By Authority.

The Quarterly

INDIAN ARMY LIST

JANUARY 1919.

Volume 1

A DISTRIBUTION LIST OF OFFICERS ON THE ACTIVE LIST OF THE ARMY IN INDIA, INDIAN DEFENCE FORCE, ETC.

TO WHICH IS ADDED, HALF-YEARLY, ORDERS OF KNIGHTHOOD, ETC., INDIAN ORDERS, FOREIGN ORDERS, NON-EFFECTIVE LISTS, AND WAR SERVICES OF OFFICERS OF THE INDIAN ARMY.

ARMY HEADQUARTERS, INDIA.

CALCUTTA
SUPERINTENDENT GOVERNMENT PRINTING, INDIA
1919

NOTE.—The Indian Army List being an Official publication, and the Press having no power to make alterations in it, all communications regarding deaths, errors, or omissions should be addressed to "Military Secretary to the Commander-in-Chief" with the words "Army List" written on the upper left-hand corner of the cover.

All correspondence regarding the War Services of Officers to be addressed to the Adjutant-General in India ***through Brigade or Divisional Commanders.*** (***Vide*** India Army Order 31 of 1911.)

CONTENTS.

	Page
Abbottabad Brigade	23
Abbreviated Telegraphic Addresses	ii
Abbreviations, Explanation of	x
Accounts Department, Military	37
Adjutant-General	8, 15
Administrative Depts. of the Army, Gazetted Civil Officers	91
Administrative Staff	15
Advisory Council	4
Agra Brigade	26
Ahmednagar Brigade	35
Aides-de-Camp to the King-Emperor	1
Aides-de-Camp (Honorary) to Viceroy	2
Air Force, Royal	577
Allahabad Brigade	45
Alphabetical List of Stations	x
Ambala Brigade	28
Amharic, Interpreters in	102
Arabic, Interpreters in	102
Army Bearer Corps	1835
" Clothing Department	79
" and Marine Department, Govt. of India	4
" Head-Quarters	5
" Hospital Corps	1836
" Medical Services	1709
" Remount Department	77
" Schools, Inspectors of	22, 24, 26, 35, 37
" Service Corps (Mechanical Transports)	1689
" Veterinary Services	1838
" " Officers with Indian Civil	
" Veterinary Department	1842
Artillery, Gradation List	612
" Inspector of Royal Garrison	7
" Royal Horse	644
" Royal Field	644
" Garrison	645
" Indian Mountain Batteries	972
" School, Quetta	67
Baluchi, Officers passed	107
Bannu Brigade	29
Bareilly Brigade	27
Bath, Order of the (*a*)	1925*
Bayonet & Physical Training Schools	67
Body Guard, Governor-General's	873
" " Governor's, Bombay	874
" " " Madras	873
" " " Bengal	874
Bombay-Deolali Brigade	40
Bombing School	68
Brigade-Majors	17
Brigades, Colonels on Staff Commanding	12
" General Officers Commanding	12
British Officers, Unattached List, Indian Army	365
Burma Division	46
Cadet College, Quetta	66
" Wellington	66
Cadet Unit, Sialkot	66
Canarese, Officers passed in	107
Cantonment Magistrates' Department	1883
Casualties	2136
Cavalry, British, Regiments of	581
Cavalry Brigade (4th)	27
" " (10th)	21
Cavalry, Indian, Regiments of	873
Cavalry School, Saugor	66
Central Bombing and Stokes Howitzer School, Mhow	68
Central Flying School, Indian	66
Chief of the General Staff	6, 13
Chinese, Interpreters in	102
Chinese, Officers passed in various dialects	107
Civil employ—Supernumerary List of Officers in	1911
" Officers, Gazetted; Administrative Depts. of the Army	93
Classified Lists, Commands and Staff	50
Clothing Department	79
Colonels' Allowances	1916*
Colonels on Staff Commanding Brigades	12
Colonial employment, Officers in	96
Commander-in-Chief in India	5
Commanding Officers of Indian Regiments—List showing dates of vacating Commands	867
Commands and Staff, Classified List	50
Delhi Brigade	27
Departmental Officers with Honorary rank—	
Army Clothing Department	1849
Barrack Department, Madras	1849
India Miscellaneous List	1850
Ordnance Department, Northern Army	1843
" " Southern Army	1844
Military Works Services, India	1848
Miscellaneous List, Madras	1852
Public Works Department, Bombay	1849
" " " India	1848
" " " Madras	1848
Supply and Transport Corps, Bengal	1845
" " " Bombay	1847
" " " Secunderabad and Burma	1846
" " " Amalgamated List	
Deputy Director, Royal Indian Marine	4
Derajat Brigade	29
Director, Army Clothing Department	79
Director of Military Operations	6
Director of Military Prisons and Detention Barracks	8
Director, Royal Indian Marine	4
Director-General, Indian Medl. Service	4, 1737
" " of Military Works	11
" " of Ordnance	10
Director of Supplies and Transport	9, 1660
" Medical Services	10, 1709
" Staff duties	6
" Military Training	6
Distinguished Service Medal, Indian (*a*)	1989
" " Order (*a*)	1936*
Divisional and Brigade Commands	12
Divisional Signal Companies	1003
Dutch, Interpreters in	102
Ecclesiastical Establishment (Bengal)	1903
" " (Bombay)	1907

* Published in January and July issues only.
(*a*) Arranged alphabetically. Not indexed.

	Page
Ecclesiastical Establishment (Madras)	1906
" " (Rangoon)	1908
Educational and Training Establishments	66
Engineers, Corps of Royal	660
Equerries to the King-Emperor	1
Farms Department, Military	81
Ferozepore Brigade	25
Field Officers of Cavalry and Infantry (Local Service), according to regimental seniority (Bengal)	169
Ditto, (Bombay)	170
Ditto, (Madras)	170
Flying School, Indian Central	66
Foreign Languages, Interpreters in	102
Foreign Orders, Officers permitted to wear (*a*)	2015*
French, Interpreters in	102
Frontier Corps, Officers employed with	95
Garrison Quarter-Master, Fort William	43
Gazetted Appointments, Administrative Depts. of the Army	93
General List—Cavalry (Bengal)	169
" " " (Bombay)	170
" " Infantry (Bengal)	169
" " " (Bombay)	170
" " " (Madras)	170
General Officers Comdg.	12
General Staff	13
German, Interpreters in	103
Good Service Pensions	1955*
Gradation List, British and Indian	118
" " British Officers of the Indian Army	145
" " Artillery	612
Graduates, Staff College	97
Gunnery, Officers obtained certificates in	101
Gunnery Schools	67
Gurkha Regiments	1595
Half Pay, List of Officers on	1910
Head-Quarters Staff of the Army	5
Hindi, Officers passed in	107
Hindustani, List of British Service Officers on the active list, under the rank of General, who have qualified by the Higher Standard in—(*a*)	2024
Honorary Rank	110
" British Rank, Retired Indian Officers holding	2101*
Honorary Physicians to the King-Emperor	1
" Surgeons to the King-Emperor	1
Imperial Govt., Troops from India serving under	96
" or Colonial Employment, Officers in	96
" Service Order (*a*)	1940*
" Service Troops, Officers with	94
Indian Army (late I. S. C.), British Officers of the	171
" " Reserve of Officers	485
" " Distinguished Service Medalists (*a*)	1989*
" Land Forces, Officers of	481
" Medical Service	1737
" Order of Merit	1970*
" " " (Civil Divn.)	1088*
" Munitions Board	1884
" Officers retired with Honorary British rank	2101*
Indian Medical Dept.	1787
" Medical Dept. (Mily. Sub-Asst. Surgeons)	1807
" Mountain Artillery	972
" Central Flying School	66
" Defence Force, Alphabetical List of Corps	674
" " " Precedence, List of	677
" " " Mounted Units	601
" " " Artillery	651
" " " Infantry	801
Infantry Regiments, British, Alphabetical List	672
" " Indian	1016
Inspector of Royal Garrison Artillery	7, 14
" Machinery, Military Works Services	11
" Infantry, North	7, 14
" " South	7
" Cavalry	7, 14
" R. H. and R. F. A.	7, 14
" R. E. & Pioneers	7, 14
" Army Signals	7, 14
Inspectors of Army Schools	22, 24, 26, 35, 37
" Ordnance Machinery	74
" Physical and Bayonet Training	19, 25
" Munitions	73
Interpreters, Modern Foreign Languages, List of	102
Italian, Interpreters in	104
Japanese, Interpreters in	104
Jhansi Brigade	34
Jhelum Brigade	23
Jubbulpore Brigade	33
Judge Advocate-General	8
Kaisar-i-Hind Medal (*a*)	1943*
Karachi Brigade	39
Kohat Brigade	29
Languages, Modern Foreign, Interpreters in	102
" appertaining to India, Officers who have passed in	107
Local Service (Bengal)	169
" " (Bombay)	170
" " (Madras)	170
Lucknow Division	43
Lucknow Brigade	45
MacGregor Memorial Silver Medalists	2023*
Machine Gun Corps	797
Machine Gun Training Centre	68
Marathi, Officers passed	108
Medical Branch, Army Head Quarters	10
" Services	1709
" Service, Indian	1737
" "	1787
" Store-keepers to Government	78
Meerut Division	26
Mhow Division	32
Military Cross, officers decorated with (*a*)	1940*
Military Accounts Department, Govt. of India	87
" Farms Department	81
" Operations, Director of	6
" Staff duties, Director of	6
" Training, Director of	6
" Works Branch	11
Mobilisation Committee	4
Modern Foreign Languages, Interpreters in	102

* Published in January and July issues only.
(*a*) Arranged alphabetically. Not indexed.

CONTENTS.

	Page
Abbottabad Brigade	23
Abbreviated Telegraphic Addresses	ii
Abbreviations, Explanation of	x
Accounts Department, Military	37
Adjutant-General	8, 15
Administrative Depts. of the Army, Gazetted Civil Officers	91
Administrative Staff	15
Advisory Council	4
Agra Brigade	26
Ahmednagar Brigade	35
Aides-de-Camp to the King-Emperor	1
Aides-de-Camp (Honorary) to Viceroy	2
Air Force, Royal	577
Allahabad Brigade	45
Alphabetical List of Stations	x
Ambala Brigade	28
Amharic, Interpreters in	102
Arabic, Interpreters in	102
Army Bearer Corps	1835
„ Clothing Department	79
„ and Marine Department, Govt. of India	4
„ Head-Quarters	5
„ Hospital Corps	1836
„ Medical Services	1709
„ Remount Department	77
„ Schools, Inspectors of	22, 24, 26, 35, 37
„ Service Corps (Mechanical Transports)	1689
„ Veterinary Services	1838
„ „ Officers with Indian Civil	
„ Veterinary Department	1842
Artillery, Gradation List	612
„ Inspector of Royal Garrison	7
„ Royal Horse	644
„ Royal Field	644
„ Garrison	645
„ Indian Mountain Batteries	972
„ School, Quetta	67
Baluchi, Officers passed	107
Bannu Brigade	29
Bareilly Brigade	27
Bath, Order of the (*a*)	1925*
Bayonet & Physical Training Schools	67
Body Guard, Governor-General's	873
„ „ Governor's, Bombay	874
„ „ „ Madras	873
„ „ „ Bengal	874
Bombay-Deolali Brigade	40
Bombing School	68
Brigade-Majors	17
Brigades, Colonels on Staff Commanding	12
„ General Officers Commanding	12
British Officers, Unattached List, Indian Army	365
Burma Division	46
Cadet College, Quetta	66
„ Wellington	66
Cadet Unit, Sialkot	66
Canarese, Officers passed in	107
Cantonment Magistrates' Department	1883
Casualties	2136
Cavalry, British, Regiments of	581
Cavalry Brigade (4th)	27
„ „ (10th)	21
Cavalry, Indian, Regiments of	873
Cavalry School, Saugor	66
Central Bombing and Stokes Howitzer School, Mhow	68
Central Flying School, Indian	66
Chief of the General Staff	6, 13
Chinese, Interpreters in	102
Chinese, Officers passed in various dialects	107
Civil employ—Supernumerary List of Officers in	1911
„ Officers, Gazetted; Administrative Depts. of the Army	93
Classified Lists, Commands and Staff	50
Clothing Department	79
Colonels' Allowances	1916*
Colonels on Staff Commanding Brigades	12
Colonial employment, Officers in	96
Commander-in-Chief in India	5
Commanding Officers of Indian Regiments—List showing dates of vacating Commands	867
Commands and Staff, Classified List	50
Delhi Brigade	27
Departmental Officers with Honorary rank—	
Army Clothing Department	1849
Barrack Department, Madras	1849
India Miscellaneous List	1850
Ordnance Department, Northern Army	1843
„ „ Southern Army	1844
Military Works Services, India	1848
Miscellaneous List, Madras	1852
Public Works Department, Bombay	1849
„ „ „ India	1848
„ „ „ Madras	1848
Supply and Transport Corps, Bengal	1845
„ „ „ Bombay	1847
„ „ „ Secunderabad and Burma	1846
„ „ „ Amalgamated List	
Deputy Director, Royal Indian Marine	4
Derajat Brigade	29
Director, Army Clothing Department	79
Director of Military Operations	6
Director of Military Prisons and Detention Barracks	8
Director, Royal Indian Marine	4
Director-General, Indian Medl. Service	4, 1737
„ „ of Military Works	11
„ „ of Ordnance	10
Director of Supplies and Transport	9, 1660
„ Medical Services	10, 1709
„ Staff duties	6
„ Military Training	6
Distinguished Service Medal, Indian (*a*)	1989
„ „ Order (*a*)	1936*
Divisional and Brigade Commands	12
Divisional Signal Companies	1003
Dutch, Interpreters in	102
Ecclesiastical Establishment (Bengal)	1903
„ „ (Bombay)	1907

* Published in January and July issues only.
(*a*) Arranged alphabetically. Not indexed.

	Page
Ecclesiastical Establishment (Madras)	1906
" " (Rangoon)	1908
Educational and Training Establishments	66
Engineers, Corps of Royal	660
Equerries to the King-Emperor	1
Farms Department, Military	81
Ferozepore Brigade	25
Field Officers of Cavalry and Infantry (Local Service), according to regimental seniority (Bengal)	169
Ditto, (Bombay)	170
Ditto, (Madras)	170
Flying School, Indian Central	66
Foreign Languages, Interpreters in	103
Foreign Orders, Officers permitted to wear (*a*)	2015*
French, Interpreters in	102
Frontier Corps, Officers employed with	95
Garrison Quarter-Master, Fort William	43
Gazetted Appointments, Administrative Depts. of the Army	93
General List—Cavalry (Bengal)	169
" " " (Bombay)	170
" " Infantry (Bengal)	169
" " " (Bombay)	170
" " " (Madras)	170
General Officers Comdg.	12
General Staff	13
German, Interpreters in	103
Good Service Pensions	1955*
Gradation List, British and Indian	118
" " British Officers of the Indian Army	145
" " Artillery	612
Graduates, Staff College	97
Gunnery, Officers obtained certificates in	101
Gunnery Schools	67
Gurkha Regiments	1595
Half Pay, List of Officers on	1910
Head-Quarters Staff of the Army	5
Hindi, Officers passed in	107
Hindustani, List of British Service Officers on the active list, under the rank of General, who have qualified by the Higher Standard in—(*a*)	2024
Honorary Rank	110
" British Rank, Retired Indian Officers holding	2101*
Honorary Physicians to the King-Emperor	1
" Surgeons to the King-Emperor	1
Imperial Govt., Troops from India serving under	96
" or Colonial Employment, Officers in	96
" Service Order (*a*)	1940*
" Service Troops, Officers with	94
Indian Army (late I. S. C.), British Officers of the	171
" " Reserve of Officers	485
" " Distinguished Service Medalists (*a*)	1989*
" Land Forces, Officers of	481
" Medical Service	1737
" Order of Merit	1970*
" " " (Civil Divn.)	1088*
" Munitions Board	1884
" Officers retired with Honorary British rank	2101*
Indian Medical Dept.	1787
" Medical Dept. (Mily. Sub-Asst. Surgeons)	1807
" Mountain Artillery	972
" Central Flying School	66
" Defence Force, Alphabetical List of Corps	674
" " " Precedence, List of	677
" " " Mounted Units	601
" " " Artillery	651
" " " Infantry	801
Infantry Regiments, British, Alphabetical List	672
" " Indian	1016
Inspector of Royal Garrison Artillery	7, 14
" Machinery, Military Works Services	11
" Infantry, North	7, 14
" " South	7
" Cavalry	7, 14
" R. H. and R. F. A.	7, 14
" R. E. & Pioneers	7, 14
" Army Signals	7, 14
Inspectors of Army Schools	22, 24, 26, 35, 37
" Ordnance Machinery	74
" Physical and Bayonet Training	19, 25
" Munitions	73
Interpreters, Modern Foreign Languages, List of	102
Italian, Interpreters in	104
Japanese, Interpreters in	104
Jhansi Brigade	34
Jhelum Brigade	23
Jubbulpore Brigade	33
Judge Advocate-General	8
Kaisar-i-Hind Medal (*a*)	1943*
Karachi Brigade	39
Kohat Brigade	29
Languages, Modern Foreign, Interpreters in	102
" appertaining to India, Officers who have passed in	107
Local Service (Bengal)	169
" " (Bombay)	170
" " (Madras)	170
Lucknow Division	43
Lucknow Brigade	45
MacGregor Memorial Silver Medalists	2023*
Machine Gun Corps	797
Machine Gun Training Centre	68
Marathi, Officers passed	108
Medical Branch, Army Head Quarters	10
" Services	1709
" Service, Indian	1737
" "	1787
" Store-keepers to Government	78
Meerut Division	26
Mhow Division	32
Military Cross, officers decorated with (*a*)	1940*
Military Accounts Department, Govt. of India	87
" Farms Department	81
" Operations, Director of	6
" Staff duties, Director of	6
" Training, Director of	6
" Works Branch	11
Mobilisation Committee	4
Modern Foreign Languages, Interpreters in	102

* Published in January and July issues only.
(*a*) Arranged alphabetically. Not indexed.

	Page
Movements Directorate, Army Head Quarters	9
Munitions, Inspectors of	73
Munitions Board, Indian	1884
Musketry Schools	68
Nasirabad Brigade	33
Non-effective Lists	2031*
Northern Command, Staff of	19
Norwegian, Interpreters in	104
Nowshera Brigade	21
Nursing Service (Military) for India, Queen Alexandra's	1734
Officers employed with Frontier Corps	95
" on the temporary half-pay list	1910
" Retired, etc., List of	2038
" in Civil employ, Supernumerary List of	1911
" Commanding Indian Regiments	857
" in Civil employment	1880
" in Civil employment, Assam	1898
" " " " Baluchistan	1898
" " " " Bengal	1893
" " " " Bihar and Orissa	1894
" " " " Bombay	1892
" " " " Burma	1894
" " " " Central India	1899
" " " " Central Provinces	1899
" " " " Hyderabad	1900
" " " " India, Govt. of	1889
" " " " Madras	1892
" " " " Mysore and Coorg	1900
" " " " N.-W. Frontier Prov.	1900
" " " " P. W. D.	1881
" " " " Punjab	1896
" " " " Railway Dept.	1881
" " " " Rajputana	1901
" " " " Survey of India	1880
" " " " Travancore	1892
" " " " U. P. of Agra and Oudh	1895
" School of Instruction, Bangalore	67
Order of the Bath (a)	925*
" British Empire (a)	1905
" British India (a)	1956*
" " Imperial Service Troops (a)	1767*
" Frontier Militia Corps and Military Police (a)	1968*
" Distinguished Service (a)	1936*
" Imperial Service (a)	1940*
" Indian Empire (a)	1931*
" Merit (Indian)	1970*
" Military Cross (a)	1940*
" Royal Red Cross (a)	1943*
" Royal Victorian (a)	1934*
" St. Michael and St. George (a)	1923*
" Star of India (a)	1928*
" Victorian Order (a)	1934*
Orders, Foreign, Officers permitted to wear	2015*
Ordnance Department	71
" Branch	10
" Manufacturing Section	71
" Factories	71
Ordnance Inspection Section	72
Ordnance Stores Section	9, 74
Pensions, Good Service	1955*
Persian, Interpreters in	104
Personal Staff, Viceroy and Governor-General	2
" " Commander-in-Chief in India	5
Peshawar Division	20
Peshawar Brigade	21
Physicians, Honorary, to the King-Emperor	1
Physical Training, Inspectors of	19, 31
Physical and Bayonet Training School	67
Poona Division	35
" Brigade	36
Precedence List, Regiments of Indian Army	866
" Warrant of, in India	2126
Presidency Brigade	43
Principal Veterinary Officer in India	9, 1838
Public Works Department	1881
Qualifications, Professional, List of references	vii
" Lingual, List of references	viii
Quarter-Master General	9, 16
Quartering Directorate, Army Head Quarters	9
Queen Alexandra's Military Nursing Service for India	1734
Quetta Division	42
Quetta 10th, 11th and 57th Infantry Brigades	42
Railway Battalion, Sappers and Miners	1001
Railway Dept., Officers in the	1881
Rangoon Brigade	47
Rawal Pindi Division	22
" " Infantry Brigade	22
Recruiting Section, Army Head Quarters	8
" Officers	82
Regimental Centres of Indian Infantry Regts.	1014
Registration Officers (Transport)	1864
Remount and Veterinary Directorates, Army Head Quarters	9
Reserve of Officers	1919*
" of Officers, Indian Army	485
Resident's Escort, Nepal	1659
Retired Indian Officers holding Honorary British rank	2101*
Retired Lists	2038*
" Officers, Indian Army on the Reserve of Officers' List	1919
Risalpur Cavalry Brigade	21
Royal Army Medical Corps	1709
" Air Force	577
" Garrison Artillery, Inspr. of	7
Russian, Interpreters in	105
Sappers and Miners, Corps of, 1st, 2nd & 3rd	985 to 1001
School of Physical and Bayonet Training, N. & S. Armies	67
" " Bombing	68
" " Signalling	68
Schools of Musketry	68
School of Instruction for Officers, Bangalore	67
" " " " Nasik	67
" Artillery	67
Secunderabad Division	37
" Cavalry Brigade	
" Infantry Brigade	
Senior Asst. Surgeons	1787
Sialkot Brigade	23

*Published in January and July issues only.
(a) Arranged alphabetically. Not indexed.

	Page
Signal Companies, Sappers and Miners	1003
Signalling Schools	68
Somali, Interpreters in	106
Southern Command, Staff of	31
„ Brigade	37
Spanish, Interpreters in	106
Staff Captains	17
„ College, Camberley	66
„ „ Graduates	97
„ „ Quetta	66
Staff School, Saugor	66
Staffs of Armies	12
Star of India, Order of the (a)	1928*
Station Staff Officers	17
St. Michael and St. George, Order of (a)	1929*
Superintendents of Physical Training	31
Supernumerary List, Officers transferred to	1911
Supply and Transport Corps	1660
„ „ „ Directorate, Army Head Quarters	9
„ „ „ Training Establishment	67
Supply and Transport Corps Reserve, Indian Officers of the—	1686
Supply and Transport Corps and Units	1660
Surgeons, Honorary, to the King-Emperor	1
Surgeons, Honorary to the Viceroy	3
„ „ Assistant, to the Viceroy	3
Survey of India	1880
Tamil, Officers passed in	108
Telegraphic Addresses, Abbreviated	xii
Transport Corps and Cadres	1681
Transport Registration Officers	1685
Transport Veterinary Assistants	1687
Troops from India serving under Imperial Govt.	96
Turkish, Interpreters in	105
Unattached List, Indian Army British Officers	365

	Page
Unemployed Supernumerary List (a)	2031*
United Service Institution, Medalists	2023*
Urdu, Officers passed in	108
Veterinary Service, Army	1838
„ Department, Indian, Civil	1841
Viceroy and Governor-General of India	9
Victoria Cross, Officers decorated with (a)	1925*
Victorian Order (a)	1933*
Warrant Officers, Army Clothing Dept.	1872
„ „ Barrack Dept., Punjab-Bengal	1870
„ „ India Miscellaneous List	1872
„ „ Indian Sub.-Medl. Dept.	1791
„ „ Indian Telegraph Dept.	1872
„ „ Indo-European Telegraph Dept.	1872
„ „ Miscellaneous List, Madras	1879
„ „ Non-dept.	1879
„ „ Ordnance Dept., Northern Army	1853
„ „ Ordnance Dept., Southern Army	1854
„ „ Ordnance Dept., General List	1855
„ „ P. W. Dept., Bombay	1868
„ „ „ „ India	1867
„ „ „ „ Madras	1868
„ „ Supply and Transport Corps (Amalgamated List)	1863
„ „ Supply and Transport Corps (Bengal)	1861
„ „ „ „ (Bombay)	1862
„ „ „ „ (Burma)	1861
„ „ „ „ (Secunderabad)	1861
Warrant of Precedence in India	2126
War Services	2146
Wesleyan Chaplains	1908

* Published in January and July issues only.
(a) Arranged alphabetically. Not indexed.

LIST OF REFERENCES—PROFESSIONAL.

12 Passed Course of Range-finding.
20 Telegraphy.
42 Ballooning Certificate.
43 Cavalry Pioneers Certificate.
45a Cavalry School, Saugor or Netheravon (Proficiency Certificate).
46 Instructor in Bombing.
47 Qualified in Hotchkiss Gun (Distinguished).
47a Qualified in Hutchkiss Gun.
48 Qualified in Lewis Gun (Distinguished).
48a Qualified in Lewis Gun.
49 Qualified in Stokes Howitzers.
50 Qualified at a School of Musketry (Extra or Distinguished).
51 Qualified at a School of Musketry.
52 Qualified in Machine Gun.
53 Qualified in Machine Gun (Extra or Distinguished).
54 Passed Musketry Course for Volunteer Officers.
56 Musketry Certificate (Indian ranks).
57 Certificated Instructor in Army Signalling.
58 Certificated Assistant Instructor in Army Signalling.
58a Certificated Assistant Instructor in Army Signalling (Distinguished).
59 Certificated Instructor in Army Signalling (Distinguished).
60a Certificate of Physical Training and Swordsmanship.
61 Certificate of Physical and Bayonet Training.
64 Army Veterinary School Certificate (with credit).
65 Army Veterinary School Certificate.
65a Officers who have passed the Supply Course at the S. and T. Schools of Instruction.
66 Officers who have passed the Course of Transport Instruction at the S. and T. Schools of Instruction.
66a Officers who have passed the Course of Transport Instruction (*vide* para. 959, A. R. I., Vol. II).
66b Officers who are in possession of "Extra Transport Certificate" referred to in para. 4, Appx. VII, A. R. I., Vol. II.
67 Transport Course, Aldershot.
67a Mechanical Transport Course, Aldershot.
71 Equitation Certificate.
72 Officers who have passed a course of Mounted Infantry training at Home.
72a Officers who have passed a course of Mounted Infantry training at a school in India.
72b Officers who have served with Mounted Infantry in the field.
74 Passed for Judge Advocate-General's Department.
79 Passed Proficiency Examination for Volunteer Officers (Para. 80, A. R. I., Vol. IX).
80 Passed I. D. F. Test.
82 Passed Artillery Course for Volunteer Officers.
82a Passed School of Instruction for Volunteer and Militia Officers.

LIST OF REFERENCES—LINGUAL.

83 Ambaric.
84 Angami.
88 Arabic (Higher Standard).
89 ,, (Lower Standard).
89a ,, (Elementary).
89b ,, (F. S. Proficiency Test).
90 Assamese (Higher Standard).
91 ,, (Lower Standard).
92 ,, (Cl. 8, G. G. O. 734 of 1864)
93 Baluchi (Interpreter).
94 ,, (Higher Standard).
95 ,, (Lower Standard).
96 Bengali (Degree of Honor).
97 ,, (High Proficiency).
98 ,, (Higher Standard).
99 ,, (Lower Standard).
100 Brahui.
101 Burmese (Degree of Honor).
102 ,, (High Proficiency).
103 ,, (Interpreter).
104 ,, (Higher Standard).
105 ,, (Lower Standard).
106 ,, (Elementary).
107 ,, (Colloquial).
108 Canarese (Degree of Honor).
109 ,, (High Proficiency).
110 ,, (Interpreter).
111 ,, (Higher Standard).
112 ,, (Colloquial).
112a ,, (Obligatory).
113 Chin (Baungshe Dialect).
114 ,, (Bokiyi ,,).
115 ,, (Chinbok ,,).
116 ,, (Chinme ,,).
117 ,, (Haka ,,).
118 ,, (Kami ,,).
118a ,, (Lai ,,).
119 ,, (Sebaung ,,).
120 ,, (Siyin ,,).
120a ,, (Talaing ,,).
121 ,, (Tashon ,,).
122 ,, (Taungzat ,,).
123 ,, (Yaklaing ,,).
124 Chinese (Final).
124a ,, (Field Service test).
125 ,, (Preliminary).
125a ,, (Yunanese Dialect).
125b ,, (Pekingese ,,).
125c ,, (Cantonese ,,).
125d Chinese, regimental test.
125e Dutch, Passed.
127 French (Passed).
128 ,, (Preliminary).
129 German (Passed).
130 Gujrati (Degree of Honor).
131 ,, (Interpreter).
132 ,, (Higher Standard).
133 ,, (Lower Standard).
134 ,, (Cl. 8, G.G.O. 734 of 1864).
135 Hindi (Degree of Honor).
136 ,, (High Proficiency).
137 ,, (Obligatory).
139 Italian (Passed).
139b Japanese (Preliminary).
140 Kachin (Chinpaw Dialect).
141 Karen (Sgau Karen).
142 ,, (Pwo Karen).
143 Karenni.
144 Khasi.
144a Khowar Dialect.
144b Lisaw.
145 Lushai.
146 Malay.
147 Malayalam (Higher Standard).
148 ,, (Colloquial).
148a ,, (Obligatory).
148b Maru Language.
149 Manipuri.
150 Marathi (Degree of Honor).
151 ,, (High Proficiency).
152 ,, (Interpreter).
153 ,, (Higher Standard).
154 ,, (Lower Standard).
155 ,, (Obligatory).
155(a) Miri.
156 Parvatiya or Khaskura (Colloquial).
157 ,, ,, (Obligatory)
161 Persian (Higher Standard).
161a ,, (Preliminary Interpreter).
162 ,, (Lower Standard).
162a ,, (Obligatory).
162b ,, (F. S. Proficiency Test).
163 ,, (Cl. 8, G.G.O. 734 of 1864).
165 Punjabi (Higher Standard).
166 ,, (Cl. 8, G.G.O. 734 of 1864)
167 ,, (Obligatory).
168 Pushtu (Interpreter).
169 ,, (Higher Standard).
170 ,, (Lower Standard).
171 ,, (Cl. 8, G.G.O. 734 of 1864).
172 ,, (Obligatory).
172a ,, (Elementary).
174 Russian (Passed).
175 ,, (Preliminary Examination).
176 Sanskrit (Degree of Honor).
177 ,, (High Proficiency).
178 ,, (Higher Standard).
179 ,, (Lower Standard).
180 Shan (Higher Standard).
181 ,, (Lower Standard).
182 ,, (Elementary).
182a ,, (Hsawng sup Dialect).
182b Shina Dialect.
183 Siamese.
183a Sikkim Bhuttia.
184 Sindhi (Interpreter).
185 ,, (Higher Standard).
186 ,, (Lower Standard).
188 Somali (Higher Standard).
188a ,, (Colloquial).
188b ,, (Elementary).
190 Spanish (Passed).
191 Tamil (Interpreter).
192 ,, (High Proficiency).
193 ,, (Higher Standard).
194 ,, (Lower Standard).
195 ,, (Obligatory).
196 Taungthu.
197 Telegu (High Proficiency).
198 ,, (Higher Standard).
199 Tibetan.
200 Turki (Eastern).
202 Turkish (Passed).
202a ,, (F. S. Proficiency Test).
203 Urdu (Degree of Honor)
204 ,, (High Proficiency).
205 ,, (Interpreter).
205a ,, (Proficiency).
206 Hindustani (Higher Standard).
207 ,, (Lower Standard).
207a ,, (Colloquial).
208 Uriya (High Proficiency).
209 ,, (Higher Standard).

EXPLANATIONS OF ABBREVIATIONS.

NOTE.—Where no corps or regiment is specified after an officer's name, it is to be understood that he belongs to the Indian Army.

A. B. C.	Army Bearer Corps.
A. C. R. E.	Assistant Commanding Royal Engineer.
A. D. C.	Aide-de-Camp General to the King.
A. D. C.	Aide-de-Camp to the King.
A. D. C.	Honorary Aide-de-Camp to the Viceroy.
A. F. C.	Air Force Cross.
A. F. M.	Air Force Medal.
A. R. R. C.	Associate of Royal Red Cross.
A. V. C.	Army Veterinary Corps.
A. T. M.	Ahmudan gaung Tazeikya Min (Title).
B.	Bengal.
Bo.	Bombay.
C. V. D.	Civil Veterinary Department.
C	Served in the Crimean Campaign.
c.	Colonel.
C. B.	Companion of the Order of the Bath.
C. B. E.	Commander of the Order of the British Empire.
C. I. E.	Companion of the Most Eminent Order of the Indian Empire.
C. M. G.	Companion of the Order of St. Michael and St. George.
C. P.	Central Provinces.
C. R. E.	Commanding Royal Engineer.
C. S. I.	Companion of the Most Exalted Order of the Star of India.
C. V. O.	Commander of the Royal Victorian Order.
DI, II, III, or IV.	Field Officers and Captains of I. D. F. units who, as Volunteer Officers, have passed examination in tactics, military law, organization, etc., or Military history.
dI, II, III, or IV.	Lieutenants of I. D. F. units who, as Volunteer Officers, have passed Examination in tactics, Military law, organization, etc., or Military history.
D. C. M.	Distinguished Conduct Medal.
D. F. C.	Distinguished Flying Cross.
D. F. M.	Distinguished Flying Medal.
D. S. C.	Distinguished Service Cross.
D. S. O.	Companion of the Distinguished Service Order.
(*e*)	Qualified in a course of Instruction at the London School of Economics.
Eq.	Equerry to the King.
(*f*)	Passed Firemasters' Course, Ordnance College.
fm.	Field Marshal.
g.	General.
(*g*)	Obtained a 1st class certificate in the Long course of Gunnery or Gunnery Staff Course prior to 1906, or the Certificate of the Gunnery Staff Course subsequent to 1905.
G. B. E.	Knight Grand Cross of the Order of the British Empire.
G. C. B.	Knight Grand Cross of the Order of the Bath.
G. C. I. E.	Knight Grand Commander of the Most Eminent Order of the Indian Empire.
G. C. M. G.	Knight Grand Cross of the Order of St. Michael and St. George.
G. C. S. I.	Knight Grand Commander of the Most Exalted Order of the Star of India.
G. C. V. O.	Knight Grand Cross of the Royal Victorian Order.
G. E.	Garrison Engineer.
G. M. I. E.	Grand Master of the Most Eminent Order of the Indian Empire.
G. M. S. I.	Grand Master of the Most Exalted Order of the Star of India.
H.	Passed Interpreter's test in Hindustani under the Madras or Bombay Regulations.
H. M.	His Majesty.
h. p.	Half Pay.
I. A.	Indian Army.
I. D. S. M.	Indian Distinguished Service Medal.
I. M. S.	Indian Medical Service.
I. O. M.	Inspector of Ordnance Machinery.
I. O. M.	Member of the Indian Order of Merit.
I. M. D.	Indian Medical Department.
I. S. O.	Companion of the Imperial Service Order.

K. B. E.	Knight Commander of the Order of the British Empire
K. C. B.	Knight Commander of the Order of the Bath
K. C. I. E.	Knight Commander of the Most Eminent Order of the Indian Empire.
K. C. M. G.	Knight Commander of the Order of St. Michael and St. George.
K. C. S. I.	Knight Commander of the Most Exalted Order of the Star of India.
K. C. V. O.	Knight Commander of the Royal Victorian Order.
K. H. C.	Hony. Chaplain to the King.
K. H. P.	Hony. Physician to the King.
K. H. S.	Hony. Surgeon to the King.
K. G.	Knight of the Order of the Garter.
K. S. M.	Kyet thaye zaung shwe Salwè ya Min (Title).
K. P.	Knight of the Order of St. Patrick.
K. T.	Knight of the Order of the Thistle.
[L.]	Qualified as 1st class Interpreter in a Modern Foreign Language.
[*l*]	Qualified as 2nd class Interpreter in a Modern Foreign Language.
l. c.	Lieutenant-Colonel.
l. g.	Lieutenant-General.
M	Served in the Indian Mutiny Campaign.
M.	Madras.
m	Major.
M. B. E.	Member of the Order of the British Empire.
M. C.	Military Cross.
m. c.	Medical Certificate.
m. g.	Major-General.
M. M.	Military Medal.
M. V. O.	Member of the 4th or 5th class of the Royal Victorian Order.
M. W. S.	Military Works Services.
N.	New Leave rules for Departmental officers with Hony. rank, Senior Asst. Surgeons and Warrant Officers of the Indian Army.
(*n*)	Officers who have attended Naval War Course.
(*o*)	Passed the Ordnance Course.
O. B. E.	Officer of the Order of the British Empire.
O. M.	Member of the Order of Merit.
P.	Passed Interpreter's Examination before District Committees.
p. a. c.	Passed the final examination of the advanced class Ordnance College.
P. C.	Passed Interpreter's test in Calcutta.
P. H.	Passed Examination in the Hindustani Language.
p. s.	Passed for Staff employ (*vide* I. A. O. No. 252 of 1907).
p. s. c.	Passed the final examination at the Staff College.
P.	Punjab.
P. W. D.	Public Works Department.
(Q).	Qualified in tactical fitness for Comd., or in the case of Officers of the R.A.M.C., for promotion to the rank of Lieutenant-Colonel.
q.	Passed field works course, Chatham, or course of Instruction in pioneering at H.Q. Corps of Sappers and Miners.
q. s.	Officer qualified for Staff employment in consequence of service on the Staff in the Field.
[R.]	Officers who are in receipt of Good Service Pension.
Ⓡ	Officers who have vacated Good Service Pensions.
R. A. M. C.	Royal Army Medical Corps.
r. d.	Regimental Depôt.
R. R. C.	Royal Red Cross.
u. p. a.	Urgent private affairs.
T. D.	Territorial Decoration.
T. D. M.	Thuye gaung ngwe Da ya Min (Title).
u. s. l.	On Unemployed Supernumerary List.
V. C.	Victoria Cross.
V.D.	Volunteer Officers' Decoration.
V. H. S.	Hony. Surgeon to the Viceroy.

ALPHABETICAL LIST OF STATIONS COMPRISED IN THE SEVERAL DIVISIONS AND BRIGADES OF THE ARMY IN INDIA.

Stations.	Page.
A	
Abazai	21
Abbottabad	23
Abu Road	33
Aden	
Adra	44
Agar	32
Agra	26
Ahmedabad	40
Ahmednagar	35
Ajmer	33
Aligarh	26
Alipore	44
Allahabad	45
Almora	27
Alwar	33
Ambala	28
Amritsar	25
Anandi	35
Arangaon	35
Asansol	44
Attock	22
Aurungabad	35
Azamgarh	45
B	
Babugarh	26
Badarpur	44
Bahawalpur	24
Bakloh	24
Baksa Duar	43
Balaghat	33
Ballygunge	44
Banda	45
Bandikui	33
Bangalore	38
Bankipore	44
Bannu	29
Bara Gali	23
Barakar	44
Bareilly	27
Barian	22
Barnagore	44
Baroda	40
Barrackpore	43
Belgaum	36
Bellary	38
Benares	45
Berhampore	44
Bethul	33
Bezwada	38
Bhagalpore	44
Bhamo	40
Bhandara	33
Bharatpur	26
Bhusawal	36
Bijapur	36
Bikaner	33
Bilaspur	33
Bolarum	37
Bombay	40
Buddertollah	44
Burhan	23
Buxar	44
C	
Calcutta	44
Calicut	38
Campbellpore	21
Cannanore	38
Cawnpore	45
Chakardharpur	44
Chakdara	20
C—*contd.*	
Chaklala	22
Chakrata	26
Chaman	40
Chamiari	23
Chaubatia	27
Cherat	21
Chickmagalur	38
Chindwara	33
Chin Hills	46
Chitral	20
Chittagong	44
Chunar	45
Chupra	44
Coorjee	44
D	
Dacca	43
Dagshai	28
Dalhousie	25
Dam Dim	44
Damoh	33
Dangiri	33
Dargai	20
Darjeeling	43
Darrang	44
Deesa	33
Dehra Dun	26
Delhi	27
Deolali	40
Deoli	33
Dera Ismail Khan	29
Dharmsala	25
Dharwar	36
Dholpur	26
Dhond	35
Dhulia	36
Dhurbanga	44
Dibrugarh	44
Dinapore	43
Drazinda	
Drosh	20
Drug	42
Dullabcherra	44
Dum Dum	44
E	
Erinpura	33
Etawah	45
F	
Fatehgarh	45
Fatehpur	45
Ferozepore	25
Fort Gloster	44
Fort Lahore	
Fort Lockhart (Samana)	29
Fort Saint George	38
Fort Sandeman	42
Fort William	43
Fyzabad	45
G	
Gadag	36
Ganjdundwara	26
Gantok	44
Garden Reach	44
Gaya	43
Gharial	22
Ghaziabad	27
Ghazipore	45
Ghoom	43
G—*contd.*	
Ghora Daka	23
Giridih	44
Girdao	42
Girni	29
Gonatea	44
Gonda	45
Goosery	44
Gooty	38
Gorakhpur	45
Gulistan	42
Guna	34
Gumbaz	42
Gungapur	32
Guntakel	38
Gwalior	34
Gyantse	44
H	
Hamirpur	45
Hangu	29
Hapur	26
Harda	33
Harnai	42
Hasan Abdal	23
Hindu Bagh	42
Hirok	42
Hollicherra	44
Howrah	44
Hubli	36
Hyderabad (Sind)	39
I	
Ichapur	44
Indore	32
J	
Jacobabad	42
Jaipur	33
Jajha	44
Jalgaon	36
Jalpaiguri	44
Jamalpur	
Jamrud	21
Jaunpur	45
Jandola	29
Jatta	29
Jhansi	34
Jhelum	23
Jherria	44
Jodhpur	33
Jorhat	44
Jubbulpore	33
Jugatdal	44
Jullundur	25
Jutogh	23
K	
Kailana	26
Kakul	23
Kalabagh	23
Kalchini	44
Kalimpong	44
Kalka	28
Kamarhatty	44
Kamptee	33
Kamrup	44
Kanchrapara	44
Kankinarrah	44
Karachi	39

Stations.	Page.
K—*contd.*	
Kasauli	28
Kasganj	26
Khandalla	36
Khandwa	32
Khanspur	23
Khardah	44
Khargpur	44
Khirgi	29
Khurda Road	44
Khyra Gali	23
Killa Saifulla	42
Kiamari	39
Kirkee	36
Kishingi	42
Kohat	29
Kohima	44
Kolar Gold Field	38
Kotah	34
Kotri	39
Kuldana	22
Kulti	44
Kurseong	44
L	
Lahore Cantonment	24
Lakhimpur	44
Lalitpur	34
Lalmonir Hat	44
Laskarpur	44
Landour	26
Lansdowne	26
Lashio	46
Laskarpur	
Lebong	44
Lilloah	44
Loralai	42
Lower Topa	22
Lower Gharial	22
Lucknow	45
Lumding	44
M	
Madras	38
Magwe	46
Mahoba	45
Malakand	
Malappuram	38
Mandalay	46
Mangaldai	44
Manipur	44
Manmad	35
Manora	39
Maratangi	42
Mardan	21
Mastung	42
Maymyo	46
Meerut	27
Mehidpore	32
Meiktila	46
Mercara	38
Mergui	47
Mhow	32
Midnapore	44
Miranshah	29
Mirzapore	45
Moghal Serai	45
Mogok	46
Mona	
Montgomery	24
Monywa	46
Moradabad	27
Motihari	44
Moulmein	47
Mount Abu	33
Mozufferpore	44
Multan	29
Murgha	42

Stations.	Page.
M—*contd.*	
Murree	22
Musa Khel	42
Mussoorie	26
Muttra	27
Myitkyina	46
Mysore	38
N	
Nagpur	33
Nagrakata	44
Naini Tal	27
Nasik	
Narainganj	44
Nasirabad	33
Nazafgarh	45
Neemuch	33
Negapatam	38
Nowgong (Assam)	44
Nowgong	34
Nowshera	21
O	
Ondal	44
Ootacamund	38
P	
Pachmarhi	33
Pahartali	44
Pallaveram	38
Panchgani	36
Parachinar	29
Pathankot	25
Patna	44
Perim	
Peshawar	20
Phillour	28
Phulera	33
Pishin	42
Pollibetta	38
Poona	36
Poonamallee	38
Port Blair	47
Purandhar	36
Purna	
Pyinmana	46
Q	
Quetta	42
R	
Raipur	33
Rajbari	44
Rajkot	40
Rampur Haut	44
Ranchi	44
Raneegunge	44
Ranikhet	27
Rangoon	47
Ratlam	32
Rawal Pindi	22
Rewat	23
Risalpur	21
Rishra	44
Rurki	26
S	
Sadiya	44
Saharanpur	26
Sahibgunge	44
Saidpore	44
Sakchi	44
Saklaspur	38
St. Thomas's Mount	38
Santa Cruz	40
Santragachi	44

Stations.	Page.
S—*contd.*	
Sara	44
Sargodha	23
Satara	36
Saugor	34
Secunderabad	37
Seebpore	44
Sehore	34
Seoni	33
Shabkadr	21
Shahjahanpur	27
Shaikh Othman	
Shelabagh	42
Shillong	44
Shimoga	38
Shingarh	42
Sholapur	36
Shwebo	46
Sialkot	23
Sibi	42
Sibsagar District	44
Sijua	44
Silchar	44
Simla	28
Sirdarpore	32
Sitabaldi	33
Sitapur	45
Sitarampur	44
Sivasamudram Power Works	38
Solon	28
Somastipore	44
Sonepore	44
Srimangal	44
Subathu	28
Sukkur	39
Surat	31
Sutna	33
T	
Takdah	44
Tank	29
Tanjore	38
Taunggyi	46
Tavoy	47
Thal	29
Thayetmyo	47
Tittaghur	44
Tonk	33
Trichinopoly	38
Trimulgherry	37
Trivandrum	
U	
Upper Topa	22
Uttarshisha	23
V	
Vizianagram	38
W	
Wana	29
Wellington	38
Whitefield	38
Y	
Yamethin	46
Yenangyaung	45
Yercaud	
Z	
Zam	

ABBREVIATED TELEGRAPHIC ADDRESSES.

Full address.	Abbreviated address.
Adjutant	Adjt.
Adjutant-General in India	A. G.
Administrative Staff Officer of the Brigade	Adminstaff Bde.
Administrative Staff Officer of the Division	Adminstaff Dvn.
Aide-de-Camp	A. D. C.
Aide-de-Camp to His Excellency the Commander-in-Chief	A. D. C. Chief.
Aide-de-Camp to His Excellency the Viceroy	A. D. C. Viceroy.
Aide-de-Camp to the General Officer Commanding	A. D. C. Gl. Comdg.
Aide-de-Camp to the General Officer Commanding Northern Command	A. D. C. Northern.
Aide-de-Camp to the General Officer Commanding Southern Command	A. D. C. Southern.
Aide-de-Camp to the Lieutenant-Governor, Burma	A. D. C. Burma.
Assistant Adjutant General	A. A. G.
Assistant Adjutant General, Record Section, New Custom House, Bombay	Records, Bombay.
Assistant Commanding Royal Engineer	Ascomenger.
Assistant Commanding Royal Engineer, Chaklala District	Chaklenger.
Assistant Controller of Wheat Products, Army Headquarters, India	Conatta.
Assistant Controller of Wood, fuel, Army Headquarters, India	Conwood.
Assistant Director of Aeronautics	Aeronautics.
Assistant Director, Dairy Farms, Northern Circle	Nordairies.
Assistant Director, Dairy Farms, Southern Circle	Soudairies.
Assistant Director of Grass Farms	Adfodder.
Assistant Director, Medical Services (British Service)	A. D. M. S. India.
Assistant Director, Medical Services (Indian Service)	
Assistant Director, or Deputy Director, Medical Services (Division)	Divmed.
Assistant Director, Medical Services (Brigade)	Brigmed.
Assistant Director of Ordnance Services, Northern Command	Ordnance, Northern.
Assistant Director of Ordnance Stores	Arsenal.
Assistant Director of Supplies and Transport	Suptrans (Divison).
Assistant Director of Supplies and Transport (Embarkation), Bombay	Suptrans Embarkation, Bombay.
Assistant Director of Supplies and Transport (Embarkation), Karachi	Suptrans Embarkation, Karachi.
Assistant Military Secretary to the Commander-in-Chief	A. M. S. Chief.
Assistant Military Secretary, Northern Command	A. M. S. Northern.
Assistant Military Secretary, Southern Command	A. M. S. Southern.
Assistant Quarter Master General	A. Q. M. G.
Assistant Secretary, Government of India, Army Department	Asst. Army.
Brigade Major	Bde. Major.

Full address.	Abbreviated address.
Brigade Major, Royal Artillery, Army Head Quarters	Bde. Major Arty.
Brigadier-General	Brigr.
Brigade Supply Officer	Supplies (Brigade).
British Units.	
British Reserve Cavalry Regiment (India), Meerut	Rescav, Meerut.
No. 1 Reserve Battalion (India), Bangalore	One Resban, Bangalore.
No. 2 Reserve Battalion (India), Bangalore	Two Resban, Bangalore.
No. 3 Reserve Battalion (India), Bangalore	Three Resban, Bangalore.
No. 4 Reserve Battalion (India), Bangalore	Four Resban, Bangalore.
No. 5 Reserve Battalion (India), Kirkee	Five Resban, Kirkee.
No. 6 Reserve Battalion (India), Kirkee	Six Resban, Kirkee.
No. 7 Reserve Battalion (India), Belgaum	Seven Resban, Belgaum.
No. 8 Reserve Battalion (India), Secunderabad	Eight Resban, Secunderabad.
No. 1 Special Bullock Battery, R. A., Allahabad	Bullock Battery, Allahabad.
British Malaria Convalescent Depôt, Dagshai	Malarcon.
Cantonment Magistrate	C. M.
Captain	Capt.
Casualty Branch, War Office	Proelicas Ave, London.
Chief Auditor Non-Military Pay and Pension Accounts, Combined Labour Depôt, Bombay	Calabour.
Chief Engineer, Northern Command / Southern Command	Chiefenger.
Chief of the General Staff	Chief Genstaff.
Chief Inspector of General Stores	Cigstor.
Colonel	Col.
Commandant, No. 1 Armoured Motor Brigade	Motorbat.
Commandant, Armoured Motor Instructional Classes	Motorclas.
Commandant	Comdt.
Commandant, Cavalry School, Saugor	Riding.
Commandant, Central School of Musketry	Musketry.
Commandants, Branch Schools of Musketry	Musketry.
Commandant, Central Bombing School, Mhow	Bombs.
Commandant, Cadet College, Quetta	Cadets, Quetta.
Commandant, Cadet College, Wellington	Cadets, Wellington.
Commandant, Depôt, Labour Cos. for France	Franzlab.
Commandant, Officers' School of Instruction, Ambala	Cadets, Ambala.
Commandant, Officer Cadet Unit, Sialkote	Cadets, Sialkote.
Commandant, Officers' School of Instruction, Bangalore	Cadets, Bangalore.
Commandant, Training School for Indian Cadets	Cadets, Indore.
Commandant, Indian Central Flying School, Sitapur	Flying.

Full address.	Abbreviated address.
Commandant, Indian Convalescent Section	Consecind.
Commandant, Machine Gun Training Centre	Maxima.
Commandant, Mechanical Transport Training School, Sohan Camp, R'Pindi	Oktrans.
Commandant, Non-Commd. Officers' School, Secunderabad	Amalgam.
Commandant, Porter Corps Depôt	Porcorps.
Commandant, Prisoners of War Camp	Prisoners.
Commandant, Works Directorate Depôt, Railway Camp, Dadar, Bombay	Workdep.
Commanding Royal Artillery	Comdg. Arty.
" Royal Engineer	Comenger.
Contractor for Printing Government of India Stock Forms.	Contractor.
Controller of Contracts, Army Headquarters, India	Concon.
Controller of Military Accounts	Cr. Mily.
Controller of Military Supply Accounts	Cr. Mily. Supply.
Controller of Printing, Stationery and Stamps, India, Calcutta	Pleximon.
Deputy Adjutant General	D. A. G.
Deputy Assistant Adjutant General Musketry, I. S. Troops	Metford.
Deputy Assistant Adjutant General	D. A. A. G.
Deputy Assistant Director, Medical Services (Sanitary)	Medsan.
Deputy Assistant Director, Medical Services (Mobilization)	Medmob.
Deputy Assistant Director of Supplies and Transport	Suptrans (Brigade).
Deputy Assistant Quarter Master General	D. A. Q. M. G.
Deputy Controller of Military Accounts	Dy. Cr. Mily.
Deputy Judge Advocate General	D. J. A. G.
Deputy Judge Advocate General, Northern Command	D. J. A. G., Northern.
Deputy Judge Advocate General, Southern Command	D. J. A. G., Southern
Deputy Quarter Master General	D. Q. M. G.
Deputy Director of Equipment, Army Head-Quarters	Equipment.
Deputy Director, Medical Services in India	Dy. D. M. S. India.
Deputy Director, or Assistant Director, Medical Services (Division)	Divmed.
Deputy Director, Medical Services, Northern Command	Northmed.
Deputy Director, Medical Services, Southern Command	Southmed.
Director, Medical Services in India	Medical, India.
Deputy Director of Stocks, Supply and Transport Corps	Stocks.
Deputy Director of Veterinary Services, Northern Command	Veterinary, Northern.
Director General, Army Remount Department	D. G. Remounts.
Director General of Indian Medical Service	Hospitals, India.
Director General of Military Works	Mily. Works.
Director General of Ordnance in India	Ordnance, India.
Director of Artillery, Army Headquarters	Dirart.
Director of Clothing	Director, Clothing.
Director of Military Farms	Dir. Farms.
Director of Ordnance Factories	Ordnance Factories.
Director of Ordnance Inspection	Ordnance Inspection.
Director of Ordnance Stores	Ordnance Stores.
Director of Veterinary Services in India	Vety. India.
Director of Supplies and Transport, Army Headquarters, India	Suptrans, India.
Deputy Director of Supplies and Transport, Northern Command	Suptrans, Northern.
Deputy Director of Supplies and Transport, Southern Command	Suptrans, Southern.
Divisional Commanding Royal Engineer	Divl. C. R. E.
Divisional Disbursing Officer, 6th (Poona) Division, Force "B" Section	Disbursing, 6th Dn Force "B," Poona.
Divisional Disbursing Officer, 6th (Poona) Division, Aden Section	Disbursing, 6th Dn., Aden, Poona.
Divisional Disbursing Officer, 6th (Poona) Division	Disbursing, 6th Division, Poona.
Divisional Troops Supply Officer	Supplies, (Brigade).
Electrical Garrison Engineer	Elgarenger.
Examiner of Accounts, Military Works Services	Correct.
Examiner of Ordnance Factory Accounts in India	Exr. Factories.
Garrison Engineer	Garenger.
General	Gl.
General Officer Commanding Brigade	Genl. Bde.
General Officer Commanding Cavalry Brigade	General Cavalry Brigade.
General Officer Commanding Division	Genl. Division.
General Officer Commanding Divisional Area	Area.
General Officer Commanding the Troops in Ceylon	General, Colombo.
General Officer Commanding Infantry Brigade	General Infantry.
General Officer Commanding Northern Command	General Northern.
General Officer Commanding Southern Command	G. O. C. Southern.
General Staff of the Command, Northern	Genstaff, Northern.
General Staff of the Command, Southern	Genstaff, Southern.
General Staff of the Brigade	Genstaff Bde.
General Staff of the Division	Genstaff Dvn.
Governor General of India	Viceroy.
His Excellency the Commander-in-Chief in India	Chief.
Hospital, Sixty-five British General—Mesopotamia	Gensifif.

Indian Defence Force.

Full address.	Abbreviated address.
1st Bihar Light Horse (I. D. F.)	1st Horse Defence.
2nd Surma Valley Light Horse (I. D. F.)	2nd Horse Defence.
3rd Calcutta Light Horse (I. D. F.)	3rd Horse Defence.
4th Bombay Light Horse (I. D. F.)	4th Horse Defence.
5th Punjab Light Horse (I. D. F.)	5th Horse Defence.
6th Assam Valley Light Horse (I. D. F.)	6th Horse Defence.

Full address.	Abbreviated address.
7th (S) United Provinces Horse (I. D. F.)	7th Horse Defence.
8th (N) United Provinces Horse (I. D. F.)	8th Horse Defence.
9th Northern Bengal Mounted Rifles (I. D. F.)	9th Mounted Rifles Defence.
10th Southern Provinces Mounted Rifles (I. D. F.) .	10th Mounted Rifles Defence.

Artillery and Electrical Engineers.

Full address.	Abbreviated address.
1st (Calcutta Port Defence) Group, Garrison Artillery (I. D. F.) . . . . Nos. 1 and 2 Companies, Garrison Artillery (I. D. F.) . No. 1 Company, Electrical Engineers (I. D. F.) . . .	1st Garrison Artillery Defence.
2nd (Madras) Group Garrison Artillery (I. D. F.) "The Duke's Own" . . . . . . Nos. 3, 4 and 5 Companies, Garrison Artillery (I. D. F.) . No. 5 Company, Electrical Engineers (I. D. F.) . . .	2nd Garrison Artillery Defence.
3rd (Rangoon Port Defence) Group, Garrison Artillery (I. D. F.) . . . . Nos. 6, 7, 8 and 9 Companies, Garrison Artillery (I. D. F.) . No. 2 Company, Electrical Engineers (I. D. F.) . . .	3rd Garrison Artillery Defence.
4th (Bombay) Group, Garrison Artillery (I. D. F.) . . . Nos. 10, 11 and 12 Companies, Garrison Artillery (I. D. F.) .	4th Garrison Artillery Defence.
No. 3 Company, Electrical Engineers (I. D. F.) . . .	3rd Electrical Defence.
No. 13 (Ahmedabad) Company, Garrison Artillery (I. D. F.) .	13th Company Garrison Artillery Defence.
F. (Kirkee) Battery, Mobile Artillery (I. D. F.) . . .	F. Mobile Battery Defence.
1st (Cossipore) Brigade, Mobile Artillery (I. D. F.) . . . "A" Battery, Mobile Artillery (I. D. F.) . . . . "B" Battery, Mobile Artillery (I. D. F.) . . . . "C" Battery, Mobile Artillery (I. D. F.) . . . . "D" Battery, Mobile Artillery (I. D. F.)	1st Mobile Artillery Defence.
2nd Karachi Brigade, Mobile Artillery (I. D. F.) . . . E. Battery, Mobile Artillery (I. D. F.) . . . . No. 4 Company, Electrical Engineers (I. D. F.) . . .	2nd Mobile Artillery Defence.
5th (Lucknow) Group, Garrison Artillery (I. D. F.) . . . Nos. 14 and 15 Companies, Garrison Artillery (I. D. F.) .	5th Garrison Artillery Defence.

Infantry.

Full address.	Abbreviated address.
1st Madras Guards (I. D. F.) .	1st Guards Defence.
2nd Nagpur Rifles (I. D. F.) .	2nd Rifles Defence.
3rd Punjab Rifles (I. D. F.) .	3rd Rifles Defence.
4th Simla Rifles (I. D. F.) . .	4th Rifles Defence.
5th Calcutta Battalion (I. D. F.)	5th Battalion Defence.
6th Bangalore, Coorg and Mysore Battalion (I. D. F.) .	6th Battalion Defence.

Full address.	Abbreviated address.
7th East Indian Railway Battalion (I. D. F.) . . .	7th Battalion Defence.
8th Allahabad Rifles (I. D. F.) .	8th Rifles Defence.
9th Mussoorie Battalion (I. D. F.)	9th Battalion Defence.
10th (Naini Tal) Company . .	10th Company Defence.
12th Eastern Bengal Railway Battalion (I. D. F.) . . .	12th Battalion Defence.
1-13th Great Indian Peninsula Railway Battalion (I. D. F.) .	1-13th Battalion Defence.
2-13th Great Indian Peninsula Railway Battalion (I. D. F.) .	2-13th Battalion Defence.
14th Tennaserim Battalion (I. D. F.)	14th Battalion Defence.
15th Bombay Battalion (I. D. F.)	15th Battalion Defence.
16th Cawnpore Rifles (I. D. F.)	16th Rifles Defence.
1-17th Bombay, Baroda and Central India Railway Battalion (I. D. F.)	1-17th Battalion Defence.
2-17th Bombay, Baroda and Central India Railway Battalion (I. D. F.)	2-17th Battalion Defence.
18th Rangoon Battalion (I. D. F.)	18th Battalion Defence.
19th Agra Company (I. D. F.) .	19th Company Defence.
20th West Coast Battalion (I. D. F.)	20th Battalion Defence.
21st Burma Railways Battalion (I. D. F.)	21st Battalion Defence.
22nd Bengal and North-Western Railway Battalion (I. D. F.)	22nd Battalion Defence.
23rd Sind Battalion (I. D. F.) .	23rd Battalion Defence.
1-24th North Western Railway Battalion (I. D. F.) . . .	1-24th Battalion Defence.
2-24th North Western Railway Battalion (I. D. F.) . . .	2-24th Battalion Defence.
25th Oudh and Rohilkhand Railway Battalion (I. D. F.) .	25th Battalion Defence.
26th Hyderabad Rifles (I. D. F.)	26th Rifles Defence.
27th Baluchistan Company (I. D. F.)	27th Company Defence.
28th South Andaman Company (I. D. F.)	28th Company Defence.
29th South Indian Railway Battalion (I. D. F.) . . .	29th Battalion Defence.
1-32nd Madras and Southern Mahratta Railway Rifles (I. D. F.)	1-32nd Rifles Defence.
2-32nd Madras and Southern Mahratta Railway Rifles (I. D. F.)	2-32nd Rifles Defence.
34th Upper Burma Battalion (I. D. F.)	34th Battalion Defence.
35th Poona Battalion (I. D. F.)	35th Battalion Defence.

Full address.	Abbreviated address.
1-36th Bengal-Nagpur Railway Battalion (I. D. F.) . . .	1-36th Battalion Defence.
2-36th Bengal-Nagpur Railway Battalion (I. D. F.) . . .	2-36th Battalion Defence.
7th Calcutta Presidency Battalion (I. D. F.)	37th Battalion Defence.
38th East Coast Battalion (I. D. F.)	38th Battalion Defence.
39th Chota Nagpur Regiment (I. D. F.)	39th Regiment Defence.
41st Eastern Bengal Company (I. D. F.)	41st Company Defence.
42nd Assam-Bengal Railway Battalion (I. D. F.) . . .	42nd Battalion Defence.
43rd Kolar Goldfields Battalion (I. D. F.)	43rd Battalion Defence.
44th Calcutta Scottish (I. D. F.)	44th Scottish Defence.
45th Aden Rifles (I. D. F.) . . .	45th Rifles Defence.
Inspecting Officer, Frontier Corps, North-West Frontier Province	Frontier Corps.
Inspecting Veterinary Officer .	I. V. O.
Inspecting Officer, Punjab Cavalry	Ghorchura.
Inspecting Officer, Punjab Infantry	Sikh.
Inspecting Officer, Sappers .	Grenade.
Inspecting Officer, Kashmir Artillery	Golandaz.
Inspecting Officer, Kashmir Infantry	Dogra.
Inspecting Officer, Rajputana Cavalry	Rathor.
Inspecting Officer, Rajputana Infantry	Rajput.
Inspecting Officer, Camel Corps	Shutar.
Inspecting Officer, Central India Cavalry	Pindhari.
Inspecting Officer, Kathiawar Cavalry	Kathi.
Inspecting Officer, Hyderabad and Mysore Cavalry	Deccani.
Inspector General, Imperial Service Troops	Imperial.
Inspector of Cavalry in India .	Cavalry.
Inspector of General Stores .	Ilantex.
Inspector of Gymnasia, Northern and Southern Command	Physical.
Inspector of Harness and Saddlery	Inspecsad.
Inspector of Mechanical Transport Vehicles, North-Western Circle, Peshawar	Inspectrans.
Inspector of Mechanical Transport Vehicles, Central Circle, Rawalpindi	Inspectrans.
Inspector of Mechanical Transport Vehicles, Southern Circle, Poona	Inspectrans.
Inspector of Royal Engineers and Pioneers	Enpy.
Inspector of Royal Garrison Artillery in India	Guncase.
Inspector of Royal Horse and Royal Field Artillery in India.	Gunlock.
Inspector of Infantry, North .	Infantry, North.
Inspector of Infantry, South .	Infantry, South.
Inspector of Stocks, Supply and Transport Corps, North-Western Area	Instocks.
Inspector of Stocks, Supply and Transport Corps, North-Eastern Area	Instocks.
Inspector of Stocks, Supply and Transport Corps, Southern Area	Instocks.
Inspector of Supply and Transport Services, Northern Command	Supply, Northern.
Inspector of Volunteers . . .	Volunteers.
Inspector of Armoured Motors .	Motorspec.
Inspector of Ammunition (No. 2 charge)	Inspector, Ammunition.
Inspector of Army Schools, 1st Circle	Army Schools, 1st Circle.
Inspector of Army Schools, 2nd Circle	Army Schools, 2nd Circle.
Inspector of Army Schools, 3rd Circle	Army Schools, 3rd Circle.
Inspector of Army Schools, 4th Circle	Army Schools, 4th Circle.
Inspector of Army Schools, 5th Circle	Army Schools, 5th Circle.
Inspector of Army Schools, 6th Circle	Army Schools, 6th Circle.
Inspector of Army Signals . .	Signalling.
Inspector of Guns and Rifles (No. 1 charge)	Inspector Guns.
Inspector of Guns, Carriages and Vehicles (No. 3 charge)	Inspector, Carriages.
Inspector of Ordnance Machinery	I. O. M.
Inspector of Signalling, I. S. Troops	Marconi.
Inspector of Territorial Artillery, "Northern Group" . . .	Artillery, Northern.
Inspector of Territorial Artillery, "Southern Group" . . .	Artillery, Southern.
Judge Advocate General in India	J. A. G.
Karachi Embarkation Commandant	Embn., Kiamar Karachi.
57th Kite Balloon Section, Aden	Balloon.
Lieutenant	Lt.
Managers of Military Grass Farms	Grass.
Managers of Military Dairy Farms	Mildairy.
Mechanical Transport.	
Mechanical Transport Coy. (No. 3), Bombay	Baytrans.
No. 5 Mechanical Transport Company	Cinqtrans.
No. 6 Mechanical Transport Company (No. 2 Motor Ambulance Convoy)	Vultrans.
1st Advanced Repair Workshops, Peshawar	Reptrans, Peshawar.
Central Mechanical Transport Workshops Section, Rawalpindi	Reptrans, Rawalpindi.
Inspector of Mechanical Transport Vehicles, North-Western Circle, Peshawar	Inspectrans.
Inspector of Mechanical Transport Vehicles, Central Circle, Rawalpindi	Inspectrans.
Inspector of Mechanical Transport Vehicles, Southern Circle, Poona	Inspectrans.
Medical Storekeeper to Government	Stores Med.
Military Accountant General	Accts. Mily.
Military Deputy Auditor General, Northern Command.	Auditor, Northern.
Military Deputy Auditor General, Southern Command.	Auditor, Southern.

Full address.	Abbreviated address.
Military Secretary to the Governor of Bengal . . .	M. S. G., Bengal.
Military Secretary to the Governor of Bombay . . .	M. S. G., Bombay.
Military Secretary to the Governor of Madras . .	M. S. G., Madras.
Military Secretary to His Excellency the Commander-in-Chief	M. S. Chief.
Military Secretary to His Excellency the Viceroy . . .	M. S. V.
Military Secretary to Resident, Hyderabad	M. S. Res.
Northern Command School of Signalling, Kasauli . . .	Northsigs.
Officer	Offr.
Officer Commanding . . .	O. C.
Officer Commanding, Aircraft Park, Royal Flying Corps .	Airpark
Officer Commanding Armoured Car Batteries	Armbat.
Officer Commanding, Base Postal Depôt, Bombay . .	Bapode.
Officer Commanding Indian Convalescent Section, Karachi	Convalescent.
Officer Commanding, Labour Corps Depôt	Labcorps.
Officer Commanding Field Veterinary Hospital	Field Vety. Hospital.
Officer Commanding, Royal Engineer Training Company (India), Bangalore . . .	Retaining.
Officer Commanding Station Hospital	O. C. Hospital.
Officer Commanding Station Veterinary Hospital . . .	Vety. Hospital.
Officer Commanding Viceroy's Body-Guard	O. C., Body-Guard.
Officer Commanding Supply Depôt Company	Sudecoy.
Officer Commanding Mule Training Depôt	Trandep.
Officer Commanding R. A. M. C. Depôt, Deolali . . .	Ramdep.
Officer in Charge "D" Section Grain Depôt, Mazagaon . .	Hay Depôt (Bombay)
Officer in Charge Detachment, South Persia Mechanical Transport Section, Bander Abbas	Spertrans, Bander Abbas.
Officer in Charge Fuel Depôt, Murmugao	Fudep, Murmugao.
Officer in Charge Military Food Laboratory, Kasauli . . .	Food, Kasauli.
Officer in Charge Supply Depôt	Supdep.
Officer in Charge Veterinary Store Depôt	Vety. Stores.
Postal Disbursing Officer, Staff, Poona	Disbursing, Staff, Poona.
Postal Disbursing Officer, B. T., Poona	Disbursing, B. T., Poona.
Postal Disbursing Officer, 13th Division, Poona	Disbursing, 13th Dn. Poona.
Postal Disbursing Officer, R. E. and Medical, Poona . . .	Disbursing, R. E., Poona.
Postal Disbursing Officer, I. T., Poona	Disbursing, I. T., Poona.
Postal Disbursing Officer, Force "A," Poona	Disbursing, "A," Poona.
Postal Disbursing Officer, Force E., Poona	Disbursing, "E," Poona.
Quarter Master	Q. M.
Quarter Master General in India	Q. M. G.
Railway Staff Officer (at) . .	Reclusus.
Railway Transport Officer . .	R. T. O.
Recruiting Officer	Recruiting.
Regiment	Regt.
Rice Commissioner for India .	Ricom, Rangoon.
Royal Artillery	R. A.
Royal Flying Corps, Aden .	Aviation.
Sanitary Officer (of a Division)	Sany. Offr. Dn.
Sanitary Officer, Burma .	Sany. Offr., Burma.
Secretary and Executive Officer, Provincial Recruitment Board, Bihar and Orissa . . .	Recsec.
Secretary to the Defence Committee	Defences.
Secretary to the Government of India, Army Department .	Army.
Secretary, War Section, General Staff Branch, Army Head Quarters, India . . .	War Section.
Senior Medical Officer (at) .	Senior Medical.
Senior Veterinary Officer of Division	Senior Vety.
Shipping Officer for Engineer and Munition Board Stores, Bombay	Muniship.
Signal Company Commander, Group I	Signals, 2nd Divn.
Signal Company Commander, Group II	Signals, 3rd Divn.
Signal Company Commander, Group III	Signals, 4th Divn.
Signal Company Commander, Group IV	Signals, 6th Divn.
Southern Command, School of Signalling	Southsigs.
Station Staff Officer . . .	Stastaff.
Sub-Divisional Officer, Military Works Services	Essdyo.
Superintendent, Ammunition Factory	Ammunition.
Superintendent, Army Clothing Factory	Clothing, Alipore.
Superintendent, Army Clothing Factory, Madras . . .	Clothing, Madras.
Superintendent, Army Remount Department, Baluchistan Circle	Studs, Baluchistan.
Superintendent, Army Remount Department, Chenab Canal Colony Circle	Studs, Chenab.
Superintendent, Army Remount Department, Deccan Circle .	Studs, Deccan.
Superintendent, Army Remount Department, Jhelum C. C. Circle	Studs, Jhelum.
Superintendent, Army Remount Department, Lahore Circle .	Studs, Lahore.
Superintendent, Army Remount Department, Rawal Pindi Circle	Studs, Rawal Pindi.
Superintendent, Army Remount Department, United Provinces Circle	Studs, U. P.
Superintendent of Government Cattle Farm	Farm.
Superintendent, Government Printing, India, Calcutta . .	Printing.
Superintendent, Gun and Shell Factory, Cossipore . . .	Shell.
Superintendent, Gun Carriage Factory	Carriage.
Superintendent of Gymnasia .	Supdt., Gymnasia.
Superintendent, Harness and Saddlery Factory . .	Harness.
Superintendent of Recruitment, Adjutant General's Branch, Army Head Quarters, India.	Personnel.
Superintendent, Remount Depôt	Remounts.
Superintendent, Rifle Factory .	Rifle.
Supply and Transport Officer on Special Duty in Kashmir	Supply, Kashmir.
Surveyor General of India . .	Surveys.

Abbreviated address.	Full address.
A. A. G.	Assistant Adjutant General.
Records Bombay.	Assistant Adjutant General, Record Section, New Custom House, Bombay.
A. A. G. Arty.	Assistant Adjutant General for Royal Artillery.
Accts. Mily.	Military Accountant General.
A. D. C.	Aide-de-Camp.
A. D. C. Chief	Aide-de-Camp to His Excellency the Commander-in-Chief.
A. D. C. Gl. Comdg.	Aide-de-Camp to the General Officer Commanding.
A. D. C. Northern	Aide-de-Camp to the General Officer Commanding Northern Command.
A. D. C. Southern	Aide-de-Camp to the General Officer Commanding Southern Command.
A. D. C. Viceroy	Aide-de-Camp to His Excellency the Viceroy.
A. D. M. S. India	Assistant Director, Medical Services (British Service). Assistant Director, Medical Services (Indian Service).
Adjt.	Adjutant.
Adfodder	Assistant Director, Grass Farms.
Adminstaff Bde.	Administrative Staff Officer of the Brigade.
Adminstaff Divn.	Administrative Staff Officer of the Division.
A. G.	Adjutant General in India.
Airpark	Officer Commanding, Aircraft Park, Royal Flying Corps.
Amalgam	Commandant, Non-Commd. Officers' School, Secunderabad.
Ammunition	Superintendent, Ammunition Factory.
A. M. S. Chief	Assistant Military Secretary to Commander-in-Chief.
A. M. S. Northern	Assistant Military Secretary, Northern Command.
A. M. S. Southern	Assistant Military Secretary, Southern Command.
A. Q. M. G.	Assistant Quarter Master General.
Aeronautics	Assistant Director of Aeronautics.
Area	General Officer Commanding Divisional Area.
Armbat	Officer Commanding Armoured Car Batteries.
Army	Secretary to the Government of India, Army Department.
Army School, 1st Circle.	Inspector of Army Schools, 1st Circle.
Army School, 2nd Circle.	Inspector of Army Schools, 2nd Circle.
Army Schools, 3rd Circle.	Inspector of Army Schools, 3rd Circle.
Army Schools, 4th Circle.	Inspector of Army Schools, 4th Circle.
Army Schools, 5th Circle.	Inspector of Army Schools, 5th Circle.
Army Schools, 6th Circle.	Inspector of Army Schools, 6th Circle.
Arsenal	Assistant Director of Ordnance Stores.
Ascomenger	Assistant Commanding Royal Engineer.
Asst. Army	Assistant Secretary, Government of India, Army Department.
Asst. Trans.	Deputy Assistant Director of Transport.
Auditor, Northern	Military Deputy Auditor General, Northern Command.
Auditor, Southern	Military Deputy Auditor General, Southern Command.
Aviation	Royal Flying Corps, Aden.
Balloon	57th Kite Balloon Section, Aden.
Bapode	Officer Commanding Base Postal Depôt, Bombay.
Baytrans	No. 3 Mechanical Transport Coy., Bombay.

Abbreviated address.	Full address.
Bde. Major	Brigade Major.
Bombs	Commandant, Central Bombing School, Mhow.
Brigmed	Asst. Director, Medical Services Brigade.
Brigr.	Brigadier-General.
British	*Units.*
Rescav, Meerut	British Reserve Cavalry Regiment, (India), Meerut.
One Resban, Bangalore.	No. 1 Reserve Battalion (India), Bangalore.
Two Resban, Bangalore.	No. 2 Ditto Ditto.
Three Resban, Bangalore.	No. 3 Ditto Ditto.
Four Resban, Bangalore.	No. 4 Ditto Ditto.
Five Resban, Kirkee	No. 5 Reserve Battalion (India), Kirkee.
Six Resban, Kirkee.	No. 6 Ditto Ditto.
Seven Resban, Belgaum.	No. 7 Reserve Battalion (India), Belgaum.
Eight Resban, Secunderabad.	No. 8 Reserve Battalion (India), Secunderabad.
Bullock Battery, Allahabad.	No. 1 Special Bullock Battery, R. A., Allahabad.
Cadets, Indore,	Commandant, Training School for Indian Cadets, Indore.
„ Quetta	Commandant, Cadet College, Quetta.
„ Wellington	Commandant, Cadet College, Wellington.
„ Ambala	Commandant, Officers' School of Instruction, Ambala.
„ Sialkote	Commandant, Officer Cadet Unit, Sialkote.
„ Bangalore	Commandant, Officers' School of Instruction, Bangalore.
Calabour	Chief Auditor, Non-Military Pay and Pension Accounts, Combined Labour Depôt, Bombay.
Capt.	Captain.
Carriage	Superintendent, Gun Carriage Factory.
Cavalry	Inspector of Cavalry in India.
Chaklenger	Assistant Commanding Royal Engineer, Chaklala District.
Chief	His Excellency the Commander-in-Chief in India.
Chief Genstaff	Chief of the General Staff.
Chiefenger	Chief Engineer, Northern / Southern Command.
Cigstor	Chief Inspector of General Stores.
Cinqtrans	No. 5 Mechanical Transport Company.
Clothing, Alipore	Superintendent, Army Clothing Factory, Alipore.
Clothing, Madras	Superintendent, Army Clothing Factory, Madras.
C. M.	Cantonment Magistrate.
Col.	Colonel.
Comdg. Arty.	Commanding Royal Artillery.
Comenger	Commanding Royal Engineer.
Comdt.	Commandant.
Consecind.	Commandant, Indian Convalescent Section, Amballa.
Conatta	Assistant Controller of Wheat Products, Army Headquarters, India.
Concon	Controller of Contracts, Army Headquarters, India.
Contractor	Contractor for Printing Government of India Stock Forms.
Convalescent	Officer Commanding Indian Convalescent Section, Karachi.
Conwood	Assistant Controller of Wood, Fuel, Army Headquarters, India.

Abbreviated address.	Full address.
Cordite	Superintendent, Cordite Factory
Correct	Examiner of Accounts, Military Works Services.
Cr. Mily.	Controller of Military Accounts.
Cr. Mily. Supply	Controller of Military Supply Accounts, Central Section, Calcutta.
D. A. A. G.	Deputy Assistant Adjutant General.
D. A. G.	Deputy Adjutant General.
D. A. Q. M. G.	Deputy Assistant Quarter Master General.
Defences	Secretary to the Defence Committee.
D. G., Remounts	Director General, Army Remount Department.
Deccani	Inspecting Officer, Hyderabad and Mysore Cavalry.
Dirart	Director of Artillery, Army Headquarters.
Director, Clothing	Director of Clothing.
Disbursing, Staff, Poona.	Postal Disbursing Officer, Staff, Poona.
Disbursing, B. T., Poona.	Postal Disbursing Officer, B. T., Poona.
Disbursing, 13th Division, Poona.	Postal Disbursing Officer, 13th Division, Poona.
Disbursing, R. E., Poona.	Postal Disbursing Officer (R.E. and Medical), Poona.
Disbursing, I. T., Poona.	Postal Disbursing Officer, I.T., Poona.
Disbursing, "A," Poona.	Postal Disbursing Officer, Force "A," Poona.
Disbursing, "E," Poona.	Postal Disbursing Officer, "E," Poona.
Disbursing, 6th Division Force "B," Poona.	Divisional Disbursing Officer, 6th (Poona) Division, Force "B" Section.
Disbursing, Aden, Poona.	Divisional Disbursing Officer, 6th (Poona) Division, Aden Section.
Disbursing, 6th Division, Poona.	Divisional Disbursing Officer, 6th (Poona) Division.
Divl. C. R. E.	Divisional Commanding Royal Engineer.
Divmed	Depy. Director, or Asst. Director, Medical Services, Division.
D. J. A. G.	Deputy Judge Advocate General.
D. J. A. G. Northern	Deputy Judge Advocate General, Northern Command.
D. J. A. G. Southern	Deputy Judge Advocate General, Southern Command.
D. Q. M G.	Deputy Quarter Master General.
Dogra	Inspecting Officer, Kashmir Infantry.
Dtmt.	Detachment.
Dtr.	Director.
Dy. D. M. S. India	Deputy Director, Medical Services in India.
Electrical, Roorkee	Electrical Engineer, Military Works Services, Roorkee.
Elgarenger	Electrical Garrison Engineer.
Enpy	Inspector of Royal Engineers and Pioneers.
Equipment	Deputy Director of Equipment A. H. Qrs.
Essdyo	Sub-Divisional Officer, Military Woriks Services.
Exr., Factories	Examner of Ordnance Factory Accounts in India.
Farm	Superintendent, Government Cattle Farm.
Flying	Commandant Indian Central Flying School, Sitapur.
Food, Kasauli	Officer in Charge, Military Food Laborat ry, Kasauli.
Franzlab	Commandant, Depôt, Labour Corps for France.
Frontier Corps	Inspecting Officeer, Frontier Corps, North-West Frontier Province.
Fudep	Officer in charge Fuel Depôt, Murmugao.
Garenger	Garrison Engineer.
General Cavalry Brigade.	General Officer Commanding Cavalry Brigade.
General Infantr Brigade.	General Officer Commanding Infantry Brigade.
Genl. Bde.	General Officer Commanding Brigade.
Genl. Divn.	General Officer Commanding Division.
Gensifif	Sixty-five British General Hospital, Mesopotamia.
Genstaff, Northern	General Staff of the Command, Northern.
Genstaff, Southern	General Staff of the Command, Southern.
Genstaff Bde.	General Staff Officer of the Brigade.
Genstaff Dvn.	General Staff Officer of the Division.
Genl.	General.
Genl. Colombo	General Officer Commanding the Troops in Ceylon.
Genl. Northern	General Officer Commanding, Northern Commend.
Genl. Southern	General Officer Commanding, Southern Command.
Ghorchura	Inspecting Officer, Punjab Cavalry.
Golandaz	Inspecting Officer, Kashmir Artillery.
Grass	Managers of Military Grass Farms.
Grenade	Inspecting Officer, I. S. Sappers.
Guncase	Inspector, Royal Garrison Artillery in India.
Gunlock	Inspector, Royal Horse and Royal Field Artillery in India.
Harness	Superintendent, Harness and Saddlery Factory.
Hay Depôt	Officer in charge "D" Section, Gram Depôt, Mazagaon.
Hospitals, India	Director General of Indian Medical Service.

Indian Defence Force.

Light Horse and Mounted Rifles.

1st Horse Defence	1st Bihar Light Horse (I. D.F.).
2nd Horse Defence	2nd Surma Valley Light Horse (I. D. F.).
3rd Horse Defence	3rd Calcutta Light Horse (I. D. F.).
4th Horse Defence	4th Bombay Light Horse (I. D. F.).
5th Horse Defence	5th Punjab Light Horse (I. D. F.).
6th Horse Defence	6th Assam Valley Light Horse (I. D. F.).
7th Horse Defence	7th (S) United Provinces Horse (I. D. F.).
8th Horse Defence	8th (N) United Provinces Horse (I. D. F.).
9th Mounted Rifles Defence.	9th Northern Bengal Mounted Rifles (I. D. F.).
10th Mounted Rifles Defence.	10th Southern Provinces Mounted Rifles (I. D. F.).

Artillery and Electrical Engineers.

1st Garrison Artillery Defence.	1st (Calcutta Port Defence) Group, Garrison Artillery (I. D. F.). Nos. 1 and 2 Companies, Garrison Artillery (I. D. F.). No. 1 Company, Electrical Engineers (I. D. F.).

Abbreviated address.	Full address.
2nd Garrison Artillery Defence.	2nd (Madras) Group Garrison Artillery (I. D. F.), "The Duke's Own." Nos. 3, 4 and 5 Companies, Garrison Artillery (I. D. F.). No. 5 Company, Electrical Engineers (I. D. F.).
3rd Garrison Artillery Defence.	3rd (Rangoon Port Defence) Group, Garrison Artillery (I. D. F.). Nos. 6, 7, 8 and 9 Companies, Garrison Artillery (I. D. F.). No. 2 Company, Electrical Engineers (I. D. F.).
4th Garrison Artillery Defence.	4th (Bombay) Group, Garrison Artillery (I. D. F.). Nos. 10, 11 and 12 Companies, Garrison Artillery (I. D. F.).
3rd Electrical Defence	No. 3 Company, Electrical Engineers (I. D. F.).
13th Company Garrison Artillery Defence.	No. 13 (Ahmadabad) Company, Garrison Artillery (I. D. F.).
F. Mobile Battery Defence.	F. (Kirkee) Battery, Mobile Artillery (I. D. F.).
1st Mobile Battery Defence.	1st (Cossipore) Brigade, Mobile Artillery (I. D. F.). "A" Battery, Mobile Artillery (I. D. F.). "B" Battery, Mobile Artillery (I. D. F.). "C" Battery, Mobile Artillery (I. D. F.). "D" Battery, Mobile Artillery (I. D. F.).
2nd Mobile Artillery Defence.	2nd Karachi Brigade Mobile Artillery (I. D. F.). E. Battery, Mobile Artillery (I. D. F.). No. 4 Company, Electrical Engineers (I. D. F.).
5th Garrison Artillery Defence.	5th (Lucknow) Group Garrison Artillery (I. D. F.). Nos. 14 and 15 Companies, Garrison Artillery (I. D. F.).
	Infantry.
1st Guards Defence .	1st Madras Guards (I. D. F.) .
2nd Rifles Defence .	2nd Nagpur Rifles (I. D. F.).
3rd Rifles Defence .	3rd Punjab Rifles (I. D. F.).
4th Rifles Defence .	4th Simla Rifles (I. D. F.).
5th Battalion Defence	5th Calcutta Battalion (I. D. F.).
6th Battalion Defence	6th Bangalore, Coorg and Mysore Battalion (I. D. F.).
7th Battalion Defence	7th East Indian Railway Battalion (I. D. F.).
8th Rifles Defence .	8th Allahabad Rifles (I. D. F.).
9th Battalion Defence	9th Mussoorie Battalion (I. D. F.).
10th Company Defence.	10th (Naini Tal) Company.
12th Battalion Defence.	12th Eastern Bengal Railway Battalion (I. D. F.).
1-13th Battalion Defence.	1-13th Great Indian Peninsula Railway Battalion (I. D. F.).
2-13th Battalion Defence.	2-13th Great Indian Peninsula Railway Battalion (I. D. F.).
14th Battalion Defence.	14th Tennaserim Battalion (I. D. F.).
15th Battalion Defence.	15th Bombay Battalion (I.D.F.).
16th Rifles Defence	16th Cawnpore Rifles (I. D. F.).
1-17th Battalion Defence.	1-17th Bombay, Baroda and Central India Railway Battalion (I. D. F.).
2-17th Battalion Defence.	2-17th Bombay, Baroda and Central India Railway Battalion (I. D. F.).
18th Battalion Defence.	18th Rangoon Battalion (I. D. F.).
19th Company Defence.	19th Agra Company (I. D. F.).
20th Battalion Defence.	20th West Coast Battalion (I. D. F.).
21st Battalion Defence.	21st Burma Railways Battalion (I. D. F.).
22nd Battalion Defence.	22nd Bengal and North-Western Railway Battalion (I. D. F.).
23rd Battalion Defence.	23rd Sind Battalion (I. D. F.).
1-24th Battalion Defence.	1-24th North Western Railway Battalion (I. D. F.).
2-24th Battalion Defence.	2-24th North-Western Railway Battalion (I. D. F.).
25th Battalion Defence.	25th Oudh and Rohilkhand Railway Battalion (I. D. F.).
26th Rifles Defence .	26th Hyderabad Rifles (I. D. F.).
27th Company Defence.	27th Baluchistan Company (I. D. F.).
28th Company Defence.	28th South Andaman Company (I. D. F.).
29th Battalion Defence.	29th South Indian Railway Battalion (I. D. F.).
1-32nd Rifles Defence	1-32nd Madras and Southern Mahratta Railway Rifles (I. D. F.).
2-32nd Rifles Defence	2-32nd Madras and Southern Mahratta Railway Rifles (I. D. F.).
34th Battalion Defence.	34th Upper Burma Battalion (I. D. F.).
35th Battalion Defence.	35th Poona Battalion (I. D. F.).
1-36th Battalion Defence.	1-36th Bengal-Nagpur Railway Battalion (I. D. F.).
2-36th Battalion Defence.	2-36th Bengal-Nagpur Railway Battalion (I. D. F.).
37th Battalion Defence.	37th Calcutta Presidency Battalion (I. D. F.).
38th Battalion Defence.	38th East Coast Battalion (I. D. F.).
39th Regiment Defence.	39th Chota Nagpur Regiment (I. D. F.).
41st Company Defence.	41st Eastern Bengal Company (I. D. F.).
42nd Battalion Defence.	42nd Assam-Bengal Railway Battalion (I. D. F.).
43rd Battalion Defence.	43rd Kolar Goldfields Battalion (I. D. F.).
44th Scottish Defence	44th Calcutta Scottish (I. D. F.)
45th Rifles Defence .	45th Aden Rifles (I. D. F.).
Ilantex	Inspector of General Stores.
Physical	Inspector of Gymnasia, Northern and Southern Command.
Imperial	Inspector General, Imperial Service Troops.
Infantry, North . .	Inspector of Infantry, North.
Infantry, South . .	Inspector of Infantry, South.
Inspecsad . . .	Inspector of Harness and Saddlery.
Inspector, Ammunition	Inspector of Ammunition (No. 2 charge).
Inspector, Carriages	Inspector of Guns, Carriages and Vehicles (No. 3 charge).
Inspector, Guns . .	Inspector of Guns and Rifles (No. 1 charge).
Inspector, Miscellaneous	Assistant Inspector, Miscellaneous Stores.
Inspectrans . . .	Inspector of Mechanical Transport Vehicles, North-Western Circle, Peshawar.
Inspectrans . . .	Inspector of Mechanical Transport Vehicles, Central Circle, Rawalpindi.
Inspectrans . . .	Inspector of Mechanical Transport Vehicles, Southern Circle, Poona.
Instocks . . .	Inspector of Stocks, Supply and Transport Corps, North-Western Area. Inspector of Stocks, Supply and Transport Corps, North Eastern Area. Inspector of Stocks, Supply and Transport Corps, Southern Area,

Abbreviated address.	Full address.
I. O. M.	Inspector of Ordnance Machinery.
I. V. O.	Inspecting Veterinary Officer.
J. A. G.	Judge Advocate General in India.
Embn. Kiamari, Karachi.	Embarkation Commandant, Karachi.
Kathi	Inspecting Officer, Kathiawar Cavalry.
Labcorps	Officer Commanding Labour Corps Depôt.
Lt.	Lieutenant.
Malarcon	British Malaria Convalescent Depôt, Dagshai.
Mildairy	Managers of Military Dairy Farms.
Mily. Works	Director General of Military Works.
Motorbat	Commandant, No. 1 Armoured Motor Brigade.
Motorclas	Commandant, Armoured Motor Instructional Classes.
Motorspec	Inspector of Armoured Motors.
M. O.	Medical Officer (at).
M. S. Chief	Military Secretary to His Excellency the Commander-in-Chief.
M. S. G., Bengal	Military Secretary to the Governor of Bengal.
M. S. G., Bombay	Military Secretary to the Governor of Bombay.
M. S. G., Madras	Military Secretary to the Governor of Madras.
M. S. V.	Military Secretary to the Viceroy.
Marconi	Inspector of Signalling, I. S. Troops.
Maxima	Commandant, Machine Gun Training Centre.
Medical, India	Director, Medical Services in India.
Medmob	Depy. Asst. Director, Medical Services (Mobilization).
Medsan	Depy. Asst. Director, Medical Services (Sanitary).
Metford	Deputy Assistant Adjutant General, Musketry, I. S. Troops.
Mideforce	1st Madras Infantry. (I. D. F.)
Muniship	Shipping Officer for Engineer and Munition Board Stores, Bombay.
Musketry	Commandant, Central School of Musketry. Commandants, Branch Schools of Musketry.
Nordairies	Assistant Director, Dairy Farms, Northern Circle.
Northmed	Depy. Director, Medical Services, Northern Command.
Northsigs	Northern Command School of Signalling, Kasauli.
O. C.	Officer Commanding.
O. C., Body-Guard	Officer Commanding Viceroy's Body-Guard.
O. C., Hospital	Officer Commanding Station Hospital.
O. C., Stn.	Officer Commanding Station.
Offr.	Officer.
Oktrans	Commandant, Mechanical Transport Training School, Sohan Camp, Rawalpindi.
Ordnance Depôt	Warrant Officer in charge Ordnance Depôt.
Ordnance Factories	Director of Ordnance Factories
Ordnance India	Director General of Ordnance in India.
Ordnance Inspection	Director of Ordnance Inspection.
Ordnance, Northern	Assistant Director of Ordnance Services, Northern Command.
Ordnance Stores	Director of Ordnance Stores.
Personnel	Superintendent of Recruitment. Adjutant General's Branch. Army Head Quarters, India.
Pleximon	Controller of Printing, Stationery and Stamps, India, Calcutta.
Pindhari	Inspecting Officer, Central India Cavalry I. S. Troops.
Porcorps	Commandant, Porter Corps Depôt.
Printing	Superintendent Government Central Printing Office. Superintendent Government Printing, India, Calcutta.
Prisoners	Commandant, Prisoners of War Camp.
Proelicas Ave, London.	Casualty Branch, War Office.
Q. M.	Quarter Master.
Q. M. G.	Quarter Master General in India.
R. A.	Royal Artillery.
Rajput	Inspecting Officer, Rajputana Infantry.
Ramdep	Officer Commanding R. A. M. C. Depot Deolali.
Rathor	Inspecting Officer, Rajputana Cavalry.
Reclusus	Railway Staff Officer (at).
Recruiting	Recruiting Officer.
Recsec	Secretary and Executive Officer, Provincial Recruitment Board, Bihar and Orissa.
Regt.	Regiment.
Remounts	Superintendent, Remount Depôt.
Reptrans, Peshawar	1st Advanced Repair Workshops, Peshawar.
Reptrans, Rawalpindi	Central Mechanical Transport Workshops Section, Rawalpindi.
Retaining	Officer Commanding, Royal Engineer Training Company (India), Bangalore.
Ricom, Rangoon	Rice Commissioner for India.
Riding	Commandant, Cavalry School, Saugor.
Rifles	Superintendent, Rifle Factory.
R. O.	Registration Officer.
R. T. O.	Railway Transport Officer.
Sany. Offr. Dn.	Sanitary Officer (of a Division).
Senior Medical	Senior Medical Officer (at)
Senior Vety.	Senior Veterinary Officer of Division.
Shutar	Inspecting Officer, Camel Corps, I. S. Troops.
Signalling	Inspector of Army Signals.
Signals, 2nd Divn.	Signal Company Commander, Group I.
Signals, 3rd Divn.	Signal Company Commander, Group II.
Signals, 4th Divn.	Signal Company Commander, Group III.
Signals, 6th Divn.	Signal Company Commander, Group IV.
Sikh	Inspecting Officer, Punjab Infantry.
Soudairies	Assistant Director of Dairy Farms, Southern Circle.
Southmed	Deputy Director, Medical Services, Southern Command.
South Sigs	Southern Command, School of Signalling.
Spertrans, Bander Abbas.	Officer in Charge Detachment, South Persia Mechanical Transport Section, Bander Abbas.
Stastaff	Station Staff Officer.
Stocks	Deputy Director of Stocks, Supply and Transport Corps.
Stores Med.	Medical Storekeeper to Government.
Supdt., Gymnasia	Superintendent of Gymnasia.
Supply Kashmir	Supply and Transport Officer on special duty in Kashmir.

Abbreviated address.	Full Address.	Abbreviated address.	Full address.
Sudecoy	Officer Commanding Supply Depôt Company.	Vety. Hospital	Officer Commanding Station Veterinary Hospital.
Supdep	Officer in charge Supply Depôt.	Vety. India	Director of Veterinary Services in India.
Supplies (Brigade)	Brigade Supply Officer. Divisional Troops Supply Officer.	Veterinary, Northern	Deputy Director of Veterinary Services, Northern Command.
Suptrans Embarkation Bombay, Karachi.	Assistant Director of Supplies and Transport, Bombay. Assistant Director of Supplies and Transport, Karachi.	Vety. School	Principal, Army Veterinary School.
		Vety. Stores	Officer in charge Veterinary Store Depôt.
Suptrans (India)	Director of Supplies and Transport, Army Headquarters.	Viceroy	Governor General of India.
Suptrans (Northern)	Deputy Director of Supplies and Transport, Northern Command.	Vols.	Inspector of Volunteers in India.
		Vultrans	No. 6 Mechanical Transport Company. (No. 2 Motor Ambulance Convoy).
Suptrans (Southern)	Deputy Director of Supplies and Transport, Southern Command.	Warsection	Secretary, War Section, General Staff Branch, Army Head Quarters, India.
Suptrans (Division)	Assistant Director of Supplies and Transport.		
Suptrans (Brigade)	Deputy Assistant Director of Supplies and Transport.	Workdep	Commandant, Works Directorate Depôt, Railway Camp, Dadar, Bombay.
Surveys	Surveyor General of India.		
Trandep	Officer Commanding Mule Training Depôt.		

The KING-EMPEROR.

Field-Marshal . . . . 7 May 1910

COL.-IN-CHIEF, 1st Duke of York's Own Lancers (Skinner's Horse), 6th King Edward's Own Cavalry, 11th King Edward's Own Lancers (Probyn's Horse), 18th King George's Own Lancers, 26th King George's Own Light Cavalry, 38th and 39th King George's Own Central India Horse, 1st King George's Own Sappers and Miners, 14th King George's Own Ferozepore Sikhs, 61st King George's Own Pioneers, 102nd King Edward's Own Grenadiers, 130th King George's Own Baluchis (Jacob's Rifles), 1st King George's Own Gurkha Rifles (The Malaun Regiment), 2nd King Edward's Own Gurkha Rifles (The Sirmoor Rifles).

Personal Aide-de-Camp.

Field Marshal *His Royal Highness* Arthur William Patrick Albert, *Duke of* **Connaught and Strathearn**, K.G., K.T., K.P., G.C.B., G.C.S.I., G.C.M.G., G.C.I.E., G.C.V.O., G.B.E., *COL.-IN-CHIEF*, 13th Duke of Connaught's Lancers (Watson's Horse), 31st Duke of Connaught's Own Lancers, 7th Duke of Connaught's Own Rajputs, 129th Duke of Connaught's Own Baluchis, Supply and Transport Corps . . . . 26 May 76

Aides-de-Camp General.

Barrett, Genl. *Sir* A. A., G.C.B., K.C.S.I., K.C.V.O. . . . . 3 Nov. 17
Birdwood, Genl. *Sir* W. R., K.C.B., K.C.S.I., K.C.M.G., C.I.E., D.S.O., *q.s.* . . . . 3 Nov. 17
Monro, Genl. *Sir*, C. C., G.C.B., G.C.M.G., *p.s.c.* . . . . 1 July 18

Aides-de-Camp.

Goodwin, Col. F., C.I.E., V.D., 17th Bombay, Baroda and Central India Ry. Bn. . . . . . . 27 May 12
Edwards, Col. Fitz J. M., C.B., C.M.G., D.S.O., *p.s.c.* . . . . 1 Jan. 17
Jacob, Col. A. LeG., C.M.G., C.I.E., D.S.O. . . . . 3 June 18
Norman, Bt.-Col. (*temp. Brig. Genl.*) C. L., D.S.O., M.V.O., *p.s.c.* . . . . 3 June 18
Tarver, Bt.-Col. A. L., D.S.O., *p.s.c.* . . . . 3 June 18

Honorary Aides-de-Camp.

Hony. Lt.-Genl. *His Highness Sir* Madho Rao Scindia, *Bahadur, Maharaja* of **Gwalior**, G.C.S.I., G.C.V.O., G.B.E., 1st D. Y. O. Lrs. . . . . 29 Nov. 00
Hony. Lieut.-Genl. *His Highness Maharaja Bahadur Sir* **Pratap Singh** (of Jodhpur), G.C.B., G.C.S.I., G.C.V.O., 34th Horse . . . . 11 Jan. 02
Hony. Maj.-Genl. *His Highness Maharaja Sir* Ganga Singh, *Bahadur, Maharaja* of **Bikaner**, G.C.S.I., G.C.I.E., K.C.B., 2nd Lrs. . . . . 3 June 10
Hony. Col. *His Highness Nawab Sir* Muhammad Hamid Ali Khan, *Bahadur, Nawab* of **Rampur**, G.C.I.E., G.C.V.O., 9th Horse . . . . 3 June 10

Equerries to the King.

Wigram, Bt.-Lt.-Col. C., C.B., C.V.O., C.S.I., 18th Lrs., *p.s.c.* (*Asst. Pte. Secy.*) . . . . . . 10 June 10
Campbell, Lt.-Col. C. F., C.I.E., Ret. (*Gent.-at-Arms*) (*Extra*) . . . . 10 June 10
VC, **Probyn**, Genl. *Rt. Hon. Sir* D. M., G.C.B., G.C.S.I., G.C.V.O., I.S.O., 11th Lrs. (*Extra*) . 10 June 10
Watson, Col. (*tempy.* Maj.-Genl.) H. D., C.B., C.M.G., C.I.E., M.V.O. (*Extra*) . . . . 10 June 10

Honorary Physicians to the King.

McLeod, Hony. Col. K., M.D., ret. . . . . 2 May 06
Lyons, Maj.-Genl. R. W. S., M.D., I.M.S. . . . . 31 Dec. 14
VC **Crimmin**, Col. J., C.B., C.I.E., V.D. . . . . 8 Feb. 16
Edwards, *The Hon'ble* Maj.-Genl. W. R., C.B., C.M.G., M.D., I.M.S. [R] . . . . 18 Feb. 17

Honorary Surgeons to the King.

Cleghorn, Maj.-Genl. J., C.S.I., M.D., ret. . . . . 5 Oct. 98
Turnbull, Maj.-Genl. P. S., M.D., ret. . . . . 20 Sep. 02
Hooper, Hony. Col. *Sir* W. R., K.C.S.I., ret. . . . . 6 Aug. 04
Hendley, Maj.-Genl. H., M.D., I.M.S. . . . . 23 Sep. 15
Banatvala, *Hon'ble* Col. H. E., I.M.S. . . . . 22 Oct. 17

Honorary Chaplain to the King.

Wickins, *Ven. Archdeacon* W. J., M.A. . . . . 23 Aug. 12

VICEROY AND GOVERNOR-GENERAL OF INDIA.

His Excellency the Right Honourable FREDERIC JOHN NAPIER, *Baron* CHELMSFORD, P.C., G.M.S.I., G.C.M.G., G.M.I.E., G.B.E. . . 5 April 1916.

	Name, Rank and Corps.	Date of appointment.	Date of completion of tenure.
Private Secretary .	**Maffey**, J. L., Esq., C.I.E., Ind. Civil Service	5 Apl. 16	4 Apl. 21(*c*)
Asst. Private Secretary .	**Mackenzie**, D. G., Esq., I.C.S.	1 Nov. 18	. . . .
Military Secretary	**Verney**, Maj. (*temp. Lt.-Col.*) R., Rif. Brig.	5 Apl. 16	4 Apl. 21(*c*)
Persl. Asst. . . .	**Parsons**, R., Esq.	9 Jan. 09	
Comptroller of the Household .	**Mackenzie**, Maj. J., C.I.E., 35th Sikhs	15 Mar. 07(*a*)	31 Mar. 19
Aides-de-Camp . .	**Denny**, Lt. J. A., Gr. Gds. (Spec. Res.)	1 Apl. 17	4 Apl. 21(*c*)
	Carnegie, Capt. C. A., *Lord*, Scots Gds.	16 June 17	4 Apl. 21(*c*)
	Alexander, Maj. R. D., 3rd G. R.	25 Feb. 18	4 Apl. 21(*c*)
	Gordon-Ives, Lt. C. M. G., Scots Gds.	......	.....
	Muhi-ud-din Khan, Risaldar-Maj., C.I.E., *Sardar Bahadur*, 31st Lrs.	19 Apl. 14	5 Dec. 18
	Muhammad Akbar Ali Khan, Risaldar-Maj., *Sardar Bahadur*, 7th Lrs.	1 Dec. 17	
Surgeon	**Smith**, Lt.-Col. H. A., C.I.E., M.B., I. M. S.	9 Oct. 16	4 Apl. 21 (*c*)
Honorary Aides-de-Camp.	**Warburton**, Lt.-Col. (*Hony. Col., late Ind. Volrs.*) H. G., 5th Luck. Gp. Garr. Arty.	14 Nov. 13(*a*)	13 Nov. 18
	Pugh, Lt.-Col. (*Hony. Col., late Ind. Volrs*) A. J., V.D., O.B.E., 3rd Calcutta L. H.	1 Jan. 14(*a*)	31 Dec. 18
	Reed, Maj. (*Hony. Lt. Col., late Ind. Volrs.*) *Sir* H. Stanley, *Kt.*, 4th Bo. L. H.	2 May. 14(*a*)	1 May 19
	Clark, Lt.-Col. (*Hony. Col., late Ind. Volrs.*) A. M., V.D., 2-36th B. N. Rly. Bn.	5 Apl. 16	4 Apl. 21(*c*)
	Evans, Lt.-Col. (*Hony. Col., late Ind. Volrs.*) G. H., C.I.E., 3rd Rangoon P. D. Gp. Garr. Arty.	5 Apl. 16	4 Apl. 21(*c*)
	Lumsden, Commodore W., C.V.O., C.I.E., R.N.	29 Aug. 16	4 Apl. 21(*c*)
	Mitchell, Lt.-Col. (*Hony. Col., late Ind Volrs.*) H. H. G., 2nd Madras Gp. Garr. Arty.	1 Feb. 17	31 Jan. 22
	Hickley, Lt.-Col. (*Hony. Col., late Ind. Volrs.*) V. N., C.I.E., V.D., 1st Bihar L. H.	7 July 17	6 July 22
	Constable, Lt.-Col. (*Hony. Col., late Ind. Volrs.*) E. A. (R. N., retd.), 1st Calcutta Port Defence Gp. Garr. Arty. . .	25 Nov. 17	24 Nov. 22
	Wright, Lt.-Col. (*Hony. Col., late Ind. Volrs.*) W. T., 3rd Punj. R.	9 Mar. 18	8 Mar. 23
	Grimston, Lt.-Col. L. A., V.D., 6th Assam V. L. H. . . .	22 Feb. 18	21 Feb. 23
	Rana Jodha Jung, *Bahadur*, M.C., Lt., 3rd Brahmans, Commdt., Tehri Imperial Service Sappers	5 Apl. 16	4 Apl. 21(*c*)
	Raja Sir **Hari Singh**, K.C.I.E., Hony. Capt. General in, and Commander-in-Chief of, the Jammu and Kashmir State Forces	18 Jan. 17	17 Jan. 22
	Ganpat Rao Raghunath Rajwade, Colonel in Gwalior State Army	18 Jan. 17	17 Jan. 22
	Haji Hafiz Obeidullah Khan, Hony. Major, *Nawabzada*, C.S.I., Brig.-General in, and Commander-in-Chief of, the Bhopal State Forces	1 Jan. 18	
	Wali Muhammad, Risaldar-Maj. (Hony. Capt.), *Sardar Bahadur, late* Govr.-Genl.'s Body-Guard	1 June 01(*b*)	
	Abdul Aziz, Risaldar-Maj. (Hony. Capt.), *Sardar Bahadur, late* 5th Cav.	9 Apl. 09(*b*)	
	Madho Singh Rana, Subadar-Major (Hony. Capt.), *Sardar Bahadur, late* 4th G.R.	10 June 10(*b*)	
	Abdul Karim Khan, Risaldar-Major (Hony Capt.), *Sardar Bahadur, late* Govr.-Genl.'s Body-Guard	19 Apl. 14(*b*)	
	Mit Singh, I.O.M., Subdr-Maj. (*Hony. Capt.*) *Sardar Bahadur*, late 53rd Sikhs	10 June 15(*b*)	
	Karm Singh, Risaldar-Maj., *Bahadur*, late 13th Lrs. . .	1 Dec. 17(*b*)	

(*a*) Date of re-appointment, 5th April 1916.
(*b*) For life.
(*c*) On termination of present Viceroyalty.

	Name, Rank and Corps.	Date of appointment.	Date of completion of tenure.
Honorary Surgeons.	**Treherne,** Maj.-Genl. *Sir* F. H., K.C.M.G., F.R.C.S., EDIN., Brit. Ser.	5 Nov. 13(*a*)	4 Nov. 18
	Matthews, Maj. E. A. C., D.S.O., M.B., I.M.S.	18 Mar. 14(*a*)	17 Mar. 19
	Granger, Lt.-Col. T. A., C.M.G., M. B., I.M.S.	16 Sep. 14(*a*)	15 Sep. 19
	Wanhill, Lt.-Col. C.F., R.A.M.C.	4 Feb. 15(*a*)	2 Feb. 20
	Rugg, Maj. G. F., R.A.M.C.	5 Apl. 16	4 Apl. 21
	MacIlwaine, Capt. A.G. J., C.I.E., R.A.M.C.	5 Apl. 16	4 Apl. 21
	Hugo, Lt.-Col. E. V., C.M.G., M.D., F.R.C.S., I.M.S.	8 May 16	7 May 21
	Molesworth, Col. W., C.I.E., C.B.E., M.B., I.M.S.	8 May 16	7 May 21
	Beyts, Bt.-Col. W. G., R.A.M.C.	14 Feb. 17	13 Feb. 22
	Heard, Bt.-Col. R., M.D., I.M.S.	1 Apl. 18	31 Mar. 23
	Smith, Col. J. B., C.B., M.B., I.M.S.	1 Oct. 18	30 Sep. 23
Honorary Assistant Surgeons.	**Sorabji, Fardunji Gandhi,** L.M. and S. (Bombay)	25 Dec. 13(*a*)	24 Dec. 18
	Pandit Balkishan Kaul, *Rai Bahadur* (Punjab)	16 Oct. 15(*a*)	15 Oct. 20
	Barada Kanta Ray, *Rai Bahadur* (Bihar & Orissa)	30 Sep. 15(*a*)	29 Sep. 20
	Rama Rao, *Rao Sahib* C. B., B.A., L.M.S., Madras M. D. (Dub.) (Madras)	5 Apl. 16	4 Apl. 21
	Satis Chandra De, *Rai Bahadur,* M.A., M.B. (Bengal)	5 Apl. 16	4 Apl. 21
	Isa Charan, Mr., L.R.C.P. and S. (Edin.), D.P.H. (Dub.), L.M. (Rot.) (Burma)	5 Apl. 16	4 Apl. 21
	Lakshmi Narayan Chaudhri, *Rai Bahadur,* L.M. and S. (Central Provinces)	5 Apl. 16	4 Apl. 21
	Rai Baij Nath Vyas, *Bahadur* (United Provinces)	27 July 18	26 July 23

(*a*) Date of re-appointment 5 April 1916.

ARMY AND MARINE DEPARTMENT, GOVERNMENT OF INDIA.

MEMBER OF COUNCIL IN CHARGE.

His Excellency General *Sir* Charles Carmichael Monro, G.C.M.G., K.C.B., A. D. C., *p.s.c., Commander-in-Chief in India.*

Extraordinary Member of the Council of the Governor General in Council 1 Oct. 1916

Secy.	**Bingley,** *The Hon'ble* Maj.-Genl. *Sir* A.H., K.C.I.E., C.B.	6 July 16
Dy. Secy.	**Spence,** Lt.-Col. (*temp. Col.*) A. H. O., C.I.E., 39th Horse [*l*]	12 Dec. 14
Dy. Secy. (*Addl.*)	**Sharp,** Maj. (*temp. Lt.-Col.*) A., C.M.G., S. & T. Corps [*l*]	1 Apl. 17
Asst. Secy.	*Whelan, A. A., Esq.* (On deputation)	*12 Dec. 14*
Asst. Secy.	**Ogilvie,** Maj. G. D., I.A.	5 Jan. 15
Asst. Secy.	**Gwynne,** 2nd-Lt. (*temp. Maj.*) C. W., I. A. R. O.	23 Feb. 18
Asst. Secy.	**Moore,** Maj. F. G., 6th Jat., L. I.	16 May 18
Asst. Secy. (*Addl.*)	**Marshall,** Capt. (*local Maj.*) G. W., I.S.O., V.D., 4th Simla Rif., I. D. F.	26 Mar. 17
Registrar	**Tharle-Hughes,** Lt. (*temp. Capt.*) R., M.B.E., I. A. R O.	12 Dec. 14
Officer in charge of Medal Distribution	**Debenham,** W. C., Esq.	20 Mar. 18

Dir.-Genl., Indian Medical Service	**Edwards,** The Hon'ble Maj.-Genl. W. R., C.B., C.M.G., M.D., I.M.S., K.H.P. [R]	8 Jan. 18
Dy. Dir.-Genl., Indian Medical Service	**Needham,** Maj. R. A. (*temp. Lt.-Col.*), D.S.O., M.B., I.M.S.	25 Oct. 18
Asst. ditto (*Stores*)	**Ross,** Maj. H. (*temp. Lt.-Col.*), O.B.E., M.B., F.R.C.S.I., I.M.S.	25 Apl. 18
Asst. ditto (*Sany.*)	**Stevenson,** Maj. W. D. H., C.I.E., M.D., I.M.S.	

Dir., R. I. M.	**Wilson,** Capt. N. F. J., C.M.G., C.B.E., R.I.M. (*offg.*)	12 June 17
Dy. Dir., R. I. M.	**Jones,** Capt. B. H., R.I.M. (*offg.*)	12 June 17
Capt. Supdt., Kidderpore Dockyard	Farewell, Capt. M. W., C.I.E., R. I. M. (*offg.*)	1 July 18
Asst. Dir., R. I. M. (*Admin.*)	**Whelan,** A. A., Esq.	24 Feb. 18

ADVISORY COUNCIL.

President	His Excellency the Army Member.
Members	The Chief of the General Staff. The Secretary to the Government of India in the Army Department. The Financial Adviser, Military Finance. The Adjutant-General. The Quartermaster-General. The Director-General of Military Works. The Director-General of Ordnance. The Director, Medical Services.
Associate Member	The Director-General, Indian Medical Service.
Secretary	The Deputy Secretary, Army Department.

MOBILISATION COMMITTEE.

President	Director of Organisation and Recruiting.
Members	Representative from General Staff Branch. " Adjutant General's Branch. " Quartermaster General's Branch. " Military Secretary's Branch. " Medical Branch. " Ordnance Branch. " Military Works Branch.
Associate Member	Representative from Telegraph Department.
Secretary	D. A. A. G. (Mob. I), Adjutant General's Branch.

COMMANDER-IN-CHIEF.

His Excellency General *Sir* CHARLES CARMICHAEL MONRO, G.C.B., G.C.M.G., A.D.C., *p.s.c.* 1 Oct. 1916.

HEAD-QUARTERS STAFF OF THE ARMY IN INDIA.

Military Secretary	**Scott,** Maj.-Genl. T. E., C.B., C.I.E., D.S.O.,	1 Oct. 16
Dy. Mily. Secretary	**Maclean,** Maj. (*temp. Lt.-Col.*) G. G. C., 104th Rif.	25 Jan. 18
Asst. Mily. Secretaries	**Miles,** Capt. (*temp. Maj.*) *Sir* C. W., *Bart.*, Som. L. I.	1 Oct. 16
	Trail, Capt. (*temp. Maj.*) W. S., M.C., 57th Rif.	8 Mar. 17
	Wilberforce-Bell, Capt. (*temp. Maj.*) H., I. A.	4 May 18
	Addison, Capt. (*temp. Maj.*) A. W. N., 1-6th Hamps. R.	20 June 18
Dy. Asst. Mily. Secretaries	**Priestman,** Capt. H. E., L'pool. R.	15 Apl. 18
	Hillman, Lt. (*temp. Capt.*) A. J. H. S., R. F. A.	23 June 18
	Norris, Capt. W. A., I.A.R.O.	8 Oct. 18
	Morrow, Capt. C. T., I.A.R.O.	8 Oct. 18
Aides-de-Camp	**Muir,** Maj. W. W., 15th Sikhs	1 Oct. 16
	Troutbeck, 2nd-Lt. J. M., 12th Bn., Lond. R.	1 Oct. 16
	Macartney, Maj. F. A., M.C., 53rd Sikhs	29 Aug. 17
	Dost Muhammad Khan, I.O.M., Ressaidar, 18th Lrs.	1 Oct. 16
	Bahadur Singh, Risaldar Major, *Sardar Bahadur,* I.O.M., Q. V. O. Corps of Guides	1 Jan. 17
Surgeon	**Bridges,** Capt. A. B. H., R.A.M.C.	1 Oct. 16
Honorary Aides-de-Camp	**Richards,** Lt.-Col. C. H., 43rd Kolar G. F. R. Bn. I. D. F.	8 Mar. 14 1 Oct. 16
	Crice, Lt.-Col., W. T., C.I.E., V.D., 5th Calcutta Bn. I. D. F.	26 Jan. 18
	Anderson, Lt.-Col. C. S., V.D., 23rd Sind Bn. I. D. F.	18 July 18
	Judbir Thapa, Hony. Capt., *Sardar Bahadur, late* 2nd K. E. O., G. R.	4 May 99 1 Oct. 16
	Sher Singh, Hony. Capt., *Sardar Bahadur, late* 13th Lrs.	22 Oct. 04 1 Oct. 16
	Bidhi Chand, Hony. Capt., *Sardar Bahadur, late* 38th Dogras	1 Apl. 09 1 Oct. 16
	Santbir Gurung, I.O.M., Hony. Capt., *Sardar Bahadur, late* 2nd G. R.	1 Oct. 10 1 Oct. 16
	Prem Singh, I.O.M., Hony. Capt., *Sardar Bahadur, late* 32nd Prs.	1 July 12 1 Oct. 16
	Devasahayam, Hony. Capt., M.B.E., *Sardar Bahadur, late* 2nd S. & M.	9 Apl. 13 1 Oct. 16
	Yar Muhammad, Hony. Capt., *Sardar Bahadur, late* 127th Infy.	15 Sep. 13 1 Oct. 16
	Mir Jafar Khan, Hony. Capt., I.O.M., *Sardar Bahadur, late* 9th Horse	1 Oct. 16

GENERAL STAFF BRANCH.

CHIEF OF THE GENERAL STAFF.	**Kirkpatrick**, Lieut.-Genl. (*tempy.*) *Sir* G. M., K.C.B., K.C.S.I., Brit. Ser., *p.s.c.* . . . 13 Jan. 16
Persl. Asst.	**Campbell**, L. S., *Esq.* (*Lv. ex I., p. a., from 19 June 16*) 11 Sep. 08
Do. (*Offg.*)	**Taylor**, Capt. T., I. A. R. O. . . . 19 June 16
DY. CHIEF OF THE GENL. STAFF.	**Skeen**, Maj.-Genl. A., C.M.G., *p.s.c.* . . . 7 Nov. 17
General Staff Officer, 1st Grade	**Villiers-Stuart**, Bt. Lt.-Col. J. P., D.S.O., 55th Rif. *p.s.c.* . . . 1 Nov. 17
DIRECTOR OF MILITARY OPERATIONS	**Moberly**, Brig.-Genl. F. J., D.S.O., 25th Punj., *p.s.c.* . 7 Nov. 17
General Staff Officers, 1st Grade	**Twiss**, Bt. Lt.-Col. W. L. O., M.C., 9th G. R., *p.s.c.* [L] 7 Nov. 17
	Pepys, Maj. (*temp. Lt.-Col.*) G. L., D.S.O., 57th Rif., *p.s.c.* [L] . . . 17 Nov. 17
	Rogers, Maj. (*temp. Lt.-Col.*) G. H., 11th Rajputs, *p.s.c.* 21 Jan. 18
	Jeffery, Maj. (*temp. Lt.-Col.*) W. H., C.I.E., 73rd Infy. 24 June 18
	Muspratt, Bt. Lt.-Col. S. F., D.S.O., 12th. Cavy., *p.s.c.* 27 Nov. 18
General Staff Officers, 2nd Grade.	**Finlay**, Maj. R. F., D.S.O., 58th Rif. (*Special Intell., Peshawar*) . . . 1 Apl. 16
	Barnes, Maj. L. E., R. E., *p. s.c.* . . . 11 July 16
	Halland, Lt. (*temp. Maj.*) G. H. R., O.B.E., I. A. R. O. 1 Nov. 17
	Williamson, Maj. A. G., 7th Lrs. . . . 18 Jan. 18
	Paige, Maj. C. P., C.I.E., 109th Infy. . . . 26 Aug. 18
	Hills, Maj. A. E., 1-5th R. W. Kents. . . . 1 Sep. 18
	Symns, 2nd-Lt. (*temp. Maj.*) J. M., I. A. R. O. . 28 Oct. 18
	Gompertz, Capt. (*temp. Maj.*) A.V., M.C., R.E., . 28 Nov. 18
	Bennett, Maj. E. E. S., 66th Punj., *p.s.c.* . . . Dec. 18
	Harper, Maj. A. F., 84th Punj . . . 8 Dec. 18
General Staff Officers, 3rd Grade.	**James**, Capt. W. R., M. C., 89th Punj. (*Special Intell., Burma*) . . . 23 July 17
	Penny, 2nd-Lt. (*temp. Capt.*) J. D., I. A. R. O. . 29 Sep. 18
	Jefferys, 2nd. Lt. (*temp. Capt.*) R. C., I. A. R. O. . 18 Oct. 18
	Marshall, Capt. P. H., I. A. R. O. . . . 1 Nov. 18
	Parks, Capt. H. C., I. A. R. O. . . .
DIRECTOR OF STAFF DUTIES	**Norman**, Brig.-Gen. C. L., D.S.O., M.V.O., A.D.C., Guides, *p.s.c.* . . . 11 July 17
General Staff Officers, 1st Grade	**Mennie**, Col. J. O., I. A. . . . 17 Apl. 15
	Moberly, Bt. Lt.-Col. B. R., D.S.O., 56th Rif., *p.s.c.* . . . 2 July 16
	Gwyer, Bt. Lt.-Col. (*temp. Lt. Col.*) V. E., 10th Jats, *p.s.c.* . . . 25 Nov. 17
	Ellis, Maj. (*temp. Lt.-Col.*) A. C. S. B., 121st Prs., *p.s.c.* (*e*) . . . 24 May 18
General Staff Officers, 2nd Grade.	**Harbord**, Capt. (*temp. Maj.*) L. B., M.C., 44th Infy. 1 Nov. 17
	Walker, Maj. C. W. G., D.S.O., *p.s.c.* . . .
	Slim, Lt. (*temp. Maj.*) W. J., M.C., W. I. Regt. . 2 Nov. 18
General Staff Officers, 3rd Grade.	**Brewer**, Capt. J. C., M.C., 90th Punj. . . . 14 Feb. 18
	Hill, Capt. A. J., I. A. R. O. . . . 2 Nov. 18
DIRECTOR OF MILITARY TRAINING	**Isacke**, Brig.-Genl. H., C.M.G., R. W. Kent R., *p.s.c.* [*l*] 8 July 16
General Staff Officers, 1st Grade	**Molloy**, Maj. (*temp. Lt.-Col.*) G. M., 34th Horse, *p.s.c.* 21 Sep. 17
	Wyatt, Maj. (*temp. Lt.-Col.*) E. R. C., D.S.O., 106th Prs., *p.s.c.* . . . 6 Aug. 18

GENERAL STAFF BRANCH—*contd.*

General Staff Officers, 2nd Grade	**Lumby**, Capt. (*temp. Maj.*) A. F. R., 69th Punj.	7 July 17
	Walker, Capt. (*temp. Maj.*) E. J. P. T., 32nd Lrs.	6 Aug. 18
General Staff Officer, 3rd Grade	**Ruegg**, Capt. A. J., I. A. R. O.	14 Nov. 18
Asst. Dir. of Army Signals (graded as G. S. O. 1).	**Day**, Maj. (*temp. Lt.-Col.*) J., D.S.O., R. E.	1 Jan. 18
Dy. Asst. Dir. of Army Signals (graded as G. S. O. 2).	**Kidner**, Maj. W. E., M. C., R. E.	1 Jan. 18

ATTACHED.

Inspr. of Cavy.	**Leader**, Maj.-Genl. H. P., C.B. Brit. Ser.	16 Dec. 16
Staff Offr. to Do.	**Oswald**, Capt. F., 20th Horse	8 Oct. 18
Inspr. R. H. & R. F. A.	**Lecky**, Maj.-Genl. T. St. C., C.B., C.M.G., R.A.	4 Nov. 18
Staff Offr. to Do.	**Mason MacFarlane**, Capt. F. N., M.C., R.F.A.	4 Nov. 18
Ord. Offr. to Do.	**Butler**, Lt. J. L., M.C., R. F. A.	23 Apl. 18
Inspr., Royal Garrn. Arty.	**Vincent**, Brig.-Gen. H. O., C.B., C.M.G.	21 Sep. 18
Staff Offr. to Do.	**Ellis**, Lt. G. A., R. G. A.	4 Oct. 18
Inspr. of Infantry, North	**Villiers-Stuart**, Brig.-Genl. W. D., 5th G. R.	20 Sep. 18
Staff Offr. to Do.	**Ketley**, Lt. E. W., 1st G. B., R. Scots.	3 Dec. 18
Inspr. of Infantry, South	**Walton**, Brig.-Genl. W. C., C.B., C.M.G.	15 June 18
Staff Offr. to Do.	**Houghton**, 2nd-Lt. A. T., 2-5th Som. L. I.	23 Sep. 18
Inspr. of R. E. & Pioneers	**Stockley**, Brig. Genl., E. R., C.I.E., R. E.	28 Oct. 18
Staff Offr. to Do.	**Bell**, Lt. N. W., 106 Prs.	
Officer for Translation Duties	**Wigley**, Capt. F. G. M., 88th Infy.	9 May 18
Intelligence Officer, Persian Gulf	**Wallace**, 2nd-Lt. (*temp. Capt.*) C. L., I. A. R. O.	29 Mar. 18
Asst. Intelligence Offr. (Peshawar)	**Clarke**, Capt. F. S., 46th Punj.	28 Sep. 18

Specially attached	**Kaye**, Lt.-Col. C., C.S.I., C.I.E., 20th Infy.	9 Oct. 14
	Sanford, Lt.-Col. G. B., Punjab Commission	6 Feb. 15
	Burke, Maj. R. J. C., *Supy. List*	23 Aug. 15
	Gaskell, Maj. T. K., I.A.	20 Oct. 16
	Wilson, 2nd-Lt. G. E., I.S.O., I. A. R. O.	1 Dec. 17
	Gould, 2nd-Lt. W. G., I. A. R. O.	18 Feb. 18
	Bagley, 2nd-Lt. A. A., I. A. R. O.	22 Feb. 18
	Fill, 2nd-Lt. H. H. G., I. A. (*T. C.*)	3 Apl. 18
	Clark, Lt. C., S. Gds.	24 July 18
	Higgins, Lt. G. L., 9th Midd'x R.	26 Aug. 18
	Macnabb, Lt. H., Lovats Scouts	28 Aug. 18
	Andrews, Lt. H. W., 18th Rif. Brig.	
	Smith, Maj. O. F., 9th Horse	28 Sep. 18
	Whitehead, Capt. G., 1st G. B., Yorks R.	28 Sep. 18
	Tester, Lt. A. F., 1-4th R. W., Kent R.	
	Hill, 2nd-Lt. C. J. H., 1st G. B., Bedford R.	
	Allen, 2nd-Lt. A., 1st G. B., R. Scots. Fus.	
	Cross, 2nd-Lt. A. B. R., R. E., I. W. T.	
	Gavin, 2nd-Lt. W. L., 4th Gord. Highrs.	
	Dundas, Capt. R. H., 3rd R. Highrs.	28 Oct. 18
	Danskins, Capt. J. W., I. A. R. O.	1 Nov. 18
	Leslie, Lt. W. D., I. A. R. O.	23 Nov. 18
	Marks, Capt. F. C., I. A. R. O.	28 Nov. 18
	Cheers, Capt. J. McG., I. A. R. O.	8 Dec. 18
	Fulton, Lt. E. A. C., 4th G. R.	
	Shaddock, Lt. J., 6th Hants	
	Young, Lt. W. D., A. S. G.	19 Dec. 18
	Engledue, Maj. R. S., 89th Punj.	
	Muhammad Feroz Khan, Subadar, 31st Punj.	1 Apl. 11
	Ghafur Shah, Jemadar, Guides Infy.	1 Sep. 15
	Nur Akhmad, Jemadar, 58th Rif.	1 Sep. 15

ADJUTANT-GENERAL'S BRANCH.

ADJUTANT-GENERAL	**Hudson**, Lt.-Genl. *Sir* H., K.C.B., C.I.E.	5 Feb. 17
Personal Assistant	**Knowles**, C. J., Esq., M.B.E.	28 May 15
Dy. Adjt.-Genl.	**Cooke**, Maj.-Genl. (*temp.*) H. F., D.S.O., 32nd Prs.	8 Oct. 17
Financial Officer	**Rolfe**, R. H., Esq.	1 June 17

DIRECTORATE OF ORGANIZATION AND RECRUITING.

Dir. of Orgn. & Recg.	**Cooke**, Maj.-Genl. (*temp.*) H. F., D.S.O., 32nd Prs.	8 Oct. 17
Asst. Adjutant-General	**Rind**, Bt. Lt.-Col. (*temp. Lt.-Col.*) G. B. A., 124th Infy. *p.s.c.*	8 Feb. 17
Do.	**Bannatyne**, Maj. (*temp. Lt.-Col.*) N. C., 128th Prs	6 Nov. 17
Do.	**Duncan**, Maj. (*temp. Lt.-Col.*) H. C., D.S.O., 9th G.R., *p. s. c*	...
Dy. Asst. Adjutant-Genl.	**Maxwell**, Maj. W. F., D.S.O., R. E.	...
Do.	**Wilson**, Maj. T. A. A., 4th Cav.	1 Sep. 17
Do.	**Hassells-Yates**, Maj. G. A., R.A.	4 Apl. 18
Do.	**Adams**, Lt. (*temp. Maj.*) K. L., R. E.	1 Feb. 18
Do.	**Gordon**, Maj. J. G. B., M.C., 52nd Sikhs (F. F.)	...
Do.	**Lamb**, Capt. (*temp. Maj.*) D. O. W., 10th Lrs.	28 Nov. 18
Do.	**McWatters**, Maj. (*Bt. Lt. Col.*) H. C., D.S.O., *p. s. c.*, 24th Punj.	3 Dec. 18
Attached	**Whiffen**, Maj. G. G., 1-5th E. Surr. R.	16 Aug. 18

DIRECTORATE OF RECRUITING.

Director	**Harrison**, Brig.-Genl. A. H. P., 33rd Punj.	1 Dec. 16
Asst. Adjt.-Genl. (*addtl.*) (*Technical*)	**Kirkhope**, Lt. (*temp. Lt.-Col.*) K. M., I. A. R. O.	26 July 17
Asst. Adjt.-Genl. (*addtl.*) (*non-combatant*).	**Nevill**, Lt.-Col. H. R., 7th U. P. Horse, I. D. F.	17 Nov. 17
Dy. Asst. Adjutant-Genl.	**Johnston**, Maj. R., D.S.O., 4th Cav.	7 Feb. 18
Do.	**Dobbs**, Lt. (*temp. Maj.*) W. E. J., I. A. R. O.	22 Aug. 18
Do. (*Technical*)	**Cairns**, Capt. (*temp. Maj.*) J., 3rd E. Lan. R.	22 Apl. 18
Do. (*Insptg. offr. Labour and Porter Corps*)	**Murphy**, Lt. (*temp. Maj.*) A., I. A. R. O.	18 Aug. 18
Attached	**Feiling**, Lt. (*temp. Capt.*) K. G., 1st G. B. R. Sco. Fus.	6 June 17
Do.	**Thomas**, 2nd-Lt. (*temp. Capt.*) E. F., I. A. R. O.	3 June 18
Do.	**Porrett**, 2nd-Lt. (*temp. Capt.*) C. R., I. A. R. O.	10 June 18
Do. (*Technical*)	**Harvey-Kelly**, Lt. (*temp. Capt.*) G. I., I. A. R. O.	1 Nov. 17
Do.	**Doherty**, Lt. J. M. (*Spl. List*), Attd. I. W. T.	5 July 18
Do.	**Matthews**, 2nd-Lt. A., I. A. R. O.	11 Nov. 18

DIRECTORATE OF PERSONAL SERVICES.

Dir. of Persl. Services	**Barstow**, Brig.-Genl. H., 38th Dogras	8 Feb. 17
Asst. Adjutant-General	**Shea**, Maj. (*temp. Lt.-Col.*) A. G., D.S.O., 51st Sikhs	19 Feb. 18
D. A. A. G.	**Carr**, Lt. (*temp. Maj.*), C.T., 2-4th Wilts. R.	1 July 18
Attached	**Walker**, Lt. (*temp. Capt.*) W. R., 2-4th Bord R.	2 Feb. 17
Do.	**Forde**, Capt. A. J., I. A. R. O.	8 Oct. 18

DIRECTORATE OF ADMINISTRATION.

Dir. of Administration	**Luke**, Brig.-Genl. T. M., D.S.O., O.B.E., R.A.	12 Oct. 16
Asst. Adjutant-General		
Dy. Asst. Adjutant-General	**Cotton**, Capt. (*temp. Maj.*) P. V., 1-9th Middx. R.	22 June 17
Do.	**Sankey**, Capt. (*temp. Maj.*) G. B., 1-1 Kent Bn.	24 Feb. 18
Staff Capt. (*Addl.*)	**Murray**, Capt. C. J., I. A. R. O.	4 Oct. 18
Attached	**Collis**, Capt. J. G., R. W. Surr. R.	30 Aug. 18
Do.	**Creak**, Capt. E. V., I. A. R. O.	4 Oct. 18
Do.	**Hussey**, Capt. H., Inspr. of Army Schools	...
Do.	**Ginns**, Capt. L.B., I. A. R. O.	1 Nov. 18

JUDGE ADVOCATE-GENERAL'S SECTION.

Judge Advoc.-Genl.	**Caruana**, Col. A. J., C.I.E.	1 May 13
Asst. Judge Advoc.-Genl.	**Nicholas**, Lt.-Col. S. H. E., I. A.	1 May 13
Attached	**Osmond**, Lt. W. R. F., R. A.	20 Oct. 17
Legal Asst. to J. A. G.	**Haldipur**, S. V., Esq., B.A., LL.B.	1 Feb. 18

ADJUTANT-GENERAL'S BRANCH—(*contd.*)

RECORDS SECTION (BOMBAY).

A. A. G.	**Wheelwright**, Maj. T. H., Genl. List.	26 Oct. 18
D. A. A. G.	**McPherson**, Capt. A. B., M.C., 6th Jats.	10 Dec. 18
Staff Captains	**Cotter**, Capt. P., I. A. R. O.	20 Nov. 18
	Sturdy, Lt. J. R., M.C., 3-5th G. R.	25 Nov. 18
Adjt. & Qr. Mr.	**Ambrose**, Capt. J., 25th Lond. R.	30 Nov. 18
Cipher Officer	**Orwin**, Capt. W., E. Kent R.	28 Nov. 18
Record Officers	**Edwards**, Capt. W. H., I.A.R.O.	29 Nov. 18
	Coward, Capt. A. J., I.A.R.O.	28 Nov. 18
	Wilson, Capt. T. G., I.A.R.O.	28 Nov. 18
	Cutler, Lt. A. C. H., R.F.A.	26 Nov. 18
	Adams, Lt. R. P., Beckroek Bn.	30 Nov. 18
	Hulbert, Lt. H. B., 9th Worc. R.	30 Nov. 18
	Haggett, 2nd-Lt. H. G., I.A.R.O.	30 Nov. 18
S. S. O., Simla	**Pierce**, Lt.-Col. F. G., 102nd Grs.	15 June 18
Dir. of Mily. Prisons and Det. Barracks	**Morse**, Lt.-Col. C. J., Brit. Ser.	11 Mar. 14

QUARTERMASTER-GENERAL'S BRANCH.

QUARTERMASTER-GENERAL	**Altham**, Lt.-Genl. *Sir* E. A., K.C.B., C.M.G., *p.s.c.* Brit. Ser.	18 Feb. 17
Personal Asst.	**Lloyd**, Hony. Capt. T. W.	18 Dec. 15
Dy. Quartermaster-General	**Richardson**, Maj.-Genl. C. W. G., C.S.I., *p.s.c.* [L].	1 Oct. 17
Assistant Quartermaster-General	**Moody**, Maj. (*temp. Lt.-Col.*) F. H., M.C., 13th Lrs.	11 July 18
Staff Captain	**Lang**, Capt. W. F., I. A. R. O.	8 Nov. 18
Financial Officer	**Cumber**, H. S., Esq.	30 July 17

DIRECTORATE OF QUARTERING.

Dir. of Quartering	**Holdich**, Brig.-Genl. H. A., D.S.O., 5th G. R., *p.s.c.*	19 Nov. 17
Asst. Quartermaster-General	**Muspratt**, Maj. (*tempy. Lt.-Col.*) V. E., 30th Lrs., *p.s.c.*	27 Feb. 18
Dy. Assistant Quartermaster-General	**Wilson**, Maj. W. H. D., 1st Brah.	7 May 17
Ditto	**Clarke**, Maj. B. G. S., 10th G. R. [*i*]	23 Mar. 18
Ditto	**Green-Armytage**, Maj. P. D., 117th Mahrattas	21 Aug. 18
Ditto	**Meiklejohn**, Capt. (*temp. Maj.*) J. F., 33rd Cav.	30 Nov. 18
Staff Captain.	**Taylor**, Lt. (*temp. Capt.*) W. R., I. A. R. O.	4 Mar. 18
Ditto	**Menzies**, Capt. F. A., M.C., I. A. R. O.	21 May 18
Ditto	**Jackson**, Capt. C. G., I. A. R. O.	11 Oct. 18
Attached	**Woodward**, Maj. C. P.	16 Oct. 18
Ditto	**Bruce-Kingsmill**, Lt.-Col., J. C. de K., R.F.A. (*T*)[*l*]	20 Mar. 18

MOVEMENTS DIRECTORATE.

Deputy Director, Movements	**Ross**, Bt. Lt.-Col. (*tempy. Lt.-Col.*) R. C., D.S.O., 6th Jats	14 Feb. 18
Dy. Dir. I. W. T.	**Maxwell**, Tempy. Maj. (*Actg. Lt.-Col.*) B.C.D'A., R.E.	26 Aug. 16
Dy. Assistant Quarter-Master General	**Hall**, Maj. E. G., 117th Mah.	20 June 17
Ditto	**Phillips**, Capt. (*temp. Maj.*) G. A., I. A. R. O., V.D.	9 Mar. 18
Ditto	**Oakes**, Maj T. T., 102nd Grs.	7 Aug. 18
Staff Captain	**Ferrier**, Lt. (*Temp. Capt.*) C. D., I. A. R. O.	26 Feb. 18
Attd. I. W. T.	**Evans**, 2nd-Lt. D. J., R. E.	29 Aug. 17

SUPPLY AND TRANSPORT DIRECTORATE.

Director of Supplies and Transport	**Vaughan**, Brig.-Gen. R. E., C.B., S. & T. Corps	5 Oct. 16
Asst. Director of Supplies & Transport	**Douglas**, Maj. (*temp. Lt.-Col.*) H. A., S. & T. Corps	8 Mar. 18
Deputy Director of Supplies	**Young**, Maj. (*temp. Lt.-Col.*) H. N., S. & T. Corps	9 Mar. 18
Assistant Director of Supplies	**Twiss**, Maj. (*temp. Lt.-Col.*) H. W. F., D.S.O., S. & T. Corps	9 Mar. 18
Dy. Asst. Dir. of Supplies	**Wright**, Capt. (*temp. Maj.*) A. St. J., S. & T. Corps	27 Feb. 18
Staff Captain (Supplies)	**Chesney**, Capt. E. C., S. & T. Corps	14 Feb. 18
Ditto	**Hunt**, Capt. G. E. D., 1st G. B., S. Staff. R.	9 May 18
Deputy Director of Transport	**Codrington**, Lt.-Col. H. de B., S. & T. Corps.	11 Nov. 18
Dy. Asst. Director of Transport	**Mayo**, Maj. G. L. L., S. & T. Corps	20 Feb. 18
Staff Captain (Transport)	**Markwick**, Capt. H. W., S. & T. Corps	4 Mar. 18
Dy. Dir. of Transport (M. T.)	**Robinson**, Maj. (*temp. Lt.-Col.*) W. P., D.S.O., A. S. C.	1 Sep. 18
Dy. Asst. Dirs. Trans. (M. T.)	**Phillips**, Maj. H. A., 1-4th Wilts. R.	1 Apl. 17
Ditto (M. T.)	**Barrett**, Capt. (*temp. Maj.*) B. H. H., A. S. C.	8 Jan. 18
Ditto (M. T.)	**Stevens**, Lieut. (*temp. Maj.*) H. L., A. S. C.	1 Sep. 18
Staff Captain (M.T.)	**Bartlett**, Lt. (*temp Capt.*) D. F., I. A. R. O.	1 Sep. 18
Attached (Supplies)	**Judge**, Lt. (*temp. Capt.*) C. E. M., M.B.E., I. A. R. O.	28 Feb. 18
Ditto (Transport)	**Cohen**, Lt. N. H., I. A. R. O.	9 Dec. 18
Ditto (Personnel)	**Watson**, Capt. J., S. and T. Corps	18 Mar. 18
Ditto	**Ross**, Lt. J. W., A. S. C.	29 Oct. 18

QUARTERMASTER-GENERAL'S BRANCH—*concld.*

EQUIPMENT AND ORDNANCE STORES DIRECTORATE.

Dir. of Equipment and Ordnance Stores	**Perry**, Col. (*Temp. Maj.-Genl.*). *Sir* H. W., K.C.M.G., C.B., C.S.I., Army Ordce. Dept.	1 Apl. 18
Dy. Ditto	**Douglas**, Lt.-Col. A. P., C.M.G., R. A.	1 Apl. 17
Asst. Ditto	**Carey**, Lt.-Col. W. L. J., R. A.	1 Apl. 18
Ditto	**Tyrrell**, Lt.-Col. J. F., R. A.	1 Apl. 18
Dy. Asst. Dir. Ditto	**Johnston**, Maj. C. E. L., R. A., (L)	1 June 18
Addl. Ordce. Offr.	**Smith**, Comsy. & Hony. Maj. B. E., Ind. Ordce. Dept.	1 Apl. 18
Depy. Dir. of Clothing	**Kennedy**, Lt.-Col. A. C., D.S.O., R. A.	25 Feb. 18
Ditto	**Robertson**, Maj. (*temp. Lt.-Col.*) R. A. H., 30th Punj.	21 Aug. 17
Dy. Asst. ditto	**Ventris**, Maj. E. F. V., R.A.	21 Aug. 17
Ditto ditto	**Dawson**, Capt. (*temp. Maj.*) B. N., A. S. C.	14 Feb. 18
Attached	**Chamberlain**, Lt. J. A., A. O. D.	14 Aug. 18
Ditto	**Palairet**, Capt. (*temp. Maj.*) R. C. N., Devon. R.	7 Sep. 18

REMOUNT DIRECTORATE.

Dir.-Genl., Remt. Dept.	**Templer**, Brig.-Genl. C. F., C.M.G., C.I.E.	5 Sep. 15
Dy. Dir., Remounts	**Gillett**, Maj. (*temp. Lt.-Col.*) E. S., C.I.E., A.V.C.	1 May 18
Asst. Dir., Remounts	**Gordon**, Maj. (*temp. Lt.-Col.*) H. F., A. R. D.	26 Aug. 18
Dy. Asst. Dir., Remounts	**Kennedy**, 2nd-Lt. (*temp. Maj.*) W. D., I. A. R. O.	29 May 18

VETERINARY DIRECTORATE.

Dir., Vety. Services in India	**Nuthall**, Brig.-Genl. C. E., C.B., C.M.G., A. V. S.	26 Jan. 18
Dy. Dir., Vety. Services	**Sawyer**, Lt.-Col. H. T., D.S.O., A. V. C.	12 Dec. 18
Dy. Asst. Dir., Vety. Services	**Sikes**, Maj. F. G., A. V.C.	17 Dec. 17

FARM DIRECTORATE.

Dir. of Farms	**Hallowes**, Col. F. W., C.I.E., S. & T. Corps.	1 Apl. 12
Dy. Asst. Dir. of Farms	**Airy**, Maj. (*temp. Lt.-Col.*) G., Mily. Farms Dept.	31 Jan. 18
Staff Captain	**Prior**, Capt. C. J., I. A. R. O.	11 Oct. 18

Controller of Contracts	**Bray**, Brig.-Genl. *Sir* E. H., *Kt.*	2 Apl. 18
Military Adviser	**Walter**, Maj. E., C.I.E., S. & T. Corps	20 Mar. 18
Asst. Controller of Contracts	**Gwynn**, Lt. (*temp. Maj.*) C. H., I. A. R. O.	1 Dec. 17
Ditto Ditto	**Parker**, Lt. (*temp. Maj.*) R. N., I. A. R. O.	20 Dec. 17
Attached	**Foskett**, Lt. (*temp. Capt*). R. C., I. A. R. O.	31 May 18
Ditto	**Humphreys**, *Hony.* Lt. A. F., S. & T. Corps.	23 Mar. 18
Ditto	**Keeling**, Mr. R. F.	1 July 18

Inspector of Q. M. G.'s services	**Jennings**, Brig.-Genl. H. A. K., C.I.E., Ordce. Dept.	1 Apl. 18

Adviser M. T.	**Montagu**, Col. (*temp. Brig.-Genl.*) W. E., *Lord*, C.S.I., V.D., F. Res.	7 Apl. 15
Asst. to Adviser M. T.	**Burgoyne**, Maj. A. H.	17 June 17

MEDICAL BRANCH.

DIRECTOR, MEDICAL SERVICES	**O'Donnell**, Lieut.-Genl. (*temp.*) *Sir* T. J., K.C.I.E., C.B., D.S.O., M.B. ret. A. M. S.	13 July 16
Dy. ditto	**Molesworth**, Col. W., C.I.E., C.B.E., *V.H.S.*, I.M.S.	1 Nov. 16
Inspector of Medical Services	**Heard**, Bt.-Col. R., *V.H.S.*, I.M.S.	23 Nov. 15
Asst. Dir. Medical Services	**Granger**, Lt.-Col., T. A., C.M.G., V.H.S., M.B., I.M.S.	1 July 18
Ditto	**Langstaff**, Lt.-Col. J. W., D.S.O., R.A.M.C.	14 Sep. 18
Asst. ditto	**Fleming**, Maj. (*temp. Lt.-Col.*) J. K. S., I.M.S.	2 Dec. 15
Ditto (Sanitary)	**Browne**, Maj. (*temp. Lt.-Col.*) W. W., R.A.M.C.	3 July 17
Asst. ditto	**Dwyer**, Maj. (*temp. Lt.-Col.*) P., M.C., R.A.M.C.	14 July 16
Dy. Asst. Dir., Medical Services	**McNeight**, Maj. A. A., I.M.S. [L]	13 Oct. 17
Ditto	**McCombe**, Capt. (*temp. Maj.*) J. S., D.S.O., R.A.M.C.	23 Apl. 17
Ditto	**Armstrong**, Capt. (*temp. Maj.*) E. R., I.M.S.	15 June 16
Chief Lady Supdt.	*Waterhouse, Miss A. M., R.R.C., Q.A.M.N.S.I.*	*22 Feb. 16*
Ditto	**Rait**, Miss H. A. M., R.R.C., Q.A.M.N.S.I.	18 Sep. 17
Attached	**Hodge**, Capt. E. H. V., I. M. S.	20 Jun. 18
Ditto	**Pitt**, Capt. W. O., R. A. M. C. (*T. C.*)	20 July 18
Ditto	**Stevens**, Capt. & Qr.-Mr. S., R. A. M. C.	1 Oct. 17

ORDNANCE BRANCH.

DIRECTOR-GENERAL OF ORDNANCE.	**Renny,** Maj.-Gen. (*temp.*) S. M., C.I.E., Brit. Ser.	16 Feb. 17
Persl. Asst.	**Villiers,** Comsy. & Hony. Maj. A.	16 Feb. 17

DIRECTORATE OF ARTILLERY.

Dy. Director of Artillery	**Rich,** Col. H. H., Brit. Ser.	30 Sep. 18
Asst. Director-Genl. of Artillery	**Ogg,** Col. G. S., C.I.E., Brit. Ser.	30 Sep. 18

DIRECTORATE OF ORDNANCE INSPECTION.

Director of Ordnance Inspection	**Lawrence-Archer,** Lt.-Col. (*tempy. Col.*) J. H., R.A.	17 Feb. 17
Asst. Dir. of Ordnance Inspection	**Waller,** Maj. (*temp. Lt.-Col.*) R. S. de W., R.A.	10 Apl. 18
Attached	**Wortham,** Maj. O. O., A. O. D.	10 Sep. 18

MILITARY WORKS BRANCH.

DIRECTOR-GENERAL OF MILITARY WORKS.	**Williams,** Maj.-Genl. *Sir* G., K.C.I.E., C.B., Brit. Ser.	26 Apl. 13
Dy. Directors-Genl. of Military Works	**Coffin,** Col. C., C.M.G., *p.s.c.*	2 Sep. 17
	Cumberlege, Lt.-Col. (*temp. Col.*) A. F., R.E.	2 Apl. 17
Inspr. of Machinery	**Nugent,** Lt.-Col. C. H. H., C.I.E., R.E.	17 Feb. 11
Asst. Dirs.-Genl. of Military Works	**LeMesurier,** Maj. (*temp. Lt.-Col.*) H. G., C.I.E., R.E.	21 Oct. 14
	Fenton, Maj. (*temp. Lt.-Col.*) G. C. V., D.S.O., R.E.	3 Jan. 17
	Bocquet, Maj. (*temp. Lt.-Col.*) R. N., R.E.	30 Sep. 17
Dy. Asst. Dirs.-Genl. of Military Works	**Stagg,** Capt. (*temp. Maj.*) M., R.E.	2 Oct. 16
	Hounsell, Capt. (*temp. Maj.*) R. H. S., R. E.	16 Oct. 18
Staff Captains	**Parselle,** Lt. (*temp. Capt.*) J., R. E.	5 May 17
	Lemon, Capt. G. T., I. A. R. O.	1 June 17
	Raven, 2nd-Lt. (*temp. Capt.*) T. F., I. A. R. O.	19 Aug. 18
	Richardson, Capt. F. L., I. A. R. O.	25 Oct. 18
Attached	**Holmes,** Lieut. H. G., R. E.	27 Oct. 18

HEAD QUARTERS STAFF OF ROYAL AIR FORCE IN INDIA.

Dir of Aeronautics	**Tennant,** Lt.-Col. J. E. D.S.O., M.C., R. A. F. (O.C., R.A.F. India)	2 July 18
S. O. II (Air)	**Howell,** Capt. (*Actg. Maj.*) R. H., R. A. F.	18 Nov. 18
S. O. II (T)	**Bruce,** Capt. (*temp. Maj.*) W. R., R. A. F.	26 Oct. 16
Attached	**Scowcroft,** Lt. H. D. (O), R. A. F.	19 May 18

ARMY, DIVISIONAL AND BRIGADE COMMANDS.

Names, Rank and Corps.	Date of present appointment.	Date, completion tenure.	Remarks.
General Officers Commanding Commands.			
Barrett, Gen. *Sir* A. A., G.C.B., K.C.S.I., K.C.V.O., A.D.C.	31 May 16	....	Northern Command.
Anderson, Lt.-Gen. *Sir* C. A., K.C.B., Brit. Ser.	6 Apl. 17	....	Southern Command.
General Officers Commanding Divisions.			
Campbell, Lt.-Gen. *Sir* F., K.C.B., D.S.O.	26 June 15	25 June 19	1st Peshawar.
Barratt, Maj.-Gen. W. C., C.B., D.S.O.	5 Dec. 16	....	9th Secunderabad.
Wapshare, Lieut.-Gen. R., C.B., C.S.I.	6 Dec. 16	....	4th Quetta.
Kemball, Maj-Gen. *Sir* G. V., K.C.M.G., C.B., D.S.O., *p.s.c.* (*g.*) Brit. Ser.	1 June 17	....	5th Mhow.
Tighe, Maj.-Gen. *Sir* M. J., K.C.M.G., C.B., C.I.E., D.S.O.	22 July 17	....	Poona.
Dobell, Maj.-Gen. *Sir* C. M., K.C.B., C.M.G., D.S.O., *p.s.c.* [*l*] Brit. Ser.	6 Aug. 17	....	2nd Rawal Pindi.
Keary, Lt.-Gen. *Sir* H. D'U., K.C.B., K.C.I.E., D.S.O.	16 Dec. 17	....	Burma.
Beynon, Maj.-Gen. *Sir* W. G. L., K.C.I.E., C.B., D.S.O.	15 June 18	....	16th Dn.
Nugent, Maj.-Gen. O. S. W., C.B., D.S.O., *p.s.c.* [*l*]	3 Aug. 18	....	Meerut.
Scott, Maj.-Gen. *Sir* A. B., K.C.B., D.S.O., Brit. Ser.	5 Aug. 18	....	8th Lucknow.
General Officers Commanding Brigades.			
Young, Maj.-Gen. C. F. G.	20 June 12	19 June 16*	Rangoon.
Poore, Brig.-Gen. R. M., C.I.E., D.S.O., Brit. Ser.	7 Aug. 15	....	Jhansi.
Hardy, Maj.-Gen. T. H.	15 Oct. 15	....	11th Infy.
Strange, Brig.-Gen. R. G., C.I.E., Brit. Ser.	15 Dec. 15	....	8th Lucknow.
Eustace, Maj.-Gen. A. H., C.B., D.S.O.	1 Feb. 16	....	Kohat.
Roberts, Brig.-Gen. H. L., C.B., M.V.O.	29 Mar. 16	....	4th Cavy.
Knight, Maj.-Gen. W. C., C.B., C.S.I., D.S.O., *p.s.c.* [L]	31 Mar. 16	....	Bombay.
Peterson, Brig.-Gen. F. H., C.B., D.S.O.	31 Mar. 16	....	5th Infy.
Betham, Brig.-Gen. R. M.	31 Mar. 16	....	Garhwal.
Crocker, Brig.-Gen. G. D., Brit. Ser.	23 Apl. 16	....	Peshawar.
Commeline, Brig.-Gen. F. H. B.	3 Sep. 16	....	Secunderabad Cav.
Hendley, Maj.-Gen. C. E., C.B.	6 Mar. 17	....	Ambala.
Climo, Maj.-Genl. S. H., C.B., D.S.O.	8 May 17	....	Nowshera.
Fowler, Maj.-Gen. F. J., C.B., D.S.O.	27 July 17	....	Karachi.
Fowler, Maj.-Gen. C. A., C.B., D.S.O.	29 July 17	....	Abbottabad.
Wolley-Dod, Brig.-Gen. O. C., C.B., D.S.O., *p.s.c.*, [*l*], Brit. Ser.	26 Nov. 17	....	Bangalore.
Prendergast, Brig.-Gen. C. G., C.B.	18 Jan. 18	....	Jubbulpore.
Peebles, Brig.-Gen. E. C., C.B., C.M.G., D.S.O., Brit. Ser.	11 Mar. 18	....	Rawalpindi Infy.
Miles, Brig.-Gen. P. J., C.B.	13 May 18	....	Multan.
Woodyatt, Maj.-Gen. N. G., C.B.	15 June 18	....	Derajat.
Sladen, Brig.-Genl. D. R., C.M.G., D.S.O.	27 Aug. 18	....	Southern.
Berners, Brig.-Genl. R. A., D.S.O.	1 Sep. 18	....	Rawalpindi Addl.
Clarke, Brig. Genl. J. L. J., E. York R.	28 Sep. 18	....	43rd Bde.
Christian, Brig.-Genl. G., C.B., D.S.O. [*l*], Brit. Ser.	3 Oct. 18	....	6th Infy.
Dunsterville, Maj-Genl. L. C., C.B., (L)	5 Dec. 18	....	Agra.
Colonels on the Staff Commanding Brigades.			
Baldwin, Brig.-Gen. G. M., D.S.O.	15 Feb. 16	....	10th Cav.
Davies, Brig.-Gen. F. G. H.	18 June 16	....	Risalpur Cav.
Smith, Brig.-Gen. F. A.	19 June 16	....	Secunderabad Infy.
Bruce, Brig.-Gen. *Hon.* C. G., C.B., M.V.O.	28 Aug. 16	....	Bannu.
Cadell, Brig.-Genl. A.	27 Dec. 16	....	Allahabad.
Grimston, Brig.-Genl. S. B., C.M.G.	6 Mar. 17	....	Delhi.
Southey, Brig.-Gen. W. M., C.M.G.	31 Mar. 17	....	44th Infy.
Norie, Brig.-Genl. C. E. de M., C.B., C.M.G., D.S.O., *p.s.c.* [*l*]	30 June 17	....	Poona.
Andrew, Brig.-Genl. D. C. A.	15 Oct. 17	....	Nasirabad.
Dyer, Brig.-Genl. R. E. H., C.B., *p.s.c.*	12 Mar. 18	....	Jullundur.
Campbell, Brig.-Gen. L. W. Y., C.M.G.	15 Apl. 18	....	Sialkot.
Gray, Brig.-Gen. F. W. B., C.M.G., D.S.O.	13 May 18	....	Bareilly.
Wood, Brig.-Gen. E. J. M., D.S.O.	21 May 18	....	Ahmednagar.
Adye, Brig.-Genl. D. R., 98th Infy. [L]	14 June 18	....	59th Infy.
Tanner, Brig.-Genl. C. O. O.	25 June 18	....	58th Infy.
Birdwood, Brig.-Genl. H. B.	27 June 18	....	12th Mtd.
Richardson, Brig.-Genl. H. L., 47th Sikhs	6 July 18	....	Jhelum.
Burton, Brig.-Genl. R. G. [L.]	16 Aug. 18	....	Madras.
Beatty, Brig.-Genl. G. A. H., D.S.O., 9th Hors.	1 Sep. 18	....	13th Mtd. Bde.
Gordon, Brig.-Genl. J. L. R., C.B., 15th Sikhs	16 Oct. 18	....	57th Infy. Bde.
O'Grady, Brig-Genl. H. De C., *p.s.c.*	31 Oct. 18	....	10th Infy. Bde.
Pritchard, Brig.-Genl. A. G., C.M.G., 2nd Lrs. (Q)	27 Nov. 18	....	Lucknow.

* Tenure extended.

THE STAFF.

Names, Rank and Corps.	Date, commencement tenure.	Date, completion tenure.	Remarks.
Chief of the General Staff.			
Kirkpatrick, Maj.-Gen. (*tempy.* Lt.-Genl.) *Sir* G. M., K.C.B., K.C.S.I., Brit. Ser., *p.s.c.*	13 Jan. 16		
Deputy Chief of the General Staff.			
Skeen, Maj.-Gen. A., C.M.G., *p.s.c.*	7 Nov. 17	. . . .	Dy. C. G. S. India.
Brigadier-General, General Staff.			
Loch, Brig.-Gen. S. G., D.S.O., *p.s.c.*	7 July 16	. . . .	Northern Command.
Isacke, Brig.-Gen. H., C. M. G., *p.s.c.*	8 July 16	. . . .	Dir. of Mily. Training.
Norman, Brig.-Gen. C. L., D.S.O., M.V.O., A. D. C., Guides, *p.s.c.*	11 July 17	. . . .	Dir. of Staff Duties.
Browne, Brig.-Gen. H. J. P., 5th G. R., *p.s.c.*	19 Oct. 17	. . . .	Southern Command.
Moberly, Brig.-Gen. F. J., D.S.O., 25th Punj., *p.s.c.*	7 Nov. 17	. . . .	Dir. of Military Operations.
General Staff Officers, 1st grade.			
Mennie, Col. J. O.	17 Apl. 15	. . . .	Army Hdqrs.
Tarver, Bt.-Col. A. L., D.S.O., A. D. C. 124th Infy	1 Dec. 15	. . . .	Peshawar.
Moberly, Bt.-Lt.-Col. (*temp. Lt.-Col.*) B. R., D.S.O., 56th Rif., *p.s.c.*	2 July 16	. . . .	Army Hdqrs.
Moberly, Lt.-Col. H. S., 66th Punj.	10 Nov. 16	. . . .	Lucknow.
Tod, Col. J. K., I.A.	20 Feb. 17	. . . .	Burma.
Conway-Gordon, Maj. (*temp. Lt.-Col.*) E. C. W., 3rd Horse, *p.s.c.*	13 Aug. 17	. . . .	Poona.
Rivett-Carnac, Maj. (*temp. Lt.-Col.*) J. S., 14th Lrs, *p.s.c.*	8 Sep. 17	. . . .	Secunderabad.
Molloy, Maj. (*temp. Lt.-Col.*) G. M., 34th Horse, *p.s.c.*	21 Sep. 17	. . . .	Army Hdqrs.
Casson, Bt.-Lt.-Col. (*temp. Lt.-Col.*) W. F. S., D.S.O., 27th Cav., *p.s.c.*	22 Sep. 17	. . . .	16th Dn.
Ward, Bt.-Lt.-Col. H. C. S., 2nd Lrs., *p.s.c.* (*e*)	6 Nov. 17	. . . .	Rawalpindi.
Villiers-Stuart, Bt.-Lt.-Col. J. P., D.S.O., 55th Rif., *p.s.c.*	1 Nov. 17	. . . .	Army Hdqrs.
Twiss, Bt.-Lt.-Col. W. L. O., M.C., 9th G. R., *p.s.c.* [L]	7 Nov. 17	. . . .	Army Hdqrs.
Pepys, Maj. (*temp. Lt.-Col.*) G. L., D.S.O., 57th Rif., *p.s.c.* [L]	17 Nov. 17	. . . .	Army Hdqrs.
Gwyer, Bt.-Lt.-Col. (*temp. Lt.-Col.*) V. E., 10th Jats, *p.s.c.*	25 Nov. 17	. . . .	Army Hdqrs.
Rogers, Maj. (*temp. Lt.-Col.*) G. H., 11th Rajputs, *p.s.c.*	21 Jan. 18	. . . .	Army Hdqrs.
Ellis, Maj. (*temp. Lt.-Col.*) A. C. S. B., 121st Prs., *p.s.c.* (*e*)	24 May 18	. . . .	Army Hdqrs.
Jeffery, Maj. (tempy. Lt.-Col.) W. H., C.I.E., 73rd Infy.	24 June 18	. . . .	Army Hdqrs.
Wyatt, Maj. (*temp. Lt.-Col.*) E. R. C., D.S.O., 106th Prs., *p.s.c.*	6 Aug. 18	. . . .	Army Hdqrs.
Murray, Maj. (*temp. Lt.-Col.*) K. D. B., D.S.O., 59th Rif., *p.s.c.*	. . . .	. . . .	Meerut.
Baird, Bt.-Lt.-Col. H. B. D., C.M.G., D.S.O., 12th Cav., *p.s.c.*		. . . .	Quetta.
Muspratt, Bt.-Lt.-Col. S. F., D.S.O., 12th Cavy., *p.s.c.*	27 Nov. 18	. . . .	Army Hdqrs.
Orr, Bt. Col. G. M., D.S.O., 11th Lrs., *p.s.c.*	. . . .	. . . .	Mhow.
General Staff Officers, 2nd grade.			
Marshall, Maj. A., D.S.O., 28th Cav.	21 Feb. 16	. . . .	Quetta.
Finlay, Maj. R. F., D.S.O., 58th Rif.	1 Apl. 16	. . . .	Special Intelligence.
Ogle, Bt.-Lt.-Col. N., D.S.O., *p.s.c.*, 67th Punj.	1 Oct. 18	. . . .	Cadet College, Quetta.
Macleod, Maj. W. M., 31st Lrs., *p.s.c.*	9 Apl. 17	. . . .	Northern Command.
Meadows, Maj. C. A. G. P., 7th Rajputs	20 July 1[illegible]	. . . .	Staff School, Saugor.
Lumby, Capt. (*temp. Maj.*) A. F. R., 69th Punj.	7 July 17	. . . .	Army Hdqrs.
Craster, Maj G., D.S.O., *p.s.c.*, 6th Cav.	. . . .	. . . .	Bannu Bde.

Names, Rank and Corps.	Date, commencement tenure.	Date, completion tenure.	Remarks.
General Staff Officers, 2nd grade—*concld.*			
Carruthers, Maj. J. H. deW., M.C., 39th Garh. Rif., *p.s.c.*	24 Aug. 17	. . . . .	Rawalpindi.
Hogg, Maj. C. M. T., D.S.O., 4th G. R., *p.s.c.*	15 Sep. 17	. . . . .	Mhow.
VC Wylly, Maj. G. G. E., D.S.O., Guides, *p.s.c.*	21 Sep. 18	. . . . .	Kohat.
Compton, Maj. C. W. McG., 69th Punj.	17 Oct. 17	. . . . .	Secunderabad.
Pott, Maj. E. H., D.S.O., 1st Lrs.	17 Oct. 17	. . . . .	Poona.
Davis, Maj. C. T., 107th Prs.	12 Nov. 17	. . . . .	Burma.
Dick-Lauder, Maj. J. N. D., 16th Cav.	13 Dec. 17	. . . . .	Lucknow.
Muir, Capt. (*temp. Maj.*) A. H. H., 15th Sikhs	5 Sep. 18	. . . . .	Cadet College, Quetta.
Harbord, Capt. (*temp. Maj.*) L. B., M.C., 44th Infy.	1 Nov. 17	. . . . .	Army Hdqrs.
Yearsley, Capt. (*temp. Maj.*) R. A., 28th Punj.	. . . .	. . . . .	Cadet Coll., Wellington.
Halland, Lt. (*temp. Maj.*) G. H. R., O.B.E., I.A.R.O.	1 Nov. 17	. . . . .	Army Hdqrs.
Trench, Capt. (*temp. Maj.*) C. F., D.S.O., 7th Lrs.	9 Dec. 17	. . . . .	Derajat.
Paige, Maj. C. P., C.I.E., 109th Infy.	. . . .	. . . . .	Army Hdqrs.
Williamson, Maj. A. G., 7th Lrs.	. . . .	. . . . .	Army Hdqrs.
Buckland, Maj. G. C. B., D.S.O., M.C., 8th G. R.	23 May 18	. . . . .	Quetta.
Marshall, Bt.-Maj. J. S., D.S.O., 35th Sikhs	2 July 18	. . . . .	Presidency Bde.
Walker, Capt. (*temp. Maj.*) E. J. P. T., 32nd Lrs.	6 Aug. 18	. . . . .	Army Hdqrs.
Smyth, Maj. C. A. W., 1st Brahmans	. . . .	. . . . .	Rawalpindi.
DeBurgh, Lt.-Col. E., D.S.O., 9th Horse	. . . .	. . . . .	Peshawar.
Wilson, Capt. H. M., 1st Lrs.	. . . .	. . . . .	Meerut.
Symns, 2nd-Lt. (temp. Maj.) J. M., I. A. R. O.	28 Oct. 18	. . . . .	Army Hdqrs.
Weeks, Maj. H. E., 10th G. R.	12 Nov. 18	. . . . .	Ambala Bde.
Plinston, Capt G. H., 11th Rajputs	. . . .	. . . . .	Bangalore Bde.
Moberly, Maj C. N. 15th Bombay Bw., I. D. F.	. . . .	. . . . .	Bombay.
Hallett, Maj P., 1-6th E. Surrey R.	. . . .	. . . . .	16th Dn.
Walker, Maj. C. W. G., D.S.O., 37th Dogras, *p. s. c.*	30 Oct. 18	. . . . .	Army Hdqrs.
Slim, Lt. (*temp. Maj.*) W. J., M.C., W. I. Regt.	2 Nov. 18	. . . . .	Army Hdqrs.
Gompertz, Capt. (*temp. Maj.*) A. V., M.C., R.E.	28 Nov. 18	. . . . .	Army Hdqrs.
Harper, Maj. A. F., 84th Punj.	8 Dec. 18	. . . . .	Army Hdqrs,
Bennett, Maj. E. E. S., 66th Punj., *p.s.c.*	1 Dec. 18		
Gray, Capt. (*temp. Maj.*) G. C. G., 3rd Horse	6 Nov. 18	. . . . .	Southern Command.
Tobin, Maj. H. W., D.S.O., 128th Prs., *p.s.c.*	. . . .	. . . . .	Karachi Bde.
General Staff Officers, 3rd grade.			
James, Capt. W. R., M.C., 89th Punj.	23 July 17	. . . . .	Special Intelligence.
Hamilton, Lt. (*temp. Capt.*) P. S., 2-4th Bord. R.	29 Nov. 17	. . . . .	Peshawar.
McIntosh, Capt. J. A., 1st G. B., Sco. Rif.	. . . .	. . . . .	Secunderabad.
Brewer, Capt. J. C., M.C., 90th Punj.	14 Feb. 18	. . . . .	Army Hdqrs.
Trought, Capt. T. R. W., Kent. R.	6 May 18	. . . . .	Northern Command.
Whigham-Teasdale, Lt. (*temp. Capt.*) D. G., M.C., I. A. R. O.	. . . .	. . . . .	
Smith, Lt. (*temp. Capt.*) V. C. E., Hamps. R.	. . . .	. . . . .	Southern Command.
Malet, Capt. W. De C., 14th Lrs.	. . . .	. . . . .	Mhow.
Penny, 2nd-Lt. (temp. Capt.) J. D., I. A. R. O.	29 Sep. 18	. . . . .	Army Hdqrs.
Jefferys, 2nd-Lt. (*temp. Capt.*) R. C., I.A.R.O.	18 Oct. 18	. . . . .	Army Hdqrs.
Marshall, Capt. P. H., I.A.R.O.	1 Nov. 18	. . . . .	Army Hdqrs.
Hill, Capt. A. J., I. A. R. O.	2 Nov. 18	. . . . .	
Ruegg, Maj. A. J., I. M. L.	14 Nov. 18	. . . . .	Army Hdqrs.
Parks, Capt. H. C., I.A.R.O.	8 Dec. 18	. . . . .	Army Hdqrs.
Attached to the General Staff.			
Leader, Maj.-Genl. H. P., C.B., Brit. Ser.	16 Dec. 16	. . . . .	Inspector of Cavalry.
Lecky, Maj.-Genl. R. St. C., C.B., C.M.G., R. A.	4 Nov. 18	. . . . .	Inspector of R. H. and R. F. A.
Wigley, Capt. F. G. M., 88th Infy.	9 May 18	. . . . .	Officer for Translation Duties.
Vincent, Brig.-Genl. H. O., C. B., C.M.G.	. . . .	. . . . .	Inspector of R. G. A.
Stockley, Brig.-Genl. H. R., R. E.	. . . .	. . . . .	Inspector of R.E. Pioneers.
Villiers-Stuart, Brig-Genl. W. D., 5th G. R.	. . . .	. . . . .	Inspector of Infantry (North).
Walton, Brig.-Genl. W. C., C. B., C.M.G.	. . . .	. . . . .	Inspector of Infantry (South.)
Brigadier Generals, Indian Depôts.			
Badcock, Brig. Genl. F. F., D.S.O.		. . . . .	Rawalpindi.
Codrington, Brig. Genl. H. W.	23 Feb. 18	. . . . .	16th Dn.
Twigg, Brig Genl. R. H.	25 July 17	. . . . .	Meerut.
Drake-Brockman, Brig. Genl. D. H., C.M.G.	13 Mar. 18	. . . . .	Meerut.
Lane, Brig.-Genl. H. A.	10 July 18	. . . . .	Multan & Ambala Ades.
Herbert, Brig.-Genl. L. N., C.B.	23 July 18	. . . . .	Mhow.
McCrea, Brig.-Genl. A. C., C.M.G.	26 Aug. 18	. . . . .	Mhow.
Buchanan, Brig.-Genl. K. J., C.B.	27 Sep. 17	. . . . .	Poona.
Somerset, Brig.-Genl. C. W., C.B., C.M.G., M.V.O.	27 Sep. 17	. . . . .	Secunderabad.
Money, Brig.-Genl. E. D., C.I.E.	8 July 18	. . . . .	Quetta.
Lowis, Brig.-Genl. H. E.	30 Sep. 17	. . . . .	Lucknow.
Boualer, Brig.-Genl. E. W.	11 Mar. 18	. . . . .	Lucknow.

Names, Rank and Corps.	Date, commencement tenure.	Date, completion tenure.	Remarks.
Adjutant General.			
Hudson, Lt.-Genl. *Sir* H., K.C.B., C.I.E.	5 Feb. 17	....	
Deputy Adjutant General.			
Cooke, Col. (*temp. Maj.-Genl.*) H.F., D.S.O., 32nd Prs.	8 Oct. 17	....	Army Hdqrs.
Directors.			
Luke, Brig.-Genl. T. M., D.S.O., O.B.E., R. A.	12 Oct. 16	....	Dir. of Admn.
Harrison, Brig.-Genl. A. H. P., 33rd Punj.	1 Dec. 16	....	Dir. of Rectg.
Barstow, Brig.-Genl. H., 38th Dogras	8 Feb. 17	....	Dir. of Personal Services.
Assistant Adjutant General.			
Rind, Bt.-Lt.-Col. (*temp. Lt.-Col.*) G.B.A., 124th Infy., *p.s.c.*	8 Feb. 17	....	Army Hdqrs.
Irving, Capt. (*temp. Lt.-Col.*) M., I. A. R. O.	20 Feb. 17	....	Southern Comd.
Bannatyne, Maj. (*temp. Lt.-Col.*) N. C., 128th Prs.	8 Apl. 17	....	Army Hdqrs.
Kirkhope, Lt. (*temp. Lt.-Col.*) K. M., I. A. R. O.	26 July 17	....	Army Hdqrs.
Nevill, Lt.-Col. H. R., 7th U. P. Horse, I. D. F.	17 Nov. 17	....	Army Hdqrs.
Shea, Maj. (*temp. Lt.-Col.*) A. G., D.S.O., 51st Sikhs	....	....	Army Hdqrs.
Leslie-Jones, 2nd-Lt. (*temp. Lt.-Col.*) L. H., I.A.R.O.	....	....	Northern Commd.
Cook, Lt.-Col. C. C., D.S.O.	....	....	Mhow.
Powell, Bt.-Lt.-Col. D. H., M.C., 105th Mah. L. I.	....	....	Secunderabad.
Wheelwright, Maj. T. H., Genl. List	26 Oct. 18	....	A. Gs. Bch. (Bombay).
Duncan, Maj. (temp. Lt.-Col.) H. C., D.S.O., 9th G.R. *p.s.c.*	....	....	Army Hdqrs.
Crombie, Maj. D. C., 23rd Cav., *p.s.c.* (n)	....	....	Poona.
Assistant Adjutant and Quartermaster Generals.			
Chitty, Lt.-Col. A. W., 126th Bal., *p.s.c.*	25 Dec. 16	....	Peshawar.
Bickford, Lt.-Col. W. W., C.I.E., 107th Prs.	5 Dec. 16	....	16th Dn.
Fetherstonhaugh, Bt.-Lt.-Col. W. A., D.S.O., 8th Cavy., *p.s.c.*	23 Dec. 17	....	Quetta.
Rice, Maj. (*temp. Lt.-Col.*) S. M., C.I.E., 64th Punj., *p.s.c.*	11 Feb. 18	....	Northern Command.
Tyrrell, Maj. (*temp. Lt.-Col.*) A. C. L., 25th Cav., *p.s.c.*	1 Feb. 18	....	Southern Command.
Lindesay, Maj. (*temp. Lt.-Col.*) F. S., I. A.	18 Jan. 18	....	Burma.
Binny, Maj. (*temp. Lt.-Col.*) A. J. M., 1st Lrs., *p.s.c.*	....	....	Derajat.
McConaghey, Maj. J. G. D. S. O., 25th Cav.	....	....	Rawalpindi.
Robb, Maj. (*temp. Lt.-Col.*) C. B., 1-4th R. W. Kent R	....	....	Meerut Dn.
Deputy Assistant Adjutant Generals.			
Carr, Lt. (*temp. Maj.*) C. T., 2-4th Wilts. R.	30 Dec. 15		Army Hdqrs.
Goodland, Capt. (*temp. Maj.*) C. H., 1-5th Som. L. I.	23 Oct. 16	....	Burma.
Cotton, Capt. (*temp. Maj.*) P. V., 1-9th Middx. R.	22 June 17	....	Army Hdqrs.
Wilson, Maj. T. A. A., 4th Cav.	1 Sep. 17	....	Army Hdqrs.
Bowden, Capt. (*temp. Maj.*) F. H., Mad. Gds.	....	....	Secunderabad.
Griffin, Capt. (*temp. Maj.*) J. B., 1-10th Midd'x. R.	3 Dec. 17	....	Lucknow.
Rogers, Lt. (*temp. Maj.*) E. W., W. Rid. R.	1 Dec. 17	....	Rawalpindi.
Niven, Maj. D. S., 25th Cav.	28 Feb. 17	....	Northern Command.
Grant, Maj. E. P., 25th Cav.	6 Feb. 18	....	Peshawar.
Porteous, Capt. (*temp. Maj.*) D. G., 90th Punj.	....	....	Southern Command.
Adams, Lt. (*temp. Maj.*) K. L., R. E.	....	....	Army Hdqrs.
Johnston, Maj. R., D.S.O., 4th Cav.	7 Feb. 18	....	Army Hdqrs.
Sankey, Capt. (*temp. Maj.*) G. B., 1-1 Kent Bn.	....	....	Army Hdqrs.
Hassell-Yates, Maj. G. A., R. A.	4 Apl. 18	....	Army Hdqrs.
Cairns, Capt. (*temp. Maj.*) J., 3rd E. Lan. R. attd. R.E.	22 Apl. 18	....	Army Hdqrs.
Dawson, Maj. H. H., 75th Infy.	8 June 18	....	Northern Command.
Steele, Bt.-Maj. C., 13th Hrs.	....	....	Poona Division.
Murphy, Lt. (*temp. Maj.*) A., I. A. R. O.	18 Aug. 18	....	Army Hdqrs.
Gillies, Maj. H., 1st Lrs.	16 Sep. 18	....	Mhow.
Murland, Maj. H. F., 64th Prs.	....	....	Secunderabad.
Taylor, Capt. (*temp. Maj.*) G. C. R., 1-8th Essex R.	....	....	16th Dn.
Maxwell, Maj. D. G., R. E.	....	....	Army Hdqrs.
Dobbs, Lt. (*temp. Maj.*) W. E. J., I. A. R. O.	22 Aug. 18	....	Army Hdqrs.
Mackenzie, Maj. H. J., D.S.O., 51st Sikhs	....	....	Rawalpindi.
Ball, Capt. (*temp. Maj.*) C. F., I. D. F.	....	....	Meerut Dn.
Gordon, Maj. J. G. B., M.C., 52nd Sikhs (F. F.)	....	....	Army Hdqrs.
Lamb, Capt. (*temp. Maj.*) D. O. W., 10th Lrs.	28 Nov. 18	....	Army Hdqrs.
McWatters, (Maj. Bt.-Lt.-Col.) H.C., D. S. O., *p. s. c.*, 24th Panj.	3 Dec. 18	....	Army Hdqrs.
McPherson, Capt. A. B., M. C., 6th Jats	10 Dec. 18	....	A. Gs. Bch. (Bombay).
Foster, Maj. H. M., 1-4th Hamps. R.	....	....	Quetta.

Names, Rank and Corps.	Date, commencement tenure.	Date, completion tenure.	Remarks.
Quartermaster General.			
Altham, Lt.-Genl. *Sir* E. A., K.C.B., C.M.G., *p.s.c.*, Brit. Ser.	18 Feb. 17	. . . .	Army Hdqrs.
Deputy Quartermaster General.			
Richardson, Maj.-Genl. C. W. G., C.S.I., *p.s.c.* [*L*]	1 Oct. 17	. . . .	Army Hdqrs.
Mahon, Maj. (*temp. Col.*) H. J., C.I.E., 23rd Sind Bn. I.D.F.	30 Jan. 17	. . . .	Karachi (Embarkation).
Brigadier Generals.			
Templer, Brig.-Genl. C. F., C.M.G., C.I.E.	5 Sep. 15	. . . .	D. G., A. R. D., Army Hdqrs.
Vaughan, Brig.-Genl. R. E., C.B., S. & T. Corps	5 Oct. 16	. . . .	D. S. & T., Army Hdqrs.
Holdich, Brig.-Genl. H. A., D.S.O., 5th G. R., *p.s.c.*	19 Nov. 17	. . . .	D. Q., Army Hdqrs.
Nuthall, Brig.-Genl. C. E., C.B., C.M.G., A. V. S.	26 Jan. 18	. . . .	D. V. S., Army Hdqrs.
Assistant Quartermaster Generals.			
Ward, Maj. (*temp. Lt.-Col.*) B. S., 99th Infy.	. . . .	. . . .	Bombay.
Ross, Bt.-Lt.-Col. (*temp. Lt.-Col.*) R. C., D.S.O., 6th Jats	17 Feb. 17	. . . .	Army Hdqrs.
Sherer, Lt.-Col. J. C., 37th Lrs.	. . . .	. . . .	Bombay.
Muspratt, Maj. (*temp. Lt.-Col.*) V. E., 30th Lrs. *p.s.c.* [*l*]	27 Feb. 18	. . . .	Army Hdqrs.
Moody, Maj. (*temp. Lt.-Col.*) F. H., M.C., 13th Lrs.	11 July 18	. . . .	Army Hdqrs.
Twiss, Maj. (*temp. Lt.-Col.*) E. K., D.S.O., 10th Jats	. . . .	. . . .	Karachi.
Calvert, Lt.-Col. R. T. C., 122nd Infy, *p.s.c.*	. . . .	. . . .	Mhow.
Kitson, Bt.-Lt.-Col. C.E., D.S.O., R. W. Kent R.	. . . .	. . . .	Poona.
Cox, Bt. Lt.-Col. R. O'B. C.M.G., C.I.E., 19th Lrs, *p.s.c.* [L]	. . . .	. . . .	Secunderabad.
Lloyd, Lt.-Col. J. H., 6th G. R.	. . . .	. . . .	Lucknow.
Deputy Assistant Quartermaster Generals.			
Mason, Capt. (*temp. Maj.*) A. S., 1-6th Devon. R.	27 June 16	. . . .	16th Dn.
Wigley, Maj. A. B., R. Suss. R.	14 Mar. 17	. . . .	Peshawar.
Farquharson, Capt. (*temp. Maj.*) C. W., 10th Jats.	. . . .	. . . .	Bombay.
Wilson, Maj. W. H. D., 1st Brah.	7 May 17	. . . .	Army Hdqrs.
Betham, Capt. (*temp. Maj.*) G. L., 15th Sikhs	. . . .	. . . .	Karachi.
Hall, Maj. G. E., 117th Mah.	20 June 17	. . . .	Army Hdqrs.
Channer, Maj. B. G., 54th Sikhs	23 Aug. 17	. . . .	Southern Command.
Williams, Capt. (*temp. Maj.*) S. C., S. & T. Corps	25 Feb. 18	. . . .	Quetta.
Brewis, Capt. (*temp. Maj.*) R. D., 1-5th R. W. Kent R.	28 Nov. 17	. . . .	Meerut.
Clarke, Maj. B. G. S., 10th G. R. [*l*]	23 Mar. 18	. . . .	Army Hdqrs.
Phillips, Capt. (*temp. Maj.*) G. A., I.A. R.O., V.D.	9 Mar. 18	. . . .	Army Hdqrs.
Humphrey, Capt. (*temp. Maj.*) J., I. A. R.O.	. . . .	. . . .	Karachi.
Haig, Lt. (*temp. Maj.*) H. G., I. A. R. O.	. . . .	. . . .	Burma.
Mackenzie, Capt. (*temp. Maj.*) J. D., 41st Dogras	. . . .	. . . .	Poona.
Oakes, Maj. T. T., 102nd Grs.	7 Aug. 18	. . . .	Army Hdqrs.
Green-Armytage, Maj. P. D., 117th Mahrattas.	21 Aug. 18	. . . .	Army Hdqrs.
Foster, Maj. C. S., 37th Lrs.	. . . .	. . . .	Secunderabad.
Hammersley-Smith, Maj. R. H., 14th Lrs.	. . . .	. . . .	Mhow.
Leapman, Maj. H. M., 13th Rajputs	. . . .	. . . .	Rawalpindi.
Meiklejohn, Capt. (*temp. Maj.*) J. F., 33rd Cav.	30 Nov. 18	. . . .	Army Hdqrs.
Deputy Assistant Adjutant and Quartermaster Generals.			
Story, Capt. (*temp. Maj.*) H. F., 35th Sikhs	24 Feb. 17	. . . .	Bannu.
Ball, Capt. (*temp. Maj.*) E. P., 110th Mah. L. I.	25 Feb. 17	. . . .	Bombay.
Cobb, Capt. (*temp. Maj.*) C., S. and T. Corps	30 Aug. 17	. . . .	Derajat.
Bannatyne, Maj. F. C., 109th Infy.	. . . .	. . . .	Kohat.
Raymond, Maj. E. D., D.S.O., M.C., 30th Lrs.	12 Jan. 18	. . . .	Presidency.
Lucas, Capt. (*temp. Maj.*) H. de N., 30th Lrs.	5 Feb. 18	. . . .	Northern Command.
Curry, Capt. (*temp. Maj.*) F. R., R. Dub. Fusrs.	20 Apl. 18	. . . .	Bangalore.
Kenworthy, Capt. (*temp. Maj.*) W., 33rd Cav.	15 Apl. 18	. . . .	Ambala.
Goldingham, Capt. D. D., 6th Middx. R.	. . . .	. . . .	Karachi.
Cave-Browne, Capt. H., 23rd Cav.	. . . .	. . . .	Jullundur Bde.

Names, Rank and Corps.	Date, commencement tenure.	Date, completion tenure.	Remarks.

Administrative Staff, General Officers.

Names, Rank and Corps.	Date, commencement tenure.	Date, completion tenure.	Remarks.
Walter, Maj.-Genl. J. MacN., C.B., C.S.I., D.S.O., Brit. Ser.	5 Feb. 17	. . . . .	Northern Command.
James, Maj.-Genl. W. B., C.B., C.I.E., M.V.O.	18 Feb. 17	. . . . .	Southern Command.
Drew, Brig.-Genl. A. B. H., C.I.E.	. . . .	. . . . .	5th Dn.
Glanville, Brig.-Genl. F., D.S.O., Brit. Ser.	. . . .	. . . . .	8th (Lucknow) Dn.
Lukin, Brig.-Genl. R. C. W., D.S.O., 9th Horse	. . . .	. . . . .	2nd (Rawalpindi) Dn.
Jackson, Brig.-Genl. H. W. D.S.O., 89th Punj.	30 Sep. 18	. . . . .	4th (Quetta) Dn.
Boddam, Lt.-Col. E. B. C. 5th G. R., *p.s.c.*	. . . .	. . . . .	Poona Dn.
Cox, Col. F. W. H., C. B., I.A.	. . . .	. . . . .	9th (S'.bad) Dn.

Brigade Majors.

Names, Rank and Corps.	Date, commencement tenure.	Date, completion tenure.	Remarks.
Evans, Capt. B. L., 1-4th R. W. Surr. R.	13 Nov. 16	. . . .	Allahabad Bde.
Griffith, Maj. L., 107th Prs.	14 Dec. 16	. . . .	Abbottabad.
Edward-Collins, Capt. G., M.C., 33rd Cav.	. . .	. . . .	4th Cav. Bde.
Williams, Maj. R. C. B., 35th Sikhs.	17 Apl. 17	. . . .	Ferozepore Bde.
Manners, Maj. R. H. H., 106th Prs.	19 Aug. 17	. . . .	Jubbulpore Bde.
Nelson, Capt. C., D.S.O., 15th Hrs.	. . .	. . . .	10th Cav. Bde.
B.C. Smyth, Capt. J. G., 15th Sikhs	. . .	. . . .	43rd Bde.
Young, Capt. K. deL., M.C., 8th Cav.	21 Sep. 17	. . . .	Secunderabad Cav. Bde.
Satterthwaite, Capt. R. E., 1-5th R. W. Kent R.	12 Oct. 17	. . . .	11th Inf. Bde.
Whitworth, Maj. T. S., 72nd Punj.	2 Nov. 17	. . . .	Nowshera Bde.
Scoones, Capt. G. A. P., D.S.O., M.C., 2nd G. R.	27 Oct. 17	. . . .	10th Infy. Bde.
Pogson, Capt. C. A., M.C., 117th Mah.	1 Dec. 17	. . . .	Delhi Bde.
Mackenzie, Capt. R. H. T., 1-10th Middx. R.	. . . .	. . . .	
Haughton, Maj. E. J. H., D.S.O., 105th Infy.	. . . .	. . . .	Peshawar Bde.
Macleod, Maj. N., D.S.O., 15th Sikhs, *p.s.c.*	28 Jan. 18	. . . .	Rangoon Bde.
Bertin, Capt. H., 1st G. B., E. Yorks. R.	23 Nov. 17	. . . .	Lucknow Bde.
deCourcy, Maj. *Hon.* M. W. R., D.S.O., 32nd Prs.	. . . .	. . . .	6th Infy. Bde.
Reader, Capt., F. E. M., 1-9th Hamp. R.	. . . .	. . . .	Southern Bde.
Mundy, Capt. H. A. G., 2-6th R. Suss. R.	. . . .	. . . .	Poona Bde.
Green, Capt. E. C., 1-4th E. Kent R.	18 Apl. 18	. . . .	Garhwal Bde.
Nugent, Capt. J. F. H., D.S.O., 28th Punj.	. . . .	. . . .	57th Infy. Bde.
Kellie, Capt. G. J. D., M.C., 30th Punj.	. . . .	. . . .	Jhelum Bde.
Sherston, Capt., J. R. V., D.S.O., M.C., 11th Lrs.	. . . .	. . . .	1st Cav. Bde.
Maynard, Capt., P. A., 1-5th E. Kent R.	. . . .	. . . .	Nasirabad Bde
Harbord, Capt. R. E., 1-4th E. Kent R.	26 Oct. 16	. . . .	Ahmednagar Bde.
Foster, Maj. W. M. A., D.S.O., 10th Lrs.	4 Sep. 18	. . . .	Rawalpindi.
Nicolls, Capt. B. E., 7th N. Staff. R.	. . . .	. . . .	Bareilly Bde.
Briggs, Capt. F. C. C., D.S.O., L'pool R.	. . . .	. . . .	45th Infy. Bde.
Wilson, Capt. A. P., 1-4th Bord. R.	. . . .	. . . .	Multan Bde.
Birch, Maj. C. F. M., 1st G. R.	. . . .	. . . .	Jhansi Bde.
Mallock, Capt. A. R. O., 42nd R.	. . . .	. . . .	58th Infy. Bde.
Erskine, Capt. K. C. S., M.C., 5th G. R.	. . . .	. . . .	59th Infy. Bde.
Gray, Maj. E. St. C., M.C., 34th Horse.	. . . .	. . . .	12th Mtd. Bde.
Morris, Maj. G. P., 30th Lrs.	. . . .	. . . .	13th Mtd. Bde.
White, Maj. S. H., 1-10th Middx. R.	15 Nov. 18	. . . .	Ambala Bde.
Bourne, Capt. F. C., 1-4th R. W. Kent. R.	. . . .	. . . .	5th Infy. Bde.
Vallance, Lt. (*temp. Capt.*) G. A. G., 2-5th Som. L. I.	. . . .	. . . .	Sialkot Bde.

Staff Captains.

Names, Rank and Corps.	Date, commencement tenure.	Date, completion tenure.	Remarks.
Angier, Lt. (*temp. Capt.*) C. T., 24 Rif. Bde.	20 Sep. 16	. . . .	Bombay Bde.
Ryder, 2nd-Lt. (*temp. Capt.*) E. N. 21st Lrs.	9 Oct. 16	. . . .	Risalpur Cav. Bde.
Reed, Capt. H. S., 2-4th Devon R.	22 Jan. 17	. . . .	Bangalore.
Duffy, Capt. L., S., Lanc. R.	. . .	. . . .	Madras.
Broadmead, Lt. (*temp. Capt.*) H. H., 2-5th Som. L.I.	24 Apl. 17	. . . .	43rd Bde.
Marshall, Capt. H. G., 2-4th Bord. R.	1 June 17	. . . .	Peshawar Bde.
Fisher, 2nd-Lt. (*temp. Capt.*) M. G., 2-4th Bord R.	6 Aug. 17	. . . .	Jubbulpore Bde.
Stracihey, Maj. *Hon.* E., 1-4th Som. L. I.	14 Sep. 17	. . . .	Karachi Bde.
Phillips, Capt. C. S., 1-1st Kent Bn.	. . .	. . . .	Delhi Bde.
Jackson, Lt. (*temp. Capt.*) A. H. C., I.A.R.O.	. . .	. . . .	Bombay Bde.
Orchard, Capt. J. W., 1-4th Devon R.	. . .	. . . .	Madras.
Coburn, Lt. (*temp. Capt.*) F. R., Gen. List	24 Sep. 17	. . . .	Bangalore.
Tregoning, Capt. E. A., 2nd G. B., North'd. Fus.	1 Oct. 17	. . . .	6th Infy. Bde.
Horsfield, Lt. (*temp. Capt.*) H. E., I. A. R. O.	30 Nov. 17	. . . .	Allahabad Bde.
Lee, Capt. G. B., R. Ir. Rif.	12 Dec. 17	. . . .	59th Infy. Bde.
Hobday, Lt. (*Temp. Capt.*) W. E., I. A. R. O.	15 Jan. 17	. . . .	Ambala Bde.
Cressall, 2nd-Lt. (*Temp. Capt.*) H. A., 1-10th Midd'x. R.	23 Nov. 17	. . . .	Lucknow Bde.
Trusted, Lt. H. H., 2-4th D. of Corn. L. I.	27 Dec. 17	. . . .	Derajat Bde.

Staff Captains—*concld.*

Names, Rank and Corps.	Date, commencement tenure.	Date, completion tenure.	Remarks.
Taylor, Lt. (*temp. Capt.*) W. R., I. A. R. O.	4 Mar. 18		Army Hdqrs.
Ferrier, Lt. (*temp. Capt.*) C. D., I. A.R.O.	26 Feb. 18		Army Hdqrs.
Horner, Capt., B. S, 1-6th E. Surr. R.			45th Bde.
Joseph, Capt. A. F., I.A.R.O.			Bareilly Bde.
Emery, Lt. (*temp. Capt.*) T. S., M.C., 1-76th Punj.			Jhansi Bde.
John, Lt. (*temp. Capt.*) W. R., I. A. R. O.			Rawalpindi Infy. Bde.
Oliver, 2nd-Lt. (*temp. Capt.*) St. J. L., I. A. R. O.			5th Infy. Bde.
Tweedy, Capt. R. J., 19th Lrs.			11th Infy. Bde.
Banks, Capt. T. M., Bord. R.			Northern Command.
Gould, Lt. (*temp. Capt.*) H. R., I.A.R.O.			Bangalore.
Buck, 2nd-Lt. (*temp. Capt.*) H., 5th Suff. R., Attd. 4th Devon. R.	19 May 18		Nowshera Bde.
Braund, Capt. H. B. L., 5th Wilts. R.			10th Infy. Bde.
Menzies, Capt. F. A., M.C., I.A.R.O.	21 May 18		......
Dymott, Capt. B. K., 31st Punj.			57th Infy. Bde.
Gladstone, Capt. T. H., 1st Dn. Gds.			44th Infy. Bde.
Bostock, Lt. (*temp. Capt.*) J. A., I. A. (*on probn.*)			45th Infy. Bde.
Langford, Lt. (*temp. Capt.*) W., 1st Dn. Gds.			4th Cav. Bde.
Binning, Lt. (*temp. Capt.*) H. A., I.A.R.O.			12th Mtd. Bde.
Ibbotson, Lt. (*temp. Capt.*) A. W., M.C., I.A.R.O.			13th Mtd. Bde.
Fear, Capt. F. R., 1-5th Hamps R.			Rangoon Bde.
Cambier, Capt. V., 32nd Prs.			Jhelum Bde.
Maxwell, 2nd-Lt. W. R. Attd. Arg. & Suth'd Highrs.			Abbottabad Bde.
Paterson, Lt. G. S. V., I.A.R.O.			Garhwal Bde.
Gomperty, Lt. M. C. T., 8th Rajputs			Multan Bde.
Kent, Lt. C. E., L'pol. R.			Ahmednagar Bde.
Vousley, Capt. F., I.A.R.O.			16th Dn.
Prior, Capt. C. J., I.A.R.O.	11 Oct. 18		Army Hdqrs.
Jackson, Capt. C. G., I.A.R.O.	11 Oct. 18		Army Hdqrs.
Lang, Capt. W. F., I.A.R.O.	8 Nov. 18		Army Hdqrs.
Blair, Lt. (*temp. Capt.*) J. R., I.A.R.O.			10th Cav. Bde.
Cotter, Capt. P., I.A.R.O.	20 Nov. 18		A. Gs. Bch. (Bombay).
Sturdy, Capt. J. R., M.C., 3-5th G.R.	25 Nov. 18		A. Gs. Bch. (Bombay).
Parselle, Lt. (*temp. Capt.*) J., R. E.	5 May 17		Army Hdqrs.
Lemon, Capt. G. T., I.A.R.O.	1 June 17		Army Hdqrs.
Raven, 2nd-Lt. (*temp. Capt.*) L.F., I.A.R.O.	19 Aug. 18		Army Hdqrs.
Richardson, Capt. F. L., I.A.R.O.	25 Oct. 18		Army Hdqrs.
Whittington, Capt. L., 1-5th R. W., Surr. R.			58th Infy. Bde.
Twiddy, Lt. A. A., 2-9th Delhi Infy.			Nasirabad Bde.
Sheppard, 2nd-Lt. (*temp. Capt.*) W. H., attd. 14th Hrs.			Poona Bde.
Hall, Lt. (*temp. Capt.*) F. G., Dorset R.			Karachi Bde.
Stuart, Lt. D. S., attd. 112th Infy.			Bannu Bde.

Station Staff Officers, 1st class.

Names, Rank and Corps.	Date, commencement tenure.	Date, completion tenure.	Remarks.
Walker, Capt. T., Midd'x R.	15 Apl. 15		Darjeeling.
Parke, Capt. L. S., 1-7th Hamps. R.	4 Nov. 15		Meerut.
Sinclair, Capt. W. C. B., 18th Rif. Brig.			Rangoon.
Tanner, Lt. (*temp. Capt.*) G., 1st G. B., Bedf. R.			Delhi.
Ralph, Lt. (*temp. Capt.*) M. M., 1 G. B., Sco. Rif.			Pachmarhi.
Armour, Capt. J. S. D. M., 1. G. B., Sco. Rif.	27 July 17		Bombay.
Bird, 2nd-Lt. (*temp. Capt.*) R., 1-4th Devon R.			Secunderadad.
Gillett, Capt. N. C., 1-25th Lond R.	4 Dec. 17		Multan.
Woodliffe, Maj. H. D., Brecknock Bn.			Dalhousie.
Crump, Lt. (*temp. Capt.*) D. J., 1-9th Middx. R.	1 Mar. 18		Jhelum.
Smith, 2nd-Lt. (*temp. Capt.*) J., I.A.R.O.			Quetta.
Walmsley, Capt., E. S., 1st Kent Bn.			Ambala
Henderson, Lt. (*temp. Capt.*) J. T., 1st G. B. Sco. Rif			Landour.
Williams, Capt. J., 1-1st Brecknock Bn.			Mhow.
Jones, Capt. V. S., 2-4th Bord. R.			Peshawar.
D'Ambruninel, Lt. W., 2-4th D. of Corn. L. I.			Chakrata.
Robinson, Lt. (*temp. Capt.*) F. W. O., 29th Lrs.	23 Nov. 18		Poona.

(*a*) Tenure extended.

NORTHERN COMMAND.

General Officer Commanding.	**Barrett,** Genl. *Sir* A. A., G.C.B., K.C.S.I., K.C.V.O. A.D.C.	31 May 16
A.-M. S.	**Cornish,** Maj. A. W. D., D.S.O., M.C., 6th G. R.	25 Oct. 17
A.-D.-C.	**Hinds,** Capt. W. A. G., M.C., 103rd Infy.	12 Feb. 17
A.-D.-C.	**Khwaja Muhammad Khan,** Risaldar, *Sardar Bahadur.* Guides.	19 June 16
GENERAL STAFF— *Brig.-Genl.*	**Looh,** Brig.-Gen. S. G., D.S.O., Brit. Ser., *p. s. c.*	7 July 16
Genl. Staff. Offr., 2nd gr.	**Macleod,** Maj. W. M., 31st Lrs., *p. s. c.*	20 June 17
Genl. Staff Offr., 3rd gr.	**Trought,** Capt. T., R. W. Kent. R.	6 May 18
ADMINISTRATIVE STAFF— *General Officer*	**Walter,** Maj.-Genl. J. MacN., C.B., C.S.I., D.S.O., Brit. Ser.	5 Feb. 17
A. A. G.	**Leslie-Jones,** 2nd-Lt. (*temp. Lt.-Col.*) L. H., I. A. R. O.	...
A. A. & Q. M. G.	**Rice,** Maj. (*temp. Lt.-Col.*) S. M., C.I.E., 64th Prs. *p.s.c.*	11 Feb. 18
D. A. A. G.	**Dawson,** Maj. H. H., 75th Infy.	8 June 18
D. A. A. & Q. M. G.	**Lucas,** Capt. (*temp. Maj.*) H. de N., 30th Lrs.	5 Feb. 18
Staff Capt.	**Banks,** Capt. T. M., Bord. R.	...
Asst. Dir. Ord. Service	**Donovan,** Lt.-Col. C. C., R. A.	24 May 18
ATTACHED— *Chief Engineer*	**Sorsbie,** Brig.-Genl. R. F., C.I.E., Brit. Ser.	21 Feb. 17
Deputy Chief Engineer	**Bond,** Lt.-Col. R.F.G., R.E.	...
Staff Offr. R.E.	**Phipps,** Capt. C. C., M.C.	...
Asst. Inspr. of Machinery	**Young,** Lt. (*temp. Maj.*) J., I. A. R. O.	15 Mar. 18
D. D., Vety. Ser.	**Edwards,** Lt.-Col. W. B., A. V. C.	8 Dec. 17
Inspr. of Musky. & Machine Guns.	**Burt,** Capt. H. V., 37th Lrs.	18 Apl. 18
Inspr. of Physical and Bayonet Training.	**Coulon,** Lt. (*temp. Maj.*) F., O.B.E.,	...
Inspr. of S. & T. Services	**Chamier,** Lt.-Col. W. St. G., S. & T. Corps	3 Mar. 17
D. D., S. & T.	**Jones,** Lt.-Col. (*temp. Col.*) H. H., S. &. T. Corps	9 Mar. 18
D. A. D., S. & T.	**Steward,** Maj. E. M., S. & T. Corps	23 Feb. 18
D. D. M. S.	**Hehir,** Maj.-Genl. P., C.B., C.M.G., C.I.E., M.D.	13 June 17
D. A. D. M. S.	**Popham,** Maj. R. L., R.A.M.C.	22 Feb. 17
D. J. A.-G.	**Beatson-Bell,** Lt.-Col. J.	1 Nov. 10
A. J. A. G.	**Smallpeice,** Maj. F. W., 1-5th R. W. Surr. R.	26 Sep. 17
Inspr. of Bombing	**Burbidge,** 2nd-Lt. (*temp. Capt.*) W. H. M., R. Suss. R.	...
Comdt., School of Physical and Bayonet Training.	**Watson,** Lt. (*temp. Capt.*) P., High L.I.	16 June 15

1st (*PESHAWAR*) DIVISION.

COMMANDER	**Campbell**, Lt.-Genl. *Sir* F., K.C.B., D.S.O.
A.-D.-C.	**Hill**, 2nd-Lt. H. B. C., I. A. R. O.
GENERAL STAFF—	
Gen. Staff Offr., 1st gr. . .	**Tarver**, Bt.-Col. A. L., D.S.O., A. D. C., 124th Infy. *p.s.c.*
Gen. Staff Offr., 2nd gr. . .	**De Burgh**, Lt-Col. E., D.S.O., 9th Horse.
Gen. Staff Offr., 3rd gr. . .	**Hamilton**, Lt. (*temp. Capt.*) P. S., 2-4th Bord. R.
ADMINISTRATIVE, TECHNICAL, AND DEPARTMENTAL STAFF—	
A. A. & Q. M.-G. . . .	**Chitty**, Lt.-Col. A. W., 126th Baluchis, *p.s.c.*
D. A. A. G.	**Grant**, Maj. E. P., 25th Cav.
D. A. Q. M.-G.	**Blood**, Capt. W. H., Guides Infy.
Comdg. R. A.	**McLeod**, Brig.-Genl. W. K., Brit. Ser. (*Also C. R. A. Frontier Bdes.*).
Orderly Offr.	
C. R. E.	**Fraser**, Col. H. A. D., R.E.
Asst. Dir. of Supplies & Tranport.	**Roddy**, Lt.-Col. H. H., C.B., S. & T. Corps.
D. A. D., Supplies . . .	**Muscroft**, Bt. Lt.-Col. W. St. C., C.B., S. & T. Corps.
D. A. D., Transport . . .	**Wildes**, Maj. C. S. D., S. & T. Corps.
Offg. in charge Supply Depôts	**Webb**, Lt.-Col. A. E., S. & T. Corps.
Asst. Dir. of Grass Farms	**Obbard**, Lt.-Col. E. N.
Asst. Dir. Medl. Services	**Condon**, Col. E. H., A. M. S.
Dy. Asst. Dir. Medl. Services (Sany.).	**Bridges**, Capt. (*temp. Maj.*) R. F., R.A.M.C. (*Offg.*).
Dy. Asst. Dir. Medl. Services (Mobn.).	**Doyle**, Capt. (*temp. Maj.*) E. E., I.M.S.
D. A. D. Ord. Stores . .	**Whitaker**, Bt.-Maj., H. C., A.O.D.
Dy. Asst. Dir., Vety. Ser. .	**Davis**, Maj. T. J., A. V. C.
Contr., Mily. Accts. (Under Finance Dept.) .	**Jackson**, Lt.-Col. K. H.

Chitral and Drosh.

Comdg.	**Samborne-Palmer**, Lt.-Col. F. C. S., 11th Rajputs.
S. O. (3rd cl.)	**Scudamore**, Capt. P. C., 11th Rajputs.
Offr. in charge Supplies & Transport.	**Sulley**, Lt. (*temp. Capt.*) J. P., I. A. R. O.
G. E.	**Riccomini**, Ty. 2nd-Lt. J. E., R. E.

Malakand, (*including Chitral and Dargai*).

Comdg. (1st cl.) . . .	**Oldfield**, Lt.-Col. G. E. B., 2-89th Punj.
S. O. (3rd cl.) . . .	**Clinton**, Lt. A. G., 2-2nd Rajputs.
G. E.	**Heenan**, Lt. J. R. E.

Chakdara.

Dargai.

S. O. (3rd cl.) . . .	**Wodehouse**, Capt. C. G., 2-89th Punj.

1st (*PESHAWAR*) DIVISION—*concld.*

PESHAWAR BRIGADE—

Comdg. . . . . . . **Baldwin**, Brig.-Genl. G. M., D.S.O.
S.O. (1st cl.) . . **Jones**, Capt. V. S., 2-4th Bord. R.
A. C. R. E. . . . . . **Scott**, Capt. M. A. H., M.C., R. E. (*offg.*)
Offr. in charge Station Supply Depôt. **Green**, Maj. W. W., S. & T. Corps.
G. E. . . . . . . **Battman**, Capt. A. R.
Cant. Magte. . . . . **Wimberley**, Lt.-Col. F. G. A., I.A.

1st INFANTRY BRIGADE—

Bde.-Comdr. . . . . . **Crocker**, Brig.-Genl. G. D., Brit. Ser.

Bde.-Maj. . . . . **Haughton**, Maj. E. J. H., D.S.O., 105th Infy.

Staff Capt. . . . . . **Marshall**, Capt. H. G., 2-4th Bord. R.

10th CAVALRY BRIGADE—
Bde.-Comdr. . . . . . **Baldwin**, Brig.-Genl. G. M., I. A., D.S.O.
Bde.-Maj. . . . . . . **Nelson**, Capt. C., D.S.O., 15th Hrs.
Staff Captn. . . . . **Blair**, Lt. (*temp. Capt.*) J. R., I. A. R. O.

Abazai.

Cherat.

Comdg. . . . . . . **Carey**, Lt.-Col. M. D., 1st York. R.
S. O. and Cant. Magte. **Ball**, Lt. G. H., 1st York. R.

Jamrud.

Comdg. . . . . . **Clark**, 2nd-Lt. J., I. A. R. O.

Shabkadar.

Comdg. . . . . . . **Tancock**, Maj. A. C., 31st Punj.
G. E. . . . . **Brown**, 2nd-Lt. G., R. E.

3rd INFANTRY BRIGADE—
Bde.-Comdr. . . . . .
Bde.-Maj. . . . . . .
Staff. Capt. . . . . .

Campbellpore.

Comdg. . . . . **Munn**, Capt. F. L. R., M.C., 26th Punj.
S.O. (2nd cl.) . . . . . **Reeves**, 2nd-Lt. S. G., D.C.M., R. F. A.

NOWSHERA BRIGADE—
Bde.-Comdr. . . . . **Climo**, Maj.-Genl. S. H., C.B., D.S.O.
Bde.-Maj. . . . . **Whitworth**, Maj. T. S., 72nd Punj.
Staff Capt. . . . . **Buck**, 2nd-Lt. (*temp. Capt.*) H., 5th Suff. R. attd. 4th Devon. R.

Nowshera.

S. O. (2nd. cl.) . . . **Crickett**, Lt. G. W., 2nd N. Staffs. R.
A. C. R. E. . . . . . **Gervers**, Maj. F. R. S., R.E. (on deputation.)
A. C. R. E. . . . . . **Evans**, Maj. A. S., R.E (*Offg.*)
Station Transport Officer **Lancaster**, Lt., C. E. A., S. and T. Corps.
Offr. in charge Station Supply Depôt. **Jones**, Lt. J. E., S. & T. Corps.
G. E. . . . .
G. E. Electrical . . . **Hopps**, Lt. O. D., I. A. R. O.
Asstt. G. E. . . . . . **Webster**, 2nd-Lt. A., I. A. R. O.
Cant. Magte. . . . . . **Gordon**, 2nd-Lt. N. H., I. A. R. O.

RISALPUR (CAVALRY) BRIGADE
Bde.-Comdr. . . . . . **Davies**, Brig.-Genl. F. G. H., I. A.
Bde.-Maj. . . . . . **Sherston**, Capt. J. R. V., D.S.O., M.C., 11th Lrs.
Staff Capt. . . . . . **Ryder**, Lt. (*temp. Capt.*) E. N., D. C. M., 21st Lrs.

Risalpur.

G. E. . . . . **Tayler**, Lt. H. J. D., R. E.
S.O. (2nd cl.) . . . . **Baddeley**, 2nd-Lt. H., 21st Lrs.

Mardan.

Comdg. . . . . **Sproule**, Maj. (*Actg. Lt.-Col.*) H. H., 4th Cav.
S. O. (3rd cl.) and Cant. Magte. **Eldon**, Lt. A., 1st Guides Infy.

2nd (*RAWAL PINDI*) DIVISION.

COMMANDER . . . . . . **Dobell**, Maj.-Genl. *Sir* C. M., K.C.B., C.M.G., D.S.O., *p.s.c.* [*l*] Brit. Ser.
A.-D.-C. . . . . . **Gordon**, Lt. C. F., 24th Rif. Brig.
Orderly Offr. . . . . .
GENERAL STAFF.
Gen. Staff Offr., 1st gr. . . **Ward**, Bt.-Lt.-Col. H. C. S., 2nd Lrs., *p.s.c.* (*e*)
Gen. Staff Offr., 2nd gr. . . **Carruthers**, Maj. J. H. deW., M.C., 39th Garh. Rif., *p.s.c.*
Gen. Staff Offr., 2nd gr. **Smyth**, Maj. C. A. W., 1st Brahmans.

ADMINISTRATIVE, TECHNICAL AND DEPARTMENTAL STAFF.

General Officer . . . **Lukin**, Brig.-Genl. R. C. W., D.S.O., 35th Horse.
A. A., & Q. M. G. . . **McConaghy**, Maj. J. G., D.S.O., 25th Cav.
D. A. Q. M. G. . . . **Leapman**, Maj. H. M., 13th Rajputs.
D. A. A. G. . . . . **Rogers**, Lt. (*temp. Maj.*) E. W., W. Rid. R.
D. A. A. G. . . . . . **Mackenzie**, Maj. H. J., D.S.O, 51st Sikhs.
Comdg. R. A. . . . . . **Caulfeild**, Brig.-Genl. C. T., C.M.G.
Orderly Offr. . . . . . **Hawker**, Lt. G. C., R. F. A.
Brig.Genl., Indian Depôts **Badcock**, Brig.-Genl. F. F., D.S.O.
C. R. E. . . **Duff**, Col. G. M., R.E.
Asst. Dir. of Supplies & Transport. **Lindsay**, Bt.-Col. H. A. P., C.M.G., S. & T. Corps [L].
D. A. D. Supplies . . . **Morris**, Maj. C. O., D.S.O., S. & T. Corps.
D. A. D. Transport . . **Bryson**, Capt. A. B., S. & T. Corps.
Asst. Dir. of Grass Farms **Obbard**, Lt.-Col. E. N.
Dy. Asst. Dir. of Grass Farms. **Wheeler**, Maj. E. S., Mily. Farms Deptt.
Asst. Dir., Medl. Services **Mawhinny**, Lt.-Col. R. J. W., R. A. M. C.
Dy. Asst. Dir., Medl. Services (Sany.) . . **Ryles**, Capt. (*temp. Maj.*) C., R. A. M. C. (*Offg.*)
Personal Asst. to D. A. D. M. S. (Sany). **Cullen**, Capt. J. A. W., R. A. M. C.
Dy. Asst. Dir. Medl. Services (Mobn.) **Hood**, Capt. (*temp. Maj.*) A., R.A.M.C.
D. A. D. of Ord. Ser. . **Neale**, Maj., M. R., M.C., A.O.D.
Dy. Asst. Dir., Vety. Ser. **Hayes**, Capt. (*temp. Maj.*) F. B., A.V.C.
Inspr. of Army Schools . **Parish**, Capt. A. H. Inspr., Army Schools (1st Circle).
Contr., Mily. Accts. (Under Finance Dept.) **Daldy**, Maj. A. W.

4th INFANTRY BRIGADE—

Bde. Comdr. . . . . . **Peebles**, Brig.-Genl. E. C., C.B., C.M.G., D.S.O., Brit. Ser.
Bde.-Maj. . . . . . **Foster**, Maj. W. M. A., D.S.O., 10th Lrs.
Staff Capt. . . . . . **John**, Lt. (*temp. Capt.*) W. R., I. A. R. O.

RAWALPINDI (ADDITIONAL) BRIGADE—
Bde. Comdr. . . . . . **Berners**, Brig.-Genl. R. A., D.S.O.
Bde. Maj. . . . . . . **Mackenzie**, Capt. R. H. T., 1-10th Midd'x. R.
Staff Capt. . . . . . .

Rawal Pindi.

Comdg. . . . . . . **Berners**, Brig.-Genl. R. A., D.S.O.
S. O. (1st class) . . .
A. C. R. E. . . . . . . **Murray**, Lt.-Col. J. H. S., R.E.
Offr. in charge Supply Depôt **Synge**, Maj., M., D.S.O., (*e*) S. & T. Corps.
Offr. in charge Supplies. (Murree Hills).
G. E., North . . . . **Faulkner**, Lt., (*Actg. Capt.*) E. A., attd. R. E.
G. E., South . . . . **Hall**, Lt. T. C., M. W. S.

Cant. Magte. . . . . . **Browning**, Lt.-Col. T. C. Supy. L 1st.

Attock.

Comdg. . . . . . .
S. O. (3rd cl.) and Cant. Magte. . . . . . . } **Wheatley**, Capt. T. P., 1st G. B., York. R.

Murree (Sanitarium).

Comdt . . . . . **Cooper**, Lt.-Col. J. H.
S. O. (3rd cl.) . . . . **Carson**, Lt. C. G., 1st G. B., Som. L. I.
G. E. . . . . **Richardson**, Lt. (*temp. Capt.*) R. J. D., I. A. R. O.

Barian. Kuldana. Upper Topa. Lower Topa. Gharial. Lower Gharial.

6th INFANTRY BRIGADE.
Comdg. . . . . . . **Christian**, Brig.-Genl. G., C.B., D.S.O., [*l*] Brit. Ser.
Bde. Maj. . . . . . . **deCourcy**, Maj. *Hon.* M. W. R., D.S.O., 32nd Prs.
Staff Capt. . . . . . **Tregoning**, Capt. E. A., 2nd G. B., North'd Fus.

Chaklala.

2nd (*RAWAL PINDI*) DIVISION—*concld.*

5TH INFANTRY BRIGADE—

Bde. Comdr. . . . **Peterson,** Brig.-Genl. F. H., C.B., D.S.O.
Bde.-Maj. . **Bourne,** Capt. F. C., 1-4th R. W. Kent R.
Staff Capt. . . **Oliver,** Lt. (*temp. Capt.*) St. J. L., I.A.R.O.

Uttarshisha.

JHELUM BRIGADE—

Bde. Comdr . . **Richardson,** Brig.-Genl. H. L.
Bde.-Maj. . . . **Kellie,** Capt. G. J. D., M.C., 30th Punj.
Staff Capt . . . **Cambier,** Capt. V. 32nd Prs.
S. O. (1st cl.) . . **Crump,** Lt. (*temp. Capt.*) D. J., 9th Midx R.
Cant. Magis. . . **Murphy,** Lieut. R. S., I.A.R.O.

Chamiara, Jhelum, Mona, Rewat, Sargodha.

58TH INFANTRY BRIGADE—

Bde. Comdr. . . . . **Tanner,** Brig.-Genl. C. O. O.
Bde. Maj. . . . . **Mallock,** Capt. A. R. O., 42nd R.
Staff Capt. . . . . **Whittington,** Capt. L., 1-5th R. W. Surr. R.

Burhan.

ABBOTTABAD BRIGADE—

Bde. Comdr. . . . **Fowler,** Maj.-Genl. C. A., C.B., D.S.O.
Bde.-Maj. . . . **Griffith,** Maj. L., 107th Prs.
Staff Capt. . . **Maxwell,** 2nd-Lt. W. R., attd. Arg. & Suth'd Highrs.
Asst. Dir., Medl. Services. **Davidson,** Lt.-Col. J., D.S.O., I.M.S. (*also A. D. M. S., Campbellpore*).

59TH INFANTRY BRIGADE—

Bde. Comdr. . . . **Adye,** Brig.-Genl. D. R., 98th Infy. [L]
Bde. Maj. . . . . **Erskine,** Capt. K. C. S., M.C., 5th G. R.
Staff Capt. . . . . **Lee,** Capt. G. B., R. Ir. Rif.

Abbottabad.

S. O. (2nd cl.) . . . **Thomas,** Lt. T. L., 1st G. B., Som. L. I.
A. C. R. E. . . . **Kelsall,** Lt.-Col. H. W., R.E.
Cant. Magte. . . **Vaughan,** Lt.-Col. R. R., Supy. List.

Kakul. Hasan Abdal.

Bara Gali. Ghora Daka. Kalabagh. Khanspur. Khyra Gali.

SIALKOT BRIGADE—

Bde.-Comdr. . . . **Campbell,** Brig.-Genl. L. W. Y., C.M.G.
Bde.-Maj. . . . **Vallance,** Lt. (*temp. Capt.*) G. A. G., 2-5th Som. L. I.
Staff Capt. . . . **McCay,** Lt. R. C., D.S.O., attd. 17th Cav.
Asst. Dir., Medl. Services. **Smith,** Col. J. B., I. M. S. (*also A. D. M. S., Jhelum Bdc.*)

Sialkot.

Commdg. . . . . **Campbell,** Brig.-Genl. L. W. Y., C.M.G.
S. O. (2nd cl.) . . **McCrossan,** Lt. A. D., 1st G. B., Yorks. R.
Bde. Supply Officer. **Hartles,** Lt. W. E., S. & T. Corps.

Cant. Magte. . . **Lincoln,** 2nd-Lt. (*Temp. Lt.*) A. E., I. A. R. O.

16th DIVISION.

COMMANDER	**Beynon**, Maj.-Genl. *Sir* W. G. L., K.C.I.E., C.B., D.S.O.
A. D. C.	**Hog**, 2nd-Lt. R. D., 11th Lrs.
GENERAL STAFF.	
Gen. Staff Offr., 1st gr.	**Casson**, Bt.-Lt.-Col. (*temp. Lt.-Col.*) W. F. S., D.S.O., *p.s.c.*, 27th Cav.
Gen. Staff Offr., 2nd gr.	**Hallett**, Maj P., 1-6th E. Surr. R.
Gen. Staff Offr., 3rd gr.	**Whigham-Teasdale**, Lt. (*temp. Capt.*) D. G., M.C., I.A.R.O.
ADMINISTRATIVE, TECHNICAL AND DEPARTMENTAL STAFF.	
A. A. & Q. M. G. . .	**Bickford**, Lt.-Col. W. W., C.I.E., 107th Prs.
D. A. A. G.	**Taylor**, Capt. (*temp. Maj.*) G. C. R., 1-8th Essex R.
D. A. Q. M. G. . . .	**Mason**, Capt. (*temp. Maj.*) A. S., 1-6th Devon R.
Staff Capt.	**Vousley**, Capt. F., I.A.R.O.
Staff Capt. I. D. F. . .	**Bolster**, Maj. R. C., 5th Punj. L. H.
Comdg. R. A. . . .	**Cartwright**, Brig.-Genl. G. N., C.M.G., D.S.O.
Orderly Offr. . . .	**Barford**, Lt. (*temp. Capt.*) G. E., R.A.
Brig.-Genl., Indian Depôts.	
C. R. E.	**Campbell**, Col. G. P., Brit. Ser.
Asst. Dir. of Supplies & Transport.	**Burlton**, Lt.-Col. A. R., S. & T. Corps.
D. A. D. Supplies . .	**Harington**, Lt.-Col. A. W. D., S. & T. Corps.
D. A. D. Transport .	**Trelawny**, Maj. L. C., M.C., S. & T. Corps.
Asst. Dir. of Grass Farms	**Marriott**, Maj. (*temp. Lt.-Col.*) A. S.
Asst. Dir., Medl. Services	**Cleveland**, Bt.-Col. H. F., C.I.E., I.M.S.
Dy. Asst. Dir., Medical Services (Mobn.) .	**Reynolds**, Capt. (*temp. Maj.*) D., R.A.M.C.
Dy. Asst. Dir., Medl. Services (Sanitation)	**Mackenzie**, Maj. J., R.A.M.C.
D. A. D. of Ord. Ser. .	**Mills**, Lt. (*temp. Maj.*) H. A., A. O. D.
Dy. Asst. Dir., Vety. Ser.	**Lang**, Capt. W. W., A.V.C.
Inspr. of Army Schools.	**Bacon**, Capt. W. F., Inspr., Army Schools (2nd Circle).
Contr., Mily. Accts. (Under Finance Dept.)	**Wylde**, Maj. L. F. G. S.

Lahore.

43RD BRIGADE—

Bde. Comdr.	**Clarke**, Br. Genl. J.L.J., C.M.G., E. York.
Bde. Maj.	V.C. **Smyth**, Capt. J. G., 15th Sikhs.
Staff Capt.	**Broadmead**, Lt. (*temp. Capt.*) H. H., 2-5th Som. L. I.
Bde. Supply Officer .	**Logan**, Lt. A. T., S. & T. Corps.
Bde. Trans. Officer .	**Harvey**, Lt. D. V., S. & T. Corps.

Montgomery.

Comdg.	
S. O. (3rd cl.) . . .	**Halloran**, Lt. H., 1-124th Baluch. Infy.

Bahawalpur.

Lahore Cantonment

S. O. (2nd cl.) . . .	**Gray**, Capt. J. K., 2-6th R. Suss. R.
G. E.	**Beck**, Maj. T., M. W. S.
Cant. Magte. . . .	**Spencer**, F. R. B., Esq.

16th DIVISION—*concld.*

Ferozepore.

44TH INFANTRY BRIGADE—

Bde. Comdr.	**Southey,** Brig.-Genl. W. M., C.M.G.
Bde. Maj.	**Williams,** Maj. R. C. B., 35th Sikhs.
Staff Capt.	**Gladstone,** Capt. T. H., 1st Dn. Gds.
A. C. R. E.	**Gillespie,** Maj. R. St. J., C.I.E., R.E.
O.C. No. 6 Supply Depôt Section.	
Bde. Supply Offr.	**Recton,** Lt. T., I. A. R.O.
Offr. in charge Transpt.	**Deane,** 2nd Lt. N. G.
G. E.	
Elec. G. E.	**Pottle,** Lt. A. E., I. A. R. O.
S. O.	**Hancock,** Lt. F. B., 1-1st Kent R.
Cant. Magte.	**Trafford,** E. A., Maj.

JULLUNDUR BRIGADE (*45th Infy. Bde.*)

Bde. Comdr.	**Dyer,** Brig.-Gen. R. E. H., C.B., *p.s.c.*
Bde. Maj.	**Briggs,** Capt. F. C. C., D.S.O., L'pool R.
D. A. A. & Q. M. G.	**Cave-Browne,** Capt. H., 23rd Cav.
Staff Capt.	**Bostock,** Capt. J. A., I. A.

Jullundur.

Comdg.	
S. O. (2nd cl.)	**De la Coze,** 2nd-Lt. F. M., 1-25th Lond. R.
A. C. R. E.	**Burn,** Lt.-Col. E. M. J., R. E.
G. E.	**Fox,** 2nd-Lt. J. McD., I. A. R. O.
Bde. Supply Offr.	**Powell,** Lt. F. J., S. & T. Corps.
Cant. Magte.	**Knowles,** Maj. J. A., I. A.

Dalhousie.

Comdg. *S. O. (1st cl.) and Cant. Magte.*	**Woodliffe,** Maj. H. D., Brecknock Bn., S. W. Bord.

Amritsar

Comdg.	**Massey,** Capt. J. W., 1st G. B. Som. L. I.
S. O. (3rd cl.)	**Thompson,** 2nd-Lt. R., 23 Rif. Brig.

Dharmsala.

Comdg.	**Shakespear,** Maj. W. B., 1st G. R.
S. O. (3rd cl.)	**Cowell,** 2nd Lt. F. J., 1st G. R.

Bakloh.

Comdg.	**Maxwell,** Maj. D. W., 4th G. R.
S. O. (3rd cl.) and Cant. Magte.	**Atkinson,** Lt. D. J., 4th G. R.

Pathankote.

MEERUT DIVISION.

COMMANDER	**Nugent,** Maj.-Genl. O. S. W., C.B., D.S.O.
A. D. C.	**Campbell,** Maj. G. C., K. R. Rif.
GENERAL STAFF.	
G. S. O., 1st grade	**Murray,** Maj. (*temp. Lt.-Col.*) K. D. B., D.S.O., *p.s.c.*, 59th Rif.
G. S. O., 2nd gr.	**Wilson,** Capt. H. M., 1st Lrs.
ADMINISTRATIVE TECHNICAL AND DEPARTMENTAL STAFF.	
General Officer	**Codrington,** Brig-Genl. H. W., I. A.
A. A. & Q. M. G.	**Robb,** Maj. (*temp. Lt.-Col.*) C. B., 1-4th R. W. Kent. R.
D. A. A. G.	**Ball,** Capt. (*temp. Maj.*) C. F., I.D.F.
D. A. Q. M. G.	**Brewis,** Capt. (*temp. Maj.*) R. D., 1-5th R. W. Kent. R.
Comdg. R. A.	**Gordon,** Brig.-Genl. L. G. F., C.B., D.S.O. (*g*) (Brit. Ser. (*Also C. R. A., 8th Dn.*)
Orderly Offr.	**Moconnochie,** Lt. J. S., R. G. A.
Brig.-Genl., Indian Depôts No. 1.	**Twigg,** Brig.-Genl. R. H., I.A.
Brig.-Genl., Indian Depots. No. 2.	**Drake-Brockman,** Brig.-Genl. D. H., C.M.G., I.A.

MEERUT DIVISION—*contd*

C. R. E.	**Dealy,** Col. J. A., C.M.G., D.S.O.
Asst. Dir. of Supplies and Transport.	**Thomason,** Lt.-Col. A. F., S. & T. Corps.
D. A. D. Supplies . . .	**Shaw,** Maj. H. P. J., S. & T. Corps.
D. A. D. Transport . .	**Hall,** Maj. H. C., S. and T. Corps.
Asst. Dir. of Grass Farms	**Mellor,** Maj. (*temp. Lt.-Col.*) G. F.
A. D. M. S.	**Barratt,** Col. H. J., C.I.E., Brit. Ser.
Persnl. Asst. to A. D. M. S.	**Hirst,** Maj. G. G., I. M. S.
Dy. Asst. Dir., Medl. Services (Sany.)	**Hingston,** Capt. (*temp. Maj.*) J. C. L., R.A.M.C.
Dy. Asst. Dir., Medl. Services (Mobn.)	**Penny,** Capt. E. A., I.M.S.
D. A. D. Ord. Ser. . .	**Deighton,** Lt. H., (*temp. Capt.*) A. O. D.
Dy. Asst. Dir., Vetv. Ser.	**Ludgate,** Maj. W., A. V. C.
Inspector of Army Schools	**Sugden,** Capt. A. S., Inspr. Army Schools, 3rd Circle (*Naini Tal*).
Contr., Mily. Accts. (Under Finance Dept.)	**Anderson,** Maj. E. S. J.
Offr. a/c War Fodder, Northern Circle.	**Campbell,** Lt. (*temp. Capt.*) T. J., I. A. R. O.

Rurki.

Comdg.	**Stiffe,** Lt.-Col. A. F. E., R.G.A.
S. O. (2nd cl.) and Cant. Magte.	**Falkner,** Capt. J. M., I. A. R. O.
O. C. Bde. 19th Supply Sub-Section.	

Babugarh, Kashganj, Hapur, Granjdundwara.

AGRA BRIGADE.

Bde. Comdr.	**Dunsterville,** Maj.-Genl. L. C., C.B.
Bde. Maj.	**MacCausland,** Capt. R. A., 45th Sikhs.
Staff Capt.	**Williams,** Lt. T. L., 2-4th D. of Corn. L. I.

Aligarh Bharatpur Dholpur.

Agra.

Comdg.	**Drayson,** Lt.-Col. A. P. O.B.E. 1-6th E. Surr. R.
S. O. (3rd cl.).	
G. E.	
Bde. Supply Offr. . . .	**Allen,** 2nd-Lt., C. E.
Cant. Magte.	**O'Neill,** 2nd-Lt. (*temp. Capt.*) F. O., I. A. R. O.

Muttra.

Comdg.	**Rogers,** Lt.-Col., P. H.
S. O. (3rd cl.) and Cant. Magte. . . .	

GARHWAL BRIGADE—

Bde. Comdr.	**Betham,** Brig.-Genl. R. M.
Bde. Maj.	**Green,** Capt. E. C., 1-4th E. Kent. R.
Staff Capt.	**Patterson,** Lt. G. S. V., I. A. R. O.
Asst. Dir., Med. Ser. . .	**Fallon,** Col. J., Brit. Ser. (*Also A. D. M. S., Delhi & Bareilly Bdes.*)
A. C. R. E.	**Pilcher,** Lt-Col. A. J., R. E.

Lansdowne.

Comdg.	**Cassels,** Lt.-Col. K. S., 8th G. Rif.
S.O. (2nd cl.) and Cant. Magte.	**Grille,** 2nd-Lt. F. L., attd. 2-3rd G. R.
G. E.	

Saharanpur.

Dehra Dun.

Comdg.	**Betham,** Brig. Genl. R. M.
S. O. (2nd cl.)	**Templeton,** Lt. R. B., 222nd M. G. Coy.
O. C. Supply Depôt. . .	**Palin,** Col. G. W., C.B., C.I.E., S. & T. Corps.
O. C. 19th Bde. Supply Section.	**Davies,** Lt. A. F., A. S. C. attd. S. & T. Corps.
G. E.	
Cant. Magte.	**Leslie,** Lt.-Col. W. C. C., Supy. List.

Chakrata.

Comdg.	**Haig,** Lt.-Col. D. P., 1st G. B. Sco. Rif.
S. O. (1st class) . .	**D'Ambrunenil,** Lt. W., 2-4th D. of Corn. L. I.
Cant Magte. . . .	**Cowie,** Lt.-Col. W. A. L., *Supy. List.*

Landour.

Comdg. & Cant. Magte.	**Leslie,** Lt.-Col. W. C. C., I.A.
S. O. (1st class) . .	**Henderson,** Lt. (*temp. Capt.*) J. T., 1st G. B. Sco. Rif.

Mussoorie. Kailana.

MEERUT DIVISION—*concld.*

DELHI BRIGADE—

Bde.-Comdr. . . . **Grimston,** Brig.-Genl. S. B., C.M.G.
Bde.-Major . . . **Pogson,** Capt. C. A., M.C., 117th Mah.
Staff Capt. . . . **Phillips,** Capt. C. S., 1-1st Kent. Bn.
Asst. Dir. Med. Ser. . **Fallon,** Col. J., Brit Ser. (*Also A. D. M. S., Bareilly & Garhwal Bdes.*)

Delhi.

Comdg. . . . . . . **Grimston,** Brig.-Genl. S. B., C.M.G.
S. O. (1st cl.) . . . **Tanner,** Lt. (*Actg. Capt.*) G., 1st. G. B., Bedf. R.
Cant. Magte. . . .
A. C. R. E. . . . . . **Bond,** Lt.-Col. R. F. G., R.E.
G. E. (New Cantonments) . . . . **Ricardo,** Lt. W. H.
Cant. Magte. (New Cantonments) . . **Coffey,** Lt. E. P., I. A. R. O.

Ghaziabad.

Meerut.

Commanding . . . . **Roberts,** Brig.-Genl. H. L., C.B., M.V.O.
S. O. (1st cl.) . . . . **Parke,** Capt. L. S. 1-7th Hamps. R.
A. C. R. E. . . . . . **Wildeblood,** Capt. (*Actg. Maj.*) J. P., I. A. R. O.
Bde. Supply Offr. . . . **Pow,** 2nd Lt. J. S.
Offr. i/c Transport . . . **Handcock,** 2nd-Lt. W. G., Guides Cav.
Cant. Magte. . . . **Peart,** Maj. D. G., I. A. (retd.).

4th CAVALRY BRIGADE

Bde. Comdr. . . . **Roberts,** Brig.-Genl. H. L., C.B., M.V.O.
Bde.-Maj. . . . . **Edward Collins,** Capt. G., M.C., 33rd Cav.
Staff Capt. . . . . **Langford,** Lt. W., 1st Dn. Gds.

BAREILLY BRIGADE—

Bde.-Comdr. . . . . **Gray,** Brig.-Genl. F. W. B., C.M.G., D.S.O.
Bde.-Major . . . **Nicolls,** Capt. B. E., 7th N. Staff. R.
Staff Capt. . . . . **Joseph,** Lt. (*temp. Capt.*) A. F., I. A. R. O.
Asst. Dir. Medl. Ser. . **Fallon,** Col. J., Brit. Ser. (*Also A. D. M. S., Delhi & Garhwal Bdes.*)

Bareilly.

Comdg. . . . . . . **Gray,** Brig.-Genl. F. W. B., C.M.G., D.S.O.
S. O. (2nd cl.) . . . **Thomson,** Capt. R. St. C., 1-5th Hamps. R.
A. C. R. E. . . . . . **Gardner,** Capt. J., R. E. (*Offg.*)
Offr. in charge Supplies **King,** 2nd-Lt. C. A., S. & T. Corps.
Cant. Magte. . . . **Reid,** Maj. W. D.

Almora.

Comdg., *S. O. (3rd cl.) and Cant. Magte.* . . } **Mead,** Maj. H. R. C., 1-50th Rif.

Ranikhet.

Comdg. . . . . .
Cant. Magte. . . . . **Herbert,** Lt.-Col. C., D.S.O., Supy. List.
S. O. (2nd cl.) . . .

Shahjehanpur.

Comdg., *S. O. (3rd cl.) and Cant. Magte.* . . . } **Corbett,** Maj. B. A., 13th Rajputs.

Naini-Tal (Sanitarium).

Comdg., *S. O. (2nd cl.) and Cant. Magte.* . . } **Bacon,** Maj. R. J., 1-7th Hamps. R.
Asst. Provost Marshal .

Moradabad.

Comdg. . . . . . . **Warner,** Lt.-Col. H. R., 1-131st Regt.

Chaubattia.

Comdg. . . . .

AMBALA BRIGADE.

Bde-Comdr.	**Hendley,** Maj.-Genl. C. E., C.B.
D. A. A. & Q. M. G.	**Kenworthy,** Capt. (*temp. Maj.*) W., 33rd Cav.
Gen. Staff Offr. 2nd Gr.	**Weekes,** Maj. H. E., 10th G. R.
Bde. Maj.	**White,** Maj. S. H., 1-10th Middx. R.
Staff Capt.	**Hobday,** Lt. (*temp. Capt.*) W. E., I. A. R. O.
A. C. R. E.	**Barton,** Lt.-Col. H. J., R.E. [*l*]
Brig.-Genl., Indian Depots.	**Lane,** Brig.-Genl. H. A. (*Also Multan Bde.*).
D. A. D., S. & T.	**Chanter,** Maj. A. D., S. & T. Corps.
A. D. M. S.	**Carr-White,** Col. P., I.M.S.
D. A. D. M. S.	**Stevenson,** Capt. (*temp. Maj.*) W. B., R.A.M.C.
D. A. D. V. S.	**Kerr,** Capt. (*temp. Maj.*) P. J., I.A.R.O.

Ambala.

Commdg.	**Hendley,** Maj. Genl. C. E., C.B.
S. O. (1st cl.)	**Walmsley,** Capt. E. S., 1-1st Kent Bn.
Bde. Supply Officer.	**Rivett,** Lt. E. G., I.A.R.O., S. & T. Corps.
Offr. in charge Transport.	**Kemp,** Capt. W. J., 60th Mule Corps.
Offr. in charge Supplies	**Manson,** Maj. J. A. P., S. & T. Corps.
G. E.	**Pearson,** Lt. A. F., R.E.
Cant. Magte.	**Bosworth-Smith,** B. N., Esq., I.C.S.

Kasauli.

Comdg.	**Talbot,** Lt.-Col. G. R., R.F.A.
S. O. (2nd cl.)	**Davies,** Lt. R. T., 1-9th Middx. R.
Cant. Magte.	**Hodgson,** Lt.-Col. W. G., Supy. List.
Supply Officer	**Flemington,** Lt. (*temp. Capt.*) R. R., I. A. R. O.

Kalka.

Comdg.	**Oxenham,** Lt. S. H., I. A. R. O.

Jutogh.

Comdg. *S. O. (3rd cl.) & Cant. Magte.*	**O'Rorke,** Capt. C. H., R. A. M. C.

Solon.

Sabathu.

Comdg.	**Gompertz,** Capt. (*temp. Lt.-Col.*) M. L. A., 108th Infy.
S. O. (3rd cl.)	**Austin,** Lt. A. McN., 1st G. B., Som. L. I.
Asst. Cant. Magte.	**Gibson,** Maj. A. W., R.A.M.C.

Dagshai.

Comdg.	**Bellhouse,** Lt. A. P., 8th Rajputs.
S. O. (3rd cl.) & Cant. Magte.	**Stevenson,** Lt. J. F. B., M.C., 44th Infy.

Simla.

S. O.	**Pierce,** Lt.-Col. F. G., 102nd Grs.

Phillour.

Comdg.	**Mercer,** 2nd-Lt. (*Actg. Lt.-Col.*) A. W., 4-30th Punj.

KOHAT BRIGADE.

Bde. Comdr. . . . . . **Eustace**, Maj.-Genl. A. H., C. B., D.S.O.
Gen. Staff Offr., 2nd gr. **VC Wylly**, Major G. G. E., D.S.O., Guides, *p.s.c.*
D. A. A. and Q. M. G. . **Bannatyne**, Maj. F. C., 109th Infy.
D. A. D., S. & T. . . . **Scott**, Capt. (*temp. Maj.*) G. C., R. I. Ri.
A. D. M. S. . . . . . . **VC Crimmin**, Col. J., C.B., C.I.E., V.D., K.H.S., I.M.S.
Asst. Dir. Vety. Ser. . . **Kelly**, Capt. G. A., A.V.C. (*Also Bannu & Derajat Bdes.*)

Kohat.

A. C. R. E. . . . . . . **Haswell**, Bt.-Lt.-Col. C. H., R.E.
G. E. . . . .
S. O. (3rd cl.) . . . . . **Lethbridge**, Lt. M. H. B., I. A. R. O.

Fort Lockhart.

Commdg. . . . . **Hill**, Maj. R. D. O., 3-8th G. R.
S. O. . . . .

Thal.

Commdg. . . . . **Fowler**, Maj. E. K.

Hangu.

Commdg. . . . . . . **Giffard**, Maj. (*temp. Lt.-Col.*) D. H. R., 3-8th G. R.
S. O. (3rd cl.) . . . . **Haynes**, Lt. S. J., 3-8th G. R.
G. E. . . . . **Harrison**, Lt. (*temp. Capt.*) A. F. L. H., I.A.R.O.

Parachinar.

G. E. (Samana and Kurram)

BANNU BRIGADE.

Bde. Comdr. . . . . . . **Bruce**, Brig.-Genl. *Hon.* C. G., C.B., M.V.O.
Gen. Staff Offr., 2nd gr. . **Craster**, Maj. G., D.S.O., *p.s.c.* 6th Cav.
D. A. A. and Q. M. G. . **Story**, Capt. (*temp. Maj.*) H. F., 35th Sikhs.
Staff Capt. . . . . . **Stuart**, Lt. D. S., Attd. 112th Infy.
D. A. D. Supplies and Transport. **Rebsch**, Capt. (*temp. Maj.*) W. K., D.S.O.
A. D. M. S. . . . . . **Stodart**, Col. T., I. M. S.
D. A. D. Ord. Ser. . . **Robinson**, Capt. F. J.
Asst. Dir. Vety. Ser. . . **Kelly**, Capt. (*temp. Maj.*) G. A., A.V.C. (*Also Kohat & Derajat Bdes.*)

Bannu.

A. C. R. E. . . . . . . **Bell**, Maj. A. H., D.S.O., R.E. [L].
G. E. (Bannu) . . . . **Byrne**, Lt. (*Actg. Capt.*) H. S., R.E. (*T. C.*)
G. E. (Tochi) . . . . . **Toller**, Lt. (*temp. Capt.*) G. G. T., I.A.R.O.
S. O. (*3rd cl.*)
A. G. E. (Tochi) . . . **Macrostie**, E. J., 2nd-Lt., 1-25th Lond.

Miranshah.

DERAJAT BRIGADE.

Bde. Comdr. . . . . . **Woodyatt**, Maj.-Gen. N. G., C.B.
Gen. Staff Offr., 2nd gr. . **Trench**, Capt. (*temp. Maj.*) C. F., D.S.O., 7th Lrs.
A. A. & Q. M. G. . . . **Binny**, Maj. (*temp. Lt.-Col.*) A. J. M., 1st Lrs., *p.s.c.*
D. A. A. and Q. M. G. . **Cobb**, Capt. (*temp. Maj.*) C., S. & T. Corps.
Staff Capt. . . . . . **Trusted**, Lt. (*temp. Capt.*) H. H., 2-4th D. of Corn. L. I.
D. A. D., S. & T. . . . **Duke**, Capt. (*temp. Maj.*) A. A. G., S. & T. Corps.
A. D. M. S. . . . . . **Close**, Lt.-Col. J. K., M.D., I.M.S.
Dy. Asst. Dir. Vety. Ser. . **Kelly**, Capt. (*temp. Maj.*) G. A., A.V.C. (*Also Kohat & Bannu Bdes.*)

Dera Ismail Khan.

A. C. R. E. . . . . . **Fox**, Maj. B. H., R.E.
G. E. . . . . **Corkery**, Lt. A. T. M., I.A.R.O.
Offr. in charge Supplies .

Manzai.

G. E. . . . . **Rochfort**, 2nd-Lt. H. G., I.A.R.O.

Tank.

A. C. R. E. . . . . **Bigsby**, Maj. S. H., I.A.R.O. (Attd. R. E.)
G. E. . . . . **Richards**, Lt. W. H., R.E.
G. E. (Mechanical) . . **Field**, Capt. C. D., Attd. R. E.

Wana.

G. E. . . . . **Wauchope**, Capt. R. S., Attd. R. E.

Jandola. Girni. Khirgi. Jatta.

MULTAN BRIGADE.

Multan.

Bde. Comdr. . . . . **Miles**, Brig.-Genl. P. J., C.B.
Bde. Maj. . . . . . **Wilson**, Capt. A. P., 1-4th Bord. R.
Staff Capt. . . . . . **Gomperty**, Lt. M. C. T., 8th Rajputs.
Brig.-Genl., Indian Depots. **Lane**, Brig.-Genl. H. A. (*Also Ambala Bde.*)
S. O. (1st cl.) . . . **Gillett**, Capt. N. C., 1-25th Lond. R.
Cant. Magte. . . . . **Deeks**, Lt. T. B., I. A.R.O.
G. E. . . . . **Sawyer**, 2nd-Lt. N. H., I.A.R.O.
Elec. G. E. . . . . . **Watson**, 2nd-Lt. R. M., I.A.R.O.

SOUTHERN COMMAND.

Headquarters.

General Officer Commanding.	**Anderson**, Lt.-Genl. *Sir* C. A., K.C.B., Brit. Ser.	6 Apl. 17
A. M. S.	**Brock**, Maj. B. deL., M.C., 126th Infy.	24 Aug. 17
A.-D.-C.	**Anderson**, Lt. N. M., Leic. R.	6 Apl. 17
A.-D.-C.	**Ghaus Muhammad Khan**, *Bahadur*, Risdr. Major, *late* 21st Cav.	31 July 17
General Staff—		
Brig.-Genl.	**Browne**, Brig.-Genl. H. J. P., 5th G. R., *p.s.c.*	19 Oct. 17
Genl. Staff Offr., 2nd gr.	**Gray**, Capt. (*temp. Maj.*) G. C. G., 3rd Horse	6 Nov. 18
Genl. Staff Offr., 3rd gr.	**Smith**, Lt. (*temp. Capt.*) V. C. E., Hamps. R.	16 Sep. 18
Administrative Staff—		
General Officer	**James**, Maj.-Genl. W. B., C.B., C.I.E., M.V.O.	18 Feb. 17
A. A. G.	**Irving**, Capt. (*temp. Lt.-Col.*) M., I.A.R.O.	18 July 18
A. Q. M. G.	**Tyrrell**, Maj. (*temp. Lt.-Col.*) A. C. L., 25th Cav., *p.s.c.*	1 Feb. 18
D. A. A. G.	**Porteous**, Capt. (*temp. Maj.*) D. G., 90th Punj.	11 Feb. 18
D. A. Q. M. G.	**Channer**, Maj. B. G., 54th Sikhs	25 Mar. 18
Staff Capt.	**Windeler**, Capt. W. T., 63rd Infy.	6 Feb. 18
Asst. Dir. of Ord. Services.	**Smallwood**, Lt.-Col. F. G., C.V.O., R. A.	27 May 18
Attached—		
Chief Engineer	**Nanton**, Brig.-Genl. H. C., C.B., Brit. Ser.	14 Jan. 18
Dy. Chief Engineer	**Walker**, Maj. E. C., R.E. [*l*]	1 Feb. 18
Staff Offr. R. E.	**Pullar**, Capt. E. J., I.A.R.O.	28 June 18
Asst. Inspr. of Machinery.	**Brazier**, Lt. (*temp. Maj.*) C. C. H., I.A.R.O.	7 Aug. 18
Inspr. of Musky. & Machine Guns.		
Inspr. of Physical and Bayonet Training.	**Chambers**, Capt. (*temp. Maj.*) G. H., M.C., 123rd Rif.	31 Mar. 17
Inspr. of Bombing	**Paterson**, Capt. (*temp. Maj.*) G. K., 3rd Arg. & Suth'd. Highrs.	1 June 18
Inspr. of S. & T. Services.	**Lushington**, Lt.-Col. W. N., S. & T. Corps	25 May 18
D. D. M. S.	**Anderson**, Maj.-Genl. L. E., C.B., Brit. Ser.	14 Mar. 18
D. A. D. M. S.	**Stanger-Leathes**, Maj. H. E., I.M.S.	2 Oct. 18
D. D. S. & T.	**Moore**, Lt.-Col. C. H. G., C.M.G., D.S.O., S. & T. Corps	18 Nov. 18
D. A. D. S. & T.	**Hume-Spry**, Maj. W. E., D.S.O., S. & T. Corps	23 Feb. 18
D. D. Vety. Ser.	**Larnder**, Col. E. W., A.V.S.	
D. J. A. G.	**Onslow**, Col. R. C.	3 Nov. 17
A. J. A. G.	**Macfie**, Lt. (*temp. Maj.*) J., I.A.R.O.	26 Sep. 17
Commdt., School of Physical and Bayonet Training.	**Ranger**, Lt. (*temp. Capt.*) W. J., Som. L. I.	25 June 15

KOHAT BRIGADE.

Bde. Comdr.	**Eustace**, Maj.-Genl. A. H., C.B., D.S.O.
Gen. Staff Offr., 2nd gr.	VC **Wylly**, Major G. G. E., D.S.O., Guides, *p.s.c.*
D. A. A. and Q. M. G.	**Bannatyne**, Maj. F. C., 109th Infy.
D. A. D., S. & T.	**Scott**, Capt. (*temp. Maj.*) G. C., R. I. Ri.
A. D. M. S.	VC **Crimmin**, Col. J., C.B., C.I.E., V.D., K.H.S., I.M.S.
Asst. Dir. Vety. Ser.	**Kelly**, Capt. G. A., A.V.C. (*Also Bannu & Derajat Bdes.*)

Kohat.

A. C. R. E.	**Haswell**, Bt.-Lt.-Col. C. H., R.E.
G. E.	
S. O. (3rd cl.)	**Lethbridge**, Lt. M. H. B., I. A. R. O.

Fort Lockhart.

Commdg.	**Hill**, Maj. R. D. O., 3-8th G. R.
S. O.	

Thal.

Commdg.	**Fowler**, Maj. E. K.

Hangu.

Commdg.	**Giffard**, Maj. (*temp. Lt.-Col.*) D. H. R., 3-8th G. R.
S. O. (3rd cl.)	**Haynes**, Lt. S. J., 3-8th G. R.
G. E.	**Harrison**, Lt. (*temp. Capt.*) A. F. L. H., I.A.R.O.

Parachinar.

G. E. (Samana and Kurram)

BANNU BRIGADE.

Bde. Comdr.	**Bruce**, Brig.-Genl. *Hon.* C. G., C.B., M.V.O.
Gen. Staff Offr., 2nd gr.	**Craster**, Maj. G., D.S.O., *p.s.c.* 6th Cav.
D. A. A. and Q. M. G.	**Story**, Capt. (*temp. Maj.*) H. F., 35th Sikhs.
Staff Capt.	**Stuart**, Lt. D. S., Attd. 112th Infy.
D. A. D. Supplies and Transport.	**Rebsch**, Capt. (*temp. Maj.*) W. K., D.S.O.
A. D. M. S.	**Stodart**, Col. T., I. M. S.
D. A. D. Ord. Ser.	**Robinson**, Capt. F. J.
Asst. Dir. Vety. Ser.	**Kelly**, Capt. (*temp. Maj.*) G. A., A.V.C. (*Also Kohat & Derajat Bdes.*)

Bannu.

A. C. R. E.	**Bell**, Maj. A. H., D.S.O., R.E. [L].
G. E. (Bannu)	**Byrne**, Lt. (*Actg. Capt.*) H. S., R.E. (*T. C.*)
G. E. (Tochi)	**Toller**, Lt. (*temp. Capt.*) G. G. T., I.A.R.O.
S. O. (*3rd cl.*)	
A. G. E. (Tochi)	**Macrostie**, E. J., 2nd-Lt., 1-25th Lond.

Miranshah.

DERAJAT BRIGADE.

Bde. Comdr.	**Woodyatt**, Maj.-Gen. N. G., C.B.
Gen. Staff Offr., 2nd gr.	**Trench**, Capt. (*temp. Maj.*) C. F., D.S.O., 7th Lrs.
A. A. & Q. M. G.	**Binny**, Maj. (*temp. Lt.-Col.*) A. J. M., 1st Lrs., *p.s.c.*
D. A. A. and Q. M. G.	**Cobb**, Capt. (*temp. Maj.*) C., S. & T. Corps.
Staff Capt.	**Trusted**, Lt. (*temp. Capt.*) H. H., 2-4th D. of Corn. L. I.
D. A. D., S. & T.	**Duke**, Capt. (*temp. Maj.*) A. A. G., S. & T. Corps.
A. D. M. S.	**Close**, Lt.-Col. J. K., M.D., I.M.S.
Dy. Asst. Dir. Vety. Ser.	**Kelly**, Capt. (*temp. Maj.*) G. A., A.V.C. (*Also Kohat & Bannu Bdes.*)

Dera Ismail Khan.

A. C. R. E.	**Fox**, Maj. B. H., R.E.
G. E.	**Corkery**, Lt. A. T. M., I.A.R.O.
Offr. in charge Supplies	

Manzai.

G. E.	**Rochfort**, 2nd-Lt. H. G., I.A.R.O.

Tank.

A. C. R. E.	**Bigsby**, Maj. S. H., I.A.R.O. (Attd. R. E.)
G. E.	**Richards**, Lt. W. H., R.E.
G. E. (Mechanical)	**Field**, Capt. C. D., Attd. R. E.

Wana.

G. E.	**Wauchope**, Capt. R. S., Attd. R. E.

Jandola. Girni. Khirgi. Jatta.

MULTAN BRIGADE.

Multan.

Bde. Comdr.	**Miles**, Brig.-Genl. P. J., C.B.
Bde. Maj.	**Wilson**, Capt. A. P., 1-4th Bord. R.
Staff Capt.	**Gomperty**, Lt. M. C. T., 8th Rajputs.
Brig.-Genl., Indian Depots.	**Lane**, Brig.-Genl. H. A. (*Also Ambala Bde.*)
S. O. (1st cl.)	**Gillett**, Capt. N. C., 1-25th Lond. R.
Cant. Magte.	**Deeks**, Lt. T. B., I. A.R.O.
G. E.	**Sawyer**, 2nd-Lt. N. H., I.A.R.O.
Elec. G. E.	**Watson**, 2nd-Lt. R. M., I.A.R.O.

SOUTHERN COMMAND.

Headquarters.

General Officer Commanding.	**Anderson,** Lt.-Genl. *Sir* C. A., K.C.B., Brit. Ser.	6 Apl. 17
A. M. S.	**Brock,** Maj. B. deL., M.C., 126th Infy.	24 Aug. 17
A.-D.-C.	**Anderson,** Lt. N. M., Leic. R.	6 Apl. 17
A.-D.-C.	**Ghaus Muhammad Khan,** *Bahadur,* Risdr. Major, *late* 21st Cav.	31 July 17

General Staff—

Brig.-Genl.	**Browne,** Brig.-Genl. H. J. P., 5th G. R., *p.s.c.*	19 Oct. 17
Genl. Staff Offr., 2nd gr.	**Gray,** Capt. (*temp. Maj.*) G. C. G., 3rd Horse	6 Nov. 18
Genl. Staff Offr., 3rd gr.	**Smith,** Lt. (*temp. Capt.*) V. C. E., Hamps. R.	16 Sep. 18

Administrative Staff—

General Officer	**James,** Maj.-Genl. W. B., C.B., C.I.E., M.V.O.	18 Feb. 17
A. A. G.	**Irving,** Capt. (*temp. Lt.-Col.*) M., I.A.R.O.	18 July 18
A. Q. M. G.	**Tyrrell,** Maj. (*temp. Lt.-Col.*) A. C. L., 25th Cav., *p.s.c.*	1 Feb. 18
D. A. A. G.	**Porteous,** Capt. (*temp. Maj.*) D. G., 90th Punj.	11 Feb. 18
D. A. Q. M. G.	**Channer,** Maj. B. G., 54th Sikhs	25 Mar. 18
Staff Capt.	**Windeler,** Capt. W. T., 63rd Infy.	6 Feb. 18
Asst. Dir. of Ord. Services.	**Smallwood,** Lt.-Col. F. G., C.V.O., R. A.	27 May 18

Attached—

Chief Engineer	**Nanton,** Brig.-Genl. H. C., C.B., Brit. Ser.	14 Jan. 18
Dy. Chief Engineer	**Walker,** Maj. E. C., R.E. [*l*]	1 Feb. 18
Staff Offr. R. E.	**Pullar,** Capt. E. J., I.A.R.O.	28 June 18
Asst. Inspr. of Machinery.	**Brazier,** Lt. (*temp. Maj.*) C. C. H., I.A.R.O.	7 Aug. 18
Inspr. of Musky. & Machine Guns.		
Inspr. of Physical and Bayonet Training.	**Chambers,** Capt. (*temp. Maj.*) G. H., M.C., 123rd Rif.	31 Mar. 17
Inspr. of Bombing	**Paterson,** Capt. (*temp. Maj.*) G. K., 3rd Arg. & Suth'd. Highrs.	1 June 18
Inspr. of S. & T. Services.	**Lushington,** Lt.-Col. W. N., S. & T. Corps	25 May 18
D. D. M. S.	**Anderson,** Maj.-Genl. L. E., C.B., Brit. Ser.	14 Mar. 18
D. A. D. M. S.	**Stanger-Leathes,** Maj. H. E., I.M.S.	2 Oct. 18
D. D. S. & T.	**Moore,** Lt.-Col. C. H. G., C.M.G., D.S.O., S. & T. Corps	18 Nov. 18
D. A. D. S. & T.	**Hume-Spry,** Maj. W. E., D.S.O., S. & T. Corps	23 Feb. 18
D. D. Vety. Ser.	**Larnder,** Col. E. W., A.V.S.	
D. J. A. G.	**Onslow,** Col. R. C.	3 Nov. 17
A. J. A. G.	**Macfie,** Lt. (*temp. Maj.*) J., I.A.R.O.	26 Sep. 17
Commdt., School of Physical and Bayonet Training.	**Ranger,** Lt. (*temp. Capt.*) W. J., Som. L. I.	25 June 15

5th (*MHOW*) DIVISION.

COMMANDER	**Kemball**, Maj.-Genl. *Sir* G. V., K.C.M.G., C.B., D.S.O., *p.s.c.* (*g*) Brit. Ser.
A.-D.-C.	**Wall**, Capt. C. E., 1st G. B., Norf. R.

GENERAL STAFF—

Gen. Staff Offr., 1st gr.	**Orr**, Bt.-Col. G. M., D.S.O., 11th Lrs. *p.s.c.*
Gen. Staff Offr., 2nd gr.	**Hogg**, Maj. C. M. T., D.S.O., 4th G. R., *p.s.c.*
Gen. Staff Offr., 3rd gr.	**Malet**, Capt. W. DeC., 14th Lrs.

ADMINISTRATIVE, TECHNICAL AND DEPARTMENTAL STAFF—

General Officer . . .	**Drew**, Brig.-Genl. A. B. H., C.I.E.
A. A. G.	**Cook**, Col. C. C., D.S.O., I. A.
A. Q. M. G.	**Calvert**, Lt.-Col. R. T. C., 122nd Infy., *p.s.c.*
D. A. A. G.	**Gillies**, Maj. H., 1st Lrs.
D. A. Q. M. G. . . .	**Hammersley-Smith**, Maj. R. H., 14th Lrs.
Brig. Genl., Indian Depôts.	**Herbert**, Brig.-Genl. L. N., C.B. [*l*].
Do. do.	**McCrea**, Brig.-Genl. A. C., C.M.G., I. A.
C. R. E.	**Kirby**, Col. N., C.I.E., Brit. Ser.
Comdg. R. A.	**Quinton**, Brig.-Genl. F. W. D., Brit. Ser. (*Also C.R.A., Poona 9th Dn.*)
Orderly Offr.	**Lewis**, Capt. H., R.F.A.
Asst. Dir. of Supplies & Tpt.	**Dallas**, Lt.-Col. A. E., C.M.G., S. & T. Corps.
D. A. D. S. & T. . .	**Bond**, Maj. G. W., D.S.O., S. & T. Corps.
Asst. Director of Grass Farms.	**Elliott**, Maj. (*temp. Lt.-Col.*) H. H.
Asst. Dir. Medl. Services.	**Tate**, Col. A. E., C.M.G., I.M.S.
Dy. Asst. Dir. Medl. Services (Sany.).	**Clarke**, Maj. F. A. H., R.A.M.C.
Dy. Asst. Dir. Medl. Services (Mobn.).	**Fawcett**, Maj. C. E. W. S., R.A.M.C.
D. A. D. of Ord. Ser. .	**Mowels**, Capt., (*temp. Maj.*) J. H., A.O.D.
D. A. D. Vety. Ser. . .	**Deacon**, Maj. A. F., A.V.C.
Contr., Mily. Accts. (Under Finance Dept.)	**Anderson**, Lt.-Col. R. H. B.
Supvg. Offr. Physical and Bay. Training.	**Johnston**, 2nd-Lt. F., 2/99th Infy.

Mhow.

Comdg.	**Martin**, Col. R. H., C.B., C.M.G., 1st G. B., Norf. R.
S. O. (1st cl.) . . .	**Williams**, Capt. J., 1-1st Brecknock Bn.
A. C. R. E.	**Noble**, Maj. W. J. W., R. E.
Cant. Magte. . . .	**Burn**, Lt.-Col. G. C., *Supy. List.*
Cav. Bde. Supply Officer	**Morley**, Lt. M. H. A., I.A.R.O.
G. E.	**Perfect**, 2nd-Lt. S. C., I.A.R.O.

Agar.

Comdg.	**Abbott**, Bt. Maj. R. S., M.C., 38th Horse.

Indore.

Comdg.	**Talbot**, Lt. W. L.
S. O. (3rd cl.) . .	**Hughes**, Lt., H. D., 1-1st Brecknock Bn.

Gungapur. Khandwa. Mehidpore. Ratlam. Sirdarpore.

5th (*MHOW*) DIVISION—*contd.*

NASIRABAD BRIGADE—

Bde.-Comdr. . . . . **Andrew**, Brig.-Genl. D. C. A.
Bde.-Maj . . . . **Maynard**, Capt. P. A., 1-5th E. Kent R.
Staff Capt. . . . . **Twiddy**, Lt. A. A., 2-9th Delhi Infy.

Nasirabad.

Comdg. . . . . . . **Andrew**, Brig.-Genl. D.C.A.
G. E. . . . . **Connor**, Lt. G. H., 1-33rd Punj.
Offr. in charge Supplies **Williams**, Lt. A. C., I.A.R.O.
Bde. Supply Offr. . . **Boynton-Lee**, Lt. F., I.A.R.O.
Cant. Magte. . . . **Riley-Irving**, 2nd-Lt. J. H., 3rd G. R.

Ajmer.

Comdg. . . . . . **Fenton**, Capt. J. N., 2-43rd R.
S. O. (3rd cl.) . . .

Deesa.

Comdg. . . . . . **Lane**, Maj. (*actg. Lt.-Col.*) H. V., 2-3rd Brahmans.
Cant. Magte. . . .

Deoli.

Comdg. . . . . **George**, Maj. D. S., 1-42nd Deoli R.
Cant. Magte. . . . **Waterfield**, S., Esq.

Erinpura.

Comdg. . . . . . **Cobban**, Maj. C. L. [L]., 43rd R.

Mount Abu (Sanitarium).

Comdt. . . . . . **Macaulay**, Maj. J., 1st G. B., R. Sco. Fus

Neemuch.

Comdg. . . . . . **Biddulph**, Col. S. F.
S. O. (3rd cl.) . . . **Beale**, Capt. G. H., 6th Dns.
G. E. . . . . . .
S. and T. Officer . . .
Cant. Magte. . . .

Bikaner.

Comdg. . . . . . **Van der Gucht**, Maj. G. T., 1-141st Infy.

Jodhpur.

Comdg. . . . . . **Money**, Maj. E. F. D., D.S.O., 1-142nd Infy.

Alwar.

Comdg. . . . . . **Cruickshank**, Maj. J. E., 1-145th Infy.

Abu Road, Bandikui, Bikaner, Jaipur, Phulera, Tonk.

JUBBULPORE BRIGADE—

Bde. Comdr. . . . **Prendergast**, Brig.-Genl. C. G., C.B.
Bde.-Maj. . . . **Manners**, Maj. E. H. H., 106th Prs.
Staff Capt. . . . **Fisher**, Lt. (*temp. Capt.*) M. G., 2-4th Bord. R.

Jubbulpore.

S. O. (2nd cl.) . . . **Smithers**, Lt. K. O., 1-5th R. W. Kent R.
A. C. R. E. . . . . **Haig**, Maj. D.de H., R. E.
Offr. in charge Supply Depôt. **Jarrett-Kerr**, Maj. H. J., S. & T.
Bde. Supply Offr. . . **Laythorpe**, Lt. F. W. H., I. A. R. O.
G. E. . . . . . . { **Baxter**, Lt. E. E., R. E. (T. C.) / **Hudson**, 2nd-Lt. R. J. H., I. A. R. O. }
Cant. Magte. . . . **Carwithen**, Lt.-Col., E. T., I. A.
A. D. M. S. . . . . **Davidson**, Col. J. S., A. M. S. (*also A. D. M. S., Jhansi Bde.*)

Kamptee.

Comdg. . . . . . . **Carey**, Brig.-Genl. O. W., C.M.G.
S. O. (2nd cl.) . . . **Sinclair**, Capt. W. V. S., 2nd G. B., Essex R.
Cant. Magte. . . . **Slaney**, W. G., Esq.

Pachmarhi.

Comdt. . . . . . . **Grove**, Maj. (*temp. Col.*) P. L., Cheshire, R.
S. O. (1st cl.) . . . **Ralph**, Lt. (*temp. Capt.*) M. M., 1 G. B., Sco. Rif.
Cant. Magte. . . . .

Balaghat, Bethul, Bhandara, Bilaspur, Chindwara, Damoh, Dangiri, Harda, Nagpur, Raipur Seoni, Sitabaldi, Sutna.

5th (*MHOW*) DIVISION—*concld.*

JHANSI BRIGADE—

Bde. Comdr. . . .	**Poore,** Brig.-Genl. R. M., C.I.E., D.S.O., Brit. Ser.
Bde.-Maj.	**Birch,** Maj. C. F. M., 1st G. R.
Staff Capt.	**Emery,** Lt. (*temp. Capt.*) T. S., M.C., 1-76th Punj.
Asst. Dir., Medl. Services	**Davidson,** Col. J. S., A.M.S. (*also A. D. M. S., Jubbulpore Bde.*)

Jhansi.

S. O. (2nd cl.) . . .	**Hubble,** Lt. (*Actg. Capt.*) L. U., 2nd G. B., Essex R.
A. C. R. E.	**Davies,** Lt. (*Actg. Maj.*) J. F., I. A. R. O.
Offr. in charge Stn. Supply Depôt.	**Gordon,** Hony. Lt. W. D., S. & T. Corps.
Bde. Supply Offr. . .	**Harris,** Lt. L., S. & T. Corps.
G. E.	
Cant. Magte. . . .	**Walker,** Lt.-Col. F. S., Supy. List.

Guna.

Comdg.	**Capper,** Lt.-Col. A. S., D.S.O., 39th Horse.

Nowgong.

Comdg.	**Harding,** Lt.-Col. M. F., C. M. G.
Cant. Magte. . . .	**Lothian,** A. G., Esq., I.C.S.

Saugor.

Comdg.	**Dunsford,** Lt.-Col. F. P. S., D.S.O., 1-2nd Rajputs.
S. O. (3rd cl.) . . .	**Maunder,** Lt. R. A., I. A. R. O.
Cant. Magte. . . .	**Tucker,** W. A., Esq.

Sehore.

Comdg.	**Willis,** Maj. G. C. W., 31st Lrs.

Kotah.

Gwalior. Lalitpur.

POONA DIVISION.

COMMANDER . . .	**Tighe**, Maj. Gen. *Sir* M. J., K.C.M.G., C.B., C.I.E., D.S.O.
A. D. C.	**Nicholson**, Lt. A. R., I. A. R. O.
Hony. A. D. C.	*Nawab Sidi* **Ibrahim, Muhammad Yakub Khan,** Nawab of Sachin Hony. Capt.
GENERAL STAFF . . .	
Gen. Staff Offr., 1st Gde.	**Conway-Gordon,** Maj. (*temp. Lt.-Col.*) E. C. W., 3rd Horse, *p.s.c.*
Gen. Staff Offr., 2nd Gde.	**Pott,** Maj. E. H., D.S.O., 1st Lrs.

ADMINISTRATIVE, TECHNICAL AND DEPARTMENTAL STAFF—

General Officer . . .	**Boddam,** Lt.-Col., E. B. C., 5th G. R., *p.s.c.*
A. Q. M. G.	**Kitson,** Bt.- Lt.-Col. C. E., D.S.O., R. W. Kent R.
A. A. G.	**Crombie,** Maj. D. C., 23rd Cav., *p.s.c.* (n).
D. A. Q. M. G. . .	**Mackenzie,** Capt. (*temp. Maj.*) J. D., 41st Dogras.
D. A. A. G.	**Steele,** Bt.-Maj. C., 13th Hrs.
Comdg. R. A. . .	**Quinton,** Brig.-Genl. F. W. D., Brit. Ser. (*Also C. R. A., 5th & 9th Dns.*)
Brig. Genl., Indian Depots.	**Buchanan,** Brig. Genl. K. J., C.B., I.A.
C. R. E.	**Brewin,** Col. J. P., Brit. Ser.
Asst. Dir. of Supplies & Transport.	**Fulton,** Col. F. C., S. & T. Corps.
D. A. D. Supplies . .	**Fox,** Lt.-Col. L. W., S. & T. Corps.
Offr. i/c Supply Depots .	**Nicholson,** Lt.-Col. R. W., S. & T. Corps.
Dy. Asst. Dir. of Grass Farms.	**Finch,** Lt.-Col. C.
A. D. M. S.	**Jennings,** Col. W. E., I.M.S.
Depy. Asst. Dir., Medl. Services (Sany.).	**Bignold,** Capt. (*temp. Maj.*) C. A., R.A.M.C. (S.R.)
Depy. Asst. Dir., Medl. Services (Mobn.).	**Barry,** Capt. (*temp. Maj.*) S. J., R.A.M.C.
D. A. D. of Ord. Ser. .	**Oldbury,** Lt. (*temp. Capt.*) A. F., A.O.D.
Dy. Asst. Dir. Vety. Ser. .	**Stewart,** Capt. (*temp. Maj.*) H. C., A. V. C.
Inspr. of Army Schools .	**Ruthven,** Capt. W. J., Inspr. of Army Schools, 4th and 6th Circles (*Poona*).
Contr., Mily. Accts. (Under Finance Dept.)	**MacDonnell,** Lt.-Col. H. N. F.

AHMEDNAGAR BRIGADE—

Bde. Comdr. . . .	**Wood,** Brig.-Genl. E. J. M., D.S.O.
Bde. Maj.	**Harbord,** Capt. R.E., 1-4th E. Kent R.
Staff Capt.	**Kent,** Lt. C. E., L'pool R.
A. D. M. S.	**Lumsden,** Lt.-Col. P. J., M.B., I.M.S.
D. A. D. M. S.	**Cragg,** Maj. F. W., M.D., I.M.S.

Ahmednagar.

S. O. (2nd class) . .	**French,** Capt. H. J., Highl. Cyclist Bn.
A. C. R. E.	**Festing,** Maj. F. E. G., R.E.
Offr. in charge Supply Depot.	**Sayers,** Maj. C., S. & T. Corps.
Bde. Supply Officer .	**Fawcett,** Lt. H. E. M., I.A.R.O.
Cant. Magte. . . .	**C. W. A. Turner,** Esq., I.C.S.

Aurungabad.

Comdg.	**Moore,** Lt.-Col. A. G. W., I.A.
S.O. (3rd cl.) . . .	**McDougall,** Lt. D. M., 101st Grenadiers.
Cant. Magte. . . .	**Chatfield,** Lt. H. S., attd. 1-76th Punj.

Arangaon.

Comdg.	
Staff Capt.	

Anandi, Manmad, Dhond.

POONA DIVISION—*concld.*

POONA BRIGADE—

Commander . . . . **Norie,** Brig.-Genl. C. E. deM., C.B., C.M.G., D.S.O., *p.s.c.* [*l*].
Bde.Maj. . . . . . . **Mundy,** Capt. H. A. G., 2-6th R. Suss. R.
Staff Capt. . . . . **Sheppard,** 2nd-Lt. (*temp. Capt.*) W. H., attd. 14th Hrs.

Poona.

S. O. (*1st cl.*) . . . **Robinson,** Lt. F. W. O., 29th Lrs. (*Offg.*)

A.C.R.E. . . . . . **Lathbury,** Lt.-Col. H. O., R.E.
Bde. Supply Officer . **Dilke,** 2nd-Lt. T. T., A. S. C.
G. E. . . . . **Lord,** Capt. R. C., M.C., R.E.
Cant. Magte. . . . **Peck,** Lt.-Col. H., I.A.

Kirkee.

G. E. . . . .
Cant. Magte. . . . **Corbould,** Capt. P. S., I.A.R.O.

Purandhar (Sanitarium).

Comdg. & Cant. Magte. **Roper** Capt. J., 1-4th Dorset R.

Khandalla (Sanitarium).

Satara.

Comdg. . . . . . . **Hulbert,** Maj. T. E., 3rd Horse.

Bhusawal. Bijapur. Dhulia. Jalgaon. Panchgani. Sholapur.

Belgaum.

Comdg. . . . . . . **Mead,** Col. H. R., I.A.
S. O. (*2nd cl.*) . . . **Frere,** Capt. R. T., Norf. R.
Bde. Supply Offr. . . **Thomas,** Lt. A. R., I.A.R.O.
G. E. . . . . . . **Stewart,** 2nd-Lt. H. W., I.A.R.O.
Cant. Magte. . . . **Davies,** Lt. J. P., I.A.R.O.

Dharwar. Gadag. Hubli.

9th (*SECUNDERABAD*) DIVISION.

COMMANDER	**Barratt**, Maj.-Genl. W. C., C.B., D.S.O.
A.-D.-C.	**John**, Capt. A. W. U., V.D., I.A.R.O.

GENERAL STAFF—

Gen. Staff Offr., 1st gr. .	**Rivett-Carnac**, Maj. (*temp. Lt.-Col.*) J. S., 14th Lrs., *p.s.c.*
Gen. Staff Offr., 2nd gr. .	**Compton**, Maj. C. W. McG., 69th Punj.
Gen. Staff Offr., 3rd gr. .	**McIntosh**, Lt. (*temp. Capt.*) J. A., 1st G. B., Sco. Rif.

ADMINISTRATIVE, TECHNICAL AND DEPARTMENTAL STAFF—

General Officer	**Cox**, Col. F. W. H., C.B.
A. Q. M. G.	**Taylor**, Bt.-Lt.-Cl. R. O'B., C.M.G., C.I.E., 19th Lrs. *p.s.c.* [L.]
A. A. G.	**Powell**, Bt.-Lt.-Col., D. H., M.C., 105th Mah. L. I.
D. A. A. G.	**Murland**, Maj. H. F., 64th Prs.
Ditto	**Bowden**, Capt. (*temp. Maj.*) F. H., Mad. Gds., I. D. F.
D. A. Q. M. G. . . .	**Foster**, Maj. C. S., 37th Lrs.
Comdg. R. A. . . .	**Quinton**, Brig.-Genl. F. W. D., Brit. Ser. (*Also C. R. A., 5th Poona Dn.*)
Orderly Offr.	
Brig.-Genl., Indian Depôts	**Somerset**, Brig.-Genl. C. W., C.B., C.M.G., M.V.O., I.A.
C. R. E. .	**Picton**, Col. R. E. (*Offg.*)
Asst. Dir. of Supplies & Transport.	**Dickson**, Lt.-Col. J. H., C.I.E., S. & T. Corps.
D. A. D. Supplies . .	**Watts**, Maj. H. D., S. & T. Corps.
Asst. Dir. of Grass Farms	**Home**, Maj. (*temp. Lt.-Col.*) D. C.
Dy. Dir., Medl. Services	**Manifold**, Maj.-Genl. C. C., C.B., C.M.G., M.B., I.M.S.
Dy. Asst. Dir., Medl. Services (Sany.).	**Wallace**, Maj. G. S., R.A.M.C.
Dy. Asst. Dir., Medl. Services (Mobn.).	**Dutton**, Maj. H. R., I.M.S.
Dy. Asst. Dir., Medl. Services (Addl.)	**Fretz**, Maj. W. L. E.
D.A.D. of Ord. Ser. . .	**Dallimore**, Capt. A. G., A.O.D.
Dy. Asst. Dir., Vety. Ser.	**Jackson**, Maj. G. T. T., A.V.C.
Inspr. of Army Schools	**Connell**, Capt. E. S., Inspr. of Army Schools, 4th Circle (*Bangalore*).
Contr. Mily. Accts. (Under Finance Dept.)	**Szczepanski**, Maj. H. C.

SECUNDERABAD (CAVALRY) BRIGADE.

Bde.-Comdr.	**Commeline**, Brig.-Genl. F. H. B.
Bde.-Maj. . . .	**Young**, Capt. K. deL., M.C., 8th Cav.

SECUNDERABAD (INFANTRY) BRIGADE.

Bde.-Comdr.	**Smith**, Brig.-Genl. F. A.
Bde.-Maj.	**Magnay**, Capt. A. D., 1st Lrs.

SECUNDERABAD.

Comdg.	**Commeline**, Brig.-Genl. F. H. B.
S. O. (1st cl.)	**Bird**, 2nd-Lt. (*temp. Capt.*) R., 1-4th Devon. R.
Offr. in charge Supplies .	**Lincoln**, Lt.-Col. S. W., I.A.
Asst. Dir., Medl. Services	**Pridmore**, Lt.-Col. W. G., C.M.G., M.B., I.M.S.
A. C. R. E.	**Greenstreet**, Bt.-Lt.-Col. C. B. L., R.E. (*Offg.*)

Trimulgherry, Bolarum.

SOUTHERN BRIGADE.

Bde.-Comdr.	**Sladen**, Brig.-Genl. D. R., C.M.G., D.S.O.
Bde.-Maj.	**Reader**, Capt. F. E. M., 1-9th Hamp. R.
Staff Capt.	**Kirkpatrick**, Capt. H., 20th Nilgiri Malabar Bn. I.D.F.
Asst. Dir., Medl. Services	**O'Reilly**, Lt.-Col. H. W. H., R.A.M.C.
Offr. in charge of Station Supply Depot.	**Watling**, Maj. L. D., S. & T. Corps.
A. C. R. E.	**Lathbury**, Lt.-Col. H. O., R. E.

9th (*SECUNDERABAD*) DIVISION—*concld.*

SOUTHERN BRIGADE—*contd.*

Fort St. George and Madras.

Comdg. . . . . **Burton,** Brig.-Genl. R. G., (L)
Staff Capt. . . . . . **Orchard,** Capt. J. W., 1-4th Devon. R.
Do. . . . . **Duffy,** Capt. L., S. Lancs. R.
G. E. . . . . **Farrington,** Lt. F. L., I.A.R.O.
Staff Surgn. . . . . **D'Avray,** Capt. A. D., R.A.M.C.

Calicut.

Comdg. . . . . . . **Ward,** Lt.-Col. V. D., 20th Nilgiri-Malabar Bn.

Cannanore.

Comdg. . . . . **Mouat,** Maj. G. E. D., 2-63rd Infy.
S. O. . . . . **Chamberlain,** Lt. W. G., 2-63rd Infy.

Malappuram.

Comdg. . . . . . . **Kerr,** 2nd-Lt. D. H., High. L. I.

Poonamallee (Sanitarium).

Comdt. . . . . . . **Power,** Maj. P., R.A.M.C.

St. Thomas' Mount and Pallaveram.

Comdg. . . . . **Mackie,** Lt.-Col. R., 2-81st Prs.

S. O. (3rd cl.) & Cant. Magte. . . . **McNaughton,** Lt. J., 1st Sea. Highrs.

Trichinopoly.

Comdg. . . . . **LeMessurier,** Maj. G. D. S., 83rd Infy.

Wellington.

Comdg. . . . . **Sladen,** Brig.-Genl. D. R., C.M.G., D.S.O.

S. O. (2nd cl.) . . . . **Bone,** Capt. C. B., 2-4th Devon. R.

Offr. in charge Station Supply Depot. **Watling,** Maj. L. D., S. and T. Corps.
Bde Supply Officer . . **Moran,** Maj. J., Attd. S. & T. Corps.
G. E. . . . . **Abbott,** Capt. T. W., R. E.

Vizianagram.

Bezwada. Negapatam. Ootacamund. Tanjore.

BANGALORE BRIGADE—

Bde. Comdr. . . . . . **Wolley-Dod,** Brig.-Gen. O. C., C.B., D.S.O., *p.s.c.* [*l*] Brit. Ser.
D. A. A. & Q. M. G. . **Curry,** Capt. (*temp. Maj*) F. R., R. Dub. Fus.
Gen. Staff Offr.-2nd gr. **Plinston,** Capt. G. H., 11th Rajputs.
Staff Capt. . . . . **Coburn,** Lt. (*temp. Capt.*) F. R. Genl. List.
Staff Capt. . . . . **Gould,** Lt. (*temp. Capt.*) H. R., I.A.R.O.

Bangalore.

Comdg. . . . . . . **Wolley-Dod,** Brig.-Genl. O. C. C.B., D.S.O., *p.s.c.* (*l*) Brit. Ser.
S. O. (2nd cl.) . . . **Reynolds,** Capt. B. O., 2nd (*Madras*) Group, Arty. I. D. F.

A. C. R. E. . . . . . **MacGeorge,** Maj. J. B., R.E.
A. D. M. S. . . . . . **More,** Col. L. P., A. M. S.

Offr. in charge Stn. Supply Depôt. **Carter,** Lt.-Col. E. J., S. and T. Corps.
Bde. Supply Offr. . . . **Bell,** Lt. H., I.A.R.O.
Bde. Transport Offr. . . **Goodhall,** Comsy. & Hon. Maj. E. J., U. L., S. & T. Corps.
Staff Surgeon . . . **Maunsell,** Maj. E. C. C., I.M.S.
G. E. . . . . **Burnett,** Lt. A. S., I.A.R.O.
Asst. Provost Marshal . **Cummings,** 2nd-Lt. (*Temy. Capt.*) W. H., Manch. R.
Ditto . . . . . . **Peden,** 2nd-Lt. (*temp. Capt.*) W., 1st Bn. High. L. I.

Bellary. Chickmagalur. Gooty. Guntakal. Kolar Gold Field. Mercara. Mysore. Pollibetta. Saklaspur. Sivasamudram Power Works. Shimoga. Whitefield.

KARACHI BRIGADE.

Bde. Comdr. . . . **Fowler**, Maj.-Genl. F. J., C.B., D.S.O.

Gen. Staff Offr., 2nd gr. **Tobin**, Maj. H. W., D.S.O., 128th Prs., *p.s.c.*

D. A. A. and Q. M. G. **Goldingham**, Capt. D. D., 6th Middx. R.

Staff Capt. . . . . **Hall**, Lt. (*temp. Capt.*) F. G., Dorset R.

Do. . . . . **Lloyd**, Lt. (*temp. Capt.*) G. T., I.A.R.O.

Intell. Offr. . . . . . **Alford**, Capt. E. N., 2nd G. B. Bedf. R.

Comdg. R. A. . . . **Cook**, Brig.-Genl. H. R., Brit. Ser. [L] (*Also C. R. A., 4th Dn.*)

A. C. R. E. . . . **Harston**, Maj. J., Attd. R. E. (*Offg.*)

D. A. D. S. & T. . . **Ditmas**, Lt.-Col. A. R., S. & T.

Bde. Supp. Offr. . . **Dunn**, Capt. F. E., I.A.R.O.

Asst. Dir. Medl. Services (Distribution). **Smith**, Lt.-Col. S. B., C.M.G., I.M.S. (*Offg.*)

D. A. D. M. S. (*Santy.*) **Rennie**, Maj. P. M., I.M.S.

Karachi.

S. O. (*2nd cl.*) . . **Kearsey**, Lt. S. H., I.A.R.O.

G. E. . . . . . **Roland-Price**, Lt. (*temp. Capt.*) A., I.A.R.O.
Davies, 2nd-Lt. W. B., I. A. R. O.

Cant. Magte. . . . . **Richardson**, W. N., Esq.

Kiamari.

Embn. Comdt. . . . **Mahon**, Maj. (*temp. Col.*) H. J., C.I.E.
Asst. Embn. Comdt . **Twiss**, Maj. (*temp. Lt.-Col.*) E. K., D.S.O., 10th Jats.
A. D. S. & T. Emb. . **Preston**, Lt.-Col. E. E., S. & T. Corps.
D. A. Q. M. G. . . **Humphrey**, Capt. (*temp. Maj.*) J., I.A.R.O.
D. A. Q. M. G. . . . **Ruble**, Capt. (*temp. Maj.*) C. B., Lan. Fus.
D. A. D. M. S. (*Dis*). . **Spong**, Capt. (*temp. Maj.*) W. A., R A.M.C.
D. A. D. M. S. (*Emb.*) **Levack**, Capt. J. S., (*tempy. Maj.*) R. A. M. C.
D. A. D. V. S. . . . **Mackie**, Capt. C., A. V. C.

Manora.

G. E. . . . . . . **Mattocks**, Lt. F. S., R.E.

Hyderabad.

Comdg. . . . . **Parrington**, Maj. (*Acg. Lt.-Col.*) J. W., R. F. A.

S. O. (*3rd cl.*) . . **Steedman**, Lt. C. D., R.F.A.

G. E. . . . . . . **Chivers**, Lt. E., R.E.

Cant. Magte. . . **Mr. Wahid Baksh.**

Kotri. Sukkur.

BOMBAY BRIGADE.

Bde. Comdr.	**Knight,** Maj.-Genl. W. C., C.B., C.S.I., D.S.O., *p.s.c.* [L]
A.Q. M. G.	**Sherer,** Lt.-Col. J. C., 37th Lrs.
Gen. Staff Offr., 2nd Gde.	**Moberly,** Maj. C. N., 15th Bombay Bn., I. D. F.
D. A. A. and Q. M. G. .	**Ball,** Capt. (*temp. Maj.*) E. P., 110th Mah. L. I.
Staff Capt.	**Angler,** Lt. (*temp. Capt.*) C. T., 24th Rif. Brig.
Ditto	**Jackson,** Lt. (*temp. Capt.*) A H. C., I. A. R. O.
Ditto (Registration) .	**Sheppard,** Capt. S. T., 15th Bombay Bn., I. D. F.
Intelligence Offr. . . .	**Kinnear,** Capt. N. B., 15th Bombay Bn., I. D. F.
A.C.R.E.	**Clarke,** Maj. C. J., D.S.O., R. E.
Chief Ord. Offr. . . .	**Clarke,** Maj. H., R. A.
Offr. i/c Ord. Depôt . .	**Harding,** Maj. G., I.O.D.
D. A. D. S. & T.	**Rendall,** Maj. J. C. S., S. & T. Corps.
Offr. in charge Supply Depôt.	**Mathias,** Lt.-Col. L. J., S. & T. Corps.
Bde. Supply Officer . .	**Carroll,** Lt. F. H., S. & T. Corps.
A. D. M. S.	**Beyts,** Col. W. G., *V.H.S.*, R.A.M.C.
D. A. D. M. S. (Sany.) .	**Stewart,** Capt. A. D., I.M.S.

Embn. Comdt. . . .	**St. John,** Brig.-Genl. R. S., D.S.O., 20th Punj. *p. s. c.*
Asst. Embn. Comdt. . .	**Maitland,** Maj. (*temp. Lt.-Col.*) G. R., D.S.O., 14th Lrs.
A.Q.M.G. (Addl.) . . .	**Ward,** Maj. (*temp. Lt.-Col.*) B. S., 99th Infy. [*l*]
D.A.Q.M.G.(Embarkation)	**Farquharson,** Capt. (*temp. Maj.*) C. W., 10th Jats.
A. D. S. & T.	**Geoghegan,** Lt.-Col. F. E., C.I.E., S. & T. Corps.
Embn. Staff Offr.	**Salusbury,** Maj. R. T. G., 26th Punj.
D. A. D. R. T.	**Matthews,** Capt. (*temp. Maj.*) F. M., 22nd Punj.
E. & M. B. Shipping Offr.	
A. D. M. S. (Distn.) .	**Barnardo,** Bt. Lt.-Col. F. A. F., C.I.E., I.M.S.
D. A. D. Vety. Ser. .	**Hewlett,** Lt. (*temp. Maj.*) K., Vety. Bch., I. D. F.
Offr. in charge, Clo. Depôt	**Burlton,** Lt.-Col. W. E. F., S. & T. Corps.
Asst. Dir., Docks . . .	**Cherry,** Maj. (*temp. Lt.-Col.*), J. A., I.A.R.O.
Asst. Dir., Railways . .	**Reynolds,** Maj. (*temp. Lt.-Col.*) J. R., I.A.R.O.
R. A. F. Embn. & Shipping Offr.	**Solomon,** 2nd-Lt. E. V., R. A. F.
Embn. Supply Offr. . .	**Wardell,** Maj. G. F. E., S. & T. Corps.

Bombay.

S. O. (1st cl.) . . .	**Armour,** Capt. J. S. D. M., I. G. B., Sco. Rif.
Provost Marshal . .	**Claridge,** Lt.-Col. P. S. F., 28th Cav.
Asst. Provost Marshal.	**Everton,** 2nd-Lt. R. F., R.G.A.
Ditto	**Newman,** Lt. R. A., 9th Worc. R.

Deolali.

Comdg.	**MacGeorge,** Col. H. K., C.B.
S. O. (2nd cl.) . . .	**Barber,** Capt. V. J., 1st Yorks R.
Cant. Magte. . . .	**Molony,** Maj. O. H. C., 94th Infy.
A. C. R. E.	**Ainslie,** Lt.-Col. C., R.E.
Asst. Provost Marshal .	**Howell,** Capt. C. J., 1-4th The Buffs.

Santa Cruz.

Comdg.	**Loudon,** Lt.-Col. J. A., 2-43rd Erinpura R.

Ahmedabad.

Comdg.	**Frazer,** Col. G. S., C.M.G., I.A.
Bde. Maj.	
Staff Capt.	**Pearson,** Lt. (*temp. Capt.*) H. C., 1-2nd G. R.
S. O. (3rd cl.)	**Butterfield,** Lt. B. F., 2-97th Infy.
Cant. Magte.	**Fitzpatrick,** Lt. J. T., I. A. R. O.

Baroda.

Comdg.	**Lawrenson,** Lt.-Col. W. R., 2-73rd Infy.
S. O. (3rd cl.) . . .	**Rowcliffe,** Capt. N. H. B., 2-73rd Infy.

Rajkot.

Comdg.	**Tate,** Maj. W. M., 125th Rif.

4th (*QUETTA*) DIVISION.

COMMANDER . . . . . . Wapshare, Lieut.-Gen. R., C.B., C.S.I.
A.-D.-C. . . . .

GENERAL STAFF—
Gen. Staff Offr., 1st gr. . Baird, Bt.-Col. H. B. D., C.M.G., D.S.O. 12th Cav. *p.s.c.*
Gen. Staff Offr., 2nd gr. . Buckland, Maj. G.C.B., D.S.O., M.C., 2-8th G.R.
Gen. Staff Offr., 2nd gr. . Marshall, Maj. A., D.S.O., 28th Cav.
ADMINISTRATIVE TECHNICAL AND DEPARTMENTAL STAFF—
General Officer . . . . Jackson, Brig.-Genl. H. W., D.S.O., 89th Punj.
A. A. and Q. M. G. . . Fetherstonhaugh, Bt.-Lt.-Col. W. A., D.S.O., 8th Cav., *p.s.c.*
D. A. Q. M. G. . . . Williams, Capt. (*temp. Maj.*) S. C., S. & T. Corps.
D. A. A. G. . . . . . . Foster, Maj. H. M., 1-4th Hamps. R.
Comdg. R. A. . . . . . Cook, Brig.-Genl. H. R., Brit. Ser. [L] (*Also C. R. A., Karachi Bde.*
Orderly Offr. . . . . . Blomenstok, Lt. L. M., M.C., 102nd Batty., R. F. A.
Brig.-Genl., Indian Depôts Money, Lt.-Col. (*tempy. Brig.-Genl.*) E. D., C.I.E., D.S.O., 1st G. R.
C. R. E. . . . . . . Maclagan, Col. R. S., C. B., C.S.I., Brit. Ser.
Asst. Dir. of Supply and Transport. Brooke, Lt.-Col. H. M. M., S. & T. Corps.
D. A. D. Supplies . . . Stewart, Lt.-Col. R. J. T., I.A.
D. A. D. Transports . . Walsh, Capt. G. G., I.A.R.O.
O. C., "G." Supply Depôt Coy. Steele, Maj. C. E. B., S. & T. Corps.
Dy. Dir., Medl. Services . Hendley, Maj.-Genl. H., M.D., I.M.S.
Dy. Asst. Dir., Medl. Services (Sany.). Stevenson, Maj. W. D. H., C.I.E., I.M.S.
Dy. Asst. Dir., Medl. Services (Mobn.). Babonau, Capt. (*temp. Maj.*) A. F., I.M.S.
D. A. D. of Ord. Ser. . . Farquharson-Roberts, Capt. K. F., A. O. D.
Dy. Asstt. Dir., Vety. Ser. Dalgleish, Maj. A. J., A. V. C.
Contr. Mily. Accts. . .
(*Under Finance Dept.*) . . Shairp, Maj. H. F.

Quetta.

Comdg. . . . . Hardy, Maj.-Genl. T. H.
S. O. (1st cl.) . . . . Smith, 2nd-Lt. (*temp. Capt.*) J., I.A.R.O.
A. C. R. E. . . . . . Ogilvie, Col. E. C., C.M.G., R.E.
A. C. R. E., Civil . . . Saunders, Lt.-Col. F. W., R.E.
Cant. Magte. . . Acheson, J. G., Esq., I.C.S.
Asst. Provost Marshal .

Chaman.

Comdg. . . . . . . Gale, Lt.-Col. E. S., 2-10th G. R.
S. O. (3rd cl.) . . .

Fort Sandeman.

Comdg. . . . . . . Ryder, Lt.-Col. W. L., 3-1st G. R.
G. E. . . . . Darling, Lt. (*temp. Capt.*) H. J., I.A.R.O.

Loralai.

Comdg. . . . . . . Sealy, Lt.-Col. A. E., 3-7th G. R.
S. O. (3rd cl.) and Cant. Magte. Russell, Lt. A. M. H., 3rd Horse.
A. C. R. E. . . . . . Murray, Lt.-Col. J. H. S., R. E.
G. E. . . . . Hodgson, Lt. H. L.

Drug, Girdao, Gulistan, Gumbaz, Harnai, Hindubagh, Hirok, Marataugi, Mastung, Murgha, Musa Khel, Pishin, Jacobabad, Killa Saifulla, Kishingi, Shelabagh, Shingarh, Sibi.

10th (INFANTRY) BRIGADE—
Bde. Comdr. . . . . . O'Grady, Brig.-Genl. Hde C., 52nd Sikhs, *p.s.c.*
Bde.-Maj. . . . . Scoones, Capt. G. A. P., D.S.O., M.C., 2nd G. R.
Staff Capt. . . . . . Braund, Capt. H. B. L., 5th Wilts. R.

11th (INFANTRY) BRIGADE—
Bde. Comdr. . . . Hardy, Maj.-Genl. T. H., I.A.
Bde.-Maj. . . . Satterthwaite, Capt. R. E., 1-5th R. W. Kent. R.
Staff Capt. . . . Tweedy, Capt. R. J., 19th Lrs.
57th (INFANTRY) BRIGADE—
Bde. Comdr. . . . Gordon, Brig.-Genl. J. L. R., C.B., 15th Sikhs.
Bde.-Maj. . . . . Nugent, Capt. J. F.H., D.S.O., 28th Punj.
Staff Capt. . . . . Dymott, Capt. B. K., 31st Punj.
12th MOUNTED BRIGADE—
Bde. Comdr. . . . Birdwood, Brig.-Genl. H. B.
Bde. Maj. . . . Gray, Maj. E. St. C., M.C., 34th Horse.
Staff Capt. . . . Binning, Lt. (*temp Capt.*) H. A., I.A.R.O.
13th MOUNTED BRIGADE—
Bde. Comdr. . . . .
Bde. Maj. . . . . . Morris, Maj. G. P., 30th Lrs.
Staff Capt. . . . Ibbotson, Lt. (*temp. Capt.*) A. W., M.C., I.A.R.O.

8th (*LUCKNOW*) DIVISION.

COMMANDER	**Scott**, Maj.-Genl. *Sir* A. B., K.C.B., D.S.O., Brit. Ser.
A.-D.-C.	**Young**, Capt. G. A. G., 1-6th Jats.
GENERAL STAFF—	
Gen. Staff Offr., 1st gr. .	**Moberly**, Lt.-Col. H. S., 66th Punj. (*Q*), *p.s.c.*
Gen. Staff Offr., 2nd gr. .	**Dick-Lauder**, Maj. J. N. D., 16th Cav.
ADMINISTRATIVE, TECHNICAL AND DEPARTMENTAL STAFF—	
General Officer	**Glanville**, Brig.-Genl. F., D.S.O., Brit. Ser.
A. Q. M. G.	**Lloyd**, Lt.-Col. J. H., 6th G. R.
D. A. A. G.	**Griffin**, Capt. (*temp. Maj.*) J. B., 1-10th Middx. R.
D. A. A. G. (*R.*) . . .	**Phillips**, Capt. (*temp. Maj.*) H. S., 27th L. Cav.
Comdg. R. A.	**Gordon**, Brig.-Genl. L. G. F., C.B., D.S.O. (*g*), Brit. Ser. (*Also C. R. A., Meerut Dn.*)
Orderly Offr.	
Brig.-Genl., Indian Depôts	**Lowis**, Brig.-Genl. H. E., I.A.
Do.	**Boudler**, Brig.-Genl. E. W., I.A.
C. R. E.	**Weedon**, Col. F. F., Brit. Ser. [L].
Asst. Dir. of Supplies and Transport.	**Forteath**, Lt.-Col. F. W. H., S. & T. Corps.
D. A. D. Supplies . . .	**Western**, Maj. C. M. T., S. & T. Corps.
Offr. i/c. Supply Depôts .	
Asst. Dir. of Grass Farms	**Hawks**, Maj. (*temp. Lt.-Col.*) G. A.
Dy. Dir. Medl. Services	**Cree**, Maj.-Genl. G., C.B., C.M.G., Brit. Ser.
Dy. Asst. Dir., Medl. Services (Sany.)	**Cairns**, Maj. J., R.A.M.C.
Dy. Asst. Dir., Medl. Services (Mobn.).	**Kerans**, Maj. G. C. L., D.S.O., I.M.S.
D. A. D. of Ord. Ser. .	**Hopkins**, Capt. (*temp. Maj.*) A. C., A.O.D.
Dy. Asst. Dir., Vety. Ser. .	**Lewis**, Capt. (*temp. Maj.*) H. D., A.V.C.
Inspr. of Army Schools .	**Connell**, Capt. E. S. 5th Circle (*Darjeeling*)
Contr., Mily. Accts. (Under Finance Dept.) .	**Will**, Maj. D. A. E.
PRESIDENCY BRIGADE—	
Bde. Comdr.	**Strange**, Brig.-Genl. R. G., C.I.E., Brit. Ser.
Gen. Staff Offr., 2nd gr.	**Marshall**, Bt.-Maj. J. S., D.S.O., 35th Sikhs.
D. A. A. and Q. M. G.	**Raymond**, Maj. E. D., D.S.O., M.C., 30th Lrs.

Fort William.

Asst. Dir. Medl. Services	**Daly**, Col. J. T., I.M.S.
A. C. R. E.	**Kealy**, Maj. P. H., R.E.
Offr. in charge Supplies .	**Brunker**, Maj. J. R., S. & T. Corps.
Garrn. Qr.-Mr. . . .	**Studd**, Capt. E., I. A. (T. C.).
G. E.	**Lloyd**, 2nd-Lt. H. S. R.E.
Provost Marshal, Calcutta	**Pollard**, Capt. E. A., 5th W. Yorks. R.
Asst. Embarkation Staff Offr. & Railway Transport Offr.	**Waller**, Capt. G. D., I.A.R.O.

Buksa Duar.

Comdg.	

Barrackpore.

Comdg.	**Kirkpatrick**, Maj. K. E., 3rd Brahmans.
S. O. (*3rd cl.*)	**Dowling**, Lt. G., 3rd Brahmans.
A. C. R. E.	
Cant. Magte.	**Stevens**, C. G. B., Esq., I.C.S.

Darjeeling.

Comdg.	
S. O. (*1st cl.*) *and Cant. Magte.*	**Walker**, Capt. T., 1-10th Midd'x. R.
S. M. O.	**Leslie**, Maj. T. C. C., D.S.O., R.A.M.C.
Provost Marshal, Darjeeling.	

Dacca.

Comdg.	
Cant. Magte.	

Dinapore.

Comdg.	**Paull**, Lt.-Col. J. R., 2-5th Som. L. I.
S.O. (*3rd cl.*)	**Kerridge**, 2nd-Lt. T. H., 2-5th Som. L. I.
Cant. Magte.	**O'Sullivan**, Capt. K. L., R.A.M.C.

Gaya, Ghoom.

8th (*LUCKNOW*) DIVISION—*contd.*

Dum Dum.

Comdg. . . . . . . **Cook**, Capt. J., 2-5th Som. L. I.
Cant. Magte. . . . . . . **Stevens**, C. G. B., Esq., I.C.S.

Gantok.

Comdg. . . . . . .
S.O. (*3rd cl.*) . . . . . .

Gyantse.

Comdg. . . . . **Roberts**, Capt. M. R., 2-113th Infy.

Lebong.

Comdg. . . . .
S. O. . . . . } **Walker**, Capt. T., 1-10th Middx. R.

Takdah.

S. O. (*3rd cl.*) . . . .
Cant. Magte. . . . . } **Walker**, Capt. T., 1-10th Middx. Regt.

Alipore
Asansol
Ballygunge
Bankipore
Barakar, Buxar, Patna
Ranchi
Narainganj
Chittagong
Calcutta*
Mozufferpore
Somastipore
Jamalpore
Giridih, Jajha, Ondal, Rampur Haut, Raneegunge, Sahibgunge, Sitarampur
Coorjee

Adra, Chakardharpur
Khargpur
Berhampore, Midnapore
Bhagalpur, Chupra, Dhurbanga, Motihari
Barnagore, Buddertollah, Fort Gloster, Goosery, Howrah, Ichapur, Jugatdal, Kamarhatty, Kankinarrah, Khardah, Rishra, Seebpore, Tittaghur
Seebpore, Lillooah, Howrah
Garden Reach, Khurda Road
Santragachi
Sakchi, Kulti
Jherria, Sijua
Dam Dim
Kurseong, Jalpaiguri, Nagrakata, Kalchini, Kalimpong
Gonatea, Sara
Lalmonir Hat, Kanchrapara, Rajbari, Saidpore
Sonepore

* The troops at Alipore and Calcutta form part of the garrison of Fort William.

Shillong.

Comdg. . . . . . . **Roosmale-Cocq**, Lt.-Col. C. A., 1-8th G. R.
S.O. (*3rd cl.*) *and Cant. Magte.* **Powell**, Lt. L. F., 1-17th Infy.

Dibrugarh.

S. O. (*3rd cl.*) *& Cant. Magte.*

Kohima.

Comdg. and S. O. (*3rd cl.*)

Manipur.

Comdg. . . . . **Row**, Lt.-Col. G., I.A. (retd.).
S. O. (*3rd cl.*) . . . .

Sadiya.

Dullabcherra.
Hollicherra.
Srimangal.
Laskarpur.

Pahartali.
Lumding.
Badarpur.

Silchar.

Lakhimpur Dist.
Sibsagar Dist.
Jorhat Dist.
Darrang Dist.

Nowgong.
Mangaldai.
Kamrup.

8th (*LUCKNOW*) DIVISION—*concld.*

ALLAHABAD BRIGADE—

Bde. Comdr.	**Cadell**, Brig.-Gen. A., I. A.
Bde.-Maj.	**Evans**, Capt. B. L., 1-4th R. W. Surr. R.
Staff Capt.	**Horsfield**, Lt. (*temp. Capt.*) H. E., I.A.R.O.
Asst. Dir., Medl. Services	**Garvie**, Col. J., I. M. S.

Allahabad.

A. C. R. E.	**Greer**, Maj. R. E., R.E.
Offr. in charge Supply Depôt.	**Gibb**, Lt. J. B., I. A. R. O.
Bde. Supply Officer	**Angelo**, Lt. M. O., I.A.R.O.
G. E.	**Sykes**, Lt. J., P.W.D.
Cant. Magte	**Hamilton**, Capt., A. H. C., I.A.R.O.
S. O. (2nd cl.)	**Dehn**, 2nd-Lt. C. G., 2-4th Wilts. R.

Benares.

Comdg.	**Fenton**, Maj. L. S., 1-113th Infy.
S. O. (3rd cl.) & Cant. Magte.	**Vere de Matthews**, Lt. R., 1-113th Infy.

Cawnpore.

Comdg. (1st cl.)	**Wood**, Col. A. H., C.B.E., 1st G B., R. Ir. Rif.
S. O. (2nd cl.)	**Bell**, Temp. Lt. N. G., 1st G.B., R. Ir. Rif.
Offr. Comdg. "J." Supply Depôt. Coy.	**Home**, Maj. J. E., S. and T. Corps.
G. E.	**McDonough**, Lt. A. D.
Cant. Magte.	**Lawrenson**, Lt.-Col. T. G. P., Supy. List.

Fatehpur.
Ghazipore.
Hamirpur.
Azamgarh.
Banda.
Jaunpur.
Etawah.
Mahoba.
Nazafgarh.
Chunar.
Mirzapore.
Moghal Serai.

Fatehgarh.

Comdg.	**Hirst**, Maj. F. C., 1-1st Brahmans.
S. O. (3rd cl.) / *Cant. Magte.*	**Stowell**, Lt. P. C., I.A.R.O.

LUCKNOW BRIGADE—

Bde. Comdr.	**Pritchard**, Brig.-Genl. A. G., C.M.G., 2nd Lrs.
Bde. Maj.	**Bertin**, Capt. H., 1st G. B., E. Yorks R.
Staff Capt.	**Cressall**, Lt. (*temp. Capt.*) H. A., 1-10th Middx. R.
S. O. (1st cl.)	**Goode**, Capt. W. W., 2-5th Som. L. I.
*Asst. Dir., Medl. Services.**	

Lucknow.

A. C. R. E.	**Craster**, Tempy. Maj. E. E., R.E.
Offr. in charge Supply Depôt.	**Parsons-Smith**, Maj. S. H., S. & T. Corps.
Bde. Supply Officer	**Clack**, Lt. G. H., I.A.(T.C.)
G. E.	
Cant. Magte.	**Prince**, Lt.-Col. W. C. S., (*Supy. List*) [L].

Fyzabad.

Comdg.	**Sutton**, Col. H. G., 21st Ref. Brig.
S. O. (2nd cl.)	**Hart**, Capt. A. E., 1-10th Midd'x. R.
Offr. in charge Supplies	
G. E.	
Cant. Magte.	**Hart**, Capt. A. E., 1-10th Middx. R.

Gonda. **Gorakhpur.**

Sitapur.

Comdg.	**Mactavish**, Lt.-Col. A., 2-1st Brahmans.
S. O. (3rd cl.)	**Finch**, 2nd-Lt. H. C., 2-1st Brahmans.
Cantt. Magte.	**Horner**, Capt. W. J., 2-66th Punj.

BURMA DIVISION.

COMMANDER	**Keary**, Lt.-Genl. *Sir* H. D'U., K.C.B., K.C.I.E., D.S.O.
A.-D.-C.	**Lushington**, Lt. A.N., I.A.R.O.
GENERAL STAFF.	
Gen. Staff Offr., 1st gr. . .	**Tod**, Col. J. K.
Gen. Staff Offr., 2nd gr. . .	**Davis**, Maj. C. T., 107th Prs.
ADMINISTRATIVE, TECHNICAL AND DEPARTMENTAL STAFF.	
A. A. and Q. M. G. . . .	**Lindesay**, Maj. (*temp. Lt.-Col.*) F. S., I. A.
D. A. A. G.	**Goodland**, Capt. (*temp. Maj.*) C. H., 2-5th Som. L. I.
D. A. Q. M. G.	**Haig**, Lt. (*temp. Maj.*) H.G., I.A.R.O.
Asst. Dir. of Supplies and Transport.	**Swan**, Lt.-Col. J. S., S. & T. Corps.
O. C. "E" Supply Depôt Coy.	**Vaughan**, Lt.-Col. E. G., D.S.O., S. & T. Corps.
Asst. Dir., Medl. Services	**Fooks**, Col. H., I.M.S.
Dy. Asst. Dir., Medl. Services (Sany.)	**Stack**, Maj. H. T., R.A.M.C.
D. A. D. of Ord. Ser. .	
D. A. D., Vety. Ser. . . .	**Stevenson**, Lt. (*temp. Capt.*) W., I.A.R.O.
Contr. Mily. Accts. (under Finance Dept.) .	**Chandler**, Lt.-Col. H. G. W.

Mandalay.

Comdg.	**Richardson**, Maj. G. G., 85th Burma Rif.
S. O. (2nd cl.) . . .	**Basset**, Lt. W. S., 1st G. B., R. Ir. Fus.
Cant. Magte.	**Carey**, J., Esq.

Bhamo.

Comdg.	**Biggs-Atkinson**, Capt. T. B.
S. O. (3rd cl.) . . .	**Wayte**, Lt. V. W., 86th Infy.

Maymyo.

Comdg.	**Robinson**, Lt.-Col. G. W., C.B. (I.A.), 1st G. B., R.Ir. Fus.
S. O. (2nd cl.)	**Morley**, Capt. B. E., 18th Rif. Bde.

Shwebo.

Comdg.	**O'Reilly**, Lt.-Col. C. M., 86th Infy.
S. O. (3rd cl.)	**Felton**, 2nd-Lt. G. H., 86th Infy.

Chin Hills, Lashio, Magwe, Myitkyina, Monywa, Mogok, Pyinmana, Taunggy, Yamethin, Yenangyaung.

Meiktila.

Comdg.	**Armstrong**, Lt.-Col. W. McG., 3rd G. B. Bedf. R.
S. O. (3rd cl.)	**Martin**, Lt. R. McG., 3rd G. B., Bedf. R.
Offr. in charge Supplies .	**Wilkinson**, Lt. E. J., S. & T. Corps.

BURMA DIVISION—*concld.*

RANGOON BRIGADE—

Bde. Comdr.	**Young,** Maj.-Genl. C. F. G.
Bde.-Maj.	**Macleod,** Maj. N., D.S.O., 15th Sikhs, *p.s.c.*
Staff Capt.	**Fear,** Capt. F. R., 1-5th Hamps. R.

Rangoon.

Stn. Staff Offr.	**Sinclair,** Capt. W. C. B., 18th Rif. Brig.
Offr. in charge Supply Depôt.	**Jones,** Maj. L. A. M., S. & T. Corps.
G. E.	**Gill,** Capt. A. G., R.E.
Cant. Magte.	**Kirkwood,** Lt.-Col. A. T., Supy. List.

Port Blair.

Comdg.	**Ricketts,** Capt. F. W., 1st G. B., R. I. Fus.
S. O. (3rd cl.)	**Gillon,** 2nd-Lt. J. B., 63rd Infy.

Thayetmyo.

Comdg.	**Holmes,** Lt.-Col. G. V., 43rd Infy.
S. O. (3rd cl.) . . .	**Lewis,** Lt. C. B., I.A.R.O.
Offr. in charge Supplies .	**Routleff,** Lt. J. I., S. and T. Corps.

Moulmein.

Mergui
Tavoy } . . . .

BRIGADE COMMANDERS.

Watson, Maj.-Genl. (*tempy.*) H. D., C.B., C.M.G., C.I.E., M.V.O.	11 Oct. 14
Delamain, Maj.-Genl. *Sir* W. S., K.C.M.G., C.B., D.S.O.	13 Oct. 14
VC **Melliss**, Maj.-Genl. *Sir* C. J., K.C.B. K.C.M.G. [L]	6 Apl. 15
Douglas, Brig.-Genl. J. A., C.M.G., C.I.E. [L]	16 Sep. 15
Hamilton, Brig.-Genl. W. G., C.B., C.S.I., D.S.O., Brit. Ser., *p.s.c.*	25 Oct. 15
Tidswell, Brig.-Genl. E. C., C.B., D.S.O., Brit.Ser.	2 Dec. 15
Austin, Brig.-Genl. H. H., C.B., C.M.G., D.S.O., R. E., *p.s.c.* [*l*]	28 Jan. 16
Edwardes, Brig.-Genl. S. M., C.B., C.M.G., D.S.O.	19 Feb. 16
Evans, Brig.-Genl. U. W., C.B., C.M.G., Brit. Ser.	13 Apl. 16
Gregory, Brig.-Genl. C. L., C.B., *p.s.c.*, 19th Lrs.	1 May 16
Lucas, Col. (*temp. Brig.-Genl.*) F. G., C. B., 5th G. R.	1 Apl 16
Davidson, Brig.-Genl. S. R., C.M.G., 47th Sikhs	12 May 16
Stewart, Maj.-Genl. J. M., C.B.	1 June 16
Davies, Brig.-Genl. C. H., C.M.G., D.S.O.	6 July 16
Andrus, Brig.-Genl. T. A., C.M.G., Brit. Ser.	11 July 16
Banbury, Brig.-Genl. W. E., C.M.G., I.A.	27 July 16
Lock, Brig.-Genl. F. R. E., D.S.O.	19 Sep. 16
Legard, Brig.-Genl. D. A., 17th Lrs. *p.s.c.* [*l*]	31 Oct. 16
Bethell, Brig.-Genl. H. K., C.M.G., D.S.O. 7th Hrs. A.-D. C.	16 Oct. 16
Younghusband, Brig.-Genl. L. N., C.B., C.M.G.	1 Nov. 16
Holland-Pryor, Brig.-Genl. P., C.M.G., D.S.O., M.V.O.	10 Feb. 17
Leslie, Brig.-Genl. W. B., R. E.	27 Jan. 17
Vincent, Brig.-Genl. B., 6th Dns.	. . .
Murray, Brig.-Genl. E. R. B., 90th Punj	15 Mar. 17
Norton, Brig.-Genl. C. E. G., 7th Hrs.	24 Mar. 17
Maclachlan, Brig.-Genl. T. R., C.M.G., 92nd Punj.	11 May 17
Wauchope, Brig.-Genl. A. G., C.M.G., D.S.O., Ryl. Highrs.	11 May 17
Luard, Brig.-Genl. C. C., C.M.G., Brit. Ser.	30 July 17
Bateman-Champain, Brig.-Genl. H. F., C.M.G., 9th G. R. [*l*].	6 Aug. 17
Hogg, Brig.-Genl. R. E. T., C.I.E., 38th Horse.	22 June 17
Gwyn-Thomas, Brig.-Genl. G., C.M.G., D.S.O., 2nd Lrs.	20 July 17
Andrew, Brig.-Genl. F. A., D.S.O., 128th Prs.	15 July 17
Andrew, Brig.-Genl. A. W., C.M.G.	20 Aug. 17
Hildyard, Brig.-Genl. R. J. T., C.M.G., D.S.O., *p.s.c.*, R. W. Kent R.	12 Sep. 17
Pryce, Brig.-Genl. H. E. apR., C.M.G. D.S.O., 18th Infy.	17 Oct. 17
Cassels, Brig.-Genl. R. A., C.B., D.S.O., 32nd Lrs. *p.s.c.*	14 Nov. 17
Sanders, Brig.-Genl. G. A. F., R. E., *p.s.c.*	19 Nov. 17
Beatty, Brig.-Genl. L. N.	20 Nov. 17
Kemball, Brig.-Genl. A. G., 31st Punj.	16 Dec. 17
Nightingale, Brig.-Genl. M. R. W., C.M.G., D.S.O., 5th G. R. [L]	20 Dec. 17
Dale, Brig.-Genl. G. A., C.M.G., 19th Punj.	1 Jan. 18
Clery, Brig.-Genl. C.B. L., C.B.	6 Jan. 18
Coleridge, Brig.-Genl. J.F. S. D., C.M.G., D.S.O., 8th G. R., *p.s.c.*	31 Oct. 17
Worgan, Brig.-Genl. R. B., D.S.O., 20th Horse.	2 Oct. 17
Fagan, Brig.-Genl. E. A., D.S.O., 36th Horse.	29 Nov. 17
Pitcher, Brig.-Genl. G. leD., C.M.G., 39th Horse.	27 Dec. 17
Godwin, Brig.-Genl. C. A. C., D.S.O., 23rd Cav.	5 Jan. 18
Morris, Brig.-Genl. G. M., D.S.O., 62nd Punj.	2 Mar. 18
VC **Costello**, Brig.-Genl. E. W., C.M.G., D.S.O., 22nd Punj., *p.s.c.*	4 May 18
Elsmie, Brig.-Genl A. M. S., C.M.G., *p.s.c.* [*l*]	...

PERSONAL STAFF.

Asstt. Mily. Secy.

Cahusac, Capt. C. F., D.S.O., 36th Horse	23 Aug. 16
Goldie, Maj. K. O., 10th Lrs.	14 Oct. 16
Williams, Bt.-Lt.-Col. L. G., D.S.O., 5th Cav.	1 Sept. 17

Aide-de-Camp.

Khan Muhammad Akbar Khan, C.I.E., Capt.	24 Aug. 14
Khawja Muhammad Khan, Risaldar, Guides.	30 Sep. 14
Kennedy, Lt. S. V., 13th Hrs.	1 July 15
Hon. Lt.-Col. Muhammad Ali Beg, Nawab Afsar-i-Jang, Afsar-ud-Daula *Bahadur, Sir*, K.C.I.E., M.V.O.	1 Nov. 15
Wilson, Capt. R. E., D.S.O., R.A. . .	
Craik, Lt. D., I. A. R. O.	10 Dec. 15
Woodhouse, 2nd-Lt. R. A. G., 14th Hrs.	
Clifton, Lt. H. A., E. Surr. R.	28 Jan. 16
Cardew, Maj. F. G. (retd.)	1 Apl. 16
Bowen, 2nd-Lt. H. C., Lanc. Fus. . .	29 July 16
Morshead, 2nd-Lt. V. H., R.A. . . .	5 Jan. 17
Wynne-Jones, Lt. C. L., 17th Lrs. . .	31 Oct. 16
Colman, Lt. B. A., I. A. R. O. . . .	12 Feb. 17
Brooke, 2nd-Lt. H. A., 14th Hrs. . . .	23 Dec. 16
Milne, Lt. E. O., 1-4th Som. L. I. . .	12 Apl. 17
Dixon, Lt., M. J., I. A. R. O.	
Birdwood, 2nd-Lt. C. B., I. A. . . .	10 Mar. 18
Prescott, Lt. J. A., G. Gds. (S. R.).	
Ewart, Lt. J., I. A. R. O.	16 Jan. 18

GENERAL STAFF.

Brig.-Genl., General Staff.

Charles, Brig.-Genl. (*temp. Maj.-Genl.*) J. R.E., C.B., D.S.O., R. E., *p.s.c.*	1 Aug. 15
DePree, Bt.-Col. (*temp.* Brig.-Genl.) H. D., C.B., R. A., *p.s.c.* [*l*] .	5 Feb. 16
Beach, Brig.-Genl. W. H., C.M.G., D.S.O., R.E.	12 Oct. 16
Wigram, Brig.-Genl. K., C.B., D.S.O., 2nd G. R.	4 Feb. 17
MacMullen, Brig.-Genl. C. N., 15th Sikhs	26 Sep. 17
Pope-Hennessy, Brig.-Genl. L. H. R., D.S.O., Oxf. and Bucks L. I., *p.s.c.* [*l*]	5 Sep. 17
Dick, Brig.-Genl., R. N., R. Suss. R., *p.s.c.*	14 Nov. 17

G. S. O., 1st Grade.

Western, Col. J. S. E. (retd.), *p.s.c.* . .	29 Oct. 14
Freeland, Bt. Lt.-Col. J. C., 35th Sikhs, *p.s.c.*	23 Dec. 15
Young, Col. F. de B., C.M.G., *p.s.c.* . .	3 Aug. 16
Hill-Whitson, Lt.-Col. T. E. L., 14th Hrs.	1 Nov. 16
Forster, Bt.-Col. (*temp. Lt.-Col.*) D., D.S.O., R.E. .	13 Sep. 16
Steel, Bt.-Col. R. A., C.I.E., 17th Cav. .	1 Nov. 16
Stokes, Bt. Lt.-Col., C. B., C.I.E., 3rd Horse, *p.s.c.* [L]	10 Feb. 17
Leslie, Bt.-Lt.-Col. W.S., D.S.O., 31 Punj., *p.s.c.*	28 Feb. 17
Drew, Bt.-Lt.-Col. B.C.H., 23rd Prs., *p.s.c.*	28 Feb. 17
Shuttleworth, Bt.-Lt.-Col. (*temp. Lt.-Col.*) D. I., 3rd G. R., *p.s.c.*	27 June 17
Norman, Bt.-Col. W. H., D.S.O., 11th Lrs., *p.s.c.*	6 Aug. 17
Grant, Maj. (*temp. Lt.-Col.*) G. P., D.S.O., 106th Prs.	1 Aug. 17
Muirhead, Maj. (*temp. Lt.-Col.*) J. A., D.S.O., 1st Lrs., *p.s.c.*	6 June 17
de Burgh, Bt.-Maj. (*temp. Lt.-Col.*) E., D.S.O., 9th Horse., *p.s.c.*	22 June 17
Allanson, Bt.-Lt.-Col. (*temp. Lt.-Col.*) C. J. L., C.I.E., D.S.O., 6th G.R., *p.s.c.* .	9 Oct. 17
Morgan-Owen, Bt. Lt.-Col., L. I. G., C.M.G., D.S.O., S. Wales Bord., *p.s.c.*	11 Sep. 17
Moens, Bt. Lt.-Col., A. W. H. M., D.S.O., 52nd Sikhs, *p.s.c.*	14 Sep. 17
Hay, Bt. Lt.-Col., C. J. B., D.S.O., Guides, *p.s.c.* [*l*]	4 Nov. 17
Storr, Maj. (*temp. Lt.-Col.*) C. L., 54th Sikhs, *p.s.c.*	30 Oct. 17
Dent, Bt.-Lt.-Col. (*temp. Lt.-Col.*) W., D.S.O., 103rd Infy., *p.s.c.*	25 Jan. 18
Bambridge, Lt.-Col. W. F., C.M.G., D.S.O., 54th Sikhs, *p.s.c.*	24 Oct. 17
Rowlandson, Bt. Lt.-Col. M. G. D., D.S.O., 38th Dogras [L] *p.s.c.*	18 Jan. 18
Waterfield, Lt.-Col. A. C. M., M.V.O. .	29 Dec. 17
Keen, Maj. (*temp. Lt.-Col.*) F. S., D.S.O., 45th Sikhs. *p.s.c.*	20 Apl. 18
Milward, Bt. Lt.-Col. C. A., D.S.O., 53rd Sikhs, *p.s.c.*	4 May 18
Wilson-Johnston, Bt. Lt.-Col., W. E., D.S.O., 36th Sikhs, *p.s.c.*	5 Dec. 17
Stapleton, Bt. Lt.-Col. F. H., Oxf. & Bucks. L. I.	4 Sep. 17
Thomson, Maj. (*temp. Lt.-Col.*) A. G., D.S.O., *p.s.c.*, 58th Rif.	30 Mar. 18
Barge, Capt. (*temp. Lt.-Col.*) K., D.S.O., M.C., 17th Cav.	3 June 18
Hopwood, Bt.-Col. H. R., I. A. . . .	13 Apl. 18

G. S. O., 2nd Grade.

Peel, Maj. B. G., D.S.O., 81st Prs. . .	7 June 15
Julius, Maj. S. de V. A., R. Suss. R., *p.s.c.* [*l*]	18 June 15
More, Maj. J. C., D.S.O., 51st Sikhs [L].	...
Pope, Maj. S. B., D.S.O., 58th Rif., *p.s.c.*	4 Aug. 16
Eadie, Maj. J. I., D.S.O., 97th Infy. [L] .	1 Aug. 16
Kelly, Maj. E. H., R.E.	17 Sep. 16
Garforth, Maj. W., D.S.O., M.C., R.E. .	10 Oct. 16
Boyd-Rochfort, Capt. H., D.S.O., 21st Lrs.	15 Oct. 16
Stone, Capt. A. G., 3rd G. R.	26 Oct. 16
Craster, Maj. G., D.S.O., 6th Cav., *p.s.c.*	1 Dec. 16
Hamilton, Bt.-Lt.-Col. E.G., D.S.O., M.C., Conn. Rang	18 Nov. 16
McMurrough-Kavanagh, Capt. A.T., M.C., 7th Hrs.	17 Feb. 17
O'Rorke, Capt. J. M. W., D.S.O., 25th Cav.	14 Feb. 17
Kirkpatrick, Bt. Lt.-Col. C., Guides, *p.s.c.*	2 Apl. 17
Stoney, Maj. P. S., 26th Punj. *p.s.c.*	27 June 17
Shakespear, Bt.-Maj. G. F. C., D.S.O., M.C., 25th Punj.	9 Sep. 17
Gabriel, Ty. Lt.-Col. E. V., C.S.I., C.M.G., C.V.O., Genl. List	20 Sep. 17
Thomson, Maj. A. G., D.S.O., I. A. . .	...
Ross, Maj. A. C., D.S.O., 20th Cav. . .	16 Nov. 17
Mason Bt. Maj. K., M.C., R.E. . . .	20 Nov. 17
Cruickshank, Capt. A. J., D.S.O., R.E. .	20 Nov. 17
Chambers, Maj. P. R., D.S.O., 10th Lrs., *p. s. c.*	5 Dec. 17
Kennedy-Craufurd-Stuart, Maj. C., D.S.O., 127th Infy. .	15 Jan. 18
Gordon, Lt.-Col. P. J. (retd.)	16 Jan. 18
Blaker, Maj. W. F., R. F. A.	25 Jan. 18
Wagstaff, Maj. L. C., 2nd Rajput L. I.	13 Mar. 18
Radcliffe-Smith, Maj. N. M. R., M. G., 44th Infy.	25 Aug. 18

G. S. O., 3rd Grade.

Morland, Capt. W. E. T., D.S.O., M.C., Oxf. and Bucks L. I.	20 June 15
Pigou, Maj. F. H., *retd.*	23 July 15
Leith-Ross, Capt. W., M.C., 55th Rif. .	16 May 16
Hickman, Capt. H. T. D., M.C., 34th Prs.	1 June 16
Lang, Bt. Maj. W. H., 16th Cav. . .	1 July 16
Lumley, Capt. J. N., 13th Hrs. . . .	6 Sep. 16
Lepper, Capt. J. G., R. W. Kent R. . .	7 Nov. 16
Darley, Capt. C. G., D.S.O., 14th Hrs. .	23 Dec. 16
Dewing, Lt. (*temp. Capt.*) R. H., D.S.O., R.E.	10 July 16
Bray, Capt. N. N. E., M.C., 18th Lrs. . .	20 Dec. 16
Scott, Capt. K. R., M.C., 31st Punj. . .	23 Jan. 17
Bovill, Capt W. J., 43rd Regt.	...
Batten, Capt. E. F., I. A. R. O. . . .	21 Jan. 17
Chetwynd-Stapylton, Maj. G.B., 1-5th E. Surr. R.	...
Hawes, Maj. C. M., 20th Infy. . . .	12 Sep. 17
Evans, Capt. J. W. D., 21st Lrs. . . .	1 Oct. 17
Whittall, Capt. J. M., 6th G. R. . . .	5 Oct. 17
Wagstaff, Capt. H. W., M.C., R.E. .	13 Nov. 17
Guise, Lt. (*temp. Capt.*) H.R.C., I.A.R.O.	...
Wheatley, Capt. A. H. S., 6th Cav. . .	27 Dec. 17
Saunders, Capt. M., D.S.O., 36th Sikhs.	18 Jan. 18
Harvey-Kelly, Maj. C. H. G. H., 127th Baluch. L.I.	22 June 18
Hornsby, Capt. (*temp. Maj.*) R. W., 19th Punj.	30 June 18

ADMINISTRATIVE STAFF.

Dy. Adj. & Q. M. Genl.

Carruthers, *Brig.-Genl.* R. A., C.B.	12 Dec. 14
Holman, Maj.-Genl. H. C., C.B., C.M.G., D.S.O. *p.s.c.* [L]	14 Nov. 15
Rattray, Brig.-Genl. C., C.B., *p.s.c.*	12 Mar. 16
Willoughby, Brig.-Genl. M. E., C.B., C.M.G.	8 Nov. 16
Robinson, Brig.-Genl. S. W., R. A.	28 Oct. 16
Abbott, Brig.-Genl. L. H., C.M.G., 8th Rajputs, *p.s.c.*	8 Nov. 17
Peck, Bt.-Col. (*temp. Maj.-Genl.*) A. W., C.M.G., 25th Cav.	15 Feb. 18

D. A. G.

Campbell, Brig.-Genl. W. N., C.S.I., C.M.G., D.S.O., I. A. (retd.)	28 June 17

A. A. G.

Thornhill, Lt.-Col. *Sir* H. B., K.C.I.E. (*retd.*)	22 May 15
Whitehead, Bt.-Lt.-Col. (*temp. Lt.-Col.*) J., D.S.O., 1st Brah. *p.s.c.*	27 Feb. 16
Morphett, Bt.-Lt.-Col. (*temp. Lt.-Col.*) G. C., C.M G., D.S.O., R. Suss. R.	28 June 17
Ward, Lt.-Col. A., D.S.O.	19 July 17
Wilson, Bt. Lt.-Col. R. C., D.S.O., M.C., 114th Mah. *p.s.c.*	11 Sep. 17
Robertson, Maj. (*temp. Lt.-Col.*) D. E., 11th Lrs.	8 Feb. 18

D. A. A. G.

Davie, Lt.-Col. J. H. M., 34th Horse	6 Nov. 14
Berkeley, Maj. R. E. G., 86th Infy.	26 Sep. 16
Duffy, Hony. Maj. (*temp. Capt.*) T. A. (I.M.L.)	22 Apl. 16
Dunbar, Maj. J. C., D.S.O., R. A.	4 May 16
Shute, Capt. C. A., 107th Prs.	19 May 16
Carpendale, Bt.-Maj. F. M., 42nd R.	27 July 16
Audain, Maj. G. M. (*retd.*)	4 July 16
Yates, Maj. J. A., C.I.E., D.S.O., 103rd Infy.	4 Aug. 16
Kettlewell, Maj. A. M., 39th Horse, *p.s.c.*	10 Nov. 16
Foster, Bt.-Maj. T. F. V., M.C., Conn. Rang.	4 Mar. 17
Lane, Maj. F. B., 8th Cav.	11 Oct. 17
Jacob, Capt. R. M., 130th Baluchis	23 Feb. 17
Haig, Bt.-Maj. (*temp. Maj.*) W. de H., D.S.O., R.E.	25 Feb. 18
Baldwin, Maj. F. E. W., 121st Prs.	10 July 17
Tyrer, Maj. R., 84th Punj.	19 May 17
Greenshields, Maj. D. J., M.C., R.A.	16 Sep. 17
Goldie, Maj. K. O., 10th Lrs.	7 Sep. 18

A. Q. M. G.

Chitty, Col. W. W.	12 June 15
Cobbe, Lt.-Col. H. H., C.M.G., D.S.O., 13th Lrs.	18 Aug. 15
Winter, Bt.-Col. C. B., C.M.G., D.S.O., 112th Infy. [L]	9 Nov. 15
Parr, Bt.-Col. H. O., 7th Rajputs	...
Russell, Lt.-Col. F. D., 1st Lrs., *p.s.c.* [L]	12 Mar. 16
Belgrave, Bt.-Lt.-Col. H. D., D.S.O., R. W. Kent R.	15 Nov. 15
Catty, Bt. *Lt.-Col.* T. C., D.S.O., 69th Punj.	4 Mar. 17
Barron, Bt. Lt.-Col., F. W., R. A., *p.s.c.* (*g*)	4 May 17
Senior, Col. H. W. R., D.S.O., 10th G. R., *p.s.c.* [L]	8 June 17
Duncan, Maj. H. C., (*temp. Lt.-Col.*) D.S.O., 9th G. R., *p. s. c.*	11 Jan. 18
Lakin, Maj. (*temp. Lt.-Col.*) J. H. F., 7th G. R., *p.s.c.*	24 Jan. 18
Butler, Bt.-Maj. (*temp. Lt.-Col.*) R.B., M.C., 30th Lrs.	24 Jan. 18

D. A. Q. M. G.

Haworth, Maj. L. B. H., Supy. List	18 Dec. 14
Langhorne, Major J. A. D., D.S.O., R.A.	14 Jan. 15
Gilchrist, Maj. W. F. C., C.I.E., 52nd Sikhs	13 Aug. 15
Ingram, Maj. C. R., R. W. Kent R.	9 Dec. 15
Vicary, Bt.-Maj. E. S., 16th Lrs.	8 Sep. 16
Pope, Lt. (*temp. Capt.*) S. F., 2-6th Devon Regt.	22 Sep. 16
Horsley, Capt. (*temp. Maj.*) C. D., M.C., 21st Lrs.	30 Dec. 16
Hamilton, Maj. (*temp. Lt.-Col.*) F. A., 8th Cav.	24 Feb. 17
Scott, Maj. A. E. S., 37th Lrs.	27 Feb. 17
Peart, Maj. (*temp. Lt.-Col.*) C. L., 106 Prs. *p.s.c* [L]	...
McSwiney, Capt. H. F. C., M.C., 3rd G. R.	3 July 17
Graham, Bt.-Lt.-Col. R. B., 33rd Punj.	22 July 17
Lloyd-Evans, Lt. (*temp. Capt.*) H. J., M.C., I. A. R. O.	24 Jan. 18
May, Lt. (*temp. Capt.*) E., R.W. Kent R.	21 Feb. 18
Clarke, 2nd-Lt. (*temp. Capt.*) A.J., M.C., Norf. R.	25 Feb. 18
Holliday, Capt. J. C. H., 122nd Infy.	8 Dec. 17
Walker, Maj. C. W. G., 37th Dogras	20 Jan. 18
Biscoe, Maj. H. V., I. A.	23 Apl. 18
Wingfield, Maj. W. J. R., 7th Hrs.	24 Aug. 18
Stork, 2nd-Lt. ((*temp. Capt.*) H. G., I. A. R. O.	...

A. A. & Q. M. G.

Wingate, Col. A. W. S., C.M.G. (retd.)	12 Oct. 14
Cartwright, Col. C. M., C.M.G.	5 Jan. 15
Mathewes, Col. J. R.	17 May 15
Brockman, Lt.-Col. P. W. D. (retd.) [L]	12 June 15
Home, Bt.-Col. J. M., I.A.	1 Nov. 16
Medlicott, Maj. (*temp. Lt.-Col.*) H. E., D.S.O., 3rd Horse	4 Sep. 16
Jacob, Col. A. LeG., C.I.E., C.M.G., D.S.O., A. D. C., I. A.	21 Mar. 17
Hocken, Lt.-Col. C. A. F., 35th Horse, *p.s.c.* [L]	10 July 17
Hoskyn, Maj. (*temp. Lt.-Col.*) J. C. M., D.S.O., 44th Infy.	7 Sep. 17
Marjoribanks, Maj. (*temp. Lt.-Col.*) R. D., 107th Prs. *p.s.c.*	17 Sep. 17
MacMahon, Maj. (*temp. Lt.-Col.*) H. F. E., M.C., S. & T. Corps.	12 Dec. 17
Nisbet, Bt. Lt.-Col. T., 28th Cav.	23 Mar. 18
Dodd, Maj. A. H. R., (*temp. Lt.-Col.*) 17th Cav.	23 Sep. 18
Cochran, Bt. Lt.-Col. G. W., D.S.O., 81st Prs.	17 Nov. 18

D. A. A. & Q. M. G.

Turner, Lt.-Col. F. C., Northd. Fus., *p.s.c.* [L]	15 Nov. 14
Hibbert, Capt. O. Y., R. W. Kent R.	9 Nov. 15
Loch, Maj. J. C., 3rd G. R.	25 Feb. 17
Hobson, Capt. J. W. S., 16th Rajputs	10 Nov. 17
Mocatta, Bt.-Maj. V. E., 14th Hrs.	15 Nov. 16
Catterson-Smith, Capt. T. M. O., 12th Prs	24 Feb. 17
Raymond, Capt. M. C., M.C., 34th Horse	23 Aug. 17
Spence, Maj. R. B., 96th Infy.	3 Aug. 17
Gray, Capt. (*temp. Maj.*), G. C. G., 3rd Horse	30 June 18

BRIGADE-MAJORS.

Dunn, Maj. E. G., R. Ir. Rif., *p.s.c.*	6 Nov. 14
Hawes, Lt.-Col. C. H., D.S.O., M.V.O., 23rd Cav.	26 July 15
Worsley, Maj. C. F. M., 21st Cav.	
Foster, Maj. W. M. A., D.S.O., 10th Lrs.	2 Jan. 16
McKenna, Maj. J. C., D.S.O., 16th Rajputs, *p.s.c.*	20 Mar. 16
Willoughby, Maj. D. V., D.S.O., 1st Brah.	10 May 16
Dayrell, Maj. W. S., D.S.O., 72nd Punj.	9 Aug. 16
Howes, Capt. S., M.C., 31st Lrs.	25 Oct. 16
Tree, Capt. H. B. LeD., 112th Infy.	16 Nov. 16
Harvey, Capt. C. O., M.C., 38th Horse.	2 Jan. 17
Stocker, Capt. E. H., 13th Hrs.	20 Dec. 16
Beresford, Capt. G. de la P., M.C., 10th Lrs.	24 Feb. 17
Herapath, Capt. D. K. MacK., 7th D. Gds.	5 Mar. 17
Daniels, Bt.-Maj. A. M., 3rd Horse	8 Mar. 17
Crowdy, Maj. J. D., D.S.O., 5th G. R.	22 Apl. 17
Acworth, Capt. D. H., M.C., 55th Rif.	9 Apl. 17
Stoddart, Capt. G., 104th Rif.	30 July 17
Howell, Capt. G. B., 19th Lrs.	26 June 17
Gretton, Maj. J. C., 126th Infy.	12 Nov. 17
Macdonald, Bt.-Maj. H., D.S.O., 11th Lrs.	14 Nov. 17
Duncan, Bt. Maj. R. C., 2-5th G. Rif. F. F.	17 Nov. 17
Palmer, Capt. (*temp. Maj.*) J. H. G., 21st Punj.	9 Oct. 17
Lane, Capt. C. R. C., M.C., 19th Lrs.	27 Dec. 17
Lilly, Capt. A. N. I., 7th Rajputs	15 Jan. 18
Hornsby, Capt. R. W., 19th Punj.	1 Feb. 18
Crick, Capt. G. C., 109th Infy.	7 May 18
Hemsley, Maj. C., D.S.O., 64th Prs.	19 Aug. 18
Berryman, Bt. Maj. E. R. P., 2-39th Garh. Rif.	19 Aug. 18
Exham, Cap. H., 2-7th G. R.	21 Sep. 18

STAFF CAPTAINS.

Hunt, Capt. (*temp. Maj.*) C. E., M.C., 34th Prs. *p.s.c.*	5 Aug. 14
Boswell, Capt. W. R., 28th Punj.	19 Jan. 15
Wagstaff, Maj. L. C., C.I.E., 2nd Rajputs [L]	4 May 15
Mair, Capt. B. V., M.C., Manch. R.	16 July 15
Johnson, Maj. M. E. S., 48th Prs., *p.s.c.*	13 Aug. 15
Wilson, Bt. Maj. N. M., D.S.O., 7th G. R.	23 Nov. 15
Burnett, Capt. A. H., D.S.O., 72nd Punj.	29 Dec. 15
Griffin, Lt. (*temp. Capt.*) T., I. A. R. O.	7 Feb. 16
Bartlett, Commy. and Hony. Maj. E. L., I. M. L	10 Feb. 16
L'Estrange, Maj. E. B., 79th Infy.	23 Feb. 16
Clarke, Depy. Commy. & Hony. Capt. J. E., I. M. L.	10 Apl. 16
Miller, Capt. G. N., 90th Punj.	9 May 16
Inskip, Capt. R. D., D.S.O., M.C., 59th Rif.	11 May 16
Bell, Capt. J. G., I. A. R. O.	19 May 16
Hackett, Capt. H. H. M., M.C., 6th G. R.	1 Apl. 16
Chesney, Capt. D. P. 23rd Prs. [L]	25 June 16
Ward, Maj. G. L. S. (I. A. retd.)	1 July 16
Dickinson, Lt. (*temp. Capt.*) W. M. K. (hp. list)	14 June 16
Gouldsbury, Bt.-Maj. C. A., 5th G. R.	24 Aug. 16
Patterson, Lt. (*temp. Capt.*) C., 1-5th Wilts. R.	1 Sep. 16
Tresawna, 2nd-Lt. (*temp. Capt.*) H., 2-4th D. of Corn. L. I.	27 Oct. 16
Anderson, Lt. C. E. L., I. A. R. O.	4 Dec. 16
Mabbett, Capt. P. W., I. A. R. O.	21 June 16
Yeld, Lt. (*temp. Capt.*) R. K., I. A. R. O.	4 Dec. 16
Pullman, Capt. G. C., E. Surr. R.	28 Feb. 17
Robinson, 2nd-Lt. (*temp. Capt.*) G. D., Genl List	21 Feb. 18
Beaver, Lt. (*temp. Capt.*) G. D. C., I. A. R. O.	24 Feb. 18
Smith, Lt. (*temp. Capt.*) R. W., Manch. Regt.	3 Mar. 18
Woodfall, Lt. (*temp. Capt.*) H. A. J., E. Surr. R.	11 Jan. 18
Hart, Lt. (*temp. Capt.*) A. C., R. W. Kent R.	26 Feb. 18
France, Lt. (*temp. Capt.*) E., I.A.R.O.	1 Mar. 18
Collingwood, Lt. (*temp. Capt.*) F. W. N., Middx. R.	23 Mar. 18

STAFF CAPTAINS—*concld.*

Malyon, Capt. F. H., 21st Punj.	17 Aug. 16
Durand, Capt. E. P. M., 8th Rajputs	2 June 16
Edwards, Maj. D. B., 39th Horse	27 Nov. 16
Weir, Capt. D. L., D.S.O., Leic. R.	23 July 16
Parminter, Capt. R. H. R., D.S.O., Manch. R.	4 Mar. 17
Saunders, Lt. (*temp. Capt.*) A., I. A. R. O.	4 Feb. 17
Munden, 2nd-Lt. S. E., I. A. R. O.	10 Nov. 16
Duncan, 2nd-Lt. J. G., I. A. R. O.	11 Jan. 16
Sassoon, Lt. (*temp. Capt.*) A. M., 13th Hrs	8 Mar. 17
Cuningham, Capt. A. P., 127th Infy.	22 Mar. 17
Boles, Capt. D. C., 17th Lrs.	30 Mar. 17
Brown, Capt. E. C., 6th G. R.	2 Apl. 17
Dennison, Capt. G. D, 1-9th Hamps. R.	7 Nov. 17
Mangin, Capt. E. B., M.C., 107th Prs.	. . .
Bacon, Capt. C. G., 29th Lrs.	. . .
Weatherall, Capt. N. E., 7th Hrs.	14 Nov. 17
Moore, Capt. C. F. F., M.C., 123rd Rif.	3 July 17
Gwatkin, Maj. F., M.C., 18th Lrs.	8 June 17
Ashe, Lt. (*temp. Capt.*) H. D., attd I. A.	3 Sep. 17
Slater, Capt. O., R. E.	1 July 17
Todd, Capt. E. T. T., 121st Prs.	6 Oct. 17
Bradford, Capt. W. H., 4th Som. L. I.	19 Nov. 17
Mellersh-Jackson, Capt. L., 1-9th Middx. R.	9 Oct. 17
Mockett, Lt. (*temp. Capt.*) V., E. Kent R.	23 July 17
Gwyn-Williams, Lt. (*temp. Capt.*) R. H., R. W. Fus.	30 Aug. 17
Welchman, Lt. (*temp. Capt.*) G. de. V., R. A.	1 Sep. 17
de Fonblanque, Capt. E. B., R. A.	1 Sep. 17
Stewart, 2nd-Lt. (*temp. Capt.*) C. K., R. A.	6 Sep. 17
Hardy, 2nd-Lt. (*temp. Capt.*) V., Hamps. R.	8 Sep. 17
Vizard, 2nd-Lt. (*temp. Capt.*) W. G., Dorset R.	11 Sep. 17
Bradley, Lt. (*temp. Capt.*) V. M., R. A.	15 Sep. 17
Mead, Lt. (*temp. Capt.*) R., Hamps. R.	28 Sep. 17
Barnes, Lt. (*temp. Capt.*) D. H., I. A. R.O.	3 Oct. 17
Sayers, 2nd-Lt (*temp. Capt.*) W. W., R. Suss. R.	6 Oct. 17
Glynne-Jones, Capt. R. R., 1-1st Brecknock Bn.	. . .
Booth, 2nd-Lt. A. N., R. F. A.	27 Sep. 18

STAFF LIEUTENANTS.

Ellis, Temp. 2nd-Lt. S., attd. 13th Rajputs	1 Oct. 18

SPECIAL APPOINTMENTS.

Graded as Brig. General.

Nepean, Col. (*temp. Brig.-Genl.*) H. E. C. B., C.M.G., I. A.	11 Feb. 18

(Graded as Assistant Adjutant Generals.)

SECTION COMMANDER.

Palin, Maj.-Genl. P. C., C. B., C.M.G.	4 Feb. 17

(Graded for purposes of pay as a Chief Engineer.)

Panet, Brig.-Genl. A. E., C.M.G., R.E.	17 Jan. 17

(Graded for purposes of pay as G. S. O., 1st Grade.)

Kaye, Lt.-Col. C., C.S.I., C.I.E., 20th Infy.	9 Oct. 14

(Graded for purposes of pay as G. S. O., 2nd Grade.)

Branson, Maj. L. H., 110th Infy.	1 Apl. 15
Wake, Lt.-Col. E. St. A. (retd.)	13 Jan. 16
Bishop, Bt.-Col. E. B., *u. s. l.*	2 Aug. 14
Ostrehan, Lt.-Col. F. G. R., *u. s. l.*	9 Sep. 14
Vivian, Bt.-Col. F. G. (retd.)	16 Dec. 14
Hunter, Maj. F. F., D.S.O., I.A.	30 Apl. 16
Cameron, Maj. D. H. (retd.)	7 Aug. 17
Creagh, Lt.-Col. E. C. (retd.)	1 Oct. 17
Clutterbuck, Maj. (*temp. Lt.-Col.*) C. H., 125th Rif.	9 Nov. 17
Applin, Maj. (*temp. Lt.-Col.*) R. V. K., D.S.O., 14th Hrs.	16 Oct. 17
Christian, Lt. (*temp. Capt.*) A. J.	
O'Bryen, Lt.-Col. C. W., I.A.	8 Mar. 18

(Graded as General Staff Officers, 3rd Grade.)

Burne, Col. K. P. (retd.)	9 Aug. 14
Dick, Col. A. R., C.B. [L]	15 Aug. 14
Goodenough, Lt.-Col. H. L., I.A.	9 Sep. 14
Hooper, 2nd-Lt. C. A., Leic. R.	7 July 16
Reed, Lt. G. S., M.C., I. A. R. O.	19 Aug. 16
Routh, 2nd-Lt. C. F. R., Middx. R.	21 Aug. 16
Clauson, Lt. G. L. M., Som. L. I.	26 Aug. 16
Geary, 2nd-Lt. A. B., Gen. List	23 Sep. 16
Lander, 2nd-Lt. A. Spec. List	18 Oct. 16
Bettlheim, Capt. H., Spec. List	19 Oct. 16
Teversham, Bt.-Col. R. K., D.S.O., I.A. (*u. s. l.*)	20 Feb. 17
Presgrave, Col. E. R. J., C.B., D.S.O.	26 Mar. 17
Abdul Samad Shah, Hony. Lt.	23 Nov. 14
Harvey, 2nd-Lt. R. O., Res. R. Cav.	1 Apl. 14
Macpherson, Lt. C. F., Res. R. Cav.	13 May 15
Williams, Lt. (*temp. Maj.*) N. M., Glouc. R.	25 Feb. 16
Wyles, 2nd-Lt. C. D., 1-4th Hamps. R.	2 Mar. 16
Marshall, Lt. J., M.C., I.A.R.O.	1 Apl. 16
Moore, Lt. E. C. F., M.C., I.A.R.O.	1 Apl. 16
Watts, 2nd-Lt. W. H., Glouc. R.	17 July 16
Archer, Lt.-Col. C., C.S.I., C.I.E. (retd.)	5 Apl. 17
King, Maj. J. St. A., 62nd Punj	7 Aug. 17
Atkinson, Lt.-Col. G. C. (retd.)	3 Dec. 17
Van Someren, Maj. W. W., D.S.O., 45th Sikhs	1 Mar. 18
Hutchesson, Lt. (*temp.Capt.*) J.F., R.A.	10 Mar. 18
Hutson, Lt. (*temp. Capt.*) H. W. P., D.S.O. M.C., R.E.	25 June 17
Levy, 2nd-Lt. R., Spec. List	7 July 17
Houy, Lt. (*temp. Capt.*) H. C., Spec. List	16 July 17

(Graded as A. Q. M. G.)

Browne, Maj. (*temp. Lt.-Col.*) H. H. G., D.S.O., Som. L. I.	13 Aug. 17
Battye, Maj. (*temp. Lt.-Col.*) B.C., D.S.O., R. E.	28 Oct. 17

(Graded as A. A. & Q. M.G.)

Hunter, Maj. (*temp. Lt.-Col.*) F. F., D.S.O., I. A.	1 Nov. 16
Farran, Maj. (*temp. Lt.-Col.*) G. L., D.S.O., M.C., 4th Cav.	23 Feb. 17
Oakes, Maj. (*temp. Lt.-Col.*) H. C., 28th L. Cav.	24 July 17
Williams, Maj. (*temp. Lt.-Col.*) V. P. B., 4th Cav.	24 Oct. 17
Hamilton, Maj. (*temp. Lt.-Col.*) F. A., 8th Cav.	1 Apl. 18

(Graded as Brigade-Major.)

Pollok, Capt. (*temp. Maj.*) A. B., 7th Hrs.	1 Oct. 16
Hay, Maj. A. S., 31st Lrs.	15 Jan. 17
Fitzgerald, Maj. (*actg. Lt.-Col.*) C. R. L., D.S.O., 126th Infy	15 Jan. 17
Boyd, Capt. O. T., M.C., I. A.	25 Nov. 17

(Graded as Staff Captains.)

Thompson, Ty. Capt. R. C.	2 Jan. 15
Purcell-Gildea, Lt. (*temp. Capt.*) E. J., I. A. R. O.	9 Jan. 17
Fleming, Maj. W. E., M.C., 41st Dogras.	1 June 16
Shewen, Col. (*Hony. Br. Gen.*) M. T. (retd.)	16 May 17
Stowell, Lt. (*temp.Capt.*) G.C., I.A. R. O.	17 Apl. 17
Moore, Capt. R. J., Notts. & Derby. R.	15 Nov. 17
Dunning, Lt. (*temp. Capt.*) H. G., I.A. R.O.	30 Aug. 17
Reith, Lt. (*temp. Capt.*) C. E. W., I.A. R.O.	10 Sep. 17
Whitamore, Lt. (*temp. Capt.*) A. G., I.A.R.O.	10 Sep. 17
Ebden, Lt. (*temp. Capt.*) J. W., I.A.R.O.	13 Oct. 17
Kirkpatrick, Lt. (*temp. Capt.*) W., I.A.R.O.	15 Oct. 17
Ezra, Lt. (*temp. Capt.*) E., I.A.R.O.	21 Jan. 18
Harrison, 2nd-Lt. (*temp. Capt.*) C., High L. I.	3 Feb. 18
Donald, Lt. (*temp. Capt.*) C. D., attd. Maxim Gun Corps	7 Feb. 18
Duggan, Lt. (*temp. Capt.*) R. D., I.A.	21 Feb. 18
Fitzpatrick, Lt. (*temp. Capt.*) G. V., I.A.R.O.	1 Apl. 18
Ord, 2nd-Lt. (*temp. Capt.*) B., Devon R.	9 Apl. 18

(Graded as Staff Lieutenants.)

Burton, Col. F. C. (retd.)	21 Sep. 14
Hobday, Col. J. R. (retd.)	21 Sep. 14
Ferris, Lt.-Col. W. B., *u. s. l.*	7 Oct. 14
Renny, Bt.-Col. G. B. (retd.)	20 Oct. 14
Hunter, Hony. Capt. J. (retd.)	30 Oct. 14
Mein, Bt.-Col. J. E. (retd.)	31 Oct. 14
Lowry, Lt.-Col. H. W. (retd.)	31 Oct. 14
Ashby, Lt.-Col. J. S. (retd.)	2 Nov. 14
Gwatkin, Col. F. S., C.B. (retd.)	4 Nov. 14
Allen, Lt.-Col. B. M. (retd.)	4 Nov. 14
Davies, Lt.-Col. J. (retd.)	4 Nov. 14
Plowden, Lt.-Col. W. F. C. (retd.)	9 Nov. 14
Pears, Lt.-Col. T. C., *u. s. l.*	16 Dec. 14
Browning, Lt.-Col. H. A. (retd.)	20 Nov. 15
Lowe, Lt. F. G., I.A.R.O.	22 Oct. 17

(Graded as Staff Lieutenants, 1st class.)

Craster, Lt.-Col. J. C. B. (retd.)	6 Nov. 14
Beresford, Maj. J. H. B. de la P. (retd.)	24 Jan. 15
Giles, Lt.-Col. W. (retd.)	4 Feb. 15
Sparling, Lt.-Col. J. P. (retd.)	15 Feb. 15
Stevens, Maj. C. (retd)	10 Mar. 15
Lewin, Lt.-Col. W. H. (retd.)	11 Mar. 15
Hughes, Capt. A. P. (retd.)	23 Mar. 15
Watson, Lt.-Col. G. H. (retd.)	25 Mar. 15
Grey, Lt.-Col. A., V.D., late Ind. Vols.	25 Mar. 15
Welchman, Bt.-Col. E. W. St. G., C.B. (retd.)	14 Apl. 15

(*Graded as Staff Lieutenants, 1st class*)—contd.

Loch, Lt.-Col. G. H., C.I.E. (retd.)	14 Apl. 15
Martindale, Maj. C. S de B. (retd.)	14 Apl. 15
Taylor, Lt.-Col. R. E. S. (retd.) [*l.*]	20 Apl. 15
Gott, Bt.-Col. G. A.	19 May 15
Macey, Hony. Maj. T. Y. (retd.)	22 May 15
Meade, Lt.-Col. J. W. B., *u. s. l.*	25 May 15
Hume-Wright, Capt. G., Bangalore Coorg and Mysore Bn., I. D. F.	15 July 16
Bouchier, Maj. A. C. F. (retd.)	30 May 15
Chapman, Maj. and Hon. Lt.-Col. A. C., V.D.	31 May 15
Atkinson, Hon. Brig.-Genl. F. G., C.B. (retd.)	16 June 15
Dempster, Maj. F. F., C.I.E., late Ind. Vols.	7 July 15
Yale, Col. F. H.	13 July 15
Clark-Kennedy, Col. A. H. (retd.)	21 July 15
Dawes, Lt.-Col. W. M. (retd.)	25 Nov. 15
Gastrell, Col. E. T.	2 Mar. 16
Phillips, Col. A. L. (retd.)	2 Mar. 16
Herbert, Maj.-Genl. L., C.V.O., C.B.(retd.)	3 Mar. 16
Hatch, Col. A. V. (retd.)	3 Mar. 16
Burton, Col. E. B., C.B. (retd.)	15 May 17
Sexton, Maj. J. J. O. B. (retd.)	23 May 17
Blyth, Lt.-Col. F. A. (retd.)	10 Mar. 16
MacIntyre, Maj.-Genl. D. C. F., C.B.	13 Mar. 16
Wimberley, Lt.-Col. C. I. (retd.)	11 Mar. 16
Dobbs, Lt.-Col. A. H. (retd.)	16 Mar. 16
Rynd, Lt.-Col. F. C. (retd.)	1 May 16
Godfrey, Lt.-Col. S. H., C.I.E. (retd.)	4 May 16
Kenny, Col. H. T. (retd.) *p.s.c.* [L.]	15 July 16
Hamilton, Maj. C. (retd.)	13 July 16
Arbuthnot, Lt.-Col. G. H. (retd.)	29 Aug. 16
Ward, Lt.-Col. G. A.	1 Sep. 16
Retallick, Lt.-Col. J. M. A. (retd.)	14 Dec. 16
Turner, Maj.-Gen. J. G.	18 Jan. 17
McKay, Lt.-Col. F. (retd.)	4 Feb. 17
Rose, Col. H. (retd.)	4 Apl. 17
Barry, Lt.-Col. J. F., I. A.	4 Apl. 17
Campbell, Maj. I. H. (retd.)	4 Apl. 17
Grey, Bt.-Col. E. (retd.)	1 Feb. 17
Turner, Brig.-Genl. G. H. (retd.)	14 July 17
Renny, Bt.-Col. A. Mc W. (retd.)	13 June 17
Curll, Bt.-Col. C. E. (retd)	17 June 17
Swayne, Bt.-Col. *Sir* E. J. E., K.C.M.G., C.B. (retd.)	17 June 17
Stainforth, Bt.-Col. L. C. H., (retd.)	17 June 17
Beddek, Maj. E. E. (retd.)	26 July 17
Jermyn, Bt.-Col. T. (retd.)	8 Aug. 17
Rainy Robinson, Col. R. M., C.B., C.M.G. (retd.).	15 July
Ffrench, Lt.-Col. T. (retd.)	23 Aug. 17
Grant, Col. D. St. J. D., M.B., I. M. S.	24 Aug. 17
Hunter, Maj.-Genl. J. G., C.B., I. A.	21 Sep. 17
How, Lt.-Col A. P., (retd.)	21 Sep. 17
Beale, Bt.-Col. A (retd.)	29 Aug. 17
Johnston, Lt.-Col. H. W., 67th Punj.	10 Oct. 17
Widdicombe, Lt.-Col. G. T., C. B.	10 Oct. 17
Murray, Maj. H. B. (retd.).	22 Oct. 17
Garret, Lt.-Col. R. V. (retd.)	25 Oct. 17
Holland, Lt.-Col. G. L., M.V.O., I. A.	7 Nov. 17
Searle, Bt.-Col. A. E. S. (retd.)	14 Nov. 17
Radcliff, Bt.-Col. S. G. (retd.)	21 Nov. 17
Tregear, Lt.-Col. V. F. W., 98th Infy.	21 Nov. 17
Elliot, Lt.-Col. C. R., I. A.	6 Dec. 17
Johnson, Maj. C. A. K. (retd.)	31 Jan. 18
Priestley, Lt.-Col., G. W. (retd.)	1 Feb. 18
Strong, Lt.-Col., N. B. G. (retd.)	14 Feb. 18
Jollie, Lt.-Col. F. I. A.	4 June 18
Burn, Lt.-Col. A. G., I. A. (retd.)	27 May 18

(*Graded as Staff Lieutenant, 2nd class.*)

Fisher, Col. J., C.B., I. A.	15 May 17
Stotherd, Lt.-Col. E. A. W. (retd.)	16 May 17
Rainsford, Lt.-Col. J. A. G., I. A.	17 June 17
Warden, Lt.-Col. A. W. (retd.)	25 July 17

(*Graded as Staff Lieutenant, 2nd class*)—contd.

Grey, Capt., W. A. S. (h. p. List)	27 Dec. 17
Lang, Lt.-Col. A. G. B. (retd.)	1 Feb. 18
Medley, Lt.-Col. A. G., I. A.	1 Feb. 18
Gabbett, Maj. (Hon. Lt.-Col.) A. C. (retd.)	8 Feb. 18
Luck, Col. C. A (retd.).	8 Feb. 18
Ransom, Bt.-Col., J. M. (retd.)	8 Feb. 18

(*Graded as Staff Lieutenant, 3rd class.*)

Campbell, Lt. L. S.	4 Oct. 16
Mangin, 2nd-Lt. P. C., I. A. R. O.	22 Mar. 17
Saulez, Maj. E. (retd.)	28 Nov. 17

(*Graded as A. A. G.*)

Fellows, Bt.-Maj. (*temp. Lt.-Col.*) B. C. (retd.).	9 Aug. 17
Becher, Lt.-Col. G. A., D.S.O., 8th Cav.	1 Nov. 17
Turner, Lt.-Col. A. G. B., D.S.O., I.A.	1 Nov. 17
Davy, Maj. (*temp. Lt.-Col.*) R. M. M., Glouc. R.	1 Nov. 17

(*Graded as D. A. A. G.*)

Harley, Maj. A. B., D.S.O., 121st Prs.	14 Mar. 17
Scott Elliot, Lt.-Col. C. R., 61st Prs.	2 Sep. 17
Williams, Maj. V. P. B., 4th Cav.	1 Aug. 17
Bruce, Maj. S. M., 37th Lrs.	25 Oct. 17
Stewart, Maj. A. E., M.C., 124th Infy.	5 Dec. 17

(*Graded as A. Q. M. G.*)

Bartholomew, Maj. A. G., 1-9th Middx. R.	9 Nov. 17

OFFR. I/C. OF LABOUR RECORDS.

Formby, Lt.-Col. R. F. R. (retd.)	1 Apl. 17

LIAISON OFFICERS.

Wilson, Capt. (*temp. Maj.*) D. W., C.I.E., Bo. L. H. I. D. F.	4 Nov. 16

ASST. MILITARY ATTACHÉ.

Thornhill, Bt.-Lt.-Col. (*temp. Lt.-Col.*) C. J. M., D.S.O., 6th Jats.	16 July 16

DEPY. ASST. DIR. OF LABOUR.

Aspinall, H. H. H., Lt.-Col. *u.s.l.*, I. A.	28 May 17

LABOUR CORPS.

Edwards, Col. J. B., D.S.O. (retd.)	13 Apl. 17
Thring, Col. R. H. D.	13 Apl. 17
Balfour, Bt.-Col. J. H.	13 Apl. 17
Burne, Lt.-Col. N. A. K. (retd.)	13 Apl. 17
Talbot, Lt.-Col. J.	13 Apl. 17
Playfair, Lt.-Col. A. (Supy. List).	31 Aug. 17
Barnes, Lt. (*temp. Lt.-Col.*) H. C., I. A. R. O.	16 July 17
Bell, Lt.-Col. R. C., D.S.O.	10 Mar. 18

AREA QUARTERING COMMITTEES.

Price, Col. C. H. U., C.B., D.S.O., I. A.	29 May 17

DY. JUDGE ADVOCATE GENERALS.

Palmer, Bt.-Lt.-Col. G. S., 97th Infy.	27 July 16
Crawford, Capt. S. C. R., 1-5th E. Surr. R.	12 Aug. 16
Thomson, Col. W. D., C.M.G., I. A.	28 Feb. 17

ATTACHED TO HEADQUARTER UNITS.

Cobham, Brig.-Genl. H. W., I. A. (retd.)	14 Jan. 17

BRIG.-GENERALS, ROYAL ARTILLERY.

Lecky, Maj.-Genl. R. St. C., C.B., C.M.G., Brit. Ser.	18 Dec. 14
Smith, Brig.-Genl. G. B., C.B.	30 Apl. 15
Kirby, Brig.-Genl. A. D., C.M.G.	16 Nov. 15
Tyler, Brig.-Genl. J. A., C.M.G.	16 Nov. 15
Grier, Brig.-Genl. H. D., D.S.O.	23 Feb. 16
White, Brig.-Genl. G. F., C.B., C.M.G.	14 Mar. 16
Ouseley, Brig.-Genl. R. G., C.B., C.M.G., D.S.O.	22 Apl. 17
Bright, Brig.-Genl. R. A., C.B., *p.s.c.*	5 Jan. 17
Delaforce, Brig.-Genl. E. F., C.M.G.	25 Oct. 16
Hardy, Lt.-Col. (*temp. Brig.-Genl.*) W. K., R.A.	15 Jan. 17
Smith, Lt.-Col. (*temp. Br.-Genl.*) L. A., R. A.	16 Sep. 17

C. R. A.

Maule, Lt.-Col. H. N. St. J., C.M.G., R.A.	7 May 16
Patch, Lt.-Col. (*temp. Brig.-Genl.*), F.R.	
Mackey, Maj. (*temp. Brig.-Genl.*) H. V. A.	22 May 16
Davies, Bt.-Col. (*temp. Brig.-Genl.*) W. P. L., C.M.G., D.S.O., R.A.	11 Dec. 16

BRIG.-MAJ., ROYAL ARTILLERY.

Wilmer, Bt.-Lt.-Col. E. R. G., D.S.O., R.A.	14 Mar. 16
Forbes, Capt. R. M. N., M.C., R. A.	27 Sep. 16
Kenyon, Capt. H. E., R. A.	15 Jan. 17
Latham, Capt. H. B., R. F. A.	13 Sep. 17
Anderson, Capt. T. R., M.C. R. A.	1 Sep. 17

STAFF CAPT., ROYAL ARTILLERY.

Wingate, Capt. G. F. R., R.A.	20 May 16
Marriott, Lt. (*temp. Capt.*) W. G., R.A.	14 Mar. 16
Rawson, Maj. C. D., D.S.O., R.A.	7 May 16
Cotton, Lt. D. C., R.A.	12 May 16
Harding, Lt. R. A. C., R. F. A. (T. F.)	12 Sep. 17

COMMANDERS.

COLUMN COMMANDERS.

BRIG.-GENL., ROYAL ENGINEERS.

ASST. ENGINEER-IN-CHIEF.

Ewbank, Col. W., C.I.E., Brit. Ser.	3 Mar. 17

CHIEF ENGINEER.

Swiney, Brig.-Genl. A. J. H., C.B., C.M.G., Brit. Ser.	25 Jan. 16
Bremner, Brig.-Genl. A. G., C.M.G., R. E.	4 Feb. 17
Johnson, Brig.-Genl. E. P., C.B., R. E.	11 Jan. 17
Waghorn, Brig.-Genl. W. D., C.M.G., Brit. Ser.	20 Nov. 17
Powell, Brig.-Genl. S. H., C.B., Brit. Ser.	23 June 17
Rushton, Brig.-Genl. H. W., D.S.O., R.E.	23 Sep. 17
Atkinson, Maj. Genl. (Temp.) E. H. deV., C.B., C.M.G., C.I.E., R.E.	1 Nov. 17
Marshall, Brig.-Genl. H. J. M., C.M.G., R. E.	22 Nov. 17

C. R. E.

Wilson, Lt.-Col. F. A., C.B., R.E.	19 July 15
Stack, Bt.-Lt.-Col. G. H., D.S.O., R. E.	28 Mar. 17
Hill, Bt.-Lt.-Col. E. F. J., M.C., R. E.	11 Feb. 17
Turner, Maj. J. F., D.S.O., R. E.	29 Aug. 17
Loch, Maj. G. C. B., R. E., *p.s.c.*	

BDE.-MAJ., ROYAL ENGINEERS.

Squires, Capt. E. K., D.S.O., M.C., R.E.	2 Feb. 16

DIRECTOR OF WORKS.

Roe, Brig.-Genl. C. H., C.I.E., Brit. Ser.	5 Jan. 18

DEPUTY DIRECTOR OF WORKS.

Pearson, Maj. (*temp. Col.*) H. L., D.S.O., R. E.	27 Mar. 17
Edgar, Maj. (*temp. Col.*) D.K., D.S.O., R.E.	15 Feb. 18
Pitkeathly, Maj. (*temp. Col.*) J. S., C.V.O., Spec. List	5 Dec. 17

ASSISTANT DIRECTOR OF WORKS.

Mann, Capt. (*actg. Maj.*) E. A., R.E.	18 Sep. 17
Pryor, Capt. (*actg. Maj.*) T., D.S.O., I.A.R.O.	27 July 18
Trench, Maj. A. C., R.E.	15 Nov. 17
Loch, Maj. G. C. B., R.E.	8 Feb. 18
Archbold, Maj. F. H. W., R.E.	23 May 18
Bridcut, Lt. (*actg. Maj.*) S. H., R.E. (*T.C.*)	24 Jan. 18
Norrie, Capt. (*actg. Maj.*) C. M., D.S.O., R.E. (*T.C.*)	4 Jan. 18
Borton, Maj. W. N., R. E.	4 Sep. 18

DEPUTY ASST. DIRECTORS OF WORKS.

Harrison, Lt. (*actg. Capt.*) W. N.	9 July 17
Theobald, 2nd-Lt. (*actg. Capt.*) J. M., R.E. (*T.C.*)	26 Feb. 18
Mason, Lt. (*actg. Capt.*) H. C., I.A.R.O.	5 July 18
Stevens, Lt. (*actg. Capt.*) T. H. G., I.A.R.O.	25 Feb. 18
Bampton, Lt. (*actg. Capt.*) J. A. H., I.A.R.O.	20 Dec. 17
Brown, 2nd-Lt. (*actg. Capt.*) R. E. (*T.C.*)	28 Nov. 17
Lightfoot, Capt. K., R.E. (*T.C.*)	7 Apl. 18
Donkin, 2nd-Lt. H. J., R.E. (*T.C.*)	28 Nov. 17

Provost Marshal.

Harding, Commsy. and Hon. Maj. G., I.O.D.	10 Sep. 15
Percy-Smith, Bt. Lt.-Col. E. S., I.A.	2 July 16
Brookes, Lt. *temp. Capt.*) A., I. A. R. O.	21 Aug. 16

Asst. Provost Marshal.

(Graded for purposes of pay as D.A.G.)

Anderson, Col. R. E. H. (retd.)	6 Mar. 16
Payne, Lt. (*temp. Capt.*) J. H. Dorset R.	9 Jan. 16

(Graded for purposes of pay as D. A. A. G.)

Maxwell, Maj. (*Temp. Lt.-Col.*) A. G., I. A. (retd.)	15 Nov. 16

(Graded for purposes of pay as Staff Capt.)

Pinney, Maj. J. C. P., I.A. (retd).	11 Nov. 16
Sebag-Montefiore, Lt. (*temp. Capt.*) G. E., 21st Lrs.	12 Nov. 16
Lumley-Smith, Capt. T. G. L., 21st Lrs.	1 Apl. 17
Goldfrap, Capt. H. W., M.C., 103rd Infy.	28 May 15
Browne, Maj. W. S. W., 44th Infy.	8 Sep. 15
Pinney, Maj. J. C. D. (retd.)	25 Sep. 15
Cole, 2nd-Lt. J. T., 3rd Devon. R.	9 Dec. 15
Westmacott, Capt. (temp.) T. H., Gen. List.	24 Dec. 15
Hooton, Capt. A. C., York. R.	
Thane, Capt C. S., I. A. R. O.	4 Sep. 16
Hudson, 2nd-Lt. J., Conn. Rang.	26 Nov. 16
Lee, Lt. A.K., 2-4th Som. L. I.	1 June 17
Aikman, Lt. (*temp. Capt.*) E. H., I.A. R. O.	10 Apl. 17
Wood, Lt. (*temp Capt.*) R. B. M., I.A. (retd.)	20 Aug. 17
Hardwick, Lt. (*temp. Capt.*) N. de C., I. A. R. O.	4 Feb. 18
Butler, 2nd-Lt. (*temp. Capt.*) F. H. C., Hamps. R.	9 Apl. 17
Kennedy, Lt. (*temp. Capt.*) B., Sea. Highrs.	10 Sep. 17
Cunningham, 2nd-Lt. (*temp. Capt.*) A. W., Som. L. I.	9 Oct. 17
Parker, 2nd-Lt. (*temp. Capt.*) G., Hamps. R.	26 Nov. 17
Walsh, Lt. (*temp. Capt.*) G. J. T.	27 Jan. 18
Goyder, Lt. (*temp. Capt.*) G.B., I.A.R.O.	1 Feb. 18

APPOINTMENTS ON LINES OF COMMUNICATION.

Commands.

Boome, Bt.-Col.(*temp. Brig.-Genl.*) E.H., C.B., C.M.G., I.A.	3 Jan. 18
Peyton, Col. A. G. (retd.)	12 May 16
McCalmont, Bt.-Maj. (*temp. Maj.*) D. H. B., 7th Hrs.	23 Dec. 16
Waterfield, Maj. B. C. (retd.)	17 Mar. 16

Inspector General of Communications.

Dickson, Br.-Genl. W. E. R., C.I.E., R.E. [*l*]	...

Administrative Comdt.

Graded as A.Q.M.G.

Commandants, Mobile Columns.

(Graded for pay as A. Q. M. G.'s)

Staff Officers, 1st Class.

Bloxham, Capt. H. F., I. A. R. O.	1 Apl. 16
Tredaway, Lt. G. H., I. A. R. O.	23 Apl. 16

Staff Officers, 2nd Class.

Morley, Lieut. H. G., M.C., R.F.A.	29 May 16
Hill, Lt. A. L., M.C., I.A.R.O.	25 Aug. 16
Dockrell, Capt. (*temp. Maj.*) M., 116th Mah.	19 Jnly 18

Staff Officers, 3rd Class

Lentaigne, Capt. E. C., D.S.O., 4th G. R.	9 Nov. 16
Rivett-Carnac, Capt. H. G., 79th Infy.	9 Nov. 16
Hutchison, Lt. (*temp. Capt.*) B. O., 7th Hrs.	19 Jluy 18

Base Commandants.

Lang, Col. E. B.	16 Sep. 15
Chenevix-Trench, Lt.-Col. G. F., C.I.E. (Ret.)	15 Jan. 16
Brownlow, Lt.-Col. (*temp. Brig.-Genl.*) d' A. C., C.M.G., C.I.E.	7 Feb. 16
Swifte, Lt.-Col. F. L., 57th W. Rif.	23 Feb. 18
Bramley, Lt. (*temp. Maj.*) P. B., I. A. R. O.	23 Feb. 18
Colvin, Capt. (*temp. Maj.*) E. J. D., I.A.	18 Oct. 17
White, Maj. (*temp. Lt.-Col.*) W. B., 38th Horse	19 June 18

Asst. Base Commandant.

Hall, Lt. (*temp. Capt.*) F. G., Dorset R.	4 May 17

Camp Commandants.

Bremner, Capt. H. N. K., 114th Mah.	28 Apl. 16
Peek, Capt. *Sir* W., Ryl. 1st Devon Yeo.	1 Jul. 16
Jones, Maj. T. W. H. (retd.)	2 Oct. 16
Stephens, Capt. (*temp. Maj.*) R., Oxf. & Bucks L. I.	1 Apl. 16

HEAD-QUARTERS OF ADMINISTRATIVE SERVICES AND DEPARTMENTS.

Director of Army Signals.

(*Graded for purposes of pay as a Brig.-Genl.*)

Franklin, Bt. Lt.-Col. (*temp. Br.-Genl.*) H. S. E., D.S.O., 14th Sikhs.	29 Oct. 17

Assistant Director of Army Signals.

Knight, Bt. Maj. (*temp. Lt.-Col.*) J. H., R. A.	
Allen, Capt. (*temp. Lt.-Col.*) H. I., D.S.O., N. Staff R.	

Director of Supplies and Transport.

Dy. Director of Supplies and Transport.

Assistant Director of Supply and Transport.

Babington, Maj. A. H., S. & T. Corps .	4 Sep. 15
Sherrard, Bt. Lt.-Col. (*temp. Lt.-Col.*) J. W., S. & T. Corps.	1 Nov. 16
Hart, Maj. (*temp. Lt.-Col.*) E. G., D.S.O., S. & T. Corps.	1 Nov. 16
Porch, Maj. (*temp. Lt.-Col.*) E. A., M.C., S. & T. Corps.	1 Nov. 16
Brown, Bt.-Lt.-Col. (*temp. Lt.-Col.*) W., S. & T. Corps.	1 Nov. 16
Pearse, Maj. (*temp. Lt.-Col.*) G., S. & T. Corps.	1 Nov. 16
Kemble, Maj. P. B., S. & T. Corps . .	20 Sept. 17
Wardell, Maj. G. F. E., S. & T. Corps .	8 May 17
McEnery, Maj. (*temp. Lt.-Col.*) R. T., S. & T. Corps.	26 Apl. 18

D. A. D. of S. and T.

Davies, Maj. H. W., S. & T. Corps . .	24 Nov. 15
Manson, Maj. J. A. P., S. & T. Corps .	29 Jan. 17
Hurst, Lt. (*temp. Capt.*) F. W., I.A.R.O.	26 Apl. 18

Assistant Director of Supplies.

Annesley, Bt.-Col. A. S. R., S. & T. Corps.	16 Apl. 15

Assistant Director of Transport.

Forbes, Maj. E. E., S. & T. Corps . .	29 May 15
Haydon, Lt.-Col. W. P., D.S.O., S. & T. Corps	

Base Transport Officer.

Smith, Maj. S. H. P., S. & T. Corps.

Deputy Director of Traffic.

Sparke, Capt. (*temp. Lt.-Col.*) H. C., I.A.R.O.	5 Feb. 17

Depy. Director Genl. of Transportation.

Crookshank, Brig.-Genl. S. D'A., C.B., C.I.E., D.S.O., M.V.O., R.E.	3 Mar. 17

Asst. Dir. Genl. of Transportation.

Craik, Maj. (*temp. Lt.-Col.*) J. (retd.)	9 May 17

Depy. Asst. Dir.-Genl. of Transportation.

O'Connor, Maj. M. S. S., I.A.R.O. . .	21 July 17

Director of Transportation.

Murray, Col. V., C. B., C.M.G., Brit. Ser.	23 Feb. 17

Deputy Directors of Transportation.

Russell, Maj. (*temp. Col.*) W. K., D.S.O., R.E.	23 Nov. 16
Birney, Bt. Lt.-Col. (*temp. Col.*) C. F., D.S.O., R.E.	1 Dec. 16
Magniac, Bt. Lt.-Col. (*temp. Col.*) C. L., C.M.G., R.E.	1 Dec. 16

Asst. Director of Transportation.

Manton, Capt. (*temp. Lt.-Col.*) L., D.S.O., R.E.	1 Dec. 16

Depy. Asst. Director of Transportation.

Robertson, 2nd-Lt. (*temp. Capt.*) F. E., I. A. R.O.	1 Dec. 16
Lynch, Maj. C. St. J., D.S.O., R.E. . .	16 Jan. 17
Moody, Lt. (*temp. Capt.*) B., I.A.R.O.	8 Jan. 18

Asst. Dir. of Docks.

Cherry, Lt. (*temp. Lt.-Col.*) J. A., I.A.R.O.	6 Feb. 17

Director of Railways.

Lubbock, Bt.-Col. Brig.-Genl. G., C.M.G., R. E.	25 Aug. 16

Asst. Dir. of Railways.

Reynolds, Lt. (*temp. Lt.-Col.*) J. R., I.A.R.O.	6 Feb. 17

Director of Light Railways.

Deputy Director of Light Railways.

Asst. Dir. of Railway Construction.

Mount, Maj. (*temp. Lt.-Col.*) A. H. L., R. E.	22 Aug. 16

Deputy Director of Railway Transport.

Cowie, Col. C. H., C.I.E.

Assistant Director of Railway Transport.

(*Graded for purposes of pay as A. Q. M. G.*)

Colvin, Capt. (*temp. Lt.-Col.*) G. L., late Ind. Vols.	27 Aug. 16
Stallard, Capt. (*temp. Lt.-Col.*) R. H., R.E.	1 May 17

Deputy Assistant Director of Railway Transport.	
(Graded for purposes of pay as D.A.Q.M.G.)	
Matthews, Capt. (*temp. Maj.*) F. M., 22nd Punj.	23 Sept. 16
Rickell, Capt. (*temp. Maj.*) G. G., 1-6th E. Surr. R.	
Railway Transport Officers.	
Baxter, Lt. R. H. N., I. A. R. O. .	28 Oct. 15
Cross, Lt. (*temp. Capt.*) A. B., I. A. R. O.	13 Jan. 16
Roseveare, Capt. E., 5th Bn. Devon. .	14 Feb. 16
Faithfull, Lt. F. H., I. A. R. O. . .	14 Aug. 16
Rosher, Capt. E. M., M.C., Welsh R. .	26 Dec. 16
Bradley, Lt.-Col. F. G., C.M.G. (retd.) .	24 Apl. 17
Ashton, 2nd-Lt. R. W., I. A. R. O. .	16 May 17
Anderson, 2nd-Lt. (*Temp. Capt.*) F. E., Oxf. & Bucks L I.	23 Oct. 17
Deputy Director of Inland Water Transport.	
(Graded for purpose of pay as A.D. S. & T.)	
Greenhaugh, Temp. Maj.) (*temp. Col.*) F. H., D.S.O., R. E.	17 Mar. 18
Assistant Director of Inland Water Transport.	
Southey, *temp.* Lt.-Col. B. I., R. E. .	17 Mar. 18
Pennant, *temp.* Lt.-Col. W. A. T., M.C. R.E.	17 Mar. 18
Daniels, *temp.* Lt.-Col. L. S., R. E. .	17 Mar. 18
Tribe, Capt. C. J., Brit. Ser.	

Embarkation Commdt.	
(Graded for purposes of pay as a Brig.-General.)	
Forbes, Brig.-Genl. A. W., C.B. . .	22 Jan. 16
(Graded for purposes of pay as A. Q. M. G.)	
Dundas, Lt.-Col. P. H., D.S.O., 6th Jats.	17 July 16
Embarkation Staff Officers. *(Graded for purposes of pay as D. A. Q.M.G.)*	
Whittall, Bt. Maj. G. E., M.C., Oxf. & Bucks L.I.	12 May 15
Maxwell, Maj. P. A., 3rd Brah. . .	1 Jan. 16
Parsley, 2nd-Lt. (*temp. Capt.*) W., M.C., Norf. R.	21 Feb. 18
Asstt. Embarkation Staff Officer. *(Graded for purposes of pay as Staff Captain.)*	
Loverock Capt. R. C., D.S.O., Oxf. & Bucks. L.I.	July 15
Ashton, Lt. (*temp. Capt.*) F. J., I.A. R. O.	1 Dec. 15
Wylde, Lt. E. W., I. A. R. O. . .	1 Jan. 16
Borissow, Lt. (*temp. Capt.*) F. A., I.A. R. O.	19 July 16
Ecott, 2nd-Lt. (*temp. Capt.*) H., 14th Hrs.	1 Nov. 16
Morgan, Lt. H. H., Manch. R. . .	24 Nov. 16
Padfield, Lt. (*temp. Maj.*) R. C. H., I.A. R. O.	10 Oct. 14
Brewin, Capt. T. J., Poona Bn., I. D. F.	3 Sep. 15
Baird, Lt. (*temp. Capt.*) A. E., I. A. R. O.	13 Mar. 16
Chapman, Lt. (*temp. Capt.*) C. W., I. A. R. O.	12 Aug. 17
Musto, 2nd-Lt. (*temp. Capt.*) A. A., I. A. R. O.	5 Mar. 17
Biscoe, Lt. (*temp. Capt.*) W. T., N. W. Ry. Bn.	18 Apl. 17
Winn, 2nd-Lt. (*temp. Capt.*) C. W., Devon R.	18 May. 17
Stewart, 2nd-Lt. (*temp. Capt.*) W. H., R.W. Fus.	24 Oct. 17
Ashton, Lt. (*temp. Capt.*) G. R. I.A.R.O.	17 Oct. 17
Womersley, 2nd-Lt. (*temp. Capt.*) F. E., R. W. Fusrs.	26 Jan. 18
Carter, Lt. (*temp. Capt.*) C. H., Yorks. R.	20 Feb. 18
Paterson, 2nd-Lt. (*temp. Capt.*) S., R. Lan. R.	21 Feb. 18
Cayford, 2nd-Lt. (*temp. Capt.*) W., Som. L. I.	10 Mar. 18
deClermont, Lt. (*temp. Capt.*) A. C., I. A. R. O.	25 Mar. 18
Rodger, 2nd-Lt. F. Mc. G., 4th R. Scots.	
Train Conducting Officers.	
(Graded as Staff Captains.)	
Moore, Lt.-Col. F. L.	1 Feb. 15

APPOINTMENTS CONNECTED WITH MEDICAL, VETERINARY, ORDNANCE, ETC.

Deputy Director of Medical Services.

Robinson, Maj.-Genl. W. H. B., C.B., I.M.S.	16 Jan. 15
Clemesha, Lt.-Col. (*temp. Col.*) W. W., M.D., I.M.S.	1 July 17

Assistant Director of Medical Services.

Burden, Lt.-Col. (*Actg. Col.*) H., C.I.E., I.M.S.	
Gee, Lt.-Col. F. W., C.I.E., M.B.	
White, Col. W. W., I. M. S.	...
Johnston, Lt.-Col. (*temp. Col.*) C. A., C. B., I. M. S.	
Wall, Bt.-Lt.-Col. F., C.M.G., I.M.S.	14 Dec. 14
Graham, Bt. Lt.-Col. J. D., M.B., I. M. S.	15 Jan. 17
Hamilton, Bt.-Lt.-Col. W. H., D.S.O., F.R.C.S., I. M. S.	1 Apl. 15

Deputy Assistant Director, Medical Services.

Browne-Mason, Lt.-Col. H. O. B., D.S.O., R.A.M.C.	
Croly, Maj. W. C., R.A.M.C.	
Horton, Bt. Lt.-Col. J. H., D.S.O., M.B., I. M. S.	6 Nov. 14
Harvey, Lt.-Col. F., R.A.M.C.	21 Apl. 15
Jones, Maj. S. W., I. M. S.	
Hudson, Major C., D.S.O., F.R.C.S.E.,	23 Apl. 15
I. M. S.	20 May 15
Cruddas, Major H. M., C.M.G., I. M. S.	...
Shettle, Maj. F. B., I. M. S.	
Christophers, Maj. S. R., C.I.E., M.B.	
Sloan, Bt.-Col. J. M., C.M.G., D.S.O., M.B., R.A.M.C.	...
Harvey, Bt. Lt.-Col. A. W. M., I. M. S.	...
Williams, Capt. W. E. R., I.M.S.	
Holmes, Maj. J. M., I.M.S.	
Paterson, Maj. T. G. F., D.S.O., I.M.S.	6 Mar. 16
	9 Jan. 17
	29 Mar. 17

Director of Veterinary Services.

Williams, Lt.-Col. (*temp. Col.*) G. M.	

Deputy Director of Veterinary Services.

Assistant Directors of Veterinary Services.

Williams, Maj. (*Actg. Lt.-Col.*) A. J., D.S.O.	
Harrison, Capt. (*temp. Maj.*) J., A. V. C.	28 Feb. 18

Deputy Assistant Directors of Veterinary Services.

Nicholl, Capt. (*Actg. Maj.*) E. Mc K.	5 Nov. 17
Thwaytes, Capt. (*temp. Maj.*) R. C. G.	
Shea, Capt. (*temp. Maj.*) T. L.	
Meadows, Lt. (*temp. Maj.*) D.	
James, Capt. (*temp. Maj.*) W. H. (*T. C.*)	
Beck, Capt. (*Temp. Maj.*) D. A. S.	...
Ward, Capt. (*temp. Maj.*) J. A.	...

Deputy Director of Remounts.

Anderson, Bt. Lt.-Col. J. F. H., D.S.O.	12 Jan. 16

Deputy Asst. Dir. of Remounts.

Griffiths, Lt.-Col. C. (Retd.)	12 Feb. 15

Base Remount Depôt.

Hughes, Maj. J. E.	...

Deputy Director of Ordnance Services.

Bowen, Lt.-Col. (*temp. Col.*). H. W., D.S.O., R. A.	14 Dec. 17

Assistant Director of Ordnance Services.

Riccard, Maj. (*temp. Lt.-Col.*) J. S., R. A.	
Finnis, Maj. (*Actg. Lt.-Col.*) F. A., R.A.	
Reed, Bt. Lt.-Col. C., D.S.O., R.A.	
Murray, Maj. (*temp. Lt.-Col.*) F. M., R.A.	

ESTABLISHMENTS.

EDUCATIONAL AND TRAINING.

STAFF COLLEGES.

CAMBERLEY.
QUETTA. } *Closed temporarily.*

CAVALRY SCHOOL, Saugor.—*Estbd. 1st Apl. 1910.*

(Closed temporarily.)

INDIAN CENTRAL FLYING SCHOOL.

(SITAPUR). *(Closed on Mobilisation.)*

CADET COLLEGE, QUETTA.

Commdt.	**Boileau**, Lt.-Col. E. R. P., 2nd G. R., [L]	6 Feb. 17
Asst. Commdt.	**Arthur**, Maj. L. F., 26th Cav. D.S.O., *p. s. c.* [L]	13 Oct. 17
G.S.O., 2nd Gr.	**Ogle**, Bt. Lt.-Col. N., D.S.O., *p.s.c.*, 67th Punj.	1 Oct. 18
Do.	**Muir**, Capt. (*temp. Maj.*) A. H. H., 15th Sikhs	5 Sep. 18
Instr. in Urdu	**Harcourt**, Maj. E. S., M.C., 7th G. R.	5 Sep. 18
Adjt. & Q.M.	**Beanland**, Capt. D., 22nd Punj.	10 Nov. 18
Compy. Commdr.	**Heath**, Capt. L. M., M.C., 19th Punj.	5 Sep. 18
Compy. Officer	**Borrowman**, Capt. C. G., 4th G. R.	28 Sep. 17
Do.	**Molesworth**, Capt. G. E. N., 69th Punj.	26 Sep. 18
Medical Offr.	**Murphy**, Capt. (*temp.*) R. W., R.A.M.C.	23 June 16
Riding Master	**Naylor**, Maj. H. W., R.F.A.	1 May 15

CADET COLLEGE, WELLINGTON.

Commdt.	**Maxwe**, Bt.-Col. L. L., C.M.G., 2nd Lrs.	15 Aug. 18
Asst. Commdt.	**Hutchinson**, Col., C. A. R., D.S.O., *p. s. c.*, 41st Dogras	31 Mar. 17
G. S. O., 2nd Gr.	**Yearsley**, Capt. (*temp. Maj.*) R. A., 28th Punj.	5 Mar. 18
Do.	**Skinner**, Capt. (*temp. Maj.*) G. E. M., L'pool R.	15 Nov. 18
Instr. in Urdu	**Hunt**, Lt.-Col. V. de V., I. A	25 Oct. 15
Adjt. and Q. M.	**Johnston**, Capt. E. N. W., *I. A. R. O.*	1 Apl. 18
Company Comdr.	**Finnis**, Capt. H., M.C., 53rd Sikhs (*F.F.*)	15 Mar. 18
Compy. Cfficer	**James**, Capt B. S., attd. 25th Cav. (F.F.)	15 Mar. 18
Do.	**Gore**, Capt. C. H. C., 39th Garh. R.	10 Sep. 18
Medical Offr.	**Anthonisz**, Capt. C. F., R.A.M.C. (T. C.)	14 Sep. 18
Riding Master	**Wark**, Capt. R. G., R.F.A.	7 Oct. 15

CADET UNIT, CAMPBELLPORE.

Commdt.	**Yates**, Bt.-Lt.-Col. H. P., D.S.O., S. W. Bord. R.	7 June 18
Asst. Commdt.	**Guinness**, Capt. (*temp Maj.*) W., 4th Manch. R.	16 Oct. 17
Coy. Comdr.	**Wace**, Maj. M., 14th Sikhs	16 Nov. 18
Platoon Comdr.	**Thompson**, Capt. N. R., 1st G. B., R. Ir. Rif.	15 Jan. 18
Do.	**Helps**, Capt. M. D., 1-4th R. W. Surr. R.	7 May 17
Do.	**Cormocy**, Lieut. (*temp. Capt.*) E. K. R. Suss. R.	18 Jan. 17
Do.	**Howard**, 2nd-Lieut. (*temp. Capt.*) R., S. Lancs. R.	12 May 17
Do.	**Napier**, Lt. (*temp. Capt.*) M., 1-5th Hamps. R.	2 Dec. 17
Do.	**Street**, Lieut. (*temp. Capt.*) A. R., 2nd Som. L. I.	18 Mar. 17
Do.	**Noad**, Lt. (*temp. Capt.*) C. K., 1-9th G. R.	1 Nov. 18
Do.	**Butcher**, Lt. (*temp. Capt.*) F. G., 2-2nd Rajputs	30 Oct. 18
Adjt.	**Bury**, Capt. E. P., S. W. Bord. R.	25 Sep. 18
Qr. Mr.	**Knight**, Lt. (*temp.*) H., Gen. List.	15 Apl. 18

ESTABLISHMENTS.

OFFICERS SCHOOL OF INSTRUCTION, BANGALORE.

Commandant	**Skinner,** Maj. (*temp. Lt.-Col.*) A. B., D.S.O., 5th Cav.	...
Asst. Comdt.	**Watts,** Capt. (*temp. Maj.*) N. H. L., 92nd Punj.	...
Cavy. Instr.	**Ward,** Lt. (*temp. Capt.*) B. M., 1st Dn. Gds.	...
Do.	**Johnson,** Capt. H. A. B., 26th L. Cav.	...
Compy. Comdr.	**Gardner,** Capt. C. P., 73rd Infy.	...
Platoon Comdr.	**Robertson,** Capt H. E., 107th Prs.	...
Do.	**Scott,** 2nd-Lt. (*temp. Capt.*) I. J., 3rd S. W. Bord.	...
Do.	**Webb-Johnson,** Capt. S., 1-6th E. Surr. R.	...
Do.	**Ellson,** Lt. (*temp. Capt.*) P., 9th Middx. R.	...
Do.	**Gordon,** Capt. D. H., D.S.O., 34th Prs.	...
Do.	**Gamble,** Lt. W. M. F., M.C., I. A. (*on probn.*)	...
Do.	**Jimby,** 2nd Lt. (*temp. Capt.*) C.-W., Som. L.I.	...
Adjutant	**Woods,** Lt. (*Actg. Capt.*) J., 2nd Som. L. I.	...
Qr. Master	**Ticehurst,** Lt. W., Genl. List	...

OFFICERS' SCHOOL OF INSTRUCTION, NASIK.

Commandant	**Dease,** Maj. (*temp. Lt.-Col.*) C. E., 91st Punj.	19 Feb. 18
Asst. Comdt.	**Errington,** Capt. (*temp. Maj.*) J. P., 1-4th Bord. R.	2 Mar. 18
Adjt.	**Newman,** 2nd-Lt. (*Actg. Capt.*) H., 3rd Hamps. R.	12 Nov. 18
Compy. Comdr.	**Layng,** Capt. T.M., M.C.	4 Nov. 18
Do.	**Abraham,** Capt. A. C., Devon. R.	26 Apl. 18
Platoon Comdr.	**Dunn,** Lt. (*temp. Capt.*) N. E. J., Sea. Highrs.	19 Feb. 18
Do.	**McEntire,** Lt. (*temp. Capt.*) H. S. M., 123rd Rif.	29 Oct. 18
Do.	**Fischer,** Lt. (*temp. Capt.*) T. G. T., 9th G. R.	...
Do.	**Moody,** Lt. (*temp. Capt.*) C. G., Middx. R.	8 Mar. 18
Do.	**Curphey,** 2nd-Lt. (*temp. Capt.*) R. L., E. Lancs. R.	...
Do.	**Parry,** 2nd-Lt. (*temp. Capt.*) E. A., Notts & Derby. R.	20 July 18
Do.	**Carter,** 2nd-Lt. (*temp. Capt.*) F. J., M.M., Yorks. L. I.	12 Nov. 18
Do.	**Sharpe,** 2nd-Lt. (*temp. Capt.*) P., W. Rid. R.	12 Oct. 18
Supy. Platoon Comdr.	**Rowbotham,** 2nd-Lt. (*Actg. Capt.*) A. S., Glouc. R.	11 Nov. 18
Do.	**Simpson,** 2nd-Lt. (*Actg. Capt.*) T. W., Welsh R.	12 Nov. 18
Qr. Master	**Lloyd,** Hony. Lt. E., S. W. Bords.	25 July 18

SUPPLY AND TRANSPORT CORPS TRAINING ESTABLISHMENT.

Rawalpindi.

Comdt. & Chief Instructor.	**Amesbury,** Lt.-Col. W. L. R., S. & T. Corps.	19 Sep. 18
Instructor	**Lyster,** Maj. C. J. H., M.C., S. & T. Corps	15 Feb. 18
Do.	**Aylmer,** Maj. G., M.C., S. & T. Corps	16 Apl. 18
Do.	**Reyner,** Maj. B., A. S. C.	15 May 18

CENTRAL SCHOOL OF PHYSICAL AND BAYONET TRAINING, NORTHERN COMMAND.

(Ambala-Kasauli.)

Commandant.	**Watson,** Lt. (*temp. Capt.*) P., High. L. I.	

CENTRAL SCHOOL OF PHYSICAL AND BAYONET TRAINING, SOUTHERN COMMAND.

(Poona.)

Commandant	**Ranger,** Lt. (*temp. Capt.*) W. J., Som. L. I.	1 July 17
Adjt. & Qr. Mr.	**Calder,** 2nd-Lt. J. F., attd. 18th Bn. Rif. Bde.	

ARTILLERY SCHOOL, QUETTA.

Comdt. & Chief Instructor.	**Smith,** Lt.-Col. H. R. W. M., D.S.O., R. A.	21 Aug. 16
Asst. Comdt.	**Batten,** Capt. J. F., M.C., R. A. (*temp. Maj.*)	20 Dec. 16
Asst. Instructor	**Addis,** Lt. (*temp. Capt.*) W. H., R.A.	8 Aug. 16
Camp Adjt. & Qr. Mr.	**Theaker,** Lt. H., R. A.	- Jan. 17

ESTABLISHMENTS.

MACHINE GUN TRAINING CENTRE, MHOW.

Comdt.	**Howson,** Maj. (*temp. Lt.-Col.*) G., M.C., 4th Cav.	22 Dec. 17
Instructor	**Slaughter,** Capt. (*temp. Maj.*) G. A., M.C., Worc. R., and M. G. Corps	17 Nov. 16
Asst. Instr.	**Lecky,** Maj. J. G., 119th Infy.	21 Jan. 18
Adjutant	**Martin,** Lt. (*temp. Capt.*) W. H., M. G. Corps.	17 Sep. 17
Offr. i/c Indian Depôt.	**Green,** Lt. (*temp. Capt.*) P. J., R. F. A.	1 Feb. 17
Attd. Indian Depôt.	**Avery,** Lt. A. H., I. A. R. O.	1 Sep. 17
Extra Asst. Instr.	**Humphries,** Lt. A., I. A R. O.	5 Feb. 18
Qr. Master.	**Tanner,** Lt. G., Brit. Ser.	12 June 13

SCHOOLS OF MUSKETRY.

CENTRAL SCHOOL, PACHMARHI.

Commandant	**Grove.** Bt. Lt.-Col (*temp. Col.*) P. L., Ches. R.	...
Instructor	**Smith,** Maj. M. C., 64th Prs.	...
Ditto	**Moss-Blundell,** Maj. R. S., M.C., 27th Cav.	...
Asst. Instructor	**Thorburn,** Capt. M. M., M.C., 3rd R. Highrs.	...
Ditto	**Brunton,** Lt. (*temp. Capt.*) F. H., 1st G. B., E. Yorks. R.	...
Ditto	**Mahoney,** Lt. (*temp, Capt.*) A., J., I.A.R.O.	...
Extra Asst. Instr.	**Wise,** 2nd-Lt. H., R. Suss. R.	...
Ditto	**Field,** Lt. F. C., 32nd Lrs.	...
Ditto	**Carter,** Lt. A. B., 53rd Sikhs	...
Ditto	**Webb,** Lt. F. B., 3-8th G. Rif.	...
Adjt & Qr. Mr.	**Hare.** Lt. S. St. G., 2-76th Punj.	...

BRANCH SCHOOL, SATARA.

Commandant	**Hulbert,** Maj. T. E., 3rd Horse	...
Instructor	**Bingham,** Capt. C. D' A., 109th Infy.	...
Asst. Instructor	**Guthrie, Lt.** (*temp. Capt.*) W., R. Suss. R.	...
Ditto	**Mudie,** Capt.. R. F., I A. R. O.	...
Extra do.	**Hawke,** Lt. (*temp. Capt.*) A. W., Durh. L.I.	...
Do. do.	**Haigh,** Lt. P. B., I. A. R. O.	...
Do. do.	**Copeman,** Lt. D. H., M.C., 8th N. Lan. R.	...
Do. do.	**Delafens,** Lt. R. A, 3-7th G. R.	...
Do. do.	**Watson,** Capt. H. B., 3rd Horse	...
Adjt. & Qr. Mr.	**Woolgar,** Lt. C., W. Rid. R.	...

BRANCH SCHOOL, RAWALPINDI.—CHANGLA GALI.

Commandant	**Sterndale-Bennett,** Maj. J., 107th Prs.	...
Instructor	**MacDougall,** Capt. R. McI., T. F. Res.	...
Asst. Instructor	**Pearce,** Capt. W. K., 1-5th Hamps. R.	...
Do. do.	**Meyer,** Lt. B. (*temp. Capt.*) 2-39th Garh. Rif.	...
Extra do.	**Platt,** Lt. E. P., 41st Dogras	...
Do. do.	**Webb,** Lt. G. S. R., M.C., 37th Lrs.	...
Do. do.	**Irwin,** 2nd-Lt. C., 2nd N. Staff. R.	...
Adjt. & Qr. Mr.	**Tye,** 2nd-Lt. (*temp. Capt.*) H. J., 19th Lrs.	...

CENTRAL BOMBING & (STOKES) TRENCH HOWITZER SCHOOL, MHOW.

Commandant	**Lee.** Maj. (*temp. Lt-Col.*) G., D.S.O., M.C., 1st B., E. Kent R.	...
Asst. Commandant	**Michell,** 2nd-Lt. (*temp. Maj.*) R. J. V., 13th Hrs.	...
Adjt. & Qr. Mr.	**Dardier,** Capt. C. E., 1-10th Midd'x. R.	...
Instructor	**Brewer,** Lt. (*temp. Capt.*) G. L., 1-4th E. Kent R.	...
Ditto	**Hide,** Lt. (*temp.-Capt.*) T. G. S., 29th Punj.	...
Ditto	**Gillies-Reyburn,** Lt. (*temp. Capt.*) D., 93rd Infy.	...
Ditto	**Edwards,** Lt. (*temp. Capt.*) A. L., M.C., 2nd R. Highrs.	...
Ditto	**Shallow,** Lt. (*temp. Capt.*) C. R., 74th Punj.	...
Ditto	**Lynch,** Lt. (*temp. Capt.*) J. A., 35th Sikhs.	...
Ditto	**Williams,** Lt. J. H., 1-6th Devon. R.	...

SIGNALLING SCHOOLS.

NORTHERN COMMAND SIGNALLING SCHOOL, LUCKNOW.

Commandant	**Foot,** Capt. P. B. D-H., Middx. R. (*31st Divl. Sig. Coy.*)	...
Instructor	**Hasluck,** Lt. R. E., I.A.R.O.	...

Bangalore.

Commandant	**Stebbing,** Capt. H. T., R. G. A. (*31st Divl. Sig. Coy.*)	...
Instructor	**Birtwistle,** Lt. J., 2-10th Jats	...

Kasauli.

Commandant & Chief Instr.	**Shoulder,** Lt. (*temp. Capt.*) H., L'pool. R.	2 Oct. '5

ESTABLISHMENTS.

TRAINING SCHOOL FOR INDIAN CADETS, INDORE.

Commandant	**Black**, Brig.-Gen. W. C., 12th Prs., *p.s.c.*	5 July 18
Senior Instructor	**Haughton**, Maj. H. L., 36th Sikhs	1 Sept. 18
Instructor	**Barstow**, Capt. A. E., M.C., 15th Sikhs	1 Oct. 18
Ditto	**Meade**, Capt. M. E. M., 125th Rif.	10 Oct. 18
Adjt. & Qr. Mr. & Riding Master.	**Harris**, Capt. A. E. C., M.C., 7th Lrs.	15 Sept. 18
Medical Officer	**Smith**, Lt.-Col. F. A., M.D., I.M.S.	16 Sep. 18

DEPARTMENTS OF THE ARMY.

[NOTE.—An asterisk * prefixed to an officer's name indicates that he has been selected for continuous service in India.]

ORDNANCE DEPARTMENT IN INDIA.

Names and Rank.	Appointments.	First appt. to Dept.	Present appointment.	Remarks.
***Renny**, Maj.-Gen. (*temp.*) S. M., C.I.E., Brit. Ser.	Director-General of Ordnance in India.	19 Dec. 87	16 Feb. 17	Simla.
***Ogg**, Col. G. S., C.I.E., Brit. Ser.	Assistant Director of Artillery.	17 Jan. 93	30 Sep. 18	Do.
Villiers, Comsy. and Hy. Maj. A.	Personal Asst. to D. G. O.	25 May 89	16 Feb. 17	Do.
**Campbell, Col. M. S. C., C.I.E., Brit. Ser. (ff).*	*Ordnance Consulting Officer for India.*	*10 June 88*	*6 Apl. 11*	India Office, London.
***Foote**, Lt.-Col., H. B., R.A. . .	Assistant Ordnance Consulting Officer for India.	1 Mar. 93	11 Oct. 17	Do. (Temp.)

ORDNANCE FACTORIES.

(UNDER CONTROL OF THE INDIAN MUNITIONS BOARD.)

Names and Rank.	Appointments.	First appt. to Dept.	Present appointment.	Remarks.
***Palmer**, Lt.-Col. C. C., R.A., *p.a.c.*	Military Assistant (Ordnance Factories).	1 Mar. 02	1 Jan. 18	Indian Munitions Board, Simla.
***Morrison**, Maj. J., R.A. . . .	2nd Mily. Asst. (Ordnance Factories).	4 July 07	11 July 18	Do.
***Young**, Lt.-Col. (*temp. Brig. Genl.*) H. A., C.I.E., R.A. (*ff*)	Director of Ordnance Factories.	26 June 92	17 Feb. 17	Calcutta.
Kenyon, Bt.-Col. (temp. Brig. Genl.) L. R., C.B., R.A., p.a.c.	*Deputy Director of Ordnance Factories.*	*1 Jan. 11*	*1 Jan. 11*	*On duty England.*
***Sturrock**, Lt.-Col, G. C., R. A. (*g*).	Do.	27 Dec. 97	14 Apl. 16	Calcutta.
Fraser, Maj. T. C., R.A. . .	Asst. Dir. of Ord. Factories	8 Apl. 18	8 Apl. 18	Do.

Ammunition Factories, Dum Dum and Kirkee.

Names and Rank.	Appointments.	First appt. to Dept.	Present appointment.	Remarks.
Cowie, Maj. D. G., R.A. . . .	Superintendent . .	21 Feb. 02	1 June 18	Dum Dum.
***Sitwell**, Maj. N. S. H., R.A. .	Do. .	17 Apl. 01	12 May 17	Kirkee.
Brown, Maj. B. H., R.A., *p.a.c.* .	Assistant Superintendent .	31 Mar. 14	14 Apl. 16	Dum Dum.
Hare, Maj. G. A., R. A. (c) (o) . .	Do. . . .	*25 Apl. 09*	...	*On depn. England.*

Cordite Factory, Aruvankadu, Nilgiri Hills.

Names and Rank.	Appointments.	First appt. to Dept.	Present appointment.	Remarks.
***Babington**, Col. D. M., C.I.E., Brit. Ser.	Superintendent . .	7 May 91	15 May 95	Aruvankadu.
***Urquhart**, Maj. A. M., R.A. .	Assistant Superintendent .	20 May 07	14 Oct. 16	Do.
Wiles, Lt. H. H., I. A. R. O. .	Danger Building Officer . .	27 Apl. 16	27 Apl. 16	Do.

Gun Carriage Factory, Jubbulpore.

Names and Rank.	Appointments.	First appt. to Dept.	Present appointment.	Remarks.
***Tomkins**, Lt.-Col. E. L., R.A. (*ff*) (*g*).	Superintendent . . .	11 Oct. 97	7 Jan. 17	Jubbulpore.
Tute, Maj. C. S., R.A. (*a*) (*o*) .	Assistant Superintendent .	22 June 10	18 Nov. 18	Jubbulpore.

(*a*) Tenure extended to 21st June 1920.
(*ff*) Passed Firemaster's Course with honours.
(*c*) Tenure extended to 24 Apl. 19.
(*o*) Passed Ordnance Course.

Names and Rank.	Appointments.	First appt. to Dept.	Present appointment.	Remarks.
ORDNANCE FACTORIES—*concld.*				
Gun and Shell Factory, Cossipore (Branch at Ishapore).				
*Parbury, Maj. E., R.A.	Superintendent	2 Dec. 02	3 Apl. 18	Cossipore.
*Vines, Maj. C. E., R. A., *p.a.c.*	Assistant Superintendent	25 Apl. 08	13 Nov. 18	Ishapore.
*Lenfesty, Maj. L. d'E., R. A.	Ditto	5 Oct. 06	12 Nov. 18	Cossipore
Wheeler, Lt. P. J., R. F. A.	Ditto	5 Aug. 16	5 Aug. 16	Ishapore.
Harness and Saddlery Factory, Cawnpore.				
*Lane, Lt.-Col. F. C., R.A.	Superintendent	23 July 96	1 Oct. 08	Cawnpore.
Larmour, Maj. L. C., R.A. (*o*)	Assistant Superintendent	6 Oct. 14	7 Nov. 18	Do.
Blank, Lt. A. L., I. A. R.O.	Ditto	...	16 Apl. 18	Do.
Rifle Factory, Ishapore.				
Kinloch, Capt. R. I. H., R.A., *p.a.c.* (*l*)†	Assistant Superintendent	27 Apl. 13	22 July 18	Ishapore.
ORDNANCE INSPECTION STAFF.				
Under control of the Dir. Genl. of Ord. in India.				
*Lawrence-Archer, Lt.-Col. (*temp. Col.*) J. H., R.A.	Director of Ordnance Inspection.	18 July 96	17 Feb. 17	Simla.
Waller, Maj. (*temp. Lt.-Col.*) R.S. de W., R.A. (*b*).	Assistant Director of Ordnance Inspection.	7 Dec. 09	10 Apl. 18	Do.
Inspectors. (6)				
*Schrottky, Maj. C. E. G., R.A. (*o*).	Inspector of Ammunition	7 Feb. 08	22 Dec. 14	Dum Dum.
*Spalding, Maj. W. B., R.A.	Chief Inspector of General Stores.	17 Mar. 02	16 Sep. 18	Cawnpore.
De la Bère, Lt.-Col. H. P., Brit. Ser.	Inspector of Guns and Rifles.	17 Aug. 15	17 Aug. 15	Ishapore.
*Thomas, Maj. C. S. W., R.A.	Inspector of Gun Carriages and Vehicles.	14 May 01	4 Oct. 15	Jubbulpore.
Forrester, Capt. H. W.	Inspector of General Stores	2 May 17	16 Sep. 18	Cawnpore.
Reavell, Capt. H. S.	Inspector of Harness and Saddlery.	9 Aug. 17	16 Sep. 18	Do.
Assistant Inspectors. (10)				
*Reeves, Maj. S. R., R.A.	Assistant Inspector of Guns and Rifles.	14 Mar. 04	13 June 17	Ishapore.
Drake-Brockman, Maj. V., R.A., *p.a.c.* [*l*].	Assistant Inspector of Ammunition.	8 Mar. 14	17 Apl. 16	Dum Dum.
Hewitt, C. R. L., Esqr.	Asst. Inspec. of Gun Carriages.	25 Apl. 16	25 Apl. 16	Jubbulpore.
Marsh, Lt. T., R.F.A.	Assistant Inspector of Guns and Rifles.	14 Sep. 17	14 Sep. 17	Ishapore.
Teague, Lt. S. W. M., R.F.A.	Asst. Inspector of Ammunition.	4 Jan. 18	4 Jan. 18	Dum Dum.
Curtis, Lt. R. J., Genl. List.	Asst. Inspector of General Stores.	3 Oct. 18	3 Oct. 18	Cawnpore.
Wilton, Lt. J. S., Genl. List.	Asst. Inspector of Harness & Saddlery.	3 Oct. 18	3 Oct. 18	Do.
Prendergast, Lt. R. V., R. A.	Asst. Chief Inspector of General Stores.	6 June 17	17 Oct. 18	Do.
Drew, Lt. A.	Asst. Inspector of General Stores.	24 May 18	14 Oct. 18	Do.
Howard, Maj. C.	Asst. Inspector of Harness & Saddlery.	...	1 Nov. 18	Do.
Proof and Experimental Establishment, Balasore.				
Freeth, Lt.-Col. C. J. D., R. A. (*g*).	Proof Officer	25 Mar. 12	28 July 15	Balasore.
Sewell, Lt. S.	Assistant Proof Officer	18 Sep. 17	18 Sep. 17	Do.
Central Chemical Laboratory, Naini Tal.				
Marshall, A., Esq., F.I.C., F.C.S.	Chemical Inspector	4 Oct. 08	4 Oct. 08	Naini Tal.

(*b*) Tenure extended to 6th December 1919.
(*o*) Passed Ordnance Course.
† Appointed subject to passing the examination laid down by Government.

ORDNANCE INSPECTION STAFF—*concld.*

Inspectors of Munitions. (Temporary.)

Names.	Station.	Date of joining appointment.	REMARKS.
Barnard, A. A., Esq.	Bombay	10 Mar. 16	
Brown, 2nd-Lt. N. L., R. F. A.	Madras	2 May 18	
Drew, 2nd-Lt. J., 1st K. D. Gds.	Cawnpore	24 May 18	
Evans, 2nd-Lt. C., 5th Res. Bn.	Agra	28 Aug. 18	
Hughes, Lt. F., 8th Res. Bn.	Ferozepore	4 Sep. 18	
Jenkins, Lt. W. M., I. A. R. O.	Cawnpore	31 May 18	
Jones, 2nd-Lt. S. C., 1st G.B., R. I. Fus.	Cawnpore	9 Sep. 18	
Leighton, Lt. J. H. L., I. A. R. O.	Bombay	22 May 18	
Maidlow, Lt. J. F., 1st G. B. Linc. R.	Calcutta	17 Apl. 18	
Maitland, 2nd-Lt. (*temp. Capt.*) A., I.A.R.O.	Rawal Pindi	28 May 17	
Ogilvie, Lt. (*temp. Capt.*) P. G., I. A. Res.	Delhi	13 May 16	
Outerson, Lt. (*tempt. Cap.*) W., 1st G. B. R. Scots. Fus.	Agra	13 July 17	
Pennington, Lt. C. E. R.	Cawnpore	27 May 18	
Preston, 2nd-Lt. E. A., R. F. A.	Jubbulpore	19 July 18	
Prior, Lt. E. G., Ox. and Bucks. L. I.	Ferozepore	8 June 17	
Pullen, Lt. A. P., 1st G. B. Manch. R.	Lahore	24 June 18	
Roskilly, Lt. S. T. H., 1st G. B., Linc. R.	Allahabad	18 Jan. 18	
Silvester, C. J., Esq.	Calcutta	3 July 15	
Smith, 2nd-Lt. C., 5th Res. Bn.	Cawnpore	27 Nov. 18	
Smith, Lt. F. N., 1st G. B., Linc. R.	Cawnpore	21 Feb. 18	
Todd, Lt. (*temp. Capt.*) A. S.	Madras	17 Apl. 17	
Treanor, R. F. S., Esq.	Calcutta	23 Mar. 16	
Tunaley, Lt. A. E., 1st G. B., Linc. R.	Bombay	17 Jan. 18	
Uridge, Lt. H. J., I. A. R. O.	Calcutta	22 July 16	
Workman, Lt. D. J. G., 1st G. B., Bedf. R.	Rangoon	14 Jan. 18	
Wortham, Maj. O. O., A. O. D.	Simla	10 Sep. 18	
Wright, 2nd-Lt. G., 1st G. B. R. Sc. Fus.	Jubbulpore	23 Sep. 18	

INDIAN ORDNANCE DEPARTMENT.
(Under the Quartermaster General in India.)

Names and Rank.	Appointments.	First appt. to Dept.	Present appointment.	Remarks.
Perry, Maj.-Genl., *Sir* H. W., K.C.M.G., C.B., C.S.I., A. O. D.	Dir. of Eqt. & Ord. Stores.	1 Apl. 18	1 Apl. 18	Delhi
***Douglas,** Lt.-Col. A. P., C. M. G., R. A. (*h*).	Dy. Dir. of Eqt. & Ord. Stores.	10 June 95	1 Apl. 18	Simla.
***Carey,** Lt.-Col. W.L.J., R.A. (*l*) (*o*).	Asst. Dir. of Eqt. & Ord. Stores.	12 Oct. 98	1 Apl. 18	Do.
***Tyrrell,** Lt.-Col. J. F., R.A.	Asst. Dir. of Eqt. & Ord. Stores.	11 July 99	1 Apl. 18	Delhi.
Johnston, Maj. (*temp. Lt.-Col.*) C. E. L., R. A [L.]	Dy. Asst. Dir. of Eqt. & Ord. Stores.	1 Oct. 09	1 June 18	Simla.
Palairet, Capt. (*temp. Maj.*) R. C. N., Devon R.	Addl. Dy. Asst. Dir. of Eqt. & Ord. Stores.	7 Sep. 18	7 Sep. 18	Do.
Smith, Maj. B. E.	Additional Ord. Offr.	2 July 85	1 Apl. 18	Delhi.

Names and Rank.	First appt. to Dept.		Present appointment.		Remarks.
	Ordnance Officers.				
**Campbell, Col. M. S. C.*, C.I.E., *Brit. Ser.* (*ff*).	*10 June*	*88*	*6 Apl.*	*11*	*Lond. Ord. Consulting Offr.*
***Smallwood,** Lt.-Col. F. G., C.V.O., R. A. (*f*)	26 Apl.	91	27 May	18	Poona, A.D.O.S., Southern Command.
***Radcliffe,** Lt.-Col. W. C. A., R.A.	7 May	92	1 Jan.	11	Quetta, C. O. O.
**Foote, Lt-Col. H. B., R. A.*	*1 Mar.*	*93*	*11 Oct.*	*17*	*London, Asst. C. O. O. (Tempy.)*
***Lethbridge,** Lt.-Col. S., R.A.	23 Mar.	94	28 Oct.	17	Ferozepore, C. O. O.
**Douglas, Lt.-Col. A. P.*, C.M.G., *R.A.* (*h*)	*10 June*	*95*	*1 Apl.*	*18*	*Simla, D. D. E. O. S.*
***Woods,** Lt.-Col. G. G., R.A. (*h*)	1 June	95	7 Nov.	17	Rawalpindi, C. O. O.
***Donovan,** Lt.-Col. C. C., R.A. (*h*).	22 Jan.	96	24 May	18	Rawalpindi, A. D. O. S., Northern Command.
***Cooper,** Lt.-Col. P. T., R.A. (*g*)(*h*)	25 May	96	5 Mar.	17	Allahabad, C. O. O.
***Carter,** Lt.-Col. E. P., R.A. (*h*)	17 May	97	23 Aug.	17	Kirkee, C. O. O.
***Bowen,** Lt.-Col. (*temp. Col.*) H.W., D.S.O., R. A.	27 June	98	...		
**Kennedy, Lt.-Col. A. C.*, D.S.O., *R.A.* (*h*)	*26 Sep.*	*98*	*25 Feb.*	*18*	*Simla, D. D., Clothing.*
***Garstin,** Lt.-Col. H. E., D.S.O., R.A. (*h*)	28 Apl.	99	...		
**Carey, Lt.-Col. W. L. J., R. A.* [*l*] (*o*)	*12 Oct.*	*98*	*1 Apl.*	*18*	*Simla, A. D. E. O. S.*
**Tyrrell, Lt.-Col. J. F., R.A.*	*11 July*	*99*	*1 Apl.*	*18*	*Simla, A. D. E. O. S.*
***Kemble,** Lt.-Col. F. A., R.A.	10 Nov.	00	8 Aug.	18	Madras, C. O. O.
***Gardiner,** Maj. H.W., R.A. (*o*)	25 May	01	...		
***Walker,** Maj. W. B., R.A (*h*)	21 July	00	23 Apl.	18	Rangoon, C. O. O.
***Riccard,** Maj. (*temp. Lt.-Col.*) J. S., R.A. (*h*)	11 May	01	...		
***Wilkes,** Maj. G. S. R. A. (*o*)	30 June	02	20 Nov.	18	Ferozepore O. O.
***Clarke,** Maj. H., R. A.	20 Sep.	02	2 Nov.	18	Bombay, C. O. O.
***Fanshawe,** Maj., (*Actg. Lt.-Col.*) L.A., D.S.O., R.A. (*o*).	1 Oct.	04	8 Sep.	18	
*** Mecredy,** Maj. H., R.A. (*o*)	20 Nov.	05	5 May	18	Kirkee.
***Finnis,** Maj. (*Actg. Lt.-Col.*) F. A., R.A. (*o*)	3 Feb.	06	...		
***Dickins,** Maj. F., R. A.	2 Oct.	06	...		
***Routh,** Maj. G. M., D.S.O., R.A. [L]	11 Jan.	08	...		
Reed, Maj. *Bt. Lt.-Col.* C., D.S.O., R.A.	10 Aug.	08	...		
Murray, Maj. Bt. *Lt.-Col.* F. M., R.A. (*i*).	5 Dec.	08	...		
Robertson-Glasgow, Maj. N., R.A. (*b*)	11 May	09	...		
Johnston, *Maj. C. E. L., R.A.* [*L*] (*c*)	*1 Oct.*	*09*	*1 June*	*18*	*Simla, D. A. D. E. O. S.*
Jenkin, Bt.-Lt.-Col. F. C., D.S.O., R.A. (*p*)	21 May	11	7 Aug.	18	Rawalpindi, O. O.
Long, Capt. W. E. L., R. A. (*o*) (*d*)	1 Aug.	11	...		
Barron, Maj. A. F. N., R.A. (*f*) (*e*)	15 Oct.	11	8 July	18	
Bryant, Maj. J. M., R.A.	1 Apl.	12	...		
Beamish, Maj., W. S., R.A. (*o*)	26 Apl.	13	...		
Temple, Maj. T., R.A. (*o*)	9 Apl.	15	...		

(*b*) Tenure extended to 10 May 1919.
(*c*) Ditto 30 September 1919.
(*d*) Ditto 31 July 1921.
(*e*) Ditto 14 October 1921.
(*f*) Passed Firemaster's Course.
(*h*) Passed Ordnance Course (excepting A. O. duties).
(*j*) Tenure extended to 14 October 1921.
(*k*) Ditto 20 May 1921.
(*o*) Passed Ordnance Course.
(*p*) Tenure extended to 20th May 1921.

INDIAN ORDNANCE DEPARTMENT—*contd.*

Attached for duty in the Ordnance Department.

Names and Rank.	First appt. to Dept.	Present appointment.	Remarks.
Whitaker, Bt. Maj. H. C., A.O.D.	7 July 18	7 July 18	D. A. D. O. S., 1st Divn.
Neale, Capt. (*temp. Maj.*) M. R., M.C., A. O. D.	4 July 18	4 July 18	D. A. D. O. S., 2nd Divn.
Farquharson-Roberts (*temp. Maj.*), K. F., A.O.D.	8 July 18	8 July 18	D. A. D. O. S., 4th Divn.
Mowels, *Temp. Maj.* J. H. B., A.O.D.	29 June 18	29 June 18	D. A. D. O. S., 5th Divn.
Dallimore, *Temp. Maj.* A. G., A.O.D.	1 July 18	1 July 18	D. A. D. O. S., 9th Divn.
Hopkins, *Temp. Maj.* A. C., A.O.D.	28 June 18	28 June 18	D. A. D. O. S., 8th Divn.
Mills, *Temp. Maj.* H. A., A.O.D.	3 July 18	3 July 18	D. A. D. O. S., 16th Divn.
Gilbert, Capt. B. J., 1st G. B., Oxf. & Bucks. L. I.	22 Nov. 17	...	
Robinson, *Temp. Capt.* F. J., A.O.D.	6 July 18	6 July 18	D. A. D. O. S., Frontier Bde.
Oldbury, *Temp. Capt.* A. E., A.O.D.	29 June 18	29 June 18	D. A. D. O. S., Poona Divn.
Deighton, *Temp. Capt.* H., A.O.D.	2 July 18	2 July 18	D. A. D. O. S., Meerut Divn.
Mawdesley, Actg. Capt. J. L., A.O.D.	29 June 18	29 June 18	Kirkee.
Nixon, Actg. Capt. A. I., A.O.D.	1 July 18	Dec. 18	Army Hd.qrs., Q.M. G.'s Br.
Chester, Lt. S. C. R., A.O.D.	4 July 18	...	
Henderson, Lt. L. D., A.O.D.	15 Aug. 18	15 Aug. 18	Ferozepore.
Fitz-Williams, Lt. M. R., A.O.D.	30 June 18	...	
Gillham, Lt. J. C. E., A.O.D.	29 June 18	29 June 18	Kirkee.
Toy, Lt. L., A.O.D.	27 June 18	27 June 18	Bombay.
Mullins, Lt. C. W., A.O.D.	6 May 18	30 Oct. 18	Rawal Pindi.
Tranter, Lt. F. G., I.A.R.O.	23 Apl. 17	...	
Bettison, Lt. L. W., A.O.D.	30 June 18	Dec. 18	Rangoon.
Echlin, Lt. G. F. W., A.O.D.	1 July 18	1 July 18	Rawal Pindi.
Dawson, Lt. W. C., A.O.D.	1 July 18	1 July 18	Rawal Pindi.
Thompson, Lt. H. W., A.O.D.	13 Aug. 18	13 Aug. 18	Kirkee.
Roberts, Lt. C., A.O.D.	1 July 18	1 July 18	Madras.
Ellis, Lt. E. R., R.A.	31 May 18	31 May 18	Allahabad.
Townsend, Lt. R. R., A.O.D.	13 Nov. 18	13 Nov. 18	Jubbulpore.
Dee, 2nd-Lt. J., Essex R.	12 Apl. 18	4 Oct. 18	Quetta.
Campbell, 2nd-Lt. C., 1st W. Rid. R.	27 Nov. 18	27 Nov. 18	Allahabad.
Davies, 2nd-Lt. C. W., Welsh R.	14 Oct. 18	14 Oct. 18	Madras.
Dickson, 2nd Lt. T., R. Berks. R.	18 Oct. 18	18 Oct. 18	Rangoon.
Honebon, 2nd-Lt. H., 1-4th R. W. Kent R.	17 Aug. 18	17 Aug. 18	Quetta.

Inspectors of Ordnance Machinery (A. O. D.). (5)

Names and Rank.	Regimental Rank.	Remarks.
Shipman, Maj. H. J.	11 Nov. 09	Bombay.
White, Lt. (*Actg. Capt.*) W.	1 Nov. 16	Quetta.
Marsh, Lt. (*Actg. Capt.*) W. R.	1 Nov. 16	Rawalpindi.
Roberts, Lt. (*Actg. Capt.*) E. F.	18 Aug. 17	Allahabad.
Kemp, Lt. A. H.	23 Sep. 15	Ferozepore.
(Attached.)		
Craigen, Lt. A.	19 Nov. 17	Aden.

Date of 1st Comn.	Names and Rank.	First appt. to Dept.	Date of present appt.	Remarks.
	ARMY CLOTHING DEPARTMENT. (Under the Quartermaster General in India.)			
23 Feb. 81	Perry, Brig.-Genl. *Sir* H. W., K.C.M.G., C.B., C.S.I., Brit. Ser.	1 Apl. 18	1 Apl. 18	Dir. of Eqt. and Ord. Stores, A. H. Q.
4 Nov. 91	Kennedy, Lt.-Col. A. C., D.S.O., R.A.	25 Feb. 18	25 Feb. 18	Dy. Dir. of Clothing, A. H. Q.
28 July 00	Robertson, Maj. (*temp. Lt.-Col.*) R. A. H., 30th Punj.	16 June 16	21 Aug. 17	Asst. Dir. of Clothing, A. H. Q.
15 July 03	Ventris, Maj. E. F. V., R.A.	21 Aug. 17	21 Aug. 17	Dy. Asst. Dir. of Clothing, A. H. Q.
25 Jan. 11	Dawson, Capt. (*temp. Maj.*) B. N., A.S.C.	14 Feb. 18	14 Feb. 18	Dy. Asst. Dir. of Clothing, A. H. Q.
20 Nov. 14	Chamberlain, Lt. J. A., A.O.D.	14 Aug. 18	14 Aug. 18	Attd. Offr., A. H. Q.
17 Jan. 91	O'Meara, Lt.-Col. C. A. E., C.I.E.	1 Aug. 06	21 Apl. 10	*Lv., ex I.*, extended to 12 Sept. 18.
	Assistant Superintendents.			
28 July 00	*Robertson, Maj. (temp. Lt.-Col.) R. A. H., 30th Punj.*	*16 June 16*	....	*Asst. Dir., Clothing A. H. Q.*
	Attached (Temporary).			
23 May 85	Burlton, Lt.-Col. W. E. F., I.A.	....	....	Bombay.
1 Jan. 16	Cameron, Hony. Lieut. (*temp. Capt.*) A. J., I.M.L.	....	....	Rawalpindi.
12 Jan. 17	Campbell, Lt. (*temp. Capt.*) T. P., I.A.R.O.	....	....	Quetta.
27 Dec. 16	Chalmers, Lieut. (*tempt Cap.*) T. A., I.A.R.O.	....	....	Kiamari.
....	Cooper, 2nd-Lieut. (*temp. Capt.*) W. P., 7th Suff. R.	....	....	Rangoon.
11 Dec. 14	Griffin, Capt. T., I.A.R.O.	....	....	Madras.
....	Hird, 2nd-Lieut. S., R. W. Kent R.	....	....	Rawalpindi.
....	Knock, Lieut. (*temp. Capt.*) J., R.F.A.	....	....	Peshawar.
....	Lloyd-Jones, 2nd-Lieut. (*temp. Capt.*) R. W. A., 4th E. Kent R.	....	....	Agra.
....	Lougheed, Lieut. (*temp. Capt.*) A., 5th Wilts. R.	....	....	Lahore.
11 Aug. 80	Melville, Bt.-Col. J. S., I.A.	....	....	Poona.
24 Mar. 15	Plant, Capt. J. A., 24th Bn. Rif. Brig.	....	....	Allahabad.
1 July 17	Rix, 2nd-Lieut. C. H. B., I.A.R.O. (Ty. Capt.)	....	....	Bombay.
2 May 18	Rodgers, 2nd-Lieut. (*temp. Capt.*) J. F., I.A.R.O.	....	....	Poona.
25 Mar. 15	Rogers, Lieut. (*temp. Capt.*) J. R. D., I.A.R.O.	....	....	Jubbulpore.
....	Tapp, 2nd-Lieut. A. V., R. War. R.	....	....	Lahore.
1 Mar. 17	Winter, Lt. H. B., I.A.R.O.	....	....	Madras.
	Attached (Serving with Army Clothing Factories, Clothing Contracts, Munitions and Boot Depôts under control of the Indian Munitions Board).			
	Superintendent.			
23 Dec. 93	Bowden, Maj. W. H.	1 Apl. 09	....	I. M. B.
	Assistant Superintendent.			
25 Jan. 99	Eliot, Maj. (*temp. Lt.-Col.*) M., I.A.	1 June 12	....	Dir. of A. C. Factories, Calcutta.
29 May 17	*Beasant, Lt. C. A., I A.R.O.*	....	....	
20 June 16	*Bowler, Lt. H. M., I.A.R.O.*	....	....	
....	*Bryer, 2nd-Lt. P. J., I.A.R.O.*	....	....	
....	*Buck, 2nd-Lieut A. G.*	....	....	
....	*Colquhoun, Major A. R., 3rd Arg. & Suth. Highrs.*	....	....	
1 Jan. 17	*Cooper, Lieut. R. E., I.A.R.O.*	....	....	
26 Aug. 16	*Davidson, Lieut. D. J., I.A.R.O.*	....	....	
18 Jan. 11	*Gabbett, Capt. K. J., 23rd Prs.*	....	....	*I. M. B., Simla.*
1 Dec. 16	*Galvin, Lieut. Q. H., I.A.R.O.*	....	....	
29 Mar. 15	*Garson, Lieut. I. G. M., 1st G. B., R. Sc. Fus.*	....	....	
....	*Greenway, Major C. D. K., Brit. Ser.*	....	....	
21 Apl. 17	*Holt, Lieut. E., I.A.R.O.*	....	....	*I. M. B., Simla.*
21 Aug. 15	*Horton-Bennett, Capt. A. G., I.A.R.O.*	....	....	*I. M. B., Simla.*
15 June 16	*Lemon, Lieut. W. S., I.A.R.O.*	....	....	
....	*Ling, Lieut. F. G., I.A.R.O.*	....	....	
24 July 15	*Loewe, 2nd-Lieut. H. M. J., 1st G.B. Linc. R.*	....	....	

ARMY CLOTHING DEPARTMENT—*concl'd.*

Date of 1st Comn.	Names of Rank.	First appt. to Dept.	Date of present appt.	Remarks.
1 July 16	*McDonald, Lieut. S. H., I. A. R. O.*	. . . . .	. . . . .	
10 Feb. 17	*Manduell, Lieut. J. S., I. A. R. O.*	. . . . .	. . . . .	
16 May 17	*Nichols, Lieut. J. E., I. A. R. O.*	. . . . .	. . . . .	
27 Dec. 16	*Nickels, Lieut. T., I. A. R. O.*	. . . . .	. . . . .	
14 Aug. 16	*Paige, Lieut. R. H., I. A. R. O.*	. . . . .	. . . . .	
8 Jan. 15	*Riley, Lieut. (temp. Capt.) D., I. A. R. O.*	. . . . .	. . . . .	
27 Dec. 16	*Rockley, Lieut. F. M., I. A. R. O.*	. . . . .	. . . . .	
9 Dec. 14	*Sellers, Capt. W. L., 1st G. B. Linc. R.*	. . . . .	. . . . .	
10 June 16	*Spragge-White, Lieut. T. E., I. A. R. O.*	. . . . .	. . . . .	
18 Mar. 15	*Stuart-Prince, Lieut. (temp. Capt.) D., I.A. R. O.*	. . . . .	. . . . .	
7 Oct. 15	*Vessey, Lieut.(temp. Capt.) S. J., I. A. R. O.*	. . . . .	. . . . .	
15 June 16	*Watkins, Lieut. F. J., I. A. R. O.*	. . . . .	. . . . .	
4 June 15	*Webb, 2nd-Lieut. H. G., 1st G. B. E. York R.*	. . . . .	. . . . .	
9 Sep. 15	*Williams, Lieut. C. A. L., I. A. R. O.*	. . . . .	. . . . .	
4 Dec. 17	Aitken, 2nd-Lieut. W. G., I. A. R. O., attd. S. & T. Corps	. . . . .	. . . . .	
. . . .	Anderson, 2nd-Lieut. J.-H., Notts. & Derby. R.	. . . . .	. . . . .	
5 Apl. 17	Anderson, Capt. R. W. A., I. A. R. O.	. . . . .	. . . . .	
15 July 17	Bell, 2nd-Lieut. T. W., 1st G. B., R. Sco. Fus.	. . . . .	. . . . .	
17 June 17	Botterill, 2nd-Lieut. P. W., 1st G. B., R. Ir. Rif.	. . . . .	. . . . .	
20 Oct. 15	Denison, 2nd-Lieut. O. G., Worc. R.	. . . . .	. . . . .	
5 Oct. 17	Harvey, 2nd-Lieut. H., attd. 2-99th Infy.	. . . . .	. . . . .	
26 Aug. 17	Jacobs, 2nd-Lieut. E. J., Yorks. L. I.	. . . . .	. . . . .	
. . . .	Levy, Lieut. M. P., No. 6 Res. Bn.	. . . . .	. . . . .	
. . . .	Macmillan, 2nd-Lieut. H. A., 5th Bn. High. L. I.	. . . . .	. . . . .	
26 Apl. 17	McKee, 2nd-Lieut. R., K. O. S. Bord.	. . . . .	. . . . .	
14 May 15	Petters, Lieut. H. D., I. A. R. O.	. . . . .	. . . . .	
27 June 18	Phillips, 2nd-Lieut. P. F., I. A. R. O.	. . . . .	. . . . .	
. . . .	Rose, 2nd-Lt. A. S., I. A. R. O.	. . . . .	. . . . .	
26 Sep. 17	Savage, 2nd-Lieut. C. T., attd. 1st S. Lan. R.	. . . . .	. . . . .	
. . . .	Walsh, 2nd-Lieut. F. T., E. Lan. R.	. . . . .	. . . . .	
29 May 17	Archer, 2nd-Lt. A. S., 48th Prs.	. . . . .	. . . . .	
8 Nov. 17	Belcher, 2nd-Lt. A. R., 1st G. B., Norf. R.	. . . . .	. . . . .	
. . . .	Collinge, 2nd-Lt. J., 2 G. B., North'd Fus.	. . . . .	. . . . .	I. M. B.
. . . .	Cothay, Capt. H. W., 4th North'd Fus.	. . . . .	. . . . .	
29 Mar. 18	Edwards, 2nd-Lt. D. J., 2-69th Punj.	. . . . .	. . . . .	
28 Mar. 17	Gray, 2nd-Lt. T., No. 3 Res. Bn. (India)	. . . . .	. . . . .	
15 Apl. 17	Hirst, 2nd-Lt. J., 1. G. B., Suff. R.	. . . . .	. . . . .	
30 May 17	Mackenzie, 2nd-Lt. D., No. 3 Res. Bn. (India)	. . . . .	. . . . .	
. . . .	Meese, Lt. W. A. B., 127nd Infy.	. . . . .	. . . . .	
. . . .	Miller, 2nd-Lt. T. S., 8th Worc. R.	. . . . .	. . . . .	
9 Apl. 15	Murphy, Lt. D. W., R. F. A.	. . . . .	. . . . .	
25 Mar. 16	Noble, Capt. T. L., I. A. R. O.	. . . . .	. . . . .	
. . . .	Paterson, 2nd-Lt. A. H., No. 7 Res. Bn.	. . . . .	. . . . .	
. . . .	Reynolds, 2nd-Lt. N., 1st W. Rid. R.	. . . . .	. . . . .	
19 Mar. 18	Richards, 2nd-Lt. C. G., 4-9th Bhopal Infy.	. . . . .	. . . . .	
11 Mar. 17	Sharman, 2nd-Lt. E. P., 1 G. B., Sco. Rif.	. . . . .	. . . . .	
3 Oct. 14	Wemyss, Capt. A. B., 2-4th Dorset R.	. . . . .	. . . . .	
. . . .	Wilson, 2nd-Lt. A. T., 1 G. B., R. Sco. Fus.	. . . . .	. . . . .	
. . . .	Woods, 2nd-Lt. L. H. W., 1 G. B., S. Staff. R.	. . . . .	. . . . .	

ARMY REMOUNT DEPARTMENT IN INDIA.

Director General.

Date of 1st Comn.	Names and Rank.	First appt. to Dept.	Date, present apptt.	Remarks.
15 Feb. 89	**Templer,** Brig.-Genl. C. F., C.M.G., C.I.E.	28 Mar. 01	5 Sept. 15	Delhi.

Superintendents.

Date of 1st Comn.	Names and Rank.	First appt. to Dept.	Date, present apptt.	Remarks.
7 Jan. 91	**Peacocke,** Col. T. G., C.I.E.	2 May 96	17 Apl. 14	Remount Depôt, Ahmednagar.
5 Aug. 96	**Bruce,** Maj. (*temp. Lt.-Col.*) J.	12 Aug. 02	13 Aug. 18	On depn. to Australia.
28 July 00	**Vanrenen,** Maj. D. H.	21 Sep. 03	12 Aug. 14	Lahore Circle.
21 Jan. 93	**Loch,** Maj. E. C.	11 Mar. 04	2 Aug. 18	Remt. Depôt, Saharanpur.
25 Jan. 99	**Mayne,** Maj. C. E. M.	11 Dec. 04	15 Feb. 18	Sargodha Depôt.
19 May 00	**Hewitt,** Maj. D. R.	1 Jan. 05	1 Apl. 14	Remt. Depôt, Mona.
18 Jan. 02	**Scott,** Maj. R. S.	13 July 06	10 Mar. 16	Remt. Depôt, Hosur.
24 June 99	**Hagger,** Maj. E. R.	10 Mar. 07	18 Feb. 18	Rawalpindi Circle.
25 Jan. 99	**Gordon,** Maj. (*temp. Lt.-Col.*) H. F.	2 Apl. 08	25 Aug. 18	Asst. Dir. Remts., A. Hdqrs.
11 Feb. 99	**Knaggs,** Maj. C. F.	21 June 09	16 June 17	On depn. N. China.
8 May 01	**Passy,** Maj. F. H. B.	12 July 09	9 Oct. 15	Baluchistan Circle.
12 Aug. 99	**Anderson,** Maj. (*temp. Lt.-Col.*) W. H.	5 Oct. 13	20 July 18	
27 Aug. 02	**Anderson,** *Bt. Lt.-Col.* J. F. H., D.S.O.	16 Oct. 13	24 Aug. 18	Remt. Depôt, Babugarh.
29 July 03	**Hotham,** Maj. J. C.	1 Apl. 16	26 Feb. 18	Remt. Depôt, Mona.
22 Oct. 02	**Polley,** Maj. S. T., 3rd Horse	21 Sep. 16	12 Feb. 18	Chenab Canal Colony Circle.
19 Jan. 07	**Jobson,** Capt. T. S., 27th Cav.	5 Nov. 15	23 Nov. 16	Jhelum Canal Colony Circle.
9 Sep. 08	**Hickson,** Capt. B. E., 27th Cav.	6 Sep. 17	6 Sep. 17	Remt. Depôt, Sargodha.
22 Mar. 03	**Young,** Maj. M. S., 14th Lrs.	13 June 18	13 June 18	Remt. Depôt, Babugarh.
Feb. 05	**McTavish,** Capt. F. H. C.	....	24 Sep. 18	Saharanpur.
27 June 18	**Staughton,** Lt. (*temp. Capt.*) A. J., I.A.R.O.	....	11 Sep. 18	Ahmednagar.

Commandants, Temporary Remount Depôts.

Date of 1st Comn.	Names and Rank.	First appt. to Dept.	Date, present apptt.	Remarks.
5 Sep. 77	**Biddulph,** Col. S. F.	....	5 Dec. 14	Neemuch.
28 Feb. 85	**Moore,** Lt.-Col. A. G. W., I.A.	....	1 Dec. 15	Aurangabad.
25 Aug. 86	**Turner,** Lt.-Col. F. W. C., I.A.	....	2 Dec. 16	Bangalore (I.C.T.D.).
25 Aug. 86	**Vincent,** Lt.-Col. F. L., I.A.	....	24 Nov. 16	Saugor.
10 Nov. 86	**Ricketts,** Lt.-Col. H. C., I.A.	....	23 Jan. 17	Mhow.
30 Jan. 89	**Tennant,** Lt.-Col. E.	....	3 Feb. 17	Bellary.
23 July 87	**Skipwith,** Maj. (*temp. Lt.-Col.*) P. A., R.A.	....	7 Oct. 14	Bangalore (Br.).
15 Feb. 88	**Hodgkinson,** Maj. G. H., R. F. A. (T.F.)	....	21 Feb. 16	Meerut.
29 Nov. 90	**Teed,** Lt.-Col. A. S. H.	....	31 Oct. 11	Deolali.
5 Feb. 87	**Wikely,** Lt.-Col. J. M.	....	4 June 18	Gujrat.
4 May 01	**Willis,** Maj. G. C. W., D.S.O., 31st Lrs.	....	2 Nov. 16	Sehore.
10 Sep. 16	**Cameron,** Maj. C. J. N., I.A.R.O.	....	1 Oct. 16	Fyzabad.
22 Oct. 02	**Gannon,** Maj. J. C. R., 23rd Cav.	....	10 June 18	Muttra.
6 Nov. 14	**Porter,** Capt. (*temp. Maj.*) L. L., I.A.R.O.	....	18 Aug. 18	Bulandshahr.

[Attached.]

Date of 1st Comn.	Names and Rank.	First appt. to Dept.	Date, present apptt.	Remarks.
28 Mar. 00	**Egerton,** Maj. S. W., 7th Lrs.	....	7 Dec. 16	Bangalore (I.C.T.D.).
16 Sept. 05	**Hudson,** Capt. B. R., 8th Cav.	....	2 June 18	Saugor.
21 Aug. 15	**Temperley,** Lt. (*actg. Capt.*) H. R.	....	16 Sep. 16	Deolali.
13 Nov. 14	**Hunt,** Capt. J. V., I.A.R.O.	....	28 Jan. 17	On depn. N. China.
20 Nov. 14	**Thomas,** Lt. D. E., I. A. R.O.	....	27 Jan. 15	Muttra.
22 Jan. 15	**Cavell,** Lt. R. G., I. A. R. O.	....	22 Dec. 16	Bangalore (I.C.T.D.).
5 Mar. 15	**Parmer,** Lt. S. W., I. A. R. O.	....	23 Feb. 18	Muttra.
30 July 15	**Symonds,** Lt. (*actg. Capt.*) A. H. B., I.A.R.O.	....	31 July 15	Meerut.
18 Aug. 15	**Shaw,** Lt. (*actg. Capt.*) R. B., R. F. A.	....	22 Feb. 17	Mhow.
15 Dec. 15	**Defries,** Capt. A. G., I. A. R. O.	....	27 Dec. 16	On depn. N. China.
18 Dec. 14	**Duff,** Capt. K. C., I.A.R.O.	....	14 Aug. 17	Saugor.
12 Oct. 16	**Frost,** Lt. (*actg. Capt.*) F. J. T., I. A. R. O.	....	25 Feb. 18	Aurangabad.
24 Mar. 17	**Maunder,** Lt. R. A., I. A. R. O.	....	5 May 18	Saugor.
16 June 17	**Meatyard,** Lt. W. E., I. A. R. O.	....	27 June 17	On depn. N. China.
30 June 17	**Batchelor,** 2nd-Lt. A. E. F., R. F. A.	....	20 June 17	Gujrat.
22 June 17	**Aitchison,** Lt. W., I. A. R. O.	....	1 July 17	Fyzabad.
12 Nov. 17	**Hudson,** 2nd-Lt. P. B., I.A.R.O.	....	26 Nov. 17	Bangalore (Br.).
23 Nov. 17	**Satchell,** 2nd-Lt. (*actg. Capt.*) W.	....	23 Nov. 17	Neemuch.
3 Feb. 94	**Fleming,** Riding Master & Major F., R. F. A.	....	11 Apl. 18	Mhow.
16 Oct. 01	**Bastyan,** Riding Master & Major B. J., R. H. A.	....	....	Sehore.
6 May 06	**Beale,** Riding Master & Capt. G. H. J., 6th Dns.	....	....	Neemuch.
18 June 10	**Lynch,** Riding Master & Capt. J., R.F.A.	....	....	Bangalore (Br.).
1 June 18	**Pugh,** 2nd-Lt. R. A. C., I.A.R.O.	....	....	Sehore.
1 July 18	**Hemsted,** Lt. F., I. A. R. O.	....	2 July 18	Sehore.
15 Nov. 17	**Dickinson,** Lt. W. G., I. A. R. O.	....	7 Apl. 18	Bellary.
15 June 18	**Hartley,** 2nd-Lt. E. R., I. A. R. O.	....	2 July 18	Bangalore (Br.).
10 Sep. 18	**Bald,** 2nd-Lt. A., I. A. R. O.	....	9 Oct. 18	Aurangabad.
....	**Battishill,** 2nd-Lt. J. H.	....	27 Nov. 18	Bellary.
1 July 15	**Daunt,** Lt. E. W. F. T.	....	29 Aug. 18	On depn. N. China.

Date of 1st Comn.	Names and Rank.	First appt. to Dept.	Date, present apptt.	Remarks.

Remount Purchasing Officers.

Date of 1st Comn.	Names and Rank.	First appt. to Dept.	Date, present apptt.	Remarks.
5 Feb. 87	**Roome,** Col. R. E.	. . . .	17 June 17	Lahore.
22 Aug. 88	**Henslowe,** Lt.-Col. F. J. D.	. . . .	19 Nov. 17	Meerut.
21 Sep. 89	**Young,** Lt.-Col. W. E.	. . . .	29 May 18	Ajmere.
25 Aug. 86	**Daunt,** Lt.-Col. W. D.	. . . .	16 Aug. 18	Lucknow.
4 May 98	**Rennick,** Maj. A. deC.	. . . .	18 June 18	Kalimpong.
22 Jan. 98	**Coaker,** Maj. V. A.	. . . .	12 June 17	Baluchistan and Sind.
13 Aug. 92	**Bennett,** Maj. J. T.	. . . .	2 Nov. 18	Lahore.
15 May 97	**Meynell,** Maj. H.	. . . .	20 June 18	Ahmedabad.
15 Mar. 17	**Thomson,** Lt. J. L., I. A. R. O.	. . . .	31 May 18	Meerut.
7 Sep. 16	**Dodd,** Lt. J. J., I. A. R. O.	. . . .	14 May 18	Lahore.
12 Aug. 18	**Elliot,** 2nd-Lt. (*temp. Capt.*) A. O., I.A.R.O.	. . . .	21 Aug. 18	Kalimpong.
3 Oct. 14	**Ward-Jackson,** Capt. C.	. . . .	10 Oct. 18	Meerut.
21 Sep. 18	**Cargill,** 2nd-Lt. J. R., I.A.R.O.	. . . .	12 Oct. 18	Lahore.
19 Sep. 18	**Tyler,** 2nd-Lt. E. W., I.A.R.O.	. . . .	1 Oct. 18	Mona.

Veterinary Officers.

Date of 1st Comn.	Names and Rank.	First appt. to Dept.	Date, present apptt.	Remarks.
3 Feb. 92	**Hunt,** Lt.-Col. F. C., A.V.C.	1 Feb. 97	1 Apl. 03	Remt. Depôt, Calcutta.
2 Nov. 01	**Gillett,** Lt.-Col. E. S., C.I.E., A.V.C.	6 Feb. 04	1 May 18	Dy. Dir. of Remts.
15 Apl. 02	**Roberts,** Maj. N. d'E., A.V.C.	20 Mar. 06	6 Sep. 15	Remt. Depôt, Hosur.
6 Feb. 04	**Russell,** Maj. E. C., F.R.C.V.S., A.V.C.	14 Mar. 07	13 Oct. 15	Remt. Depôt, Hapur and United Provinces Circle.
16 May 03	**Allen,** Maj. H., A.V.C.	14 Oct. 07	1 Mar. 16	Remt. Depôt, Sargodha, Chenab Canal Colony Circle and half Rawalpindi Circle.
1 Nov. '14	**Barry,** Capt. G.	. . . .	10 Jan. 17	Remt. Depôt, Saharanpur.
1 Aug. 16	**Howie,** Capt. G., A. V. C.	20 Apl. 18	20 Apl. 18	Remt. Depôts, Aurangabad & Deolali.
4 Aug. 17	**Arnold,** Lt. T. F., A. V. C.	15 Mar. 18	15 Mar. 18	Jhelum Canal Colony Circle.
. . . .	**Isherwood,** Capt. R., A. V. C.	30 Oct. 18	30 Oct. 18	Remt. Depôt, Mona & half Rawalpindi Circle.
25 July 16	**Power,** Capt. J. A., A.V.C.	. . . .	Nov. 18	Baluchistan Circle.
11 June 17	**Tay,** Capt. B., A.V.C.	. . . .	2 Oct. 18	Ahmednagar.

Indian Officers.

Date of 1st Comn.	Names and Rank.	First appt. to Dept.	Date, present apptt.	Remarks.
5 Oct. 03	**Gunpat Rao Umre,** Risaldar-Major	5 Oct. 03	4 Jan. 10	Rawal Pindi Circle.
20 Mar. 95	**Sultan Jan,** Risaldar-Major	1 Nov. 03	9 Jan. 09	Lahore Circle.
7 Nov. 03	**Muhammad Ali Khan,** Risaldar.	7 Nov. 03	7 Nov. 03	Baluchistan Circle.
27 Sep. 04	**Ewaz Khan,** Ressaidar	27 Sep. 04	27 Sep. 04	United Provs. Circle.
1 Nov. 06	**Murad Ali Khan,** Ressaidar	1 Nov. 06	1 Nov. 06	Chenab Canal Colony Circle, Lyallpur.
28 Feb. 85	**Muhammad Khan,** Jemadar	1 Oct. 13	1 Oct. 13	Jhelum Canal Colony Circle.

MEDICAL STORE DEPARTMENT.

Medical Store-keepers to Government.

Date of 1st Comn.	Names and Rank.	First appt. to Dept.	Date, present apptt.	Remarks.
27 July 92	**Swinton,** Lt.-Col. F. E., I.M.S.	18 May 01	13 Mar. 18	Madras.
28 July 94	**Hayward,** Lt.-Col. W. D., M.B., I.M.S.	27 Apl. 10	1 Dec. 18	Calcutta.
29 July 95	**Gibbs,** Lt.-Col. A. A., I.M.S. (*e*)	18 Nov. 11	18 Nov. 11	Lahore Cantt.
27 June 01	**Marr,** Maj. C. F., M.B., I.M.S.	14 Apl. 13	16 Aug. 16	Bombay.
1 Feb. 06	**Fit Gerald,** Maj. D.C.V., M.C., I.M.S.	23 Jan. 17	23 Jan. 17	Dy. M. S. K., Bombay (*temp.*).
28 Jan. 11	**Tate,** Capt. G., I.M.S.	14 May 17	14 May 17	Dy. M. S. K., Madras (*temp.*).

Date of 1st Comn.	Names and Rank.	First appt. to Dept.	Present appointment.	Remarks.

MILITARY FARMS DEPARTMENT.

Director.

5 Aug. 86	**Hallowes,** Col. F. W., C.I.E. (S. and T. Corps)	15 May 90	1 Apl. 12	Simla.

Assistant Directors, Grass Farms, and Assistant Officers.

Appointed under I. A. O., clause 82, of 1900, and I. A. O. 473 of 1905.

30 Jan. 84	**Finch,** Lt.-Col. C.	13 May 97	13 May 14	Poona.
10 Oct. 94	**Hawks,** Maj. (*temp. Lt.-Col.*) G. A.	20 Oct. 99	20 Oct. 16	Lucknow.
14 Sep. 87	**Obbard,** Lt.-Col. E. N.	4 Mar. 01	4 Mar. 13	Rawal Pindi.
19 May 00	*Matson, Maj. (temp. Lt.-Col.) J.*	*10 July 02*	*10 July 14*	*Ambala.*
14 Sep. 01	**Mellor,** Maj. (*temp. Lt.-Col.*) G. F.	21 Jan. 04	21 Jan. 16	Meerut.

(Formerly graded in S. and T. Corps or appointed subsequent to 1st February 1912.)

26 Apl. 93	**Elliott,** Maj. (*temp. Lt.-Col.*) H. H.	10 June 06	4 June 17	Mhow.
20 Feb. 97	**Marriott,** Maj. (*temp. Lt.-Col.*) A. S.	14 June 04	20 Feb. 14	Lahore.
5 July 99	**Kirkwood,** Maj. A. S., O.B.E.	3 Nov. 06 6 Sep. 12	5 July 16	Quetta.
28 July 00	**Airy,** Maj. (*temp. Lt.-Col.*) G.	21 Apl. 06	28 July 17	Asst. Dir Farms, A. H. Qrs.
8 Jan. 01	**Home,** Maj. (*temp. Lt.-Col.*) D. C.	12 July 07	8 Jan. 14	Bangalore.
4 July 03	**Wheeler,** Maj. E. S.	20 June 11 28 Mar. 13	5 Oct. 15	Rawalpindi.
27 Jan. 04	**Budden,** Capt. J. A.	28 May 10	27 Jan. 17	Saugor.
29 July 04	**Crawford,** Capt. (*temp. Maj.*) A.E., M.C.	20 July 12	29 July 17	With R. A.
5 Aug. 05	**Brett,** Capt. V. W.	27 Apl. 12	5 Aug. 18	Secunderabad.
9 Sep. 08	**Staples,** Capt. E. C., M.C.,	10 Jan. 17	10 Jan. 17	Lahore Cantt.
2 Jan. 15	**Watson,** Lt. O.	....	....	Mhow.
25 July 15	**Allen,** Lt. B.	....	14 Sep. 17	On probn. Poona.
1 June 16	**Malcolm,** Lt. J. F.	9 Oct. 17	9 Oct. 17	On probn. Ambala.
19 May 15	**Lester,** Lt. H. N., 2nd G. B., Essex. R.	....	....	On probn. Poona.

Assistant Directors, Dairy Farms.

19 May 00	**Matson,** Maj. (*temp. Lt.-Col.*) J.	10 July 02	10 July 14	Ambala.
...	**Smith,** W., Esq.	1 Feb. 12	1 Feb. 12	Poona.

RECRUITING OFFICERS.

Centre.	Name, Rank and Corps.	Date of apptt.	Appointment.
Director, Recruiting Section, A.H. Qrs.	**Harrison,** Brig.-Genl. A. H. P., 33rd Punj	(a) 12 Mar. 12	...
	Nevill, Lt.-Col. H. R., 7th U. P. Horse, I. D. F.	17 Nov. 17	A. A. G.
	Johnston, Maj. R., D.S.O., 4th Cavy.	7 Feb. 18	D. A. A. G.
	Dobbs, Lt. (*temp. Maj.*) W. E. J., I. A. R. O.	1 May 17	D. A. A. G.
	Kirkhope, Lt. (*temp. Lt.-Col.*) K. M., I. A. R. O.	26 July 17	A. A. G.
	Cairns, Capt. (*temp. Maj.*) J., 3rd E. Lan. R., employed R. E.	22 Apl. 18	D. A. A. G.
	Murphy, Lt. (*temp. Maj.*) A., I. A. R. O.	18 Aug. 18	D. A. A. G.
	Harvey-Kelly, Lt. (*temp. Capt.*) G., I. A. R. O.	21 Sep. 17	Attached.
	Doherty, Lt. J. M., I.W.T., Spl. List	5 Aug. 18	Attached.
	Thomas, Lt. (*temp. Capt.*) E. F., I. A. R. O.	3 June 18	...
	Porrett, 2nd-Lt. (*temp. Capt.*) C. R., I. A. R. O.	10 June 18	Attached.
	Matthews, 2nd-Lt. A., I. A. R. O.	11 Nov. 18	Attached.
Divl. Recruiting Officer, PESHAWAR.	**Falcon,** Lt.-Col. R. W., I.A. (retd.)	...	D. R. O.
	Simpson, Capt. C. N., I. A. R. O.	...	A. R. O.
	Falcon, Capt. R. R. B., I. A. R. O.	...	A. R. O.
	Holt, Lt. B. G.	1 Nov. 18	A. R. O.
Divl. Rectg. Officer, RAWALPINDI	**Batten,** Maj. H. H., Attd. 46th Punj.	5 Jan. 15	D. R. O.
	Malik Mumtaz Muhammad Khan, Capt., 4th Cav.	3 Apl. 17	A. R. O.
	Malik Sardar Khan	19 Sep. 18	A. R. O.
	Cunningham, 2nd-Lt. E.C.J., I. A. R. O.	7 June 18	A. R. O.
	Gagliardi, 2nd-Lt. F. B., I. A. R. O.	7 Aug. 18	A. R. O.
Divl. Rectg. Officer, LAHORE	**Barnes,** Maj. E. C., 19th Punj.	15 Aug. 17	D. R. O.
	Herdon, Lt. (*temp. Capt.*) S. P. N., I.A. R. O.	25 Feb. 16	A. R. O.
	Jarrad, Lt. (*temp. Capt.*) E. D. D., I. A. R. O.	20 Ma . 17	A. R. O.
Divl. Rectg. Officer, MULTAN	**Savi,** Lt.-Col. R. J. T., I.A.	21 Sep. 18	D. R. O.
Divl. Rectg. Officer, JULLUNDUR	**Wolley,** Maj. H. S. L., 56th Rif.	...	D. R. O.
	Branch, 2nd-Lt. (*temp. Capt.*) W. G., I. A. R. O.	3 Aug. 18	A. R. O.
	Garrett, Lt. (*temp. Capt.*) H. L. O., I. A. R. O.	1 Oct. 17	A. R. O.
	Stokes, 2nd-Lt. (*temp. Capt.*) S. E., I. A. R. O.	23 Oct. 17	A. R. O.
	Preston, 2nd-Lt. (*temp. Capt.*) E. B. C., I. A. R. O.	8 July 18	A. R. O.
	Derwent, 2nd-Lt. F. R.	...	A. R. O.
Divl. Rectg., Officer, DELHI	**Greig,** Maj. W. B., I.-A. (retd.)	28 Sep. 15	A. R. O.
	Warburton, 2nd-Lt. (*temp. Capt.*) R. P., I. A. R. O.	7 June 17	A. R. O.
	Puech, (*temp. Capt.*) A. G., O.B.E., V.D., I. A.	29 Jan. 17	A. R. O.
Divl. Rectg. Officer, BAREILLY	**Willcocks,** 2nd-Lt. (*temp. Maj.*) W. G., I. A. R. O.	17 June 18	D. R. O.
	Sandford, Capt. W. S. B., I. A. R. O.	13 Aug. 17	A. R. O.
	Slater, 2nd-Lt. (*temp. Capt.*) A. E., I.A.R.O.	30 July 18	A. R. O.
	Smyth, Lt. (*temp. Capt.*) G. H., I. A. R. O.	14 June 18	A. R. O.
Divl. Rectg. Officer, LUCKNOW	**Wills,** Capt. (*temp. Maj.*) E. V., 9th Infy.	20 Dec. 15	D. R. O.
	Clifford, Lt. (*temp. Capt.*) W. H. R., I.A. R. O.	23 Mar. 16	A. R. O.
	Storey, 2nd-Lt. F. G. F., I. A. R. O.	23 Apl. 18	A. R. O.
	O'Gorman, 2nd-Lt. C. J., I. A. R. O.	11 Oct. 18	A. R. O.
Divl. Rectg. Officer, FYZABAD	**Forbes,** 2nd-Lt. (*temp.-Maj.*) B. C., I. A. R. O.	20 June 18	D. R. O.
	Durant, 2nd-Lt. (*temp. Capt.*), E. J., I. A. R. O.	13 Aug. 17	A. R. O.
Divl. Rectg. Officer, AJMER	**Cole,** Maj. B. L., 13th Rajputs	1 Nov. 16	D. R. O.
	Hodgen, Lt. (*temp. Capt.*) G. W., 123rd Rif.	2 June 16	A. R. O.

(a) Tenure extended.

RECRUITING OFFICERS—*contd.*

Centre.	Name, Rank and Corps.	Date of apptt.	Appointment.
Divl. Rectg. Officer, JUBBULPORE	**Crosthwaite,** 2nd-Lt. (*temp. Lt.-Col.*), H. R., I. A. R. O.	23 Oct. 17	D. R. O.
	Lowrie, 2nd-Lt. (*temp. Capt.*) A. E., I. A. R. O.	17 Oct. 17	A. R. O.
	O'Grady, Lt. (*temp. Capt.*) W. J., I.A.R.O.	1 Dec. 17	A. R. O.
Divl. Rectg. Officer, KIRKEE	**Johnstone,** Maj. F. A. B., 101st Grs.	30 Nov. 18	D. R. O.
	Kerr, 2nd-Lt. (*temp. Capt.*) W. H., I. A. R. O.	29 Apl. 18	A. R. O.

RECRUITING OFFICERS—*concld.*

Centre.	Name, Rank and Corps.	Date of apptt.	Appointment.
	Casling, Lt. (*temp. Capt.*) D. M., I. A. R. O.	31 Mar. 17	A. R. O.
	Clabby, 2nd-Lt. (*temp. Capt.*) D. N. J., I. A. R. O.	31 Oct. 17	A. R. O.
Divl. Rectg. Officer, SECUNDERABAD.	**Haslehust**, Lt. (*temp. Maj.*) H. M., I. A. R. O.	4 Feb. 18	D. R. O.
	Wright, 2nd-Lt. (*tempt. Capt.*) C. R., S. & T. Corps	13 Aug. 18	A. R. O.
Divl. Rectg. Officer, BANGALORE.	**Graham**, Col. S. B.	15 June 17	D. R. O.
	King, Lieut. (*temp. Capt.*) E. O., I. A. R. O.	4 Aug. 16	A. R. O.
	Breithaupt, 2nd-Lt. (*temp. Capt.*) G. W., I. A. R. O.	7 Nov. 17	A. R. O.
	Doral, Hony. 2nd-Lt. R. B., Raja, I. L. Forces	21 May 18	A. R. O.
Divl. Rectg. Officer, CALCUTTA	**Oakes**, Lt.-Col. G. R., I. A.	3 Sep. 18	D. R. O.
	Gregory, Lt. (*temp. Capt.*) G. M., I. A. R. O.	16 Apl. 17	A. R. O.
	Taylor, Lt. S. G., 49th Bengalis	..	A. R. O.
Divl. Rectg. Officer, BANKIPORE	**Russell**, Lt. (*temp. Maj.*) R. E., I.A.R.O.	...	D. R. O.
	Hervey, 2nd-Lt. N. L., I. A. R. O.	23 Oct. 18	A. R. O.
Recruiting Officer, GORAKHPUR	**Stewart**, Major J. M. V., 10th G. R.	30 Mar. 18	D. R. O.
	Smyth-Osbourne, Lt. D. G., I. A. R. O.	5 Oct. 16	A. R. O.
	Flyter, 2nd-Lt. M. J., I. A. R. O.	...	A. R. O.
	Partridge, 2nd-Lt. E. C., I. A. R. O.	..	A. R. O.
Divl. Rectg. Officer, MEIKTILA (BURMA).	**Lowry**, Maj.	14 Oct. 18	D. R. O.
	Gray, 2nd-Lt. F. W.	24 Sep. 18	A. R. O.
	Leach, Lt. A. G.	17 Oct. 18	A. R. O.
	O'Brien, Lt. B. P. T.	15 Oct. 18	A. R. O.
	Maung Ba Tin, 2nd-Lt.	17 Sep. 18	A. R. O.
Asst. Rectg. Officer, LANSDOWNE	**Martin**, Lt. (*temp. Capt.*) A. S. W., I. A. R. O.	...	A. R. O.
Asst. Rectg. Officer, SIMLA	**Pierce**, Lt.-Col. F. G.	...	A. R. O.

Date of 1st commn.	Names and rank.	First appt. to Dept.	Present appointment.	Remarks.
	MILITARY ACCOUNTS DEPARTMENT IN INDIA.			
	(UNDER THE FINANCE DEPARTMENT OF THE GOVERNMENT OF INDIA.)			
2 Nov. 84	**Marlow**, Col. B. W., C.S.I., C.I.E.	16 Aug. 91	13 Apl. 08(*a*)	**Military Accountant General.**
7 Feb. 85	**Harrison**, Col. T. A., C.S.I.	18 Dec. 92	1 Apl. 06(*a*)	**Senior Contr., Mily. Supply and Marine Accts.**
29 Dec. 88	**Sheppard**, Bt.-Col. G. S., C.M.G. (*temp. Col.*).	20 June 95	10 Aug. 15(*b*)	**Mily. Dy. Auditor Genl., Southern Command.**
19 Sep. 88	**Baker**, Lt.-Col. C. N.	28 May 96	29 Feb. 16	**Mily. Dy. Auditor Genl., Northern Command.**
22 Aug. 88	**Bagshawe**, Lt.-Col. F. W., C.I.E. [L].	10 Feb. 97	3 Oct. 15	Senior Contr., Mily. Supply and Marine Accts., s.*p.t.*
21 Sep. 89	**Payne**, Lt.-Col. E. H.	24 Feb. 97	29 Feb. 16	Dy. Mily. Acct. Genl.
	Military Accountants, 1st class. (3)			
4 Dec. 89	**Chandler**, Lt.-Col. H. G. W.	16 Nov. 98	10 Aug. 15	Contr., Burma Dn.
4 Mar. 91	**Hudson**, Lt.-Col. J. H. [L]	30 May 99	3 Oct. 15	Junior Contr., Mily. Supply and Marine Accts., s.*p.t.*
12 Dec. 94	**Carr-Hall**, Maj. (*temp.* Lt.-Col.) R. E., C.I.E.	8 May 00	1 Apl. 17	Field Contr., Poona.
10 Sep. 75	*Ramsden, Col. H. F. S., I. A. (retd.).*	1 Feb. 18	...	*Temply. re-employed. F. C. M. A. Ind. Cavy. Corps, Force "A."*
	Military Accountants, 2nd class. (7)			
10 Oct. 91	**Anderson**, Lt.-Col. R. H. B.	23 Oct. 99	16 Sep. 10	Contr., 5th (Mhow) Dn.
7 Mar. 94	**Will**, Maj. D. A. E.	4 Feb. 01	30 Nov. 14	Contr., 8th (Lucknow) Dn., s.*p.t.*, 1st class.
28 Jan. 93	**Peacock**, Maj. (*temp. Lt.-Col.*) E. B.	1 Dec. 02	1 Oct. 15	Dy. Finl. Adviser, Mily. Finance, s. *p. t.*, 1st class.
5 Aug. 96	**Daldy**, Maj. A. W.	28 Jan. 03	1 Apl. 17	Contr., 2nd (Rawal Pindi) Dn., s. *p. t.*, 1st class.
17 June 91	**Jackson**, Lt.-Col. K. H.	17 Aug. 03	27 Sep. 17	Contr., 1st (Peshawar) Dn.
13 Aug. 92	**Bruce**, Lt.-Col. M. E. L.	1 Aug. 03	1 July 18	Contr., 9th (Secunderabad) Divn.
23 Nov. 92	**deLabilliere**, Lt.-Col. E. G. D.	20 Aug. 03	1 July 18	Asst. Audit Offr., 5th (Mhow) Dn.
	Military Accountants, 3rd class. (7)			
5 Aug. 96	**Sharp**, Maj. H. F.	9 May 03	1 Oct. 15	Contr., 4th (Quetta) Dn., s.*p.t.*, 2nd class.
1 Dec. 97	**Anderson**, Maj. E. S. J.	11 June 03	3 Oct. 15	Contr., 7th (Meerut) Dn., s.*p.t.*, 2nd class.
8 Jan. 98	**Szczepanski**, Maj. H. C.	20 July 03	1 Apl. 17	s.*p.t.*, 2nd class.
22 Jan. 98	**Ross**, Maj. G. W., D.S.O.	20 July 03	1 Apl. 17	Field Contr.'s Office, Poona.
27 July 98	**MacDonnell**, Bt.-Lt.-Col. H. N. F.	3 Sep. 03	1 Apl. 17	Contr., 6th (Poona) Dn.
7 May 98	**Hardinge**, Maj. H. R. von Donop	18 Dec. 03	1 Apl. 17	Fd. Contr.'s Office, Poona.
20 Feb. 97	**Richards**, Maj. W. V.	14 Apl. 05	30 Apl. 18	Field Contr.'s Office, Poona.
	Military Accountants, 4th class. (10)			
20 July 98	**Wylde**, Maj. L. F. G. S.	5 Apl. 05	5 Apl. 15	Controller, 3rd (Lahore) Dn., s.*p.t.*, 3rd class.
27 July 98	**Bird**, Maj. C. J. G.	10 Mar. 05	10 Mar. 15	Fd. Contr.'s Office, Poona, s.*p.t.*, 3rd class.
20 May 99	**Ellis**, Maj. S. G. V., D.S.O.	1 Feb. 05	1 Feb. 15	Fd. Contr.'s Office, Poona, s.*p.t.*, 3rd class.
15 Nov. 99	**Butler**, Maj. C. W.	28 Mar. 05	28 Mar. 15	Divl. Disbrsg. Offr., 4th (Quetta) Dn., s.*p.t.*, 3rd class.
19 May 00	**Murray**, Maj. (*temp. Lt.-Col.*) H.	5 Oct. 06	5 Oct. 16	s.*p.t.*, 3rd class.
12 Aug. 99	**Prince**, Maj. (*temp. Lt.-Col.*) R.	3 Jan. 07	3 Jan. 17	s.*p.t.*, 3rd class.
6 Dec. 99	**Murray**, Maj. A. G.	15 Apl. 07	15 Apl. 17	Divl. Disbrsg. Offr., 2nd (Rawal Pindi) Dn., s.*p.t.*, 3rd class.
[illegible] Sep. 99	**Morgan**, Maj. G. H.	28 Nov. 07	28 Nov. 17	Divl. Disbrsg. Offr., 3rd (Lahore) Dn., s.*p.t.*, 3rd class.
17 Feb. 00	**Whitchurch**, Maj. R. H. S.	15 Jan. 08	15 Jan. 18	Fd. Contr.'s Office, Poona, s.*p.t.*, 3rd class.
28 July 00	**Ashfield**, Maj. P.	14 Sep. 08	14 Sep. 18	Field Contr.'s Office, Poona, s.*p.t.*, 4th class.
	Assistant Military Accountants, 1st class. (3)			
20 Jan. 00	**Allen**, Maj. J. F.	4 Aug. 09	4 Aug. 14	Divl. Disbrsg. Office, 7th (Meerut) Dn., s. *p. t.*, 4th class.
18 Jan. 02	**Watson**, Maj. H. D. [L]	2 Aug. 09	2 Aug. 14	Field Contr.'s Office, Poona, s.*p.t.*, 4th class.
30 Apl. 02	**Graham**, Maj. J. S.	2 Aug. 09	2 Aug. 14	s. *p. t.*, 4th class.

(*a*) Tenure extended until further orders.
(*b*) Tenure extended to 9 Aug. 1920.

MILITARY ACCOUNTS DEPARTMENT IN INDIA—*contd.*

(UNDER THE FINANCE DEPARTMENT OF THE GOVERNMENT OF INDIA)—*contd.*

MILITARY ACCOUNTANTS, 2nd* CLASS. (23)

Names.	Date of appointment.	Remarks.
(i) Officers appointed in England.		
Playne, H. F., Esq., B.A.	1 Dec. 09	Field Accts. Offr., Bushire.
Tyndale, A. C., Esq., M.A.	1 Dec. 09	Divl. Disbrsg. Offr., 5th (Mhow) Dn.
Slater, A., Esq., B.A.	1 Dec. 09	Field Contr's Office, Poona.
O'Dowd, G. R., Esq., B.A.	10 Dec. 09	Asst. Audit Offr., 9th (Sec'bad) Dn.
Coates, W. J., Esq., B.A.	21 Jan. 10	Divl. Disbrsg. Offr., 8th (Lucknow) Dn.
Livingstone, H. L., Esq.	4 Mar. 10	
Coburn, M. R., Esq.	11 Mar. 10	Dy. Controller of War Accounts.
Harper, E., Esq., M.A., B.SC.	28 Oct. 10	Field Contr.'s Office, Poona.
Hastings, A. W., Esq.	2 Feb. 12	Asst. Audit Offr., 1st (Peshawar) Dn.
Mason, N., Esq., B.A.	31 Jan. 13	(*Lv., ex Ind.*)
Parlby, J., Esq., B.SC.	31 Oct. 13	On depn. For. & Pol. Dept.
Sanders, B. G. T., Esq.	31 Oct. 13	Divl. Disbrsg. Offr., 6th (Poona) Dn.
Stott, R., Esq., M.A.	28 May 15	Depy. Contr. of War Accounts.
Hinchliffe, *Hony.* 2nd-Lt. A. H. S.	11 Oct. 16	Office of M. D. A. G., S. Command.
(ii) Officers specially appointed in India.		
Leys, W. G., Esq.	16 Nov. 12	Asst. Audit Offr., 7th (Meerut) Dn.
Rolfe, R. H., Esqr.	24 July 16	Finl. Officer, A.G.'s Bch., A. Hdqrs.
Cookman, G. H., Esq., I.S.O.	26 Aug. 15 (*e*)	M. A. G.'s Office.
Dover, R. W., Esq.	26 Aug. 15	Depy. Contr. of War Accounts.
Cooke, T. C., Esq.	1 July 16 (*c*)	Asst. Audit Offr., Mily. Supply Accts.
Wells, W. G., Esq.	1 July 16	Field Contr.'s Office, Poona.
O'Brian, H. E. W., Esq.	1 July 16	Field Contr.'s Office, Poona.
Crawford, J. C., Esq.	21 Aug. 16	Asst. Audit Offr., 8th (Lucknow) Divn.
Shaw, 2nd-Lt. (*Hony. Lt.*) C. H., T. F. Res.	4 Dec. 16	Depy Contr. of War Accts.
Officer appointed by the Secy. of State under special terms.		
Matley, C. A., D.Sc., F. G. S.	28 Mar. 13	Dy. Controller of War Accounts.

GAZETTED APPOINTMENTS.

Names.	Date of appointment.	Remarks.
Assistant Military Accountant General. (1)		
* * *		
Depy. Examrs., 1st grade (18), including 12 tempy. and supernumerary.		
Smith, T. G., Esq.	1 Oct. 08	Office of Contr., Mily. Supply Accts.
Grinnol, H. R., Esq.	14 Dec. 13 (*a*)	Office of Contr., Mily. Supply Accts.
O'Hara, A. E., Esq.	26 Aug. 15	Divl. Audit Office, 8th (Lucknow) Divn.
Rice, T. W., Esq.	1 July 16	
deSouza, E., Esq.	1 July 16	Divl. Disbrsg. Office, 9th (Sec'bad) Dn.
Gracias, H. D., Esq.	1 July 16	Asst. Contr. of War Accts.
Vengu Aiyar, P. R., *Rao Bahadur*	1 July 16	(Temp. & Supy.) Divl. Audit Office, 6th (Poona) Dn.
Clarke, F. I. L., Esq.	1 July 16	(Temp. & Supy.) Assistant Controller of War Accounts.
Smith, H. M., Esq.	1 June 17 (*d*)	(Temp. & Supy.) Divl. Audit Office, 3rd (Lahore) Dn.
Earle, E. J., Esq.	15 June 17	(Temp. & Supy.) Divl. Audit Office, 2nd (Rawalpindi) Dn.
Cumber, H. S., Esq.	30 July 17	(Temp. & Supy.) Finl. Officer, Q. M. G.'s Bch., A. H. Qrs.
King, J. R., Esq.	1 Nov. 17 (*h*)	(Temp. & Supy.) Divl. Audit Office, 8th (Lucknow) Dn.
Lewis, J. W., Esq.	1 Nov. 17	(Temp. & Supy.) Treasure Chest Office, Bombay.
Sterling, E., Esq.	1 Nov. 17	(Temp. & Supy.) Divl. Disbrsg. Office, 2nd (Rawalpindi) Dn.
Gonsalves, E. F., Esq.	1 Nov. 17	(Temp. & Supy.) M. A. G.'s Office.
Lichmore, B. C., *Khan Saheb*	1 Nov. 17 (*b*)	(Temp. & Supy.) Divl. Disbrsg. Office, 5th (Mhow) Dn.
Fairhurst, A., Esq.	1 Nov. 17	(Temp. & Supy.)
Pereyra, A., Esq., B.A.	26 Aug. 18	(Temp. & Supy.) Field Acct. Offr., Aden.
Deputy Examiners, 2nd grade (20, including 1 tempy. and supy.).		
Shepherd, W. S. W., Esq.	1 Oct. 13	Divl. Disbrsg. Office, 4th (Quetta) Dn.
Gonsalves, C., Esq.	26 Aug. 15	*s.p.t.*, 1st grade.
Kalyanasundram, V. R., *Rao Sahib*	1 July 16	Office of Contr., Mily. Supply Accts.
Windsor, H. A., Esq.	1 July 16	M. A. G.'s Office, Simla.
Rozier, C., Esq.	1 July 16	Divl. Disbrsg. Office, 5th Dn.
Kali Charan Dutt, *Rai Sahib*, B.A.	10 July 16	M. A. G.'s Office, Simla.

* Under the reorganisation scheme of 1910.
(*a*) Service extended 13 Sep. 1919.
(*b*) " " to 10th Nov. 1919.
(*c*) " " " 16th Dec. 1919.
(*d*) " " " 2nd Jan. 1920.
(*e*) " " " 23rd Oct. 1919.

MILITARY ACCOUNTS DEPARTMENT IN INDIA—*contd.*
(UNDER THE FINANCE DEPARTMENT OF THE GOVERNMENT OF INDIA)—*contd.*

GAZETTED APPOINTMENTS—*contd.*

Names.	Date of appointment.	Remarks.
Court, A. E., Esq.	15 June 17	Field Contr.'s Office, Poona.
Paulie, E., Esq., B.A.	1 July 16	(Temp. & Supy.) Field Contr.'s Office, Poona.
Browne, J., Esq.	10 July 16	(Temp. & Supy.) Divl. Disbrsg. Office, 9th Divn.
Satyendra Nath Aditya, Rai Sahib	10 July 16	(Temp. & Supy.) Office of Contr., Mily. Supply, Accts.
Glancey, J. W., Esq.	1 June 17	(Temp. & Supy.) Divl. Audit Office, 4th (Quetta) Dn.
deRozario, N., Esq., B.A.	15 June 17	(Temp. & Supy.) *Offg. Chief Acct., Bombay Dockyard.*
Allen, W. J., Esq.	30 July 17	(*Temp. & Supy.*)
Mordecai Joseph, Khan Sahib	13 Aug. 14	(*Temp.*) Field Contr.'s Office, Poona.
Browne, W. S., Esq.	8 *Feb.* 15	(*Temp.*) Field Contr.'s Office, Poona.
Anantanarayana Aiyar, C. P., Esq.	3 May 15	(*Temp.*) Asstt. Contr. of War Accts.
Duckworth, G. P., Esq.	1 July 16	(Temp.) Divl. Disbrsg. Office, 6th (Poona) Dn.
Colkers, A. B., Esq.	10 July 16	(Temp.) Office of Contr. of Mily. Supply Accts.
Grimson, G. R., Esq.	24 Aug. 16	(Temp.) Field Contr.'s Office, Poona.
Greenaway, W. J., Esq.	4 Sep. 16	(Temp.) Field Contr.'s Office, Poona.
Shaikh Ahmed Sahib, Esq.	18 Sep. 16	(Temp.) Divl. Audit Office, 6th (Poona) Dn.
Johnston, W. H., Esq.	20 Sep. 16	(Temp.) Office of Contr., 1st (Peshawar) Dn.
Leatham, H. H., Esq.	14 Oct. 16	(Temp.) Divl. Audit Office, 5th (Mhow) Dn.
Hellier, J., Esq.	30 Oct. 16	(Temp.)
Framji, A., Khan Sahib	29 Nov. 16(*f*)	(Temp.) Office of Contr., 4th (Quetta) Dn.
Hari Das Gupta, Rai Sahib	12 Mar. 17	(Temp.) temp. attd. to Kidderpore Dockyard.
Davies, W., Esq.	24 May 17	(Temp.) Divl. Disbrsg. Office, 6th (Poona) Divn.
Hardingham, H., Esq.	11 Nov. 17	(Temp.) Attd. Bombay Dkyd.
Richards, W. E., Esq.	9 July 17	(Temp.) Field Contr.'s. Office, Poona.
Sorabjee Pallanjee, B.A., *Khan Sahib*	1 Aug. 17	(Temp.) Divl. Disbrsg. Office, 6th (Poona) Dn.
Raghavulu Chettiar, S., Rao Sahib	15 Aug. 17(*g*)	(Temp.) Office of Contr., Mily. Supply Accts.
Whitehead, J., Esq.	21 Aug. 17	(Temp.) Divl. Disbrsg. Office, 7th (Meerut) Dn.
Rodrigues, H. A., Esq.	12 Jan. 18	(Temp.) Attd Bombay Dockyard.
Royal, A., Esq.	1 Apl. 18	(Temp.) Office C. M. A., 9th (Sec'bd.) Dn.
Ramjee Dass, Esq.	25 Apl. 18	(Temp.) D. D. Office, 3rd (Lahore) Dn.
Jeste, D. G., Esq.	8 May 18	(*Temp.*), Office of Contr., 4th (Quetta) Dn.
Mahmood Mian Sahib, S., Khan Sahib	19 June 18	(Temp.) Office of Contr., Mily. Supply Accts.
Brisley, H. F., Esq.	21 Oct. 18	(Temp.) Field Contr's. Office, Poona.
Ramaswamy Iyengar, R., Esq., B.A.	1 Nov. 18	(Temp.) Office of Chief Auditor Non-Mily. Pay and Pension Accts.

(*f*) Service extended to 10th Sep. 1919.
(*g*) Ditto 20th Aug. 1919.
(*h*) Ditto 18th May 1919.
(*i*) Ditto 14th Jan. 1920.

British Service Officers temporarily employed with the Military Accounts Department.

Names.	Office in which employed.
Trevenen, Maj. W. B., 1-6th Devon. R.	Field C. M. A., Poona.
Burridge, Capt. W. T., 1-5th Som. L. I.	Mily. Dy. Auditor Gl., N Command.
Campbell, Capt. A., 1st G. B. Som. L. I.	Field C. M. A., Poona.
Borissow, Capt. R. W., 2nd G. B., Essex R.	C. M. A., 3rd (Lahore) Dn.
Hibbert, Capt. W. J. C., 6th R. Lanc. R.	Field C. M. A., Poona.
Rimmer, Capt. H. A., 23rd Bn., Rif. Brig.	C. M. A., 2nd (Rawalpindi) Dn.
Gerard, Capt. M. F. 2nd G. B., Essex R.	Field C. M. A., Poona.
Bendall, Capt. J., 3rd Bn., High L. I.	Financial Advisor, Baghdad.
Robertson, Capt. R. H., 1st G. B., Gord. Highrs.	C. M. A., 4th (Quetta) Dn.
Cockburn, Capt. J. G. H., 2-6th R. Suss. R.	C. M. A., 6th (Poona) Dn.
Wade-Evans, Capt. H., 1-1st Brecknock Bn.	Treasure Chest, Karachi.
Wilson, Capt. A. H., 1st G. B., R. Ir. Fus.	C. M. A., 1st (Peshawar) Dn.
Wicks, Capt. W. M., 1-10th Middx. R.	Field C. M. A., Poona.
Shortt, Capt. G. A. No. 6 Res. Bn.	C. M. A., 5th (Mhow) Divn.
Hobkirk, Capt. R., 1st Manch. R.	Field C. M. S. A., Basra.
Hargreaves, Capt. R., Bord. R.	Paymaster, Amara.
Richards, Capt. A. P., Durh. L. I.	C. M. A., 9th (Sec'bad.) Dn.
Young, Capt. J. H., 1-5th R. W. Surr. R.	Field C. M. S. A., Basra.
Thomas, Lt. (*temp. Capt.*) A. E., 1-6th Devon. R.	Paymaster, Baghdad.
Thornton, Lt. (*temp. Capt.*) A. M., Manch. R.	Base Paymaster, Basra.
Symonds, Lt. (*temp. Capt.*) W. F., 2-4th Dorset. R.	Field C. M. S. A., Basra.
Triphook, Lt. (*temp. Capt.*) O. L., R. F. A.	Field C. M. S. A., Basra.
Auger, Lt. A. E., 24th Bn., Rif. Brig.	C. M. A., 1st (Peshawar) Dn.
Turrell, Lt. J. W., 1st G. B., Manch. R.	Field C. M. A., Poona.
McLellan, Lieut. J. S., 24th Bn., Rif. Brig.	C. M. A., 1st (Peshawar) Dn.

MILITARY ACCOUNTS DEPARTMENT IN INDIA—*contd.*
(UNDER THE FINANCE DEPARTMENT OF THE GOVERNMENT OF INDIA)—*contd.*

GAZETTED APPOINTMENTS—*contd.*

British Service Officers temporarily employed with the Military Accounts Department—*contd.*

Names.	Office in which employed.
Moore, Lt. H. E., 2nd G. B., Essex R.	C. M. A., 7th (Meerut) Dn.
Martin, Lt. E. L., 1-4th Bn., The Buffs	C. M. A., 7th (Meerut) Dn.
Wright, Lt. J. H., 4th Bn., Lond. R.	Field C. M. A., Poona.
Beaumont, Lt. H. S., 6th L. N. Lancs. R.	Ditto.
Radford, Lt. E. J., 23rd Bn., Rif. Brig.	C. M. A., 6th (Poona) Dn.
Seal, Lt. J. R., 1st (G. B.) E. Yorks R.	C. M. A., 2nd (Rawalpindi) Dn.
Kite, Lt. P., 2-7th Hamps. R.	Field C. M. A., Poona.
Platt, Lt. W. P., 2nd G. B., North'd. Fusrs.	C. M. A., 4th (Quetta) Dn.
White, Lt. R. S., 2-4th Som. L. I.	C. M. A., 4th (Quetta) Dn.
Johnson, Lt. V. A., 18th Bn. Rif. Brig.	C. M. S. A., Calcutta.
Tremaine, Lt. J., 8th Welsh Prs.	Paymaster, Hillan.
Page, Lt. L. A., 5th Wilts. R.	Paymaster, Dewani.
James, Lt. E. J. T. 6th S. Lan. R.	Paymaster, Bagubah.
Carpenter, Lt., D. B. B., R. G. A.	Field C. M A., Poona.
Knowles, Lt. C. G., 9th R. War. R.	Field C. M. S. A., Basra.
Robotham, Lt. W. D. N., 1st W. Rid. R.	C. M. A., 2nd (Rawal Pindi) Dn.
Fisk, Lt. C. L., 6th R. Lanc. R.	Field C. M. A., Poona.
Hurburgh, Lt. L. H., 1-25th Bn., Lond. R.	C. M. A., 2nd (Rawal Pindi) Dn.
Dickie, Lt. D. H., 1-4th Bord. R.	Field C. M. A., Poona.
Sandison, Lt. J., 5th R. Scots.	C. M. A., 3rd (Lahore) Dn.
Barry-Briant, Lt. G. S. B., R. F. A.	Field C. M. A., Poona.
Blount, Lt. C., 1-4th Dorset R.	C. M. A., Burma Dn.
Drake, Lt. F. H., 2-4th D. of Corn. L. I.	C. M. A. 3rd (Lahore) Dn.
O'Brien, Lt. J. G., 6th S. Lanc. R.	Field C. M. A., Poona.
Drayton, Lt. P., York. Dgns. (T. F.)	C. M. A., Burma Dn.
Nuttall, Lt. B. A., Yorks R.	C. M. A. 9th (Sec'bad) Dn.
Sheppard, Lt. S. V., 1st G. B. E., Yorks R.	C. M. A., 3rd (Lahore) Dn.
Dyer, Lt. C. H. F., 1st G. B., Linc. R.	C. M. A., 8th (Lucknow) Dn.
Fraser, 2nd-Lt. J., 1st (G. B.) Oxf. & Bucks. L. I.	Field C. M. S. A., Basra.
Wells, 2nd-Lt. C. Mackenzie, 1-4th The Buffs	C. M. A., 3rd (Lahore) Dn.
Beckworth, 2nd-Lt. G., 4th Dorset R.	Field C. M. A., Poona.
Facon, 2nd-Lt. W. S., 9th Midd'x R.	Field C. M. A., Poona.
Barchard, 2nd-Lt. L. B., 1st G. B., E. York. R.	C. M. A., 1st (Peshawar) Dn.
Scoble, 2nd-Lt. A., 2nd R. Highrs.	Field F. C. O., Egypt.
Sanders, 2nd-Lt. F. T., 9th Midd'x R.	C. M. A., 6th (Poona) Dn.
Frith, 2nd-Lt. N. R. C., Notts. and Derby. R.	C. M. A., 6th (Poona) Dn.
Taylor, 2nd-Lt. E. J., K. R. Rif. C.	Field C. M. A., Poona.
Combe, 2nd-Lt. H. W., 5th R. W. Kent. R.	Field C. M. A., Poona.
Keywood, 2nd-Lt., H. A., 1st Bn., S. Lan. R.	C. M. A., 4th (Quetta) Dn.
McNaught, 2nd-Lt. D., 2-10th G. Rif.	F. C. M. A., Cairo.
Naylor, 2nd-Lt. T., 1st G. B. R. Ir. Rif.	C. M. A., 8th Lucknow) Dn.
Floyd, 2nd-Lt. H., 3rd Bn. The Welsh R.	F. C. M. A., Poona.
Campbell-Smith, 2nd-Lt. 1st G. B., R. Ir. Rif.	C. M. A., 3rd (Lahore) Dn.
Bird, 2nd-Lt. T. E. H., 1-4th R. W. Kent R.	C. M. A., 4th (Quetta) Dn.
Indian Army Reserve Officers temporarily employed with the Military Accounts Department.	
Swainson, Lt.-Col. C. A.	Finl. Adviser to the G. O. C., S. Persia Rif.
O'Brien, Capt. J.	Field C. M. S. A., Basra.
Willison, Capt. W. L. P.	Field C. M. A., Poona.
Hewetson, Lt. (*temp. Capt.*) F. E. A.	Field C. M. A., Poona.
Hepworth, Lt. (*temp. Capt.*) N.	Field C. M. A., Poona.
Rountree, Lt. (*temp. Capt.*) A. F.	Field C. M. S. A., Basra.
Wilts, Lt. G. S. F.	F. C. M. A., Poona.
Green, Lt. (Actg. Capt.) E. J.	F. C. M. S. A., Basra.
Anger, Lt. E. A.	C. M. A., 1st (Peshawar) Dn.
Knott, Lieut. W. J.	C. M. A., 4th (Quetta) Dn.
Martin, Lt. J.	C. M. A., Burma Dn.
Husband, Lt. R.	Field C. M. A., Poona.
Mare, Lt. E. J	C. M. A., 9th (S'bad) Dn.
Dawson, Lt. (*temp. Capt.*) T. F.	Field C. M. S. A., Basra.
Eden, Lt. B.	C. M. A., 5th (Mhow) Dn.
Neill, Lt. J. F. R.	Field C. M. A., Poona.
Askew, Lt. T. E.	C. M. A., 5th (Mhow) Dn.
Hamer, Lt. R. G.	Field C. M. A., Poona.
Griffie-Williams, Lt. (*temp. Capt.*) J. H.	Field C. M. S. A., Basra.
Rushforth, Lt. F. V.	Controller of War Accounts.
Finlayson, Lt. C. S. M.	Field C. M. A., Egypt.
Luck. Lt. J.	C. M. A., 2nd (Rawalpindi) Dn.
Young, Lt. E. L.	C. M. A., 2nd (Rawalpindi) Dn.
Cox, Lt. A. J.	Field C. M. A., Poona.
Ellis, Lt. T. S.	C. M. A., 7th (Meerut) Dn.
McCall, Lt. B.	F. C. M. A., Cairo.
Jones, Lt. E. O.	F. C. M. A., Cairo.
Wadeson, Lt. F. F.	F. C. M. A., Poona.
Edwards, Lt. F. B.	C. M. A., 2nd (Rawalpindi) Dn.
Bateman, Lt. C. W.	F. C. M. A., Poona.
Saunders, 2nd-Lieut. E. J.	C. M. A., 6th (Poona) Dn.
Allan, 2nd-Lt. J. M.	C. M. A., 6th (Poona) Dn.
Powell, Lt. (*temp. Capt.*) W.	Field C. M. S. A., Basra.
Blick, 2nd-Lt. L. H.	C. M. A., 9th (S'bad) Dn.

MILITARY ACCOUNTS DEPARTMENT IN INDIA—*concld.*
(UNDER THE FINANCE DEPARTMENT OF THE GOVERNMENT OF INDIA)—*concld.*
Civil Department Officers who are temporarily employed in the Military Accounts Department.

Names.	Office in which employed.
Conibear, 2nd-Lt. W.	F. C. M. A., Poona.
Wickham, Lt. W. C.	Field C. M. A., Poona.
Capstick, 2nd-Lt. F. H.	C. M. A., 2nd (Rawalpindi) Dn.
Morton, 2nd-Lt. W.	Chief Auditor, Non-Military Pay and Pension Accts.
Walker, 2nd-Lt. L. H. C.	Chief Auditor, Non-Military Pay and Pension Accts.
Green, 2nd-Lt. G. H. B.	C. M. A., 7th (Meerut) Dn.
Storrow, 2nd-Lt. S. L.	Field T. C., Egypt.
Ord, 2nd-Lt B. K.	Do. Do.
Bernard, 2nd-Lt. P.	Chief Auditor, Non-Military, Pay and Pension Accts.
Turner. 2nd-Lt. J. M.	C. M. A., 6th (Poona) Dn.
Johnson, 2nd-Lt. W. J.	F. C. M. A., Cairo.
Shepherd, 2nd-Lt. F. A.	F. C. M. A., Cairo.
Daniels, 2nd-Lt. A.	C. M. A., 9th (S'bad) Dn.
Spence, 2nd-Lt. W. J. F.	F. C. M. A., Poona.
Higham, 2nd-Lt. L. A.	F. C. M. A., Poona.
Wilson, 2nd-Lt. J.	F. C. M. A., Poona.
Cronan, 2nd-Lt. C. P.	Chief Auditor, Non.-Mily. Pay and Pension Accts.
Officers deputed from the Indian Finance Department.	
Waterfield, R., Esq., B.A., (Oxon.)	Financial Adviser to the G. O. C. South Persia Rif., Shiraz.
Barr, A. G., Esq.	Senior Audit Officer, South Persia Rif., Shiraz.
Mitra, B. N., Esq., C.I.E., O.B.E., M.A.	Controller of War Accounts.
Hartley, J. M., Esq.	Financial Adviser to D. R. I. M.
Deane, L. S., Esq.	C. M. A. 3rd (Lahore) Dn.
James, C. H., Esq.	C. M. A., 3rd (Lahore) Dn.
Gupta, A. C., Esq, M.A.	Controller of War Accounts.
Currie, A. J., Esq.	On Field Service.
Sundarachari, R., Esq., B. A., F. R. E. S., F. S. S.	Chief Auditor, Non-Military Pay and Pension Accounts.
Ewing, S R., Esq.	Senior C. M. S. A., Calcutta.
Bayley, A. S. B., Esq.	C. M. A., 3rd (Lahore) Dn.
Sarma, T. R. V., Esq., B. A.	C M. A., 6th (Poona) Dn.
Sadashiva Rau, K. R., Esq., B.A. (Cantab.)	C. M. A., 7th (Meerut) Dn.
Harris, B. A., Esq.	C. M. A., 6th (Poona) Dn.
Young, J. W., Esq.	Controller of War Accounts.
Lacey, R. H. D., Esq.	Field C. M. A , Poona.
Gupta, S. D., M.A., F.R.E.S., F.S.S.	Controller of War Accounts.
Watts, M. E., Esq., B. A.	C. M. A., 1st (Peshawar) Dn.
Reilly, C., Esq.	C. M. A., Burma Dn.
Sheikh Taj Mohammad, M.A., M.Sc.	C. M. A., 2nd (Rawal Pindi) Dn.
Iyer, T. S. S., Esq., B.A., B. E.	C. M. A., 6th (Poona) Dn.
Chanter, E. D., Esq.	C. M. A., 6th (Poona) Dn.
Officers deputed by the Provincial Governments.	
Marshall, W. S., Esq.	C. M. A., 5th (Mhow) Dn.
Gill, H. W., Esq.	C. M. A., 8th (Lucknow) Dn.
D'Abreo, F., Esq.	C. M. A., 6th (Poona) Dn.
Grant, Allan, Esq.	Field C. M. A., Poona.
Banerjie, Raj Narain, Esq.	C. M. A., 8th (Lucknow) Dn.
Stephens, E. J., Esq.	C. M. A., 4th (Quetta) Dn.
Retired Officers who are temporarily re-employed.	
Mole, P. C., Esq.	Chief Auditor, Non-Military Pay and Pension Accounts.
Wolfe, G. C., Esq.	Senior C. M. S. A., Calcutta.
O'Brien, J. A., Esq.	Field C. M. A., Poona.
Cross, H. L., Esq.	Field C. M. A., Poona.
Gopalan Nayar, C., *Diwan Bahadur*	C. M. A., 9th (S'bad) Dn.
Bhattacharjie, Debendra Nath, *Rai Sahib*	C. M. A., 6th (Poona) Dn.
Chatterjee, Bhutnath, *Rai Sahib*	C. M. A., 8th Dn.
Asstt. Accounts Officers, Class II, Indian Finance Department and Supdts. and Acctts. of that Deptt. deputed as gazetted officers.	
George, A. B., Esq.	Chief Auditor, Non-Military Pay and Pension Accounts.
Ricketts, C. H., Esq.	Chief Auditor, Non-Military Pay and Pension [Accounts.
Singh, Hukum, Esq.	C. M. A., 8th (Lucknow) Dn.
Balkishen, Esq.	C. M. A., 3rd (Lahore) Dn.
Davids, W. J., Esq., B.A	C. M. A., 6th (Poona) Dn.
Young, W. A., Esq.	C. M. A., 7th (Meerut) Dn.
Clark, A. G. N., Esq.	Controller of War Accounts.
Roy, J. N., Esq., M.A.	Senior C. M. S. A., Calcutta.
Iyer, A. Rangaswami, Esq.	Contr. of War Accounts.
Lamb, A. R., Esq.	Chief Auditor, Non-Military Pay and Pension Accounts.
Singha, N. L., Esq., B.A.	Controller of War Accounts.
Iyengar, C. N. Ramaswami, Esq., M. A.	Controller of War Accts.

Names.	Office in which employed.
Officers who did not previously hold any appointment under Government.	
Poulter, H. W., Esq.	Accounts officer to the Dir. of S. & T. Corps, Baghdad.
Medland, J. W., Esq.	Accounts officer to the Dir. of I. W. T., Basra.
Other officers who are temporarily attached.	
Harvey, E. E. Esq., I.S.O., Chief Acctt., Dir. Ord. Factories.	Controller of War Accounts.

Names.	Date of apptt	Remarks.
GAZETTED APPOINTMENTS.		
Bayne, A., Esq.	5 Jan. 14	Textile Expert, Army Clothing Department.
Biswas, *Rai Sahib* S. C.	13 Dec. 14	Supt., Army Dept.
Campbell, L. S., Esq.	11 Sep. 08	Persl. Asst. to Chief of the Gen. Staff. *Lv.*, ex I., P.A., 19th June 1916.
Daniel, E. A., Esq.	20 Mar. 18	Supt., Army Dept.
Debenham, W. C., Esq.	1 Dec. 09	Offr. i-c Medal Distbn., Army Dept.
Haldipur, S. V., Esq., B. A., L L. B.	1 Feb. 18	Legal Asst. to the J. A. G., in India.
Hypher, P. P., Esq.	1 Jan. 14	Supt., Army Dept.
Knowles, C. J., Esq., M.B.E.	28 May 15	Personal Asst. to Adjt. General in India.
Kunning, A. B., Esq.	1 May 11	Supt., Army Dept.
Leslie, J. C. R., Esq.	1 Nov. 15	Supt., Army Dept.
Lloyd. *Hony.* Capt. T. W., S. & T. Corps	18 Dec. 15	Personal Asst. to the Quarter Master General.
Villiers, Commy. & *Hony.* Maj. A.	17 Feb. 17	Persl. Asst. to Dir.-Genl. of Ordnance.

Note.—The names of Gazetted Civil Officers of the Ordnance Department, etc., are not shown on this list at they are already borne on their respective Departmental Civil Lists.

OFFICERS EMPLOYED WITH IMPERIAL SERVICE TROOPS.

Names, Rank and Corps.	Date of appointment.	Date of completion of tenure.	Remarks.
Inspector General.			
Rose, Brig.-Genl. J. L., C.I.E., 1st G. R.	20 Oct. 05	. . . . .	Simla.
Inspecting Officers.			
Kerr, Maj. A. N., M.C., 38th Dogras	13 Sep. 12	. . . .	Satwari Cant. *s.p.t.*
Dickinson, Maj. R. C., 28th L. C.	7 Dec. 14	. . . .	Jaipur. Offg. I. O., Rajputana Cav. and Transport.
Grimshaw, Maj. R. W. W., 34th Horse.	3 Nov. 18	. . . .	
Dening, Maj. H., 25th Cav.	11 Oct. 14	. . . .	Bahawalpur, 10 Camel Corps. Offg. D. A. A. G. M., I. S. troops, in addition to duty.
Barnard, Lt.-Col. A. E., I.A.	8 Sep. 16	. . . .	Ambala. Punjab I. S. Cav. *s.p.t.*
Cotter, Maj. V. R., R.A.	24 Nov. 16	. . . .	Satwari. I. O., Kashmir I. S. Arty. *s.p.t.*
MacBrayne, Maj. R. J., M.C., 15th Sikhs.	28 Feb. 14	. . . .	Ambala. I. O. Punjab I. S. Infy. *s.p.t.*
Logan, Capt. A. F., M.C., 21st (F. F.)	28 Jan. 18	. . . .	Bangalore. L. O. Hyd. and Mysore Cy. *s.p.t.*
Chaldecott, Maj. W. H., R. E.	27 Sep. 17	. . . .	Roorkee. I. O., I. S. Sappers. *s.p.t.*
Ouseley, Capt. J. W. G., I. A. R. O.	15 May 15	. . . .	Gwalior Residency. I. O., Central India I. S. Cav. Offg. I. O. Kathiawar States I. S. Cav. in addition to duties. *s.p.t.*
Hughes, Maj. G. R., 43rd R.	1 Oct. 18	. . . .	Gwalior Resy. I. O., Rajputana I. S. Infy. *s.p.t.* in addition to duties as Offg. Inspr. of Signalling, I. S. Troops.
Dy. Asst. Adjt. Genl. for Musketry.			
Dening, Maj. H., 25th Cav.	1 June 16	. . . .	Meerut. Offg. in addition to duties as I. O., I. S. Camel Corps.
Inspector of Signalling.			
Hughes, Maj. G. R., 43rd R.	1 Oct. 18	. . . .	Meerut. Offg. in addition to duties as I. O., Rajputana I. S. Infy.
Assistant Inspecting Officers.			
Bickford, Capt. M. H., 38th Dogras	15 Nov. 16	. . . .	Kashmir I. S. Infy., Gilgit. *s.p.t.*
Liston, Capt. R. C. C., 102nd Grs.	27 Nov. 17	. . . .	Kashmir I. S. Infy., Gilgit. *s.p.t.*
Voelcker, Capt. G. G., 1-3rd G. R.	11 May 18	. . . .	Kashmir I. S. Infy., Satwari Cant. *s.p.t.*
Steane, Lt. S. W., R. F. A.	9 Nov. 16	. . . .	Kashmir I. S. Arty., Gilgit. *s.p.t.*
Norman, Capt. H. J., 102 Grs.	13 Apl. 18	. . . .	Punjab Infy., Amballa. *s.p.t.*
Hammick, Capt. S. T. V. F., 13th Rajputs	1 Apl. 18	. . . .	Bharatpur City. Rajputana Infy. *s.p.t.*

OFFICERS EMPLOYED WITH FRONTIER CORPS.

Name, Rank and Corps.	Date of appointment to Militia.	Date of completion of tenure.	Remarks.
Inspecting Officer, Frontier Corps.			
Kemball, Lt.-Col. J. S., 29th Punj.	16 May 14	15 May 17	Lv. m.c. 17 yr. from 16 Oct. 18.
Tancock, Maj. A. C., 31st Punj.	6 Dec. 11	5 Dec. 16(*a*)	Offg.
Khyber Rifles.			
Roos-Keppel, *the Hon.* Lt.-Col. *Sir* G. O., G.C.I.E., K.C.S.I. (L)	28 Sep. 09	. . .	Hony. Colonel.
Campbell, Maj. C. H., Guides	21 Apl. 12	20 Apl. 17(*a*)	Comdt. *s.p.t.*
Jackson, Capt. C. H., 21st Punj.	13 Sep. 14	12 Sep. 19	2nd-in-Comd.
Barton, Capt. L. E., 30th Punj.	4 Apl. 14	3 Apl. 19	Wing Comdr.
Stuart, Lt. (*temp. Capt.*), P. P., I. A. R. O.	22 Apl. 16	21 Apl. 19	Adjt.
Grant, Lt. (*temp. Capt.*) I., I. A. R. O.	28 Apl. 16	27 Apl. 19	Qr.-Mr.
McCracken, Lt. (*temp. Capt.*) A., I. A. R. O.	28 Feb. 17	27 Feb. 20	Wing Offr.
Payton, Lt. (*temp. Capt.*) W. H., I. A. R. O.	2 Mar. 17	1 Mar. 20	Wing Offr.
Carter, Lt. R. F. F., 15th Sikhs	3 Mar. 18	2 Mar. 21	Wing Offr. (*Attd. K. M.*)
Kurram Militia.			
Davis, Lt.-Col. G. J., D.S.O., 22nd Punj.	8 Oct. 16	7 Oct. 19	Comdt. Mily. Secy. to Ch. Commr.
Dodd, Capt. P. C. R., 31st Lrs.	7 Aug. 10	6 Aug 15 (*a*)	*Offg.* Comdt.
Wilson, Capt. R. H., 82nd Punj.	25 Mar. 13	24 Mar. 18(*a*)	*Offg.* Comdt. *s.p.t.*
Higgs, Lt. (*temp. Capt.*), P. R., I. A. R. O.	17 Mar. 16	16 Mar. 19	Adjt. Qr. Mr. *s.p.t.*
Champion, Lt. (*temp. Capt.*) F. W., I. A. R. O.	8 Mar. 17	7 Mar. 20	Wing Offr.
Beamish, Lt. R. P., 21st Punj.	28 Nov. 17	27 Nov. 20	Adjt. & Qr. Mr. *s.p.t.*
Farwell, Lt. A. E., 2-129th Baluchis	17 Aug. 18	16 Aug. 21	Wing Offr.
North Waziristan Militia.			
Scott, Maj. G. B., D.S.O., 27th Punj.	29 June 11	8 June 16(*a*)	Comdt. *Lv., m. c., 12 mos. fm. 30 Dec. 17.*
Russell, Maj. G. H., 126th Infy.	1 June 14	31 May 19	Comdt. *s.p.t.*
Prendergast, Capt. N. H., Guides	1 June 14	31 May 17	2nd-in-Comd.
Puckle, Lt. (*temp. Capt.*), F. H., I. A. R. O.	20 Apl. 16	19 Apl. 19	Wing Offr. (*Attd. Mohmand Militia.*)
Petrie, Lt. J. F., Guides	30 Sep. 17	29 Sep. 20	Adjt.
Graham, Lt. J. P., attd. 2-35th Sikhs	2 Feb. 18	1 Feb. 21	Qr.-Mr.
Poulton, Lt. H. M., attd. 1-35th Sikhs	1 Feb. 18	31 Jan. 21	Wing Offr.
Lynch, Lt. R. J., attd. 126th Baluchis	7 Feb. 18	6 Feb. 21	Wing Offr.
South Waziristan Militia.			
Davis, Maj. G. E. P., 106th Prs.	25 Jan. 12	24 Jan. 17	Comdt.
Parsons, Capt. A. E. B., 52nd Sikhs	3 July 13	2 July 16(*a*)	
Burn-Murdoch, Capt. C. T., 20th Punj.	17 Mar. 17	16 Mar. 20	Wing Comdr.
Cotton, Lt. (*temp. Capt.*) W. B., I. A. R. O.	15 Mar. 16	14 Mar. 19	Wing Offr.
Wylie, 2nd-Lt. (*temp. Capt.*) F. V., I. A. R. O.	4 Mar. 17	3 Mar. 20	Adjt. & Qr.-Mr.
Cade, Capt. C. M. D., I. A. R. O.	10 Apl. 17	9 Apl. 20	Wing Offr.
Cumming, Lt. W. J., I. A.	11 Sep. 17	10 Sep. 20	Wing Offr.
Barker, Lt. H. R.	23 Aug. 18	22 Aug. 21	Wing Offr.
Mohmand Militia.			
Bennett, Maj. O. D., 26th Punj.	22 Feb. 12	21 Feb. 17(*a*)	2nd-in-Comd. Offg. Comdt.
Campbell, Capt. F. C. G., M.C., 40th Pathans	4 June 17	3 June 20	Wing Comdr.
Almond, 2nd-Lt. (*temp. Capt.*) J., I. A. R. O.	2 Mar. 17	1 Mar. 20	Adjt. & Qr. Mr.
Wood, Lt. (*temp. Capt.*) C. F., I. A. R. O.	9 Mar. 17	8 Mar. 20	Wing Offr.
Johnston, Lt. R. H. G., 21st Punj.	23 Oct. 17	22 Oct. 20	Wing Offr.
Attfield, Lt. C. J., 24th Punj.	18 Oct. 17	17 Oct. 20	Wing Offr.
Chitral Scouts.			
Reilly, Maj. N. E., I. A.	20 Oct. 15	. . .	Comdt.
Crimmin, Capt. C. C., I. A. R. O.	19 Mar. 18	18 Mar. 21	2nd-in-Command.
Zhob Militia.			
Paul, Lt.-Col. R. S. 126th Infy.	16 July 17	15 July 20	Comdt.
Turner, Lt. (*temp. Capt.*) J. T., I. A. R. O.	29 Oct. 17	28 Oct. 20	2nd-in-Comd., Cavalry.
Slesser, Lt. (*temp. Capt.*) W. D. V., I. A. R.	24 Aug. 16	23 Aug. 19	Adjutant, Infantry.
Reilly, Lt. A. F. (*temp. Capt.*) I. A. R. O.	17 Dec. 16	16 Dec. 19	Adjt. Cav.
Swift, Lt. G. A., Bedf. R.	19 May 18	18 May 21	Reserve Officer.
Mekran Levy Corps.			
Hales, Lt. G., I. A. R. O.	23 Nov. 17	. . . .	2nd-in-Comd.
Maclachlan, Lt. D., I A. R. O.	28 Sep. 17	. . . .	Adjt.
Gilgit Scouts.			
Miller-Hallet, Lt. J. L., I.A.	23 Oct. 17	. . . .	Asst. Comdt.

(*a*) Tenure extended.

OFFICERS AND TROOPS IN IMPERIAL OR COLONIAL EMPLOYMENT.

Name, Rank and Corps.	Date of appointment.	How employed.
Blanford, Capt. P. T., M.C., 84th Punj. . .	29 Mar. 14	Malay States Guides.
Curtis, Lt. R. L., I. A. R. O.	. . . .	With Somaliland Indian Contgt., K. A. Rif.
Fraser, Lt. A. H., I. A. R. O.	Oct. 16	A. D. C. to H. E. the Governor of Ceylon.
Haymes, 2nd-Lt. W. R., I. A. R. O. . . .	. . . .	With Somaliland Camel Corps.
Howard, Capt. C. A. L., 32nd Lrs.	11 Apl. 13	With Somaliland Indian Contgt., K. A. Rif.
Ismay, Capt. (*temp. Maj.*) H. L., 21st Cav. . .	. . . .	With Somaliland Indian Contgt., K. A. Rif.
Lees, Lt.-Col. C. H. B., C.M.G., 53rd Sikhs, *p.s.c.*	Mar. 14	Malay States Guides.
Lewis, Capt. T. G., 10th Jats	. . . .	With Somaliland Indian Contgt. K. A. Rif.
Marsh, Bt.-Lt.-Col. (*temp. Col.*) F. G., C.M.G., 9th G.R., *p.s.c.* [L].	24 Nov. 13	Mily. Attaché, Meshed. (*G. S. O., 2nd gr.*)
Minniken, Capt. T. B., 76th Punj.	23 Apl. 13	Malay States Guides.
Moll, Capt. W. R., 9th Infy.	. . . .	With Somaliland Indian Contgt. K. A. Rif.
Steel, Bt.-Col. R. A., C.I.E., 17th Cav., *p.s.c.* [*l*].	16 Dec. 13	Mily. Attaché, Teheran. (*G. S. O., 2nd gr.*)
Summers, Bt.-Maj. G. H. (*temp. Lt.-Col.*), 26th Cav.	10 Aug. 12	With Somaliland Indian Contgt., K. A. Rif.
Turner, Capt. G. F., 82nd Punj.	14 Nov. 11	Malay States Guides.
Voysey, Lt. C. C. A., attd. 35th Sikhs . . .	. . . .	Malay States Guides.
Wigram, Bt.-Lt.-Col. C., C.B. C.V.O., C.S.I., 18th Lrs., *p.s.c.*, Eq.	8 Apl. 08	Asst. Pte. Secy. to the King.
Worlledge, Capt. F. H., 26th Cav. . . .	. . . .	With Somaliland Indian Contgt.

OFFICERS NOW SERVING WHO ARE STAFF COLLEGE GRADUATES.

NOTE.—Officers whose names are preceded by an asterisk (*) have passed with Honours.

Names.	Corps.	Appointment.
GENERAL.		
Monro, *Sir* C. C. G.C.B., G.C.M.G., A.D.C.	Brit. Ser.	Comdr.-in-Chief in India.
LIEUT.-GENERALS.		
Altham, *Sir* F. A., K.C.B., C.M.G.	Brit. Ser.	Q.M.G. in India.
MAJOR-GENERALS.		
Barrow, G. deS., K.C.M.G., C.B. [*l*]	...	......
Bunbury, W. E., C.B.	I. A.	......
VC Cobbe, A. S., C.B., D.S.O.	...	......
Kemball, *Sir* G. V., K.C.M.G., C.B., D.S.O. (*g*) (206).	Brit. Ser.	G. O. C., 5th (Mhow) Dn.
Kirkpatrick, *Sir* G. M., K.C.B., K.C.S.I., (206) (*c*).	Brit. Ser.	Ch. of the Genl. Staff in India.
Knight, W. C., C.B., C.S.I., D.S.O. [L]	I. A.	G. O.C., Bombay Bde.
Nugent, O.S.W., C.B., D.S.O. [*l*].	Br. Ser.	G. O. C., Meerut Dn.
Richardson, C. W. G., C.S.I. [L]	...	D. Q. M. G. A. Hd. Qrs.
Shea, J. S. M., C.B., C.M.G., D.S.O., *q.s.*	I. A.	......
Skeen, A., C.M.G.	...	Depy. C. G. Staff, India.
Walter, J. MacN., C.B., C.S.I., D.S.O. (206).	Brit Ser.	Maj.-Genl., Admn. N. Comd.
Watson, W. A., C.B., C.M.G., C.I.E.	I. A.	
Younghusband, *Sir* G. J., K.C.M.G., K.C.I.E., C.B.	I. A.	......
COLONELS.		
Austin, H. H., C.B., C.M.G., D.S.O. [*l*]. (206) (*temp. Br.-Genl.*)	R. E.	......
Bell-Smyth, J. A. (*temp. Br.-Genl.*)	1st Dn. Gds.	......
Black, W. C. [L] (*temp. Br.-Genl.*)	12th Prs.	Cadet School, Indore.
Bright, R. A., C.B., *q.s.* (*temp. Br. Genl.*)	R. A.	...
Browne, H.J.P. (*temp. Br.-Genl.*)	5th G. R.	Brig.-Genl., G.S. S. Comd.
Cassels, R. A., C.B., D.S.O.	32nd Lrs.	....
Charles, (*Temp. Maj. Genl.*) J. R. E., C.B., D.S.O.	R. E.	......
Clay, C. H. [*l*]	I. A.	......
Coffin, C., C.M.G. (206)	R. E.	Dy. D. G., M. W. Army Hdqrs.
Cummins, H. A. V.	I. A.	......
Davies, W. P. L., C.M.G., D.S.O. (*temp. Br. Genl.*) (206).	R. A.	......
Dyer, R. E. H., C.B. (*temp. Br.-Genl.*)	I. A.	G. O. C., Jullundur Bde.
Edwards, FitzJ. M., C.B., C.M.G., D.S.O.	I. A.	......
Elsmie, A. M. S., C.M.G. [L] (*temp. Br.-Genl.*)	56th Rif.	......
Ewbank, W., C.I.E. (206)	Brit. Ser.	......
Fraser, T., C.B., C.S.I., C.M.G. R.E., *p.s.c.* (*n*) [L] (*temp. Br. Genl.*) (50, 206)	R. E.	......
Godwin, C. A. C., D.S.O.	23rd Cav.	......

Names.	Corps.	Appointment.
COLONELS—*contd.*		
Gregory, C. L., C.B. (*temp. Br.-Genl.*)	19th Lrs.	......
Hamilton, W. G., C.B., C.S.I., D.S.O. (206). (*temp. Br.-Genl.*)	Brit. Ser.	......
Holloway, B., C.I.E. [L] (*temp. Br.-Genl.*)	I. A.	......
Holman, H.C., C.B., C.M.G., D.S.O. [L] (*temp. Maj.-Genl.*).	16th Cav.	......
Home, J. M. [L].	...	......
Hopwood, H. R., C.S.I.	33rd Cav.	......
Hutchinson, C. A. R., D.S.O.	41st Dogras	Cadet Coll., W'ton.
Isacke, H., C.M.G. [*l*] (*temp. Br.-Genl.*)	...	Dir. of Mily. Tg. A. Hd. Qrs.
Knox, A. W. F., C.B. (*temp. Br.-Genl.*) [L]	58th Rif.	......
Jones, L. C., C.M.G., M.V.O. [*l*] (*temp. Maj. Genl.*)	7th Lrs.	......
Loch, H. F.	I. A.	......
Loch, S. G., D.S.O. (206) (*temp. Br.-Genl.*)	R. E.	Brig.-Genl., Gen. Staff, N. Comd.
Macmullen, C. N., D.S.O. (*temp. Bg.-Genl.*)	15th Sikhs	......
Norman, C. L., D.S.O., M.V.O., A.D.C.	Guides	Dir. of Staff duties, India.
Norie (*temp. Br.-Genl.*) C. E. deM., C.B., C.M.G. D.S.O. [*l*]	2nd G. R.	Comdg. Poona Bde.
O'Grady, H. de C.	52nd Sikhs	Bde. Comdr., 10th Infy. Bde.
Orr, G. M., D.S.O.	11 Lrs.	G. S. O., 1st gr. Mhow Dn.
Selwyn, C. H., C.B. [L]	I. A.	......
Senior, H. W. R. [L].	...	......
Sheppard, S. H., C.B., C.M.G., D.S.O. (206) (*temp. Br.-Genl.*)	R. E.	......
Shore, O. B. S. F., C.B., C.I.E., D.S.O. [L]. (*temp. Br.-Genl.*).	I. A.	......
Steel, R. A., C.I.E. [*l*].	17th Cav.	......
Stewart, J. H. K., D.S.O.	39th Garh. Rif.	......
Symons, A., C.M.G. [*l*].	Brit. Ser.	......
Tarver, Bt.-Col. A. L., D.S.O.	124th Inf.	......
Tod, J. K. [L]	I. A.	G. S. O., 1st gr., Burma.
Vaughan, L. R., C.B., D.S.O.	7th G. R.	......
Wigram, K., C.B., D.S.O. (*temp. Br.-Genl.*)	2nd G. R.	......
Willoughby, M. E., (*temp. Br.-Genl.*) C.B., C.M.G. [*l*]	I. A.	......
Wolley-Dod, O.C., C.B. D.S.O. [*l*] (206). (*temp. Br-Genl.*)	Brit. Ser.	G. O. C. Bangalore Bde.
Young, F. DeB., C.M.G.	I. A.	......
LIEUT.-COLONELS.		
Abbott, L. H., C.M.G. (*temp. Brig.-Genl.*)	8th Rajputs.	......
Alexander, E. C., D.S.O.	55th Rif.	......
Allanson, C. J. L., C.I.E., D.S.O.	6th G. R.	
Baird, H. B. D., C.M.G.	12th Cav.	G. S. O., 1st gr., Quetta.
Bainbridge, W. F., C.M.G., D.S.O.	54th Sikhs	......
Battine, R. St. C., D.S.O. [L].	21st Cav.	......

GRADUATED AT STAFF COLLEGE—*contd.*

Names.	Corps.	Appointment.
LIEUT.-COLONELS—*contd.*		
Belgrave, H. D., D.S.O.	R. W. Kent R.	......
Bethell, H. K., C.M.G., D.S.O. (*temp. Br.-Genl.*)	7th H.	......
Boddam, E. B. C.	5th G. R.	B. G. Adm., Poona.
Bourne, W. K.	2nd Lcrs.	......
Bridges, A. H., D.S.O.	112th Mah.	On deputation, Australia.
Bruce, G. D.	61st Prs.	......
Cameron, A.R., C.M.G.	R. Highrs.	......
Capper, A. S., D.S.O.	39th Horse	......
Cassels, G. R.	35th Sikhs	......
Catty, T. C., D.S.O.	69th Punj.	......
Chitty, A. W.	126th Infy.	A. A. & Q. M. G., Rawalpindi.
Clarke, L. J. L., C.M.G. [*l*].	E. York R.	......
Coleridge, J. F. S. D., C.M.G., D.S.O. (*temp. Brig.-Genl.*)	8th G. R.	......
VC Costello, E. W., C.M.G., D.S.O.	22nd Punj.	......
Deane, D., D.S.O. [*l*]	12th Cav.	......
DePree, H. D., C.B. (*temp. Br.-Genl.*). [L] (206).	R. A.	......
Dick, R. N. (*temp. Br.-Genl.*).	R. Suss. R.	......
Dobbs, C. F., D.S.O.	94th Infy.	......
Dundas, P. H., D.S.O.	6th Infy.	......
Drew, B. C. H.	23rd Prs.	......
Fetherstonhaugh, W. A., D.S.O.	8th Cav.	A. A. & Q. M. G., Quetta.
Fisher, B. D., D.S.O.	17th Lrs.	......
Fordham, W. M.	20th Infy.	......
Forster, D., D.S.O.	R. E.	......
Geoghegan, S., C.B., [L]	62nd Punj.	......
Giles, E. D., D.S.O.	35th Horse	......
Glasfurd, A. I. R., C.M.G., D.S.O.	46th Punj.	......
Gwyer, V. E. (*temp. Lt.-Col.*)	10th Jats	G.S.O., 1st gde., Army Hd.Qrs.
Gwyn-Thomas, G., C.M.G., D.S.O. (*temp. Br.-Genl.*)	2nd Lrs.	......
Hay, C. J. B., D.S.O. [L]	Guides	......
Hay, W. N., C.I.E., D.S.O. [L]	129th Baluchis.	......
Hocken, C. A. F. [L]	35th Horse	......
Holdich, H. A., D.S.O. (*tempy. Brig.-Genl.*)	5th G. R.	Dir. of Q. A. Hd. Qrs.
Lees, C. H. B., C.M.G. (*temp Lt.-Col.*)	53rd Sikhs	Comdt., Malay States Guides.
Legard, D' A. [*l*]. (206) (*temp. Bg.-Genl.*)	17th Lrs.	......
Leslie, W. S., D.S.O.	31st Punj.	......
Longhurst, A. L. [L]	7th G. R.	......
McCulloch, A. J. (206)	7th D. Gds.	......
Marsh, F. G., C.M.G. [L] (*temp. Col.*)	9th G. R.	......
Moberly, F. J., D.S.O. (*temp. Br.-Genl.*)	25th Punj.	Dir. Mily. Operations.
Moberly, H. S.	66th Punj.	G. S. O., 1st gde., Lucknow.
Moens, A. W. H. M., D.S.O. (*n*).	52nd Sikhs	......
Murray, W. G. P. (*temp. Lt.-Col.*) (*n*).	21st Punj.	......
Muspratt, S. F., D.S.O.	12th Cav.	G. S. O., 1st gr. A. H. Qrs.
Newnham, C. C., D.S.O.	6th Cav.	......
Norman, Bt.-Col. W. H., D.S.O.	11th Lrs.	......
Ogle, N., D.S.O.	67th Punj.	Cadet Coll., Quetta.
LIEUT.-COLONELS—*concld.*		
Pryce, H. E. apR., C.M.G., D.S.O. (*temp. Brig.-Genl.*).	18th Infy.	......
Rattray, C., C.B. (*temp. Bg.-Genl.*)	20th Infy.	......
Ricketts, R. L.	10th Lrs.	......
Rind, G. B. A.	124th Infy.	A. A. G., A. Hdqrs.
Rowlandson, M. G. D., D.S.O. [L].	38th Dogras	G. S. O., 1st Gr.
Russell, F. D. [L]	1st Lrs.	......
St. John, R. S., D.S.O. (*temp. Brig. Gen.*)	20th Infy.	Embn. Comdt., Bombay.
Sanders, G. A. F. (*temp. Br.-Genl.*) (206).	R. E.	...
Shuttleworth, D. I. (*temp. Lt.-Col.*)	3rd G. R.	......
Stapleton, F. H.	Oxf & Bucks L. I.	G. S. O., 1st gr.
Talbot, F. E. G.	14th Sikhs	......
Taylor, R. O. B., C.M.G., C.I.E. [L].	19th Lrs.	A. Q. M. G., S'bad. Dn.
Thompson, A. G.	58th Rif.	......
Tennant, H. L. (*e*)	R. F. A	......
Turner, F. C. [L] (206)	N. Fus.	D. A. A. and Q. M. G.
Twiss, W. L. O., M.C. [L].	9th G. R.	G. S. O., 1st gr., A. Hdqrs.
Villiers Stuart, J. P., D.S.O.	55th Rif.	G. S. O., 1st gr., A. Hdqrs.
Vincent, B. (206)	6th Dns.	......
Wagstaff, C. M., C.M.G., C.I.E., D.S.O. (*temp. Brig.-Genl.*) (206)	R. E.	......
Ward, H. C. S.	2nd Lrs.	G. S. O., 1st gde., R' Pindi.
Waterfield, A. C. M., M.V.O.	....	......
Watson, L. A. [L]	74th Punj.	......
Watt, D. M., D.S.O.	2nd G. R.	......
Whitehead, J., D.S.O.	1st Brah.	......
Wigram, C., C.V.O., C.S.I., Eq.	18th Lrs.	Asst. Pte. Secy. to the King.
Wilson, R. C., D.S.O., M.C.	114th Mah.	......
Herdon, H. E.	55th Rif.	......
Casson, Maj. W. F. S., D.S.O.	27th Cav.	G. S.O., 1st gr., 16th Divn.
MAJORS.		
Anderson, B. E., D.S.O. [L].	59th Rif.	......
Arthur, L. F., D.S.O. [L]	26th Cav.	Asst. Comdt., Cadet College, Quetta.
Bannatyne, N. C.	128th Prs.	A. A. G., Army Hdqrs.
Barnes, L. E. (206)	R. E.	G. S. O., 2nd gr., A. Hdqrs.
Bennett, E. E. S.	66th Punj.	G. S. O., 2nd gr., A. Hdqrs.
Binny, A. J. M. (*temp. Lt.-Col.*)	1st Lrs.	A. A. & Q. M. G., Derajat.
Bradshaw, C. R. [*l*]	9th G. R.	......
Cadell, J. G., D.S.O.	45th Sikhs	G.S.O., 2nd gr., S. Comd.
Calvert, R. T. C.	122nd Infy.	A. Q. M. G., Mhow.
Cameron, A. R. (*temp. Bg.-Genl.*), C.M.G.	R. Highrs.	......
Carruthers, J. H. de W., M.C.	39th G. R.	G.S.O., 2nd gr., Rawalpindi.
Chambers, P. R., D.S.O.	10th Lcrs.	......
Cloete, L. B.	37th Dogras	......
VC Colvin, J. M. C., *q.s.* (206).	R. E.	......

GRADUATED AT STAFF COLLEGE—*contd.*

Names.	Corps.	Appointment.
MAJORS—*contd.*		
Coningham, R. E.	10th G. R.	G. S.O., 2nd gr., A. Hdqrs.
Conway-Gordon, E.C.W.	3rd Horse	G. S. O., 1st gr., Poona.
Craster, G., D.S.O.	6th Cav.	......
Crombie, D. C. (*n*)	23rd Cav.	G. S.O., 2nd gr., Bannu.
Crowdy, J. D., D.S.O.	5th G. R.	......
Cunningham, C. C.	12th Prs.	......
de Burgh, E., D.S.O. (*temp. Lt.-Col.*)	9th Horse	......
Dent, W., D.S.O. (*temp. Lt.Col.*) (Bt.-Lt.-Col.)	103th Infy.	......
Duncan, H. C., D.S.O.	9th G. R.	A. A. G. A. Hdqrs.
Dunn, F. G (206)	R. Ir. Rif.	.....
Elias Wynn, A. H. W.	1st G. R.	......
Ellis, A. C. S. B. (*e*) (*temp. Lt.-Col.*)	121st Prs.	G. S. O., 1st gr., A. Hdqrs.
Etheridge, F., D.S.O.	7th Raj.	D. A. A. G., 4th Dn.
Freeland, Bt.-Lt.-Col. J. C.	35th Sikhs	......
Gardiner, R. [*l*]	53rd Sikhs	......
Gilchrist, W. F. C., C.I.E. (*n*).	52nd Sikhs	......
Gillies, F. G.	33rd Cav.	......
Glynton, G. M.	3rd G. R.	......
Grant, G. P., D.S.O. (*temp. Lt.-Col.*)	106th Prs.	......
VC Grant, J. D.	8th G. Rif.	......
Greene, A. C. R. (206)	R. G. A.	......
Hamer, M. A., D.S.O., M.C.	129th Baluchis.	......
Haslehust, G. W. (*temp. Lt.-Col.*)	29th Punj.	......
Henderson, K., D.S.O.	39th Garh. Rif.	......
Hill-Whitson, T. E. L.	14th Hrs.	......
Hogg, C. M. T., D.S.O.	4th G. R.	G. S. O., 2nd gr., Mhow.
Holmes, C. W. A.	116th Mah.	......
Hunt, H. R. A.	25th Punj.	......
Hutchinson, A. G. C.	23rd Prs.	......
Hutchison, G.	122nd Infy.	......
Johnson, M. E. S.	48th Prs.	......
Julius, S. deV. A. [*l*]	R. Suss. R.	......
Keen, F. S., D.S.O.	45th Sikhs.	......
Kensington, E. C., M.C.	130th Baluchis.	......
Kirkpatrick, C. (Bt.-Lt.-Col.)	Guides	......
Kettlewell, A. M.	39th Horse	......
Lakin, J. H. F. (*temp. Lt.-Col.*)	7th G. R.	......
Loch, G. C. B. (206)	R. E.	......
McKenna, J. C., D.S.O.	16th Rajputs	......
McLaughlin, W.P.M.D.	107th Prs.	......
McRae, H. St. G. M.	45th Sikhs	G. S. O., 2nd gr., 16th Dn.
McWatters, H. C., D.S.O.	24th Punj.	......
Macleod, N., D.S.O.	15th Sikhs	Bde. Maj., Rangoon.
Macleod, W. M.	31st Lrs.	G. S. O., 2nd gr., N. Com'd.
Macrae, J. C.	19th Punj.	Staff Sch., Saugor.
Maitland, G. R., D.S.O.	14th Lrs.	......
Marchant, G.	29th Lrs.	......
Marjoribanks, R. D. (*temp. Lt.-Col.*)	107th Prs.	......
Maxwell, D. W.	4th G. R.	......
Meiklejohn, W. L. S.	106th Prs.	

Names.	Corps.	Appointment.
MAJORS—*concld.*		
Milward, C. A., D.S.O. (Bt. Lt.-Col.)	53rd Sikhs	......
Moberly, B. R., D.S.O. (*temp. Lt.-Col.*) (Bt. Lt.-Col.)	56th Rif.	G. S. O., 1st gr., A. H. Qrs.
Molloy, G. M. (*temp. Lt.-Col.*)	34th Horse	G. S. O., 1st gr., A. Hdqrs.
Muirhead, J. A.	1st Lrs.	......
Murray, K. D. B., D.S.O.	59th Rif.	
Muspratt, V. E. [*l*] (*temp. Lt.-Col.*)	30th Lrs.	A. Q. M. G., A. Hdqrs.
Nangle, M. C. [*l*]	92nd Punj.	
Ogg, A. C., D.S.O.	7th Rajputs	......
Orton, E. F. (Bt. Col.)	37th Lrs.	......
Orton, S. B.	39th G. R.	......
Peart, C. L. [L]	106th Prs.	......
Penton, B. C.	25th Punj.	......
Pepys, G. L., D.S.O. [L] (*temp Lt.-Col.*)	57th Rif.	G. S. O., 1st gr., A. Hdqrs.
Pope, S. B., D.S.O.	58th Rif.	......
Rice, S. M., C.I.E. (*temp. Lt.-Col.*)	64th Prs.	A. A. & Q. M. G., Northn. Comd.
Ridgeway, D. G. (*temp. Lt.-Col.*)	3rd G. R.	Staff Sch., Saugor.
Rivett-Carnac, J. S. (*temp. Lt.-Col.*)	14th Lrs.	G. S. O., 1st gr., S'bad.
Robinson, D. G., D.S.O.	46th Punj.	G. S. O., 2nd gr., A. Hdqrs.
Rogers, G. H. (*temp. Lt.-Col.*)	11th Rajputs	G. S. O., 1st gr., A. Hdqrs.
Scott, H. L., D.S.O., M.C.	1st G. R.	......
Seymour, C. R. F. (*temp. Lt.-Col.*) [L]	13th Rajputs	......
Sherlock, G. W. S.	6th G. R.	
Shuttleworth, A. R. B.	S. and T. Corps.	......
Shuttleworth, B. W.	45th Sikhs.	......
Skipwith, F.	5th G. R.	......
Stansfeld, C. G.	8th G. R.	......
Stokes, C. B., C.I.E. (*temp. Lt.-Col.*) [L] (Bt. Lt.-Col.)	3rd Horse	......
Stoney, P. S.	26th Punj.	......
Storr, C. L. (*temp. Lt.-Col.*)	54th Sikhs	......
Taylor, E. M.	22nd Cav.	......
Tobin, H. W., D.S.O.	128th Prs.	G. S. O., 2nd gr., Karachi Bde.
Tyrrell, A. C. L. (*temp. Lt.-Col.*)	25th Cav.	A. A. and Q. M. G., S. Com'd.
Underhill, G. E. C.	62nd Punj.	......
Walker, C. W. G., D.S.O.	37th Dogras	......
Williams, A. F. C., D.S.O. (*temp. Lt.-Col.*)	31st Lrs.	......
Wilson, D. D., M.C.	17th Cav.	......
Wilson-Johnston, W. E., D.S.O. (Bt. Lt.-Col.)	36th Sikhs	......
Withers, C. McG.	87th Punj.	.....
Wyatt, E. R. C., D.S.O. (*temp. Lt.-Col.*)	106th Prs.	G.S. O., 1st gr., A. Hdqrs.
VC Wylly, G. G. E., D.S.O.	Guides	G. S. O., 2nd gr., Kohat.

GRADUATED AT STAFF COLLEGE—*concld.*

Names.	Corps.	Appointment.	Names.	Corps.	Appointment.
CAPTAINS.			CAPTAINS—*concld.*		
Hibbert, O. Y., D.S.O. .	R. W. Kent R.	....	Thompson, *Sir* T. R. L., *Bt.* M.C.	7 Hrs. .	......
			Tollemache, *Hon.* D. P. (*temp. Maj.*).	7 Hrs. .	......

OFFICERS NOW SERVING WHO ARE QUALIFIED FOR STAFF EMPLOYMENT IN CONSEQUENCE OF SERVICE ON THE STAFF IN THE FIELD.

Names.	Corps.	Appointment.	Names.	Corps.	Appointment.
GENERAL.			COLONEL.		
Birdwood, *Sir* W. R., K.C.B., K.C.S.I., K.C.M.G., C.I.E., D.S.O., A.D.C.	Ind. Army	......	Bright, R. A., *p.s.c.*	R. H. A.	......
MAJ.-GENERALS.			LIEUT.-COLONEL.		
Macandrew, H. J. M., C.B., D.S.O.	Ind. Army.	......	V.C. Colvin, J. M. C., *p.s.c.*	R. Eng.	......
Shea, J. S. M., C.B., C.M.G., D.S.O., *p.s.c.*	Ind. Army.	......			

OFFICERS NOW SERVING WHO HAVE PASSED A COURSE OF INSTRUCTION AT THE LONDON SCHOOL OF ECONOMICS.

Names, Rank and Corps.	Names, Rank and Corps.	Names, Rank and Corps.
Alexander, Maj. H. M., D.S.O., S. & T. Corps.	Fraser, Lt.-Col. H. F., C.M.G., D.S.O., 21 Lrs.	Smith, Maj. A. A., 58 Rif.
Bright, Bt.-Col. R. A., C.B., R.A., *p.s.c., q.s.*	Geoghegan, Maj. O. A., S. & T. Corps.	Smith, Lt.-Col. W. F., S. & T. Corps.
deBerry, Lt.-Col. P. P. E., R.F.A.	Gibbs, Lt.-Col. A. A., I.M.S.	Synge, Maj. M., D.S.O., S. & T. Corps.
Duxbury, Maj. C. D., S. & T. Corps.	Grove, Bt. Lt.-Col. P. L., Ches. R.	Tennant, Lt.-Col. H. L., R.F.A., *p.s.c.*
Ellis, Maj. A. C. S. B., 121 Prs. *p.s.c.*	Maule, Lt.-Col. H. N. St. J., C.M.G., R.F.A.	Timbrell, Maj. T., D.S.O., S. & T. Corps.
Fitzgerald, Capt. A., S. & T. Corps.	Plummer, Lt.-Col. E. W., R.F.A.	Ward, Bt.-Lt.-Col. H. C. S., 2 Lrs., *p.s.c.*
	Robinson, Maj. E. L., 75 Infy.	

OFFICERS NOW SERVING WHO HAVE PASSED THE FINAL EXAMINATION OF THE ADVANCED CLASS, ORDNANCE COLLEGE.

Brown, Maj. B. H., R.A.	Harvey, Maj. A. F. B., R.A.	Palmer, Lt.-Col. C. C., R.A.
Cruickshank, Maj. H. St. J., R.A.	Kenyon, Bt.-Col. L. R., C.B., R.A.	Vines, Maj. C. E., R.A.
Drake-Brockman, Capt. V., R.A. [*l*]	Kinloch, Capt. R.I. H., R.A. [*l*]	

OFFICERS NOW SERVING WHO HAVE PASSED THE FIRST YEAR'S COURSE AT THE ORDNANCE COLLEGE.

Beamish, Maj. W. S., R.A.	Hare, Maj. G. A., R.A.	Marshall, Maj. F. H., R.A.
Carey, Lt.-Col. W. L. J., R.A. [*l*]	Kirke, Maj. K. St. G., D.S.O., R.A.	Schrottky, Maj. C. E. G., R.A.
Fanshawe, Maj. L. A., D.S.O., R.A.	Lane, Capt. H. E., R.A.	Snowdon, Maj. H. S. K., R.A.
Finnis, Maj. F. A., R.A.	Larmour, Maj. L. C., R.A.	Temple, Maj. T., R.A.
Gardiner, Maj. H. W., R.A.	Long, Capt. W. E. L., R.A.	Tute, Capt. C. S., R. A.
	Mecredy, Maj. H., R.A.	Wilkes, Maj. G. S., R.A.

OFFICERS NOW SERVING WHO HAVE OBTAINED A FIRST CLASS CERTIFICATE IN THE LONG COURSE OF GUNNERY OR GUNNERY STAFF COURSE PRIOR TO 1906, OR HAVE OBTAINED THE CERTIFICATE OF THE GUNNERY STAFF COURSE SUBSEQUENT TO 1905.

Alexander, Maj. W. P., R.A.	Fergusson, Lt.-Col. A. C., C.M.G., D.S.O., R.A.	Ouseley, Col. R. G., C.M.G., D.S.O., R.A. (*f*)
Anderson, Lt.-Col., A. T., R.A.	Foote, Lt.-Col. H. B., R.A.	Parrington, Maj. J. W., R.F.A.
Barton, Maj. P., D.S.O., R. A.	Freeth, Lt.-Col. C. J. D., R.A. [R.A	Potts, Col. F., C.M.G., R.A.
Brewster, Lt.-Col., R. F., R.A.	Gordon, Col. L. G. F., C.B., D.S.O.,	Preston, Maj. P. H. H., R.A.
Campbell, Maj. K. G., D.S.O., R.A.	Halford, Capt. E. S., R.A.	Pullen, Col. A. F., Brit. Ser. [*l*]
Cooper, Lt.-Col. P. T., R.A.	Kemball, Maj.-Genl. *Sir* H. V., K.C.M.G., C.B., D.S.O., Brit. Ser., *p.s.c.*	Stokes, Col. A., C.B., C.M.G., D.S.O., Brit. Ser.
Corrie, Maj. W. F. T., R.A., *p.a.c.*	Langhorne, Maj. J. A. D., R. A.	Sturrock, Lt.-Col. G. C., R.A.
Davidson, Lt.-Col. F. M., R.A.	Mackintosh, Lt.-Col. J. B., R.A.	Swann, Maj. C. J. H., R.A. (*e*)
Duhan, Brig.-Gen. W. W. A.		Tomkins, Lt.-Col. E. L., R.A. (*f*)
Elliott, Capt. B. H., R.A.		

OFFICERS NOW SERVING WHO HAVE QUALIFIED (OR ARE CLASSED) AS INTERPRETERS IN MODERN FOREIGN LANGUAGES.

Amharic.

1st Class.

Name, Rank and Corps.	Date of Qualification.
Eadie, Capt. J. I., 97 Infy.	Jan. 14

Arabic.

1st Class.

Name, Rank and Corps.	Date of Qualification.
§*Boyle, Capt. C. A., 11 Lrs.	Apl. 16
Cook, Lt.-Col. H. R., R.A.	Apl. 97
§Daly, Capt. C. K., S. & T. Corps	July 11
§Eadie, Maj. J. I., D.S.O., 97 Infy.	July 10
Fowle, Capt. T. C. W., 40 Pathans	June 11
Grey, Lt.-Col. W. G., Supy. List	Aug. 02
‡Johnson, Col. F. E., C.M.G., D.S.O., R.A.	Jan. 90
Mackenzie, Maj. C. F., C.I.E.	July 07
Malyon, Capt. F. H., 21 Punj.	Apl. 12
More, Maj. J. C., D.S.O., 51 Sikhs	Feb. 10
Morton, Col. W. R., Brit. Ser.	Oct. 00
Murphy, Bt.-Lt.-Col. C. C. R., 30 Punj.	Apl. 11
Norbury, Maj. P. Fitz G., 34 Horse	Oct. 13
Pulley, Maj. H. C., 12 Prs.	Dec. 12
§Routh, Maj. G. M., D.S.O., R. A.	July 10
Young, Capt. H. W., 116 Mah.	June 12

2nd Class.

Name, Rank and Corps.	Date of Qualification.
Pullen, Col. A. F., Brit. Ser. (*g*)	Oct. 95

Cape Dutch.

1st Class.

Name, Rank and Corps.	Date of Qualification.
Rynd, Maj. F. F., D.S.O., R.G.A.	Jan. 09
White, Capt. R. W., R.F.A.	June 12

2nd Class.

Name, Rank and Corps.	Date of Qualification.
Gordon, Capt. C. C., R.F.A.	Jan. 10
Mintoft, Capt. T. C., York. R.	Jan. 09

Chinese.

1st Class.

Name, Rank and Corps.	Date of Qualification.
Worrall, Maj. E. W., Som. L. I.	Dec. 08
Cardew, Maj. A. M., R. E.	June 13

2nd Class.

Name, Rank and Corps.	Date of Qualification.
Campbell, Maj. W. L., C.I.E. (Supy. List) (A. O. 150 of 1899).	Feb. 03
Poole, Maj. F. G., D.S.O., Midd'x. R. (A. O. 150 of 1899).	July 02

French.

1st Class.

Name, Rank and Corps.	Date of Qualification.
Anderson, Bt.-Maj. B. E., D.S.O., 59 Rif., *p.s.c.*	July 07
Anderson, Bt.-Lt.-Col. C. F., D.S.O., R.E.	July 05
†Anderson, Capt. M. G., 7 G. R.	Oct. 09
Battine, Lt.-Col. R. St. C., *p.s.c.*, 21 Cav.	Oct. 02
Biscoe, Maj. H. V., Supy. List	Oct. 04
Burn, Lt.-Col. E. M. J., R.E.	Apl. 00
Cheke, Lt.-Col. E. G., R. A.	Apl. 02
Chesney, Capt. E. C., 38 Dogras	Oct. 05
De Pree, Bt.-Col. H. D., *p.s.c.*, R.A.	Apl. 04
dePret, 2nd-Lt. J., 21 Lrs.	Jan. 14
Dick, Col. A. R., C.B.	July 09
Farebrother, Capt. F. H., 128 Prs.	Oct. 01
Fraser, Lt.-Col. T., C.B., C.S.I., C.M.G., *p.s.c.*, R.E.	Apl. 01
Giffard, Maj. J. F., N. Staff. R.	Jan. 11
Giles, Bt.-Lt.-Col. V., D.S.O., R. E.	Oct. 08
Goldman-Monk, Capt. M. E. de B., 91 Punj.	Oct. 13
Hay, Bt.-Lt.-Col. C. J. B., D.S.O., *p.s.c.*, Guides.	Apl. 04
Hocken, Lt.-Col. C. A. F., *p.s.c.*, 35 Horse.	Apl. 04
Holman, Bt.-Col. H. C., C.B., C.M.G., D.S.O., *p.s.c.*, 16 Cav.	Oct. 97
Johnston, Maj. C. E. L., R. A.	July 06
Knight, Maj.-Genl. W. C., C.B., C.S.I., D.S.O., *p.s.c.*	Oct. 98
Lee, Capt. S. S., D.S.O. R. A.	Jan. 11
Macaulay, Maj. D. I. M., 1 Lrs.	July 05
Pepys, Maj. G. L., D.S.O., 57 Rif., *p.s.c.*	Jan. 11
Preston, Lt.-Col. E. E., S. & T. Corps	Jan. 13
Redl, Maj. E. A. F., C.I.E., 113 Infy.	Oct. 91
Richardson, Maj.-Genl. C. W. G., C.S.I., *p.s.c.*	Oct. 99
Roos-Keppel, Lt.-Col. *Sir* G. O., G.C.I.E., K.C.S.I., Supy. List.	Oct. 90
Russell, Lt.-Col. F. D., *p.s.c.*, 1 Lrs.	Oct. 99
Scovell, Lt. A. M., Sea. Highrs.	July 06
Seys, Capt. R. C., D.S.O., R.A.	Oct. 02
Taylor, Lt.-Col. R.O'B., C.M.G., C.I.E., *p.s.c.*, 19 Lrs.	Apl. 01
Turner, Lt.-Col. F. C., *p.s.c.*, North'd Fus.	Apl. 99
Twiss, Maj. H. W. F., D.S.O., S. and T. Corps.	June 11
Twiss, Bt.-Lt.-Col. W. L. O., M.C., *p.s.c.*, 9 G. R.	Apl. 06
Walker, Maj. A. D., R. E.	Oct. 99
Willoughby, Col. M. E., C.B., C.M.G., *p.s.c.*, 2 Lrs.	Oct. 97
Young, Capt. H. W., 116 Mah.	July 07

2nd Class.

Name, Rank and Corps.	Date of Qualification.
Alexander, Lt.-Col. C. H.	Oct. 04
Allen, Maj. W. T., 23 Cav.	Apl. 13
Barrow, Maj.-Genl. *Sir* G. deS., K.C.M.G., C.B., *p.s.c.*	Oct. 97

* Official Interpreter in Arabic.
† Official Interpreter in Russian.
‡ Degree of Honour, 2nd Division.
§ High Proficiency.

OFFICERS NOW SERVING WHO HAVE QUALIFIED (OR ARE CLASSED) AS INTERPRETERS IN MODERN FOREIGN LANGUAGES—*contd.*

French—*contd.*

2nd Class—contd.

Name, Rank and Corps.	Date of Qualification.
	June 10
Bell, Maj. A. H., D.S.O., R. E.	Jan. 09
Benson-Cooke, Maj. P. G., 80 Infy.	Jan. 14
Bernard, Capt. R. P. St. V., D.S.O., M. 6 G. R.	July 05
Bingham, Maj. W. H., 69 Punj.	Jan. 12
‡Blair, Maj. G. L., 36 Sikhs	Jan. 13
Bodley, Lt. R. V. C., K. R. Rif. C.	Oct. 96
Bowker, Maj. W. J., D.S.O., Som. L.I.	Oct. 04
Bradshaw, Maj. C. R., *p.s.c.*, 9 G. R.	Apl. 03
Bridges, Bt.-Lt.-Col. A. H., D.S.O., *p.s.c.*, 116 Mah.	June 14
Brodie, 2nd-Lt. R. H., Midd'x. R.	June 11
Carey, Maj. W. L. J. (*o*) R. A.	Jan. 11
Carlisle, Maj. T. R. M., D.S.O., R. A.	Apl. 89
Christian, Col. G., D.S.O.	Jan. 12
Clarke, Maj. B. G. S., 10 G. R.	Oct. 06
Cooper, Maj. G. C., M.C., 121 Prs.	Oct. 91
Dale, Lt.-Col. G. A., C.M.G.	Apl. 05
Deane, Bt.-Lt.-Col. D., D.S.O., *p.s.c.*, 12 Cav.	Jan. 11
Drake-Brockman, Capt. V., *p.a.c.*, R.A.	June 14
Duke, Capt. O. A., 22 Cav.	Apl. 02
Elsmie, Col. A. M. S., C.M.G., *p.s.c.*	Oct. 05
Etherton, Maj. P. T., 39 Garh. Rif.	June 14
Fulton, Capt. C. M., 33 Cav.	Oct. 88
Gompertz, Capt. A. V., M.C., R. E.	Jan. 11
Gompertz, Capt. M. L.A., 108 Infy.	Jan. 05
Harding, Lt.-Col. M. F., C.M.G.	Oct. 01
Hay, Capt. A., R. A.	Jan. 12
Hay, Lt.-Col. W. N., C.I.E., D.S.O., 129 Baluchis, *p.s.c.*	Apl. 02
Herbert, Bt. Col. L. N.	Oct. 95
Hogg, Capt. D. McA., M.C., R. E.	July 07
Holme, Maj. A. S., R. E.	Apl. 96
Home, Lt.-Col. J. M., *p.s.c.*	Oct. 99
Hudson, Lt.-Col. J. H.	June 13
Hunt, Maj. H. R. A., D.S.O., 25 Punj.	June 14
Isacke, Bt.-Col. H., C.M.G., *p.s.c.*, R. W. Kent. R.	Apl. 05
Jones, Maj. A. N. G., D.S.O., 36 Horse	Oct. 09
Jones, Bt.-Col. L. C., C.M.G., M.V.O., *p.s.c.*, 7 Lrs.	Oct. 04
Julius, Maj. S. de V.A., *p.s.c.*, R. Suss. R.	Oct. 04
Kinloch, Capt. R. I. H., R.A., *p.a.c.*	Oct. 07
Knox, Bt.-Col. A. W. F., C.B., *p.s.c.*, 58 Rif	Apl. 04
Loch, Capt. P. G., 97 Infy.	Jan. 12
Luxmore, Maj. C. de J., 99 Infy.	Oct. 94
McHarg, Bt.-Lt.-Col. A. A., D.S.O., R. E.	Oct. 96
Mocatta, Bt.-Maj. V. E., 14 Hrs.	Apl. 04
Muspratt, Maj. V. E., 30 Lrs.	June 14
Norie, Col. C. E. de M., C.B., C.M.G., D.S.O., *p.s.c.*	Apl. 01
Palmer, Maj. J. H. G., 21 Punj.	Jan. 09
Povah, Lt. F., R. Scots	Apl. 05
Powell, Col. S. H., C.B., Brit. Ser.	Jan. 14
Quayle, Capt. P. R., 127 Infy.	Oct. 09
Ritchie, Maj. C. Macl., R. A.	Apl. 97
Ryder, Lt.-Col. C. H. D., C.I.E., D.S.O., R.E.	July 06
Sartorius, Capt. G., 6 Cav.	June 14
Shairp, Maj. A., C.M.G., S. and T. Corps.	Oct. 06
Shakeshaft, Capt. A. J., Norf. R.	Jan. 14
Spence, Lt.-Col. A. H. O., C.I.E., 39 Horse	Oct. 98
Steel, Bt.-Col. R. A., C.I.E., *p.s.c.*, 17 Cav.	Oct. 04

French—*concld.*

2nd Class—concld.

Name, Rank and Corps.	Date of Qualification.
Stokes, Bt. Lt.-Col. C. B., C.I.E., *p.s.c.*, 3 Horse	Apl. 01
Stuart-Menteth, Maj. A. H., 112 Infy.	July 05
Symons, Col. A., C.M.G., *p.s.c.*	Apl. 01
Trevor, Lt.-Col. A. P., C.O.I., C.I.E., Supy. List.	Apl. 07
Van Someren, Maj. W. W. D.S.O., 45 Sikhs	Oct. 02
Walker, Maj. E. C., R.E.	Apl. 08
Waterfield, Lt.-Col. A. C. M., M.V.O., *p.s.c.*	Oct. 99
Watson, Lt.-Col. L.A., *p.s.c.*	Apl. 03
Weir, Maj. J. L.-R.	July 07
Willis, Capt. S. G. R., D.S.O., R. A.	Oct. 03
Wilson, Bt. Lt.-Col. A. T., C.M.G., C.I.E., D.S.O.	Oct. 03
Woodward, Capt. R. F., 130 Baluchis	Apl. 13

German.

1st Class.

Name, Rank and Corps.	Date of Qualification.
Battine, Lt.-Col. R. St. C., D.S.O., *p.s.c.*,	Oct. 01
21 Cav.	Apl. 99
Burn, Lt.-Col. E. M. J., R. E.	July 08
Chesney, Capt. D. P., 23 Prs.	Jan. 12
Delme-Radcliffe, Maj. A., 105 Infy.	Apl. 99
Giles, Bt. Lt.-Col. V., D.S.O., R. E.	Jan. 14
Green-Armytage, Maj. P. D., 117 Mah.	June 10
Harcourt, Capt. E. S., M.C., 7 G. R.	Oct. 98
Holman, Bt.-Col. H. C., C. B., C.M.G., D.S.O., *p.s.c.*, 16 Cav.	Jan. 07
Holme, Maj. A. S., R. E.	Oct. 98
Home, Bt.-Col. J M., *p.s.c.*	June 13
Longhurst, Bt.-Lt.-Col. A.L., *p.s.c.*, 7 G. R.	Oct. 09
Quayle, Lt. P. R., 127 Infy.	Apl. 02
Quin, Maj. R. P. W., 41 Dogras	Oct. 93
Selwyn, Col. C. H., C.B., *p.s.c.*, 28 Cav.	Apl. 02
Senior, Col. H. W. R., D.S.O., *p.s.c.*	Jan. 11
Trotter, Maj. E. G. S.	Apl. 02
Turner, Lt.-Col. F. C., *p.s.c.*, North'd Fus.	Apl. 06
Twiss, Bt.-Lt.-Col. W. I. O., M.C., *p.s.c.*, 9 G. R.	

2nd Class.

Name, Rank and Corps.	Date of Qualification.
Austin, Bt.-Col. H. H., C.B., C.M.G., D.S.O., *p.s.c.*, R. E.	Apl. 04
Barton, Lt.-Col. H. J., R. E.	Apl. 88
Bennett, Maj. F. L., R. A.	Oct. 99
Bingham, Maj. W. H., 69 Punj.	Oct. 10
Campbell, Maj. A., D.S.O., R. E.	July 06
Daly, Lt. C. K., S. & T. Corps.	Oct. 08
Dunsterville, Maj.-Genl. L. C., C.B.	Apl. 96
Manton, Capt. L., D.S.O., R.E.	Jan. 11
Gardiner, Lt.-Col. R., *p.s.c.*, 53 Sikhs	Apl. 95
Gibson, Capt. L. G., R.A.M.C.	Apl. 07
Hay, Capt. A., R. A.	Jan. 12

‡ Official Interpreter in German.

OFFICERS NOW SERVING WHO HAVE QUALIFIED (OR ARE CLASSED) AS INTERPRETERS IN MODERN FOREIGN LANGUAGES—*contd.*

German—*concld.*

2nd Class—concld.

Name, Rank, and Corps.	Date of Qualification.
Hay, Bt.-Lt.-Col. C. J. B., D.S.O., *p.s.c.*, Guides.	Apl. 04
Hay, Lt.-Col. W. N., C.I.E., D.S.O., *p.s.c.*, 129 Baluchis.	Apl. 02
Hume, Maj. A. H. B., R. E.	Jan. 08
Knox, Bt.-Col. A. W. F., C.B., *p.s.c.*, 58 Rif.	Apl. 04
Lee, Capt. S. S., D.S.O., R.A.	June 13
Legard, Lt.-Col. D'A., *p.s.c.*, 17 Lrs.	Apl. 06
Montgomery, Capt. J., 7 Dgn. Gds.	July 07
Newell, Lt.-Col. H. A., S. & T. Corps.	Apl. 96
Northey, Capt. W. B., M.C., 1 G. R	June 11
O'Bryen, Lt.-Col. C. W.	Apl. 02
Ozzard, Lt.-Col. F. R., I.M.S.	Jan 14
Pepys, Maj. G. L., D.S.O., *p.s.c.*, 57 Rif.	Jan. 11
Reinhold, Maj. H. E., M.C., 27 Punj.	Oct. 10
Roberts, Maj. W. B., D.S.O., 101 Grs.	Jan. 12
Shakeshaft, Capt. A. J., Norf. R.	June 10
Spence, Lt.-Col. A. H. O., C.I.E., 39 Horse.	Apl. 99
Steel, Bt.-Lt.-Col. R. A., *p.s.c.*, 17 Cav.	Apl. 03
Ward, Capt. B.S., 99 Infy.	Oct. 08
Wright-Warren, Capt. C. T., 80 Infy.	Jan. 10

Dutch.

2nd Class.

Name, Rank, and Corps.	Date of Qualification.
Trotter, Maj. E. G. S.	Apl. 01

Italian.

1st Class.

Name, Rank, and Corps.	Date of Qualification.
Burn, Lt.-Col. E. M. J., R.E.	Oct. 08
Holme, Maj. A. S., R.E.	Jan. 07

2nd Class.

Name, Rank, and Corps.	Date of Qualification.
Dunolly, Lt.-Col. K. J. G., 119 Infy.	Apl. 92
Yeats-Brown, Capt. F. C. C., 17 Cav.	July 09

Japanese.

1st Class.

Name, Rank, and Corps.	Date of Qualification.
* Cardew, Maj. A. M., R. E.	May 10
* McNeight, Maj. A. A., I.M.S.	June 14

2nd Class.

Name, Rank, and Corps.	Date of Qualification.
Barr, Maj. J. H. V., 109 Infy.	Apl. 12
Bateman-Champain, Lt.-Col. H. F., C.M.G. 9 G. R.	Feb. 08
‡ Blair, Maj. G. L., 36 Sikhs	Apl. 11
Cheyne, Lt.-Col. R. E., 8 Cav.	Feb. 08
* Combe, Bt.-Lt.-Col. S. B., M.C., 47 Sikhs	Mar. 12
Dopping-Hepenstal, Lt.-Col. M. E., D.S.O., 1 G. R.	Feb. 09

Japanese.

2nd Class—contd.

Name, Rank, and Corps.	Date of Qualification.
Hart-Synnot, Maj. A. H. S., D.S.O., *p.s.c.*, E. Surr. R.	Dec. 05
Hogge, Maj. L. R., 47 Sikhs	June 10
Jolly, Maj. G. A., I.M.S.	...
Kirby, Bt. Lt.-Col., J. T., D.S.O., S. & T. Corps.	Apl. 13
Nangle, Maj. M. C., *p.s.c.*, 92 Punj.	Apl. 13
Steel, Bt.-Col. R. A., C.I.E., *p.s.c.*, 17 Cav.	Feb. 08
Wallace, Maj. E. C. L., D.S.O., 30 Punj.	Feb. 09
Walter, Maj. E., C.I.E., S. and T. Corps.	Apl. 12
Woulfe Flanagan, Lt.-Col. R. J., R. W. Kent R.	Mar. 06

Norwegian.

2nd Class.

Name, Rank, and Corps.	Date of Qualification.
Paterson, Lt.-Col. P. J., D.S.O., R.A.	Jan. 11

Persian.

1st Class.

Name, Rank, and Corps.	Date of Qualification.
†Ferrar, Maj. M. L., Supy. List (*a*)	Apl. 11
Grey, Lt.-Col. W. G. (*a*)	Apl. 11
Haig, Lt.-Col. T. W., C.M.G., Supy. List (*a*)	Oct. 95
Hoskyn, Maj. J. C. M., D.S.O., 44 Infy. (*a*)	Oct. 10
Johnson, Brig.-Genl. F. E., C.M.G., D.S.O., R.A. (*b*)	Oct. 86
Peart, Maj. C. L., *p.s.c.*, 106 Prs. (*a*)	Nov. 09
Stokes, Bt. Lt.-Col. C. B., *p.s.c.*, 3 Horse (*c*)	Oct. 04

Name, Rank, and Corps.	Date of Qualification.
Andrew, Col. A. W., C.M.G. (*c*)	Apl. 99
Andrews, Col. L. J. (*c*)	July 92
Bagshawe, Lt.-Col. F. W., C.I.E. (*c*)	July 93
Banbury, Bt.-Col. W. E., C.M.G. (*c*)	July 92
Bhola Nauth, Lt.-Col., C.I.E., I.M.S. (*c*) (*c*)	Apl. 02
Bigg-Wither, Maj. F., Supy. List (*c*)	Apl. 89
Boyle, Capt. C. A., 11 Lrs. (*c*)	Mar. 13
Brooking, Maj.-Genl. *Sir* H. T., K.C.B., K.C.M.G. (*c*)	Jan. 92
Bullock, Lt. R. S., 27 Punj.	Apl. 13
Cleveland, Col. H. F., C.I.E., I.M.S. (*c*)	July 99
Cobban, Maj. C. L., 43 Erin. R. (*c*)	Oct. 06
Daly, Capt. C. K., 119 Infy. (*c*)	Oct. 12
Durham, Maj. W. G. W., 27 Lt. Cav. (*c*)	Oct. 08
Eadie, Maj. J. I., D.S.O., 97 Infy. (*c*)	Jan. 97
Evans, Col. U. W., C.B., C.M.G., Brit. Ser. (*c*).	... 93
Fraser, Bt.-Maj. W. A. K., M.C., 39 Horse	Sep. 13
Haworth, Maj. L. B. H. (*c*)	Mar. 06
Hudson, Lt.-Col. J. H. (*c*)	July 98
Keogh, Lt.-Col. J. H., R.A. (*c*)	Oct. 99
Kirkwood, Maj. C. H. K., 23 Cav. (*c*)	Mar. 14
Lane, Capt. C. R. C., M.C., 19 Lrs.	Apl. 14
Lindsay, Bt.-Col. H. A. P., C.M.G., S. and T. Corps. (*c*)	July 85

(*a*) Degree of Honour, 1st Divn.
(*b*) Degree of Honour, 2nd Divn.
(*c*) High Proficiency.

* Official Interpreter in Japanese.
† Official Interpreter in Persian.
‡ Official Interpreter in German.

OFFICERS NOW SERVING WHO HAVE QUALIFIED (OR ARE CLASSED) AS INTERPRETERS IN MODERN FOREIGN LANGUAGES—*contd.*

Persian.

1st Class—concld.

Name, Rank and Corps.	Date of Qualification.
Maturin, Capt. H. G., 61 Prs. (*c*)	Apl. 11
Mitter, Lt.-Col. R. K., I.M.S. (*c*)	July 99
Nixon, Genl. *Sir* J. E., K.C.B. (*c*)	Oct. 85
Noel, Capt. E. W. C., C.I.E., 44 Infy. (*c*)	Feb. 10
O'Brien, Lt.-Col. A. J., C.I.E., Supy. List (*c*)	Sep. 94
Oliver, Maj. D. R. G. (*c*)	...
Rai, Maj. D. G., I.M.S. (*c*)	July 05
Rodwell, Brig.-Genl. E. H., C.B., *p.s.c.* (*b*)	July 87
Routh, Maj. G. M., D.S.O., R.A. (*c*)	Feb. 13
Sheringham, Maj. A. T., D.S.O., 121 Prs. (*c*)	July 13
Stephenson, Lt.-Col. J., M.B., F.R.C.S., I.M.S. (*c*)	Oct. 02
Turner, Maj.-Genl. J. G., C.B. (*c*)	Apl. 88
Weedon, Col. F. F., Brit. Ser. (*c*)	Jan. 94
Wickham, Lt. E. T. R., 39 Horse	Apl. 14
Young, Maj. H. N., S. and T. Corps (*c*)	Jan. 98
Young, Capt. H. W., 116 Mah. (*c*)	Oct. 11

2nd Class.

Name, Rank and Corps.	Date of Qualification.
Elgood, Lt. C., 14 Lrs.	Apl. 17
Fowle, Capt. T. C., 40 Pathans	Apl. 14
Hart, Maj. E. G., D.S.O., S. and T. Corps	Apl. 14
Meiklejohn, Capt. W. L. S., *p.s.c.*, 106 Prs.	Sep. 14
Wadia, Capt. M. D., I.M.S.	Nov. 15
Wagstaff, Capt. L. C., 2 Raj. L. I.	Jan. 14

Russian.

1st Class.

Name, Rank and Corps.	Date of Qualification.
Adye, Bt.-Col. D. R., 98 Infy.	Oct. 99
Anderson, Capt. M. G., 7 G. R.	June 13
Aplin, Bt.-Col. P. J. H., D.S.O.	Oct. 89
Arthur, Maj. L. F., D.S.O., *p.s.c.*, 26 Cav.	Apl. 06
Battine, Lt.-Col. R. St. C., D.S.O., *p.s.c.*, 21 Cav.	Apl. 05
Bell, Maj. A. H., D.S.O., R.E.	Jan. 07
* Bennett, Maj. J. H., 3 Horse	Apl. 06
*Bingham, Maj. W. H., 69 Punj.	July 05
Black, Col. W. C., *p.s.c.*	Oct. 94
‡ Blair, Maj. G. L., 36 Sikhs	Oct. 04
* Bodkin, Maj. L. F., 113 Infy.	Apl. 10
Boileau, Lt.-Col. E. R. P., 2 G. R.	Oct. 02
Browning, Maj. E. R. L., 1 Brah.	Oct. 11
Burton, Col. R. G.	Oct. 93
Clutterbuck, Maj. C. H., 125 Rif.	June 11
Cobban, Maj. C. L., 43 R.	Oct. 11
Collum, Capt. J. A., D.S.O., 26 Cav.	Oct. 10
Cooper, Maj. G. C., M.C., 121 Prs.	June 11
Corlett, Lt-Col. J. S., 15 Lrs.	Jan. 09
Davidson, Lt.-Col. J., D.S.O., M.D., I.M.S.	Oct. 03
Douglas, Col. J. A., C.M.G., C.I.E.	Oct. 90
Duke, Capt. O. A., 22 Cav.	Oct. 10
Dunsterville, Maj.-Genl. L. C., C.B.	Oct. 97
Eastmead, Lt.-Col. C. S., 3 G. R.	Apl. 04
Fisher, Col. J., C.B.	Oct. 95
* Ford, Capt. G. N., 105 Infy.	Apl. 13
Geoghegan, Lt.-Col. S., C.B., *p.s.c.*, 62 Punj.	Apl. 04

Russian—*contd.*

1st Class—concld.

Name, Rank and Corps.	Date of Qualification.
Goldfrap, Capt. H. W., M.C., 103 Infy.	Oct. 12
Grover, Lt.-Genl. *Sir* M. H. S., K.C.B., K.C.I.E.	Apl. 89
* Haines, Capt. J. B., 10 Jats	Apl. 14
Hay, Lt.-Col. W. N., C.I.E., D.S.O., *p.s.c.*, 129 Baluchis	Oct. 05
Holloway, Col. B., C.I.E., *p.s.c.*	Apl. 89
Holman, Bt.-Col. H. C., C.B., C.M.G., D.S.O., *p.s.c.*, 16 Cav.	Apl. 95
Home, Bt.-Col. J. M., *p.s.c.*	Apl. 01
Hutchinson, Col. F. P., C.B.	Oct. 89
Keyes, Bt. Lt.-Col. T. H., C.I.E. Supy. List	Oct. 02
Knox, Bt.-Col. A. W. F., C.B., *p.s.c.*, 58 Rif.	Aug. 07
Koebel, Maj. F. E., 51 Sikhs	Oct. 10
*Leslie, Capt. W., R. A.	July 07
Lord, Maj. H. P. M., Supy. List	Apl. 04
McNeile, Lt.-Col. D. H., 19 Lrs.	Oct. 02
Macaulay, Maj. D. I. M., 1 Lrs.	Apl. 05
Macdonald, Lt.-Col. F. W. P., Supy. List.	Oct. 96
* Marsh, Bt.-Lt.-Col. F. G., C.M.G., *p.s.c.*, 9 G. R.	Mar. 11
Mears, Lt.-Col. A., Supy. List.	Oct. 96
Medley, Lt.-Col. A. G., 19 Lrs.	Apl. 95
*‡ Melliss, Maj.-Gen. *Sir* C. J., K.C.B., K.C.M.G.	Apl. 89
Mockler, Lt.-Col. G. H. G.	Apl. 93
*Morgan, Maj. M. H. L., D.S.O., 62 Punj.	Oct. 11
Morris, Bt.-Col. G. M., D.S.O. 62 Punj.	Apl. 98
Newcombe, Maj. G. H., Supy. List	Oct. 04
Nightingale, Maj. M. R. W., C.M.G., D.S.O., 5 G.R.	Oct. 04
Noel, Capt. E. W. C., C.I.E.	Oct. 12
Oakes, Lt.-Col. H. C., 28 Lt. Cav.	Apl. 00
Palmer, Bt Lt.-Col. G. S., 97 Infy.	Apl. 04
Patrick, Capt. R. M. F., M.C., 42 R.	Oct. 10
Phayre, Bt.-Maj. R. B., M.C., 4 G. R.	Oct. 11
Powell, Maj.-Genl. C. H., C.B. [R]	Apl. 89
Prince, Lt.-Col. W. C. S., Supy. List	Apl. 94
Redl, Maj. E. A. F., C.I.E., 113 Infy.	Oct. 08
Reinhold, Maj. H. E., M.C., 27 Punj.	Oct. 08
Richardson, Maj.-Genl. C. W. G., C.S.I., *p.s.c.*	Oct. 97
Robertson, Lt.-Col. G. A.	Oct. 90
Robson, Lt.-Col. C. G., 10 Jats.	Jan. 09
Rowlandson, Maj. H.W., 82 Punj.	Oct. 11
Rowlandson, Bt. Lt.-Col. M. G. D., D.S.O., *p.s.c.*, 38 Dogras*	Apl. 06
Saunders, Capt. M., D.S.O., 36 Sikhs.	Nov. 11
Scale, Maj. J. D., D.S.O., 87 Punj.	June 13
Searight, Capt. A. K., R. W. Kent. R.	Oct. 08
Selwyn, Col. C. H., C. B., *p.s.c.*, 28 Cav.	Oct. 93
*Seymour, Maj. C. R. F., *p.s.c.*, 13 Raj.	Oct. 13
Shore, Brig.-Gen. O. B. S. F., C.B., C.I.E., D.S.O., *p.s.c.*	Oct. 88
Steele, Col. St. G. L., C. B.	Oct. 88
Stevens, Lt.-Col. N. M. C., C.M.G.	Apl. 90
Tennant, Lt.-Col. E., 20 Horse.	Apl. 00
Thornhill, Bt. Lt.-Col. C. J. M., D.S.O., 6 Jats.	Oct. 11
Tod, Col. J. K., *p.s.c.*	Oct. 21
Van Someren, Maj. W. W., D.S.O., 45 Sikhs.	Oct. 00
Wagstaff, Maj. L. C., C.I.E., 2 Raj. L. I.	Oct. 11
Watson, Maj. H. D., Mily. Accts. Dept.	Oct. 08
Watson, Lt.-Col. L. A., *p.s.c.*, 74 Punj.	Apl. 06
Winter, Bt.-Col. C. B., C.M.G., D.S.O., 112 Infy.	Apl. 01
Worlledge, Bt.-Col. J. F.	Oct. 89

* Official Interpreter in Russian. ‡ Official Interpreter in German.
(*b*) Degree of Honour, 2nd Divn. (*c*) High Proficiency.

OFFICERS NOW SERVING WHO HAVE QUALIFIED (OR ARE CLASSED) AS INTERPRETERS IN MODERN FOREIGN LANGUAGES—*concld.*

Name, Rank and Corps.	Date of Qualification	Name, Rank and Corps.	Date of Qualification.
Russian—*concld.*		**Spanish.**	
2nd Class.		*1st Class.*	
Blair, Maj. R. W. C., 123 Rif.	Jan. 07	Burn, Lt.-Col. E. M. J., R.E.	Oct. 03
Bone, Capt. A. F., 76 Punj.	Oct. 12		
Boome, Bt.-Col. E. H., C.B., C.M.G., 112 Infy.	Oct. 02	*2nd Class.*	
Clay, Col. C. H., *p.s.c.*	Apl. 90		
Crookshank, Maj. C. K., 34 Prs.	Oct. 09	Scovell, Lt. A. M., Sea. Highrs.	Jan. 10
Dickson, Bt.-Col. W. E. R., C.I.E., R. E.	Apl. 91		
Haldane, Lt.-Col. C. L., C.M.G., D.S.O., 7 G.R.	Apl. 00		
Hynes, Capt. H. G. C., 4 G. R.	Aug. 11	**Turkish.**	
Maynard, Maj. F. H., M.C., 125 Rif.	Jan. 14		
Peterson, Col. F. H., C.B., D.S.O.	Oct. 03	*1st Class.*	
Pogson, Capt. F. V., S. & T. Corps	Jan. 10		
Roberts, Lt.-Col. H. L., C.B., M.V.O.	Oct. 07	Batchelor, Capt. V. A., R. F. A.	Jan. 12
Souter, Lt.-Col., A. B., 21 Cav.	Apl. 00	Morton, Col. W. R., Brit. Ser.	Oct. 00
Wilkins, Lt.-Col. E. G. R.	Oct. 06		
Somali.			
2nd Class.		*2nd Class.*	
		Prince, Lt.-Col. W. C. S., Supy. List	Apl. 93
Pritchard, Lt.-Col. C. H., Supy. List	...		
Trotter, Maj. E. G. S.	May 04		

OFFICER NOW SERVING WHO HAS QUALIFIED IN COLLOQUIAL CANTONESE AT HONG KONG OR THE STRAITS SETTLEMENTS.

Name, Rank and Corps.	Date of Qualification.	Name, Rank and Corps.	Date of Qualification.
Bingham, Maj. W. H., 69 Punj. [L]	Aug. 02		

OFFICER NOW SERVING WHO HAS QUALIFIED AT THE CHINESE REGIMENT TEST EXAMINATION IN THE CHINESE LANGUAGE.

Name, Rank and Corps.	Date of Qualification.	Name, Rank and Corps.	Date of Qualification.
Watson, Col. W. M., Brit. Ser.	Apl. 01		

OFFICERS NOW SERVING WHO HAVE QUALIFIED IN VARIOUS CHINESE DIALECTS.

Name, Rank, and Corps.	Date of Qualification.
Pekingese Dialect.	
(*Final.*)	
Bingham, Maj. W. H., 69 Punj. [L]	Aug. 01.
Campbell, Maj. W. L., C.I.E. (Supy. List)	Feb. 03.
Douglas, Lt.-Col. A. P., C.M.G., R.A.	Aug. 07.
Edwardes, Lt.-Col. J. G., 3 G. R.	Nov. 07.
Grey, Maj. A. J. H.	...
Heron, Maj. D., C.I.E., I.M.S.	Mar. 08.
Hogge, Maj. L. R., 47 Sikhs.	Feb. 07.
Jeffery, Maj. W. H., C.I.E., 73 Infy.	Aug. 03.
Kaye, Lt.-Col. C., C.S.I., C.I.E.	Apl. 04.
Mockler, Lt.-Col. G. H. G., 46 Punj. [L].	Sep. 99.
Needham, Maj. R. A. M.B., I.M.S.	Mar. 08.
Peart, Maj. C. L., *p.s.c.*, 106 Prs.	Mar. 00.
Ralston, Maj. W.H., D.S.O., M.C., 47 Sikhs	Feb. 07.
Webster, Maj. C. D., 30 Punj	May 03.
Willoughby, Col. M. E., C.B., C.M.G., *p.s.c.*, 2 Lrs. [L]	Mar. 07.
Woods, Lt.-Col. G. G., R.A.	Apl. 04.
Cantonese Dialect.	
Bingham, Maj. W. H., 69 Punj.	Aug. 02.
Campbell, Capt. W. L., C.I.E.	July 01.
Kelsall, Maj. H. W., R.E.	Oct. 04.
Yunnanese Dialect.	
Abbey, Lt.-Col. W. B. T., Supy. List	Apl. 02.
Broome, Capt. W. G. K., 89 Punj.	Nov. 10.
Browne, Capt. St. J. A., 53 Sikhs	June 14.
Cloete, Maj. H. D., M.C., 92 Punj.	June 08.
Crawford, Bt. Maj. J. D., D.S.O., M.C., 89 Punj.	May 08.
Yunnanese Dialect—contd.	
Crossfield, Capt. E. A. K., 10 G. R.	...
Dening, Maj. H., 25 Cav.	Apl. 08.
Dening, Capt. R., 18 Lanc.	...
Digges La Touche, Maj. C. H., 16 Cav.	May 11.
Egerton, Maj. S. W., 7 Lrs.	Oct. 10.
Engledue, Bt.-Maj. R. S., 89 Punj.	Aug. 11.
Farfan, Maj. A. J. T., D.S.O., R.A.	Jan. 09.
Forster, Bt.-Lt.-Col. D., D.S.O., R.E., *p.s.c.*	...
Fox-Male, Maj. W. C. E. T., M.C., 93 Infy.	Feb. 12.
Geoghegan, Bt.-Lt.-Col., N. M., 89 Punj.	Oct. 10.
Goldman-Monk, Capt. M. E. de B., 91 Punj.	Feb. 15.
Hastings, Maj. W. H., D.S.O., 92 Punj.	May 09.
Haughton, Maj. H. L., 92 Punj.	Aug. 08.
Henderson, Maj. C. G. H., 29 Lrs.	May 08.
Hensley, Maj. F.K., Guides	Nov. 09.
James, Capt. W. R., M.C., 89, Punj.	Jan. 09.
Johnston, Maj. C. E. L., R.A.	Mar. 06.
Kemble, Maj. R. B., S. and T. Corps.	Feb. 12.
Kemmis, Maj. A. W. M., D.S.O., 10 Lrs.	Mar. 06.
Kennedy-Craufurd-Stuart, Maj. C., D.S.O., 127 Infy.	Sep. 13.
Lakin, Maj. J. H. F., 7 G. R.	Sep. 05.
Landon, Maj. C. R. H. P., 35 Horse	June 11.
Mathew, Lt.-Col. C. M., I.M.S.	Jan. 09.
Murray, Lt.-Col. A. B.	Mar. 03.
Orman, Lt.-Col. F. L., 10 G. R.	Oct. 13.
Pelly, Capt. H. R., 89 Punj.	Aug. 10.
Percy-Smith, Bt. Lt.-Col. E. S., Supy. List	Feb. 05.
Phythian-Adams, Capt. E. G., 80 Infy.	June 14.
Rebsch, Capt. W. K., D.S.O., S. & T. Corps.	...
Scruby, Maj. W. S. J., 89 Punj.	Dec. 09.
Simpson, Maj. J., 10 G.R.	Nov. 08.
Watts, Capt. N. H. L., 92 Punj.	June 08.
Wilkinson, Maj. T. O., 91 Punj	June 07.
Williams, Capt H. R., 10 G. R.	...
Wright-Warren, Capt. C. T., 80 Infy	...

OFFICERS NOW SERVING WHO HAVE QUALIFIED IN LANGUAGES APPERTAINING TO INDIA.

NOTE—Officers whose names are preceded by an asterisk (*) have passed in the First Division and have been awarded Gold Medals.

Name, Rank and Corps.	Date of Qualification.
Baluchi.	
Interpreter's Test.	
Jacob, Lt.-Genl. *Sir* C. W., K.C.B.	Apl. 89.
Kelly, Maj.-Genl. F.H., C.B., Brit. Ser.	July 90.
Leslie, Bt.-Col. G. A. J., C.B., R. E.	Apl. 93.
Wadeson, Maj.-Genl. F. W. G., C.B.	Sep. 87.
Watling, Bt.-Col. G., 11 Raj.	Jan. 89.
Canarese.	
(*Interpreter.*)	
Bannerman, Maj.-Genl. W. B., C.S.I., M.D., I.M.S., K.H.P.	1892
Hindi.	
(*High Proficiency.*)	
Eadie, Maj. J. I., D.S.O., 97 Infy.	
Pulley, Maj. H. C., 12 Prs. [L]	Oct. 11.
Robertson-Glasgow, Capt. N., R.A.	Oct. 05.
Seconde, Capt. E. C., 16 Raj.	Mar. 15.
Wigley, Capt. P. J. R., M.C., 90 Punj.	Oct. 10.

OFFICERS NOW SERVING WHO HAVE QUALIFIED IN LANGUAGES APPERTAINING TO INDIA—*contd.*

Marathi.

(High Proficiency.)

Name, Rank and Corps.	Date of Qualification.
Wilberforce-Bell, Capt. H.	Sep. 11.

(Interpreter's Test.)

Name, Rank and Corps.	Date of Qualification.
Andrew, Col. A. W., C.M.G.	Mar. 95.

Tamil.

(High Proficiency.)

Name, Rank and Corps.	Date of Qualification.
Roe, Col. C. H., C.I.E., Brit. Ser.	Jan. 06.
Sorsbie, Col. R. F., C.I.E., R. E.	July 94.

Urdu.

(Honors.)

Name, Rank and Corps.	Date of Qualification.
Bhola Nauth, Lt.-Col., C.I.E., I.M.S.	July 07.
*Boyle, Capt. C. A., 11 Lrs. [L]	Mar 17.
*Ferrar, Maj. M. L., Supy. List	Apl. 15.
*Haig, Lt.-Col. T. W., C. M. G., Supy. List. [L]	Oct. 96.
*Harcourt, Capt. E. S., M.C., 7 G. R. [L]	July 12.
Hoskyn, Maj. J. C. M., D.S.O., 44 Infy.	Sep. 14.
*Lindsay, Bt.-Col. H. A. P., C.M.G., S. and T. Corps.	July 11.
*Peart, Maj. C. L., *p.s.c.*, 106 Prs.	Dec. 08.
Vazifdar, Capt. S. S., I.M.S.	July 12.

(High Proficiency.)

Name, Rank and Corps.	Date of Qualification.
Andrews, Col. L. J.	Jan. 92.
Anderson, Major A. T., R.F.A.	Oct. 08.
Bagshawe, Lt.-Col. F. W., C.I.E.	Jan. 92.
Baldwin, Col. G. M., D.S.O.	Oct. 05.
Banbury, Bt.-Col. W. E., C.M.G.	Jan. 94.

Urdu—*contd.*

(High Proficiency)—contd.

Name, Rank and Corps.	Date of Qualification.
Bigg-Wither, Maj. F., Supy. List	Apl. 97.
Birch, Maj. V. K., 29 Lrs.	Apl. 05.
Branson, Maj. L. H., 110 Lt. Infy.	Oct. 11.
Brown, Lt.-Col. W. H., C.B.	July 00.
Chopra, Capt. R. N., I.M.S.	Mar. 12.
Cleveland, Col. H. F., C.I.E., I.M.S.	July 06.
Condon, Lt.-Col. J. K.,	July 98.
Condon, Capt. W. E. H., 17 Infy.	Apl. 14.
Corbett, Capt. T. W., M.C., 9 Horse.	July 14.
Doveton, Capt. A. C., 21 Punj.	Oct. 12.
Durham, Maj. W. G. W., 27 Lt. Cav.	Apl. 06.
Eadie, Maj. J. I., D.S.O., 97 Infy.	Jan. 11.
Elgood, Lt.-Col. C., 14 Lrs.	Nov. 17.
Gompertz, Capt. M. L. A., 108 Infy. [*l*]	Apl. 14.
Greenhouse, Capt. F. S., 94 Infy.	Oct. 13.
Grey, Lt.-Col. W. G., Supy. List	Apl. 96.
Harvest, Lt.-Col. H. deV., Supy. List.	Apl. 97.
Heale, Maj. R. J. W., O.B.E., Supy. List	Oct. 09.
Hudson, Lt.-Col. J. H.	July 99.
Hunt, Lt.-Col. V. deV.	Oct. 97.
Jacob, Lt.-Col. H. F., C.S.I., Supy. List.	Apl. 96.
Jamal-ud-din, Lt., M.D., I.M.S.	Mar. 15
Jardine, Bt.-Lt.-Col., C. H., 96 Infy.	Apl. 12.
Kanwar Pirthi Singh. Capt., 5 Cav.	July 12.
Khan, Capt. A. S., I.M.S.	Oct 11.
Kirkwood, Maj. C. H. K., 23 Cav. [L]	Mar. 13.
Kureishi, Capt. M. D. A., I. M. S.	Sep. 14.
Maturin, Capt. H. G., 61 Prs. [L]	July 14.
Mehta, Capt. S. B., I.M.S.	Apl. 13.
O'Brien, Lt.-Col. A. J., C.I.E., Supy. List.	Oct. 69.
Palmer, Bt. Lt.-Col. G. S., 97 Infy. [L]	Apl. 04.
Parsons, Capt. A. E. B., 52 Sikhs	July 14.
Poole, Maj. I. M. C., D.S.O., S. and T. Corps.	Mar. 10.
Prince, Lt.-Col. W. C. S., Supy. List. [L]	July 92.
Pulley, Maj. H. C., 12 Prs. [L]	July 09.
Rai, Maj. D. G., I.M.S.	July 04.
Rai, Capt. D. H., M.C., I. M. S.	Jan. 11.
Robertson-Glasgow, Maj. N., R.A.	Apl. 12.
Ross, Bt. Lt.-Col. R. C., D.S.O., 6 Infy.	Oct. 10.
Routh, Maj. G. M., D.S.O., R.A.	Sep. 10.
Ruck, Capt. R. C., 95 Infy.	July 13.
Secondé, Capt. E. C., 16 Raj.	July 12.
Sheringham, Maj. A. T., D.S.O., 121 Prs. [L]	Jan. 12.
Sodhi, Capt. N. S., M.C., I.M.S.	Oct. 09.
Stokes, Bt. Lt.-Col. C. B., C.I.E., 3 Horse, *p.s.c.* [L].	Oct. 99.
Szczepanski, Maj. H. C.	Jan. 06.
Turner, Maj.-Genl. J. G., C.B.	Jan. 89.
Walker, Maj. A.M.E.H.T.	Jan. 11.
Watts, Capt. N. H. L., 92 Punj.	
Wigley, Capt. P. J. R., M.C., 90 Punj.	July 10.
Williamson, Maj. A. G., 7 Lrs.	Oct. 13.
Winter, Bt.-Col. C.B., C.M.G., D.S.O., 112 Infy. [L]	Oct. 09.
Young, Bt.-Lt.-Col., E. deL., R.E.	Oct. 11.

* 1st Divn. with gold medals.

HONORARY RANK.

Names.	Date of rank.	Remarks.
OFFICERS WITH HONORARY RANK IN THE ARMY.		
Lieutenant-Generals.		
Nepal, *His Excellency Maharaja Sir* Chandra Shumshere Jung, *Rana Bahadur,* G.C.B., G.C.S.I., G.C.V.O., *Prime Minister, Marshal* of—	11 Nov. 15	Hon. Col., 4th G. R.
Jammu and Kashmir, *His Highness Maharaja Sir* Partab Singh, *Bahadur,* G.C.S.I., G.C.I.E., G.B.E., *Maharaja* of—	3 June 16	Hon. Col., 37th Dogras.
Pratap Singh, *His Highness Maharaja Bahadur, Sir,* G.C.B., G.C.S.I., G.C.V.O., A.-D.-C. (of Jodhpur)	3 June 16	Hon. Col., 34th Horse.
Gwalior, *His Highness Sir* Madho Rao Scindia, *Bahadur,* G.C.S.I., G.C.V.O., G.B.E., A.-D.-C., *Maharaja* of—	1 Jan. 18	Hon. Col., 1st Lancers.
Hyderabad, *His Exalted Highness Nawab Mir Sir* Usman Ali Khan, *Bahadur, Fateh Jang,* G.C.S.I., G.B.E., *Nizam* of—	1 Jan. 18	Hon. Col., 20th Deccan Horse.
Major-Generals.		
Jaipur, *His Highness Maharajadhiraja Sir* Sawai Madho Singh *Bahadur,* G.C.S.I., G.C.I.E., G.C.V.O., G.B.E., *Maharaja* of—	12 Dec. 11	Hon. Col., 13th Rajputs.
Bikaner, *His Highness Maharaja Sir* Ganga Singh *Bahadur,* G.C.S.I., G.C.I.E., K.C.B., A.-D.-C., *Maharaja* of—	1 Aug. 17	
Patiala, *His Highness Maharajadhiraja Sir* Bhupindar Singh Mahindar *Bahadur,* G.C.I.E., G.B.E., *Maharaja* of—	15 June 18	Hon. Col., 15th Sikhs.
Colonels.		
Rampur, *His Highness Nawab Sir* Muhammad Hamid Ali Khan *Bahadur,* G.C.I.E., G.C.V.O., A.-D.-C., *Nawab* of—	3 June 10	9th Hodson's Horse.
Mysore, *His Highness Maharaja* Sri *Sir* Krishnaraja Wadiyar *Bahadur,* G.C.S.I., G.B.E., *Maharaja* of—	24 June 10	Hon. Col., 26th Cavalry.
Kolhapur, *His Highness Sir* Shahu Chhatrapati Maharaj, G.C.S.I., G.C.I.E., G.C.V.O., LL.D., *Maharaja* of—	1 Jan. 15	Hon. Col., 103rd Infantry.
Ratlam, *His Highness Raja Sir* Sajjan Singh, K.C.S.I., *Raja* of—	1 Jan. 18	39th Horse.
Muhammad Nasrulla Khan, *Nawab of* Bhopal	1 Jan. 18	Hon. Col., 9th Bhopal Infantry.
Lieutenant-Colonels.		
Muhammad Ali Beg, *Nawab Afsar i-Jang, Afsar-ud daula, Bahadur, Sir* K.C.I.E., M.V.O.	9 Aug. 02	20th Deccan Horse.
Sirmur, *Kumar* Bir Bikram Sing, C.S.I., C.I.E., of—	24 June 10	1st S. & M.
Lambagraon, *Raja Sir* Jai Chand, K.C.I.E., C.S.I., *Raja* of—	25 Jan. 11	37th Dogras.
Rewa, *His Highness Maharaja Sir* Venkat Raman Singh *Bahadur,* G.C.S.I., *Maharaja* of—	1 Jan. 15	
Kotah, *His Highness Maharao Sir* Umed Singh *Bahadur,* G.C.S.I., G.C.I.E., G.B.E., *Maharao* of—	1 Jan. 15	42nd Deoli Regt
Alwar, *His Highness Maharaja Sir* Jey Singh *Bahadur,* K.C.S.I., K.C.I.E., *Maharaja* of—	1 Jan. 15	
Kishangarh, *His Highness Maharajadhiraj Sir* Madan Singh *Bahadur,* K.C.S.I., K.C.I.E., *Maharaja* of—	1 Aug. 17	
Desraj Urs, C.I.E., M.V.O., Col., Mysore Impl. Serv. Lancers	11 Aug. 17	
Rose, Horace Arthur	19 Oct. 17	
Idar, *His Highness Maharaja* Daolat Singhji, *Maharaja* of—	1 Jan. 18	
Navanagar, *His Highness Maharaja Shri Sir* Ranjitsinji Vibhaji, K.C.S.I., *Maharaja* of—	1 Jan. 18	
Khairpur, *His Highness Mir Sir* Imam Baksh Khan, *Talpur,* G.C.I.E., *Mir* of—	1 Jan. 18	
Bhavnagar, *His Highness Maharaja Sir* Bhavsinhji Takhtsinhji, K.C.S.I., *Maharaja* of—	1 Jan. 18	
Jind, *His Highness Maharaja Sir* Ranbir Singh, *Rajendra Bahadur,* G.C.I.E., K.C.S.I., *Maharaja* of—	1 Jan. 18	
Kapurthala, *His Highness Maharaja Sir* Jagatjit Singh *Bahadur,* G.C.S.I., *Maharaja* of—	1 Jan. 18	
Sirmur, *His Highness Maharaja Sir* Amar Parkash *Bahadur,* K.C.S.I., *Maharaja* of—	1 Jan. 18	
Majors.		
Jaora, *His Highness Nawab Sir* Muhammad Iftikar Ali Khan *Bahadur,* K.C.I.E., *Nawab* of—	12 Dec. 11	39th Horse.
Haji Hafiz Obeidullah Khan, *Nawabzada,* C.S.I., A.-D.-C. Brig.-Gen., & Commander-in-Chief, Bhopal State Forces	12 Dec. 11	
Maler Kotla, *His Highness Nawab Sir* Ahmad Ali Khan *Bahadur,* K.C.S.I., *Nawab* of—	14 June 16	3rd S. and M.
Malik Umar Hayat Khan, *The Hon'ble Sir,* K.C.I.E., M.V.O., Tiwana, of Kalra, Shahpur District	3 June 17	18th Lancers.
Burgoyne, Alan Hughes	17 June 17	(temp.)
Faridkot, *His Highness Maharaja* Brij Indar Singh *Bahadur, Raja* of—	1 Jan. 18	
Shah Mirza Beg, Hyderabad Imperial Service Lancers	8 Mar. 18	
Gurnam Singh, Patiala State Mil. Forces	4 May 18	(temp.)

HONORARY RANK.

Names.	Date of rank.	Remarks.
OFFICERS WITH HONORARY RANK IN THE ARMY—*contd.*		
Captains.		
Holmes, Bernard Willoughby, E. I. R. Vol. Rifles	12 Jan. 01	
Sidey, Charles Lyon, Surma Val. L. H.	12 Jan. 01	
Powell, Surgn.-Capt. Samuel Arthur, M.D., Bo. L. H.	12 Jan. 01	
Skinner, Stanley Edgar	12 Oct. 10	1st Lancers.
Malik Muhammad Mubariz Khan, *Nawab*, O.B.E., Tiwana, of Shahpur *late* 9th Bengal Lancers	24 Apl. 12	9th Hodson's Horse.
Ahmed Nawaz Khan, *Saddozai*, M.B.E., *Nawab*, of Dera	7 Dec. 12	35th Horse.
Barwani, *His Highness, Rana, Sir* Ranjit Singh, K.C.I.E., *Rana* of—	18 Feb. 15	
Wren, Percival Christopher	1 Nov. 15	
Vankaner, *Raj Saheb Sir* Amarsinhji Banesinhi, K.C.I.E., Raj Saheb of—	9 Jan. 16	
Llewhellin, Neville Gerald Richard	10 Apl. 16	
Tale Muhammad Khan, *Nawabzada*	29 Oct. 16	
Hodges, Robert William	9 May 17	
Thakur Dip Singh, M.C., *Rao Bahadur*	9 Aug. 17	
Abdul Samad Shah	2 Sep. 17	(temp.) 31st Lancers.
Fernandes, Cajetan	20 Oct. 17	(temp.)
Jamkhandi, *Meherban Sir* Parashramrav Ramchandrarav (Bhausaheb Patwardhan), K.C.I.E., *Chief* of—	1 Jan. 18	
Sachin, *Nawab Sidi* Ibrahim Mohamed Yakub Khan, *Nawab of*—	1 Jan. 18	Hon. A.-D.-C. to G.O.C., [Poona.
Savanur, *Meherban* Abdul Majid Khan, *Nawab of*—	1 Jan. 18	29th Lancers. (Lt., I. L. Forces.)
Akalkot, *Meherban Shrimant* Fatehsinhrav Shahaji Raje Bhonsle, Bapusaheb, *Raje Saheb* of—	1 Jan. 18	
Barya, *Maharawal Shri* Ranjitsinghji Mansinghji, *Raja of*—	1 Jan. 18	
Hari Singh, *Raja Sir*, K.C.I.E.	27 Jan. 18	
Ganpat Rao Raghunath Rajwade, A.-D.-C., of Gwalior	27 Jan. 18	
Cope, Alexander Arthur Richard	26 Mar. 18	
Bilaspur, *His Highness Raja Sir* Bije Chand, K.C.I.E., C.S.I., Raja of—	7 Aug. 18	(temp.)
Veterinary Captain.		
Stevenson, William, *late* Lumsden's Horse	12 June 01	
Lieutenants.		
Pugh, Harbet Owain, D.S.O., *late* Lumsden's Horse	12 Jan. 01	
Abdul Rahman Khan, *Sardar, Effendi*	17 Sep. 13	5th Cavalry.
Baba Kartar Singh Bedi	7 Feb. 14	38th Horse.
Panna, Hira Singh, *Rajkumar, of*—	15 Oct. 14	
Cooch Behar, Hitendra Narayan, *Maharaj Kumar, of*—	15 Oct. 14	(temp.)
Tikari, Gopal Saran Narain Singh, *Maharaj Kumar, of*—	15 Oct. 14	(temp.)
Jogendra Singh, *Sirdar, of* Patiala	15 Oct. 14	(temp.)
Qadir Bakhsh Khan	3 Feb. 15	(temp.)
Mehr Mohammad Khan *Bahadur*, of Maler Kotla	3 July 15	
Tehri, *His Highness Raja* Narendra Sah, *Raja of*—	14 June 16	1/39th Garh. Rif.
Younghusband, Cyril	4 Sep. 16	
Ker, Edward Cameron	7 Sep. 16	
Gwatkin, Willougby Levinge Stapleton	19 Oct. 16	
Robertson, Frederick Arthur deVere	22 Oct. 16	
Morrison, William Hay	24 Nov. 16	
Arnott, H. W.	25 Nov. 16	
Durand, Percy Fisher, M. C.	28 Dec. 16	
Ward, Horace Futcher	5 Feb. 17	
Ghulam Ali Khan, *Mirzada Mir*, of Khairpur	9 Apl. 17	
Anderson, Archie Ripon	14 June 17	
Harker, John Thwaite	4 July 17	
Ruthven, Hugh Peter Lansdale	25 July 17	
Shelverton, St. John Alfred	28 July 17	
McConnell, Clarence Henry Quentin	1 Aug. 17	
Radcliffe, Cyril Elphinstone St. John	4 Aug. 17	
Winckworth, William Norman	8 Sep. 17	
Reid, Claud Lestock	21 Sep. 17	
Clark, Henry Agar	14 Oct. 17	
Macdonald, Ian Pendlebury	18 Oct. 17	
Wace, Ernest William Cornish	20 Oct. 17	
Marshall, William David Lucas	2 Nov. 17	
Humfress, Harold Tunmer	8 Nov. 17	
Tomlinson, Ernest William	13 Nov. 17	
Peto, Valentine Frank Cane	29 Nov. 17	
Gillon, Walter Douglas	. . . .	
Stoddard, Arthur Askwith	. . . .	
Manisty, George Stewart	. . . .	
Cox, Cuthbert Eustace Connop	. . . .	
Price, Edmund William Arthur	22 Dec. 17	
Pinniger, Reginald Charles	26 Dec. 17	
Rayneau, Jules Robert	28 Dec. 17	

HONORARY RANK.

OFFICERS WITH HONORARY RANK IN THE ARMY—*contd.*

Lieutenants—*contd.*

Names.	Date of rank.	Remarks.
Bharatpur, *His Highness Maharaja Sri* Brajindra Sawai Kishan Singh, *Bahadur, Maharaja* of—	1 Jan. 18	
Evans, Cyril Herbert	1 Jan. 18	
Burton, Hugh Ashby	8 Jan. 18	
McColl, James Surherland	8 Jan. 18	
Tresler, Karl Theodore Berthold	1 Apl. 18	
Aitkenhead, James Dykes, M.C.	1 Apl. 18	
Yates, Kenneth	3 Apl. 18	
Apcar, Thomas Seth	8 Apl. 18	
Tilley, William Ernest Cedric	15 May 18	
Stedman, Charles Richard	21 June 18	
Rodwell, E. H.	6 July 18	
Bikaner, *Maharaj Kumar Sri* Sadul Singh, *Sahib Bahadur,* of—	7 Aug. 18	

2nd-Lieutenants.

Names.	Date of rank.	Remarks.
Hemsted, Frederick	24 Dec. 15	
Krishna Chandra Kalubha, *Kumar Shri,* of Bhavnagar	9 Jan. 16	
Shivsingji Karan Sinhji, *Kumar Shri, Jhala,* of Vánkáner	9 Jan. 16	
Cahiran Jivabhai Mahinetura, *Secretary,* L.R.C.P., L.R.C.S., L.F.P.S. (Glas.), Chief Medl. Offr. of Vánkáner State	9 Jan. 16	(temp.
Mudhol, *Meherban* Malojirao Vyankatrav Raje Ghorpade (Nanasaheb), *Chief of*—	27 Apl. 16	
Sagat Singh, *Rao-Raja,* of Jodhpur	12 July 16	(temp.)
Hanut Singh, *Rao-Raja,* of Jodhpur	12 July 16	(temp.)
Savantvadi, Raje Bahadur, Shrimant Khem Savant Bhonsle (Bapu Saheb), Sar Desai of—	24 June 17	(temp.)
Stayner, Joscelyn Winstowe	5 July 17	
Cameron, Alan	21 Oct. 17	
Hurlbatt, Douglas Gray	. . . .	
Bear, Richard Mason	. . . .	
Watkis, Gerard Linton	. . . .	
Benson, George Falls	31 Jan. 18	
Norman, Frank	2 July 18	

OFFICERS OF THE BRITISH SERVICE SERVING IN INDIA, AND OF THE INDIAN ARMY (FIELD MARSHAL TO LIEUT.-COLONELS).

Field-Marshal.	Date of rank.
Egerton, *Sir* C. C., G.C.B., D.S.O. [R]	17 Mar. 17
Generals.	
Barrow, *Sir* E. G., G.C.B., G.C.S.I. [R]	27 July 09
Monro, *Sir* C. C., G.C.B., G.C.M.G., Brit. Ser., *p.s.c.*, A.-D.-C.	1 Oct. 16
Barrett, *Sir* A. A., G.C.B., K.C.S.I., K.C.V.O., A.-D.-C. [R]	1 Aug. 17
Birdwood, *Sir* W. R., K.C.B., K.C.S.I., K.C.M.G., C.I.E., D.S.O., A.-D.-C. *q.s.*	23 Oct. 17
Lieutenant-Generals.	
Anderson, *Sir* C. A., K.C.B., Brit. Ser. (206)	12 May 14
VC Aylmer, *Sir* F. J., K.C.B., Brit. Ser. (206) [R]	11 June 15
Phayre, *Sir* A., K.C.B.	30 Oct. 15
Altham, *Sir* E. A., K.C.B., C.M.G., Brit. Ser. *p.s.c.*	1 Jan. 17
Campbell, *Sir* F., K.C.B., D.S.O.	1 Jan. 17
Cox, *Sir* H. V., K.C.B., K.C.M.G., C.S.I	1 Jan. 17
Jacob, *Sir* C. W., K.C.B.	3 June 17
Keary, *Sir* H. D'U., K.C.B., K.C.I.E., D.S.O.	1 Aug. 17
Egerton, *Sir* R. G., K.C.B., K.C.I.E.	27 Aug. 17
Hudson, *Sir* H., K.C.B., C.I.E.	23 Oct. 17
Wapshare, R., C.B., C.S.I.	15 June 18
Temporary. [Brit. Ser.,	
O'Donnell, *Sir* T. J., K.C.I.E., C.B., D.S.O.	21 Sep. 16
Major-Generals.	
Powell, *Sir* C. H., K.C.B. [L] [R]	15 Nov. 07
Hastings, E. S., C.B., D.S.O.	15 Nov. 07
Mackenzie-Kennedy, E. C. W., C.B.	9 June 08
Drummond, *Sir* F. H. R., K.C.I.E., C.B.	27 July 09
Kelly, F. H., C.B., Brit. Ser. (206)	4 Dec. 09
Wallace, A., C.B.	3 May 11
Bell, *Sir* J. A., K.C.V.O.	6 May 11
Davison, K. S., C.B.	3 June 11
Townshend, *Sir* C. V. F., K.C.B., D.S.O., Brit. Ser. (206)	19 June 11
Bannerman, *Hon.* W. B., C.S.I., M.D., I.M.S., K.H.P.	26 July 11
Younghusband, *Sir* G. J., K.C.M.G., K.C.I.E., C.B., *p.s.c.*	29 Aug. 11
Cookson, G. A., C.B., C.M.G.	23 Oct. 11
VC Melliss, *Sir* C. J., K.C.B., K.C.M.G. [L]	19 Mar. 12
Bunbury, W. E., C.B., *p.s.c.*	1 May 12
Anderson, L. E., C.B., Brit. Ser.	4 May 12
Carnegy, P. M., C.B.	1 Sep. 12
Raitt, H. A., C.B., Brit. Ser. (206)	4 Sep. 12
Young, C. F. G.	21 Sep. 12
Lyons, *Hon.* R. W. S., M.D., I.M.S., K.H.P.	11 Jan. 14
Pirie, C. P. W., C.B.	30 Jan. 14
Turner, J. G., C.B. [L]	5 Feb. 14
Fry, C. I., C.B.	11 Mar. 14
Shaw, D. G. L.	20 Mar. 14
Kemball, *Sir* G. V., K.C.M.G., C.B., D.S.O., Brit. Ser., *p.s.c.* (*g*) (206)	8 Apl. 14
Macintyre, D. C. F., C.B.	4 May 14
Treherne, *Sir* F. H., K.C.M.G., F.R.C.S., Edin., Brit. Ser.	1 Mar. 15
‡Kirkpatrick, *Sir* G. M., K.C.B., K.C.S.I., Brit. Ser. *p.s.c.* (206)	3 June 15
Tighe, *Sir* M. J., K.C.M.G., C.B., C.I.E., D.S.O.	3 June 15
Delamain, *Sir* W. S., K.C.M.G., C.B., D.S.O.	3 June 15
Stewart, J. M., C.B.	3 June 15
‡Marshall, *Sir* W. R., K.C.B., K.C.S.I., Brit. Ser.	3 June 15
Bingley, *Hon. Sir* A. H., K.C.I.E., C.B.	3 June 15
Dobell, *Sir* C. M., K.C.B., C.M.G., D.S.O., Brit. Ser., *p.s.c.* [*l*]	3 June 15
Cowper, M., C.B., C.I.E.	27 Aug. 15
Stuart, *Sir* R. C. O., K.C.S.I., Brit. Ser. (206)	26 Oct. 15
Wadeson, F. W. G., C.B.	28 Oct. 15

Major-Generals—*contd.*	Date of rank.
Scott, *Sir* A. B., K.C.B., D.S.O., Br. Ser.	1 Jan. 16
Nugent, O. S. W., C.B., D.S.O., *p.s.c.*	1 Jan. 16
Edwards, W. R., C.B., C.M.G., M.D., I.M.S. K.H.P. [R]	1 Apl. 16
‡ Cobbe, *Sir* A. S., K.C.B., C.S.I., D.S.O., *p.s.c.*	3 June 16
Rimington, J. C., C.B., C.S.I., Br. Ser. (206)	3 June 16
‡Barrow, *Sir* G. deS., K.C.M.G., C.B., *p.s.c.* [*l*]	3 June 16
Hunter, J. G., C.B.	30 Aug. 16
Malleson, W., C.B., C.I.E.	20 Sep. 16
Leader, H. P., C.B., Br. Ser.	1 Jan. 17
Williams, *Sir* G., K.C.I.E., C.B., Br. Ser. (207)	1 Jan. 17
Fane, *Sir* V. B., K.C.I.E., C.B.	1 Jan. 17
Barratt, W. C., C.B., D.S.O.	1 Jan. 17
Ewart, *Sir* R. H., K.C.M.G., C.B., C.I.E., D.S.O.	1 Jan. 17
Beynon, *Sir* W. G. L., K.C.I.E., C.B., D.S.O.	1 Jan. 17
Brooking, *Sir* H. T., K.C.B., K.C.M.G., [L]	1 Jan. 17
Woodyatt, N. G., C.B.	1 Jan. 17
Eustace, A. H., C.B., D.S.O.	1 Jan. 17
Knight, W. C., C.B., C.S.I., D.S.O., *p.s.c.* [L]	1 Jan. 17
Macandrew, H. J. M., C.B., D.S.O., *q.s.*	1 Jan. 17
Twining, P. G., C.B., C.M.G., M.V.O., Br. Ser.	1 Jan. 17
Shea, J. S. M., C.B., C.M.G., D.S.O., *p.s.c.*, *q.s.*	1 Jan. 17
Lecky, R. St. C., C.B., C.M.G., Br. Ser. (206)	3 June 17
Fowler, F. J., C.B., D.S.O.	23 Oct. 17
Hardy, T. H.	3 Dec. 17
Manifold, *Hon.* C. C., C.B., C.M.G., M.B., I.M.S.	29 Dec. 17
Carr, H., C.B., M.D., Br. Ser.	26 Dec. 17
Robinson, W. H. B., C.B., I. M. S.	8 Jan. 18
Hehir, P., C.B., C.M.G., C.I.E., M.D., F.R.C.S., Edin., I.M.S.	13 Mar. 18
Fowler, C. A., C.B., D.S.O.	6 Apl. 18
Dunsterville, L. C., C.B.	3 June 18
Hendley, C. E., C.B.	3 June 18
Cayley, *Sir* W. deS., K.C.M.G., C.B., Br. Ser.	3 June 18
James, W. B., C.B., C.I.E., M.V.O.	3 June 18
Climo, S. H., C.B., D.S.O.	3 June 18
Hill, J., C.B., D.S.O.	3 June 18
Skeen, A., C.M.G., *p.s.c.*	3 June 18
Palin, P. C., C.B., C.M.G.	3 June 18
Scott, T. E., C.B., C.I.E., D.S.O.	3 June 18
Richardson, C. W. G., C.S.I., *p.s.c.* [L]	3 June 18
Walter, J. MacN., C.B., C.S.I., D.S.O., Br. Ser., *p.s.c.* (206)	16 June 18
Hendley, H., M.D., I.M.S., K.H.S.	21 June 18
Giffard, G. G., C.S.I., I.M.S.	6 July 18
Temporary.	
Ravenshaw, H. S. L., C.M.G., Br. Ser.	23 Sep. 16
Holman, H. C., C.B., C.M.G., D.S.O.	13 Nov. 16
Ritchie, A. B., C.M.G., Br. Ser.	5 Dec. 16
Renny, S. M., C.I.E., Br. Ser.	10 Feb. 17
Johnson, F. E., C.M.G., D.S.O., Br. Ser.	22 Feb. 17
Jones, L. C., C.M.G., M.V.O., *p.s.c.* [*l*]	10 Mar. 17
Vaughan, L. R., C.B., D.S.O.	21 May 17
Thuillier, H. F., C.B., C.M.G., R. E.	17 June 17
Freeland, H. F. E., C.B., M.V.O., D.S.O., R. E. (206)	27 Oct. 17
Atkinson, E. H. deV., C.B., C.M.G., C.I.E., Br. Ser.	1 Nov. 17
Thomson, W. M., C.B., M.C., Sea. Highrs.	22 Nov. 17
Leslie, G. A. J., C.B., C.M.G., R.E.	17 Dec. 17
Watson, H. D., C.B., C.M.G., C.I.E., M.V.O., *Eq.*	7 Jan. 1
Peck, A. W., C.M.G., 17 Cav.	15 Feb. 1
Cooke, H. F., D.S.O.	1 Apl. 1
Bethell, H. K., D.S.O., 7th Hrs.	31 Mar. 18
Charles, J. R. E., C.B., D.S.O., R.E., *p.s.c.* (206)	4 Aug. 18

‡ Temporary Lieut.-General.

Brigadier-Generals.	Date of rank.
Forbes, A. W., C.B.	26 Jan. 16
Nanton, H. C., C.B., Brit. Ser	14 Jan. 18
Hamilton, W. G., C.B., C.S.I., D.S.O., Brit. Ser., (206)	6 Oct. 12
Johnstone, A. A. J.	22 Apl. 13
Smith, G. B., C.B., Brit. Ser. (206)	21 Dec. 11
Perry, *Sir* H. W., K.C.M.G., C.B., C.S.I., Brit. Ser.	2 Nov. 15
Stokes, A., C.B., C.M.G., D.S.O., Brit. Ser. (*g*) (206)	...
Strange, R. G., C.I.E., Brit. Ser. (206)	23 Sep. 14
Twigg, R. H., C.B.	25 July 17
Glanville, F., D.S.O., (206)	4 Feb. 16
Herbert, L. N., C.B., [*l*]	23 July 17
Iggulden, H. A., Brit. Ser. (206)	22 Jan. 17
Nuthall, C. E., C.B., C.M.G., Brit. Ser.	11 Oct 16
Buchanan, K. J., C.B.	26 Apl. 16
Campbell, A. A. E.	26 Dec. 14
Tidswell, E.C., C.B., D.S.O., Brit. Ser. (206)	25 Nov. 14
McLeod, W. K., Brit. Ser.	25 Apl. 16
Wolley-Dod, O.C., C.B., D.S.O., *p.s.c.* [*l*] (206)	3 June 15
Holloway, B., C.I.E., *p.s.c.*	10 July 16
Caulfeild, C. T., C.M.G., Brit. Ser.	3 Nov. 16
Muspratt, F. C., C.B., C.M.G.	3 Sep. 17
Roe, C. H., C.I.E., Brit. Ser. (206)	5 Jan. 18
Duhan, W. W. T. (*g*), Brit. Ser.	6 Oct. 16
Peterson, F. H., C.B., D.S.O.	9 Aug. 15
Carruthers, R. A., C.B., C.M.G.	12 Dec. 14
Betham, R. M.	26 June 15
Commeline, F. H. B.	14 Apl. 15
Miles, P. J. C.B.	25 July 17
Colomb, G. H. C., C.M.G.	1 Dec. 14
Walton, W. C., C.B., C.M.G.	15 June 18
Powell, S. H., C.B. [L] (206)	23 June 17
Dealy, J. A., C.M.G., D.S.O., Brit. Ser.	7 Dec. 15
Soady, G. J. Fitz M., C.M.G.	4 Oct. 17
Evans, U. W., C.B., C.M.G., Brit. Ser.	13 Apl. 16
Sorsbie, R. F., C.I.E. Brit. Ser.	14 May 17
Grier, H. D., C.B., D.S.O., Brit. Ser. (206)	23 Feb. 16
Burton, R. G. [L]	. . .
Roberts, H. L., C.B., M.V.O.	21 Nov. 14
Douglas, J. A., C.M.G., C.I.E. [L]	16 Sep. 15
Edwardes, S. M., C.B., C.M.G., D.S.O.	6 Sep. 15
Prendergast, C. G., C.B.	16 Dec. 15
Somerset, C. W., C.B., C.M.G., M.V.O.	27 Sep. 17
Cadell, A.	27 Dec. 16
Willoughby, M. E., C.B., C.M.G., *p.s.c.* [L]	8 Nov. 16
Jennings, H. A. K., C.I.E., Brit. Ser.	23 Nov. 17
Parker, J. L., C.M.G., Brit. Ser.	21 Sep. 15
White, G. F., C.B., C.M.G., Brit. Ser.	4 Jan. 15
Christian, G., C.B., D.S.O., Brit. Ser. [*l*]	28 Dec. 16
Grimston, S. B., C.M.G.	6 Mar. 14
Johnson, E. P., C.B., Brit. Ser.	10 Dec. 16
Poore, R. M., C.I.E., D.S.O., Brit. Ser.	14 Oct. 15
Swiney, A. J. H., C.B., Brit. Ser.	25 Jan. 16
Smith, F. A.	19 June 16
Norie, C. E. de M., C.B., C.M.G., D.S.O., *p.s.c.* [*l*]	16 July 15
Dyer, R. E. H., C.B., *p.s.c.*	12 Mar. 18
Richardson, H. L.	7 July 18
Sullivan, E. L., C.B., C.M.G.,	25 Feb. 17
Gordon, L. G. F., C.B., D.S.O., R. A. (*g*)	21 June 17
Boudier, E. W.	11 Mar. 18
Lowis, H. E.	30 Sep. 17
Baldwin, G. M., D.S.O.	15 Feb. 16
Southey, W. M., C.M.G., 129th Baluchis	31 Mar. 17
Vaughan, R. E., C.B., S. & T. Corps	5 Oct. 16
Peebles, E.C., C.B., C.M.G., D.S.O., Norf. R.	26 Jan. 16
Clery, C. B. L., C.B., 104th Rif.	6 Jan. 18
Sheppard, S.H., C.B., C.M.G., D.S.O., R.E., *p.s.c.*	1 Feb. 16
Bruce, *Hon.* C. G., C.B., M.V.O.	3 Sep. 16
Kenyon, L. R., C.B., R. A., *p.a.c.*	3 Jan. 18
Cook, H. R., Brit. Ser.	2 Feb. 15
Drew, A. B.H. C.I.E.	8 May 18
Willis, E. H., C.M.G., R.A.	30 Jan. 16
Cameron, A.R., C.B., C.M.G., *p.s.c.*	21 Oct. 15
Watson, W. M., Brit. Ser. (207)	1 Jan. 16
Crooker, G. D., Brit. Ser.	23 Apl. 16
Luard, C. C., C.M.G. Brit. Ser.	10 June 16
Younghusband, L. N., C.B., C.M.G.	1 Nov. 16

Brigadier-Generals—*contd.*	Date of rank.
Carey, O. W., C.M.G.	23 Sep. 16
Elsmie, A.M.S., C.M.G., *p.s.c.*, [*l*]	7 Apl. 16
Davies, C. H., C.M.G., D.S.O., 53rd Sikhs	6 July 16
Boome, E. H., C.B., C.M.G. [*l*]	26 Sep. 17
Rattray, C., C.B., *p.s.c.*	12 Mar. 16
Anderson, W. C., C.M.G., I.G.R.	11 Mar. 18
Davidson, S. R., C.M.G.	12 May 16
Browne, H. J. P., 5th G. R., *p.s.c.*	19 Oct. 17
Musgrave, A. D., R.A.	14 Mar. 16
Potts, F., C.M.G., R.A. (207)	17 Sep. 15
Fraser, T., C.B., C.S.I., C.M.G., Brit. Ser. (*n*)	23 Oct. 16
Wauchope, A.G., C.M.G., D.S.O., R. Highrs.	20 May 17
Quinton, F. W. D., R. A.	15 June 15
Look, F. R. E., D.S.O.	19 Sep. 16
Davies, F. G. H.	18 June 16
Nepean, H. E. C. B., C.M.G., 123rd Rif.	25 Nov. 16
Andrew, D. C. A.	14 Mar. 17
Black, W. C., *p.s.c.* [L]	10 Oct. 17
Adye, D. R, 98th Infy. [L]	14 June 18
Holland-Pryor, P., C.M.G., D.S.O., M.V.O.	24 Feb. 16
Isacke, H., C.M.G., *p.s.c.*, R.W. Kent. R. (*l*)	8 July 16
Waghorn, W. D., C.M.G., R. E.	22 Nov. 16
Rotton, J. G., C.B., C.M.G., R.A.	26 July 15
DePree, H. D., C.B., C.M.G., *p.s.c.*, R. A., (206)	5 Feb. 16
Knox, A. W. F., C.B., 58th Rif., *p.s.c.* [L]	27 Apl. 17
Gregory, C. L., C.B., 19th Lrs., *p.s.c.*	1 May 16
Maclachlan, T. R., C.M.G.	11 May 17
Davies, W. P. L., C.M.G., D.S.O., R. A., *p.s.c.* (206)	1 Dec. 16
Stewart, J. H. K., D.S.O., 39th Garh. Rif.	13 Mar. 17
Cassels, R. A., C.B., D.S.O., 32nd Lrs., *p.s.c.*	16 Aug. 17
Banbury, W. E., C.M.G.	27 July 16
Austin, H. H., C.B., C.M.G., D.S.O., R.E. *p.s.c.* [*l*]	1 Apl. 15
Lucas, F. G., C.B., D.S.O.	1 Apl. 16
Campbell, L. W. Y., C.M.G.	16 Apl. 18
Bright, R. A., C.B., *p.s.c.q.s.*, R.A. (*e*) (207)	23 Mar. 16
Pritchard, A. G., 2nd Lrs.	14 Apl. 17
Loch, S. G., D.S.O., *p.s.c.*, R.E.	7 July 16
Dickson, W. E. R., C.I.E., R. E. [*l*]	24 Aug. 17
Wood, E. J. M, D.S.O., 97th Infy.	4 July 17
Tanner, C. O. O.	9 June 17
Gray, F. W. B., C.M.G., D.S.O.	13 May 18
Tyler, J. A., C.M.G., R.A.	16 Nov. 15
Bell-Smyth, J. A., C.M.G., *p.s.c.*, 1 Dn. Gds.	16 Oct. 15
Kirby, A. D., C.M.G., R.A.	16 Nov. 15
Delaforce, E. F., C.M.G., R. A.	25 Oct. 16
Lamont, J. W. F., C.M.G., D.S.O., R. A.	28 Aug. 16
Macmullen, C. N., D.S.O., 15 Sikhs, *p.s.c.*	26 Sep. 17
Lesslie, W. B., C.M.G., R. E.	5 Oct. 15
Wigram, K., C.B., D.S.O., *p.s.c.*, 2nd G. R.	4 Feb. 17
Lubbock, G., D.S.O., R. E.	8 Nov. 16
Crookshank, S. D'A., C.B., C.I.E., M.V.O., D.S.O., R.E.	15 Jan. 17
Birdwood, H. B.	9 Mar. 18
Andrew, A. W., C.M.G. [L]	20 Aug. 17
Codrington, H. W.	. . .
Beatty, L. N.	20 Nov. 17
Maxwell, L. L., C.M.G., 2 Lrs.	6 Jan. 18
Norton, C.E.G., A.-D.-C., 7 Hrs.	24 Apl. 17
Norman, C. L., D.S.O., M.V.O., A.-D.-C. *p.s.c.*, Guides.	11 July 17
Beach, W. H., C.M.G., D.S.O., R. E. (206)	12 Oct. 16
Robinson, S. W., D.S.O., R. A. (206)	28 Oct. 16
Rankin, C. H., C.M.G., D.S.O., 7 Hrs.	15 Apl. 16
Godwin, C. A. C., D.S.O., *p.s.c.* 23 Cav.	30 Sep. 17
Morris, G. M., D.S.O., 62 Punj.	2 Mar. 18
Drake-Brockman, D. H., C.M.G.	13 Mar. 16
Knatchbull, G. W. C., C.M.G.	10 Aug. 15
Stanton, F. W. S., D.S.O., R. A.	1 Feb. 15
Hardy, W. K., R.A.	14 Mar. 16
Dale, G. A., C.M.G.	11 Jan. 18
Moberly, F. J., D.S.O., *p.s.c.*	7 Nov. 16
Lane, H. A.	10 July 18
Dunsford, F.P.S., D.S.O.	17 June 16
Rose, J. L., C.I.E., 1st G.R.	8 Apl. 16
Marshall, H. J. M., C.M.G., R. E.	22 Nov. 17
Murray, E.R.B., 90th Punj.	15 Mar. 17
Sykes, *Sir* P. M., K.C.I.E., C.M.G.	11 Mar. 16
Young, H. A., C.I.E., R. A. (206)	8 Mar. 18

Brigadier-Generals—*concld.*

	Date of rank.
Money, E. D., C.I.E., D.S.O.	8 July 18
Cartwright, G. N., C.M.G., D.S.O., R. A.	15 Oct. 15
Rettie, W. J. K., D.S.O., R. A.	20 Feb. 17
Rushton, H. W., D.S.O., R. E. (207)	23 Sep. 17
Sykes, C. A., D.S.O., R. A.	13 July 17
Smith, L. A., R. A.	16 Sep. 17
Goring-Jones, M. D., C.M.G., Brit. Ser.	12 Apl. 18
Patch, F. R., C.M.G., D.S.O., R. A.	7 May 16
Templer, C. F., C.M.G. C.I.E.	26 Dec. 16
Legard, D'A., *p.s.c.*, 17 Lrs.	31 Oct. 16
Andrew, F. A., D.S.O., 128 Prs.	15 July 17
Sanders, G. A. F., *p.s.c.*, R.E.	11 Aug. 16
Holdich, H. A., D.S.O.	19 Nov. 17
Bateman-Champain, H. F., C.M.G., 9 G. R.	6 Aug. 17
Brownlow, d'A. C., C.M.G., C.I.E.	8 Sep. 16
Harrison, A. H. P., 33 Punj.	1 Dec. 16
Luke, T. M., D.S.O., O.B.E., R. A.	12 Oct. 16
Hildyard, R. J. T., D.S.O.	12 Sep. 17
Fagan, E. A., C.M.G., D.S.O., 36 Horse	29 Nov. 17
Nightingale, M. R. W., D.S.O., 5 G. R.	20 Dec. 17
Kemball, A. G., 31 Punj.	17 Dec. 17
Lukin, R. C. W., D.S.O., 35th Horse	25 Sep. 18
Watt, D. M., D.S.O., *p.s.c.*, 2 G. R.	28 Nov. 16
VC **Costello, E. W.**, C.M.G., D.S.O. *p.s.c.*	4 May 18
Andrus, T. A., C.M.G., N. Staff. R.	11 July 16
Pryce, H. E. apR., C.M.G., D.S.O., *p.s.c.*	17 Oct. 17
St. John, R. S., D.S.O., *p.s.c.*, 20 Infy.	21 June 17
Brough, A., D.S.O., R. E.	10 July 17
Franklin, H. S. E., D.S.O., 15 Sikhs	29 Oct. 17
Dick, R. N., R. Suss. R.	14 Nov. 17
Barstow, H., 38 Dogras	8 Feb. 17
Wagstaff, C. M., C.M.G., C.I.E., D.S.O., *p.s.c.*, R. E.	7 Jan. 18
Pitcher, D. LeG., C.M.G., 39 Horse	1 Apl. 16
Coleridge, J. F. S. D., C.M.G., D.S.O., *p.s.c.*, 8 G. R.	31 Oct. 17
Bethell, H. K., C.M.G., D.S.O., 7 Hrs.	
Mackey, H. J. A., D.S.O., M.V.O., R. A.	2 May 16
Gwyn-Thomas, G., C.M.G., D.S.O., 2 Lrs.	20 July 17
Abbott, L. H., C.M.G., 81 Raj., *p.s.c.*	8 Nov. 17
Taylor, R. O'B., C.M.G., C.I.E., *p.s.c.*, 19 Lrs.	22 May 16
Elliott, C. A., D.S.O., R. E.	17 Feb. 18
Hogg, R. E. T., C.I.E., 38 Horse	22 June 17
Harbord, C. R., D.S.O., 30 Lrs.	16 Apr. 17
Hoare, C. G., 39 Horse	1 Aug. 17
Pope-Hennessy, L. H. R., D.S.O., Oxf. & Bucks, L.I.	5 Sep. 17
Worgan, R. B., D.S.O., 20 Horse	2 Oct. 17
Newall, C. L. N., 2-2nd G. R.	28 Dec. 17

Colonels

	Date of rank.
Blyth, F. A.	13 June 04
Gott, G. A.	23 Feb. 05
Maxwell, G. W.	28 Oct. 05
†*Forbes, A. W.*, C.B.	11 Sep. 06
Cowie, C. H., C.I.E., Brit. Ser. (207)	29 Nov. 06
Renny-Tailyour, T. F. B., C.S.I., Brit Ser.	29 Nov. 06
Westlake, A. P., D.S.O.	29 Nov. 06
Browne, C. A. R., Brit. Ser. (207)	3 Jan. 07
Gastrell, E. T.	30 Jan. 07
Whistler, A. E.	30 Jan. 07
Rodwell, E. H., C.B., *p.s.c.* [L]	30 Jan. 07
†*Nanton, H. C.*, C.B., Brit. Ser.	1 Feb. 07
†*Hamilton, W. G.*, C.B., C.S.I., D.S.O., Brit. Ser., *p. s. c.* (206)	12 Mar. 07
Hodgson, G. B.	1 May 07
Thackwell, C. G. R., C.B., D.S.O.	1 May 07
Thomson, W. D., C.M.G.	1 May 07
Hutchinson, F. P., C.B. [L]	1 May 07
Aplin, P. J. H., D.S.O. [L]	1 May 07
Dobbie, H. H.	1 May 07
Richardson, F. B. W.	11 May 07
Justice, C. Le G.	11 May 07
Steele, St. G. L., C.B. [L]	1 June 07
Yate, F. H.	1 June 07
Dennys, W. A. B.	6 June 07

Colonels—*contd.*

	Date of rank.
Dobbie, W. H., C.B.	6 June 07
Aitken, A. E.	9 June 07
Loch, H. F., *p.s.c.*	10 June 07
Johnson, C. E.	11 June 07
†*Johnstone, A. A. J.*	11 June 07
Unwin, G. B., D.S.O.	15 Sep. 07
Templer, H.	28 Oct. 07
†*Smith, G. B.*, C.B., Brit. Ser. (206)	7 Nov. 07
Burrard, *Sir* S. G., K.C.S.I., Brit. Ser. (207)	15 Nov. 07
Price, C. H. U., C.B., D.S.O.	1 Mar. 08
Balfour, J. H.	23 Mar. 08
Dallas, J., Brit. Ser. (206)	22 Apl. 08
†*Perry*, Sir *H. W.*, K.C.M.G., C.B., C.S.I., Brit. Ser.	21 June 08
Dick, A. R., C.B. [L]	16 July 08
Dundee, W. J. D., C.I.E., Brit. Ser. (207)	16 July 08
Maclagan, R. S., C.B., C.S.I., Brit. Ser. (206)	12 Aug. 08
Leonard, A. W.	2 Oct. 08
Baddeley, C. E., C.M.G., Brit. Ser. (206)	22 Oct. 08
†*Stokes, A.*, C.B., C.M.G., D.S.O., Brit. Ser. (*g*) (206)	14 Nov. 08
**Johnson, F. E.*, C.M.G., D.S.O., Br. Ser.	20 Dec. 08
Marlow, B. W., C.S.I., C.I.E.	23 Dec. 08
†*Strange, R. G.*, C.I.E. Brit. Ser. (206)	15 Jan. 09
†*Twigg, R. H.*, C.B.	4 Feb. 09
Lindesay, A. L.	13 Feb. 09
†*Glanville, F.*, D.S.O., Brit. Ser. (206)	1 Apl. 09
Ellis, W. M., C.I.E., Brit. Ser. (207)	1 Apl. 09
Shakespear, L. W. C.B.	2 Apl. 09
†*Herbert, L. N.*, C.B., [*l*]	4 Apl. 09
Palin, G. W., C.B., C.I.E.	1 May 09
Walton, E. W., Brit. Ser. (207)	25 July 09
†*Iggulden, H. A.*, Brit. Ser. (206)	11 Aug. 09
Craster, S. L., C.B., C.I.E., Brit. Ser. (206)	13 Aug. 09
Harrison, T. A, C.S.I.	1 Apl. 10
†*Nuthall, C. E.*, C B., C.M.G., Brit. Ser.	4 June 10
Shore, O. B. S. F., C.B., C.I.E., D.S.O., *p.s.c.* [L]	30 July 10
†*Buchanan, K. J.*, C. B.	27 Sep. 10
Dowell, G. C., C.M.G., R.A. (206)	10 Oct. 10
†*Campbell, A. A. E.*	29 Oct. 10
Cleveland, H. F., C.I.E., I.M.S. [L]	1 Jan. 11
Woodward, J. A. H.	30 June 11
Houston, E., Brit. Ser. (207)	19 July 11
Fisher, J., C. B. [L]	29 Aug. 11
†*Tidswell, E. C.*, C.B., D.S.O., Brit. Ser. (206)	30 Aug. 11
Mainwaring, C. V.	3 Oct. 11
†*McLeod, W. K.*, Brit. Ser.	4 Oct. 11
†*Wolley-Dod, O. C.*, C.B., D.S.O., *p.s.c.* [*l*] (206)	4 Oct. 11
Rose, C. S., Brit Ser. (207)	6 Dec. 11
Millar, W. H.	26 Dec. 11
Cartwright, C. M., C.B., C.M.G.	1 Jan. 12
†*Holloway, B.*, C.I.E., *p.s.c.* [L]	14 Jan. 12
Edwards, Fitz J. M., C.B., C.M.G., D.S.O., *p.s.c.*, **A.-D.-C.**	14 Jan. 12
Morton, W. R., Brit. Ser. [L] (206)	7 Mar. 12
†*Caulfeild, C. T.*, C.M.G., Brit. Ser.	7 Mar. 12
†*Muspratt, F. C.*, C.B., C.M.G.	1 Apl. 12
†*Roe, C. H.*, C.I.E., Brit. Ser. (206)	6 June 12
**Renny, S. M.*, C.I.E., Brit. Ser. (206)	6 June 12
†*Duhan, W. W. T.* (*g*) Brit. Ser.	6 June 12
†*Peterson, F. H.*, C.B., D.S.O. [*l*]	6 June 12
Andrews, L. J. [L]	26 June 12
†*Carruthers, R. A.*, C.B., C.M.G.	29 June 12
Newbold, A. W.	11 Aug. 12
Mead, H. R.	19 Aug. 12
†*Betham, R. M.*	1 Sep. 12
Savage, W. H., C.M.G.	1 Sep. 12
Cox, F. W. H., C.B.	21 Sep. 12
Mathewes, J. R.	21 Sep. 12
Johnston, H. M.	5 Oct. 12
Fasken, W. H., C.B.	12 Oct. 12
Thomas, F. H. S., C.B.	23 Oct. 12
Blakeney, W. E. A.	23 Oct. 12
Brewin, J. P., Brit. Ser.	30 Oct. 12
Rich, H. H., Brit. Ser.	4 Dec. 12
Kennedy, H., C.B.	30 Dec. 12
†*Commeline, F. H. B.*	31 Dec. 12

* Ty. Maj.-Genl. † Brigadier-General.

Colonels—*contd.*

	Date of rank.
†*Miles, P. J.*, C.B.	31 Jan. 13
Lang, E. B.	14 Feb. 13
Sullivan, R. E.	21 Feb. 13
Cripps, A. W., C.B., C.M.G.	13 Mar. 13
†*Colomb, G. H. C.*, C.M.G.	24 Mar. 13
Brown, R. J. R.	24 Mar. 13
†*Walton, W. C.*, C.B., C.M.G.	31 Mar. 13
Crocker, S. F., C.B.	31 May 13
†*Powell, S. H.*, C.B., Brit. Ser. [l] (206)	2 June 13
†*Dealy, J. A.*, C.M.G., D.S.O., Brit. Ser. (207)	2 June 13
Ewbank, W., C.I.E., Br. Ser., *p.s.c.* (206)	2 June 13
Clay, C. H., *p.s.c.* [l]	30 Sep. 13
VC Crimmin, J., C.B., C.I.E., I.M.S., K.H.P.	1 Oct. 13
†*Soady, G. J. FitzM.*, C.M.G.	22 Oct. 13
†*Evans, U. W.*, C.B., C.M.G., Brit. Ser. [L] (206)	16 Dec. 13
†*Sorsbie, R. F.*, C.I.E., Brit. Ser. (206)	16 Dec. 13
†*Grier, H. D.*, C.B., D.S.O., R.A. (206)	16 Dec. 13
†*Burton, R. G.* [L]	19 Dec. 13
Cole, E. H., C.B., C.M.G.	2 Jan. 14
†*Roberts, H. L.*, C.B., M.V.O., [l]	8 Jan. 14
Dyson, T. E., M.B., I.M.S.	10 Jan. 14
Selwyn, C. H., C.B., *p.s.c.* [L]	20 Jan. 14
†*Douglas, J. A.*, C.M.G., C.I.E. [L]	28 Jan. 14
†*Edwardes, S. M.*, C.M.G., D.S.O.	5 Feb. 14
†*Prendergast, C. G.*, C.B.	12 Feb. 14
†*Somerset, C. W.*, C.B., C.M.G., M.V.O.	11 Mar. 14
†*Cadell, A.*	1 Apl. 14
Banatvala, *Hon.* H. F., C.S.I., I.M.S., K.H.S.	2 Apl. 14
**Watson, H. D.*, C.B., C.M.G., C.I.E., M.V.O., *Eq.*	19 Apl. 14
†*Willoughby, M. E.*, C.B., C.M.G., *p.s.c.* [L]	4 May 14
Tod, J. K., *p.s.c.* [L]	10 May 14
Strickland, P. C. H., I.M.S.	15 May 14
†*Jennings, H. A. K.*, C.I.E., (*f*) Brit. Ser. (206)	15 June 14
Lenox-Conyngham, G. P., Brit. Ser.	15 June 14
Rouse, H., C.B., D.S.O., R. A. (206)	15 June 14
**Atkinson, E. H. deV.*, C.B., C.M.G., C.I.E., Brit. Ser.	15 June 14
Askwith, H. F., C.M.G., Brit. Ser.	15 June 14
†*Parker, J. L.*, C.B., C.M.G., Brit. Ser. (206)	15 June 14
Babington, D. M., C.I.E., Brit. Ser.	15 June 14
†*White, G. F.*, C.B., C.M.G., Brit. Ser.	15 June 14
†*Christian, G.*, C.B., D.S.O., Brit. Ser. [l]	15 June 14
Scott, B.	1 Sep. 14
Roche, H. J., C.B.	4 Nov. 14
†*Grimston, S. B.*, C.M.G.	8 Dec. 14
†*Johnson, E. P.*, C.B., Brit. Ser. (206)	15 Dec. 14
Campbell, M. S. C., C.I.E., Brit. Ser. (206)	15 Dec. 14
†*Poore, R. M.*, C.I.E., D.S.O., Brit. Ser.	15 Dec. 14
Scudamore, W. V., Brit. Ser. (206)	15 Dec. 14
Williamson Oswald, O. C., C.B., C.M.G., R. A. (206)	15 Dec. 14
†*Swiney, A. J. H.*, C.B., C.M.G., Brit. Ser. (207)	15 Dec. 14
Graham, S. B.	4 Feb. 15
Mennie, J. O.	10 Feb. 15
VC Boisragon, G. H.	17 Feb. 15
†*Smith, F. A.*	18 Feb. 15
†*Norie, C. E. deM.*, C.B., C.M.G., D.S.O., *p.s.c.* [Z]	18 Feb. 15
Adamson, H. M., C.B., M.B., Brit. Ser.	1 Mar. 15
Davidson, J. S., M.B., Brit. Ser.	1 Mar. 15
Fallon, J., Brit. Ser.	1 Mar. 15
Eckersley, E., M.B., Brit. Ser.	1 Mar. 15
Burnside, E. A., Brit. Ser.	1 Mar. 15
Inniss, B. J., Brit. Ser.	4 Mar. 15
†*Dyer, R. E. H.*, C.B., *p.s.c.*	5 Mar. 15
Mactaggart, *Hon.* C., C.I.E., M.B., I.M.S.	29 Mar. 15

Colonels—*contd.*

	Date of rank.
Daly, J. T., M.B., I.M.S.	1 Apl. 15
†*Richardson, H. L.*	15 Apl. 15
†*Sullivan, E. L.*, C.B., C.M.G.	2 May 15
†*Gordon, L. G. F.*, C.B., D.S.O., R. A. (*g*) (206)	17 May 15
White, W. W., C.B., M.D., I.M.S.	3 June 15
†*Boudier, E. W.*	3 June 15
Frazer, G. S., C.M.G.	3 June 15
Batten, F. G., D.S.O.	3 June 15
Bailey, C.	3 June 15
†*Lewis, H. E.*	3 June 15
Chitty, W. W.	3 June 15
†*Baldwin, G. M.*, D.S.O.	3 June 15
†*Southey, W. M.*, C.M.G.	3 June 15
†*Vaughan, R. E.*, C.B., S. & T. Corps	3 June 15
†*Peebles, E. C.*, C.B., C.M.G., D.S.O., Brit. Ser. (207)	3 June 15
Fordyce, H. L. D., S. & T. Corps	3 June 15
Harward, A. J. N.	3 June 15
**Ritchie, A. B.*, C.M.G., Brit. Ser.	3 June 15
†*Clery, C. B. L.*, C.B., 104 Rif.	3 June 15
Annesley, A. S. R., S. & T. Corps	3 June 15
Molesworth, R. P., R. A.	3 June 15
Bell, G. J. H., C.I.E., M.B., I.M.S.	21 June 15
Fooks, H., I.M.S.	12 July 15
†*Sheppard, S. H.*, C.B., C.M.G., D.S.O., R. E. *p.s.c.* (206)	24 July 15
Duff, G. M., Brit. Ser. (207)	25 July 15
Joly de Lotbiniere, A. C. Del, C.B., C.S.I., C.I.E., Brit. Ser. (207)	6 Aug. 15
Picton, R. E., Brit. Ser. (207)	13 Aug. 15
Kirby, N., C.I.E., Brit. Ser. (207)	18 Aug. 15
Lane, H. P.	27 Aug. 15
Pullen, A. F., Brit. Ser. (*g*) [l] (207)	29 Sep. 15
Young, F. DeB., C.M.G., *p.s.c.*	9 Oct. 15
Luck, C. A.	28 Oct. 15
†*Bruce*, Hon. *C. G.*, C.B., M.V.O.	8 Nov. 15
Heard, R., M.B., I. M. S.	29 Nov. 15
†*Kenyon, L. R.*, C.B., R.A., *p. a. c.*	29 Nov. 15
†*Cook, H. R.*, Brit. Ser. [L] (206)	4 Dec. 15
Jennings, E., I.M.S.	1 Jan. 16
Macnab, A. J., C.B., F.R.C.S., I.M.S.	1 Jan. 16
Sanders, G. L'H., S. & T. Corps	1 Jan. 16
Caruana, A. J., C.I.E.	1 Jan. 16
†*Drew, A. B. H.*, C.I.E.	1 Jan. 16
**Holman, H. C.*, C.B., C.M.G., D.S.O., 16 Cav., *p.s.c.* (L)	1 Jan. 16
**Ravenshaw, H. S. L.*, C.M.G., Brit. Ser.	1 Jan. 16
†*Willis, E. H.*, C.M.G., R. A.	1 Jan. 16
Sloan, J. M., C.M.G., D.S.O., M.B., R.A.M.C.	1 Jan. 16
†*Cameron, A. R.*, C.B., C.M.G., *p.s.c.*, Brit. Ser.	1 Jan. 16
Peacocke, T. G., C.I.E., A. Remt. Dept.	6 Jan. 16
†*Watson, W. M.*, Brit. Ser. (207)	3 Feb. 16
†*Crocker, G. D.*, Brit. Ser.	17 Feb. 16
†*Luard*, C. C., C.M.G., Brit. Ser. (207)	29 Feb. 16
†*Younghusband, L. N.*, C.B., C.M.G.	31 Mar. 16
Campbell, A. J., D.S.O.	3 June 16
Wright, E. G.	3 June 16
†*Carey O. W.*, C.M.G.	3 June 16
†*Elsmie, A. M. S.*, C.M.G., *p.s.c.*, [L]	3 June 16
Lindsay, H. A. P., C.M.G., S. & T. Corps [L]	3 June 16
†*Davies, C. H.*, C.M.G., D.S.O., 53 Sikhs	3 June 16
Campbell, G. P., R. E. (207)	3 June 16
†*Boome, E. H.*, C.B., C.M.G., 102 Grs. [l]	3 June 16
Parr, H. O., 7 Raj.	3 June 16
Ogilvie, E. C., C.M.G., R. E. (207)	3 June 16
†*Rattray, C.*, C.B., *p.s.c.*	3 June 16
Fulton, F. C. R., S. & T. Corps	3 June 16
**Leslie, G. A. J.*, C.B., C.M.G., R. E. (206)	3 June 16
Sheppard, G. S., C.M.G., Mily. Accts. Dept.	3 June 16
†*Anderson, W. C.*, C.M.G., 1 G. R.	3 June 16
†*Davidson, S. R.*, C.M.G., 47 Sikhs	3 June 16

* Ty. Maj.-Genl. † Brigadier-General.

Colonels—contd.	Date of rank.
*Charles, J. R. E., C.B., D.S.O., p.s.c., R. E.	3 June 16
Collins, D. J., M.D., R.A.M.C.	3 June 16
Cummins, H. A. V., p.s.c., 24 Punj.	3 June 16
Hopwood, H. R., C.S.I., p.s.c., 33 Cav.	3 June 16
†Browne, H. J. P., 5 G. R., p.s.c.	3 June 16
†Musgrave, A. D., R. A.	3 June 16
Carr-White, P., M.B., F.R.C.S.E., I.M.S.	7 June 16
Seddon, T. Y.	9 June 16
Murray, V., C.B., C.M.G., Brit. Ser. (206)	22 June 16
Weedon, F. F., Brit. Ser. [L] (206)	1 July 16
Tytler, H. C., D.S.O.	2 July 16
Creagh, A. H. D., C.M.G., M.V.O.	30 Aug. 16
Jennings, W. E., M.D., I.M.S.	1 Oct. 16
Blakeway, J. P., C.M.G., Brit Ser. (207)	1 Oct. 16
Coffin, C., C.M.G., p.s.c., R. E., (206)	1 Oct. 16
Green, C. R. M., M.D., F.R.C.S., I.M.S.	11 Nov. 16
†Potts, F., C.M.G., R. A. (g) (207)	20 Nov. 16
†Fraser, T., C.B., C.S.I., C.M.G., R.E., p.s.c. (n) [L.] (206)	19 Dec. 16
†Wauchope, A. G., C.M.G., D.S.O., R. Highrs	23 Dec. 16
*Thomson, W. M., C.B., M.C., Sea. Highrs.	23 Dec. 16
†Quinton. F. W. D., R.A. (206)	28 Dec. 16
Carnegy, C. G., M.V.O.	1 Jan. 17
†Lock, F. R. E., D.S.O.	1 Jan. 17
†Davies, F. G. H.	1 Jan. 17
†Nepean, H.E., C. B., C.M.G.	1 Jan. 17
Jacob, A. LeG., C.M.G., C.I.E., D.S.O., A.-D.-C.	1 Jan. 17
Lucas, F. G., C.B., D.S.O.	1 Jan. 17
†Andrew, D. C. A.	1 Jan. 17
Wall, E. W., C.M.G.	1 Jan. 17
Campbell, L. W. Y., C.M.G.	1 Jan. 17
Home, J. M., p.s.c., [L]	1 Jan. 17
†Black, W. C., p.s.c., [L]	1 Jan. 17
†Adye, D. R., 98 Infy. [L].	1 Jan. 17
†Holland Pryor, P., C.M.G., D.S.O., M.V.O.	1 Jan. 17
†Isacke, H., C.M.G., p.s.c., R. W. Kent. R. [L]	1 Jan. 17
Price, C. U., C.M.G., 130 Baluchis	1 Jan. 17
†Waghorn, W. D., C.M.G., R.E.	1 Jan. 17
†Rotton, J. G., C.B., C.M.G., R.A.	1 Jan. 17
*Thuillier, H. F., C.B., C.M.G., R.E.	1 Jan. 17
†De Pree, H.D., C.B., C.M.G., p.s.c., R.A. (206)	1 Jan. 17
*Peck, A. W., C.M.G., 17 Cav.	1 Jan. 17
More, L. P., M.B., R. A. M. C.	1 Jan. 17
†Knox, A. W. F., C.B., 58 Rif., p.s.c. [L]	1 Jan. 17
†Gregory, C. L., C.B., p.s.c., 19 Lrs.	1 Jan. 17
*Vaughan, L. R., C.B., D.S.O., 7 G. R.	1 Jan. 17
Molesworth, W., C.I.E., C.B.E., M.B., I.M.S. V.H.S.	22 Jan. 17
Lane, C. A., M.B., R.A.M.C.	24 Jan. 17
Ogg, G. S., C.I.E., Brit. Ser.	15 Feb. 17
†Maclachlan, T. R., C.M.G., 92 Punj.	10 Mar. 17
†Davies, W. P. L., C.M.G., D.S.O., R. A., p.s.c. (206)	10 Mar. 17
†Stewart, J. H. K., D.S.O., p.s.c., 39 G.	10 Mar. 17
†Cassels, R. A., C.B., D.S.O., p.s.c., 32 Lrs.	10 Mar. 17
Garvie, J., M.B., I.M.S.	25 Mar. 17
Fraser, H. A. D., Brit. Ser.	16 Apl. 17
Cook, C. C., D.S.O.	18 Apl. 17
MacGeorge, H. K., C.B.	22 Apl. 17
Hallowes, F. W., C.I.E.	3 May 17
Ouseley, R. G., C.B., C.M.G., D.S.O., R.A. (g) (206)	5 May 17
†Banbury, W. E., C.M.G., [L]	3 June 17
†Austin, H. H., C.B., C.M.G., D.S.O., R. E., p.s.c.	3 June 17
†Lucas, F. G., C.B., D.S.O.	3 June 17
†Campbell, L. W. Y., C.M.G.	3 June 17
*Cooke, H. F., D.S.O., 32 Prs.	3 June 17
Symons, A., C.M.G., 13 Hrs., p.s.c. [L]	3 June 17
*Jones, L. C., C.M.G., M.V.O., p.s.c. 7 Lrs. [l]	3 June 17
†Bright, R. A., C.B., p.s.c. q.s. (e) R. A. (207)	3 June 17
†Pritchard, A. G., C.M.G., 2 Lrs.	3 June 17
Farmer, J., C.I.E., F.R.C.V.S.	20 July 17
Hutchinson, C. A. R., D.S.O., p.s.c.	31 Aug. 17
†Loch, S. G., D.S.O., R. E., p.s.c. (206)	17 Sep. 17
†Dickson, W. E. R., C.I.E. [l] R. E.	17 Sep. 17

Colonels—concld.	Date of rank.
Badcock, F. F., D.S.O.	20 Oct. 17
Senior, H. W. R., D.S.O., p.s.c. [L]	1 Dec. 17
†Wood, E. J. M., D.S.O.	3 Dec. 17
Condon, E. H., M.B., Brit. Ser.	26 Dec. 17
Grattan, H. W., Brit. Ser.	26 Dec. 17
†Tanner, C. O. O.	1 Jan. 18
Eassie, F., C.M.G., D.S.O., A.V.C.	1 Jan. 18
†Gray, F. W. B., C.M.G., D.S.O.	1 Jan. 18
†Tyler, J. A., C.M.G., R.A.	1 Jan. 18
†Bell-Smyth, J. A., C.M.G., 1 Dgn. Gds., p.s.c.	1 Jan. 18
†Kirby, A. D., C.M.G., R. A.	1 Jan. 18
†Delaforce, E. F., C.M.G., R. A. (206)	1 Jan. 18
Perreau, A. M., C.M.G., R.A.	1 Jan. 18
†Lamont, J. W. F., C.M.G., D.S.O., R A.	1 Jan. 18
†Macmullen, C. N., D.S.O., 15 Sikhs, p.s.c.	1 Jan. 18
Massie, R. H., R. A. (206)	1 Jan. 18
†Lesslie, W. B., C.M.G., R. E. (206)	1 Jan. 18
†Wigram, K., C.B., D.S.O., p.s.c., 2 G. R.	1 Jan. 18
Bremner, A. G., C.M.G., R.E. (207)	1 Jan. 18
Steel, R. A., C.I.E., 17th Cavy. p.s.c. [l]	1 Jan. 18
†Lubbock, G., C.M.G., D.S.O., R.E.	1 Jan. 18
†Crookshank, S. D' A., C.B., C.I.E., M.V.O., D.S.O., R.E.	1 Jan. 18
Macwatt, R. C., C.I.E., M.B., I.M.S.	8 Jan. 18
†Birdwood, H. B.	6 Apl. 18
Platt, E. H., R. C. R., Som. L. I. (206)	21 Apl. 18
†Andrew, A. W., C.M.G. [L]	18 May 18
Dunsterville, E. L., Brit. Ser. (206)	20 May 18
McCrea, A. C., C.M.G.	21 May 18
†Codrington, H. W.	25 May 18
†Beatty, L. N.	3 June 18
Ridgway, R. T. I., C.B.	3 June 18
Cobbe, H. H. C.M.G., D.S.O., 13th Lrs.	3 June 18
†Maxwell, L. L., C.M.G., 2nd Lrs.	3 June 18
†Norton, C.E.G., A.-D.-C. 7th Hrs.	3 June 18
†Norman, C. L., D.S.O., M.V.O., A.-D.-C., p.s.c., Guides	3 June 18
†Beach, W. H. C.M.G., D.S.O., R.E. (206)	3 June 18
†Robinson, S. W., D.S.O., R. A. (206)	3 June 18
†Rankin, C. H., C.M.G., D.S.O., 7th Hrs.	3 June 18
†Godwin, C.A.C., D.S.O., 23rd Cav., p.s.c.	3 June 18
Winter, C. B., C.M.G., D.S.O., 112th Infy	3 June 18
Tarver, A. L., D.S.O., A.-D.-C., p.s.c., 124th Infy.	3 June 18
Orr, G. M., D.S.O., 11th Lrs., p.s.c.	3 June 18
Leslie, W. S., D.S.O., p.s.c., 31st Punj.	3 June 18
†Morris, G. M., D.S.O., [L] 62nd Punj.	3 June 18
Norman, W. H., D.S.O., p.s.c., 11th Lrs.	3 June 18
Panet, A. E., C.M.G., D.S.O., R. E.	3 June 18
Vincent, B., C.M.G., 6th Dgns. p.s.c., (206)	3 June 18
O'Grady, H.de C., 54th Rif.	3 June 18
Henderson, M. H., 26th Lt. Cavy.	3 June 18
†Drake-Brockman, D. H., C.M.G.	16 June 18
Close, J. K., M.D., I.M.S.	15 Aug. 18
Hennessy, J. P. C., C.M.G.	27 Aug. 18
Smith, J. B., C.B., M.B., I.M.S., V.H.S.	3 Sep. 18
Stodart, T., M.B., I.M.S.	12 Oct. 18
Lieutenant-Colonels.	
Plumer, T. H.	1 June 04
Henderson, S. H., M.B., I.M.S.	30 Sep. 06
Castor, R. H., I.M.S.	30 Sep. 06
Morwood, J., M.D., I.M.S.	30 Sep. 06
Hall, E. A. W., M.B., I.M.S.	30 Sep. 06
Giles, A.	8 Dec. 06
O'Bryen, C. W., [l]	10 Mar. 07
Sunder, C. E., M.B., I.M.S.	31 Mar. 07
Buchanan, A., I.M.S.	31 Mar. 07
Fischer, L. G., I.M.S.	31 Mar. 07
Vost, W., M.B., I.M.S.	31 Mar. 07
Stirling, W. G. H.	2 Apl. 07
Davidson, D. MacD., M.B., I M.S.	1 Oct. 07
Maynard, F. P., M.B., F.R.C.S., I.M.S.	1 Oct. 07
Nott, A. H., M.D., I.M.S.	1 Oct. 07
Coleman, A., M.B., I.M.S.	1 Oct. 07
Lane, D. T., M.D., C.I.E., I.M.S.	1 Oct. 07
Woodwright, W. H.E., F.R.C.S.I., I.M.S.	1 Oct. 07

*Ty. Maj.-Genl. † Brigadier-General.

Lieutenant-Colonels—*contd.*	Date of rank.
Buchanan, *Sir* W. J., K.C.I.E., M.D., I.M.S.	1 Oct. 07
Dewes, F. J., I.M.S.	1 Oct. 07
Stewart, T. W., I.M.S.	1 Oct. 07
Wynch, F. J. H.	19 Oct. 07
Erck, J. C. W.	26 Oct. 07
Roberts, *Sir* J. R., *Kt.*, C.I.E., M.B., F.R.C.S., I.M.S.	31 Mar. 08
Hojel, J. G., C.I.E., M.B., I.M.S.	31 Mar. 08
Gee, F. W., C.I.E., M.B., I.M.S.	31 Mar. 08
O'Gorman, P. W., C.M.G., I.M.S.	31 Mar. 08
Holland, G. L., M.V.O.	30 Apl. 08
Nicholson, R. W.	2 Aug. 08
Henderson, T. H.	11 Aug. 08
Jones, H. J., D.S.O.	25 Aug. 08
Drake-Brockman, H. E., F.R.C.S.E., I.M.S.	29 Sep. 08
Lane, W. B., C.I.E., I.M.S.	29 Sep. 08
Lumsden, P. J., I.M.S.	29 Sep. 08
Prall, S. E., M.B., I.M.S.	29 Sep. 08
Henegan, J., D.S.O.	3 Jan. 09
Whiffin, H. E.	27 Jan. 09
Young, W. B.	27 Jan. 09
Baynes, C. E.	14 Feb. 09
Dann, H. C. B.	28 Feb. 09
Radcliffe, W. C. A., R.A. (206)	5 Mar. 09
†*Knatchbull, G. W. C.*, C.M.G.	10 Mar. 09
Loudon, J. A.	10 Mar. 09
Ozzard, F. R., I.M.S. [*l*]	30 Mar. 09
Calvert, J. T., M.B., I.M.S.	30 Mar. 09
Hudson, C. T., C.M.G., I.M.S.	30 Mar. 09
Williams, A. H.	28 Apl. 09
Thomas, H. St. G., D.S.O.	16 May 09
Muspratt-Williams, C. A., C.I.E., R.A. (206)	25 May 09
Rainsford, J. A. G.	5 July 09
Wilkins, E. G. R. [*l*]	31 July 09
Grant, A.	3 Aug. 09
Melville, H. B., M.B., I.M.S.	30 Sep. 09
Vaughan, J. C. S., M.B., I.M.S.	30 Sep. 09
Walker, M., R. A. (206)	1 Oct. 09
Pritchard, H. E.	11 Nov. 09
Palmer, H. I. E., C.M.G.	5 Dec. 09
Finch, C.	30 Jan. 10
Pollard, J. H.	6 Feb. 10
Medley, A. G., [L]	6 Feb. 10
Murray, C. F. T., C.B., M.V.O., S. and T.	6 Feb. 10
Barnes, J. P.	6 Feb. 10
Hilliard, H. N., S. and T. Corps	18 Feb. 10
Rea, F. W.	26 Mar. 10
Jackson, *Hon* J., M.B., C.I.E., I.M.S.	31 Mar. 10
Smith, H., C.I.E., M.D., I.M.S.	31 Mar. 10
Wimberley, C. N. C., C.M.G., M.B., I.M.S.	31 Mar. 10
Street, A., M.B., F.R.C.S., I.M.S.	31 Mar. 10
Sutherland, W. D., M.D., I.M.S.	31 Mar. 10
Wright, E. H., I.M.S.	31 Mar. 10
Johnston, C. A., C.B., D.S.O., M.B., I.M.S.	31 Mar. 10
Laing, F. C.	30 Apl. 10
Dennys, A. H.	12 May 10
Hill, J. R.	26 May 10
Ward, G. A.	2 July 10
Maxwell, J. McC., C.B., D.S.O., R.A.	20 July 10
Ryder, W. I.	9 Aug. 10
Budd, N. A. H.	11 Aug. 10
Harvey, A. B.	19 Aug. 10
Gordon, R. F. C., C.I.E.	23 Aug. 10
Goodenough, H. L.	23 Aug. 10
Roome, H. N.	3 Sep. 10
Dobbie, C. F.	8 Oct. 10
Barry, J. F.	25 Oct. 10
Battye, A. H.	12 Nov. 10
Lowther, H.	22 Jan. 11
Crawford, J. M., O.B.E., M.B., I.M.S.	31 Jan. 11
Singh, B. J., C.I.E., I.M.S.	31 Jan. 11
James, C. H., C.I.E., I.M.S.	31 Jan. 11
O'Kinealy, F., M.S.	31 Jan. 11

Lieutenant-Colonels—*contd.*	Date of rank.
Buist, A. W. T., M.B., I.M.S.	31 Jan. 11
Low, R. B., D.S.O.	7 Feb. 11
Moore, F. L.	7 Feb. 11
Ditmas, A. R., S. and T. Corps	7 Feb. 11
Stuart, D. F.	7 Feb. 11
Cooper, P. T., R. A. (*g*) (206)	11 Feb 11
Moore, A. G. W.	28 Feb. 11
Foote, H. B., R.A. (*g*) (206)	1 Apl. 11
Robin, N. E.	6 May 11
Bell, L. M.	9 May 11
Burlton, W. E. F., S. and T. Corps	23 May 11
White, W. E.	12 June 11
Deare, B. H., I.M.S.	28 July 11
Smith, S. B., C.M.G., I.M.S.	28 July 11
Frost, G. H., M.B., I.M.S.	28 July 11
Wood, H. S., M.B., I.M.S.	28 July 11
Irvine, T. W., M.B., F.R.C.S.E., I. M. S.	28 July 11
Entrican, J., M.D., I.M.S.	28 July 11
Pridmore, W. G., C.M.G., M.B., I.M.S.	28 July 11
Donovan, C., M.D., I.M.S.	28 July 11
Penny, J., I.M.S.	28 July 11
Graves, D. H. McD., M.B., I.M.S.	28 July 11
Trydell, B., 7 G. R.	29 Aug. 11
Beville, C. H., C.B., S. and T. Corps	29 Aug. 11
Dennys, H. T., C.I.E.	29 Aug. 11
Widdicombe, F. S., 116 Mah.	
Geoghegan, S., C.B. [L]	29 Aug. 11
Ward, T. M.	29 Aug. 11
Swan, C. T.	17 Sep. 11
Browne, A. J. W.	18 Sep. 11
Wilson, F. A., C.B., R.E. (207)	25 Sep. 11
Bolton, A. H.	13 Nov. 11
†*Stanton, F. W. S.*, D.S.O., R.A.	22 Nov. 11
Bell, R. M., S. and T. Corps	25 Nov. 11
Dunlop, H. H.	25 Nov. 11
Heycock, C. H., R.E. (206)	18 Jan. 12
Mitter, R. K., M.B., I.M.S. [L]	30 Jan. 12
Johnston, H. W., 67 Punj.	30 Jan. 12
Elliot, C. R.	30 Jan. 12
Jollie, F.	30 Jan. 12
Arnold, A. S., 26 Lt. Cav.	30 Jan. 12
Widdicombe, G. T., C.B., 9 G. R.	30 Jan. 12
Magrath, H. A. F.	1 Feb. 12
Hurly, M. R.	1 Feb. 12
Taylor, A. W. N., C.S.I., 7 G. R.	13 Feb. 12
Morse, C. J., Brit. Ser.	20 Mar. 12
Thacker, H. J., S. and T. Corps	28 Apl. 12
Brown, W. H., C.B.	10 May 12
Bertie-Clay, N. S., R. A. (206)	18 May 12
Hawkins, G. A.	19 May 12
Burlton, A. R., S. and T. Corps	19 May 12
Stone, C. A., M.D., R. A. M. C.	21 May 12
Dunolly, K. J. G., 122 Infy. [*l*]	30 May 12
Tylden-Pattenson, E. C., D.S.O., R.E. (207)	6 June 12
Way, L., R. A. M. C. (*temp. Col.*) (206)	8 June 12
Waterfield, A. C. M., M.V.O., *p.s.c.* [*l*]	10 July 12
Tennant, H. L., R. A., *p.s.c.*, (*e*)	21 July 12
Haig, P. B., C.B., M.B., I. M. S.	27 July 12
Maddox, R. H., M.B., I.M.S.	27 July 12
Hugo, E. V., C.M.G., M.D., F.R.C.S., I.M.S., *V.H.S.*	27 July 12
Smith, H. A., C.I.E., M.B., I.M.S.	27 July 12
Smith, G. McI. C., M.B., I.M.S.	27 July 12
Swinton, F. E., I.M.S.	27 July 12
Burnett, S. H., M.B., I.M.S.	27 July 12
Jackson, T., M.B., I.M.S.	27 July 12
Macrae, J. L., M.B., I.M.S.	27 July 12
Hall, E. S.	25 Aug. 12
Turner, F. W. C., 20 Horse	25 Aug. 12
Daunt, W. D.	25 Aug. 12
Vincent, F. L.	25 Aug. 12
Moore, H. A., 36 Sikhs	25 Aug. 12
Watson, A. C.	25 Aug. 12
Murray, J. H. S., R. E. (206)	26 Aug. 12

†Brigadier-General.

Lieutenant-Colonels—*contd.*	Date of rank.
Mawhinny, R. J. W., C.B., R.A.M.C.	11 Oct. 12
Mockler, G. H. G.	19 Oct. 12
Malcolm, W. L., 31 Punj.	27 Oct. 12
Ricketts, H. C.	10 Nov. 12
Forster, P. B. B.	10 Nov. 12
Stewart, F. T.	10 Nov. 12
Jamieson, A. J.	10 Nov. 12
Newsom, A. C., C.M.G., A.V.C.	16 Nov. 12
Warren, W. L., R. A.	22 Nov. 12
Hickley, A. C., 3 G. R.	8 Dec. 12
Denne, R. W. A., 79 Infy.	8 Dec. 12
Kirkwood, T. M.	8 Dec. 12
Pilcher, A. J., R.E. (206)	25 Dec. 12
Roome, R. E.	14 Jan. 13
†*Hardy, W. K.*, R.A. (207)	29 Jan. 13
Milne, C., M.B., I.M.S.	30 Jan. 13
Young, W., M.B., F.R.C.S.E., I.M.S.	30 Jan. 13
Bourke, J. J., M.B., I.M.S.	30 Jan. 13
Chatterton, B. R., M.D., F.R.C.S., I.M.S.	30 Jan. 13
Williams, C. E., M.D., I.M.S.	30 Jan. 13
Ogilvie, W. H., C.M.G., M.B., I.M.S.	30 Jan. 13
Parry, E. R., M.B., I.M.S.	30 Jan. 13
More, P. St. C., M.B., I.M.S.	30 Jan. 13
Bidie, G., M.D., F.R.C.S.E., I.M.S.	30 Jan. 13
Stevens, C. R., M.D., F.R.C.S., I.M.S.	30 Jan. 13
Rogers, *Sir* L., Kt., C.I.E., M.B., F.R.C.P., F.R.C.S., I.M.S.	30 Jan. 13
Newman, E. A. R., C.I.E., M.D., I.M.S.	30 Jan. 13
Birdwood, G. T., M.D., I.M.S.	30 Jan. 13
Turner, R. G., C.M.G., D.S.O., F.R.C.S., I.M.S.	30 Jan. 13
Davidson, J., D.S.O., M.D., I.M.S. [L]	30 Jan. 13
Bhola Nauth, C.I.E., I.M.S.	30 Jan. 13
Foulkes, T. H., F.R.C.S., I.M.S.	30 Jan. 13
Chitty, E. R. I.	5 Feb. 13
Priestly, C. E. N., 42 R.	5 Feb. 13
Twigg, E. F., 124 Infy.	5 Feb. 13
Holmes, G. V.	5 Feb. 13
Wikeley, J. M.	5 Feb. 13
Browne, A. P., D.S.O.	5 Feb. 13
Roddy, H. H., C.B., S. and T. Corps	5 Feb. 13
Tweddell, H.	5 Feb. 13
†*Dale, G. A.*, C.M.G., [*l*]	5 Feb. 13
Turner, A. G. B., D.S.O.	5 Feb. 13
†*Moberly, F. J.*, D.S.O., 25 Punj. *p.s.c.*	13 Feb. 13
Faunce, R. DeL.	15 Feb. 13
Saunders, A. R., 29 Lrs.	16 Feb. 13
Pearson, H. F. A.	21 Feb. 13
Harding, M. F., C.M.G.	10 Mar. 13
Wilson, J. A., D.S.O., 8 G.R.	27 Mar. 13
Cunningham, P. H., 101 Grs.	12 Apl. 13
Oakes, G. R.	4 May 13
Anderson, A. M.	4 May 13
Maxwell, H. G.	7 May 13
Savi, R. J. T.	7 June 13
Ainslie, C., R.E. (206)	10 July 13
Ryder, C. H. D., C.I.E., D.S.O., R.E. [*l*] (206)	14 July 13
Wooldridge, W. H.	20 July 13
Bennett, H., M.B., F.R.C.S., I.M.S.	29 July 13
Barry, C. C. S., C.I.E., I.M.S.	29 July 13
Gould, J., C.B.E., M.B., I.M.S.	29 July 13
Mulvany, J., I.M.S.	29 July 13
Burden, H., C.I.E., F.R.C.S., I.M.S. (*temp. Col.*)	29 July 13
Bowle-Evans, C. H., C.M.G., M.B., I.M.S.	29 July 13
Berry, A. E., I.M.S.	29 July 13
MacLeod, E. C., I.M.S.	29 July 13
Burn, E. M. J., R.E. [L] (207)	1 Sep. 13
Hardy, W. E., R. A. M. C.	13 Sep. 13
Wintour, E.	13 Sep. 14
Robinson, G. W., C.B.	14 Sep. 13
Keays, R. W. C., 83 Infy.	14 Sep. 13

Lieutenant-Colonels—*contd.*	Date of rank.
ffrench-Mullen, J. L. W., C.I.E.	14 Sep. 13
†*Lane, H. A.*	14 Sep. 13
Powell, W. B., C.M.G., 7 G.R.	14 Sep. 13
Watling, G., 11 Raj.	14 Sep. 13
Thornton, C. E., C.M.G.	14 Sep. 13
Obbard, E. N.	14 Sep. 13
Talbot, J.	14 Sep. 13
Jones, H. H., S. and T. Corps (*temp. Col.*)	14 Sep. 13
Bayley, E. C., C.I.E., 15 Lrs.	14 Sep. 13
Robertson, C. L., C.M.G., R.E. (207)	25 Sep. 13
Smith, W. D., C.M.G., D.S.O., A. V. C.	15 Oct. 13
Mussenden, F. W., 8 Hrs.	19 Oct. 13
Boddam, E. B. C., 5 G. R., *p.s.c.*	19 Oct. 13
Lethbridge, E. A. E., C.M.G., D.S.O., Oxf. & Bucks. L. I. (207)	25 Oct. 13
Chaplin, R. E., 8 Cav.	2 Nov. 13
Watson, L. A., *p.s.c.* [L]	13 Nov. 13
Chichester, A. G. deV.	16 Nov. 13
Thomason, A. F., S. and T. Corps	16 Nov. 13
†*Dunsford, F. P. S.*, D.S.O.	16 Nov. 13
Nisbett, W. G., 92nd Punj.	16 Nov. 13
Sutton, F. G. H., 10th G. R.	16 Nov. 13
†*Rose, J. L.*, C.I.E.	16 Nov. 13
Porteous, C. McL.	16 Nov. 13
Robson, C. G., 10th Jats [L]	16 Nov. 13
Webb, A. E., S. and T. Corps	16 Nov. 13
Barton, H. J., R.E., (*l*) (206)	5 Dec. 13
Ward, A., D.S.O.	12 Dec. 13
Jourdain, C. E. A., D.S.O., N. Lan. R.	17 Dec. 13
Granger, T. A., C.M.G., M.B., I.M.S., *V.H.S.*	1 Jan. 14
†*Marshall, H. J. M.*, C.M.G., R.E. (207)	5 Jan. 14
Bainbridge, W. F., C.M.G., D.S.O., 54th Sikhs, *p.s.c.*	7 Jan. 14
Muscroft, W. St. C., S. and T. Corps	7 Jan. 14
Wooldridge, H. C.	21 Jan. 14
Davidson, F. M., R. A. (*g*) (207)	28 Jan. 14
Fisher, J., D.S.O., I.M.S.	29 Jan. 14
Harriss, S. A., M.B., I.M.S.	29 Jan. 14
Sutherland, D. W., C.I.E., M.D., I.M.S.	29 Jan. 14
Bennett, V. B., M.B., F.R.C.S., I.M.S.	29 Jan. 14
Pearce, C. R., M.B., I.M.S.	29 Jan. 14
†*Murray, E. R. B.*, 90th Punj.	1 Feb. 14
Macquoid, C. E. E. F. K., D.S.O., 4th Cav.	11 Feb. 14
Stack, C. S., C.M.G., 33rd Cav.	11 Feb. 14
Dickson, J. H., C.I.E., S. and T. Corps (*Ty. Col.*)	11 Feb. 14
Lloyd, W. E. E., 97th Infy.	11 Feb. 14
Pennington, A. W., M.V.O., 9th Horse	11 Feb. 14
Longden, A. B., 38th Dogras	11 Feb. 14
Eastmead, C. S., 3rd G. R. [L]	11 Feb. 14
Vaughan, E. G., D.S.O., S. and T. Corps	11 Feb. 14
Vanrenen, G. R., 16th Rajputs	11 Feb. 14
Stevens, N. M. C., C.M.G.	11 Feb. 14
Heffernan, H. W., 86th Infy.	11 Feb. 14
Major, F. F., 95th Infy.	11 Feb. 14
Brooke, H. M. M., S. and T. Corps	11 Feb. 14
Sherer, J. C.	11 Feb. 14
Beatson Bell, J.	11 Feb. 14
Lloyd-Jones, F. L., 113th Infy.	11 Feb. 14
Hitchins, H. E., 3rd Brah.	11 Feb. 14
Barnard, A. E., 30th Lrs.	11 Feb. 14
Pierce, F. G., 102nd Grs.	11 Feb. 14
Walker, W. R.	11 Feb. 14
Kauntze, C. R.	11 Feb. 14
Moore, C. H. G., C.M.G., D.S.O., S. and T. Corps	11 Feb. 14
Kemball, J. S.	11 Feb. 14
Baker, H. R., 73rd Inf.	15 Feb. 14
Murray, A. B.	17 Feb. 14
†*Young, H. A.*, C.I.E., R.A. (*temp. Col.*)(206)	21 Feb. 14
Young, W. E.	8 Mar. 14
Ford, H. B., 29th Punj.	14 Mar. 14
Dallas, J. S., 83rd Lt. Infy.	9 May 14
Hutchinson, C. R. M., 4th G. R.	9 May 14
†*Money, E. D.*, C.I.E., D.S.O.	9 May 14
Lathbury, H.O., R.E. (207)	16 May 14
Wall, F., C.M.G., I.M.S.	31 May 14

† Brigadier-General.

Lieutenant-Colonels—*contd.*	Date of rank.
Chatterton, G. D. L., D.S.O.	1 June 14
Hennessy, J., C. B., M.B., R.A.M.C.	7 July 14
Cobbe, H. H., C.M.G., D.S.O., 13th Lrs.	27 July 14
Bamfield, H., J K., D.S.O., I.M.S.	28 July 14
Grant, J. W., M.B., I.M.S.	28 July 14
Hayward, W. D., M.B., I.M S.	28 July 14
Evans, S., M.B., I.M.S.	29 July 14
Hamilton, J. A., C.M.G., M.B., F.R.C.S.E., I.M.S.	29 July 14
McDonald, J. H., M.B., I.M.S.	29 July 14
Bagshawe, F. W., C.I.E., Mily. Accts. Dept. [L]	22 Aug. 14
Henslowe, F. J. D.	22 Aug. 14
Gordon, J. L. R., C.B., 15th Sikhs	22 Aug. 14
Cooper, W. G.	22 Aug. 14
Ralph, A. C., D.S.O., 4th Rajputs	22 Aug. 14
Stewart, J. A., 13th Rajputs	22 Aug. 14
Humfrey, F. G. C., 12th Cav.	22 Aug. 14
Ainslie, H. P.	22 Aug. 14
Basevi, W. H. F., 91st Punj.	22 Aug. 14
Fraser, F. J.	12 Sep. 14
Baker, C. N.	19 Sep. 14
O'Meara, C. A. E., C.I.E.	23 Oct. 14
Clarkson, B. St. J., C.M.G., D.S.O., Dorset. R.	25 Oct. 14
Bond, R. F. G., R.E. (206)	27 Oct. 14
†*Cartwright, G. N.*, C.M.G., D.S.O., R.A.	30 Oct. 14
†*Rettie, W. J. K.*, D.S.O., R.A.	30 Oct. 14
Tancock, O. K., C.M.G., R.A. (206)	30 Oct. 14
Brewster, R. F., R.A. (206)	30 Oct. 14
Smallwood, F. G., C.V.O., R.A. (206)	30 Oct. 14
Grove, E. W., D.S.O., R.A. (207)	30 Oct. 14
Marsden, R. T., R.A.	30 Oct. 14
Bowring, A. H., R.A.	30 Oct. 14
Rivett-Carnac, S. G., R.E.	30 Oct. 14
†*Rushton, H. W.*, D.S.O., R.E. (207)	30 Oct. 14
Douglas, A. P., C.M.G., R.A. (206)	30 Oct. 14
Stiffe, A. F. E., R.A. (207)	30 Oct. 14
†*Sykes, C. A.*, D.S.O., R.A.	30 Oct. 14
Stockley, H. R., C.I.E., R.E.	30 Oct. 14
Winsloe, A. R., D.S.O., R.E.	30 Oct. 14
Smyth, W. C., R.E.	30 Oct. 14
Elderton, M. T., 108th Infy.	10 Nov. 14
Haldane, C. L., C.M.G., D.S.O., 7th G.R. [l]	10 Nov. 14
Amesbury, W. L. R., S. and T. Corps	10 Nov. 14
Codrington, H. deB., S. and T. Corps	28 Nov. 14
Coke, A. H., 112th Infy.	5 Dec. 14
Lillingston, W. E. G.	8 Dec. 14
Strahan, G. A., 90th Punj.	8 Dec. 14
Carter, G. L., C.I.E., 106th Prs.	8 Dec. 14
Firth, R. A., D.S.O., 8th G. R.	8 Dec. 14
†*Smith, L. A.*, R.A.	17 Dec. 14
Sparrow, R., C.M.G., 7th Dgn. Gds.	24 Dec. 14
†*Goring Jones, M. D.*, C.M.G., Brit. Ser.	7 Jan. 15
Lane, F. C., R.A. (206)	23 Jan. 15
Stevens, A. F., I.M.S.	29 Jan. 15
Bensley, C. H., I.M.S.	29 Jan. 15
Watling, F. H., M.B., I.M.S.	29 Jan. 15
Mathew, C. M., I.M.S.	29 Jan. 15
Stephenson, J., M.B., F.R.C.S. [L] I.M.S.	29 Jan. 15
Windsor, F. N., M.B., I.M.S.	29 Jan. 15
Waters, E. E., M.D., I.M.S.	29 Jan. 15
Leventon, A., F.R.C.S.I., I. M. S.	29 Jan. 15
King, A. F. W., F.R.C.S.E., I.M.S.	29 Jan. 15
Standage, R. F., M.R.C.P., I.M.S.	29 Jan. 15
Gibbs, A. A., I.M.S. (e)	29 Jan. 15
Illington, E. M., F.R.C.S.E., I.M.S.	29 Jan. 15
Lee, A. W H., 7th G. R.	30 Jan. 15
Douglas, W. B., 1st Brah.	30 Jan. 15
Tennant, E. [L]	30 Jan. 15
Paul, R. S., 126th Infy.	30 Jan. 15
†*Patch, F. R.*, C.M.G., D.S.O., R.A.	1 Feb. 15
†*Templer, C. F.*, C.M.G., C.I.E., A. Remt. Dept.	15 Feb. 15
Hill-Whitson, T. E. L., 14 Hrs.	22 Feb. 15
†*Legard, D'A.*, *p.s.c.*, 17 Lrs.	24 Feb. 15

Lieutenant-Colonels—*contd.*	Date of rank.
O'Reilly, H. W. H., M.B., R.A.M.C.	1 Mar. 15
Read, H. W. K., R.A.M.C.	1 Mar. 15
Mangin, F. M., R.A.M.C.	1 Mar. 15
Longhurst, B. W., R.A.M.C.	1 Mar. 15
Clark, E. S., M.B., R.A.M.C.	1 Mar. 15
Staddon, H. E., R.A.M.C.	1 Mar. 15
Killery, St. J. B., R.A.M.C.	1 Mar. 15
Addams-Williams, L., R.A.M.C.	1 Mar. 15
McCarthy, J. McD., M.B., R.A.M.C.	1 Mar. 15
Young, A. H. O., R.A.M.C.	1 Mar. 15
Ross, N. H., M.B., R.A.M.C.	1 Mar. 15
Hall, S. O., R.A.M.C.	1 Mar. 15
Cowan, J., M.B., R.A.M.C.	1 Mar. 15
O'Flaherty, A. R., R.A.M.C.	1 Mar. 15
Walton, H. B. G., R.A.M.C.	1 Mar. 15
Mainprise, C. W., R.A.M.C.	1 Mar. 15
Scott, A. L., R.A.M.C.	1 Mar. 15
Winkfield, W. B., R.A.M.C.	1 Mar. 15
Houghton, J. W. H., R.A.M.C.	1 Mar. 15
Lloyd, R. H., R.A.M.C.	1 Mar. 15
Packer, H. D., R.A.M.C.	1 Mar. 15
Gill, J. G., R.A.M.C.	1 Mar. 15
Lauder, T. C., R.A.M.C.	1 Mar. 15
Palmer, F. J., R.A.M.C.	1 Mar. 15
Falkner, P. H. E., F.R.C.S.I., R.A.M.C.	1 Mar. 15
Langstaff, J. W., D.S.O., R.A.M.C.	1 Mar. 15
Cunningham, R. A., M.B., R.A.M.C.	1 Mar. 15
Falkner, P. H., F.R.C.S.I., R.A.M.C.	1 Mar. 15
Harvey, F., R.A.M.C.	1 Mar. 15
Hime, H. C. R., M.B., R.A.M.C.	1 Mar. 15
Woodside, W. A., R.A.M.C.	1 Mar. 15
Seeds, A. A., M.D., R.A.M.C.	1 Mar. 15
Bond, J. H. R., D.S.O., R.A.M.C.	1 Mar. 15
Kaye, C., C.S.I., C.I.E.	6 Mar. 15
Plummer, E. W., D.S.O., R.A. (e) (206)	15 Mar. 15
†*Andrew, F. A.*, D.S.O., 128th Prs.	23 Mar. 15
Munn, R. G., 36th Sikhs	23 Mar. 15
Alexander, H. S., D.S.O., 44 Infy.	23 Mar. 15
Travers, R. E.	23 Mar. 15
Geoghegan, F. E., C.I.E., S. & T. Corps	23 Mar. 15
Cullen, E. H. S., C.M.G., D.S.O., M.V.O., 34th Prs.	24 Mar. 15
Clarke, J. L. J., C.M.G., *p.s.c.*, E. York R. [l]	27 Mar. 15
Hume, A. H. B., R.E. [l] (207)	29 Mar. 15
Bannerman, W. P., 31st Lrs	24 Apl. 15
Cassels, G. R., 35th Sikhs	24 Apl. 15
Becher, G. A., D.S.O., 8th Cavalry	27 Apl. 15
Charlton, C. E. C. G., D.S.O., R.A.	9 May 15
Byers, C. B., S. & T. Corps.	15 May 15
Wigram, C., C.B., C.S.I., C.V.O., *p.s.c.*, Eq., 18th Lrs.	3 June 15
Smith, W. F., S. & T. Corps (e)	3 June 15
†*Sanders, G. A. F.*, C.M.G., *p.s.c.*, R. E.	3 June 15
Boileau, E. R. P., 2nd G. R.	3 June 15
Tomkins, H. L., C.M.G., D.S.O., 28 Punj.	3 June 15
Leeds, T. L., C.M.G., D.S.O., 59th Rif.	3 June 15
**Freeland, H. F. E.*, C.B., M.V.O., D.S.O., R.E. (206)	3 June 15
Murphy, C. C. R., 30th Punj.	3 June 15
†*Holdich, H. A.*, D.S.O., 5th G. R.	3 June 15
Forteath, F. W. H., S. & T. Corps	8 June 15
Moore, A., 66th Punjabis	6 July 15
Roosmale-Cocq, C. A., 8th G.R.	6 July 15
Darley, J. R., C.I.E., D.S.O., 119th Punj.	6 July 15
Cotton, W. L., 5th Infy.	8 July 15
Maule, H. N. St. J., C.M.G., R.A. (e)	20 July 15
Chapman, P. F., M.B.	29 July 15
Hooton, A., I.M.S.	29 July 15
Blenkinsop, F. L., M.B.	29 July 15
Knapton, H. A. F., I.M.S.	29 July 15
Cochrane, A. W. R., M.B., F.R.C.S.	29 July 15
Clemesha, W. W., M.D. (*temp. Col.*)	29 July 15
Black, J. A., M.B.	29 July 15
Wilson, R. P., F.R.C.S.	29 July 15
Lindesay, V. E. H., M.B.	29 July 15

* Ty. Maj.-Genl. † Brigadier-General.

Lieutenant-Colonels—*contd.*	Date of rank.
Robertson, J. C., C.M.G., C.I.E., M.B.	29 July 15
Rainier, N. R. J., I.M.S.	29 July 15
Perry, E. L., D.S.O., I.M.S.	29 July 15
Niblock, W. J., M.B., F.R.C.S.I.	29 July 15
Rost, E. R., I.M.S.	29 July 15
Riach, A. H. D., R.E. (206)	12 Aug. 15
Close, L. H., C.M.G., R.E. (207)	12 Aug. 15
Mackenzie, E. L., D.S.O., R. Suss. R.	12 Aug. 15
Garstin, H. E., D.S.O., R.A.	3 Sep. 15
Fraser, H. F., C.M.G., D.S.O., 21st Lrs. (*e*)	6 Sep. 15
Carter, E. P., R.A.	7 Sep. 15
deBerry, P. P. E., C.M.G., R.A. (*e*)	13 Sep. 15
Cheyne, R. E., 8th Cav.	21 Sep. 15
Payne, E. H., Mily. Accts. Dept.	21 Sep. 15
†*Bateman-Champain, H. F.*, C.M.G., 9th G. R.	21 Sep. 15
Dallas, A. E., C.M.G., S. & T. Corps	21 Sep. 15
†*Brownlow, d'A. C.*, C.M.G., C.I.E., J. A. G.'s Deptt.	21 Sep. 15
Prentis, W. S., 72nd Punj.	21 Sep. 15
Nugent, C. H. H., C.I.E., R.E. (207)	4 Oct. 15
Coldstream, W. M., R.E. (207)	22 Oct. 15
Samborne-Palmer, F. C S., 11th Raj.	23 Oct. 15
Cassels, K. S., 8th G. R.	23 Oct. 15
Bayley, L. S., R. A.	8 Nov. 15
Garratt, H. S., 103th Infy.	9 Nov. 15
Lloyd, J. H., 6th G. R.	9 Nov. 15
Johnson, C. R., 5th G. R.	9 Nov. 15
Hay, L. T., 5th Cav.	9 Nov. 15
Ravenhill, C., R.A.	16 Nov. 15
Saunders, F. W., R. E.	22 Nov. 15
Ritchie, C. Macl., R.A. [*l*] (207)	24 Nov. 15
†*Harrison, A. H. P.*, 33rd Punj.	2[illegible] Nov. 15
Palmer, C. C., *p.a.c.*, R.A. (206)	29 Nov. 15
†*Luke, T. M.*, D.S.O., O.B.E., R.A. (206)	29 Nov. 15
Chandler, H. G. W., Mily. Accts. Deptt.	4 Dec. 15
Hunt, F. D., A. V. C.	15 Dec. 15
Alexander, A. V., 74th Punj.	21 Dec. 15
Beatty, G. A. H., D.S.O., 9th Horse	21 Dec. 15
Ward, C. H., 120th Infy.	21 Dec. 15
Nicholson, W. C., 17th Infy.	21 Dec. 15
Rawlins, A. K., C.I.E., D.S.O., 24th Punj.	21 Dec. 15
Mackenzie, C., D.S.O., 13th Lrs.	21 Dec. 15
Swan, J. S., S. and T. Corps	21 Dec. 15
Anderson, S., M. B., I.M.S.	1 Jan. 16
Newnham, C. C., D.S.O., 6th Cav., *p.s.c.*	1 Jan. 16
Lushington, W. N., S. & T. Corps	1 Jan. 16
Cumberlege, A. F., R.E. (206) (*temp. Col.*)	1 Jan. 16
Boulton, H., M.B.	1 Jan. 16
Dobbs, C. F., D.S.O., 94th Infy., *p.s.c.*	1 Jan. 16
Hunter, C. G. W., D.S.O., R.E.	1 Jan. 16
Browse, G., D.S.O., M.B., I.M.S.	1 Jan. 16
Fergusson, A. C., C.M.G., D.S.O., R.A.	1 Jan. 16
Kitson, C. E., D.S.O., R. W. K. Regt.	1 Jan. 16
Cox, C. E. S., 67th Punjabis	1 Jan. 16
Woods, H. K., S. Lan. R.	1 Jan. 16
Browne, C. M., C.M.G., D.S.O., R.E.	1 Jan. 16
†*Hildyard, R. J. T.*, D.S.O., R. W. Kent R.	1 Jan. 16
Tomkins, E. L., R.A. (206)	6 Jan. 16
Lethbridge, S., R.A. (206)	6 Jan. 16
Bakhle, C. R., I.M.S. (206)	29 Jan. 16
Kukday, K. V., I.M.S. (206)	29 Jan. 16
Dawes, C. D., I.M.S.	29 Jan. 16
Symons, T. H., O.B.E., I.M.S.	29 Jan. 16
Walton, H. J., M.D., F.R.C.S. (206)	29 Jan. 16
Ainsworth, H., M.B., F.R.C.S. (206)	29 Jan. 16
Smith, F. A., M.D.	29 Jan. 16
Cornwall, J. W., M.D.	29 Jan. 16
Miller, A., M.B.	29 Jan. 16
Hammond, F. A. L., I.M.S.	29 Jan. 16
Brown, H. R., I.M.S.	29 Jan. 16
Fleming, A. N., D.S.O., M.B., F.R.C.S.E.	29 Jan. 16
Dee, P., M.B.	29 Jan. 16
Mell, F. O. N., C.I.E., M.B.	29 Jan. 16
Bourne, W. Fitz G., 6th Jats	31 Jan. 16
Maffett, R. E., W. Rid. R.	3 Feb. 16
Stallard, S. F., C.M.G., D.S.O., R.A.	2 Feb. 16

Lieutenant-Colonels—*contd.*	Date of rank.
†*Fagan, E. A.*, C.M.G., D.S.O., 36th Horse	61 Feb. 16
Mackintosh, J. B., R.A. (206)	68 Feb. 16
Willoughby-Osborne, A. de V., S. Lan. R.	28 Feb. 16
Field, C. D., 75th Infy.	1 Mar. 16
Hocken, C.A.F., 35th Horse, *p.s.c.* (*temp. Col.*)	1 Mar. 16
†*Nightingale, M. R. W.*, C.M.G., D.S.O., 5th G. R.	1 Mar. 16
Davis, G. J., D.S.O., 22nd Punj.	1 Mar. 16
Kirkpatrick, E., 59th Rifles	1 Mar. 16
Redl, E. A. F., C.I.E., 113th Infy.	1 Mar. 16
Spence, A. H. O., C.I.E., 39th Horse (*Ty. Col.*)	1 Mar. 16
†*Kemball, A. G.*, 31st Punj.	1 Mar. 16
Woulfe Flanagan, R. J., R. W. Kent R.	14 Mar. 16
Glasfurd, A. I. R., C.M.G., D.S.O., 40th Pathans	29 Mar. 16
Alexander, C. H.	29 Mar. 16
Hogg, J. Mc K. T., 9th G. R.	29 Mar. 16
Wilkinson, C. W., C.M.G., D.S.O., R.E. (207)	1 Apl. 16
McCormick, A. L. C., R.E. (207)	1 Apl. 16
Davidson, F. D., 99th Infy.	1 Apl. 16
Ricketts, P. E., D.S.O., M.V.O., (*temp. Col.*)	3 Apl. 16
Farmer, W. C. R., R.A (206)	10 Apl. 16
Smith, H. B., D.S.O., R.A.	19 Apl. 16
Woods, G. G., R.A. (206)	20 Apl. 16
Prevost, G. H., 87th Punj	22 Apl. 16
Barton, P., D.S.O., R.A. (206)	24 Apl. 16
Carter, R. M., C.B., F.R.C.S., I.M.S.	26 Apl. 16
Kelsall, H. W., R. E.	30 Apl. 16
O'Reilly, C. M., 86th Infy.	3 May 16
Bliss, J. A., D.S.O., M.V.O., 64th Prs.	3 May 16
Colan, W. R. B., D.S.O., 67th Punj.	3 May 16
†*Lukin, R. C. W.*, D.S.O., 3[illegible]th Horse	3 May 16
Hole, H. L., 80th Infy.	3 May 16
Orman, F. L., 10th G. R.	3 May 16
Fellows, C. S., 109th Infy.	3 May 16
Archdale, T. M., D.S.O., R.A.	20 May 16
Sealy, A. E., 4th G. R.	24 May 16
Wynter, C. P., 52nd Sikhs	24 May 16
Ross, H., 103rd Infy.	31 May 16
Kirke, K. St. G., D.S.O., R.A.	1 June 16
Pennefather-Evans, G., 19th Punj.	3 June 16
Mears, C. D., M.V.O., 16th Cav.	3 June 16
Davie, J. H. M., 34th Horse	3 June 16
†*Watt, D. M.*, D.S.O., *p.s.c.*, 2nd G. R.	3 June 16
Edwardes, A. C., 102nd Grs.	3 June 16
Boileau, G. H., C.M.G., D.S.O., R.E.	3 June 16
Walton, G., 46th Punj.	3 June 16
Keelan, H. P., 121st Prs.	3 June 16
Hamilton, W. A., Conn. Rang.	3 June 16
†VC *Costello, E. W.*, C.M.G., D.S.O., *p.s.c.*, 22nd Punj.	3 June 16
Scott-Elliott, C. R., 61st Prs.	3 June 16
Greenstreet, C. B. L., R. E.	3 June 16
Deacon, H. R. G., D.S.O., Conn. Rang.	3 June 16
†*Andrus, T. A.*, C.M.G., N. Staff. R.	3 June 16
Goodbody, C. M., C.I.E., D.S.O., I.M.S.	3 June 16
Moberly, H. S., *p.s.c.*, 66th Punj.	3 June 16
Russell, F. D., *p.s.c.*, [L] 1st Lrs.	3 June 16
Longhurst, A. L., *p.s.c.*, 7th G. R. [L]	3 June 16
Anthony, W. S., A.V.C.	3 June 16
Furney, J. L., 22nd Punj.	3 June 16
Foster, J. G., R.A.M.C.	3 June 16
Winsloe, H. E., D.S.O., R.E. (206)	3 June 16
Young, A. W. C., M.B.	3 June 16
Anderson, C. F., D.S.O., R.E. [L] (206)	3 June 16
Watson, S. B., 64th Prs.	3 June 16
†*Pryce, H. E. ap. R.*, C.M.G., D.S.O., *p.s.c.*, 18th Infy.	3 June 16
Bourne, W. K., *p.s.c.*, 2nd Lrs.	3 June 16

† Brigadier-General.

Lieutenant-Colonels—*contd.*	Date of rank.
Fisher, B. D., D.S.O., *p.s.c.*, 17th Lrs.	3 June 16
†*St. John, R. S.* D.S.O., *p.s.c.*, 20th Infy.	3 June 16
Marsh, F. G., C.M.G., *p.s.c.*, 9th G. R.	3 June 16
Cramer-Roberts, W. E., Norf. R.	3 June 16
Crossley, H. J., R.A.M.C.	3 June 16
Geoghegan, N. M., 89th Punj.	3 June 16
†*Brough, A.*, D.S.O., R.E.	3 June 16
Queripel, L. H., D.S.O., R. A.	3 June 16
Beddy, P. L., 51st Sikhs	3 June 16
Hardcastle, R. N., D.S.O., Manch. R.	3 June 16
†*Franklin, H. S. E.*, D.S.O., 14th Sikhs	3 June 16
Lewis, F., D.S.O., Leic. R.	3 June 16
Stapleton, F. H., Oxf. and Bucks L. I.	3 June 16
Joicey, *Hon.* H. E., 14th Hrs.	3 June 16
Corbould-Warren, E., R.A.	3 June 16
Belgrave, H. D., D.S.O., R. W. Kent R.	3 June 16
†*Dick, R. N.*, R. Suss. R.	3 June 16
Leachman, G. E., C.I.E.	3 June 16
Rowlandson, M. G. D., D.S.O., *p.s.c.*	3 June 16
Case, H. A., D.S.O., Dorset. R.	3 June 16
Carwithen, E. T., Cant. Magte. Dept.	4 June 16
Fisher, F.	5 June 16
Sherer, J. D., D.S.O., R.A.	18 June 16
Kemble, F. A., R.A. (206)	18 June 16
Strange, A. J., 6th G. R.	28 June 16
Bidwell, R. F., 101st Grs.	28 June 16
Battine, R. St. C., D.S.O., *p.s.c.*, [L] 21st Cav.	23 July 16
Woolley, J. M., M. D., I.M.S.	29 July 16
Hugo, J. H., D.S.O., M. B., I.M.S.	29 July 16
Kenrick, W. H., I.M.S.	29 July 16
Price, R. H., M. B., F.R.C.S.E., I.M.S.	29 July 16
Bryson, R., F.R.C.S.E., I.M.S.	29 July 16
Adams, F., D.S.O., 20th Horse	29 July 16
Sangster, P. B., D.S.O., 29 Lrs.	7 Aug. 16
Earle, E. S., 14th Sikhs	8 Aug. 16
Westropp, F. M., D.S.O., R.E.	13 Aug. 16
Cameron, H. A., R.E. (207)	18 Aug. 16
Kelly, F. P. C., C.M.G., D.S.O., 125th Rif.	24 Sept. 16
Powell, E. E., R. A. M. C.	26 Sept. 16
Young, H. G., D.S.O., 22nd Cav.	8 Oct. 16
Nangle, K. E., 88th Infy.	8 Oct. 16
Nicholas, S. H. E.	8 Oct. 16
Coningham, F. E., D.S.O., 10th G. R.	8 Oct. 16
Hill, H. C., D.S.O., 110th Infy.	29 Oct. 16
Haydon, W. P., D.S.O., S. & T. Corps.	29 Oct. 16
McLeod, T. J., D.S.O., 7th G. R.	29 Oct. 16
Cox, A. D., 26th Punj.	29 Oct. 16
Crawford, J. H., 32nd Lrs.	1 Nov. 16
Lough, A. T., R.A.	10 Nov. 16
Buist, A. H., M.V.O., Guides	14 Nov. 16
Souter, H. M. W., C.M.G., D.S.O., 14th Lrs.	14 Nov. 16
Edwardes, J. G., 3rd G. R.	29 Nov. 16
Newell, H. A., S. & T. Corps.	29 Nov. 16
Barclay, P. C. R., 122nd Infy.	29 Nov. 16
Teed, A. S. H., 14th Lrs.	29 Nov. 16
Sarel, G. B. M., 34th Horse	29 Nov. 16
Fenn, E. H. McB.	29 Nov. 16
Crawford, A. G., 84th Punj.	29 Nov. 16
Knox, E. F., D.S.O., 36th Sikhs	29 Nov. 16
Cheyne, A. Y., 15th Lrs.	29 Nov. 16
Smith, A. Le F., 2nd Rajputs	3 Dec. 16
Souter, A. B., 21st Cav.	17 Dec. 16
Tillard, A. B., D.S.O., 2nd G. R.	24 Dec. 16
Bridges, A. H., D.S.O., *p.s.c.* (*l*)	1 Jan. 17
McClintock, R. L., D.S.O., (206)	1 Jan. 17
Seton-Browne, C. L. S., D.S.O.	1 Jan. 17
Jackson, H. W., D.S.O., 89th Punj.	1 Jan. 17
Sweet, E. H., D.S.O., 2nd G. R.	1 Jan. 17
Dyke, P. H., D.S.O., 130th Baluchis.	1 Jan. 17
Wikham, H. R. 1st Dn. Gds.	1 Jan. 17
Lawrence-Archer, J. H., R. A. (206) (*temp. Col.*)	1 Jan. 17
Bowen, H. W., D.S.O., R. A. (206) (*temp. Col.*)	1 Jan. 17
Bruce, G. D., D.S.O., *p.s.c.*, 81st Prs.	1 Jan. 17
Deane, D., D.S.O., *p.s.c.*, 12th Cav. [*l*]	1 Jan. 17
†*Barstow, H.*, 38th Dogras	1 Jan. 17

Lieutenant-Colonels—*contd.*	Date of rank.
Herbert, G. M., D.S.O., Dorset. R.	1 Jan. 17
Barnardo, F. A. F., C.I.E., M.D., F.R.C.S.	1 Jan. 17
Malan, L. N., R. E.	1 Jan. 17
†*Wagstaff, C. M.*, C.M.G., C.I.E., D.S.O., *p.s.c.* R. E. (206)	1 Jan. 17
Baird, H. B. D., C.M.G., D.S.O., *p.s.c.* 12th Cav.	1 Jan. 17
Hay, C. J. B., D.S.O., *p.s.c.* Guides	1 Jan. 17
†*Pitcher, D. Le G.*, C.M.G., 39th Horse	1 Jan. 17
Neumann, C. W., 48th Prs.	1 Jan. 17
†*Coleridge, J. F. S. D.*, C.M.G., D.S.O., *p.s.c.*, 8th G. R.	1 Jan. 17
Talbot, F. E. G., *p.s.c.* 14th Sikhs	1 Jan. 17
Moens, A. W. H. M., D.S.O., *p.s.c.* 52nd Sikhs	1 Jan. 17
Drew, B. C. H., 23rd Prs., *p.s.c.*	1 Jan. 17
MacDonnell, H. N. F., M. Accts. Dept.	1 Jan. 17
Whitehead, J., D.S.O., *p.s.c.*, 1 Brah.	1 Jan. 17
Hawtrey, H. C., C.M.G., D.S.O., R.E.	1 Jan. 17
†*Bethell, H. K.*, C.M.G., D.S.O., 7th Hrs.	1 Jan. 17
†*Mackey, H. J. A.*, M.V.O., D.S.O., R.A.	3 Jan. 17
Swifte, E. L. 57th Rif.	17 Jan. 17
VC **Colvin**, J. M. C., *p.s.c.*, *q.s.* R.E.	18 Jan. 17
Lane, C. A., M. D.	28 Jan. 17
Kelly, T. B., D.S.O., F.R.C.S.E.	28 Jan. 17
Watson, C. H., D.S.O., I.M.S.	28 Jan. 17
Baines, E. F. E., I.M.S.	28 Jan. 17
Sealy, G. O. F., F.R.C.S.	28 Jan. 17
Murray, J. G. P., M.D., F.R.C.S.E.	28 Jan. 17
Hutchinson, F. H. G., M.D.	28 Jan. 17
Marjoribanks, J. L., M.D.	28 Jan. 17
Fenton, A., M.B.	28 Jan. 17
Knox, R. W., D.S.O., M.B., F.R.C.S.E.	28 Jan. 17
Butler, A. H., 44th Infy.	4 Feb. 17
McNeile, D. H., 19th Lrs. [L]	13 Feb. 17
Capper, A. S., D.S.O., *p.s.c.*, 39th Horse	15 Feb. 17
Milne, J. W., 82nd Punj.	4 Mar. 17
Hudson, J. H., Mily. Accts. Dept. [L]	4 Mar. 17
Henderson, M. H., 26th Cav.	4 Mar. 17
Tregear, F. C., 7th Rajputs	4 Mar. 17
Brackenbury, W. J. W., 43rd R.	4 Mar. 17
†*Gwyn-Thomas, G.*, C.M.G., D.S.O., 2nd Lrs.	4 Mar. 17
Anstruther, P. G., Sea. Highrs.	10 Mar. 17
Jenkin, F. C., D.S.O., R.A.	10 Mar. 17
Pocock, P. F., D.S.O., 120th Infy.	25 Mar. 17
Donovan, C. C., R.A.	1 May 17
Cotter, H. J., C.I.E., D.S.O., R.A.	1 May 17
Easton, F. A., R.A.	1 May 17
Lowis, P. S., D.S.O., R.A.	1 May 17
Kennedy, A. C., R.A.	1 May 17
Molesworth, H. E., D.S.O., R.A.	1 May 17
Wyatt, F. O., M.V.O., R.A.	1 May 17
Freeth, C. J. D., R.A.	1 May 17
Sturrock, G. C., R.A.	1 May 17
Corrie, G. G. W., R.A.	1 May 17
Riddell, R. B., R.A.	1 May 17
Fox, L. W., S. & T. Corps.	2 May 17
Hawkes, R., 3rd Brah.	10 May 17
Tregear, V. F. W., 98th Infy	23 May 17
Cunningham, A. H., R.E.	1 June 17
Dundas, P. H., D.S.O., *p.s.c.*, 6th Lt. Infy.	3 June 17
Watts, G. A. R., 25th Cavalry	3 June 17
Prissick, C., 56th Rifles	3 June 17
Graham, J. D., M.B., I.M.S.	3 June 17
Deas, L. J. M., M.B., F.R.C.S.E., I.M.S.	3 June 17
Cuming, R. J., D.S.O., 48th Pioneers	3 June 17
Gunter, C. P., R.E.	3 June 17
Williams, T. S. B., M.B., I.M.S.	3 June 17
Alexander, E. C., D.S.O., *p.s.c.*, 55th Rifles	3 June 17
Simpson, J. C., 16th Rajputs	3 June 17
†*Abbott, L. H.*, C.M.G., *p.s.c.*, 8th Rajputs	3 June 17
Melhuish, H. M. H., I.M.S.	3 June 17
Fetherstonhaugh, W. A., D.S.O., *p.s.c.*, 8th Cav.	3 June 17
Shuttleworth, D. I., *p.s.c.*, 3rd G. R.	3 June 17
Wintle, C. E. H., 114 Mah.	3 June 17
Wilmer, E. R. G., D.S.O., R.A.	3 June 17
Sherrard, J. W., S. & T. Corps	3 June 17

† Brigadier-General.

Lieutenant-Colonels—*contd.*	Date of rank.
Stack, G. H., R. E. (206)	3 June 17
Jardine, C. H., 96th Infy.	3 June 17
Hill, E. F. J., M.C., R. E.	3 June 17
Harvey, A. W. M., M.B., I.M.S.	3 June 17
Gem, G., R. Suss. R.	3 June 17
Brock, A. W. S., D.S.O., Leic. R.	3 June 17
Combe, S. B., M.C., 47th Sikhs	3 June 17
Twiss W. L. D., M.C., *p.s.c.*, 9th G. R.	3 June 17
Muspratt, S. F., D.S.O., *p.s.c.*, 12th Cav.	3 June 17
McWatters, H. C., D.S.O., *p.s.c.*, 24th Punj.	3 June 17
Ward H. C. S., *p.s.c.*, (*e*) 2nd Lrs.	3 June 17
Kensington, E. C., M.C., *p.s.c.*, 130th Baluchis.	3 June 17
Goodfellow, N. G. B., 128th Prs.	3 June 17
Gwyer, V. E., 10th Jats.	3 June 17
Young, E. deL., R. E.	3 June 17
Hewlett, E. A., 14th Lrs.	3 June 17
Williams, L. G., D.S.O., 5th Cav.	3 June 17
Wilson, R. C., D.S.O., M.C., *p.s.c.*, 114th Mah.	3 June 17
Woodhouse, P. St. J. R.	3 June 17
Hamilton, E. G., D.S.O., M.C., Conn. Rang.	3 June 17
McCaskill, J. C. H., 88th Inf.	11 June 17
Jackson, K. H., Mily. Accts. Dept.	17 June 17
Stewart, R. J T., Cantt. Magte.'s Dept.	17 June 17
Barry-Smith, R. C., 17th Cav.	19 June 17
Carey, W. L. J., R. A.	13 July 17
Walker, G. K., C.I.E., O.B.E., F.R.C.V.S., I. C.V.D.	20 July 17
Hay, W. N., C.I.E., D.S.O., *p.s.c.* 129th Baluchis.	24 July 17
Hunt, V. de V., Cantt. Magte.'s Dept.	24 July 17
Churchill, G. R. D., 19th Punjabis	24 July 17
Preston, E. E., S. & T. Corps	24 July 17
Rait, J. W. F., M.B., I.M.S.	28 July 17
O'Meara, E. J., F.R.C.S., I.M.S.	28 July 17
Tate, G., M.B., I.M.S.	28 July 17
Baird, R. F., I.M.S.	28 July 17
Gage, A. T., M.B., I M.S.	28 July 17
McPherson, G., C.I.E., M.B., F.R.C.S.E., I.M.S.	28 July 17
Hunt, S., M.B., I.M.S.	28 July 17
Condon, deV., M.D., I.M.S.	28 July 17
Lethbridge, W., M.B., I.M.S.	28 July 17
Ellwood, B. P., 31st Lrs.	29 July 17
McVean, D. A. D., D.S.O., 45th Sikhs	29 July 17
Keogh, J. B., D.S.O., 32nd Lrs.	23 Aug. 17
Bredin, A., 93rd Infy.	28 Aug. 17
Dent, W., D.S.O., *p.s.c.*, 103rd Infy.	28 Aug. 17
Tyrrell, J. F., R.A.	8 Sep. 17
Preston, G. A., 6th G. R.	9 Sep. 17
Chamier, W. St. G., S. & T. Corps	9 Sep. 17
Hulseburg, H., D.S.O., 127th Infy	9 Sep. 17
Bickford, W. W., C.I.E., 107th Prs.	17 Sep. 17
Edwards, W. B., A. V. C.	23 Sep 17
Turner, H.H., R.E.	1 Oct. 17
Anderson, R. H.B , Mily. Accts. Dept.	10 Oct. 17
Mitchell, W. J., 124th Infy.	28 Oct. 17
Hawes, C.H., M.V.O., D.S.O., 23rd Cav.	7 Nov. 17
Scott, W., 104th Rif.	7 Nov. 17
Keighley, V. A. S., M.V.O., 18th Lrs	2 Dec. 17
Battye, A.B., 6th Cav.	5 Dec. 17
Houston, A. M., 69th Punj.	5 Dec. 17
Oakes, H. C., 28th Cav.	17 Dec. 17
Beazeley, G. A., R.E.	19 Dec. 17
Rundle, F. P., D.S.O., R.E.	1 Jan. 18
†*Taylor, R. O'B.*, C.M.G., C.I.E., 19th Lrs.	1 Jan. 18
†*Elliott, C. A.*, D.S.O., R.E.	1 Jan. 18
Hearn, G. R., D.S.O., R.E.	1 Jan. 18
Leicester, J. C. H., M.D., F.R.C.S., I.M.S.	1 Jan. 18
Gordon, G. A. S., 13th Lrs.	1 Jan. 18
Barratt, A. L., 126th Infy.	1 Jan. 18
Mackelvie, M., M.B., F R.C.S. Edin., I.M.S.	1 Jan. 18
Wood, H., R.E.	1 Jan. 18
McCarrison, R., M.B., I.M.S.	1 Jan. 18
Leonard, W. H., F.R.C.S., I.M.S.	1 Jan. 18
Palmer, G. S., 97th Infy.	1 Jan. 18
McHarg, A. A., D.S.O., R.E.	1 Jan. 18

Lieutenant-Colonels—*contd.*	Date of rank.
Barron, R.M., I.M.S.	3 June 18
Hughes Hallett, R. L., S. & T. Corps	3 June 18
†*Hogg, R. E. T.*, C.I.E., 38th Horse	3 June 18
Magniac, C. L., C.M.G., R.E.	3 June 18
West, A. H. D., D.S.O., R.A.	3 June 18
Oldham, G. M., D.S.O., R.E.	1 Jan. 18
Giles, V., D.S.O., R.E.	1 Jan. 18
Walton, C., D.S.O., R.E.	1 Jan. 18
Casson, W. F. S., D.S.O., *p.s.c.*, 27th Cav.	1 Jan. 18
Dunlop, W. B., D.S.O., S. & T. Corps	1 Jan. 18
Kirkpatrick, C., *p.s.c.*, Corps of Guides	1 Jan. 18
Villiers-Stuart, J. P., D.S.O., 55th Rif.	1 Jan. 18
Porch, E. A. M.C., S. & T. Corps	1 Jan. 18
McMahon, H. F. E. M.C., S. & T. Corps	1 Jan. 18
Worgan, R. B , D.S.O., 20th Horse	1 Jan. 18
Brown, W., S. & T. Corps	1 Jan. 18
Haswell, C. H., R. E.	1 Jan. 18
Hamilton, W. H. D.S.O., F.R.C.S., I.M.S.	1 Jan. 18
Hewlett, A., 38th Horse	15 Jan. 18
Murray, W. G. P., 21st Punj.	22 Jan. 18
Fordham, W. M., 20th Inf.	22 Jan. 18
Pye, W. E., C.I.E., 98th Inf.	27 Jan. 18
Stewart, A. C., Guides	27 Jan. 18
Harington, A. W. D., S. & T. Corps	27 Jan. 18
Waller, A. A. P., 105th Lt. Inf.	27 Jan. 18
Cattell, G. L.	27 Jan. 18
Lees, C. H. B., C.M.G., 51st Sikhs	27 Jan. 18
Kirkpatrick, H., M.B., I.M.S.	28 Jan. 18
Fayrer, F. D. S., I.M.S.	28 Jan. 18
Chitale, P. K., I.M.S.	28 Jan. 18
Hunter, T., M.D., I.M.S.	28 Jan. 18
Battye, W. R., D.S.O., M.B., F.R.C.S., I.M.S.	28 Jan. 18
Hutcheson, G., M.B., I.M.S.	28 Jan. 18
Liston, W. G., M.D., C.I.E., I.M S.	28 Jan. 18
Anthony, R. W., M.B., F.R.C.S.E., I.M.S.	28 Jan. 18
Tucker, E. F. G. M.B., I.M.S.	28 Jan. 18
Stewart, G. E., M.B., F.R.C.S.E., I.M.S.	28 Jan. 18
Thompson, F. S. C., M.B., I.M.S.	28 Jan. 18
Novis, F. S., F.R.C.S., I.M.S.	28 Jan. 18
Watson, J. W., I.M S.	28 Jan. 18
Rowcroft, C. H., 9th Horse	19 Feb. 18
Wall, A. C., 8th G. R.	20 Feb. 18
Gaussen, J. R., C.M.G., D.S.O. 3rd Horse	20 Feb. 18
Ottley, W. J., 23rd Prs.	21 Feb. 18
Keogh, J. H., R.A.	1 Mar. 18
Herdon, H. E., 5 th Rif.	2 Mar. 18
Ricketts, R L., 10th Lrs.	8 Mar. 18
Willans, T. J., D.S.O., 57th Rif.	8 Mar. 18
Gardiner, R., *p.s.c.*, 53rd Sikhs	10 Mar. 18
Dopping-Hepenstal, M. E., D.S.O., 1st G. R.	11 Mar. 18
James, A. A.	12 Mar. 18
Rice, S. K. B., 35th Sikhs	14 Mar. 18
Carnegy, G. P. O., D.S.O., 7th Lrs.	15 Mar. 18
Hunter, W. J. H., 117th Mah.	18 Mar. 18
Strong, H. P., 101st Grs.	30 Mar. 18
Ommanney, C. V., 109th Infy.	29 May 18
Thomson, A. G., D.S.O., 58th Rif.	3 June 18
Graham, R. B., 46th Punjabis	3 June 18
Strong, A. D., D.S.O., 10th Lrs.	3 June 18
Stokes, C. B., C.I.E., *p.s.c.*, 3rd Horse	3 June 18
Hodgson, P. E., D.S.O., R.E.	3 June 18
Hogg, C. C. H., C.M.G., R. E.	3 June 18
Reed, C., D.S.O., R. A.	3 June 18
Birney, C. F., D.S.O., R.E.	3 June 18
Allanson, C. J. L., D.S.O., 6th G.R.	3 June 18
Barron, F. W., *p.s.c.*, R. A.	3 June 18
Popham, E.L., D.S.O., 26th Lt. Cavy.	3 June 18
Freeland, J. C., *p.s.c.*, 35th Sikhs	3 June 18
Moberly, B. R., D.S.O., *p.s.c.*, 56th Rif.	3 June 18
Milward, C. A., D.S.O., 53rd Sikhs	3 June 18
Smith, H. H., D.S.O., 9th Infy.	3 June 18
Wilson-Johnston, W. E., D.S.O., *p.s.c.*, 36th Sikhs.	3 June 18
Giles, E. D., D.S.O., 35th Horse	3 June 18
Catty, T. C., D.S.O., *p.s c.*, 69th Punj.	3 June 18
Rowan Robinson, J. R., D.S.O., S. & T. Corps.	3 June 18

† Brigadier-General.

Lieutenant-Colonels—*contd.*	Date of rank.
Orton, S. B., *p.s.c.*, 39th Grah. Rif.	3 June 18
Gill, C. A., I.M.S.	3 June 18
Nisbet, T., D.S.O., 28th Lt. Cavy.	3 June 18
Lloyd, C. R., D.S.O., S. & T. Corps	3 June 18
Powell, D. H., M.C., 105th Lt. Infy.	3 June 18
Kirby, J. T., D.S.O., S. & T. Corps	3 June 18
Cochran, G. W., D.S.O., 81st Prs.	3 June 18
Ogle, N., D.S.O., 67th Punj.	3 June 18
North, H. N., D.S.O., R. E.	3 June 18
Thornhill, C. J. M., D.S.O., 6th Lt. Infy.	3 June 18
Anderson, J. F. H., D.S.O., A. Remt, Dept.	3 June 18
Little, W. C., 6th G. R.	11 June 18
Browne, L. S., 37th Dogras	15 June 18
Cheape, G. R. H., 1 Dn. Gds.	15 June 18
Hodgkinson, C., 6th Cav.	18 June 18
Hodgkinson, J., 5th Cav.	18 June 18
Bradshaw, C. R., *p.s.c.*, 9th G. R.	22 June 18
Johnson, J. E. B., 21st Cav.	1 July 18
Gale, E. S., 10th G. R.	3 July 18
Hume, S. G. W., 7th Rajputs	6 July 18
Chitty, A. W., 126th Infy.	13 July 18
Calvert, R. T. C., *p.s.c.*, 120 Infy.	13 July 18
Harcourt, J. S. M., 2nd G. R.	13 July 18
Innes, H., M.D., I. M. S.	28 July 18
Willmore, W. S., M.D., I. M. S.	28 July 18
Walter, A. E., O.B.E., I. M. S.	28 July 18
Hudson, C., D.S.O., F.R.C.S.E., I.M.S.	28 July 18
Hutchinson, L. T. R., M.D., I. M. S.	28 July 18

Lieutenant-Colonels—*concld.*	Date of rank.
Fleming, A. M., M.B., F.R.C.S.E., I. M. S.	28 July 18
Ward, E. L., I. M. S.	28 July 18
Walker, J. N., I. M. S.	28 July 18
Roberts, V. H., F.R.C.S.E., I. M. S.	28 July 18
Ross, T. S., I. M. S.	28 July 18
Pearson, W. M., M.B., I. M. S.	28 July 18
Kemp, D. C., I. M. S.	28 July 18
Kauntze, B. C., 79th Infy.	31 July 18
Nicolay, B. U., 4th G. R.	3 Aug. 18
Bruce, M. E. L.	13 Aug. 18
Stansfeld, C. G., 8th G. R.	13 Aug. 18
Bruce, G. D., D.S.O., *p.s.c.*, 81st Prs.	13 Aug. 18
Waring, E. H., 63rd Infy.	19 Aug. 18
Campbell, H. B. D., R. E.	27 Aug. 18
Preston, W. J., D.S.O., 97th Infy.	31 Aug. 18
Harbord, C. R., D.S.O., 30th Lancers	1 Sep. 18
Corlett, J. S., 15th Lrs.	3 Sep. 18
Stewart, A. F., 114th Mah.	3 Sep. 18
Quin, R. P. W., 41 Dogras.	12 Sep. 18
VC **Costello**, E. W., C.M.G., D.S.O., 22nd Punj.	22 Sep. 18
Orton, E. F., 37th Lrs.	26 Sep. 18
Hawkes, W. C. W., D.S.O., 12th Prs.	20 Oct. 18
Saville, A. R. C., 31 Lrs.	16 Nov. 18
deLabilliere, E. G. D.	23 Nov. 18
Villiers-Stuart, W. D., 5 G. R.	19 Dec. 18
Grove, P. L., Ches. R.	19 Dec. 18
Rind, G. B. A., 124 Infy.	19 Dec. 18
Ross, R. C., D.S.O., 6 Lt. Infy.	19 Dec. 18

GRADATION LIST OF OFFICERS OF THE INDIAN ARMY.

(FIELD-MARSHAL TO LIEUTENANT-COLONEL.)

No.	Names.	Date of birth.	Date of first Commission.	Lieut.	Captain.	Major.	Lieut.-Colonel.	Colonel.	Major-Genl.	Lt.-Genl.	General.	Ceased to be employed.
	Field-Marshal.											
	Egerton, *Sir* Charles Comyn, G.C.B., D.S.O., [R] (Fd. Ml., 17 Mar. 17).	10 Nov. 48	8 June 67	19 Oct. 69	8 June 79	8 June 87	1 Sep. 91	1 Sep. 95	1 Apl. 02	28 Oct. 03	28 Oct. 06	. . . .
	Generals—3.											
1	**Barrow,** *Sir* Edmund George, G.C.B., G.C.S.I. [R]	28 Jan. 52	. . . .	30 Dec. 71	30 Dec. 83	16 Feb. 87	16 Jan. 05	16 Jan. 08	1 Apl. 02	16 July 05	27 July 09	. . . .
	**Nixon,* Sir *John Eccles,* K.C.B. [L] [R]	*16 Aug. 57*	. . . .	*10 Sep. 75*	*10 Sep. 86*	*10 Sep. 95*	*22 Jan. 96*	*12 Mar. 99*	*12 Jan. 04*	*15 Feb. 09*	*4 May 14*	. . . .
	**Pearson,* Sir *Alfred Astley,* K.C.B., [R]	*5 June 50*	*8 Feb. 70*	*28 Oct. 71*	*8 Feb. 82*	*1 July 87*	*8 Feb. 96*	*8 Feb. 00*	*27 June 04*	*3 May 11*	*1 Mar. 17*	*3 Oct. 12*
	**Woon,* Sir *John Blaxell,* K.C.B., [R]	*24 Feb. 56*	. . . .	*1 Oct. 73*	*8 Jan 87*	*1 Oct. 93*	*20 May 98*	*29 Nov. 00*	*10 June 05*	*3 June 11*	*7 June 17*	*15 June 14*
	**Scallon,* Sir *Robert Irvin,* G.C.B., K.C.I.E., D.S.O., [R]	*3 Apl. 57*	. . . .	*12 Feb. 76*	*12 Feb. 87*	*12 Feb 96*	*12 Feb. 02*	*7 June 02*	*1 Dec. 06*	*3 Oct. 11*	*30 June 17*	. . . .
	Barrett, *Sir* Arthur Arnold, G.C.B., K.C.S.I., K.C.V.O., A.D.C. [R]	3 June 57	. . . .	10 Sep. 75	10 Sep. 86	10 Sep. 95	20 May 98	11 Oct. 02	1 Dec. 06	23 Oct. 11	1 Aug. 17	. . . .
	**Grover,* Sir *Malcolm Henry Stanley,* K.C.B., K.C.I.E. [L] [R]	*19 July 58*	*11 Sep. 76*	*11 Sep. 76*	*11 Sep. 87*	*11 Sep. 96*	*20 May 98*	*27 Nov. 02*	*1 June 07*	*21 Sep. 12*	*27 Aug. 17*	. . . .
	**Watkis,* Sir *Henry Bulckley Burlton,* K.C.B., *p.s.c.* [R]	*4 June 60*	*11 May 78*	*2 June 80*	*11 May 89*	*11 May 98*	*20 May 98*	*4 July 03*	*1 June 07*	*4 May 14*	*1 Oct. 17*	*8 Sep. 15*
3	**Birdwood,** *Sir* William Riddell, K.C.B., K.C.S.I., K.C.M.G., C.I.E., D.S.O., A.D.C., *q.s.*	13 Sep. 65	. . . .	9 May 85	9 May 96	29 Nov. 00	26 June 02	26 June 05	3 Oct. 11	28 Oct. 15	23 Oct. 17	. . . .
	Lieutenant-Generals—6.											
	**Phayre, Sir Arthur,* K.C.B. . . .	*23 Feb. 66*	*1 May 78*	*2 June 80*	*1 May 89*	*1 May 98*	*29 Nov 00*	*10 Feb. 04*	*7 Apl. 08*	*30 Oct. 15*	. . . .	. . . .
1	**Campbell,** *Sir* Frederick, K.C.B., D.S.O.	25 Feb. 60	13 Aug. 79	15 Mar. 80	13 Aug. 90	22 Jan. 96	1 June 04	10 Nov. 04	29 June 11	1 Jan. 17	. . . .	. . . .
	Cox, *Sir* Herbert Vaughan, K.C.B., K.C.M.G., C.S.I.	12 July 60	14 Jan. 80	1 July 81	14 Jan. 91	14 Jan. 00	1 June 04	9 June 07	17 Feb. 15	1 Jan. 17	. . . .	. . . .
3	***Jacob,** *Sir* Claud William, K.C.B. .	21 Nov. 63	. . . .	9 Sep. 82	9 Sep. 93	10 July 01	1 Oct. 04	1 Oct. 08	1 Jan. 16	3 June 17	. . . .	. . . .

* Supernumerary.

4	**Keary,** Sir Henry D'Urban, K.C.B., K.C.I.E., D.S.O.	28 Apl. 57	†10 Sep. 75	10 Sep. 77	10 Sep. 88	29 Dec. 93	10 Sep. 03	10 Sep. 06	1 Dec. 11	1 Aug. 17	. . . .
	Egerton, Sir Raleigh Gilbert, K.C.B., K.C.I.E.	25 Sep. 60	13 Aug. 79	19 Feb. 81	13 Aug. 90	18 Nov. 96	13 Aug. 05	6 June 07	14 Nov. 14	27 Aug. 17	. . . .
	Hudson, Sir Havelock, K.C.B., C.I.E.	26 June 62	. . . .	22 Oct. 81	22 Oct. 92	29 Nov. 00	13 May 07	17 Feb. 11	18 Feb. 15	23 Oct. 17	. . . .
7	**Wapshare,** Richard, C.B., C.S.I.	6 Jan. 60	. . . .	1 Feb. 80	1 Feb. 91	1 Feb. 00	1 June 04	10 June 07	5 Mar. 15	15 June 18	. . . .

Major-Generals—24.

	**Powell,* Sir *Charles Herbert,* K.C.B. [L] [R]	23 Oct. 57	. . . .	27 Jan. 76	27 Jan. 87	27 Jan. 96	20 May 98	10 Feb. 04	15 Nov. 07	. . . .	20 June 12
1	**Hastings,** Edward Spence, C.B., D.S.O.	17 Jan. 56	†28 Feb. 74	28 Feb. 74	28 Feb. 85	28 Feb. 94	28 Feb. 00	10 Feb. 04	15 Nov. 07	. . . .	12 Apl. 12
	**Mackenzie-Kennedy, Edward Charles William,* C.B.	6 July 60	13 Aug. 79	30 Mar. 81	13 Aug. 92	13 Aug. 99	29 Nov. 00	10 Feb. 04	9 June 08	. . . .	7 Jan. 13
	**Drummond,* Sir *Francis Henry Rutherford,* K.C.I.E., C.B.	9 Sep. 57	†1 Feb. 75	11 Feb. 75	11 Feb. 86	16 Feb. 87	11 Feb. 01	11 Feb. 04	27 July 09	. . . .	17 Feb. 12
	**Wallace, Alexander,* C.B.	22 Aug. 58	†11 Sep. 76	11 Sep. 76	11 Sep. 87	11 Sep. 96	11 Sep. 02	7 Sep. 04	3 May 11	. . . .	17 Oct. 13
	**Bell* Sir *James Alexander,* K.C.V.O.	28 Jan. 56	†10 Sep. 75	10 Sep. 75	10 Sep. 86	10 Sep. 95	10 Sep. 01	10 Sep. 04	6 May 11	. . . .	14 Nov. 14
	**Davison, Kenneth Stewart,* C.B.	25 Dec. 56	. . . .	19 Aug. 75	19 Aug. 86	19 Aug. 95	19 Aug. 01	3 Oct. 04	3 June 11	. . . .	. . . .
	Younghusband, Sir George John, K.C.M.G., K.C.I.E., C.B., p.s.c.	9 July 59	1 May 78	15 Mar. 80	1 May 89	22 Jan. 95	29 Nov. 00	14 Feb. 05	29 Aug. 11	. . . .	. . . .
	Cookson, George Arthur, C.B., C.M.G.	6 Aug. 60	11 Sep. 80	1 July 81	11 Sep. 91	11 Sep. 00	29 Nov. 00	30 July 05	23 Oct. 11	. . . .	. . . .
	VC **Melliss,** Sir Charles John, K.C.B., K.C.M.G. [L]	12 Sep. 62	. . . .	9 Sep. 82	9 Sep. 93	10 July 01	11 July 01	15 Jan. 07	19 Mar. 12	. . . .	. . . .
	**Bunbury, William Edwin,* C.B., p.s.c.	5 Apl. 58	30 Jan. 78	14 Sep. 80	30 Jan. 89	30 Jan. 98	30 Jan. 04	30 Jan. 07	1 May 12	. . . .	. . . .
	**Carnegy, Philip Mainwaring,* C.B.	12 Dec. 58	30 Jan. 78	28 Feb. 79	27 July 89	30 Jan. 98	30 Jan. 04	30 Jan. 07	1 Sep. 12	. . . .	7 Sep. 15
5	**Young,** Charles Frederic Gordon	12 Nov. 59	1 May 78	7 Jan. 80	1 May 89	1 May 98	1 May 04	1 May 07	21 Sep. 12	. . . .	. . . .
	**Pirie, Charles Patrick William,* C.B.	17 Aug. 59	11 May 78	8 Dec. 80	11 May 89	11 May 89	11 May 04	1 June 07	30 Jan. 14	. . . .	. . . .
	**Turner, James Gibbon,* C.B., [L]	24 Aug. 59	. . . .	18 Dec. 78	18 Dec. 89	2 Jan. 96	18 Dec. 04	1 June 97	5 Feb. 14	. . . .	15 Sep. 14
	**Fry, Charles Irwin,* C.B.	2 Sep. 58	22 Jan 79	1 Dec. 80	22 Jan. 90	22 Jan. 99	1 June 04	4 June 07	11 Mar. 14	. . . .	. . . .
	Shaw, David George Levinge	1 July 60	21 June 79	29 Apl. 81	21 June 90	21 June 99	1 June 04	5 June 07	20 Mar. 14	. . . .	5 Aug. 16
	**Macintyre, Donald Charles Frederick,* C.B.	17 Apl. 59	13 Aug. 79	1 July 81	13 Aug. 90	20 May 98	1 June 04	6 June 07	4 May 14	. . . .	14 Oct. 14
	Tighe, Sir Michael Joseph, K.C.M.G., C.B., C.I.E., D.S.O.	21 May 64	. . . .	25 Aug. 83	25 Aug. 94	25 Aug. 01	25 June 07	8 Dec. 07	3 June 15	. . . .	. . . .
	Delamain, Sir Walter Sinclair, K.C.M.G., C.B., D.S.O.	18 Feb. 62	. . . .	22 Oct. 81	22 Oct. 92	10 July 01	12 Jan. 05	12 Jan. 08	3 June 15	. . . .	. . . .
10	**Stewart,** James Marshall, C.B.	9 Aug. 61	. . . .	22 Oct. 81	22 Oct. 92	10 July 01	20 Mar. 05	20 Mar. 08	3 June 15	. . . .	. . . .
	Bingley, *The Hon. Sir* Alfred Horsford, K.C.I.E., C.B.	28 May 65	. . . .	7 Feb. 85	7 Feb. 96	7 Feb. 03	19 Mar. 06	19 Mar. 10	3 June 15	. . . .	. . . .
	**Cowper, Maitland,* C.B., C.I.E.	24 Dec. 59	11 Aug. 80	29 Mar. 81	11 Aug. 91	11 Aug. 00	1 Jan. 06	10 June 07	27 Aug. 15	. . . .	. . . .
	**Watson, William Arthur,* C.B., C.M.G., C.I.E., p.s.c.	25 Sep. 60	11 Aug. 80	1 July 81	11 Aug. 91	11 Aug. 00	1 Apl. 06	10 June 07	7 Sep. 15	. . . .	. . . .
	**Wadeson, Frederick William George,* C.B.	12 July 60	17 Apl. 80	1 July 81	17 Apl. 91	17 Apl. 00	17 Apl. 00	10 June 07	28 Oct. 15	. . . .	. . . .
	*VC *Walker William George,* C.B.	28 May 63	. . . .	29 Aug. 85	29 Aug. 96	29 Aug. 03	7 Sep. 04	8 May 10	1 Jan. 16	. . . .	. . . .
12	VC **Cobbe,** Sir Alexander Stanhope, K.C.B., C.S.I., D.S.O., p.s.c. (*temp. Lt.-Genl.*	5 June 70	21 Sep. 89	4 Mar. 92	21 Sep. 00	21 Sep. 07	22 Sep. 07	1 Dec. 11	3 June 16	. . . .	. . . .

* Supernumerary. † Date of Sub-Lieutenancy.

Lieutenant-Generals—*contd.*

No.	Names.	Date of birth.	Date of first Commission.	Lieut.	Captain.	Major.	Lieut.-Colonel.	Colonel.	Major-Genl.	Lt.-Genl.	Ceased to be employed.
13	***Barrow,** *Sir* George deSymons, K.C.M.G., C.B., *p.s.c.* [L] (*temp. Lt.-Genl.*)	25 Oct. 64	1 May 78	24 Sep. 79	1 May 89	1 May 98	1 May 04	1 May 07	3 June 16	. . . .	. . . .
	**Hunter, John Gunning*, C.B. (*a*) . . .	1 Nov. 59	. . . .	23 Aug. 84	23 Aug. 95	23 Aug. 02	23 Aug. 10	19 Aug. 15	30 Aug. 16	. . . .	. . . .
	Malleson, Wilfd, C.B., C.E. . . .	8 Sep. 66	. . . .	17 Feb. 86	28 Sep. 96	9 Oct. 01	20 Oct. 04	20 Oct. 08	20 Sep. 16	. . . .	. . . .
15	**Fane,** *Sir* Vere Bonamy, K.C.I.E., C.B. (*Leave without pay in extension for 11 Mar. 03.*)	16 June 63	. . . .	12 Nov. 84	12 Nov. 95	12 Nov. 02	1 Jan. 03	20 Oct. 08	1 Jan. 17	. . . .	. . . .
	Barratt, William Cross, C.B., D.S.O. .	2 June 62	. . . .	5 Dec. 83	5 Dec. 94	25 Jan. 99	25 July 05	25 July 09	1 Jan. 17	. . . .	. . . .
	Ewart, *Sir* Richard Henry, K.C.M.G., C.B., C.I.E., D.S.O.	26 Dec. 64	25 Aug. 83	25 Aug. 83	25 Aug. 94	25 Aug. 01	1 May 06	1 May 10	1 Jan. 17	. . . .	. . . .
	Beynon, *Sir* Wm. Geo. Lawrence, K.C.I.E., C.B., D.S.O.	5 Nov. 66	5 Feb. 87	11 Dec. 89	5 Feb. 98	20 May 98	10 Nov. 04	10 Nov. 10	1 Jan. 17	. . . .	. . . .
	Brooking, *Sir* Harry Triscott, K.C.B., K.C.M.G.	13 Jan. 64	. . . .	6 Feb. 84	6 Feb 95	29 Nov. 00	21 Nov. 06	21 Nov. 10	1 Jan. 17	. . . .	. . . .
20	**Woodyatt,** Nigel Gresley, C.B. . . .	30 Mar. 61	. . . .	12 May 83	12 May 94	10 July 01	20 Sep. 07	25 Oct. 11	1 Jan. 17	. . . .	. . . .
	Eustace, Alexander Henry, C.B., D.S.O.	1 June 63	. . . .	6 May 85	6 May 96	6 May 03	22 Sep. 07	1 Dec. 11	1 Jan. 17	. . . .	. . . .
	***Knight,** Wyndham Charles., C.B., C.S.I., D.S.O., *p.s.c.*, [L]	30 Nov. 63	. . . .	25 Aug. 83	25 Aug. 94	29 Nov. 00	25 Aug. 09	19 Mar. 12	1 Jan. 17	. . . .	. . . .
	***Macandrew,** Henry John Milnes, C.B., D.S.O., *q.s.*	7 Aug. 66	. . . .	10 Nov. 85	10 Nov. 97	10 Nov. 04	10 Nov. 12	3 June 15	1 Jan. 17	. . . .	. . . .
	***Shea,** John Stuart Mackenzie, C.B., C.M.G., D.S.O., *p.s.c.*, *q.s.*	17 Jan. 69	11 Feb. 88	18 Feb. 90	11 Feb. 99	22 Aug. 02	10 July 12	1 Jan. 16	1 Jan. 17	. . . .	. . . .
25	**Fowler** Francis John, C.B., D.S.O. . .	31 July 64	. . . .	25 Aug. 83	25 Aug. 94	4 Oct. 99	28 Jan. 07	6 May 11	23 Oct. 17	. . . .	. . . .
	Hardy, Thomas Henry	25 Nov. 63	. . . .	25 Aug. 83	25 Aug. 94	25 Aug. 01	1 July 07	25 Sep. 11	3 Dec. 17	. . . .	. . . .
	Fowler, Charles Astley, C.B., D.S.O. .	9 Nov. 65	. . . .	7 Feb. 85	7 Feb. 96	7 Feb. 03	22 Sep. 07	12 Nov. 11	6 Apl. 18	. . . .	. . . .
	Dunsterville, Lionel Charles, C.B., [L]	9 Nov. 65	. . . .	23 Aug. 84	23 Aug. 95	23 Aug. 02	21 Jan. 03	1 May 12	3 June 18	. . . .	. . . .
	***Hendley,** Charles Edward, C.B. . .	26 Dec. 63	. . . .	12 Nov. 84	12 Nov. 95	12 Nov. 02	29 Dec. 08	23 Apl. 13	3 June 18	. . . .	. . . .
30	***James,** William Bernard, C.B., C.I.E., M.V.O.	8 Feb. 65	. . . .	30 Jan. 86	30 Jan. 97	30 Jan. 04	30 Jan. 12	3 June 15	3 June 18	. . . .	. . . .
	***Climo,** Skipton Hill, C.B., D.S.O. . .	24 Dec. 63	11 Feb. 88	11 July 89	11 Feb. 99	12 Feb. 99	13 Feb. 12	3 June 15	3 June 18	. . . .	. . . .
	***Hill,** John, C.B., D.S.O., ...	14 Jan. 66	5 Feb. 87	11 Feb. 88	5 Feb. 98	5 Feb. 05	5 Feb. 13	3 June 15	3 June 18	. . . .	. . . .
	***Skeen,** Andrew, C.M.G., *p.s.c.* . . .	20 Jan. 73	5 Dec. 91	2 Oct. 93	10 July 01	7 Sep. 04	7 Jan. 14	3 June 15	3 June 18	. . . .	. . . .
	***Pain,** Philip Charles, C.B., C.M.G. .	8 Aug. 64	. . . .	28 Apl. 86	28 Apl. 97	28 Apl. 04	28 Apl. 12	8 Nov. 15	3 June 18	. . . .	. . . .
35	***Scott,** Thomas Edwin, C.B., C.I.E., D.S.O.	6 Mar. 67	9 May 88	18 Dec. 89	9 May 99	10 May 99	10 May 12	29 Nov. 15	3 June 18	. . . .	. . . .
36	***Richardson,** Charles William Grant, C.S.I. *p.s.c.* [L]	27 Oct. 68	11 Feb. 88	23 Apl. 90	11 Feb. 99	13 Feb. 05	10 July 12	1 Jan. 17	3 June 18	. . . .	. . . .

*Supernumerary.

No.	Names.	Date of rank.	Ceased to be employed.
	Colonels—110.		
	Gastrell, E. T.	30 Jan. 07	
	Rodwell, E. H., C.B., *p.s.c.* [L]	30 Jan. 07	30 Jan. 16
	Thackwell, C. G. R., C.B., D.S.O.	1 May 07	1 Apl. 17
	Thomson, W. D., C.M.G.	1 May 07	30 Apl 09
	Hutchinson, F. P., C.B. [L]	1 May 07	
	Richardson, F. B. W.	11 May 07	
	Steele, St. G. L., C.B. [L]	1 June 07	1 Sep. 13
	Yate, F. H.	1 June 07	
	Dobbie, W. H., C.B.	6 June 07	
	Aitken, A. E.	9 June 07	
	Loch, H. F., *p.s.c.*	10 June 07	
	Johnson, C. E.	11 June 07	
	Johnstone, A. A. J.	11 June 07	
	Price, C. H. U., C.B., D.S.O.	1 Mar. 08	
	Dick, A. R., C.B. [L]	16 July 08	
	Marlow, B. W., C.S.I., C.I.E.	23 Dec. 08	
	Twigg, R. H., C.B.	4 Feb. 09	
	Lindesay, A. L.	13 Feb. 09	
	Shakespear, L. W., C.B.	2 Apl. 09	
5	Herbert, L. N., C.B.	4 Apl. 09	
	Palin, G. W., C.B., C.I.E.	1 May 09	
	Harrison, T. A., C.S.I.	1 Apl. 10	
	Shore, O. B. S. F., C.B., C.I.E., D.S.O., *p.s.c.* [L]	30 July 10	
	Buchanan, K. J., C.B.	27 Sep. 10	
	Campbell, A. A. E.	29 Oct. 10	
10	Woodward, J. A. H.	30 June 11	
	Fisher, J., C.B. [L]	29 Aug. 11	
	Mainwaring, C. V.	3 Oct. 11	
	Millar, W. H.	26 Dec. 11	19 Oct. 12
	Cartwright, C. M., C.B., C.M.G.	1 Jan. 12	
	Holloway, B., C.I.E., *p.s.c.* [L]	14 Jan. 12	
	Edwards, Fitz J. M., C.B., C.M.G., D.S.O., A.-D.-C., *p.s.c.*	14 Jan. 12	
	Thring, R. H. D.	23 Feb. 12	
15	Muspratt, F. C., C.B., C.M.G.	1 Apl. 12	
	Peterson, F. H., C.B., D.S.O. [*l*]	6 June 12	
	Andrews, L. J. [L]	26 June 12	
	Carruthers, R. A., C.B., C.M.G.	29 June 12	
	Newbold, A. W.	11 Aug. 12	
20	Mead, H. R.	19 Aug. 12	
	Betham, R. M.	1 Sep. 12	
	Savage, W. H., C.M.G.	1 Sep. 12	
	Cox, F. W. H. C.B.	21 Sep. 12	
	Mathewes, J. R.	21 Sep. 12	
	Johnston, H. M.	5 Oct. 12	
25	Fasken, W. H., C.B.	12 Oct. 12	
	Thomas, F. H. S., C.B.	23 Oct. 12	
	Blakeney, W. E. A.	23 Oct. 12	
	Kennedy, H., C.B.	30 Dec. 12	
	Commeline, F. H. B.	31 Dec. 12	
30	Miles, P. J., C.B.	31 Jan. 13	
	Lang, E. B.	14 Feb. 13	
	Sullivan, R. E.	21 Feb. 13	
	Cripps, A. W., C.B.	13 Mar. 13	
	Colomb, G. H. C., C.M.G.	24 Mar. 13	
35	Brown, R. J. R.	24 Mar. 13	13 Nov. 13
	Walton, W. C., C.B. C.M.G.	31 Mar. 13	11 Nov. 17
	Crocker, S. F., C.B.	31 May 13	
	Clay, C. H., *p.s.c.* [*l*]	30 Sep. 13	
	Soady, G. J. Fitz M., C.M.G.	22 Oct. 13	
40	Burton, R. G. [L]	19 Dec. 13	
	Cole, E. H., C.B., C.M.G.	2 Jan. 14	
	Roberts, H. L., C.B., M.V.O. [*l*]	8 Jan. 14	
	Selwyn, C. H., C.B., *p.s.c.* [L]	20 Jan. 14	
	Douglas, J. A., C.M.G., C.I.E. [L]	28 Jan. 14	
45	Edwardes, S. M., C.B., C.M.G., D.S.O.	5 Feb. 14	
46	Prendergast, C. G., C.B.	12 Feb. 14	

No.	Names.	Date of rank.	Ceased to be employed.
	Colonels—*contd.*		
47	Somerset, C. W., C.B., C.M.G., M.V.O.	11 Mar. 14	
	Cadell, A.	1 Apl. 14	
	Watson, H. D., C.B., C.M.G., C.I.E., M.V.O., *Eq.*	19 Apl. 14	
50	Willoughby, M. E., C.B., C.M.G., *p.s.c.* [L]	4 May 14	
	Tod, J. K., *p.s.c.* [L]	10 May 14	
	Scott, B.	1 Sep. 14	
	Roche, H. J., C.B.	4 Nov. 14	
	Grimston, S. B., C.M.G.	8 Dec. 14	
55	Graham, S. B.	4 Feb. 15	
	Mennie, J. O.	10 Feb. 15	
	VC Boisragon, G. H.	17 Feb. 15	
	Smith, F. A.	18 Feb. 15	
	Norie, C. E. deM., C.B., C.M.G., D.S.O., *p.s.c.* [*l*]	18 Feb. 15	
60	Dyer, R. E. H., C.B., *p.s.c.*	5 Mar. 15	
	Richardson, H. L.	15 Apl. 15	
	Sullivan, E. L., C.B., C.M.G.	2 May 15	
	Boudier, E. W.	3 June 15	
	Frazer, G. S., C.M.G.	3 June 15	
65	Batten, F. G., D.S.O.	3 June 15	
	Bailey, C.	3 June 15	
	Lowis, H. E.	3 June 15	
	Chitty, W. W.	3 June 15	
	Baldwin, G. M., D.S.O.	3 June 15	
70	Southey, W. M., C.M.G.	3 June 15	
	Vaughan, R. E., C.B.	3 June 15	
	Lane, H. P.	27 Aug. 15	
	Young, F. De B., C.M.G., *p.s.c.*	9 Oct. 15	
	Luck, C. A.	28 Oct. 15	
75	Sanders, G. L. H.	1 Jan. 16	
	Caruana, A. J., C.I.E.	1 Jan. 16	
	Drew, A. B. H., C.I.E.	1 Jan. 16	
	Younghusband, L. N., C.B., C.M.G.	31 Mar. 16	
	Campbell, A. J., D.S.O.	3 June 16	
80	Wright, E. G.	3 June 16	
	Carey, C. W., C.M.G.	9 June 16	
	Elsmie, A. M. S., C.M.G., *p.s.c.* [L]	3 June 16	
	Seddon, T. V.	3 June 16	
	Tytler, H. C., D.S.O.	2 July 16	
85	Creagh, A. H. D., C.M.G., M.V.O.	30 Aug. 16	
	Lock, F. R. E., D.S.O.	1 Jan. 17	
	Davies, F. G. H.	1 Jan. 17	
	Nepean, H. E. C. B., C.M.G.	1 Jan. 17	
	Jacob, A. Le G., C.M.G., C.I.E., D.S.O. A.-D.-C.	1 Jan 17	
90	Lucas, F. G., C.B., D.S.O.	1 Jan. 17	
	Andrew, D. C. A.	1 Jan. 17	
	Wall, E. W., C.M.G.	1 Jan. 17	
	Campbell, L. W. Y., C.M.G.	1 Jan. 17	
	Black, W. C., *p.s.c.* [L]	1 Jan. 17	
95	Cook, C. C., D.S.O.	18 Apl. 17	
	MacGeorge, H. K., C.B.	22 Apl. 17	
	Hallowes, F. W., C.I.E.	3 May 17	
	Cooke, H. F., D.S.O.	3 June 17	
	Hutchinson, C. A. R., D.S.O., *p.s.c.*	31 Aug. 17	
100	Badcock, F. F., D.S.O.	20 Oct. 17	
	Senior, H. W. R., D.S.O., *p.s.c.* [L]	1 Dec. 17	
	Wood, E. J. M., D.S.O.	3 Dec. 17	
	Tanner, C. O. O.	1 Jan. 18	
	Birdwood, H. B.	6 Apl. 18	
105	Andrew, A. W., C.M.G., [L]	14 May 18	
	McCrea, A. C., C.M.G.	21 May 18	
	* * *		
	Codrington, H. W.	25 May 18	
	Beatty, L. N.	3 June 18	
	Drake-Brockman, D. H., C.M.G.	16 June 18	
110	Hennessy, J. P. C., C.M.G.	27 Aug. 18	

Brevet-Colonels.

Names.	Date of rank.	Ceased to be employed.
Blyth, F. A.	13 June 04	31 May 07
Gott, G. A.	23 Feb. 05	9 June 07
Maxwell, G. W.	28 Oct. 05	. . . .
Forbes, A. W., C.B.	11 Sep. 06	10 Feb. 08
Westlake, A. P., D.S.O.	29 Nov. 06	26 Oct. 15
Whistler, A. E.	30 Jan. 07	11 Sep. 09
Hodgson, G. B.	1 May 07	28 Oct. 13
Aplin, P. J. H., D.S.O. [L]	1 May 07	. . . .
Dobbie, H. H.	1 May 07	12 Dec. 07
Justice, C. Le G.	11 May 07	. . . .
Dennys, W. A. B.	6 June 07	. . . .
Unwin, G. B., D.S.O.	15 Sep. 07	. . . .
Templer, H.	28 Oct. 07	26 Apl. 10
Balfour, J. H.	23 Mar. 08	9 Apl. 09
Leonard, A. W.	2 Oct. 08	2 Oct. 11
Fordyce, H. L. D.	3 June 15	. . . .
Harward, A. J. N.	3 June 15	. . . .
Clery, C. B. L., C.B.	3 June 15	. . . .
Annesley, A. S. R.	3 June 15	. . . .
Bruce, *Hon.* C. G., C.B., M.V.O.	8 Nov. 15	. . . .
Holman, H. C., C.B., C.M.G., D.S.O., *p.s.c.*	1 Jan. 16	. . . .
Lindsay, H.A.P., C.M.G. [L]	3 June 16	. . . .
Davies, C. H., C.M.G., D.S.O.	3 June 16	. . . .
Boome, E. H., C.B., C.M.G.	3 June 16	. . . .
Parr, H. O.	3 June 16	. . . .
Rattray, C., C.B., *p.s.c.*	3 June 16	. . . .
Fulton, F. C. R.	3 June 16	. . . .
Sheppard, G. S., C.M.G.	3 June 16	. . . .
Anderson, W. C., C.M.G.	3 June 16	. . . .
Davidson, S. R., C.M.G.	3 June 16	. . . .
Cummins, H. A. V., *p.s.c.*	3 June 16	. . . .
Hopwood, H. R., C.S.I., *p.s.c.*	3 June 16	. . . .
Browne, H. J. P., *p.s.c.*	3 June 16	. . . .
Carnegy, C. G., M.V.O.	1 Jan. 17	. . . .
Home, J. M., *p.s.c.* [L]	1 Jan. 17	. . . .
Adye, D. R. [L]	1 Jan. 17	. . . .
Holland-Pryor, P., C.M.G., D.S.O., M.V.O.	1 Jan. 17	. . . .
Price, C. U., C.M.G.	1 Jan. 17	. . . .
Peck, A. W., C.M.G.	1 Jan. 17	. . . .
Knox, A. W. F., C.B., *p.s.c.* [L]	1 Jan. 17	. . . .
Gregory, C. L., C.B., *p.s.c.*	1 Jan. 17	. . . .
Vaughan, L. R., C.B., D.S.O., *p.s.c.*	1 Jan. 17	. . . .
Maclachlan, T. R., C.M.G.	10 Mar. 17	. . . .
Stewart, J. H. K., D.S.O., *p.s.c.*	10 Mar. 17	. . . .
Cassels, R. A., C.B., D.S.O., *p.s.c.*	10 Mar. 17	. . . .
Banbury, W. E., C.M.G. [L]	3 June 17	. . . .
Jones, L. C., C.M.G., M.V.O., *p.s.c.* [L]	3 June 17	. . . .
Pritchard, A. G., C.M.G.	3 June 17	. . . .
Gray, F. W. B., C.M.G., D.S.O.	1 Jan. 18	. . . .
Macmullen, C.N., D.S.O., *p.s.c.*	1 Jan. 18	. . . .
Wigram, K., D.S.O., *p.s.c.*	1 Jan. 18	. . . .
Steel, R. A., C.I.E., *p.s.c.* [*l*]	1 Jan. 18	. . . .
Ridgway, R. T. I., C.B.	3 June 18	. . . .
Cobbe, H. H., C.M.G., D.S.O.	3 June 18	. . . .
Maxwell, L. L., C.M.G.	3 June 18	. . . .
Norman, C. L., D.S.O., M.V.O., A.-D.-C., *p.s.c.*	3 Jnne 18	. . . .
Godwin, C.A.C., D.S.O., *p.s.c.*	3 June 18	. . . .
Winter, C.B., C.M.G., D.S.O.	3 June 18	. . . .
Tarver, A. L., D.S.O., A. D.-C., *p.s.c.*	3 June 18	. . . .
Orr, G.M., D.S.O., *p.s.c.*	3 June 18	. . . .
Leslie, W. S., D.S.O., *p.s.c.*	3 June 18	. . . .
Morris, G. M., D.S.O., [L]	3 June 18	. . . .
Norman, W. H., D.S.O., *p.s.c.*	3 June 18	. . . .
O'Grady, H. de C.	3 June 18	. . . .
Henderson, M. H.	3 June 18	. . . .

Lieut.-Colonels.

(In this list Brevet Lieut.-Colonels are shown in italic type and arranged in order of Army seniority. The second date after an officer's name is the date of his substantive rank.)

Names.	Date of rank.	Ceased to be employed.
Plumer, T. H.	1 June 04	5 Aug. 05
Giles, A.	8 Dec. 06	. . . .
O'Bryen, C. W. [*l*]	10 Mar. 07	9 Mar. 12
Stirling, W. G. H.	2 Apl. 07	. . . .
Wynch, F. J. H.	10 Oct. 07	. . . .
Erck, J. C. W.	26 Oct. 07	26 Oct. 12
Holland, G. L., M.V.O.	30 Apl. 08	. . . .
Nicholson, R. W.	2 Aug. 08	23 May 12
Henderson, T. H.	11 Aug. 08	13 Apl. 14
Jones, H. J., D.S.O.	25 Aug. 08	. . . .
Henegan, J., D.S.O.	3 Jan. 09	3 Jan. 14
Whiffin, H. E.	27 Jan. 09	. . . .
Young, W. B.	27 Jan. 09	. . . .
Baynes, C. E.	14 Feb. 09	28 May 14
Dann, H. C. B.	28 Feb. 09	28 Feb. 14
Knatchbull, G. W. C., C.M.G.	10 Mar. 09	. . . .
Loudon, J. A.	10 Mar. 09	. . . .
Williams, A. H.	28 Apl. 09	. . . .
Thomas, H. St. G., D.S.O.	16 May 09	. . . .
Rainsford, J. A. G.	5 July 09	. . . .
Wilkins, E. G. R. [*l*]	31 July 09	. . . .
Grant, A.	3 Aug. 09	3 Aug. 14
Pritchard, H. E.	11 Nov. 09	. . . .
Palmer, H. I. E., C.M.G.	5 Dec. 09	. . . .
Finch, C.	30 Jan. 10	. . . .
Pollard, J. H.	6 Feb. 10	. . . .
Medley, A. G. [L]	6 Feb. 10	. . . .
Murray, C. F. T., C.B., M.V.O.	6 Feb. 10	4 Jan. 18
Barnes, J. P.	6 Feb. 10	1 May 14
Hilliard, H. N.	18 Feb. 10	. . . .
Rea, F. W.	26 Mar. 10	. . . .
Laing, F. C.	30 Apl. 10	. . . .
Dennys, A. H.	12 May 10	. . . .
Hill, J. R.	26 May 10	. . . .
Ward, G. A.	2 July 10	. . . .
Ryder, W. I.	9 Aug. 10	. . . .
Budd, N. A. H.	11 Aug. 10	11 Aug. 15
Harvey, A. B.	23 Aug. 10	. . . .
Gordon, R. F. C., C.I.E.	23 Aug. 10	. . . .
Goodenough, H. L.	23 Aug. 10	. . . .
Roome, H. N.	3 Sep. 10	. . . .
Dobbie, C. F.	8 Oct. 10	. . . .
Barry, J. F.	25 Oct. 10	. . . .
Battye, A. H.	12 Nov. 10	. . . .
Lowther, H.	22 Jan. 11	. . . .
Low, R. B., D.S.O.	7 Feb. 11	. . . .
Moore, F. L.	7 Feb. 11	. . . .
Ditmas, A. R.	7 Feb. 11	. . . .
Stuart, D. F.	7 Feb. 11	. . . .
Moore, A. G.W.	28 Feb. 11	. . . .
Robin, N. E.	6 May 11	23 Jan. 15
Bell, L. M.	9 May 11	. . . .
Burlton, W. E. F.	23 May 11	. . . .
White, W. E.	12 June 11	. . . .
Trydell, B.	29 Aug. 11	. . . .
Beville, C. H., C.B.	29 Aug. 11	. . . .
Dennys, H. T., C.I.E.	29 Aug. 11	. . . .
Widdicombe, F. S.	29 Aug. 11	. . . .
Geoghegan, S., C.B., *p.s.c.* [L]	29 Aug. 11	27 Dec. 17
Ward, T. M.	29 Aug. 11	. . . .
Swan, C. T.	17 Sep. 11	. . . .
Browne, A. J. W.	18 Sep. 11	. . . .
Bolton, A. H.	13 Nov. 11	. . . .
Bell, R. M.	25 Nov. 11	. . . .
Dunlop, H. H.	25 Nov. 11	. . . .
Johnston, H.W.	30 Jan. 12	. . . .
Elliot, C. R.	30 Jan. 12	. . . .
Jollie, F.	30 Jan. 12	. . . .
Arnold, A. S.	30 Jan. 12	. . . .
Widdicombe, G. T., C.B.	30 Jan. 12	. . . .
Magrath, H. A. F.	1 Feb. 12	12 June 16
Hurly, M. R.	1 Feb. 12	. . . .

In this list Brevet Lieut.-Colonels are shown in italic type.

Names.	Date of rank.	Ceased to be employed.
Lieut.-Colonels—*contd.*		
Taylor, A. W. N., C.S.I.	13 Feb. 12	....
Thacker, H. J.	28 Apl. 12	....
Brown, W. H., C.B.	10 May 12	15 Feb. 14
Hawkins, G. A.	19 May 12	....
Burlton, A. R.	19 May 12	....
Dunolly, K. J. G. [l]	30 May 12	....
Waterfield, A. C. M., M.V.O., *p.s.c.* [l]	10 July 12 5 Feb. 13	....
Hall, E. S.	25 Aug. 12	....
Turner, F. W. C.	25 Aug. 12	....
Daunt, W. D.	25 Aug. 12	....
Vincent, F. L.	25 Aug. 12	....
Moore, H. A.	25 Aug. 12	....
Watson, A. C.	25 Aug. 12	....
Mockler, G. H. G. [L]	19 Oct. 12	....
Malcolm, W. L.	27 Oct. 12	....
Ricketts, H. C.	10 Nov. 12	....
Forster, P. B. B.	10 Nov. 12	....
Stewart, F. T.	10 Nov. 12	....
Jamieson, A. J.	10 Nov. 12	....
Hickley, A. C.	8 Dec. 12	....
Denne, R. W. A.	8 Dec. 12	....
Kirkwood, T. M.	8 Dec. 12	....
Roome, R. E.	14 Jan. 13	....
Chitty, E. R. I.	5 Feb. 13	....
Priestley, C. E. N.	5 Feb. 13	....
Twigg, E. F.	5 Feb. 13	....
Holmes, G. V.	5 Feb. 13	....
Wikeley, J. M.	5 Feb. 13	....
Browne, A. P., D.S.O.	5 Feb. 13	....
Roddy, H. H., C.B.	5 Feb. 13	....
Tweddell, H.	5 Feb. 13	....
Dale, G. A. C.M.G. [l]	5 Feb. 13	....
Turner, A. G. B., D.S.O.	5 Feb. 13	20 May 16
Moberly, F. J., D.S.O., *p.s.c.*	13 Feb. 13	....
Faunce, R. De L.	13 Feb. 13	....
Saunders, A. R.	16 Feb. 13	....
Pearson H. F. A.	21 Feb. 13	....
Harding, M. F., C.M.G. [l]	10 Mar. 13	....
Wilson, J. A., D.S.O.	27 Mar. 13 16 Nov. 13	....
Cunningham, P. H.	12 Apl. 13	....
Oakes, G. R.	4 May 13	....
Anderson, A. M.	4 May 13	....
Maxwell, H. G.	7 May 13	....
Savi, R. J. T.	7 June 13	....
Wooldridge, W. H.	20 July 13	....
Wintour, E.	14 Sep. 13	....
Robinson, G. W., C.B.	14 Sep. 13	....
Keays, R. W. C.	14 Sep. 13	....
ffrench-Mullen, J. L. W., C.I.E.	14 Sep. 13	....
Lane, H. A.	14 Sep. 13	....
Powell, W. B., C.M.G.	14 Sep. 13	....
Watling, G.	14 Sep. 13	....
Thornton, C. E., C.M.G.	14 Sep. 13	....
Obbard, E. N.	14 Sep. 13	....
Talbot, J.	14 Sep. 13	....
Jones, H. H. (*temp.* Col.)	14 Sep. 13	....
Bayley, E. C., C.I.E.	14 Sep. 13	....
Boddam, E. B. C., *p.s.c.*	19 Oct. 13	....
Chaplin, R. E.	2 Nov. 13	....
Watson, L. A., *p.s.c.* [L]	13 Nov. 13	....
Chichester, A. G. deV.	16 Nov. 13	....
Thomason, A. F.	16 Nov. 13	....
Dunsford, F. P. S., D.S.O.	16 Nov. 13	....
Nisbett, W. G.	16 Nov. 13	....
Sutton, F. G. H.	16 Nov. 13	....
Rose, J. L., C.I.E.	16 Nov. 13	....
Porteous, C. McL	16 Nov. 13	....
Robson, C. G. [L.]	16 Nov. 13	....

Names.	Date of rank.	Ceased to be employed.
Lieut.-Colonels—*contd.*		
Webb, A. E.	16 Nov. 13	....
Ward, A., D.S.O.	12 Dec. 13	....
Bainbridge, W. F., C.M.G., D.S.O., *p.s.c.*	7 Jan. 14 24 May 16	....
Muscroft, W. St. C.	7 Jan. 14	....
Wooldridge, H. C.	21 Jan. 14	....
Murray, E. R. B.	1 Feb. 14	....
Macquoid, C. E. E. F. K., D.S.O.	11 Feb. 14	....
Stack, C. S., C.M.G.	11 Feb. 14	....
Dickson, J. H., C.I.E.	11 Feb. 14	....
Lloyd, W. E. E.	11 Feb. 14	....
Pennington, A. W., M.V.O.	11 Feb. 14	....
Longden, A. B.	11 Feb. 14	....
Eastmead, C. S., [L]	11 Feb. 14	....
Vaughan, E. G., D.S.O.	11 Feb. 14	....
Vanrenen, G. R.	11 Feb. 14	....
Stevens, N. M. C., C.M.G [L]	11 Feb. 14	....
Heffernan, H. W.	11 Feb. 14	....
Major, F. F.	11 Feb. 14	....
Brooke, H. M. M.	11 Feb. 14	....
Sherer, J. C.	11 Feb. 14	....
Beatson Bell, J.	11 Feb. 14	....
Lloyd-Jones, F. L.	11 Feb. 14	....
Hitchins, H. E.	11 Feb. 14	....
Barnard, A. E.	11 Feb. 14	....
Pierce, F. G.	11 Feb. 14	....
Walker, W. R.	11 Feb. 14	....
Kauntze, C. R.	11 Feb. 14	....
Moore, C. H. G., C.M.G., D.S.O.	11 Feb. 14	....
Kemball, J. S.	11 Feb. 14	....
Baker, H. R.	15 Feb. 14	....
Murray, A. B.	17 Feb. 14	....
Young, W. E.	8 Mar. 14	....
Ford, H. B.	14 Mar. 14	....
Dallas, J. S.	9 May 14	....
Hutchinson, C. R. M.	9 May 14	....
Money, E. D., C.I.E., D.S.O.	9 May 14	....
Chatterton, G. D. L., D.S.O.	1 June 14	....
Cobbe, H. H., C.M.G., D.S.O.	27 July 14	....
Bagshawe, F. W., C.I.E. [L]	22 Aug. 14	....
Henslowe, F. J. D.	22 Aug. 14	....
Gordon, J. L. R., C.B.	22 Aug. 14	....
Cooper, W. G.	22 Aug. 14	....
Ralph, A. C., D.S.O.	22 Aug. 14	....
Stewart, J. A.	22 Aug. 14	....
Humfrey, F. G. C.	22 Aug. 14	....
Ainslie, H. P.	22 Aug. 14	....
Basevi, W. H. F.	22 Aug. 14	....
Fraser, F. J.	12 Sep. 14	....
Baker, C. N.	19 Sep. 14	....
O'Meara, C. A. E., C.I.E.	23 Oct. 14	....
Elderton, M. T.	10 Nov. 14	....
Haldane, C. L., C.M.G., D.S.O. [l]	10 Nov. 14	....
Amesbury, W. L. R.	10 Nov. 14	....
Codrington, H. de B.	28 Nov. 14	....
Coke, A. H.	5 Dec. 14	....
Lillingston, W. E. G.	8 Dec. 14	....
Strahan, G. A.	8 Dec. 14	....
Carter, G. L., C.I.E.	8 Dec. 14	....
Firth, R. A., D.S.O.	8 Dec. 14	....
Lee, A. W. H.	30 Jan. 15	....
Douglas, W. B.	30 Jan. 15	....
Tennant, E. [L]	30 Jan. 15	....
Paul, R. S.	30 Jan. 15	....
Templer, C. F., C.M.G., C.I.E.	15 Feb. 15	....
Kaye, C., C.S.I., C.I.E.	6 Mar. 15	....
Andrew, F. A., D.S.O.	23 Mar. 15	....
Munn, R. G.	23 Mar. 15	....
Alexander, H. S., D.S.O.	23 Mar. 15	....
Travers, R. E.	23 Mar. 15	....

In this list Brevet Lieut.-Colonels are shown in italic type.

Lieut.-Colonels—*contd.*

Names.	Date of rank.	Ceased to be employed.
Geoghegan, F. E., C.I.E.	23 Mar. 15	
Cullen, E. H. S., C.M.G., D.S.O., M.V.O.	24 Mar. 15	
Bannerman, W. P.	24 Apl. 15	
Cassels, G. R., *p.s.c.*	24 Apl. 15	
Becher, G. A., D.S.O.	27 Apl. 15	
Byers, C. B.	15 May 15	
Wigram, C., C.B., C.S.I., C.V.O., *p.s.c.*, Eq.	3 June 15	
Smith, W. F. (*e*)	3 June 15 21 Sep. 15	
Boileau, E. R. P.	3 June 15 2 Apl. 16	
Tomkins, H. L., C.M.G., D.S.O.	3 June 15 29 Oct. 16	
Leeds, T. L., C.M.G., D.S.O.	3 June 15 20 Feb. 17	
Murphy, C. C. R.	3 June 15	
Holdich, H. A., D.S.O.	3 June 15	
Forteath, F. W. H.	8 June 15	
Moore, A.	6 July 15	
Roosmale-Cocq, C. A.	6 July 15	
Darley, J. R., C.I.E., D.S.O.	6 July 15	
Cotton, W. L.	8 July 15	
Cheyne, R. E.	21 Sep. 15	
Payne, E. H.	21 Sep. 15	
Bateman-Champain, H. F., C.M.G.	21 Sep. 15	
Dallas, A. E., C.M.G.	21 Sep. 15	
Brownlow, d'A. C., C.M.G., C.I.E.	21 Sep. 15	
Prentis, W. S.	21 Sep. 15	
Samborne-Palmer, F. C. S.	23 Oct. 15	
Cassels, K. S.	23 Oct. 15	
Garratt, H. S.	9 Nov. 15	
Lloyd, J. H.	9 Nov. 15	
Johnson, C. R.	9 Nov. 15	
Hay, L. T.	9 Nov. 15	
Harrison, A. H. P.	29 Nov. 15	
Chandler, H. G. W.	4 Dec. 15	
Alexander, A. V.	21 Dec. 15	
Beatty, G. A. H., D.S.O.	21 Dec. 15	
Ward, C. H.	21 Dec. 15	
Nicholson, W. C.	21 Dec. 15	
Rawlins, A. K., C.I.E., D.S.O.	21 Dec. 15	
Mackenzie, C., D.S.O.	21 Dec. 15	
Swan, J. S.	21 Dec. 15	
Newnham, C. C., D.S.O., *p.s.c.*	1 Jan. 16 4 Feb. 16	
Lushington, W. N.	1 Jan. 16	
Dobbs, C. F., D.S.O., *p.s.c.*	1 Jan. 16 4 Feb. 17	
Cox, C. E. S.	1 Jan. 16	
Bourne, W. Fitz G.	31 Jan. 16	
Fagan, E. A., C.M.G., D.S.O.	18 Feb. 16	
Field, C. D.	1 Mar. 16	
Hocken, C. A. F., *p.s.c.*	1 Mar. 16	
Nightingale, M. R. W., C.M.G., D.S.O.	1 Mar. 16	
Davis, G. J., D.S.O.	1 Mar. 16	
Kirkpatrick, E.	1 Mar. 16	
Redl, E. A. F., C.I.E.	1 Mar. 16	
Spence, A. H. O., C.I.E.	1 Mar. 16	
Kemball, A. G.	1 Mar. 16	
Glasfurd, A. I. R., C.M.G., D.S.O.	29 Mar. 16	
Alexander, C. H.	29 Mar. 16	
Hogg, J. McK. T.	29 Mar. 16	
Davidson, F. D.	1 Apl. 16	
Ricketts, P. E., D.S.O., M.V.O.	3 Apl. 16	
Prevost, G. H.	22 Apl. 16	
O'Reilly, C. M.	3 May 16	
Bliss, J. A., D.S.O., M.V.O.	3 May 16	
Colan, W. R. B., D.S.O.	3 May 16	
Lukin, R. C. W., D.S.O.	3 May 16	

Lieut.-Colonels—*contd.*

Names.	Date of rank.	Ceased to be emloyed.
Hole, H. L.	3 May 16	
Orman, F. L.	3 May 16	
Fellows, C. S.	3 May 16	
Sealy, A. E.	24 May 16	
Wynter, C. P.	24 May 16	
Ross, H.	31 May 16	
Pennefather Evans, G.	3 June 16 16 July 16	
Mears, C. D., M.V.O.	3 June 16 9 Sep. 17	
Davie, J. H. M.	3 June 16 7 Nov. 17	
Watt, D. M., D.S.O., *p.s.c.*	3 June 16 7 Nov. 17	
Edwardes, A. C.	3 June 16 12 Oct. 17	
Walton, G.	3 June 16 20 Oct. 17	
Keelan, H. P.	3 June 16 14 Aug. 16	
VC *Costello, E. W.*, C.M.G., D.S.O., *p.s.c.*	3 June 16	
Scott-Elliott, C. R.	3 June 16	
Moberly, H. S., *p.s.c.*	3 June 16	
Russell, F. D., *p.s.c.* [L]	3 June 16 7 July 16	
Longhurst, A. L., *p.s.c.* [L]	3 June 16	
Furney, J. L.	3 June 16	
Watson, S. B.	3 June 16	
Pryce, H. E. ap R., C.M.G., D.S.O., *p.s.c.*	3 June 16 17 Aug. 18	
Bourne, W. K., *p.s.c.*	3 June 16	
St. John, R. S., D.S.O., *p.s.c.*	3 June 16	
Marsh, F. G., C.M.G., *p.s.c.*	3 June 16	
Geoghegan, N. M.	3 June 16	
Beddy, P. L.	3 June 16	
Franklin, H. S. E., D.S.O.	3 June 16	
Rowlandson, M. G. D., D.S.O., *p.s.c.* [L]	3 June 16	
Carwithen, T.	4 June 16	
Fisher, F.	5 June 16	
Strange, A. J.	28 June 16	
Bidwell, R. F.	28 June 16	
Battine, R. St. C., D.S.O., *p.s.c.* [L]	23 July 16	
Adams, F., D.S.O.	29 July 16	
Sangster, P. B., D.S.O.	7 Aug. 16	
Earle, E. S.	8 Aug. 16	
Keily, F. P. C., C.M.G., D.S.O.	24 Sep. 16	
Young, H. G., D.S.O.	8 Oct. 16	
Nangle, K. E.	8 Oct. 16	
Nicholas, S. H. E.	8 Oct. 16	
Coningham, F. E., D.S.O.	8 Oct. 16	
Hill, H. C., D.S.O.	29 Oct. 16	
Haydon, W. P., D.S.O.	29 Oct. 16	
McLeod, T. J., D.S.O.	29 Oct. 16	
Cox, A. D.	29 Oct. 16	
Crawford, J. H.	1 Nov. 16	
Buist, A. H., M.V.O.	14 Nov. 16	
Souter, H. M. W., C.M.G., D.S.O.	14 Nov. 16	
Edwardes, J. G.	29 Nov. 16	
Newell, H. A.	29 Nov. 16	
Barclay, P. C. R.	29 Nov. 16	
Teed, A. S. H.	29 Nov. 16	
Sarel, G. B. M.	29 Nov. 16	
Fenn, E. H. McB.	29 Nov. 16	
Crawford, A. G.	29 Nov. 16	
Knox, E. F., D.S.O.	29 Nov. 16	
Cheyne, A. Y.	29 Nov. 16	
Smith, A. LeF.	3 Dec. 16	
Souter, A. B.	17 Dec. 16	
Tillard, A. B.	24 Dec. 16	
Bridges, A. H., D.S.O., *p.s.c.* [*l.*]	1 Jan. 17 25 Mar. 17	
Seton-Browne, C. L. S., D.S.O.	1 Jan. 17	

In this list Brevet Lieut.-Colonels are shown in italic type.

Lieut.-Colonels—*contd.*

Names.	Date of rank.	Ceased to be employed.
Jackson, H. W., D.S.O.	1 Jan. 17 5 Dec. 17	
Sweet, E. H., D.S.O.	1 Jan. 17	
Dyke, P. H., D.S.O.	1 Jan. 17 18 May 18	
Bruce, G. D., D.S.O., *p.s.c.*	1 Jan. 17 13 Aug. 18	
Deane, D., D.S.O. *p.s.c.* [*l*]	1 Jan. 17	
Barstow, H.	1 Jan. 17	
Baird, H. B. D., C.M.G., D.S.O., *p.s.c.*	1 Jan. 17	
Hay, C. J. B., D.S.O., *p.s.c.*	1 Jan. 17	
Pitcher, D. Le G., C.M.G.	1 Jan. 17	
Neumann, C. W.	1 Jan. 17	
Coleridge, J. F. S. D., C.M.G., D.S.O., *p.s.c.*	1 Jan. 17	
Talbot, F. E. G., *p.s.c.*	1 Jan. 17	
Moens, A. W. H. M., D.S.O., *p.s.c.*	1 Jan. 17	
Drew, B. C. H., *p.s.c.*	1 Jan. 17	
MacDonnell, H. N. F.	1 Jan. 17	
Whitehead, J., D.S.O., *p.s.c.*	1 Jan. 17	
Swifte, E. L.	17 Jan. 17	
Butler, A. H.	4 Feb. 17	
McNeile, D. H. [L]	13 Feb. 17	
Capper, A. S., D.S.O., *p.s.c.*	15 Feb. 17	
Milne, J. W.	4 Mar. 17	
Hudson, J. H. [L]	4 Mar. 17	
Henderson, M. H.	4 Mar. 17	
Tregear, F. C.	4 Mar. 17	
Brackenbury, W. J. W.	4 Mar. 17	
Gwyn-Thomas, G., C.M.G., D.S.O., *p.s.c.*	4 Mar. 17	
Pocock, P. F., D.S.O.	25 Mar. 17	
Fox, L. W.	2 May 17	
Hawkes, R.	10 May 17	
Tregear, V. F. W.	23 May 17	
Dundas, P. H., D.S.O., *p.s.c.*	3 June 17 3 Sep. 18	
Watts, G. A. R.	3 June 17 3 Sep. 18	
Prissick, C.	3 June 17 22 Sep. 18	
Cuming, R. J., D.S.O.	3 June 17 24 June 18	
Alexander, F. C., D.S.O., *p.s.c.*	3 June 17	
Simpson, J. C.	3 June 17	
Abbott, L. H., C.M.G., *p.s.c.*	3 June 17	
Fetherstonhaugh, W. A., D.S.O., *p.s.c.*	3 June 17	
Shuttleworth, D. I., *p.s.c.*	3 June 17	
Wintle, C. E. H.	3 June 17 29 Mar. 18	
Sherrard, J. W.	3 June 17	
Jardine, C. H.	3 June 17	
Combe, S. B., M.C.	3 June 17	
Twiss, W. L. O., M.C., *p.s.c.*	3 June 17	
Muspratt, S. F., D.S.O., *p.s.c.*	3 June 17	
McWatters, H. C., D.S.O., *p.s.c.*	3 June 17	
Ward, H. C. S., *p.s.c.* (*e*)	3 June 17	
Kensington, E. C., M.C., *p.s.c.*	3 June 17	
Goodfellow, N. G. B.	3 June 17	
Gwyer, V. E.	3 June 17	
Hewlett, E. A.	3 June 17	
Williams, L. G., D.S.O.	3 June 17	
Wilson, R. C., D.S.O., M.C., *p.s.c.*	3 June 17	
Woodhouse, P. St. J. R.	3 June 17	
McCaskill, J. C. H.	11 June 17	
Jackson, K. H.	17 June 17	
Stewart, R. J. T.	17 June 17	
Barry-Smith, R. C.	19 June 17	
Hay, W. N., C.I.E., D.S.O., *p.s.c.*	24 July 17	
Hunt, V. de V.	24 July 17	
Churchill, G. R. D.	24 July 17	
Preston, E. E.	24 July 17	
Ellwood, B. P.	29 July 17	

Lieut.-Colonels—*contd.*

Names.	Date of rank.	Ceased to be employed.
McVean, D. A. D., D.S.O.	29 July 17	
Keogh, J. B., D.S.O.	23 Aug. 17	
Bredin, A.	28 Aug. 17	
Dent, W., D.S.O., *p.s.c.*	28 Aug. 17	
Preston, G. A.	9 Sep. 17	
Chamier, W. St. G.	9 Sep. 17	
Hulseburg, H., D.S.O.	9 Sep. 17	
Bickford, W. W., C.I.E.	17 Sep. 17 5 Dec. 17	
Anderson, R. H. B.	10 Oct. 17	
Mitchell, W. J.	28 Oct. 17	
Hawes, C. H., D.S.O., M.V.O.	7 Nov. 17	
Scott, W.	7 Nov. 17	
Keighley, V. A. S., M.V.O.	2 Dec. 17	
Battye, A. B.	5 Dec. 17	
Houston, A. M.	5 Dec. 17	
Oakes, H. C.	17 Dec. 17	
Taylor, R. O'B., C.M.G., C.I.E.	1 Jan. 18 13 July 18	
Gordon, G. A. S.	1 Jan. 18	
Barrett, A. L.	1 Jan. 18	
Palmer, G. S.	1 Jan. 18	
Hughes Hallett, R. L.	1 Jan. 18	
Hogg, R. E. T., C.I.E.	1 Jan. 18	
Casson, W. F. S., D.S.O., *p.s.c.*	1 Jan. 18	
Dunlop, W. B., D.S.O.	1 Jan. 18	
Kirkpatrick, C.	1 Jan. 18	
Villiers Stuart, J. P., D.S.O.	1 Jan. 18	
Porch, E. A., M.C.	1 Jan. 18	
McMahon, H. F. E., M.C.	1 Jan. 18	
Worgan, R. B., D.S.O.	1 Jan. 18	
Brown, W.	1 Jan. 18	
Hewlett, A.	15 Jan. 18	
Murray, W. G. P.	22 Jan. 18	
Fordham, W. M.	22 Jan. 18	
Pye, W. E., C.I.E.	27 Jan. 18	
Stewart, A. C.	27 Jan. 18	
Harrigton, A. W. D.	27 Jan. 18	
Waller, A. A. P.	27 Jan. 18	
Cattell, G. L.	27 Jan. 18	
Lees, C. H. B., C.M.G.	27 Jan. 18	
Rowcroft, C. H.	19 Feb. 18	
Wall, A. C.	20 Feb. 18	
Gaussen, J. R., C.M.G., D.S.O.	20 Feb. 18	
Ottley, W. J.	21 Feb. 18	
Herdon, H. E.	2 Mar. 18	
Ricketts, R. L.	8 Mar. 18	
Willans, T. J., D.S.O.	8 Mar. 18	
Gardiner, R., *p.s.c.*	10 Mar. 18	
Dopping-Hepenstal, M. E., D.S.O.	11 Mar. 18	
James, A. A.	12 Mar. 18	
Rice, S. K. B.	14 Mar. 18	
Carnegy, G. P. O., D.S.O.	15 Mar. 18	
Hunter, W. J. H.	18 Mar. 18	
Strong, H. P.	30 Mar. 18	
Ommanney, C. V.	3 June 18	
Thomson, A. G., D.S.O.	3 June 18	
Graham, R. B.	3 June 18	
Strong, A. D., D.S.O.	3 June 18	
Allanson, C. J. L., D.S.O.	3 June 18	
Stokes, C. B., C.I.E. *p.s.c.*	3 June 18	
Popham, E. L., D.S.O.	3 June 18	
Freeland, J. C., *p.s.c.*	3 June 18	
Moberly, B. R., D.S.O., *p.s.c.*	3 June 18	
Milward, C. A., D.S.O.	3 June 18	
Smith, H. H., D.S.O.	3 June 18	
Willson-Johnston, W. E., D.S.O., *p.s.c.*	3 June 18	
Giles, E. D., D.S.O.	3 June 18	
Catty, T. C., D.S.O., *p.s.c.*	3 June 18	
Rowan Robinson, J. R., D.S.O.	3 June 18	
Orton, S. B., *p.s.c.*	3 June 18	
Nisbet, T., D.S.O.	3 June 18	
Lloyd, C. R., D.S.O.	3 June 18	
Powel, D. H., M.C.	3 June 18	
Kirby, J. T., D.S.O.	3 June 18	

In this list Brevet Lieut.-Colonels are shown in italic type.

Names.	Date of rank.	Ceased to be employed.	Names.	Date of rank.	Ceased to be employed.
Lieut.-Colonels—*contd.*			**Lieut.-Colonels**—*concld.*		
Cochran, G. W., D.S.O.	3 June 18	....	Stansfeld, C. G.	13 Aug. 18	....
Ogle, N., D.S.O.	3 June 18	....	Bruce, G. D., D.S.O., *p.s.c.*	13 Aug. 18	....
Thornhill, C. J. M., D.S.O.	3 June 18	....	Waring, E. H.	13 Aug. 18	....
Anderson, J. F. H., D.S.O.	3 June 18	....	Preston, W. J., D.S.O.	31 Aug. 18	....
Little, W. C.	11 June 18	....	Harbord, C. R., D.S.O.	1 Sep. 18	....
Browne, L. S.	15 June 18	....	Corlett, J. S.	3 Sep. 18	....
Hodgkinson, C.	18 June 18	....	Stewart, A. F.	3 Sep. 18	....
Hodgkinson, J.	18 June 18	....	Quin, R. P. W.	12 Sep. 18	....
Bradshaw, C. R., *p.s.c.*	22 June 18	....	VC Costello, E. W., C.M.G., D.S.O.	22 Sep. 18	....
Johnson, J. E. C.	1 July 18	....	Orton, E. F.	26 Sep. 18	....
Gale, E. S.	3 July 18	....	Hawkes, W. C. W., D.S.O.	20 Oct. 18	....
Hume, S. G. W.	6 July 18	....	Saville, A. R. C.	16 Nov. 18	....
Chitty, A. W.	13 July 18	....	deLabilliere, E. G. D.	23 Nov. 18	....
Calvert, R. T. C.	13 July 18	....	*Villiers-Stuart, W. D.*	19 Dec. 18	....
Harcourt, J. S. M.	13 July 18	....	*Rind, G. B. A.*	19 Dec. 18	....
Kauntze, B. C.	31 July 18	....	*Ross, R. C.*, D.S.O.	19 Dec. 18	....
Nicolay, B. U.	3 Aug. 18	....			
Bruce, M. E. L.	13 Aug. 18	....			

OFFICERS OF THE LOCAL SERVICE AND GENERAL LIST CAVALRY AND INFANTRY ON THE UNEMPLOYED SUPERNUMERARY LIST.

BENGAL.

LOCAL SERVICE.

CAVALRY.

First Commn.	Names.		Date as Colonel.		Army rank.	Remarks.
30 Dec. 54	𝔐 Mackenzie, *Sir* Alfred Robert Davidson, K.C.B. Ⓡ	PH	. . .	*c.*	23 July 82	
20 Oct. 49	VC 𝔐 Probyn, *Rt. Hon. Sir* Dighton Macnaghten, G.C.B., G.C.S.I., G.C.V.O., I.S.O., *Eq.* Ⓡ . .	PH	1 Apl. 93	*g.*	1 Dec. 88	Col., 11 Lancers.

INFANTRY.

First Commn.	Names.				Army rank.	Remarks.
3 June 51	𝔐 Cockburn, Henry Alexander . . .	PH	6 Mar. 80	*g.*	1 Apl. 94	
20 Dec. 52	𝔐 Baillie, Jas. Cadogan Parkison (50, 99) .	P	6 May 81	*g.*	1 Apl. 94	
4 Oct. 55	𝔐 Franks, William Astell	PH	1 Jan. 82	*m.g.*	1 Apl. 94	
4 Feb. 58	𝔐 Howey, William	PH	9 Aug. 82	*m.g.*	1 Apl. 94	
25 Aug. 59	Jones, Griffith Turner	PH	11 Sep. 84	*l.g.*	1 Jan. 93	
13 Dec. 56	𝔐 Dalmahoy, Patrick Carfrae (57)	PH	9 Sep. 85	*m.g*	20 Aug. 90	
26 July 56	𝔐 McNair, Edward John (30, 50, 57) . . .	PH	1 Jan. 86	*l.g.*	20 Dec. 91	
30 Oct. 58	Stewart, John Mackie (50)	PH	1 Apl. 87	*l.g.*	10 Feb. 94	
11 Dec. 58	Woodhouse, Harvey	PH	16 July 88	*c.*	16 July 88	
16 Sep. 57	𝔐 Bartleman, Woodburn Francis (50, 57)	PH	12 Mar. 89	*m.g.*	5 Oct. 93	

GENERAL LIST.

CAVALRY.

First Commn.	Names.		Date as Lt.-Col.		Army rank.	Remarks.
20 May 59	Jennings, *Sir* Robert Melvill, K.C.B. . . .	PH	20 May 85	*g.*	28 Jan. 05	Col., 6 Cavalry.

INFANTRY.

First Commn.	Names.				Army rank.	Remarks.
20 Feb. 59	Beckett, William Henry	PH	20 Feb. 85	*c.*	20 Feb. 89	
4 Mar. 59	Quin, Thomas James	PH	4 Mar. 85	*m.g.*	6 June 98	
20 May 59	Marsh, Frank Hale Berwick	PH	20 May 85	*m.g.*	12 Aug. 97	
11 June 59	Tregear, *Sir* Vincent William, K.C.B. . . .	PH	11 June 85	*m.g.*	12 Aug. 97	
9 July 59	Marriott, Edwin Metcalfe Leslie	PH	9 July 85	*c.*	9 July 89	
20 Sep. 59	Chatterton, Frank William, C.I.E. . . .	PH	20 Sep. 85	*c.*	20 Sep. 89	
20 Dec. 59	Mainwaring, Edward Phillipson (57) . . .	PH	20 Dec. 85	*c.*	20 Dec. 89	
8 June 60	Carruthers, Joseph Geo. Thomson . . .	PH	8 June 86	*c.*	8 June 90	
4 Nov. 60	Bruce, Andrew McCrae, C.B. Ⓡ . . .	PH	4 Nov. 86	*c.*	4 Nov. 90	

OFFICERS UNDER THE I. A. (LATE STAFF CORPS) RULES OF PROMOTION,

First Commn.	Names.		Col. Allce.		Army rank.	Remarks.
20 Dec. 58	Halliday, George Thomas	PH	20 Dec. 96	*l.g.*	4 Nov. 00	Col., 16 Cavalry.
20 Nov. 58	Currie, Fendall	PH	20 Nov. 96	*m.g.*	6 June 98	

OFFICERS OF THE LOCAL SERVICE AND GENERAL LIST CAVALRY AND INFANTRY ON THE UNEMPLOYED SUPERNUMERARY LIST—*contd.*

First Commn.	Names			Army rank.	Remarks.

MADRAS.

LOCAL SERVICE.

INFANTRY.

First Commn.	Names		Date as Col.	Army rank.	Remarks.
9 Sep. 38	Halliday, John Gustavus		8 Jan. 76	*g.* 1 Dec. 88	
20 Aug. 55	ℳ Elliot, Henry Riversdale		5 Oct. 86	*m.g.* 13 June 91	
10 Oct. 57	Berkeley, James Cavan, C.I.E.		21 Aug. 87	*m.g.* 1 Jan. 93	
5 May 57	Tyrrell, Francis Hardinge (204)		1 Nov. 87	*l.g.* 1 Aug. 94	Col., 74th Punj.

GENERAL LIST.

INFANTRY.

First Commn.	Names		Date as Lt.-Col.	Army rank.	Remarks.
23 Feb. 59	Hutchins, Alfred George		20 Feb. 85	*c.* 28 Nov. 88	
4 Apl. 59	Begbie, Elphinstone Waters, C.B., D.S.O., ⊛ (50, 57, 191) .		4 Apl. 85	*m.g.* 6 June 98	
4 Aug. 59	Weston, George Edward (50)		4 Aug. 85	*c.* 4 Aug. 89	

OFFICERS UNDER THE I. A. (LATE STAFF CORPS) RULES OF PROMOTION.

First Commn.	Names		Col. Allce.	Army rank.	Remarks.
4 Oct. 58	ℳ Walker, James Grant Duff		14 Oct. 96	*m.g.* 12 Sep. 97	

BOMBAY.

LOCAL SERVICE.

CAVALRY.

First Commn.	Names		Date as Colonel.	Army rank.	Remarks.
20 Feb. 53	ℳ Stack, Charles Edward (57)	PH	. . .	*c.* 31 Oct. 82	
4 Mar. 54	ℳ Macnaghten, Arthur William	PH	. . .	*c.* 4 Mar. 84	
4 Nov. 56	ℳ Moore, Charles Alfred (57) ⊛	H	2 July 97	*m.g.* 25 Apl. 95	

INFANTRY.

First Commn.	Names		Date as Colonel.	Army rank.	Remarks.
1 Jan. 44	Burrows, George Reynolds Scott . . .	H	16 July 76	*g.* 17 Aug. 90	
10 Dec. 54	ℳ Iredell, Francis Shrubb	H	31 Mar. 85	*l.g.* 1 Jan. 93	Col., 114th Mah.
23 Dec. 53	ℳ Phelps, Arthur (152)	PH	1 Jan. 86	*l.g.* 24 Dec. 92	
4 Oct. 57	Hunt, Robert Augustus Carew	PH	30 July 80	*m.g.* 6 Feb. 91	Col., 105th Infy.
31 July 58	ℳ Chambers, Robert Macdonald		5 Oct. 86	*m.g.* 1 July 91	

GENERAL LIST.

INFANTRY.

First Commn.	Names			Army rank.	Remarks.
7 Dec. 61	Stevens, Malcolm Wilkinson	PH	. . .	*l.c.* 27 Dec. 87	

OFFICERS UNDER THE I. A. (LATE STAFF CORPS) RULES OF PROMOTION.

First Commn.	Names		Col. Allce.	Army rank.	Remarks.
27 Oct. 58	Hogg, Geo. Crawford, C.B. (100) ⊛ . .	H	27 Oct. 96	*m.g.* 20 Sep. 97	Col., 34th Horse.
27 Oct. 58	Currie, Algernon (152)	H	27 Oct. 96	*c.* 27 Oct. 88	

BRITISH OFFICERS OF THE INDIAN ARMY.

(Late Indian Staff Corps.)

Vide *Royal Warrant, 15 December 1902.* (*G. G. O. 51 of 1903.*)

Field-Marshal 1—Generals 3—Lieut.-Generals 6—Major-Generals 24—Colonels and Lt.-Colonels—Majors—Captains—Lieutenants—2nd-Lieutenants.

NOTE I.—As all officers of the Indian Army must qualify in Hindustani, by the Higher Standard, the reference "206" denoting that qualification is omitted opposite the names of all Officers above the rank of Lieutenant. The Furlough Rules for the Indian Army which apply to all officers whose admission to the I. A. bears date subsequent to the 30th June 1885, have also been removed.

NOTE II.—Service for Indian pension and for promotion reckons from 1st Commission except as otherwise shown.

No.	Former Corps.	Names.	Date of birth.	Date of first commission.	Date of appointment to Indian Army	Date of rank.	Date of Colonel's allowance.	Fr. rule.	Remarks.
		Field-Marshal—1.							
	6 F.	**Egerton,** *Sir* Charles Comyn, G.C.B., D.S.O., [R] (94, 162)	10 Nov. 48	8 June 67	30 May 71	16 Mar. 17	. . .	1A	Col. 23 Cavalry.
		Generals—3.							
	2 *Eur.*	M *Aitchison, Charles Terrington,* C.B.	*28 May* 23	*10 June* 41	*18 Feb.* 61	*22 Jan.* 89	*10 June* 80	—	*u. s. l.*
	60 N. I.	M *Lumsden,* Sir *Peter Stark,* G.C.B., C.S.I. . PH	*9 Nov.* 29	*10 Dec.* 47	*18 Feb.* 61	*17 Aug.* 90	*10 Dec.* 85	68	*u. s. l.*
	28 N. I.	VC M *Watson,* Sir *John,* G.C.B. . . . . . H	*4 Sep.* 29	*3 Feb.* 48	*18 Feb.* 61	*1 Mar.* 91	*3 Feb.* 86	—	*u. s. l.* Col., 13 Lancers.
	49 N. I.	*Gordon, William,* C.I.E. (58) . . . . . PH	*10 Feb.* 24	*11 June* 41	*18 Feb.* 61	*1 Apl.* 94	*11 June* 79	68	*u. s. l.*
	54 N. I.	M *Anderson, Harry Cortlandt,* C.B. . . P	*28 May* 26	*2 Sep.* 43	*16 Feb.* 61	*1 Apl.* 94	*2 Sep* 80	68	*u. s. l.* Col., 4 Rajputs.
	104 F.	*Evans,* Sir *Horace Moule,* K.C.B. (50, 57) [(50) [R]	*8 Dec.* 41	*20 June* 58	*7 May* 68	*9 Nov.* 01	*20 June* 96	—	*u. s. l.* Col., 8 G. R.
1	89 F.	**Barrow,** *Sir* Edmund George, G.C.B., G.C.S.I.	28 Jan. 52	30 Dec. 71	17 July 77	27 July 09	. . . .	1A	Col., 7 Rajputs.
	25 F.	**Nixon,* Sir *John Eccles,* K.C.B., (161, 166, 169) [L] [R]	16 Aug. 57	10 Sep. 75	2 May 78	4 May 14	. . . .	1A	
	107 F.	**Pearson,* Sir *Alfred Astley,* K.C.B. [R]	5 June 50	8 Feb. 70	4 Nov. 75	1 Mar. 17	. . . .	68	*Resdg. in India* - Col. 124 Infy.
	R.M.A.	**Woon,* Sir *John Blaxell,* K.C.B., (57, 169)[R] . .	*24 Feb.* 56	*1 Oct.* 73	*23 June* 80	*7 June* 17	. . . .	1A	*Resdg. ex India.*
	72 F.	**Scallon,* Sir *Robert Irvin,* G.C.B., K.C.I.E., D.S.O. [R]	*3 Apl.* 57	*12 Feb.* 76	*3 Sep* 77	*30 June* 17	. . . .	1A	
	44 F.	**Barrett,** *Sir* Arthur Arnold, G.C.B., K.C.S.I., K.C.V.O. A.D.C. (50, 169) [R]	3 June 57	10 Sep. 75	17 Jan. 79	1 Aug. 17	. . . .	1A	Comdg., Northern Comd.
	89 F.	**Grover,* Sir *Malcolm Henry Stanley,* K.C.B., K.C.I.E. (170),[L] [R]	*19 July* 58	*11 Sep.* 76	*4 Mar.* 78	*27 Aug.* 17	. . . .	1A	*Resdg. ex India.*
	6 F.	**Watkis,* Sir *Henry Buckley Burlton,* K.C.B., *p.s.c.* (50, 162) [R]	*4 June* 60	*11 May* 78	*22 Apl.* 81	*1 Oct.* 17	. . . .	1A	*Resdg. ex India.*—Col., 31 [Punjabis.
3	12 Lancers	**Birdwood,** *Sir* William Riddell, K.C.B., K.C.S.I., K.C.M.G., C.I.E., D.S.O., A.D.C. (50, 64, 156, 162, 166, 170), *q.s.*	13 Sep. 65	9 May 85	20 Dec. 86	23 Oct. 17	. . . .	1A	

* Supernumerary.

No.	Former Corps.	Names.	Date of birth.	Date of first commission.	Date of appointment to Indian Army.	Date of rank.	Date of Colonel's allowance.	F. rule.	Remarks.
		Lieutenant-Generals—6.							
	72 N. I.	*𝔙𝔆 𝔐 Lyster, Harry Hammon,* C.B. ● . . *PH*	*24 Dec. 30*	*20 Sep. 48*	*12 Sep. 66*	*1 Sep. 91*	*20 Sep. 86*	68	*u. s. l.*
	48 N. I.	*𝔐 Smith, Octavius Ludlow* . . . . . . *PH*	*30 July 28*	*12 June 47*	*18 Feb. 61*	*1 July 92*	*12 June 85*	—	*u. s. l.*
	24 M. I.	***Hankin, Edward Lewis*** . . . . *P*	*12 Nov. 31*	*10 Dec. 47*	*18 Feb. 61*	*23 Sep. 92*	*10 Dec. 83*	68	*u. s. l.*
	30 N. I.	*Morton, Boye William Dunlop* (152) . . . *P*	*24 Aug. 29*	*9 June 48*	*18 Feb. 61*	*10 Dec. 92*	*9 June 86*	68	*u. s. l.*
	4 N. I.	*Willoughby, Michael Weekes* C.S.I. (152) ● . H	*18 Mar. 33*	*9 June 49*	*18 Feb. 61*	*9 Dec. 93*	*9 June 87*	68	*u. s. l.*
	38 N. I.	*𝔐 Clifford, Robert Cecil Richard,* C.B. ● . *PH*	*30 June 39*	*13 Dec. 56*	*12 Sep. 66*	*7 Feb. 99*	*3 Dec. 94*	—	*u. s. l.* Col., 22 Cavalry.
	43 N. I.	*Handcock, Arthur Gore,* C.B. (30, 57) ● . . *PH*	*24 Feb. 40*	*11 Dec. 58*	*12 Sep. 66*	*11 Feb. 00*	*11 Dec. 96*	—	*u. s. l.* Col., 6 Lt. Infantry.
	Devon. R.	**Phayre,* Sir *Arthur,* K.C.B. . . . [170]	*23 Feb 50*	*1 May 78*	*8 Dec 81*	*30 Oct. 15*	. . . .	IA	
1	S. Lanc. R.	**Campbell,** *Sir* Frederick, K.C.B., D.S.O. (30, 50, 57, [170]	25 Feb. 60	13 Aug. 79	18 Sep. 82	1 Jan. 17	. . . .	IA	Divsl. Comdr.—Col., 40 Pathans.
	K. O. S. B.	**Cox,** *Sir* Herbert Vaughan., K.C.B., K.C.M.G., C.S.I. (50, 57, 66*a*, 124*a*, 161, 167) .	12 July 60	14 Jan. 80	25 Aug. 82	1 Jan. 17	. . . .	IA	Col., 69 Punj.—Mily Secy., India Office.
	Worc. R.	***Jacob,** *Sir* Claud William, K.C.B. (50, 93, 169) . .	21 Nov. 63	9 Sep. 82	16 Dec. 84	3 June 17	. . . .	IA	Col., 106 Pioneers.
	12 7.	**Keary,** *Sir* Henry D'Urban, K.C.B., K.C.I.E., D.S.O. (54, 72*b*, 124 *a*, 162).	18 Apl. 57	10 Sep. 75	**7 May 77**	1 Aug. 17	. . . .	IA	Divl. Comdr.—Col., 91 **Punjabis.**
5	Leic. R.	**Egerton,** *Sir* Raleigh Gilbert, K.C.B., K.C.I.E., (50, 71, 170)	25 Sep. 60	13 Aug. 7?	1 Nov. 81	27 Aug. 17	. . . .	IA	
	North'n R.	**Hudson,** *Sir* Havelock, K.C.B., C.I.E. (162, 169) .	26 June 62	22 Oct. 81	4 Jan. 85	23 Oct. 17	. . . .	IA	A. G. in India.
7	R. Mar.	**Wapshare,** Richard, C.B., C.S.I. (65)	6 Jan. 60	1 Feb. 80	18 Nov. 82	15 June 18	. . . .	IA	Divl. Comdr.
		Major-Generals—24.							
	22 M. I.	*Magrath, Beauchamp Henry Whittingham* ● *H*	*4 Feb. 32*	*20 Dec. 49*	*18 Feb. 61*	*31 May 90*	*20 Dec. 86*	68	*u. s. l.*
	9 N. I.	*Coape Smith, Henry* . . . . *PH*	*24 Sep. 29*	*20 Dec. 48*	*18 Feb. 61*	*31 May 90*	*20 Dec. 86*	—	*u. s. l.*
	41 N. I.	*Rowcroft, George Cleland* (50) . . . . . . *PH*	*17 Sep. 31*	*10 Aug. 49*	*18 Feb. 61*	*17 Aug. 90*	*10 Aug. 87*	—	*u. s. l.*
	32 M. I.	*Searle, Arthur Thaddes*	*29 Dec. 30*	*12 Dec. 49*	*12 Sep. 66*	*8 Oct. 90*	*12 Dec. 87*	—	*u. s. l.*
	4 N. I.	*𝔐 Chamier, Francis Edward Archibald,* C.B., C.I.E. *P*	*13 May 33*	*20 Feb. 50*	*18 Feb. 61*	*8 Oct. 90*	*20 Feb. 88*	68	*u. s. l.*
	5 E. L. C.	*𝔐 Prendergast, Guy Annesley* . . . . . *PH*	*11 Mar. 34*	*20 Mar. 50*	*12 Sep. 66*	*3 Jan. 91*	*20 Mar. 88*	68	*u. s. l.* Col., 15 Lancers.
	12 M. I.	*Church, Thomas Ross,* C.I.E. . . . . . . *H*	*5 Jan. 31*	*14 June 50*	*18 Feb. 61*	*13 June 91*	*14 June 88*	—	*u. s. l.*
	61 N. I.	*𝔐 Campbell, Archibald Edwards* . . . . . *PH*	*18 Mar. 34*	*20 Jan. 51*	*12 Sep. 66*	*1 July 92*	*20 Jan. 89*	—	*u. s. l.*
	3 L. C.	*𝔐 LeGeyt, Philip Harrison* . . . . . . *H*	*22 Dec. 34*	*20 Feb. 51*	*18 Feb. 61*	*23 Sep. 92*	*20 Feb. 89*	68	*u. s. l.*
	33 N. I.	*Rutherfurd, Thomas Walter* . . . . . . *P*	*4 Sep. 32*	*14 June 51*	*18 Feb. 61*	*23 Sep. 92*	*14 June 89*	68	*u. s. l.*
	28 N. I.	*𝔐 Gurdon, Evelyn Pulteney* . . . . *P*	*4 June 33*	*20 Jan. 52*	*18 Feb. 61*	*10 Dec. 92*	*20 Jan. 90*	—	*u. s. l.*
		𝔐 Maclean, Charles Smith, C.B., C.I.E. (60) ● . *PH*	*15 Dec. 36*	*14 Dec. 53*	*18 Feb. 61*	*2 Sep. 93*	*14 Dec. 91*	68	*u. s. l.* Col., 21 Cavalry.

	49 N. I.	M Chapman, Hamilton (57) . . . . PH	11 Oct. 35	15 July 56	18 Feb. 61	9 Dec. 93	15 July 94	—	u. s. l. Col., 8 Cavalry.
	30 N. I.	M Boyd, Julius Middleton . . . . H	1 Oct. 37	20 Feb. 54	6 July 63	9 June 94	20 Feb. 92	—	u. s. l.
	46 N. I.	M Upperton, John, C.B. . . . . P	10 June 38	10 June 54	18 Feb. 61	20 Sep. 94	10 June 92	—	u. s. l.
	19 M. I.	M Jacob, William (57) . . . . H	14 Dec. 37	20 Jan. 55	18 Feb. 61	22 Oct. 94	20 Jan. 93	—	u. s. l.
	49 N. I.	Smith, Joseph Barnard . . . . PH	12 June 39	26 June 56	12 Sep. 66	3 Mar. 95	26 June 94	—	u. s. l.
	6 E. R.	Stewart, James Calder, C.B. ● . . . . . . PH	20 June 40	4 Sep. 57	12 Sep. 66	29 Oct. 95	4 Sep. 95	—	u. s. l. Col., 7 Lancers.
	58 N. I.	Sym, Sir John Munro, K.C.B. (57) ● . . . . PH	15 Feb. 39	11 Dec. 58	23 Dec. 62	4 Oct. 96	11 Dec. 96	—	u. s. l. Col., 5 G. R.
	G. L. I.	Toker, Sir Alliston Champion, K.C.B. ● (50, 57, 88, (97, 136, 178, 203)	10 Dec. 43	4 May 60	29 Dec. 62	2 May 97	4 May 98	—	u. s. l. Col., 18 Infantry.
	21 Hus.	Lawrence, William Alexander . . . . PH	5 July 43	4 Feb. 61	20 Dec. 64	8 Nov. 97	4 Feb. 99	—	u. s. l. Col., 17 Cavalry.
	40 N. I.	M Szczepanski, Henry Charles Antony (155) . . PH	21 Jan. 41	13 June 57	18 Feb. 61	10 Dec. 97	13 June 95	—	u. s. l.
	23 N. I.	M Gatacre, Sir John, K.C.B. (57) . . . . H	14 Feb. 41	13 June 57	12 Sep. 66	10 Dec. 97	13 June 95	—	u. s. l. Col., 123 Rifles.
	15 M. I.	Elton, Henry Strachan ●	30 Dec. 41	20 Oct. 58	12 Sep. 66	10 June 98	20 Oct. 99	—	u. s. l. Col., 76 Punjabis.
	28 N. I.	Cuningham, Charles Alexander (57) . . . . . H	29 Jan. 42	20 June 58	19 Nov. 63	12 Jan. 99	20 June 96	—	u. s. l. Col., 127 Lt. Infy.
	16 N. I.	M Westmacott, Sir Richard, K.C.B., D.S.O. (131) ● .	16 Mar. 41	19 Mar. 59	18 Feb. 63	20 June 99	19 Mar. 97	68	u. s. l. Col., 128 Pioneers.
	R. A. (B).	Waterhouse, James . . . . PH	24 July 42	10 June 59	27 June 66	27 June 99	10 June 97	68	u. s. l.
	G. L.	Fagan, James Lawtie (110, 131, 152) . . . . H	4 Feb. 43	11 June 59	9 Nov. 62	6 Sep. 99	11 June 97	—	u. s. l.
	20 Hus.	Shakespear, George Robert James ● . . . . PH	14 Oct. 42	7 Sep. 60	17 Apl. 65	27 Sep. 99	7 Sep. 98	68	u. s. l.
	G. L.	Hughes, Charles Frederick, C.B. ●	19 Mar. 44	7 June 61	1 Jan. 64	21 Feb. 00	7 June 99	—	u. s. l.
	R. Mar.	*Powell, Sir Charles Herbert, K.C.B. (50, 156) [L] [R] .	23 Oct 57	27 Jan. 76	23 June 80	15 Nov. 07	. . . . .	1A	Resdg. ex India.—Col. 1 G. R.
1	100 F.	Hastings, Edward Spence, C.B., D.S.O. (64, 71, 105)	17 Jan. 56	28 Feb. 74	17 Aug. 76	15 Nov. 07	. . . . .	1A	Resdg. ex India.—Col., 92 Punjabis.]
	York R.	*Mackenzie-Kennedy, Edward Chas. Wm., C.B. (50)	6 July 60	13 Aug. 79	3 May 82	9 June 08	. . . . .	1A	Resdg. ex India—Col., 61 Punjabis.]
	109 F.	*Drummond, Sir Francis Henry Rutherford, K.C.I.E., C.B.	9 Sep. 57	11 Feb. 75	17 Aug. 77	21 July 09	. . . . .	—	Resdg. ex India.
	62 F.	*Wallace, Alexander, C.B. (127, 129, 162, 169, 174) .	22 Aug. 58	11 Sep. 76	28 Aug. 79	3 May 11	. . . . .	1A	Col., 27 Punjabis.
	44 F.	*Bell, Sir James Alexander, K.C.V.O.	28 Jan. 56	10 Sep. 75	22 Mar. 79	6 May 11	. . . . .	1A	Col., 42 Deoli Regt.
	R. A.	*Davison, Kenneth Stewart, C.B. (50, 150, 161, 169) .	25 Dec. 56	19 Aug. 75	14 Aug. 82	3 June 11	. . . . .	1A	
	Leic. R.	Younghusband, Sir George John, K.C.M.G., K.C.I.E., C.B., p.s.c.	9 July 59	1 May 78	5 Oct. 80	29 Aug. 11	. . . . .	1A	Resdg. ex India.
	Sco. Rif.	Cookson, George Arthur, C.B., C.M.G. (50, 64, 72b, 162)	6 Aug. 60	11 Sep. 80	26 Jan. 84	23 Oct. 11	. . . . .	1A	Col., 29 Lancers.
	E. York. R.	VC Melliss, Sir Charles John, K.C.B., K.C.M.G. (162) [L]	12 Sep. 62	9 Sep. 82	13 June 84	19 Mar. 12	. . . . .	1A	
	2 F.	*Bunbury, William Edwin, C.B., p.s.c. (50, 162, 165, 170)	5 Apl. 58	30 Jan. 78	14 Sep. 80	1 May 12	. . . . .	1A	Resdg. in India.
	Hamps R.	*Carnegy, Philip Mainwaring, C.B. (50, 71, 156) .	12 Dec. 58	30 Jan. 78	23 Oct. 84	1 Sep. 12	. . . . .	1A	Resdg. ex India.
5	Wilts. R.	Young, Charles Frederic Gordon (50)	12 Nov. 59	1 May 78	14 July 80	21 Sep. 12	. . . . .	1A	Bde. Comdr.
	Devon. R.	*Pirie, Charles Patrick William, C.B. (50, 57, 65, 162) .	17 Aug. 59	11 May 78	1 July 82	30 Jan. 14	. . . . .	1A	
	R. A.	*Turner, James Gibbon, C.B. (50, 64, 72b, 169, 204 [L]	24 Aug 59	18 Dec. 78	27 Mar. 81	5 Feb. 14	. . . . .	1A	Resdg. ex India.—Col., 4 Cav.
	R. Ir. Fus.	*Fry, Charles Irwin, C.B. (50, 71, 162)	2 Sep. 58	22 Jan. 79	13 Feb. 83	11 Mar. 14	. . . . .	1A	Resdg. ex India—Col., 125 Rifles.
	Hamps. R.	Shaw, David George Levinge (50, 66a, 170) . . .	1 July 60	21 June 79	29 Sep. 82	20 Mar. 14	. . . . .	1A	Resdg. ex India.
	Sea. Higrs.	*Macintyre, Donald Chas. Fredk., C.B. (144, 153, 156, 170)	17 Apl. 59	13 Aug. 79	1 Jan. 82	4 May 14	. . . . .	1A	
	Leins. R.	Tighe, Sir Michael Joseph, K.C.M.G., C.B., C.I.E., D.S.O. (72b, 162, 170)	21 May 64	25 Aug. 83	27 June 85	3 June 15	. . . . .	1A	Divl. Comdr.
	R. Berks R.	Delamain, Sir Walter Sinclair, K.C.M.G., C.B., D.S.O. (50, 71)	18 Feb. 62	22 Oct. 81	1 Feb. 85	3 June 15	. . .	1A	
	Glouc. R.	Stewart, James Marshall, C.B. (50, 157, 162, 169) .	9 Aug. 61	22 Oct. 81	28 Feb. 83	3 June 15	. . . . .	1A	Bde. Comdr.
10	Leins. R.	Bingley, The Hon. Sir Alfred Horsford, K.C.I.E., C.B. (162)	28 May 65	7 Feb. 85	9 Aug. 87	3 June 15	. . . . .	1A	Secy., Army Dept.

* Supernumerary.

Major-Generals—*contd.*

No	Former Corps.	Names.	Date of birth.	Date of first commission.	Date of appointment to Indian Army.	Date of rank.	Date of Colonel's allowance.	ℱ. rule.	Remarks.
	Leins. R.	**Cowper, Maitland*, C.B., C.I.E. (65, 161, 166, 170) .	24 *Dec.* 59	11 *Aug.* 80	13 *Oct.* 84	27 *Aug.* 15	. . . .	IA	Col., 39 Horse.
	Worc. R.	**Watson, William Arthur*, C.B., C.M.G., C.I.E., *p.s.c.* (162)	25 *Sep* 60	11 *Aug.* 80	23 *June* 82	7 *Sep.* 15	. . . .	IA	Col., 10 Lancers.
	K. O. S. B.	**Wadeson, Frederick William George*, C.B. (50, 93) .	12 *July* 60	17 *Apl.* 80	17 *June* 82	28 *Oct.* 15	. . . .	IA	
	Suff. R.	*VC *Walker, William George*, C.B. (50, 156) . .	28 *May* 63	29 *Aug.* 85	28 *May* 87	1 *Jan.* 16	. . . .	IA	
11	S. W. Bord.	VC **Cobbe,** *Sir* Alexander Stanhope, K.C.B., C.S.I., D.S.O., *p.s.c.* (50, 52) (*tempy. Lt.-Genl.*)	5 June 70	21 Sep. 89	12 Aug. 92	3 June 16	. . . .	—	
	Conn. Rang.	***Barrow,** *Sir* George deSymons, K.C.M.G., C.B., *p.s.c.* (50) [L] (*temp. Lt.-Genl.*)	25 Oct. 64	23 Aug. 84	28 Apl. 86	3 June 16	. . . .	—	Col., 25 Horse.
	E. Surr. R.	**Hunter, John Gunning*, C.B. (162)	1 *Nov.* 59	1 *May* 78	20 *Oct.* 80	30 *Aug.* 16	. . . .	IA	
	R. A.	**Malleson,** Wilfrid, C.B., C.I.E (162, 166, 170) . .	8 Sep. 66	17 Feb. 86	1 Jan. 04	20 Sep. 16	. . . .	—	
	Manch. R.	**Fane,** *Sir* Vere Bonamy, K.C.I.E., C.B. (50, 170) . . *Leave without pay in extension for 11 Mar. 03.*	16 June 63	12 Nov. 84	30 Mar. 88	1 Jan. 17	. . . .	IA	Col., 21 Punjabis.
15	R. Berks R.	**Barratt,** William Cross, C.B., D.S.O. (50, 57, 71, 169)	1 June 62	5 Dec. 83	11 Nov. 86	1 Jan. 17	. . . .	IA	Divl. Comdr.
	Hamps. R.	**Ewart,** *Sir* Richard Henry, K.C.M.G., C.B., C.I.E., D.S.O. (165, 170)	26 Dec. 64	25 Aug. 83	27 Mar. 86	1 Jan. 17	. . . .	IA	
	R. Suss. R.	**Beynon,** *Sir* Wm. Geo. Lawrence, K.C.I.E., C.B., D.S.O. (50, 53, 72*b*, 156)	5 Nov. 66	5 Feb. 87	11 Dec. 89	1 Jan. 17	. . . .	—	Divl. Comdr.
	S. W. Bord.	**Brooking,** *Sir* Harry Triscott, K.C.B., K.C.M.G. (50, 57, 195) [L]	13 Jan. 64	6 Feb. 84	27 Oct. 85	1 Jan. 17	. . . .	IA	Col., 81 Pioneers.
	Ches. R.	**Woodyatt,** Nigel Gresley, C.B. (50, 57, 71, 157) .	30 Mar. 61	12 May 83	22 Mar. 86	1 Jan. 17	. . . .	—	Bde. Comdr. Col., 7 Gurkha Rifles.
20	E. Surr. R.	**Eustace,** Alexander Henry, C.B., D.S.O. (50, 53, 57, 170.	1 June 63	6 May 85	13 July 87	1 Jan. 17	. . . .	IA	Bde. Comdr.
	Worc. R.	***Knight,** Wyndham Chas., C.B., C.S.I., D.S.O., *p.s.c.* (50, 65, 66*a*, 72*b*, 162) [L]	30 Nov. 63	25 Aug. 83	29 Apl. 85	1 Jan. 17	. . . .	—	Bde. Comdr.
	Linc. R.	***Macandrew,** Henry John Milnes, C.B., D.S.O., *q.s.* (50, 71)	7 Aug. 66	10 Nov. 86	30 Mar. 88	1 Jan. 17	. . . .	—	
	R. Ir. R.	***Shea,** John Stuart Mackenzie, C.B., C.M.G., D.S.O., *p.s.c.*, *q.s.* (50, 162, 170)	17 Jan. 69	11 Feb. 88	24 Nov. 91	1 Jan. 17	. . . .	—	
	N. Lan. R.	**Fowler,** Francis John, C.B., D.S.O. (50, 71, 169) . .	31 July 64	25 Aug. 83	13 Nov. 85	23 Oct. 17	. . . .	IA	Bde. Comdr.
25	Worc. R.	**Hardy,** Thomas Henry (50, 71)	25 Nov. 63	25 Aug. 83	24 Nov. 88	3 Dec. 17	. . . .	IA	Bde. Comdr.

26	Devon. R.	**Fowler,** Charles Astley, C.B., D.S.O. (50, 162, 165, 170)	9 Nov. 65	7 Feb. 85	16 Aug. 86	6 Apl. 18	. . . .	IA	Bde. Comdr.
	R. Suss. R.	**Dunsterville,** Lionel Charles, C.B. [L] (50, 53, 124*a*, 162, 166, 169)	9 Nov. 65	23 Aug. 84	18 Feb. 87	3 June 18	. . . .	IA	
	Welsh R.	***Hendley,** Charles Edward, C.B. (50, 53, 65, 66*a*, 71, 195)	26 Dec. 63	12 Nov. 84	15 Oct. 86	3 June 18	. . . .	IA	Bde. Comdr.
	Derby. R.	***James,** William Bernard, C.B., C.I.E., M.V.O. (50, 162, 170)	8 Feb. 65	30 Jan. 86	5 Dec. 87	3 June 18	. . . .	—	
	Bord. R.	***Climo,** Skipton Hill, C.B., D.S.O. (50, 53, 162, 170) .	24 Dec. 68	1 Feb. 88	14 Sept. 89	3 June 18	. . . .	—	Bde Comdr.
30	Oxf. L. I.	***Hill,** John, C.B., D.S.O., (167)	14 Jan. 66	5 Feb. 87	11 Feb. 88	3 June 18	. . . .	—	Dy. Chief of the Genl. Staff.
	K. O. Sco. Bord.	***Skeen,** Andrew, C.M.G., *p.s.c.* (50. 53, 66*a*, 72*b*, 162, 166, 169)	20 Jan. 73	5 Dec. 91	2 Oct. 93	3 June 18	. . . .	—	
	Ches. R.	***Palin,** Philip Charles, C.B., C.M.G. (50, 53, 167) . .	8 Aug. 64	28 Apl. 86	15 June 88	3 June 18	. . . .	—	Mily. Secy. to C.-in C.
	R. Ir. Fus	***Scott,** Thomas Edwin, C.B., C.I.E. D.S.O., (50, 162, 170)	6 Mar. 67	9 May 88	17 Apl. 90	3 June 18	. . . .	—	
35	Norf. R.	***Richardson,** Charles William Grant, C.S.I., *p.s.c.*, (50, 72*b*, 170) [L]	27 Oct. 68	11 Feb. 88	6 Oct. 93	3 June 18	. . . .	—	Dy. Quartermaster-General.

* Supernumerary.

Colonels and Lieutenant-Colonels in receipt of Colonel's Allowances.

No.	Former Corps.	Names.		Date of birth.	Date of first commission.	Date of appointment to Indian Army.	Date of Rank. Substantive Lt.-Colonel.	Date of Rank. Army.	Date of Colonel's Allowance.	F. rule.	Remarks.
	31 L. I.	𝔐 St. Maur-Wynch, Henry St. Maur		23 Dec. 33	12 June 52	12 Sep. 66	12 June 78	c. 12 June 82	12 June 90	—	u. s. l.
	2 M. I.	Griffith, Richard (50)		1 Feb. 33	13 June 52	4 Nov. 61	13 June 78	c. 13 June 82	13 June 90	—	u. s. l.
	23 N. I.	𝔐 Ross, William Henry (65) ●		23 Mar. 35	20 Dec. 52	18 Feb. 61	20 Dec. 78	c. 20 Dec. 82	20 Dec. 90	68	u. s. l.
	21 M. I.	Clarke, Tredway George		10 June 35	26 Jan. 53	18 Feb. 61	26 Jan. 79	c. 26 Jan. 83	26 Jan. 91	—	u. s. l.
	1 Lan.	𝔐 Anderson, Charles John ●		2 Sep. 35	20 May 53	18 Feb. 61	20 May 79	c. 20 May 83	20 May 91	—	u. s. l.
	47 M. I.	Pearse, Charles John		20 July 34	20 June 53	18 Feb. 61	20 June 79	c. 20 June 83	20 June 91	68	u. s. l.
	48 N. I.	deKantzow, Charles Adolphus	PH	27 June 35	14 Sep. 53	28 Feb. 61	14 Sep. 79	c. 14 Sep. 83	14 Sep. 91	—	u. s. l.
	R. A. (B)	𝔐 Sconce, James	PH	2 Apl. 36	9 Dec. 53	8 Mar. 69	9 Dec. 79	c. 9 Dec. 83	9 Dec. 91	68	u. s. l.
	R. A. (B)	𝔐 Malloch, Henry Archibald	PH	9 Apl. 34	9 Dec. 53	17 Oct. 67	9 Dec. 79	c. 9 Dec. 83	9 Dec. 91	68	u. s. l.
	47 M. I.	Lugard, Henry Ibbetson	H	8 Jan. 37	14 Dec. 53	18 Feb. 61	14 Dec. 79	c. 14 Dec. 83	14 Dec. 91	—	u. s. l.
	19 M. I.	𝔐 Plowden, Charles Henry	H	22 Feb. 37	14 Feb. 54	12 Sep. 66	14 Feb. 80	c. 14 Feb. 84	14 Feb. 92	68	u. s. l.
	2 E. B. F.	𝔙ℭ 𝔐 Cadell, Thomas, C.B.	PH	5 Sep. 35	17 Apl. 54	18 Feb. 61	17 Apl. 80	c. 17 Apl. 84	17 Apl. 92	—	u. s. l.
	63 N. I.	Gordon, Henry Wilson (52)	PH	21 May 37	6 May 54	18 Feb. 61	6 May 80	c. 6 May 84	6 May 92	—	u. s. l.
	2 E. L. I.	Preston, Barnard Hughes		13 Feb. 36	10 June 54	12 Sep. 66	10 June 80	c. 10 June 84	10 June 92	68	u. s. l.
	54 N. I.	𝔐 Vibart, Edward Daniel Hamilton	PH	20 Aug. 37	4 Sep. 54	12 Sep. 66	4 Sep. 80	c. 4 Sep. 84	4 Sep. 92	—	u. s. l.
	46 N. I.	𝔐 Boyle, Hon. Robert Elphinstone	PH	3 June 37	10 Dec. 54	12 Sep. 66	13 Dec. 80	c. 10 Dec. 84	10 Dec. 92	—	u. s. l.
	Bo. S. C.	𝔐 Barnes, Osmond, C.B. (57)	H	23 Dec. 34	4 Mar. 55	18 Feb. 61	4 Mar. 81	c. 4 Mar. 85	4 Mar. 93	—	u. s. l.
	19 M. I.	𝔐 McInroy, Charles, C.B.		3 Mar. 38	17 Mar. 55	12 Sep. 66	17 Mar. 81	c. 17 Mar. 85	17 Mar. 93	68	u. s. l.
	56 N. I.	Matthews, Charles Richard (74)	PH	11 Nov. 38	20 Mar. 55	18 Feb. 61	20 Mar. 81	c. 20 Mar. 85	20 Mar. 93	—	u. s. l.
	42 M. I.	𝔐 Wright, Henry Charles (191, 198)	H	30 May 35	4 Apl. 55	30 Sep. 64	4 Apl. 81	c. 4 Apl. 85	4 Apl. 93	—	u. s. l.
	12 M. I.	Smith, John MacDonald		9 Aug. 37	20 Apl. 55	8 Nov. 61	20 Apl. 81	c. 20 Apl. 85	20 Apl. 93	—	u. s. l.
	56 F.	Swinhoe, Charles	H	29 Aug. 38	27 July 55	1 Oct. 63	27 July 81	c. 27 July 85	27 July 93	—	u. s. l.
	3 L. I.	Crawfurd, Hugh Patrick Rolston Francis (60)		27 Aug. 35	1 Sep. 55	19 Feb. 63	1 Sep. 81	c. 1 Sep. 85	1 Sep. 93	—	u. s. l.
	47 M. I.	Weldon, Francis		30 July 36	26 Sep. 55	12 Sep. 66	26 Sep. 81	c. 26 Sep. 85	26 Sep. 93	—	u. s. l.
	45 M. I.	Ewing, John		13 Mar. 36	7 Dec. 55	18 Feb. 61	7 Dec. 81	c. 7 Dec. 85	7 Dec. 93	—	u. s. l.
	26 M. I.	𝔐 McNeill, Duncan		25 June 36	8 Dec. 55	12 Sep. 66	8 Dec. 81	c. 8 Dec. 85	8 Dec. 93	—	u. s. l.
	1 E. B. F.	𝔐 Chapman, Arthur Robert	PH	26 Oct. 36	4 Jan. 56	18 Feb. 61	4 Jan. 82	c. 4 Jan. 86	4 Jan. 94	—	u. s. l.

16 N. I.	Young, George Gordon	PH	5 Jan. 37	13 June 56	18 Feb. 61	13 June 82	c. 13 June 86	13 June 94	—	u. s. l.
13 N. I.	𝔐 Tweedie, Maurice		18 Mar. 38	20 Sep. 56	18 Feb. 61	20 Sep. 82	c. 20 Sep. 86	20 Sep. 94	68	u. s. l.
40 F.	Humfrey, Frederick Thomas (57) ●		25 Jan. 36	26 Sep. 56	16 May 68	26 Sep. 82	c. 22 Nov. 83	26 Sep. 94	—	u. s. l.
29 N. I.	Jopp, John, C.B. (Hony. Brig.-Genl.)	H	13 Mar. 40	20 Oct. 56	7 May 63	20 Oct. 82	c. 20 Oct. 86	20 Oct. 94	68	u. s. l.
3 E. R.	𝔐 Jarrett, Henry Sullivan, C.I.E. (169, 203)		17 June 39	13 Dec. 56	18 Feb. 61	13 Dec. 82	c. 13 Dec. 86	13 Dec. 94	68	u. s. l.
16 N. I.	𝔐 Grey, Leopold John Herbert, C.S.I. (98)	PH	1 July 40	13 Dec. 56	18 Feb. 61	13 Dec. 82	c. 13 Dec. 86	13 Dec. 94	68	u. s. l.
51 N. I.	𝔐 Pratt, Henry Marsh, C.B. (162) ●	PH	24 Oct. 38	13 Dec. 56	18 Feb. 61	13 Dec. 82	c. 2 Mar. 85	13 Dec. 94	—	u. s. l.
2 E. B. F.	𝔐 Cantor, Charles Henry	H	21 Nov. 36	13 Dec. 56	12 Sep. 66	13 Dec. 82	c. 13 Dec. 86	13 Dec. 94	—	u. s. l.
22 N. I.	Jacob, George Adolphus (152, 177)		21 Aug. 40	13 Dec. 56	27 May 61	13 Dec. 82	c. 13 Dec. 86	13 Dec. 94	—	u. s. l.
27 M. I.	𝔐 Porteous, Charles Arkcol, C.I.E.		27 July 39	13 Dec. 56	12 Sep. 66	13 Dec. 82	c. 13 Dec. 86	13 Dec. 94	68	u. s. l.
21 N. I.	‡ 𝔐 Adam, Fred. John Stuart (17)		22 Sep. 36	20 Dec. 56	19 Feb. 61	20 Dec. 82	c. 20 Dec. 86	20 Dec. 94	—	u. s. l.
18 N. I.	𝔐 Wilson, William Henry (131, 152)	H	13 Sep. 39	20 Dec. 56	18 Feb. 61	20 Dec. 82	c. 20 Dec. 86	20 Dec. 94	68	u. s. l.
47 M. I.	𝔐 Furlong, Malcolm	H	11 Aug. 39	4 Jan. 57	6 Dec. 62	4 Jan. 83	c. 4 Jan. 87	4 Jan. 95	—	u. s. l.
53 N. I.	Beddy, Edwin	PH	11 Jan. 37	20 Feb. 57	18 Feb. 61	20 Feb. 83	c. 20 Feb. 87	20 Feb. 95	—	u. s. l.
8 M. I.	Hesketh, Robert Walter		8 Sep. 37	20 Mar. 57	12 Sep. 66	20 Mar. 83	c. 20 Mar. 87	20 Mar. 95	—	u. s. l.
46 M. I.	Gordon, James Henry, C.B., D.S.O.		25 Jan. 39	20 Apl. 57	12 Sep. 66	20 Apl. 83	c. 20 Apl. 87	20 Apl. 95	—	u. s. l.
14 N. I.	Codrington, George Henry Forbes (110)		29 May 37	13 June 57	18 Feb. 61	13 June 83	c. 13 June 87	13 June 95	—	u. s. l.
28 N. I.	Scott, Thomas Augustus	PH	29 Aug. 39	13 June 57	14 Sep. 65	13 June 83	c. 13 June 87	13 June 95	68	u. s. l.
52 M. I.	Kilgour, Frederick		9 Apl. 41	11 July 57	12 Sep. 66	11 July 83	c. 11 July 87	11 July 95	68	u. s. l.
1 M. F.	𝔐 Luxmoore, Coryndon Thomas Putt		16 Aug. 39	4 Aug. 57	12 Sep. 66	4 Aug. 83	c. 4 Aug. 87	4 Aug. 95	—	u. s. l.
33 F.	Philips, James (57)		4 Dec. 37	27 Nov. 57	23 May 64	27 Nov. 83	c. 27 Nov. 87	27 Nov. 95	—	u. s. l.
30 Hus.	Hogg, Theodore William	PH	30 July 41	4 Dec. 57	28 Dec. 64	4 Dec. 83	c. 4 Dec. 87	4 Dec. 95	—	u. s. l.
1 M. F.	𝔐 Tabuteau, Thomas Rooke	PH	7 Apl. 38	11 Dec. 57	12 Sep. 66	11 Dec. 83	c. 11 Dec. 87	11 Dec. 95	68	u. s. l.
R. A.	Grierson, John	PH	24 Aug. 39	12 Dec. 57	10 Feb. 64	12 Dec. 83	c. 12 Dec. 87	12 Dec. 95	—	u. s. l.
R. A. (B)	Wilkins, William Henry		6 Oct. 41	12 Dec. 57	11 May 64	12 Dec. 83	c. 12 Dec. 87	12 Dec. 95	68	u. s. l.
10 M. I.	Shaw, Elphinstone		28 Sep. 41	12 Dec. 57	14 June 64	12 Dec. 83	c. 12 Dec. 87	12 Dec. 95	—	u. s. l.
3 M. E. R.	𝔐 Chrystie, George		7 Sep. 41	12 Dec. 57	12 Sep. 66	12 Dec. 83	c. 12 Dec. 87	12 Dec. 95	—	u. s. l.
65 N. I.	𝔐 Macpherson, James Duncan	PH	16 May 41	19 Dec. 57	7 Aug. 62	19 Dec. 83	c. 19 Dec. 87	19 Dec. 95	68	u. s. l.
24 N. I.	𝔐 Jacob, Herbert Bruce (110, 131, 152, 204)		30 July 41	20 Dec. 57	19 Feb. 61	20 Dec. 83	c. 20 Dec. 87	20 Dec. 95	—	u. s. l.
38 M. I.	Eyre, Edmund Henry, C.B. (50) ●		26 Nov. 38	4 Mar. 58	10 Dec. 64	4 Mar. 84	c. 1 July 87	4 Mar. 96	—	u. s. l.
37 M. I.	Michell, James William Abbot		22 Nov. 40	4 Apl. 58	12 Sep. 66	4 Apl. 84	c. 4 Apl. 88	4 Apl. 96	68	u. s. l.
Artv.	Trevor, George Herbert, C.S.I.		29 Jan. 40	12 June 58	25 June 62	12 June 84	c. 12 June 88	12 June 96	68	u. s. l.
34 L.	Hatchell, David Thompson		2 Apl. 39	12 June 58	12 Sep. 56	12 June 84	c. 12 June 88	12 June 96	—	u. s. l.
66 N. I.	Young, George (50, 60)	PH	12 Aug. 38	20 Aug. 58	28 Oct. 65	20 Aug. 84	c. 20 Aug. 88	20 Aug. 96	68	u. s. l.
41 M. I.	Rowlandson, Michael Alexander		12 Nov. 41	20 Aug. 58	21 Nov. 61	20 Aug. 84	c. 20 Aug. 88	20 Aug. 96		u. s. l.
20 M. I.	Goldie, James Ord		8 May 41	4 Sep. 58	12 Sep. 66	4 Sep. 84	c. 4 Sep. 88	4 Sep. 96	—	u. s. l.
107 F.	Willis, James L. N. (57, 60)		29 Apl. 41	13 Nov. 58	24 Dec. 66	13 Nov. 84	c. 13 Nov. 88	13 Nov. 96	—	u. s. l.
R. A.	Major, Francis Ward (110, 152)		12 Aug. 40	10 Dec. 58	23 Apl. 65	10 Dec. 84	c. 10 Dec. 88	10 Dec. 96	—	u. s. l.
16 N. I.	Lane, Clayton Turner, C.I.E.	PH	20 Feb. 42	11 Dec. 58	14 Apl. 62	11 Dec. 84	c. 11 Dec. 88	11 Dec. 99	68	u. s. l.
65 N. I.	Thomas, Robert Mosely Bryce	PH	14 Sep. 40	6 Jan. 59	12 Sep. 66	6 Jan. 85	c. 6 Jan. 89	6 Jan. 97	—	u. s. l.
37 M. I.	Bruce, Elliott Armstrong (50) ●	H	9 Aug. 40	6 Jan. 59	26 Aug. 63	6 Jan. 85	c. 1 July 87	6 Jan. 97	—	u. s. l.
91 F.	Bromhead, Sir Benjamin Parnell, Bart., C.B.		24 Oct. 38	18 Jan. 59	23 June 66	18 Jan. 85	c. 18 Jan. 89	18 Jan. 97	—	u. s. l.
58 F.	Bredin, Alexander		10 Oct. 37	11 Mar. 59	23 Feb. 69	11 Mar. 85	c. 11 Mar. 89	11 Mar. 97	—	u. s. l.
R. A.	Hancock, George Edward (131, 152)	H	28 Jan. 41	10 June 59	12 Oct. 63	10 June 85	c. 10 June 89	10 June 97	68	u. s. l.
R. A.	Wilmer, John Randal		20 Jan. 41	22 June 59	24 Aug. 69	22 June 85	c. 22 June 89	22 June 97	68	u. s. l.
G. L. I.	Peart, George Richard		17 July 40	12 Aug. 59	31 May 62	12 Aug. 85	c. 12 Aug. 89	12 Aug. 97	—	u. s. l.
21 Hus.	Pitcher, Duncan George		2 Nov. 39	3 Sep. 59	30 June 68	3 Sep. 85	c. 3 Sep. 89	3 Sep. 97	68	u. s. l.
38 F.	Skinner, George John, D.S.O. (50)	PH	16 June 41	16 Sep. 59	2 Mar. 65	16 Sep. 85	c. 16 Sep. 89	16 Sep. 97	—	u. s. l.
G. L. I.	Marshall, Charles Henry Tilson	PH	8 July 41	12 Oct. 59	19 Oct. 67	12 Oct. 85	c. 12 Oct. 89	12 Oct. 97	68	u. s. l.
G. L. I.	Campbell, John Edward (161)	PH	3 Sep. 42	10 Dec. 59	16 Apl. 67	10 Dec. 85	e. 10 Dec. 89	10 Dec. 97	68	u. s. l.
G. L.	Scott, Walter (50, 131, 152)		3 Jan. 43	12 Mar. 60	1 Jan. 64	12 Mar. 86	c. 12 Mar. 90	12 Mar. 98	—	u. s. l.
R. A. (B)	Rivaz, Vincent, C.B. (161, 171)	PH	26 June 42	8 June 60	12 Sep. 70	8 June 86	c. 8 June 90	8 June 98	—	u. s. l.

Colonels and Lieutenant-Colonels in receipt of Colonel's Allowance—*concld.*

No.	Former Corps.	Names.	Date of birth.	Date of first commission.	Date of appointment to Indian Army.	Date of Rank. Substantive Lt.-Colonel.	Date of Rank. Army.	Date of Colonel's Allowance.	F. rule.	Remarks.
	5 F.	*Money, Elliott Alexander,* C.B. (57)	16 *Aug.* 43	16 *July* 60	22 *Sep.* 69	16 *July* 86	c. 16 *July* 90	16 *July* 98	—	*u. s. l.*
	R. A.	*King-Harman, Montague Jocelyn*	1 *Mar.* 43	2 *Oct.* 60	3 *Oct.* 71	2 *Oct.* 86	c. 2 *Oct.* 90	2 *Oct.* 98	—	*u. s. l.*
	G. L. I.	*Shepherd, Charles Edward* . . . PH	22 *Mar.* 43	19 *Dec.* 60	8 *May* 66	19 *Dec.* 86	c. 19 *Dec.* 90	19 *Dec.* 98	68	*u. s. l.*
	G. L. I.	*Temple, John Alexander* . . . PH	4 *Apl.* 42	19 *Dec.* 60	15 *Sep.* 66	19 *Dec.* 86	c. 19 *Dec.* 90	19 *Dec.* 98	68	*u. s. l.*
	G. L.	*Tandy, Henry Stratford* ⊛	30 *June* 43	12 *June* 61	13 *Sep.* 67	12 *June* 87	c. 29 *Aug.* 93	12 *June* 99	—	*u. s. l.*
	106 F.	*McNair, Alfred Lionel* . . . (204)	12 *Oct.* 43	27 *Aug.* 61	9 *Oct.* 64	27 *Aug.* 87	c. 29 *Aug.* 93	27 *Aug.* 99	—	*u. s. l.*
	6 F.	*Carey, Horace Richard Le Marchant* (136, 204)	28 *June* 42	15 *Jan.* 61	6 *Dec.* 70	15 *Jan.* 87	. . . .	13 *Feb.* 00	68	*u. s. l.*
	54 E.	*Bishop, Edward Barry* . . . PH	28 *Sep.* 43	2 *Dec.* 62	2 *Dec.* 65	2 *Dec.* 88	c. 29 *Aug.* 93	2 *Dec.* 00	IA	*u. s. l.*
	79 F.	*Turner, Augustus Henry,* C.B. (57) ⊛	13 *Oct.* 42	24 *May* 61	7 *May* 68	24 *May* 87	c. 1 *Sep.* 91	17 *Nov.* 00	IA	*u. s. l.*
	82 F.	*Hutchinson, James Bird,* C.S.I. . . . PH	22 *Mar.* 44	17 *July* 63	17 *July* 66	17 *July* 89	. . .	17 *July* 01	68	*u. s. l.*
	R. A.	*Hunter, John Muir,* C.S.I. (131)	31 *Mar.* 44	12 *Apl.* 64	15 *May* 69	12 *Apl.* 90	. . .	12 *Apl.* 02	68	*u. s. l.*
	56 F.	*Leckie, Fred. William Vans* ⊛	4 *Apl.* 47	5 *July* 04	31 *Oct.* 77	5 *July* 90	c. 26 *July* 94	5 *July* 02	IA	*u. s. l.*
	59 F.	*Jones, William Henry Dawes*	4 *Nov.* 44	26 *July* 64	7 *June* 71	16 *July* 90	. . .	7 *May* 03	IA	*u. s. l.*
	G. L. I.	V.C. *Hammond,* Sir *Arthur George,* K.C.B., D.S.O. ⊛ *H.*	28 *Sep.* 43	7 *June* 61	10 *June* 67	7 *June* 87	c. 12 *Feb.* 90	8 *Sep.* 08	IA	*u. s. l.*
	R. A.	*Swinton-Skinner, Evelyn* (50)	29 *Jan.* 43	4 *Jan.* 60	22 *June* 69	4 *Jan.* 86	c. 4 *Jan.* 90	7 *May* 09	68	*u. s. l.*
	G. L. I.	*Sandeman, John Edward* . . . PH	3 *Sep.* 43	7 *June* 61	9 *May* 67	7 *June* 87	c. 29 *Aug.* 93	28 *Aug.* 09	IA	*u. s. l.*
	G. L. I.	*Brownlow, Celadon Charles,* C.B. ⊛ . . PH	20 *Nov.* 43	8 *June* 61	8 *Mar.* 67	8 *June* 87	c. 1 *Sep.* 91	27 *June* 10	IA	*u. s. l.*
	19 Hus.	M *Biddulph, John* . . . PH	25 *July* 40	20 *Jan.* 58	16 *Aug.* 77	22 *Oct.* 87	c. 24 *Feb.* 8[illegible]	26 *June* 14	IA	*u. s. l.*
	12 F.	*Little, John Albert*	7 *Feb.* 40	18 *Mar.* 62	2 *Mar.* 67	18 *Mar.* 88	. . . .	22 *Mar.* 15	68	*u. s. l.*
	59 F.	*Forjett, Frederick Hutchinson* (131, 152)	28 *Jan.* 42	10 *June* 65	30 *Mar.* 70	10 *June* 91	. . . .	31 *May* 16	IA	*u. s. l.*
	21 F.	*Melliss,* Sir *Howard,* K.C.S.I. ⊛	2 *Apl.* 47	30 *June* 65	31 *Mar.* 69	30 *June* 91	c. 30 *June* 95	23 *Nov.* 16	IA	*u. s. l.*
	2 D. G.	*Newill, Joseph Henry*	12 *Sep.* 44	4 *July* 65	7 *Nov.* 68	4 *July* 91	. . . .	7 *Dec.* 16	68	*u. s. l.*
	21 F.	*Robertson,* Sir *Donald,* K.C.S.I.	24 *June* 47	8 *Aug.* 65	8 *Aug.* 67	8 *Aug.* 91	. . . .	28 *Apl.* 17	68	*u. s. l.*
	19 F.	*Rolland Stuart Erskine* (50, 57, 65) (*Hony. Brig.-Genl.*)	7 *Nov.* 46	9 *Nov.* 15	9 *Dec.* 68	9 *Nov.* 92	c. 29 *Aug.* 93	8 *May* 18	IA	*u. s. l.*—Col., 86 Infantry.

Colonels and Lieutenant-Colonels—602.

(The names are arranged in order of seniority according to substantive rank in the Indian Army.)

94 F.	Kelly, James Graves, C.B. (50, 57, 170)	28 Nov. 43	11 Sep. 63	30 Aug. 67	11 Sep. 89	c. 10 July 95	IA	u. s. l
108 F.	Cooke, Crawford Boyd	17 Mar. 44	8 Mar. 64	3 Apl. 67	8 Mar. 90	. . . .	68	u. s. l.
3 F.	Rennick, Albert de Claney (136)	2 May 44	16 July 64	18 July 68	6 July 90	. . . .	68	u. s. l.
71 F.	Martelli, Norton Charles	13 May 43	31 Oct. 64	1 Oct. 70	11 Oct. 90	. . . .	68	u. s. l.
46 F.	Grant, James	17 Sep. 47	30 June 65	4 Aug. 68	30 June 91	c. 30 June 95	IA	u. s. l.
95 F.	Kirkwood, James Nicholson Soden (59)	4 Jan. 46	1 Sep. 65	10 Aug. 70	1 Sep. 91	c 1 Sep. 95	IA	u. s. l
25 F.	Cuninghame, David Stanley (50, 57, 136)	5 June 47	1 Dec. 65	1 Oct. 69	1 Dec. 91	c. 29 Oct. 98	IA	u. s l.
109 F.	Salmon, William Henry	2 Feb. 48	12 Jan. 66	8 Sep. 71	12 Jan. 92	c. 12 Jan. 96	IA	u. s. l.
49 F.	Wetherall, William Alexander	8 Apl. 47	30 Jan. 66	8 Dec. 69	30 Jan. 92	c. 30 Jan. 96	IA	u. s. l
19 F.	Butler, James	4 Mar. 46	20 Feb. 66	9 July 68	20 Feb. 92	. . . .	68	u. s. l.
41 F.	Money-Simons, John James (50, 74, 136)	26 Sep. 44	11 Jan. 67	2 Jan. 71	11 Jan. 93	. . . .	IA	u. s. l.
R. A.	Talbot, Sir Adelbert Cecil, K.C.I.E.	3 June 45	15 Jan. 67	1 Aug. 69	15 Jan. 93	. . . .	68	u. s. l.
19 Hus.	Aberigh-Mackay, Jas. Livingstone (50, 57, 161)	6 June 44	13 Feb. 67	10 Sep. 69	13 Feb. 93	. . . .	IA	u. s. l.
8 F.	Macbay, William George Wolfe (94, 200)	23 Feb. 45	23 Mar. 67	13 July 70	23 Mar. 93	. . . .	IA	u. s. l.
55 F.	Walker, John Newman	22 Aug. 47	20 July 67	17 Aug. 70	20 July 93	c. 20 July 97	IA	u. s. l.
49 F.	Yate, Chas. Edward, C.S.I., C.M.G.	28 Aug. 49	9 Nov. 67	22 Mar. 71	9 Nov. 93	c. 13 Jan. 01	IA	u. s. l.
109 F.	Gordon, Robert	18 Jan. 50	8 Jan. 68	3 July 74	8 Jan. 94	c. 8 Jan. 98	IA	u. s. l.
37 F.	Abbott, Henry Alexius, C.B. (50, 57, 166) (Hony. Brig.-Genl.)	22 Jan. 49	15 Feb. 68	18 Nov. 70	15 Feb. 94	c. 15 Feb. 98	IA	u. s. l.—Col., 15 Sikhs
24 F.	Pryce, Douglas Davidson (50)	23 May 48	8 July 68	15 July 71	8 July 94	. . . .	IA	u. s. l.
106 F.	Maxwell, Henry St. Patrick, C.S.I. (92)	17 Mar. 50	6 Feb. 69	27 Apl. 71	6 Feb. 95	. . . .	68	u. s. l.
76 F.	Thomas, Henry Robert Dacres	9 Dec. 48	10 Feb. 69	3 Mar. 73	10 Feb. 95	c. 10 Feb. 99	IA	u. s. l.
62 F.	Clibborn, John, C.I.E.	8 Dec. 47	14 July 69	20 Mar. 72	14 July 95	. . . .	68	u. s. l.
104 F.	Ferris, William Butler (152)	29 Oct. 51	18 Aug. 69	7 Oct. 71	18 Aug. 95	. . . .	IA	u. s. l.
5 Lanc.	Garbett, Charles Henry Vincent (166, 171)	19 Mar. 49	17 Nov. 69	14 Oct. 73	17 Nov. 95	c. 11 July 01	IA	u. s. l.
5 Lanc.	Fancourt, St. John Michell, C.B.	30 Sep. 47	1 Dec. 69	10 Nov. 73	1 Dec. 95	c. 26 Mar. 95	—	u. s. l.
19 F.	Hawkes, George (54, 156)	10 Oct. 51	19 Feb. 70	12 Feb. 74	19 Feb. 96	c. 19 Feb. 00	IA	u. s. l.
65 F.	Wheatley, Henry Spencer, C.B. (50, 20)	16 May 51	3 Aug 70	3 July 74	3 Aug. 96	c. 24 Oct. 00	IA	u. s. l.
17 F.	Maisey, Frederick Charles (170)	7 July 51	23 Sep. 71	25 Sep. 74	23 Sep. 97	. . . .	IA	u. s. l.
19 F.	Wingate, George, C.I.E. (50, 57)	21 Nov. 52	23 Sep. 71	7 Sep. 75	23 Sep. 97	c. 23 Sep. 01	IA	u. s. l.
11 Hus.	Macmullen, William Henry Fothergill (57, 74, 161)	13 June 48	28 Oct. 71	19 Apl. 75	28 Oct. 97	. . . .	IA	u. s. l.
14 F.	Hewett, William Setwood (30, 50, 57, 98)	30 Dec. 51	30 Dec. 71	25 June 78	30 Dec. 97	. . . .	IA	u. s. l.
49 F.	Rurdall, Frank Montagu, C.B., D.S.O. (156, 170)	18 May 51	9 Mar. 72	26 Nov. 74	9 Mar. 98	c. 8 June 02	IA	u. s. l.
70 F.	Pears, Thomas Caldwell (161, 169)	27 June 51	29 May 72	11 Feb. 76	29 May 98	. . . .	IA	u. s. l.
65 F.	Lean, William Walter (50)	19 June 53	11 Sep. 72	4 Aug. 76	11 Sep. 98	. . . .	IA	u. s. l.
8 F.	Money, George Alfred (57)	24 Apl. 53	11 Sep. 72	17 Sep. 75	11 Sep. 98	c. 20 Nov. 02	IA	u. s. l.
83 F.	Read, Hastings, C.B. (Hony. Brig.-Genl.)	19 Aug. 52	15 Jan. 73	28 Aug. 76	15 Jan. 99	c. 15 Jan. 03	IA	u. s. l.
51 F.	Gartside-Tipping, Robert Francis, C.B.	17 Aug. 52	3 Aug. 73 3 Aug. 72*	15 June 76	3 Aug. 99	. . . .	IA	u. s. l.
43 F.	Lawford, Eustace Edward Melville	11 Mar. 56	9 Aug. 73	17 Oct. 76	9 Aug. 99	c. 9 Aug. 02	IA	u. s. l.
83 F.	Cadell, Hew Francis	24 Nov. 52	12 Feb. 74 12 Feb. 73*	5 July 75	12 Feb. 00	. . . .	IA	u. s. l.
98 F.	Preston, Frederick George Service for Indian pension reckons from 10 June 74.	18 Nov. 55	28 Feb. 74	24 May 79	28 Feb. 00	. . . .	IA	u. s. l.
45 F.	Worlledge, John Franklin (57, 170, 204) [L]	16 Apl. 56	28 Feb. 74	2 Jan. 78	28 Feb. 00	c. 10 Feb. 04	IA	u. s. l.
70 F.	Jones, Alfred Edwin, p.s.c. (57, 127) [L]	6 Sep. 55	28 Feb. 74	13 Oct. 76	28 Feb. 00	. . . .	IA	u. s. l.
E. Sur. R.	Juxon-Jones, Juxon Henry (50)	1 Feb. 55	10 June 74	8 Mar. 79	10 June 00	. . . .	IA	u. s. l.
108 F.	Aslett, William Charles	12 Jan. 56	13 June 74	12 Sep. 76	13 June 00	. . . .	IA	u. s. l.

Colonels and Lieutenant-Colonels—*contd.*

No.	Former Corps.	Names.	Date of birth.	Date of first commission.	Date of appointment to Indian Army.	Date of Rank. Substantive Lt.-Colonel.	Date of Rank. Army.	F. rule.	Remarks.
	Glouc. R.	*Beckham, Lionel Grafton* (50)	14 Jan. 56	13 June 74	7 Apl. 79	13 June 00	. . . .	75	u. s. l.
	S. York R.	*Preston, Jenico Edward*, C.B., D.S.O.	30 Nov. 55	21 Sep. 74	20 Sep. 80	21 Sep. 00	. . . .	1A	u. s. l.
	8 F.	*Hawkins, Frederick* (57, 162)	18 Sep. 56	21 Sep. 74	22 Jan. 78	21 Sep. 00	c. 10 Feb. 04	1A	u. s. l.
	F.	*Presgrave, Edward Robert John*, C.B., D.S.O.	29 June 55	11 Feb. 75	19 Dec. 78	11 Feb. 01	c. 11 Feb. 04	1A	u. s. l.
1	R. Highrs.	Blyth, Frederick Augustus (57, 162)	20 Nov. 57	11 Feb. 75	26 Jan. 81	11 Feb. 01	c. 13 June 04	1A	Resdg. ex India.
	67 F.	*Jones, Richard Godfrey*, C.I.E.	26 May 55	13 June 75	7 Jan. 73	13 June 01	. . . .	1A	u. s. l.
	1 W. I. R.	*Lawson, Herbert* (104, 169)	31 Jan. 56	12 Aug. 75 12 Aug. 74*	15 Mar. 80	12 Aug. 01	. . . .	1A	u. s. l.
	89 F.	*Cook, Walter* (50, 107, 169)	19 Nov. 57	10 Sep. 75	5 Mar. 77	10 Sep. 01	. . . .	1A	u. s. l.
	21 F.	*Tewersham, Richard Kinlock*, D.S.O. (50)	6 Sep. 56	10 Sep. 75	3 July 77	10 Sep. 01	c. 10 Sep. 04	1A	u. s. l.
	1 W. I. R.	*Peirse, Charles Edward* (64, 109)	6 June 54	13 Nov. 75	19 Dec. 79	13 Nov. 01	c. 7 Mar. 05	1A	u. s. l.
	R. Mar.	*Baillie, Robert*	28 Jan. 57	27 Jan. 76	3 Oct. 80	27 Jan. 02	c. 27 Jan. 05	1A	u. s. l.
	R. Mar.	Gott, George Arthur	16 Sep. 57	23 Feb. 76	8 Dec. 80	23 Feb. 02	c. 23 Feb. 05	1A	Resdg. ex India.
	E. Surr. R.	*Lumsden, Henry Richmond William*	9 Sep. 53	6 Sep. 76	24 Sep. 79	6 Sep. 02	. . . .	1A	u. s. l.
	51 F.	*Monck-Mason, Henry Espine* (57, 106)	11 Feb. 56	10 Sep. 76 10 Sep. 75*	3 July 77	10 Sep. 02	. . . .	1A	u. s. l.
	12 F.	*Onslow, Richard Cranley* (50, 57, 74, 161, 169)	26 July 57	11 Sep. 76	7 June 79	11 Sep. 02	c. 11 Sep. 05	1A	u. s. l.
	1 W. I. R.	*Yate, Arthur Campbell* (72b, 95, 169)	18 Feb. 53	22 Sep. 76 22 Sep. 75*	10 Feb. 79	22 Sep. 02	. . . .	1A	u. s. l.
	8 F.	Balfour, Joseph Hume (50)	9 Apl. 57	6 Oct. 76 6 Oct. 75*	11 Mar. 80	6 Oct. 02	c. 23 Mar. 08	1A	Resdg. ex India.
4	40 F.	Maxwell, George William (50)	31 Dec. 57	11 Oct. 76	2 July 77	11 Oct. 02	c. 28 Oct. 05	1A	
	R. Ken. R.	*Middleton, Herbert John James*	31 Mar. 56	28 Oct. 76	1 Oct. 08	28 Oct. 02	c. 28 Oct. 05	1A	u. s. l.
	R. A.	*Meade, John William Babington* (162)	30 Dec. 56	25 Jan. 77	2 Apl. 81	25 Jan. 03	. . . .	1A	u. s. l.
	15 F.	*Aspinall, Hugh Harry Haworth* (50)	24 Dec. 56	22 Mar. 77 25 Mar. 76*	20 July 77	22 Mar. 03	. . .	1A	u. s. l.
	59 F.	*Gordon, Louis Augustus, C.B.*, (50, 64, 71)	28 May 57	14 June 77 14 June 76*	3 Dec. 77	14 June 03	c. 23 Oct. 06	1A	u. s. l.
		Service for pension from 27 Mar. 77; for promotion from 14 June 77.							
	63 F.	*Vansittart, Eden* (125, 156)	19 Apl. 56	10 Sep. 77 10 Sep. 75*	20 July 77	10 Sep. 03	c. 10 Sep. 06	1A	u. s. l.
	54 F.	*Jones, Frank Lewis* (50, 162) [L]	5 Aug. 56	10 Sep. 77 10 Sep. 75*	20 Apl. 77	10 Sep. 03	. . . .	1A	u. s. l.
		Service for pension from 6 Dec. 76; for promotion from 10 Sep. 77.							

5	65 F.	**Forbes**, Arthur William, C.B. (50)	17 Sep. 58	11 Sep. 77 / 11 Sep. 76*	16 Aug. 78	11 Sep. 03	c. 11 Sep. 06	IA	
	R. Mar.	*Stevens, George Borlase* (72b)	*5 Feb. 59*	*19 Sep. 77*	*3 Oct. 80*	*19 Sep. 03*	. . . .	IA	*u. s. l.*
		Forfeits 1 day for pension.							
	70 F.	**Westlake**, Almond Paul, D.S.O.	2 Mar. 58	29 Nov. 77 / 29 Nov. 76*	2 Sep. 78	29 Nov. 03	c. 29 Nov. 06	IA	
	Leins. R.	**Gastrell**, Everard Thuillier (50, 72b, 107, 165, 170)	9 Oct. 58	30 Jan. 78	29 Mar. 81	30 Jan. 04	c. 30 Jan. 07	IA	*Resdg. ex India.*
	Wilts. R.	**Whistler**, Albert Edward (50, 57)	27 Dec. 57	30 Jan. 78	18 Dec. 80	30 Jan. 04	c. 30 Jan. 07	IA	
	E. Surr. R.	**Rodwell**, Ernest Hunter, C.B., *p.s.c.* (50, 57, 65, 169, 175) [L]	10 Feb. 58	30 Jan. 78	28 May 80	30 Jan. 04	c. 30 Jan. 07	IA	
	9 F.	*Georges, Henry Wallace Edgcome* (50, 125)	*1 Sep. 56*	*26 Feb. 78 / 26 Feb. 76**	*20 July 77*	*26 Feb. 04*	*c. 26 Feb. 07*	IA	*u. s. l.*
	Durh. L. I.	*Purdon, David William*, C.M.G. (50, 161)	*17 July 53*	*1 May 78*	*17 Mar. 80*	*1 May 04*	. . . .	IA	*u. s. l.*
10	Ches. R.	**Hodgson**, Godfrey Beckett (156)	28 Oct. 58	1 May 78	15 Sep. 80	1 May 04	c. 1 May 07	75	*Resdg. ex India.*
	York L. I.	**Thackwell**, Colquhoun Grant Roche, C.B., D.S.O. (171)	8 Dec. 57	1 May 78	23 Feb. 81	1 May 04	c. 1 May 07	IA	A.D.S. & T. 4th Dn.
	R. Ir. Fus.	**Thomson**, William David, C.M.G. (50, 57, 74, 162, 169)	4 Feb. 58	1 May 78	13 Sep. 80	1 May 04	c. 1 May 07	IA	*Resdg. ex India.*
	Suff. R.	**Hutchinson**, Francis Patrick, C.B. (50, 162, 173) [L]	15 Apl. 58	1 May 78	26 Nov. 80	1 May 04	c. 1 May 07	IA	
	15 F.	**Aplin**, Philip John Hanham, D.S.O. (66 a, 107, 162, 170) [L]	18 Dec. 59	1 May 78	22 May 80	1 May 04	c. 1 May 07	IA	
15	Som. L. I.	**Templer**, Henry (50, 170)	19 June 58	1 May 78	20 Nov. 80	1 May 04	c. 28 Oct. 07	IA	*Resdg. ex India.*
	103 F.	*King, Henry Thomas* (50, 57, 71)	*20 Feb. 56*	*1 May 78*	*16 Aug. 80*	*1 May 04*	*c. 22 May 07*	IA	*u. s. l.*
	R. Ir. Fus.	**Dobbie**, Herbert Hugh (50, 57, 161, 169)	19 Apl. 59	1 May 78	29 May 81	1 May 04	c. 1 May 07	IA	
	E. York R.	*Piers, William Barrington*	*4 Apl. 59*	*1 May 78*	*29 Mar. 81*	*1 May 04*	. . . .	IA	*u. s. l.*
	Linc. R.	**Richardson**, Francis Bernard Walter (50, 149)	9 June 59	11 May 78	16 Aug. 80	11 May 04	c. 11 May 07	IA	
	N. Staff. R.	**Steele**, St. George Loftus, C.B. (50, 57, 66a, 156, 161, 170) [L]	31 Mar. 59	11 May 78	10 Mar. 80	11 May 04	c. 1 June 07	IA	
	Wilts. R.	**Justice**, Charles Le Gendre (50)	5 Aug. 59	11 May 78	16 June 80	11 May 04	c. 11 May 07	IA	
	16 F.	*Ostrehan, Francis George Rodney* (50, 71, 157, 161, 170)	*20 Aug. 59*	*11 May 78*	*1 May 80*	*11 May 04*	. . . .	IA	*u. s. l.*
	1 D.G.	*Wright, Ernest Leonard* (57)	*28 Oct. 58*	*11 May 78*	*12 May 81*	*11 May 04*	. . . .	IA	*u. s. l.*
20	13 F.	**Unwin**, Gaston Bouverie, D.S.O. (170)	9 July 59	25 May 78	28 Apl. 81	25 May 04	c. 15 Sep. 07	IA	
	Wilts. R.	**Dennys**, William Annesley Burton (50, 170)	26 June 59	13 Aug. 79	30 Oct. 80	1 June 04	c. 6 June 07	IA	*Resdg. ex India.*
	Dorset. R.	**Dobbie**, William Hugh, C.B. (50)	9 May 59	13 Aug. 79	10 Mar. 82	1 June 04	c. 6 June 07	IA	*Resdg. ex India.*
	28 F.	**Plumer**, Thomas Hall	30 Dec. 60	13 Aug. 79	25 Oct. 80	1 June 04	. . . .	IA	*Resdg. ex India.*
	Worc. R.	**Aitken**, Arthur Edward (50, 71, 169)	25 May 61	14 Jan. 80	31 Dec. 82	1 June 04	c. 9 June 07	75	*Resdg. ex India.*
25	S. York. R.	**Johnson**, Charles Edward (50, 57, 167)	4 Aug. 61	22 Jan. 81	20 Oct. 82	1 June 04	c. 11 June 07	IA	Attd. A. G.'s Branch.
	Leins. R.	**Johnstone**, Augustus Anderson Jervis (50, 170)	30 May 62	22 Jan. 81	17 Aug. 83	1 June 04	c. 11 June 07	57	
	62 F.	*Fryer, George Willoughby Smith* (50)	*16 Sep. 58*	*21 Aug. 78*	*31 Mar. 80*	*21 Aug. 04*	. . . .	IA	*u. s. l.*
	Essex R.	**Leonard**, Alfred Wilberforce (51, 105)	3 June 62	10 Mar. 83	13 Nov. 85	2 Oct. 04	c. 2 Oct. 08		*Resdg. India.*
	Glouc. R.	**Marlow**, Benjamin William, C.S.I., C.I.E.	1 Dec. 63	12 Nov. 84	5 Nov. 86	23 Dec. 04	c. 23 Dec. 08		Mily. Acct. Genl.
	R. Suss. R.	**Yate**, Frederick Herbert (170)	2 May 58	22 Jan. 79	24 Mar. 80	22 Jan. 05	c. 1 June 07		*Resdg. ex India.*
30	W. Rid R.	**Lindesay**, Alexander Lumsdaine (50, 170)	5 July 62	10 Mar. 83	24 May 86	13 Feb. 05	c. 13 Feb. 09		
	Welsh. R.	**Price**, Chas. Henry Uvedale, C.B., D.S.O. (50, 93, 170, 185)	16 June 62	22 Oct. 81	16 June 83	1 Mar. 05	c. 1 Mar. 08	IA	*Resdg. ex India.*
	R. Ir. R.	**Herbert**, Lionel Norton, C.B., (50, 127, 170) [*l*]	18 Feb. 63	9 Sep. 82	24 June 84	4 Apl. 05	c. 4 Apl. 09	75	Inspr. of Depôts, 5th Divn.
	N. Staff. R.	**Palin**, Gilbert Walter, C.B., C.I.E	29 Sep. 62	12 May 83	3 Nov. 85	1 May 05	c. 1 May 09	IA	S. and T. Corps.
	R. Ir. R.	**Loch**, Harry Frere, *p.s.c.* (50, 104, 161)	20 Oct. 60	11 Aug. 80	25 Aug. 82	7 May 05	c. 10 June 07	IA	*Resdg. ex India.*
35	War. R.	**Twigg**, Robert Henry, C.B. (50, 162, 169)	4 Apl. 60	13 Aug. 79	26 Dec. 80	13 Aug. 05	c. 4 Feb. 09	IA	
	Dorset. R.	**Harrison**, Thomas Aylet C.S.I. (50)	26 Mar. 65	7 Feb. 85	12 Jan. 88	1 Apl. 06	c. 1 Apl. 10	IA	Mily. Accts. Dept.
	L'pool R.	**Shakespear**, Leslie Waterfield, C.B. (50, 157, 170)	5 Sep. 60	22 Jan. 81	27 Jan. 85	2 Apl. 06	c. 2 Apl. 09	IA	
38	W. York. R.	**Shore**, Offley Bohun Stovin Fairless, C.B., C.I.E., D.S.O., *p.s.c.* (67, 127, 162, 165) [L]	9 Aug. 63	10 May 82	27 Apl. 84	1 June 06	c. 30 July 10	IA	

Colonels and Lieutenant-Colonels—*contd.*

No.	Former Corps.	Names.	Date of birth.	Date of first commission.	Date of appointment to Indian Army.	Date of rank. Substantive Lt.-Colonal.	Date of rank. Army.	F. rule.	Remarks.
39	R. Mar.	**Buchanan,** Kenneth James, C.B. (50, 170)	7 Mar. 63	1 Feb. 83	2 Jan. 87	27 Sep. 06	c. 27 Sep. 10	1A	
40	Derby. R.	**Campbell, Alexander Augustus Elphinstone** (50, 161, 170)	14 May 61	10 May 82	23 Nov. 84	20 Oct. 06	c. 29 Oct. 10.	1A	Bde. Comdr.
	E. Lan. R.	**Giles,** Albert (50, 175)	3 Mar. 62	9 Sep. 82	1 Jan. 86	8 Dec. 06		1A	
	Oxf. L. I.	**Thring,** Richard Henry Dugdale (57, 107)	23 June 61	22 Jan. 81	12 Oct. 83	22 Jan. 07	c. 23 Feb. 12	1A	
	Ches. R.	**O'Bryen,** Charles William (50, 72b, 106, 162, 166, 169)[*l*]	25 Feb. 62	9 Sep. 82	20 Aug. 84	10 Mar. 07	. . . .	1A	Postal Censor, Karachi.
	6 D. G.	**Stirling,** William George Hay (50, 104, 195)	21 Apl. 61	9 Sep. 82	29 Nov. 84	2 Apl. 07	. . . .	1A	
45	Bord. R.	**Woodward,** John Alfred Hudson (50, 165, 170)	14 Oct. 62	9 Sep. 82	18 Dec. 84	1 Apl. 07	c. 30 June 11	75	
	Norf. R.	**Fisher,** John, C.B. (50, 53, 57, 157) [L]	9 Mar. 62	10 Mar. 83	10 Dec. 84	26 May 07	c. 29 Aug. 11	75	
	R. Innis. F.	**Mainwaring,** Charles Vaughan (50, 71, 72b, 72c)	7 May 62	10 Mar. 83	21 June 85	22 July 07	c. 3 Oct. 11	1A	
	R. A.	*Stratton, Wallace Christopher Ramsay,* C.I.E. (*161*)	*19 Aug. 63*	*30 July 81*	*7 Aug. 83*	*26 July 07*	. . . .	75	*Supy. List.*
	R. W. S.	**Millar,** William Henry (50, 57, 72b, 162, 170)	29 Sep. 62	23 Aug. 83	19 Aug. 87	19 Oct. 07	c. 26 Dec. 11	1A	*Resdg. ex India.*
	Midd'x. R.	**Wynch,** Frederick John Henry (50, 71)	28 Aug. 62	14 May 84	11 Mar. 86	19 Oct. 07	. . . .	1A	
	R. Ir. R.	*Macdonald, Fredk. Weston Peile* (*50, 94, 161, 165, 169*) [L]	*19 Apl. 63*	*22 Oct 81*	*17 June 83*	*22 Oct. 07*	. . . .	75	*Supy. List.*
50	R. Ir. F.	**Cartwright,** Charles Marling, C.B., C. M. G. (50, 53, 65)	25 May 62	22 Oct. 81	8 Apl. 83	22 Oct. 07	c. 1 Jan. 12	1A	
	Oxf. L. I.	**Holloway,** Benjamin, C.I.E., *p.s.c.* (50) [L]	27 Aug. 61	22 Oct. 81	11 May 83	22 Oct. 07	c. 14 Jan. 12	1A	Bde. Comdr.
	R. W. S. R.	**Edwards,** FitzJames Maine, C.B., C.M.G., D.S.O., **A.D.C.,** *p.s.c.* (124a)	21 May 61	22 Oct. 81	8 Nov. 83	22 Oct. 07	c. 14 Jan. 12	1A	
	R. Sc. F.	**Erck,** John Caillard Wentworth (50, 57, 71, 107, 124a, 140, 181, 195)	6 May 62	9 Sep. 82	5 Mar. 84	26 Oct. 07	. . . .	1A	*Resdg. ex India.*
	W. York. R.	**Muspratt,** Francis Clifton, C.B., C.M.G. (50, 53, 167, 170)	21 Feb. 64	7 Feb. 85	26 July 86	13 Dec. 07	c. 1 Apl 12	1A	
55	Linc. R.	**Andrews,** Louis James (50, 71, 204) [L]	6 Oct. 62	10 Mar. 83	14 Aug. 84	29 Feb. 08	c. 26 June 12	1A	
	Devon. R.	**Carruthers,** Robert Alexander, C.B., C.M.G. (50, 107, 170)	20 Feb. 62	10 Mar. 83	8 Nov. 85	2 Mar. 08	c. 29 June 12	1A	
	R. Suss. R.	**Newbold,** Ambrose William (107)	23 Sep. 63	23 Aug. 84	19 Jan. 87	2 Mar. 08	c. 11 Aug. 12	1A	
	Durh. L. I.	**Mead,** Harold Richard (50, 155)	15 Sep. 63	10 Mar. 83	9 June 87	26 Mar. 08	c. 19 Aug. 12	1A	
	Worc. R.	**Betham,** Robert Mitchell (50, 71, 153)	2 May 64	10 Mar. 83	15 Feb. 85	12 Apl. 08	c. 1 Sep. 12	1A	Bde. Comdr.
60	N. Lan. R.	**Savage,** William Henry, C.M.G., (50, 71, 157, 162)	18 Apl. 63	9 Sep. 82	6 Nov. 83	15 Apl. 08	c. 1 Sep. 12	1A	
	Oxf. L. I.	**Cox,** Francis William Henry, C.B., (72b, 106) *Forfeits 1 day for pension.*	29 Oct. 63	6 Feb. 84	27 Nov. 86	30 Apl. 08	c. 21 Sep. 12	1A	
	L'pool. R.	**Holland,** Guy Lushington, M.V.O. (50, 57, 71)	21 Feb. 61	27 Jan. 83	15 Apl. 85	30 Apl. 08	. . . .	1A	
63	Oxf. L. I.	**Mathewes,** John Rocke (50, 53)	16 June 61	10 May 82	11 Aug. 84	10 May 08	c. 21 Sep. 12	1A	

	R. Sco. F.	**Johnston**, Harley Macalpine (50, 57, 170)	1 Aug. 62	10 May 82	6 Sep. 85	10 May 08	*c.* 5 Oct. 12	1A	
65	Linc. R.	**Fasken**, William Henry, C.B., (50, 64, 170)	9 May 63	10 May 82	15 Oct. 85	10 May 08	*c.* 12 Oct. 12	1A	
	R. War. R.	**Thomas**, Francis Herbert Sullivan, C.B., (156)	2 Oct. 62	10 May 82	5 Nov. 85	10 May 08	*c.* 23 Oct. 12	1A	S. and T. Corps.
	Bedf. R.	*Minchin, Charles Frederick*, D.S.O. (50, 94, 161, 165, 169, 175)	*22 Sep. 62*	*10 May 82*	*3 Feb. 85*	*10 May 08*	. . . .	1A	*Supy. List.*
	R. Ir. R.	*Leslie, William Clarence Colebrook (161)*	*13 Mar. 61*	*10 May 82*	*3 Mar. 85*	*10 May 08*	. . . .	1A	*Supy. List.*
	K. O. S. B.	**Blakeney**, William Edward Albemarle (50, 57, 66*a*, 162, 170)	5 Dec. 62	10 May 82	6 Dec. 85	10 May 08	*c.* 23 Oct. 12	1A	
	R. A.	**Kennedy**, Hugh, C.B. (50, 162, 170)	16 Nov. 62	28 Feb. 83	14 Sep. 86	10 May 08	*c.* 30 Dec. 12	1A	*Resdg. ex India.*
	S. Lan. R.	**Commeline**, Francis Henry Bagot (50, 65)	7 June 63	10 Mar. 83	27 Apl. 85	10 May 08	*c.* 31 Dec. 12	1A	
70	R. A.	**Lang**, Elliot Brownlow (50, 71)	10 Aug. 62	25 July 82	14 May 86	25 July 08	*c.* 14 Feb. 13	1A	
	19 Huss.	**Nicholson**, Ralph Wise (162)	23 May 60	2 Aug. 82	29 Jan. 85	2 Aug. 08	. . . .	1A	*Resdg. ex India.*
	R. Innis. Fus.	**Scott**, Bernard (50, 161)	1 Aug. 66	9 May 88	26 Aug. 89	10 Aug. 08	*c.* 1 Sep. 14	1A	*Resdg. ex India.*
	Som L. I.	**Sullivan**, Rupert Edward (106)	12 Sep. 62	27 Jan. 83	4 June 84	11 Aug. 08	*c.* 21 Feb. 13	1A	*Resdg. in India.*
	N. Lanc. R.	**Henderson**, Thomas Hume (50)	20 May 64	6 Feb. 84	13 July 85	11 Aug. 08	. . . .	1A	*Resdg. ex India.*
75	F. Lan. R.	**Jones**, Herbert John, D.S.O. (124*a*, 162, 166, 169)	23 Dec. 65	23 Aug. 84	16 June 86	25 Aug. 08	. . . .	1A	
	Devon. R.	*Gurdon*, the Hon. *Philip Richard Thornhagh*, C.S.I. (162)	*2 Feb. 63*	*9 Sep. 82*	*12 July 84*	*9 Sep. 08*	. . . .	75	*Supy. List.*
	Leins. R.	**Cripps**, Arthur William, C.B., (170)	16 Jan. 62	9 Sep. 82	21 Nov. 85	9 Sep. 08	*c.* 13 Mar. 13	1A	S. and T. Corps.
	L'pool. R.	**Colomb**, George Henry Cooper C.M.G. (50, 53, 71, 88, 156)	8 Dec. 62	6 Feb. 84	16 Aug. 86	19 Sep. 08	*c.* 24 Mar. 13	1A	
	R. Suss. R.	**Brown**, Robert James Reid (71, 72*b*, 104, 141, 181)	24 Apl. 63	10 Mar. 83	7 June 86	13 Nov. 08	*c.* 24 Mar. 13	1A	
	R. Sco. F.	**Walton**, William Crawford, C.B., C.M.G. (50, 53, 137, 162, 170)	15 Nov. 64	6 Feb. 84	15 June 86	17 Nov. 08	*c.* 31 Mar. 13	1A	
80	S. W. Bord	**Henegan**, John, D.S.O.	29 Jan. 65	6 Feb. 84	30 Dec. 85	3 Jan. 09	. . . .	1A	*Resdg. ex India.*
	E. Surr. R.	**Whiffin**, Henry Edward (50, 71, 107, 120, 162, 169)	1 Nov. 62	27 Jan. 83	7 July 85	27 Jan. 09	. . . .	1A	
	Essex R.	**Young**, William Bensley (50, 71)	12 Oct. 60	27 Jan. 83	7 Aug. 85	27 Jan. 09	. . . .	8A	
	Hamps. R.	**Banbury**, Walter Edward, C.M.G. (50, 53, 147, 159, 170, 204) [L]	25 Dec. 63	25 Aug. 83	7 July 85	1 Feb. 09	*c.* 3 June 17	1A	
	R. A.	**Baynes**, Charles Edward (50, 65, 71)	17 July 63	14 Feb. 83	6 July 88	14 Feb. 09	. .	1A	
85	E. York. R.	**Carnegy**, Charles Gilbert, M.V.O. (50, 57, 155)	29 June 64	23 Aug. 84	23 Mar. 86	27 Feb. 09	*c.* 1 Jan. 17	1A	
	W. I. R.	**Dann**, Henry Churnside Beaumont (50, 71, 74, 155, 162, 169)	26 May 63	20 Oct. 83	2 Aug. 87	28 Feb. 09		1A	*Resdg. ex India.*
	Norf. R.	VC *Smith, John Manners*, C.V.O., C.I.E. (50)	*30 Aug. 64*	*10 Mar. 83*	*25 Mar. 85*	*10 Mar. 09*		75	*Supy. List.*
	Leic. R.	**Crocker**, Sydney Francis, C.B. (50, 65, 162, 169)	5 Aug. 64	10 Mar. 83	15 Feb. 87	10 Mar. 09	*c.* 31 May 13	1A	*Resdg. ex India.*
	Welsh R.	**Knatchbull**, George Wyndham Chichester, C.M.G. (50, 65, 71)	3 Mar. 62	10 Mar. 83	13 July 85	10 Mar. 09	. . . .	1A	
	Leic. R.	**Loudon**, John Archer (50, 71)	16 Oct. 63	10 Mar. 83	15 Nov. 85	10 Mar. 09	. . . .	1A	43 Erin. R.
90	Bord. R.	**Dick**, Arthur Robert, C.B. (57) [L] *Half pay from 25 Sep. 91 to 20 Nov. 94 (1 yr. to count.)*	22 Apl. 60	22 Jan. 81	29 July 84	20 Mar. 09	*c.* 16 July 08	A1	
	R. Ir. Fus.	**Clay**, Charles Herbert, *p.s.c.* (50, 53, 71, 156, 162, 170)	24 Jan. 65	6 Feb. 84	11 Apl. 86	31 Mar. 09	*c.* 30 Sep. 13	1A	
	Midd'x. R.	**Williams**, Arthur Henry (50, 71, 106, 172)	13 Dec. 65	7 Feb. 85	30 Apl. 86	28 Apl. 09	. . . .	1A	
	R. Suss.	**Thomas**, Hubert St. George, D.S.O. (50, 53, 71)	23 Apl. 62	14 May 84	2 Nov. 86	16 May 09	. . . .	1A	
	Devon. R.	**Soady**, George Joseph FitzMaurice, C.M.G. (50, 59, 170)	5 June 63	6 Feb. 84	10 Sep. 87	1 June 09	*c.* 22 Oct. 13	1A	*Lv. ex. I., m.c., 6 m.*
95	W. I. R.	**Burton**, Reginald George (50, 52, 66*a*, 71) [L]	8 July 64	23 Aug. 84	9 July 87	17 June 09	*c.* 19 Dec. 13	1A	
96	Derby. R.	**Cole**, Edward Hearle, C.B., C.M.G. (50, 57, 71, 162, 169, 174)	27 July 63	19 Dec. 83	17 Apl. 86	25 June 09	*c.* 2 Jan. 14	1A	*Resdg. in India.*

*Date of Sub-Lieutenancy.

Colonels and Lieutenant-Colonels—*contd.*

No.	Former Corps.	Names.	Date of birth.	Date of first commission.	Date of appointment to Indian Army.	Date of Rank. Substantive Lt.-Colonel.	Date of Rank. Army.	F. rule.	Remarks.
97	S. Lan. R.	**Rainsford**, John Augustus Godfrey (51, 57, 65, 106, 140, 167)	29 Sep. 62	22 Nov. 84	10 Nov. 86	5 July 09	. . . .	IA	
	Midd'x. R.	**Wilkins**, Ernest George Rule (50, 71, 153, 162) [*l*]	30 May 62	7 Feb. 85	29 Apl. 86	31 July 09	. . . .	75	
	Dorset R.	**Grant**, Arthur (156)	18 Feb. 65	7 Feb. 85	13 June 87	3 Aug. 09	. . . .	IA	2nd G. R.
100	R. Muns. F.	**Roberts**, Hereward Llewelyn, C.B., M.V.O. (50, 169, 174) [*l*]	3 Oct. 64	25 Aug. 83	27 Oct. 85	25 Aug. 09	*c.* 8 Jan. 14	IA	Bde. Comdr.
	Conn. Rang.	**Selwyn**, Chas. Henry, C.B., *p.s.c.* (64, 71, 124*a*, 162, 165, 170) [L]	1 Mar. 63	25 Aug. 83	3 Dec. 86	25 Aug. 09	*c.* 20 Jan. 14	IA	
	S. W. Bord.	*Field, Charles William* (74, 169)	*30 Jan. 63*	*25 Aug. 83*	*27 May 85*	*25 Aug. 09*	. . . .	IA	*Supy. List.*
	Dorset. R.	**Douglas**, James Archibald, C.M.G., C.I.E. (50, 65, 71, 162, 170) [L]	21 Oct. 62	25 Aug. 83	24 Dec. 85	25 Aug. 09	*c.* 28 Jan. 14	IA	
	R. M. L. I.	*Aplin, Stephen Lushington*, C.S.I. (65, 104)	*17 Oct. 63*	*1 Sep. 83*	*13 Jan. 87*	*1 Sep. 09*	. . . .	IA	*Supy. List.*
	R. Mar.	**Miles**, Philip John, C.B. (50, 53, 66*a*, 162, 170)	23 Dec. 64	1 Feb. 85	31 Dec. 87	9 Sep. 09	*c.* 31 Jan. 13	IA	
	N. Lan. R.	**Edwardes**, Stanley Malcolm, C.B., C.M.G., D.S.O. (50)	18 Nov. 63	23 Aug. 84	24 Jan. 86	17 Sep. 09	*c.* 5 Feb. 14	IA	Bde. Comdr.
105	Oxf. L.	**Prendergast**, Charles Gordon, C.B. (50, 169)	18 Apl. 64	23 Aug. 84	29 Apl. 86	8 Oct. 09	*c.* 12 Feb. 14	IA	Bde. Comdr.
	R. Suss. I. R.	**Somerset**, Charles Wyndham, C.B., C.M.G., M.V.O. (50, 52, 72*b*, 144, 162)	24 June 62	30 Jan. 84	28 June 86	2 Nov. 09	*c.* 11 Mar. 14	IA	
	R. Ir. R.	**Pritchard**, Henry Edward (106)	11 Sep. 62	12 Nov. 84	2 Apl. 87	11 Nov. 09	. . . .	IA	
	Worc. R.	**Palmer**, Henry Ingham Evered, C.M.G. (50, 53, 57, 65, 71, 170)	5 Apl. 62	5 Dec. 83	6 Dec. 85	5 Dec. 09	. . . .	IA	
	York. L. I.	**Sanders**, Gerhardt L'Honneux (162, 170)	22 Sep. 62	5 Dec. 83	22 Nov. 85	5 Dec. 09	*c.* 1 Jan. 16	IA	S. and T. Corps.
110	York. L. I.	**Peterson**, Frederick Hopewell, C.B., D.S.O. (50, 52, 167) [*l*]	5 Sep. 64	25 Nov. 85	4 Sep. 87	31 Dec. 09	*c.* 6 June 12	IA	Bde. Comdr.
	Manch. R.	**Cadell**, Alexander (50, 52, 162, 167)	27 Jan. 67	29 Aug. 85	10 May 87	4 Jan. 10	*c.* 1 Apl. 14	IA	Bde. Comdr.
	Dorset. R.	**Watson**, Harry Davis, C.B., C.M.G., C.I.E., M.V.O. (50, 157) *Eq.*	18 July 66	7 Feb. 85	2 June 86	6 Jan. 10	*c.* 19 Apl. 14	IA	
	R. Suss. R.	**Finch**, Colin (50)	18 Jan. 63	30 Jan. 84	8 Sep. 87	30 Jan. 10	. . . .	IA	Mily. Farms Dept.
	North'd. F.	**Willoughby**, Michael Edward, C.B., C.M.G., *p.s.c.* (50, 65, 124, 125*b*, 127, 162) [L]	18 June 64	6 Feb. 84	23 Nov. 85	6 Feb. 10	*c.* 4 May 14	IA	
	Sea Higrs.	*Haig, Thomas Wolseley* C.M.G. (106, 155, 170, 203) [L]	*7 Aug. 65*	*6 Feb. 84*	*14 July 87*	*6 Feb. 10*	. . . .	IA	*Supy. List.*
115	R. Ir. Fus.	**Tod**, John Kelso, *p.s.c.* (50, 57, 64, 71, 161, 169) [L]	3 July 64	6 Feb. 84	17 Dec. 85	6 Feb. 10	*c.* 10 May 14	IA	
	Ches. R.	**Roche**, Henry John, C.B. (50)	12 Aug. 64	6 Feb. 84	25 Apl. 86	6 Feb. 10	*c.* 4 Nov. 14	IA	*Resdg. ex India.*
	L'pool. R.	**Grimston**, Sylvester Bertram, C.M.G. (50, 165, 170)	27 Nov. 64	6 Feb. 84	12 May 86	6 Feb. 10	*c.* 8 Dec. 14	IA	
	Glouc. R.	**Pollard**, James Herbert (50, 65, 66*a*, 162)	25 May 64	6 Feb. 84	20 Mar. 87	6 Feb. 10	. . . .	IA	*Resdg. ex India.*
	Sco. Rif.	*Cox, Sir Percy Zachariah*, G.C.I.E., K.C.S.I., (88, 153, 161) (*Hon. Maj.-Genl.*, *17 May 17*)	*21 Nov. 64*	*6 Feb. 84*	*19 Nov. 89*	*6 Feb. 10*	. . . .	IA	*Supy. List.*
	Suff. R.	**Medley**, Allen George (161, 179) [L]	14 Sep. 63	6 Feb. 84	14 Apl. 86	6 Feb. 10	. . . .	IA	
120	R. Suss. R.	**Graham**, Stuart Bruce (71, 195)	7 June 65	6 Feb. 84	26 Mar. 86	6 Feb. 10	*c.* 4 Feb. 15	IA	
	Worc. R.	**Mennie**, John Oliver (66*a*, 71, 127, 172)	28 Dec. 64	6 Feb. 84	1 June 85	6 Feb. 10	*c.* 10 Feb. 15	IA	G.S.O., 1st gr., A. Hd. Qrs.
122	S. Staff R.	**Murray**, Cyril Francis Tyrell, C.B., M.V.O. (162)	19 Jan. 63	6 Feb. 84	24 Jan. 86	6 Feb. 10	. . . .	IA	S. and T. Corps.
	N. Staff. R.	*Douglas, Montagu William*, C.I.E. (162, 165)	*23 Nov. 63*	*6 Feb. 84*	*24 Mar. 87*	*6 Feb. 10*	. . . .	IA	*Supy. List*

	W. Rid. R.	*Bayley, Stuart Farquharson* (*161*)	*14 Aug. 63*	*6 Feb. 84*	*25 Mar. 86*	*6 Feb. 10*	. . . .	IA	*Supy. List.*
	E. Lan. R.	**Barnes**, James Petit (65)	21 Aug. 63	6 Feb. 84	31 May 86	6 Feb. 10	. . . .	IA	
	Linc. R.	**de Boisragon**, Guy Hudleston (157, 170)	5 Nov. 64	7 Feb. 85	29 Sep. 86	11 Feb. 10	c. 17 Feb. 15	IA	Bde. Comdr.
125	York. L. I.	**Smith**, Frederic Alexander (50, 53, 71, 137)	18 Mar. 64	7 Feb. 85	1 Mar. 88	13 Feb. 10	c. 18 Feb. 15	IA	S. and T. Corps.
	R. War. R.	**Hilliard**, Harold Nash	10 Oct. 63	12 Nov. 84	21 July 86	18 Feb. 10	. . . .	IA	
	R. W. S. R.	**Dyer**, Reginald Edward Harry, C.B., *p. s. c.* (50, 162, 167, 170)	9 Oct. 64	29 Aug. 85	30 Aug. 87	25 Feb. 10	c. 5 Mar. 15	IA	
	York. L. I.	**Richardson**, Herbert Lance (162, 165, 170)	6 Sep. 65	7 Feb. 85	19 June 88	23 Mar. 10	c. 15 Apl. 15	IA	*Resdg. ex India.*
	Bord. R.	**Rea**, Frederic William (50, 71)	15 Oct. 65	29 Aug. 85	11 Jan. 89	26 Mar. 10	. . . .	IA	
130	Bord. R.	**Laing**, Frederick Charles (50, 53, 71, 162)	5 Dec. 65	8 Dec. 86	21 Feb. 88	30 Apl. 10	. . . .	IA	
	Leic. R.	**Sullivan**, Edward Langford, C.B., C.M.G. (50, 72*b*, 167, 170)	8 Jan. 65	25 Nov. 85	5 Sep. 89	5 May 10	c 2 May 15	IA	
	R. Mar.	**Dennys**, Alexander Harry (50, 71)	1 Aug. 65	1 Feb. 85	18 Dec. 88	12 May 10	. . . .	IA	
	Ches. R.	**Hill**, Joseph Robert (50, 53, 57)	20 Oct. 62	12 Nov. 84	23 Oct. 86	26 May 10	. . . .	IA	
	Durh. L. I.	**Ward**, George Alexander (50, 66*a*, 71, 167)	7 June 61	2 July 84	12 Nov. 85	2 July 10	. . . .	IA	
135	R. A.	**Boudier**, Edward William (50, 57, 65, 169)	20 Jan. 65	5 July 84	19 Feb. 87	5 July 10	c. 3 June 15	75	*Resdg. ex India.*
	Suff. R.	**Ryder**, Wilfred Ironside (71, 72*b*, 156)	2 June 66	23 Aug. 84	12 Apl. 86	9 Aug. 10	. . . .	IA	
	Oxf. L. I.	**Budd**, Norman Arthur Hay (51, 65, 66*a*, 170)	25 Apl. 64	9 May 85	19 Oct. 86	11 Aug. 10	. . . .	IA	
	Leic. R.	**Harvey**, Ambrose Blacklock (50, 53, 71, 162)	4 July 64	29 Aug. 85	12 Feb. 87	19 Aug. 10	. . . .	IA	*Supy. List.*
	Devon. R.	*McConaghey, Allen*, C.I.E. (*57, 94, 161, 169*)	*31 Mar. 64*	*23 Aug. 84*	*18 Apl. 87*	*23 Aug. 10*	. . . .	IA	*Supy. List.*
	H. L. I.	*Crichton, Richmond Trevor*, C.I.E. (*50, 162*)	*8 Mar. 65*	*23 Aug. 84*	*4 June 88*	*23 Aug. 10*	. . . .	IA	
	Hamps. R.	**Gordon**, Ramsay Frederick Clayton, C.I.E. (107, 167, 170)	17 Nov. 64	23 Aug. 84	2 Jan. 88	23 Aug. 10	. . . .	IA	*Supy. List*
	H. L. I.	*Macnabb, Donald John Campbell*, C.S.I. (*104*)	*24 June 64*	*23 Aug. 84*	*9 Mar. 86*	*23 Aug. 10*	. . . .	IA	*Supy. List*
	E. York. R.	*Young*, Sir *Frank Popham*, K.B.E., C.I.E.	*24 Dec. 63*	*23 Aug. 84*	*19 Sep. 86*	*23 Aug. 10*	. . . .	IA	*Resdg. ex India.*
140	Bord. R.	**Goodenough**, Herbert Lane (102)	26 Mar. 65	23 Aug. 84	6 June 86	23 Aug. 10	. . . .	IA	*Resdg. ex India.*
	York. L. I.	**Roome**, Henry Napier (172)	2 July 64	23 May 85	12 June 87	3 Sep. 10	. . .	IA	
	E. York. R.	**Frazer**, George Stanley, C.M.G. (57, 64, 65, 66*a*, 71)	9 Feb. 65	29 Aug. 85	6 Mar. 87	1 Oct. 10	c. 3 June 15	IA	
	York. L. I.	**Dobbie**, Charles Francis (50, 52, 71, 162)	28 May 63	28 Feb. 85	10 Aug. 87	8 Oct. 10	. . . .	IA	*Resdg. ex India.*
	4 Hus.	**Barry**, John Frederick	28 Nov. 62	25 Oct. 84	1 Aug. 85	25 Oct. 10	. . .	IA	
		Service for pension; In Ranks from 1 Mar. 80 to 24 Oct. 84 (half to count).							
145	Midd'x R.	**Batten**, Frederick Græme, D.S.O. (50, 57, 65, 71, 107, 195)	23 Oct. 63	6 May 85	10 June 87	5 Nov. 10	c. 3 June 15	IA	*Supy. List*
	E. York. R.	*Herbert, Claude*, D.S.O. (*156*)	*20 Oct. 62*	*12 Nov. 84*	*30 Sep. 86*	*12 Nov. 10*	.	IA	
	R. Sc. Fus.	**Bailey**, Charles (50)	5 Dec. 62	12 Nov. 84	5 Aug. 86	12 Nov. 10	c. 3 June 15	IA	
	Ches. R.	**Battye**, Arthur Henry (50, 157)	2 July 63	12 Nov. 84	12 July 86	12 Nov. 10	. . . .	IA	
	K.O.S. Bord.	**Lane**, Horace Powys (50, 66*a*, 71)	18 Sep. 65	29 Aug. 85	10 Jan. 89	12 Nov. 10	c. 27 Aug. 15		*Resdg. ex India.*
		Half pay from 2 Apl. to 31 July 88 (to count).							
	R. A.	**Young**, Frederick DeBudé, C.M.G., *p.s.c.* (71, 72*b*, 162)	31 Mar. 65	9 Dec. 84	7 Feb. 89	9 Dec. 10	c. 9 Oct. 15		
150	R. Mar.	**Luck**, Churchill Arthur (50, 170)	1 Mar. 66	1 Sep. 85	23 Nov. 87	21 Dec. 10	c. 28 Oct. 15		J. A. Genl.
	E. Lan. R.	**Caruana**, Alfred Joseph, C.I.E. (50, 57, 74, 169)	13 Dec. 65	14 Jan. 85	15 Sep. 88	14 Jan. 11	c. 1 Jan. 16		
	North'n R.	**Lowther**, Henry (52)	11 Oct. 62	23 May 85	25 May 87	22 Jan. 11	. . . .		
153	W. I. R.	**Lowis**, Harry Elliot (50, 57, 71, 169)	9 Sep. 64	9 May 85	1 Apl. 89	28 Jan. 11	c. 3 June 15		

Colonels and Lieutenant-Colonels—*contd.*

No.	Former Corps.	Names.	Date of birth.	Date of first commission.	Date of appointment to Indian Army.	Date of Rank. Substantive Lt.-Colonel.	Date of Rank. Army.	F. rule.	Remarks.
154	W. York. R.	**Younghusband,** Leslie Napier, C.B., C.M.G. (50, 57, 65, 170)	5 Aug. 66	7 Feb. 85	1 June 87	7 Feb. 11	c. 31 Mar. 16		
155	Conn. Rang.	**Moore,** Frederick Lewis (50, 66a, 170)	22 Sep. 65	7 Feb. 85	24 Apl. 87	7 Feb. 11			
	E. York. R	**Chitty,** Walter Willis (50, 53, 64, 71, 72b)	11 June 66	7 Feb. 85	12 July 86	7 Feb. 11	c. 3 June 15		
	R. W. S. R.	**Ditmas,** Alfred Robertson	19 Nov. 65	7 Feb. 85	6 Aug. 85	Feb. 11			S. and T. Corps.
	Bord. R.	*Waller, Edmund (159. 170)* *Forfeits period 11 May to 1 June 03 for pension.*	*18 Aug. 65*	*7 Feb. 85*	*3 Apl. 88*	*7 Feb. 11*			*Supy. List.*
	North'n. R.	**Drew,** Arthur Blanshard Hawley, C.I.E. (50)	17 June 65	7 Feb. 85	28 Nov. 87	7 Feb. 11	c. 1 Jan. 16		
	R. Ir. R.	**Low,** Robert Balmain, D.S.O. (66a, 71, 170)	7 Oct. 64	7 Feb. 85	12 Oct. 86	7 Feb. 11			
160	Leic. R.	**Stuart,** Donald Forbes (50, 53, 57, 162)	28 Aug. 65	7 Feb. 85	30 July 85	7 Feb. 11			
	Welsh. R.	**Moore,** Arthur Guy Welman (65)	29 Apl. 63	28 Feb. 85	8 Mar. 87	28 Feb. 11			Remount Dept.
	York. L. I.	**Norie,** Chas. Edwd. deManley, C.B., C.M.G., D.S.O., *p.s.c.* (50, 57, 66a, 156)[*l*]	7 Aug. 66	9 May 85	31 July 87	2 Apl. 11	c. 18 Feb. 15		Bde. Comdr.
	W. York. R.	**Robin,** Nicholas Edmund (50, 53)	23 Jan. 63	6 May 95	4 Oct. 86	6 May 11			
	Derby. R.	*Burlton, Philip Sykes Murphy (50, 170)*	*28 Nov. 65*	*9 May 85*	*9 Dec. 87*	*9 May 11*			*Supy. List*
	E. Lan. R.	**Campbell,** Alan James, D.S.O. (51)	10 May 65	9 May 85	11 Jan. 89	9 May 11	c 3 June 16		*Resdg. ex India.*
	Ches. R.	*Hughes, Francis Townshend Cunynghame (50, 161)*	*8 Aug. 66*	*9 May 85*	*2 Feb. 87*	*9 May 11*			*Supy. List.*
165	Bedf. R.	**Wright,** Ernest Granville (50, 72b, 137, 162)	20 June 65	9 May 85	2 Aug. 86	9 May 11	c 3 June 16		18 Infantry.
	Worc. R.	**Seddon,** Thomas Young (50, 53, 165)	23 Nov. 64	9 May 85	27 Apl. 87	9 May 11	c. 9 June 16		
	Linc. R.	**Bell,** Lindsay Maxwell	25 June 65	25 Aug. 86	7 Dec. 87	9 May 11			*Resdg. ex India.*
	Manch. R.	**Tytler,** Harry Christopher, D.S.O.	26 Sep. 67	30 Jan. 86	9 Nov. 87	22 May 11	c. 2 July 16		*Resdg. ex India.*
	Hamps. R.	**Burlton,** William Erle Ferdinand	29 Dec. 63	23 May 85	3 Aug. 86	23 May 11			S. and I. Corps.
	R. Dub. Fus.	*Cowie, William Alexander Lomer (107).* *Forfeits one day for pension.*	*19 May 62*	*23 May 85*	*19 Nov. 86*	*23 May 11*			*Supy. List.*
170	W. Rid. R.	**White,** William Edward (50)	4 Dec. 65	16 Nov. 87	25 Nov. 89	12 June 11			

171	Worc. R.	**Creagh**, Arthur Henry Dopping, C.M.G., M.V.O. (50, 66*a*, 162, 170)	29 Apl. 66	5 Feb. 87	23 Jan. 88	9 July 11	*c* 30 Aug. 16	
	Glouc. R.	**Look**, Frederick Robert Edward, D.S.O. (51, 52, 71, 72*a*, 94, 106, 120, 169 175)	21 Apl. 67	25 Aug. 86	2 June 89	23 July 11	*c*. 1 Jan. 17	
	W. I. R.	**Trydell**, Botet (50, 72*b*, 94, 157, 169)	15 Apl. 64	29 Aug. 85	11 Apl. 88	29 Aug. 11		*Resdg. ex India.*
	Wilts. R.	**Davies**, Frederick George Hugh (50, 52, 71, 169)	7 July 66	29 Aug. 85	20 Feb. 87	29 Aug. 11	*c*. 1 Jan. 17	
	H. L. I.	*Bannerman, Arthur D'Arcy Gordon*, C.I.E., C.V.O. (*161*)	*30 Feb. 66*	*29 Aug. 85*	*30 Nov. 88*	*29 Aug. 11*	. . .	*Supy. List.*
	Derby. R.	*Spence, Philip Trevor Augustine* (*161, 169*)	*8 Aug. 65*	*29 Aug. 85*	*26 Aug. 87*	*29 Aug. 11*	. . .	*Supy. List.*
175	Worc. R.	**Beville**, Charles Hamilton, C.B.	18 June 65	29 Aug. 85	1 Dec. 86	29 Aug. 11	. . .	S. and T. Corps.
	Manch. R.	**Dennys**, Hector Travers, C.I.E.	9 Mar. 64	29 Aug. 85	30 May 87	29 Aug. 11	. . .	Civil employ.
	Worc. R.	**Widdicombe**, Frederick Servington (50, 57, 155)	21 Aug. 66	29 Aug. 85	26 Oct. 86	29 Aug. 11	. . .	116 Mahrattas.
	Wilts. R.	*Thompson, Cyril Powney* (*94, 169*)	*21 Aug. 64*	*29 Aug. 85*	*11 Apl. 88*	*29 Aug. 11*	. . .	*Supy. List.*
	York. L. I.	*Barton, Arthur Elliott*	*12 Oct. 64*	*29 Aug. 85*	*12 Oct. 88*	*29 Aug. 11*	. . .	*Supy. List.*
	Leic. R.	*Obbard, Owen James* (*104*)	*7 Mar. 65*	*29 Aug. 85*	*22 Mar. 87*	*29 Aug. 11*	. . .	*Supy. List.*
	North'd. Fus.	*Cole, Henry Walter George*, C.S.I. (*57*)	*8 Mar. 66*	*29 Aug. 85*	*20 May 87*	*29 Aug. 11*	. . .	*Supy. List.*
	R. Ir. R.	**Geoghegan**, Stannus, C.B., *p.s.c.* (50, 71, 127) [L]	18 Feb. 66	29 Aug. 85	22 Nov. 89	29 Aug. 11	. . .	*Resdg. ex India.*
	W. Kent. R.	**Nepean**, Herbert Evan Chas. Bayley, C.M.G. (50, 71, 105, 125*a*, 148*a*, 161, 185)	10 Oct. 65	29 Aug. 85	23 Jan. 88	29 Aug. 11	*c*. 1 Jan. 17	
180	R. Innis. F.	**Ward**, Trimnell Martin (50, 52, 57, 66*a*, 71)	24 Oct. 66	29 Aug. 85	24 Oct. 88	29 Aug. 11	. . .	97 Infantry.
	R. Berks. R.	**Swan**, Charles Tarrant (50, 71)	20 Aug. 64	25 Aug. 86	26 Jan. 88	17 Sep. 11	. . .	
	R. Innis. F.	**Browne**, Alexander John Wogan (51, 52, 66*a*) [94, 169)	7 Nov. 64	16 Dec. 85	24 Oct. 87	18 Sep. 11		
	N. Lan. R.	**Jacob**, Arthur LeGrand, C.M.G., C.I.E., D.S.O. A.D.C. (50, 57, 71,	19 Feb. 67	30 Jan. 86	11 June 87	1 Oct. 11	*c*. 1 Jan. 17	
	R. War. R.	**Andrew**, David Clapham Adrian (57, 72*b*, 107, 170)	3 Jan. 66	30 Jan. 86	17 Apl. 87	22 Oct. 11	*c*. 1 Jan. 17	
185	G. Lan. R.	**Lucas**, Frederic George, C.B., D.S.O. (50, 53, 57, 71, 157, 162, 169)	20 Oct. 66	25 Aug. 86	1 Aug. 88	22 Oct. 11	*c*. 1 Jan. 17	
	Dub. Fus.	**Bolton**, Arthur Hely (59)	3 July 64	10 Nov. 86	1 Mar. 88	13 Nov. 11	. . .	*Resdg. ex India.*
	Dorset R.	**Bell**, Ralph Maitland (106)	31 July 63	25 Nov. 85	8 June 87	25 Nov. 11	. . .	S. and T. Corps.
	Devon. R.	**Dunlop**, Henry Hinton (50, 53)	16 May 64	25 Nov. 85	22 Feb. 87	25 Nov. 11		
	Linc. R.	*Herbert, Douglas*	*11 July 65*	*25 Nov. 85*	*14 June 87*	*25 Nov. 11*	. . .	*Supy. List.*
	Oxf. L. I.	**Johnston**, Herbert William (50, 52, 195)	27 Nov. 65	30 Jan. 86	14 Apl. 87	30 Jan. 12	. . .	
	N. Staff. R.	*Beville, Francis Granville*, C.I.E. (*161*)	*24 Mar. 67*	*30 Jan. 86*	*21 Jan. 88*	*30 Jan. 12*	. . .	*Supy. List.*
190	Middx. R.	**Elliot**, Charles Riversdale (50)	10 Oct. 65	30 Jan. 86	16 Mar. 87	30 Jan. 12	. . .	*Resdg. ex India.*
191	Devon. R.	**Wall**, Edward Watkin, C.M.G. (51, 66*a*, 71, 170)	20 Dec. 66	30 Jan. 86	22 Sep. 89	30 Jan. 12	*c*. 1 Jan. 17	

Colonels and Lieutenant-Colonels—*contd.*

No.	Former Corps.	Names.	Date of birth.	Date of first commission.	Date of appointment to Indian Army.	Date of Rank. Substantive Lt.-Colonel.	Date of Rank. Army.	Remarks.
192	Derby. R.	**Campbell**, Leslie Warner Yule, C.M.G. (50, 52, 71, 195)	24 Aug. 67	30 Jan. 86	22 Feb. 88	30 Jan. 12	*c.* 1 Jan. 17	
	20 Huss.	**Jollie**, Francis (50, 167)	30 Mar. 65	30 Jan. 86	30 Jan. 86	30 Jan. 12	. . . .	*Resdg. ex India.*
	N. Lan. R.	**Baldwin**, Guy Melfort, D.S.O. (50, 52, 71, 161, 165, 169, 204)	22 Mar. 65	30 Jan. 86	22 Mar. 88	30 Jan. 12	*c.* 3 June 15	
195	R. Muns. Fus.	**Carey**, Octavius William, C.M.G. (50, 71, 162, 169)	7 Sep. 65	30 Jan. 86	13 Oct. 87	30 Jan. 12	*c.* 3 June 16	
	York. L. I.	*Halliday, Hugh Maclean* (57, 90, 98)	*12 Apl. 66*	*30 Jan. 86*	*1 Apl. 91*	*30 Jan. 12*	. . . .	*Supy. List.*
	18 Huss.	**Arnold**, Arthur Seymour (50, 162)	24 Apl. 65	30 Jan. 86	19 Oct. 88	30 Jan. 12	. . . .	
	Leins. R.	**Widdicombe**, George Templer, C.B. (50, 53, 157)	13 Oct. 67	30 Jan. 86	10 June 88	30 Jan. 12	. . . .	
	D. C. L. I.	**Southey**, William Melvill, C.M.G. (50, 57, 169)	31 Aug. 66	30 Jan. 86	4 Jan. 90	30 Jan. 12	*c.* 3 June 15	
	Lan. Fus.	*Wodehouse, Frederic William*, C.I.E. (89, 133, 153)	*7 Apl. 67*	*30 Jan. 86*	*17 June 87*	*30 Jan. 12*	. . . .	*Supy. List.*
	R. Mar.	**Magrath**, Henry Augustus Frederick (66*a*, 71, 170)	20 Oct. 67	1 Feb. 86	3 Jan. 89	1 Feb. 12	. . . .	
200	R. Mar.	**Hurly**, Maurice Randall (50, 71, 104, 167)	12 June 64	1 Feb. 86	27 Dec. 88	1 Feb. 12	. . . .	*Resdg. ex India.*
	R. Scots.	**Taylor**, Arthur William Neufville, C.S.I. (50, 95, 120, 156)	11 May 63	13 Feb. 86	20 June 87	13 Feb. 12	. . . .	
		Service for pension;—*In ranks from 16 Oct. 82 to 12 Feb. 86 (1 yr. 245 ds. to count).*						
	R. A.	**Cook**, Charles Chesney, D.S.O. (57, 66*a*, 66*b*, 71, ?, 100, 137, 162, 169)	23 Aug. 66	18 Feb. 86	30 Mar. 88	18 Feb. 12	*c.* 18 Apl. 17	
	R. A.	**Home**, Jas. Murray, *p.s.c.* (50, 53, 71, 157,) [L]	12 Feb. 66	18 Feb. 86	3 Dec. 89	18 Feb. 12	*c.* 1 Jan. 17	
	7 D. G.	**MacGeorge**, Henry King, C.B. (57, 71, 155, 15)	10 May 65	3 Mar. 86	26 Feb. 90	3 Mar. 12	*c.* 22 Apl. 17	
205	R. War. R.	**Thacker**, Henry Joseph (90)	16 Jan. 65	28 Apl. 86	23 Feb. 88	28 Apl. 12	. . . .	*Resdg. ex India.*
	W. Kent R.	**Brown**, Walter Henry, C.B. (51, 52, 71, 147, 162, 193, 204)	27 Jan. 67	25 Aug. 86	2 Dec. 89	10 May 12	. . . .	
	Glouc. R.	**Hawkins**, George Arthur	17 Feb. 63	19 May 86	9 Mar. 87	19 May 12	. . . .	*Resdg. ex India.*
		Service for pension.—In Ranks from 17 Jan. 82 to 18 May 86 (half to count).						
	R. A.	**Burlton**, Alfred Ralph (64)	27 June 67	24 July 86	18 Feb. 89	19 May 12	. . . .	S. and T. Corps.
	Bord. R.	**Hallowes**, Francis William C.I.E. (94, 162, 169)	28 May 66	25 Aug. 86	20 Sep. 88	19 May 12	*c.* 3 May 17	S. and T. Corps.
210	Essex. R.	**Dunolly**, Kenneth James Grant (50, 53, 57, 71, 139, 172*a*, 195) [*l*]	27 Dec. 64	3 Dec. 86	29 Feb 88	30 May 12		119 Infantry.
	Wilts. R.	**Badcock**, Francis Frederick, D.S.O. (50, 66*a*, 157, 170)	13 Sep. 67	25 Aug. 86	23 Dec. 87	11 July 12	*c.* 20 Oct. 17	
	Norf. R.	**Vaughan**, Robert Edward, C.B. (172*a*)	12 Aug. 66	5 Feb. 87	27 Oct. 88	30 July 12	*c.* 3 June 15	S. and T. Corps.
	R. Sco. F.	*Roos-Keppel*, Hon. Sir *George Olof*, G.C.I.E., K.C.S.I. (106, 162, 199) [*L*]	*7 Sep. 66*	*25 Aug. 86*	*19 Sep. 97*	*25 Aug. 12*	. . . .	*Supy. List.*
	Leins. R.	**Senior**, Hy. Wm. Richard, D.S.O., *p.s.c.* (50, 53, 66*a*, 127, 156, 69) [L]	13 Sep. 66	25 Aug. 86	5 July 88	25 Aug. 12	*c.* 1 Dec. 17	
	Wilts. R.	**Hall**, Ernest Stanley (50, 52, 57, 167)	9 Apl. 67	25 Aug. 86	1 Jan. 88	25 Aug. 12	. . . .	
215	Midd'x R.	**Turner**, Frederick William Charles (50, 71, 106)	2 Sep. 67	25 Aug. 86	11 Dec. 87	25 Aug. 12	. . . .	20 Horse.
	E. Surr. R.	**Wood**, Ernest Joseph Macfarlane, D.S.O. (50, 66*a*, 74)	5 Feb. 67	25 Aug. 86	28 Jan. 88	25 Aug. 12	*Ty. c.* 25 Feb. 17 *c.* 3 Dec. 17	97 Infantry.
	7. D. G.	**Daunt**, Walter Dickens (50, 57, 65, 66*a*, 74)	2 Jan. 67	25 Aug. 86	13 Jan. 89	25 Aug. 12	. . . .	
	York. L. I.	**Vincent**, Frank Lloyd (50, 57)	22 Oct. 65	25 Aug. 86	1 Aug. 88	25 Aug. 12	. . . .	A. A. and Q. M. G.
	Bord. R.	**Moore**, Harry Alexander (50, 52, 166)	30 Dec. 66	25 Aug. 86	25 Feb. 88	25 Aug. 12	. . . .	36 Sikhs.
	Worc. R.	*Hodgson, Walter George* (50, 57, 195)	*7 Feb. 67*	*25 Aug. 86*	*14 Jan. 88*	*25 Aug. 12*	. . . .	*Supy. List.*
220	Wilts. R.	**Watson**, Alfred Carnac (50, 53)	26 Nov. 66	25 Aug. 86	28 Jan. 89	25 Aug. 12	. . . .	

	Leins. R.	*Merewether, John Walter Beresford,* C.I.E. *(50, 88, 106, 154)*	*24 June 67*	*25 Aug. 86*	*26 Mar 92*	*25 Aug. 12*	. . . .	*Supy. List.*
	S. Lan. R.	*Grey, William George (169, 204)* [L]	*12 Oct. 66*	*25 Aug. 86*	*23 Oct. 89*	*25 Aug. 12*	. . .	*Supy. List.*
	R. Mar.	*Windham, Charles Joseph,* C.I.E.	*7 Aug. 67*	*1 Sep. 86*	*6 Jan. 90*	*1 Sep. 12*	. . .	*Supy. List.*
	Devon. R.	**Andrew,** Albert William, C.M.G. (152) [L]	29 Mar. 66	8 Sep. 86	7 Feb. 89	8 Sep. 12	*c.* 14 May 18	
221	R. Ir. F.	**McCrea,** Alfred Coryton, C.M.G. (50, 107, 137)	6 Sep. 64	19 Jan. 87	22 Apl. 89	9 Sep. 12	*c.* 21 May 18	
	L'pool R.	**Codrington,** Hubert Walter (50, 52, 169)	17 Nov. 64	30 Mar. 87	9 Apl. 88	20 Sep. 12	*c.* 25 May 18	
		Service for pension;—In ranks from 17 Mar. 84 to 29 Mar. 87, (half to count).						
	Worc. R.	**Beatty,** Lionel Nicholson	4 May 67	4 May 87	1 Nov. 89	6 Oct. 12	*c.* 3 June 18	
	7 Huss.	*Peacock, Henry Barnes (161)*	*19 Oct. 66*	*13 Oct. 86*	*9 May 88*	*13 Oct. 12*	.	*Supy. List.*
225	Dorset. R.	**Mockler,** Guy Henry Gaston (105, 125*b*) [L]	9 Sep. 66	14 Sep. 87	22 Feb. 88	19 Oct. 12	. . . .	
	R. Sco. F.	**Malcolm,** William Leith (50, 57, 167, 169)	1 Dec. 66	27 Oct. 86	27 Aug. 90	27 Oct. 12	. . . .	21 Punjabis
	R. Ir. F.	**Ricketts,** Henry Carew (50, 66*a*, 167, 170)	22 Oct. 65	10 Nov. 86	9 June 88	10 Nov. 12		
	N. Staff R.	**Tanner,** Charles Oriel Oliphant (50, 162, 169)	4 Feb. 67	10 Nov. 86	14 Mar. 88	10 Nov. 12	*c.* 1 Jan. 18	
	Middl'x R.	**Forster,** Philip Byron Bohun (52, 170)	24 May 64	10 Nov. 86	10 Apl. 88	10 Nov. 12		
230	H. L. I.	**Stewart,** Francis Thornton (50, 167)	29 July 65	10 Nov. 86	2 Jan. 90	10 Nov. 12	.	
	N. Lan. R.	**Jamieson,** Aubrey James (50)	7 May 65	10 Nov. 86	18 Feb. 89	10 Nov. 12	.	
	R. Mar.	**Drake-Brockman,** David Henry, C.M.G. (50, 59, 157, 162, 170)	2 Apl. 68	1 Feb. 87	23 Oct. 91	22 Nov. 12	*c.* 16 June 18	
	W. Rid. R.	**Hickley,** Alfred Charles (52, 66*a*, 71, 157)	2 June 65	8 Dec. 86	10 May 88	8 Dec. 12	. . . .	
	R. Inns. F.	**Denne,** Richard William Arthur (51, 71, 195)	2 Oct. 65	8 Dec. 86	17 Apl. 89	8 Dec. 12	. .	79 Infantry.
235	Som. L. I.	**Kirkwood,** Thomas Moore (50, 52, 57, 66*a*, 71, 162, 170)	4 Feb. 65	8 Dec. 86	29 Dec. 87	8 Dec. 12	. . .	
	R. Sco. F.	**Roome,** Reginald Eckford (50, 53, 169)	12 July 66	5 Feb. 87	20 Apl. 88	14 Jan. 13	. . . .	
	S. Staff R.	*Carter, John Robert Blackhall Graham (50, 133)*	*10 Aug. 67*	*5 Feb. 87*	*29 Jan. 88*	*5 Feb. 13*	. . .	*Supy. List.*
	Worc. R.	**Chitty,** Ernest Richard Inglis (50, 52, 66*a*, 71, 155, 162, 169)	11 Oct. 67	5 Feb. 87	1 Feb. 88	5 Feb. 13	. . . .	
	Welsh R.	**Priestley,** Charles Eccles Nixon (51, 52, 66*a*, 167)	24 Apl. 67	5 Feb. 87	21 Feb. 88	5 Feb. 13	. . . .	42 Deoli Regt.
	York. R.	**Twigg,** Edward Francis (66*a*, 94, 100, 162, 169)	21 Feb. 68	5 Feb. 87	27 Feb. 88	5 Feb. 13	.	124 Infantry.
240	S. W. Bord.	**Hennessey,** John Patrick Cumberlege, C.M.G. (65, 71, 94, 162, 169)	24 Sep. 67	5 Feb. 87	22 Mar. 88	5 Feb. 13	{ *Ty. c.* 9 Mar. 18 / { *c.* 27 Aug. 18	} S. and T. Corps.
	Essex R.	**Holmes,** Gilbert Vallentin (50, 53, 57, 74, 172*a*, 202*a*)	14 Mar. 68	5 Feb. 87	12 Apl. 88	5 Feb. 13	. . . .	
	R. War. R.	**Wikeley,** James Masson (50, 169)	2 Oct. 67	5 Feb. 87	4 May 88	5 Feb. 13	. . . .	
	6 D. G.	**Browne,** Alfred Percy, D.S.O. (51)	15 Jan. 68	5 Feb. 87	10 May 88	5 Feb. 13	. . . .	
	1 D. G.	**Waterfield,** Arthur Chas. Malleson, M.V.O., *p.s.c.* (50, 71, 88, 161, 169) [*l*]	4 July 66	5 Feb. 87	12 May 88	5 Feb. 13	10 July 12	
245	R. Dub. Fus.	**Roddy,** Henry Hugh, C.B. (57)	11 Oct. 66	5 Feb. 87	13 May 88	5 Feb. 13	. . . .	S. and T. Corps.
	K.O.S. Bord.	*Kennion, Roger Lloyd,* C.I.E. *(161, 169)*	*16 Dec. 66*	*5 Feb. 87*	*9 June 88*	*5 Feb. 13*	. . .	*Supy. List.*
	W. York R.	**Tweddell,** Henry (50, 172*a*)	4 Sep. 66	5 Feb. 87	2 Aug. 88	5 Feb. 13	. . .	
	Worc. R.	*Vaughan, Richard Randal (170)*	*15 Dec. 66*	*5 Feb. 87*	*25 Feb. 89*	*5 Feb. 13*	. . .	*Supy. List.*
	Bedf. R.	**Dale,** George Arthur, C.M.G. (50, 53, 105, 162, 169) [*l*]	19 Nov. 66	5 Feb. 87	30 July 89	5 Feb. 13	. . . .	
	W. I. R.	**Lindsay,** Henry Arthur Peyton, C.M.G. (57, 203) [L]	29 Jan. 68	5 Feb. 87	29 Nov. 89	5 Feb. 13	*c.* 3 June 16	S. and T. Corps
	Devon. R.	**Turner,** Alfred Granville Burne, D.S.O. (50, 71, 162, 169)	20 May 66	5 Feb. 87	6 May 89	5 Feb. 13	.	
250	Ches. R.	**Davies,** Charles Henry, C.M.G., D.S.O. (162)	20 Nov. 67	5 Feb. 87	1 Sep. 89	5 Feb. 13	*c.* 3 June 16	
	York L. I.	*Benn, Robert Arthur Edward,* C.I.E. *(94, 169)*	*4 Feb. 67*	*5 Feb. 87*	*5 Feb. 90*	*5 Feb. 13*	. . . .	*Supy. List.*
	Oxf. L. I.	**Moberly,** Frederick James, D.S.O., *p.s.c.* (162, 165, 170)	15 Sep. 67	22 Aug. 88	4 Oct. 91	13 Feb. 13	. . . .	25 Punjabis.
252	Oxf. L. I.	**Faunce,** Robert DeLaune (50, 148*a*)	25 Sep. 67	17 Sep. 87	27 Aug. 89	13 Feb. 13	. . . .	

Colonels and Lieutenant-Colonels—*contd.*

No.	Former Corps.	Names.	Date of birth.	Date of first commission.	Date of appointment to Indian Army.	Date of Rank. Substantive Lt.-Colonel.	Date of Rank. Army.	Remarks.
253	R. A.	**Fordyce**, Henry Lawrence Dingwall	12 July 67	16 Feb. 87	28 June 89	16 Feb. 13	*c.* 3 July 15	S. and T. Corps.
	R. A.	**Saunders**, Arthur Rivers (12, 50, 66*a*, 71)	6 Oct. 67	16 Feb. 87	23 June 90	16 Feb. 13	....	
255	Manch. R.	**Pearson**, Hugh Frederic Archie (50, 53, 71, 162, 165)	19 June 69	22 Aug. 88	6 Nov. 89	21 Feb. 13	....	
	R. Mar.	**Harding**, Maynard Ffolliott, C.M.G. (66*a*, 71, 162, 195) [*l*]	10 Jan. 69	1 Feb. 88	6 Nov. 93	10 Mar. 13		
	Conn. Rang.	**Ridgway**, Richard Thomas Incledon, C.B. (50, 169)	17 Aug. 68	14 Sep. 87	3 Mar. 89	22 Mar. 13	*Ty. c.* 10 July 17 *c.* 3 June 18	
	R. Sco. F.	**Cunningham**, Percy Henry (66*a*, 95, 155, 162, 169)	15 Nov. 63	13 Apl. 87	29 Feb. 88	12 Apl. 13	....	
		Service for pension;—In ranks, 3 yrs. 57ds. (half to count); for promotion from 13 Apl. 87.						
	Bedf. R.	**Oakes**, George Richard (50, 195)	29 Apl. 65	4 May 87	29 Feb. 88	4 May 13	....	
260	Oxf. L. I.	**Bruce**, *Hon.* Chas. Granville, C.B., M.V.O. (51, 156, 170)	7 Apl. 66	4 May 87	29 Oct. 88	4 May 13	*c.* 8 Nov. 15	
	A. & S. H.	**Anderson**, Allan Meyrick (51, 52, 57, 71, 195)	13 Aug. 66	4 May 87	4 Apl. 89	4 May 13	....	
	Conn. Rang.	**Maxwell**, Hamilton George (72*b*, 107, 167)	7 July 63	7 May 87	29 Jan. 88	7 May 13	....	
		Service for pension;—In ranks, 3 yrs. 89ds. (half to count); for promotion from 7 May 87						
	Dorset. R.	**Savi**, Robert James Taylor (50, 53, 66*a*, 71, 167)	26 Mar. 68	14 Sep. 87	25 Feb. 88	7 June 13	....	*Resdg. ex India.*
	R. Berks. R.	**Wooldridge**, William Hugh (50, 57, 71, 94, 169)	19 Mar. 68	16 Nov. 87	25 Mar. 90	20 July 13	....	
	Linc. R.	*Jacob, Harold Fenton*, C.S.I. (162, 175, 204)	*22 Dec. 66*	*14 Sep. 87*	*22 Feb. 88*	*14 Sep. 13*	....	*Supy. List.*
265	Glouc. R.	**Wintour**, Evelyn (50)	12 Dec. 64	14 Sep. 87	23 Feb. 88	14 Sep. 13	....	
	W. I. R.	**Robinson**, Godfrey Walker, C.B. (50, 66*a*, 71, 170)	28 Oct. 63	14 Sep. 87	29 Feb. 88	14 Sep. 13	....	
	Leic. R.	**Adye**, Daniel Richard (50, 66*a*, 71, 161, 167) [L]	29 Aug. 67	14 Sep. 87	10 Mar. 88	14 Sep. 13	*c.* 1 Jan. 17	98 Infantry.
	Som. L. I.	**Keays**, Robert William Clairmonte (50, 71, 162, 195)	8 June 68	14 Sep. 87	31 Aug. 88	14 Sep. 13	....	83 Infantry.
	7 D. G.	**ffrench-Mullen**, John Lawrence Wm., C.I.E. (57, 105, 125*a*, 140, 148*b*, 162, 169, 180)	24 Mar. 68	14 Sep. 87	30 Jan. 89	14 Sep. 13	....	
270	3 D. G.	**Lane**, Henry Arthur (66*a*, 71)	6 Oct. 68	14 Sep. 87	9 Feb. 89	14 Sep. 13	....	
	North'n R.	**Powell**, William Bowen, C.M.G. (50, 57, 71, 95, 157)	28 Feb. 68	14 Sep. 87	10 Mar. 89	14 Sep. 13	....	7 G. R.
	R. Ir. F.	*Harvest, Herbert de Vere* (104, 153, 162, 204)	*9 May 68*	*14 Sep. 87*	*17 Dec. 90*	*14 Sep. 13*	....	*Supy. List.*
	Linc. R.	*Pirrie, Francis William*, C.M.G., C.I.E. (170)	*23 Aug. 67*	*14 Sep. 87*	*25 Oct. 89*	*14 Sep. 13*	*c. 3 June 16*	*Supy. List.*
	R. Lanc. R.	**Watling**, Gordon (51, 52, 57, 71, 93, 162, 170)	14 Sep. 67	14 Sep. 87	12 May 89	14 Sep. 13	....	11 Rajputs.
	N. Lanc. R.	**Thornton**, Charles, Edward, C.M.G. (51, 167)	10 Nov. 67	14 Sep. 87	22 Mar. 90	14 Sep. 13	....	
	Leic. R.	**Obbard**, Edward Naismith (50, 57)	14 Nov. 67	14 Sep. 87	10 June 89	14 Sep. 13	....	Mily. Farms Dept.
275	R. Muns. Fus.	**Talbot**, John (51, 169)	4 Feb. 68	14 Sep. 87	24 Feb. 90	14 Sep. 13	....	
	Devon. R.	**Jones**, Harry Harvey	9 Dec. 67	14 Sep. 87	7 Mar. 90	14 Sep. 13	*Ty. c.* 9 Mar. 18	S. and T. Corps.
	Sco. Rif.	**Bayley**, Edward Charles, C.I.E. (50, 167, 170)	13 Dec. 67	14 Sep. 87	25 Oct. 90	14 Sep. 13	....	
278	Bord. R.	**Elsmie**, Alexr. Montagu Spears, C.M.G., *p.s.c.* (50, 53, 129, 161, 166, 169) [*l*]	2 Nov. 69	30 Jan. 89	3 May 99	22 Sep. 13	*c.* 3 June 16	

	W. York. R.	*DesVœux*, the Hon. *Herbert* (104)	*27 Dec. 63*	*28 Sep. 87*	*5 Feb. 88*	*28 Sep. 13*	. . . .	*Supy. List.*
		Service for Pension.—In Ranks—3 yrs. 335 ds. (Half to count.)						
279	21 Huss.	**Harward**, Arthur John Netherton (66*a*, 74, 162, 170)	20 Apl. 67	5 Oct. 87	1 Sep. 88	5 Oct. 13	*c.* 3 June 15	
280	W. I. R.	**Boddam**, Ernest Berkeley Chas., *p.s.c.* (50, 53, 66*a*, 71, 157, 170)	29 Dec. 66	19 Oct. 87	5 Apl. 90	19 Oct. 13	. . . .	5 G. R.
	Leic. R.	**Chaplin**, Ronald Eustace (50)	19 Feb. 66	16 Nov. 87	31 Mar. 88	2 Nov. 13	. . . .	8 Cavalry.
	Leic. R.	**Watson**, Lionel Arthur, *p.s.c.* (51, 66*a*, 94, 129, 162, 165, 169) [L]	24 Aug. 68	22 Aug. 88	11 Nov. 90	13 Nov. 13	. . . .	
	Manch. R.	**Chichester**, Alfred Godfrey deVaud (50, 57, 71, 72*a*, 170)	7 Nov. 66	16 Nov. 87	1 Apl. 88	16 Nov. 13	. . . .	
	R. Sco. F.	**Thomason**, Archibald Fawcett (170)	31 Mar. 66	16 Nov. 87	30 Oct. 88	16 Nov. 13	. . . .	S. and T. Corps.
285	Durh. L. I.	**Dunsford**, Francis Pearson Shaw, C.I.E. (50, 53, 71, 137, 161)	20 Oct. 66	16 Nov. 87	11 Dec. 88	16 Nov. 13	. . . .	
	R. W. S. R.	**Nisbett**, William Gibbon (50, 107, 167)	24 Sep. 65	16 Nov. 87	4 Jan. 89	16 Nov. 13	. . . .	92 Punjabis.
	Lancs. Fus.	*Stone, William Richard* (104)	*8 Aug. 66*	*16 Nov. 87*	*21 Jan. 89*	*16 Nov. 13*	. . . .	*Supy. List.*
	S. Lan. R.	**Boome**, Edward Herbert, C.B., C.M.G. (51, 52, 66*a*, 71, 195) [*l*]	31 July 65	16 Nov. 87	11 Feb. 89	16 Nov. 13	*c.* 3 *June 16*	
	Leins. R.	*Bowen, Charles Eustace* (50, 104)	*3 Aug. 65*	*16 Nov. 87*	*26 Feb. 89*	*16 Nov. 13*	. . . .	*Supy. List.*
	D. C. L. I.	*Ware, Frank Cooke Webb*, C.I.E. (57, 94, 170)	*10 Nov. 66*	*16 Nov. 87*	*28 Feb. 89*	*16 Nov. 13*	. . . .	*Supy. List.*
	R. Ir. R.	**Sutton**, Francis Goodwin Hill (50, 107, 120, 157)	20 Feb. 68	16 Nov. 87	17 Mar. 89	16 Nov. 13	. . . .	10 G. R.
	H. L. I.	**Rose**, John Latham, C.I.E. (50, 57, 66*a*, 124*a*, 127, 156)	24 Apl. 67	16 Nov. 87	6 May 89	16 Nov. 13	. . . .	
290	L'pool R.	**Porteous**, Charles McLeod (51, 66*a*, 71, 72*a*, 156, 162, 169)	15 Aug. 65	16 Nov. 87	30 May 89	16 Nov. 13	. . . .	
	Hamps. R.	**Robson**, Charles Græme (50, 52, 57, 60, 71, 162) [L]	5 Oct. 67	16 Nov. 87	1 Sep. 89	16 Nov. 13	. . . .	10 Jats.
	Sea. Highrs.	**Wilson**, James Alban, D.S.O. (50, 53, 57, 71, 157)	12 Feb. 65	16 Nov. 87	1 Nov. 89	16 Nov. 13	27 Mar. 13	8 G. R.
	E. Sur. R.	**Parr**, Harington Owen (50, 53, 72*b*, 157, 162)	21 Apl. 67	16 Nov. 87	26 Mar. 90	16 Nov. 13	*c.* 3 *June 16*	7 Rajputs.
	Midd'x R.	*Prince, Wm. Chas. Stuart* (50, 106, 154, 162, 169, 195, 202, 204) [L]	*22 Nov. 67*	*16 Nov. 87*	*20 Apl. 90*	*16 Nov. 13*	. . . .	*Supy. List.*
	North'd. Fus.	**Webb**, Algernon Edward (161)	13 Feb. 65	16 Nov. 87	14 May 90	16 Nov. 13	. . . .	S. and T. Corps.
295	3 D. G.	**Holland-Pryor**, Pomeroy, C.M.G., D.S.O., M.V.O., (50, 66*a*, 66*b*, 71, 162, 169)	7 July 66	16 Nov. 87	14 Jan. 91	16 Nov. 13	*c.* 1 Jan. 17	Bde. Comdr.
	Leic. R.	**Ward**, Arthur, D.S.O. (65, 162, 165, 169)	1 Feb. 66	4 May 87	21 Mar. 90	12 Dec. 13	. . . .	30 Punjabis.
		Half pay from 29 Aug. 94 to 11 Feb. 97 (1 yr. to count).						
	Worc. R.	**Wooldridge**, Herbert Cheselden (50, 71)	30 Jan. 70	26 Apl. 90	6 Apl. 92	21 Jan. 14	. . . .	
	Devon. R.	**Rattray**, Charles, C.B., *p.s.c.* (50, 53, 57, 127, 165, 169	26 July 68	11 Feb. 88	27 Feb. 90	22 Jan. 14	*c.* 3 June 16	
	R. Mar.	**Murray**, Edward Rushworth Blakiston (50, 52, 57, 71, 167)	7 Dec. 68	1 Feb. 88	21 Oct. 92	1 Feb. 14	. . . .	90 Punjabis.
300	R. Ir. Rif.	**Clery**, Carleton Buckley Laming C.B. (50, 57, 66*a*, 71, 107)	6 Sep. 69	24 Apl. 89	2 Apl. 90	6 Feb. 14	*c.* 3 June 15	104 Rifles.
	L'pool. R.	**Macquoid**, Chas. Edwd. Every Francis Kirwan, D.S.O. (50, 71, 125, 162)	2 Aug. 69	11 Feb. 88	26 Aug. 88	11 Feb. 14	. . . .	4 Cavalry.
	6 D. G.	**Stack**, Charles Spottiswoode, C.M.G. (50, 53, 66*a*, 124*a*, 162, 169)	28 Sep. 68	11 Feb. 88	13 Sep. 88	11 Feb. 14	. . . .	33 Lt. Cavalry.
	R. Lanc. R.	**Dickson**, John Herbert, C.I.E. (65)	10 Apl. 67	11 Feb. 88	2 Apl. 89	11 Feb. 14	*Ty. c.* 20 Feb. 17	S. and T. Corps.
	Durh. L. I.	**Lloyd**, William Edmund Eyre (50)	10 Nov. 67	11 Feb. 88	26 June 89	11 Feb. 14	. . . .	97 Infantry.
305	Bord. R.	**Pennington**, Arthur Watson, M.V.O. (50, 57, 71, 170)	20 Mar. 67	11 Feb. 88	12 Jan. 90	11 Feb. 14	. . . .	
	R. War. R.	**Longden**, Arthur Berridge (50, 72*a*, 170)	10 Mar. 68	11 Feb. 88	25 Dec. 91	11 Feb. 14	. . . .	38 Dogras.
	Manch. R.	**Eastmead**, Charles Sidney (50, 53, 66*a*, 71, 157, 162) [L]	30 Sep. 69	11 Feb. 88	18 July 89	11 Feb. 14	. . . .	3 G. R.
	Bedf. R.	**Vaughan**, Edward Gyles, D.S.O. (65, 162, 170)	21 Nov. 67	11 Feb. 88	26 July 89	11 Feb. 14	. . . .	S. and T. Corps.
	R. War. R.	**Vanrenen**, George Rainer (50, 52, 166)	24 Nov. 68	11 Feb. 88	13 July 90	11 Feb. 14	. . . .	16 Rajputs.
310	Bedf. R.	**Stevens**, Nathaniel Melhuish Comins, C.M.G. (50, 53, 71, 94, 169, 193) [L]	10 Aug. 68	11 Feb. 88	11 Nov. 89	11 Feb. 14	. . . .	
	R. Ir. F.	**Annesley**, Arthur Stephen Robert (170)	12 Jan. 69	11 Feb. 88	1 Aug. 89	11 Feb. 14	*c.* 3 June 15	S. and T. Corps.
	Norf. R.	**Heffernan**, Herbert William (50, 53, 71, 195)	16 Oct. 67	11 Feb. 88	10 Aug. 89	11 Feb. 14	. . . .	
	Linc. R.	**Donnan**, William (136, 174, 203) [L]	21 Sep. 67	11 Feb. 88	12 Aug. 89	11 Feb. 14	. . . .	
		On pension establishment from 9 Mar. 16 to 11 Jan. 18.						
		Reinstated in the Indian Army with effect from 12 Jan. 18.						
	Dorset. R.	**Major**, Francis Forbes (50, 66)	23 Dec. 67	11 Feb. 88	10 Oct. 89	11 Feb 14	. . . .	95 Infantry.
315	R. Essex.	**Brooke**, Harry Morris Mitchelson (50)	6 Dec. 68	11 Feb. 88	22 Oct. 89	11 Feb. 14	. . . .	S. and T. Corps.
	Dorset. R.	**Sherer**, John Corrie (50, 94, 162, 169)	26 Sep. 68	11 Feb. 88	3 Nov. 89	11 Feb. 14	. . . .	
	W. York. R.	**Beatson Bell**, John (50, 74)	12 June 66	11 Feb. 88	15 Jan. 90	11 Feb. 14	. . . .	D. J. A.-G., N. Army.
318	York. L. I.	**Lloyd-Jones**, Frederic Llewellyn (50, 53)	12 Sep. 67	11 Feb. 88	24 Nov. 89	11 Feb. 14	. . . .	113 Infantry.

Colonels and Lieutenant-Colonels—*contd.*

No.	Former Corps.	Names.	Date of birth.	Date of first commission.	Date of appointment to Indian Army.	Date of Rank. Substantive Lt.-Colonel.	Date of Rank. Army.	Remarks.
319	W. York. R.	**Hitchins**, Harold Edmund (50, 57, 66*a*)	6 Feb. 68	11 Feb. 88	6 May 90	11 Feb. 14		
320	Derby. R.	**Barnard**, Andrew Edward (66*a*, 107, 156)	30 Jan. 67	11 Feb. 88	15 Oct. 91	11 Feb. 14		
	R. Dub. Fus.	**Price**, Cyril Uvedale, C.M.G. (50, 71, 94, 169)	13 May 68	11 Feb. 88	4 Jan. 90	11 Feb. 14	*c.* 1 Jan. 17	130 Baluchis.
	Derby. F.	**Black**, Walter Clarence, *p.s.c.* (50, 53, 162, 170) [L]	29 Jan. 67	11 Feb. 88	19 Apl. 90	11 Feb. 14	*c.* 1 Jan. 17	
	R. Dub. Fus.	**Pierce**, Frederick George (52, 71, 195)	9 Apl. 67	11 Feb. 88	1 Mar. 92	11 Feb. 14		102 Grenadiers.
	Derby. R.	**Walker**, William Robert (50, 57, 66*a*)	21 Apl. 69	11 Feb. 88	6 Oct. 91	11 Feb. 14		
325	Suff. R.	**Kauntze**, Cedric Richard (51, 66*a*, 71, 148*a*, 195)	20 Nov. 67	11 Feb. 88	2 Apl. 90	11 Feb. 14		
	W. York R.	**Moore**, Charles Hesketh Grant, C.M.G., D.S.O. (170)	5 July 68	11 Feb. 88	11 Sep. 91	11 Feb. 14		S. and T. Corps.
	Linc. R.	**Kemball**, John Shaw (50, 161, 167, 169)	18 Nov. 67	11 Feb. 88	7 Oct. 92	11 Feb. 14		
	W. I. R.	**Baker**, Harold Robert (50, 53, 57, 71, 195)	5 Feb. 68	15 Feb. 88	1 Feb. 91	15 Feb. 14		73 Infantry.
	R. A.	**Murray**, Alan Beville (50, 71, 105, 125*a*, 140, 167)	18 Sep. 68	17 Feb. 88	2 Feb. 90	17 Feb. 14		
	R. A.	*Morris, Donald Ogilvy*	*22 Oct. 68*	*17 Feb. 88*	*19 Oct. 91*	*17 Feb. 14*		*Supy. List.*
	2 D. G.	*Sykes*, Sir *Percy Molesworth*, K.C.I.E., C.M.G.	*28 Feb. 67*	*29 Feb. 88*	*22 Nov. 02*	*1 Mar. 14*		*Supy. List.*
	W. Rid. R.	*Beale, Willy (89, 132)*	*8 May 67*	*29 Feb. 88*	*31 Oct. 89*	*1 Mar. 14*		*Supy. List.*
330	R. Berks. R.	**Gray**, Frederick William Barton, C.M.G., D.S.O. (50, 170)	7 Feb. 67	22 Aug. 88	22 Jan. 90	8 Mar. 14	*c.* 1 Jan. 18	Bde. Comdr.
	7 D. G.	**Young**, Wilfred Edward (50, 52, 170)	12 July 68	21 Sep. 89	24 May 92	8 Mar. 14		
	R. W. Fus.	**Ford**, Harry Burroughes (72*a*, 72*b*, 162, 165, 169)	6 June 68	14 Mar. 88	4 Apl. 90	14 Mar. 14		29 Punjabis.
	R. W. Kent R.	**Dallas**, Julian Stuart (50, 53, 71, 195)	21 June 67	9 May 88	18 Nov. 89	9 May 14		83 Lt. Infantry.
	W. York. R.	**Hutchinson**, Charles Roderick Maclagan (71, 125, 157)	29 Sep. 67	9 May 88	21 Jan. 92	9 May 14		*Resdg. ex India.*
335	Linc. R.	**Money**, Ernest Douglas, C.I.E., D.S.O. (50, 57, 157, 170)	11 Mar. 66	9 May 88	4 Mar. 91	9 May 14		
	Shrops. L. I.	**Chatterton**, Geo. Davenport Latham, D.S.O. (30, 57, 66*a*, 71, 170)	22 Dec. 67	8 Dec. 88	13 Mar. 90	1 June 14		
	R. A.	**Cobbe**, Henry Hercules, C.M.G., D.S.O. (71, 107, 170, 181)	26 Feb. 69	27 July 88	11 July 91	27 July 14	*c.* 3 June 18	13 Lancers.
	Oxf. L. I.	**Bagshawe**, Frederick William, C.I.E. (50, 57, 66*a*, 204) [L]	6 Mar. 68	22 Aug. 88	19 Sep. 89	22 Aug. 14		Mily. Accts. Dept.
	E. Surr. R.	**Henslowe**, Francis John Deloraine (50, 53, 162, 169)	29 Nov. 68	22 Aug. 88	24 Oct. 89	22 Aug. 14		
340	S. Staff R.	**Fulton**, Frederick Charles Rampini	4 July 68	22 Aug. 88	13 Mar. 91	22 Aug. 14	*c.* 3 June 16	S. and T. Corps.
	R. Ir. Rif.	*Playfair, Alan (50, 66, 145, 161)*	*15 June 68*	*22 Aug. 88*	*3 Feb. 90*	*22 Aug. 14*		*Supy. List.*
	Worc. R.	**Gordon**, John Lewis Randolph, C.B. (50, 165, 170)	17 Nov. 67	22 Aug. 88	19 July 91	22 Aug. 14		15 Sikhs.
	Leic. R.	**Cooper**, William George (65, 66*a*, 71)	31 Dec. 68	22 Aug. 88	9 Apl. 90	22 Aug. 14		
	L'pool R.	*Blakeway*, the Hon. *Denys Brooke*, C.I.E. (*161, 165, 169*)	*29 May 70*	*22 Aug. 88*	*14 June 90*	22 Aug. 14		*Supy. List.*
	L'pool R.	**Ralph**, Alfred Colyer, D.S.O. (50)	3 May 69	22 Aug. 88	11 Aug. 90	*22 Aug. 14*		4 Rajputs.
	Manch. R.	**Stewart**, John Alexander (50, 52, 66*a*, 162)	5 June 68	22 Aug. 88	1 June 93	22 Aug. 14		13 Rajputs.
	Manch. R.	*Condon, James Knighton, (104, 133, 161, 204)*	*29 Aug. 67*	*22 Aug. 88*	*30 July 90*	*22 Aug. 14*		*Supy. List.*
345	7 D. G.	**Humfrey**, Fredk. Geo. Clinton (50, 65, 66*a*, 106, 117, 162, 167, 170)	31 July 69	22 Aug. 88	20 Nov. 90	22 Aug. 14		12 Cavalry.
	N. Staff R.	**Ainslie**, Henry Perceval (51, 52, 71, 195)	2 Dec. 68	22 Aug. 88	17 Nov. 90	22 Aug. 14		
	S. W. Bord	**Basevi**, William Henry Francis (66*a*, 105, 121, 140, 167)	16 May 68	22 Aug. 88	12 Apl. 92	22 Aug. 14		91 Punjabis.
348	L'pool R.	**Fraser**, Frederick John (50, 172) *Forfeits 17 Jan. 03 for pension*	6 May 67	12 Sep. 88	24 Apl. 92	12 Sep. 14		

349	Linc. R.	**Baker**, Cecil Norris	29 Jan. 69	19 Sep. 88	16 Aug. 91	19 Sep. 14	. . . .	Mily. Accts. Dept.
	Wilts. R.	*Knox, Stuart George*, C.S.I., C.I.E. (50, 88, 94, 100, 127, 162, 169)	*7 Oct. 69*	*19 Sep. 88*	*14 Apl. 91*	*19 Sep. 14*	.	*Supy. List.*
350	4 D. G.	**Sheppard**, George Sidney, C.M.G.	1 Nov. 67	29 Dec. 88	31 Oct. 90	23 Oct. 14	c. 3 June 16	Mily. Accts. Dept.
	E. Lan. R.	**O'Meara**, Charles Albert Edmond, C.I.E. (162, 170)	7 Aug. 68	17 Jan. 91	16 July 93	23 Oct. 14	. .	Army Clothing Dept.
	R. In. Fus.	*Dew, Armine Brereton*, C.S.I., C.I.E. (50, 94, 100, 156, 161, 169)	*27 Sep. 67*	*10 Nov. 88*	*16 Dec. 89*	*10 Nov. 14*	. . . .	*Supy. List.*
	E. Surr. R.	**Elderton**, Murray Trent (50, 71, 155)	26 Dec. 66	10 Nov. 88	2 July 91	10 Nov. 14	. . . .	108 Infantry
	W. York. R.	**Haldane**, Charles Levenax, C.M.G., D.S.O. (50, 57, 66a, 71, 94, 107, 157, 162, 172a 174) [*l*]	23 Nov. 66	10 Nov. 88	21 Apl. 91	10 Nov. 14	. . . .	7 G. R.
	E. Lan. R.	**Amesbury**, Wheaton Lipyeatt Raleigh (50, 57, 64, 71, 105)	4 Mar. 68	10 Nov. 88	30 Apl. 91	10 Nov. 14	. . . .	S. and T. Corps.
355	W. York. R.	**Codrington**, Harry de Burgh (64, 67a) *Service for pension;—In Ranks, 10 Mar. 86 to 27 Nov. 88 (half to count).*	10 Feb. 67	28 Nov. 88	10 May 90	26 Nov. 14	. . . .	S. and T. Corps.
	E. Surr. R.	**Coke**, Augustus Hodson (50, 71, 105)	21 May 66	5 Dec. 88	14 Jan. 91	5 Dec. 14	. . . .	96 Infantry.
	W. Rid. R.	**Lillingston**, William Edward Gordon (50, 57, 65, 71)	15 June 67	8 Dec. 88	8 May 90	8 Dec. 14	. . . .	*Resdg. ex India.*
	Midd'x R.	**Strahan**, Geo. Aubrey (50, 52, 71, 167)	16 Mar. 66	8 Dec. 88	11 Sep. 90	8 Dec. 14	. . . .	90 Punjabis.
	R. Ir. Rif.	**Carter**, Godfrey Lambert, C.I.E. (95, 162a, 169)	2 Nov. 68	8 Dec. 88	5 Oct. 90	8 Dec. 14	. . . .	106 Pioneers.
360	H. L. I.	**Anderson**, William Christian, C.M.G. (50, 72a, 72b, 157)	20 Feb. 67	8 Dec. 88	4 May 91	8 Dec. 14	c. 3 June 16	1 G. R.
	E. York R.	**Firth**, Richard Anson, D.S.O. (50, 125a, 157, 199)	17 Nov. 66	8 Dec. 88	2 Apl. 91	8 Dec. 14	.	8 G. R.
	R. Lanc. R.	*Mears, Arthur* (50, 162) [L]	*15 June 69*	*30 Jan. 89*	*3 May 90*	*30 Jan. 15*	. . . .	*Supy. List.*
	R. Ir. R.	*Nethersole, Frederick Ralph*	*17 Aug. 65*	*30 Jan. 89*	*15 Aug. 90*	*30 Jan. 15*	. . . .	*Supy. List.*
	G. Highrs.	**Lee**, Alexander William Henry (50, 66a, 71, 94, 105, 140, 157, 169)	17 Oct. 67	30 Jan. 89	28 Aug. 90	30 Jan. 15	*Ty. l. c.* 6 Oct. 14	7 G. R.
	Glouc. R.	**Douglas**, Walter Binny (50, 66a, 71)	7 June 68	30 Jan. 89	4 May 92	30 Jan. 15	. . . .	1 Brahmans.
	R. Lanc. R.	**Tennant**, Edward (50, 57, 64, 71, 174) [L]	21 Feb. 69	30 Jan. 89	17 Nov. 90	30 Jan. 15	. . . .	
365	Bord. R.	**Paul**, Robert Sears (50, 66a, 94, 100, 162, 169)	23 Sep. 68	30 Jan. 89	3 July 91	30 Jan. 15	. . . .	126 Infantry.
	R. Muns. Fus.	*Oldfield, Thomas Allan Francis Ross* (153) *Half pay from 30 June 93 to 1 Dec. 96.*	*21 Sep. 66*	*29 Aug. 85*	*18 Oct. 88*	*31 Jan. 15*	. . . .	*Supy. List.*
	R. A.	**Templer**, Cyril Frank, C.M.G., C.I.E. (50, 161)	3 Oct. 69	15 Feb. 89	17 May 91	15 Feb. 15	*Ty. c.* 5 Sep. 15	D. G., Army Remt. Dept.
	Derby. R.	**Kaye**, Cecil, C.S.I., C.I.E. (50, 53, 88, 125b, 161, 169)	27 May 68	6 Mar. 89	26 June 92	6 Mar. 15	*Ty. l. c.* 6 Oct. 14	
	Lanc. Fus.	**Maxwell**, Laurence Lockhart, C.M.G. (50, 162, 170)	23 Oct. 68	23 Mar. 89	20 Sep. 91	23 Mar. 15	c. 3 June 18	2 Lancers.
	R. Ir. R.	**Andrew**, Frederick Annesley, D.S.O. (50)	30 May 68	23 Mar. 89	26 Aug. 91	23 Mar. 15	. . . .	128 Pioneers.
370	Derby R.	**Munn**, Reginald George (50, 52, 66a, 162, 165, 169)	20 Aug. 69	23 Mar. 89	28 Sep. 91	23 Mar. 15	. . . .	36 Sikhs.
	Wilts. R.	**Alexander**, Henry Stirling, D.S.O. (50, 53, 57, 162)	21 Apl. 70	23 Mar. 89	13 Nov. 90	23 Mar. 15	. . . .	44 Infy.
	Lans. Fus.	*Burton, Richard Watkins*	*5 Nov. 68*	*23 Mar. 89*	*16 Dec. 90*	*23 Mar. 15*	. . . .	*Supy. List.*
	R. W. S. R.	*Plowden, Trevor Chichele* (50, 170)	*20 July 69*	*23 Mar. 89*	*6 Apl. 92*	*23 Mar. 15*	. . . .	*Supy. List.*
	R. W. S. R.	**Travers**, Robert Eaton (50, 66a, 71, 157)	13 Dec. 69	23 Mar. 89	10 Feb. 92	23 Mar. 15	. . . .	
	Glouc. R.	**Geoghegan**, Francis Edward, C.I.E.	14 Aug. 69	23 Mar. 89	12 Feb. 91	23 Mar. 15	. . . .	S. and T. Corps.
	Bord. R.	**Maclachlan**, Thomas Robertson, C.M.G. (50, 169)	8 Apl. 70	23 Mar. 89	1 Feb. 93	23 Mar. 15	c. 10 Mar. 17	92 Punjabis.
375	W. York R.	**Cullen**, Ernest Henry Scott, C.M.G., D.S.O., M.V.O. (50, 52, 107)	16 Nov. 69	1 Mar. 90	2 July 92	24 Mar. 15	. .	34 Pioneers.
	Glouc. R.	**Bannerman**, Wyndham Philip (66a, 172a)	16 Feb. 70	24 Apl. 89	29 Oct. 90	24 Apl. 15	. . . .	31 Lancers.
	Worc. R.	**Cassels**, Gilbert Robert, *p.s.c.* (50, 53, 66a, 162, 167)	4 Apl. 70	24 Apl. 89	20 Feb. 92	24 Apl. 15	*Ty. l. c.* 23 Sep. 14	123 Rifles.
	R. Dub. Fus.	*Peck, John Herbert* (66a, 94, 107, 170)	*23 Nov. 68*	*24 Apl. 89*	*10 Aug. 91*	*24 Apl. 15*	. . . .	*Supy. List.*
	Norf. R.	**Becher**, George Arthur, D.S.O. (50, 57)	15 Mar. 69	27 Apl. 89	14 Mar. 92	27 Apl. 15	. . . .	8 Cavalry.
	Linc. R.	**Byers**, Cyril Bertram (66a, 155)	1 Nov. 69	15 May 89	30 Mar. 92	15 May 15	. . . .	S. and T. Corps.
380	H. L. I.	**Forteath**, Frederick William Hughes (75)	6 Oct. 67	8 June 89	30 Aug. 90	8 June 15	. . . .	S. and T. Corps.
	Conn. Rang.	**Moore**, Alexander (50, 71, 145, 161, 167)	5 Jan. 68	6 July 89	4 Apl. 91	6 July 15	. . . .	66 Punjabis.
	Leic. R.	**Roosmale-Cocq**, Charles Alexander (50, 53, 60a, 71, 95, 170)	10 July 68	6 July 89	13 Dec. 90	6 July 15	. . . .	8 G. R.
383	Glouc. R.	**Darley**, James Russel, C.I.E., D.S.O. (50, 53, 66a, 71, 107, 117, 140, 149, 180)	5 Jan. 68	6 July 89	20 Dec. 90	6 July 15	. . . .	119 Infantry.
	Leins. R.	*Lyon, John Wilmot Harley* (50, 60, 148a, 195)	*14 Oct. 66*	*6 July 89*	*22 Aug. 91*	*6 July 15*	. . . .	*Supy. List.*

Colonels and Lieutenant-Colonels—*contd.*

No.	Former Corps.	Names.	Date of birth.	Date of first commission.	Date of appointment to Indian Army.	Date of rank. Substantive Lt.-Colonel.	Date of rank. Army.	Remarks.
384	York. L. I.	Cotton, William Lewis (50, 57, 66a)	15 Sep. 71	27 Jan. 92	12 July 93	8 July 15		5 Infantry.
	R. A.	*Pottinger, Robert Southey* (107, 162)	*31 Jan. 70*	*27 July 89*	*4 Feb. 92*	*27 July 15*		*Supy. List.*
385	R. Ir. F.	Gregory, Charles Levinge, C.B., *p. s. c.* (50, 53, 105, 170, 181)	22 Aug. 70	9 Nov. 89	1 July 92	28 Aug. 15	c. 1 Jan. 17	10 Lancers.
	Oxf. L. I.	Payne, Edward Henry (50, 66a)	24 Dec. 68	21 Sep. 89	8 Nov. 90	21 Sep. 15		Mily. Accts. Dept.
	W. York. R.	Bateman-Champain, Hugh Fredk., C.M.G. (30, 50, 53, 157, 170) [*l*]	6 Apl. 69	21 Sep. 89	3 Jan. 91	21 Sep. 15		9 G. R.
	R. Fus.	Cummins, Harry Ashley Vane, *p.s.c.* (50, 53, 66a, 107, 127, 172)	9 Aug. 70	21 Sep. 89	15 Jan. 91	21 Sep. 15	*c* 3 *June 16* *Ty. l.c.* 27 Dec. 14	24 Punjabis.
	L'pool R.	Cheyne, Reginald Edmonstone (50, 65, 66a, 161, 170) [*l*]	3 Sep. 70	21 Sep. 89	28 Apl. 92	21 Sep. 15		8 Cavalry.
390	L'pool R.	Dallas, Alexander Egerton, C.M.G. (50, 169)	25 Feb. 69	21 Sep. 89	4 Apl. 91	21 Sep. 15	*Ty. c.* 10 Dec. 15	S. and T. Corps.
	Devon. R.	Holman, Herbert Campbell, C.B., C.M.G., D.S.O., *p.s.c.* (50, 65, 162, 165, 169) [L]	3 May 69	21 Sep. 89	3 May 92	21 Sep. 15	*c. 1 Jan. 16*	16 Cavalry.
	R. W. Fus.	Smith, William Frank (165, 170) [*e*]	28 Jan. 70	21 Sep. 89	6 June 92	21 Sep. 15	3 June 15 *Ty. c.* 5 Dec. 17	S. and T. Corps.
	Worc. R.	*Harold, Charles Frederick* (50, 132, 154, 170)	*26 Dec. 68*	*21 Sep. 89*	*22 Aug. 91*	*21 Sep. 15*		*Supy. List.*
	K.O.S. Bord	*Cumberlege, Charles John* (50, 61, 71, 133, 169)	*1 Apl. 69*	*21 Sep. 89*	*29 Aug. 91*	*21 Sep. 15*		*Supy. List.*
	Leins. R.	Brownlow, d'Arcy Charles, C.M.G., C.I.E. (66a, 74, 162, 170)	30 Oct. 69	21 Sep. 89	2 Dec. 91	21 Sep. 15		
	Oxf. L. I.	Prentis, Walter Sladen (50, 52, 71, 105, 124a, 167)	5 Aug. 70	21 Sep. 89	28 June 92	21 Sep. 15		72 Punjabis.
395	Bedf. R.	Samborne-Palmer, Fredk. Carey Stukeley (50, 57, 74, 121, 127)	18 Aug. 68	23 Oct. 89	19 Dec. 91	23 Oct. 15		11 Rajputs.
	Hamps. R.	Cassels, Kenneth Scougall (50, 53, 71, 157)	5 June 69	23 Oct. 89	12 July 92	23 Oct. 15		8 G. R.
	Bedf. R.	Garratt, Harry Sumner (50, 53, 66a, 95, 153, 162, 169)	30 Oct. 68	9 Nov. 89	10 July 91	9 Nov. 15		103 Lt. Infantry.
	Conn. Rang.	*Roe, Bernhard Oswald*	*29 Dec. 70*	*9 Nov. 89*	*20 Oct. 91*	*9 Nov. 15*		*Supy. List.*
	Norf. R.	Lloyd, John Herbert (50, 66a, 151, 157, 162)	28 July 68	9 Nov. 89	12 May 92	9 Nov. 15	*Ty. l.c.* 13 Aug. 15	6 G. R.
	R. War. R.	Johnson, Charles Richard (50, 53, 57, 71, 157, 170)	22 Oct. 68	9 Nov. 89	18 Dec. 91	9 Nov. 15		5 G. R.
400	Bord. R.	Hay, Lenox Theobald (50, 56, 71)	15 Apl. 69	9 Nov. 89	5 Feb. 92	9 Nov. 15		5 Cavalry.
	R. Innis. F.	*Lawrenson, Thomas George Pasley* (65, 71, 105, 169)	*16 May 71*	*27 Nov. 89*	*8 Mar. 91*	*27 Nov. 15*		*Supy. List.*
	Middx. R.	Chandler, Hubert George William (51, 71, 162)	6 Sep. 69	4 Dec. 89	11 June 94	4 Dec. 15		Mily. Accts. Dept.
	R. Ir. R.	Alexander, Arthur Vickers (50, 52, 71, 195)	28 June 69	21 Dec. 89	5 Apl. 91	21 Dec. 15		74 Punjabis.
	R. Ir. R.	Beatty, Guy Archibald Hastings, D.S.O. (51, 66a, 71, 162, 170)	22 June 70	21 Dec. 89	27 May 92	21 Dec. 15		9 Horse.
	Conn. Rang.	Pritchard, Aubrey Gordon, C.M.G. (50, 170)	8 Oct. 69	21 Dec. 89	13 Oct. 91	21 Dec. 15	*c.* 3 June 17	2 Lancers.
405	S. Lan. R.	Ward, Cyril Harding (50, 53, 71, 169, 172a)	14 Apl. 68	21 Dec. 89	13 Oct. 91	21 Dec. 15		120 Infy.
	Conn. Rang.	Nicholson, William Cumberland (157)	16 June 68	21 Dec. 89	31 Oct. 91	21 Dec. 15		17 Infantry.
	E. Surr. R.	Rawlins, Arthur Kennedy, C.I.E., D.S.O. (50, 170)	15 May 68	21 Dec. 89	19 May 92	21 Dec. 15		24 Punjabis.
	Bedf. R.	Mackenzie, Charles, D.S.O. (50, 71, 170, 172)	25 July 69	21 Dec. 89	5 Apl. 92	21 Dec. 15		13 Lancers.
409	K.O.S. Bord.	Swan, John Stewart	24 Jan. 68	21 Dec. 89	21 Jan. 93	21 Dec. 15		S. and T. Corps.

410	R.Muns.Fus.	**Bourne,** Walter FitzGerald (50, 52, 71, 170)	20 May 72	29 Oct. 90	21 June 93	31 Jan. 16	. . . .	6 Jats
	Sco. Rif.	**Newnham,** Charles Cowan, D.S.O., *p.s.c.* (50, 66a, 72b)	15 June 71	25 Mar. 91	4 Jan. 95	4 Feb. 16	*Ty.l.c.* 26 Dec. 14 / 1 Jan. 16	6 Cavalry.
	R. A.	*O'Connor, William Fredk. Travers,* C.I.E. *(65, 125, 156, 162, 169, 174, 199) Service for pension from 27 Feb. 90.*	*30 July 70*	*14 Feb. 90*	*18 June 05*	*14 Feb. 16*	. . . .	*Supy. List.*
	Som. L. I.	**Davidson,** Sisley Richard, C.M.G. (50, 52, 57, 66a, 167)	26 Aug. 69	16 Apl. 90	3 June 92	18 Feb. 16	*c.* 3 June 16	47 Sikhs.
	R.W.Kent.R	**Fagan,** Edward Arthur, C.M.G., D.S.O. (50, 52, 162, 172)	27 Nov. 71	2 May 91	3 July 94	18 Feb. 16	.	36 Horse.
	Devon R.	**Field,** Charles Douglas (50, 53, 66a, 71, 72b, 107, 147, 162, 193)	6 Apl. 70	1 Mar. 90	22 Sep. 91	1 Mar. 16	. . . .	75 Infantry.
415	Sco. Rif.	**Hocken,** Chas. Augustus Fredk., *p.s.c.* (50, 57, 162, 169) [L]	4 Aug. 70	1 Mar. 90	1 Sep. 91	1 Mar. 16	*Ty. l.c.* 3 Mar. 15	35 Horse.
	Ches. R.	**Nightingale,** Manners Ralph Willmot, C.M.G., D.S.O. (50, 53, 59, 66a, 72a, 157, 160, 169) [L].	15 Apl. 71	1 Mar. 90	23 Dec. 91	1 Mar. 16	. . .	5 G. R.
	Sco. Rif.	**Davis,** Gronow John, D.S.O. (50, 94, 169)	9 Nov. 69	1 Mar. 90	15 Feb. 92	1 Mar. 16	. . . .	22 Punjabis.
	Linc. R.	**Peck,** Arthur Wharton, C.M.G. (50, 53, 170)	23 Apl. 69	1 Mar. 90	10 Mar. 92	1 Mar. 16	*c.* 1 Jan. 17	25 Cavalry.
	Linc. R.	**Winter,** Clifford Boardman, C.M.G., D.S.O. (65, 94, 161, 165, 169, 185, 204) [L]	16 Jan. 69	1 Mar. 90	9 June 92	1 Mar. 16	*Ty.l.c.* 29 Oct. 15 / *c.* 3 June 18	112 Infantry.
	Derby. R.	*Beadon, Henry Cecil (169)*	*28 Nov. 69*	*1 Mar. 90*	*1 May 92*	*1 Mar. 16*		*Supy. List.*
420	Bedf. R.	**Jones,** Leslie Cockburn, C.M.G., M.V.O., *p.s.c.* (50, 129, 162) [*l*]	14 Jan. 70	1 Mar. 90	7 Jan. 93	1 Mar. 16	*c.* 3 June 17	7 Lancers.
	Som. L. I.	**Kirkpatrick,** Edwad (50, 37, 66a, 169)	16 Nov. 69	1 Mar. 90	14 June 92	1 Mar. 16	. . . .	59 Rifles.
	R. Lan. R.	**Redl,** Ernest Arthur Frederick, C.I.E. (50, 53, 88, 129, 161) [L]	1 Aug. 69	1 Mar. 90	7 Oct 93	1 Mar. 16	. . . .	113 Infantry.
	K.O.S. Bord.	**Spence,** Alexander Hieron Ogilvy, C.I.E. (50, 162) [*l*]	23 May 69	1 Mar. 90	26 Aug. 93	1 Mar. 16	*Ty. c.* 11 Dec. 17	39 C. I. Horse.
	Conn. Rang.	**Kemball,** Alick Gurdon (50, 162, 165, 170)	16 Oct. 69	1 Mar. 90	8 Oct. 92	1 Mar. 16	. . . .	31 Punjabis.
425	N. Lanc. R.	**Glasfurd,** Alexander Inglis Robertson, C.M.G., D.S.O., *p.s.c.* (50, 53, 66a, 71)	4 Dec. 70	29 Mar. 90	17 June 91	29 Mar. 16	. . . .	40 Pathans.
	North'n R.	**Alexander,** Charles Henry (50, 53, 65, 66a, 72b, 94, 127, 162, 169) [*l*.]	28 Aug. 69	29 Mar. 90	12 Oct. 91	29 Mar. 16	. . . .	
	R. Muns.Fus	*Walker, Francis Spring (57, 169)*	*8 Feb. 70*	*29 Mar. 90*	*11 Apl. 92*	*29 Mar. 16*	. . . .	*Supy. List.*
	R.Muns.Fus	*Browning, Theodore Charles (50, 59)*	*8 Oct. 68*	*29 Mar. 90*	*5 Apl. 92*	*29 Mar. 16*	. . . .	*Supy. List.*
	R. Berks R.	**Hogg,** John McKenzie Trower (50, 53, 57, 66a, 71, 157, 169)	20 May 69	29 Mar. 90	12 Nov. 93	29 Mar. 16	. . . .	9 G. R.
	Wilts. R.	**Davidson,** Francis Dean (51, 66a, 71)	17 Feb. 72	13 July 92	11 Aug. 90	1 Apl. 16	. . . .	90 Infantry.
	E. Lan. R.	**Boileau,** Etienne Ronald Partridge (53, 57, 157) [L]	8 Apl. 70	28 June 90	10 Oct. 95	2 Apl. 16	3 June 15	2 G. R.
430	Devon. R.	**Ricketts,** Percy Edward, M.V.O., D.S.O. (50, 64, 167, 170) *Half-pay from 27 Mar. 01 to 6 Nov. 03 (1 yr. to count).*	5 Mar. 68	22 Aug. 88	17 June 90	3 Apl. 16	*Ty. c.* 5 May 17	
	R. Suss. R.	*Elliott, Alfred Charles,* C.B.E. *(169)*	*20 Sep. 70*	*9 Apl. 90*	*31 Dec. 91*	*9 Apl. 16*	. . . .	*Supy. List.*
	K. R. Rif. C.	**Prevost,** George Herbert (50, 66a, 71)	12 Apl. 68	28 June 90	7 Feb. 93	22 Apl. 16	. . . .	87 Punjabis.
	R. Ir. R.	**O'Reilly,** Charles Myles (66a, 71, 195)	2 Jan. 69	3 May 90	13 Oct. 92	3 May 16	. . . .	85 Infantry.
	Midd'x R.	**Bliss,** James Arthur, D.S.O., M.V.O. (50, 66a, 71, 72b, 162, 195)	15 Feb. 70	3 May 90	4 Dec. 91	3 May 16	. . . .	64 Pioneers.
	W. York. R.	**Colan,** William Robert Boyle, D.S.O. (50, 52, 57, 66a, 71, 95, 100, 167, 195)	25 May 70	3 May 90	14 Dec. 91	3 May 16	. . . .	67 Punjabis.
435	York L. I.	**Lukin,** Robert Clarence Wellesley, D. S. O. (50, 66a, 71, 170)	22 Dec. 70	3 May 90	16 June 92	3 May 16	*Ty. l.c.* 22 Sep. 15	35 Horse.
	Midd'x R.	*Prideaux, Francis Beville,* C.S.I., C.I.E. *(162, 170)*	*26 Nov. 71*	*3 May 90*	*16 Sep. 92*	*3 May 16*	. . . .	*Supy. List.*
	Shrops. L.I.	**Hole,** Henry Lewis (51, 71, 195)	2 Jan. 70	3 May 90	2 Nov. 92	3 May 16	. . . .	80 Infantry.
	D. C. L. I.	**Orman,** Frank Leslie (50, 66a, 105, 125a, 140, 148b, 157, 181)	17 Jan. 69	3 May 90	5 Sep. 93	3 May 16	. . . .	10 G. R.
	S. Lanc. R.	**Fellows,** Christopher Stewart (50, 52, 71, 155)	21 Nov. 70	3 May 90	8 Nov. 94	3 May 16	. . . .	109 Infantry.
	Manch. R.	**Sealy,** Alfred Edward (50, 53, 71, 157)	27 June 70	8 Oct. 90	1 Dec. 95	24 May 16	. . . .	7 G. R.
440	Suff. R.	**Wynter,** Charles Philip (50, 53, 59, 71, 72a, 94, 142, 169)	12 July 69	17 Jan. 91	19 Sep. 93	24 May 16	. . . .	52 Sikhs.
	Sco. Rif.	**Bainbridge,** Wm. Frank, C.M.G., D.S.O., *p.s.c.* (50, 53, 71, 162, 169)	15 Jan. 73	7 Nov. 91	21 May 95	24 May 16	7 Jan. 14	54 Sikhs.
	Som. L. I.	**Ross,** Harry (51, 53, 155)	17 Apl. 69	31 May 90	24 Apl. 92	31 May 16	. . . .	103 Lt. Infantry
	Ches. R.	**Cooke,** Herbert Fothergill, D.S.O. (50, 52, 167)	13 Nov. 71	27 Jan. 92	24 Nov. 93	3 June 16	*c.* 3 June 17	32 Pioneers.
	Wilts. R.	*Gough, Hugh Augustus Keppel,* C.I.E. *(94, 156, 169)*	*5 Mar. 71*	*4 June 90*	*31 Aug. 91*	*4 June 16*	*Ty. l.c.* 16 *Apl.* 16	*Supy. List.*
444	K.O.S.Bord.	**Carwithen,** Edward Terry (64, 66a, 71, 162, 170)	8 Jan. 69	4 June 90	26 Nov. 91	4 June 16	. . . .	Cant. Magte.'s Dept.

No.	Former Corps.	Names.	Date of birth.	Date of first commission.	Date of appointment to Indian Army.	Date of Rank. Substantive Lt.-Colonel.	Date of Rank. Army.	Remarks.
		Colonels and Lieutenant-Colonels—*contd.*						
445	Essex R.	**Fisher**, Frederick (50, 52, *66a*, 71, 169) *Forfeited from 11 Feb. 06 to 4 June 08, for promotion.*	17 Sep. 67	11 Feb. 88	20 Oct. 89	5 June 16		
	Worc. R.	**Birdwood**, Halhed Brodrick (50)	29 May 70	7 Nov. 91	26 Dec. 95	20 June 16	29 Aug. 12 / c. 16 Apl. 18	
	Devon. R.	**Morris**, George Mortimer, D.S.O. (50, 53, 57, 71, 167, 195) [L]	17 Mar. 68	28 June 90	7 Sep. 92	8 June 16	c. 3 June 18	62 Punjabis.
	R. Innis. F.	*Kennedy, the Hon. William Magill*, C.I.E. (50)	*21 Aug. 68*	*28 June 90*	*31 July 92*	*28 June 16*	. . . .	*Supy. List.*
	North'd Fus.	*Drummond, Arthur Berkeley* (94, 156, 162, 169)	*27 Nov. 69*	*28 June 90*	*18 Aug. 92*	*28 June 16*	. . . .	*Supy. List.*
	Bedf. R.	**Strange**, Arthur James (50, 57, 71, 157, 170)	11 Mar. 68	28 June 90	4 Mar. 93	28 June 16	. . . .	6 G. R.
	York & L.R.	**Bidwell**, Reginald Frank (50, 71, 105)	6 Mar. 68	28 June 90	31 Oct. 92	28 June 16	. . . .	101 Grenadiers.
450	Unattd. List	**Stewart**, John Henry Keith, D.S.O., *p.s.c.* (50, 52, 157, 162)	31 Aug. 72	3 Sep. 92	28 Dec. 93	1 July 16	3 June 16 / c. 10 Mar. 17	39 Garhwal Rifles
	R. War. R	**Norman**, Walter Henry, D.S.O., *p.s.c.* (50, 71, 162, 169)	14 June 71	8 Oct. 90	22 Dec. 91	4 July 16	c. 3 June 18	11 Lancers.
	Unattd. List	**Russell**, Francis Deane, *p.s.c.* (50, 66a, 170) [L]	4 July 73	30 Aug. 93	23 Jan. 95	7 July 16	3 June 16 / *Ty. l.c.* 31 Oct. 15	1 Lancers.
	S. W. Bord.	**Pennefather Evans**, Granville (107, 120, 170)	9 Dec. 68	16 July 90	20 July 93	16 July 16	3 June 16	19 Punjabis.
	16 Lanc. R.	**Battine**, Reginald St. Clair, C.B., D.S.O., *p.s.c.* (50, 53, 71, 162, 169) [L]	4 Sep. 69	23 July 90	31 July 94	23 July 16		21 Cavalry.
455	R. Ir. Rif.	**Knox**, Alfred William Fortescue, C.B., *p.s.c.* (57, 162, 169) [L]	30 Oct. 70	2 May 91	18 Nov. 93	24 July 16	c. 1 Jan. 17	58 Rifles.
	Unattd. List	**Adams**, Francis, D.S.O. (50, 167)	17 Apl. 74	2 June 94	5 Oct. 95	29 July 16	. . . .	20 Horse.
	Unattd. List	**Sangster**, Patrick Barclay, D.S.O. (45a, 50, 66a, 137)	28 Aug. 72	28 Nov. 94	22 Feb. 96	7 Aug. 16	. . . .	29 Lancers.
	Bord. R.	**Earle**, Edward Savi (50, 53, 124a, 165, 170)	8 Apl. 71	5 Dec. 91	24 July 96	8 Aug. 16	. . . .	14 Sikhs.
	Lanc. Fus.	**Keelan**, Henry Percival (50, 162, 172)	26 Feb. 72	18 June 92	14 Sep. 96	14 Aug. 16	3 June 16	121 Pioneers.
460	R. Lanc. R.	**Kelly**, Frederick Peter Charles, C.M.G., D.S.O. (50, 71, 66a, 155, 170)	25 Dec. 70	25 July 91	9 Aug. 93	24 Sep. 16	. . . .	125 Rifles.
	Wilts. R.	*Sanford, George Batthyany* (50, 57, 106, 165)	*19 July 71*	*8 Oct. 90*	*17 Jan. 92*	*8 Oct. 16*	. . . .	*Supy. List.*
	R. Fus.	**Young**, Henry George, D.S.O. (50, 52, 71, 170)	7 Mar. 70	8 Oct. 90	25 Nov. 92	8 Oct. 16	. . . .	22 Cavalry.
	R. Innis. Fus.	**Nangle**, Kenlis Edward (50, 52, 66a)	5 Feb. 71	8 Oct. 90	5 Mar. 92	8 Oct. 16	. . . .	96 Infantry.
	K.O.S. Bord.	*Ormiston, Thomas Lane* (50, 76, 104, 162)	*30 May 67*	*8 Oct. 90*	*5 Apl. 92*	*8 Oct. 16*	. . . .	*Supy. List.*
	Midd'x R.	*Berthon, Henry Warwick* (133)	*7 Jan. 70*	*8 Oct. 90*	*14 Apl. 92*	*8 Oct. 16*	. . . .	*Supy. List.*
	E. Surr. R.	**Nicholas**, Stephen Henry Edmund (50, 52, 74, 137, 162)	26 Dec. 70	8 Oct. 90	4 Aug. 92	8 Oct. 16	. . . .	J. A. G.'s Deptt.
	Bord. R.	**Coningham**, Frank Evelyn, D.S.O. (137, 157, 162, 170)	6 June 70	8 Oct. 90	6 Jan. 96	8 Oct. 16	. . . .	10 G. R.
465	R. Muns. Fus.	**Hill**, Harold Charles, D.S.O. (50, 66a, 71, 155)	17 Sep. 69	29 Oct. 90	20 Feb. 93	29 Oct. 16	. . . .	110 Lt. Infantry.
	E. Surr. R.	**Haydon**, William Pitt, D.S.O.	20 Oct. 70	29 Oct. 90	14 Apl. 93	29 Oct. 16	. . . .	S. and T. Corps.
	R. W. Fus.	**Tarver**, Alexander Leigh, D.S.O., A.D.C., *p.s.c.* (66a, 94, 169)	30 Aug. 71	29 Oct. 90	12 Apl. 93	29 Oct. 16	3 June 16 / c. 14 Aug. 18	124 Infantry.
	K.R. Rif. C.	**McLeod**, Torquil John, D.S.O. (50, 66a, 156, 170)	30 June 70	29 Oct. 90	26 Aug. 92	29 Oct. 16	. . . .	7 G. R.
	North'n R.	**Cox**, Arthur Ditmas (50, 53, 66a, 72a, 162, 167, 169, 195)	21 Mar. 70	29 Oct. 90	29 May 93	29 Oct. 16		26 Punjabis.
470	R. Lanc. R.	**Tomkins**, Harry Leith, C.M.G., D.S.O. (50, 120, 172)	25 Jan. 70	29 Oct. 90	14 Oct. 94	29 Oct. 16	3 June 15	28 Punjabis.
	R. A.	**Crawford**, John Halket (12, 50, 57, 71, 137)	28 Sep. 68	1 Nov. 90	3 Mar. 93	1 Nov. 16	.	32 Lancers.
	R. Sco. Fus.	*Hamilton, Robert Edward Archibald*, C.I.E. (57, 157, 162, 169)	*8 Apl. 71*	*8 Nov. 90*	*19 June 93*	*8 Nov. 16*	. . . .	*Supy. List.*
	R. Sco. Fus.	**Buist**, Arthur Hunter, M.V.O. (50, 52, 172)	3 June 71	9 Sep. 91	20 Aug. 96	14 Nov. 16	. . . .	Guides.
	Manch. R	**Souter**, Hugh Maurice Wellesley, D.S.O. (50, 66a, 72b, 162, 72a)	30 Oct. 72	18 May 92	19 Dec. 96	14 Nov. 16	. . . .	14 Lancers.
	R. Ir. R.	**Edwardes**, John Grahame (50, 52, 66a, 71, 125b, 156)	9 Sep. 69	29 Nov. 90	10 Aug. 92	29 Nov. 16	*Ty. l.c.* 8 Aug. 16	3 G. R.
475	R. Muns. Fus.	**Newell**, Herbert Andrews (57, 162, 170) [*l*]	23 Aug. 69	29 Nov. 90	5 Apl. 92	29 Nov. 16	. . . .	S. and T. Corps.

	Derby. R.	*Leslie, Philip Norman* (50, 66*a*)	*4 Aug. 68*	*29 Nov. 90*	*10 May 93*	*29 Nov. 16*	. . . .	*Supy. List.*
476	Norf. R.	**Barclay**, Patrick Charles Robertson (50, 53, 57, 71, 162, 169)	9 Apl. 72	29 Nov. 90	2 Sep. 02	29 Nov. 16	*Ty.l.c.* 14 May 15	122 Infantry.
	L'pool R.	**Teed**, Arthur Seymour Holland (50, 53, 94, 100, 161, 165, 169, 185)	7 Apl. 72	29 Nov. 90	13 Sep. 02	29 Nov. 16	. . . .	
	R. Sco. Fus.	**Sarel**, George Benedic Molyneux (50, 71, 170)	21 Mar. 71	29 Nov. 90	20 Mar. 94	29 Nov. 16	. . . .	34 Horse.
	Norf. R.	**Fenn**, Edwin Harold McBride (57, 107, 125)	18 May 71	29 Nov. 90	24 Apl. 93	29 Nov. 16	. . . .	
480	North'n R.	**Crawford**, Arthur Gosset (51, 52, 71, 167, 195)	23 Aug. 60	29 Nov. 90	15 May 93	29 Nov. 16	. . . .	84 Punjabis
	Norf. R.	**Knox**, Ernest Francis, D.S.O. (50, 52, 59, 72, 163, 167)	27 July 71	29 Nov. 90	30 Nov. 96	29 Nov. 16	. . . .	36 Sikhs.
	R. Scots.	*Coghill, Norman Sinclair* (76, 89, 132)	*1 Aug. 69*	*29 Nov. 90*	*30 July 94*	*29 Nov. 16*	. . . .	*Supy. List.*
	Welsh R.	**Cheyne**, Archibald Ythen (50, 64, 66*a*, 71, 167)	18 June 72	29 Nov. 90	26 Nov. 96	29 Nov. 16	. . . .	15 Lancers.
	7 Dn. Gds.	**Smith**, Alick Le Fleming (50, 53, 66*a*, 71, 137)	23 June 68	3 Dec. 90	22 Feb. 93	3 Dec. 16	. . . .	2 Lt. Infantry.
	Devon. R.	**Souter**, Arthur Bourke, C.M.G. (71, 172) [*l*]	8 Feb. 71	17 Dec. 90	28 Jan. 94	17 Dec. 16	. . . .	21 Cavalry.
485	Hamps. R.	**Tillard**, Arthur Basil, D.S.O. (50, 53, 57, 71, 157)	10 June 70	24 Dec. 90	30 Jan. 92	24 Dec. 16	. . . .	4 G. R.
	Wilts. R.	**Seton-Browne**, Clement Laurence Seton, D.S.O. (50, 53, 172)	9 Mar. 69	10 Oct. 91	7 Mar. 94	1 Jan. 17	. . . .	
	R. Ir. R.	**Sweet**, Edward Herbert, D.S.O. (50, 53, 66*a*, 71, 157, 170)	1 June 71	9 Apl. 92	10 July 06	1 Jan. 17	. . . .	2 G. R
	Lanc. Fus.	**Dobbs**, Charles Fairlie, *p.s.c.* (50, 52, 57, 71, 162)	1 July 72	18 June 92	11 Oct. 94	4 Feb. 17	1 Jan. 16	94 Infantry.
	Devon. R.	**Swifte**, Edmund Lenthal (50, 66*a*, 170)	25 May 60	17 Jan. 91	22 Mar. 04	17 Jan. 17	*Ty.l.c.* 26 Nov. 14	57 Rifles.
	R. W. S. P	*Langtry, Henry Vivian Montague* (52, 164)	*12 Nov. 69*	*4 Feb. 91*	*21 May 92*	*4 Feb. 17*	. . . .	*Supy. List.*
490	Ches. R.	**Butler**, Alexander Hazelwood (50, 53, 66*a*, 195)	17 Aug. 69	4 Feb. 91	28 Aug. 92	4 Feb. 17	. . . .	44 Infantry.
	R. A.	**McNeile**, Donald Hugh (51, 65, 162, 165, 169) [L]	6 Feb. 71	13 Feb. 91	4 Sep. 95	13 Feb. 17		19 Lancers
	R. A.	**Capper**, Alfred Stewart, D.S.O., *p.s.c.* (50, 71, 170)	2 Mar. 71	24 July 91	12 Aug. 93	15 Feb. 17	*Ty.l.c.* 4 June 16	39 C. I. Horse
	Derby. R.	**Leeds**, Thomas Louis, C.M.G., D.S.O. (51, 66*a*, 169)	25 July 69	9 Sep. 91	13 July 94	20 Feb. 17	3 June 15 *Ty.l.c.* 18 Apl. 15	59 Rifles.
	Glouc. R.	*Jacob, Arthur Leslie*, C.I.E., O.B.E. (50, 161, 169)	*20 Dec. 70*	*4 Mar. 91*	*21 Apl. 92*	*4 Mar. 17*		*Supy. List.*
		Forfeits 3 days for pension.						
	Sco. Rif.	**Milne**, James William (50, 57, 66*a*, 162, 165, 169, 193, 198)	24 Mar. 70	4 Mar. 91	31 July 92	4 Mar. 17	. . . .	82 Punjabis.
495	R. Suss. R.	**Hudson**, Josiah Howard (50, 159, 192, 204) [L]	26 June 70	4 Mar. 91	28 Oct. 94	4 Mar. 17	. . . .	Mily. Accts. Dept
	7 Dn. Gds.	**Henderson**, Maxwell Hume (50, 162, 170)	17 Dec. 69	4 Mar. 91	17 Nov. 92	4 Mar. 17	*c.* 3 June 18	26 Lt. Cavalry.
	W. I. R.	**Tregear**, Frederick Charles (50, 52, 137)	2 May 71	4 Mar. 91	11 Dec. 94	4 Mar. 17	. . . .	7 Rajputs.
	Bedf. R.	*Minchin, Alfred Beckett*, C.I.E. (94, 162, 165, 169)	*23 Dec. 70*	*4 Mar. 91*	*14 Nov. 93*	*4 Mar. 17*	. . . .	*Supy. List.*
	Suff. R.	**Brakenbury**, Walter John William (50, 53, 72*a*, 155, 161, 169)	14 Nov. 70	4 Mar. 91	21 July 96	4 Mar. 17	. . . .	43 Regiment.
	Devon. R.	**Gwyn-Thomas**, Gwyn, C.M.G. D.S.O., *p.s.c.* (50, 57, 65, 72, 161, 170)	19 July 71	4 Mar. 91	25 May 95	4 Mar. 17	*Ty.l.c.* 27 Sep. 16	2 Lancers
	R. Ir. R.	*McConaghey, Frank* (94, 170)	*27 Apl. 71*	*25 Mar. 91*	*28 Dec. 93*	*25 Mar. 17*	. . . .	*Supy. List.*
500	L'pool R.	**Pocock**, Philip Frederick, D.S.O. (50, 53, 66, 71, 170)	5 Dec. 71	25 Mar. 91	6 July 92	25 Mar. 17	. . . .	120 Infantry.
	Durh. L. I.	**Bridges**, Arthur Holroyd, D.S.O., *p.s.c.* (50, 155) [*l*]	21 Apl. 71	25 Mar. 91	8 July 92	25 Mar. 17	Jan. 17	116 Mahrattas.
	R. Suss. R.	*Buck, Cecil Henry*	*15 Jan. 72*	*25 Mar. 91*	*31 May 93*	*25 Mar. 17*	. . . .	*Supy. List.*
	Worc. R.	**Fox**, Lionel Whitelaw (162)	12 July 70	2 May 91	10 July 93	2 May 17	. . . .	S. and T. Corps.
	Unattd. List	**Scott-Elliot**, Charles Reginald (51, 52, 57, 105, 124*a*, 162, 195)	19 Sep. 73	28 Jan. 93	5 Apl. 94	5 May 17	*Ty.l.c.* 28 Jan. 15 3 June 16	61 Pioneers.
	Unattd. List	**Hawkes**, Robert (51, 66*a*, 170)	26 Sep. 74	28 Jan. 93	13 July 94	10 May 17	. . . .	3 Brahmans
505	R. Ir. R.	**Lushington**, William Nelson (50)	5 Sep. 70	23 May 91	17 June 96	23 May 17	1 Jan. 16 *Ty.l.c.* 16 Oct. 15	S. and T. Corps
	D. C. L. I.	**Tregear**, Vincent Francis William (50, 52, 71)	1 Jan. 70	23 May 91	24 Sep. 92	23 May 17	. . . .	98 Infantry.
	R. Ir. R.	**McCaskill**, John Charles Henry (50, 53, 66*a*, 71, 170)	25 Feb. 72	27 Jan. 92	27 Mar 94	11 June 17	. . . .	88 Infantry.
	Linc. R.	*Irvine, Andrew Alexander* (162)	*28 Sep. 71*	*17 June 91*	*12 Oct. 92*	*17 June 17*	. . . .	*Supy. List.*
	L'pool R.	**Jackson**, Keith Hungerford (106, 113, 170)	3 Dec. 70	17 June 91	6 Nov. 92	17 June 17	. . . .	Mily. Accts. Dept.
	North'd Fus.	**Stewart**, Robert Joseph Tucker (66*a*, 94, 100, 162, 165, 169)	18 Aug. 72	17 June 91	27 Oct. 94	17 June 17	. . . .	Cant. Magte.'s Dept.
510	Glouc. R.	**Barry-Smith**, Raymond Coape (50, 53, 57, 66*a*, 71, 169)	31 May 74	20 May 93	7 July 98	19 June 17	. . . .	17 Cavalry.
	R. A.	**Hay**, Westwood Norman, C.I.E., D.S.O., *p.s.c.* (50, 94, 162, 169) [L]	7 Oct. 71	24 July 91	4 Aug. 93	24 July 17	*Ty.l.c.* 18 Jan. 17	120 Baluchis.
	R. A.	*Trevor, Arthur Prescott*, C.S.I., C.I.E. (161, 169) [*l*]	*17 Mar. 72*	*24 July 91*	*22 Aug. 97*	*24 July 17*	. . . .	*Supy. List.*
512	R. A.	**Hunt**, Vere De Vere (51, 161, 165, 169, 204)	26 Dec. 70	24 July 91	11 Oct. 95	24 July 17	. . . .	Cant. Magte.'s Dept.
	Glouc. R.	*Burn, George Callander* (105)	*22 Oct. 70*	*25 July 91*	*11 Oct. 92*	*25 July 17*	. . . .	*Supy. List.*

Colonels and Lieutenant-Colonels—*contd.*

No.	Former Corps.	Names.	Date of birth.	Date of first commission.	Date of appointment to Indian Army.	Date of Rank. Substantive.	Date of Rank. Army.	Remarks.
	N. Lan. R.	*O'Brien, Aubrey John*, C.I.E. (*161, 204*) [L]	*5 Dec. 70*	*25 July 91*	*17 May 95*	*25 July 17*	. . . .	*Supy. List.*
513	E. Surr. R.	**Churchill**, George Ross Deas (50, 53, 162, 170)	2 July 70	25 July 91	9 Feb. 96	25 July 17	. . . .	19 Punjabis.
	R.Innis.Fus.	**Preston**, Eyre Evans (71, 105, 162, 169) [L]	12 Mar. 72	25 July 91	16 May 93	25 July 17	. . . .	S. and T. Corps.
515	York. L. I.	**Ellwood**, Bertram Price (50, 66*a*, 155, 169, 172*a*)	30 Sep. 72	18 May 92	30 Oct. 93	29 July 17	. . . .	31 Lancers.
	Manch. R.	**McVean**, Donald Archibald Dugald, D.S.O. (50, 127, 129, 162)	22 July 70	7 Nov. 91	8 Feb. 96	29 July 17	. . . .	45 Sikhs.
	R. Lanc. R.	**Keogh**, James Blair (50, 71, 137, 170)	25 Jan. 71	9 Sep. 91	10 Nov. 95	23 Aug. 17	. . . .	32 Lancers.
	York. R.	**Bredin**, Alexander (51, 53, 57, 105, 157, 169, 172)	22 Feb. 71	9 Jan. 92	9 July 96	28 Aug. 17	. . . .	93 Infantry.
	R. Mar.	**Hopwood**, Herbert Reginald, C.S.I., *p.s.c.* (50, 53, 66*a*, 162)	23 Aug. 71	1 Sep. 91	22 Oct. 94	1 Sep. 17	*c.* 3 June 16	33 Lt. Cavalry.
520	R. Scots.	**Preston**, George Allen (50, 57, 71, 169)	14 Jan. 68	9 Sep. 91	14 Jan. 93	9 Sep. 17	. . . .	6 G. R.
	S.W Bord.	*Wimberley, Francis Gordon Arabin*	*22 Aug. 69*	*9 Sep. 91*	*24 Oct. 93*	*9 Sep. 17*	. . . .	*Supy. List.*
	Leins. R.	**Chamier**, William St. George (51, 66*a*, 69, 71, 75)	3 Aug. 70	9 Sep. 91	11 Oct. 94	9 Sep. 17	. . . .	S. and T. Corps.
	E. Lan. R.	**Mears**, Cecil Delarue, M.V.O. (50, 162)	26 July 71	9 Sep. 91	18 June 95	9 Sep. 17	3 June 16	16 Cavalry.
	Leins. R.	*Kirkwood, Andrew Torton (104, 167)*	*29 Aug. 71*	*9 Sep. 91*	*15 June 93*	*9 Sep. 17*	. . . .	*Supy. List.*
	Ches. R.	**Hulseberg**, Herbert, D.S.O. (51, 60, 95, 162, 169)	16 Oct. 70	9 Sep. 91	17 June 96	9 Sep. 17	. . . .	127 Lt. Infantry.
	Leins. R.	*Wood, Wyndham Madden Pierpoint*, C.I.E. (*154, 172*)	*10 Sep. 69*	*10 Oct. 91*	*21 Aug. 94*	*10 Oct 17*	. . . .	*Supy. List.*
	Durh. L. I.	**Anderson**, Robert Holme Bankes (57)	28 Aug. 71	10 Oct. 91	10 June 94	10 Oct. 17	. . . .	Mily. Accts. Dept.
525	Oxf. L. I.	**Edwardes**, Alexander Coburn (50, 52, 59, 71, 88, 137, 162)	15 Jan. 73	27 Jan. 92	22 Oct. 96	12 Oct. 17	3 June 16	} 102 Grenadiers.
	R. War. R.	**Walton**, Graham (50, 53, 172)	8 May 70	20 Feb. 92	3 Nov. 94	20 Oct. 17	*Ty. l.c. 10 Sep. 16*	}
	N. Lan. R.	**Mitchell**, Wilfrid James (50, 66*a*, 72*a*, 169)	5 Mar. 71	28 Oct. 91	12 Apl. 93	28 Oct. 17	{ 3 June 16	} 46 Punjabis.
	R. Scots.	**Davie**, James Henry Morison (50, 65, 66*a*, 72)	13 Aug. 70	7 Nov. 91	13 Jan. 07	7 Nov. 17	{ *Ty.l.c. 7 May 16*	}
	W. York. R.	**Hawes**, Charles Howard, D.S.O., M.V.O. (50, 59, 66*a*, 94, 100, 170)	31 Aug. 71	7 Nov. 91	7 July 95	7 Nov. 17	3 June 16	124 Infantry. 34 Horse.
530	R. Ir. R.	**Hutchinson**, Chas. Alex. Robert, D.S.O., *p.s.c.* (50, 53, 66*a*, 71, 162, 169)	31 May 27	7 Nov. 91	7 Apl. 96	7 Nov. 17	{ *c.* 31 Aug. 17	23 Cavalry.
	Midd'x R.	*Macpherson, Archibald Duncan (161, 169)*	*14 Jan. 72*	*7 Nov. 91*	*6 Apl. 93*	*7 Nov. 17*	{ 6 June 12	} 41 Dogras.
	Welsh. R.	**Scott**, William (50, 53, 71)	21 Nov. 71	7 Nov. 91	24 May 93	7 Nov. 17	. . . .	*Supy. List.*
	Gord. Highrs	**Watt**, Donald Munro, D.S.O., *p.s.c.* (50, 53, 57, 72, 157)	18 June 71	7 Nov. 91	17 Aug. 96	7 Nov. 17 }	*Ty.l.c.* 28 May 16	104 Rifles.
	Essex. R.	*Smith, Charles Aitchison (144a, 162, 169)*	*12 Sep. 71*	*7 Nov. 91*	*18 Jan. 96*	*7 Nov. 17*	3 June 16	} 2 G. R.
	Sco. Rif.	**Keighley**, Vernon Aubrey Scott, M.V.O. (71, 167, 170)	15 Dec. 74	28 Dec. 95	5 June 99	2 Dec. 17	. . . .	*Supy. List.* 18 Lancers.
534	K.O.S.Bord	**Browne**, Herbert José Pierson, *p.s.c.* (50, 53, 157, 170)(*c*)	21 Jan. 72	5 Dec. 91	14 Apl. 93	5 Dec. 17 {	*c.* 3 June 16 *Ty. l. c.* 1 *Apl. 15*	} 5 G. R.

	Derby. R.	*Leslie, Philip Norman* (50, 66a)	*4 Aug. 68*	*29 Nov. 90*	*10 May 93*	*29 Nov. 16*	. . . .	*Supy. List.*
476	Norf. R.	Barclay, Patrick Charles Robertson (50, 53, 57, 71, 162, 169)	9 Apl. 72	29 Nov. 90	2 Sep. 02	29 Nov. 16	*Ty.l.c.* 14 May 15	122 Infantry.
	L'pool R.	Teed, Arthur Seymour Holland (50, 53, 94, 100, 161, 165, 169, 185)	7 Apl. 72	29 Nov. 90	13 Sep. 02	29 Nov. 16	. . . .	
	R. Sco. Fus.	Sarel, George Benedic Molyneux (50, 71, 170)	21 Mar. 71	29 Nov. 90	20 Mar. 94	29 Nov. 16	. . . .	34 Horse.
	Norf. R.	Fenn, Edwin Harold McBride (57, 107, 125)	18 May 71	29 Nov. 90	24 Apl. 93	29 Nov. 16	. . . .	
480	North'n R.	Crawford, Arthur Gosset (51, 52, 71, 167, 195)	23 Aug. 60	29 Nov. 90	15 May 93	29 Nov. 16	. . . .	84 Punjabis.
	Norf. R.	Knox, Ernest Francis, D.S.O. (50, 52, 59, 72, 163, 167)	27 July 71	29 Nov. 90	30 Nov. 96	29 Nov. 16	. . . .	36 Sikhs.
	R. Scots.	*Coghill, Norman Sinclair* (76, 89, 132)	*1 Aug. 69*	*29 Nov. 90*	*30 July 94*	*29 Nov. 16*	. . . .	*Supy. List.*
	Welsh R.	Cheyne, Archibald Ythen (50, 64, 66a, 71, 167)	18 June 72	29 Nov. 90	26 Nov. 96	29 Nov. 16	. . . .	15 Lancers.
	7 Dn. Gds.	Smith, Alick Le Fleming (50, 53, 66a, 71, 137)	23 June 68	3 Dec. 90	22 Feb. 93	3 Dec. 16	. . . .	2 Lt. Infantry.
	Devon. R.	Souter, Arthur Bourke, C.M.G. (71, 172) [*l*]	8 Feb. 71	17 Dec. 90	28 Jan. 94	17 Dec. 16	. . . .	21 Cavalry.
485	Hamps. R.	Tillard, Arthur Basil, D.S.O. (50, 53, 57, 71, 157)	10 June 70	24 Dec. 90	30 Jan. 92	24 Dec. 16	. . . .	4 G. R.
	Wilts. R.	Seton-Browne, Clement Laurence Seton, D.S.O. (50, 53, 172)	9 Mar. 69	10 Oct. 91	7 Mar. 94	1 Jan. 17	. . . .	
	R. Ir. R.	Sweet, Edward Herbert, D.S.O. (50, 53, 66a, 71, 157, 170)	1 June 71	9 Apl. 92	10 July 06	1 Jan. 17	. . . .	2 G. R.
	Lanc. Fus.	Dobbs, Charles Fairlie, *p.s.c.* (50, 52, 57, 71, 162)	1 July 72	18 June 92	11 Oct. 94	4 Feb. 17	1 Jan. 16	91 Infantry.
	Devon. R.	Swifte, Edmund Lenthal (50, 66a, 170)	25 May 60	17 Jan. 91	22 Mar. 04	17 Jan. 17	*Ty.l.c.* 26 Nov. 14	57 Rifles.
	R. W. S. F.	*Langtry, Henry Vivian Montague* (52, 164)	*12 Nov. 69*	*4 Feb. 91*	*21 May 92*	*4 Feb. 17*	. . . .	*Supy. List.*
490	Ches. R.	Butler, Alexander Hazelwood (50, 53, 66a, 195)	17 Aug. 69	4 Feb. 91	28 Aug. 92	4 Feb. 17	. . . .	44 Infantry.
	R. A.	McNeile, Donald Hugh (51, 65, 162, 165, 169) [L]	6 Feb. 71	13 Feb. 91	4 Sep. 95	13 Feb. 17		19 Lancers
	R. A.	Capper, Alfred Stewart, D.S.O., *p.s.c.* (50, 71, 170)	2 Mar. 71	24 July 91	12 Aug. 93	15 Feb. 17	*Ty.l.c.* 4 June 16	39 C. I. Horse
	Derby. R.	Leeds, Thomas Louis, C.M.G., D.S.O. (51, 66a, 169)	25 July 69	9 Sep. 91	13 July 94	20 Feb. 17	3 June 15 *Ty.l.c.* 18 Apl. 15	59 Rifles.
	Glouc. R.	*Jacob, Arthur Leslie*, C.I.E., O.B.E. (50, 161, 169) *Forfeits 3 days for pension.*	*20 Dec. 70*	*4 Mar. 91*	*21 Apl. 92*	*4 Mar. 17*		*Supy. List.*
	Sco. Rif.	Milne, James William (50, 57, 66a, 162, 165, 169, 193, 198)	24 Mar. 70	4 Mar. 91	31 July 92	4 Mar. 17	. . . .	82 Punjabis.
495	R. Suss. R.	Hudson, Josiah Howard (50, 159, 192, 204) [L]	26 June 70	4 Mar. 91	28 Oct. 94	4 Mar. 17		Mily. Accts. Dept
	7 Dn. Gds.	Henderson, Maxwell Hume (50, 162, 170)	17 Dec. 69	4 Mar. 91	17 Nov. 92	4 Mar. 17	*c.* 3 June 18	26 Lt. Cavalry.
	W. I. R.	Tregear, Frederick Charles (50, 52, 137)	2 May 71	4 Mar. 91	11 Dec. 94	4 Mar. 17	. . . .	7 Rajputs.
	Bedf. R.	*Minchin, Alfred Beckett*, C.I.E. (94, 162, 165, 169)	*23 Dec. 70*	*4 Mar. 91*	*14 Nov. 93*	*4 Mar. 17*	. . . .	*Supy. List.*
	Suff. R.	Brackenbury, Walter John William (50, 53, 72a, 155, 161, 169).	14 Nov. 70	4 Mar. 91	21 July 96	4 Mar. 17		43 Regiment.
	Devon. R.	Gwyn-Thomas, Gwyn, C.M.G. D.S.O., *p.s.c.* (50, 57, 65, 72, 161, 170)	19 July 71	4 Mar. 91	25 May 95	4 Mar. 17	*Ty.l.c.* 27 Sep. 16	2 Lancers
	R. Ir. R.	*McConaghey, Frank* (94, 170)	*27 Apl. 71*	*25 Mar. 91*	*28 Dec. 93*	*25 Mar. 17*	. . . .	*Supy. List.*
500	L'pool R.	Pocock, Philip Frederick, D.S.O. (50, 53, 66, 71, 170)	5 Dec. 71	25 Mar. 91	6 July 92	25 Mar. 17		120 Infantry.
	Durh. L. I.	Bridges, Arthur Holroyd, D.S.O., *p.s.c.* (50, 155) [*l*]	21 Apl. 71	25 Mar. 91	8 July 92	25 Mar. 17	Jan. 17	116 Mahrattas.
	R. Suss. R.	*Buck, Cecil Henry*	*15 Jan. 72*	*25 Mar. 91*	*31 May 93*	25 Mar. 17	. . . .	*Supy. List.*
	Worc. R.	Fox, Lionel Whitelaw (162)	12 July 70	2 May 91	10 July 93	2 May 17		S. and T. Corps.
	Unattd. List	Scott-Elliot, Charles Reginald (51, 52, 57, 105, 124a, 162, 195)	19 Sep. 73	28 Jan. 93	5 Apl. 94	5 May 17	*Ty.l.c.* 28 Jan. 15 3 June 16	61 Pioneers.
	Unattd. List	Hawkes, Robert (51, 66a, 170)	26 Sep. 74	28 Jan. 93	13 July 94	10 May 17	. . . .	3 Brahmans
505	R. Ir. R.	Lushington, William Nelson (50)	5 Sep. 70	23 May 91	17 June 96	23 May 17	1 Jan. 16 *Ty.l.c.* 16 Oct. 15	S. and T. Corps
	D. C. L. I.	Tregear, Vincent Francis William (50, 52, 71)	1 Jan. 70	23 May 91	24 Sep. 92	23 May 17	. . . .	98 Infantry.
	R. Ir. R.	McCaskill, John Charles Henry (50, 53, 66a, 71, 170)	25 Feb. 72	27 Jan. 92	27 Mar 94	11 June 17	. . . .	88 Infantry.
	Linc. R.	*Irvine, Andrew Alexander* (162)	*28 Sep. 71*	*17 June 91*	*12 Oct. 92*	*17 June 17*	. . . .	*Supy. List.*
	L'pool R.	Jackson, Keith Hungerford (106, 113, 170)	3 Dec. 70	17 June 91	6 Nov. 92	17 June 17	. . . .	Mily. Accts. Dept.
	North'd Fus.	Stewart, Robert Joseph Tucker (66a, 94, 100, 162, 165, 169)	18 Aug. 72	17 June 91	27 Oct. 94	17 June 17	. . . .	Cant. Magte.'s Dept.
510	Glouc. R.	Barry-Smith, Raymond Coape (50, 53, 57, 66a, 71, 169)	31 May 74	20 May 93	7 July 98	19 June 17	. . . .	17 Cavalry.
	R. A.	Hay, Westwood Norman, C.I.E., D.S.C., *p.s.c.* (50, 94, 162, 169) [L]	7 Oct. 71	24 July 91	4 Aug. 93	24 July 17	*Ty.l.c.* 18 Jan. 17	129 Baluchis.
	R. A.	*Trevor, Arthur Prescott*, C.S.I., C.I.E. (161, 169) [*l*]	*17 Mar. 72*	*24 July 91*	*22 Aug. 97*	*24 July 17*	. . . .	*Supy List.*
512	R. A.	Hunt, Vere De Vere (51, 161, 165, 169, 204)	26 Dec. 70	24 July 91	11 Oct. 95	24 July 17	. . . .	Cant. Magte.'s Dept.
	Glouc. R.	*Burn, George Callander* (105)	*22 Oct. 70*	*25 July 91*	*11 Oct. 92*	*25 July 17*	. . . .	*Supy. List.*

Colonels and Lieutenant-Colonels—*contd.*

No.	Former Corps.	Names.	Date of birth.	Date of first commission.	Date of appointment to Indian Army.	Date of Rank. Substantive.	Date of Rank. Army.	Remarks.
	N. Lan. R.	*O'Brien, Aubrey John,* C.I.E. *(161, 204)* [L]	*5 Dec. 70*	*25 July 91*	*17 May 95*	*25 July 17*	. . . .	*Supy. List.*
513	E. Surr. R.	**Churchill**, George Ross Deas (50, 53, 162, 170)	2 July 70	25 July 91	9 Feb. 96	25 July 17	. . . .	19 Punjabis.
	R.Innis.Fus.	**Preston**, Eyre Evans (71, 105, 162, 169) [L]	12 Mar. 72	25 July 91	16 May 93	25 July 17	. . . .	S. and T. Corps.
515	York. L. I.	**Ellwood**, Bertram Price (50, 66a, 155, 169, 172a)	30 Sep. 72	18 May 92	30 Oct. 93	29 July 17	. . . .	31 Lancers.
	Manch. R.	**McVean**, Donald Archibald Dugald, D.S.O. (50, 127, 129, 162)	22 July 70	7 Nov. 91	8 Feb. 96	29 July 17	. . . .	45 Sikhs.
	R. Lanc. R.	**Keogh**, James Blair (50, 71, 137, 170)	25 Jan. 71	9 Sep. 91	10 Nov. 95	23 Aug. 17	. . . .	32 Lancers.
	York. R.	**Bredin**, Alexander (51, 53, 57, 105, 157, 169, 172)	22 Feb. 71	9 Jan. 92	9 July 96	28 Aug. 17	. . . .	93 Infantry.
	R. Mar.	**Hopwood**, Herbert Reginald, C.S.I., *p.s.c.* (50, 53, 66a, 162)	23 Aug. 71	1 Sep. 91	22 Oct. 94	1 Sep. 17	c. 3 June 16	33 Lt. Cavalry.
520	R. Scots.	**Preston**, George Allen (50, 57, 71, 169)	14 Jan. 68	9 Sep. 91	14 Jan. 93	9 Sep. 17	. . . .	6 G. R.
	S.W Bord.	*Wimberley, Francis Gordon Arabin*	*22 Aug. 69*	*9 Sep. 91*	*24 Oct. 93*	*9 Sep. 17*	. . . .	*Supy. List.*
	Leins. R.	**Chamier**, William St. George (51, 66a, 69, 71, 75)	3 Aug. 70	9 Sep. 91	11 Oct. 94	9 Sep. 17	. . . .	S. and T. Corps.
	E. Lan. R.	**Mears**, Cecil Delarue, M.V.O. (50, 162)	26 July 71	9 Sep. 91	18 June 95	9 Sep. 17	3 June 16	16 Cavalry.
	Leins. R.	*Kirkwood, Andrew Torton (104, 167)*	*29 Aug. 71*	*9 Sep. 91*	*15 June 93*	*9 Sep. 17*	. . . .	*Supy. List.*
	Ches. R.	**Hulseberg**, Herbert, D.S.O. (51, 60, 95, 162, 169)	16 Oct. 70	9 Sep. 91	17 June 90	9 Sep. 17	. . . .	127 Lt. Infantry.
	Leins. R.	*Wood, Wyndham Madden Pierpoint,* C.I.E. *(154, 172)*	*10 Sep. 69*	*10 Oct. 91*	*21 Aug. 94*	*10 Oct 17*	. . . .	*Supy. List.*
	Durh. L. I.	**Anderson**, Robert Holme Bankes (57)	28 Aug. 71	10 Oct. 91	10 June 94	10 Oct. 17	. . . .	Mily. Accts. Dept.
525	Oxf. L. I.	**Edwardes**, Alexander Coburn (50, 52, 59, 71, 88, 137, 162)	15 Jan. 73	27 Jan. 92	22 Oct. 96	12 Oct. 17	3 June 16 Ty.l.c. 10 Sep. 16	102 Grenadiers.
	R. War. R.	**Walton**, Graham (50, 53, 172)	8 May 70	20 Feb. 92	3 Nov. 94	20 Oct. 17	3 June 16 Ty.l.c. 7 May 16	46 Punjabis.
	N. Lan. R.	**Mitchell**, Wilfrid James (50, 66a, 72a, 169)	5 Mar. 71	28 Oct. 91	12 Apl. 93	28 Oct. 17	. . . .	124 Infantry.
	R. Scots.	**Davie**, James Henry Morison (50, 65, 66a, 72)	13 Aug. 70	7 Nov. 91	13 Jan. 07	7 Nov. 17	3 June 16	34 Horse.
	W. York. R.	**Hawes**, Charles Howard, D.S.O., M.V.O. (50, 59, 66a, 94, 100, 170)	31 Aug. 71	7 Nov. 91	7 July 95	7 Nov. 17	. . . .	23 Cavalry.
530	R. Ir. R.	**Hutchinson**, Chas. Alex. Robert, D.S.O., *p.s.c.* (50, 53, 66a, 71, 162, 169)	31 May 27	7 Nov. 91	7 Apl. 96	7 Nov. 17	c. 31 Aug. 17 6 June 12	41 Dogras.
	Midd'x R.	*Macpherson, Archibald Duncan (162, 169)*	*14 Jan. 72*	*7 Nov. 91*	*6 Apl. 93*	*7 Nov. 17*	. . . .	*Supy. List.*
	Welsh. R.	**Scott**, William (50, 53, 71)	21 Nov. 71	7 Nov. 91	24 May 93	7 Nov. 17	. . . .	104 Rifles.
	Gord. Highrs	**Watt**, Donald Munro, D.S.O., *p.s.c.* (50, 53, 57, 72, 157)	18 June 71	7 Nov. 91	17 Aug. 96	7 Nov. 17	Ty.l.c. 28 May 16 3 June 16	2 G. R.
	Essex. R.	*Smith, Charles Aitchison (144a, 162, 169)*	*12 Sep. 71*	*7 Nov. 91*	*18 Jan. 96*	*7 Nov. 17*	. . . .	*Supy. List.*
	Sco. Rif.	**Keighley**, Vernon Aubrey Scott, M.V.O. (71, 167, 170)	15 Dec. 74	28 Dec. 95	5 June 99	2 Dec. 17	. . . .	18 Lancers.
534	K.O.S.Bord	**Browne**, Herbert José Pierson, *p.s.c.* (50, 53, 157, 170)(c)	21 Jan. 72	5 Dec. 91	14 Apl. 93	5 Dec. 17	c. 3 June 16 Ty. l. c. 1 Apl. 15	5 G. R.

535	E. Lan. R.	**Battye**, Arthur Baldwyn (65, 170)	2 Nov. 72	5 Dec. 91	31 Oct. 96	5 Dec. 17	. . . .	6 Cavalry.
	Midd'x R.	**Jackson**, Herbert William, D.S.O. (105, 120, 167)	9 June 72	5 Dec. 91	23 Aug. 93	5 Dec. 17	1 Jan. 17 *Ty.l.c.* 7 *Aug.* 16	89 Punjabis
	Manch R.	**Houston**, Arthur Manson (50, 53, 66*a*, 169)	16 Oct. 70	5 Dec. 91	15 July 96	5 Dec. 17	.	69 Punjabis.
	R.W.S.R.	**Bickford**, William Wilfrid, C.I.E. (51, 71, 72*a*, 137, 162*a*, 169)	22 Sep. 71	5 Dec. 91	22 July 96	5 Dec. 17	17 Sep. 17	107 Pioneers
	Durh. L. I.	**Oakes**, Horace Charles (50, 71, 161) [L]	3 May 71	12 Mar. 92	21 May 96	17 Dec. 17	. . .	28 Lt. Cavalry.
	R. Ir. R.	*O'Brien, Edward (88, 111, 132, 153, 162, 169)*	*20 Sep. 72*	*30 Dec. 91*	*2 July 94*	*30 Dec. 17*	. . . .	*Supy. List.*
	N. Lanc. R.	*Warren-Codrington, Hopton Osbert (50, 66a, 153, 162)*	*11 Feb. 71*	*9 Jan. 92*	*5 Aug. 93*	*9 Jan. 18*	. . . .	*Supy. List.*
540	Unattd. List	**Hewlett**, Alan (50, 65, 71, 162)	10 Nov. 74	28 Jan. 93	23 Mar. 94	15 Jan. 18	. . . .	38 C. I. Horse.
	Unattd. List	**Murray**, Walter Godfrey Patrick, *p.s.c.* (50, 170) (*n*)	30 May 74	28 Jan. 93	5 Apl. 94	22 Jan. 18	. . . .	21 Punjabis.
	Unattd. List	**Fordham**, William Marshall, *p.s.c.* (50, 53, 162, 166, 169)	9 Aug. 75	14 Aug. 95	15 Oct. 96	22 Jan. 18	. . . .	20 Infantry.
	Derby. R.	**Pye**, William Edmund, C.I.E. (66*a*, 71, 124*a*, 162)	29 Apl. 72	27 Jan. 92	26 Feb. 98	27 Jan. 18	. . . .	98 Infantry.
	Oxf. L. I.	**Stewart**, Archibald Campbell (51, 52, 71, 169)	27 Dec. 72	27 Jan. 92	21 Mar. 93	27 Jan. 18	. . . .	Guides.
545	Glouc. R.	**Harington**, Arthur William Donnelly (50)	7 Jan. 73	27 Jan. 92	8 June 93	27 Jan. 18	. . . .	S. and T. Corps.
	Glouc. R.	**Waller**, Alfred Alexander Phillips (50, 52, 71, 72*a*, 155)	5 Sep. 70	27 Jan. 92	14 Aug. 93	27 Jan. 18	. . . .	105 Lt. Infy.
	Midd'x R.	**Cattell**, Gilbert Landale (50, 95, 106, 148*a*, 157, 195)	23 Oct. 71	27 Jan. 92	4 Aug. 96	27 Jan. 18	. . . .	Cant. Magte.'s Dept.
	Wilts. R.	**Lees**, Charles Henry Brownlow, C.M.G., *p.s.c.* (50, 53, 162, 16)	17 Oct. 71	27 Jan. 92	25 Nov. 95	27 Jan. 18	*Ty. l.c.* 16 Mar. 14	51 Sikhs.
	R. A.	**Rowcroft**, Claude Harold (51, 94, 162, 169)	2 Mar. 72	19 Feb. 92	15 Nov. 96	19 Feb. 18	. . . .	9 Horse.
550	Midd'x R.	**Wall**, Allan Copinger (12, 50, 53, 66*a*, 71, 157, 162, 170)	1 Nov. 71	20 Feb. 92	28 Jan. 96	20 Feb. 18	*Ty. l.c.* 13 Nov. 15	8 G. R.
	Hamps. R.	**Gaussen**, James Robert, C.M.G., D.S.O. (50, 66*a*, 125, 137)	20 Mar. 71	20 Feb. 92	21 June 94	20 Feb. 18	.	3 Horse.
	Leins. R.	*Owens, Francis Charles (94, 103, 113, 162)*	*7 July 72*	*20 Feb. 92*	*22 Aug. 94*	*20 Feb. 18*	*Ty.l.c.* 21 *June* 17	*Supy. List.*
	R.Muns.Fus	**Ottley**, William John (50, 52, 71, 72*a*, 161, 167, 170, 195)	17 Mar. 70	9 Apl. 92	7 Nov. 94	21 Feb. 18	. . . .	23 Pioneers.
	Wilts. R.	**Herdon**, Hugh Edward, *p.s.c.* (57, 66*a*, 72*a*, 94, 162, 165, 169)	28 Oct. 71	13 July 92	1 Dec. 96	2 Mar. 18	*Ty.l.c.* 10 *July* 17	55 Rifles.
	Unattd. List	**Ricketts**, Robert Lumsden, *p.s.c.* (52, 57, 170)	9 July 72	3 Sep. 92	24 Dec. 93	8 Mar. 18	*Ty.l.c.* 25 *Aug.* 16	10 Lancers.
555	R. Ir. R.	**Willans**, Thomas James, D.S.O. (169)	2 Sep. 72	21 Oct. 93	2 Sep. 97	8 Mar 18	. . . .	57 Rifles.
	S. Lan. R.	**Gardiner**, Richard, *p.s.c.* (50, 53, 57, 71, 72*a*, 94, 165, 169), [*l*]	31 Oct. 74	6 Mar. 95	28 Oct. 97	10 Mar. 18	. . . .	53 Sikhs.
	Worc. R.	**Dopping-Hepenstal**, Maxwell Edward, D.S.O. (50, 52, 59, 71, 72*a*, 157) [*l*]	7 Mar. 72	18 June 92	25 July 96	11 Mar. 18	. . . .	1 G. R.
	K. O. S. B.	**James**, Alfred Albert (50, 167)	27 July 69	12 Mar. 92	9 Apl. 94	12 Mar. 18	. . . .	
	Unattd. List	**Rice**, Sheridan Knowles Brownlow (51, 53, 66*a*, 162, 165, 170)	29 July 72	3 Sep. 92	26 Dec. 93	14 Mar. 18	. . . .	35 Sikhs.
560	S. Staff. R.	**Carnegy**, Gerald Patrick Ogilvy, D.S.O. (51, 57, 66*a*, 71, 137)	25 Nov. 73	7 Dec. 95	24 Oct. 97	15 Mar. 18	. . . .	7 Lancers.
	Glouc. R.	**Hunter**, Walter James Henry (51, 52, 153)	25 Jan. 71	13 Aug. 92	2 Apl. 95	18 Mar 18	. . . .	117 Mahrattas.
	A. S. C.	**Wintle**, Charles Edmund Hunter (51, 52, 66*a*, 71) *Service for pension from 24 Aug. 94.*	4 May 71	16 Dec. 96	18 Nov. 10	29 Mar. 18	*l.c.* 3 June 17	96 Infantry.
	Bedfc. R.	**Strong**, Herbert Percy (51, 52, 66*a*, 69, 71, 155)	10 June 74	12 Dec. 94	23 July 98	30 Mar 18	. . . .	101 Grenadiers.
	Devon. R.	*Anderson, William Cortlandt (51, 170, 195)*	*22 Sep. 69*	*9 Apl. 92*	*17 Sep. 94*	*9 Apl. 18*	. . . .	*Supy. List.*
	Worc. R.	**Dyke**, Percyvall Hart, D.S.O. (50, 66*a*, 162, 169)	24 Aug. 72	18 May 92	4 Sep. 96	18 May 18	1 Jan. 17 *Ty. c.* 15 May 17.	130 Baluchis.
	W.York R.	*Luard, Charles Ekford (71)*	*11 Oct. 69*	*18 May 92*	*26 Sep. 94*	*18 May 18*	. . . .	*Supy. List.*
	York L. I.	*Harrison, Hugh (50, 65, 133, 162, 165)*	*27 May 70*	*18 May 92*	*21 Nov. 95*	*18 May 18*	. . . .	*Supy. List.*
565	Suff. R.	**Ommanney**, Charles Vernon (50, 195)	29 Dec. 72	7 Dec. 95	1 Mar. 98	29 May 18	. . . .	109 Infantry.
	Unattd. List	**Moberly**, Hugh Stephenson, *p.s.c.* (50, 71, 124*a*, 167)	25 Aug. 73	30 Aug. 93	25 Jan. 95	1 June 18	3 June 16 *Ty.l.c.* 10 Nov. 16	66 Punjabis.
	Unattd. List	**Little**, William Campbell (50, 157) *Service for promotion from 1 Mar. 95.*	26 Feb. 75	1 Sep. 93	23 Mar. 99	11 June 18	*Ty.l.c.* 1 Mar. 17	6 G. R.
	R. Ir. R.	**Browne**, Leslie Swinton (50, 52, 162, 167)	3 Dec. 72	19 Nov. 92	27 Feb. 94	15 June 18	. . . .	37 Dogras.
	5 D. G.	**Hodgkinson**, Charles (50, 66*a*)	14 Apl. 70	18 June 92	9 May 95	18 June 18	. . . .	6 Cavalry.
570	5 Lrs.	**Hodgkinson**, John (50, 66*a*)	14 Apl. 70	18 June 92	17 Jan. 95	18 June 18	. . . .	5 Cavalry.

No.	Former Corps.	Names.	Date of birth.	Date of first commission.	Date of appointment to Indian Army.	Date of Rank. Substantive Lt.-Colonel.	Date of Rank. Army.	Remarks.
		Colonels and Lieutenant-Colonels—*concld.*						
571	R. W. Kent. R.	**Bradshaw**, Charles Richard, *p.s.c.* (50, 129, 157, 162, 169) [*l*]	18 Jan. 73	19 July 93	8 May 97	22 June 18	. . . .	9 G. R.
	Suff. R.	**Cuming**, Robert John D.S.O., (51, 52, 66*a*, 137)	28 Aug. 72	7 Mar. 94	2 Aug. 97	24 June 18	3 June 17	48 Pioneers.
	Unattd. List	**Johnson**, John Ernest Blois (71, 170)	10 June 72	30 Aug. 93	24 Jan. 95	1 July 18	. . . .	21 Cavalry.
	R. Dub. Fus.	**Gale**, Ernest Septimus (51, 61, 145, 157, 162)	14 Oct. 72	28 June 93	11 Mar. 97	3 July 18	. . . .	10 G. R.
575	Bord. R.	**Hume**, Seignelay Gustavus Willoughby (51, 52, 137)	31 Aug. 71	29 Nov. 90	6 Dec. 04	6 July 18	. . . .	7 Rajputs.
	Glouc. R.	**Chitty**, Arthur Wilson, *p.s.c.* (50, 57, 66*a*, 71, 94, 100, 162, 169)	7 May 72	13 July 92	24 Aug. 94	13 July 18	*Ty. l.c.* 25 *Dec.* 16	126 Infantry.
	Worc. R.	*Ramsay, Arthur Dennys Gilbert*, C.I.E., O.B.E. (*88, 161, 196*)	*9 Oct. 72*	*13 July 92*	*14 Dec. 96*	*13 July 18*	.	*Supy. List.*
	Wilts. R.	**Taylor**, Reginald O'Bryan, C.M.G., C.I.E., *p.s.c.* (51, 53, 69, 157, 170) [L]	6 Feb. 72	13 July 92	13 Aug. 96	13 July 18	1 Jan. 18	19 Lancers.
	Worc. R.	**Calvert**, Robert Thomas Christopher, *p.s.c.* (50, 57, 66*a*, 71, 137, 162)	20 Feb. 73	13 July 92	23 Oct. 96	13 July 18	. . . .	122 Infantry
	R. Dub. Fus.	**Harcourt**, John Stewart Mortimer (51, 53, 66*a*, 71, 157)	7 Oct. 72	13 July 92	5 Jan. 97	15 July 18	. . . .	2 G. R.
580	E. York. R.	**Kauntze**, Bertram Charles (12, 50, 53, 59, 66*a*, 71, 193)	17 Oct. 73	7 Mar. 94	23 Aug. 98	31 July 18	. . . .	86 Infantry.
	Hamps. R.	**Nicolay**, Bernard Underwood (50, 124*a*, 157)	23 Dec. 73	7 Mar. 94	27 Mar. 98	3 Aug. 18	. . . .	4 G. R.
	Glouc. R.	*Friselle, Julian* (*162, 170*)	*30 July 71*	*31 Aug. 92*	*23 Apl. 95*	*4 Aug. 18*	. . . .	*Supy. List.*
	Derby. R.	**Bruce**, Malcolm Edward Lloyd (50, 137)	12 Feb. 72	13 Aug. 92	18 July 96	13 Aug. 18	*Ty. l.c. 11 June 15*	Mily. Accts. Dept.
	R. W. S. R.	*Patterson, Stewart Blakely Agnew* (*66a, 169*)	*18 Mar. 72*	*13 Aug. 92*	*31 Dec. 93*	*13 Aug. 18*	. . . .	*Supy. List.*
	Manch. R.	**Stansfeld**, Cyril Grey, *p.s.c.* (50, 53, 72, 157, 172*a*, 199)	17 July 71	13 Aug. 92	24 Aug. 96	13 Aug. 18	. . . .	8 G. R.
	R. Fus.	**Bruce**, George David, D.S.O., *p.s.c.* (50, 53, 71, 185, 195)	7 Oct. 72	20 May 93	27 Feb. 97	13 Aug. 18	*Ty. l.c.* 31 May 16 / 1 Jan. 17	81 Pioneers.
585	Unattd. List	**Pryce**, Henry Edward ap Rhys., C.M.G., D.S.O., *p.s.c.*, (50, 64, 162, 170)	30 Nov. 74	16 Jan. 95	2 Apl. 96	17 Aug. 18	3 June 16 / *Ty l.c.* 19 Apl. 15	18 Infantry.
	Unattd. List	**Waring**, Edmund Henry (50, 57, 66*a*, 71)	25 Mar. 75	14 Aug. 95	14 Oct. 96	19 Aug. 18	. . . .	63 Infantry.
	Mdd'x R.	**Preston**, William John Phælim, D.S.O. (50, 52, 57, 66*a*, 71)	11 June 73	31 Aug. 92	24 Oct. 96	31 Aug. 18	. . . .	97 Infantry.
	Unattd. List	**Harbord**, Cyril Rodney, D.S.O. (66*a*, 71)	2 Dec. 73	3 Sep. 92	26 Dec. 93	1 Sep. 18	. . . .	30 Lancers.
	Unattd. List	**Dundas**, Patrick Henry, D.S.O., *p.s.c.* (50)	17 Oct. 71	3 Sep. 92	5 Jan. 94	3 Sep. 18	*Ty. l.c.* 17 *July* 16 / 3 June 17	6 Lt. Infantry.
	Unattd. List	*Abbey, Walter Bulmer Tait* (*104, 117, 125a, 140, 148b, 162, 181, 202*)	*8 Aug. 72*	*3 Sep. 92*	*25 Dec. 93*	*3 Sep. 18*	. . . .	*Supy. List.*
590	Unattd. List	**Corlett**, John Stebbing (162, 165, 169) [L]	18 June 72	3 Sep. 92	28 Dec. 93	3 Sep. 18	. . . .	15 Lancers.
	Unattd. List	**Watts**, George Annesley Ross (94, 162, 169)	3 June 73	3 Sep. 92	24 Dec. 93	3 Sep. 18	*l.c.* 3 June 17	25 Cavalry.
	R. Suss. R.	**Quin**, Richard Percival Wemyss (50, 53, 57, 71, 127, 162, 165, 170 [L]	10 Feb. 74	15 Feb. 93	31 July 97	12 Sep. 18	. . . .	41 Dogras.
	W. York. R.	**V. C. Costello**, Edmond William, C.M.G., D.S.O., *p.s.c.* (50, 162, 165, 169)	7 Aug. 73	19 Nov. 92	21 Mar. 94	22 Sep. 18	3 June 16	22 Punjabis.

594	Unattd. List	**Prissick,** Cuthbert (50, 53, 162, 169)	15 Sep. 71	23 Dec. 93	27 Mar. 95	22 Sep. 18	3 June 17	56 Rifles.
595	R. Dub. Fus.	**Orton,** Ernest Frederick, *p.s.c.* (50, 59, 66*a*, 94, 127, 129, 161, 169) .	27 Apl. 74	7 Mar. 94	4 June 97	26 Sep. 18	*Ty. l. c.* 21 Feb. 16 *Ty. c.* 29 Oct. 17	37 Lancers.
	Durh. L. I.	**Stewart,** Archibald Francis (66*a*, 155)	12 Sep. 71	28 Sep. 92	21 Mar. 97	28 Sep. 18	. . .	114 Mahrattas.
	R. Ir. Fus.	**Steel,** Richard Alexander, C.I.E., *p.s.c.* (45*a*, 50, 53, 127, 129, 172 (l)	6 Oct. 73	5 Oct. 92	18 Nov. 96	5 Oct. 18	*c.* 1 Jan. 18 *Ty. l.c.* 1 *Nov.* 16	17 Cavalry.
	Unattd. List	**Hawkes,** William Cotter Williamson, D.S.O. (50, 66*a*, 71, 95, 162*a*, 169)	13 June 72	23 Dec. 93	27 Mar. 95	20 Oct. 18	*Ty. l.c.* 20 Feb. 16	12 Pioneers.
	Unattd. List	**Harrison,** Arthur Howarth Pryce (165, 170)	14 July 71	28 June 93	20 Oct. 94	10 Nov. 18	29 Nov. 15,	33 Punjabis.
600	R. W. S. R.	**Savile,** Albany Robert Cecil (155)	13 Sep. 71	16 Nov. 92	1 Jan. 96	16 Nov. 18	. . . .	31 Lancers.
	R. W. Fus.	***Keen,** William John,* C.I.E. (*169*)	*24 Mar. 73*	*19 Nov. 92*	*19 Dec. 93*	*19 Nov. 18*	. .	*Supy. List.*
	Norf. R.	**Abbott,** Leonard Henry, C.M.G., *p.s.c.* (50, 53, 57, 137, 162). . . .	10 Sep. 75	29 Feb. 96	3 May 98	21 Nov. 18	*Ty. l.c.* 17 Aug. 15 3 June 17	8 Rajputs.
602	Unattd. List	**deLabilliere,** Edward Gardiole Delacour	10 July 72	23 Nov. 92	2 Mar. 94	23 Nov. 18	. . . .	Mily. Accts. Dept.

No.	Former Corps.	Names.	Date of birth.	Date of first commission.	Date of appointment to Indian Army.	Date of Rank. Substantive.	Date of Rank. Army.	Remarks.
		Majors—1,115.						
1	R. Ir. R.	**Hill**, William Alfred (66*a*, 170) *Forfeited from 24 Apl. to 13 Dec. 07 for promotion.*	1 Feb. 66	24 Apl. 89	24 Jan. 91	14 Dec. 07	. . . .	
	S. Staff. R.	**Home**, Pelham Maitland (50, 52, 66*a*, 72, 157)	6 Oct. 70	25 Mar. 91	21 Dec. 94	25 Mar. 09	. . . .	39 Garhwal Rifles.
	R. W. Fus.	**Blair**, Richard William Creighton (50, 66*a*, 71, 137, 162, 170) [*l*]. *Half pay from 29 Apl. 07 to 26 Dec. 07 (to count).*	10 Oct. 70	2 May 91	30 Oct. 96	2 May 09	. . . .	123 Rifles.
	R. Muns. Fus.	**Haughton**, Hugh Latimer (50, 125*a*, 102, 105)	18 Sep. 70	17 June 91	27 Jan. 96	17 June 09	. . . .	93 Infy.
5	18 Huss.	**Harbottle**, Robert Cecil (50, 65, 71, 137)	23 Nov. 71	10 Oct. 91	20 Oct. 96	10 Oct. 09	. . . .	9 Lancers.
	18 Huss.	**Claridge**, Percy Shelley Freese (50, 52)	15 May 68	5 Dec. 91	12 May 93	5 Dec. 09	. . . .	28 Lt. Cavalry.
	Leins. R.	**Cates**, William Johnstone (50, 94, 169, 185)	2 Sep. 71	20 Feb. 92	14 June 95	20 Feb. 10	. . . .	S. and T. Corps.
	R. Dub. Fus.	**Reece**, John Deane (195)	11 Oct. 71	20 Feb. 92	27 July 95	20 Feb. 10	. . . .	88 Infantry.
	Glouc. R.	**Bennett**, John Francis (51, 66*a*, 71, 137)	25 Oct. 72	13 Aug. 92	5 Oct. 97	13 Aug. 10	. . . .	2 Lancers.
10	Unattd. List	**Pratt**, Ernest George Wilberforce (50, 52, 57, 66*a*)	6 Aug. 73	3 Sep. 92	9 Nov. 93	3 Sep. 10	. . . .	95 Infantry.
	Derby. R.	**Keene**, Charles William (50, 66*a*, 72, 94, 169)	5 Dec. 70	28 Sep. 92	26 July 96	28 Sep. 10	. . . .	28 Punjabis.
	Durh. L. I.	**Martin**, Frank (50, 162, 170)	22 May 70	28 Sep. 92	23 July 96	28 Sep. 10	. . . .	25 Punjabis.
	North'n R.	**Faunce**, Alured De Laune (66*a*, 71, 107, 140, 172)	3 July 70	19 Oct. 92	9 Oct. 94	19 Oct. 10	. . . .	9 Infantry
	R. Fus.	**Gaskell**, Thomas Kershaw (50, 66*a*, 71, 167) *Half pay from 1 Apl. to 10 Oct. 16.*	1 Oct. 70	19 Oct. 92	12 Jan. 96	19 Oct. 10	. . . .	
15	North'n R.	**McGill**, Harry Robertson (51, 52, 66*a*, 105, 167)	21 July 72	9 Nov. 92	16 Sep. 96	9 Nov. 10	. . . .	90 Punjabis.
	Midd'x. R.	**Haslehust**, George Williams, *p. s. c.* (51, 57, 72*b*) *Service for pension from 8 Apl. 93.*	19 Oct. 73	19 Nov. 92	8 Feb. 00	19 Nov. 10	*Ty.l.c.* 10 Apl. 17	29 Punjabis.
	N. Lanc. R.	**Greig**, John Glennie, C.I.E. (66, 71)	24 Oct. 71	19 Nov. 92	26 Oct. 95	22 Sep. 18	. . . .	121 Pioneers.
	Unattd. List	**Bogle**, John Savile (50, 52, 59, 156, 169)	25 Sep. 72	23 Nov. 92	2 Mar. 94	23 Nov. 10	. . . .	Guides.
19	Lanc. Fus.	**Lloyd**, Kenrick Horace (50, 53, 66*a*, 162, 167, 170, 195)	1 Jan. 74	17 Dec. 92	4 Aug. 97	17 Dec. 10	. . . .	62 Punjabis.

20	Midd'x R.	Muscroft, William St. Clair (75, 124a, 162)	14 June 73	17 Dec. 92	19 May 97	17 Dec. 10	l. c. 7 Jan. 14	S. and T. Corps.
	R. A.	Gordon, George Alexander Steuart (59, 71, 162, 167, 170)	6 Nov. 72	2 Jan. 93	19 Dec. 95	2 Jan. 11	l.c. 1 Jan. 18	13 Lancers.
	High. L. I.	Loch, Edward Campbell (162, 170)	15 Dec. 71	21 Jan. 93	9 Apl. 94	21 Jan. 11	. . . .	Army Remt. Dept.
	R. Fus.	*Bigg-Wither, Ferdinand* (88, 100, 103, 127, 1129, 169, 180, 185, 204)[L]	*29 Aug. 72*	*21 Jan. 93*	*20 Feb. 97*	*21 Jan. 11*	. . . .	*Supy. List.*
	Unattd. List	Rice, Sidney Mervyn, C.I.E., p.s.c. (50, 71, 162, 195)	4 Mar. 73	28 Jan. 93	3 Apl. 94	28 Jan. 11	Ty. l.c. 9 Nov. 16	64 Pioneers.
	Unattd. List	Hope, Adrian Victor Webley (50, 53, 66a, 71, 165, 195)	16 Feb. 73	28 Jan. 93	3 Apl. 94	28 Jan. 11	Ty. l.c. 20 Nov. 15	32 Pioneers.
	Unattd. List	*Palin, Randle Harry* (57, 94, 169)	*20 Feb. 73*	*28 Jan. 93*	*1 Apl. 94*	*28 Jan. 11*	. . .	*Supy. List.*
	Unattd. List	*Knollys, Robert Walter Edmund* (50, 144a, 162, 169)	*10 Nov. 72*	*28 Jan. 93*	*28 Mar. 94*	*28 Jan. 11*	Ty. l.c. 29 May 15	*Supy. List.*
25	Unattd. List	Taylor, Edward Moore, p.s.c. (50, 53, 66a, 162, 169)	1 Sep. 72	28 Jan. 93	28 Mar. 94	28 Jan. 11	Ty. l.c. 9 July 16	22 Cavalry.
	Unattd. List	Peacock, Edward Barnes (88, 161, 167, 169)	28 July 73	28 Jan. 93	25 Mar. 94	28 Jan. 11	Ty. l.c. 31 Mar. 17	Mily. Accts. Dept.
	R. W. Kent R.	*Stewart, Hugh*, C.I.E. (50, 66a, 161, 169)	*5 Nov. 72*	*15 Mar. 93*	*4 Apl. 97*	*15 Mar. 11*	. . .	*Supy. List.*
	R. Suss. R.	Elliott, Herbert Henry (50, 72)	30 Dec. 69	26 Apl. 93	25 Nov. 04	26 Apl. 11	Ty. l.c. 25 Feb. 18	Mily. Farms Dept.
	Unattd. List	Holbrooke, Bernard Frederick Roper (66a, 72b, 94, 170)	7 Oct. 71	28 June 93	20 Oct. 94	28 June 11	. . . .	124 Infantry.
	E. Surr. R.	Sproule, Harry Hatton (51, 52, 65, 66a, 124a)	27 June 73	28 June 93	10 Dec. 97	28 June 11	. . . .	4 Cavalry.
30	R. A.	Kettlewell, Archibald Middleton p.s.c. (12, 54, 57, 94, 162, 169)	7 Oct. 73	28 June 93	11 Oct. 97	28 June 11	. . . .	39 C. I. Horse.
	S. Staff. R.	Corbett, Basil Andrew (52, 72, 137, 162)	11 Nov. 73	19 July 93	29 Sep. 97	19 July 11	. . . .	13 Rajputs.
	Derby. R.	Knowles, George, D.S.O. (50, 137, 169)	16 Oct. 73	19 July 93	6 Aug. 97	19 July 11	. . . .	2 Lancers.
	R. Ir. Rif.	Molloy, Gerald Macleay, p.s.c. (50, 71, 137, 188a)	29 Sep. 73	19 July 93	28 Oct. 97	19 July 11	Ty. l.c. 21 Sep. 17	34 Horse.
	Unattd. List	O'Grady, Henry DeCourcy, p.s.c. (51, 144a, 169)	26 Aug. 73	30 Aug. 93	25 Jan. 95	30 Aug. 11	Ty. l.c. 15 Feb. 16 l.c. 1 Jan. 17 c. 3 June 18	52 Sikhs.
	Unattd. List	*Coldstream, John Glayton* (94, 166, 169)	*16 July 72*	*30 Aug. 93*	*27 Jan. 95*	*30 Aug. 11*	. . .	*Supy. List.*
35	Unattd. List	Storr, Charles Lancelot, p.s.c. (71, 172)	18 Jan. 74	30 Aug. 93	26 Jan. 95	30 Aug. 11	Ty. l.c., 30 Oct. 17	54 Sikhs.
	Unattd. List	Barrett, Arthur Lennard (50, 66a, 71, 94, 169)	9 Nov. 72	30 Aug. 93	24 Jan. 95	30 Aug. 11	l.c. 1 Jan. 18	126 Infantry.
	Unattd. List	*Nicolas, Francis Carminowe* (57, 65, 94, 165)	*16 Nov. 74*	*30 Aug. 93*	*25 Jan. 95*	*30 Aug. 11*	. . . .	*Supy. List.*
	Unattd. List	Ames, Launcelot Hope Rix (51, 57, 71, 106, 195)	31 July 74	30 Aug. 93	25 Jan. 95	30 Aug. 11	. . . .	75 Infantry.
	Unattd. List	*St. John, Henry Beauchamp*, C.I.E. (94, 162 170)	*26 Aug. 74*	*30 Aug. 93*	*27 Jan. 95*	*30 Aug. 11*	. . . .	*Supy. List.*
	Unattd. List	Dalrymple-Hay, Stair Francis Barton (50, 53, 66a, 71, 157, 172a)	21 Jan. 72	30 Aug. 93	25 Jan. 95	30 Aug. 11	. . . .	43 Regiment.
	Unattd. List	Heyland, Arthur Kyffin, D.S.O. (67, 75)	22 Apl. 74	30 Aug. 93	24 Jan. 95	30 Aug. 11	. . . .	S. and T. Corps.
40	Bord. R.	Longhurst, Arthur Lyster, p.s.c. (50, 59, 71, 94, 105, 157, 162, 169) [L]	1 Apl. 72	9 Sep. 93	25 Feb. 97	9 Sep. 11	l.c. 3 June 16	7 G. R.
	R. A.	Deans, Dennis, D.S.O. p.s.c. (50, 72a, 167) [l]	14 Sep. 74	14 Sep. 93	22 Sep. 97	14 Sep. 11	Ty. l.c. 1 Apl. 16 l.c. 1 Jan. 17	12 Cavalry.
	R. A.	Wigram, Clive, C.B., C.S.I., C.V.O., p.s.c., Eq. (51, 65, 72b, 162, 165, 169)	5 July 73	4 Oct. 93	14 Sep. 97	4 Oct. 11	l.c. 3 June 15	18 Lancers.
	Unattd. List	Steen, Hunter Carmichael (71, 155)	10 Aug. 72	11 Oct. 93	11 Jan. 95	11 Oct. 11	. . . .	114 Mahrattas.
	North'n R.	Thomson, Alexander Guthrie, D.S.O., p.s.c. (50, 71, 169)	11 Feb. 73	21 Oct. 93	6 May 97	21 Oct. 11	Ty. l.c., 30 Mar. 18 l.c. 3 June 18	58 Rifles.
45	Suff. R.	Bowden, William Harvey (50, 71, 106)	9 May 71	23 Dec. 93	16 Nov. 07	23 Dec. 11	. . . .	Army Clothing Dept.
	Unattd. List	Furney, John Leared (170)	14 Jan. 72	23 Dec. 93	1 Apl. 95	23 Dec. 11	l.c. 3 June 16	22 Punjabis.
	Unattd. List	*Fenning, William Maxwell* (162)	*29 June 73*	*23 Dec. 93*	*26 Mar. 95*	*23 Dec. 11*	. . . .	*Supy. List.*
	Suff. R.	Murphy, Charles Cecil Rowe (50, 53, 71, 105, 161, 167, 170) [L]	2 Mar. 72	23 Dec. 93	22 May 09	23 Dec. 11	l. c. 3 June 15	30 Punjabis.
	York. L. I.	*Haworth, Lionel Berkeley Holt* (94, 165, 169) [L]	*30 Sep. 73*	*23 Dec. 93*	*22 May 98*	*23 Dec. 11*	. . . .	*Supy. List.*
	Unattd. List	Babington, Alfred Harcourt (95)	18 Mar. 73	31 Jan. 94	16 Apl. 95	31 Jan. 12	. . . .	S. and T. Corps.
	Unattd. List	*Strong, Henry Stuart* (132, 153)	*30 Oct. 73*	*31 Jan. 94*	*10 Apl. 95*	*31 Jan. 12*	. . . .	*Supy. List.*
49	Unattd. List	Keen, Frederick Stewart, D.S.O., p.s.c. (50, 66a, 167, 170)	22 June 74	31 Jan. 94	11 Oct. 95	31 Jan. 12	. . . .	45 Sikhs.

No.	Former Corps.	Names.	Date of birth.	Date of first commission.	Date of appointment to Indian Army.	Date of Rank. Substantive.	Date of Rank. Army.	Remarks.
		Majors—*contd.*						
50	A.&S.Highrs	**Graham**, Robert Blackall (107, 162, 170)	2 Mar. 74	7 Mar. 94	28 Aug. 08	7 Mar. 12	*l.c.* June 18	46 Punjabis.
	R. Ir. R.	**Will**, Duncan Alured Elmsly	23 Aug. 72	7 Mar. 94	27 Apl. 97	7 Mar. 12	. . .	Mily. Accts. Dept.
	S. Lanc. R.	**Grant**, William Ogilvie (51, 59, 66*a*, 71, 172)	18 Mar. 74	7 Mar. 94	17 Oct. 97	7 Mar. 12	*Ty.l.c.* 27 Oct. 14	127 Lt. Infantry.
	E. Lanc. R.	**Hilson**, Robert James (50, 66*a*, 167)	9 Feb. 74	7 Mar. 94	5 Nov. 97	7 Mar. 12	. . .	91 Punjabis.
	E. Lanc. R.	**Lambert**, Walter John, D.S.O. (66*a*, 71, 161)	2 Apl. 74	7 Mar. 94	19 Nov. 97	7 Mar. 12	*Ty. l.c.* 5 Aug. 18	29 Lancers.
55	Unattd. List	**Villiers-Stuart**, William Desmond (50, 53, 157, 170)	11 Apl. 72	2 June 94	10 Oct. 95	2 June 12	*Ty. l.c.* 8 *Aug.* 16 / *l.c.* 19 Aug. 18	5 G. R.
	E. Surr. R.	*Barry, William Thurburn* (50, 56*a*, 71, 137)	*10 May 74*	*2 June 94*	*24 July 97*	*2 June 12*		*Supy. List.*
	Sco. Rif.	**Holdich**, Harold Adrian, D.S.O., *p.s.c.* (50, 53, 56*a*, 71, 157 169)	20 Mar. 74	10 Oct. 94	9 Oct. 97	10 Oct. 12	*Ty. l.c.* 10 Dec. 15 / *l.c.* 3 June 15	5 G. R.
	R. Ir. R.	**Luxmoore**, Charles de Joncourt (50, 52, 57, 66*a*, 72*a*) [*l*]	4 Sep. 73	10 Oct. 94	12 Nov. 97	10 Oct. 12	. . .	99 Infantry.
	Norf. R.	**Shoubridge**, Charles Alban Grevis (50, 53, 66*a*, 130, 137, 169)	6 Aug. 74	10 Oct. 94	19 Apl. 98	10 Oct. 12	. . .	112 Infantry.
	Unattd. List	**Watson**, Spencer Burton (50, 71, 162, 193)	11 Aug. 74	10 Oct. 94	20 Dec. 95	10 Oct. 12	*l.c.* 3 June 16	64 Pioneers.
60	Unattd. List	**Henderson**, Robert Wynne, D. S. O. (51, 52, 71, 72*b*, 161, 165, 169)	20 Aug. 74	10 Oct. 94	10 Jan. 96	10 Oct. 12	. . .	17 Cavalry.
	Unattd. List	**Macaulay**, Denzil Ibbetson Michael (51, 66, 66*b*, 162, 170) [L]	18 Sep. 72	10 Oct. 94	28 Dec. 95	10 Oct. 12	. . .	1 Lancers.
	Unattd. List	**Young**, Harry Norman (50, 124*a*, 159) [L]	2 Jan. 74	10 Oct. 94	23 Dec. 95	10 Oct. 12	*Ty. l.c.*, 14 Feb. 18.	S. and T. Corps.
	Unattd. List	**Cox**, Thomas Sands (65, 66*a*, 71, 124*a*, 162*a*, 167, 170, 175)	14 Mar. 74	10 Oct. 94	9 Jan. 96	10 Oct. 12	. . .	37 Dogras.
	Unattd. List	**Hogg**, Allan Gilbert Mayhew (50, 53, 66*a*, 94, 153, 162, 169)	24 July 75	10 Oct. 94	17 Dec. 95	10 Oct. 12	. . .	42 Regiment.
65	Unattd. List	**Bateman-Champain**, Claude Edward (51, 52, 59, 71, 157, 162, 170)	30 Mar. 75	10 Oct. 94	10 Jan. 96	10 Oct. 12	. . .	5 G. R.
	Unattd. List	**Carey**, Cecil William (50, 52, 71, 169)	11 Sep. 74	10 Oct. 94	5 Jan. 96	10 Oct. 12	. . .	Guides.
	Unattd. List	**Hawks**, George Augustus (56, 66*a*, 195)	27 May 74	10 Oct. 94	5 Jan. 96	10 Oct. 12	*Ty. l.c.* 25 Feb. 18	Mily. Farms Dept.
	York L. I.	**Lane**, William Horsburgh (65, 107, 121)	2 Aug. 74	10 Oct. 94	18 Aug. 97	10 Oct. 12	. . .	94 Infantry.
	Som. L. I.	**Nangle**, Montague Claude, *p.s.c.* (51, 104, 167) [*l*]	22 July 74	10 Oct. 94	4 Sep. 97	10 Oct. 12	. . .	92 Punjabis.
70	E. Sur. R.	**Davies**, Hugh Walter (50, 172*a*)	15 July 73	10 Oct. 94	14 Apl. 98	10 Oct. 12	. . .	S. and T. Corps.
	Midd'x R.	*Trotter, Everard Graham Stanley* (51, 71, 125*a*, 127 137) [*L*]	*22 Jan 73*	*10 Oct. 94*	*21 Jan. 98*	*10 Oct. 12*	. . .	*Supy. List.*
71	R. A.	**Lennox**, George Montagu (50, 71, 104, 113, 165, 170)	74 Aug. 22	17 Nov. 94	3 Aug. 98	17 Nov. 12	. . .	117 Mahrattas.

72	Unattd. List	**Shairp**, Alexander, C.M.G. (71, 127, 139) [*l*]	17 Sep. 73	28 Nov. 94	20 Feb. 96	28 Nov. 12	*Ty. l.c.* 1 Apl. 17	S. and T. Corps.
	Unattd. List	**Synge**, Mark, D.S.O. (162, 170) (*e*)	25 Apl. 71	28 Nov. 94	21 Feb. 96	28 Nov. 12	. . . .	S. and T. Corps.
	Unattd. List	**Hesketh**, Walter, D.S.O. (51, 167) (*actg. l.c. 3 Aug. 16*)	1 Nov. 72	28 Nov. 94	22 Feb. 96	28 Nov. 12	. . . .	12 Cavalry.
	Som. L. I.	*James, Edmund Henry Salt* (50, 57, 94, 162, 169)	*26 June* 74	*12 Dec. 94*	*2 May 97*	*12 Dec.* 12	. . . .	*Supy. List.*
75	Durh. L.I.	**Carr-Hall**, Ralph Ellis, C.I.E. (162, 170)	30 Oct. 73	12 Dec. 94	18 May 96	12 Dec. 12	*Ty. l.c.* 24 Mar. 16.	Mily. Accts. Dept.
	Ches. R.	**Thomas**, Reginald Seymour (66*a*, 67*a*, 172)	7 Mar. 73	12 Dec. 94	11 July 97	12 Dec. 12	. . . .	S. and T. Corps.
	S. Staff. R.	**Lang**, Cecil Frederick Grant	15 Feb. 74	12 Dec. 94	17 Nov. 98	12 Dec. 12	. . . .	S. and T. Corps.
	Unattd. List	**Kidd**, Bertram Graham Balfour (50, 71, 72, 137) *Half pay from 15 Feb. to 30 Apl. 14 (to count).*	6 Dec. 75	16 Jan. 95	27 Mar. 96	16 Jan. 13	. . . .	125 Rifles.
	Unattd. List	**Rivett-Carnac**, John Stirling, *p.s.c.* (59, 167, 170)	25 Feb. 75	16 Jan. 95	8 Mar. 96	16 Jan. 13	*Ty. l. c. 8 Sep. 17*	14 Lancers.
80	Unattd. List	**Hoskyn**, John Cunningham Moore, D.S.O. (51, 137, 139, 204) [L]	20 Dec. 75	16 Jan. 95	27 Mar. 96	16 Jan. 13	*Ty. l. c.* 7 Sep. 17	44 Infantry.
	Unattd. List	**Channer**, George Kendall, D.S.O. (51, 53, 71, 157)	5 Oct. 73	15 Jan. 95	3 Apl. 96	16 Jan. 13	. . . .	3 G. R.
	Unattd. List	**Palmer**, Gerald Steuart (50, 52, 71, 74, 161, 204) [L]	4 June 75	15 Jan. 95	28 Mar. 96	16 Jan. 13	*l.c.* 1 Jan. 18	97 Infantry.
	Unattd. List	**Godwin**, Charles Alexander Campbell, D.S.O., *p.s.c.* (51, 66*a*, 94, 161, 169)	28 Oct. 73	16 Jan. 95	28 Mar. 96	16 Jan. 13	*c.* 3 June 18 *t.c.* 1 Jan. 16 *Ty.l.c.*3 Aug. 15	23 Cavalry.
	Unattd. List	**Tyndall**, Henry Stuart, D.S.O. (51, 59, 169)	16 July 75	16 Jan. 95	29 Mar. 96	16 Jan. 13	. . . .	40 Pathans.
85	Unattd. List	**Conway-Gordon**, Esme Cosmo William, *p.s.c.* (50, 137, 162, 170)	29 Aug. 75	15 Jan. 95	6 Apl. 96	16 Jan. 13	*Ty.l.c.* 13 Aug. 17	3 Horse.
	Unattd. List	**Goodfellow**, Robert Charles (57, 65, 66*a*, 162, 170)	15 May 75	16 Jan. 95	2 Apl. 96	16 Jan. 13	. . . .	38 C. I. Horse.
	Unattd. List	**Lance**, Frederick FitzHugh (51, 52, 59, 172)	8 Dec. 73	16 Jan. 95	7 Mar. 96	16 Jan. 13	. . . .	19 Lancers.
	Unattd. List	**Birch**, Valentine Kingston (50, 57, 162, 204)	4 July 75	16 Jan. 95	29 Mar. 96	16 Jan. 13	. . . .	29 Lancers.
	Unattd. List	**Battye**, Ivan Urmston, D.S.O. (50, 53, 72*a*, 162, 169)	5 Mar. 75	16 Jan. 95	1 Apl. 96	16 Jan. 13	. . . .	Guides.
90	Unattd. List	**Strong**, Addington Dawsonne, D.S.O. (50, 52, 71, 172)	13 Mar. 75	16 Jan. 95	27 July 96	16 Jan. 13	*l.c.* 3 June 18	10 Lancers.
	Unattd. List	**Smith**, John Lindsay (162, 170)	19 Apl. 74	16 Jan. 95	6 Apl. 96	16 Jan. 13	. . . .	S. and T. Corps.
	Manch. R.	*Hirst, Frederick Christian* (170)	*24 Dec. 74*	*20 Feb. 95*	*10 Nov. 97*	*20 Feb. 13*	. . . .	*Supy. List.*
	H. L. I.	**Maxwell**, Denis Wellesley, *p.s.c.* (50, 53, 60, 71, 72*a*, 157, 162, 172*a*)	24 Mar. 75	20 Feb. 95	19 Jan. 99	20 Feb. 13	. . . .	4 G. R.
	Norf. R.	**Wheatley**, Guy Rutherford Prescott (50, 53, 71)	12 May 75	20 Feb. 95	30 Oct. 97	20 Feb. 13	. . . .	27 Lt. Cavalry.
	Hamps. R.	**Hennell**, Alfred Montagu (50, 57, 71, 137)	16 Aug. 74	20 Feb. 95	9 Oct. 97	20 Feb. 13	. . . .	32 Lancers.
	E. York. R.	**Davis**, George Henville (50, 71, 94, 137, 162, 169)	11 Oct. 74	20 Feb. 95	23 June 98	20 Feb. 13	. . . .	S. and T. Corps.
95	R. Muns. Fus.	**Fagan**, Bernard Joseph, D.S.O. (51, 59, 71)	8 Oct. 74	6 Mar. 95	23 Nov. 98	6 Mar. 13	. . . .	17 Infantry.
	R. Innis. Fus.	**Arbuthnot**, Archibald Hugh (50, 71, 94, 145, 157, 162, 169)	7 Dec. 74	6 Mar. 95	6 Dec. 99	6 Mar. 13	. . . .	7 G. R.
	S. Staff. R.	**Wardell**, Geoffrey Francis Evans (50)	23 Nov. 74	6 Mar. 95	15 Sep. 99	6 Mar. 13	*Ty. l.c.* 8 May 17	S. and T. Corps.
	Unattd. List	*Garratt, Richard* (51, 157, 161, 169)	*3 Oct. 74*	*15 May 95*	*24 Jan. 98*	*15 May 13*	. . . .	*Supy. List.*
	A. S. C.	**Douglas**, Herbert Archibald	2 Apl. 74	29 May 95	23 Jan. 09	29 May 13	*Ty. l.c.*, 8 Mar. 18.	S. and T. Corps.
100	Unattd. List	**Pratt**, Henry Roger Evelyn, D.S.O. (51, 52, 59, 66*a*, 167, 170)	8 Dec. 75	14 Aug. 95	13 Dec. 96	14 Aug. 13	. . . .	35 Sikhs.
	Unattd. List	**Forbes**, Edward Ernest (50, 59, 66*a*, 144, 162)	16 Jan. 75	14 Aug. 95	10 Dec. 96	14 Aug. 13	. . . .	S. and T. Corps.
	Unattd. List	**Dundas**, Walter Leslie (50, 53, 59, 156, 170)	22 Mar. 76	14 Aug. 95	13 Dec. 96	14 Aug. 13	*Ty.l.c. 14 July 16*	3 G. R.
	Unattd. List	**Vaughan**, Louis Ridley, C.B., D.S.O., *p.s.c.* (50, 71, 147, 16, 195)	7 Aug. 75	14 Aug. 95	11 Dec. 96	14 Aug. 13	*c.* 1 Jan. 17	7 G. R.
	Unattd. List	**Clarke**, George Stanley (50, 167)	26 Apl. 75	14 Aug. 95	17 Dec. 96	14 Aug. 13	. . . .	34 Pioneers.
105	Unattd. List	**Wallace**, Edward Charles Lloyd, D.S.O. (50, 52, 124*a*, 167) [*l*]	14 Oct. 75	14 Aug. 95	14 Dec. 96	14 Aug. 13	. . . .	30 Punjabis.
	Unattd. List	**Henderson**, Kenneth, D.S.O., *p.s.c.* (50, 72, 72*b*, 157)	14 Apl. 75	14 Aug. 95	26 Oct. 96	14 Aug. 13	. . . .	39 Garhwal Rifles.
	Unattd. List	**Ryves**, Benjamin Harvey (162, 170)	14 Nov. 75	14 Aug. 95	13 Dec. 96	14 Aug. 13	. . . .	S. and T. Corps.
	Unattd. List	**McWatters** George Alfred (50, 53, 66*a*, 66*b*, 195)	20 Dec. 75	14 Aug. 95	14 Dec. 96	14 Aug. 13	. . . .	81 Pioneers.
109	Unattd. List	**Duxbury**, Charles Douglas (50, 55, 66*a*, 195) (*e*)	30 Oct. 75	14 Aug. 95	15 Dec. 96	14 Aug. 13	. . . .	S. and T. Corps.

Majors—*contd.*

No.	Former Corps.	Names.	Date of birth.	Date of first commission.	Date of appointment to Indian Army.	Date of Rank. Substantive.	Date of Rank. Army.	Remarks.
110	Unattd. List	Magniac, Francis Arthur (51, 52, 107, 140, 172)	2 Aug. 75	14 Aug. 95	13 Dec. 96	14 Aug. 13	. . . .	22 Punjabis.
	Unattd. List	Manson, James Arthur Prendergast (50, 72, 94, 137, 172*a*) . .	12 Feb. 76	14 Aug. 95	6 Mar. 98	14 Aug. 13	. . . .	S. and T. Corps.
	E. Kent. R.	Stokes, Claude Bayfield, C.I.E., *p.s.c.* (51, 57, 71, 137, 160, 165, 160, 204) [L]	Oct. 75	28 Sep. 95	7 Oct. 97	28 Sep. 13	*Ty.l.c.* 10 *Feb.* 17. *l.c.* 3 June 18	3 Horse.
	R.W. Kent. R.	Muspratt, Vivian Edward, *p.s.c.* (51, 66*a*, 71, 170) [*I*]	28 June 75	28 Sep. 95	7 Oct. 99	28 Sep. 13	*Ty. l.c.* 14 Feb. 18	30 Lancers.
	Dorset R.	Alexander, Edward Currie, D.S.O., *p.s.c.* (51, 52, 66*a*, 72*a*, 169) .	15 Sep. 75	28 Sep. 95	5 Aug. 98	28 Sep. 13	*l.c.* 3 June 17 *Ty. l.c.* 5 May 16	55 Rifles.
115	R. A.	Bignell, Bertram Harold (51, 52, 71, 157)	25 Mar. 76	2 Nov. 95	22 Aug. 00	2 Nov. 13	. . . .	9 G. R.
	Unattd List	Bourne, Walter Kemp, *p.s.c.* (50, 53, 137, 162, 170).	29 June 72	13 Nov. 95	20 Jan. 97	13 Nov. 13	*Ty.l.c*, 18 Dec. 14 *l.c.* 3 June 16	2 Lancers.
	Wilts. R.	Chauncy, Charles Henry Kemble (66*a*, 71, 94, 162, 169) . .	21 Dec. 73	7 Dec. 95	13 Aug. 98	7 Dec. 13	. . . .	124 Infantry.
	E. York. R.	Green, Wilfrith Gerald Key, D.S.O. (50, 52, 71, 137, 162, 165, 169)	3 Nov. 72	7 Dec. 95	3 Oct. 97	7 Dec. 13	*Ty.l.c.* 21 Dec. 15	36 Horse.
	R. Sco. Fus.	*Williams, George Robert Keppel* (59, 104, 170)	22 *July* 73	28 *Dec.* 95	25 *July* 98	28 *Dec.* 13	. . . .	*Supy. List.*
	Unattd. List	St. John, Richard Stukeley, D.S.O., *p.s.c.* (50, 53, 124*a*, 161, 169)	15 Jan. 76	22 Jan. 96	11 Apl. 97	22 Jan. 14	*l.c.* 3 June 16 *Ty.l.c.* 8 Mar. 16	20 Infantry.
120	Unattd. List	Peart, Charles Lube, *p.s.c.* (50, 59, 124, 125*b*, 158, 169, 203) [L]	2 Apl. 76	22 Jan. 96	30 Mar. 97	22 Jan. 14	. . . .	106 Pioneers.
121	Unattd. List	Norman, Claude Lumsden, D.S.O., M.V.O., A.D.C., *p.s.c.* (50, 53, 64, 71, 162, 169)	19 Feb. 76	22 Jan. 96	14 Mar. 97	22 Jan. 14	*l.c.* 29 Nov. 15 *c.* 1 Aug. 18	

	Unattd. List	*Chenevix-Trench, Richards Henry* (*127, 162, 169*).	*4 Apl. 76*	*22 Jan. 96*	*12 Apl. 97*	*22 Jan. 14*	. . . .	*Supy. List.*
122	Unattd. List	**Barstow**, Henry (50, 52, 165, 169)	14 Nov. 76	22 Jan. 96	31 Mar. 97	22 Jan. 14	*Ty. l.c.* 12 Oct. 16; *l.c.* 1 Jan. 17	38 Dogras.
	Unattd. List	**Wigram**, Kenneth, C.B., D.S.O., *p.s.c.* (51, 53, 157, 169)	5 Dec. 75	22 Jan. 96	29 Mar. 97	22 Jan. 14	*c.* 1 Jan. 18; *l.c.* 3 June 16; *Ty.l.c.* 4 Feb. 16	2 G. R.
	Unattd. List	**Orr**, Gerald Maxwell, D.S.O., *p.s.c.* (51, 71, 169)	9 June 76	22 Jan. 96	9 Apl. 97	22 Jan. 14	*l.c.* 3 June 16; *Ty. c.* 15 May 17; *Ty. l.c.* 1 Feb. 16; *c.* 3 June 18	11 Lancers
25	Unattd. List	**Marsh**, Frank Graham, C.M.G., *p.s.c.* (51, 52, 127, 145, 157, 162, 195) [L]	20 Sep. 75	22 Jan. 96	30 Mar. 97	22 Jan. 14	*Ty. c.* 13 Mar. 17; *l.c.* 3 June 16; *Ty. l.c* 12 May 16	9 G. R.
	Unattd. List	**Woodhouse**, Claude Gregory (51, 72*a*, 95, 162, 169)	12 Dec. 75	22 Jan. 96	3 Apl. 97	22 Jan. 14	. . . .	126 Infantry.
	Unattd. List	**Bingham**, William Henry (50, 72*b*, 124*a*, 125*b*, 125*c*, 137, 162, 167, 169) [L]	19 Mar. 77	22 Jan. 96	29 Mar. 97	22 Jan. 14	. . . .	69 Punjabis.
	Unattd. List	**Macpherson**, James, D.S.O. (50, 66*a*, 72*a*, 162, 169).	20 June 76	22 Jan. 96	30 Mar. 97	22 Jan. 14	. . . .	124 Infantry.
	Unattd. List	**Barlow**, Cecil Popham (51, 52, 57, 66*a*, 157, 162).	23 Aug. 75	22 Jan. 96	3 Apl. 97	22 Jan. 14	. . . .	8 G. R.
130	Unattd. List	**Scott**, Gerald Bassett, D.S.O. (51, 59, 72*a*, 162, 169)	27 Oct. 75	22 Jan. 96	1 Apl. 97	22 Jan. 14	. . . .	27 Punjabis.
	Unattd. List	**Cassels**, Robert Archibald, C.B., D.S.O., *p.s.c.* (50, 71, 137). . .	15 Mar. 76	22 Jan. 96	30 Mar. 97	22 Jan. 14	*c.* 10 Mar. 17; *l.c.* 18 Aug. 16	32 Lancers.
	Unattd. List	*Ferrar, Michael Lloyd* (*57, 94, 165, 169, 203*) [L]	*16 Apl. 76*	*22 Jan. 96*	*6 May 97*	*22 Jan. 14*	. . . .	*Supy. List.*
	Unattd. List	**Simpson**, John Chalmers (51, 52, 66*a*, 71, 137, 169)	19 Sep. 76	22 Jan. 96	11 Oct. 97	22 Jan. 14	*l.c.* 3 June 17	16 Rajputs.
	Unattd. List	**Coke**, Algernon John Parker (51, 53, 59, 64, 71, 172)	10 Mar. 76	22 Jan. 96	3 Apl. 97	22 Jan. 14	. . . .	22 Cavalry.
	Wilts. R. .	**D'Oyly**, John (51, 52, 57, 137)	17 Apl. 76	29 Feb. 96	4 July 98	1 Mar. 14	. . . .	119 Infantry.
135	R.W.KentR.	**Leslie**, Walter Stewart, D.S.O., *p.s.c.* (50, 66*a*, 72*a*, 165, 170) . .	23 Mar. 76	29 Feb. 96	22 Sep. 98	1 Mar. 14	*l.c.* 3 June 16; *c.* 3 June 18	31 Punjabis.
	Shrops L.I.	**Slingsby**, Thomas William (51, 53, 71, 172)	6 Mar. 76	29 Feb. 96	3 Feb. 99	1 Mar. 14	. . . .	22 Cavalry.
	Wilts. R.	**Cox**, Charles Ernest Somers (51, 52, 57, 66*a*, 71, 94, 100, 162, 167, 169)	21 Aug. 75	25 Mar. 96	23 Apl. 98	25 Mar. 14	*l.c.* 1 Jan. 16	67 Punjabis.
	W. Rid. R.	**Bally**, Henry George (51, 52, 71, 72*b*, 137)	9 Jan. 77	25 Mar. 96	1 Sep. 00	25 Mar. 14	. . . .	10 Jats.
	Dorset. R.	**Woodward**, Charles Frederick (50, 72, 172)	11 Mar. 75	6 June 96	4 Feb. 00	6 June 14	. . . .	54 Sikhs.
140	York. L. I.	**Shuttleworth**, Allen Robert Betham, *p. s. c.* (50, 66*a*) . .	20 Feb. 73	6 June 96	16 Feb. 98	6 June 14	. . . .	S. and T. Corps.
	R. War. R.	**Jackson**, Freeman Astley, D.S.O. (51, 52, 71, 162, 169)	11 Mar. 74	24 June 96	10 Mar. 99	24 June 14	. . . .	21 Cavalry.
	R. Suss. R.	**Bazett**, Seymour (104, 175)	13 Apl. 74	24 June 96	28 Mar. 99	24 June 14	. . . .	Civil employ.
	R. Scots.	**Wildes**, Charles Stuart Dudlow (66, 70)	17 Dec. 75	22 July 96	26 Sep. 00	22 July 14	. . . .	S. and T. Corps.
	4th Huss.	**Williams**, Arthur Frederick Carlisle, D.S.O., *p.s.c.* (50, 155) . .	19 June 76	5 Aug. 96	20 Dec. 97	5 Aug. 14	*Ty. l.c.* 11 Mar. 18	31 Lancers.
145	Unattd. List	**Geoghegan**, Norman Meredith (51, 52, 72, 105, 125*a*, 140, 148*b*, 167)	7 Sep. 76	5 Aug. 96	22 Oct. 97	5 Aug. 14	*l.c.* 3 June 16	89 Punjabis.
	Unattd. List	*Lorimer, David Lockhart Robertson*, C.I.E. (*161, 169*)	*24 Dec. 76*	*5 Aug. 96*	*20 Nov. 97*	*5 Aug. 14*	. . . .	*Supy. List.*
	Unattd. List	**Arthur**, Lionel Francis, D.S.O., *p.s.c.* (50, 66, 162) [L] . . .	14 Mar. 76	5 Aug. 96	22 Oct. 97	5 Aug. 14	. . . .	26 Lt. Cavalry.
	Unattd. List	**Shairp**, Henry Frank (59, 195)	13 Mar. 76	5 Aug. 96	25 Nov. 97	5 Aug. 14	*Ty.l.c.* 30 July 15	Mily. Accts. Dept.
	Unattd. List	**Ashburner**, Harley Wentworth, D.S.O. (57, 95, 162, 169) . . .	13 Dec. 75	5 Aug. 96	20 Oct. 97	5 Aug. 14	. . . .	106 Pioneers.
	Unattd. List	**Fetherstonhaugh**, William Albany, D.S.O., *p.s.c.* (50, 64, 66*a*, 71, 94, 162, 169)	24 Oct. 76	5 Aug. 96	4 Nov. 97	5 Aug. 14	*Ty.l.c.* 19 Mar. 16; *l.c.* 3 June 17	8 Cavalry.
50	Unattd. List	**Reid**, William David (50, 53, 71, 193)	14 Nov. 76	5 Aug. 96	8 Oct. 97	5 Aug. 14	. . . .	Cantt. Magte.'s Deptt.

No.	Former Corps.	Names.	Date of birth.	Date of first commission.	Date of appointment to Indian Army.	Date of Rank. Substantive.	Date of Rank. Army.	Remarks.
		Majors—*contd.*						
151	5 Lancers.	**Bruce,** John (71)	22 Jan. 76	5 Aug. 96	11 Feb. 98	5 Aug. 14	. . . .	Army Remt. Dept.
	Unattd. List	**Dunsford,** Evelyn Henry (51, 72a, 169)	15 Sep. 77	5 Aug. 96	22 Nov. 97	5 Aug. 14	. . . .	69 Punjabis.
	Unattd. List	**Bald,** John Arthur, C.M.G. (50, 53, 59, 71, 193)	20 July 76	5 Aug. 96	19 Oct. 97	5 Aug. 14	. . . .	61 Pioneers.
	Unattd. List	**Daldy,** Alexander William (51, 155)	3 July 76	5 Aug. 96	20 Nov. 97	5 Aug. 14	. . . .	Mily. Accts. Dept.
155	Unattd. List	**Crichton,** McVeagh (51, 52, 66a, 72a, 155, 156)	3 Oct. 76	5 Aug. 96	20 Nov. 97	5 Aug. 14	. . . .	103 Lt. Infantry.
	Unattd. List	**Shuttleworth,** Digby Inglis, *p.s.c.* (50, 53, 155, 157)	23 Aug. 76	5 Aug. 96	3 Oct. 97	5 Aug. 14	*Ty. l.c.* 27 June 17 *l.c.* 3 June 17	3 G. R.
	Unattd. List	**Mackenzie,** John, C.I.E. (51, 53, 59, 167, 170)	21 Sep. 76	5 Aug. 96	21 Oct. 97	5 Aug. 14	. . . .	35 Sikhs.
	Unattd. List	**Jewett,** Ashley Ernest (51, 52, 104, 141, 142, 144)	27 Oct. 76	5 Aug. 96	19 Nov. 97	5 Aug. 14	. . . .	8 G. R.
	Unattd. List	**Van Someren,** William Weymouth, D.S.O. (51, 52, 107, 127, 146, 162) [L]	17 Sep. 76	5 Aug. 96	8 Oct. 97	5 Aug. 14	. . . .	45 Sikhs.
	Unattd. List	***Anderson,* *Gordon Hay*** (100, 162, 169)	22 *Jan.* 77	5 *Aug.* 96	18 *Nov.* 97	5 *Aug.* 14	. . . .	*Supy. List.*
160	Unattd. List	**Ames,** Cecil Gardner (51, 52, 144 a, 169, 172, 182b)	2 Oct. 76	5 Aug. 96	19 Oct. 97	5 Aug. 14	. . . .	52 Sikhs.
	Unattd. List	**Ellis,** Alfred Charles Samuel Burdon, *p.s.c.* (50, 53, 57, 71, 137, 155) (*e*)	9 June 76	5 Aug. 96	16 Nov. 97	5 Aug. 14	*Ty. l.c.* 24 May 18	121 Pioneers.
	Unattd. List	**Jarrett,** Hubert Cecil Delacour (51, 52, 57, 169)	22 July 76	5 Aug. 96	10 Dec. 97	5 Aug. 14	. . . .	19 Punjabis.
	Unattd. List	***Hutchinson,* *William Gordon*,** O.B.E. (51, 94, 162, 167, 170)	15 *June* 76	5 *Aug.* 96	15 *Dec.* 97	5 *Aug.* 14	. . . .	*Supy. List.*
	Unattd. List	**Watson,** John Hugh (50, 71, 162, 167, 170)	28 Sep. 76	5 Aug. 96	1 Jan. 98	5 Aug. 14	. . . .	13 Lancers.
	Unattd. List	**Taylor,** Percy Henry Mitchell (51, 52, 64, 71, 137)	28 Aug. 76	5 Aug. 96	5 Oct. 97	5 Aug. 14	. . . .	32 Lancers.
165	Unattd. List	**Rybot,** Norman Victor Lacey, D.S.O. (51, 59, 71, 125, 162, 167, 170)	7 Jan. 76	5 Aug. 96	19 Nov. 97	5 Aug. 14	. . . .	76 Punjabis.
	4th Hrs.	**Dening,** Lewis Eales (51, 52, 66a, 137)	10 Mar. 76	5 Aug. 96	2 May 99	5 Aug. 14	. . . .	33 Lt. Cavalry.
	Unattd. List	**Cheyne,** Henry Baliol (51)	28 May 76	5 Aug. 96	12 Oct. 97	5 Aug. 14	. . . .	1 Lancers.
	Unattd. List	*Lyall, Robert-Adolphus,* D.S.O. (162, 169)	20 *Oct.* 76	5 *Aug.* 96	31 *Dec.* 97	5 *Aug.* 14	. . . .	*Supy. List.*
	Unattd. List	**Keene,** Geoffrey Norman Stewart (51, 52, 64, 106, 167, 170, 181, 196)	2 Jan. 76	5 Aug. 96	16 Dec. 97	5 Aug. 14	. . . .	30 Punjabis.
	E. York. R.	**Wilford,** Edmund Ernest, D.S.O. (50, 71, 72)	13 Jan. 76	5 Sep. 96	19 Dec. 98	5 Sep. 14	*Ty. l.c.* 11 Nov. 15	30 Lancers.
170	R. Ir. Rif.	**Hughes Hallett,** Richard Lionel	19 Dec. 75	5 Sep. 96	3 Dec. 00	5 Sep. 14	*l.c.* 1 Jan. 18	S. and T. Corps.
	Midd'x. R.	**Tyrer,** Reginald (50, 53, 66a, 72a, 107, 167)	27 Aug. 76	5 Sep. 96	27 Sep. 12	5 Sep. 14	. . . .	84 Punjabis.
	Som. L. I.	**Rice,** Brinsley Alexander McHenry, D.S.O. (66a, 157, 170) (*a*)	7 Feb. 76	5 Sep. 96	26 Apl. 98	5 Sep. 14	. . . .	6 G. R.
	Lanc. Fus.	*Bruce, Charles Edward* (72, 162, 169)	23 *Mar.* 76	5 *Sep.* 96	7 *Dec.* 97	5 *Sep.* 14	. . . .	*Supy. List.*
	Bord. R.	**Grant,** George Patrick, D.S.O., *p.s.c.* (94, 162, 169)	22 Sep. 76	5 Sep. 96	27 Aug. 01	5 Sep. 14	*Ty. l.c.* 1 Aug. 17	106 Pioneers.
	R. A.	**Hogg,** Rudolph Edward Trower, C.I.E. (51, 169)	19 Jan. 77	21 Sep. 96	22 July 01	21 Sep. 14	*Ty. c.* 1 Jan. 17 *l.c.* 1 Jan. 18	38 C. I. Horse.
175	Derby. R.	**Hallowes,** John Hope (51, 59, 71, 167) *Service for pension—In ranks 2 yrs. 131 days (half to count).*	15 Feb. 73	23 Sep. 96	12 July 99	23 Sep. 14	. . . .	15 Lancers.
	Unattd. List	**Hardie,** Grant Espie (50, 59, 62, 71, 137)	14 May 74	23 Sep. 96	20 Dec. 97	23 Sep. 14	. . . .	123 Rifles.
	R. Sco. Fus.	**George,** Douglas Stewart (50) *Service for pension from 12 Mar. 98.*	19 Dec. 74	9 Dec. 96	26 Apl. 13	9 Dec. 14	. . . .	42 Regiment.
178	Manch. R.	**Routh,** Gordon Stewart Floyd (51, 53, 57, 162, 169) *Service for pension from 13 May 94.*	8 Apl. 73	16 Dec. 96	26 Feb. 98	16 Dec. 14	. . . .	56 Rifles.

179	Unattd. List	**Cutler,** Howard Challoner (51, 52, 66*a*, 137)	29 Mar. 77	20 Jan. 97	15 Mar. 98	20 Jan. 15	....	8 Rajputs.
180	Unattd. List	**Hodgson,** George Cecil, D.S.O. (50, 52, 59, 66*a*, 162, 167, 169)	10 July 76	20 Jan. 97	26 Mar. 98	20 Jan. 15	....	23 Pioneers.
	Unattd. List	**Sherrard,** James William (106, 169, 195)	12 June 77	20 Jan. 97	20 Mar. 98	20 Jan. 15	*Ty. l.c. 1 Nov. 16* *l.c.* 3 June 17	S. and T. Corps.
	Unattd. List	**Popham,** Edward Leyborne, D.S.O. (50)	14 Dec. 76	20 Jan. 97	22 Mar. 98	20 Jan. 15	*l.c.* 3 June 18	26 Lt. Cavalry.
	Unattd. List	*Keyes, Terence Humphrey,* C.I.E. *(94, 162, 165, 167, 170)* [*L*]	*28 May 77*	*20 Jan. 97*	*22 Mar. 98*	*20 Jan. 15*	*l.c.* 3 June 18	*Supy. List.*
	Unattd. List	**Moore,** Frederick Grattan (51, 66*a*, 72*a*, 137)	26 Apl. 77	20 Jan. 97	20 Mar. 98	20 Jan. 15	....	6 Light Infantry.
	Unattd. List	**Mactavish,** Archibald (51, 52, 66*a*, 137)	31 Dec. 76	20 Jan. 97	20 Mar. 98	20 Jan. 15	....	3 Brahmans.
185	Unattd. List	**Wilkinson,** Charles Robert (50, 53, 172)	9 May 77	20 Jan. 97	16 Mar. 98	20 Jan. 15	....	52 Sikhs.
	Unattd. List	*Thuillier, Leslie Cardew*	*7 Oct. 77*	*20 Jan. 97*	*22 Mar. 98*	*20 Jan. 15*	....	*Supy. List.*
	Unattd. List	**Murray,** Francis Stewart James (51, 71, 162, 169)	6 June 77	20 Jan. 97	15 Mar. 98	20 Jan. 15	....	S. and T. Corps.
	Unattd. List	**Hutchison,** George, *p.s.c.* (51, 71, 66*a*, 95, 137, 162, 170)	9 Apl. 77	20 Jan. 97	15 Mar. 98	20 Jan. 15	....	122 Infantry.
	Unattd. List	**Beddy,** Percy Langdon (51, 52, 71, 169)	3 Oct. 76	20 Jan. 97	28 Mar. 98	20 Jan. 15	*l.c.* 3 June 16	51 Sikhs.
	Unattd. List	**Baird,** Harry Beauchamp Douglas, C.M.G., D.S.O., *p.s.c.* (51, 71, 162, 167, 169)	4 Apl. 77	20 Jan. 97	18 Mar. 98	20 Jan. 15	*l.c.* 1 Jan. 17	12 Cavalry.
190	Unattd. List	**Sherlock,** George Wright Stratford, *p.s.c.* (51, 66*a*, 72, 157, 170)	16 Feb. 77	20 Jan. 97	13 Mar. 98	20 Jan. 15	....	6 G. R.
	Unattd. List	**Gunning,** George Hamilton, D.S.O., (51, 52, 71, 172)	3 Nov. 76	20 Jan. 97	15 Mar. 98	20 Jan. 15	....	21 Cavalry.
	Unattd. List	**Rennick,** Robert William de Clanay (52)	5 June 78	20 Jan. 97	22 Feb. 98	20 Jan. 15	....	95 Infantry.
	Unattd. List	**Browne,** Horace Fairfax Moresby (50, 104 124, 148*b*, 162, 165, 172, 175)	14 Dec. 76	20 Jan. 97	24 Mar. 98	20 Jan. 15	....	Civil, Burma.
	Unattd. List	**Butler,** Henry Maurice (50, 53, 137)	26 Sep. 76	20 Jan. 97	19 Mar. 98	20 Jan. 15	....	104 Rifles.
195	Unattd. List	**Innes-Lillingston,** Frederick Francis (50)	1 June 77	20 Jan. 97	22 Feb. 98	20 Jan. 15	....	S. and T. Corps.
	Unattd. List	**Campbell,** Hector, D.S.O., M.V.O. (51, 52, 172)	24 Oct. 77	20 Jan. 97	31 Mar. 98	20 Jan. 15	....	Guides.
	Unattd. List	**Withers,** Charles McGregor, *p.s.c.* (50, 59, 167, 195)	29 Nov. 76	20 Jan. 97	8 Sep. 98	20 Jan. 15	....	87 Punjabis.
	Unattd. List	**Middlemass,** Robert William Hume	29 Nov. 77	20 Jan. 97	7 Mar. 98	20 Jan 15	....	S. and T. Corps.
	R. War. R.	**Nicolas,** Charles Percy (51, 52, 71, 172)	30 Oct. 73	19 May 00	9 Mar. 07	3 Feb. 15	....	
		Service for pension from 21 Oct. 93 to 2 Feb. 97; In Ranks from 25 Nov. 99 to 18 May 00 (half to count); for promotion from 21 Oct. 93 to 2 Feb. 97.						
200	Bedf. R.	**Sexton,** Edmund George (50 59, 64, 66*a* 71, 172*a*) ***Forfeits from 28 Sep. 13 to 18 Feb. 15, for promotion.***	24 Apl. 75	28 Sep. 95	4 Jan. 98	19 Feb. 15	....	1 Lancers.
	Bedf. R.	**Marriott,** Alfred Sinclair (51)	3 July 76	20 Feb. 97	8 Apl. 00	20 Feb. 15	*Ty. l.c.* 25 Feb. 18	Mily. Farms Dept.
	Wilts. R.	**Steele,** Charles Edward Beevor (66, 161, 169)	6 Nov. 76	20 Feb. 97	9 July 98	20 Feb. 15	....	S. and T. Corps.
	Bedf. R.	**Richards,** William Vellacott (51)	1 Dec. 76	20 Feb. 97	26 Sep. 98	20 Feb. 15	....	Mily. Accts. Dept.
	Bord. R.	**James,** Charles Ashton (50, 72*a*, 94, 162, 169)	10 Oct. 76	20 Feb. 97	19 Oct. 98	20 Feb. 15	*Ty. l.c.* 23 July 16	126 Infantry.
205	Durh. L. I.	**Duncan,** Henry Clare, D.S.O., *p.s.c.* (50, 53, 59, 157, 172*a*)	12 July 76	20 Feb. 97	2 Nov. 00	20 Feb. 15	*Ty. l.c.*, 11 Jan. 18	9 G. R.
	E. Surr. R.	**Oldfield,** Guy Edmund Bladensberg (50, 52, 66*a*, 71, 107)	24 Nov. 74	3 Mar. 97	27 July 99	3 Mar. 15	....	74 Punjabis.
	R. A.	**Allanson,** Cecil John Lyons, C.I.E., D.S.O., *p. s. c.* (12, 51, 157)	2 Apl. 77	23 Mar. 97	21 Mar. 02	23 Mar. 15	*Ty. l.c.* 9 Oct. 17 *l. c.* 3 June 18	6 G. R.
	Norf. R.	**White,** Harold Philip Muir (51, 67*a*, 167, 170)	28 Nov. 74	24 Mar. 97	12 Aug. 99	24 Mar. 15	....	S. and T. Corps.
	Suff. R.	**Meynell,** Hugo (50, 65, 66*a*, 95, 170)	1 July 75	15 May 97	10 Dec. 07	15 May 15	....	29 Lancers.
210	York R.	**Morris,** Charles Temple (50, 66*a*, 66*b*, 72*a*, 95, 155, 162, 167, 172*a*)	9 Jan. 76	15 May 97	12 Apl. 00	15 May 15	....	82 Punjabis.
	Unattd. List	**Freeland,** John Cavendish, *p.s.c.* (51, 53, 162, 167)	22 Jan. 77	4 Aug. 97	8 Nov. 98	4 Aug. 15	*l.c.* 3 June 18	35 Sikhs.
	Unattd. List	**Nicolas,** Robert (50, 53, 59, 167)	3 Dec. 76	4 Aug. 97	8 Nov. 98	4 Aug. 15	....	23 Pioneers.
213	Unattd. List	**Cloete,** Laurence Balfour, *p.s.c.* (50, 66*a*, 72*a*, 137, 162, 170)	3 July 77	4 Aug. 97	7 Nov. 98	4 Aug. 15	....	37 Dogras.

(*a*) Passed in Hausa (Colloquial).

No.	Former Corps.	Names.	Date of birth.	Date of first commission.	Date of appointment to Indian Army.	Date of Rank. Substantive.	Date of Rank. Army.	Remarks.
		Majors—*contd.*						
214	Unattd. List	**Casson**, Wilfrid Francis Seymour, D.S.O., *p. s. c.* (50, 53, 71)	4 Oct. 77	4 Aug. 97	8 Nov. 98	4 Aug. 15	*Ty.l.c.* 14 June 17; *l.c.* 1 Jan. 18	27 Lt. Cavalry.
215	Unattd. List	**Moberly**, Bertrand Richard, D.S.O., *p.s.c.* (172)	15 Oct. 77	4 Aug. 97	7 Nov. 98	4 Aug. 15	*Ty.l.c.* 2 July 16; *l.c.* 3 June 18	56 Rifles.
	Unattd. List	**Webber**, Mervyn Robert Howe (51, 57, 71, 137, 162, 165, 170)	4 Nov. 76	4 Aug. 97	22 Nov. 98	4 Aug. 15	. . . .	3 Horse.
	Unattd. List	**Mackie**, Ralph (50, 59, 71, 195)	12 Apl. 77	4 Aug. 97	21 Nov. 98	4 Aug. 15	. . . .	64 Pioneers.
	Unattd. List	**Hay**, Charles John Bruce, D.S.O., *p.s.c.* (50, 53, 59, 66*a*, 156, 162, 167, 169) [L]	18 May 77	4 Aug. 97	24 Nov. 98	4 Aug. 15	*l.c.* 1 Jan. 17	Guides.
	Unattd. List	**Macmullen**, Cyril Norman, D.S.O., *p.s.c.* (50, 95, 107, 140, 165, 169)	13 Dec. 77	4 Aug. 97	29 Nov. 98	4 Aug. 15	*c.* 1 Jan. 18; *l.c.* 1 Jan. 16; *Ty.l.c.* 4 Oct. 15	15 Sikhs.
220	Unattd. List	**Sargent**, William Patrick Miller (51, 59, 66*a*, 72*b* 125*a*, 137, 140, 162, 170)	2 Dec. 77	4 Aug. 97	23 Oct. 98	4 Aug. 15	. . . .	104 Rifles.
	Unattd. List	, Patrick Neil (51, 59, 137)	11 Feb. 77	4 Aug. 97	20 Oct. 98	4 Aug. 15	. . . .	6 Cavalry.
	Unattd. List	**Anderson**, Reginald Hugh (51, 71, 172)	12 Mar. 77	4 Aug. 97	29 Oct. 98	4 Aug. 15	. . . .	11 Lancers.
	Unattd. List	*Crosthwaite, Charles Gilbert* (162, 169)	*3 June* 78	*4 Aug* 97	*5 Dec.* 98	*4 Aug.* 15	. . . .	*Supy. List.*
	Unattd. List	**Hawley**, William George Broughton Ischia (51, 52, 64, 71, 172*a*)	16 Aug. 77	4 Aug. 97	25 Nov. 98	4 Aug. 15	. . . .	28 Lt. Cavalry.
	Unattd. List	**Lumb**, Frederick George Edward, M.C. (50, 157, 162, 169)	10 Jan. 77	4 Aug. 97	21 Nov. 98	4 Aug. 15	. . . .	39 Garhwal Rifles.
	Unattd. List	*Mosse, Arthur Henry Eyre* (89, 132)	*28 Sep.* 77	*4 Aug.* 97	*20 Nov.* 98	*4 Aug* 15	. . . .	*Supy. List.*
225	Unattd. List	**Franklin**, Harold Scott Erskine, D.S.O. (50, 59, 124*a*, 125*b*, 165)	25 Nov. 78	4 Aug. 97	22 Oct. 98	4 Aug. 15	*l.c.* 3 June 16	14 Sikhs.
	Unattd. List	**Graham**, Bazett Colvin (51, 71, 157)	16 Feb. 77	4 Aug. 97	25 Feb. 99	4 Aug. 15	. . . .	125 Rifles.
	Unattd. List	**Loch**, John Carysfort (50, 53, 59, 137, 157)	25 Dec. 77	4 Aug. 97	26 Oct. 98	4 Aug. 15	. . . .	3 G. R.
	Unattd. List	**Milward**, Clement Arthur, D.S.O., *p.s.c.* (51, 53, 59, 66*a*, 72, 162, 169)	20 May 77	4 Aug. 97	20 June 99	4 Aug. 15	*l.c.* 3 June 18	53 Sikhs.
	York. R.	**Hogg**, Willoughby Lugard (50, 72*b*, 106)	29 Dec. 81	11 Aug. 00	13 Aug. 03	†11 Aug. 15	. . . .	3 Brahmans.
230	R. A.	**Hart**, Eric George, D.S.O. (71, 88, 161, 188) [*l*]	6 Feb. 78	1 Sep. 97	26 Feb. 09	1 Sep. 15	*Ty. l.c. 1 Nov.* 16	S. and T. Corps.
	Hamps. R.	**Clementi**, John (51, 59, 71, 107, 172)	27 Sep. 77	8 Sep. 97	14 Dec. 99	1 Sep. 15	. . . .	Guides.
	Wilts. R.	*Burne, Lindsay Eliott Lumley* (104, 120, 140, 167, 170)	*1 July* 77	*8 Sep.* 97	*14 Nov.* 99	*1 Sep.* 15	. . . .	*Supy. List.*
	Glouc. R.	**Maunsell**, Eustace Ball, *p. s. c.* (51, 57, 64, 77, 172)	11 Oct. 78	8 Sep. 97	30 Aug. 99	1 Sep. 15	. . . .	35 Horse.
	R. Ir. Rif.	**Saunders**, Ernest Howie, D.S.O. (50, 57, 72*b*)	18 Nov. 77	8 Sep. 97	24 Nov. 08	1 Sep. 15	. . . .	S. and T. Corps.
	S. W. Bord.	**Syer**, Herbert Hingston (66*a*, 162)	15 Nov. 76	15 Sep. 97	11 Apl. 99	1 Sep. 15	. . . .	S. and T. Corps.
235	R. A.	**Finlay**, Robert Francis, D.S.O. (50, 53, 161, 169)	31 Jan. 77	22 Oct. 97	7 Nov. 00	1 Sep. 15	. . . .	58 Rifles.
	R. Scots.	**Gledstanes**, Ambrose Upton (51, 71)	5 July 76	1 Dec. 97	4 July 01	1 Sep. 15	. . . .	30 Lancers.
	Linc. R.	**Woodcock**, Ernest Elborough (50, 53, 59, 71, 137, 162)	19 Nov. 76	1 Dec. 97	4 Aug. 00	1 Sep. 15	. . . .	4 Rajputs.
	Durh. L. I.	**Watts**, Herbert Ponsonby	6 Dec. 76	1 Dec. 97	21 Jan. 01	1 Sep. 15	. . . .	S. and T. Corps.
	N. Staff. R.	**Anderson**, Edward Sambrooke John (71)	22 Aug. 76	1 Dec. 97	12 Nov. 00	1 Sep. 15	. . . .	Mily. Accts. Dept.
240	Y. & Lanc. R.	**Lind**, Alexander Gordon, D.S.O. (51, 52, 59, 66*a*, 71, 95, 169)	15 June 77	1 Dec. 97	16 Sep. 00	1 Sep. 15	*Ty. l.c.* 20 Feb. 17	58 Rifles.
	E. Lan. R.	**Smith**, Alfred Aquila (51, 52, 66*a*, 71, 72*b*, 169, 188*a*) (*e*)	2 Jan. 77	1 Dec. 97	8 Feb. 01	1 Sep. 15	. . . .	58 Rifles.
	W. Rid. R.	**Wright**, Cron Hope Baillie (51, 167)	9 Mar. 75	15 Dec. 97	27 June 99	1 Sep. 15	. . . .	62 Punjabis.
	R. A.	**Szczepanski**, Hugh Clayton (12, 66*a*, 94, 162, 165, 169, 204)	4 Dec. 77	8 Jan. 98	12 Aug. 02	1 Sep. 15	. . . .	Mily. Accts. Dept.
	Unattd. List	**Turnbull**, George Oliver, D.S.O. (50, 59, 169)	21 July 77	22 Jan. 98	26 Mar. 99	1 Sep. 15	. . . .	26 Punjabis.
245	Unattd. List	**VC Grant**, John Duncan, *p.s.c.* (51, 66*a*, 72, 157)	28 Dec. 77	22 Jan. 98	23 Mar. 99	1 Sep. 15	. . . .	8 G. R.
246	Unattd. List	**Skipwith**, Frederick, *p.s.c.* (51, 52, 65, 66*a*, 72*b*, 125*b*, 157, 167, 169)	23 Dec. 77	22 Jan. 98	1 Apl. 99	1 Sep. 15	. . . .	5 G. R.

247	Unattd. List	**Kay**, Henry Clifford (137)	16 June 78	22 Jan. 98	9 Apl. 99	1 Sep. 15	. . .	8 Cavalry.
	Unattd. List	*Burke, Richard John Charles (132, 153)*	*5 May 78*	*22 Jan. 98*	*23 Mar. 99*	*1 Sep. 15*	. . .	*Supy. List.*
	Unattd. List	**Combe**, Sydney Boyce, M.C. (51, 52, 51, 167) [I]	13 June 78	22 Jan. 98	23 Mar. 99	1 Sep. 15	*l.c.* 3 June 17	47 Sikhs.
	Unattd. List	**Hamer**, Meredith Ashton, D.S.O., M.C., *p.s.c.* (5) 59, 66*a*, 162, 170)	11 Aug. 77	22 Jan. 98	20 Mar. 99	1 Sep. 15	. . . .	129 Baluchis.
250	Unattd. List	**Dyce**, Hugh Lennox, M.C. (51, 71, 172)	24 Dec. 77	22 Jan. 98	4 Apl. 99	1 Sep. 15	. . . .	9 Horse.
	Unattd. List	**Rowlandson**, Michael George Dobbie, D.S.O., *p.s.c.* (51, 52, 162, 167, 169) [L]	3 Aug. 78	22 Jan. 98	29 Mar. 99	1 Sep. 15	*l.c.* 3 June 16 *Ty. l.c.* 18 Jan. 18.	38 Dogras.
	Unattd. List	**Watling**, Leonard Douglas (66, 172)	3 Apl. 78	22 Jan. 98	2 Apl. 99	1 Sep. 15	. . . .	S. and T. Corps.
	Unattd. List	**Malet**, Robert James (50, 64, 66*a*, 71, 162, 195)	15 Nov. 77	22 Jan. 98	28 Mar. 99	1 Sep. 15	. . . .	64 Pioneers.
	Unattd. List	**Merriman**, Athol Bulkley (51, 107, 170)	19 Nov. 78	22 Jan. 98	28 Mar. 99	1 Sep. 15	. . . .	127 Lt. Infantry.
255	Unattd. List	**Crombie**, David Campbell, *p.s.c.* (45*a*, 51, 59, 71, 172) (*n*)	19 Nov. 77	22 Jan. 98	22 Mar. 99	1 Sep. 15	. . . .	23 Cavalry.
	Unattd. List	**Ross**, George Whitehill, D.S.O. (71, 137)	6 June 78	22 Jan. 98	27 Mar. 99	1 Sep. 15	. . . .	Mily. Accts. Dept.
	Unattd. List	**Woodham**, James Francis (51, 167, 195)	20 Oct. 77	22 Jan. 98	27 Mar. 99	1 Sep. 15	. . . .	76 Punjabis.
	Unattd. List	**Poole**, Ivan Maxwell Conway, D.S.O. (95, 137, 161, 165, 169, 204)	23 Feb. 78	22 Jan. 98	29 Mar. 99	1 Sep. 15	. . . .	S. and T. Corps.
	Unattd. List	**Twiss**, William Louis Oberkirch, M.C., *p.s.c.* (50, 52, 66*a*, 157, 167, 169) [L]	18 Jan. 79	22 Jan. 98	28 Mar. 99	1 Sep. 15	*Ty. l. c.* *13 June 16* *l. c.* 3 June 17	9 G. R.
260	Unattd. List	**Clifford**, Richard (50, 52, 107, 140, 172)	4 Sep. 77	22 Jan. 98	28 Mar. 99	1 Sep. 15	. . . .	22 Punjabis.
	Unattd. List	**Dunlop**, William Bruce, D.S.O. (*Ty. Lt.-Col. 20 Mar. 16*)	2 Nov. 77	22 Jan. 98	5 Mar. 99	1 Sep. 15	*l.c.* 1 Jan. 18	S. and T. Corps.
	Unattd. List	*Lord, Hugh Probyn Malet (104, 137, 170) [L]*	*26 Aug. 78*	*22 Jan 98*	*7 Apl. 99*	*1 Sep. 15*	. . . .	*Supy. List.*
	Unattd. List	**Muspratt**, Sydney Frederick, D. S. O., *p.s.c.* (71, 162, 167, 169)	11 Sep. 78	22 Jan. 98	23 Sep. 99	1 Sep. 15	*Ty. l. c.* *23 Dec. 15* *l. c.* 3 June 17	12 Cavalry.
	Unattd. List	**Macleod**, William Myers, *p.s.c.* (51, 57, 64, 66*a*, 71, 155)	16 Mar. 78	22 Jan. 98	2 Apl. 99	1 Sep. 15	. . . .	31 Lancers.
	Unattd. List	**Marchant**, Godfrey, *p. s. c.* (51)	17 Apl. 78	22 Jan. 98	22 Mar. 99	1 Sep. 15	. . . .	29 Lancers.
265	Unattd. List	**Coaker**, Vere Arthur (51, 66*a*, 137)	4 Jan. 78	22 Jan. 98	30 Mar. 99	1 Sep. 15	. . . .	3 Horse.
	K. O. S. B.	**Stewart**, James Montgomery Vansittart (50, 157)	12 July 77	16 Feb. 98	11 Dec. 07	1 Sep. 15	. . . .	10 G. R.
	R. Suss. R.	**Hughes**, Charles Frederic Watson, M.C. (50, 66*a*, 167, 172*a*)	28 Sep. 77	16 Feb. 98	15 Sep. 02	1 Sep. 15	. . . .	15 Sikhs.
	Bedf. R.	**Binny**, Allan James Murray, *p.s.c.* (50, 65, 72)	9 Nov. 78	16 Feb. 98	18 Sep. 00	1 Sep. 15	*Ty. l. c.* *20 Feb. 17*	1 Lancers.
	S. W. Bord.	**Pitcher**, Duncan Le Geyt, C.M.G. (50 66*a*, 71)	31 Aug. 77	16 Feb. 98	16 Oct. 01	1 Sep. 15	*l.c.* 1 Jan. 17	39 C. I. Horse.
270	North'r. R.	**Baily**, Frederick Kennedy (51, 71, 137, 172)	29 Sep. 75	16 Feb. 98	27 Nov. 00	1 Sep. 15	. . . .	14 Lancers.
	York. R.	**Keen**, Patrick Houston (51, 53, 172)	19 Sep. 77	16 Feb. 98	12 Aug. 00	1 Sep. 15	29 Nov. 15	56 Rifles.
	Hamps. R.	**McCormick**, Gregory Day (51, 53, 172)	13 Sep. 75	23 Feb. 98	8 June 99	1 Sep. 15	. . . .	72 Punjabis.
	Norf. R.	**Worsley**, Charles Fleetwood Mant (50, 53, 66*a*, 71, 169) *Service for pension:—In Ranks, 30 Aug. 94 to 5 Apl. 98 (half to count) less 2days forfeited*	30 Dec. 75	6 Apl. 98	20 Aug. 99	1 Sep. 15	. . . .	21 Cavalry.
	W. Rid R.	**Brunker**, James Robert (71, 162)	29 Apl. 78	4 May 98	25 June 09	1 Sep. 15	. . . .	S. and T. Corps.
275	Linc. R.	**Dyke**, Oswald Muirhead (51, 52, 59, 71, 72, 172)	11 Apl. 79	4 May 98	17 Nov. 02	1 Sep. 15	. . . .	21 Cavalry.
	Suff. R.	*Knowles, James Keilly*	*8 Oct. 77*	*4 May 98*	*20 Aug. 00*	*1 Sep. 15*	. . . .	*Supy. List.*
	North'n. R.	**Mackinnon**, Alexander Charles Broughton (51, 52, 157)	1 Feb. 78	4 May 98	13 July 01	1 Sep. 15	. . . .	9 G. R.
	S. Staff. R.	**Milford**, Edward (50, 59, 157, 167)	28 Sep. 78	4 May 98	15 Nov. 12	1 Sep. 15	. . . .	76 Punjabis.
	Hamps. R.	**Hardinge**, Henry Reginald von Donop (59)	20 May 78	7 May 98	16 July 01	1 Sep. 15	. . . .	Mily. Accts. Dept.
	S. Lanc. R.	**Jennings**, Robert Daniel (51, 66*a*, 71, 94, 162, 165, 169)	15 Aug. 77	7 May 98	1 Oct. 00	1 Sep. 15	. . . .	13 Lancers.
280	Suff. R.	**Robinson**, Daniel George, D.S.O., *p.s.c.* (51, 52, 95, 107, 140, 162, 169, 181)	25 May 78	7 May 98	30 Aug. 00	1 Sep. 15	. . . .	46 Punjabis.
281	Glouc. R.	**Smith**, Hubert Hamilton, D.S.O.† (51, 52, 66, 71, 137)	5 Mar. 78	7 May 98	16 Apl. 02	1 Sep. 15	*l.c.* 3 June 18	9 Infantry.

†Without pay and allowances prior to 1 Sep. 16.

No.	Former Corps.	Name.	Date of birth.	Date of first commission.	Date of appointment to Indian Army.	Date of Rank. Substantive.	Date of Rank. Army.	Remarks.
		Majors—*contd.*						
282	R. Muns. Fus.	**Tyrrell**, Augustus Charles Lionel, *p.s.c.* (51, 52, 71, 162, 169, 172)	15 June 79	7 May 98	11 Oct. 02	1 Sep. 15	*Ty. l. c.* 1 Feb. 18	25 Cavalry.
	Dorset. R.	**Ogg**, Arthur Charles, D.S.O., *p.s.c.* (50, 52, 137, 162, 205*a*)	16 Sep. 78	7 May 98	16 Jan. 02	1 Sep. 15	. . . .	7 Rajputs.
	North'n. R.	**Morris**, Charles Oxley, D.S.O. (51, 162, 170)	13 Jan. 79	7 May 98	19 Dec. 01	1 Sep. 15	. . . .	S. and T. Corps.
285	York. R.	**Craster**, William Shafto (51, 65, 66*a*, 162, 169)	17 Apl. 73	18 May 98	13 Feb. 04	1 Sep. 15	. . . .	8 Cavalry.
	R. Scots.	*Scott, Robert* (58, 71, 169)	*17 Sep. 75*	*28 May 98*	*12 Sep. 00*	*1 Sep. 15*	. . . .	*Supy List.*
	R. A.	**Benn**, John Piercy (51, 169)	25 July 79	23 June 98	20 July 05	1 Sep. 15	*Ty. l.c.* 29 Sep. 17	37 Lancers.
	R. A.	**Gretton**, George Foster (12, 51, 65, 71, 137)	11 Apl. 78	24 June 98	17 Sep. 01	1 Sep. 15	. . . .	7 Lancers.
	Unattd. List	**Watts**, Arthur Hugo (59, 162, 165, 169)	2 June 79	20 July 98	13 Oct. 99	1 Sep. 15	. . . .	S. and T. Corps.
	Unattd. List	**Wylde**, Lennard Francis George Stovin (162, 169)	24 Aug. 78	20 July 98	3 Oct. 99	1 Sep. 15	. . . .	Mily. Accts. Dept.
290	Unattd. List	**McWatters**, Herbert Claude, D.S.O., *p.s.c.* (50, 53, 57, 165, 172)	19 Oct. 78	20 July 98	5 Oct. 99	1 Sep. 15	*l.c.* 3 June 17	24 Punjabis.
	Unattd. List	**Holmes**, Cecil William Archibald, *p.s.c.* (50, 155, 188*b*)	16 Jan. 78	20 July 98	13 Oct. 99	1 Sep. 15	. . . .	116 Mahrattas.
	Unattd. List	**Watson**, Henry Rowsell (50, 53, 71, 195)	27 Aug. 78	20 July 98	5 Oct. 99	1 Sep. 15	. . . .	81 Pioneers.
	Unattd. List	**McLaughlin**, Harry Dalzell (157, 169)	10 Apl. 78	20 July 98	13 Oct. 99	1 Sep. 15	. . . .	6 G. R.
	Unattd. List	**Meiklejohn**, Walter Lloyd Sinclair, *p.s.c.* (51, 66*a*, 71, 172*a*, 195) [*l*]	31 Jan. 78	20 July 98	5 Oct. 99	1 Sep. 15	. . . .	106 Pioneers.
295	Unattd. List	**Craster**, George, D.S.O., *p.s.c.* (51, 59, 64, 137)	27 Dec. 78	20 July 98	16 Oct. 99	1 Sep. 15	. . . .	6 Cavalry.
	Unattd. List	**Smithers**, Leonard Sueton Hirsch (51, 59)	26 Jan. 79	20 July 98	7 Oct. 99	1 Sep. 15	. . . .	17 Infantry.
	Unattd. List	**Shaw**, Harold Middleton Drury (51, 52, 157)	5 June 78	20 July 98	19 Oct. 99	1 Sep. 15	. . . .	1 G. R.
	Unattd. List	**Evans**, Walter James (51, 52, 157)	3 May 78	20 July 98	3 Oct. 99	1 Sep. 15	. . . .	1 G. R.
	Unattd. List	**Timbrell**, Thomas, D.S.O. (167) (*e*)	2 Nov. 78	20 July 98	9 Oct. 99	1 Sep. 15	*Ty. l.c.* 1 Nov. 16	S. and T. Corps.
300	Unattd. List	**Lakin**, John Henry Foster, *p.s.c.* (50, 105, 125*a*, 140, 167, 169)	29 Nov. 78	20 July 98	5 Oct. 99	1 Sep. 15	*Ty. l.c.* 4 Jan. 8	7 G. R.
	Unattd. List	**Money**, Ernle Francis Drummond, D.S.O. (51, 66*a*, 157, 170)	18 June 78	20 July 98	28 Oct. 99	1 Sep. 15	. . . .	4 G. R.
	Unattd. List	**Lucas**, Guy William Caswell (51, 71, 137)	23 Apl. 79	20 July 98	10 Oct. 99	1 Sep. 15	. . . .	34 Horse.
	Unattd. List	**Neumann**, Charles William (51, 52, 137)	4 Aug. 78	20 July 98	6 Oct. 99	1 Sep. 15	*l.c.* 1 Jan. 17	48 Pioneers.
	Unattd. List	*Roberts, Arthur Salisbury Browne* (71, 104)	*24 Sep. 78*	*20 July 98*	*10 Oct. 99*	*1 Sep. 15*	. . . .	*Supy. List.*
	Unattd. List	**Coleridge**, John Francis Stanhope Duke, C.M.G., D.S.O., *p.s.c.* (51, 52, 71, 157) (*Ty. l.c. 21 Apl. 16.*)	25 Apl. 78	20 July 98	16 Apl. 00	1 Sep. 15	*l.c. 1 Jan. 17*	8 G. R.
305	Unattd. List	**Home**, John Elton (59, 167)	27 Oct. 78	20 July 98	12 June 00	1 Sep. 15	*Ty. l. c.* 1 Nov. 16	S. and T. Corps.
	Unattd. List	*Sandeman, Alfred Patrick* (104, 113)	*17 Mar. 78*	*20 July 98*	*12 Apl. 00*	*1 Sep. 15*	. . . .	*Supy. List.*
	Unattd. List	**Vander Gucht**, George Tristram (51, 57, 71, 137)	2 June 79	20 July 98	7 Oct. 99	1 Sep. 15	. . . .	3 Horse.
	Unattd. List	**Finnis**, Benjamin Hubert (51, 107, 172)	11 Feb. 78	20 July 98	6 Oct. 99	1 Sep. 15	. . . .	20 Infantry.
	Unattd. List	**Dening**, Harold (50, 53, 57, 71, 125*a*, 169) *Service for promotion from 21 July 98.*	18 July 79	1 Sep. 97	26 Nov. 02	1 Sep. 15	. . . .	25 Cavalry.
	Unattd. List	**Johnston**, Robert, D.S.O. (51, 53, 71, 137, 162, 170)	9 Apl. 79	27 July 98	19 Nov. 99	1 Sep. 15	. . . .	4 Cavalry.
310	Unattd. List	**Talbot**, Frank Eustace George, *p.s.c.* (50, 53, 57, 167)	6 Oct. 79	27 July 98	13 Oct. 99	1 Sep. 15	*l.c. 1 Jan. 17*	14 Sikhs.
311	Unattd. List	**Jeffery**, Walter Hugh, C.I.E. (71, 124, 125*b*, 195)	15 Dec. 78	27 July 98	12 Oct. 99	1 Sep. 15	*Ty. l.c.*, 24 June 18.	73 Infantry.

312	Unattd. List	**Gilchrist,** Walter Fellowes Cowan, C.I.E., *p.s.c.* (51, 52, 66*a*, 169) (*n*)	1 Apl. 79	27 July 98	12 Oct. 99	1 Sep. 15	. . . .	52 Sikhs.
	Unattd. List	**Burton,** Harold Franklin (50, 57, 71, 167)	11 July 78	27 July 98	12 Oct. 99	1 Sep. 15	. . . .	66 Punjabis.
	Unattd. List	**Barr,** John Henry Vere (51, 57, 71, 155) [*l*]	19 Oct. 78	27 July 98	11 Oct. 99	1 Sep. 15	. . . .	109 Infantry.
	Unattd. List	*Grey, Arthur James Herbert* (94, 124, 125*b*, 161, 172, 175)	*26 May 80*	*27 July 98*	*10 Oct. 99*	*1 Sep. 15*	. . . .	*Supy. List.*
315	Unattd. List	**Molony,** Arthur Deane (51, 95, 107, 146, 157)	7 Apl. 79	27 July 98	17 Oct. 99	1 Sep. 15	. . . .	7 G. R.
	Unattd. List	**McCleverty,** Anson Hugh, *p.s.c.* (51, 52, 71, 72*b*, 137)	26 Jan. 79	27 July 98	14 Oct. 99	1 Sep. 15	. . . .	13 Rajputs.
	Unattd. List	**Johnson,** Maurice Eustace Stanley, *p.s.c.* (51, 52, 137, 162, 165, 169)	21 Apl. 79	27 July 98	18 Oct. 99	1 Sep. 15	. . . .	48 Pioneers.
	Unattd. List	**Johnson,** Allen Edwin (51, 59, 157)	2 Oct. 78	27 July 98	14 Oct. 99	1 Sep. 15	. . . .	1 G. R.
	Unattd. List	**Pepys,** Gerald Leslie, D.S.O., *p.s.c.* (51, 72*a*, 169) [L]	30 Jan. 79	27 July 98	17 Oct. 99	1 Sep. 15	*Ty. l.c.* 17 Nov. 17	57 Rifles.
320	Unattd. List	**Cadell,** John George, D.S.O., *p.s.c.* (50, 66*a*, 72*a*, 137, 162, 165, 169, 205*a*)	7 Mar. 79	27 July 98	2 Nov. 99	1 Sep. 15	. . . .	45 Sikhs.
	Unattd. List	**Ward,** Henry Charles Swinburne, *p.s.c.* (51, 64, 65, 71, 137, 169) (*e*)	12 June 79	27 July 98	12 Oct. 99	1 Sep. 15	*Ty. l.c.* 9 Feb. 17; *l.c.* 3 June 17	2 Lancers.
	Unattd. List	**Wilson-Johnston,** Walter Edward, D.S.O., *p.s.c.* (50, 52, 162, 165, 169)	27 Oct. 78	27 July 98	19 Oct. 99	1 Sep. 15	*l.c.* 3 June 18	36 Sikhs.
	Unattd. List	**Kensington,** Edgar Claude, M.C., *p.s.c.* (50, 71, 95, 155, 169)	2 June 79	27 July 98	15 Oct. 99	1 Sep. 15	*l.c.* 3 June 17	130 Baluchis.
	Unattd. List	**McConaghy,** John Gerald, D.S.O. (51, 52, 71, 169)	19 Apl. 79	27 July 98	22 Oct. 99	1 Sep. 15	. . . .	25 Cavalry.
325	Unattd. List	**Wilson,** Geoffrey Edward Holt (50, 52, 167)	17 Apl. 79	27 July 98	17 Oct. 99	1 Sep. 15	. . . .	34 Pioneers.
	Unattd. List	**Hogg,** George Elliot Minchin (71, 170)	23 July 79	27 July 98	25 Oct. 99	1 Sep. 15	. . . .	S. and T. Corps.
	Unattd. List	**Moens,** Arthur William Hamilton May, D.S.O., *p.s.c.* (51, 52, 59, 95, 162, 169) (*n*)	20 Dec. 79	27 July 98	13 Mar. 00	1 Sep. 15	*Ty. l.c.* 5 Nov. 16; *l.c.* 1 *Jan.* 17	52 Sikhs.
	Unattd. List	**Drew,** Bertie Clephane Hawley, *p.s.c.* (50, 53, 71, 167, 172)	9 Feb. 80	27 July 98	14 Oct. 99	1 Sep. 15	*l.c.* 1 *Jan.* 17	23 Pioneers.
	Unattd. List	**Kirkpatrick,** Charles, *p.s.c.* (51, 52, 72*a*, 95, 162, 169)	18 May 79	27 July 98	20 Oct. 99	1 Sep. 15	*l.c.* 1 Jan. 18	Guides.
330	Unattd. List	**Bird,** Charles John Gwynne (162, 167, 170)	18 June 78	27 July 98	28 Oct 99	1 Sep. 15	. . . .	Mily. Accts. Dept.
	Unattd. List	**Goodfellow,** Napier George Barras (51, 162, 172)	9 Sep. 78	27 July 98	16 Oct. 99	1 Sep. 15	*l.c.* 3 June 17	128 Pioneers.
	Unattd. List	**Hogge,** Leared Row (51, 72*a*, 125*b*, 140*b*, 162, 167, 169) [*l*]	23 Feb. 80	27 July 98	14 July 00	1 Sep. 15	. . . .	47 Sikhs.
	Unattd. List	**MacDonnell,** Hercules Neville Francis (71)	29 May 79	27 July 98	19 Oct. 99	1 Sep. 15	*l.c.* 1 *Jan* 17	Mily. Accts. Dept.
	Camn. H.	**Duff-Sutherland-Dunbar,** *Sir* George, *Bart.* (162, 167)	29 May 78	3 Aug. 98	21 June 02	1 Sep. 15	. . . .	19 Punjabis.
335	Bedf. R.	**Muir,** Wingate Wemyss (50, 66*a*, 72*a*, 162, 165, 169)	12 June 79	3 Aug. 98	3 Nov. 00	1 Sep. 15	. . . .	15 Sikhs.
	Bedf. R.	**Ridgeway,** David Græme, *p.s.c.* (50, 53, 72*a*, 157)	12 Dec. 79	3 Aug. 98	20 Dec. 01	1 Sep. 15	*Ty. l.c.* 4 *Aug.* 17	3 G. R.
	R. Ir. R.	**Fraser,** Alexander Charles (162, 172*a*)	2 Sep. 78	3 Aug. 98	28 Oct. 00	1 Sep. 15	. . . .	S. and T. Corps.
	Hamps. R.	**Noake,** Arthur Stratford (75, 170)	25 June 79	3 Aug. 98	21 Apl. 02	1 Sep. 15	. . . .	S. and T. Corps.
	Wilts. R.	**Marjoribanks,** Robert Douglas, *p.s.c.* (51, 95, 162, 169)	9 June 79	3 Aug. 98	29 Nov. 00	1 Sep. 15	4 Aug. 07; *Ty. l.c.* 7 Dec. 16	107 Pioneers.
340	R. Innis. F*R*.	**Hutchinson,** Alan George Caldwell, *p.s.c.* (50, 66*a*, 167)	25 July 79	3 Aug. 98	11 Sep. 00	1 Sep. 15	. . . .	23 Pioneers.
	E. Surr.	*Yule, Robert Arbercromby* (51, 137, 162, 169)	*4 Nov. 79*	*3 Aug. 98*	*9 Jan 01*	*1 Sep. 15*	. . . .	*Supy. List.*
	Suff. R.	**Torkington,** Gerard Stukeley (51, 50, 72*a*, 167)	24 Sep. 78	3 Aug. 98	18 Aug. 00	1 Sep. 15	. . . .	69 Punjabis.
	S. Lanc. R.	**Delme-Radcliffe,** Alfred (51, 52, 66*a*, 71, 155) [L]	14 June 79	3 Aug. 98	8 Aug. 00	1 Sep. 15	. . . .	105 Lt. Infantry.
	S. Lanc. R.	**Hogg,** Cyril Mindon Trower, D.S.O., *p.s.c.* (50, 53, 129, 157, 162) *Service for pension from 10 Jan. 00.*	25 Jan. 79	3 Aug. 98	23 Aug. 00	1 Sep. 15	. . . .	4 G. R.
	Suff. R.	**Hay,** Arthur Sidney (51, 57, 155, 162, 169)	21 Feb. 79	3 Aug. 98	2 Oct. 00	1 Sep. 15	. . . .	31 Lancers.
345	Essex R.	**Day,** Arthur Duncan (51, 105, 140, 167)	24 Dec. 78	3 Aug. 98	24 Dec. 99	1 Sep. 15	. . . .	91 Punjabis.
	Bord. R.	**McKenna,** James Charles, D.S.O., *p.s.c.* (51, 137, 156)	4 Aug. 79	3 Aug. 98	11 Sep. 01	1 Sep. 15	. . . .	16 Rajputs.
	Bedf. R.	**Rogers,** Guy Hamilton, *p.s.c.* (50, 53, 59, 95, 137)	29 Nov. 77	10 Aug. 98	25 Sep. 00	1 Sep. 15	*Ty. l.c.* 21 Jan. 18	11 Rajputs.
	Unattd. List	**Hunter,** Frederick Fraser, D.S.O. (124*a*, 169, 172*a*)	7 Aug. 76	10 Aug. 98	18 Nov. 99	1 Sep. 15	*Ty. l.c.* 1 *Nov.* 16	Survey Dept.
349	E. Kent R.	**Rennick,** Alexander deClanay (51, 52, 66, 71, 172) *Forfeited from 4 May to 13 Aug. 07, for promotion.*	2 Oct. 78	4 May 98	2 Nov. 00	1 Sep. 15	. . . .	11 Lancers.

No.	Former Corps.	Names.	Date of birth.	Date of first commission.	Date of appointment to Indian Army.	Date of Rank. Substantive.	Date of Rank. Army.	Remarks.
		Majors—*contd.*						
350	Unattd. List	**Nixon**, Edmund Malet (51, 52, 59, 94, 169) *Forfeited from 27 July to 17 Dec. 07, for promotion.*	7 May 79	27 July 98	11 Oct. 99	1 Sep. 15	. . . .	36 Horse.
	R. A.	*Campbell, William Lachlan*, C.I.E. (*125b, 125c, 169, 199*) [*l*]	*10 Mar. 80*	*23 Dec. 98*	*19 Nov. 05*	*1 Sep. 15*	. . . .	*Supy. List.*
	R. A.	**Tobin**, Harry Walter, D.S.O., *p.s.c.* (51, 94, 162, 169)	17 Sep. 79	23 Dec. 98	9 Jan. 03	1 Sep. 15	3 June 16	128 Pioneers.
	Hamps. R.	**Skinner**, Henry Terence, D.S.O. (51, 59, 66a, 167, 170)	15 Apl. 77	23 Nov. 98	19 Nov. 00	1 Sep. 15	. . . .	29 Punjabis.
	R. Ir. R.	**McRae**, Henry St. George Murray, D.S.O., *p.s.c.* (57, 72a, 124a, 167, 170)	10 Nov. 78	4 Jan. 99	12 Sep. 00	1 Sep. 15	. . . .	45 Sikhs.
	Wilts. R.	**Rigg**, Henry Masson (107, 167, 170)	22 June 78	4 Jan. 99	31 Jan. 01	1 Sep. 15	. . . .	92 Punjabis.
355	N. Staff. R.	**Baddeley**, Robert John Halkett, M.C. (51, 65, 71, 167)	19 Aug. 77	4 Jan. 99	12 Sep. 00	1 Sep. 15	. . . .	15 Lancers.
	E. Lan. R.	**Renny**, George Sidney (51, 52, 66a, 167, 172a)	21 Aug. 78	4 Jan. 99	9 Apl. 03	1 Sep. 15	. . . .	38 Dogras.
	S. W. Bord.	**Doveton**, Hector James (51, 52, 169)	18 May 78	4 Jan. 99	15 Feb. 02	1 Sep. 15	. . . .	21 Punjabis.
	Suff. R.	**Ball**, Thomas Leeson (66a, 72)	10 Sep. 75	25 Jan. 99	21 Feb. 09	1 Sep. 15	. . . .	S. and T. Corps.
	Unattd. List	**Stoney**, Patrick Sinclair, *p.s.c.* (51, 66a, 72a, 161, 165, 160, 205a)	9 Jan. 80	25 Jan. 99	20 Apl. 00	1 Sep. 15	. . . .	26 Punjabis.
360	Unattd. List	**Wyatt**, Ernest Robert Caldwell, D.S.O., *p.s.c.* (51, 59, 162a, 169)	27 Apl. 80	25 Jan. 99	15 Apl. 00	1 Sep. 15	*Ty. l.c.* 6 Aug. 18.	106 Pioneers.
	Unattd. List	**Hunt**, Henry Richard Augustus, D.S.O., *p.s.c.* (51, 162, 169) [*l*]	24 July 79	25 Jan. 99	14 Apl. 00	1 Sep. 15	1 Jan. 16	25 Punjabis.
	Unattd. List	**Tancred**, John Yule (50, 53, 66a, 72a, 169)	16 Apl. 79	25 Jan. 99	14 Apl. 00	1 Sep. 15	. . . .	19 Punjabis.
	Unattd. List	**Villiers-Stuart**, John Patrick, D.S.O., *p.s.c.* (51, 52, 66a, 72b, 169)	1 Mar. 79	25 Jan. 99	14 Apl. 00	1 Sep. 15	*Ty. l.c.* 1 Nov. 17 *l.c.* 1 Jan. 18	55 Rifles.
	Unattd. List	**Mayne**, Charles Eric Mosley (94, 100, 137)	4 July 79	25 Jan. 99	13 Apl. 00	1 Sep. 15	. . . .	Army Remt. Dept.
365	Unattd. List	**Hume-Spry**, William Edmund, D.S.O.	20 July 79	25 Jan. 99	17 Apl. 00	1 Sep. 15	. . . .	S. and T. Corps.
	Unattd. List	**Penton**, Bertie Cyril, *p.s.c.* (51, 59, 95, 161, 165, 169, 205a)	1 July 80	25 Jan. 99	9 Apl. 00	1 Sep. 15	. . . .	25 Punjabis.
	Unattd. List	**Glynton**, Gerard Maxwell, *p.s.c.* (51, 53, 59, 65, 66, 72a, 157)	18 Feb. 79	25 Jan. 99	16 Apl. 00	1 Sep. 15	. . . .	3 G. R.
	Unattd. List	**Gordon**, Huntly Fleetwood (51, 71)	21 Mar. 80	25 Jan. 99	12 Mar. 00	1 Sep. 15	*Ty. l.c.* 26 Aug. 18.	Army Remt. Dept.
	Unattd. List	**Skinner**, Alexander Baird D.S.O. (51, 71, 137)	13 Oct. 79	25 Jan. 99	15 Apl. 00	1 Sep. 15	*Ty. l.c.* 31 Oct. 17	5 Cavalry.
	Unattd. List	*Newcombe, George Harley* (*104, 157, 162, 170*) [*L*]	*30 Nov. 79*	*25 Jan. 99*	*12 Apl. 00*	*1 Sep. 15*	. . . .	*Supy. List.*
370	Unattd. List	**Eliot**, Mark (50, 53, 66a, 167, 170)	24 June 79	25 Jan. 99	14 Apl. 00	1 Sep. 15	*Ty. l. c.* 1 June 18	Army Clo. Dept.
	Unattd. List	**Meadows**, Charles Alfred Gordon Pierrepont (51, 53, 137)	17 Jan. 79	25 Jan. 99	15 Apl. 00	1 Sep. 15	. . . .	7 Rajputs.
	Unattd. List	**Bayley**, Lytton Cecil Lambert (51, 162)	9 Apl. 79	25 Jan. 99	17 Apl. 00	1 Sep. 15	. . . .	6 Cavalry.
	Unattd. List	**Macleod**, Norman, D.S.O., *p.s.c.* (51, 167)	8 Jan. 79	25 Jan. 99	11 Apl. 00	1 Sep. 15	. . . .	15 Sikhs.
	Unattd. List	**Roberts**, William Bradley, D.S.O. (51, 52, 95, 129, 155) [*l*]	10 Mar. 79	25 Jan. 99	10 Apl. 00	1 Sep. 15	. . . .	101 Grenadiers.
375	Unattd. List	**Landon**, Charles Richard Henry Palmer (66a, 104, 125a, 162, 172, 180, 196)						
	Unattd. List	**Hewlett**, Evelyn Alfred (51, 52, 71, 137, 172a)	6 June 79	25 Jan. 99	22 Mar. 00	1 Sep. 15	. . . .	35 Horse.
	Unattd. List	**Stewart**, Alexander Frederick (95)	6 July 79	25 Jan. 99	15 Apl. 00	1 Sep. 15	*l.c.* 3 June 17	14 Lancers.
	Unattd. List	**Shea**, Alexander Gallwey, D.S.O. (50, 53, 169)	29 Apl. 79	25 Jan. 99	12 Apl. 00	1 Sep. 15	. . . .	S. and T. Corps.
			23 June 80	25 Jan. 99	21 Apl. 00	1 Sep. 15	*Ty. l.c.* 4 Feb. 18.	51 Sikhs.
	Unattd. List	**Beyts**, Roger Fleetwood Sconce (51, 66a, 137, 162) *Service for promotion from 26 Jan. 99.*	26 Feb. 79	1 Jan. 98	30 Nov. 02	1 Sep. 15	. . . .	Cantt. Magt.'s. Dept.
380	Unattd. List	**Morgan**, Henry Arthur (51, 72a, 148a) *Service for promotion from 26 Jan. 99.*	2 Mar. 79	1 Jan. 98	27 Nov. 02	1 Sep. 15	. . . .	18 Infantry.

381	E. York. R.	**Dickson**, Egbert, C.M.G. (51, 5², 57, 66*a*, 71, 167)	19 Aug. 79	11 Feb. 99	12 Mar. 03	1 Sep. 15	. . . .	67 Punjabis.
	Devon. R.	**Twiss**, Horace William Francis, D.S.O. [L]	20 Nov. 79	11 Feb. 99	10 Feb. 09	1 Sep. 15	*Ty. l.c.* 9 Mar. 18.	S. and T. Corps.
	Wilts. R.	**Kendall**, Sydney Robert Gordon (51, 66*a*, 170)	18 Apl. 79	11 Feb. 99	11 June 03	1 Sep. 15	. . . .	S. and T. Corps.
	Devon. R.	**Willis**, Evelyn Stuart Constantine, D.S.O. (51, 52, 95, 162, 169)	9 Oct. 79	11 Feb. 99	10 Dec. 03	1 Sep. 15	. . . .	58 Rifles.
385	R. War. R.	**Torrie**, Claud Jameson, D.S.O. (50, 53, 59, 167)	20 Mar. 79	11 Feb. 99	14 Aug. 00	1 Sep. 15	. . . .	30 Punjabis.
	R. Innis. Fus.	**Knaggs**, Charles Frederick (51, 66*a*, 71, 137, 162, 170)	21 Mar. 79	11 Feb. 99	11 Aug. 02	1 Sep. 15	. . . .	Army Remt. Dept.
	Norf. R.	**Trueman**, Henry John Hamilton (51, 52, 66*a*, 71)	28 Dec. 78	11 Feb. 99	14 Oct. 04	1 Sep. 15	. . . .	43 Regiment.
	R. Muns. Fus.	**Branson**, Lionel Hugh (51, 71, 155, 161, 170, 204)	8 Apl. 79	11 Feb. 99	17 Feb. 01	1 Sep. 15	. . . .	110 Lt. Infantry.
	R. Muns. Fus.	**Murray**, Kenelm Digby Bold, D.S.O., *p.s.c.* (50, 52, 169)	30 Oct. 79	11 Feb. 99	18 June 03	1 Sep. 15	*Ty. l.c.* 29 July 18.	59 Rifles.
390	York. R.	**Smith**, Eustace Clementi (51, 52, 59, 72*a*, 169)	8 Nov. 79	11 Feb. 99	29 Sep. 00	1 Sep. 15	. . . .	22 Punjabis.
	Unattd. List	**Lawrenson**, William Ralph (51, 52, 57, 71, 195)	8 Dec. 77	20 July 98	5 Oct. 99	1 Sep. 15	. . . .	86 Infantry.
		Forfeited from 20 July 07 to 12 Feb. 08, for promotion.						
	Unattd. List	**Mouat**, George Elliott Dundas, D.S.O. (51, 52, 66, 71, 195)	14 Feb. 79	25 Jan. 99	10 Apl. 00	1 Sep. 15	. . . .	63 Lt. Infantry.
		Forfeited from 25 Jan. to 12 Feb. 08, for promotion.						
	Midd'x. R.	**Jardine**, Charles Herbert (50, 53, 62, 72, 204)	1 June 73	11 Mar. 99	6 Sep. 12	1 Sep. 15	*l.c.* 3 June 17	96 Infantry.
		Service for pension from 13 June 00.						
	Y. & L. R.	**Boxwell**, Ambrose, C.I.E. (50, 53, 66*a*, 72*b*, 88, 137, 169)	9 Nov. 76	17 May 99	16 Mar. 09	1 Sep. 15	. . . .	119 Infantry.
395	D. C. L. I.	**Vickers**, Aubrey (51)	23 Sep. 78	20 May 99	18 Nov. 00	1 Sep. 15	. . . .	48 Pioneers.
	R. Ir. R.	**Grimshaw**, Roland William Wrigley (51, 57, 71, 72, 137)	14 Aug. 80	20 May 99	8 Dec. 00	1 Sep. 15	. . . .	34 Horse.
	Conn. Rang.	**Walker**, Herbert Thomas (51, 57, 71, 107, 120, 140, 148*b*)	8 June 79	20 May 99	19 Jan. 03	1 Sep. 15	. . . .	30 Lancers.
	W. I. R.	**Ellis**, Sherman Gordon Venn, D.S.O.	20 Feb. 80	20 May 99	28 Jan. 04	1 Sep. 15	. . . .	Mily. Accts. Dept.
	Conn. Rang.	**Wallis**, Henry Jocelyn Kennedy (51, 52, 57, 64, 71, 162, 174)	20 Jan. 78	20 May 99	17 Feb. 01	1 Sep. 15	. . . .	20 Horse.
400	S. Lan. R.	**Marsh**, Cunliffe Hebbert, D.S.O. (50, 71, 167) (*Ty. l. c. 13 May 16*)	30 Oct. 78	20 May 99	6 Sep. 02	1 Sep 15	. . . .	18 Lancers.
	Welsh R.	**Morgan**, Bernard Evan, D.S.O. (50, 57, 104, 167)	17 June 78	20 May 99	11 Feb. 01	1 Sep. 15	. . . .	93 Infantry.
	W. York. R.	**Gretton**, John Cunliffe (51, 71, 72*b*, 169)	7 May 80	20 May 99	7 Dec. 04	1 Sep. 15	. . . .	126 Infantry.
	R. A.	**King-Mason**, Alan Richard Leopold (51, 52, 71, 155, 162)	9 Aug. 78	20 May 99	28 July 04	1 Sep. 15	. . . .	108 Infantry.
	W. York. R.	**Porch**, Edward Albert, M.C.	23 June 79	20 May 99	22 Jan. 12	1 Sep. 15	*l.c.* 1 Jan 17	S. and T. Corps.
	Ches. R.	*Pritchard, Hugh Robert Norman (71, 169, 188a)*	*11 Apl. 79*	*20 May 99*	*10 Oct. 00*	1 Sep. 15	. . . .	*Supy. List.*
405	W. York. R.	**Lyster**, Cecil Julius Hamilton, M.C. (66, 72*b*)	1 Dec. 77	20 May 99	1 Aug. 09	1 Sep. 15	. . . .	S. and T. Corps.
	R. A.	**Ross**, Alan Campbell, D.S.O. (12, 50, 52, 71)	4 May 78	4 Mar. 99	3 May 03	1 Sep. 15	. . . .	20 Horse.
		Forfeited from 4 Mar. to 20 May 08, for promotion.						
	Som. L. I.	**Hagger**, Edgar Robert (51, 170)	2 Sep. 79	24 June 99	26 May 04	1 Sep. 15	. . . .	Army Remt. Dept.
	R. A.	*Mackenzie, Charles Fraser*, C.I.E. (137, 161) (*L*)	*21 Sep. 80*	*25 June 99*	*12 Jan. 03*	1 Sep 15	. . . .	*Supy. List.*
	R. A.	**Cole**, Brian Lytton (51, 52, 72*a*, 137)	26 May 80	25 June 99	23 Apl. 01	1 Sep. 15	. . . .	13 Rajputs.
	R. Ir. Rif.	**Kirkwood**, Andrew Samuel (51, 66*a*, 71)	6 June 75	5 July 99	2 Nov. 09	1 Sep. 15	. . . .	Mily. Farms Dept.
410	Unattd. List	**Bannatyne**, Neil Charles, *p.s.c.* (50, 59, 172)	24 Jan. 80	19 July 99	30 Sep. 00	1 Sep. 15	*Ty. l.c.* 8 Apl. 17	128 Pioneers.
	R. A.	**Gourlie**, James (51, 65, 66*a*)	8 Apl. 76	26 July 99	29 Mar. 04	1 Sep. 15	. . . .	38 C. I. Horse.
	Unattd. List	**Peters**, James (50, 52, 71, 172)	18 July 78	2 Aug. 99	25 Nov. 00	1 Sep. 15	. . . .	10 Lancers.
	Unattd. List	**Sweeny**, Rogers Lewis Campbell, D.S.O., M.C. (64)	21 Nov. 78	2 Aug. 99	28 Nov. 00	1 Sep. 15	. . . .	S. and T. Corps.
	R. Suss. R.	**Hill**, Henry Walter Dunlop (51, 52, 59, 71, 167)	2 Mar. 79	9 Aug. 99	28 Feb. 02	1 Sep. 15	. . . .	16 Cavalry.
415	R. Dub. Fus.	**Lane**, Frank Bernard (51, 58, 71, 137, 170)	16 Aug. 79	12 Aug. 99	20 Aug. 02	1 Sep. 15	. . . .	8 Cavalry.
	L'pool R.	**Prince**, Robert (72*a*)	10 Feb. 80	12 Aug. 99	7 Jan. 05	1 Sep. 15	*Ty. l. c.* 6 Apl. 18.	Mily. Accts. Dept.
	E. York R.	**Burd**, Edmund (66*a*, 104, 140, 167)	4 Feb. 80	12 Aug. 99	23 June 04	1 Sep. 15	. . . .	93 Infantry.
	E. York R.	**Gordon**, Geoffrey Seton (51, 66*a*, 72*a*, 172)	2 Nov. 80	12 Aug. 99	10 Jan. 05	1 Sep. 15	. . . .	35 Horse.
419	S. Staff. R.	**Porter**, Fortescue Geoffrey (71, 155, 170)	21 Dec. 80	12 Aug. 99	12 Oct. 01	1 Sep. 15	. . . .	17 Infantry.

No.	Former Corps.	Names.	Date of birth.	Date of first commission.	Date of appointment to Indian Army.	Date of Rank. Substantive.	Date of Rank. Army.	Remarks.
		Majors—*contd.*						
420	Shrops. L.I.	**Giles**, Edward Douglas, D.S.O., *p.s.c.* (51, 59, 95, 169)	13 Oct. 79	12 Aug. 99	17 Jan. 01	1 Sep. 15	*Ty. l. c.* 7 May 17; *l. c.* 3 June 18	35 Horse.
	Yorks. L. I.	*Walker, Arthur Mills Elloughton Henry Travis* (204)	*26 Mar. 80*	*12 Aug. 99*	*10 Oct. 03*	*1 Sep. 15*	. . . .	*Supy. List.*
	Norf. R.	**Anderson**, William Henniker (51, 52, 66*a*, 172*a*)	17 Apl. 80	12 Aug. 99	12 June 03	1 Sep. 15	*Ty. l. c.* 25 Feb. 18.	Army Remt. Dept.
	R. Ir. R.	**Norbury**, Paul Fitz Gerald (51, 71, 88, 137) [L]	28 Sep. 79	12 Aug. 99	24 June 01	1 Sep. 15	. . . .	34 Horse.
	N. Staff. R.	**Salmon**, William Harry Broome	4 Nov. 79	12 Aug. 99	5 Apl. 02	1 Sep. 15	. . . .	S. and T. Corps.
	Y. & L.R.	*Daukes, Clendon Turberville* (162, 169)	*24 Dec. 79*	*12 Aug. 99*	*22 Mar. 01*	1 Sep. 15	. . . .	*Supy. List.*
	S. Staff. R.	**Geoghegan**, Oswald Alan (51) (*e*)	9 Oct. 79	12 Aug. 99	6 Apl. 09	1 Sep. 11	*Ty. l. c.* 28 Mar. 17	S. and T. Corps.
425	W. York R.	**Catty**, Thomas Claude, D.S.O., *p.s.c.* (50, 66*a*, 162, 166, 195)	6 Nov. 79	12 Aug. 99	6 July 01	1 Sep. 15	*Ty. l. c.* 4 Mar. 17; *l.c.* 3 June 18.	69 Punjabis.
	Linc. R.	**Dunlop**, Charles Andrew Montague, M.C. (51, 72*a*, 137)	22 Aug. 79	12 Aug. 99	23 July 01	1 Sep. 15	. . . .	37 Dogras.
	S. W. Bord.	*Wilson, Denis Grey* (51, 162)	*6 Jan. 80*	*12 Aug. 99*	*23 Aug. 01*	*1 Sep. 15*	. . . .	*Supy. List.*
	R. Suss. R.	**Tancock**, Alexander Charles (50, 162, 167, 169)	27 Dec. 79	12 Aug. 99	31 July 01	1 Sep. 15	. . . .	31 Punjabis.
	R. W. K. R.	**Whitehead**, James, D.S.O., *p.s.c.* (51, 95, 137)	18 Jan. 80	12 Aug. 99	29 Oct. 01	1 Sep. 15	*Ty. l.c.* 27 Feb. 16; *l.c.* 1 Jan. 17	1 Brahmans.
	20 Huss.	*Percy-Smith, Eric Sydney* (51, 104, 125*a*, 140, 181) *Forfeited from 3 Aug. 07 to 6 Oct. 08, for promotion.*	*26 Jan. 79*	*3 Aug. 98*	*20 Aug. 06*	1 Sep. 15	*l.c.* 3 June 18	*Supy. List.*
	R. Innis. Fus.	**Pott**, Henry Percy (51, 66)	17 May 79	18 Oct. 99	24 Sep. 03	1 Sep. 15	. . . .	5 Light Infantry.
430	H. L. I.	**Blair**, Douglas Alexander (51, 52, 157)	10 July 79	18 Oct. 99	1 Nov. 03	1 Sep. 15	. . . .	39 Garhwal Rifles.
	Som. L. I.	**Harrison**, Louis George Beresford (51, 52, 137)	12 Nov. 78	18 Oct. 99	15 Sep. 03	1 Sep. 15	. . . .	125 Rifles.
	Conn. Rang	**Hutcheson**, Steele (51, 72*b*)	1 Mar. 80	18 Oct. 99	26 Nov. 03	1 Sep. 15	. . . .	3 Brahmans.
	Suff. R.	**Butler**, Charles Walter, (71, 72*a*)	17 Feb. 80	15 Nov. 99	10 June 04	1 Sep. 15	. . . .	Mily. Accts. Dept.
	Suff. R.	**Stuart**, William (51, 53, 71, 72*a*, 106, 125*a*, 140, 162, 172)	14 Jan. 78	20 May 99	14 Jan. 03	1 Sep. 15	. . . .	22 Cavalry.
435	R. A.	**Gray**, Evelyn St. Clair, M.C. (50, 66, 137)	22 Nov. 80	22 Nov. 99	15 Sep. 03	1 Sep. 15	. . . .	34 Horse.
	Y. & L. R.	**Wilson**, Denis Daly, M.C., *p.s.c.* (50, 53, 59, 64, 71, 169) (*Ty.l.c.* 29 *Mar.* 16)	22 Oct. 78	29 Nov. 99	29 July 02	1 Sep. 15	. . . .	17 Cavalry.
	Y. & L. R.	**Bethell**, Leonard Arthur (57, 72*b*, 157, 170)	6 Dec. 79	6 Dec. 99	17 Oct. 02	1 Sep. 15	. . . .	10 G. R.
	W. York R.	**Shuttleworth**, Betham Wilkins, *p. s. c.* (50, 72*a*, 95, 165, 172*a*)	16 Mar. 80	6 Dec. 99	24 June 03	1 Sep. 15	. . . .	45 Sikhs.
	Dorset. R.	**Murray**, Arthur Gaussen (51, 137)	3 May 80	6 Dec. 99	28 May 01	1 Sep. 15	. . . .	Mily. Accts. Dept.
440	L'pool R.	**Browne**, Walter Stephen Wilfrid (51, 53, 137) *Service for pension from 10 July 03.*	9 May 77	9 Dec. 99	29 Feb. 12	1 Sep. 15	. . . .	44 Infantry.
	R. A.	**Crowdy**, James Dunscomb, D.S.O., *p. s. c.* (50, 157, 170)	11 Jan. 81	6 Jan. 00	7 Feb. 05	1 Sep. 15	. . . .	5 G. R.
	R. A.	**Henderson**, Clarence Gordon Hume (12, 43, 51, 125*a*, 140, 148*b*, 162)	21 Nov. 80	6 Jan. 00	12 July 03	1 Sep. 15	. . . .	29 Lancers.
	R. A.	**Broadbent**, John Theodore Crofton (12, 50, 71, 170, 195)	10 May 81	6 Jan. 00	2 May 04	1 Sep. 15	. . . .	88 Infantry.
	R. A.	**Eliot**, Harold Selwyn (94, 169)	20 Apl. 81	6 Jan. 00	1 July 09	1 Sep. 15	. . . .	S. and T. Corps.
445	R. A.	**Walter**, Edmund, C.I.E. (94, 162, 169, 205*a*) [*l*]	29 Jan. 81	6 Jan. 00	7 May 09	1 Sep. 15	. . . .	S. and T. Corps.

446	S. Lanc. R.	**Wilkinson**, George Hamilton, D.S.O. *Service for pension;—Embd. Mila. from 1 Jan. 00 to 3 Apl. 00; for promotion from 11 Jan. 00 to 3 Apl. 00.*	14 Apl. 78	4 Apl. 00	26 Jan. 12	1 Sep. 15	. . . .	S. and T. Corps.
	3 D. G.	**Dyer**, Hyde Ridgway D.S.O., (172) *Service for pen son;—Attd. Reglrs. from 10 Jan. to 9 Nov. ;00 for promotion from 15 Jan. 00.*	29 June 80	10 Nov. 00	7 Oct. 02	1 Sep. 15	. . . .	35 Horse.
	Manch. R.	**Gordon**, James Guy Birnie, M.C. (51, 52, 66a, 172) *Service for pension;—Embd. Mila. from 13 Dec. 99 to 3 May 01; for promotion from 16 Jan. 00.*	14 Oct. 81	4 May 01	27 May 05	1 Sep. 15	. . . .	52 Sikhs.
	Unattd. List	**Cochran**, Ale xr. Chas. urves (51, 169)	17 Sep. 79	17 Jan. 00	13 Apl. 01	1 Sep. 15	. . . .	40 Pathans.
450	Unattd. List	**Gwyer**, Vernon Edward, *p.s.c.* (50, 52, 59, 95, 137)	19 Sep. 79	17 Jan. 00	14 Apl. 01	1 Sep. 15	*l.c.* 3 June 17 *Ty. l. c.* 25 Nov. 17	10 Jats.
	Unattd. List	**Bennett**, Ernest Edward Sebastian, *p.s.c.* (50, 162, 167)	3 Nov. 79	17 Jan. 00	10 Apl. 01	1 Sep 15	. . . .	66 Punjabis.
	Unattd. List	**Chambers**, Philip Roper, D.S.O., *p.s.c.* (51, 52, 59, 66a, 71, 95, 162, 172)	21 June 81	17 Jan. 00	18 Apl. 01	1 Sep. 15	3 June 16	10 Lancers.
	Unattd. List	*Oliver, Douglas Roderic Giffard* (167, 169) [L]	*17 Aug. 80*	*17 Jan. 00*	*16 Apl. 01*	*1 Sep. 15*	. . . .	*Supy. List.*
	Unattd List	*Rae, Maitland Easton* (51, 169)	*6 Mar. 80*	*17 Jan 00*	*10 Apl. 01*	*1 Sep 15*	. . . .	*Supy. List.*
	Unattd. List	**Burrowes**, Patrick William (51, 52, 71, 72, 162, 169)	16 Jan. 81	17 Jan. 00	20 Apl. 01	1 Sep. 15	. . . .	25 Cavalry.
	Unattd. List	**Scott**, Arthur Elliott Somerville (59, 169)	13 Oct. 79	17 Jan. 00	13 Apl. 01	1 Sep. 15	. . . .	37 Lancers.
455	Unattd. List	**Faris**, John George (51, 59, 95, 147, 157)	9 Feb. 80	17 Jan. 00	9 Apl. 01	1 Sep. 15	. . . .	7 G.R.
	Unattd. List	*Biscoe, Hugh Vincent* (71, 88, 94, 100, 161, 169) [L]	*27 June 81*	*17 Jan. 00*	*17 Apl. 01*	*1 Sep. 15*	. . . .	*Supy. List.*
	Unattd. List	**Kemble**, Paul Berthon (66, 125a, 137)	2 Sep. 79	17 Jan. 00	10 June 01	1 Sep. 15	. . . .	S. and T. Corps.
	Unattd. List	*Prescott, Cyril William* (162, 172)	*22 Jan. 80*	*17 Jan 00*	*15 Apl 01*	*1 Sep. 15*	. . . .	*Supy. List.*
	Unattd. List	**Harvey-Kelly**, Harvey St. George Hume (50, 105, 129, 143, 181, 195)	10 Oct. 80	17 Jan. 00	26 Mar. 01	1 Sep. 15	. . . .	64 Pioneers.
	Unattd. List	**Davis**, George Edward Parry (162a, 169)	2 May 80	17 Jan. 00	15 Apl. 01	1 Sep. 15	. . . .	106 Pioneers.
	Unattd. List	**Howell**, Robert Hugh (59, 64, 66, 71, 167, 169)	20 Mar. 80	17 Jan. 00	10 Apl. 01	1 Sep. 15	. . . .	15 Lancers.
460	Unattd. List	**Bailey**, William Bryan (51, 66a)	23 July 79	17 Jan. 00	12 Apl. 01	1 Sep. 15	. . . .	2 G. R.
	Unattd. List	**Sanders**, Geoffrey Percival (50, 53, 59, 157, 169)	28 Feb. 80	17 Jan. 00	12 Apl. 01	1 Sep. 15	. . . .	5 G. R.
	Unattd. List	**Greenaway**, Hugh (51, 66, 71, 167, 169)	21 Mar 80	17 Jan. 00	15 June 01	1 Sep. 15	. . . .	69 Punjabis.
	Unattd. List	**Scott**, Arthur Napier De Vere (50, 57, 71, 137)	3 Aug. 80	17 Jan. 00	12 Apl. 01	1 Sep. 15	. . . .	2 Lancers.
	Unattd. List	**Lane**, Joscelyn Theodore Horsburgh, M.C. (50, 59, 72a, 157, 205a)	7 Feb. 80	17 Jan. 00	14 Apl. 01	1 Sep. 15	. . . .	39 Garhwal Rifles.
	Unattd. List	**Reed**, Henry Robert Baynes, M.C. (51, 52, 59, 157)	3 Jan. 80	17 Jan. 00	12 Apl. 01	1 Sep. 15	. . . .	39 Garhwal Rifles.
465	Unattd. List	**Robertson**, Donald Elphinston (51, 59, 71, 167, 172)	22 Dec. 79	17 Jan. 00	28 Mar. 01	1 Sep. 15	*Ty. l. c.* 8 Feb. 18.	11 Lancers.
	Unattd. List	**Davson**, Harold John Hunter (50, 72a, 107, 129, 167, 172a) (n) *Service for promotion from 18 Jan. 00.*	15 Jan. 80	1 Sep. 98	27 Nov. 02	1 Sep. 15	. . . .	82 Punjabis.
	Unattd. List	**McLaughlin**, Wm. Pulteney Michael Dalzell, *p.s.c.* (50, 172a) *Service for promotion from 18 Jan. 00.*	3 Nov. 80	1 Sep. 98	29 Nov. 02	1 Sep. 15	. . . .	107 Pioneers.
	Unattd. List	**Lindesay**, Frederick Sinclair (95, 105, 162, 172, 185) *Service for promotion from 18 Jan. 00.*	16 Jan. 80	1 Sep. 98	17 Jan. 04	1 Sep. 15	*Ty. l.c.* 18 Jan. 18	Cant. Magte.'s Deptt.
	Essex R.	**Saunders**, Patrick Cornelius (51, 66)	13 Sep. 79	20 Jan. 00	13 Oct. 02	1 Sep. 15	. . . .	S. and T. Corps.
470	Glouc. R.	**Hamilton**, Claude Alex Brewer (50, 53, 72b, 157)	29 Aug. 79	20 Jan. 00	25 Aug. 02	1 Sep. 15	. . . .	3 G. R.
	E. Kent R.	**Ransford**, Francis Metcalfe (51, 72b, 157, 161, 169, 188b)	16 Oct. 80	20 Jan. 00	27 Sep. 02	1 Sep. 15	. . . .	6 G. R.
	Norf. R.	**Lyon**, Clive Wentworth (50, 53, 66a, 71, 165)	30 June 80	20 Jan. 00	11 Feb. 03	1 Sep. 15	. . . .	42 Regiment.
	R. Ir. R.	**Kemmis**, Arthur William Mersh, D.S.O. (51, 52, 71, 107, 125a, 140, 172, 181)	7 July 81	20 Jan. 00	21 July 01	1 Sep. 15	. . . .	10 Lancers.
475	R. Ir. R.	**Dease**, Conly Edward (50, 72b, 167)	20 Nov. 80	20 Jan. 00	28 Sep. 03	1 Sep. 15	*Ty. l. c.* 19 Feb. 18.	91 Punjabis.
	Durh. L. I.	**Whitby**, Henry Frank (50, 53, 71, 94, 169)	15 Mar. 80	20 Jan. 00	18 Aug. 04	1 Sep. 15	. . . .	19 Lancers.
	R.W.Kent R.	**Henderson**, Kenneth Douglas Murray, M.C. (51)	29 Mar. 81	20 Jan. 00	19 Nov. 02	1 Sep. 15	. . . .	17 Infantry.
	Y. & L. R.	**Ruel**, William George, D.S.O.	12 Jan. 80	20 Jan. 00	11 Apl. 04	1 Sep. 15	. . . .	S. and T. Corps.
479	Y. & L. R.	**Corbyn**, Malcolm McLeod (51, 59, 66, 172)	5 Jan. 80	20 Jan. 00	5 Jan. 03	1 Sep. 15	. . . .	S. and T. Corps.

No.	Former Corps.	Names.	Date of birth.	Date of first commission.	Date of appointment to Indian Army.	Date of Rank. Substantive.	Date of Rank. Army.	Remarks.
		Majors—*contd.*						
480	Board R.	**Sutherland**, Arthur Orr (51, 52, *66a*, 72, 172)	21 Dec. 79	20 Jan. 00	17 July 03	1 Sep. 15	. . . .	22 Punjabis.
	R. W. S. R.	**Lancaster**, Edmund Henry (71, 72)	13 Mar. 81	20 Jan. 00	31 May 03	1 Sep. 15	. . . .	S. and T. Corps.
	R. Sco. Rif.	**Grant**, Euan Patrick (51, 65, 71, 72*b*, 172)	24 Apl. 81	20 Jan. 00	25 Nov. 03	1 Sep. 15	. . . .	25 Cavalry.
	Unattd. List	**Elias Wynn**, Alfred Hearst Wynn, *p.s.c.* (50, 57, 157, 170)	20 Jan. 81	20 Jan. 00	21 Apl. 01	1 Sep. 15	. . . .	1 G. R.
	Unattd. List	**Allen**, John Frederick (57, 137)	11 May 80	20 Jan. 00	21 Apl. 01	1 Sep. 15	. . . .	Mily. Accts. Dept.
485	Unattd. List	**MacMahon**, Hugh Francis Edward, M.C. (169)	13 Oct. 80	20 Jan. 00	21 Apl. 01	1 Sep. 15	*Ty. l.c.* 12 Dec. 17 *l.c.* 1 Jan. 18	S. and T. Corps.
	Unattd. List	**Wheeler**, George Campbell, (50, 59, 71, 157)	7 Apl. 80	20 Jan. 00	18 Apl. 01	1 Sep. 15	. . . .	9 G. R.
	Unattd. List	**Gillies**, Frederick George, *p.s.c.* (50, 53, 65, 71, 137, 162)	28 Jan. 81	20 Jan. 00	20 Apl. 01	1 Sep. 15	*Ty. l. c.* 26 Apl. 18.	33 Lt. Cavalry.
	Unattd. List	**Lyall**, Charles Ross (50, 52, 167)	3 Oct. 81	20 Jan. 00	19 Apl. 01	1 Sep. 15	. . . .	36 Sikhs.
	Unattd. List	**Apthorpe**, Evelyn Hugh Foyer (51, 52, 57, 105, 140, 167)	31 Aug. 80	20 Jan. 00	18 Apl. 01	1 Sep. 15	. . . .	90 Punjabis.
	Unattd. List	*Butterfield, Edward*, D.S.O. (*51, 59, 104, 167*)	*9 Apl. 80*	*20 Jan 00*	*18 Apl. 01*	*1 Sep. 15*	. . . .	*Supy. List.*
490	Unattd. List	**Reinhold**, Hermann Edward, M.C. (51, 172) [L]	17 Nov. 81	20 Jan. 00	23 Apl. 01	1 Sep. 15	. . . .	27 Punjabis.
	Unattd. List	**Bignell**, Richard Lowis, D.S.O. (50, 53, 162, 167)	4 June 81	20 Jan. 00	27 Apl. 01	1 Sep. 15	3 June 16	41 Dogras.
	Unattd. List	**McEnery**, Reginald Thomas (50, 53, 59, 66, 172)	20 Feb. 80	20 Jan. 00	18 Apl. 01	1 Sep. 15	*Ty. l.c.* 26 Apl. 18	S. and T. Corps.
	Unattd. List	**Coleridge**, Percy Lovel (50, 66, 195)	20 July 80	20 Jan. 00	18 Apl. 01	1 Sep 15	. . . .	80 Infantry.
	12 Lrs.	**Connop**, Harry Ernest (51, 172*a*)	5 July 80	20 Jan. 00	22 May 11	1 Sep. 15	. . . .	35 Horse.
495	Unattd. List	**Hewett**, George, D.S.O. (51, 52, 59, 137)	18 Mar. 81	20 Jan. 00	22 Apl. 01	1 Sep. 15	. . . .	48 Pioneers.
	Unattd. List	**Graham**, Dudley Stevens (51, 94, 162, 169)	28 Nov. 80	20 Jan. 00	30 Apl. 01	1 Sep. 15	. . . .	128 Pioneers.
	Unattd. List	**Strettell**, Chauncy Batho Dashwood (50, 71, 140, 162, 169)	6 Aug. 81	20 Jan. 00	22 Apl. 01	1 Sep. 15	. . . .	23 Cavalry.
	Unattd. List	**Alexander**, Robert Dundas (50, 53, 157)	28 Aug. 80	20 Jan. 00	27 Apl. 01	1 Sep. 15	. . . .	3 G. R.
	Unattd. List	**Luck**, Tylden (51, 52, 66, 94, 100, 167, 169)	3 Apl. 81	1 Jan. 99	10 Oct. 05	1 Sep. 15	. . . .	67 Punjabis.
		Service for promotion from 21 Jan 00.						
500	Hamps. R.	**Hooker**, Joseph Symonds (51, 52, 71, 72*b*, 137)	14 Dec. 77	24 Jan. 00	6 Dec. 02	1 Sep. 15	. . . .	12 Pioneers.
	R.W.Kent.R.	**Bennett**, John Humphrey (51, 137) [L]	1 Dec. 78	24 Jan. 00	2 Aug. 02	1 Sep. 15	. . . .	3 Horse.
	Manch. R.	**Kennedy-Craufurd-Stuart**, Charles, D.S.O. (51, 52, 71, 106, 115, 125*a*, 140, 162, 169)	29 Aug. 79	4 Apl. 00	3 Aug. 03	1 Sep. 15		127 Infantry.
		Service for pension.—Embd. Mila. from 13 Dec. 99 to 3 Apl. 00; for promotion from 17 Jan. 00 and forfeited 17 Jan. to 26 Jan. 09 for promotion.						
	Som. L. I.	**Peel**, Basil Gerard, D. S. O. (51, 52, 66*a*, 71, 195)	26 Dec. 81	18 Apl. 00	31 Dec. 07	1 Sep. 15	. . . .	81 Pioneers.
		Service for pension.—Embd. Mila. & Attd. Reglrs. from 14 Dec. 99 to 17 Apl. 00; for promotion from 8 Feb. 00.						
	Unattd. List	**Webster**, Charles David (50, 53, 72*a*, 124, 125*b*, 165, 169, 205*a*)	19 July 81	14 Feb. 00	10 May 01	1 Sep. 15	. . . .	129 Baluchis.
505	Unattd. List	**Simpson**, James (50, 53, 106, 125*a* 140, 157)	17 July 80	14 Feb. 00	10 May 01	1 Sep. 15	. . . .	10 G. R.
	Unattd. List	**Davidson**, Hugh Middleton (50, 52, 64, 172)	1 Sep. 80	14 Feb. 00	10 May 01	1 Sep. 15	. . . .	36 Horse.
	Unattd. List	**Young**, George Herbert (51, 52, 107, 167, 170)	30 July 80	14 Feb. 00	19 Apl. 01	1 Sep. 15	. . . .	38 Dogras.
508	Unattd. List	**Farran**, George Lambert, D.S.O., M.C., (45*a*, 51, 52, 66*a*, 71, 137)	30 Jan. 81	14 Feb. 00	12 May 01	1 Sep. 15	*Ty. l. c.* 23 Feb. 17	4 Cavalry.

509	N. Staff R.	**Steward**, Edward Merivale (66a) *Service for pension.—Embd. Mila. from 17 Jan. 00 to 3 May 01; for promotion from 14 Feb. 00.*	5 Feb. 81	4 May 01	6 July 11	1 Sep. 15	. . . .	S. and T. Corps.
510	N. Lan. R.	**Salusbury**, Robert Thelwall Gwlym (50, 59, 66a, 169) *Service for pension.—Embd. Mila. from 23 Jan. 00 to 13 Sep. 01; for promotion from 14 Feb. 00.*	27 June 81	14 Sep. 01	5 Dec. 03	1 Sep. 15	. . . .	26 Punjabis.
	Lanc. Fus.	**Marriott**, John Harold Goodwyn (51, 59, 66) *Service for pension.—Embd. Mila. & Attd. Reglrs. from 24 Jan. to 18 Sep. 00; for promotion from 16 Feb. 00.*	3 Sep. 78	19 Sep. 00	17 Feb. 03	1 Sep. 15	. . . .	12 Pioneers.
	E. Kent. R.	**Baldwin**, Francis Edward Wm. (51, 53, 65a, 66, 66b, 94, 155, 169)	8 Jan. 80	17 Feb. 00	9 July 04	1 Sep. 15	. . . .	121 Pioneers.
	Bedf. R.	**Whitchurch**, Reginald Henry Sedgwick (51, 71)	27 Feb. 79	17 Feb. 00	27 Oct. 01	1 Sep. 15	. . . .	Mily. Accts. Dept.
	W. Kent R.	**Hulbert**, Thomas Ernest (50, 137)	13 July 79	17 Feb. 00	16 May 02	1 Sep. 15	. . . .	3 Horse.
515	Hamps. R.	**Seymour**, Charles Richard Foster, *p.s.c.* (50, 52, 137) [L]	9 Apl. 80	17 Feb. 00	21 Nov. 02	1 Sep. 15	*Ty. l. c.* 1 June 18	13 Rajputs.
	Oxf. L. I.	**Norman**, Alfred Cecil (51, 71, 137)	1 Apl. 79	17 Feb. 00	10 Jan. 02	1 Sep. 15	. . . .	5 Cavalry.
	R. War. R.	**Price**, Henry William (51, 65, 66, 137) *Service for pension—Emod. Mila. from 22 Jan. 00 to 3 May 01; for promotion from 19 Feb. 00.*	27 May 80	4 May 01	13 Jan. 04	1 Sep. 15	. . . .	S. and T. Corps.
	Midd'x. R.	**Phillips**, Jas. Charles (50, 165, 172) *Service for pension;—Embd. Mila. 1 yr. 190 ds. from 27 Jan. 00; for promotion from 19 Feb. 00.*	23 July 81	27 July 01	24 May 04	1 Sep. 15	. . . .	28 Punjabis.
	Essex. R.	**Abbay**, Bryan Norman (51, 52, 66a, 71, 72a)	6 June 81	21 Feb 00	5 Feb. 05	1 Sep. 15	. . . .	27 Lt. Cavalry.
520	A. S. C.	**Worgan**, Rivers Berney, D.S.O. (50, 53, 67, 71, 162)	4 Mar. 81	21 Feb. 00	17 Oct. 05	1 Sep. 15	*l.c.* 1 Jan. 18; *Ty. l. c.* 30 Jan. 16; 3 June 16	20 Horse.
	Ches. R.	**Ramsay**, Alexander Fitz Gerald (51, 53, 71, 172)	29 Nov. 80	21 Feb. 00	3 July 04	1 Sep. 15	. . . .	22 Cavalry.
	Durh. L. I.	**Phillips**, Lionel (51, 52, 66, 71, 195)	25 June 80	21 Feb. 00	28 June 05	1 Sep. 15	. . . .	63 Infantry.
	Oxf. L. I.	**Thorne**, Arthur Charles Drought (51, 52, 155)	8 Oct. 81	21 Feb. 00	8 Aug 02	1 Sep. 15	. . . .	103 Lt. Infantry.
	Midd'x R.	**Harcourt**, Cecil Beckham (51, 66a, 169) *Service for pension;—Embd. Mila. from 25 Jan. to 17 Apl. 00; for promotion from 25 Feb. 00.*	25 Feb. 80	18 Apl. 00	10 Jan. 04	1 Sep. 15	. . . .	28 Punjabis.
525	Ches. R.	**Bean**, James Allen (51, 52, 71, 155) *Service for pension;—Embd. Mila. from 22 Jan. 00 to 4 Jan. 01; for promotion from 25 Feb. 00;*	29 Nov. 81	5 Jan. 01	10 Sep. 05	1 Sep. 15	. . . .	101 Genadiers.
	Ches. R.	**Orchard**, Albert John (51, 52, 72, 137) *Service for pension;—Embd. Mila. from 22 Jan. 00 to 4 Jan. 01; for promotion from 25 Feb. 00.*	5 May 81	5 Jan. 01	22 July 10	1 Sep. 15	. . . .	8 Rajputs.
	Berks. R.	**Chamberlayne**, Athelstan (51, 66a, 71) *Forfeited from 27 Jan. to 1 Mar. 09 for promotion.*	16 June 78	27 Jan. 00	2 Aug. 02	1 Sep. 15	. . . .	1 Lancers.
	R. Scots.	**Collingridge**, Herbert Frederic (50, 66a, 156) *Service for pension;—Embd. Mila. from 5 Dec. 99 to 4 Jan. 01; for promotion from 2 Mar. 00.*	24 July 81	5 Jan. 00	19 Dec. 02	1 Sep. 15	. . . .	9 G. R.
	Devon. R.	**Edward-Collins**, Charles Edward (51)	28 May 81	3 Mar. 00	14 Oct. 09	1 Sep. 15	6 June 12; *Ty. l.c.* 24 Sep. 18.	S. and T. Corps.
530	Manch. R.	**Kirkpatrick**, Kenneth Eric	29 July 79	7 Mar. 00	29 Mar. 03	1 Sep. 15	. . . .	3 Brahmans.
	R. Fus.	**Ivens**, Harold Thomason Carew (50, 169)	27 June 80	7 Mar. 00	10 Jan. 07	1 Sep. 15	. . . .	26 Punjabis.
532	R. Ir. Fus.	**Mahon**, Alfred Ernest, D.S.O. (51, 169) *Service for pension;—Embd. Mila. from 6 Dec. 99 to 6 Mar. 00.*	11 Dec. 78	7 Mar. 03	12 Feb. 03	1 Sep. 15	. . . .	55 Rifles.

Majors—*contd.*

No.	Former Corps.	Names.	Date of birth.	Date of first commission.	Date of appointment to Indian Army.	Date of Rank. Substantive.	Date of Rank. Army.	Remarks.
533	K. O. S. B.	**Jameson**, Alexander McCulloch (51, 71, 167) *Service for pension;—Embd. Mila. from 27 Feb. 00 to 4 Jan. 01; for promotion from 9 Mar. 00.*	2 May 81	5 Jan. 01	21 Aug. 05	1 Sep. 15	. . . .	13 Lancers.
	York R.	**Sanderson**, Francis Grey Oke (51, 66*a*, 71, 95, 169) *Service for pension;—Embd. Mila. from 14 Dec. 99 to 17 Apl. 00; for promotion from 16 Mar. 00.*	26 Sep. 7	18 Apl. 00	10 Oct. 03	1 Sep. 15	. . .	126 Infantry.
535	R. A.	**Koebel**, Frederick Ernest, D.S.O. (50, 169, 174, 205*a*) [L]	20 Sep. 81	17 Mar. 00	12 Jan. 04	1 Sep. 15	. . . .	51 Sikhs.
	R. A.	**Martin**, Jasper, M.C. (51, 52)	20 July 79	17 Mar. 00	16 Aug. 04	1 Sep. 15	. . . .	94 Infantry.
	Leic. R.	**Hunt**, Rochfort Vere (51, 72, 137) *Service for pension;—Embd. Mila. & Attd. Reglrs. from 5 Dec. 99 to 10 May 00; for promotion from 17 Mar. 00.*	11 May 80	11 May 00	2 Dec. 04	1 Sep. 15	. . . .	123 Rifles.
	Devon. R.	**Pearse**, John Cecil (50, 53, 57, 71, 72*b*, 137) *Service for pension;—Embd. Mila. & Attd. Reglrs. from 4 Dec. 99 to 3 Apl. 00; for promotion from 24 Mar. 00.*	13 Mar. 80	4 Apl. 00	5 Mar. 04	1 Sep. 15	. . . .	4 Cavalry.
	R. Dub. Fus.	**Marsh**, Roland Henry (50, 71, 167)	25 Mar. 79	24 Mar. 00	30 Mar. 03	1 Sep. 15	. . . .	18 Lancers.
540	Dorset. R.	**Fell**, Ernes, Arthur Hunter (51, 162, 167) *Service for pension;— Embd. Mila. & Attd. Reglrs. from 19 Nov. 99 to 3 May 01; for promotion from 24 Mar. 00.*	9 Aug. 80	4 May 01	17 June 04	1 Sep. 15	. . . .	12 Cavalry.
	R. A.	**Smith**, Clement Arthur Johnstone (12, 51, 52, 71) *Forfeited from 17 Mar. to 25 Mar. 09 for promotion.*	1 Feb. 80	17 Mar. 00	24 Jan. 04	1 Sep. 15	. . . .	109 Infantry.
	R. A.	**Egerton**, Samuel William (57, 64, 66*a*, 71, 94, 106, 125*a*, 137, 140, 162, 165, 170)	4 Oct. 77	28 Mar. 00	27 Mar. 02	1 Sep. 15	. . . .	7 Lancers.
	R. A.	*Lawrence, Henry Rundle* (169)	*20 July 78*	*28 Mar. 00*	*13 Apl. 03*	1 Sep. 15	. . . .	*Supy. List.*
	R. War. R.	**Bond**, George Weston, D.S.O. (51, 65*a*, 66, 72, 95, 162, 170) *Service for pension;—In Ranks from 25 Mar. 96 to 10 Apl. 00.*	25 Mar. 76	11 Apl. 00	7 Oct. 09	1 Sep. 15	. . . .	S. and T. Corps.
	Essex R.	**Holderness**, Hardwicke (50, 53, 59, 157)	10 Jan. 79	11 Apl. 00	21 Nov. 01	1 Sep. 15	. . . .	1 G. R.
545	Worc. R.	**Wood**, Henry Oswald Bell (50, 53, 66*a*, 137, 172*a*) (*Actg. Maj. 22 Apl. 16*)	28 Oct. 81	18 Apl. 00	17 Nov. 03	1 Sep. 15	3 June 15	8 Rajputs.
	L'pool. R.	**Gaskell**, Reginald Williamson (51, 52, 66*a*, 71, 72, 172*a*)	29 June 80	18 Apl. 00	4 Oct. 03	1 Sep. 15	. . . .	9 Infantry.
	Derby. R.	**Sargent**, Arthur Edward Every, D.S.O., M.C. (66, 72)	22 May 79	18 Apl. 00	30 Jan. 03	1 Sep. 15	. . . .	S. and T. Corps.
	Lan. Fus.	**Gatherer**, Ronald Gordon Thompson (106, 140, 157)	25 Sep. 79	18 Apl. 00	13 June 04	1 Sep. 15	. . . .	10 G. R.
	R. Muns. Fus.	**Strong**, Reginald Godfrey (51, 52, 72*a*, 137)	16 Nov. 79	18 Apl. 00	2 Dec. 01	1 Sep. 15	. . . .	11 Rajputs.
550	Essex R.	**Raymond**, Evelyn Dalrymple, D.S.O., M.C. (51, 66*a*, 71)	2 May 81	18 Apl. 00	12 Mar. 07	1 Sep. 15	. . . .	30 Lancers.
	R. Fus.	**Carey**, Thomas de Beauvoir (51, 72*a*, 127)	10 Oct. 81	18 Apl. 00	25 Oct. 13	1 Sep. 15	. . . .	119 Infantry.
	North'n R.	**Giffard**, Douglas Henry Randall (50, 53, 57, 72*a*, 157)	7 Mar. 80	18 Apl. 00	17 Apl. 04	1 Sep. 15	. . . .	8 G. R.
	Sco. Rif.	**Cameron**, Charles Shaw (51, 57, 71, 95, 172) *Service for pension;— Embd. Mila. from 13 Dec. 99 to 17 Apl. 00.*	11 May 80	18 Apl. 00	16 Apl. 04	1 Sep. 15	. . . .	23 Cavalry.
	Manch. R.	**Carroll**, Herbert Owen (57)	7 Aug. 79	21 Apl. 00	5 Jan. 03	1 Sep. 15	. . . .	S. and T. Corps.
555	Manch. R.	**Gell**, Philip Francis (51, 71, 105, 113, 137, 169, 181) *Service for pension;—Embd. Mila. from 13 Dec. 99 to 20 Apl. 00.*	6 Mar. 80	21 Apl. 00	12 Apl. 02	1 Sep. 15	. . . .	14 Lancers.

556	Manch. R.	**Ardoino**, Joseph (51, 52, 57, 66a, 167, 172a, 175)	20 May 79	21 Apl. 00	9 June 02	1 Sep. 15	. . . .	74 Punjabis.
	S. Lan. R.	**De Wylly**, Guy George Egerton, D.S.O., *p.s.c.* (59, 72b, 169) . . . *Service for pension;—Tasm. I. Bushmen from 26 Apl. to 4 Dec. 00; for promotion from 26 Apl. 00.*	17 Feb. 80	5 Dec. 00	1 Oct. 02	1 Sep. 15	. . . .	Guides.
	R. A.	**Stewart**, James St. Clair Drysdale (51, 59, 162, 169)	30 Aug. 81	2 May 00	4 Oct. 03	1 Sep. 15	. . . .	37 Lancers.
	R. A.	**Hext**, George Trevor Barkley, D.S.O.	23 Apl. 81	2 May 00	1 Dec. 03	1 Sep. 15	. . . .	S. and T. Corps.
560	R. A.	**Medlicott**, Henry Edward, D.S.O. (51, 137)	24 July 82	2 May 00	1 July 02	1 Sep. 15	{ *Ty. l.c.* 4 Sep. 16 3 June 16 }	3 Horse.
	Bedf. R.	**Barnes**, Edward Christian (51, 52, 72b, 172) . . . . *Service for pension;—In Ranks from 27 Mar. 97 to 4 May 00 (Half to count).*	27 Sep. 78	5 May 00	12 Dec. 07	1 Sep. 15	. . . .	19 Punjabis.
	Manch. R.	**Waters**, Robert Sidney (51, 66a, 72b, 140, 161, 169)	24 Aug. 78	5 May 00	27 Aug. 02	1 Sep. 15	. . . .	40 Pathans.
	Bedf. R.	**Compton**, Clifton Wm. McGrath (50, 59)	20 July 78	5 May 00	24 Sep. 03	1 Sep. 15	. . . .	69 Punjabis.
	N. & D. R.	**Hext**, John Edward (51, 66, 172a)	21 Dec. 79	5 May 00	1 Dec. 03	1 Sep. 15	. . . .	S. and T. Corps.
565	Leic. R.	**Rowan-Robinson**, John Rowan, D.S.O. (51, 66, 71, 72, 72b) . .	25 Jan. 76	5 May 00	30 Sep. 09	1 Sep. 15	*l.c.* 3 June 18	S. and T. Corps.
	R. War. R.	**Lane**, Henry Vials (50, 66a, 137)	7 Sep. 78	5 May 00	10 May 02	1 Sep. 15	. . . .	1 Brahmans.
	Leic. R.	**Smith**, Samuel Henry Parsons (51, 66a, 72)	8 Mar. 78	5 May 00	4 Oct. 09	1 Sep. 15	. . . .	S. and T. Corps.
	R. Fus.	**Adler**, Benjamin Isidore Herbert (51, 52, 66a, 137)	28 July 78	5 May 00	11 Apl. 02	1 Sep. 15	. . . .	113 Infantry.
	E. Surr. R.	**Messenger**, Herbert Thomas Knapp (50, 65, 66, 71)	24 Aug. 75	5 May 00	9 June 11	1 Sep. 15	. . . .	S. and T. Corps.
570	R. Scots.	**Mollison**, Edwin James, C.I.E. (51, 59, 137, 161)	11 Aug. 76	5 May 00	4 Feb. 05	1 Sep. 15	. . . .	125 Rifles.
	R. Berks. R.	**Crosleigh**, Evan Leigh (51, 52, 72b, 162, 167, 170) . . . . . *Service for pension;—Embd. Mila. from 9 Feb. 00 to 18 Oct. 01; for promotion from 8 May 00.*	30 Oct. 80	19 Oct. 01	25 Apl. 03	1 Sep. 15	. . . .	23 Pioneers.
	Devon. R.	**Hext**, Charles Wilfred (51, 65a, 66a, 72b) . . . . *Service for pension;—Embd. Mila. & Attd. Reglrs. from 4 Dec. 99 to 4 Jan. 01; for promotion from 17 May 00.*	14 Nov. 80	5 Jan. 01	18 Aug. 11	1 Sep. 15	. . .	S. and T. Corps.
	N. Staff. R.	**Matson**, John (72) . . . . *Service for pension;—In Ranks—Colonial Contgt. from 19 Oct. 99 to 18 May 00 (half to count).*	14 Apl. 76	19 May 00	10 Jan. 03	1 Sep. 15	*Ty. l. c.* 25 Feb. 18.	Mily. Farms Dept.
	K. O. S. B.	**Douglas**, George Stuart (51, 66a, 72a, 172a)	11 Feb. 79	19 May 00	12 June 03	1 Sep. 15	. . . .	18 Infantry.
575	Midd'x. R.	**Mitchell**, Hay Stewart (50, 52, 167)	5 Feb. 79	19 May 00	8 Feb. 03	1 Sep. 15	. . . .	32 Pioneers.
	Bedf. R.	**Murray**, Howard	20 May 76	19 May 00	21 Sep. 04	1 Sep. 15	*Ty. l. c.* 1 Nov. 17.	Mily. Accts. Dept.
	Bedf. R.	**McCleverty**, Patrick Hope (51, 52, 169)	30 Oct. 80	19 May 00	26 Feb. 02	1 Sep. 15	. . . .	20 Infantry.
	E. Lan. R.	**Hewitt**, Dudley Ridisford, (71, 72b, 172)	30 Mar. 77	19 May 00	29 Mar. 02	1 Sep. 15	. . . .	Army Remt. Dept.
	Leins. R.	**Bredin**, Andrew Noble (51, 172a) . . . . *Service for pension from 4 Nov. 02.*	9 Aug. 75	19 May 00	27 Oct. 11	1 Sep. 15	. . . .	6 G. R.
580	R. Scots.	**Morton-Marshall**, Frank Walter (50, 66a, 66b, 172)	8 July 79	19 May 00	16 Dec. 02	1 Sep. 15	19 May 16	130 Baluchis.
	S. Staff. R.	**Reynolds**, Alban John (51, 57, 72b, 89a, 172) . . . . . . *Service for pension;—In Ranks from 20 Oct. 99 to 18 May 00 (half to count, i.e., from 3 Feb. 00).*	19 Jan. 81	19 May 00	12 Apl. 03	1 Sep. 15	. . . .	37 Lancers.
	W. Rid. R.	**Lewis**, Gilbert (47a, 50, 53, 71, 72b, 172a) . . . . *Service for pension;—In Ranks—South Australian Bushmen from 1 May 00 to 18 May 00 (half to count).*	11 Apl. 79	19 May 00	9 Aug 03	1 Sep. 15	. . .	38 C. I. Horse.
	A. S. C.	**Brown**, William (205a) . . . . *Service for pension from 9 Mar. 04.*	18 Apl. 73	22 May 00	9 Mar. 12	1 Sep. 15	*l. c.*, 1 Jan. 18	S. and T. Corps.
	A. S. C.	**Rennison**, Arthur John, D.S.O. (94)	3 Nov. 76	22 May 00	18 Apl. 09	1 Sep. 15	3 June 16	S. and T. Corps.
585	Manch. R.	**Moran**, John (137)	14 May 80	23 May 00	10 Nov. 02	1 Sep. 15	. . . .	Cant. Magte.'s Dept.
	Worc. R.	*Humphrys, Francis Henry* (161, 166, 169)	*24 Apl. 79*	*23 May 00*	*18 Oct. 02*	*1 Sep. 15*	. . . .	*Supy. List.*

No.	Former Corps.	Names	Date of birth.	Date of first commission.	Date of appointment to Indian Army.	Date of Rank. Substantive.	Date of Rank. Army.	Remarks.
		Majors—*contd.*						
586	Worc. R.	**Sterndale-Bennett,** John (50, 66*a*, 99)	10 Sep. 79	23 May 00	3 July 02	1 Sep. 15		107 Pioneers.
	Worc. R.	**Harrison,** Arthur Pemberton (51, 155)	5 Sep. 81	23 May 00	13 Jan. 04	1 Sep. 15		116 Mahrattas.
	R. War. R.	**Dawson,** Henry Hubert (51)	26 Dec. 77	23 May 00	24 Dec. 02	1 Sep. 15		73 Infantry.
	R. A.	**Stewart,** Harold Souther (51, 66*a*, 172)	5 Sep. 81	23 May 00	2 Aug. 06	1 Sep. 15		17 Cavalry.
590	N. Staff. R.	**Shine,** Eugene Percy Forrest	1 May 75	23 May 00	6 Nov. 03	1 Sep. 15		S. and T. Corps.
	E. York. R.	**Swayne,** Francis Gerald (51, 57, 167)	15 Jan. 76	23 May 00	29 Oct. 12	1 Sep. 15		35 Sikhs.
	S. Staff. R.	*Heale, Robert John Wingfield,* O.B.E. (*162, 165, 169,* **204**)	*24 Sep. 76*	*23 May 00*	*21 Nov. 01*	*1 Sep. 15*		*Supy. List.*
	E. York. R.	**Bodkin,** Leo Francis (51, 52, 137) [L]	13 Oct. 79	23 May 00	7 Feb. 04	1 Sep. 15		113 Infantry.
	R. A.	**Cobban,** Clement Lee (51, 52, 174, 205*a*) [L]	24 Mar. 79	26 May 00	12 Jan. 03	1 Sep. 15		43 Regiment.
	R. A.	*Brett, John Aloysius (12, 66a, 94, 161, 169)*	*20 June 79*	*26 May 00*	*11 Oct. 02*	*1 Sep. 15*		*Supy. List.*
	R. A.	**Mayo,** George Lawrence Lacy	20 Jan. 77	26 May 00	11 June 11	1 Sep. 15		S. and T. Corps.
595	R. Ir. Fus.	**Flood,** John Duncan McIntyre (51, 52, 66*a*, 95, 162, 169) *Service for pension;—Embd. Mila. & Attd. Reglrs. from 8 May 00; for promotion from 26 May 00.*	24 Jan. 81	23 Apl. 02	26 Jan. 04	1 Sep. 15		58 Rifles.
	S. Lan. R.	**Ley,** Arthur Edwin Hale (51, 52, 71) *Service for pension—Rl. Navy from 15 Apl. 99 to 17 May 00.*	18 June 79	13 June 00	8 July 02	1 Sep. 15		20 Horse.
	E. York. R.	**Grant-Smith,** Herbert George Henry (51, 52, 72*b*, 137, 162, 172*a*) *Service for promotion from 13 June 00.*	5 Feb. 78	19 Jan. 01	1 Apl. 03	1 Sep. 15		104 Rifles.
	R. Highrs.	**Stewart,** Arthur Easdale, M.C. (50, 56, 66*b*, 172) *Service for pension;—Attd. Reglrs. from 23 Apl. 00 to 18 June 01; for promotion from 10 May 00 to 18 May 01.*	18 May 78	26 June 01	9 Nov. 02	1 Sep. 15		124 Infantry.
	R. Lan. R.	**Maude,** Eustace Arthur (51, 72*b*, 172) *Service for pension;—Embd. Mila. & Attd. Reglrs. from 18 Feb. 00 to 7 Feb. 02; for promotion from 6 July 00.*	10 July 82	8 Feb. 02	17 Jan. 04	1 Sep. 15		26 Punjabis.
600	R. War. R.	**Norris,** Hubert (50, 193) *Service for pension;—Embd. Mila. from 14 Dec. 99 to 3 June 01; for promotion from 11 Jan. 00 to 1 June 01.*	18 Aug. 80	4 Dec. 01	15 June 05	1 Sep. 15		80 Infantry.
	R. A.	**Muirhead,** John Arthur, D.S.O. *p. s. c.* (65, 66*a*, 98)	24 May 79	18 July 00	9 Apl. 02	1 Sep. 15	*Ty. l. c. 6 June 17*	1 Lancers.
	R. A.	**Ewart,** Robert Frederick, D.S.O.	8 May 75	25 July 00	2 May 10	1 Sep. 15		S. and T. Corps.
	Wilts. R.	**Ommanney,** Arthur Hammond (162, 169) *On pension estab. from 21 Feb. 06 to 12 Oct. 14. Re-instated in the Indian Army with effect from the 13th October 1914*	12 Aug. 71	5 Dec. 91	13 Feb. 93	1 Sep. 15		Attached 88 Infantry.
	Unattd. List	**Blair,** Geoffrey Leigh (51, 102, 107, 170) [L]	29 Sep. 80	28 July 00	6 Oct. 01	1 Sep. 15		36 Sikhs.
605	Unattd. List	**Norris,** Arthur Kenneth (51, 140, 147)	13 Dec. 81	28 July 00	9 Apl. 02	1 Sep. 15		123 Rifles.
	Unattd. List	*Anson, George Frank Wemyss (51, 137, 161)*	*22 Mar. 81*	*28 July 00*	*6 Oct. 01*	*1 Sep. 15*		*Supy. List.*
	Unattd. List	**Latham,** Alan (50, 53, 157) (*Ty. Maj., 2 June 16*)	15 Feb. 81	28 July 00	7 Oct. 01	1 Sep. 15		1 G. R.
	Unattd. List	**Carter,** John Fillis Carre (105, 140, 167, 170)	11 Jan. 82	28 July 00	6 Oct. 01	1 Sep. 15		Civil, Burma.
608	Unattd. List	**Clarke,** Gunning Campbell Charles (66*a*, 137)	1 Sep. 80	28 July 00	6 Oct. 01	1 Sep. 15		S. & T. Corps.
	Unattd. List	*MacDonald, Roderick William,* D.S.O. (*51, 104, 120, 120a, 146, 166, 169*)	*22 Apl. 81*	*28 July 00*	*6 Oct. 01*	*1 Sep. 15*		*Supy. List.*

	Unattd. List	**Robertson**, Robert Arthur Harvey (50, 53, 155, 167)	10 Nov. 80	28 July 00	6 Oct. 01	1 Sep. 15	*Ty. l. c.* 25 Feb. 18	30 Punjabis.
610	Unattd. List	**Hathronthwaite**, James Carnegy (51, 52, 59, 155)	29 Nov. 80	28 July 00	3 Oct. 01	1 Sep. 15	. . . .	121 Pioneers.
	Unattd. List	**Harper**, Alexander Forrest (51, 52, 72a, 167, 172a)	11 Apl. 81	28 July 00	8 Oct. 01	1 Sep. 15	. . . .	84 Punjabis.
	Unattd. List	**Tyrrell**, Cecil Herbert (51, 71, 95, 162, 170)	27 Jan. 82	28 July 00	10 Oct. 01	1 Sep. 15	. . . .	25 Punjabis.
	Unattd. List	**Airy**, George (66a, 71, 162, 172)	25 May 81	28 July 00	7 Oct. 01	1 Sep. 15	*Ty. l. c.* 25 Feb. 18	Mily. Farms Dept.
	Unattd. List	**LeMessurier**, George Darel Senhouse (51, 57, 167)	29 Oct. 81	28 July 00	11 Oct. 01	1 Sep. 15	. . . .	85 Infantry.
615	Unattd. List	**Ashfield**, Percy (162, 170)	19 June 82	28 July 00	8 Oct. 01	1 Sep. 15	. . . .	Mily. Accts. Dept.
	Unattd. List	**Patrickson**, Dudley Mostyn (51, 52, 66, 162, 195)	13 Oct. 81	28 July 00	10 Oct. 01	1 Sep. 15	. . . .	86 Infantry.
	Unattd. List	**Turnbull**, Bruce (51, 52, 66a, 167)	4 Nov 80	28 July 00	9 Oct. 01	1 Sep. 15	*Ty. l. c.* 12 July 17.	23 Pioneers.
	Unattd. List	**Green-Armytage**, Paul Dayrell (51, 52, 65, 71, 155) [L] *Service for pension from 1 Oct. 99; for promotion from 28 July 00*	28 Mar. 81	1 Sep. 99	1 Oct. 05	1 Sep. 15	. . . .	117 Mahrattas.
	Unattd. List	**Morgan**, Godfrey Howard *Service for pension from 1 Oct. 99; for promotion from 28 July 00.*	26 May 81	1 Sep. 99	4 Oct. 05	1 Sep. 15	. . . .	Mily. Accts. Dept.
620	Unattd. List	**Salkeld**, Harold Yorke, D.S.O. (30, 51, 52, 57, 66a, 71) *Service for promotion from 28 July 00.*	22 Oct. 81	1 Sep. 99	28 Sep. 05	1 Sep. 15	. . . .	2 Lancers.
	Unattd. List	*Ogilvie, George Drummond (161, 169)*	*18 Feb. 82*	*28 July 00*	*10 Oct. 01*	*1 Sep 15*	. . . .	*Supy. List.*
	Unattd. List	**Smith**, Colin Walter Johnstone, D.S.O. (51, 66a, 167)	15 Dec. 81	28 July 00	9 Oct. 01	1 Sep. 15	. . . .	29 Punjabis.
	Unattd. List	**Garstin**, William Arthur MacDonell (157, 161)	14 May 82	28 July 00	7 Oct. 01	1 Sep. 15	. . . .	Poll. employ.
	Unattd. List	**Vanrenen**, Deneys Henry (71) [205a]	12 Sep. 82	28 July 00	21 Oct. 01	1 Sep. 15	. . . .	Army Remt. Dept.
	Unattd. List	**Carruthers**, John Harvey DeWiederhold, M.C., *p.s.c.* (50, 157, 162, 169,	2 Oct. 82	28 July 00	12 Oct. 01	1 Sep. 15	. . . .	39 Garhwal Rifles.
	Unattd. List	*Field, Donald Moyle (66a, 161, 165, 169, 205a)*	*19 Nov. 81*	*28 July 00*	*10 Oct. 01*	*1 Sep. 15*	. . . .	*Supy. List.*
625	Unattd. List	**Crosthwait**, Leland George (51, 59, 137)	29 Aug. 81	28 July 00	18 Oct. 01	1 Sep. 15	. . . .	Survey Dept.
	Unattd. List	*Bailey, Frederick Marshman,* C.I.E. *(72b, 162, 199)*	*3 Feb. 82*	*28 July 00*	*26 Oct. 01*	*1 Sep. 15*	*Ty. l. c.,* 1 Apl. 18.	*Supy. List.*
	Unattd. List	**Rind**, George Burnet Abercrombie, *p.s.c.* (51, 66a, 71, 172)	30 Aug. 80	28 July 00	7 Oct. 01	1 Sep. 15	*Ty. l. c.* 8 Feb. 17 *l.c.* 19 Dec. 18.	124 Infantry.
	Unattd. List	**Ebden**, Francis Thomas Powney (50, 53, 65, 71, 137, 161, 195)	15 Jan. 82	28 July 00	7 Oct. 01	1 Sep. 15	. . . .	73 Infantry.
	Unattd. List	**Cunningham**, Cuthbert Colpoys, *p.s.c.* (50, 53, 71, 137)	13 Oct. 81	28 July 00	11 Oct. 01	1 Sep. 15	3 June 16	12 Pioneers.
	Unattd. List	**Pearse**, Godfrey (71)	28 Oct. 81	28 July 00	10 Oct. 01	1 Sep. 15	. . . .	S. and T. Corps.
630	Unattd. List	**Hewett**, Murray Selwood, D.S.O. (50, 137)	22 Nov. 81	28 July 00	11 Oct. 01	1 Sep. 15	. . . .	S. and T. Corps.
	Unattd. List	**Sandiford**, Victor Valentine Vincent (50, 167)	7 Apl. 81	28 July 00	7 Oct. 01	1 Sep. 15	. . . .	62 Punjabis.
	Unattd. List	**Morgan**, Mountiford Hickman Llewellyn, D.S.O. (50, 57, 165, 175) [L]	18 Nov. 81	28 July 00	10 Feb. 03	1 Sep. 15	. . . .	62 Punjabis.
	Unattd. List	**Coates**, Victor (51, 71, 167)	14 Mar. 82	28 July 00	5 Oct. 01	1 Sep. 15	. . . .	87 Punjabis.
	Wilts R.	**Murray**, Harold Ashe (51, 52, 66a, 165, 169)	26 Mar. 81	4 Aug. 00	12 Mar. 05	1 Sep. 15	. . . .	30 Punjabis. [Dept.
635	E. Kent R.	**Woodward**, Charles Palgrave (74, 162, 167)	17 Apl. 81	11 Aug. 00	10 Dec. 02	1 Sep. 15	. . . .	Cant. Magte.'s
	Ches. R.	**Byrne**, Edward Gavin Johnstone (50, 52, 60, 71, 137)	11 May 81	11 Aug. 00	8 Apl. 03	1 Sep. 15	3 June 15	104 Rifles. Dept.
	Wilts. R.	**Anderson**, Kenneth Edward (51, 162, 169)	3 May 81	11 Aug. 00	13 Sep. 02	1 Sep. 15	. . . .	Cant. Magte.'s
	E. Kent R.	**Oakes**, Thomas Trevenen (51, 52, 66a, 72a)	13 Aug. 80	11 Aug. 00	22 Dec. 05	1 Sep. 15	. . . .	102 Grenadiers.
	Suff. R.	**Seaton**, Francis Daniel Ross (51, 52, 66a, 137, 172a)	29 Mar. 81	11 Aug. 00	25 Aug. 03	1 Sep. 15	. . . .	112 Infantry.
640	R. Lanc. R.	**Cargill**, Ronald Jasper (51, 66a, 72a, 172)	27 Apl. 81	11 Aug. 00	24 Aug. 03	1 Sep. 15	. . . .	26 Punjabis.
	Shrops. L. I.	**Johnston**, Percy Douglas Campbell, D.S.O. (51, 66a, 72, 137, 161)	8 July 82	11 Aug. 00	23 Jan. 05	1 Sep. 15	. . . .	8 Rajputs.
	Hamps. R.	**Macrae**, John Cecil, D.S.O., *p.s.c.* (66a, 72a, 169)	1 Sep. 81	11 Aug. 00	16 Sep. 04	1 Sep. 15	*Ty. l. c.* 8 Sep. 16	19 Punjabis.
	York. R.	**Whitaker**, Charles Lionel David Herbert (51)	10 Mar. 81	11 Aug. 00	29 May 03	1 Sep. 15	. . . .	18 Infantry.
644	E. Surr. R.	**Ross**, Ronald Campbell, D.S.O. (51, 71, 72b, 137, 161, 165, 204)	28 June 82	11 Aug. 00	3 June 03	1 Sep. 15	*Ty. l. c.* 7 Feb. 17 *l.c.* 19 Dec. 18.	6 Lt. Infantry.

Majors—*contd.*

No.	Former Corps.	Names.	Date of birth.	Date of first commission.	Date of appointment to Indian Army.	Date of rank. Substantive.	Date of rank. Army.	Remarks.
645	R. Suss. R.	**Howell**, Christopher Horace (45a, 50, 59, 71, 167)	22 Oct. 81	11 Aug. 00	20 Feb. 02	1 Sep. 15	. . . .	18 Lancers.
	Som. L. I.	**Alexander**, Heber Maitland, D.S.O. (50, 53, 65a, 66, 71, 162, 170)(e)	1 Feb. 81	11 Aug. 00	19 June 11	1 Sep. 15	*Ty.' l. c.* 25 Feb. 18	S. and T. Corps.
	Welsh R.	**Dickinson**, Reginald Charles (51, 52, 72, 95)	31 Dec. 81	11 Aug. 00	22 Jan. 13	1 Sep. 15	. . . .	28 Lt. Cavalry,
	20 Huss.	**Underwood**, George Frederick (51, 167)	23 Mar. 81	11 Aug. 00	22 Feb. 02	1 Sep. 15	. . . .	15 Lancers.
	S. Lan. R.	**Johnston**, William, M.C. (51, 52, 72a, 147, 157)	29 Sep. 81	11 Aug. 00	19 Feb. 02	1 Sep. 15	. . . .	7 G. R.
650	Midd'x. R.	**Pulley**, Harry Cuthbert (51, 71, 88, 136, 153, 161, 165, 169, 204) [L]	3 Nov. 82	11 Aug. 00	8 Feb. 02	1 Sep. 15	. . . .	12 Pioneers.
	Bord. R.	**Orton**, Sidney Bernard, *p.s.c.* (51, 59, 95, 157, 169)	21 Apl. 81	11 Aug. 00	25 Feb. 02	1 Sep. 15	*l.c.* 3 June 18	39 Garhwal Rifles.
	R. Sco. Fus.	**Willoughby**, Douglas Vere, D.S.O. (50, 57, 137, 162)	6 Feb. 82	11 Aug. 00	19 Dec. 02	1 Sep. 15	. . . .	1 Brahmans.
	W. I. R.	**Weekes**, Harold Ernest (50, 53, 71, 156) *Service for pension;—In Ranks from 1 Jan. to 17 Aug. 00 (half to count).*	27 Sep. 80	18 Aug. 00	22 May 04	1 Sep. 15	. . . .	10 G. R.
	R. A.	**Browning**, Edward Richard Lyndale (51, 137, 174) [L]	27 Aug. 81	18 Aug. 00	8 Apl. 03	1 Sep. 15	. . . .	1 Brahmans.
655	R. A.	**Underhill**, Guilford Edgar Charlton, *p.s.c.* (50, 52, 71, 165, 172a)	2 Feb. 82	18 Aug. 00	18 Oct. 03	1 Sep. 15	. . . .	62 Punjabis.
	R. A.	**Cotter**, Vere Rogerson (12, 162, 170)	18 Feb. 82	18 Aug. 00	29 Feb. 04	1 Sep. 15	. . . .	Survey Dept.
	R. A.	**Wace**, Sidney Lawrie (50, 66, 162, 169)	1 Mar. 82	18 Aug. 00	12 Oct. 03	1 Sep. 15	. . . .	S. and T. Corps.
	R. A.	**Maitland**, George Ramsay, *p.s.c.* (51, 137, 162, 169)	20 Dec. 82	18 Aug. 00	25 Apl. 05	1 Sep. 15	. . . .	14 Lancers.
	Midd'x. R.	**Watney**, Charles Wanford (50, 53, 65a, 66, 71, 72b, 167) *Service for pension; —In Ranks for 2 yrs. 188 ds. (half to count).*	22 Dec. 80	29 Aug. 00	23 July 04	1 Sep. 15	. . . .	S. and T. Corps.
660	R. Dub. Fus.	**Molony**, Otho Hugh Chartres (72b) *Forfeited from 18 Apl. to 31 Aug. 09, for promotion.*	28 Aug. 80	18 Apl. 00	21 Nov. 05	1 Sep. 15	. . . .	94 Infantry.
	R. A.	**Clark**, Arthur Sydney (65a, 66, 167) *Service for pension;—Embd. Mila. 124 ds.*	25 Jan. 79	8 Sep. 00	27 Jan. 03	8 Sep. 15	. . . .	S. and T. Corps.
	R. War R.	**Jones**, Leopold Aloysius Matthew (94, 162, 169) *Service for pension;—Embd. Mila. & Attd. Reglrs. from 3 May 00 to 17 Jan. 01 and 4 Apl. to 3 May 01; for promotion from 26 May to 21 Nov. 00 and from 6 Apl. 01.*	30 Mar. 79	4 May 01	4 Sep. 02	8 Oct. 15	. . . .	S. and T. Corps.
	N. Staff R.	**Williams**, Leslie Gwatkin, D.S.O. (50, 52, 71, 72b) *Service for pension;—In Ranks from 19 Feb. to 16 Oct. 00 (half to count).*	12 Oct. 78	17 Oct. 00	14 Apl. 02	17 Oct. 15	*l. c.* 3 June 17.	5 Cavalry.
	Leic. R.	**Sealy**, Henry George (51, 169) *Service for pension;—Embd. Mila. from 8 May to 16 Oct. 00.*	30 June 79	17 Oct. 00	21 Mar. 03	17 Oct. 15	. . . .	130 Baluchis.
665	Norf. R.	**Tuck**, Charles Harold Amys (51, 52, 66a, 71, 157) *Service for pension;—Embd. Mila. from 20 July 00 to 18 Oct. 01; for promotion from 17 Oct. 00.*	6 Dec. 80	19 Oct. 01	20 Mar. 05	17 Oct. 15	. . . .	3 G. R.

666	R. War R.	**Redman**, Harry Eastland (51, 57, 72*b*) . . . . *Service for pension;—Embd. Mila. from 14 Dec. 99 to 4 Jan. 01; for promotion from 11 Jan. 00; forfeited from 11 Jan. to 19 Oct. 09, for promotion.*	9 Feb. 80	5 Jan. 01	15 May 05	20 Oct. 15	. . . .	99 Infantry.
	Manch. R.	**Hastings**, Geoffrey Francis (51, 172) . . . . *Service for pension;—Embd. Mila. from 17 Oct. 00 to 18 Oct. 01; for promotion from 17 Nov. 00.*	17 May 82	19 Oct. 01	14 Aug. 05	17 Nov. 15	. . . .	22 Cavalry.
	R. A.	**Stuart**, Donald Neil (57, 104) . . . . *Service for pension;—Embd. Mila. from 8 May to 3 Oct. 00.*	8 Oct. 81	19 Dec. 00	12 Mar 04	19 Dec. 15	. . . .	Civil employ.
	R. A.	**Hackett**, Thomas Dalby Hutchison (71, 72*b*, 104, 137, 172*a*)	19 Sep. 81	19 Dec. 00	17 Sep 07	19 Dec. 15	. . . .	Civil, Burma.
670	R. A.	**Taylor**, Norman Callender, D.S.O. (65, 71, 167, 174, 205*a*)	2 Dec. 82	19 Dec. 00	3 July 05	19 Dec. 15	. . . .	92 Punjabis.
	W. I. R.	**Sneringham**, Arthur Thos., D.S.O. (51, 52, 66, 89, 95, 155, 161, 169, 204) [L] *Service for pension. Embd. Mila. and Attd. Reglrs. from 1 May 00 to 4 Oct. 02; for promotion from 25 Aug. 00 to 3 Oct. 02.*	19 Aug. 81	28 Jan. 03	11 Aug. 05	†19 Dec. 15	. . . .	121 Pioneers.
	R. A.	*Weir, Jas. Leslie Rose, (71, 94, 162, 170, 199) [l]*	*29 Jan. 83*	*21 Dec. 00*	*16 Aug. 04*	*21 Dec. 15*	. . . .	*Supy. List.*
	R. A.	**Mackintosh**, Gordon Hathaway Teesdale (51, 59, 65*a*, 66, 66*a*, 66*b*, 95, 162)	22 Apl. 83	21 Dec. 00	21 Apl. 05	21 Dec. 15	. . . .	5 Cavalry.
	R. A.	**Christie**, Henry Geo. Fredk. (67*a*, 137)	22 Aug. 83	21 Dec. 00	21 Dec. 03	21 Dec. 15	. . . .	S. and T. Corps.
	19 Hrs.	**Mansfield**, Charles Bertram Heyworth (51, 52, 66*a*, 71)	24 Feb. 79	29 Dec. 00	6 May 02	29 Dec. 15	. . . .	8 Cavalry.
675	Leins. R.	**L'Estrange**, Edward Bomford (51, 52, 59, 71, 195) . . . . . . *Service for pension;—Embd. Mila. & Attd. Reglrs. from 23 Jan. 00 to 4 Jan. 01; for promotion from 29 June to 4 Nov. 00, forfeited from 29 Aug. to 30 Dec. 09, for promotion.*	20 Feb. 82	5 Jan. 01	5 Aug. 05	31 Dec. 15	. . . .	79 Infantry.
	Worc. R.	**Collins**, Lionel Peter, D.S.O. (51, 165) . . . . *Service for pension;—Embd. Mila. from 8 May to 19 Oct. 00.*	27 Nov. 78	5 Jan. 01	14 Nov. 02	5 Jan. 16	3 June 16	4 G. R.
	Midd'x. R.	**Murray**, Graham Bell (162) . . . . *Service for pension;—Embd. Mila. from 5 Dec. 99 to 4 Dec. 00.*	29 July 81	5 Jan. 01	31 Oct. 02	5 Jan. 16	. . . .	S. and T. Corps.
	Suff. R.	**Moore**, Herbert Durie, D.S.O. (51, 52, 59, 172) . . . . *Service for pension;—Embd. Mila. from 9 May 00 to 4 Jan. 01.*	24 May 82	5 Jan. 01	29 July 03	5 Jan. 16	. . . .	130 Baluchis.
	Dorset. R.	**Mitford**, John Philip, D.S.O. (50, 52, 59, 72*a*, 162, 170) . . . . . *Service for pension;—Embd. Mila. from 29 Jan. to 17 Oct. 00 & Attd. Reglrs. from 18 Oct. to 31 Dec. 00.*	12 June 80	5 Jan. 01	15 Dec. 03	5 Jan. 16	. . . .	98 Infantry.
680	W. Kent R.	**Tugwell**, Wm. Basil Pope (50, 53, 57, 71, 94, 155, 162) . . . . . . *Service for pension;—Embd. Mila. from 1 May to 11 Oct. 00.*	13 Jan. 81	5 Jan. 01	24 Dec. 03	5 Jan. 16	. . . .	121 Pioneers.
	Conn. Rang.	**Peyton**, Alfred Henry (51, 57, 66, 137, 162, 172*a*)	7 May 82	5 Jan. 01	21 Nov. 02	5 Jan. 16	. . . .	S. and T. Corps.
	Conn. Rang.	**Hall**, Hugh Constable (51, 66, 67, 69, 71) . . . . *Service for pension from 1 May 00.*	13 Oct. 80	5 Jan. 01	23 Apl. 05	5 Jan. 16	. . . .	S. and T. Corps.
683	R. Suss. R.	**Drinan**, Gerald Chas. (51, 52, 66*a*, 66*b*, 72*b*, 137, 170) . . . . *Service for pension;—Embd. Mila. from 12 May to 15 Oct. 00 & 29 Oct. 00 to 4 Jan 01.*	13 June 80	5 Jan. 01	17 Dec. 04	5 Jan. 16	. . . .	48 Pioneers.

† Without pay and allowances prior to 1 Sep. 16

Majors—*contd.*

No.	Former Corps.	Names.	Date of birth.	Date of first commission.	Date of appointment to Indian Army.	Date of rank. Substantive.	Date of rank. Army.	Remarks.
684	S. Lan. R.	**Brander**, Edward Hepburn Clay (51, 71, 72*b*, 137)	19 Aug. 81	8 Jan. 01	19 June 03	8 Jan. 16	. . . .	2 Lt. Infantry.
685	R. War. R.	**Rowlandson**, Herbert Wynyard (51, 57, 167, 172, 175) [L]	19 June 82	8 Jan. 01	9 Sep. 03	8 Jan. 16	. . . .	82 Punjabis.
	R. War. R.	**King**, Jack St. Aubyn (51, 71)	7 Jan. 82	8 Jan. 01	4 June 03	8 Jan 16	. . . .	62 Punjabis.
	Dorset. R.	**Conry**, John deLisle (51, 52, 71, 107)	14 Apl. 82	8 Jan. 01	28 Nov. 03	8 Jan. 16	. . . .	96 Infantry.
	E Surr. R.	**Davis**, Thomas Arnoll (51, 137, 162, 167, 172*a*, 205*a*)	5 Jan. 82	8 Jan. 01	24 Mar. 06	8 Jan. 16	. . . .	8 Rajputs.
	E. Surr. R.	**Mortimer**, Philip (51, 95)	22 Apl. 82	8 Jan. 01	7 May 03	8 Jan. 16	. . . .	S. and T. Corps.
690	Unattd. List	**Simonds**, Maurice Hayes (51, 71, 137)	6 Nov. 82	8 Jan. 01	15 Apl. 02	8 Jan. 16	. . . .	7 Lancers.
	Unattd. List	**Davis**, Charles Thomas (50, 52, 170)	29 Apl. 81	8 Jan. 01	8 Apl. 02	8 Jan. 16	. . . .	107 Pioneers.
	Unattd. List	**Lyell**, Alfred George (50, 52, 66*a*, 157)	20 June 82	8 Jan. 01	8 Apl. 02	8 Jan. 16	. . . .	39 Garhwal Rifles.
	R. Scots.	**Home**, Duncan Charles (51, 167)	23 Oct. 81	8 Jan. 01	4 Apl. 03	8 Jan. 16	*Ty. l. c.* 25 Feb. 18	Mily. Farms Dept.
	Unattd. List	**Hensley**, Frederick King (51, 52, 84, 105, 125*a*, 140, 145, 156, 169)	29 July 81	8 Jan. 01	12 Apl. 02	8 Jan. 16	. . . .	Guides.
695	Unattd. List	**Shakespear**, Henry Talbot (51, 59, 66*a*, 71, 162, 167, 170)	2 May 82	8 Jan. 01	23 Mar. 02	8 Jan. 16	. . . .	12 Cavalry.
	Unattd. List	**Channer**, Bernard Gordon (51, 172)	7 Oct. 81	8 Jan. 01	9 Apl. 02	8 Jan. 16	. . . .	54 Sikhs.
	Unattd. List	*Crosthwaite, John. Symons (94, 162, 169)*	*25 May 82*	*8 Jan. 01*	*12 Apl. 02*	*8 Jan. 16*	. . . .	*Supy. List.*
	Unattd. List	**Clutterbuck**, Cecil Hulton (51, 52, 137, 174) [L]	21 Oct. 81	8 Jan. 01	5 Apl. 02	8 Jan. 16	* *Ty. l. c.* 9 Nov. 17	125 Rifles.
	High. L. I.	**McLeod**, Kenneth (51, 72, 95, 162, 167, 169)	5 Dec. 81	8 Jan. 01	8 Oct. 07	8 Jan. 16	. . . .	31 Punjabis.
	Essex. R.	**Hamilton**, Frederick Arthur (50, 53, 66*a*, 71, 72*a*, 95, 137)	16 Dec. 80	8 Jan. 01	22 Jan. 13	8 Jan. 16	*Ty. l. c.* 1 Apl. 18.	8 Cavalry.
	W. York R.	*Harris, Thomas Guy Marriott (51, 94, 169)*	*7 Mar. 82*	*8 Jan. 01*	*3 May 03*	*8 Jan. 16*	. . . .	*Supy. List.*
700	Bedf. R.	**Molesworth**, Alick Lindsay Mortimer (50, 53, 66*a*, 72, 157)	24 Mar. 81	8 Jan. 01	27 Mar. 05	8 Jan. 16	. . . .	8 G. R.
	Unattd. List	**McCudden**, John Hugh, D.S.O., M.C. (51, 52, 57, 65, 71, 94, 162, 172)	31 Jan. 81	8 Jan. 01	7 Apl. 02	8 Jan. 16	. . . .	21 Cavalry.
	Durh. L. I.	**Hartley**, Alan Fleming, D.S.O. (50, 57, 65, 71, 72*a*, 172)	24 Oct. 82	8 Jan. 01	10 May 05	8 Jan. 16	. . . .	11 Lancers.
	Dorset. R.	**Rendall**, John Charles Shuttleworth (51, 65, 66*a*, 162, 170)	19 Mar. 82	8 Jan. 01	9 Jan. 04	8 Jan. 16	. . . .	S. and T. Corps.
	Unattd. List	**Morice**, Ronald Charles Llewellyn, M.C. (51, 57, 66*a*, 137)	25 May 81	8 Jan. 01	11 Apl. 02	8 Jan. 16	. . . .	37 Dogras.
705	Wilts. R.	**Brett**, Francis Lionel Seymour (50, 57, 66*a*, 157, 172*a*)	11 Mar. 81	8 Jan. 01	23 Sep. 03	8 Jan. 16	. . . .	7 G. R.
	Unattd. List	**Skene**, John Gordon (51, 52, 107, 157)	1 June 82	8 Jan. 01	9 Apl. 02	8 Jan. 16	. . . .	8 G. R.
	Unattd. List	**Wardell**, Charles George Vaughan Morgan (46, 51, 53, 172)	13 Sep. 82	1 Jan. 01	6 Apl. 02	8 Jan. 16	. . . .	21 Punjabis.
	York. R.	**Massy**, Seaton Dunham, D.S.O. (51, 52, 57, 167)	25 Apl. 82	8 Jan. 01	5 Sep. 03	8 Jan. 16	3 June 15	29 Punjabis.
	Unattd. List	**Williams**, Vivian Pericles Barrow (51, 52, 66*a*, 71, 107, 125*a*, 137, 140, 144*b*, 148*b*, 162)	30 Oct. 81	8 Jan 01	15 Apl. 02	8 Jan. 16	*Ty. l. c.* 24 Oct. 17	4 Cavalry.
710	Suff. R.	**Massy**, Francis Seaton (50, 95)	14 July 81	8 Jan. 01	23 Apl. 05	8 Jan. 16	. . . .	4 G. R.
	Bedf. R.	**Smith**, Oswald Fairbanks (50, 71, 172)	1 Apl. 81	8 Jan. 01	30 Sep. 02	8 Jan. 16	. . . .	9 Horse.
	E. Kent. R.	**Harley**, Arthur Bertram, D.S.O. (50, 53, 155)	30 Sep. 81	8 Jan. 01	12 Apl. 06	8 Jan. 16	. . . .	121 Pioneers.
	Norf. R.	**Keppel**, Thos. Robt. Harvey (50, 52, 57, 66*a*, 144*a*, 167, 169)	27 Sep. 81	8 Jan. 01	18 July 05	8 Jan. 16	. . . .	30 Punjabis.
	R. Sco. Fus.	**Kempson**, Edward Chad. Randall (51, 71, 137)	24 Nov. 81	8 Jan. 01	19 Feb. 04	8 Jan. 16	. . . .	1 Brahmans.
715	E. Lan. R.	**Jarrad**, Frederick Launcelot Donne (50, 66, 66*b*, 71, 172)	31 Dec. 80	8 Jan. 01	21 Oct. 04	† 8 Jan. 16	. . . .	126 Infantry.
716	Unattd. List	**Flint**, Alban Lothair Arthur (66, 95, 137, 162, 170)	29 Oct. 81	8 Jan. 01	9 Apl. 02	8 Jan. 16	. . . .	S. and T. Corps.

717	R. Ir. Fus.	**Pott,** Edward Helme, D.S.O. (50, 72)	17 Dec. 80	8 Jan. 01	23 Apl. 05	8 Jan. 16	. . . .	1 Lancers.
	Worc. R.	**Mainwaring-White,** Edmund Perceval (50, 95, 161, 172) *Forfeited from 21 Apl. 09 to 10 Jan. 10, for promotion.*	26 May 78	21 Apl. 00	19 Oct. 02	11 Jan. 16	*Ty. l. c.* 23 Dec. 17.	120 Infantry.
	Leic. R.	**Ferris,** Wm. Angelo Tate (51, 66*a*, 71, 137) *Forfeited from 8 to 21 Jan. 10, for promotion.*	3 Mar. 82	8 Jan. 01	18 Nov. 04	22 Jan. 16	. . . .	33 Lt. Cavalry.
720	R. Dub. Fus.	**Scott,** Henry St. George Stewart, D.S.O. (50, 52, 59, 156) *Service for pension;—Embd. Mila. from 4 May to 18 Oct. 00 & Attd. Reglrs. from 5 Nov. 00 to 25 June 01; for promotion from 6 Feb. 01.*	26 July 80	26 June 01	20 Feb. 03	6 Feb. 16	. . . .	4 G. R.
	A. S. C.	**Aylmer,** Gerald, M.C. *Service for pension from 9 Mar. 04; for promotion from 6 Feb. 01.*	14 Feb. 79	23 Oct. 01	22 Nov. 13	6 Feb. 16	. . . .	S. and T. Corps.
	Unattd. List	**Murphy,** James Lloyd (51, 66*a*, 71, 105) *Forfeited from 17 Jan. 09 to 6 Feb. 10, for promotion.*	18 Nov. 79	17 Jan. 00	7 Apl. 01	7 Feb. 16	. . . .	79 Infantry.
	A. S. C.	**Hammersley-Smith,** Ralph Henry (51, 57, 67, 71, 137) *Service for pension;—In Ranks—Impl. Yeo. from 23 Feb. 00 to 13 Feb. 01 (half to count) & Attd. Reglrs. from 14 Feb. to 22 Oct. 01 for promotion from 14 Feb. 01.*	7 June 80	23 Oct. 01	8 Feb. 05	14 Feb. 16	. . . .	14 Lancers.
	K. O. S. B.	**Cruickshank,** John Eustace (52, 148*b*, 157) *Service for pension;—Embd. Mila. from 28 Nov. 00 to 13 Sep. 01; for promotion from 15 Feb. 01.*	23 Jan. 83	14 Sep. 01	6 Dec. 05	15 Feb. 16	. . . .	2 G. R.
725	Worc. R.	**Lee,** Maurice George (51, 59, 172) *Service for pension;—Embd. Mila. and Attd. Reglrs. from 1 May to 20 Oct. 00, 27 Oct. 00 to 24 Feb. 01, 25 Feb. to 8 Dec. 01, 9 Dec. 01 to 29 Apl. 02; for promotion from 25 Feb. 01.*	12 Nov. 80	30 Apl. 02	19 Mar. 05	25 Feb. 16	. . . .	40 Pathans.
	Middx. R.	**Hill,** Herbert Augustus (51, 162, 172, 195) *Service for pension;—Embd. Mila. from 26 Jan. to 18 Oct. 01; for promotion from 28 Feb. 01. Half pay from 1 Aug. 14 to 31 July 1915, (to count).*	2 Dec. 81	19 Oct. 01	5 Oct. 05	28 Feb. 16	. . . .	
	5 Lanc. R.	**Mowat,** James Scott (51, 59, 66, 72*b*, 137) *Service for pension;—In Ranks, from 10 Jan. 00 to 28 Feb. 01 (half to count)—Impl. Yeo. from 1 Mar. to 25 June 01; for promotion from 1 Mar. 01.*	9 Sep. 80	26 June 01	22 May 03	1 Mar. 16	. . . .	26 Lt. Cavalry.
	Shrops. L. I.	**Mellor,** George Frederick *Service for pension;—In Ranks—Impl. Yeo. from 4 Jan. 00 to 28 Feb. 01 (half to count), Impl. Yeo. (Comm'd.) from 1 Mar. to 13 Sep. 01; for promotion from 1 Mar. 01.*	12 Nov. 79	14 Sep. 01	9 Dec. 03	1 Mar. 16	*Ty. l. c.* 25 Feb. 18	Mily. Farms Dep't.
	N. Lan. R.	**Jackson,** David Robert Henry (51, 72*b*, 157) *Service for pension;—Embd. Mila. and Attd. Reglrs. from 9 July 00 to 29 Apl. 02; for promotion from 2 Mar. 01.*	8 July 83	30 Apl. 02	8 Aug. 04	2 Mar. 16	. . . .	6 G. R.
730	R. Muns. Fus.	**Powell,** Dacre Hamilton, M.C. (51, 52, 66*a*, 72*a*, 155) *Service for pension;—Embd. Mila. from 30 Dec. 00 to 18 Oct. 01; for promotion from 3 Mar. 01.*	21 Aug. 83	19 Oct. 01	26 Sep. 04	8 Jan. 16	*Ty. l.c.* 14 Oct. 17. *l.c.* 3 June 18	105 Lt. Infantry.

* Without the pay of the rank. † Without pay and allowances prior to 1 Sep. 16.

Majors—*contd.*

No.	Former Corps.	Names	Date of birth.	Date of first commission.	Date of appointment to Indian Army.	Date of Rank. Substantive.	Date of Rank. Army.	Remarks.
731	Suff. R.	**Fenton,** Lionel Schofield (51, 52) *Forfeited from 8 Jan. to 9 Mar. 10, for promotion.*	19 Apl. 81	8 Jan. 01	28 Apl. 03	10 Mar. 16	. . . .	113 Infantry.
	5 Lanc.	**Johnstone,** Fitzroy Augustus Beauclerk (51, 52, 71, 155) *Service for pension from 20 Feb. 00; forfeited from 15 Aug. 09 to 22 Mar. 10, for promotion.*	4 July 78	15 Aug. 00	18 Nov. 02	23 Mar. 16	. . . .	101 Grenadiers.
	High. L. I.	**Campbell,** William, D.S.O. (66a, 72b) *Service for pension;—Impl. Yeo. from 24 Mar. to 13 Sep. 01; for promotion from 24 Mar. 01.*	19 Jan. 79	14 Sep. 01	29 Oct. 02	24 Mar. 16	. . . .	S. and T. Corps.
	Unattd. List	**Dela Motte,** Reginald Barker, D.S.O. (51, 66, 67, 71, 155) *Service for pension;—Embd. Mila. from 2 May to 18 Oct. 00, Attd. Reglrs. from 12 Nov. 00 to 10 Mar. 01 & Impl. Yeo. from 11 Mar. 01 to 15 Aug. 02, i.e., from 26 May 00; for promotion from 28 Mar. 01.*	19 Aug. 78	16 Aug. 02	31 Dec. 05	28 Mar. 16	. . . .	S. and T. Corps.
735	Devon. R.	**Mills,** Arthur Mordaunt, D.S.O. (50, 66a, 71, 72b, 94, 167) *Service for pension;—Embd. Mila. from 24 May 00 to 3 May 01; for promotion from 29 Mar. 01.*	13 Aug. 79	4 May 01	10 Oct. 02	29 Mar. 16	*Ty. l. c.* 20 *Apl.* 16.	18 Lancers.
	Leic. R.	**Fennell,** Alfred Ernest Slater (51, 52, 105, 140, 148b, 195) *Forfeited from 8 Jan. to 30 Mar. 10 for promotion.*	25 Apl. 81	8 Jan. 01	24 Nov. 02	31 Mar. 16	. . . .	81 Pioneers.
	Som. L. I.	**Nisbet,** Thomas, D.S.O. (51, 52, 66a, 71, 72b, 94, 162, 167, 172a) *Service for pension;—In Ranks from 28 Jan. to 31 Mar. 01 (half to count) and Impl. Yeo. from 1 Apl. to 13 Sep. 01; for promotion from 1 Apl. 01.*	1 Dec. 82	14 Sep. 01	16 Oct. 03	1 Apl. 16	*Ty. l. c.* 17 *Aug.* 17. *l. c.* 3 June 18.	28 Lt. Cavalry.
738	W. Rid. R.	**Goût,** Percival James, M.C. (66a, 72b) *Service for pension;—Embd. Mila. from 15 May to 17 Oct. 00, Attd. Reglrs. from 24 Oct. 00 to 29 Mar. 01 and Impl. Yeo. from 30 Mar. to 13 Sep. 01; for promotion from 6 Apl. 01.*	26 Jan. 80	14 Sep. 01	28 Aug. 04	6 Apl. 16	. . . .	94 Infantry.

739	High. L. I.	**Stevenson-Hamilton,** Olmar Charles John (51, 65*a*, 66 72*b*, 162)	15 Sep. 75	24 July 01	7 Nov. 13	15 Apl. 16	. . . .	S. and T. Corps.
		Service for promotion from 15 Apl. 01.						
740	R. Scots.	**Rombulow-Pearse,** Aubrey Bewicke (51, 59, 71, 157)	20 Mar. 82	30 Apl. 02	17 Aug. 07	16 Apl. 16	. . . .	6 G. R.
		Service for pension;—Embd. Mila. from 3 May 00 to 29 Apl. 02; for promotion from 16 Apl. 01.						
	A. S. C.	**Parkin,** John Francis, D.S.O. 51, 52, 66*a*, 71, 137, 162)	29 Apl. 82	1 May 01	9 Mar. 04	1 May 16	. . . .	113 Infantry.
	A. S. C.	**Lloyd,** Charles Robert, D.S.O. (172*a*)	7 Oct. 82	1 May 01	22 Jan. 09	1 May 16	*l.c.* 3 June 18	S. and T Corps.
	Midd'x. R.	**Wright,** George Darling (51, 52, 71)	13 Oct. 79	18 Apl. 00	29 Jan. 04	4 May 16	. . . .	98 Infantry.
		Forfeited from 18 Apl. 09 to 3 May 10 for promotion.						
	Ches. R.	**Pritchard,** Philip Allan Raymond (107, 140, 148*b*, 162, 167, 169)	18 Mar. 78	4 May 01	29 Dec. 02	4 May 16	. . . .	Civil employ.
		Service for pension from 21 Nov. 00.						
745	R. War. R.	**Kirby,** Joseph Thomas, D.S.O. [*l*] (170)	2 Aug. 80	4 May 01	8 Apl. 03	4 May 16	*l.c.* 3 June 18	S. and T. Corps.
		Service for pension;—Embd. Mila. from 28 May 00 to 3 May 01.						
	R. War. R.	**Skinner,** George Summerson (51, 52, 72*a*, 172)	20 May 81	4 May 01	3 Jan. 03	4 May 16	. . .	56 Rifles.
		Service for pension;—Embd. Mila. & Attd. Reglrs. 273 days.						
	R. Innis. Fus.	**Forteath,** Alister Mackenzie (51, 52, 107, 140, 172)	8 Sep. 80	4 May 01	30 Mar. 03	4 May 16	. . . .	36 Horse.
	R. A.	**Willis,** Gerald Charles Wale, D.S.O. (45*a*, 50, 05, 172*a*)	26 Feb. 82	4 May 01	17 Oct. 06	4 May 16	. . . .	31 Lancers.
	R. In. R.	**Pope,** Sydney Buxton, D.S.O., *p.s.c.* (50, 53, 59, 66*a*, 71, 72, 172)	9 Feb. 79	4 May 01	8 Apl. 03	4 May 16	. . . .	58 Rifles.
		Service for pension;—Embd. Mila. from 12 July to 4 Dec. 00.						
750	Hamps. R.	**MacMahon,** George Dudley Ruadh (50, 162, 165, 169)	23 Jan. 81	4 May 01	12 Aug. 03	4 May 16	. . . .	124 Infantry.
		Service for pension;—Embd. Mila. from 19 Feb. to 17 Oct. 00 and Attd. Reglrs. from 2 Nov. 00 to 14 Feb. 01.						
	Norf. R.	**Etheridge,** Frank, D.S.O., *p. s. c.* (51, 52, 137, 162, 169, 205*a*)	17 July 80	4 May 01	7 Apl. 03	4 May 16	. . . .	7 Rajputs.
	Derby R.	**Slater,** Alfred Frank Murray (51, 66*a*, 105, 137)	26 Sep. 82	4 May 01	5 Sep. 03	4 May 16	. . . .	Civil employ.
	R. A.	**Jones,** Archibald Nelson Gavin, D.S.O. (51, 52, 64, 66*a*, 127, 169, 205*a*) [*l*]	11 Nov. 82	4 May 01	9 Oct. 03	4 May 16	*Ty. l. c.* 11 July 16.	36 Horse.
		Service for pension;—Embd. Mila. 152 *ds.*						
	Hamps. R.	**MacLeod,** Raoul Donald Carnegy, D.S.O. (51, 53)	16 Feb. 79	24 July 01	24 June 03	4 May 16	. . . .	19 Lancers.
		Service for pension;—In Ranks from 20 Oct. 99 to 9 Nov. 00 (half to count) and Impl. Yeo. from 10 Apl. to 23 July 01; for promotion from 4 May 01.						
755	Devon. R.	**Britton,** John Alfred Hamilton (51, 71, 195)	22 Sep. 82	4 May 01	17 Oct. 06	4 May 16	. . . .	83 Lt. Infantry.
	W. I. R.	**Hughes,** Gerald Rickards (51, 52, 57)	28 Feb. 82	8 May 01	28 Aug. 04	8 May 16	. . . .	43 Regiment.
	Midd'x R.	**Passy,** Frederic Harry Burn (51, 57)	29 Jan. 83	8 May 01	30 Sep. 03	8 May 16	. . . .	Army Remt. Dept.
	Midd'x R.	**Griffith,** Ralph Edwin Hotchkin, C.I.E. (51, 162, 169)	4 Mar. 82	8 May 01	1 Apl. 03	8 May 16	. . . .	Poll. employ.
	Midd'x R.	**Cloete,** Harry Durie, M.C. (51, 57, 71, 105, 125*a*, 167)	13 Jan. 82	8 May 01	27 Sep. 03	8 May 16	. . . .	92 Punjabis.
760	Midd'x R.	**Thompson,** Geoffrey Norton (51, 52, 71)	3 July 81	8 May 01	21 Apl. 04	8 May 16	. . . .	43 Regiment.
	Unattd. List	**Wilson,** Thomas Arthur Atkinson (51, 53, 71, 137, 162)	17 June 82	8 May 01	6 Nov. 02	8 May 16	. . . .	4 Cavalry.
	Unattd. List	**Morris,** George Philip (51, 64, 66*a*, 71)	12 Mar. 82	8 May 01	7 Nov. 02	8 May 16	. . . .	30 Lancers.
	Unattd. List	**Batten,** Henry Hallett (51, 59, 104, 172)	10 Oct. 81	8 May 01	4 Nov. 02	8 May 16	. . . .	Civil Burma.
	Bedf. R.	**Webb,** William Francis Richmond, D.S.O. (50, 53, 165, 169) (*temp. Lt. Col.* 16 May 17.	12 Mar. 83	8 May 01	11 Oct. 02	8 May 16	3 June 16	22 Punjabis.
765	North'n. R.	**Ralph,** Stanley Lush (51, 52, 66*a*, 137, 162, 172)	19 June 82	8 May 01	19 Sep. 07	8 May 16	. . . .	8 Rajputs.
	Unattd. List	**Clarke,** Brinsley Guise Sheridan (51, 52, 66*a*, 157, 161, 167, 169) [*l*]	20 Oct. 81	8 May 01	13 Nov. 02	8 May 16	. . . .	10 G. R.
	Unattd. List	**Cochran,** George Walker, D.S.O. (50, 53, 66*a*, 195)	13 Jan. 82	8 May 01	3 Nov. 02	8 May 16	*l.c.* 3 June 18	81 Pioneers.
	Unattd. List	**MacBrayne,** Robert Jaffray, M.C. (51, 59, 66*a*, 95, 167, 170)	12 Apl. 82	8 May 01	8 Nov. 02	8 May 16	. . . .	15 Sikhs.
	Unattd. List	**Anderson,** Barton Edward, D.S.O., *p. s. c.* (51, 52, 57, 162, 169) [L]	15 Sep. 81	8 May 01	4 Nov. 02	8 May 16	18 Feb. 15	59 Rifles.
770	Som. L. I.	**Robarts,** Charles Stuart Hamilton (51, 52, 71, 162)	6 Feb. 82	8 May 01	7 June 03	8 May 16	. . . .	75 Infantry.

Majors—*contd.*

No.	Former Corps.	Names.	Date of birth.	Date of first commission.	Date of appointment to Indian Army.	Date of rank. Substantive.	Date of rank. Army.	Remarks.
771	Unattd. List	**Brown,** Cecil Arthur (51, 52, 71, 162, 165, 169	4 May 82	8 May 01	6 Nov. 02	8 May 16	. . . .	84 Punjabis.
	E. Surr. R.	**Gould,** George, D.S.O. (51, 66*a*, 137)	20 June 82	8 May 01	2 Oct. 06	8 May 16	. . . .	2 Lancers.
	Midd'x R.	**Anderson,** Robert Heath (51, 167)	4 Aug. 82	8 May 01	18 Nov. 02	8 May 16	. . . .	45 Sikhs.
	Unattd. List	**Baker,** Wilfred Bertram (50, 59, 107, 157, 181)	6 Apl. 82	8 May 01	6 Nov. 02	8 May 16	. . . .	10 G. R.
775	R. War. R.	**Scale,** John Dymoke, D.S.O. (51, 66*a*, 165, 172*a*, 175) [L]	27 Dec. 82	8 May 01	15 June 03	8 May 16	. . . .	87 Punjabis.
	N. Lan. R.	**Warton,** Charles Percival Fenwick (51, 52, 66*a*, 71, 137)	8 Jan. 82	8 May 01	20 Apl. 03	8 May 16	. . . .	102 Grenadiers.
	Unattd. List	**Venning,** Francis Esmond Wingate, D.S.O. (51, 59, 104, 113, 162, 167, 170)	26 Jan. 82	8 May 01	9 Nov. 02	8 May 16	. . . .	31 Punjabis.
	Unattd. List	**Foster,** Rodney (51, 172)	27 May 82	8 May 01	6 Nov. 02	8 May 16	. . . .	Survey Dept.
	Unattd. List	**Cooper** George Chamberlain, M.C. (51, 52, 95, 127, 155, 175) [L]	7 Oct. 81	8 May 01	7 Nov. 02	8 May 16	. . . .	121 Pioneers.
780	4 Huss.	**White,** William Blomfield (50)	13 May 82	8 May 01	9 Mar. 10	8 May 16	*Ty. l. c.* 19 June 18	39 C. I. Horse.
	Unattd. List	**Stuart-Menteth,** Arthur Hastings (51, 52, 66*a*, 161, 167, 172) [*I*].	13 Feb. 82	8 May 01	5 Nov. 02	8 May 16	. . . .	112 Infantry.
	Unattd. List	**Milne,** Thomas (51, 52, 57, 66*a*, 169)	24 Feb. 82	8 May 01	6 Nov. 02	8 May 16	. . . .	55 Rifles.
	Unattd. List	**Thompson,** John Pickering (51, 172)	14 Dec. 82	8 May 01	20 Nov. 02	8 May 16	. . . .	35 Horse.
	N. Staff R.	**Boyce,** Clement James (51, 66, 162, 172*a*)	20 June 82	8 May 01	4 June 03	8 May 16	. . . .	S. and T. Corps.
785	Worc. R.	**Hoare,** Cuthbert Gurney (50, 162)	21 Jan. 83	8 May 01	13 Feb. 03	†8 May 16	*Ty. l. c.* 2 *Apl.* 16.	39 C. I. Horse.
	Unattd. List	**Barnett,** William Henry (51, 66*a*, 107, 118*a*, 167)	2 Aug. 81	8 May 01	26 Nov. 02	8 May 16	. . . .	91 Punjabis.
	Unattd. List	**Kirkwood,** Charles Home Kingston (51, 94, 161, 165, 169, 204) [L]	1 Dec. 82	8 May 01	25 Nov. 02	8 May 16	. . . .	23 Cavalry.
	Ches. R.	**Moir,** John Stewart Sutherland (50, 64, 107, 137, 140, 148*b*)	11 Jan. 82	8 May 01	27 Apl. 03	8 May 16	. . . .	26 Lt. Cavalry.
	Hamps. R.	**Macnabb,** Rawdon James (51, 71, 162)	7 Feb. 83	8 May 01	29 June 03	8 May 16	. . . .	Poll. employ.
790	N. Staff. R.	**Wells,** Frederick Walter Arbuthnot, M.C. (51, 57, 107, 137)	8 Mar. 82	8 May 01	24 Sep. 03	8 May 16	. . . .	6 Lt. Infantry.
	Unattd. List	**Goldie,** Kenneth Oswald (51, 52, 172)	19 Sep. 82	8 May 01	5 Nov. 02	8 May 16	. . . .	10 Lancers.
	N. Lan. R.	**Kunhardt,** Frederick George (51, 52, 71, 107, 167)	29 July 82	8 May 01	9 Nov. 03	8 May 16	. . . .	74 Punjabis.
	Ches. R.	**Carey,** John Edward (50, 52, 65, 72*a*, 172)	2 Aug. 82	8 May 01	31 Mar. 04	8 May 16	. . . .	28 Punjabis.
	Unattd. List	**Connor,** Albemarle Dare (51, 52, 66*a*)	23 Nov. 82	8 May 01	24 Nov. 02	8 May 16	. . . .	42 Regiment.
795	Unattd. List	**Lorimer,** Edward (51, 71, 169)	4 Apl. 82	8 May 01	7 Nov. 02	8 May 16	. . . .	33 Lancers.
	Gord. Highrs	**Dayrell,** Wilfred Stuart, D.S.O. (50, 157, 170)	23 Dec. 82	8 May 01	11 Feb. 04	8 May 16	. . . .	72 Punjabis.
	N. Staff. R.	**Livesay,** John (51, 52, 66*a*, 137)	14 Jan. 83	8 May 01	19 Feb. 04	8 May 16	. . . .	102 Grenadiers.
	Unattd. List	**DeGruchy,** Francis Arthur Labey, *p.s.c.* (51, 52, 161)	14 June 83	8 May 01	18 Nov. 02	8 May 16	. . . .	95 Infantry.
	Essex. R.	**Hanson,** Grenville John George, *p. s. c.* (51, 167, 169)	17 May 82	8 May 01	18 Apl. 04	8 May 16	. . . .	69 Punjabis.
800	Ches. R.	**Wilson,** Roger Cochrane, D.S.O., M.C., *p. s. c.* (51, 52, 155)	26 Dec. 82	8 May 01	19 Apl. 04	8 May 16	*l. c.* 3 June 17.	114 Mahrattas.
	N. Lan. R.	**Yates,** Robert James Burton, D.S.O. (51, 53, 71, 95, 169)	4 Mar. 83	8 May 01	29 June 05	8 May 16	*Ty. m.* 15 *Apl.* 16.	22 Cavalry.
	Lan. Fus.	**Radcliffe-Smith,** Noel Mervyn Radcliffe, M.C. (51, 52, 137, 161, 205*a*)	31 May 82	8 May 01	8 June 04	8 May 16	. . . .	44 Infantry.
	Yorks L. I.	**Daniell,** John Acheson Staines, D.S.O. (51, 52, 94, 167)	7 May 82	8 May 01	22 Dec. 05	8 May 16	. . . .	14 Sikhs.
804	Unattd. List	**Ogle,** Nathaniel, D.S.O., *p. s. c.* (51, 52, 94, 167)	17 Sep. 82	8 May 01	3 Nov. 02	8 May 16	*l.c.* 3 Jnue 18	67 Punjabis.

805	Ches. R.	**Connell-McDowell**, Arnot Edward Joseph (51, 57, 155, 162, 172*a*, 204)	13 Mar. 83	8 May 01	1 Feb. 06	8 May 16	. . . .	Civil employ.
	Unattd. List	**Wylle**, Macleod (51, 52, 72*a*, 157)	26 Sep. 81	8 May 01	8 Nov. 02	8 May 16	. . . .	4 G. R.
	Unattd. List.	**Stockley**, Charles Hugh, M.C. (50, 51, 72*b*, 167, 170)	12 Feb. 82	8 May 01	6 Nov. 02	8 May 16	. . . .	66 Punjabis.
	D. C. L. I.	**Hammond**, Wm. Philip (51)	20 Oct. 82	8 May 01	11 Apl. 05	8 May 16	. . . .	18 Infantry.
	Bedf. R.	**Benson-Cooke**, Philip Geoffrey (50, 66*a*, 72*b*, 155, 162, 205*a*) [*l*]	18 June 82	11 May 01	8 Nov. 09	11 May 16	. . . .	80 Infantry.
810	R. A.	**Hutchison**, John Robert (57, 65, 66, 71) *Service for pension—Embd. Mila. from 4 May to 12 Oct. 00.*	1 Feb. 83	15 May 01	3 Dec. 03	15 May 16	. . . .	38 C. I. Horse.
	R. A.	**Paterson**, Richard Harold O'Donnell (50, 57, 137, 162) *Service for pension;—Embd. Mila. and Attd. Reglrs. from 8 May 00 to 10 July 01; for promotion from 13 July 00 to 8 July 01.*	14 Nov. 80	24 May 02	7 Apl. 06	27 May 16	. . . .	34 Horse.
	R. Fus.	**Wolley**, Hugh Seymour Lamplugh (51, 66*a*, 72*a*, 172) *Service for pension;—Embd. Mila. and Attd. Reglrs. from 24 Apl. to 16 Oct. 00, 18 Dec. 00 to 23 Apl. 01, 6 May 01 to 29 Apl. 02; for promotion from 4 June 01.*	7 July 82	30 Apl. 02	29 Dec. 04	4 June 16	. . . .	56 Rifles.
	E. Lan. R.	**Galbraith**, Edgar David, D.S.O. (51, 52, 169) *Service for pension;—Embd. Mila. from 3 May to 4 Dec. 00.*	4 Feb. 81	19 June 01	9 June 04	19 June 16	1 Jan. 17	55 Rifles.
	S. Staff. R.	**Brock**, Basil deLisle, M.C. (51, 71, 172) *Service for pension;—Embd. Mila. from 3 May to 4 Dec. 00, 6 May 01 to 20 Apl. 02; for promotion from 22 June 01.*	13 Nov. 79	30 Apl. 02	25 Sep. 04	22 June 16	. . . .	126 Infantry.
815	S. W. Bord.	**Praed**, Geoffrey Arthur Mackworth (51, 57, 72*b*) *Service for pension;—In Ranks, 2 yrs. 167 ds. (half to count).*	31 Dec. 78	26 June 01	21 Sep. 03	26 June 16	. . . .	5 Lt. Infantry.
	K. O. S. B.	**Mackenzie**, Herbert John, D.S.O. (51, 52, 66*a*, 72*b*, 169) *Service for pension;—In Ranks—Impl. Yeo. from 1 Mar. 00 to 15 June 01 (half to count).*	19 Sep. 78	26 June 01	10 May 03	26 June 16	. . . .	51 Sikhs.
	R. Scots.	**Bannatyne**, Frederic Cooke (51, 52, 71, 72*a*, 72*b*) *Service for pension;—In Ranks, from 15 Jan. 00 to 11 June 01 (half to count).*	16 May 78	26 June 01	12 May 03	26 June 16	3 June 15	109 Infantry.
	R. W. Fus.	**Frere**, Allan Gray (51, 52, 71) *Service for pension;—In Ranks—3 yrs. 144 ds. (half to count).*	8 Jan. 80	26 June 01	10 Apl. 04	26 June 16	. . . .	86 Infantry.
	S. Staff. R.	**Tyndall**, John Wm. Henry Darby (50, 66*a*, 71, 72*b*) *Service for pension;—In Ranks, Impl. Yeo. from 29 Jan. 00 to 16 May 01 and Attd. Reglrs. from 17 May 01 to 23 July 02; for promotion from 17 May 01 to 23 July 02.*	28 June 79	3 Sep. 02	1 Jan. 06	27 June 16	. . . .	5 Lt. Infantry.
820	N. & D. R.	**St. John**, Oliver (51, 52, 71) *Service for pension;—Embd. Mila. from 14 May to 15 Oct. 00 and Attd. Reglrs. from 27 Oct. 00 to 7 Feb. 02; for promotion from 30 Mar. to 24 Dec. 01; forfeited from 14 May to 28 June 10 for promotion.*	30 July 81	8 Feb. 02	30 June 05	29 June 16	. . . .	94 Infantry.
	Midd'x. R.	**Trafford**, Edward Armstrong (52) *Forfeited from 8 May to 5 July 10 for promotion.*	3 Sep. 82	8 May 01	4 June 04	6 July 16	. . . .	Cant. Magte.'s Deptt.
	E. Lan. R.	**Maclean**, Geo. Gilbert Creswick (50, 53, 72*b*, 137) *Service for pension;—Embd. Mila. & Attd. Reglrs. from 16 Feb. 01 to 22 Apl. 02; for promotion from 10 July 01.*	28 Aug. 80	23 Apl. 02	20 Dec. 04	10 July 16	{ 3 June 16 *Ty. l. c.* 25 Jan. 18 }	104 Rifles.
	R. A.	**Holmes**, Hubert Lee (51, 137, 172*a*)	22 Mar. 83	23 July 01	21 Jan. 09	23 July 16	. . . .	122 Infantry.
	R. A.	**Burnett**, Robert Francis Day, M.C. (50, 53)	1 Nov. 82	23 July 01	3 Oct. 06	23 July 16	. . . .	42 Regiment.
825	R. A.	**Waller**, Robert de Warrenne (51, 52, 155) *Service for pension from 18 May 04; for promotion from 23 July 01.*	28 May 83	23 July 01	22 May 09	23 July 16	. . . .	108 Infantry.

† Without pay and allowances prior to 1 Sep. 16.

No.	Former Corps.	Names.	Date of birth.	Date of first commission.	Date of appointment to Indian Army.	Date of Rank. Substantive.	Date of Rank. Army.	Remarks.
		Majors—*contd.*						
826	R. A.	**Risley**, Crescent Gebhard, D.S.O. (51, 71, 167) (*actg. Maj. 6 Oct. 16.*)	8 Oct. 80	23 July 01	4 Feb. 05	23 July 16	. . . .	18 Lancers.
	Shrops L. I.	**Forbes**, Leonard (50, 59, 169)	13 Jan. 80	24 July 01	4 June 03	24 July 16	. . . .	57 Rifles.
	Linc. R.	**Kirkwood**, William Clarke (51, 58, 66a, 71)	12 Mar. 80	24 July 01	19 Mar. 03	24 July 16	. . . .	97 Infantry.
	Devon. R.	**Twiss**, Edward Kemble, D.S.O. (51, 52, 57, 72b, 95, 137)	9 Nov. 82	24 July 01	19 Apl. 05	24 July 16	3 June 16 *Ty. l.c.* 29 Apl. 18.	10 Jats.
830	Dorset R.	**Smyth**, Chas. Alfred Watson (51, 72, 137) *Service for pension from 12 Nov. 00.*	27 Nov. 80	24 July 01	19 Oct. 04	24 July 16	. . . .	1 Brahmans.
	Manch. R.	**Colan**, Harry Norman, D.S.O. (51, 71, 72b, 167)	15 June 80	27 July 01	3 Mar. 03	27 July 16	. . . .	67 Punjabis.
	R. Ir. R.	**Hogan**, Edward Michael Angelo John (72b, 162) *Service for pension;—In Ranks from 28 Dec. 98 to 26 July 01 (half to count).*	28 Aug. 80	27 July 01	22 Jan. 12	27 July 16	. . . .	S. and T. Corps.
	R. Sco Fus.	*Hodgson, Wm. Hayhurst (50, 57, 98, 137)* *Service for pension from 22 Jan. 00.*	*20 July 80*	*27 July 01*	*21 Jan. 04*	*27 July 16*	. . . .	*Supy. List.*
	R. Sco. Fus.	**Strong**, James Douglas (51, 52, 167) *Service for pension;—Embd. Mila. & Attd. Reglrs. from 8 May to 3 Oct. 00, 19 Dec. 00 to 28 Feb. 01 and 1 Mar. 01 to 26 July 01;*	12 Jan. 81	27 July 01	4 Oct. 04	27 July 16	. . . .	90 Punjabis.
	Manch. R.	**Mackrell**, Ashton (72b, 107) *Service for pension from 26 July 01; forfeited from 24 July to 8 Aug. 10, for promotion.*	17 May 80	24 July 01	16 Mar. 05	9 Aug. 16	. . . .	109 Infantry.
835	Unattd. List	**Robertson**, Charles Arthur (51, 52, 195)	17 July 81	28 Aug. 01	13 Dec. 02	28 Aug. 16	. . . .	61 Pioneers.
	S. W. Bord.	**Prescott**, Henry Cecil (72a, 105, 167) *Service for pension;—Embd. Mila. and Attd. Reglrs. from 16 Apl. 01 to 29 Apl. 02; for promotion from 31 Aug. 01.*	1 Mar. 82	30 Apl. 02	23 Nov. 03	31 Aug. 16	. . . .	Civil, Burma.
	S. Lan. R.	**Inglis**, Alexander Francis (50, 53, 157 172a, 182b)	16 Oct. 81	14 Sep. 01	12 Nov. 02	14 Sep. 16	. . . .	1 G. R.
	R. Ir. R.	**Apthorp**, Shirley East, D.S.O. (51, 52, 71, 72b) *Service for pension;—Embd. Mila., Attd. Reglrs. & Impl. Yeo. from 5 Dec. 99 to 8 Oct. 02; for promotion from 4 May 01 to 8 Sep. 02.*	2 Jan. 82	28 Jan. 03	15 July 05	22 Sep. 16	. . . .	96 Infantry.
	W. Rid. R.	**Skinner**, Thomas Burrell (51, 52, 57, 66a, 71, 162, 195) *Service for pension;—Embd. Mila. and Attd. Reglrs. from 15 Feb. 01 to 8 Oct. 02; for promotion from 1 Jan. 01 to 8 Oct 02.*	5 Feb. 84	4 July 03	11 May 05	26 Sep. 16	*Ty. m.* 17 *Aug. 16.*	64 Pioneers.
840	Norf. R.	**Robertson**, Kenneth (50, 66a, 71, 137) *Forfeited from 8 Jan. to 3 Oct. 10, for promotion.*	26 Oct. 81	8 Jan. 01	22 Sep. 05	4 Oct. 16	. . . .	2 Lancers.
	Y. and L. R.	**Smart**, Reginald Walter Louis DeBernier (51, 52, 71, 72b, 172)	17 Oct. 80	5 Oct. 01	27 Jan. 04	5 Oct. 16	. . . .	21 Cavalry.
	S. W. Bord.	**Wilson**, Wm. Henry Deane (51, 72, 137) *Service for pension;—Embd. Mila. and Attd. Reglrs. from 3 May to 5 Dec. 00 and from 6 Oct. 01; for promotion from 6 Oct. 01.*	21 May 81	23 Apl. 02	14 June 05	6 Oct. 16	. . . .	1 Brahmans.
843	Essex. R.	**Carroll**, Brian Maurice (51, 66a, 71, 167) *Service for pension;—Embd. Mila. from 6 May to 18 Oct. 00 and 6 May 01 to 30 Sep. 02; for promotion from 17 June 01 to 30 Sep. 02*	2 Aug. 81	28 Jan. 03	23 Apl. 04	14 Oct. 16	. . . .	87 Punjabis.

844	R. Lanc. R.	**Temple**, Percy DeLuders (12, 51, 172). *Forfeited from 14 Sep. to 17 Oct. 10 for promotion.*	17 Aug. 81	14 Sep. 01	30 Mar. 07	18 Oct. 16	. . . .	37 Lancers.
845	R. Muns. Fus.	**Rice**, Gerald Dominick, D.S.O. (51, 75) *Service for pension;—Embd. Mila. from 12 June 00 to 8 Oct. 01.*	12 June 83	19 Oct. 01	4 Apl. 03	19 Oct. 16	. . . .	S. and T. Corps.
	Unattd. List	**Maynard**, Francis Herbert, M.C. (50, 52, 57, 137, 175) [*l*].	21 Dec. 81	30 Oct. 01	30 Jan. 03	30 Oct. 16	. . . .	125 Rifles.
	R. A.	**Chanter**, Arthur Douglas (66, 137)	24 June 83	6 Nov. 01	9 Dec. 07	6 Nov. 16	. . . .	S. and T. Corps.
	E. Lan. R.	**Jackson**, Vivian Leonard Frank Postlethwaite (45*a*, 51, 52, 66*a*, 71, 72, 137) *Service for pension;—Embd. Mila. from 5 Sep. 00 to 20 Jan. 01; forfeited from 11 May to 7 Nov. 10 for promotion.*	18 Dec. 80	11 May 01	20 Oct. 03	8 Nov. 16	. . . .	27 Lt. Cavalry.
	W. I. R.	**Hancock**, Eric Sortain (51, 52, 167) *Service for pension;—Embd. Mila. from 3 Apl. 99 to 3 Dec. 01;*	26 Mar. 82	4 Dec. 01	10 Nov. 04	4 Dec. 16	. . . .	90 Punjabis.
850	W. I. R.	**Hutcheson**, Wellesley Cathcart (51, 66*a*, 137) *Service for promotion;—Embd. Mila. from 30 Apl. to 16 Oct. 00 and 2 and 3 Dec. 01.*	8 Apl. 83	4 Dec. 01	26 Aug. 04	4 Dec. 16	. . . .	3 Brahmans.
	W. I. R.	**Greville**, Brooke Southwell Algernon Faulke (51, 52, 137) (*Actg. Maj., 21 Dec. 15*)	2 June 82	4 Dec. 01	11 Oct. 04	4 Dec. 16	. . . .	16 Rajputs.
	Midd'x R.	**Wheatley**, William Prescott Ross, D.S.O. (51, 72*a*, 72*b*)	6 Nov. 78	4 Dec. 01	10 July 11	4 Dec. 16	. . .	S. and T. Corps.
	S. W. Bord.	**Berkeley**, Rupert Edric Gifford (51, 52, 71, 195) *Service for pension;—In Ranks from 1 Mar. 00 to 24 June 01 (half to count).*	29 Mar. 80	4 Dec. 01	14 Sep. 03	4 Dec. 16	. . . .	86 Infantry.
	Shrops. L. I.	**Fleming**, William Ernest, M.C. (51, 52, 167) *Service for pension;—Embd. Mila. from 22 Sep. 00 to 11 July 01.*	12 Aug. 81	4 Dec 01	11 Sep. 03	4 Dec. 16	. . . .	41 Dogras.
855	Hamps. R.	**Morris**, Charles Edward (50, 53, 66*a*, 106, 127, 140, 169, 181)	16 Oct. 81	4 Dec. 01	23 May 03	4 Dec. 16	. . . .	Guides.
	R. Scots.	**Manners**, Richard Henry Hedges (50, 66, 72*a*, 94, 100, 155, 162*a*, 169, 205*a*).	10 July 80	4 Dec. 01	18 July 03	4 Dec. 16	. . . .	106 Pioneers.
	Worc. R.	**Pollock**, Ralph Chas. Geoffrey (51, 52, 57, 169)	16 Aug. 81	4 Dec. 01	2 July 04	4 Dec. 16	. . . .	51 Sikhs.
	Ches. R.	**Beaumont**, Seymour Jas. Gordon (51, 52, 57, 71, 72*b*, 175)	2 Jan. 82	4 Dec. 01	23 Nov. 05	4 Dec. 16	. . . .	101 Grenadiers.
	Norf. R.	**Sutherland**, Robert Orr, D.S.O. (51) *Service for pension;—Embd. Mila. from 1 May 00 to 17 July 01.*	1 Oct. 81	4 Dec. 01	1 Nov. 05	4 Dec. 16	. . . .	14 Lancers.
	A. S. C.	*Gabriel, Cecil Hamilton* (162, 169)	*16 Feb. 79*	*7 Dec. 01*	*10 Feb. 03*	*7 Dec. 16*	. . . .	*Supy. List.*
860	L'pool R.	**Brook**, Thomas Fleetwood *Service for pension;—Embd. Mila. from 2 Dec. 01 to 2 Sep. 02; for promotion from 16 Dec. 01 to 2 Sep. 02.*	22 Nov. 81	3 Sep. 02	14 May 15	16 Dec. 16	. . . .	S. & T. Corps.
	Leic. R.	**Yates**, James Ainsworth, C.I.E., D.S.O. (51, 52, 60, 155) *Service for pension;—Embd. Mila. from 9 Dec. 01 to 2 Sep. 02; for promotion from 21 Dec. 01.*	24 Nov. 83	3 Sep. 02	9 Oct. 07	21 Dec. 16	. . . .	103 Lt. Infantry.
	Y. and L. R.	**Biscoe**, John Vincent Meetkerke (137, 172) *Service for pension;—Embd. Mila. and Attd. Reglrs. from 18 Mar. to 4 Dec. 00, 12 Apl. 01 to 29 Apl. 02; for promotion from 24 Dec. 01.*	20 July 80	30 Apl. 02	8 July 06	24 Dec. 16	. . . .	Cant. Magtes.' Dept.
	S. Staff. R.	**Saxton**, Percy Douglas, M.C. (51, 52, 66*a*, 169) *Service for pension;—Embd. Mila. from 6 May 01 to 19 July 02; for promotion from 17 June 01 to 19 July 02.*	15 Jan. 83	28 Jan. 03	17 Feb. 08	25 Dec. 16	. . . .	20 Infantry.
864	Y. & L. R.	**Plumer**, Chas. Geo. Marshall (51, 52, 72*b*, 193). *Service for pension from 13 July 01 to 7 July 02 (With irreglrs.); for promotion from 9 Oct. 01 to 9 Aug. 02.*	16 Sep. 78	12 Nov. 02	20 Sep. 03	11 Jan. 17	. . . .	61 Pioneers.

No.	Former Corps.	Names.	Date of birth.	Date of first commission.	Date of appointment to Indian Army.	Date of Rank. Substantive.	Date of Rank. Army.	Remarks.
		Majors—*contd.*						
865	W. I. R.	Garratt, Bertram Francis (51, 52, 137, 162)	29 Jan. 81	18 Jan. 02	29 May 04	18 Jan. 17	. . . .	16 Rajputs.
	Unattd. List	Ross, Donald Brackenbury (50, 53, 66a, 169)	22 Aug. 82	18 Jan. 02	9 Apl. 03	18 Jan. 17	. . . .	19 Punjabis.
	Unattd. List	Eadie, John Inglis, D.S.O. (50, 53, 66a, 83, 86, 132, 136, 172a, 202, 204) [L]	6 June 83	18 Jan. 02	6 Apl. 03	18 Jan. 17	. . . .	97 Infantry.
	Unattd. List	Hall, Edward George (50, 53, 59, 71, 155)	13 Sep. 82	18 Jan. 02	23 Mar. 03	18 Jan. 17	. . . .	117 Mahrattas.
	Unattd. List	Elliott, Percy William (50, 53, 161, 165, 169, 205a)	19 June 83	18 Jan. 02	27 Mar. 03	18 Jan. 17	. . . .	20 Infantry.
870	Unattd. List	Coningham, Robert Evelyn, *p.s.c.* (51, 66a, 137, 157)	23 Jan. 83	18 Jan. 02	11 Apl. 03	18 Jan. 17	1 Jan. 16	10 G. R.
	Unattd. List	Beadle, Robert Denis (50, 53, 162, 169)	24 June 82	18 Jan. 02	26 Mar. 03	18 Jan. 17	. . . .	46 Punjabis.
	Unattd. List	Ovens, Thomas Leslie (48, 51, 57, 66a, 137)	29 Oct. 83	18 Jan. 02	11 Apl. 03	18 Jan. 17	. . . .	11 Rajputs.
	Unattd. List	Scott, Raymond Somerville (50, 71, 162, 167, 169)	4 Dec. 82	18 Jan. 02	31 Mar. 03	18 Jan. 17	. . . .	Army Remt. Dept.
	Unattd. List	Chaldecott, Oswald Arthur (50, 59, 94, 100, 162, 169)	9 Oct. 82	18 Jan. 02	11 Apl. 03	18 Jan. 17	. . . .	124 Infantry.
875	Durh. L. I.	Woodhouse, Percy St. John Rance (51, 61, 66, 169)	25 Aug. 83	18 Jan. 02	30 Sep. 06	18 Jan. 17	*l.c.*, 3 June 17	S. and T. Corps.
	Unattd. List	Kerr, Alexander Nairne, M.C. (50, 52, 66, 167, 170)	7 June 82	18 Jan. 02	8 Apl. 03	18 Jan. 17	. . . .	38 Dogras.
	Unattd. List	Watson, Hugh Darvill (51, 162, 169) (L)	16 Nov. 83	18 Jan. 02	25 Mar. 03	18 Jan. 17	. . . .	Mily. Accts. Dept.
	Wilts. R.	Macdonald, William (50, 95, 167, 205a)	25 Nov. 82	18 Jan. 02	26 Oct. 03	18 Jan. 17	. . . .	31 Punjabis.
	Unattd. List	Black, Geo. Cumine Strahan (50, 53, 59, 137)	3 Jan. 82	18 Jan. 02	8 Apl. 03	18 Jan. 17	. . . .	8 Rajputs.
880	Dorset. R.	Wagstaff, Lewis Cecil, C.I.E., (51, 52, 71, 137, 162, 174, 205a) (L)	23 Jan. 82	18 Jan. 02	20 Sep. 03	18 Jan. 17	. . . .	2 Lt. Infantry.
	Shrops. L. I.	Bell, Henry Corbould Forbes (51, 52, 195)	8 Sep. 82	18 Jan. 02	12 Mar. 08	18 Jan. 17	. . . .	73 Infantry.
	Unattd. List	Breithaupt, Edward Avenel (50, 53, 66a, 71, 195)	17 July 83	18 Jan. 02	6 Apl. 03	18 Jan. 17	. . . .	63 Lt. Infantry.
	Unattd. List	Williams, Raymond Carol Bridgewater (51, 52, 57, 167, 170)	15 Oct. 82	18 Jan. 02	8 Apl. 03	18 Jan. 17	. . . .	35 Sikhs.
	Unattd. List	Bunbury, Gerald Bruce St. Pierre (51, 52, 72a, 137)	21 Mar. 83	18 Jan. 02	11 Apl. 03	18 Jan. 17	. . . .	13 Rajputs.
885	Essex R.	Riddell, Archibald, D.S.O. (51, 52, 72, 167)	2 Apl. 82	18 Jan. 02	4 Sep. 07	18 Jan. 17	. . . .	74 Punjabis.
	Y. & L. R.	Thornhill, Cudbert John Massy, D.S.O. (50, 137, 161) (L)	4 Oct. 83	18 Jan. 02	19 Dec. 02	18 Jan. 17	1 Jan. 17 *Ty. m.* 12 Jan. 16. *Ty. l. c.*, 5 Jan. 18. *l.c.*, 3 June 18.	5 Lt. Infantry.
	R. Sco. Fus.	More, James Carmichael, D.S.O. (51, 52, 87, 161, 169, 205a) [L]	13 Feb. 83	18 Jan. 02	28 Jan. 04	18 Jan. 17	. . . .	51 Sikhs.
	North'n. R.	Champion de Crespigny, Henry, M.C. (50, 53, 172)	11 July 82	18 Jan. 02	11 Aug. 07	18 Jan. 17	. . . .	56 Rifles.
	Linc. R.	Scott, Henry Lawrence, D.S.O., M.C., *p.s.c.* (51, 52, 57, 157)	6 Apl. 82	18 Jan. 02	8 July 03	18 Jan. 17	. . . .	1 G. R.
890	E. York. R.	Smith, Yorick Edgar (51, 59, 66, 71, 167)	29 July 82	18 Jan. 02	31 Oct. 03	18 Jan. 17	. . . .	66 Punjabis.
	S. Lan. R.	Whitworth, Thomas Stanley (51, 52, 57, 169)	29 July 82	18 Jan. 02	7 Sep. 03	18 Jan. 17	. . . .	72 Punjabis.
	S. Lan. R.	Maxwell, Percy Alexander (51, 137)	11 May 83	18 Jan. 02	19 Nov. 03	18 Jan. 17	. . . .	3 Brahmans.
	Unattd. List	Molloy, Henry Townsend (51, 52, 66a, 157)	2 Sep. 82	18 Jan. 02	14 Apl. 03	18 Jan. 17	. . . .	5 G. R.
	Linc. R.	Robinson, Ernest Leef (51, 65a, 52, 66, 71, 162, 169, 195) (e)	19 Sep. 83	18 Jan. 02	24 Apl. 03	18 Jan. 17	. . . .	75 Infantry.
895	Midd'x. R.	Lucas, Malcolm Hugh (169)	2 June 82	18 Jan. 02	22 May 03	18 Jan. 17	. . . .	37 Lancers.
	R. Fus.	Carey, Percival Gaussen (50, 66a, 167)	11 Dec. 82	18 Jan. 02	7 Feb. 04	18 Jan. 17	. . . .	31 Punjabis.
	Linc. R.	Clarke, Gerald Anderson (51, 66a, 71, 137, 167)	11 Nov. 83	18 Jan. 02	15 Oct. 03	18 Jan. 17	. . . .	12 Pioneers.
	Unattd. List	Scruby, William Sidney James (51, 52, 57, 125a, 167)	1 Apl. 82	18 Jan. 02	31 Mar. 03	18 Jan. 17	. . . .	89 Punjabis.
	R. War. R.	Haughton, Edward Juxon Henry, D.S.O. (51, 52, 148a)	19 Jan. 83	18 Jan. 02	13 Nov. 03	18 Jan. 17	. . . .	105 Lt. Infantry.
900	York. R.	Curtis, Frederick (51, 52, 71, 169)	3 Dec. 82	18 Jan. 02	11 Nov. 03	18 Jan. 17	. . . .	19 Lancers.
	Midd'x. R.	Jarvis, Cecil, M.C. (51, 52, 71, 72a)	1 Jan. 83	18 Jan. 02	6 Dec. 04	18 Jan. 17	. . . .	20 Horse.
	Unattd. List	*Reilly, Bernard Rawdon (66, 88, 137, 153)*	*5 Mar. 82*	*18 Jan. 02*	*5 Apl. 03*	*18 Jan. 17*	. . . .	*Supy. List.*
902	Unattd. List	Ralston, William Henry, D.S.O., M.C. (51, 52, 124, 125b, 162, 165, 170)	27 Mar. 83	18 Jan. 02	26 Mar. 03	18 Jan. 17	. . . .	47 Sikhs.

903	North'd.Fus.	**Marshall**, Anthony, D.S.O. (45*a*, 51, 71, 167)	4 Apl. 82	18 Jan. 02	10 May 04	18 Jan. 17	. . . .	28 Lt. Cavalry.
	R. Ir. R.	**Roe**, Charles Douglas, D.S.O. (50, 53, 57, 66*a*, 71, 100, 157, 167, 172*a*)	26 Apl. 82	18 Jan. 02	18 Aug. 04	18 Jan. 17	. . . .	4 G. R.
905	York L. I.	**Fox-Male**, Wm. Cecil Edward Trevan, M.C. (51, 106, 125*a*, 167)	15 May 83	18 Jan. 02	5 Mar. 05	18 Jan. 17	. . . .	93 Burma Infy.
	Unattd. List	**Ames**, William Rix (51, 52, 107, 125*a*, 137, 140, 148*b*)	4 Dec. 82	18 Jan. 02	7 Apl. 03	18 Jan. 17	. . . .	4 Rajputs.
	Ches. R.	**McGann**, Henry Herbert (51, 107, 167)	16 Aug. 82	18 Jan. '02	23 Aug. 04	18 Jan. 17	. . . .	69 Punjabis.
	E. Lan. R.	**Hardcastle**, John (51, 66*a*)	1 Aug. 83	18 Jan. 02	17 Sep. 04	18 Jan. 17	. . . .	46 Punjabis.
	Unattd. List	**Massy**, Wm. Ingoldsby Justice (51, 52, 66*a*, 71, 107, 120, 137)	16 Feb. 83	18 Jan. 02	21 Oct 03	18 Jan. 17	. . . .	12 Pioneers.
910	R. Suss. R.	**Atkins**, Geoffrey Willoughby, M.C. (51, 169)	25 Jan. 83	18 Jan. 02	30 Nov. 04	18 Jan. 17	. . . .	25 Punjabis.
	Unattd. List	**Heyland**, George Dominic (51, 66*a*, 71, 107, 155)	30 Nov. 82	18 Jan. 02	6 Apl. 03	18 Jan. 17	. . . .	121 Pioneers.
	Hants. R.	**Renny**, Hugh Blakiston, M.C. (51, 53, 167)	8 Jan. 83	18 Jan. 02	8 Apl. 05	18 Jan. 17	. . . .	23 Pioneers.
	R. Sco. Fus.	**Slater**, Duncan McLauchlan, (51, 52, 72*a*, 137)	21 May 83	18 Jan. 02	9 May 05	18 Jan. 17	. . . .	11 Rajputs.
	Unattd. List	**Whitwell**, Harry William (51, 52, 72*a*, 137)	25 Mar. 82	18 Jan. 02	24 Mar. 03	18 Jan. 17	. . . .	13 Rajputs.
915	Y. & L. R.	**Miller**, Francis (51, 52, 55, 57, 67, 153, 155)	7 Nov. 81	29 Jan. 02	24 Mar. 04	29 Jan. 17	. . . .	108 Infantry.
	Bedf. R.	**Orchard**, Donald Stuart (52, 66, 156) *Service for pension;—Embd. Mila. from 25 May to 18 Oct. 00.*	8 Jan. 83	29 Jan. 02	21 Sep. 04	29 Jan 17	. . . .	8 G. R.
	R. Fus.	**Colenso**, John Eric (50, 53, 66, 66*b*, 71, 72, 95, 107, 120, 157) *Service for pension;—Embd. Mila. from 20 Sep. to 18 Oct. 00 and Attd. Reglrs. from 22 Mar. to 7 June 01.*	28 Nov. 80	29 Jan. 02	6 Mar. 04	29 Jan. 17	. . . .	7 G. R.
	Linc. R.	**Robertson**, Edward Douglas Struan (51, 52, 64, 71, 72*a*, 162, 172) *Service for pension;—Embd. Mila. and Attd. Reglrs. from 6 Mar. to 17 Oct. 00, 1 May 01 to 28 Jan. 02.*	18 Nov. 82	29 Jan. 02	24 July 03	29 Jan. 17	. . .	21 Cavalry.
	Leic. R.	**Hildebrand**, Henry Algernon (51, 65, 66*a*, 71, 72*a*, 137)	12 July 82	29 Jan. 02	12 Oct. 06	29 Jan. 17	. . . .	34 Horse.
920	D. C. L. I.	**Paige**, Cyril Penroses, C.I.E. (50, 57, 65, 170)	7 July 82	29 Jan. 02	30 Sep. 05	29 Jan. 17	. . . .	109 Infantry.
	North'd.Fus.	**FitzGerald**, Chas. Robert Lewis, D.S.O. (51, 95, 161, 172) *Service for pension;—Embd. Mila. from 18 May to 15 Oct. 00 and 21 Feb. 01 to 9 June 02, for promotion from 1 May 01 to 29 Apl. 02.*	17 May 83	28 Jan. 03	14 Dec. 05	29 Jan. 17	. . . .	126 Infantry.
	R. War. R.	**Baille**, Sydney Mitford Hope (51, 52, 66*a*, 71, 72*b*, 167, 172*a*) *Service for pension;—In Ranks from 27 Nov. 99 to 10 Mar. 00, 15 Mar. to 15 Dec. 00 and 29 Dec. 00 to 4 July 01 (half to count).*	14 July 81	8 Feb. 02	11 July 05	8 Feb. 17	. . . .	87 Punjabis.
	2 L. G	**Gregson**, Charles Dalton (51, 52, 57, 71, 162)	22 Feb. 81	15 Feb. 02	13 Apl. 04	15 Feb. 17	. . . .	20 Horse.
	Devon R.	*Neale, Walter Gordon (66a, 38, 161, 169)*	*2 Oct. 80*	*15 Feb. 02*	*1 Feb. 04*	*15 Feb. 17*	. . . .	*Supy. List.*
	7 D. G.	**Potter**, Richard James Keath (51, 52, 71)	25 May 82	20 Mar. 02	26 Aug. 00	20 Mar. 17	. . . .	117 Mahrattas.
925	L'pool. R.	**Wilcox**, Hugh Berkeley Frederick (51, 167) *Service for pension;—In Ranks from 11 Jan. 99 to 11 Feb. 02 (half to count); forfeited from 12 Feb. to 5 Apl. 11 for promotion.*	11 Oct. 79	12 Feb. 02	19 Nov. 03	6 Apl. 17	. . . .	91 Punjabis.
	Bedf. R.	**Ruck**, Oliver Laurence (50, 72*b*, 169)	6 Sep. 81	23 Apl. 02	8 July 03	23 Apl. 17	. . . .	54 Sikhs.
	S. Lan. R.	**Ball**, Lionel Plomer, D.S.O., (50, 57, 72*b*)	22 Jan. 82	23 Apl. 02	2 Mar. 03	23 Apl. 17	. . . .	5 Lt. Infantry.
	R. Innis. Fus.	**Bradley**, Chas. Raymond Strathearn (50, 52, 72, 137)	21 Jan. 82	23 Apl. 02	17 June 05	23 Apl. 17	*Ty. m. 27 Mar. 16.*	4 Cavalry.
	E. York R.	**Walker**, Charles Clarke (71, 94, 162, 167)	20 June 83	7 May 02	6 Mar. 05	7 May 17	. . . .	28 Lt. Cavalry.
930	E. Lan R.	**Hickie**, Cecil Corbett (51, 66*a*, 105, 137) *Service for pension;—Embd. Mila. from 21 Jan. 01 to 6 May 02.*	13 Dec. 83	7 May 02	11 Jan. 04	7 May 17	. . . .	Civil employ.

No.	Former Corps.	Names.	Date of birth.	Date of first commission.	Date of appointment to Indian Army.	Date of Rank. Substantive.	Date of Rank. Army.	Remarks.
		Majors—*contd.*						
931	Dorset. R.	**Edwards**, Ernest (51, 167). *Service for pension;—Embd. Mila. from 10 May 01 to 6 May 02.*	24 Nov. 81	7 May 02	24 June 04	7 May 17	. . . .	29 Punjabis.
	E. Lan. R.	**Scrafton**, Joseph Elwin Bishop (50)	29 June 82	7 May 02	2 Aug. 04	7 May 17	. . . .	Civil employ.
	Midd'x. R.	**Thackwell**, Arthur Charles Austen. M.C. (51, 52, 71, 195) *Service for pension;—Embd. Mila. from 3 May to 5 Dec. 00 and 16 Jan. to 6 Sep. 02; for promotion from 23 Jan. to 6 Sep. 02.*	17 Nov. 82	28 Jan. 03	5 Nov. 04	15 June 17	. . . .	81 Pioneers.
	Manch. R.	**de Burgh**, Eric, D.S.O., *p.s.c.* (51, 52, 57, 66*a*, 71, 72, 172) *Service for pension;—Attd. Reglrs. from 25 Nov. 01 to 16 Sep. 02; for promotion from 7 Feb. to 15 Sep. 02.*	10 May 81	28 Jan 03	28 July 04	21 June 17	3 June 16. *Ty. l. c.* 22 June 17.	9 Horse.
935	Manch. R.	**Graham**, John Scot (51, 155) *Service for pension;—Embd. Mila. & Attd. Reglrs. from 8 May 00 to 9 June 01 and 2 Sep. 01 to 29 Apl. 02; for promotion from 10 Jan. to 7 Apl. 01; forfeited from 1 Feb. to 9 July 11 for promotion.*	24 Aug. 80	30 Apl. 02	24 Dec. 05	10 July 17	. . . .	Mily. Accts. Dept.
	S. Lanc. R.	**Ayscough**, Geoffrey Martin (51) *Service for pension;—Embd. Mila. from 10 Mar. to 6 Oct. 02; for promotion from 27 Mar. to 6 Oct. 02.*	16 Dec. 83	28 Jan. 03	2 Dec. 06	18 July 17	. . . .	116 Mahrattas.
	R. A.	**Digges-La Touche**, Cecil Huntington (71, 106, 125*a*, 127, 162, 167, 172*a*, 175, 188*a*) (*g*)	4 Apl. 83	31 July 02	26 July 05	31 July 17	. . . .	16 Cavalry.
	Dorset. R.	**Ross**, Eglintoune Frederick (51, 52, 72, 72*a*) *Service for pension;—Embd. Mila. from 22 Nov. 00 to 30 June 01; Attd. Reglrs. 1 July 01 to 11 Aug. 02.*	20 June 83	12 Aug. 02	12 Dec. 09	12 Aug. 17	. . . .	43 Infantry.
	Unattd. List	**Walker**, Charles Wm. Garne, D.S.O., *p.s.c.* (51, 57, 137, 162, 170)	2 July 82	27 Aug. 02	20 Dec. 03	27 Aug. 17	. . . .	37 Dogras.
940	Unattd. List	**James**, Frank Harley, M.C. (51, 52, 71, 137)	28 Feb. 83	27 Aug. 02	9 Jan. 04	27 Aug. 17	. . . .	104 Rifles.
	Unattd. List	**Russell**, Guy Hamilton (51, 59, 94, 100, 162, 169)	22 Oct. 82	27 Aug. 02	9 Jan. 04	27 Aug. 17	. . . .	126 Infantry.
	Unattd. List	**Henderson**, George (45*a*, 51, 52, 57, 71)	26 Dec. 82	27 Aug. 02	8 Jan. 04	27 Aug 17	. . . .	38 C. I. Horse.
	Unattd. List	**Carpendale**, Maxwell Montague (51, 52, 162, 169)	12 July 83	27 Aug. 02	9 Jan. 04	27 Aug. 17	. . . .	36 Horse.
	Unattd. List	**Niven**, Douglas Scott (50, 53, 71, 172)	5 May 82	27 Aug. 02	6 Jan. 04	27 Aug. 17	. . . .	25 Cavalry.
945	Unattd. List	**Chamier**, John Adrian, D.S.O. (51, 165, 169)	26 Dec. 83	27 Aug. 02	11 Jan. 04	27 Aug. 17	*Ty. l. c.* 29 *Apl.* 17.	33 Punjabis.
	Unattd. List	**Ford**, Geoffry Noel (50, 52, 59, 127, 155, 162, 170) [L]	25 Dec. 83	27 Aug. 02	10 Jan. 04	27 Aug. 17	. . . .	105 Lt. Infantry.
	Unattd. List	**Hodson**, Vernon Charles Paget (50, 52, 59, 71, 95, 162, 172)	17 May 83	27 Aug. 02	7 Jan. 04	27 Aug. 17	. . . .	10 Lancers.
	Unattd. List	**Murland**, Howard Ferguson (51, 52, 140, 193)	20 Dec. 82	27 Aug. 02	7 Jan. 04	27 Aug. 17	. . . .	64 Pioneers.
	Unattd. List	**Moody**, Francis Hyde, M.C. (45*a*, 50, 71, 72*b*, 95, 167, 169)	21 Feb. 83	27 Aug. 02	18 Jan. 04	27 Aug. 17	*Ty. l. c.* 11 July 18	13 Lancers.
								96 Infantry.
950	Unattd. List	**Coningham**, Meynell Evelyn (51, 52, 57, 66*a*, 162)	19 Mar. 84	27 Aug. 02	15 Jan. 04	27 Aug. 17	. . . .	99 Infantry.
	Unattd. List	**Ward**, Basil Seth (51, 52) [*l*] (*Ty. l. c., 1 Feb. 17*)	1 Mar. 83	27 Aug. 02	15 Jan. 04	27 Aug. 17	1 Jan. 17	*Supy. List.*
	Unattd. List	*Barrett, Cyril Charles Johnson*, C.I.E. (66, 66*a*, 88, 94, 132, 153, 170	*26 May 84*	*27 Aug. 02*	*19 Jan 04*	*27 Aug. 17*	. . . .	
952	Unattd. List	**Anderson**, John Faulkener Henniker, D.S.O. (51, 169) (*Ty. Maj., 12 Jan. 16*)	10 Dec. 83	27 Aug. 02	16 Jan. 04	27 Aug. 17	3 June 17 *Ty. l.c* 25 Feb. 18. *l.c.* 3 June 18.	Army Remt. Dept.

953	Unattd. List	**Haughton,** Henry Lawrence (50, 52, 57, 162, 167, 170, 182*b*)	1 Nov. 83	27 Aug. 02	10 Jan. 04	27 Aug. 17	. . . .	36 Sikhs.
	Unattd. List	**Campbell,** Charles Hay (50, 53, 66*a*, 169)	14 June 83	27 Aug. 02	21 Jan. 04	27 Aug. 17	*Ty. m.* 19 June 16.	Guides.
955	Unattd. List	**Cook,** Walter Kingsmill (51, 52, 65*a*, 66, 125, 172)	12 Oct. 83	27 Aug. 02	20 Jan. 04	27 Aug. 17	. . . .	S. and T. Corps.
	Unattd. List	**Reilly,** Noel Edmund (94)	30 Dec. 83	27 Aug. 02	18 Jan. 04	27 Aug. 17	. . . .	Poll. employ.
	Unattd. List	**Masters,** John (51, 52, 137, 145, 149)	5 Oct. 83	27 Aug. 02	26 Jan. 04	27 Aug. 17	. . . .	16 Rajputs.
	Unattd. List	**Meade,** Walter Lambert (51, 52, 71, 94, 161, 169, 205*a*)	5 Dec. 82	27 Aug. 02	5 Jan. 04	27 Aug. 17	. . . .	95 Infantry.
	Unattd. List	**Alexander,** Christopher, M.C. (51, 52, 66*a*, 195)	24 Dec. 82	27 Aug. 02	15 Jan. 04	27 Aug. 17	. . . .	81 Pioneers.
	Unattd. List	**Oldham,** Joseph Stransham (51, 57, 66*a*, 71, 137)	17 Dec. 82	27 Aug. 02	15 Jan. 04	27 Aug. 17	. . . .	6 Cavalry.
960	Unattd. List	**Shepherd,** Claude Innes, D.S.O. (50, 53, 66*a*, 161, 165, 169, 205*a*)	14 Jan. 84	27 Aug. 02	19 Jan. 04	27 Aug. 17	. . . .	53 Sikhs.
	Unattd. List	**Gwyn,** Aylmer Hamond (137)	13 Feb. 83	27 Aug. 02	28 Jan. 04	27 Aug. 17	. . . .	Survey Dept.
	Unattd. List	**Reid,** Walter Clarke (50, 71, 107, 137)	12 Apl. 81	27 Aug. 02	28 Jan. 04	27 Aug. 17	. . . .	32 Lancers.
	Unattd. List	**Dent,** Wilkinson, D.S.O., *p.s.c.* (50, 53, 59, 153, 169)	11 June 83	27 Aug. 02	19 Jan. 04	27 Aug. 17	*Ty. l. c.* 25 Jan. 18 *l. c.* 28 Aug. 17.	103 Lt. Infantry.
965	Unattd. List	**Hume,** Ernest George (51, 57, 66*a*, 71, 161, 165, 169)	14 Nov. 83	27 Aug. 02	20 Feb. 04	27 Aug. 17	. . . .	7 Lancers.
	Unattd. List	**Dodd,** Arthur Harvey Russell (50, 53, 71, 172)	5 Oct. 83	27 Aug. 02	19 Jan. 04	27 Aug. 17	. . . .	17 Cavalry.
	Unattd. List	**Monteith,** Montagu Edmund Cameron Von Stein (51, 52, 66*a*, 155)	20 Nov. 83	27 Aug. 02	28 Jan. 04	31 Aug. 17	. . . .	105 Lt. Infantry.
	Unattd. List	**Bailey,** Percival Henry Havelock, M.C. (51, 66*a*, 71, 167)	25 June 83	27 Aug. 02	21 Jan. 04	27 Aug. 17	. . . .	34 Pioneers.
	Bedf. R.	**Paterson,** Henry Francis Wm. (50, 52) *Forfeited from 4 May 10 to 7 Sep. 11 for promotion.*	28 Aug. 80	4 May 01	9 Apl. 04	8 Sep. 17	. . . .	Cant. Magte.'s Deptt.
970	Leins. R.	**Thomson,** Alexander Jas. Douglas (50, 57, 105) *Service for pension;—Embd. Mila. from 5 June to 17 Oct. 00 & Attd. Reglrs. from 24 Apl. 01 to 6 May 02; forfeited from 7 May to 23 Sep. 11 for promotion.*	4 June 82	7 May 02	1 Jan. 04	24 Sep. 17	. . . .	Civil employ.
	Bedf. R.	**Wheeler,** Edward Somerville (66*a*, 137, 162) *Service for pension;—Embd. Mila. from 13 Dec. 01 to 15 Sep. 02; for promotion from 16 Dec. 01 to 13 Sep. 02.*	21 Aug. 83	4 July 03	5 Nov. 04	5 Oct. 17	. . . .	Mily. Farms Dept.
	Sco. Rif.	**Campbell,** John, M.C. (50, 95, 167, 172*a*) *Service for pension;—Embd. Mila. & Attd. Reglrs. from 10 Sep. to 9 Nov. 01 and 12 Nov. 01 to 19 Sep. 02; for promotion from 23 Dec. 01 to 19 Sep. 02.*	29 Dec. 82	4 July 03	19 Sep. 05	6 Oct. 17	. . . .	29 Punjabis.
	3 Huss.	**Foster,** Claude Seymour (51, 162, 172) *Service for pension;—In Ranks—Impl. Yeo. from 20 Feb. 00 to 8 Aug. 01 (half to count, i.e., from 21 Jan. 02).*	27 June 80	15 Oct. 02	16 Jan. 05	15 Oct. 17	. . . .	37 Lancers.
	4 Huss.	**de Gale,** Wilfred Arthur Selwyn (51, 57) *Service for pension;—Embd. Mila. from 2 July to 5 Dec. 00 and 21 Aug. 01 to 20 Sep. 02; for promotion from 24 Au~. 01 to 20 Sep. 02, forfeited from 1 Aug. to 16 Oct. 11 for promotion.*	22 Mar. 81	28 Jan. 03	7 Aug. 06	17 Oct. 17	. . . .	5 Cavalry.
975	R. Lanc. R.	**Hunt,** Frederick Eckstein, D.S.O. (71) *Service for pension—Embd. Mila. 296 days, 3 Sep. 00 to 25 June 01. Forfeited from 26 June 10 to 17 Oct. 11 for promotion.*	15 Dec. 79	26 June 01	15 Dec. 05	18 Oct. 17	. . . .	29 Lancers.

No.	Former Corps.	Names.	Date of birth.	Date of first commission.	Date of appointment to Indian Army.	Date of rank. Substantive.	Date of rank. Army.	Remarks.
		Majors—*contd.*						
976	S. W. Bord.	**Hankin**, David Lewis (51) *Service for pension;—Embd. Mila. from 24 Jan. 00 to 14 Jan. 02; for promotion from 14 Feb. 00, forfeited from 15 Jan. to 18 Oct. 11 for promotion.*	15 Apl. 82	15 Jan. 02	15 Oct. 05	19 Oct. 17	. . . .	18 Infantry.
	Bord. R.	**Engledue**, Ralph Sandwith (50, 52, 125a, 167)	27 Apl. 84	22 Oct. 02	15 May 04	22 Oct. 17	8 Nov. 15	89 Punjabis.
	Bedf. R.	**Paterson**, Francis Wm. Joseph. M.C. (51, 167)	23 Nov. 83	22 Oct. 02	23 Jan. 05	22 Oct. 17	. . . .	66 Punjabis.
	W. I. R.	**Polley**, Sydney Trepess (66, 66b, 137)	7 Dec. 82	22 Oct. 02	14 Aug. 05	22 Oct. 17	. . . .	3 Horse.
980	W. I. R.	**Coates**, Sydney Baron, M.C. (51, 52, 57, 137, 162)	11 Nov. 83	22 Oct. 02	11 Aug. 05	22 Oct. 17	. . . .	102 Grenadiers.
	Midd'x. R.	**Western**, Charles Maximilian Thomas (65a, 66, 95, 169)	15 Dec. 82	22 Oct. 02	16 Jan. 05	22 Oct. 17	. . . .	S. and T. Corps.
	Conn. Rang.	**DeCourcy**, *Hon.* Michael William Robert, D.S.O. (50, 52, 162, 165, 205a)	26 Sep. 82	22 Oct. 02	29 July 04	22 Oct. 17	. . . .	32 Pioneers.
	Manch. R.	**Martin**, George Dodd, M. C. (51, 52, 59, 66, 71)	9 Apl 83	22 Oct. 02	10 Dec. 03	22 Oct. 17	. . . .	9 Infantry.
	E. York. R.	**Cornish**, Arthur Wm. Dauncey, D.S.O., M.C. (51, 57, 157, 172)	24 Apl. 84	22 Oct. 02	7 Mar. 04	22 Oct. 17	. . . .	6 G. R.
985	Leic. R.	**Vernon**, Herbert Bowater, M. C. (51, 148a, 172)	14 Mar. 83	22 Oct. 02	15 Feb. 04	22 Oct. 17	. . . .	27 Punjabis.
	Wilts. R.	**Tayler**, Francis Lionel (50, 53, 57, 95)	31 July 83	22 Oct. 02	16 July 04	22 Oct. 17	. . . .	42 Regiment.
	W. York R.	**Auret**, Alan Sauer (51, 66, 169)	20 Feb. 83	22 Oct. 02	25 Mar. 04	22 Oct. 17	. . . .	127 Lt. Infantry.
	Linc. R.	**Clarke**, Richard Charles, M.C. (51, 66)	3 Feb. 83	22 Oct. 02	13 July 04	22 Oct. 17	. . . .	S. and T. Corps.
	S. Staff R.	**Gannon**, Jack Compton Rose (51, 95, 172)	1 Nov. 82	22 Oct. 02	4 Oct. 06	22 Oct. 17	. . . .	23 Cavalry.
990	York. R.	**Rollo**, Leonard Duncan (50, 52, 57, 170)	18 June 83	22 Oct. 02	7 Feb. 04	22 Oct. 17	. . . .	96 Infantry.
	Som. L. I.	**Foster**, William Montague Arnaud, D.S.O. (51, 52, 172)	7 Apl. 83	22 Oct. 02	5 Oct. 06	22 Oct 17	. . . .	10 Lancers.
	Ches. R.	**Leapman**, Henry Martin (51, 52, 72a, 95, 137)	14 Mar. 83	22 Oct. 02	3 Aug. 07	22 Oct 17	. . . .	13 Rajputs.
	S. Staff. R.	**Macpherson**, Duncan Iver (51, 95, 137)	17 June 83	22 Oct. 02	2 July 07	22 Oct 17	. . . .	10 Jats.
	Ir. Gds.	**Beck**, Ralph Coker (60, 66, 71, 172)	27 June 83	22 Oct. 02	18 Nov. 06	22 Oct. 17	. . . .	S. and T. Corps.
995	E. York. R.	**Wilkinson**, Thomas Owen (51, 66, 125a, 140, 167)	24 May 83	22 Oct. 02	10 Feb. 06	22 Oct. 17	. . . .	91 Punjabis.
	Som. L. I.	**Williamson**, Andrew Geo. (51, 66a, 71, 137, 162, 169, 204)	15 Jan. 84	22 Oct. 02	27 Mar. 05	22 Oct. 17	. . . .	7 Lancers.
	S. Lan. R.	**Bird**, Walter Frederick (51, 52, 65a, 66, 162, 172)	13 Feb. 83	22 Oct. 02	17 Apl. 05	22 Oct. 17	. . . .	S. and T. Corps.
	S. Lan. R.	**Wilson**, Wm. Kenneth Pollexfen (66, 167)	6 June 83	22 Oct. 02	30 July 05	22 Oct. 17	. . . .	30 Punjabis.
	S. Lan. R.	**Martin**, Percival Henry (51, 71, 195)	23 Apl. 84	22 Oct. 02	18 Aug. 05	22 Oct. 17	. . . .	83 Infantry.
1000	2 D. G.	**Gillies**, Herbert (45a, 51) *Service for pension;—Embd. Mila. from 6 Jan. to 30 Sep. 02; for promotion from 13 Feb. to 29 Sep. 02.*	19 Feb. 81	22 July 03	11 Mar. 06	12 Dec. 17	. . . .	1 Lancers.
	R. Suss. R.	**Monck-Mason**, Godfrey Noel Grey (51, 162, 167) *Service for pension;—Embd. Mila. from 17 Feb. 01 to 29 Apl. 02; for promotion from 25 Mar. 01, forfeited from 25 Mar. 10 to 23 Dec. 11 for promotion.*	24 Oct. 82	30 Apl. 02	3 Oct. 04	24 Dec. 17	. . . .	84 Punjabis.
	R. A.	**Benton**, William Burgess, M.C. (51, 52, 153, 161)	17 Sep. 83	24 Dec 02	9 Nov. 06	24 Dec. 17	. . . .	105 Lt. Infantry.
1003	R. A.	**Durham**, Wm. Glynn Wakeman (64. 66a, 71, 104, 137, 161, 204) [L]	24 July 84	24 Dec. 02	28 July 05	24 Dec. 17	. . . .	27 Lt. Cavalry.

1004	Worc. R.	**Skinner**, Russell Herbert (51, 57, 71, 167)	14 Sep. 81	*July 03*	18 Nov. 05	11 Jan. 18	. . . .	91 Punjabis.
		Service for pension;—Embd. Mila. from 9 May 00 to 5 July 01 & Attd. Reglrs. from 1 Dec. to 31 Jan. 02, Embd. Mila. from 17 Feb. 02 to 6 Oct. 02; for promotion from 21 Mar. to 5 Oct. 02.						
1005	Leins. R.	**Jarrett-Kerr**, Herbert Jarrett (51, 66, 72b)	26 Jan. 83	9 Jan. 04	11 Apl. 14	12 Jan. 18	. . . .	S. and T. Corps.
		Service for pension;—Embd. Mila. from 3 Oct 00 to 25 Feb. 02; for promotion from 28 Feb. 01 to 24 Feb. 02.						
	Sco. Rif.	**Barge**, Kenneth, D.S.O., M.C. (57, 64, 71, 172)	27 May 83	9 Sep. 03	6 Jan. 05	18 Jan. 18	*Ty. l. c.* 3 June 18.	17 Cavalry.
		Service for pension;—Embd. Mila. from 6 Jan. to 23 Sep. 02; for promotion from 1 Feb. to 22 Sep. 02.						
	Unattd. List	**Bagnall**, Ralph (50, 53, 167)	19 Jan. 84	21 Jan. 03	2 Apl. 02	21 Jan. 18	. . . .	74 Punjabis.
	Unattd. List	**Griffith**, Wm. Robt. Clavey, M.C. (51, 52, 59, 71, 137, 162, 170, 205a)	26 Mar. 84	21 Jan. 03	31 Mar. 04	21 Jan. 18	. . . .	2 Lt. Infantry.
	Unattd. List	**Sandeman**, Donald George (50, 52, 53, 59, 162, 169)	14 Oct. 84	21 Jan. 03	7 Apl. 04	21 Jan. 18	. . . .	Guides.
1010	Unattd. List	**Laing**, Stanley van Buren, M.C. (51, 52, 59, 167)	21 Apl. 84	21 Jan. 03	6 Apl. 04	21 Jan. 18	. . . .	76 Punjabis.
	Unattd. List	**Gall**, Lawrence, M. C. (45a, 51, 52, 57, 66, 71, 172)	10 Sep. 84	21 Jan. 03	9 Apl. 04	21 Jan. 18	. . . .	25 Cavalry.
	Unattd. List	**Rae**, James Gordon (50, 71, 137)	2 June 83	21 Jan. 03	10 Apl. 04	21 Jan. 18	. . . .	123 Rifles.
	Unattd. List	**Wellborne**, Cyril deMontfort (57, 94, 105, 155, 162, 170)	2 July 84	21 Jan. 03	10 Apl. 04	21 Jan. 18	. . . .	Civil employ.
	Unattd. List	**Channer**, Guy (50, 53, 94, 162, 165, 169)	12 Nov. 84	21 Jan. 03	8 Apl. 04	21 Jan. 18	. . . .	14 Sikhs.
1015	Unattd. List	**Thompson**, Claude Matthew (50, 94, 162, 167, 205a)	21 Sep. 83	21 Jan. 03	7 Apl. 04	21 Jan. 18	. . . .	Survey Dept.
	Unattd. List	**Hickie**, Henry Whitten (50, 53, 66a, 71, 165, 169)	18 Aug. 84	21 Jan. 03	7 Apl. 04	21 Jan. 18	. . . .	84 Punjabis.
	Unattd. List	**Hawes**, Claude Mackinnon (50, 53, 59, 169)	20 May 84	21 Jan. 03	14 Apl. 04	21 Jan. 18	. . . .	20 Infantry.
	Unattd. List	*Meek, Arthur Stanley (57, 132, 195)*	*7 Oct. 83*	*21 Jan. 03*	*17 Apl. 04*	*21 Jan. 18*	*Ty. m.* 22 Aug. 17	*Supy. List.*
	Unattd. List	**Dick-Lauder**, John North Dalrymple (50, 52, 71, 167)	22 July 83	21 Jan. 03	7 Apl. 04	21 Jan. 18	1 Jan. 16	16 Cavalry.
	Unattd. List	**Hastings**, William Holland, D.S.O. (51, 65, 66a, 71, 72a, 125a, 167)	17 Mar. 84	21 Jan. 03	6 Apl. 04	21 Jan. 18		92 Punjabis.
1020	Unattd. List	**Lecky**, John Gage (51, 52, 95, 137, 139b)	14 Nov. 83	21 Jan. 03	11 Apl. 04	21 Jan. 18	. . . .	119 Infantry.
	Unattd. List	**Howson**, Geoffrey, M. C. (51, 52, 59, 71, 137)	14 Aug. 83	21 Jan. 03	16 Apl. 04	21 Jan. 18	*Ty. m.* 17 Jan. 17. *Ty. l. c.* 22 Dec. 17.	4 Cavalry.
	Unattd. List	**Leman**, Frederic Baring (51, 52, 71, 195)	10 July 83	21 Jan. 03	8 Apl. 04	21 Jan. 18	. . . .	86 Infantry.
	Unattd. List	**Auchinleck**, Claud John Eyre, D.S.O. (51, 52, 71, 165, 169)	21 June 84	21 Jan. 03	10 Apl. 04	21 Jan. 18	. . . .	62 Punjabis.
	Unattd. List	**Stainton**, Thomas Henry (51, 65a, 66, 94, 137, 161, 205a)	4 Oct. 83	21 Jan. 03	11 Apl. 04	21 Jan. 18	. . . .	S. and T. Corps.
1025	Unattd. List	**Radford**, Oswald Harvey (50, 53, 95, 172)	21 May 84	21 Jan. 03	9 Apl. 04	21 Jan. 18	*Ty. l. c.* 14 Jan. 18.	57 Rifles.
	Unattd. List	**Boyce**, Reginald Baron (51, 66a, 94, 137, 169)	4 May 84	21 Jan. 03	12 Apl. 04	21 Jan. 18	. . . .	12 Pioneers.
	Unattd. List	**Bruce**, George Eyre, M. C. (59, 165, 169)	2 Sep. 83	21 Jan. 03	8 Apl. 04	21 Jan. 18	. . . .	53 Sikhs.
	Unattd. List	**Kirkwood**, Thomas Wm. (51, 66a, 71, 169)	13 Feb. 84	21 Jan. 03	25 Apl. 04	21 Jan. 18	. . . .	17 Cavalry.
	Unattd. List	**Hill**, Robert Daubeney Ottley (157, 172)	17 Mar. 84	21 Jan. 03	8 Mar. 04	21 Jan. 18	. . . .	2 G. R.
1030	Unattd. List	**Harenc**, Roger Edward (45a, 50, 53, 71, 137)	31 Oct. 83	21 Jan. 03	18 Jan. 05	21 Jan. 18	. . . .	4 Cavalry.
	Unattd. List	**Bromilow**, David George, D.S.O. (45a, 51, 66a, 66b, 71, 137, 172a)	8 Feb. 84	21 Jan. 03	10 Apl. 04	21 Jan. 18	. . . .	14 Lancers.
	Bord. R.	**Burnett**, Allan Harrington, D.S.O. (51, 53, 57, 66, 169)	9 June 84	28 Jan. 03	10 Jan. 05	28 Jan. 18	. . . .	72 Punjabis.
		Service for pension;—Embd. Mila. from 23 Sep. 01 to 1 Oct. 02.						
	Conn. Rang.	**Dickson**, Harold Richard Patrick, C.I.E. (50, 60a, 66, 71)	4 Feb. 81	28 Jan. 03	3 Feb. 07	28 Jan. 18	. . . .	29 Lancers.
	Bedf. R.	**Seymour**, Montague Henry (50, 53, 157)	29 Sep. 79	25 Feb. 03	19 Dec. 13	25 Feb. 18	. . . .	10 G. R.
1035	Manch. R.	**Young**, Montague Sheridan (51, 52, 66, 137)	6 Aug. 83	22 Mar. 03	3 June 06	22 Mar. 18	. . . .	14 Lancers.
	W. I. R.	**Kreyer**, John Arthur Claude (50, 57, 66, 71, 94, 162, 167)	7 June 83	22 Apl. 03	6 July 05	22 Apl. 18	. . . .	28 Lt. Cavalry.
	W. I. R.	**Ralston**, William John Hopkins (137)	16 Dec. 82	22 Apl. 03	17 June 06	22 Apl. 18	. . . .	4 Rajputs.
	W. York R.	**Davies**, Ronald Despard (51, 66a, 71, 95, 162, 170)	1 Oct. 83	22 Apl. 03	31 Mar. 09	22 Apl. 18	. . . .	127 Infantry.
1039	R. Muns. Fus.	**Henry**, Wilfrid Richardson Peacock (45a, 59, 162, 165, 170)	21 June 83	22 Apl. 03	14 Oct. 06	22 Apl. 18	. . . .	12 Cavalry.

No.	Former Corps.	Names.	Date of birth.	Date of first commission.	Date of appointment to India Army.	Date of Rank. Substantive.	Date of Rank. Army.	Remarks.
		Majors—*contd.*						
1040	W. Rid. R.	**Reynolds**, Henry Victor (162, 167, 172*a*)	22 Aug. 83	22 Apl. 03	26 Sep. 08	22 Apl. 18	....	82 Punjabis.
	Welsh R.	**Griffith**, Llewelyn (50, 52, 59, 172)	13 Nov. 84	22 Apl. 03	25 Oct. 06	22 Apl. 18	....	107 Pioneers.
	Essex. R.	**Cooper**, Kenneth Edward, M C. (51, 66, 155)	13 July 83	22 Apl. 03	16 Nov. 07	22 Apl. 18	....	110 Lt. Infantry.
	Suff. R.	**Trelawny**, Lancelot Conyers, M. C. (66*a*, 66, 167)	3 Oct. 84	22 Apl. 03	10 Dec. 06	22 Apl. 18	....	S. and T. Corps.
	W. Rid. R.	**Turner**, Horace Gough (51, 66*a*, 94, 162*a*, 169)	15 Sep. 84	22 Apl. 03	18 June 05	22 Apl. 18	....	106 Pioneers.
1045	Manch. R.	**Sutton**, Henry Godwyn (52, 66*a*)	29 Feb. 84	22 Apl. 03	30 June 05	22 Apl. 18	....	42 Regiment.
	W. Rid. R.	**Chope**, Arthur John Herbert, D.S.O. (50, 53, 157, 170)	23 Aug. 84	22 Apl. 03	29 Jan. 06	22 Apl. 18	....	2 G. R.
	York. L. I.	**Allen**, Wm. Townsend (51, 52, 66*a*, 172]	11 June 84	2 May 03	6 July 05	2 May 18	....	3 Cavalry.
	R. Berks. R.	**Simonds**, James Napier (45*a*, 51, 52, 66*a*, 72)	23 July 83	28 Nov. 03	5 Apl. 09	2 May 18	....	35 Horse.
		*Service for pension;—**Embd. Mila. from 8 Jan. to 17 Sep. 02**, for promotion from **23 Jan. to 20 Aug. 02**.*						
	Leic. R.	**Morris**, Allan Leslie (51)	5 July 83	2 May 03	11 Apl. 06	2 May 18	....	17 Infantry.
1050	S. Staff. R.	**Duke**, Arthur Alexander Geo. (51, 65*a*, 66*a*, 66)	29 Oct. 81	16 May 03	18 June 06	16 May 18	*Ty. m.* 25 Feb. 18.	S. & T. Corps.
		Service for pension from 21st November 1901.						
	W. Surr. R.	**Wise**, Alan Douglas (50, 53, 71)	12 June 84	27 May 03	17 Oct. 05	27 May 18	....	27 Lt. Cavalry.
	North'n. R.	**Smith**, Maurice Castle (50, 53, 195)	8 June 82	16 Dec. 03	25 June 05	3 July 18	....	64 Pioneers.
		Service for pension;—Embd. Mila. from 17 Mar. to 20 Sep. 02; for promotion from 8 Apl. to 20 Sep. 02.						
	North'n. R.	**Clive**, Henry William Fitzroy (66*a*, 72*a*, 172)	17 Mar. 84	4 July 03	18 Mar. 09	4 July 18	....	129 Baluchis.
	North'n. R.	**Etherton**, Percy Thos. (50, 51, 72*b*, 127, 157, 162, 169, 205*a*) [*I*]	4 Sep. 79	4 July 03	4 Nov. 04	4 July 18	*Ty. m.* 1 Apl. 18.	39 Garhwal Rifles.
		Service for pension;—In Ranks from 17 Oct. 00 to 16 July 01 and 19 July 01 to 23 Jan. 02 (half to count).						
1055	R. A.	**Fuller**, Herbert Foster (66, 162)	31 Oct. 83	15 July 03	29 Nov. 12	15 July 18	....	S. and T. Corps.
	R. A.	**Holliday**, John Cecil Hamilton (12, 50, 53, 59, 71, 72*c*, 137)	4 May 84	15 July 03	22 Sep. 08	15 July 18	....	122 Infantry.
	R. A.	**Roughton**, Francis Arthur Guy (51, 52, 137)	6 July 84	15 July 03	3 Mar. 08	15 July 18	....	113 Infantry.
	R. A.	**Shaw**, Heeton Percy John (60*a*, 172*a*)	6 Apl. 84	15 July 03	10 May 15	15 July 18	....	S. and T. Corps.
	R. A.	**Chapman**, William Leslie Bedell (50, 52, 167)	13 July 84	15 July 03	24 Mar. 07	15 July 18	....	89 Punjabis.
1060	R. A.	**Tucker**, Hngh Boswell. (71, 74, 94, 100, 161 169, 205*a*)	2 Aug. 84	15 July 03	11 June 08	15 July 18	....	98 Infantry.
	S. W. Bord.	**Stone**, Alan Gething, M.C. (51, 52, 57, 66*a*, 157)	4 Nov. 83	27 Jan. 04	4 Sep. 05	19 July 18	....	3 G. R.
		*Service for pension;—**Embd. Mila. from 26 Nov. 01 to 28 July 02**; for promotion from 16 Jan. to 26 July 02.*						
	15 Huss.	**Wetherall**, Geo. Augustus Champagne (50)	24 July 83	22 July 03	29 Sep. 06	22 July 18	....	1 Lancers.
1063	W. I. R.	**Shaw**, William Western (51, 52, 66*a*, 137)	18 May 80	29 July 03	27 June 05	29 July 18	....	S. and T. Corps.
		*Service for pension;—Embd. Mila. **from 5 June 00 to 1 Nov. 00 and Attd. Reglrs. from 12 Nov. 00 to 31 Oct. 02.***						

1064	E. York R.	**Finnie,** Andrew McLean (53, 53, 105, 167)	18 Oct. 79	29 July 03	10 Mar. 05	29 July 18		Civil employ.
		*Service for pension;—In Ranks from 18 Oct. 97 to **28 July 03** (half to count.)*						
1065	Lan. Fus.	**Pope,** Philip Fell (51, 66a, 71, 137)	17 Aug. 81	29 July 03	7 June 05	29 July 18		1 Brahmans
	Lan. Fus.	**Spence,** Richard Bennett (51, 52, 71, 172a)	8 Mar. 82	29 July 03	29 July 05	29 July 18		96 Infantry.
	Suff. R.	**Hotham,** John Clarence (51, 52, 71, 175, 172a)	25 Mar. 82	29 July 03	5 Oct. 06	29 July 18		4 Cavalry.
	Unattd. List	**Wilson,** Arnold Talbot, C.M.G., C.I.E., D.S.O. (57, 66a, 162, 165, 169) (a) [l]	18 July 84	19 Aug. 03	18 Dec. 04	19 Aug. 18	*l.c.* 3 June 18	Poll. employ.
	Unattd. List	**Palmer,** John Harald Gore (51. 52, 66, 161, 169, 205a) [l]	11 July 84	19 Aug. 03	31 May 05	19 Aug. 18	*Ty. m.* 19 July 18.	21 Punjabis.
1070	Unattd. List	**Knollys,** Denis Erskine (50, 53, 59, 172)	20 Feb. 84	19 Aug. 03	18 Dec. 04	19 Aug. 18		19 Punjabis.
	Unattd. List	**Gwatkin,** Frederick, M.C. (45a, 50, 57, 167)	12 Apl. 85	19 Aug. 03	19 Dec. 04	19 Aug. 18		18 Lancers.
	Unattd. List	**McLeod,** Donald Kenneth, D.S.O. (50, 53, 59, 71, 169)	19 June 85	19 Aug. 03	29 Nov. 04	19 Aug. 18	*Ty. m.* 25 Feb. 18	Guides.
	Unattd. List	**Farquhar,** Francis Robert, M. C. (51, 53, 57, 172)	11 June 84	19 Aug. 03	19 Dec. 04	19 Aug. 18	*Ty. m.* 12 Oct. 14	36 Horse.
	Unattd. List	**Thompson,** Alec. (50, 53, 65a, 137)	28 Aug 83	19 Aug. 03	18 Dec. 04	19 Aug. 18		112 Infantry.
1075	Unattd. List	**Bennett,** Ogilvie David (50, 59, 66a, 169)	10 June 85	19 Aug. 03	17 Dec. 04	19 Aug. 18		26 Punjabis.
	Unattd. List	**Slacke,** George Cave (51, 104, 137, 161, 165, 205a,)	26 Feb. 84	19 Aug. 03	18 Oct. 04	19 Aug. 18		Civil, Burma.
	Unattd. List	**Buckland,** Gerald Chas. Balfour, D.S.O., M.C. (50, 53, 157)	18 May 84	19 Aug 03	9 Nov. 04	19 Aug. 18	*Ty. m.* 23 May 18	8 G. R.
	Unattd. List	**Lee,** Wm. Cuppaidge Norris (51, 52, 60a, 64, 66a, 71, 94, 169)	6 July 84	19 Aug. 03	19 Dec. 04	19 Aug. 18		37 Lancers.
	Unattd. List	**Barrow,** Ronald Edmund (51, 52, 57, 66, 167)	10 June 85	19 Aug. 03	1 Jan. 05	19 Aug. 18		38 Dogras.
1080	Unattd. List	**Hemsley,** Charles, D.S.O. (51, 52, 57, 162, 193)	6 Feb. 84	19 Aug. 03	10 Nov. 04	19 Aug. 18		64 Pioneers
	Unattd. List	**Harcourt,** Edric Sherwood, M.C. (51, 52, 65a, 94, 100, 157, 161, 169, 203) [L]	2 June 85	19 Aug. 03	9 Nov. 04	19 Aug. 18	*Ty. m.* 25 Feb. 18	7 G. R.
	Unattd. List	**Cheeke,** Arthur Marcus Mosseley (57, 71, 125a, 137)	17 Feb 85	19 Aug. 03	9 Nov. 04	19 Aug. 18		104 Rifles.
	Unattd. List	**Pemberton,** Eric Harry (51, 52, 169)	6 Sep. 83	19 Aug. 03	9 Nov. 04	19 Aug. 18		19 Punjabis.
	Unattd. List	**Beadon,** Vernon (104)	24 Apl. 84	19 Aug. 03	9 Nov 04	19 Aug. 18		Civil employ.
1085	Unattd. List	**Hobday,** Eric Maitland (51, 52, 125a, 140, 167, 172a)	12 May 84	19 Aug. 03	24 Oct. 04	19 Aug. 18		41 Dogras.
	Unattd. List	**Richardson,** Geoffrey Gordon (51, 52, 107, 140, 148b, 155)	27 Aug. 83	19 Aug. 03	22 Dec. 04	19 Aug. 18		105 Lt. Infantry.
	S. Lan. R.	**Moss-Blundell,** Reginald Seymour, M.C. (51. 52, 71, 72b)	26 Oct. 82	4 June 04	31 July 07	26 Aug. 18		27 Lt. Cavalry.
		Service for pension;—Embd. Mila. from 1 May 00 to 3 Oct. 00 and Impl. Yeo. from 7 Feb. 02 to 15 Nov. 02.						
	R. Ir. R.	**Langrishe,** Richard Bellingham (65a, 66, 167)	9 Sep. 84	10 Oct. 03	9 Apl. 06	10 Oct. 18		S. and T. Corps.
	E. Surr. R.	**Cardew,** Henry Spencer, M.C. (50, 59, 157) 71, 72a, 167, 170)	2[illegible] Mar. 84	10 Oct. 03	28 Oct. 06	10 Oct. 18		39 Garhwal Rifles.
1090	Wilts. R.	**Clarke,** Arthur Calvert Keir Stewart, M.C. (51, 65a, 71, 72a, 167, 170)	26 Jan. 84	10 Oce. 03	10 Jan. 07	10 Oct. 18		15 Lancers.
		Service for pension;—Embd. Mila. from 28 Jan. to 10 June 01, i.e., from 29 May 03.						
	Y. and L. R.	**Exham,** Harold (50, 52, 57, 94, 157, 172a)	13 Oct. 84	10 Oct. 03	23 July 07	10 Oct. 18		7 G. R.
	Ches. R.	**Taylor,** John McLeod Grigor (51, 52, 57, 137, 161)	24 July 84	10 Oct. 03	3 Aug. 04	10 Oct. 18		119 Infantry.
	E. Surr. R.	**Dennys,** Guy Tullock (51, 57, 66a, 165, 169)	14 May 84	10 Oct. 03	1 Aug. 05	10 Oct. 18		31 Punjabis.
	North'n. R.	**Steel,** Cyril Nisbet (51, 162, 172)	30 Apl. 84	10 Oct. 03	25 Oct. 06	10 Oct. 18	3 June 17	25 Punjabis.
1095	Suff. R.	**Wace,** Montague (51, 52, 57, 66, 167)	22 Oct. 84	10 Oct. 03	30 July 05	10 Oct. 18		14 Sikhs.
	North'n. R.	**Chesney,** Harold George (50, 53, 106, 137, 140, 169)	8 Jan. 85	10 Oct. 03	25 June 05	10 Oct. 18		4 Rajputs.
1097	R. Suff.	**Comyn,** Geoffrey Valence, M.C. (51, 52, 66)	16 June 85	10 Oct. 03	9 Aug. 05	10 Oct. 18		S. and T. Corps.

(a) Passed Persian (C. S. Commrs., London).

Majors—*concld.*

No.	Former Corps.	Names.	Date of birth.	Date of first commission.	Date of appointment to Indian Army.	Date of Rank. Substantive.	Date of Rank. Army.	Remarks.
1098	The Queen's	**Campbell**, Roy Neil Boyd (51, 52, 57, 66*a*, 165, **170**)	24 Sep. 84	10 Oct. 03	25 Mar. 05	10 Oct. 18	. . . .	23 Pioneers.
	W. York R.	**Cosens**, Francis Richard (51, 52, 57, 72*a*, 127, **129**, **137**)	5 Sep. 84	10 Oct. 03	14 July 07	10 Oct. 18	. . . .	4 Rajputs.
1100	R. W. Surr. R.	**Enriquez**, Colin Metcalfe Dallas (50, 53, 105, 125*a*, 140, **162**, **172**)	3 Oct. 84	10 Oct. 03	3 Apl. 05	10 Oct. 18	. . . .	21 Punjabis.
	R. War. R.	**Woodward**, Robert Francis (51, 71, 94, 161, 169) [*l*]	21 Oct. 84	10 Oct. 03	7 Aug. 07	10 Oct. 18	. . . .	130 Baluchis.
	S. Lan. R.	**Baldwin**, Montagu Chenevix, M.C. (51, 52, 157)	2 Mar. 84	10 Oct. 03	9 July 05	10 Oct. 18	. . . .	1 G. R.
	R. War. R.	**Macartney**, Frank Alan, M.C. (51, 53, 64, 72*a*, 95, 169)	19 Dec. 84	10 Oct. 03	20 Sep. 08	10 Oct. 18	. . . .	53 Sikhs.
	L'pool. R.	**Pottinger**, John Ashton, M.C. (51, 52, 66*a*, 167)	28 July 84	10 Oct. 03	20 July 07	10 Oct. 18	. . . .	30 Punjabis.
1105	E. York. R.	**Malet**, William de Carteret (51, 52)	7 Apl. 84	10 Oct. 03	9 Feb. 06	10 Oct. 18	. . . .	14 Lancers.
	Midd'x. R.	**Mullaly**, Herbert (50, 53, 71, 66*a*, 157)	27 May 84	4 Nov. 03	20 June 05	4 Nov. 18	. . . .	9 G. R.
	Midd'x. R.	**Cooper**, Geoffrey Beauchamp Astley (65*a*, 66)	5 Feb. 84	4 Nov. 03	6 Aug. 08	4 Nov. 18	. . . .	S. and T. Corps.
	Midd'x. R.	**Byrch**, Arthur Edward Berry (66, 67*a*, 71, 162)	17 Feb. 84	4 Nov. 03	21 Aug. 05	4 Nov. 18	. . . .	S. and T. Corps.
	R. War. R.	**Stevenson**, Harry Daniel Mühldorff, M.C.	9 Oct. 84	18 Nov. 03	27 Apl. 15	18 Nov. 18	. . . .	S. & T. Corps.
1110	R. W. Surr. R.	**Nicholson**, Francis Lothian, D.S.O., M.C. (51, 52, 137, 140)	16 Jan. 84	18 Nov. 03	23 June 05	18 Nov. 18	. . . .	37 Dogras.
	Wilts. R.	**Birch**, Colvin Frederic Montagu (51, 52, 66, 157)	3 Nov. 84	18 Nov. 03	2 July 05	18 Nov. 18	. . . .	1 G. R.
	Midd'x. R.	**Hinde**, Henry Thos. Langford (51, 53)	31 Mar. 84	18 Nov. 03	22 Mar. 06	18 Nov. 18	. . . .	99 Infantry.
	R. Suss. R.	**Paske**, Chas. Alfred Montanaro (172)	19 Nov. 84	28 Nov. 03	15 Aug. 05	28 Nov. 18	. . . .	23 Cavalry.
	Welsh. R.	**Lethbridge**, Alfred (50, 172)	29 Sep. 84	12 Dec. 03	11 Aug. 05	12 Dec. 18	. . . .	54 Sikhs.
1115	R. Ir. R.	**Patrick**, Richard Martin Forsythe, M.C. (51, 52, 66*a*, 72*b*, 95, 107, 169, 175) [L] *Service for pension;—In Ranks from 3 Nov. 98 to 15 Dec. 03 (half to count).*	3 Feb. 80	16 Dec. 03	26 Jan. 06	16 Dec. 18	. . . .	42 Regiment.

Captains—979.

1	Unattd. List	**Yates,** Robert Charles Byam (51, 57, 157, 170)	9 June 78	4 Aug. 97	26 Oct. 98	4 Aug. 06	. . . .	4 G. R.
	R. Highrs.	**Grant,** Charles John Patrick MacAlpine (51, 52, 66*a*, 162)	23 Apl. 80	4 Apl. 00	13 Sep. 04	18 Jan. 09	. . . .	99 Infantry.
		Service for pension ;— Embd. Mila. from 13 Dec. 99 to 3 Apl. 00 ; for promotion from 18 Jan. 00.						
	R. A.	**Todd,** James Farquhar (45*a*, 51)	24 Dec. 79	26 May 00	21 Apl. 04	26 May 09	. . . .	39 C. I. Horse.
	Unattd. List	**Hunt,** Cecil Edwin, M.C., *p. s. c.* (50, 162, 165)	13 Oct. 80	28 July 00	4 Oct. 01	28 July 09	. . . .	34 Pioneers.
5	R. Muns. Fus.	**Maydwell,** Henry Somers Lawrence (51, 66*a*, 71)	8 Feb. 81	5 Jan. 01	9 Sep. 03	1 Mar. 10	. . . .	
		Service for pension ;—Embd. Mila. from 5 Dec. 99 to 4 Jan. 01 ; for promotion from 23 Feb. 00 ; forfeited from 23 Feb. 09 to 28 Feb. 10, for promotion.						
	R. Muns. Fus.	**Stoddart,** Hugh (72*b*, 137, 162, 170, 199)	8 May 79	28 Jan. 03	15 Jan. 04	12 July 10	. . . .	11 Rajputs.
		Service for pension ;—Embd. Mila. from 12 Mar. 01. to 29 Sep. 02; for promotion from 12 Mar. 01 to 27 Sep. 02.						
	E. Surr. R.	**Maunsell,** Ivan O'Grady (51, 71, 106, 167)	28 Dec 81	18 Jan. 02	18 June 04	18 Jan. 11	. . . .	92 Punjabis.
8	W. I. R.	**Wayman,** Walter Lancelot (51, 66*a*, 155)	7 Mar. 82	4 May 01	1 June 04	10 Mar. 11	. . . .	94 Infantry.
		Forfeited from 4 May 10 to 9 Mar. 11 for promotion.						
	Unattd. List	**Mackenzie,** Chas. Alexr. Chandos (45*a*, 51, 52, 59, 71)	21 Feb. 84	21 Jan. 03	7 Apl. 04	21 Jan. 12	. . . .	20 Horse.
10	Suff. R.	**Standbridge,** Charles Haswell (66*a*)	9 May 82	18 Jan 02	22 May 04	16 Mar. 12	. . . .	117 Mahrattas.
		Forfeited from 18 Jan. 11 to 15 Mar. 12 for promotion.						
	Y. & L. R.	**Montagu-Smith,** Edmund (45*a*, 51, 71, 172)	10 Sep. 84	10 Oct. 03	1 Feb. 09	10 Oct. 12	. . . .	35 Horse.
	R. W. Surr. R.	**Nixon,** Chas. Howard Featherstonehaugh, D.S.O. (51, 57, 157, 167)	14 Aug. 84	4 Nov. 03	26 Aug. 05	4 Nov. 12	. . . .	91 Punjabis.
	W. I. R.	**Minniken,** Thomas Brutton (51, 71, 167)	11 Mar. 84	18 Nov. 03	22 July 05	18 Nov. 12	*Ty. m.* 2 Aug. 17.	76 Punjabis.
	4 D. G.	**Brooke,** Augustine Henry (51, 167)	25 Feb. 84	28 Nov. 03	8 Oct. 05	28 Nov. 12	. . . .	18 Lancers.
15	R. A.	**Saunders,** Macan, D.S.O. (12, 50, 52, 66*a*, 71, 162, 167, 174) [L]	9 Nov. 84	23 Dec. 03	12 Nov. 07	23 Dec. 12	. . . .	36 Sikhs.
	R. A.	**Gray,** David Benjamin (12, 51, 52, 137, 162, 205*a*)	19 July 84	23 Dec. 03	7 Oct. 05	23 Dec. 12	. . . .	48 Pioneers.
	R. A.	**Blanford,** Philip Thomas, M.C. (51, 52, 165)	14 Sep. 83	23 Dec. 03	30 Mar. 09	23 Dec. 12	. . . .	84 Punjabis.
	Unattd. List	**Anderson,** Mark Gordon (50, 59, 94, 157, 161, 169, 175) [L]	24 Feb. 85	9 Jan. 04	8 Apl. 05	9 Jan. 13	. . . .	7 G. R.
	Unattd. List	**Lamb,** David Ogilvy Wight (45*a*, 51, 52, 57, 66*a*, 71, 172)	29 Jan. 85	9 Jan. 04	21 Mar. 05	9 Jan. 13	. . . .	10 Lancers.
20	Unattd. List	**Trevor,** Arthur Cuthbert Henry (50, 53, 71, 161, 169, 172)	6 Mar. 84	9 Jan. 04	22 Mar. 05	9 Jan. 13	. . . .	24 Punjabis.
	Unattd. List	**Budgen,** Hugh Vincent (51, 52, 66*a*, 137)	21 May 85	9 Jan. 04	24 June 05	9 Jan. 13	. . . .	112 Infantry.
	Unattd. List	**Anderson,** Guy Willoughby (50, 169)	3 Apl. 85	9 Jan. 04	2 Mar. 05	9 Jan. 13	. . . .	26 Punjabis.
	Unattd. List	**Lang,** Walter Hamilton (45*a*, 51, 59, 66*a*, 71, 167)	30 Aug 84	9 Jan. 04	28 Mar. 05	9 Jan. 13	*m.* 3 June 17	16 Cavalry.
	Unattd. List	**Walker,** Hugh Reginald Oriel (50, 59, 172)	1 Nov. 83	9 Jan. 04	8 Mar. 05	9 Jan. 13	. . . .	25 Punjabis.
25	Unattd. List	**Drake-Brockman,** Gerald Travers (51, 59)	9 Jan. 85	9 Jan. 04	28 Mar. 05	9 Jan. 13	. . . .	18 Infantry.
26	Unattd. List	**Bird,** Wm. Arthur Henry (50, 53, 71, 165, 170)	6 Nov. 84	9 Jan. 04	11 Mar. 05	9 Jan. 13	. . . .	23 Pioneers.

Captains—*contd.*

No.	Former Corps.	Names.	Date of birth.	Date of first commission.	Date of appointment to Indian Army.	Date of rank: Substantive.	Date of rank: Army.	Remarks.
27	Unattd. List	**Watts**, Nicholas Hugh Lawrence (51, 57, 66a, 125a, 140, 162, 167, 170, 204)	8 June 85	9 Jan. 04	24 Mar. 05	9 Jan. 13	*Ty. m.* 20 June 17.	92 Punjabis.
	Unattd. List	**Wallace**, Charles William (51, 53, 172)	24 Nov. 84	9 Jan. 04	2 Apl. 05	9 Jan. 13	. . . .	22 Punjabis.
	Unattd. List	**Burrows**, Hollis Martin (51, 59, 161, 169)	15 Sep. 84	9 Jan. 04	9 Mar. 05	9 Jan. 13	. . . .	51 Sikhs.
30	Unattd. List	**Kenworthy**, Walter (45a, 50, 52, 57, 71, 95, 137)	1 Sep. 84	9 Jan. 04	28 Mar. 05	9 Jan. 13	*Ty. m.* 15 Apl. 18	33 Lt. Cavalry.
	Unattd. List	**Haviland**, Leonard Proby (45a, 51, 57, 66, 71, 167)	3 Jan. 84	9 Jan. 04	26 Mar. 05	9 Jan. 13	. . . .	13 Lancers.
	Unattd. List	**Quinan**, Edward Pellew (51, 52, 169)	9 Jan. 85	9 Jan. 04	25 Mar. 05	9 Jan. 13	*m.* 1 Jan. 18	27 Punjabis.
	Unattd. List	**Ross**, Edward Johnson, M.C. (50, 53, 157)	2 Mar 84	9 Jan. 04	24 Mar. 05	9 Jan. 13	. . . .	8 G. R.
	Unattd. List	**Paterson**, George Fredrick Joseph, M.C. (50, 53, 167)	26 Mar. 85	9 Jan. 04	25 Mar. 05	9 Jan. 13	. . . .	34 Pioneers.
35	Unattd. List	**Leith-Ross**, William, M.C. (50, 53, 66a, 169)	14 Nov. 84	9 Jan. 04	8 Mar. 05	9 Jan. 13	. . . .	55 Rifles.
	Unattd. List	**Christopher**, Charles de Lona (66, 71, 167, 172a)	16 May 85	9 Jan 04	1 Mar. 05	9 Jan. 13	*m.* 1 Jan. 18	S. and T. Corps.
	Unattd. List	**Gould**, Standish Charles (57, 104 167)	3 Jan. 85	9 Jan. 04	27 Mar. 05	9 Jan 13	. . . .	Civil, Burma.
	Unattd. List	**Ricketts**, Edward Wallace Claud (51, 52, 71, 167, 170)	1 Apl. 84	9 Jan. 04	24 Mar. 05	9 Jan. 13	. . . .	23 Pioneers.
	Unattd. List	**Summers**, Gerald Henry (51, 65, 66). (*Ty. l.c.* 20 Jan. 16)	12 Oct. 85	9 Jan. 04	24 Mar. 05	9 Jan. 13	*m.* 7 Jan. 14	26 Lt. Cavalry.
40	Unattd. List	**Pogson**, Frederick Vere (51, 52, 65a, 66, 71) [*l*]	13 Apl. 85	9 Jan. 04	21 Mar. 05	9 Jan. 13	. . . .	S. and T. Corps.
	Unattd. List	**Manderson**, Robert Wardlaw (51, 57, 137)	2 Feb. 85	9 Jan. 04	21 May 05	9 Jan. 13	*Ty. m.* 20 July 18	3 Horse.
	Unattd List	**Burn**, Alexander Henderson (51, 161, 169)	5 Feb. 85	9 Jan. 04	24 Jan. 06	9 Jan. 13	. . . .	123 Rifles.
	Unattd. List	**Smith**, Edmund **Denman** Dallas (162, 170)	20 Nov. 83	9 Jan. 04	9 Mar. 05	9 Jan. 13	. . . .	27 Punjabis.
	Unattd. List	**Dodd**, Percy Chas. Russell (50, 59, 169)	15 Nov. 85	9 Jan. 04	2 Sep. 05	9 Jan. 13	. . . .	31 Lancers.
45	Unattd. List	**Lake**, Morice Challoner (51, 52, 59, 71, 170)	28 May 85	9 Jan. 04	3 Apl. 05	9 Jan. 13	. . . .	109 Infantry.
	Unattd. List	**Cobb**, Charles (51, 65a, 66, 95, 167, 169)	20 Aug. 84	9 Jan. 04	10 Mar. 05	9 Jan. 13	*Ty. m.* 25 Feb. 18	29 Punjabis.
	Unattd. List	**Thyne**, Walter (66a, 107, 148b, 167)	28 Mar. 84	9 Jan. 04	3 Apl. 05	9 Jan. 13	. . . .	90 Punjabis.
	Unattd. List	**Ring**, Jas. Sinclair Henry (172)	4 Jan. 84	9 Jan. 04	24 Mar. 05	9 Jan. 13	. . . .	46 Punjabis.
	Unattd. List	**Mallock**, Arthur Richard Ogilvie (50, 53, 71, 205a)	21 Oct. 85	9 Jan. 04	29 Mar. 05	9 Jan. 13	. . . .	42 Regiment.
50	Unattd. List	**Crawford**, John Douglas, D.S.O., M.C. (50, 52, 125a, 148b, 167)	21 Jan. 85	9 Jan. 04	11 Mar. 05	9 Jan. 13	*m.* 1 Jan. 18 *Ty. c.* 24 May 18	89 Punjabis.
	Unattd. List	**Grose**, Edward (137)	11 Feb. 85	9 Jan. 04	10 Sep. 05	9 Jan. 13	. . . .	16 Rajputs.
	S. W. Bord.	**Phythian-Adams**, Edward Gwynne Phythian (50, 125a, 170)	3 Aug. 83	27 Jan. 04	23 Aug. 05	27 Jan. 13	. . . .	80 Infantry.
	Welsh R.	**Williams**, Severn (66a, 94, 162, 169)	14 Feb. 84	27 Jan. 04	11 Oct. 05	27 Jan. 13	. . . .	Pol. Dept.
	Bedf. R.	**Place**, Norbert Dumayne (51, 52, 66, 137, 162)	18 Sep. 81	27 Jan. 04	3 Aug. 07	27 Jan. 13	. . . .	8 Rajputs.
55	Berks. R.	**McCausland** Edmund Thos. Wm. (51, 52, 71, 157)	11 Dec. 83	27 Jan. 04	9 Mar. 09	27 Jan. 13	. . . .	3 G. R.
56	Suss. R.	**Story**, Henry Francis (51, 52, 167) *Service for pension;—Attd. Reglrs. & Embd. Mila. from 27 Jan. to 27 Sep. 02.*	23 Dec. 83	27 Jan. 04	9 Dec. 05	27 Jan. 13	*Ty. m.* 25 Feb. 18	35 Sikhs.

57	R. War. R.	**Budden**, John Austin (167, 172a)	27 May 84	27 Jan. 04	2 Nov. 05	27 Jan. 13	. . . .	Mily. Farms Dept.
		Service for pension;—Embd. Mila. from 7 Apl. to 1 Oct. 02.						
	R. Scots	**Brickmann**, Francis Ivan Oscar, M.C. (50, 53, 57, 66a, 71, 136, 161)	19 Aug. 84	27 Jan. 04	19 Mar. 08	27 Jan. 13	. . . .	119 Infantry.
	Bedf. R.	**Hamilton**, Archibald Francis, M.C. (50, 53, 71, 195)	12 Mar. 83	27 Jan. 04	20 Aug. 05	27 Jan. 13	. . . .	61 Pioneers.
60	E. Surr. R.	**Power**, John Edward (95, 107, 140, 148b, 170)	5 Sep. 82	27 Jan. 04	28 June 05	27 Jan. 13	. . . .	19 Lancers.
		Service for pension;—Austln. Horse from 12 May to 13 Aug. 02.						
	Midd'x. R.	**Tree**, Harry Ben Le Despencer (51, 52, 57, 66a, 137)	3 Mar. 82	10 Feb. 04	14 Oct. 05	10 Feb. 13	. . . .	112 Infantry.
	Unattd. List	**Greig**, Robert Moore (51, 52, 66a)	7 Nov. 83	19 Aug. 03	7 Nov. 04	13 Feb. 13	. . . .	42 Regiment.
		Forfeited from 19 Aug. 12 to 12 Feb. 13 for promotion.						
	York. R.	**Bovill**, William John (50, 53, 66, 66b, 71, 157)	14 May 85	2 Mar. 04	6 Feb. 06	2 Mar. 13	. . . .	43 Regiment.
	Linc. R.	**Lentaigne**, Edward Charles, D.S.O. (51, 52, 66a, 157)	16 June 84	2 Mar 04	13 Aug. 05	2 Mar. 13	. . . .	4 G. R.
65	North'n R.	**Raymond**, Maurice Claud, M.C. (51, 71, 137, 162)	2 Dec. 84	2 Mar. 04	27 Aug. 05	2 Mar. 13	. . . .	34 Horse.
	York. R.	**Gore**, Frederick Lawrence (51, 52, 57, 66a, 137)	18 Jan. 84	2 Mar. 04	9 Apl. 06	2 Mar. 13	. . . .	113 Infantry.
	E. Surr. R.	**Allan**, Patrick Alexander (51, 167)	3 July 84	2 Mar. 04	14 Dec. 05	2 Mar. 13	. . . .	66 Punjabis.
	Som. L. I.	**Williams**, Harold Radcliffe (50, 66a, 125a, 157)	30 Aug. 84	16 Mar. 04	16 Mar. 09	16 Mar. 13	. . . .	10 G. R.
	Manch. R.	**Grace**, Humphrey Gilbert, M.C. (57, 71, 169)	30 July 83	27 Jan. 04	13 Nov. 05	11 Apl. 13	. . . .	21 Cavalry.
		Forfeited for promotion from 27 Jan. to 10 Apl. 13.						
70	R. W. Fus.	**Wilson**, Nigel Maitland, D.S.O. (50, 59, 157)	6 Apl. 84	20 Apl. 04	27 Nov. 09	20 Apl. 13	m. 1 Jan. 17	7 G. R.
	Devon. R.	**Marshall**, John Stuart, D.S.O. (51, 52, 95, 162, 167) *Service for pension;—Embd. Mila. from 27 Feb. to 10 Oct. 02; for promotion from 21 Mar. to 9 Oct. 02.*	24 Feb. 83	3 Dec. 04	25 Sep. 07	14 May 13	Ty. m. 2 July 18 m. 3 June 18	35 Sikhs.
	Essex R.	**Wills**, Edgar Vernon (51, 71, 72, 105)	1 Aug. 83	4 June 04	26 July 09	4 June 13	Ty. m. 30 July 18	9 Infantry.
	Manch R.	**Pilnston**, George Herbert (51, 57, 71, 137)	26 Feb. 85	4 June 04	29 June 06	4 June 13	. . . .	11 Rajputs.
	R. A.	**Abbott**, Reginald Stuart, M. C. (50, 53, 71, 161, 172a)	12 Nov. 82	4 June 04	17 Apl. 09	4 June 13	m. 3 June 17	38 C. I. Horse.
75	R. A.	**Bridge**, Richard Edwyn Athol, D.S.O. (66, 71, 205a)	13 Apl. 82	4 June 04	29 Sep. 07	4 June 13	. . . .	S. and T. Corps.
	R. A.	**Burmester**, Zante Gower (50, 53, 172a)	23 May 85	29 July 04	10 Nov. 08	29 July 13	. . . .	31 Lancers.
	R. A.	**Howlett**, Edmund George (51, 172)	7 Dec. 84	29 July 04	19 Mar. 09	29 July 13	. . . .	72 Punjabis.
	R. A.	**Pengelley**, George Hastings (66, 169)	26 Feb. 85	29 July 04	17 Jan. 14	29 July 13	Ty. m. 25 Feb. 18.	S. and T. Corps.
	Devon. R.	**Gunn**, Alistair Dudley (51, 107, 155) *Forfeited from 2 Mar. to 1 Aug. 13, for promotion.*	17 Apl. 84	2 Mar. 04	11 July 05	2 Aug. 13	. . . .	110 Lt. Infantry.
80	Unattd. List	**Hackett**, Harold Monck Mason, M.C. (50, 66a, 157)	2 June 83	3 Aug. 04	28 Oct. 05	3 Aug. 13	. . . .	6 G. R.
	Unattd. List	**Hay**, John Basil Whitmore (167)	30 July 85	13 Aug. 04	11 Nov. 05	13 Aug. 13	. . . .	36 Sikhs.
	Unattd. List	**Tuite-Dalton**, Edward George Thomas, M.C. (50, 53, 57, 157)	2 July 84	13 Aug. 04	28 Oct. 05	13 Aug. 13	. . . .	3 G. R.
	Midd'x. R.	**White**, Archibald Geo. (51, 172)	1 Oct. 84	13 Aug. 04	15 Jan. 06	13 Aug. 13	. . . .	126 Infantry.
	Unattd. List	**Farrell**, Gilbert Valentine (51, 52, 71, 172)	30 Sep. 84	13 Aug. 04	29 Oct. 05	13 Aug. 13	. . . .	99 Infantry.
85	Unattd. List	**Lowry-Corry**, Walter (51, 172)	6 Mar. 85	13 Aug. 04	30 Oct. 05	13 Aug. 13	. . . .	23 Cavalry.
	Unattd. List	**Gover**, Arnold Charles, M.C. (51, 155)	11 Mar. 84	13 Aug. 04	1 Nov. 05	13 Aug. 13	. . . .	121 Pioneers.
87	Unattd. List	**Middlemass**, Geoffrey de Saumarez Hume (51, 52, 137)	18 July 84	13 Aug. 04	29 Oct. 05	13 Aug. 13	. . . .	122 Infantry.

Captains—*contd.*

No	Former Corps.	Names.	Date of birth.	Date of first commission.	Date of appointment to Indian Army.	Date of rank. Substantive.	Date of rank. Army.	Remarks.
88	Unattd. List	**Murray,** Stewart George Cromartie C.I.E. (51, 59, 71, 95, 137, 172*a*)	21 Apl. 84	13 Aug. 04	28 Oct. 05	13 Aug. 13	. . . .	12 Pioneers.
	Unattd. List	**Collum,** James Alfred, D.S.O. (51, 65, 66, 137) [L]	13 Aug. 84	13 Aug. 04	28 Oct. 05	13 Aug. 13	. . . .	26 Lt. Cavalry.
90	Unattd. List	**Tilly,** Robert (57)	3 June 84	13 Aug. 04	3 Nov. 05	13 Aug. 13	. . . .	123 Rifles.
	Unattd. List	**Berryman,** Edward Rolleston Palmer (51, 52, 57, 157)	10 July 83	13 Aug. 04	29 Oct. 05	13 Aug. 13	*m.* 1 Jan. 18	39 Garhwal Rifles.
	Unattd. List	**Allan,** Keith Douglas (51, 52, 165)	26 Sep. 84	13 Aug. 04	29 Oct. 05	13 Aug. 13	. . . .	35 Sikhs.
	Unattd. List	**Turner,** John (52, 71)	2 June 84	13 Aug. 04	30 Oct. 05	13 Aug. 13	. . . .	114 Mahrattas.
	Unattd. List	**Maturin,** Hugh Geoffrey (51, 195, 204) [L]	26 Mar. 85	13 Aug. 04	29 Oct. 05	13 Aug. 13	. . . .	61 Pioneers.
95	Unattd. List	**Duke,** Oriel Augustus (45*a*, 162, 165, 169) [L]	18 Aug. 84	13 Aug. 04	28 Oct. 05	13 Aug. 13	. . . .	22 Cavalry.
	Unattd. List	**Harrison,** Christopher Heathfield (65, 65*a*, 66, 172)	1 Sep. 84	13 Aug. 04	29 Oct. 05	13 Aug. 13	. . . .	S. & T. Corps.
	Unattd. List	**Abbott,** Frank Berkeley, D.S.O. (51, 52, 57)	2 Aug. 85	13 Aug. 04	30 Oct. 05	13 Aug. 13	. . . .	6 G. R.
	Unattd. List	**Rebsch,** William Knowles, D.S.O. (65*a*, 66*a*, 66, 125*a*, 167)	29 Aug. 85	13 Aug. 04	29 Oct. 05	13 Aug. 13	*Ty. m.* 25 Feb. 18	S. and T. Corps.
	Unattd. List	**Bird,** Austin Carlos (66, 172*a*)	26 Jan. 84	13 Aug. 04	31 Oct. 05	13 Aug. 13	. . . .	S. and T. Corps.
100	Unattd. List	**James,** Frank (45*a*, 50, 71, 1*fg*)	10 May 85	13 Aug. 04	29 Oct. 05	13 Aug. 13	. . . .	28 Lt. Cavalry.
	Unattd. List	**Pulley,** Owen Christopher (137)	8 Sep. 85	13 Aug. 04	22 Oct. 06	13 Aug. 13	*Ty. m.* 10 Feb. 18.	16 Rajputs.
	Unattd. List	**Nicoll,** Noel Fiennes (51, 52, 66, 172)	22 Dec. 84	13 Aug. 04	29 Oct. 05	13 Aug. 13	. . . .	54 Sikhs.
	Unattd. List	**Montefiore,** Claude Emanuel (51, 52, 66*a*, 155)	17 Aug. 83	13 Aug. 04	29 Oct. 05	13 Aug. 13	. . . .	110 Lt. Infantry.
	Unattd. List	**Mangin,** Etienne Bessonnet, M.C. (51, 172)	4 Sep. 84	13 Aug 04	29 Oct. 05	13 Aug. 13	. . . .	107 Pioneers.
105	R. Suss. R.	**Wilson,** Thomas Kenneth (51, 52, 71)	20 Apl. 84	13 Aug. 04	14 July 09	13 Aug. 13	. . . .	20 Horse.
	Unattd. List	**Benton,** Hugh Gordon (45*a*, 51, 52, 71, 137)	12 June 85	13 Aug. 04	29 July 06	13 Aug. 13	. . . .	4 Cavalry.
	Essex R.	**Lloyd,** Charles Geoffrey, M.C. (94, 100)	12 Mar. 84	3 Sep. 04	26 Oct. 15	3 Sep. 13	. . . .	S. & T. Corps.
	Unattd. List	**Doveton,** Arnold Chadwick (51, 57, 66, 136, 162, 165, 169, 204) *Forefeited from 9 Jan. to 18 Nov. 13, for promotion.*	20 Mar. 84	9 Jan. 04	23 May 05	19 Nov. 13	. . . .	21 Punjabis.
	R. Suss. R.	**Whittell,** Hubert Massey, (65*a*, 66, 162, 169)	24 Mar. 83	3 Dec. 04	28 June 07	3 Dec. 13	. . . .	56 Rifles.
110	E. Lan. R.	**Whitaker,** Sidney Stratton, M. C. (66, 95, 157)	2 Dec. 83	3 Dec. 04	9 July 07	3 Dec. 13	. . . .	7 G. R.
	R. Suss. R.	**Logan-Home,** William Miles (51, 52, 137)	6 June 84	3 Dec. 04	1 July 08	3 Dec. 13	. . . .	112 Infantry.
	York. R.	**Gompertz,** Martin Louis Alan (50, 52, 57, 71, 127, 153, 162, 204) [*l*]	23 Feb. 86	3 Dec. 04	6 Apl. 06	3 Dec. 13	*m.* 21 Dec. 16. *Ty. l. c.* 24 Apl. 18.	108 Infantry.
113	Wilts. R.	**Ralston,** Norman Hopkins Hargreaves (51, 52, 66*a*, 71, 72*a*, 106)	25 July 84	3 Dec. 04	10 Aug. 07	3 Dec. 13	. . . .	9 Infantry.

114	Linc. R.	**Gardner,** Charles Philipps (51, 66, 71, 105)	30 July 83	3 Dec. 04	21 Mar. 09	3 Dec. 13	. . . .	73 Infantry.
115	R. A.	**Young,** Hubert Winthrop (50, 52, 155) [L]	6 July 85	21 Dec. 04	24 Dec. 08	21 Dec. 13	. . . .	116 Mahrattas.
	R. A.	**Noyes,** Cyril Duprè, M.C. (51, 137)	3 Feb. 85	21 Dec. 04	23 Sep. 08	21 Dec. 13	. . . .	2 Lt. Infy.
	R. A.	**Noel,** Edward William Charles, C.I.E. (51, 52, 59, 175) [L]	14 Apl. 86	21 Dec. 04	27 Nov. 08	21 Dec. 13	. . . .	Poll. Dept.
	R. A.	**Scott,** Francis Burgess	3 Mar. 85	21 Dec. 04	16 Feb. 09	21 Dec. 13	. . . .	Survey Dept.
	Unattd. List	**Tate,** John Creery (95, 132, 153, 172)	14 Aug. 84	18 Jan. 05	25 Mar. 06	18 Jan. 14	*m.* 10 Mar. 17	Poll., Bombay.
120	Unattd. List	**Daniels,** Arthur Marston (45*a*, 51, 57, 94, 137, 162, 172)	4 Aug. 85	18 Jan. 05	7 Mar. 06	18 Jan. 14	. . . .	3 Horse.
	Unattd. List	**Dormer,** James Garrad (51, 52, 57, 71, 137)	19 Dec. 84	18 Jan. 05	19 Mar. 06	18 Jan. 14	. . . .	104 Rifles.
	Unattd. List	**Carpendale,** Trevor Maxwell (45*a*, 51, 52, 57, 71, 137, 172*a*)	18 Sep. 85	18 Jan. 05	16 Mar. 06	18 Jan. 14	. . . .	8 Cavalry.
	Unattd. List	**Finnis,** Herbert Cobb. (52, 95, 167, 169, 172*a*)	6 May 85	18 Jan. 05	15 Mar. 06	18 Jan. 14	. . . .	Poll. Dept.
	Unattd. List	**Bell-Kingsley,** Harold Evelyn William, D.S.O. (50, 53, 157)	23 Dec. 85	18 Jan. 05	16 Mar. 06	18 Jan. 14	. . . .	4 G. R.
125	Unattd. List	**Ostrehan,** Malcolm (51, 52, 59, 157)	18 May 86	18 Jan. 05	14 Mar. 06	18 Jan. 14	. . . .	9 G. R.
	Unattd. List	**Lane,** Hugh Robert Charles (51, 52, 59, 66*a*, 155*a*, 157, 170)	20 Dec. 85	18 Jan. 05	14 Mar. 06	18 Jan. 14	. . . .	5 G. R.
	Unattd. List	**Haig,** Arthur Brodie (50, 53, 59, 172)	31 Jan. 86	18 Jan. 05	17 Mar. 06	18 Jan. 14	. . . .	24 Punjabis.
	Unattd. List	**Greenhouse,** Frank Stewart (51, 52, 57, 66*a*, 163, 169, 204)	24 Dec. 85	18 Jan. 05	19 Mar. 06	18 Jan. 14	. . . .	94 Infantry.
	Unattd. List	**Muir,** Archibald Huleatt Huntly (50, 53, 59, 167)	20 Apl. 86	18 Jan. 05	16 Mar. 06	18 Jan. 14	*Ty. m.* 5 Sep. 18	15 Sikhs.
130	Unattd. List	**Blackwell,** Henry (52, 66*a*, 71, 94, 137, 162*a*, 169)	5 Aug. 84	18 Jan. 05	16 Mar. 06	18 Jan. 14	. . . .	106 Pioneers.
	Unattd. List	**Ogilvie,** Jas. Farquhar White (45*a*, 51, 167, 172)	16 Mar. 85	18 Jan. 05	17 Mar. 06	18 Jan. 14	. . . .	Guides.
	Unattd. List	**Heath,** Lewis Macclesfield, M.C. (50, 53, 59, 165, 169)	23 Nov. 85	18 Jan. 05	15 Mar. 06	18 Jan. 14	. . . .	19 Punjabis.
	Unattd. List	**Deane,** Chas. Herbert Stanley (50, 57, 137, 161)	26 Mar. 86	18 Jan. 05	17 Mar. 06	18 Jan. 14	. . . .	44 Infantry.
	Unattd. List	**Paterson,** Thomas Schomberg (45*a*, 51, 52, 57, 162, 172)	29 May 85	18 Jan. 05	16 Mar. 06	18 Jan. 14	. . . .	9 Lancers.
135	Unattd. List	**Reynolds,** Eric William (51, 52, 57, 162)	27 June 85	18 Jan. 05	14 Mar. 06	18 Jan. 14	. . . .	96 Infantry.
	Unattd. List	**Howell,** Geoffrey Bulmer (45*a*, 51, 52, 169)	15 Nov. 85	18 Jan. 05	17 Mar. 06	18 Jan. 14	. . . .	19 Lancers.
	Unattd. List	**Rainsford Hannay,** Donald (50, 53, 57, 170)	16 Oct. 85	18 Jan. 05	17 Mar. 06	18 Jan. 14	*Ty. m.* 5 *Dec.* 16.	53 Sikhs.
	Unattd. List	**Burn-Murdoch,** Jan (50, 52, 57, 165, 172*a*)	1 Nov. 85	18 Jan. 05	10 Mar. 06	18 Jan. 14		32 Pioneers.
	Unattd. List	**Moss,** Thomas (50, 53, 167)	1 Sep. 84	18 Jan. 05	8 Mar. 06	18 Jan. 14	*Ty. m.* 25 Feb. 18.	30 Punjabis.
140	Unattd. List	**Northey,** William Brook, M.C. (51, 59, 157) [*l*]	29 Dec. 84	18 Jan. 05	10 Mar. 06	18 Jan. 14	. . . .	1 G. R.
	Unattd. List	**Goldfrap,** Harold Wyn, M.C. (53, 155) [L]	8 Apl. 84	18 Jan. 05	8 Mar. 06	18 Jan. 14	. . . .	103 Lt. Infantry.
	Unattd. List	**Eden,** John Lancelot (51, 52, 57, 66)	24 Jan. 85	18 Jan. 05	6 Mar. 06	18 Jan. 14	. . . .	43 Regiment.
	Unattd. List	**Manners,** Charles Molyneux Sandys, D.S.O. (51, 52, 66, 71, 137, 185)	24 Jan. 85	18 Jan. 05	6 Mar. 06	18 Jan. 14	. . . .	104 Rifles.
	Unattd. List	**Browne,** St. John Arthur (51, 52, 66, 106, 118*a*, 125*a*, 148*b*, 165, 162, 172)	21 Feb. 85	18 Jan. 05	6 Mar. 06	18 Jan. 14	. . . .	53 Sikhs.
145	Unattd. List	**Harvey-Kelly,** Chas. Hamilton Grant Hume (50, 66*a*, 94, 169)	15 June 85	18 Jan. 05	10 Mar. 06	18 Jan. 14	m. *3 June 17*	127 Lt. Infantry.
	Unattd. List	**Warden,** William Reginald (51, 52, 57, 71, 94, 100, 172*a*, 195)	27 Dec. 85	18 Jan. 05	7 Mar. 06	18 Jan. 14	. . . .	75 Infantry.
	Unattd. List	**Cunningham,** William Bruce (50, 52, 66*a*, 165)	22 Mar. 85	18 Jan. 05	6 Mar. 06	18 Jan. 14	. . . .	38 Dogras.
	Unattd. List	**Trench,** Chas. Frederick, D.S.O. (45*a*, 51, 59, 71, 137)	29 July 85	18 Jan. 05	6 Mar. 06	18 Jan. 14	*Ty. m.* 25 Feb. 18.	7 Lancers.
	Unattd. List	**Waller,** John Edmond (51, 57, 167)	21 Sep. 84	18 Jan. 05	6 Mar. 06	18 Jan. 14	. . . .	45 Sikhs.
150	Unattd. List	**Mulloy,** Noel Frank Coote, M. C. (50, 66*a*, 71, 137)	28 Feb. 86	18 Jan. 05	6 Mar. 06	18 Jan. 14	. . . .	32 Lancers.
	Unattd. List	**Finnis,** Steriker William), 95, 162, 169)	3 Sep. 85	18 Jan. 05	7 Mar. 06	18 Jan. 14	. . . .	25 Punjabis.
	Unattd. List	**Lang,** Lionel Edward, M. C. (94, 132, 162*a*, 172*a*)	1 Nov. 85	18 Jan. 05	24 Mar. 06	18 Jan. 14	*Ty. m.* 25 Sep. 17.	Poll. employ.
	Unattd. List	**Broome,** Wm. George King (51, 52, 66*a*, 107, 125*a*, 167)	17 Dec. 84	18 Jan 05	8 Mar. 06	18 Jan. 14	. . . .	89 Punjabis.
	Unattd. List	**Money,** Wigram Seymour Elliot (45*a*, 51, 53, 57, 169)	7 July 85	18 Jan. 05	9 Mar. 06	18 Jan. 14	. . . .	22 Cavalry.
155	Unattd. List	**Stoddart,** Guy (51, 52, 71)	1 Sep. 84	18 Jan. 05	8 Mar. 06	18 Jan. 14	. . . .	104 Rifles.
156	Unattd. List	**O'Rorke,** John Marcus William, D.S.O. (45*a*, 50, 52, 71, 162, 169)	12 Nov. 84	24 Jan. 05	12 Mar. 08	24 Jan. 14	. . . .	25 Cavalry.
		Service for pension from 19 Jan. 07.						

No.	Former Corps.	Names.	Date of birth.	Date of first commission.	Date of appointment to Indian Army.	Date of Rank. Substantive.	Date of Rank. Army.	Remarks.
		Captains—*contd.*						
157	R. Suss. R.	**Ryan,** Denis George Jocelyn, D.S.O. (50, 52, 157)	6 Apl. 85	28 Jan. 05	11 July 07	28 Jan. 14	. . . .	6 G. R.
	Bedf. R.	**Daly,** Christopher Edward (50, 57, 167)	18 Dec. 84	28 Jan. 05	29 July 07	28 Jan. 14	. . . .	66 Punjabis.
	Wilts. R.	**Edwards,** Douglas Bruce (51, 66a, 71, 162 .)	15 Oct. 84	28 Jan. 05	6 Mar. 07	28 Jan. 14	. .	39 C. I. Horse.
160	4 D. G.	**Vicary,** Ernest Samuel (45a. 51, 65, 71) *Forfeited from 27 Jan. 13 to 19 Apl. 14 for promotion.*	10 Dec. 84	27 Jan. 04	10 Nov. 05	20 Apl. 14	*m.* 3 *June* 17	16 Cavalry.
	R. A.	**Hogan,** Eric Oakley (65a, 66a, 95, 162, 169, 205a)	8 Nov. 84	17 May 05	27 Nov. 16	17 May 14	. . . .	S. & T. Corps.
	Essex R.	**Wright,** Arthur St. John (65a, 66, 72)	27 Apl. 83	20 May 05	28 Aug. 14	20 May 14	*Ty. m.* 27th Feb. 18.	S. and T. Corps.
	North'n. R.	**Miller,** George Nigel (51, 52, 167)	6 July 83	20 May 05	4 Nov. 08	20 May 14	. . . .	90 Punjabis.
	Y.&Lanc.R.	**Philips,** James Washington (51, 65a, 66, 137)	29 Jan. 84	20 May 05	27 Oct. 08	20 May 14	. . . .	S. and T. Corps.
165	W. Rid. R	**Campbell,** Colin (50, 52, 57, 66a, 162, 167, 169)	27 Nov. 82	20 May 05	9 Aug. 08	20 May 14	. . . .	74 Punjabis.
	Midd'x R.	**Fitz Gerald,** Arthur (65, 66)(e) *Service for pension—In Ranks from 1 Feb. 98 to 20 June 05 (half to count)*	31 July 81	21 June 05	13 Aug. 11	21 June 14	. . . .	S. and T. Corps.
	Sco. Rif.	**Campbell,** Frederick Robert Maxwell (167, 175)	27 Jan. 84	21 June 05	30 Sep. 09	21 June 14		5 G. R.
	Unattd. List	**Jukes,** Andrew Henry, D.S.O. (51, 52, 57, 157, 172a)	11 Nov. 85	22 June 05	6 Dec. 06	22 June 14	*m.* 1 Jan. 17	9 G. R.
	Unattd. List	**Colvin,** Elliot James Dowell (161, 172a) *Forfeited from 9 Jan. 13 to 8 July 14 for promotion.*	27 July 85	9 Jan. 04	23 Mar. 05	9 July 14	*Ty. m.* 28 Sept. 18.	Poll. Dept.
170	L'pool. R.	**Jackson,** Mansel Halket, M.C. (43, 45a, 51, 71) *Service for pension—In Ranks from 24 Oct. 02 to 18 July 05 (half to count)*	24 Nov. 82	19 July 05	1 Mar. 07	19 July 14	. . .	29 Lancers.
	R. A.	**Brown,** Eric Carmichael (157)	17 Jan. 85	27 July 05	12 July 10	27 July 14	. . . .	6 G. R.
	Suff. R.	**Clarke,** Herbert	15 Jan. 83	2 Aug. 05	10 Jan. 08	2 Aug. 14	. . . .	4 Rajputs.
	Unattd. List	**Græme,** Ninian Francis (51, 52, 71, 157)	28 Oct. 84	5 Aug. 05	3 Dec. 06	5 Aug. 14	. . . .	1 G. R.
	Unattd. List	**Welch,** Frederick Ernest (57, 148)	13 Jan. 84	5 Aug. 05	6 Dec. 06	5 Aug. 14	. . . .	110 Lt. Infantry.
175	Unattd. List	**Hertz-Smith,** Geoffrey Vidal, M.C. (172)	19 Feb. 85	5 Aug. 05	21 Oct. 06	5 Aug. 14	. . . .	27 Punjabis.
	Unattd. List	**Smith,** Alexander Dallas, M.C. (65a, 157)	17 Oct. 85	5 Aug. 05	6 Dec. 06	5 Aug. 14	. . . .	2 G. R.
	Unattd. List	**Jopp,** Charles Hector Keith (50, 53, 137, 172a)	1 Sep. 85	5 Aug. 05	8 Dec. 06	5 Aug. 14	. . . .	14 Lancers.
	Unattd. List	**Ford,** Evelyn George (51, 52, 66a, 155)	18 Sep. 84	5 Aug. 05	3 Nov. 06	5 Aug. 14	. . . .	108 Infantry.
	Unattd. List	**Brett,** Vernon Wilford (51, 52, 65a, 66, 157, 172a)	23 July 84	5 Aug. 05	18 Nov. 06	5 Aug. 14	. . . .	Mily. Farms Dept.
180	Unattd. List	**Western,** Charles Edward Murray (45a, 51, 52, 172)	22 Mar. 85	5 Aug. 05	18 Nov. 06	5 Aug. 14	. . . .	37 Lancers.
181	Unattd. List	**Harbord,** Lionel Berkeley, M.C. (51, 52, 66a)	17 May 86	5 Aug. 05	18 Nov. 06	5 Aug. 14	*Ty. m.* 25 Feb. 18.	44 Infantry.

182	Unattd. List	**Goad**, Cecil Elliot Godfrey Boileau, M.C. (51, 52, 172)	5 Jan. 85	5 Aug. 05	16 Nov. 06	5 Aug. 14	. . . .	128 Pioneers.
	Unattd. List	**Dockrell**, Morgan (66a, 162) *Service for pension from 10 Jan. 06.*	13 May 83	5 Aug. 05	25 Feb. 08	5 Aug. 14	. . . .	116 Mahrattas.
	Unattd. List	**Grylls**, William Michell (51, 52, 59, 66a, 169, 172)	9 Jan. 85	5 Aug. 05	23 Oct. 06	5 Aug. 14	. . . .	52 Sikhs.
185	Unattd. List	**Acworth**, Douglas Harry, M.C. (50, 53, 66a, 169)	27 Oct. 85	5 Aug. 05	18 Nov. 06	5 Aug. 14	*Ty. m.* 25 Feb. 18.	55 Rifles.
	Unattd. List	**Shirley**, Sewallis Robert, M.C. (50, 172)	31 Jan. 85	5 Aug. 05	18 Nov. 06	5 Aug. 14	. . . .	54 Sikhs.
	Unattd. List	**Bray**, Norman Napier-Evelyn, M.C. (50, 52, 71, 167)	17 Aug. 85	5 Aug. 05	18 Nov. 06	5 Aug. 14	. . . .	18 Lancers.
	Unattd. List	**Yearsley**, Ralph Algernon (172)	13 Sep. 86	5 Aug. 05	18 Nov. 06	5 Aug. 14	*Ty. m.* 6 Mar. 18.	28 Punjabis.
	Unattd. List	**Beaman**, Ardern Arthur Hulme, D.S.O. (45a, 51, 57, 66, 71, 162, 169)	18 Aug. 86	5 Aug. 05	30 Sep. 06	5 Aug. 14	. . . .	1 Lancers.
190	Unattd. List	**Farebrother**, Francis Hughes (50, 53, 57, 94, 169) [L]	23 July 86	5 Aug. 05	8 Dec. 06	5 Aug. 14	. . . .	128 Pioneers.
	Unattd. List	**Inskip**, Roland Debenham, D.S.O., M.C. (51, 59, 172)	17 Sep. 85	5 Aug. 05	8 Dec. 06	5 Aug. 14	. . . .	59 Rifles.
	Unattd. List	**Trall**, William Samuel, M.C. (50, 53, 172)	13 Nov. 85	5 Aug. 05	22 Nov. 06	5 Aug. 14	*Ty. m.* 25 Feb. 18.	57 Rifles.
	Unattd. List	**Malet**, Alexander Wyndham (51, 52, 59, 107, 125a, 129, 140, 144b, 148b, 162, 165, 169, 205a)	28 Feb. 86	5 Aug. 05	30 Oct. 06	5 Aug. 14	. . . .	38 Dogras.
	Unattd. List	**Morse**, Lionel Henry, M.C. (59, 172a)	12 Sep. 85	5 Aug. 05	30 Oct. 06	5 Aug. 14	. . . .	62 Punjabis.
195	Unattd. List	**Childers**, Hugh Francis Eardley (71, 107, 137, 172a)	17 July 86	5 Aug. 05	6 Dec. 06	5 Aug. 14	. . . .	32 Lancers.
	Unattd. List	**James**, William Rhodes, M.C. (51, 52, 66, 125a, 167, 205a)	28 Aug 86	5 Aug. 05	18 Nov. 06	5 Aug. 14	. . . .	89 Punjabis.
	Unattd. List	**Wynter**, John Rawson, D.S.O. (52, 59, 172)	6 Oct. 85	5 Aug. 05	7 Dec. 06	5 Aug. 14	*Ty. m.* 28 July 18.	52 Sikhs.
	Unattd. List	**Bevington**, Leonard Farquhar (51, 52, 57, 66, 71, 72a, 155, 172a)	14 Jan. 85	5 Aug. 05	6 Dec. 06	5 Aug. 14	. . . .	121 Pioneers.
	Unattd. List	**Reyne**, Gerard van Rossum, D.S.O. (51, 52, 66a, 167)	29 Mar. 86	5 Aug. 05	23 Oct. 06	5 Aug. 14	. . . .	76 Punjabis.
200	Unattd. List	**Cookson**, Stewart Macdonald (65a, 66, 172)	8 May 86	5 Aug. 05	6 Dec. 06	5 Aug. 14	. . . .	21 Punjabis.
	Unattd. List	**Woodhouse**, Frank Saltoun (50, 53, 65a, 66, 172a)	5 Oct. 84	5 Aug. 05	23 Oct. 06	5 Aug. 14	. . . .	9 G. R.
	Unattd. List	**Story**, John Alexis, M.C. (51, 52, 195)	7 Feb. 86	5 Aug. 05	8 Dec. 06	5 Aug. 14	. . . .	61 Pioneers.
	Unattd. List	**Broughton**, Gilbert (162, 165, 169, 205a)	13 Oct. 86	5 Aug. 05	9 Oct. 06	5 Aug. 14	. . . .	33 Punjabis.
	Unattd. List	**Barbour**, Kenneth Durand (45a, 51, 52, 57, 66, 71)	22 Dec. 85	5 Aug. 05	18 Nov. 06	5 Aug. 14	. . . .	15 Lancers.
205	Unattd. List	**Phayre**, Robert Bernard, M.C. (50, 53, 157, 175) [L]	2 Jan. 86	5 Aug. 05	3 Dec. 06	5 Aug. 14	*Ty. m.* 23 *Oct. 16.* *m.* 3 June 18	4 G. R.
	Unattd. List	**Binny**, Alexander Chrystie Murray (51)	13 July 86	5 Aug. 05	24 Oct. 06	5 Aug. 14	. . . .	18 Infantry.
	Unattd. List	**Duncan**, Ronald Cardew (51, 52, 66a, 66, 157, 170, 172a)	9 Sep. 86	5 Aug. 05	4 Dec. 06	5 Aug. 14	*m.* 3 June 18	5 G. R.
	Unattd. List	**Simpson**, Alan Fraser (51, 52, 57, 66a)	6 Mar. 85	5 Aug. 05	9 May 07	5 Aug. 14	. . . .	95 Infantry.
	Unattd. List	**Butler**, Richard Bolger, M.C. (45a, 50, 52, 57, 71) *Service for pension from 13 June 06.*	26 June 84	5 Aug. 05	14 Dec. 07	5 Aug. 14	*m.* 1 Jan. 18 *Ty. l. c.* 24 Jan. 18.	30 Lancers.
210	Unattd. List	**Beazley**, Walter Edwin, M.C. (59, 72, 167, 169)	26 June 86	5 Aug. 05	6 Dec. 06	5 Aug. 14	*m.* 1 Jan. 18	54 Sikhs.
	Unattd. List	**Ismay**, Hastings Lionel (51, 52, 71, 162, 169)	21 June 87	5 Aug. 05	11 Dec. 06	5 Aug. 14	*Ty. m.* 31 Oct. 17.	21 Cavalry.
	Unattd. List	**Fraser**, Wm. Archibald Kenneth, M.C. (45a, 50, 66a, 71) [L]	19 Dec. 86	5 Aug. 05	7 Dec. 06	5 Aug. 14	*Ty. l. c.* 1 Oct. 18 *m.* 1 Jan. 18	39 C. I. Horse.
	Unattd. List	**Reilly**, Hugh Lambert, D.S.O. (162, 167, 169, 205a)	18 Oct. 86	5 Aug. 05	23 Oct. 06	5 Aug. 14	*Ty. m.*, 9 Apl. 15. 3 June 15	82 Punjabis.
	Unattd. List	**Morrell**, Harold George (57, 71, 72a, 162, 169, 195)	19 Nov. 85	5 Aug. 05	22 Oct. 06	5 Aug. 14	. . . .	119 Infantry.
215	Unattd. List	**Brooke**, Arthur (45a, 50, 52, 66, 71, 167)	13 Feb. 86	5 Aug. 05	21 Dec. 06	5 Aug. 14	. . . .	18 Lancers.
	Unattd. List	**Cave-Browne**, Horace (45a, 51, 52, 57, 66a, 66, 95, 172)	1 Jan. 86	5 Aug. 05	30 Oct. 06	5 Aug. 14	. . . .	23 Cavalry.
17	Unattd. List	**Thompson**, Paul Tempest Lambert (50, 52, 71, 195)	15 Oct. 85	5 Aug. 05	28 Oct. 06	5 Aug. 14	. . . .	79 Infantry.

No.	Former Corps.	Names.	Date of birth.	Date of first commission.	Date of appointment to Indian Army.	Date of Rank. Substantive.	Date of Rank. Army.	Remarks.
		Captains—*contd.*						
218	Unattd. List	**Segar,** Edward (46, 51, 52, 172)	13 Feb. 85	5 Aug. 05	20 Dec. 06	5 Aug. 14	. . . .	40 Pathans.
	Unattd. List	**Johnson,** Henry Allen Beaumont (45*a*, 51, 57, 137)	3 Jan. 87	5 Aug. 05	24 Oct. 06	5 Aug. 14	. . . .	26 Lt. Cavalry.
220	Unattd. List	**Lake,** Ernest Atwell Winter (51, 53, 107, 172)	5 Aug. 86	5 Aug. 05	28 Oct. 06	5 Aug. 14	. . . .	127 Lt. Infantry.
	Unattd. List	**Blood,** William Holcroft (45*a*, 51, 66, 162, 169)	29 May 87	5 Aug. 05	22 Oct. 06	5 Aug. 14	*Ty. m* 7 Oct 18.	Guides.
	Unattd. List	**Chesney,** Ellis Campbell (65*a*, 66, 161, 165, 169, 205*a*) [L]	3 May 86	5 Aug. 05	8 Dec. 06	5 Aug. 14	. . . .	S. & T. Corps.
	Unattd. List	**Yeats-Brown,** Francis Charles Claypon (45*a*, 51, 52, 71, 139, 162, 169) [*l*]	15 Aug. 86	5 Aug. 05	6 Dec. 06	5 Aug. 14	. . . .	17 Cavalry.
	Unattd. List	**Spankie,** Charles George (51, 52, 66*a*)	31 May 86	5 Aug. 05	24 Oct. 06	5 Aug. 14	. . . .	42 Regiment.
225	Unattd. List	**Clark,** William Hartley (45*a*, 51, 127, 161, 169)	12 Jan. 87	5 Aug. 05	24 Oct. 06	5 Aug. 14	. . . .	5 Cavalry.
	Unattd. List	**Poole,** Evelyn Jas. Evered (51, 172)	2 Jan. 86	5 Aug. 05	23 Oct. 06	5 Aug. 14	. . . .	46 Punjabis.
	Unattd. List	**Loch,** Percy Gordon (51, 52, 88, 127, 162) [*l*]	16 Jan. 87	5 Aug. 05	7 Dec. 06	5 Aug. 14	. . . .	Pol. Dept.
	Unattd. List	**Elphinston,** William Graham (45*a*, 50, 52, 57, 66*a*, 71, 137)	5 June 86	5 Aug. 05	24 Oct. 06	5 Aug. 14	. . . .	34 Horse.
	Unattd. List	**Lucas,** Hugh de Neufville (51, 71)	6 Dec. 86	5 Aug. 05	19 Aug. 07	5 Aug. 14	*Ty. m.* 25 Feb. 18	30 Lancers.
230	Unattd. List	**Mainwaring,** Guy Rowland (157)	5 Aug. 85	5 Aug. 05	30 Oct. 06	5 Aug. 14	. . . .	39 Garhwal Rifles.
	North'n R.	**Lynch-Staunton,** Henry Victor Cormick (51, 172)	17 June 85	16 Aug. 05	18 June 08	16 Aug. 14	. . . .	124 Baluchis.
	Wilts R.	**Mulliken,** Edward Morris (52, 65*a*, 66, 155)	14 Jan. 85	16 Aug. 05	7 Apl. 09	16 Aug. 14	. . . .	S. & T. Corps.
	R. Ir. Rif.	**Kenny,** Edward Courtenay (50, 52, 107)	20 Nov. 85	16 Aug. 05	24 Sep. 07	16 Aug. 14	. . . .	98 Infantry.
	W. Rid. R.	**Smythe,** Arthur Clifford Gordon (51, 52, 127, 129, 137, 140, 148*b*, 169)	15 Jan. 85	16 Aug. 05	14 Mar. 08	16 Aug. 14	. . . .	5 Cavalry.
235	E. Yorks. R.	**Haynes,** William Feilden (51, 52, 106, 172)	8 Oct. 84	16 Aug. 05	23 Aug. 08	16 Aug. 14	. . . .	21 Punjabis.
	R. War. R.	**Newall,** Cyril Louis Norton (50, 53, 95, 157, 169) (*b*)	15 Feb. 86	16 Aug. 05	16 Sep. 09	16 Aug. 14	*Ty. l. c.,* 1 Feb. 16 *m.* 1 Jan. 18	2 G. R.
	E. Surr. R.	**Porter,** Alexander Barr (45*a*, 51, 66*a*, 107, 140, 148*b*)	24 Oct. 85	16 Aug. 05	30 June 07	16 Aug. 14	. . . .	6 Cavalry.
	Bedf. R.	**McCleverty,** Guy Massy, M.C. (57, 156)	17 Sep. 85	16 Aug. 05	29 July 07	16 Aug. 14	. . . .	2 G. R.
	Leic. R.	**Hornsby,** Robert William (57, 66*a*, 95, 162, 169)	7 Aug. 85	16 Aug. 05	2 Aug. 07	16 Aug. 14	*Ty. m.* 30 June 18	19 Punjabis.
240	North'n R.	**Roberts,** Francis Lane (72*a*, 107, 137, 162)	24 Apl. 86	16 Aug. 05	17 Aug. 07	16 Aug. 14	. . . .	7 Rajputs.
	R. War. R.	**Franks,** Kendal Ferguson (50, 53, 66*a*)	22 Sep. 86	16 Aug. 05	11 Mar. 08	16 Aug. 14	. . . .	117 Infantry.
	W. Yorks. R.	**Ponsonby,** Douglas Gordon, D.S.O. (51, 52, 165, 179*a*)	12 Apl. 86	16 Aug. 05	29 Jan. 11	16 Aug. 14	. . . .	34 Pioneers.
	Midd'x. R.	**Williams,** Stanley Price (59, 66*a*, 100, 162, 169, 199)	18 June 85	16 Aug. 05	4 Aug. 07	16 Aug. 14	. . . .	51 Sikhs.
	R. Suss. R.	**Hall,** William Draper, M.C. (57, 66*a*, 162, 205*a*)	13 Sep. 85	16 Aug. 05	26 July 09	16 Aug. 14	. . . .	5 Lt. Infantry.
245	Suff. R.	**Wightwick,** Herbert Milner (88, 132, 153, 172)	17 Apl. 85	16 Aug. 05	30 Nov. 07	16 Aug. 14	. . . .	Poll. employ.
	North'd Fus.	**Staples,** Edward Henry (51, 52, 66*a*, 137)	12 Sep. 85	16 Aug. 05	20 June 08	16 Aug. 14	. . . .	122 Infantry.
	Durh. L. I.	**Shute,** Cyril Aveling (50, 53, 66*a*, 172)	30 Jan. 86	16 Aug. 05	11 Mar. 08	16 Aug. 14	. . . .	107 Pioneers.
	Ches. R.	**Frost,** Frank Dutton, M.C. (51, 65, 66, 72*a*, 162) *Service for pension—Embd. Mila, 6 Jan. 02 to 4 Oct. 02—Imp. Yao., 20 Feb. 00 to 24 June 01 (Half to count); for promotion from 23 Jan. 02 to 3 Oct. 02.*	17 Jan. 82	23 May 06	21 Sep. 14	11 Sep. 14	*Ty. m.,* 6 Nov. 16 *m.* 1 Jan. 18.	S. and T. Corps.
249	Dragoons	**Hudson,** Brereton Rowland (45*a*, 52, 66*a*)	31 July 85	16 Sep. 05	15 Oct. 08	16 Sep. 14	. . . .	8 Cavalry.

251	E. Lan. R.	**Francis,** Reginald Frankland, M.C. (88, 165, 205*a*)	18 May 86	27 Sep. 05	2 Nov. 07	27 Sep. 14		47 Sikhs.
	Leic. R.	**Falkland,** Henry Louis Francis (52, 71, 137)	22 Sep. 82	8 Nov. 05	4 Aug. 07	8 Nov. 15		13 Rajputs
		Service for pension ;—In Ranks 3 yrs. 23ds. (half to count.)						
	R. Muns. Fus.	**Fowle,** Trenchard Craven Wm. (162, 169) [L].	28 Nov. 84	29 Nov. 05	14 Sep. 07	29 Nov. 14		40 Pathans.
	Devon. R.	**Kingdon,** Jack (51, 52, 57, 72*b*, 188*a*, 195)	2 Oct. 83	29 Nov. 05	2 Nov. 07	29 Nov. 14		86 Infantry.
255	Suff. R.	**Daly,** Claud Dangar (65*a*, 66, 167)	25 Feb. 84	29 Nov. 05	20 Oct. 07	29 Nov. 14		S. and T. Corps.
	Som. L. I.	**Burlton-Bennet,** John Arthur (65*a*, 66, 71)	26 Mar. 86	29 Nov. 05	26 Oct. 07	29 Nov. 14		Do.
	Leic. R.	**Masters** Oswald (51, 66*a*, 71)	12 Aug. 86	29 Nov. 05	9 Nov. 08	29 Nov. 14		99 Infantry.
	R. Sco. Fus.	**Brickmann,** John Macdonald (51, 66*a*, 137)	26 June 86	29 Nov. 05	23 May 08	29 Nov. 14		3 Brahmans.
	Leic. R.	**Cole,** Henry Alick (51, 65*a*, 66, 66*b*)	19 Aug. 85	29 Nov. 05	31 Aug. 08	29 Nov. 14		44 Infantry.
260	Durh L. I.	**Coote,** Mervyn Chidley (72)	5 June 85	29 Nov. 05	17 Sep. 09	29 Nov. 14		107 Pioneers.
	Conn. Rang.	**Wilberforce-Bell,** Harold (132, 137, 151, 157, 161*a*)	17 Nov. 85	29 Nov. 05	25 Mar. 08	29 Nov. 14	*Ty. m.* 4 May 18.	Poll. employ.
	Suff. R.	**Atkins,** Basil Sydney (137)	4 Apl. 85	13 Dec. 05	4 Nov. 08	13 Dec. 14		11 Rajputs.
	E. York. R	**Goldsmith,** Horace Armstrong, M.C. (53)	28 July 86	20 Dec. 05	24 July 08	20 Dec. 14	*m.* 3 June 18	95 Infantry.
	Unattd. List	**Davies,** George Bevan (51, 137)	24 July 83	24 Jan. 06	1 Dec. 07	24 Jan. 15		48 Pioneers.
		Service for pension from 15 Sept. 06.						
265	Unattd. List	**Beattie,** John Osborne (57, 137, 162)	3 Nov. 84	24 Jan. 06	12 Mar. 08	24 Jan. 15		S & T. Corps.
		Service for pension from 19 Jan. 07.						
	Unattd. List	**Beresford,** George de la Poer, M.C. (51, 52, 95, 169)	21 July 85	24 Jan. 06	10 Mar. 07	24 Jan. 15		10 Lancers.
	Unattd. List	**Cadell,** Arthur Patrick Hamilton (51, 52, 65, 66, 162, 175)	9 Aug. 86	24 Jan. 06	11 Mar. 07	24 Jan. 15		109 Infantry.
	Unattd. List	**Jacob,** Rupert Montague, M.C. (50, 53, 57, 94, 100, 162, 169)	31 July 86	24 Jan. 06	17 Mar. 07	24 Jan. 15		130 Baluchis.
	Unattd. List	**Christie,** Robert Cecil, M.C. (51, 52, 66*a*, 71, 137, 162)	12 Dec. 85	24 Jan. 06	10 Mar. 07	24 Jan. 15		102 Grenadiers.
270	Unattd. List	**Gardiner,** William Alexander (50, 53, 59, 157)	2 July 86	24 Jan. 06	13 Feb. 08	24 Jan. 15		4 G. R.
	Unattd. List	**Miskin,** William Lancelot (51)	4 Apl. 86	24 Jan. 06	12 Mar. 07	24 Jan. 15		120 Infantry.
	Unattd. List	**Young,** Richard (51, 52, 66*a*, 167, 172*a*)	22 July 86	24 Jan. 06	10 Mar 07	24 Jan. 15		62 Punjabis.
	Unattd. List	**Conder,** Hugh (51, 52, 57, 66*a*, 137, 172*a*)	17 Sep. 86	24 Jan. 06	10 Mar. 07	24 Jan. 15		112 Infantry.
	Unattd. List	**Hartwell,** John Redmond, D.S.O. (50, 53, 157)	7 May 87	24 Jan. 06	18 Mar. 07	24 Jan. 15		4 G. R.
275	Unattd. List	**Prendergast,** Norman Henry (50, 52, 66, 71, 169)	7 Nov. 86	24 Jan. 06	6 Apl. 07	24 Jan. 15		Guides.
	Unattd. List	**Bignell,** Guy Newcome, M.C. (66*a*, 95, 167)	3 Dec. 86	24 Jan. 06	17 Mar. 07	24 Jan. 15		29 Punjabis.
	Unattd. List	**Stranack,** Harington Rivers (51, 52, 57, 167)	8 Feb. 86	24 Jan. 06	17 Mar. 07	24 Jan. 15		29 Punjabis.
	Unattd. List	**Smith,** Victor Henry Sherwood (51, 71, 137, 169)	24 May 87	24 Jan. 06	15 Mar. 07	24 Jan. 15		32 Lancers.
	Unattd. List	**Vigors,** Mervyn Doyne, M.C. (45*a*, 50, 53, 71, 172)	19 Nov. 86	24 Jan. 06	12 Apl. 07	24 Jan. 15		9 Horse.
280	Unattd. List	**Jacomb,** Fredk. Basil Wood (51, 52, 66*a*, 155)	18 Sep. 85	24 Jan. 06	15 Mar. 07	24 Jan. 15		S. & T. Corps.
	Unattd. List	**Merritt,** Bertam Thomas Barnett (157)	24 Sep. 85	24 Jan. 06	27 Mar. 07	24 Jan. 15		10 G. R.
	Unattd. List	**Wilson,** Hugh Maitland (50, 52, 59, 65, 66*a*, 163, 169, 205*a*)	6 Apl. 86	24 Jan. 06	19 Mar. 07	24 Jan. 15		1 Lancers.
	Unattd. List	**Bremmer,** Harold Noel Keble (51, 52, 66)	3 Mar. 86	24 Jan. 06	10 Mar. 07	24 Jan. 15		114 Mahrattas.
	Unattd. List	**Evans-Gordon,** Kenmure Alick Garth (57, 71, 162)	20 Aug. 85	24 Jan. 06	8 Apl. 07	24 Jan. 15		39 C. I. Horse.
285	Glouc R.	**Rae,** Gustavus Rochfort (51, 52, 65*a*, 66, 137 172*a*)	1 Oct. 86	24 Jan. 06	16 Aug. 08	24 Jan. 15		7 Rajputs.
	Ches. R.	**Wheatly,** Arthur Haldane Spencer (45*a*, 51, 57, 137)	9 Sep. 86	24 Jan. 06	7 July 07	24 Jan. 15		6 Cavalry.
	S. W. Bord.	**Cahusac,** Chas. Fitzroy, D.S.O. (50, 52, 66*a*, 71, 196)	0 Sep. 85	24 Jan. 06	19 Sep. 07	24 Jan. 15		36 Horse.
	R. Lanc. R.	**Blackett,** Clement Parker (50, 52, 66*a*, 66*b*, 167)	4 Apl. 87	24 Jan. 06	9 Nov. 08	24 Jan. 15		6 G. R.
	Unattd. List	**Palmer,** Arthur Cecil (51, 66*a*, 72*a*)	28 June 87	24 Jan. 06	18 Mar. 07	24 Jan. 15		94 Infantry.
290	R. War. R.	**Lorimer,** John Malcolm (137)	7 Oct. 86	24 Jan. 06	17 Sep. 09	24 Jan. 15	*Ty. m.* 20 July 18	3 Horse.
	Unattd. List.	**Burridge,** Thos. Francis Stannard (107, 162, 167)	3 June 87	24 Jan. 06	9 Mar. 07	24 Jan. 15		35 Sikhs.
	Manch. R.	**Cates,** Kenneth George Hyde	21 July 85	24 Jan. 06	5 Mar. 09	24 Jan. 15		45 Sikhs.
	Unattd. List.	**Sartorius,** Gerald (51, 57, 95, 169) [*l*]	12 Nov. 87	24 Jan. 06	11 Mar. 07	24 Jan. 15	*Ty. m.* 14 Oct. 16.	6 Cavalry.
	Essex R.	**Fenton,** James Newmarch (95)	18 Apl. 87	24 Jan. 06	25 Sep. 11	24 Jan. 15		43 Infantry.
295	R. Sus. R.	**Atkins,** Reginald Fenner (50, 52, 57, 66*a*, 95, 167)	8 Mar. 87	24 Jan. 06	4 Aug. 08	24 Jan. 15		67 Punjabis.

No.	Former Corps.	Names.	Date of birth.	Date of first commission.	Date of appointment to Indian Army.	Date of Rank. Substantive.	Date of Rank. Army.	Remarks.
		Captains—*contd.*						
296	Unattd. List	**Pughe,** Owen Llewellyn (51, 57, 107, 140, 162, 172)	30 May 87	24 Jan. 06	18 Mar. 07	24 Jan. 15		72 Punjabis.
	Unattd. List	**McPherson,** Alan Bruce, M.C. (51, 52, 57, 71, 137, 162)	6 Jan. 87	24 Jan. 06	18 Mar. 07	24 Jan. 15		6 Lt. Infantry.
	Unattd. List	**Chesney,** Duncan Patrick (66 167) [L]	17 July 85	24 Jan. 06	10 Mar 07	24 Jan. 15		23 Pioneers.
	Unattd. List	**Yule,** Harry Victor (51, 65a, 66, 71, 169, 172)	4 Feb. 87	24 Jan. 06	12 Mar. 07	24 Jan. 15		25 Cavalry.
300	Norf. R.	**Pogson,** Charles Aubrey, M.C. (59, 66a, 95)	27 June 85	24 Jan. 06	23 Apl. 12	24 Jan. 15		117 Mahrattas.
	N. & D. R.	**Parsons,** Arthur Edward Broadbent (161, 165, 169, 204)	5 Aug. 84	4 Apl. 06	30 Sep. 08	4 Apl. 15		52 Sikhs.
	Ches. R.	**Huxford,** Harold James (51, 52, 72, 72a, 137, 175)	19 Oct. 84	25 Apl. 06	31 Mar. 09	25 Apl. 15		125 Rifles.
		*Service for pension ;—**In Ranks from 19 Nov. 03 to 24 Apl. 06** (half to count)*						
	N. Staff. R.	**Stone,** Burnett Sidney (66a)	25 Apl. 86	23 May 06	19 Mar. 09	23 May 15		88 Infantry.
	E. Surr. R.	**Smith,** Christopher Thos. Marshall (51, 52, 165, 72a)	7 July 85	23 May 06	16 July 09	23 May 15	*m.* 3 June 18	87 Punjabis.
305	R.W.Surr.R.	**Matthews,** Frank Melvin (172)	11 June 85	23 May 06	16 July 09	23 May 15	*Ty. m.* 25 Feb. 18.	22 Punjabis.
	R.W.Surr.R.	**Nixon,** Brian George Michael Freder(51, 52, 65, 66a, 71, 107, 167)	5 June 86	23 May 06	29 Oct. 08	23 May 15		41 Dogras.
	R. Sco. Fus.	**Nairne,** Robert Gordon (51, 107, 195)	2 Apl. 84	16 June 06	16 Aug. 08	16 June 15		61 Pioneers.
	R. A.	**Bacon,** Charles Gilmore (12, 162, 167, 172a, 205a)	19 Oct. 85	25 July 06	4 Sep. 11	25 July 15		29 Lancers.
	R. A.	**Daunt,** Charles O'Brien, M.C. (71, 172a)	3 Sep. 86	25 July 06	25 Sep. 09	25 July 15		38 C. I. Horse.
310	R. A.	**Todd,** Eric Ludor Tallent (50, 94, 169)	17 May 86	25 July 06	19 June 10	25 July 15		121 Pioneers.
	Conn. Rang.	**Markwick,** Harold William (65a, 66, 169)	15 June 83	15 Aug. 06	23 Sep. 14	15 Aug. 15		S. and T. Corps.
	Unattd. List	**Proctor,** George Norman (51, 66a, 195)	22 Aug. 86	29 Aug. 06	12 Nov. 08	29 Aug. 15		83 Infantry.
		Service for pension from 3 Aug. 07.						
	Unattd. List	**Barker,** Ronald William (104, 167)	20 Sep. 83	29 Aug. 06	13 Feb. 09	29 Aug. 15		Civil, Burma.
		Service for pension from 3 Aug. 07.						
	Unattd. List	**Blandy,** Raleigh, M.C. (120, 157)	22 May 84	29 Aug. 06	12 Nov. 08	29 Aug. 15		9 G. R.
		Service for pension from 3 Aug. 07.						
315	Unattd. List	**Secondé,** Emile Charles (57, 136, 162, 172a, 204)	17 June 84	29 Aug. 06	30 Sep. 08	29 Aug. 1		16 Rajputs.
		Service for pension from 3 Aug. 07.						
	Unattd. List	**Ball,** Eric Percy, M.C. (50, 155)	26 June 87	29 Aug. 06	25 Oct. 07	29 Aug. 15	*Ty. m.* 25 Feb. 18.	110 Lt. Infantry.
	Unattd. List	**Wells,** Thomas Ralph, M.C. (59, 173)	11 Mar. 87	29 Aug. 06	13 Nov. 07	29 Aug. 15		33 Punjabis.
	Unattd. List	**Erskine,** Claude Ernest Torin (51, 52, 59, 66, 169)	1 May 87	29 Aug. 06	13 Oct. 07	29 Aug. 15		Guides.
	Unattd. List	**Walker,** Edward John Patteson Travis (45a, 51, 52, 59, 71, 137)	10 Apl. 88	29 Aug. 06	12 Nov. 07	29 Aug. 15	*Ty. m.* 6 Aug. 18.	32 Lancers.
320	Unattd. List	**Haines,** John Bertram (51, 52, 71, 95, 137, 161, 174, 205a) [L]	31 July 86	29 Aug. 06	27 Oct. 07	29 Aug. 15		10 Jats.
	Unattd. List	**Robson,** Henry Wm. Cumine (53, 94, 157, 162, 169)	8 Dec. 86	29 Aug. 06	22 Oct. 07	29 Aug. 15		7 G. R.
	Unattd. List	**Evans,** Dudley Mac Neil, M.C. (51, 52, 59, 66a, 169)	11 Dec. 86	29 Aug. 06	27 Oct. 07	29 Aug. 15		55 Rifles.
	Unattd. List	**Coffin,** Alexander Sebright (65a, 66, 172)	7 Jan. 87	29 Aug. 06	26 Jan. 08	29 Aug. 15		28 Punjabis.
	Unattd. List	**Dyce,** Alan George (51, 52, 57, 95,)	24 Oct. 86	29 Aug. 06	13 Nov. 07	29 Aug. 15		S. & T. Corps.
325	Unattd. List	**Connell,** Victor John Alexander (45a, 51, 65, 71, 167, 172a)	18 Aug. 87	29 Aug. 06	13 Nov. 07	29 Aug. 15		13 Lancers.
	Unattd. List	**Dane,** Richard, M.C. (51, 52, 66, 71, 169, 205a)	21 Aug. 87	29 Aug. 06	12 Nov. 07	29 Aug. 15		21 Cavalry.
	Unattd. List	**Macdonald,** Harry, D.S.O. (45a, 51, 53, 71, 162, 172)	3 May 86	29 Aug. 06	27 Oct. 07	29 Aug. 15	3 June 17	111 Lancers.
	Wilts. R.	**Eberhardie,** Rudolph Charles (51, 52, 58, 155, 161, 169, 172a, 205a)	23 June 87	29 Aug. 06	16 June 08	29 Aug. 15		16 Mahrattas.
329	E. Lan. R.	**Porteous,** David Guy (50, 66a, 72a, 167)	6 Sep. 86	29 Aug. 06	2 Nov. 11	29 Aug. 15	*Ty. m.* 25 Feb. 18.	90 Punjabis.

251	E. Lan. R.	**Francis,** Reginald Frankland, M.C. (88, 165, 205*a*)	18 May 86	27 Sep. 05	2 Nov. 07	27 Sep. 14	. . . .	47 Sikhs.
	Leic. R.	**Falkland,** Henry Louis Francis (52, 71, 137)	22 Sep. 82	8 Nov. 05	4 Aug. 07	8 Nov. 15	. . . .	13 Rajputs
		Service for pension ;—In Ranks 3 yrs. 23ds. (half to count.)						
	R. Muns. Fus.	**Fowle,** Trenchard Craven Wm. (162, 169) [L].	28 Nov. 84	29 Nov. 05	14 Sep. 07	29 Nov. 14	. . . .	40 Pathans.
	Devon. R.	**Kingdon,** Jack (51, 52, 57, 72*b*, 188*a*, 195)	2 Oct. 83	29 Nov. 05	2 Nov. 07	29 Nov. 14	. . . .	86 Infantry.
255	Suff. R.	**Daly,** Claud Dangar (65*a*, 66, 167)	25 Feb. 84	29 Nov. 05	20 Oct. 07	29 Nov. 14	. . . .	S. and T. Corps.
	Som. L. I.	**Burlton-Bennet,** John Arthur (65*a*, 66, 71)	26 Mar. 86	29 Nov. 05	26 Oct. 07	29 Nov. 14	. . . .	Do.
	Leic. R.	**Masters** Oswald (51, 66*a*, 71)	12 Aug. 86	29 Nov. 05	9 Nov. 08	29 Nov. 14	. . . .	99 Infantry.
	R. Sco. Fus.	**Brickmann,** John Macdonald (51, 66*a*, 137)	26 June 86	29 Nov. 05	23 May 08	29 Nov. 14	. . . .	3 Brahmans.
	Leic. R.	**Cole,** Henry Alick (51, 65*a*, 66, 66*b*)	19 Aug. 85	29 Nov. 05	31 Aug. 08	29 Nov. 14	. . . .	44 Infantry.
260	Durh L. I.	**Coote,** Mervyn Chidley (72)	5 June 85	29 Nov. 05	17 Sep. 09	29 Nov. 14	*Ty. m.* 4 May 18.	107 Pioneers.
	Conn. Rang.	**Wilberforce-Bell,** Harold (132, 137, 151, 157, 161*a*)	17 Nov. 85	29 Nov. 05	25 Mar. 08	29 Nov. 14		Poll. employ.
	Suff. R.	**Atkins,** Basil Sydney (137)	4 Apl. 85	13 Dec. 05	4 Nov. 08	13 Dec. 14	. . . .	11 Rajputs.
	E. York. R.	**Goldsmith,** Horace Armstrong, M.C. (53)	28 July 86	20 Dec. 05	24 July 08	20 Dec. 14	*m.* 3 June 18	95 Infantry.
	Unattd. List	**Davies,** George Bevan (51, 137)	24 July 83	24 Jan. 06	1 Dec. 07	24 Jan. 15	. . . .	48 Pioneers.
		Service for pension from 15 Sept. 06.						
265	Unattd. List	**Beattie,** John Osborne (57, 137, 162)	3 Nov. 84	24 Jan. 06	12 Mar. 08	24 Jan. 15	. . . .	S & T. Corps.
		Service for pension from 19 Jan. 07.						
	Unattd. List	**Beresford,** George de la Poer, M.C. (51, 52, 95, 169)	21 July 85	24 Jan. 06	10 Mar. 07	24 Jan. 15	. . . .	10 Lancers.
	Unattd. List	**Cadell,** Arthur Patrick Hamilton (51, 52, 65, 66, 162, 175)	9 Aug. 86	24 Jan. 06	11 Mar. 07	24 Jan. 15	. . . .	109 Infantry.
	Unattd. List	**Jacob,** Rupert Montague, M.C. (50, 53, 57, 94, 100, 162, 169)	31 July 86	24 Jan. 06	17 Mar. 07	24 Jan. 15	. . . .	130 Baluchis.
	Unattd. List	**Christie,** Robert Cecil, M.C. (51, 52, 66*a*, 71, 137*a*, 162)	12 Dec. 85	24 Jan. 06	10 Mar. 07	24 Jan. 15	. . . .	102 Grenadiers.
270	Unattd. List	**Gardiner,** William Alexander (50, 53, 59, 157)	2 July 86	24 Jan. 06	13 Feb. 08	24 Jan. 15	. . . .	4 G. R.
	Unattd. List	**Miskin,** William Lancelot (51)	4 Apl. 86	24 Jan. 06	12 Mar. 07	24 Jan. 15	. . . .	120 Infantry.
	Unattd. List	**Young,** Richard (51, 52, 66*a*, 167, 172*a*)	22 July 86	24 Jan. 06	10 Mar. 07	24 Jan. 15	. . . .	62 Punjabis.
	Unattd. List	**Conder,** Hugh (51, 52, 57, 66*a*, 137, 172*a*)	17 Sep. 86	24 Jan. 06	10 Mar. 07	24 Jan. 15	. . . .	112 Infantry.
	Unattd. List	**Hartwell,** John Redmond, D.S.O. (50, 53, 157)	7 May 87	24 Jan. 06	18 Mar. 07	24 Jan. 15	. . . .	4 G. R.
275	Unattd. List	**Prendergast,** Norman Henry (50, 52, 66, 71, 169)	7 Nov. 86	24 Jan. 06	6 Apl. 07	24 Jan. 15	. . . .	Guides.
	Unattd. List	**Bignell,** Guy Newcome, M.C. (66*a*, 95, 167)	3 Dec. 86	24 Jan. 06	17 Mar. 07	24 Jan. 15	. . . .	29 Punjabis.
	Unattd. List	**Stranack,** Harington Rivers (51, 52, 57, 167)	8 Feb. 86	24 Jan. 06	17 Mar. 07	24 Jan. 15	. . . .	29 Punjabis.
	Unattd. List	**Smith,** Victor Henry Sherwood (51, 71, 137, 169)	24 May 87	24 Jan. 06	15 Mar. 07	24 Jan. 15	. . . .	32 Lancers.
	Unattd. List	**Vigors,** Mervyn Doyne, M.C. (45*a*, 50, 53, 71, 172)	19 Nov. 86	24 Jan. 06	12 Apl. 07	24 Jan. 15	. . . .	9 Horse.
280	Unattd. List	**Jacomb,** Fredk. Basil Wood (51, 52, 66*a*, 155)	18 Sep. 85	24 Jan. 06	15 Mar. 07	24 Jan. 15	. . . .	S. & T. Corps.
	Unattd. List	**Merritt,** Bertam Thomas Barnett (157)	24 Sep. 85	24 Jan. 06	27 Mar. 07	24 Jan. 15	. . . .	10 G. R.
	Unattd. List	**Wilson,** Hugh Maitland (50, 52, 59, 65, 66*a*, 163, 169, 205*a*)	6 Apl. 86	24 Jan. 06	19 Mar. 07	24 Jan. 15	. . . .	1 Lancers.
	Unattd. List	**Bremmer,** Harold Noel Keble (51, 52, 66)	3 Mar. 86	24 Jan. 06	10 Mar. 07	24 Jan. 15	. . . .	114 Mahrattas.
	Unattd. List	**Evans-Gordon,** Kenmure Alick Garth (57, 71, 162)	20 Aug. 85	24 Jan. 06	8 Apl. 07	24 Jan. 15	. . . .	39 C. I. Horse.
285	Glouc R.	**Rae,** Gustavus Rochfort (51, 52, 65*a*, 66, 137 172*a*)	1 Oct. 86	24 Jan. 06	16 Aug. 08	24 Jan. 15	. . . .	7 Rajputs.
	Ches. R.	**Wheatly,** Arthur Haldane Spencer (45*a*, 51, 57, 137)	9 Sep. 86	24 Jan. 06	7 July 07	24 Jan. 15	. . . .	6 Cavalry.
	S. W. Bord.	**Cahusac,** Chas. Fitzroy, D.S.O. (50, 52, 66*a*, 71, 196)	0 Sep. 85	24 Jan. 06	19 Sep. 07	24 Jan. 15	. . . .	36 Horse.
	R. Lanc. R.	**Blackett,** Clement Parker (50, 52, 66*a*, 66*b*, 167)	4 Apl. 87	24 Jan. 06	9 Nov. 08	24 Jan. 15	. . . .	6 G. R.
	Unattd. List	**Palmer,** Arthur Cecil (51, 66*a*, 72*a*)	28 June 87	24 Jan. 06	18 Mar. 07	24 Jan. 15	. . . .	94 Infantry.
290	R. War. R.	**Lorimer,** John Malcolm (137)	7 Oct. 86	24 Jan. 06	17 Sep. 09	24 Jan. 15	*Ty. m.* 20 July 18	3 Horse.
	Unattd. List.	**Burridge,** Thos. Francis Stannard (107, 162, 167)	3 June 87	24 Jan. 06	9 Mar. 07	24 Jan. 15	. . . .	35 Sikhs.
	Manch. R.	**Cates,** Kenneth George Hyde	21 July 85	24 Jan. 06	5 Mar. 09	24 Jan. 15	. . . .	45 Sikhs.
	Unattd. List.	**Sartorius,** Gerald (51, 57, 95, 169) [*l*]	12 Nov. 87	24 Jan. 06	11 Mar. 07	24 Jan. 15	*Ty. m.* 14 Oct. 16.	6 Cavalry.
	Essex R.	**Fenton,** James Newmarch (95)	18 Apl. 87	24 Jan. 06	25 Sep. 11	24 Jan. 15	. . . .	43 Infantry.
295	R. Sus. R.	**Atkins,** Reginald Fenner (50, 52, 57, 66*a*, 95, 167)	8 Mar. 87	24 Jan. 06	4 Aug. 08	24 Jan. 15	. . . .	67 Punjabis.

No.	Former Corps.	Names.	Date of birth.	Date of first commission.	Date of appointment to Indian Army.	Date of Rank. Substantive.	Date of Rank. Army.	Remarks.
		Captains—*contd.*						
296	Unattd. List	**Pughe,** Owen Llewellyn (51, 57, 107, 140, 162, 172)	30 May 87	24 Jan. 06	18 Mar. 07	24 Jan. 15	. . . .	72 Punjabis.
	Unattd. List	**McPherson,** Alan Bruce, M.C. (51, 52, 57, 71, 137, 162)	6 Jan. 87	24 Jan. 06	18 Mar. 07	24 Jan. 15	. . . .	6 Lt. Infantry.
	Unattd. List	**Chesney,** Duncan Patrick (66 167) [L]	17 July 85	24 Jan. 06	10 Mar 07	24 Jan. 15	. . . .	23 Pioneers.
	Unattd. List	**Yule,** Harry Victor (51, 65a, 66, 71, 169, 172)	4 Feb. 87	24 Jan. 06	12 Mar. 07	24 Jan. 15	. . . .	25 Cavalry.
300	Norf. R.	**Pogson,** Charles Aubrey, M.C. (59, 66a, 95)	27 June 85	24 Jan. 06	23 Apl. 12	24 Jan. 15	. . . .	117 Mahrattas.
	N. & D. R.	**Parsons,** Arthur Edward Broadbent (161, 165, 169, 204)	5 Aug. 84	4 Apl. 06	30 Sep. 08	4 Apl. 15	. . . .	52 Sikhs.
	Ches. R.	**Huxford,** Harold James (51, 52, 72, 72a, 137, 175) *Service for pension;—In Ranks from 19 Nov. 03 to 24 Apl. 06 (half to count)*	19 Oct. 84	25 Apl. 06	31 Mar. 09	25 Apl. 15	. . . .	125 Rifles.
	N. Staff. R.	**Stone,** Burnett Sidney (66a)	25 Apl. 86	23 May 06	19 Mar. 09	23 May 15	. . . .	88 Infantry.
	E. Surr. R.	**Smith,** Christopher Thos. Marshall (51, 52, 165, 72a)	7 July 85	23 May 06	16 July 09	23 May 15	*m.* 3 June 18	87 Punjabis.
305	R.W.Surr.R.	**Matthews,** Frank Melvin (172)	11 June 85	23 May 06	16 July 09	23 May 15	*Ty. m.* 25 Feb. 18.	22 Punjabis.
	R.W.Surr.R.	**Nixon,** Brian George Michael Freder(51, 52, 65, 66a, 71, 107, 167)	5 June 86	23 May 06	29 Oct. 08	23 May 15	. . . .	41 Dogras.
	R. Sco. Fus.	**Nairne,** Robert Gordon (51, 107, 195)	2 Apl. 84	16 June 06	16 Aug. 08	16 June 15	. . . .	61 Pioneers.
	R. A.	**Bacon,** Charles Gilmore (12, 162, 167, 172a, 205a)	19 Oct. 85	25 July 06	4 Sep. 11	25 July 15	. . . .	29 Lancers.
	R. A.	**Daunt,** Charles O'Brien, M.C. (71, 172a)	3 Sep. 86	25 July 06	25 Sep. 09	25 July 15	. . . .	38 C. I. Horse.
310	R. A.	**Todd,** Eric Ludor Tallent (50, 94, 169)	17 May 86	25 July 06	19 June 10	25 July 15	. . . .	121 Pioneers.
	Conn. Rang.	**Markwick,** Harold William (65a, 66, 169)	15 June 83	15 Aug. 06	23 Sep. 14	15 Aug. 15	. . . .	S. and T. Corps.
	Unattd. List	**Proctor,** George Norman (51, 66a, 195) *Service for pension from 3 Aug. 07.*	22 Aug. 86	29 Aug. 06	12 Nov. 08	29 Aug. 15	. . . .	83 Infantry.
	Unattd. List	**Barker,** Ronald William (104, 167) *Service for pension from 3 Aug. 07.*	20 Sep. 83	29 Aug. 06	13 Feb. 09	29 Aug. 15	. . . .	Civil, Burma.
	Unattd. List	**Blandy,** Raleigh, M.C. (120, 157) *Service for pension from 3 Aug. 07.*	22 May 84	29 Aug. 06	12 Nov. 08	29 Aug. 15	. . . .	9 G. R.
315	Unattd. List	**Secondê,** Emile Charles (57, 136, 162, 172a, 204) *Service for pension from 3 Aug. 07.*	17 June 84	29 Aug. 06	30 Sep. 08	29 Aug. 1	. . . .	16 Rajputs.
	Unattd. List	**Ball,** Eric Percy, M.C. (50, 155)	26 June 87	29 Aug. 06	25 Oct. 07	29 Aug. 15	*Ty. m.* 25 Feb. 18.	110 Lt. Infantry.
	Unattd. List	**Wells,** Thomas Ralph, M.C. (59, 173)	11 Mar. 87	29 Aug. 06	13 Nov. 07	29 Aug. 15	. . . .	33 Punjabis.
	Unattd. List	**Erskine,** Claude Ernest Torin (51, 52, 59, 66, 169)	1 May 87	29 Aug. 06	13 Oct. 07	29 Aug. 15	. . . .	Guides.
	Unattd. List	**Walker,** Edward John Patteson Travis (45a, 51, 52, 59, 71, 137)	10 Apl. 88	29 Aug. 06	12 Nov. 07	29 Aug. 15	*Ty. m.* 6 Aug. 18.	32 Lancers.
320	Unattd. List	**Haines,** John Bertram (51, 52, 71, 95, 137, 161, 174, 205a) [L]	31 July 86	29 Aug. 06	27 Oct. 07	29 Aug. 15	. . . .	10 Jats.
	Unattd. List	**Robson,** Henry Wm. Cumine (53, 94, 157, 162, 169)	8 Dec. 86	29 Aug. 06	22 Oct. 07	29 Aug. 15	. . . .	7 G. R.
	Unattd. List	**Evans,** Dudley Mac Neil, M.C. (51, 52, 59, 66a, 169)	11 Dec. 86	29 Aug. 06	27 Oct. 07	29 Aug. 15	. . . .	55 Rifles.
	Unattd. List	**Coffin,** Alexander Sebright (65a, 66, 172)	7 Jan. 87	29 Aug. 06	26 Jan. 08	29 Aug. 15	. . . .	28 Punjabis.
	Unattd. List	**Dyce,** Alan George (51, 52, 57, 95,)	24 Oct. 86	29 Aug. 06	13 Nov. 07	29 Aug. 15	. . . .	S. & T. Corps.
325	Unattd. List	**Connell,** Victor John Alexander (45a, 51, 65, 71, 167, 172a)	18 Aug. 87	29 Aug. 06	13 Nov. 07	29 Aug. 15	. . . .	13 Lancers.
	Unattd. List	**Dane,** Richard, M.C. (51, 52, 66, 71, 169, 205a)	21 Aug. 87	29 Aug. 06	12 Nov. 07	29 Aug. 15	. . . .	21 Cavalry.
	Unattd. List	**Macdonald,** Harry, D.S.O. (45a, 51, 53, 71, 162, 172)	3 May 86	29 Aug. 06	27 Oct. 07	29 Aug. 15	3 June 17	111 Lancers.
	Wilts. R.	**Eberhardie,** Rudolph Charles (51, 52, 58, 155, 161, 169, 172a, 205a)	23 June 87	29 Aug. 06	16 June 08	29 Aug. 15	. . . .	16 Mahrattas.
329	E. Lan. R.	**Porteous,** David Guy (50, 66a, 72a, 167)	6 Sep. 86	29 Aug. 06	2 Nov. 11	29 Aug. 15	*Ty. m.* 25 Feb. 18.	90 Punjabis.

0	Unattd. List	**Thornhill**, Charles Massy, D.S.O., M.C. (51, 66a, 172)	5 Feb. 85	29 Aug. 06	22 Oct. 07	29 Aug 15	. . . .	24 Punjabis.
	R. Highrs.	**Murray**, Eric Mackay, M.C. (45a, 71, 127, 172a)	24 Sep. 86	29 Aug. 06	25 July 09	29 Aug 15	. . . .	Guides.
	Unattd. List	**Corse-Scott**, Erule James, M.C. (50, 53, 59, 157)	16 Nov. 86	29 Aug. 06	17 Oct. 07	29 Aug. 15	. . . .	2 G. R.
	Suff. R.	**Crossfield**, Eric Arthur Kaffin (51, 52, 125a, 140, 157, 162)	1 Feb. 87	29 Aug. 06	16 Mar. 09	29 Aug. 15	. . . .	10 G. R.
	Unattd. List	**Carpendale**, Frederic Maxwell (45a, 51, 52)	9 July 87	29 Aug. 06	24 Sep. 08	29 Aug 15	*m.* 1 Jan. 18	42 Regiment.
335	Dorset. R.	**Moore**, Charles Francis Fitzgerald, M.C. (62, 137, 140)	18 June 86	29 Aug. 06	13 Mar. 09	29 Aug. 15	. . . .	123 Rifles.
	Unattd. List	**Oswald**, Frank (45a, 51, 52, 71)	27 May 87	29 Aug. 06	16 Oct. 07	29 Aug. 15	. . . .	20 Horse.
	S. Staff. R.	**Skinner**, Edmund Morland (50, 52, 59, 137)	16 Jan. 87	29 Aug. 06	1 Aug. 08	29 Aug. 15	. . . .	125 Rifles.
	Unattd. List	**Thomson**, Alister Ralph, M.C. (51, 52, 57, 66a, 137)	24 Oct. 86	29 Aug. 06	11 Nov. 07	29 Aug. 15	. . . .	7 Rajputs.
	Essex. R.	**Waller**, Alfred George (51, 137, 162)	20 Apl. 86	29 Aug. 06	25 July 09	29 Aug. 15	. . . .	44 Infantry.
340	Unattd. List	**Todd**, Reginald (172)	23 Oct. 86	29 Aug. 06	14 Nov. 07	29 Aug. 15	. . . .	27 Punjabis.
	R. Ir. R.	**Gordon**, John de la Hay, M.C. (88, 153, 162, 167, 169)	31 Mar. 87	29 Aug. 06	2 Aug. 08	29 Aug. 15	*Ty. m.*, 19 Sep. 16	Poll. employ.
	Ches. R.	**Baker**, Harry Norman (51, 52, 72a, 195)	16 June 87	29 Aug. 06	12 Nov. 08	29 Aug. 15	. . . .	75 Infantry.
	Unattd. List	**Leslie-Smith**, Colin (50, 57, 66, 107, 140, 172)	11 July 87	29 Aug. 06	15 Oct. 07	29 Aug. 15	. . . .	22 Punjabis.
	Suff. R.	**Chapman**, Eric Haldane (51, 169)	31 Aug. 87	29 Aug. 06	23 Apl. 12	29 Aug. 15	. . . .	107 Pioneers.
345	R. Ir. Rif.	**Waller**, Edmund de Warrenne (51, 52, 72a, 172)	20 Jan. 88	29 Aug. 06	22 July 09	29 Aug. 15	. . . .	72 Punjabis.
	Unattd. List	**Digby**, Thomas Arthur (162, 167, 172a)	5 Oct. 86	29 Aug. 06	12 Nov. 07	29 Aug. 15	. . . .	30 Punjabis.
	Unattd. List	**Hickman**, Henry Temple Devereux, M.C. (51, 66, 165, 172a)	24 Feb. 88	29 Aug. 06	22 Oct. 07	29 Aug. 15	. . . .	34 Pioneers.
	Y. & Lanc. R	**Wilson**, Richard Henry (51, 52, 59, 71, 167, 169)	10 June 86	29 Aug. 06	22 Sep. 08	29 Aug. 15	. . . .	82 Punjabis.
	R. Ir. R.	**Thomson-Glover**, John William (59, 161, 165, 172a)	30 July 87	29 Aug. 06	17 Feb. 09	29 Aug 15	. . . .	35 Sikhs.
350	Wilts. R.	**Fowler**, Ernest Kynaston, M.C. (51, 52, 59, 169)	13 Nov. 86	29 Aug. 06	19 Mar. 09	29 Aug. 15	*m.* 1 Jan. 18	57 Rifles.
	E. Lan. R.	**Kemble**, Thomas Aubrey (51, 52, 195)	10 Dec. 85	29 Aug. 06	26 Sep. 08	29 Aug. 15	. . . .	81 Pioneers.
	Unattd. List	**O'Neill**, Heremon Jas. Drummond, M.C. (51, 57, 172)	18 Mar. 88	29 Aug. 06	18 Oct. 07	29 Aug. 15	. . . .	127 Baluch. Lt. Infy.
	Norf. R.	**Duberly**, Montagu Richard William (57, 66, 137, 162, 167)	30 June 87	29 Aug. 06	2 July 12	29 Aug. 15	. . . .	23 Pioneers.
	Conn. Rang.	**Pelly**, Hutchison Raymond (57, 94, 104, 125a, 162, 167, 205a)	22 Sep. 87	29 Aug. 06	2 Nov. 09	29 Aug. 15	. . . .	Civil, Burma.
355	Unattd. List	**Bennett**, Alexander Dumaresq, M.C. (51, 52, 95, 169)	16 Apl. 87	29 Aug. 06	13 Nov. 07	29 Aug. 15	. . . .	19 Punjabis.
	Ches. R.	**Hume-Wright**, Francis Joseph Owen (51, 52, 52, 137)	3 Mar. 87	29 Aug. 06	8 Sep. 11	29 Aug. 15	*m.* 3 June 18	16 Rajputs.
	R. Highrs.	**Campbell**, Wentworth Edward Dallas (59, 71, 137, 162)	6 Mar. 87	29 Aug. 06	12 Sep. 09	29 Aug. 15	. . . .	34 Horse.
	Unattd. List	**Ommanney**, Walter Montagu (51, 52, 66a, 167)	17 Aug. 85	29 Aug. 06	16 Nov. 07	29 Aug. 15	. . . .	32 Pioneers.
	1 Dgns.	**Watson**, Trevor Newall, M.C. (45a, 50, 167)	18 Apl. 86	29 Aug. 06	11 June 10	29 Aug. 15	*m.* 3 June 18	12 Cavalry.
360	Unattd. List	**Lucas**, Charles Alfred (51, 66, 137)	9 July 84	29 Aug. 06	17 Oct. 07	29 Aug. 15	. . . .	S. & T. Corps.
	Unattd. List	**McSwiney**, Herbert Frederick Cyril, M.C. (59, 157)	8 Nov. 86	29 Aug. 06	2 Nov. 07	29 Aug. 15	. . . .	3 G. R.
	R. Muns. Fus.	**Ireland**, Gilbert (51, 52, 59, 66a)	20 May 88	29 Aug. 06	12 Mar. 10	29 Aug. 15	. . . .	13 Rajputs.
	R. Muns. Fus	**LeCornu**, Charles James Seward, M.C. (57, 59, 72a, 94, 172)	12 Nov. 87	29 Aug. 06	16 Sep. 09	29 Aug. 15	. . . .	56 Rifles.
	Unattd. List	**Wauchope**, Robert Stuart (162, 172)	20 Mar. 87	29 Aug. 06	11 Mar. 08	29 Aug. 15	. . . .	Survey Dept.
365	Unattd. List	**Hailes**, Walter Leslie (51, 137, 162, 172a)	21 June 88	29 Aug. 06	22 Oct. 07	29 Aug. 15	*Ty. m.* 23 Aug. 16.	6 Lt. Infantry.
	Unattd. List	**Strover**, Wyndham George (50, 52, 66a)	10 Dec. 87	29 Aug. 06	6 June 08	29 Aug. 15	. . . .	5 Lt. Infantry.
	Midd'x. R.	**Phillips**, Herbert Stewart (51, 52, 66, 137)	9 Apl. 86	6 Oct. 06	20 Sep. 09	1 Sep. 15	*Ty. m.* 19 June 18.	27 Cavalry.
	R. Muns. Fus.	**Pott**, Douglas, D.S.O., M.C. (45a, 51, 57, 71, 95, 167)	24 Feb. 88	6 Oct. 06	26 Apl. 09	1 Sep. 15	. . . .	13 Lancers.
	Bord. R.	**Mylne**, Ronald Heathcote (50, 157, 162)	28 June 87	6 Oct. 06	4 Oct. 06	1 Sep. 15	. . . .	10 Gurkhas.
370	Worc. R.	**Burroughs**, Shirley Victor Gore (72a, 137, 172a)	20 Apl. 87	6 Oct. 06	22 July 09	1 Sep. 15	. . . .	4 Rajputs.
	S. Lans. R.	**Howard**, Charles Alfred Lowray (45a, 51, 52, 66a, 71, 137)	15 Mar. 86	7 Nov. 06	24 Nov. 08	1 Sep. 15	. . . .	32 Lancers.
	Worc. R.	**Solano**, Bernard Austin (51, 63, 162, 169)	5 May 87	7 Nov. 06	25 Mar. 09	1 Sep. 15	. . . .	46 Punjabis.
	W. Rid. R.	**Ford**, John Meredith Randle, M.C. (59, 162, 195)	17 Nov. 87	7 Nov. 06	20 Sep. 08	1 Sep. 15	. . . .	61 Pioneers.
	S. Lan. R.	**Hammick**, St. Vincent Frederick (52)	11 Aug. 86	7 Nov. 06	8 Dec. 09	1 Sep. 15	. . . .	13 Rajputs.
375	Shrops. L. I.	**Bickford**, Maurice Herbert (50, 52, 57, 167, 172a)	2 Sep. 87	21 Nov. 06	30 Oct. 08	1 Sep. 15	. . . .	38 Dogras.
376	R. A.	**Goldman-Monk**, Max Emile de Baillon (59, 106, 125a, 140, 162, 167, 169, 205a) [L].	6 Aug. 88	20 Dec. 06	10 Sep. 11	1 Sep. 15	. . . .	91 Punjabis.

No.	Former Corps.	Names.	Date of birth.	Date of first commission.	Date of appointment to Indian Army.	Date of Rank. Substantive.	Date of Rank. Army.	Remarks.
		Captains—*contd.*						
377	R. A.	**Taylor,** Aveline Thompson Waters (59)	26 Feb. 88	20 Dec. 06	21 Feb. 14	1 Sep. 15	. . . .	43 Regiment.
	R. A.	**Condon,** William Edmund Hunt (94, 161, 204)	28 Oct. 86	20 Dec. 06	28 Sep. 11	1 Sep. 15	. . . .	17 Infantry.
	E. Surr. R.	**Boswell,** William Roy (51, 66*a*, 172)	24 June 85	16 Jan. 07	3 Nov. 08	1 Sep. 15	. . . .	28 Punjabis.
380	S. Lan. R.	**Edward-Collins,** Gerald, M.C. (51, 57, 71, 137)	23 Feb. 85	16 Jan. 07	11 Mar. 09	1 Sep. 15	. . . .	33 Lt. Cavy.
	Notts. & Derby R.	**Simpson,** Richard Charles (51, 52, 66*a*, 72*a*, 172)	7 Mar. 86	16 Jan. 07	6 Oct. 11	1 Sep. 15	. . . .	130 Baluchis.
	Unattd. List	**Gardner,** Robert Geoffrey (51, 172)	25 Oct. 85	19 Jan. 07	31 Mar 09	1 Sep. 15	. . . .	25 Punjabis.
		Service for pension from 8 Feb. 08.						
	Unattd. List	**Strover,** Ernest James (51, 137, 162)	23 May 85	19 Jan. 07	31 Mar. 09	1 Sep. 15	. . . .	3 Brahmans.
		Service for pension from 8 Feb. 08.						
	Unattd. List	**Sheepshanks,** Richard Hassell, D.S.O. (167)	19 Feb. 85	19 Jan. 07	18 Dec. 09	1 Sep. 15	. . . .	12 Cavalry.
		Service for pension from 8 Feb. 08.						
385	Unattd. List	**Corbett,** Thomas William, M.C. (4, 51, 52, 71, 172, 204)	2 June 88	19 Jan. 07	15 Mar. 08	1 Sep. 15	. . . .	9 Horse.
	Unattd. List	**Campbell,** Frederick Chas. Gunning, M.C. (51, 52, 66*a*, 162, 169, 175, 205*a*)	11 Aug. 87	19 Jan. 07	15 Mar. 08	1 Sep. 15	. . . .	40 Pathans.
	Unattd. List	**Munton,** Vincent Robert (57, 153)	1 May 86	19 Jan. 07	18 Mar. 08	1 Sep. 15	. . . .	103 Lt. Infantry
	Unattd. List	**Minchinton,** Henry Darrell (51, 66*a*, 157)	15 Nov. 87	19 Jan. 07	2 Mar. 08	1 Sep. 15	. . . .	1 G. R.
	Unattd. List	**Walker,** Geoffrey Beresford, M.C. (51, 94, 161, 167, 169, 205*a*)	23 Apl. 87	19 Jan. 07	16 May 08	1 Sep. 15	. . . .	Pol. Dept.
390	Unattd. List	**Blacker,** Latham Valentine Stewart (57, 169)	1 Oct. 87	19 Jan. 07	15 Mar. 08	1 Sep. 15	*Ty. m.*, 1 Apl. 18.	Guides.
	Unattd. List	**Somerville,** Edward Annesley (45*a*, 51, 65, 71, 169)	4 June 87	19 Jan. 07	15 Mar. 08	1 Sep. 15	. . . .	22 Cavalry.
	Unattd. List	**Nethersole,** John, M.C. (45*a*, 51, 52, 59, 66*a*, 71, 172)	12 July 88	19 Jan. 07	18 Mar. 08	1 Sep. 15	. . . .	25 Cavalry.
	Unattd. List	**Turner,** Guy Fisher, M.C., (162, 167, 170)	28 Sep. 86	19 Jan. 07	18 Mar. 08	1 Sep. 15	. . . .	82 Punjabis.
	Unattd. List	**Ponting,** Theophilus John, M.C. (50, 167)	15 Sep. 86	19 Jan. 07	13 Mar. 08	1 Sep. 15	. . . .	92 Punjabis.
395	Unattd. List	**Boyle,** Cecil Alexander (71, 88, 161, 165, 169, 203) [L]	28 Mar. 88	19 Jan. 07	13 Mar. 08	1 Sep. 15	. . . .	11 Lancers.
	Unattd. List	**Cripps,** Ernest John (66, 162, 205*a*)	24 Feb. 87	19 Jan. 07	11 Mar. 08	1 Sep. 15	. . . .	S. & T. Corps.
	Unattd. List	**Arnott,** Allan Maxwell (51, 52, 157)	3 Oct. 86	19 Jan. 07	13 Mar. 08	1 Sep. 15	. . . .	4 G. R.
	Unattd. List	**Welman,** Wm. Harvey Loftus John (51, 57, 172)	7 June 88	19 Jan. 07	12 Mar. 08	1 Sep. 15	. . . .	3 G. R.
	Unattd. List	**Farquharson,** Christopher William (65*a*, 66, 71, 95, 137)	27 Aug. 87	19 Jan. 07	9 Mar. 08	1 Sep. 15	*Ty. m.* 25 Feb. 18.	10 Jats.
400	Unattd. List	**Belli-Bivar,** Herbert Frederick (51, 52, 66*a*, 172)	16 Nov. 88	19 Jan. 07	17 Mar. 08	1 Sep. 15	. . . .	127 Lt. Infantry.
	Unattd. List	**Hopkinson,** George Hamilton (155)	28 Feb. 87	19 Jan. 07	13 Mar. 08	1 Sep. 15	. . . .	116 Infantry.
	Unattd. List	**Baddeley,** Sydney Ernest Lodington (45*a*, 162, 169)	19 Aug. 86	19 Jan. 07	14 Mar. 08	1 Sep. 15	. . . .	19 Lancers.
	Unattd. List	**Ogilvy,** John Donald (51, 52, 66*a*, 157, 172*a*)	23 Mar. 87	19 Jan. 07	8 Mar. 08	1 Sep. 15	. . . .	1 G. R.
	Unattd. List	**Jobson,** Thomas Smith (45*a*, 51, 52, 71)	26 Jan. 87	19 Jan. 07	1 Mar. 08	1 Sep. 15	. . . .	27 Lt. Cavalry.
405	Unattd. List	**Pearson,** Henry Sim Ibbetson (51, 52)	16 June 88	19 Jan. 07	2 Mar. 08	1 Sep. 15	. . . .	114 Mahrattas.
	Unattd. List	**Eales,** Arthur Cecil Herbert Adrian (51, 66*a*, 137)	27 July 87	19 Jan. 07	1 Mar. 08	1 Sep. 15	. . . .	119 Infantry.
	Unattd. List	**Hinde,** Reginald Graham (51, 52, 172)	19 May 87	19 Jan. 07	4 Mar. 08	1 Sep. 15	. . . .	124 Infantry.
	Unattd. List	**Atkinson,** Eric Garnett (45*a*, 71, 102, 172)	9 Apl. 87	19 Jan. 07	16 Mar. 08	1 Sep. 15	. . . .	17 Cavalry.
	Unattd. List	**Norman,** Herbert John (57, 71)	30 Jan. 87	19 Jan. 07	8 Mar. 08	1 Sep. 15	. . . .	102 Grenadiers.
410	Dorset R.	**Wigley,** Frederic Gwyn Mackay (51, 162, 165, 169, 195, 205*a*)	23 Dec. 85	16 Feb. 07	4 Oct. 09	1 Sep. 15	. . . .	88 Infantry.
	R. I. Fus.	**Herapath,** Conrad Boehm (65*a*, 66, 71, 95, 137, 162, 172*a*)	25 July 86	16 Feb. 07	25 July 09	1 Sep. 15	. . . .	16 Rajputs.
412	R. I. Fus.	**Gardner,** Charles Frederick (51, 52, 137, 162, 165, 169)	22 Sep. 84	16 Feb. 07	21 Feb. 09	1 Sep. 15	. . . .	35 Sikhs.

413	R. Innis. Fus.	**Rudkin,** Benjamin Arthur (51, 71, 72*a*)	4 Sep. 85	16 Feb. 07	3 Dec. 12	1 Sep. 15	80 Infantry.
	Unattd. List	**Plin,** George Adrien (51, 52, 66*a*, 71, 95, 172) *Service for pension from 6 Feb. 09.*	7 Feb. 88	21 Feb. 07	2 Apl. 10	1 Sep. 15	130 Baluchis.
415	R. I. Fus.	**Blacker,** William Frederick (57, 71, 72*a*, 162, 172)	14 Nov. 85	2 Mar. 07	12 July 09	1 Sep. 15	36 Horse.
	Leins. R.	**Ferguson,** Geo. Chichester Smythe (45*a*, 51, 172*a*)	26 Aug. 85	20 Mar. 07	17 Sep. 09	1 Sep. 15	6 Cavalry.
	Devon R.	**Addington,** *Hon.* Raymond Anthony	24 Jan. 87	4 May 07	2 July 12	1 Sep. 15	26 Cavalry.
	R. Lanc. R.	**Lane,** Frank Nowell (50, 53, 66*a*, 71, 156, 162, 172*a*)	4 Jan. 88	4 May 07	12 Sep. 11	1 Sep. 15	7 G. R.
	E. York. R.	**May,** Christopher Delabère (51, 52, 66*a*, 172*a*)	3 Aug. 86	4 May 07	21 Sep. 11	1 Sep. 15	5 Lt. Infantry.
420	E. Surr. R.	**Durand,** Edward Percy Marion	11 July 84	29 May 07	21 June 10	1 Sep. 15	8 Rajputs.
	R. Ir. Regt.	**Collings,** Ernest Gardiner (57, 66, 95, 167)	7 Dec. 85	12 June 07	15 July 09	1 Sep. 15	35 Sikhs.
	R. A.	**Hankin,** Hugh Mordaunt (58, 64, 66*a*, 172)	26 Dec. 86	23 July 07	30 June 10	1 Sep. 15	Guides.
	Unattd. List	**Gibson,** John Gordon, M. C. (45*a*, 52, 66, 71)	10 Sep. 86	7 Aug. 07	8 Nov. 08	1 Sep. 15	14 Lancers
	Unattd. List	**Drought,** John Victor (51, 52, 172) *Service for pension from 13 June 08.*	9 Jan. 87	17 Aug. 07	26 Sep. 09	1 Sep. 15	124 Infantry.
425	Unattd. List	**Empson,** Arthur Hugh Ackland (59, 71, 94, 137, 162) *Service for pension from 15 Aug. 08.*	26 July 86	17 Aug. 07	30 Oct. 09	1 Sep. 15	8 Cavalry.
	Unattd. List	**Metcalfe,** Edward Dudley, M.C. (45*a*, 51, 52, 57, 137) *Service for pension from 15 Aug. 08.*	16 Jan. 87	17 Aug. 07	8 Nov. 09	1 Sep. 15	3 Horse.
	Unattd. List	**Boulton,** Clive Essington (51) *Service for pension from 15 Aug. 08.*	28 July 84	17 Aug. 07	25 Oct. 09	1 Sep. 15	5 Lt. Infantry.
	Unattd. List	**Paget,** Harold Edmund George (172) *Service for pension from 3 Oct. 08.*	6 July 86	17 Aug. 07	27 Dec. 09	1 Sep. 15	46 Punjabis.
	Unattd. List	**Meredith,** Arthur Elsdale, M.C. (59, 155*a*) *Service for pension from 15 Aug. 08.*	30 Nov. 85	17 Aug. 07	25 Oct. 09	1 Sep. 15	114 Mahrattas
430	Unattd. List	**Bernard,** Ronald Playfair St. Vincent, D.S.O., M.C. (51, 52, 57, 157, 172*a*)[*l*]	25 Apl. 88	17 Aug. 07	11 Nov. 08	1 Sep. 15	6 G. R.
	Unattd. List	**Burne,** Frank Owen Newdigate, M.C. (51, 52, 66*a*, 89*b*, 162, 193)	20 Oct. 87	17 Aug. 07	10 Nov. 08	1 Sep. 15	64 Pioneers.
	Unattd. List	**Peet,** Lionel Meredith (51, 137, 161, 205*a*)	6 July 87	17 Aug. 07	9 Apl. 09	1 Sep. 15	6 Lt. Infantry.
	Unattd. List	**Davidson,** Henry Betham (557, 157)	26 Nov. 87	17 Aug. 07	11 Nov. 08	1 Sep. 15	10 G. R.
	Unattd. List	**Neville Rolfe,** Arthur (51, 2, 157)	15 Jan. 88	17 Aug. 07	13 Nov. 08	1 Sep. 15	5 G. R.
435	Unattd. List	**King-Salter,** Nigel Henry (51, 66, 157)	9 May 88	17 Aug. 07	28 Oct. 08	1 Sep. 15	6 G. R.
	Unattd. List	**Burt,** Henry Vere (45*a*, 51, 52, 59, 172)	2 May 88	17 Aug. 07	28 Oct. 08	1 Sep. 15	37 Lancers.
	Unattd. List	**Manisty,** George Jennings Eldon (51, 66, 66*b*, 95, 167, 169)	9 Nov. 87	17 Aug. 07	26 Oct. 08	1 Sep. 15	67 Punjabis.
	Unattd. List	**Sherston,** John Reginald Vivian, D.S.O., M. C. (71, 172)	2 Oct. 88	17 Aug. 07	12 Nov. 08	1 Sep. 15	11 Lancers.
	Unattd. List	**Lewis,** Roland Ford (45*a*, 51, 71, 169)	8 May 88	17 Aug. 07	11 Nov. 08	1 Sep. 15	21 Cavalry.
440	Unattd. List	**Baxter,** Darell St. John (51, 52, 71, 167)	12 Dec. 88	17 Aug. 07	4 Nov. 08	1 Sep. 15	91 Punjabis.
	Unattd. List	**De Butts,** Frederick Cromie, D.S.O., M. C. (51, 52, 57, 59, 169)	6 Nov. 87	17 Aug. 07	27 Oct. 08	1 Sep. 15	55 Rifles.
	Unattd. List	**Jackson,** Cecil Hungerford (57, 169)	7 Nov. 87	17 Aug. 07	10 Nov. 08	1 Sep. 15	21 Punjabis.
	Unattd. List	**Burdett,** Ernes Wyndham, M. C. (59, 137)	5 Sep. 87	17 Aug. 07	12 Nov. 08	1 Sep. 15	48 Pioneers.
	Unattd. List	**Macausland,** Redmond Alexander (66*a*)	6 Dec. 86	17 Aug. 07	11 Nov. 08	1 Sep. 15	45 Sikhs.
445	Unattd. List	**Messum,** Leo Ruston (51, 52, 57, 172)	2 Dec. 87	17 Aug. 07	10 Nov. 08	1 Sep. 15	21 Punjabis.

No.	Former Corps.	Names.	Date of birth.	Date of first commission.	Date of appointment to Indian Army.	Date of Rank. Substantive.	Date of Rank. Army.	Remarks.
		Captains—*contd.*						
446	Unattd. List	**Griffith-Griffin**, Forrester Metcalfe, M. C. (51, 52, 66a, 169)	3 Apl. 89	17 Aug. 07	10 Nov. 08	1 Sep. 15	m. 1 Jan. 17	129 Baluchis.
	Unattd. List	**Davison**, Douglas Stewart (45a, 50, 52, 137)	17 Feb 88	17 Aug. 07	12 Nov. 08	1 Sep. 15	. . . .	2 Lancers
	Unattd. List	**Fraser**, Stannus Grant Gordon (51, 52, 57, 195)	22 May 88	17 Aug. 07	11 Nov. 08	1 Sep. 15	. . . .	81 Pioneers.
	Unattd. List	**Money**, George Wigram Pocklington (51, 157)	25 Sep. 88	17 Aug. 07	11 Nov. 08	1 Sep. 15	. . . .	3 G. R.
450	Wilts. R.	**Hogan**, Patrick Thomas (51, 52, 66a)	13 Dec. 85	4 Sep. 07	17 Sep. 09	1 Sep. 15	. . . .	96 Infantry.
		Service for pension;—In Ranks from 13 June 05 to 3 Sep. 07 (half to count).						
	Leins. R.	**Mitchell**, Roland Peto Johnstone (12, 47a, 48a, 51, 52, 57, 95, 172)	30 Jan. 87	25 Sep. 07	18 Oct. 12	1 Sep. 15	. . . .	37 Lancers.
	R. Lanc. R.	**Wigley**, Philip John Reynell, M. C. (132, 136, 161, 167, 172a, 204)	10 Mar. 87	9 Oct. 07	7 Oct. 11	1 Sep. 15	. . . .	90 Punjabis.
	17 Lanc.	**Willoughby**, Michael Guy Percival, M. C. (45a, 51, 52, 172)	21 Oct. 87	9 Oct. 07	11 Oct. 09	1 Sep. 15	. . . .	10 Lancers.
	R. Ir. Regt.	**Torrie**, Lawrence Jameson (172)	23 Mar. 89	12 Oct. 07	20 Sep. 09	1 Sep. 15	. . . .	26 Punjabis.
455	W. York. R.	**Malyon**, Frank Hailstone (161, 169) [L]	20 Mar. 85	11 Dec. 07	21 Sep. 09	1 Sep. 15	. . . .	21 Punjabis.
	E. Surr. R.	**Hornsby**, Henry Fyshe-Palmer (45a, 51, 52, 65, 71)	7 July 84	11 Dec. 97	16 Sep. 10	1 Sep. 15	. . . .	5 Cavalry.
	Welsh. R.	**Jones**, Bertram Lowther (65a, 66, 172a)	18 Jan. 87	11 Dec. 07	10 Aug. 12	1 Sep. 15	. . . .	S. and T. Corps.
	W. York. R.	**Foottit**, James Augustus Warwick (51, 71, 172)	26 Aug. 85	11 Dec. 07	9 Oct. 09	1 Sep. 15	. . . .	17 Cavalry.
	Bedf. R.	**Semini**, Edward Jerome (71, 162)	1 Mar. 86	11 Dec. 07	3 May 12	1 Sep. 15	. . . .	79 Carnatic Infy.
460	R. A.	**Daly**, Clive Kirkpatrick (71, 128, 137, 161, 202, 205a) [L]	3 Apl. 88	18 Dec. 07	14 Oct. 11	1 Sep. 15	. . . .	S. and T. Corps.
	R. A.	**Dening**, Roland, M.C. (45a, 57, 125a)	13 Sep. 88	18 Dec. 07	24 Jan. 11	1 Sep. 15	. . . .	18 Lancers.
	3 Huss.	**Mackenzie**, Thomas Chalmers (51, 66a)	30 Dec. 85	20 Dec. 07	6 June 12	1 Sep. 15	. . . .	6 Cavalry.
	Unattd. List	**Barstow**, Arthur Edward, M.C. (59, 60a, 66a, 95, 167, 172a)	17 Mar. 88	25 Jan. 08	11 Mar. 09	1 Sep. 15	. . . .	15 Sikhs.
	Unattd. List	**Crick**, Charles Churchill (51, 71, 162)	9 Nov. 88	25 Jan. 08	31 July 09	1 Sep. 15	. . . .	109 Infantry.
465	Unattd. List	**Wetherall**, Frederic George Bonney (51, 52, 57, 66a, 95, 169)	2 May 88	25 Jan 08	13 Mar. 09	1 Sep. 15	. . . .	127 Infantry.
	Unattd. List	**Mackenzie**, Eric Alistcan Hay (57, 137)	4 Mar. 88	25 Jan. 08	22 July 09	1 Sep. 15	. . . .	3 Brahmans.
	Unattd. List	**Seton**, Henry Winton (157)	1 Aug. 87	25 Jan. 08	18 Apl. 09	1 Sep. 15	. . . .	9 G. R.
	Unattd. List	**Murphy**, Gerald Patrick (167, 172a)	8 May 88	25 Jan. 08	15 Mar. 09	1 Sep. 15	. . . .	Poll. Bombay.
	Unattd. List	**Leslie**, Francis Downie (137)	21 Apl. 88	25 Jan. 08	1 Apl. 09	1 Sep. 15	. . . .	48 Pioneers.
470	Unattd. List	**Hornor**, Francis Harold Fardell, M. C.	30 Jan. 88	25 Jan. 08	18 May 11	1 Sep. 15	. . . .	107 Pioneers.
	Unattd. List	**Hare**, Ivan Lancelot O'Hara (195)	16 Jan. 88	25 Jan. 08	28 Mar. 09	1 Sep. 15	. . . .	64 Pioneers.
	Unattd. List	**Greenway**, Charles Kelvynge (45a, 65, 137)	24 Mar. 88	25 Jan. 08	21 Mar. 09	1 Sep. 15	. . . .	26 Lt. Cavalry.
	Unattd. List	**Darell**, Robert Darell Elford (51, 57, 167)	5 Aug. 87	25 Jan. 08	28 Mar. 09	1 Sep. 15	. . . .	76 Punjabis.
	Unattd. List	**Spencer**, Laurence (51, 137)	19 Feb. 88	25 Jan. 08	14 Mar. 09	1 Sep. 15	. . . .	8 Rajputs.
475	Unattd. List	**Nance**, Walter Joseph (51, 52, 172a, 195)	21 Apl. 88	25 Jan. 08	13 Mar. 09	1 Sep. 15	. . . .	21 Pioneers.
	Unattd. List	**Tennant**, Claude Alwyne Muriel (59, 66a, 167)	9 June 87	25 Jan. 08	20 Mar. 09	1 Sep. 15	m. 1 Jan. 18	S. & T. Corps.
477	Unattd. List	**Copeland**, Alik Gerald Joseph (59, 65, 71, 90, 167, 204)	13 June 87	25 Jan. 08	14 Mar. 09	1 Sep. 15	. . . .	5 Cavalry.

478	Unattd. List	**Dent,** Robert Boisragon (51, 52, 57, 157)	25 Sep. 87	25 Jan. 08	28 Mar. 09	1 Sep. 15	. . . .	3 G. R.
	Unattd. List	**Macdonald,** Donald Theodore, M.C. (51, 59, 137)	9 Mar. 88	25 Jan. 08	13 Mar. 09	1 Sep. 15	. . . .	125 Rifles.
480	Unattd. List	**Bacon,** Roger Gilmore (94, 172)	21 Mar. 87	25 Jan. 08	13 Mar. 09	1 Sep. 15	. . . .	71 Punjabis.
	Unattd. List	**Molony,** Claud Weldon (51, 52, 59, 172)	27 May 87	25 Jan. 08	15 July 09	1 Sep. 15	. . . .	57 Rifles.
	Unattd. List	**Barron,** Albert Henry Darley, M.C. (51)	13 Jan. 88	25 Jan. 08	13 Mar. 09	1 Sep. 15	. . . .	17 Infantry.
	Unattd. List	**Cobbold,** Montagu Francis Davy (51, 57, 157)	8 Mar. 88	25 Jan. 08	16 Mar. 09	1 Sep. 15	. . . .	1 G. R.
	Unattd. List	**Harington,** Hastings Roy (51, 52, 157)	20 Sep. 88	25 Jan. 08	10 Apl. 09	1 Sep. 15	. . . .	8 G. R.
485	Unattd. List	**Seymour,** Edward Victor Francis (45a, 71, 172)	5 Aug. 87	25 Jan. 08	19 Mar. 09	1 Sep. 15	. . . .	9 Horse
	Unattd. List	**Dawson,** Arthur Annesley Francis Charles Hutton (51, 52)	8 Dec. 86	25 Jan. 08	11 Apl. 09	1 Sep. 15	. . . .	20 Infantry.
	Unattd. List	**Deighton,** Francis John (95, 172) *Service for pension from 3 Feb. 09.*	30 Oct. 84	22 Feb. 08	31 Mar. 10	1 Sep. 15	. . . .	S. & T. Corps.
	Essex R.	**Carter,** Eric Nelson (51, 52, 94, 137, 162, 169)	22 Jan. 88	22 Feb. 08	5 Nov. 11	1 Sep. 15	. . . .	112 Infantry.
	Manch R.	**Fox,** Frederick Noel, M.C. (51, 52)	11 Feb. 88	22 Feb. 08	15 Feb. 14	1 Sep. 15	. . . .	39 Garhwal Rifles.
490	R. I. Regt.	**Donovan,** Charles Hugh (57, 172a)	12 Oct. 87	27 May 08	9 Sep. 11	1 Sep. 15	. . . .	94 Infantry.
	Leins. R.	**Fletcher,** Wilfrid Taylor	21 Sep. 85	27 May 08	6 Sep. 11	1 Sep. 15	. . . .	97 Infantry.
	R. Scots.	**Guthrie,** Ivan Douglas, M.C. (71)	6 Aug. 86	27 May 08	8 Oct. 11	1 Sep. 15	. . . .	17 Cavalry.
	R.F.A.	**Spain,** Granville Alaric Richard (155, 162, 169)	19 Dec. 88	26 June 08	7 Sep. 13	1 Sep. 15	. . . .	103 Infantry.
	Unattd. List	**Lewis,** Harold Victor, D.S.O., M.C. (50, 52, 59, 169)	6 Sep. 87	9 Sep. 08	15 Nov. 09	1 Sep. 15	. . . .	129 Baluchis.
495	Unattd. List	**Fulton,** James Deas, M.C. (172)	23 Aug. 88	9 Sep. 08	15 Nov. 09	1 Sep. 15	. . . .	26 Punjabis.
	Unattd. List	**Andrewes,** Claude Lancelot, M.C. (45a, 59, 169)	6 Oct. 87	9 Sep. 08	9 Nov. 09	1 Sep. 15	. . . .	31 Lancers.
	Unattd. List	**Battye,** Thomas Howard (50, 59, 157, 172a)	12 Nov. 87	9 Sep. 08	8 Nov. 09	1 Sep. 15	. . . .	10 G. R.
	Unattd. List	**Armstrong,** Arthur Godfrey (45a, 51, 137)	4 May 89	9 Sep. 08	25 Nov. 09	1 Sep. 15	. . . .	26 Cavalry.
	Unattd. List	**Gray,** Gerald Cecil Gibson (51, 52, 59, 137)	11 Jan. 88	9 Sep. 08	28 Nov. 09	1 Sep. 15	*Ty. m.*, 1 Nov. 18.	3 Horse.
500	Unattd. List	**Gatherer,** Arthur Henry Norman, M.C. (57, 172)	8 Oct. 87	9 Sep. 08	13 Dec. 09	1 Sep. 15	. . . .	46 Punjabis.
	Unattd. List	**Chamier,** Richard Outram (51, 57, 155, 162)	24 Aug. 88	9 Sep. 08	8 Nov. 09	1 Sep. 15	. . . .	110 Lt. Infantry.
	Unattd. List	**Ruck,** Richard Conyers (51, 57, 161, 169, 204)	24 Aug. 87	9 Sep. 08	7 Nov. 09	1 Sep. 15	. . . .	95 Infantry.
	Unattd. List	**Grant,** Ronald Alister Percival, M.C. (51, 57, 137)	10 Dec. 87	9 Sep. 08	7 Nov. 09	1 Sep. 15	. . . .	112 Infantry.
	Unattd. List	**Henstock,** Roy Parnell (51, 94, 157)	4 Oct. 88	9 Sep. 08	8 Nov. 09	1 Sep. 15	. . . .	7 G. R.
505	Unattd. List	**Macpherson,** Alexander Kilgour (51, 52, 71, 137, 169, 172a)	6 Mar. 88	9 Sep. 08	13 Feb. 10	1 Sep. 15	. . . .	12 Pioneers.
	Unattd. List	**Catterson-Smith,** Thos. Mervyn Osborne (51, 52, 71, 95, 137, 169, 172a)	1 June 88	9 Sep. 08	9 Nov. 09	1 Sep. 15	. . . .	12 Pioneers.
	Unattd. List	**Coats,** Norman George Rogers (57, 95, 165, 169)	17 Oct. 88	9 Sep. 08	7 Nov. 09	1 Sep. 15	. . . .	25 Punjabis.
	Unattd. List	**Moore-Lane,** Vernon Wm. Corbyn (51, 52, 57, 94, 162a, 172a)	4 Oct. 88	9 Sep. 08	7 Nov. 09	1 Sep. 15	. . . .	106 Pioneers.
	Unattd. List	**Farwell,** Walter Lamond (157)	1 Aug. 88	9 Sep. 08	8 Nov. 09	1 Sep. 15	. . . .	4 G. R.
510	Unattd. List	**Brodrick,** William Le Couteur (45a, 51, 52, 71, 137)	27 Jan. 88	9 Sep. 08	23 Oct. 09	1 Sep. 15	. . . .	S. & T. Corps.
	Unattd. List	**Staples,** Edmund Cyril, M.C. (51, 52, 137)	27 Oct. 88	9 Sep. 08	15 Nov. 09	1 Sep. 15	. . . .	Mily. Farms Dept.
	Unattd. List	**Dawes,** Alfred Myers (195)	1 Dec. 87	9 Sep. 08	25 Nov. 09	1 Sep. 15	. . . .	61 Pioneers.
	Unattd. List	**Watts,** Edward Ponsonby (172)	15 June 87	9 Sep. 08	2 Dec. 09	1 Sep. 15	. . . .	53 Sikhs.
	Unattd. List	**Hughes,** Gerald Vincent, M.C. (52, 137)	30 July 88	9 Sep. 08	7 Nov. 09	1 Sep. 15	. . . .	S. & T. Corps.
515	Unattd. List	**Curtis,** Oscar Owen (51, 169)	8 Jan. 89	9 Sep. 08	8 Nov. 09	1 Sep. 15	. . . .	56 Rifles.
	Unattd. List	**Garstin,** Horace Anthony, M.C. (45a, 52, 57, 71, 172)	18 Nov. 88	9 Sep. 08	9 Nov. 09	1 Sep. 15	. . . .	21 Cavalry.
	Unattd. List	**Bainbridge,** Douglas, M.C. (66a, 172)	2 June 88	9 Sep. 08	28 Nov. 09	1 Sep. 15	. . . .	57 Rifles.
	Unattd. List	**Pearson,** Hugh Guy Astley (66, 172)	2 Sep. 88	9 Sep. 08	9 Nov. 09	1 Sep. 15	. . . .	53 Sikhs.
	Unattd. List	**Hickson,** Bertram Evelyn (45a, 71) *Service for pension from 11 Aug. 09.*	25 Sep. 85	9 Sep. 08	29 Oct. 10	1 Sep. 15	. . . .	27 Lt. Cavalry.
520	Unattd. List	**Rogers,** Geoffrey Swarbreck *Service for pension from 11 Aug. 09.*	5 Mar. 88	9 Sep. 08	10 Oct. 10	1 Sep. 15	. . . .	39 Garhwal Rifles.
	Unattd. List	**Fisher,** Gerald Thomas (157, 162, 169, 172a, 205a) *Service for pension from 11 Aug. 09.*	27 Aug. 87	9 Sep. 08	1 Nov. 10	1 Sep. 15	. . . .	3 G. R.
522	Unattd. List	**Harvey,** Charles Offle, M.Cy. (45a, 71, 72a, 172a)	16 July 88	9 Sep. 08	28 Nov. 09	1 Sep. 15	. . . .	38 C. I. Horse.

Captains—*contd.*

No.	Former Corps.	Names.	Date of birth.	Date of first commission.	Date of appointment to Indian Army.	Date of Rank. Substantive.	Date of Rank. Army.	Remarks.
523	Unattd. List	**Brooke,** Arthur Septimus, M.C. (50, 53, 59)	28 Sep. 88	9 Sep. 08	8 Nov. 09	1 Sep. 15	. . . .	18 Infantry.
	Unattd. List	**Erskine,** Kenneth Charles Stuart, M.C. (50, 52, 59, 157)	22 Oct. 88	9 Sep. 08	14 Nov. 09	1 Sep. 15	. . .	5 G. R.
525	Unattd. List	**Smith,** Lawrence (46, 50, 53, 94, 162, 172*a*)	21 Dec. 88	9 Sep. 08	7 Nov. 09	1 Sep. 15	. . . .	6 Cavalry.
	Unattd. List	**Mitchell-Carruthers,** Nigel Laurie (59, 71, 162, 167)	12 Nov. 88	9 Sep. 08	7 Nov. 09	1 Sep. 15	*Ty. m.* 18 July 17.	30 Punjabis.
	Unattd. List	**Mockler,** Edward Claude (51, 52, 157)	13 Mar. 89	9 Sep. 08	8 Nov. 09	1 Sep. 15	. . .	1 G. R.
	Unattd. List	**Leslie,** *Sir* Norman Roderick Alexander David, *Bart.* (45*a*, 51, 52, 71, 162)	10 Jan. 89	9 Sep. 08	15 Nov. 09	1 Sep. 15	*m.* 1 Jan. 18	38 C. I. Horse.
	Unattd. List	**Trail,** Reginald Magnus (52, 66*a*, 162, 167, 169, 205*a*)	6 June 88	9 Sep. 08	12 Dec. 09	1 Sep. 15	. . . .	38 Dogras.
530	Unattd. List	**Shakespear,** Wyndham Biddulph (157)	9 Nov. 88	9 Sep. 08	14 Dec. 09	1 Sep. 15	. . . .	2 G. R.
	Unattd. List	**Mayne,** Ashton Gerard Oswald Mosley, D.S.O. (51, 52, 71, 167)	24 Apl. 89	9 Sep. 08	9 Dec. 09	1 Sep. 15	. . . .	13 Lancers.
	Unattd. List	**Yeates,** Edwin Percival (51, 52, 57, 66*a*, 71, 95, 137, 169, 172*a*)	7 Nov. 88	9 Sep. 08	30 Nov. 09	1 Sep. 15	. . . .	12 Pioneers.
	Unattd. List	**Wright,** Rawdon Douglas (45*a*, 57, 71)	29 May 89	9 Sep. 08	11 Dec. 09	1 Sep. 15	. . . .	29 Lancers.
	Unattd. List	**MacGregor,** Raymond Guy (45*a*, 71, 137)	2 Jan. 89	9 Sep. 08	31 Dec. 09	1 Sep. 15	. . . .	34 Horse.
535	Unattd. List	**Pulvermann,** Alfred William (51, 59, 169)	1 Nov. 88	9 Sep. 08	11 Dec. 09	1 Sep. 15	. . . .	128 Pioneers.
	Unattd. List	**Walker,** James Coulthard (45*a*, 51, 52)	7 Mar. 89	9 Sep. 08	23 Oct. 09	1 Sep. 15	. . . .	5 Cavalry.
	Unattd. List	**Andrewes,** Hugh William (51, 52, 165, 167 205*a*)	24 Feb. 89	9 Sep. 08	11 Dec. 09	1 Sep. 15	. . . .	32 Pioneers.
	Unattd. List	**Tate,** William Moresby (52, 169, 172)	6 Nov. 87	9 Sep. 08	11 Dec. 09	1 Sep. 15	. . . .	55 Rifles.
	Unattd. List	**Laird,** Harold Gordon Canny (50, 52, 71, 155)	6 Jan. 89	9 Sep. 08	10 Dec. 09	1 Sep. 15	. . . .	101 Grenadiers.
540	Unattd. List	**Skinner,** Arthur Lennox (50, 59, 137)	12 Jan. 89	9 Sep. 08	10 Dec. 09	1 Sep. 15	. . . .	122 Infantry.
	Unattd. List	**Mather,** Oswald (50)	17 Oct. 88	9 Sep. 08	11 Dec. 09	1 Sep. 15	. . . .	109 Infantry
	Unattd. List	**Meade,** Malcolm Ellis Morton (66*a*, 137, 162)	23 Feb. 89	9 Sep. 08	5 Dec. 09	1 Sep. 15	. . . .	125 Rifles.
	Unattd. List	**Liston,** Robert Charles Crawford (57)	5 July 88	9 Sep. 08	11 Dec. 09	1 Sep. 15	. . . .	102 Grenadiers
	Unattd. List	**Masters,** Edwin (51, 52, 155)	24 Oct. 88	9 Sep. 08	1 Dec. 09	1 Sep. 15	. . . .	108 Infantry.
545	Unattd. List	**Brewer,** John Cartwright, M. C. (66*a*, 167)	11 Apl. 89	9 Sep. 08	11 Dec. 09	1 Sep. 15	. . . .	90 Punjabis.
	Unattd. List	**Buckland,** Lionel Lindsay (57, 71, 162, 169)	14 June 88	9 Sep. 08	14 Dec. 09	1 Sep. 15	. . . .	25 Cavalry.
	Unattd. List	**Metcalfe,** Eric Debonnair Theophilus, M.C. (51, 53, 66, 157, 172*a*)	2 Mar. 89	9 Sep. 08	30 Nov. 09	1 Sep. 15	. . . .	S. & T. Corps.
	R. Ir. R.	**Wallis,** Beresford Herbert, M.C. (57, 169)	14 Nov. 88	19 Sep. 08	7 Oct. 11	1 Sep. 15	. . . .	107 Pioneers.
	R. Ir. Rif.	**Ffrench,** Raymont Patrick Thomas (51, 52, 60*a*)	17 Mar. 86	19 Sep. 08	18 Mar. 14	1 Sep. 15	. . . .	123 Rifles.
550	York. & L.R.	**Harris,** Arthur Edward Campbell, M.C. (45*a*, 52, 71, 137)	25 July 88	19 Sep. 08	4 Nov. 10	1 Sep. 15	. . . .	7 Lancers.
	N.Lan. R.	***Brotherton,** Leslie Palmer 207	25 Mar. 88	19 Sep. 08	16 Mar. 14	1 Sep. 15	. . . .	3 Brahmans.
	R. Fus.	***Gouldsbury,** Cyril Arkwright 207	20 Mar. 88	14 Oct. 08	17 Mar. 14	1 Sep. 15	*m.* 1 Jan. 18	5 G. R.
	W. Rid. R.	***Bald,** Eric Alfred (71) 207	18 Oct. 88	14 Oct. 08	26 Oct. 14	1 Sep. 15	. . . .	10 G. R.
	R. Suss. R.	**Hamilton,** Owen Philip (66*a*, 167, 172*a*)	27 Feb. 89	14 Oct. 08	4 Mar. 14	1 Sep. 15	. . . .	74 Punjabis.
555	L'pool R.	**Gilbert,** Henry Hugh	12 Oct. 87	14 Oct. 08	22 May 17	1 Sep. 15	. . . .	S. & T. Corps.
	Manch. R.	**Hickie,** Frederick Corbett (60, 65*a*, 66*a*, 162, 172*a*)	8 May 88	16 Dec. 08	5 May 14	1 Sep. 15	. . . .	108 Infantry.
	Unattd. List	**Christie,** Benjamin (51, 52, 57, 66*a*, 167) *Service for pension from 20 Oct. 09.*	17 Nov. 85	20 Oct. 08	6 Jan. 11	1 Sep. 15	. . . .	32 Pioneers.
558	Unattd. List	**Scroope,** Charles Frederick (57, 60*a*, 167) *Service for pension from 15 Dec. 09.*	14 Feb. 86	20 Jan. 09	27 Feb. 11	1 Sep. 15	*Ty. m.* 6 June 17	66 Punjabis

559	Unattd. List	**Strahan,** Geoffrey Cartaret (52, 71, 157) *Service for pension from 2 Feb. 10.*	17 Sep. 86	20 Jan. 09	7 Mar. 11	1 Sep. 15	. . . .	6 G. R.
560	Unattd. List	**Donaldson,** Alastair (167) *Service for pension from 2 Feb. 10.*	7 Feb. 86	20 Jan. 09	6 Mar. 11	1 Sep. 15	. . . .	34 Pioneers.
	Unattd. List	**Crompton,** James (50, 57, 137) *Service for pension from 2 Feb. 10.*	19 July 86	20 Jan. 09	7 Mar. 11	1 Sep. 15	. . . .	10 Jats.
	Unattd. List	**Baker,** William Henry Goldney, D.S.O. (45a, 50, 169)	17 Dec. 88	20 Jan. 09	29 Mar. 10	1 Sep. 15	m. 1 Jan. 18	31 Lancers.
	Unattd. List	**Mackenzie,** Colin Forbes (51, 52, 66a, 169)	10 Nov. 89	20 Jan. 09	11 Mar. 10	1 Sep. 15	. . . .	21 Punjabis
	Unattd. List	**Betham,** Geoffrey Lawrence (66a, 94, 161, 165, 169, 185, 205a)	8 Apl. 89	20 Jan. 09	11 Mar. 10	1 Sep. 15	Ty. m. 25 Feb. 18	15 Sikhs.
565	Unattd. List	**Ferrier,** Ilay (65, 137)	28 Apl. 89	20 Jan. 09	11 Mar. 10	1 Sep. 15	. . . .	48 Pioneers.
	Unattd. List	**Faithfull,** Geo. Ferdinand Hay (172) (*Ty. Maj. 30 Aug. 17 & without pay and allowances from 5 Dec. 16*).	13 Dec. 88	20 Jan. 09	9 Mar. 10	1 Sep. 15	. . . .	126 Infantry.
	Unattd. List	**Gulland,** John Perfect (51, 52, 167, 172a)	3 June 88	20 Jan. 09	12 Mar. 10	1 Sep. 15	. . . .	69 Punjabis.
	Unattd. List	**Scott,** Robert Noel Girling (162, 169, 205a)	25 Dec. 88	20 Jan. 09	6 May 10	1 Sep. 15	. . . .	28 Punjabis.
	Unattd. List	**Windeler,** Wilfrid Templeman (51, 59, 195)	3 Jan. 89	20 Jan. 09	16 Mar 11	1 Sep. 15	. . . .	63 Infantry.
570	Unattd. List	**Spurgin,** Eric Wingrove, M.C. (43, 45a, 51, 71)	3 July 89	20 Jan. 09	11 Mar. 10	1 Sep. 15	. . . .	29 Lancers.
	Unattd. List	**Boyd,** Owen Tudor, M.C. (59, 172a)	30 Aug. 89	20 Jan 09	11 Mar. 10	1 Sep. 15	. . . .	
	Unattd. List	**Fraser,** Denholme DeMontalte Stuart (57, 88, 162, 205a)	5 Oct. 89	20 Jan. 09	28 Mar. 10	1 Sep. 15	. . . .	39 C. I. Horse.
	Unattd. List	**Nugent,** John Fagan Henslowe, D.S.O. (172)	1 July 89	20 Jan. 09	28 Mar. 10	1 Sep. 15	. . . .	28 Punjabis.
	Unattd. List	**Hobson,** John William Sydney (51, 52, 95, 137)	19 June 89	20 Jan. 09	27 Mar. 10	1 Sep. 15	. . . .	16 Rajputs.
575	Unattd. List	**Webb,** Archibald Wilfrid Tisdal (132, 153, 169)	3 Aug. 89	20 Jan. 09	26 Mar. 10	1 Sep. 15	. . . .	Poll. Employ.
	Unattd. List	**Grant,** Francis Robert Elliott (51, 57, 66a, 137, 172a)	14 Mar. 89	20 Jan. 09	11 Mar. 10	1 Sep. 15	. . . .	122 Infantry.
	Unattd. List	**Taylor,** Neville Scott (50, 165)	21 Mar. 89	20 Jan. 09	12 Mar. 10	1 Sep. 15	. . . .	32 Pioneers.
	Unattd. List	**Murray,** Robert George Hamilton, M.C. (157)	11 Nov. 88	20 Jan. 09	17 Mar. 10	1 Sep. 15	. . . .	9 G. R.
	Unattd. List	**Munn,** Frederick Legh Richmond, M.C. (172)	20 July 89	20 Jan. 09	13 Mar. 10	1 Sep. 15	. . . .	46 Punjabis
580	Unattd. List	**Johnson,** George Thomas (172a)	14 Nov. 89	20 Jan. 09	12 Mar. 10	1 Sep. 15	. . . .	22 Punjabis.
	Unattd. List	**Young,** Barclay North (51, 52, 66a, 71, 137, 161, 163, 172a, 205a)	19 Oct. 88	20 Jan. 09	26 Mar. 10	1 Sep. 15	. . . .	12 Pioneers.
	Unattd. List	**Drummond,** James Geoffrey Powys (51, 52, 172)	27 July 89	20 Jan. 09	29 Mar 10	1 Sep. 15	. . . .	19 Punjabis.
	Unattd. List	**Martin,** Norman Macdonald (57, 71, 169) (*Ty. Maj. 13 May 17*)	25 July 89	20 Jan. 09	30 Mar. 10	1 Sep. 15	. . . .	38 C. I. Horse.
	Unattd. List	**Cassidy,** Vivian Clemons (195)	3 Oct. 89	20 Jan. 09	28 Mar. 10	1 Sep. 15	. . . .	S. & T. Corps.
585	14 Hussars	* **Watson,** George Newall (51, 52)	29 Dec. 88	27 Jan. 09	25 July 14	1 Sep. 15	. . . .	13 Lancers.
	Unattd. List	**Williamson,** Andrew Arnold *Service for pension from 25 June 11.*	3 Nov. 88	29 Jan. 09	31 Mar. 12	1 Sep. 15	. . . .	75 Infantry.
	E. Surr. R.	***Mackenzie,** James Dunbar	19 July 89	6 Feb. 09	14 Aug. 14	1 Sep. 15	Ty. m. 11 May 18	41 Dogras.
	S. Lan. R.	* **Warner,** Charles Thornton (51, 52)	29 Aug. 89	6 Feb. 09	19 Aug. 14	1 Sep. 15	. . . .	22 Punjabis.
	N. Staff. R.	**Gompertz,** Arthur William Bean (52, 53, 66, 172a)	8 Dec. 88	22 May 09	22 Aug. 14	1 Sep. 15	. . . .	94 Infantry.
590	W. York R.	**Anderson,** George Douglas *Service for pension from 23 April 08.*	28 July 87	16 June 09	13 June 12	1 Sep. 15	. . . .	62 Punjabis.
	R. A.	§* **Bryson,** Alexander Buchanan (66, 169, 172a)	30 Aug. 88	23 July 09	2 May 14	1 Sep. 15	. . . .	S. & T. Corps.
	R. F. A.	* **Buist,** Claude Napier (172a)	9 July 88	23 July 09	8 July 14	1 Sep. 15	. . . .	51 Sikhs.
	Unattd. List	**Hutton,** Guy Maurice (57)	2 Aug. 87	25 Aug. 09	1 Nov. 10	1 Sep. 15	. . . .	22 Cavalry.
	Unattd. List	**Leslie,** Thomas Leslie (45a) *Service for pension from 24 Aug. 10.*	26 Sep. 87	8 Sep. 09	1 Oct. 11	1 Sep. 15	. . . .	35 Horse.
595	Unattd. List	**Fraser-Tytler,** William Kerr, M.C. (45a, 51, 52, 71, 169, 172) *Service for pension from 24 Aug. 10.*	26 Dec. 86	8 Sep. 09	30 Oct. 11	1 Sep. 15	. . . .	25 Cavalry.
596	Unattd. List	**Tweedy,** Roger John (172, 205a) *Service for pension from 24 Aug. 10.*	16 Nov. 87	8 Sep. 09	30 Oct. 11	1 Sep. 15	. . . .	19 Lancers.

* Has not passed the final examination. § Admitted subject to H. M.'s approval.

No.	Former Corps.	Names.	Date of birth.	Date of first commission.	Date of appointment to Indian Army.	Date of Rank. Substantive.	Date of Rank. Army.	Remarks.
		Captains—*contd.*						
597	Unattd. List	**Williams,** Austin Henry (45*a*, 71, 162, 172*a*)	11 Feb. 90	8 Sep. 09	4 Dec. 10	1 Sep. 15	. . . .	38 C. L. Horse.
	Unattd. List	**Barton,** Leslie Eric (50, 66*a*, 71, 162, 167, 169)	3 Feb. 89	8 Sep. 09	4 Dec. 10	1 Sep. 15	. . . .	30 Punjabis.
	Unattd. List	**Channer,** George Osborne De Renzy (50, 52, 59, 95, 157, 162)	25 Feb. 90	8 Sep. 09	24 Dec. 10	1 Sep. 15	. . . .	7 G. R.
600	Unattd. List	**Kennedy,** Angus Menzies (50, 157)	29 Oct. 89	8 Sep. 09	1 Nov. 10	1 Sep. 15	. . . .	8 G. R.
	Unattd. List	**Pitman,** Charles Robert Senhouse, D.S.O., M.C. (51, 162, 172)	19 Mar. 90	8 Sep. 09	23 Dec. 10	1 Sep. 15	. . . .	27 Punjabis.
	Unattd. List	**Anderson,** James McKenzie, M.C. (57)	15 Dec. 89	8 Sep. 09	24 Dec. 10	1 Sep. 15	. . . .	109 Infantry.
	Unattd. List	**Raynor,** Cyril Arthur, M.C. (137)	6 Dec. 89	8 Sep. 09	24 Oct. 10	1 Sep. 15	. . . .	48 Pioneers.
	Unattd. List	**Quayle,** Pietro Rudolph (94, 127, 162, 169) [L]	22 Mar. 90	8 Sep. 09	5 Dec. 10	1 Sep. 15	. . . .	127 Infantry.
605	Unattd. List	**Deane,** Henry Harold Rookhurst (162, 172*a*, 205*a*)	14 Dec. 89	8 Sep. 09	31 Dec. 10	1 Sep. 15	. . . .	64 Pioneers.
	Unattd. List	**Scratton,** Guy Walsh Blackburn, M.C. (172)	5 Apl. 90	8 Sep. 09	7 Jan. 11	1 Sep. 15	*Ty. m.*, 27 Dec. 18	19 Lancers.
	Unattd. List	**Ogilvy,** Abercromby Graham (51, 52, 66*a*, 167, 172*a*)	1 Oct. 89	8 Sep. 09	31 Dec. 10	1 Sep. 15	. . . .	38 Dogras.
	Unattd. List	**Drysdale,** Alexander Edward, D.S.O., M.C. (167)	7 May 89	8 Sep. 09	3 Nov. 10	1 Sep. 15	. . . .	47 Sikhs.
	Unattd. List	**Ozanne,** Guy Durand, M.C. (59, 66*a*, 162, 205*a*)	2 Apl. 89	8 Sep. 09	5 Dec. 10	1 Sep. 15	. . . .	99 Infantry.
610	Unattd. List	**Jackson,** Lawrence Hungerford (59)	27 Sep. 89	8 Sep. 09	1 Nov. 10	1 Sep. 15	. . . .	13 Rajputs.
	Unattd. List	**Mein,** Desbrisay Blundell, M.C. (12, 51, 52, 172)	7 Dec. 89	8 Sep. 09	2 Nov. 10	1 Sep. 15	. . . .	55 Rifles.
	Unattd. List	**Skipwith,** Charles Gray Yule (51, 59, 71, 162, 172)	20 Apl. 90	8 Sep. 09	23 Dec. 10	1 Sep. 15	. . . .	17 Cavalry.
	Unattd. List	**Brunskill,** Bertie Arthur Stephen (51, 52, 57)	14 Apl. 90	8 Sep. 09	30 Oct. 10	1 Sep. 15	. . . .	79 Infantry.
	Unattd. List	**Shakespear,** George Frederick Corlandt, D.S.O., M.C. (51, 161)	31 Oct. 89	8 Sep. 09	31 Oct. 10	1 Sep. 15	*m.* 1 Jan. 17	88 Infantry.
615	Unattd. List	**O'Neill,** William Hickley Lovell, M.C. (57, 172*a*)	7 Feb 90	8 Sep. 09	3 Jan. 11	1 Sep. 15	. . . .	51 Sikhs.
	Unattd. List	**Finnis,** Henry, M.C. (59, 95, 172)	21 Apl. 90	8 Sep. 09	30 Oct. 10	1 Sep. 15	. . . .	53 Sikhs.
	Unattd. List	**Wilson,** William Clement Francis Allan (57, 66*a*, 66, 137, 162)	25 Mar. 00	8 Sep. 09	3 Nov. 10	1 Sep. 15	*m.* 3 June 18	104 Rifles.
	Unattd. List	**Meiklejohn,** John Forbes (51, 52, 57, 71, 95, 137)	10 Oct. 89	8 Sep. 09	30 Oct. 10	1 Sep. 15	. . . .	33 Lt. Cavalry.
	Unattd. List	**Ozanne,** Edward Harry Brock (51, 95, 71, 172)	29 Dec. 89	8 Sep. 09	23 Dec. 10	1 Sep. 15	. . . .	121 Pioneers.
620	Unattd. List	**Perram,** Henry Charles (51, 167, 172*a*)	13 Dec. 89	8 Sep. 09	4 Dec. 10	1 Sep. 15	. . . .	84 Punjabis.
	Shrops. L. I.	***Johnston,** Allen Leigh (50, 52, 60*a*, 71, 161, 205*a*)	7 May 90	18 Sep. 09	20 Aug. 14	1 Sep. 15	. . . .	126 Infantry.
	W. Rid. R.	**Russell,** Raymond Wycliffe (51, 66*a*, 71, 72, 157)	29 July 89	18 Sep. 09	19 Sep. 14	1 Sep. 15	. . . .	9 G. R.
	R. W. Kent. R	**Bonham-Carter,** Brian Hulbert	15 May 89	18 Sep. 09	16 Aug. 13	1 Sep. 15	. . . .	40 Pathans.
	Notts & Derby R.	**Happell,** William Horatio (50, 52, 162)	15 Apl. 90	18 Sep. 09	28 Mar. 14	1 Sep. 15	. . . .	99 Infantry.
625	E. York. R.	***Eccles,** Malcolm, M.C. (60*a*) 207	4 Jan. 90	18 Sep. 09	13 Aug. 11	1 Sep. 15	. . . .	119 Infantry.
	N. Staff. R.	**Young,** Keith de Lorentz, M.C. (45*a*, 51, 66, 71, 172*a*)	5 Aug. 89	18 Sep. 09	5 June 12	1 Sep. 15	. . . .	8 Cavalry.
	Leic. R.	**Williams,** Sydney Cecil (65*a*, 66, 162)	5 Jan. 89	18 Sep. 09	15 May 17	1 Sep. 15	*Ty. m.* 25 Feb. 18	S. & T. Corps.
	Leins. R.	**McLean,** Frederick George Swan (172*a*)	25 Jan. 89	18 Sep. 09	2 May. 14	1 Sep. 15	. . . .	67 Punjabis.
	L. Ir. Regt.	**Chisholm,** Douglas Kerr Joseph 207	23 Jan. 89	6 Nov. 09	27 Feb. 14	1 Sep. 15	. . . .	114 Mahrattas.
630	R. A.	**Urmson,** Denis George Severn (65)	26 Feb. 88	11 Dec. 09	20 Feb. 14	1 Sep. 15	. . . .	64 Pioneers.
	R. Ir. Fus.	***Kellie,** George Jerome Dunlop, M.C. (51, 52, 66*a*, 95, 172*a*)	28 Sep. 88	11 Dec. 09	6 Sep. 14	1 Sep. 15	. . . .	30 Punjabis.
	R. Ir. Fus.	**Jones,** Basil Ivor, M.C. (60*a*, 94, 169)	25 Apl. 89	25 Dec. 09	14 Oct. 14	1 Sep. 15	. . . .	38 Dogras.
633	Unattd. List	**Kane,** John Leonard Kirkpatrick *Service for pension from 25 Jan. 11.*	7 Oct. 89	29 Jan. 10	27 Mar. 12	1 Sep. 15	. . . .	109 Infantry.

634	Unattd. List	**Cahusac,** Stuart Douglas Nugent (45*a*, 51, 52, 95, 172)	5 Dec. 86	29 Jan. 10	23 Mar. 12	1 Sep. 15	. . . .	10 Lancers.
		Service for pension from 25 Jan. 11.						
635	Unattd. List	**Meade,** Henry Richard Caine (88, 94, 155*a*, 157, 162, 172*a*)	9 July 90	29 Jan. 10	11 Mar. 11	1 Sep. 15	. . . .	8 G. R.
	Unattd. List	**Vickers,** Wilmot Gordon Hilton (50, 53, 169)	8 June 90	29 Jan. 10	13 Mar. 11	1 Sep. 15	. . . .	31 Lancers.
	Unattd. List	**Lilly,** Alfred Noel Irvine (57, 95, 137, 161)	22 Dec. 89	29 Jan. 10	5 Mar. 11	1 Sep. 15	. . . .	7 Rajputs.
	Unattd. List	**Standen,** James Douglas (50, 71, 95, 137, 172*a*)	5 July 88	29 Jan. 10	9 Mar. 11	1 Sep. 15	. . . .	10 Jats.
	Unattd. List	**Barstow,** Thomas Clement Erskine (157)	30 June 90	29 Jan. 10	4 Apl. 11	1 Sep. 15	. . . .	6 G. R.
640	Unattd. List	**Dawe,** Thomas Reed (50, 71, 195)	7 July 90	29 Jan. 10	6 Mar. 11	1 Sep. 15	. . . .	79 Infantry.
	Unattd. List	**Whitford-Hawkey,** Edward Horace (45*a*, 66*a*, 71, 167)	11 Aug. 90	29 Jan. 10	28 June 11	1 Sep. 15	. . . .	16 Cavalry.
	Unattd. List	**Robinson,** Dudley Clare, M.C. (50, 52, 59, 167)	1 Oct. 90	29 Jan. 10	31 Mar. 11	1 Sep. 15	. . . .	36 Sikhs.
	Unattd. List	**Corke,** Rupert Mortimer (50, 52)	22 Oct. 89	29 Jan. 10	21 Mar. 11	1 Sep. 15	. . . .	94 Infantry.
	Unattd. List	**Deedes,** Ralph Bouverie, M.C. (50, 52, 59, 66*a*, 167)	17 Oct. 90	29 Jan. 10	5 Mar. 11	1 Sep. 15	. . . .	31 Punjabis.
645	Unattd. List	**Sanders,** Claude William (51, 53, 57, 137, 162, 172*a*)	10 Oct. 90	29 Jan. 10	6 Mar. 11	1 Sep. 15	. . . .	1 Brahmans.
	Unattd. List	**O'Sullivan,** Justin (51, 52, 137)	13 June 90	29 Jan. 10	6 Mar. 11	1 Sep. 15	. . . .	4 Rajputs.
	Unattd. List	**Dennys,** Lancelot Ernest (50, 66*a*, 71, 169)	10 May 90	29 Jan. 10	7 Mar. 11	1 Sep. 15	. . . .	54 Sikhs.
	Unattd. List	**Whitworth,** Dysart Edward, M.C. (45*a*, 51, 52, 71, 137)	13 July 90	29 Jan. 10	6 Mar. 11	1 Sep. 15	. . . .	2 Lancers.
	Unattd. List	**Crew,** Robert Douglas (51, 52, 57)	10 Aug. 90	29 Jan. 10	7 Mar. 11	1 Sep. 15	. . . .	52 Sikhs.
650	Unattd. List	**FitzGerald,** Gerald Michael, M.C. (169)	28 Aug. 89	29 Jan. 10	6 Mar. 11	1 Sep. 15	. . . .	19 Lancers.
	Unattd. List	**Matheson,** Bertram Heylyn (46, 50, 52, 57, 59, 162, 169)	24 July 90	29 Jan. 10	7 Mar. 11	1 Sep. 15	. . . .	54 Sikhs.
	Unattd. List	**Henderson,** George Burton (51, 52, 66*a*, 95, 165, 167, 169, 205*a*)	6 July 90	29 Jan. 10	7 Mar. 11	1 Sep. 15	. . . .	15 Sikhs.
	Unattd. List	**Langhorne,** Cuthbert Charles (52, 66, 172)	7 Nov. 90	29 Jan. 10	11 Mar. 11	1 Sep. 15	. . . .	24 Punjabis.
	Unattd. List	**Wright-Warren,** Christopher Thomas (51, 57, 125*a*, 129) [*l*]	4 Sep. 89	29 Jan. 10	11 Mar. 11	1 Sep. 15	. . . .	80 Infantry.
655	Unattd. List	**Hungerford,** Sidney Arthur Hodder	25 Feb. 90	29 Jan. 10	7 Mar. 11	1 Sep. 15	. . . .	116 Mahrattas.
	Unattd. List	**Bingham,** Charles D'Arcy (50, 51, 52, 162, 205*a*)	16 Aug. 90	29 Jan. 10	6 Mar. 11	1 Sep. 15	. . . .	109 Infantry.
	Unattd. List	**Irwin,** Hubert Richard Benjamin Hinde (51, 52, 59)	25 Feb. 90	29 Jan. 10	16 Mar. 11	1 Sep. 15	. . . .	105 Infantry.
	Unattd. List	**Harward,** Robert Blake (51, 52)	12 Apl. 90	29 Jan. 10	31 Mar. 11	1 Sep. 15	. . . .	73 Infantry.
	Unattd. List	**Wickham,** Edward Thomas Ruscombe (94, 163) [L]	4 May 90	29 Jan. 10	7 Mar. 11	1 Sep. 15	. . . .	39 C. I. Horse.
660	Unattd. List	**Davison,** Kenneth Herbert Metcalfe (51, 52, 66*a*, 95, 162*a*)	20 Jan. 90	29 Jan. 10	7 Mar. 11	1 Sep. 15	. . . .	106 Pioneers.
	Manch. R.	***Mardall,** Francis Hay	2 July 90	20 Apl. 10	26 Feb. 14	1 Sep. 15	. . . .	41 Dogras.
	R. Fus.	**Schofield,** William Hugh Samuel	27 Feb. 90	28 May 10	6 Dec. 12	1 Sep. 15	. . . .	18 Infantry.
	Leic. R.	**Longbotham,** Cyril Murgatroyd (51, 95, 169)	3 July 87	28 May 10	5 Oct. 12	1 Sep. 15	. . . .	72 Punjabis.
	Unattd. List	**Ferguson,** John Allan (50, 52)	22 May 89	2 Sep. 10	8 Oct. 12	1 Sep. 15	. . . .	13 Rajputs.
		Service for pension from 16 Aug. 11.						
665	Unattd. List	**Montford,** Douglas Raymond (51, 52, 66*a*)	18 Nov. 89	2 Sep. 10	7 Oct. 12	1 Sep. 15	. . . .	98 Infantry.
		Service for pension from 16 Aug. 11.						
	Unattd. List	**Stevenson,** Maurice Macaw (45*a*)	15 Apl. 89	2 Sep. 10	8 Oct. 12	1 Sep. 15	. . . .	1 Lancers.
		Service for pension from 16 Aug. 11.						
	Unattd. List	**Mathias,** Leonard William Henry, D.S.O.	31 Oct. 90	3 Sep. 1•	26 Oct. 11	1 Sep. 15	. . . .	128 Pioneers.
	Unattd. List	**Graham,** Lancelot Cecil Torbock, M.C. (57, 66, 71)	9 Apl. 90	3 Sep. 10	12 Dec. 11	1 Sep. 15	. . . .	9 Horse.
	Unattd. List	**Shearer,** Johnston Evelyn, M.C. (172*a*)	26 Dec. 90	3 Sep. 10	1 Nov. 11	1 Sep. 15	. . . .	26 Punjabis.
670	Unattd. List	**Morrogh Bernard,** John Alexander Dermot (51, 52, 95)	8 Dec. 90	3 Sep. 10	27 Jan. 12	1 Sep. 15	. . . .	22 Cavalry.
	Unattd. List	**Drake-Brockman,** Francis Travers (71, 137, 162)	4 Dec. 90	3 Sep. 10	27 Oct. 11	1 Sep. 15	. . . .	7 Lancers.
	Unattd. List	**Whittall,** James Mance (157)	18 Sep. 90	3 Sep. 10	10 Dec. 11	1 Sep. 15	. . . .	6 G. R.
	Unattd. List	**Lane,** Charles Reginald Cambridge, M.C. (172, 205*a*) [L]	4 Dec. 90	3 Sep. 10	12 Dec. 11	1 Sep. 15	. . . .	19 Lancers.
	Unattd. List	**O'Donnell,** Brian Henry (51, 52, 137)	24 Jan. 91	3 Sep. 10	13 Dec. 11	1 Sep. 15	. . . .	26 Cavalry.
675	Unattd. List	**Bunbury,** Noel Louis St. Pierre (50, 53, 66, 95, 162*a*, 169)	25 Dec. 90	3 Sep. 10	1 Nov. 11	1 Sep. 15	. . . .	106 Pioneers.
	Unattd. List	**Laville,** Theodore Thompson (71, 137, 172*a*)	8 Apl. 91	3 Sep. 10	11 Dec. 11	1 Sep. 15	. . . .	11 Rajputs.
	Unattd. List	**Lunn,** Arthur John, M.C. (51, 52, 60*a*, 66*a*, 167)	3 Mar. 91	3 Sep. 10	12 Dec. 11	1 Sep. 15	. . . .	82 Punjabis.
678	Unattd. List	**Andrews,** Harold Arthur, M.C. (137)	25 Apl. 91	3 Sep. 10	28 Nov. 11	1 Sep. 15	. . . .	37 Dogras.

* Has not passed the final examination.

No.	Former Corps.	Names.	Date of birth.	Date of first commission.	Date of appointment to Indian Army.	Date of Rank. Substantive.	Date of Rank. Army.	Remarks.
		Captains—*contd.*						
679	Unattd. List	**Mosse,** Charles Oliver Robins, M.C. (51, 52, 172a)	31 May 91	3 Sep. 10	27 Nov. 11	1 Sep. 15	. . . .	120 Infantry.
680	Unattd. List	**Rivett-Carnac,** Percy Kenneth	3 May 90	3 Sep. 10	28 Nov. 11	1 Sep. 15	. . . .	17 Infantry.
	Unattd. List	**Maltby,** Christopher Michael, M.C. (46, 50, 53)	13 Jan. 91	3 Sep. 10	12 Dec. 11	1 Sep. 15	. . . .	95 Infantry.
	Unattd. List	**Lemon,** Roy Lancelot (167)	15 Oct. 90	3 Sep. 10	26 Oct. 11	1 Sep. 15	. . . .	10 G. R.
	Unattd. List	**Penrose-Welsted,** Reginald Hugh (51, 137)	23 Apl. 91	3 Sep. 10	28 Nov. 11	1 Sep. 15	. . . .	4 Rajputs.
	Unattd. List	**Scott,** William Richard Annesley (66a, 137)	4 Dec. 90	3 Sep. 10	27 Nov. 11	1 Sep. 15	. . . .	S. & T. Corps.
685	Unattd. List	**Gell,** Humphrey Vickers (57, 167)	11 Feb. 91	3 Sep. 10	27 Nov. 11	1 Sep. 15	. . . .	69 Punjabis.
	Unattd. List	**Elkington,** Edward Henry Ommanney, M.C. (71, 167, 205a)	23 Nov. 90	3 Sep 10	27 Nov. 11	1 Sep. 15	. . . .	36 Sikhs.
	Unattd. List	**Bird,** William Trevor (57, 71, 172a, 205a)	1 Sep. 90	3 Sep. 10	1 Dec. 11	1 Sep. 15	. . . .	30 Lancers.
	Unattd. List	**Lawrence,** Richard Travers (45a, 51, 52)	7 July 90	3 Sep. 10	28 Nov. 11	1 Sep. 15	. . . .	10 Lancers.
	Unattd. List	**Abbott,** James Fraser, M.C. (50, 157)	2 July 90	3 Sep. 10	29 Oct. 11	1 Sep. 15	. . . .	8 G. R.
690	Unattd. List	**Grey,** Percy	29 Apl. 91	3 Sep. 10	26 Nov 11	1 Sep. 15	. . . .	S. & T. Corps.
	Unattd. List	**Mullaly,** John Clive Doveton (51, 52, 172a)	4 Jan. 91	3 Sep. 10	27 Nov. 11	1 Sep. 15	. . . .	9 Infantry.
	Unattd. List	**Hebbert,** Percival Benbow	24 Mar. 90	3 Sep. 10	28 Nov. 11	1 Sep. 15	. . . .	44 Infantry.
	Unattd. List	**Fulton,** Charles Mathew (47a, 51, 71) [l]	3 Oct. 90	3 Sep. 10	26 Oct. 11	1 Sep. 15	. . . .	33 Lt. Cavalry.
	Unattd. List	**MacLean,** Lachlan Loudoun	23 July 90	3 Sep. 10	29 Oct. 11	1 Sep. 15	. . . .	8 G. R.
695	Unattd. List	**Minter,** John Paul Surtees (172)	24 Jan. 91	3 Sep. 10	29 Oct. 11	1 Sep. 15	. . . .	130 Baluchis.
	R. Berks. R.	**Handley,** Leonard Mourant, M.C. (57)	9 Dec. 90	5 Oct. 10	8 Oct. 13	1 Sep. 15	. . . .	26 Cavalry.
	N. & D. R.	* **Chambers,** George Harold, M.C. (60a) 207	30 June 90	5 Oct. 10	22 Aug. 14	1 Sep. 15	*Ty. m.* 1 Feb. 18	123 Rifles.
	Unattd. List	**McElwaine,** Erick James Dalby (50, 167) *Service for pension from 11 Nov. 11.*	13 Apl. 87	17 Jan. 11	7 Jan. 13	1 Sep. 15	. . . .	76 Punjabis.
	Unattd. List	**Wallace,** Alexander James Lumsden (169) *Service for pension from 7 Feb. 12.*	8 Aug. 89	17 Jan. 11	25 Mar. 13	1 Sep. 15	. . . .	52 Sikhs.
700	Unattd. List	**Malone,** Edmond Mortimer *Service for pension from 7 Feb. 12.*	6 Sep. 90	17 Jan. 11	24 Mar. 13	1 Sep. 15	. . . .	5 Lt. Infantry.
	Unattd. List	**Brandon,** Roy Victor (46, 66a) *Service for pension from 7 Feb. 12.*	15 Oct. 89	17 Jan. 11	24 Mar. 13	1 Sep. 15	. . . .	18 Infantry.
	Unattd. List	**Thompson,** Sidney Lionel (50, 52) *Service for pension from 21 Feb. 12.*	29 July 88	17 Jan. 11	23 Mar. 13	1 Sep. 15	. . . .	113 Infantry.
	Unattd. List	**Pollard,** George Leslie Grove (57, 95, 162a, 172a)	1 Sep. 91	18 Jan. 11	25 Mar. 12	1 Sep. 15	. . . .	106 Pioneers.
	Unattd. List	**Cuddeford,** Roy Bathurst Gore (50, 52, 95)	7 Apl. 91	18 Jan. 11	25 Mar. 12	1 Sep. 15	. . . .	127 Infantry.
705	Unattd. List	**Haycraft,** William Church Stacpoole, M.C. (167)	24 Oct. 91	18 Jan. 11	25 Mar. 12	1 Sep. 15	. . . .	93 Infantry.
	Unattd. List	**Taylor,** Valentine Colebrooke Lilly (57, 106)	9 Oct. 91	18 Jan. 11	25 Mar. 12	1 Sep. 15	. . . .	93 Infantry.
	Unattd. List	**Brunskill,** George Stephen, M.C. (57, 167)	26 Aug. 91	18 Jan. 11	6 Mar. 12	1 Sep. 15	. . . .	47 Sikhs.
	Unattd. List	**Watkis,** Harold Lithgow, M.C. (66, 95, 167, 169)	23 Sep. 91	18 Jan. 11	6 Mar. 12	1 Sep. 15	. . . .	37 Lancers.
	Unattd. List	**Geard,** Reginald Cheniston (71, 95, 167, 172a, 205a)	5 Oct. 91	18 Jan. 11	6 Mar. 12	1 Sep. 15	. . . .	15 Lancers.
710	Unattd. List	**Brander,** Ralph Whitley (66a)	13 May 91	18 Jan. 11	5 Mar. 12	1 Sep. 15	. . . .	S. & T. Corps.
	Unattd. List	**Shewen,** Douglas Gordon Pigou Mansel	2 July 91	18 Jan. 11	29 Mar. 12	1 Sep. 15	. . . .	27 Punjabis.
	Unattd. List	**North,** Kenneth Amber, M.C. (52, 57, 167)	12 July 91	18 Jan. 11	24 Mar. 12	1 Sep. 15	. . . .	32 Pioneers.
713	Unattd. List	**Gabbett,** Kenneth Joseph (51, 52, 66a, 167, 172a)	2 Sep. 91	18 Jan. 11	6 Mar. 12	1 Sep. 15	. . . .	23 Pioneers.

714	Unattd. List	**Hughes,** Lionel William (137)		30 Oct. 91	18 Jan. 11	4 Mar. 12	1 Sep. 15		125 Rifles.
715	Unattd. List	**Rich,** Henry Hampden (51, 95)		30 Mar. 91	18 Jan. 11	23 Mar. 12	1 Sep. 15		120 Infantry.
	Unattd. List	***Gaisford,** Philip (94. 167, 169, 172*a*)		28 Nov. 91	18 Jan. 11	27 Mar. 12	1 Sep. 15		31 Punjabis.
	Unattd. List	**Steward,** Cecil Walter d'Alterac (51, 66*a*)	207	1 Oct. 91	18 Jan. 11	25 Mar. 12	1 Sep. 15		66 Punjabis.
	Unattd. List	**Radley,** Hugh Poynton, M.C. (59,95,172)		14 Nov. 91	18 Jan. 11	6 Mar. 12	1 Sep. 15		72 Punjabis.
	Unattd. List	**Middleton,** Montagu (51, 52, 57, 71, 165)		11 June 91	18 Jan. 11	25 Mar. 12	1 Sep. 15		87 Punjabis.
720	Unattd. List	**Booth,** Edward Robert Courtenay (51, 52)		2 Aug. 90	18 Jan. 11	25 Mar. 12	1 Sep 15		35 Horse.
	Unattd. List	**Ricketts,** Edwin Leslie (50, 60*a*, 71)		16 July 91	18 Jan. 11	6 Mar. 12	1 Sep. 15		96 Infantry.
	Unattd. List	**Craig,** Herbert Teesdale (51, 52)		26 Aug. 91	18 Jan. 11	25 Mar. 12	1 Sep. 15		43 Regiment.
	Unattd. List	**Landale,** Charles Hugh (59, 71, 167)		8 Oct. 91	18 Jan. 11	5 Mar. 12	1 Sep. 15		16 Cavalry.
	Unattd. List	**Scudamore,** Philip Cowie (137)		11 Aug. 91	18 Jan. 11	7 Mar. 12	1 Sep. 15		11 Rajputs.
725	Unattd. List	**Bampfield,** George Charles (51, 167)		1 Sep. 91	18 Jan. 11	5 Mar. 12	1 Sep. 15		90 Punjabis.
	Unattd. List	**Brownlow,** John (66*a*, 71)		31 Jan. 91	18 Jan. 11	6 Mar. 12	1 Sep. 15		28 Cavalry.
	Unattd. List	**Branfoot,** Denis Charles (51, 52)		18 July 91	18 Jan. 11	8 Mar. 11	1 Sep. 15		22 Cavalry.
	Essex R.	***Bremner,** Claude Edward Urquhart, M.C. (95, 169)	205*a*	30 Aug. 91	4 Mar. 11	11 Feb. 14	1 Sep. 15		29 Punjabis.
	Unattd. List	***Silver,** Harold Edward	207	24 Sep. 91	30 Aug. 11	6 Nov. 12	1 Sep. 15		18 Infantry.
730	Unattd. List	***Pim,** Hugh Maxwell, M.C.		10 Oct. 92	24 Aug. 12	7 Oct. 14	1 Sep. 15		24 Punjabis.
		Service for pension from 14 Jan. 14.							
	A. S. C.	*Stubbs, John Bushby	207*a*	2 Nov. 91	4 Feb. 11	26 Oct. 18	1 Sep. 15		13 Lancers.
	Unattd. List	**Lumby,** Arthur Friedrich Rawson (51) ‡		13 Aug. 90	5 Sep. 11	5 Oct. 13	5 Sep. 15	*Ty.* *m* 25 Feb. 17	59 Punjabis.
		Service for pension from 7 Aug. 12.							
	Unattd. List	***Batten,** William Douglas Grant (52)		9 July 90	5 Sep. 11	21 Mar. 14	5 Sep. 15		3 G. R.
		Service for pension from 18 Jan. 13.							
	Unattd. List	**Ewing,** Robert Ross (107, 172*a*)		22 Feb. 91	5 Sep. 11	11 Oct. 13	5 Sep. 15		61 Pioneers.
		Service for pension from 7 Aug. 12.							
735	Unattd. List	***Mackay,** Alexander Sutherland, M.C. (51, 52, 66*a*)	207	18 June 90	5 Sep. 11	21 Nov. 13	5 Sep. 15		7 G. R.
		Service for pension from 7 Aug. 12.							
	Unattd. List	***Randall,** Basil Fitz Herbert (95)	207	2 Nov. 90	5 Sep 11	9 Mar. 14	†5 Sep. 15		3 Horse.
		Service for pension from 20 Aug. 13.							
	Unattd. List	***Armstrong,** Francis Harold Courtenay (71)		20 Oct 90	5 Sep. 11	8 Mar. 14	5 Sep. 15		67 Punjabis.
		Service for pension from 18 Jan. 13.							
	Unattd. List	**Harding,** Cuthbert Neil (51, 52)		31 Dec. 90	6 Sep 11	5 Nov. 12	6 Sep. 15		88 Infantry.
	Unattd. List	**Channer,** Keith Francis (95, 172, 169)		18 Feb. 92	6 Sep. 11	3 Dec. 12	6 Sep. 15		36 Horse.
740	Unattd. List	**Marshall,** Harold Hamilton, M.C. (172*a*)		4 July 92	6 Sep. 11	28 Oct. 12	6 Sep. 15		48 Pioneers.
	Unattd. List	**Geldt,** Edward Wollaston		5 Dec. 91	6 Sep. 11	6 Nov. 12	6 Sep. 15		107 Pioneers.
	Unattd. List	**Rodgerson,** Alan Patrick D.S.O., (52, 137, 172*a*)		11 May 92	6 Sep. 11	3 Dec. 12	6 Sep. 15		39 Garh. R.
	Unattd. List	**Rule,** Donald Gordon (172)		29 Nov. 91	6 Sep. 11	3 Dec. 12	6 Sep. 15		27 Punjabis.
	Unattd. List	**Simson,** Gerard Oliver (71, 137)		18 Nov. 91	6 Sep. 11	4 Nov. 12	6 Sep. 15		34 Horse.
745	Unattd. List	**Cubbon,** Francis Richard, M.C. (95, 169)		29 Nov. 92	6 Sep. 11	1 Dec. 12	6 Sep. 15		72 Punjabis.
	Unattd. List	**Wellesley,** Henry Augustus, M.C. (157)		25 Sep. 91	6 Sep. 11	4 Dec. 12	6 Sep. 15		5 G. R.
	Unattd. List	**Hind,** Neville Godfray (157, 172*a*)		8 Jan. 92	6 Sep. 11	3 Dec. 12	6 Sep. 15		2 G. R.
	Unattd. List	***Shearer,** Eric James, M.C. (50, 52)		23 Nov. 92	6 Sep. 11	3 Dec. 12	6 Sep. 15		31 Lancers.
	Unattd. List	**Hammond,** Arthur Verney, (51, 52, 71)		16 Oct. 92	6 Sep. 11	3 Dec. 12	6 Sep. 15		Guides.
750	Unattd. List	**Tinley,** Francis Benton Newpot, M.C. (69, 172)		23 Aug. 92	6 Sep. 11	3 Dec. 12	6 Sep. 15		20 Horse.
	Unattd. List	**Prinsep,** Evelyn Seigfried MacLeod (71, 95, 172)		23 Apl. 92	6 Sep. 11	7 Nov. 12	6 Sep. 15		11 Lancers.
	Unattd. List	* **Graham,** Roderic Arthur (167)		2 July 91	6 Sep. 11	4 Nov. 12	6 Sep. 15		31 Punjabis.
	Unattd. List	**Campbell,** Ian (51, 52, 71, 95, 172, 172*a*)		27 Oct. 91	6 Sep. 11	6 Nov. 12	6 Sep. 15		11 Lancers.
	Unattd. List	**Rivett-Carnac,** Herbert Gordon (51, 52, 71)		13 Feb. 92	6 Sep. 11	4 Nov. 12	6 Sep. 15		79 Infantry.
755	Unattd. List	* **Watson,** Eric Edward (205)	207	17 Sep. 92	6 Sep. 11	4 Dec. 12	6 Sep. 15		62 Punjabis.

* Has not passed the final examination. † Without pay and allowances prior to 1 Sept. 16. ‡ Official Interpreter in German.

Captains—*contd.*

No.	Former Corps.	Names.	Date of birth.	Date of first commission.	Date of appointment to Indian Army.	Date of Rank. Substantive.	Date of Rank. Army.	Remarks.
756	Unattd. List	* **Cuningham,** Alured Phayre (172)	24 Sep. 92	6 Sep. 11	4 Nov. 12	6 Sep. 15		27 Infantry.
	Unattd. List	**Mullaly,** Brian Reginald (156)	1 Aug. 92	6 Sep. 11	3 Dec. 12	6 Sep. 15		10 G. R.
	Unattd. List	**Stevens,** Charles Francis Linnæus, M.C.	26 Jan. 92	6 Sep 11	2 Dec. 12	6 Sep. 15		10 Lancers.
	Unattd. List	* **Royston,** Cyril George (172*a*) 207	4 Nov. 92	6 Sep. 11	3 Dec. 12	6 Sep. 15		18 Lancers.
760	Unattd. List	* **Marriott,** Charles Frere 207	5 May 92	6 Sep. 11	4 Dec. 12	6 Sep. 15		10 Jats.
	R. W. Fus.	* **Scott,** Kenneth Robert, M.C. (50, 52, 95, 167, 172*a*) 207	23 Nov. 91	20 Sep. 11	26 Feb. 14	20 Sep. 15		31 Punjabis.
	Midd'x R.	* **Gordon,** Andrew Douglas	21 Nov 92	20 Sep. 11	9 Aug. 18	20 Sep. 15		16 Cavalry.
	Unattd. List	**Borrowman,** Charles Gordon (51, 95, 172*a*) *Service for pension from 18 Jan. 13.*	30 May 92	19 Jan. 12	11 Mar. 14	19 Jan. 16		4 G. R.
	Unattd. List	**Gillan,** George Van Baerle (51, 52, 157, 162, 172*a*)	3 Sep. 90	19 Jan. 12	7 Oct. 14	19 Jan. 16		9 G. R.
765	Unattd. List	* **Firth,** Francis John Waller (205*a*) *Service for pension from 20 Aug. 13.*	28 Aug. 91	19 Jan. 12	30 Sept. 14	19 Jan. 16		6 Cavy.
	Unattd. List	**Bruce,** Robert Mason, M.C. *Service for pension from 20 Aug. 13.*	3 May 91	19 Jan. 12	8 Oct. 14	19 Jan. 16		5 G. R.
	Unattd. List	**Edwards,** Noel Hyde (57, 94, 157)	10 Dec. 91	20 Jan. 12	9 Mar. 13	20 Jan. 16		7 G.R.
	Unattd. List	**Crowther,** Arthur Henry (66*a*, 71, 95)	17 Apl. 93	20 Jan. 12	9 Mar. 13	20 Jan. 16		67 Punjabis.
	Unattd. List	**Turner,** Christopher John Radcliffe (52, 59, 95, 165, 167, 172*a*)	20 July 92	20 Jan. 12	24 Mar. 13	20 Jan. 16		15 Sikhs.
770	Unattd. List	**Carr-White,** Reginald Alfred (51, 172*a*)	20 Aug. 92	20 Jan. 12	1 Apl. 13	20 Jan. 16		31 Lancers
	Unattd. List	**Hindmarsh,** John Herbert Livingston (51, 52, 71, 170)	14 Oct. 92	20 Jan. 12	8 Mar. 13	20 Jan. 16		8 Rajputs.
	Unattd. List	**Clarke,** Frank Scamander (66, 169, 205*a*)	15 Jan. 93	20 Jan. 12	9 Mar. 13	20 Jan. 16		46 Punjabis.
	Unatt'd. List	* **Rundle,** Cubitt Archer Grant, M.C.	10 Jan. 93	20 Jan. 12	26 Mar. 13	20 Jan. 16		128 Pioneers.
	Unattd. List	**Grant,** Patrick (51, 52, 156)	10 Jan. 93	20 Jan. 12	8 Mar. 13	20 Jan. 16		Guides.
775	Unattd. List	**Guise,** Francis Edward Boissian (51, 52, 172*a*)	21 Dec. 92	20 Jan. 12	8 Mar. 13	20 Jan. 16		42 Regiment.
	Unattd. List	* **Dickinson,** Thomas Malcolm (71)	13 Feb. 93	20 Jan. 12	26 Mar. 13	20 Jan. 16		16 Cavalry.
	Unattd. List	**Poynder,** Frederick Sinclair (71)	8 Apl. 93	20 Jan. 12	8 Mar. 13	20 Jan. 16		9 G. R.
	Unattd. List	* **Browne,** Raleigh Barton (71*a*) 207	26 Jan. 93	20 Jan. 12	8 Mar. 13	20 Jan. 16		34 Pioneers.
	Unattd. List	**Gray,** Eric Hubert (167)	30 Mar. 92	20 Jan. 12	8 Mar. 13	20 Jan. 16		74 Punjabis.
780	Unattd. List	**Carpendale,** William St. John	26 July 92	20 Jan. 12	9 Mar. 13	20 Jan. 16		1 G. R.
	Unattd. List	**Williams,** James Hubert (31, 52, 66*a*, 172*a*)	18 Aug. 91	20 Jan. 12	8 Mar. 13	20 Jan. 16		10 G. R.
	Unattd. List	**Clarke,** Arthur Emlyn	14 Sep. 92	20 Jan. 12	9 Mar. 13	20 Jan. 16		39 Garhwal Rifles.
	Unattd. List	* **Bradley,** Rayer Oswald	14 Mar. 93	20 Jan. 12	8 Mar. 13	20 Jan. 16		29 Lancers.
	Unattd. List	**Scoones,** Geoffry Allen Percival, D.S.O., M.C.	25 Jan. 93	20 Jan. 12	8 Mar. 13	20 Jan. 16		2 G. R.
785	Unattd. List	**Burn-Murdoch,** Colin Thomas (172)	2 Feb. 92	20 Jan. 12	25 Mar. 13	20 Jan. 16		20 Infantry.
	Unattd. List	**Fasken,** Kenneth Llewellyn (52, 51, 172*a*)	28 Jan. 93	20 Jan. 12	9 Mar. 13	20 Jan. 16		95 Infantry.
	Unattd. List	* **Massy,** Derrick Fitzgerald (71, 95)	30 Jan. 93	20 Jan. 12	28 Feb. 13	20 Jan. 16		33 Lt. Cavalry.
	Unattd. List	**Hogge,** Arthur Herbert Fountaine (167)	13 Dec. 91	20 Jan. 12	8 Mar. 13	20 Jan. 16		23 Pioneers.
	Unattd. List	* **Luttman-Johnson,** Henry William (51, 52)	23 July 92	20 Jan. 12	6 Feb. 13	20 Jan. 16		9 Horse.
790	Unattd. List	* **Scott,** James Bruce, M.C. (95) 207	25 Dec. 92	20 Jan. 12	12 Mar 13	20 Jan. 16		33 Punjabis.
	Unattd. List	* **Candy,** Harrison Rolfe	8 Aug. 92	20 Jan. 12	8 Mar. 13	20 Jan. 16		73 Infantry.
	Unattd. List	**Passy,** Wilton DeLacy (51)	26 July 92	20 Jan. 12	13 Mar. 13	20 Jan. 16		113 Infantry.
	Unattd. List	**DeBrath,** Neville Stanley	15 Dec. 92	20 Jan. 12	9 Mar. 13	20 Jan. 16		15 Lancers.

794	Unattd. List	* **Brett,** Edric Walter		11 May 93	20 Jan. 12	10 Mar. 13	20 Jan. 16	. . . .	18 Infantry.
795	Unattd. List	**Kennedy,** Ernest Elliott, M.C. (51)		21 July 91	20 Jan. 12	9 Mar. 13	20 Jan. 16	. . . .	35 Horse.
	Unattd. List	**Rogers,** George Gordon		9 Apl. 93	20 Jan. 12	8 Mar. 13	20 Jan. 16	. . . .	1 G. R.
	Unattd. List	* **Boggs,** Arther Beaumont (172a)		28 July 91	3 Aug. 12	20 Nov. 13	23 Aug. 16	. . . .	29 Lancers.
	Unattd. List	* **Johanson,** John Laurits	207	18 Sep. 92	24 Aug. 12	2 Oct. 14	24 Aug. 16	. . . .	7 G.R.
		Service for Pension from 14 Jan. 14.							
	Unattd. List	* **Johnson,** Gordon Saffery, M.C.		4 Nov. 90	24 Aug. 12	1 Oct. 14	24 Aug. 16	. . . .	61 Pioneers.
		Service for pension from 14 Jan. 14.							
800	Unattd. List	* **Gracey,** Hugh Malcolm Kirkwood		10 Mar. 92	24 Aug. 12	1 Oct 14	24 Aug. 16	. . . .	101 Grenadiers.
	Unattd. List	**Kemp,** Thomas Norman Cameron, M.C. (50)		21 June 93	24 Aug. 12	4 Nov. 13	24 Aug. 16	. . . .	9 G. R.
	Unattd. List	**Magnay,** Arthur Douglas (51, 52)		14 May 93	24 Aug. 12	4 Nov. 13	24 Aug. 16	. . . .	1 Lancers.
	Unattd. List	* **Jones,** Mervyn Stockton Harvey (50, 53)	207	18 July 93	24 Aug. 12	3 Nov. 13	24 Aug. 16	. . . .	39 C. I. Horse.
	Unattd. List	* **StCSmyth,** John George (57, 167, 172a)		24 Oct. 93	24 Aug. 12	5 Nov. 13	24 Aug. 16	. . . .	15 Sikhs.
805	Unattd. List	* **De Wilton,** Julian Gerald Barnes (137, 172a)		5 Aug. 93	24 Aug. 12	20 Nov. 13	24 Aug. 16	. . . .	8 Cavalry.
	Unattd. List	**Trehane,** Charles Hony (51, 71, 172)		12 May 93	24 Aug. 12	10 Nov. 13	24 Aug. 16	. . . .	25 Cavalry.
	Unattd. List	* **Spooner,** Frank Vivian (47, 51, 52, 71)		10 July 93	24 Aug. 12	14 Oct. 13	24 Aug. 16	. . . .	30 Lancers.
	Unattd. List	* **Weld,** Charles Joseph		4 Feb. 93	24 Aug. 12	5 Nov. 13	24 Aug. 16	. . . .	56 Rifles.
	Unattd. List	* **Cochrane,** Richmond Pelham Inglis		18 Aug. 93	24 Aug. 12	1 Nov. 13	24 Aug.	. . . .	17 Cavalry.
810	Unattd. List	* **May-Somerville,** John Alexander Colquhoun (51, 52, 71, 95)		28 June 93	24 Aug. 12	13 Oct. 13	24 Aug. 16	. . . .	11 Lancers.
	Unattd. List	**Thomas,** Herbert Prichard (161, 169, 172)		22 Dec. 92	24 Aug. 12	15 Oct. 13	24 Aug 16	. . . .	126 Infantry.
	Unattd. List	**McCalmont,** William Vernon (57)		24 Mar. 93	24 Aug. 12	15 Oct. 13	24 Aug. 16	. . . .	10 Jats.
	Unattd. List	**Mosse,** Claude Boileau		8 Oct. 92	24 Aug. 12	2 Nov. 13	24 Aug. 16	. . . .	116 Mahrattas.
	Unattd. List	* **Fowle,** John Le Clerc		13 Sep. 93	24 Aug. 12	3 Nov. 13	24 Aug. 16	. . . .	15 Lancers.
815	Unattd. List	* **Morrison,** Charles Colquhoun (51, 52, 57)		25 Mar. 93	24 Aug. 12	21 Nov. 13	24 Aug. 16	. . . .	35 Horse.
	Unattd. List	**Meade,** Paul Anthony (137, 172a)		29 Feb. 92	24 Aug. 12	1 Nov. 13	24 Aug. 16	. . . .	112 Infantry.
	Unattd. List	* **Hinds,** Walter Augustus Gore, M.C.	207	5 Sep. 93	24 Aug. 12	1 Nov. 13	24 Aug. 16	. . . .	103 Infantry.
	Unattd. List	* **Welsh,** Ralph Henry Wellesley (71, 172a)		24 July 93	24 Aug. 12	12 Oct. 13	24 Aug. 16	. . . .	16 Cavalry
	Unattd. List	* **Dane,** Henry (57, 71)		9 Sep. 93	24 Aug. 12	18 Nov. 13	24 Aug 16	. . . .	Guides.
820	Unattd. List	* **Campbell,** Malcolm Hay Alexander		2 Aug. 92	24 Aug. 12	15 Dec. 13	24 Aug. 16	. . . .	89 Punjabis.
	Unattd. List	* **Williamson,** Cyril Philip Francis (51)	207	6 Dec. 92	24 Aug. 12	5 Nov. 13	24 Aug. 16	. . . .	95 Infantry.
	Unattd. List	**Plowden,** Byan Edward Chicheley (167, 169)		16 June 92	24 Aug. 12	12 Oct. 13	24 Aug. 16	. . . .	82 Punjabis.
	Unattd. List	* **Keene,** Brian Gerald	207	18 Mar. 93	24 Aug. 12	12 Oct. 13	24 Aug. 16	. . . .	92 Punjabis.
	Unattd. List	* **Worlledge,** Franklin Herbert		23 Aug. 92	24 Aug. 12	14 Oct. 13	24 Aug. 16	. . . .	26 Cavalry.
825	Unattd. List	* **Vansittart,** Guy Nicholas		8 Sep. 93	21 Jan. 13	4 Oct. 14	21 Jan. 17	. . . .	39 C. I. Horse.
		Service for pension from 14 Jan. 14.							
	Unattd. List	* **Layng,** Thomas Malcolm, M.C.		23 Apl. 92	21 Jan. 13	29 Aug. 15	21 Jan. 17	. . . .	Attd. 10 Jats.
		Service for pension from 5 Aug. 14.							
	Unattd. List	* **Mackenzie,** Daniel Barton		15 Oct. 92	21 Jan. 13	28 Aug. 15	21 Jan. 17	. . . .	58 Rifles.
	Unattd. List	* **Blaxland,** Alan Bruce		17 Oct. 92	21 Jan. 13	10 Sep. 15	21 Jan. 17	. . . .	Attd. 16 Rajputs.
		Service for pension from 5 Aug. 14.							
	Unattd. List	**Stable,** Robert Henry, D.S.O. (95)		2 Aug. 93	22 Jan. 13	11 Mar. 14	22 Jan. 17	. . . .	122 Infantry.
830	Unattd. List	* **Lockhart,** Rob MacGregor Macdonald, M.C.	207	23 June 93	22 Jan. 13	8 Mar. 14	22 Jan. 17	. . . .	51 Sikhs.
	Unattd. List	**Cambier,** Valentine (51, 52, 71)		5 Feb. 94	22 Jan. 13	7 Mar. 14	22 Jan. 17	. . . .	32 Pioneers.
	Unattd. List	* **Forbes,** Arthur Francis Gordon	207	19 July 93	22 Jan. 13	15 Mar. 14	22 Jan. 17	. . . .	18 Lancers
	Unattd. List	**Lennox,** Gilbert (50, 52)		19 Feb. 94	22 Jan. 13	9 Mar. 14	22 Jan. 17	*Ty. m.* 15 Oct. 18	37 Dogras.
	Unattd. List	* **Ashmore,** Edwin James Caldwell, M.C.		11 July 93	22 Jan. 13	8 Mar. 14	22 Jan. 17	. . . .	10 G. R.
835	Unattd. List	**Norris,** Reginald John Nelson		27 Jan. 94	22 Jan. 13	8 Mar. 14	22 Jan. 17	. . . .	7 Rajputs.
836	Unattd. List	**Minchin,** Hugh Charles Stephens (53, 169, 172)		5 Jan. 93	22 Jan. 13	16 Mar. 14	22 Jan. 17	. . . .	52 Sikhs.

* Has not passed the final examination.

No.	Former Corps.	Names.	Date of birth.	Date of first commission.	Date of appointment to Indian Army.	Date of Rank. Substantive.	Date of Rank. Army.	Remarks.
		Captains—*contd.*						
837	Unattd. List	**Stuart,** Lionel Arthur (52)	22 Apl. 92	22 Jan. 13	16 Mar. 14	22 Jan. 17		123 Rifles.
	Unattd. List	* **Weir,** Ian Everest Rose	5 Jan. 94	22 Jan. 13	8 Mar. 14	22 Jan. 17		126 Infantry.
	Unattd. List	**Lemarchand,** Frank Edward	9 Sep. 93	22 Jan. 13	8 Mar. 14	22 Jan. 17		5 G. R.
840	Unattd. List	* **Beanland,** Douglas (95)	4 Mar. 93	22 Jan. 13	13 Oct. 14	22 Jan. 17		22 Punjabis.
	Unattd. List	* **Dallas,** Douglas Arbuthnot Grant (46, 51, 52, 71) .207	2 Mar. 94	22 Jan. 13	8 Mar. 14	22 Jan. 17		36 Horse.
	Unattd. List	* **Robinson,** Guy St. John (46, 51, 52, 95)	4 Aug. 93	22 Jan. 13	11 Mar. 14	22 Jan. 17		88 Infantry.
	Unattd. List	**Pegg,** Albert Cecil (95)	20 June 93	22 Jan. 13	11 Mar. 14	22 Jan. 17		93 Infantry.
	Unattd. List	**Uloth,** Gerald (50, 71, 95, 100, 172*a*)	27 Dec. 93	22 Jan. 13	11 Mar. 14	22 Jan. 17		28 Cavalry.
845	Unattd. List	* **Messervy,** Frank Walter .207	9 Dec. 93	22 Jan. 13	8 Mar. 14	22 Jan. 17		9 Horse.
	Unattd. List	* **Baines,** George Douglas (71) .207	5 Mar. 93	22 Jan. 13	7 Mar. 14	22 Jan. 17		33 Cavalry.
	Unattd. List	**Hall,** Ian Francis Gordon (52, 71)	19 May 93	22 Jan. 13	7 Mar. 14	22 Jan. 17		25 Cavalry.
	Unattd. List	* **Bird,** Edward Kenelm .207	26 Mar. 93	22 Jan. 13	8 Mar. 14	22 Jan. 17		29 Punjabis.
	Unattd. List	* **Tindall,** John William Brooke (46) .207	30 July 93	22 Jan. 13	29 Sep. 14	22 Jan. 17		7 Rajputs.
850	Unattd. List	* **Stuart-Prince,** Charles Edward .207	12 July 93	22 Jan. 13	8 Mar. 14	22 Jan. 17		59 Rifles.
	Unattd. List	* **Harvey,** George Bourke (51)	6 Apl. 94	22 Jan. 13	8 Mar. 14	22 Jan. 17		5 Cavalry.
	Unattd. List	* **Vaughan,** Edward William Drummond	12 Mar. 94	22 Jan. 13	13 Mar. 14	22 Jan. 17		2 Lancers
	Unattd. List	**Payne,** George Drummond (52)	12 Aug. 92	22 Jan. 13	8 Mar. 14	22 Jan. 17		98 Infantry
	I. A. R. O.	§* **Hudson-Heaven,** Leonard Gyde *Service for promotion from 14 July 13 under I. A. O. 510 of Oct. 1917.*	. . . .	14 Oct. 12	26 Sep. 18	14 July 17		Attd. 62 Punjabis.
855	Unattd. List	* **MacIvor,** Robert Sandeman Peisley (50, 52)	7 Oct. 91	26 July 13	8 Oct. 14	26 July 17		129 Baluchis.
	Unattd. List	**Zorawar Singh,** M.C.	1 Apl. 83	25 Aug. 17	25 Aug. 17	25 Aug. 17		1 Lancers.
	Unattd. List	**Kanwar Amar Singh** (162)	24 July 78	25 Aug. 17	25 Aug. 17	25 Aug. 17		2 Lancers.
	Unattd. List	**Aga Cassim Shah** (47*a*, 51, 65, 88, 207)	13 Nov. 83	25 Aug. 17	25 Aug. 17	25 Aug. 17		3 Cavalry.
	Unattd. List	**Khan Muhammad Akbar Khan,** C.I.E.	27 Feb. 85	25 Aug. 17	25 Aug. 17	25 Aug. 17		1 Brahmans.
860	Unattd. List	**Malik Mumtaz Muhammad Khan**	14 Aug. 85	25 Aug. 17	25 Aug. 17	25 Aug. 17		4 Cavalry.
	Unattd. List	**Kanwar Pirthi Singh** (66*a*, 165, 172*a*, 204)	15 Apl. 85	25 Aug. 17	25 Aug. 17	25 Aug. 17		5 Cavalry.
	Unattd. List	**Bala Sahib Daphle** (66*a*, 132)	21 Sep. 83	25 Aug. 17	25 Aug. 17	25 Aug. 17		2 Rajput Lt. Infy.
	Unattd. List	* **Lindsay-Young,** Evelyn (51, 52, 57, 207*a*) *Service for pension from 5 Aug. 14.*	25 Nov. 93	2 Sep. 13	29 Aug. 15	2 Sep. 17		Attd. 19 Punjabis.
	S. W. Bord	* **Dawes,** Herbert Douglas 207*a*	9 Mar. 94	3 Sep. 13	24 May 18	3 Sep. 17		Attd. 32 Pioneers.
865	Unattd. List	* **Kidd,** Gerald Ralph *Service for pension from 26 Aug. 14*	9 July 94	13 Jan. 14	17 Sep. 15	13 Jan. 18		Attd. 20 Punjabis.
	Unattd. List	* **Pratt,** Mervyn Palles 207*a* *Service for pension from 26 Aug. 14.*	22 Oct. 93	13 Jan. 14	26 Aug. 15	13 Jan. 18		Attd. 121 Pioneers.
	Unattd. List	* **McConkey,** Alan Ivor Grey (57) *Service for pension from 26 Aug. 14.*	12 Nov. 96	13 Jan. 14	12 Sep. 15	13 Jan. 18		58 Rifles.
	Unattd. List	* **Lane,** Geoffrey Horsburgh (50) .207	5 Mar. 95	14 Jan. 14	10 Oct. 14	14 Jan. 18		103 Lt. Infantry.
	Unattd. List	* **Christian,** Henry Noel (51, 52) .207	22 July 94	14 Jan. 14	12 Oct. 14	14 Jan. 18		120 Infantry.
870	Unattd. List	* **Savory,** Reginald Arthur, M. C.	26 July 94	14 Jan. 14	21 Oct. 14	14 Jan. 18		14 Sikhs.
	Unattd. List	* **Trevelyan,** Walter Raleigh Fetherstonhaugh	16 Oct. 93	14 Jan. 14	10 Oct. 14	14 Jan. 18		93 Infantry.
	Unattd. List	**Tuker,** Francis Ivan Simms (51, 52)	14 July 94	14 Jan. 14	11 Oct. 14	14 Jan. 18		2 G. R.
	Unattd. List	* **Penfold,** Hugh de Lisle	17 Feb. 95	14 Jan. 14	10 Oct. 14	14 Jan. 18		9 G. R.
874	Unattd. List	* **White,** Benjamin Going (50, 52) .206	31 Aug. 94	14 Jan. 14	13 Oct. 14	14 Jan. 18		7 G. R.

875	Unattd. List	* **Broadfoot,** William Alexander (51)		7 Oct. 94	14 Jan. 14	10 Oct. 14	14 Jan. 18	. . . .	3 Horse.
	Unattd. List	* **Fletcher,** Roger Cornell		16 Feb. 94	14 Jan. 14	11 Oct. 14	14 Jan. 18	. . . .	23 Pioneers.
	Unattd. List	**Campbell,** Walter Fendall (51, 57, 169)	.206	20 May 94	14 Jan. 14	11 Oct. 14	14 Jan. 18	. . . .	52 Sikhs.
	Unattd. List	* **Nepean,** Herbert Dryden Home Yorke, D.S.O.	.206	27 Nov. 93	14 Jan. 14	10 Oct. 14	14 Jan. 18	. . . .	5 G. R.
	Unattd. List	* **Garrett,** Kenneth Alfred, M.C.	.207	12 Nov. 94	14 Jan. 14	11 Oct. 14	14 Jan. 18	. . . .	51 Sikhs.
880	Unattd. List	* **Lewis,** Trevor Gordon		25 Feb. 94	14 Jan. 14	12 Oct. 14	14 Jan. 18	. . . .	10 Jats.
	Unattd. List	* **Frazer,** Donald Stewart	.207	26 Sep. 94	14 Jan. 14	12 Oct. 14	14 Jan. 18	. . . .	18 Lancers.
	Unattd. List	* **Gore,** Charles Humphrey Casamaijor (48a, 51)	207	7 July 94	14 Jan. 14	19 Dec. 13	14 Jan. 18	. . . .	39 Garhwal Rifles.
	Unattd. List	* **Brunlees,** Leslie Howard (51)	.207	18 July 93	14 Jan. 14	13 Oct. 14	14 Jan. 18	. . . .	4 G. R.
	Unattd. List	* **Bellers,** Eric Vernon Robert		30 June 94	14 Jan. 14	16 Oct. 14	14 Jan. 18	. . . .	1 G. R.
885	Unattd. List	* **Dowland,** George Roy, M.C. (51, 52, 172a)	.207	23 Mar. 93	14 Jan. 14	15 Oct. 14	14 Jan. 18	. . . .	58 Rifles.
	Unattd. List	* **Gordon,** Harold Somerville (50)	.206	19 Sep. 93	14 Jan. 14	13 Oct. 14	14 Jan. 18	. . . .	8 G. R.
	Unattd. List	* **Robertson,** Henry Everton		27 July 93	14 Jan. 14	10 Oct. 14	14 Jan. 18	*Ty. Capt.* 18 July 17.	107 Pioneers.
	Unattd. List	* **Glegg,** John Allan (172a)	.207	13 Sep. 93	14 Jan. 14	11 Oct. 14	14 Jan. 18	. . . .	57 Rifles.
	Unattd. List	* **Taylor,** Jack Hulme (51, 52, 71)	.207	11 Dec. 94	14 Jan. 14	12 Oct. 14	14 Jan. 18	. . . .	11 Lancers.
890	Unattd. List	* **Moll,** William Rudolf		22 Jan. 94	14 Jan. 14	13 Oct. 14	14 Jan. 18	. . . .	9 Infantry.
	Unattd. List	* **Hamilton,** Harry Montgomery	.206	4 July 94	14 Jan. 14	14 Oct. 14	14 Jan. 18	. . . .	59 Rifles.
	Unattd. List	* **Fullerton,** John Parke (51)		8 Sep. 94	14 Jan. 14	12 Oct. 14	14 Jan. 18	. . . .	41 Dogras.
	Unattd. List	* **Dening,** John Pitt (51, 52)	.207	21 Nov. 94	14 Jan. 14	10 Oct. 14	14 Jan. 18	. . . .	21 Cavalry.
	Unattd. List	* **Fulton,** Edmund James		8 Feb. 95	14 Jan. 14	17 Oct. 14	14 Jan. 18	. . . .	1 Lancers.
895	Unattd. List	* **Bidie,** Allan George Charles (51, 52, 57)		3 Apl. 94	14 Jan. 14	14 Oct. 14	14 Jan. 18	. . . .	10 Lancers.
	Unattd. List	* **Maltby,** Francis Norman (51, 52)	.207	28 July 93	14 Jan. 14	13 Oct. 14	14 Jan. 18	. . . .	125 Rifles.
	Unattd. List	* **Gordon,** Douglas Hamilton, D.S.O.		27 Jan. 95	14 Jan. 14	11 Oct. 14	14 Jan. 18	. . . .	34 Pioneers.
	Unattd. List	* **Walker,** Bertram William George, M.C.		22 Aug. 94	14 Jan. 14	11 Oct. 14	14 Jan. 18	. . . .	23 Cavalry.
	Ches. R.	* **Law,** Charles Harold (47a)§		17 Aug. 93	24 Jan. 14	29 July 18	24 Jan. 18	. . . .	Attd. 30 Lancers.
900	Unattd. List	* **Gardner,** Jack Montfort Stanley		27 Apl. 94	{†5 Aug. 14 / 9 Oct. 15}	1 Feb. 16	5 Aug. 18	. . . .	Attd. 124 Infy.
		Service for promotion and pension from 5 Aug. 14.							
	Unattd. List	* **Fletcher,** Arthur Barry		28 May 93	{†5 Aug. 14 / 17 Oct. 15}	25 Feb. 16	5 Aug. 18	. . . .	Attd. 10 Jats.
		Service for promotion and pension from 5 Aug. 14.							
	Unattd. List	* **Barwis,** Cuthbert William Arthur (51)			7 Aug. 14	9 Nov. 15	7 Aug. 18	. . . .	54 Sikhs.
	Unattd. List	* **Filose,** Anthony Aloysius Emmanuel		19 Sep. 95	8 Aug. 14	7 Sep 15	8 Aug. 18	. . . .	Attd. 39 Horse.
	Unattd. List	* **Campbell,** Alexander Donald Powys (50, 172a)		6 July 94	8 Aug. 14	20 Aug. 15	8 Aug. 18	. . . .	106 Pioneers.
905	Unattd. List	* **Dymott,** Brian Kingsley		17 Aug. 95	8 Aug. 14	18 Aug. 15	8 Aug. 18	*Ty.Capt.* 21 June 18.	Attd. 31 Punjabis.
	Unattd. List	* **Roberts,** Michael Roukherst		24 Oct. 94	8 Aug. 14	10 Sep. 15	8 Aug. 18	. . . .	Attd. 112 Infy.
	Unattd. List	* **Hamber,** Lancelot George Werge (51, 59)		23 Aug. 94	8 Aug. 14	9 Sep. 15	8 Aug. 18	. . . .	1 G. R.
	Unattd. List	* **Powell,** Frank, M.C. (57)		7 July 93	8 Aug. 14	8 Aug. 15	8 Aug. 18	. . . .	39 Garh. Rif.
	Unattd. List	* **Churchill,** Alec Palmer ffleetwood		5 Feb. 95	8 Aug. 14	28 Aug. 15	8 Aug. 18	. . . .	Attd. 37 Dogras.
910	Unattd. List	* **Searle,** Carleton Stuart, M.C.		29 Nov. 95	8 Aug. 14	9 Aug. 15	8 Aug. 18	. . . .	Attd. 97 Infy.
	Unattd. List	* **Harding,** Kenneth O'Brien (51, 52)		28 Aug. 94	8 Aug. 14	21 Sep. 15	8 Aug. 18	. . . .	Attd. 13 Lancers.
	Unattd. List	* **Palmer,** Stephen Vaughan (50, 52)		15 Jan. 95	8 Aug. 14	13 Oct. 14	8 Aug. 18	. . . .	84 Punjabis.
	Unattd. List	* **Thwaytes,** Lancelot Lawrence	.207	10 Aug. 95	8 Aug. 14	24 Oct. 14	8 Aug. 18	. . . .	22 Cavy.
915	Unattd. List	* **Lambert,** William Robert (46)		29 July 94	8 Aug. 14	12 Oct. 14	8 Aug. 18	. . . .	97 Infantry.
	Unattd. List	* **Whistler,** Aubery Rivett		7 July 94	8 Aug. 14	1 Sep. 15	8 Aug. 18	. . . .	Attd. 10 Lancers.
	Unattd. List	* **Gordon,** Duncan St. Vincent	.207	22 June 95	8 Aug. 14	9 Sep. 15	8 Aug. 18	. . . .	Attd. 39 C.I.Horse
	Unattd. List	* **Palin,** Charlton Walter, M.C. (52)		14 Apl. 94	8 Aug. 14	15 Oct. 14	8 Aug. 18	. . . .	129 Baluchis.
918	Unattd. List	**Curtis,** Alfred Cyril, M.C.	207a	2 Nov. 94	8 Aug. 14	8 Aug. 15	8 Aug. 18	. . . .	Attd. 45 Sikhs.

* Has not passed the final examination. † Date from which seniority counts. § Admitted subject to H. M.'s approval.

No.	Former Corps.	Names	Date of birth.	Date of first commission.	Date of appointment to Indian Army.	Date of Rank. Substantive.	Date of Rank. Army.	Remarks.
		Captains—*concld.*						
919	Unattd. List	* **Eustace**, Rowland (46, 51)	11 Aug. 95	8 Aug. 14	19 Aug. 15	8 Aug. 18	. . . .	39 Garhwal Rifles.
920	Unattd. List	* **Wilson**, Robert Hamilton Birch (52, 157, 172*a*)	20 May 95	8 Aug. 14	13 Oct. 14	8 Aug. 18	. . . .	9 G. R.
	Unattd. List	* **Prinsep**, David Walter McLeod (51) 207	16 June 95	8 Aug. 14	13 Oct. 14	8 Aug. 18	. . . .	3 Horse.
	Unattd. List	* **Marriott**, Neville Edward (71)	July 94	8 Aug. 14	13 Oct. 14	8 Aug. 18	. . . .	25 Cavalry.
	Unattd. List	* **Fausset-Baker**, Arthur Amys (57)	29 Sep. 94	8 Aug. 14	13 Oct. 14	8 Aug. 18	. . . .	21 Punjabis.
	Unattd. List	* **Guise**, John William (51, 52)	20 Feb. 94	8 Aug. 14	15 Oct. 14	8 Aug. 18	. . . .	45 Sikhs.
925	Unattd. List	* **Johnson**, James Clyde	10 Sep. 94	15 Aug 14	10 Sep. 15	14 Aug. 18	. . . .	Attd. 97 Infantry.
	Unattd. List	* **Dods**, Eric Raymond Sandars, M.C.	25 Nov. 95	15 Aug. 14	5 Sep. 15	15 Aug. 18	. . . .	102 Grenadiers.
	Unattd. List	* **Miller-Stirling**, Arthur Eustace Stirling	15 July 95	15 Aug. 14	26 Aug. 15	15 Aug. 18	. . . .	
	Unattd. List	* **Gracey**, Douglas David, M.C.	3 Sep. 94	15 Aug. 14	15 Sep. 15	15 Aug. 18	. . . .	1 G. R.
	Unattd. List	* **Greenhouse**, Ronald Eric (51, 52) .207	26 Jan. 96	15 Aug. 14	14 Oct. 14	15 Aug. 18	. . . .	83 Infantry.
930	Unattd. List	* **Logan**, Arthur Fancourt, M.C. .207	23 Mar. 94	15 Aug. 14	20 Aug. 15	15 Aug. 18	. . . .	Attd. 21 Cavy.
	Unattd. List	* **Young**, Gerald Antony Gordon (50) .207	9 Nov. 95	15 Aug. 14	9 Sep. 15	15 Aug. 18	. . . .	6 Jats.
	Unattd. List	* **Briggs**, Harold Rawden	14 July 94	15 Aug. 14	16 Sep. 15	15 Aug. 18	. . . .	Attd. 31 Punjabis.
	Unattd. List	* **Minchin** Reginald Humphrey Loder	18 May 96	15 Aug. 14	24 Aug. 15	15 Aug. 18	. . . .	,, 45 Sikhs.
	Unattd. List	* **Loch**, Humphrey Norman	11 Nov. 95	15 Aug. 14	11 Sep. 15	15 Aug. 18	. . . .	,, 5 G. R.
935	Unattd. List	* **Miller-Hallet**, John Lionel	10 June 95	15 Aug. 14	12 Sep. 15	15 Aug. 18	. . . .	,, 2 G. R.
	Unattd. List	* **LeFleming**, Roger Eustace	20 Apl. 95	15 Aug. 14	18 Aug. 15	15 Aug. 18	. . . .	102 Grenadiers.
	Unattd. List	* **Upton**, Robert Babington Everard (46)	25 Apl. 96	15 Aug. 14	20 Aug. 15	15 Aug. 18	. . . .	1 G. R.
	Unattd. List	* **Christie**, Douglas Albert (50)	20 Dec. 94	15 Aug. 14	25 Aug. 15	15 Aug. 18	. . . .	14 Sikhs.
	Unattd. List	* **Bucher**, Francis Robert Roy 207	31 Aug. 95	15 Aug. 14	5 Sep. 15	15 Aug. 18	. . . .	31 Lancers.
940	Unattd. List	* **Fraser**, James Donald 207	16 Jan. 94	15 Aug. 14	29 Aug. 15	15 Aug. 18	. . . .	Attd. 32 Lancers.
	Unattd. List	* **Davidson**, John Waugh	11 Jan. 95	15 Aug. 14	9 Sep. 15	15 Aug. 18	. . . .	12 Cavalry.
	Unattd. List	* **Bunbury**, George Francis 207	9 Nov. 95	15 Aug. 14	30 Aug. 15	15 Aug. 18	. . . .	14 Sikhs.
	Unattd. List	* **Teversham**, Mark Symonds	5 Apl. 95	15 Aug. 14	28 Aug. 15	15 Aug. 18	. . . .	32 Pioneers.
	Unattd. List	* **Eales**, Charles Herbert Harberton, M.C. .207	16 Jan. 95	15 Aug. 14	25 Aug. 15	15 Aug. 18	. . . .	Guides.
945	Unattd. List	* **Money**, Harold Douglas Kyrle (51, 52)	3 Sep. 95	15 Aug. 14	9 Sep. 15	15 Aug. 18	. . . .	Guides.
	Unattd. List	* **Mercer**, Leslie Faithfull	25 June 93	15 Aug 14	26 Aug. 15	15 Aug. 18	. . . .	6 G. R.
	Unattd. List	* **Hodgen**, Gordon West	2 Feb. 94	16 Sep. 14	30 Sep. 15	16 Sep. 18	. . . .	Attd. 123 Rifles.
	Unattd. List	* **Ackland**, Heber Jackson	8 Nov. 95	1 Oct. 14	15 Dec. 14	1 Oct. 18	. . . .	1 Brahmans.
	Unattd. List	* **Key**, Berthold Wells	19 Dec. 95	1 Oct. 14	18 Dec. 14	1 Oct. 18	. . . .	45 Sikhs.
950	Unattd. List	* **Raynor**, Vernon Hamilton 207	6 Dec. 95	1 Oct. 14	22 Dec. 14	1 Oct. 18	. . . .	87 Punjabis.
	Unattd. List	* **Ansell**, John Wolseley (57) 206	18 Apl. 95	1 Oct. 14	15 Dec. 14	1 Oct. 18	. . . .	8 Rajputs.
	Unattd. List	* **Farwell**, Gerald Alexander Leith	11 Sep. 95	1 Oct. 14	16 Dec. 14	1 Oct. 18	. . . .	26 Punjabis.
	Unattd. List	* **Voelcker**, George Geeoffrey	7 May 95	1 Oct. 14	19 Dec. 14	1 Oct. 18	. . . .	3 G. R.
	Unattd. List	* **Williams**, William Reginald Bridgewater (52)	1 Apl. 96	1 Oct. 14	19 Dec. 14	1 Oct. 18	. . . .	7 G. R.
955	Unattd. List	* **Percy-Jones**, Cydril Muncaster 207	12 Dec. 95	1 Oct. 14	17 Dec. 14	1 Oct. 18	. . . .	21 Punjabis.
	Unattd. List	* **Carr-White**, Gerald	31 Aug. 96	1 Oct. 14	15 Dec. 14	1 Oct. 18	. . . .	23 Cavalry.
	Unattd. List	* **Reiche**, Alan David Eugene	17 July 96	1 Oct. 14	19 Dec. 14	1 Oct. 18	. . . .	38 Dogras.
	Unattd. List	* **Savage**, Eric Giles	11 Nov. 95	1 Oct. 14	23 Dec. 14	1 Oct. 18	. . . .	112 Infantry.
	Unattd. List	* **Clarke**, Geoffrey Raymond	11 July 95	1 Oct. 14	17 Dec. 14	1 Oct. 18	. . . .	72 Punjabis.
960	Unattd. List	* **Light**, Hilton	20 Feb. 95	1 Oct. 14	16 Dec. 14	1 Oct. 18	. . . .	84 Punjabis.
961	Unattd. List	* **Spottiswoode**, Arthur Ulric Molyneux	18 Jan. 95	1 Oct. 14	17 Dec. 14	1 Oct. 18	. . . .	5 G. R.

962	Unattd. List	* **Traill,** Henry Rauthwells		15 Nov. 95	1 Oct. 14	17 Dec. 14	1 Oct. 18	. . . .	87 Punjabis.
	Unattd. List	* **Flemming,** Robert Louis Victor (52)		27 Aug. 94	1 Oct. 14	18 Dec. 14	1 Oct. 18	. . . .	116 Mahrattas.
	Unattd. List	* **Shute,** George Eric FitzGerald (51, 52, 57)		5 June 96	1 Oct. 14	19 Dec. 14	1 Oct. 18	. . . .	19 Punjabis.
965	Unattd. List	* **Henry,** John Rochfort Armstrong (52) (*temp. Capt. 3 Jan. 18*)		20 Nov. 96	1 Oct. 14	18 Dec. 14	1 Oct. 18	. . . .	10 Jats.
	Unattd. List	* **Ekin,** Roger Gillies		18 Nov. 95	1 Oct. 14	16 Dec. 14	1 Oct. 18	. . . .	55 Rifles.
	Unattd. List	* **Kennedy,** Edward Richard		7 June 96	1 Oct. 14	16 Dec. 14	1 Oct. 18	. . . .	18 Infantry.
	Unattd. List	* **Smith,** Maurice Henry Howard		17 Mar. 96	1 Oct. 14	17 Dec. 14	1 Oct. 18	. . . .	6 G. R.
	Unattd. List	* **Feltham,** William Ivor Parsons, M.C. (52)		27 June 96	1 Oct. 14	18 Dec. 14	1 Oct. 18	. . . .	52 Sikhs.
970	Unattd. List	* **Keene,** Mervyn Robert		23 Sep. 96	1 Oct. 14	16 Dec. 14	1 Oct. 18	. . . .	97 Infantry.
	Unattd. List	* **Walton,** Lewine Charles James Bendyshe		18 Aug. 95	1 Oct. 14	16 Dec. 14	1 Oct. 18	. . . .	4 G. R.
	Unattd. List	* **Hanwell,** Joseph Osborne (71)		15 Oct. 96	1 Oct. 14	15 Dec. 14	1 Oct. 18	. . . .	11 Lancers.
	R.Muns.Fus.	**O'Reilly,** Harold Richard Hutchinson		. . . .	1 Oct. 14	21 Aug. 18	1 Oct. 18	. . . .	Attd. 10 Jats.
	Unattd. List	* **West,** Edward Maxwell		10 June 96	11 Nov. 14	27 Jan. 15	11 Nov. 18	. . . .	3 Gurkhas.
975	Unattd. List	* **Ross,** Alfred John Hume (52)		20 Sep. 96	11 Nov. 14	30 Jan. 15	11 Nov. 18	. . . .	7 Gurkhas.
	Unattd. List	* **Powell,** Donald	.206	21 Oct. 96	11 Nov. 14	25 Jan. 15	11 Nov. 18	. . . .	30 Punjabis.
	Unattd. List	* **Herbert,** Douglas Maxwell Arthur		18 Aug. 95	11 Nov. 14	15 Mar. 15	11 Nov. 18	. . . .	82 Punjabis.
	Unattd. List	* **Gough,** William George Hugh, M.C. (12, 46, 51)		16 Feb. 97	11 Nov. 14	28 Jan. 15	11 Nov. 18	. . . .	2 G. R.
979	Unattd. List	* **Stead,** Edward Alfred, M.C.	.206	7 Feb. 95	11 Nov. 14	26 Jan. 15	11 Nov. 18	. . . .	35 Sikhs.

* Has not passed the final examination.

No.	Former Corps.	Names.	Date of birth.	Date of first commission.	Date of appointment to Indian Army.	Date of rank.	Remarks.
		Lieutenants.					
1	Unattd. List	* **Browning,** Bertram Wilson	13 May 94	22 Jan. 13	7 Mar. 14	22 Apl. 15	9 Infantry.
	Unattd. List	* **Pigot,** Geoffrey	16 Feb. 94	11 Nov. 14	30 Jan. 15	11 Nov. 15	19 Punjabis.
	Unattd. List	* **Richardson,** Roland, M.C. .207	25 Mar. 94	11 Nov. 14	26 Jan. 15	11 Nov. 15	55 Rifles.
	Unattd. List	* **Maconchy,** George Alexander	12 Nov. 94	11 Nov. 14	26 Jan. 15	11 Nov. 15	5 G. R.
5	Unattd. List	* **Lloyd,** Jasper Searle	18 May 94	11 Nov. 14	27 Jan. 15	11 Nov. 15	2 G. R.
	Unattd. List	* **Fraser,** Charles James Stewart (57)	5 Feb. 96	11 Nov. 14	26 Jan. 15	11 Nov. 15	54 Sikhs.
	Unattd. List	* **Beckett,** Reinzi Stephen Barry (*temp. Capt., 8 June 16*)	31 July 96	16 Dec. 14	11 Feb. 15	16 Dec. 15	8 G. R.
	Unattd. List	* **Hunt,** John Montgomerie .206	17 Apl. 97	16 Dec. 14	12 Feb. 15	16 Dec. 15	87 Punjabis.
	Unattd. List	* **Ford,** Charles Wilbraham Watson (59)	17 July 96	16 Dec. 14	11 Feb. 15	16 Dec. 15	35 Sikhs.
10	Unattd. List	* **Bradshaw,** John Richard Landon .207	8 Nov. 96	16 Dec. 14	13 Mar. 15	16 Dec. 15	14 Lancers.
	Unattd. List	* **Hancock,** Cyril Percy	18 Sep. 96	16 Dec. 14	18 Feb. 15	16 Dec. 15	114 Mahrattas.
	Unattd. List	* **Bagley,** Arthur Handcock (52)	27 May 97	16 Dec. 14	11 Feb. 15	16 Dec. 15	8 Rajputs.
	Unattd. List	* **Kellie,** Charles Champion Jerome (50, 57) .207	19 May 96	23 Dec. 14	25 Feb. 15	23 Dec. 15	1 Brahmans.
	Unattd. List	* **Critchley,** Drury Ronald	10 Sep. 96	23 Dec. 14	1 Mar. 15	23 Dec. 15	32 Pioneers.
15	Unattd. List	* **Weallens,** Wam Russell Willoughby	30 Aug. 96	23 Dec. 14	28 Feb. 15	23 Dec. 15	4 Gurkhas.
	Unattd. List	* **Hawkes,** Montague Hay (51, 52)	25 Apl. 96	23 Dec. 14	27 Feb. 15	23 Dec. 15	12 Pioneers.
	Unattd. List	* **Woolf,** Sydney Henry (51, 52, 94) .206	15 Aug. 96	23 Dec. 14	27 Feb. 15	23 Dec. 15	19 Punjabis.
	Unattd. List	* **Clarke,** Creemer Paul	27 Dec. 95	23 Dec. 14	24 Feb. 15	23 Dec. 15	30 Punjabis.
	Unattd. List	* **Vaughan-Jones,** Henry	. . . .	29 June 16	9 July 16	4 Jan. 16	Attd. 113 Infantry.
		Service for promotion from 12th August 1914 to 3rd February 1916—1 yr. 176 d.					
20	Unattd. List	* **Sanders,** Reginald George (57)	23 Nov. 96	13 Jan. 15	25 Feb. 15	13 Jan. 16	38 Dogras.
	Unattd. List	* **Smith,** Arthur Beaumont	28 Aug. 96	13 Jan. 15	6 Mar. 15	13 Jan. 16	42 Regiment.
	Unattd. List	* **Laville,** Frank Autone Joseph (57) 207	29 Nov. 96	13 Jan. 15	17 Mar. 15	13 Jan. 16	16 Rajputs.
	Unattd. List	* **Bookey,** John Terence Brownrigg	29 Feb. 96	13 Jan. 15	26 Feb. 15	13 Jan. 16	Attd. 52nd Sikhs.
	Unattd. List	* **Palin,** Charles Edward Elliot	13 Mar. 96	13 Jan. 15	28 Feb. 15	13 Jan. 16	43 Regiment.
25	Unattd. List	* **Paterson,** Llewellyn Mostyn, M.C.	6 Mar. 97	13 Jan. 15	27 Feb. 15	13 Jan. 16	32 Sikhs.
	Unattd. List	* **Beer,** William Robert	. . . .	{†15 Jan. 15 14 Nov. 16}	27 Nov. 16	15 Jan. 16	Attd. 7 G. R.
		Service for promotion from 5th August 1914 to 2nd June 1916—1 yr. 303d.					
	Unattd. List	* **Phillips,** Robert Stowell (57)	. . . .	29 June 16	5 July 16	16 Jan. 16	39 Garh. Rif.
		Service for promotion from 22nd August 1914 to 1st February 1916—1 yr. 164 d.					
	Unattd. List	* **Burgan,** Norman 206	. . . .	{†19 Jan. 15 14 Nov. 16}	24 Nov. 16	19 Jan. 16	4 G. R.
		Service for promotion from 5th August 1914 to 29 May 1916—1 yr. 299d.					
	Unattd. List	* **Roy,** Robert Maitland	1 Jan. 96	22 Jan. 15	22 Mar. 15	22 Jan. 16	73 Infantry.
30	Unattd. List	* **Ham,** Douglas Miller	30 Apl. 95	22 Jan. 15	21 Mar. 15	22 Jan. 16	75 Infantry.

31	Unattd. List	***Clemence**, Stanley			29 June 16	29 June 16	14 Feb. 16	Attd. 7 G. R.
		Service for promotion from 21st September 1914 to 2nd February 1916—1 yr. 135 days.						
	A. S. C.	***Field**, Ferdinand Cromwell			15 Feb. 15	4 Oct. 18	15 Feb. 16	Attached 32 Lancers.
		9 months forfeited for seniority under conditions of I. A. O. 511 of 1917.						
	Unattd. List	§***Hall**, Edwin Raymond Macpherson			{†17 Feb. 15 / 27 Oct. 16}	5 Jan. 17	17 Feb. 16	Attd. 38 Dogras.
	Unattd. List	***Philpott**, Francis Leslie (51, 57, 71)	207a		29 June 16	19 July 16	23 Feb. 16	Attd. 27 Cavalry.
		Service for promotion from 2nd Oct. 1914 to 4th Feb. 1916.—1 yr. 126d.						
35	Unattd. List	***Mold**, Gilbert Leslie (51)	207		29 June 16	3 July 16	1 Mar. 16	40 Pathans.
		Service for promotion from 9th October 1914 to 5th February 1916—1yr. 120 days.						
	Unattd. List	***Harris**, Eric Noel	206		{†11 Mar. 15 / 14 Nov. 16}	24 Nov. 16	11 Mar. 16	Attd. 4 G. R.
	Unattd. List	§***Hugh-Jones**, Noel	207a		{†13 Mar. 15 / 18 June 17}	3 July 17	13 Mar. 16	53 Sikhs.
	Unattd. List	***Combe**, Lionel Molyneux (52, 60a)	207		29 June 16	2 July 16	14 Mar. 16	Attd. 32 Pioneers.
		Service for promotion from 22nd September 1914 to 31st December 1915—1yr. 101 days.						
	Unattd. List	***Knight**, Austen Bertram	206	29 Mar. 97	17 Mar. 15	10 May 15	17 Mar. 16	4 Cavy.
40	Unattd. List	***Berkeley**, Ernest Leopold Fitzhardinge		17 Aug. 96	17 Mar. 15	3 May 15	17 Mar. 16	25 Punjabis.
	Unattd. List	***Deakin**, Robert Hartley (57)		28 Jan. 95	22 Dec. 15	2 Mar. 16	28 Mar. 16	Attd. 10 Jats.
		Service for promotion and pension 269 days from 28th November 1914 to 23rd August 1915.						
	Unattd. List	§***Tollemache**, Harold Vincent			4 Apl. 15	6 Nov. 17	4 Apl. 16	
	Unattd. List	***Conner**, Terence Seymour (57)			29 June 16	5 July 16	10 Apl. 16	Attd. 88 Infantry.
		Service for promotion and pension to count from 10 April 15.						
	Unattd. List	***Williams**, John Walter			15 Nov. 15	19 Nov. 15	11 Apl. 16	Attd. 87 Punjabis.
		Service for promotion and pension 218 days from 17th September 1914 to 22nd April 1915.						
45	Unattd. List	***Wickham**, John	207a	16 Aug. 97	17 Apl. 15	28 Oct. 15	17 Apl. 16	Attd. 21 Punjabis.
	Unattd. List	***Boileau**, Walter Lind Sherwill			{19 Apl. 15† / 30 Jan. 17}	1 Feb. 17	19 Apl. 16	7 G. R.
		Service for promotion and pension from 20 November 14 to 30 Aug. 16—1yr. 285d.						
	Unattd. List	***Clark**, Cyril Drummond Le Gros	206		{†20 Apl. 15 / 18 June 17}	24 June 17	20 Apl. 16	34 Horse.
	Unattd. List	***Carleton**, Keith Osborne	207	29 Aug. 91	22 Apl. 15	5 Nov. 15	22 Apl. 16	Attached 25 Punjabis.
	Unattd. List	***Morgan**, Kenneth Sayce	207a		{†28 Apl. 15 / 18 Jan. 17}	25 June 17	28 Apl. 16	Attached 34 Pionrs.
50	Unattd. List	***Moss**, Eric de Vere			2 May 15	2 Dec. 17	2 May 16	Attached 112 Infantry.
	Unattd. List	***Martin**, Eric Henry de Carteret (46)	207		{†3 May 15 / 14 Nov. 16}	24 Nov. 16	3 May 16	120 Infy.
		Service for promotion from 20th November 1914 to 1st June 1910—1 yr. 196d.						
52	R. W. Surr. R.	***Gabb**, James Desmond			5 May 15	6 July 18	5 May 16	Attached 74 Punjabis.
		9 months forfeited for seniority under conditions of I. A. O. 511 of 1917.						

* Has not passed the final examination. † The date from which seniority counts. § Admitted subject to H. M.'s approval.

No.	Former Corps.	Names.	Date of birth.	Date of first commission.	Date of appointment to Indian Army.	Date of rank.	Remarks.
		Lieutenants—*contd.*					
53	Ayr. Yeo.	*§ **Howatson,** Charles N. *Service for promotion from 5 May 15 under I. A. O. 510 of 1917.*	....	5 Aug. 14	15 Jan. 18	5 May 16	Attached 19 Lancers.
	Midd'x. R.	***Tooviy,** Cecil Wotton *9 months forfeited for seniority under conditions of I. A. O. 511 of 1917.*	....	5 May 15	3 Apl.	5 May 16	Attached 86 Punjabis.
55	Midd'x R.	* **Gifford,** Frederick Richard (51, 57, 170) 205a *9 months forfeited for seniority under conditions of I. A. O. 511 of 1917.*	....	5 May 15	13 June 18	5 May 16	Attached 10 Gurkha Rifles.
	Mon. R.	* **Hockaday,** Percy Holland (48a, 51) *9 months forfeited for seniority under conditions of I. A. O. 511 of 1917.*	....	5 May 15	6 May 18	5 May 16	Attached 86 Infantry.
	A. S. C.	* **Quin,** Leonard Morallee *9 months forfeited for seniority under conditions of I. A. O. 511 of 1917.*	....	5 May 15	6 May 18	5 May 16	Attd. 88 Infantry.
	R. W. Surr. R.	***Turner,** Evelyn Lennox 207a *9 months forfeited for seniority under conditions of I. A. O. 511 of 1917.*	....	5 May 15	6 Aug. 18	5 May 16	Attached 2 Lancers.
	Som. L. I.	* **Ruck,** Charles Frederick Leyson (59) *9 months forfeited for seniority under conditions of I. A. O. 511 of 1917.*	....	5 May 15	10 July 18	5 May 16	15 Divsl. Sig. Coy.
60	Lanark. Yeo.	* **Hutchison,** James Riley Holt 207a *9 months forfeited for seniority under conditions of I. A. O. 511 of 1917.*	....	5 May 15	30 June 18	5 May 16	Attd. 17 Cavalry.
	Scot. Horse	* **Lothian,** James *9 months forfeited for seniority under conditions of I. A. O. 511 of 1917.*	....	5 May 15	19 June 18	5 May 16	Attached 28 Lt. Cavalry.
	Aus. Imp. Force.	***White,** Godfrey William Warren 207a *9 months forfeited for seniority under conditions of I. A. O. 511 of 1917.*	....	5 May 15	28 July 18	5 May 16	Attached 5 Cavalry.
	R. Scots.	* **Barker,** Thomas Ludwig *9 months forfeited for seniority under conditions of I. A. O. 511 of 1917.*	....	5 May 15	29 July 18	5 May 16	Attached 107 Pioneers.
	R. W. Kent. R.	***Dowling,** Guy *9 months forfeited for seniority under conditions of I. A. O. 511 of 1917.*	....	5 May 15	3 Nov. 18	5 May 16	Attached 3 Brahmans.
65	R. Ir. Rif.	* **Morton,** Thomas Mathew *9 months forfeited for seniority under conditions of I. A. O. 511 of 1917.*	....	5 May 15	1 July 18	5 May 16	Attd. 3 Gurkha Rif.
	N. Staff. R.	***Copland,** Reginald Wallace (48a, 51) 207a *9 months forfeited for seniority under conditions of I. A. O. 511 of 1917.*	....	5 May 15	16 Sep. 18	5 May 16	Attached 1 Gurkha Rifles.
	Lond. R.	* **Smith,** Walter Stanley 207a *9 months forfeited for seniority under conditions of I. A. O. 511 of 1917.*	....	5 May 15	29 July 18	5 May 16	Attached 19 Punjabis.
	Lond. R.	***Herring,** Ralph Lenex *9 months forfeited for seniority under conditions of I. A. O. 511 of 1917.*	....	5 May 15	22 Oct. 18	5 May 16	Attached 1 Gurkha Rifles.
69	Glouc. R.	* **Wade,** Harry *9 months forfeited for seniority under conditions of I. A. O. 511 of 1917.*	....	5 May 15	21 Aug. 18	5 May 16	Attached 56 Rifles.

70	E. Surr. R.	* **Maynard,** Crichton Ivan Fleming (57) *9 months forfeited for seniority under conditions of I. A. O. 511 of 1917.*	207*a*	. . . .	5 May 15	18 Aug. 18	5 May 16	Attached 21 Punjabis.
	Manch. R.	* **Barlow,** Alfred Edmund, M.C. *9 months forfeited for seniority under conditions of I. A. O. 511 of 1917.*		. . . .	5 May 15	6 Oct. 18	5 May 16	Attached 5 Gurkha Rifles.
	Lan. Fus.	* **Woodcock,** Russell Claude (12, 47*a*, 51, 66) *9 months forfeited for seniority under conditions of I. A. O. 511 of 1917.*	207*a*	. . . .	6 May 15	16 May 18	6 May 16	Attached 37 Lancers.
	Welsh. R.	* **Bruce,** John Geoffrey *9 months forfeited for seniority under conditions of I. A. O. 511 of 1917*	207*a*	. . . .	8 May 15	29 July 18	8 May 16	Attached 3-6th Gurkha Rifles.
	Unattd. List	* **Brennan,** Cyril Gascoigne Oxley		21 May 95	12 May 15	6 Nov. 15	12 May 16	Attached 31 Punjabis.
75	Unattd. List	* **Barlow,** Cyril Bourchier		24 Dec. 96	12 May 15	29 Oct. 15	12 May 16	Attached 52 Sikhs.
	Unattd. List	* **Salvesen,** Harold Keith (51, 57, 95)	207	29 July 97	12 May 15	1 Nov. 15	12 May 16	Attached 42 Regt.
	Unattd. List	* **Ridge,** Kenneth John Thriepland, M.C.		8 Aug. 93	12 May 15	8 Nov. 15	12 May 16	Attached 99 Infantry.
	R. E.	* **Hobson,** Evelyn, M.C. (51) *9 months forfeited for seniority under conditions of I. A. O. 511 of 1917.*		. . . .	12 May 15	12 June 18	12 May 16	Attached 30 Lancers.
	Unattd. List	§* **Grundy,** Sidney Elston		18 July 95	13 May 15	25 Nov. 17	13 May 16	Attached 86 Infantry.
80	Lond. R.	* **Leonard,** Dennis James (172) *9 months forfeited for seniority under conditions of I. A. O. 511 of 1917.*	270*a*	. . . .	14 May 15	8 Aug. 18	14 May 16	Attached 57 Rifles.
	Notts. & Derby. R.	* **Holden,** Wilfrid Shuttleworth *9 months forfeited for seniority under conditions of I. A. O. 511 of 1917.*		. . . .	15 May 15	24 Oct. 18	15 May 16	Attached 9 Bhopal Infy.
	Worc. R.	* **Hurford-Jones,** Montague *9 months forfeited for seniority under conditions of I. A. O. 511 of 1917.*	207*a*	. . . .	15 May 15	28 July 18	15 May 16	Attached 63 Infantry.
	S. Staff R.	* **Fitzpatrick,** Leslie Vernon (57) *9 months forfeited for seniority under conditions of I. A. O. 511 of 1917.*		. . . .	15 May 15	28 July 18	15 May 16	Attached 19 Punjabis.
	Essex. R.	* **Blake,** Alexander Gianville *9 months forfeited for seniority under conditions of I. A. O. 511 of 1917.*		. . . .	15 May 15	22 Nov. 18	15 May 16	Attached 2 Rajput Lt. Infy.
85	Durh. L. I.	* **Robertson,** John *9 months forfeited for seniority under conditions of I. A. O. 511 of 1917.*		. . . .	15 May 15	21 Aug. 18	15 May 16	Attached 10 Jats.
	Manch R.	* **Evans,** Ivor Ballantyne *9 months forfeited for seniority under conditions of I. A. O. 511 of 1917.*		. . . .	15 May 15	17 Oct. 18	15 May 16	Attached 5 Lt. Infy.
	Hamps. R.	* **Mosely,** Basil *9 months forfeited for seniority under conditions of I. A. O. 511 of 1917.*		. . . .	1[illegible] May 15	24 Apl. 18	16 May 16	Attached 110 Lt. Infantry.
	Unattd. List	* **Collingridge,** Hugh Vivian *(Service for promotion and pension to count from 19 May 15.)*		. . . .	14 Nov. 16	22 Nov. 16	19 May 16	Attached 6 G. R.
	Som. L. I.	* **Virgin,** Arthur Herbert *9 months forfeited for seniority under conditions of I. A. O. 511 of 1917.*		. . . .	22 May 15	7 June 18	22 May 16	Attached 12 Pioneers.
90	E. Kent. R.	* **Peters,** Joseph Henry Gervese *9 months forfeited for seniority under conditions of I. A. O. 511 of 1917.*		. . . .	22 May 15	24 Oct. 18	22 May 16	Attached 51 Sikhs.
	Essex R.	* **Priestly,** Edgar Charles *9 months forfeited for seniority under conditions of I. A. O. 511 of 1917.*		. . . .	26 May 15	12 June 18	26 May 16	Attached 67 Punjabis.
92	R. Scots.	* **Murison,** Gilbert deHussey *9 months forfeited for seniority under conditions of I. A. O. 511 of 1917.*		. . . .	26 May 15	6 May 18	26 May 16	Attached 1 Gurkha Rifles.

*Has not passed the final examination. † The date from which seniority counts. § Admitted subject to H. M's. approval.

No.	Former Corps.	Names.	Date of birth.	Date of first Commission.	Date of appointment to Indian Army.	Date of rank.	Remarks.
		Lieutenants—*contd.*					
93	R. Dub. Fus.	* **Wilkinson**, Gerrard Napier *9 months forfeited for seniority under conditions of I. A. O. 511 of 1917.*		26 May 15	7 Apl. 18	26 May 16	Attached 39 Garh. Rif.
	R. W. Fus.	* **Greer**, Eric Roberts *9 months forfeited for seniority under conditions of I. A. O 511 of 1917.*		26 May 15	18 June 18	26 May 16	Attached 106 Pioneers.
95	Manch. R.	* **Heywood**, Hugh Christopher Lewpriere *9 months forfeited for seniority under conditions of I. A. O. 511 of 1917.*		26 May 15	29 July 18	26 May 16	Attached 74 Punjabis.
	Hamps. R.	***Coles**, Guy Rogers (47a, 50, 60a) *9 months forfeited for seniority under conditions of I. A. O. 511 of 1917.*		29 May 15	29 Apl. 18	29 May 16	Attached 30 Lancers.
	Unattd. List	***Ninis**, George Woodward *Service for promotion 1 yr. 246 d. from 24 Oct. 14 to 25 June 16.*		†29 May 15 30 Jan. 17	14 Feb. 17	29 May 16	Attached 9 Horse.
	Ches. R.	* **White**, Frank *9 months forfeited for seniority under conditions of I. A. O. 511 of 1917.*		29 May 15	9 Aug. 18	29 May 16	Attached 25 Punjabis.
	R. War. R.	* **Teague**, John, M.C. *9 months forfeited for seniority under conditions of I. A. O. 511 of 1917.*		29 May 15	5 Oct. 18	29 May 16	Attached 124 Infantry.
100	Oxf. & Bucks. L. I.	* **Maude**, Edwin Walter, M.C. *9 months forfeited for seniority under conditions of I. A. O. 511 of 1917.*		1 June 15	20 Aug. 18	1 June 16	Attached 55 Rifles.
	A. S. C.	***Hamlyn**, Herbert William Andrew *9 months forfeited for seniority under conditions of I. A. O. 511 of 1917.*		1 June 15	12 June 18	1 June 16	Attached 88 Infantry.
	Wilts. R.	* **Harrison**, Edmund Louis *9 months forfeited for seniority under conditions of I. A. O. 511 of 1917.*		1 June 15	10 Sep. 18	1 June 16	Attached 67 Punjabis.
	Unattd. List	***Doudney**, George William *Service for promotion from 12th December 1914 to 24th May 1916—1 yr. 165d.*		†2 June 15 14 Nov. 16	24 Nov. 16	2 June 16	Attached 38 Horse.
	North'n R.	***Crampton**, Gerald Philip 207a *9 months forfeited for seniority under conditions of I. A. O. 511 of 1917.*		2 June 15	30 June 18	2 June 16	Attached 9 Gurkha Rifles.
105	Unattd. List	***Hampton**, Alick Victor Frank (57) *Service for promotion 1 yr. 24 days.*		29 June 16	11 July 16	5 June 16	Attached 12 Pioneers.
	Midd'x R.	* **King**, Ronald Mark Eaton (59) 207a *9 months forfeited for seniority under conditions of I. A. O. 511 of 1917.*		5 June 15	16 Sep. 18	5 June 16	Attached 123 Rifles.
	Unattd. List	§***Marjoribanks**, Stewart Dudley		6 June 15	6 Nov. 17	6 June 16	
	Unattd. List	§***Pocock**, John Grahame		8 June 15	6 Nov. 17	8 June 16	
	Midd'x R.	***Challen**, Phillip *9 months forfeited for seniority under conditions of I. A. O. 511 of 1917.*		9 June 15	6 June 18	9 June 16	Attached 3-39th Gurkha Rifles.
110	W. York R.	* **Fawcett**, Walter Lindley, M.C. *9 months forfeited for seniority under conditions of I. A. O. 511 of 1917.*		10 June 15	16 Sep. 18	10 June 16	Attached 9 Gurkha Rifles.

111	R. Innis. Fus.	* **Furness**, William Stanley, M.C. *9 months forfeited for seniority under conditions of I. A. O. 511 of 1917.*		. . . .	13 June 15	6 May 18	13 June 16	Attached 83 Infantry.
	R. Ir. Fus.	* **Tudor-Craig**, Philip Carlisle *9 months forfeited for seniority under conditions of I. A. O. 511 of 1917.*		. . . .	9 June 15	20 Oct. 18	9 June 16	Attached 5 Gurkha Rifles.
	Unattd. List	* **Oldridge dela Hey**, Cyril James		28 Apl. 97	16 June 15	27 Oct. 15	16 June 16	27 Cavalry.
	Unattd. List	* **Tomlinson**, Frank (59)		2 Jan. 96	16 June 15	28 Oct. 15	16 June 16	Attached 105 Infantry.
115	Unattd. List	* **Tirebuck**, Cecil Thomas		11 Feb. 96	16 June 15	13 Feb. 16	16 June 16	4 G. R.
	Unattd. List	* **Carruthers**, Humphrey St. John		16 Aug. 97	16 June 15	30 Oct. 15	16 June 16	Attached 7 G. R.
	Unattd. List	* **Fulton**, John Oswald (50, 52, 57)		16 July 97	16 June 15	28 Oct. 15	16 June 16	Attached 1 G. R.
	Unattd. List	* **Anderson**, Philip Berkeley (57) *Service for promotion from 12th December 1914 to 24th May 1916—1 yr. 165d.*		24 Apl. 96	16 June 15	2 Jan. 16	16 June 16	Attached 116 Infantry.
	R. Ir. Fus.	* **Ledgerwood**, Frederic Hill *9 months forfeited for seniority under conditions of I. A. O. 511 of 1917.*		. . . .	16 June 15	21 Aug. 18	16 June 16	Attached 89 Panjabis.
120	R. Suss. R.	* **Campbell-Harris**, Clifford *9 months forfeited for seniority under conditions of I. A. O. 511 of 1917.*		. . . .	16 June 15	12 June 18	16 June 16	Attached 17 Cavalry.
	S. Wales Bord.	* **Taylor**, George Derbyshire *9 months forfeited for seniority under conditions of I. A. O. 511 of 1917.*		. . . .	17 June 15	5 July 18	17 June 16	Attached 7 Gurkha Rifles.
	Leins. R.	* **FitzGerald**, Lancelot Cecil *9 months forfeited for seniority under conditions of I. A. O. 511 of 1917.*	207*a*	. . . .	19 June 15	12 June 18	19 June 16	Attached 113 Infantry.
	R. Innis. Fus.	* **Ballentine**, George Cecil *9 months forfeited for seniority under conditions of I. A. O. 511 of 1917.*	207*a*	. . . .	19 June 15	1 July 18	19 June 16	Attached 66 Punjabis.
	High. L. I.	* **Edwards**, Cosmo Grant Niven *9 months forfeited for seniority under conditions of I. A. O. 511 of 1917.*	207*a*	. . . .	19 June 15	6 May 18	19 June 16	Attached Corps of Guides.
125	Linc. R.	* **Pattinson**, Henry *9 months forfeited for seniority under conditions of I. A. O. 511 of 1917.*	207*a*	. . . .	19 June 15	28 July 18	19 June 16	Attached 103 Infantry.
	E. Lanc. R.	* **Sharp**, Cecil Jervis *9 months forfeited for seniority under conditions of I. A. O. 511 of 1917.*	207*a*	. . . .	19 June 15	20 Aug. 18	19 June 16	Attached 13 Rajputs.
	Gord. Highrs.	* **Pitcairn**, Douglas Neville *9 months forfeited for seniority under conditions of I. A. O. 511 of 1917.*	207*a*	. . . .	22 June 15	12 June 18	22 June 16	Attached 3 Gurkha Rifles.
	L'pool. R.	* **Jameson**, John Henry *9 months forfeited for seniority under conditions of I. A. O. 511 of 1917.*		. . . .	22 June 15	6 May 18	22 June 16	Attached 54 Sikhs.
	Glouc. R.	* **Clarke**, Francis Edward Merley *9 months forfeited for seniority under conditions of I. A. O. 511 of 1917.*	207*a*	. . . .	22 June 15	7 July 18	22 June 16	Attached 19 Punjabis.
130	L'pool. R.	* **Jennison**, Sydney Angelo (47*a*, 51, 61) *9 months forfeited for seniority under conditions of I. A. O. 511 of 1917.*	207*a*	. . . .	22 June 15	16 May 18	22 June 16	Attached 3 Horse.

Lieutenants—*contd.*

No.	Former Corps.	Names.	Date of birth.	Date of first commission.	Date of appointment to Indian Army.	Date of rank.	Remarks.
131	Arg. & Suth'd Highrs.	* **Duff-Torrance,** William Russell . . . 207*a* *9 months forfeited for seniority under conditions of I. A. O. 511 of 1917.*	. . . .	22 June 15	24 Aug. 18	22 June 16	Attached 67 Punjabis.
	Bord. R.	* **Elliot,** Alfred Cecil *9 months forfeited for seniority under conditions of I. A. O. 511 of 1917.*	. . . .	22 June 15	9 Aug. 18	22 June 16	Attached 95 Infantry.
	R. Ir. Rif.	* **Comerford,** Thomas James *9 months forfeited for seniority under conditions of I. A. O. 511 of 1917.*	. . . .	23 June 15	22 Nov. 18	23 June 16	Attached 18 Lancers.
	A. S. C.	* **Chamberlain,** Edward Francis *9 months forfeited for seniority under conditions of I. A. O. 511 of 1917.*	. . . .	24 June 15	31 July 18	24 June 16	Attached 88 Infantry.
135	Suff. R.	* **Platts,** John Carrick *9 months forfeited for seniority under conditions of I. A. O. 511 of 1917.*	. . . .	25 June 15	6 May 18	25 June 16	Attached 17 Cavalry.
	Unattd. List	§* **Monier-Williams,** Laurence	. . . .	28 June 15	5 Jan. 17	28 June 16	Attached 1-3rd Gurkha Rifles.
	Unattd. List	* **Mackie,** Andrew Hugh (57) *Service for promotion from 20th January 1915 to 3rd June 1916—1 yr. 136d.*	. . . .	{ †1 July 15 { 14 Nov. 16	} 29 Nov. 16	1 July 16	Attached 54 Sikhs.
138	R. F. A.	* **Menzies,** William Ferguson (51) . . . 207*a* *9 months forfeited for seniority under conditions of I. A. O. 511 of 1917.*	. . . .	1 July 15	6 May 18	1 July 16	Attached 7 Lancers.

139	Unattd. List	§*Sheldon, Harold George	16 June 96	3 July 15	12 Dec. 17	3 July 16	Attached 10 Lancers.
140	N. Devon Huss.	* Collyns, Gilbert George (50) . . . 270a	. . . .	5 July 15	31 July 18	5 July 16	Attached 37 Lancers.
		9 months forfeited for seniority under conditions of I. A. O. 511 of 1917.					
	Conn. Rang.	*Bourke, Arthur John Henry	. . . .	5 July 15	21 Aug. 18	5 July 16	Attached 15 Sikhs.
		9 months forfeited for seniority under conditions of I. A. O. 511 of 1917.					
	Unattd. List	*Hearn, Walter Scott Hogg	. . . .	{†10 July 15 14 Nov. 16}	27 Nov. 16	10 July 16	Attached 67 Punj.
		Service for promotion 3rd October 1914 to 6th February 1916—1 yr. 127d.					
	I. A. R. O.	*§ Seager, Richard Linger	. . . .	12 Oct. 14	16 Jan. 18	12 July 16	Attached 18 Lancers.
		Service for promotion from 12th July 1915 under I. A. O. 510 of 1917.					
	N. Som. Yeo.	*Gradidge, John Henry	. . . .	14 July 15	24 Oct. 18	12 July 16	Attached 14 Lancers
		9 months forfeited for seniority under conditions of I. A. O. 511 of 1917.					
145	E. York R.	*Burch, Frederick Whitmore, M.C. . . . 207a	. . . .	12 July 15	24 Sep. 18	12 July 16	Attached 7 Gurkha Rifles.
		9 months forfeited for seniority under conditions of I. A. O. 511 of 1917.					
	York L. I.	*Williams, Geoffrey Baynton	. . . .	13 July 15	20 Aug. 18	13 July 16	Attached 2 Rajput Lt. Infantry.
		9 months forfeited for seniority under conditions of I. A. O. 511 of 1917.					
	Unattd. List	* Bourdillon, Francis Henry (50) . . . 207	22 May 96	14 July 15	28 Oct. 15	14 July 16	Attached 36 Sikhs.
	Unattd. List	* Cooke, Gervase Mignon Borlase	29 Sep. 96	14 July 15	28 Oct. 15	14 July 16	Attached 27 Cavy.
	Unattd. List	* Bucknal, Richard Drummond Hay	31 Mar. 96	14 July 15	27 Oct. 15	14 July 16	Attached 2 G. R.
150	Unattd. List	* Prior, Charles Geoffrey (95)	9 Dec. 96	14 July 15	28 Oct. 15	14 July 16	Attached 127 Infantry.
	Unattd. List	* Baines, Roderick Howard, M.C.	4 Aug. 97	14 July 15	27 Oct. 15	14 July 16	Attached 121 Pioneers.
	Unattd. List	* Cameron, Ewen	22 Mar. 90	14 July 15	28 Oct. 15	14 July 16	Attached 113 Infantry.
	Unattd. List	* Hasler, Harold John . . . 207	. . . .	18 Apl. 16	27 Apl. 16	16 July 16	Attached 121 Pioneers.
		Service for promotion from 9th December 1914 to 11th September 1915 or 277 days. Service for pension 277 days plus 63 days from 5th August 1914 to 8th December 1914,—half to count					
	I. A. R. O.	* Robert, Vivian Godwin, M.C.	. . . .	22 Oct. 14	5 June 17	22 July 16	Attached 129 Baluchis.
		Service for promotion from 22nd July 1915 under I. A. O. 510 of 1917.					
155	I. A. R. O.	* Gayer, John Hamilton Courtenay	22 Oct. 92	23 Oct. 14	27 July 17	23 July 16	Attached S. & T. Corps.
		Service for promotion from 23rd July 1915 under I. A. O. 510 of 1917.					
	Gen. List	*Scott, Ronald Jocelyn Leslie (50) . . . 207a	. . . .	24 July 15	29 July 18	24 July 16	Attached 63 Infantry.
		9 months forfeited for seniority under conditions of I. A. O. 511 of 1917.					
	Unattd. List	§*Underhill, William Ernest	. . . .	25 July 15	6 Nov. 17	25 July 16	
	Dorset R.	*Rowley, Brian Gabriel . . . 207a	. . . .	28 July 15	20 Aug. 18	28 July 16	Attached 21 Punjabis.
		9 months forfeited for seniority under conditions of I. A. O. 511 of 1917.					
	Suff. R.	*Grey, Ronald Cecil . . . 270a	. . . .	28 July 15	29 July 18	28 July 16	Attached 1 Gurkha Rifles.
		9 months forfeited for seniority under conditions of I. A O. 511 of 1917.					
160	Leic. Yeo.	*Marsh, Charles Kenneth	. . . .	29 July 15	20 Oct. 18	29 July 16	Attached Corps of Guides.
		9 months forfeited for seniority under conditions of I. A. O. 511 of 1917.					
	Suff. R.	*Ashtion, Percy Wyndham Bankes	. . . .	29 July 15	31 Oct. 18	29 July 16	Attached 58 Rifles.
		9 months forfeited for soniority under conditions of I. A. O. 511 of 1917.					
162	I. A. R. O.	§*Butcher, Frederick George	. . . .	30 Oct. 14	30 July 18	30 July 16	Attached 2 Rajput Lt. Inft.
		Service for promotion from 30th July 15 under I. A. O. 510 of 1917.					

* Has not passed the final examination. † Date from which seniority counts. § Admitted subject to H. M.'s approval.

Lieutenants—*contd.*

No.	Former Corps.	Names.	Date of birth.	Date of first Commission.	Date of appointment to Indian Army.	Date of rank.	Remarks.
163	E. Lan. R.	***Peak,** Cyril Ashbrooke . . . . *9 months forfeited for seniority under conditions of I. A. O. 511 of 1917.*	. . . .	2 Aug. 15	22 Nov. 18	2 Aug. 16	Attached 10 Jats.
	Unattd. List	* **Talbot,** William Edward Harper (57)	10 Dec. 96	4 Aug. 15	27 Oct. 15	4 Aug. 16	Attached 83 Infatry.
165	Unattd. List	* **Page,** Herbert Vero Shaw	31 July 97	4 Aug. 15	4 Jan. 16	4 Aug. 16	Attached 123 Rifles.
	Unattd. List	* **Matthewman**, Arthur Stanford (57) 207	3 July 95	4 Aug. 15	28 Oct. 15	4 Aug. 16	Attached 94 Infantry.
	Unattd. List	* **Matthewman,** Gerald Beverley	10 June 95	4 Aug. 15	27 Oct. 15	4 Aug. 16	Attached 113 Infantry.
	Unattd. List	* **Manuk,** Thomas Mornington . . . . *Service for promotion 329 days.*	. . . .	29 June 16	11 July 16	5 Aug. 16	Attached 22 Cavalry.
	Trng. Res. Bn.	* **Wallis,** Cedric. . . . . 207a *9 months forfeited for seniority under conditions of I. A. O. 511 of 1917.*	. . . .	5 Aug. 15	6 May 18	5 Aug. 16	Corps of Guides.
170	I. A. R. O.	* **Collin,** Edward Phillip Compton . . . . *Service for promotion from 6th August 1915 under I. A. O. 510 of 1917.*	. . . .	6 Nov. 14	6 June 17	6 Aug. 16	Attached 18 Lancers.
	I. A. R. O.	§***Grant,** Terence Guy L'Estrange . . . . *Service for promotion from 6th August 1915 under I. A. O. 510 of 1917.*	Mar. 94	6 Nov. 14	4 Apl. 18	6 Aug. 16	Attached Alwar I. S. Infantry.
	R. Suss. R.	* **Kennelly,** Donald McKeon . . . . *9 months forfeited for seniority under conditions of I. A. O. 511 of 1917.*	. . . .	7 Aug. 15	7 July 18	7 Aug. 16	Attached 117 Mahrattas.
173	S. Wales Bord.	***Welsh,** Arthur Conrad Robert, M.C. . . . . 207a *9 months forfeited for seniority under conditions of I. A. O. 511 of 1917.*	. . . .	9 Aug. 15	17 Oct. 18	9 Aug. 16	Attached 119 Infantry.

174	Unattd. List	***Curle,** Victor	29 July 96	11 Aug. 15	28 Oct. 15	11 Aug. 16	Attached 81 Pioneers.
175	Unattd. List	***Rossel,** Henry George Theodore (51) 207*a*	7 July 94	11 Aug. 15	7 Nov. 15	11 Aug. 16	*Attached 64 Pioneers.
	Unattd. List	***Roberts,** Harold Bradley (57)	9 Jan. 91	11 Aug. 15	27 Oct. 15	11 Aug. 16	Attached 75 Infantry.
	Unattd. List	* **Ridley,** Ernest Rupert (52)	. . . .	11 Aug. 15	30 Oct. 15	11 Aug. 16	Attached 80 Infantry.
	Unattd. List	* **Pritchard,** Cecil Dryden *Service for promotion and pension 1 yr. 95d.*	. . . .	†11 Aug. 15 14 Nov. 16	27 Nov. 16	11 Aug. 16	Attached 9 G. R.
	Sea. Highrs.	* **Mackenzie,** William Stuart (46, 51, 53) 207*a* *9 months forfeited for seniority under conditions of I. A. O. 511 of 1917.*	. . . .	11 Aug. 15	4 Apl. 18	11 Aug. 16	Attached 4 G. R.
180	5 Dgn. Gds	* **Hilliard,** Leo Matthew 207*a* *9 months forfeited for seniority under conditions of I. A. O. 511 of 1917.*	. . . .	11 Aug. 15	29 July 18	11 Aug. 16	Attached 7 Lancers.
	Essex R.	* **Litton,** George Edward (46) 207*a* *9 months forfeited for seniority under conditions of I. A. O. 511 of 1917.*	. . . .	11 Aug. 15	22 Nov. 18	11 Aug 16	Attached 32 Pioneers.
	I. A. R. O.	* **Bolton,** Robert Henry Dundus *Service for promotion from 13th August 1915 under I. A. O. 510 of 1917.*	. . . .	13 Nov. 14	4 June 17	13 Aug. 16	Attached 101 Grenadiers.
	I. A. R. O.	§***Galloway,** William Cuerdon *Service for promotion from 13th August 1915 under I. A. O. 510 of 1917.*	28 May 89	13 Nov. 14	28 July 17	13 Aug. 16	Attached 99 Infantry.
	Durh. L. I.	* **Pringle,** John Hedley (46) *9 months forfeited for seniority under conditions of I. A. O. 511 of 1917.*	. . . .	13 Aug. 15	12 June 18	13 Aug. 16	Attached 126 Infantry.
185	M. G. Corps.	* **Nightingale,** George Oswald *9 months forfeited for seniority under conditions of I. A. O. 511 of 1917.*	. . . .	13 Aug. 15	25 Sep. 18	13 Aug. 16	Attached 9 Delhi Infantry.
	R. Innis. Fus.	* **White,** Ormond John *9 months forfeited for seniority under conditions of I. A. O. 511 of 1917.*	. . . .	14 Aug. 15	12 June 18	14 Aug. 16	Attached 5 Gurkha Rifles.
	Unattd. List	* **Melville,** Hugh Jack	. . . .	†15 Aug. 15 18 June 17	29 June 17	15 Aug. 16	22 Cavalry.
	Unattd. List	§***Pryor,** Oswald John	. . . .	15 Aug. 15	7 Nov. 17	15 Aug. 16	Attached 1-9th Gurkha Rifles.
	R. Lanc. R.	* **Nunn,** Richard Narcissus, M.C. (51) *9 months forfeited for seniority under conditions of I. A. O. 511 of 1917.*	. . . .	16 Aug. 15	29 July 18	16 Aug. 16	Attached 66 Punjabis.
190	Scot. Rif.	* **Brown,** John *9 months forfeited for seniority under conditions of I. A. O. 511 of 1917.*	. . . .	16 Aug. 15	1 Oct. 18	16 Aug. 16	Attached 7 Gurkha Rifles.
191	Linc. R.	* **Selby,** Wallace Raymond *9 months forfeited for seniority under conditions of I. A. O. 511 of 1917.*	. . . .	18 Aug. 15	6 May 18	18 Aug. 16	Attached 98 Infantry.

* Has not passed the final examination. † The date from which seniority counts. § Admitted subject to H. M's. approval.

Lieutenants—*contd.*

No.	Former Corps.	Names.	Date of birth.	Date of first commission.	Date of appointment to Indian Army.	Date of rank.	Remarks.
192	I. A. R. O.	§F **Curnow**, Eland . . . . *Service for promotion from 20th August 1915 under I. A. O. 510 of 1917.*	7 Apl. 92	20 Nov. 14	16 Aug. 18	20 Aug. 16	Attached 67 Punjabis.
	I. A. R. O.	§* **O'Reilly**, Keith Wilfred Robert . . . . *Service for promotion from 20th August 1915 under I. A. O. 510 of 1917.*	. . . .	20 Nov. 14	2 Nov. 17	20 Aug. 16	119 Infantry.
	Unattd. List	* **Abraham**, Roy Clive . . . . *Service for promotion from 22nd January 1915 to 18th January 1916—362 days.*	16 Dec. 90	16 Aug. 16	10 Nov. 16	20 Aug. 16	Attached 109 Infantry.
195	I. A. R. O.	§* **Jones**, Reginald Maurice (51, 52) . . . . *Service for promotion from 26th August 1915 under I. A. O. 510 of 1917.*	. . . .	20 Nov. 14	26 Apl. 18	20 Aug. 16	Attached 48 Pioneers.
	I. A. R. O.	§* **Orr**, Ian Macbeth . . . . *Service for promotion from 20th August 1915 under I. A. O. 510 of 1917.*	. . . .	20 Nov. 14	26 Sep. 18	20 Aug. 16	Attached 119 Infantry.
	E. Surr. R.	* **Baker**, George Morgan . . . . 207a *9 months forfeited for seniority under conditions of I. A. O. 511 of 1917.*	. . . .	23 Aug. 15	27 June 18	23 Aug. 15	Attached 13 Rajputs.
	R. W. Fus.	* **Lewis**, Francis Herbert . . . . 207a *9 months forfeited for seniority under conditions of I. A. O. 511 of 1917.*	. . . .	24 Aug. 15	27 July 18	24 Aug. 16	Attached 3 Gurkha Rifles.
199	Leins. R.	* **O'Morchoe**, Nial Creagh . . . . 207a *9 months forfeited for seniority under conditions of I. A. O. 511 of 1917.*	. . . .	24 Aug. 15	29 July 18	24 Aug. 16	Attached 29th Punjabis.

200	I. A. R. O.	§* **Vaughan,** Harry Russell (79) *Service for promotion from 27th August 1915 under I. A. O. 510 of 1917.*	. . . .	27 Nov. 14	30 July 18	27 Aug. 16	Attached 41 Dogras.
	I. A. R. O.	§* **Dunford,** Cyril Maton Periam (51) 207a *Service for promotion from 27th August 1915 under I. A. O. 510 of 1917.*	. . . .	27 Nov. 14	5 Dec. 17	27 Aug. 16	Attached 123 Rifles.
	I. A. R. O.	§* **Lloyd Evans,** Horace John, M.C. *Service for promotion from 27th August 1915 under I. A. O. 510 of 1917.*	. . . .	27 Nov. 14	4 Apl. 18	27 Aug. 16	
	Unattd. List	§* **Smelt,** Mervyn Vincent	. . . .	27 Aug. 15	30 Nov. 17	27 Aug. 16	Attached Corps of Guides.
	Unattd. List	* **Humby,** Lancelot James *Service for promotion from 21st January 1915 to 11th September 1915 or 234 days. Service for pension 234 days plus 28 days from 26th November 1914 to 21st January 1915—half to count.*	. . . .	18 Apl. 16	25 Apl. 16	28 Aug. 16	Attached 32 Pioneers.
205	Midd'x R.	* **Noad,** Colin Kenneth (51) 207a *9 months forfeited for seniority under conditions of I. A. O. 511 of 1917.*	. . . .	28 Aug. 15	29 July 18	28 Aug. 16	Attached 9 Gurkha Rifles.
	R. War. R.	* **Jones,** John Keane *9 months forfeited for seniority under conditions of I. A O. 511 of 1917.*	. . . .	28 Aug. 15	20 Aug. 18	28 Aug. 16	Attached 6 Gurkha Rifles.
	Unattd. List	* **Hatch,** Keith 206 *Service for promotion from 16th November 1914 to 30th January 1916—1yr. 76 days.*	. . . .	14 Nov. 16	6 Dec. 16	30 Aug. 16	34 Horse.
	Spl. List	* **Willoughby,** Bernard Digby 207a *9 months forfeited for seniority under conditions of I. A. O. 511 of 1917.*	. . . .	1 Sep. 15	21 Aug. 18	1 Sep. 16	Attached 21 Punjabis.
	Leic. R.	* **Silberbauer,** Harold Lewis *9 months forfeited for seniority under conditions of I. A. O. 511 of 1917.*	. . . .	1 Sep. 15	28 July 18	1 Sep. 16	Attached 12 Pioneers.
210	Som. L. I.	* **Davis,** Harold 207a *9 months forfeited for seniority under conditions of I. A. O. 511 of 1917.*	. . . .	2 Sep. 15	24 Aug. 18	2 Sep. 16	Attached 3 Gurkha Rifles.
	Gen. List	* **Noel,** Kenneth Hugh *9 months forfeited for seniority under conditions of I. A. O. 511 of 1917.*	. . . .	2 Sep. 15	16 Sep. 18	2 Sep. 16	Attached 107 Pioneers.
	I. A. R. O.	§* **Woledge,** Frank Leonard *Service for promotion from 4th September 1915 under I. A. O. 510 of 1917.*	. . . .	4 Dec. 14	4 Apl. 18	4 Sep. 16	Attached 9 Bhopal Infantry.
	I. A. R. O.	§* **Crocker,** Wilfred James *Service for promotion from 4 September 1915 under I. A. O. 510 of 1917.*	27 Apl. 93	4 Dec. 14	5 Dec. 17	4 Sep. 17	Attached 14 Sikhs.
	I. A. R. O.	§* **Waite,** Eric Wilfed *Service for promotion from 4th September 1915 under I. A. O 510 of 1917.*	27 Oct. 94	4 Dec. 14	30 July 18	4 Sep. 16	Attached 66 Punjabis.
215	Oxf. & Bucks. L. I.	* **Barnes,** Robert Money *9 months forfeited for seniority under conditions of I. A. O. 511 of 1917.*	. . . .	7 Sep. 15	21 Aug. 18	7 Sep. 16	Attached 21 Punjabis.
216	Glouc. R.	* **Farbrother,** Arthur Alfred *9 months forfeited for seniority under conditions of I. A. O. 511 of 1917.*	. . . .	7 Sep. 15	9 Aug. 18	7 Sep. 16	Attached 86 Infantry.

* Has not passed the final examination. § Admitted subject to H. M.'s approval.

No.	Former Corps.	Names.		Date of birth.	Date of first commission.	Date of appointment to Indian Army.	Date of rank.	Remarks.
		Lieutenants—*contd.*						
217	Devon. R.	* **Lewis,** Arthur Milton		. . . .	7 Sep. 15	29 July 18	7 Sep. 16	Attached 52 Sikhs.
		9 months forfeited for seniority under conditions of I. A. O. 511 of 1917.						
	Hamps. R.	* **Balding,** Charles Dennis	207a	. . . .	9 Sep. 15	11 Apl. 18	9 Sep. 16	Attached 113 Infantry.
		9 months forfeited for seniority under conditions of I. A. O. 511 of 1917.						
	Lond. Yeo.	* **Wallington,** John Mildmay (71)	207a	. . . .	10 Sep. 15	27 Mar. 18	10 Sep. 16	33 Cavalry.
		9 months forfeited for seniority under conditions of I. A. O. 511 of 1917.						
220	I. A. R. O.	§* **Halliley,** William Sydney, M. C.		. . . .	11 Dec. 14	17 Aug. 18	11 Sep. 16	Attached 7 Rajputs.
		Service for promotion from 11th September 1915 under I. A. O. 510 of 1917.						
	I. A. R. O.	§* **O'Connor,** Roderick Lewis (51, 57)		. . . .	11 Dec. 14	5 Dec. 17	11 Sep. 16	Attached 6 Jat Lt. Infy.
		Service for promotion from 11th September 1915 under I. A. O. 510 of 1917.						
	Unattd. List	* **Mallam,** George Leslie	207	. . . .	†13 Sep. 15 14 Nov. 16	22 Nov. 16	13 Sep. 16	Punjabis.
		Service for promotion from 30th November 1914 to 30th January 1916—1yr. 62 days.						
	Unattd. List	* **Cruikshank,** James Richard Holcombe		. . . .	†14 Sep. 15 30 Jan. 17	14 Feb. 17	14 Sep. 16	
		Service for promotion 1 yr. 138 d. from 11 Dec. 14 to 26 Apl. 16.						
	Unattd. List	* **Barrow,** Richard		29 May 95	15 Sep. 15	27 Oct. 15	15 Sep. 16	Attached 86 Infantry.
225	Unattd. List	* **Conolly,** John James Pollock		7 June 96	15 Sep. 15	28 Oct. 15	15 Sep. 16	Attached 35 Sikhs.
	Unattd. List	* **Lord,** James William		31 Oct. 97	15 Sep. 15	30 Oct. 15	15 Sep. 16	Attached 124 Infantry.
	Unattd. List	* **Howitt,** Cyril Naunton (48a, 51)		20 Mar. 97	15 Sep. 15	13 Nov. 15	15 Sep. 16	Attached 108 Infantry.
	Unattd. List	* **Walmsley,** John Bankes		9 Oct. 96	15 Sep. 15	15 Sep. 16	15 Sep. 16	5 Infy.
229	Unattd. List	* **Marindin,** John Francis		22 Dec. 97	15 Sep. 15	28 Oct. 15	15 Sep. 16	Attached 5 G. R.

No.	Corps	Name					
230	Worc. R.	*Leach, Alleyn Borie Robert (48a, 51) *9 months forfeited for seniority under conditions of I. A. O. 511 of 1917.*	. . . .	15 Sep. 15	26 June 18	15 Sep. 16	Attached 98 Infantry.
	Trng. Res. Bn.	* Hibberd, Andrew Stuart *9 months forfeited for seniority under conditions of I. A. O. 511 of 1917.*	. . . .	18 Sep. 15	6 May 18	18 Sep. 16	46 Punjabis.
	I. A. R. O.	§*Irwin, Henry Neville, M.C. (51, 52) *Service for promotion from 18th September 15 under I. A. O. 510 of 1917.*	28 Oct. 95	18 Dec. 14	4 Apl. 18	18 Sep. 16	Attached 3-3rd Gurkha Rifles.
	M. G. Corps	* Carstairs, Norman Moray *9 months forfeited for seniority under conditions of I. A. O. 511 of 1917.*	. . .	23 Sep. 15	20 Oct. 18	23 Sep. 16	Attached 97 Infantry.
	I. A. R. O.	* Roxburgh, George Archibald *Service for promotion from 24th Sep. 15 under I. A. O. 510 of 1917.*	31 Dec. 90	24 Dec. 14	6 June 17	24 Sep. 16	Attached 38 C. I. H.
235	I. A. R. O	§* Lewis, Charles Cecil Courtney 207 *Service for promotion from 24 Sep. 15 under I. A. O. 510 of 1917.*	. . . .	24 Dec. 14	21 Aug. 18	24 Sep. 16	Attached S. & T. Corps.
	York. R.	* Evans, Tubervile Rowley 207a *9 months forfeited for seniority under conditions of I. A. O. 511 of 1917.*	. . . .	26 Sep. 15	25 June 18	26 Sep 16	Attached 29 Punjabis.
	Scot. Rif.	* Urquhart, Douglas David Malcolm *9 months forfeited for seniority under conditions of I. A. O. 511 of 1917.*	. . . .	29 Sep. 15	5 May 18	29 Sep. 16	Attached 63 Infantry.
	Unattd. List	* Butler, Alfred George (51) *Service for promotion from 18th December 1914 to 30th January 1916—1 yr. 44d.*	. . . .	† 1 Oct. 15 14 Nov. 16	23 Nov. 16	1 Oct. 16	104 Rifles.
	Unattd. List	*Duncan, Donald Leslie *Service for promotion 1 yr. 121 d. from 23 Apl. 15 to 21 Aug. 16.*	. . . .	†1 Oct. 15 30 Jan. 17	17 Feb. 17	1 Oct. 16	Attached 2 G. R.
240	Unattd. List	* Bevan-Petman, Bertram Hamersley *Service for promotion 271 days.*	. . . .	29 June 16	2 July 16	2 Oct. 16	Attached 21 Cavalry.
	S. Wales Bord.	* Bodenham, Albert *9 months forfeited for seniority under conditions of I. A. O. 511 of 1917.*	. . . .	3 Oct. 15	17 Sep. 18	3 Oct. 16	Attached 41 Dogras.
	Unattd. List	* Berkeley, Cyril Sackville Jocelyn (57) *Service for promotion and pension from 1st March 1915 to 11th September 1915 or 195 days.*	. . .	18 Apl. 16	26 Apl. 16	6 Oct. 16	Attached 3 G. R.
	Gen List	* Delafons, Reginald Arthur (50) *9 months forfeited for seniority under conditions of I. A. O. 511 of 1917.*	26 Jan. 93	6 Oct. 15	31 July 18	6 Oct. 16	Attached 7 Gurkha Rifles.
	I. A. R. O.	*§Taylor, Arthur Thomas *Service for promotion from 8 Oct. 15 under I. A. O. 510 of 1917.*	. . . .	8 Jan. 15	16 July 17	8 Oct. 16	Attached 17 Mahrattas.
245	I. A. R. O.	§* Watson, Owen 207 *Service for promotion from 8 Oct. 15 under I. A. O. 510 of 1917.*	. . . .	8 Jan. 15	30 July 18	8 Oct. 16	Attached Mily. Farms Dept.
	I. A. R. O.	* Jamieson, James William Eric *Service for promotion from 8 Oct. 15 under I. A. O. 510 of 1917.*	7 Aug. 95	8 Jan. 15	4 June 17	8 Oct. 16	Attached 58 Rifles.
247	Oxf. & Buck, L. I.	* Molyneux, Philip Lucas *9 months forfeited for seniority under conditions of I. A. O. 511 of 1917.*	. . . .	11 Oct. 15	5 Aug. 18	11 Oct. 16	Attached 39 Garhwal Rifles.

* Has not passed the final examination. † The date from which seniority counts. § Admitted subject to His Majesty's approval.

17

Lieutenants—*contd.*

o.	Former Corps.	Names.	Date of birth.	Date of first commission.	Date of appointment to Indian Army.	Date of rank.	Remarks.
248	R. Fus.	* **Honeyball,** Frederick Ralph . . . . 207a *9 months forfeited for seniority under conditions of I. A. O. 511 of 1917.*	. . . .	11 Oct. 15	22 Nov. 18	11 Oct. 16	Attached 124 Infantry.
	I. A. R. O.	*§**Reeves,** Arthur James Medlicott . . . . *Service for promotion from 14 Oct. 15 under I. A. O. 510 of 1917.*	. . . .	14 Jan. 15	2 Nov. 17	14 Oct. 16	62 Punjabis.
250	I. A. R. O.	* **Gillies Peyburn,** Donald (46) . . . . 207a *Service for promotion from 14 Oct. 15 under I. A. O. 510 of 1917.*	. . . .	14 Jan. 15	6 June 17	14 Oct. 16	Attached 93 Infantry.
	I. A. R. O.	§* **Scott,** Curtis Wilson Folliott . . . . *Service for promotion from 14 Oct. 15 under I. A. O. 510 of 1917.*	. . . .	14 Jan. 15	5 Dec. 17	14 Oct. 16	Attached 9th Gurkha Rifles.
	I. A. R. O.	*§**Wemyss,** Alexander Francis, M.C. . . . . *Service for promotion from 14 Oct. 15 under I. A. O. 510 of 1917.*	. . . .	14 Jan. 15	2 Nov. 17	14 Oct. 16	62 Punjabis.
	R. Lanc. R.	* **Schneider,** Alexander (59) . . . . *9 months forfeited for seniority under conditions of I. A. O. 511 of 1917.*	. . . .	15 Oct. 15	22 Nov. 18	15 Oct. 16	Attached 38 Dogras.
	Unattd. List	* **Hemsworth,** Noel Edward Campbell (46)	. . . .	†17 Oct. 15 18 June 17	24 June 17	17 Oct. 16	Attached 29 Punjabis.
255	Arg. & Suth'd Highrs.	* **Laird,** Walter Ross . . . . *9 months forfeited for seniority under conditions of I. A. O. 511 of 1917.*	. . . .	9 Oct. 15	17 Oct. 18	19 Oct. 16	Attached 43 Regiment.
	Unattd. List	* **Steer,** John Mansel Bruce (51, 52)	5 Jan. 98	20 Oct. 15	10 Dec. 15	20 Oct. 16	Attached 83 Infantry.
	Unattd. List	* **Massy,** Ingoldsby Philip	22 Oct. 96	20 Oct. 15	11 Dec. 15	20 Oct. 16	Attached 81 Pioneers.
	I. A. R. O.	§* **France,** Edwin . . . . *Service for promotion from 22 Oct. 15 under I. A. O. 510 of 1917.*	. . . .	22 Jan. 15	31 Aug. 18	22 Oct. 16	Attached S. & T. Corps.
	Sea. Highrs.	* **Black,** Robert . . . . *9 months forfeited for seniority under conditions of I. A. O. 511 of 1917.*	. . . .	22 Oct. 15	29 July 18	22 Oct. 16	Attached 38 Dogras.
260	M. G. Corps.	* **Carr,** Harold James . . . . *9 months forfeited for seniority under conditions of I. A. O. 511 of 1917.*	. . . .	24 Oct. 15	2 Oct. 18	24 Oct. 16	Attached 23 Pioneers.
	Unattd. List	* **Kenworthy,** Leslie Allan 207a	. . . .	†26 Oct. 15 18 June 17	6 July 17	26 Oct. 16	Attached 76 Punjabis.
262	R. Highrs.	* **Wallace,** Andrew Alexander . . . . *9 months forfeited for seniority under conditions of I. A. O. 511 of 1917.*	. . . .	28 Oct. 15	6 May 18	28 Oct. 16	Attached Corps of Guides.

263	R. G. A.	* **Hodgins**, John Noel Miles *9 months forfeited for seniority under conditions of I. A. O. 511 of 1917.*	29 Oct. 15	16 Apl. 18	29 Oct. 16	Attached Corps of Guides.
	Aus. Imp. Force.	* **Pender**, William Stanhope *9 months forfeited for seniority under conditions of I. A. O. 511 of 1917*	3 Nov. 15	3 Nov. 18	3 Nov. 16	Attached 11 Lancers.
265	Dorset R.	* **Gillespie**, Donald Stanley *9 months forfeited for seniority under conditions of I. A. O. 511 of 1917.*	4 Nov. 15	6 Apl. 18	4 Nov. 16	58 Rifles.
	Scot. Rif.	***Christison**, Fredrick Hay (*tem. Capt. 4 Oct. 18*) *9 months forfeited for seniority under conditions of I.A.O. 511 of 1917.*	4 Nov. 15	17 Oct. 18	4 Nov. 16	Attached 23 Pioneers.
	I. A. R. O. G C P	§***Lovat-Fraser**, William Alexander *Service for promotion from 5 Nov. 1916 under I. A. O. 510 of 1917.*	5 Feb. 15	†30 July 13	5 Nov. 16	Attached 48 Pioneers.
	E. Kent R.	***Bevan**, Alan Harry Reynolds *9 months forfeited for seniority under conditions of I. A. O. 511 of 1917.*	10 Nov. 15	18 June 18	10 Nov. 16	Attached 120 Infantry.
	R. Scots.	* **MacIntyre**, Ian Hamilton *9 months forfeited for seniority under conditions of I. A. O. 511 of 1917.*	10 Nov. 15	29 July 18	10 Nov. 16	Attached 41 Dogras.
270	I. A. R. O.	§***Angelo**, Neville Leander *Service for promotion from 12 Nov. 15 under I. A. O. 510 of 1917.*	12 Feb. 15	4 Apl. 18	12 Nov. 16	Attached 96 Berar Infantry.
	Bedf. R.	* **Bull**, Geoffrey George Coulson *9 months forfeited for seniority under conditions of I. A. O. 511 of 1917.*	12 Sep. 15	24 May 18	12 Sep. 16	Attached 98 Infantry.
	Shrops. Yeo.	* **Newill**, Wilfred Marston (47a, 50, 71) 207a *9 months forfeited for seniority under conditions of I. A. O. 511 of 1917.*	14 Nov. 15	29 July 18	14 Nov. 16	Attached 33 Cavalry.
	Devon. R.	* **Adams**, Arthur Gwyther (59) *9 months forfeited for seniority under conditions of I. A. O. 511 of 1917.*	14 Nov. 15	1 July 18	14 Nov. 16	Attached 98 Infantry.
	Unattd. List	* **Amies**, Basil Jerome (51, 52) 207	15 Nov. 15	18 Nov. 15	15 Nov 16	Attached 21 Punjabis.
275	Unattd. List	* **Mizen**, Harold Alfred	15 Nov. 15	19 Nov. 15	15 Nov. 16	Attached 8 Rajputs.
	Unattd. List	† **Bromham**, Angus Claude	15 Nov. 15	19 Nov. 15	15 Nov. 16	Attached 16 Rajputs.
	Unattd. List	* **Ranking**, Robert Philip Lancaster 206	15 Nov. 15	18 Nov. 15	15 Nov. 16	Attached 5 Cavalry.
	Unattd. List	* **Challen**, Bernard Stanley 207	15 Nov. 15	19 Nov. 15	15 Nov. 16	Attached 72 Punjabis.
	Unattd. List	* **Laughton**, Joseph Watson	15 Nov. 15	27 Nov. 15	15 Nov. 16	Attached 10 G. R.
280	Unattd. List	* **Harrison**, Alfred Marshall Langton, M.C. 207a	15 Nov. 15	18 Nov. 15	15 Nov. 16	4 G. R.
	Unattd. List	* **Robinson**, Herbert Reginald	15 Nov. 15	23 Nov. 15	15 Nov. 16	Attached 91 Punjabis.
	Unattd. List	* **Rich**, Edward Pye	15 Nov. 15	19 Nov. 15	15 Nov. 16	Attached Guides.
	Unattd. List	* **Hirst**, Joseph Edward (57)	15 Nov. 15	16 Nov. 15	15 Nov. 16	Attached 12 Pioneers.
	Unattd. List	* **Birdwood**, Felix Tolcher 207	15 Nov. 15	19 Nov. 15	15 Nov. 16	36 Sikhs.
285	Unattd. List	* **Brilliant**, Leopold	15 Nov 15	17 Nov. 15	15 Nov. 16	127 Baluchis.
	Unattd. List	* **Wilson**, Alan David (57) 207	15 Nov. 15	19 Nov. 15	15 Nov. 16	Attached 1 Lancers.
	Unattd. List	* **Laing**, Alan Daniel Mackinlay Gordon (51)	15 Nov. 15	26 Nov. 15	15 Nov. 16	Attached 79 Infantry.
	Unattd. List	* **Robinson**, Edward Christopher (51, 52)	15 Nov. 15	18 Nov. 15	15 Nov. 16	39 G. R.
	Unattd. List	* **Longbottom**, John Leslie	15 Nov. 15	17 Nov. 15	15 Nov. 16	Attached 67 Punjabis.
290	Unattd. List	* **Mackintosh**, Gordon Nasmyth (46) 207	15 Nov. 15	19 Nov. 15	15 Nov. 16	Attached 36 Sikhs.
	Unattd. List	* **Browne**, William Reginald Howe	15 Nov. 15	17 Nov. 15	15 Nov. 16	Attached 127 Infantry.
	Unattd. List	* **Stapylton**, John Miles 207	15 Nov. 15	19 Nov. 15	15 Nov. 16	Attached 1 Lancers.
293	Unattd. List	* **Hansen**, Gordon Ericksen (57)	15 Nov. 15	19 Nov. 15	15 Nov. 16	Attached 52 Sikhs.

* Has not passed the final examination. | † The date from which seniority counts. | § Admitted subject to H. M.'s approval.

Lieutenants—*contd.*

No.	Former Corps.	Names.	Date of birth.	Date of first commission.	Date of appointment to Indian Army.	Date of rank.	Remarks.
294	Unattd. List	* **Walker**, Frederick Rutley		15 Nov. 15	18 Nov. 15	15 Nov. 16	Attached 31 Punjabis.
295	Unattd. List	* **Woodhouse**, Francis Victor Reginald (57)		15 Nov. 15	24 Nov. 15	15 Nov. 16	Attached 99 Infantry.
	Unattd. List	* **Nadin**, Guy		15 Nov. 15	18 Nov. 15	15 Nov. 16	12 Cavalry.
	Unattd. List	* **Goddard**, Eric Norman .207		15 Nov. 15	18 Nov. 15	15 Nov. 16	Attached 107 Pioneers.
	Unattd. List	* **Rees**, Thomas Winford		15 Nov 15	24 Nov. 15	15 Nov. 16	Attached 73 Infantry.
	Unattd. List	* **Purves**, James Iarvie (48a, 165) .206		15 Nov. 15	18 Nov. 15	15 Nov. 16	Attached 35 Sikhs.
300	Unattd. List	* **Anderson**, Donald Clive		15 Nov. 15	18 Nov. 15	15 Nov. 16	Attached 122 Infantry.
	Unattd. List	* **Weatherly**, Edward Frederick Christopher Ford (57)		15 Nov. 15	18 Nov. 15	15 Nov. 16	37 Dogras.
	Unattd. List	* **Hayne**, Maurice Leslie (46, 51, 52) .207		15 Nov. 15	18 Nov. 15	15 Nov. 16	37 Dogras.
	Unattd. List	* **Adams**, Robert Frederick George		15 Nov. 15	16 Nov. 15	15 Nov. 16	Attached 19 Punjabis.
	Unattd. List	* **Heraper**, Joseph Clifford		15 Nov 15	18 Nov. 15	15 Nov. 16	,, 113 Infantry.
305	Unattd. List	* **Mountain**, Ronald Gervase		15 Nov. 15	17 Nov. 15	15 Nov. 16	,, 127 Infantry.
	Unattd. List	* **Barton**, Francis Valentine Goold (46)		15 Nov. 15	18 Nov. 15	15 Nov. 16	1st Brahmans.
	Unattd. List	* **Newton-King**, Richard Mowbray		15 Nov. 15	20 Nov. 15	15 Nov. 16	Attached 38 Dogras.
	Unattd. List	* **Gemmell**, Geoffrey Woodburn		15 Nov. 15	18 Nov. 15	15 Nov. 16	4 Rajputs.
	Unattd. List	* **Hawkes**, Gordon Edward, M.C. (57b, 172) 206		15 Nov. 15	18 Nov. 15	15 Nov. 16	Attached 54 Sikhs.
310	Unattd. List	* **Dawes**, Kenneth Christopher Beresford 207a		15 Nov. 15	18 Nov. 15	15 Nov. 16	24 Punjabis.
	Unattd. List	* **Titchmarsh**, Edward Arthur (*Ty. Capt.*, 19 *Feb.* 18) 207		15 Nov. 15	19 Nov. 15	15 Nov. 16	7 Rajputs.
	Unattd. List	* **Soden**, James Newton (46)		15 Nov. 15	17 Nov. 15	15 Nov. 16	Attached 127 Infy.
	Unattd. List	* **Cant**, Ernest Dudley (95)		15 Nov. 15	20 Nov. 15	15 Nov. 16	Attached 84 Punjabis.
	Unattd. List	* **Pratt**, John (51, 71)		15 Nov. 15	18 Nov. 15	15 Nov. 16	Attached 5 Cavalry.
315	Unattd. List	* **Barrett**, Michael Long (57) .207		15 Nov. 15	19 Nov. 15	15 Nov. 16	Attached Guides.
	Unattd. List	* **Snelling**, Arthur Hugh Jay		15 Nov. 15	18 Nov. 15	15 Nov. 16	4 Rajputs.
	Unattd. List	* **Slade**, Mead (48a, 50) 207a		15 Nov. 15	28 Nov. 15	15 Nov. 16	64 Pioneers.
	Unattd. List	* **Rowlinson**, Oliver Alfred (51, 52)		15 Nov. 15	23 Nov. 15	15 Nov. 16	Attached 80 Infantry.
	Unattd. List	* **Fenwicke-Clennell**, Geoffrey Edward		15 Nov. 15	16 Nov. 15	15 Nov. 16	Attached 106 Pioneers.
320	Unattd. List	* **Spaull**, Cecil Mickleburgh		15 Nov. 15	30 Nov. 15	15 Nov. 16	Attached 86 Infantry.
	Unattd. List	* **Lalor**, James Francis Victor		15 Nov 15	19 Nov. 15	15 Nov. 16	,, 94 Infantry.
	Unattd. List	* **Gazzard**, Thomas		15 Nov. 15	18 Nov. 15	15 Nov. 16	Attached 32 Pioneers.
	Unattd. List	* **Dann**, Charles Frederick M.C.		15 Nov. 15	18 Nov. 15	15 Nov. 16	Attached 25 Punjabis.
	Unattd. List	* **Russell**, Oswald Bede Plowden		15 Nov. 15	17 Nov. 15	15 Nov. 16	Attached 10 Lancers.
325	Unattd. List	* **Tinney**, George Edward (59)		15 Nov. 15	18 Nov. 15	15 Nov. 16	28 Punjabis.
	Unattd. List	* **Phillips**, Charles Edmund Vyvyan		15 Nov. 15	16 Nov. 15	15 Nov. 16	Attached 106 Pioneers.
	Unattd. List	* **Blackburn**, Basil Marcus 206		15 Nov. 15	23 Nov. 15	15 Nov. 16	61 Pioneers.
	Unattd. List	* **Connor**, George Henry		15 Nov. 15	18 Nov. 15	15 Nov. 16	33 Punjabis.
	Unattd. List	* **Dunn**, Douglas Arthur Charles		15 Nov. 15	19 Nov. 15	15 Nov. 16	Attached 8 Rajputs.
330	Unattd. List	* **Hayward**, Thomas Edward (51)		15 Nov. 15	18 Nov. 15	15 Nov. 16	Attached 32 Pioneers.
	Unattd. List	* **Cutler**, Melvin Howard Challoner 207a		15 Nov. 15	18 Nov. 15	15 Nov. 16	Attached 54 Sikhs.
	Unattd. List	* **Jepson**, Stanley		15 Nov. 15	19 Nov. 15	15 Nov. 16	40 Pathans.
	Unattd. List	* **Kilpin**, Joseph Gordon (51, 52)		15 Nov. 15	18 Nov. 15	15 Nov. 16	Attached 25 Punjabis.
	Unattd. List	* **Moore**, William Stewart		15 Nov. 15	20 Nov. 15	15 Nov. 16	Attached 38 Dogras.
335	Unattd. List	* **Plomer**, Geoffrey .207		15 Nov. 15	18 Nov. 15	15 Nov. 16	Attached 107 Pioneers.

336	Unattd. List	* **Salvesen,** Frederick Malcolm Ross (51) 207a	. . . .	15 Nov. 15	19 Nov. 15	15 Nov. 16	Attached 87 Punjabis.
	Unattd. List	* **Shaw,** Reginald Ladbrooke	. . . .	15 Nov. 15	19 Nov. 15	15 Nov. 16	" 48 Pioneers.
	Unattd. List	* **Thomas,** Godfrey Herbert	. . . .	15 Nov. 15	20 Nov. 15	15 Nov. 16	3 G. R.
	Unattd. List	* **Coates,** John Cuthbert	. . . .	15 Nov. 15	19 Nov. 15	15 Nov. 16	Attached 5 G. R.
340	Unattd. List	* **Barlow,** Walter Northey Cecil, M.C.	. . . .	15 Nov. 15	19 Nov. 15	15 Nov. 16	Attached 30 Punjabis.
	Unattd. List	* **Wilkinson,** Harold Ayton	. . . .	15 Nov. 15	18 Nov. 15	15 Nov. 16	56 Rifles.
	Unattd. List	* **Teague,** Robert Joseph Dudley (46, 51, 52)	. . . .	15 Nov. 15	16 Nov. 15	15 Nov. 16	Attached 12 Pioneers.
	Unattd. List	* **Henson,** Robert Kent (46)	. . . .	15 Nov. 15	18 Nov. 15	15 Nov. 16	Attached 25 Punjabis.
	Unattd. List	* **Troughton-Dean,** Geoffrey Percival	. . . .	15 Nov. 15	19 Nov. 15	15 Nov. 16	24 Punjabis.
345	Unattd. List	* **Devenish-Meares,** Basil John	. . . .	15 Nov. 15	18 Nov. 15	15 Nov. 16	Attached 38 Dogras.
	Unattd. List	* **Leggatt,** Henry Allan (172a) 207	. . . .	15 Nov. 15	20 Nov. 15	15 Nov. 16	Attached 11 Lancers.
	Unattd. List	* **Forrest,** Lionel Hugh (*Ty. Lt., Gen. List, 1 Sep. 17*)	. . . .	15 Nov. 15	18 Nov. 15	15 Nov. 16	47 Sikhs.
	Unattd. List	* **Walker,** Stevenson Eden 207	. . . .	15 Nov. 15	20 Nov. 15	15 Nov. 16	47 Sikhs.
	Unattd. List	* **Twinberrow,** John Dennis	. . . .	15 Nov. 15	13 Nov. 15	15 Nov. 16	59 Rifles.
350	Unattd. List	* **Pine-Coffin,** Edward Claude 207a	. . . .	15 Nov. 15	17 Nov. 15	15 Nov. 16	Attached 22 Punjabis.
	Unattd. List	* **Whitehead,** Wilfrid Arthur 207	. . . .	15 Nov. 15	18 Nov. 15	15 Nov. 16	76 Punjabis.
	Unattd. List	* **Ellison,** Douglas Heins	. . . .	15 Nov. 15	21 Nov. 15	15 Nov. 16	95 Infantry.
	Unattd. List	* **Garbett,** Dennis Adair	. . . .	15 Nov. 15	21 Nov. 15	15 Nov. 16	96 Infantry.
	Unattd. List	* **Pearse,** Theodore Franklyn	. . . .	15 Nov. 15	22 Nov. 15	15 Nov. 16	120 Infantry.
355	Unattd. List	* **Bullock,** Penry Gerard Willoughby	. . . .	15 Nov. 15	28 Nov. 13	15 Nov. 16	Attached 22 Cavalry.
	I. A. R. O.	§* **Hansard,** Arthur John St. Leger (48a) . . . . *Service for promotion from 17 Nov. 15 under I. A. O. 510 of 1917.*	. . . .	17 Feb. 15	16 July 17	17 Nov. 16	" 98th Infantry.
	I. A. R. O.	*§ **Jervis** Thomas Sanford . . . . *Service for promotion from 18 Nov. 15 under I. A. O. 510 of 1917.*	. . . .	18 Feb 15	4 Apl. 18	18 Nov. 16	96 Infantry.
	R. Dub. Fus.	* **Spankie,** Hugh Vernon . . . . *9 months forfeited for seniority under conditions of I. A. O. 511 of 1917.*	. . . .	9 Nov. 15	6 May 18	19 Nov. 16	Attached 15 Sikhs.
	Unattd. List	* **Grey,** Reginald William Hargrave	21 Oct. 97	24 Nov. 15	1 Feb. 16	24 Nov. 16	Attached 5 G. R.
360	I. A. R. O.	§* **Cullen,** Albert Bryant, M. C. . . . . *Service for promotion from 26 Nov. 15 under I. A. O. 510 of 1917.*	. . . .	26 Feb. 15	16 Aug. 18	26 Nov. 16	" 2 S. & M.
	York. & Lanc. R.	* **Watkins,** Alfred Norman Mitchell . . . . *9 months fofeited for seniority under conditions of I. A. O. 511 of 1917.*	. . . .	24 Nov. 15	22 Nov. 18	24 Nov. 16	Attached 47 Sikhs.
	I. A. R. O.	§* **Salt,** James Joseph . . . . *Service for promotion from 5 Dec. 1917 under I. A. O., 510 of 1915.*	. . . .	5 Mar. 15	20 Sep. 18	5 Dec. 16	Attached 4 Rajputs.
	I. A. R. O.	*§ **Wallace,** Edward Graingar . . . . *Service for promotion from 5 Dec. 15 under I. A. O. 510 of 1917*	. . . .	5 Mar. 15	2 Nov. 17	5 Dec. 16	96 Infantry.
	W. York R.	* **Peake-Cottam,** Harold . . . . *9 months forfeited for seniority under conditions of I. A. O. 511 of 1917.*	. . . .	5 Dec. 15	6 May 18	5 Dec. 16	Attached 88 Infantry.
365	Oxf. & Buck. L. I.	* **Craig,** Charles Crawford . . . . *9 months forfeited for seniority under conditions of I. A. O. 511 of 1917.*	. . . .	5 Dec. 15	22 Nov. 18	5 Dec. 16	Attached 2 Infantry.

* Has not passed the final examination. † The date from which seniority counts. § Admitted subject to H. M.'s approval.

Lieutenants—*contd.*

No.	Former Corps.	Names.	Date of birth.	Date of first commission.	Date of appointment to Indian Army.	Date of rank.	Remarks.
366	Wilts. R.	* **Fifth,** John Griffith . . . . 207a *9 months forfeited for seniority under conditions of I. A. O. 511 of 1917.*	. . . .	10 Dec. 15	1 July 18	10 Dec. 16	Attached 126 Infantry.
	R. W. Surr. R.	* **Smallwood,** Leonard Stephen . . . . 207a *9 months forfeited for seniority under conditions of I. A. O. 511 of 1917.*	. . . .	10 Dec. 15	21 Aug. 18	10 Dec. 16	Attached 94 Infantry.
	R. Fus.	* **Larlham,** Percival Edward, ... *9 months forfeited for seniority under conditions of I.A.O. 511 of 1917.*	. . . .	11 Dec. 15	6 June 18	11 Dec. 16	Attached 98 Infantry.
	Unattd. List	***Dawes,** Richard Mallinson (57)	. . . .	29 June 16	6 Sep. 16	13 Dec. 16	6 Jats.
370	Unattd. List	***Nicholas,** Alexander Albert . . . . *Service for promotion 199 days.*	. . . .	29 June 16	9 July 16	13 Dec. 16	27 Punjabis.
	Unattd. List	***Nicholas,** Cecil Arthur Newbold . . . . *Service for promotion 199 days.*	. . . .	29 June 16	1 July 16	13 Dec. 16	Attached 126 Infantry.
	Unattd. List	***Bird,** George Garfield . . . . *Service for pension—5th August 1914 to 19th July 1915—half to count—174d; service for promotion counts from 14th December 1915*	. . . .	29 June 16	29 July 16	14 Dec. 16	Attached 73 Infantry.
	I. A. R. O.	§* **Owen,** Arthur Trevor, John, ... *Service for promotion from 19 December 15 under I.A.O. 510 of 1917.*	. . . .	19 Mar. 15	17 July 18	19 Dec. 16	Attached 56 Rifles.
	I. A. R. O.	§***Duguld,** Thomas Coutts (51) (*temp. Capt. 16 Jan. 18.*) . . . . *Service for promotion from 19 Dec. 15 under I. A. O. 510 of 1917.*	. . . .	19 Mar. 15	16 July 17	19 Dec. 16	,, 74th Punjabis.
375	Bord. R.	* **Dedman,** John Johnstone . . . . 207a *9 months forfeited for seniority under conditions of I. A. O. 511 of 1917.*	. . . .	19 Dec. 15	22 Nov. 16	19 Dec. 16	,, 119 Infantry.
	I. A. R. O.	*§**Preston,** Kenneth Harvoy (51,52) . . . . *Service for promotion from 20 Dec. 15 under I. A. O. 510 of 1917.*	. . . .	20 Mar 15	4 Apl. 18	20 Dec. 16	,, 45 Sikhs.
	Unattd. List	* **Solly,** Alec Richard	7 Jan. 96	22 Dec. 15	2 Mar. 16	22 Dec. 16	,, 116 Mahrattas.
	Unattd. List	* **Tyson,** Eric William (51)	4 Aug. 96	22 Dec. 15	1 Mar. 16	22 Dec. 16	,, 106 Pioneers.
	Unattd. List	* **Ricketts,** Thomas Charles Donald	3 July 96	22 Dec. 15	3 Mar. 16	22 Dec. 16	,, 46 Punjabis.
380	Unattd. List	* **Mercer,** Andrew Vivian Aldie	14 Mar. 96	22 Dec. 15	29 Feb. 16	22 Dec. 16	,, 7 G. R.
	M. G. Corps.	* **Neal,** Harold Aldridge . . . . *9 months forfeited for seniority under conditions of I. A. O. 511 of 1917.*	. . . .	25 Dec. 15	29 July 18	25 Dec. 16	,, 119 Infantry.
	I. A. R. O.	* **Powell,** Leslie Francis . . . . *Service for promotion from 26 Dec. 15 under I. A. O. 510 of 1917.*	. . . .	26 Mar. 15	15 June 17	26 Dec. 16	,, 17 Infy.
383	Unattd. List	* **Tipping,** Richard Percy Macrae . . . . *Service for promotion 186 days.*	. . . .	29 June 16	3 July 16	26 Dec. 16	Attached 81 Pioneers.

336	Unattd. List	* **Salvesen,** Frederick Malcolm Ross (51) . . . 207a	. . . .	15 Nov. 15	19 Nov. 15	15 Nov. 16	Attached 87 Punjabis.
	Unattd. List	* **Shaw,** Reginald Ladbrooke	. . . .	15 Nov. 15	19 Nov. 15	15 Nov. 16	,, 48 Pioneers.
	Unattd. List	* **Thomas,** Godfrey Herbert	. . . .	15 Nov. 15	20 Nov. 15	15 Nov. 16	3 G. R.
	Unattd. List	* **Coates,** John Cuthbert	. . . .	15 Nov. 15	19 Nov. 15	15 Nov. 16	Attached 5 G. R.
340	Unattd. List	* **Barlow,** Walter Northey Cecil, M.C.	. . . .	15 Nov. 15	19 Nov. 15	15 Nov. 16	Attached 30 Punjabis.
	Unattd. List	* **Wilkinson,** Harold Ayton	. . . .	15 Nov. 15	18 Nov. 15	15 Nov. 16	56 Rifles.
	Unattd. List	* **Teague,** Robert Joseph Dudley (46, 51, 52)	. . . .	15 Nov. 15	16 Nov. 15	15 Nov. 16	Attached 12 Pioneers.
	Unattd. List	* **Henson,** Robert Kent (46)	. . . .	15 Nov. 15	18 Nov. 15	15 Nov. 16	Attached 25 Punjabis.
	Unattd. List	* **Troughton-Dean,** Geoffrey Percival	. . . .	15 Nov. 15	19 Nov. 15	15 Nov. 16	24 Punjabis.
345	Unattd. List	* **Devenish-Meares,** Basil John	. . . .	15 Nov. 15	18 Nov. 15	15 Nov. 16	Attached 38 Dogras.
	Unattd. List	* **Leggatt,** Henry Allan (172a) 207	. . . .	15 Nov. 15	20 Nov. 15	15 Nov. 16	Attached 11 Lancers.
	Unattd. List	* **Forrest,** Lionel Hugh (*Ty. Lt., Gen. List, 1 Sep. 17*)	. . . .	15 Nov. 15	18 Nov. 15	15 Nov. 16	47 Sikhs.
	Unattd. List	* **Walker,** Stevenson Eden 207	. . . .	15 Nov. 15	20 Nov. 15	15 Nov. 16	47 Sikhs.
	Unattd. List	* **Twinberrow,** John Dennis	. . . .	15 Nov. 15	13 Nov. 15	15 Nov. 16	59 Rifles.
350	Unattd. List	* **Pine-Coffin,** Edward Claude . . . 207a	. . . .	15 Nov. 15	17 Nov. 15	15 Nov. 16	Attached 22 Punjabis.
	Unattd. List	* **Whitehead,** Wilfrid Arthur 207	. . . .	15 Nov. 15	18 Nov. 15	15 Nov. 16	76 Punjabis.
	Unattd. List	* **Ellison,** Douglas Heins	. . . .	15 Nov. 15	21 Nov. 15	15 Nov. 16	95 Infantry.
	Unattd. List	* **Garbett,** Dennis Adair	. . . .	15 Nov. 15	21 Nov. 15	15 Nov. 16	96 Infantry.
	Unattd. List	* **Pearse,** Theodore Franklyn	. . . .	15 Nov. 15	22 Nov. 15	15 Nov. 16	120 Infantry.
355	Unattd. List	* **Bullock,** Penry Gerard Willoughby	. . . .	15 Nov. 15	28 Nov. 13	15 Nov. 16	Attached 22 Cavalry.
	I. A. R. O.	§* **Hansard,** Arthur John St. Leger (48a)	. . . .	17 Feb. 15	16 July 17	17 Nov. 16	,, 98th Infantry.
		Service for promotion from 17 Nov. 15 under I. A. O. 510 of 1917.					
	I. A. R. O.	*§ **Jervis** Thomas Sanford	. . . .	18 Feb 15	4 Apl. 1?	18 Nov. 16	96 Infantry.
		Service for promotion from 18 Nov. 15 under I. A. O. 510 of 1917.					
	R. Dub. Fus.	* **Spankie,** Hugh Vernon	. . . .	9 Nov. 15	6 May 18	19 Nov. 16	Attached 15 Sikhs.
		9 months forfeited for seniority under conditions of I. A. O. 511 of 1917.					
	Unattd. List	* **Grey,** Reginald William Hargrave	21 Oct. 97	24 Nov. 15	1 Feb. 16	24 Nov. 16	Attached 5 G. R.
360	I. A. R. O.	§* **Cullen,** Albert Bryant, M. C.	. . . .	26 Feb. 15	16 Aug. 18	26 Nov. 16	,, 2 S. & M.
		Service for promotion from 26 Nov. 15 under I. A. O. 510 of 1917.					
	York. & Lanc. R.	* **Watkins,** Alfred Norman Mitchell	. . . .	24 Nov. 15	22 Nov. 18	24 Nov. 16	Attached 47 Sikhs.
		9 months fofeited for seniority under conditions of I. A. O. 511 of 1917.					
	I. A. R. O.	§* **Salt,** James Joseph	. . . .	5 Mar. 15	20 Sep. 18	5 Dec. 16	Attached 4 Rajputs.
		Service for promotion from 5 Dec. 1917 under I. A. O., 510 of 1915.					
	I. A. R. O.	*§ **Wallace,** Edward Graingar	. . . .	5 Mar. 15	2 Nov. 17	5 Dec. 16	96 Infantry.
		Service for promotion from 5 Dec. 15 under I. A. O. 510 of 1917					
	W. York R.	* **Peake-Cottam,** Harold	. . . .	5 Dec. 15	6 May 18	5 Dec. 16	Attached 88 Infantry.
		9 months forfeited for seniority under conditions of I. A. O. 511 of 1917.					
365	Oxf. & Buck. L. I.	* **Craig,** Charles Crawford	. . . .	5 Dec. 15	22 Nov. 18	5 Dec. 16	Attached 2 Infantry.
		9 months forfeited for seniority under conditions of I. A. O. 511 of 1917.					

* Has not passed the final examination. † The date from which seniority counts. § Admitted subject to H. M.'s approval.

No.	Former Corps.	Names.	Date of birth.	Date of first commission.	Date of appointment to Indian Army.	Date of rank.	Remarks.
		Lieutenants—*contd.*					
366	Wilts. R.	* **Fifth**, John Griffith . . . . 207a *9 months forfeited for seniority under conditions of I. A. O. 511 of 1917.*	. . . .	10 Dec. 15	1 July 18	10 Dec. 16	Attached 126 Infantry.
	R. W. Surr. R.	* **Smallwood**, Leonard Stephen . . . . 207a *9 months forfeited for seniority under conditions of I. A. O. 511 of 1917.*	. . . .	10 Dec. 15	21 Aug. 18	10 Dec. 16	Attached 94 Infantry.
	R. Fus.	* **Larlham**, Percival Edward, ... *9 months forfeited for seniority under conditions of I.A.O. 511 of 1917.*	. . . .	11 Dec. 15	6 June 18	11 Dec. 16	Attached 98 Infantry.
	Unattd. List	***Dawes**, Richard Mallinson (57)	. . . .	29 June 16	6 Sep. 16	13 Dec. 16	6 Jats.
370	Unattd. List	***Nicholas**, Alexander Albert . . . . *Service for promotion 199 days.*	. . . .	29 June 16	9 July 16	13 Dec. 16	27 Punjabis.
	Unattd. List	***Nicholas**, Cecil Arthur Newbold . . . . *Service for promotion 199 days.*	. . . .	29 June 16	1 July 16	13 Dec. 16	Attached 126 Infantry.
	Unattd. List	***Bird**, George Garfield . . . . *Service for pension—5th August 1914 to 19th July 1915—half to count—174d; service for promotion counts from 14th December 1915*	. . . .	29 June 16	29 July 16	14 Dec. 16	Attached 73 Infantry.
	I. A. R. O.	§* **Owen**, Arthur Trevor, John, ... *Service for promotion from 19 December 15 under I.A.O. 510 of 1917.*	. . . .	19 Mar. 15	17 July 18	19 Dec. 16	Attached 56 Rifles.
	I. A. R. O.	§***Duguld**, Thomas Coutts (51) (*temp. Capt. 16 Jan. 18.*) . . . . *Service for promotion from 19 Dec. 15 under I. A. O. 510 of 1917.*	. . . .	19 Mar. 15	16 July 17	19 Dec. 16	„ 74th Punjabis.
375	Bord. R.	* **Dedman**, John Johnstone . . . . 207a *9 months forfeited for seniority under conditions of I. A. O. 511 of 1917.*	. . . .	19 Dec. 15	22 Nov. 16	19 Dec. 16	„ 119 Infantry.
	I. A. R. O.	*§**Preston**, Kenneth Harvoy (51,52) . . . . *Service for promotion from 20 Dec. 15 under I. A. O. 510 of 1917.*	. . . .	20 Mar 15	4 Apl. 18	20 Dec. 16	„ 45 Sikhs.
	Unattd. List	* **Solly**, Alec Richard	7 Jan. 96	22 Dec. 15	2 Mar. 16	22 Dec. 16	„ 116 Mahrattas.
	Unattd. List	* **Tyson**, Eric William (51)	4 Aug. 96	22 Dec. 15	1 Mar. 16	22 Dec. 16	„ 106 Pioneers.
	Unattd. List	* **Ricketts**, Thomas Charles Donald	3 July 96	22 Dec. 15	3 Mar. 16	22 Dec. 16	„ 46 Punjabis.
380	Unattd. List	* **Mercer**, Andrew Vivian Aldie	14 Mar. 96	22 Dec. 15	29 Feb. 16	22 Dec. 16	„ 7 G. R.
	M. G. Corps.	* **Neal**, Harold Aldridge . . . . *9 months forfeited for seniority under conditions of I. A. O. 511 of 1917.*	. . . .	25 Dec. 15	29 July 18	25 Dec. 16	„ 119 Infantry.
	I. A. R. O.	* **Powell**, Leslie Francis . . . . *Service for promotion from 26 Dec. 15 under I. A. O. 510 of 1917.*	. . . .	26 Mar. 15	15 June 17	26 Dec. 16	„ 17 Infy.
383	Unattd. List	* **Tipping**, Richard Percy Macrae . . . . *Service for promotion 186 days.*	. . . .	29 June 16	3 July 16	26 Dec. 16	Attached 81 Pioneers.

384	N. Staff. R.	*Salt, George (207a) . . . *9 months forfeited for seniority under conditions of I. A. O. 511 of 1917.*	. . . .	27 Dec. 15	29 Aug. 18	27 Dec. 16	Attached 73 Infantry.
385	Unattd. List	*Partridge, James Wallace . . .	. . . .	†27 Dec. 15 / 18 June 17	26 June 17	27 Dec. 16	Attached 95 Infantry.
	R. Innis. Fus.	*Oliffe, Francis (46) . . . *9 months forfeited for seniority under conditions of I. A. O. 511 of 1917.*	. . . .	27 Dec. 15	13 June 18	27 Dec. 16	Attached 76 Punjabis.
	Unattd. List	*Cumming, Walter James (52) . . . *Service for promotion 184 days.*	. . . .	29 June 16	3 July 16	28 Dec. 16	Attached 52 Sikhs.
	Unattd. List	*Cumming, Arthur Edward (57) . . . 207 *Service for promotion 184 days.*	. . . .	29 June 16	2 July 16	28 Dec. 16	58 Rifles.
	Lond. R.	*Rice, Arthur Kemp (46) . . . *9 months forfeited for seniority under conditions of I. A. O. 511 of 1917.*	. . . .	29 Dec. 15	1 July 18	29 Dec. 16	Attached 5 Gurkha Rifles.
390	Unattd. List	*Norris, William Cyril Griffin (46) . . . *Service for promotion, 317d. from 20th July 1915 to 31st May 1916 and for pension, 60 days in addition.*	. . . .	14 Nov. 16	18 Nov. 16	2 Jan. 17	„ 103 Mahrattas.
	I. A. R. O.	§*Allanson, Greorge Hubert . . . *Service for promotion from 2nd January 1916 under I. A. O. 510 of 1917.*	. . . .	2 Apl. 15	4 Apl. 18	2 Jan. 17	Attached 2-3rd Gurkha Rifles.
	Linc. R.	*Barroll, William Stacpoole . . . *9 months forfeited for seniority under conditions of I. A. O. 511 of 1917.*	. . . .	2 Jan. 16	6 May 18	2 Jan. 17	94 Infantry.
	Unattd. List	*Finlay, John Alexander . . .	. . . .	†3 Jan. 16 / 18 June 17	25 June 17	3 Jan. 17	45 Sikhs.
	Norf. R.	*Mitchley, George Ambrose (46) . . . *9 months forfeited for seniority under conditions of I. A. O. 511 of 1917.*	. . . .	3 Jan. 16	17 June 18	3 Jan. 17	Attached 74 Punjabis.
395	Gen. List.	*Woollam, James Leonard . . . *9 months forfeited for seniority under conditions of I. A. O. 511 of 1917.*	. . . .	3 Jan. 16	31 July 18	3 Jan. 17	Attached 106 Pioneers.
	E. Kent. R.	*Woodward, William James . . . *9 months forfeited for seniority under conditions of I. A. O. 511 of 1917.*	. . . .	3 Jan. 16	5 Oct. 18	3 Jan. 17	Attached 1 Gurkha Rifles.
	Unattd. List	*James, William Arthur Lester . . . *Service for promotion from 20th July 1915 to 29th May 1916—315 days.*	. . . .	14 Nov. 16	23 Nov. 16	4 Jan. 17	Attached 123 Rifles.
	M. G. Corps.	*Lloyd, Hamilton Samuel John (52, 53, 207a) . . . *9 months forfeited for seniority under conditions of I. A. O. 511 of 1917.*	. . . .	4 Jan. 16	16 Sep. 18	4 Jan. 17	Attached 1 Gurkha Rifles.
	E. Lan. R.	*Grime, Harold Riley . . . 207 *9 months forfeited for seniority under conditions of I. A. O. 511 of 1917.*	. . . .	7 Jan. 16	6 May 18	7 Jan. 17	Attached 83 Infantry.
400	1 Res. R. of Cav.	* Hobbs, Charles Eric . . . *9 months forfeited for seniority under conditions of I. A. O. 511 of 1917.*	. . . .	8 Jan. 16	20 Aug. 18	8 Jan. 17	Attached 16 Cavalry.

No.	Former Corps.	Names.	Date of birth.	Date of first commission.	Date of appointment to Indian Army.	Date of ranks.	Remarks.
		Lieutenants—*contd.*					
401	I. A. R. O.	§***Denny,** Malcolm Percy Richard (46, 51) *Service for promotion from 9 Jan. 16 under I. A. O. 510 of 1917.*	. . . . .	9 Apl. 15	26 Apl. 18	9 Jan. 17	Attached 43 Regiment.
	I. A. R. O.	***Bevis,** Frederic George (48a, 50, 172a) *Service for promotion from 9th January 1916 under I.A.O. 510 of 1917.*	. . . . .	9 Apl. 15	2 Sep. 18	9 Jan. 17	Attached 97 Infantry.
	York & Lanc. R.	* **Drake,** Douglas Harold 207a *9 months forfeited for seniority under conditions of I. A. O. 511 of 1917.*	. . . . .	9 Jan. 16	29 July 18	9 Jan. 17	Attached 2 Rajput Lt. Infy.
	R. Lanc. R.	* **Webb,** Frank Burns 207a *9 months forfeited for seniority under conditions of I. A. O. 511 of 1917.*	. . . . .	10 Jan. 16	21 Aug. 18	10 Jan. 17	Attached 3-8th Gurkha Rifles.
405	W. Rid. R.	***Graham,** Norman Ingerfield *9 months forfeited for seniority under conditions of I. A. O. 511 of 1917.*	. . . . .	17 Jan. 16	8 May 18	17 Jan. 17	Attached 91 Punjabis.
	R. Ir. Fus.	***Manning,** Jack (46) (*temp. Capt. 11th January 1918*) *9 months forfeited for seniority under conditions of I. A. O. 511 of 1917.*	. . . . .	19 Jan. 16	7 May 18	19 Jan. 17	86 Infantry.
	Welsh. R.	***Williams,** Hubert Melville *9 months forfeited for seniority under conditions of I. A. O. 511 of 1917.*	. . . . .	19 Jan. 16	29 July 18	19 Jan. 17	Attached 74 Punjabis.
	R. Innis, Fus.	***Allen,** James Shaw *9 months forfeited for seniority under conditions of I. A. O. 511 of 1917.*	. . . . .	19 Jan. 16	24 Oct. 18	19 Jan. 17	Attached 18 Infantry.
	Scot. Rif..	***Evetts,** David William *9 months forfeited for seniority under conditions of I. A. O. 511 of 1917.*	. . . . .	19 Jan. 16	25 June 18	19 Jan. 17	Attached 11 Lancers.
410	Dorset. R.	***Franklin,** Eric Angus McAlister M.C. *9 months forfeited for seniority under conditions of I.A.O. 511 of 1917.*	. . . . .	19 Jan. 16	29 July 18	19 Jan. 17	Attached 97 Infantry,
		***Moore,** Francis Malcolm *9 months forfeited for seniority under conditions of I.A.O. 511 of 1917.*	. . . . .	21 Jan. 16	6 May 18	21 Jan. 17	Attached 52 Sikhs.
	Norf. R.	***Howdle,** Thomas George. *9 months forfeited for seniority under conditions of I. A. O. 511 of 1917.*	. . . . .	23 Jan. 16	3 Apl. 18	23 Jan. 17	Attached 120 Infantry.
413	Bord. R.	***Darwell,** Guy Preston, .207a *9 months forfeited for seniority under conditions of I. A. O. 511 of 1917.*	. . . . .	24 Jan. 16	3 Apl. 17	24 Jan. 17	Attached 120 Infantry.

414	Lond. R.	*Tearoe, Arthur Tresillian	. . . .	25 Jan. 16	11 Sep. 18	25 Jan. 17	Attd. 124 Infantry.
		9 months forfeited for seniority under conditions of I. A. O. 511 of 1917.					
415	Unattd. List	*Forbes, Leonard William Mckay (57)	3 June 99	26 Jan. 16	19 Mar. 16	26 Jan. 17	Attached 3 G. R.
	Unattd. List	*Robb, Thomas Ross (51)	2 Apl. 97	26 Jan. 16	17 Mar. 16	26 Jan. 17	,, 88 Infy.
	Unattd. List	*Wooldridge, James Hugh Copleston	. . . .	26 Jan. 16	21 Mar. 16	26 Jan. 17	,, 126 Infy.
	Unattd. List	*Davies, Henry Lowrie (51)	. . . .	26 Jan. 16	18 Mar. 16	26 Jan. 17	39 Garh. R.
	Unattd. List	*Rimell, Francis John 207a	17 May 97	26 Jan. 16	20 Mar. 16	26 Jan. 17	Attached 13 Rajputs.
420	Unattd. List	*Davies, Frederick Arthur	19 Sep. 97	26 Jan. 16	19 Mar. 16	26 Jan. 17	,, Guides.
	Unattd. List	*Lunham, Thomas Maybury	. . . .	†26 Jan. 16	14 Feb. 17	26 Jan. 17	34 Horse.
		Service for promotion 1 yr. 4 d. from 25 June 15 to 28 June 16.		30 Jan. 17			
	Essex. R.	*Bayliss, Bertram	. . . .	28 Jan. 16	22 June 18	28 Jan. 17	Attd. 76 Punjabis.
		9 months forfeited for seniority under conditions of I. A. O. 511 of 1917.					
	Unattd. List	*Tucker, Bernard Henry George (48a)	. . . .	29 June 16	12 July 16	29 Jan. 17	Attached 10 G. R.
		Service for promotion from 8th September 1915 to 6th February 1916 or 152 days.					
		Service for pension 152 days plus 189 days from 25th August 1914 to 7th September 1915–half to count.					
	Unattd. List	* Mull, Alan	. . . .	29 June 16	3 July 16	3 Feb. 17	2 Rajputs.
		Service for promotion from 8th September 1915 to 1st February 1916—147 days.					
425	York. & Lanc. R.	* Clark, John 207a	. . . .	6 Feb. 16	20 Aug. 18	6 Feb. 17	Attached 39 Garhwal Rifles.
		9 months forfeited for seniority under conditions of I. A. O. 511 of 1917.					
	E. Surr. R.	*Goolden, Alexander Wood	. . . .	9 Feb. 16	25 Sep. 18	9 Feb. 17	Attd. 124 Infantry.
		9 months forfeited for seniority under conditions of I. A. O. 511 of 1917.					
	R. W. Fus.	*Sowton, Bernard Stafford	. . . .	11 Feb. 16	6 May 18	11 Feb. 17	Attd. 12 Pioneers.
		9 months forfeited for seniority under conditions of I. A. O. 511 of 1917.					
	I. A. R. O	§*Ferguson, William Alexander Macleod (172a) 206	. . . .	12 May 15	16 July 17	12 Feb. 17	Attached 4-3rd G. R.
		Service for promotion from 12th February 1916 under I. A. O. 510 of 1917.					
	Bord. R.	*Parmenter, Fred	. . . .	12 Feb. 16	24 Oct. 18	12 Feb. 17	Attd. 107 Pioneers.
		9 months forfeited for seniority under conditions of I. A. O. 511 of 1917.					
430	R. Suss. R.	*Eccles, George Jenner 270a	. . . .	13 Feb. 16	29 July 18	13 Feb. 17	Attd. 3 Horse.
		9 months forfeited for seniority under conditions of I. A. O. 511 of 1917.					
	Unattd. List	*Oates, William John Stewart	. . . .	29 June 16	2 July 16	15 Feb. 17	23 Pioneers.
		Service for promotion 135 days.					
	Aug. Imp. Force.	*Abbott, Adrian Augustus	. . . .	15 Feb. 16	22 Nov. 18	15 Feb. 17	Attd. 6 Cavalry.
		9 months forfeited for seniority under conditions of I. A. O. 511 of 1917.					
	Unattd. List	*Tulloch, Herbert Maurice (51, 71) 207	2 Aug. 93	18 Feb. 16	6 Nov. 16	18 Feb. 17	Attached 33 Cavy.
		(Service for pension from 23 May 15.)					
	R. War. R.	*Pugh, Leslie Beresford (51) 207a	. . . .	18 Feb. 16	12 June 18	18 Feb. 17	Attd. 9 Gurkha Rifles.
		9 months forfeited for seniority under conditions of I.A.O. 511 of 1917.					
435	Mont. Yeo.	*Mostyn-Owen, Herbert Louis 207a	. . . .	20 Feb. 16	24 Aug. 18	20 Feb. 17	Attd. 26 Cavalry.
		9 months forfeited for seniority under conditions of I. A. O. 511 of 1917.					
436	I. A. R. O.	§*Ingles, Henry Noel George	. . . .	21 May 15	16 Aug. 18	21 Feb. 17	Attd. 31 Punjabis.
		Service for promotion from 21 Feb. 16 under I. A. O. 510 of 1917.					

* Has not passed the final examination. † The date from which seniority counts. § Admitted subject to H. M.'s approval.

No.	Former Corps.	Names.	Date of birth.	Date of first commission.	Date of appointment to Indian Army.	Date of rank.	Remarks.
		Lieutenants—*contd.*					
437	Lond. R.	***Spencer,** Harold John *9 months forfeited for seniority under conditions of I. A. O. 511 of 1917*	. . . .	22 Feb. 16	31 Oct. 18	22 Feb. 17	Attd. 3 Brahmans.
	R. Innis. Fus.	***Porter,** Albert George *9 months forfeited for seniority under conditions of I. A. O. 511 of 1917.*	. . . .	27 Feb. 16	29 July 18	27 Feb. 17	Attd. 10 Jats.
	Unattd. List	***Retallack,** Cyrus William (48a) *Service for promotion 121 days.*	. . . .	29 June 16	3 July 16	28 Feb. 17	101 Grenadiers.
440	Lan. Fus.	***Hampson,** John Henry Japheth *9 months forfeited for seniority under conditions of I.A.O. 511 of 1917.*	. . . .	28 Feb. 16	1 Oct. 18	28 Feb. 17	Attached 1 Gurkha Rifles.
	I. A. R. O.	*§**Murray,** Victor William *Service for promotion from 1st March 1916 under I.A.O. 510 of 1917.*	. . . .	29 May 15	15 June 18	1 Mar. 17	45th Cavalry.
	Sea. Highrs.	***Daw,** Donald James, M.C. *9 months forfeited for seniority under conditions of I. A. O. 511 of 1917.*	. . . .	2 Mar. 16	16 Sep. 18	2 Mar. 17	Attd. 9 Gurkha Rifles.
	D.L.O. Yeo.	***Anson,** Edward Ellis 207 *9 months forfeited for seniority under conditions of I. A. O. 511 of 1917.*	. . . .	2 Mar. 16	24 Oct. 18	2 Mar. 17	Attd. 34 Horse.
	Manch R.	***McClenaghan,** Henry Clarke *9 months forfeited for seniority under conditions of I. A. O. 511 of 1917.*	. . . .	3 Mar. 16	21 July 18	3 Mar. 17	Attd. 43 Regt.
445	Ches. R.	***Heaver,** Lyonel George Homewood *9 months forfeited for seniority under conditions of I. A. O. 511 of 1917.*	. . . .	8 Mar. 16	29 July 18	8 Mar. 17	Attd. 41 Dogras.
	Lond. R.	***Randall,** Leslie Ernest *9 months forfeited for seniority under conditions of I. A. O. 511 of 1917.*	. . . .	8 Mar. 16	17 Sep. 18	8 Mar. 17	Attd. 41 Dogras.
	Suff. R.	***Whitcombe,** Leslie Clarence *9 months forfeited for seniority under conditions of I. A. O. 511 of 1917.*	. . . .	8 Mar. 16	29 July 18	8 Mar. 17	Attd. 74 Punjabis.
	R. Muns. Fus.	***Sheperd,** Albert Victor Wilberforce *9 months forfeited for seniority under conditions of I. A. O. 511 of 1917.*	. . . .	10 Mar. 16	6 June 18	10 Mar. 17	Attd. 6 Gurkha Rifles.
	Dorset R.	***Hawkins,** Frederic Trevor *9 months forfeited for seniority under conditions of I. A. O. 511 of 1917.*	. . . .	10 Mar. 16	1 Sep. 13	10 Mar. 17	Attd. 117 Mahrattas.
450	North'd Fus.	***Richards,** William Norman 207a *9 months forfeited for seniority under conditions of I.A.O. 511 of 1917.*	. . . .	12 Mar. 16	18 June 18	12 Mar. 17	Attd. 120 Infantry.
	Unattd. List	***Jessop,** Charles Richard *(Service for promotion and pension to count from 15 Mar. 16.)*	. . . .	14 Nov. 16	28 Nov. 16	15 Mar. 17	Attached 67 Punj.
	R. W. Fus.	***Davies,** Victor Kenneth Neville *9 months forfeited for seniority under conditions of I. A. O. 511 of 1917.*	. . . .	15 Mar. 16	28 Oct. 18	15 Mar. 17	Attd. 9 Gurkha Rifles.
	M. G. Corps	***Avent,** James George *9 months forfeited for seniority under conditions of I. A. O. 511 of 1917.*	. . . .	17 Mar. 16	28 July 18	17 Mar. 17	Attd. 117 Mahrattas.
	I. A. R. O.	*§**Jahans,** Lorraine Errol Hay (46) 207 *(Service for promotion from 18th March 1916 under I. A. O. 510 of 1917.)*	. . . .	18 June 15	3 Nov. 17	18 Mar. 17	109 Infantry.
455	I. A. R. O.	§***Wilkinson,** Robert Joseph *Services for promotion from 18 Mar. 16 under I. A. O. 510 of 1917.*	. . . .	18 June 15	22 Aug. 18	18 Mar. 17	Attd. Sig. Service.

456	Devon. R.	***Gosnell,** Kenneth Arthur	. . . .	24 Mar. 16	6 May 18	24 Mar. 17	13 Rajputs.
		9 months forfeited for seniority under conditions of I. A. O. 511 of 1917.					
	I. A. R. O.	*§ **Powell,** Lancelot Kaye Machattie	. . . .	25 June 15	26 June 18	25 Mar. 17	Attd. 19 Lancers.
		Service for promotion from 25th March 1916 under I. A. O. 510 of 1917.					
	Notts. Derby. R.	***Loup,** Louis Anthony (46) . . . 207a	. . . .	25 Mar. 16	12 June 18	25 Mar. 17	,, 10 Gurkha Rifles.
		9 months forfeited for seniority under conditions of I.A.O. 511 of 1917.					
	Liens. R.	* **Lyon,** Wallace Adelbert . . . 207a	. . . .	26 Mar. 16	16 Sep. 18	26 Mar. 17	,, 52 Sikhs.
		9 months forfeited for seniority under conditions of I. A. O. 511 of 1917.					
460	S. Staff R.	***Portas,** John	. . . .	1 Apl. 16	12 June 18	1 Apl. 17	,, 126 Infantry.
		9 months forfeited for seniority under conditions of I.A.O. 511 of 1917.					
	Sea. Highrs.	* **Walker,** Allan Robertson	. . . .	3 Apl. 16	29 July 18	3 Apl. 17	Attached 15 Sikhs.
		9 months forfeited for seniority under conditions of I. A. O. 511 of 1917.					
	Unattd. List	***Wingfield,** Richard Wiliam Digby	. . . .	14 Nov. 16	8 Dec. 16	4 Apl. 17	,, 18 Infantry.
		Service for promotion from 21st June 1915 to 30th January 1916—224 days.					
	Unattd. List	***Bain,** George Alexander (4a. 51)	15 Mar. 98	7 Apl. 16	6 Nov. 16	7 Apl. 17	,, 2 G. R.
	Unattd. List	***Adams,** Herbert Vincent (46)	6 Apl. 98	7 Apl. 16	5 Nov. 16	7 Apl. 17	Attached 13 Rajputs.
465	Unattd. List	***Corfield,** Victor Charles William (57)	7 Mar. 98	7 Apl. 16	15 Nov. 16	7 Apl. 17	,, 10 G. R.
	Unattd. List	***Johnston,** Dorian Dudley Hope (51) . . . 207 a	15 Feb. 98	7 Apl. 16	0 Nov. 16	7 Apl. 17	,, 11 Rajputs.
	Unattd. List	***Seagrim,** Cyril Bruce (46)	7 Jan. 98	7 Apl. 16	9 Nov. 16	7 Apl. 17	,, 99 Infantry.
	Unattd. List	***Wardle,** Frank Moore (46)	30 Nov. 93	7 Apl. 16	5 Nov. 16	7 Apl. 17	,, 7 Rajputs.
	N. Staff R.	* **Fitzherbert,** Rowden French	. . . .	7 Apl. 16	20 Aug. 18	7 Apl. 17	,, 19 Punjabis.
		9 months forfeited for seniority under conditions of I. A. O. 511 of 1917.					
470	Unattd. List	***Davies,** John Trevor . . . 206	. . . .	{ †8 Apl. 16 { 30 Jan. 17	14 Feb. 17	8 Apl. 17	19 Lancers.
		Service for promotion 297 d. from 5 Sep. 15 to 28 June 16.					
	I. A. R. O.	§* **Nash,** Albert Victor (71)	. . . .	9 July 15	10 Sep. 18	9 Apl. 17	Attached 67 Punjabis.
		Service for promotion from 9 April 16 under I. A. O. 510 of 1917.					
	I. A. R. O.	§***Youngman,** John Southgate (51)	. . . .	9 July 15	25 Feb. 18	9 Apl. 17	Attached 121 Pioneers.
		Service for promotion from 9 Apl. 16 under I. A. O. 510 of 1917.					
	I. A. R. O.	§***Simpson,** Allan Richard (57)	. . . .	9 July 15	19 July 17	9 Apl. 17	,, 15 Sikhs.
		Service for promotion from 9 Apl. 16 under I. A. O. 510 of 1917.					
	Scot. Horse	* **Ambler,** Richard . . . 207a	. . . .	10 Apl. 16	12 Sep. 18	10 Apl. 17	,, 2 Lancers.
		9 months forfeited for seniority under conditions of I. A. O. 511 of 1917.					
475	Sea. Highrs.	* **Jackson,** Laurence Ernest William . . . 207a	. . . .	15 Apl. 16	16 Sep. 18	15 Apl. 17	,, 8th Gurkha Rifles.
		9 months forfeited for seniority under conditions of I. A. O. 511 of 1917.					
	Unattd. List	***Spear,** Christopher Ronald (51) . . . 206	. . . .	18 Apl. 16	25 Apl. 16	18 Apl. 17	,, 58 Rifles.
	Unattd. List	***Moore,** Alleyn Cardwell	. . . .	18 Apl. 16	23 Apl. 16	18 Apl. 17	,, 92 Punjabis.
	Unattd. List	***Reed,** Ronald Robert	. . . .	18 Apl. 16	24 Apl. 16	18 Apl. 17	,, 48 Pioneers.
	Unattd. List	***Smith,** Brian Leslie Sundius	. . . .	18 Apl. 16	30 Apl. 16	18 Apl. 17	,, 126 Infantry.
480	Unattd. List	***Guy,** Norman Greenwood . . . 206	. . . .	18 Apl. 16	25 Apl. 16	18 Apl. 17	,, 57 Rifles.
	Unattd. List	***Hailstone,** Robert	. . . .	18 Apl. 16	29 Apl. 16	18 Apl. 17	,, 10 G. R.
	Unattd. List	***Townley,** Neil Patersson (51, 52) . . . 207	. . . .	18 Apl. 16	24 Apl. 16	18 Apl. 17	,, 41 Dogras.
483	Unattd. List	***Hewetson,** John Jervis . . . 207a	. . . .	18 Apl. 16	26 Apl. 16	18 Apl. 17	,, 8 G. R.

* Has not passed the final examination. § Admitted subject to H. M.'s approval.

No.	Former Corps.	Names.	Date of birth.	Date of first commission.	Date of appointment to Indian Army.	Date of rank.	REMARKS.
		Lieutenants—*contd.*					
484	Unattd. List	***Cristall**, Brian Pery .206		18 Apl. 16	15 May 16	18 Apl. 17	Attached 51 Sikhs.
485	Unattd. List	***Bridges**, Kenneth Wilfred		18 Apl. 16	23 Apl. 16	18 Apl. 17	7 Rajputs.
	Unattd. List	***Dallson**, John Bernard		18 Apl. 16	30 Apl. 16	18 Apl. 17	67 Punjabis.
	Unattd. List	***Skene**, Stuart Gray Sanders		18 Apl. 16	26 Apl. 16	18 Apl. 17	8 G. R.
	Unattd. List	***Warren**, Victor Alexander		18 Apl. 16	25 Apl. 16	18 Apl. 17	37 Dogras.
	Unattd. List	***Wallis**, Stanley Ivan (60*a*) .206		18 Apl. 16	25 Apl. 16	18 Apl. 17	Attached 23 Pioneers.
490	Unattd. List	***Tomkins** Gerard Lionel, M. C.		18 Apl. 16	1 May 16	18 Apl. 17	„ 128 Pioneers.
	Unattd. List	***MacLachlan**, Archibald		18 Apl. 16	24 Apl. 16	18 Apl. 17	„ 51 Sikhs.
	Unattd. List	***Lloyd-Jones**, William Rice (51)		18 Apl. 16	23 Apl. 16	18 Apl. 17	„ 128 Pioneers.
	Unattd. List	***Nicolson**, Donald Oliver		18 Apl. 16	22 Apl. 16	18 Apl. 17	„ 108 Infantry.
	Unattd. List	***Clark**, William Valiant		18 Apl. 16	24 Apl. 16	18 Apl. 17	„ 107 Pioneers.
495	Unattd. List	***Yarrow**, Thomas Bryden		18 Apl. 16	28 Apl. 16	18 Apl. 17	„ 24 Punjabis.
	Unattd. List	***Simpson**, Francis Cortlan (57)		18 Apl. 16	25 Apl. 16	18 Apl. 17	37 Dogras.
	Unattd. List	***Keegan**, Denis		18 Apl. 16	25 Apl. 16	18 Apl. 17	Attached 33 Punjabis.
	Unattd. List	***Gompertz**, Mervyn Clive Theodore .206		18 Apl. 16	25 Apl. 16	18 Apl. 17	„ 8 Rajputs.
	Unattd. List	***Bell**, Hugh Gilchrist (57) .207*a*		18 Apl. 16	26 Apl. 16	18 Apl. 17	„ 3 Horse.
500	Unattd. List	***Power**, Henry Ross		18 Apl. 16	25 Apl. 16	18 Apl. 17	„ 37 Dogras.
	Unattd. List	***McClenagnan**, Percy John Warren (51, 95) .207		18 Apl. 16	27 Apl. 16	18 Apl. 17	129 Baluchis.
	Unattd. List	***Millar**, Arthrles Buckingham		18 Apl. 16	26 Apl. 16	18 Apl. 17	Attached 56 Rifles.
	Unattd. List	***Malden**, Charles Melville (146)		18 Apl. 16	25 Apl. 16	18 Apl. 17	„ 59 Rifles.
	Unattd. List	***Jenkins**, Norman Leslie (46)		18 Apl. 16	25 Apl. 15	18 Apl. 17	„ 33 Punjabis.
505	Unattd. List	***Dean**, William Edward		18 Apl. 16	25 Apl. 16	18 Apl. 17	53 Sikhs.
	Unattd. List	***Winthrop**, Hugh Erskine (51) .207		18 Apl. 16	25 Apl. 16	18 Apl. 17	14 Sikhs.
	Unattd. List	***Warhurst**, Albert Edward		18 Apl. 16	25 Apl. 16	18 Apl. 17	Attd. 6 G. R.
	Unattd. List	***Ansell**, Cyril Edwin Osborne (46) 207*a*		18 Apl. 16	29 Apl. 16	18 Apl. 17	„ 8 Cavalry.
	Unattd. List	***Frankland**, Alec		18 Apl. 16	24 Apl. 16	18 Apl. 17	„ 101 Grenadiers.
510	Unattd. List	***Edward**, William Donald		18 Apl. 16	25 Apl. 16	18 Apl. 17	„ 57 Rifles.
	Unattd. List	***Neale**, Ronald Kingsley (57)		18 Apl. 16	25 Apl. 16	18 Apl. 17	„ 25 Punjabis.
	Unattd. List	***Doyle**, Patrick James (51)		18 Apl. 16	26 Apl. 16	18 Apl. 17	„ 21 Punjabis.
	Unattd. List	***Bennett**, William Gwynn .207		18 Apl. 16	24 Apl. 16	18 Apl. 17	47 Sikhs.
	Unattd. List	***Ambrose**, Robert Denis, M.C. (51)		18 Apl. 16	23 Apl. 16	18 Apl. 17	104 Rifles.
515	Unattd. List	***Gulton**, Louis Wansbrough (51,52)		18 Apl. 16	30 Apl. 16	18 Apl. 17	67 Punjabis.
	Unattd. List	***Sutcliffe**, Oliver Dudley		18 Apl. 16	26 Apl. 16	18 Apl. 17	Attached 87 Punjabis.
	Unattd. List	***Glen**, John Vivian Johnson		18 Apl. 16	24 Apl. 16	18 Apl. 17	6 Jats.
	Unattd. List	***Mason**, Douglas Richard Wordsworth 207		18 Apl. 16	24 Apl. 16	18 Apl. 17	2 Rajputs.
	Unattd. List	***Sevenoaks**, Charles Lutman (52)		18 Apl. 16	28 Apl. 16	18 Apl. 17	Attached 24 Punjabis.
520	Unattd. List	***Pybus**, Gilbert Douglas (46) 207		18 Apl. 16	24 Apl. 16	18 Apl. 17	14 Sikhs.
	Unattd. List	***Huelin**, William (51)		18 Apl. 16	20 Apl. 16	18 Apl. 17	Attd. 13 Rajputs.
	Unattd. List	***Hurly**, Maurice Francis Fitzmaurice (46)		18 Apl. 16	24 Apl. 16	18 Apl. 17	„ 34 Pioneers.
	Unattd. List	***Wingfield**, Charles Mervyan Hunt (57)		18 Apl. 16	25 Apl. 16	18 Apl. 17	2 G. R.
524	Unattd. List	***Voysey**, George Cuming Annesley (46, 57)		18 Apl. 16	26 Apl. 16	18 Apl. 17	Attd. 35 Sikhs.

525	Unattd. List	*Jackson, Harold John Ross	207	18 Apl. 16	25 Apl. 16	18 Apl. 17	„ 58 Rifles.
	Unattd. List	*Bland, George Hubert, M.C.		18 Apl. 16	26 Apl. 16	18 Apl. 17	„ 105 Infantry.
	Unattd. List	*Bragg, Harry Vyner		18 Apl. 16	25 Apl. 16	18 Apl. 17	„ 10 Jats.
	Unattd. List	*Foster, John Wilford (48a, 57, 66a, 71)	205a	18 Apl. 16	26 Apl. 16	18 Apl. 17	„ 25 Cavalry.
	Unattd. List	*Gray, Charles Edward (51, 52, 66a)		18 Apl. 16	25 Apl. 16	18 Apl. 17	„ 3 G. R.
530	Unattd. List	*Loftus-Tottenham, Frederick Joseph		18 Apl. 16	25 Apl. 16	18 Apl. 17	„ 2 G. R.
	Unattd. List	*Stewart, Charles Harvey		18 Apl. 16	30 Apl. 16	18 Apl. 17	„ 106 Pioneers.
	Unattd. List	*Leese, Francis Winton		18 Apl. 16	29 Apl. 16	18 Apl. 17	29 Punjabis.
	Unattd. List	*Baker Cedric Stansmore (57)	207	18 Apl. 16	25 Apl. 16	18 Apl. 17	Attached 5 G. R.
	Unattd. List	*Chappell, Hereward		18 Apl. 16	27 Apl. 16	18 Apl. 17	39 Garh. R.
535	Unattd. List	*Ronnfeldt, William Eric		18 Apl. 16	24 Apl. 16	18 Apl. 17	62 Punjabis.
	Unattd. List	*Hodson, Denham (46, 57)		18 Apl. 16	25 Apl. 16	18 Apl. 17	Attd. 116 Mahrattas.
	Unattd. List	*Rodwell, Edward Hunter		18 Apl. 16	25 Apl. 16	18 Apl. 17	„ 53 Sikhs.
	Unattd. List	*Wernicke, Andrew Edmund	207a	18 Apl. 16	29 Apl. 16	18 Apl. 17	29 Punjabis.
	Unattd. List	*Tester, Charles Knowler		18 Apl. 16	25 Apl. 16	18 Apl. 17	6 Jats.
540	Unattd. List	*Thomason, Archibald David Fawcett		18 Apl. 16	23 Apl. 16	18 Apl. 17	Attd. 12 Cavalry.
	Unattd. List	*Fox, Walter Basil Orlando (51)		18 Apl. 16	25 Apl. 16	18 Apl. 17	„ 5 G. R.
	Leic. R.	*Hepworth, Maurice Arthur	207a	18 Apl. 16	21 Aug. 18	18 Apl. 17	Attd. 21 Punjabis.
		9 months forfeited for seniority under conditions of I. A. O. 511 of 1917.					
	Glam. Yeo.	* Scott, Arthur Geoffrey		18 Apl. 16	29 July 18	18 Apl. 17	Attd. 94 Infantry.
		9 months forfeited for seniority under conditions of I. A. O. 511 of 1917.					
	I. A. R. O.	§*Cox, Gordon Wallace		20 July 15	2 Nov. 17	20 Apl. 17	31 Punjabis.
		(Service for promotion from 20 April 1916 under I.A.O. 510 of 1917)					
545	I. A. R. O.	§* Harrison, James Alfred	206	20 July 15	26 Apl. 18	20 Apl. 17	Attd. 1st Brahmans.
		(Service for promotion from 20 April under 1916 under I. A. O. 510 of 1917.)					
	I. A. R. O.	§*Skilliter, Walter Miles		20 July 15	10 Sep. 18	20 Apl. 17	Attd. 125 Rifles.
		Service for promotion from 20 April 1916 under I. A. O. 510 of 1917.					
	I. A. R. O.	§* Nichols, Frederick William (59)		20 July 15	26 Apl. 18	20 Apl. 17	Attd. Signal Service.
		(Service for promotion from 20 April 1916 under I. A. O. 510 of 1917.)					
	I. A. R. O.	§*Isaacs, Ralph Leicestor		20 July 15	30 July 18	20 Apl. 17	„ 116 Mahrattas.
		Service of promotion from 20 April 1916 under I. A. O. of 510 of 1917.					
	I. A. R. O.	§*Allen, Brian		22 July 15	17 July 18	22 Apl. 17	Attd. Mily. Farms Dept.
		Service for promotion from 22 April 1916 under I. A. O. 510 of 1917.					
550	I. A. R. O.	*§Commeline, Francis Tom		23 July 15	14 July 17	23 Apl. 17	Attd. 2 Lancers.
		(Service for promotion from 23rd Apl. 1916 under I.A.O. 510 of 1917.)					
	Lond. R.	*Brown, Archibald Ernest		24 Apl. 16	24 Oct. 18	24 Apl. 17	Attd. 107 Pioneers.
		9 months forfeited for seniority under canditions of I. A. O. 511 of 1917.					
	L'pool R.	* Waring, James Livingston	207a	24 Apl. 16	10 July 18	24 Apl. 17	Attd. 104 Rifles.
		9 months forfeited for seniority under conditions of I. A. O. 511 of 1917.					
553	I. A. R. O.	§* Leir, John Russell, Charles Vernon		25 July 15	16 Aug. 18	25 Apl. 17	Attd. 6 Gurkha Rifles.
		Service for promotion from 25 April 1916 under I. A. O. 510 of 1917					

* Has not passed the final examination.

§ Admitted subject to His Majesty's approval.

No.	Former Corps.	Names.	Date of birth.	Date of first commission.	Date of appointment to Indian Army.	Date of rank.	REMARKS.
		Lieutenants—*contd.*					
554	York & Lanc. R.	* **Steele,** Richard Philip . . . . *9 months forfeited for seniority under conditions of I. A. O. 511 of 1917.*	. . . . .	28 Apl. 16	29 July 18	28 Apl. 17	Attd. 97 Infantry.
555	I. A. R. O.	§***Nye,** Arthur Reginald . . . . *Service for promotion from 29 April 1916 under I. A. O. 510 of 1917.*	. . . . .	29 July 15	5 Dec. 17	29 Apl. 17	4 Gurkha Rifles.
	I. A. R. O.	§* **Evennett,** Harold Clement (52) . . . . *Service for promotion from 30 April 1916 under I. A. O. 510 of 1917.*	. . . . .	30 July 15	16 Aug. 18	30 Apl. 17	Attd. 12 Cavalry.
	Lord. R.	***Atkins,** Percy Coyney . . . . *9 months forfeited for seniority under conditions of I. A. O. 511 of 1917.*	. . . . .	1 May 15	16 Sep. 18	1 May 17	Attd. 71 Punjabis.
	Unattd. List	***Barry,** Frederick Cecil	. . . . .	14 Nov. 16	25 Nov. 16	3 May 17	" 52 Sikhs.
	Aus. Imp. Force.	***Stanfield,** Ernest Alfred Field . . . . *9 months forfeited for seniority under conditions of I. A. O. 511 of 1917.*	. . . . .	4 May 16	31 Oct. 18	4 May 17	Attd. 14 Lancers.
560	Sea. Highrs.	* **Sainter,** James Dow, M. C. . . . . *9 months forfeited for seniority under conditions of I. A. O. 511 of 1917.*	. . . . .	7 May 16	16 Sep. 18	7 May 17	Attd. 66 Punjabis.
	Bedf. R.	***Jones,** Kenneth Charles Johnston, M.C. . . . . *9 months forfeited for seniority under conditions of I. A. O. 511 of 1917.*	. . . . .	12 May 16	22 Nov. 18	12 May 17	Attd. 9 Gurkha Rifles.
	Arg. & Suth'd Highrs.	***Cowan,** David Tennant, M. C. . . . . *9 months forfeited for seniority under conditions of I. A. O. 511 of 1917.*	. . . . .	14 May 16	27 July 18	14 May 17	Attd. 3 Gurkha Rifles.
	Ayr. Yeo.	**Chalmers,** Thomas Cuthbert . . . . *9 months forfeited for seniority under conditions of I. A. O. 511 of 1917.*	. . . . .	17 May 16	16 Aug. 18	17 May 17	Attd. 26 Cavalry.
	I. A. R. O.	§* **Penn-Simkins,** Gerald . . . . *Service for promotion from 18 May 1916 under I. A. O. 510 of 1917.*	. . . . .	18 Aug. 15	26 Apl. 18	18 May 17	Attd. 28 Cavalry.
565	Unattd. List	***Oates,** Alfred Tennyson 207a	. . . . .	{†20 May 16 {18 June 17	} 25 June 17	20 May 17	10 Lancers.
	R. Innis. Fus.	***Morrow,** John Stanley . . . . *9 months forfeited for seniority under conditions of I. A. O. 511 of 1917.*	. . . . .	23 May 16	26 Sep. 18	23 May 17	Attd. 9 Gurkha Rifles
	I. A. R. O.	§* **Joyce,** Wilfred Digby (46, 51, 52) . . . . 207a *Service for promotion from 24 May 1916 under I. A. O. 510 of 1917.*	. . . . .	24 Aug. 15	26 Apl. 18	24 May 17	Attd. 22 Punjabis.
	R. W. Fus.	*VC **Barter,** Frederick, M.C., (207a) . . . . 207a *9 months forfeited for seniority under conditions of I. A. O. 511 of 1917.*	. . . . .	26 May 16	6 May 18	26 May 17	3 Gurkha Rifles.
569	R. G. A.	**Lickman,** Earnest William	. . . . .	26 May 16	6 June 18	26 May 17	Attd. 7 Lancers.

570	Suff. R.	***Richdale**, Conroy Richdale *9 months forfeited for seniority under conditions of I.A.O. 511 of 1917.*	. . . .	26 May 16	14 Sep. 18	26 May 17	Attd. 94 Infantry.
	Unattd. List	***Tomsett**, Ernest William *Services for promotion 32 days.*	. . . .	29 June 16	20 July 16	28 May 17	3 G. R.
	E. Kent. R.	***Green**, William George . . . 207*a* *9 months forfeited for seniority under conditions of I.A.O. 511 of 1917*	. . . .	30 May 16	22 Nov. 18	30 May 17	Attd. 16 Rajputs.
	R. Fus.	***Renwick**, Hew *9 months forfeited for seniority under conditions of I.A.O. 511 of 1917.*	. . . .	3 June 16	6 May 18	3 June 17	Attd. 96 Infantry.
	Unattd. List	***French**, Francis Holroyd *Service for promotion 32 days.*	. . . .	8 Sep. 15	5 Dec. 17	8 June 17	Attd. 27 Punjabis.
575	I. A. R. O.	§***Hathaway**, Cecil John *(Service for promotion from 8 June 1916 under I.A.O. 510 of 1917)*	. . . .	8 Sep. 15	2 Nov. 17	8 June 17	31 Punjabis.
	I. A. R. O.	***Tapsell**, Neville Bailey (51, 52) *Service for promotion from 8th June 1916 under I.A.O. 510 of 1917*	. . . .	8 Sep. 15	4 Apl. 18	8 June 17	Attd. 2 Gurkha Rifles.
	I. A. R. O.	§***Jeffreys**, James Frederick Donohue *Service for promotion from 8th June 1916 under I.A.O. 510 of 1917*	. . . .	8 Sep. 15	4 Apl. 18	8 June 17	Attd. 96 Infantry.
	I. A. R. O.	§***Thomson**, Christopher Gardner *Service for promotion from 8th June 1916 under I.A.O. 510 of 1917*	. . . .	8 Sep. 15	26 Sep. 18	8 June 17	Attd. 6 Jat Lt. Infantry.
	I. A. R. O.	§***Corrin**, Thomas Oswald *(Service for promotion from 8 June 1916 under I.A.O. 510 of 1917)*	. . . .	8 Sep. 15	2 Nov. 17	8 June 17	31 Punjabis.
580	Devon R.	***Ellis**, Arthur Besley *9 months forfeited for seniority under conditions of I.A.O. 511 of 1917*	. . . .	9 June 16	5 Oct. 18	9 June 17	Attd. 32 Pioneers.
	E. Lan. R.	***Birtwistle**, Joseph (57) *9 months forfeited for seniority under conditions of I.A.O. 115 of 1917*	. . . .	11 June 16	20 Aug. 18	11 June 17	Attd. 10 Jats.
	Suff R.	***Thorpe**, Charles William Gordon . . . 207*a* *9 months forfeited for seniority under conditions of I.A.O. 511 of 1917*	. . . .	12 June 16	17 Sep. 18	12 June 17	Attd. 52 Sikhs.
	Unattd. List	§***Beynon**, Albert Cyril	. . . .	18 June 16	2 Jan. 17	18 June 17	Attd. 98 Infantry.
	R. Ir. Fus.	***Loyd**, Edward Basil Kirkman *9 months forfeited for seniority under conditions of I.A.O. 511 of 1917*	. . . .	20 June 16	1 Aug. 18	20 June 17	Attd. 11 Lancers.
585	I. A. R. O.	§***Morgan**, Lewis Edyvean *(Service for promotion from 22nd June 1916 under I.A.O. 510 of 1917)*	. . . .	22 Sep. 15	5 Dec. 17	22 June 17	Attd. 56 Rifles.
	Lov. Scouts	***Huntley**, William Burney *9 months forfeited for seniority under conditions of I. A.O. 511 of 1917*	. . . .	23 June 16	4 May 18	23 June 17	Attd. 123 Rifles.
587	E. Surr. R.	***Stott**, Frank Leonard *9 months forfeited for seniority under conditions of I.A.O. 511 of 1917*	. . . .	23 June 16	30 July 18	23 June 17	Attd. 112 Infantry.

*** Has not passed the final examination. § Admitted subject to His Majesty's approval.**

Lieutenants—*contd.*

No.	Former Corps.	Names.	Date of birth.	Date of first commission.	Date of appointment to Indian Army.	Date of rank.	Remarks.
588	E. Kent. R.	* **Reynolds,** Alfred William . . . 207a *9 months forfeited for seniority under conditions of I. A. O. 511 of 1917*		22 June 16	29 July 18	22 June 17	Attd. 119 Infantry.
	Unattd. List	***Freeman-Thomas,** *Hon.* Inigo Brassey (57)		30 Jan. 17	5 Feb. 17	24 June 17	Attd. 3 Horse.
590	R. Fus.	***Landon,** Sylvanus Luke *9 months forfeited for seniority under conditions of I.A.O. 511 of 1917*		25 June 16	27 Sep. 18	25 June 17	Attd. 106 Pioneers.
	Unattd. List	***O'Connor,** Kenneth Kennedy		29 June 16	2 July 16	29 June 17	14 Sikhs.
	Unattd List	***Macintosh,** Kenneth John (46)		29 June 16	3 July 16	29 June 17	Attd. 5 G. R.
	Unattd. List	***Maconochie,** Alexander Evan Frederick		29 June 16	5 July 16	29 June 17	Attd. 12 Cavy.
	Unattd. List	***Bartlett,** Harold Victor (*Ty. Lt. 20 Oct. 16*) . . . 206		29 June 16	11 July 16	29 June 17	Attd. 91 Punjabis.
595	Unattd. List	***O'Brien,** Brian Palliser Tieque, M.C. (57)		29 June 16	7 July 16	29 June 17	Attd. 8 G. R.
	Unattd. List	***Proudlock,** Robert Victor		29 June 16	3 July 16	29 June 17	92 Punjabis.
	Unattd. List	***Johnston,** Ronald Hilary Graves (57) . . . 207a		29 June 16	2 July 16	29 June 17	Attd. 21 Punjabis.
	Unattd. List	***Ross-Hurst,** William James O'Brien . . . 206		29 June 16	10 July 16	29 June 17	58 Rifles.
	Unattd. List	***Beamish,** Richard Percival (51)		29 June 16	2 July 16	29 June 17	Attd. 21 Punjabis.
600	Unattd. List	***Thomson,** Alexander Percival Quin		29 June 16	3 July 16	29 June 17	Attd. 5 G. R.
	Unattd. List	***Mannock,** William Arthur		29 June 16	12 July 16	29 June 17	Attd. 10 G. R.
	Unattd. List	***Cruickshank,** Alexander Stanley . . . 207a		29 June 16	2 July 16	29 June 17	14 Sikhs.
	Unattd List	***Pollock-Gore,** William Alexander Joseph		29 June 16	3 July 16	29 June 17	128 Pioneers.
	Unattd. List	***Ruffell,** Bernard (52)		29 June 16	3 July 16	29 June 17	93 Infy.
605	Unattd. List	***Skrine,** Philip Roger Huntley (57)		29 June 16	8 July 16	29 June 17	Attd. 13 Rajputs.
	Unattd. List	***Cameron-Webb,** John Haseltine		29 June 16	13 Aug. 16	29 June 17	14 Sikhs.
	Unattd. List	***Mannock,** Reginald Chater (57)		29 June 16	7 July 16	29 June 17	Attd. 8 G. R.
	Unattd. List	***Hervey,** William Fitz-Gerald (51)		29 June 16	12 July 16	29 June 17	Attd. 10 G. R.
	Unattd. List	***Kenny,** Edward Joseph (51)		29 June 16	3 July 16	29 June 17	Attd. 116 Mahrattas.
610	Unattd. List	***Fischer,** Thomas Gordon Temple		29 June 16	3 July 16	29 June 17	Attd. 9 G. R.
	Unattd. List	***Lindquist,** William Herbert Horatio (46)		29 June 16	12 July 16	29 June 17	Attd. 10 G. R.
	Unattd. List	***Tayler,** Skipwith Edward		29 June 16	5 July 16	29 June 17	39 G. R.
	Unattd. List	***Woodward,** Robert Geoffrey, M. C.		29 June 16	3 July 16	29 June 17	Attd. 52 Sikhs.
	Unattd. List	***Robinson,** James Arthur (51)		29 June 16	6 July 16	29 June 17	Attd. 83 Infy.
615	Unattd. List	***Boyton-Smith,** Edward Sydney (46)		29 June 16	3 July 16	29 June 17	Attd. 10 Jats.
616	Unattd. List	***Bannerman,** Halcro D'Arcy		29 June 16	3 July 16	29 June 17	Attd. 52 Sikhs.

617	Unattd. List	***deSalis,** Rudolph Albert	207a	29 June 16	2 July 16	29 June 17	4 Cavy.
	Unattd. List	***Playfair,** Harold Robert (57)		29 June 16	2 July 16	29 June 17	Attd. 87 Punjabis.
	Unattd. List	***Darell,** Edward Marmaduke Darell (66a)		29 June 16	2 July 16	29 June 17	Attd. 87 Punjabis.
620	Unattd. List	***Moorhead,** Henry Dawson	207	29 June 16	30 June 16	29 June 17	31 Punjabis.
	Unattd. List	***McCarthy,** Terence Cormac		29 June 16	3 July 16	29 June 17	6 Jats Lt. Infy.
	Unattd. List	***Taylor,** Arthur Gilbert Vivian		29 June 16	4 July 16	29 June 17	41 Dogras.
	Unattd. List	***Raynor,** Edmund Victor		29 June 16	10 July 16	29 June 17	Attd. 55 Rifles.
	Unattd. List	***Gatehouse,** Robert Oswald Arthur (71)		29 June 16	1 July 16	29 June 17	28 Cavy.
625	Unattd. List	***Allsebrook,** Maurice Bryer		29 June 16	3 July 16	29 June 17	Attd. 9 G. R.
	Unattd. List	***D'Oyly-Hughes,** Richard		29 June 16	4 July 16	29 June 17	Attd. 1 G. R.
	Unattd. List	***Tahourdin,** Raoul John Liddon		29 June 16	13 July 16	29 June 17	47 Sikhs.
	Unattd. List	***Scotland,** Alastair Graeme		29 June 16	5 July 16	29 June 17	36 Sikhs.
	Unattd. List	***Petrie,** James Francis (57, 172a)		29 June 16	3 July 16	21 June 17	Attd. 5 G. R.
630	Unattd. List	***Murray-Shirreff,** Bruce Godwin		29 June 16	17 July 16	29 June 17	Attd. 33 Punjabis.
	Unattd. List	***Eccles-Holmes,** James Francis		29 June 16	17 July 16	29 June 17	Attd. 33 Punjabis.
	Unattd. List	***Morris,** Henry Graham		29 June 16	3 July 16	29 June 17	56 Rifles.
	Unattd. List	***Bion,** Roy Harold	207	29 June 16	4 July 16	29 June 17	17 Cavalry.
	Unattd. List	***Hewett,** William George O'Connell (51)		29 June 16	2 July 16	29 June 17	48 Pioneers.
635	Unattd. List	***Hamilton,** Leslie Steuart		29 June 16	14 Aug. 16	29 June 17	Attd. 81 Pioneers.
	Unattd. List	***Dowling,** Dermot Michael (51)		29 June 16	3 July 16	29 June 17	Attd. 108 Infy.
	Unattd. List	***Revell,** Alfred John		29 June 16	3 July 16	29 June 17	61 Pioneers.
	Unattd. List	***Oatts,** Henry Augustus	207	20 June 16	4 July 16	29 June 17	41 Dogras.
	Unattd. List	***Miller,** George Reginald Bruce		29 June 16	1 July 16	29 June 17	2 Rajputs.
640	Unattd. List	***Smithers,** Harold Edgar		29 June 16	3 July 16	29 June 17	2 Grenadiers.
	Unattd. List	***Roper,** Richard Blaney		29 June 16	3 July 16	29 June 17	27 Punjabis.
	Unattd. List	***Acheson,** Ivan St. George (51)		29 June 16	4 July 16	29 June 17	Attd. 127 Infantry.
	Unattd. List	***Barker,** Denis		29 June 16	3 July 16	29 June 17	Attd. 72 Punjabis.
	Unattd. List	***Goddard,** Frederick Charles		29 June 16	3 July 16	29 June 17	114 Mahrattas.
645	Unattd. List	***Badman,** Francis Henry		29 June 16	3 July 16	29 June 17	Attd. 46 Punjabis.
	Unattd. List	***James,** Henry Osborne Blewett	207a	29 June 16	8 July 16	29 June 17	Attd. 84 Punjabis.
	Unattd. List	***Denehy,** Francis Patrick Gordon	207	29 June 16	9 July 16	29 June 17	Attd. 32 Lancers.
	Unattd. List	***Davys,** John Victor		29 June 16	8 July 16	20 June 17	Attd. 13 Rajputs.
	Unattd. List	***Saunders,** Claude Winstanley		29 June 16	3 July 16	29 June 17	Attd. 108 Infantry.
650	Unattd. List	***Perry,** Eric Gordon (57)		29 June 16	2 July 16	29 June 17	51 Sikhs.
	Unattd. List	***Wainwright,** John Gordon		29 June 16	2 July 16	29 June 17	Attd. Guides.
	Unattd. List	***deCourcy,** Gerald Lyle (66a)	206	29 June 16	2 July 16	29 June 17	51 Sikhs.
	Unattd. List	***Douglas,** Alexander Edward Ringham		29 June 16	2 July 16	29 June 17	Attd. 105 Infantry.
	Unattd. List	***Worlledge,** Leofric Henry		29 June 16	3 July 16	29 June 17	107 Pioneers.
655	Unattd. List	***Short,** John McLoughlin (51, 57)	207a	29 June 16	5 July 16	29 June 17	47 Sikhs.
	Unattd. List	***Gordon-Powell,** David Henry (46)		29 June 16	30 June 16	29 June 17	Attd. 106 Pioneers.
	Unattd. List	***Lang,** Joseph Harold, M. C.		29 June 16	1 July 16	29 June 17	Attd. 10 Rajputs.
	I. A. R. O.	***O'Gorman,** Maurice Kevin *Service for promotion from 1 July 16 under I. A. O. 510 of 1917.*		1 Oct. 15	30 July 18	1 July 17	Attd. 116 Mahrattas.
	M. G. Corps.	***Hill,** Harold, Victor John *9 months forfeited for seniority under conditions of I. A. O. 511 of 1917.*		1 July 16	26 June 18	1 July 17	Attd. 29 Punjabis.
660	North'd Fus.	***Frost,** Cyril Charles James (48a, 51) *9 months forfeited for seniority under conditions of I. A. O. 511 of 1917*	207a	4 July 16	12 June 18	4 July 17	Attd. 10 Gurkha Rifles.
661	Yorks. L. I.	***Duggan,** Ralph Desmond *9 months forfeited for seniority under conditions of I. A. O. 511 of 1917.*		4 July 16	6 May 18	4 July 17	Attd. 25 Punjabis.

* Has not passed the final examination.

Lieutenants—*contd.*

No.	Former Corps.	Names.	Date of birth.	Date of first commission.	Date of appointment to Indian Army.	Date of rank.	Remarks.
662	M. G. Corps	***Hunter,** James Herbert Douglas	. . . .	10 July 16	20 Aug. 18	6 July 17	Attd. 94 Infantry.
		9 months forfeited for seniority under conditions of I. A. O. 511 of 1917.					
	Durh. L. I.	***Corbett,** Vincent William	. . . .	14 July 16	6 May 18	14 July 17	„ 3 Gurkha Rifles.
		9 months forfeited for seniority under conditions of I. A. O. 511 of 1917.					
665	Unattd. List	§***Gilligan,** William Boyd (46)	. . . .	14 July 17	9 Dec. 17	14 July 17	„ 88 Infantry.
	Lond. R.	***Latham,** Harold Cecil	. . . .	15 July 16	29 July 18	15 July 17	„ 19 Punjabis.
		9 months forfeited for seniority under conditions of I. A. O. 511 of 1917.					
	Unattd. List	***Molesworth,** Richard Cecil	30 May 98	19 July 16	9 Nov. 16	19 July 17	„ 1 G. R.
	Unattd. List	***Attfield,** Charles John 207*a*	18 Apl. 98	19 July 16	8 Nov. 16	19 July 17	„ 24 Punjabis.
	Unattd. List	***Allen,** Cecil Arthur 207*a*	16 Aug. 97	19 July 16	12 Nov. 16	19 July 17	„ 42 Regiment.
	Unattd. List	***Abram,** Gilbert Faulkner	23 June 97	19 July 16	12 Nov. 16	19 July 17	„ 42 Regiment.
670	Unattd. List	***Cloran,** Dudley Joseph Hutchinson	19 May 98	19 July 16	9 Nov. 16	19 July 17	„ 81 Pioneers.
	Unattd. List	***Cassell,** Alfred Duncan	28 July 97	19 July 16	16 Nov. 16	19 July 17	„ 84 Punjabis.
	Unattd. List	***Lillies,** Geoffery Leonard	14 Mar. 98	19 July 16	15 Nov. 16	19 July 17	„ 83 Infantry.
	Unattd. List	***Drake,** James Arthur	9 Sep. 98	19 July 16	16 Nov. 16	19 July 17	„ 84 Punjabis.
	Unattd. List	***McConnell,** James Kerr	23 Jan. 95	19 July 16	8 Dec. 16	19 July 17	„ 74 Punjabis.
675	Unattd. List	***Lewis,** Ernest 207*a*	21 May 98	19 July 16	9 Nov. 16	19 July 17	„ 87 Punjabis.
	Unattd. List	***Jones,** Cyril Robert	29 July 97	19 July 16	16 Nov. 16	19 July 17	„ 84 Punjabis.
	Unattd. List	***Rogers,** Timothy James (51)	24 Oct. 97	19 July 16	11 Nov. 16	19 July 17	„ 89 Punjabis.
	Unattd. List	***Farwell,** Arthur Evelyn	3 July 98	19 July 16	11 Nov. 16	19 July 17	„ 15 Sikhs.
	Unattd. List	***Smithett,** William Hamilton	17 July 98	19 July 16	11 Nov. 16	19 July 17	„ 72 Punjabis.
680	Unattd. List	***Wilsey,** Frederick Herbert Willasey (51)	25 June 98	19 July 16	16 Nov. 16	19 July 17	„ 9 G. R.
	York. L. I.	**Collingwood,** Arthur Leslie	. . . .	21 July 16	20 Aug. 18	21 July 17	„ 117 Mahrattas.
		9 months forfeited for seniority under conditions of I. A. O. 511 of 1917.					
	Trng. Res. Bn.	***Martin,** William Milles	. . . .	21 July 16	22 Nov. 18	21 July 17	„ 38 C. I. Horse.
		9 months forfeited for seniority under conditions of I. A. O. 511 of 1917.					
	Leic. R.	***Morris,** Charles John 207*a*	. . . .	22 July 16	29 July 18	22 July 17	„ 9 Gurkha Rifles.
		9 months forfeited for seniority under conditions of I.A.O. 511 of 1917.					
	Unattd. List	***Vrooman,** Percy Ham (71) 207	12 Oct. 95	27 July 16	15 Nov. 16	27 July 17	„ 33 Cavalry.
685	Lond R.	***Walsh,** Maurice Campbell	. . . .	27 July 16	17 Oct. 18	27 July 17	„ 69 Punjabis.
		9 months forfeited for seniority under conditions of I. A. O. 511 of 1917.					
	I. A. R. O.	***House,** Edward Hector	. . . .	4 Nov. 15	14 July 17	4 Aug. 17	„ 6 Cavalry.
		Service for promotion from 4 Aug. 16 under I. A. O. 510 of 1917.					
	Suff. R.	***Peacock,** Leslie Vernon	. . . .	4 Aug. 16	30 July 18	4 Aug. 17	„ 46 Punjabis.
		9 months forfeited for seniority under conditions of I. A. O. 511 of 1917.					
688	Gord. Highrs.	* **Watt,** William Riach, M.C.	. . . .	7 Aug. 16	29 July 18	7 Aug. 17	„ 5 Gurkha Rifles.
		9 months forfeited for seniority under conditions of I. A. O. 511 of 1917.					

689	Lond. Yeo.	*Calger, Arthur Clarence Everett	. . . .	8 Aug. 16	31 Oct. 18	8 Aug. 17	Attd. 36 Horse.
		9 months forfeited for seniority under conditions of I. A. O. 511 of 1917.					
690	Gord. Highrs.	*Caruth, Alexander Caruth . . . 207	. . . .	11 Aug. 16	6 Apl. 18	11 Aug. 17	Attd. 7 Gurkha Rifles.
		9 months forfeited for seniority under conditions of I. A. O. 511 of 1917.					
	N. Lanc. R.	Bigger, Theodore, Anthony Lennox	. . . .	11 Aug. 16	22 Nov. 18	11 Aug. 17	Attd. 61 Pioneer.
		9 months forfeited for seniority under conditions of I. A. O. 511 of 1917.					
	I. A. R. O.	*Waller, Bertram Crichton (51, 52)	. . . .	12 Nov. 15	23 Sep. 15	12 Aug. 17	Attd. 36 Horse.
		Service for promotion from 12 Aug. 16 under I. A. O. 510 of 1917.					
	Unattd. List	*Holmes, George Leonard	17 Sep. 97	16 Aug. 16	11 Nov. 16	16 Aug. 17	Attd. 99 Infantry.
	Unattd. List	*Hughes, Francis Edmund Charles . . . 207a		16 Aug. 16	11 Nov. 16	16 Aug. 17	Attd. 4 G. R.
695	Unattd. List	*Ballentine, John Steventon . . . 207a	10 Mar. 97	16 Aug. 16	1 Nov. 16	16 Aug. 17	Attd. 17 Cavalry.
	Unattd. List	*Carter, James Langham (46)	18 Sep. 98	16 Aug. 16	6 Nov. 16	16 Aug. 17	Attd. 15 Sikhs.
	Unattd. List	*Wace, Michael Henry (51) . . . 207a	25 Oct. 96	16 Aug. 16	10 Nov. 16	16 Aug. 17	Attd. 127 Infantry.
	Unattd. List	*Hamlyn, Edwin Athelstan	8 Dec. 96	16 Aug. 16	10 Nov. 16	16 Aug. 17	Attd. 103 Mahrattas.
	Unattd. List	*Hunter, Hugh Campbell	2 Nov. 96	16 Aug. 16	11 Nov. 16	16 Aug. 17	Attd. 116 Mahrattas.
700	Unattd. List	*Schofield, Henry William Melles (46, 57) . . . 207a	6 Mar. 98	16 Aug. 16	16 Nov. 16	16 Aug. 17	Attd. 123 Rifles
	Unattd. List	*Sansom, Henry Jonathan	26 Mar. 97	16 Aug. 16	11 Nov. 16	16 Aug. 17	Attd. 123 Rifles.
	Unattd. List	*Barry, Jack Leslie (51)	25 Nov. 98	16 Aug. 16	11 Nov. 16	16 Aug. 17	Attd. 1 G. R.
	Unattd. List	*Simms, Geoffrey Howard	18 May 96	16 Aug. 16	10 Nov. 16	16 Aug. 17	Attd. 123 Rifles.
	Unattd. List	*Jones, Louis	17 Oct. 97	16 Aug. 16	10 Nov. 16	16 Aug. 17	Attd. 123 Rifles.
705	Unattd. List	*Parker, Vivian Lewes . . . 207a	20 Aug. 96	16 Aug. 16	11 Nov. 16	16 Aug. 17	Attd. 5 Cavalry.
	Unattd. List	*Windham, John Bayntun Russell . . . 207a	27 Aug. 97	16 Aug. 16	14 Nov. 16	16 Aug. 17	18 Lancers.
	Unattd. List	*Prendergast, Lancelot Claude (51)	9 Aug 98	16 Aug. 16	12 Nov. 16	16 Aug. 17	Attached 39 Garh. R.
	Unattd. List	§*Franklin, Henry James	. . . .	16 Aug. 16	1 Dec. 17	16 Aug. 17	,, 17th Infantry.
	E. Surr. R.	*Russell, John Gniles	. . . .	16 Aug. 16	22 Nov. 18	16 Aug. 17	Attd. 2 Infantry.
		9 months forfeited for seniority under conditions of I. A. O. 511 of 1917.					
710	D. of Corn L.I.	*Blanc, Louis Cerald	. . . .	22 Aug. 16	18 June 18	22 Aug. 17	Attd. 71 Punjabis.
		9 months forfeited for seniority under conditions of I. A. O. 511 of 1917.					
	Unattd. List	Rana Jodha Jang Bahadur, M.C., A.D.C., (95, 136)	5 Jan. 90	. . . .	25 Aug. 17	25 Aug. 17	3 Brahmans.
	Unattd. List	Kunwar Saval Sinhji	5 Oct. 90	. . . .	25 Aug. 17	25 Aug. 17	4 Rajputs.
	Shrops. L. I.	*Smith, Cecil Curtis Harold	. . . .	26 Aug. 16	6 May 18	26 Aug. 17	10 Jats.
		9 months forfeited for seniority under conditions of I A.O. 511 of 1917					
	R. Scots.	*Montgomerie, Lennox John	. . . .	28 Aug. 16	29 July 18	28 Aug. 17	Attached 54 Sikhs.
		9 months forfeited for seniority under conditions of I. A. O. 511 of 1917.					
715	I. A. R. O.	*Sydney-Smith, Dermot Crawford	. . . .	1 Dec. 15	16 Aug. 18	1 Sep. 17	Attd. 64 Pioneers.
		Service for promotion from 1 Sep. 1916 under I. A. O. 510 of 1917.					
	I. A. R. O.	*§Conyers-Brown, Horace Percival (46) (*temp. Capt. 2 Jan. 1918*)	. . . .	3 Dec. 15	5 Dec. 17	3 Sep. 17	Attd. 63 Infy.
		(Service for promotion from 3 Sep. 1916 under I.A.O. 510 of 1917)					
	Camn. Highrs.	*Marshall, John Dalrymple Calder	. . . .	3 Sep. 16	1 July 18	3 Sep. 7	Attd. 39 Garh Rifles.
		9 months forfeited for seniority under conditions of I. A. O. 511 of 1917.					
	Dorset R.	*Smith, Alan Langdale	. . . .	4 Sep. 16	29 July 18	4 Sep. 17	Attd. 63 Infantry.
		9 months forfeited for seniority under conditions of I. A. O. 511 of 1917.					
	S. Lanc. R.	*Rowson, John Corlett	. . . .	9 Sep. 16	31 May 18	9 Sep. 17	Attd. 27 Punjabis.
		9 months forfeited for seniority under conditions of I. A. O. 511 of 1917.					
720	I. A. R. O.	*Money, William Greenfield Mann	. . . .	10 Dec. 15	26 Sep. 18	10 Sep. 17	Attd. 101 Grenadiers.
		Service for promotion from 10 Sep. 1916 under I. A. O. 510 of 1917.					
	Unattd. List	*Champion, Harry Claude (47a, 51)	. . . .	†11 Sep. 16 / 18 June 17	22 June 17	11 Sep. 17	Attd. 7 Lancers.
722	R. War. R.	*Hall, Russell Wykeham Morgan (46)	. . . .	16 Sep. 16	16 May 18	16 Sep. 17	Attd. 29 Punjabis.
		9 months forfeited for seniority under conditions of I. A. O. 511 of 1917					

Lieutenants—*contd.*

No.	Former Corps.	Names.	Date of birth.	Date of first commission.	Date of appointment to Indian Army.	Date of rank.	Remarks.
723	Lan. Hrs.	*__Barraclough,__ Martin Wales Robinson 207*a* *9 months forfeited for seniority under conditions of I. A. O. 511 of 1917.*		25 Sep. 16	6 June 18	25 Sep. 17	7 Lancers.
	R. W. Fus.	*__Jones,__ Glyn Frank (46) 207*a* *9 months forfeited for seniority under conditions of I.A.O. 511 of 1917*		27 Sep. 16	6 May 18	27 Sep. 17	Attd. 7 Lancers.
725	I. A. R. O.	§*__Hassell-Wright,__ Edmund Louis Cecil *(Service for promotion from 1 Oct. 1916 under I.A.O. 510 of 1917)*		1 Jan. 16	2 Nov. 17	1 Oct. 17	Attd. 92 Punjabis.
	R. Scots	*__Smith,__ Reuben Godfrey *9 months forfeited for seniority under conditions of I. A. O. 511 of 1917.*		1 Oct. 16	16 Sep. 18	1 Oct. 17	Attd. 124 Infantry.
	Trng. Res. Bn.	*__Moore,__ Harry Leon *9 months forfeited for seniority under conditions of I. A. O. 511 of 1917.*		4 Oct. 16	1 Aug. 18	4 Oct. 17	Attd. 9 Delhi Infantry.
	Som. L. I.	*__Sheppard,__ Bertram Kingston *9 months forfeited for seniority under conditions of I. A. O. 511 of 1917.*		15 Oct. 16	7 Sep. 18	15 Oct. 17	Attd. 1 Gurkha Rifles.
	Welsh R.	*__Thomas,__ Kempson Frederick Welman *9 months forfeited for seniority under conditions of I. A. O. 511 of 1917.*		22 Oct. 16	15 Sep. 18	22 Oct. 17	Attd. 1 Gurkha Rifles.
730	Manch. R.	*__Harris,__ Alfred Andrew Fraser *9 months forfeited for seniority under conditions of I.A.O. 511 of 1917.*		25 Oct. 16	12 June 18	25 Oct. 17	Attd. 25 Punjabis.
	Unattd. List	§*__Cohen,__ Nicholas Cyril		27 Oct. 16	2 Jan. 17	27 Oct. 17	Attd. 76 Punjabis.
	Unattd. List	§*__Rogers,__ William Grainge 207*a*		27 Oct. 16	4 Aug. 17	27 Oct. 17	Attd. 13 Rajputs.
	Unattd. List	§*__Barker,__ Alan Robert (51) 207*a*		27 Oct. 16	25 Sep. 17	27 Oct. 17	Attd. 124 Infantry.
	Unattd. List	§*__Hill,__ Dayrell Reed Oakley (52)		27 Oct. 16	3 Jan. 17	27 Oct. 17	Attd. 1—7th Gurkha Rifles.
735	Unattd. List	§*__Alison,__ Thornton Fredrick Alexander Robert		27 Oct. 16	3 Jan. 17	27 Oct. 17	Attd. 110 Infantry.
	Unattd. List	§*__Martin,__ Horace James		27 Oct. 16	3 Jan. 17	27 Oct. 17	Attd. 110 Infantry.
	Unattd. List	§*__Henry,__ Maurice (51)		27 Oct. 16	3 Jan. 17	27 Oct. 17	Attd. 117 Mahrattas.
	Unattd. List	§*__Widdicombe,__ John Line Templer (46,52)		27 Oct. 16	4 Jan. 17	27 Oct. 17	Attd. 1—9th Gurkha Rifles.
	Unattd. List	§*__Linney,__ William Walter (46)		27 Oct. 16	3 Jan. 17	27 Oct. 17	Attd. 120 Infantry.
740	Unattd. List	§*__Morrison,__ Brian Harford 207*a*		27 Oct. 16	25 Sep. 17	27 Oct. 17	Attd. 124 Infantry.
	Unattd. List	§*__O'Sullivan,__ Eugene Thomas		27 Oct. 16	3 Jan. 17	27 Oct. 17	Attd. 66 Punjabis.
	Unattd. List	§*__Bullock,__ Humphry		27 Oct. 16	2 Jan. 17	27 Oct. 17	Attd. 95 Infantry.
	Unattd. List	§*__Baldwin,__ Patrick Guy Melfort		27 Oct. 16	11 Apl. 17	27 Oct. 17	Attd. 11 Lancers
	Sea. Highrs.	* __Spark,__ John Archibald *9 months forfeited for seniority under conditions of I. A. O. 511 of 1917.*		27 Oct. 16	7 July 18	27 Oct. 17	Attd. 19 Punjabis.
745	R. Scots.	* __Sowler,__ James. *9 months forfeited for seniority under conditions of I. A. O. 511 of 1917.*		29 Oct. 16	18 June 18	29 Oct. 17	Attd. 113 Infantry.
	R. Innis. Fus.	*__Strong,__ Cecil Alfred, M.C. *9 months forfeited for seniority under conditions of I. A. O. 511 of 1917.*		29 Oct. 16	1 Oct. 18	29 Oct. 17	Attd. 1 Gurkha Rifles.
747	Unattd. List	*__Boucher,__ Charles Hamilton (57)		14 Nov. 16	1 Dec. 16	14 Nov. 17	Attd. 2 Gurkha Rifles.

748	Unattd. List	*Beckett, Alfred	207		14 Nov. 16	20 Nov. 16	14 Nov. 17	Attd. 8 G. R.
	Unattd. List	*Harvey, Christopher			14 Nov. 16	26 Nov. 16	14 Nov. 17	„ 29 Lancers.
750	Unattd. List	*Robinson, Norman Purkiss			14 Nov. 16	22 Nov 16	14 Nov. 17	61 Pioneers.
	Unattd. List	*Searle, William Arthur Aitken			14 Nov. 16	8 Dec. 16	14 Nov. 17	101 Grenadiers.
	Unattd. List	*Elliott, James Gordon	207		14 Nov. 16	21 Nov. 16	14 Nov. 17	Attached 55 Rifles.
	Unattd. List	*Long, Denis Thorby			14 Nov. 16	26 Nov. 16	14 Nov. 17	„ 9 Horse.
	Unattd. List	*Castells, Edmunds (46)			14 Nov. 16	22 Nov. 16	14 Nov. 17	„ 6 G. R.
755	Unattd. List	*Burch, Philif Douglas			14 Nov. 16	24 Nov. 16	14 Nov. 17	„ 1 G. R.
	Unattd. List	*Latter, Frank Austen			14 Nov. 16	1 Dec. 16	14 Nov. 17	„ 6 Cavalry.
	Unattd. List	*Gastrell, Everard Huddleston			14 Nov. 16	25 Nov. 16	14 Nov. 17	56 Punj.
	Unattd. List	*Good, Sydney Barrett			14 Nov. 16	23 Nov. 16	14 Nov. 17	Attd. 35 Sikhs.
	Unattd. List	*Williamson, George Alexander (46)			14 Nov. 16	27 Nov. 16	14 Nov. 17	43 Regiment.
760	Unattd. List	*Wells, Andrew Milton	207a		14 Nov. 16	17 Nov. 16	14 Nov. 17	Attd. 127 Infantry.
	Unattd. List	*Wilson, John Finney Stanford			14 Nov. 16	21 Nov. 16	14 Nov. 17	20 Horse.
	Unattd. List	*Campbell, Cyril Maxwell St. Albans	207a		14 Nov. 16	22 Nov. 16	14 Nov. 17	Attd. 8 G. R.
	Unattd. List	*Dagge, Richard Canning (51, 57)	207a		14 Nov. 16	18 Nov. 16	14 Nov. 17	„ 103 Mahrattas.
	Unattd. List	§Colbourne, William Anthony			14 Nov. 16	22 Nov. 16	14 Nov. 17	„ 25 Punjabis.
765	Unattd. List	*Stewart, Cyril Elliot			14 Nov. 16	22 Nov. 16	14 Nov. 17	9 Infy.
	Unattd. List	*Pecker, Claude Bertram (71)			14 Nov. 15	29 Nov. 16	14 Nov. 17	Attd. 25 Cavy.
	Unattd. List	*Stroud, George Mannering			14 Nov. 16	23 Nov. 16	14 Nov. 17	„ 36 Horse.
	Unattd. List	*Acworth, John Pelham			14 Nov. 16	27 Nov. 16	14 Nov. 17	„ 34 Horse.
	Unattd. List	*Wilson-Haffenden, Eric Lawrence (46)			14 Nov. 16	22 Nov. 16	14 Nov. 17	„ 6 Cavalry.
770	Unattd. List	*Read, James	207	11 Jan. 98	14 Nov. 16	21 July 17	14 Nov. 17	„ 16 Cavalry.
	Unattd. List	*Shaw, Brian Van Dyke Havergal (*Ty. Capt., 8 May 17*)	207		14 Nov. 16	26 Nov. 16	14 Nov. 17	„ 107 Pioneers.
	Unattd. List	*Armstrong, John Alec Egerton			14 Nov. 16	21 Nov. 16	14 Nov. 17	„ 2 G. R.
	Unattd. List	*Davis, Gilbert George			14 Nov. 16	24 Nov. 16	14 Nov. 17	40 Pathans.
	Unattd. List	*Cox, Michael			14 Nov. 16	27 Nov. 16	14 Nov. 17	Attd. 38 Horse.
775	Unattd. List	*Monies, Vernon Alfred (57)	207a		14 Nov. 16	28 Nov. 16	14 Nov. 17	„ 67 Punj.
	Unattd. List	*Fry, George Cyril			14 Nov. 16	23 Nov. 16	14 Nov. 17	36 Horse.
	Unattd. List	*Walter, Edward Fraser			14 Nov. 16	27 Nov. 16	14 Nov. 17	44 Infy.
	Unattd. List	*Dodds, Harold Christopher			14 Nov. 16	27 Nov. 16	14 Nov. 17	43 Regt.
	Unattd. List	*Oddie, Bernard Ethelbert			14 Nov. 16	18 Nov. 16	14 Nov. 17	Attd. 6 Cavy.
780	Unattd. List	*Dormer, Ernest John			14 Nov. 16	18 Nov. 16	14 Nov. 17	„ 36 Horse.
	Unattd. List	*Dobson, Eric Trist			14 Nov. 16	8 Dec. 16	14 Nov. 17	Attd. 74 Punjabis.
	Unattd. List	*Murphy, James Roney Keogh			14 Nov. 16	25 Nov. 16	14 Nov. 17	„ 9 Horse.
	Unattd. List	*Howland, Cecil Claridge			14 Nov. 16	21 Nov. 16	14 Nov. 17	„ 34 Horse.
	Unattd. List	*Hemeon, Clarence Reid	207		14 Nov. 16	27 Nov. 16	14 Nov. 17	44 Infy.
785	Unattd. List	*Fisher, Francis Arthur Braddon			14 Nov. 16	24 Nov. 16	14 Nov. 17	Attd. 5 G. R.
	Unattd. List	*Watkins, Frederick William Springett			14 Nov. 16	23 Nov. 16	14 Nov. 17	„ 36 Horse.
	Unattd. List	*Corner, Richard James	207a		14 Nov. 16	21 Nov. 16	14 Nov. 17	„ 31 Lancers.
	Unattd. List	*Chetwynd-Palmer, Edgar Richard (51, 57)	207a		14 Nov. 16	26 Feb. 17	14 Nov. 17	„ 89 Punj.
	Unattd. List	*Jewell, John Witherow (71)	207a		14 Nov. 16	24 Nov. 16	14 Nov. 17	„ 25 Cavy.
790	Unattd. List	*Woollcombe, Geoffrey Harley Douglas			14 Nov. 16	21 Nov. 16	14 Nov. 17	„ 2 G. R.
	Unattd. List	*Vosper, Eric Essington			14 Nov. 16	24 Nov. 16	14 Nov. 17	36 Horse.
	Unattd. List	*Paton, Herbert Gordon			14 Nov. 16	26 Nov. 16	14 Nov. 17	Attd. 126 Infy.
	Unattd. List	*Berkeley, Stewart Sackville Hamilton (46)	207a		14 Nov. 16	20 Nov. 16	14 Nov. 17	„ 8 G. R.
	Unattd. List	*Humfrey, Henry Brabazon			14 Nov. 16	21 Nov. 16	14 Nov. 17	20 Horse.
795	Unattd. List	*Cousens, Richard Purdon			14 Nov. 16	23 Nov. 16	14 Nov. 17	Attd. 38 C. I. Horse.
796	Unattd. List	*Cartlidge, Cecil			14 Nov. 16	21 Nov. 16	14 Nov. 17	„ 19 Lancers.

* Has not passed the final examination. † The date from which seniority counts. § Admitted subject to H. M.'s approval.

Lieutenants—*contd.*

No.	Former Corps.	Names.	Date of birth.	Date of first Commission.	Date of appointment of Indian Army.	Date of rank.	Remarks.
797	Unattd. List	***Lewis**, Percy Herbert		14 Nov. 16	21 Nov. 16	14 Nov. 17	93 Infy.
	Unattd. List	***Cashmore**, Albert Edward		14 Nov. 16	21 Nov. 16	14 Nov 17	29 Lancers.
	Unattd. List	***Pigot**, Harold		14 Nov. 16	26 Nov. 16	14 Nov. 17	Attd. 38 C. I. Horse.
800	Unattd. List	***Walker**, John Eric		14 Nov. 16	29 Nov. 16	14 Nov. 17	,, 9 Horse.
	Unattd. List	***Rennick**, Guy Barrett (57)		14 Nov. 16	21 Nov. 16	14 Nov. 17	89 Punj.
	Unattd. List	***George**, Richard		14 Nov. 16	24 Nov. 16	14 Nov. 17	Attd. 38 C. I. Horse.
	Unattd. List	***Hartigan**, Guy Edward Ross Stewart (51, 52)		14 Nov. 16	21 Nov. 16	14 Nov. 17	,, 2 G. R.
	Unattd. List	***Bloomfield**, Fitz-Maurice John Commerell 207*a*		14 Nov. 16	21 Nov. 16	14 Nov. 17	,, 1 Lancers.
805	Unattd. List	***Maffett**, Desmond Bloomfield (57) 207*a*		14 Nov. 16	4 Dec. 16	14 Nov. 17	,, 29 Punj.
	Unattd. List	***Ryan**, Beaumont Astle (51)		14 Nov. 16	24 Nov. 16	14 Nov. 17	,, 4 G. R.
	Unattd. List	***Williams**, Basil Francis		14 Nov. 16	30 Nov. 16	14 Nov. 17	38 C. I. Horse.
	Unattd. List	***Murray**, Thomas William		14 Nov. 16	24 Nov. 16	14 Nov. 17	117 Mahrattas.
	Unattd. List	***Mellor**, Leycester Andrew St. John		14 Nov. 16	30 Nov. 16	14 Nov. 17	98 Infy.
810	Unattd. List	***Frith**, James Wiltshire (46, 57)		14 Nov. 16	22 Nov. 16	14 Nov. 17	Attd. 8 G. R.
	Unattd. List	***Woolley**, John Harold Crown		14 Nov. 16	24 Nov. 16	14 Nov. 17	,, 19 Lancers.
	Unattd. List	***Gordon**, Stuart Hamilton 207*a*		14 Nov. 16	24 Nov. 16	14 Nov. 17	,, 99 Infy.
	Unattd. List	***Race**, Terence Thomas George		14 Nov. 16	8 Dec. 16	14 Nov. 17	101 Grenadiers.
	Unattd. List	***Harris**, Charles Edwin Laurence		14 Nov. 16	24 Nov. 16	14 Nov. 17	Attd. 2 Lancers.
815	R. Ir. Regt.	*** McCutcheon**, George Henry 207*a* *9 months forfeited for seniority under conditions of I. A. O. 511 of 1917*		18 Nov. 16	30 July 18	18 Nov. 17	Attd. 56 Rifles.
	I. A. R. O.	**§*Morris**, Horace Stanley, M.C. (51, 52) *(Service for promotion from 20 Nov. 16 under I. A. O. 510 of 1917.)*		20 Feb. 16	2 Nov. 17	20 Nov. 17	92 Punjabis.
	I. A. R. O.	**§* Hawksworth**, Percy (51) *Service for promotion from 20 Nov. 16 under I. A. O. 510 of 1917.*		20 Feb. 16	26 Apl. 18	20 Nov. 17	Attd. 44 Infantry.
	I. A. R. O.	**§*Rodwell**, Lindsay Hassell *Service for promotion from 26 Nov. 16 under I. A. O. 510 of 1917*		26 Feb. 16	4 Apl. 18	26 Nov. 17	,, 9 Bhopal Infantry.
	A. S. C.	***Poole**, Esmond Randolph Luscombe *9 months forfeited for seniority under conditions of I. A. O. 511 of 1917.*		29 Nov. 16	12 June 18	29 Nov. 17	Attd. 31 Lancers.
820	I. A. R. O.	**§* O'Leary**, Maurice Patrick (51) *Service for promotion from 25 Dec. 16 under I. A. O. 510 of 1917.*		25 Mar. 16	26 Apl. 18	25 Dec. 17	Attd. 122 Infantry.
	I. A. R. O.	**§*Cayley**, Philip Eastcourt (51) *Service for promotion from 25 Dec 16 under I. A. O. 510 of 1917.*		25 Mar. 16	16 Aug. 18	25 Dec. 17	Attd. 5 Gurkha Rifles.
822	S. Afr. Force	***English**, Frederick William Hawthorne, M. C. *9 months forfeited for seniority under conditions of I. A. O. 511 of 1917.*		25 Dec. 16	5 Oct. 18	25 Dec. 17	Attd. 2nd Gurkha Rifles.

823	I. A. R. O.	*§**Hargreaves**, Patrick Shiels		30 Mar. 16	4 Apl. 18	30 Dec. 17	
		Service for promotion from 30 Dec. 16 under I. A. O. 510 of 1917					
	Unattd. List	***Craig**, Samuel Maurice Mateer (57)		29 June 16	1 July 16	1 Jan. 18	Attd. 123 Rifles.
825	Worc. R.	***Knowles, Williams**, Scott., M.C.		23 Jan. 17	22 Nov. 18	23 Jan. 18	Attd. 1 Gurkha Rifles.
		9 months forfeited for seniority under conditions of I. A. O. 511 of 1917.					
	Unattd. List	§***Dalrymple-Hay**, Brian George		28 Jan. 17	6 Nov. 17	28 Jan. 18	
	Unattd. List	***Lillie**, Cecil James Wingate	207a	30 Jan. 17	3 Feb. 17	30 Jan. 18	,, 4 G. R.
	Unattd. List	***Petrie**, Arthur Gordon		30 Jan. 17	14 Feb. 17	30 Jan. 18	
	Unattd. List	***Bendle**, Maurice Scott		30 Jan. 17	14 Feb. 17	30 Jan. 18	
830	Unattd. List	***Schofield**, Alfred John Albert		30 Jan. 17	14 Feb. 17	30 Jan. 18	
	Unattd. List	***King**, Rodney Stewart		30 Jan. 17	14 Feb. 17	30 Jan. 18	
	Unattd. List	***Ryan**, Cecil Crawford Lindsay		30 Jan. 17	14 Feb. 17	30 Jan. 18	Attd. 20 Horse.
	Unattd. List	***Clarke**, Thomas Veitch	207a	30 Jan. 17	14 Feb. 17	30 Jan. 18	
	Unattd. List	***Hopkins**, William John		30 Jan. 17	14 Feb. 17	30 Jan. 18	
835	Unattd. List	***Lindsay**, Douglas Mollison		30 Jan. 17	3 Feb. 17	30 Jan. 18	59 Rifles.
	Unattd. List	***Wakeham**, Charles Eric		30 Jan. 17	14 Feb. 17	30 Jan. 18	
	Unattd. List	***Butler**, Gervase Rigby	207a	30 Jan. 17	5 Feb. 17	30 Jan. 18	Attd. 29 Punj.
	Unattd. List	***Keightley**, Maurice Frederick		30 Jan. 17	14 Feb. 17	30 Jan. 18	
	Unattd. List	***Robertson**, Harold Les Cunynghame		30 Jan. 17	3 Feb. 17	30 Jan. 18	26 Punjabis.
840	Unattd. List	***Joyce**, Alfred Henry Bertram, M. C.	207a	30 Jan. 17	10 Feb. 1	30 Jan. 18	62 Punjabis.
	Unattd. List	***Griffin**, Gerald William		30 Jan. 17	14 Feb. 17	30 Jan. 18	Attd 6 Cavalry.
	Unattd. List	***Abernethy**, Peter Paul	207a	30 Jan. 17	7 Feb. 17	30 Jan. 18	Attd. 63 Infy.
	Unattd. List	***Pollitt**, Herbet Tomlin		30 Jan. 17	14 Feb. 17	30 Jan. 18	
	Unattd. List	***Caunter**, Henry Edmund	207a	30 Jan. 17	3 Feb. 17	30 Jan. 18	40 Pathans.
845	Unattd. List	***Levingston**, Thomas Clarke		30 Jan. 17	7 Feb. 17	30 Jan. 18	119 Infantry.
	Unattd. List	***Sandeman**, Ronald Frederick		30 Jan. 17	14 Feb. 17	30 Jan. 18	
	Unattd. List	***Cumming**, Richard Hugh Rossiter		30 Jan. 17	14 Feb. 17	30 Jan. 18	
	Unattd. List	***Galbraith**, Ian William, M. C.		30 Jan. 17	14 Feb. 17	30 Jan. 18	
	Unattd. List	***Edmondson**, Brian Edward Wickham		30 Jan. 17	14 Feb. 17	30 Jan. 18	
850	Unattd. List	***Sawers**, Charles Maxwell Orr		30 Jan. 17	14 Feb. 17	30 Jan. 18	
	Unattd. List	***Meyer**, Bertie (51). (*temp. Capt., 1 Sep. 18*).		30 Jan. 17	1 Feb. 17	30 Jan. 18	39 Garh. R.
	Unattd. List	***Russell**, Alexander Alfred (52)		30 Jan. 17	5 Feb. 17	30 Jan. 18	43 Regt.
	Unattd. List	***Bowers**, Thomas Ivan		30 Jan. 17	4 Feb. 17	30 Jan. 18	11 Rajputs.
	Unattd. List	***White**, Douglas Onslow		30 Jan. 17	20 Feb. 17	30 Jan. 18	3 Brahmans.
855	Unattd. List	***Benzie**, James Henry Swan		30 Jan. 17	14 Feb. 17	30 Jan. 18	
	Unattd. List	***Hodgson**, Harold Christopher		30 Jan. 17	14 Feb. 17	30 Jan. 18	
	Unattd. List	***Armitage**, Aubrey Seymour		30 Jan. 17	10 Feb. 17	30 Jan. 18	Attd. 3 G. R.
	Unattd. List	***McCaughey**, Frederick George		30 Jan. 17	3 Feb. 17	30 Jan. 18	Attd. 74 Punj.
	Unattd. List	***Grove**, Geoffrey Reginald (51)		30 Jan. 17	5 Feb. 17	30 Jan. 18	6 G. R.
860	Unattd. List	***Spurway**, William Richard Pyne		30 Jan. 17	5 Feb. 17	30 Jan. 18	Attd. Guides.
	Unattd. List	***Hallifax**, John Staniford		30 Jan. 17	14 Feb. 17	30 Jan. 18	
	Unattd. List	***Allen**, Frederick Joshua		30 Jan. 17	14 Feb. 17	30 Jan. 18	
	Unattd. List	***Moss**, Charles Henry de Vere		30 Jan. 17	14 Feb. 17	30 Jan. 18	
	Unattd. List	***Ennis**, John Ignatius		30 Jan. 17	10 Feb. 17	30 Jan. 18	53 Sikhs.
865	Unattd. List	***Pierce**, Ian Gilbert Fielding		30 Jan. 17	9 Feb. 17	30 Jan. 18	Attd. 10 G. R.
	Unattd. List	***Deans**, Theodore John (57)	207a	30 Jan. 17	14 Feb. 17	30 Jan. 18	
	Unattd. List	***Spencer**, Egerton Charles		30 Jan. 17	3 Feb. 17	30 Jan. 18	92 Punj.
	Unattd. List	***McNeill**, Douglas Sidney Elliott		30 Jan. 17	14 Feb. 17	30 Jan. 18	45 Cavalry.
	Unattd. List	***Farrar**, William Brookes	207a	30 Jan. 17	3 Feb. 17	30 Jan. 18	5 Infy.
870	Unattd. List	***Sykes**, Walter David		30 Jan. 17	14 Feb. 17	30 Jan. 18	

* Has not passed the final examination. § Admitted subject to H. M.'s approval.

Lieutenants—*contd.*

No.	Former Corps.	Names.	Date of birth.	Date of first commission.	Date of appointment to Indian Army.	Date of rank.	Remarks.
871	Unattd. List	***Whelan**, Herbert Clive		30 Jan. 17	14 Feb. 17	30 Jan. 18	
	Unattd. List	***Pocock**, Nicholas Edgar Innes (46)		30 Jan. 17	1 Feb. 17	30 Jan. 18	4 Rajputs.
	Unattd. List	***O'Connor**, Michael Joseph		30 Jan. 17	3 Feb. 17	30 Jan. 18	6 Jats.
	Unattd. List	***Flynn**, Terence Francis Axen		30 Jan. 17	14 Feb. 17	30 Jan. 18	
875	Unattd. List	***Lawler**, Guy Feinaigle		30 Jan. 17	14 Feb. 17	30 Jan. 18	
	Unattd. List	***Holmes**, Richard Eric James (57) 207		30 Jan. 17	1 Feb. 17	30 Jan. 18	Attd. 39 Garh. R.
	Unattd. List	***Whittick**, Humphrey Donald		30 Jan. 17	14 Feb. 17	30 Jan. 18	
	Unattd. List	***Wheeler**, Gerald Arden Savill d'Erf		30 Jan. 17	14 Feb. 17	30 Jan. 18	
	Unattd. List	***Church**, Charles John Bromley (51)		30 Jan. 17	3 Feb. 17	30 Jan. 18	14 Sikhs.
880	Unattd. List	***Boxall**, Maurice William James (46) *207a*		30 Jan. 17	14 Feb. 17	30 Jan. 18	28 Punj.
	Unattd. List	***Bullock**, James Charles Critchen		30 Jan. 17	14 Feb. 17	30 Jan. 18	
	Unattd. List	***Burch**, Allen Paton		30 Jan. 17	15 Feb. 17	30 Jan. 18	Attd. 67 Punj.
	Unattd. List	***Pratt**, Arthur Reginald *207a*		30 Jan. 17	4 Feb. 17	30 Jan. 18	130 Baluchis.
	Unattd. List	***Fleming**, Robert Stevenson Tristram *207a*		30 Jan. 17	3 Feb. 17	30 Jan. 18	93 Infy.
885	Unattd. List	***Telford**, Brian Desmond *207a*		30 Jan. 17	3 Feb. 17	30 Jan. 18	93 Infy.
	Unattd. List	***Gibbs**, Egbert Willis Burnett		30 Jan. 17	4 Feb. 17	30 Jan. 18	Attd. 113 Mah. L. I.
	Unattd. List	***Hawkes**, Thomas Yerrow *207a*		30 Jan. 17	14 Feb. 17	30 Jan. 18	
	Unattd. List	***Gerrard**, Robert Joesph		30 Jan. 17	3 Feb. 17	30 Jan. 18	33 Punj.
	Unattd. List	***Stock**, Sholto Merrett		30 Jan. 17	7 Feb. 17	30 Jan. 18	Attd. 120 Infy.
890	Unattd. List	***Lane**, Philip Gerald Cambridge		30 Jan. 17	14 Feb. 17	30 Jan. 18	
	Unattd. List	***Wyrill**, Leslie Moore		30 Jan. 17	14 Feb. 17	30 Jan. 18	
	Unattd. List	***Young**, William Alfred Leonard (46)		30 Jan. 17	3 Feb. 17	30 Jan. 18	Attd. 98 Infy.
	Unattd. List	***Begg**, Patrick Hunter *207a*		30 Jan. 17	4 Feb. 17	30 Jan. 18	Attd. 110 Mahrattas.
	Unattd. List	***Curtin**, Arthur Frederick Stanley (51)		30 Jan. 17	5 Feb. 17	30 Jan. 18	44 Infy.
895	Unattd. List	***Evans**, Norman William		30 Jan. 17	3 Feb. 17	30 Jan. 18	47 Sikhs.
	Unattd. List	***Ellison**, Edward Courtenay (57)		30 Jan. 17	10 Feb. 17	30 Jan. 18	62 Punj.
	Unattd. List	***Moysey**, Reginald Abel Lewis *207a*		30 Jan. 17	1 Feb. 17	30 Jan. 18	22 Punj.
	Unattd. List	***Daldy**, Arthur John *207a*		30 Jan. 17	3 Feb. 17	30 Jan. 18	61 Pioneers.
	Unattd. List	***Foster**, Lionel Alfred		30 Jan. 17	8 Feb. 17	30 Jan. 18	1 G. R.
900	Unattd. List	***Fforde**, John Stuart Beresford		30 Jan. 17	7 Feb. 17	30 Jan. 18	119 Infy.
	Unattd. List	***Craig**, Gordon		30 Jan. 17	14 Feb. 17	30 Jan. 18	
	Unattd. List	***Carberry**, Thomas Bernard		30 Jan. 17	9 Feb. 17	30 Jan. 18	Attd. 117 Mahrattas.
	Unattd. List	***McDonnell**, Francis Alexander *207a*		30 Jan. 17	3 Feb. 17	30 Jan. 18	5 Infy.
	Unattd. List	***Cawood**, Cyril Walter (51) *107a*		30 Jan. 17	6 Feb. 17	30 Jan. 18	Attd. 80 Infy.
905	Unattd. List	***Smith**, Edward Travers		30 Jan. 17	3 Feb. 17	30 Jan. 18	Attd. 9 G. R.
906	Unattd. List	***Marshall**, Thomas Ralph		30 Jan. 17	14 Feb. 17	30 Jan. 18	Attd. 38 Horse.

907	Unattd. List	***Short**, Vere Charles Gurdon O'Borthistle	30 Jan. 17	19 Feb. 17	30 Jan. 18	2 Rajputs.
	Unattd. List	***Martin**, Stanley Frederic	30 Jan. 17	13 Feb. 17	30 Jan. 18	108 Infy.
	Unattd. List	***O'Sullivan**, Hugh Patrick	30 Jan. 17	16 Mar. 17	30 Jan. 18	3 Brahmans.
910	Unattd. List	***Frost**, Robert Lewis	30 Jan. 17	15 Feb. 17	30 Jan. 18	69 Punj.
	Unattd. List	***Edgcome**, George Edward Rupert	30 Jan. 17	14 Feb. 17	30 Jan. 18	
	Unattd. List	***Pottinger**, Eric Guy	30 Jan. 17	14 Feb. 17	30 Jan. 18	
	Unattd. List	***Dawson**, Arthur John Eardley	30 Jan. 17	9 Feb. 17	30 Jan. 18	Attd. 117 Mahrattas.
	Unattd. List	***Kilroy**, Alan James Minchin 207a	30 Jan. 17	7 Feb. 17	30 Jan. 18	122 Infy.
915	Unattd. List	***Macalister**, Sidney Cairnes	30 Jan. 17	3 Feb. 17	30 Jan. 18	105 Infy.
	Unattd. List	***Robinson**, Frederick Roland Henry	30 Jan. 17	14 Feb. 17	30 Jan. 18	
	Unattd. List	***Blakeney**, Stephen Richard St. Leger (57)	30 Jan. 17	10 Feb. 17	30 Jan. 18	82 Punj.
	Unattd. List	***Hadaway**, John Carre Snelling	30 Jan. 17	11 Feb. 17	30 Jan. 18	20 Infy.
	Unattd. List	***Silver**, George James	30 Jan. 17	14 Feb. 17	30 Jan. 18	
920	Unattd. List	***King**, Ernest St. John	30 Jan. 17	14 Feb. 17	30 Jan. 18	
	Unattd. List	***Cleave**, John Clive Tyler 207a	30 Jan. 17	11 Feb. 17	30 Jan. 18	7 Rajputs.
	Unattd. List	***Chamberlain**, Lindsay Llewellyn	30 Jan. 17	12 Feb. 17	30 Jan. 18	27 Punj.
	Unattd. List	***Macdonald**, Norman	30 Jan. 17	4 Feb. 17	30 Jan. 18	Attd. 41 Dogras.
	Unattd. List	***Wilkinson**, Joseph Harold	30 Jan. 17	14 Feb. 17	30 Jan. 18	
925	Unattd. List	***Hodge**, John Furber	30 Jan. 17	3 Feb. 17	30 Jan. 18	125 Rifles.
	I. A. R. O.	§***Hart**, Neville Vincent (46, 95) *Service for promotion from 10 Feb. 1917 under I. A. O. 510 of 1917.*	10 May 16	26 Apl. 18	10 Feb. 18	Attd. 19 Punjabis.
	I. A. R. O.	§***Long**, Walt r Herbert *Service for promotion from 10 Feb. 17 under I. A. O. 510 of 1917.*	10 May 16	16 Aug. 18	10 Feb. 18	Attd. 49 Bengalis.
	I. A. R. O.	§***Osborne**, Harold (51) *Service for promotion from 10 Feb. 1917 under I. A. O. 510 of 1917.*	10 May 16	26 Apl. 18	10 Feb. 18	Attd. 102 Grenadiers.
	I. A. R. O.	§***Gordon**, Francis Leslie, (57) *Service for promotion from 10 Feb. 17 under I. A. O. 510 of 1917.*	10 May 16	30 July 18	10 Feb. 18	Attd. Sig. Service.
930	I. A. R. O.	§***Judge**, Charles Edward Miller, M.B.E. *Service for promotion from 15 Feb. 17 under I. A. O. 510 of 1917*	15 May 16	21 Sep. 17	15 Feb. 18	Attd. S. and T. Corps.
	E. Surr. R.	***Pattinson**, Howard Alan, M.C. *9 months forfeited for seniority under conditions of I. A. O. 511 of 1917.*	9 Mar. 17	9 Aug. 18	9 Mar. 18	Attd. 9 Delhi Infantry.
	I. A. R. O.	§***Ingle**, Louis Sobaux *Service for promotion from 10 Mar. 1917 under I. A. O. 510 of 1917.*	10 June 16	26 Apl. 18	10 Mar. 18	Attd. 19 Punjabis.
	I. A. R. O.	§***Nicholl**, Edwin Mckillop *Service for promotion from 15 Mar. 17 under I. A. O. 510 of 1917*	15 June 16	15 Aug. 18	15 Mar. 18	Attd. 14 Lancers.
	I. A. R. O.	§***Sanderson**, Patrick Claud *Service for promotion from 17 Mar. 1917 under I. A. O. 510 of 1917.*	17 June 16	26 Apl. 18	17 Mar. 18	Attd. 1-4th Gurkha Rifles.
935	I. A. R. O.	§***Hennessey**, Dudley Michael Leslie 207 *Service for promotion from 27 Mar. 1917 under I. A. O. 510 of 1917.*	27 June 16	26 Apl. 18	27 Mar. 18	Attd. 9 Infantry.
	Unattd. List	§***Hewett**, James Pearce Bowen O'Connel	1 Apl. 17	5 Nov. 17	1 Apl. 18	Attd. 37 Dogras.
937	Dorset. R.	***Dawkes**, Robert Bolton *9 months forfeited for seniority under conditions of I. A. O. 511 of 1917.*	7 Apl. 17	16 Apl. 18	7 Apl. 18	Attd. 89 Punjabis.

* Has not passed the final examination.

§ Admitted subject to H. M.'s approval.

Lieutenants—*contd.*

No.	Former Corps.	Names.	Date of birth.	Date of first Commission.	Date of appointment to Indian Army.	Date of rank.	Remarks.
938	Dorset. R.	***Bullock**, Herbert Poe Story *9 months forfeited for seniority under conditions of I. A. O. 511 of 1917.*		7 Apl. 17	16 Apl. 18	7 Apl. 18	Attd. 89 Punjabis.
	W. York. R.	***Fell**, Alfred Lakeland *9 months forfeited for seniority under conditions of I. A. O. 511 of 1917.*		11 Apl. 17	24 Sep. 18	11 Apl. 18	Attd. 2 Gurkha Rifles.
940	S. Wales Bord.	***Brigham**, Guy *9 months forfeited for seniority under conditions of I. A. O. 511 of 1917.*		19 Apl. 17	9 Aug. 18	19 Apl. 18	Attd. 95 Infantry.
	Lond. Yeo.	***Harland**, Frederick William *9 months forfeited for seniority under conditions of I. A. O. 511 of 1917.*		29 Apl. 17	3 Apl. 18	29 Apl. 18	Attd. 4 Cavalry.
	Suff. Yeo	***Hudson**, Thompson (57) 207a *9 months forfeited for seniority under conditions of I. A. O. 511 of 1917.*		29 Apl. 17	6 May 18	29 Apl. 18	31 Lancers.
	Unattd. List	§***Percy-Smith**, Hubert Kendall		1 May 17	30 Nov. 17	1 May 18	Attd. 40 Pathans.
	Unattd. List	§***Milne**, Oswald John Tattersall 207a		1 May 7	30 Nov. 17	1 May 18	,, 18 Lancers.
945	Unattd. List	§***Dulcken**, Henry Ward		1 May 17	2 Dec. 1	1 May 18	,, 19 Lancers.
	Unattd. List	§***Bellamy**, Lewis Charles Montagu (46, 61)		1 May 17	28 Nov. 17	1 May 18	,, 116 Mahrattas.
	Unattd. List	§***Stephen**, Alexander Thomas		1 May 17	30 Nov. 17	1 May 18	,, 56 Rifles.
	Unattd. List	§***Craft**, Henry Taylor		1 May 17	2 Dec. 17	1 May 18	,, 120 Infantry.
	Unattd. List	§***Stanley**, Douglas Richmond		1 May 17	24 Dec. 17	1 May 18	,, 69 Punjabis.
950	Unattd. List	§***Gault**, Cecil (48a, 51) 207a		1 May 17	29 Nov. 17	1 May 18	,, 24 Punjabis.
	Unattd. List	§***Piper**, Walter John Fletcher		1 May 17	2 Dec. 17	1 May 18	,, 41 Dogras.
	Unattd. List	§***Orr**, John Henry Reginald		1 May 17	2 Dec. 17	1 May 18	,, 39 Garhwal Rifles.
	Unattd. List	§***Carter**, William Thomas		1 May 17	2 Dec. 17	1 May 18	,, 120 Infantry.
	Unattd. List	§***Leake**, Jonus Inchiquin		1 May 17	1 Dec. 17	1 May 18	,, 32 Pioneers.
955	Unattd. List	§***Orpen**, Gerald Erpingham (46)		1 May 17	5 Dec. 17	1 May 18	,, 39 Garhwal Rifles.
	Unattd. List	§***Rayne**, Percy Claude		1 May 17	4 Dec. 17	1 May 18	,, 47 Sikhs.
	Unattd. List	§***Croucher**, Francis Edward		1 May 17	30 Nov. 17	1 May 18	,, 9 Hodson's Horse.
	Unattd. List	§***Stiven**, Robert		1 May 17	1 Dec. 17	1 May 18	,, 66 Punjabis.
	Unattd. List	§***Brown**, Geoffrey Record		1 May 17	13 Dec. 17	1 May 18	,, 7 Gurkha Rifles.
960	Unattd. List	§**Evans**, John Emrys		1 May 17	1 Dec. 17	1 May 16	,, 18 Lancers.
961	Yorks. L. I.	***Beadon**, William a'Court 207a *9 months forfeited for seniority under conditions of I. A. O. 511 of 1917.*		5 May 17	1 July 18	5 May 18	Attd. 9 Infantry.

962	I. A. R. O.	§*Stubbs, Harvey Goodwin Pattinson *Service for promotion from 25 May 17 under I. A. O. 510 of 1917.*		25 Aug. 16	4 Apl. 18	25 May 18	Attd. 24 Punjabis.
	I. A. R. O.	§*Slade, Joseph William Carter *Service for promotion from 28 May 7 under I. A. O. 510 of 1917.*		28 Aug. 16	16 Aug. 18	28 May 18	Attd. 101 Grenadiers.
	E. Lanc. R.	*Ballance, Samuel Henry *0 months forfeited for seniority under conditions of I. A. O. 511 of 1917.*		5 June 17	14 Sep. 18	5 June 18	Attd. 2 Gurkha Rifles.
965	M. G. Corps	*Chapman, Charles Albert *9 months for feited for seniority under conditions of I. A. O. 511 of 1917.*		5 June 17	25 Sep. 18	5 June 18	Attd. 71 Punjabis.
	Unattd. List	*Kelly, Vivian Stanisland		30 Jan. 17	17 Feb. 17	18 June 18	130 Baluchis.
	Unattd. List	*Hallstone, Geoffery Ronald	207a	18 June 17	27 June 17	18 June 18	Attd. 8 G. R.
	Unattd. List	*Holworthy, Arthur Wilmot Wadeson (46)		18 June 17	29 June 17	18 June 1	1 G. R.
	Unattd. List	*Fox-Strangways, Vivian (48, 51)		18 June 17	27 June 17	18 June 18	Attd. Corps of Guides.
970	Unattd. List	*Macmillan, Donald MacLachlan (57)	207	18 June 17	25 June 17	18 June 18	34 Pioneers.
	Unattd. List	*Carter, Reginald Francis Fernie		18 June 17	25 June 17	18 June 18	15 Sikhs.
	Unattd. List	*Iliff, William Angus Boyd		18 June 17	·7 June 17	18 June 18	57 Rifles.
	Unattd. List	*Jackman, Herbert Ralph	207a	18 June 17	25 June 17	18 June 18	Attd. 46 Punjabis.
	Unattd. List	*Graham, John Park		18 June 17	25 June 17	18 June 18	Attd. 52 Sikhs.
975	Unattd. List	*Pouncey, George Ronald	207a	18 June 17	4 July 17	18 June 18	,, 21 Punjabis.
	Unattd. List	*Belben, Thomas Alan (46)	207a	18 June 17	29 June 17	18 June 18	,, 123 Rifles.
	Unattd. List	*Pert, Claude Ernest	207a	18 June 17	23 June 17	2· June 18	,, 17 Cavalry.
	Unattd. List	*Preston, George Allan	207a	18 June 17	25 June 17	18 June 18	41 Dogras.
	Unattd. List	*Crichton, Terence Claud		18 June 17	25 June 17	18 June 18	Attd. 5 Cavalry.
980	Unattd. List	*Sandham, William Frederick Norman		18 June 17	29 June 17	18 June 18	,, 66 Punjabis.
	Unattd. List	*Ylend, Edgar Kenelm		18 June 17	25 June 17	18 June 18	33 Punjabis.
	Unattd. List	*Middleton, Thomas Benjamin	207a	18 June 17	29 June 17	18 June 18	3 G. R.
	Unattd. List	*Platt, Eric Peregrine		18 June 17	25 June 17	18 June 18	41 Dogras.
	Unattd. List	*Bampfield, Richard	207a	18 June 17	25 June 17	18 June 18	Attd. 35 Sikhs.
985	Unattd. List	*Turner, Cecil Douglas Lovett (57)		18 June 17	25 June 17	18 June 18	Attd. 117 Mahrattas.
	Unattd. List	*Robertson, Douglas Graeme (46)	207a	18 June 17	23 June 17	18 June 18	,, 10 Jats.
	Unattd. List	*Clarke, Percy Trant	207a	18 June 17	25 June 17	18 June 18	59 Rifles.
	Unattd. List	*Prioleau, Charles Philip Jerome		18 June 17	25 June 17	18 June 18	Attd. Corps of Guides.
	Unattd. List	*Chaplin, Ernest Leslie (46)	207a	18 June 17	25 June 17	18 June 18	,, 110 Mahrattas.
990	Unattd. List	*Rerrie, Erichsen Shafton Ewen	207a	18 June 17	22 June 17	18 June 18	,, 103 Mahrattas.
	Unattd. List	*Maclaren, Colin Shaw		18 June 17	27 June 17	18 June 18	57 Rifles.
	Unattd. List	*Buck, Frederick Alfred	207a	18 June 17	29 June 17	18 June 18	Attd. 2-67 Punjabis.
	Unattd. List	*Metcalfe, Marmaduke Reginald		18 June 17	26 June 17	18 June 18	,, 42 Deoli Regt.
	Unattd. List	*Harding, William John		18 June 17	23 June 17	18 June 18	92 Punjabis.
995	Unattd. List	*McIntyre, Henry Finch		18 June 17	25 June 17	18 June 18	37 Dogras.
	Unattd. List	*Macann, Arthur Ernest Henry (172a)	207a	18 June 17	25 June 17	18 June 18	107 Pioneers.
	Unattd. List	*Lawton, Thaddeus Daniel Vincent	207a	18 June 17	24 June 17	18 June 18	Attd. 113 Infantry.
	Unattd. List	*Shippard, Seymour Sidney Ayshford		18 June 17	25 June 17	18 June 18	,, Corps of Guides.
	Unattd. List	*Mandeville, Walter	207a	18 June 17	26 June 17	18 June 18	,, 95 Infantry.
1000	Unattd. List	*Baritrop, Adrian Bazeley	207a	18 June 17	26 June 17	18 June 18	4 G. R.
	Unattd. List	*Hanmer, Charles Gordon	207a	18 June 17	29 June 17	18 June 18	21 Cavalry.
	Unattd. List	*Sansom, Albert Edward	206	18 June 17	24 June 17	18 June 18	Attd. 39 C. I. H.
	Unattd. List	*Macleod, Norman Murray		18 June 17	28 June 17	18 June 18	1 G. R.
	Unattd. List	*Carruthers, Richard John	207a	18 June 17	5 July 17	18 June 18	Attd. 10 G. R.
1005	Unattd. List	*Yates, Derek Harold Peel (46, 51)	207a	18 June 17	25 June 17	18 June 18	94 Infantry.
	Unattd. List	*Hendrick, Charles Albert		18 June 17	25 June 17	18 June 18	41 Dogras.
1007	Unattd. List	*Whitehead, Edward Croxall		18 June 17	24 June 17	18 June 18	Attd. 98 Infantry.

No.	Former Corps.	Names.		Date of birth.	Date of first commission.	Date of appointment to Indian Army.		Remarks.
		Lieutenants—*contd.*						
1008	Unattd. List	***Shuttleworth**, Edward Hugh Septimus			18 June 17	7 Sep. 17	18 June 18	53 Sikhs.
	Unattd. List	***Parkin**, John Henry Devereux (46, 51)	207a		18 June 17	24 June 17	18 June 18	Attd. 12 Pioneers.
1010	Unattd. List	***Warran**, William Simeon Edmund (51, 52)	207a		18 June 17	5 July 17	18 June 18	Attd. 10 G. R.
	Unattd. List	***Bucknall**, Hugh Livingstone			18 June 17	25 June 17	18 June 18	,, Corps of Guides.
	Unattd. List	***Perkins**, Wilfrid Harold (51)			18 June 17	20 June 17	18 June 18	,, 88 Infantry.
	Unattd. List	***Lynch**, Reginald Joseph (57)	207a		18 June 17	28 June 17	18 June 18	,, 126 Infantry.
	Unattd. List	***Woodman**, Jack Bolton (71)	207a		18 June 17	24 June 17	18 June 18	,, 33 Cavalry.
1015	Unattd. List	***Covernton**, James Waller Gargrave (59)			18 June 17	22 June 17	18 June 18	,, 103 Mah. Lt. Infy.
	Unattd. List	***Salter**, Eric			18 June 17	29 June 17	18 June 18	3 G. R.
	Unattd. List	***Stevenson**, Herbert Allan			18 June 17	28 June 17	18 June 18	1 G. R.
	Unattd. List	***Prophit**, George Craigie			18 June 17	28 June 17	18 June 18	Attd. 39 Garhwal Rifles.
	Unattd. List	***Morton**, Henry Kingsley			18 June 17	25 June 17	18 June 18	,, 38 Dogras.
1020	Unattd. List	***Sproule**, Robert Francis (71)	207a		18 June 17	23 June 17	18 June 18	,, 27 Lt. Cavalry.
	Unattd. List	***Fox**, Ronald De Vere Richardson			18 June 17	25 June 17	18 June 18	45 Sikhs.
	Unattd. List	***Ommanney**, Frederick Roy	207a		18 June 17	26 June 17	18 June 18	Attd. 120 Infantry.
	Unattd. List	***Humphreys**, Charles Edward	207a		18 June 1[illegible]	24 June 17	18 June 18	,, 74 Punjabis.
	Unattd. List	***McKernan**, Francis Patrick	207a		18 June 17	24 June 17	18 June 18	,, 74 Punjabis.
1025	Unattd. List	***Foster**, Esmond Lewis Pearce (57)	207a		18 June 17	25 June 17	18 June 18	,, 3—39 Garhwal Rifles.
	Unattd. List	***Eastmead**, Colinton (51)			18 June 17	25 June 17	18 June 18	,, 15 Sikhs.
	Unattd. List	***Reeve**, Montagu Banks Pearman (51, 207a)			18 June 17	14 July 17	18 June 18	,, 19 Punjabis.
	Unattd. List	***Davies**, Rowland Marshall (48a, 51)	207a		18 June 17	23 June 17	18 June 18	,, 104 Rifles.
	Unattd. List	***Bates**, Ronald Charles Sidney			18 June 17	25 June 17	18 June 18	Attd. 89 Punjabis.
1030	Unattd. List	***McCleery**, Robert John (61)	207a		18 June 17	24 June 17	18 June 18	Attd. 4 Cavalry.
	Unattd. List	***Hartland**, Brantford Shadwell (57)	207a		18 June 17	27 June 17	18 June 18	Attd. 9 G. R.
	Unattd. List	***Marshall**, Campbell	207		18 June 17	24 June 17	18 June 18	Attd. 37 Lancers.
	Unattd. List	***Kirk**, John Cedric Mylrea (46)			18 June 17	25 June 17	18 June 18	61 Pioneers.
	Unattd. List	***Heggart**, John Edward			18 June 17	24 June 17	18 June 18	Attd. 83 Infantry.
1035	Unattd. List	***Featherstone**, Belford Kinchant			18 June 17	25 June 17	18 June 18	Attd. 54 Sikhs.
	Unattd. List	***Richards**, Cuthbert Denison	207a		18 June 17	25 June 17	18 June 18	Attd. 8 Cavalry.
	Unattd. List	***Herbert**, Henry Arthur			18 June 17	17 Oct. 17	18 June 18	Attd. 2 Rajputs.
	Unattd. List	***Seater**, George Harold	207a		18 June 17	1 July 17	18 June 18	Attd. 2 Rajputs.
	Unattd. List	***Lovett**, Osmond de Turville			18 June 17	27 June 17	18 June 18	Attd. 9 G. R.
1040	Unattd. List	***Roxburgh**, William	207a		18 June 17	25 June 17	18 June 18	Attd. 25 Punjabis.
	Unattd. List	***Poulton**, Henry Mortimer (57)			18 June 17	9 Sep. 17	18 June 18	Attd. 35 Sikhs.
1042	Unattd. List	***Tottenham**, James Loftus	207a		18 June 17	14 July 17	18 June 18	22 Punjabis.

1043	Unattd. List	***Coventry,** Henry Francis	207a	18 June 17	22 June 17	18 June 18	Attd. 2-9th Infantry.
	Unattd. List	***Needham,** James Owen (46)	207a	18 June 17	25 June 17	18 June 18	Attd. 25 Punjabis.
1045	Unattd. List	***Morton,** Charles Edward		18 June 17	28 June 17	18 June 18	Attd. 89 Punjabis.
	Unattd. List	***Chilcott,** John Shepherd		18 June 17	28 June 17	18 June 18	Attd. 39 Garhwal Rifles.
	Unattd. List	***Bake,** Arthur Trevithick Tyack	207a	18 June 17	28 June 17	18 June 18	126 Infantry.
	Unattd. List	***Hayes-Newington,** Eric Adrian		18 June 17	24 June 17	18 June 18	Attd. 29 Punjabis.
	D. of Lanc. Yeo.	***Barlow,** John Hall(57)		26 June 17	29 July 18	26 June 18	Attd. 4 Cavalry.
		9 months forfeited for seniority under conditions of I. A. O. 511 of 1917.					
1050	Unattd. List	§***Bates,** Richard Sydney Patrick		5 July 17	30 Nov. 17	5 July 18	Attd. 119 Infantry.
	Unattd. List	§***Barr,** Montague Claude		5 July 17	12 Dec. 17	5 July 18	Attd. 72 Punjabis.
	Unattd. List	§***Forrest,** Albert Edmund George		5 July 17	1 Dec. 17	5 July 18	Attd. 1 Lancers.
	Unattd. List	§***Milling,** Robert Cranston		5 July 17	1 Dec. 17	5 July 18	Attd. 2-1st Gurkha Rifles.
	Oxf. & Buck.	***Gass,** Fenwick Temperley (48a, 50)		25 July 17	27 July 18	25 July 18	Attd. 113 Infantry.
		9 months forfeited for seniority under conditions of I. A. O. 511 of 1917.					
1055	Hamps. R.	***Beazeley,** George Edward Armitage		7 Aug. 17	1 Oct. 18	7 Aug. 18	Attd. 8 Gurkha Rifles.
		9 months forfeited for seniority under conditions of I. A. O. 511 of 1917.					
	1 Res. R. of Cov.	***Robinson,** Frank William Overton		28 Aug. 17	28 July 18	28 Aug. 18	Attd. 26 Cavalry.
		9 months forfeited for seniority under conditions of I. A. O. 511 of 1917.					
	Unattd. List	§***Bowden,** Cecil Charles Hooper		31 Aug. 17	13 Nov. 17	31 Aug. 18	Attd. 38 C. I. Horse.
	Unattd. List	§***Simpson,** Edward Skinner		11 Sep. 17	23 Sep. 18	. . . .	Attd. 73 Infantry.
	Unattd. List	§***LeMesurier,** Tom Henry Ferrers (57)		12 Sep. 17	24 Mar. 18	12 Sep. 18	Attd. Corps of Guides.
1060	Unattd. List	§***Wilton,** Arthur John Millard		12 Sep. 17	7 Dec. 17	12 Sep. 18	Attd. 116 Mahrattas.
	Unattd. List	§***Russell,** Henry Brian		12 Sep. 17	9 Dec. 17	12 Sep. 18	Attd. 107 Pioneers.
	Unattd. List	§***Fiddes,** James (46)		12 Sep. 17	9 Dec. 17	12 Sep. 18	Attd. 88 Infantry.
	Unattd. List	§***Amsden,** Brian Hennell		12 Sep. 17	9 Dec. 17	12 Sep. 18	Attd. 57 Rifles.
	Unattd. List	§***Marriott,** John Richard		12 Sep. 17	7 Dec. 17	12 Sep. 18	Attd. 99 Infantry.
1065	Unattd. List	§***Dicker,** Bertie Percival		12 Sep. 17	13 Dec. 17	12 Sep. 18	Attd. 83 Infantry.
	Unattd. List	§***Jones,** Joseph Llewellyn		12 Sep. 17	17 Dec. 17	12 Sep. 18	Attd. 91 Punjabis.
	Unattd. List	§***Burridge,** Richard Arthur		12 Sep. 17	8 Dec. 17	12 Sep. 18	Attd. 25 Cavalry.
	Unattd. List	§***O'Brien,** William James		12 Sep. 17	9 Dec. 17	12 Sep. 18	Attd. 102 Grenadiers.
	Unattd. List	§***Tapp,** Donald Hansom		12 Sep. 17	7 Dec. 17	12 Sep. 18	Attd. 114 Mahrattas.
1070	Unattd. List	§***Barton,** Victor George Joseph		12 Sep. 17	8 Dec. 17	12 Sep. 18	Attd. 75 Infantry.
	Unattd. List	§***Sale,** Wilfred George		12 Sep. 17	9 Dec. 17	12 Sep. 18	Attd. 112 Infantry.
	Unattd. List	§***Deakin,** James Stanley (57)		12 Sep. 17	9 Dec. 17	12 Sep. 18	Attd. 84 Punjabis.
	Unattd. List	§***Illingworth,** Guy Russell		12 Sep. 17	30 Dec. 17	12 Sep. 18	Attd. 91 Punjabis.
	Unattd. List	§***East,** Cecil Frederick		12 Sep. 17	9 Dec. 17	12 Sep. 18	Attd. 87 Punjabis.
1075	Unattd. List	§***Clarke,** John Arundell		12 Sep. 17	16 Dec. 17	12 Sep. 18	Attd. 33 Cavalry.
	Unattd. List	§***Rae,** Kenneth William Ross (207a)		12 Sep. 17	9 Dec. 17	12 Sep. 18	Attd. 108 Infantry.
	Unattd. List	§***Broadway,** Philip Robert		12 Sep. 17	11 Dec. 17	12 Sep. 18	Attd. 5 Gurkha Rifles.
	Unattd. List	§***Macvean,** Eric Douglas		12 Sep. 17	13 Jan. 18	12 Sep. 18	Attd. 7 Lancers.
	Unattd. List	§***Woodman,** Reginald Stanley (61)		12 Sep. 17	7 Dec. 17	12 Sep. 18	Attd. 116 Mahrattas.
1080	Unattd. List	§***Dyer,** Godfrey Maxwell		12 Sep. 17	10 Dec. 17	12 Sep. 18	Attd. 14 Lancers.
	Unattd. List	§***Brown,** Ralph		12 Sep. 17	11 Dec. 17	12 Sep. 18	Attd. 67 Punjabis.
1082	Unattd. List	§***Hervalwil-Cox,** Vernon Maurice		2 Sep. 17	10 Dec. 17	12 Sep. 18	Attd. 89 Punjabis.

* Has not passed the final examination. § Admitted subject to H. M.'s approval.

Lieutenants—*contd.*

No.	Former Corps.	Name.		Date of birth.	Date of first Commission.	Date of appointment to Indian Army.	Date of rank.	Remarks.
1083	Unattd. List	§*Archer, Wilfred Algernon (57)			12 Sep. 17	9 Dec. 17	12 Sep. 18	Attd. Corps of Guides.
	Unattd. List	§*Dean, Arthur			12 Sep. 17	17 Dec. 17	12 Sep. 18	Attd. 80 Infantry.
1085	Unattd. List	§*Johns, Reginald Charles Walmngton (46)			12 Sep. 17	8 Dec. 17	12 Sep. 18	Attd. 92 Punjabis.
	Unattd. List	§*Plowden, James Miles Binden Chicheley			12 Sep. 17	9 Dec. 17	12 Sep. 18	Attd. 39 Garhwal Rifles.
	Unattd. List	§*Larking, Henry Sloane			12 Sep. 17	8 Dec. 17	12 Sep. 18	Attd. 97 Infantry.
	Unattd. List	§*Cobbett, James Edward			12 Sep. 17	11 Dec. 17	12 Sep. 18	Attd. 76 Punjabis.
	Unattd. List	§*McMahon, Harry Anderson	207a		12 Sep. 17	9 Dec. 17	12 Sep. 18	Attd. 97 Infantry.
1090	Unattd. List	§*Dickinson, Frederick Richard			12 Sep. 17	22 Dec. 17	12 Sep. 18	Attd. 20 Horse.
	Unattd. List	§*Arbuthnott, John St. Clair			12 Sep. 17	12 Dec. 17	12 Sep. 18	Attd. 59 Rifles.
	Unattd. List	§*Smith, Leslie Evan			12 Sep. 17	9 Dec. 17	12 Sep. 18	Attd. 112 Infantry.
	Unattd. List	§*Duke, Allen Lingard Rae			12 Sep. 17	15 Dec. 17	12 Sep. 18	Attd. 82 Punjabis.
	Unattd. List	§*Brown, Leslie Charles Hutchinson			12 Sep. 17	10 Dec. 17	12 Sep. 18	Attd. 67 Punjabis.
1095	Unattd. List	§*Mullaly, Hugh Alexander			12 Sep. 17	7 Dec. 17	12 Sep. 18	Attd. 99 Infantry.
	Unattd. List	§*Walsh, Arthur Patrick			12 Sep. 17	10 Dec. 17	12 Sep. 18	Attd. 117 Mahrattas.
	Unattd. List	§*Stratton, Eric Bruce			12 Sep. 17	9 Dec. 17	12 Sep. 18	Attd. 112 Infantry.
	Unattd. List	§*Barnes, John Frederick Fortescue			12 Sep. 17	1 Jan. 18	12 Sep. 18	Attd. 2 Lancers.
	Unattd. List	§*Judd, Edwin Leeb			12 Sep. 17	11 Dec. 17	12 Sep. 18	Attd. [illegible] Horse.
1100	Unattd. List	§*Schokman, Clive Conrad	207a		12 Sep. 17	16 Jan. 18	12 Sep. 18	Attd. 25 Cavalry.
	Unattd. List	§*Lambert, William George Arthur	207a		12 Sep. 17	27 Dec. 17	12 Sep. 18	Attd. 32 Pioneers.
	Unattd. List	§*Weston, John Ruahine (48a, 51)			12 Sep. 17	9 Dec. 17	12 Sep. 18	Attd. 119 Infantry.
	Unattd. List	§*Badham, Robert Armstrong Alexander			12 Sep. 17	9 Dec. 17	12 Sep. 18	Attd. 61 Pioneers.
	Unattd. List	§*McDougall, Dugald Mitchell			12 Sep. 17	11 Dec. 17	12 Sep. 18	Attd. 101 Grenadiers.
1105	Unattd. List	§*Rundall, John Wingate			12 Sep. 17	10 Dec. 17	12 Sep. 18	Attd. 1 Gurkha Rifles.
	Unattd. List	§*Phillips, Robert Lionel (61)			12 Sep. 17	8 Dec. 17	12 Sep. 18	Attd. 105 Infantry.
	Hamps. R.	*Matthews, Frederick Jackson (57)			19 Sep. 17	5 Oct. 18	19 Sep. 18	Attd. 2 Gurkha Rifles
		9 months forfeited for seniority under conditions of I. A. O. 511 of 1917.						
	3 Dn. Gds.	*Bruce-Smith, Ronald			21 Sep. 17	1 Oct. 18	21 Sep. 18	Attd. 26 Cavalry.
		9 months forfeited for seniority under conditions of I. A. O. 511 of 1917.						
	Unattd. List	*Mearns, James Patrick			14 Nov. 16	24 Nov. 16	18 Oct. 18	2 Rajputs.
1110	R. Sc. Fus.	*Gibson, James Russell			25 Oct. 17	24 Oct. 18	25 Oct. 18	Attd. 110 Infantry.
		9 months forfeited for seniority under conditions of I.A.O. 511 of 1917.						
1111	Unattd. List	§*Godley, Brian Richard			27 Oct. 17	6 Nov. 17	27 Oct. 18	Attd. 6 Jat L. Infy.

1112	Unattd. List	§***Lewis**, Charles Benjamin (57)	27 Oct. 17	4 Nov. 17	27 Oct. 18	Attd. 2-1st Gurkha Rifles.
	Unattd. List	§***Stephenson**, John	27 Oct. 17	2 Nov. 17	27 Oct. 18	Attd. 106 Pioneers.
	Unattd. List	§***Dedman**, James Baillie	27 Oct. 17	6 Nov. 17	27 Oct. 18	Attd. 105 Lt. Infantry.
1115	Unattd. List	§***Ward**, George Ross	27 Oct. 17	5 Nov. 17	27 Oct. 18	Attd. 1-6th Gurkha Rifles.
	Unattd. List	§* **Moore**, Clement Selby	27 Oct. 17	8 Nov. 17	27 Oct. 18	Attd. 1-8th Gurkha Rifles.
	Unattd. List	§***Ries**, James Leonard	27 Oct. 17	5 Nov. 17	27 Oct. 18	Attd. 38 Dogras.
	Unattd. List	§***Edge**, Joseph	27 Oct. 17	5 Nov. 17	27 Oct. 18	Attd. 27 Punjabis.
	Unattd. List	§***Macdonald**, James Balfour	27 Oct. 17	4 Nov. 17	27 Oct. 18	Attd. 23 Pioneers.
1120	Unattd. List	§***Owens**, Theobald David Cogswell	27 Oct. 17	1 Nov. 17	27 Oct. 18	Attd. 1-4th Gurkha Rifles.
	Unattd. List	§***Winckley**, Sidney Freer	27 Oct. 17	5 Nov. 17	27 Oct. 18	Attd. 12 Cavalry.
	Unattd. List	§***Healy**, John Joseph	27 Oct. 17	5 Nov. 17	27 Oct. 18	Attd. 27 Punjabis.
	Unattd. List	§***Gibbs**, Herbert Ronald Keith	27 Oct. 17	6 Nov 17	27 Oct. 18	Attd. 3-6th Gurkha Rifles.
	Unattd. List	§***Phillips**, Percival Jerome	27 Oct. 17	5 Nov. 17	27 Oct. 18	Attd. 89 Punjabis.
1125	Unattd. List	§***Hilliard**, Percy Joseph	27 Oct. 17	4 Nov. 17	27 Oct. 18	Attd. 19 Lancers.
	Unattd. List	§***Fox**, Edward Lionel Wakefield	27 Oct. 17	1 Nov. 17	27 Oct. 18	Attd. 39 Garhwal Rifles.
	Unattd. List	§* **Upton**, Joseph Annesley Everard	27 Oct. 17	6 Nov. 17	27 Oct. 18	Attd. 1-3rd Gurkha Rifles.
	Unattd. List	§* **Blackwood**, Charles Vincent Ralph	27 Oct. 17	7 Nov. 17	27 Oct. 18	Attd. 2-7th Gurkha Rifles.
	Unattd. List	§* **Buss**, Ronald Austen	27 Oct. 17	5 Nov. 17	27 Oct. 18	Attd. 2-2nd Gurkha Rifles.
1130	Unattd. List	§* **Forsyth**, Ian Mortimer	27 Oct. 17	7 Nov. 17	27 Oct. 18	Attd. 3-8th Gurkha Rifles.
	Unattd. List	§* **Hay**, Colin Elliot	27 Oct. 17	7 Nov. 17	27 Oct. 18	Attd. 30th Lancers.
	Unattd. List	§* **Reynolds**, Charles Clavell Richardson	27 Oct. 17	3 Nov. 17	27 Oct. 18	Attd. 19th Lancers.
	Unattd. List	§* **Aspinall**, Arthur William	27 Oct. 17	6 Nov. 17	27 Oct. 18	Attd. 2-39th Gurkha Rifles.
	Unattd. List	§* **Nicholl**, Maynard Charrington	27 Oct. 17	4 Nov. 17	27 Oct. 18	Attd. 54th Sikhs.
1135	Unattd. List	§* **Pearson**, George Thomas (48a, 52)	27 Oct. 17	3 Nov. 17	27 Oct. 18	Attd. 58th Rifles.
	Unattd. List	§* **Warren**, Alan George	27 Oct. 17	2 Nov. 17	27 Oct. 18	Attd. 22nd Punjabis.
	Unattd. List	§* **Standish**, Norman	27 Oct. 17	5 Nov. 17	27 Oct. 18	Attd. 37th Dogras.
	Unattd. List	§* **Brooke-Hunt**, Robert Cecil	27 Oct. 17	4 Nov. 17	27 Oct. 18	Attd. Corps of Guides.
	Unattd. List	§* **Weatherdon**, Hugh Eric	27 Oct. 17	6 Nov. 17	27 Oct. 18	Attd. 3-2nd Gurkha Rifles.
1140	Unattd. List	§* **Cooper-Chadwick**, John Lionel	27 Oct. 17	3 Nov. 17	27 Oct 18	Attd. 55th Rifles.
	Unattd. List	§* **Shipton**, Geoffrey Leonard	27 Oct. 17	6 Nov. 17	27 Oct 18	Attd. 105th Lt. Infantry.
	Unattd. List	§* **Armstrong**, Howard Francis Cloncurry	27 Oct. 17	5 Nov. 17	27 Oct. 18	Attd. 3-2nd Gurkha Rifles.
	Unattd. List	§* **Platts**, Morris Arthur	27 Oct. 17	6 Nov. 17	27 Oct. 18	Attd. 2-5th Gurkha Rifles.
	Unattd. List	§* **Hobbs**, Athol Joseph	27 Oct. 17	4 Nov. 17	27 Oct. 18	Attd. 23rd Pioneers.
1145	Unattd. List	§* **Craig**, Archibald Douglas	27 Oct. 17	1 Nov. 17	27 Oct. 18	Attd. 2-10th Gurkha Rifles.
	Unattd. List	§* **Smyth**, Henry Malcolm	27 Oct. 17	7 Nov. 17	27 Oct. 18	Attd. 1-9th Gurkha Rifles.
	Unattd. List	§* **Rouse**, Reginald Pilkington	27 Oct. 17	1 Nov. 17	27 Oct. 18	Attd. 1-7th Gurkha Rifles.
	Unattd. List	§* **Billing**, Ernest Wilmot	27 Oct. 17	4 Nov. 17	27 Oct 18	Attd. 20th Infantry.
1150	Unattd. List	§* **Penney**, Nicholas Arthur	27 Oct. 17	7 Nov. 17	27 Oct. 18	Attd. 25th Punjabis.
	Unattd. List	§* **Cummins**, John Eric	27 Oct. 17	4 Nov. 17	27 Oct. 18	Attd. 93rd Infantry.
	Unattd. List	§* **Kershaw**, Harold Willoughby le Pelly 207a	27 Oct. 17	2 Nov. 17	27 Oct. 18	Attd. 107th Pioneers.
	Unattd. List	§* **Milliken**, Ernest Kenneth	27 Oct. 17	4 Nov. 17	27 Oct. 18	Attd. 54th Sikhs.
	Unattd. List	§* **Bannon**, John Denton	27 Oct. 17	4 Nov. 17	27 Oct. 18	Attd. 67th Punjabis.
	I. A. R. O.	§***Alfieri**, Frederick John 207a *Service for promotion from 29 Oct. 17, under I. A. O. 510 of 1917.*	29 Jan. 17	30 July 18	29 Oct. 18	Attd. 34 Pioneers.
1155	R. Innis. Fus	***Notley**, Walter, M.C. *9 months forfeited for seniority under conditions of I.A.O. 511 of 1917.*	26 Nov. 17	5 Oct. 28	26 Nov. 18	Attd. 107 Pioneers.
1156	Aus. Imp. Force.	§* **Hanna**, Arthur Leslie *9 months forfeited for seniority under conditions of I. A. O. 511 of 1917.*	3 Dec. 17	3 Nov. 18	3 Dec. 18	Attd. 11 Lancers.

* Has not passed the final examination.

§ Admitted subject to H. M.'s approval.

No.	Former Corps.	Name.	Date of birth.	Date of first Commission.	Date of appointment to Indian Army.	Date of rank.	Remarks.
		Lieutenants—*concld.*					
1157	Aus. Imp. Force.	§* **Bowe,** Roy William *9 months forfeited for seniority under conditions of I. A. O. 511 of 1917.*	. . . .	6 Dec. 17	3 Nov. 18	3 Dec. 18	Attd. 15 Lancers.
		2nd-Lieutenants—376.					
	Unattd. List	***Mitchell,** John Harris 206	. . . .	15 Nov. 15	19 Nov. 15	. . . .	Attached hs.
	Unattd. List.	§***Brooks,** Joshua Rowland	. . . .	12 Jan. 16	4 Sep. 18	. . . .	Attd. 21 Cavalry.
	W. Kent. Yeo.	***Collis,** Frederick Scrivener *9 months forfeited for seniority under conditions of I. A. O. 511 of 1917.*	. . . .	12 Apl. 16	16 June 18	. . . .	Attd. Corps of Guides.
	Unattd. List.	§***Thornton,** Charles Arnold	. . . .	26 Jan. 16	6 Sep. 18	. . . .	Attd. 39 Garhwal Rifles.
5	Unattd. List	§* **Hog,** Ralph Dundas	. . . .	30 Apl. 16	21 Mar. 18	27 Oct. 18	Attd. 11 Lancers.
	Unattd. List	***Blakeney,** Edward Charles William	. . . .	14 Nov. 16	21 Nov. 16	. . . .	Attd. 72 Punjabis.
	I. A. R. O.	***Mann,** Stephen Harold *Service for promotion from 19 Jan. 17 under I. A. O. 510 of 1917.*	. . . .	19 Apl. 16 19 Jan. 17	5 Dec. 17	. . . .	8 Rajputs.
	Unattd. List	***Fordyce,** John Frederick Dingwall	. . . .	30 Jan. 17	10 Feb. 17	. . . .	34 Pioneers.
	Unattd. List	***Larymore,** Patrick Alan Douglas	. . . .	30 Jan. 17	9 Feb. 17	. . . .	Attd. 10 G. R.
10	Unattd. List	***Dickens,** Arthur Charles	. . . .	30 Jan. 17	11 Feb. 17	. . . .	125 Rifles.
	Unattd. List	§***Tower,** Herbert George Eric, M.C.	. . . .	14 Feb. 17	7 Oct. 18	. . . .	Attd. 57 Rifles.
	I. A. R. O.	§***Bourne,** William Harvey *Service for promotion from 9 June 1917 under I. A. O. 510 of 1917.*	. . . .	9 Sep. 16 9 June 17	26 Apl. 18	. . . .	„ 9 Infantry.
	I. A. R. O.	§***Rippon,** Percy Ivor Vincent *Service for promotion from 10 June 17 under I. A. O. 510 of 1917.*	. . . .	10 Sep. 16 10 June 17	5 Dec. 17	. . . .	114 Mahrattas.
	Unattd. List	***Bebbington,** Humphrey John 207a	. . . .	18 June 17	25 June 17	18 June 18	Attd. 94 Infantry.
15	Unattd. List	***Clarke-Bell,** Donald Mary	. . . .	18 June 17	25 June 17	. . . .	Attd. 6 G. R.
	Unattd. List	§***Gray,** Lionel George	. . . .	21 June 17	6 Nov. 17		
	I. A. R. O.	§***Lindsley,** Reginald Maurice *Service for promotion from 6 July 1917 under I. A.O. 510 of 1917.*	. . . .	6 Oct. 16 6 July 17	26 Apl. 18	. . . .	Attd. 13 Rajputs.
	Unattd. List	§***Jackson,** Joseph Devoncher	. . . .	12 Sep. 17	8 Dec. 17	. . . .	Attd. 92 Punjabis.

	I. A. R. O.	§*Smallwood, Robert (207a)	1 Jan. 17 1 Oct. 17	1 Sep. 18	Attd. 154 Infantry.
		Service for promotion from 1 Oct. 17 under I. A. O. 510 of 1917.			
20	Unattd. List	§*Reynolds, John Ridgway	27 Oct. 17	6 Nov. 17	
	Unattd. List	§*Hurst, George Stanley	27 Oct. 17	6 Nov. 17	
	Unattd. List	§*Hutchings, Claude Morgan	27 Oct. 17	6 Nov. 17	
	Unattd. List	§*Hopper, Francis Cyril	27 Oct 17	4 Nov. 17	Attd. 21 Punjabis.
	Unattd. List	§*Harris, John Noel Alexander	27 Oct. 17	6 Nov. 17	
25	Unattd. List	§*Gwyn, Philip Eustace Congreve Jermy	27 Oct. 17	2 Nov. 17	Attd. 22 Punjabis.
	Unattd. List	§*Cotton, John Colson	27 Oct. 17	3 Nov. 17	Attd. 24 Punjabis.
	Unattd. List	§* Madge, Hubert Edward	27 Oct. 17	1 Nov. 17	Attd. 1-4th Gurkha Rifles.
	Unattd. List	§*Caires, Charles Albert	27 Oct. 17	6 Nov. 17	
	Unattd. List	§*Allington, Noel Stanley	27 Oct. 17	6 Nov. 17	
30	Unattd. List	§*Oswald, Richard Alexander	27 Oct. 17	6 Nov. 17	
	Unattd. List	§*Spooner, Hugh	27 Oct. 17	6 Nov. 17	
	Unattd. List	§*Stephens, John Roland Milner	27 Oct. 17	6 Nov. 17	
	Unattd. List	§*Grieves, Roy Bourne	27 Oct. 17	4 Nov. 17	Attd. 32 Pioneers.
	Unattd. List	§*Gorringe, William Pennington Hugh	27 Oct. 17	6 Nov. 17	
35	Unattd. List	§*Geddes, Thomas John Roy	27 Oct. 17	6 Nov. 17	
	Unattd. List	§*Goode, Robert Leigh	27 Oct. 17	3 Nov. 17	Attd. 56 Rifles.
	Unattd. List	§*Weigall, Geoffrey Stephen Carmac	27 Oct. 17	6 Nov. 17	
	Unattd. List	§*Allen, Rickard Francis William Kinkead	27 Oct. 17	6 Nov. 17	Attd. 6 Jat Lt. Infy.
	Unattd. List	§*Day, Eric Rouviere	27 Oct. 17	3 Nov. 17	Attd. 3 Horse.
40	Unattd. List	§*John, Kenneth Edwin	27 Oct. 17	6 Nov. 17	
	Unattd. List	§*Hossack, Ian Fraser	27 Oct. 17	6 Nov. 17	
	Unattd. List	§*A'Brook, Willie Richard Cecil	27 Oct. 17	4 Nov. 17	Attd. 54 Sikhs.
	Unattd. List	§*Buckton, George Neol	27 Oct. 17	7 Nov. 17	Attd. 10 Jats.
	Unattd. List	§*Tatham, Wilfred Meaburn	27 Oct. 17	7 Nov. 17	Attd. 16 Cavalry.
45	Unattd. List	§*Sherman, Douglas McCleod Hasted	27 Oct. 17	6 Nov. 17	
	Unattd. List	§*Thunder, Henry Francis Anthony	27 Oct. 17	8 Nov. 17	Attd. 72 Punjabis.
	Unattd. List	§*Haslett, Robert Stuart	27 Oct. 17	6 Nov. 17	
	Unattd. List	§*Francis, Mark Herbert	27 Oct. 17	6 Nov. 17	
	Unattd. List	§*Flett, Robert	27 Oct. 17	4 Nov. 17	
50	Unattd. List	§*Johnson, Edmund Cowper	27 Oct. 17	6 Nov. 17	
	Unattd. List	§*Meyrick-Jones, Philip	27 Oct. 17	5 Nov. 17	Attd. 2-2nd Gurkha Rifles.
	Unattd. List	§* Burnand, Ronald Fleming	27 Oct. 17	6 Nov. 17	
	Unattd. List	§* Heaton, Ronald Branston	27 Oct. 17	6 Nov. 17	
	Unattd. List	§* White, Cyril Wilton Geoffrey	27 Oct. 17	6 Nov. 17	
55	Unattd. List	§* Spence, William McGregor	27 Oct. 17	3 Nov. 17	Attd. 53rd Sikhs.
	Unattd. List	§* Kitson, George Hubert Wentworth	27 Oct. 17	7 Nov. 17	Attd. 10th Jats.
	Unattd. List	§* Cobb, Henry James Geoffrey	27 Oct. 17	6 Nov. 17	
	Unattd. List	§* Fowler, Hugh Oliver Wilson	27 Oct. 17	6 Nov. 17	
	Unattd. List	§* Thew, Illius Augustus	27 Oct. 17	5 Nov. 17	Attd. Corps of Guides.
60	Unattd. List	§* Jackson, George O'Neill Bayley	27 Oct. 17	6 Nov. 17	
	Unattd. List	§* Pearson, Ronald Forbes	27 Oct. 17	7 Nov. 17	Attd. 30th Lancers.
	Unattd. List	§* Garvie, Kenneth Clephen	27 Oct. 17	13 Nov. 17	Attd. 2-5th Gurkha Rifles.
	Unattd. List	§* Gee, William Henry Stuart	27 Oct. 17	5 Nov. 17	Attd. 48th Pioneers.
	Unattd. List	§* Bodle, Richard Terence	27 Oct. 17	7 Nov. 17	Attd. 10th Jats.
65	Unattd. List	§* Simpson, James Samuel Couldrey	27 Oct. 17	5 Nov. 17	Attd. 125th Rifles.
	Unattd. List	§* Cross, Stanley Graham Norwood	27 Oct. 17	3 Nov. 17	Attd. 12th Pioneers.
67	Unattd. List	§* Brett, Edward Tillotson St. Maur	27 Oct. 17	3 Nov. 17	Attd. 51st Sikhs.

2nd-Lieutenants—*contd.*

No.	Former Corps.	Names.	Date of birth.	Date of first commission.	Date of appointment to Indian Army.	Remarks.
68	Unattd. List	§* **Brooke,** Arthur Edgerton		27 Oct. 17	7 Nov. 17	Attd. 16th Cavalry.
	Unattd. List	§* **Ropner,** Geoffrey Vivian		27 Oct. 17	6 Nov. 17	Attd. 7th Rajputs.
70	Unattd. List	§* **Tweedy,** Henry Richard Francis		27 Oct. 17	4 Nov. 17	Attd. 20th Infantry.
	Unattd. List	§* **Willis,** John Henry Patrick		27 Oct. 17	2 Nov. 17	Attd. 106th Pioneers.
	Unattd. List	§* **Anderson,** Ola		27 Oct. 17	6 Nov. 17	Attd. 52nd Sikhs.
	Unattd. List	§* **Whittaker,** Eric Horton		27 Oct. 17	21 Nov. 17	Attd. 5th Cavalry.
	Unattd. List	§* **Alexander,** Leigh Arbuthnot		27 Oct. 17	6 Nov. 17	Attd. 2-5th Gurkha Rifles.
75	Unattd. List	§* **Tervitt,** John Furley Taylor		27 Oct. 17	12 Mar. 18	Attd. 1-4th Gurkha Rifles.
	Bedf. R.	* **Kent,** Henry Taylor *9 months forfeited for seniority under conditions of I. A. O. 511 of 1917.*		1 Dec. 17	22 Nov. 18	Attd. 10 Jats.
	Aus. Imp. Force.	* **Breton,** Charles George *9 months forfeited for seniority under conditions of I. A. O. 511 of 191*[illegible]		11 Dec. 17	5 Nov. 18	Attd. 69 Punjabis.
	Unattd. List	§* **Hope,** Adrian Alexander (48*a*, 50) 207*a*		21 Dec. 17	21 Mar. 18	Attd. 35 Sikhs.
	Unattd. List	§* **Sangster,** Douglas Bruce		21 Dec. 17	21 Mar. 18	Attd. 51 Shiks.
80	Unattd. List	§* **Burden,** Geoffrey Noel		21 Dec. 17	22 Mar. 18	Attd. 7 Gurkha Rifles.
	Unattd. List	§* **Hutton,** Antony Reginald		21 Dec. 17	21 Mar. 18	Attd. 2 Gurkha Rifles.
	Unattd. List	§* **Clease,** Edward Kenneth James		21 Dec. 17	21 Mar. 18	Attd. 53 Sikhs.
	Unattd. List	§* **Volt,** Reginald Austen 207*a*		21 Dec. 17	20 Mar. 18	Attd. 35 Sikhs.
	Unattd. List	§* **Roberts,** Godfrey Edward		21 Dec. 17	21 Mar. 18	Attd. Corps of Guides.
85	Unattd. List	§* **Andrews,** John Albert		21 Dec. 17	21 Mar. 18	Attd. 67 Punjabis.
	Unattd. List	§* **Bowen,** William Oswald		21 Dec. 17	22 Mar. 18	Attd. 6 Gurkha Rifles.
	Unattd. List	§* **Robinson,** Charles Patrick Whitmore		21 Dec. 17	21 Mar. 18	Attd. 54 Sikhs.
	Unattd. List	§* **Lund,** Charles		21 Dec. 17	21 Mar. 18	Attd. 8 Gurkha Rifles.
	Unattd. List	§* **Broughton,** Coote Edgar		21 Dec. 17	14 Apl. 18	Attd. 34 Horse.
90	Unattd. List	§* **Crabbe,** Cecil Brooksby		21 Dec. 17	29 Mar. 18	Attd. 129 Baluchis.
	Unattd. List	§* **Cunningham,** William Ernest		21 Dec. 17	21 Mar. 18	Attd. 52 Sikhs.
	Unattd. List	§* **Overton,** Arthur Gœ		21 Dec. 17	21 Mar. 18	Attd. 21 Cavalry.
	Unattd. List	§* **Bryceson,** Donald John		21 Dec. 17	21 Mar. 18	Attd. 54 Sikhs.
	Unattd. List	§* **Noel,** Owen Cyril		21 Dec. 17	2 Apl. 18	Attd. 1 Gurkha Rifles.
95	Unattd. List	§* **Weeks,** Courtenay Alison		21 Dec. 17	21 Mar. 18	Attd. 4 Gurkha Rifles.
	Unattd. List	§* **Kitchen,** Alfred Clarence		21 Dec. 17	28 Mar. 18	Attd. 57 Rifles.
	Unattd. List	§* **Nolan,** Michael Charles		21 Dec. 17	21 Mar. 18	Attd. 25 Punjabis.
	Unattd. List	§* **Huth,** Frederick Lancelot		21 Dec. 17	4 June 18	Attd. 14 Lancers.

	Unattd. List	§* **Danson**, Edward Towne	21 Dec. 17	20 Mar. 18	Attd. 53 Sikhs.
100	Unattd. List	§* **Kearns**, Roger Willis	21 Dec. 17	20 Mar. 18	Attd. 84 Punjabis.
	Unattd. List	§* **Jefcoate**, Horace Marcus	21 Dec. 17	22 Mar. 18	Attd. 57 Rifles.
	Unattd. List	§* **Rootes**, Augustus Hugh	21 Dec. 17	21 Mar. 18	Attd. 6 Jat Lt. Infy.
	Unattd. List	§* **Devonshire**, Norman George	21 Dec. 17	22 Mar. 18	Attd. Corps of Guides.
	Unattd. List	§* **Jones**, Hugh Iorwerth Kiffig	21 Dec. 17	22 Mar. 18	Attd. 21 Punjabis.
105	Unattd. List	§* **Orgill**, Arthur William	21 Dec. 17	21 Mar. 18	Attd. 2 Gurkha Rifles.
	Unattd. List	§* **Freston**, Thomas Anthony	21 Dec. 17	21 Mar. 18	Attd. 17 Cavalry.
	Unattd. List	§* **Davis**, Thomas Arthur Warren	21 Dec. 17	22 Mar. 18	Attd. 2 Gurkha Rifles.
	Unattd. List	§* **Radcliff**, Henry Garnett	21 Dec. 17	27 Mar. 18	Attd. 7 Gurkha Rifles.
	Unattd. List	§* **Evans**, Thomas Llewellin	21 Dec. 17	20 Mar. 18	Attd. 21 Punjabis.
110	Unattd. List	§* **Rivers**, William Elphinstone	21 Dec. 17	27 Mar. 18	Attd. 10 Gurkha Rifles.
	Unattd. List	§* **Hay**, Edward James Alan	21 Dec. 17	22 Mar. 18	Attd. 41 Dogras.
	Unattd. List	§* **Moodie**, Eric Herbert	21 Dec. 17	22 Mar. 18	Attd. 4 Cavalry.
	Unattd. List	§* **Christie**, George Hewitt	21 Dec. 17	27 Mar. 18	Attd. 10 Gurkha Rifles.
	Unattd. List	§* **Bell**, John Vincent	21 Dec. 17	21 Mar. 18	Attd. 9 Gurkha Rifles.
115	Unattd. List	§* **Harrison**, Thomas Reed	21 Dec. 17	23 Mar. 18	Attd. 5 Gurkha Rifles.
	Unattd. List	§* **Hughes**, Henry	21 Dec. 17	22 Mar. 18	Attd. 56 Rifles.
	Unattd. List	§* **Forward**, Frances Charles Miller	21 Dec. 17	21 Mar. 18	Attd. 3 Gurkha Rifles.
	Unattd. List	§* **Robinson**, John Gawden Kirkup	21 Dec. 17	21 Mar. 18	Attd. 16 Rajputs.
	Unattd. List	§* **Seller**, Charles	21 Dec. 17	22 Mar. 18	Attd. 37 Dogras.
120	Unattd. List	§* **Dalgety**, Algy Emlyn	21 Dec. 17	22 Mar. 18	Attd. 15 Sikhs.
	Unattd. List	§* **Parry**, John Keith	21 Dec. 17	20 Mar. 18	Attd. 89 Punjabis.
	Unattd. List	§* **Evans**, Samuel Frederick	21 Dec. 17	22 Mar. 18	Attd. Corps of Guides.
	Unattd. List	§* **Wright**, Cedric Cecil Vernon	21 Dec. 17	21 Mar. 18	Attd. 1 Brahmans.
	Unattd. List	§* **Norris**, Herbert Francis	21 Dec. 17	20 Mar. 18	Attd. 114 Mahrattas.
125	Unattd. List	§* **Daly**, Francis Owen William Augustin	21 Dec. 17	25 Mar. 18	Attd. 56 Rifles.
	Unattd. List	§* **Wilder**, Percival Ernest	21 Dec. 17	21 Mar. 18	Attd. 2 Gurkha Rifles.
	Unattd. List	§* **ffolkes**, Edward John Patrick Boschetti	21 Dec. 17	21 Mar. 18	Attd. 11 Lancers.
	Unattd. List	§* **Littlebury**, Arthur Theodore Baron	21 Dec. 17	22 Mar. 18	Attd. 34 Pioneers.
	Unattd. List	§* **Partridge**, Harry Victor	21 Dec. 17	22 Mar. 18	Attd. 1 Gurkha Rifles.
130	Unattd. List	§* **Vickers**, Charles Keisop	21 Dec. 7	20 Mar. 18	Attd. 8 Cavalry.
	Unattd. List	§* **Powell**, Francis Truman	21 Dec. 17	27 Mar. 18	Attd. 8 Gurkha Rifles.
	Unattd. List	§* **Long**, Harold Wallis	21 Dec. 17	21 Mar. 18	Attd. 125 Rifles.
	Unattd. List	§* **Williams**, Sidney Morgan	21 Dec. 17	27 Mar. 18	Attd. 4 Gurkha Rifles.
	Unattd. List	§* **Connolly**, Dermot Richard Vincent	21 Dec. 17	26 Mar. 18	Attd. 9 Gurkha Rifles.
135	Unattd. List	§* **Draffen**, John Edward	21 Dec. 17	22 Mar. 18	Attd. 38 Dogras.
	Unattd. List	§* **Fraser**, Thomas Frederick Glenelge	21 Dec. 17	28 Mar. 18	Attd. 3 Horse.
	Unattd. List	§* **Newland**, John Henry Frankland	21 Dec. 17	21 Mar. 18	Attd. 22 Punjabis.
	6 Res. R. Cav.	* **Bock**, Charles Cleaver *9 months forfeited for seniority under conditions of I. A. O. 511 of 1917.*	17 Jan. 18	23 Nov. 18	Attd. 30 Lancers.
	R. Scots.	* **Simpson**, George Archibald Dundas *9 months forfeited for seniority under conditions of I. A. O. 511 of 1917.*	26 Jan. 18	3 Oct. 18	Attd. 8 Gurkha Rifles.
140	Unattd. List	§* **Fullerton**, Gordon Alexander	31 Jan. 18	4 Feb. 18	Attd. 2-4th Gurkha Rifles.
	Unattd. List	§* **Roach**, James Osbert	31 Jan. 18	7 Feb. 18	Attd. 56 Rifles.
	Unattd. List	§* **Wilder**, Burnard Leslie	31 Jan. 18	6 Feb. 18	Attd. 52 Sikhs.
	Unattd. List	§* **Sinker**, Robert Carey	31 Jan. 18	7 Feb. 18	Attd. 1-6th Gurkha Rifles.
144	Unattd. List	§* **Macguckin**, Charles Edward	31 Jan. 18	7 Feb. 18	Attd. 33 Punjabis.

2nd-Lieutenants—*contd.*

No.	Former Corps.	Names.	Date of birth.	Date of first commission.	Date of appointment to Indian Army.	Remarks.
145	Unattd. List	§* **Duthie,** Kenneth James Nasmyth		31 Jan. 18	11 Feb. 18	Attd. 41 Dogras.
	Unattd. List	§* **Phelan,** Percy George . . . 207*a*		31 Jan. 18	8 Feb. 18	Attd. 103 Mah. L. I.
	Unattd. List	§* **Clery,** Percy Arthur Adamson		31 Jan. 18	6 Feb. 18	Attd. 53 Sikhs.
	Unattd. List	§* **Seccombe,** John Bertie Philip		31 Jan. 18	6 Feb. 18	Attd. 59 Rifles.
	Unattd. List	§* **Wallace,** Roland William Astell		31 Jan. 18	15 Feb. 18	Attd. 50 Rifles.
150	Unattd. List	§* **Stewart,** Frederick Gordon		31 Jan. 18	7 Feb. 18	Attd. 55 Rifles.
	Unattd. List	§* **Tatham,** Philip Robert		31 Jan. 18	6 Feb. 18	Attd. 23 Cavalry.
	Unattd. List	§* **Steed,** Frederic Charles William		31 Jan. 18	6 Feb. 18	Attd. 1 Brahmans.
	Unattd. List	§* **Bates,** Alfred Thomas Trevor		31 Jan. 18	6 Feb. 18	Attd. 2-1st Gurkha Rifles.
	Unattd. List	§* **Codrington,** Kenneth de Burgh		31 Jan. 18	9 July 18	Attd. 33 Cavalry.
155	Unattd. List	§* **Finlay,** James Robert Paul		31 Jan. 18	7 Feb. 18	Attd. 15 Sikhs.
	Unattd. List	§* **Noel,** Edward Francis Hamlyn		31 Jan. 18	6 Feb. 18	Attd. 12 Cavalry.
	Unattd. List	§* **Howman,** Ross Cosens (48*a*, 51)		31 Jan. 18	12 Feb. 18	Attd. 85 Rifles.
	Unattd. List	§* **Ker,** Thomas Menzies		31 Jan. 18	7 Feb. 18	Attd. 34 Pioneers.
	Unattd. List	§* **Wilkinson,** Robert Rede		31 Jan. 18	7 Feb. 18	Attd. Guides.
160	Unattd. List	§* **Powell,** Reginald Edward		31 Jan. 18	7 Feb. 18	Attd. 37 Dogras.
	Unattd. List	§* **Ross-Hurst,** Brian Plunkett		31 Jan. 18	7 Feb 18	Attd. 55 Rifles.
	Unattd. List	§* **Bodenham,** Konstantine Lubienski . . 207*a*		31 Jan. 18	19 Feb. 18	Attd. 1-4th Gurkha Rifles.
	Unattd. List	§* **Allison,** Gordon . . . 207*a*		31 Jan. 18	5 Feb. 18	Attd. 3-1st Gurkha Rifles.
	Unattd. List	§* **Mallinson,** Ernest Henry Powell		31 Jan. 18	6 Feb. 18	Attd. 30 Punjabis.
165	Unattd. List	§* **Hourihane,** Thomas Cormac		31 Jan. 18	8 Feb. 18	Attd. 41 Dogras.
	Unattd. List	§* **Putnam,** William Augustus (57) . . (207*a*)		31 Jan. 18	7 Feb. 18	Attd. 35 Sikhs.
	Unattd. List	§* **Welsh,** Thomas Vaughan		31 Jan. 18	6 Feb. 18	Attd. 12 Cavalry.
	Unattd. List	§* **Guise,** Christopher Probyn (61)		31 Jan. 18	7 Feb. 18	Attd. 15 Sikhs.
	Unattd. List	§* **Hardman,** Edward Fitz-Maur Chambre		31 Jan. 18	6 Feb. 18	Attd. 2-8th Gurkha Rifles.
170	Unattd. List	§* **Harrison,** Jim Willoughby Hyicee		31 Jan. 18	6 Feb. 18	Attd. 2-1st Gurkha Rifles.
	Unattd. List	§* **Fogerty,** William Perrott (48*a*, 51)		31 Jan. 18	7 Feb. 18	Attd. 57 Rifles.
	Unattd. List	§* **Saunders,** William Eric Pentland		31 Jan. 18	5 Feb. 18	Attd. 41 Dogras.
	Unattd. List	§* **Maxwell,** William Ernest		31 Jan. 18	18 Feb. 18	Attd. 127 Baluch L. I.
	Unattd. List	§* **Lewis,** Horace Ross		31 Jan. 18	6 Feb. 18	Attd. 9 Bhopal Infy.
175	Unattd. List	§* **Campbell,** Albert Gordon		31 Jan. 18	12 Feb. 18	Attd. 85 Rifles.
	Unattd. List	§* **Collins,** Malcolm Reginald		31 Jan. 18	6 Feb. 18	Attd. 19 Lancers.
	Unattd. List	§* **Bullard,** Hugh		31 Jan. 18	12 Feb. 18	Attd. 85 Rifles.
	Unattd. List	§* **Gray,** Archibald Reynolds (46)		31 Jan. 18	7 Feb. 18	Attd. 3-5th Gurkha Rifles.
	Unattd. List	§* **Price,** Cyril Jack		31 Jan. 18	6 Feb. 18	Attd. 33 Punjabis.
180	Unattd. List	§* **Hunt,** Joseph Wray Angus		31 Jan. 18	23 Feb. 18	Attd. 106 Pioneers.
	Unattd. List	§* **Gilbert,** Clarance		31 Jan. 18	6 Feb. 18	Attd. 56 Rifles.
	Unattd. List	§* **Burkett,** Richard Norman		31 Jan. 18	8 Feb. 18	Attd. 37 Lancers.
183	Unattd. List	§* **Phelan,** Patrick Joseph		31 Jan. 18	3 Feb. 18	Attd. 101 Grenadiers.

	Unattd. List	*§**Higginbotham**, Philip Pleasants		31 Jan. 18	4 Feb. 18	Attd. 70 Rifles.
185	Unattd. List	*§**Dowse**, Benjamin Thomas John		31 Jan. 18	6 Feb. 18	Attd. 30 Punjabis.
	Unattd. List	*§**Fitz-Maurice**, Wilfred Houghton		31 Jan. 18	5 Feb. 18	Attd. 21 Punjabis.
	Unattd. List	*§**Smith**, John Lithgow Vincent		31 Jan. 18	6 Feb. 18	Attd. 12 Pioneers.
	Unattd. List	*§**Bateman**, Herbert Elliott Richard		31 Jan. 18	6 Feb. 18	Attd. 54 Sikhs.
	Unattd. List	*§**Jones**, Maurice Rowland		31 Jan. 18	21 Feb. 18	Attd. 19 Punjabis.
190	Unattd. List	*§**Davy**, Herbert Nightingale		31 Jan. 18	6 Feb. 18	Attd. 12 Cavalry.
	Unattd. List	*§**Stephenson**, George Eric Burdett		31 Jan. 18	6 Feb. 18	Attd. 35 Sikhs.
	Unattd. List	*§**Berkeley**, Maurice Hamilton		31 Jan. 18	1 Mar. 18	Attd. 3-7th Gurkha Rifles.
	Unattd. List	*§**Saunders-Jones**, David Willoughby		31 Jan. 18	6 Feb. 18	Attd. 76 Punjabis.
	Unattd. List	*§**Green**, Eric Guy Hampden		31 Jan. 18	7 Feb. 18	Attd. 2-5th Gurkha Rifles.
195	Unattd. List	*§**Churchill**, John Stanton		31 Jan. 18	7 Feb. 18	Attd. 22 Punjabis.
	Unattd. List	*§**George**, Sandys Parker		31 Jan. 18	6 Feb. 18	Attd. 35 Sikhs.
	Unattd. List	*§**Peckover**, Richard Gurney		31 Jan. 18	5 Feb. 18	Attd. 72 Punjabis.
	Unattd. List	*§**Fitch**, Keith Simpson		31 Jan. 18	7 Feb. 18	Attd. 54 Sikhs.
	Unattd. List	*§**Marsh**, Francis Caldwell		31 Jan. 18	23 Feb. 18	Attd. 106 Pioneers.
	Unattd. List	*§**Pearson**, Harry Stuart		31 Jan. 18	6 Feb. 18	Attd. 34 Horse.
200	Unattd. List	*§**Blaber**, Hugh Kenneth		31 Jan. 18	4 Feb. 18	Attd. 70 Rifles.
	Unattd. List	*§**Parker**, Cecil John		31 Jan. 18	8 Feb. 18	Attd. 89 Punjabis.
	Unattd. List	*§**Curtoys**, Oswald Francis (59)		31 Jan. 18	7 Feb. 18	Attd. 34 Pioneers.
	Unattd. List	*§**Leonard**, Ronald Garvie		31 Jan. 18	7 Feb. 18	Attd. 3-5th Gurkha Rifles.
	Unattd. List	*§**Coldstream**, George Anson Probyns		31 Jan. 18	7 Feb. 18	Attd. 57 Rifles.
205	Unattd. List	*§**Hargrave**, Joshua Foster Harrison		31 Jan. 18	7 Feb. 18	Attd. 1-3rd Gurkha Rifles.
	Unattd. List	*§**Townshend**, Charles Maurice Waddington Loftus		31 Jan. 18	9 Feb. 18	Attd. 9 Horse.
	Unattd. List	*§**Cahill**, John		31 Jan. 18	20 Feb. 18	Attd. 129 Baluchis.
	Unattd. List	*§**Smythe**, Lee Cordon Fuller		31 Jan. 18	12 Feb. 18	Attd. 70 Rifles.
	Unattd. List	*§**Everett**, William Hanmer		31 Jan. 18	7 Feb. 18	Attd. 38 Horse.
210	Unattd. List	*§**Wilson**, Lesile Muir		31 Jan. 18	6 Feb. 18	Attd. 12 Pioneers.
	Unattd. List	*§**Greer**, Usher McGregor		31 Jan. 18	15 Feb. 18	Attd. 4-3rd Gurkha Rifles.
	Unattd. List	*§**Strickland**, Hugh Baring		31 Jan. 18	6 Feb. 18	Attd. 18 Lancers.
	Unattd. List	*§**Adames**, Frederick Henry		31 Jan. 18	15 Feb. 18	Attd. 50 Rifles.
	Unattd. List	*§**Lindsay**, Pooler Leeman		31 Jan. 18	6 Feb. 18	Attd. 130 Baluchis.
215	Unattd. List	*§**Francis**, Will		31 Jan. 18	1 Mar. 18	Attd. 3 Horse.
	Unattd. List	*§**Newington**, Gordon Henderson		31 Jan. 18	7 Feb. 18	Attd. 33 Punjabis.
	Unattd. List	*§**Dwane**, Leonard Milton Robson		31 Jan. 18	11 Feb. 18	Attd. 6 Jat L. I.
	Unattd. List	*§**Henshaw**, Alan George		31 Jan. 18	5 Feb. 18	Attd. 72 Punjabis.
	Unattd. List	*§**Bullard**, Daniel Ainslie (48*a*, 51)		31 Jan. 18	12 Feb. 18	Attd. 70 Rifles.
220	Unattd. List	*§**Batey**, George Henry Montague		31 Jan. 18	19 Feb. 18	Attd. 124 Infantry.
	Unattd. List	*§**Newbery**, Eric Archibald		31 Jan. 18	5 Feb. 18	Attd. 6 Jat L. I.
	Unattd. List	*§**Garrett**, Jack Trevor		31 Jan. 18	6 Feb. 18	Attd. 15 Lancers.
	Unattd. List	*§**Ryper**, George Henry		31 Jan. 18	2 Mar. 18	Attd. 3 Horse.
	Unattd. List	*§**Wells**, Harry Vincent		31 Jan. 18	7 Feb. 18	Attd. 35 Sikhs.
225	Unattd. List	*§**Barnes**, John Henry		31 Jan. 18	19 Feb. 18	Attd. 129 Baluchis.
	Unattd. List	*§**Johnson**, Alec Henry Garton		31 Jan. 18	7 Feb. 18	Attd. 38 Dogras.
	Unattd. List	*§**Phipps**, John Pedder		31 Jan. 18	7 Feb. 18	Attd. 4 Cavalry.
	Unattd. List	*§**Smith**, James		31 Jan. 18	5 Feb. 18	Attd. 3-9th Gurkha Rifles.
	Unattd. List	*§**Strump**, Reuben		31 Jan. 18	7 Feb. 18	Attd. 4 Cavalry.
230	Unattd. List	*§**Tittle**, Desmond Eric		31 Jan. 18	5 Feb. 18	Attd. 25 Punjabis.
	Unattd. List	*§**Larminie**, John Alexander		31 Jan. 18	20 Feb. 18	Attd. 129 Baluchis.
	Unattd. List	*§**Welsford**, Edward Percy		31 Jan. 18	7 Feb. 18	Attd. 1 Lancers.
	Unattd. List	*§**Maund**, John Anthony Hansby		31 Jan. 18	11 Feb. 18	Attd. 3-6th Gurkha Rifles.
234						

2nd-Lieutenants—*contd.*

No.	Former Corps.	Names.	Date of birth.	Date of first commission.	Date of appointment to Indian Army.	REMARKS.
235	A.S.C.	*Swann, Arthur Edmond. *9 months forfeited for seniority under conditions of I. A. O. 511 of 1917.*		3 Feb. 18	22 Nov. 18	Attd. 38 Horse.
	Arg.&Suth'd Highrs.	*John, David Ceredig. *9 months forfeited for seniority under conditions of I.A.O. 511 of 1917.*		28 Feb. 18	22 Nov. 18	Attd. 62 Punjabis.
	Unattd. List	§* Carmichael, William Bertram Graham		31 Aug. 18	16 Sep. 18	Attd. 123 Rifles.
	Unattd. List	§* Gordon, Ronald		31 Aug. 18	3 Sep. 18	Attd. 15 Sikhs.
	Unattd. List	§* Officer, Hugh Ronald		31 Aug. 18	4 Sep. 18	Attd. 54 Sikhs.
240	Unattd. List	§* Bryant, George Dangerfield		31 Aug. 18	5 Sep. 18	Attd. Corps of Guides.
	Unattd. List	§* Byers, Eric		31 Aug. 18	5 Sep. 18	Attd. 47 Sikhs.
	Unattd. List	§* May, Charles Richard Allen		31 Aug. 18	2 Sep. 18	Attd. 56 Rifles.
	Unattd. List	§* Williams, David Henry James		31 Aug. 18	5 Sep. 18	Attd. 35 Sikhs.
	Unattd. List	§* Thompson, William Gerald Murray		31 Aug. 18	4 Sep. 18	Attd. 18 Lancers.
245	Unattd. List	§* Darnell, William Richmond		31 Aug. 18	3 Sep. 18	Attd. 14 Sikhs.
	Unattd. List	§* Clapp, Cecil Douglas		31 Aug. 18	3 Sep. 18	Attd. 45 Sikhs.
	Unattd. List	§* Hill, Francis Ormsby Rowland		31 Aug. 18	7 Sep. 18	Attd. 37 Dogras.
	Unattd. List	§*Johnson, Samuel Holmes		31 Aug. 18	4 Sep. 18	Attd. 2 Lancers.
	Unattd. List	§* Mildon, Robert Karn		31 Aug. 18	4 Sep. 18	Attd. 124 Infantry.
250	Unattd. List	§* Ingram-Johnson, Ronald Edward Sydney		31 Aug. 18	5 Sep. 18	Attd. Corps of Guides.
	Unattd. List	§* Wilkins, Richard		31 Aug. 18	2 Sep. 18	Attd. 55 Rifles.
	Unattd. List	§* Wallace, David John Christie		31 Aug. 18	2 Sep. 18	Attd. 55 Rifles.
	Unattd. List	§* Lane, Charles Malet		31 Aug. 18	5 Sep. 18	Attd. 26 Punjabis.
	Unattd. List	§* McGregor, William Douglas		31 Aug. 18	13 Sep. 18	Attd. 11 Gurkha Rifles.
255	Unattd. List	§* Skone, Hubert Allen		31 Aug. 18	6 Sep. 18	Attd. 3 Gurkha Rifles.
	Unattd. List	§* Tyne, Cuthbert John		31 Aug. 18	5 Sep. 18	Attd. 2 Gurkha Rifles.
	Unattd. List	§* Alston, William Lowry		31 Aug. 18	4 Sep. 18	Attd. 32 Pioneers.
	Unattd. List	§* McDougall, Roy Malcolm		31 Aug. 18	5 Sep. 18	Attd. 15 Sikhs.
	Unattd. List	§* Blackden, John Franklyn		31 Aug. 18	5 Sep. 18	Attd. 31 Lancers.
260	Unattd. List	§* Clarabut, Frank Stone		31 Aug. 18	14 Sep. 18	Attd. 26 Cavalry.
	Unattd. List	§* Jones, Victor Walter		31 Aug. 18	15 Sep. 18	Attd. 48 Pioneers.
	Unattd. List	§* Murray, Shadwell Keith		31 Aug. 18	1 Sep. 18	Attd. 10 Gurkha Rifles.
	Unattd. List	§* Hindle, Malcolm Douglas		31 Aug. 18	4 Sep. 18	Attd. 35 Sikhs.
	Unattd. List	§* Stevenl, Osker Kinsman		31 Aug. 18	1 Sep. 18	Attd. 107 Pioneers.

265	Unattd. List	§* **Smith**, Terence Nelson		31 Aug. 18	5 Sep. 18	Attd. 6 Gurkha Rifles.
	Unattd. List	§* **Foskett**, Richard Maurice		31 Aug. 18	13 Sep. 18	Attd. 11 Gurkha Rifles.
	Unattd. List	§* **Worthington**, Roger Francis		31 Aug. 18	14 Sep. 18	Attd. 64 Pioneers.
	Unattd. List	§* **Tomlinson**, Robert Alexander Henry		31 Aug. 18	5 Sep. 18	Attd. 50 Rifles.
	Unattd. List	§* **Kenny**, William David		31 Aug. 18	15 Sep. 18	Attd. 11 Gurkha Rifles.
270	Unattd. List	§* **Proctor**, Percival		31 Aug. 18	4 Sep. 18	Attd. 76 Punjabis.
	Unattd. List	§* **Raper**, Charles Fraser		31 Aug. 18	4 Sep. 18	Attd. 20 Infantry.
	Unattd. List	§* **Ferguson**, Dudley Charles Haden		31 Aug. 18	5 Sep. 18	Attd. 93 Infantry.
	Unattd. List	§* **Will**, Henry St. Clair Ogilvie		31 Aug. 18	3 Sep. 18	Attd. 45 Sikhs.
	Unattd. List	§* **Bulgin**, John		31 Aug. 18	6 Sep. 18	Attd. 39 Rifles.
275	Unattd. List	§* **Pirouet**, Henry Nicolle		31 Aug. 18	4 Sep. 18	Attd. 34 Pioneers.
	Unattd. List	§* **Rundle**, Arthur Leslie de Villiers		31 Aug. 18	4 Sep. 18	Attd. 1 Gurkha Rifles.
	Unattd. List	§* **Henderson**, Edwin Graham Straton		31 Aug. 18	10 Sep. 18	Attd. 125 Rifles.
	Unattd. List	§* **Richardson-Gibson**, Frederick Joseph		31 Aug. 18	2 Sep. 18	Attd. 56 Rifles.
	Unattd. List	§* **Dallas Conte**, Norman		31 Aug. 18	15 Sep. 18	Attd. 104 Rifles.
280	Unattd. List	§* **Spurgin**, Arthur Rushworth		31 Aug. 18	5 Sep. 18	Attd. 12 Pioneers.
	Unattd. List	§* **Duesbury**, Alan Arthur		31 Aug. 18	3 Sep. 18	Attd. 2 Gurkha Rifles.
	Unattd. List	§* **Burge**, Norman Benjamin		31 Aug. 18	2 Sep. 18	Attd. 106 Pioneers.
	Unattd. List	§* **Greatwood**, Hugh Egerton		31 Aug. 18	5 Sep. 18	Attd. 123 Rifles.
	Unattd. List	§* **Price**, James Chermside		31 Aug. 18	13 Sep. 18	Attd. 117 Mahrattas.
285	Unattd. List	§* **Coleman**, Cecil Maltwood		31 Aug. 18	5 Sep. 18	Attd. 25 Cavalry.
	Unattd. List	§* **Palmer**, George Wilfred		31 Aug. 18	5 Sep. 18	Attd. 41 Dogras.
	Unattd. List	§* **Barnes**, Eric Cecil		31 Aug. 18	11 Sep. 18	Attd. 51 Sikhs.
	Unattd. List	§* **Whitley**, Archibald Pepin		31 Aug. 18	4 Sep. 18	Attd. 34 Pioneers.
	Unattd. List	§* **Ruttledge**, Robert Francis		31 Aug. 18	4 Sep. 18	Attd. 34 Horse.
290	Unattd. List	§* **Plunkett**, Humphrey Ronald Sturge		31 Aug. 18	5 Sep. 18	Attd. 5 Gurkha Rifles.
	Unattd. List	§* **Malden**, William Jack		31 Aug. 18	5 Sep. 18	Attd. Corps of Guides.
	Unattd. List	§* **White**, Edward Willoughby		31 Aug. 18	3 Sep. 18	Attd. 22 Cavalry.
	Unattd. List	§* **McOstrich**, Alexander Henry Morris		31 Aug. 18	6 Sep. 18	Attd. 39 Rifles.
	Unattd. List	§* **Etherington**, Roy Alexander		31 Aug. 18	1 Sep. 18	Attd. 4 Gurkha Rifles.
295	Unattd. List	§* **Day**, Horace Melville		31 Aug. 18	5 Sep. 18	Attd. 5 Cavalry.
	Unattd. List	§* **Story**, Francis Edward		31 Aug. 18	14 Sep. 18	Attd. 20 Horse.
	Unattd. List	§* **Power**, Francis William		31 Aug. 18	5 Sep. 18	Attd. 41 Dogras.
	Unattd. List	§* **Phelps**, John Newbery		31 Aug. 18	4 Sep. 18	Attd. 38 Dogras.
	Unattd. List	§* **Pryce**, Walter William Richard		31 Aug. 18	14 Sep. 18	Attd. 120 Infantry.
300	Unattd. List	§* **Alderson**, Marmaduke Guy		31 Aug. 18	22 Sep. 18	Attd. 10 Gurkha Rifles.
	Unattd. List	§* **Davies**, Ernest Richard Lasham		31 Aug. 18	5 Sep. 18	Attd. 8 Rajputs.
	Unattd. List	§* **Wardle**, John Leek		31 Aug. 18	14 Sep. 18	Attd. 20 Horse.
	Unattd. List	§* **Jones**, Walter Henry Clulee		31 Aug. 18	3 Oct. 18	Attd. 103 Infantry.
	Unattd. List	§* **Mather**, Robert William		31 Aug. 18	14 Sep. 18	Attd. 114 Mahrattas.
305	Unattd. List	§* **Hale**, William Godfrey	(207*a*)	31 Aug. 18	14 Sep. 18	Attd. 114 Mahrattas.
	Unattd. List	§* **Diack**, Henry James Wallace		31 Aug. 18	5 Sep. 18	Attd. 5 Gurkha Rifles.
	Unattd. List	§* **Bates**, Guy Ernest Lockington		31 Aug. 18	16 Sep. 18	Attd. 38 Horse.
	Unattd. List	§* **Grainger**, Stephen Norman		31 Aug. 18	13 Sep. 18	Attd. 112 Infantry.
	Unattd. List	§* **Knight**, Richard		31 Aug. 18	4 Sep. 18	Attd. Corps of Guides.
310	Unattd. List	§* **Dixon**, Vyvyan Hustler		31 Aug. 18	4 Sep. 18	Attd. 17 Cavalry.
	Unattd. List	§* **Price**, Christopher Salter		31 Aug. 18	5 Sep. 18	Attd. Corps of Guides.
	Unattd. List	§* **Moreton**, Basi Meredith		31 Aug. 18	4 Sep. 18	Attd. 37 Dogras.
	Unattd. List	§* **Jennings**, Lewis Wensley Bond		31 Aug. 18	24 Sep. 18	Attd. 64 Pioneers.
	Unattd. List	§* **Costello**, John Walter		31 Aug. 18	4 Sep. 18	Attd. 23 Cavalry.
315	Unattd. List	§* **Watson**, Nigel		31 Aug. 18	4 Sep. 18	Attd. 9 Horse.
316	Unattd. List	§* **Power**, William Noel		31 Aug. 18	4 Sep. 18	Attd. 52 Sikhs.

2nd-Lieutenants—*contd.*

No.	Former Corps.	Names.	Date of birth.	Date of first commission.	Date of appointment to Indian Army.	Remarks.
317	Unattd. List	§*Abercrombie, Desmond Eden		31 Aug. 18	3 Sep. 18	Attd. 10 Lancers.
	Unattd. List	§*Wilkinson, Allan Scott		31 Aug. 18	4 Sep. 18	Attd. 11 Lancers.
	Unattd. List	§*Roche, Ian Aird		31 Aug. 18	6 Sep. 18	Attd. 9 Gurkha Rifles.
320	Unattd. List	§*Mason, John Lea		31 Aug. 18	14 Sep. 18	Attd. 101 Grenadiers.
	Unattd. List	§*Donnelly, Harold Roland		31 Aug. 18	14 Sep. 18	Attd. 11 Gurkha Rifles.
	Unattd. List	§*Lawrence, Edward Daniel		31 Aug. 18	18 Sep. 18	Attd. 75 Infantry.
	Unattd. List	§*Turner, Miles Ransome		31 Aug. 18	4 Sep. 18	Attd. 25 Cavalry.
	Unattd. List	§*Thomson, William Robert		31 Aug. 18	5 Sep. 18	Attd. 109 Infantry.
325	Unattd. List	§*Warren, Peter		31 Aug. 18	4 Sep. 18	Attd. 16 Cavalry.
	Unattd. List	§*Dolphin, Harold Maximilian Burton		31 Aug. 18	2 Sep. 18	Attd. 56 Rifles.
	Unattd. List	§*Carles, John Robin		31 Aug. 18	3 Sep. 18	Attd. 15 Sikhs.
	Unattd. List	§*Davis, Henry Carlisle Monsell		31 Aug. 18	6 Sep. 18	Attd. 3 Rifles.
	Unattd. List	§*Russell, Hubert Leslie		31 Aug. 18	5 Sep. 18	Attd. 27 Cavalry.
330	Unattd. List	§* Fife, Dougalas McDuff		31 Aug. 18	14 Sep. 18	Attd. 6 Infantry.
	Unattd. List	§* Platts, Shelley Arden		31 Aug. 18	5 Sep. 18	Attd. 9 Gurkha Rifles.
	Unattd. List	§* Brett, Wills Hill		31 Aug. 18	14 Sep. 18	Attd 26 Cavalry.
	Unattd. List	§* Robinson, John Godfrey		31 Aug. 18	4 Sep. 18	Attd. 27 Punjabis.
	Unattd. List	§* Florence, George Douglas		31 Aug. 18	14 Sep. 18	Attd. 42 Deoli Regt.
335	Unattd. List	§* Collier, William James		31 Aug. 18	13 Sep. 18	Attd. 122 Infantry.
	Unattd. List	§* Geere, Alfred John Kelsall		31 Aug. 18	15 Sep. 18	Attd. 67 Punjabis.
	Unattd. List	§* Brown, William Felix		31 Aug. 18	14 Sep. 18	Attd. 10 Jats.
	Unattd. List	§* Bates, Frank Page Elton		31 Aug. 18	4 Sep. 18	Attd. 54 Sikhs.
	Unattd. List	§* Jones, Reginald John Peter		31 Aug. 18	3 Sep. 18	Attd. 22 Cavalry.
340	Unattd. List	§* Windsor-Aubrey, Wentworth Arthur		31 Aug. 18	5 Sep. 18	Attd. 32 Pioneers.
	Unattd. List	§* Meyer, William Reginald Macdonald		31 Aug. 18	4 Sep. 18	Attd. 21 Cavalry.
	Unattd. List	§* A'Brook, Donald		31 Aug. 18	5 Sep. 18	Attd. 2 Infantry.
	Unattd. List	§* Roberts, Vyvyan Prowting		31 Aug. 18	5 Sep. 18	Attd. 27 Cavalry.
	Unattd. List	§* Auton, John Howell		31 Aug. 18	4 Sep. 18	Attd. 25 Punjabis.
345	Unattd. List	§* Griffiths, Frank Julian		31 Aug. 18	5 Sep. 18	Attd. 113 Infantry.
	Unattd. List	§* Hotton, Hugh Overthwaite		31 Aug. 18	6 Sep. 18	Attd. 11 Gurkha Rifles.

	Unattd. List	§* **Blackwell,** John Norman	31 Aug. 18	5 Sep. 18	Attd. 4 Cavalry.
	Unattd. List	§* **Scott,** John Graham	31 Aug. 18	1 Sep. 18	Attd. 11 Gurkha Rifles.
	Unattd. List	§* **Eyre,** John Stephen Giles	31 Aug. 18	14 Sep. 18	Attd. 27 Punjabis.
350	Unattd. List	§* **Watts,** Lancelot Gutichen	31 Aug. 18	4 Sep. 18	Attd. 1 Gurkha Rifles.
	Unattd. List	§* **Hotham,** George Ignatius Montague	31 Aug. 18	8 Sep. 18	Attd. 4 Cavalry.
	Unattd. List	§* **Gillmon,** John Hewat	31 Aug. 18	14 Sep. 18	Attd. 74 Punjabis.
	Unattd. List	§* **Drought,** Alexander Charles	31 Aug. 18	5 Sep. 18	Attd. 113 Infantry.
	Unattd. List	§* **Alexander,** Augustus Gordon Stewart	31 Aug. 18	16 Sep. 18	Attd. 39 C. I. Horse.
355	Unattd. List	§* **Lane-Ryan,** Stanley Edgar	31 Aug. 18	4 Sen. 18	Attd. 9 Horse.
	Unattd. List	§* **Mawhood,** John Charles	31 Aug. 18	4 Sep. 18	Attd. Corps of Guides.
	Unattd. List	§* **Tennant,** John Henry	31 Aug. 18	5 Sep. 18	Attd. 50 Rifles.
	Unattd. List	§* **Hutchinson,** George Willey	31 Aug. 18	4 Sep. 18	Attd. 23 Pioneers.
	Unattd. List	§* **Dawson,** Arthur Vernon	31 Aug. 18	4 Sep. 18	Attd. 15 Lancers.
360	Unattd. List	§* **Ross,** James Rolland	31 Aug. 18	11 Nov. 18	Attd. 23 Infantry.
	Unattd. List	§* **West,** Philip Scarborough	31 Aug. 18	28 Sep. 18	Attd. 33 Punjabis.
	I. A. R. O.	*§Lang, Henry George *Service for promotion from 20 Mar. 18 under I. A. O. 510 of 1918.*	20 June 17 20 Mar. 18	26 Sep. 18	Attd. 95 Infantry.
	I. A. R. O.	*§Croft, Frederick Roy 207a *Service for promotion from 20 Mar. 18 under I. A. O. of 1918.*	20 June 17 20 Mar. 18	30 Sep. 18	Attd. 8 Rajputs.
	I. A. R. O.	*§Leonard, Patrick Joseph 207a *Service for promotion from 15 Feb. 18 under I. A. O. 510 of 1918.*	5 Mar. 17 5 Feb. 18	3 Oct. 18	Attd. 83 Lt. Infy.
365	Unattd. List	§* **Mallaby,** Aubertin Walter Southern	1 Oct. 18	8 Oct. 18	Attd. 27 Punjabis.
	Unattd. List	§* **Dent,** Robert Cecil	1 Oct. 18	7 Oct. 18	Attd. 57 Rifles.
	Unattd. List	§* **Garrow,** Robert Kenneth	1 Oct. 18	13 Oct. 18	Attd. 11 Rajputs.
	Unattd. List	§* **Pargiter,** Justin Gerrard	1 Oct. 18	12 Oct. 18	Attd. 43 Regiment.
	Unattd. List	§* **Spencer,** Leonard Algernon	1 Oct. 18	7 Oct. 18	Attd. 34 Horse.
370	Unattd. List	§* **Staecy,** Leonard Brown	1 Oct. 18	8 Oct. 18	Attd. 28 Punjabis.
	Unattd. List	§* **Broadway,** Alan Colquhoun	1 Oct. 18	7 Oct. 18	Attd. 56 Rifles.
	Unattd. List	§* **Lentaigne,** Walter David Alexander	1 Oct. 18	11 Oct. 18	Attd. 4 Gurkha Rifles.
	Unattd. List	§* **Nicholls,** Alfred William	1 Oct. 18	6 Oct. 18	Attd. 88 Infantry.
	Unattd. List	§* **Kerr,** John Norman	1 Oct. 18	8 Oct. 18	Attd. 37 Dogras.
375	Unattd. List	§* **Lowsley,** Geoffrey Lionel	1 Oct. 18	7 Oct. 18	Attd. 4 Gurkha Rifles.
	Unattd. List	§* **Gordon,** Robert David Lyons	1 Oct. 18	14 Oct. 18	Attd. 48 Pioneers.
	Unattd. List	§* **Tribe,** Albert Geoffrey	1 Oct. 18	10 Oct. 18	Attd. 107 Pioneers.
	Unattd. List	§* **Coleman,** George Vincent Leigh	1 Oct. 18	7 Oct. 18	Attd. 55 Rifles.
	Unattd. List	§* **Straker,** Frederick Eaton	1 Oct. 18	7 Oct. 18	Attd. 92 Punjabis.
380	Unattd. List	§* **Dobbin,** Fergus Le Fanu	1 Oct. 18	7 Oct. 18	Attd. 6 Gurkha Rifles.
	Unattd. List	§* **Gardner,** John Henry Dunningham	1 Oct. 18	8 Oct. 18	Attd. 8 Gurkha Rifles.
	Unattd. List	§* **O'Ferrall,** John Lawrence Francis	1 Oct. 18	7 Oct. 18	Attd. 47 Sikhs.
	Unattd. List	§* **Pollock,** Vincent David	1 Oct. 18	6 Oct. 18	Attd. 35 Sikhs.
	Unattd. List	§* **Peacocke,** Gerald Ivan Crozier	1 Oct. 18	6 Oct. 18	Attd. 108 Infantry.
385	Unattd. List	§* **Brown,** Robert Thirkell	1 Oct. 18	11 Oct. 18	Attd. 63 Infantry.
	Unattd. List	§* **Sinclair,** Malcolm Cecil	1 Oct. 18	11 Oct. 18	Attd. 129 Baluchis.
	Unattd. List	§* **Akehorst,** Charles Henry Irwin	1 Oct. 18	12 Oct. 18	Attd. 34 Pioneers.
	Unattd. List	§* **Watkins,** Bernard Springett	1 Oct. 18	6 Oct. 18	Attd. 54 Sikhs.
	Unattd. List	§* **Gostling,** Denis Erroll	1 Oct. 18	7 Oct. 18	Attd. 36 Horse.
390	Unattd. List	§* **Coole,** Arthur Evans	1 Oct. 18	12 Oct. 18	Attd. 32 Pioneers.

No.	Former Corps.	Names.	Date of birth.	Date of first commission.	Date of appointment to Indian Army.	Remarks.
		2nd-Lieutenant—*concld.*				
391	Unattd. List	§***Mitchell,** Sydney Andrew	. . . . .	1 Oct. 18	7 Oct. 18	Attd. 40 Pathans.
	Unattd. List	§***Stables,** Francis Hugh Anthony	. . . . .	1 Oct. 18	7 Oct. 18	Attd. 5 Gurkha Rifles.
	Unattd. List	§***Stevenson,** Philip Noel	. . . . .	1 Oct. 18	7 Oct. 18	Attd. 109 Infantry.
	Unattd. List	§***Lownders,** Donald George	. . . . .	1 Oct. 18	21 Oct. 18	Attd. 39 Garhwal Rifles.
395	Unattd. List	§***Keenan,** Terence Patrick Charles Seymour	. . . . .	1 Oct. 18	8 Oct. 18	Attd. 130 Baluchis.
	Unattd. List	§***Carse,** William Mitchell	. . . . .	1 Oct. 18	6 Oct. 18	Attd. 3 Brahmans.
	Unattd. List	§***Price,** Cyril Hugh	. . . . .	1 Oct. 18	7 Oct. 18	Attd. 34 Pioneers.
	Unattd. List	§***Butler,** Arthur Hedley	. . . . .	1 Oct. 18	10 Oct. 18	Attd. 127 Infantry.
398	Unattd. List	§***Alban,** Reginald George Evelin William	. . . . .	1 Oct. 18	11 Oct. 18	Attd. 7 Gurkha Rifles.

No.	Names.	Date of rank.	Remarks.

OFFICERS OF THE INDIAN ARMY HAVING TEMPORARY COMMISSIONS.

Major.

No.	Names.	Date of rank.	Remarks.
1	**Arbuthnot,** Percy Bingham (54, 79)	14 May 15	

Captains.

No.	Names.	Date of rank.	Remarks.
	Studd, Eric (D. 1) (54, 79)	11 Mar. 16	
	Guthrie, Alan	15 May 16	
	Puech, Albert Gottleib, O.B.E., V. D.	28 Jan. 17	
	Deane, Edward Stanley (D. 1) (51, 52, 54, 71, 79)	14 May 17	
5	**Pearse,** George Gerald (54, 79)	27 Feb. 18	

Lieutenants.

No.	Names.	Date of rank.	Remarks.
1	**Tedman,** Cyril, M.C. (79)	23 Apl. 15	
	Brown, Archibald	1 Sep. 18	
	Taylor, Walter (207*a*)	1 Sep. 18	
	Fishburn, Joseph	1 Sep. 18	
5	**Husband,** William Frederick	1 Sep. 18	
	Roberts, Charles Joseph	1 Sep. 18	
	Elles, Frederick William Cesari	1 Sep. 18	
	Down, Harold Richard	1 Sep. 18	
	Wicks, Frederick George	12 Sep. 18	
10	**Cockerill,** James Wilfred	13 Sep. 18	
	McCuish, Angus (71)	13 Sep. 18	
	Cotton, Trevor Charles (207*a*)	13 Sep. 18	
	Cross, Stanley David	13 Sep. 18	
	Dance, Albert James (207*a*)	13 Sep. 18	
15	**Stephen,** Arthur Alexander	13 Sep. 18	
	Crozier, David	14 Sep. 18	
	Dumper, Ernest	25 Sep. 18	
	Moore, Albert John	1 Oct. 18	
	Freeman, Ernest Charles	1 Oct. 18	
20	**Wyatt,** Cyril Jack	1 Oct. 18	
	Fox, Bertram Frederick	1 Oct. 18	
	Walton, George William (57	1 Oct. 18	
	Warwick, Thomas George	1 Oct. 18	
	Middleweek, Leonard William Samuel	1 Oct. 18	
25	**Roberts,** Samuel Norman	1 Oct. 18	
	Rhodes, Herbert Harpham	1 Oct. 18	
	Dyson, Harry Peak	1 Oct. 18	
	Jones, Thomas Percy	1 Oct. 18	
	Wood, Maurice Hiram	1 Oct. 18	
30	**Cullingford,** Frank Harold	1 Oct. 18	
	Bugler, Ernest Frank (207*a*)	1 Oct 18	
	Rose, Samuel George Stephens	1 Oct. 18	
	Mitchell, Reginald Robert	1 Oct. 18	
	Bowditch, Clement Edgar	1 Oct. 18	
35	**Guppy,** Samuel John	1 Oct. 18	
	Lambden, Charles Baker	1 Oct. 18	
	Belben, Herbert Reginald	1 Oct. 18	
	Rendall, Charles	1 Oct. 18	
	Bunce, Louis Leslie	1 Oct. 18	
40	**Leat,** Herbert George	1 Oct. 18	
	Manthel, Wilfred John (207*a*)	1 Oct. 18	
	Chapman, Edgar John	1 Oct. 18	
	Josland, George Challice	5 Oct. 18	
	Armstrong, Alfred John Eite	5 Oct. 18	
45	**Watson,** Albert Ernest	5 Oct. 18	
	Wright, Albert Alexander (48*a*)	5 Oct. 18	
	Gibson, Philip George Hammond (48*a*)	5 Oct. 18	
	Clack, George Henry	5 Oct. 18	
	Pugh, Frank Marlow (207*a*)	5 Oct. 18	
50	**Holbrook,** Harold Frederick (207*a*)	5 Oct. 18	
	Curley, Walter Langley (46, 207*a*)	5 Oct. 18	
	Hore, Archibald Tom (207*a*)	5 Oct. 18	
	Hodder, Eric Mesney	5 Oct. 18	
	Carter, Frederick Norman	5 Oct. 18	
55	**Laver,** William Ewart	5 Oct. 18	
	Pitt, Ronald Harry	5 Oct. 18	
57	**Carter,** George Hartle Horace (207*a*)	5 Oct. 18	

OFFICERS OF THE INDIAN ARMY HAVING TEMPORARY COMMISSIONS—*contd.*

Lieutenants—*contd.*

No.	Names.	Date of rank.	Remarks.
58	**Brown,** Thomas Harold Cliffe	5 Oct. 18	
	Farley, Henry James	5 Oct. 18	
60	**Ashton,** Cecil Thomas (48*a*, 51)	5 Oct. 18	
	Pook, Ernest Richard	5 Oct. 18	
	Stephen, Peter Wiseman Rettie (46)	5 Oct. 18	
	Hume, Robert	5 Oct. 18	
	Shepard, Arthur John	5 Oct. 18	
65	**Buley,** Sidney Havelock	5 Oct. 18	
	Waitt, William Edwin	5 Oct. 18	
	Newnham, Henry John	5 Oct. 18	
	Van Houten, Frederic	5 Oct. 18	
	Spencer, William Greenwood (207*a*)	5 Oct. 18	
70	**Bell,** William Francis	5 Oct. 18	
	Page, Charles	5 Oct. 18	
	Swanborough, Frederick Claud Valentine	5 Oct. 18	
	Dover, William Ernest	5 Oct. 18	
	Wykes, Edwin	5 Oct 18	
75	**Baker,** George Mallam	5 Oct. 18	
	Mullarkey, Patrick Stephen (207*a*)	5 Oct. 18	
	Woods, Henry Alexander	5 Oct. 18	
	Howitt, George	5 Oct. 18	
	Stone, Algernon Percy Grimmith	5 Oct. 18	
80	**Linell,** Charles James Bishton (48*a*)	5 Oct. 18	
	Ingram, Frederick Ernest	5 Oct. 18	
	Harmer, Stanley Hugh	5 Oct. 18	
	Linker, Lionel Leslie (207*a*)	5 Oct. 18	
	Stillman, Archibald Charles Henry	5 Oct. 18	
85	**Brett,** Clarence Edwin	5 Oct. 18	
	Higgs, John Vinson	5 Oct. 18	
	Cleaver, Norman Richard	5 Oct. 18	
	Woods, Herbert	5 Oct. 18	
	White, Percy Ralph	5 Oct. 18	
90	**Bonner,** Frederick John	5 Oct. 18	
	Wallace, Edward Henry	5 Oct. 18	
	James, William Richard	5 Oct. 18	
	Hallam, Edward Hubert (207*a*)	5 Oct. 18	
	Murrell, Arthur Frederick	5 Oct. 18	
95	**Mugford,** Thomas George (51, 207*a*)	5 Oct. 18	
	Legg, Percy Cecil (207*a*)	5 Oct. 18	
	Waight, Gordon	5 Oct. 18	
	Harvey, Herbert	5 Oct. 18	
	Hewitt, Sydney Herbert	5 Oct. 18	
100	**Brookes,** Thomas Louis (207*a*)	5 Oct. 18	
	Richmond, Montague George (57)	5 Oct. 18	
	Bethell, Charles Victor	5 Oct. 18	
	Falkenstein, Albert James	5 Oct. 18	
	Tegetmeier, Philip Anthony	5 Oct. 18	
105	**Saunders,** Walter Henry	5 Oct. 18	
	Kind, Francis (207*a*)	5 Oct. 18	
	Hughes, Samuel	5 Oct. 18	
	Conibear, William John	10 Oct. 18	
	Thornton, George Edward	1 Nov. 18	
110	**Riggs,** Alfred James	1 Nov. 18	
	Burgess, Benjamin (207*a*)	1 Nov. 18	
	Miller, Francis Beatty	1 Nov. 18	
	Cotton, Henry John	1 Nov. 18	
	Hatton, William Austin	1 Nov. 18	
115	**Taylor,** Charles William	1 Nov. 18	
	Corbould, Eric Chilton	1 Nov. 18	
	Fraser, Roy	1 Nov. 18	
	Macmillan, James (51)	1 Nov. 18	
	McNab, Andrew (207*a*)	1 Nov. 18	
120	**Deane,** Francis John (207*a*)	1 Nov. 18	
	Price, William Harold	1 Nov. 18	
	Poole, George William (46)	1 Nov. 18	
	Ellis, Stanley (207*a*)	1 Nov. 18	
	Bicker, Hedley John	1 Nov. 18	
125	**Hillier,** Leslie Oswald	1 Nov. 18	
	Jenkins, Clarence Henley	1 Nov. 18	
127	**Roberts,** Anthony Charles	1 Nov. 18	

OFFICERS OF THE INDIAN ARMY HAVING TEMPORARY COMMISSIONS—*contd*

Lieutenants—*contd.*

No.	Names.	Date of rank.	Remarks.
128	**Michel,** Thomas John	1 Nov. 18	
	Mills, Edward George	1 Nov. 18	
130	**Danagher,** Arthur Vincent	1 Nov. 18	
	Newman, James	1 Nov. 18	
	Upchurch, Leslie Charles	1 Nov. 18	
	Carter, Louis Harold	1 Nov. 18	
	Martin, John Henry Allan	1 Nov. 18	
135	**Gigg,** Frederick John (51)	1 Nov. 18	
	Bassett, Thomas Ralph	1 Nov. 18	
	Davies, William John	1 Nov. 18	
	Seeley, Alfred Charles	1 Nov. 18	
	Nicholson, Robert Lees	1 Nov. 18	
140	**Pinkerton,** George Elliott	1 Nov. 18	
	Donnell, Norman Augustus	1 Nov. 18	
	Taverner, Daniel Thomas	1 Nov. 18	
	Rice, Howard George	5 Nov. 18	
	Hargreaves, Frank Bertram	5 Nov. 18	
145	**Haynes,** Sidney John (57)	5 Nov. 18	
	Hampton, Cyril Herbert	5 Nov. 18	
	Rood, Leonard Frederick (48*a*)	5 Nov. 18	
	Rowe, Walter Henry (46)	5 Nov. 18	
	Wonnacott, Alfred John	5 Nov. 18	
150	**Barraclough,** Arnold Ewart	5 Nov. 18	
	Biddlecombe, Gordon Lewis (*temp. Capt., 13 Aug. 18*)	5 Nov. 18	
	Ford, Ronal Arthur	5 Nov. 18	
	Hasnip, Geoffrey Herbert	5 Nov. 18	
	Rayner, Thomas Austen	5 Nov. 18	
155	**Boyce,** William Arthur	5 Nov. 18	
	Sergeant, Arthur Horace (207*a*)	5 Nov. 18	
	Handy, Frank Robert	5 Nov. 18	
	Matthews, Sidney Fulk Weale	5 Nov. 18	
	Merrill, William Eric (207*a*)	5 Nov. 18	
160	**Stark,** Clarence William	5 Nov. 18	
	Hoare, Arthur William	5 Nov. 18	
	Read, Robert Stanley (207*a*)	5 Nov. 18	
	Sutherland, Joseph White	5 Nov. 18	
	Dew, Thomas Henry	5 Nov. 18	
165	**Kent,** Alfred Robert	5 Nov. 18	
	Legg, Edward Dixon Burns	5 Nov. 18	
	Mockridge, Arthur Percival (207*a*)	5 Nov. 18	
	Smart, John Nelson	5 Nov. 18	
	Sparkes, Henry Bolton	5 Nov. 18	
170	**Wood,** Cyril	5 Nov. 18	
	Fullerton, Lionel William	5 Nov. 18	
	Julian, Alfred Leslie	5 Nov. 18	
	Linker, John Edwin	5 Nov. 18	
	Jones, Robert Ivor	5 Nov. 18	
175	**Barr,** Arthur	5 Nov. 18	
	Cude, Charles John	5 Nov. 18	
	Hunt, Henry Gilbert	5 Nov. 18	
	Strange, Frank	5 Nov. 18	
	Brooks, Robbie Winter	5 Nov. 18	
180	**Fournier,** Albert Edward	5 Nov. 18	
	Musto, Clifford Roberts	5 Nov. 18	
	O'Connell, William Robert Francis (59)	5 Nov. 18	
	Wyatt, Cecil Henry	5 Nov. 18	
	Wyatt, Reginald William	5 Nov. 18	
185	**Collett,** Reginald Arthur	5 Nov. 18	
	Bellamy, Ernest Adolphus	5 Nov. 18	
	Blackmore, John Phillip (207*a*)	5 Nov. 18	
	Mills, Cyril Frank	5 Nov. 18	
	Fellingham, Frank Wesley	5 Nov. 18	
190	**Innes,** Gilbert Anderson	5 Nov. 18	
	Norman, Stuart Henry	5 Nov. 18	
	D'Arcy, Michael Angelo Aloysius (59)	5 Nov. 18	
	Tilley, Harold Rondolph	5 Nov. 18	
	Turnbull, David	5 Nov. 18	
195	**Philpott,** Gilbert Richard	5 Nov. 18	
	Philp, William Hayward Stokes	5 Nov. 18	
197	**Walter,** Charles Alexander	5 Nov. 18	

OFFICERS OF THE INDIAN ARMY HAVING TEMPORARY COMMISSIONS—*contd.*

Lieutenants—*concld.*

No.	Names.	Date of rank.	Remarks.
198	**Cross,** William	5 Nov. 18	
	Raeburn, Joseph Percy	5 Nov. 18	
200	**Taylor,** Wesley	5 Nov. 18	
	Chilcott, Wilfred John	5 Nov. 18	
	Goldsack, Arthur George Richard	5 Nov. 18	
	Creasy, Noel Edward	5 Nov. 18	
	Cude, Thomas	5 Nov. 18	
205	**Cunningham,** William James	5 Nov. 18	
	Pole, George Edward	5 Nov. 18	
	Massey, Albert Peirce (46)	5 Nov. 18	
	Mooring, Leslie Lionel	5 Nov. 18	
	Strudwick, Walter Albert Jones	5 Nov. 18	
210	**Bailey,** Walter George	5 Nov. 18	
	Cunningham, Terence Claud	5 Nov. 18	
	Parsons, Leonard John	5 Nov. 18	
	Lockstone, John	5 Nov 18	
	Abbott, Edwin James	5 Nov. 18	
215	**Bodley,** Cyril Athelstan	5 Nov. 18	
	Clarke, William Henry	5 Nov. 18	
	Newlyn, Edward William George	5 Nov. 18	
	Senior, Eric Boyd Graham	5 Nov. 18	
	Turl, Gilbert Holway	5 Nov. 18	
220	**Brown,** Edward Reginald	5 Nov. 18	
	Haymes, Thomas Fergus (207*a*)	5 Nov. 18	
	Thurston, Claude Walter	5 Nov. 18	
	Ashton, Francis Clifford	5 Nov 18	
	Flye, Thomas Henry	5 Nov. 18	
225	**Saunders,** Thomas Henry (207*a*)	5 Nov. 18	
	Whittington, Harold	5 Nov. 18	
	Kent, William Harvey	5 Nov. 18	
	Caminada, Charles Bernard	5 Nov. 18	
	Loveland, Frank	5 Nov. 18	
230	**Lamble,** Sidney Harold	5 Nov. 18	
	Tarr, Walter Govett (207*a*)	5 Nov. 18	
	Beesley, Frank	5 Nov. 18	
	Rackett, George Leopold	5 Nov. 18	
	Steele, Harold	5 Nov. 18	
235	**Baker,** Victor Albert	5 Nov. 18	
	Roughton, George Edward	10 Nov. 18	

2nd-Lieutenants.

No.	Names.	Date of rank.	Remarks.
1	**Bradshaw,** Frederick James	5 Oct. 17	
	Pitchford, Herbert	5 Nov. 17	
	Bell, William Arthur John	5 Nov. 17	
	Haymes, George Noel	5 Nov. 17	
5	**Taylor,** Sydney Thomas Ransome	5 Nov. 17	
	Muschamp, William Alfred Henuson	5 Nov. 17	
	White, Gilbert	5 Nov. 17	
	Symonds, Eneas	5 Nov. 17	
	Ball, William Valentine	5 Nov. 17	
10	**Rawlinson,** Charles Albert Walter	5 Nov. 17	
	Foster, Thomas Duncan	26 Nov. 17	
	Taggart, Michael John	7 Dec. 17	
	Smith, Stanley	11 Dec. 17	
	Stone, John Frederick	20 Dec. 17	
15	**Vanstone,** Wallace Herbert	30 Dec. 17	
	Sullivan, George Marriott Rivers	4 Jan. 18	
	Thompson, George Laidlaw	5 Jan. 18	
	Evans, Ambrose	5 Jan. 18	
	Bloxsome, William Auther	5 Jan. 18	
20	**Partridge,** Cecil	5 Jan. 18	
	Watts, Harold Edgar	5 Jan. 18	
	Collett, William Arthur	5 Jan. 18	
	Borthwick, Charles James	5 Jan. 18	
	Winter, John Herbert	5 Jan. 18	
25	**Fill,** Hugo Hugh Golding	5 Jan. 18	
	Flynn, Joseph William	5 Jan. 18	
	Clark, Howard Williams	5 Jan. 18	
	Green, Harold	5 Jan. 18	
	Smith, William Herbert	5 Jan. 18	
30	**Butcher,** Ralph Abercrombie	5 Jan. 18	
	Hawkes, John Stanley	5 Jan. 18	
	Wearing, Walter	5 Jan. 18	
	Davis, Frank Leonard	5 Jan. 18	
34	**Adkins,** Earnest Chapman	5 Jan. 18	

OFFICERS OF THE INDIAN ARMY HAVING TEMPORARY COMMISSIONS—*contd.*

2nd-Lieutenants—*contd.*

No.	Names.	Date of rank.	REMARKS.
35	**List,** Frank Joseph	5 Jan. 18	
	Halstead, John	5 Jan. 18	
	Cleator, Alfred Ewart	5 Jan. 18	
	Ashford, Ernest	5 Jan. 18	
	Kent, Jackson Brownley	5 Jan. 18	
40	**Wreford,** Arthur Percy	5 Jan. 18	
	Roberts, Arthur Cecil	5 Jan. 18	
	West, Wilfred Garrett	5 Jan. 18	
	Munro, Harry	5 Jan. 18	
	Jennings, Hartley William	5 Jan. 18	
45	**Garbett,** Robert Beaufort Lynch	5 Jan. 18	
	Sugden, John Wilfred	5 Jan. 18	
	Hambly, William Edward	5 Jan. 18	
	Wright, Ernest Walter	5 Jan. 18	
	Denness, Sidney William George	5 Jan. 18	
50	**Richardson,** John Henry Swain	5 Jan. 18	
	Coupar, John Mitchell	5 Jan. 18	
	Sabine, Harold Edward	5 Jan. 18	
	King, Frederick John Herbert	5 Jan. 18	
	Woodgate, Henry William	9 Jan. 18	
55	**Lenox,** George Hartley	12 Jan. 18	
	Thomas, Gordon Stewart	15 Jan. 18	
	Bowley, Gordon William	15 Jan. 18	
	Sheppard, Ernest Alfred	15 Jan. 18	
	Jacob, Lionel William Richard	15 Jan. 18	
60	**Clark,** George Samuel Henry	15 Jan. 18	
	Darnell, William Alfred	15 Jan 18	
	Eldred, Harold Somers	15 Jan. 18	
	Francis, Thomas Vaughan	15 Jan. 18	
	Tuffin, Regnald	15 Jan. 18	
65	**Prebble,** Bernard	15 Jan. 18	
	Thomson, William	15 Jan. 18	
	Clarke, Arthur Conard	15 Jan. 18	
	Russel, John Scott	15 Jan. 18	
	Watts, Bert	15 Jan. 18	
70	**Weston,** Allan	15 Jan. 18	
	Morgan, Reginald Walter	15 Jan. 18	
	Raines, Harold	15 Jan. 18	
	Jeffreys, William John	15 Jan. 18	
	Richardson, Gordon	15 Jan. 18	
75	**Wragg,** Walter Francis	15 Jan. 18	
	Ingham, Harold	15 Jan. 18	
	Slocombe, Walter	15 Jan. 18	
	Shaw, Mark Harold	15 Jan. 18	
	Ovenden, Edward Archibald	15 Jan. 18	
80	**Trigg,** Thomas Robert	15 Jan. 18	
	Daveport, Roland	15 Jan. 18	
	Stansfeld, William	15 Jan. 18	
	Craven, Leo	15 Jan. 18	
	Davenport, Charles Vivian	15 Jan. 18	
85	**Green,** Harold Thornton	15 Jan. 18	
	Bugler, John	15 Jan. 18	
	Libby, William Henry	15 Jan. 18	
	Lovering, Arthur James	15 Jan. 18	
	McKinney, John	15 Jan. 18	
90	**Pratt,** Stanley Sidney Truscott	15 Jan. 18	
	Forder, Hugh Nicholson	15 Jan. 18	
	Robertson, Stewart	15 Jan. 18	
	Ford, Frederick Paul	15 Jan. 18	
	Brown, William Fred	15 Jan. 18	
95	**Harrison,** Archibald Sedgwick	15 Jan. 18	
	Drudge, Herbert Austin	15 Jan. 18	
	Jones, James Sidney	15 Jan. 18	
	Ryan, Thomas	15 Jan. 18	
	Hughes, William	15 Jan. 18	
100	**Jeffreys,** Kenneth	15 Jan. 18	
	Minshall, Albert Oswald	15 Jan. 18	
	Price, Thomas Winter	15 Jan. 18	
	Barnett, William Edward	15 Jan. 18	
	Fairclough, Frederick Stafford	15 Jan. 18	
105	**Taggart,** Percy Douglas	15 Jan. 18	
	Atkinson, Leslie Richard	15 Jan. 18	
107	**Otter,** Reginald Bruce	15 Jan. 18	

No.	Names.	Date of rank.	Remarks.
	OFFICERS OF THE INDIAN ARMY HAVING TEMPORARY COMMISSIONS—*contd.*		
	2nd-Lieutenants—*contd.*		
108	**Painter,** Roland Frederic	15 Jan. 18	
	Bennion, William Owen (*temp. Capt., 15 Sep. 18*)	15 Jan. 18	
110	**Williamson,** Arthur	15 Jan. 18	
	Jones, William Edgar Richard	15 Jan. 18	
	Felton, George Horace	15 Jan. 18	
	Gayten, Chester	15 Jan. 18	
	Godly, Bernard William Turley	15 Jan. 18	
115	**Harding,** Alfred Gordon Leonard	15 Jan. 18	
	Howells, Thomas Reginald Raymond	15 Jan. 18	
	Evans, Arthur	15 Jan. 18	
	Bromley, Charles Gordon	15 Jan. 18	
	Finch, Henry Cuthbert	15 Jan. 18	
120	**Ryde,** Albert Arthur	15 Jan. 18	
	Nightingale, Victor Russell John	15 Jan. 18	
	Waddell, James Henry	15 Jan. 18	
	Wiffin, George Samuel Wilfred	15 Jan. 18	
	Allen, Charles Edwin	15 Jan. 18	
125	**Ledingham,** John James	15 Jan. 18	
	Flawn, Arthur Henry Tayler	15 Jan. 18	
	Leyshon, William	15 Jan. 18	
	Watts, Francis Henry	15 Jan. 18	
	Hollamby, William Charles	15 Jan. 18	
130	**Squires,** Archibald W. Towers	15 Jan. 18	
	Lawrie, Willie Wilfred	15 Jan. 18	
	Rodd, Neville James(207*a*)	15 Jan. 18	
	Pibworth, Arthur Henry	15 Jan. 18	
	Shuell, Cecil Anthony John	15 Jan. 18	
135	**Perry,** Ernest Leslie	15 Jan. 18	
	Bally, Philip Edwin Canning	15 Jan. 18	
	McManus, Andrew Augustus	15 Jan. 18	
	Hoggett. Francis Albert Jack	15 Jan. 18	
	Miller, Reginald Gordon	15 Jan. 18	
140	**Waite,** Walter Henry	15 Jan. 18	
	Pearson, William George	17 Jan. 18	
	Mellis, John Hampton Strachan	19 Jan. 18	
	Holyoak, Albert Henry Perkins	19 Jan. 18	
	Blockley, John Frank Grosvenor (207*a*)	21 Jan. 18	
145	**Lowe,** Thomas Whitson	12 Mar. 18	
	Price, Guy	12 Mar. 18	
	Ludlow, Herbert	19 Mar. 18	
	Snatt, Sidney George	19 Mar. 18	
	Wellings, Dennis	19 Mar. 18	
150	**Burrough,** Ian John(207*a*)	19 Mar. 18	
	Nickalls, John William Davies	19 Mar. 18	
	Walker, Jack	19 Mar. 18	
	Coldham, Percy Charles	19 Mar. 18	
	George, Aurthur Cecil	19 Mar. 18	
155	**Loveless,** George Jerred	19 Mar. 18	
	Blanchford, Montigue William	19 Mar. 18	
	Lloyd, Edwin Thomas	19 Mar. 18	
	Knott, Victor (46)	19 Mar. 18	
	James, Alfred Ernest	19 Mar. 18	
160	**Bibby,** Edwin	19 Mar. 18	
	Pollard, Charles Crossley	19 Mar. 18	
	de Coverly, Alfred Thomas Edward	19 Mar. 18	
	Harry, Hubert Teare	19 Mar. 18	
	Ridler, William	19 Mar. 18	
165	**Homan,** Ralph	19 Mar. 18	
	Obourn, Norman Arthur	19 Mar. 18	
	Muspratt, Raynor Alan Leslie	19 Mar. 18	
	Soper, William Thomas	19 Mar. 18	
	Bowen, John	19 Mar. 18	
170	**Richards,** Clifford George	19 Mar. 18	
	Shaw, William Henry	28 Mar. 18	
	Gurton-Hobbs, Arthur Gerald	28 Mar. 18	
	Harman, Frederick Arthur	29 Mar. 18	
	Clift, Arthur	29 Mar. 18	
175	**Butcher,** Eugene Andrew	29 Mar. 18	
	Cunningham, Vivian	29 Mar. 18	
	Stuart, Alan	29 Mar. 18	
	Askew, Jaccob	29 Mar. 18	
	Greenwood, Norris	29 Mar. 18	
180	**Miles,** Emlyn	29 Mar. 18	
	Phillips, Evan Curwen	29 Mar. 18	
182	**Lewis,** John Stanford	29 Mar. 18	

No.	Names.	Date of rank.	Remarks.

OFFICERS OF THE INDIAN ARMY HAVING TEMPORARY COMMISSIONS—*contd.*

2nd-Lieutenants—*contd.*

No.	Names.	Date of rank.	Remarks.
183	**Price,** Charles Edward	29 Mar. 18	
	Bray, George Frederick	29 Mar. 18	
	Deung, Martin Thomas	29 Mar. 18	
185	**Johnston,** Fraser	29 Mar. 18	
	Anscombe, Reginald Thomas John	29 Mar. 18	
	Ward, Ernest	29 Mar. 18	
	West, Charles	29 Mar. 18	
	Burchett, Victor William Kenward	29 Mar. 18	
190	**Kerley,** Sidney Walter	29 Mar. 18	
	Connell, Ernest William Heath	29 Mar. 18	
	Griffiths, Roland Wither	29 Mar. 18	
	Oakman, Alfred Stanley	29 Mar. 18	
	Kitchen, William Henry	29 Mar. 18	
195	**Lacey,** James Stanley	29 Mar. 18	
	Carey, Thomas William	29 Mar. 18	
	Webber, Cyril John	29 Mar. 18	
	Wilson, John Thomas	29 Mar. 18	
	Terry, John Henry Scrase (46)	29 Mar. 18	
200	**Casey,** John William	29 Mar. 18	
	Edwards, David John	29 Mar. 18	
	Durban, Edward Charles	29 Mar. 18	
	Pillar Edward John	29 Mar. 18	
	Pope, Horace William	29 Mar. 18	
205	**Wheatley,** Cyril Knifton	29 Mar. 18	
	Bayley, William Henry	29 Mar. 18	
	Lambert, Herbert	29 Mar. 18	
	Robinson Albert Edward (59)	29 Mar. 18	
	Stevens, Frederic William	29 Mar. 18	
210	**Clark,** Victor Percy James	29 Mar. 18	
	Dicker, Donland Edward	29 Mar. 18	
	Ford, Charles William	29 Mar. 18	
	de Beer, Esmond Samuel	29 Mar. 18	
	Brown, Frederick James	29 Mar. 18	
215	**Waller,** Leonard George	29 Mar. 18	
	Pillar, Charles Henry	29 Mar. 18	
	Lake, Arthur Ernest	29 Mar. 18	
	Burton, Victor Marshall	29 Mar. 18	
	Newman, Gilbert Joseph	29 Mar. 18	
220	**Richardson,** Francis Johnson	29 Mar. 18	
	Hoskin, John Grose	29 Mar. 18	
	Garrett, John Noel	29 Mar. 18	
	Townsend, Thomas	29 Mar. 18	
	Cooke, Clement Aubrey	29 Mar. 18	
225	**Adams,** Francis Alan	29 Mar. 18	
	Murray, Philip Frederick	29 Mar. 18	
	Webb, Thomas	29 Mar. 18	
	Mace, Philip Henry	29 Mar. 18	
	Hayes, Alfred William	29 Mar. 18	
230	**Oke,** Harry	29 Mar. 18	
	Dear, William Thomas	29 Mar. 18	
	Wicks, Eric Leslie	29 Mar. 18	
	Pleasance, Alfred George	29 Mar. 18	
	Sawford, Alfred Norman	29 Mar 18	
235	**Small,** John Henry	29 Mar. 18	
	Little, Walter James	29 Mar. 18	
	De ves, Norman Duglas	29 Mar. 18	
	Sterl, Thomas Henry	29 Mar. 18	
	Dennes, Harry Melbourne	29 Mar. 18	
240	**Logan,** James Stuart	29 Mar. 18	
	Wilkins, Horace William John	29 Mar. 18	
	Evenden, James Miller	29 Mar. 18	
	Gwinnett, Richard Charles Havard	29 Mar. 18	
	Wooff, Sidney Herbert	29 Mar. 18	
245	**Carlile,** Robert Love Hunter	6 Apl. 18	
	Logan, John	9 Apl. 18	
	Bateman, Louis Charles	22 Apl. 18	
	Weston-Foster, Aubrey Wright	25 Apl. 18	
	McNeish, Arthur William	20 Apl. 18	
250	**Stuart,** Robert	29 Apl. 18	
	Smart, Ernest Leofwin	1 May 18	
	Steel, George	1 May 18	
254	**Keogh,** Easil Hewlett	15 May 18	

OFFICERS OF THE INDIAN ARMY HAVING TEMPORARY COMMISSIONS—*contd.*

2nd-Lieutenants—*contd.*

No.	Names.	Date of rank.	Remarks.
255	**Withnell**, John	2 June 18	
	Goodwin, James	2 June 18	
	Robinson, Frederick Charles	2 June 18	
	Knight, Wilfred Norton	2 June 18	
	Westlake, Tom	2 June 18	
260	**Stokes**, Reginald George	2 June 18	
	Bickel, Thomas Harold	2 June 18	
	O'Meara, Michael Joseph	2 June 18	
	Smith, Arthur Henry	2 June 18	
	Brown, Albert William Chesterton	2 June 18	
265	**Cartwright**, William Ashton	2 June 18	
	Harrison, Herbert	2 June 18	
	Randell, John William	2 June 18	
	Richards, Reginald William	2 June 18	
	Clifton, Francis Russell	2 June 18	
270	**Bendall**, Arthur Frederic	2 June 18	
	Gillespie, James	5 June 18	
	Wrigglesworth, Sydney	7 June 18	
	Gilbert, Charles	7 June 18	
	Henley, Richard William	7 June 18	
275	**Morgan**, Leoline	7 June 18	
	Charlesworth, Fred Watson	7 June 18	
	Bishop, Frederick John	7 June 18	
	Hickmott, William Howard	7 June 18	
	Willie, Earnes Edgar	7 June 18	
280	**Price**, Arthur Woodman	7 June 18	
	Vincent, John Arthur	7 June 18	
	Wells, Charles Arthur	7 June 18	
	Pain, Sidney Albert	7 June 18	
	Wibby, Edward John	7 June 18	
285	**Michael**, William	7 June 18	
	Strawbridge, Herbery George	7 June 18	
	Sage, Harold	7 June 18	
	Gould, Percival Rowland	7 June 18	
	L[illegible]der, Leslie Norman	7 June 18	
290	**Mintey**, Arthur Harold	7 June 18	
	Wheeler, William	7 June 18	
	Mitchell, Harold Frederick	7 June 18	
	Stuart-Shepherd, Ughtred Guy Stuart	7 June 18	
	Barber, Phillip Thomas	7 June 18	
295	**Branson**, Claude Harry	7 June 18	
	White, William Henry	7 June 18	
	Tidmarsh, Charles Frederick	7 June 18	
	Castle, Eric Donald Forest	7 June 18	
	Livesay, Edward	7 June 18	
300	**O'Connor**, Maurice Joseph	7 June 18	
	Williamson, Ashley Uden	7 June 18	
	Holden, George Patrick Cecil	7 June 18	
	Moase, Wilfrid	7 June 18	
	Eden, Alfred William Burrage	7 June 18	
305	**Eden**, Herbert George Arthur	7 June 18	
	Livingstone, Douglas Alexander	7 June 18	
	Stannard, George Edward	7 June 18	
	Blanche, Frederick Edgar	7 June 18	
	Clemo, John	7 June 18	
310	**Padgham**, Thomas Herbert	7 June 18	
	Quinton, Arthur Walter	7 June 18	
	Hart, Walter	7 June 18	
	Kelly, Roland George	7 June 18	
	Wells, Cecil Alfred	7 June 18	
315	**Coleman**, Arthur James	7 June 18	
	Irving Cecil Eric	7 June 18	
	Irving, Malcolm Claude	7 June 18	
	Barnes, Arthur Edward Christopher	7 June 18	
	Bryant, Leonard	7 June 18	
320	**Marden**, John	7 June 18	
	Nichols, Dudley	7 June 18	
	Daniels, Arthur	7 June 18	
	Douglas, George Douglas Eric	7 June 18	
	Talbot, John Edward Lightfoot	7 June 18	
325	**Nicholas**, Oswald Robert	7 June 18	
326	**Green**, Arthur John	7 June 18	

No.	Names.	Date of rank.	Remarks.

OFFICERS OF THE INDIAN ARMY HAVING TEMPORARY COMMISSIONS—*contd.*

2nd-Lieutenants—*contd.*

No.	Names.	Date of rank.	Remarks.
327	**James**, Henry Pearce	7 June 18	
	Green, Victor Norman	7 June 18	
	Hopkinson, George Greenwood	7 June 18	
330	**Chater**, Albert	7 June 18	
	Hutton, Edward Arnold	7 June 18	
	Pollock, Kelvin Clayton	7 June 18	
	Swinburn, Henry Robinson	7 June 18	
	O'Reilly, James Michael	7 June 18	
335	**Pilbeam**, Frederick William	7 June 18	
	Willmore, Percy Alfred	7 June 18	
	Yare, Frederick Robert	7 June 18	
	Southgate, Edward Harry Dray	7 June 18	
	Woolsey, Charles Horace	7 June 18	
340	**Betteridge**, Raymond Arthur	7 June 18	
	Barron, William James	7 June 18	
	Bisher, Douglas Randolph	7 June 18	
	Lester, Colin	7 June 18	
	Simmons, Frederick Samuel	7 June 18	
345	**Evans**, Victor Charles	7 June 18	
	Baily, John	7 June 18	
	Hays, Walter	7 June 18	
	Lynch, Percy William	7 June 18	
	Bishop, Wilfred Ewart	7 June 18	
350	**Millin**, Dennis John	7 June 18	
	Hall, Frank Richard	7 June 18	
	Tyndale, Humphrey Earle	7 June 18	
	Williams, Arthur Johnson	7 June 18	
	Egginton, John Foley	7 June 18	
355	**Sparham**, Alan Charles	7 June 18	
	Peace, Joseph Norman	7 June 18	
	Roberts, George Brook	7 June 18	
	Ford, Albert	7 June 18	
	Ainscow, Albert	7 June 18	
360	**Duckenfield**, Walter	7 June 18	
	Hyder, Herbert Sydney	7 June 18	
	Mackey, Dennis Daniel	7 June 18	
	Wyatt, Edwin Wellington	7 June 18	
	Brown, Charles	7 June 18	
365	**Jones**, Sidney Vincent	7 June 18	
	Grover, William	7 June 18	
	Stock, Charles Albert	7 June 18	
	Simpson, Hugh Beckitt	7 June 18	
	Brunger, Bertie	7 June 18	
370	**Barnett**, Christopher Edward	7 June 18	
	Cory-Smith, Leslie Beverley	7 June 18	
	Silver, Percy William	7 June 18	
	Mount, James Thomas	7 June 18	
	O'Donnell, Irwin Aubrey	5 July 18	
375	**Manson**, William Mackenzie	10 July 18	
	Llewellyn, Ralph George	10 July 18	
	Chidley, Leonard	10 July 18	
	Thorn, George	10 July 18	
	Hall, Thomas Harold	10 July 18	
380	**Badhams**, Oswald George	10 July 18	
	Magee, John Hamilton	10 July 18	
	Scorgie, Norman Anderson	10 July 18	
	Urwin, Samuel Ernest	10 July 18	
	Bell, William Samuel Brittain	10 July 18	
385	**Thirsk**, Frederick Atkinson	10 July 18	
	Amos, Cecil	10 July 18	
	Brooks, Percy	10 July 18	
	Gunn, Alexander	10 July 18	
	Hubbard, Herbert	10 July 18	
390	**Spreadbrow**, Francis James	10 July 18	
	Cartmel, Alfred Edward	10 July 18	
	Bishop, Harold Percival	10 July 18	
	Lonsdale, Arthur John	10 July 18	
	Drever, George Douglas	10 July 18	
395	**Symons**, Henry George	10 July 18	
	Stott, Henry William Mackie	10 July 18	
397	**Thorne**, Ernest Pollett	10 July 18	

OFFICERS OF THE INDIAN ARMY HAVING TEMPORARY COMMISSIONS—*contd.*

2nd-Lieutenants—*contd.*

No.	Names.	Date of rank.	Remarks.
398	**Russell,** George Arthur	10 July 18	
	Wright, Hubert	10 July 18	
400	**Pugh,** John Churcher	12 July 18	
	Anderson, Adam	16 July 18	
	Lewis, Harold Constable	18 July 18	
	Medley, Arnold Foster	22 July 18	
	Tutton, Harold Knight	22 July 18	
405	**Pollitt,** James	22 July 18	
	Bedford, Harry	22 July 18	
	Strangeman, Clive Venn	22 July 18	
	Irwin-Carter, Ewart	22 July 18	
	Hutchings, Frederic Samuel	22 July 18	
410	**Loveless,** William George	22 July 18	
	Cruttenden, Leonard Charles	22 July 18	
	Salt, John	22 July 18	
	Wright, Harry Robert	22 July 18	
	Donaldson, Albert Harvey	22 July 18	
415	**Ward,** William Guyatt	22 July 18	
	Adams, Ernest John	22 July 18	
	Hooper, William John	22 July 18	
	Jones, Henry Harvey	22 July 18	
	Hunter, Alexander	22 July 18	
420	**Moody,** Francis Wollaston	22 July 18	
	Whitaker, William Arthur	22 July 18	
	Lowther, Albert William	22 July 18	
	Lawson, Walter George	22 July 18	
	Foster, Albert Edward	22 July 18	
425	**Coster,** George Worthy	22 July 18	
	Bee, John	22 July 18	
	Burscough, George Edward	22 July 18	
	Wardle, Frank	22 July 18	
	Watts, Henry James	22 July 18	
430	**Moore,** Philip James	22 July 18	
	Wall, Walter Howard	22 July 18	
	Cockeran, Alfred Ernest	22 July 18	
	Langley, William Victor	22 July 18	
	Plant, Ivor Thomas	22 July 18	
435	**Gerrey,** George William	22 July 18	
	Moss, Arthur George	22 July 18	
	Boswarva, Stanley John	22 July 18	
	Boniface, Charley Frank	22 July 18	
	Noakes, Stanley Winton	22 July 18	
440	**Searl,** Arthur Edward George	22 July 18	
	Pryor, John William	22 July 18	
	Trevett, Herbert Victor	22 July 18	
	Bibby, James William	22 July 18	
	Mason, John Sidney	22 July 18	
445	**Holloway,** Charles Arthur	22 July 18	
	Guest, Frank Farmer	22 July 18	
	Gordon, Douglas Bently	22 July 18	
	Davis, George Ratcliff	22 July 18	
	Bitton, David	22 July 18	
450	**Clapperton,** Harold	22 July 18	
	Price, Rueben	22 July 18	
	Smith, Arthur	22 July 18	
	Warr, Charles Hugh	22 July 18	
	Yates, George Charles Henry	22 July 18	
455	**Nankervis,** Thomas James	22 July 18	
	Runnalls, Henry George	22 July 18	
	Nicholas, Charles Frederick	22 July 18	
	Collins, Richard Edward	22 July 18	
	McPherson, William Hodgson	22 July 18	
460	**Opie,** Arthur Leslie	22 July 18	
	Coombs, Ernest William Geoffrey	22 July 18	
	Crandell, William	22 July 18	
	Young, Stuart Adolphus	22 July 18	
	Brierley, Robert Jones	22 July 18	
465	**Rayner,** Arthur George	22 July 18	
	Boughton, Albert Eward	22 July 18	
	Brewer, Arnold Abbotson	22 July 18	
468	**Cull,** Bertram Noel	22 July 18	

OFFICERS OF THE INDIAN ARMY HAVING TEMPORARY COMMISSIONS—*contd.*

2nd-Lieutenants—*contd.*

No.	Names.	Date of rank.	Remarks.
469	**Newham,** Frank George	22 July 18	
470	**Niblett,** Leslie Phillip	22 July 18	
	Pearce, George Alfred	22 July 18	
	Peacock, Ronald Percy Montague	22 July 18	
	Tregale, Percy Morrish	22 July 18	
	Cordwell, Ernest Thomas	22 July 18	
475	**Longthorn,** Joseph	22 July 18	
	Noble, Herbert Brown	22 July 18	
	Edwards, Archibald Horace Muspratt	22 July 18	
	Towner, Cecil	22 July 18	
	Pickering, Miles Bristow	22 July 18	
480	**Langdon,** Allan George	22 July 18	
	White, Norman	22 July 18	
	King, John Henry	22 July 18	
	Wigner, George Douglas	5 Aug. 18	
	Coleman, Herbert Charles	9 Aug. 18	
485	**Pelmear,** Andrew	10 Aug. 18	
	Steel, Ernest	15 Aug. 18	
	Barry, Joseph	2 Sep. 18	
	Anderson, George Kelsey	10 Sep. 18	
	Crabtree, Herbert	10 Sep. 18	
490	**Sawyer,** Arthur John	25 Sep. 18	
	Maddison, John George	3 Oct. 18	
	Bird, Harold Wallace	15 Oct. 18	
	Jones, Wallace Homeyard	15 Oct. 18	
	Grant, Alexander Rae	15 Oct. 18	
495	**Hogg,** Harold Charles George	15 Oct. 18	
	Dennett, Frank Cecil Hansford	15 Oct. 18	
	French, Leonard	15 Oct. 18	
	Upson, Percy Charles	15 Oct. 18	
	Brown, Gerald James	15 Oct. 18	
500	**Bell,** Nelson Ewart	15 Oct. 18	
	Utting, Louis Edgar	15 Oct. 18	
	Parry, Arthur	5 Oct. 18	
	Corke, Leonard Harold	15 Oct. 18	
	Kirkham, Harry	15 Oct. 18	
505	**Goodwin,** Fred Allis	15 Oct. 18	
	Davies, John Brechfa	15 Oct. 18	
	Mylon, Daniel	15 Oct. 18	
	Lewis, Henry John	15 Oct. 18	
	Dearnley, Horace Sayle	15 Oct. 18	
510	**Bryson,** Herbert Edward	15 Oct. 18	
	Webb, Horace Percy	15 Oct. 18	
	Mills, Albert Edward	15 Oct. 18	
	Butler, Wilfred Marsh	15 Oct. 18	
	Beer, Gerald Edward	15 Oct. 18	
515	**Thomson,** James	15 Oct. 18	
	Poole, Arthur Robert	15 Oct. 18	
	Britton, Thomas Rowland	15 Oct. 18	
	Webb, Ernest Alfred	15 Oct. 18	
	Higginson, John	15 Oct. 18	
520	**Newman,** William Cecil	15 Oct. 18	
	Vigar, William Arthur	15 Oct. 18	
	Meates, William Wylie	15 Oct. 18	
	Capon, Albert James	15 Oct. 18	
	Lawton, Charles Armitage	15 Oct. 18	
525	**Smith,** Samuel Charles	15 Oct. 18	
	Pibworth, Frederick	15 Oct. 18	
	Rose, Reginald Peter	15 Oct. 18	
	Rope, Cecil	15 Oct. 18	
	Blake, Stanley Brinkworth	15 Oct. 18	
530	**Davies,** Robert James	15 Oct. 18	
	Mason, James Cecil	15 Oct. 18	
	Richards, William	15 Oct. 18	
	Bugg, William Arthur	15 Oct. 18	
	Curthoys, Frederick Harold	15 Oct. 18	
535	**Richards,** Harold	15 Oct. 18	
	Fairhurst, John Aloyious	15 Oct. 18	
	Rose, Albert George	15 Oct. 18	
	Blackburn, Harold	15 Oct. 18	
	Brinkler, Charles Henry	15 Oct. 18	
540	**Swain,** George Victor	15 Oct. 18	

OFFICERS OF THE INDIAN ARMY HAVING TEMPORARY COMMISSIONS—*contd.*

2nd-Lieutenants—*contd.*

No.	Names.	Date of rank.	REMARKS.
541	**Scrivens,** Aubrey Thomas	15 Oct. 18	
	Stallard, Harol . Ernest	15 Oct. 18	
	Andrews, William	15 Oct. 18	
	Roberts, Arthur	15 Oct. 18	
545	**Atherton,** Percy	15 Oct. 18	
	Provan, Andrew	15 Oct. 18	
	Bentley, Charles William	15 Oct. 18	
	Freegard, Samuel Edward Lawrence	15 Oct. 18	
	Strellett,-David Louis	15 Oct. 18	
550	**Ostler,** Rupert Stanley	15 Oct. 18	
	Wilson, William	15 Oct. 18	
	Dart, Kenneth George	15 Oct. 18	
	Hurst, George Thomas	15 Oct. 18	
	Crowley, John Thomas	15 Oct. 18	
555	**Lindsay,** Alexander	19 Oct 18	
	Lyle, Peter	19 Oct. 18	
	Howard, Harry Edward	19 Oct. 18	
	Jennings, Albert Rudd	19 Oct. 18	
	Kirkham, Harry	19 Oct. 18	
560	**Plater,** William Edward	19 Oct. 18	
	McGuire, John	19 Oct. 18	
	Bailey, John William Stanley	19 Oct. 18	
	Ricketts, Percy	19 Oct. 18	
	Hearsey, John Theodore Moss	19 Oct. 18	
565	**Peach,** Walter Albert ..	19 Oct. 18	
	Morley, Lawrence Vincent	19 Oct. 18	
	Spivey, James Albert	19 Oct. 18	
	Payne, Edward Charles	19 Oct. 18	
	Ward, Charles Herbert	19 Oct. 18	
570	**Pickford,** Hube t Victor Fred	19 Oct. 18	
	Martin, Cecil James Simpson	19 Oct. 18	
	Stannard, Percy Victor	19 Oct. 18	
	Evans, William Kenneth	19 Oct. 18	
	Treasure, Francis Lionel	19 Oct. 18	
575	**Evans,** Robert Augustus Gordon	19 Oct. 18	
	Budd, Frank	19 Oct. 18	
	Orris, Walter Edward	19 Oct. 18	
	Haynes, Albert Harry	19 Oct. 18	
	Rainford, Herbert	19 Oct. 18	
580	**Bennett,** Arthur John	19 Oct. 18	
	Irving, Robert	19 Oct. 18	
	Choake, Frederick James	19 Oct. 18	
	Francis, William Lenard	19 Oct. 18	
	Hilton, Walter Harold	19 Oct. 18	
585	**Dalzell,** James	19 Oct. 18	
	Rudduck, William	19 Oct. 18	
	Fewery, James	19 Oct. 18	
	Levinson, Herbert	19 Oct. 18	
	Brown, George	19 Oct. 18	
590	**Sharp,** Thomas George	19 Oct. 18	
	Sudlow, Frederick Sharman	19 Oct. 18	
	Thwaites, Alfred James	19 Oct. 18	
	Francis, Arthur Stanley	19 Oct. 18	
	Redfarn, Frederick William	19 Oct. 18	
595	**Fraser,** David Francis Scott	19 Oct. 18	
	Hayter, Lionel Neville	19 Oct. 18	
	Perry, Harold Stanley James	19 Oct. 18	
	Gramson, Daniel Thomas Richard	19 Oct. 18	
	Hurley, William Charles	19 Oct. 18	
600	**Guest,** Richard	19 Oct. 18	
	Talbot, Charles Cecil	19 Oct. 18	
	Bicker, Leslie James	19 Oct. 18	
	Ford, Montague Alfred	19 Oct. 18	
	Brazier, Frederick Robert	19 Oct. 18	
605	**Steabben,** John Orchard	19 Oct. 18	
	Henderson, Cecil	19 Oct. 18	
	Hoggett, James Edward	19 Oct. 18	
	Gill, William Howell Inniss	19 Oct. 18	
	Graves, John	19 Oct. 18	
610	**Mac Phail,** Donald	20 Oct. 18	
	Down, Ernest Augustus	27 Oct. 18	
	Cooper, Frank Ledsome	27 Oct. 18	
	Pentland, Albert	27 Oct. 18	

OFFICERS OF THE INDIAN ARMY HAVING TEMPORARY COMMISSIONS—*concld.*

2nd-Lieutenants—*concld.*

No.	Names.	Date of rank.	Remarks.
	Mendoze, Alfred Lewis	27 Oct. 18	
615	**Borton,** Edwin	27 Oct. 18	
	Fraser, James William	27 Oct. 18	
	Bridgwater, Ernest Leonard	27 Oct. 18	
	Drew, Oliver William	27 Oct. 18	
	Jecks, Ernest Dudley	27 Oct. 18	
620	**Brown,** Frank Henry Joseph	27 Oct. 18	
	Harden, Henry Albert	27 Oct. 18	
	Shackell, Augustus	27 Oct. 18	
	Stiles, Joseph Haines	27 Oct. 18	
	Brown, Arthur	27 Oct. 18	
625	**Everett,** Charles Frederick	27 Oct. 18	
	Hedger, Harry	27 Oct. 18	
	Carter, Albert James	27 Oct. 18	
	Sampson, Cecil Arthur	27 Oct. 18	
	Watts, Albert Everett	27 Oct. 18	
630	**Bennett,** Samuel Bertram	27 Oct. 18	
	Hawkins, Frederick Leslie	27 Oct. 18	
	Barrow, Arthur	27 Oct. 18	
	Perrin, Francis Edward	27 Oct. 18	
	Greenwood, Harry	27 Oct. 18	
635	**Fielding** John	27 Oct. 18	
	Lord, Arthur James	27 Oct. 18	
	Vanston-Rumney, William Francis	27 Oct. 18	
	Miles, Joseph William	27 Oct. 18	
	Jeffery, Frederick Brewer	27 Oct. 18	
640	**Tumber,** William Albert	27 Oct. 18	
	Hogg, Ralph	5 Nov. 18	
	Whittome, Arthur George	5 Nov. 18	
	Butcher, Albert Joseph	5 Nov. 18	
	Smith, John Frederick Spencer	5 Nov. 18	
645	**Collishaw,** Thomas	5 Nov. 18	
	Reeves, Frederick Herbert	5 Nov. 18	
	Abbott, Albert Stanley	5 Nov. 18	
	Close, William Austin	5 Nov. 18	
	Graham, Thomas	5 Nov. 18	
650	**Leighton,** William	5 Nov. 18	
	Murray, Bernard Joseph	5 Nov. 18	
	Shuttleworth, Frederick	5 Nov. 18	
	Thatcher, Clifford	5 Nov. 18	
	Scriven, Robert Guy	5 Nov. 18	
655	**Dean,** Trevor Hicks	5 Nov. 18	
	Macfarlane, Edward Allan	5 Nov. 18	
	Hume, Francis John	5 Nov. 18	
	Collins, Joseph (95)	5 Nov. 18	
	Barrett, Fred	5 Nov. 18	
660	**Batley,** William	12 Nov. 18	
	Etchells, John Edward	12 Nov. 18	
	Barker, Harold	12 Nov. 18	
	Argyle, Victor Lewis	12 Nov. 18	
	Watts, Gilbert Sydney	12 Nov. 18	
665	**Blackburn,** Fred	12 Nov. 18	
	Dawson, John Clifford	12 Nov. 18	
	Alexander, James Lauchland	12 Nov. 18	
	Shields, Chirstopher John	12 Nov. 18	
	Birch, Arnold	12 Nov. 18	
670	**Entwistle,** James	12 Nov. 18	
	Martin, John	12 Nov. 18	
	Worth, Arthur Marshall	12 Nov. 18	
	Creasey, Alfred Charles	12 Nov. 18	
	Bennett, Charles Reginald	12 Nov. 18	
675	**Coath,** Percy Stantan	12 Nov. 18	
676	**Rice,** Reginald William	12 Nov. 18	

UNATTACHED LIST.

2nd-Lieutenants (for appointment to the Indian Army).

Names.	Date of rank.	British Corps to which attached.	Date of joining British Corps.
Johnston, Ernest Augustus	21 Dec. 17		
Birdwood, Christopher Bromhead	21 Dec. 17		
Mackenzie, David Alexander Laurence	24 Apl. 18		
Dinwiddie, Herbert William	24 Apl. 18		
Plewman, Thomas Leslie	24 Apl. 18		
Love, Robert Mellis	24 Apl. 18		
Graham, Reginald Moray	24 Apl. 18		
Aitken, Robert Gordon	24 Apl 18		
Crosbie, Frank Stephen	24 Apl. 18		
Murray, James Roberts	24 Apl. 18		
Dunley-Owen, Arthur Hugh Upcher	24 Apl. 18		
Knocker, John Bedingfield	24 Apl. 18		
Allen, William Ashwin	24 Apl. 18		
Wilson, John Hector Alexander	24 Apl. 18		
Angus, Tom Hardy	24 Apl. 18		
Warwick, Guy Nevile	24 Apl. 18		
Salmon, Frederick Henry	24 Apl. 18		
Cooper, Charles Henry	24 Apl. 18		
Desborough, John Noel Campbell	24 Apl. 18		
Simmons, John Marcus Rainsford	24 Apl. 18		
deSilva, Walter Raymond	24 Apl. 18		
vanRenen, James Rudolph Douglas	24 Apl. 18		
Benson, Charles Richard	24 Apl. 18		
Finch, Kenneth Hamilton	24 Apl. 18		
Banks, Alexander Thomas	24 Apl 18		
Bayley, John Philip	24 Apl. 18		
Brampton, Charles	24 Apl. 18		
Hartigar, Kenneth Leslie Stewart	24 Apl. 18		
Ballinger, George Alexander	24 Apl. 18		
Sadlier, James Willoughby Vickers	24 Apl. 18		
James, Leigh	24 Apl. 18		
Holder, Eric Denzil	24 Apl. 18		
Howard, Cecil Wilfred	24 Apl. 18		
Macqueen, Ronald Brodie	24 Apl. 18		
Ryley, Leslie Frederic Macpherson	24 Apl. 18		
Gomer, Leslie William	24 Apl 18		
Ekin, William John	24 Apl. 18		
Tilden, Edward Divers	24 Apl. 18		
Bryant, Eric John Aldous	24 Apl. 18		
FitzGerald, Knightly Purefoy	24 Apl. 18		
Maxwell, Frnancis Scott	24 Apl. 18		
Sykes, Hugh	24 Apl. 18		
Hastings, Percy Graham	24 Apl. 18		
Collard, Basil Hewlett Aldred	24 Apl. 18		
Holmes, Laurence Vernon	24 Apl. 18		
Walker, Wilfrip	24 Apl. 18		
Sawdy, Alfred Eustace Leon	24 Apl. 18		
Parnes, Thomas William Chadborn	24 Apl. 18		
Woodman, William Henry Lionel	24 Apl. 18		
Hughes, Allan Harwood	24 Apl. 18		
Stewart, Charles Duncan Andrew	24 Apl. 18		
Hime, Ernest Holley	24 Apl. 18		
Anderson, Francis Gordon	24 May 18		
Gilder, Tom Norman	25 June 18		
Hurst, Thomas Robert	23 Aug. 18		
Burt, Philip Jeffery	31 Aug. 18		
Bunn, Clarence Walter	31 Aug. 18		
Sampson, John Conrad Reginald	21 Aug. 18		
Dorman-Smith, Reginald Hugh	21 Aug. 18		
Buckley, John Frederic Sydney	21 Aug. 18		
Hill, Herbert Lawrence	21 Aug. 18		
Whitaker, William Benjamin	21 Aug. 18		
Westmacott, Alan Rendel	21 Aug. 18		
Bott, Francis George	21 Aug. 18		
Evans, William Arthur	21 Aug. 18		
Woods, George Bernard Calder	21 Aug. 18		
Burbury, William Hawkesley	21 Aug. 18		
Cook, Douglas Oswald	21 Aug. 18		
Telford, Carmont Owen	21 Aug. 18		
Jarman, George Standerwick	21 Aug. 18		
MacKirdy, James Joseph Lambert	21 Aug. 18		
Taylor, Robert Douglas Clague	21 Aug. 18		
Phillips, Willam Kimberley	21 Aug. 18		

UNATTACHED LIST.

2nd-Lieutenants (for appointment to the **Indian Army**).

Names.	Date of rank.	British Corps to which attached.	Date of joining British Corps.
Webber, Geoffrey Huyshe Stafford	21 Aug. 18		
Maude, Jack Carstairs	21 Aug. 18		
Echlin, Francis Charles	21 Aug. 18		
Sankey, Clifford Seddon	21 Aug. 18		
Martin, Leslie Jack	21 Aug. 18		
le Maistre, Alfred Villeneufve	21 Aug. 18		
Myers, Barnett Berlyn	21 Aug. 18		
Wood, James Lindsay	21 Aug. 18		
Whitby, Henry Francis	21 Aug. 18		
Marten, Eric Charles	21 Aug. 18		
Williams, John Whitaker	21 Aug. 18		
Sheldrake, Bevis Swinborne	21 Aug. 18		
Tobin, Ralph Walter	21 Aug. 18		
Bean, John Rowland	21 Aug. 18		
Dale, Edwin Forbes	21 Aug. 18		
Waters, Basil Edmond Hervey	21 Aug. 18		
Smith, Reginald Henry Trivess	21 Aug. 18		
Dixon, Frederick Hedley	21 Aug. 18		
Eustace, Edward Arther Rawlins	21 Aug. 18		
Wray, Daniel Leslie	21 Aug. 18		
Mitchell, Francis Arthur Frederick	21 Aug. 18		
Cartwright, Harold Eustace	21 Aug. 18		
Jarman, Cyril John Russell	21 Aug. 18		
Davidson, Andrew Forbes	21 Aug. 18		
Williams, Claude Cuthbert	21 Aug. 18		
Edwards, Ian Arther Talbot	21 Aug. 18		
Howell, Rodney Warwick	21 Aug. 18		
Edwards, Edward Leigh	21 Aug. 18		
Grinsted, Norman Angus	21 Aug. 18		
Lord, John Charles Hamilton	21 Aug. 18		
Cockaday, Leslie Charles	21 Aug. 18		
Fegan, John Ardra de Crespigny	21 Aug. 18		
Cann, William Maybery	21 Aug. 18		
Holcroft, Reginald Culcheth	21 Aug. 18		
Morris, Frederick Arthur	21 Aug. 18		
Bird, Muir Douglas Wilberforce	21 Aug. 18		
Ambler, Frank de Vismes	21 Aug. 18		
Lyons, James John	21 Aug. 18		
Watson, Kenneth Cradock Cradock	21 Aug. 18		
Dennison, Richard	21 Aug. 18		
Waterer, Michael Desmond George	21 Aug. 18		
Everett, John Henry	21 Aug. 18		
Rice, Eric Hamilton Buller	21 Aug. 18		
Coney, Guy Archibald	21 Aug. 18		
Hayes, Archibald	21 Aug. 18		
Harper, Leonard Græme	21 Aug. 18		
Travers, Temple Hugh	21 Aug. 18		
Inglis, Ernest John D'Oyly	21 Aug. 18		
Wharton, Robert Eglinton	21 Aug. 18		
Ford, Dudley Herbert	21 Aug. 18		
Henfrey, Thomas Hetherington	21 Aug. 18		
Cramer, Jocelyn Marmaduke Coghill	21 Aug. 18		
Hall, Thomas Peter Maubray	21 Aug. 18		

OFFICERS OF THE INDIAN ARMY ON PROBATION WITH A VIEW TO PERMANENT APPOINTMENT.

No.	Former Corps.	Names.	Date of birth.	Date of first commission.	Date of appointment to Indian Army.	Date of rank.	Remarks.
		Captains.					
1	L'pool R.	**Trevor,** Christopher Philip	31 Dec. 90	5 Oct. 10	25 Mar. 18	1 Sep. 15	Attached 2-2nd Gurkha Rifles.
	R. Mun's Fus	**James,** Bernard Seton	8 Sep. 89	6 Nov. 09	29 Dec. 16	1 Sep. 15	,, 25 Cavalry.
	Conn. Rang.	**Anderson,** Arthur Lewis Brunker	16 Oct. 89	28 May 10	29 June 18	1 Sep. 15	,, 37 Lancers.
	R. War. R.	**Bamber,** Claude Charles	29 July 89	18 Sep. 09	30 June 18	1 Sep. 15	,, 1-7th Gurkha Rifles.
5	R. Tr. Rif.	**Macqueen,** Edward Theodore Lindesay	. . . .	10 Feb. 15	23 Oct. 18	1 Sep. 15	,, 11 Rajputs.
		Previous service to count 4 yrs. 150 days					
	Ches. R.	**Villiers-Stuart,** Percival Grandison, M.C.	3 Apl. 92	20 Sep. 11	18 May 18	20 Sep. 15	,, 97 Infantry.
	W. I. R.	**Tinney,** Lionel Henry	22 Nov. 90	20 Sep. 11	2 Oct. 17	20 Sep. 15	,, 34 Pioneers.
	14 Huss.	**St. Maur,** Richard William Maxwell	22 Mar. 92	7 Feb. 12	3 Aug. 18	7 Feb. 16	,, 14 Lancers.
		Simpson, Vere Elliot Ward, M.C. 207a	. . . .	9 Sep. 14	24 Oct. 17	28 May 16	,, 42 Deoli Regt.
		Previous service to count 3 yrs. 3 mos. 12 days less 9 months.					
10		**Herbert,** Harold Rupert	27 Feb. 92	4 Sep. 12	12 Jan. 18	4 Sep. 16	,, 42 Deoli Regt.
	R. War. R.	**Craddock,** Alexander Bainbridge	2 Feb. 93	21 Jan. 13	20 Nov. 17	21 Jan. 17	,, 2 Infantry.
	Ches. R.	**Watson,** Gerald Thomas Colclough	24 Oct. 92	5 Feb. 13	1 Mar. 18	5 Feb. 17	,, 9 Infantry.
	A. S. C.	**Lynn,** John Arthur Goodenough	29 Oct. 94	17 Sep. 13	12 May 18	17 Sep 17	,, 25 Cavalry.
		Pinhey, Robert Alexander	. . . .	6 Jan. 15	13 Jan. 18	4 Nov. 17	,, 1 Brahmans.
		Previous service to count 1 yr. 64 days less 9 months.					
15	W. Rid. R.	**Sleigh,** Gervase Paget	9 Sep. 93	23 Jan. 14	29 July 18	23 Jan. 18	,, 17 Cavalry.
	Hamps. R.	**Watts,** Ralph Ponsonby	24 Sep. 92	23 Jan. 14	13 Aug. 18	23 Jan. 18	,, 99 Infantry.
	Conn. Rang.	**Dennys,** Cecil Hector Massy	24 Sep. 94	24 Jan. 14	18 Feb. 18	24 Jan. 18	,, 1-4th Gurkha Rifles.
	Bedf. R.	**Mills,** Stephen Douglas, M.C.	26 Nov. 92	14 Feb. 12	30 Sep. 18	14 Feb. 16	,, 43 Regiment.
	Lond. R.	**Mahon,** Brian MacMahon, D.S.O., M.C.	. . . .	26 Aug. 14	20 Feb. 18	20 Feb. 18	,, 25 Punjabis.
20	Ches R.	**Pogson,** Lewis Victor John, M.C.	19 Apl. 94	8 Aug. 14	4 July 18	8 Aug. 18	,, 107 Pioneers.
	L'pool R.	**Kynaston,** Norman Edward Vernon	12 July 92	1 Sep. 14	27 Apl. 18	14 Aug. 18	,, 55 Rifles.
	L'pool R.	**Fulton,** Charles Frederick Vernon, M.C.	28 May 96	15 Aug. 14	28 Sep. 18	15 Aug. 18	,, Mily. Farms Dept.
	19 Huss.	**Scott,** Richard Andrew	. . . .	1 Oct. 14	25 Mar. 18	1 Oct. 18	,, 34 Horse.
	Leins R.	**Godsland,** Cyril Herbert, M.C.	. . . .	7 Nov. 14	23 Nov. 17	7 Nov. 18	,, 19 Punjabi.
25	L'pool R.	**Holland,** George Walter (51, 57, 71, 95, 169)	6 Feb. 90	7 Nov. 14	12 Apl 18	7 Nov. 18	,, 21 Punjabis.
		Lieutenants.					
1		**Cay,** Cyril Stancomb	. . . .	16 Dec. 14	22 Nov. 17	16 Dec. 15	,, 33 Cavalry.
	R. Suss. R.	**Fewtrell,** Alford Hastings Heap	. . . .	16 Dec. 14	25 Mar. 18	16 Dec. 15	,, 27 Cavalry.
		Parsons, James William Arthur	. . . .	13 Jan. 15	22 Nov. 17	13 Jan. 16	,, 32 Pioneers.
	Essex R.	**Monypenny,** Robert d'Arblay	. . . .	15 Feb. 15*	29 July 18	15 Feb. 16	,, 61 Punjabis.
5	Worc. R.	**Ward,** Herbert Geoffrey Leng	Nov. 95	17 Feb. 15	29 July 18	17 Feb. 16	,, 102 Grenadiers.
	E. Surr. R.	**Bousfield,** Hambleton Neville	. . . .	14 Mar. 15	25 Mar. 18	14 Mar. 16	,, 69 Punjabis.
	Midd'x R.	**Milne,** Annesley Henry	27 Jan. 95	17 Mar. 15	6 Jan. 17	17 Mar. 16	,, 42 Deoli Regt.
	Midd'x. R.	**Thorne,** Malcolm	9 Nov. 90	17 Mar. 15	11 Oct. 18	17 Mar. 16	,, 52 Sikhs.
	Worc. R.	**Milward,** Henry Harding	. . . .	20 Mar. 15	7 Feb. 18	20 Mar. 16	,, 4-39th Rifles.
10	A. S. C.	**Moss,** Lionel	1 May 90	17 Apl. 15	30 June 18	17 Apl. 16	,, 53 Sikhs.
		Lester-Garland, Arthur Garland	. . . .	5 May 15*	22 May 17	5 May 16	,, 76 Punjabis.
		Boss, Thomas Henry 207a	. . . .	5 May 15*	24 Oct. 17	5 May 16	,, 107 Pioneers.

10		Stewart, Ian	5 May 15*	23 Nov. 17	5 May 16	,,	38 Dogras.
		Curtler, Walter Laurence	5 May 15*	22 Nov. 17	5 May 16	,,	1-4th Gurkha Rifles.
		Leeward, Richard Boyd	5 May 15*	29 Aug. 17	5 May 16	,,	38 Dogras.
		Emery, Thomas Smythe, M.C. (*temp Capt., 16 Apl. 16.*) 207a	5 May 15*	31 July 17	5 May 16	,,	76 Punjabis.
		Huthwaite Henry Yoole, M.C.	5 May 15*	22 Nov 17	5 May 16	,,	56 Rifles.
15		Davies, Kenneth Cuthbert Johnson	5 May 15*	30 Nov. 17	5 May 16	,,	2-10th Gurkha Rifles.
		Evans, Arthur Llewelyn (48a, 51)	5 May 15*	24 Nov. 17	5 May 16	,,	9 Bhopal Infantry.
		Forshaw, William Thomas	5 May 15*	25 Nov 17	5 May 16	,,	76 Punjabis.
		Colnett, Richard Daunteshey	5 May 15*	22 Nov. 17	5 May 16	,,	80 Punjabis.
		Portal, Gervas Edward	5 May 15*	31 July 17	5 May 16	,,	5 Cavalry.
20		Leach, Richard Blaylock, M.C.	5 May 15*	12 Jan. 18	5 May 16	,,	35 Sikhs.
		Magner, John	5 May 15*	17 Jan. 18	5 May 16	,,	16 Cavalry.
		Donaldson, Arthur Leslie	5 May 15*	22 Jan. 18	5 May 16	,,	2-2nd Gurkha Rifles.
		O'Brien, John Joseph (46)	5 May 15*	22 Jan. 18	5 May 16	,,	112 Infantry.
		Adams, Douglas Raymond (48a, 50)	5 May 15*	22 Jan. 18	5 May 16	,,	35 Sikhs.
25		Ham, Paul Sison	5 May 15*	22 Jan. 18	5 May 16	,,	15 Sikhs.
		Greaves, John	5 May 15*	22 Jan. 18	5 May 16	,,	10 Jats.
		Stubbings, Robert Arthur	5 May 15*	9 Aug. 17	5 May 16	,,	30 Punjabis.
	Lan. Fus.	Holden, George Green	5 May 15*	11 Apl. 18	5 May 16	,,	19 Lancers.
		Turner, John Walter	5 May 15*	25 Mar. 18	5 May 16	,,	27 Cavalry.
30		Robertson, Eric	5 May 15*	25 Mar. 18	5 May 16	,,	43 Regiment.
		Applewhaite, Gerald Richard	5 Mar. 15*	26 Mar. 18	5 May 16	,,	17 Infantry.
	N. Zea. Force.	Free, Cuthbert William, M.C.	5 May 15*	5 Dec. 17	5 May 16	,,	Corps of Guides.
	R. F. A.	Gribble, Dudley Stewart	5 May 15*	23 Jan. 18	5 May 16	,,	3-9th Gurkha Rifles.
	Notts. & Derby. R.	Kirby, Hugh	5 May 15*	18 May 18	5 May 16	,,	15 Lancers.
35	Lond. R.	Clarke, Frederick Arthur Stanley, D.S.O.	5 May 15*	22 Apl. 18	5 May 16	,,	57 Rifles.
	Glas. Yeo	Larkin, Edwin Percy	5 May 15*	14 Aug. 17	5 May 16	,,	20 Horse.
	L'pool. R.	Bell, Joseph	5 May 15*	6 June 18	5 May 16	,,	1-9th Bhopal Infantry.
	E. Surr. R.	Paterson, Charles Ramsay Stuart	5 May 15*	29 Aug. 17	5 May 16	,,	32 Pioneers.
	Devon R.	Jervis, Walter Willson	5 May 15*	22 Jan. 18	5 May 16	,,	2-2nd Gurkha Rifles.
40	R. W. Surr. R.	Jefferis, Arthur Reginald	5 May 15*	20 Dec. 17	5 May 16	,,	3 Skinner's Horse.
	Hamps. R.	Lisby, Vincent Charles	5 May 15*	1 Jan. 18	5 May 16	,,	7 Rajputs.
	R. War. R.	Whitehill, Robert Douglas	5 May 15*	26 June 18	5 May 16	,,	121 Pioneers.
	Notts. and Derb. R.	Milward, Roderic Gwynne	5 May 15*	27 Apl. 18	5 May 16	,,	14 Sikhs.
	Bedf. R.	Richardson, Archibald James Colquhoun	5 May 15*	26 Mar. 18	5 May 16	,,	8 Cavalry.
45	R. W. Surr. R.	Falcon, Robert Rex Berney	5 May 15*	22 July 18	5 May 16	,,	117 Mahrattas.
	Welsh R.	Dart, Leonard Victor, M.C.	5 May 15*	12 Feb. 18	5 May 16	,,	Corps of Guides.
	E. Kent.	Witts, Charles Sydney Forster	5 May 15*	3 Sep. 18	5 May 16	,,	Mily. Accts. Dept.
	W. York. R.	Whitley, Henry Cyril	5 May 15*	31 Aug. 18	5 May 16	,,	54 Sikhs.
	Midd'x. R.	Templer, Harold Edward	5 May 15*	18 Sep 18	5 May 16	,,	11 Gurkha Rifles.
50	Ches. R.	Watkin, Hugh Geoffrey	5 May 15*	30 Sep. 18	5 May 16	,,	28 Punjabis.
	E. York. R.	Bickersteth, Ernest Henry	5 May 15*	24 Sep. 18	5 May 16	,,	103 Infantry.
	Norf. Yeo.	Walsh, Lawrence Waddington	5 May 15*	10 Oct. 18	5 May 16	,,	21 Punjabis.
	N. Staff. R.	Paget, George Howard	5 May 15*	10 Oct. 18	5 May 16	,,	3 Gurkha Rifles.
54	York. L. I.	Muirhead, James Ingram, M.C.	5 May 15*	29 Oct. 18	5 May 16	,,	28 Cavalry.

* 9 months forfeited for seniority under conditions of India Army Order 511, dated 14th May 17.

No.	Former Corps.	Names.		Date of birth.	Date of first commission.	Date of appointment to Indian Army.	Date of rank.	Remarks.
		Lieutenants—*contd.*						
55		**Chaundler,** Oliver Stephen	207*a*		6 May 15*	29 July 17	6 May 16	Attached 2 Rajput Infantry.
		Snell, William John Bennett			6 May 15*	3 Mar. 18	6 May 16	,, 58 Rifles.
		Harry, Frank Leslie			12 May 15	22 Nov. 17	12 May 16	,, 20 Infantry.
		Cox, Sydenham John Somers, M.C.	207*a*		12 May 15	8 Dec. 17	12 May 16	,, 97 Infantry.
		Macpherson, Herbert Alexander			12 May 15	9 Jan. 18	12 May 16	,, 70 Rifles.
60		**Morton,** Arthur Frederick			12 May 15	2 Jan. 18	12 May 16	,, 8 Rajputs.
		Gillespie, James Knox			12 May 15	8 Feb. 18	12 May 16	,, 25 Horse.
	Yorks L. J.	**Jones,** Basil Mathwin		24 May 95	12 May 15	28 Nov. 17	12 May 16	,, 2-8th Gurkha Rifles.
	Midd'x R.	**Lightfoot,** Gordon Shelley			12 May 15	23 Nov. 17	12 May 16	,, 101 Grenadiers.
	Ches. R.	**Johnson,** Harold Brodie		8 Sep. 97	12 May 15	16 Apl. 18	12 May 16	,, 3-8 Gurkha Rifles.
65	Suff. R.	**Taylor,** John Norton		17 Nov. 97	12 May 15	24 July 18	12 May 16	,, 82 Punjabis.
	Sco. Rif.	**Brown,** Archibald Francis Wise		16 Aug. 94	12 May 15	26 Oct. 17	12 May 16	,, 9 Gurkha Rifles.
		Atkinson, Norman Evans			15 May 15*	14 Aug. 17	15 May 16	,, 104 Rifles.
		D'Albani, Charles Louis			15 May 15*	3 Dec. 17	15 May 16	,, 123 Rifles.
		Newland, Cecil Dunbar			15 May 15*	3 Dec. 17	15 May 16	,, 2-1st Gurkha Rifles.
70		**Merrylees,** Stewart Bates, M.C.			15 May 15*	12 Jan. 18	15 May 16	,, 75 Panjabis.
	N. Lanc. R.	**Barker,** Ronald Auriol (50)	207*a*	13 Sep. 92	15 May 15*	22 Jan. 18	15 May 16	,, 4-3rd Gurkha Rifles.
	Conn. Rang.	**Lyons,** Henry Harvey, M.C.			15 May 15*	17 Jan. 18	15 May 16	,, 3 Horse.
		Morris, Henry Matthew			15 May 15*	29 July 17	15 May 16	,, 89 Punjabis.
		Chenevix-Trench, Hugo			15 May 15*	24 Feb. 18	15 May 16	,, 3-5th Gurkha Rifles.
75		**O'Leary,** Herbert Homfray			15 May 15*	26 Mar. 18	15 May 16	,, 76 Punjabis.
	Aus. Imp. Force.	**Holmes,** Basil			15 May 15*	8 Feb. 18	15 May 16	,, 38 Horse.
	R. Dub. Fus.	**Jameson,** Charles Edward Eustace						
	5 Dgn. Gds.	**Chard,** Alexander Richard			15 May 15*	12 May 18	15 May 16	,, 17 Infantry.
	R. Ir. Fus.	**Colhoun,** William Albert			15 May 15*	12 May 18	15 May 16	,, 38 C. I. Horse.
80	W. York. R	**Ferguson,** Fritz Eberhard			15 May 15*	1 June 18	15 May 16	,, 29 Punjabis.
	Dorset. R.	**Sinton,** James Scott			15 May 15*	1 June 18	15 May 16	,, 1-6th Jat. Lt. Infantry.
	R. Ir. Rif.	**Anderson,** David Mitchell			15 May 15*	20 Aug. 17	15 May 16	,, 97 Infantry.
	R. Scots	**Barton,** Ralph Firebrace			15 May 15*	6 Sep. 18	15 May 16	,, 3 Brahmans.
	R. Ir. Fus.	**Henry,** Thomas Richard			15 May 15*	29 Sep. 18	15 May 16	,, 56 Rifles.
85		**Haines,** Maurice Arthur			15 May 15*	18 Oct. 18	15 May 16	,, 9 Infantry.
	4 Dgn. Gds.	**Dudley,** Stewart	207		18 May 15*	21 Aug. 17	18 May 16	,, 39 Garhwal Rifles.
		Hedding, John Edward			19 May 15*	6 Oct. 17	19 May 16	,, 7 Rajputs.
	Oxf. and Bucks. L.I.	**Fenwick,** John Stanhope			22 May 15*	3 Dec. 17	22 May 16	,, 35 Sikhs.
					22 May 15*	12 Apl. 18	22 May 16	,, 104 Rifles.
	Aus. Imp. Force.	**Catron,** Joseph Edward Thomas, M.C.			24 May 15*	5 Feb. 18	24 May 16	,, 50 Rifles.
90	Midd'x. R.	**Elgood,** Cyril Lloyd (57, 161, 204)		14 Oct. 92	26 May 15*	4 July 18	26 May 16	,, 14 Lancers.
		Wordsworth, Robert Harley			27 May 15*	3 Nov. 17	27 May 16	,, 16 Cavalry.
		Powell, William Neilson, M.C.			28 May 15*	15 Apl. 17	28 May 16	,, 26 Cavalry.
	Oxf. and Bucks. L.I	**Evans,** Eric Agard			28 May 15*	4 July 18	28 May 16	,, 40 Pathans.
		Shuldham, Walter Frank Quantock			29 May 15*	3 Apl. 17	29 May 16	,, 104 Rifles.

95		**Roper,** Thomas Percy			29 May 15*	29 July 17	29 May 16	,,	11 Rajputs.
		Kirby, Norman Conyers			29 May 15*	25 Mar. 18	29 May 16	,,	35 Sikhs.
	Lond. R.	**Barham,** William Cecil			29 May 15*	12 May 18	29 May 16	,,	11 Rajputs.
	High L. I.	**Martin,** John, M.C.			29 May 15*	1 Oct. 17	29 May 16	,,	123 Rifles.
	E. Surr. R.	**Churcher,** Walter Donald			29 May 15*	16 Aug. 18	29 May 16	,,	67 Punjabis.
100	18 Huss	**Young,** George Murdoch Kirkpatrick			30 May 15*	1 June 18	30 May 16	,,	27 Cavalry.
		Argles, Edward Venables (*temp. Capt., 25 Aug. 18*)			31 May 15*	4 Dec. 17	31 May 16	,,	2 Rajput Infantry.
	R. W. Fus.	**Gibby,** David			31 May 16*	14 Dec. 17	31 May 16	,,	2-3rd Gurkha Rifles.
	A. S. C.	**Cornell,** William Ernest Leonard			31 May 15*	25 Feb. 18	31 May 16	,,	33 Cavalry.
		Hartnoll, Forrest Henry (46)			1 June 15*	14 Sep. 17	1 June 16	,,	36 Gurkha Rifles.
105		**Gibson,** Arthur Blair			1 June 15*	26 Oct. 17	1 June 16	,,	56 Rifles.
	A. S. C.	**Kirkbride,** George			1 June 15*	17 Oct. 17	1 June 16	,,	27 Punjabis.
	M. G. Corps	**Harvey,** James Steel			1 June 15*	1 June 18	1 June 16	,,	127 Baluch Lt. Infy.
	R. Ir. Fus.	**Gordon-Smith,** William			1 June 15*	29 Jan. 18	1 June 16	,,	129 Baluchis.
	F. Scots	**Macpherson,** James Ewan			1 June 15*	3 Nov. 18	1 June 16	,,	90 Punjabis.
110		**Woodyer,** John Mason	207*a*		2 June 15*	2 Apl. 17	2 June 16	,,	104 Rifles
		Morgan, Paul Robert James	207*a*		2 June 15*	25 Sep. 17	2 June 16	,,	3-8 Gurkha Rifles.
	Gord Highrs.	**Calder,** John MacLaren			2 June 15*	16 Sep. 17	2 June 16	,,	111 Mahars.
	Essex R.	**Greenhow,** Edward Leicester			3 June 15*	5 July 18	3 June 16	,,	32 Lancers.
	R. Fus.	**Martin,** William James Royle, M.C.			3 June 15*	29 Aug. 18	3 June 16	,,	50 Rifles.
115		**Gordon,** Thomas James, M.C.			5 June 15*	5 Dec. 17	5 June 16	,,	10 Jats.
		Coles, Nicolas Robert			9 June 15*	3 Dec. 17	9 June 16	,,	123 Rifles.
	Ches. R.	**Marsh,** Bertie Cecil			9 June 15*	17 Dec. 17	9 June 16	,,	6[illegible] Punjabis.
	High. L. I.	**Carruthers,** Walter, Ord			10 June 15*	17 June 17	10 June 16	,,	19 Lancers.
	Devon R.	**Molesworth,** Gilbert Edmund Nassau			10 June 15*	16 Aug. 18	10 June 16	,,	69 Punjabis.
120		**O'Brien,** Edwin Cecil			12 June 15*	4 Dec. 17	12 June 16	,,	3-8th Gurkha Rifles.
		Norton, Charles Edgar			12 June 15*	24 Jan. 18	12 June 16	,,	123 Rifles.
		Stubbs, Robert Hindmarsh			12 June 15*	8 Feb. 18	12 June 16	,,	2 Rajput L. I.
	Conn. Rang.	**Charlton,** Frederick Johnston			12 June 15*	12 May 18	12 June 16	,,	57 Rifles.
	Norf. R.	**Dallas,** Hugh Waldron			12 June 15*	24 Dec. 17	12 June 16	,,	82 Punjabis.
125	A. Cyclist Bn	**Willans,** Norman Jeune			12 June 15*	30 June 18	12 June 16	,,	106 Pioneers.
	York. R.	**Hawkes,** Lawrence Vincent Clare			12 June 15*	30 Sep. 18	12 June 16	,,	63 Infantry.
		Smith, Lawder Benjamin Sandys			14 June 15*	25 Mar. 18	14 June 16	,,	119 Infantry.
	R. Ir. R.	**Claridge,** Donovan Percy			14 June 15	25 Feb. 18	14 June 16	,,	27 Cavalry.
		Southgate, Trevor Davidson			16 June 15	25 Nov. 17	16 June 16	,,	19 Lancers.
130		**Farrell,** Joseph Keith O'Moore		15 Jan. 95	16 June 15*	22 Nov. 17	16 June 16	,,	2 Lancers.
	Midd'x R.	**Longden,** Derek John			16 June 15	25 Mar. 18	16 June 16	,,	23 Pioneers.
	Yorks L. I.	**Ramsden,** Alfred Reginald			16 June 15*	25 Mar. 18	16 June 16	,,	30 Lancers.
		Underwood, Herbert John	207*a*		17 June 15*	6 Apl. 17	17 June 16	,,	29 Punjabis.
		Manlove, John Edward Davis			17 June 15*	12 Jan. 18	17 June 16	,,	10 Jats.
135		**Webb,** Ronald Alfred			17 June 15*	16 Jan. 18	17 June 16	,,	98 Infantry.
	R. Ir. Rif.	**Hanna,** Herbert			17 June 15*	11 Apl. 18	17 June 16	,,	10[illegible] Lt. Infantry.
	E. York R.	**Hickey,** Humphrey Gilbert			17 June 15*	15 June 18	17 June 16	,,	18 Pioneers.
		Poole, Richard Lionel Guy	207*a*		19 June 15*	6 May 17	19 June 16	,,	3 Horse.
		Blair-White, John Henry			19 June 15*	8 Feb. 18	19 June 16	,,	1 Brahmans.
140	N. Staff R.	**Lea,** Edward Heath			19 June 15*	12 May 18	19 June 16	,,	102 Grenadiers.
	N. Staff R.	**Langlands,** Eric Wilfred			19 June 15*	9 Jan. 18	19 June 16	,,	2-3rd Gurkha Rifles.
	Linc. R.	**White,** Reginald Strelley Moresby			19 June 15*	17 Apl. 18	19 June 16	,,	9 Bhopal Infy.
143	North'd Fus	**Southgate,** Compton, M.C.			19 June 15*	29 Aug. 18	19 June 16	,,	81 Pioneers.

* 9 months forfeited for seniority under conditions of India Army Order 511, dated 4th May 17.

Lieutenants—*contd.*

No.	Former Corps.	Names.	Date of birth.	Date of first commission.	Date of appointment to Indian Army.	Date of rank.	Remarks.
144	L'pool R.	**Knight,** Lionel Rankin		19 June 15*	25 Feb. 18	19 June 16	Attached Corps of Guides.
145		**Lloyd,** Wilfrid Lewis, M.C.		20 June 15*	25 Nov. 17	20 June 16	„ 4-39th Garhwal Rifles.
		Banks, Russell Everard		21 June 15*	6 Apl. 17	21 June 16	„ 15 Sikhs.
		Brewis, Ronald Stanley, M.C. 207a		21 June 15*	14 Aug. 17	21 June 16	„ 72 Punjabis.
	R. Muns. Fus.	**Thornton,** Ernest Alfred		21 June 15*	30 June 18	21 June 16	„ 56 Rifles.
		Mackarness, Roger Seymour Patteson		22 June 15*	6 May 17	22 June 16	„ 35 Sikhs.
150		**Carter,** Edward Maurice 207a		22 June 15*	21 Aug. 17	22 June 16	„ 120 Infantry.
		Hillman Alfred Kammeyer		22 June 15*	22 Nov. 17	22 June 16	„ 111 Mahars.
		Russell, Dudley (46)		22 June 15*	25 Nov. 17	22 June 16	„ 97 Infantry.
		Hingston, Alan Trevillion (52) 207a	6 June 96	22 June 15*	20 Aug. 17	22 June 16	„ 3-7th Gurkha Rifles.
		Warry, Harbin Edward		22 June 15*	9 Jan. 18	22 June 16	„ 106 Infantry.
155		**Whitamore,** Edward Osmond		22 June 15*	22 Jan. 18	22 June 16	„ 16 Cavalry.
	Wilt. R.	**Cockayne,** Marcus Henry Douglas		22 June 15*	11 Sep. 17	22 June 16	„ 1 Brahmans.
	L'pool R.	**Farrance,** Sydney Charles Ernest		22 June 15*	25 Feb. 18	22 June 16	„ 8 Cavalry.
	Leic. R.	**Tanner,** George Walter, M.C.		22 June 15*	8 Feb. 18	22 June 16	„ 54 Sikhs.
		Cross, Frank Noel		22 June 15*	5 Apl. 18	22 June 16	„ 33 Cavalry.
160		**Dodd,** Arthur Campbell Beadon		22 June 15*	11 Apl. 18	22 June 16	„ 33 Punjabis.
	R.Ir.Regt.	**Slater,** Reginald Anderson		22 June 15*	13 May 18	22 June 16	„ 37 Punjabis.
	Devon R.	**Freeland,** Rowan Arthur Bayfield, M.C.		22 June 15*	5 Oct. 17	22 June 16	„ 35 Sikhs.
	W. Rid. R.	**Bolland,** Leslie Guy Stewart		22 June 15*	1 June 18	22 June 16	„ 8 Rajputs.
	R. W. Surr. R.	**Taylor,** Harold Duncan		22 June 15*	5 Oct. 17	22 June 16	„ 66 Punjabis.
165	Hamps. R.	**Read,** John Edward		22 June 15*	12 Apl. 18	22 June 16	„ 1-4th Gurkha Rifles.
	Arg.&Suth'd. Highrs.	**Innes,** John Hay, M.C.		22 June 15*	28 Nov. 17	22 June 16	„ 3-8th Gurkha Rifles.
	Notts. & Derby. R.	**Fyldes,** George Borford, M.C.		22 June 15*	27 Aug. 18	22 June 16	„ 45 Sikhs.
	North'd Fus.	**Lake,** Frederick Thomas Macdonald		22 June 15*	6 Sep. 18	22 June 16	„ 2 Rajput L. I.
	Durh. L. I.	**Smith,** Noel Vincent		22 June 15*	30 Sep. 18	22 June 16	„ 34 Pioneers.
170		**Greer,** Roderick Denis		23 June 15*	25 Mar. 18	23 June 16	„ 3-7th Gurkha Rifles.
	L'pool R.	**Pattison,** Fitzgerald Ralph		23 June 15*	18 May 18	23 June 16	„ 54 Sikhs.
	R. Muns. Fus.	**Eagar,** Frederick Maitland		23 June 15*	27 Apl. 18	23 June 16	„ 102 Grenadiers.
		Chambers, Henry Maurice		25 June 15*	3 Dec. 17	25 June 16	„ 117 Mahrattas.
	S. Staff R.	**Hildick-Smith,** John Norman, M.C.		25 June 15*	3 Nov. 18	25 June 16	„ 8 Rajputs.
175		**Steele,** James Stuart		26 June 15*	5 Dec. 17	26 June 16	„ 123 Rifles.
	A. S. C.	**Gahan,** John		26 June 15*	8 Feb. 17	26 June 16	„ 35 Horse.
	A. S. C.	**Stuart-William,** Wilfred Lawrence		26 June 15*	26 June 18	26 June 16	„ 13 Lancers.
		Hastings, John Maurice 207a		28 June 15*	20 Aug. 17	28 June 16	„ 31 Lancers.
	R. Ir. Fus.	**Blood,** Jeffrey Armstrong		30 June 15*	17 Oct. 17	30 June 16	„ 54 Sikhs.
180		**Sheppard-Capurro,** Robert Michael Frederick Francis de Sales		1 July 15*	25 Nov. 17	1 July 16	„ 22 Cavalry.
		Williams, Arthur Hugh Kay 207a		2 July 15*	18 June 17	2 July 16	„ 39 C. I. Horse.
		White, Bernard Kerr		3 July 15*	8 Feb. 18	3 July 16	„ 3 Guides Infy.
183	Dorset. R.	**Hay,** William Rupert		3 July 15*	23 Jan. 18	3 July 16	„ 24 Punjabis.

184	Dorset. R.	**Stable,** Hugh Huntington			3 July 15*	27 Aug. 18	3 July 16	,,	39 C. I. Horse.
185	Som. L. I.	**Luscombe,** Arthur Maitland Popham			3 July 15*	18 Sep. 18	3 July 16	,,	17 Infantry.
	Bedf. R.	**Nokes,** Graham Brunel Ingle			5 July 15*	12 Jan. 18	5 July 16	,,	35 Scinde Horse.
	M. G. Corp	**Crawshaw,** Harry Heaton			6 July 15*	1 Oct. 17	6 July 16	,,	87 Punjabis.
		Chappel, Brian Herbert			7 July 15*	15 Aug. 17	7 July 16	,,	107 Pioneers.
	Som. L. I.	**Thatcher,** Allan Frank Beauchamp			7 July 15*	20 Dec. 17	7 July 16	,,	1-10th Gurkha Rifles.
190	Life Gds.	**Roark,** Charles Thomas Irvine			7 July 15*	13 May 18	7 July 16	,,	34 Horse.
	M. G. Corps.	**Irvin,** David			7 July 15*	18 May 18	7 July 16	,,	102 Grenadiers.
		Fenwicke-Clennell, Edward Clennell, M.C.	207*a*		8 July 15*	22 Nov. 17	8 July 16	,,	107 Pioneers.
		Burch, Gerald Walter			10 July 15*	8 Feb. 18	10 July 16	,,	8 Cavalry.
	R. Fus.	**Rawbone,** Douglas Bertram Murray	207*a*		11 July 15*	1 June 18	11 July 16	,,	5 Lt. Infantry.
195		**Vetch,** William Hamilton			12 July 15*	9 Jan. 18	12 July 16	,,	2-8th Gurkha Rifles.
		Eales, Thomas Francis John			13 July 15*	14 Nov. 17	13 July 16	,,	97 Infantry.
		Keary, Henry Plumer			13 July 15*	26 Feb. 18	13 July 16	,,	30 Punjabis.
		Watts, Harold Christopher	207*a*		14 July 15*	25 Nov. 17	14 July 16	,,	108 Infantry.
		Rayner, Ralph Herbert			14 July 15*	12 Jan. 18	14 July 16	,,	41 Dogras.
200	Yorks. L.I.	**Cameron,** Ronald Thomas		19 Apl. 96	14 July 15*	28 Nov. 17	14 July 16	,,	1-5th Gurkha Rifles.
	3 Dgn. Gds.	**Briggs,** George			14 July 15*	6 Jan. 18	14 July 16	,,	7 Lancers.
	Wilts. R.	**Bearne,** Lewis Henry		23 Oct. 97	14 July 15*	12 Oct. 18	14 July 16	,,	151 Infantry.
	Derb. Yeo.	**Rodgers,** Joseph			15 July 15*	14 Apl. 18	15 July 16	,,	27 Punjabis.
		Denny, Thomas Hamilton			17 July 15*	15 July 17	17 July 16	,,	7 Gurkha Rifles.
205		**Ridley,** Gerard Arthur			17 July 15*	16 May 17	17 July 16	,,	30 Lancers.
	Essex. R.	**Scott,** Sidney Clermont, M.C.			17 July 15*	25 Feb. 18	17 July 16	,,	109 Infantry.
		Wilson, Lionel Earnest			18 July 15*	9 Aug. 17	18 July 16	,,	27 Punjabs.
		Scott, William			19 July 15*	8 Feb. 18	19 July 16	,,	21 Punjabis.
	L'pool R.	**Taylor,** Geoffrey Lyon			19 July 15*	6 Sep. 18	19 July 16	,,	7 Gurkha Rifles.
210	Hamps. R.	**Pike,** Wilfrid Holton			20 July 15*	12 Dec. 17	20 July 16	,,	7 Lancers.
	Hamps. R.	**Jenner,** Charles Herbert			20 July 15*	30 Sep. 17	20 July 16	,,	23 Pioneers.
		Goddard, Arthur Gerald Ivo Ayling			21 July 15*	12 Jan. 18	21 July 16	,,	41 Dogras.
	North'd Fus.	**Swan,** John Harold			21 July 15*	27 Apl. 18	21 July 16	,,	42 Deoli Regt
		Cairn-Duff, John Carlyle			22 July 15*	18 Aug. 17	22 July 16	,,	54 Sikhs.
215	Devon R.	**Tayleur,** Guy			22 July 15*	12 May 18	22 July 16	,,	33 Punjabis.
		Snelling, Cyril Grey (*temp. Capt.*, *6 Sep. 18*)	207*a*		23 July 15*	12 July 17	23 July 16	,,	104 Rifles.
		Crittall, Philip Charles			23 July 15*	12 Jan. 18	23 July 16	,,	16 Cavalry.
		Adderley, Alfred Herbert			23 July 15*	1 Mar 18	23 July 16	,,	35 Sikhs.
	High. L. I.	**Bowers,** John Frederick Farmar			23 July 15*	1 June 18	23 July 16	,,	107 Pioneers.
220	Genl. List	**Paton,** John Esplen Falconer			23 July 15*	28 Feb. 18	23 July 16	,,	31 Lancers.
		Gelston, John Epworth Hubbard			24 July 15*	8 Feb. 18	24 July 16	,,	1 Lancers.
		Gilmore, Horatio Frederick			24 July 15*	29 July 17	24 July 16	,,	11 Rajputs.
	W. Rid. R.	**Longbottom,** Arthur William			24 July 15*	13 May 18	24 July 16	,,	67 Punjabis.
	A. S. C.	**Fearnside-Speed,** Kenneth Andrew San Germano Nelson			24 July 15*	29 Aug. 18	24 July 16	,,	8 Cavalry
225	K. R. Rif. C.	**Compston,** Paul			27 July 15*	25 Sep. 17	27 July 16	,,	3-9th Gurkha Rifles.
	Conn. Rang.	**Stileman,** Cecil Everard Cheere			29 July 15*	12 May 18	29 July 16	,,	55 Rifles.
	M. G. Corps	**Milligan,** Henry Cecil			29 July 15*	1 June 18	29 July 16	,,	129 Baluchis.
	Lond. R.	**Hind,** Hugh William, D.S.O., M.C.			29 July 15*	18 Oct. 18	29 July 16	,,	22 Punjabis.
	Derby. Yeo.	**Gilbin,** Eric Lethbridge Paul Purcell			31 July 15*	6 Sep. 18	31 July 16	,,	31 Lancers.
230	Bord. R.	**Jones,** Rees Stanley			1 Aug. 15*	20 Aug. 17	1 Aug. 16	,,	3-9th Gurkha Rifles.
	S. W. Bord	**Lochner,** Conrad Bertie, M.C.			1 Aug. 15*	29 July 17	1 Aug. 16	,,	31 Lancers.
	S. Afr. Force	**Nicholson,** Cyril Francis Shaw			1 Aug. 15*	5 Dec. 17	1 Aug. 16	,,	3 Gurkha Rifles.
	S. Staff R.	**Pritchard,** Alexander John Loft, M.C.			1 Aug. 15*	25 Feb. 18	1 Aug. 16	,,	Corps of Guides.

* 9 months forfeited for seniority under conditions of I. A. O. 511, dated 14th May 1917.

No.	Former Corps.	Names.	Date of birth.	Date of first commission.	Date of appointment to Indian Army.	Date of rank.	Remarks.
		Lieutenants—*contd.*					
234		**Stewart**, Malcolm John (48*a*, 51)		2 Aug. 15*	4 Dec. 17	2 Aug. 16	Attached 42 Deoli Regiment.
235		**Osborne**, Charles Edward Frank		3 Aug. 15*	22 Aug. 17	3 Aug. 16	,, 63 Infantry.
		Hunt, Randall Evelyn		3 Aug. 15*	22 Nov. 17	3 Aug. 16	,, 31 Punjabis.
		Blomeley, Cyril		3 Aug. 15*	25 Mar. 18	3 Aug. 16	,, 33 Punjabis.
	A. S. C.	**Stansfeld**, Cecil Alfred		3 Aug. 15*	4 Sep. 18	3 Aug. 16	,, 18 Infantry.
		Jones, Thomas Norman Baskerfeild		4 Aug. 15*	12 Jan. 18	4 Aug. 16	,, 71 Punjabis.
240		**Duncan**, Torquil Macleod		4 Aug. 15*	8 Feb. 18	4 Aug. 16	,, 18 Infantry.
	Dorset R.	**Dyer**, Ivon Reginald		4 Aug. 15*	27 Aug. 18	4 Aug. 16	,, 82 Punjabis.
		Allen, Arthur Emilio		5 Aug. 15*	8 Feb. 18	5 Aug. 16	,, 3 Horse.
		Brigg, Cecil John 207*a*		6 Aug. 15*	24 May 17	6 Aug. 16	,, 8 Cavalry.
		Guy, Kenneth		7 Aug. 15*	25 Mar. 18	7 Aug. 16	,, 123 Rifles.
245	K. O. Sco. Bord.	**Brewitt**, Arthur Claude		7 Aug. 15*	13 Aug. 18	7 Aug. 16	,, 8 Rajputs.
		Gordon, Michael (48*a*, 51)		9 Aug. 15*	25 Nov. 17	9 Aug. 16	,, 76 Punjabis.
		Bremner, Stanley		9 Aug. 15*	25 Mar. 18	9 Aug. 16	,, 2 Lancers.
	Midd'x R.	**Garrow**, Vernon James		9 Aug. 15*	29 Aug. 18	9 Aug. 16	,, 7 Rajputs.
	North'd Fus.	**Curry**, Arthur Vincent (46)		10 Aug. 15*	11 Apl. 18	10 Aug. 16	,, 2 Rajput Lt. Infy.
250		**Cole**, Ronald St. George (57)		11 Aug. 15	23 Nov. 17	11 Aug. 16	,, 42 Regt.
		Diggory, Percival		11 Aug. 15*	23 Jan. 18	11 Aug. 16	,, 3-9th Regiment.
		Trow, Vivian Harold (*tempt. Capt. 14 Oct. 18*)		11 Aug. 15*	8 Feb. 18	11 Aug. 16	,, 120 Infantry.
		Carpenter, Hedley James		11 Aug. 15*	8 Feb. 18	11 Aug. 16	,, 3 Horse.
	Aus. Imp. Force.	**Pulling**, Hugh Douglas (46) (*temp. Capt. 9 June 18*)		11 Aug. 15*	25 Mar. 18	11 Aug. 16	,, 3-6th Gurkha Rifles.
255	S. Lan. R.	**MacKenzie**, Herbert Cedric		11 Aug. 15*	13 Feb. 17	11 Aug. 16	,, 7 Rajputs.
	Hamps. R.	**Harvey**, Cecil Walter Lewery		12 Aug. 15*	17 Feb. 18	12 Aug. 16	,, 2-1st Gurkha Rifles.
	A. S. C.	**Hall**, William Henry		12 Aug. 15*	12 May 18	12 Aug. 16	,, 98 Infantry.
	Buck. Yeo.	**Charles**, Percy John		12 Aug. 15*	27 Aug. 18	12 Aug. 16	,, 10 Lancers.
	R. Highrs.	**Telfer**, James William		13 Aug. 15*	11 Apl. 18	13 Aug. 16	,, 3-9 Gurkha Rifles.
260	R. E.	**Shute**, Percy Burgess		13 Aug. 15*	12 May 18	13 Aug. 16	,, 64 Pioneers.
	Scot. Rif.	**McLaren**, John Archibald		13 Aug. 15*	3 Nov. 18	13 Aug. 16	,, 151 Infantry.
		Phelps, Albert Edwin (46)		14 Aug. 15*	3 Dec. 17	14 Aug. 16	,, 56 Rifles.
	Sco. Rif.	**Imrie**, William		14 Aug. 15*	1 June 18	14 Aug. 16	,, 55 Rifles.
	5 Res. R. of Cav.	**Davis**, Cecil Kenneth		14 Aug. 15*	26 June 18	14 Aug. 16	,, 29 Lancers.
265	Welsh R.	**James**, Herbert Cyril		14 Aug. 15*	24 Sep. 18	14 Aug. 16	,, 117 Mahrattas.
		Seed, John Andrew Theodore (51) 207*a*		18 Aug. 15*	29 July 17	18 Aug. 16	,, 110 Infantry.
		Butler, Herbert Marshall		19 Aug. 15*	8 Feb. 18	19 Aug. 16	,, 7 Lancers.
	M. G. Corps	**Martin**, Vivian John		19 Aug. 15*	2 Oct. 17	19 Aug. 16	,, 23 Pioneers.
	R. Fus.	**Sheehan**, Austin Francis Campbell		19 Aug. 15*	25 Nov. 17	19 Aug. 16	,, 2-3rd Gurkha Rifles.
270		**Weaver**, John Yardley		20 Aug. 15*	22 Jan. 18	20 Aug. 16	,, 13 Lancers.
	A. S. C.	**Goodall**, Lionel Mackenzie		20 Aug. 15*	16 Oct. 17	20 Aug. 16	,, 2-5th Gurkha Rifles.
	Suff. R.	**Down**, Francis Denis		23 Aug. 15*	11 Apl. 18	23 Aug. 16	,, 124 Infantry.
	Shrops. L. I.	**Lawrence**, Arthur Edward, M.C.		23 Aug. 15*	21 Jan. 18	23 Aug. 16	,, 4 Gurkha Rifles.
	Lan. Fus.	**Brear**, Bernard Haigh		24 Aug. 15*	3 Dec. 17	24 Aug. 16	,, 66 Punjabis.
275	Oxf. & Bucks. L. I.	**Rayner**, Charles Sydney William		27 Aug. 15*	22 Nov. 17	27 Aug. 16	,, 95 Infantry.

	R. Berks. R.	**Elphick,** Hugh Clifford		27 Aug. 15*	8 Dec. 17	27 Aug. 16	„	15 Sikhs.
		Hide, Thomas George Seymour (46 (†*temp. Capt.*, 1 *May 18*)	207*a*	28 Aug. 15*	24 May 17	28 Aug. 16	„	29 Punjabis.
		Turner, Roland Valford		28 Aug. 15*	20 Feb. 17	28 Aug. 16	„	6 Cavalry.
	13 Cav.	**Booth,** Richard		28 Aug. 15*	12 Jan. 18	28 Aug. 16	„	37 Lancers.
	Dorset R.	**Sawyer,** Claude Patrick Bevan		28 Aug. 15*	30 June 18	28 Aug. 16	„	10 Lancers.
	A. S. C.	**Henrotin,** Herbert Victor		28 Aug. 15*	27 Aug. 18	28 Aug. 16	„	31 Lancers.
	A Cyclist Corps.	**Booty,** Francis Charles		30 Aug. 15*	15 Mar. 18	30 Aug. 16	„	89 Punjabis.
	Welsh R.	**Howell,** Vernon		30 Aug. 15*	30 Sep. 18	30 Aug. 16	„	26 Punjabis.
		Todd, Fred Walton		1 Sep. 15*	22 Jan. 18	1 Sep. 16	„	107 Pioneers.
285	War. Yeo.	**King,** Frank Harold		1 Sep. 15*	14 Apl. 18	1 Sep. 16	„	29 Lancers.
	Bord. R.	**Hardy,** Edwin Pearce		1 Sep. 15*	1 Jan. 18	1 Sep. 16	„	72 Punjabis.
	Som. L. I.	**Pearce,** Charles Frederick Byrde		1 Sep. 15*	5 Oct. 18	1 Sep. 16	„	17 Infantry.
	S. W. Bord.	**Taylor,** John Nigel Caryl		2 Sep. 15*	12 Jan. 18	2 Sep. 16	„	15 Sikhs.
		Thompson, Robert Mayne	207*a*	3 Sep. 15*	6 June 17	3 Sep. 16	„	2-9th Delhi Infantry
290	R. Muns. Fus.	**Furney,** Simon Kenny, M.C.		3 Sep. 15*	29 Oct. 18	3 Sep. 16	„	54 Sikhs.
	L'pool R.	**Shier,** Morton Barker		3 Sep. 15*	4 Nov. 18	3 Sep. 16	„	11 Gurkha Rifles.
		Campbell, William Balfour	207*a*	4 Sep. 15*	3 Dec. 17	4 Sep. 16	„	113 Infantry.
		Hasted, Arthur William Ord		5 Sep. 15*	19 Aug. 17	5 Sep. 16	„	113 Infantry.
		Herrick, John Riversdale Warren		5 Sep. 15*	12 Jan. 18	5 Sep. 16	„	3-2nd Gurkha Rifles.
295	R. W. Surr. R.	**Faulkner,** Roland, M.C.		5 Sep. 15*	31 Aug. 18	5 Sep. 16	„	6 Gurkha Rifles.
	R. Suss. R.	**Best,** Ernest		7 Sep. 15*	22 Apl. 18	7 Sep. 16	„	25 Punjabis.
		Jones, Philip Perkyns		9 Sep. 15*	19 Jan. 18	9 Sep. 16	„	85 Rifles.
	Devon R.	**Tweed,** John Reginald Howard		9 Sep. 15*	28 July 17	9 Sep. 16	„	94 Infantry.
	Hamps. R.	**McEntire,** Henry Sandys Mecredy		9 Sep. 15*	24 July 18	9 Sep. 16	„	123 Rifles.
300		**Columbine,** Roland Herbert		10 Sep. 15*	29 July 17	10 Sep. 16	„	39 Rifles.
	Manch. R.	**Wood,** Ernest, M.C.		10 Sep. 15*	13 Aug. 18	10 Sep. 16	„	2 Rajput L. I.
	W. Rid. R.	**Charlesworth,** John Henry		10 Sep 15*	27 Aug. 18	10 Sep. 16	„	38 Dogras.
	W. Rid. R.	**Faulder,** George Bertram		11 Sep. 15*	30 June 18	11 Sep. 16	„	94 Infantry.
	R. Fus.	**Williams,** Ronald George		11 Sep. 15*	26 Sep. 18	11 Sep. 16	„	154 Infantry.
305		**Ogier,** Lionel Leverett	207*a*	12 Sep. 15*	22 Nov. 17	12 Sep. 16	„	124 Infantry.
	A. S. C.	**Sharpe,** Alfred George		12 Sep. 15*	5 Feb. 18	12 Sep. 16	„	4 Cavalry.
	Ches. R.	**Ellis,** Henry Carthew (46)		14 Sep. 15*	9 Apl. 18	14 Sep. 16	„	40 Pathans.
		Evanson, Eric Ashley		15 Sep. 15*	8 Feb. 18	15 Sep. 16	„	41 Dogras.
		Valentine, William Stainer		15 Sep. 15*	8 Feb. 18	15 Sep. 16	„	3-9th Gurkha Rifles.
310	R. Berk. R.	**Turner,** Reginald Hamilton Fyers		15 Sep. 15*	23 Apl. 18	15 Sep. 16	„	2 Lancers.
	Welsh R.	**Bellhouse,** Alfred Panter		16 Sep. 15*	28 Jan. 18	16 Sep. 16	„	2 Rajput L. I.
	Manch. R.	**Assig,** Harold Leslie, M.C.		16 Sep. 15*	31 July 18	16 Sep. 16	„	S. & T. Corps.
		Bletsoe, William Bruce		17 Sep. 15*	21 Feb. 17	17 Sep. 16	„	18 Lancers.
		Driver, Rupert Arthur		18 Sep. 15*	6 May 17	18 Sep. 16	„	19 Punjabies.
315	Glouc. R.	**Toogood,** Cyril George, D.S.O.		18 Sep. 15*	16 Aug. 17	18 Sep. 16	„	2 Gurkha Rifles.
		Beaton, Dickson James		18 Sep. 15*	23 Nov. 17	18 Sep. 16	„	8 Rajputs.
		Tonkin, Reginald Sebert		18 Sep. 15*	24 Jan. 18	18 Sep. 16	„	111 Mahars.
	L'pool R.	**Fawkner,** Harold Stanley		18 Sep. 15*	22 Apl. 18	18 Sep. 16	„	28 Lt. Cavy.
	R. Dub. Fus.	**Stobart,** Robert McIlroy		18 Sep. 15*	4 Sep. 18	18 Sep. 16	„	2 Rajput L. I.
		Aitken, Charles Fleming	207*a*	20 Sep. 15*	29 July 17	20 Sep. 16	„	39 C. I. Horse.
320	Welsh R.	**Chapman,** Herbert Reginald		21 Sep. 15*	12 May 18	21 Sep. 16	„	23 Cavalry.
		Dempster, Douglas Erskine		22 Sep. 15*	14 Sep. 17	22 Sep. 16	„	54 Sikhs.

* 9 months forfeited for seniority under conditions of I. A. O. 511, dated 14th May 1917. † With pay and allowances as for Lieutenant.

No.	Former Corps.	Names.	Date of birth.	Date of first commission.	Date of appointment to Indian Army.	Date of rank.	Remarks.
		Lieutenants—*contd.*					
		Leys, John Farquharson 207*a*		22 Sep. 15*	22 Nov. 17	22 Sep. 16	Attached 2 Infantry.
	L'pool R.	**Baker**, John Norman Leonard		22 Sep. 15*	13 May 18	22 Sep. 16	" 119 Infantry.
325	York. R.	**Wilkinson**, William Dale		22 Sep. 15*	1 June 18	22 Sep. 16	" 9 Infantry.
	Worc. R.	**Marsh**, Courtney Cyril Allen		22 Sep. 15*	15 Dec. 17	22 Sep. 16	" 47 Sikhs.
		Stranger, Frank Railton		23 Sep. 15*	25 Mar. 18	23 Sep. 16	" 33 Punjabis.
	York. R.	**Brown**, Henry Noel		23 Sep. 15*	4 Jan. 18	23 Sep. 16	" 107 Pioneers.
	Sea. Highrs.	**Duncan**, Herbert Cecil		23 Sep. 15*	29 Aug. 18	23 Sep. 16	" 98 Infantry.
		Bennett, Norman Butlar		24 Sep. 15*	23 Jan. 18	24 Sep. 16	" 31 Lancers.
330	S. Lanc. R.	**Abraham**, Clement Erwin Simon, M.C.		24 Sep. 15*	12 Dec. 17	24 Sep. 16	" 127 Baluch Lt. Infy.
	M. G. Corps.	**Cockrill**, Stanley Robert		24 Sep. 15*	13 Aug. 18	24 Sep. 16	" 3 Gurkha Rifles.
	W. York. R.	**Rogers**, Thomas Paterson Walker, M.C.		26 Sep. 15*	15 June 18	26 Sep. 16	" 124 Infantry.
	E. Lanc. R.	**Durrant**, Oliver Torlesse		26 Sep. 15*	12 Dec. 17	26 Sep. 16	" 103 Mah. Lt. Infy.
		Rea, Alfred		28 Sep. 15*	11 Sep. 17	28 Sep. 16	" 7 Rajputs.
335		**Delap**, Alexander (48*a*, 51)		28 Sep. 15*	24 Oct. 17	28 Sep. 16	" 3-1st Gurkha Rifles.
	Nav. Vol.	**Webb**, Wilfrid Francis		29 Sep. 15*	12 June 17	29 Sep. 16	" 104 Rifles.
	Suff. R.	**Johnson**, Richard Arthur		29 Sep. 15*	1 June 18	29 Sep. 16	" 66 Punjabis.
	Bord. R.	**Brownlie**, John Kerr Cross		29 Sep. 15*	1 Oct. 17	29 Sep. 16	" 42 Deoli Regiment.
340		**Atkinson**, Denis Jackson		30 Sep. 15*	12 Jan. 18	30 Sep. 16	" 2-4th Gurkha Rifles.
		Mackay, Douglas Sween (48*a*)		30 Sep. 15*	21 Jan. 18	30 Sep. 16	" 15 Lancers.
		Moritz, Francis Edward Binny 207*a*		30 Sep. 15*	8 Feb. 18	30 Sep. 16	" 71 Punjabis.
	North'd. Fus.	**Simons**, John Guy		30 Sep. 15*	30 June 18	30 Sep. 16	" 2-1st Gurkha Rifles.
	S. Staff R.	**Fussell**, Henry Stephen		2 Oct. 15*	25 Sep. 17	2 Oct. 16	" 111 Mahars.
345	Lanc. Huss.	**Robinson**, Clive Whateley		3 Oct. 15*	26 June 18	3 Oct. 16	" 5 Cavalry.
	North'n. R.	**Mackay**, Richard Johnstone		4 Oct. 15*	5 Oct. 17	4 Oct. 16	" 112 Infantry.
	R. W. Fus.	**Ford**, Charles Bell		4 Oct. 15*	22 Oct. 17	4 Oct. 16	" 103 Mah. Lt. Infy.
		Corcon, Cecil Ilderton, M.C.		6 Oct. 15*	26 Feb. 18	6 Oct. 16	" 22 Punjabis.
		Tankard, Brian Lincey		9 Oct. 15*	29 Nov. 17	9 Oct. 16	" 116 Mahrattas.
350	Ches. R.	**Powell**, Harold Norman		9 Oct. 15*	28 Sep. 18	9 Oct. 16	" 39 C. I. Horse.
	Linc. R.	**Thorn**, Arthur Percy		9 Oct. 15*	1 Mar. 18	9 Oct. 16	" 21 Punjabis.
	Suff. R.	**Sturdy**, John Rowland, M.C. (*temp. Capt., 28 June 18*)	25 Oct. 89	10 Oct. 15*	18 Oct. 17	10 Oct. 16	" 3-5th Gurkha Rifles.
	Bedf. R.	**Taitt**, Noel Rothwell		10 Oct. 15*	27 Aug. 18	10 Oct. 16	" 5 Lt. Infy.
		Nicholas, Charles Henry 207*a*		11 Oct. 15*	11 Sep. 17	11 Oct. 16	" 18 Infantry.
355		**Mocatta**, Cyril Herbert		12 Oct. 15*	23 Jan. 18	12 Oct. 16	" 2-3 Gurkha Rifles.
	Gen. List.	**Quest**, Thomas Percival		13 Oct. 15*	11 Apl. 18	13 Oct. 16	" 119 Infantry.
	Linc. R.	**Wright**, George Alexander		14 Oct. 15*	1 June 18	14 Oct. 16	" 15 Lancers.
		Adams, Arthur Wilfred		15 Oct. 15*	28 July 17	15 Oct. 16	" 107 Pioneers.
	York. L. I.	**Russell**, Alec Gaudie		16 Oct. 15*	16 Sept. 17	16 Oct. 16	" 89 Punjabis.
360	Manch. R.	**Sykes**, Frank Clifford	27 Mar. 96	20 Oct. 15*	13 Aug. 18	20 Oct. 16	" 29 Punjabis.
	York. & Lanc. R.	**Thompson**, John Stanley		21 Oct. 15*	30 Sep. 18	21 Oct. 16	" 7 Rajputs.
		Ewen, Alan Charles		22 Oct. 15*	19 Sep. 17	22 Oct. 16	" 3-9th Gurkha Rifles.
	A. S. C.	**Raven-Hill**, Arthur Leonard		23 Oct. 15*	11 Apl. 18	23 Oct. 16	" 22 Punjabis.
		Stevenson, Alexander Macneil		24 Oct. 15*	24 Jan. 18	24 Oct. 16	" 76 Punjabis.
	Manch. R.	**Widdows**, Frank Mason		24 Oct. 15*	6 Sep. 18	24 Oct. 16	" 22 Punjabis.
365		**Butcher**, Alfred Leopold		25 Oct. 15*	5 Dec. 17	25 Oct. 16	" 52 Sikhs.

		Pickard, Bernard de Hailes			25 Oct. 15*	19 Jan. 18	25 Oct. 16	,,	35 Horse.
	Linc. R.	Hurst, Robert Haslewood			25 Oct. 15*	4 Mar. 18	25 Oct. 16	,,	4-3 Gurkha Rifles.
		Cornish, Harold Denson			26 Oct. 15*	5 Dec. 17	26 Oct. 16	,,	2-2nd Gurkha Rifles.
370		Lowe, Edward Arthur Howard			26 Oct. 15*	17 Sep. 17	26 Oct. 16	,,	16 Rajputs.
	Glouc. R.	Jackson, Norman Carstairs			26 Oct. 15*	16 Dec. 17	26 Oct. 16	,,	33 Punjabis.
	Midd'x. R.	Bennett, William Harris			26 Oct. 15*	22 Jan. 18	26 Oct. 16	,,	1 Brahmans.
	Scot. Rif.	Charter, William Forster			26 Oct. 15*	12 Oct. 18	26 Oct. 16	,,	21 Punjabis.
		McCoy-Hill, Lancelot McCoy			27 Oct. 15*	3 Dec. 17	27 Oct. 16	,,	57 Rifles.
375		Moyes, Eric Milne			27 Oct. 15*	25 Mar. 18	27 Oct. 16	,,	76 Punjabis.
		Teape, Cyrud Charles Tulloh	207a		28 Oct. 15*	24 May 17	28 Oct. 16	,,	21 Punjabis.
	Linc. Yeo.	Steward, John Reginald O'Bryan			28 Oct. 15*	2 June 17	28 Oct. 16	,,	6 Cavalry.
		Jones, Philip James Desprez			30 Oct. 15*	22 Jan. 18	30 Oct. 16	,,	11 Rajputs.
		Brothers, John Arnold	207a		30 Oct. 15*	8 Feb. 18	30 Oct. 16	,,	21 Punjabis.
380	M. G. Corps.	Barstow, Henry Lionel	207a		1 Nov. 15*	7 Sep. 17	1 Nov. 16	,,	2-4 Gurkha Rifles.
	Arg.&Suth'd Highrs.	Tickell, George Warren			1 Nov. 15*	30 Sep. 18	1 Nov. 16	,,	36 Sikhs.
	Bedf. R.	Wilson, Eric Chilver			3 Nov. 15*	11 Apl. 18	3 Nov. 16	,,	2-1 Gurkha Rifles.
	Norf. R.	Beckerson, Robert Ernest			4 Nov. 15*	31 May 18	4 Nov. 16	,,	54 Sikhs.
		Regan, Graford Henry			5 Nov. 15*	12 Jan. 18	5 Nov. 16	,,	110 Mah. L. I.
385	Trg.Res.Bn.	Topping, Walter Darling			6 Nov. 15*	18 May 18	6 Nov. 16	,,	3 Brahmans.
		Porral, Francis Joseph William			7 Nov. 15*	12 Jan. 18	7 Nov. 16	,,	119 Infantry.
		Rowlands, Ioan Barry Hughes			8 Nov. 15*	6 Feb. 18	8 Nov. 16	,,	109 Infantry.
		Irwin, Stephen Fenemore			9 Nov. 15*	8 Feb. 18	9 Nov. 16	,,	76 Punjabis.
		Grell, Louis George Nevile			9 Nov. 15*	23 Jan. 18	9 Nov. 16	,,	27 Punjabis.
390		Salt, Robert Francis			10 Nov. 15*	12 Jan. 18	10 Nov. 16	,,	1-39th Garhwal Rifles.
	R. Fus.	Hussey, Glen Owen Gilbert			10 Nov. 15*	18 Mar. 18	10 Nov. 16	,,	4 Cavalry.
	Lond. R.	Gain, Joseph Charles			12 Nov. 15*	11 Apl. 18	12 Nov 16	,,	41 Dogras.
	Lond. R.	Richmond, John Philip			12 Nov. 15*	14 Apl. 18	12 Nov. 16	,,	33 Punjabis.
	L'pool R.	Jones, Gryffydh Watcyn			15 Nov. 15*	11 Apl. 18	15 Nov. 16	,,	61 Pioneers.
395	Som. L. I.	Gilmore-Ellis, Patrick Douglas			15 Nov. 15*	17 Oct. 17	15 Nov. 16	,,	54 Sikhs.
	Lond. R.	Thorogood, Percival Walter			21 Nov. 15*	20 Oct. 17	21 Nov. 16	,,	10 Jats.
		Pearson, Hugh Castell, M.C. (*temp. Capt., 1 June 18*)			23 Nov. 15*	25 Sep. 17	23 Nov. 16	,,	1-2nd Gurkha Rifles.
		Brady, Wiliam Alfred Henry (48a, 51)			23 Nov. 15*	26 Oct. 17	23 Nov. 16	,,	35 Sikhs.
		Kelly, Peirse Ignatious			23 Nov. 15*	25 Mar. 18	23 Nov. 16	,,	86 Infantry.
400	Worc. R.	Carter, Sydney Noel			23 Nov. 15*	12 May 18	23 Nov. 16	,,	2 Raj. L. I.
	S. W. Bord.	Morgan, William Gordon Curtis			23 Nov. 15*	18 May 18	23 Nov. 16	,,	81 Pioneers.
	Northn. R.	Leslie-Smith, Brian Russell			23 Nov. 15*	1 Feb. 18	23 Nov. 16	,,	129 Baluchis.
		Matthews, Reginald			24 Nov. 15*	13 Jan. 18	24 Nov. 16	,,	39 C. I. Horse.
	Scot. Rif.	Torrance, Ian Symington	207a		24 Nov. 15	1 Oct. 17	24 Nov. 16	,,	42 Regt.
405	Lanc. Fus.	Myres, Norman Garth		12 Apl. 97	24 Nov. 15	29 July 18	24 Nov. 16	,,	119 Infantry.
	Sco. Rif.	Rusby, Lionel Stuart			24 Nov. 15*	9 Jan. 18	24 Nov. 16	,,	3rd Gurkha Rifles.
	E. Kent R.	Atteridge, Laurence Foley			26 Nov. 15*	24 Sep. 18	26 Nov. 16	,,	15 Sikhs.
		Maunder, George Henry	207a		25 Nov. 15*	26 Mar. 18	25 Nov. 16	,,	107 Pioneers.
	High. L. I.	Fulton, Edward Arthur Craig			25 Nov. 15*	1 June 18	25 Nov. 16	,,	1-4th Gurkha Rifles.
410	Lan. Fus.	Rudd, Clement Hugh Chadd			26 Nov. 15*	8 Dec. 17	26 Nov. 16	,,	41 Dogras.
		Buist, Andrew Augustine (51)	207a		26 Nov. 15*	6 May 17	26 Nov. 16	,,	4 Gurkha Rifles.
	M. G. Corps.	Davenport, Thomas			26 Nov. 15*	27 Apl. 18	26 Nov. 16	,,	40 Pathans.
	M. G. Corps	Hayes, Ernest James Thomas			26 Nov. 15*	3 Feb. 18	26 Nov. 16	,,	50 Rifles.
414	Oxf. & Buck. R.	Rowe, William Campbell			27 Nov. 15*	26 June 18	27 Nov. 16	,,	72 Punjabis.

* 9 months forfeited for seniority under conditions of I. A. O. 511, dated 14th May 1917.

No.	Former Corps.	Names.	Date of birth.	Date of first commission.	Date of appointment to Indian Army.	Date of rank.	Remarks.
		Lieutenants—*contd.*					
415	R. Ir. Rif.	**Gault,** Joseph		27 Nov. 15*	27 Aug. 18	27 Nov. 16	Attached 35 Sikhs.
		Morris, Herbert Edwin Abrahall		1 Dec. 15*	26 Feb. 18	1 Dec. 16	" 30 Punjabis.
	A. S. C.	**Brown,** John Ray		1 Dec. 15*	35 Feb. 18	1 Dec. 16	" 16 Cavalry.
	Glouc. R.	**Thacker,** Noel, M.C. (71)		2 Dec. 15*	27 Apl. 18	2 Dec. 16	" 15 Sikhs.
	A. S. Corps	**Smith,** Arthur Alborough		2 Dec. 15*	29 July 18	2 Dec. 16	" 31 Lancers.
420	Lord. R.	**Noakes,** Eric John Wickham		3 Dec. 15*	18 May 18	3 Dec. 16	" 26 Cavalry.
		O'Callaghan, Cornelius Howden Denis		4 Dec. 15*	8 July 17	4 Dec. 16	" 66 Punjabis.
	Lond. R.	**Tooth,** Hugh Cyril Artemus		9 Dec. 15*	19 Apl. 18	9 Dec. 16	" 124 Infantry.
		Charnley, Frederick Harold (48*a*, 51) 207*a*	19 Dec. 92	10 Dec. 15*	10 Sep. 17	10 Dec. 16	" 3-7th Gurkha Rifles.
		Donaldson, John		10 Dec. 15*	9 Jan. 18	10 Dec. 16	" 99 Infantry.
425		**Chapman,** Ralph Vincent		10 Dec. 15*	3 Dec. 17	10 Dec. 16	" 2nd Guides Infantry.
		Perrett, Wilfred Arthur		10 Dec. 15*	22 Jan. 18	10 Dec. 16	" 76 Punjabis.
		Duchesne, Rupert Leslie		10 Dec. 15*	26 Feb 18	10 Dec. 16	" 2 Lancers.
	R. H. A.	**Heath,** Henry Norman		10 Dec. 15*	30 June 18	10 Dec. 16	" 26 Cavalry.
	R. W. Fus.	**Tewson,** Reginald George		10 Dec. 15*	4 Sep. 18	10 Dec. 16	" 11 Gurkha Rifles.
430		**Booth,** Allan Hodgson		11 Dec. 15*	25 Nov. 17	11 Dec. 16	" 14 Sikhs.
		Kane, Gordon Blackburne		11 Dec. 15*	3 Dec. 17	11 Dec. 16	" 41 Dogras.
	Suff. R.	**Jay,** Cyril Everard		11 Dec. 15*	13 Aug. 18	11 Dec. 16	" 1 Brahmans.
	A. S. C.	**Stevenson,** John		12 Dec. 15*	12 Oct. 17	12 Dec. 16	" 97 Infantry.
	Scot. H.	**Hardie,** John David		12 Dec. 15*	19 Mar. 18	12 Dec. 16	" 3 Horse.
435		**Uff,** Gordon Thomas Rogers		13 Dec. 15*	22 Nov. 17	13 Dec. 16	" 54 Sikhs.
		Secombe, Herbert Stanley (57)		13 Dec. 15*	22 Nov. 17	13 Dec. 16	" 43 Regiment.
		Fetherstonhaugh, Mark Rupert		13 Dec. 15*	22 Jan. 18	13 Dec. 16	" 33 Cavalry.
		Pigott, Fredrick Michael		13 Dec. 15*	7 Feb. 18	13 Dec. 16	" 30 Punjabis.
		Henderson, William Smith Ramsay (59)		13 Dec. 15*	25 Mar. 18	13 Dec. 16	" 103 Mah. L. I.
440	R. Berk R.	**Lawrence,** Thomas Bernard		13 Dec. 15*	1 June 18	13 Dec. 16	" 88 Infantry.
	Lond. Yeo.	**Bostock,** Joseph Anthony		13 Dec. 15*	12 May 18	13 Dec. 16	" 11 Lancers.
		Farrow, John Kenneth		15 Dec. 15*	8 Feb. 18	15 Dec. 16	" 2-8th Gurkha Rifles.
	M.G. Corps.	**Scott,** William		15 Dec. 15*	12 May 18	15 Dec. 16	" 52 Sikhs.
		Bond, Kenneth Hills, M.C.		16 Dec. 15*	25 Mar. 18	16 Dec. 16	" 129 Baluchis.
445	R. Suss. R.	**Williams,** Ernest William		16 Dec. 15*	30 Sep. 18	16 Dec. 16	" 28 Punjabis.
		McMahon-O'Byrne, Dermott		17 Dec. 15*	2 July 17	17 Dec. 16	" 25 Cavalry.
	Kent Cyclist Bn.	**Raven-Hill,** Lucien Ennar		17 Dec. 15*	31 Dec. 17	17 Dec. 16	" 61 Pioneers.
	E. Kent. R.	**Saunders,** Gilbert Norman		19 Dec. 15*	22 Feb. 18	19 Dec. 16	" 30 Punjabis.
	E. Kent. R.	**Fulton,** Roderick		19 Dec. 15*	30 Dec. 17	19 Dec. 16	" 53 Sikhs.
450	R. W. Fus.	**Irwin,** Norman Leslie Crozier		19 Dec. 15*	5 Aug. 18	19 Dec. 16	" 8th Gurkha Rifles.
		Stepto, Edward		20 Dec. 15*	6 June 17	20 Dec. 16	" 5 Cavalry.
		Greenop, Maurice Ewdin		20 Dec. 15*	22 Nov. 17	20 Dec. 16	" 18 Infantry.
		Bassett, Frederick Sidney (57)		20 Dec. 15*	25 Nov. 17	20 Dec. 16	" 67 Punjabis.
	Aus. Imp. Force.	**Sydenham,** Alwyn Francis		20 Dec. 15*	25 Mar. 18	20 Dec. 16	" 17 Cavalry.
455	Scot. Rif.	**Stevens,** Cyril Walter		20 Dec. 15*	14 Apl. 18	20 Dec. 16	" 30 Punjabis.
	Gord. Highrs.	**Priday,** Arthur Kenneth		20 Dec 15*	11 Apl. 18	20 Dec. 16	" 25 Punjabis.
457	York & Lans. R.	**Sheppard,** Frank Eric		20 Dec. 15*	6 Oct. 17	20 Dec. 16	" 123 Rifles.

457	Midd'x. R.	**Folkard**, William Francis Douglas	. . . . .	20 Dec. 15*	5 Jan. 18	20 Dec. 16	,,	97 Infantry.
	R. W. Surr. R.	**Wheeler**, Geoffrey Edleston	. . . . .	20 Dec. 15*	31 Aug. 18	20 Dec. 16	,,	6 Gurkha Rifles.
	S. Staff. R.	**Bull**, Leslie Colyn, M.C.	7 Oct. 95	22 Dec. 15	27 Apl. 18	23 Dec. 16	,,	79 Infantry.
460	S. Wales B.	**Manley**, Walter James (46)	20 Sep. 97	22 Dec. 15	1 June 18	22 Dec. 16	,,	21 Punjabis.
		Maclachlan, Alastan Banks	. . . . .	24 Dec. 15*	26 Oct. 17	24 Dec. 16	,,	8 Rajputs.
	Aus. Imp. F.	**Alderson**, Victor Clarkson	. . . . .	24 Dec. 15*	25 Feb. 18	24 Dec. 16	,,	120 Baluchis.
	R. Sc. Fus.	**Laurie**, James	. . . . .	24 Dec. 15*	3 Nov. 18	24 Dec. 16	,,	42 Deoli Regt.
		Thompson, Richard Herbert	. . . . .	25 Dec. 15*	22 Jan. 18	25 Dec. 16	,,	10 Jats.
465		**Carey**, Joseph	. . . . .	25 Dec. 15*	8 Feb. 18	25 Dec. 16	,,	22 Punjabis.
	R. Fus.	**Davis**, Allan Thomas	. . . . .	25 Dec. 15*	30 Sep. 18	25 Dec. 16	,,	18 Infantry.
		Lavender, Sydney Salter	. . . . .	26 Dec. 15*	12 Jan. 18	26 Dec. 16	,,	46 Punjabis.
		Rich, Gregory	. . . . .	26 Dec. 15*	8 Feb. 18	26 Dec. 16	,,	107 Pioneers.
	Linc. R.	**Loudoun**, Robert Franklin	. . . . .	26 Dec. 15*	12 Feb. 18	26 Dec. 15	,,	2-9th Infantry.
470	R. Innis. Fus.	**Armstrong**, Albert Herbert Howard	. . . . .	26 Dec. 15*	13 May 18	26 Dec. 16	,,	29 Lancers.
	War. R.	**Jenkins**, Arthur George	. . . . .	26 Dec. 15*	3 Feb. 18	26 Dec. 16	,,	50 Rifles.
		Clarabut, Reginald Blaxland (48a)	. . . . .	27 Dec. 15*	25 Nov. 17	27 Dec. 16	,,	67 Punjabis.
	R. Sc. Fus.	**Kerr**, Neil	. . . . .	27 Dec. 15*	13 July 18	27 Dec. 16	,,	82 Punjabis.
	Devon. R.	**Watts**, Reginald Bryan	. . . . .	28 Dec. 15*	22 Mar. 18	28 Dec. 16	,,	15 Sikhs.
475	R. Innis. Fus.	**McNiel**, William Alexander	. . . . .	29 Dec. 15*	22 Apl. 18	29 Dec. 16	,,	25 Punjabis.
	Conn. Rang.	**Reed**, Samuel Percy Keppel John Hendy	. . . . .	31 Dec. 15*	5 Mar. 18	31 Dec. 16	,,	124 Infantry.
	North'd Fus.	**Rutledge**, Crawford Jordan	. . . . .	31 Dec. 15*	31 Aug. 18	31 Dec. 16	,,	126 Infantry.
	R. Suss. R.	**Spencer**, Wilfred Edward Chetwode	. . . . .	31 Dec. 15*	30 Sep. 18	31 Dec. 16	,,	89 Punjabis.
		Watkins, Vere Clifford	. . . . .	1 Jan. 16*	9 Jan. 18	1 Jan. 17	,,	23 Cavalry.
480		**Walkinshaw**, Ivon Curzon	. . . . .	2 Jan. 16*	8 Feb. 18	2 Jan. 17	,,	10 Jats.
		Anderson, Cecil Winterton	. . . . .	2 Jan. 16*	9 Mar. 18	2 Jan. 17	,,	9 Infantry.
	S. Lan. R.	**Batchelor**, Arnold James	. . . . .	2 Jan. 16*	14 July 18	2 Jan. 17	,,	28 Punjabis.
	R. E.	**Round**, George Edward Guy	. . . . .	2 Jan. 16*	27 Apl. 18	2 Jan. 17	,,	Sig. Unit Depôt.
		Matthews, Leslie Gard	. . . . .	3 Jan. 16*	1 Aug. 17	3 Jan. 17	,,	2-9th Delhi Infantry.
485	R. Ir. Fus.	**Brennan**, Reginald, M.C.	. . . . .	3 Jan. 16*	18 Oct. 17	3 Jan. 17	,,	3-5th Gurkha Rifles.
	E. Lanc. R.	**Buckley**, Fred	. . . . .	3 Jan. 16*	4 July 18	3 Jan. 17	,,	45 Sikhs.
	A. S. C.	**Sutton**, Gilbert Francis	. . . . .	3 Jan. 16*	30 June 18	3 Jan. 17	,,	122 Infantry.
		Dagg, Alan Victor	. . . . .	5 Jan. 16*	25 Nov. 17	5 Jan. 17	,,	66 Punjabis.
	Leins. R.	**Renshaw**, Charles Mitford	. . . . .	6 Jan. 16*	16 Jan. 18	6 Jan. 17	,,	9 Bhopal Infantry.
490	Lan. Fus.	**Openshaw**, Sydney Noel	. . . . .	6 Jan. 16*	11 Oct. 18	6 Jan. 17	,,	66 Punjabis.
		Simpson, Herbert James Attwood (48a, 51)	. . . . .	7 Jan. 16*	5 Dec. 17	7 Jan. 17	,,	2-3rd Gurkha Rifles.
	N. Staff. R.	**North**, Leslie Vivian	. . . . .	7 Jan. 16*	12 Jan. 18	7 Jan. 17	,,	39 Garh. Rifles.
	Manch. R.	**Humfrey**, Bruce	. . . . .	7 Jan. 16*	27 May 18	7 Jan. 17	,,	39 (Divsl.) Sig. Coy.
	Sea. Highrs.	**Reid**, Dennis Whitehorn, M.C.	. . . . .	7 Jan. 16*	27 Aug. 18	7 Jan. 17	,,	103 Mah. L. I.
495	M. G. Corps.	**Rowland**, Cyril Justice	. . . . .	7 Jan. 16*	6 Sep. 18	7 Jan. 17	,,	102 Grenadiers.
		Hull, Bernard	. . . . .	8 Jan. 16*	5 Dec. 17	8 Jan. 17	,,	130 Baluchis.
		Rutledge, Laurence Hugh Nesbitt	. . . . .	8 Jan. 16*	22 Jan. 18	8 Jan. 17	,,	1-8th Gurkha Rifles.
		Thomson, Robert Francis	. . . . .	9 Jan. 16*	2 July 17	9 Jan. 17	,,	28 Lt. Cavalry.
	Lan. Fus.	**Hodkin**, Wilfred	. . . . .	9 Jan. 16*	14 Apl. 18	9 Jan. 17	,,	109 Infantry.
500		**Palk**, Leonard Cayme	. . . . .	9 Jan. 16*	22 Jan. 18	10 Jan. 17	,,	14 Sikhs.
		Clyde, William Thomas	. . . . .	10 Jan. 16*	25 Mar. 18	10 Jan. 17	,,	94 Infantry.
	York. L. I.	**Ricketts**, Gordon Archibald McGeorge	. . . . .	10 Jan. 16*	1 June 18	10 Jan. 17	,,	5 Lt. Infy.
	Conn. Rang.	**Tyte**, Alfred Norman	. . . . .	11 Jan. 16*	7 Mar. 18	11 Jan. 17	,,	47 Sikhs.
504	A. S. Corps	**Maguire**, James Joseph	. . . . .	12 Jan. 16*	16 Sep. 17	12 Jan. 17	,,	124 Infantry.

* 9 months forfeited for seniority under conditions of I. A. O. 511, dated 14 May 1917.

Lieutenants—*contd.*

No.	Former Corps.	Names.	Date of birth.	Date of first commission.	Date of appointment to Indian Army.	Date of rank.	Remarks.
505		**Axtell,** Leslie Harry Westwood (46)		13 Jan. 16*	22 Nov. 17	13 Jan. 17	Attached 2-9th Delhi Infy.
		Maclaren, Walter de Bois 207*a*		13 Jan. 16*	24 Sep. 17	13 Jan. 17	,, 112 Infantry.
		Whittington, George William		13 Jan. 16*	10 Mar. 18	13 Jan. 17	,, 7 Lancers.
	Som. L. I.	**Wall,** Roland Hampden		14 Jan. 16*	1 June 18	14 Jan. 17	,, 129 Baluchis.
	13 Huss.	**Sheil,** Patrick Henry Joseph		14 Jan. 16*	2 Mar. 18	14 Jan. 17	,, 32 Lancers.
510	Hamps. R.	**Capes,** Gerard Alphonsus		16 Jan. 16*	16 Aug. 17	16 Jan. 17	,, 1-4th Gurkha Rifles.
		Hare, Sydney St. George (†*Temp. Capt., 26 Nov. 18*)		18 Jan. 16*	25 Mar. 18	18 Jan. 17	,, 76 Punjabis.
	R. W. Surr. R.	**Burdon,** Randal Matthews, M.C.		18 Jan. 16*	16 Oct. 18	18 Jan. 17	,, 81 Pioneers.
		Bell, Norman Keith 207*a*		19 Jan. 16*	5 Dec. 17	19 Jan. 17	,, 106 Pioneers.
		Young, John William (48*a*, 51)	13 Nov. 88	19 Jan. 16*	3 Dec. 17	19 Jan. 17	,, 3-7th Gurkha Rifles.
515		**Bigwood,** Wilfrid, M.C.		19 Jan. 16*	12 Jan. 18	19 Jan. 17	,, 16 Cavalry.
	R. Innis. Fus.	**Adams,** Thomas		19 Jan. 16*	12 May 18	19 Jan. 17	,, 3-1st Gurkha Rifles.
	Suff. R.	**Dann,** Charles Walter		19 Jan. 16*	17 Oct. 17	19 Jan. 17	,, 112 Infantry.
	Aus. Lt. H.	**Shaw,** John Frederick		19 Jan. 16*	12 Dec. 17	19 Jan. 17	,, 122 Infantry.
		Phillips, Charles Arthur 207*a*		20 Jan. 16*	22 Jan. 18	20 Jan. 17	,, 22 Punjabis.
520	Lan. Fus.	**Leigh,** Arthur Podeus		20 Jan. 16*	4 July 18	20 Jan. 17	,, 25 Cavalry.
		Christie, Gordon		21 Jan. 16*	30 July 17	21 Jan. 17	,, 10 Jats.
		Girling, John Brough		22 Jan. 16*	4 Dec. 17	22 Jan. 17	,, 12 Pioneers.
		Johnson, George Macness Whiteway		23 Jan. 16*	25 Nov. 17	23 Jan. 17	,, 83 Infantry.
	R. W. Fus.	**Morgan,** William Marychurch		23 Jan. 16*	13 May 18	23 Jan. 17	,, 42 Regiment.
525	R. Ir. Fus.	**Colquhoun,** John Workman Coates		23 Jan. 16*	1 June 18	23 Jan. 17	,, 24 Punjabis.
	Midd'x R.	**Roberts,** Harold Charles West, M.C.		23 Jan. 16*	6 Sep. 18	23 Jan. 17	,, Corps of Guides.
		Peel, Wilmot Reginald Bloomfield (71)		25 Jan. 16*	22 Nov. 17	25 Jan. 17	,, 33 Cavalry.
	L'pool R.	**Cockram,** Leslie Harding		25 Jan. 16*	12 Oct. 18	25 Jan. 17	,, 7 Gurkha Rifles.
	R. Sc. Fus.	**Fawcett,** Robert Beresford		26 Jan. 16	25 Mar. 18	26 Jan. 17	,, 1-9th Gurkha Rifles.
530	R. Ir. Regt.	**Pilkington,** Thomas Lionel		26 Jan. 16	12 May 18	26 Jan. 17	,, 128 Pioneers.
	York. L. I.	**Mitchley,** Edward Charles Thomas		26 Jan. 16*	27 Nov. 17	26 Jan. 17	,, 1-1st Gurkha Rifles.
	A. S. C.	**Barron,** Arthur Sydney Noel		26 Jan. 16*	25 Feb. 18	26 Jan. 17	,, 1-4th Gurkha Rifles.
		Watson, Aubery Caiger		27 Jan. 16*	25 Nov. 17	27 Jan. 17	,, 94 Infantry. [N. Circle.
		Malcolm, James Fitzgerald		27 Jan. 16*	12 Dec. 17	27 Jan. 17	,, Mily. Farms Dept,
535		**Qualtrough,** Edward Fleming		27 Jan. 16*	8 Feb. 18	27 Jan. 17	,, 3-6th Gurkha Rifles.
		Hale, Ernest Cecil		28 Jan. 16*	6 May 17	28 Jan. 17	,, 11 Rajputs.
	R. E.	**Farrar,** Cyril Beresford		28 Jan. 16*	24 Jan. 18	29 Jan. 17	,, 4 Cavalry.
	Aus. Imp. Force.	**Lodge,** Augustine Bernard, D.S.O.		28 Jan. 16*	29 July 18	28 Jan. 17	,, 2 Corps of Guides.
	Lond. R.	**Frankish,** William Martin		29 Jan. 16*	11 Apl. 18	29 Jan. 17	,, 9 Bhopal Infy.
540	Aus. Imp. Force.	**Oates,** Augustus, D.S.O.		29 Jan. 16*	25 Mar. 18	29 Jan. 17	,, 67 Punjabis.
	W. York. R.	**Miller,** Luke Henry		29 Jan. 16*	18 May 18	29 Jan. 17	,, 3-23rd Sikkh Infantry.
	Aus. Imp. Force.	**Bastin,** Hector Ernest, M.C.		29 Jan. 16*	29 July 18	29 Jan. 17	,, 15 Lancers.
		Walters, Edgar Cyril		30 Jan. 16*	6 Nov. 17	30 Jan. 17	,, 28 Punjabis.
		Andrews, John Spencer Trevor		30 Jan. 16*	22 Nov. 17	30 Jan. 17	,, 13 Rajputs.
545	R. Ir. Fus.	**Phillips,** William Joseph Christopher		30 Jan. 16*	25 June 17	30 Jan. 17	,, 116 Mahrattas.

546	R. W. Fus.	**Davies,** Ernest Llewellyn			30 Jan. 16*	10 Oct. 18	30 Jan. 17	,,	3 Gurkha Rifles.
	Ans. Imp. F.	**Lloyd,** John Edward, M.C.			1 Feb. 16*	25 Mar. 18	1 Feb. 17	,,	35 Sikhs.
	R. Fus.	**Roth,** Ernest John Harold			3 Feb. 16*	13 May 18	3 Feb. 17	,,	30 Lancers.
	Worc. R.	**Burn,** Ronald Francis			2 Feb. 16*	4 Sep. 18	2 Feb. 17	,,	40 Cavalry.
550	Sea. Highrs.	**Gore,** Gecil Harry			3 Feb. 16*	27 Aug. 18	3 Feb. 17	,,	42 Cavalry.
		Mayne, Paul (71)			4 Feb. 16*	12 Jan. 18	4 Feb. 17	,,	33 Cavalry.
	Aus. Imp. Force.	**Brown,** Montague Ambrose			4 Feb. 16*	3 Nov. 17	4 Feb. 17	,,	9 Horse.
		Gowland, Cyril James (59, 157)	207*a*	19 Aug. 95	5 Feb. 16*	6 May 17	5 Feb. 17	,,	7 Gurkha Rifles.
		Duncan, Alan Wilson			5 Feb. 16*	3 Dec. 17	5 Feb. 17	,,	2-8th Gurkha Rifles.
555	York. R.	**Finch,** Percy Wagner (95)			5 Feb. 16*	20 Feb. 18	5 Feb. 17	,,	107 Pioneers.
	Aus. Imp. Force.	**Anderson,** Robert Christian	207*a*		5 Feb. 16*	8 Feb. 18	5 Feb. 17	,,	76 Punjabis.
	Wilts. R.	**Atkinson,** Claude Milnes			5 Feb. 16*	13 May 18	5 Feb. 17	,,	116 Mahrattas.
		Dove, Maurice	207*a*		6 Feb. 16*	23 Nov. 17	6 Feb. 17	,,	16 Rajputs.
	R.W.Surr. R	**Abigail,** Reginald Philip			6 Feb. 16*	18 May 18	6 Feb. 17	,,	2-7th Gurkha Rifles.
560	Welsh. R.	**James,** Harold Vernon			6 Feb. 16*	12 May 18	6 Feb. 17	,,	23 Pioneers.
	Notts. & Derby R.	**Robinson,** James McVynn			6 Feb. 16*	30 June 18	6 Feb. 17	,,	113 Infantry.
	S. Wales Bord.	**Jones,** Roderick Idrisyn			6 Feb. 16*	25 Aug. 18	6 Feb. 17	,,	88 Infantry.
	E. York. R,	**Holbrook,** Sam Daykin			6 Feb. 16*	29 Oct. 18	6 Feb. 17	,,	153 Punjabis.
		Cinnamond, Hubert Park, M.C.			7 Feb. 16*	12 Jan. 18	7 Feb. 17	,,	43 Regiment.
565		**Jardine.** Arthur			7 Feb. 16*	12 Jan. 18	7 Feb. 17	,,	54 Sikhs.
	North'n R.	**Turner,** Harold Fredrick George			7 Feb. 16*	30 Jan. 18	7 Feb. 17	,,	28 Cavalry.
	Aus. Imp. F.	**Bland,** Ewen MacKay			7 Feb. 16*	26 Feb. 18	7 Feb. 17	,,	120 Baluchis.
		Warburton, John France			8 Feb. 16*	5 Nov. 17	8 Feb. 17	,,	48 Pioneers.
		Charles, Frank Dallas			8 Feb. 16*	25 Mar. 18	8 Feb. 17	,,	40 Pathans.
570	R. Innis. Fus.	**Kirkman,** Harry Gayton			8 Feb. 16*	1 June 18	8 Feb. 17	,,	107 Pioneers.
		Pennington, Charles Eric Rainy			10 Feb. 16*	5 Dec. 17	10 Feb. 17	,,	33 Punjabis.
	Essex R.	**Shallow,** Charles Rushworth (†*temp. Capt.*, *10 May 18*)			10 Feb. 16*	9 Jan. 18	10 Feb. 17	,,	74 Punjabis.
		Allan, Ranold James			11 Feb. 16*	9 Jan. 18	11 Feb. 17	,,	2-8 Gurkha Rifles.
		Jackson, Wilfred Edward			11 Feb. 16*	6 Apl. 17	11 Feb. 17	,,	120 Infantry.
575		**Johnson,** Ernest Claud			11 Feb. 16*	6 May 17	11 Feb. 17	,,	35 Sikhs.
	Lond. R.	**Aron,** David Julius			11 Feb. 16*	27 Apl. 18	11 Feb. 17	,,	2-9th Gurkha Rifles,
		Donald, George (46)			12 Feb. 16*	23 Nov. 17	12 Feb. 17	,,	124 Infantry.
	6 Dgn. Gds.	**Tunrer,** Fric			12 Feb. 16*	25 Mar. 18	12 Feb. 17	,,	Guides Cavalry.
	Gen. List.	**Pettman,** Frank Leslie			12 Feb. 16*	30 July 18	12 Feb. 17	,,	99 Infantry.
580	Shrops. L. I.	**Hughes.** Hugh Thomas			13 Feb. 16*	14 Apl. 18	13 Feb. 17	,,	22 Punjabis.
		Wilson, Jarvie Webb			14 Feb. 16*	22 Jan. 18	14 Feb. 17	,,	28 Punjabis.
		Brown, Denis Sidney			15 Feb. 16*	17 Sep. 17	15 Feb. 17	,,	39 Garhwal Rifles.
	Can. Force	**Ellis,** Albert Henry James			15 Feb. 16*	25 Mar. 18	15 Feb. 17	,,	57 Rifles.
		Parish, George Nowell Arnott			16 Feb. 16*	29 July 17	16 Feb. 17	,,	10 Jats.
585	Aus. Imp. Force.	**Clarke,** Mathew John			15 Feb. 16*	12 Jan. 18	16 Feb. 17	,,	70 Rifles.
		Gowing, Dudley Sayers			17 Feb. 16*	25 Sep. 17	17 Feb. 17	,,	12 Pioneers.
		Trechmann, Albert David			17 Feb. 16*	12 Jan. 18	17 Feb. 17	,,	67 Punjabis.
		Heath, Hugh Charles Spencer			18 Feb. 16*	6 May 17	18 Feb. 17	,,	52 Sikhs.
		Tollit, William Ernest			18 Feb. 16*	20 Aug. 17	18 Feb. 17	,,	112 Infantry.
590		**Mackay,** Alexander John, M.C.			18 Feb. 16*	12 Jan. 18	18 Feb. 17	,,	54 Sikhs.

*** 9 months forfeited for seniority under conditions of I. A. O. 511, dated 14th May 1917. | † With pay and allowances as for Lieutenant.**

No.	Former Corps.	Names.	Date of birth.	Date of first Commission.	Date of appointment to Indian Army.	Date of rank.	Remarks.
		Lieutenants—*contd.*					
585		**Hartmann,** Werner Hugo (47*a*, 51)		18 Feb. 16*	22 Jan. 18	18 Feb. 17	Attached 4 Cavalry.
	R. Suss. R.	**Lester,** Humphrey Noel		19 Feb. 16*	4 Oct. 18	19 Feb. 17	„ Mily. Farms Dept.
		Stevenson, John Francis Barr, M.C.		20 Feb. 16*	22 Nov. 17	20 Feb. 17	„ 44 Infantry.
		Pyper, Hector McDonald		20 Feb. 16*	29 July 17	20 Feb. 17	„ 3-2nd Gurkha Rifles.
	W. York. R.	**Shields,** Frederick Trevor		20 Feb. 16*	25 Mar. 18	20 Feb. 17	„ 109 Infantry.
590	Sea. Highrs.	**Campbell,** William James		20 Feb. 16*	7 July 17	20 Feb. 17	„ 74 Punjabis.
	A. S. C.	**Makenzie,** James Ferguson Ogilvie Park		20 Feb. 16*	19 Mar. 18	20 Feb. 17	„ 89 Punbjabis.
	Devon. R.	**Robinson,** Frank Calvert		20 Feb. 16*	26 Sep. 18	20 Feb. 17	„ 123 Rifles.
		Lawrence, Joseph Ashton		21 Feb. 16*	25 Nov. 17	21 Feb. 17	„ 84 Panjabis.
		Craster, Reginald Fenwick 207*a*		22 Feb. 16*	5 May 17	22 Feb. 17	„ 39 Garhwal Rifles.
595		**Ganniclifft,** Albert Leslie 207*a*		22 Feb. 16*	29 July 17	22 Feb. 17	„ 42 Regiment.
		Webber, Cyril Donald		22 Feb. 16*	9 Jan. 18	22 Feb. 17	„ 20 Infantry.
	R. Ir. Fus.	**Reid,** William Fitzgerald		22 Feb. 16*	12 Oct. 18	22 Feb. 17	„ 31 Punjabis.
	R. Ir. Rif.	**Rainier,** William Harvey Cecil (46)		23 Feb. 16*	26 Mar. 18	23 Feb. 17	„ 57 Rifles.
		Hoskins, Cecil John		26 Feb. 16*	27 Aug. 17	26 Feb. 17	„ 46 Punjabis.
600	E. York R.	**Jenkins,** Frederick Arthur Montagu Bertram		26 Feb. 16*	6 May 17	26 Feb. 17	„ 72 Punjabis.
		Shaw, Frederick Charles		27 Feb. 16*	22 June 17	27 Feb. 17	„ 19 Punjabis.
		Porter, William Dunbar		27 Feb. 16*	29 July 17	27 Feb. 17	„ 10 Jats.
	M. G. Corps	**Shaw,** Vincent Eric		28 Feb. 16*	25 Mar. 18	28 Feb. 17	„ 18 Infantry.
	Manch. R.	**Atkins,** Walter Vincent		28 Feb. 16*	25 Jan. 18	28 Feb. 17	„ 35 Sikhs.
605	M. G. Corps	**Shaw,** Vincent Eric		28 Feb. 16*	26 Mar. 18	28 Feb. 17	„ 18 Infanfry.
	R. Scots.	**Hay,** Thomas William Denholm		28 Feb. 16*	24 Sep. 18	28 Feb. 17	„ 72 Punjabis.
		Reed, John Hastings		1 Mar. 16*	6 May 17	1 Mar. 17	„ 38 Dogras.
	Sea. Highrs.	**Proctor,** Walter James Evans		3 Mar. 16*	28 July 17	3 Mar. 17	„ 35 Sikhs.
	Suff. R.	**Hamill,** George English		3 Mar. 16*	6 Sep. 18	3 Mar. 17	„ 63 Lt. Infy.
610	A. S. C.	**Battye,** Rowland Dawson		3 Mar. 16*	30 Sep. 18	3 Mar. 17	„ 96 Infantry.
		Riley, Charles Lloyd		4 Mar. 16*	12 Jan. 18	4 Mar. 17	„ 41 Dogras.
	Can. Engrs.	**Smith,** Daniel Richard		4 Mar. 16*	5 Oct. 17	4 Mar. 17	„ 34 Pioneers.
		McCarthy, William Ogle		5 Mar. 16	9 Feb. 18	5 Mar. 17	„ 76 Punjabis.
	R. Dub. Fus.	**Devoy,** Joseph Charles		5 Mar. 16	12 May 18	5 Mar. 17	„ 117 Mahrattas.
615		**Wilkinson,** Frederick Richardson 207*a*		7 Mar. 16*	15 Sep. 17	7 Mar. 17	„ 12 Pioneers.
		Michell, John Henry		7 Mar. 16*	10 Mar. 18	7 Mar. 17	„ 39 Horse.
		Hart, Lionel Warren		8 Mar. 16*	22 Jan. 18	8 Mar. 17	„ 12 Pioneers.
	Glouc. R.	**Grieves,** Arthur James Mansfiled		8 Mar. 16*	26 June 18	8 Mar. 17	„ 108 Infantry.
		Warren, Dermot Frederick William 207*a*		10 Mar. 16*	6 May 17	10 Mar. 17	„ 12 Pioneers.
620		**Bratton,** Victor Wilson 207*a*		10 Mar. 16*	20 Aug. 17	10 Mar. 17	„ 86 Infantry.
		Leawood, Reginald Reed		10 Mar. 16*	22 Nov. 17	10 Mar. 17	„ 3-2nd Gurkha Rifles.
	North'd Fus.	**Hobley,** Alfred Jabez		10 Mar. 16*	28 Feb. 18	10 Mar. 17	„ 25 Punjabis.
	Oxf.&Buck's L. I.	**Carew-Hughes,** Cyril Francis		10 Mar. 16*	11 July 18	10 Mar. 17	„ 31 Punjabis.
	M. G. Corps.	**Brooks,** Frederick Lewis		10 Mar. 16*	31 Aug. 18	10 Mar. 17	„ 74 Punjabis.
625	R. Lanc. R.	**Ashley-Smith,** Ronald		10 Mar. 16*	16 Aug. 18	10 Mar. 17	„ 97 Infantry.
	R. Fus.	**St. Aubyn,** John Henry Arundel Godolphin		10 Mar. 16*	30 Sep. 18	10 Mar. 17	„ 156 Infantry.
	North'd Fus.	**Robson,** Ernest		10 Mar. 16*	24 July 18	10 Mar. 17	„ 2 Rajput Lt. Infy.
628	E. Kent R.	**Williams,** William Theophilus, M.C.		11 Mar. 16*	12 May 18	11 Mar. 17	„ 112 Infantry.

635	Hamps. R.	**King,** James Henry Lawrence			11 Mar. 16*	2 July 18	11 Mar. 17	,,	39 Garh. Rifles.
	E. Kent. R.	**Macaulay,** Archibald Duncan Campell			12 Mar. 16*	24 Oct. 17	12 Mar. 17	,,	2-7th Gurkha Rifles.
		Shuker, Harold			14 Mar. 16*	5 Dec. 17	14 Mar. 17	,,	3-6th Gurkha Rifles.
		Pulleine, Humphrey			14 Mar. 16*	8 Feb. 18	14 Mar. 17	,,	8 Cavalry.
	R. Ir. Fus.	**Best,** William James			15 Mar. 16*	13 May 18	15 Mar. 17	,,	25 Punjabis.
640		**Carling,** Thomas Edward			17 Mar. 16*	24 Jan. 18	17 Mar. 17	,,	3-5th Gurkha Rifles.
	R. Muns. Fus.	**Baldwin,** James Welply			17 Mar. 16*	1 June 18	17 Mar. 17	,,	28 Punjabis
	R. Muns. Fus.	**Wetherall,** Francis Gyrth Johnston			18 Mar. 16*	5 Oct. 17	18 Mar. 17	,,	72 Punjabis.
	Suff. R.	**Rea,** Arthur Noel (46)			19 Mar. 16	25 Mar. 18	19 Mar. 17	,,	2 Rajput Lt. Infy.
	L'pool R.	**Cleveland,** Eric Albert Mendham			21 Mar. 16*	29 Oct. 18	21 Mar. 17	,,	3 Gurkha Rifles.
645	R. Lanc. R.	**Wesson,** Reginald Arthur			22 Mar. 16*	27 Aug. 18	22 Mar. 17	,,	98 Infantry.
	S. Staff R.	**Briggs,** Geoffrey			22 Mar. 16*	28 Sep. 18	22 Mar. 17	,,	1 Brahmans.
		Burrell, Kenneth			23 Mar. 16*	3 Dec. 17	23 Mar. 17	,,	2-8th Gurkha Rifles.
		Crocombe, Frederick Richard			26 Mar. 16*	6 June 17	26 Mar. 17	,,	76 Punjabis.
		Armstrong, Herbert (46)			26 Mar. 16*	21 July 17	26 Mar. 17	,,	110 Mahratta Lt. Infy.
650	M. G. Corps	**Belford,** Harry David			29 Mar. 16*	14 Apl. 18	29 Mar. 17	,,	11 Rujputs.
	R. Berk R.	**Webb,** George Sholto Ripley			29 Mar. 16*	18 May 18	29 Mar. 17	,,	37 Lancers.
	R. W. Surr. R.	**Fison,** Cornell Henry			3 Apl. 16*	6 Mar. 18	3 Apl. 17	,,	19 Punjabis.
	R. Scots.	**Burnett,** Robert Richardson			3 Apl. 16*	1 Oct. 18	3 Apl. 17	,,	27 Punjabis.
		Marshall, William Dobie	207*a*		4 Apl. 16*	20 Aug. 17	4 Apl. 17	,,	103 Mahratta Lt. Infy.
655		**Barnes,** Thomas James			4 Apl. 16*	26 Nov. 17	4 Apl. 17	,,	64 Pioneers.
		Kevill-Davies, Ernest Francis William Humphrey			7 Apl. 16	26 Nov. 17	7 Apl. 17	,,	3-6th Gurkha Rifles.
		Boyle, Hugh MacCormac, M.C.			7 Apl. 16	8 Feb. 18	7 Apl. 17	,,	31 Lancers.
		O'Conner, Terence Joseph			7 Apl 16	24 Jan. 18	7 Apl 17	,,	31 Lancers.
		Jacob, George Webb Butler			7 Apl. 16	8 Feb. 18	7 Apl. 17	,,	35 Horse.
660	Hamps. R.	**Armstrong,** Robert Henry			7 Apl. 16	25 Mar. 18	7 Apl. 17	,,	75 Infantry.
	A. S. C.	**Hargreaves,** Harold			7 Apl. 16	25 Mar. 18	7 Apl. 17	,,	124 Infantry.
	Ches. R.	**Meers,** Philip James			7 Apl. 16	26 Mar. 18	7 Apl. 17	,,	56 Rifles.
	S. Lanc. R.	**Dixon,** William			7 Apl. 16	27 Feb. 18	7 Apl. 17	,,	35 Sikhs.
	Suff. R.	**Howell,** Harold Rodolph			7 Apl. 16	12 May 18	7 Apl. 17	,,	2-9th Gurkha Rifles.
665	E. Lanc. R.	**Bradshaw,** Alan Thomas Rothney		19 Feb. 98	7 Apl. 16	17 Apl. 18	7 Apl. 17	,,	57 Rifles.
	R. W. Fus.	**Hughes,** Charles Robert			7 Apl. 16*	13 Feb. 18	7 Apl. 17	,,	Corps of Guides.
	Dorset R.	**Murray,** Nigel Trenly George		4 Apl. 98	7 Apl. 16	22 Oct. 18	7 Apl. 17	,,	29 Punjabis.
		Dilks, Arthur Charles			8 Apl. 16*	23 Dec. 17	8 Apl. 17	,,	30 Punjabis
	Lond. R.	**Heaton,** Arthur Douglas			8 Apl. 16*	11 Apl. 18	8 Apl. 17	,,	6 Jat Lt. Infy.
670	R. Lanc. R.	**Hall,** Gordon Civil Oswald			8 Apl. 16*	12 May 18	8 Apl. 17	,,	99 Infantry.
	Scot. H.	**Taylor,** James Scott			10 Apl. 16*	12 May 18	10 Apl. 17	,,	8 Cavalry.
	Scot. H.	**Miller,** Archibald Bryce			10 Apl. 16*	13 Aug. 18	10 Apl. 17	,,	47 Sikhs.
	Durh. L. I.	**Johnson,** Reginald Sidney			11 Apl. 16*	31 May 18	11 Apl. 17	,,	4 Rajputs.
		Grant, John Crawford (46, 48*a*, 51)	207*a*		12 Apl. 16*	22 Jan. 18	12 Apl. 17	,,	1-9th Bhopal Infantry.
675		**Pickersgill,** Richard			12 Apl. 16*	22 Jan. 18	12 Apl. 17	,,	35 Sikhs.
	Dorc. R.	**Weldhen,** Robert Brown			12 Apl. 16*	12 Feb. 18	12 Apl. 17	,,	124 Infantry.
	M. G. Corps.	**Doig,** Edward Willim			12 Apl. 16*	1 June 18	12 Apl. 17	,,	61 Pioneers.
	Worc. R.	**Anstey,** Daniel			12 Apl. 16*	16 Dec. 17	12 Apl. 17	,,	31 Punjabis.
		Rendall, Francis Cuthbert Eugene			13 Apl. 16*	8 Feb. 18	13 Apl. 17	,,	63 Lt. Infantry.
680	Lond. R.	**Lancaster,** Alexander Stalker (48*a*, 51)			13 Apl. 16*	20 Oct. 17	13 Apl. 18	,,	66 Punjabis.
		Hey, Benjamin Stuart			15 Apl. 16*	29 July 17	15 Apl. 17	,,	43 Regiment.
682	Gord. Highrs	**Morrison,** George Alexander			15 Apl. 16*	12 May 18	15 Apl. 17	,,	91 Punjabis.

* 9 months forfeited for seniority under conditions of I. A. O. 511, dated 14th May 1917.

No.	Former Corps.	Names.	Date of birth.	Date of first commission.	Date of appointment to Indian Army.	Date of rank.	Remarks.
		Lieutenants—*contd.*					
683		**Parsons,** John Henry (51)		19 Apl. 16*	19 Jan. 18	19 Apl. 17	Attached 31 Lancers.
	Midd'x. R.	**Lesster,** Laban		19 Apl. 16*	30 Sep. 18	19 Apl. 17	,, 39 Garhwal Rifles.
685	E. York. R.	**O'Dowd,** James Henry Barratt		20 Apl. 16*	11 Apl. 18	20 Apl. 17	,, 1 Brahmans.
	Leins. R.	**Watson,** James Harvey Long		20 Apl. 16*	18 May 18	20 Apl. 17	,, 3-6th Gurkha Rifles.
	Aus. Imp. Force.	**Hobbs,** John Mervyn		20 Apl. 16*	27 Feb. 18	20 Apl. 17	,, 33 Punjabis.
	Sea. Higrs.	**Low,** John		21 Apl. 16*	14 Aug. 18	21 Apl. 17	,, 114 Mahrattas.
		Grosvenor, Chetwynd John Pershall		22 Apl. 16*	8 Feb. 18	22 Apl. 17	,, 18 Lancers.
690		**Elliot,** Charles Denton		22 Apl. 16*	9 Aug. 17	22 Apl. 17	,, 46 Punjabis.
	R. Suss. R.	**Tutton,** Percy Charles		22 Apl. 16*	5 June 18	22 Apl. 17	,, 90 Punjabis.
	Hamps. R.	**Long,** Percival		22 Apl. 16*	22 Aug. 18	22 Apl. 17	,, Sig. Service.
	York R.	**Bayliss,** Clive Lane		22 Apl. 16*	30 Sep. 18	22 Apl. 17	,, 35 Shiks.
	R. Ir. Rif.	**Rouche,** George Patrick (51)		23 Apl. 16*	13 May 18	23 Apl. 17	,, 32 Pioneers.
695		**Hyde,** George Leslie		24 Apl. 16*	6 May 17	24 Apl. 17	,, 76 Punjabis.
	Dorset. R.	**Brindley,** William Walter		24 Apl. 16*	12 Feb. 18	24 Apl. 17	,, 3-9th Gurkha Rifles.
	A. Cyclist Corps.	**Jones,** Cyril Dover		24 Apl. 16*	6 Sep. 18	24 Apl. 17	,, 41 Dogras.
	S. Lanc. R.	**Glover,** Malcolm		24 Apl. 16*	23 Sep. 18	24 Apl. 17	,, 22 Punjabis.
	Hamps. R.	**Hoare,** Harold Percival Brindley		25 Apl. 16*	20 Mar. 18	25 Apl. 17	,, 66 Punjabis.
700	E. Surr. R.	**Tucker,** Bryant Burgess		25 Apl. 16*	20 Dec. 17	25 Apl. 17	,, 2-2nd Gurkha Rifles.
	E. Lanc. R.	**Gamble,** Lawrance		26 Apl. 16*	25 Sep. 17	26 Apl. 17	,, 1-4th Gurkha Rifles.
	North'd. Fus.	**Welton,** James John Graham		26 Apl. 16*	30 June 18	26 Apl. 17	,, 110 Mah. L. I.
	A. S. C.	**Goldthorp,** Henry Douglas		26 Apl. 16*	1 Aug. 18	26 Apl. 17	,, 22 Cavalry.
		Lewis, David Godfrey 207a		28 Apl. 16*	9 Jan. 18	28 Apl. 17	,, 37 Lancers.
705		**Meehan,** Cecil Frank 207a		29 Apl. 16*	29 July 17	29 Apl. 17	,, 42 Regiment.
		White, Stanley Edward Chapman		29 Apl. 16*	8 Feb. 18	29 Apl. 17	,, 41 Dogras.
	M. G. Corps.	**Hind-Fletcher,** Thomas Ralph		30 Apl. 16*	27 Aug. 18	30 Apl. 17	,, 112 Infantry.
	High. L. I.	**Stevenson,** William Hugh		1 May 16*	14 Apl. 18	1 May 17	,, 33 Punjabis.
		Powell, Frank Newman Howard		4 May 16*	26 Nov. 17	4 May 17	,, 66 Punjabis.
710		**Philpott,** Douglas Arthur		4 May 16*	25 Nov. 17	4 May 17	,, 97 Infantry.
	Midd'x. R.	**Andrew,** Herbert Wayte		6 May 16	14 Apl. 18	6 May 17	,, 35 Sikhs.
		Briggs, Robert Andrew (12, 57) 207a		7 May 16*	25 Nov. 17	7 May 17	,, 1-5th Gurkha Rifles.
	Essex R.	**Jones,** Harry		7 May 16*	26 Sep. 18	7 May 17	,, 49 Bengalis.
		Allen, Geoffrey Carl 207a		9 May 16*	28 July 17	9 May 17	,, 29 Punjabis.
715	Lond. R.	**Jeffreys,** Herbert Frederick		10 May 16*	1 June 18	10 May 17	,, 95 Infantry.
	Notts & Derby R.	**Owen,** Arthur Wynne		10 May 16*	29 Dec. 17	10 May 17	,, 70 Burma Rifles.
		Carr, Douglas Gerald (57)		11 May 16*	25 Sep. 17	11 May 17	,, 120 Infantry.
	Welsh R.	**Gibson,** Frederick Corbett		11 May 16*	8 July 18	11 May 17	,, 122 Infantry.
		Warmington, William Ernest		12 May 16*	22 Nov. 17	12 May 17	,, 36 Sikhs.
720		**McCarthy,** Charles Leopold Joseph (48a, 51)		12 May 16*	25 Nov. 17	12 May 17	,, 88 Infantry.
		Stokes, Henry Alban		12 May 16*	9 Feb. 18	12 May 17	,, 1 Brahmans.
	Essex R.	**Bird,** Mervyn		12 May 16*	16 Apl. 18	12 May 17	,, 121 Pioneers.
	R. Dub. Fus.	**Hurst,** Nicholas		12 May 16*	27 Apl. 18	12 May 17	,, 2-9th Gurkha Rifles
724	Norf. R.	**Crozier,** William Henry		12 May 16*	27 Nov. 17	12 May 17	,, 2-6th Gurkha Rifles.

724		**Cotton-Swanston,** William Lawrence		14 May 16*	8 Feb. 18	14 May 17	,,	4-39th Kumaon Rifles.
725		**Leeming,** Alfred (207)		14 May 16*	16 Sep. 17	14 May 17	,,	106 Pioneers.
		Babtie, John William Alexander		14 May 16*	30 Nov. 17	14 May 17	,,	58 Rifles.
	R. Fus.	**Lalore,** Denis Cawthorne		14 May 16*	26 Mar. 18	14 May 17	,,	39 Horse.
	M. G. Corps.	**Bartleman,** Edmund Leslie		14 May 16*	25 Mar. 18	14 May 17	,,	111 Mahars.
	R. Lanc. R.	**Flood,** Frederick William Shine Patrick		14 May 16*	21 Jan. 18	14 May 17	,,	35 Horse.
730	Som. L. I.	**Lee,** Tom Reginald		14 May 16*	5 July 18	14 May 17	,,	21 Punjabis.
	E. Surr. R.	**Dodd,** Basil Feilding		14 May 16*	12 Oct. 18	14 May 17	,,	107 Pioneers.
	R. Muns. Fus.	**Roe,** Kingsley Thomas		15 May 16*	5 Oct. 17	15 May 17	,,	1-3rd Gurkha Rifles.
		Wilson, William Maxson Collingwood (46, 50)		16 May 16*	22 Nov. 17	16 May 17	,,	62 Punjabis.
		Godfrey, Richard Edgar		16 May 16*	8 Feb. 18	16 May 17	,,	94 Infantry.
735	N. Staff. R.	**Davison,** Harold Thomas		16 May 16*	18 May 18	16 May 17	,,	91 Punjabis.
		Plaford, Robert Ballantyne	207a	17 May 16*	22 Nov. 17	17 May 17	,,	106 Pioneers.
		Bullock-Webster, Aubrey Philson		18 May 16*	23 Nov. 17	18 May 17	,,	2 Lancers.
		Esse, Frank Adolphus		18 May 16*	26 Feb. 18	18 May 17	,,	2-2nd Gurkha Rifles.
	Aus. Imp. Force.	**Cummins,** William		18 May 16*	25 Mar. 18	18 May 17	,,	4 Cavalry.
740		**Stanley,** Philip Timbrell		19 May 16*	12 Jan. 18	19 May 17	,,	94 Infantry.
	Hamps. R.	**Prince,** Huburt		19 May 16*	26 Mar. 18	19 May 17	,,	39 Garh. Rifles.
	Corn. L. I.	**Lander,** John Henry		19 May 16*	26 Apl. 18	19 May 17	,,	Signal Service.
	R. Lanc. R.	**Oglethorpe,** Ralph Stuart		19 May 16*	3 Feb. 18	19 May 17	,,	50 Rifles.
		Green, William		20 May 16*	22 Jan. 18	20 May 17	,,	3-9th Delhi Regt.
745		**Campbell,** James		21 May 16*	25 Nov. 17	21 May 17	,,	74 Punjabis.
	Midd'x R.	**Smith,** Richard Skilbeck		21 May 16*	14 Feb. 18	21 May 17	,,	1 Brahmans.
		McKinstry, Herbert (51)		23 May 16*	25 Nov. 17	23 May 17	,,	124 Infantry.
		Winnall, Eric Ivan		23 May 16*	8 Feb. 18	23 May 17	,,	33 Punjabis.
	M. G. Corps	**D'Este,** Hubert Adrian Middleton		23 May 16*	25 Mar. 18	23 May 17	,,	94 Infantry.
750	North'n R.	**Bower,** Samuel William		23 May 16*	4 Sep. 18	23 May 17	,,	7 Rajputs.
		Hartshorn, Erik Ponsonby	207a	24 May 16*	29 July 17	24 May 17	,,	3-8th Gurkha Rifles.
	R. I. Rif.	**Gilligan,** Cecil Herbert		24 May 16*	25 Feb. 18	24 May 17	,,	30 Punjabis.
		Allott, Arthur Duncan	207a	26 May 16*	29 July 17	26 May 17	,,	7 Lancers.
		Morphy, Eric Charles Owen		26 Mar. 16*	12 Jan. 18	26 Mar. 17	,,	63 Infantry.
755		**Harvey,** Ernest Fitzroy		26 May 16*	12 Jan. 18	26 May 17	,,	107 Pioneers.
		Saxton, Arther Cecil		27 May 16*	8 Feb. 18	27 May 17	,,	3-9th Delhi Infantry.
	R. Suss. R.	**Holdsworth,** Maurice Joseph		27 May 16*	30 Sep. 18	27 May 17	,,	111 Mahars.
	R. E.	**Murcott,** Stanley Howard Dowling	207a	28 May 16*	11 Apl. 18	28 May 17	,,	107 Pioneers.
	1 Res. R. of Cav.	**O'Keeffe,** Alan Richard William Mary		28 May 16*	1 Oct. 17	28 May 17	,,	27 Cavalry.
	Midd'x R.	**Woodliffe,** Lawrence Frederick		28 May 16*	22 Apl. 18	28 May 17	,,	9 Infantry.
760		**Hutton,** Gordon Campbell		29 May 16*	25 Sep. 17	29 May 17	,,	119 Infantry.
		Paterson, Kenneth		2 June 16*	3 Dec. 17	2 June 17	,,	2-9th Gurkha Rifles.
	R. Berk. R.	**Hutchings,** Dudley Alfred		2 June 16*	18 May 18	2 June 17	,,	99 Infantry.
	York. R.	**Luigi,** Parisotti		2 June 16*	28 Nov. 17	2 June 17	,,	2-3rd Gurkha Rifles.
	Bedf. R.	**Smith,** Harry Marchant		2 June 16*	26 June 18	2 June 17	,,	1-2nd Gurkha Rifles.
765	Lond. R.	**Baird,** Christopher James		3 June 16*	29 Nov. 17	3 June 17	,,	2-8th Gurkha Rifles.
		Bell, Leonard Charles		5 June 16*	13 Jan. 18	5 June 17	,,	12 Pioneers.
		Wells, Wilfred George Holmes (51)		6 June 16*	5 Dec. 17	6 June 17	,,	1-6th Gurkha Rifles.
		Spencer, Geoffrey Wareing		8 June 16*	8 Feb. 18	8 June 17	,,	30 Punjabis.
770		**Edwards,** Frederick Oscar		9 June 16*	5 Dec. 17	9 June 17	,,	125 Rifles.
771	Lan. Fus.	**Oliphant,** Ian Maurice Bruce		9 June 16*	19 Feb. 18	9 June 17	,,	8 Rajputs.

* 9 months forfeited for seniority under conditions of I. A. O. 511, dated 14th May 1917.

No.	Former Corps.	Names.	Date of birth.	Date of first commission.	Date of appointment to Indian Army.	Date of rank.	Remarks.
		Lieutenants—*contd.*					
772		**Hoare**, John George		10 June 16*	22 Nov. 17	10 June 17	Attached 31 Punjabis.
	Shrops. L. I.	**Walsh**, Percy George		10 June 16*	14 Apl. 18	10 June 17	" 123 Rifles.
	Can. Forces	**Baird**, Robert Patrick Hay		11 June 16*	26 Mar. 18	11 June 17	" 43 Regiment.
775		**Hunter**, James 207a		13 June 16*	22 Nov. 17	13 June 17	" 71 Punjabis.
	Bedf. R.	**Brickell**, Herbert James		13 June 16*	3 Dec. 17	13 June 17	" 86 Infantry.
	R. Sc. Fus.	**Nimmo**, Andrew		13 June 16*	31 May 18	13 June 17	" 54 Sikhs.
	York. R.	**James**, John Russell		13 June 16*	27 Aug. 18	13 June 17	" 5 Lt. Infy.
	M. G. Corps	**Kendrick**, Victor John Marshall		14 June 16*	27 Apl. 18	14 June 17	" 18 Infantry.
780	A. S. C.	**Clarke**, Joseph Robert Rice		14 June 16*	6 July 18	14 June 17	" 17 Infantry.
	R. Lanc. R.	**Dixon**, Thomas, M.C.		16 June 16*	30 Sep. 18	16 June 17	" 28 Punjabis.
	R. War. R.	**Thornton**, Walter William		18 June 16*	13 May 18	18 June 17	" 2 Raj. L. I.
	Bord. R.	**Johnson**, Harry Hall		19 June 16	1 June 18	19 June 17	" 52 Sikhs.
	L'pool R.	**Jardine**, John Douglas		19 June 16*	30 Sep. 18	19 June 17	" 124 Infantry.
785		**Hopkins**, Frederick		20 June 16*	11 Sep. 17	20 June 17	" 1 Brahmans.
		Parker, Eric		21 June 16*	8 Feb. 18	21 June 17	" Guides Infantry.
	K. R. Rif. C.	**Stevenson**, Thomas Ivor (51)		21 June 16*	25 Mar. 18	21 June 17	" 2-8th Gurkha Rifles.
	A. S. Corps	**Haynes**, Alfred Ernest		21 June 16*	1 June 18	21 June 17	" 11 Rajputs.
	E. York. R.	**Desborough**, George		21 June 16*	29 Oct. 18	21 June 17	" 62 Punjabis.
790	M. G. Corps	**Jordan**, Ernest George		23 June 16*	13 May 18	23 June 17	" 17 Infantry.
	6 Dgns.	**Auber**, Joseph		24 June 16*	14 Apl. 18	24 June 17	" 16 Cavalry.
		Douglas, Leslie Norman		25 June 16*	26 Nov. 17	25 June 17	" 2-1st Gurkha Rifles.
	Aus. Imp. Force.	**Copplestone**, Frank Eugene		25 June 16*	25 Mar. 18	25 June 17	" 63 Infantry.
	Can. Forces	**Stuart**, Douglas		25 June 16*	12 Jan. 18	25 June 17	" 112 Infantry.
795	M. G. Corps	**Cormack**, Robert Kerr Weir		25 June 16*	27 Aug. 18	25 June 17	" 5 Gurkha Rifles.
	Oxf. & Buck. L. I.	**Rutherford**, John Sanders		25 June 16*	29 Oct. 18	25 June 17	" 15 Sikhs.
	R. E.	**Cheverton**, Francis Penn		26 June 16*	27 Apl. 18	26 June 17	" 9 Infantry.
	M. G. Corps	**Caldwell**, Lindsay Roy		27 June 16*	1 Oct. 17	27 June 17	" 124 Infantry.
800	Tank Corps	**Rayner**, John		27 June 16*	1 June 18	27 June 17	" 33 Punjabis.
	M. G. Corps	**Couldrey**, Oscar, M.C.		27 June 16*	30 Sep. 18	27 June 17	" 23 Infantry.
	Aus. Imp. Forces.	**Bowtell-Harris**, James Frederic, M.C.		28 June 16*	29 July 18	28 June 17	" 45 Sikhs.
	R. Suss. R.	**Carden**, Frederick George		28 June 16*	13 Aug. 18	28 June 17	" 84 Punjabis.
	Lond. R.	**Alsop**, Reginald Philip		29 June 16*	26 Mar. 18	29 June 17	" 3-8th Gurkha Rifles.
805		**Newcomb**, Harry Reed		30 June 16*	6 May 17	30 June 17	" 12 Pioneers.
	R. Fus.	**Skead**, Bernard		30 June 16*	27 Aug. 18	30 June 17	" 7 Gurkha Rifles.
		Reed, H. James Frederick		1 July 16*	17 Sep. 17	1 July 17	" 34 Pioneers.
	Man. R.	**Dore**, Gilbert Bennet		1 July 16*	15 Feb. 18	1 July 17	" 123 Rifles.
	Aus. Imp. Forces.	**Pulling**, Guy Harris		1 July 16*	4 July 18	1 July 17	" 6 Gurkha Rifles.
		Stirling, James Morrison		2 July 16*	5 Dec. 17	2 July 17	" 59 Rifles.
810		**Bartlett**, Percy James Austin		4 July 16*	9 Jan. 18	4 July 17	" 13 Lancers.
811	Leins. R.	**Toomey**, Waldron Gosselin (46)		4 July 16*	12 May 18	4 July 17	" 31 Punjabis.

	R. Ir. Rif.	**Crawford,** Charles Oliver, M.C.			4 July 16*	6 Sep. 18	4 July 17	,,	67 Punjabis.
	E. York R.	**Hearring,** Thomas Charles, M.C.			4 July 16*	27 Aug. 18	4 July 17	,,	123 Rifles.
	E. Kent R.	**Tustin,** Charles Victor			5 July 16*	27 Sep. 18	5 July 17	,,	40 Cavalry.
815		**Macdonald,** Somerled	207*a*		7 July 16*	9 Aug. 17	7 July 17	,,	54 Sikhs.
		Lewis, Howard Allan Cornewell (46)			7 July 16*	29 July 17	7 July 17	,,	113 Infantry.
		Birkin, William Neville, M.C.			7 July 16*	22 Jan. 18	7 July 17	,,	110 Infantry.
	Aus. Imp. Force.	**Trott,** William Augustus, M.C.			7 July 16*	25 Feb. 18	7 July 17	,,	3-9th Bhopal Infy.
	R. Ir. Fus.	**Waters,** William Haviland			7 July 16*	13 Aug. 18	7 July 17	,,	99 Infantry.
820	R. Ir. Fus.	**Hill,** Douglas Nel			7 July 16*	24 July 18	7 July 17	,,	36 Sikhs.
	Bedf. R.	**Winters,** Leo Patrick			7 July 16*	29 Sep. 18	7 July 17	,,	39 Garhwal Rifles.
	E. York. R.	**Fielder,** George Harold			8 July 16*	25 Mar. 18	8 July 17	,,	3-7th Gurkha Rifles.
	R. Inns. Fus	**Metcalfe,** Charles James			8 July 16*	12 May 18	8 July 17	,,	8 Rajputs.
	Midd'x. R.	**Longley,** Arthur Harry			8 July 16*	4 July 18	8 July 17	,,	6 Jat. L. I.
825	5 Lanc.	**Armstrong,** Kevill			8 July 16*	28 Oct. 18	8 July 17	,,	124 Infantry.
		Hunt, James Louis			9 July 16*	5 Nov. 17	9 July 17	,,	37 Dogras.
	Dorset R.	**Chamberlain,** William Harlow			9 July 16*	29 Oct. 18	9 July 17	,,	1 Brahmans.
	Herts. Yeo.	**Gripper,** Harold Basil			12 July 16*	14 Nov. 17	12 July 17	,,	22 Cavalry.
		Golden, Harold Arthur			13 July 16*	6 June 17	13 July 17	,,	32 Pioneers.
830	York R.	**Fagan,** Herbert Archer (46)			13 July 16*	12 May 18	13 July 17	,,	2-9th Gurkha Rifles.
	Aus. Imp. Force.	**Stewart,** Alastair Moray			14 July 16*	26 Mar. 18	14 July 17	,,	35 Horse.
		Souter, James Henry			16 July 16*	16 Jan. 18	16 July 17	,,	109 Infantry.
	R. Highrs.	**Bett,** Frederick John Hunter			16 July 16*	11 Apl. 18	16 July 17	,,	3-6th Gurkha Rifles.
	W. York R.	**Skinner,** Lewis Sammuel Corbett			16 July 16*	25 Mar. 18	16 July 17	,,	39 Horse.
835	Lond. R.	**Martin,** Geoffrey St. John			17 July 16*	1 June 18	17 July 17	,,	127 Baluch L. I.
	Bedf. R.	**Edward,** Malcolm Frederick			17 July 16*	29 July 18	17 July 17	,,	15 Sikhs.
		Burnley, William Douglas			18 July 16*	12 Jan. 18	18 July 17	,,	31 Lancers.
	Gen. List	**Whitaker,** Samuel Whiteley			18 July 16*	4 Sep. 18	18 July 17	,,	16 Cavalry
		Butterfield, Basil Frederick	207*a*		19 July 16	12 Jan. 18	19 July 17	,,	97 Infantry.
840		**Oldham,** Arthur Wilfred			19 July 16	22 Jan. 18	19 July 17	,,	12 Cavalry.
		Little, Edward Gerald			19 July 16*	16 Jan. 18	19 July 17	,,	1 Brahmans.
	Devon. R.	**Old,** Reginald Eustace			19 July 16*	12 May 18	19 July 17	,,	81 Pioneers.
	Leins. R.	**Jameson,** Gerald Grey		4 July 98	19 July 16	1 Mar. 18	19 July 17	,,	30 Punjabis.
	Durh. L. I.	**Smith,** Vernon Thomas			19 July 16*	17 Oct. 17	19 July 17	,,	34 Pioneers.
845		**Ballantine,** Richard Waverly Head	207*a*		20 July 16*	31 July 17	20 July 17	,,	81 Pioneers.
	Linc. R.	**Spencer,** John Charles			20 July 16*	11 Apl. 18	20 July 17	,,	72 Punjabis.
		Quin, Richard Berchmans (46)			21 July 16*	6 May 17	21 July 17	,,	2 Gurkha Rifles.
		Allen, Norman Charles			21 July 16*	5 Dec. 17	21 July 17	,,	119 Infantry.
	North'd. Fus.	**Thomas,** Lancelot Arthur Smith			21 July 16*	1 June 18	21 July 17	,,	129 Baluchis.
850	A Cyclist Corps.	**Barfield,** Frank Harold			21 July 16*	18 Oct. 18	21 July 17	,,	43 Regt.
		Cruickshank, James Gray			22 July 16*	10 Mar. 18	22 July 17	,,	18 Infantry.
	Camn. Hghrs	**Sutherland,** Alexander Watts			22 July 16*	5 Mar. 18	22 July 17	,,	110 Infantry.
		Jones, Cecil Ynyr			24 July 16*	8 Feb. 18	24 July 17	,,	30 Punjabis.
	E. Sur. R.	**Bates,** Henry Vivian Ross			25 July 16*	27 Apl. 18	25 July 17	,,	27 Punjabis.
855	R. Highrs	**Brown,** James			25 July 16*	30 Sep. 18	25 July 17	,,	40 Pathans.
	Nott. Yeo.	**Wing,** Richard			25 July 16*	26 Sep. 18	25 July 17	,,	16 Cavalry.
857	N. Zea. Forces.	**Garden,** Robert Leslie			26 July 16*	5 Dec. 17	26 July 17	,,	130 Baluchis.

* 9 months forfeited for seniority under conditions of I. A. O. 511, dated 14th May 1917.

No.	Former Corps.	Names.	Date of birth.	Date of first commission.	Date of appointment to Indian Army.	Date of rank.	Remarks.
		Lieutenants—*contd.*					
857	R. Scots.	**Simpson,** Joseph Norman		26 July 16*	27 Aug. 18	25 July 17	Attached 20 Horse.
		Dunbar-Sutherland, Cyril Francis		27 July 16*	27 July 17	28 July 17	,, 4 Cavalry.
	Worc. R.	**Garrod,** William Francis		29 July 16*	11 Apl. 18	29 July 17	,, 25 Punjabis.
860	Linc. R.	**Theak,** Leslie Fredric		29 July 16*	14 Apl. 18	29 July 17	,, 94 Infantry.
	Bedf. R.	**Liswell,** John Alfred		29 July 16	1 Jan. 18	29 July 17	,, 41 Dogras.
		Gilbert, Eric Clair		30 July 16*	23 Jan. 18	30 July 17	,, 6 Infantry.
		Sommerville, Robert		30 July 16*	8 Feb. 18	30 July 17	,, 18 Infantry.
		Kersey, Arthur Oliver (46) 207*a*		1 Aug. 16*	5 Dec. 17	1 Aug. 17	,, 49 Bengalis.
865		**Hawken,** Gerald Humphrey		1 Aug. 16*	12 Jan. 18	1 Aug. 17	,, 27 Cavalry.
		Hawken, Bernard Roger Philip 207*a*		1 Aug. 16*	8 Feb. 18	1 Aug. 17	,, 113 Infantry.
	R. Highrs.	**Schmidt,** Cecil Trant		1 Aug. 16*	26 June 18	1 Aug. 17	,, 13 Lancers.
	Essex. R.	**Willoughby,** John Jeston		1 Aug. 16*	29 Aug. 18	1 Aug. 17	,, 27 Punjabis.
	E. Surr. R.	**Beard,** Francis		2 Aug. 16*	29 July 18	2 Aug. 17	,, 54 Sikhs.
870	Devon R.	**Rains-Bath,** John Gardiner		2 Aug. 16*	28 Sep. 18	2 Aug. 17	,, 25 Punjabis.
	Scot. H.	**Orr,** Harry Douglas		3 Aug. 16*	13 May 18	3 Aug. 17	,, 37 Lancers.
		Black, John Mayes		4 Aug. 16*	3 Dec. 17	4 Aug. 17	,, 121 Pioneers.
	North'd.Fus.	**Green,** James Henry		7 Aug. 16*	18 May 18	7 Aug. 17	,, 85 Rifles.
	Midd'x. R.	**Taylor,** Arthur Charles, D.S.O. (46)		7 Aug. 16*	6 Sep. 18	7 Aug. 17	,, 2 Infantry.
875	S. Lan. R.	**Cunningham,** Allan Proud		8 Aug. 16*	1 June 18	8 Aug. 17	,, 24 Punjabis.
		Butler, Alfred Home		9 Aug. 16*	6 May 17	9 Aug. 17	,, 16 Cavalry.
		Downie, Archibald		9 Aug. 16*	5 Dec. 17	9 Aug. 17	,, 17 Infantry.
	Aus. Imp. Force.	**Cawthorn,** Walter Joseph		9 Aug. 16*	25 Mar. 18	9 Aug. 17	,, 3-5th Gurkha Rifles.
	Mance R.	**Barton,** John Edward Broadbent		10 Aug. 16*	6 Sep. 18	10 Aug. 17	,, 109 Infantry.
880	R. Berk R.	**Gibson,** Dudley Robert		11 Aug. 16*	4 July 18	11 Aug. 17	,, 83 Infantry.
	R. Innis. Fus.	**Naye,** Rowland Alexander		11 Aug. 16*	29 Oct. 18	11 Aug. 17	,, 95 Infantry.
		Dossett, Albert Edward, M.C.		12 Aug. 16*	25 Nov. 17	12 Aug. 17	,, 74 Punjabis.
		Marshall, Robert Husband		12 Aug. 16*	8 Feb. 18	12 Aug. 17	,, 4 Cavalry.
	R. W. Surr. R.	**McDermott,** Francis		14 Aug. 16*	27 Apl. 18	14 Aug. 17	,, 42 Regiment.
885	Linc. R.	**Burrows,** Lionel Beeley		14 Aug. 16*	2 Mar. 18	14 Aug. 17	,, 26 Punjabis.
	Aus. Imp. Force.	**Hirst,** Arthur Havelock		14 Aug. 16*	25 Mar. 18	14 Aug. 17	,, 98 Infantry.
		Goss, John Eric Sabine 207*a*		15 Aug. 16*	22 Jan. 18	15 Aug. 17	,, 7 Rajputs.
		Comins, Richard Innes		16 Aug. 16	22 Nov. 17	16 Aug. 17	,, 5 Lt. Infantry.
		Sivary, Herbert Sydney		16 Aug. 16	23 Nov. 17	16 Aug. 17	,, 32 Pioneers.
890	R. Ir. R.	**MacIvor,** Robert James		16 Aug. 16*	25 Mar. 18	16 Aug. 17	,, 2.10th Gurkha Rifles.
	R. Ir. Rif.	**Darling,** Owen Richard		16 Aug. 16*	11 Apl. 18	16 Aug. 17	,, 18 Infantry.
	R. E.	**Bentham,** Vincent Harry		16 Aug. 16*	2 Mar. 18	16 Aug. 17	,, 3-5th Gurkha Rifles.
	A. Cyclist Corps.	**Harrison,** Henry Leonard		16 Aug. 16*	12 May 18	16 Aug. 17	,, 9 Bhopal Infantry.
	E. Surr. R.	**Bussell,** John Chiles		16 Aug. 16*	23 Nov. 17	16 Aug. 17	,, 2 Raj. L. I.
895	Manch. R.	**Greengrass,** Reginald Percival William		16 Aug. 16*	4 Sep. 18	16 Aug. 17	,, 128 Pioneers.
	Midd'x. R.	**Baker,** Arthur Hillman		16 Aug. 16*	10 Oct. 18	16 Aug. 17	,, 107 Pioneers.

	Glous. Yeo.	**Lawson**, Joseph Frank		17 Aug. 16*	25 Feb. 18	17 Aug. 17	,,	39 C. I. Horse.
	Dorset R.	**Strudwicke**, Montague Alexander		18 Aug. 16*	6 Sep. 18	18 Aug. 17	,,	105 Mah. L. I.
	Linc. Yeo.	**Coates**, Howard Lewis		19 Aug. 16*	12 Oct. 18	19 Aug. 17	,,	73 Infantry.
900		**Monell**, Douglas Waldo (46)		20 Aug. 16*	22 Jan. 18	20 Aug. 17	,,	15 Sikhs.
	R. Scots.	**Murray**, Randolph George		20 Aug. 16*	12 May 18	20 Aug. 17	,,	26 Cavalry.
	York & Lanc. R.	**Bailey**, John Howard		21 Aug. 16*	13 Apl. 18	21 Aug. 17	,,	86 Infantry.
	Gord. Highrs.	**Wiseman**, William		22 Aug. 16*	1 June 18	22 Aug. 17	,,	11 Rajputs.
		Thompson, David	207a	26 Aug. 16*	29 Sep. 17	26 Aug. 17	,,	15 Lancers.
905	Lvt. Scouts	**Dore**, Archibald Mervyn		26 Aug. 16*	25 Mar. 18	26 Aug. 17	,,	35 Horse.
	L'pool R.	**Foot**, Howard Owen		27 Aug. 16*	18 May 18	27 Aug. 17	,,	3-2nd Gurkha Rifles.
	North'd. Fus.	**Steedman**, Leonard		29 Aug. 16*	18 May 18	29 Aug. 17	,,	25 Cavalry.
	R. War R.	**Eastabrook**, Ivor		29 Aug. 16*	5 July 18	29 Aug. 17	,,	1-5 Gurkha Rif.
		Carter, Harold Leslie Mark		30 Aug. 16*	29 July 17	30 Aug. 17	,,	28 Cavalry.
910	York R.	**Beall**, John Samuel		30 Aug. 16*	12 May 18	30 Aug. 17	,,	33 Punjabis.
		Jones, George Frederick		1 Sep. 16*	9 Jan. 18	1 Sep. 17	,,	103 Mah. L. I.
	A. S. C.	**Tait**, Thomas Rennie		1 Sep. 16*	13 Aug. 18	1 Sep. 17	,,	102 Grenadiers.
		Pitt, Benno Irmangarda		2 Sep. 16*	8 Feb. 18	2 Sep. 17	,,	124 Infantry.
	R. Scots.	**Langlands**, Andrew Romilly		2 Sep. 16*	1 June 18	2 Sep. 17	,,	113 Infantry.
915	Gord. Highrs	**Fleming**, James George Grant, D.S.O., M.C.		3 Sep. 16*	25 Mar. 18	3 Sep. 17	Attached	21 Punjabis.
	Notts. & Derby. R.	**Kelk**, Basil Milner		4 Sep. 16*	18 May 18	4 Sep. 17	,,	81 Pioneers.
	M. G. Corps.	**Lewis**, Thomas Meredith		4 Sep. 16*	30 Oct. 18	4 Sep. 17	,,	79 Infantry.
		Southern, Willis (59)		6 Sep. 16*	12 Jan. 18	6 Sep. 17	,,	112 Infantry.
		Curtis, Arthur Gilbert Clements		6 Sep. 16*	9 Aug. 17	6 Sep. 17	,,	28 Punjabis.
920	High L. I.	**Lewellin**, David John		6 Sep. 16*	15 June 18	6 Sep. 17	,,	2 Raj L. I.
		Standing, Herbert (57)		7 Sep. 16*	6 Jan. 18	7 Sep. 17	,,	30 Lancers.
	S. Lan. R.	**Griffiths**, Charles Percy		7 Sep. 16*	12 Feb. 18	7 Sep. 17	,,	Corps of Guides.
	Welsh L.	**Underwood**, William Reynish		7 Sep. 16*	29 Dec. 17	7 Sep. 17	,,	70 Rifles.
		Craven, Arthur Gerald William Scott	207a	8 Sep. 16*	24 Oct. 17	8 Sep. 17	,,	17 Infantry.
925	Aus. Imp. Force.	**Boileau**, Cecil Mackintosh		8 Sep. 16*	15 Feb. 18	8 Sep. 17	,,	37 Lancers.
	Aog. Imp. Force.	**Roach**, Leonard Nairn		8 Sep. 16*	1 June 18	8 Sep. 17	,,	129 Baluchis.
		Breething, William Joseph (172a)	207a	9 Sep. 16*	6 May 17	9 Sep. 17	,,	81 Pioneers.
		Simms, James		9 Sep. 16*	25 Nov. 17	9 Sep. 17	,,	99 Infantry.
		Baker, Henry Leslie	207a	9 Sep. 16*	5 Dec. 17	9 Sep. 17	,,	86 Infantry.
930		**Williams**, David Montygue		9 Sep. 16*	9 Jan. 18	9 Sep. 17	,,	73 Infantry.
		Fanning, Robert James		9 Sep. 16*	9 Feb. 18	9 Sep. 17	,,	39 Garhwal Rif.
		Holmes, Stanley Brereton		9 Sep. 16*	9 Mar. 18	9 Sep. 17	,,	1 Brahmans.
	L'pool R.	**Jones**, Reginald Rowland		9 Sep. 16*	30 Sep. 18	9 Sep. 17	,,	87 Punjabis.
	K. E. Horse.	**Moffat**, Leslie Palmer		10 Sep. 16*	29 Aug. 18	10 Sep. 17	,,	44 Cavalry.
935		**Fagnani**, Henry Hutchinson		13 Sep. 16*	11 Apl. 17	13 Sep. 17	,,	Corps of Guides.
		Warrener, John Victor		13 Sep. 16*	12 Jan. 18	13 Sep. 17	,,	97 Infantry.
		King, Stanley Jesse		14 Sep. 16*	11 Sep. 17	14 Sep. 17	,,	22 Punjabis.
		Birney, Herbert		14 Sep. 16*	12 Jan. 18	14 Sep. 17	,,	123 Rifles.
	R. Ir. R.	**Hall**, Frederick Ambrose		14 Sep. 16*	13 May 18	14 Sep. 17	,,	96 Infantry.
940		**Johnson**, Edward George		15 Sep. 16*	6 May 17	15 Sep. 17	,,	30 Garwal Rifles.
	E. Surr. R.	**Recknell**, Dickson Lynn		15 Sep. 16*	17 June 18	15 Sep. 17	,,	12 Pioneers.

* 9 months forfeited for seniority under conditions of I. A. O. 511, dated 14th May 1917.

Lieutenants—*contd.*

No.	Former Corps.	Names.		Date of birth.	Date of first commission.	Date of appointment to Indian Army.	Date of rank.	Remarks.
	Bord. R.	**Crawford,** Gavin Alexander			16 Sep. 16*	18 May 18	16 Sep. 17	Attached 102 Grenadiers.
	E. Lan. R.	**Foster,** Reginald Frank			17 Sep. 16*	26 Mar. 18	17 Sep. 17	,, 70 Rifles.
	Notts. & Derby R.	**Overton,** Harold			19 Sep. 16*	30 Sep. 18	19 Sep. 17	,, 3 Brahmans.
945	Cam'n Highrs.	**McIntyre,** John McDonald			19 Sep. 16*	17 Sep. 17	19 Sep. 17	,, 43 Regiment.
		Tyrrell, Harry Ernest			20 Sep. 16*	16 Jan. 18	20 Sep. 17	,, 30 Punjabis.
		De la Fargue, Dudley Gordon Heriot			21 Sep. 16*	21 Feb. 17	21 Sep. 17	,, 6 Cavalry.
		Vines, Ronald Gordon	207a		22 Sep. 16*	3 July 17	22 Sep. 17	,, 28 Cavalry.
		Poynor, Maurice Pugh			22 Sep. 16*	23 Nov. 17	22 Sep. 17	,, 94 Infantry.
950		**Savary,** James Richard			23 Sep. 16*	23 Jan. 18	23 Sep. 17	,, 3-2nd Gurkha Rifles.
	Worc. Yeo.	**Griffiths,** William Stanley			23 Sep. 16*	1 Mar. 18	23 Sep. 17	,, 17 Cavalry.
		George, Alexander			24 Sep. 16*	12 Jan. 18	24 Sep. 17	,, 66 Punjabis.
		Preddle, Arthur Reginald			24 Sep. 16*	9 Feb. 18	24 Sep. 17	,, 15 Sikhs.
		James, Cyril Merton			24 Sep. 16*	8 Feb. 18	24 Sep. 17	,, 117 Mahrattas.
955		**Osborne,** Coles Alexander (46)			26 Sep. 16*	12 Jan. 18	26 Sep. 17	,, 35 Sikhs.
		Voyle, Stanley George			27 Sep. 16*	5 Dec. 17	27 Sep. 17	,, 2-6th Gurkha Rifles.
	R. Mar.	**Woods,** John Mc Birney			27 Sep. 16*	13 May 18	27 Sep. 17	,, 14 Sikhs.
	W. York. R.	**Hanson,** Frank Allen			28 Sep. 16*	29 Aug. 18	28 Sep. 17	,, 61 Pioneers.
		Seaton, Oscar Clark (46)			30 Sep. 16*	22 Nov. 17	30 Sep. 17	,, 42 Regiment.
960		**Hodgshon,** Frederick William			30 Sept. 16*	8 Feb. 18	30 Sep. 17	,, 3-2nd Gurkha Rifles.
	North'd Fus.	**Burton,** John Alexander, M.C.			30 Sep. 16*	31 Aug. 18	30 Sep. 17	,, 79 Infantry.
	Lond. R.	**Sendall,** Walter Harold			30 Sep. 16*	4 Sep. 18	30 Sep. 17	,, 96 Bearar Ify.
		Hay, Cecil Milnes			1 Oct. 16*	12 Jan. 18	1 Oct. 17	,, 35 Sikhs.
	K. R. Rif. C.	**Peach,** Eric Noel			1 Oct. 16*	1 Oct. 17	1 Oct. 17	,, 111 Mahars.
965	Berks. Yeo.	**Simons,** Francis Henry			1 Oct. 16*	4 Oct. 17	1 Oct. 17	,, 27 Cavalry.
	W. York. R.	**Davies,** Cuthbert Collin			1 Oct. 16*	30 June 18	1 Oct. 17	,, 2-1st Gurkha Rifles.
	Manch. R.	**Reeve,** Alfred John			2 Oct. 16*	25 Mar. 18	2 Oct. 17	,, 30 Punjabis.
		Thomas, Arthur Frank Friend			3 Oct. 16*	23 Nov. 17	3 Oct. 17	,, 46 Punjabis.
		Patton, William Francis	207a		6 Oct. 16*	6 May 17	6 Oct. 17	,, 1 Gurkha Rifles.
970		**Storrow,** Samuel Loram			6 Oct. 16*	6 May 17	6 Oct. 17	,, 54 Sikhs.
	Aus. Imp. Force.	**Iceton,** Edward Arthur			6 Oct. 16*	26 Mar. 18	6 Oct. 17	,, 1-9th Gurkha Rifles.
	Lond R.	**Pettigrew,** Frederick Charles			6 Oct. 16*	12 May 18	6 Oct. 17	,, 2-8th Gurkha Rifles.
	Manch. R.	**Law,** Edward Lionel			6 Oct. 16*	13 Aug. 18	6 Oct. 17	,, 26 Punjabis.
	N. Ir. Horse	**Mcmullen,** William Kenneth			6 Oct. 16*	30 Sep. 18	6 Oct. 17	,, 41 Cavalry.
975		**Morris-Jones,** John	207a		7 Oct. 16*	29 July 17	7 Oct. 17	,, 39 C. I. Horse.
		Steel, Edward Gerald	207a		7 Oct. 16*	8 Feb. 18	7 Oct. 17	,, 129 Baluchis.
	Manch. R.	**Carruthers,** Francis William Walter			7 Oct. 16*	30 Sep. 18	7 Oct. 17	,, 17 Infantry.
		Kirkham, Joseph Lionel (51)			8 Oct. 16*	17 Oct. 17	8 Oct. 17	,, 10 Jats.
	Welsh. R.	**Threle,** Cecil Edward			8 Oct. 16*	30 June 18	8 Oct. 17	,, 96 Infantry.
980	Denbig. Yeo	**Weaver,** Moses George			8 Oct. 16*	22 Apl. 18	8 Oct. 17	,, 21 Cavalry.
	R. War. R.	**Tyers,** William Frederick			8 Oct. 16*	31 Aug. 18	8 Oct. 17	,, 27 Punjabis.
		Cotton, James Thomas (61)			10 Oct. 16*	9 Feb. 18	10 Oct. 17	,, 41 Dogras.
	Midd'x R.	**Fowler,** Gordon Harry Ashton			10 Oct. 16*	25 Mar. 18	10 Oct. 17	,, Corps of Guides.

984	North'd Fus.	**Carlile**, Thomas Alexander			10 Oct. 16*	30 Sep. 18	10 Oct. 17	,,	7 Rajputs.
		Blyde, Wilfrid Arthur			11 Oct. 16*	29 July 17	11 Oct. 17	,,	1 Lancers.
985	Glouc. R.	**Edwards**, Gerald Alexander			11 Oct. 16*	1 June 18	11 Oct. 17	,,	26 Cavalry.
	Devon. R.	**Porteous**, Hugh Aubrey			11 Oct. 16*	24 Sep. 18	11 Oct. 17	,,	7 Rajputs.
		Bradford, Francis Kysall			13 Dec. 16*	8 Feb. 18	13 Dec. 17	,,	1-4th Gurkha Rifles.
		Bygate, Charles Mountford			20 Oct. 16*	22 Jan. 18	20 Oct. 17	,,	22 Punjabis.
990	Aus. Imp. Force.	**Barclay**, Harry Lewis			20 Oct. 16*	25 Feb. 18	20 Oct. 17	,,	109 Infantry.
	North'd Fus.	**Kerridge**, Sidney Charles, M.C.			20 Oct. 16*	29 Aug. 18	20 Oct. 17	,,	26 Punjabis.
	Conn. Rang.	**Hyland**, James Patrick			20 Oct. 16*	30 Oct. 18	20 Oct. 17	,,	10 Jats.
		Mitchell, Leonard	207a		21 Oct. 16*	10 Aug. 17	21 Oct. 17	,,	110 Infantry.
		Lane, Alfred Donald			21 Oct. 16*	10 Mar. 18	21 Oct. 17	,,	1 Brahmans.
995	R. Highrs.	**Cowie**, James MacDonald			21 Oct. 16*	27 Mar. 18	21 Oct. 17	,,	33 Punjabis.
		Jones, Rignald Austin			22 Oct. 16*	19 Jan. 18	22 Oct. 17	,,	3-9th Gurkha Rifles.
	M. G. Corps.	**Hitchock**, Lawrence William			23 Oct. 16*	25 Feb. 18	23 Oct. 17	,,	22 Punjabis.
	R. Highrs.	**Blackmore**, Edwin Wallace Henry			26 Oct. 16*	27 Apl. 18	26 Oct. 17	,,	11 Rajputs.
	Lond. R.	**Weston**, Edwin Bernard			26 Oct. 16*	1 June 18	26 Oct. 17	,,	117 Mahrattas.
1000	W.Rid. R.	**Laughton**, Francis Seymour		5 Oct. 98	27 Oct. 16	18 May 18	27 Oct. 17	,,	4 Rajputs.
		Power, William Albeus			28 Oct. 16*	5 Nov. 17	28 Oct. 17	,,	20 Infantry.
		Hunt, Louis Frederick			28 Oct. 16*	12 June 17	28 Oct. 17	,,	120 Infantry.
		Thomas, John Ernest			29 Oct. 16*	22 Jan. 18	29 Oct. 17	,,	26 Punjabis.
	R. Ir. R.	**Smith**, Frederick George			29 Oct. 16*	12 May 18	29 Oct. 17	,,	124 Infantry.
1005		**Hudson**, Alexander Robert			29 Oct. 16*	6 May 17	29 Oct. 17	,,	8 Cavalry.
	Manch. R.	**Foskett**, Walter, M.C.			29 Oct. 16*	26 June 18	29 Oct. 17	,,	73 Infantry.
	Ches. R.	**Edgson**, Theodore John			29 Oct. 16*	12 May 18	29 Oct. 17	,,	5 Lt. Infantry.
		Currie, James			31 Oct. 16*	5 Dec. 17	31 Oct. 17	,,	124 Infantry.
	Gord. Highrs	**Mills**, John Edward			31 Oct. 16*	12 Oct. 18	31 Oct. 17	,,	150 Infantry.
1010	Lond. R.	**Belcher**, Douglas Walter, V. C.			10 Nov. 16*	30 Oct. 18	10 Nov 17	,,	6 Gurkha Rifles.
	Scot. Rif.	**Page**, Arthur Jutting			17 Nov. 16*	12 Feb. 18	17 Nov. 17	,,	21 Cavalry.
	R. Ir. Rif.	**Booth**, Arthur Ronald			18 Nov. 16*	18 Oct. 17	18 Nov. 17	,,	31 Punjabis.
	A. Cyclist Corps.	**Cooke**, Charles William			18 Nov. 16*	3 Dec. 17	18 Nov. 17	,,	12 Pioneers.
	Aus. Imp. Force.	**McKenzie-Muirson**, Kenneth Basil, M.C.			20 Nov. 16*	8 Feb. 18	20 Nov. 17	,,	1 Lancers.
1015	Aus. Imp. Force.	**Carter**, Alan Victor			20 Nov. 16*	25 Mar. 18	20 Nov. 17	,,	112 Infantry.
	Aus. Imp. Force.	**Wrigley**, Hugh, M.C.			20 Nov. 16*	1 Oct. 17	20 Nov. 17	,,	123 Rifles.
	Aus. Imp. Force.	**Bursey**, Thomas Frederick McClean M.C.			20 Nov. 16*	6 Sep. 18	20 Nov. 17	,,	125 Rifles.
	R. Mar. Arty.	**Sienesi**, Alan Russell			28 Nov. 16*	11 Apl. 18	28 Nov. 17	,,	1-4th Gurkha Rifles.
	Linc. R.	**Clark**, Ernest William			3 Dec. 16*	14 Apl. 18	3 Dec. 17	,,	111 Mahars.
1020		**Balch**, Charles Harry			5 Dec. 16*	12 Jan. 18	5 Dec. 17	,,	7 Lancers.
	Welsh R.	**Figgins**, Arthur Percival			5 Dec. 16*	30 June 18	5 Dec. 17	,,	31 Punjabis.
	Life Gds.	**Anderson**, George			6 Dec. 16*	13 May 18	6 Dec. 17	,,	3 Horse.
	Aus. Imp. Force.	**Cameron**, Frank Osborne			6 Dec. 16*	1 June 18	6 Dec. 17	,,	107 Pioneers.
		Webb, Stuart Davies	207a		10 Dec. 16*	23 Nov. 17	10 Dec. 17	,,	124 Infantry.
1025		**Mellor**, John Russell			12 Dec. 16*	22 Nov. 17	12 Dec. 17	,,	33 Punjabis.
1026	Aus. Imp. Force.	**McLean**, Douglas Allan			12 Dec. 16*	8 Feb. 18	12 Dec. 17	,,	34 Pioneers.

* 9 months forfeited for seniority under conditions of I. AO. 511, dated 14th May 1917.

No.	Former Corps.	Names.	Date of birth.	Date of first commission.	Date of appointment to Indian Army.	Date of rank.	Remarks.
		Lieutenants—*contd.*					
1027	Aus. Imp. Force	**Lang**, Joe Tilson, M.C.		12 Dec. 16*	9 Feb. 18	12 Dec. 17	Attached 2 Raj. L. I.
		Simpson, William Arthur		1 Jan. 17	12 Jan. 18	1 Jan. 18	„ 4-3rd Gurkha Rifles.
	Ches. R.	**Johnstone**, Edward (57)		1 Jan. 17	1 Mar. 18	1 Jan. 18	„ 89 Punjabis.
1030		**Wickham**, John Joseph Onslow		2 Jan. 17*	8 Feb. 18	2 Jan. 18	„ 129 Baluchis.
	M. G. Corps	**Pepperall**, Robert Augustus		2 Jan. 17*	30 June 18	2 Jan. 18	„ 67 Punjabis.
	Aus. Imp. Force.	**Gore**, Reginald Malpas		5 Jan. 17*	1 June 18	5 Jan. 18	„ 88 Infantry.
	Lond R.	**Clappen**, Stanley Graham		15 Jan. 17*	1 June 18	15 Jan. 18	„ 55 Rifles.
		Forth, Charles Ogilvie		18 Jan. 17*	29 July 17	18 Jan. 18	„ 34 Pioneers.
1035	York. L. I.	**Jennings**, Carl Kaberry		26 Jan. 17*	14 Apl. 18	26 Jan. 18	„ 31 Punjabis.
		O'Donnell, Frank George		30 Jan. 17*	22 Jan. 18	30 Jan. 18	„ 25 Punjabis.
	Glouc. R.	**Stenson**, Lawrence Henry Edward		30 Jan. 17*	27 Aug. 18	30 Jan. 18	„ 45 Cavalry.
	R. Ir. Rif.	**O'Halloran**, Henry		2 Feb. 17*	12 May 18	2 Feb. 18	„ 124 Infantry.
	W. York R.	**Melhuish**, Harold		7 Feb. 17*	30 Sep. 18	7 Feb. 18	„ 3 Brahmans.
1040	Notts. & Derby. R.	**Gregory**, Samuel	3 June 91	12 Feb. 17	1 June 18	12 Feb. 18	„ 26 Cavalry.
		Fennell, William George		16 Feb. 17*	16 Jan. 18	16 Feb. 18	„ 94 Infantry.
	L'pool R.	**Cangley**, Frederick George, D.S.O.		23 Feb. 17*	13 Aug. 18	23 Feb. 18	„ 89 Punjabis.
	York. L. I.	**Clews**, Lionel Charles		26 Feb. 17*	11 Apl. 18	26 Feb. 18	„ 32 Lancers.
		Robinson, Harold		2 Mar. 17*	5 Dec. 17	2 Mar. 18	„ 2-6th Gurkha Rifles.
1045	L'pool R.	**Lunt**, Ralph Marshall		2 Mar. 17*	30 Sep. 18	2 Mar. 18	„ 87 Punjabis.
	Aus. Imp. Force.	**Wagstaffe**, Hector		9 Mar. 17*	5 Feb. 18	9 Mar. 18	„ 50 Rifles.
		Witts, Cecil Arthur		10 Mar. 17*	17 Sept. 17	10 Mar. 18	„ 2 Raj. L. I.
		Carr, William Robert		13 Mar. 17*	3 Nov. 17	13 Mar. 18	„ 16 Cavalry.
	Lond. R.	**Field**, Reginald Thomas		15 Mar. 17*	26 Mar. 18	15 Mar. 18	„ 112 Infantry.
1050		**Watters**, George Herbert		16 Mar. 17	9 Jan. 18	16 Mar. 18	„ 96 Infantry.
	R. F. A.	**Gerrard**, Henry James Frederick		17 Mar. 17*	5 Sep. 18	17 Mar. 18	„ 21 Cavalry.
	Lond. R.	**Crawley**, Norman Leslie		22 Mar. 17*	26 Mar. 18	22 Mar. 18	„ 30 Punjabis.
		Davis, Ralph		23 Mar. 17*	26 Feb. 18	23 Mar. 18	„ 1-9th Gurkha Rifles.
	R. Berk. R.	**Rolinson**, Wilfred James		27 Mar. 17*	27 Aug. 18	27 Mar. 18	„ 90 Punjabis.
1055		**Baxter**, Elliott Esmond		5 Apl. 17*	8 Feb. 18	5 Apl. 18	„ 107th Pioneers.
		Keenan, Felix William		7 Apl. 17*	28 July 17	7 Apl. 18	„ 72 Punjabis.
		Shearman, Walter West. 207a		7 Apl. 17*	7 May 17	7 Apl. 18	„ 3-8th Gurkha Rifles.
	R. Ir. Rif.	**Gordon**, John Patrick		7 Apl. 17*	1 Oct. 17	7 Apl. 18	„ 3-2nd Gurkha Rifles.
	Dorset R.	**Butler**, Charles Theophilus		7 Apl. 17*	12 May 18	7 Apl. 18	„ 119 Infantry.
1060	Cam'n Highlanders.	**Murray**, Vivian Royston		7 Apl. 17*	27 Nov. 17	7 Apl. 18	„ 2-2nd Gurkha Rifles.
	Notts. and Derby. R.	**White**, George Frederick, M.C.		7 Apl. 17*	30 June 18	7 Apl. 18	„ 13 Lancers.
	Dorset R.	**Doherty**, Patrick Joseph		7 Apl. 17*	30 June 18	7 Apl. 18	„ 19 Punjabis.
	R. War. R.	**Palmer**, William John		7 Apl. 17*	7 June 18	7 Apl. 18	„ 3-8th Gurkha Rifles.
	Scot. Rif.	**Woodford**, Archibald Stanley		7 Apl. 17*	26 July 18	7 Apl. 18	„ 110 Mah. Lt. Infy.

1065	R. Ir. Rif.	**Pemberton,** Joseph Desmond		7 Apl. 17*	28 Aug. 18	7 Apl. 18	,,	28 Punjabis.
	R. W. Fus.	**Fellix-Williams,** Arthur		7 Apl. 17*	10 Oct. 18	7 Apl. 18	,,	55 Rifles.
	A. S. C.	**MacCulloch,** Hugh Shedden		8 Apl. 17*	29 July 18	8 Apl. 18	,,	10 Lancers.
		Flitcroft, Ernest Albert Seth		9 Apl. 17*	12 Jan. 18	9 Apl. 18	,,	32 Pioneers.
		Haylock, Stanley William (46)		11 Apl. 17*	25 Nov. 17	11 Apl. 18	,,	82 Punjabis.
1070		**Cooke,** Archibald Barton . . . 207a		11 Apl. 17*	12 Jan. 18	11 Apl. 18	,,	119 Infantry.
	Bord R.	**Borwick,** Robert Hartley		11 Apl. 17*	4 July 18	11 Apl. 18	,,	43 Regiment.
	Lond. R.	**Evans,** Douglas William		11 Apl. 17*	3 Nov. 18	11 Apl. 18	,,	37 Dogras.
		Beatson, John		15 Apl. 17*	23 Nov. 17	15 Apl. 18	,,	13 Rajputs.
		Scott, Morrison		17 Apl. 17*	8 Feb. 18	17 Apl. 18	,,	16 Cavalry.
1075	Cam'n Highrs.	**Souter,** William Dewar		18 Apl. 17*	30 Sep. 18	18 Apl. 18	,,	44 Infantry.
	A. S. C.	**Bishop,** Eric Cuthbert Gabriel		24 Apl. 17*	29 June 18	24 Apl. 18	,,	5 Cavalry.
	1 Dgn. Gds.	**Norris,** James William		24 Apl. 17*	1 Aug. 18	24 Apl. 18	,,	14 Lancers.
		Straker, William Rider		29 Apl. 17*	8 Feb. 18	29 Apl. 18	,,	28 Cavalry.
	Notts. Yeo.	**Williams,** Godfrey Edgar Mostyn		29 Apl. 17*	14 Apl. 18	29 Apl. 18	,,	31 Lancers.
1080		**Corke,** Francis Sinclair		1 May 17*	21 Nov. 18	1 May 18	,,	30 Punjabis.
		Brookman, John (46)		1 May 17	3 Dec. 17	1 May 18	,,	1-1st Gurkha Rifles.
		Hill, Frank Reginald		1 May 17	12 Jan. 18	1 May 18	,,	7 Lancers.
	R. Lanc. R.	**Shelley,** Percy		1 May 17	25 Mar. 18	1 May 18	,,	6 Jat L. I.
	R. Dub. Fus.	**Murphy,** John James Christopher		1 May 17	18 May 18	1 May 18	,,	98 Infantry.
1085	K. R. Rif. C	**Seeber,** Harold Charles Rundle		1 May 17	13 May 18	1 May 18	,,	1-2nd Gurkha Rifles.
	York R.	**Hannam,** Herbert Lovel	28 Feb. 98	1 May 17	1 June 18	1 May 18	,,	66 Punjabis.
	R. Innis. Fus.	**Lindsay,** Henry Norman		4 May 17*	12 Oct. 18	4 May 18	,,	59 Rifles.
		Cremer, George Oliver Rowland		5 May 17*	29 July 18	5 May 18	,,	27 Cavalry.
		Moreton, Ernest Albert (51, 59)		5 May 17*	12 Jan. 18	5 May 18	,,	41 Dogras.
1090		**Carter,** Tom Lawrence		5 May 17*	16 Jan. 18	5 May 18	,,	8 Cavalry.
		Cumming, John Fleming		5 May 17*	22 Jan. 18	5 May 18	,,	18 Infantry.
		Dawson, Thomas		5 May 17*	24 Jan. 18	5 May 18	,,	34 Pioneers.
		Finlayson, Thomas		5 May 17*	23 Jan. 18	5 May 18	,,	35 Sikhs.
		Weir, George		5 May 17*	19 Jan. 18	5 May 18	,,	85 Rifles.
1095		**Mallet,** William Harrison		5 May 17*	9 Jan. 18	5 May 18	,,	2-2nd Gurkha Rifles.
		Furminger, Horace John		5 May 17*	8 Feb. 18	5 May 18	,,	56 Rifles.
		White, Luke William Newbold		5 May 17*	22 Nov. 17	5 May 18	,,	113 Infantry.
		Strachan, George Ross		5 May 17*	26 Mar. 18	5 May 18	,,	129 Baluchis.
		Norrish, Frank		5 May 17*	25 Mar. 18	5 May 18	,,	98 Infantry.
1100		**Forrest,** James Dickson Douglas		5 May 17*	26 Mar. 18	5 May 18	,,	6 Jat L. I.
		Holmes, Ernest (46)		5 May 17*	26 Mar. 18	5 May 18	,,	70 Rifles.
	R. Ir. Fus.	**Clark,** William James		5 May 17*	27 Apl. 18	5 May 18	,,	95 Infantry.
	S. Lanc. R.	**Neale,** Ernest Albert Charles		5 May 17*	14 Apl. 18	5 May 18	,,	3-39th Garh. Rifles.
	High. L. I.	**Walker,** Charles Augustus Percival		5 May 17*	13 May 18	5 May 18	,,	44 Infantry.
1105	Aus. Imp. Force.	**Pinches,** Arthur George		5 May 17*	14 Apl. 18	5 May 18	,,	3-5th Gurkha Rifles.
	Lov. Scouts	**Colborne,** Thomas Louis Desvages		5 May 17*	1 Oct. 17	5 May 18	,,	3-9th Gurkha Rifles.
	Notts. & Derby R.	**Lilley,** William Galpin		5 May 17*	30 June 18	5 May 18	,,	4 Rajputs.
	Lond. R.	**Henwood,** Fredrick Henry		5 May 17*	31 Aug. 18	5 May 18	,,	47 Punjabis.
	M. G. Corps.	**Pill,** Gordon Cecil Robert		5 May 17*	31 Aug. 18	5 May 18	,,	120 Infantry.
1110	Arg. & Suth'd. Highrs.	**Dron,** Charles William		5 May 17*	16 Oct. 18	5 May 18	,,	10 Jats.

* 9 months forfeited for seniority under conditions of I. A. O. 511, dated 14th May 1917.

No.	Former Corps.	Names.	Date of birth.	Date of first commission.	Date of appointment to Indian Army.	Date of rank.	Remarks.
		Lieutenants—*contd.*					
	N. Lanc. R.	**Hamilton,** Douglas Stewart		5 May 17*	30 Sep. 18	5 May 18	Attached 150 Infantry,
	Aus. Imp. Force.	**Snelling,** Charles		5 May 17*	6 Sep. 18	5 May 18	,, 81 Pioneers.
		Topham, Sidney 207*a*		6 May 17*	6 May 17	6 May 18	,, 15 Sikhs.
		Relief, Paul Johannes		6 May 17*	12 Jan. 18	6 May 18	,, 76 Punjabis.
	North'd Fus.	**Noble,** George		9 May 17*	11 Apl. 18	9 May 18	,, 33 Cavalry.
1115	Aus. Imp. Force.	**Gamble,** Walter Morris Felix (*temp. Capt. 26 Oct. 18*)		9 May 17*	25 Mar. 18	9 May 18	,, 86 Infantry.
	Aus. Imp. Force.	**Lundy,** Clarence Robeat Collie		9 May 17	11 Apl. 18	9 May 18	,, 12 Pioneers.
	Suff. R.	**Knights,** Robert		10 May 17*	9 Dec. 17	10 May 18	,, 2-8th Gurkha Rifles.
	Aus. Imp. Force.	**White,** Harold Stephen		16 May 17*	25 Mar. 18	16 May 18	,, 76 Punjabis.
		Mathews, Geoffrey Lister		14 May 17*	23 Sep. 17	14 May 18	,, 88 Infantry.
1120	North'd Fus.	**Finlayson,** Ronald Edward Maclennan		14 May 17*	27 Aug. 18	14 May 18	,, 38 Dogras
	Glouc. R.	**Tucker,** John Darley		16 May 17*	25 Feb. 18	16 May 18	,, 70 Burma Rifles.
		Delamain, John Herbert		19 May 17*	12 June 17	19 May 18	,, 106 Pioneers.
		Munckton, Frank Ernest		20 May 17*	25 Mar. 18	20 May 18	,, 7 Lancers.
	Aus Imp. Force.	**Lodge,** Edmund Roy		1 June 17*	30 Sep. 18	1 June 18	,, 80 Infantry.
1125		**Skinner,** Frank Hollamby		5 June 17*	26 Sep. 17	5 June 18	,, 12 Pioneers.
		Coventon, John Adams (51) 207*a*		5 June 17*	2 Aug. 17	5 June 18	,, 8 Cavalry.
		Clarke, John Arthur (46, 71)		5 June 17*	29 July 17	5 June 18	,, 66 Punjabis.
		Burnell, Reginald George 207*a*		5 June 17*	12 Jan. 18	5 June 18	,, 29 Punjabis.
		Webster, Harold Pearson		5 June 17*	8 Feb. 18	5 June 18	,, 32 Pioneers.
1130		**Moorcroft,** Richard		5 June 17*	8 Feb. 18	5 June 18	,, 57 Rifles.
		Meadmore, Thomas Bennett Robert		5 June 17*	25 Mar. 18	5 June 18	,, 67 Punjabis.
		Wilson, William Lee		5 June 17*	25 Mar. 18	5 June 18	,, 40 Pathans.
		Shaw, Leonard Bilton		5 June 17*	25 Mar. 18	5 June 18	,, 6 Jat L. I.
	R. Ir. Regt.	**Turner,** William Tyrie		5 June 17*	27 Apl. 18	5 June 18	,, 94 Infantry.
1135	E. York. R.	**Gregory,** Thomas		5 June 17*	13 May 18	5 June 18	,, 9 Bhopal Infantry.
	Devon R.	**Jones,** Leslie William		5 June 17*	12 May 18	5 June 18	,, 33 Punjabis.
	R. Ir. Fus.	**Collins,** Frederick Charles		5 June 17*	24 Dec. 17	5 June 18	,, 91 Punjabis.
	Sea. Highrs.	**Spence,** James, M.C.		5 June 17*	4 July 18	5 June 18	,, 89 Punjabis.
	W. York. R.	**Wilton,** Vincent Raine		5 June 17*	13 Aug. 18	5 June 18	,, 43 Regiment.
1140	M. G. Corps.	**Pilcher,** Vernon Doyle		5 June 17*	13 Aug. 18	5 June 18	,, 1st Gurkha Rifles.
	Sea. Highrs.	**Cooper,** Leslie Gordon		5 June 17*	6 Sep. 18	5 June 18	,, 105 Mah L. I.
	L. Lanc. R.	**Jewell,** Edward Collins		5 June 17*	19 Sep. 18	5 June 18	,, 46 Punjabis.
	Devon R.	**Dening,** Samuel John Claude		5 June 17*	3 Nov. 18	5 June 18	,, 23 Pioneers.
	R. Ir. Rif.	**Kelly,** Michael Gabriel		8 June 17	12 May 18	8 June 18	,, 98 Infantry.
1145	Aus. Imp. Force.	**Howlett,** Alexander Harold James		15 June 17*	4 July 18	15 June 18	,, 48 Pioneers.
	Wilt. R.	**Illingworth,** John Richardson		19 June 17*	6 Sep. 18	19 June 18	,, 125 Rifles.
	Welsh. R.	**Bowen,** George Lloyd	16 June 93	19 June 17*	4 Sep. 18	19 June 18	,, 3 Brahmans.
	A. S. C.	**Shipman,** Harry Leonard		23 June 17*	6 Oct. 17	23 June 18	,, 124 Infantry.

	Lan. Fus.	*V.C.*, **Grimshaw**, John Elisha		13 Feb. 93	24 June 17*	4 Sep. 18	24 June 18	,,	105 Mah. Lt. Infy.
1150		**Constable**, Wilfred Nevile Brown			26 June 17*	25 July 17	26 June 18	,,	94 Infantry.
		Ballard, Victor	207a		26 June 17*	3 Nov. 17	26 June 18	,,	20 Infantry.
		Watson, Percy Charles (48a, 51)			26 June 17*	22 Jan 18	26 June 18	,,	98 Infantry.
		Davis, Douglas Arthur			26 June 17*	4 Mar. 18	26 June 18	,,	70 Rifles
		Harper, Reginald Kirman			26 June 17*	25 Feb. 18	26 June 18	,,	7 Lancers.
1155		**Smith**, Frederick Miller			26 June 17*	25 Mar. 18	26 June 18	,,	70 Rifles.
		Steele, Lindsay			26 June 17*	26 Mar. 18	26 June 18	,,	37 Lancers.
	Som. L. I.	**Battle**, John William			26 June 17*	14 Apl. 18	26 June 18	,,	71 Punjabis.
	Leic. R.	**Nunn**, Howard			26 June 17*	27 Apl. 18	26 June 18	,,	31 Punjabis.
	Sea Highrs.	**King**, Lawrence Charles, M.C.			26 June 17*	27 Apl. 18	26 June 18	,,	27 Punjabis.
1160	Som. L. I.	**Bleathman**, William Edward Percy			26 June 17*	3 Jan. 18	26 June 18	,,	70 Rifles.
	Lan. Fus.	**Westoby**, Charles Brewerton			26 June 17*	8 July 18	26 June 18	,,	36 Sikhs.
	Manch. R.	**Wayte**, Victor William			26 June 17*	22 June 18	26 June 18	,,	86 Infantry.
	R. Dub. Fus.	**Mathews**, Alfred Henry			26 June 17*	13 Aug. 18	26 June 18	,,	28 Punjabis.
	N. Lanc. R.	**Leggat**, Wilson			26 June 17*	31 Sep. 18	26 June 18	,,	150 Infantry.
1165	R. W. Fus.	**Baker**, Reginald Augustus Charles		8 Feb. 94	27 June 17	31 Aug. 18	27 June 18	,,	8 Rajputs.
	R. W. Fus.	**Norris**, William John		11 Aug. 94	27 June 17	4 Sep. 18	27 June 18	,,	41 Dogras.
	S. Afr. Infy.	**Carstens**, John Eustace			6 July 17*	29 Aug. 18	6 July 18	,,	3 Gurkha Rifles.
	R. G. A.	**Thomas**, Frederick Gordon Sherriff			12 July 17*	26 June 18	12 July 18	,,	97 Infantry.
	R. Fus.	**Tomkins**, Charles Jocelyn Deane			25 July 17*	24 May 18	25 July 18	,,	38 Dogras.
1170		**Ogden**, Howard Leonard (*temp Capt., 23 Nov. 18*)			25 July 17*	22 Jan. 18	25 July 18	,,	7 Lancers.
		Austin, Lawrence Augustus, D.C.M.			25 July 17*	22 Nov. 17	25 July 18	,,	120th Delhi Infantry.
		Adams, Hugh Robert			25 July 17*	9 Feb. 18	25 July 18	,,	7[illegible] Punjabis.
		Hartley, Clifford Reginald Bernard			25 July 17*	24 Jan. 18	25 July 18	,,	41 Dogras.
		Martyr, Harold			25 July 17*	26 Feb. 18	25 July 18	,,	1-9th Gurkha Rifles.
1175		**Averill**, John Eric (46)			25 July 17*	8 Feb. 18	25 July 18	,,	1-10th Gurkha Rifles.
		Parke, Arthur Henry			25 July 17*	2 Mar. 18	25 July 18	,,	33 Cavalry.
		Butler, Frederick Harvey			25 July 17*	25 Mar. 18	25 July 18	,,	57 Rifles.
		Green, Percy Arnold			25 July 17*	25 Mar. 18	25 July 18	,,	Guides.
	North'n R.	**Merriman**, Sydney Thomas			25 July 17*	11 Apl. 18	25 July 18	,,	76 Punjabis.
1180	W. York, R	**Jermyn**, Keppel			25 July 17*	27 Apl. 18	25 July 18	,,	31 Punjabis.
	D. of Corn. L. I.	**Dodd**, Arthur Vernon			25 July 17*	27 Apl. 18	25 July 18	,,	106 Pioneers.
	Scot. Horse	**McDougall**, Andrew			25 July 17*	14 Apl. 18	25 July 18	,,	30 Lancers.
	Lord. R.	**Moriarty**, Dennis			25 July 17*	18 May 18	25 July 18	,,	27 Punjabis.
	Ches. R.	**Seal**, Reginald Charles (48a, 51)			25 July 17*	1 Oct. 17	25 July 18	,,	124 Infantry.
1185	M. G. Corps	**Gregson**, Harry			25 July 17*	1 June 18	25 July 18	,,	12[illegible] Infantry.
	Manch R.	**Sutton**, Jack Cyril			25 July 17*	1 June 18	25 July 18	,,	8[illegible] Punjabis.
	M. G. Corps	**Roberts**, Gwylym Henry			25 July 17*	1 Oct. 17	25 July 18	,,	111 Mahars.
	M. G. Corps	**Laidler**, Thomas			25 July 17*	30 June 18	25 July 18	,,	126 Infantry.
	L'pool R.	**Lee**, Wilfred Millson			25 July 17*	30 June 18	25 July 18	,,	10 Jats.
1190	M. G. Corps	**Sutcliffe**, James Edward			25 July 17*	31 Aug. 18	25 July 18	,,	96 Infantry.
	W. Rid. R.	**Selway**, Herbert			25 July 17*	4 Sep. 18	25 July 18	,,	7 Rajputs.
	N. Staff. R.	**Oswell**, Gordon Cecil			25 July 17*	6 Sep. 18	25 July 18	,,	50 Kumaon Rifles.
	R. Berk. R.	**Crampin**, Ronald William			25 July 17*	30 Sep. 18	25 July 18	,,	150 Infantry.
		Groocock, George Hughes			1 Aug. 17*	25 Mar. 18	1 Aug. 18	,,	76 Punjabis.
1195	Aus. Imp. Force.	**Hutton**, Ernest Daniel			1 Aug. 17*	26 Mar. 18	1 Aug. 18	,,	33 Cavalry.
1196	Aus. Imp. Force.	**Seager**, John Eden Baldwin			1 Aug. 17*	19 Apl. 18	1 Aug. 18	,,	12 Pioneers.

* 9 months forfeited for seniority under conditions of I. A. O. 511, dated 14th May 1917.

No.	Former Corps.	Names.	Date of birth.	Date of first commission.	Date of appointment to Indian Army.	Date of Rank.	Remarks.
		Lieutenants—*contd.*					
1197	S.Afr.Force.	**Ashworth**, Harold Gwyn	. . . .	7 Aug. 17*	26 June 18	7 Aug. 18	Attached 119 Infantry.
	S.Afr. Force.	**Kirkpatrick**, Athol Roy	. . . .	7 Aug. 17*	29 July 18	7 Aug. 18	,, 3 Gurkha Rifles.
		Thurgood, William James	. . . .	12 Aug. 17*	25 Mar. 18	12 Aug. 18	,, 3-9th Delhi Regt.
1200		**Pertwee**, Lionel	. . . .	13 Aug. 17*	5 Dec. 17	13 Aug. 18	,, 3-6th Gurkha Rifles.
	R. Innis Fus.	**Whiting**, Walter Cecil	. . . .	13 Aug. 17*	20 Aug. 17	13 Aug. 18	,, 3-1st Gurkha Rifles.
	Lan. Fus.	**McVay**, Stephen	. . . .	14 Aug. 17*	4 Oct. 17	14 Aug. 18	,, 3-6th Gurkha Rifles.
		West, Edwin	. . . .	22 Aug. 17*	23 Nov. 17	22 Aug. 18	,, 47 Sikhs.
		Barden, Edward Albert (57)	. . . .	22 Aug. 17*	25 Nov. 17	22 Aug. 18	,, 66 Punjabis.
1205		**Lees**, Leslie	. . . .	22 Aug 17*	25 Nov. 17	22 Aug. 18	,, 84 Punjabis.
		Sallal, Charles Greenslade	. . . .	22 Aug. 17*	8 Feb. 18	22 Aug. 18	,, 7 Lancers.
	Leic. R.	**Connell**, Justin Harold	. . . .	22 Aug. 17	25 Mar. 18	22 Aug. 18	,, 57 Rifles.
	Lond. Yeo.	**Taylor**, George Frederick	. . . .	22 Aug. 17*	27 Apl. 18	22 Aug. 18	,, 31 Lancers.
	Lond. Yeo.	**Fleming**, Francis	. . . .	22 Aug. 17*	27 Apl. 18	22 Aug. 18	,, 31 Lancers.
1210	R. Dub. Fus.	**Aue**, Werner Charles Rudolph	. . . .	22 Aug. 17*	29 Apl. 18	22 Aug. 18	,, 70 Rifles.
	Rif. Bde.	**Day**, Edward Victor Grace (46)	. . . .	22 Aug. 17*	23 Jan. 18	22 Aug. 18	,, 2-10 Gurkha Rifles.
	Dorset. R.	**Worden**, Geoffrey Mawdsley	. . . .	22 Aug. 17*	12 May 18	22 Aug. 18	,, 45 Sikhs.
	R. Ir. R.	**Lewis**, James Albert, M.C.	. . . .	22 Aug. 17*	25 Sep. 17	22 Aug. 18	,, 110 Infantry.
	E. Kent. R.	**Harman**, Brian Relton	. . . .	22 Aug. 17*	5 Oct. 17	22 Aug. 18	,, 110 Mah. L. Infy.
1215	R. Ir. Fus.	**Beaty**, Arthur Patrick	. . . .	22 Aug. 17*	10 Oct. 18	22 Aug. 18	,, 53 Sikhs.
		Lane, Vivian Charles Thomas (57)	. . . .	23 Aug. 17*	22 Jan. 18	23 Aug. 18	,, 33 Cavalry.
		Macintyre, John Taylor	. . . .	23 Aug. 17*	17 Sep. 17	23 Aug. 18	,, 43 Regiment.
		Raynbird, Herbert	. . . .	23 Aug. 17*	25 Mar. 18	23 Aug. 18	,, 117 Mahrattas.
		Graham, Frederick	. . . .	23 Aug. 17*	25 Mar. 18	23 Aug. 18	,, 33 Punjabis.
1220	High.L. I.	**Brown**, Frederick Logan	. . . .	23 Aug. 17*	5 Oct. 17	23 Aug. 18	,, 112 Infantry.
	N. Lanc R.	**Seed**, Richard Burton	. . . .	23 Aug. 17*	30 June 18	23 Aug. 18	,, 73 Infantr
	Glouc. R.	**Chivers**, Norman Melville, M.C.	. . . .	23 Aug. 17*	3 Nov. 18	23 Aug. 18	,, 80 Infantry.
	R. Berks. R.	**Murray**, Peter	. . . .	23 Aug. 17*	3 Nov. 18	23 Aug. 18	,, 25 Punjabis.
		Falconer, George Arthur Leslie	. . . .	28 Aug. 17*	28 Aug. 17	28 Aug. 18	,, 28 Cavalry.
1225		**Bullmore**, Francis Edward	. . . .	28 Aug. 17*	27 Jan. 18	28 Aug. 18	,, 27 Cavalry.
		Moorhouse, Eric Godfrey	. . . .	28 Aug. 17*	26 Feb. 18	28 Aug. 18	,, 30 Lancers.
	Tank Corps.	**Pritchard**, Thomas Archer	. . . .	10 Sep. 17*	31 Aug. 18	10 Sep. 18	,, 32 Pioneers.
	Tank Corps.	**Pitt**, Geoffrey	. . . .	10 Sep. 17*	6 Sep. 18	10 Sep. 17	,, 72 Punjabis.
	E. York. R.	**Penn**, Claude Leslie	. . . .	11 Sep. 17*	13 May 18	11 Sep. 18	,, 128 Pioneers.
1230	Manch. R.	**Rudolph**, William	14 Feb. 97	12 Sep. 17	26 Mar. 18	12 Sep. 18	,, 56 Rifles.
	York. R.	**Shannon**, Edmund Cecil Boyd	11 Sep. 98	12 Sep. 17	27 Apl. 18	12 Sep. 18	,, 15 Lancers.
	Wilts. R.	**Cockburn**, Robert Claude Radcliffe	12 Jan. 99	12 Sep. 17*	30 June 18	12 Sep. 18	,, 36 Sikhs.
	Midd'x. R.	**Wilson**, Ronald Eric	. . . .	12 Sep. 17	11 Oct. 18	12 Sep. 18	,, 59 Rifles.
		Rennie, William Ledingham	. . . .	19 Sep. 17*	16 Sep. 17	19 Sep. 18	,, 89 Punjabis.
		Reilly, Thomas Joseph	. . . .	19 Sep. 17*	8 Feb. 18	19 Sep. 18	,, 67 Punjabis.
1235	W. York. R.	**Bradshaw-Smith**, Brian	. . . .	19 Sep. 17*	11 Apl. 18	19 Sep. 18	,, 3-2nd Gurkha Rifles.
	Bord. R.	**Haswell**, Francis William	. . . .	19 Sep. 17*	18 May 18	19 Sep. 18	,, 107 Pioneers.
	R. Dub. Fus.	**McClelland**, William	. . . .	19 Sep. 17*	13 May 18	19 Sep. 18	,, 99 Infantry.
		McCarthy, John William	. . . .	19 Sep. 17*	27 Apl. 18	19 Sep. 18	,, 44 Infantry.

	R. Ir. Fus.	**McCall,** Robert George		19 Sep. 17*	29 July 18	19 Sep. 18	,,	7 Rajputs.
1240	Lond. R.	**Arnold,** Leonard Frank Cecil		19 Sep. 17*	13 Aug. 18	19 Sep. 18	,,	82 Punjabis.
	R. Innis. Fus.	**Paterson,** George Frederick		29 Sep. 17*	29 July 18	19 Sep. 18	,,	5 Gurkha Rifles.
	Scot. Rif.	**Purvis,** Robert Graham		19 Sep. 17*	10 Oct. 18	19 Sep. 18	,,	10 Jats.
	Manch. R.	**Clarke,** John William Alphonsus		19 Sep. 17*	30 Sep. 18	19 Sep. 18	,,	21 Punjabis.
1245	R. W. Fus.	**Storey-Cooper,** Edward Samuel	207*a*	19 Sep. 17*	31 May 18	19 Sep. 18	,,	6 Jat. Lt. Infy.
	R. Dub. Fus.	**Anderson,** Robert Cecil		19 Sep. 17*	25 Feb. 18	19 Sep. 18	,,	98 Infantry.
	Manch R.	**Leonard,** Herbert William		21 Sep. 17*	25 Mar. 18	21 Sep. 18	,,	30 Lancers.
	5 Dgn. Gds.	**Peate,** Watkin Earnest		21 Sep. 17*	18 May 18	21 Sep. 18	,,	27 Cavalry.
	5 Dgn. Gds.	**Salt,** Alfred Cyril Josiah		21 Sep. 17*	4 July 18	21 Sep. 18	,,	12 Cavalry.
1250	E. Kent. R.	**Northfield,** Edward Walter		26 Sep. 17	13 May 18	26 Sep. 18	,,	26 Cavalry.
	Welsh, Gds.	**Baness,** Horace Edward		26 Sep. 17*	29 Oct. 18	26 Sep. 18	,,	1 Gurkha Rifles.
	York R.	**Thornton,** Harry Eden, M.C.		28 Sep. 17*	6 Sep. 18	28 Sep. 17	,,	96 Infantry.
	R. Dub. Fus.	**Hill,** Charles, M.C.		1 Oct. 17*	11 Apl. 18	1 Oct. 18	,,	89 Punjabis.
	Suff. R.	**Robins,** Richard Terence, M.C.		9 Oct. 17*	1 June 18	9 Oct. 18	,,	6 Jat Lt. Infy.
1255	Aus. Imp. Force.	**Kelly,** Wallscourt Steen		17 Oct. 17*	25 Mar. 18	17 Oct. 18	,,	94 Infantry.
		Hawthorn, Douglas Cyril		25 Oct. 17*	8 Feb. 18	25 Oct. 18	,,	3 Horse.
	Corn. L. I.	**Piper,** Albert Benjamin Charles		25 Oct. 17*	1 June 18	25 Oct. 18	,,	120 Infantry.
	Mon. R.	**Smith,** Donald Frank		25 Oct. 17*	25 Feb. 18	25 Oct. 18	,,	50 Rifles.
1260	Wilt. R.	**Gosden,** William Ivon		25 Oct. 17*	4 July 18	25 Oct. 18	,,	102 Grenadiers.
	Welsh R.	**Lewis-Lewis,** Sydney		25 Oct. 17*	6 Sep. 18	25 Oct. 18	,,	128 Pioneers.
	N. Lanc. R.	**Scholes,** William		25 Oct. 17*	31 Aug. 18	25 Oct. 18	,,	106 Pioneers.
	R. Fus.	**Adams,** Alfred Edward St. John		25 Oct. 17*	30 Sep. 18	25 Oct. 18	,,	3 Gurkha Rifles.
	Lan. Fus.	**Greeves,** Stuart		25 Oct. 17*	29 Oct. 18	25 Oct. 18	,,	152 Punjabis.
	Tank Corps.	**Mackenzie** Fleming		26 Oct. 17*	18 May 18	26 Oct. 18	,,	30 Punjabis.
1265	High. L. I.	**Stone,** Thomas William		27 Oct. 17*	9 June 18	27 Oct. 18	,,	107 Pioneers.
		Clapp, John Henry		28 Oct. 17*	5 Dec. 17	28 Oct. 18	,,	2 Lancers.
	E. York. R.	**Parks,** Herbert James		28 Oct. 17	9 Jan. 18	28 Oct. 18	,,	121 Pioneers.
	Tank Corps.	**Stevens,** Daniel Thomas		31 Oct. 17*	13 May 18	31 Oct. 18	,,	31 Lancers.
	R. W. Fus.	**Coy,** Francis Caleb		11 Nov. 17*	3 Dec. 17	11 Nov. 18	,,	63 Infantry.
	North'd Fus.	**McCluskey,** Francis John		13 Nov. 17*	30 June 18	13 Nov. 18	,,	18 Infantry.
1270		**Kerr,** Horatio Nelson		17 Nov. 17*	22 Jan. 18	17 Nov. 18	,,	16 Cavalry.
		Mee, Edwin Arthur		17 Nov. 17*	8 Feb. 18	17 Nov. 18	,,	8 Cavalry.
	Buck. Yeo.	**Chapman,** Arthur Reginald		17 Nov. 17*	25 Oct. 18	17 Nov. 18	,,	3 Horse.
	R. Innis. Fus.	**Boyce,** Thomas Walker		20 Nov. 17*	12 Oct. 18	20 Nov. 18	,,	111 Mahars.
	Glouc R.	**Trotman,** Ewart Washington		26 Nov. 17*	30 June 18	26 Nov. 18	,,	104 Rifles.
1275		**Peffers,** John		1 Dec. 17*	22 Jan. 18	1 Dec. 18	,,	19 Punjabis.
		Magenty, Ernest Ross		1 Dec. 17*	26 Feb. 18	1 Dec. 18	,,	35 Sikhs.
		Wilton, Christopher Eric Pleydell		1 Dec. 17*	25 Mar. 18	1 Dec. 18	,,	40 Pathans.
	York and Lanc. R.	**Davidson,** Angus		1 Dec. 17*	27 Apl. 18	1 Dec. 18	,,	79 Infantry.
	Dorset. R.	**Pye,** Howard Bartle		1 Dec. 17*	20 Feb. 18	1 Dec. 18	,,	129 Baluchis.
1280	R. Dub. Fus.	**Burke,** Antony Owen Leo		1 Dec. 17*	18 May 18	1 Dec. 18	,,	2-7th Gurkha Rifles.
	E. York. R.	**Eccles,** John Gerald Turton		1 Dec. 17*	1 June 18	1 Dec. 18	,,	1 Brahmans.
	North'd Fus.	**Donald,** Hugh		1 Dec. 17*	30 June 18	1 Dec. 18	,,	80 Infantry.
1283	R. Muns. Fus.	**Baily,** Oswin Hellier		1 Dec. 17*	23 June 18	1 Dec. 18	,,	7 Rajputs.

* 9 months forfeited for seniority under conditions of I. A. O. 511, dated 14th May 1917.

No.	Former Corps.	Nomes.	Date of birth.	Date of fi st commisssion.	Date of appointment to Indian Army.	Date of rank.	Rrks
		Lieutenants—*concld.*					
1284	E. Lanc. R.	**Cowling**, Charles Stuart Cornish		1 Dec. 17*	29 July 18	1 Dec. 18	Attached 7th Gurkha Rifles.
1285	High. L.I.	**Lawson**, John		1 Dec. 17*	5 Sep. 18	1 Dec. 18	" 27 Lt. Cavy.
	Leins. R.	**Butler**, Robert O'Neill		1 Dec. 17*	13 Aug. 18	1 Dec. 18	" 112 Infantry.
	Suff. R.	**Bowden**, Spencer Albert		1 Dec. 17*	30 Sep. 18	1 Dec. 18	" 32 Pioneers.
	Durh. L. I.	**Holmes**, Joseph Andrew		1 Dec. 17*	30 Sep. 18	1 Dec. 18	" 43 Regt.
	R. W. Fus.	**Griffiths**, Creighton Edmund		1 Dec. 17*	3 Nov. 18	1 Dec. 18	" 119 Infantry.
1290	W. I. R.	**Hamer**, Frederick Robert		6 Dec. 17*	6 Sep. 18	6 Dec. 18	" 88 Infantry.
	R. W. Fus.	**McCarthy**, Eugene Daniel		11 Dec. 17*	22 Nov. 17	11 Dec. 18	" 12 Pioneers.
		2nd-Lieutenants.					
	Aus. Imp. Force.	**Anderson**, Gordon		6 June 17*	25 Mar. 18	6 June 17	" 35 Sikhs.
		Cooke, Richard Lindsay		1 Dec. 17*	22 Nov. 17	1 Dec. 17	" 107 Pioneers.
		Neale, Frederick Claud Lawrence		15 Dec. 17*	5 Dec. 17	15 Dec. 17	" 11 Lancers.
5	M. G. Corps.	**Platt**, Howard Chappell		27 Dec. 17*	1 June 18	27 Dec. 17	" 5 Lt. Infy.
		Guinee, James Aloysius		28 Dec. 17*	9 Jan. 18	28 Dec. 17	" 70 Rifles.
		Halloran, George Ballantine		28 Dec. 17*	24 Jan. 18	28 Dec. 17	" 94 Infantry.
		White, Frederick James		28 Dec. 17*	8 Feb. 18	28 Dec. 17	" 2-10 Gurkha Rifles.
	R. Dub. Fus.	**Cranwill**, Guy Butler		28 Dec. 17*	10 Mar. 18	28 Dec. 17	" 6 Jat L. I.
	Sea. Highrs.	**Munro**, Alexander		28 Dec. 17*	12 May 18	28 Dec. 17	" 95 Infantry.
10	R. Fus.	**Lamble**, James		28 Dec. 17*	13 May 18	28 Dec. 17	" 4 Rajputs.
	North'n R.	**Dean**, George Henry		28 Dec. 17*	1 Feb. 18	28 Dec. 17	" 35 Horse.
	K. R. Rif. C.	**Jones**, Oscar Cenric (46)		28 Dec. 17*	20 Feb. 18	28 Dec. 17	" 4-3rd Gurkha Rifles.
	Lond. R.	**Alberga**, Edward Moriee		28 Dec. 17*	26 May 18	28 Dec. 17	" 3-2nd Gurkha Rifles.
	W. York. R.	**Thornon**, Cecil Albert Forbes		28 Dec. 17*	31 May 18	28 Dec. 17	" 99 Infantry.
15	North'd Fus.	**Hall**, Ridley Martin		28 Dec. 17*	30 June 18	28 Dec. 17	" 112 Infantry.
	Lond. R.	**Arnold**, Lawrence Patrick		28 Dec. 17*	30 June 18	28 Dec. 17	" 127 Infantry.
	R. Ir. Fus.	**Powell**, Edward Hawkshaw		28 Dec. 17*	30 June 18	28 Dec. 17	" 84 Punjabis.
	North'd Fus.	**Bowden**, Walter Reginald		28 Dec. 17*	30 June 18	28 Dec. 17	" 1-2nd Gurkha Rifles.
	North'd Fus.	**Dwelley**, William Edwin		28 Dec. 17*	29 July 18	28 Dec. 17	" 62 Punjabis.
20	N. Lanc. R.	**Smith**, William George		28 Dec. 17*	16 July 18	28 Dec. 17	" 99 Infantry.
	York. & Lanc. R.	**Ralphs**, Frederic George		28 Dec. 17*	4 Sep. 18	20 Dec. 17	" 81 Pioneers.
	High. L. I.	**Meikle**, William Allan		20 Dec. 17*	1 June 18	28 Dec. 17	" 55 Rifles.
	Staff R.	**Bree**, Philip George		28 Dec. 17*	23 Oct. 18	28 Dec. 17	" 66 Punjabis.
	Arg. & Suth'd Highrs.	**Dunn**, David		28 Dec. 17*	29 Oct. 18	28 Dec. 17	" 112 Infantry.

25		**Moore**, Gilbert		13 Jan. 18*	25 Mar. 18	13 Jan. 18	„	72 Punjabis.
	Suff. R.	**Cane**, Sydney Phillips		15 Jan. 18*	27 Apl. 18	15 Jan. 18	„	70 Infantry.
		Lightbody, James Gordon		17 Jan. 18*	22 Nov. 17	17 Jan. 18	„	36 Horse.
		McCleery, Henry William Frederick		17 Jan. 18*	22 Jan. 18	17 Jan. 18	„	35 Horse.
		Kelso, Walter Duncan (57)		17 Jan. 18*	22 Jan. 18	17 Jan. 18	„	31 Lancers.
30		**Kidd**, Robert John		26 Jan. 18*	3 Dec. 17	26 Jan. 18	„	1-1st Gurkha Rifles.
		Mackinon, Robert Archibald	207*a*	26 Jan. 18*	12 Jan. 18	26 Jan. 18	„	112 Infantry.
		Kirkhope, John Love		26 Jan. 18*	22 Jan. 18	26 Jan. 18	„	1 Brahmans.
		McKay, Thomas Alexander	(57)	26 Jan. 18*	22 Jan. 18	26 Jan. 18	„	123 Rifles.
	E. York. R.	**Cranswick**, William		26 Jan. 18*	4 Sep. 18	26 Jan. 18	„	8 Rajputs.
35		**Moreland**, James Alexander		26 Jan. 18*	25 Mar. 18	26 Jan. 18	„	8 Cavalry.
		Lea, William		26 Jan. 18*	26 Mar. 18	26 Jan. 18	„	Brahmans.
		Wards, George Thexton		26 Jan. 18*	25 Mar. 18	26 Jan. 18	„	84 Punjabis.
		Arnold, Victor Douglas		26 Jan. 18*	26 Mar. 18	26 Jan. 18	„	6 Jat L. I.
	York. L. I.	**Head**, Henry Reginald		26 Jan. 18*	11 Apl. 18	26 Jan. 18	„	30 Punjabis.
40	M. G. Corps.	**Danby**, John Alexander Macduff		26 Jan. 18*	14 Apl. 18	26 Jan. 18	„	76 Punjabis.
	R. Fus.	**Green**, Arthur William		26 Jan. 18*	13 May 18	26 Jan. 18	„	95 Infantry.
	Lan. Fus.	**Collett**, Norman Temple		26 Jan. 18*	11 Apl. 18	26 Jan. 18	„	35 Sikhs.
	R. Ir. Rif.	**Somerset**, Isaiah		26 Jan. 18*	12 May 18	26 Jan. 18	„	105 Mah. L. I.
	R. Scots	**Macfarlane**, Henry		26 Jan. 18*	4 Oct. 17	26 Jan. 18	„	111 Mahars.
45	R. W. Fus.	**Evans**, Roy Meyrick		26 Jan. 18*	1 June 18	26 Jan. 18	„	61 Pioneers.
	R. W. Surr. R.	**Ellen**, Walter		26 Jan. 18*	1 June 18	26 Jan. 18	„	55 Rifles.
	Sea. Highrs.	**Geddie**, Douglas Walton		26 Jan. 18*	1 July 18	26 Jan. 18	„	102 Grenadiers.
	Gord. Highrs	**Grant**, Arther James		26 Jan. 18*	13 Aug. 18	25 Jan. 18	„	10 Lancers.
	S. Staff. R.	**David**, John		26 Jan. 18*	13 Aug. 18	26 Jan. 18	„	89 Punjabis.
50	L'pool R.	**Pearcey**, Reginald Arthur		26 Jan. 18*	13 Aug. 18	26 Jan. 18	„	62 Punjabis.
	S. Staff. R.	**Law**, George, Henry Major		26 Jan. 18*	31 Aug. 18	26 Jan. 18	„	112 Infantry.
	Lond. R.	**Elmer**, Philip Steggall		26 Jan. 18*	12 Oct. 18	26 Jan. 18	„	50 Infantry.
	Lanc. Regt.	**Snock**, Charles Gorell		26 Jan. 18*	Nov. 18	26 Jan. 18	„	43 Regiment.
		Ednie, Lingard		27 Jan. 18*	2 Jan. 18	27 Jan. 18	„	117 Mahrattas.
55	Breck. Bn.	**Smith**, John Coleman		28 Jan. 18*	23 Mar. 18	28 Jan. 18	„	61 Pioneers.
	Breck. Bn.	**Colsey**, John William Thomas		28 Jan. 18*	2 July 18	28 Jan. 18	„	40 Bengalis.
	Can. Forces.	**Boughton**, Eric James		28 Jan. 18*	29 July 18	28 Jan. 18	„	84 Punjabis.
	Wilt. R.	**Davies**, Wilfred Vaughan		28 Jan. 18*	20 Sep. 18	28 Jan. 18	„	34 Pioneers.
		Mooney, Herbert Francis		1 Feb. 18*	25 Nov. 17	1 Feb. 18	„	12 Cavalry.
60		**Ferguson**, John Mackey		1 Feb. 18*	25 Nov. 17	1 Feb. 18	„	11 Lancers.
	6 Res. Cav.	**Collings**, John Stanley		1 Feb. 18*	27 Apl. 18	1 Feb. 18	„	18 Lancers.
	S. Ir. Horse.	**Aylward**, Thomas Benjamin		1 Feb. 18*	10 Apl. 18	1 Feb. 18	„	39 Central India Horse
	4 Res. R. of Cav.	**Bradshaw**, George Thomas		1 Feb. 18*	26 Sep. 18	1 Feb. 18	„	121 Pioneers.
		Arnold, Ralph Thomas		2 Feb. 18*	5 Dec. 17	2 Feb. 18	„	110 Infantry.
65	M. G. Corps	**Whelton**, John Henry		26 Feb. 18*	27 Aug. 18	26 Feb. 18	„	1 Brahman.
		Bennett, Maurice Chesterton (57)		28 Feb. 18*	25 Nov. 17	28 Feb. 18	„	66 Punjabis.
		French, Cyril Robert		28 Feb. 18*	12 Jan. 18	28 Feb. 18	„	94 Infantry.
	Tank Corps	**McCluskey**, James Joseph Patrick		28 Feb. 18*	11 Apl. 18	28 Feb. 18	„	76 Punjabis.
		Roberts, John Rees		28 Feb. 18*	10 Mar. 18	28 Feb. 18	„	33 Punjabis.
70	Suff. R.	**White**, Thomas Nicholas		28 Feb. 18*	8 Feb. 18	28 Feb. 18	„	35 Sikhs.
71	Oxf. & Buck. R.	**Marigold**, Arthur Joseph		28 Feb. 18*	14 Apl. 18	28 Feb. 18	„	14 Sikhs.

*9 months forfeited for seniority under conditions of **I. A. O.** 511, dated 14th May 1917.

No.	Former Corps.	Names.	Date of birth.	Date of first commission.	Date of appointmen to Indian Army.	Date of rank.	Remarks.
		2nd-Lieutenants—*contd.*					
72	High. L. I.	**Martin,** David Donald (46)		28 Feb. 18*	18 May 18	28 Feb. 18	Attached 3-2nd Gurkha Rifles.
	E. Kent R.	**Prebble,** Charlie Edmund		28 Feb. 18*	18 May 18	28 Feb. 18	„ 17 Infantry.
	York. L. I.	**Burton,** George Herbert		28 Feb. 18*	1 June 18	28 Feb. 18	„ 107 Pioneers.
75	E. Surr. R.	**Whittle,** Horace Edwin		28 Feb. 18*	11 Feb. 18	28 Feb. 18	„ 112 Infantry.
	R. Ir. Fus.	**McGowan,** Thomas Edmond		28 Feb. 18*	14 June 18	28 Feb. 18	„ 36 Sikhs.
	R. W. Kent. R.	**Wright,** Percy		20 Feb. 18*	30 Sep. 18	20 Feb. 18	„ 117 Mahrattas.
	Cam's Highrs.	**Gourlie,** Norman Elliot John		28 Feb. 18*	24 Sep. 18	28 Feb. 18	„ 34 Pioneers.
	R. Fus.	**Toms,** Reginald Gillingham		28 Feb. 18*	30 Sep. 18	28 Feb. 18	„ 39 Rifles.
80	Lond R.	**Weeks,** Horace Augustine		28 Feb. 18*	11 Oct. 18	28 Feb. 18	„ 48 Pioneers.
	R. Scots.	**Pattullo,** Henry Alexander		28 Feb. 18*	3 Nov. 18	28 Feb. 18	„ 6th Jats. L. I.
	L'pool. R.	**Gemmell,** Alexander		28 Feb. 18*	3 Nov. 18	28 Feb. 18	
	R. F. A.	**Tye,** Harry James († *temp. Capt. 18 Nov. 18*)		28 Feb. 19*	4 July 18	28 Feb. 18	
		Autv, Stanley Critchley		23 Mar. 18*	4 July 18	23 Mar. 18	„ 19 Lancers.
85	M. C. Corps	**Elson,** Edgar John		26 Mar. 18*	25 Mar. 18	26 Mar. 18	„ 3-9th Delhi Regt.
		Glenie, Julian William		26 Mar. 18*	30 Sep. 18	26 Mar. 18	„ 111 Mahars.
		Cook, William George		27 Mar. 18*	8 Feb. 18	27 Mar. 18	„ 70 Rifles.
		Westrupp, Percy		27 Mar. 18*	22 Jan. 18	27 Mar. 18	„ 1-6th Gurkha Rifles.
		Trutch, Horace Frank		27 Mar. 18*	6 Dec. 17	27 Mar. 18	„ 3-7th Gurkha Rifles.
90	Dorset. R.	**Hartman,** Vincent John		27 Mar. 18*	25 Mar. 18	27 Mar. 18	„ 40 Pathans.
	Bord. R.	**Marsh,** John James		27 Mar. 18*	8 Feb. 18	27 Mar. 18	„ 3-9th Gurkha Rifles.
	R. Suss. R.	**Cole,** Arthur Hubert		27 Mar. 18*	18 May 18	27 Mar. 18	„ 2-8th Gurkha Rifles.
	Welsh R.	**Hughes,** Thomas		27 Mar. 18*	12 May 18	27 Mar. 18	„ 2-23rd Sikh Infy.
	R. W. Fus.	**Pollicott,** Harold Ernest		27 Mar. 18*	1 June 18	27 Mar. 18	„ 26 Punjabis.
95	R. Innis. Fus	**Reade,** Patrick Valentine Joseph		27 Mar. 18*	1 June 18	27 Mar. 18	„ 124 Infantry.
	Suff. R.	**Pearson,** Cyril Vietor Bolton		27 Mar. 18*	1 June 18	27 Mar. 18	„ 120 Infantry.
	Can. Force.	**Laing,** Horold James Gowans		27 Mar. 18*	30 June 18	27 Mar. 18	„ 1 Lancers.
	S. Staff. R.	**Berrisford,** Ernest Alfred		27 Mar. 18*	30 June 18	27 Mar. 18	„ 117 Mahrattas.
	R. W. Fus.	**Young,** John William Francis		27 Mar. 18*	13 Aug. 18	27 Mar. 18	„ 62 Punjabis.
100	Norf. R.	**Brawn,** John Agutta		27 Mar. 18*	4 Sep. 18	27 Mar. 18	„ 15 Sikhs.
	Glouc. R.	**Dickens,** Harold Edington	13 Sep. 95	27 Mar. 18*	11 Oct. 18	27 Mar. 18	„ 66 Punjabis.
	R. Fus.	**Parry-Jones,** Wynn Hughes		27 Mar. 18*	30 Sep. 18	27 Mar. 18	„ 98 Infantry.
	R. Ir. Regt.	**Kennedy,** Alphonsus Patrick		27 Mar. 18*	25 Feb. 18	27 Mar. 18	„ 9 Gurkha Rifles.
	Hamps. R	**Fisher,** Archibald Robert George		27 Mar. 18*	26 June 18	27 Mar. 18	„ 125 Rifles.
105	Hamps. R.	**Brannon,** Alfred Charles		27 Mar. 18*	3 Nov. 18	27 Mar. 18	„ 47 Sikhs.
	6 Res. R. of Cav.	**Molyneux,** Herbert James		27 Mar. 18*	3 Nov. 18	27 Mar. 18	„ 69 Punjabis.
				29 Mar. 18*	19 Apl. 18	29 Mar. 18	„ 7 Lancers.
	Can. Forces	**Jack,** Herbert Alexander		1 Apl. 18*	26 Feb. 18	1 Apl. 18	„ 30 Punjabis.
	Res. R. Cav.	**Sloper,** Horace Charles		8 Apl. 18*	2 Feb. 18	8 Apl. 18	„ 8 Cavalry.
	War. Yeo.	**Smitn,** Robert Henry		8 Apl. 18*	22 Apl. 18	8 Apl. 18	„ 20 Horse.
110	R. Fus.	**Sale,** Claude Arthur		15 Apl. 18*	30 Oct. 18	15 Apl. 18	„ 8 Gurkha Rifles.

	Aus. Imp. Force.	**Crooke,** Frederick Montague **Warren**		23 Apl. 18*	11 Apl. 18	23 Apl. 18	„	3/1st Gurkha Rifles.
		Brooke, Lionel Hugh		26 Apl. 18*	12 Jan. 18	26 Apl. 18	„	124 Infantry.
	Aus. Imp. Force.	**Keeley,** William Aloysius		28 Apl. 18*	29 July 18	28 Apl. 18	„	3 Brahmans.
	R. E.	**Egan,** Eric Martin		30 Apl. 18*	6 June 18	30 Apl. 18	„	10 Lancers.
115	York and Lanc. R.	**Pilling,** Reginald Westby		1 May 18*	26 Feb. 18	1 May 18	„	16 Cavalry.
	Leic. R.	**Huddleston,** John Henry		1 May 18*	26 Mar. 18	1 May 18	„	2-8th Gurkha Rifles.
	R. Lanc. R.	**Duffy,** Samuel Charles Rhodes		1 May 18*	18 May 18	1 May 18	„	95 Infantry.
	R. W. Kent R.	**Jordan,** Charles		1 May 18*	1 June 18	1 May 18	„	7 Rajputs.
	R. Fus.	**Norman,** Francis Eglesfield Bathurst		1 May 18*	3 Nov. 18	1 May 18	„	2 Rajput L. I.
120	Res. R. Cav.	**English,** Henry Human Ernest		25 May 18*	12 May 18	25 May 18	„	5 Cavalry.
	Res. R. Cav.	**Hawkins,** Maurice		25 May 18*	12 May 18	25 May 18	„	29 Lancers.
	M. G. Corps.	**Ballantyne,** William John		26 May 18*	3 Nov. 18	26 May 18	„	2 Rajput L. I.
	Hamps. R.	**Dolman,** Horace George		28 May 18*	26 Sep. 18	28 May 18	„	95 Infantry.
	Ches. R.	**Pocklington,** John Frederic Jelf		29 May 18*	27 Apl. 18	29 May 18	„	1-2nd Gurkha Rifles.
125	Leic. R.	**King,** Reginald Bertram		29 May 18*	26 Mar. 18	29 May 18	„	89 Punjabis.
	York. R.	**Dyson,** Frank		29 May 18*	†14 Apl. 18	29 May 18	„	27 Cavalry.
	Linc. R.	**Jones,** Ernest Henry		29 May 18*	13 May 18	29 May 18	„	117 Mahrattae.
	Norf. R.	**Shaw,** Alfred Athol St. John		29 May 18*	29 Aug. 18	29 May 18	„	117 Mahrattas.
	Hamps. R.	**Preiss,** Russell Bridgman		29 May 18*	20 Sep. 18	29 May 18	„	18 Infantry.
130	North'n. R.	**Kempster,** Albert Cecil		29 May 18*	30 Sep. 18	29 May 18	„	81 Pioneers.
	Suff. R.	**Edmeades,** Frederick Duncan		29 May 18*	12 Oct. 18	29 May 18	„	69 Punjabis.
	K. O. Sco. Bord.	**Little,** David Duncan		29 May 18*	30 Sep. 18	29 May 18	„	32 Pioneers.
	R. R. Rif. Corps.	**Hebron,** Charles William		29 May 18*	29 Oct. 18	29 May 18	„	129 Baluchis.
	R. G. A.	**Hall,** Harry		2 June 18*	1 June 18	2 June 18	„	107 Pioneers.
135	Can. Force.	**Lowson,** James McEwan		9 June 18*	1 June 13	9 June 18	„	2 Rajputs Lt. Infantry.
	1 Res. R. of Cav.	**Dignam,** Jack Robert		25 June 18*	1 June 18	25 June 18	„	25 Cavalry.
	Glouc. Yeo.	**Radcliffe,** Clifford		25 June 18*	30 June 18	25 June 18	„	26 Cavalry.
	M. G. Corps.	**Latimer,** Robert Anderson		26 June 18*	13 May 18	26 June 18	„	117 Mahrattas.
	L'pool R.	**Nicholson,** George Albert		26 June 18*	18 May 18	26 June 18	„	98 Infantry.
140	Midd'x R.	**Langley,** Leonard Charles		26 June 18*	1 June 18	26 June 18	„	20 Punjabis.
	Lan. Fus.	**Garner,** Edgar Myles		26 June 18*	1 June 18	26 June 18	„	61 Pioneers.
	R. Scots.	**Dow,** George		26 June 18*	5 Oct. 17	26 June 18	„	94 Infantry.
	R. Dub. Fus.	**Peniston-Bird,** Edward **Louis**		26 June 18*	14 Apl. 18	26 June 18	„	39 C. I. Horse.
	Midd'x. R.	**Clay,** Norman Cyril		26 June 18*	16 Sep. 18	26 June 18	„	105 Lt. Infy.
145	R. Highrs.	**Campbell,** James		26 June 18*	30 Sep. 18	26 June 18	„	130 Baluchis.
	R. Scots.	**Johnston,** Albert James Scott		26 June 18*	29 Oct. 18	26 June 18	„	16 Rajputs.
	M. G. Corps.	**Rahles-Rahbula,** Esmond Herbert		26 June 18*	3 Nov. 18	26 June 18	„	12 Pioneers.
	M. G. Corps.	**Creffield,** George		27 July 18*	18 May 18	27 July 18	„	17 Infantry.
	M. G. Corps.	**Bartley,** Emyr Samuel		27 July 18*	18 May 18	27 July 18	„	116 Mahrattas.
150	M. G. Corps	**Beard,** William Horace **Gladstone**		27 July 18*	1 June 18	27 July 18	„	13 Rajputs.
	R. War. R.	**Dobney,** Thomas Arthur		31 July 18*	27 Apl. 18	31 July 18	„	24 Punjabis.
	Som L. I.	**Hill,** Bernard Beaumont		31 July 18*	6 June 18	31 July 18	„	1-3rd Gurkha Rifles.
153	Suff. R.	**Durrant,** George		31 July 18*	4 Sep. 18	31 July 18	„	34 Pioneers.

* 9 months forfeited for seniority under conditions of I. A. O. 511, dated 14th May 1917. † with pay and allowances as for a Lieut.

No.	Former Corps.	Names.	Date of birth.	Date of first commission.	Date of appointment to Indian Army.	Date of rank.	Remarks.
		2nd-Lieutenants—*concld.*					
154	Lond. R.	**Wallers**, Reginald William		31 July 18*	27 Aug. 18	31 July 18	Attached 34 Pioneers.
155	R. Ir. Rif.	**Abbott**, Charles Pryce		31 July 18*	30 Sep. 18	31 July 18	„ 109 Infantry.
	North's Fus.	**Dance**, Ernest		31 July 18*	29 Oct. 18	31 July 18	„ 89 Punjabis.
	R. Ir. Fus.	**Hourihane**, Gerald O'Brien		31 July 18*	29 Oct. 18	31 July 18	„ 71 Punjabis.
	Buck. Yeo.	**Tully**, Eric Johnathan Shand		24 Aug. 18*	4 Sep. 18	24 Aug. 18	„ 44 Cavalry.
	3 Res. R. Cav.	**Jacob**, Henry Hier		24 Aug. 18*	11 Oct. 18	24 Aug. 18	„ 152 Infantry.
160	High. L. I.	**Beveridge**, John Wynford Phillipps		28 Aug. 18*	13 Aug. 18	28 Aug. 18	„ 38 Dogras.
	Essex R.	**Sisley**, Frederick Ward		28 Aug. 18*	30 Sep. 18	28 Aug. 18	„ 86 Infantry.
	High. L. I.	**Steedman**, William McQueen		28 Aug. 18*	12 Oct. 18	28 Aug. 18	„ 47 Sikhs.
	Bord. R.	**Miller**, Andrew Douglas		28 Aug. 18*	3 Nov. 18	28 Aug. 18	„ 86 Infantry.
	R. W. Kent. R.	**Callow**, James Lewis		8 Sep. 18*	27 Aug. 18	8 Sep. 18	„ 40 Cavalry.
165	Glouc. R.	**Butt**, Gilbert Fryer		18 Sep. 18*	4 July 18	18 Sep. 18	„ 11 Lancers.
	High. L. I.	**Forrest**, George		18 Sep. 18*	13 Aug. 18	18 Sep. 18	„ 1 Gurkha Rifles.
	Linc. R.	**Christal**, Frederick Augustus		18 Sep. 18*	4 July 18	18 Sep. 18	„ 74 Punjabis.
	Shrops. L. I.	**Haworth**, Gilbert Mallalieu		18 Sep. 18*	24 Sep. 18	18 Sep. 18	„ 109 Infantry.
	Aus. Imp. F. Force.	**O'Collins**, Patrick Francis		19 Sep. 18*	6 Sep. 18	19 Sep. 18	„ 25 Punjabis.
170	A. S. C.	**Potts**, William		30 Sep. 18*	20 Aug. 18	30 Sep. 18	„ 105 Lt. Infy.
	M. G. Corps.	**Moorhouse**, Frederick Harry		26 Oct. 18*	29 July 18	26 Oct. 18	„ 7 Rajputs.
	M. G. Corps.	**Barrett**, Edwin Dudley Ernest		26 Oct. 18*	29 Aug. 18	26 Oct. 18	„ 17 Infantry.
	Linc. R.	**Guyton**, George		30 Oct. 18*	31 June 18	30 Oct. 18	„ 97 Infantry.
	Shrops. L. I.	**Welch**, John Francis		30 Oct. 18*	13 Aug. 18	30 Oct. 18	„ 42 Regiment.
175	Linc. R.	**Burgess**, Frank Harold, D.C.M.		30 Oct. 18*	6 Sep. 18	30 Oct. 18	„ 102 Grenadiers.
	York. L. I.	**Waterfall**, Johan		30 Oct. 18*	30 Sep. 18	30 Oct. 18	„ 4 Rajputs.
	R. Dub. Fus.	**Kirwan**, James William Kiernan		30 Oct. 18*	28 Oct. 18	30 Oct. 18	„ 110 Infantry.
	3 Res. R. Cav.	**Brockliss**, Arthur Albert		23 Nov. 18*	29 Aug. 18	23 Nov. 18	„ 41 Cavalry.
179	Welsh R.	**Sawle**, John Frazer		27 Nov. 18*	30 Sep. 18	27 Nov. 18	„ 21 Punjabis.

* 9 months forfeited for seniority under conditions of I. A. O. 511, dated 14th May 1917.

INDIAN LAND FORCES.

Names.	Date of birth.	Date of first commission.	Date of rank.	Remarks.
Lieutenants.				
Aga Murtaza Khan, *q.*	30 May 80	1 Sep. 11	1 Dec. 13	
Savanur, Abdul Majid Khan, *Nawab* of	. . . .	24 July 17	24 July 17	Hon. Capt. in Army.
Muhammad Fakhruddin Khan	. . . .	5 Oct. 17	5 Oct. 18	(Tempy.)
2nd-Lieutenants.				
Kashi Nath, Pandit	. . . .	5 Oct. 17	5 Oct. 17	(Tempy.)
Muhammad Jamiluddin	. . . .	5 Oct. 17	5 Oct. 17	(Tempy.)
Maungpe	. . . .	5 Oct. 17	5 Oct. 17	(Tempy.)
Sirdar Jasjit Singh, Ahluwalia		. . . .	. . . .	(Tempy. and Hony.)
Rajindar Singh, *Sirdar Bahadur*		. . . .	. . . .	(Tempy. and Hony.)
Bobbili, *the Hon. Raja Sri* Ravu Venkata Kumara Krishna Ranga, *Rao Bahadur*, Raja of		. . . .	. . . .	(Tempy. and Hony.)
Venkatagiri, Raja of		. . . .	. . . .	(Tempy. and Hony.)
Telprole, Meka Sobhanadri Appa Rao Bahadur Garu, Zamindar of		. . . .	. . . .	(Tempy. and Hony.)
Ramnad, *the Hon. B. Raja* Rajeswara Setupati *alias* Mutturamalinga Setupati Avargal, Raja of		. . . .	. . . .	(Tempy. and Hony.)
Parlakimedi, Sri Krishna Chandra Gajapati Narayana Deo Garu, Zamindar of	. . . .	. . . .	. . . .	(Tempy. and Hony.)
K. K. Raman Kavalappara Muppil Nayar, *the Hon.*		. . . .	. . . .	(Tempy. and Hony.)
R. B. Dorai Raja (**tempy. Capt.* 6 *Sep.* 18)	. . . .	. . . .	. . . .	(Tempy. and Hony.)
Jeypore, the Kumara Raja of	. . . .	. . . .	. . . .	(Tempy. and Hony.)
Mandasa, Zamindar of	. . . .	. . . .	. . . .	(Tempy. and Hony.)
Bhupendra Narayan Choudhury, Srijut	. . . .	. . . .	. . . .	(Tempy. and Hony.)
Gunindranath Barua, Srijut	. . . .	. . . .	. . . .	(Tempy. and Hony.)
Nagendra Nath Choudhury, *Rai Bahadur*	. . . .	. . . .	. . . .	(Tempy. and Hony.)
Abdul Lais Saaduddin Muhammad, Maulvi	. . . .	. . . .	. . . .	(Tempy. and Hony.)
Maung Ba Tin, (3), Myook	. . . .	. . . .	12 Feb. 18	(Tempy.)
Durga Narain Singh, Raja	. . . .	. . . .	. . . .	(Tempy. and Hony.)
Hukm Tej Partab Singh, Raja	. . . .	. . . .	. . . .	(Tempy. and Hony.)
Lokendra Sah, Raja	. . . .	. . . .	. . . .	(Tempy. and Hony.)
Bhagwant Singh, Lala	. . . .	. . . .	. . . .	(Tempy. and Hony.)
Ravi Partab Narain Singh, Sahibzada	. . . .	. . . .	. . . .	(Tempy. and Hony.)
Ibrahim Ali Khan, Nawab	. . . .	17 Apl. 18	17 Apl. 18	(Tempy. and Hony.)
Balbir Singh, *Rao Bahadur*	. . . .	17 Apl. 18	17 Apl. 18	(Tempy. and Hony.)
Malik Mohamad Akbar Khan, Jodhra, *Khan Sahib*	. . . .	17 Apl. 18	17 Apl. 18	(Tempy. and Hony.)
Khan Sikandar Hayat Khan, Khattar	. . . .	17 Apl. 18	17 Apl. 18	(Tempy. and Hony.)
Chaudhri Damodar Sing	. . . .	17 Apl. 18	17 Apl. 18	(Tempy. and Hony.)
Malik Sardar Khan, Nun	. . . .	17 Apl. 18	17 Apl. 18	(Tempy. and Hony.)
Malik Khizar Mohammad Hayat	. . . .	17 Apl. 18	17 Apl 18	(Tempy. and Hony.)
Malik Mohammad Sher Khan, Tiwana	. . . .	17 Apl. 18	17 Apl. 18	(Tempy. and Hony.)
Sardar Raghbir Singh, Sindhanwala	. . . .	17 Apl. 18	17 Apl. 18	(Tempy. and Hony.)
Saileswar Singh Roy, Babu	. . . .	19 Apl. 18	19 Apl. 18	(Tempy. and Hony.)
Bejoy Prasad Singh Roy, B.A., Babu	. . . .	19 Apl. 18	19 Apl. 18	(Tempy. and Hony.)
Nalini Mohan Roy Chaudhuri, B.A., Babu	. . . .	19 Apl. 18	19 Apl. 18	(Tempy. and Hony.)
	. . . .	19 Apl. 18	19 Apl. 18	(Tempy and Hony.)
Syed Mohammad Hossain, Maulvi	. . . .	19 Apl. 18	19 Apl. 18	(Tempy. and Hony.)
Nour Ahmed Hossain, Maulvi	. . . .	19 Apl. 18	19 Apl. 18	(Tempy. and Hony.)
Madhaves varendra Sahi, Babu	. . . .	19 Apl. 18	19 Apl. 18	(Tempy. and Hony.)
Ram Prashhad Narayan Sahi, Babu	. . . .	19 Apl. 18	19 Apl. 18	(Tempy. and Hony.)
Kameshvar Narayan Singh, Babu	. . . .	19 Apl. 18	19 Apl. 18	(Tempy. and Hony.)
Mahendra Man Singh, Raja	. . . .	. . . .	. . . .	(Tempy. and Hony.)
Malojirao Mudhojirao Naik Nimbalkar, Meherban Chief of Phaltan	. . . .	24 May 18	24 May 18	(Tempy. and Hony.)

* Without pay and allowances of the rank.

INDIAN LAND FORCES.

Names.	Date of birth.	Date of first commission.	Date of rank.	Remarks.
		2nd-Lieutenants—*contd.*		
Bhimrao Nagojirao Patankar, Meherban, Sardar	. . . .	24 May 18	24 May 18	(Tempy. and Hony.)
Ramchandrarao, *alias* Abasaheb Hanmantrao-Nimbalkar (or Khardekar), Meherban	. . . .	24 May 18	24 May 18	(Tempy. and Hony.)
Vikramshah, *alias* Martandrao, Raja	. . . .	24 May 18	24 May 18	(Tempy. and Hony.)
Shankarrao B. Ingle	. . . .	24 May 18	24 May 18	(Tempy. and Hony.)
Saiyad Ghulam Ali, *Khan Bahadur* Nawab	. . . .	26 July 18	26 July 18	(Tempy. and Hony.)
Kunwar Daulat Singh	. . . .	26 July 18	26 July 18	(Tempy. and Hony.)
Girindra Sircar	. . . .	9 Aug. 18	9 Aug. 18	(Tempy. and Hony.)
Muhammad Ahmad Said Khan, Nawab	. . . .	30 Aug. 18	30 Aug. 18	(Tempy. and Hony.)
Kunwar Nand Lal	. . . .	30 Aug. 18	30 Aug. 18	(Tempy. and Hony.)
Rustom Day	. . . .	4 Oct. 18	4 Oct. 18	(Tempy.)
Thakur Karan Singh	. . . .	4 Oct. 18	4 Oct. 18	(Tempy.)
Reyazuddin Ahmed Khan . . .	. . . .	4 Oct. 18	8 Oct. 18	(Tempy.)
Narendra Nath Ghatak, M.B.E. .	. . . .	4 Oct. 18	4 Oct. 18	(Tempy. and Hony.)
Khagendra Chandra Nag . . .	. . . .	4 Oct. 18	4 Oct. 18	(Tempy. and Hony.)
Surendra Nath Sen Gupta . . .	. . . .	4 Oct. 18	4 Oct. 18	(Tempy. and Hony.)
Muazzim Husein Muhamed Furrokh, *Maulvi*	. . . .	4 Oct. 18	4 Oct. 18	(Tempy. and Hony.)
Ratansinh Sambhusinh	. . . .	4 Oct. 18	4 Oct. 18	(Tempy. and Hony.)
Ram Gopal Singh, Raja Bahadur .	. . . .	8 Oct. 18	8 Oct. 18	(Tempy. and Hony.)
Bishnath Saran Singh, Raja Bahadur	. . . .	8 Oct. 18	8 Oct. 18	(Tempy. and Hony.)
Chandra Chur Singh, The Hon'ble Raja	. . . .	8 Oct. 18	8 Oct. 18	(Tempy. and Hony.)
Amar Singh Sardar	. . . .	8 Oct. 18	8 Oct. 18	(Tempy. and Hony.)
Sheikh Imtiaz Rasul Khan	. . . .	8 Oct. 18	8 Oct. 18	(Tempy. and Hony.)
Sheikh Shahid Husain	. . . .	8 Oct. 18	8 Oct. 18	(Tempy. and Hony.)
Chaudhri Balwant Gujar Singh . .	. . . .	8 Oct. 18	8 Oct. 18	(Tempy. and Hony.)

(Appendix III, A. R. I., Vol. II.)

NOTE —1. Officers of the Indian Army Reserves should report their addresses to the Military Secretary to the Commander-in-Chief on the 1st January and 1st July of each year. If no reports are received for a year, it will be assumed that the officer is non-effective, and his name will be removed from the list.

2. The second date against an officer's name denotes the date of his appointment to the Indian Army Reserve of Officers.

3. Cavalry—633, Infantry—4,980, Veterinary Bch.—10 & Tempy. Commn.—84. Total—5,707

CAVALRY.

Majors—5.

No.	Date of First Commn.	Names.	Date of birth.	Date of rank.	Remarks.
1	4 Mar. 98	O'Connor, Malcolm Stuart Scott	4 July 70	1 Sep. 15	
	8 Jan. 16	Filgate, Townley Richard, C.I.E., V.D. . . .	. . . .	8 Jan. 16	
	10 Sep. 16	Cameron, Colin John Norman (79)	. . . .	10 Sep. 16	
	17 Nov. 17	Nevill, Henry Rivers (*temp. Lt.-Col., 17 Nov. 17*)	. . . .	17 Nov. 17	
5	25 Sep. 03	Trench, John Stewart	9 Apl. 79	25 Sep. 18	

Captains—92.

No.	Date of First Commn.	Names.	Date of birth.	Date of rank.	Remarks.
1	22 Apl. 04	Young, Ernest Meredyth	10 Mar. 76	12 Sep. 14	
	6 Nov. 14	Porter, Leonard Lachlan (54, 79) (*temp. Maj., 15 Oct. 18*)	21 Oct. 68	6 Nov. 14	
		Service for promn.—Emd. Volrs., 28 Oct. 14 to 6 Nov. 14.			
	18 Feb. 15	Ouseley, Joseph William Glynn (54, 79) . . .	. . . .	18 Feb. 15	
	18 Feb. 15	Llewhellin, George Elliott	. . . .	18 Feb. 15	
5	13 Nov. 14	Watson, James	25 July 80	25 June 15	
	7 Aug. 14	Hamley, William Studholme Clarence . . .	3 Nov. 84	1 Sep. 15	
		Service for promn.—4 yrs. 278d. from 29th Nov. 05 to 3 Sep. 10.			
	21 Apl. 11	Gordon, Alexander Huntley	31 Oct. 76	12 Sep. 15	
	28 Sep. 15	Little, John Oswald (59, 79, 206)	. . . .	28 Sep. 15	
	27 Nov. 15	Macwhirter, Dudley (79)	. . . .	27 Nov. 15	
10	1 Mar. 16	Byres, Edward Byres Moir (79)	. . . .	1 Mar. 16	
	6 Mar. 14	Westmacott, Richard	. . . .	26 Mar. 17	
	10 May 04	Gregson, Edward G., C.M.G., C.I.E. (*temp. Lt.-Col., 1 Apl. 17.*)	10 June 77	1 Apl. 17	
	16 Aug. 14	Wilson, Duncan William, C.I.E. (54, 79) (*temp. Maj., 25 Feb. 18*)	30 Sep. 81	16 Aug. 17	
	23 Jan. 14	Banister, Harry	12 Apl. 82	25 Sep. 17	
15	14 Oct. 14	Wallace, Archibald Malcolm	. . . .	14 Oct. 17	
	4 Dec. 14	Ranking, William Lancaster (79)	. . . .	8 Nov. 17	
		Service for promn.—Emb. Volrs. 1 Nov. 14 to 26 Nov. 14.			
	4 Dec. 14	Henderson, Walter Lewis (79)	. . . .	4 Dec. 17	
	18 Dec. 14	Duff, Kenneth Cyril	3 July 86	14 Feb. 18	
	13 Nov. 14	Hills, Archie Chalmers	. . . .	16 Feb. 18	
20	20 Nov. 14	Horne, Charles Walter	. . . .	17 Feb. 18	
	27 Nov. 14	Belcher, Edward Brymer (*Actg. Maj., 26 Sep. 18*)	4 Dec. 87	17 Feb. 18	
	20 Nov. 14	Wylde, Eric William	9 June 88	18 Feb. 18	
	13 Nov. 14	Murray, Walter William Macdonald . .	. . . .	25 Feb. 18	
	13 Nov. 14	Forbes, Charles Alexander	6 July 77	28 Feb. 18	
25	13 Nov. 14	Falkner, James Maurice	16 Mar. 80	28 Feb. 18	
	13 Nov. 14	Hunt, James Vere	. . . .	28 Feb. 18	
	20 Nov. 14	Shand, George	18 Nov. 77	2 Mar. 18	
	20 Nov. 14	Donald, Colin Dunlop (*Serv. for prom.—Emb. Volr. 5 Aug. 14 to 20 Nov. 14*) . . .	11 Sep. 79	2 Mar. 18	
	20 Nov. 14	Little, Robert	. . . .	28 Mar. 18	
30	22 Mar. 15	Dickinson, Henry Baron Humphrey . . .	. . . .	6 Apl. 18	
	27 Nov. 14	Sandford, William Sarsfield Beach . . .	13 Aug. 81	8 Apl. 18	
	20 Nov. 14	Fraser, Arkyl Donald	. . . .	9 Apl. 18	
	9 Apl. 15	Goyder, George Barrett (162)	. . . .	9 Apl. 18	
	20 Nov. 14	Cooper, Thomas	. . . .	10 Apl. 18	
35	26 Feb. 15	Harrison, Archibald William	. . . .	16 Apl. 18	
	22 Mar. 15	Morris, Noel Meeson	. . . .	16 Apl. 18	
	22 Mar. 15	Phillips, Gerald Courtenay	. . . .	29 Apl. 18	
	22 Mar. 15	Pilcher, Hugh Robarts	. . . .	30 Apl. 18	
	7 May 15	Haig, Harry Graham (79)(*temp. Maj. 22 Apl. 18*)	. . . .	7 May 18	
40	4 Dec. 14	Taylor, Harry (57)	. . . .	10 May 18	
	4 Dec. 14	Gray, Stuart	. . . .	10 May 18	
	27 Nov. 14	Biss, James Cuthbert de Vere	4 June 92	30 May 18	
	19 Mar. 15	Rose-Smith, William	. . . .	18 June 18	
	20 Nov. 14	Rice, Roy Alfred, M.C.	. . . .	8 July 18	
45	3 Nov. 14	Reid, Claud Lestock	29 Aug. 88	30 July 18	
	20 Nov. 14	Ferguson, Donald Harry, M.C.	8 Feb. 90	30 July 18	
	20 Nov. 14	Benfield, Claud William Collyes	. . . .	30 July 18	
48	27 Nov. 14	Nicholl, Charles Kenneth	19 Sep. 88	30 July 18	

CAVALRY—*contd.*

Captains—*contd.*

No.	Date of First Commn.	Names.	Date of birth.	Date of rank.	Remarks.
49	20 Nov. 14	Martin, Theobald Seymour Stewart		6 Aug. 18	
50	19 Sep. 14	Courthope-Munroe, Caryl Henry		13 Aug. 18	
	5 Feb. 15	Sheldrick, Alfred Henry William		13 Aug. 18	
	22 Jan. 15	Richards, Benjamin William (172*a*)		15 Aug. 18	
	13 Nov. 14	Nixon, Charles Norman	8 June 91	17 Aug. 18	
	27 Nov. 14	O'Connor, Fergus Patrick		17 Aug. 18	
55	20 Nov. 14	Lang, Hugh Francis	17 Nov. 87	20 Aug. 18	
	27 Nov. 14	Johnston, Edward Norman Wybrants		20 Aug. 18	
	16 Aug. 14	Thacker, Walter Lionel Hubert	20 July 82	20 Aug. 18	
	23 Aug. 14	Shout, William Gerdes		23 Aug. 18	
	29 Aug. 14	May, Francis William		29 Aug. 18	
60	29 Aug. 14	Watson, Harold Boyes		29 Aug. 18	
	30 Oct. 14	Rayneau, Lesnee Armand	10 Apl. 91	30 Aug. 18	
	13 Nov. 14	Fairweather, Harold	2 Dec. 86	30 Aug. 18	
	20 Nov. 14	Herrick, Robert Lysle Warren	26 May 95	30 Aug. 18	
	20 Nov. 14	Danby, Arthur Lloyd	3 Mar. 90	30 Aug. 18	
65	20 Nov. 14	MacKenzie, James McAuslan, M.C.		30 Aug. 18	
	20 Nov. 14	Sheard, Clifford		30 Aug. 18	
	27 Nov. 14	Rust, Frederic Hugh (207*a*)		30 Aug. 18	
	27 Nov. 14	Falconer John Fattes		30 Aug. 18	
	4 Dec. 14	Robertson, Frederick Ewart		30 Aug. 18	
70	4 Dec. 14	Hindley, Douglas Colin Rowland		30 Aug. 18	
	18 Dec. 14	Bradshaw, John William Hardin		30 Aug. 18	
	27 Nov. 14	Douetil, Philip Victor	7 Feb. 93	30 Aug. 18	
	18 Dec. 14	Hawkins, Robert Phillip	20 Feb. 81	30 Aug. 18	
	31 Dec. 14	Wilson, Newell Budd		30 Aug. 18	
75	8 Jan. 15	Walter, Garnet Beauchamp		30 Aug. 18	
	8 Jan. 15	Watson, Harold Reginald West		30 Aug. 18	
	21 May 15	Herbert, Vaughan Augustus, M.C.		30 Aug. 18	
	5 Aug. 14	Raikes, Marcus Hamilton		5 Aug. 18	
	20 July 15	Heaton-Armstrong, John Dunamace		8 Sep. 18	
80	23 Apl. 15	Collett, Charles Morden		9 Sep. 18	Released from Army Service.
	21 Sep. 14	Garle, John Aubrey Batten		21 Sep. 18	
	22 Sep. 14	Nash, John Alexander (79)		22 Sep. 18	
	23 Sep. 14	Persse, Sydney Henry		23 Sep. 18	
	20 Nov. 14	Liddell, James Alexander (51)	13 Mar. 86	1 Oct. 18	
85	27 Oct. 14	Defries, Alfred Gordon (52)		27 Oct. 18	
	27 Nov. 14	Fremlin, Ralph William		27 Oct. 18	
	30 Oct. 14	Carpenter, John Howard		30 Oct. 18	
	9 Nov. 14	Byrne, George Robert		9 Nov. 18	
	27 Nov. 14	Edkins, Reginald Herbert (71)		11 Nov. 18	
90	22 Feb. 15	Landale, Henry Russell		13 Nov. 18	
	20 Nov. 14	Thomas, David Edward	6 June 90	20 Nov. 18	
92	4 Dec. 14	Whitakar, Basil (51)		4 Dec. 18	

Lieutenants—469.

No.	Date of First Commn.	Names.	Date of birth.	Date of rank.	Remarks.
1	27 Nov. 14	Lawford, Edward Enfield		4 Jan. 15	
	13 Nov. 14	Shortridge, Guy Chester		26 Feb. 15	
	4 Dec. 14	Tayler, Charles Hopewell		2 Mar. 15	
	18 Dec. 14	Smallwood, Harold St. Clair (*temp. Capt. 19 July 17*).		16 Apl. 15	
5	12 Oct. 14	Barford, Luther Holden (*temp., Capt. 25 Dec. 17*)		22 May 15	
	18 Jan. 15	Allum, Hilery Gilbert		26 July 15	
	10 Sep. 14	Booth, Albert Edwin		10 Sep. 15	
	20 Nov. 14	Monks, Harold Gordon		2 Oct. 15	
	20 Nov. 14	Floyer, Ernest Ayscoghe, M.C.	30 June 88	3 Oct. 15	
10	10 Oct. 14	Padfield, Robert Charles Hillman (*temp. Maj. 10 July 18*)	5 Sep. 74	10 Oct. 15	
	23 Oct. 14	Bird, Reuben James		23 Oct. 15	
	8 Jan. 15	Chubb, William	24 Mar. 82	24 Oct. 15	
	18 Jan. 15	Iles, Alfred Gordon		24 Oct. 15	
	13 Nov. 14	Guthrie, Frederick Charles	3 Aug. 92	13 Nov. 15	
15	20 Nov. 14	Kennedy, John Pringle	7 Mar. 79	20 Nov. 15	
	20 Nov. 14	Glasspool, Leonard Alfred		20 Nov. 15	
	20 Nov. 14	Lacey, Frederick Arthur (47*a*, 51)		20 Nov. 15	
	20 Nov. 14	Bapty, Philip Sydney, M.C.		20 Nov. 15	
	27 Nov. 14	Hemsted, Frederick		27 Nov. 15	Released from Army Service.
20	27 Nov. 14	Cumberlege, John Henry	24 Jan. 85	27 Nov. 15	
	28 May 15	Trousdell, John Ponsonby		28 Nov. 15	
	18 Feb. 15	Ogley, Ernest Frederick, M.C.		2 Dec. 15	
	4 Dec. 14	Fulcher, George William		2 Dec. 15	
24	4 Dec. 14	Innes, Richard Joseph	8 Dec. 75	4 Dec. 15	

CAVALRY—*contd.*

Lieutenants—*contd.*

No.	Date of First Commn.	Names.	Date of birth.	Date of rank.	Remarks.
25	4 Dec. 14	Barrow, Reginald Lousada		4 Dec. 15	
	4 Dec. 14	Homer, Nicholas Howard Thomas		4 Dec. 15	
	9 Dec. 14	McEntire, Henry Sandys Mecredy (207a) (*temp. Capt. 30 Oct. 18*)		9 Dec. 15	
	11 Dec. 14	Fremlin, Henery James Leigh		11 Dec. 15	
	9 Apl. 15	Turner, John Trench (*Serv. for prom.—Emb. Volr. 16 Dec. 14 to 9 Apl. 15*) (*temp. Capt. 29 Oct. 17*)		16 Dec. 15	
30	16 Dec. 14	Williams, Charles Glare		16 Dec. 15	
	18 Dec. 14	Powell, Vivian Norman ffolliott		18 Dec. 15	
	18 Dec. 14	Livesey, Thomas Reginald		18 Dec. 15	
	18 Dec. 14	Robinson, Lancelot Roden Claud		18 Dec. 15	
	18 Dec. 14	McGarvey, Bernard Valentine		18 Dec. 15	
35	23 Dec. 14	Husband, Robert		23 Dec. 15	
	24 Dec. 14	Shelverton, Evelyn Manfred (51)	28 Aug. 85	24 Dec. 15	
	24 Dec. 14	Peppe, George Tusco Vivian	19 Nov. 87	24 Dec. 15	
	24 Dec. 14	Carmichael, Thomas	8 July 91	24 Dec. 15	
	24 Dec. 14	Murphy, Edward Lawrence	24 Sep. 87	24 Dec. 15	
40	24 Dec. 14	Llewellyn, Basil		24 Dec. 15	
	24 Dec. 14	Fenwick, Francis		24 Dec. 15	
	24 Dec. 14	Allen, Jeffrey Hugh Summers		24 Dec. 15	
	8 Jan. 15	Leigh-Clare, Harry James Leigh, M.C.		6 Jan. 16	
	8 Jan. 15	Smith, Victor Wallace, M.C.		8 Jan. 16	
45	8 Jan. 15	Marsh, Norman Oswald Cyril		8 Jan. 16	
	8 Jan. 15	Audsley, Albert Edward Mac Auley	1 Feb. 75	8 Jan. 16	
	8 Jan. 15	Marriott Edward Francis	6 Oct. 88	8 Jan. 16	
	5 Feb. 15	Maxwell, Robert Muir		19 Jan. 16	
	22 Jan. 15	Harvey-Hurst, Allan James		22 Jan. 16	
50	22 Jan. 15	Cavell, Reginald George		22 Jan. 16 *22 Dec. 16	
	27 Jan. 15	Tucker, Norman Paterson		27 Jan. 16	
	16 Apl. 15	Roger, Kenneth Butterworth		31 Jan. 16	
	5 Feb. 15	Ewart, James Alan, M.C.		5 Feb. 16	
	5 Feb. 15	Streatfield, Newton	29 Aug. 88	5 Feb. 16	
55	5 Feb. 15	Harrold, John Henry		5 Feb. 16	
	5 Feb. 15	Davenport, Edward Charles		5 Feb. 16	
	5 Feb. 15	Charriol, Phillip		5 Feb. 16	
	12 Feb. 15	Ives, Charles St. John		12 Feb. 16	
	12 Feb. 15	Burton, Eric Ray	21 Oct. 91	12 Feb. 16	
60	12 Feb. 15	Rust, Lancelot Doudney		12 Feb. 16	
	12 Feb. 15	Denning, John Du Pré		12 Feb. 16	
	18 Feb. 15	Abercrombie, John Robertson		18 Feb. 16	
	18 Feb. 15	Webster, Frederick Garnet		18 Feb. 16	
	23 Feb. 15	MacWilliam, Thomas Alexander		23 Feb. 16	
65	26 Feb. 15	Brown, James Alexander		26 Feb. 16	
	26 Feb. 15	Coombs, Milbourne Edward (207a)		26 Feb. 16	
	5 Mar. 15	Parmer, Stanley William (*temp. Capt. 1 Mar. 17*)		5 Mar. 16	
	5 Mar. 15	Keatinge, Gilbert		5 Mar. 16	
	5 Mar. 15	Walsh, Gordon James Turnbull		5 Mar. 16	
70	5 Mar. 15	Wynyard-Wright, Vaughan Lascelles, (79).		5 Mar. 16	
	5 Mar. 15	Mangin, Purefoy Colles		5 Mar. 16	
	7 May 15	Emerson, Donald Bomford		11 Mar. 16	
	19 Mar. 15	Stuart, Philip Price (*temp. Capt.*)		19 Mar. 16	
	19 Mar. 15	Swayne, Reginald Percy Nugent		19 Mar. 16	
75	19 Mar. 15	Newington, Malcolm Theodore Lysaught		19 Mar. 16	
	19 Mar. 15	Thompson, John Stuart		19 Mar. 16	
	19 Mar. 15	Maxwell, James Ferris		19 Mar. 16	
	19 Mar. 15	Cotton, William Bensley (*temp. Capt. 29 Dec. 17*)		19 Mar. 16	
	23 Apl. 15	Hewett, Harold Pearse		22 Mar. 16	
80	23 Mar. 15	Christie, William Alexander Kynoch		23 Mar. 16	
	7 May 15	Ibbotson, Archie William, M.C. (*temp. Capt. 29 Dec. 17.*)		25 Mar. 16	
	26 Mar. 15	Bayldon, Richard Cresswell		26 Mar. 16	
	26 Mar. 15	King, George Edward		26 Mar. 16	
	2 Apl. 15	Boyd, Henry Sidney (*temp. Capt.*)		2 Apl. 16	
85	2 Apl. 15	O'Connor, Joseph Cornelius Jerome		2 Apl. 16	
	9 Apl. 15	Deacon, Charles Gordon (51, 207a)		9 Apl. 16	
	30 Apl. 15	Lloyd-Williams, James Evan		10 Apl. 16	
	16 Apl. 15	Guise, Harry Rivett Cecil (*temp. Capt. 8 Jan. 18*)		16 Apl. 16	
	16 Apl. 15	Murray, Colin Robert Baillie (169)		16 Apl. 16	
90	16 Apl. 15	Puckle, George Hale (207)		16 Apl. 16	
91	19 Apl. 15	Dacres, Leonard Seymour Lambert		19 Apl. 16	

* Date of transfer to the Cavalry Branch.
† Without pay and allowance of the rank.

CAVALRY—*contd.*

Lieutenants—*contd.*

No.	Date of First Commn.	Names.	Date of birth.	Date of rank.	Remarks.
92	20 Apl. 15	Eyton, John Seymour (52)		20 Apl. 16	
	23 Apl. 15	Shee, John Robert Lloyd (207*a*)		23 Apl. 16	
	30 Apl. 15	Garratt, Geoffrey Theodore (52)		30 Apl. 16	
95	30 Apl. 15	Ashcroft, John Myddleton		30 Apl. 16	
	30 Apl. 15	Carpenter, Percy Henry (95) (*temp. Capt. 1 Mar. 17*)		30 Apl. 16	
	30 Apl. 15	Durack, Jeremiah Joseph		30 Apl. 16	
	7 May 15	Jones, Hubert Cecil		7 May 16	
	7 May 15	Fremantle, Alan Frederick (**temp. Capt. 22 Apl. 18.*)		7 May 16	
100	7 May 15	O'Connor, Roderick Augustus (**temp. Capt. 1 Nov. 17*)		7 May 16	
	7 May 15	Reed, Marmaduke Wilfred (51, 52)		7 May 16	
	7 May 15	Dixon, George		7 May 16	
	7 May 15	Grant, Charles Frederick (46)		7 May 16	
	28 May 15	Shepherd, William Skinner (52)		10 May 16	
105	14 May 15	Willis, Richard Dunn		14 May 16	
	14 May 15	Wood, Arthur Edward Frances (170)		14 May 16	
	14 May 15	Milner, Charles Edward		14 May 16	
	4 June 15	Fitzpatrick, Joseph O'Connor		4 June 16	
	4 June 15	Wiles, Bernard Harper		4 June 16	
110	11 June 15	Sargisson, William Fred (*temp. Capt. 1 Apl. 17*)		11 June 16	
	2 July 15	Lloyd, Claude Pemberton Wilkinson		2 July 16	
	2 July 15	Jackson, Lawrence Tait		2 July 16	
	10 July 15	Common, Frank Bell		10 July 16	
	10 July 15	Cook, Henry Olliff		10 July 16	
115	13 July 15	Peck, Leonard Jasper, M.C.		13 July 16	
	17 July 15	Moore, Edward Cecil Frederick, M.C.		17 July 16	
	18 July 15	Hannah, George Kissam (207*a*)		18 July 16	
	21 July 15	Gosselin, Hugh de Vere (207*a*)		21 July 16	
	23 July 15	Smith, Douglas Edwin		23 July 16	
120	23 July 15	Wilson, George		23 July 16	
	23 July 15	Case, John Bullen		23 July 16	
	30 July 15	McBrien, Charles Hugh		30 July 16	
	30 July 15	Symonds, Archibald Hugh		30 July 16	
	2 Aug. 15	Beattie, Jeremy Greef		2 Aug. 16	
125	7 Aug. 15	Knox, James Hay Graham (207*a*) (**temp. Capt. 1 Oct. 17*)		7 Aug. 16	
	11 Aug. 15	Danby, Edward Clement		11 Aug. 16	
	16 Aug. 15	Arrindell, Cecil Dudley Stewart May, M.C. (51, 71)		16 Aug. 16	
	20 Aug. 15	Johnson, Thomas Francis		20 Aug. 16	
	20 Aug. 15	Kinloch, William		20 Aug. 16	
130	20 Aug. 15	Blackwood, John Richardson (51)		20 Aug. 16	
	20 Aug. 15	Lovell, William Day		20 Aug. 16	
	20 Aug. 15	Parr, Alfred Ernest		20 Aug. 16	
	24 Aug. 15	Matthews, Stuart Leslie		24 Aug. 16	
	3 Sep. 15	Russell, Archibald Miles Holme		3 Sep. 16	
135	13 Sep. 15	Lambert, Dryden Henry Charles		13 Sep. 16	
	14 Sep. 15	Williams, Arther Basil		14 Sep. 16	
	16 Sep. 15	Forbes, Archibald Douglas Coldham		16 Sep. 16	
	20 Sep. 15	Williams, James Harold (52)		20 Sep. 16	
	23 Sep. 15	Bell, Henry Charles (52)		23 Sep. 16	
140	29 Sep. 15	Reilly, Charles Oliver Calcott (207)		29 Sep. 16	
	2 Oct. 15	Blakiston, John Francis		2 Oct. 16	
	2 Oct. 15	Cameron, William Wyllie Leslie		2 Oct. 16	
	7 Oct. 15	Haseler, Alan Roy		7 Oct. 16	
	8 Oct. 15	Swanzy, Robert Edward		8 Oct. 16	
145	8 Oct. 15	Elmes, Julius Paul Jonathan		8 Oct. 16	
	13 Oct. 15	Wilson, James Morton		13 Oct. 16	
	17 Oct. 15	Irving, James Henry Riley (51)		17 Oct. 16	
	11 Nov. 15	Uridge, Harry James (*temp. Capt.*)		11 Nov. 16	
149	12 Nov. 15	Owden, John Scambler		12 Nov. 16	

* Without pay and allowances of the rank.

CAVALRY—*contd.*

Lieutenants—*contd.*

No.	Date of First Commn.	Names.	Date of birth.	Date of rank.	Remarks.
152	13 Nov. 15	Smith, Malcolm Haynes		13 Nov. 16	
	4 Mar. 15	Morris, Hubert Charles (*207*)		16 Nov. 16	
	18 Nov. 15	Clifford, Henry Hebberd		18 Nov. 16	
155	20 Nov. 15	Grant, Robert (*207a*)		20 Nov. 16	
	26 Nov. 16	Munden, Sydney Edward (*temp. Capt.*, 10 *Nov.* 16)		26 Nov. 16	
	27 Nov. 15	Murray, Michael John (*72b*)		27 Nov. 16	
	1 Dec. 15	Cowan, Gavan John (*temp. Capt.* 15 *May* 18)		1 Dec. 16	
	2 Dec. 15	Hearsey, George Arthur Canning		2 Dec. 16	
160	4 Dec. 15	Pike, Albert (51)		4 Dec. 16	
	4 Dec. 15	Morris, Llewellyn John		4 Dec. 16	
	4 Dec. 15	Parker, George Thomas More (51)		4 Dec. 16	
	5 Dec. 15	Rorison, Wilfred Moncrieffe		5 Dec. 16	
	11 Dec. 15	Pegge, Frank Henry (52)		11 Dec. 16	
165	13 Dec. 15	Marshall, John, M.C. (*temp. Capt.*, 8 *Mar.* 17.)		13 Dec. 16	
	15 Dec. 15	Aucutt, Wilfred		15 Dec 16	
	23 Dec. 15	Danby, Gerald Philip		23 Dec. 16	
	31 Dec. 15	Gavaghan, Edward Laurence (51, 52)		31 Dec. 16	
	16 Jan. 16	Shaw, Frank McIlwraith		16 Jan. 17	
170	18 Jan. 16	Heath, Frederick Samuel		18 Jan. 17	
	27 Jan. 16	Stanton, Cyril Arthur (57)		27 Jan. 17	
	27 Jan. 16	Campbell, George Riddock		27 Jan. 17	
	12 Feb. 16	Wells, Richard Leonard Percy		12 Feb 17	
	12 Feb. 16	Gulliland, Colin Campbell (207*a*)		12 Feb. 17	
175	12 Feb. 16	Macmillan, Eric John Ashton (51, 52)		12 Feb. 17	
	19 Feb. 16	Onraët, Pierre Trevor		19 Feb. 17	
	20 Feb. 16	Foxlee, Frederick John		20 Feb. 17	
	20 Feb. 16	Pont, Leon Victor (51)		20 Feb. 17	
	26 Feb. 16	Harvey, Godfrey Thomas Benedict		26 Feb. 17	
180	1 Mar. 16	Cazalet, Robert George (51)		1 Mar. 17	
	4 Mar. 16	Rothera, George Howitt (51, 52)		4 Mar. 17	
	4 Mar. 16	Smith, Arthur Kirke (51, 207*a*)		4 Mar. 17	
	10 Mar. 16	Hearsey, Lionel Douglas William (79)		10 Mar. 17	
	13 Mar. 16	Baird, Alexander Edwin (§ *temp. Capt.*, 13 *Mar.* 16)		13 Mar. 17	
185	22 Mar. 16	Wilks, Errick (51)		22 Mar. 17	
	30 Mar. 16	Small, Gerald Agnew (57) (**temp. Capt.* 1 *Oct.* 17)		30 Mar. 17	
	5 Apl. 16	Reiss, Horace Baggalay (51)		5 Apl. 17	
	10 Apl. 16	Borissow, Francis Augustine (*temp. Capt.* 17 *July* 16.)		10 Apl. 17	
	13 Apl. 16	McManus, Edwin Alexander		13 Apl. 17	
190	19 Apl. 16	Sherwood, William Oscar		19 Apl. 17	
	24 Apl. 16	Larymore, Montgomery Alexander (12, 207*a*)		24 Apl. 17	
	27 Apl. 16	Rose, William Arthur (*temp. Capt.*, 8 *Aug.* 16.)		27 Apl. 17	
	29 Apl. 16	Moore, Alexander Norman FitzGerald, M.C.		29 Apl. 17	
	1 May 16	Collet, Arthur Preston (172) (*temp. Capt.*, 9 *Mar.* 17)		1 May 17	
195	12 May 16	Ostler, Clifford (51)		12 May 17	
	12 May 16	Morrison, Robert Gordon		12 May 17	
	10 June 16	Atkins, Edward Davis (*Actg. Maj.*, 26 *Sep.* 18		10 June 17	
	14 June 16	Oliver, Frank Plummer (51)		14 June 17	
	15 June 16	Smith, Guy Assheton		15 June 17	
200	15 June 16	Elsworth, Charles William Tyndale		15 June 17	
	15 June 16	Cormack Frederic Eunson		15 June 17	
	18 June 16	Sullivan, Joseph Hubert Baron		18 June 17	
	18 June 16	Simmons, Frederick (47*a*, 51)		18 June 17	
	24 June 16	Threlkeld, Stanley		24 June 17	
205	28 June 16	Latham, Philip Denys		28 June 17	
	28 June 16	Horwill, Lionel Clifford (51)		28 June 17	
	10 July 16	Williams, John Fortescue Hugh		10 July 17	
	17 July 16	Lory, Frederick Burton Pendarves (79)		17 July 17	
	22 July 16	Pope, James Alister		22 July 17	
210	22 July 16	Ziman, Solomon Nethein (51) (*temp. Capt.*, 3 *Dec.* 17)		22 July 17	
	22 July 16	Sanders, John Herne		22 July 17	Released from Army Service.
	22 July 16	Dalley, Richard Pomeroy		22 July 17	Released from Army Service.
213	22 July 16	Hoyle, Arthur Lascelles (51) (**temp. Capt.*, 27 *Mar.* 17)		22 July 17	

* Without pay and allowances of the rank.
§ Without pay and allowances prior to 10 Apl. 17.

CAVALRY—*contd.*

Lieutenants—*contd.*

No.	Date of First Commn.	Names.	Date of birth.	Date of rank.	REMARKS.
214	22 July 16	Bell, Alec Jeffrey (79) (*temp. Capt., 1 May 18*)	20 May 85	22 July 17	
215	22 July 16	Ellis, Reginald	....	22 July 17	
	28 July 16	Brigg, William Llewellyn	....	28 July 17	
	29 July 16	Butterfield, James (207*a*)	....	29 July 17	
	29 July 16	Meller, Frank Reginald (207*a*)	....	29 July 17	
	29 July 16	Jaques, Arthur Basil	....	29 July 17	
220	31 July 16	Hood, Hugh Meggison (207*a*)	....	31 July 17	
	1 Aug. 16	Askwith, Arthur Vivian (*temp. Capt., 12 Nov. 17*)	....	1 Aug. 17	
	1 Aug. 16	Tate, Thomas Bailey (**temp. Capt., 22 June 17*)	....	1 Aug. 17	
	3 Aug. 16	Slesser, William Duncan Vivian (*temp. Capt. 3 Aug. 17*)	....	3 Aug. 17	
	4 Aug. 16	McIntosh, Eric Prideaux	....	4 Aug. 17	
225	7 Aug. 16	Rivett-Carnac, John Claude Thurlow (51, 79) (**temp. Capt., 24 Sep. 17*)	....	7 Aug. 17	
	7 Aug. 16	Cones, Hubert Michael Alfred (*temp. Capt., 11 Jan. 18*)	....	7 Aug. 17	
	14 Aug. 16	Martin, Olans Macleod	....	14 Aug. 17	
	14 Aug. 16	Lethbridge, Montagu Henry Bowater	....	14 Aug. 17	
	14 Aug. 16	Drummond, John Graham (207*a*)	....	14 Aug. 17	
230	14 Aug. 16	Fawcus, Louis Reginald	....	14 Aug. 17	
	14 Aug. 16	Reid, Robert Niel (51, 52, 172*a*)	....	14 Aug. 17	
	14 Aug. 16	Rawlins, Eric Ford	....	14 Aug. 17	
	14 Aug. 16	Mainwaring, Albert James	....	14 Aug. 17	
	14 Aug. 16	Thrupp, Wilfred Graham	....	14 Aug. 17	
235	14 Aug. 16	Bavin, Lancelot Neville (**tempy. Capt. 23 Feb. 18.*)	....	14 Aug. 17	
	14 Aug. 16	Wallace, Alexander Ross	....	14 Aug. 17	
	14 Aug. 16	Reilly, Arthur Frederick (169, 206) (*temp. Capt. 14 Aug. 17*)	....	14 Aug. 17	
	18 Aug. 16	Clark, Robert Napier (*temp. Maj., 25 Feb. 18*)	....	18 Aug. 17	
	21 Aug. 16	Robinson, Edward Charles	....	21 Aug. 17	
240	21 Aug. 16	Gude, Niels Christian	....	21 Aug. 17	
	21 Aug. 16	Champion, Frederick Walter (169) (*temp. Capt., 8 Mar. 17*)	....	21 Aug. 17	
	21 Aug. 16	Pyne, Herbert Percy	....	21 Aug. 17	
	21 Aug. 16	Nicholson, John Wilfred (51)	....	21 Aug. 17	
	21 Aug. 16	Walker, Norman	....	21 Aug. 17	
245	21 Aug. 16	Clough, Thomas (51)	....	21 Aug. 17	
	21 Aug. 16	Stevens, Thomas Morton (2 7*a*)	....	21 Aug. 17	
	21 Aug. 16	Vaughan-Stevens, Dudley Lewis (51)	....	21 Aug. 17	
	24 Aug. 16	Quinlan, Patrick Bagot	....	24 Aug. 17	
	1 Sep. 16	Palmes, William Percy (51)	....	1 Sep. 17	
250	1 Sep. 16	Dreyfus, Bertram Edward (**temp. Capt. 1 Oct. 17*)	....	1 Sep. 17	
	1 Sep. 16	Vickers, Harold James	....	1 Sep. 17	
	1 Sep. 16	Avery, Arthur Henry St. John	....	1 Sep. 17	
	7 Sep. 16	Elliot, Thomas	....	7 Sep. 17	
	7 Sep. 16	Dodd, Jasper John	....	7 Sep. 17	
255	7 Sep. 16	Heslop, George Dudley (207*a*)	....	7 Sep. 17	
	8 Sep. 16	Jollye, Humphrey Cyprian Beverly (**temp. Capt., 10 Dec. 17*)	....	8 Sep. 17	
	9 Sep. 16	Tayler, Edward Reginald Noel a'Wood	....	9 Sep. 17	
	10 Sep. 16	Buckler, Francis William (51, 52)	....	10 Sep. 17	
	15 Sep. 16	Wood, Christopher Francis (46, 169) (*temp. Capt. 9 Mar. 17*)	....	15 Sep. 17	
260	15 Sep. 16	Bion, Arnold Eustace (*temp. Capt., 31 Oct. 17.*)	....	15 Sep. 17	
	15 Sep. 16	Oaten, Edward Farley (26)	....	15 Sep. 17	
	15 Sep. 16	Lloyd, Stanley (51, 207*a*)	....	15 Sep. 17	
	15 Sep. 16	Child, Thomas Hadkinson	....	15 Sep. 17	
	15 Sep. 16	Tranter, Francis Godfrey (51)	....	15 Sep. 17	
265	21 Sep. 16	Wood, Ernest Clement (51)	....	21 Sep. 17	
	25 Sep. 16	Kennedy, Kenneth Milman	....	25 Sep. 17	
	25 Sep. 16	Maxwell, Herbert Bryden	....	25 Sep. 17	
	28 Sep. 16	Warren, William John	....	28 Sep. 17	
	1 Oct. 16	Macintyre, Alexander	....	{ 1 Oct. 17 { 9 Feb. 18	
270	5 Oct. 16	Rogers, Robert Hastings (52)	....	5 Oct. 17	
271	6 Oct. 16	Horsefield, Herbert Ellis (*temp. Capt., 30 Nov. 17*)	....	6 Oct. 17	

* Without pay and allowances of the rank.

CAVALRY—*contd.*

Lieutenants—*contd.*

No.	Date of First Commn.	Names.	Date of birth.	Date of rank.	REMARKS.
272	6 Oct. 16	Smith, Stanley Brabant		6 Oct. 17	
	6 Oct. 16	Connell, Cyril Martin (51)		6 Oct. 17	
	7 Oct. 16	Wilson, Cecil Claude (51)		7 Oct. 17	
275	10 Oct. 16	Boswell, Robert Bruce		10 Oct. 17	
	11 Oct. 16	Reed, William Louis Lindsay, V.D. (*temp. Capt., 11 Oct. 16*).		11 Oct. 17	
	12 Oct. 16	Frost, Frederick James Tabor		12 Oct. 17	
	13 Oct. 16	George, Robert Esmonde Gordon		13 Oct. 17	
	20 Oct. 16	Horsman, Henry		20 Oct. 17	
280	3 Nov. 16	Nissen, Ferdinand Henry Christian (*temp. Capt., 3 Nov. 16*).		3 Nov. 17	
	13 Nov. 16	Shuttleworth, John Thornton		13 Nov. 17	
	15 Nov. 16	Dickinson, William Graham (51)		15 Nov. 17	
	22 Nov. 16	Muirhead, James Walter		22 Nov. 17	
	24 Nov. 16	Salter, Malcolm Gurney		24 Nov. 17	
285	28 Nov. 16	Bartleet, Thomas Edwin		28 Nov. 17	
	28 Nov. 16	Henderson, David Thomson		28 Nov. 17	
	28 Nov. 16	Treleaven, Conrad Vaughan		28 Nov. 17	
	9 Dec. 16	Collinson, Cyril Danby Hearn (47*a*, 51)		9 Dec. 17	
	12 Dec. 16	Rideout, John Edward's Wood		12 Dec. 17	
290	21 Dec. 16	Riddle, Thomas Ewart (137, 207*a*)		21 Dec. 17	
	21 Dec. 16	Owen, David Price		21 Dec. 17	
	1 Jan. 17	Sayres, Alexander Callander		1 Jan. 18	
	8 Jan. 17	Irwin, Cyril James		8 Jan. 18	
	22 Jan. 17	Dixon, Martin Joseph (172*a*)		22 Jan. 18	
295	25 Jan. 17	Lyell, George Drummond (207*a*)		25 Jan. 18	
	25 Jan. 17	Durant, Eric George (207*a*) (**temp. Capt. 1 Oct. 17*)		25 Jan. 18	
	29 Jan. 17	Johnson, Edward Joseph (51)		29 Jan. 18	
	6 Feb. 17	Innes, Edward Arthur Robert		6 Feb. 18	
	7 Feb. 17	King, Cuthbert (161, 204) (**temp. Capt. 22 Apl. 18*)		7 Feb. 18	
300	7 Feb. 17	Newman, Ernest John (51)		7 Feb. 18	
	13 Feb. 17	Humphrise, Eusebe Edward (*temp. Capt., 30 June 17*).		13 Feb. 18	
	17 Feb. 17	Smith, Lawrence Winsland		17 Feb. 18	
	17 Feb. 17	King, Ernest Roy		17 Feb. 18	
	17 Feb. 17	Chapman, Arthur Reginald		17 Feb. 18	
305	17 Feb. 17	Henderson, Cecil Henry		17 Feb. 18	
	26 Feb. 17	Keston, Thomas Henry		26 Feb. 18	
	1 Mar. 17	Foreman, Frank Clelend		1 Mar. 18	
	5 Mar. 17	Eaton, Herbert		5 Mar. 18	
	5 Mar. 17	Radcliffe-Barrett, Charles Lawless		5 Mar. 18	
310	5 Mar. 17	Thomson, John Lambert		5 Mar. 18	
	6 Mar. 17	House, Edgar Nelson (207*a*)		6 Mar. 18	
	7 Mar. 17	Powis, Harry Baden		7 Mar. 18	
	15 Mar. 17	Gillespie, Andrew Dollar (51, 207*a*)		15 Mar. 18	
	15 Mar. 17	Ellis, Tasseau Sandeman (207*a*)		15 Mar. 18	
315	15 Mar. 17	Heape, Charles Herbert		15 Mar. 18	
	25 Mar. 17	Simpson, Leonard Charles		15 Mar. 18	
	21 Mar. 17	Bodley, Leonard Owen (*temp. Capt 16 June 18*)		21 Mar. 18	
	21 Mar. 17	Wills, Herbert George (52, 207*a*)		21 Mar. 18	
	21 Mar. 17	McIntyre, Daniel Patrick		21 Mar. 18	
320	24 Mar. 17	Vanrenen, Louis Frederick Ernest		24 Mar. 18	
	24 Mar. 17	Allmand, Thomas Reginald		24 Mar. 18	
	24 Mar. 17	Oliver, St. John Louis (51) (*temp. Capt. 15 July 18*)		24 Mar. 18	
	24 Mar. 17	Maunder, Reginald Arthur (47*a*, 206)		24 Mar. 18	
	24 Mar. 17	Tatam, Arthur Cecil		24 Mar. 18	
325	24 Mar. 17	Hulatt, Claude		24 Mar. 18	
	24 Mar. 17	Maxwell, Charles Walter		24 Mar. 18	
	2 Apl. 17	Tapsell, Eustace Bailey		2 Apl. 18	
	4 Apl. 17	Druitt, Theodore Russell (51, 207*a*)		4 Apl. 18	
	5 Apl. 17	Cocker, Robert Whitehead (207*a*)		5 Apl. 18	
330	5 Apl. 17	Sayres, Edward Hamilton (207*a*)		5 Apl. 18	
	5 Apl. 17	Beswick, Leonard (207*a*)		5 Apl. 18	
	5 Apl. 17	Lowndes, Richard Charles (57)		5 Apl. 18	
	5 Apl. 17	Craddock, Henry Ernest Frank		5 Apl. 18	
	6 Apl. 17	Williams, Vaughan Salusbury		6 Apl. 18	
335	11 Apl. 17	Leslie, Donald Wilfrid Russell (57)		11 Apl. 18	
	11 Apl. 17	Pratt, Edmond Benard (207*a*)		11 Apl. 18	
	11 Apl. 17	McGlashan, Eric George		11 Apl. 18	
	11 Apl. 17	Bois, Harold Edward		11 Apl. 18	
	11 Apl. 17	Ormsby, Montagu Henry (*temp. Maj., 15 Aug. 18*)		11 Apl. 18	
340	12 Apl. 17	King, Ernest Henry		12 Apl 18	

* Without pay and allowances of the rank.

CAVALRY—*contd.*

Lieutenants—*contd.*

No.	Date of First Commn.	Names.	Date of birth.	Date of rank.	REMARKS.
341	16 Apl. 17	Huggins, Fredrick Mumford (*temp. Capt. 13 Oct. 17*)		16 Apl. 18	
	17 Apl. 17	Todd, Herbert John		17 Apl. 18	
	17 Apl. 17	Craufurd-Lindsay, Hugh William		17 Apl. 18	
	17 Apl. 17	Morel, Henry James (207*a*) (*temp. Capt. 19 June 18*)		17 Apl. 18	
345	17 Apl. 17	Whinyates, Stanley		17 Apl. 18	
	18 Apl. 17	Woolley, James, (57, 79)		18 Apl. 18	
	18 Apl. 17	Binning, Harold Angus (207*a*) (*temp. Capt. 17 June 18.*)		18 Apl. 18	
	19 Apl. 17	Lowe, Charles Graham		19 Apl. 18	
	19 Apl. 17	Jardine, Douglas Peter		19 Apl. 18	
350	20 Apl. 17	Clayton, John Hatch (207*a*)		20 Apl. 18	
	23 Apl. 17	Horne, George Goodman (51)		23 Apl. 18	
	23 Apl. 17	Tozer, Donald Carles Essery (207*a*)		23 Apl. 18	
	24 Apl. 17	Hill, Harold Brian Cunningham		24 Apl. 18	
	24 Apl. 17	Turle, Henry Bernard		24 Apl. 18	
355	26 Apl. 17	Clarke, Hugh		26 Apl. 18	
	26 Apl. 17	Hawkeswood, Henry William (207*a*)		26 Apl. 18	
	27 Apl. 17	Munckton, Victor Charles Alexander (207*a*)		27 Apl. 18	
	1 May 17	Bowles, Walter Eton		1 May 18	
	5 May 17	McKerrow, James Arthur		5 May 18	
360	11 May 17	Mortimer, William George (207*a*)		11 May 18	
	11 May 17	Mitchell, William		11 May 18	
	11 May 17	Wilson, James Parry		11 May 18	
	11 May 17	Allen, William Born (52, 207*a*)		11 May 18	
	11 May 17	Carter, Arthur Charles (57, 207*a*)		11 May 18	
365	12 May 17	Burgess, Jesse Fletcher		12 May 18	
	12 May 17	Drinkwater, Charles Reginald (207*a*)		12 May 18	
	14 May 17	Meyrick, Reginald		14 May 18	
	14 May 17	Richardson, Alan James		14 May 18	
	14 May 17	Stewart, James		14 May 18	
370	15 May 17	Bagg, William Frank		15 May 18	
	19 May 17	Preston, Thomas		19 May 18	
	21 May 17	Engeldow, George Robert (207*a*)		21 May 18	
	21 May 17	Wallis, Herbert (51)		21 May 18	
	23 May 17	Pitt, William Henry		23 May 18	Released from Army Service.
375	23 May 17	Harper, John Kenneth (51)		23 May 18	
	25 May 17	Cameron, William (207*a*)		25 May 18	
	25 May 17	Foster, Harold Byatt (172*a*, 207*a*)		25 May 18	
	1 June 17	Boyce, Carlyle Oswald (207*a*)		1 June 18	
	1 June 17	Gruar, Roland Foster (57)		1 June 18	
380	9 June 17	Wilkinson, Robert Cecil		9 June 18	
	12 June 17	Thornley, Joseph Herbert		12 June 18	
	12 June 17	Leach, Harold John		12 June 18	
	15 June 17	Dawson, Andrew Thomas (207*a*)		15 June 18	
	16 June 17	Meatyard, William Edward		16 June 18	
385	18 June 17	Odling, Tom Murdoch (207*a*)		18 June 18	
	21 June 17	Grummitt, William Clarke (207*a*)		21 June 18	
	22 June 17	Tanner, Jocelyn Theodore (207*a*)		22 June 18	
	22 June 17	Smellie, Charles Stewart (207*a*)		22 June 18	
	22 June 17	Aitchison, William (207*a*)		22 June 18	
390	23 June 17	Mockler, George Stainbank		23 June 18	
	23 June 17	Calvert, Geoffrey Renshaw		23 June 18	
	23 June 17	Hayes-Watkins, Alan Stanley (207*a*)		23 June 18	
	25 June 17	Studdy, Henry		25 June 18	
	26 June 17	Ker, John Stanley		26 June 18	
395	26 June 17	Munns, Ferrers Augustus Collyer		26 June 18	
	29 June 17	Bettles, Alfred Leonard		29 June 18	
	29 June 17	Hathaway, James Rex		29 June 18	
	29 June 17	Woods, Arthur Fletcher		29 June 18	
	29 June 17	Witchell, Harry		29 June 18	
400	1 July 17	Rhodes, Basil Berhard (207*a*)		1 July 18	
	2 July 17	Herald, Arthur Tarrant Ashby		2 July 18	
	2 July 17	Manley, William Bernard		2 July 18	
	4 July 17	Turner, Gathorne Fenn		4 July 18	
	8 July 17	Chetwynd, Charles William George		8 July 18	
405	9 July 17	Grigson, Pawlet St. John Baseley (207*a*)		9 July 18	
	9 July 17	Masterman, Christopher Hughes (47*a*, 51, 207*a*)		9 July 18	
	13 July 17	Morgan, Henry Stafford		13 July 18	
	16 July 17	Lipscomb, Leslie		16 July 18	
	24 July 17	Dove, Ernest Percy		24 July 18	
410	25 July 17	Corley, Joseph William		25 July 18	
	26 July 17	Alexander, Harold Bryce		26 July 18	
412	26 July 17	Luke, Oliver Edward		26 July 18	

CAVALRY—*contd.*

Lieutenants—*concld.*

No.	Date of First Commn.	Names.	Date of birth.	Date of rank.	Remarks.
411	10 Aug. 17	Cameron, Ewen Grant (207*a*)		10 Aug. 18	
	11 Aug. 17	Richardson, Richard Kenneth (207*a*)		11 Aug. 18	
	18 Aug. 17	Anderson, Frederick MacKay		18 Aug. 18	
	18 Aug. 17	Williams, William Denison Clephane (207*a*)		18 Aug. 18	
415	18 Aug. 17	Drysdale, James Kent		18 Aug. 18	
	22 Aug. 17	Lushington, Harry Ernest Cordue (207*a*)		22 Aug. 18	
	25 Aug. 17	Bosworth, Frederick James		25 Aug. 18	
	25 Aug. 17	Emeney, Arthur Michael		25 Aug. 18	
	25 Aug. 17	Stewart, Angus Wardlaw		25 Aug. 18	
420	25 Aug. 17	Hamilton, Arthur Sidney, M.M.		25 Aug. 18	
	25 Aug. 17	Weitzel, Reginald Harben		29 Aug. 18	
	3 Sep. 17	Blackham, John Colvill (207*a*)		3 Sep. 18	
	4 Sep. 17	Godfrey, Frederick		4 Sep. 18	
	4 Sep. 17	Medwell, Percy Reginald		4 Sep. 18	
425	4 Sep. 17	Croucher, Edward William		4 Sep. 18	
	11 Sep. 17	Napier-Ford, Geoffrey Selwyn (207*a*)		11 Sep. 18	
	11 Sep. 17	Broom, John Thomas		11 Sep. 18	
	12 Sep. 17	Messum, Hugh Ruston (207*a*)		12 Sep. 18	
	13 Sep. 17	Vivian, Graham Linsell (**temp. Capt.*, 3 *Dec.* 17)		13 Sep. 18	
430	13 Sep. 17	Harvey, William Naegelé		13 Sep. 17	Released from Army Service.
	14 Sep. 17	Douglas, Daniel		14 Sep. 18	
	20 Sep. 17	Warren, Edward Thomas (**Ty. Capt.*, 14 *Nov.* 17)		20 Sep. 18	
	22 Sep. 17	Morde, Roland Westphal		22 Sep. 18	Released from Army Service.
	24 Sep. 17	Sheepshanks, Edward		24 Sep. 18	Released from Army Service.
435	26 Sep. 17	Whitehead, Thomas Alex (207*a*)		26 Sep. 18	
	26 Sep. 17	Beatty, Henry (**Ty. Capt.* 1 *Nov.* 17)		26 Sep. 18	
	26 Sep. 17	Finch, Colin Hales		26 Sep. 18	
	26 Sep. 17	Palmer, Andrew Robert Barrett		26 Sep. 18	
	1 Oct. 17	Nichols, Harold Lacy		1 Oct. 18	
440	1 Oct. 17	Blomfield, Henry George		1 Oct. 18	Released from Army Service.
	1 Oct. 17	Newsam, William Oswald (207*a*)		1 Oct. 18	
	5 Oct. 17	Bromley, Ralph Edward		5 Oct. 18	
	6 Oct. 17	Hay, Eric Falconar Alston		6 Oct. 18	
	11 Oct. 17	Rigold, Bernard Francis		11 Oct. 18	
445	11 Oct. 17	Yorke, Robert Langdon		11 Oct. 18	
	11 Oct. 17	Crofton, Richard Marsh		11 Oct. 18	
	11 Oct. 17	Goldsworthy, Leonard James		11 Oct. 18	
	15 Oct. 17	Grigg, Edward Clifford		15 Oct. 18	
	17 Oct. 17	Latimer, Hugh (172*a*		17 Oct. 18	
450	18 Oct. 17	Butler, Harold Edwin		18 Oct. 18	
	18 Oct. 17	Jacob, John Reginald (207*a*)		18 Oct. 18	
	22 Oct. 17	Bles, David Geoffrey (207*a*)		22 Oct. 18	
	24 Oct. 17	Lynch-Blosse, Cecil Eagles (207*a*) (*temp. Capt.* 4 *Mar.* 18)		24 Oct. 18	
	31 Oct. 17	Cowie, William Patrick		31 Oct. 18	
455	31 Oct. 17	Martin, William Drummond		31 Oct. 18	
	31 Oct. 17	Harrison, Harold Stuart		31 Oct. 18	
	31 Oct. 17	Austin, James Bruce Gordon		31 Oct. 18	
	31 Oct. 17	Dasent, Lawrence Hastings		31 Oct. 18	
	1 Nov. 17	Hewett, Douglas Pearse (47*a*, 51, 207*a*)		1 Nov. 18	
460	3 Nov. 17	Clear, Thomas		3 Nov. 18	
	7 Nov. 17	Taffs, Leslie Hugh		7 Nov. 18	
	8 Nov. 17	Rutherford, Thomas George		8 Nov. 18	
	12 Nov. 17	Hudson, Percy Brereton		12 Nov. 18	
	14 Nov. 17	Robertson, John		14 Nov. 18	
465	15 Nov. 17	Fraser, Donald Stuart (172)		15 Nov. 18	
	20 Nov. 17	Collis, Maurice Stewart		20 Nov. 18	
	21 Nov. 17	Powell, Walter Samuel		21 Nov. 18	
	25 Nov. 17	Brown, Ernest Hugh		25 Nov. 18	
469	28 Nov. 17	Franks, Stanley Harold		28 Nov. 18	

2nd-Lieutenants—67.

No.	Date of First Commn.	Names.	Date of birth.	Date of rank.	Remarks.
1	15 June 06	Copeland, Theodore	7 Oct. 78	15 June 06	
	21 June 07	Maffey, John Loader, C.I.E.	1 July 77	21 June 07	
	2 Apl. 15	Goulding, Christopher Charles (*temp. Capt.*, 28 *May* 16)		2 Apl. 15	
	21 Aug. 16	Richard, Charles Millar		21 Aug. 16	
5	31 Oct. 17	Bromham, Arthur William		31 Oct. 17	

* Without pay and allowances of the rank.

CAVALRY—*concld.*

2nd-Lieutenants—*contd.*

No.	Date of First Commn.	Names.	Date of birth.	Date of rank.	Remarks.
6	1 Dec. 17	Leach, Frank Burton (207*a*)		1 Dec. 17	
	19 Dec. 17	Walker, Frank Leitch		19 Dec. 17	
	22 Dec. 17	Lloyd, John Thomas		22 Dec. 17	
	25 Dec. 17	Huggins, Harold Douglas		25 Dec. 17	
10	25 Dec. 17	Hall, Robert George		25 Dec. 17	
	28 Dec. 17	Brown, James Birch		28 Dec. 17	
	1 June 18	Pugh, Roland Anthony Corderon		1 June 18	
	3 Jan. 18	Digby, Everard (207*a*)		3 Jan. 18	
	16 Jan. 18	Mackay, Donald Henry		16 Jan. 18	
15	1 Feb. 18	Story, Harry Wakefield		1 Feb. 18	
	1 Feb. 18	Skinner, James Richard Ross (207*a*)		1 Feb. 18	
	5 Feb. 18	Armstrong, Charles Frederick		5 Feb. 18	
	15 Feb. 18	Ball, Arthur Harry		15 Feb. 18	
	19 Feb. 18	Pearson, Harold Frederick		19 Feb. 18	
20	19 Feb. 18	Reilly, Ernest Fredrick James		19 Feb. 18	
	16 Mar. 18	McDougall, Frederick Alexander		16 Mar. 18	
	10 Apl. 18	Bellairs, Seaforth Stephen		10 Apl. 18	
	25 Apl. 18	Preston, Arthur Kerrison		25 Apl. 18	
	13 May 18	Watson, Edward Leslie		13 May 18	
25	16 May 18	Pratt, Charles Thomas Edward		16 May 18	
	27 May 18	Edmondson, Harold Collmann		27 May 18	
	27 May 18	Carter, Edwin John		27 May 18	
	27 May 18	Ross, John Herbert Chisholm		27 May 18	
	29 May 18	Kennedy, William Denis (*temp. Maj.*, 29 *May* [18)		29 May 18	
30	29 May 18	Barrington, Arthur Harry Manliffe		29 May 18	
	31 May 18	Kemp, Walter Noel		31 May 18	
	1 June 18	Wilmer, Douglas Horsford		1 June 18	
	4 June 18	Ingen, Carl Trevor Van		4 June 18	
	7 June 18	Tait, John		7 June 18	
35	10 June 18	Barry, Charles Bissell		10 June 18	
	11 June 18	Gratton, John Stuart		11 June 18	
	15 June 18	Hartley, Ernest Richard		15 June 18	
	27 June 18	Staughton, Arthur John		27 June 18	
	24 Aug. 18	Sturdy, Edwin Carlile		24 Aug. 18	
40	27 Aug. 18	Forrester, George Douglas		27 Aug. 18	
	28 Aug. 18	Waddell, John William		28 Aug. 18	
	31 Aug. 18	Mather, William Mitchell		31 Aug. 18	
	1 Sep. 18	Drake, John Collard Bernard		1 Sep. 18	
	4 July 18	Marsden, Percy		4 July 18	
45	13 July 18	Casson, Lionel Bell		13 July 18	
	15 July 18	Bennett, John Waller		15 July 18	
	15 July 18	Kidd, Hugh Beresford		15 July 18	
	24 July 18	Milton, Alfred Martin (*temp. Capt.* 24 *July* 18)		24 July 18	
	27 July 18	Badham, John Rupert Campbell		27 July 18	
50	10 Sep. 18	Bald, Alexander		10 Sep. 18	
	11 Sep. 18	Fanshawe, Wilfred		11 Sep. 18	
	14 Sep. 18	Hunt, Cecil Vere		14 Sep. 18	
	18 Sep. 18	Albert, Edward Francis		18 Sep. 18	
	19 Sep. 18	Scott, Walter Seton		19 Sep. 18	
55	19 Sep. 18	Hackett, Harold Frederick		19 Sep. 18	
	19 Sep. 18	Tyler, Edward William		19 Sep. 18	
	20 Sep. 18	Matthews, William Harwood		20 Sep. 18	
	21 Sep. 18	Cargill, James Robert		21 Sep. 18	
	21 Sep. 18	Bingemann, Leslie Steel		27 Sep 18	
60	8 Oct. 18	Munns, Philip Geoffrey		8 Oct. 18	
	19 Oct. 18	Cockrell, Gerald Harcourt		19 Oct. 18	
	25 Oct. 18	Blackburne, Herbert Henshaw		25 Oct. 18	
	31 Oct. 18	Spencer, William		31 Oct. 18	
	31 Oct. 18	Wearn, Alfred William		31 Oct. 18	
65	12 Aug. 18	Withinshaw, Leslie		12 Aug. 18	
	12 Aug. 18	Elliot, Archibald Campbell		12 Aug. 18	
67	14 Aug. 18	Rolfe, John Shirley		14 Aug. 18	

INFANTRY.

Majors—8.

No.	Date of First Commn.	Names.	Date of birth.	Date of rank.	Remarks.
1	24 Dec. 14	Dawson, Ernest (51, 72*b*, 79, 104)	. . . .	24 Dec. 14	
	24 Dec. 14	Coen, William Whelan, V.D. (79)	. . . .	24 Dec. 14	
	5 Jan. 15	Tyndale-Biscoe, Edward Carey	. . . .	5 Jan. 15	
	22 Jan. 97	Miller, William Charles Walmer	19 Nov. 70	1 Sep. 15	Released from Army Service.
5	15 Feb 97	Morin, Archibald Henry, D.S.O., V.D. (d.l) (79)	4 Mar. 70	1 Sep. 15	
	18 Feb. 98	Proes, Ernest Marinus	28 Oct. 71	1 Sep. 15	
	30 Dec. 15	Warburton, Harris Grant (54, 79)	. . . .	30 Dec. 15	
8	7 July 16	Iles, Francis William	. . . .	7 July 16	

Captains—203.

No.	Date of First Commn.	Names.	Date of birth.	Date of rank.	Remarks.
1	3 Sep. 09	Wood, Arthur W. (79)	1 Aug. 83	3 Sep. 09	
	16 Aug. 13	Mellis, Atherley Renfrew M.C. (51, 52, 57, 71, 95, 137, 162)	2 Oct. 85	16 Aug. 13	
	11 Dec. 03	Phillips, George Alexander, V.D. (*Ty. Maj., 27 Sep. 17*)	23 June 73	14 Sep. 14	
	9 May 05	White, Henry Geo. Norman (79, 207)	17 Feb. 76	19 Sep. 14	
5	12 Oct. 00	Tobit, Percy Keymer	21 Sep. 70	19 Sep. 14	
	30 Dec. 04	Egerton, William	19 May 78	23 Sep. 14	
	13 Nov. 14	Arthur, Edmond John (*Ty. Maj.*) *Service for promn. 9 yrs. 65d. from 25 Feb. 93 to 1 May 02.*	. . . .	13 Nov. 14	
	27 Nov. 14	Brock, Charles Henry (54, 79) (*Ty. Maj., Gen. List*) *Serv., for promn.—Emb. Volrs. 28 Oct. 14 to 27 Nov. 14.*	31 Dec. 78	27 Nov. 14	
	26 Jan. 15	Berkeley, Arthur Mowbray, C.I.E. (71, 79) (*Ty. Maj., 24 Oct. 16*)	. . . .	26 Jan. 15	
10	18 Feb. 15	John, Anthony William Ulysses, V.D. (54, 79)	. . . .	18 Feb. 15	
	5 Mar. 15	Lakin, Colin Arthur	. . . .	5 Mar. 15	
	4 Dec. 14	Pope, Hugh Lethbridge (57) *Service for promn. 11 yrs. 14d. from 24 Dec. 02 to 7 Jan. 14.*	4 May 84	9 Apl. 15	
	23 Apl. 15	Mabbett, Percy Whereat (79)	. . . .	23 Apl. 15	
	15 June 15	Ray, Henry Maurice (51, 54, 79)	24 Mar. 78	15 June 15	
15	2 July 15	Davies, Henry Merson, (D. I., 54, 71, 79, 207)	. . . .	2 July 15	
	9 July 15	Walsh, George Gould	. . . .	9 July 15	
	22 July 15	Tulloch, Sylvester Stephen Gregorie, D.I. (54, 79)	. . . .	22 July 15	
	12 Aug. 15	Giffin, Roland Alexander	. . . .	12 Aug. 15	
	7 Apl. 11	Tofts, Crosier Fullerton (79)	17 Nov. 84	1 Sep. 15	
20	7 Apl. 11	Huddleston, George Reginald Graham (54, (79)	5 June 86	1 Sep. 15	
	15 Sep. 10	Corbould, Pelham Stewart (d.I) (79)	28 Sep. 80	1 Sep. 15	
	4 Nov. 10	Tayler, Henry Holroyd (50, 79, 80)	9 Nov. 84	1 Sep. 15	
	2 Feb. 11	O'Brien, Joseph (79) (*temp. Maj., 13 Sep. 8*)	7 Apl. 79	1 Sep. 15	
	5 Aug. 14	Grant, John Francis (207*a*)	. . . .	1 Sept. 15	
25	12 Sep. 14	Mottram, Piercey John McQueen	. . . .	1 Sep. 15	
	3 Apl. 15	Lang, Wynyard Freeling (71) *Service for promn. 7 yrs. 176d. from 9 Aug. 93 to 1 Feb. 01.*	. . . .	1 Sep. 15	
27	15 June 16	Smith, Cuthbert Mervyn *Service for promn. 6 yrs. 336d. from 19 May 00 to 20 Apl. 07.*	. . . .	†1 Sep. 15 15 June 16	

Without pay and allowance prior to 1 Sep. 16.

INFANTRY—*contd.*

Captains—*contd.*

No.	Date of First Commn.	Names.	Date of birth.	Date of rank.	REMARKS.
28	19 Oct. 15	Graham, Charles Townley (79, 82)	. . . .	19 Oct. 15	
	25 Oct. 15	Dunn, Frederick Eustace, V.D. (54, 79)	. . . .	25 Oct. 15	
30	1 Jan. 16	Toohey, John	. . . .	1 Jan. 16	
	4 June 15	Wildeblood, John Peake (*Actg. Maj., 11 July 18*)	. . . .	14 Mar. 16	
	30 Oct. 14	Evans, Reginald John	. . . .	16 Mar. 16	
	12 Feb. 15	Roger, George Bennett	. . . .	5 May 16	
	11 Dec. 14	MacDougall, Donald (79)	. . . .	1 July 16	
35	21 July 16	Russell, John Bertram	. . . .	21 July 16	
	2 Aug. 12	Arbuthnot, Clifford William Ernest (46, 79, 207*a*)	13 Feb. 85	2 Aug. 16	
	18 June 15	Inglis, Thomas Archibald *Serv. for pension and promotion counts from 2nd Aug. 12.*	. . . .	† 2 Aug. 16	
	27 Nov. 14	Ives, George Marquis Paterson, M.C.	30 Jan. 81	19 Sep. 16	
	13 Nov. 14	Lemon, Guy Talbot (79)	29 Jan. 86	19 Sep. 16	
40	20 Nov. 14	Venables, Arthur Vincent, M.C.	. . . .	19 Sep. 16	
	27 Nov. 14	Alderman, Robert Edward, C.I.E.	. . . .	19 Sep. 16	
	8 Jan. 15	Sargon, Arthur Irons	. . . .	19 Sep. 16	
	14 Jan. 15	Cosby, Noel Robert Charles, M.C. (57)	. . . .	19 Sep. 16	
	18 Feb. 15	Phelps, Arthur Austin	. . . .	19 Sep. 16	
45	26 July 15	Bland, Ethelbert George	. . . .	19 Sep. 16	
	4 Oct. 12	Bennet, William Ryland Moore, M.C. (79)	22 May 82	4 Oct. 16	
	1 Nov. 12	Inkson, Norman Leopold, D.S.O., (51, 79)	31 Oct. 82	1 Nov. 16	
	22 Jan. 15	Carter, Frederick George	. . . .	14 Nov. 16	
	19 Mar. 15	Townsend, Frederick Owen, D.S.O. (*Ty. Maj., 9 Dec. 16*)	. . . .	9 Dec. 16	
50	19 Dec. 16	Thirkell-White, Eric Howard	. . . .	19 Dec. 16	
	5 Feb. 17	Sparke, Harry Charles, V.D. (79, 206) (*Ty. Lt.-Col. 5 Feb. 17*).	. . . .	5 Feb. 17	
	20 Feb. 17	Irving, Miles, D. I.(79), (*Ty.Lt.-Col. 20 Feb. 17.*)	21 Aug. 76	20 Feb. 17	
	2 Mar. 17	Stephenson, Frank Harry Albert, V.D. (54, 79, 207*b*)	. . . .	2 Mar. 17	
	13 Apl. 17	Fox, Henry Benedict	. . . .	13 Apl. 17	
55	25 Apl. 13	Travers, Otho Algernon (66*a*, 71, 79, 105)	16 Feb. 82	25 Apl. 17	
	10 May 17	Tainsh, Joseph Ramsay, V.D. (79) (*Ty. Maj., 2 Apl. 17.*)	. . . .	10 May 17	
	20 May 17	Chrystall, Andrew	. . . .	20 May 17	
	20 June 13	Wigfall, Reginald Horace (54, 79)	29 Oct. 82	20 June 17	
	27 June 17	Pullar, Edgar John	. . . .	27 June 17	
60	18 July 17	Crump, Ethelbert William Ernest	. . . .	18 July 17	
	5 Mar. 15	Pryor, Thomas, D.S.O. (*Ty. Maj., 20 July 17*).	. . . .	20 July 17	
	24 July 17	Waddington, Charles Willoughby, C.I.E., M.V.O. (*Ty. Lt.-Col., 24 July 17.*)	. . . .	24 July 17	
	13 Nov. 14	Gammon, John Charles	. . . .	31 July 17	
	5 Aug. 14	Jones, Reginald Trevor	. . . .	5 Aug. 17	
65	5 Aug. 14	Claridge, John Watson Lawson	. . . .	5 Aug. 17	
	23 July 15	Cooper, Henry Lionel	. . . .	18 Aug. 17	
	18 Aug. 14	Nicholas, Richmond Horace (52, 54, 79)	. . . .	18 Aug. 17	
	2 Sep. 14	Stokes, Cyril Lovesy Lawrence	. . . .	2 Sep. 17	
	1 Oct. 17	Buzzard, Archibald Dougan	. . . .	1 Oct. 17	
70	2 Oct. 14	Hornsby, Norman (79, 172*a*)	. . . .	2 Oct. 17	
	29 Oct. 14	Williamson, Frederick	. . . .	29 Oct. 17	
	9 Apl. 15	Neild Ralph (52)	. . . .	1 Nov. 17	
	5 Aug. 14	Rooker, Guy (57)	. . . .	1 Nov. 17	
	20 Nov. 14	Woodhead, Alexander William (54, 79)	27 Apl. 89	20 Nov. 17	
75	20 Nov. 14	McLean, Robert Reginald Balfour	. . . .	20 Nov. 17	
	28 May 15	Knowland, George Henry	. . . .	23 Nov. 17	
	27 Nov. 14	Burne, Rainald Hugo, M.C.	. . . .	27 Nov. 17	
	27 Nov. 14	Burroughs, Geoffrey Herbert Gore (79)	15 Apl. 88	27 Nov. 17	
	27 Nov. 14	Scovell, George Vance Fielding	. . . .	27 Nov. 17	
80	4 Dec. 14	Cheshire, Hugh Stanley (79)	25 Mar. 75	4 Dec. 17	
	19 Mar. 15	Hotson, John Ernest Buttery, O.B.E. (79, 94, 132)	. . . .	16 Dec. 17	
	14 May 15	Kemp, Kenneth Mcintyre *Serv. for promn.—Emb. volrs., 16 Dec. 14 to 14 May 15.*	. . . .	16 Dec. 17	
	24 Dec. 14	Saunders, Alan	26 Aug. 87	24 Dec. 17	
	1 Dec. 15	David, Thomas	. . . .	1 Jan. 18	
85	8 Jan. 15	Johnstone, William Wallace, V.D. (54, 79, 95)	. . . .	8 Jan. 18	

† Without pay and allowances prior to 1 Sep. 16.

INFANTRY—*contd.*

Captains—*contd.*

No.	Date of First Commn.	Names.	Date of birth.	Date of rank.	REMARKS.
86	13 Nov. 14	Hance, Henry Malkin, D.S.O., M.C.		13 Jan. 18	
	24 Dec. 14	Tyler, Bertram Maurice Cohen (54, 66, 82*a*)	14 Jan. 86	20 Jan. 18	
	24 Dec. 14	Nesbitt, Edward Arthur		20 Jan. 18	
	27 Mar. 14	Hodson, Shirley Samuel	17 July 79	27 Mar. 18	
90	12 Feb. 15	Brunton, John Evelyn Cardet (79)		12 Feb. 18	
	12 Feb. 15	Mossop, John Craig Christie		12 Feb. 18	
	12 Mar. 15	Newman, Henry		12 Mar. 18	
	12 Mar. 15	Wootten, Eric Herbert Wootten (51, 99)		12 Mar. 18	
	4 Dec. 14	Ralph, Sidney Arthur		26 Mar. 18	
95	29 Mar. 18	Thomas, Robert Shand		29 Mar. 18	
	15 Apl. 15	MacKay, George Sutherland		15 Apl. 18	
	23 Apl. 15	Ebden, James Wylde		23 Apl. 18	
	30 Apl. 15	Dowley, Francis Michael (207*a*)		30 Apl. 18	
	30 Apl. 15	Courthope, Edward Arthur (79, 206)		30 Apl. 18	
100	30 Apl. 15	Stapleton, Henry Ernest		30 Apl. 18	
	14 May 15	Corlett, Maceolm James John		14 May 18	
	21 May 15	Lace, Ernest William		21 May 18	
	21 May 15	Potter, Lawrence Robert		21 May 18	
	13 Nov. 14	Culverwell, Joseph Stanley (79)	27 Apl. 89	3 June 18	
105	20 Nov. 14	Rooth, Albert Victor, M.C.		3 June 18	
	18 Dec. 14	Duncan, Leonard Fenton (46)	3 June 86	3 June 18	
	9 July 15	Robinson, Arthur Edgar Watson		3 June 18	
	7 Aug. 15	Newitt, Dudley Maurice		3 June 18	
	15 Dec. 15	Alexander, Colin Gibbon		3 June 18	
110	1 Jan. 16	Grant, Alan Gordon		3 June 18	
	17 Apl. 16	Fisk, Ernest Arthur		3 June 18	
	6 June 15	Watson, Hugh (79)		6 June 18	
	17 June 15	Rodger, Alexander James, M.C.		17 June 18	
	21 June 18	Hadow, Douglas Scott		21 June 18	
115	1 July 15	Bowen, Charles Henry Croasdaile		1 July 18	
	2 July 18	Houldey, John Emeris		2 July 18	
	15 July 15	Kennedy, Kenneth Hartley Heathcote, M.C.		15 July 18	
	31 Dec. 14	Hodges, Horace Sydney	25 July 91	16 July 18	
	24 Nov. 14	Thornewill, Allan Stafford		22 July 18	
120	5 Aug. 14	Wadeson, Henry Harman (60*a*, 207)		5 Aug. 18	
	5 Aug. 14	Fletcher, Howard		5 Aug. 18	
	2 Dec. 14	Keeling, Edward Herbert (79) (*Serv. for prom.—Emb. volr., 5 Aug. 14 to 21 Dec. 14.*)		5 Aug. 18	
	27 Apl. 15	Doyle, John Patrick (*Serv. for prom.—Emb. volr., 8 Aug. 14 to 27 Apl. 15*).		5 Aug. 18	
	5 Aug. 14	Cade, Charles Maurice Dakeyne		5 Aug. 18	
125	5 Aug. 14	Bloxham, Hugh Fitzherbert		5 Aug. 18	
	5 Aug. 14	Rawlings, Geoffrey Nares		5 Aug. 18	
	5 Aug. 14	Boxer, Frederick Philip		5 Aug. 18	
	5 Aug. 14	Flemington, Robert Russell		5 Aug. 18	
	5 Feb. 15	Bell, John Golden		6 Aug. 18	
130	5 Aug. 14	Falcon, Robert Rex Berney (57, 172)		5 Aug. 18	
	8 Aug. 14	Brodie, William Henry, Haddon Chambers (57)		8 Aug. 18	
	16 Aug. 14	Steuart, Finlay Graeme, M.C.	4 Feb. 83	16 Aug. 18	
	1 Nov. 14	Frankau, George Neuberg (66)		19 Aug. 18	
	20 Aug. 14	Humphrey, John (*temp. Maj., 7 Mar. 18*)		20 Aug. 18	
135	21 Aug. 14	Horton-Bennett, Alfred George Ranelagh		21 Aug. 18	
	22 Aug. 14	Shelswell, Oscar Henry Clive		22 Aug. 18	
	23 May 16	Furze, Thomas Edwin		25 Aug. 18	
	26 Aug. 14	Mudie, Robert Francis (51)		26 Aug. 18	
	2 Sep. 14	Willison, William Leonard Pollard		2 Sep. 18	
140	29 Aug. 14	Sellick, Hurbert Drayton (57, 206)		29 Aug. 18	
	29 Aug. 14	Westmacott, Peter Germanus (207)		29 Aug. 18	
	28 Aug. 14	Huss, Sydney Arthur		28 Aug. 18	
	29 Aug. 14	Matthews, Bernard Frank (*Ty. Maj., 1 Aug. 18*)		29 Aug. 18	
	4 Sep. 14	Spink, Harold Hans Marshall	18 Feb. 87	4 Sep 18	
145	6 Sep. 14	Thomas, Cecil Ernest		6 Sep. 18	
	7 Sep. 14	Armitage, Kenneth Lionel Frederick		7 Sep. 18	
	10 Sep. 14	Millar, Arthur Scoular		10 Sep. 18	
	10 Sep. 18	Paxton, Walter Macleod		10 Sep. 18	
	12 Sep. 14	Mackenzie-Rogan, Henry		12 Sep. 18	
150	14 Sep. 14	Palmer, Henry Alleyn		14 Sep. 18	

INFANTRY—*contd.*

Captains—*concld.*

No.	Date of First Commn.	Names.	Date of birth.	Date of rank.	REMARKS.
151	16 Sep. 14	Marwood, Sidney Lionel (57, 172*a*, 206)	....	16 Sep. 18	
	16 Sep. 14	Gibbs, Frederick Josiah	....	16 Sep. 18	
	22 Sep. 14	Kinns, Percy Charles	....	22 Sep. 18	
	22 Sep. 14	Dawson, William Harrington Hulton, M.C.	....	22 Sep. 18	
155	14 Sep. 14	Smith, Basil St. John (51, 54, 79, 161)	....	24 Sep. 18	
	25 Sep. 14	Macallum, Angus Hamilton	....	25 Sep. 18	
	26 Sep. 14	Kittoe, Marmaduke Claude Sutton Hamilton (207*a*)	....	26 Sep. 18	
	26 Sep. 14	Livesey-Wardle, Reginald Martin	....	26 Sep. 18	
	1 Oct. 14	Osborne, Frank William	....	1 Oct. 18	
160	4 Dec. 14	Brierley, Richard	....	1 Oct. 18	
	3 Oct. 14	Burnett, Andrew Stuart	....	3 Oct. 18	
	3 Oct. 14	Wemyss, Alan Binfield	10 Dec. 86	3 Oct. 18	
	7 Oct. 14	Priestley-Evans, Mansfield	...	7 Oct. 18	
	13 Nov. 14	Foreman, Cecil Lennox	30 Dec. 86	9 Oct. 18	
165	13 Oct. 14	Hargrave, Carroll Featherstone	....	13 Oct. 18	
	16 Oct. 14	Clifford, William Henry Richard	26 July 83	16 Oct. 18	
	16 Oct. 14	Cook, Albert Frederick	....	16 Oct. 18	
	27 Nov. 14	Fuge, Thomas Llewellyn Marshall. *Serv. for promn. to count from 17 Oct. 14.*	31 Mar. 91	17 Oct. 18	
	19 Oct. 14	Judkins, Bernerd Eimer Heygate	....	19 Oct. 18	
170	21 Oct. 14	Hearn, Edmund Risley	...	21 Oct. 18	
	21 Oct. 14	Rawlingson, Hugh George	....	21 Oct. 18	
	21 Oct. 14	Williams, Vyvyan Stephen Michael	....	21 Oct. 18	
	23 Oct. 14	Edelston, Edward, M.C.	....	23 Oct. 18	
	5 Mar. 14	Hickie, John Corbett	...	29 Oct. 18	
175	30 Oct. 14	Richards, Albert	1 Nov. 89	30 Oct. 18	
	30 Oct. 14	Hurst, Frank William	....	30 Oct. 18	
	30 Oct. 14	Crawford, Hugh	29 Mar. 85	30 Oct. 18	
	30 Oct. 14	Dunning, Herbert Guy	....	30 Oct. 18	
	30 Oct. 14	Keeling, Edward Philip (172*a*)	12 Mar. 92	30 Oct. 18	
180	30 Oct. 14	MacDermott, Michael Joseph, C.I.E.	11 Feb. 86	30 Oct. 18	
	30 Oct. 14	Menzies, Frederic Auber, M.C.	8 Feb. 83	30 Oct. 18	
	30 Oct. 14	Earp, Arthur Clifford	....	30 Oct. 18	
	30 Oct. 14	Jones, Frank Herbert	....	30 Oct. 18	
	1 Nov. 14	Alexander, George (*Ty. Maj.*, 17 *Sep.* 18)	....	1 Nov. 18	
185	5 Nov. 14	Beney, George Eric	....	5 Nov. 18	
	13 Nov. 14	Bennett, Leslie Arthur	....	13 Nov. 18	
	13 Nov. 14	Marshall, William Malcolm Campbell	26 May 95	13 Nov. 18	
	13 Nov. 14	lePatourel, Edward Claude, M.C.	15 Jan. 93	13 Nov. 18	
	13 Nov. 14	Ubsdell, Arnold Roy	12 Feb. 91	13 Nov. 18	
190	13 Nov. 14	Joseph. Arthur Frederick (52, 79)	3 Oct. 85	13 Nov. 18	
	13 Nov. 14	Hart, Edward John Alexander	....	13 Nov. 18	
	20 Nov. 14	Gray, Sidney Thomas	4 Apl. 90	20 Nov. 18	
	20 Nov. 14	Batten, Edward Fetherstonhaugh, M.B.E.	....	25 June 18	
	20 Nov. 14	MacGilivray, Duncan Grant (107, 207)	....	20 Nov. 18	
195	20 Nov. 14	Preston, Evan Bertram Charles, M.C.	17 June 83	20 Nov. 18	
	20 Nov. 14	Eiloart, Bernard Hastings, M.C. (207*a*)	....	20 Nov. 18	
	5 Mar. 15	O'Connor, Gerald Butler	....	20 Nov. 18	
	27 Nov. 14	Stone, Arthur Percival (*Ty. Capt.*, 2 *Nov.* 18)	....	27 Nov. 18	
	27 Nov. 14	Shove, Norman Arthur	3 Feb. 86	27 Nov. 18	
200	27 Nov. 14	Nicolle, Arthur Villeneuve	22 Aug. 91	27 Nov. 18	
	29 Nov. 14	Nathan, Stanley Lewin (**temp. Capt.*, 21 *Dec.* 17)	....	29 Nov. 18	
	4 Dec. 14	Treloar, Richard Martin	....	4 Dec. 18	
203	18 Dec. 14	Dunning, Alexis George Ashburton, M.C.	....	18 Dec. 18	

*Without pay and allowances of the rank.

INFANTRY—*contd.*

Lieutenants

No.	Date of First Commn.	Names.	Date of birth.	Date of rank.	REMARKS.
	5 Oct. 14	Gribble, Dudley Stewart	. . . .	5 Oct. 14	
	18 Dec. 14	Vaughan-Sheehan, George Martin Reginald	. . . .	19 Mar. 15	
	26 Mar. 15	Darby, George Scott	. . . .	26 Mar. 15	
	28 Apl. 15	Lee, George Jonston	. . . .	28 Apl. 15	
5	11 June 15	Oakley, Charles Benjamin	. . . .	11 June 15	
	19 Aug. 15	Daniels, Charles James Frank Walter	. . . .	1 Sep. 15	
	29 Aug. 14	Churcher, Walter Donald	. . . .	1 Sep. 15	
	[illegible]0 Oct. 14	Phillips, Charles Henry Walpole	29 July 89	17 Sep. 15	Released from Army Service.
	1 Oct. 14	Tilley, William Ernest Cedric	. . . .	1 Oct. 15	
10	3 Oct. 14	Hay, William Rupert (162, 172*a*, 206)	. . . .	3 Oct. 15	
	3 Oct. 14	Peele, Charles Richard DeCourcy (51, 207)	. . . .	3 Oct. 15	
	9 Oct. 14	Flaxman, Hubert James Marlowe (*Ty. Capt., 28 Jan. 18.*)	. . . .	9 Oct. 15	
	9 Oct. 14	Holmes, Arthur Stanislaus	. . . .	9 Oct. 15	
	10 Oct. 14	Moffatt, James	. . . .	10 Oct. 15	
	22 Sep. 14	Ede, Harold Stanley	. . . .	22 Sep. 15	
	5 Aug. 14	Lee, Adam Edward	. . . .	5 Aug. 15	
	2 Nov. 14	Tunnicliffe, Felix Duncan	. . . .	2 Nov. 15	
15	14 Oct. 14	Wilcox, Francis John	. . . .	14 Oct. 15	
	23 Oct. 14	Anderson, Charles Edwin Lindsay (79). (*temp. Capt., 4 Dec. 16*) (*Serv. for prom.—Emb. volr., 26 Sep. 14 to 3 Oct. 14*).	. . . .	15 Oct. 15	
20	16 Oct. 14	Macnee, Eustace Alberic (79)	11 Mar. 85	16 Oct. 15	
	17 Oct. 14	Rollo, Andrew Duncan	. . . .	17 Oct. 15	
	21 Oct. 15	Bates, Harry Stuart	. . . .	21 Oct. 15	
	24 Dec. 14	Haslam, Reginald Ernest	. . . .	24 Oct. 15	
	25 Oct. 14	England, Reginald Simson	. . . .	25 Oct 15	
25	19 Mar. 15	Venn, Tom Walters (79) (*Ty. Capt., 6 Nov. 17*) (*Serv. for prom.—Emb. volr. 26 Oct. 14 to 19 Mar. 15*).	. . . .	26 Oct. 15	
	30 Oct. 14	Powell-Price, John Cadogan (*Actg. Maj., 26 Sep. 18*)	. . . .	30 Oct. 15	
	30 Oct. 14	Hayfield, Bertram (*temp. Capt., 19 Feb. 18*)	. . . .	30 Oct. 15	
	4 Nov. 14	Finglas, Noel Howard	. . . .	4 Nov. 15	
	5 Nov. 14	Towerton, Henry John	. . . .	5 Nov. 15	
30	6 Nov. 14	O'Dea, Mathew Leo Brien Patrick	. . . .	6 Nov. 15	
	6 Nov. 14	Keatinge, Claude Albany (12, 51)	. . . .	6 Nov. 15	
	12 Feb. 15	Midgley, Ernest Arthur (*Serv. for prom.—Emb. volr., 28 Oct. 14 to 29 Jan. 15.*)	. . . .	10 Nov. 15	
	11 Nov. 14	Cooke, Sydney Frank	. . . .	11 Nov. 15	
	13 Nov. 14	Brotherton, John Basil (*Ty. Capt., 4 Feb. 18*)	28 May 89	13 Nov. 15	
35	13 Nov. 14	Lock, Herbert George (51)	. . . .	13 Nov. 15	
	13 Nov. 14	Purves, Percival Francis	7 Mar. 84	13 Nov. 15	
	13 Nov. 14	Turner, Dinis Mahon (51)	. . . .	13 Nov. 15	
	13 Nov. 14	Kelly, Walter Aloysius Doyle	21 May 85	13 Nov. 15	
	13 Nov. 14	Ronaldson, Edward Charles S'John.	5 July 85	13 Nov. 15	
40	13 Nov. 14	Craik, David	. . . .	13 Nov. 15	
	27 Nov. 14	Backhouse, Philip Brockman	4 Feb. 87	16 Nov. 15	
	24 Dec. 14	O'Hagan, Reginald (207*a*)	. . . .	16 Nov. 15	
43	16 Nov. 15	Evans, Stuart, M.C.	. . . .	16 Nov. 15	

INFANTRY—*contd.*

Lieutenants—*contd.*

No.	Date of First Commn.	Names.	Date of birth.	Date of rank.	REMARKS.
44	20 Nov. 14	Covell, Beverley Carthew (*Ty. Capt., 19 Dec. 17*)	. . . .	20 Nov. 15	
45	20 Nov. 14	Brookes, Albert (*Ty. Capt., 21 Aug. 16*)	. . . .	20 Nov. 15	
	20 Nov. 14	Clarke, Percy Scamander (*Actg. Maj., 26 Sep. 18*)	. . . .	20 Nov. 15	
	20 Nov. 14	Jones, Frank Balcombe (46, 207)	10 Sep. 95	20 Nov. 15	
	20 Nov. 14	Shrager, Archibald Lewis	6 Feb. 90	20 Nov. 15	
	20 Nov. 14	Giles, Herbert Edwin (207*a*)	. . . .	20 Nov. 15	
50	20 Nov. 14	Aikman, Edward Henry (*temp. Capt. 4 Feb. 18*)	. . . .	20 Nov. 15	
	20 Nov. 14	Ker, Tom Gillespie	. . . .	20 Nov. 15	
	20 Nov. 14	Goldsmith, William Septimus	. . . .	20 Nov. 15	
	20 Nov. 14	Miles, William Hull	. . . .	20 Nov. 15	
	20 Nov. 14	Hurst, Ivor (*temp. Capt., 26 Apl. 18*)	. . . .	20 Nov. 15	
55	20 Nov. 14	Davies, Arthur	. . . .	20 Nov. 15	
	20 Nov. 14	Stearns, Hubert Dudley	. . . .	20 Nov. 15	
	20 Nov. 14	McFadyen, Archie	. . . .	20 Nov. 15	
	20 Nov. 14	Cardew, John Horace	. . . .	20 Nov. 15	
	20 Nov. 14	Stedman, Howard, M.C. (107)	. . . .	20 Nov. 15	
60	20 Nov. 14	Ingram, George Edward	. . . .	20 Nov. 15	
	20 Nov. 14	Hendry, Frank Coutts, M.C. (*temp. Capt.*)	. . . .	20 Nov. 15	
	20 Nov. 14	Fisk-Aldis, Alfred	22 Feb. 89	20 Nov. 15	
	20 Nov. 14	Davies, John Francis (*temp. Maj., 4 Jan. 18*)	. . . .	20 Nov. 15	
	20 Nov. 14	Galvin, Hugh Fitz-Maurice	11 July 92	20 Nov. 15	
65	20 Nov. 14	Galvin, Leo Evelyn	21 Jan. 95	20 Nov. 15	
	20 Nov. 14	Wright, Douglas Gordon	. . . .	20 Nov. 15	
	20 Nov. 14	Wright, Stanley Melville	. . . .	20 Nov. 15	
	20 Nov. 14	Collins, Lionel	. . . .	20 Nov. 15	
	20 Nov. 14	O'Brien, Maurice Conry	. . . .	20 Nov. 15	
70	20 Nov. 14	Halliday, Hugh Ingle	2 Jan. 92	20 Nov. 15	
	20 Nov. 14	Nicklin, George Norman (207)	2 Dec. 99	20 Nov. 15	
	20 Nov. 14	Gordon, Charles Algernon, M.C.	. . . .	20 Nov. 15	
	20 Nov. 14	Bampton, John Augustus Hamilton	. . . .	20 Nov. 15	
	20 Nov. 14	Horwood, John Collingwood	. . . .	20 Nov. 15	
75	18 Feb. 15	Crealock, William Etienne Wyse	3 Aug. 90	20 Nov. 15	
	2 Apl. 15	Humphrey, David Alexander Reid	. . . .	20 Nov. 15	
	30 Apl. 15	Walker, Arthur William	1 Jan. 84	20 Nov. 15	
	4 Dec. 14	Nugent, Thoms Domenic (*Serv. for promn.—Emb. Volr. 28 Oct. 14 to 8 Nov. 14.*)	. . . .	22 Nov. 15	
	23 Nov. 14	Tanner, Humphrey	. . . .	23 Nov. 15	
80	24 Nov. 14	Hill, Alfred Lyon, M.C. (*temp. Capt., 1 Apl. 17*)	. . . .	23 July 18	
	24 Nov. 14	Fleming, Gerard Ross	. . . .	24 Nov. 15	
	12 Feb. 15	Lloyd, Arthur Cresswell	. . . .	25 Nov. 15	
	27 Nov. 14	Pook, John de Courcy	30 Dec. 87	27 Nov. 15	
	27 Nov. 14	Wilkinson, Harold Arthur (51)	. . . .	27 Nov. 15	
85	27 Nov. 14	Coxe, Knightley Holler	. . . .	27 Nov. 15	
	27 Nov. 14	Duke, Walter Stanley	. . . .	27 Nov. 15	
	27 Nov. 14	O'Reilly, Donald Owen	. . . .	27 Nov. 15	
	27 Nov. 14	Wilson, Kenneth Bain	. . . .	27 Nov. 15	
	27 Nov. 14	Wimbush, Norman Cameron	. . . .	27 Nov. 15	
90	27 Nov. 14	Allan, Andrew	10 June 92	27 Nov. 15	
	27 Nov. 14	Jacobson, Eric John Hart	23 Mar. 94	27 Nov. 15	
	27 Nov. 14	McDermott, Cyril Hugh	. . . .	27 Nov. 15	
	27 Nov. 14	Dobbs, Alexander Francis	29 Dec. 90	27 Nov. 15	
	27 Nov. 14	Venables, Edward Percy	13 Sep. 84	27 Nov. 15	
95	27 Nov. 14	Tredaway, George Herbert	14 Nov. 81	27 Nov. 15	
	27 Nov. 14	Winn, Herbert Ewart	3 May 91	27 Nov. 15	
	27 Nov. 14	Allan, John Caldwell	. . . .	27 Nov. 15	
	27 Nov. 14	Toller, George Gordon Taylor	. . . .	27 Nov. 15	
	27 Nov. 14	Goodall, George Montagu, M.C. (51)	. . . .	27 Nov. 15	
100	27 Nov. 14	Halley, Matthew White	. . . .	27 Nov. 15	
101	27 Nov. 14	Halford-Watkins, John Francis, (*temp. Capt., 26 Apl. 18*)	. . . .	27 Nov. 15	

INFANTRY—*contd.*

Lieutenants—*contd.*

No.	Date of First Commn.	Names.	Date of birth.	Date of rank.	Remarks.
212	27 Nov. 14	Angelo, Sydney Seymour (*temp. Capt., 15 Dec. 17*)	. . . .	27 Nov. 15	
	27 Nov. 14	Duxbury, Ernest Leslie	. . . .	27 Nov. 15	
	27 Nov. 14	Bampton, Ronald Wilson	. . . .	27 Nov 15	
215	27 Nov. 14	Willson, Charles Boswell	5 Jan. 91	27 Nov. 15	
	27 Nov. 14	Mildmay, Arthur Bernard Cardozo	. . . .	27 Nov. 15	
	27 Nov. 14	Fraser, Arthur Havelock Hugh	. . . .	27 Nov. 15	
	27 Nov. 14	Fitzpatrick, Gerald Victor (206) (*temp. Capt., 1 Apl. 18*)	. . . .	27 Nov. 15	
	27 Nov. 14	Barclay, Francis	. . . .	27 Nov. 15	
220	28 Nov 14	Savage, Ettore Gordon	. . . .	28 Nov. 15	
	29 Nov. 14	Jacob, Edwin (*Serv. for prom.—Emb. Volr. 28 Oct. 14 to 4 Jan. 15*).	. . . .	29 Nov. 15	
	1 Dec. 14	Harrison, William Norman	. . . .	1 Dec. 15	
	4 Dec. 14	Green, Ernest John	. . . .	4 Dec. 15	
	4 Dec. 14	McCallum, Daniel Vincent	. . . .	4 Dec. 15	
225	4 Dec. 14	Banon, Henry Maurye	. . . .	4 Dec. 15	
	4 Dec. 14	Widdicombe, Edward Parnell	17 July 76	4 Dec. 15	
	4 Dec. 14	Beck, Homer Clarence	15 Jan. 89	4 Dec. 15	
	4 Dec. 14	Helm, Frederick	. . . .	4 Dec. 15	
	4 Dec. 14	Field, George Lewis	. . . .	4 Dec. 15	
230	4 Dec. 14	May, Burney Henry	. . . .	4 Dec. 15	
	4 Dec. 14	Siedle, Frederick Julian	6 Oct. 89	4 Dec. 15	
	4 Dec. 14	Syer, Leslie Bell	29 Dec. 89	4 Dec. 15	
	4 Dec. 14	Hearn, Edward Stephen (*temp. Capt.*)	. . . .	4 Dec. 15	
	4 Dec. 14	Macdonald, Lauchlan Toremore	17 Jan. 93	4 Dec. 15	
235	4 Dec. 14	Wishart, George	10 July 87	4 Dec. 15	
	4 Dec. 14	Hardwick, Noel DeCourcy (**Ty. Capt. 10 Apl. 17*)	. . . .	4 Dec. 15	
	4 Dec. 14	Head, Francis Carey	. . . .	4 Dec. 15	
	4 Dec. 14	Head, Harold Kenneth	. . . .	4 Dec. 15	
	4 Dec. 14	Lincoln Norman	. . . .	4 Dec. 15	
240	4 Dec 14	Pender, George Tresilian, M.C.	. . . .	4 Dec. 15	
	4 Dec. 14	Blake, William	. . . .	4 Dec. 15	
	4 Dec. 14	Bartholomew, Hugh Lionel	4 Apl. 90	4 Dec. 15	
	4 Dec. 14	Panckridge, Hugh Rahere (*Ty. Capt., 7 Oct. 16*)	2 Oct. 85	4 Dec. 15	
	4 Dec. 14	Crossfield, Cedric Robert Kyffun (207*a*)	29 Apl. 91	4 Dec. 15	
245	5 Dec. 14	Beatty, Harvey Patrick	. . . .	5 Dec. 15	
	7 Dec 14	Caulfield-Giles, Henry Robert	. . . .	7 Dec. 15	
	9 Dec. 14	Balding, Charles Dennis (51, 52, 206)	. . . .	9 Dec. 15	
	9 Dec 14	Carroll, Eric Fawcus	. . . .	9 Dec. 15	
	11 Dec. 14	Higgin, Harold Sinclair (52)	. . . .	11 Dec. 15	
250	11 Dec. 14	Griffin, Thomas (*temp. Capt.*)	28 Aug 64	11 Dec. 15	
	11 Dec. 14	Smyth-Osbourne, Douglas Gregory	29 Dec. 90	11 Dec. 15	
	11 Dec 14	Lowrie, Alfred Edmund (*Ty. Capt., 6 Apl. 17*)	. . . .	11 Dec. 15	
	11 Dec. 14	Powell, Arthur Edward, M.C.	27 Oct. 71	11 Dec. 15	
	11 Dec. 14	Cousins, Montague Arthur Stephen	. . . .	11 Dec. 15	
255	11 Dec. 14	Milligan, John Richard	. . . .	11 Dec. 15	
	11 Dec. 14	Mair, Robert, M.C. (207*a*)	16 July 90	11 Dec. 15	
	11 Dec. 14	Cross, Alfred Butler (*Ty. Capt., 13 Jan. 17*)	. . . .	11 Dec. 15	
	11 Dec. 14	Benson, James Wilson Herbert	. . . .	11 Dec. 15	
	11 Dec. 14	Higginson, Guy St. George	. . . .	11 Dec. 15	
260	1 Dec. 14	Williams, Roland George	. . . .	11 Dec. 15	
	13 Dec. 14	Fagnani, Henry Hutchinson	. . . .	13 Dec. 15	
	11 Dec. 14	Walpole, Percy William	. . . .	15 Dec. 15	
	15 Dec. 14	Martin, Robert Fisher	. . . .	15 Dec. 15	
	15 Dec. 14	Mather, Oswald Harvey	. . . .	15 Dec. 15	
265	15 Dec. 14	Kent, Herbert Burns	. . . .	15 Dec. 15	
	2 Apl. 15	Yates, Donald Russell Martin (*Ty. Capt., 1 Dec. 16*) (*Serv. for prom.—Emb. Volr., 16 Dec. 14 to 2nd Apl. 15*).	. . . .	16 Dec. 15	
	17 Dec. 14	Fortescue, Lionel Stanhope	. . . .	17 Dec. 15	
	18 Dec. 14	Aylward, Douglas, M.C. (*Ty. Capt., 17 Aug. 16.*)	. . . .	18 Dec. 15	
	18 Dec. 14	Thomson, Rowley Stuart	. . . .	18 Dec. 15	
270	18 Dec. 14	Cuscaden, Robert Lyons	14 Mar. 93	18 Dec. 15	
	18 Dec. 14	Jarrad, Evelyn Digby Donne (*Ty. Capt., 10 Aug. 17*)	. . . .	18 Dec. 15	
	18 Dec. 14	Barnes, Douglas Hodson (*Ty. Capt., 3 Oct. 17*) (207*a*)	11 July 89	18 Dec. 15	
273	18 Dec. 14	Jenkins, Douglas Liltington	. . . .	18 Dec. 15	

* Without pay and allowances of the rank

INFANTRY—*contd.*

Lieutenants—*contd.*

	Date of First Commn.	Names.	Date of birth.	Date of rank.	Remarks.
274	18 Dec. 14	Walling, Arthur Henry Boyl	2 June 91	18 Dec. 15	
	18 Dec. 14	Broad, Robert Norman Dymoke (51)	20 Jan. 87	18 Dec. 15	
	18 Dec. 14	Matthews, John Marcus	. . . .	18 Dec. 15	
	18 Dec. 14	Dubois, Harold Anthony (*temp. Capt.*, *24 Dec. 17*) (206)	11 Feb. 92	18 Dec. 15	
	18 Dec 14	Walker, Edward Melville	8 Oct. 91	18 Dec. 15	
	18 Dec. 14	Faithfull, Ferdinand Hastings	31 Mar. 92	18 Dec. 15	
280	18 Dec. 14	Dunning, Cedric William Ashburton	. . . .	18 Dec. 15	
	18 Dec. 14	Barton, Reginald Maurice Searle	. . . .	18 Dec. 15	
	18 Dec. 14	Hill, Henry Lawrence	16 Mar. 94	18 Dec. 15	
	18 Dec. 14	Pigot, Walter Thomas (*Actg. Maj. 26 Sep. 18*)	11 Nov. 87	18 Dec. 15	
	18 Dec. 14	Johnson, Edward Power	7 Apl. 89	18 Dec. 15	
285	18 Dec. 14	Biden, Saden Williams	1 Feb. 82	18 Dec. 15	
	18 Dec. 14	Thorn, Terence Conrad, M.C.	15 Jan. 88	18 Dec. 15	
	18 Dec. 14	Duncan, James Grant (*Ty. Capt., 17 Nov. 17*).	12 May 81	18 Dec. 15	
	18 Dec. 14	Aitken, Allan Beacon, M.C. (*Ty. Capt., 14 July 17*)	. . . .	18 Dec. 15	
	18 Dec. 14	Rogers, Wilfred Percival	5 June 86	18 Dec. 15	
290	22 Dec. 14	Pedly, John Edward	. . . .	22 Dec. 15	
	23 Dec. 14	Hervey, Eric George	. . . .	23 Dec. 15	
	23 Dec. 14	Barry, Eric Hayden	. . . .	23 Dec. 15	
	24 Dec. 14	Hughes, Cuthbert Evans	. . . .	24 Dec. 15	
	24 Dec. 14	Fawcett, Arthur Henry	18 Sep. 92	24 Dec. 15	
295	24 Dec. 14	Mellish, Augustin (*Ty. Capt.*)	. . . .	24 Dec. 15	
	24 Dec. 14	Baxter, Eric Gordon (*temp. Capt.*)	8 May 84	24 Dec. 15	
	24 Dec. 14	Copeland, John Hugh	. . . .	24 Dec. 15	
	24 Dec. 14	Smyth, Guy Charles Gee	. . . .	24 Dec. 15	
	24 Dec. 14	Burke, Maurice	12 Sep. 92	24 Dec. 15	
300	24 Dec. 14	Geary, Herbert Valentine, M.C.	21 May 94	24 Dec. 15	
	24 Dec. 14	Manley, Herbert (172*a*)	. . . .	24 Dec. 15	
	24 Dec. 14	Field, Walter Henry	. . . .	24 Dec. 15	
	24 Dec. 14	Lindop, Adrian Horatius Erskine, M.C.	. . . .	24 Dec. 15	
	24 Dec. 14	Brandon, Francis James	. . . .	24 Dec. 15	
305	24 Dec. 14	Mears, Hubert	7 June 81	24 Dec. 15	
	24 Dec. 14	Cooke, Gilbert Petre	4 Sep. 92	24 Dec. 15	
	24 Dec. 14	Stewart, Robert Harold	1 Sep. 81	24 Dec. 15	
	24 Dec. 14	Lester, Ralph	. . . .	24 Dec. 15	
	24 Dec. 14	Milne, Ewan Mitchell (207*a*)	. . . .	24 Dec. 15	
310	24 Dec. 14	Hayne, Christopher William	. . . .	24 Dec. 15	
	24 Dec. 14	McGurk, Niall Austin (*Ty. Capt.*)	. . . .	24 Dec. 15	
	24 Dec. 14	Slater, Alexander	. . . .	24 Dec. 15	
	24 Dec. 14	Stonehewer, Edwin Horton	. . . .	24 Dec. 15	
	24 Dec. 14	Bracher, Henry Courtney	. . . .	24 Dec. 15	
315	24 Dec. 14	Hughes, James	15 Feb. 86	24 Dec. 15	
	24 Dec. 14	Piggford, Henry Grounds (*Ty. Capt., 25 June 18*)	16 Jan. 80	24 Dec. 15	
	24 Dec. 14	Simon, Clement Erwin	. . . .	24 Dec. 15	
	24 Dec. 14	Crace, John Henry (*temp. Capt. 18 July 17*)	. . . .	24 Dec. 15	
	29 Dec. 14	Swansten, Charles Brian Rowally	. . . .	29 Dec. 15	
320	31 Dec. 14	Flashman, Herbert Charles	. . . .	31 Dec. 15	
	31 Dec. 14	Morgan, Herbert William	. . . .	31 Dec. 15	
	31 Dec. 14	Ashton, Gerald Rossi (*Ty. Capt., 23 Aug. 16.*)	Apl. 88	31 Dec. 15	
	31 Dec. 14	Griffin, Seymour Fortescue Hassel	. . . .	31 Dec. 15	
	31 Dec. 14	Gladstone, Albert Charles	. . . .	31 Dec. 15	
325	31 Dec. 14	Hill, Lawrence Ellison	. . . .	31 Dec. 15	
	31 Dec. 14	Baird, Roland Veitch	. . . .	31 Dec. 15	
	31 Dec. 14	Ward, Francis Kingdon	. . . .	31 Dec. 15	
	1 Jan. 15	Cole, John Trevor	. . . .	1 Jan. 16	
	3 Jan. 15	Edwards, Francis Hinton, (140)	. . . .	3 Jan. 16	
330	4 Jan. 15	Gamble Victor Felix	. . . .	4 Jan. 16	
	5 Jan. 15	Colchester, Maurice Travers	. . . .	5 Jan. 16	
	7 Jan. 15	Wadeson, Frederick Fitzgerald	. . . .	7 Jan. 16	
	8 Jan. 15	Heron-Jones, Cyril Vincent, M.C.	. . . .	8 Jan. 16	
	8 Jan. 15	Stevens, Richard Henry	. . . .	8 Jan. 16	
335	8 Jan. 15	Armitage, Charles Cyrus	. . . .	8 Jan. 16	
	8 Jan. 15	Rogers, Gilbert Noel	. . . .	8 Jan. 16	
	8 Jan. 15	Harman, Sidney Henry (207)	. . . .	8 Jan. 16	
	8 Jan 15	Moriarty, John, M.C.	. . . .	8 Jan. 16	
	8 Jan. 15	Galvin, Alfred	. . . .	8 Jan. 16	
340	8 Jan. 15	Spowart, William Charles	. . . .	8 Jan. 16	
	8 Jan. 15	Flynn, George Ernest Crosby	. . . .	8 Jan. 16	
	8 Jan. 15	Salberg, Frank James (79)	. . . .	8 Jan. 16	
	8 Jan. 15	Hunnybun, Charles	. . . .	8 Jan. 16	
	8 Jan. 15	Maltby, Gerald	. . . .	8 Jan. 16	
345	8 Jan. 15	Bousfield, Henry Thomas Wishart	29 May 91	8 Jan. 16	

INFANTRY—*contd.*

Lieutenants—*contd.*

No.	Date of First Commn.	Names.	Date of birth.	Date of rank.	Remarks.
348	8 Jan. 15	Riley, Denis (*Ty. Capt., 22 Oct. 17*)		8 Jan. 16	
	8 Jan. 15	Smith, Mathew Laurance Corley		8 Jan. 16	
350	9 Jan. 15	Macgregor, Lewis Eric		9 Jan. 16	
	14 Jan. 15	Adams, Gerald (57)		14 Jan. 16	
	14 Jan. 15	Baker, David Mitchell		14 Jan. 16	
	14 Jan. 15	Barker, Henry Edward Raymond		14 Jan. 16	
	14 Jan. 15	Bayly, Philip Francis Beard, M.C.		14 Jan. 16	
355	14 Jan. 15	Bolster, Robert Crofts (46) (*Ty. Capt., 23 Oct. 17*).		14 Jan. 16	
	14 Jan. 15	Buchanan, Donald Seafield		14 Jan. 16	
	14 Jan. 15	Chester, Anthony James Bagot, M.C.		14 Jan. 16	
	14 Jan. 15	Clayton-Cowell, Henry Cecil Flemming		14 Jan. 16	
	14 Jan. 15	Hearsey, Norman Noel (51)		14 Jan. 16	
360	14 Jan. 15	Horsfall, Charles Michael (51)		14 Jan. 16	
	14 Jan. 15	Inglis, Harold Gordon, M.C.		14 Jan. 16	
	14 Jan. 15	Johnson, Eric		14 Jan. 16	
	14 Jan. 15	Knowles, George Johnstone (57)		14 Jan. 16	
	14 Jan. 15	Lushington, Alexander Nelson		14 Jan 16	
365	14 Jan. 15	Martyn, William Waddon (46)		14 Jan. 16	
	14 Jan 15	Mills, Alan William Fenwick		14 Jan. 16	
	14 Jan. 15	Mitchell, Frederick Mardies		14 Jan. 16	
	14 Jan. 15	Morris, Colin Maesmore (46)		14 Jan. 16	
	14 Jan. 15	Nicholas, John William		14 Jan. 16	
370	14 Jan. 15	Postance, John Fredrick Arthur		14 Jan. 16	
	14 Jan. 15	Phillips, Charles Henry Kerr		14 Jan. 16	
	14 Jan. 15	Phillips, William Watt		14 Jan. 16	
	14 Jan. 15	Thornton, Reginald Trelawny, M.C.		14 Jan. 16	
	14 Jan. 15	Reith, Charles Edward William (*Ty. Capt., 10 Sep. 17*)		14 Jan. 16	
375	14 Jan. 15	Rolt, Philip James		14 Jan. 16	
	14 Jan. 15	Thompson, Reginald Dempsey (46, 207*a*)	22 May 92	14 Jan. 16	
	14 Jan. 15	Wilkinson, Douglas Allix		14 Jan. 16	
	14 Jan. 15	Wreford, Reginald George		14 Jan. 16	
	14 Jan. 15	Williams, Kenneth Lawson		14 Jan. 16	
380	15 Jan. 15	Lesmond, Cyril Aloysius		15 Jan. 16	
	15 Jan. 15	Rowland, James Walter (137)		15 Jan. 16	
	15 Jan. 15	Ogilvie, George Hamilton (57)		15 Jan. 16	
	15 Jan. 15	Haslehust, Harold Maitland (137) (*temp. Maj., 30 July 18*)		15 Jan. 16	
	15 Jan. 15	Trotter, Hugh		14 May 16	
385	16 Jan. 15	Kendle, William Ernest Collingwood		16 Jan. 16	
	16 Jan. 15	Nicholas, Wilfred Alban		16 Jan. 16	
	25 Jan. 15	Campbell-Shaw, Herbert Henry *Serv. for prom.—Emb. Volr.—26 Sep. 14 to 3 Oct. 14.*		17 Jan. 16	
	29 Jan. 15	Phillips, Hunter Innis		18 Jan. 16	
	18 Jan. 15	Wilson, Robert Barron (*Ty. Lt.-Col., 20 Aug. 17*)		18 Jan. 16	
390	20 Jan. 15	Copeland, David Patrick, O.B.E. (*Ty. Capt., 19 Apl. 17*)		20 Jan. 16	
	22 Jan. 15	Woosnam, Arthur George (54, 79)		22 Jan. 16	
	22 Jan. 15	Mapp, Charles Lachmere		22 Jan. 16	
	22 Jan. 15	Lorrayne-Wadley, Gordon Cedric		22 Jan. 16	
	22 Jan. 15	McSwiny, Philip Myles		22 Jan. 16	
395	22 Jan. 15	Rouch, Eustace Arthur		22 Jan. 16	
	22 Jan. 15	Bowles, Ranulph Francis		22 Jan. 16	
	22 Jan. 15	Hart, James Hugh Barbour Berry		22 Jan. 16	
	22 Jan. 15	Rogerson, John Alexander (51, 57)		22 Jan. 16	
	22 Jan. 15	Davies, John Pitt		22 Jan. 16	
400	25 Jan. 15	Bramall, Kenneth George McLeod		25 Jan. 16	
	25 Jan. 15	Brett, Douglas Alexander		25 Jan. 16	
	26 Jan. 15	Hardinge, *Hon.* Alexander Henry Louis		26 Jan. 16	
	29 Jan. 15	Webb, Alfred Brooke		29 Jan. 16	
	29 Jan. 15	Whitaker, Charles Francis (*temp. Capt., 5 Oct. 18*).		29 Jan. 16	
405	29 Jan. 15	Crimmin, Charles Clarence (172*a*) *temp. Cap. 19 Mar. 18*		29 Jan. 16	
	29 Jan. 15	Raschen, George Hermann (*temp. Capt., 5 Aug. 18*).	3 Feb. 89	29 Jan. 16	
	29 Jan. 15	Greenfield, Arthur Cyril (*Actg. Maj 26 Sep. 18*)		29 Jan. 16	
	29 Jan. 15	Lecky, Reynolds Lament (79)		29 Jan. 16	
409	29 Jan. 15	Walters, Thomas James Grey Stewart (52, 79).		29 Jan. 16	

INFANTRY—*contd.*

Lieutenants—*contd.*

No.	Date of First Commn.	Names.	Date of birth.	Date of rank.	Remarks.
410	29 Jan. 15	Reid, Donald Eric (207)	....	29 Jan. 16	
	29 Jan. 15	Keatinge, Frederck William George, M.C.	....	29 Jan. 16	
	3 Feb. 15	Pollaco, Robert Henry (*Ty. Capt., 9 Mar. 17*)	....	3 Feb. 16	
	2 Apl. 15	Nissen, Trevor Augustus (*temp. Capt., 4 Feb. 18*)	....	3 Feb. 16	
	28 May 15	Rodgers, Joseph John (57) (*Ty. Capt., 16 Mar. 17*)	....	3 Feb. 16	
415	3 Feb. 15	Davis, William	....	3 Feb. 16	
	4 Feb. 15	Cattell, Arthur Skeels	....	4 Feb. 16	
	5 Feb. 15	Viney, John Elliott	....	5 Feb. 16	
	5 Feb. 15	deClermont, Alec Cecil (*Ty. Capt., 28 Jan. 18*)	6 May 85	5 Feb. 16	
	5 Feb. 15	Keely, Isaac Egbert	....	5 Feb. 16	
420	5 Feb. 15	Jackson, Howard	11 Nov. 86	5 Feb. 16	
	5 Feb. 15	Lee, John Fitzgerald, V.D. (94, 162) (*temp. Capt., 20 Aug. 18*)	....	5 Feb. 16	
	5 Feb. 15	Garrett, Edward Christopher McNevin. (*temp. Capt., 27 Mar. 17*)	3 June 91	5 Feb. 16	
	5 Feb. 15	Bishop, Harry Coghill Watson	15 Apl. 85	5 Feb. 16	
	5 Feb. 15	Buswell, Gordon	27 Feb. 85	5 Feb. 16	
425	5 Feb. 15	Simpson, Carleton Napier (**temp. Capt., 22 Apl. 19*)	8 June 91	5 Feb. 16	
	5 Feb. 15	Robertson, Robert Alick	....	5 Feb. 16	
	5 Feb. 15	Thorburn, Robert, M.C.	....	5 Feb. 16	
	5 Feb. 15	Adams, Percy Tidswell	....	5 Feb. 16	
	5 Feb. 15	Price, Augustus Robert Petley	4 May 85	5 Feb. 16	
430	5 Feb. 15	Hadden, Alfred John (*Ty. Capt., 24 Oct. 17*)	....	5 Feb. 16	
	5 Feb. 15	Clark, Aubrey Matthew	....	5 Feb. 16	
	5 Feb. 15	Ford, Leo Dorey	13 Dec. 86	5 Feb. 16	
	5 Feb. 15	Daley, Frederick Michael	27 June 91	5 Feb. 16	
	5 Feb. 15	Inglis, William Morley	....	5 Feb. 16	
435	8 Feb. 15	Stevens, James Reginald	....	8 Feb. 16	
	10 Feb. 15	Hussey, Glen Owen Gilbert	....	10 Feb. 16	
	10 Feb. 15	Mileham, Cyril Arthur	....		
	12 Feb. 15	Hollands, Ernest William (*Ty. Capt., 16 Jan. 17*)	....	12 July 16	
			....	12 Feb. 16	
	12 Feb. 15	Modget, Leslie Donald	....	12 Feb. 16	
440	12 Feb. 15	Browne, Herbert	....	12 Feb. 16	
	12 Feb. 15	Browne, Charles Ernest Meredith	....	12 Feb. 16	
	12 Feb. 15	Hayes, Lawrence, M.C.	....	12 Feb. 16	
	12 Feb. 15	Farley, Graham Payne (*temp. Capt., 28 Nov. 18*)	....	12 Feb. 16	
	12 Feb. 15	Fane, Arthur George Cecil, M.C.	....	12 Feb. 16	
445	12 Feb. 15	Snodgrass, John (51)	....	12 Feb. 16	
	12 Feb. 15	Perch, Howard Leonard	....	12 Feb. 16	
	12 Feb. 15	Hart, Norman Basset	....	12 Feb. 16	
	12 Feb. 15	Martyn, Joseph King, M.C. (*temp. Capt., 19 June 18*)	24 Apl. 86	12 Feb. 16	
	12 Feb. 15	Hodson, John Stewart Falconer	....	12 Feb. 16	
450	12 Feb. 15	Smith, Gerald Robert	....	12 Feb. 16	
	12 Feb. 15	Shortt, Norman Robert	14 July 83	12 Feb. 16	
	12 Feb. 15	Harvey, John Frederick Barnard (95)	....	12 Feb. 16	
	12 Feb. 15	Gimson, William Morgan	25 Apl. 91	12 Feb. 16	
	12 Feb. 15	Cox, Oswald	....	12 Feb. 16	
455	12 Feb. 15	Ogilvie, Philip Gordon (*Ty. Capt., 1 Mar. 17*)	....	12 Feb. 16	
	12 Feb. 15	Dacre-Tyler, John Lowndes	....	12 Feb. 16	
	12 Feb. 15	Maitland, Alexander (*temp. Capt., 28 Mar. 18*)	....	12 Feb. 16	
	12 Feb. 15	Gardiner, Robert Cecil	....	12 Feb. 16	
	12 Feb. 15	Bell, Howard (207*a*)	....	12 Feb. 16	
460	14 May 15	Tice, Charles Allen (66*a*)	....	12 Feb. 16	
	23 July 15	Chevallier, Francis Temple	....	12 Feb. 16	
	15 Feb. 15	Jellicoe, John Andrea	....	15 Feb. 16	
	17 Feb. 15	Rea, Frederick Arthu Reginald Riviere	....	17 Feb. 16	
	17 Feb. 15	Whittome, John Eric	....	17 Feb. 16	
465	18 Feb. 15	Gordon, John Woodburn Semphill	....	18 Feb. 16	
	18 Feb. 15	Spiller, Thomas Reubon	....	18 Feb. 16	
	18 Feb. 15	Lutter, Walter Frederick (172*a*)	....	18 Feb. 16	
	18 Feb. 15	Clarke, Walter Douglas Montgomery	....	18 Feb. 16	
	18 Feb. 15	Bampton, Richard Ewsters Fitz Symons.	....	18 Feb. 16	
470	18 Feb. 15	Wilkinson, Ronald (51)	....	18 Feb. 16	
	18 Feb. 15	Dempster, Harry Stewart (207*a*)	6 Aug. 86	18 Feb. 16	
	18 Feb. 15	Hearn, John (*temp. Capt.*)	....	18 Feb. 16	
	18 Feb. 15	McLellan, Archibald John	....	18 Feb. 16	
474	18 Feb. 15	Rix, Howard Vivian	....	18 Feb. 16	

* Without pay and allowances of the rank.

No.	Date of First Commn.	Names.	Date of birth.	Date of rank.	REMARKS.

INFANTRY—*contd.*

Lieutenants—*contd.*

No.	Date of First Commn.	Names.	Date of birth.	Date of rank.	REMARKS.
480	18 Feb. 15	Sorby, William John (207*a*)	....	18 Feb. 16	
	18 Feb. 15	Morbey, Ely Harry (*Ty. Capt. 20 Aug. 17*)	....	18 Feb. 16	
	18 Feb. 15	Muir, James Logan	26 July 91	18 Feb. 16	
	13 Feb. 15	McClumpha, Cyril Dudley	28 Apl. 94	18 Feb. 16	
	18 Feb. 15	Mulgrue, James Arthur	20 Jan. 77	18 Feb. 16	
485	18 Feb. 15	Weathrall, Ian Alister MacDonald	10 June 93	18 Feb. 16	
	18 Feb. 15	Such, John	30 Apl. 86	18 Feb. 16	
	18 Feb. 15	Leedham, Walter	....	18 Feb. 16	
	18 Feb. 15	Borlase, John	22 Aug. 90	18 Feb. 16	
	18 Feb. 15	Bennett, Henry Benjamin (*Ty. Capt., 31 Oct. 17*)	16 Mar. 83	18 Feb. 16	
490	18 Feb. 15	Scott, George Duncan (*temp. Capt., 23 May 18*)	6 Dec. 17	18 Feb. 16	
	18 Feb. 15	Beaver, Guy Duncan Campbell (*Ty. Capt., 15 May 16*)	....	18 Feb. 16	
	18 Feb. 15	Herdon, Sydney Philip Norman (*Ty. Capt., 2 Aug. 16*)	....	18 Feb. 16	
	18 Feb. 15	Hoade, John Robert	....	18 Feb. 16	
	18 Feb. 15	Auger, Augustus Edward	....	18 Feb. 16	
495	19 Feb. 15	Wiles, Humphrey Halsey	....	19 Feb. 16	
	21 Feb. 15	Thorogood, Percival Walker	....	21 Feb. 16	
	22 Feb. 15	Rudd, Noel Bateman	....	22 Feb. 16	
	30 Apl. 15	Coles, Charles Erskinner Hennessey	....	23 Feb. 16	
	25 Feb. 15	Morison, Hector McDonald	....	25 Feb. 16	
500	26 Feb. 15	Wilson, Terence Carroll (51, 52) (*Ty. Capt., 21 Apl. 17*)	....	26 Feb. 16	
	26 Feb. 15	Templeton, William Greenwood Hastings (52)	....	26 Feb. 16	
	26 Feb. 15	Craig-Adams, Robert Logan	....	26 Feb. 16	
	26 Feb. 15	Gray, James Wensley, M.C.	....	26 Feb. 16	
	26 Feb. 15	Phillips, Henry John Lawrence	....	26 Feb. 16	
505	26 Feb. 15	Elliot, Clement	....	26 Feb. 16	
	27 Feb. 15	Steane, Samuel William	....	27 Feb. 16	
	27 Feb. 15	Maxwell, Robert	....	27 Feb. 16	
	29 Feb. 15	McKeag, Hugh Theodore Alexander	....	29 Feb. 16	
	4 Mar. 15	Mangin, Paul Henry	....	4 Mar. 16	
510	4 Mar. 15	Martin, Joseph (207*a*)	....	4 Mar. 16	
	5 Mar. 15	Boyes, William Reginald	....	5 Mar. 16	
	5 Mar. 15	Jerram, Martin Ralph Knight, M.C	....	5 Mar. 16	
	5 Mar. 15	Scott-Ruffle, Clifford Fredrick (207*a*)	....	5 Mar. 16	
	5 Mar. 15	Graveston, Henry Blakeman	....	5 Mar. 16	
515	5 Mar. 15	Roland-Price, Arthur (*Ty. Capt.*)	....	5 Mar. 16	
	5 Mar. 15	Stowell, Philip Charles	....	5 Mar. 16	
	5 Mar. 15	Archdale, Fulbert Audley (172*a*) (*temp. Capt.*)	8 Aug. 90	5 Mar. 16	
	5 Mar. 15	Hobbs, Charles Christie Arthur (79)	....	5 Mar. 16	
	5 Mar. 15	O'Connor, Archibald John	....	5 Mar. 16	
520	5 Mar. 15	Hales, George (169) (*Ty. Capt., 23 Nov. 17*)	28 May 87	5 Mar. 16	
	5 Mar. 15	Elliott, George Frederick (*Ty. Capt., 9 Sep. 16.*)	5 Apl. 93	5 Mar. 16	
	5 Mar. 15	Bright, Frederick Charles (*Ty. Capt., 1 Apl. 17*)	....	5 Mar. 16	
	5 Mar. 15	Whigham-Teasdale, Dennis George, M.C. (*temp. Capt., 5 Oct. 18*)	....	5 Mar. 16	
	5 Mar. 15	Gilbert, William Gilbert Lyon (*Ty. Capt., 11 June 17*)	....	5 Mar. 16	
525	5 Mar. 15	Jefferis, Charles Edward (*Ty. Capt., 8 July 16*)	....	5 Mar. 16	
	5 Mar. 15	Duncan, John Grant (*Ty. Capt., 11 Jan. 17*)	....	5 Mar. 16	
	5 Mar. 15	Fennell, Charles Dennis (*Ty. Capt., 14 Apl. 17*)	....	5 Mar. 16	
	8 Mar. 15	Horner, William John	....	8 Mar. 16	
	10 Mar. 15	Scholefield, Maurice Theodore	....	8 Mar. 16	
530	10 Mar. 15	Smith, Vernon Sampson	....	10 Mar. 16	
	10 Mar. 15	Fletcher, James Allen	....	10 Mar. 16	
	10 Mar. 15	Haigh, Percy Barnes (51)	....	11 Mar. 16	
	12 Mar. 15	Lyster, Anthony St. George	....	12 Mar. 16	
	12 Mar. 15	McDermott, John William, C.I.E. (*Ty. Capt.*)	....	12 Mar. 16	
535	12 Mar. 15	Farquharson, Francis Archibald, M.C.	....	12 Mar. 16	
	12 Mar. 15	Kiernander, Octavius Gelhard	....	12 Mar. 16	
	12 Mar. 15	Storrs, Gerald Noel, M.C.	....	12 Mar. 16	
	12 Mar. 15	Field, Frank Moyle Shuttleworth	....	12 Mar. 16	
	12 Mar. 15	Scruby, Joseph (79)	....	12 Mar. 16	
540	12 Mar. 15	Cooper, Alfred Victor	....	12 Mar. 16	
	12 Mar. 15	Craker, Phelip William (206)	....	12 Mar. 16	
	12 Mar. 15	Young, William Hoare Hatchell (79)	....	12 Mar. 16	
543	12 Mar. 15	Gregory, Marcar Sheridan, M.C. (79)	....	12 Mar. 16	

* Without pay and allowances of the rank.

INFANTRY—*contd.*

Lieutenants—*contd.*

No.	Date of First Commn.	Names.	Date of birth.	Date of rank.	Remarks.
544	12 Mar. 15	Woakes, Russell Beaumont, M.C.	. . . . .	12 Mar. 16	
545	12 Mar. 15	Moody, Basil (*temp. Capt., 8 Jan. 18*)	. . . . .	12 Mar. 16	
	13 Mar. 15	Hambly, Richard Luskey	. . . . .	13 Mar. 16	
	15 Mar. 15	Adie, Thomas Dunbar	. . . . .	15 Mar. 16	
	15 Mar. 15	Mills, William Stead	. . . . .	15 Mar. 16	
	17 Mar. 15	Blair, James Richard (*temp. Capt., 15 Oct. 18*)	. . . . .	17 Mar. 16	
550	17 Mar. 15	Macartney, Harry James	. . . . .	17 Mar. 17	
	18 Mar. 15	Stuart-Prince, Dudley (206) (*Ty. Capt., 28 Feb. 18*)	. . . . .	18 Mar. 16	
	19 Mar. 15	Stowell, George Christopher (*Ty. Capt., 17 Apl. 17*)	. . . . .	19 Mar. 16	
	19 Mar. 15	Sturrock, Laurence Anderson	. . . . .	19 Mar. 16	
	19 Mar. 15	Yeld, Richard Kingsley (*Ty. Capt., 4 Dec. 16*)	. . . . .	19 Mar. 16	
555	19 Mar. 15	Gill, George Theodore (51) (*Ty. Capt., 31 Oct. 17*)	. . . . .	19 Mar. 16	
	19 Mar. 15	Watling, Montague Clements Garden	. . . . .	19 Mar. 16	
	19 Mar. 15	Sneyd, Robert Thomas Stuart M.C.	. . . . .	19 Mar. 16	
	19 Mar. 15	Smith, Malcolm Innes Lewis (*Ty. Capt., 30 June 17*)	. . . . .	19 Mar. 16	
	19 Mar. 15	Watkins, Harry	. . . . .	19 Mar. 16	
560	19 Mar. 15	Evans, William Arthur Avenel (51)	. . . . .	19 Mar. 16	
	19 Mar. 15	Ransford, William Reginald (51)	. . . . .	19 Mar. 16	
	19 Mar. 15	Creedy, Harold Douglas	. . . . .	19 Mar. 16	
	20 Mar. 15	De Brath, Eric Ernest (51)	. . . . .	20 Mar. 16	
	20 Mar. 15	Young, Arthur Donovon	. . . . .	20 Mar. 16	
565	22 Mar. 15	Fagan, Arthur William (207) (*temp. Capt., 10 Nov. 18*)	. . . . .	22 Mar. 16	
	25 Mar. 15	Rogers, Joseph Richard Denzil (*temp. Capt. 16 May 18*)	. . . . .	25 Mar. 16	
	25 Mar. 15	Lambert, Edward William (207)	. . . . .	25 Mar. 16	
	25 Mar. 15	Chowns, William Pharoah (95)	. . . . .	25 Mar. 16	
	26 Mar. 15	Wills, Wilfred Ridout (*Ty. Capt.*)	. . . . .	26 Mar. 16	
570	26 Mar. 15	Moylan, William Drummond	. . . . .	26 Mar. 16	
	26 Mar. 15	Crawford, Colin Grant	. . . . .	26 Mar. 16	
	26 Mar. 15	Baxter, Robert Hugh Neville (*temp. Maj., 1 June 18*)	. . . . .	26 Mar. 16	
	26 Mar. 15	Harris, George Wilfred	. . . . .	26 Mar. 16	
	26 Mar. 15	Reid, John Brown (57)	. . . . .	26 Mar. 16	
575	26 Mar. 15	Folkard, Frank Tyhurst	. . . . .	26 Mar. 16	
	26 Mar. 15	Ranger, Cecil Argill	. . . . .	26 Mar. 16	
	27 Mar. 15	Morgan, William Leslie	. . . . .	27 Mar. 16	
	2 Apl. 15	Haigh, Frank Fraser (51)	. . . . .	2 Apl. 16	
	2 Apl. 15	Harper, Richard	. . . . .	2 Apl. 16	
580	2 Apl. 15	Hardwicke, Clement Baldry	. . . . .	2 Apl. 16	
	2 Apl. 15	Salusbury-Trelawny, John Maitland, M.C.	. . . . .	2 Apl. 16	
	2 Apl. 15	Faithfull, Francis Edward Noel (51)	. . . . .	2 Apl. 16	
	2 Apl. 15	Cooper, Charles Herbert (57)	. . . . .	2 Apl. 16	
	2 Apl. 15	Sherlock, Woodford Wright	30 Oct. 82	2 Apl. 16	
585	2 Apl. 15	Sale, John Lewis (95, 165)	. . . . .	2 Apl. 16	
	2 Apl. 15	Pearce, Thomas	. . . . .	2 Apl. 16	
	2 Apl. 15	Ward, William Joseph	. . . . .	2 Apl. 16	
	2 Apl. 15	Green, Charles William Tandy	. . . . .	2 Apl. 16	
	2 Apl. 16	Souter, Leslie Fowler	. . . . .	2 Apl. 16	
590	2 Apl. 15	Rose, George Wright Norton (*Ty. Capt., 1 Dec. 16*)	. . . . .	2 Apl. 16	
	2 Apl. 15	O'Callaghan, Timothy Patrick Moriarty	. . . . .	2 Apl. 16	
	2 Apl. 15	Crisp, Robert Henry	. . . . .	2 Apl. 16	
	2 Apl. 15	Mercer, Arthur Warren	. . . . .	2 Apl. 16	
	2 Apl. 15	Cameron, George Symington, M.C.	. . . . .	2 Apl. 16	
595	2 Apl. 15	Harper, Gerald	. . . . .	2 Apl. 16	
	3 Apl. 15	Darling, Henry John	. . . . .	3 Apl. 16	
	3 Apl. 15	Kendall, Harold John	. . . . .	3 Apl. 16	
	10 July 17	Wailling, William	. . . . .	{ 5 Apl. 16 { 10 July 18	
	8 Apl. 15	Pickmere, Clement Elliots	. . . . .	8 Apl. 16	
600	9 Apl. 15	Riley, Reginald Conduit (51)	. . . . .	9 Apl. 16	

INFANTRY—*contd.*

Lieutenants—*contd.*

No.	Date of First Commn.	Names.	Date of birth.	Date of rank.	REMARKS.
603	9 Apl. 15	Rice, Philip Howard		9 Apl. 16	
	9 Apl. 15	Gibb, James Ballantine		9 Apl. 16	
605	9 Apl. 15	Wise, John Humphrey		9 Apl. 16	
	9 Apl. 15	Atkinson, Edward Dawson (*temp. Capt., 25 Jan. 17*)		9 Apl. 16	
	9 Apl. 15	Clive, John McClintock		9 Apl. 16	
	9 Apl. 15	Cook, Stanley		19 Apl. 16	
610	10 Apl. 15	Ferrier, Charles Duncan (*temp. Capt., 26 Feb. 18*)		10 Apl. 16	
	12 Apl. 15	Bostock-Wilson, Eustace Leslie		12 Apl. 16	
	13 Apl. 15	English, Bertram Mansfield		13 Apl. 16	
	14 Apl. 15	Gerrard, Frederick Wernham (*temp. Capt. 1 June 17*)		14 Apl. 16	
	16 Apl. 15	Ashton, Erick Mossop	16 Oct. 90	16 Apl. 16	
615	17 Apl. 15	Hoggan, George William (51,52)		17 Apl. 16	
	17 Apl. 15	Buchanan, Harry		17 Apl. 16	
	11 June 15	Sulley, James Penderell (*temp. Capt., 31 Dec. 17*)		18 Apl. 16	
	18 Apl. 15	Steele, James Torrance		18 Apl. 16	
	19 Apl. 15	Pickthall, Cecil Marmaduke		19 Apl. 16	
620	19 Apl. 15	Morris, John Claude Tuke	20 Apl. 88	19 Apl. 16	
	19 Apl. 15	Shaw, John Frederick (46, 207*a*)	23 June 93	19 Apl. 16	
	26 July 15	Cook, Thomas Page		19 Apl. 16	
	19 Apl. 15	Scrivener, Arthur Herbert		19 Apl. 16	
	19 Apl. 15	Gibson, Douglas Vincent		19 Apl. 16	
	19 Apl. 15	Evans, John Lewis		19 Apl. 16	
	22 Apl. 15	Coulson, Raymond Arthur		22 Apl. 16	
	22 Apl. 15	Jones, Eric Glanville (48*a*, 51)		22 Apl. 16	
	22 Apl. 15	Seal, John Robert (59, 206)		22 Apl. 16	
625	23 Apl. 15	West, Albert Percy Joynson (95)		23 Apl. 16	
	23 Apl. 15	Rackstraw, George, M.C.		23 Apl. 16	
	23 Apl. 15	Prideaux, Claud William Austice (*temp. Capt. 8 Aug. 16*)		23 Apl. 16	
	23 Apl. 15	Hills, George Gregory	Mar. 82	23 Apl. 16	
	23 Apl. 15	Grant, Charles Alexander (207*a*)		23 Apl. 16	
630	23 Apl. 15	Bristow, Charles Holditch		23 Apl. 16	
	23 Apl. 15	Puckle, Frederick Hale (*temp. Capt.*)		23 Apl. 16	
	23 Apl. 15	Stuart, John Matthew Blackwood (*temp. Capt., 22 Feb. 17*)		23 Apl. 16	
	23 Apl. 15	Higgs, Philip Robert (*temp. Capt.*)		23 Apl. 16	
	15 Jan. 15	Sothers, Donald Bevan		23 Apl. 16	Released from Army Service.
635	23 Apl. 15	Pollard, Armell Richard (*temp. Capt., 25 May 17*)		23 Apl. 16	
	23 Apl. 15	Greany, Maurice Peter Mawe	15 July 94	23 Apl. 16	
	23 Apl. 15	Brown, Leslie Nanscawon		23 Apl. 16	
	23 Apl. 15	Snow, Charles Alfred		23 Apl. 16	
	23 Apl. 15	Blandy, Richard Denis		23 Apl. 16	
640	23 Apl. 15	Davis, Theodore William Godwin (*temp. Maj. 4 Mar. 18*)		23 Apl. 16	
	23 Apl. 15	Taylor, Basil Churton		23 Apl. 16	
	23 Apl. 15	Clayton-Jones, Robert (51) (52)		23 Apl. 16	
	23 Apl. 15	Owen, Vaughan Edward O'Neill		23 Apl. 16	
	24 Apl. 15	Milward, Robert Cecil (206)		24 Apl. 16	
	25 Apl. 15	Jenkin, Reginald Trevor		25 Apl. 16	
645	26 Apl. 15	Mungavin, Henry St. John		26 Apl. 16	
	27 Apl. 15	Oppenheimer, Kurt Michael		27 Apl. 16	
	30 Apl. 15	Fieldsend, Charles Edward, M. C.		30 Apl. 16	
	30 Apl. 15	Hamilton, Robert		30 Apl. 16	
650	30 Apl. 15	Shilstone, William Charles Nelson (*temp. Capt.*)		30 Apl. 16	
	30 Apl. 15	Gregory, John Roger		30 Apl. 16	
	30 Apl. 15	Barnes, William Constantine		30 Apl. 16	
	30 Apl. 15	Bunting, Sheldon Arthur Steward		30 Apl. 16	
	30 Apl. 15	Martin, Ernest Burdsall		30 Apl. 16	
655	30 Apl. 15	Nightingale, Arthur Winfield		30 Apl. 16	
	30 Apl. 15	Lattey, Henry Patrick Tabor		30 Apl. 16	
	30 Apl. 15	Learmond, Alexander William		30 Apl. 16	
	30 Apl. 15	Marriott, Robert Ecklin		30 Apl. 16	
	30 Apl. 15	Knyvett, William Alexander (*temp. Cpt.*)		30 Apl. 16	
660	30 Apl. 15	Sheridan, Francis Sidney		30 Apl. 16	
	30 Apl. 15	Birt, Austin George	23 Jan. 84	30 Apl. 16	
	30 Apl. 15	Macdonald, Lewis John, M.C. (57)		30 Apl. 16	
	30 Apl. 15	Tipper, George Howlett		30 Apl. 16	
	30 Apl. 15	Walters, Edgar Cyril		30 Apl. 16	
	30 Apl. 15	Eden, Edwin Arthur		30 Apl. 16	
665	5 May 15	Smith, Harry Stanford		5 May 16	

INFANTRY—*contd.*

Lieutenants—*contd.*

No.	Date of First Commn.	Names.	Date of birth.	Date of rank.	Remarks.
666	1 Dec. 15	Nicholson, Eustace Marmion Ferrers (79) *Ser. for prom.—Emb. Volr. 6 May 15 to 1 Dec. 15*	. . . .	6 May 16	
	7 May 15	Cantlie, Keith (51, 52, 162)	. . . .	7 May 16	
	7 May 15	Thorp, Maurice	8 July 87	7 May 16	
	7 May 15	Stewart, John Alexander, M. C.	. . . .	7 May 16	
670	7 May 15	Twynam, Henry Joseph (51)	. . . .	7 May 16	
	7 May 15	Dewar, Francis	. . . .	7 May 16	Released from Army Service.
	7 May 15	Eastmond, Archibald, M. C.	. . . .	7 May 16	
	7 May 15	Walton, John Hugh	. . . .	7 May 16	
	7 May 15	Thomson, Herbert James	. . . .	7 May 16	
675	7 May 15	Graham, Harold	. . . .	7 May 16	
	7 May 15	Rushforth, Frank Victor	. . . .	7 May 16	
	11 May 15	Knott, William James	. . . .	11 May 16	
	12 May 15	Pettman, Frank Leslie	. . . .	12 May 16	
	12 May 15	Shepherd, Hugh Vincent (207)	. . . .	12 Mar. 16	
	12 May 15	Seymour-Higgins, Allan Charles Price	. . . .	12 May 16	
680	12 May 15	Bottrill, Edward Richard	. . . .	12 May 16	
	12 May 15	Fisher, Percy	. . . .	12 May 16	
	13 May 15	Chouler, Reay	. . . .	13 May 16	
	13 May. 15	Down, Gordon Oswald	. . . .	13 May 16	
	14 May 15	White, Colin Robert M.C.	. . . .	14 May 16	
685	14 May 15	Wilson, Herbert William Dundas	. . . .	14 May 16	
	14 May 15	Barrack, John Charles	. . . .	14 May 16	
	14 May 15	Walmesly, Herman Joseph	. . . .	14 May 16	
	14 May 15	Glanville, Ernest Alfred	. . . .	14 May 16	
	14 May 15	Petters, Harry Danvers (66*a*, 207)	. . . .	14 May 16	
690	14 May 15	Gildea, Richard Bryan	. . . .	14 May 16	
	14 May 15	Leigh-Bennet, Percy Raymond	. . . .	14 May 16	
	14 May 15	Robertson, Alexander MacLeod, M.C.	. . . .	14 May 16	
	4 May 15	Clifford, Cecil Herbert Windsor (207*a*)	. . . .	14 May 16	
	14 May 15	Paterson, George Stanley Vaughan	. . . .	14 May 16	
695	14 May 15	Gill, Eric Henry Nevile	. . . .	14 May 16	Released from Army Service.
	14 May 15	Pascoe, Edwin Hall	. . . .	14 May 16	
	14 May 15	Carter, William Herbert (57)	. . . .	14 May 16	
	14 May 15	Hicks, Lancelot Joseph (*Ty. Capt. 29 May 17*)	. . . .	14 May 16	
	14 May 15	Worster, Alexander Howett	. . . .	14 May 16	
700	14 May 15	Baldwin, Arthur Charles John	. . . .	14 May 16	
	15 May 15	Dempster, Hamish Frank	. . . .	15 May 16	
	16 May 15	Smyth, Sidney (207*a*)	. . . .	16 May 16	
	18 May 15	Hill, Ernest Douglas	. . . .	18 May 16	
	20 May 15	Young, James (*Ty. Maj. 15 Mar. 18*)	. . . .	20 May 16	
705	20 May 15	Robinson, Frank Calvert	. . . .	20 May 16	
	21 May 15	Richardson, Rosslyn James Dilyell (207) (*Ty. Capt.*)	. . . .	21 May 16	
	21 May 15	Rhodes, Charles Kenneth (52)	5 May 89	21 May 16	
	21 May 15	Gladstone, Stephen Deiniol, M.C. (157)	. . . .	21 May 16	
710	21 May 15	Dickins, Edwin Alfred, V. D. (*Ty. Capt., 9 Sep. 18*)	. . . .	21 May 16	
	21 May 15	Wilson, Charles Henry (52)	. . . .	21 May 16	
	20 Oct. 14	Pike, Wilfred Holton	. . . .	22 May 16	
	22 May 15	Rees, Idwal Desmond	. . . .	22 May 16	
	25 May 15	Hamlet, Ralph Oliver	. . . .	25 May 16	
	28 May 15	Nicholson, Alexander Richard (51)	. . . .	28 May 16	
715	28 May 15	Hayne, Arthur Robert Sinclair	. . . .	28 May 16	
	28 May 15	Criper, Stanley Fuller	. . . .	28 May 16	
	28 May 15	Keddie, Charles Mackay	. . . .	28 May 16	
	28 May 15	Tate, James M.C.	. . . .	28 May 16	
	28 May 15	Abbott-Young, Edric Charle	. . . .	28 May 16	
720	28 May 15	Pilgrim, Henry Guy Ellcocks	. . . .	28 May 16	
	28 May 15	Robson, Matthew	. . . .	28 May 16	
	28 May 15	Lee-Warner, James	. . . .	28 May 16	
	29 May 15	Johnstone, Noel William Wallace	. . . .	29 May 16	
	29 May 15	Wodehouse, Charles Gilbert (51)	. . . .	29 May 16	
725	31 May 15	Richardson Roland Alexander	. . . .	31 May 16	
	1 June 15	Smith, William Aird	29 Aug. 93	1 June 16	
	1 June 15	Hewett, Harry Selwood	. . . .	1 June 16	
	2 June 15	Dunn, David Tullis	. . . .	2 June 16	
729	3 June 15	Price, John Duckworth	. . . .	3 June 16	

Without pay and allowances of the rank.

INFANTRY—*contd.*

Lieutenants—*contd.*

No.	Date of First Commn.	Names.	Date of birth.	Date of rank.	REMARKS.
730	3 June 15	Weston, Norman Alfred		3 June 16	
	4 June 15	Spring, George Conrad (*Ty. Capt., 15 Jan. 17*)		4 June 16	
	4 June 15	Macfie, James (*Ty. Maj. 25 Feb. 18*)		4 June 16	
	10 June 15	Ashley-Smith, Roland		10 June 16	
	10 June 15	Mare, Edward Joseph		10 June 16	
735	11 June 15	Turner, Ralph Lilley		11 June 16	
	11 June 15	Angelo, Michael Oldfied		11 June 16	
	11 June 15	Goepel, John Alfred		11 June 16	
	11 June 15	Cruickshank, Fletcher Cecil Macdonald (*temp. Maj., 17 Sep. 18*)		11 June 16	
	27 Aug. 15	George, Norman Macdonald		11 June 16	
740	12 June 15	Macaulay, Archibald Duncan Campbell		12 June 16	
	13 June 15	Limby, Shirley Onslow		13 June 16	
	14 June 15	Broadbent, Henry Edward		14 June 16	
	17 June 15	Strachan, Arthur William (*For promotion from 11 Dec. 14 to 10 Feb. 16*)		17 June 16	
	18 June 15	Hickey, Patrick Francis Bourke, D.S.O. (*Ty. Capt. 8 Dec. 16.*)		18 June 16	
745	18 June 15	Moone, Albert John (*Ty. Capt., 4 Feb. 17*)		18 June 16	
	18 June 15	Harman, Oswald Aubrey, M.C.		18 June 16	
	18 June 15	Flower, Arthur Coleman		18 June 16	
	18 June 15	Jahans, Lorraine Errol (207)		18 June 16	I. A. (*on probn.*)
	18 June 15	Wainwright, Victor Paul (95, 172*a*)		18 June 16	
750	18 June 15	Millar, Alfred John Flude		18 June 16	
	18 June 15	Berkeley, Harold Fitzhardinge		18 June 16	
	18 June 15	Davies, Felix Carr (51)		18 June 16	
	18 June 15	Williamson, Philip Ralph Clements (51)		18 June 16	
	18 June 15	Harrop, Henry Robert		18 June 16	
755	18 June 15	Hickie, Allan Darrell		18 June 16	
	18 June 15	McCombe, Robert (51)		18 June 16	
	18 June 15	Tresham, Basil Francis		18 June 16	
	18 June 15	Smith, Reginald Newland Dawes		18 June 16	
	18 June* 15	McPherson, Hugh		18 June 16	Released from Army Service.
760	18 June 15	Fraser, Hugh John		18 June 16	
	22 June 15	Bowers, Percy Lloyd, M.C. (79) (*Ty. Lt.-Col., 6 Feb. 18*)		22 June 16	
	25 June 15	Maclear, Geoffrey D'Olier (46)		25 June 16	
	25 June 15	Macdonald, Sutherland Robert		25 June 16	
	29 June 15	Lee, Alexander William		29 July 17	
765	29 June 15	Twiddy, Albert Arthur		29 June 16	
	30 June 15	Halland, Gordon Herbert Ramsay, O.B.E. (*Ty. Maj., 1 Dec. 17*)		30 June 16	
	1 July 15	Daunt, Elliott Wallace Frederick Townsend		1 July 16	
	1 July 15	Ashton, Frederick James (*temp. Capt., 7 Dec. 17*)		1 July 16	
	2 July 15	Jeffs, Horace Silvanus		2 July 16	
770	8 July 15	Flewett, William Edward		8 July 16	
	8 July 15	Baumber, Joseph (46, 57, 72, 76, 170, 207)		8 July 16	
	9 July 15	Walsh, John		9 July 16	
775	9 July 15	Lawson, Robert Allison		9 July 16	
	9 July 15	Fitz-Gibbon, Leonard Burton		9 July 16	
	9 July 15	Sexton, Cecil Bernard		9 July 16	
	9 July 15	Thrupp, Guy Aveline (207)		9 July 16	
	9 July 15	Thomas, Montague Powell (52)		9 July 16	
780	9 July 15	Calvert, Frank Ernest Rivers, (51)		9 July 16	
	9 July 15	Pearson, Charles McMullen (95)		9 July 16	
	9 July 15	Wyatt, Randolph Maurice		9 July 16	
	9 July 15	Day, Louis Frederick Lerway		9 July 16	
785	9 July 15	Kempson, Carteret John		9 July 16	
	9 July 15	Austin, Arthur Charles (207)		9 July 16	
	9 July 15	Kiddle, John Arthur Copleston, M.C.		9 July 16	
	9 July 15	Davy, Basil Garnham	13 Mar. 92	9 July 16	
	9 July 15	Hillier, Victor Florian		9 July 16	
790	9 July 15	Blanford, Arthur William, M.C. (*Ty. Capt.*)	13 Aug. 85	9 July 16	
	9 July 15	Sitzler, Edward Albert		9 July 16	
792	9 July 15	Fitz Gibbon, Allen Fitz-Gerald (51, 52)		9 July 16	

* With pay and allowances as for a Lieutenant.

INFANTRY—*contd.*

Lieutenants—*contd.*

No.	Date of First Commn.	Names.	Date of birth.	Date of rank.	REMARKS.
793	9 July 15	Steel, John Gwatkin, M.C. (52)	11 Sep. 86	9 July 16	
	9 July 15	Fitzpatrick, Leo Jerome		9 July 16	
795	10 July 15	Lupton, Frank Ernest		12 July 16	
	12 July 15	Christie, Harold		10 July 16	
	12 July 15	McBride, John		12 July 16	
	13 July 15	Coulthard, George Robert		13 July 16	
	14 July 15	Berry, William James		14 July 16	
800	15 July 15	Casson, Randal (79) (*Ty. Capt.*)		15 July 16	
	16 July 15	Harrison, Edward Philip (*Ty. Capt.*)		16 July 16	
	16 July 15	Norrish, Edward Hugh Bent		16 July 16	
	19 July 15	Moore, Herbert Henry		19 July 16	
	20 July 15	Harris, Aubrey Walter		20 July 16	
805	20 July 15	Mahoney, Arthur John (207*a*) (*temp. Capt.*)		20 July 16	
	20 July 15	Fawcett, John Hope		20 July 16	
	20 July 15	Hepworth, Noel (51, 52) (*Ty. Capt. 1 Feb. 18*)		20 July 16	
	20 July 15	French, Frederick George		20 July 16	
	20 July 15	Sanders, William George (51)		20 July 16	
810	20 July 15	Thorne, Harman		20 July 16	
	20 July 15	Hocken, Norman Alfred		20 July 16	
	20 July 15	Dawe, Henry Raymond (51, 52)		20 July 16	
	20 July 15	Reed, Leslie Northcott, M.C. (52)		20 July 16	
	20 July 15	Crockwell, Leslie Horace William		20 July 16	
815	20 July 15	Doubt, Earnest Alfred		20 July 16	
	20 July 15	Hannah, Alfred James (52)		20 July 16	
	20 July 15	Hauser, Sidney Bucklee		20 July 16	
	20 July 15	Bateman, Celred William		20 July 16	
820	20 July 15	Stables, James Howard		20 July 16	
	20 July 15	Pottle, Alfred Edwin		20 July 16	
	20 July 15	Young, Edward Lionel (52)		20 July 16	
	20 July 15	Heathcote, Crawford Napier		20 July 16	
825	20 July 15	Granger, Henry Horace		20 July 16	
	20 July 15	Drew, John Denys (207 *a*)		20 July 16	
	20 July 15	Eastmond, Reginald Copleston e		20 July 16	
	20 July 15	Colman, Bernard Arnold		20 July 16	
	20 July 15	Teviot-Kerr, Albert Evelyn King		20 July 16	
830	22 July 15	Solly, Percy Robert		22 July 16	
	22 July 15	Yonge, Gustavus Edward Husedred		22 July 16	
	22 July 15	Marchant, Denzil Robert		29 July 16	
	22 July 15	Midelton, Ernest Walter		22 July 16	
	23 July 16	MacCulloch, Andrew Francis		23 July 16 16 Aug. 18	
835	24 July 15	Abel, William Henry		24 July 16	
	24 July 15	Comper, George Quintyn		24 July 16	
	24 July 15	Smalley, Skelton		24 July 16	
	24 July 15	Rowe, Stephen Gilby		24 July 16	
	25 July 15	Marley, Walter Hammerton		25 July 16	
840	25 July 15	Horton, Kenneth (51)		25 July 16	
	25 July 15	Lynes, Reginald Howard		25 July 16	
	25 July 15	Ryder, Robert Cecil		25 July 16	
	25 July 15	March, Joseph Cyril		25 July 16	
	25 July 15	Lawrence, Nevil Sherbrooke (52, 57)		15 July 16	
	26 July 15	Roberts, Frederick Charles, M. C.		26 July 16	
845	26 July 15	Hobday, Wilfred Ernest (*Ty. Capt. 15 Jan. 18*)		26 July 16	
	29 July 15	Hardman, Philip Leycester		29 July 16	
	29 July 15	Pain, Harold West, M. C.		29 July 16	
	30 July 15	Zobel, Martin		30 July 16	
	30 July 15	Hitchins, Charles Sidney		30 July 16	
850	30 July 15	Orgill, Tyrrell Churton (51, 52)		30 July 16	
	30 July 15	Cave, Cecil Roderick		30 July 16	
	30 July 15	Ritter, James Connell		30 July 16	
	30 July 15	McGavin, Alan Eugene Cecil (*Ty. Capt., 12 May 18*)		30 July 16	
	30 July 15	Augier, Douglas Eric		30 July 16	
855	30 July 15	Dennehy, Harold George (207)		30 July 16	
	30 July 15	Bell, Leslie		30 July 16	
	30 July 15	Izat, John, C.I.E.		30 July 16	
859	30 July 15	Todd, Frank Ritchie Price		30 July 16	

INFANTRY—*contd.*

Lieutenants - *contd.*

No.	Date of First Commn.	Names.	Date of birth.	Date of rank.	REMARKS.
860	30 July 15	McLean, Arthur Macdonald, M.C.	....	30 July 16	
	30 July 15	Jamieson, George	....	30 July 16	
	30 July 15	Robinson, Donald Guthrie (*Ty. Capt., 8 May. 17*)	....	30 July 16	
	30 July 15	McGeorge, David Jackson	....	30 July 16	
	30 July 15	Cuninghame, Thomas Bradley (51, 52)	....	30 July 16	
865	30 July 15	Sykes, Maurice Gaskell	....	30 July 16	
	30 July 15	DeBoinville, Gustave Alfred Chastel	....	30 July 16	
	30 July 15	Nicholson, Walter Henry (207)	....	30 July 16	
	30 July 15	Douse, Francis Walter	....	30 July 16	
	31 July 15	Haysom, Reginald Andrew	....	31 July 16	
870	1 Aug. 15	Donaldson, Samuel	....	1 Aug 16	
	2 Aug. 15	George, Arthur Basil	9 Sep. 85	2 Aug. 16	
	2 Aug. 15	Harris, Frank Oswald, M.C. (51)	....	2 Aug. 16	
	2 Aug. 15	Fleming, Thomas	....	2 Aug. 16	
	3 Aug. 15	Ashworth, Cecil Henry	....	3 Aug. 16	
875	4 Aug. 15	Dash, Arthur Jules (207)	....	4 Aug. 16	Released from Army Service.
	4 Aug. 15	King, Edward Oakes (79) (*Ty. Capt., 10 July 17*)	....	4 Aug. 16	
	4 Aug. 15	Price, Harry	....	4 Aug. 16	
	6 Aug. 15	Stewart, David Middleton (51)	....	6 Aug. 16	
	6 Aug. 15	Peake, Albert Johnstone	....	6 Aug. 16	
880	6 Aug. 15	Howland, Charles	....	6 Aug. 16	
	6 Aug. 15	Ravenhill, Montagu Hamer (*temp. Capt. 9 Apl. 18*)	....	6 Aug. 16	
	6 Aug. 15	Garthwaite, Edward James Keevil	....	6 Aug. 16	
	6 Aug. 15	Moore, Wentworth William Thompson (52, 57)	....	6 Aug. 16	
	6 Aug. 15	McFarlane, Archibald Howard	....	6 Aug. 16	
885	7 Aug. 15	Oddie, Wilfred Philip, M.C.	....	7 Aug. 16	
	7 Aug. 15	Crawford, Norman Stuart (*Ty. Capt., 4 June 18*)	....	7 Aug. 16	
	7 Aug. 15	Warren, Douglas Selby	....	7 Aug. 16	
	7 Aug. 15	Banks, Reginald Mitchell	....	7 Aug. 16	
	7 Aug. 15	Kearsey, Stephen Henry	....	7 Aug. 16	
890	7 Aug. 15	Jones, Harry	....	7 Aug. 16	
	7 Aug. 15	Coltman, Arthur Cranfield	....	7 Aug. 16	
	13 Aug. 15	Harris, Claude Pickering	....	13 Aug. 16	
	14 Aug. 15	Furneaux, Thomas Edington	....	14 Aug. 16	
	14 Aug. 15	James, Walter Ethelbert (161)	....	14 Aug. 16	
895	14 Aug. 15	Paige, Reginald Herbert	....	14 Aug. 16	
	16 Aug. 15	Harricks, Norman	....	16 Aug. 16	
	17 Aug. 15	Ramsbotham, Richard Bury (79)	....	17 Aug. 16	
	18 Aug. 15	Nichols, Herbert James (51, 207*a*)	....	18 Aug. 16	
	21 Aug. 15	Prideaux, Guy Vyvyan Austice	....	21 Aug. 16	
900	25 Aug. 15	Moor, John Alexander	....	25 Aug. 16	
	25 Aug. 15	Young, Edward George Dinsdale	....	25 Aug. 16	
	26 Aug. 15	Lind, Emile Norman Arnold	....	26 Aug. 16	
	27 Aug. 15	Cove, Vivian Douglas	....	27 Aug. 16	
	27 Aug. 15	Evans, Arthur Gordon	....	27 Aug. 16	
905	27 Aug. 15	Pearson, Leonard Clive	....	27 Aug. 16	
	28 Aug. 15	Thompson, James Stanislaus (*Ty. Capt.*)	....	28 Aug. 16	
	29 Aug. 15	Janson, Sven Jukkes	....	29 Aug. 16	
	31 Aug. 15	Oatway, Stanley Herbert (51, 52)	....	31 Aug. 16	
	31 Aug. 15	Keighley, Frederick Walter, M.C.	....	31 Aug. 16	
910	2 Sep. 15	Toovey, Ernest Hamilton	....	2 Sep. 16	
	2 Sep. 15	Bradley, Gerald Graves	....	2 Sep. 16	
	2 Sep. 15	Walton, Harry (*Ty. Capt., 10 Oct. 16*)	....	2 Sep. 16	
	2 Sep. 15	Turner, Francis Roland	....	2 Sep. 16	
	3 Sep. 15	Milne, William Graham	....	3 Sep. 16	
915	3 Sep. 15	Brockes, Edward Herbert (52)	....	3 Sep. 16	
	6 Sep. 15	Hennel, Frad John	12 May 88	6 Sep. 16	
	7 Sep. 15	Gunn, John Peterson	....	7 Sep. 16	
	7 Sep. 15	Dunn, Cyril Leslie	....	7 Sep. 16	
	8 Sep. 15	Davidson, James Ramsay, M.C. (*Ty. Capt., 8 Aug. 16*)	....	8 Sep. 16	
91	8 Sep. 15	Hosking, Frederick Mervyn	....	8 Sep. 16	

INFANTRY—*contd.*

Lieutenants—*contd.*

No.	Date of First Commn.	Names.	Date of birth.	Date of rank.	REMARKS.
920	8 Sep. 15	Jonas, Lewis Gordon		8 Sep. 16	
	8 Sep. 15	Unsworth, Norval (207*a*)		8 Sep. 16	
	8 Sep. 15	Fleming, Victor Joseph Paterson		8 Sep. 16	
	8 Sep. 15	Grant, Robert		8 Sep. 16	
	8 Sep. 15	Munsey, Sydney Thomas Hensman		8 Sep. 16	
925	8 Sep. 15	Greig, Stanley		8 Sep. 16	
	8 Sep. 15	Abrahall, Frederick Thomas Ernest (51)		8 Sep. 16	
	8 Sep. 15	Hosking, Bertrand Charles		8 Sep. 16	
	8 Sep. 15	O'Riordan, Patrick Michael (95)		8 Sep. 16	
	8 Sep. 15	Bream, John William (51, 52)		8 Sep. 16	
930	8 Sep. 15	Anderson, Alfred Thomas Duncan		8 Sep. 16	
	8 Sep. 15	Swan, John Cyril Kingsley		8 Sep. 16	
	8 Sep. 15	Robathan, Kenneth Minshull		8 Sep. 16	
	8 Sep. 15	Wilkinson, John Bertram		8 Sep. 16	
	8 Sep. 15	Lister-Jackson, Arthur		8 Sep. 16	
935	8 Sep. 15	Kirwan, Noel Gerald Brudenell		8 Sep. 16	
	8 Sep. 15	Curtis, Roi Linton		8 Sep. 16	
	8 Sep. 15	Jones, Owen Edward (51, 52)	4 Feb. 92	8 Sep. 16	
	8 Sep. 15	Mathew, William Charles		8 Sep. 16	
940	9 Sep. 15	Williams, Charles Alfred Leonard		9 Sep. 16	
	9 Sep. 15	Masters, Hugh Frederick		9 Sep. 16	
	9 Sep. 15	Latham, Edward Bryan		9 Sep. 16	
	10 Sep. 15	Durnan, John Andrew (*temp. Capt. 9 Nov. 16*) (50, 53, 58)		10 Sep. 16	
	10 Sep. 15	Hele, George Clive		10 Sep. 16	
945	15 Sep. 15	Francies, Joseph Valentine (79, 107, 207) (*Ty. Maj. 17 Jan. 18*)		15 Sep. 16	
	20 Sep. 15	Dunt, Reginald Beethoven		20 Sep. 16	
	21 Sep. 15	Rowell, Joseph Wilson (51)		21 Sep. 16	
	22 Sep. 15	Fox, Edwin Blamey (*Ty. Capt., 25 July 16*)		22 Sep. 16	
	22 Sep. 15	Jones, Elias Henry		22 Sep. 16	
950	23 Sep. 15	Butcher, William Henry		23 Sep. 16	
	23 Sep. 15	Mackay, Frank Forbes (*temp. Capt., 10 Nov. 16*)		23 Sep. 16	
	23 Sep. 15	Carr, Frank John Bedford		23 Sep. 16	
	24 Sep. 15	Thomas, Alan Robert		24 Sep. 16	
	24 Sep. 15	Brewin, George Graham		24 Sep. 16	
955	24 Sep. 15	Hope, Alexander Cunningham		24 Sep. 16	
	24 Sop. 15	Sedgley, Charles Henry		24 Sep. 16	
	24 Sep. 15	Rendell, Clifford Beard		24 Sep. 16	
	25 Sep. 15	Sellier, Phillip Joseph (*Ty. Capt., 16 July 17*)		25 Sep. 16	
	25 Sep. 15	Webb, John Elias Nanfan		25 Sep. 16	
960	25 Sep. 15	Pryor, Elmer Vela		25 Sep. 16	
	25 Sep. 15	Fischer, Norman Winther		25 Sep. 16	
	26 Sep. 15	Stevenson, John		26 Sep. 16	
	27 Sep. 15	Joyes, Archibald		27 Sep. 16	
	29 Sep. 15	Finlayson, Colin Scott Moncrieff		29 Sep. 16	
	30 Sep. 15	Finch, Reginald Harold		30 Sep. 15	
965	30 Sep. 15	Boddy, Sidney Melville		30 Sep. 16	
	1 Oct. 15	Casling, Cyril Thomas		1 Oct. 16	
	1 Oct. 15	McGuire, Thomas Beesley (*Ty. Capt., 1 Nov. 16*)		1 Oct. 16	
	1 Oct. 15	Fisher, Frederic Browell		1 Oct. 16	
	1 Oct. 15	Wright, Reginald William Barton (51)		1 Oct. 16	
970	1 Oct. 15	Milburne, William		1 Oct. 16	
	1 Oct. 15	Butterfield, Leslie Hope (51) (*Ty. Capt. 17 Feb. 17*)		1 Oct. 16	
	1 Oct. 15	Tyler, Henry Harry Francies Macdonald		1 Oct. 16	
	1 Oct. 15	Breithaupt, Guy Wallace (51) (**temp. Capt., 22 May 18*)		1 Oct. 16	
974	1 Oct. 15	Smith, Ronald William Holmes (52)		1 Oct. 16	

* Without pay and allowances of the rank.

INFANTRY—*contd.*

Lieutenants—*contd.*

No.	Date of First Commn.	Names.	Date of birth.	Date of rank.	REMARKS.
975	1 Oct. 15	Poingdestre, Charles Le Geyt	. . . .	1 Oct. 16	
	1 Oct. 15	Chihsolm, Donald Malcolm Francis (51, 52)	. . . .	1 Oct. 16	
	1 Oct. 15	Bell, John Aiton	. . . .	1 Oct. 16	
	1 Oct. 15	Hawkes, Frank Roxburgh	. . . .	1 Oct. 16	
	1 Oct. 15	Smith, Ralph Eugene	. . . .	1 Oct. 16	
980	1 Oct 15	Donald, Andrew (52)	. . . .	1 Oct. 16	
	1 Oct. 15	Norris, Roland Victor	. . . .	1 Oct. 16	
	1 Oct 15	Priestley, Eric Charles (207a)	. . . .	1 Oct 16	
	1 Oct. 15	Cooke, Gerald Hallam (*Actg. Maj. 26 Sep. 18*)	. . . .	1 Oct. 16	
	1 Oct. 15	Taylor, William Robinson, (207a) (*temp. Capt. 4 Mar. 18*)	. . . .	1 Oct. 16	
985	4 Oct. 15	Andrews, David Ernest	. . . .	4 Oct. 16	
	6 Oct. 15	Cohen, Norman Henry	. . . .	6 Oct. 16	
	7 Oct. 15	Harrison, Frank Seymour (46, 51, 52)	. . . .	7 Oct. 16	
	7 Oct. 15	Corfield, Henry Ruxton	. . . .	7 Oct. 16	
	7 Oct. 15	Baker, William Sydney, M.C.	. . . .	7 Oct. 16	
990	7 Oct. 15	Vessey, Stanley John (*Ty. Capt., 22 Oct. 17*)	. . . .	7 Oct. 16	
	12 Oct. 15	Simmons, Charles Edward	. . . .	12 Oct. 16	
	12 Oct. 15	Considine, William Joseph	. . . .	12 Oct. 16	
	12 Oct. 15	Whitamore, Arthur Gerald (*Ty. Capt., 27 Mar. 17*)	. . . .	12 Oct. 16	
	12 Oct. 15	O'Grady, William James (**Ty. Capt., 8 Dec. 17*)	. . . .	15 Oct. 16	
995	15 Oct. 15	Murphy, Alfred (*Ty. Maj., 18 Aug. 18*)	. . . .	15 Oct. 16	
	15 Oct. 15	Ainslie, Kenneth	. . . .	15 Oct. 16	
	16 Oct. 15	Yeates, Robert Montford Michaelson (*Ty. Capt.*)	. . . .	16 Oct. 16	
	16 Oct. 15	Thompson, William Jameson (*Ty. Capt., 27 July 17*)	. . . .	16 Oct. 16	
	17 Oct. 15	Crawford, Ernest Louis	. . . .	17 Oct. 16	Released from Army service.
	17 Oct. 15	Terry, Leslie Ryder	. . . .	17 Oct. 16	
1000	18 Oct. 15	Rivett, Edward Godfrey (79)	. . . .	18 Oct. 16	
	18 Oct. 15	Webster, George William	. . . .	18 Oct. 16	
	19 Oct. 15	Cox, Harold Ranyard (79)	. . . .	19 Oct. 16	
	21 Oct. 15	Centeno, John Bell	. . . .	21 Oct. 16	
1005	22 Oct. 15	Croad, Albert	. . . .	22 Oct. 16	
	23 Oct. 15	Bartlett, George Donald Newman	. . . .	23 Oct. 16	
	23 Oct. 15	Crowther, Harold Oakes	. . . .	23 Oct. 16	
	25 Oct. 15	Cooke, Joseph Charles Cameron	. . . .	25 Oct. 16	
	27 Oct. 15	Davis, Mathew Arnold	. . . .	27 Oct. 16	
	27 Oct. 15	Tuckey, Robert Eyre Shaw	. . . .	27 Oct. 16	
1010	28 Oct. 15	Cooper, Henry Guy, M.C. (*temp. Capt., 10 July 18*)	16 Sep. 90	28 Oct. 16	
	29 Oct. 15	Chater, Cecil William (207a)	. . . .	29 Oct. 16	
	29 Oct. 15	Galvin, Percy Harold (79)	. . . .	29 Oct. 16	
	29 Oct. 15	Prior, Henry Carlos (52)	. . . .	29 Oct. 16	
	29 Oct. 15	Russell, Robert Edwin (51, 52, 169, 172a) (*temp. Maj., 30 July 18*)	. . . .	29 Oct. 16	
1015	29 Oct. 15	Marshall, John Goodfellow M.C.	. . . .	29 Oct. 16	
	29 Oct. 15	Hirst, Felix Henry (207a)	. . . .	29 Oct. 16	
	29 Oct. 15	Byron, George (51, 52)	. . . .	29 Oct. 16	
	29 Oct. 15	Tirney, Harry	. . . .	29 Oct. 16	
	29 Oct. 15	Godfrey, Edward Kenneth Macgregor (46)	19 May 90	29 Oct. 16	
1020	29 Oct. 15	Campbell, Thorn Edward	. . . .	29 Oct. 16	
	29 Oct. 15	Murray, Arthur Eveyard (51)	. . . .	29 Oct. 16	
	29 Oct. 15	Simeon, Geoffrey Nelthorpe (52)	. . . .	29 Oct. 16	
	29 Oct. 15	Davies, Andrew Taylor (46, 207a)	. . . .	29 Oct. 16	
	29 Oct. 15	Smith, Stanley Harvey Melville	. . . .	29 Oct. 16	
1025	29 Oct. 15	MacFadden, Richard Read (51)	. . . .	29 Oct. 16	
	29 Oct. 15	Minnitt, George Chappell	. . . .	29 Oct. 16	
	29 Oct. 15	Hall, William Aeneas	. . . .	29 Oct. 16	
	29 Oct. 15	Quilter, Charles Frederic Grant	. . . .	29 Oct. 16	
1030	29 Oct. 15	Ezra, Ellice (*temp. Capt., 21 Jan. 18*)	. . . .	29 Oct. 16	
	30 Oct. 15	Marks, Roland Maurice	. . . .	30 Oct. 16	
1032	1 Nov. 15	Hewetson, Francis Edward Arthur	. . . .	1 Nov. 16	

*Without pay and allowances of the rank.

INFANTRY—*contd.*

Lieutenants—*contd.*

No.	Date of First Commn.	Names.	Date of birth.	Date of rank.	REMARKS.
1033	1 Nov. 15	Hall, George Hanney		1 Nov. 16	
	3 Nov. 15	Jennison, Frederick Geoffrey		3 Nov. 16	
1035	7 Nov. 15	Jenkins, Evan Merndith		7 Nov. 16	
	8 Nov. 15	Eliot, Osmond Herbert (*temp. Capt.*, 23 *Nov.* 17)		8 Nov. 16	
	8 Nov. 15	Stanbury, Sydney James (52)		8 Nov. 16	
	11 Nov. 15	Shallow, George (*temp. Capt.*, 1 *Jan.* 18)		11 Nov. 16	
	11 Nov. 15	Carter, Albert Bernard (51)		11 Nov. 16	
1040	23 Nov. 15	Gemmell, George		23 Nov. 16	
	23 Nov. 15	Martin, Thomas		23 Nov. 16	
	24 Nov. 15	Cates, Bruce Hyde (51, 52)		24 Nov. 16	
	25 Nov. 15	Peel, William		25 Nov. 16	
	27 Nov. 15	Skinner, William Edward		27 Nov. 16	
	27 Nov. 15	Leighton, John Herbert Leslie		27 Nov. 16	
1045	30 Nov. 15	Harman, Charles William Theo.		30 Nov. 16	
	1 Dec. 15	Young, Alexander Drysdale		1 Dec. 16	
	1 Dec. 15	Mole, James Anthony		1 Dec. 16	
	1 Dec. 15	Lindsay, Harold Cedric Craufurd		1 Dec. 16	
	1 Dec. 15	Izat, James Rennie		1 Dec. 16	
1050	1 Dec. 15	Cousins, Cyril Walter		1 Dec. 16	
	1 Dec. 15	Noble, Arthur Guy		1 Dec. 16	
	1 Dec. 15	Young, Frank		1 Dec. 16	
	1 Dec. 15	Bateman, Norman George		1 Dec. 16	
1055	1 Dec. 15	Humphrys, Basil Hardress		1 Dec. 16	
	1 Dec. 15	Findlay, Alfred James		1 Dec. 16	
	1 Dec. 15	Partridge, Edward Charles		1 Dec. 16	
	1 Dec. 15	Graham, Malcohn Tudor		1 Dec. 16	
	1 Dec. 15	Mackarness, Cuthbert George Milford (*temp. Capt.*, 10 *Aug.* 17)		1 Dec. 16	
1060	1 Dec. 15	Barron, William George (*temp. Capt.*, 1 *Apl.* 18)		1 Jan. 18	
	1 Dec. 15	Lawford, Norman Bernard		1 Dec. 16	
	1 Dec. 15	Ross, Kenneth Maxwell		1 Dec. 16	
	2 Dec. 15	Harvey, Bertram Lionel (*temp. Maj.*, 14 *May* 17)		2 Dec. 16	
	2 Dec. 15	O'Toole, Neil Donovan Granville		2 Dec. 16	
1065	2 Dec. 15	Hopkins, Herbert Ernest Evelyn Jesse		2 Dec. 16	
	3 Dec. 15	Saunders, Raymund Alexander Richard		3 Dec. 16	
	4 Dec. 15	Loughman, Geoffrey		4 Dec. 16	
	5 Dec. 15	Hurst, Victor de a Motte		5 Dec. 16	
	6 Dec. 15	McWhinnie, William		6 Dec. 16	
1070	6 Dec. 15	Campbell, Collin		6 Dec. 16	
	9 Dec. 15	Eden, Basil		9 Dec. 16	
	8 Dec. 15	Cleaver, Lionel		8 Dec. 16	
	8 Dec. 15	Hearn, George Sidney		8 Dec. 16	
	9 Dec. 15	Miller, William James Burns		9 Dec. 16	
	9 Dec. 15	Watkins, Alan Wynne		9 Dec. 15	
1075	7 Nov. 15	MacGrigor, Gregor Norman		10 Dec. 16	
	10 Dec. 15	Hasluck, Reginald Ffolliott		10 Dec. 16	
	12 Dec. 15	Platts, Cyril Stanley		12 Dec. 16	
	14 Dec. 15	Hogan, Edmund James Patrick		14 Dec. 16	
	14 Dec. 15	Hughes, James Emlyn		14 Dec. 16	
1080	14 Dec. 15	Luck, James		14 Dec. 16	
	14 Dec. 15	Jephcott, Archibald Humphrey		14 Dec. 16	
	15 Dec. 15	Moran, Terence		15 Dec. 16	
	15 Dec. 15	Smart, Colin Neil Boddington		15 Dec. 16	
1085	15 Dec. 15	Cox, Harry		15 Dec. 16	
	16 Dec. 15	Lowis, Frank Currie, C.I.E. (*temp. Maj.*, 1 *Aug.* 17.)		16 Dec. 16	
1087	23 Dec. 15	Baldrey, Charles Herbert		23 Dec. 16	

No.	Date of First Commn.	Names.	Date of birth.	Date of rank.	REMARKS.

INFANTRY—*contd.*

Lieutenants—*contd.*

No.	Date of First Commn.	Names.	Date of birth.	Date of rank.	REMARKS.
1086	23 Dec. 15	Harding, William Francis		23 Dec. 16	
	23 Dec. 15	Moir, Gordon		23 Dec. 16	
	23 Dec. 15	Brisley, Frank deBock		23 Dec. 16	
	24 Dec. 15	Hull, Joseph Reginald		24 Dec. 16	
	30 Dec. 15	Brady, Reginald Malcolm Ellis		30 Dec. 16	
1090	30 Dec. 15	Dunckley, Percy John (207*a*)		30 Dec. 16	
	1 Jan. 16	Mould, Arthur Ernest (79)		1 Jan. 17	
	1 Jan. 16	Wilson, James McLean		1 Jan. 17	
	1 Jan. 16	Graves, Ernest (52)		1 Jan. 17	
	1 Jan. 16	Elliot, Robert Arthur Douglas (51)		1 Jan. 17	
1095	1 Jan. 16	Morley, William Roy Stone (72*b*)		1 Jan. 17	
	1 Jan. 16	Morris, William Gerald		1 Jan. 17	
	1 Jan. 16	Pepper, George Herbert Neil		1 Jan. 17	
	1 Jan. 16	Daintith, Edward Michael St. George (*Ty. Capt., 4 Mar. 17*)		1 Jan. 17	
	1 Jan. 16	Pawson, Frederick Alexander (51)		1 Jan. 17	
1100	1 Jan. 16	Read, Walter Thomas		1 Jan. 17	
	1 Jan. 16	Tilley, James Stanley		1 Jan 17	
	1 Jan. 16	Feilmann, Percival Michael Julius		1 Jan. 17	
	1 Jan. 16	Stewart, Francis William		1 Jan. 17	
	1 Jan. 16	Winckler, Cecil Percy (51)		1 Jan. 17	
1105	1 Jan. 16	Lentaigne, Joseph Ignatius Nugent (51)		1 Jan. 17	
	1 Jan. 16	Hodding, Charles (51, 52) (*Ty. Capt., 12 Aug. 17*)		1 Jan. 17	
	1 Jan. 16	Banister, Fred (54)		1 Jan. 17	
	1 Jan. 16	Torrens, Reginald Geoffrey (51, 79)		1 Jan. 17	
1110	1 Jan. 16	Thorpe, John Aymas		1 Jan. 17	
	1 Jan. 16	Boynton-Lee, Francis		1 Jan. 17	
	1 Jan. 16	Jervis, Cyril Walter Joseph		1 Jan. 17	
	1 Jan. 16	Longdin, Walter Gardner (52, 79) (*Ty. Capt., 2 Jan. 17*)		2 Jan. 17	
	5 Jan. 16	Collet, Arthur Lowe (54, 79)		5 Jan. 17	
	6 Jan. 16	McMorran, James (51)		6 Jan. 17	
	8 Jan. 16	Mack, Cyril Hugh		8 Jan. 17	
	8 Jan. 16	Browne, Charles Leonard		8 Jan. 17	
1115	12 Jan. 16	Campbell, Thomas Patrick (100) (*temp. Capt., 21 Dec. 17*)		12 Jan. 17	
	13 Jan. 16	Cousins, Edward Charles	21 Nov. 91	13 Jan. 17	
	14 Jan. 16	Ross, Terence Ryan		14 Jan. 17	
	14 Jan. 16	Croft, Tristram de Carne		14 Jan. 17	
	14 Jan. 16	Mumford, Stanley Erroll Aubrey		14 Jan. 17	
	15 Jan. 16	Helme, Frederick Stanley		15 Jan. 17	
1120	30 Aug. 15	Russell, Bernard Hammond		18 Jan. 17	
	18 Jan. 16	Determes, Maurice Joseph		18 Jan. 17	
	20 Jan. 16	Dolphin, Wilfred Herbert (*temp. Lt., Gen. List*)		20 Jan. 17	
	22 Jan. 16	Lee, Gerald St. Clair (*Ty. Capt., 15 Aug. 16.*)		22 Jan. 17	
	24 Jan. 16	Kent, Harold Stanley		24 Jan. 17	
1125	25 Jan. 16	Page, Edmund Robertson		25 Jan. 17	
	26 Jan. 16	Lowe, Francis Gordon (51)		26 Jan. 17	
	27 Jan. 16	Kelly, Patrick Anselm (51) (*Ty. Capt., 10 June 17*)		27 Jan. 17	
	27 Jan. 16	Harrison, Douglas Hugh		27 Jan. 17	
	27 Jan. 16	Hely, Terence Francis		27 Jan. 17	
1130	27 Jan. 16	Fitzpatrick, Edward John (51)		27 Jan. 17	
	27 Jan. 16	Bazett, Lancelot Young		27 Jan. 17	
	27 Jan. 16	Gouldsbury, Cecil Philip		27 Jan. 17	
	27 Jan. 16	Cliff, Wilfred Pearse		27 Jan. 17	
	27 Jan. 16	Roberts, Richard Frederick (*Ty. Capt., 29 July 16*)		27 Jan. 17	
1135	27 Jan. 16	Fell, Richard James Hadwin		27 Jan. 17	
	27 Jan. 16	Fraser, Herbert Windham Matheson		27 Jan. 17	
	27 Jan. 16	Greenwood, Jack Arthur		27 Jan. 17	
	27 Jan. 16	Leftwich, William Martin		27 Jan. 17	
	27 Jan. 16	Seymour, Hugh Cuthbert		27 Jan. 17	
1140	27 Jan. 16	Johnston, James Halero (207)		27 Jan. 17	
	27 Jan. 16	Fraser, Ebenezer		27 Jan. 17	
	27 Jan. 16	Perdue, Cecil Graham		27 Jan. 17	
	27 Jan. 16	Channell, Edgar Cecil		27 Jan. 17	
	28 Jan. 16	Atkins, Edvin Alan		28 Jan. 17	
1144	29 Jan. 16	Lattey, John Tabor		29 Jan. 17	

INFANTRY—*contd.*

Lieutenants—*contd.*

No.	Date of First Commn.	Names.	Date of birth.	Date of rank.	REMARKS.
1145	1 Feb. 16	Kirwan, Lionel Edward (72*b*, 79) (*Temp. Capt., 1 Feb. 17*)		1 Feb. 17	
	5 May 16	Stafford, Horace Christmas Cook		1 Feb. 17	
	4 Feb. 16	Kavanagh, Courtney (*Tempy. Capt.*)		4 Feb. 17	
	6 Feb. 16	Cherry, John Arnold (*Temp. Lt.-Col., 25 Feb. 18*) (79)		6 Feb. 17	
	6 Feb. 16	Reynolds, John Richardson (*Temp. Lt.-Col., 25 Feb. 18*) (79, 207)		6 Feb. 17	
1150	7 Feb. 16	Jones, John Henry		7 Feb. 17	
	8 Feb. 16	Hewer, John Arthur, M.C.		8 Feb. 17	
	23 June 15	Jones, Thomas Joseph William		10 Feb. 17	
	19 Sep. 15	Wigfall, Cecil Stuart		10 Feb. 17	
	10 Feb. 16	Harwood, William Albert		10 Feb. 17	
1155	10 Feb. 16	Bailey, Francis Robert		10 Feb. 17	
	11 Feb. 16	Shaw, Cyril Hay		11 Feb. 17	
	12 Feb. 16	Dixon, Ralph Claiborne		12 Feb. 17	
	12 Feb. 16	Cooper, Philip Alan (79)		12 Feb. 17	
	15 Feb. 16	Flatt, Leslie Neeve (54, 79)		15 Feb. 17	
	15 Feb. 16	Thompson, Bernard Nathan		15 Feb. 17	
1160	15 Feb. 16	Morrison, Alister		15 Feb. 17	
	17 Feb. 16	Gordon, John Melvin		17 Feb. 17	
	18 Feb. 16	Richards, Thomas John Reginald		18 Feb. 17	
	19 Feb. 16	Breslauer, Leo		19 Feb. 17	
	20 Feb. 16	Coffey, Edward Powell		20 Feb. 17	
1165	20 Feb. 16	Siedle, Charles Edward		20 Feb. 17	
	20 Feb. 16	McLean, William (51)		20 Feb. 17	
	20 Feb. 16	Lake, Cecil Herbert Frank Logan (51, 52)		20 Feb. 17	
	20 Feb. 16	Dunsmure, Charles Thistlethwayth Welman		20 Feb. 17	
	20 Feb. 16	Dawson, Thomas Ernest (**Temp. Capt., 17 Feb. 18*)		20 Feb. 17	
1170	20 Feb. 16	Goldberg, Gordon Henry Archibald		20 Feb. 17	
	20 Feb. 16	Bell, Herbert Jasper (51)		20 Feb. 17	
	20 Feb. 16	Murray, Maurice James		20 Feb. 17	
	20 Feb. 16	Mould, George Brown (52)		20 Feb. 17	
1175	20 Feb. 16	Marshall, Arthur Douglas		20 Feb. 17	
	20 Feb. 16	Young, John Mervyn (52)		20 Feb. 17	
	20 Feb. 16	O'Brien, Dennis Patrick		20 Feb. 17	
	20 Feb. 16	Higby, Harold Vienna		20 Feb. 17	
	20 Feb. 16	Littledale, Anthony Godfrey		20 Feb. 17	
1180	20 Feb. 16	Perrin, Frank Hubert		20 Feb. 17	
	20 Feb. 16	Murphy, Robert Stanley		20 Feb. 17	
	20 Feb. 16	Preece, Clarance Burran		20 Feb. 17	
	20 Feb. 16	Priestley, Hugh Edgar Basil		20 Feb. 17	
	20 Feb. 16	Griffie-Williams, John Henry (**Temp. Capt., 4 Mar. 18*)		20 Feb. 17	
1185	20 Feb. 16	Austin, Joseph Crane		20 Feb. 17	
	20 Feb. 16	Pearce, John Cyprian (*Ty. Capt., 10 Aug. 17*) (207 *a*)		20 Feb. 17	
	20 Feb. 16	Normand, Charles William Blythe (79)		20 Feb. 17	
	21 Feb. 16	Dooley, James Arthur		21 Feb. 17	
1190	22 Feb. 16	Punchard, Frank Norris		22 Feb. 17	
	24 Feb. 16	Mackay, Thomas Ian Sutherland		24 Feb. 17	
	28 Feb. 16	Herbert, Reginald Lloyd		28 Feb. 17	
	28 Feb. 16	Harrison, David James (*Ty. Capt., 5 July 17*)		28 Feb. 17	
	1 Mar. 16	Russell, John Maurice		1 Mar. 17	
1195	1 Mar. 16	Paice, Alfred Hubert		1 Mar. 17	
	4 Mar. 16	Patrick, Paul Joseph		4 Mar. 17	
	8 Mar. 16	Clements, George Weston, (207*a*)		8 Mar. 17	
	8 Mar 16	Parry, David Griffith (51)		8 Mar. 17	
	8 Mar. 16	Boden, John Francis Worsley		8 Mar. 17	
1200	8 Sep. 15	Thomas, Robert Douglas		8 Mar. 17	
	26 Mar. 15	Bradley, Sydney Minta		9 Mar. 17	
	11 Mar. 16	Banon, Hugh Lavalan (165)		11 Mar. 17	
	12 Mar. 16	Drake, Arthur John Alexander (207*a*)		12 Mar. 17	
	12 Mar. 16	Hofford, Robert McVitie		12 Mar. 17	
1205	14 Mar. 16	Davidson, John, (*Temp. Maj., 29 July 17*)		14 Mar. 17	
	15 Mar. 16	Waddingham, Charles Alfred		15 Mar. 17	
	17 Mar. 16	Heathcote, Frederick Cyril		17 Mar. 17	
	18 Mar. 16	Donovan, Edward (51)		18 Mar. 17	
	20 Mar. 16	Rae, Cecil (*Ty. Capt., 18 Mar. 17*)		20 Mar. 17	
1210	20 Mar. 16	Myers, Nathan Coleman		20 Mar. 17	

* Without pay and allowances of the rank.

INFANTRY—*contd.*

Lieutenants—*contd.*

No.	Date of First Commn.	Names.	Date of birth.	Date of rank.	Remarks.
1211	19 Mar. 16	Moore, Arthur James	....	20 Mar. 17	
	20 Mar. 16	Hartnoll, Frederic Brewse (95, 207*a*)	17 Mar. 82	20 Mar. 17	
	24 Mar. 16	Oliphant, Donald Aden (79)	....	24 Mar. 17	
	25 Mar. 16	Rice, Eustace Meredith (79)	....	25 Mar. 17	
1215	25 Mar. 16	Stokes, Hugh Gabriel (*Ty. Capt., 14 Dec. 17*).	....	25 Mar. 17	
	25 Mar. 16	Modget, Herbert Raymond	....	25 Mar. 17	
	25 Mar. 16	Duffus, Alan Harold Henry	....	25 Mar. 17	
	25 Mar. 16	Hill, Donald George	....	25 Mar. 17	
	25 Mar. 16	Wilson, James Mollison (*Ty. Capt. 24 July 17*)	....	25 Mar. 17	
1220	25 Mar. 16	Noble, Tom Lawson (*Tempy. Capt., 1 June 17*)	....	25 Mar. 17	
	25 Mar. 16	Delahunty, Richard Alphonsus (207*a*)	....	25 Mar. 17	
	25 Mar. 16	Bender, William Edward Gustave (79)	....	25 Mar. 17	
	25 Mar. 16	Smith, George Makins (51)	....	25 Mar. 17	
	25 Mar. 16	Holmes, Alfred Henry Robert (*Ty. Capt.*)	....	25 Mar. 17	
1225	25 Mar. 16	White, Maurice	....	25 Mar. 17	
	25 Mar. 16	Templeton, Clifford Arthur Sheridan (51)	....	25 Mar. 17	
	25 Mar. 16	Baikie, Hugh McAndrew (*Tempy. Capt., 7 Feb. 17*)	....	25 Mar. 17	
	25 Mar. 16	Rich, Clive Noble (79)	....	25 Mar. 17	
	25 Mar. 16	Powis, Leonard John (79, 207)	....	25 Mar. 17	
1230	25 Mar. 16	Hopkins, Edward	....	26 Mar. 17	
	26 Mar. 16	Percy, Aubrey Cecil (51, 95) (*Ty. Capt., 26 Feb. 18*).	....	26 Mar. 17	
	28 Mar. 16	Marshall, Donald Frederick	....	28 Mar. 17	
	29 Mar. 16	Chapman, Clement William, D.C.M.(*Ty. Capt., 9 Nov. 16*)	....	29 Mar. 17	
	29 Mar. 16	Dudley, Robert Samuel	....	29 Mar. 17	
1235	30 Mar. 16	Irwin, George William (46, 57)	....	30 Mar. 17	
	1 Apl. 16	Manson, Charles Ernest Fellowes	....	1 Apl. 17	
	2 Apl. 16	Kempe, Edward Michael	....	2 Apl. 17	
	3 Apl. 16	Hitchcock, Ronald Victor (79)	....	3 Apl. 17	
	8 Apl. 16	Griffiths, Herbert	....	8 Apl. 17	
1240	14 Apl. 16	Stevenson, Robert Campbell (*Ty. Capt., 1 Dec. 16*).	....	14 Apl. 17	
	14 Apl. 16	Butt-Gow, Herbert Frederick Steuart	....	14 Apl. 17	
	14 Apl. 16	Turner, Noel Theodor Barwell	....	14 Apl. 17	
	15 Apl. 16	Grant, Ian (*Ty. Capt.*)	....	15 Apl. 17	
	17 Apl. 16	Williams, Albert	....	17 Apl. 17	
1245	17 Apl. 16	MacGregor, Donald Roy	....	17 Apl. 17	
	17 Apl. 16	Bispham, George	....	17 Apl. 17	
	17 Apl. 16	Nierses, Wilfred Leonard Macleod	....	17 Apl. 17	
	17 Apl. 16	Cole, William Johnson	....	17 Apl. 17	
	17 Apl. 16	McCorquodale, Alexander Robertson	....	17 Apl. 17	
1250	17 Apl. 16	Gilbert, Rupert Sampson	....	17 Apl. 17	
	17 Apl. 16	Hope, John William (207*a*)	....	17 Apl. 17	
	17 Apl. 16	Boyd, William Stuart Jackson	....	17 Apl. 17	
	17 Apl. 16	Ricketts, Edward Montgomery	....	17 Apl. 17	
	18 Apl. 16	Davies, Percy Meyrick Morris (46)	....	18 Apl. 17	
1255	21 Apl. 16	Rose, Perey Thomas	....	21 Apl. 17	
	24 Apl. 16	Splane, Edgar George	....	24 Apl. 17	
	28 Apl. 16	Turner, Herbert John Chater (*Ty. Capt. 24 Mar. 17*)	....	28 Apl. 17	
	28 Apl. 16	Proctor, David Paris	....	28 Apl. 17	
	28 Apl. 16	Jackson, Henry Stewart	....	28 Apl. 17	
1260	28 Apl. 16	Barckley, Duguidichton (51)	....	28 Apl. 17	
	28 Apl. 16	Allen, Francis	....	28 Apl. 17	
	28 Apl. 16	McRoberts, Percy Frederick (*Ty. Capt. 30 Aug. 17*)	....	28 Apl. 17	
	28 Apl. 16	Rees, Roderic (*Ty. Capt. 28 Apl. 17*)	....	28 Apl. 17	
	29 Apl. 16	Marriott, Alfred Edward	....	29 Apl. 17	
1265	29 Apl. 16	Rennie, James Ogilvie (*Ty. Capt. 5 May 17*)	....	29 Apl. 17	
1266	30 Apl. 16	Lomas, Leopold	....	30 Apl. 17	

INFANTRY—*contd.*

Lieutenants—*contd.*

No.	Date of First Commn.	Names.	Date of birth.	Date of rank.	Remarks.
1267	5 May 16	Pigott, Thomas Francis		5 May 17	
	5 May 16	Dewar, Hubert Stephen Lowry		5 May 17	
	9 May 16	Lethbridge, William Besnard		9 May 17	
1270	10 May 16	Muggridge, Herbert George (54, 79) (*Ty. Capt., 10 May 17.*)		10 May 17	
	10 May 16	Burge, James Thomas		10 May 17	
	10 May 16	Knowles, James Frank (51)		10 May 17	
	10 May 16	Musto, William Reid (*Ty., Capt. 5 Mar. 17*)		10 May 17	
	10 May 16	Forbes, Gordon Harold Norman (*Ty. Capt., 10 Oct. 16*)		10 May 17	
1275	10 May 16	Hales, George		10 May 17	
	10 May 16	Hargreaves, Frank		10 May 17	
	10 May 16	Snowden, Wallace Crawford		10 May 17	
	10 May 16	Cook, Hubert Graeme		10 May 17	
	10 May 16	Berey, Charles Eric		10 May 17	
1280	10 May 16	Gilman, Henry Charles Randon		10 May 17	
	10 May 16	Baker, Sydney		10 May 17	
	10 May 16	Clerici, Maru		10 May 17	
	10 May 16	Graham, Percy Herbert		10 May 17	
	10 May 16	Walsh, Leslie Edward (207*a*)		10 May 17	
1285	10 May 16	deBourbel, Godfrey		10 May 17	
	10 May 16	Hutcheson, John Henderson		10 May 17	
	10 May 16	Russell, Robert Tor, D.S.O.		10 May 17	
	10 May 16	Wyles, John Wallace		10 May 17	
	12 May 16	Henderson, Robert Sharp		12 May 17	
1290	12 May 16	Corin, Edward John Ronald		12 May 17	
	15 May 16	McIntosh, Alexander (79)		15 May 17	
	15 May 16	Miller, Alexander Dunbar (*Tempy. Capt., 17 Nov. 16*)		15 May 17	
	16 May 16	Harris, Christopher Ernest		16 May 17	
	17 May 16	Cullen, James Goodsir		17 May 17	
1295	18 May 16	Cory, Edgar Myrie (79)		18 May 17	
	19 May 16	Pooley, James Price Favell		19 May 17	
	22 May 16	Fricker, Ernest Stanley Gibbs (51)		22 May 17	
	26 May 16	Hughes, Edward Walter		26 May 17	
	26 May 16	Deed, Laurence Henry (57)		26 May 17	
1300	26 May 16	Kissane, Richard (207*a*)		26 May 17	
	29 May 16	Thurley, Mervyn (*Ty. Capt. 1 May 17*)		29 May 17	
	29 May 16	Lyons, Arthur Jacob (*temp. Capt., 13 Dec. 17*)		29 May 17	
	30 May 16	Wilson, Francis Woodman (*Ty. Capt. 30 May 17*).		30 May 17	
	1 June 16	List, John Napier (46)		1 June 17	
1305	1 June 16	Edwards, Osborne George (*Ty. Capt. 1 June 17*) *Service for promotion from 6 Nov. 16 to 31 May 17.*		1 June 17	
	1 June 16	Williams, Joseph Clarence		1 June 17	
	1 June 16	Annely, Francis Richard Charles		1 June 17	
	1 June 16	Lewis, George Wesley		1 June 17	
	2 June 16	Simmonds, Vivian Bruton		2 June 17	
1310	2 June 16	Shaw, George Frederick		2 June 17	
	4 June 16	Britts, Malcolm Herbert Gordon		4 June 17	
	6 June 16	Hamill, Harry		6 June 17	
	10 June 16	Mont-Thomas, Emlyn Robert		10 June 17	
	10 June 16	Lowe, John Sanderson (57)		10 June 17	
1315	10 June 16	Cable, Robert William		10 June 17	
	10 June 16	Gerrard, Edwin Frederick (*temp. Capt. 8 July 18*)		10 June 17	
	10 June 16	Hodgkins, Norman Francis		10 June 17	
	10 June 16	Spragge-White, Thomas Edwin		10 June 17	
	10 June 16	Lucas, Frederick		10 June 17	
1320	10 June 16	Muggeridge, Harold Chisman (79, 207)		10 June 17	
	10 June 16	Parker, David Ralston		10 June 17	
	10 June 16	Waymouth, Oswald Smith		10 June 17	
	10 June 16	Neal, Guy Trevor St. John		10 June 17	
	10 June 16	Walker, Frederick Kennedy (51)		10 June 17	
1325	10 June 16	Ewens, George Perryman (57)		10 June 17	
	10 June 16	Glen, Robert James Grange		10 June 17	
	10 June 16	Angus, Harry Carteledge		10 June 17	
1328	10 June 16	Herne, Reginald Valentine		10 June 17	

INFANTRY—*contd.*

Lieutenants—*contd.*

No.	Date of First Commn.	Names.	Date of birth.	Date of rank.	Remarks.
1334	12 June 16	Dempster, James		12 June 16	
1335	13 June 16	Rigg, Joseph Harold		13 June 17	
	13 June 16	Walker, Frank Hugh (46)		13 June 17	
	13 June 16	Adams, George Garnet Price Domel		14 June 17	
	14 June 16	Barrow, George Easton		14 June 17	
	14 June 16	Ward, Thomas (51)		14 June 17	
1340	15 June 16	Watkins, Frederick James		15 June 17	
	15 June 16	Lemon, William Solomon		15 June 17	
	16 June 16	Auston, George Tabor		16 June 17	
	17 June 16	Lambert, Arthur George (51, 207*a*)		17 June 17	
	19 June 16	Ward, Leonard Harris (46)		19 June 17	
1345	19 June 16	Wakeham, Arnold Rupert		19 June 17	
	20 June 16	Rowson, William Stanley		20 June 17	
	20 June 16	Bowler, Hugh Martin		20 June 17	
	20 June 16	Cole, William John Ives		20 June 17	
	20 June 16	Evelegh, Piers Bonham Carter (*Ty. Capt. 1 May 17*)		20 June 17	
1350	23 June 16	Young, James (*Ty. Capt., 23 June 16*)		23 June 17	
	23 June 16	Teehan, Patric		23 June 17	
	23 June 16	Lawson, Edward Ferguson		23 June 17	
	24 June 16	Tregale, John William Walter		24 June 17	
	25 June 16	Bramley, Percy Brooke (*Ty. Maj., 25 June 16*)		25 June 17	
1355	28 June 16	James, William Thomas		28 June 17	
	29 June 16	Watson, George Lawrence		29 June 17	
	1 July 16	Ashton, Alexander (51, 52, 79)		1 July 17	
	1 July 16	Jessop, Frederick Thomas		1 July 17	
	1 July 16	Blackwood, Neol Pinkstan O'Reilly (79)		1 July 17	
1360	1 July 16	Delmege, Claude Henry		1 July 17	
	1 July 16	Mather, James Jamieson		1 July 17	
	1 July 16	Macmillan, Kenneth Wallace		1 July 17	
	1 July 16	Havelock, William Arthur Chaplin		1 July 17	
	1 July 16	Lakeland, William John		1 July 17	
1365	1 July 16	Mcdonald, Sydney Henry		1 July 17	
	1 July 16	Stewart, Charles Wilfred (51)		1 July 17	
	1 July 16	Daly, Denis James		1 July 17	
	1 July 16	Aspinwall, William Hugh Elliot		1 July 17	
	1 July 16	Wells, Reginald Wallis		1 July 17	
1370	1 July 16	Brodie, Walter Scott Campbell		1 July 17	
	1 July 16	Tapsell, Clarence Bailey		1 July 17	
	1 July 16	Maurice, John Oliver Fiennes (51)		1 July 17	
	1 July 16	Godfrey, William Ewart		1 July 17	
	3 July 16	Wharton, Frederick		3 July 17	
1375	4 July 16	Hill, Arthur Halkett (*Ty. Capt., 26 July 17*)		4 July 17	
	5 July 16	Herdon, Albert Hampton		5 July 17	
	5 July 16	O'Leary, Michael George		5 July 17	
	5 July 16	Jaquet, Joseph Medworth		5 July 17	
	5 July 16	Hone, Arnold Charles, M.C.		5 July 17	
1380	6 July 16	Smith, Kenneth Crowley (*Ty. Capt., 4 Apl. 17*)		6 July 17	
	7 July 16	Mowforth, Haydn		7 July 17	
	7 July 16	Shearman, Wallet West		7 July 17	
	7 July 16	Harder, John Reedie		7 July 17	
	7 July 16	Cameron, Alan William		7 July 17	
1385	7 July 16	Hay, Edward Bruce		7 July 17	
	7 July 16	Nicholls, Reginald Latham		7 July 17	
	7 July 16	Collinge, William Chisholm		7 July 17	
	7 July 16	Stuart, Thomas Charles		7 July 17	
	8 July 16	Davies, John Hunter		8 July 17	
1390	13 July 16	Crawshaw, Percy William		13 July 17	
	14 July 16	Courtenay, John Matthew		14 July 17	
	14 July 16	Slaughter, Henry Laurence		14 July 17	
	14 July 16	Walters, Richard (57)		14 July 17	
	15 July 16	Booth, Walter Martin		15 July 17	
1395	16 July 16	Cox, Arthur James		16 July 17	

INFANTRY—*contd.*

Lieutenants—*contd.*

No.	Date of First Commn.	Names.	Date of birth.	Date of rank.	Remarks.
1398	17 July 16	Berry, Ernest Stanley		17 July 17	
	17 July 16	Browne, Denis Robert Howe (*temp. Maj., 25 Feb. 18*)		17 July 17	
1400	20 July 16	Stirling, Frederick		20 July 17	Released from Army Service.
	22 July 16	Williams, John Kynastan		22 July 17	
	22 July 16	Bourke, Dermot Richard Southwell (79)		22 July 17	
	22 July 16	Crossle, Henry Corry		22 July 17	
	22 July 16	Ludlow, Frank		22 July 17	
	22 July 16	Jolly, Evelyn Hugh Parker		22 July 17	
1405	22 July 16	Kirkpatrick, William (*Ty. Capt., 15 Oct. 17*)		22 July 17	
	22 July 16	Gould, Herbert Ross (*temp. Capt., 20 June 18*)		22 July 17	
	22 July 16	Collins, Godfrey Ferdinando Stratford (206)		22 July 17	Released from Army Service.
1410	22 July 16	Grant, John Leslie		22 July 17	
	22 July 16	Evershed, Wallis Albury		22 July 17	
	22 July 16	Bushby, Walker Edwin		22 July 17	
	22 July 16	Adam, Colin Gurdon		22 July 17	
	22 July 16	MacLachlan, Duncan (*temp. Capt., 28 Sep. 17*)		22 July 17	
1415	22 July 16	Norman, Douglas Reay (51, 207)		22 July 17	Released from Army Service.
	22 July 16	Farrell, Nelson Norbert		22 July 17	
	22 July 16	Furness, Horace Geoffrey (48*a*, 51, 207*a*)		22 July 17	
	22 July 16	Haughton, Wilfred Benjamin (79)		22 July 17	
	22 July 16	Scroggie, John Adam (206) (**Ty. Capt., 7 Nov. 11*)		22 July 17	
1420	22 July 16	Scott, James Alexander		22 July 17	
	12 July 16	Harding, George Trevor Hamilton		22 July 17	
	22 July 16	Wylie, Francis Verner (169) (*temp. Capt., 4 Mar. 17*)		22 July 17	
	22 July 16	Bartholomew, Geoffrey		22 July 17	
	22 July 16	Bishop, Lawrence Arthur		22 July 17	Released from Army Service.
1425	22 July 16	Digby, Kenelm George		22 July 17	
	22 July 16	Musto, Arnold Albert		22 July 17	
	23 July 16	Ellis, William Jamieson		23 July 17	
	26 July 16	Kirkhope, Kenneth Macleay (*Ty. Lt.-Col., 26 July 17*)		26 July 17	
	27 July 16	Burnett, George (172*a*, 207*a*)		27 July 17	
1430	28 July 16	Cox, Edward Orme (*Ty. Capt., 21 Oct. 16*)		28 July 17	
	28 July 16	McNair, George Douglas		28 July 17	
	29 July 16	Crocker, Arthur		29 July 17	
	29 July 16	Tinsley, William Stewart		29 July 17	
	30 July 16	English, Alexander		30 July 17	
1435	31 July 16	Gwynn, Charles Herbert (*Ty. Maj., 1 Dec. 17*)		31 July 17	
	1 Aug. 16	Parsons, Frank Farleigh		1 Aug. 17	
	1 Aug. 16	Knowles, George Bentley		1 Aug. 17	
	1 Aug. 16	Marshall, Frederick		1 Aug. 17	
	1 Aug. 16	Rowcliffe, Norman Henderson Barrington		1 Aug. 17	
1440	1 Aug. 16	Fitzpatrick, Patrick William		1 Aug. 17	
	2 Aug. 16	Pollen, Harry Stewart		2 Aug. 17	
	2 Aug. 16	Smith, John Hubert		2 Aug. 17	
	4 Aug. 16	Willett, Humfrey Gilbert (*Ty. Capt., 1 Apl. 17*)		4 Aug. 17	
	4 Aug. 16	Vyall, Frederick Stanley		4 Aug. 17	
1445	5 Aug. 16	Maloy, Thomas Henry		5 Aug. 17	
	5 Aug. 16	Stevens, Thomas Harry Goldsworthy (105, 180)		5 Aug. 17	
	5 Aug. 16	Tulloch, David		5 Aug. 17	
	5 Aug. 16	Lawson, Harold Arthur Clint		5 Aug. 17	
	5 Aug. 16	Birkett, Stanley		5 Aug. 17	
	5 Aug. 16	Gower, John Herbert Noel		5 Aug. 17	
	5 Aug. 16	Sinclair, Norman		5 Aug. 17	
	5 Aug. 16	Marston, Edward Noel		5 Aug. 17	
1450	7 Aug. 16	Quilter, George John Henry (125*a*)		7 Aug. 17	
	10 Aug. 16	Mitchell, Reginald St. John		10 Aug. 17	
	10 Aug. 16	Payton, Wilfred Hugh (169) (*Ty. Capt., 2 Mar. 17*)		10 Aug. 17	
	10 Aug. 16	McCracken, Andrew (169) (*Ty. Capt., 28 Feb. 17*)		10 Aug. 17	
	10 Aug. 16	Beyts, Mario Stafford (57)		10 Aug. 17	
1455	10 Aug. 16	Brown, Donald James (172*a*, 207*a*)		10 Aug. 17	
	10 Aug. 16	Cameron, Geoffrey St. Quentin		10 Aug. 17	
1457	10 Aug. 16	LeGeyt, Philip Sydney		10 Aug. 17	

* Without pay and allowances of the rank.

INFANTRY—*contd.*

Lieutenants—*contd.*

No.	Date of First Commn.	Names.	Date of birth.	Date of rank.	REMARKS.
1458	10 Aug. 16	Stubbs, Sydney Thomas	. . . .	10 Aug. 17	
	10 Aug. 16	Moxon, Rupert	. . . .	10 Aug. 17	
1460	10 Aug. 16	Perry, Arthur Stanley Harris	. . . .	10 Aug. 17	
	10 Aug. 16	Gorst, Robert	. . . .	10 Aug. 17	
	10 Aug. 16	Brown, Alexander (51) (*temp. Capt., 1 Feb. 18*)	. . . .	10 Aug. 17	
	10 Aug. 16	Davis, Stanley	. . . .	10 Aug. 17	
	11 Aug. 16	Baker, Frederick Clarence (95)	. . . .	11 Aug. 17	
1465	11 Aug. 16	Jones, Edward Osborne	. . . .	11 Aug. 17	
	11 Aug. 16	Routleff, Joseph Ives	. . . .	11 Aug. 17	
	11 Aug. 16	Knox, Alexander Edmund	. . . .	11 Aug. 17	
	12 Aug. 16	Murray, Granville	. . . .	12 Aug. 17	
	12 Aug. 16	Barnett, Alfred Henry (*Ty. Capt., 11 Jan. 17*)	. . . .	12 Aug. 17	
1470	12 Aug. 16	Haskew, Lindsay Clifford	. . . .	12 Aug. 17	
	12 Aug. 16	Cheesman, Robert Ernest	. . . .	12 Aug. 17	
	14 Aug. 16	Peerman, Thomas Cowderoy	. . . .	14 Aug. 17	
	14 Aug. 16	Hyde-Smith, Valentine Gardner (*Ty. Capt., 18 Oct. 16*)	. . . .	14 Aug. 17	
	14 Aug. 16	Jackson, James Duncan (*Ty. Capt., 22 Sep. 16*)	. . . .	14 Aug. 17	
1475	14 Aug. 16	Hannah, John Stewert (51, 52)	. . . .	14 Aug. 17	
	14 Aug. 16	Douglas, Robert	. . . .	14 Aug. 17	
	14 Aug. 16	Almond, James (169) (*Ty. Capt., 2 Mar. 17*)	. . . .	14 Aug. 17	
	14 Aug. 16	Parry, Nevill Edward	. . . .	14 Aug. 17	
	14 Aug. 16	Shaw, Edward Brian (207)	. . . .	14 Aug. 17	Released from Army Service.
1480	14 Aug. 16	Jackson, Arthur Herbert Conway	. . . .	14 Aug. 17	
	14 Aug. 16	Thomson, Harold Harry (169)	. . . .	14 Aug. 17	
	14 Aug. 16	Campbell, Donald Lennox (57)	. . . .	14 Aug. 17	
	14 Aug. 16	Blank, Abraham Lewis	. . . .	14 Aug. 17	
	14 Aug. 16	Blomfield, Douglas John	. . . .	14 Aug. 17	
1485	14 Aug. 16	Stork, Herbert Cecil (*Ty. Capt., 26 Apl. 17*)	. . . .	14 Aug. 17	
	14 Aug. 16	Kemm, Archibald Henry (46, 207)	. . . .	14 Aug. 17	
	14 Aug. 16	Scott, Thomas Terrance (46)	. . . .	14 Aug. 17	
	14 Aug. 16	Grose-Hodge, Humfrey (207)	. . . .	14 Aug. 17	
	15 Aug. 16	Smith, Robert Hamilton	. . . .	15 Aug. 17	
1490	17 Aug. 16	Wolters, Nicholas Charles (207*a*)	. . . .	17 Aug. 17	
	17 Aug. 16	Smyth, Gilbert Henry (*Ty. Capt., 17 Aug. 17*)	. . . .	17 Aug. 17	
	17 Aug. 16	O'Neill-Ready, Charles William	. . . .	17 Aug. 17	
	18 Aug. 16	Maclean, Henry Gordon	. . . .	18 Aug. 17	
	19 Aug. 16	Reed, George Stanley M.C. (*Ty. Capt., 30 Aug. 17*)	. . . .	19 Aug. 17	
1495	21 Aug. 16	Nevill, John Benjamin (**Ty. Capt., 24 Sep. 17*)	. . . .	21 Aug. 17	
	21 Aug. 16	Campbell, Alexander, M.C.	. . . .	21 Aug. 17	
	21 Aug. 16	Birley, Frank, D.C.M. (*Ty. Capt., 10 Jan. 18*)	. . . .	21 Aug. 17	
	21 Aug. 16	Carrick, Henry Ernest	. . . .	21 Aug. 17	
	21 Aug. 16	Anderson, Ernest Dennett (207)	. . . .	21 Aug. 17	
1500	21 Aug. 16	Clapp, William Maurice (207*a*)	. . . .	21 Aug. 17	
	21 Aug. 16	Dobbs, William Evelyn Joseph (*Ty. Maj. 22 Aug. 18*)	. . . .	21 Aug. 17	
	21 Aug. 16	Smith, Egerton	. . . .	21 Aug. 17	
	21 Aug. 16	Rowbotham, Claude John	. . . .	21 Aug. 17	
	21 Aug. 16	Dow, Thomas Miller (207*a*)	. . . .	21 Aug. 17	
1505	21 Aug. 16	Blizard, William Ewart	. . . .	21 Aug. 17	
	21 Aug. 16	Forbes, Barre Cassels (*temp. Maj., 30 July 18*)	. . . .	21 Aug. 17	
	22 Aug. 16	Bentley, Tom	. . . .	22 Aug. 17	
	22 Aug. 16	Willis, Reginald Harry (*Ty. Capt., 20 Feb. 17*)	. . . .	22 Aug. 17	
	22 Aug. 16	Todd, William	. . . .	22 Aug. 17	
1510	22 Aug. 16	Stein, John Alan	. . . .	22 Aug. 17	
	22 Aug. 16	Hogshaw, Francis Henry	. . . .	22 Aug. 17	
	22 Aug. 16	Melvin, William George	. . . .	22 Aug. 17	
	22 Aug. 16	Kalberer, John Fraser (51)	27 Dec. 89	22 Aug. 17	
	24 Aug. 16	Lake William George	. . . .	24 Aug. 17	
1515	25 Aug. 16	Beaufort, Percy Mostyn	. . . .	25 Aug. 17	
	25 Aug. 16	Tiffen, William Losh	. . . .	25 Aug. 17	
	25 Aug. 16	Pearson, James Bruce	. . . .	25 Aug. 17	
	25 Aug. 16	Higginson, Norman, M.C.	. . . .	25 Aug. 17	
	25 Aug. 16	Roan, William	. . . .	25 Aug. 17	
1520	25 Aug. 16	Gibson, David Arthur	. . . .	25 Aug. 17	
	25 Aug. 16	Simon, Alfred Owen	. . . .	25 Aug. 17	
1522	26 Aug. 16	Davidson, David James	. . . .	26 Aug. 17	

* Without pay and allowances of the rank.

INFANTRY—*contd.*

Lieutenants—*contd.*

No.	Date of First Commn.	Names.	Date of birth.	Date of rank.	Remarks.
1523	26 Aug. 16	McAlpin, Duncan Francis Claude (207*a*)	....	26 Aug. 17	
	28 Aug. 16	Barber, Charles Gordon	....	28 Aug. 17	
1525	28 Aug. 16	Smith, Charles Tysoe	....	28 Aug. 17	
	28 Aug. 16	Lee, Edward Owen (172*a*)	....	28 Aug. 17	
	28 Aug. 16	Beal, Hastings Evelyn John (*Ty. Capt*, *20 Aug. 17*)	....	28 Aug. 17	
	28 Aug. 16	Owen, Llewelyn Gordon 207*a*	....	28 Aug. 17	
	28 Aug. 16	Rust, Neville Arthur	....	28 Aug. 17	
1530	28 Aug. 16	Roxburgh, Thomas James Young (*temp. Capt.*, *12 Apl. 18*)	....	28 Aug. 17	
	28 Aug. 16	Saxton, Walter Theodore	....	28 Aug. 17	
	28 Aug. 16	Inder, Robert Wilfred	....	28 Aug. 17	
	28 Aug. 16	Barclay, Archibald George Frant	....	28 Aug. 17	
	28 Aug. 16	Davis, William John	....	28 Aug. 17	
1535	28 Aug. 16	Wace, Ferdinand Blyth (205*a*)	....	28 Aug. 17	
	29 Aug. 16	Chance, Percival Vincent	....	29 Aug. 17	
	29 Aug. 16	Staunton, James William †(*temp. Capt.*, *14 Aug. 18*)	....	29 Aug. 17	
	1 Sep. 16	Weston, Francis Henry (51, 71)	....	1 Sep. 17	
	1 Sep. 16	Stent, Percy John Hodsoll (*temp. Capt.*, *19 Apl. 18*)	....	1 Sep. 17	
1540	1 Sep. 16	Way, John Dover	....	1 Sep. 17	
	1 Sep. 16	Hart, Norman Foley	....	1 Sep. 17	
	1 Sep. 16	Fido, Clarence Brewin	....	1 Sep. 17	
	1 Sep. 16	Malcolm, James Dogi	....	1 Sep. 17	
	2 Sep. 16	Blaikie, Harold Mergler (95)	....	2 Sep. 17	
1545	2 Sep. 16	Moberly, Ronald Stuart (54, 79)	....	2 Sep. 17	
	2 Sep. 16	Brazier, Clifford Claude Hendy (*temp. Maj. 7 Aug. 18*)	....	2 Sep. 17	
	2 Sep. 16	Stubbs, Henry Hubert	....	2 Sep. 17	
	5 Sep. 16	Cuerdon, Frederick James	....	5 Sep. 17	
	5 Sep. 16	Ritchie, John Hall	....	5 Sep. 17	
1550	5 Sep. 16	Blackburn, Maurice Edgar Musgrav (*temp. Capt.*, *23 Mar. 18*)	....	5 Sep. 17	
	5 Sep. 16	Wade-Evans, Dinnie Gwynne (51, 207*a*)	....	5 Sep. 17	
	5 Sep. 16	Startin, Hugh Adkin	....	5 Sep. 17	
	6 Sep. 16	Badcock, William Burton	....	6 Sep. 17	
	6 Sep. 16	Fry, Wallace John	....	6 Sep. 17	
1555	6 Sep. 16	Wardle, Frederick McLean	....	6 Sep. 17	
	7 Sep. 16	Clapp, Leslie John (51)	....	7 Sep. 17	
	7 Sep. 16	Davies, William	....	7 Sep. 17	
	8 Sep. 16	Harlow, Christopher Millward (46)	....	8 Sep. 17	
	8 Sep. 16	Northey, Henry Stewart	....	8 Sep. 17	
1560	8 Sep. 16	Anderson, Gerald Alfrey (46)	....	8 Sep. 17	
	8 Sep. 16	Keates, Joseph Alfred	....	8 Sep. 17	
	9 Sep. 16	Keenan, Thomas James	....	9 Sep. 17	
	9 Sep. 16	Hanrahan, Maurice Webb	....	9 Sep. 17	
	9 Sep. 16	Harrison, Albert Francis Louis Helfrick (*temp. Capt.*, *15 Jan. 18*)	....	9 Sep. 17	
1565	10 Sep. 16	Godwin, Walter Edward (57)	....	10 Sep. 17	
	11 Sep. 16	Munro, Donald (46, 57)	....	11 Sep. 17	
	12 Sep. 16	Steedman, James Alexander	....	12 Sep. 17	
	13 Sep. 16	Yerbury, Stuart Browning	....	13 Sep. 17	
	13 Sep. 16	Dobson, Frank Edward (207*a*)	....	13 Sep. 17	
1570	14 Sep. 16	Joyce, Thomas Howell	....	14 Sep. 17	
	14 Sep. 16	Byron, Francis Harvey	....	14 Sep. 17	
	15 Sep. 16	Taylor, Stanley Grisewood	....	15 Sep. 17	
	15 Sep. 16	Elmslie, Cyril Stewart	....	15 Sep. 17	
	15 Sep. 16	Trusler, William Jeffrey	....	15 Sep. 17	
1575	18 Sep. 16	Bill, Edgar Llewellyn	....	18 Sep. 17	
	18 Sep. 16	Greswell, Ernest (*Ty. Capt.*, *30 Jan. 17*)	....	18 Sep. 17	
	18 Sep. 16	Slater, Arthur Cecil Norman (*Ty. Capt.*, *27 Feb. 17*)	....	18 Sep. 17	
	19 Sep. 16	Vernieux, Charles Osmond	....	19 Sep. 17	
	20 Sep. 16	Goodwin, Arthur Ernest	....	20 Sep. 17	
1580	20 Sep. 16	O'Hara, Clarence Havelock (*Ty. Capt.*, *9 Aug. 17*)	....	20 Sep. 17	
1581	20 Sep. 16	Neave, Louis Harold	....	20 Sep. 17	

† With pay and allowances as for Lieutenant.

No.	Date of First Commn.	Names.	Date of birth.	Date of rank.	REMARKS.

INFANTRY—*contd.*

Lieutenants—*contd.*

No.	Date of First Commn.	Names.	Date of birth.	Date of rank.	REMARKS.
1582	20 Sep. 16	Crichton, Alexander Baillie (*Ty. Capt., 30 Oct. 17*).	. . . .	20 Sep. 17	
	21 Sep. 16	Fordham, Harry	. . . .	21 Sep. 17	
	21 Sep. 16	Cargill, Angus Keith	. . . .	21 Sep. 17	
1585	21 Sep. 16	Deane, Victor Murray (*temp. Capt.*)	. . . .	21 Sep. 17	
	21 Sep. 16	Mooney, William James	. . . .	21 Sep. 17	
	21 Sep. 16	Corlett, Dudley Stuart	. . . .	21 Sep. 17	
	25 Sep. 16	Beams, Reginald Hervey	. . . .	25 Sep. 17	
	25 Sep. 16	Hodson, Charles	. . . .	25 Sep. 17	
1590	25 Sep. 16	Higgens, Charles Raymond	. . . .	25 Sep. 17	
	26 Feb. 16	Raynor, William Downing	. . . .	26 Sep. 17	
	26 Sep. 16	Parkinson, Herbert Leslie	. . . .	26 Sep. 17	
	26 Sep. 16	Groombridge, Kenneth Corry	. . . .	26 Sep. 17	
	28 Sep. 16	Woods, Percy William Henry (*temp. Capt., 6 June 18*)	. . . .	28 Sep. 17	
	28 Sep. 16	Smith, Richard Hugh	. . . .	28 Sep. 17	
1595	30 Sep. 16	Unger, Valentine Carl Whybrow (79)	. . . .	30 Sep. 17	
	30 Sep. 16	Crowley, Hugh Stanislaus	. . . .	30 Sep. 17	
	1 Oct. 16	Milchem, Hugh George	. . . .	1 Oct. 17	
	1 Oct. 16	Williams, Evan Powell	. . . .	1 Oct. 17	
	5 Oct. 16	Williams, Alfred Clement	. . . .	5 Oct. 17	
1600	5 Oct. 17	Craw, Henry Hewat	. . . .	5 Oct. 17	
	5 Oct. 17	Drury, John	. . . .	5 Oct. 17	
	6 Oct. 16	Monies, Sidney Oliver	. . . .	6 Oct. 17	
	6 Oct. 16	Oliver, Thomas William (46, 207*a*)	. . . .	6 Oct. 17	
	6 Oct. 16	Cambridge, Albert Frederick	. . . .	6 Oct. 17	
1605	6 Oct. 16	Wilkinson, Kenneth William (207*a*)	. . . .	6 Oct. 17	
	6 Oct. 16	Fox, Harold Vivian	. . . .	6 Oct. 17	
	6 Oct. 16	Blair, Frank Ritchie (207*a*)	. . . .	6 Oct. 17	
	6 Oct. 16	Cowan, John Macqueen	. . . .	6 Oct. 17	
	6 Oct. 16	Cooper, Lionel Kingsley	. . . .	6 Oct. 17	
1610	6 Oct. 16	Cooper, Robert Francis	. . . .	6 Oct. 17	
	6 Oct. 16	McLean, Kenneth	. . . .	6 Oct. 17	
	6 Oct. 16	Carrol, Samuel Robert (207*a*)	. . . .	6 Oct. 17	
	6 Oct. 16	Lloyd, George Thomas (*temp. Capt., 16 Apl. 18*)	. . . .	6 Oct. 17	
	6 Oct. 16	Grieve, William (207*a*)	. . . .	6 Oct. 17	
1615	6 Oct. 16	Hussey, Dudley Charles, (207*a*)	. . . .	6 Oct. 17	
	6 Oct. 16	Corry, Percy Newton	. . . .	6 Oct. 17	
	6 Oct. 16	Millar, John Garnet	. . . .	6 Oct. 17	
	6 Oct. 16	Bailey, William Alfred	. . . .	6 Oct. 17	
	6 Oct. 16	Power, Oswald Ernest Samuel	. . . .	6 Oct. 17	
1620	6 Oct. 16	Cuvelier, Edward Augustus	. . . .	6 Oct. 17	
	6 Oct. 16	Wheeler, Charles Dunn	. . . .	6 Oct. 17	
	6 Oct. 16	Milwood, Alfred Mawson	. . . .	6 Oct. 17	Released from Army Service.
	6 Oct. 16	Lyall, John Henry	. . . .	6 Oct. 17	
	6 Oct. 16	Thomas, Frederick Arthur Schomberg (*Ty. Capt., 21 June 17*)	. . . .	6 Oct. 17	
1625	6 Oct. 16	Galvin, Cyril Michael	. . . .	6 Oct. 17	
	7 Oct. 16	Finlay, John	. . . .	7 Oct. 17	
	7 Oct. 16	Baldwin, George William	. . . .	7 Oct. 17	
	7 Oct. 16	Adlard, Frank Addison, M.C.	. . . .	7 Oct. 17	
	10 Oct. 16	Clapham, Stanley	. . . .	10 Oct. 17	
1630	10 Oct. 16	Bridgewater, Albert Gerald Benjamin	. . . .	10 Oct. 17	
	10 Oct. 16	Roberts, Samuel Arthur	. . . .	10 Oct. 17	
	10 Oct. 16	Preston, Kenneth (*temp. Capt., 27 Feb. 18*)	. . . .	10 Oct. 17	
	10 Oct. 16	Haycroft, James Seymour (*Ty. Capt., 10 Oct. 16*)	. . . .	10 Oct. 17	
	11 Oct. 16	Armour, John Stuart	. . . .	11 Oct. 17	
1635	11 Oct. 16	Anderson, Matthew Longfellow	. . . .	11 Oct. 17	
	11 Oct. 16	Elkins, Walter Henry	. . . .	11 Oct. 17	
	11 Oct. 16	Lawrence, Edwin	. . . .	11 Oct. 17	
	13 Oct. 16	Burns, William	. . . .	13 Oct. 17	
	14 Oct. 16	Wilkinson, Joseph Frederick	. . . .	14 Oct. 17	
1640	15 Oct. 16	Biddle, William Charles	. . . .	15 Oct. 17	
	15 Oct. 16	Chapman, Arthur John Broadwith (206)	. . . .	15 Oct. 17	
	15 Oct. 16	Meares, Willie King	. . . .	15 Oct. 17	
	15 Oct. 16	Walker, George David (51)	. . . .	15 Oct. 17	
	17 Oct. 16	Lucas, Cedric St. John	. . . .	17 Oct. 17	
1645	18 Oct. 16	Norton, George Alfred	. . . .	18 Oct. 17	
	18 Oct. 16	Kemp, William George John (50, 52)	. . . .	18 Oct. 17	
1648	20 Oct. 16	McDonald, Harry Patrick	. . . .	20 Oct. 17	

No.	Date of First Commn.	Names.	Date of birth.	Date of rank.	REMARKS.
		INFANTRY—*contd.*			
		Lieutenants—*contd.*			
1649	20 Oct. 16	Barlow, Luther	....	20 Oct. 17	
1650	20 Oct. 16	Meares, George Charles King	....	20 Oct. 17	
	20 Oct. 16	O'Brien, Richard William	....	20 Oct. 17	
	20 Oct. 16	Fielding, Henry	....	20 Oct. 17	
	20 Oct. 16	Fitzpatrick, Clement	....	20 Oct. 17	
	20 Oct. 16	Lownes, Joseph Reginald (*Ty. Capt., 28 Nov. 17*)	....	20 Oct. 17	
1655	21 Oct. 16	Kay, Thomas Albert	....	21 Oct. 17	
	21 Oct. 16	Sutton, Stanley Hubert	....	21 Oct. 17	
	21 Oct. 16	Marsh, Cecil Henry (51)	....	21 Oct. 17	Released from Army Service.
	21 Oct. 16	Christie, Thomas Murray	....	21 Oct. 17	
	21 Oct. 16	Recton, Theodore	....	21 Oct. 17	
1660	21 Oct. 16	Morley, Michael Henry Ausbro	....	21 Oct. 17	
	25 Oct. 16	Langer, William Colgan	....	25 Oct. 17	
	25 Oct. 16	Morgan, Eric John	....	25 Oct. 17	
	25 Oct. 16	Cortwright, Thomas Henry	....	25 Oct. 17	
	27 Oct. 16	Dawson, Ralph Currey	....	27 Oct. 17	
1665	27 Oct. 16	Bartlett, Daniel Frederick (*temp. Capt., 1 Sep. 18*)	....	27 Oct. 17	
	28 Oct. 16	Roberts, Cecil Richard John	....	28 Oct. 17	
	28 Oct. 16	Trusler, Charles Henry	....	28 Oct. 17	
	28 Oct. 16	Edge, Felix Lionel	....	28 Oct. 17	
	28 Oct. 16	Shaw, Frederick Charles	....	28 Oct. 17	
1670	28 Oct. 16	Bothwell, Percival James (207*a*)	....	28 Oct. 17	
	1 Nov. 16	Barclay, John Buchanan (*Ty. Capt., 5 Feb. 17*)	....	1 Nov. 17	
	1 Nov. 16	Bull, Cecil Herbert	....	1 Nov. 17	
	1 Nov. 16	Erskine, William Robert	1 Dec. 93	1 Nov. 17	
	4 Nov. 16	Foddy, Arthur Henry	....	4 Nov. 17	
1675	6 Nov. 16	Webb, Charles Archibald	....	6 Nov. 17	
	6 Nov. 16	Harvey, Robert Harold Mansel (79)	....	6 Nov. 17	
	6 Nov. 16	Quick, Norman	....	6 Nov. 17	
	10 Nov. 16	Scott, Owen Rodway	....	10 Nov. 17	
	10 Nov. 16	Berkeley, Velters Cornwalle Fitzhardinge	....	10 Nov. 17	
1680	10 Nov. 16	Scott-Brown, Egerton Gartside (95)	....	10 Nov. 17	
	10 Nov. 16	Cleary, Cyril Joseph (51)	....	10 Nov. 17	
	10 Nov. 16	Pelham-Clay, Edward Cecil Granby	....	10 Nov. 17	
	10 Nov. 16	McKenzie, Robert Duncan	....	10 Nov. 17	
	10 Nov. 16	Wallace, James Robert Kenneth	....	10 Nov 17	
1685	10 Nov. 16	Beveridge, Ronald	....	10 Nov. 17	
	10 Nov. 16	Stevens, Arthur	....	10 Nov. 17	
	10 Nov. 16	Cornish, Harry	....	10 Nov. 17	
	10 Nov. 16	Hunter, Harry Claude Arnele	....	10 Nov. 17	
	10 Nov. 16	Harper, Cuthbert Sutcliffe	....	10 Nov. 17	
1690	10 Nov. 16	Maguire, Patrick John	....	10 Nov. 17	
	10 Nov. 16	Whitby, Robert Bertram (46, 207*a*)	....	10 Nov. 17	
	10 Nov. 16	Saunders, Ernest Edward Leslie (52)	....	10 Nov. 17	
	10 Nov. 16	Dickens, William Samuel	....	10 Nov. 17	
	10 Nov. 16	Thorne, Stanley George	....	10 Nov. 17	
1695	11 Nov. 16	Bateman, Cecil George	....	11 Nov. 17	
	11 Nov. 16	Davidson, George Stainforth (79)	....	11 Nov. 17	
	15 Nov. 16	Hyde, James Arthur	....	15 Nov. 17	
	16 Nov. 16	Smith, Hector Angus (67)	....	16 Nov. 17	
	18 Nov. 16	Harvey, Denis Victor	....	18 Nov. 17	
1700	20 Nov. 16	McCall, Ben	....	20 Nov. 17	
	21 Nov. 16	Palmer, Frederick (207*a*)	....	21 Nov. 17	
	21 Nov. 16	Phillips, John Humphrey	....	21 Nov. 17	
	21 Nov. 16	Gipps, Walter Francis	....	21 Nov. 17	
	21 Nov. 16	Davies, Hugh	....	21 Nov. 17	
1705	21 Nov. 16	Lyons, Charles Patrick	....	21 Nov. 17	
	21 Nov. 16	Fisher, Stanley	....	21 Nov. 17	
	21 Nov. 16	Hopkins, Joseph Archibald Streatfield	....	21 Nov. 17	
	21 Nov. 16	Nielson, Eric Gerald	....	21 Nov. 17	
	21 Nov. 16	Hargreaves, Harold (54, 79, 206) (*temp. Capt., 11 June 18*)	....	21 Nov. 17	
1710	22 Nov. 16	Beaumont, Harold	....	22 Nov. 17	
	22 Nov. 16	Digby, Percival Reginald	....	22 Nov. 17	
	24 Nov. 16	Barnard, William	....	24 Nov. 17	
	24 Nov. 16	Soar, Samuel Frederick	....	24 Nov. 17	
	24 Nov. 16	Wood, David Grammie (*temp. Capt., 4 June 18*)	....	24 Nov. 17	
1715	25 Nov. 16	Maher, Cyril Benjamin	....	25 Nov. 17	
1716	25 Nov. 16	Williams, Arthur deCoetlogan	....	25 Nov. 17	

INFANTRY—*contd.*

Lieutenants—*contd.*

No.	Date of First Commn.	Names.	Date of birth.	Date of rank.	REMARKS.
1722	25 Nov. 16	Gompertz, George Henry Montague		25 Nov. 17	
	27 Nov. 16	Johnston, Albert Hall (79, 207)		27 Nov. 17	
	29 Nov. 16	Tayler, Reginald Godfroy a'Wood		29 Nov. 17	
1725	1 Dec. 16	Galvin, Oswald Henry		1 Dec. 17	
	1 Dec. 16	Fforde, George Fetherstonhaugh		1 Dec. 17	
	1 Dec. 16	Mayles, Richard William John		1 Dec. 17	
	1 Dec. 16	Presswell, Thomas George (*temp. Capt., 1 Apl. 18*)		1 Dec. 17	
	3 Dec. 16	Forbes, Alister Esme Buchan (*Ty. Maj., 3 Dec. 17*)		3 Dec. 17	
1730	4 Dec. 16	Purcell-Gildea, Edward James (*Ty. Capt., 9 Jan. 17*)		4 Dec. 17	
	4 Dec. 16	Betteridge, George William		4 Dec. 17	
	5 Dec. 16	Platts, Arnold		5 Dec. 17	
	11 Dec. 16	Blackie, Frank Burnett (51)		11 Dec. 17	
	12 Dec. 16	Fitzpatrick, Cyril James		12 Dec. 17	
1735	13 Dec. 16	Blackett, Walter Scott		13 Dec. 17	
	16 Dec. 16	Howard, James George		16 Dec. 17	
	16 Dec. 16	Hogan, Percival Clement		16 Dec. 17	
	16 Dec. 16	Mansfield, Montague George		16 Dec. 17	
	16 Dec. 16	Garstin, Edward Lewis Longmore		16 Dec. 17	
1740	19 Dec. 16	Provis, Frank		16 Dec. 17	
	19 Dec. 16	Richardson, Harold		19 Dec. 17	
	19 Dec. 16	Tregaskis, Oswald		19 Dec. 17	
	19 Dec. 16	Drake, Percy Charles		19 Dec. 17	
	19 Dec. 16	Butt, Herbert		19 Dec. 17	
1745	19 Dec. 16	Peace, Alfred Creswell		19 Dec. 17	
	21 Dec. 16	Vaughan, Sidney George (207*a*)		21 Dec. 17	
	21 Dec. 16	Hebberd, Edward St. John		21 Dec. 17	
	23 Dec. 16	Charriol, Richard Alexandre		23 Dec. 17	
	27 Dec. 16	Chalmers, Thomas Andrew (*temp. Capt., 19 Nov. 17*)		27 Dec. 17	
1750	27 Dec. 16	Nickels, Thomas		27 Dec. 17	
	27 Dec. 16	Rockley, Frank Melville		27 Dec. 17	
	27 Dec. 16	Campbell, Thomas John (*temp. Capt., 4 Apl. 18*)		27 Dec. 17	
	28 Dec. 16	Fitz-Herbert, Gerald Allingham		28 Dec. 17	
	1 Jan. 17	Evans, Ivor Augustus (296)		1 Jan. 18	
1755	1 Jan. 17	Mackrow, George Henry		1 Jan. 18	
	1 Jan. 17	Cooper, Richard Elmore		1 Jan. 18	
	1 Jan. 17	Vernal, Mervyn Stanley		1 Jan. 18	
	5 Jan. 17	Macdonald, Arthur St. John		5 Jan. 18	
	5 Jan. 17	Stoddart, John Stewart Parnell (207*a*)		5 Jan. 18	
1760	5 Jan. 17	Howarth, Herbert Sutcliffe		5 Jan. 18	
	5 Jan. 17	Howe, Edgar		5 Jan. 18	
	5 Jan. 17	Clark, George Allington (57)		5 Jan. 18	
	5 Jan. 17	Muspratt, Gilbert (46)		5 Jan. 18	
	5 Jan. 17	Cooper, Herbert Reginald		5 Jan. 18	
1765	5 Jan. 17	Casling, Daniel Melville		5 Jan. 18	
	5 Jan. 17	Saunders, Wilfrid		5 Jan. 18	
	5 Jan. 17	Alderson, Ronald Cyril		5 Jan. 18	
	5 Jan. 17	Ford, Frank O'Neill		5 Jan. 18	
	5 Jan. 17	Walmsley, Denis Carlton		5 Jan. 18	
1770	5 Jan. 17	Perrott, Lionel Clifford Phair		5 Jan. 18	
	5 Jan. 17	Cornwell, Richard Barry		5 Jan. 18	
	5 Jan. 17	Wotherspoon, William Scott (207 *a*)		5 Jan. 18	
	5 Jan. 17	Ord, Bertram Knox		5 Jan. 18	
	5 Jan. 17	Hudson, James Eric		5 Jan. 18	
1775	5 Jan. 17	Hutchens, Frank Kentsbeer (57)		5 Jan. 18	
	5 Jan. 17	Robinson, Joseph Clark		5 Jan. 18	
	5 Jan. 17	Crosby-Browne, Dennis		5 Jan. 18	
	5 Jan. 17	Neil, John Alfred		5 Jan. 18	
	5 Jan. 17	Slaughter, Charles Montague (46, 51, 95)		5 Jan. 18	
1780	5 Jan. 17	Farrington, Francis Leslie		5 Jan. 18	
	5 Jan. 17	Pirie, Douglas George Littlejohn (207*a*)		5 Jan. 18	
	5 Jan. 17	Willmot, Neil Reginald		5 Jan. 18	
	5 Jan. 17	Farr, Cecil Jaffrey (207*a*)		5 Jan. 18	
	5 Jan. 17	Cole, Robert David		5 Jan. 18	
1785	5 Jan. 17	Walker, Henry Lawrence		5 Jan. 18	
	5 Jan. 17	Meager, Charles Joseph		5 Jan. 18	
	5 Jan. 17	Hibberd Ernest		5 Jan. 18	
	5 Jan. 17	James, Thomas Albert (57)		5 Jan. 18	
	5 Jan. 17	Kelly, Clarence John (207*a*)		5 Jan. 18	
1790	5 Jan. 17	Macdonald, Alexander (79)		5 Jan. 18	
1792	5 Jan. 17	Bradbury, Arthur Lyle (*Ty. Capt., 13 Jan. 18*)		5 Jan. 18	

INFANTRY—*contd.*

Lieutenants.—*contd.*

No.	Date of First Commn.	Names.	Date of birth.	Date of rank.	Remarks.
1793	8 Jan. 17	Shaw, William (51)		8 Jan. 18	
	8 Jan. 17	Haswell, Percy Leonard		8 Jan. 18	
	8 Jan. 17	Bedell, George William (207*a*)		8 Jan. 18	
1795	8 Jan. 17	Speed, William Austin		8 Jan. 18	
	8 Jan. 17	Walker, Lawrence William		8 Jan. 18	
	5 Jan. 17	Blake, Leslie Fredrick Medon (57)		5 Jan. 18	
	10 Jan. 17	Tryhorne, Mathew William		10 Jan. 18	
	10 Jan. 17	Seale, Colin Ian		10 Jan. 18	
1800	11 Jan. 17	Fawcett, Hugh Edward Morris		11 Jan. 18	
	12 Jan. 17	Hayes, Ernest Richmond Corbett		12 Jan. 18	
	12 Jan. 17	Turner, John Marsden		12 Jan. 18	
	15 Jan. 17	Stoddart, Roger Austen		15 Jan. 18	
	16 Jan. 17	Fogarty, Philip Christopher		16 Jan. 18	
1805	18 Jan. 17	Hoff, Stanley Rose		18 Jan. 18	
	20 Jan. 17	Reid, Harold Alexander		20 Jan. 18	
	20 Jan. 17	Perry, Arthur Cecil		20 Jan. 18	
	20 Jan. 17	Hainsworth, Horace Mann (46)		20 Jan. 18	
	20 Jan. 17	Griffin, Albert Marshall		20 Jan. 18	
1810	20 Jan. 17	Elton, Anthony Conway		20 Jan. 18	
	20 Jan. 17	Colwill, Frank		20 Jan. 18	
	25 Jan. 17	Richardson, John Stewart (*Ty. Capt., 21 Jan. 18*)		25 Jan. 18	
	25 Jan. 17	Todman, Arthur McNish		25 Jan. 18	
	25 Jan. 17	Russell, Edward Powys		25 Jan. 18	
1815	25 Jan. 17	Bartlett, Albert Charles (207*a*)		25 Jan. 18	
	25 Jan. 17	Barbar, Alan James		25 Jan. 18	
	26 Jan. 17	Sloane, William Henry St. John		26 Jan. 18	
	27 Jan. 17	Curry, George Richard		27 Jan. 18	
	29 Jan. 17	Morford, Eric St. Clair		29 Jan. 18	
1820	29 Jan. 17	Falgar, Richard Joseph		29 Jan. 18	
	30 Jan. 17	Lewis, Alfred Elliott (*Ty. Maj., 30 Jan. 17*)		30 Jan. 18	
	1 Feb. 17	Bush, Sydney Poynty		1 Feb. 18	
	1 Feb. 17	White, William Thomas Harvey		1 Feb. 18	
	1 Feb. 17	Swainson, Charles Anthony (79) (*Ty. Capt., 1 Aug. 17*)		1 Feb. 18	
1825	1 Feb. 17	Morrison, Arnold O'Hanlon		1 Feb. 18	
	1 Feb. 17	Good, William Jonathon (*temp. Maj. 4 Oct. 18*)		1 Feb. 18	
	1 Feb. 17	Hussey, Neville George Edmund (207*a*) (*Ty. Capt., 23 Apl. 17*)		1 Feb. 18	
	1 Feb. 17	Smethurst, Samuel James (207*a*)		1 Feb. 18	
	1 Feb. 17	Rae, Frank Langford		1 Feb. 18	
	1 Feb. 17	Blake, Hedley William		1 Feb. 18	
1830	1 Feb. 17	Leonard, Gwilon Thurnell Ingledew (51, 52, 207*a*)		1 Feb. 18	
	1 Feb. 17	Oglethorpe, John Atkinson (48*a*, 50, 207*a*)		1 Feb. 18	
	1 Feb. 17	Harrison, Kenneth Hodgson (207*a*)		1 Feb. 18	
	1 Feb. 17	Atkinson, Alan Wilson		1 Feb. 18	
	1 Feb. 17	Dunning, Angus Ashburton John		1 Feb. 18	
1835	1 Feb. 17	Hampton, Lewis Colin		1 Feb. 18	
	1 Feb. 17	Daley, John Foulstone		1 Feb. 18	
	1 Feb. 17	Kearsey, Frederick Wilfred		1 Feb. 18	
	1 Feb. 17	Elloy, Lourence Melville		1 Feb. 18	
	1 Feb. 17	Keyte, Percy Edwin		1 Feb. 18	
1840	1 Feb. 17	Lawton, Alexander Crombi (51)		1 Feb. 18	
	1 Feb. 17	Greener, William James (207*a*)		1 Feb. 18	
	1 Feb. 17	Nissen, George Ayrton (*Ty. Capt., 8 Nov. 17*)		1 Feb. 18	
	1 Feb. 17	Groom, John Philip (52)		1 Feb. 18	
	1 Feb. 17	Tarran, Frederick Percy (79)		1 Feb. 18	
1845	1 Feb. 17	Edwards, Thomas (51, 100, 207*a*)		1 Feb. 18	
	1 Feb. 17	Metcalf, Frank		1 Feb. 18	
	1 Feb. 17	Taylor, Harry Wilfred Dick Colebrooke		1 Feb. 18	
	1 Feb. 17	Vickery, Raymond Milton		1 Feb. 18	
	2 Feb. 17	Cumming, Ronaleyn Charles Rossiter		2 Feb. 18	
1850	2 Feb. 17	Crawshaw, Harold Edgar John		2 Feb. 18	
	2 Feb. 17	Batty, Archibald Douglas George Staunton		2 Feb. 18	
	31 Dec. 14	Pilcher, Arthur Hope	25 June 81	3 Feb. 18	
	3 Feb. 17	Morgan, Edwin John Lewis		3 Feb. 18	
1855	3 Feb. 17	King, James (46, 207*a*)		3 Feb. 18	
	3 Feb. 17	Phillips, Charles Edward (52, 79)		3 Feb. 18	
1860	3 Feb. 17	London, Herbert Ruthven		3 Feb. 18	
	5 Feb. 17	Aston, Robert William (*Ty. Capt., 5 Dec. 17*)		5 Feb. 18	
	5 Feb. 17	Wentworth-Hyde, Raymond Robert		5 Feb. 18	
1863	5 Feb. 17	Fulton, Thomas		5 Feb. 18	

INFANTRY—*contd.*

Lieutenants—*contd.*

No.	Date of First Commn.	Names.	Date of birth.	Date of rank.	REMARKS.
1864	5 Feb. 17	Bensly, Harold Foster		5 Feb. 18	
	5 Feb. 17	Humphries, Alfred		5 Feb. 18	
	6 Feb. 17	Shepherd, Walter Isaac		6 Feb. 18	
	6 Feb. 17	Inglefield-Dall, Bertram Maurice		6 Feb. 18	
1865	10 Feb. 17	Manduell, John Steele		10 Feb. 18	
	10 Feb. 17	Thorpe, Arthur John (207*a*)		10 Feb. 18	
	15 Feb. 17	Eldon, Arthur		15 Feb. 18	
	15 Feb. 17	Greening, Arthur Henry		15 Feb. 18	
	15 Feb. 17	Holme, Hugh Basil		15 Feb. 18	
1870	18 Feb. 17	Oxenham, Samuel Herbert		18 Feb. 18	
	20 Feb. 17	Gibb, Frederick George (207*a*)		20 Feb. 18	
	21 Feb. 17	Evatt, Charles Alexander		21 Feb. 18	
	21 Feb. 17	Hindley, Charles Edward Hugh		21 Feb. 18	
	21 Feb. 17	Holland, George Reginald		21 Feb. 18	
1875	21 Feb. 17	Keir, George		21 Feb. 18	
	21 Feb. 17	Pearman, Horace Alfred		21 Feb. 18	
	21 Feb. 17	Hopkins, Charles John Wykham		21 Feb. 18	
	21 Feb. 17	Knight, Maynard Mansfield		21 Feb. 18	
	27 Feb. 17	MacLarron, Charles Montague (207*a*)		27 Feb. 18	
1880	1 Mar. 17	Cowie, Frank		1 Mar. 18	
	1 Mar. 17	Cooper, Roland Edgar (57, 207*a*)		1 Mar. 18	
	1 Mar. 17	Jack, Louis Alexander		1 Mar. 18	
	1 Mar. 17	Brown, George Lawrence		1 Mar. 18	
	1 Mar. 17	Fowler, George Victor (207)		1 Mar. 18	
1885	1 Mar. 17	Lewis, Francis William Hugh		1 Mar. 18	
	1 Mar. 17	Jones, Jenkyn David		1 Mar. 18	
	1 Mar. 17	Walsh, Hugh (54)		1 Mar. 18	
	1 Mar. 17	Winter, Harold Berningham		1 Mar. 18	
	1 Mar. 17	Lander, Kenneth Joseph Macgladdery		1 Mar. 18	
1890	1 Mar. 17	Steele, Frank Robert		1 Mar. 18	
	1 Mar. 17	Parker, Harold Rudolph Carson		1 Mar. 18	
	1 Mar. 17	Maloney, Joseph Clifford (207*a*)		1 Mar. 18	
	1 Mar. 17	Haines, William Edward (46)		1 Mar. 18	
	1 Mar. 17	Proctor, Norman Calderbank		1 Mar. 18	
1895	1 Mar. 17	Frattini, Gerado Norman		1 Mar. 18	
	1 Mar. 17	Robertson, Richard Bruce		1 Mar. 18	
	1 Mar. 17	Stacey, Arthur Frederick		1 Mar. 18	
	1 Mar. 17	Burrell, Thomas Henry		1 Mar. 18	
	1 Mar. 17	Scott, Eric Henry		1 Mar. 18	
1900	1 Mar. 17	Roach, Joseph Percival (52, 207*a*)		1 Mar. 18	
	1 Mar. 17	Moncrieff, Robert Lucian (207*a*)		1 Mar. 18	
	1 Mar. 17	Alder-Peters, Herbert (51)		1 Mar. 18	
	1 Mar. 17	O'Donovan, Denis Fitzmaurice		1 Mar. 18	
	1 Mar. 17	Campbell, James Whittaker		1 Mar. 18	
1905	1 Mar. 17	Simpson, John Alexander		1 Mar. 18	
	1 Mar. 17	Ross-Lewin, Francis Henry Westrop		1 Mar. 18	
	1 Mar. 17	Hutchinson, Ernest		1 Mar. 18	
	30 May 16	Jesse, William		2 Mar. 17	
	3 Mar. 17	Butcher, Robin Ernest		3 Mar. 18	
1910	4 Mar. 17	Burlingham, Donald		4 Mar. 18	
	5 Mar. 17	Mason, Harold Clayforth		5 Mar. 18	
	5 Mar. 17	Young, Erick Herbert		5 Mar. 18	
	5 Mar. 17	Robinson, James Golder		5 Mar. 18	
	5 Mar. 17	Macniven, George Archibald (46)		5 Mar. 18	
1915	5 Mar. 17	Fookes, Thomas George (51, 207*a*)		5 Mar. 18	
	6 Mar. 17	Hay, Frank Stuart		6 Mar. 18	
	6 Mar. 17	Murray, Gerald Edward (46)		6 Mar. 18	
	8 Mar. 17	Sheehan, Timothy Powell		8 Mar. 18	
	8 Mar. 17	Chamier, Anthony		8 Mar. 18	
1920	8 Mar. 17	Rooke, Ingram (46, 52, 207*a*)		8 Mar. 18	
	8 Mar. 17	Entwisle, George Ralph		8 Mar. 18	
	10 Mar. 17	Wise, Arthur Arnold		10 Mar. 18	
	10 Mar. 17	McGarry, Joseph		10 Mar. 18	
	10 Mar. 17	Thomson, William Roddie (207*a*)		10 Mar. 18	
1925	10 Mar. 17	Pell, Terence Owen		10 Mar. 18	
	10 Mar. 17	Jackson, Charles		10 Mar. 18	
	10 Mar. 17	Wise, Anthony Forster (*temp. Capt.*, 26 *Mar.* 18)		10 Mar. 18	
	11 Mar. 17	Askew, Thomas Edward		11 Mar. 18	
	12 Mar. 17	Mitchell, George Greig		12 Mar. 18	
1930	14 Mar. 17	Amey, Frederick George		14 Mar. 18	
	14 Mar. 17	Murdoch-Smith, Mungo		14 Mar. 18	
	14 Mar. 17	Rowley, John Thomas		14 Mar. 18	
1937	14 Mar. 17	Baldwin, David Frederick		14 Mar. 18	

INFANTRY—*contd.*

Lieutenants—*contd.*

No.	Date of First Commn.	Names.	Date of birth.	Date of rank.	REMARKS.
1938	14 Mar. 17	Duncan, David John (*temp. Capt., 2 June 17*)		14 Mar. 18	
	14 Mar. 17	Hughes, Reginald Raymond		14 Mar. 18	
1940	15 Mar. 17	Rountree, Arthur FitzGerald (**temp. Capt., 1 Nov. 17*)		15 Mar. 18	
	15 Mar. 17	Damerell, William James		15 Mar. 18	
	15 Mar. 17	Hosie, Alexander Lindsay (59, 79)		15 Mar. 18	
	15 Mar. 17	Temple, Bernard		15 Mar. 18	
	15 Mar. 17	Davis, Claude John McNaught		15 Mar. 18	
1945	15 Mar. 17	Lawrie, George		15 Mar. 18	
	15 Mar. 17	Wilkinson, Eric John		15 Mar. 18	
	15 Mar. 17	Pogson, Charles Albion		15 Mar. 18	
	16 Mar. 17	Sampson, Richard Woodward Kendall		16 Mar. 18	
1950	16 Mar. 17	Drew, Francis John		16 Mar. 18	
	16 Mar. 15	Mitchell, Eric John Reinhold (207*a*)		16 Mar. 18	
	17 Mar. 17	Bell, Alfred Richard		17 Mar. 18	
	19 Mar. 17	Tharle-Hughes, Reginald, M.B.E. (*temp. Capt., 12 Sep. 18*)		19 Mar. 18	
	19 Mar. 17	Meehan, Michael (*temp. Capt., 25 Jan. 18*)		19 Mar. 18	
1955	20 Mar. 17	Jones, Frank Lewis		20 Mar. 18	
	20 Mar. 17	Gadsdon, Henry Arnold		20 Mar. 18	
	20 Mar. 17	Fraser, Alan		20 Mar. 18	
	20 Mar. 17	Ambler, Charles Taaffee Murray		20 Mar. 18	
	22 Mar. 17	Ball, William Charles		22 Mar. 18	
	23 Mar. 17	Fitzhardinge-Berkeley, Guy		23 Mar. 18	
1960	24 Mar. 17	Rendle, John Morgan		24 Mar. 18	
	24 Mar. 17	Laythorpe, Frederick William Haswell		24 Mar. 18	
	27 Mar. 17	Webster, Percival Stanley (*temp. Capt., 1 Nov. 17*)		27 Mar. 18	
	28 Mar. 17	Miller, Thomas		28 Mar. 18	
	28 Mar. 17	Parkinson, Arthur George		28 Mar. 18	
1965	28 Mar. 17	Forester, Frank John		28 Mar. 18	
	28 Mar. 17	Logan, Ernest Williamson		28 Mar. 18	
	28 Mar. 17	Brown, David Hill		28 Mar. 18	
	28 Mar. 17	Castle, Joseph Percy, D.S.O.		28 Mar. 18	
	1 Apl. 17	Johnston, Meredith St. George Corbett		1 Apl. 18	
	1 Apl. 17	Allen, Basil Copleston (*Ty. Lt.-Col., 1 Apl. 17*)		1 Apl. 18	
	1 Apl. 17	Black, Robert Greenhill (*Ty. Capt., 1 Apl. 17*)		1 Apl. 18	
	1 Apl. 17	Hardy, Henry De Lisle (*Ty. Capt., 1 Apl. 17*).		1 Apl. 18	
1970	2 Apl. 17	Henderson, Adrian Noe (*temp. Capt., 9 Jan. 18*)		2 Apl. 18	
	2 Apl. 17	Breadon, George William Dienley (*Ty. Capt., 17 Jan. 18*)		2 Apl. 18	
	3 Apl. 17	Gore, Francis William Baldock (*Ty. Capt., 2 Apl. 17*)		2 Apl. 18	
	5 Apl. 17	Booth, Francis Napier		5 Apl. 18	
1975	5 Apl. 17	Morris, Frank		5 Apl. 18	
	5 Apl. 17	Wilson, Charles Henry		5 Apl. 18	
	5 Apl. 17	Beaty, Percival Charles Dudley		5 Apl. 18	
	5 Apl. 17	Barber, Ernest		5 Apl. 18	
	5 Apl. 17	Anderson, Robert William Alexander (*Ty. Capt., 11 Apl. 17*)		5 Apl. 18	
1980	6 Apl. 17	Quilter, Thomas Howlet (*Ty. Capt., 16 Apl. 17*)		6 Apl. 18	
	7 Apl. 17	Wilkins, John Frederick (*Ty. Capt., 2 Oct. 17*)		7 Apl. 18	
	7 Apl. 17	Roberts, William Henry		7 Apl. 18	
	7 Apl. 17	Mahony, Charles		7 Apl. 18	
	7 Apl. 17	Connolly, Edward Thomas Patrick		7 Apl. 18	
1985	8 Apl. 17	Fossey, Robert Henry		8 Apl. 18	
	8 Apl. 17	Mulcahy, William James Patrick		8 Apl. 18	
	8 Apl. 17	Carmichael, Charles		8 Apl. 18	
	9 Apl. 17	Casey, Thomas Hill		9 Apl. 18	
	9 Apl. 17	Neill, John Francis Bernard		9 Apl. 18	
1990	9 Apl. 17	Hudson, Lovell		9 Apl. 18	
	9 Apl. 17	Whiteley, Francis Norman		9 Apl. 18	
	10 Apl. 17	Giles, Leonard		10 Apl. 18	
	10 Apl. 17	Jenkins, Stanley Evan		10 Apl. 18	
	10 Apl. 17	Brewer, Valentine (207*a*)		10 Apl. 18	
1995	11 Apl. 17	Jackson, Herbert		11 Apl. 18	
	13 Apl. 17	Lamb, Philip		13 Apl. 18	
	14 Apl. 17	Greenwood, Henry Collingwood		14 Apl. 18	
1998	14 Apl. 17	Petts, Harold John		14 Apl. 18	

* Without pay and allowances of the rank.

INFANTRY—*contd.*

Lieutenants—*contd.*

No.	Date of First Commn.	Names.	Date of birth.	Date of rank.	Remarks.
1999	16 Apl. 17	Teague-Jones, Reginald		16 Apl. 18	
2000	16 Apl. 17	Gregory, Galiston Marcar (**Ty. Capt.*, 7 *Nov.* 17)		16 Apl. 18	
	17 Apl. 17	Hogan, Robert James		17 Apl. 18	
	17 Apl. 17	Barnes, Herbert Charles		17 Apl. 18	
	17 Apl. 17	Bigsby, Sydney Herbert (*Ty. Capt.*, 17 *Apl.* 17)		17 Apl. 18	
	17 Apl. 17	Lancaster, Charles Edwin Albert		17 Apl. 18	
2005	17 Apl. 17	Phillips, Emile		17 Apl. 18	
	17 Apl. 17	Macgregor, Alexander		17 Apl. 18	
	18 Apl. 17	Griffiths, Harry Melville (*Ty. Capt.*, 18 *Apl.* 17)		18 Apl. 18	
	18 Apl. 17	Pipe, Henry William		18 Apl. 18	
	18 Apl. 17	Pain, Simon Christopher Alfred John		18 Apl. 18	
2010	18 Apl. 17	Nute, John Francis Allen		18 Apl. 18	
	18 Apl. 17	Johnston, John Alexander		18 Apl. 18	
	19 Apl. 17	Davies, George Patrick		19 Apl. 18	
	19 Apl. 17	Longstaffe, John Walter		19 Apl. 18	
	19 Apl. 17	Saunders, Ernest John		19 Apl. 18	
2015	20 Apl. 17	Harris, Leonard		20 Apl. 18	
	20 Apl. 17	Griffith, Edward Stanhope		20 Apl. 18	
	20 Apl. 17	Redmond, Charles Joseph		20 Apl. 18	
	20 Apl. 17	Beaton, James Stuart Cameron		20 Apl. 18	
	20 Apl. 17	Mills, Reginald Theodore (207*a*)		20 Apl. 18	
2020	20 Apl. 17	Needham, John (*Ty. Capt.*, 20 *Apl.* 17)		20 Apl. 18	
	21 Apl. 17	Holt, Edmund		21 Apl. 18	
	25 Apl. 17	Livingstone, Lionel (*Ty. Capt.*, 25 *Apl.* 17)		25 Apl. 18	
	25 Apl. 17	Lang, John		25 Apl. 18	Released from Army Service.
	26 Apl. 17	Burgess, William George Aveston		26 Apl. 18	
2025	26 Apl. 17	Watson, Alexander George		26 Apl. 18	
	26 Apl. 17	Firth, Herbert Victor		26 Apl. 18	
	26 Apl. 17	Chapman, Gavin Percy		26 Apl. 18	
	26 Apl. 17	Barker, Henry Arthur Eric		26 Apl. 18	
	26 Apl. 17	Rostron, Harry		26 Apl. 18	
2030	26 Apl. 17	Lyon, Henry Norcutt		26 Apl. 18	
	27 Apl. 17	Faulconer, Edwin Abraham		27 Apl. 18	
	20 Apl. 17	Savege, Canon Richmond		28 Apl. 18	
	1 May 17	Perrott, Henry Ronald-Taylor Simpson (*Ty. Lt.-Col.*, 1 *May* 17)		1 May 18	
	1 May 17	Yearsley Fletcher Allen		1 May 18	
2035	1 May 17	Bradshaw, George Thomas		1 May 18	
	1 May 17	Barton, Edwin Richard		1 May 18	
	2 May 17	Lewis, Frederick Dowding (46, 207*a*)		2 May 18	
	2 May 17	Lumsden, Ernest Stephen		2 May 18	
	2 May. 17	Fowler, Edward Oliver (*Bt.-Maj.*, 3 *June* 18) (*Ty. Capt.*, 2 *May* 17).		2 May 18	
2040	5 May 17	McGavin, Denzil Gordon Kean (207*a*)		5 May 18	
	5 May 17	Hoskins, George Douglas		5 May 18	
	5 May 17	Carter, Theodore Frederic		5 May 18	
	5 May 17	Ferguson, Ronald Henry		5 May 18	
	5 May 17	Backhouse, John Reginald Charles		5 May 18	
2045	5 May 17	Wilkinson, Frederick Basil (59, 207*a*)		5 May 18	
	5 May 17	Gosling, John Evans		5 May 18	
	5 May 17	Hamilton, Arthur James		5 May 18	
	5 May 17	Vickery, Lawrence Mostyn		5 May 18	
	5 May 17	Green, Hugh Stanley		5 May 18	
2050	5 May 17	Bracey, Sidney Walter		5 May 18	
	5 May 17	Martin, Alan (207*a*)		5 May 18	
	5 May 17	Robertson, William Vincent		5 May 18	
	5 May 17	Berry, Owen Daniel Whittle (46)		5 May 18	
	5 May 17	Lintott, Percy Joseph (207*a*)		5 May 18	
2055	5 May 17	Stewart, Arther Reginald (48*a*, 51, 207*a*)		5 May 18	
	5 May 17	Hamer, Robert Gouing (207*a*)		5 May 18	
	5 May 17	Burrows, Albert Craymer (207*a*)		5 May 18	
	5 May 17	Cronan, Lubin Bertram		5 May 18	
	5 May 17	Nougerede, Vincent Adair Dela		5 May 18	
2060	5 May 17	Collett, John William (207*a*)		5 May 18	
	5 May 17	Fitzpatrick, Maurice John Joseph		5 May 18	
	5 May 17	Paterson, Archibald Morton (207*a*)		5 May 18	
	5 May 17	Sullivan, Arthur Stevens		5 May 18	
	5 May 17	Abbott, John William (207*a*)		5 May 18	
2065	5 May 17	Dewey, Cecil Arthur Robert (207*a*)		5 May 18	
	5 May 17	Dixon, Guy		5 May 18	
2067	5 May 17	Gagliardi, Francis Biaggio (207*a*)		5 May 18	

* Without pay and allowances of the rank.

No.	Date of First Commn.	Names.	Date of birth.	Date of rank.	REMARKS.
		INFANTRY—*contd.*			
		Lieutenants—*contd.*			
2068	5 May 17	Green, Donald Clement (50, 207*a*)		5 May 18	
	5 May 17	Baxter, William George Edwin		5 May 18	
2070	5 May 17	Collins, Hermann William		5 May 18	
	5 May 17	Harper, Arthur Henry (207*a*)		5 May 18	
	5 May 17	Stevenson, Lawrence James		5 May 18	
	5 May 17	Cambridge, Sydney John		5 May 18	
	5 May 17	Twigg, Charles Hanbury (207*a*)		5 May 18	
2075	5 May 17	Brady, John Pemberton (207)		5 May 18	
	5 May 17	Houston, Charles (207*a*)		5 May 18	
	5 May 17	Bishop, Herbert Compton		5 May 18	
	5 May 17	Wheeler, Arthur George		5 May 18	
	5 May 17	Collett, Harold Stephen (207*a*)		5 May 18	
2080	5 May 17	Freemantle, Hugh Elliott		5 May 18	Released from Army Service.
	5 May 17	Papworth, Frederick Arthur George (207*a*)		5 May 18	
	5 May 17	Thompson, Walter Stuart		5 May 18	
	5 May 17	Amor, Lucien Ernest (207*a*)		5 May 18	
	5 May 17	McIlwaine, William Arthur St. John		5 May 18	
2085	5 May 17	Pitts, George		5 May 18	
	5 May 17	Frend, George Victor Royle (207*a*)		5 May 18	
	5 May 17	McMillan, Thomas Bankhed (207*a*)		5 May 18	
	5 May 17	Asquith, Edwin John (207*a*)		5 May 18	
	5 May 17	Maker, Leonard (207*a*)		5 May 18	
2090	5 May 17	Cooter, Harry		5 May 18	
	5 May 17	Gray, Frederick William		5 May 18	
	5 May 17	Andrews, William Matthew (207*a*)		5 May 18	
	5 May 17	Austin, John Mein		5 May 18	
	5 May 17	Dodd, Joseph Richard (207*a*)		5 May 18	
2095	5 May 17	Fry, Hugh Robert (207*a*) (*temp. Capt.*, 4 *Oct. 18*).		5 May 18	
	5 May 17	Kerr, Thomas (46)		5 May 18	
	5 May 17	Lynch, John Alexander (46, 207*a*) (*temp. Capt.*, *1 May 18*).		5 May 18	
	5 May 17	Rogers, Harry Buston		5 May 18	
	5 May 17	Wilks, David James (207*a*)		5 May 18	
2100	5 May 17	Anderson, Herbert John (207*a*)		5 May 18	
	5 May 17	Devonshire, Charles Harris (207*a*)		5 May 18	
	5 May 17	Caird, Donald Renny (207*a*)		5 May 18	
	5 May 17	Fraser, David (207*a*)		5 May 18	
	5 May 17	Mayer, George Hugh		5 May 18	
2105	5 May 17	Noyle, Henry Wilbert		5 May 18	
	5 May 17	Osborne, Cyril Charles George		5 May 18	
	5 May 17	Scott, Reginald Ellis (207*a*)		5 May 18	
	5 May 17	Clarke, Arthur Alexander MacKenzie (207*a*)		5 May 18	
	5 May 17	Hollis, Frederick Stanley (207*a*)		5 May 18	
2110	5 May 17	Hogan, Herbert Francis		5 May 18	
	5 May 17	Wide, Leonard Alfred Churchill (207*a*)		5 May 18	
	5 May 17	Duff, Steven Harold (46)		5 May 18	
	5 May 17	Joblin, George William (59)		5 May 18	
	5 May 17	Phillips, William Henry (207*a*)		5 May 18	
2115	5 May 17	Gilmore, Arthur Lewis (207*a*)		5 May 18	
	5 May 17	Jeffries, Hugh Randall		5 May 18	
	5 May 17	Thomas, Enoch		5 May 18	
	5 May 17	Cooling, Charles Ernest		5 May 18	
	5 May 17	Cuthbert, Leonard Arthur (59, 207*a*)		5 May 18	
2120	5 May 17	Heath, Clement John (207*a*)		5 May 18	
	5 May 17	Jackson, William Ewart (207*a*)		5 May 18	
	5 May 17	Owens, Douglas William Byrne (207*a*)		5 May 18	
	5 May 17	Trinkle, Percy (46, 59)		5 May 18	
	5 May 17	Childs, Francis Dwyer		5 May 18	
2125	5 May 17	Cox, Frederick Basil		5 May 18	
	5 May 17	Porter, George Henry Jolly		5 May 18	
	5 May 17	Stamp, Robert Guy (207*a*)		5 May 18	
	5 May 17	Herdon, Bertram Laurence (207*a*)		5 May 18	
	5 May 17	Rose, Charles Granville		5 May 18	
2130	5 May 17	Black, Henry (207*a*)		5 May 18	
	5 May 17	Koch, Edwin Miller		5 May 18	
	5 May 17	Woodman, William Frederick		5 May 18	
	5 May 17	Dand, John		5 May 18	
	5 May 17	Donaldson, Herbert Vivian		5 May 18	
2135	5 May 17	Fraser, Kenneth Mills (51, 207*a*)		5 May 18	
	5 May 17	Harrison, Rupert St. Clair (207*a*)		5 May 18	
	5 May 17	Penn, Christian Erick		5 May 18	
	5 May 17	Bramwell, William Henry (207*a*)		5 May 18	
2139	5 May 17	Goss, Bertie Graham (207*a*)		5 May 18	

† With pay and allowances as for a Lieut.

No.	Date of First Commn.	Names.	Date of birth.	Date of rank.	REMARKS.
		INFANTRY—*contd*			
		Lieutenants—*contd.*			
2138	5 May 17	Nicachi, John Constantine	. . . . .	5 May 18	
	5 May 17	Reiner, Noel Andrew (51, 207*a*)	. . . . .	5 May 18	
2140	5 May 17	Duke, William Falcon	. . . . .	5 May 18	
	5 May 17	Fritchley, George Bowen (207*a*)	. . . . .	5 May 18	
	5 May 17	Walsh, Leslie William	. . . . .	5 May 18	
	5 May 17	Pollard, Ivan Eustace (57)	. . . . .	5 May 18	
2145	5 May 17	Mitchell, Charles Hilary (46, 207*a*)	. . . . .	5 May 18	
	5 May 17	Leamon, Howard Whitwell (207*a*)	. . . . .	5 May 18	
	5 May 17	Bateman, Frank William (51, 207*a*)	. . . . .	5 May 18	
	5 May 17	Doubt, Herbert Edwin	. . . . .	5 May 18	
	5 May 17	Pearson, Herbert Grayhurst (207*a*)	. . . . .	5 May 18	
2150	5 May 17	Wilken, William	. . . . .	5 May 18	
	5 May 17	Banham, Harry (*Ty. Capt., 15 Dec. 17*)	. . . . .	5 May 18	
	5 May 17	Armstrong, Wilfrid Lawson (51)	. . . . .	5 May 18	
	5 May 17	Davey, Charles Chrismas	. . . . .	5 May 18	
	5 May 17	Hawley, Norman Francis (51, 207*a*)	. . . . .	5 May 18	
2155	5 May 17	Kidson, Reginald Ainslie Forbes (207*a*)	. . . . .	5 May 18	
	5 May 17	Gibbs, Arthur Ronald	. . . . .	5 May 18	
	5 May 17	McDougall, Peter	. . . . .	5 May 18	
	5 May 17	Medley, Frank James	. . . . .	5 May 18	
	5 May 17	Ferguson, John Hugh	. . . . .	5 May 18	
2160	5 May 17	Humphreys, Douglas William George (207*a*)	. . . . .	5 May 18	
	5 May 17	Maflin, Alfred John Godwyn	. . . . .	5 May 18	
	5 May 17	Croxson, John Arthur Frederick (*temp. Capt., 10 July 18*)	. . . . .	5 May 18	
	5 May 17	Hockin, Frederic William	. . . . .	5 May 18	
	5 May 17	Mill, George Scott (207*a*)	. . . . .	5 May 18	
2165	5 May 17	Robinson, Daniel Bartlett	. . . . .	5 May 18	
	5 May 17	Shaw, Roderick (207*a*)	. . . . .	5 May 18	
	5 May 17	Beauchamp, Kenneth George	. . . . .	5 May 18	
	5 May 17	Brant-Smith, Charles	. . . . .	5 May 18	
	5 May 17	Minter George (207*a*)	. . . . .	5 May 18	
2170	5 May 17	Sutton, Vernon Colston	. . . . .	5 May 18	
	5 May 17	Griffen, Frank William Bailey (207*a*)	. . . . .	5 May 18	
	5 May 17	Honnor, Albert John (207*a*)	. . . . .	5 May 18	
	5 May 17	Gamon, Richard Benjamin	. . . . .	5 May 18	
2175	5 May 17	Allen, George Herbert (*Ty. Capt., 5 Dec. 17*)	. . . . .	5 May 18	
	5 May 17	Binstead, Victor John	. . . . .	5 May 18	
	5 May 17	O'Hara, Edward Hamilton	. . . . .	5 May 18	
	5 May 17	Orr, James	. . . . .	5 May 18	
2180	5 May 17	Parry, John Graham	. . . . .	5 May 18	
	5 May 17	Partridge, Thomas George	. . . . .	5 May 18	
	5 May 17	Seymour, Charles Leslie (207*a*)	. . . . .	5 May 18	
	5 May 17	Stedman, John Frederick (207*a*)	. . . . .	5 May 18	
	5 May 17	Lines, Reginald Edward	. . . . .	5 May 18	
2185	5 May 17	Monk, Francis Fitzhugh (207*a*)	. . . . .	5 May 18	
	7 May 17	Cann, Mark	. . . . .	7 May 18	
	7 May 17	Hooper, Edgar John	. . . . .	7 May 18	
	7 May 17	Prance, Basil Camden (*Ty. Capt., 7 May 17*)	. . . . .	7 May 18	
	8 May 17	Connell, Harold William (207*a*)	. . . . .	8 May 18	
2190	10 May 17	Graham, Thomas Charles Valentine	. . . . .	10 May 18	
	11 May 17	Metcalfe, John Henry	. . . . .	11 May 18	
	12 May 17	Swaffer, Harry Percival	. . . . .	12 May 18	
	12 May 17	Sweeney, Sydney (207*a*)	. . . . .	12 May 18	
	12 May 17	Henderson, Horace William Woodness (*Ty. Capt., 20 Novr. 17*)	. . . . .	12 May 18	
2195	12 May 17	Holmes, Henry Robert (*Ty. Capt., 12 May 17*)	. . . .	12 May 18	
	14 May 17	Jeffery, Charles Frank (*Ty. Capt., 14 May 17*)	. . . .	14 May 18	
	16 May 17	Adams, Edward Francis	. . . .	16 May 18	
	16 May 17	Nichols, John Edwin	. . . .	16 May 18	
	17 May 17	Graham, Gilbert Duncan (*temp. Maj., 9 Oct. 18*)	. . . .	17 May 18	
2200	17 May 17	Cumming, Donald Brodby	. . . .	17 May 18	
	17 May 17	Robertson, David Kinloch	. . . .	17 May 18	
	17 May 17	Holman-Hunt, Hilary Lushington (*Ty. Capt., 17 May 17*)	. . . .	17 May 18	
	18 May 17	Middelton, Lionel William	. . . .	18 May 18	
	18 May 17	Burns, Arthur Richard (*Ty. Capt., 22 May 17*)	. . . .	18 May 18	
2205	18 May 17	Underdown, Arthur Henry	. . . .	18 May 18	
	18 May 17	Hoerder, Paul Eugene Robert	. . . .	18 May 18	
	18 May 17	Briggs, Robert Chapman (*Ty. Capt., 18 May 17*)	. . . .	18 May 18	
	22 May 17	Howard, William Henry Ker, O.B.E.	. . . . .	22 May 18	
2209	23 May 17	Miller, Alfred Roderic Crawford	. . . . .	23 May 18	

INFANTRY—*contd.*

Lieutenants—*contd.*

No.	Date of First Commn.	Names.	Date of birth.	Date of rank.	REMARKS.
2210	23 May 17	Atkins, Cyril Gregory		23 May 18	Released from Army Service.
	24 May 17	Bellairs, Robert George (*Ty. Lt.-Col.*, 24 *May*, 17)		24 May 18	
	26 May 17	Turner, Vere Everard		26 May 18	
	28 May 17	Allan, James Malcolm		28 May 18	
	28 May 17	Lyell, Thomas Reginald Guise		28 May 18	
2215	29 May 17	Beasant, Charles Aubrey		29 May 18	
	29 May 17	Welby, Bertie Conroy (*Ty. Capt.*, 29 *May* 17)		29 May 18	
	29 May 17	Watkins, Lawrence Theodore		29 May 18	
	29 May 17	Cooper, Chares Robert Plant		29 May 18	
	30 May 17	Gillett, Bernard George (*Ty. Capt.*, 30 *May* 17)		30 May 18	
2220	30 May 17	Shepherd, Frederick Arthur		30 May 18	
	30 May 17	Arridge, Herbert Chadwick		30 May 18	
	30 May 17	Woods, Nicholas		30 May 18	
	30 May 17	Cameron, Alexander Ansell		30 May 18	
	30 May 17	Game, Stanley Wilfrid		30 May 18	
	30 May 17	Perraton, Charles Whitfield		30 May 18	
	30 May 17	Coles, Thomas Henry		30 May 18	
	31 May 17	Corbett, James Edward (*Ty. Capt.*, 31 *May* 17)		31 May 18	
	1 June 17	Crayden, Claude Charles		1 June 18	
2225	1 June 17	Newquist, Herbert Henry		1 June 18	
	1 June 17	Murley, Percy Arnold		1 June 18	
	1 June 17	Eldridge, Francis Layton		1 June 18	
	1 June 17	Robb, Arthur Angus (**Ty. Capt.*, 5 *Oct.* 17)		1 June 18	
	1 June 17	Jones, Arthur Gerard		1 June 18	
2230	1 June 17	Berkeley, Maurice Fitzhardinge		1 June 18	
	1 June 17	McDonald, Donald William		1 June 18	
	1 June 17	Cooper, William Harold		1 June 18	
	1 June 17	Scott, James Edward (*Ty. Lt.-Col.*, 1 *June* 17)		1 June 18	
	1 June 17	Robinson, Charles (58, 207)		1 June 18	
	1 June 17	Craven, Charles Henry James (*Ty. Lt.-Col.*, 1 *June* 17)		1 June 18	
2235	2 June 17	Gough, Francis Berkeley (207*a*)		2 June 18	
	2 June 17	Eberhardie, Robert Edward		2 June 18	
	4 June 17	Warburton, Robert Paul (**Ty. Capt.*, 25 *Nov.* 17)		4 June 18	
	4 June 17	Kirkman, William Arthur (207*a*)		4 June 18	
	4 June 17	Williams, Arthur Tudor (*Ty. Capt.*, 4 *June* 17)		4 June 18	
2240	6 June 17	Stewart, Henry William		6 June 18	
	6 June 17	Hitchcock, Frederick Charles		6 June 18	
	6 June 17	Wright, Charles Rupert		6 June 18	
	6 June 17	Ling, Frank Girling		6 June 18	
	6 June 17	Page, William Herbert		6 June 18	
2245	8 June 17	Thompson, Eric Charles Ashington		8 June 18	
	10 June 17	McPherson, James (*Ty. Capt.*, 10 *June* 17)		10 June 18	
	12 June 17	Milne, Alexander Simpson (206)		12 June 18	
	12 June 17	Browne, Cecil Charles (207*a*)		12 June 18	
	13 June 17	Hoernle, Edward Selwyn (*Ty. Capt.*, 13 *June* 17)		13 June 18	
2250	13 June 17	Kelly, Walter George (*Ty. Capt.*, 13 *June* 17)		13 June 18	
	14 June 17	Doyle, Francis Joseph Bernard (*Ty. Capt.*, 14 *June* 17)		14 June 18	
	15 June 17	Vaughan, Richard Wyndham		15 June 18	
	17 June 17	Inrig, Leo		17 June 18	
	18 June 17	Pendrigh, Thomas Goodfellow		18 June 18	
2255	20 June 17	Neate, Walter Henry (207*a*)		20 June 18	
	20 June 17	Madge, Kenneth Alfred (207*a*)		20 June 18	
	20 June 17	Cooper, Frederick Alexander		20 June 18	Released from Army Service.
	20 June 17	Elton-Williams-Jervis, Archibald		20 June 18	
	20 June 17	Hodgson, Harry Leonard		20 June 18	
2260	20 June 17	Tippetts-Aylmer, George Alfred		20 June 18	
	20 June 17	Bush, Alfred John (57)		20 June 18	
	20 June 17	Bush, Wenden Ray (48*a*, 50)		20 June 18	
	20 June 17	Nickalls, James William		20 June 18	
2265	20 June 17	Nicolson, Harry Gow (207*a*)		20 June 18	
	20 June 17	White, William John		20 June 18	
2267	20 June 17	Goulstone, William Harrold (207*a*)		20 June 18	

* Without pay and allowances of the rank.

INFANTRY—*contd.*

Lieutenants—*contd.*

No.	Date of First Commn.	Names.	Date of birth.	Date of rank.	Remarks.
2268	20 June 17	Clinton, Arthur Gilbert (207*a*)		20 June 18	
	20 June 17	Hill, Cyril Rowland (207*a*)		20 June 18	
2270	20 June 17	Daly, Douglas Austin		20 June 18	Released from Army Service.
	20 June 17	Williams, Cecil Walter (207*a*)		20 June 18	
	20 June 17	Cutting, Charles Edward		20 June 18	
	20 June 17	Atkins, George Frederick Boissonade		20 June 18	
	20 June 17	Forsyth-Forrest, Robert Septimus		20 June 18	
2275	20 June 17	Bazett, Harold Mancle (207*a*)		20 June 18	
	20 June 17	Ferguson, Herbert John (207*a*)		20 June 18	
	20 June 17	Willard, Frederick William		20 June 18	
	20 June 17	Nuttall, Norman (51)		20 June 18	
	20 June 17	Highley, William Morgan Burlington		20 June 18	
2280	20 June 17	Hislop, Robert Gilbert George		20 June 18	
	20 June 17	Peppin, Sydenham Henry		20 June 18	
	20 June 17	Knowles, Herbert Pryce		20 June 18	
	20 June 17	Hill, Walter Eric (207*a*)		20 June 18	
2285	20 June 17	King, John Rennie		20 June 18	
	20 June 17	Storey, Frank George Foster		20 June 18	
	20 June 17	Cullis, Leonard (48, 51, 56, 207*a*)		20 June 18	
	20 June 17	Batterbury, George Henry (207*a*)		20 June 18	
2290	20 June 17	Tobin, Edward James		20 June 18	
	20 June 17	Gapper, John Edward Bristow		20 June 18	
	20 June 17	Smart, John Henry		20 June 18	
	20 June 17	Carew, Thomas Palk		20 June 18	
	20 June 17	DeBrunner, Walter Eugene (207*a*)		20 June 18	
2295	20 June 17	Lund, Richard		20 June 18	
	20 June 17	Rogers, Harry		20 June 18	
	20 June 17	Batterbury, Frank Herbert (207*a*)		20 June 18	
	20 June 17	Simpson, Clement Edward (46, 51, 207*a*)		20 June 18	
	20 June 17	Hastie, Henry Russell (207*a*)		20 June 18	
2300	20 June 17	Henderson, David Samson		20 June 18	
	20 June 17	Lett, Jonas William Henry		20 June 18	
	20 June 17	Allbrook, Walter Reginald (48*a*, 51)		20 June 18	
	20 June 17	Chalk, James		20 June 18	
	20 June 17	Fraser, Donald Hugh		20 June 18	
2305	20 June 17	Happell, Alexander John		20 June 18	Released from Army Service.
	20 June 17	Lyon, Henry Brockbank		20 June 18	
	20 June 17	Ritson, Fletcher		20 June 18	
	20 June 17	Barker, Robert		20 June 18	
2310	20 June 17	Paterson, Alexander Douglas (51, 207*a*)		20 June 18	
	20 June 17	Johnston, William Court		20 June 18	
	20 June 17	Moore, William George Thomas		20 June 18	
	20 June 17	Powell, Arthur Lenthall Harcourt		20 June 18	
	20 June 17	Perry, Frank William Robert		20 June 18	
2315	20 June 17	Crilly, Alfred Joseph		20 June 18	
	20 June 17	Marlow, Thomas (207*a*)		20 June 18	
	20 June 17	Shelverton, Hubert Melville	2 Nov. 82	20 June 18	
	20 June 17	Howe, George		20 June 18	
	20 June 17	Willcock, Frederick William		20 June 18	
2320	20 June 17	Johnston, James Stanley		20 June 18	
	20 June 17	Leggs, William Edward (46)		20 June 18	
	20 June 17	Mathews, Vere		20 June 18	
	20 June 17	Findlay, Matthew Alexander		20 June 18	
	20 June 17	McCarthy, Reginald William		20 June 18	
2325	20 June 17	Ray, Gerald George		20 June 18	
	20 June 17	Anson, Clifford		20 June 18	
	20 June 17	John, William Rees (*temp. Capt., 15 July 18*)		20 June 18	
	20 June 17	Jones, David John (57)		20 June 18	
	20 June 17	Evans, Ewart Gwyn Haydn		20 June 18	
2330	20 June 17	Martin, Arthur Stanley William (207*a*) (*Ty. Capt., 25 Mar. 18*)		20 June 18	
	20 June 17	Wright, Eric McWilliam		20 June 18	
	20 June 17	Clarke, Geoffrey		20 June 18	
	20 June 17	Dunn, Owen Henry Bourne		20 June 18	
	20 June 17	Hopps, Oswald Deacon		20 June 18	
2335	20 June 17	Forbes, Ernest Reginald Craig		20 June 18	
	20 June 17	Swithenbank, Robert Schofield		20 June 18	
	20 June 17	Pryde, Arthur Walter (48*a*, 51)		20 June 18	
2338	20 June 17	Jennings, John Martin (207*a*)		20 June 18	
2339	20 June 17	Stoddart, Walter Thomas (207*a*) (*Ty. Capt., 6 Mar. 18*)		20 June 18	

INFANTRY—*contd.*

Lieutenants—*contd.*

No.	Date of First Commn.	Names.	Date of birth.	Date of rank.	REMARKS.
2340	20 June 17	LaBouchardiere, Maurice Prince (207*a*)		20 June 18	
	20 June 17	Seal, Llewellyn Stanley		20 June 18	
	20 June 17	Bailey, Dudley Cedric		20 June 18	
	20 June 17	Fisk, Lawrence Edward		20 June 18	
2345	20 June 17	Naismith, William Norman		20 June 18	
	20 June 17	Curd, Alfred Thomas (207*a*)		20 June 18	
	20 June 17	Hughes, Henry		20 June 18	
	20 June 17	Dodds, Benjamin William Pedley		20 June 18	
	20 June 17	Audley, William Dundonald (207*a*)		20 June 18	
2350	20 June 17	Fisher, Harold Alfred		20 June 18	
	20 June 17	MacMahon, James Joseph		20 June 18	
	20 June 17	Dawson, Christopher		20 June 18	
	20 June 17	Fraser, William Colquhoun (207*a*)		20 June 18	
	20 June 17	Jenkins, William Morgan		20 June 18	
2355	20 June 17	Gaul, Alexander Duthie (207*a*)		20 June 18	
	20 June 17	Webb, William Henry		20 June 18	
	20 June 17	Ogilvie, William Napier		20 June 18	
	20 June 17	Duncan, Alistair William (207*a*)		20 June 18	
	20 June 17	Denness, Richard Bernard (207*a*) (*Ty. Capt., 15 Oct. 17*)		20 June 18	
2360	21 June 17	Chetwynd-Palmer, Robert (*temp. Capt., 14 Aug. 18*)		21 June 18	
	21 June 17	Morton, Walter David		21 June 18	
	22 June 17	Dalrymple-Waller, George Herbert Desmond (*Ty. Capt., 14 Sept. 17*)		22 June 18	
	22 June 17	Henke, Frederick Ernest		20 June 18	
	23 June 17	Gething, Leonard Rudolph		23 June 18	
	23 June 17	Good, Oliver Noel Hooper		23 June 18	
2365	23 June 17	Turton, Walter Herbert		23 June 18	
	25 June 17	Donnithorne-Lutter, Charles Edwin (*Ty. Capt., 12 July 17*)		25 June 18	
	25 June 17	Rogerson, James Frederick		25 June 18	
	25 June 17	Morris, Hedley Lloyd		25 June 18	
	25 June 17	Rivett, Walter Charles		25 June 18	
2370	26 June 17	Sportt, Frank Herbert (*Ty. Capt., 10 Dec. 17*)		26 June 18	
	27 June 17	Stevens, Robert Patrick		27 June 18	
	27 June 17	Watson, James Gordon		27 June 18	
	27 June 17	Watson, Joseph		27 June 18	
	27 June 17	Gidden, John Arthur		27 June 18	
	27 June 17	Billingham, William Frederick		27 June 18	
	27 June 17	Fuller, Albert George		27 June 18	
	27 June 17	Goode, Leonard Charles Hardy, M.C.		27 June 18	
	27 June 17	Jeynes, Alban George Hardy		27 June 18	
	27 June 17	Reddick, Gordon Philip		27 June 18	
	27 June 17	Anderson, William Henry		27 June 17	
2375	28 June 17	Coad, Thomas Herbert Edgar (*Ty. Capt., 28 June 17*)		28 June 18	
	28 June 17	Stirling, Stirling Westrop		28 June 18	
	28 June 17	Scotter, Henry Harold Gordon Dill		28 June 18	
	29 June 17	Powell, Wallace (* *temp. Capt., 27 Feb. 18*)		29 June 18	
	29 June 17	Farmer, John Cotton		29 June 18	Released from Army Service.
2380	29 June 17	Quarry, Philippe Stowell (**Ty. Capt., 7 Nov. 17*)		29 June 18	
	29 June 17	Mannooch, Geoffrey Herbert (**Ty. Capt., 7 Nov. 17*)		29 June 18	
	29 June 17	Nicholls, Ernest Maxwell (51)		29 June 18	
	29 June 17	Colwill, Richard Freeman (207*a*)		29 June 18	
	29 June 17	Harry, William Thomas		29 June 18	
2385	30 June 17	Riches, Charles Alfred (*Ty. Capt., 8 July 17*)		30 June 18	
	1 July 17	Birkett, Alfred Nicholson		1 July 18	
	2 July 17	Winters, Arthur Edward		2 July 18	
	2 July 17	Swales, Thomas		2 July 18	
	2 July 17	McMillan, William (*Ty. Capt., 14 July 17*)		2 July 18	
2390	2 July 17	Rix, Charles Harold Burton (*Ty. Capt., 18 July 17*)		2 July 18	
	2 July 17	Nesbitt, Harry Kerr (*Ty. Capt., 14 July 17*)		2 July 18	
	2 July 17	Campbell, George Edward Fredrick		2 July 18	
	2 July 17	Haywood, John Spencer		2 July 18	
	2 July 17	Mudie, Peter Urquhart		2 July 18	
2395	2 July 17	Porritt, Arthur Herbert		2 July 18	
	2 July 17	Savage, Robert		2 July 18	
	4 July 17	Woods, Harold Starmer		4 July 18	
	5 July 17	Robertson, Harold Baron		5 July 18	
2399	6 July 17	Evans, Charles Frederick		6 July 18	

* Without pay and allowances of the rank.

INFANTRY—*contd.*

Lieutenants—*contd.*

No.	Date of First Commn.	Names.	Date of birth.	Date of rank.	REMARKS.
2400	6 July 17	Goby, James Henry		6 July 18	
	7 July 17	Farrington, Hubert Richard		7 July 18	
	7 July 17	Reeks, Frederick		7 July 18	
	7 July 17	Pentreath, George Artis		7 July 18	
	7 July 17	Jaggar, John Joseph		7 July 18	
2405	8 July 17	Surrey, Henry Philip		8 July 18	
	8 July 17	Philbrick, Henry Alfred (207*a*)		8 July 18	
	8 July 17	Hansford, Edwin Victor		8 July 18	
	8 July 17	Quillinan, Lawrence Patric		8 July 18	
	8 July 17	Williams, Ernest Charles		8 July 18	
	9 July 17	Howie, Harry Campbell		9 July 18	
	10 July 17	Scott-Oldfield, John Ignatius		10 July 18	
2410	10 July 17	Griffiths, William Stephen (*Ty. Capt. 14 Sep. 17*).		10 July 18	
	10 July 17	Ottley, Vernon William		10 July 18	
	11 July 17	Smith, Thomas Purrier (**Ty. Capt. 7 Nov. 17*).		11 July 18	
	13 July 17	Ray, Reginald Edwin Anthony		13 July 18	Released from Army Service.
2415	16 July 17	Hubbard, Archibald Edward (**Ty. Capt. 1 Oct. 17*)		16 July 18	
	16 July 17	Crick, William Latta		16 July 18	
	17 July 17	Dominy, Arthur George		17 July 18	
	18 July 17	Northam, Reginald Meek		18 July 18	
	18 July 17	Elliott, Frederick Percy (*Ty. Capt. 15 Jan. 18*)		18 July 18	
2420	18 July 17	Whitburn, William Wilfred (*Ty. Capt. 15 Jan. 18*)		18 July 18	
	18 July 17	Franklin, Frederick James (*Ty. Capt. 5 Mar. 18*)		18 July 18	
	18 July 17	Cornish, Archibald Searles (*Ty Capt 14 Jan 18*)		18 July 18	
	18 July 17	Clarke, Samuel Henry		18 July 18	
	18 July 17	Norris, Charles Warmer		18 July 18	
2425	18 July 17	Kearsey, Edward Samuel		18 July 18	
	18 July 17	Stark, James Dearden		18 July 18	
	18 July 17	Deighton, Arthur Archer		18 July 18	
	18 July 17	Robinson, Charles Thomas		18 July 18	
	18 July 17	Willis, Alfred Charles		18 July 18	
2430	18 July 17	Lee, Hardie		18 July 18	
	18 July 17	Bradlaugh, John Howard		18 July 18	
	18 July 17	Carter, Gilmore Themlett		18 July 18	
	18 July 17	Thompson, Edward		18 July 18	
	18 July 17	Burrow, Sidney George		18 July 18	
2435	18 July 17	Hodgkins, Charles John		18 July 18	
	18 July 17	Watson, Alexander Wilfred		18 July 18	
	18 July 17	McAdam, Jack		18 July 18	
	18 July 17	Robertson, Edwin Adolphus		18 July 18	
	18 July 17	Aitken, Charles Stirling		18 July 18	
2440	18 July 17	Paddon-Row, Donald Alexander Thomas		18 July 18	
	19 July 17	Evanson, Gilbert Alleyne		19 July 18	
	20 July 17	Dufall, John Baker		20 June 18	
	21 July 17	Thomas, Edward Francis (*temp. Capt. 3 June 18*)		21 July 18	
	22 Aug. 17	Bregman, Theodore Montague		22 July 18	
2445	24 July 17	Pirie, George Wilson		24 July 18	
	24 July 17	Trevaskis, Hugh Kennedy*		24 July 18	Released from Army Service.
	24 July 17	Jack, William		24 July 18	
	25 July 17	Dickie, Robert Andrew		25 July 18	
2450	25 July 17	Bowder, Norman Charles		25 July 18	
	25 July 17	Smith, William Townsend (*Ty. Capt. 25 July 17*).		25 July 18	
	25 July 17	Everard, Thomas Arthur		25 July 18	
	25 July 17	Gilbert, William Henry		25 July 18	
	25 July 17	Knight, William Frederic		25 July 18	
2455	25 July 17	Inman, Harry Dent		25 July 18	
	25 July 17	Biss, Evan Ebenezer		25 July 18	
	26 July 17	Vogwill, John Wright		26 July 18	
	26 July 17	Christian, John William		26 July 18	
	26 July 17	Sare-Soar, Morris William Robert		26 July 18	
	26 July 17	Marks, Gerald Andrew		26 July 18	
2460	27 July 17	Norwood, William Stephenson		27 July 18	
	27 July 17	Duff, Baron Beatson Hamilton		27 July 18	
	27 July 17	Barker, Walter Daniel		27 July 18	
	27 July 17	Morgan, Frank		27 July 18	
	27 July 17	Harris, Harry Fredrick Ensor		27 July 18	
2465	27 July 17	Mould, Blair Stirling		27 July 18	
2466	27 July 17	Edwards, Sidny James		27 July 18	

* Without pay and allowances of the rank

No.	Date of First Commn.	Names.	Date of birth	Date of rank.	Remarks.
		INFANTRY—*contd.*			
		Lieutenants—*contd.*			
2467	27 July 17	Patterson, Andrew Alexander Claude		27 July 18	
	28 July 17	Deacon, Hector		28 July 18	
	28 July 17	Tyndall, Charles Thomas Alexander		28 July 18	
2470	29 July 17	O'Doherty, Terence Edward		29 July 18	
	29 July 17	King, James Archibald		29 July 18	
	1 Aug. 17	Connelly, Joseph Patrick		1 Aug. 18	
	1 Aug. 17	Tregaskis, George		1 Aug. 18	
	2 Aug. 17	Petrie, John Cambell Eggar (*Ty. Capt.* 2 *Aug.* 17.)		2 Aug. 18	
2475	4 Aug. 17	Pow, George Scott (207*a*)		4 Aug. 18	
	4 Aug. 17	Leigh, Philip Charles James (46)		4 Aug. 18	
	5 Aug. 17	Betteridge, Roland Clifford		5 Aug. 18	
	6 Aug. 17	Marshall, Edward Charles		6 Aug. 18	
	6 Aug. 17	Shaw, Julian Arthur		6 Aug. 18	
2480	7 Aug. 17	Shillitoe, Godfery Rukman		7 Aug. 18	
	7 Aug. 17	Coulson, William Bennie		7 Aug. 18	
	8 Aug. 17	Faulkner, William George (51, 207*a*)		8 Aug. 18	
	8 Aug. 17	Jones, George Hoel		8 Aug. 18	
	8 Aug. 17	Chambers, Thomas Brooke Winsley		8 Aug. 18	
2485	8 Aug. 17	Staveley, James Plankett		8 Aug. 18	
	8 Aug. 17	Donovan, Bartholmew David (207*a*)		8 Aug. 18	
	8 Aug. 17	Sime, Ernest Alfred		8 Aug. 18	
	8 Aug. 17	Williams, Griffith (207*a*)		8 Aug. 18	
	8 Aug. 17	Dawn, Arthur (207*a*)		8 Aug. 18	
2490	8 Aug. 17	Irving, Martin		8 Aug. 18	
	8 Aug. 17	Lewis, Lionel Raymond		8 Aug. 18	
	8 Aug. 17	Dobbie, Thomas Brown (207*a*)		8 Aug. 18	
	8 Aug. 17	Pettersen, Sverre (48, 207*a*)		8 Aug. 18	
	8 Aug. 17	Rowley, William John Duncan		8 Aug. 18	
2495	8 Aug. 17	McKenzie, Harry Neave		8 Aug. 18	
	8 Aug. 17	Haslam, Roy Havelock (207*a*)		8 Aug. 18	Released from Army Service.
	8 Aug. 17	Scott-Coward Christopher William (207*a*)		8 Aug. 18	
	8 Aug. 17	Collison, Stephen Stanley		8 Aug. 18	
	8 Aug. 17	Thewlis, Noel Eastwood (48*a*, 51, 207*a*)		8 Aug. 18	
2500	8 Aug. 17	Crowther, Harry John		8 Aug. 18	
	8 Aug. 17	Wade, Ernest George (207*a*)		8 Aug. 18	
	8 Aug. 17	Hurley, Frank Shannahan (207*a*)		8 Aug. 18	
	8 Aug. 17	Terrell, Fredrick John Alexander (51, 52, 207*a*)		8 Aug. 18	
	8 Aug. 17	Elliott, Ernest Thomas		8 Aug. 18	
2505	8 Aug. 17	Wickham, William Charles (207*a*)		8 Aug. 18	
	8 Aug. 17	Yorke, Harold Branson		8 Aug. 18	
	8 Aug. 17	Mortimer, Douglas Cruickshank		8 Aug. 18	
	8 Aug. 17	Chapman, Arthur Frederic (207*a*)		8 Aug. 18	
	8 Aug. 17	Robertson, David		8 Aug. 18	
2510	8 Aug. 17	Clarke, Arthur William (207*a*)		8 Aug. 18	
	8 Aug. 17	Gregory, Herbert George		8 Aug. 18	
	8 Aug. 17	Cox, Gordon Wilkinson		8 Aug. 18	
	8 Aug. 17	Hilliard, William Arthur (46, 207*a*)		8 Aug. 18	
	8 Aug. 17	Foster, Robert		8 Aug. 18	
2515	8 Aug. 17	Pugh, Thomas Edwin (207*a*)		8 Aug. 18	
	8 Aug. 17	George, Hugh Shaw (207*a*)		8 Aug. 18	
	8 Aug. 17	Chatfield, Hugh Stanton (207*a*)		8 Aug. 18	
	8 Aug. 17	Greenacre, George Alexander		8 Aug. 18	
	8 Aug. 17	Routh, Claude Ismal (207*a*)		8 Aug. 18	
2520	8 Aug. 17	Beauchamp, Willoughby Greaves (207*a*)		8 Aug. 18	
	8 Aug. 17	Cowell, Francis John (207*a*)		8 Aug. 18	
	8 Aug. 17	Brand, Arthur Reginald		8 Aug. 18	
	8 Aug. 17	Westland, Douglas		8 Aug. 18	
	8 Aug. 17	Leete, Benjamin Tainer		8 Aug. 18	
2525	8 Aug. 17	Woodcock, William Dennis Grant		8 Aug. 18	
	8 Aug. 17	Griffiths, Thomas Alfred (207*a*)		8 Aug. 18	
	8 Aug. 17	Mackay, John (207*a*)		8 Aug. 18	
	8 Aug. 17	Ross, Cyril Egbert William (207*a*)		8 Aug. 18	
	8 Aug. 17	Browne, George De La Rue (207*a*)		8 Aug. 18	
2530	8 Aug. 17	Pitt, Horace Frederick George (207*a*)		8 Aug. 18	
	8 Aug. 17	Martin, Stephen Alexander		8 Aug. 18	
	8 Aug. 17	Rayner, George Eric		8 Aug. 18	
	8 Aug. 17	Mann, Alexander (48*a*, 51, 207*a*)		8 Aug. 18	
	8 Aug. 17	Stampe, William Leonard (207*a*)		8 Aug. 18	
2535	8 Aug. 17	Kelly, Philip Desmond Lorne		8 Aug. 18	
	8 Aug. 17	Grierson, Gilbert Vedast		8 Aug. 18	
2537	8 Aug. 17	Littlejohn, Alexander		8 Aug. 18	

*Without pay and allowances of the rank.

INFANTRY—*contd.*

Lieutenants—*contd.*

No.	Date of First Commn.	Names.	Date of birth.	Date of rank.	REMARKS.
80	8 Aug. 17	Park, Ernest Reginald (207*a*)		8 Aug. 18	
	8 Aug. 17	Brown, Leslie Clifford (207*a*)		8 Aug. 18	
	8 Aug. 17	Rogers, Arthur Lionel (51)		8 Aug. 18	
	8 Aug. 17	Mayes, Fredrick Arthur		8 Aug. 18	
	8 Aug. 17	McDougall, John (207*a*)		8 Aug. 18	
85	8 Aug. 17	Winton, Charles		8 Aug. 18	
	8 Aug. 17	Sherard, Arthur Frederick Lancelot (207*a*)		8 Aug. 18	
	8 Aug. 17	Pattinson, Ronald Ismay		8 Aug. 18	
	8 Aug. 17	Bateman, George Oliver (207*a*)		8 Aug. 18	
90	8 Aug. 17	Jelley, Godfrey Charles Alfred (207*a*)		8 Aug. 18	
	8 Aug. 17	Hambling, Victor Frederick (207*a*)		8 Aug. 18	
	8 Aug. 17	Club, Alexander		8 Aug. 18	
	8 Aug. 17	Marriott, Leslie John		8 Aug. 18	
95	10 Aug. 17	Crump, Robert Harold		10 Aug. 18	Released from Army service.
	10 Aug. 17	Clarke, Walter Harold (46)		10 Aug. 18	
	10 Aug. 17	Emmott, Richard St. Barbe (207*a*)		10 Aug. 18	
	11 Aug. 17	Youett, Authur Noel		11 Aug. 18	
	11 Aug. 17	Evans, Edger William		11 Aug. 18	
	12 Aug. 17	Nunn, James Nester		12 Aug. 18	
	12 Aug. 17	Butcher, Charlie Robert		12 Aug. 18	
100	12 Aug. 17	Hurley, James Martin		12 Aug. 18	
	13 Aug 17	Wilkins, John Patrick Kenelem		13 Aug. 18	
	14 Aug. 17	Stobbart, John		14 Aug. 18	
	15 Aug. 17	Thomas, George Ernest Venning		15 Aug. 18	
	15 Aug. 17	Gerrard, Errol Forester Hilgrove		15 Aug. 18	
105	15 Aug. 17	Dickins, Ernest Karl		15 Aug. 18	
	15 Aug. 17	Beck, John Percy		15 Aug. 18	
	15 Aug 17	Clarke, Arthur Bennet,		15 Aug. 18	
	15 Aug. 17	Landeman, Hugh Gabriel		15 Aug. 18	
	15 Aug. 17	Melvin, John Wilson		15 Aug. 18	
110	18 Aug. 17	Daniel, Annesley Roy		18 Aug. 18	
	27 Apl. 17	Weston, Maxwell		18 Aug. 18	
	19 Aug. 17	Brandt, Henry Gustave		19 Aug. 18	
	21 Aug. 17	Lewis, Clement Lester		21 Aug. 18	
	21 Aug. 17	Parsons, Frank Stewart		21 Aug. 18	
	22 Aug. 17	Whitlock, Percy Oddie		22 Aug. 18	
	22 Aug. 17	Ward, Kenneth Edward		22 Aug. 18	
115	23 Aug. 17	Blacker, Harold Alfred Cecil		23 Aug. 18	Released from Army service.
	23 Aug. 17	Davidson, Henry Hutcheon		23 Aug. 18	
	24 Aug. 17	Ring, Harry James (*Ty. Lt. 1 Sep. 17*)		24 Aug. 18	
	25 Aug. 17	Moseley, Frank Samuel Hughes (*Ty. Lt., 25 Aug. 17*)		25 Aug. 18	
	25 Aug. 17	Gravett, George William		25 Aug. 18	
120	26 Aug. 17	White, Mark John		26 Aug. 18	
	26 Aug. 17	Taylor, Roland Charles		26 Aug. 18	
	26 Aug. 17	Rose, Clifford		26 Aug. 18	
124	29 Aug. 17	Davies, Stanley Clifford		29 Aug. 18	
	29 Aug. 17	Perks, Harry Burton (*Ty. Lt. 30 Aug. 17*)		29 Aug. 18	
125	29 Aug. 17	Hawkins, Horace William		29 Aug. 18	
	29 Aug. 17	Jeffery, Hubert Edmund		29 Aug. 18	
	29 Aug. 17	Riddell, Robert		29 Aug. 18	
	29 Aug. 17	Cater, Leonard James		29 Aug. 18	
	29 Aug. 17	Mills, Francis Charles Raymond		29 Aug. 18	
130	29 Aug. 17	Daniel, Francis Leslie		29 Aug. 18	
	29 Aug. 17	Wallace, Dixon		29 Aug. 18	
	29 Aug. 17	Faddy, Francis Leopold		29 Aug. 18	
	29 Aug. 17	Brewster, John Alexander		29 Aug. 18	
	29 Aug. 17	Clubbe, Cedric Henchman (*temp. Capt. 15 Sep. 18.*)		29 Aug. 18	
135	29 Aug. 17	Howes, Harry Charles		29 Aug. 18	
	29 Aug. 17	Binns, Bernard Ottwell		29 Aug. 18	
	29 Aug. 17	Pearson, Thomas Francis		29 Aug. 18	
	29 Aug. 17	Flower, Laurence Trevillion		29 Aug. 18	
	29 Aug. 17	Howard, Claud		29 Aug. 18	
140	29 Aug. 17	Gardiner, Douglas Ivan		29 Aug. 18	
	29 Aug. 17	Galliene, John William Horrabine		29 Aug. 18	

INFANTRY—*contd.*

Lieutenants—*contd.*

No.	Date of First Commn.	Names.	Date of birth.	Date of rank.	REMARKS.
	29 Aug. 17	Harris, Albert Thomas Robert		29 Aug. 18	
	29 Aug. 17	Long, Howard Arthur		29 Aug. 18	
	29 Aug. 17	Blyth, James Cameron		29 Aug. 18	
	29 Aug. 17	Darling, John Singleton		29 Aug. 18	
	29 Aug. 17	McCombe, Andrew		29 Aug. 18	
	29 Aug. 17	McQueen, James Alexander		29 Aug. 18	
	29 Aug. 17	Winn, Frank Blancher		29 Aug. 18	
	29 Aug. 17	Ellery, Albert Charles		29 Aug. 18	
	29 Aug. 17	Druett, Horace Charles		29 Aug. 18	
	29 Aug. 17	Harris, Donald Bruce Bowe		29 Aug. 18	
	29 Aug. 17	Bullocke, John Greville		29 Aug. 18	
	29 Aug. 17	Austen, Edward Joseph		29 Aug. 18	
	29 Aug. 17	Williams, Arthur		29 Aug. 18	
	29 Aug. 17	Corbould-Warren, William		29 Aug. 18	
	29 Aug. 17	Bell, Harold Crespo		29 Aug. 18	
	29 Aug. 17	Swan, Sealy Rollo		29 Aug. 18	
	30 Aug. 17	Liddell, Charles Oliver (*temp. Capt. 10 Oct. 18*)		30 Aug. 18	
	31 Aug. 17	Fitch, Walter Ogle		31 Aug. 18	
145	1 Sep. 17	Fitzpatrick, James Thomas		1 Sep. 18	
	1 Sep. 17	Farnon, Laurence Redmond St. John		1 Sep. 18	
	1 Sep. 17	Jackson, John		1 Sep. 18	
	3 Sep. 17	Smith, Russell Owen		3 Sep. 18	
	3 Sep. 17	Bester, George Cyril		3 Sep. 18	
150	4 Sep. 17	Clarke, Miles Fincastle Campbell		4 Sep. 18	
	4 Sep. 17	Barrington, Arthur		4 Sep. 18	
	5 Sep. 17	Park, William James		5 Sep. 18	
	5 Sep. 17	Wiggins, Douglas George		5 Sep. 18	
	5 Sep. 17	Ballantyne, John Russell		5 Sep. 18	
155	5 Sep. 17	Grant, George Nelson		5 Sep. 18	
	5 Sep. 17	Smith, Harry McMorran (*Ty. Capt. 1 May 18*)		5 Sep. 18	
	5 Sep. 17	Rose, Richard Enderwick		5 Sep. 18	
	5 Sep. 17	Turner, Charles Algernon Guy (*Ty. Capt., 29 Apl. 18*)		5 Sep. 18	
	5 Sep. 17	Manook, John Sarkies		5 Sep. 18	
160	6 Sep. 17	Gordon, Norman Horatio (172*a*)		6 Sep. 18	
	6 Sep. 17	Sargent, Hugh Hamilton		6 Sep. 18	
	6 Sep. 17	Corkery, Albert Thamas Moore		6 Sep. 18	
	6 Sep. 17	Johnson, Henry Edward		6 Sep. 18	
	7 Sep. 17	Jackson, James Wotherspoon Clarke		7 Sep. 18	Released from Army Service.
165	8 Sep. 17	Hawkes, William Blackburns (*Ty. Capt. 18 Jan. 18*)		8 Sep. 18	
	9 Sep. 17	Duke, William Victor		9 Sep. 18	
166	10 Sep. 17	Gordon, Sidney Victor Leopold deMontmorency (*Ty. Capt.,10 Sep. 17*)		10 Sep. 18	
167	11 Sep. 17	Bedingfeld, Richard Henry Norman		11 Sep. 18	
	11 Sep. 17	Shove, Harry Leonard		11 Sep. 18	
	11 Sep. 17	Dickerson, Albert Howard		11 Sep. 18	
	11 Sep. 17	Welchmen, Gerald Harington, (*Ty. Capt. 11 Sep. 17*)		11 Sep. 18	
170	12 Sep. 17	Brotherston, George Thomas(**Ty. Capt.8 Mar. 18.*)		12 Sep. 18	
	14 Sep. 17	Guider, Michael Edward		14 Sep. 18	
	14 Sep. 17	Nicoll, William Fisher		14 Sep. 18	
	15 Sep. 17	Wilkinson, James Vere Stewart		15 Sep. 18	
175	16 Sep. 17	Moss, Edgar Mark		16 Sep. 18	
	16 Sep. 17	Mitchell, Herbert Richard Wellsman		16 Sep. 18	
	20 Sep. 17	Orr, William (**Ty. Capt., 4 Jan. 18*)		20 Sep. 18	
	20 Sep. 17	Stephenson, Thomas		20 Sep. 18	
	21 Sep. 17	Fraser, William Laidlaw		21 Sep. 18	
180	21 Sep. 17	Harvey-Kelly, Gilbert Isdell (*temp. Capt. 10 July 18*)		21 Sep. 18	
	25 Sep. 17	Mullin, Reginald Henry		25 Sep. 18	
	26 Sep. 17	Covernton, Stephen Harris		26 Sep. 18	Released from Army Service.
	26 Sep. 17	Garrett, Herbert Leonard Offley (* *Ty. Capt., 1 Nov. 17*)		26 Sep. 18	
	26 Sep. 17	Rowse, Thomas Arthur		26 Sep. 18	
185	26 Sep. 17	Mitchell, Percival Wilkinson		26 Sep. 18	
	26 Sep. 17	Allen, George Mortimer Hugh		25 Sep. 18	

* Without pay and allowances of the rank.

INFANTRY—*contd.*

Lieutenants—*contd.*

No.	Date of First Commn.	Names.	Date of birth.	Date of rank.	REMARKS.
	26 Sep. 17	Odling, Richard Ernest		26 Sep. 18	
	26 Sep. 17	Craig, William		26 Sep. 18	
	26 Sep. 17	Macbeth, Charles James Savory		26 Sep. 18	
190	26 Sep. 17	Keith, Neville Yorke		26 Sep. 18	
	26 Sep. 17	Barker, Gordon Mears		26 Sep. 18	
	26 Sep. 17	Hobson, Thomas Ernest		26 Sep. 18	
	26 Sep. 17	Cattanach, Alistair		26 Sep. 18	
	26 Sep. 17	Forrester, Harry Evelyn		26 Sep. 18	
195	26 Sep. 17	Anderson, Francis Sheed		26 Sep. 18	
	26 Sep. 17	Ridlington, Harry Thomas		26 Sep. 18	
	26 Sep. 17	Breton, Basil Ogilvie		26 Sep. 18	
	26 Sep. 17	Francis, William Donald		26 Sep. 18	
	26 Sep. 17	Crockett, Norton Stephen		26 Sep. 18	
200	26 Sep. 17	Addinall, Carl Rupert		26 Sep. 18	
	26 Sep. 17	Waumsley, Thomas Harold		26 Sep. 18	
	26 Sep. 17	Hingston, Francis Perceval Randolph		26 Sep. 18	
	26 Sep. 17	Duckworth, Melancthon Bertwistle		26 Sep. 18	
	28 Sep. 17	Kearns, William Irving, (*Ty. Maj., 28 Sep. 17*)		28 Sep. 18	
205	29 Sep. 17	Lanktree, Henry Richard		29 Sep. 18	
	1 Oct. 17	Wenmoth, Harold Hocking (46)		1 Oct. 18	
	1 Oct. 17	Gadsden, Cecil Holroyd (207*a*)		1 Oct. 18	
	1 Oct. 17	Williams, Horace George Hyde		1 Oct. 18	
	1 Oct. 17	Buttress, Ewart Skelliet (207*a*)		1 Oct. 18	
210	1 Oct. 17	Hicks, Hubert Ernest		1 Oct. 18	
	1 Oct. 17	Goss, William Adams		1 Oct. 18	
	1 Oct. 17	Bundy, Frederick Claude, M.M.		1 Oct. 18	
	1 Oct. 17	Thomas, Percy Edmund		1 Oct. 18	
	1 Oct. 17	Holdcroft, George Ernest		1 Oct. 18	
215	1 Oct. 17	Phillips, Horace William		1 Oct. 18	
	1 Oct. 17	Stewert, Reginald Ronald (207*a*)		1 Oct. 18	
	1 Oct. 17	Evans, Edward Macey		1 Oct. 18	
	2 Oct. 17	Malcolm, Frederick Arthur		2 Oct. 17	
	3 Oct. 17	Kilgour, Robert		3 Oct. 18	
220	3 Oct. 17	Cutler, George James		3 Oct. 18	
	4 Oct. 17	Burgess, Wilford Moy		4 Oct. 18	
	4 Oct. 17	Albrecht, Carl Leo		4 Oct. 18	
	4 Oct. 17	Durston, Robert Edgar		4 Oct. 18	
	4 Oct 17	Thomas, Max Llewelyn		4 Oct. 18	
225	5 Oct. 17	Marriott, Bertie Millar (207*a*)		5 Oct. 18	Released from Army Service.
	5 Oct. 17	Collins, Terence Mostyn		5 Oct. 18	
	5 Oct 17	Greenwood, Alfred Bertram		5 Oct. 18	
	5 Oct. 17	MacArthur, Robert Greay		5 Oct. 18	
	5 Oct. 17	Morris, Morgan Griffiths		5 Oct. 18	
	5 Oct. 17	Cooke, Eric Robert Machin		5 Oct. 18	
	5 Oct. 17	Smith, Sydney Albert (57)		5 Oct. 18	
	5 Oct. 17	Webster, William Henry Albert		5 Oct. 18	
231	5 Oct. 17	L'Estrange, Terence William		5 Oct. 18	
	5 Oct. 17	Harding, Archibald Walter		5 Oct. 18	
	5 Oct. 17	Hayward, Russell Dallas		5 Oct. 18	
	5 Oct. 17	Nottage, John Bowtell		5 Oct. 18	
	5 Oct. 17	Gillon, John Bruce		5 Oct. 18	
	5 Oct. 17	Horne, George Henry		5 Oct. 18	
	5 Oct. 17	Sanders, Hubert Wilfred		5 Oct. 18	
	5 Oct. 17	Laurie, George		5 Oct. 18	
	5 Oct. 17	Whitcombe, Charles Edward		5 Oct. 18	
	5 Oct. 17	Young, Percival Reay		5 Oct. 18	
	5 Oct. 17	Kilner, Frederick William		5 Oct. 18	
	5 Oct. 17	Pearcy, Frederick Gladstone		5 Oct. 18	
	5 Oct. 17	Johnson, James Russell (207*a*)		5 Oct. 18	
	5 Oct. 17	Chisholm, Kenneth McKenzie		5 Oct. 18	
232	5 Oct. 17	Purcell-Gildea, Edward Thomas Harcourt		5 Oct. 18	
	5 Oct. 17	Griffin, Gerald (46)		5 Oct. 18	
	5 Oct. 17	O'Neil-Shaw, John Charles Liddell		5 Oct. 18	
235	5 Oct. 17	Bairsto, Frederick James		5 Oct. 18	
	5 Oct. 17	Leitch, Ronald Andrew (51)		5 Oct. 18	
	5 Oct. 17	Gamble, Melvilla Frederick		5 Oct. 18	
	5 Oct. 17	Miller, Charles Burns (207*a*)		5 Oct. 18	
	5 Oct. 17	Pattinson, Ernest		5 Oct. 18	
240	5 Oct. 17	Lindsay, John Gordon		5 Oct. 18	
	5 Oct. 17	Caldar, Oliver Vassall		5 Oct. 18	
	5 Oct. 17	Benson, Alexander (46)		5 Oct. 18	
	5 Oct. 17	Martin, Horace		5 Oct. 18	

INFANTRY—*contd.*

Lieutenants—*contd.*

No.	Date of First Commn.	Names.	Date of birth.	Date of rank.	REMARKS.
	5 Oct. 17	Didcock, Richard Stephen Henry		5 Oct. 18	
245	5 Oct. 17	Partridge, Ernest		5 Oct. 18	
	5 Oct. 17	Birtwistle, Robert Edward (48*a*, 50, 206)		5 Oct. 18	
	5 Oct. 17	Ross, Donald Alexander		5 Oct. 18	
	5 Oct. 17	Fitz-Patrick, Horace James		5 Oct. 18	
	5 Oct. 17	Weeks, Victor Herbert		5 Oct. 18	
250	5 Oct. 17	Williams, William Howard		5 Oct. 18	
	5 Oct. 17	Bell, Harry Hubert		5 Oct. 18	
	5 Oct. 17	Parsons, William George Lanyon		5 Oct. 18	
	5 Oct. 17	Lacey, Gerald		5 Oct. 18	
	5 Oct. 17	Ainsworth, Richard		5 Oct. 18	Released from Army Service.
255	5 Oct. 17	Turner, Gerald Knight		5 Oct. 18	
	5 Oct. 17	Mitchell, Robert Aitken		5 Oct. 18	
	5 Oct. 17	Cowie, Adam (*207a*)		5 Oct. 18	
	5 Oct. 17	Shore, Noel Beilby Porteus		5 Oct. 18	
	5 Oct. 17	Inglis, Hugh Arbuthnot		5 Oct. 18	
260	5 Oct. 17	Armstrong, Philip Andrew		5 Oct. 18	
	5 Oct. 17	Lemon, Robert Ernest		5 Oct. 18	
	5 Oct. 17	Macdonald, Malcolm		5 Oct. 18	
	5 Oct. 17	Bain, Graham Ward		5 Oct. 18	
265	5 Oct. 17	Thurston, Edward Thomas		5 Oct. 18	
	5 Oct. 17	Anderson, James Drummond		5 Oct. 18	Released from Army Service.
	5 Oct. 17	Maconachie, Richard Roy		5 Oct. 18	
	5 Oct. 17	Parker, Richard Neville (*Ty. Maj., 20 Dec. 17*)		5 Oct. 18	
	5 Oct. 17	Grille, Frederick Louis		5 Oct. 18	
270	5 Oct. 17	Sheehy, John Francis		5 Oct. 18	
	5 Oct. 17	Hartnoll, Eric Stewart		5 Oct. 18	
	5 Oct. 17	Taylor, Joseph Andrew		5 Oct. 18	
	5 Oct. 17	Richardson, Reginald Charles		5 Oct. 18	Released from Army Service.
	5 Oct. 17	McGill, Donald Scott (*207a*) (*Ty. Capt., 12 Mar. 18*)		5 Oct. 18	
275	5 Oct. 17	Farquhar, Dean		5 Oct. 18	
	5 Oct. 17	Zaiman, Barnet Abraham		5 Oct. 18	
	5 Oct. 17	Hill, Edwin William		5 Oct. 18	
	5 Oct. 17	Benton, Gordon William		5 Oct. 18	
	5 Oct. 17	Leach, Arthur Gordon		5 Oct. 18	
280	5 Oct. 17	Hitchcock, Richard Howard		5 Oct. 18	
	5 Oct. 17	Fox, Charles Baylis		5 Oct. 18	
	6 Oct. 17	Howard, George Blake		6 Oct. 18	
	6 Oct. 17	Inglet, Bertram Armstrong		6 Oct. 18	
	6 Oct. 17	Stokes, Samuel Evans (**Ty. Capt., 1 Nov. 17*)		6 Oct. 18	
285	6 Oct. 17	Edwards, Frederick Dillon		6 Oct. 18	
	7 Oct. 17	Green, Edwin Wilfred Killocksley		7 Oct. 18	
	7 Oct. 17	Page, Herbert James		7 Oct. 18	
	7 Oct. 17	Barnard, Leslie George		7 Oct. 18	
	7 Oct. 17	Culley, Ronald Harry		7 Oct. 18	
290	8 Oct. 17	Dunham, Herbert Henry		8 Oct. 17	
	8 Oct. 17	Brian, William George		8 Oct. 18	
	9 Oct. 17	Matthews, Harry		9 Oct. 18	
294	9 Oct. 17	Simmonds, Harold		9 Oct. 18	
	9 Oct. 17	Weir, Archibald Dargue		9 Oct. 18	
	9 Oct. 17	Oliver, Edgar Simpson		9 Oct. 18	
295	10 Oct. 17	McKenzie, Alexander Thomas		10 Oct. 18	
	11 Oct. 17	Rochfort, Hume Guilford		11 Oct. 18	
	11 Oct. 17	Hall, Charles Frank (**Ty. Capt., 3 Dec. 17*)		11 Oct. 18	
	11 Oct. 17	Blick, Leonard Hammond		11 Oct. 18	
	11 Oct. 17	Jones, Cyril Edger		11 Oct. 18	Released from Army Service.
	27 June 17	Kingett, Arthur Edward		11 Oct. 18	
300	13 Oct. 17	Grant-Peterkin, Malcolm Henry		13 Oct. 18	
	15 Oct. 17	Flyter, Malcolm John		15 Oct. 18	
	16 Oct. 17	Watson, Dennis George		16 Oct. 18	Released from Army Service.
	18 Oct. 17	Rule, William Norman Prichard		18 Oct. 18	
	18 Oct. 17	Matthews, Vivian		18 Oct. 18	
305	29 Aug. 17	Parmenter, Aubrey Simon Piper		18 Oct. 18	
	19 Oct. 17	D'Cruz, Harold Joseph		19 Oct. 18	
	19 Oct. 17	Oliphant, Ronald Wynne		19 Oct. 18	
	22 Oct. 17	Littlewood, Ronald Vernon		22 Oct. 18	

* Without pay and allowances of the rank.
† With pay and allowances as for a Lieut.

INFANTRY—*contd.*

Lieutenants—*contd.*

No.	Date of First Commn.	Names.	Date of birth.	Date of rank.	REMARKS.
309	23 Oct. 17	Crosthwaite, Henry Robert, C.I.E. (**Ty. Lt.-Col. 23 Oct. 17*)		23 Oct. 18	
310	23 Oct. 17	Armstrong, Arthur Campbell		23 Oct. 18	
	23 Oct. 17	Evans, Geoffery (*Ty. Capt., 23 Oct. 17*)		23 Oct. 18	
	23 Oct. 17	Watson, Robert Montador		23 Oct. 18	
	24 Oct. 17	Norris, Charles Edward (*Ty. Capt., 6 Feb. 18*)		24 Oct. 18	
	24 Oct. 17	Madden, William John		24 Oct. 18	
315	25 Oct. 17	Williams, David		25 Oct. 18	
	25 Oct. 17	Gilbert, Edwin Osborne		25 Oct 18	
	25 Oct. 17	Tancred, George Marshall Wilson		25 Oct. 18	
	25 Oct. 17	Gerard, Ernest Lewis		25 Oct. 18	
	25 Oct. 17	Clabby, Dermot Nicolas Joseph (**Ty. Capt., 6 Feb. 18*)		25 Oct 18	
320	26 Oct. 17	O'Neill, Frank Owen (**Ty. Capt., 26 Oct. 17*)		26 Oct. 18	
	28 Oct. 17	Hulbert, Leonard		28 Oct. 18	
	28 Oct. 17	Street, Walter		28 Oct. 18	
	28 Oct. 17	Dudley, Frank		28 Oct. 18	
	29 Oct. 17	Head, William Raleigh		29 Oct. 18	
	29 Oct. 17	Rundle, Henry		29 Oct. 18	
325	30 Oct. 17	Rodgers, Edmund William Frederick		30 Oct. 18	
	31 Oct. 17	Duff, Archibald Craigmile		31 Oct. 18	
	31 Oct. 17	West, George Frederic		31 Oct. 18	
	31 Oct. 17	deGale, Hugh Otway		31 Oct. 18	
330	31 Oct. 17	Abbey, John Barwick		31 Oct. 18	
	31 Oct. 17	Bapty, Patrick Aubrey		31 Oct. 18	
	31 Oct. 17	Cook, Edward Heron		31 Oct. 18	
	31 Oct. 17	Culley, Arthur		31 Oct. 18	
	31 Oct. 17	Deakin, Charles Cecil		31 Oct. 18	
335	31 Oct. 17	Fitzgerald, John Jellett		31 Oct. 18	
	31 Oct. 17	Gill, Robert Bruce		31 Oct. 18	
	31 Oct. 17	Machin, Farnham Wilbraham		31 Oct. 18	
	31 Oct. 17	Murphy, John		31 Oct. 18	
	31 Oct. 17	Nash, George Frederick		31 Oct. 18	
340	31 Oct. 17	Parry, Roy Edgardo		31 Oct. 18	
	31 Oct. 17	Pocock, Leslie James		31 Oct. 18	
	31 Oct. 17	Sharpe, George Frederick		31 Oct. 18	
	31 Oct. 17	Shaw, Reginald		31 Oct. 18	
	31 Oct. 17	Smart, Richard Lynds		31 Oct. 18	
345	31 Oct. 17	Smith, Charles Arthur Bernard		31 Oct. 18	
	31 Oct. 17	Stringer, Brian de Haviland		31 Oct. 18	
	31 Oct. 17	Townley, Claude Frank		31 Oct. 18	
	31 Oct. 17	Wright, Reginald Arthur		31 Oct. 18	
	31 Oct. 17	Wroe, Herbert		31 Oct. 18	
350	31 Oct. 17	Harrison, William Edgar Pringle		31 Oct. 18	
	31 Oct. 17	Campbell, Duncan		31 Oct. 18	
	31 Oct. 17	Weeks, Sidney Charles		31 Oct. 18	
	31 Oct. 17	McNaught, Douglas		31 Oct. 18	
	31 Oct. 17	Billson, John Vincent		31 Oct. 18	
355	31 Oct. 17	Stewart, Murdo McDonald		31 Oct. 18	
	31 Oct. 17	Stutter, John Eric		31 Oct. 18	
	1 Nov. 17	Cox, William James		1 Nov. 18	
	1 Nov. 17	Brunton, Theodore Stuart		1 Nov. 18	
	1 Nov. 17	Pine, Frank Cecil		1 Nov. 18	
360	1 Nov. 17	Frost, Howard Richard		1 Nov. 18	
	1 Nov. 17	Bailey, Albert Henry		1 Nov. 18	
	1 Nov. 17	Gould, William George		1 Nov. 18	
	1 Nov. 17	Frost, Fred Hyde		1 Nov. 18	
	2 Nov. 17	Hildyard-Marris, Reginald Willows (*Ty. Maj., 2 Nov. 17*)		2 Nov. 18	
365	2 Nov. 17	Lynn, Guy (*Ty. Capt., 2 Nov. 17*)		2 Nov. 18	
	3 Nov. 17	Shepton, Henry McIntire		3 Nov. 18	
	5 Nov. 17	Mannes-Abbott, William Somerset Erskine		5 Nov. 18	
	5 Nov. 17	Rowe, William Thomas Matthew		5 Nov. 18	
	5 Nov. 17	Hunter, George Gunning Ratcliffe (*207a*)		5 Nov. 18	
370	5 Nov. 17	Jennings, Sydney Patrick		5 Nov. 18	
	5 Nov. 17	Jones, Alfred Herbert		5 Nov. 18	
	5 Nov. 17	Brewin, Neil Alfred Patrick		5 Nov. 18	
	5 Nov. 17	Tate, Joseph Wallingford		5 Nov. 18	
	5 Nov. 17	Handscomb, Leslie George		5 Nov. 18	
375	5 Nov 17	Edwards, Edwin		5 Nov. 18	
	5 Nov. 17	Smith, Harold Plassey		5 Nov. 18	
	5 Nov. 17	Keelan, Arthur Pryor		5 Nov. 18	
	5 Nov. 17	Finnigan, Timothy Francis (*207a*)		5 Nov. 18	
380	5 Nov. 17	Dale, Charles Hamilton		5 Nov. 18	
	5 Nov. 17	Kilburn, John Maurice		5 Nov. 18	

* Without pay and allowances of the rank.

INFANTRY—*contd.*

Lieutenants—*contd.*

No.	Date of First Comm n.	Names.	Date of birth	Date of rank.	REMARKS.
3 1	5 Nov. 17	Thompson, Leslie Malcolm		5 Nov. 18	
	5 Nov. 17	Gordon, Gilbert		5 Nov. 18	
	5 Nov. 17	Shenton. Edgar Walter		5 Nov. 18	
	5 Nov. 17	Battle, William Aubrey Aylmer		5 Nov. 18	
385	5 Nov. 17	Egan-Wyer, Owen Joseph		5 Nov. 11	
	5 Nov. 17	Virgo, Frederick Raoul		5 Nov. 18	
	5 Nov. 17	Wisher, John Douglas		5 Nov. 18	
	5 Nov. 17	Hannay, Reginald James		5 Nov. 18	
	5 Nov. 17	Dirks, Alexander James (207a)		5 Nov. 18	
390	5 Nov. 17	Gordon, Hector Vivian		5 Nov. 18	
	5 Nov. 17	Nash, Stanley Gilbert Charles		5 Nov. 17	
	5 Nov. 17	Rogers, John Samuel (*207a*)		5 Nov. 17	
	5 Nov. 17	Godley, Edward Leslie		5 Nov. 17	
	5 Nov. 17	Balfour, Andrew		5 Nov. 17	
395	5 Nov. 17	Plumpton, Reginal Roy		5 Nov. 17	
	5 Nov. 17	Wilks Bertie St. Clair (207a)		5 Nov. 17	
	5 Nov. 17	Savage Walter Charles		5 Nov. 11	
	5 Nov. 17	Harper-Collins, Richard Sharman (207a)		5 Nov. 17	
	5 Nov. 17	Kelly, Richard William Lewis		5 Nov. 17	
400	5 Nov. 17	Smith, Leonard Arthur Hinton (57)		5 Nov. 17	
	5 Nov. 17	Pannett, William Joseph		5 Nov. 17	
	5 Nov. 17	Wilson, Arthur Keith		5 Nov. 17	
	5 Nov. 17	Moffatt, Sydney Hugh		5 Nov. 17	
	5 Nov. 17	Pippet, George Reginald		5 Nov. 17	
405	5 Nov. 17	Cook, Ernest John Alfred		5 Nov. 17	
	5 Nov. 17	Brodie, John Wallace		5 Nov. 17	
	5 Nov. 17	Dickson, Walter Colin		5 Nov. 17	
	5 Nov. 17	Spargo, Frederick Wilson		5 Nov. 17	
	5 Nov. 17	Dallmeyer, Harry Somerville Thomes (48a)		5 Nov. 17	
410	5 Nov. 17	Inman, William Foster		5 Nov. 17	
	5 Nov. 17	Watson, Edward Chayter (207a)		5 Nov. 17	
	5 Nov. 17	Girling, George Frederick Horace		5 Nov. 17	
	5 Nov. 17	Yorke, Roland Hudson		5 Nov. 17	
	5 Nov. 17	Anderson, Archibald Robert		5 Nov. 17	
415	5 Nov. 17	Lowry, Alexander		5 Nov. 17	
	5 Nov. 17	Roach, Leonard James		5 Nov. 17	
	5 Nov 17	Rose, Francis Upton		5 Nov. 17	
	5 Nov. 17	Robertson John McIntyre (207a)		5 Nov. 17	
	5 Nov. 17	Sawyer, Percy Seymour		5 Nov. 17	
420	5 Nov. 17	Searle, Gilbert Alfred Hamilton (207a) († *temp. Capt.*, 26 *Sep.* 18).		5 Nov. 17	Released from Army Service.
	5 Nov. 17	Gasson, Henry John		5 Nov. 17	
	5 Nov 17	Casling Whyte		5 Nov. 17	
	5 Nov. 17	Jeffreys, Reginald Castelman (207a) (*Temp. Capt.*, 18 *Oct.* 18.)		5 Nov. 17	
	5 Nov. 17	Blake, Dennis		5 Nov. 17	
425	5 Nov. 17	Connor, George Powell		5 Nov. 17	
	5 Nov. 17	Brooks, John Birthwhistle		5 Nov. 17	
	5 Nov. 17	Perry, Edward William (207)		5 Nov. 17	
	5 Nov. 17	Milward, Alfred George		5 Nov. 17	
	5 Nov 17	Smith, Herbert Cecil		5 Nov. 17	
430	5 Nov. 17	Taunton, Ivon Hope (207a)		5 Nov. 17	
	5 Nov. 17	Barter, Elder Gaul		5 Nov. 17	
	5 Nov 17	Holdsworth, Benjamin George		5 Nov. 17	
	5 Nov. 17	Teague, Lionel Elles Stanley (207a)	12 Feb. 82	5 Nov. 17	
	5 Nov. 17	Everard, Bertram Godard		5 Nov. 17	
435	5 Nov. 17	Hallahan, Richard Joseph		5 Nov. 17	
	6 Nov. 17	Dracott, Frank Herbert		6 Nov. 17	
	6 Nov. 17	Johnson, Arthur		6 Nov. 17	
	6 Nov. 17	Robinson, Arthur Wesley		6 Nov. 17	
	6 Nov. 17	Threlfall, Vincent Robert Seed		6 Nov. 17	
440	6 Nov. 17	Campbell, Charles Stewart (*Ty. Lt.-Col.*, 6 *Nov.* 17)		6 Nov. 17	
	6 Nov. 17	Gasper, Horace Le Geyt		6 Nov. 18	
	7 Nov. 17	Graham, Robert James Douglas		7 Nov. 17	
	8 Nov. 17	Staley, George		8 Nov. 17	
	10 Nov. 17	Oldall, Claude Herbert		10 Nov. 17	
445	11 Nov. 17	Johnson William Joseph		11 Nov. 17	
	12 Nov. 17	Hill, Frederick George Alexander		12 Nov 17	
	12 Nov. 17	Chappell, Frank Sidney		12 Nov. 17	
	12 Nov. 17	Gruer, Harold George		12 Nov 17	
	12 Nov. 17	Mitchell, Bertram Edward		12 Nov. 17	
450	15 Nov. 17	Jackson, William Joseph		15 Nov. 17	

† With pay and allowances as for a Lieut.

INFANTRY—*contd.*

Lieutenants—*contd.*

No.	Date of First Commn.	Names.	Date of birth.	Date of rank.	Remarks.
451	19 Nov. 17	Byrnes, William Francis (*Ty. Capt. 19 Nov. 17*)		19 Nov. 17	
	19 Nov. 17	Allen, Ernest Raymond		19 Nov. 17	
	19 Nov. 17	Fleming, Edward Gibson		19 Nov. 17	
	19 Nov. 17	Cockman, George Robinson		19 Nov. 17	
455	20 Nov. 17	Boucher, Alfred Sydney Hannah		20 Nov 17	
	20 Nov. 17	Holmes, James Frederick		20 Nov. 17	
	20 Nov. 17	Toomey, Stephen Patrick		20 Nov. 17	
	21 Nov. 17	Street, Walter Frank		21 Nov. 17	
	21 Nov. 17	Kingston, Frank		21 Nov. 18	
460	22 Nov. 17	Day, Richard Francis O'Hara Fitzgerald (*207a*)		22 Nov 17	
	22 Nov. 17	Brewin, Archibald Patrick (*207a*)		22 Nov 18	
	22 Nov. 17	Burns, Louis Davidson		22 Nov 18	
	22 Nov. 17	Earwaker, Graham Parsons		22 Nov. 18	
	22 Nov. 17	Walsh, John Edward		22 Nov. 18	
465	23 Nov. 17	Rappoport, Frederick Gerald		23 Nov. 18	
	24 Nov. 17	Bull, Harold		24 Nov. 18	
	25 Nov. 17	Herridge, William John		25 Nov. 18	
	26 Nov. 17	Linskill, Harry		26 Nov. 18	
	27 Nov. 17	Williams, Albert Henry (*Ty. Capt., 27 Nov. 17*)		27 Nov. 18	
470	28 Nov. 17	Dawes, Frederick Leopold		28 Nov. 18	
	28 Nov. 17	Adams, John Alan		28 Nov. 18	
	28 Nov. 17	Richer, Roland Admiral		28 Nov. 18	
	28 Nov. 17	Read, James		28 Nov. 18	
	28 Nov. 17	Mayers, Eric Williams		28 Nov. 18	
475	28 Nov. 17	Gow, Henry		28 Nov. 18	
	28 Nov. 17	Jacob, Garrard		28 Nov. 18	
	28 Nov. 17	Worden, Charles Norman		28 Nov. 18	
	28 Nov. 17	Crapp, William Vivian		28 Nov. 18	
	28 Nov. 17	Liston, John		28 Nov. 18	
480	28 Nov. 17	Birnie, Alexander		28 Nov. 18	
	28 Nov. 17	Moore, Ronald Nicholas		28 Nov. 17	
	28 Nov. 17	Angel, Bernard Gorwyne		28 Nov. 18	
	3 Dec. 17	James, Richard Farrer		3 Dec. 18	
	19 Dec. 17	Robins, Edward		19 Dec. 18	
485	19 Dec. 17	Cunliffe, Richard George		19 Dec. 18	
	19 Dec. 17	Walmisley, Herbert Edward		19 Dec. 18	
	19 Dec. 17	Walmisley, Arthur Lloyd		19 Dec. 18	
	19 Dec. 17	Logan, Thomas		19 Dec. 18	
	19 Dec. 17	Mercer, Alexander Robertson		19 Dec. 18	
490	19 Dec. 17	Pickard, Leonard Alfred		19 Dec. 18	
	21 Dec. 17	King, Ernest Walter		21 Dec. 18	
492	21 Dec. 17	Stewart, Charles		21 Dec. 18	

2nd-Lieutenants.

No.	Date of First Commn.	Names.	Date of birth.	Date of rank.	Remarks.
1	28 Nov. 17	Mair, Alexander		28 Nov. 17	
	16 Mar. 17	Davies, Gilbert Sinclair		16 Mar. 17	
	26 July 17	Marks, Gerelt Andrew		26 July 17 9 Aug. 17	
	30 Apl. 97	Harnett, William Falkiner, V.D.	1 May 69	30 Apl. 97	
5	25 Aug. 99	Willans, John Frederick	23 Mar. 69	25 Aug. 99	
	21 May 15	Croudace, Christopher Gardner		21 May 15	
	11 May 16	Prussia, Harry Bates		11 May 16	
	8 Aug. 17	Johnstone, Claude Walter Huntingdon		8 Aug. 16	
	5 Nov 17	Shannon, Gerald Cairns (*207a*)		5 Nov. 17	
10	5 Nov. 17	Henriques, Cyril Quixano (*207a*)		5 Nov. 17	
	26 Sep. 17	Bull, Arthur Herbert		26 Sep. 17	
	5 Oct. 17	Fairlie, John William		5 Oct. 18	
	5 Oct. 17	Gibbons, Arthur James		5 Oct. 17	
	31 Oct. 17	Herridge, William John			
15	1 Nov. 17	Balloch, Arthur Cecil		1 Nov. 17	
	5 Nov. 17	Sullivan, Robert Timothy		5 Nov. 17	
	5 Nov. 17	Buchanan-Meikle, Alan		5 Nov. 17	
	19 Nov. 17	Bruce, Edwin		19 Nov. 17	Released from Army Service.
	1 Dec. 17	Wilson, Gerald Eyre, I.S.O.		1 Dec 17	
20	1 Dec. 17	Mathew, Harold Bertram		1 Dec. 17	
	4 Dec. 17	Murphy, John Joseph		4 Dec. 17	

INFANTRY—*contd.*

2nd-Lieutenant—*contd.*

No.	Date of First Commn.	Names.	Date of birth.	Date of rank.	REMARKS.
	4 Dec. 17	Aitken, William Norman Gordon		4 Dec. 17	
	4 Dec. 17	Rushtom, John Pomfret		4 Dec. 17	
	7 Dec. 17	Lesser, Samuel		7 Dec. 17	
	8 Dec. 17	McAuliffe, John Patrick Thomas		8 Dec. 17	
	8 Dec. 17	Ashworth, Sam Lord		8 Dec. 17	
25	10 Dec. 17	Russell, Henry George		10 Dec. 17	
	10 Dec. 17	Perry, Percy Edward		10 Dec. 17	
	10 Dec. 17	Walker, Aldworth Marnock (*Ty. Lt., 10 Dec. 17*)		10 Dec. 17	
	10 Dec. 17	Saunders, John		10 Dec. 17	
	11 Dec. 17	Byles, George Frederick		11 Dec. 17	
30	11 Dec. 17	Stevens, Harold Coster		11 Dec. 17	
	14 Dec. 17	Edwards, William		14 Dec. 17	
	14 Dec. 17	Saunders, Hugh Abercrombie Childers		14 Dec. 17	
	15 Dec. 17	Elliot, George		15 Dec. 17	
	16 Dec. 17	Harridence, Robert Treslove		16 Dec. 17	
35	20 Dec. 17	Marsh, Godfrey Sutcliffe		20 Dec. 17	
	21 Dec. 17	Porrett, Cecil Robert (*161 days' embodied Volunteer Service to count for seniority*) († *temp. Capt., 4 Nov. 18*).		21 Dec. 17 31 May 18	
	21 Dec. 17	Greenwood, William Lockhart		21 Dec. 17	
	21 Dec. 17	Collins, Francis Patrick		21 Dec. 17	
	21 Dec. 17	Hays, Harold		21 Dec. 17	
40	22 Dec. 17	Syddall, John		22 Dec. 17	
	22 Dec. 17	Edge, George Percy		22 Dec. 17	
	23 Dec. 17	Berrill, Eric Francis		23 Dec. 17	
	25 Dec. 17	Corbett, Henry Reginald		25 Dec. 17	
45	26 Dec. 17	Symon, George		26 Dec. 17	
	26 Dec. 17	McGrath, William Henry (*Ty. Capt. 26 Dec. 17*)		26 Dec. 17	
	28 Dec. 17	Judah, David Alroy		28 Dec. 17	
	28 Dec. 17	Beckwith, Leonard Kildin		28 Dec. 17	
	28 Dec. 17	Walker, Louis		28 Dec. 17	
	28 Dec. 17	Morris, John Samuel		28 Dec. 17	
50	29 Dec. 17	Sykes, Harold John (*Ty. Capt. 29 Dec. 17*)		29 Dec. 17	
	29 Dec. 17	Fenstone, John		29 Dec. 17	
	31 Dec. 17	Mills, Henry Charles		31 Dec. 17	
	1 Jan. 18	Bayley, Victor, C.I.E.		1 Jan. 18	
55	4 Jan. 18	Bruce, Earnest Douglas Beresford		4 Jan. 18	
	5 Jan. 18	Lee-Mitchell, Harold Arthur		5 Jan. 18	
	5 Jan. 18	Johnstone, Archibald Meredith		5 Jan. 18	
	5 Jan. 18	Extance, Thomas John		5 Jan. 18	
	5 Jan. 18	Bird, Alfred George (46)		5 Jan. 18	
60	5 Jan. 18	Hewitt, Douglas James		5 Jan. 18	
	5 Jan. 18	Kemplay, Wilfred		5 Jan. 18	
	6 Jan. 18	Oggier, Sydney Leonard		6 Jan. 18	
	7 Jan. 18	Bowden, Norman Henry Martin		7 Jan. 18	
	7 Jan. 18	Corridon, Alfred Francis		7 Jan 18	
65	7 Jan 18	Morrow, William James Angus		7 Jan. 18	
	10 Jan. 18	Beck, Fredrick George		10 Jan. 18	
	12 Jan. 18	Bradbury, Cecil Howard		12 Jan. 18	
	15 Jan. 18	Murphy, Lawrence Arthur		15 Jan. 18	
	15 Jan. 18	Riley, Edward Charles		15 Jan. 18	
70	15 Jan. 18	Barker, Frederick Raymond		15 Jan. 18	
	15 Jan. 18	Osborne, Henry Mallett		15 Jan. 18	
	15 Jan. 18	Aikin, Ralph William Doran		15 Jan. 18	
	15 Jan. 18	Spooner, John Douglas		15 Jan. 18	
	15 Jan. 18	Turner, Joseph Henry Patrick		15 Jan. 18	
75	15 Jan. 18	McDermott, Neville William Ewart		15 Jan. 18	
	15 Jan. 18	Dickson, William André		15 Jan. 18	
	15 Jan. 18	Howard, Henry Hazel Grover (*temp. Capt. 1 Aug. 18*)		15 Jan. 18	
	15 Jan. 18	Franks, Harry Shillaker		15 Jan. 18	
	15 Jan. 18	Whiffin, Lloyd Dundas		15 Jan. 18	
80	15 Jan. 18	Campbell, Archibald Alexander		15 Jan. 18	
	15 Jan. 18	Betterton, Frederick Archibald		15 Jan. 18	
	15 Jan. 18	Cattermole, Frank Charles		15 Jan. 18	
	15 Jan. 18	Armstrong, Michael Hopson		15 Jan. 18	
84	26 July 17	Marks, Gerald Andrew		26 July 18 9 Aug. 18	

† With pay and allowances as for a Lieut.

No.	Date of First Commn.	Names.	Date of birth.	Date of rank.	REMARKS.

INFANTRY—*contd.*

2nd-Lieutenants—*contd.*

No.	Date of First Commn.	Names.	Date of birth.	Date of rank.	REMARKS.
511	15 Jan. 18	Bailey, Ronald Joseph		15 Jan. 18	
	15 Jan. 18	Diacono, Hugo Victor		15 Jan. 18	
	15 Jan. 18	Cook, Harry Winfield		15 Jan. 18	
	15 Jan. 18	Dale, Raymond		15 Jan. 18	
515	15 Jan. 18	Wilkinson, Stanley Percy		15 Jan. 18	
	15 Jan. 18	Ward, Frederick Herbert Tetley		15 Jan. 18	
	15 Jan. 18	Miller, Herbert		15 Jan. 18	
	15 Jan. 18	Roblin, Maurice Leslie		15 Jan. 18	
	15 Jan. 18	Grant, Arthur Edward		15 Jan. 18	
520	15 Jan. 18	Rayner, Arthur James		15 Jan. 18	
	15 Jan. 18	Fordyce-Harvey, Charles Arthur		15 Jan. 18	
	15 Jan. 18	Field, Walter John		15 Jan. 18	
	15 Jan. 18	Towell, Arthur Frederick Herbert		15 Jan. 18	
	15 Jan. 18	Athol-Liddell, Ralph Edmund		15 Jan. 18	
525	15 Jan. 18	Topham, Leslie Harland Charles		15 Jan. 18	
	15 Jan. 18	Tunstall, John Eric		15 Jan. 18	
	15 Jan. 18	Atkinson, Arthur John Moulton		15 Jan. 18	
	15 Jan. 18	Reynolds, Cyril George (*temp. Capt.*, 1 *July* 18)		15 Jan. 18	
	15 Jan. 18	McGregor, Alexander		15 Jan. 18	
530	15 Jan. 18	Harmer, Frederick Llewellin		15 Jan. 18	
	15 Jan. 18	Sanders, Henry Harold		15 Jan. 18	
	15 Jan. 18	Lane, Walter Edward		15 Jan. 18	
	15 Jan. 18	Fox, Thomas Francis		15 Jan. 18	
535	15 Jan. 18	Meredith, David Francis		15 Jan. 18	
	15 Jan. 18	Robb, Alexander Henry Denby		15 Jan. 18	
	15 Jan. 18	Cameron, Henry Hay		15 Jan. 18	
	15 Jan. 18	Langford, George Watson		15 Jan. 18	
	15 Jan. 18	Cahill, James (207*a*)		15 Jan. 18	
540	15 Jan. 18	Hosgood, John Winsborrow		15 Jan. 18	
	15 Jan. 18	Fitzpatrick, John Joseph		15 Jan. 18	
	15 Jan. 18	Short, Reginald Lawrence Brend		15 Jan. 18	
	15 Jan. 18	Prebble, Cecil Charles		15 Jan. 18	
	15 Jan. 18	Wilson, Edwin Henry Edmonds		15 Jan. 18	
545	15 Jan. 18	Clark, Harold		15 Jan. 18	
	15 Jan. 18	Holt, George Harold		15 Jan. 18	
	15 Jan. 18	Jackson, Arthur William James		15 Jan. 18	
	15 Jan. 18	Norton, Wallace Oswald		15 Jan. 18	
	15 Jan. 18	Thomas, Richard Harold		15 Jan. 18	
550	15 Jan. 18	Ball, John William		15 Jan. 18	
	15 Jan. 18	Corrie, Arthur Byrom		15 Jan. 18	
	15 Jan. 18	Aston, Herbert Reid		15 Jan. 18	
	15 Jan. 18	Joaquim, Basil John Patrick		15 Jan. 18	
	15 Jan. 18	Kirkby, Roland Arthur		15 Jan. 18	
555	15 Jan. 18	Butler, Stanley Gordon		15 Jan. 18	
	15 Jan. 18	Gawthorne, Charles Hugh		15 Jan. 18	
	15 Jan. 18	Eade, Richard Oswald		15 Jan. 18	
	15 Jan. 18	Gordon, Evan Ramsay MacIntosh		15 Jan. 18	
	15 Jan. 18	Sullivan, Cecil Stevens		15 Jan. 18	
560	15 Jan. 18	Clifford, Edmond King		15 Jan. 18	
	15 Jan. 18	Determes, Arthur Charles		15 Jan. 18	
	15 Jan. 18	Storm, William		15 Jan. 18	
	15 Jan. 18	Mendes, Maurice		15 Jan. 18	
	15 Jan. 18	Minhinnick, John Bertrand		15 Jan. 18	
565	15 Jan. 18	Bagley, Albert Anthony		15 Jan. 18	
	15 Jan. 18	Williamson, Alec		15 Jan. 18	
	15 Jan. 18	Woledge, Cecil George Charles		15 Jan. 18	
	15 Jan. 18	Croban-Lucas, Arthur John Lionel		15 Jan. 18	
	15 Jan. 18	Patton, Arnold Gordon		15 Jan. 18	
570	15 Jan. 18	Boddington, Eric Stanley St. Clair		15 Jan. 18	
	15 Jan. 18	Sultano, Patrick Sidney		15 Jan. 18	
	15 Jan. 18	Bowden, Colin Guy		15 Jan. 18	
	15 Jan. 18	Kelly, William James		15 Jan. 18	
	15 Jan. 18	Searle, Humphrey Frederic		15 Jan. 18	
575	15 Jan. 18	Webster, Thomas George Clarkson		15 Jan. 18	
	15 Jan. 18	Edge, Eric Harold (46)		15 Jan. 18	
	15 Jan. 18	Hannagan, Lawrence		15 Jan. 18	
	15 Jan. 18	Richardson, Worsley		15 Jan. 18	
	15 Jan. 18	Batchelor, George Ferrier Fraser		15 Jan. 18	
580	15 Jan. 18	Fuge, Richard Douglas (46)		15 Jan. 18	
	15 Jan. 18	Herman, Colin Roy		15 Jan. 18	
	15 Jan. 18	Beard, Norman Theodore		15 Jan. 18	
	15 Jan. 18	Carey, Cedric O'Donoghue		15 Jan. 18	
	15 Jan. 18	Haffenden, Edward Dalton		15 Jan. 18	
585	15 Jan. 18	Meiklejohn, William		15 Jan. 18	

No.	Date of First Commn.	Names.	Date of birth.	Date of rank.	REMARKS.
		INFANTRY—*contd.*			
		2nd-Lieutenants—*contd.*			
586	15 Jan. 18	Bedford, Edward Walter		15 Jan. 18	
	15 Jan. 18	Ruthven, John Leighton		15 Jan. 18	
	15 Jan. 18	Conder, John Marmaduke		15 Jan. 18	
	15 Jan. 18	Moylan, Thomas Joseph		15 Jan. 18	
590	17 Jan. 18	Phibbs, Harry		17 Jan. 18	
	17 Jan. 18	Forbes, Thomas Chirnside		17 Jan. 18	
	18 Jan. 18	Smyth, Edward George		18 Jan. 18	
	18 Jan. 18	Smith, Alexander Russell		18 Jan. 18	
	18 Jan. 18	Perry, Charles Alfred Stanley		18 Jan. 18	
595	19 Jan. 18	Wimshurst, Charles Rodney		19 Jan. 18	
	21 Jan. 18	Knyvett-Hough, Charles Attwood		21 Jan. 18	
	23 Jan. 18	Sievwright, Andrew George Hume (*Ty. Capt., 28 Jan. 18.*)		23 Jan. 18	
	23 Jan. 18	Ringrow, Noel Vandyke (*Ty. Lt., 23 Jan. 18*).		23 Jan. 18	
	23 Jan. 18	Hoogwerf, Vivian Horace (*Ty. Lt., 23 Jan. 18.*)		23 Jan. 18	
600	23 Jan. 18	Guest, Stanley Vowler (*Ty. Lt., 23 Jan. 18.*)		23 Jan. 18	
	23 Jan. 18	Burt, Ernest Henry		23 Jan. 18	
	23 Jan. 18	Hilton, John George Maxwell		23 Jan. 18	
	28 Jan. 18	Lynn, Thomas Valentine		28 Jan. 18	
	28 Jan. 18	Lacey, Christopher William (*Ty. Capt., 28 Jan. 18.*)		28 Jan. 18	
605	30 Jan. 18	Meudell, Edgar		30 Jan. 18	
	30 Jan. 18	Grimes, Cecil John		30 Jan. 18	
	1 Feb. 18	Carrington, Squire Sydney		1 Feb. 18	
	1 Feb. 18	Webster, Sydney (*Ty. Capt., 1 Feb. 18.*)		1 Feb. 18	
	1 Feb. 18	O'Shea, Patrick John		1 Feb. 18	
610	1 Feb. 18	Wreford, Cyril Frederick		1 Feb. 18	
	4 Feb. 18	Dowson, Valentine Hugh Wilfred		4 Feb. 18	
	5 Feb. 18	Stone, Horatio O'Dell (*207a*).		5 Feb. 18	
	5 Feb. 18	Harkness, Edwin Berg (*207a*).		5 Feb. 18	
	6 Feb. 18	Mackaness, Samuel George		6 Feb. 18	
615	7 Feb. 18	Gimson, Christopher		7 Feb. 18	
	9 Feb. 18	Davidson, Matthew Denholm		9 Feb. 18	
	11 Feb. 18	Cuningham, George Edmund (*temp. Maj. 11 Feb. 18*).		11 Feb. 18	
	11 Feb. 18	Pyett, Claud Felix		11 Feb. 18	
	11 Feb. 18	Bennison, Arthur John		11 Feb. 18	
620	11 Feb. 18	Carruthers, Angus Lennon		11 Feb. 18	
	16 Feb. 18	Woods, Charles Francis		16 Feb. 18	
	16 Feb. 18	Cooper, Frank Alexander (**temp. Capt. 30 Apl. 18*)		16 Feb. 18	
	17 Feb. 18	Graham, Cyril George		17 Feb. 18	
	18 Feb. 18	Orchard, Douglas George		18 Feb. 18	Released from Army Service.
625	19 Feb. 18	McKenna, Curtin		19 Feb. 18	
	19 Feb. 18	MacKay, Robert		19 Feb. 18	
	20 Feb. 18	Slater, Samuel Henry (*temp. Capt., 20 Feb. 18*)		20 Feb. 18	
	20 Feb. 18	Smith, John (*temp. Capt., 20 Feb. 18*)		20 Feb. 18	
	21 Feb. 18	Tyte, William Leonard		21 Feb. 18	
630	23 Feb. 18	Gwynne, Clement Wansbrough (*temp. Maj., 23 Feb. 18*)		23 Feb. 18	
	26 Feb. 18	Fuller, Robert William		26 Feb. 18	
	26 Feb. 18	Finnigan, Lawrence Desmond		26 Feb. 18	
	26 Feb. 18	Saw, George Arthur (*temp. Lt., 12 Mar. 18*).		26 Feb. 18	
	27 Feb. 18	McKernan, Robert Lucian		27 Feb. 18	Released from Army Service.
635	1 Mar. 18	Baldwin, Robert Henry		1 Mar. 18	
	1 Mar. 18	Macgregor, William Lewis		1 Mar. 18	
	1 Mar. 18	English, Eric Erle		1 Mar. 18	
	1 Mar. 18	St. Romaine, Eugene Chandos		1 Mar. 18	
	1 Mar. 18	Carey, Archie Bertram		1 Mar. 18	
640	1 Mar. 18	Campbell, Colin		1 Mar. 18	
	1 Mar. 18	Mackay, John William		1 Mar. 18 3 July 18	
	1 Mar. 18	Finnigan, William Bernard (*temp. Maj., 7 May 18*)		1 Mar. 18 22 July 18	
	7 Mar. 18	Gorrie, George		7 Mar. 18	
	8 Mar. 18	Lamacraft, Harry (*temp. Capt., 12 June 18*)		8 Mar. 18	
645	8 Mar. 18	Hodges, John Edwin Carter		8 Mar. 18	
	8 Mar. 18	Phillips, Percy Brydone		8 Mar. 18	
	8 Mar. 18	Wallace, Charles Lindsay (*temp. Capt., 27 Mar. 18*)		8 Mar. 18	
	8 Mar. 18	Davis, John Thomas (207)		8 Mar. 18	Released from Army Service.
	11 Mar. 18	Williams, Charles Beynon †		11 Mar. 18	
650	12 Mar. 18	Walford, Henry Gordon		12 Mar. 18	

* Without pay and allowances of the rank.
† With pay and allowances as for a Lieut.

INFANTRY—*contd.*

2nd-Lieutenants—*contd.*

No.	Date of First Commn.	Names.	Date of birth.	Date of rank.	REMARKS.
650	12 Mar. 18	Brayham, Edward Frederick		12 Mar. 18	
	12 Mar. 18	Thurlow, Benjamin William		12 Mar. 18	
	12 Mar. 18	Johnson, Finlayson Huson		12 Mar. 18	
	14 Mar. 18	Pettigrew. William (*temp. Capt. 9 Oct. 18*)		14 Mar. 18	
	15 Mar. 18	Edwards, William Cecil		15 Mar. 18	
655	15 Mar. 18	Dean George Edmondstone		15 Mar. 18	
	15 Mar. 18	Capstick, Frederick Herbert		15 Mar. 18	
	15 Mar. 18	Sargeant, James William		15 Mar. 18	
	15 Mar. 18	Tipple, Alan		15 Mar. 18	
	15 Mar. 18	Harris, Herbert John		15 Mar. 18	
	16 Mar. 18	Long, Clifford Arthur		16 Mar. 18 24 June 18	
660	16 Mar. 18	Purchase, Robert George		16 Mar. 18 6 July 18	
	16 Mar. 18	Beynon, Gwilym		16 Mar. 18 11 July 18	
	16 Mar. 18	Fraser, Robert Albert John		16 Mar. 18 5 Aug. 18	
	18 Mar. 18	Cross, Percy Leon		18 Mar. 18	
665	19 Mar. 18	Luscombe, Frederick Wallis		19 Mar. 18	
	19 Mar. 18	Orchard, Herbert Claude White		19 Mar. 18	Released from Army Service.
	21 Mar. 18	Crewe, Edward Neville		21 Mar. 18	
	26 Mar. 18	Holland, Thomas George (*temp. Capt., 19 Mar. 18*).		26 Mar. 18	
	19 Mar. 18	Gibson, Alexander John		19 Mar. 18	
670	19 Mar. 18	Hicks, Ethelbert Samuel		19 Mar 18	
	19 Mar. 18	Cardwell, William Albert		19 Mar. 18	
	19 Mar. 18	Small, Charles George		19 Mar. 18	
	19 Mar. 18	Foot, William John Hawken		19 Mar. 18	
	19 Mar. 18	Beer, Earnest Edger		19 Mar. 18	
675	19 Mar. 18	Dodge, Montague William		19 Mar. 18	
	19 Mar. 18	Gibbs, Albert Edward Thomas		19 Mar. 18	
	19 Mar. 18	Krapman, Earnest Hart		19 Mar. 18	
	28 Mar. 18	Smith, Walter Percival (†*temp. Capt. 10 June 18*)		28 Mar. 18	
	28 Mar. 18	Howell, Heber William Henry (*temp. Capt. 28 Mar. 18*)		28 Mar. 18	
680	29 Mar. 18	Ross, Walter James		29 Mar. 18	
	29 Mar. 18	Rixon, Frederick		29 Mar. 18	
	29 Mar. 18	Bryer, Percy James		29 Mar. 18	
	29 Mar. 18	McDonald, Donald Charles		29 Mar. 18	
	29 Mar. 18	Gillie, Kenneth Mackenzie		29 Mar. 18	
685	29 Mar. 18	Burke, Patrick Harland		29 Mar. 18	
	29 Mar. 18	Pryde, Robert William		29 Mar. 18	
	29 Mar. 18	Beard, Richard Herbert Frith		29 Mar. 18	
	29 Mar. 18	Heselton, Rodolphus Thomas		29 Mar. 18	
	29 Mar. 18	Fish, William Henry		29 Mar. 18	
690	29 Mar. 18	Smith, Walter Sherrard		29 Mar. 18	
	29 Mar. 18	Woods, George Henry		29 Mar. 18	
	29 Mar. 18	Davis, Arthur Isidore		29 Mar. 18	
	29 Mar. 18	Hudson, Edwin James Muddle		29 Mar. 18	
	29 Mar. 18	Calogreedy, Lewis Henry		29 Mar. 18	
695	29 Mar. 18	Shortland, Ralph Bertram		29 Mar. 18	
	29 Mar. 18	Holt, Denis Ralph William		29 Mar. 18	
	29 Mar. 18	Young, Stewart		29 Mar. 18	
	29 Mar. 18	Cobble, Edward William		29 Mar. 18	
	29 Mar. 18	Cabot, Cecil Edward		29 Mar. 18	
700	29 Mar. 18	Allen, Robyn Alan		29 Mar. 18	
	29 Mar. 18	Jackson, Ralph Cecil		29 Mar. 18	
	29 Mar. 18	Edwards, Hubert Bruce		29 Mar. 18	
	29 Mar. 18	Moore, John		29 Mar. 18	
	29 Mar. 18	Priddey, Alfred Henry Peter		29 Mar. 18	
705	29 Mar. 18	Cullen, Stanley Edgar		29 Mar. 18	
	29 Mar. 18	Soper, Owen Norman		29 Mar. 18	
	29 Mar. 18	Downing, Leonard David		29 Mar. 18	
	29 Mar. 18	Shortt, William Henry Oscar		29 Mar. 18	
	29 Mar. 18	Nelson, Harold		29 Mar. 18	
710	29 Mar. 18	Kingsmill, Wilfred John		20 Mar. 18	
	29 Mar. 18	Hyde, Harry Stanley		29 Mar. 18	
	29 Mar. 18	Driver, George Leslie Stewart		29 Mar. 18	
	29 Mar. 18	Gurney, Albert		29 Mar. 18	
	29 Mar. 18	Owen, William		29 Mar. 18	
715	29 Mar. 18	Williams, Samuel Nicholas		29 Mar. 18	
	29 Mar. 18	Bandey, Arthur Edgar		29 Mar. 18	
717	29 Mar. 18	Hewett, Harry		29 Mar. 18	

† With pay and allowances as for a Lieutenant.

No.	Date of First Commn.	Names.	Date of birth.	Date of rank.	Remarks.
		INFANTRY—*contd.*			
		2nd-Lieutenants—*contd.*			
718	29 Mar. 18	Hubball, John Lewis		26 Mar. 18	
	29 Mar. 18	Traylen, Harold Edward		29 Mar. 18	
720	29 Mar. 18	Tremenheere, Hugh Seymour		29 Mar. 18	
	29 Mar. 18	Killick, Charles Fredrick Rowe		29 Mar. 18	
	29 Mar. 18	Bentham, Charles Percy		29 Mar. 18	
	29 Mar. 18	Shawe, Reginald Arthur		29 Mar. 18	
	29 Mar. 18	Le Brooy, Herbert Terrence		29 Mar. 18	
725	29 Mar. 18	Meston, Alexander		29 Mar. 18	
	29 Mar. 18	Moore, Eric Richmond		26 Mar. 18	
	29 Mar. 18	Phillips, George Henry		29 Mar. 18	
	29 Mar. 18	Scutt, Charles Clyde		29 Mar. 18	
	29 Mar. 18	Johnson, Percy Edward		29 Mar. 18	
730	29 Mar. 18	Venton, William Fredrick		29 Mar. 18	
	29 Mar. 18	Coldwell, James Andre		29 Mar. 18	
	29 Mar. 18	Clarke, Percy Thomas		29 Mar. 18	
	29 Mar. 18	Rushbrooke, Albert Victor		29 Mar. 18	
	29 Mar. 18	Scott, Alexander Henry		29 Mar. 18	
735	29 Mar. 18	Adams, Perry		29 Mar. 18	
	29 Mar. 18	Burt, William George		29 Mar. 18	
	29 Mar. 18	Creasy, Robert Leslie		29 Mar. 18	
	29 Mar. 18	Winstantley, Bernard Joseph		29 Mar. 18	
	29 Mar. 18	Gomme, David Edmond		29 Mar. 18	
740	29 Mar. 18	Swynnerton, Frederic Richard		29 Mar. 18	
	29 Mar. 18	Beedham, John William		28 Mar. 18	
	29 Mar. 18	McMurray, Frederic Charles		29 Mar. 18	
	29 Mar. 18	Hobson, Robert John		29 Mar. 18	
	29 Mar. 18	George, William James		29 Mar. 18	
745	29 Mar. 18	Fairweather, Alexander		29 Mar. 18	
	29 Mar. 18	Thompson, Aiden Kenneth		29 Mar. 18	
	29 Mar. 18	Carless, Henry Alfred		29 Mar. 18	
	29 Mar. 18	Hisco, Ralph		29 Mar. 18	
	29 Mar. 18	Teal, Frank Harold Dunn		29 Mar. 18	
750	29 Mar. 18	Reeves, William Henry		29 Mar. 18	
	29 Mar. 18	Roblin, Cyril Herbert		29 Mar. 18	
	29 Mar. 18	Tritton, George Christopher		29 Mar. 18	
	29 Mar. 18	Macleod, Ian Norman		29 Mar. 18	
	30 Mar. 18	Fitzpatrick, Lawrence Francis Jerome		30 Mar. 18	
755	30 Mar. 18	Showers, Frank		30 Mar. 18	
	1 Apl. 18	Fothergill, William		1 Apl. 18	
	3 Apl. 18	Hoskings, Rutuerford Vincent Theodore		3 Apl. 18	
	5 Apl. 18	Rhodes, George Edward		5 Apl. 18	
	5 Apl. 18	James, Frederick		5 Apl. 18	
760	5 Apl. 18	Beart, Purl Douglas		5 Apl. 18	
	5 Apl. 18	Greer-Spencer, Benjamin		5 Apl. 18	
	6 Apl. 18	Rose, George Harry		6 Apl. 18	
	6 Apl. 18	Le Feuvre, Clifford Owen		6 Apl. 18	
	8 Apl. 18	White, Francis George Ignatius Marie		8 Apl. 18	
765	8 Apl. 18	King, Eric Travor		8 Apl. 18	
	9 Apl. 18	Watson, Arthur Osbourne		9 Apl. 18	
	9 Apl. 18	Hayes, Charles Robert		9 Apl. 18	
	9 Apl. 18	Knight, Herbert		9 Apl. 18	
	9 Apl. 18	Lewis, Lloyd		9 Apl. 18	
770	9 Apl. 18	Brocklebank, Herbert James		9 Apl. 18	
	9 Apl. 18	Johnson, John Frederick		9 Apl. 18	
	9 Apl. 18	McIver, Andrew Cochrane		9 Apl. 18	
	12 Apl. 18	Keer, William Horace (†*temp. Capt. 25 June 18*)		12 Apl. 18	
	13 Apl. 18	Heyne, George Arnold Frederick		13 Apl. 18	
775	17 Apl. 18	Bradley-Birt, Francis Bradley (*temp. Capt. 17 Apl. 18*)		17 Apl. 18	
	18 Apl. 18	Lysons, Kenneth Cambel		18 Apl. 18	
	18 Apl. 18	Gordon, Bruce Marie Melville		18 Apl. 18	
	18 Apl. 18	Weston, James Samuel Newman		18 Apl. 18	
	20 Apl. 18	Gilbertson, John Grigor		20 Apl. 18	
780	20 Apl. 18	Webster, Alexander		20 Apl. 18	
	20 Apl. 18	Mortimer, Hugh		20 Apl. 18	
	20 Apl. 18	Folds, James		20 Apl. 18	
	22 Apl. 18	Smith, Eric Seton		22 Apl. 18	
	23 Apl. 18	Mason, Thomas William		23 Apl. 18	
785	23 Apl. 18	Parsons, James Parkes Dudley		23 Apl. 18	
	25 Apl. 18	Adams, Percival Henry		25 Apl. 18	
	26 Apl. 18	Veysey, George		26 Apl. 18	
	27 Apl. 18	Todd, Alfred Edward		27 Apl. 18	
	27 Apl. 18	Titterington, James Hugh		27 Apl. 18	
790	29 Apl. 18	Hardinge, George William Percy		29 Apl. 18	
791	29 Apl. 18	Newman, John Herbert		29 Apl. 18	

† With pay and allowances as for a Lieutenant.

INFANTRY—*contd.*

2nd-Lieutenants—*contd.*

No.	Date of First Commn.	Names.	Date of birth.	Date of Rank.	Remarks.
791	1 May 18	Fyfe, William Smith		1 May 18	
	1 May 18	Love, George William		1 May 18	
	1 May 18	Gardner, Oswald James		1 May 18 25 June 18	
	1 May 18	Walker, John		1 May 18	
795	1 May 18	Mansfield, Philip Theodore		1 May 18 16 July 18	
	2 May 18	Box, Edward George		2 May 18	
	2 May 18	Rodgers, John Frederick		2 May 18	
	2 May 18	Webbe, Denzil Brudenell		2 May 18	
	7 May 18	MacArthur, Donald Fraser		7 May 18	
800	7 May 18	Collis, Walter Marsh		7 May 18	
	8 May 18	Mathalone, Maurice		8 May 18	
	10 May 18	Davis, Arnold Percivale		10 May 18 27 June 18	
	10 May 18	Couper, Thomas		10 May 18 3 July 18	
	10 May 18	Collings, Francis William		10 May 18 4 July 18	
805	10 May 18	Killick, Alexander Edward Hilton		10 May 18 4 July 18	
	10 May 18	Clague, John		10 May 18	
	13 May 18	Lambe, Geoffrey		13 May 18	
	13 May 18	Peters, Arthur Stanley Colin		13 May 18	
	13 May 18	Clark-Kennedy, Hew Desbrough George Armistead		13 May 18	
810	13 May 18	Huffton, Percy Norman		13 May 18	
	13 May 18	Miller, Colin Hay		13 May 18	
	13 May 18	Taylor, James Frank Stevenson		13 May 18	
	13 May 18	Ritchie, Adam Alexander		13 May 18 27 June 18	
	14 May 18	Wilson, Frederick Hamlyn		14 May 18	
815	14 May 18	Pearse, Francis Napier		14 May 18	
	15 May 18	Grant, Edmund Robert		15 May 18	
	15 May 18	Dunlop, Hope Lyle		15 May 18	
	15 May 18	Furguson, William Edward		15 May 18	
	15 May 18	Callaway, Frederick Maynard (*temp. Lt., 19 May 18*).		15 May 18	
820	15 May 18	Ramsay, William Fyfe Craik		15 May 18	
	15 May 18	Cornish-Trestrail, Arthur George		15 May 18	
	16 May 18	McNichol, William Mathew (†*temp. Capt., 18 July 18*).		16 May 18	
	16 May 18	Staig, Bertie Munro		16 May 18 27 June 18	
	16 May 18	Macpherson, Donald		16 May 18	
825	17 May 18	Rees, Colin George Ashley		17 May 18	
	17 May 18	Russell, Edward Thomas		17 May 18	
	20 May 18	Hill, Kenneth MacKenzie Ross		20 May 18	
	20 May 18	Levy, Reginald Michael		20 May 18	
	21 May 18	Roper, George Eliot		21 May 18	
830	21 May 18	Hawker, Gilbert Victor		21 May 18	
	22 May 18	FitzPatrick, Richard Henry Joseph		22 May 18	
	22 May 18	Elliot, Walter		22 May 18	
	22 May 18	Bell, John Nairn		22 May 18	
	22 May 18	Byrne, Frederick (†*temp. Capt., 31 July 18*)		22 May 18	
835	22 May 18	Teale, Douglas Brisbane Shipley		22 May 18	
	22 May 18	Wrench, John Mervyn Dallas		22 May 18	
	23 May 18	Randall, Bernard Thomas		23 May 18	
	23 May 18	Walker, Lionel Harold Chadwick		23 May 18	
	24 May 18	Critchell, Douglas Edgar Ralph		24 May 18	
840	24 May 18	Booker, George Aston		24 May 18	
	24 May 18	Mills, George Roycroft		24 May 18	
	25 May 18	Eyre, Montague Henry, (*Actg Maj., 12 July 18*)		25 May 18 12 July 18	
	27 May 18	Sharpe, Julian Norris (*temp. Capt., 27 May 18*)		27 May 18	
	27 May 18	Kingham, Patrick		27 May 18	
845	28 May 18	Watson, Harold Argyll		28 May 18	Released from Army Service.
	29 May 18	Hopewell, George Morse		29 May 18	
	29 May 18	Ottley, Adam Rupert Cyril		29 May 18	
	29 May 18	Deane, Nelson Guy		29 May 18	
	30 May 18	McCord, James Coromandel		30 May 18	
850	30 May 18	Evans, John David		30 May 18	
	30 May 18	Maynard, Nicholas Crofton		30 May 18	
	30 May 18	Clinch, Harold Gordon		30 May 18	Released from Army Service.
853	31 May 18	Foskett, Ralph Cavan		31 May 18	

† With pay and allowances as for a Lieutenant.

No.	Date of First Commn.	Names.	Date of birth	Date of Rank.	Remarks.
		INFANTRY—*contd.*			
		2nd-Lieutenants—*contd.*			
854	1 June 18	Barty, Joseph Edward		1 June 18	
855	1 June 18	Vivian, Valentine Patrick Terrel, O.B.E. (*temp. Maj., 1 June 18*)		1 June 18	
	1 June 18	Walmsley, Alfred Moss		1 June 18	
	1 June 18	Farrell, Joseph		1 June 18	
	1 June 18	Grant, David Mudie		1 June 18	
	1 June 18	Ledwich, George		1 June 18	
860	1 June 18	McCall, William Charles		1 June 18	
	1 June 18	Proust Rene		1 June 18	
	2 June 18	Marchant, Lavinius Brice		2 June 18	
	2 June 18	Dunne, Eric Leslie		2 June 18	
	2 June 18	Thomas, Harvey		2 June 18	
865	2 June 18	Pignon, Henry Leonard		2 June 18	
	2 June 18	Johnson, Ernest		2 June 18	
	2 June 18	Turner, Percival Heathcote		2 June 18	
	2 June 18	Willis, Harold Augustus		2 June 18	
	2 June 18	Boult, Nelson Cross		2 June 18	
870	2 June 18	Lowes, Gregson Wooler		2 June 18	
	2 June 18	Friend, Ernest Alfred		2 June 18	
	2 June 18	Goldsmith, Malcolm Lawrence		2 June 18	
	2 June 18	Payne, William		2 June 18	
	2 June 18	Browning, John Morgan		2 June 18	
875	2 June 18	Wimpey, Samuel George		2 June 18	
	3 June 18	Searle, Henry George		3 June 18	
	3 June 18	Rees, Henry Vaughan		3 June 18	
	3 June 18	Crowhurst, John Donald James		3 June 18	
	3 June 18	Hardaker, Harold Lawrie Walwin		3 June 18	
880	3 June 18	Bartlett, Allan Henry Stephen		3 June 18	
	3 June 18	Lewis, Richard Norman		3 June 18	
	4 June 18	Ross, John Brackenridge		4 June 18	
	4 June 18	Goodall, Francis Ernest		4 June 18	
	4 June 18	Hollingbery, Geoffery Robert Vaughan		4 June 18	
85	4 June 18	Parsons, Percival James		4 June 18	
	5 June 18	Stone, Robert Charles		5 June 18	
	5 June 18	Brohier, Clarence Percival		5 June 18	
	5 June 18	De Vos, James Percival		5 June 18	
	5 June 18	Woods-Scawen, Charles Edmund		5 June 18	
890	7 June 18	Haye, Noel		7 June 18	
	7 June 18	De Bretton, Duncan William		7 June 18	
	7 June 18	Allum, Richard Warner		7 June 18	
	7 June 18	Randall, Harold Victor		7 June 18	
	7 June 18	Foweraker, Harold Austen		7 June 18	
895	7 June 18	Young, Lyle Elliott		7 June 18	
	7 June 18	Hardwicke, Sebastien More Baldry		7 June 18	
	7 June 18	Hanley, Maurice Patrick		7 June 18	
	7 June 18	Rivett, Harold Henry		7 June 18	
	7 June 18	Brown, Henry Whitby		7 June 18	
900	7 June 18	Handley, Percy		7 June 18	
	7 June 18	Brewin, Arthur Sydney		7 June 18	
	7 June 18	Fleming-Madge, David		7 June 18	
	7 June 18	Fry, Edgar Noel		7 June 18	
	7 June 18	Ford, Guy Harold		7 June 18	
905	7 June 18	Tee, Stanley Edwin		7 June 18	
	7 June 18	Haden, William Richard		7 June 18	
	7 June 18	Fraser, Arthur Byrne		7 June 18	
	7 June 18	Lottus, George Merrick		7 June 18	
	7 June 18	Monk, Frank Harold		7 June 18	
910	7 June 18	Howe, William Gordon Theodore		7 June 18	
	7 June 18	Bond, Oscar Henry		7 June 18	
	7 June 18	Jones, Ashley Norman Postance		7 June 18	
	7 June 18	Taylor, Clarence Henry		7 June 18	
	7 June 18	Rowe, Stanley Albert		7 June 18	
915	7 June 18	Shannon, Alfred George Leonard		7 June 18	
	7 June 18	Ryper, Harvey Joseph		7 June 18	
	7 June 18	Williams, Stephen Joseph		7 June 18	
	7 June 18	Pelly, Charles Thorne		7 June 18	
	7 June 18	Brick, Patrick Joseph		7 June 18	
920	7 June 18	Hill, Albert Edward		7 June 18	
	7 June 18	Lee, Alfred John		7 June 18	
	7 June 18	Park, William Black		7 June 18	
	7 June 18	Gerard-Smith, Vincent		7 June 18	
	7 June 18	Culpeck, Thomas Charles		7 June 18	
925	7 June 18	Finelli, Charles Willing Francis		7 June 18	
	7 June 18	Robertson, Robin Haskew		7 June 18	
	7 June 18	O'Leary, Dermot Francis Kyrle		7 June 18	
	7 June 18	Glover, Cyril Arthur Wood		7 June 18	
	7 June 18	Lockhart, Walter		7 June 18	
930	7 June 18	Clark, Leonard Arthur Frank		7 June 18	

No.	Date of First Commn.	Names.	Date of birth.	Date of Rank.	REMARKS.
		INFANTRY—*contd.*			
		2nd-Lieutenants—*contd.*			
507	7 June 18	Shead, Cyril George		7 June 18	
	7 June 18	Baskett, Oswald William Bennett		7 June 18	
	7 June 18	Donald, Arthur Alexander		7 June 18	
510	7 June 18	Fraser, Cyril Douglas		7 June 18	
	7 June 18	Cannell, Michael Douglas		7 June 18	
	7 June 18	Watts, William Walter		7 June 18	
	7 June 18	Wilkinson, William Edward		7 June 18	
	7 June 18	Milner, Frank Derek		7 June 18	
515	7 June 18	Parkinson, John Warren		7 June 18	
	7 June 18	Vaughan, Cuthbert Cyril		7 June 18	
	7 June 18	Barber, Alfred Herbert James		7 June 18	
	7 June 18	Hands, Patrick Lancelot John		7 June 18	
	7 June 18	Cheshire, Walter Keenan		7 June 18	
520	7 June 18	Gemmell, Andrew		7 June 18	
	7 June 18	Blaylock, Joseph		7 June 18	
	7 June 18	Chirgwin, Samuel Herbert		7 June 18	
	7 June 18	Green, joseph		7 June 18	
	7 June 18	Purser, Charles Randolph		7 June 18	
525	7 June 18	Tyrrell, Abbey		7 June 18	
	7 June 18	Veneer, Frederick Arthur		7 June 18	
	7 June 18	Matthews, Frederick Charles		7 June 18	
	7 June 18	Bristow, Albert Charles		7 June 18	
	7 June 18	Leonard, Denis William		7 June 18	
530	7 June 18	Arthur, William George Aylmer		7 June 18	
	7 June 18	Maxwell, Herbert		7 June 18	
	7 June 18	Goff, Herbert Hope		7 June 18	
	7 June 18	Johnston, Matthew		7 June 18	
	7 June 18	Riley, Errol Samuel		7 June 18	
535	7 June 18	Read, Edgar William		7 June 18	
	7 June 18	Duncan, Sydney Fairweather		7 June 18	
	7 June 18	Egan, Vere Gerald		7 June 18	
	7 June 18	Ireland, Bertram James		7 June 18	
	7 June 18	Lichfield, Sidney john		7 June 18	
540	7 June 18	Arthur, John Richard Henry		7 June 18	
	7 June 18	Strip, Roland Alexander		7 June 18	
	7 June 18	Smith, Frederick		7 June 18	
	7 June 18	Cotton, Cyril William		7 June 18	
	7 June 18	Abbott, Percy Douglas		7 June 18	
545	7 June 18	Crawford, Frederick Reginald Ralph		7 June 18	
	7 June 18	Middleton-Stewart, John Crawford		7 June 18	[Army Service
	8 June 18	Galvin, William Hexter		8 June 18	Released from
	8 June 18	Clarke, Frank Hawkins		8 June 18	
	10 June 18	Clay-Inman, Henry		10 June 18	
550	10 June 18	Sanderson, George		10 June 18	
	10 June 18	Dalrymple, William Charles		10 June 18	
	11 June 18	Smith, Reginald Christopher		11 June 18	
	11 June 18	Reilly, Stanislaus George		11 June 18	
	11 June 18	Hill-Murray, Henry John		11 June 18	
555	12 June 18	Cameron, Bruce Lachlan		12 June 18	
	12 June 18	Parsick, Charles Edmund		12 June 18	
	12 June 18	Clery, James Austin		12 June 18	
	13 June 18	Skinner, Francis Edward		13 June 18	
	14 June 18	Occam, Conrad Kenneth		14 June 18	
560	14 June 18	Hall, Reginald Sydenham		14 June 18	
	14 June 18	Turner, Wilfred Charles		14 June 18	
	14 June 18	Wontersz, Eugene Percival Terence		14 June 18	
	14 June 18	Morgan, Eric		14 June 18	
	15 June 18	Beath, Kenneth Cameron		15 June 18	
565	15 June 18	Beale, William Francis Bennett		15 June 18	
	15 June 18	Hatton, Sydney Osborne		15 June 18	
	15 June 18	Linberry-Kerr, George Wilfred		15 June 18	
	17 June 18	Penney, James Downing		17 June 18	
	17 June 18	Shillingford, Cyril Alexander Norman		17 June 18	
570	17 June 18	Rawlands, William Shaw		17 June 18	
	19 June 18	Kingston, Gerald Eden		19 June 18	
	19 June 18	Harrison, Arthur Sidney		19 June 18	
	19 June 18	Rogers, George Denzil		19 June 18	
	19 June 18	Hordern, Richard De Laune		19 June 18	
575	20 June 18	FitzGerald, Seymour Gonne Vesey (48*a*, 50)		20 June 18	
	20 June 18	Hullock, Eric Read		20 June 18	
	20 June 18	Cox, Theodore William		20 June 18	
	20 June 18	Burke, William George		20 June 18	
	20 June 18	Walcroft, Thomas Clarence		20 June 18	
580	21 June 18	Mercer, Alexander Warren (*Actg. Lt.-Col.*)		21 June 18	
	21 June 18	Molony, Henry James Creighton		21 June 18	
	21 June 18	Isemonger, Frederick Charles (*Actg. Lt.-Col.*)		21 June 18	
583	21 June 18	Smith, John Augustus		21 June 18	

No.	Date of First Commn.	Names.	Date of birth.	Date of Rank.	REMARKS.
		INFANTRY—*contd.*			
		2nd-Lieutenants—*contd.*			
584	21 June 18	Hartless, Leonard Charles		21 June 18	
585	21 June 18	Adams, Reginald Johnston		21 June 18	
	21 June 18	Bell, Roland Stanley		21 June 18	
	21 June 18	O'Riordan, Kevin Moran		21 June 18	
	21 June 18	Horton, Ralph Albert		21 June 18	
	21 June 18	Stronach, John Lindsay		21 June 18	
590	21 June 18	Weymouth, James Charles		21 June 18	
	21 June 18	Crawford, Andrew George		21 June 18	
	21 June 18	Elms, Cecil Elrington		21 June 18	
	21 June 18	Nott-Bower, John Reginald Hornby		21 June 18	
	21 June 18	Coulton, Gordon Francis		21 June 18	
595	21 June 18	Pilditch, Denys		21 June 18	
	21 June 18	Blunden, Alfred Charles		21 June 18	
	21 June 18	Waddell, Gilbert		21 June 18	
	21 June 18	Luck, Charles Pyne		21 June 18	
	21 June 18	Field, Geoffrey Glennie		21 June 18	
600	21 June 18	Boylette, William Richard Davie		21 June 18	
	21 June 18	Hickmott, Reginald Ernest Joseph		21 June 18	
	21 June 18	Murphy, Hugh William		21 June 18	
	22 June 18	Hetherington, Carleton		22 June 18	
	22 June 18	Leighton, Walter William Michael Ferrier		22 June 18	
605	24 June 18	Carlton, Robert Frederick Cyril Lascelles Carlton.		24 June 18	
	24 June 18	Collins, Henry William Shields		24 June 18	
	24 June 18	Mathews, James Sidney		24 June 18	
	24 June 18	McCreadie, David		24 June 18	
	24 June 18	Ross, John Donald		24 June 18	
610	24 June 18	Vale, Harry		24 June 18	
	25 June 18	Cole, Owen Wynne		25 June 18	
	25 June 18	Rylands, Joseph		25 June 18	
	25 June 18	Spooner, Archibald Franklin		25 June 18	
	26 June 18	Barker, Percy		26 June 18	
615	26 June 18	Botterill, Wilfrid Reginald		26 June 18	
	26 June 18	Hilliard, Robert Bruce		26 June 18	
	26 June 18	Jarvis, Harold Lucius		26 June 18	
	26 June 18	Lawton, Samuel Peter		26 June 18	
	26 June 18	O'Leary, Ralph Frederick		26 June 18	
620	26 June 18	Saunders, Howard Fauntleroy		26 June 18	
	26 June 18	Mackenzie, James		26 June 18	
	27 June 18	Alexander, William		27 June 18	
	27 June 18	Price, George Winstanley		27 June 18	
	27 June 18	Bernard, Peter		27 June 18	
625	27 June 18	Phillips, Phillip Frank		27 June 18	
	28 June 18	Bruce, Robert William		28 June 18	
	28 June 18	Collister, Harold James		28 June 18	
	28 June 18	Bryce, William		28 June 18	[Army Service.
	28 June 18	Jefford, Alexander		28 June 18	Released from
630	28 June 18	Lamont, James (*temp. Lt., 28 July 18*)		28 June 18	
	28 June 18	Morton, John Inglis		28 June 18	
	28 June 18	Shawe, Edward Barron		28 June 18	
	28 June 18	Rees, Thomas Owen		28 June 18	
	29 June 18	Honeywell, Benjamin		29 June 18	
635	29 June 18	Kirk, Richard Thomas Francis		29 June 18	
	29 June 18	Tudor-Owen, William Courtenay		29 June 18	
	30 June 18	Philpot, Hamlet Cunningham Vachell (*temp. Capt., 2 Oct. 18*).		30 June 18	
	30 June 18	Mitchell, Kenneth Grant		30 June 18	
	30 June 18	Rampezzotti, Ernest Alfred		30 June 18	
640	1 July 18	Smith, Sidney John		1 July 18	
	1 July 18	Berridge, Samuel Caunter (*temp. Maj., 1 July 18*)		1 July 18	
	1 July 18	Shaw, James		1 July 18	
	1 July 18	Vivian, Owen Hussey		1 July 18	[Army Service.
	1 July 18	Hey, Norman Heather		1 July 18	Released from
645	1 July 18	Walze, Charles Ernest		1 July 18	
	1 July 18	Francis, George Norman		1 July 18	
	2 July 18	Waller, Frederick Crighton		2 July 18	
	2 July 18	Bennet, Albert William		2 July 18	
	2 July 18	Fox, Joseph McDermott		2 July 18	
650	2 July 18	Giddens, Harold Leslie		2 July 18	
	2 July 18	Wilson, George Balfour		2 July 18	
	2 July 18	Boyd-Parselle, Frederick George		2 July 18	
	2 July 18	McDermott, Maurice Desmond		2 July 18	
	2 July 18	Stanford, Walter Halstead Cortis		2 July 18	
655	3 July 18	Aikin, Vivian John Doran		3 July 18	[Army Service.
	3 July 18	Hay-Jahans, Philip Frederick		3 July 18	Released from
	3 July 18	Webster, Cyril Heathcock		3 July 18	Do.
	3 July 18	Nicoll, William Atkins		3 July 18	Do.
659	4 July 18	Fermer, Edward James		4 July 18	

No.	Date of First Commn.	Names.	Date of birth.	Date of Rank.	REMARKS.
		INFANTRY—*contd.*			
		2nd-Lieutenants—*contd.*			
660	4 July 18	Coombs, William Henry		4 July 18	
	4 July 18	Ball, George Wilfred Morton		4 July 18	
	4 July 18	Burnett, Edwin Collingwood		4 July 18	
	4 July 18	Sawyer, William Henry		4 July 18	
	4 July 18	Townsend, George		4 July 18	
665	5 July 18	Turner, Charles Lockhart		5 July 18	
	5 July 18	Focken, Frederick John Charles		5 July 18	
	5 July 18	Kelly, Robert James		5 July 18	
	5 July 18	O'Gorman, Andrew Reginald		5 July 18	
	5 July 18	Shaw, Herbert Huntley (†*temp. Capt.*, *13 Sep. 18*).		5 July 18	
670	5 July 18	Peal, Bernard Ashley		5 July 18	
	6 July 18	Cox, Gerald William Sherbrock		6 July 18	
	6 July 18	Keegan, Bryan Albert		6 July 18	
	6 July 18	Pantlin, Henry Charles		6 July 18	
	6 July 18	Phipps, Albert Henry		6 July 18	
675	6 July 18	Stevens, George Chesney Nevill (*temp. Maj.*, *6 July 18*)		6 July 18	
	7 July 18	Gould, Ernest Farrer		7 July 18	
	7 July 18	Howe, Thomas Morris		7 July 18	
	8 July 18	Blood, Charles Daniel		8 July 18	
	8 July 18	Dewey, Herbert Alexander Ross		8 July 18	
680	8 July 18	Plumb, Walter Cyril		8 July 18	
	9 July 18	Bean, Thomas Lawrence Lahey		9 July 18	
	9 July 18	Friel, Ralph		9 July 18	
	9 July 18	Pym, Frank William		9 July 18	
	9 July 18	White, Garrick Hanbury		9 July 18	
685	10 July 18	Field, Horace		10 July 18	
	10 July 18	Robert, Vernon Austen		10 July 18	
	10 July 18	Bullock, John Wilfrid Hollies		10 July 18	
	10 July 18	Ward, George Snell La Cloche		10 July 18	
	10 July 18	Hegarty, Ernest		10 July 18	
690	10 July 18	Martin, George Henry		10 July 18	
	10 July 18	Allan, Thomas Lawson		10 July 18	
	10 July 18	Green, Frank		10 July 18	
	10 July 18	Haggett, Hedley George		10 July 18	
	10 July 18	Jeans, Frank		10 July 18	
695	10 July 18	Hodgson, John Latimer		10 July 18	
	10 July 18	Toon, Harold Edgar		10 July 18	
	11 July 18	Butterfield, Arnold Robin		11 July 18	
	11 July 18	Turner, Herbert Ashmore		11 July 18	
	11 July 18	Willmott, Moreton Joseph		11 July 18	
700	12 July 18	Bridgnell, Eric		12 July 18	
	12 July 18	Curtis, Allan		12 July 18	
	12 July 18	Darvoll, George Gordon		12 July 18	
	12 July 18	Hayter, Lear Conrad David		12 July 18	
	12 July 18	Minto, Archibald Henry		12 July 18	
705	12 July 18	Mitchell, Philip Thomas		12 July 18	
	12 July 18	Latham, Charles Reginald Christopher		12 July 18	
	13 July 18	Barber, John Fredrick Archer		13 July 18	
	13 July 18	Ebert, William Edward Ross Bell		13 July 18	
	13 July 18	Curtis, Colin Medway		13 July 18	
710	13 July 18	Rose, Algernon Stradley		13 July 18	
	15 July 18	Carroll, Edward William		15 July 18	
	15 July 18	Cooper, Harry Lyster		15 July 18	
	15 July 18	Crowhurst, Alfred Davidson		15 July 18	
	15 July 18	Henry, Wilson		15 July 18	
715	15 July 18	Hodson, Eric		15 July 18	
	15 July 18	McArthur, Alexander Gordon		15 July 18	
	15 July 18	Marklew, Ernest George		15 July 18	
	15 July 18	Monks, Edgar Wallace		15 July 18	
	15 July 18	Shirreff, Alexander Grierson		15 July 18	
720	15 July 18	Hilabrand-Todd, Henry Vivian		15 July 18	
	15 July 18	Van Langenberg, James Frederick		15 July 18	
	15 July 18	Vaughan, Ernest		15 July 18	
	15 July 18	Keir, Charles McHardy		15 July 18	[Army Service.
	15 July 18	Marrs, Francis William		15 July 18	Released from
725	16 July 18	Judge, Charles Henry		16 July 18	
	16 July 18	LeFeuvre, Charles Edward		16 July 18	[Army Service.
	16 July 18	Nicholson, John Patrick		16 July 18	Released from
	16 July 18	Rees, Frederick Arthur		16 July 18	
	16 July 18	Rigby, James Prescott		16 July 18	
730	16 July 18	Stubbs, Shirley Graham		16 July 18	
	17 July 18	Allen, Guy Oldfield		17 July 18	
	17 July 18	Allsop, James Joseph Whittlesea		17 July 18	
733	17 July 18	Bennison, John Jennings		17 July 18	

† With pay and allowances as for a Lieutenant.

INFANTRY—*contd.*

2nd-Lieutenants—*contd.*

No.	Date of First Commn.	Names.	Date of birth.	Date of Rank.	REMARKS.
734	17 July 18	Bowen, Hugh Henry		17 July 18	
735	17 July 18	Chalk, William George		17 July 18	
	17 July 18	Freeman, Alston Barkeley		17 July 18	
	17 July 18	Gillies, John Ernest		17 July 18	
	17 July 18	Hansen, Henrick Ericksen		17 July 18	
	17 July 18	Hunter, Charles Frederick		17 July 18	
740	17 July 18	Hunter, Douglas Campbell		17 July 18	
	17 July 18	Jenkins, Walter Allen		17 July 18	
	17 July 18	Luce, Gordon Hannington		17 July 18	
	17 July 18	Olney, James William Charles		17 July 18	
	17 July 18	Parkinson, Charles Edward		17 July 18	
745	17 July 18	Saunders, James St. Clair		17 July 18	
	17 July 18	Stewart, Thomas Alexander		17 July 18	
	17 July 18	Symes, Francis Stuart		17 July 18	
	17 July 18	Tovey, Gerald Arthur Campbell		17 July 18	
	17 July 18	Lyle, Thomas McElderry (*Actg. Capt, 17 July 18*)		17 July 18	
750	18 July 18	Davies, William Alan		18 July 18	
	18 July 18	Doherty, Jack Aloysius		18 July 18	
	18 July 18	Hogg, Gilbert Pitcairn		18 July 18	
	18 July 18	Mackay, Charles Hammond		18 July 18	
	18 July 18	Whetham, Lewis Charles		18 July 18	
755	19 July 18	Hargreaves, Charles Kerrison		19 July 18	
	19 July 18	Mortimer, Arthur Cundy		19 July 18	
	19 July 18	Shirley, Geoffrey Stanhope		19 July 18	
	19 July 18	Deeks, Thomas Bloss		19 July 18	[Army Service.
	19 July 18	Beckett, James		19 July 18	Released from
760	19 July 18	Kennedy, Hugh Kay		19 July 18	
	20 July 18	Brockbanke, Arthur Cecil		20 July 18	
	20 July 18	Seaton, William Alexander		20 July 18	
	15 Mar. 18	Ross-Counsell, Lionel		15 Mar. 18 §21 July 18	
	22 July 18	White, Leslie Stuart		22 July 18	
765	22 July 18	Kitching, Bertram		22 July 18	
	22 July 18	Petersen, Richard George		22 July 18	
	22 July 18	Walker, Albert		22 July 18	
	22 July 18	King, Alfred		22 July 18	
	22 July 18	Hay, James Arthur Coulson		22 July 18	
770	22 July 18	Kellett, Walter Rennie		22 July 18	
	22 July 18	Strange, Frank		22 July 18	
	22 July 18	Osborne, Stanley		22 July 18	
	22 July 18	Martin, James Elliot Leslie		22 July 18	
	22 July 18	Craig, Samuel Ferguson		22 July 18	
775	22 July 18	Thompson, Percy		22 July 18	
	22 July 18	Brewer, James Richard		22 July 18	
	22 July 18	Jodrell, Francis Joseph		22 July 18	
	22 July 18	Lavin, Jack Finlayson		22 July 18	
	22 July 18	McCracken, George		22 July 18	
780	22 July 18	Milne, Cecil Allan		22 July 18	
	22 July 18	Barnes, Thomas		22 July 18	
	22 July 18	Dow, John Newton		22 July 18	
	22 July 18	Irving, Sydney Edward		22 July 18	
	22 July 18	Henderson, John Wing		22 July 18	
785	22 July 18	Thomas, Frank Griffith		22 July 18	
	22 July 18	Baker, Reginald Horace		22 July 18	
	22 July 18	Case, Cecil Ernest		22 July 18	
	22 July 18	Eastland, William Henry		22 July 18	
	22 July 18	Eaton, Reginald Warren		22 July 18	
790	22 July 18	Hozier, Francis Rupert		22 July 18	
	22 July 18	Hamblen, George		22 July 18	
	22 July 18	Cooper, Archibald		22 July 18	
	22 July 18	Bellamy, Frederick Henry Howard		22 July 18	
	22 July 18	Peck, Herbert Edwin		22 July 18	
795	22 July 18	Lewis, Richard Morgan		22 July 18	
	22 July 18	Brown, Herbert		22 July 18	
	22 July 18	Browne, James Frederick		22 July 18	
	22 July 18	Waldron, Charles Bernard		22 July 18	
	22 July 18	Greig, Adam Ritchie		22 July 18	
800	22 July 18	Willis, Percival George		22 July 18	
	22 July 18	Hotz, Roland		22 July 18	
	22 July 18	Over, John		22 July 18	
	22 July 18	Wilby, Vivian Gladstone		22 July 18	
	22 July 18	Wilkinson, Terence William		22 July 18	
805	22 July 18	Yuill, Alexander Ferguson		22 July 18	
806	22 July 18	Green, Samuel Arnold Collier		22 July 18	

§ To take rank and precedence from 21 July 18.
With pay and allowances as for a Lieutenant.

No.	Date of First Commn.	Names.	Date of birth.	Date of Rank.	REMARKS.
		INFANTRY—*contd.*			
		2nd-Lieutenants—*contd.*			
807	22 July 18	Shaw, George Reid		22 July 18	
	22 July 18	Shattock, Earnest Keith		22 July 18	
	23 July 18	Ashe, Clare Harold		23 July 18	
810	23 July 18	Neal, John Hales		23 July 18	
	23 July 18	Vokes, Frank Leonard		23 July 18	
	23 July 18	Hodding, Frederick		23 July 18	
	23 July 18	Gasper, Basil Counsell		23 July 18	
	23 July 18	Gurnett, Willie Vaux		23 July 18	
815	23 July 18	Holland, George Mervyn		23 July 18	
	23 July 18	Levett, Percy		23 July 18	
	23 July 18	Paramor, Cedric Herbert		23 July 18	
	23 July 18	Radford, Edward Alan		23 July 18	
	23 July 18	Yelverton, Dudley		23 July 18	
820	24 July 18	Crapper, Norman		24 July 18	
	24 July 18	Emeny, Leslie Henry		24 July 18	
	24 July 18	Warner, Harold Rudyard (*Actg. Lt.-Col., 24 July 18*)		24 July 18	
	24 July 18	Atkinson, Charles Vincent McCarthy		24 July 18	[Army Service. Released from
	25 July 18	Fraser, William Gordon		25 July 18	
825	25 July 18	Gilbanks, Cecil Bernard		25 July 18	
	25 July 18	Howden, George Henry		25 July 18	
	25 July 18	Macpherson, Alasdair Iain		25 July 18	
	25 July 18	Ryves, Cyril Harry		25 July 18	
	25 July 18	Slater, Arthur Erskine		25 July 18	
830	25 July 18	St. John, Philip George		25 July 18	
	26 July 18	Irvine, Archibald		26 July 18	
	26 July 18	Rundlett, Raymond Curtiss		26 July 18	
	26 July 18	Wootton, William John George		26 July 18	
	27 July 18	Johnston, Wilfrid Harold		27 July 18	
835	27 July 18	Bowen, Dennis Herbert		27 July 18	
	27 July 18	Corcoran, Bernard Joseph		27 July 18	
	27 July 18	Dee, Lionel James		27 July 18	
	27 July 18	Gill, Cyril Alfred Goddard		27 July 18	
	27 July 18	Gurney, Angus Hugh		27 July 18	
840	27 July 18	Illman, Sydney George		27 July 18	
	27 July 18	Jefford, Percival Eugene		27 July 18	
	27 July 18	Korb, Percy Alfred		27 July 18	
	27 July 18	McMahon, Francis Alban		27 July 18	
	27 July 18	Mathie, William Alfred Samuel		27 July 18	
845	27 July 18	Mecredy, Wills Hill		27 July 18	
	27 July 18	Nelson, Stephen		27 July 18	
	27 July 18	Sesres, Ernest Dominic		27 July 18	
	27 July 18	Williams, Thomas Taliesin		27 July 18	
	27 July 18	Bone, William Richard		27 July 18	
850	29 July 18	Kay-Mouat, Charles George		29 July 18	
	29 July 18	McQuade, Joseph Louis		29 July 18	
	29 July 18	Curran, Patrick Joseph Emmet		29 July 18	
	29 July 18	Caffyn, Sidney Charles		29 July 18	
	30 July 18	Patterson, David Clarke (†*temp. Maj., 31 Aug. 18*).		30 July 18	
855	30 July 18	Alderson, Fred Houliston		30 July 18	
	30 July 18	Baker, Robert Henry		30 July 18	
	30 July 18	Craven-Thomson, Charles Augustus		30 July 18	
	30 July 18	Demorgan, Arthur Claude		30 July 18	
	30 July 18	Duncan, Richard Adolphus		30 July 18	
860	30 July 18	Fleming, James Hamilton		30 July 18	
	30 July 18	Foot, Archibald Ballantine		30 July 18	
	30 July 18	Harding, Leslie O'Brien		30 July 18	
	30 July 18	Heughan, John Knox		30 July 18	
	30 July 18	Johnson, John Nesbitt Gordon		30 July 18	
865	30 July 18	McHaffie, James Perrie		30 July 18	
	30 July 18	Nelson, John Inglis		30 July 18	
	30 July 18	Pusey, Albert George		30 July 18	
	30 July 18	Rawlings, Donald Arthur Russell		30 July 18	
	30 July 18	Skempton, Arthur William		30 July 18	
870	30 July 18	West, Michael Philip		30 July 18	
	31 July 18	Baldwin, Eric Samuel		31 July 18	
	31 July 18	Claridge, Edward Grey Woodhouse		31 July 18	
	31 July 18	Eden, Albert Edmund		31 July 18	
	31 July 18	Forshaw, Harry William		31 July 18	
875	31 July 18	Harris, Urban Arthur		31 July 18	
	31 July 18	Havelock, Lionel Acton		31 July 18	
	31 July 18	Hill, Harold Henry		31 July 18	
	31 July 18	Ross, Herbert Stewart		31 July 18	
	31 July 18	Surrey, Herbert Rutherford		31 July 18	
880	31 July 18	Goffin, Raymond Cullis		31 July 18	
881	1 Aug. 18	Branch, William George		1 Aug. 18	

† With pay and allowances as for a Lieutenant.

No.	Date of First Commn.	Names.	Date of birth.	Date of Rank.	REMARKS.
		INFANTRY—*contd.*			
		2nd Lieutenants—*contd.*			
882	1 Aug. 18	Lynch, Joseph Victor		1 Aug. 18	
	1 Aug. 18	Mahony, Bertram Cecil		1 Aug. 18	
	1 Aug. 18	Brown, William Scott		1 Aug. 18	
885	1 Aug. 18	Curtis, Kenneth Samuel		1 Aug. 18	
	1 Aug. 18	Dickinson, Richard Stanley		1 Aug. 18	
	1 Aug. 18	Hargrave, Leonard Oswald		1 Aug. 18	
	1 Aug. 18	Lynsdale, Archibald Herbert		1 Aug. 18	
	1 Aug. 18	Madeley, William Yorke		1 Aug. 18	
890	1 Aug. 18	Parkinson, Ernest		1 Aug. 18	
	1 Aug. 18	Purcell-Gildea, John		1 Aug. 18	
	1 Aug. 18	Robertson, David		1 Aug. 18	
	1 Aug. 18	Austin, Thomas		1 Aug. 18	
	1 Aug. 18	Child, Harold William		1 Aug. 18	
895	1 Aug. 18	Clarke, Ernest Wilfrid		1 Aug. 18	
	1 Aug. 18	Jackson, Ralph		1 Aug. 18	
	1 Aug. 18	Meikle, Douglas Mathew		1 Aug. 18	
	1 Aug. 18	Mitchell, Roland Harry		1 Aug. 18	
	1 Aug. 18	Nichol, Donald Stewart Evans		1 Aug. 18	
900	1 Aug. 18	Perrin, Lancelot Charles		1 Aug. 18	
	1 Aug. 18	Smith, Howard		1 Aug. 18	
	1 Aug. 18	Warmington, James		1 Aug. 18	
	2 Aug. 18	Anding, James Barlow		2 Aug. 18	
	2 Aug. 18	Chitts, Mervyn Hugh Gordon		2 Aug. 18	
905	2 Aug. 18	Hodges, Walter Harold		2 Aug. 18	
	2 Aug. 18	Mackenzie, Kenneth Seaforth		2 Aug. 18	
	2 Aug. 18	Moreton, Tom		2 Aug. 18	[Army Service.
	2 Aug. 18	Saywell, Ronald		2 Aug. 18	Released from
	2 Aug. 18	Wakefield, Arthur Gibbon		2 Aug. 18	
910	2 Aug. 18	Tellery, Jack Anton		2 Aug. 18	
	3 Aug. 18	Berg, Clement Leonard		3 Aug. 18	
	3 Aug. 18	Grose, John Stanley		3 Aug. 18	
	3 Aug. 18	Tomkyns, Richard Alexander George		3 Aug. 18	
	4 Aug. 18	Willis, Herbert Arthur		4 Aug. 18	
915	5 Aug. 18	Clifford, Gerald Aynsley		5 Aug. 18	
	5 Aug. 18	Blaquiere, Norman Arthur		5 Aug. 18	
	5 Aug. 18	De la Harpe, Shirley St. John		5 Aug. 18	
	5 Aug. 18	Hill, William Agnew		5 Aug. 18	
	5 Aug. 18	Lewis, Alred George Llewellyn		5 Aug. 18	
920	5 Aug. 18	McLean, Thomas Dalgleish		5 Aug. 18	
	5 Aug. 18	Oliver, Frederick James		5 Aug. 18	
	5 Aug. 18	Farley, Albert Thomas		5 Aug. 18	
	6 Aug. 18	Willoughby-Harris, Arthur John		6 Aug. 18	
	7 Aug. 18	Gotch, Charles William		7 Aug. 18	
925	7 Aug. 18	Walden, George Percival		7 Aug. 18	
	7 Aug. 18	Barnett, Douglas Alfred		7 Aug. 18	
	7 Aug. 18	Fairlie, William Speirs		7 Aug. 18	
	7 Aug. 18	McKenzie, Robert Ernest		7 Aug. 18	
	7 Aug. 18	Philipsz, Bertram Stanislaus		7 Aug. 18	
930	7 Aug. 18	Vlieland, Cornelis Pieter		7 Aug. 18	
	7 Aug. 18	Roberts, Ernest Cecil		7 Aug. 18	
	7 Aug. 18	Pryde, James († *temp. Capt.*, *22 Oct. 18*)		7 Aug. 18	
	8 Aug. 18	Mecredy, George Wills		8 Aug. 18	
	8 Aug. 18	Ross, Donald James		8 Aug. 18	
935	8 Aug. 18	Devlin, Dennis George		8 Aug. 18	
	8 Aug. 18	Gilbert, Arthur Steuart Ahluwalia Stronge		8 Aug. 18	
	8 Aug. 18	Grant, Patrick William		8 Aug. 18	
	8 Aug. 18	Hart, Walter Cecil		8 Aug. 18	
	8 Aug. 18	Jahans, Frederick Rothschild		8 Aug. 18	
940	8 Aug. 18	Johnston, James Allison		8 Aug. 18	
	8 Aug. 18	Kiernander, Charles Dixie		8 Aug. 18	
	8 Aug. 18	Pentham, James		8 Aug. 18	
	8 Aug. 18	Philip, Charles Lyall		8 Aug. 18	
	8 Aug. 18	Stow, Vincent Aubery Stewart		8 Aug. 18	
945	8 Aug. 18	Sutherland, Stevens, Robert Glenton		8 Aug. 18	
	8 Aug. 18	Upshon, Llewellyn Hector		8 Aug. 18	
	8 Aug. 18	Browne, Thomas Garnet Henry (*temp. Capt.*, *8 Aug. 18*).		8 Aug. 18	
	9 Aug. 18	Allen, Alexander Lucas		9 Aug. 18	
	9 Aug. 18	Anstiss, Leslie Alfred		9 Aug. 18	
950	9 Aug. 18	Hamilton, Alan Woodforde Ball		9 Aug. 18	
	9 Aug. 18	Havelock, Robert Acton Stransham		9 Aug. 18	
	9 Aug. 18	Lowry, John		9 Aug. 18	
	9 Aug. 18	O'Connor, Arthur Brooking		9 Aug. 18	
	9 Aug. 18	Pymm, John Melville		9 Aug. 18	
955	9 Aug. 18	Chartres, Bertie Jeremiah		9 Aug. 18	
	10 Aug. 18	Gammon, Frederick George		10 Aug. 18	
957	10 Aug. 18	Tulloch, Donald Arthur		10 Aug. 18	

† With pay and allowances as for a Lieutenant.

INFANTRY—*contd.*

2nd-Lieutenants—*contd.*

No.	Date of First Commn.	Names.	Date of birth.	Date of Rank.	Remarks.
958	10 Aug. 18	Beveridge, Edmund Walter St. Clair (*temp. Capt., 10 Aug. 18*).		10 Aug. 18	
	10 Aug. 18	Chits, Terence Aubrey Oswald		10 Aug. 18	
960	12 Aug. 18	Miller, Bernard Leo St. Clair		12 Aug. 18	
	12 Aug. 18	Simpson, Gerald Hugh		12 Aug. 18	
	12 Aug. 18	Woods, Arthur		12 Aug. 18	
	12 Aug. 18	Buchanan, John Patrick		12 Aug. 18	
	12 Aug. 18	Forbes, Keppel Duncan Morrison		12 Aug. 18	
965	12 Aug. 18	McEwen, Henry Caldow		12 Aug. 18	
	12 Aug. 18	Waugh, James Nicol		12 Aug. 18	
	12 Aug. 18	Wright, Frederic Noel		12 Aug. 18	
	13 Aug. 18	Clarke, Norman St. John Sydenham		13 Aug. 18	
	13 Aug. 18	Moss, Benjamin Hugh		13 Aug. 18	
970	13 Aug. 18	Palmer, Walter Attwell Lynch		13 Aug. 18	
	13 Aug. 18	Rees, Charles Henry		13 Aug. 18	
	13 Aug. 18	Vaughan, Ernest Alfred		13 Aug. 18	
	13 Aug. 18	Ferguson, Robert Huntley		13 Aug. 18	
	13 Aug. 18	Gillon, George Kingsley		13 Aug. 18	
975	13 Aug. 18	Kalkhoven, Frederic Gordon		13 Aug. 18	
	13 Aug. 18	Mitchell, Christopher Robert Stanley		13 Aug. 18	
	14 Aug. 18	Champion, Oscar Dennis		14 Aug. 18	
	14 Aug. 18	Powell, George Augustine		14 Aug. 18	
	14 Aug. 18	Frost, George Henry (*tempy. Capt., 14 Aug. 18*)		14 Aug. 18	
980	14 Aug. 18	Blenkinsop, Bernard Arthur		14 Aug. 18	
	14 Aug. 18	Codd, Herbert Charles		14 Aug. 18	
	14 Aug. 18	Dillon, Henry Vere Shirley		14 Aug. 18	
	14 Aug. 18	Douglas, Valentine Edmund		14 Aug. 18	
	14 Aug. 18	Duncan, Charles Auld		14 Aug. 18	
985	14 Aug. 18	Heron, Edmund Franklin		14 Aug. 18	
	14 Aug. 18	Leembruggen, Bryerley Claude Fretz		14 Aug. 18	
	14 Aug. 18	Pentony, Llewellyn Richard		14 Aug. 18	
	14 Aug. 18	Pugh, Harold Valentine		14 Aug. 18	
	14 Aug. 18	Wright, Frederick Alweyne		14 Aug. 18	
990	14 Aug. 18	Wright-Neville, Vivian Russell		14 Aug. 18	
	14 Aug. 18	Cameron, Ewen Melville		14 Aug. 18	
	14 Aug. 18	Wilkinson, William John		14 Aug. 18	
	15 Aug. 18	Brown, John Alexander Henstridge		15 Aug. 18	
	15 Aug. 18	Baldrey, Walter George		15 Aug. 18	
995	15 Aug. 18	Eveling, Walter Robert		15 Aug. 18	[Army Service.
	15 Aug. 18	Gill, Archibald Whitlock Richard		15 Aug. 18	Released from
	15 Aug. 18	Hickie, John Michael		15 Aug. 18	
	15 Aug. 18	Hogshaw, Bertie William		15 Aug. 18	
	15 Aug. 18	MacKenzie, Walter Raymond		15 Aug. 18	
1000	15 Aug. 18	Pittar, Alan Herbert		15 Aug. 18	
	15 Aug. 18	Wardley, Arthur Norton		15 Aug. 18	
	15 Aug. 18	Wilson, James Lane Donald		15 Aug. 18	
	15 Aug. 18	Cameron, Donald Aytoun		15 Aug. 18	[Army Service.
	15 Aug. 18	Perfect, Stanislaus Randolph		15 Aug. 18	Released from
1005	16 Aug. 18	Bond, Francis Selwyn		16 Aug. 18	
	16 Aug. 18	Alexander, William Lucas		16 Aug. 18	
	16 Aug. 18	Browne, Thomas Lyster		16 Aug. 18	
	16 Aug. 18	Davidson, Alfred James		16 Aug. 18	
	16 Aug. 18	Ewing, James Alexander		16 Aug. 18	
1010	16 Aug. 18	Grant, Charles Taylor		10 Aug. 18	
	16 Aug. 18	Gray, Walter Dodgson		16 Aug. 18	
	16 Aug. 18	Hare-Duke, Arthur Robert Aubrey		16 Aug. 18	
	16 Aug. 18	Leach, Alfred Henry Lionel		16 Aug. 18	
	16 Aug. 18	Leggat, James Wilson		16 Aug. 18	
1015	16 Aug. 18	Mackenzie, Charles Adam		16 Aug. 18	
	16 Aug. 18	Morris, Albert Eric Camroux		16 Aug. 18	
	16 Aug. 18	Pearson, Lionel Maitland		16 Aug. 18	
	16 Aug. 18	Roberts, David Edward		16 Aug. 18	
	16 Aug. 18	Scorey, John Burnaby		16 Aug. 18	
1020	16 Aug. 18	Stork, Cornelis Hendrik		16 Aug. 18	
	16 Aug. 18	Jepson, Leonard William		16 Aug. 18	
	16 Aug. 18	Whitaker, Frederick Norman		16 Aug. 18	
	16 Aug. 18	Lincoln, Edward Henning (*temp. Lt., 16 Aug. 18*).		16 Aug. 18	
	17 Aug. 18	Eldridge, George Thomas		17 Aug. 18	
1025	17 Aug. 18	Ward, Thomas Edward		17 Aug. 18	
	19 Aug. 18	Bennett, William James		19 Aug. 18	
	19 Aug. 18	Higson, William		19 Aug. 18	
	19 Aug. 18	Humphrys, Elwin Frank Adams		19 Aug. 18	
	19 Aug. 18	Le Frank, Guy Antoine		19 Aug. 18	
1030	19 Aug. 18	Morris, Alfred Rosslyn		19 Aug. 18	

No.	Date of First commn.	Names.	Date of birth.	Date of Rank.	REMARKS.

INFANTRY—*contd.*

2nd-Lieutenants—*contd.*

No.	Date of First commn.	Names.	Date of birth.	Date of Rank.	REMARKS.
1031	19 Aug. 18	Wardle, Walter Frederick		19 Aug. 18	
	19 Aug. 18	Price, Colin Anthony		19 Aug. 18	
	19 Aug. 18	Raven, Thomas Frederick (*temp. Capt., 19 Aug. 18*).		19 Aug. 18	
	20 Aug. 18	Alder, William George		20 Aug. 18	
1035	20 Aug. 18	Belton, Peter		20 Aug. 18	
	20 Aug. 18	Hepburn, William Smith		20 Aug. 18	
	20 Aug. 18	Howison, George		20 Aug. 18	
	20 Aug. 18	Major, Charles Richard		20 Aug. 18	
	20 Aug. 18	Somerville, Aubrey Percival		20 Aug. 18	
1040	20 Aug. 18	Stuart, John		20 Aug. 18	
	20 Aug. 18	White, Hamilton Robey		20 Aug. 18	
	20 Aug. 18	Williamson, Ulric Gladwin		20 Aug. 18	
	21 Aug. 18	Allen, Edward Charles		21 Aug. 18	
	21 Aug. 18	Chapman, John Joseph		21 Aug. 18	
1045	21 Aug. 18	Darwin, John Henry		21 Aug. 18	
	21 Aug. 18	Ennis, Charles		21 Aug. 18	
	21 Aug. 18	Harris, Basil Blenman		21 Aug. 18	
	21 Aug. 18	Naughton, Fred Alexander		21 Aug. 18	
	21 Aug. 18	O'Connor, Arthur Timothy		21 Aug. 18	
1050	21 Aug. 18	Rieu, Emile Victor		21 Aug. 18	
	21 Aug. 18	Travers-Drapes, Ernest Higginson Ardagh		21 Aug. 18	
	21 Aug. 18	Woodhouse-Adolphus, Wilfrid Harold		21 Aug. 18	
	21 Aug. 18	Bishop-Smith, James		21 Aug. 18	
	22 Aug. 18	Hunsworth, Ernest		22 Aug. 18	
1055	22 Aug. 18	Coleman, Percy William		22 Aug. 18	
	22 Aug. 18	Frew, Jeffrey Park		22 Aug. 18	
	22 Aug. 18	Hall, Francis William		22 Aug. 18	
	22 Aug. 18	Hereford, Reginald James		22 Aug. 18	
	22 Aug. 18	King, William Bernard		22 Aug. 18	
1060	22 Aug. 18	Pollacco, Rene Gaston		22 Aug. 18	
	22 Aug. 18	Symns, John Montfort (*temp. Maj., 28 Oct. 18*)		22 Aug. 18	
	22 Aug. 18	Toomey, Arthur Thomas		22 Aug. 18	
	23 Aug. 18	Davis, Gerald Alfred		23 Aug. 18	
	23 Aug. 18	Helme, Charles George Gordon		23 Aug. 18	
1065	23 Aug. 18	Harborow, James Edward William Percy Ashton.		23 Aug. 18	
	23 Aug. 18	McErvel, Duncan Stephen		23 Aug. 18	
	23 Aug. 18	Morley, Edward Austin		23 Aug. 18	
	23 Aug. 18	Stewart, William		23 Aug. 18	
	23 Aug. 18	Robey, Clarence Algernon		23 Aug. 18	
1070	23 Aug. 18	Flashman, Thomas Charles (*temp. Capt., 23 Aug. 18*).		23 Aug. 18	
	24 Aug. 18	Fowler, Cecil William		24 Aug. 18	
	24 Aug. 18	Knowles, Leonard Thomas		24 Aug. 18	
	24 Aug. 18	Lee, Joseph Standish Searchfield		24 Aug. 18	
	24 Aug. 18	Saubolle, Cyril Edward		24 Aug. 18	
1075	25 Aug. 18	Liddell, William Temple		25 Aug. 18	
	26 Aug. 18	Barker, Thomas Garner		26 Aug. 18	
	26 Aug. 18	Collyer, James		26 Aug. 18	
	26 Aug. 18	Harvey, Charles William		26 Aug. 18	
	26 Aug. 18	Thompson, Gilbert Vernon Leslie		26 Aug. 18	
1080	26 Aug. 18	Wood-Scawen, Philip Neri		26 Aug. 18	
	26 Aug. 18	Carn, Alec Thomas, M.M.		26 Aug. 18	
	26 Aug. 18	McDonald, Colin Keith		26 Aug. 18	
	26 Aug. 18	Thomas, Allan Hugh		26 Aug. 18	
	26 Aug. 18	Thomas, Donald Victor		26 Aug. 18	
1085	26 Aug. 18	Townsend, William Albert		26 Aug. 18	
	27 Aug. 18	Blyth, Ernest William		27 Aug. 18	
	27 Aug. 18	Audsley, Charles Edward		27 Aug. 18	
	27 Aug. 18	DeSilva, Edmund Harold		27 Aug. 18	
	27 Aug. 18	Bain, Leslie Terrence		27 Aug. 18	
1090	27 Aug. 18	Beale, Edwin James		27 Aug. 18	
	27 Aug. 18	Binge, Felix George Henry		27 Aug. 18	
	27 Aug. 18	Brodie, Robert		27 Aug. 18	
	27 Aug. 18	Elliot, Samuel Louis Howell		27 Aug. 18	
	27 Aug. 18	Hinton, Frederick Ernest		27 Aug. 18	
1095	27 Aug. 18	Hubbard, Douglas Stanhope		27 Aug. 18	
	27 Aug. 18	Joyce, Clive Eardley		27 Aug. 18	
	27 Aug. 18	Keeling, Frederick George		27 Aug. 18	
	27 Aug. 18	Lampard, Leonard Ashmall		27 Aug. 18	
	27 Aug. 18	Miller, Arthur Congreve		27 Aug. 18	
1100	27 Aug. 18	Morton, Edgar Hector		27 Aug. 18	
	27 Aug. 18	Perry, Henry Augustus Palmer		27 Aug. 18	
	27 Aug. 18	Rose, George Charles Griffiths Forbes		27 Aug. 18	
	27 Aug. 18	Tait, Arthur		27 Aug. 18	
1104	28 Aug. 18	Bickers, Sydney Merypn		28 Aug. 18	

No.	Date of First Commn.	Names.	Date of birth.	Date of Rank.	REMARKS.
		INFANTRY—*contd.*			
		2nd-Lieutenants—*contd.*			
1105	28 Aug. 18	Cowper, George Godwin St. John		28 Aug. 18	
	28 Aug. 18	Harper, George Milne		28 Aug. 18	
	28 Aug. 18	MacBey, Stephen Gordon		28 Aug. 18	
	28 Aug. 18	Mallis, James Ernest		28 Aug. 18	
	28 Aug. 18	Seaman, Charles Kenningale		28 Aug. 18	
1110	28 Aug. 18	Tostevin, Fred Percy		28 Aug. 18	
	28 Aug. 18	Watson, John James		28 Aug. 18	
	28 Aug. 18	Davies, William Barron		28 Aug. 18	
	29 Aug. 18	Cloy, Frank Alexander		29 Aug. 18	
	29 Aug. 18	Cox, Aubrey Paul		29 Aug. 18	
1115	29 Aug. 18	Edwards, Frank		29 Aug. 18	
	29 Aug. 18	Kewley, John		29 Aug. 18	
	29 Aug. 18	Kingsley, Charles Herbert		29 Aug. 18	
	29 Aug. 18	Nichols, James Arthur		29 Aug. 18	
	29 Aug. 18	Simpson, Alexander Allan		29 Aug. 18	
1120	29 Aug. 18	Smith, William Henry		29 Aug. 18	
	30 Aug. 18	Force-Jones, Rowland		30 Aug. 18	
	30 Aug. 18	Lloyd, Robert Howard		30 Aug. 18	
	30 Aug. 18	Atkinson, Albert Allan		30 Aug. 18	
	30 Aug. 18	Bertie, Thomas Paterson		30 Aug. 18	
1125	30 Aug. 18	Blewitt, Albert Ernest		30 Aug. 18	
	30 Aug. 18	Chorley, Richard Cleasby		30 Aug. 18	
	30 Aug. 18	Curtis, Patrick Henry		30 Aug. 18	
	30 Aug. 18	Danter, Arthur William		30 Aug. 18	
	30 Aug. 18	Gidley, Douglas John		30 Aug. 18	
1130	30 Aug. 18	Graham, Daniel		30 Aug. 18	
	30 Aug. 18	Maher, Frederick Michael		30 Aug. 18	
	30 Aug. 18	Mills, Seymour Herbert Hatten		30 Aug. 18	
	30 Aug. 18	Patton, Sydney Leonard Scott		30 Aug. 18	
	30 Aug. 18	Rebertson, Harold Edward		30 Aug. 18	
1135	30 Aug. 18	Roy, Charles Barclay		30 Aug. 18	
	30 Aug. 18	St. Pourcain, Eric Albert		30 Aug. 18	
	30 Aug. 18	Taylor, Ralph Norman		30 Aug. 18	
	30 Aug. 18	Teeling, Arthur Robert Bayly		30 Aug. 18	
	30 Aug. 18	Wood, William Leslie		30 Aug. 18	
1140	30 Aug. 18	Henderson, Donald Raoul		30 Aug. 18	
	30 Aug. 18	Mackay, Ronald David		30 Aug. 18	
	30 Aug. 18	King, Frank Harvey		30 Aug. 18	
	30 Aug. 18	Allen, Arthur James Leverstone		30 Aug. 18	
	31 Aug. 18	Tribble, Norman Vautin		31 Aug. 18	
1145	31 Aug. 18	Beale, George McLachlan		31 Aug. 18	
	31 Aug. 18	Borebank, Frederick William		31 Aug. 18	
	31 Aug. 18	Carreras, Lester Alfred		31 Aug. 18	
	31 Aug. 18	Collins, John Laville		31 Aug. 18	
	31 Aug. 18	Fay, Archibald Thomas		31 Aug. 18	
1150	31 Aug. 18	George, Arthur Gregory		31 Aug. 18	
	31 Aug 18	O'Brien, William Gilbert		31 Aug. 18	
	31 Aug. 18	Wilkins, George Alfred William		31 Aug. 18	
	31 Aug. 18	Master, Alfred		31 Aug. 18	
	2 Sep. 18	Wilkins, William Samuel		2 Sep. 18	
1155	2 Sep. 18	Baird, Thomas Walter		2 Sep. 18	
	2 Sep. 18	Bunch, John Benson Shorthouse		2 Sep. 18	
	2 Sep. 18	Burns, George Thomas Abercromby		2 Sep. 18	
	2 Sep. 18	Cheadle, George		2 Sep. 18	
	2 Sep. 18	Fleming, George Gillan		2 Sep. 18	
1160	2 Sep. 18	Inglis, James Paul		2 Sep. 18	
	2 Sep. 18	Mackinnon, Alexander		2 Sep. 18	
	2 Sep. 18	Rundlett, Leslie St. Clare		2 Sep. 18	
	2 Sep. 18	Hamilton, Arthur Harry Claude (*temp. Capt., 2 Sep. 18*).		2 Sep. 18	
	3 Sep. 18	Butchart, Alexander Davidson		3 Sep. 18	
1165	3 Sep. 18	Collinson, Charles Percy		3 Sep. 18	
	3 Sep. 18	Cullen, Leonard		3 Sep. 18	
	3 Sep. 18	Earle, Cyril Oswald		3 Sep. 18	
	3 Sep. 18	Gardener, George Oscar Thompson		3 Sep. 18	
	3 Sep. 18	Harrison, Allen Arthur		3 Sep. 18	
1170	3 Sep. 18	Mackey, Dermot Aubrey		3 Sep. 18	
	3 Sep. 18	Moffat, John		3 Sep. 18	
	3 Sep. 18	O'Connor, Terence Gordon		3 Sep. 18	
	3 Sep. 18	Parry, Bernard King		3 Sep. 18	
	3 Sep. 18	Pescod, Robert Scott		3 Sep. 18	
1175	3 Sep. 18	Warren, Ernest Harry		3 Sep. 18	
	3 Sep. 18	James, Harold Reginald		3 Sep. 18	
	4 Sep. 18	Sugden, Harold		4 Sep. 18	
	4 Sep. 18	Sawtel, William George		4 Sep. 18	
	4 Sep. 18	Vaughan, John		4 Sep. 18	
1180	4 Sep. 18	Battersby, Richard Shalcross		4 Sep. 18	

INFANTRY—*contd.*
2nd-Lieutenants—*contd.*

No.	Date of First Commn.	Names.	Date of birth.	Date of Rank.	Remarks.
1181	4 Sep. 18	Cleary, George William		4 Sep. 18	
	4 Sep. 18	Goutiers, George Francis Clement de Marquetière.		4 Sep. 18	
	4 Sep. 18	Pim, Gerald Edward		4 Sep. 18	
	5 Sep. 18	Cotton, Stanley Charles Neville		5 Sep. 18	
1185	5 Sep. 18	Dupree, Harry		5 Sep. 18	
	5 Sep. 18	Falloon, Percy Howard		5 Sep. 18	[Army Service.
	5 Sep. 18	Harper, Kenneth Gordon		5 Sep. 18	Released from
	5 Sep. 18	Keegan, Robert James		5 Sep. 18	
	5 Sep. 18	Langley, Eric Erskine		5 Sep. 18	
1190	5 Sep. 18	Sinclair, William (*temp. Capt.*, *26 Sep. 18*)		5 Sep. 18	
	5 Sep. 18	Sinclair-Day, Albert Richard		5 Sep. 18	
	5 Sep. 18	Swales, Fredrick William		5 Sep. 18	
	5 Sep. 18	Lawrence, Albert Gordon		5 Sep. 18	
	5 Sep. 18	Minshull, William Travers		5 Sep. 18	
1195	5 Sep. 18	Anstey, John Charles Knight		5 Sep. 18	
	6 Sep. 18	Andrews, Dennis Carew		6 Sep. 18	
	6 Sep. 18	Clair-Smith, Eric Inman		6 Sep. 15	
	6 Sep. 18	Corbett, Eugene Christopher		6 Sep. 18	
	6 Sep. 18	Fleming, Michael Joseph		6 Sep. 18	
1200	6 Sep. 18	Grieve, Cyril Charles Morton		6 Sep. 18	
	6 Sep. 18	Guthrie, David Topping Goodwin		6 Sep. 18	
	6 Sep. 18	Hill, Claude Foster		6 Sep. 18	
	6 Sep. 18	Kernick, John Eddy		6 Sep. 18	
	6 Sep. 18	Morgan, Walter Duncan		6 Sep. 18	
1205	6 Sep. 18	Petch, Philip Bernard		6 Sep 18	
	6 Sep. 18	Smith, Frank Gwyther		6 Sep. 18	
	6 Sep. 18	Trimmings, Arthur Ernest		6 Sep. 18	
	6 Sep. 18	Harper, James Percival		6 Sep. 18	
	6 Sep. 18	Hatchman, Gerald Hugh Stinton		6 Sep. 18	
1210	7 Sep. 18	Bickers, Austin Algernon Daniel		7 Sep. 18	
	7 Sep. 18	Graveston, George James Blakeman		7 Sep. 18	
	7 Sep. 18	Otto, Edward Clarence		7 Sep. 18	
	7 Sep. 18	Gratton, Cyril Keith		7 Sep. 18	
	7 Sep. 18	James, William Henry		7 Sep. 18	
1215	7 Sep. 18	Laws, Wilfred		7 Sep. 18	
	9 Sep. 18	Emmett, Cecil Henry		9 Sep. 18	
	9 Sep. 18	Hartle, Julian Walter Leslie		9 Sep. 18	
	9 Sep. 18	McLean, William Adam Gardner		9 Sep. 18	
	10 Sep. 18	Ardagh, Edmund Vernon		10 Sep. 18	
1220	10 Sep. 18	Lecky-Thompson, William		10 Sep. 18	
	10 Sep. 18	Roy, Malcolm Hughes Trevor		10 Sep. 18	
	10 Sep. 18	Summers, Ernest Warren		10 Sep. 18	
	10 Sep. 18	Munn, Albert Ernest		10 Sep. 18	
	10 Sep. 18	Brander, Andrew Archibald		10 Sep. 18	
1225	10 Sep. 18	Bruce, Edward		10 Sep. 18	
	10 Sep. 18	Evans, John William Percival		10 Sep. 18	
	10 Sep. 18	Farrow, George Ulfrid		10 Sep. 18	
	10 Sep. 18	Ireland, Cecil Ernest		10 Sep. 18	
	10 Sep. 18	Johnston, Robert Burns		10 Sep. 18	
1230	10 Sep. 18	Lake, Charles Harry Wynne		10 Sep. 18	
	10 Sep. 18	Minahan, Darrel John		10 Sep. 18	
	10 Sep. 18	Smith, Ernest		10 Sep. 18	
	10 Sep. 18	Williams, John Melville		10 Sep. 18	
	11 Sep. 18	Walden, Harold Leslie		11 Sep. 18	
1235	11 Sep. 18	Cook, William Marsden		11 Sep. 18	
	12 Sep. 18	Hardman, Robert George		12 Sep. 18	
	12 Sep. 18	Dickinson, Leonard Duncan		12 Sep. 18	
	12 Sep. 18	Johnstone, Robert Barton		12 Sep. 18	
	12 Sep. 18	Parsons-Smerdon, Charles		12 Sep 18	
1240	13 Sep. 18	Heath, Leonard James		13 Sep. 18	
	13 Sep. 18	Robertson, Alexander Lindsay		13 Sep. 18	
	13 Sep. 18	Brito, Edward		13 Sep. 18	
	13 Sep. 18	Buchanan, Andrew Thomas		13 Sep. 18	
	13 Sep. 18	Dunlop, Robert William Barr		13 Sep. 18	
1245	13 Sep. 18	Elliott, Anthony Blaker		13 Sep. 18	
	13 Sep. 18	Graham, John Stevenson		13 Sep. 18	
	13 Sep. 18	Harrison, Henry Lewis		13 Sep. 18	
	13 Sep. 18	Hudson, Richard John Harrington		13 Sep. 18	
	13 Sep. 18	Jackson, Arthur Leslie		13 Sep. 18	
1250	13 Sep. 18	Laws, Maurice William Parry		13 Sep. 18	
	13 Sep. 18	Murray, Herbert Lanyan		13 Sep. 18	
	13 Sep. 18	Newson, William		13 Sep. 18	
	13 Sep. 18	Spence, William John Forrest		13 Sep. 18	
	13 Sep. 18	Wambeek, Christopher Leonard		13 Sep. 18	
1255	13 Sep. 18	Kenderdine, William Norman		13 Sep. 18	

INFANTRY—*contd.*

2nd-Lieutenants—*contd.*

No.	Date of First Commn.	Name.	Date of birth.	Date of Rank.	REMARKS.
1256	13 Sep. 18	Sinclair, Albert Vernon		13 Sep. 18	
	14 Sep. 18	Hogan, William Edward		14 Sep. 18	
	14 Sep. 18	Pettitt, Nelson		14 Sep. 18	
	14 Sep. 18	Phillips, Carl Archibald		14 Sep. 18	
1260	14 Sep. 18	Prosser, Patrick Hepburn		14 Sep. 18	
	14 Sep. 18	Meredith, Felix Bathridge Stark		14 Sep. 18	
	14 Sep. 18	Mitchell, Herbert James		14 Sep. 18	
	14 Sep 18	Dalling, William Edwin		14 Sep. 18	
	16 Sep. 18	Bion, Clarence Raymond Field		16 Sep. 18	
1265	16 Sep. 18	Clayton, John		16 Sep. 18	
	16 Sep. 18	Cobble, Victor George		16 Sep. 18	
	16 Sep. 18	Wells-Smith, Felix Tom		16 Sep. 18	
	16 Sep. 18	Wright, Frederick Clarence		16 Sep. 18	
	17 Sep. 18	Gain, Charles Thomas		17 Sep. 18	
1270	17 Sep. 18	Madden, Samuel Fitzgerald		17 Sep. 18	
	17 Sep. 18	Carstaris, George		17 Sep. 18	
	17 Sep. 18	Treeby, Henry John		17 Sep. 18	
	18 Sep. 18	Dempster, Eric Richard		18 Sep. 18	
	18 Sep. 18	Frater, William Malcolm		18 Sep. 18	
1275	18 Sep. 18	Gard, Herbert Morris		18 Sep. 11	
	18 Sep. 18	Hollingbery, Neville Horace		18 Sep. 18	
	18 Sep. 18	Hynes, William		18 Sep. 18	
	18 Sep. 18	Leonard, John Joseph		18 Sep. 18	
	18 Sep. 18	MacLeod, Donald Gordon		18 Sep. 18	
1280	18 Sep. 18	Moncur, John Oliver		18 Sep. 18	
	18 Sep. 18	Rae, Langford Frank Dennis		18 Sep. 18	
	18 Sep. 18	Shepherd, Ronald		18 Sep. 18	
	18 Sep. 18	Whelan, Joseph		18 Sep. 18	
	19 Sep. 18	Engley, Ernest Richard		19 Sep. 18	
1285	19 Sep. 18	Elliott, John Joseph		19 Sep. 18	
	19 Sep. 18	Webb, Arthur Chisholm		19 Sep. 18	
	19 Sep. 18	Wilson, Frederick		19 Sep. 18	
	19 Sep. 18	Northcote, Charles Carrington		19 Sep. 18	
	19 Sep. 18	Page, Thomas Slade		19 Sep. 18	
1290	19 Sep. 18	Walsh, Edward		19 Sep. 18	
	19 Sep. 18	Munro, Alexander Stewart		19 Sep. 18	
	20 Sep. 18	Cooke, Harry Lewis		20 Sep. 18	
	20 Sep. 18	Fleming, Donald James		20 Sep. 18	
	20 Sep. 18	Gainsford, Dundonald		20 Sep. 18	
1295	20 Sep. 18	Glen, John Charles Grange		20 Sep. 18	
	20 Sep. 18	Grassby, Cyril George		20 Sep. 18	
	20 Sep. 18	Hinch, Thomas William		20 Sep. 18	
	20 Sep. 18	McCann, Wilfred Maurice		20 Sep. 18	
	20 Sep. 18	Shaw, Clark John		20 Sep. 18	
1300	20 Sep. 18	Stevens, Charles William		20 Sep. 18	
	20 Sep. 18	Steward, Alexander William		20 Sep. 18	
	20 Sep. 18	Walden, Albert Edward		20 Sep. 18	
	20 Sep. 18	Watson, John Vernon		20 Sep. 18	
	20 Sep. 18	Guilford, Edmund (†*temp. Capt.*, *1 Oct. 18*)		20 Sep. 18	
1305	21 Sep. 18	Campbell, Roland		21 Sep. 18	
	21 Sep. 18	Collins, Thomas Charles		21 Sep. 18	
	21 Sep. 18	Losasso, Mario Raffaele		21 Sep. 18	
	21 Sep. 18	Prescott, John Edward		21 Sep. 18	
	21 Sep. 18	Scully, Henry		21 Sep. 18	
1310	21 Sep. 18	Watson, George Arthur		21 Sep. 18	
	21 Sep. 18	Blackburn, Desmond Allan		21 Sep. 18	
	21 Sep. 18	Miller, Ewart Vivian		21 Sep. 18	
	21 Sep. 18	Miller, Otwell Grenville		21 Sep. 18	
	23 Sep. 18	Cornfield, Charles James		23 Sep. 18	
1315	23 Sep. 18	Dukes, Marcus Braden		23 Sep. 18	
	23 Sep. 18	Lucas, Pierre Noel Arthur		23 Sep. 18	
	23 Sep. 18	Niblett, Robert Howard		23 Sep. 18	
	23 Sep. 18	Burnett, Alfred Harding		23 Sep. 18	
	23 Sep. 18	Bayly, Henry		23 Sep. 18	
1320	23 Sep. 18	Hutchinson, Frederick Heap		23 Sep. 18	
	23 Sep. 18	Young, Harold Percy Frederick		23 Sep. 18	
	24 Sep. 18	Cary, William Ambrose		24 Sep. 18	
	24 Sep. 18	Heppolette, Frederick Harold		24 Sep. 18	
	24 Sep. 18	Warren, Stanley		24 Sep. 18	
1325	24 Sep. 18	Beard, Hugh		24 Sep. 18	
	24 Sep. 18	Bennet, Lawrence George		24 Sep. 18	
	24 Sep. 18	Berriff, Alfred William		24 Sep. 18	
	24 Sep. 18	Brown, Herbert Haddon		24 Sep. 18	
	24 Sep. 18	Dickson, Eric James		24 Sep. 18	
1330	24 Sep. 18	Furey, Thomas Francis		24 Sep. 18	
	24 Sep. 18	Gregory, Vernon Winter		24 Sep. 18	
	24 Sep. 18	Hayes-Flanagan Joseph		24 Sep. 18	
1333	24 Sep. 18	Holmes, Henry William James		24 Sep. 18	

† With pay and allowances as for a Lieutenant.

INFANTRY—*contd.*

2nd-Lieutenants—*contd.*

No.	Date of First Commn.	Name.	Date of birth.	Date of Rank.	Remarks.
1334	24 Sep. 18	Marshal, Arnold Earnest		24 Sep. 18	
1335	24 Sep. 18	Morrison, Marco Dennis		24 Sep. 18	
	24 Sep. 18	Perkin, Charles Rundle		24 Sep. 18	
	24 Sep. 18	Priestley, Gerald William		24 Sep. 18	
	24 Sep. 18	Slark, James		24 Sep. 18	
	24 Sep. 18	Spencer, Harry Young		24 Sep. 18	
1340	25 Sep. 18	Kay, Arnold Innes		25 Sep. 18	
	25 Sep. 18	Scott, William David		25 Sep. 18	
	25 Sep. 18	Cockburn James		25 Sep. 18	
	25 Sep. 18	Hardie, William Alexander		25 Sep. 18	
	25 Sep. 18	Higham, Lewis Aldis		25 Sep. 18	
1345	25 Sep. 18	Willoughby, Frank Archibald		25 Sep. 18	
	25 Sep. 18	Mackenzie, Stephen Nairne		25 Sep. 18	
	26 Sep. 18	Gidney, Lemuel Arthur		26 Sep. 18	
	26 Sep. 18	Hitchcock, Walter Henry Herbert		26 Sep. 18	
	26 Sep. 18	O'Brien, Terence		26 Sep. 18	
1350	26 Sep. 18	Sargeant, William		26 Sep. 18	
	26 Sep. 18	Simons, Arthur Lionel		26 Sep. 18	
	26 Sep. 18	Girsham, William Samuel		26 Sep. 18	
	26 Sep. 18	Smith, Swithin		26 Sep. 18	
	26 Sep. 18	Clogg, Carl Tresilian		26 Sep. 18	
1355	27 Sep. 18	Christie, John Sutherland		27 Sep. 18	
	27 Sep. 18	Congdon, Joseph Clemens		27 Sep. 18	
	27 Sep. 18	Gateley, Terrence Patrick		27 Sep. 18	
	27 Sep. 18	Langshaw, Claude Melbourne		27 Sep. 18	
	27 Sep. 18	McCarthy, Michael Phillips		27 Sep. 18	
1360	27 Sep. 18	O'Flynn, Antony James		27 Sep. 18	[Army Service.
	28 Sep. 18	Carter, George Edward Lovelace		28 Sep. 18	Released from
	28 Sep. 18	Garratty, Michael Collingwood		28 Sep. 18	
	28 Sep. 18	Grahamslaw, John George		28 Sep. 18	
	28 Sep. 18	Jones, Frank Leslie		28 Sep. 18	
1365	28 Sep. 18	Sampson, John Ernst		28 Sep. 18	
	28 Sep. 18	Dyton, Albert Henry		28 Sep. 18	
	28 Sep. 18	Lucey, Francis Edward		28 Sep. 18	
	28 Sep. 18	Mahoney, Rowland Shawnasy		28 Sep. 18	
	30 Sep. 18	Farren, Daniel		30 Sep. 18	
1370	30 Sep. 18	Madan, Francis Falconer		30 Sep. 18	
	30 Sep. 18	Wilson, James		30 Sep. 18	
	1 Oct. 18	Prichard, Henry Montague		1st Oct. 18	
	1 Oct. 18	Grange, Denis Goodman		1 Oct. 18	
	1 Oct. 18	Bown, Joseph Edgar		1 Oct. 18	
1375	1 Oct. 18	Brett, Cyril Templeton		1 Oct. 18	
	1 Oct. 18	Butterfield, Philip Henry Anthoney		1 Oct. 18	
	1 Oct. 18	Warn, Joseph Edward		1 Oct. 18	
	1 Oct. 18	Tyler, William		1 Oct. 18	
	1 Oct. 18	Arratoon, John		1 Oct. 18	
1380	1 Oct. 18	Donne, Archibald James		1 Oct. 18	
	1 Oct. 18	MacArthur, Donald Frederick Swan		1 Oct. 18	
	1 Oct. 18	Montgomery, Henry		1 Oct. 18	
	1 Oct. 18	Rushton, Ernest Harcourt		1 Oct. 18	
	1 Oct. 18	Travers-Drapes, Vernon Lamphier		1 Oct. 18	
1385	2 Oct. 18	Leslie, William Henry Hamilton		2 Oct. 18	
	2 Oct. 18	Winter, George William		2 Oct. 18	
	2 Oct. 18	Barrie, William George		2 Oct. 18	
	2 Oct. 18	Cotton, Alfred Christopher		2 Oct. 18	
	2 Oct. 18	Fitzherbert, Richard Anthoney		2 Oct. 18	
1390	2 Oct. 18	Hunneybun, Bertie		2 Oct. 18	
	2 Oct. 18	Marchant, Harry Leonard MacLeod		2 Oct. 18	
	2 Oct. 18	Cotton, Alfred Christopher		2 Oct. 18	
	2 Oct. 18	Rowton, John		2 Oct. 18	
	2 Oct. 18	Stewart, James Ronald		2 Oct. 18	
1395	2 Oct. 18	Wells, Aelian		2 Oct. 18	
	3 Oct. 18	Barron, Francis Harry		3 Oct. 18	
	3 Oct. 18	Goode, Harold Hayden		3 Oct. 18	
	3 Oct. 18	Hayling, Arthur Allen		3 Oct. 18	
	3 Oct. 18	Rodgers, Frank Donald		3 Oct. 18	
1400	3 Oct. 18	Scarff, Alexander		3 Oct. 18	
	3 Oct. 18	Tait, Robert		3 Oct. 18	
	3 Oct. 18	Jansz, William Henry Scheffer		3 Oct. 18	
	4 Oct. 18	Ash, Lindsay		4 Oct. 18	
	4 Oct. 18	Coventry, Walter Altred (*temp. Capt., 10 Oct. 18*).		4 Oct. 18	
1405	4 Oct. 18	Danter, Jamens Edward		4 Oct. 18	
	4 Oct. 18	Lynsdale, Cyril Beaufort Fitzgerald		4 Oct. 18	
	4 Oct. 18	Morgans, Clarence Gladwin		4 Oct. 18	
1408	4 Oct. 18	Warburton, Julian Durani MacLean Ives		4 Oct. 18	

No.	Date of First Commn.	Name.	Date of birth.	Date of Rank.	REMARKS.
		INFANTRY—*contd.*			
		2nd Lieutenants—*contd.*			
1409	4 Oct. 18	Christophers, Harry Ainslie Shannon		4 Oct. 18	
1410	4 Oct. 18	Glover, Edward Herbert		4 Oct 18	
	4 Oct. 18	Large, Osbert		4 Oct. 18	
	4 Oct. 18	Shelton-Palmer Arthur Lucas		4 Oct. 18	
	4 Oct. 18	Tibbs, Dudley Muriel		4 Oct. 18	
	4 Oct. 18	Whitworth, Herbert John		4 Oct. 18	
1415	4 Oct. 18	James, Reginald Valpy		4 Oct. 18	
	5 Oct. 18	Halsey, George Edward Clarence		5 Oct. 18	
	5 Oct. 18	Stall, Charles John		5 Oct. 18	
	5 Oct. 18	Whistondale, Frederick		5 Oct. 18	
	5 Oct. 18	Bolan, William		5 Oct. 18	
1420	7 Oct. 18	Foyster, Frederick		7 Oct. 18	
	7 Oct. 18	Jones, Alexander Edward		7 Oct. 18	[Army Service.
	7 Oct. 18	Lescher, Robin		7 Oct. 18	Released from
	7 Oct. 18	Sykes, Edward Francis		7 Oct. 18	
	7 Oct. 18	Hamilton, Claude Baker		7 Oct. 18	
1425	7 Oct. 18	Williams, Edward Pritchett		7 Oct. 18	
	7 Oct. 18	Bowman, William		7 Oct. 18	
	7 Oct. 18	Croghan, Samuel Culling		7 Oct. 18	
	7 Oct. 18	Hawkins, Ernest James		7 Oct. 18	
	7 Oct. 18	Nicholson, James		7 Oct. 18	
1430	7 Oct. 18	Holmes, William George		7 Oct. 18	
	7 Oct. 18	Morgan, Ivor Montague		7 Oct 18	
	7 Oct. 18	Goodger, Arthur		7 Oct. 18	
	7 Oct. 18	Hazzard, Percy Hugh		7 Oct. 18	
	7 Oct. 18	Pouncey, Thomas Cyril		7 Oct. 18	
1435	7 Oct 18	Howson, William George		7 Oct. 18	
	7 Oct. 18	Teire, Daniel		7 Oct. 18	
	7 Oct. 18	Bonniface, Bettram Harry		7 Oct. 18	
	7 Oct. 18	Price, Ernest John		7 Oct. 18	
	7 Oct. 18	John, William Henry		7 Oct. 18	
1440	7 Oct. 18	Kniveton, Frank		7 Oct. 18	
	7 Oct. 18	Dillingham, Charles Stanley Denbigh		7 Oct. 18	
	7 Oct. 18	Nicholas, Thomas		7 Oct. 18	
	7 Oct. 18	Phillips, Herbert Theodore		7 Oct. 18	
	Oct. 18	Donne, William John Burnett		7 Oct. 18	
1445	7 Oct. 18	Foster, Ernest Charles Henry		7 Oct. 18	
	7 Oct. 18	Taylor, Edward Richard Louis		7 Oct. 18	
	7 Oct. 18	Hitchin, Frank Ernest		7 Oct. 18	
	7 Oct. 18	Collins, Frederick Tom		7 Oct. 18	
	7 Oct. 18	Tutte, Edwin		7 Oct. 18	
1450	7 Oct. 18	Rutherford, Howard Richard		7 Oct. 18	
	7 Oct. 18	Edlin, Charles Arthur Edgar		7 Oct. 18	
	8 Oct. 18	Allan, David Sinclair		8 Oct. 18	
	8 Oct. 18	Archar, Clement John		8 Oct. 18	
	8 Oct. 18	Banyard, Edward Francis		8 Oct. 18	
1455	8 Oct. 18	Barry, Frederick		8 Oct. 18	
	8 Oct. 18	Cochrane, James Alexander		8 Oct. 18	
	8 Oct. 18	Finch, Charles		8 Oct. 18	
	8 Oct. 18	Cleeson, John		8 Oct. 18	
	8 Oct. 18	Grant, Norman Maling		8 Oct. 18	
1460	8 Oct. 18	Spinks, John Henry Charles		8 Oct. 18	
	8 Oct. 18	Christian, Herschel Dear		8 Oct. 18	
	9 Oct. 18	Bredin, John Arthur		9 Oct. 18	
	9 Oct. 18	Grant, James		9 Oct. 18	
	9 Oct. 18	O'Gorman, Charles James		9 Oct. 18	
1465	9 Oct. 18	Hogh, Allan Leslie Von		9 Oct. 18	
	10 Oct. 18	Atkinson, Henry Oscar		10 Oct. 18	
	10 Oct. 18	Hampson, Albert		10 Oct. 18	
	10 Oct. 18	Haygarth, Roy Neville		10 Oct. 18	
	10 Oct. 18	Mahoney, Thomas		10 Oct. 18	
1470	10 Oct. 18	Shubart, Stanley David		10 Oct. 18	
	10 Oct. 18	Slatter, Charles Beaumont		10 Oct. 18	
	10 Oct 18	Whalley, Claude Albert		10 Oct. 18	
	10 Oct. 18	Corner, Edward Seaton		10 Oct. 18	
	10 Oct. 18	Cook, Charles Wellstood		10 Oct. 18	
1475	11 Oct. 18	Best, George Henry		11 Oct. 18	
	11 Oct. 18	Holmes, Arthur Clement		11 Oct. 18	
	11 Oct. 18	Korb, George Lionel		11 Oct. 18	
	11 Oct. 18	Priestley, William Hessel		11 Oct. 18	
	11 Oct. 18	Schrader, George Justus		11 Oct. 18	
1480	11 Oct. 18	Speldewinde, Charles George Oliver		11 Oct. 18	
	11 Oct. 18	Toussaint, Julian Basil		11 Oct. 18	
	11 Oct. 18	Wharton, Robert Francis Harold		11 Oct. 18	
	11 Oct. 18	Wilson, James Charles Montgomery		11 Oct. 18	
	11 Oct 18	Brown, Edward James		11 Oct. 18	
1485	12 Oct. 18	Breeze, Charles Leslie		12 Oct. 18	

No.	Date of First Commn.	Names.	Date of birth.	Date of Rank.	REMARKS.
		INFANTRY—*contd.* 2nd-Lieutenants—*contd.*			
1486	12 Oct. 18	Hastings, James Everard		12 Oct. 18	
	12 Oct. 18	Holton, William Ernest		12 Oct. 18	
	14 Oct. 18	Booley, George Horrace		14 Oct. 18	
	14 Oct. 18	Crisp, Robert Ernest		14 Oct. 18	
1490	14 Oct. 18	Porter, Nelson		14 Oct. 18	
	14 Oct. 18	Standish, Alexander		14 Oct. 18	
	14 Oct. 18	Tingley, Albert		14 Oct. 18	
	14 Oct. 18	Lowry, William Edward (*temp. Maj.* 14 *Oct.* 18.)		14 Oct. 18	
	14 Oct. 18	McClay, Arthur Duncan Glynnly		14 Oct. 18	
1495	15 Oct. 18	Vivian, James Francis Gordon		15 Oct. 18	
	15 Oct. 18	Stephenson, Percival William		15 Oct. 18	
	15 Oct. 18	Oakeshott, Douglas Benjamin		15 Oct. 18	
	15 Oct. 18	O'Brien, Michael John		15 Oct. 18	
	15 Oct. 18	Stonor, Wilfred Thomas		15 Oct. 18	
1500	15 Oct. 18	Willmer, Eric John		15 Oct. 18	
	15 Oct. 18	Donald, James		15 Oct. 18	
	15 Oct. 18	Pinker, Frederick Rashleigh		15 Oct. 18	
	15 Oct. 18	Wright, Harry Walter		15 Oct. 18	
	15 Oct. 18	Netherwood, Edgar		15 Oct. 18	
1505	15 Oct. 18	O'Brien, John Michael Denis		15 Oct. 18	
	15 Oct. 18	Gamble, Harold Charles		15 Oct. 18	
	15 Oct. 18	Kingsbury, Walter George		15 Oct. 18	
	15 Oct. 18	Hamil, William James Wallace		15 Oct. 18	
	15 Oct. 18	Sutton, William Reginald		15 Oct. 18	
1510	15 Oct. 18	Gaunt, Alfred Frederick		15 Oct. 18	
	15 Oct 18	Smith, Ernest		15 Oct. 18	
	15 Oct. 18	Taylor, Walter Frank		15 Oct. 18	
	15 Oct. 18	Carter, John Henry		15 Oct. 18	
	15 Oct. 18	Edmondson, William Henry		15 Oct. 18	
1515	15 Oct. 18	Gouge, William George		15 Oct. 18	
	15 Oct. 18	Kingdon, Sydney Charles Kent		15 Oct. 18	
	15 Oct. 18	Smith, Arthur Sidney David		15 Oct. 18	
	15 Oct. 18	Thompson, John		15 Oct. 18	
	15 Oct. 18	Harris, Arthur Marshall		15 Oct. 18	
1520	15 Oct. 18	Birch, Charles		15 Oct. 18	
	15 Oct. 18	Powell, Frank Edwin Thomas		15 Oct. 18	
	15 Oct. 18	Harwood, Walter Frank		15 Oct. 18	
	15 Oct. 18	Morgan, George James		15 Oct. 18	
	15 Oct. 18	Virgin, Carlton Dean		15 Oct. 18	
1525	15 Oct 18	Coney, Harold Greatrex		15 Oct. 18	
	15 Oct. 18	Fernbach, Herbert Helmer		15 Oct. 18	
	15 Oct. 18	Mumford, Ernest Cyril		15 Oct. 18	
	15 Oct. 18	Renton, Henry Darling		15 Oct. 18	
	15 Oct. 18	Selley, Wallace Alfred		15 Oct. 18	
1530	15 Oct. 18	Pickering, Frederick Charles		15 Oct. 18	
	15 Oct. 18	Sales, John James		15 Oct. 18	
	15 Oct. 18	Bassett, Bert Jesse		15 Oct. 18	
	15 Oct. 18	Cummings, Charles George		15 Oct. 18	
	16 Oct. 18	Brown, Cecil Jermyn		16 Oct. 18	
1535	16 Oct. 18	Tottle, John Bews		16 Oct. 18	
	16 Oct. 18	Barber, Frederick Farrel		16 Oct. 18	
	16 Oct. 18	Frisby, William Thomas		16 Oct. 18	
	16 Oct. 18	Loughlin, James Arthur		16 Oct. 18	
	16 Oct. 18	Maybury, Melville Alfred Kingston		16 Oct. 18	
1540	17 Oct. 18	Edwards, John William		17 Oct. 18	
	17 Oct. 18	McWhinnie, Robert Smith		17 Oct. 18	
	17 Oct. 18	Sinclair, George		17 Oct. 18	
	17 Oct. 18	Woods-Jennings, Harold William		17 Oct. 18	
	17 Oct. 18	Cole, John Edward Lyall		17 Oct. 18	
1545	17 Oct. 18	Gibson, Herbert Arthur		17 Oct. 18	
	17 Oct. 18	Ford, Daniel O'Connell		17 Oct. 18	
	17 Oct. 18	Dorrell, James Okey		17 Oct. 18	
	18 Oct. 18	Armstrong, Frederick Edmund John		18 Oct. 18	
	18 Oct. 18	Court, Edwin Thomas		18 Oct. 18	
1550	18 Oct. 18	Lister, Charles Edward Langley		18 Oct. 18	
	18 Oct. 18	Mander, Thomas James		18 Oct. 18	
	18 Oct. 18	Rabbitt, Joseph Barnard		18 Oct. 18	
	19 Oct. 18	Symes, Leonard		19 Oct. 18	
	19 Oct. 18	Annear, William Caxton		19 Oct. 18	
1555	19 Oct. 18	Elliott, Joseph Edward		19 Oct. 18	
	19 Oct. 18	Treloar, Cyril Conrad Martyn		19 Oct. 18	
	19 Oct. 18	Cotterell, Robert John		19 Oct. 18	
	19 Oct. 18	Spence, Arthur Leslie		19 Oct. 18	
	19 Oct. 18	King, Reginald Maurice		19 Oct. 18	
1560	19 Oct. 18	Lenton, Henry Thomas Francis		19 Oct. 18	
	19 Oct. 18	Powell, Arthur		19 Oct. 18	
1562	19 Oct. 18	MacLeod, Leslie Charles		19 Oct. 18	

INFANTRY—*contd.*

2nd-Lieutenants—*contd.*

No.	Date of First Common.	Names.	Date of birth.	Date of rank.	Remarks.
1563	19 Oct. 18	Hart, Herbert Arthur		19 Oct. 18	
	19 Oct. 18	Meiklejohn, Robert		19 Oct. 18	
1565	19 Oct. 18	Powers, George Comen		19 Oct. 18	
	19 Oct. 18	Harding, Fred		19 Oct. 18	
	19 Oct. 18	Williamson, Allan Boak		19 Oct. 18	
	19 Oct. 18	Green, William		19 Oct. 18	
	19 Oct. 18	Colthup, Bertam William		19 Oct. 18	
1570	19 Oct. 18	Reid, John		19 Oct. 18	
	19 Oct. 18	Wyndham, Arnold Edwin		19 Oct. 18	
	19 Oct. 18	Gray, John Maxwell		19 Oct. 18	
	19 Oct. 18	Turner, David Victor		19 Oct. 18	
	19 Oct. 18	McArdle, Raymond Harold		19 Oct. 18	
1575	19 Oct. 18	Jackson, Thomas Lloyd		19 Oct. 18	
	19 Oct. 18	Amers, Thomas		19 Oct. 18	
	19 Oct. 18	Charlesworth, Frederick		19 Oct. 18	
	19 Oct. 18	Jones, George Edward		19 Oct. 18	
	19 Oct. 18	Legge, Albert Grenville		19 Oct. 18	
1580	19 Oct. 18	Waite, Harry Spacey		19 Oct. 18	
	19 Oct. 18	Taylor, Henry Smithson		19 Oct. 18	
	19 Oct. 18	Cougle, Magnus I.		19 Oct. 18	
	19 Oct. 18	Cole, Alfred Charles		19 Oct 18	
	19 Oct. 18	Fair, Frederick William		19 Oct. 18	
1585	19 Oct. 18	Cross, Cecil Herbert Mark		19 Oct. 18	
	19 Oct. 18	Hancock, Harold Frank		19 Oct. 18	
	19 Oct. 18	Hanna, John		19 Oct. 18	
	19 Oct. 18	Richardson, Charles		19 Oct. 18	
	19 Oct. 18	Scotcher, Walter James		19 Oct. 18	
1590	19 Oct. 18	Tate, John		19 Oct. 18	
	19 Oct. 18	Sherman, Frederick John		19 Oct. 18	
	19 Oct. 18	Brightmore-Armour, Charles James		19 Oct. 18	
	19 Oct. 18	Brooks, Cyril William		19 Oct. 18	
	19 Oct. 18	Cox, Charles John Bede		19 Oct. 18	
1595	19 Oct. 18	Dorrington, George Herbert		19 Oct. 18	
	19 Oct. 18	Gilroy, Alexander William		19 Oct. 18	
	19 Oct. 18	Graburn, Nelson Edmund Bromehead		19 Oct. 18	
	19 Oct. 18	Hayes, Hugh Cameron		19 Oct. 18	
	19 Oct. 18	Hervey, Ney Lindsay		19 Oct. 18	
1600	19 Oct. 18	Jackson, Robert John		19 Oct. 18	
	19 Oct. 18	Park, Herbert Hugh		19 Oct. 18	
	19 Oct. 18	Sherson, Dudley Berkeley Nowell		19 Oct. 18	
	20 Oct. 18	Patmore, Samuel Harold		20 Oct. 18	
	20 Oct. 18	Gunning, Cardwell Sinclair		20 Oct. 18	
1605	20 Oct. 18	Bogaerde, Leon Alidor Francois Van den		20 Oct. 18	
	21 Oct. 18	Donald, John		21 Oct. 18	
	21 Oct. 18	Haworth, Richard		21 Oct. 18	
	21 Oct. 18	Somerville, Thomas		21 Oct. 18	
	21 Oct. 18	Grandison, Robert McKinlay		21 Oct. 18	
1610	21 Oct. 18	Webster, Cecil George Gordon		21 Oct. 18	
	22 Oct. 18	Cracknell, Harry William		22 Oct. 18	
	22 Oct. 18	Preston, Louis Charles Edward Wharton		22 Oct. 18	
	22 Oct. 18	Thompson, Charles Barnard		22 Oct. 18	
	22 Oct. 18	Brown, John Robertson		22 Oct. 18	
1615	23 Oct. 18	Clen-Murphy, Ernest William		23 Oct. 18	
	23 Oct. 18	Netscher, Edward George		23 Oct. 18	
	23 Oct. 18	Tobin, James		23 Oct. 18	
	23 Oct. 18	Day, John Henry		23 Oct. 18	
	24 Oct. 18	Arndt, Carl Evan		24 Oct. 18	
1620	24 Oct. 18	Cronan, Cyril Preston		24 Oct. 18	
	24 Oct. 18	Grant, Donald Stuart		24 Oct. 18	
	24 Oct. 18	Keats, Albert William		24 Oct. 18	
	24 Oct. 18	Mack, Frank Cruikshank		24 Oct. 18	
	24 Oct. 18	Rodda, William John		24 Oct. 18	
1625	24 Oct. 18	Hayward, Herbert Hayward		24 Oct. 18	
	24 Oct. 18	Jones, Harry Richar		24 Oct. 18	
	24 Oct. 18	Nesh, Harry		24 Oct. 18	
	24 Oct. 18	Sykes, Arthur William		24 Oct. 18	
	24 Oct. 18	Middleton-Stewart, Charles Edward (*temp. Capt., 24 Oct. 18*).		24 Oct. 18	
1630	25 Oct. 18	Egan-Wyear, Terence Michael		25 Oct. 18	
	25 Oct. 18	Lockwood, Reginald Ughtred		25 Oct. 18	
	25 Oct. 18	Mitchell, James		25 Oct. 18	
	25 Oct. 18	Sackett, Frank Colyer, (*temp. Capt., 25 Oct. 18*)		25 Oct. 18	
	25 Oct. 18	Charmarette, Robert Douthwaite Camaron		25 Oct. 18	
1635	26 Oct. 18	Harkess, Albert Ernest		26 Oct. 18	
	26 Oct. 18	Utting, Frederick Albert Ker		26 Oct. 18	
1637	27 Oct. 18	Lancaster, George Cursham		27 Oct. 18	

No.	Date of First Commn.	Names.	Date of birth.	Date of rank.	REMARKS.
		INFANTRY—*contd.*			
		2nd-Lieutenants—*contd.*			
1638	27 Oct. 18	Swinton, Thomas Swinton		28 Oct. 18	
	27 Oct. 18	Lawson, Allexander George		27 Oct. 18	
1640	27 Oct. 18	Rule, Dennis Duncan Pritchard		27 Oct. 18	
	27 Oct. 18	Wright, Harold Charles		27 Oct. 18	
	28 Oct. 18	Watts, Percival John		28 Oct. 18	
	28 Oct. 18	Millett, Thomas William		28 Oct. 18	
	28 Oct. 18	Kettleband, Herbert William Bernard		28 Oct. 18	
1645	28 Oct. 18	Lowcock, Douglas Raymond		27 Oct. 18	
	28 Oct. 18	Prebble, John		28 Oct. 18	
	28 Oct. 18	Soar, Joseph		28 Oct. 18	
	28 Oct. 18	Wallace, Henry Charles		28 Oct. 18	
	28 Oct. 18	Fenn, Charles Deane		28 Oct. 18	
1650	29 Oct. 18	Clark, Allan Cunningham		29 Oct. 18	
	29 Oct. 18	Clarke, John William		29 Oct. 18	
	29 Oct. 18	Thomson, Samuel Harold		29 Oct. 18	
	29 Oct. 18	Crumblehulme, William		29 Oct. 18	
	29 Oct. 18	Curtis, Morley		29 Oct. 18	
1655	29 Oct. 18	Edwards, Henry James		29 Oct. 18	
	29 Oct. 18	Gilchrist, Thomas Ernest		29 Oct. 18	
	29 Oct. 18	O'Sullivan, Owen Donal		29 Oct. 18	
	29 Oct. 18	Wattom, Ivor Charles		29 Oct. 18	
	29 Oct. 18	Bell, Charles Francis John		29 Oct. 18	
1660	30 Oct. 18	Ford, Basil Henry		30 Oct. 18	
	30 Oct. 18	George, James Goulder		30 Oct. 18	
	30 Oct. 18	Gibson, Robert James		30 Oct. 18	
	30 Oct. 18	Mayers, Henry Sylvester		30 Oct. 18	
	30 Oct. 18	Mayhew, Alfred George		30 Oct. 18	
1665	30 Oct. 18	Parker, Leonard		30 Oct. 18	
	30 Oct. 18	Stanes, Frederick James		30 Oct. 18	
	31 Oct. 18	Thrupp, Harold Frank		31 Oct. 18	
	31 Oct. 18	Challice, John		31 Oct. 18	
	31 Oct. 18	Graham, Archibald Cyril		31 Oct. 18	
1670	31 Oct. 18	Mawson, Lawrence Rivers		31 Oct. 18	
	31 Oct. 18	Webb, William		31 Oct. 18	
	1 Nov. 18	Molesworth, Anthony Oliver		1 Nov. 18	
	1 Nov. 18	Ayckbourne, Frederick Sydney		1 Nov. 18	
	1 Nov. 18	Labanti, Vincent Wilbert		1 Nov. 18	
1675	1 Nov. 18	Benny, Henry Flamank		1 Nov. 18	
	1 Nov. 18	Harrison, Albert John		1 Nov. 18	
	1 Nov. 18	Patchett, Ephraim		1 Nov. 18	
	2 Nov. 18	Blomfield, Christopher		2 Nov. 18	
	2 Nov. 18	Lewis, Wynne Hill		2 Nov. 18	
1680	4 Nov. 18	Doran, Henry Francis		4 Nov. 18	
	4 Nov. 18	Hampton, Henry Verner		4 Nov. 18	
	4 Nov. 18	Scott, Edwin Algernon		4 Nov. 18	
	4 Nov. 18	Weinman, Aubrey Neil		4 Nov. 18	
	4 Nov. 18	Wise, Harold Hardy		4 Nov. 18	
1685	4 Nov. 18	Wyncoll, Claude Horace Albert		4 Nov. 18	
	5 Nov. 18	Reeves, Harold MacDonald		5 Nov. 18	
	5 Nov. 18	Swords, William John		5 Nov. 18	
	5 Nov. 18	Hewitt, John		5 Nov. 18	
	5 Nov. 18	Boorman, George Henry Frederick		5 Nov. 18	
1690	5 Nov. 18	Leeman, John Albert		5 Nov. 18	
	5 Nov. 18	Robinson, William Arthur		5 Nov. 18	
	5 Nov. 18	Brandon, Rudolph William		5 Nov. 18	
	5 Nov. 18	Ascroft, Philip Thomas		5 Nov. 18	
	5 Nov. 18	Greene, Ernest Arnold Jeffries		5 Nov. 18	
1695	5 Nov. 18	Sellick, William Henry		5 Nov. 18	
	5 Nov. 18	Major, Eric		5 Nov. 18	
	5 Nov. 18	Sellers, Stanley Riley		5 Nov. 18	
	5 Nov. 18	Bestford, Thomas		5 Nov. 18	
	5 Nov. 18	Ledgerwood, Thomas		5 Nov. 18	
1700	5 Nov. 18	Carpenter, Ernest Clement		5 Nov. 18	
	5 Nov. 18	Nickolds, Arthur		5 Nov. 18	
	5 Nov. 18	Batty, Ralph Edward		5 Nov. 18	
	5 Nov. 18	Sloan, Robert		5 Nov. 18	
	5 Nov. 18	Beeson, Reginald George		5 Nov. 18	
1705	5 Nov. 18	Skellon, Paul Edward		5 Nov. 18	
	5 Nov. 18	Verner, Evelyn Aubrey		5 Nov. 18	
	5 Nov. 18	Thomas Samuel		5 Nov. 18	
	5 Nov. 18	Boddy, Percy Charles		5 Nov. 18	
	5 Nov. 18	Rowson, Claude Montague		5 Nov. 18	
1710	5 Nov. 18	Marsden, Gilbert		5 Nov. 18	
	5 Nov. 18	Andrews, William		5 Nov. 18	
	5 Nov. 18	Cawthron, Edwin		5 Nov. 18	
1713	5 Nov. 18	Lofthouse, Arthur		5 Nov. 18	

INFANTRY—*contd.*

2nd-Lieutenants—*contd.*

No.	Date of first Common.	Name.	Date of birth.	Date of rank.	REMARKS.
1714	5 Nov. 18	Shipp, George Richard Gordon		5 Nov. 18	
1715	5 Nov. 18	Cooper, Albert Percival		5 Nov. 18	
	5 Nov. 18	Newman, Tom Douglas		5 Nov. 18	
	5 Nov. 18	Huthinson, Alexander McIntyre		5 Nov. 18	
	5 Nov. 18	Russell, William Henry		5 Nov. 18	
	5 Nov. 18	Smith, Tredgar Sidney		5 Nov. 18	
1720	5 Nov. 18	Corkery, Joseph Samuel Waters		5 Nov. 18	
	5 Nov. 18	Enticknapp, Arthur William		5 Nov. 18	
	5 Nov. 18	Moir, John Horton		5 Nov. 18	
	5 Nov. 18	Riley-Irving, Charles Terence		5 Nov. 18	
	5 Nov. 18	Burn, Walter		5 Nov. 18	
1725	6 Nov. 18	Heath, Arthur		6 Nov. 18	
	6 Nov. 18	Moss, William Henry		6 Nov. 18	
	6 Nov. 18	Stewart, John Jackson		6 Nov. 18	
	6 Nov. 18	Terry, Edwin		6 Nov. 18	
	6 Nov. 18	Levick, Henry		6 Nov. 18	
1730	6 Nov. 18	Walden, William Charles		6 Nov. 18	
	7 Nov. 18	Jackson, William Frederick Barry		7 Nov. 18	
	7 Nov. 18	Smith, Horace Vivion		7 Nov. 18	
	7 Nov. 18	Almond, Howard Raymond		7 Nov. 18	
	7 Nov. 18	Childs, Henry		7 Nov. 18	
1735	7 Nov. 18	Jackson, Walter George		7 Nov. 18	
	7 Nov. 18	Jansz, Egerton Oliver		7 Nov. 18	
	7 Nov. 18	Lewis, Reginald Elliot		7 Nov. 18	
	7 Nov. 18	MacLean, Charles Hector MacKenzie		7 Nov. 18	
	7 Nov. 18	Mathews, Ernest		7 Nov. 18	
1740	7 Nov. 18	Plommer, George Frederick		7 Nov. 18	
	7 Nov. 18	Sharp, John Michael Curran		7 Nov. 18	
	8 Nov. 18	Whittaker, Alfred Patrick		8 Nov. 18	
	8 Nov. 18	Dawson, Raymund Mowbray		8 Nov. 18	
	8 Nov. 18	Gooch, Frank Crawford		8 Nov. 18	
1745	8 Nov. 18	Montgomery, Hugh		8 Nov. 18	
	8 Nov. 18	O'Sullivan, Anthony George Joseph		8 Nov. 18	
	8 Nov. 18	Parsons, James		8 Nov. 18	
	8 Nov. 18	Quinton, Leslie Mark		8 Nov. 18	
	8 Nov. 18	Stallard, Godfrey Bradley		8 Nov. 18	
1750	8 Nov. 18	Vermont, Claude Logan		8 Nov. 18	
	8 Nov. 18	Williams, Philip Martin		8 Nov 18	
	8 Nov. 18	Biggs, William		8 Nov. 18	
	8 Nov. 18	Goff, Cyril Arthur		8 Nov. 18	
	9 Nov. 18	Murcutt, Frederick Andrew		9 Nov. 18	
1755	9 Nov. 18	Sco'man, Robert		9 Nov. 18	
	11 Nov. 18	Franks, Shirley Wilfred		11 Nov. 18	
	11 Nov. 18	Matthews, Arthur		11 Nov. 18	
	11 Nov. 18	Woods, Dermot Frederick St. John		11 Nov. 18	
	11 Nov. 18	Farmer, Trevor Mervyn		11 Nov. 18	
1760	12 Nov. 18	Elliott, Walter		12 Nov. 18	
	12 Nov. 18	Handley, Herbert George		12 Nov. 18	
	12 Nov. 18	Austen, William George		12 Nov. 18	
	12 Nov. 18	Allen, Ernest Frank		12 Nov. 18	
	12 Nov. 18	Reed, Francis William		12 Nov. 18	
1765	12 Nov. 18	Fogwell, Leonard Roberts		12 Nov. 18	
	12 Nov. 18	Goddard, Leonard Arthur		12 Nov. 18	
	12 Nov. 18	McLennan, George Lawrence		12 Nov. 18	
	12 Nov. 18	Strong, George		12 Nov. 18	
	12 Nov. 18	Collins, Alfred Henry Thomas		12 Nov. 18	
1770	12 Nov. 18	Boorman, Thomas		12 Nov. 18	
	12 Nov. 18	Blake, William John		12 Nov. 18	
	12 Nov. 18	Smith, Henry Charles Edward		12 Nov. 18	
	12 Nov. 18	Turner, George Rivers		12 Nov. 18	
	12 Nov. 18	Henderson, George Paterson		12 Nov. 18	
1775	12 Nov. 18	Grover, George Henry		12 Nov. 18	
	12 Nov. 18	Simpson, Hector McKinnon		12 Nov. 18	
	13 Nov. 18	Johnson-Coombs, Kenneth Smith		13 Nov. 18	
	13 Nov. 18	Grimmond, John Alexander		13 Nov. 18	
	15 Nov. 18	Calder, Charles Cumming		15 Nov. 18	
1780	15 Nov. 18	Kinneen, James Francis		15 Nov. 18	
	15 Nov. 18	Hickman, Walter Thomas		15 Nov. 18	
	21 Nov. 18	McGregor, Kenneth Stuart		21 Nov. 18	
	23 Nov. 18	Ramsay, Alick Hector Winson		23 Nov. 18	
1784	25 Nov. 18	Morby, Joseph Charles Henry		25 Nov. 18	

OFFICERS WITH TEMPORARY COMMISSIONS.

Lieutenant-Colonel.

No.	Date of First Commn.	Names.	Date of birth.	Date of rank.	Remarks.
1	13 May 18	Leslie-Jones, Leycester Hudson		28 Aug. 18	

Captains.

No.	Date of First Commn.	Names.	Date of birth.	Date of rank.	Remarks.
1	6 July 17	Shaw, G. W.		15 July 18	
	4 Oct. 18	Murray, Charles John	2 June 71	4 Oct. 18	
	4 Oct. 18	Norris, William Albert	8 Oct. 78	4 Oct. 18	
	4 Oct. 18	Forde, Andrew	5 Oct. 80	4 Oct. 18	
5	4 Oct. 18	Morrow, Charles Thomas	27 Nov. 76	4 Oct. 18	
	4 Oct. 18	Creak, Evan Victor	15 June 79	4 Oct. 18	
	4 Oct. 18	Vousley, Frederick	25 Mar. 69	4 Oct. 18	
	11 Oct. 18	Prior, Charles James	10 Apl. 75	11 Oct. 18	
	11 Oct. 18	Jackson, Charles Gowan	9 Nov. 77	11 Oct. 18	
10	25 Oct. 18	Richardson, Francis Lionel	12 Nov. 76	25 Oct. 18	
	1 Nov. 18	Ginns, Lewis Bertrand	2 June 81	1 Nov. 18	
	1 Nov. 18	Marks, Frederick Charles	26 Dec. 70	1 Nov. 18	
	1 Nov. 18	Danskin, John William	13 May 75	1 Nov. 18	
	1 Nov. 18	Marshall, Percy Harold	16 June 78	1 Nov. 18	
15	2 Nov. 18	Hill, Alfred John	29 Apl. 73	2 Nov. 18	
	14 Nov. 18	Ruegg, Alfred James	17 Apl. 65	14 Nov. 18	
	20 Nov. 18	Cotter, Parr	22 Aug. 73	20 Nov. 18	
	26 Nov. 18	Edwards, William Henry	23 Mar. 73	26 Nov. 18	
	26 Nov. 18	Coward, Arthur John	11 Jan. 72	26 Nov. 18	
20	26 Nov. 18	Wilson, Thomas George	16 Dec. 73	26 Nov. 18	

Lieutenants.

No.	Date of First Commn.	Names.	Date of birth.	Date of rank.	Remarks.
1	5 Apl. 17	Smith, T. S.		5 Apl. 18	
	6 Apl. 17	Whitten, A. M.		6 Apl. 18	
	11 Apl. 17	More, J. S.		11 Apl. 18	
	13 Apl. 17	Dorward, W.		13 Apl. 18	
5	17 Apl. 17	Bourn, H. (207a)		17 Apl. 18	
	19 Apl. 17	Wilkins, F. B.		19 Apl. 18	Released from Army Service.
	20 Apl. 17	Knight, Alfred Frederick John		20 Apl. 17	
	28 Apl. 17	Doubell, P. S.		28 Apl. 18	
	28 Apl. 17	Stephen, R. L.		28 Apl. 18	
10	30 Apl. 17	Le Patourel, H. J. B.		30 Apl. 18	
	1 May 17	Alderson, H. W.		1 May 18	
	1 May 17	Baddeley, G. T. B.		1 May 18	
	7 May 17	Woolmer, H. E.		7 May 18	
	19 May 17	Fern, E. A.		19 May 18	
15	22 May 17	Booth, J. J.		22 May 18	
	22 May 17	Pettifer, F. G.		22 May 18	
	23 May 17	Thompson, A.		23 May 18	
	2 June 17	Crowe, J. A.		2 June 18	
19	5 Oct. 17	Williamson, H.		12 Apl. 18	

2nd-Lieutenants.

No.	Date of First Commn.	Names.	Date of birth.	Date of rank.	Remarks.
1	22 Mar. 17	Terry, P. E.		22 Mar. 17	
	1 Apl. 17	Davies, D. S.		1 Apl. 17	
	5 Apl. 17	Borrett, T. L.		5 Apl. 17	
	6 Apl. 17	Benson, G. W.		6 Apl. 17	
5	15 Apl. 17	Lucas, E. W.		15 Apl. 17	
	15 Apl. 17	Reid, W. F.		15 Apl. 17	
	29 Apl. 17	Gibbons, J. H.		29 Apl. 17	
	22 May 17	Walker, E. C.		22 May 17	
	1 June 17	Metcalfe, T.		1 June 17	
10	11 June 17	Oates, S. J.		11 June 17	
	15 June 17	Place, W.		15 June 17	
	18 June 17	Kingsley, W. E. (*Actg. Cap., 9 June 18*)		18 June 17	
	5 Sep. 17	Bourdillon, Bernard Henry		5 Sep. 17	
	2 Jan. 18	Henderson, Alan Gerald Russell		2 Jan. 18	
15	2 Jan. 18	Cooper, George Milne		2 Jan. 18	
	25 Jan. 18	Cattell, James George		25 Jan. 18	
	1 Mar. 18	MacKay, John William		1 Mar. 18	
	1 Mar. 18	Steinhoff, Bertram Godwin		1 Mar. 18	
	1 Apl. 18	Wroughton, Francis Harold		1 Apl. 18	
20	1 May 18	Edye, Ernest Henry Huish		1 May 18	
21	1 May 18	Willcocks, William Gabriel (*temp. Maj., 20 July 18*)		1 May 18	

OFFICERS WITH TEMPORARY COMMISSIONS—*contd.*

2nd-Lieutenants—*contd.*

No.	Date of First Commn.	Names.	Date of birth.	Date of rank.	Remarks.
22	25 May 18	Eyre, Montague Henry	....	25 May 18	
	1 June 18	Hare, Thomas	....	1 June 18	
	1 July 18	Metcalfe, James Shepherd	....	1 July 18	
25	1 July 18	Hyde, Gilbert William Thomas	....	1 July 18	
	1 July 18	Jones, Hugh Erskine	....	1 July 18	
	15 Aug. 18	Gantzer, Joseph Francis	....	15 Aug. 18	
	17 Aug. 18	Lancashire, John Kenneth	....	17 Aug. 18	
	1 Sep. 18	Senneck, Francis Henry	....	1 Sep. 18	
30	5 Sep. 18	Barry, Dennis Thurburn	....	5 Sep. 18	
	5 Sep. 18	Chatterton, William Henry Hart	....	5 Sep. 18	
	5 Sep. 18	Clift, Frank Charles Lechmere	....	5 Sep. 18	
	5 Sep. 18	MacQueen, Percy	....	5 Sep. 18	Released from Army Service.
	5 Sep. 18	Parnell, Frederick Richard	....	5 Sep. 18	
35	15 Sep. 18	Cammiade, Louis Aimé	....	15 Sep. 18	
	16 Sep. 18	Crampton, Ernest Arthur	....	16 Sep. 18	
	24 Sep. 18	D'Cruz, George Lawrence	....	24 Sep 18	
	1 Oct. 18	Wilkinson, Algernon Charles Legge.	....	1 Oct. 18	
	1 Oct. 18	Hawes, Ernest Howard	....	1 Oct. 18	Released from Army Service.
40	6 Oct. 18	Davis, Godfrey	....	6 Oct. 18	
	10 Oct. 18	Harris, Frederick William	....	10 Oct. 18	
	15 Oct. 18	Hesketh, Marsh	....	15 Oct. 18	
	8 Nov. 18	Thompson, William Hugh	....	8 Nov. 18	
	10 Nov. 18	Younie, John	....	10 Nov. 18	
45	15 Nov. 18	Tremenheere, Charles Wyyn	....	15 Nov. 18	

VETERINARY BRANCH.

Captains.

No.	Date of First Commn.	Names.	Date of birth.	Date of rank.	Remarks.
1	1 Nov. 14	Weathrall, Charles Hoffman George	....	1 Nov. 15	
	4 Dec. 14	Stevenson, William (*Ty. Maj., 25 Feb. 18*)	....	4 Dec. 15	
	1 July 15	Kerr, Percival James (*Ty. Maj., 25 Feb. 18*)	....	1 July 16	
	10 Feb. 16	MacGregor, Andrew Dawson	....	10 Feb. 17	
5	27 Feb. 17	Assheton-Smith, Thomas (*Ty. Maj., 13 May 18*)	....	27 Feb. 18	
	28 Feb. 17	Pool, William Arthur	....	28 Feb. 18	
	19 Apl. 17	Meadows, Daunt	....	19 Apl. 18	
8	13 Dec. 17	Hickey, Sidney George Middleton	....	13 Dec. 18	

Lieutenants.

No.	Date of First Commn.	Names.	Date of birth.	Date of rank.	Remarks.
1	9 July 15	Farbrother, Earnest Samuel	....	9 July 15	
2	15 June 17	Cross, Henry Eustace (*tempy. Capt., A.V.C., 3 July 18*)	....	15 June 17	

OFFICERS OF THE INDIAN ARMY RESERVE OF OFFICERS ON PROBATION.

Regiment.	Date of first Commission.	Names and rank.	Remarks.
N. Z. Rif. Bde.	4 Mar. 16	Abernethy, 2nd-Lt. R. C.	
Gen. List	29 Oct. 17	Aird, 2nd-Lt. M.	
Gen. List	....	Alderton, 2nd-Lt. F. W.	
Bedf. R.	....	Alleguen, 2nd-Lt. E. J.	
Gen. List	....	Allen, 2nd-Lt. H. K. S.	
Gen. List	31 Oct. 17	Ambler, 2nd-Lt. L. E.	
Gen. List	....	Antrobus, 2nd-Lt. P. C. T.	
K. O. Sco. Bord.	15 Mar. 15	Anderson, 2nd-Lt. C. B.	
R. Ir. Rif.	4 Oct. 15	Andrews, 2nd-Lt. D. E.	
Oxf. and Bucks. L. I.	29 Aug. 17	Andrews, 2nd-Lt. P. J.	
West and Cumb. Yeo.	10 July 15	Appleyard, 2nd-Lt. J. R.	
Notts. and Derby R.	30 May 15	Archer, 2nd-Lt. A. S.	
Gen. List	28 Nov. 17	Arnold, 2nd-Lt. F.	
R. W. Fus.	8 July 16	Arthur, 2nd-Lt. G. J.	
Gen. List	31 Oct. 17	Atkins, 2nd-Lt. C. R.	
E. Surr. R.	3 Apl. 15	Atkins, 2nd-Lt. E. R. H.	
Gen. List	26 Sep. 17	Bacon, 2nd-Lt. W. D.	
Innis. Dns.	6 Sep. 16	Baillie, 2nd-Lt. P.	
Gen. List	....	Baldwin, 2nd-Lt. R. A. C.	
Gen. List	....	Bance, 2nd-Lt. W. J.	
Gen. List	....	Bannister, 2nd-Lt. W. C.	
Gen. List	26 Aug. 17	Bancroft, 2nd-Lt. G. R. C.	
Gen. List	9 Oct. 17	Banwell, 2nd-Lt. J. A.	
R. N. V. Res.	....	Barclay,-Brown, 2nd-Lt. R.	
Gen. List	....	Barkway, 2nd-Lt. A. F. C.	
R. Fus.	2 Dec. 15	Barker, 2nd-Lt. H. A.	
R. Rif. C.	26 Sep. 17	Barker, 2nd-Lt. W. D.	
S. Staff. R.	17 Sep. 15	Barton, 2nd-Lt. A. G. A.	
Hussars	29 June 17	Barton, 2nd-Lt. J. W.	
Norf. R.	21 July 15	Beales, Lt. W. L.	
Gen. List	....	Beaumont, 2nd-Lt. R.	
Gen. List	26 Sep. 17	Belfield, 2nd-Lt. G. H.	
Gen. List	29 Aug. 17	Bell, 2nd-Lt. W. C.	
R. Fus.	30 May 17	Berry, 2nd-Lt. A.	
Gen. List	....	Berry, 2nd-Lt. A.	
Lond. R.	13 Aug. 15	Betts, 2nd-Lt. R. O.	
Gen. List	19 Oct. 17	Binns, 2nd-Lt. B. O.	
Gord. Highrs.	26 Sep. 17	Bissett, 2nd-Lt. J. D.	
Gen. List	....	Bishop, 2nd-Lt. H. G. S.	
Sco. Horse	11 Jan. 16	Black, Lt. J. S. C.	
A. S. C.	16 Aug. 15	Black, Lt. A. S.	
R. Scots	27 June 17	Bladworth, 2nd-Lt. H. M.	
Gen. List	31 Oct. 17	Blair, 2nd-Lt. J. M.	
Gen. List	....	Bonney, 2nd-Lt. G. S.	
Gen. List	30 Dec. 17	Bower, 2nd-Lt. F.	
R. N. V. Res.	....	Boyd, Sub-Lt. N. R.	
Cam'n. Highrs.	3 July 15	Boyle, 2nd-Lt. J.	
R. Ir. Rif.	27 Mar. 17	Brady, 2nd-Lt. H. K.	
Gen. List	26 Nov. 17	Braimbridge, 2nd-Lt. R. G.	
Gen. List	31 Oct. 17	Brand, 2nd-Lt. F. T.	
Gen. List	28 Nov. 17	Bray, 2nd Lt. N. D.	
Gen. List	....	Brereton, 2nd-Lt. H.	
Gen. List	....	Brett, 2nd-Lt. A. L.	
Gen. List	19 Dec. 18	Brewer, 2nd-Lt. C. E.	
Cav.	1 May 17	Brewster, 2nd-Lt. F.	
Gen. List	....	Bristow, 2nd-Lt. W. H.	
Gen. List	19 Dec. 17	Brocklebank, 2nd-Lt. G. S.	
Hamps. R.	24 Apl. 15	Brodie, 2nd-Lt. R. P.	
Gen. List	31 Oct. 17	Bromby, 2nd-Lt. T. A.	
Gen. List	29 July 17	Bromley, 2nd-Lt. B. E.	
R. W. Fus.	22 Nov. 16	Brommage, 2nd-Lt. J. C.	
Gen. List	5 Jan. 18	Browne, 2nd-Lt. V. S. T.	
W. Rid. R.	1 Aug. 17	Brown, 2nd-Lt. W. R.	
S. A. Forces	1 Oct. 14	Brown, Lt. S. B.	
Gen. List	....	Brown, 2nd-Lt. A. D.	
Gen. List	....	Brown, 2nd-Lt. W.	
Gen. List	....	Brown, 2nd-Lt. K. S.	
Gen. List	28 Feb. 18	Bruce, 2nd-Lt. W.	
R. F. A.	11 Aug. 17	Bryant, Lt. R. J. C.	
Gen. List	31 Oct. 17	Buckley, 2nd-Lt. P. W.	
R. N. Dn.	...	Buckland, Sub-Lt. L. J. E.	
Gen. List	26 Mar. 18	Buckner, 2nd-Lt. C. S.	
R. F. A.	31 Jan. 15	Bulley, Lt. H. R.	
Gen. List	29 Aug. 17	Bullocke, 2nd-Lt. J. G.	

Regiment.	Date of first Commission.	Names and rank.	Remarks.

OFFICERS OF THE INDIAN ARMY RESERVE OF OFFICERS ON PROBATION—*contd.*

Regiment.	Date of first Commission.	Names and rank.	Remarks.
Norf. R.	19 Nov. 14	Bunn, Lt. W.	
Gen. List	...	Bunt, 2nd-Lt. H.	
Gen. List	27 June 17	Burder, 2nd-Lt. M. N.	
Lancs. R.	...	Burns, 2nd-Lt. F.	
R. N. V. Res.	27 June 17	Burns, Sub-Lt. J.	
Notts. & Derby.	19 Sep. 14	Burrows, Lt. G. T.	
Gen. List	29 Aug. 17	Burrows, 2nd-Lt. S. J.	
L'pool. R.	...	Burr ws, 2nd-Lt. E.	
Ches. R.	14 Nov. 14	Burton, Lt. R. A.	
R. Ir. Rif.	25 Apl. 15	Buser, Lt. C. F.	
Wilts. R.	11 June 15	Butler, Lt. M.	
Gen. List	31 Oct. 17	Cadman, 2nd-Lt. A. L.	
Derby. Yeo.	16 Apl. 15	Calder, Lt. J. T.	
Gen. List	18 Dec. 17	Campbell, 2nd-Lt. W. B.	
R. Ir. Rif.	31 Mar. 15	Carson, 2nd-Lt. J. C.	
W. York R.	30 May 17	Carter, 2nd-Lt. J. T.	
Gen. List	...	Cave, 2nd-Lt. H.	
Gen. List	26 Sep. 17	Chadwick, 2nd-Lt. W. H.	
L'pool R.	5 Aug. 16	Chamberlain, 2nd-Lt. W. G.	
D. of Corn. L. I.	19 Aug. 15	Champion, 2nd-Lt. H. S.	
Gen. List	...	Chamings, 2nd-Lt. R. G.	
Gen. List	...	Chanin, 2nd-Lt. P. R.	
R. N. V. Res.	...	Chapman, Sub-Lt. D. J.	
Gen. List	28 Nov. 17	Chenery, 2nd-Lt. A. F.	
Sco. Horse	22 Nov. 16	Christie, 2nd-Lt. I. F.	
York. L. I.	...	Christopherson, 2nd-Lt. D.	
Can. I.	26 Jan. 18	Clarke, Lt. D. G.	
Gen. List	...	Clarke, 2nd-Lt. H.	
High. L. I.	26 Oct. 15	Clarke, 2nd-Lt. A. W. C.	
Gen. List	...	Clark, Lt. A. H.	
Gen. List	19 Dec. 17	Clark, 2nd-Lt. J. D.	
High. L. I.	13 Jan. 18	Clarkson, Lt. R. M.	
High. L. I.	30 Jan. 18	Clarkson, 2nd-Lt. A. S.	
Notts. & Derby. R.	5 Nov. 15	Clay, Lt. R. D.	
Hamps. R.	26 Aug. 14	Clement, Lt. J.	
R. Ir. Rif.	23 Aug. 15	Clendenning, 2nd-Lt. H.	
Gen. List	25 Dec. 17	Cochrane, 2nd-Lt. D. W.	
Can. Infy.	6 Apl. 15	Colley, Lt. T. B.	
Gen. List	21 Dec. 17	Collins, 2nd-Lt. F. J.	
Gen. List	28 Nov. 17	Conway, 2nd-Lt. H.	
Gen. List	31 Oct. 17	Cook, 2nd-Lt. R. H.	
R. N. Dn.	30 May 17	Cooper, 2nd-Lt. A. C.	
R. W. Kent. R.	1 Aug. 17	Cooper, 2nd-Lt. N. E.	
R. Ir. Fus.	6 Jan. 16	Cooper, 2nd-Lt. R. G.	
R. Lanc. R.	27 Oct. 15	Copleston, 2nd-Lt. R. T.	
Gen. List	31 Oct. 17	Corser, 2nd-Lt. H. D.	
Gen. List	19 Dec. 17	Cornwell, 2nd-Lt. V. J. S.	
Gen. List	26 Sep. 17	Cottam, 2nd-Lt. G. H.	
S. Wales Bord.	1 May 18	Cottrell, 2nd-Lt. H.	
Can. Contgt.	...	Coulter, Lt. A. W.	
R. Innis. Fus.	...	Courley, 2nd-Lt. J. Mc. C.	
Notts. & Derby. R.	...	Cowan, Lt. S. N.	
Gen. List	...	Cowper-Smith, 2nd-Lt. S. C.	
Gen. List	1 May 18	Cox, 2nd-Lt. H. G.	
Gen. List	26 Sep. 17	Craig, 2nd-Lt. J. W.	
Gen. List	...	Craigie, 2nd-Lt. J. H.	
Tr. Res. Bn.	29 Jan. 15	Crichton, Lt. W. A.	
Gen. List	...	Crisp, 2nd-Lt. M. F.	
R. Ir. Fus.	16 Nov. 15	Crothers, Lt. A. V.	
Gen. List	...	Crowe, 2nd-Lt. H. W.	
Gen. List	20 Sep. 17	Crowley, 2nd-Lt. B. E. S.	
R. Lanc. R.	30 Jan. 18	Cruise, 2nd-Lt. W.	
Gen. List	25 Dec. 17	Cunningham, 2nd-Lt. E. C. J.	
R. Dub. Fus.	22 Apl. 15	Dade, 2nd-Lt. R. T.	
A. S. C.	11 Nov. 14	Dale, Lt. D. H.	
Lond. R.	26 Feb. 15	Dalton, M.C., Lt. R. C.	
Gen. List	23 June 17	Darby, 2nd-Lt. I. C. L.	
A. S. C.	7 Apl. 15	Davey, Lt. H. R.	
R. W. Fus.	6 Mar. 15	Davies, Lt. G. C.	
Gen. List	9 Aug. 17	Davies, 2nd-Lt. J. F.	
S. Wales Bord.	27 Feb. 18	Davies, 2nd-Lt. R. T.	
Gen. List.	....	Davis, 2nd-Lt. N. E.	
Gen. List.	....	Dawson, 2nd-Lt. E. J.	
Glouc. Yeo.	23 Mar. 15	deFreville, Lt. G. V.	

OFFICERS OF THE INDIAN ARMY RESERVE OF OFFICERS ON PROBATION—*contd.*

Regiment.	Date of first Commission.	Names and rank.	Remarks.
Gen. List	....	Denyer, 2nd-Lt. P. H.	
Tr. Res. Bn.	21 Aug. 15	Derwent, 2nd-Lt. F. R.	
Gen. List	19 Dec. 17	Diack, 2nd-Lt. W. J.	
Gen. List	7 Nov. 17	Dickens, 2nd-Lt. C. E.	
Gen. List	....	Dick, 2nd-Lt. H. N. H.	
Gen. List	....	Dixon, 2nd-Lt. R.	
Gen. List	....	Donald, 2nd-Lt. J. W.	
Gen. List	25 Sep. 17	Drake, 2nd-Lt. A. V.	
Essex Yeo.	Apl. 14	Dudding, Lt. M.	
Gen. List	28 Nov. 17	Dumbreck, 2nd-Lt. W. V.	
Gen. List	31 Oct. 17	Dunwell, 2nd-Lt. C. N.	
Gen. List	29 Aug. 17	Du Pre, 2nd-Lt. M. D. A. G.	
R. Suss. R.	1 Mar. 17	Dutton, 2nd-Lt. W. H. R.	
Gen. List	....	Dybell, 2nd-Lt. C. W.	
Tr. Res. Bn.	27 July 15	Dyer, Lt. A. J.	
Gen. List	26 Sep. 17	Eaden, 2nd-Lt. C. L.	
Gen. List	26 Sep. 17	Eaton, 2nd-Lt. R. S.	
D. of Corn. L. I.	22 Apl. 15	Eckel, Lt. P.	
Gen. List	....	Edmiston, 2nd-Lt. R. W.	
Worc. R.	9 Sep. 15	Edwards, Lt. F. O.	
S. Staff. R.	2 Oct. 15	Edwards, 2nd-Lt. S. W.	
N. Lan. R.	29 Aug. 17	Edwards, 2nd-Lt. H.	
Mon. R.	1 Aug. 17	Edwards, 2nd-Lt. A. W.	
R. Fus.	....	Ellis, 2nd-Lt. A. W.	
Gen. List	....	Elliot, 2nd-Lt. W. J.	
Gen. List	29 Aug. 17	Elson, 2nd-Lt. W.	
Dns.	1 May 17	Elton, 2nd-Lt. C. B.	
Lan. Fus.	25 Oct. 16	Evans, 2nd-Lt. F. L.	
Gen. List	26 Sep. 17	Evans, 2nd-Lt. J.	
Gen. List	....	Evans, 2nd-Lt. H. R.	
Lan. Fus.	25 Oct. 16	Evans, 2nd-Lt. F. L.	
Gen. List	19 Dec. 17	Everett, 2nd-Lt. W. M.	
Lond. R.	7 Oct. 14	Eve, Lt. H. E., M.C.	
R. W. Fus.	13 Nov. 16	Ewin, 2nd-Lt. H. L. H.	
Gen. List	....	Fairclough, 2nd-Lt. J. E. B.	
Gen. List	28 Nov. 17	Farley, 2nd-Lt. F. A.	
Gen. List	31 Oct. 17	Finglas, 2nd-Lt. U. C. N.	
Gen. List	....	Fitch, 2nd-Lt. S. A.	
York and Lanc. R.	18 Feb. 15	Fitton, Lt. C. V.	
Som. L. I.	1 Dec. 14	Flower, Lt. H. G.	
Gen. List	....	Forbes, Lt. A.	
Innis. Fus.	8 Nov. 15	Fortescue, Lt. A.	
Gen. List	....	Forster, 2nd-Lt. H. R.	
Sco. Horse	11 June 15	Forsyth, Lt. C. S.	
Gen. List	....	Foster, 2nd-Lt. H. E.	
Gen. List	31 Oct. 17	Fox, 2nd-Lt. R. J.	
Essex R.	1 Aug. 17	Fraser, 2nd-Lt. E. R. D.	
Lond. R.	20 Oct. 16	Franklin, 2nd-Lt. S. S.	
Bord. Horse	6 May 16	Friend, 2nd-Lt. M.	
R. Innis. Fus.	30 Jan. 18	Frizzel, 2nd-Lt. W. G.	
Lond. R.	26 Sep. 16	Froy, Lt. W. A.	
Gen. List	....	Fuller, 2nd-Lt. B. H.	
Gen. List	....	Gale, 2nd-Lt. H. C.	
Norf. R.	25 Jan. 17	Gardiner, 2nd-Lt. A. L.	
Lab. Corps	....	Gemmell, 2nd-Lt. J. T. I.	
Ches. R.	1 Aug. 17	Gerrard, 2nd-Lt. D. M.	
Gen. List	....	Giblett, 2nd-Lt. J. V.	
Gen. List	....	Gibson, 2nd-Lt. P. R.	
Lond. R.	8 Aug. 15	Giffen, Lt. L. E.	
R. Fus.	....	Giland, 2nd-Lt. J. G.	
Gen. List	....	Gilbert, 2nd-Lt. J. R.	
Gen. List	26 Sep. 17	Gilkes, 2nd-Lt. C. H.	
Gen. List	....	Gilliatt, 2nd-Lt. L.	
Gen. List	19 Dec. 17	Gillman, 2nd-Lt. B. B.	
Gen. List	31 Oct. 17	Ginns, 2nd-Lt. R.	
Gen. List	....	Gordon, 2nd-Lt. A. R.	
Cav.	....	Gordon, 2nd-Lt. W.	
Lond. R.	28 Mar. 17	Gosling, 2nd-Lt. E. A. C.	
E. Surr. R.	30 May 17	Goulden, 2nd-Lt. C. F.	
Gen. List	....	Grabham, 2nd-Lt. H. M.	
C. Gds.	16 Sep. 14	Graham, Lt. M. D.	
M. G. Corps	24 Oct. 14	Grant, Lt. A. G. C.	
Gen. List	26 Nov. 17	Grant, 2nd-Lt. A. O.	
High. L. I.	22 Nov. 16	Gray, Lt. W.	

OFFICERS OF THE INDIAN ARMY RESERVE OF OFFICERS ON PROBATION—*contd.*

Regiment.	Date of first Commission.	Names and ranks.	Remarks.
Notts. and Derby. R.	5 Aug. 16	Gray, 2nd-Lt. A.	
Lan. R.	27 Mar. 18	Gray, 2nd-Lt. F.	
Gen. List	....	Gray, 2nd-Lt. G. E.	
Gen. List	26 Sep. 17	Green, 2nd-Lt. G. H. B.	
Manch. R.	30 May 17	Green, 2nd-Lt. T. W.	
Gen. List	....	Gregson, 2nd-Lt. D. A.	
Gen. List	....	Gregory, 2nd-Lt. C. J. E.	
W. Rid. R.	11 Feb. 15	Greenwood, Lt. R. S.	
Gen. List	31 Oct. 17	Gronow, 2nd-Lt. I. A.	
Linc. R.	29 Aug. 17	Groom, 2nd-Lt. C. O.	
Gen. List	29 Aug. 17	Gudgeon, 2nd-Lt. P. E.	
Gen. List	26 Sep. 17	Gunson, 2nd-Lt. G. H.	
Yeo.	28 Nov. 17	Hadfield, 2nd-Lt. H. W.	
Gen. List.	31 Oct. 17	Hall, 2nd-Lt. R. F.	
Gen. List	25 Dec. 17	Hall, 2nd-Lt. R. G.	
R. E.	19 Dec. 15	Hamilton, Lt. G.	
Gen. List	29 Aug. 17	Hancock, 2nd-Lt. A.	
A. S. C.	25 Mar. 15	Handcock, 2nd-Lt. W. G.	
R. Innis. Fus.	14 July 15	Hanna, 2nd-Lt. J. R. M.	
Gen. List	....	Hannigan, 2nd-Lt. A.	
Gen. List	3 Oct. 17	Hanson, 2nd-Lt. H. A.	
R. N. V. Res.	26 June 15	Hanvey, Sub-Lt. M. P.	
S. Wales Bord.	27 Mar. 18	Hardie, 2nd-Lt. J. J.	
Gen. List	....	Hardman, 2nd-Lt. A.	
R. Highrs.	11 Aug. 15	Harley, 2nd-Lt. N.	
Gen. List	26 Sep. 17	Harley, 2nd-Lt. W.	
Gen. List	31 Oct. 17	Harris, 2nd-Lt. E. J.	
L'pool R.	19 May 15	Harris, Lt. H.	
Lan. Fus.	29 Aug. 17	Harrison, 2nd-Lt. W. H.	
Gen. List	....	Harrison, 2nd-Lt. M. K.	
Gen. List	31 Oct. 17	Harrison, 2nd-Lt. J. S.	
Gen. List	26 Sep. 17	Harrison, 2nd-Lt. G. E.	
Bord. R.	29 Jan. 16	Harold, 2nd-Lt. P. H.	
L'pool R.	8 July 15	Hartles, 2nd-Lt. W. E.	
Ches. R.	29 Aug. 17	Hartley, 2nd-Lt. G. C.	
Gen. List	19 Oct. 17	Harvey, 2nd-Lt. A. J.	
R. N. Dn.	25 Oct. 16	Harwood, 2nd-Lt. G. E.	
Gen. List	....	Haughton, 2nd-Lt. H.	
N. Staff. R.	21 Sep. 14	Hawkes, M. C., Lt. H.	
Gen. List	....	Hawkins, 2nd-Lt. F. H.	
Gen. List	29 Aug. 17	Hay, 2nd-Lt. R. W. M.	
Gen. List	....	Haymes, 2nd-Lt. W. R.	
Gen. List	....	Hayward, 2nd-Lt. H. B.	
Gen. List	....	Hayward, 2nd-Lt. C. D.	
A. Cyclist Corps	15 Oct. 14	Heath, Lt. W. A. H.	
Gen. List	29 Jan. 18	Helmore, 2nd-Lt. G. D.	
Cav.	1 May 17	Herbison, 2nd-Lt. L. J.	
Arg. & Suth'd Highrs.	21 Sep. 15	Hewison, Lt. A.	
War. Yeo.	17 Feb. 17	Hicks, 2nd-Lt. W. A.	
Gen. List	31 Oct. 17	Hickson, 2nd-Lt. J. C. A.	
R. War. R.	1 May 18	Hill. 2nd-Lt. H. L.	
Gen. List	20 Apl. 17	Hind, 2nd-Lt. F. M. J.	
Gen. List	21 Mar. 18	Hipsey, 2nd-Lt. A. W. L.	
Gen. List	....	Hodgman, 2nd-Lt. S. F. J.	
Gen. List	29 Aug. 17	Hogbin, 2nd-Lt. C. J.	
Bord. R.	20 Nov. 14	Holdsworth, Lt. P. B. C.	
Gen. List	....	Holland, 2nd-Lt. A. P.	
Gen. List	....	Hollwell, 2nd-Lt. P. F.	
R. F. A.	8 Mar. 15	Holman, 2nd-Lt. J. C. L.	
Notts. & Derby, R.	1 July 17	Holmes, 2nd-Lt. C.	
Cav.	6 Apl. 18	Holt. 2nd-Lt. E. M.	
Norf. R.	30 July 15	Hood, Lt. F. W.	
Cam'n Highrs.	22 June 17	Hood, 2nd-Lt. D. A.	
S. Lan. R.	25 Sep. 15	Hopewell, 2nd-Lt. T.	
Cav.	21 Dec. 16	Hopkins, 2nd-Lt. F. L.	
Gen. List	....	Horncastle, 2nd-Lt. J. A.	
Uganda Police	1 Dec. 15	Houchen, Lt. J. B.	
Gen. List	....	Houghton, 2nd-Lt. H.	
Glouc. R.	....	Houghton, 2nd-Lt. W. H.	
Cav.	28 Nov. 16	Howe, 2nd Lt. G. E.	
Cav.	1 May 17	Howell, 2nd-Lt. W. E.	
Gen. List	....	Howells, 2nd-Lt. D. R.	
Gen. List	....	Howfield, 2nd-Lt. J. A. J.	
R. F. A.	15 Aug. 17	Howison, 2nd-Lt. T. A.	

Regt.	Date of first Commission.	Names and Rank.	Remarks.

OFFICERS OF THE INDIAN ARMY RESERVE OF OFFICERS ON PROBATION—*contd.*

Regt.	Date of first Commission.	Names and Rank.	Remarks.
Gen. List	29 Aug. 17	Hughes, 2nd-Lt. F. E.	
Gen. List	31 Oct. 17	Hughes, 2nd-Lt. S.	
D. of Corn. L.	10 July 15	Hughesdon, M. C. Lt., R. H.	
N. Dev. Hrs.	7 Mar. 15	Hulbert, Lt. M. L.	
Gen. List	. . .	Hu bert, 2nd-Lt. R. C.	
E. Kent, R.	. . .	Humphries, Lt. E. H.	
Midd x. R.	3 Sep. 15	Hunter, 2nd-Lt. T. C. L.	
R. Rif. C.	15 May 15	Ingman, 2nd-Lt. G. E.	
War. Yeo.	28 Nov. 16	Innocent, 2nd-Lt. A.	
R. Ir. Fus.	20 July 15	Jack, Lt. A. J.	
Gord. Highrs.	9 Apl. 15	Jack, Lt. M. M.	
Gen. List	11 Oct. 17	Jacks, 2nd-Lt. W. R.	
Lond. R.	31 Oct. 17	Jackson, 2nd-Lt. H. L.	
Gen. List	. . .	Jackson, 2nd-Lt. T.	
S. Notts. Hrs.	18 Dec. 14	Jacques, Lt. G. P. R.	
Tr. Res. Bn.	31 May 15	James, Lt. H.	
Dns.	17 Apl. 17	Jaques, 2nd-Lt. B. W.	
R. W. Fus.	26 Sep. 17	Jenkins, 2nd-Lt. J. W. P. J.	
A. S. C.	8 Sep. 15	Johnson, Lt. K. H.	
R. Innis. Fus.	11 Dec. 14	Johnson, Lt. T. G	
Gen. List	26 Sep. 17	Johnson, 2nd-Lt. F. L.	
A. S. C.	4 May 15	Jones, Lt. R. E.	
L'pool R.	22 Jan. 16	Jones, Lt. T. E. S.	
Gen. List	. . .	Jones, 2nd-Lt. G. E.	
Gen. List	. . .	Jones, 2nd-Lt. A.	
Gen. List	. . .	Jones, 2nd-Lt. C.	
R. W. Fus.	7 Feb. 16	Jones, 2nd-Lt. J. H.	
Genl. List	28 Nov. 17	Jones, 2nd-Lt. B. C.	
Norf. R.	29 Aug. 17	Jones, G. P.	
Genl. List	. . .	Jones, 2nd Lt. J. L.	
M. G. Corps	26 Nov. 17	Joshua, 2nd-Lt. W. J.	
Yorks L. I.	25 Jan. 17	Joyce, 2nd-Lt. B. J. M. R.	
A. S. C.	15 Nov. 15	Keay, Lt. A. J.	
Dorset R.	19 Dec. 16	Kellaway, 2nd-Lt. G. L.	
Gen. List	25 Dec. 17	Kelly, 2nd-Lt. A.	
A. S. C.	31 Dec. 16	Kelly, 2nd-Lt. A. G.	
Gen. List	31 Oct. 17	Kenward, 2nd-Lt. P. E.	
Arg. & Suth'd. Highrs.	1 Mar. 17	Kerr, 2nd-Lt. J. C.	
Gen. List	8 Oct. 17	Kincaid, 2nd-Lt. J.	
Gen. List	31 Oct. 17	King, 2nd-Lt. T. A.	
Gen. List	26 Sep. 17	Kinnaird, 2nd-Lt. C. H.	
Gen. List	28 Nov. 17	Knatchbull, 2nd-Lt. R. E.	
Gen. List	. . .	Knapton, 2nd-Lt. J. H.	
Gen. List	31 Oct. 17	Knight, 2nd-Lt. P. H.	
Gord. Highrs.	30 May 17	Kydd, 2nd-Lt. W. A.	
Gen. List	. . . .	Laird, 2nd-Lt. A. B.	
Lond. Sco.	27 Apl. 17	Laird, 2nd-Lt. E. H.	
Gen. List	8 July 17	Laming, 2nd-Lt. S. A.	
Gen. List	31 Oct. 17	Lancaster, 2nd-Lt. M. P.	
Gen. List	26 Sep. 17	Lange, 2nd-Lt. P. M. A.	
A. S. C.	31 Aug. 14	Larkin, Lt. A. S.	
Gen. List	26 Sep. 17	Lawrence, 2nd-Lt. H. A.	
R. N. V. Res.	. . .	Lawrie, Sub-Lt. T. M.	
R. Innis. Fus.	26 Sep. 17	Layfield, 2nd-Lt. F.	
Hrs.	24 Nov. 17	Leader, 2nd-Lt. S.	
A. S. C.	. . . .	Leake, 2nd-Lt. H.	
W. York. R.	28 Dec. 15	Lee, Lt. D. H.	
W. York. R.	1 Aug. 17	Lee, 2nd-Lt. F. A.	
Gen. List	26 Sep. 17	Lennox, 2nd-Lt. A.	
R. Lan. R.	27 Feb. 18	Leonard, 2nd-Lt. R. E.	
Gen. List	. . . .	Lethern, 2nd-Lt. W. E.	
R. Fus.	. . . .	Libby, 2nd-Lt. H. J.	
R. Ir. Rif.	22 Nov. 16	Lewis, 2nd-Lt. J. A.	
R. Suss. R.	17 Nov. 16	Lewis, 2nd-Lt. J. D.	
R. W. Fus.	30 May 17	Lewis, 2nd-Lt. E. S.	
Gen. List	. . . .	Limb, 2nd-Lt. H. E.	
Gen. List	31 Oct. 17	Lindsay, 2nd-Lt. A. R.	
Gen. List	28 Nov. 17	Lintott, 2nd-Lt. L. E.	
R. W. Fus.	15 July 15	Lloyd, 2nd-Lt. J. H.	
Gen. List	. . . .	Lloyd, 2nd-Lt. J. W.	
Lan. R.	27 Feb. 18	Lomax, 2nd-Lt. G. W.	
Cav.	27 Feb. 18	Lordan, 2nd-Lt. J. P.	
Gen. List	31 Oct. 17	Low, 2nd-Lt. W. H.	
Leic. R.	28 Mar. 15	Lowe, 2nd-Lt. H. A.	

OFFICERS OF THE INDIAN ARMY RESERVE OF OFFICERS ON PROBATION—*contd.*

Regiment.	Date of first Commission.	Name and rank.	Remarks.
Essex R.	6 Apl. 5	Lunnon, Lt. J. D. K.	
Notts. & Derby. R.	21 Jan. 15	Lurcott, Lt. R. G.	
R. Ir. Rif.	21 May 15	Lynch, 2nd-Lt. H. E. H.	
Gen. List	25 Nov. 17	Lyte, 2nd-Lt. C. F. N.	
Gen. List	29 Aug. 17	Macdonald, 2nd-Lt. D.	
R. N. Dn.	6 June 15	Macdonald, Sub-Lt. P. S.	
Gen. List	....	Macgregor, 2nd-Lt. J. S.	
Welsh. R.	13 June 13	Machon, Capt. G. C. S.	
Gen. List	28 Nov. 17	Mac Intyre, 2nd-Lt. W.	
Gen. List	26 Nov. 17	Mackis, 2nd-Lt. J.	
A. S. C.	....	Mackenzie, R. H. H.	
Lan. R.	25 Mar. 18	MacQueen, 2nd-Lt. S. Y.	
Gen. List	31 Oct. 17	Madley, 2nd-Lt. J.	
Gen. List	21 Dec. 17	Male, 2nd-Lt. E.	
R. Berk. R.	26 Jan. 15	Manders, Capt. H. C.	
Norf. R.	28 Nov. 14	Manning, 2nd-Lt. T. W.	
Gen. List	....	Manning Press, 2nd-Lt. C. A.	
Gen. List	....	Mantle, 2nd-Lt. W. E.	
Gen. List	11 Oct. 17	Margerison, 2nd-Lt. L. A.	
R. N. V. Res.	....	Marlow, W.	
North'd. Fus.	16 Nov. 14	Marple, 2nd Lt. R. G.	
Gen. List	29 Aug. 17	Marsh-Smith, 2nd-Lt. E. C	
R. N. Dn.	26 Apl. 17	Marsland, Sub-Lt. K. D.	
S. Lan. R.	26 Jan. 15	Martin, 2nd-Lt. S. F.	
Gen. List	....	Martin, 2nd-Lt. G. F.	
Gen. List	....	Martin, 2nd-Lt. H. L.	
Oxf. & Bucks L. I.	25 Sep. 15	Mason, Lt. H. C. E.	
Gen. List	....	Mason, 2nd-Lt. E. W. R.	
R. N. Dn.	....	Matcham, Sub-Lt. L. C.	
R. F. A.	....	Maunder, 2nd-Lt. G. H. O.	
Gen. List	....	Mawson, 2nd-Lt. J. E.	
Gen. List	19 Dec. 7	May, 2nd-Lt. A. V.	
York, L. I.	28 Mar. 17	Maycock, 2nd-Lt. H. W.	
Gen. List	....	Maynard, 2nd-Lt. R.	
Gen. List	....	Mc Andrew, 2nd-Lt. A.	
Gen. List	29 Aug. 17	Mc Combe, 2nd-Lt. A.	
Bord R.	7 May 15	Mc Crea, 2nd-Lt. J.	
Gen. List	....	Mc Evoy, 2nd-Lt. J. F.	
R. Scots	6 Aug. 15	Mc Farland, 2nd-Lt. A. H.	
Gen. List	29 Aug. 17	Mc Gorry, 2nd-Lt. F. E.	
Bedf. R.	29 May 18	Mc Intosh, 2nd-Lt. R. W.	
Gord. Highrs.	2 Oct. 14	Mc Kerron, Lt. P. A. B.	
Gen. List	1 Aug. 17	Mc Kinnon, 2nd-Lt. J. A.	
Lan. Fus.	1 Aug. 17	Mc Larty, 2nd-Lt. A.	
A. S. C.	11 Oct. 15	Mc Laurin, Lt. D. J.	
R. Rif. C.	27 Nov. 14	Mc Lean, 2nd-Lt. A. E. J.	
R. Ir Rif.	May 15	Mc Neill, 2nd-Lt. M. D.	
Gen. List	...	Medlicott, 2nd-Lt. S. J. E.	
R. Fus.	4 Aug. 15	Meeks, 2nd-Lt. F. S.	
R. War. R.	29 May 17	Meese, 2nd-Lt. W. A. B.	
Gen. List	21 Dec. 17	Mehew, 2nd-Lt. R. S.	
Surr. Yeo.	29 Aug. 14	Mellor, 2nd-Lt. A. R. I.	
Suff. R.	23 Apl. 16	Meredith, 2nd-Lt. H. T.	
Norf. R.	15 Sep. 15	Merry, Lt. R.	
Midd'x. R.	30 Jan. 18	Merry, 2nd-Lt. N. Mc I.	
Sco. Bord.	25 Jan. 16	Michie, 2nd-Lt. J. B.	
Gen. List	31 Oct. 17	Millar, 2nd-Lt. R. S. C.	
Gen. List	...	Mills, 2nd-Lt. S. W.	
Hrs.	25 Aug. 17	Mitchem, 2nd-Lt. A. H.	
Gen. List.	...	Mitchell, 2nd-Lt. J.	
Devon R.	10 Oct. 14	Monk, Lt. G. A. W.	
R. Ir. Fus.	28 Feb. 17	Montgomery, 2nd-Lt. R. M.	
R. Innis. Fus.	...	Montray, Cap. F. C.	
Leins. R.	11 Nov. 14	Moran, Lt. E. W.	
Gen. List.	25 Dec. 17	Morgan, 2nd-Lt. D. T.	
Gen. List	29 Aug. 17	Moriarty, 2nd-Lt. P.	
Dns.	17 Apl. 17	Morrison, 2nd-Lt. R. M. S.	
York. L. I.	30 Jan. 18	Morrison, 2nd-Lt. J. W.	
Bedf. R.	27 Mar. 18	Morris, 2nd-Lt. J. J.	
Norf. Yeo.	22 Nov. 16	Muirhead, 2nd-Lt. H. W.	
R. Lan. R.	29 May 18	Muirhead 2nd-Lt. J. M.	
Gen. List.	...	Murray, 2nd-Lt. W. T. W.	
Gen. List.	...	Murray, 2nd-Lt. J. W.	
R. G. A.	20 Dec. 16	Myles, 2nd-Lt R. H. S.	

Regiment.	Date of first Commission.	Name and Rank.	Remarks.

OFFICERS OF THE INDIAN ARMY RESERVE OF OFFICERS ON PROBATION—*contd.*

Regiment.	Date of first Commission.	Name and Rank.	Remarks.
E. Kent. R.	5 Aug. 16	Neal, 2nd-Lt. G. E.	
L'pool R.	14 Nov. 14	Neale, Lt. A. K. H.	
Gen. List	...	Neale, 2nd-Lt. R. T.	
S. Staff. R.	1 Aug. 17	Newbold, 2nd-Lt. A. W.	
Gen. List	...	Newman, 2nd-Lt. C. E. T.	
Gen. List	...	Newsam, 2nd-Lt. E. S.	
Ches. R.	22 Jan. 15	Newton, Lt. J. E.	
R. W. Fus.	27 Mar. 18	Nicholas, 2nd-Lt. J. R.	
Durh. L. I.	8 Nov. 15	Nicholson, Capt. E. A.	
Lond. R.	15 July 15	Nichol, Lt. F. C.	
Glouc. R.	5 Sep. 16	Norman, 2nd-Lt. C. A.	
W. Rid. R.	30 July 15	Norris, 2nd-Lt. W. T.	
Lond. R.	26 Oct. 16	North, 2nd-Lt. A. E.	
Gen. List	...	North, 2nd-Lt. F. E.	
Gen. List	28 Nov. 17	Obee, 2nd-Lt. L. E.	
Gen. List	31 Oct. 17	O'Calloghan, 2nd-Lt. F. J.	
Ir. Horse	21 Dec. 16	Oecken, 2nd-Lt. C. A. J.	
W. Surr. R.	1 May 18	Olden, 2nd-Lt. S. M.	
Gen. List	...	Oldroyd, O. E.	
Arg. & Suth'd Highrs	25 Aug. 15	Oliver, Lt. J. S.	
Gen. List	26 Sep. 17	Oliver, 2nd-Lt. N. E.	
A. S. C.	19 July 15	O'Meara, Lt. A. J.	
Gen. List	...	O'Reilly, 2nd-Lt. R.	
R. Ir. Rif.	21 May 15	Orr, 2nd-Lt. C. H. H.	
Gen. List	28 Nov. 17	Padfield, 2nd-Lt. C. A.	
Dns.	1 Sep. 15	Pallant, 2nd-Lt. H. C.	
Dorset R.	2 June 15	Park, 2nd-Lt. L.	
Manch. R.	Feb. 13	Park, Lt. D.	
Gen. List	...	Parsons, 2nd-Lt. J.	
W. Rid. R.	29 Aug. 17	Patch, 2nd-Lt. V. L.	
Gen. List	...	Peacock, 2nd-Lt. A.	
Midd'x R.	12 Oct. 15	Peattie, 2nd-Lt. G. D.	
R. Innis. Fus.	22 Jan. 16	Peddie, Lt. R. M.	
Gen. List	...	Penadon, 2nd-Lt. F. L.	
York. L. I.	30 May 17	Penny, 2nd-Lt. C. E.	
Norf. R.	25 Oct. 16	Perks, 2nd-Lt. G. A. C.	
Gen. List	...	Perrins, 2nd-Lt. H. G.	
Suff. R.	27 June 17	Peters, 2nd-Lt. S. A.	
Bedf. R.	29 May 18	Peters, M. M., 2nd-Lt. W.	
Notts. & Derby. R.	5 Sep. 16	Petts, 2nd-Lt. C. E.	
Gen. List	31 Oct. 17	Petty, 2nd-Lt. H. G.	
Durh. L. I.	20 July 15	Philipson, 2nd-Lt. H.	
Hrs.	30 July 15	Phillips, 2nd-Lt. E. I.	
Gen. List	27 July 17	Phillips, 2nd-Lt. N. R. G.	
Gen. List	9 Oct. 17	Pierce, 2nd-Lt. H. J. J.	
Manch. R.	16 July 16	Pitt., M. C., 2nd-Lt. F.	
Hrs.	24 Nov. 17	Polland, 2nd-Lt. R. C.	
Gen. List	28 Nov. 17	Pollard, 2nd-Lt. C. A.	
Gen. List	...	Powell, 2nd-Lt. C. T.	
Gen. List	17 Feb. 18	Power, 2nd-Lt. C. C.	
Shrop. L. I.	30 Nov. 14	Preece, Lt. D. E. G.	
Gen. List	...	Preston, Lt. N.	
Tr. Res. Bn.	20 Apl. 15	Prestwich, Lt. E.	
L'pool. R.	7 Aug. 15	Pumfrey, Lt. A. I.	
Gen. List	...	Putnam, 2nd-Lt. G. W.	
Gen. List	27 Oct. 17	Putnam, 2nd-Lt. T. C.	
Lan. Fus.	22 Nov. 16	Radcliffe, 2nd-Lt. H.	
Gen. List	...	Rae, 2nd-Lt. R. J.	
R. War. R.	10 July 15	Rayner, 2nd-Lt. G. H.	
Gen. List	15 Sep. 17	Redfern, 2nd-Lt. T.	
Sco. Horse	19 Dec. 16	Reekie, Lt. J.	
R. W. Fus.	...	Rees, Lt. H. W.	
R. W. Fus.	22 May 15	Rees, 2nd-Lt. I. D.	
Norf. R.	27 Feb. 18	Reeve, 2nd-Lt. H. L.	
R. Ir. Rif.	9 Jan. 15	Reid. Lt. W. E.	
Gen. List	29 Aug. 17	Reiph, 2nd-Lt. S. D.	
S. Wales. Bord.	26 Aug. 15	Reynolds, 2nd-Lt. J. W.	
Gord. Highrs.	22 Oct. 14	Rice, Lt. J.	
Gen. List	...	Riches, 2nd-Lt. C. F. h.	
Gen. List	31 Oct. 17	Rickard, 2nd-Lt. J. H.	
Sco. Horse	5 Aug. 16	Riddock, 2nd-Lt. J. D.	
R. W. Fus.	19 May 15	Roberts, 2nd-Lt. C. R.	
Gen. List	...	Roberts 2nd-Lt. W. C.	
A. Cyclist C.	15 June 15	Robertson, Lt. H. K.	

Regiment.	Date of first Commission.	Name and Rank.	Remarks.

OFFICERS OF THE INDIAN ARMY RESERVE OF OFFICERS ON PROBATION—*contd.*

Regiment.	Date of first Commission.	Name and Rank.	Remarks.
R. N. Dn.	27 June 17	Robertson, Sub-Lt. A. J.	
Gord. Highrs.	23 Nov. 16	Robertson Lt. R.	
E. Kent. R.	26 June 18	Robertson, 2nd-Lt. R.	
Sco. Horse	22 Nov. 14	Robertson, 2nd-Lt. J.	
Lan. R.	28 Nov. 17	Robertson, 2nd-Lt. J. P. W.	
R. Lan. R.	30 Jan. 18	Robinson, 2nd-Lt. H.	
Gen. List	...	Rogers, 2nd-Lt. J. W.	
W. Surr. R.	1 May 18	Rogers, 2nd-Lt. R. L.	
Seaf. Highrs.	5 Sep. 16	Roger, Lt. T. T.	
Gen. List	26 Sep. 17	Roe, 2nd-Lt. F. W.	
E. Surr. R.	25 Oct. 16	Rollinson, 2nd-Lt. R. G.	
E. Lan. R.	...	Rollinson, 2nd-Lt. A. M.	
Gen. List	...	Rose, 2nd-Lt. W.	
R. Ir. Rif.	29 Aug. 17	Rose, 2nd-Lt. F. W.	
Gen. List	...	Ross, 2nd-Lt. C.	
Gen. List	19 Dec. 17	Ross, 2nd-Lt. R. P.	
Gen. List	29 Aug. 17	Rowbotham, 2nd-Lt. A. W.	
Gen. List	29 Aug. 17	Rowlinson, 2nd-Lt. P. J.	
Leins. R.	5 Aug. 16	Roycroft, 2nd-Lt. B. C.	
Leins. R.	5 Aug. 16	Roycroft, 2nd-Lt. B. F.	
North'd Fus.	3 Apl. 15	Russell, 2nd-Lt. C. S. C.	
Norf. R.	27 Mar. 18	Russell, 2nd-Lt. C. R.	
North'd Fus.	31 Mar. 15	Rutledge, Lt. C. J.	
Gen. List	29 Aug. 17	Ryall, 2nd-Lt. H. H.	
Norf. R.	1 May 18	Sales, 2nd-Lt. H. D.	
Gen. List	...	Salmon, 2nd-Lt. F.	
R. Ir. Rif.	15 Dec. 14	Samuels, 2nd-Lt. C. J. H.	
R. N. Dn.	..	Sanday, Sub-Lt. W. J. W.	
Gen. List	...	Savage, 2nd-Lt. P. G.	
Dorset R.	3 June 13	Saywell, Capt. J. R.	
Linc. R.	1 Mar. 17	Scott, 2nd-Lt. K. D.	
Gen. List	29 Aug. 17	Segar, 2nd-Lt. W. J.	
W. Surr. R.	29 May 18	Seth, 2nd-Lt. E. R.	
Gen. List	...	Sevenoaks, 2nd-Lt. P. L.	
Gen. List	...	Severn, 2nd-Lt. F.	
R. A. F.	...	Seward, 2nd-Lt. J. B.	
D. of Corn. L. I.	29 Aug. 14	Seymour, Lt. A. H.	
R. Fus.	21 June 15	Seymour, 2nd-Lt. W. W.	
Gen. List	...	Shallcroft, 2nd-Lt. L. E.	
Sco. Rif.	14 Dec. 14	Sharp, Lt. J. C.	
Gen. List	29 Aug. 17	Shaw, 2nd-Lt. W.	
Gen. List	29 Aug. 17	Shearer, 2nd-Lt. F. C.	
Gen. List	26 Nov. 17	Shearer, 2nd-Lt. W.	
Gen. List	15 June 18	Sheffield, 2nd-Lt. J. C.	
W. York. R.	12 Nov. 15	Shelley, Lt. C. E.	
Gen. List	...	Shepherd, 2nd-Lt. H. C. B.	
Gen. List	29 Aug. 17	Sheridan, 2nd-Lt. F. L.	
Gen. List	26 Nov. 17	Shipley, 2nd-Lt. H.	
Lond. R.	1 Nov. 15	Short, 2nd-Lt. S. E.	
Bord. R.	22 Dec. 14	Sim, 2nd-Lt. D.	
R. War. R.	13 May 15	Simms, Lt. F. H.	
Dorset R.	26 Sep. 16	Simmons, Lt. F. D.	
R. F. A.	16 Sep. 15	Simpson, Lt. S. R.	
E. Kent. R.	17 Nov. 14	Skey, Lt. L. G.	
Gen. List	28 Nov. 17	Slatter, 2nd-Lt. J. C.	
Gord. Highrs.	24 Oct. 14	Lleigh, Lt. G. R.	
Gen. List	...	Smallman, 2nd-Lt. W. F.	
R. Rif. C.	22 Nov. 16	Smerdon, 2nd-Lt. H. E.	
York and Lancs.	13 Nov. 14	Smith, Lt. G. W.	
Gen. List	28 Nov. 17	Smith, 2nd-Lt. R. H.	
Tr. Res. Bn.	3 Aug. 15	Smith, 2nd-Lt. R.	
Gen. List	29 Aug. 17	Smith, 2nd-Lt. R. C. K.	
Gen. List	13 Nov. 17	Smith, 2nd-Lt. A. F.	
Norf. R.	18 Aug. 15	Smith, 2nd-Lt. H. H.	
Gen. List	27 Feb. 18	Smith, 2nd-Lt. H. N.	
Gen. List	...	Smith, 2nd-Lt. E. H.	
Gen. List	19 Nov. 17	Smithson, 2nd-Lt. E. A.	
Sco. Bord.	26 Jan. 15	Smyth, Lt. J.	
A. S. C.	26 June 16	Soden, 2nd-Lt. H. L. B.	
Gen. List	28 Nov. 17	Sparkes, 2nd-Lt. H. W.	
Suff. R.	26 June 17	Spurgeon, 2nd.Lt. F. W.	
Oxf. and Bucks. L. I.	7 Nov. 15	Squire, 2nd-Lt. L. G.	
Ches. R.	10 Dec. 14	Stafford, 2nd-Lt. W.	
W. York. R.	12 Oct. 15	Stanley, Lt. S. P.	

OFFICERS OF THE INDIAN ARMY RESERVE OF OFFICERS ON PROBATION.—*contd.*

Regiment.	Date of first Commission.	Name and rank.	Remarks.
Essex R.	25 Jan. 17	Stanley, 2nd-Lt. C. H.	
Lond. Yeo.	5 Jan. 18	Stanley, 2nd-Lt. G. C.	
Bedf. R.	...	Storr, 2nd.Lt. W. G.	
Gen. List	...	Stearn, 2nd-Lt. C.	
R. Fus.	14 Mar. 17	Steel, 2nd-Lt. J.	
S. Lan. R.	3 Apl. 15	Steele, 2nd-Lt. G. B. E.	
R. Suss. R.	27 Mar. 18	St. John, 2nd-Lt. C. R.	
Lond. R.	18 Dec. 17	Steward, 2nd-Lt. D. H.	
Newfoundland R.	1 July 16	Stick, 2nd-Lt. L. T.	
Oxf. and Bucks L. I.	25 Oct. 16	Straith, 2nd-Lt. J. D. M.	
R. N. V. Res.	5 Dec. 15	Strang, Sub.-Lt. D. A.	
North'd Fus.	6 Sep. 15	Stuart, Lt. J. P.	
Tr. Res. Bn.	7 July 16	Stuart, 2nd-Lt. T. C.	
Gen. List	26 Sep. 17	Sutherland-Hawes, 2nd-Lt. H. V.	
Lond. R.	...	Sussex, 2nd-Lt. H. G. V.	
A. S. C.	15 May 15	Sutton, Lt. R. W. E.	
A. S. C.	26 June 16	Syme, Lt. K.	
A. S. C.	17 Dec. 13	Symonds, Capt. J. D. B.	
R. E.	6 Oct. 15	Symonds, 2nd-Lt. P. R.	
Gen. List	4 Dec. 17	Tams, 2nd-Lt. G. E.	
Gen. List	26 Aug. 17	Tancred, 2nd-Lt. N. G.	
Shrop. L. I.	26 Feb. 18	Tansen, 2nd-Lt. H. A.	
Gen. List	...	Tatem, 2nd-Lt. E. G.	
Ches. R.	15 Mar. 15	Tatham, Lt. C. W.	
A. S. C.	28 Sep. 15	Taylor, Lt. S. A.	
Gen. List	19 Feb. 18	Taylor, 2nd-Lt. S.	
Gen. List	31 Oct. 17	Taylor, 2nd-Lt. A. H. S.	
Gen. List	28 Nov. 17	Taylor, 2nd-Lt. B.	
Gen. List	. . . .	Taylor, 2nd-Lt. W. D.	
R. Dub. Fus.	11 Aug. 17	Telfer, 2nd-Lt. A. F.	
Gen. List	26 Sep. 17	Tennant, 2nd-Lt. F. O.	
Suff. R.	6 Oct. 15	Tennant, 2nd-Lt. A. W.	
A. Cyclist C.	9 Dec. 16	Tetley, Lt. J. A.	
R. E.	9 Nov. 14	Thomas, Lt. J. V.	
Gen. List	. . . .	Thomas, 2nd-Lt. C. A.	
R. W. Kent R.	1 Aug. 17	Thomas, 2nd-Lt. E. A.	
Norf. R.	27 Mar. 18	Thompson, 2nd-Lt. T. R.	
Bedf. R.	25 Oct. 16	Thompson, 2nd-Lt. F. C. A.	
Midd'x	29 May 18	Thompson, 2nd-Lt. N. L.	
Gen. List	. . . .	Thompson, 2nd-Lt. B.	
R. W. Fus.	12 Oct. 15	Thompson, 2nd-Lt. J.	
Hrs.	8 July 17	Thorne, 2nd-Lt. A. M.	
Gen. List	. . . .	Thornley, 2nd-Lt. V.	
Yorks. R.	28 Dec. 16	Thornton, M.C., 2nd-Lt. H. E.	
Hrs.	8 July 17	Tinkler, 2nd-Lt. G. de L. B.	
Hrs.	8 July 17	Toogood, 2nd-Lt. L. G. C.	
R. Rif. C.	5 Sep. 16	Towndrow, 2nd-Lt. F. E.	
Gen. List	26 Nov. 17	Trent, 2nd-Lt. A. R.	
W. Rid. R.	29 Aug. 17	Trowsdale, 2nd-Lt. H. J.	
Lond. R.	24 Mar. 15	Tucker, Lt. V.	
Gen. List	. . . .	Tucker, 2nd-Lt. K. D.	
Gen. List	31 Oct. 17	Tuckniss, 2nd-Lt. C. S.	
Gen. List	19 Dec. 17	Tungate, 2nd-Lt. W. C.	
Lond. R.	25 Jan. 17	Turner, 2nd-Lt. A. W.	
R. Lan. R.	25 Jan. 17	Turney, 2nd-Lt. H. M.	
R. Fus.	30 May 17	Twelftree, 2nd-Lt. A. E.	
R. Ir. Rif.	16 Nov. 15	Unsworth, 2nd-Lt. V.	
Gen. List	28 Nov. 17	Vaughan, 2nd-Lt. F. B.	
Lab. Corps	18 Dec. 17	Veale, 2nd-Lt. R. de S.	
A. S. C.	24 May 15	Vernon, Lt. T. F.	
Gen. List	. . . .	Vernon-Browne, 2nd-Lt. E. C.	
R. E.	1 Apl. 17	Vickers, 2nd-Lt. T. B.	
E. Kent. R.	27 Feb. 18	Walsh, 2nd-Lt. V. P. F.	
Gen. List	. . .	Walton, 2nd-Lt. G.	
Norf. R.	. . . .	Ward, 2nd-Lt. E.	
Devon R.	26 Sep. 14	Wardle, Lt. R. M. L.	
Gen. List	21 Dec. 17	Wardle, 2nd-Lt. R. P.	
E. Surr. R.	28 Sep. 14	Watney, Lt. R. W.	
Gen. List	26 Nov. 17	Watson, 2nd-Lt. W. B.	
Lan. R.	29 May 18	Watson, 2nd-Lt. H. S.	
R. Fus.	1 Aug. 17	Watts, 2nd-Lt. J. W.	
Gen. List	. . . .	Watts, F. C.	
R. Fus.	7 Nov. 15	Webb, 2nd-Lt. S. R.	
Gen. List	9 Oct. 17	Webber, 2nd-Lt. W. G.	

OFFICERS OF THE INDIAN ARMY RESERVE OF OFFICERS ON PROBATION—*contd.*

Regiment.	Date of first Commission.	Name and Rank.	Remarks.
Gen. List	....	Webster, J.	
Norf. R.	....	Weeden, 2nd-Lt. H. H.	
Gen. List	....	Welman, 2nd-Lt. E. J.	
Gen. List	7 Oct. 17	Welsman, 2nd-Lt. C. H.	
Gen. List	29 Aug. 17	Westgarth, 2nd Lt. S. H.	
Can. Contgt.	21 Aug. 17	White, Lt. E. W.	
Gen. List	....	White, 2nd-Lt. C J.	
Gen. List	....	Whitelock, 2nd-Lt. T.	
Gen. List	....	Whiteside, 2nd-Lt. A.	
Manch. R.	29 June 15	Whitley, 2nd-Lt. L. G. M.	
Gen. List	....	Whitwam, 2nd-Lt. R.	
Gen. List	....	Wigden-Jenks, 2nd-Lt. J. E.	
R. E.	1 Dec. 15	Wigley, Lt. A. J	
Gen. List	31 Oct. 17	Wilcock, 2nd-Lt. H. A.	
Sco. Bor.	30 Oct. 15	Wiley, 2nd-Lt. W. J. G.	
R. Innis. Fus.	24 Aug. 15	Wilkinson, 2nd-Lt. E. M.	
Gen. List	....	Will, 2nd-Lt. G. D.	
S. Wales Bord.	15 July 15	Williams, Lt. B. T.	
Ches. R.	26 Apl. 17	Williams, 2nd-Lt. G. H.	
D. of Corn L. I.	30 Nov. 15	Williams, 2nd-Lt. L. J.	
R. N. V. Res.	....	Williams, Sub-Lt. R. F.	
Gen. List	....	Williams, 2nd-Lt. J. T.	
Gen. List	....	Williamson, 2nd-Lt. A. F.	
A. S. C.	11 Nov. 14	Wilson, Lt. E. L. B.	
High. L. I.	28 May 17	Wilson, D.C.M., 2nd-Lt. W.	
Devon. R.	13 Aug. 09	Wilton, Capt. T.	
R. War. R.	28 Mar. 17	Wing, 2nd-Lt. H. J. T.	
Notts and Derby R.	25 Oct. 15	Wing, Lt. R.	
Gord. Highrs.	26 Apl. 17	Winton, 2nd-Lt. W. C.	
Midd'x R.	17 Aug. 15	Withers, 2nd-Lt. F. W.	
Gen. List	....	Wood, 2nd-Lt. J. E.	
A. S. C.	16 Aug. 15	Wood, Lt. E. H. M.	
Camb. R.	8 Aug. 14	Wood, Lt. E. R.	
Gen. List	....	Woodcock, T.	
Gen. List	....	Wooldridge, 2nd-Lt. L. E.	
Gen. List	....	Woolner, 2nd Lt. R.	
Gen. List	28 Nov. 17	Wootten, 2nd-Lt. R. W.	
Gen. List	....	Worsfold, 2nd-Lt. J. E.	
A. S. C.	....	Worram, Lt. G.	
Gen. List	26 Nov. 17	Wright, 2nd-Lt. R. J.	
Cav.	21 Dec. 16	Wrigley, 2nd-Lt. F. H.	
Gen. List	....	Wycherley, 2nd.Lt. R. H.	
Lab. Corps	18 Dec. 17	Wykes, 2nd-Lt. H J.	
Gen. List	....	Yealand, 2nd-Lt. W. C.	
Gen. List	29 Aug. 17	Yeomans, 2nd-Lt. R. A.	
Gen. List	26 Sep. 17	Young, 2nd-Lt. J. C.	

ROYAL AIR FORCE.

" Per Ardua ad Astra "

Colonel-in-Chief . . . . The King.

N. B.—(i) The date after an officer's name is that of his Army, etc., Rank.
(ii) (O) Denotes that the officer is a qualified Observer gazetted to the Royal Air Force.
(iii) All officers serving in the Royal Flying Corps on the 31st March 1918, were granted temporary commissions in the Royal Air Force with effect from 1st April 1918, in ranks which appear in the Royal Air Force List.

52nd WING, R. A. F.

(Headquarters.)

Lt.-Col. Flying (A)	**Heathcote,** Lt.-Col. J. R. C., Cam'n. Highrs. & R. A. F.	1 Apl. 18
Capt. Admins. (A)	**Tilley,** Capt. H., 1 Bn., Durh. L. I. & R. A. F.	1 Apl. 18
2nd-Lt. Admin. (S)	**Waymouth,** 2nd-Lt. O. S., R. A. F. *Hony. Lt.*	1 Apl. 18
Capt. Technical (W)	**Stuart,** (*Actg. Capt., 10th Apl. 18*) L. J., R. A. F.	1 Apl. 18
Lieut. Tech. (Amrt.)	**Hooper,** S. (*Hon. Capt.*)	1 Apl. 18
2nd-Lieut. Tech. (Photo.)	**Goudge,** B. N. (*Hon. Lt.*)	1 Apl. 18

(No. 31 SQUADRON.)

Major Flying (A)	**Millar,** E. L., M.B.E.	1 Apl. 18
Capt. Flying (A)	**Hughes,** Temp. Capt. O., R. A. F.	1 Apl. 18
Do. (A)	**Shears,** Capt. R. H., R. A. F.	1 Apl. 18
Do. (A)	**Travis,** (*Actg. Capt., 10th Apl. 18*) G. R., R. A. F.	1 Apl. 18
Lieut. Flying (A)	**Boyd,** Lt. L. C., R. A. F.	1 Apl. 18
Do.	**Gompertz,** Lieut. A. V. H., R. A. F.	1 Apl. 18
Do. (O)	**Scowcroft,** Lieut. H. D. R. A. & R. A. F.	1 Apl. 18

(Temply. attached R. A. F. Hdqrs.)

Lieut. Flying (A)	**King,** Lt. E. H., R. A. F.	1 Apl. 18
Do. (A)	**Rankin,** Lieut. R. M., R. A. F.	1 Apl. 18
Do. (A)	**Toulmin,** Lieut. V. F., R. A. F.	1 Apl. 18
Do. (A)	**Thompson,** Lieut. C. N., R. A. F.	1 Apl. 18
Do. (A)	**Bellamy,** 2nd-Lieut. C. S., R. A. F.	1 Apl. 18
Do. (A)	**Barker,** Lieut. C. G., R. A. F.	1 Apl. 18
Do.	**Devonshire,** Lieut. F. V.	1 Apl. 18
Do. (A)	**Whiteley,** Lieut. F. N., R. A. F.	1 Apl. 18
Do. (A)	**Ashton,** Lieut. (*Hon. Capt.*) E. M., R. A. F.	1 Apl. 18
Do.	**Cosslett,** Lieut. E. T., R. A. F.	1 Apl. 18
Do. (A)	**Kirby,** Lieut. J. L., R. A. F.	1 Apl. 18
Do. (A)	**Longfellow,** Lieut. G. D., R. A. F.	1 Apl. 18
Do. (A)	**Dodkins,** Lieut. L. C., R. A. F.	1 Apl. 18
Do. (A)	**Neville,** Lieut. W. E., R. A. F.	1 Apl. 18
Do. (A)	**Lang,** Lieut. D. W., R. A. F.	1 Apl. 18
2nd-Lieut. (O)	**Hutton,** D. J.	1 Apl. 18
Lieut. Admin. (A)	**Rumney,** Lieut. T.	1 Apl. 18
2nd-Lieut. Admin. (T.)	**White,** 2nd-Lieut. P. G.	1 Apl. 18
2nd-Lieut. Technical (W).	**Mullinger,** 2nd-Lieut. R. O., R. A. F.	1 Apl. 18

ROYAL AIR FORCE.

Attached.

(Qualified Observers.)

2nd-Lieut.	**Hayward,** R., 2 Bn., N. Staff R.	18 Oct. 18
Do.	**Hoare,** L. J., 1 Bn., York. R.	18 Oct. 18
Do.	**Plait,** J., R. Dub. Fus.	18 Oct. 18
Lieut.	**Liston,** W. D. F., I. A. R. O.	18 Oct. 18
2nd-Lieut.	**Thomson,** D., 9 Bn., Manch. R.	3 Nov. 18
Do.	**Burke,** N., Oxf. & Bucks. L. I.	3 Nov. 18
Do.	**Body,** G. A., 1-5 Bn., E. Kent. R.	3 Nov. 18
Do.	**Williams,** R. C., 3 Bn., S. Wales. Bord.	3 Nov. 18
Do.	**Moyles,** S., R. Dub. Fus.	3 Nov. 18

(Under instruction as Observer.)

2nd-Lieut.	**Stewart,** W. F., R. Sc. Fus.	

(No. 114 SQUADRON, R. A. F.)

Major Flying (A)	**Stodart,** Maj. D. E., D.S.O., D.F.C., R. A. F.	1 Apl. 18
Capt. Flying (A)	**Mackrell,** Capt. G., R. A. F.	1 Apl. 18
Do. (A)	**Massy,** Capt. D. F., 33rd Cavalry & R. A. F.	1 Apl. 18
Capt. Flying (A)	**Gove,** (*Act'g. Capt. 10 Aug. 18*) R. V., R. A. F.	1 Apl. 18
Lieut. Flying (A)	**Ellis,** Lieut. E. T. H., R. A. F.	1 Apl. 18
Do. (A)	**Gunning,** Lt. (*Hon. Capt.*) H. R. (F.), R. A. F.	1 Apl. 18
Do. (O)	**Owden,** Lieut. J. S., R. A. F.	1 Apl. 18
Do. (A)	**Horne,** Lieut. A. W. M., R. A. F.	1 Apl. 18
Do. (A)	**Antelme,** Lieut. P. F., R. A. F.	1 Apl. 18
Do. (A)	**Parsons,** Lieut. A. H., R. A. F.	1 Apl. 18
Do. (A)	**Gilbert,** Lieut. T., R. A. F.	1 Apl. 18
Do. (A)	**Brown,** Lieut. K. H., R. A. F.	1 Apl. 18
Do. (O)	**Gilpin,** Lieut. F. E., R. G. A., R. A. F.	1 Apl. 18
Do. (A)	**Low,** Lieut. J. H., R. A. F.	1 Apl. 18
Do. (A)	**Taylor,** Lieut. D. E. D., M.C., R. A. F.	1 Apl. 18
Do. (A)	**Huggard,** Lieut. G. C., R. A. F.	1 Apl. 18
Do. (A)	**Bernard,** Lieut. J. L., R. A. F.	1 Apl. 18
Capt. Admins. (A)	**Renny,** Capt. S. C., 1st K. O. Sco. Bord. & R. A. F.	1 Apl. 18
2nd-Lieut. Admins. (S.)	**Leonard,** 2nd-Lieut. J. A., R. A. F.	1 Apl. 18
Lieut. Technical (W.)	**Dowling,** Lieut. W. H., R. A. F.	1 Apl. 18
2nd-Lieut. Tech. (W)	**Dinnage,** 2nd-Lt. H. A.	1 Apl. 18

Attached.

(Qualified Observers.)

2nd-Lieut.	**Lawson,** J. E., R. Highrs.	18 Oct. 18
Do.	**Manson,** W. A., M.M., Gord. Highrs.	18 Oct. 18
Do.	**Porter,** W. J., D.C.M., M.M., Gord. Highrs.	18 Oct. 18
Do.	**Simister,** G. H., Welsh. R.	18 Oct. 18
Do.	**Saunderson,** R. G., R. F. A.	18 Oct. 18

(Under instruction as Observers.)

2nd-Lieut.	**Parsons,** W. G. L., I. A. R. O.	...
Do.	**MacDonald,** W. C., 1 Bn., R. Suss. R.	...
Lieut.	**Smith,** R. J., 2-4 Bn., Bord. R.	...
Do.	**Anson,** R. R., 13 Hrs.	...
2nd-Lieut.	**Roberts,** V. L., 1-5 Bn., E. Kent. R.	...

ROYAL AIR FORCE.

(R. A. F., DETACHED FORCE.)

Captain Flying (A)	**Sherwin,** Capt. C. E., M.C., R. A. F.	1 Apl. 18
Lieut. Flying (A)	**Thomas,** Lieut. R. L., R. A. F.	1 Apl. 18
Do. (A)	**Knight,** Lieut. H. F., R. A. F.	1 Apl. 18
Do. (A)	**Morrell,** Lieut. S. J., R. A. F.	1 Apl. 18
Lieut. Flying (O)	**Harris,** 2nd-Lieut. A. F., 1-9 Hamps. R.	...
Do. (O)	**Pereira,** 2nd-Lieut. J. E., R. A. F.	...
Lieut. Admins. (A)	**Barr,** Temp. Lieut. G., R. A. F.	1 Apl. 18
2nd-Lieut. Technical	**Smyth,** Lieut. F. E., Genl. List, R. A. F.	1 Apl. 18
2nd-Lieut. Technical (W)	**Featherstone,** 2nd-Lieut. W. A. E., R. A. F.	1 Apl. 18

AIRCRAFT PARK.

Major Technical	**Rumbold,** Major, R. S., Som. L. I. & R. A. F.	1 Apl. 18
		1 Apl. 18
Lieut. Flying (A)	**Rooke,** Lieut. D. M., R. A. F.	1 Apl. 18
Lieut. Admins. (S)	**Bennet,** 2nd-Lieut. (*Hon. Lieut.*) G. L., R. A. F.	22 June 18
Lieut. Technical (Engine).	**Wale,** 2nd-Lieut. (*Hon. Lieut.*) N. W., R. A. F.	9 May 18
2nd-Lieut. Technical	**Hilsdon,** 2nd-Lieut. H., R. A. F.	1 Apl. 18

Attached.

Hill-Workman, Capt. J., 3 Bn., R. Scots. . . . . . 1 June 15

(*For trade testing of transfers.*)

R. A. F. DEPOT, BOMBAY.

2nd-Lieut. Tech.	**Solomon,** Temp. 2nd-Lieut. E. V., R. A. F.	1 Apl. 18

AIRCRAFT MANUFACTURE.

Superintendent (Capt. Technical).	**Robertson,** Capt. J. H., R. A. F.	23 Mar. 18

1st (King's) Dragoon Guards.

Arrived in India, 28th November 1907, from England.

THE ROYAL CYPHER WITHIN THE GARTER.
"Blenheim" "Ramillies" "Oudenarde" "Malplaquet" "Dettingen" "Warburg" "Beaumont" "Waterloo"
"Sevastopol" "Taku Forts" "Pekin, 1860" "South Africa, 1879, 1901-02."

Uniform—Scarlet. *Facings*—Blue. *Plume*—Red.

COLONEL—Major-General W. V. Brownlow, C.B. ret. pay . 13 June 08.

First Commn.	Names.	Regimental rank.	Army rank.	Remarks.
	Lieutenant-Colonel.			
4 Mar. 91	Williams, H. J., D.S.O. (Q., 43, 51, 57, 65, 161, 172a, 205a)	14 June 18		
	Majors.			
4 July 96	Wickham, H. F. (*temp. Lt.-Col., 30 June 15*) (Q. 51, 57)	14 June 10	*l. c.*, 1 Jan. 17	
15 May 97	Hunt, R. S., D.S.O., (Q., 43, 50, 207)	31 Mar. 11	14 July 10	Empld. Tr. Res. Bn.
4 Mar. 00	Turner, A. M., D.S.O., (43, 45a, 51, 52, 57, 65)	9 Sep. 14	. . . .	G. S. O.
1 Dec. 01	Cheape, G. R. H., D.S.O., M.C. (*temp. Br.-Genl.*) (43, 45a, 51, 52, 57, 65, 67)	14 June 18	*l.c.*, 15 June [18	
	Captains.			
28 Jan. 03	Spurrier, R. S. (*temp. Maj., 10 July 16*) (51, 52)	9 Apl. 10		
8 May 01	Wienholt, E. A. (*temp. Maj.* (51, 52)	17 Dec. 12		
29 Nov. 05	*Harvey, S. E.* (51)	23 Dec. 13	. . . .	*Staff Capt.*
15 Aug. 03	Cooper, W. R. F., M.C. (51, 52) .	14 June 14		
24 Jan. 06	Fleming, H. M. (45a, 57) .	20 Nov. 14		
22 Jan. 08	Alexander, L. W. (45a) . . .	23 June 16		
19 Sep. 08	*Sprot, E. W. H.* (45a, 66a) . .	23 June 16	. . . .	*Adjt.*
27 Jan. 09	Gladstone, T. H.	7 Jan. 17	. . . .	Staff Capt., 44th Bde.
16 Mar. 09	*Read, W. R.*, M.C. (*temp. Maj.*), (51, 52, 66a, 207)	19 Aug. 17	. . . .	*R. A. F.*
4 Aug. 09	Hatfeild, H. S.	31 Aug. 17		
	Lieutenants.			
27 May 08	*Fox, R. G.* (51, 207) . . .	13 May 10	. . . .	*4th Res. Regt. of Cav.*
7 Feb. 12	*Ward, B. M.* (*temp. Capt., 9 Sep. 18*) (50)	10 Sep. 14	. . . .	Cavy. Instr., Offrs'. Schl. of Instn., Bangalore.
7 Feb. 12	*Carleton-Smith, D. L. G.* (*act'g. Capt.*)	1 Nov. 14		
1 Sep. 12	Card, F. W. F. (47a, 51) . . .	20 Nov. 14		
2 Sep. 13	Wilson, C.E. (*act'g. Capt., 21 Sep. 18*)	2 Feb. 16		
25 Feb. 14	Muir, W. H., M.C.	23 June 16		
12 Aug. 14	Richardson, A. B. H. N. . . .	23 June 16		
15 Aug. 14	*Holt, B. G.*	20 Oct. 16	. . . .	*Bde. M. G. Coy.*
1 Sep. 14	Parker, W. S. H.	20 Oct. 16		
28 Jan. 15	Jacques, A.	7 Jan. 17		
	Murray-Johnson, F. K. (*temp.*) .	7 Jan. 17	. . . .	Embn. Staff, Bombay.
17 Mar. 15	Moreton, J. P., M.C.	1 July 17		
4 May 15	Greenshields, R. L.	1 July 17		
12 May 15	Tiarks, J. G. E.	1 July 17		
12 May 15	*Jaffray, Sir W. E., Bart.* . .	1 July 17		
16 June 15	Waggett, J. L.	1 July 17		
5 July 15	Leicester, C. B. W.	1 July 17		
15 July 15	*Chalmers, T. H.* (*temp. Capt.*)	1 July 17		
23 July 15	Langford, W. (temp. Capt., 28 Sep. 18)	1 July 17	. . . .	Staff Capt., 4th Cav. Bde.
3 Dec. 15	Peacocke, L. G. L. (57) . . .	1 July 17		
3 Dec. 15	Ward, R. I.	1 July 17		
19 July 16	Bogle-Smith, J. A. (61) . . .	19 Jan. 18		
19 July 16	Barnes, B. D. J.	9 Jan. 18		
. . . .	*Norris, J. W.*	24 Jan. 18	. . . .	*I. A.* (*on probn.*).
	2nd-Lieutenants.			
6 Nov. 15	Osborne, F. W.	6 Nov. 15		
6 Feb. 17	*Rigby, W.* (*temp. Capt.*) . . .	6 Feb. 17	. . . .	*Adjt., Punj. L. H.*
24 Apl. 18	Spurrier, J. R.	24 Apl. 18		

1st (King's) Dragoon Guards—*contd.*

First Commn.	Names.	Regimental rank.	Army rank.	Remarks.
	SPECIAL RESERVE.			
	Captains.			
	Brown, D. C. (*Capt. ret. pay*)	2 Feb. 10		
	Crossley, A. A.	5 Aug. 14		
	Lieutenants.			
	Brown, A., M. C.	7 Jan. 17		
	Cumberlege, C. F.	7 Jan. 17		
	McCollough, R. L.	7 Jan. 17		
	Mott, G. E. T.	1 July 17		
	2nd-Lieutenants.			
	Walker, N.	30 Apl. 15	16 Feb. 17	
19 Sep. 08	Sprot, Capt. E. W. H.	20 Feb. 17	. . . .	Adjt.
22 Sep. 03	Wells, Maj. W. T. L.	3 June 17	. . . .	Qr.-Mr.
. . . .	Haws, *Hony.* Capt. A. H., 2-2nd Sco. Horse	24 Sep. 17	. . . .	Qr.-Mr.

7th (Princess Royal's) Dragoon Guards.

Arrived in India, 12th October 1910, from Egypt.

IN THE CENTRE THE CORONET OF HER LATE MAJESTY THE EMPRESS AND QUEEN FREDERICK GERMANY AND PRUSSIA AS PRINCESS ROYAL OF GREAT BRITAIN AND IRELAND.
"Blenheim" "Ramillies" "Oudenarde" "Malplaquet" "Dettingen" "Warburg" "South Africa, 1846-7." "Tel-el-Kebir" "Egypt, 1882" "South Africa, 1900—02."

Uniform—Scarlet. *Facings*—Black. *Plume*—Black and White.

COLONEL-IN-CHIEF—*H. R. H. the* Princess Royal.
COLONEL—Major-General *Sir* H. P. Ewart, *Bt.*, G.C.V.O., K.C.B., ret. pay, *Eq.* 10 Feb. 00.

First Commn.	Names.	Regimental Rank.	Army Rank.	Remarks.
	Lieutenant-Colonel.			
12 Mar. 92	Sparrow, R., C.M.G. (43, 51)	24 Dec. 14		
	Majors.			
28 June 90	Dyer, J. E. F. (Q. 51, 57)	6 June 05	29 Nov. 00	
31 Oct. 90	Clay, B. G., C.M.G., D.S.O. (*temp. Br.-Genl.*) (Q. 51, 57, 67, 207)	6 Feb. 09	6 Mar. 07 / *l.c.* 3 June 17	Bde. Comdr.
20 Feb. 97	Campbell, N. D. H. (51, 57, 71)	18 Mar. 11	. . . .	G. S. O., 2nd gr.
7 May 98	Chappell, W. F. (51, 71)	24 Dec. 14	. . . .	With 4th Res. Regt. of Cav.
	Captains.			
10 Jan. 00	Mansel, J. L. (43, 45*a*, 51, 65, 71)	14 Mar. 06		
28 Mar. 00	Lindsay, M. E., D.S.O. (*temp. Lt.-Col.*, 43, 51, 52, 65, 71, 72*b*).	6 Apl. 07	. . . .	With 13 Bn., Durh. L. I.
15 Aug. 00	Greene, J., D.S.O. (*temp. Br.-Genl.*) (43, 45*a*, 51, 52, 57, 65, 71)	9 Nov. 10	. . . .	With 13 Bn., Midd'x. R.
8 Feb. 02	*Watson, E. C.* (51, 60*a*)	9 Nov. 10	. . . .	*Comdg., Bde. M. G. Sqdn.*
4 Aug. 00	*McCulloch, A. J., p.s.c.* (50, 52, 72*b*, 162, 226)	18 Mar. 11	*m.*, 3 June 16 / *l.c.* 3 June 17	*With 9 Bn., York. L. I.*
9 Aug. 06	*Cairnes, T. A. E.*, D.S.O. (43, 45*a*, 51, 65, 71)	22 Apl. 11	. . . .	*R. A. F.*
2 Aug. 99	*Montgomery, J.* [*l*] (51, 129)	16 Dec. 11	18 July 08	*Spec. Appt.*
13 Aug. 04	*Herapath, D. K. MacK*	10 June 14	6 Mar. 12	*Bde. Maj.*
15 Sep. 06	*Wood, R. G. P.* (45*a*, 51)	24 Dec. 14	. . .	*G. S. O., 3rd gr.*
2 Feb. 07	*Scott, F. J.*, M.C. (43, 45*a*, 51, 57, 65, 71)	8 May 15	. . . .	*Bde. Maj.*
18 Mar. 08	Friend, A. L. I. (51, 57, 67, 72)	17 May 15		
28 May 10	*Noble, R.* (57, 60*a*)	16 Nov. 15	. . . .	*With Sig. Ser.*
4 Feb. 11	Ansdell, R. C. (43, 45*a*, 51, 65, 71)	16 Nov. 15		
4 Feb. 13	*Roberts, R. G.*	10 July 16	. . . .	*Adjt.*
22 Jan. 13	Sanford, S. A.	13 Apl. 17		
10 Dec. 13	Mellor, J. S.	13 Apl. 17		
8 Aug. 14	*Anson, T. G.*	22 Aug. 17	. . . .	*M. G. Sqdn.*
14 Aug. 14	Malone, H. A. P.	22 Aug. 17		
	Lieutenants.			
3 Sep. 12	Mann, R. L.	12 Nov. 13		
15 Aug. 14	*Mauduit, R. F. S.*, M.C.	9 Aug. 15	. . . .	*R. A. F.*
15 Aug. 14	Watson, L. D.	16 Nov. 15		
15 Aug. 14	Hastings, G. L., M.C.	27 Nov. 15		
11 Nov. 14	*Johns, A. B.*	2 Feb. 16	. . . .	*With Sig. Ser.*
6 Dec. 14	Pope, H. W., M.C. (*temp. Capt.*)	10 July 16	. . . .	With N. Som. Yeo.
6 Dec. 14	*Fraser, J. A.*, D.S.O., D.C.M. (*temp. Lt.-Col.*)	15 July 16	. . . .	*With 5th Bn. R. Scots*
12 Dec. 14	Sunderland, W.	15 July 16		
13 Dec. 14	Penny, E.	4 Sep. 16		
16 Dec. 14	Adair, D. H.	13 Apl. 17		
16 Dec 14	Carr, P. G.	25 Apl. 17		
17 Sep. 14	Struben, L. F.	1 July 17	17 Sep. 14	
7 Apl. 16	Bolckow, H. C. R.	7 Oct. 17		
7 Apl. 16	Murray, T. H.	7 Oct. 17		
14 July 16	*Chase, H. L.*	14 Jan. 18	. . . .	*With Sig. Ser.*
27 Sep. 16	Wood, T.	27 Mar. 18		
26 Mar. 16	Thompson, A.	12 Apl. 18	. . . .	With 6th Res. Regt. of Cav.
	2nd-Lieutenants.			
4 Feb. 13	Roberts Capt., R. G.	13 Apl. 17	. . . .	Adjt.
17 June 03	Huxtable, Maj. A. E. V. (71, 207)	1 July 17	. . . .	Rdg. Mr. With 7th Res. Regt. of Cav.
6 June 06	Hiatt, Capt. A.	6 June 16	. . .	Qr.-Mr.

Agents:—MESSRS. COX & CO.

6th (Inniskilling) Dragoons.

Arrived in India, 8th October 1908, from Egypt.

THE CASTLE OF INNISKILLING, WITH THE ST. GEORGE'S COLOURS, AND THE WORD "INNISKILLING" UNDERNEATH.
"Dettingen" "Warburg" "Willems" "Waterloo" "Balaklava" "Sevastopol"
"South Africa, 1899—1902."

Uniform—Scarlet. *Facings*—Primrose. *Plume*—White.

COLONEL-IN-CHIEF—Field Marshal *His Royal Highness* ARTHUR WILLIAM PATRICK ALBERT, *Duke of* CONNAUGHT AND STRATHEARN, K.G., K.T., K.P., G.C.B., G.C.S.I.; G.C.M.G., G.C.I.E., G.C.V.O., G.B.E., A.D.C., 2 June 97.
COLONEL—Maj.-General M. F. Rimington, C.B., C.V.O. 17 June 12.

First Commn.	Names.	Regimental rank.	Army rank.	Remarks.
	Lieutenant-Colonel.			
9 Sep. 93	Paterson, E., D.S.O. (43, 51, 57, 65)	2 July 16		
	Majors.			
24 July 91	Vincent, B., C.M.G., *p.s.c.* Q. (*temp. Br.-Genl.*) 12, 65, 72*b*, 206	14 Oct. 08	*l.c.* 1 Jan. 17; *c.* 3 June 18	Bde. Comdr.
9 Dec. 99	Terrot, C. R., D.S.O. (*temp. Lt.-Col.*, *18 Nov. 16*) (51, 57)	12 Aug. 11	. . . .	Comdt., Schl. of Instn.
7 July 00	FitzGerald, E. A. (50, 53, 59, 65)	2 July 12		
18 Aug. 00	Nixon, F. B., D.S.O. (45*a*, 51, 59)	2 July 16	. . . .	Bde.-Maj.
	Captains.			
18 Apl. 00	Ritson, R. G. (51, 57)	14 Oct. 08	8 Jan. 07; *m.* 1 Jan. 17; *l.c.* 1 Jan. 18	G. S. O., 2nd gr.
6 Dec. 99	*Moncrieff, A. R.*, M.C., (43, 45*a*, 51, 57, 65)	25 June 10	. . . .	*G. S. O., 3rd gr.*
15 Aug. 03	Fleury Teulon, C. A. (51, 65)	2 July 12		
13 Aug. 04	Bowen, E. C., M.C., (43, 45*a*, 51)	15 July 12		
13 Oct. 06	Burnham, G. LeR	11 Feb. 13	. . . .	
2 Feb. 05	Sassoon, S. J. (43, 45*a*, 65)	11 Feb. 13		
20 May 05	Brooke, J. A. (43, 45*a*, 50, 52, 59, 65, 66, 162, 206)	2 Dec. 17		
9 Sep. 08	*Humfrey, J. C.*, (*14*) (*51, 52*)	*2 Dec. 17*	. . . .	*M. G. C.*
27 Jan. 09	*Wootten, R. M.* (*45a, 51, 66a*)	*2 Dec. 17*	. . . .	*Adjt.*
4 Feb. 11	*Dudgeon, M. F.*, M.C. (*57, 66a*)	*2 Dec. 17*		
	Lieutenants.			
23 Feb. 10	Barton, N. F. (45*a*) (*temp. Capt.*)	2 July 12		
4 Feb. 11	Rimington, R. G. W. (65)	6 Nov. 12		
9 Sep. 11	Freer-Smith, D. F.	26 Feb. 13		
7 Feb. 12	Kemmis, W. D. O., M.C. (66*a*)	10 May 13		
3 Sep. 13	Macintyre, D. I.	24 Dec. 15		
10 Aug. 12	Usher, I. T.	19 Apl. 16		
24 Jan. 14	*Windham, A. W. G.*	*26 June 17*	. . . .	*M. G. C.*
10 June 14	Graham, J. M.	26 June 17	. . . .	
1 Oct. 14	Birch-Reynardson, M. H.	26 June 17		
7 Nov. 14	Totterdell, J. L.	1 July 17	. . . .	With Sig. Ser.
16 Dec. 14	Shelton, R. C. M.	1 July 17		
16 June 15	Young, G. D.	1 July 17		
16 June 15	Dames, T. D. L.	1 July 17		
16 June 15	Hilder, T. M. M.	1 July 17		
25 Aug. 15	Barraclough, W.	1 July 17		
20 Oct. 15	*Knott, H.*	*1 July 17*	. . . .	*Adjt., 7th (Southern)* ***U. P. Horse***
20 Oct. 15	Elliot, R. H.	1 July 17	. . . .	Res. Regt. of Cav.
7 Apl. 16	Anstice, J. H.	7 Oct. 17		
9 July 16	Pease, C. N.	19 Jan. 18		
27 Oct. 16	Pratt, C.	27 Apl. 18		
12 Nov. 16	Lupton, G.	12 May 18		
	2nd-Lieutenants.			
17 Jan. 09	Wootten, Capt. R. M.	25 July 15	. . . .	Adjt.
16 May 06	Beale, Capt. G. H.	16 May 16	. . . .	Rdg. Mr., S.S.O., Neemuch.
16 Aug. 99	Siddall, Maj. J. R.	16 Aug. 14	. . . .	*r. d.* Qr.-Mr.
25 Nov. 11	Macpherson, Capt. W.	1 July 17	. . . .	Qr.-Mr.

Agents :—MESSRS. COX & CO

7th (Queen's Own) Hussars.

Arrived in India, 2nd November 1911, from England.

THE LETTERS "Q. O." INTERLACED, WITHIN THE GARTER.
"Dettingen" "Warburg" "Beaumont" "Willems" "Orthes" "Peninsula" "Waterloo" "Lucknow" "South Africa, 1901-02."

Uniform—Blue. *Busby-bag*—Scarlet. *Plume*—White.

COLONEL—Major-General *Sir* H. McCalmont, K.C.B., C.V.O., ret. pay. . 13 May 07.

First Commn.	Names.	Regimental rank.	Army rank.	Remarks.
	Lieutenant-Colonel.			
9 Oct. 89	Norton, C. E. G. A.-D.-C. (*temp. Br.-Genl.*, 50, 52, 206)	26 June 15	*c.* 3 June 18	
	Majors.			
6 Feb. 90	Dalgety, H. B., (*act'g. Lt.-Col.*, 8 *Apl.* 17) (Q. 43, 51, 52)	26 Feb. 08		
26 Nov. 90	Fryer, J. (Q. 43, 50, 57, 65)	12 Feb. 10	. . . .	
21 Oct. 93	Rankin, C. H., C.M.G., D.S.O. (*temp. Brig.-Gen.* Q. 50, 57)	26 June 11	*l.c.* 1 Jan. 16 *c.* 3 June 18	
12 Aug. 99	Leyland, F. B. (45*a*, 51, 57)	26 June 15	. . . .	Asst. Provost Marshal.
	Captains.			
17 Feb. 00	Gibbs, W. (43, 45*a*, 51, 65)	20 Apl. 07		
17 Feb. 00	*Pollok, A. B.* (*temp. Maj.*) (43, 45*a*, 51)	1 Nov. 07	. . . .	*Comdg. Cadet Sqdn.*
19 Sep. 00	Paget-Tomlinson, W. (43, 45*a*, 50, 52, 59, 65).	12 Apl. 10		
22 Oct. 02	*Tollemache*, Hon. *D. P.*, *p.s.c.* (*act'g. Maj.*, (43, 45*a*, 51, 52, 57)	9 Oct. 11	*m.* 3 June 16	
14 Jan. 03	*Watson, A. C.*, D.S.O. (*act'g. Maj.*) (43, 45*a*, 51).	11 Oct. 11	. . . .	
19 Aug. 03	Walker, P. L. E. (*act'g. Maj.*) (51, 52)	10 Mar. 13	. . . .	With R. War. R.
29 Aug. 06	Thornton, T. A. (*act'g. Maj.* (45*a*, 51)	9 Aug. 13		
8 May 07	*McCalmont, D. H. B.*, M.C. (51, 67)	9 Aug. 13	*m.* 1 Jan. 18	
4 Sep. 07	*McMurrough Kavanagh, A. T.*, M.C. (51, 52)	1 Nov. 13	. . . .	*G. S. O., 2nd gr.*
24 Dec. 02	Bethell, H. K., C.M.G., D.S.O., *p.s.c.* (*temp. Maj.-Genl.*) (12, 51, 71, 157, 206)	14 Feb. 14	24 Dec. 11 *m.* 3 June 16 *l.c.* 1 Jan. 17	
22 Nov. 99	*Thompson*, Sir *T. R. L.*, Bt., M.C., *p.s.c.*	18 Feb. 14	22 Nov. 08	
27 May, 02	*Colmore, H.* (64, 66*a*)	15 July 14	31 Aug. 12	*R. A. F.*
8 Jan. 05	*Bryant, C. E.*, D.S.O.	27 July 14	. . . .	*R. A. F.*
4 Feb. 11	Breitmeyer, G. C. A. (51)	2 Dec. 14		
9 Dec. 11	Evans, R. (57)	2 Dec. 14		
4 Mar. 11	Weatherall, N. E. (51, 57)	14 Dec. 14	. . . .	
	Lieutenants.			
9 Sep. 11	*Hill, G. D.* (*temp. Capt.*) (51)	10 Mar. 13	. . . .	*R. A. F.*
9 Sep. 11	*Creagh, D. V.*, M.C. (*temp. Capt.*).	11 Mar. 13	. . . .	*Staff Capt.*
9 Dec. 11	Hutchison, B. O. (50, 52, 57, 207).	9 Aug. 13	. . . .	With Sig. Serv.
3 Sep. 12	Scott, J. B. (*act'g. Capt.*) (51, 52, 207)	25 May 14		
21 Jan. 13	Payne-Gallwey, L. P. (51, 52)	25 May 14		
12 Jan. 13	Chandos-Pole, G. R. (51, 52, 60*a*)	25 May 14	. . . .	Attd. 1st Bn., York R.
12 Sep. 13	*Pelly, H. A* (*temp. Capt.*).	2 Dec. 14	. . . .	*With Household Bn.*
27 Sep. 13	Hallowes, J. C. (51, 52, 57)	14 Dec. 14		
5 Feb. 14	Fullerton, R. A. D. (51, 52, 57)	21 Oct. 15		
5 Aug. 14	Dixson, H. F. (51, 52, 59)	1 July 17	. . . .	With Sig. Ser.
5 Aug. 14	Fielden, G. (51, 52).	1 July 17	. . . .	*R. A. F.*
21 Aug. 14	*Stevens, A. E.* (*act'g. Capt.*)	1 July 17	. . . .	Adjt.
7 Nov. 14	Bone, J. H.	1 July 17	. . . .	Adjt., Kailana Camp.
7 Nov. 14	Ashcroft, W. (51, 52)	1 July 17	. . . .	
7 Nov. 14	Rennie, C. A. (51, 53)	1 July 17	. . . .	With M. G. Sqdn.
7 Nov. 14	Perry, W. H. G. (57)	1 July 17		
15 Sep. 15	Bell, J. G. A. (51, 52)	1 July 17		
19 Jan. 16	Potter, G. E., M.C.	19 July 17	. . . .	
4 Mar. 16	Devonshire, F. V.	4 Sep. 17	. . . .	*R. A. F.*
27 Oct. 16	Porchester, *Lord* H. G. A. M. V. F.	27 Apl. 18		
31 Jan. 17	Chapman, J. J. (51)	31 July 18	. . . .	With M. G. Sqdn.
	2nd-Lieutenants.			
1 May 17	Roberts, A. W.	1 May 17		
24 Oct. 17	Dibble, H. R.	24 Oct. 17		
24 Oct. 17	Thomas, D.	24 Oct. 17		
5 Nov. 17	Russell, F.	5 Nov. 17		
19 Nov. 17	Taylor, R.	5 Nov. 17		
5 Dec. 17	Jones, E. H.	19 Dec. 17		

Agents:—MESSRS. COX & CO.

7th (Queen's Own) Hussars—*contd.*

First Commn.	Names.	Regimental rank.	Army rank.	REMARKS.
	Attached.			
6 Nov. 14	Buckwell, temp. 2nd-Lt. H. L. (*temp. Capt.*)	6 Nov. 14	. . . .	
	Jagger, Lt. R.	24 Mar. 15	. . . .	York. Dns.
	Spittal, Lt. J. (51)	20 Sep. 15	. . . .	Sco. Horse.
4 Nov. 15	Gibbons, 2nd-Lt. W. R.	4 Nov. 15	. . . .	Gen. Ser. Cav.
12 May 16	Tonkin, 2nd-Lt. O. P.	12 May 16	. . . .	Sco. Horse.
29 July 16	Cox, 2nd-Lt. H. T.	29 July 16	. . . .	Notts. Yeo.
29 July 16	Day, 2nd-Lt. C. M.	29 July 16	. . . .	North'd. Huss.
16 Apl. 17	Bowskill, 2nd-Lt. H. V.	16 Apl. 17	. . . .	Gen. Ser. Cav.
1 May 17	Taylor, 2nd-Lt. C. A.	1 May 17	. . . .	Gen. Ser. Cav.
21 Aug. 14	Stevens, 2nd-Lt. (*act'g. Capt.*) A. E.	19 Oct. 17	. . . .	Adjt.
14 Sep. 04	Andrews, Maj. F. L. (71)	14 Sep. 17	. . . .	Rdg. Mr.
30 Aug. 11	Paterson, *Hony.* Lt. E. A. . . .	30 Aug. 11	. . . .	Qr.-Mr. *Lv., ex I., m.c.,* *6 mos.*
25 Aug. 14	Akhurst, Capt. W. H..	25 Aug. 17	. . . .	*Remount Service, England.*
4 June 18	Beardall, Lt. C. H.	4 June 18	. . . .	Qr.-Mr.
10 Sept. 14	Wilbraham, *Hony.* Lt. A.E. (temp.)	10 Sep. 14	. . . .	7th Res. Regt. of Cav.

8th (King's Royal Irish) Hussars.

Arrived in India, 19th October 1909, from England.

THE HARP AND CROWN. *"Pristinæ virtutis memores."*

"Leswarree" "Hindoostan" "Alma" "Balaklava" "Inkerman" "Sevastopol" "Central India" "Afghanistan, 1879—80" "South Africa, 1900—02."

Uniform—Blue. *Busby-bag*—Scarlet. *Plume*—Red and White.

COLONEL—Lt.-General *Rt. Hon. Sir* B. T. Mahon, K.C.B., K.C.V.O., D.S.O. . . 7 Mar. 10.

First Commn.	Name.	Regimental rank.	Army rank.	Remarks.
	Lieutenant-Colonel.			
11 June 90	Mussenden, F. W. (43, 51, 52, 65, 127)	19 Oct. 13		
8 Sep. 97	Mort, G. M. (51, 52, 127)	19 Oct. 17		
	Majors.			
11 July 94	*McClellan, H. T.* (*Q*, 50, 53, 59) . .	7 Nov. 16	6 May 05 *l.c.*, 3 June 18.	
7 May 98	Van der Byl, J., D.S.O. (*temp. Lt.-Col.*) (51, 52, 127)	19 Oct. 13	. . . .	Comdt., Sch. of Instn.
11 Aug. 00	Curell, A. (43, 45*a*, 51, 52, 65, 71, 127)	21 Dec. 15		
21 Dec. 01	*Broadbent, E. R.*, M.C. (43, 45*a*, 51, 52, 65, 71)	19 Oct. 17	. . . .	*Staff Capt.*
	Captains.			
3 Sep. 02	*Blakiston Houston, J. E.* (*temp. Lt.-Col.*) (45*a*, 51, 52, 66*a*, 71) .	2 July 10	. . . .	*A. A. & Q. M. G.*
19 Nov. 02	Alexander, P. S. (45*a*, 51, 52, 57, 71, 206)	10 June 11		
12 Sep. 03	Weldon, E. G. (*temp. Maj.*) (43, 45*a*, 51, 65, 71)	16 Oct. 12	. . . .	Comdt., Sch. of Instn.
29 Aug. 06	Cairnes, W. E. P. (45*a*, 51, 52) . .	3 Dec. 13		
10 Oct. 03	Adlercron, G. R. L. (43, 45*a*, 51, 52, 65, 71)	4 Apl. 14		
29 Aug. 06	*Regnart, N. S.*, M.C. (51, 57, 127, 129) .	5 Feb. 16	. . . .	*Army Sig. Service.*
23 Feb. 10	Leech, J. C.	24 July 16		
	Lieutenants.			
28 May 10	Mulliner, A. R., M.C. (45*a*) . . .	21 Oct. 11		
3 Sep. 10	*Wathen, L. W. D.* (66, 127) . . .	3 May 12	. . . .	*Brig. M. G. Coy.*
18 Sep. 09	Pope, D., M.C. (*act'g. Maj.*) (51, 127)	16 Oct. 12	. . . .	With Sig. Ser.
2 Sep. 10	Hornby, R. F., (*act'g. Capt.*) (51, 52)	3 Dec. 13		
7 Dec. 10	*Atkinson-Willes, G. A.*	5 Aug. 14	. . . .	*Adjt.*
4 Feb. 11	*King, B. C.*	21 Dec. 15	. . . .	M. G. Corps (Cav.)
9 Sep. 11	*Baldwin, J. E. A.* (*temp. Maj.*) .	2 Feb. 16	. . . .	*R. A. F.*
9 Dec. 11	*Rowley, G. S.*	2 Feb. 16		
5 Feb. 13	Clifford, G. G. J.	2 Feb. 16		
5 Feb. 13	*Kennedy, R. L.*	2 Feb. 16		*R. A. F.*
17 Sep. 13	*Staniland, E. A.*	5 Feb. 16	. . . .	*With Sig. Service.*
11 Nov. 14	Daly, D. W.	19 Apl. 16		
. . . .	Mansfield, W. S.	24 July 16		
16 Dec. 14	Bradish-Ellames, J. E. M. . . .	12 Sep. 16		
16 Dec. 14	Watson, T. G.	6 Mar. 17		
16 Dec. 14	Knight, R. A. W.	1 July 17		
23 Dec. 14	Stourton, *Hon.* W. M.	1 July 17		
15 Feb. 15	Armstrong, J. R. B.	1 July 17		
17 Apl. 15	Collett-Mason, W. W. C.	1 July 17	. . . .	With 2nd Res. Regt. of Cav.
15 May 15	Dent, W. E.	1 July 17		
11 Aug. 15	*Clowes, W. P.*	1 July 17	. . . .	*M. G. Corps (Cav.)*
4 Dec. 15	Paton, J. D.	1 July 17		
11 Aug. 15	Chirnside, P. E. F.	1 July 17	. . . .	With Res. Regt. of Cav.
16 May 16	Robinson, J. P.	16 Nov. 17		
16 Aug. 16	House, B. H.	16 Feb. 18		
22 Aug. 16	*Turton, A. E.* (*temp. Capt.*) (12, 51, 52)	22 Feb. 18	. . . .	*Adjt., 8 (Northern) Horse. U. P.*
27 Oct. 16	Evans-Lombe, T. B. A.	27 Apl. 18		
27 Oct. 16	Goodson, D. F. J.	27 Apl. 18		
27 Oct. 16	Montgomery, S. M.	27 Apl. 18		
27 Oct. 16	Graham, F. R. V.	27 Apl. 18		
15 Dec. 17	*Hodgetts, W.*	15 June 18	. . . .	*M. G. Corps.*
	2nd-Lieutenants.			
13 May 17	*Dyball, E. E.* (*temp. Capt.*)	13 May 17	. . . .	*Adjt., Surma V. L. H.*
17 May 17	Hounsell, T. H.	17 May 17	. . . .	
29 May 17	Greenstreet, W., M.C.	29 May 17		
7 June 17	Roff, C. W.	7 June 17	6 Sept. 15	
7 Dec. 10	Atkinson-Willes, Lieut. G. A. . . .	5 Apl. 17	. . . .	Adjt.
17 Jan. 06	Turnbull, Capt. J. (71)	17 Jan. 16	. . . .	Riding Master.—With 5 Res. Regt. of Cav.
24 Jan. 03	Hampton, Maj. R. C.	1 July 17	. . . .	*r. d.*—Qr.-Mr.
25 Sep. 14	Newling, Capt. A. R.	25 Sep. 17	. . . .	Qr.-Mr.

Agents:—MESSRS. COX & Co.

13th Hussars.

Arrived in India, 1st October 1904, from England.

" Viret in Æternum."
" Albuhera " " Vittoria " " Orthes " " Toulouse " " Peninsula " " Waterloo "
" Alma " " Balaklava " " Inkerman " " Sevastopol "
" Relief of Ladysmith" " South Africa, 1899-1902 "

Uniform—Blue. *Collars*—Buff. *Busby-bag*—Buff. *Plume*—White.

COLONEL—Lt.-General *Sir* R. S. S. Baden-Powell, K.C.B., K.C.V.O. ret. pay 26 Nov. 11.

First Commn.	Names.	Regimental rank.	Army rank.	Remarks.
	Lieutenant-Colonel.			
5 Oct. 92	Richardson, J. J., D.S.O. (43, 51, 57, 65a, 66)	1 July 17		
	Majors.			
10 Oct. 91	Cox, R. F. (Q. 50, 57, 66a, 162, 175, 206) .	6 Nov. 07	17 July 05	
4 May 98	Twist, E. F. (51, 57)	28 Sep. 12		
3 July 98	Marchant, T. H. S., D.S.O. (43, 51, 57)	1 July 13	. . .	With 2-4th S. Lan. R.
26 Mar. 02	*Carter, E. J.* (51)	1 July 17	3 June 17	*M. G. Corps.*
	Captains.			
21 Mar. 00	Kennard, W. A., D.S.O. (51) (*temp. Maj.*)	10 Nov. 09	13 Dec. 06	
12 Sep. 03	*Stocker, E. H.* (51, 57)	23 June 09	. . .	*Staff Capt.*
13 Aug. 04	Steele, C. (51, 52).	13 July 10	*m.* 1 Jan. 18	D. A. A. G., Poona.
16 Aug. 05	Hind, J. H. (45a, 50, 53). . . .	25 Sep. 12		
27 May 08	*Lumley, J. N.* (45a, 51, 52) . . .	6 Dec. 13	. . .	*G. S. O., 3rd gr.*
10 Jan. 08	Huntingfield, W. C. A., *Lord* . .	26 Mar. 17	6 Dec. 13	
15 Sep. 08	Oakes, J. O. (66a) . . .	28 Dec. 15		
22 Jan. 09	Jones, H. L., D.S.O. (*actg. Lt.-Col.*) (45a, 57)	1 Jan. 16	*m.* 1 Jan. 18	With E. Lanc. R.
27 Jan 09	*Sassoon, A. M.*, M.C. (45a, 51, 52)	17 Apl. 18	. . .	*Staff Capt.*
27 Jan. 09	*Kennedy, S. V.*, M.C. (43, 51) . .	17 Apl. 18	. . .	*D. A. A. & Q. M. G.*
15 Sep. 09	*Dawson, J. V.* (57)	17 Apl 18		
3 Sep. 10	Jeffrey, J. A., M.C.	17 Apl. 18		
	Lieutenants.			
5 Feb. 13	*Gore, R.*	12 Mar. 15	. . .	*R. A. F.*
24 Jan. 14	Kennedy, M. C.	3 Dec. 15		
25 Feb. 14	Barrett, J.L.M. (*actg. Capt.*) .	5 Dec. 15		
12 Aug. 14	Wingfield, C. A. F.	28 Dec. 15		
7 Nov. 14	Annett, A. E.	1 Jan. 16		
7 Nov. 14	*Gowan, C. H.* (*actg. Capt.*) . . .	2 Feb. 16	. . .	*Adjt.*
7 Nov. 14	*Chrystall, J. I.*, M.C.	1 July 17	. . .	*Brig. M. G. Coy.*
16 Dec. 14	Norton, D. J. E., M.C. (57) . .	1 July 17		
16 Dec. 14	Hirsch, J. H.	1 July 17		
17 Feb. 15	Stirling, D. A.	1 July 17		
25 Feb. 15	Pedder, G. R.	1 July 17		
9 June 15	Ormrod, L. A.	1 July 17		
4 July 15	Williams-Taylor, T.	1 July 17		
18 July 15	*Blyth, W.*	1 July 17		
6 Nov. 15	*Williams, A.*	1 July 17		
19 Mar. 16	Hill, R. C.	19 Sep. 17		
7 Apl. 16	Welstead, G. L. M.	7 Oct. 17		
1 June 16	*Shanley, T.*	1 Dec. 17	. . .	*K. Afr. Rif.*
28 Aug. 16	Hampson, J..	28 Feb. 18	. . .	Arab Labour Corps.
	2nd-Lieutenants.			
27 Oct. 16	Hawkins, H. R.	27 Oct. 16		
12 Sep. 17	Wood, A. H.	12 Sep. 17	. . .	With Res. Regt. of Cav.
12 Sep. 17	Constable, R. C.	12 Sep. 17	. . .	Do.
12 Sep. 17	Collins, C. F.	12 Sep. 17	. . .	Do.
12 Dec. 17	Sibson, A. J. R. M.	21 Dec. 17		
21 Jan. 18	Barrington, A. C.	21 Jan. 18		
21 Jan. 18	Wrigley, G. F.	21 Jan. 18		
	Attached.			
8 Feb. 08	Methven, Capt. D. M., 21st Lrs. .	11 Sep. 14		
7 Nov. 14	**Gowan, Lt.** (*actg. Capt.*) **C. H.** . .	28 Dec. 16	. . .	Adjt.
26 Nov. 10	Cooke, Maj. A. (51)	3 June 18	. . .	Qr.-Mr.

Agents :—MESSRS. COX & CO.

14th (King's) Hussars.

Arrived in India, 27th September 1906, from England.

THE ROYAL CREST WITHIN THE GARTER.

"Douro" "Talavera" "Fuentes d'Onor" "Salamanca" "Vittoria" "Pyrenees" "Orthes" "Peninsula" "Chillianwallah" "Goojerat" "Punjaub" "Persia" "Central India" "Relief of Ladysmith" "South Africa, 1900-02"

Uniform—Blue. *Busby-bag*—Yellow. *Plume*—White.

COLONEL—Major-General B. A. Combe, C.B., ret. pay . 9 July 04.

First Commn.	Names.	Regimental rank.	Army rank.	Remarks.
	Lieutenant-Colonel.			
6 June 96	Hill-Whitson, T. E. L., *p.s.c.* (50, 57, 65, 127)	22 Feb. 15	. . . .	*Lv., ex I., m. c.*
	Majors.			
16 Aug. 02	Applin, R. V. K., D.S.O. (51, 52, 146, 206)	3 June 11		
4 Nov. 99	Browne, J. G., C.M.G., D.S.O., *p.s.c.* (43, 51, 52, 65) . .	3 May 12	*l.c.*, 3 June 17	G. S. O., 1st gr.
3 Oct. 00	Joicey, *Hon.* H. E. (*temp. Lt.-Col.*) (51, 57)	22 Feb. 15	*l.c.*, 3 June 16	With Suff. R.
13 Sep. 02	Bridges, E. J., M.C. (*actg. Lt.-Col., 15 Oct. 17*) (43, 51, 57) .	1 Oct. 17	. . . .	Comdg.
	Captains.			
27 July 01	Secker, V. H. (*actg. Maj.*) (51, 52, 127, 104)	14 Dec. 07	. . . .	Adjt., Ind. Cavy. Depôt, Bangalore.
11 Feb. 03	Woodhouse, J. D. F. (*actg. Maj., 15 Oct. 17*) (43, 45*a*, 51, 52, 59, 65, 161, 206).	6 Apl. 08		
10 Oct. 03	*Mocatta, V. E.* [*l*] (51, 52, 205)	17 Nov. 11	*m.* 3 June 17	*D. A. A. & Q.-M. G.*
19 Aug. 06	*Mason, G. K. M.*, D.S.O. (43, 45*a*, 50, 59, 60*a*, 65, 161, 206)	3 May 12	. . . .	*With Notts. Yeo.*
19 Sep. 06	Fetherstonhaugh, A. E. H. (43, 45*a*, 51, 65)	24 Dec. 12		
18 Jan. 05	Hunter-Jones, L. (51, 52, 57, 127, 129)	22 Feb. 13	. . . .	War Office.
16 Jan. 07	*Darley, C. G.*, D.S.O. (51, 52, 66*a*)	8 July 16	. . . .	*G. S. O., 3rd gr.*
16 Jan. 07	*Ambler, M. J.* (45*a*, 50, 203) . .	15 Nov. 16		
18 Dec. 07	*Miller, J. A. T.* (51, 52, 207) . .	23 Dec. 16	. . . .	*Adjt.*
16 Dec. 08	Pope, A. V. (51, 52)	9 Feb. 17		
27 Jan. 09	*Fooks, O. J. F.* (51, 53, 57, 161, 206)	6 Mar. 17		
9 Sep. 11	*Cropper, P. G.*	6 Mar. 17		
8 Nov. 11	deWend-Fenton, J. D. L.	6 Mar. 17		
20 Dec. 11	Moule, G. G., M.C. (66*a*) . .	1 Oct. 17		
	Lieutenants.			
29 Jan. 12	Hamer, G. M. (51, 52)	25 July 14		
17 Feb. 12	Whidborne, C. S. L. . .	17 June 15		
7 Feb. 12	*St Maur, R W. M.* (51, 52) . .	17 Oct. 16	. . . .	*I. A.* (*on probn.*).
4 Sep. 12	*Macintyre, F P.* (51, 52) . .	15 Nov. 16	. . . .	*A. D. C.*
7 Sep. 13	Woodhouse, R. A. G. . . .	23 Dec. 16		
1 Oct. 14	Stanley, D. M.	9 Feb. 17		
7 Nov. 14	Taylor, D...	16 Mar. 17		
. . .	Shard, C. B.	1 July 17		
4 Jan. 17	Whadcoat, C. C.	1 July 17		
15 Sep. 15	Burt, R. F.	1 July 17		
6 Nov. 15	Ecott, H. (*temp. Capt.*)	1 July 17	. . . .	Asst. Embkn. Staff Offr.
6 Nov. 15	*Brooke, H. A.*	1 July 17	. . . .	*A. D. C.*
7 Apl. 16	Petre, A. H. C.	7 Oct. 17		
7 Apl. 16	Wolfenden, J. C.	7 Oct. 17	. . . .	*Lv., ex I., m. c.*
19 July 16	Brown, E. B.	19 Jan. 18		
16 Aug. 16	Jones, V. H.	16 Feb. 18		
	2nd-Lieutenants.			
1 May 17	Forster, R. W. H.	1 May 17		
1 May 17	James, R. E.	1 May 17		
19 Dec. 17	Brown, J. W.	19 Dec. 17		
18 Dec. 07	Miller, Capt J. A. T.	23 Dec. 16	. . . .	*Adjt.*
16 Jan. 01	Pridgeon, Maj. A. F., D.S.O., D.C.M. (*temp. Maj. in Army*) .	16 Jan. 16	. . . .	Qr.-Mr.
25 Aug. 14	Best, Capt. J. F.	25 Aug. 17	. . . .	Qr.-Mr.

Agents:—MESSRS. COX & CO.

17th (Duke of Cambridge's Own) Lancers.

Arrived in India, 28th September 1905, from England.

DEATH'S HEAD "*Or Glory.*"

"Alma" "Balaklava" "Inkerman" "Sevastopol" "Central India"
"South Africa, 1879, 1900–02."

Uniform—Blue. *Facings and Plume*—White.

COLONEL—Field-Marshal *Sir* Douglas Haig, K.T., G.C.B., G.C.V.O., K.C.I.E., *p.s.c.*, . 14 May 12

First Commn.	Names.	Regimental rank.	Army rank.	Remarks.
	Lieutenant-Colonel.			
7 Dec. 95	Legard, D'A., *p.s.c.* [*l*] (*temp. Br.-Genl.*) (43, 50, 53, 66*a*, 206)	24 Feb. 15		
	Majors.			
1 Dec. 97	Fletcher, A. F., D.S.O., M.V.O., (*temp. Lt.-Col.*) (51, 57)	15 May 12	. . . .	Asst. Mily. Secy.
23 May 00	Fisher, B. D., D.S.O., *p.s.c.* (43, 51, 59, 206)	7 Jan. 14	*l.c.* 3 June 16	
14 Oct. 96	Melvill, T. P., D.S.O., (*act'g. Lt.-Col.*) (50, 53, 59, 206)	24 Feb. 15		
11 Aug. 00	Bruce, R. W. V. (51, 52) . . .	12 July 16		
	Captains.			
8 Jan. 01	*Talbot, D. H.*, D.S.O., M.C. (51, 64)	9 May 06	. . . .	*G.S.O., 2nd gr.*
2 Aug. 00	Lockett, V. N. (50, 53) .	4 Feb. 07		
6 Mar. 02	Nutting, *Sir* H. S. *Bart.* (51, 52, 66)	14 Aug. 13		
16 Aug. 05	*Sopper, E.*, M. C. (50, 53) . . .	16 Sep. 13	. . . .	*With City of Lond. Yeo.*
16 Aug. 05	Micholls, M. G. (51, 52) . . .	13 Oct. 13		
29 Nov. 05	Parbury, H. F. (66*a*) . .	15 Apl. 14		
12 May 06	*Boles, D. C.* (45*a*, 51, 52) . . .	2 May 14	. . . .	*Staff Capt.*
29 Aug. 06	Tutnor, H. B., M.C. (45*a*) . .	2 May 14		
29 Aug. 06	*Dubs, C. E. D.* (59) . . .	18 Dec. 14	. . . .	*With Sig. Serv.*
29 Aug. 10	Plowden, H. R. H. (51) . . .	27 Feb. 15		
27 Jan. 09	Eckstein, H. F. (57)	12 July 16		
	Lieutenants.			
29 Aug. 06	*Muir, J. H.* (*act'g. Maj.*) (45*a*, 57, 66*a*, 169)	16 Sep. 13 22 Dec. 15	. . .	With K. O. Sco. Bord.
3 Sep. 10	*Clements-Finnerty, H.* (66*a*) . .	16 Sep. 13	. . . .	*R. A. F.*
3 Sep. 10	*Venn, E. N. L.* (*temp. Capt.*) . .	13 Oct. 13	. . . .	Instr., R. Mil. Coll., Sandhurst.
3 Sep. 10	*Wynne-Jones, C. L.* (*act'g. Capt.*) (66*a*)	2 May 14	. . . .	*A.-D.-C.*
24 Oct. 11	Black, G. B. (*temp. Capt.*) (64)	12 Sep. 14		
25 Oct. 11	Fowler, R. St. L., M.C. (*act'g. Capt.*)	18 Dec. 14		
19 Jan. 12	*Cornwallis, F. W. M.* (*temp. Capt.*)	27 Feb. 15	. . . .	*M. G. Sqdn.*
4 Sep. 12	Butler, J. A. N. (*temp. Capt.*) (66*a*)	2 Feb. 16		
21 Jan. 13	Hedley, I. M. (*temp. Capt.*) .	2 Feb. 16		
21 Jan. 13	*Stanyforth, R. T.*	18 Aug. 16	. . . .	*Adjt.*
21 Jan. 13	Lees, G. W. M.	2 Sep. 16		
5 Nov. 13	Hay, C. E. E.	31 Oct. 16		
1 Oct. 14	Harris, P. V.	1 July 17		
1 Oct. 14	*Egerton, F.*	1 July 17	. . . .	
1 Oct. 14	Verney, *Hon.* J. H. P., M.C. .	1 July 17		*R. A. F.*
14 Dec. 14	Souchon, G. H. . . .	1 July 17		
16 Dec. 14	Killeen, O. J. H. P., M.C., *Lord*	1 July 17		
23 Dec. 14	Daniel, H. C.	1 July 17		
25 Jan. 15	Baker, D. G.	1 July 17		
20 Oct. 15	*Bury, J.*	1 July 17		
4 May 15	Gossage, A. F. W.	1 July 17		
14 July 15	Brooke, B. G.	1 July 17		
15 July 15	Kirk, C. G. P.	1 July 17		
16 Feb. 16	*Honywood, W. W.*, M.C. . . .	1 July 17		
13 Oct. 15	Perpe, J. H. H.	1 July 17		
27 Oct. 16	Smith, R. H. V.	27 Apl. 18		
27 Oct. 16	Heron, G. C. H.	27 Apl. 18		
7 Mar. 16	Malcolm, A. A.	17 May 18		
14 Jan. 17	Jacobs, A.	14 July 18	. . . .	With 1st Res. Regt. of Cav.
	2nd-Lieutenants.			
1 May 17	Dighton, P. C.	1 May 17	. . . .	With Res. Regt. of Cav.
1 May 17	Mitford, W. S.	1 May 17		
1 May 17	Dereham, H. G.	1 May 17		
4 May 17	Hargrave, J. D.	4 May 17	. . . .	Ac'g. Qr.-Mr.
5 July 17	Birley, N. A.	5 July 17	. . . .	With Res. Regt. of Cav.
5 July 17	Misa, J. M.	5 July 17	. . . .	With Res. Regt. of Cav.
12 Sep. 17	Alford, L. H. McN.	12 Sep. 17	. . . .	With Res. Regt. of Cav.
	Attached.			
5 Sep. 14	Andrews, 2nd-Lt. G. O., Spec. Res.	5 Jan. 17	5 Sep. 14	
9 Nov. 14	Stevenson, 2nd-Lt. A. F. M., Spec. Res.	18 Mar. 17	9 Nov. 14	
26 Nov. 16	Roughead, 2nd-Lt. F. A., Spec. Res.	10 Aug. 17	26 Nov. 16	
21 Jan. 13	Stanyforth, Lt. (*act'g. Capt.*) R. T.	24 Feb. 18	. . . .	Adjutant.

Agents:—MESSRS. COX & CO.

21st (Empress of India's) Lancers.

Arrived in India, 10th October 1912, from Egypt.

" Khartoum."

Uniform—Blue. *Facings*—French grey. *Plume*—White.

COLONEL—Hon. Maj.-Genl. A. H. Taylor, ret. pay . . . . . . 20 Aug. 16.

First Commn.	Names.	Regimental rank.	Army rank.	Remarks.
	Lieutenant-Colonel.			
7 Dec. 95	Fraser, H. F., C.M.G., D.S.O. (*e*) (43, 50, 59)	6 Sep. 15	. . . .	
	Majors.			
1 Dec. 97	Brinton, O. W. (*act'g. Lt.-Col.*) (50, 57)	7 Sep. 10	. . . .	Comdg., Br. Res. Cav. Regt., Muttra.
18 Jan. 99	Cecil, R. E., D.S.O. (*temp. Lt.-Col.*) (51, 57)	10 Dec. 13	. . .	G. S. O., 1st gr.
12 Aug. 99	Pilcher, A. S. (45*a*, 51, 65, 72*b*) .	7 Sep. 14		
18 Apl. 00	Delmege, C. H. (43, 45*a*, 50, 52, 57, 64)	6 Sep. 15	. . . .	1st Res. Cav. Regt.
	Captains.			
5 May 00	*Wheler, G. B. H.* (43, 51, 57, 65, 89, 130)	12 Oct. 07	. . . .	*Bde. Maj.*
23 May 00	*Horsley, C. D.*, M.C. (*temp. Maj., 25 Feb. 18.*) (51, 127)	1 Nov. 10	. . . .	D. A. Q. M. G.
23 May 00	*Lumley-Smith, T. G. L.*, D.S.O. (*temp. Maj.*) (43, 51, 65)	16 Jan. 12	. . . .	*Asst. Provost Marshal.*
23 May 00	Reynolds, G. N. M.C. (43, 45*a*, 51, 59)	4 Apl. 12		
5 Jan. 01	*Boyd-Rochfort, H.* D.S.O., M.C., (51, 60*a*, 64)	4 Apl. 12	. . . .	*G. S. O., 2nd gr.*
8 May 01	Godfree, D. W. (45*a*, 51, 60*a*, 65)	4 Apl. 12	. . . .	With 13th Hrs.
18 Jan. 02	Lister, C. C. (45*a*, 51, 52, 65) . .	26 May. 17	1 Oct. 12	
30 Apl. 02	Cotesworth, C. H. (43, 51) . .	4 May 14	. . . .	Attd. 10 Bn., R. W. Kent R.
22 Jan. 08	Evans, J. W. D. (45 *a*, 51, 52) .	5 Aug. 14	. . . .	G. S. O., 3rd gr.
8 Feb. 08	Methven, D. M. (51, 52, 64) . .	11 Sep. 14	. . . .	With 13th Hussars.
29 Nov. 05	Nicholas, B. G. (45*a*, 51, 52) . .	3 Nov. 14	. . . .	With R. H. A.
10 July 09	*Howes, S.*, M.C. (*temp. Maj.*)	31 Jan. 15	. . . .	*G. S. O., 2nd gr.*
	Lieutenants.			
19 Jan. 12	Brierley, R. W. (*act'g. Capt.*) (12, 51, 52)	4 May 14	. . . .	*Lv., ex I., m.c., 6 mos.*
4 Sep. 12	*Holland, W. T. F.* (*temp. Capt.*)	9 July 14	. . . .	*R. A. F.*
14 Sep. 12	dePret, J., M.C. (*temp. Capt.*) [L] (127)	11 Sep. 14	. . . .	Spec. Apptt.
17 Apl. 13	*Conran, E. L.*, M.C. (*temp. Lt.Col.*)	23 Sep. 14	. . . .	*R. A. F.*
3 Sep. 13	*Kekewich, S.* (*temp. Capt.*) . .	23 Sep. 14	. . . .	*Staff Capt., War Office.*
23 Jan. 14	Lloyd, L. W. (*act'g. Capt.*)	1 Nov. 14	. . . .	Adjt. Comd. Depot.
24 Jan. 14	*Sebag-Montefiore, G. E.*, M.B.E. (*temp. Capt.*)	31 Jan. 15	. . . .	*Asst. Provost Marshal.*
15 Aug. 14	Barker, G. G. (52, 57)	6 Sep. 15	. . . .	With Sig. Ser.
15 Aug. 14	Hoare, E. R. (*act'g. Capt., 23 Oct. 18*) (51, 52)	5 Oct. 15		
25 Oct. 14	Pigott, *Sir* B., *Bart.* (*act'g. Capt., 13 Nov. 18*) (47*a*, 51, 170)	29 July 16		
7 Nov. 14	*Marriette, F. W.* (12, 47*a*, 50, 53)	10 Oct. 16	. . . .	*Adjt.*
7 Nov. 14	*Murphy, E. J.* (47*a*, 50, 59, 170)	1 June 17	. . . .	*1st Cav. Bde. Sig. Troop.*
23 Dec. 14	*St. Clair, L. F. C.*	1 July 17	. . . .	*1st. Res. Cav. Regt.*
9 Jan. 15	Baddeley, H. (51, 172*a*) . . .	1 July 17	. . . .	S. S. O., Risalpur.
17 Feb. 15	Norman, G. R. (51, 57, 61) . .	1 July 17		
13 Mar. 15	*Few, G.* (51, 52)	1 July 17	. . . .	*M. G. Corps.*
17 Mar. 15	Keppel-Palmer, C. M. . . .	1 July 17	. . . .	*Lv., ex I., m. c., 6m., 24 Apl. 18.*
23 Sep. 15	Boothby, R. E. (47*a*, 51) . . .	1 July 17		
10 Oct. 15	Gilholme, E. (*act'g. Capt.*) . .	1 July 17	. . . .	With Glouc. Yeo.
30 Oct. 15	Ryder, E. N., D.C.M. (*temp. Capt.*)	1 July 17	. . . .	Staff Capt., Risalpur Cavy. Bde.
19 Nov. 15	Hill, F. (95, 170)	1 July 17	. . . .	With S. Persian R.
7 Apl. 16	La Terriere, F. J. B. deS. . .	7 Oct. 17	. . . .	1st Res. Cav. Regt.
4 Nov. 16	Hathaway, H. (*act'g. Capt.*) (45*a*, 51, 65)	4 May 18	. . . .	Base Comdt.
24 Nov. 16	Melville, D. W. L., M.C.	24 May 18	. . . .	1st Res. Cav. Regt.
12 Jan. 17	Miller, F. N. (45*a*, 172*a*) . . .	12 July 18		
12 Jan. 17	Hill, F. (50, 52)	12 July 18		
	2nd-Lieutenants.			
7 Feb. 17	*Jennings, P. W.* (50, 53) . .	7 Feb. 17	. . . .	*M. G. Corps.*
1 May 17	Wyatt, W. P.	1 May 17		
1 May 17	Harford, G. A.	1 May 17		
12 Sep. 17	Aveideick, G. U	12 Sep. 17		
12 Sep. 17	Benjamin, A. L.	12 Sep. 17		
12 Sep. 17	Macintosh, I. D.	12 Sep. 17		
21 Dec. 17	Wilson, F. G.	21 Dec. 17		
24 Apl. 18	Ballantine, C. L.	24 Apl. 18		

21st (Empress of India's) Lancers—*contd.*

First Commn.	Names.	Regimental rank.	Army rank.	Remarks.
	Special Reserve.			
	Lieutenants.			
16 Sep. 14	Drake, E. T. T.	29 July 16		
7 Jan. 15	Lyon, C. B. (12, 47*a*, 51)	1 July 17		A.-D.-C. to H.E. the Govr. of Bengal (Extra).
27 Mar. 15	Fraser, H. A. H. (51, 57)	1 July 17		
7 Apl. 15	Jones, W. A. (46, 169)	29 May 18		
21 Dec. 16	Lowndes, W. G.	21 June 18		
	2nd-Lieutenants.			
16 Dec. 17	Prynne, H. F. (46, 47*a*, 57)	16 Dec. 17		
	Attached.			
19 Dec. 00	Carr-Ellison, Capt. H. G. C.	19 Aug. 14	....	North'd Hrs. Attd. Hdqrs., 2nd (Rawalpindi) Dn.
14 Oct. 14	St. Maur, Lt. F. P. (12, 47*a*, 51)	1 June 16	....	1st Royal Devon Yeo.
8 Jan. 15	Storey, Lieut. T.M.M. (46, 47*a*, 51)	8 Oct. 16	....	Linc. Yeo.
10 July 15	Rae, Lieut. R. (46, 61)	1 July 17	....	Sco. Horse.
20 Aug. 16	Rutherford, 2nd-Lt. E.V. (46, 51)	20 Aug. 16	....	North'd Hrs.
21 Dec. 16	Cheesman, Temp. 2nd-Lt. G. W. (47*a*, 51)	21 Dec. 16	....	3rd Res. Regt. of Cav.
15 Dec. 17	Stirling, Temp. 2nd-Lt. R.	15 Dec. 17	....	2nd Res. Regt. of Cav.
7 Nov. 14	Marriette, Lt. F. W.	15 May 18	....	Adjutant.
18 June 04	Laughton, Maj. J. V. (60*a*, 65, 207)	1 July 17	....	*r. d.*-Qr.-Mr.
24 Feb. 12	Ward, Capt. R.	1 July 17	....	Qr.-Mr.

Agents :—Messrs. Cox & Co.

British Reserve Cavalry Regiment. (India.)

Formed in India.

First Commission.	Name.	Regimental rank.	Army rank.	Remarks.
	Major.			
1 Dec. 97	Brinton, O. W. (*actg. Lt.-Col.*) (50, 57).	7 Sep. 10	. . . .	Comdg.
	Captains.			
27 July 01	Secker, V. H. (51, 52, 127, 194).	14 Dec. 07	. . . .	Adjt., Remt. Depôt, Bangalore.
13 Aug. 04	Steele, C. (51, 52)	13 July 10	. . . .	D. A. A. G., Poona Dn.
8 May 01	Wienholt, E. A. (*actg. Maj.*) (51, 52).	17 Dec. 12	. . . .	2nd-in-Command.
15 Sep. 08	Oakes, J. O. (66*a*)	28 Dec. 15		
22 Jan. 08	Alexander, L. W. (45*a*)	23 June 16		
	Lieutenants.			
	Foster, C. R.	1 May 09		
29 Jan. 12	Hamer, G. M. (51, 52)	25 July 14		
3 Oct. 14	Lock, C. F.	16 June 16		
	Ryder, C. F.	14 Nov. 14	28 Aug. 17	
	Riley-Smith, W.	13 Nov. 14	19 Sep. 17	
	Cohn, D. M.	30 Dec. 16	. . . .	Stn. Staff Office, Meerut.
1 Oct. 14	Stanley, D. M.	9 Feb. 17		
22 Mar. 15	Lord, J. A. (*actg. Capt.*)	11 Mar. 17	. . . .	Actg. Adjt.
. . . .	Stevens, A. E. (*actg. Capt.*)	1 July 17		
. . . .	Ormord, L. A.	1 July 17		
. . . .	Goodman, E.	1 July 17		
21 Aug. 16	Scott, E. J. W. (*temp.*)	3 Aug. 17		
. . . .	Reddington, D. C. (*temp.*)	20 Nov. 17		
	2nd-Lieutenants.			
14 July 14	Gillanders, C. F.	14 July 14		
6 Oct. 14	Mitchell, J. R. V. (*temp. Maj.*)	6 Oct. 14	. . . .	Comdt. C. Bomb. Sch., Mhow.
7 Oct. 15	Anson, R. R.	7 Oct. 15		
1 Nov. 15	Hewetson, E. F.	1 Nov. 15	. . . .	Mily. Accts. Dept., Poona.
28 Aug. 16	Hampson, J.	28 Aug. 16		
27 Sep. 16	Cullum, A. E.	27 Sep. 16		
28 Nov. 16	Mc'Lagan, C. A. G.	28 Nov. 16		
28 Nov. 16	Woodhouse, G.	28 Nov. 16		
12 Dec. 16	Moore, H. A.	12 Dec. 16		
16 Dec. 16	Webley E. A.	16 Dec. 16		
1 May 17	Hill-Forster, R. W.	1 May 17		
1 May 17	Still, C. H.	1 May 17		
1 May 17	James R. E.	1 May 17		
17 May 17	Hounsell, T. H.	17 May 17		
29 June 17	Hollingdale, E. T.	29 June 17		
1 July 17	Bone, J. H.	1 July 17		
8 July 17	Morrell, W. T.	8 July 17		
24 Aug. 17	O'Connor-George, J. M.	24 Aug. 17		
24 Aug. 17	Jack, W. B.	24 Aug. 17		
24 Aug. 17	Evans. D. R.	24 Aug. 17		
24 Aug. 17	Crease, E. H.	24 Aug. 17		
14 Sep. 17	Gosnell, A. G.	14 Sep. 17		
23 Sep. 17	Fossett, F. C.	23 Sep. 17		
25 Sep. 17	Nordon, R. W.	25 Sep. 17		
8 Oct. 17	Longstreet, R. W.	8 Oct. 17		
20 Nov. 17	Redington, D. C.	20 Nov. 17		
. . . .	Harvey, R. L.			
19 Dec. 17	Brown, J. W.	19 Dec. 17	. . . .	Mily. Accts. Dept., Poona.
22 Mar. 15	Lord, Lieut. (*actg. Capt.*) J. A.	25 Mar. 18	. . . .	Adjutant.
14 Sep. 04	Andrews, Major F. L., D.C.M.	14 Sep. 17	. . . .	Riding Master.
. . . .	Beard, Lieut. R.	. . . .	. . . .	Actg. Quartermaster.

Uniform.—Full dress—(Officers only) Blue cloth.
Working dress—Khaki drill.

1st Bihar Light Horse.

Formed as Bihar Mtd. Rifles by Govt. of Bengal Notfn., dated 8th December 1862. Designation changed to Bihar Light Horse by Mily. Dept. Notfn. No. 120, dated 29th February 1884, and reconstituted as an I. D. F. Unit with present designation 1st Apl. 1917.

Establishment—2 Combined Active & Reserve Sqdns.
2 do. do. Coys
2 Machine Gun Detts.

Troops at Motihari, Chupra, Bankipore, Darbhanga and Bhagalpur.

HONORARY COMMANDANT—*The Hon'ble Sir* EDWARD ALBERT GAIT, K.C.S.I., C.I.E. 1 Apl. 17.

First Commn.	Names.	Date of rank.	Date of appointment to corps.	Remarks.
	Lieutenant-Colonels.			
1 Apl. 17	Hickley, Victor North, C.I.E., V.D., A.D.C. (70).	1 Apl. 17	1 Apl. 17	Comdt. (*Hon. Col. late Ind. Volrs.*)
1 Apl. 17	*Filgate, Townley Richard,* C.I.E., V.D.	1 Apl. 17	1 Apl. 17	*Adjutant. I. A. R. O.*
	Majors.			
1 Apl. 17	Wilson, John Arthur Maclean, V.D. (79)	1 Apl. 17	1 Apl. 17	2nd-in-Comd. Hony. A.D.C. to the Lieut.-Govnr., Bihar and Orissa.
1 Apl. 17	*Llewhellin, George Elliott* (79)	1 Apl. 17	1 Apl. 17	*I. A. R. O.*
1 Apl. 17	Hubback, John Austen (79)	1 Apl. 17	1 Apl. 17	
1 Apl. 17	Lees, Cecil George (79)	1 Apl. 17	1 Apl. 17	
	Captains.			
1 Apl. 17	*Cheshire, Hugh Stanley* (79)	1 Apl. 17	1 Apl. 17	*I. A. R. O.*
1 Apl. 17	Whitty, John Tarlton (79)	1 Apl. 17	1 Apl. 17	
1 Apl. 17	Leslie, Stephen (54, 79)	1 Apl. 17	1 Apl. 17	
1 Apl. 17	McNeil, Robert (54, 79)	1 Apl. 17	1 Apl. 17	Machine Gun. Det., Monghyr.
	Lieutenants.			
1 Apl. 17	Temple, Frederick Charles	1 Apl. 17	1 Apl. 17	
1 Apl. 17	Gale, Horace Charles Macleod	1 Apl. 17	1 Apl. 17	
1 Apl. 17	Henry, John (79)	1 Apl. 17	1 Apl. 17	
1 Apl. 17	Mackenzie, William	1 Apl. 17	1 Apl. 17	
	2nd-Lieutenants.			
1 Apl. 17	*Reed, William Louis Lindsay,* V.D. (79)	1 Apl. 17	1 Apl. 17	*I. A. R. O.*
1 Apl. 17	*Lyall, Frank Frederick,* C.I.E. (79)	*1 Apl. 17*	*1 Apl. 17*	*I. A. R. O.*
1 Apl. 17	Caldwell, Kenneth Somerville	1 Apl. 17	1 Apl. 17	
1 Apl. 17	Shaw, Frederick John Freshwater	1 Apl. 17	1 Apl. 17	
1 Apl. 17	Fielder, Hubert Gordon	1 Apl. 17	1 Apl. 17	
1 Apl. 17	*Armstrong, C.F.*	*1 Apl. 17*	*1 Apl. 17*	*I. A. R. O.*
1 Apl. 17	Brebner, 2nd-Lt. Alexander	1 Apl. 17	1 Apl. 17	
	Supernumerary.			
1 Apl. 17	*Hervey, Ney Lindsay (79)*	*1 Apl. 17*	*1 Apl. 17*	*I. A. R. O.*
1 Apl. 17	Morshead, Leonard Frederick (79)	1 Apl. 17	1 Apl. 17	Lv. England.
	Staff.			
1 Apl. 17	*Filgate, Maj. T. R.,* C.I.E., V.D.	*1 Apl. 17*	*1 Apl. 17*	*Adjt. I. A. R. O.*

Uniform—Full dress—Blue serge.
Working dress—Khaki drill.

2nd Surma Valley Light Horse.

Formed 6th Apl. 83. Designation altered by Mily. Dept. Notfn. No. 557, dated 13th Aug. 1886, and reconstituted as an I. D. F. Unit with present designation 1st Apl. 1917.

"Non Sibi Sed Patriæ."

Establishment—3 Combined Active & Reserve Sqdns.

Troops at Dullabcherra, Hollicherra, Srimangal and Laskarpur.

HONORARY COMMANDANT—*The Hon'ble Mr.* NICHOLAS DODD BEATSON-BELL, C.S.I., C.I.E. . . . . . 1 July 18.

(*Chief Commissioner of Assam.*)

First Commn	Names.	Date of rank.	Date of appointment to corps.	Remarks.
	Lieutenant-Colonels.			
1 Apl. 17	Knowles, John George, C.I.E., V.D.	1 Apl. 17	1 Apl. 17	Comdt. Hon. Col. late Ind. Volrs.
	Majors			
1 Apl. 17	Mason, Walter, V.D.	1 Apl. 17	1 Apl. 17	Lt.-Col. late Ind. Volrs.
17 Apl. 17	Hickman, Robert St. John, V.D.	1 Apl. 17	1 Apl. 17	[2nd-in-Comd.
1 Apl. 17	*Fox, Henry Benedict*, V.D.	*1 Apl. 17*	*1 Apl. 17*	*I. A. R. O.*
1 Apl. 17	Marshall, Hannath Douglas, V.D.	1 Apl. 17	1 Apl. 17	
	Captains.			
1 Apl. 17	Fraser, Robert Thomson, V.D.	1 Apl. 17	1 Apl. 17	
1 Apl. 17	Girling, Harry Morton, V.D.	1 Apl. 17	1 Apl. 17	
1 Apl. 17	*Rees, Roderick*	*1 Apl. 17*	*1 Apl. 17*	*I. A. R. O.*
1 Apl. 17	*Macwhirter, Dudley* (*I. A. Res.*) (54)	*1 Apl. 17*	*1 Apl. 17*	*I. A. R. O.*
1 Apl. 17	Carslaw, Frank Henderson	1 Apl. 17	1 Apl. 17	
	Lieutenants.			
1 Apl. 17	Pullen, Frank (54)	1 Apl. 17	1 Apl. 17	
1 Apl. 17	MacLeod, Charles	1 Apl. 17	1 Apl. 17	
1 Apl. 17	*Mortimore, Ralph*	*1 Apl. 17*	*1 Apl. 17*	*Temp. Commn. in Army.*
1 Apl. 17	*Smith, William Townsend*, V.D.	*1 Apl. 17*	*1 Apl. 17*	*I. A. R. O.*
1 Apl. 17	James, Herbert Mark	1 Apl. 17	1 Apl. 17	
1 Apl. 17	MacKnight, John	1 Apl. 17	1 Apl. 17	
	2nd-Lieutenants.			
1 Apl. 17	Pringle, Robert	1 Apl. 17	1 Apl. 17	
1 Apl. 17	*Doubell, Philip Stanley*	*1 Apl. 17*	*1 Apl. 17*	*I. A. R. O.*
1 Apl. 17	Green, Thomas Walter	1 Apl. 17	1 Apl. 17	
1 Apl. 17	Boyle, Reginald George	1 Apl. 17	1 Apl. 17	
1 Apl. 17	Baker, Edward Banham	1 Apl. 17	1 Apl. 17	
18 June 18	Stuart, Alasdair Francis	18 June 18	18 June 18	
18 June 18	MacIver, Peter	18 June 18	18 June 18	
	Staff.			
13 May 17	Dyball, 2nd-Lt. (*temp. Capt.*) E. E., 8 Huss.	13 May 17	7 July 17	Adjutant.
1 Apl. 17	White, Rev. William Ebenezer, V.D.	1 Apl. 17	1 Apl. 17	Hony. Chaplain.

Uniform—Khaki drill.

3rd Calcutta Light Horse.

Formed 7th Oct. 81. Reorganised 14th Sep. 86. Reconstituted, G. G. O. 437 of 1898, Central Bengal Light Horse amalgamated by G. G. O. No. 1000, dated 1st November 1901, and reconstituted as an I. D. F. Unit with present designation 1st Apl. 1917.

Establishment—1 Res. Sqdn.
1 Combined Active & Reserve Sqdn.
2 Machine Gun Detts.

HONORARY COMMANDANT—*H. E. the Rt. Hon.* FREDERIC JOHN NAPIER, *Baron* CHELMSFORD, P.C., G.M.S.I., G.C.M.G., G.M.I.E., G.B.E. . . 1 Apl. 17.
(*Viceroy and Governor-General of India.*)

First Commn.	Names.	Date of rank.	Date of appointment to corps.	Remarks.
	Lieutenant-Colonel.			
1 Apl. 17	Pugh, Archibald John, O.B.E., V.D., A.D.C.	1 Apl. 17	1 Apl. 17	Comdt., Hony. Col. late Ind. Volrs.
1 Apl. 17	*Westmacott, Richard*	*1 Apl. 17*	*1 Apl. 17*	*I. A. R. O.*
	Majors.			
1 Apl. 17	*Westmacott, Thomas Horatio* . .	*1 Apl. 17*	*1 Apl. 17*	(*Temp. Capt. in Army 24 Dec. 15*).
1 Apl. 17	Beadel, Cecil Frank (54) . . .	1 Apl. 17	1 Apl. 17	2nd-in-Comd.
1 Apl. 17	Crum, Walter Erskine, O.B.E. (D.I.) (50) .	1 Apl. 17	1 Apl. 17	
1 Apl. 17	Pickford, Alfred Donald . . .	1 Apl. 17	1 Apl. 17	
	Captains.			
1 Apl. 17	Harvey, Gourlay	1 Apl. 17	1 Apl. 17	
1 Apl. 17	Law, Robert Oswald	1 Apl. 17	1 Apl. 17	
1 Apl. 17	*Studd, Eric* (D. I.) (54) . . .	*1 Apl. 17*	*1 Apl. 17*	*I. A. (T. C.)*
1 Apl. 17	*Arthur, Charles Gordon* . . .	*1 Apl. 17*	*1 Apl. 17*	(*Temp. Lt. in Army, 25 Feb. 16.*)
1 Apl. 17	*Deane, Edward Stanley* (D. I.) (51, 52, 54, 71.)	*1 Apl. 17*	*1 Apl. 17*	*I. A. (T. C.) Adjutant.*
	Lieutenants.			
	2nd-Lieutenants.			
1 Apl. 17	*Allen, Jeffery Hugh Summers* . .	*1 Apl. 17*	*1 Apl. 17*	*I. A. R. O.*
1 Apl. 17	Moinet, Louis Vincent	1 Apl. 17	1 Apl. 17	
1 Apl. 17	Craik, William Ross (D. I.) . . .	1 Apl. 17	1 Apl. 17	
	Supernumerary.			
1 Apl. 17	Steele-Perkins, Lieut. Charles Steele .	1 Apl. 17	1 Apl. 17	
1 Apl. 17	Birkmyre, The Hon. *Sir* Archibald, KT., C. B. E.	1 Apl. 17	1 Apl. 17	
	Staff.			
5 Feb. 15	Deane, temp. 2nd-Lt. (*temp. Capt.*) E. S. (D. I.) I. A. (51, 52, 54, 7 .)	5 Feb. 15	8 May 15	Adjutant.
1 Apl. 17	Houseman, Maj. Edward Albert . .	1 Apl. 17	. . .	Medl. Offr.
1 Apl. 17	Caddy, Maj. Adrian, M.D., F.R.C.S.	1 Apl. 17	. . .	Ditto.

Uniform—Full dress—Blue serge
Working dress—Khaki drill.

4th Bombay Light Horse.

Formed 28th May 85. Reorganised 28th Mar. 87 and re-constituted as an I.D.F. Unit with present designation, 1st Apl. 17.

Establishment—1 Active Sqdn.

HONORARY COMMANDANT—*H. E. the Rt. Hon.* FREEMAN FREEMAN-THOMAS, *Baron* WILLINGDON OF RATTON, G.C.S.I., G.C.I.E., G.B.E. . . . . . 1 Apl. 17
(*Governor of Bombay.*)

First Commn.	Names.	Date of rank.	Date of appointment to corps.	Remarks.
	Major.			
1 Apl. 17	Reed, *Sir* Herbert Stanley, *Kt.*, A. D. C.	1 Apl. 17	1 Apl. 17	Comdt. (Hon. Lt.-Col., late Ind. Volrs.)
	Captain.			
Apl. 17	*Wilson, Duncan William,* C. I. E. .	*1 Apl. 17*	*1 Apl. 17*	*I. A. R. O.*
Apl. 17	Annesley, Francis Charles (50, 80) . .	1 Apl. 17	1 Apl. 17	
	Lieutenants.			
1 Apl. 17	Spence, Reginald Arthur (51, 80) . . .	1 Apl. 17	1 Apl. 17	
1 Apl. 17	*Baird, Alexander Edwin*	*1 Apl. 17*	*1 Apl. 17*	*I. A. R. O.*
	2nd-Lieutenants.			
1 Apl. 17	*Smith, Arthur Kirke* (51)	*1 Apl. 17*	*1 Apl. 17*	*I. A. R. O.*
1 Apl. 17	Pearson, Edmund George (51, 80) . . .	1 Apl. 17	1 Apl. 17	Actg. Adjt.
21 Nov. 17	Hearson, Henry Francis Pengelly (51, 80)	21 Nov. 17	21 Nov. 17	
22 Nov. 17	Partridge, Charles Henry (50, 80) . .	22 Nov. 17	22 Nov. 17	

Staff.

1 Apl. 17	Hewlett, Kenelm Lt. (*temp. Maj.*,) *2 Oct. 17* (Vety. Corps).	1 Apl. 17	1 Apl. 17	Vety. Embarkation Offr., Bombay.

Uniform—Blue.
Facings—White.
Working dress—Khaki drill.

5th Punjab Light Horse.

Formed by G. G. O. No. 460, dated 12th May 1893, and reconstituted as an I. D. F. Unit with present designation, 1 Apl. 17.

"De Bon Vouloir Servir Le Roy."

Establishment—2 Combined Active and Reserve Sqdns. 2 Machine Gun Detts.

Troops at Lahore, Ambala, Rawal Pindi and Delhi.

HONORARY COMMANDANT—*The Hon'ble Sir* MICHAEL FRANCIS O'DWYER, G.C.I.E., K.C.S.I., 1 Apl. 17.

(*Lieutenant-Governor of the Punjab.*)

First Commn.	Names.	Date of rank.	Date of appointment to corps.	Remarks.
	Lieutenant-Colonel.			
1 Apl. 17	Barron, Claud Alexander, C.I.E. (80) (169)	17 Feb. 18	1 Apl. 17	Comdt.
	Majors.			
1 Apl. 17	*Bolster, Reginald Charles* (51, 52, 80, 94, 165, 205a)	*1 Apl. 17*	*1 Apl. 17*	*2nd-in-Command. Staff Capt., 16 Ind. Dn.*
1 Apl. 17	*Crookshank, Sidney D'Aquilar,* C.I.E., M.V.O., D.S.O. (Bt.-Col., R.E.)	*1 Apl. 17*	*1 Apl. 17*	
1 Apl. 17	Wilson, William	1 Apl. 17	1 Apl. 17	R' Pindi.
1 Apl. 17	*Hadow, Douglas Scott* (169)	*1 Apl. 17*	*1 Apl. 17*	*I. A. R. O.*
1 Apl. 17	Mitchell, John (54) (80)	1 July 18	1 Apl. 17	Amritsar.
	Captain.			
1 Apl. 17	*Rivett, Edwin Godfrey*	*1 Apl. 17*	*1 Apl. 17*	*I. A. R. O.*
	Lieutenants.			
1 Apl. 17	Duncan, George (54)	1 Apl. 17	1 Apl. 17	R' Pindi.
1 Apl. 17	*Howell, Gerard Charles Lisle*	*1 Apl. 17*	*1 Apl. 17*	(*Actg. Capt. in Army*).
1 Apl. 17	Blascheck, Arthur David (54)	1 Apl. 17	1 Apl. 17	R' Pindi.
1 Apl. 17	Wilson-Johnston, Joseph, C.B.E. (80, 94, 165, 169)	1 Apl. 17	1 Apl. 17	R' Pindi.
1 Apl. 17	Parker, Henry Edmond (51)	1 Apl. 17	1 Apl. 17	Delhi.
	2nd-Lieutenants.			
1 Apl. 17	*Carter, Frederick George*	*1 Apl. 17*	*1 Apl. 17*	*I. A. R. O.*
1 Apl. 17	Bowles, Charles William	1 Apl. 17	1 Apl. 17	Patiala.
1 Apl. 17	*Pitkeathly, James Scott*	*1 Apl. 17*	*1 Apl. 17*	*Temp. Col. in Army.*
1 Apl. 17	Ferguson, John Alexander (165)	1 Apl. 17	1 Apl. 17	Lahore.
1 Apl. 17	Cornforth, Ernest	1 Apl. 17	1 Apl. 17	Lahore.
	Supernumerary.			
1 Apl. 17	Donald, Capt. Charles Hilliard (51, 54, 72b)	1 Apl. 17	1 Apl. 17	Mesopotamia.
1 Apl. 17	Metcalfe, Lt. Herbert Aubrey Francis	1 Apl. 17	1 Apl. 17	Peshawar.
1 Apl. 17	Thompson, Lt. John Perronet (79)	. . . . .	1 Apl. 17	Lahore.
	Staff.			
6 Feb. 17	Rigby, 2nd-Lt. (temp. Capt.) W., 1st (King's) Dragoon Gds.	6 Feb. 17	6 Feb. 17	Adjutant.
1 Apl. 17	Stephenson, Revd. H. S.	1 Apl. 17	1 Apl. 17	Hony. Chaplain.

Uniform—Khaki drill.
Mess Uniform—Blue with white facings.

6th Assam Valley Light Horse.

The Sibsagar Mtd. Rifles, Lakhimpur Mtd. Rifles, Darrang Mtd. Rifles and Nowgong Mtd. Rifles amalgamated and the Corps designated the Assam Valley Mounted Rifles on 6th November 1891. Reorganised and designated the Assam Valley Light Horse by G. G. O. No. 1086, dated 25th September 1896. Shillong Volunteer Rifles amalgamated as a dismounted detachment 1st May 1906, G. G. O. 100 of 1906 and reconstituted as an I. D. F. Unit with present designation, 1st Apl. 1917.—"Semper Paratus."

Estabilshment—3 Combined Active & Reserve Sqdns.
1 do. do. Coy.
1 Machine Gun Dett.

Troops at Lakhimpur, Sibsagar, Nowgong, Kamrup, Shillong and Darrang.

HONORARY COMMANDANT—*The Hon'ble* Mr. Nicholas Dodd Beatson-Bell, C.S.I., C.I.E. *(Chief Commissioner of Assam).*

First Commn.	Names.	Date of rank.	Date of appointment to corps.	Remarks.
	Lieutenant-Colonel.			
1 Apl. 17	Grimston, Lionel Augustus, V.D., A.D.C. (80).	1 Apl. 17	1 Apl. 17	Comdt. *3 mos. Lv. in I. from 10 Nov. 18.*
	Majors.			
1 Apl. 17	Garbett, Hubert Champion, V. D. (80)	1 Apl. 17	1 Apl. 17	2nd-in-Comd.
1 Apl. 17	*Chrystall, Andrew,* V. D.	*1 Apl. 17*	*1 Apl. 17*	*I. A. R. O.*
1 Apl. 17	Mucklow, John, V. D. (80)	1 Apl. 17	1 Apl. 17	
1 Apl. 17	Abbott, Osborne Alan, V. D. (80)	1 Apl. 17	1 Apl. 17	
1 Apl. 17	Lovell, William Cecil Lennox (80)	1 Apl. 17	1 Apl. 17	
	Captains.			
1 Apl. 17	Chambers, Cyril Claridge (80)	1 Apl. 17	1 Apl. 17	
1 Apl. 17	*Ebden, James Wylde*	*1 Apl. 17*	*1 Apl. 17*	*I. A. R. O.*
1 Apl. 17	Hawkins, Francis Henry (80)	1 Apl. 17	1 Apl. 17	
1 Apl. 17	Allen, Charles Wilfred (80)	1 Apl. 17	1 Apl. 17	
	Lieutenants.			
1 Apl. 17	Elliot, James Stenhouse, V. D. (80)	1 Apl. 17	1 Apl. 17	
1 Apl. 17	Stewart, George (80)	1 Apl. 17	1 Apl. 17	
1 Apl. 17	*Story, Harry Wakefield*	*1 Apl. 17*	*1 Apl. 17*	*I. A. R. O.*
1 Apl. 17	Roffey, Leslie Arden (80)	1 Apl. 17	1 Apl. 17	Offg. Adjt.
1 Apl. 17	*Gore, Francis William*	*1 Apl. 17*	*1 Apl. 17*	*I. A. R. O.*
1 Apl. 17	Hannen, Frederick Barlow (80)	1 Apl. 17	1 Apl. 17	
1 Apl. 17	*Perry, Charles Alfred Stanley*	*1 Apl. 17*	*1 Apl. 17*	*I. A. R. O.*
1 Apl. 17	*Nevill, Edward Parry*	*1 Apl. 17*	*1 Apl. 17*	*I. A. R. O.*
	2nd-Lieutenants.			
1 Apl. 17	Newbery, John Everard (80)	1 Apl. 17	1 Apl. 17	
1 Apl. 17	White, Gerald Fox (80)	1 Apl. 17	1 Apl. 17	
1 Apl. 17	*Reid, William Peter*	*1 Apl. 17*	*1 Apl. 17*	*I. A. R. O. (T. C.)*
1 Apl. 17	Lane, Stuart Thomas Henry (80)	1 Apl. 17	1 Apl. 17	
1 Apl. 17	Cockburn, Robert Charles Radcliffe (80)	1 Apl. 17	1 Apl. 17	
1 Apl. 17	Moran, John Percy (80)	1 Apl. 17	27 June 18	
1 Apl. 17	Ashe, Wellesley St. George (80)	1 Apl. 17	27 June 18	
1 Apl. 17	Bower, James Anderson (80)	1 Apl. 17	27 June 18	
1 Apl. 17	Roffey, Edgar Stuart (80)	1 Apl. 17	27 June 18	
	Supernumerary.			
1 Apl. 17	Fairley, Capt. John (80)	1 Apl. 17	1 Apl. 17	
	Staff.			
1 Apl. 17	*Byres, Capt. E. B. M.*	*1 Apl. 17*	*1 Apl. 17*	*Adjt. I. A. R. O. 6 mos. Lv. in I. 19, Sep. 18.*
1 Apl. 17	Harris, Capt. William (80)	1 Jan. 18	1 Apl. 17	Vety. Offr.
1 Apl. 17	McCombie, Lieut. Frederick Champ (80)	1 Apl. 17	....	Med. Offr.
28 Feb. 18	Brain, Hony. Lt. W. T.	28 Feb. 18	27 Mar. 18	Qr.-Mr.
1 Apl. 17	Forsyth, Lieut. Charles Edwin Percy	1 Apl. 17	....	Med. Offr.

Uniform—Khaki.

7th (S) United Provinces Horse.

Formed as U. P. Light Horse 1st Apl. 1904—G. G. O. 158 of 1904. Reorganised 1st Apl. 1909, by Army Dept. Notfns. No. 188 and 309 of 1909, and reconstituted as an I. D. F. Unit with present designation 1st Apl. 1917.

Establishment—3 Combined Active & Reserve Sqdns.
1 Motor Cycle Machine Gun Dett.

Troops at Allahabad, Ghazipur, Lucknow and Gorakhpur.

HONORARY COMMANDANT—

First Commn.	Names.	Date of rank.	Date of appointment to corps.	Remarks.
	Lieutenant-Colonels.			
1 Apl. 17	*Faunthorpe, John Champion,* M.C., V.D.	*1 Apl. 17*	*1 Apl. 17*	*Temp. Lt.-Col. in Army.*
1 Apl. 17	*Nevill, Henry Rivers* (DII)	*1 Apl. 17*	*1 Apl. 17*	*A. A. G., Army Hd. Qrs.*
1 Apl. 17	Watson, Alexander (DII)	1 Apl. 17	1 Apl. 17	Comdt.
	Majors.			
1 Apl. 17	Harris, Charles Gordon (80)	1 Apl. 17	1 Apl. 17	2nd-in-Comd., *17* Nov. 17.
1 Apl. 17	Silver, James Robert Robertson (80)	1 Apl. 17	1 Apl. 17	
1 Apl. 17	Taylor, Samuel Henry	1 Apl. 17	1 Apl. 17	
1 Apl. 17	Oliver, Edward William	1 Apl. 17	1 Apl. 17	
	Captains.			
1 Apl. 17	Williamson, James	1 Apl. 17	1 Apl. 17	
1 Apl. 17	Turner, Maurice (80)	1 Apl. 17	1 Apl. 17	
1 Apl. 17	Leake, Hugh Martin (52) (80)	1 Apl. 17	1 Apl. 17	
	Lieutenants.			
1 Apl. 17	*Cotton, William Benjamin*	*1 Apl. 17*	*1 Apl. 17*	*I. A. R. O.*
1 Apl. 17	Pearson, James Rae (80)	1 Apl. 17	1 Apl. 17	
1 Apl. 17	*O'Neill, Frank Owen* (80)	*1 Apl. 17*	*1 Apl. 17*	*I. A. R. O.*
1 Apl. 17	*Nash, John Alexander*	*1 Apl. 17*	*1 Apl. 17*	*I. A. R. O.*
1 Apl. 17	*Bender, William Edward Gustave*	*1 Apl. 17*	*1 Apl. 17*	*I. A. R. O.*
1 Apl. 17	*Ibbotson, Archie William,* M.C.	*1 Apl. 17*	*1 Apl. 17*	*I. A. R. O.*
1 Apl. 17	*Hearsey, Lionel Douglas William,* M.C.	*1 Apl. 17*	*1 Apl. 17*	*I. A. R. O.*
1 Apl. 17	Wild, Charles Edmund (80)	1 Apl. 17	1 Apl. 17	
1 Apl. 17	*Bourdillon, Bernard Henry*	*1 Apl. 17*	*1 Apl. 17*	*I. A. R. O.*
	2nd-Lieutenant.			
1 Apl. 17	*Anderson, Robert William*	*1 Apl. 17*	*1 Apl. 17*	*I. A. R. O.*
	Supernumerary.			
1 Apl. 17	Gavin-Jones, Lt.-Col. Tracy French (80) (DII)	17 Nov. 17	25 June 18	
1 Apl. 17	Rigg, Capt. Harry (80)	1 Apl. 17	1 Apl. 17	
1 Apl. 17	Rix, Capt. Harry Sidney (80)	1 Apl. 17	1 Apl. 17	
1 Apl. 17	Smart, Lt. Arthur Wernicke	1 Apl. 17	1 Apl. 17	
1 Apl. 17	de Hoxar, Lt. Arthur Reginald	1 Apl. 17	1 Apl. 17	
1 Apl. 17	Harris, Lt. Frederick William	1 Apl. 17	1 Apl. 17	
	Staff.			
20 Oct. 15	Knott, Lt. H., 6th Dgns.	1 July 17	20 Oct. 15	Adjutant.
1 Apl. 17	Fuller, Lt.-Col. Charles Arthur, M.B.	1 Apl. 17	. . . .	Medl. Offr.

Uniform—Khaki drill.
Full dress—Blue, Facings, white.

8th (N.) United Provinces Horse.

Formed 1st Apl. 1909 from Dehra Dun Mtd. Rifles and Mtd. Coy., Naini Tal Vol. Rifles, by Army Dept. Notfns. No. 188 and 309 of 1909. Reconstituted as an I. D. F. Unit with present designation 1st Apl. 1917.

Establishment—*1 Combined Active and Reserve Sqdn.*
2 Machine Gun Detts.

HONORARY COMMANDANT—

First Commn.	Names.	Date of rank.	Date of appointment to corps.	Remarks.
	Majors.			
1 Apl. 17	*Ouseley, Joseph William Glynn*, V.D. (54)	*1 Apl. 17*	*1 Apl. 17*	*I. A. R. O.*
1 Apl. 17	Lomas, Hugh Arthur	19 May 18	1 Apl. 17	Comdt.
	Captains.			
1 Apl. 17	*Courthope, Edward Arthur*	*1 Apl. 17*	*1 Apl. 17*	*I. A. R. O.*
1 Apl. 17	*Johnstone, William Wallace*, V.D. (54, 95)	*1 Apl. 17*	*1 Apl. 17*	*I. A. R. O.*
1 Apl. 17	Billson, Herbert George (54)	*1 Apl. 17*	*1 Apl. 17*	
	Lieutenants.			
1 Apl. 17	*fforde, Arthur Brownlow*	*1 Apl. 17*	*1 Apl. 17*	
1 Apl. 17	*Dacres, Leonard Seymour Lambert*	*1 Apl. 17*	*1 Apl. 17*	*I. A. R. O.*
	2nd-Lieutenants.			
1 Apl. 17	Wimbush, Anthony	1 Apl. 17	1 Apl. 17	Lt., late Ind. Vols.
1 Apl. 17	Douglas-Hamilton, James Alexander Gordon	1 Apl. 17	1 Apl. 17	Lt., late Ind. Vols.
1 Apl. 17	Grant, Charles William	1 Apl. 17	1 Apl. 17	
13 June 18	Hall, Richard Ferguson	13 June 18	13 June 18	
	Supernumerary.			
1 Apl. 17	Johnston, Capt. David Landale	1 Apl. 17	1 Apl. 17	
1 Apl. 17	Quarry, Capt. Frederick George	1 Apl. 17	1 Apl. 17	
	Staff.			
22 Aug. 16	Turton, Lt. (*temp. Capt.*) A. E., 8 Huss.	22 Feb. 18	22 Aug. 16	Adjutant.

Uniform—Full dress—Scarlet. *Facings*—White.
Working dress—Khaki drill.

9th Northern Bengal Mounted Rifles.

Formed 6th Aug. 73. Reorganised 5th Feb. 89. Reconstituted as an I. D. F. Unit with present designation 1 Apl. 1917.

Establishment—3 Combined Active and Reserve Sqdn.
1 Do. do. Coy.
1 Cadet Coy.
1 Machine Gun Dett.

Squadrons and Companies at Kurseong, Mal, Nagrakata, Kalchini and Kalimpong.

HONORARY COMMANDANT—*His Excellency the Right Hon'ble* LAWRENCE JOHN LUMLEY DUNDAS, *Earl of* Ronaldshay, G.C.I.E. . . . . . . 1 May 17.

(*Governor of Bengal.*)

HONORARY LIEUTENANT—*His Highness Maharaja Sir* JITENDRA NARAYAN BHUP BAHADUR, K.C.S.I., of Cooch Behar . . . .

First Comm.	Names.	Date of rank.	Date of appointment to corps.	Remarks.
	Lieutenant-Colonel.			
1 Apl. 17	Wood, Frederick Evan	1 Apl. 17	17 Oct. 18	Comdt.
	Majors.			
1 Apl. 17	Gollan, Robert, V.D.	1 Apl. 17	1 Apl. 17	Pillans Hat. 2nd-in-Comd.
1 Apl. 17	Drysdale, Lawrance, V.D.	1 Apl. 17	1 Apl. 17	Mal.
1 Apl. 17	Webb, George Augustus (54)	1 Apl. 17	1 Apl. 17	Kurseong.
	Captains.			
1 Apl. 17	*Little, John Oswald* (59, 206) (D. I, II, III, IV)	1 Apl. 17	1 Apl. 17	*I. A. R. O.*
1 Apl. 17	*Woodhead, Alexander William* (54) .	1 Apl. 17	1 Apl. 17	*I. A. R. O.*
1 Apl. 17	Travers, Walter Lancelot	1 Apl. 17	1 Apl. 17	Mal.
1 Apl. 17	Ansell, Arthur Molloy (54)	1 Apl. 17	1 Apl. 17	Cadet Coy., Kurseong.
1 Apl. 17	Ripley, Thornton	1 Apl. 17	1 Apl. 17	Cadet Coy., Kalimpong.
1 Apl. 17	Luard, Bernard Eustace	1 Apl. 17	1 Apl. 17	Mal.
1 Apl. 17	Macalister, Alexander Stormonth . .	1 Apl. 17	1 Apl. 17	*Lv. ex I. 9 mos., 1 Mar. 18.*
1 Apl. 17	Addis, Robert Bawn	1 Apl. 17	1 Apl. 17	
	Lieutenants.			
1 Apl. 17	*Ward, Horace Futcher*	1 Apl. 17	1 Apl. 17	*I. A. R. O.*
1 Apl. 17	*Pook, John deCourcy* (54)	1 Apl. 17	1 Apl. 17	*I. A. R. O.*
1 Apl. 17	Macbean, James Alexander	1 Apl. 17	1 Apl. 17	Mal.
1 Apl. 17	Polwhele, John Archibald	1 Apl. 17	1 Apl. 17	Domohani.
1 Apl. 17	Luard, Godfrey Ernest	1 Apl. 17	1 Apl. 17	Telepara.
1 Apl. 17	Wrangham-Hardy, George (54)	1 Apl. 17	1 Apl. 17	Darjeeling.
	2nd-Lieutenants.			
Apl. 17	*Winn, Herbert Ewart*	1 Apl. 17	1 Apl. 17	*I. A. R. O.*
Apl. 17	Webb, Norman Gotliebe	1 Apl. 17	1 Apl. 17	Kalchini. Lieut. late Ind. Volrs.
1 Apl. 17	*Harvey, Robert Harold Mansel* . . .	1 Apl. 17	1 Apl. 17	*I. A. R. O.*
1 Apl. 17	Candler, Leo Meighan	1 Apl. 17	1 Apl. 17	Cadet Coy., Kurseong.
1 Apl. 17	*Cuvelier, Edward Augustus*	1 Apl. 17	1 Apl. 17	*I. A. R. O.*
1 Apl 17	Renwick, William Barr Kushtia . . .	1 Apl. 17	1 Apl. 17	Paksey.
1 Apl. 17	Davis, William John	1 Apl. 17	1 Apl. 17	Kurseong.
1 Apl. 17	Webb, Alexander O'Brien	1 Apl. 17	1 Apl. 17	Kurseong.
20 May 18	Addis, Norman Frank	20 May 18	20 May 18	Kurseong.
20 May 18	Davis, Albert William	20 May 18	20 May 18	Mal.
	Supernumerary.			
1 Apl. 17	Edwards, Maj. William Harold, V.D. .	1 Apl. 17	1 Apl. 17	
1 Apl. 17	Mackenzie, Capt. Kenneth Lauchlan .	1 Apl. 17	1 Apl. 17	
	Staff.			
15 Mar. 18	James, Lt. F. A.	15 Mar. 18	15 Mar. 18	Qr.-Mr. and Actg. Adjt.
1 Apl. 17	Seal, Maj. Charles Edward Baldwin, V.D.	1 Apl. 17	. . .	Medl. Offr.
1 Apl. 17	McCutcheon, Maj. Oliver Edward, M.B. .	1 Apl. 17	. . .	Ditto.
1 Apl. 17	Stone, Lieut. Charles Ainslie	1 Apl. 17	. . .	Ditto.

Uniform—Khaki.

10th Southern Provinces Mounted Rifles.

Formed by G. G. O. No. 157, dated 12th February 1904. Change of Hd.-Qrs. G. G. O. 1064 of 1905. Re-constituted as an I.D.F. Unit, with present designation, 1 Apl. 17.

Establishment—2 Active Sqdns. 1 Res. Sqdn.

Troops at Coimbatore, Waltair, Munaar, Tinnevelly and Madras.

HONORARY COMMANDANT—*H. E. The Rt. Hon.* JOHN, *Baron* PENTLAND of LYTH, P.C., G.C.I.E. (*Governor of Madras*) 1st Apl. 17.

First Commn.	Names.	Date of rank.	Date of appointment to corps.	Remarks.
	Lieutenant-Colonels.			
1 Apl. 17	Buckley, Llewellyn Eddison, C.S.I.	1 Apl. 17	1 Apl. 17	Comdt.
	Majors.			
1 Apl. 17	Phillips, William Watkin	1 Apl. 17	1 Apl. 17	
1 Apl. 17	Jackson, Arthur Bushe	1 Apl. 17	1 Apl. 17	
1 Apl. 17	Barber, *Hon. Sir* Edward Fairless, Kt.	1 Apl. 17	1 Apl. 17	
7 Apl. 17	***Morin, Archibald Henry,*** D.S.O., V.D.	*1 Apl. 17*	*1 Apl. 17*	*I. A. R. O.*
	Captains.			
1 Apl. 17	*Ranking, William Lancaster*	*1 Apl. 17*	*1 Apl. 17*	*I. A. R. O.*
1 Apl. 17	Carless, John Elliott	1 Apl. 17	1 Apl. 17	
1 Apl. 17	*Cox, Stephen*	*1 Apl. 17*	*1 Apl. 17*	
1 Apl. 17	*King, Edward Oakes*	*1 Apl. 17*	*1 Apl. 17*	*I. A. R. O.*
1 Apl. 17	Turnbull, William Joseph Ugo	1 Apl. 17	1 Apl. 17	
1 Apl. 17	Lacey, Colin Warner	1 Apl. 17	1 Apl. 17	
	Lieutenants.			
1 Apl. 17	*Ormsby, Montagu Henry*	*1 Apl. 17*	*1 Apl. 17*	*I. A. R. O.*
1 Apl. 17	*Mackay, Frank Forbes*	*1 Apl. 17*	*1 Apl. 17*	*I. A. R. O.*
1 Apl. 17	John, Arthur Wood	1 Apl. 17	1 Apl. 17	
1 Apl. 17	*Lampard, Leonard Ashmall*	*1 Apl. 17*	*1 Apl. 17*	*I. A. R. O.*
7 Feb. 18	Richmond, Robert Daniel	7 Feb. 18	7 Feb. 18	
	2nd-Lieutenants.			
1 Apl. 17	Minchin, Alfred Alyson Fennel	1 Apl. 17	1 Apl. 17	
1 Apl. 17	*Hodding, Frederic*	*1 Apl. 17*	*1 Apl. 17*	*I. A. R. O.*
1 Apl. 17	Powell, Stratford Hamilton	1 Apl. 17	1 Apl. 17	
1 Apl. 17	Leech, Arthur John (51)	1 Apl. 17	1 Apl. 17	
7 Feb. 18	Ward, Frank Robinson	7 Feb. 18	7 Feb. 18	
1 Apl. 18	Pirrie, James Voce	1 Apl. 18	1 Apl. 18	
19 Mar. 18	Dixson, Archibald Hulke	19 Mar. 18	19 Mar. 18	
	Supernumerary.			
1 Apl. 17	Malcolm, Lt.-Col., Bernard Edward	1 Apl. 17	1 Apl. 17	
1 Apl. 17	Campbell, Major Archibald Young Gipps, C.I.E.	1 Apl 17	1 Apl. 17	
1 Apl. 17	Campbell, Capt. Herbert Neil Colquhoun	1 Apl. 17	1 Apl. 17	
1 Apl. 17	Chatterton., Capt. Francis	1 Apl. 17	1 Apl. 17	
	Staff.			
4 Dec. 14	Ranking, Capt. W. L.	4 Dec. 14	. . . . .	*I. A. R. O.*, Adjutant (*tempy.*)
19 Apl. 18	McGivren, Lt. F. G.	19 Apl. 18	19 *Apl.* 18	Qr. Mr.

ROYAL REGIMENT OF ARTILLERY.

THE ROYAL ARMS AND SUPPORTERS, WITH A GUN.
"*Ubique*" (above). "*Quo Fas et Gloria ducunt*" (below the Gun).

Uniform—Blue. *Facings*—Scarlet. *Busby-bag*—Scarlet. *Plume*—White.

COLONEL-IN-CHIEF—THE KING.

First Commn.	Names.	Army rank.	Remarks.

OFFICERS OF THE ROYAL REGIMENT OF ARTILLERY SERVING IN INDIA

NOTE.—An asterisk * prefixed to an officer's name indicates that he has been selected for continuous service in the Ordnance Department in India.

REMOVED FROM THE REGIMENT AND STILL ON THE ACTIVE LIST.

General Officers.

2 Feb. 76	Anderson, *Sir* C. A., K.C.B. (65, 162, 170, 206)	*l.g.* 12 May 14	Comdg. Southern Army.
18 Dec. 78	Kemball, *Sir* G. V., K.C.M.G., C.B., D.S.O., *p.s.c.* (*g*)	*m.g.* 8 Apl. 14	Comdg., 5th (Mhow) Dn.
19 May 80	Stuart, *Sir* R. C. O., K.C.S.I. (*f*) (161, 206)	*m.g.* 26 Oct. 15	Member, Indian Munitions Board.
23 Feb. 81	Scott, *Sir* A. B., K.C.B., D.S.O.	*m.g.* 1 Jan. 16	Comdg., 8th (Lucknow) Dn.
25 July 82	Lecky, R. St. C., C.B., C.M.G. (206)	*m.g.* 3 June 17	Inspr. of R.H. & R.F.

Colonels.

6 Apl. 79	Smith, G. B., C.B. (*temp. Brig-Gen.*, 93, 184, 206)	7 Nov. 07	Brig.-Gen., Royal Arty.
8 Feb. 80	Stokes, A., C.B., C.M.G., D.S.O. (*g*) (*temp. Brig.-Gen.*, 206)	14 Nov. 08	Brig.-Gen., Royal Arty.
27 July 80	Johnson, F. E., C.M.G., D.S.O.	20 Dec. 08	
7 July 80	Strange, R. G., C.I.E. (*temp. Brig.-Gen.*, 65, 162, 206)	15 Jan. 09	Comdg. Presdy. Bde.
22 Feb. 82	McLeod, W. K. (*temp. Brig.-Gen.*)	4 Oct. 11	Comdg. R. A., 1st Dn. & Frontier Bde.
25 July 82	Caulfeild, C. T., C.M.G. (*temp. Brig.-Gen.*)	7 Mar. 12	C. R. A., 2nd R'Pindi Dn.
*27 July 80	Renny, S. M., C.I.E., (*temp. Maj.-Gen.*) (206)	6 June 12	Dir.-Genl. of Ordce. in India.
27 July 80	Duhan, W. W. T. (*g*) (*temp. Br.-Gen.*)	6 June 12	
23 Feb. 81	Rich, H. H.	4 Dec. 12	Dy. Dir. of Equip., Ord. Bch.
*25 July 82	Jennings, H. A. K., C.I.E. (*temp. Brig.-Gen.*) (*f*) (206)	15 June 14	Inspr. of Q.M.G.'s Services.
	Rouse, H., C.B., D.S.O. (12, 206)	15 June 14	
5 July 84	Babington, D. M., C.I.E. (206)	15 June 14	*Ordce. Dept.*
*1 Oct. 82	Parker, J. L., C.M.G. (*temp. Brig.-Gen.*) (165, 169, 206)	15 June 14	Brig.-Gen., Royal Arty.
1 Oct. 82			
9 Dec. 84	White, G. F., C.B., C.M.G. (*temp. Brig.-Gen.*)	15 June 14	Brig.-Gen., Royal Arty.
1 Oct. 82	Vincent, H. O., C.B., C.M.G. (*temp. Br.-Genl.*)	15 Dec. 14	Inspr. of R. G. Arty.
4 Feb. 83	Campbell, M. S. C., C.I.E. (*f*) (169, 206)	15 Dec. 14	*Ordce. Dept.*
1 Aug. 83	*Gordon, L. G. F.*, C.B., D.S.O. (*temp. Brig.-Gen.*) (*g*) (206)	17 May 15	Comdg. R. A., Meerut and Lucknow Dns.
14 Feb. 83	Pullen, A. F. (*g*) [*l*] (95, 207)	29 Sep. 15	
14 Feb. 83	Cook, H. R. (*temp. Brig.-Gen.*) [L] (105, 162, 169, 206)	4 Dec. 15	Comdg. R.A., 4th Dn. & Karachi Bde.
16 Sep. 85	*Potts, F.*, C.M.G. (*temp. Brig.-Gen.*) (*g*) (207)	20 Nov. 16	*Brig.-Gen., R. A.*
16 Sep. 85	*Quinton, F. W. D.* (*temp. Brig.-Gen.*) (162, 206)	28 Dec. 16	Comdg. R.A., 5th Poona and 9th Dns.
29 Apl. 85	*Ogg, G. S.*, C.I.E., (206)	15 Feb. 17	*Ordce. Dept.*
15 Feb. 84	Maxwell, J. McC., C.B., D.S.O.	5 Jan. 18	Comdg. XI. Bde., R F. A.

ROYAL HORSE AND ROYAL FIELD ARTILLERY IN INDIA.

Lieutenant-Colonels.

First Commn.	Batty.	Names.	Regimental rank.	Army rank.	Remarks.
*1 Oct. 82	—	*Radcliffe, W. C. A.* (12, 169, 206)	5 Mar. 09	. . . .	*Ordce. Dept.*
*9 Dec. 84	—	*Cooper, P. T.* (*g*) (*h*) (170, 206)	11 Feb. 11	. . . .	*Ordce. Dept.*
29 Apl. 85	—	Tennant, H. L., *p.s.c.* (*e*)	21 July 12	. . . .	*Lv., ex I., m.c., 4m.*
17 Feb. 86	—	*Ouseley, R. G.*, C.B., C.M.G., D.S.O., (*temp. Brig.-Gen.*) (*g*) (*f*)	5 May 13	*s.c.* 5 May 17	Brig.-Gen., R. A.
24 July 86	—	*Tyler, J. A.*, C.M.G. (*temp. Brig.-Gen.*)	16 May 14	*c.* 1 Jan. 18	*Brig.-Gen., R. A.*
23 July 87	—	*Kirby, A. D.*, C.M.G. (*temp. Brig.-Gen.*)	15 Oct. 14	*c.* 1 Jan. 18	*Brig.-Gen., R. A.*
17 Feb. 88	—	*Rotton, J. G.*, C.B., C.M.G. (12) (*temp. Brig.-Gen.*)	30 Oct. 14	*c.* 1 Jan. 17	*Brig.-Gen., R. A.*
17 Feb. 88	—	Cartwright, G. N., C.M.G., D.S.O. (*temp. Brig.-Gen.*)	30 Oct. 14	. . . .	C. R. A., Lahore Dn.
17 Feb. 88	—	*Rettie, W. J. K.*, D.S.O. (*temp. Brig.-Gen.*) (12)	*30 Oct. 14*	. . . .	*Brig-Gen., R. A.*
27 July 89	—	Grove, E. W., D.S.O. (207)	30 Oct. 14		
27 July 89	—	Marsden, R. T. (65) . .	30 Oct. 14		
27 July 89	—	Bowring, A. H. (57, 65) . .	30 Oct. 14		
27 July 89	—	*Bright, R.A.*, C.B., *p.s.c., q.s.* (*temp. Brig.-Gen.*) (*e*)(207	30 Oct. 14	*c.* 3 June 17	*Brig.-Gen., R. A.*
27 July 89	—	*Sykes, C. A.*, D.S.O. (*temp Brig.-Gen.*)	30 Oct. 14	. . . .	Brig.-Gen., R. A.
14 Feb. 90	—	*Willis, E. H.*, C.M.G. (*temp. Brig.-Gen.*) (57)	30 Oct. 14	*c.* 1 Jan. 16	*Brig.-Gen., R. A.*
14 Feb. 90	—	Smith, L. A. (*temp. Brig.-Gen.*) (2, 57)	17 Dec. 14	. . . .	*Comdg. R. A.*
25 July 90	—	*Delaforce, E. F.*, C.M.G. (*temp. Br.-Gen.*) (57, 170, 206)	7 Jan. 15	*c.* 1 Jan. 18	*Brig.-Gen., R. A.*
25 July 90	—	*DePree, H. D.*, C.B., *p.s.c.* [L] (*temp. Brig.-Gen.*) (169, 206)	11 Jan. 15	*c.* 1 Jan. 17	*Brig.-Gen., Gen. Staff.*
*25 July 90	—	*Lane, F. C.* (206) . . .	23 Jan. 15	. . . .	*Govt. Inspr., Army Boots, Cawnpore.*
25 July 90	—	*Perreau, A. M.*, C.M.G. (*temp. Brig.-Gen.*), (206)	23 Jan. 15	*c.* 1 Jan. 18	*Brig-Gen., R. A.*
13 Feb. 91	—	Plummer, E. W., D.S.O. (*e*) (57, 206)	15 Mar. 15		
24 July 91	R.H.A. I Bde.	Charlton, C. E. C. G., (*temp. Brig.-Gen.*), D.S.O.	9 May 15		
24 July 91	—	Dennistoun, J. G., D.S.O., (Q) (57, 72 *b*, 162, 206)	10 May 15	. . . .	Comdg. VII Bde., R. F. A.
12 Feb. 92	R.F.A. X Bde.	Maule, H. N. St. J., C.M.G (*e*)	20 July 15	. . . .	Comdg. X Bde., R. F. A.
*1 Apl. 92	—	*Garstin, H. E.*, D.S.O., (*h*) (162, 206)	3 Sep. 15	. . . .	*Ordce. Dept.*
27 July 92	—	de Berry, P.P.E., C.M.G. (*e*)	13 Sep. 15		
1 Jan. 93	—	*Lamont, J. W. F.*, C.M.G., D.S.O.	6 Oct. 15	*c.* 1 Jan. 18	*Brig.-Gen., R. A.*
1 Feb. 93	—	*Davies, W. P. L.*, C.M.G., D.S.O., *p.s.c.* (*temp. Brig.-Gen.*) (12, 59, 206)	7 Nov 15	*c.* 10 Mar. 17	*Brig.-Gen., R. A.*
25 July 90	—	Ravenhill, C.	16 Nov. 15		
14 June 93	—	Ritchie, C. MacI. [*l*] (12, 57, 207).	24 Nov. 15	. . . .	*Lv., ex I., m. c., 6 mos.*
11 Aug. 93	—	*Stallard, S. F.*, C.M.G., D.S.O.	6 Feb. 16	. . . .	*Comdg. R. A.*
21 Oct. 93	—	Musgrave, A. D. (*temp. Brig.-Gen.*)	1 Mar. 16	*c.* 3 June 16	Brig.-Gen., R. A.
17 Nov. 94	—	Smith, H. B., D.S.O. (129)	19 Apl. 16		
17 Nov. 94	—	Barton, P., D.S.O. (*g*) (57, 206)	24 Apl. 16		
12 Dec. 94	—	Archdale, T. M., D.S.O. (127, 129)	20 May 16		
15 June 95	—	Kirke, K. St. G., D.S.O., (*o*)	1 June 16		
*15 June 95	—	*Kemble, F. A.* (162, 206) .	18 June 16	. . . .	*Ordce. Dept.*
21 Mar. 96	14	Lough, A. T. (Q) . . .	10 Nov. 16	. . . .	*Lv., ex I., m. c., 4 m.*
21 Mar. 96	—	*Mackey, H. J. A.*, C.M.G., D.S.O., M.V.O., (*temp. Brig.-Gen.*)	3 Jan. 17	. . . .	*Comdg. R. A.*
14 July 96	—	Paterson, P. J., D.S.O.	8 May 17		

(*h*) Passed Ordce. Course (except A. O. duties).

First Commn.	Batty.	Names.	Date of rank.	Army rank.	Remarks.
		ROYAL HORSE AND ROYAL FIELD ARTILLERY—*contd.*			
		Lieutenant-Colonels—*contd.*			
23 Mar. 97	—	Willis, S. G. R., D.S.O. [*l*] . .	1 Oct. 17		
1 Sep. 97	—	Shewell, E. F., C.M.G., D.S.O. . .	20 Mar. 18		
19 Jan. 98	N	West, A. H. D., D.S.O.	24 July 14	1 Jan. 18.	
		Majors.			
23 June 98	—	*Schrottky, C. E. G.* (*o*) (206)	30 Oct. 14	. . . .	*Ordce. Dept.*
22 Dec. 98	—	Harris, O. M., D.S.O. (*act'g. Lt.-Col.*) (12)	30 Oct. 14	*l.c.*, 1 Jan. 18.	
23 Dec. 98	...	*Heslet, R. K.*, D.S.O., *p. a. c.* .	30 Oct. 14	*l.c.*, 1 Jan. 18	*Member, Ordce. Commtee.*
23 Dec. 98	A	Mellor, A., D.S.O. (*temp. Lt.-Col.*)	30 Oct. 14		
25 June 99	...	*Lascelles, R. H.*, D.S.O. (*act'g. Lt.-Col.*)	30 Oct. 14	. . . .	*Bde.-Maj.*
25 June 99	...	*Murray, F. M.* (*temp. Lt.-Col.*) (51, 207)	30 Oct. 14	. . . .	*Ordce. Dept.*
18 Oct. 99	76	Lloyd, O. S. (57)	30 Oct. 14		
22 Nov. 99	...	*Queripel, L. H.*, C.M.G., D.S.O. (*temp. Lt.-Col.*) (59, 206)	30 Oct. 14	*l.c.*, 3 June 16.	
22 Nov. 99	—	*Hare, G. A.* (*o*) (206) .	30 Oct. 14	. . .	*Ordce. Dept.*
22 Nov. 99	—	*Duthie, A. M.*, D.S.O. (*act'g. Lt.-Col.*	30 Oct. 14		
6 Jan. 00	—	Ingram, J. M. (*act'g. Lt.-Col.*) (51)	30 Oct. 14		
17 Mar. 00	—	*Dunbar, J. C.*, D.S.O. (*act'g. Lt.-Col.*, *19 July 17*)	30 Oct. 14	. . . .	*G. S. O., 2nd gr.*
17 Mar. 00	—	Younger, A. A. S., D.S.O. (*act'g Lt.-Col.*, *23 May 17*) (51, 57, 95).	30 Oct. 14		
28 Mar. 00	V	Philips, J. L., D.S.O. . .	30 Oct. 14	*l.c.*, 1 Jan. 18	
17 Mar. 00	—	Eden, A. G. (*act'g. Lt.-Col., Dec. 17*) (57)	30 Oct. 14		
2 May 00	—	Renny-Tailyour, J. W., D.S.O. (*act'g. Lt.-Col., 22 Aug. 17*).	30 Oct. 14	. . . .	Comdg. XVI Bde., R.F.A.
2 May 00	94	Raikes, L. T., D.S.O. (*act'g. Lt.-Col., 20 Sept. 17*) (206)	30 Oct. 14		
2 May 00	—	Marshall, F. H. (*o*) (71)	30 Oct. 14		
2 May 00	—	Warren, W. R. (*act'g. Lt.-Col., 13 Oct. 17*)	30 Oct. 14		
2 May 00	—	Cowan, J. De B. (206) . .	30 Oct. 14		
2 May 00	..	*Harrison, J. M. R.*, D.S.O.	30 Oct. 14		
12 May 00	Q	Paynter, W. P. (*act'g. Lt.-Col., 18 Nov. 17*) (51)	30 Oct. 14		
12 May 00	—	Parrington, J. W. (*act'g. Lt.-Col., 26 Jan. 17*) (*g*) (65)	30 Oct. 14	. . . .	Comdg. XXI. Bde., R.F.A.
2 May 00	59	Nornabell, H. M., D.S.O. . . .	30 Oct. 14		
12 May 00	7 *a.c.*	Cullerne, C. P.	30 Oct. 14		
16 May 00	20	Gidley, C deB., D.S.O. (*act'g. Lt.-Col., 19 July 16*)	30 Oct. 14		
15 May 00	—	Waycott, E. W. J. (51, 57, 65, 71)	30 Oct. 14		
*26 May 00	...	*Mecredy, H.* (206)	30 Oct. 14	. . . .	*Ordce. Dept.*
27 June 00	—	Hunt, C. O. C. (57, 127, 129, 190)	30 Oct. 14		
18 Aug. 00	66	Crofton, M. R. H., D.S.O. . .	18 Dec. 14	. . . .	*Lv., ex I., m.c., 6 m.*
18 Aug. 00	28	Wilmer, E. R. G., D.S.O. (59, 207)	5 Jan. 15	*l.c.*, 3 June 17	
18 Aug. 00	—	McGowan, T., D.S.O. (51) . .	13 Feb. 15		
8 Sep. 00	—	Twisleton-Wykeham-Fiennes, N. I. E. (12, 65, 207)	10 Apl. 15		
6 Oct. 00	—	Theobald, A. C. L., D.S.O. (64) .	23 Apl. 15		
*6 Oct. 00	—	*Parbury, E.* (170, 206)	5 May 15	. . .	*Ordce Dept.*
19 Dec. 00		Coldwell-Horsfall, J. H. (*act'g. Lt.-Col., 1 July 17*) (57.	10 May 15		
19 Dec. 00	82	Corbould-Warren, E.	15 May 15	*l.c.*, 3 June 16	
19 Dec. 00	—	Turnbull, P. (51)	23 July 15		
2 May 00	—	Mascall, F. E.	20 Aug 15		
21 Dec. 00	—	Batchelor, V. A., D.S.O. [L] (45*a*, 51)	31 Aug. 15		
1 Dec. 00	—	*Harvey, A. F. B.*, *p.a.c.* . .	6 Sep. 15	. . . .	*Adjt., X Bde., R. F. A*
1 Dec. 00	—	*Bolitho, E. H. W.* (57)	13 Sep. 15	. . . .	*Lv., ex I.*
4 May 01	—	Povah, J. W. (57, 207) . .	5 Oct. 15		
4 May 01	—	Ward, B. S. (129, 206) . .	8 Oct. 15		
23 July 01	82	Baylay, E. J. L. (45*a*) . .	10 Nov. 15		
16 Nov. 01	102	Clayton, W. A.	16 Nov. 15	*l.c.*, 3 June 17	
4 Dec. 01	—	*Cooper, G. S.*, D.S.O. (59, 207) . .	23 Nov. 15	. . . .	*Bde.-Maj.*
4 Dec. 01	—	Halliday, W. J. F., D.S.O. . .	21 Dec. 15		
24 Dec. 01	63	Carlisle, T. R. M., D.S.O. (*l*) (51, 57	17 Jan. 16		
21 Dec. 01	—	Gordon, J. K. (51, 65) . . .	27 Jan. 16		
24 May 02	—	Hornsby, F. H. (12, 51, 129)	9 Mar. 16	. . .	*Lv., ex I., m. c., 6 m.*
4 May 01	94	D'Esterre, J. C. E.	3 Apl. 16	. . . .	*Murree.*
31 July 02	—	*Warton, R. B.*, D.S.O. (57)	9 Apl. 16		
31 July 02	X	Walker, H. W., D.S.O. . . .	8 Apl. 16		
24 June 03	—	*Temple, T* [*o*]	13 June 16	. . . .	*Ordce. Dept.*
25 Dec. 03	76	Peebles, R. E. (127, 129) . .	22 July 16		
23 Dec. 03	S	Harrison, P. F. (57)	10 July 16		
29 July 04	—	Murray, G. (95)	26 Aug. 16	8 May 17	
21 Dec. 04	—	Burne, S. F., D.S.O. (59, 162, 169, 205*a*)	1 Oct. 16		

ROYAL HORSE AND ROYAL FIELD ARTILLERY—*contd.*

Majors—*contd.*

First Comm'n	Batty.	Names.	Date of rank.	Army rank.	Remarks.
27 July 05	—	Wilson, D. C., D.S.O.	23 Nov. 16		
20 Dec. 05	S	Greenshields, D. J., D.S.O., M. C. (206)	30 Dec. 16		
23 May 06	2	Beckley, T. H., D. S. O.	7 Jan. 17		
23 May 06	—	*Waller, H. W. L.*, M. C.	17 Jan. 17	. . . .	*Bde.-Maj.*
25 July 06	*6 a.c.*	*Martin, H. G.* (12, 59, 95)	12 Feb. 17	. . . .	*Bde.-Maj.*
25 July 06	—	Mullings, J. R	17 Mar. 17		
4 Dec. 01	428	Watson, E. V. (71)	1 May 17		
23 July 07	—	Allen, R. F. C., M.C.	9 July 17		
23 July 07	—	Welsh, W. M. M. O.D., M.C. (95, 207)	12 July 17		
23 July 07	*9 a.c.*	Lambert, A. F., M. C.	1 Aug. 17		
18 Dec. 07	—	Pyne, F. S., D.S.O.	22 Oct. 17	. . . .	*Lv., ex I., m. c.*
18 Dec. 07	—	Lucas, J. deB. T., M.C.	22 Oct. 17		
18 Dec. 07	—	Harford, H. H. (57)	26 Nov. 17		
18 Dec. 07	84	Green, W. W., M.C.	2 Dec. 17		
18 Dec. 07	83	McMaster, W., M.C. (51)	12 Dec. 17		
29 July 08	*7 a.c.*	Benson, G. R. (57)	20 Feb. 18		
29 July 08	—	Akerman, W. P. J., D.S.O.	22 Mar. 18	. . . .	*Lv., ex I., m. c., 4 m.*
29 July 08	—	Colfox, W. P. (51)	24 Mar. 18		

Captains.

First Comm'n	Batty.	Names.	Date of rank.	Army rank.	Remarks.
19 Dec. 00	—	Millar, J. H. B. (57)	26 July 12		
18 Dec. 08	*4 a.c*	Martin, E. T. (95)	18 Dec. 14		
18 Dec. 08	—	Crewdson, W. T. O. (*act'g. Maj., 22 Jan. 17*).	18 Dec. 14		
13 July 09	81	Bayne-Jardine, C. W., D.S.O., M.C. (*act'g. Maj., 25 Sep. 16*), (51, 65)	23 July 15		
23 July 09	—	Burrard, G., D.S.O.	23 July 15		
23 July 09	S	Fleet, A. L. E. (65)	23 July 15		
23 July 09	*7 a. c.*	Freer, R. C., M.C.	23 July 15		
23 July 09	—	Eden, H. C. H., (*act'g. Maj., 9 Feb. 17*) M.C.	23 July 15		
23 July 09	—	*Mason MacFarlane, F. N.*, M.C.	23 July 15	. . . .	Staff Offr. to Inspr. of R. H. & R. F. A.
23 July 09	—	Moorhead, E. J.	23 July 15		
23 July 09	—	Kempe, H. F. C. (172*a*) (*act'g. Maj., 1 Apl. 17*).	23 July 15		
23 July 09	—	Cotter, R. K. (59, 66*a*, 206)	23 July 15		
23 July 09	28	Dorling, L. H. G. (12, 207)	23 July 15		
23 Dec. 09	—	Elles, P. G. M., M.C. (95)	23 Dec. 15		
23 Dec. 09	64	Glendinning, H. J. (*Bt.-Maj., 1 Jan. 17*)	23 Dec. 15		
23 Dec. 09	—	Cammell, G. A., D.S.O.	23 Dec. 15		
23 July 10	82	Wilson, F. W. B. (65)	23 July 16		
23 July 10	M	Lee, S. S., D.S.O. (*act'g. Maj., 27 Aug. 17*) [L.] (95).	23 July 16		
23 Dec. 10	—	*Adam, N. M.* (*act'g. Maj., 25 Sep. 17*)	*8 Aug. 16*		
23 Dec. 10	—	Kemp, G. C., M.C.	8 Aug. 16		
20 July 11	44	Hart, S. (*act'g. Maj., 23 Oct. 17*)	8 Aug. 16		
20 July 11	—	Latham, E., M.C. (65, 172*a*, 207)	8 Aug. 16		
20 July 11	S	Roney-Dougal, C. N. (51)	8 Aug. 16		
20 July 11	429	Kennedy, A. S. C. (*act'g Maj.*) (95, 207)	8 Aug. 16		
20 July 11	—	Beckett, C. T., M C. (*act'g. Maj., 27 Apl. 17*) (95, 172*a*)	8 Aug. 16		
. . . .	—	Arkright, E. H. (*temp. Maj., 2 Mar. 16*)	8 Aug. 16		
23 Dec. 11	—	Harriott, W. G., M.C.	8 Aug. 16		
23 Dec. 11	—	Macleod, R. (*act'g. Maj., 15 Apl. 17*), M.C.	8 Aug. 16		
23 Dec. 11	—	Maxwell, A. (*act'g. Maj., 25 Apl. 17*), M.C. (95)	8 Aug. 16		
23 Dec. 11	A	McKay, C. I. (*act'g Maj., 26 Jan. 17*).	8 Aug. 16		
23 Dec. 11	74	Stillman, R. C. B. (*act'g. Maj., 1 July 17*)	8 Aug. 16		
23 Dec. 11	—	*Anderson, T. R.*, M. C. (*act'g. Maj., 22 Aug., 16*)	8 Aug. 16	. . . .	*Bde.-Maj.*
23 Dec. 11	—	Staveley, R., D.S.O	8 Aug. 16		
23 Dec. 11	8	Harrison, H. T., M.C.	8 Aug. 16		
22 May 12	—	Orr-Ewing, M. R., M.C. (59)	8 Aug. 16		
22 May 12	S	MacIlwaine, A. H. (*act'g. Maj., 5 Sep. 17*) M.C.	8 Aug. 16		
22 May 12	V	Wingfield Digby, K. E.	8 Aug. 16		
19 July 12	U	Huggins, H. W. (*act'g. Maj., 4 June 17*) D.S.O.	8 Aug. 16		
19 July 12	—	Dale, E. C. B.	8 Aug. 16		
19 July 12	N	Martin, G. N. C. (*act'g. Maj., 12 Aug. 17*)	8 Aug. 16		
19 July 12	2	Harbord, G. M., M.C.	8 Aug. 16	*m.* 3 June 18	
19 July 12	Q	Simpson, G. P., M.C. (*act'g. Maj.*)	8 Aug. 16		

First Commn.	Batty.	Names.	Date of rank.	Army rank.	Remarks.

ROYAL HORSE AND ROYAL FIELD ARTILLERY—*contd.*

Captains—*contd.*

First Commn.	Batty.	Names.	Date of rank.	Army rank.	Remarks.
19 July 12	63	*Latham, H. B.* (169)	8 Aug. 16	....	*Bde. Maj.*
19 July 12	—	Crowe, W. H. F., M.C. (*act'g. Maj., 15 Oct. 16*)	8 Aug. 16		
19 July 12	—	Caudle, H. C. R.	8 Aug. 16		
19 July 12	—	Temple, E.	8 Aug. 16		
20 Dec. 12	—	Batten, J. F., M.C.	20 Dec. 16		
20 Dec. 12	—	Thomas, G. I., M.C. (*act'g. Maj., 15 June 17*) (64)	20 Dec. 16		
20 Dec. 12	—	Gordon, R. E. (*act'g. Maj., 30 June 17*)	20 Dec. 16	....	*Lv., ex I., m.c., 6 m.*
20 Dec. 12	—	Reynolds, B. T.	20 Dec. 16		
20 Dec. 12	—	*Carne E.F.* (*act'g. Maj., 11 Apl. 17*), M.C. (57)	20 Dec. 16	....	*Adjt., XI Bde., R. F. A.*
18 Dec. 13	89	Mallalieu, W.	3 Nov. 17		
10 June 14	89	Hollwey, James B., M.C.	3 Nov. 17		
17 July 14	N	Anderson, F. S., M.C.	3 Nov. 17		
11 Aug. 14	77	Buckley, E. S.	3 Nov. 17		
11 Aug. 14	X	Gottwaltz, R. L.	3 Nov. 17		
12 Aug. 14	U	Campbell, C. F.	3 Nov. 17		
12 Aug. 14	—	*Howard, E. S. G.*	3 Nov. 17	....	*Adjt., I Bde., R. H. A.*
16 Sep. 14	W	Payne, H. L. H.	8 Feb. 18		
	90	Little, A. G.	10 Mar. 18		

Lieutenants.

First Commn.	Batty.	Names.	Date of rank.	Army rank.	Remarks.
10 Oct. 14	—	Murison, C. A. P.	10 Mar. 11	....	No. 7 Mtn. Bde., R.G.A
30 Oct. 14	—	Westrope, F. E.	9 June 15	....	Orderly Offr. to G. O. C. [1st Bde. R. H. A.
7 Nov. 14	—	Jackson, L.	9 June 15		
12 Nov. 14	—	Walker, R. Y. K., M.C.	9 June 15		
15 Nov. 14	19	Rankin, H. W. (*act'g. Capt., 7 Nov. 16*)	9 June 15		
	—	Heron, R.	13 July 15		
17 Nov. 14	N.	Fenton, G. A.	22 Nov. 15		
23 Nov. 14	—	Ferguson, G., M.C., (*act'g. Maj., 16 June 17*)	3 Dec. 15		
24 Nov. 14	—	Reynolds, C. (57)	16 Dec. 15	....	With Sig. Serv.
3 Dec. 14	14	Fraser, D. F.	23 Dec. 15		
10 Dec. 14	222Bde a.c.	Harrop, T. G. (*act'g. Capt., 9 Oct. 17*)	29 Dec. 15		
10 Dec. 14	—	*McCarthy, C. E. A.*	1 Jan. 16	....	*Adjt., XVIII Bde., R.F.*
10 Dec. 14	—	*Mitchell, J.*, M.C.	1 Jan. 16	....	*A. R. A. F.*
12 Dec. 14	19	Norman, G.	11 Jan. 16		
14 Dec. 14	20	Turner, H.	27 Feb. 16		
14 Dec. 14	—	Hayes, F.	2 Mar. 16		
18 Dec. 14	—	Humphrey, H. P.	23 July 16		
20 Dec. 14	—	Grace, F.	26 July 16		
20 Dec. 14	—	Birtles, H.	8 Aug. 16	....	O C.R. A Depôt, Kirkee.
20 Dec. 14	38	Stoyle, E. G.	8 Aug. 16		
7 Jan. 15	81	Roberts, H. A.	8 Aug. 16		
7 Jan. 15	90	White, R. E.	8 Aug. 16		
12 Jan. 15	—	Adams, C.	8 Aug. 16		
12 Jan. 15	—	Keating, J. W. (20) (*act'g. Capt., 11 Oct. 16*)	8 Aug. 16		
12 Jan. 15	W	Payne, P. (*act'g. Capt., 1 Oct. 17*)	8 Aug. 16		
12 Jan. 15	78	Collison, A. E. (*act'g. Capt., 1 Feb. 17*)	8 Aug. 16		
12 Jan. 15	4	Addis, W. H. (172a) (*temp. Capt.*)	8 Aug. 16	....	Asst. Instr., School of [Gunnery.
12 Jan. 15	—	Spinks, W.	8 Aug. 16		
12 Jan. 15	--	Humphris, T. H. (*act'g. Maj., 23 July 18*)	8 Aug. 16	....	Comdg. R. A. Depot.
12 Jan. 15	M.	Carnochan, D. S. (95, 172a)	8 Aug. 16		
12 Jan. 15	—	Burgess, R. A.	8 Aug. 16		
12 Jan. 15	76	Devereux, W., M.C.	8 Aug. 16		
10 Feb. 15	14	Jungblüth, V. A., M.C.	8 Aug. 16		
10 Feb. 15	1064	O'Beirne, J. J.	8 Aug. 16		
7 Oct. 16	X	West (*act'g.* Capt.) C.	7 Oct. 16		
12 Feb. 15	375	Ealand, V. F.	1 Nov. 16		
14 Feb. 15	—	Knock, J. (*temp. Capt., 10 Apl. 18*)	4 Dec. 16	....	A. C. Factory, Alipore.
15 Feb. 15	—	*Gooch, K. T.*, M.C.	10 Dec. 16	....	*Staff Capt.*
9 Apl. 15	77	Murphy, D. W.	19 Mar. 17		
1 June 17	—	Bradley, V. M.	1 June 17		
22 Apl. 15	64	Weigall, E. T.	1 July 17		
22 Apl. 15	73	Villiers, S. A.	1 July 17		
22 Apl. 15	—	Lumsden, E.	1 July 17		
18 May 15	—	Green, P. J.	1 July 17	....	S. S. O., Campbellpore.
26 May 15	64	Hayward, R. H. P.	1 July 17		
	1098	Robinson, H. N.	1 July 17		
	—	Gittings, T. A. (*act'g. Maj., 23 July 18*)	1 July 17	....	R. A. Depôt, Jubbulpoe.
	215	*Wilkins, T.*, M.C.	1 July 17	....	*Adjt., V Bde., R. F A.*
	Bde. a.c.	Langford (*act'g.* Capt.) R.	1 July 17		

ROYAL HORSE AND ROYAL FIELD ARTILLERY—*contd.*

Lieutenants—*contd.*

First Commn.	Batty.	Names.	Date of rank.	Army Rank.	Remarks.
	—	*Durant, R. M.*	1 July 17		*Adjt., XXI Bde., R. F. A.*
	78	MacKichan, K. W. (57, 79)	1 July 17		
	—	*Browning, L. H.*	1 July 17		*R. A. F.*
15 May 15	—	Osmond, W. R. F.	1 July 17		
3 Aug. 15	—	Graham-Brown, E. D.	1 July 17		
	4	Griggs, A.	1 July 17		R. A. Depôt, Jubbulpore.
	—	Andrews, A. C.	1 July 17		
	14	Cunningham, W., M.C.	1 July 17		
	—	Conlon, H.	1 July 17		
	—	*Moojen, L. G.* D.C.M., *(temp. Capt.)* (51)	1 July 17		*Adjt., Cal. P. D. Vols.*
	—	Oatey, G. H.	1 July 17		
	Q	Lenanton, G. F. R.	1 July 17		
	U	Kemm, St. J. G.	1 July 17		
	N	Mitchell, E. A.	1 July 17		
	"B" *a.c.*	Robinson, G.	1 July 17		
	V	Sutton, F. K.	1 July 17		
	—	Tudor, W. J. E. (59)	1 July 17		S. S. Depôt.
	—	Harrison, H. H.	1 July 17		
	11 *a.c.*	Seeley, B. J.	1 July 17		
	—	Fahey, W.	1 July 17		
	375	Howe, W.	1 July 17		
4 Oct. 16	—	Parkes, R. S.	1 July 17		
1 Feb. 16	90	Shepard, R. W.	1 Aug. 17		
4 Apl. 16	M	Waller, C. *(actg. Capt., 27 Aug. 17)*	4 Oct. 17		
10 May 16	X	Eastwood, G. A.	10 Nov. 17		
	X	Phipps, H. C.	10 Nov. 17		
17 June 16	2 *a.c.*	Hill, J.	17 Dec. 17		
4 July 16	38	Williamson, J. R.	4 Jan. 18		
11 July 16	—	Eales, R. A. G.	11 Jan. 18		No. II B. M. A. Bde.
	—	Dobson, F. S. A.	28 Mar. 18		
6 Oct. 16	1086	Wynne, C. H. R.	6 Apl. 18		
12 Oct. 16	—	Fraser, J.	12 Apl. 18		
13 Oct. 16	1 *a.c.*	Mills, G. A.	13 Apl. 18		
15 Oct. 16	—	Prior, E. *(act'g. Capt., 19 Sep. 17)*	15 Apl. 18		
16 Oct. 16	—	Lewis, W.	16 Apl. 18		
16 Oct. 16	—	Hobbs, H. R., D.C.M.	16 Apl. 18		
16 Oct. 16	428	Terry, H. J.	16 Apl. 18		
5 Nov. 16	—	Woodhouse, S.	5 May 18		
5 Nov. 16	—	McDermott, W. *(actg. Capt., 5 Nov. 17)*	5 May 18		
5 Nov. 16	—	Wakeling, A.	5 May 18		
8 Nov. 16	M	Mills, G.	8 May 18		
8 Nov. 16	—	Hearnden, H. J.	8 May 18		
10 Nov. 16	"A" *a.c.*	Lamb, G.	10 May 18		
11 Nov. 16	W	Beckett, H. J. (57)	11 May 18		
11 Nov. 16	W	Spencer, H. A. (57, 59)	11 May 18		
14 Nov. 16	W	Simmonds, W. G.	14 May 18		
17 Nov. 16	—	Hallett, S. C. *(act'g. Maj. 23 July 18)*	17 May 18		
23 Nov. 16	X	Tod, A. F.	23 May 18		
11 Dec. 16	77	Price, F.	11 June 18		
13 Dec. 16	—	Dupe, T. G.	13 June 18		
13 Dec. 16	—	Dodds, H. S., M.C.	13 June 18		
16 Dec. 16	—	Jones, W. E.	16 June 18		
18 Dec. 16	M	Hoile, G. W. C. H.	18 June 18		
20 Dec. 16	—	Warner, J.	20 June 18		Frontier Garrison Artillery.
23 Dec. 16	—	Gleig, D. D. W.	23 June 18		*Lv. ex I., m. c., 6 mos.*
23 Dec. 16	—	Mace, F. C.	23 June 18		
25 Dec. 16	—	Carter, J.	25 June 18		
27 Dec. 16	—	Flight, W. E. *(act'g. Maj., 23 July 18)*	27 June 18		Comdg., R. A. Depôt.
28 Dec. 16	79	Stanley, A.	28 June 18		
29 Dec. 16	—	Collard, E.	29 June 18		
31 Dec. 16	—	White, G.	1 July 18		
31 Dec. 16	—	Steward, M. E. C.	1 July 18		
1 Jan. 17	—	Lawson, C.	1 July 18		
2 Jan. 17	—	Goodman, R. H. B.	2 July 18		
17 Jan. 17	—	Woodhouse, C. H.	17 July 18		
20 Jan. 17	221 *Bde. a.c.*	Johnson, J.	20 July 18		

2nd-Lieutenants.

First Commn.	Batty.	Names.	Date of rank.	Army Rank.	Remarks.
	77	Haynes, W.	16 June 16		
	—	*Desmond, A. E.*	14 Jan. 17		
	102	Moss, C. E. R.	18 Dec. 16		
	—	Darts, W.	16 Feb. 17		
	—	Tranthen, J. R.	22 Feb. 17		

ROYAL HORSE AND ROYAL FIELD ARTILLERY—*contd.*

2nd-Lieutenants.—*contd.*

First Comm n.	Batty.	Names.	Date of rank.	Army Rank.	Remarks.
	221 Bde. a.c.	Hall, C. E.	25 Feb. 17	. . .	
	—	*Winslow, L. C.*	4 Mar. 17	. . . .	*Adjt., 216th Bde.*
	—	Doy, E. P.	28 Mar. 17	. . . .	
	—	Miles, H.	31 Mar. 17	. . . .	
	222 Bde. a c.	Kennard, W. G. L.	31 Mar. 17	. . . .	
	—	Malcolm, A. R. H.	3 Apl. 17	. . . .	
	—	Rolfe, F. W.	3 Apl. 17	. . . .	
	—	Grant, J. D.	4 Apl. 17	. . . .	
	—	Davies, J. A.	15 Apl. 17	. . . .	
	—	Higgins, F. M. P.	22 Apl. 17	3 . . .	
	—	*Heslop, W. T. B.*	*4 May* 17	. . . .	*Adjt., 222nd Bde., R. F. A.*
	—	Batchelor, A. E. F.	30 Jun. 17	. . . .	Attd. Army Remount Depôt, Aurungabad.
	No. 189 a.c.	Clements, C.	9 Jul. 17	. . .	
	4	Woodrow, T. W.	13 Sep. 17	. . . .	
	—	Newstead, F.	18 Sep. 17	. . . .	
	79	Maidment, W. T.	21 Sep. 17	. . . .	*Lv., ex I., m. c., 6 m.*
	—	Ayres, G.	3 Oct. 17	. . . .	
	—	Meaburn, M.	3 Oct. 17	. . . .	
	—	Browning, G.	4 Oct. 17	. . . .	
	—	Dootson, H.	4 Oct. 17	. . . .	
	—	Holland, G.	5 Oct. 17	. . . .	
	—	Greatorex, J.	5 Oct. 17	. . . .	
	89	Butler, A. E.	3 Nov. 17	. . . .	
	12 a.c.	Dawes, W.	4 Nov. 17	. . . .	
	12 a.c.	Warner, E. P.	11 Nov. 17	. . . .	
	—	Satchell	23 Nov. 17	. . . .	
	4 a.c.	Lewis, W. J.	28 Nov. 17	. . . .	
	—	Chalker, J. R.	26 Dec. 17	. . . .	
	—	Messenger, W.	26 Dec. 17	. . . .	
	—	Rettie, A. H.	26 Dec. 17	. . . .	
	—	Ferst, W. J.	27 Dec. 17	. . . .	
	—	Gooch, H. W.	27 Dec. 17	. . . .	
	—	Kentish, E. G.	27 Dec. 17	. . . .	
	—	Honour, W.	27 Dec. 17	. . . .	
	—	Marshall, A.	27 Dec. 17	. . . .	
	—	Stiff, R.	27 Dec. 17	. . . .	
	—	Hinton, N. R.	28 Dec. 17	. . . .	
	78	Smith, W. J.	30 Dec. 17	. . . .	
	1064	Wakeling, H. E.			

Temporary Majors.

77 Knyvett, F. B., D.S.O., D.C.M. . . . . 25 May 15

Temporary Captains.

First Comm n.	Batty.	Names.	Date of rank.	Army Rank.	Remarks.
	375	McKenna, J. W., M.C. (*act'g. Major*)	15 Oct. 14	. . . .	
	—	Fagan, L. E. (*act'g. Maj., 1 Oct. 18*)	1 Dec. 14	. . .	No. 33rd Mtn. By.
	—	Senior, Capt. H. F.	30 Jan. 15	. . . .	*Lv., ex I., m. c., 6 m.*
	—	Cameron-Smith, G.	13 Feb. 15	. . . .	
	38	Archbold, R.	7 May 15	. . . .	
	1098	Priestley, W.	20 May 15	. . . .	
	101	Wilson, C. H. E. (*act'g. Major*)	5 June 15	. . . .	
	1089	Coffin, D. M.	22 Dec. 15	. . . .	
	89	Cavenagh, O. (*act'g. Maj., 18 Apl. 17*)	24 Dec. 15	. . . .	*Lv., ex I., m.c., 6 mos.*
	85	Denman, F. C.	4 May 16	. . . .	
	7	Knight, E. E., M.C.	18 Sep. 17	. . . .	
	93	Grant, W. C.	18 Sep. 17	. . . .	
	59	Warwick, D. V.	18 Sep. 17	. . . .	
13 Nov. 14	81	Carbery, M., M.C.	12 Sep. 15	. . . .	
17 Nov. 14	83	Boulter, F. C., M.C.	12 Sep. 15	. . . .	
	94	Abell, N.	18 Sep. 17	. . . .	
27 Feb. 15	84	Busk, E. W., M.C. (*act'g. Maj., 27 June 17*)	18 Sep. 17	. . . .	

Temporary Lieutenants.

First Comm n.	Batty.	Names.	Date of rank.	Army Rank.	Remarks.
	84	Meysey-Thompson, A. de C. C.	2 Apl. 15	. . . .	
	78	Allman, R.	8 Apl. 15	. . . .	

First Commn.	Batty.	Names.	Date of rank.	Army rank.	Remarks.
		ROYAL HORSE AND ROYAL FIELD ARTILLERY—*contd.*			
		Temporary Lieutenants—*contd.*			
	1093	Herschorn, D. F.	22 Apl. 15		
	2	Mein, J. J.	9 May 15		*Lv. in I., m.c., 4 m. 10 d.*
	44	Battrick, E. V. B.	12 Sep. 15		
30 Oct. 14	94	Hewetson, R. P.	12 Sep. 15		
17 Nov. 14	73	Hepburn, C. A.	12 Sep. 15		
	—	*Cotton, D. C.*	*12 Sep. 15*		*Adjt., XIII Bde., R. F. A.*
	—	*Mims, C. C.*	*12 Sep. 15*		*Adjt., IV Bde., R. F. A.*
6 Feb. 15	93	Walker, R. G.	12 Sep. 15		
	89	Ambler, H. L., M.C.	1 Nov. 15		
26 Aug. 14	59	Poynter, H. G.	8 Dec. 15		
11 Dec. 14	81	Dening, D. M.	8 Dec. 15		
	102	Steedman, C. D.	8 Dec. 15		S.S.O., Hyderabad.
	38	Davy, T. A. L., (*act'g.* Capt.)	8 Dec. 15		
28 Oct. 14	1091	Smith, R. (*act'g. Capt., 14 July 18*)	19 Dec. 15		217th Bde. (T. F.)
	—	*Begbie, G. (act'g. Capt., 18 Sep. 17)*	*1 Jan. 16*		*Adjt., XVI Bde., R. F. A.*
16 Jan. 15	83	Dare, N. F., M.C. (*act'g. Maj., 14 Oct. 17*)	31 Jan. 16		
	—	Williams, B. W.	25 Feb. 16		No. 2 Mtn. Bde., R. G. A.
	-	Blomenstock, L. M., M. C. (*Actg. Capt., 16th June 18.*)	4 Mar. 16		
	A	Evans, P. R.	4 Mar. 16		
7 Dec. 14	85	Wickham-Jones, E.	15 Mar. 16		
29 Jan. 15	8 *a.c.*	Leckie, A.	15 Mar. 16		
	—	Lingeman, A. A.	15 Mar. 16		Orderly Offr., XI Bde., R. F. A.
	—	*Wheeler, P. J.*	*28 Mar. 16*		*Asst. Supdt., Gun & Shell Factory, Cossipore.*
	20	Locks, G.M., M.C.	28 Mar. 16		
	—	Brinkley, E. G.	28 Mar. 16		No. 94 Coy. R. G. A.
	2	Fawcett, A. M., M.C.	28 Mar. 16		
	—	*Guiness, J. F. G.*	*28 Mar. 16*		*Adjt., IX Bde., R. F. A.*
	—	Baines, J.	28 Mar. 16		
	38	Mackay, G. S.	28 Mar. 16		
	19	Sewell, H. K.	28 Mar. 16		
	9 *a.c.*	Gordon, G.	28 Mar. 16		
	73	Stapleton, J., M.C.	20 May 16		
	—	Nairn, A. M.	1 June 16		
	—	McGill, E. A.	1 June 16		
	21 Mn.	*Wells, R. H.*	1 Aug. 16		
	—	Bodilly, G. R.	3 Aug. 16		
	4 *a.c.*	Hill, W. D.	6 Aug. 16		
	—	Barry, C. A.	6 Aug. 16		
	76	Leeds, H. J.	6 Aug. 16		
	—	Thierry, E. H.	6 Aug. 16		
	77	Cunningham, W.	6 Aug. 16		
	—	Gate, H. J.	18 Aug. 16		
	79	Barnes, W. A.(*act'g. Capt., 11 June 18*)	15 Nov. 16		Adjt., 217th Bde.
	4	Joseph, E. V.	24 Oct. 16		
	44	Leggatt, G.	1 July 17		
	1091	Coles, F. H.	1 July 17		
	—	Rivers-Smith, P. E.	1 July 17		29 Mtn. By.
	—	Hunter, J. P.	1 July 17		
	74	Florance, D. C. H.	1 July 17		
	63	Lewis, P. S.	1 July 17		*Lv., ex I., m. c., 4 m.*
	101	Maltby, G. B.	1 July 17		
6 Oct. 14	101	Samuel, C. M.	1 July 17		
1 Feb. 15	1 *a.c.*	Holford Walker, L.	1 July 17		
19 Feb. 15	102	Ziegler, J. H.	1 July 17		
17 Mar. 15	—	Vears, L.	1 July 17		*Lv. ex I., 10 m.*
	79	Blackburn, J.	1 July 17		
7 Oct. 14	—	King, A. J. J.	1 July 17		
	76	Mousley, E. O.	1 July 17		
	—	Davies, R.M.	1 July 17		
	—	Teague, S. W. M.	1 July 17		
	—	Watkins, E. W.	1 July 17		
	—	Eldred, W. (*act'g. Capt., 8 Aug. 17*)	1 July 17		
	6 *a. c.*	Trafford, W. E.	1 July 17		
	33 *Mn.*	*Willis, C. R. (act'g. Capt.)* (207*a*)	11 Aug. 17		
	—	*Limpenny, F.R. (act'g. Capt., 4 Sep. 17)*	26 Aug. 17		*Adjt., 221st Bde., R. F. A.*
	—	Colman, E.	2 Nov. 17		
	77	Harraway, C. G. (57)	2 Nov. 17		
2 May 16	1096	Holliday, J. M.	2 Nov. 17		
	—	Lattey, W. T.	2 Nov. 17		
	1088	Ardron, J. S.	20 Dec. 17		
	74	Bransden, W. J.	10 Jan. 18		
		Temporary 2nd-Lieutenants.			
	—	Hume-Spry, F.	17 Nov. 14		
	22 Mn.	Carringan, R. W., M.C.	11 Dec. 14		
	94	Gamble, V. F.	4 Jan. 15		No. 64 Coy., R. G. A.
	—	Greatbatch, G.	5 Jan. 15		
	6 *a. c.*	Edmonds, P. N.	19 Jan. 15		

ROYAL HORSE AND ROYAL FIELD ARTILLERY—*contd.*

Temporary 2nd-Lieutenants—*contd.*

First Commn.	Batty.	Names.	Date of rank.	Army rank.	Remarks.
	81	Graham, R. L.	26 Jan. 15		
	—	Greenwood, E. M.	5 Mar. 15		
	12 *a.c.*	Lockett, G. B.	10 Mar. 15		*Lv., ex I., m. c., 6 m.*
	—	*MacMahon-O'Byrne, D.* (207*a*)	*17 Mar. 15*		*I. A.* (*on probn.*)
	102	Brook, L. A.	13 May 15		
	4	Boling, L. C.	4 June 15		
	8	Forrester, W. R., M. C.	12 June 15		
	—	Green, H. L.	21 June 15		
	—	Ker, W. E.	3 July 15		*Lv., ex I., m. c., 3 m.*
	—	Hulme, H.	22 July 15		
	—	Trowbridge, B. B.	28 July 15		No. 52 Coy., R. G. A.
	—	Burleigh, H.	18 Aug. 15		
	—	Parkin, R. H.	2 Sep. 15		10-pr. Batty.
	101	Umfreville, A. T. (*act'g. Capt., 27 May 18*)	6 Sep. 15		
13 Oct. 15	1089	Harries, D. G.	13 Oct. 15		
	74	Williams, J. G.	7 Nov. 15		"M" Batty., R. H. A.
	—	Morfey, R. J.	7 Nov. 15		
	—	Holman, A. T.	7 Nov. 15		
	—	Pedder, J. W. R.	7 Dec. 15		
	—	Mawer, E. S.	7 Apl. 16		
	—	Gamblin, W. H.	2 May 16		
	—	Bonser, A.	2 May 16		
	—	Joynes, A.	2 May 16		
	7	Hobbs, A. W.			
	2	Cowie, H. K.			
	1107	Wright, A.	23 Apl. 17		
	90	Sexton, D. C.	{ 7 July 17 24 Oct. 14		
	—	Paterson, H.	10 July 17		

Special Reserve.

Captain.

First Commn.	Batty.	Names.	Date of rank.	Army rank.	Remarks.
	—	Creasy, E. C. (*act'g. Maj., 1 Dec. 17*)	30 Oct. 14		
17 Nov. 09	90	Pease, G. B. R. (*act'g. Maj., 20 Aug. 17*)	8 Aug. 16		
	—	Furlong, G. O.	3 Nov. 17		

Lieutenants.

First Commn.	Batty.	Names.	Date of rank.	Army rank.	Remarks.
	—	Roberts, C. F., M.C.	9 June 15		Orderly officer. V. Bde., R. F. A.
	—	Hawker, G. C.	12 Sep. 15		Orderly Offr. to C. R. A., 2nd (R'pindi) Dn.
	—	Storrs, B. St. J. (57, 175)	28 Nov. 15		
	—	*Powell, W. N.* (57)	26 Feb. 16		*With Sig. Serv.*
	101	Pettitt, A. H.	{ 1 Jan. 16 22 June 16		
	90	Duns, G. S. S.	1 Jan. 17		
	82	Macpherson, H. R.	12 Feb. 17		
	89	Hart, F. A. A.	12 Feb. 17		
	77	Prendergast, R. V.	17 Feb. 17		With I. O. Dept.
	—	Jarvis, E. D. (*act'g. Capt., 6 Sep. 17*)	11 June 17		
	89	Gibbon, S. R.	11 June 17		
	—	Barker, F.	1 July 17		
	79	Cowan, A. H.	1 July 17		*Lv., ex I., m.c., 6 m.*
	—	McFarlane, F. N.	1 July 17		
	93	Cunningham, N. V.	1 July 17		
	64	Roseveare, R. V. H.	1 July 17		
18 Aug. 15	12 *a.c.*	Gibson, L. B.	1 July 17		*Lv., ex I., m.c., 5 m.*
	79	Ormond, E. C.	1 July 17		
8 Sep. 15	8 *a.c.*	Pye, L. K.	1 July 17		
	—	Wilson, J. C. J.	1 July 17		
	4	Tredinnick, N. W., M.C.	1 July 17		
	79	Friend, O. F.	1 July 17		
	84	Hillier, F. N.	1 July 17		
	66	Shepherd, F. M., (*Act'g., 26 May 18*)	1 July 17		
	A	Cowling, E. J. S.	1 July 17		
	—	Dickson, A.	1 July 17		
	4	Arundell, W. R. H.	1 July 17		
	77	Roche, J. M.	1 July 17		
	1 *a.c.*	Crawford, T. A.	1 July 17		*Lv., ex I., m.c., 6 m.*
	—	Barr, E. N. D.	1 July 17		*R. A. F.*
	—	Cook, P. R.	1 July 17		
	64	Turner, G. H., M.C. (*tempy. Lt.*)	1 July 17		
	Q	Wood, F. M. A. (*act'g. Capt., 16 Oct. 17*)	1 July 17		
	—	Aldous, T. R.	1 July 17		
	—	Corfield, R. Fitz H. (*act'g. Capt., 17 Sep. 17*)	1 July 17		
	—	Bailey, H. E.	1 July 17		Ind. Mun. Board.

ROYAL HORSE AND ROYAL FIELD ARTILLERY—*contd.*

Special Reserve—*contd.*

Lieutenants—*contd.*

First Commn.	Batty.	Names.	Date of rank.	Army rank.	REMARKS.
	—	Waring, H. S.	1 July 17		
	—	Hopkinson, W. D.	1 July 17		*Lv., ex I., m.c., 4 m.*
	—	Bowden, C. A. (95)	1 July 17		28 Mn. Batty.
	—	Barnet, D. G.	1 July 17		
	85	Rigby, G. H.	1 July 17		
	A	Holmes, W. K.	1 July 17		
	—	Turner, H. R., M.C.	1 July 17		
	8 *a. c.*	Falcke, J. (*actg.* Capt.)	1 July 17		
	73	Mann, A. R.	1 July 17		
	85	Pakes, W. C.	1 July 17		
	3 *a. c.*	Maxwell, M. T.	1 July 17		
	—	Gorman, T. S.	1 July 17		
	102	Hotblack, H. C.	1 July 17		
	38	Shaw, R. B.	1 July 17		
	1 *a.c.*	Howard, H. G., M.C.	1 July 17		
	77	Saunderson, R. G.	1 July 17		
	—	Trousdell, C. F.	1 July 17		
	90	Harbord, G.	1 July 17		
	4	Bowman, R. S.	1 July 17		R. A. Depôt, Ambala.
	—	Reid, L. H.	1 July 17		
1 Nov. 15	1093	du Vallon, G. C. de J.	11 Nov. 15		
	Q	Thornhill, C. H.	1 July 17		
	22MB	Bastow, H.	1 July 17		
	—	Malone, A.	1 July 17		
	—	*Richardson, S. H. (act'g. Capt., 16 Sep. 18)*	*1 July 17*		*Adjt. No. II Br. Mtn.* [*Arty. Bde.*
	74	Cattell, E. E.	1 July 17		
	—	Winn, W. J.	1 July 17		*Lv., ex I., m.c., 6 m.*
8 Jan. 16	27 Mn	Smith, R. A.	8 July 17		
19 Jan. 16	25Mn.	Hawes, E. J. (207*a*)	19 July 17		
	16 Bde.	Bosher, F.	19 July 17		
	—	*Duncan, T. G.*	*19 July 17*		
	—	Smart, G. M.	31 July 17		Mtn. Arty.
	—	*Hughes, A. S., (act'g. Capt., 11 Aug. 17)*	*18 Aug. 17*		*Adjt. IV Mn. Batty.,* [*R.G.A.*
	—	Bright, W. J.	10 Sep. 17		
	215 Bde. *a.c.*	Mappin, J.	11 Nov. 17		
	—	Rothe, C. W.	11 Nov. 17		No. 74 Coy., R. G. A.
	—	Englefield, R. H.	17 Dec. 17		
	74	Behrens, D. H.	19 Feb. 18		

2nd-Lieutenants.

First Commn.	Batty.	Names.	Date of rank.	Army rank.	REMARKS.
	—	Roosmale-Cocq, C. W. V. (*temp. Lieut.*)	5 July 15		
	90	Haworth-Booth, M. (57)	2 Aug. 15		
	—	Bradfield, L. G.	22 Jan. 16		
	—	Sykes, A. A.	4 June 16		
	—	*Gerrard, H. J. F.*	*17 June 16*		
	78	Stanton, H. M. A.	17 Dec. 16		*I. A. (on probn.)*
	—	Sweeney, M.	17 June 16		
	—	Spark, A. J.	6 Aug. 16		
	101	Evans, D. M.	6 Aug. 16		
	79	Young, A. V. W. (57)	6 Aug. 16		
6 Aug. 16	2 *a.c.*	Campbell, E. C. M.	6 Aug. 16		
6 Aug. 16	1072	Fish, F. M.	6 Aug. 16		
2 Aug. 16	77	Moor, G. W.	6 Aug. 16		
2 Sep. 16	1086	Douglas, J. S.	2 Sep. 16		
2 Sep. 16	89	Stephenson, N. A. (57)	2 Sep. 16		
14 Dec. 16	89	Metcalfe, H. J.	4 Dec. 16		
23 Jan. 17	102	Miller, H.	23 Jan. 17		
23 Dec. 16	79	Butcher, G. W. K.	23 Dec. 16		
21 Jan. 17	—	Jones, F. M. L.	21 Jan. 17		
	79	Duncan, D. D.	28 Jan. 17		Barrackpore.
	79	Freeman, K. C.	28 Jan. 17		
	—	McGee, G. L.	4 Feb. 17		
	—	Dodd, F.	4 Feb. 17		

ROYAL HORSE AND ROYAL FIELD ARTILLERY—*concld.*

Special Reserve—*contd.*

2nd-Lieutenants—*contd.*

First Commn.	Batty.	Names.	Date of rank.	Army rank.	Remarks.
	—	Davies, D. M.	11 Feb. 17		
	102	Cole, L. B.	27 Mar. 17		
8 Apl. 17	90	Chambers, G. W.	8 Apl. 17		
8 Apl. 17	1091	Bishop, T. C.	8 Apl. 17		
24 Apl. 17	74	Young, D. (57)	24 Apl. 17		
6 May 17	—	Shurey, H. R.	6 May 17		
	90	Thomson, A. A. McD.	20 May 17		
	38	Sopwith, R. J.	20 May 17		Attached.
26 May 17	1107	Madland, C.	26 May 17		
23 June 17	1094	Knight, L. C. (57)	23 June 17		
23 June 17	102	Everard, K. V.	23 June 17		
23 June 17	1091	Holland, D. M.	23 June 17		
23 June 17	1093	Irvine, W. D.	23 June 17		
2 July 17	1104	Martin, W. J.	2 July 17		
21 July 17	2 *a.c.*	Sanderson, R. J.	21 July 17		
5 Aug. 17	1094	Nussey, H. S.	5 Aug. 17		
12 Aug. 17	102	Smyth, H. A.	12 Aug. 17		
12 Aug. 17	74	Empson, C.	12 Aug. 17		
12 Aug. 17	1107	Perrin, N. J.	12 Aug. 17		
19 Aug. 17	51	Friedlander, W.	19 Aug. 17		
28 Aug. 17	90	Hicks, A. N.	28 Aug. 17		
31 Aug. 17	—	Moody, J.	31 Aug. 17		34 (Res.) Mn. Batty.
3 Sep. 17	89	Grantham, G. J.	3 Sep. 17		
9 Sep. 17	"M"	Trobridge, F. M. (57)	9 Sep. 17		
	38	Barlow, R. F.	21 Sep. 17		
21 Sep. 17	38	Hepburn, R.	21 Sep. 17		
27 Sep. 17	29 M B.	Gilbert, V. J.	27 Sep. 17		
5 Oct. 17	12 *a.c.*	Farley, L. G.	5 Oct. 17		
1 Oct. 17	4	Reeves, S.G., D.C.M.	1 Oct. 17		
7 Oct. 17	1091	Harrison, H.	7 Oct. 17		
7 Oct. 17	1093	Powe, H.	7 Oct. 17		
20 Oct. 17	27Mn.	Moulding, E. J.	20 Oct. 17		
20 Oct. 17	4	Storey, B. L.	20 Oct. 17		
28 Oct. 17	1093	Myles, G. S.	28 Oct. 17		
28 Oct. 17	38	Barrett, R. C.	28 Oct. 17		
4 Nov. 17	90	Ballenden, H. McN.	4 Nov. 17		
4 Nov. 17	77	Willcock, A.	4 Nov. 17		
18 Nov. 17	1096	Hill, H. G.	18 Nov. 17		
2 Dec. 17	1096	Collier, E.	2 Dec. 17		
2 Dec. 17	1089	Hamer, R. F.	2 Dec. 17		
1 Dec. 17	74	Laing, H.	1 Dec. 17		
1 Dec. 17	4	Hollebone, H. C.	1 Dec. 17		
21 Dec. 17	4	White, R. Y.	21 Dec. 17		

Riding Masters.

First Commn.	Batty.	Names.	Date of rank.	Army rank.	Remarks.
	R.F.A.	Fleming, Maj. F.	3 Feb. 09		
	R.H.A.	Connell, Maj. T. (65, 71, 207)	23 Feb. 13		
	R.H.A.	Bastyan, Maj. S. J.	16 Oct. 16		
	R.F.A.	Naylor, Maj. H. W. (71)	28 Mar. 15		
	R.H.A.	Lynch, Capt. J.	1 July 17		
	R.F.A.	Wark, Capt. R. G.	1 July 17		

District Officers.

First Commn.	Batty.	Names.	Date of rank.	Army rank.	Remarks.
	—	Bolton, Lieut. F. (*temp. Capt., 30 July 16*) (71)	3 Feb. 12		
	9 *a.c.*	Cockaday, Lieut. W.G.C., M.C.	3 Feb. 12		
	10 *a.c.*	Davies, Lieut. T. W. E (*act'g. Capt., 3 Aug. 17 in R. F. A.*)	3 Feb. 12		
	8 *a.c.*	Kershaw, Lieut. (*temp. Capt., 13 Sep. 15*) C. H.	3 Feb. 12		
	7 *a.c.*	Martin, Lieut. R. (*temp. Capt., 21 Nov. 14*)	3 Feb. 12		
	—	Farrow, Lieut. A. E. (71, 95)	27 Mar. 12		*Lv., ex I., m.c., 6 m.*
	11 *a.c.*	Harris, L. C. (*temp. Capt., 2 Sep. 15*)	23 Nov. 12		
	—	Syrett. Lieut. F. G.	23 Nov. 12		
	3 *a.c.*	Walker, Lieut. T. H., M.C. (*temp. Capt., 11 Apl. 15*)	25 Dec. 12		
	1 *a.c.*	Bell, Lieut. W. J.	25 Dec. 12		
	—	Ramsay, Lieut. J. (*temp. Capt., 16 Nov. 15*)	3 Sep. 15		

First Commn.	Batty. or Coy.	Names.	Regimental rank.	Remarks.
		ROYAL HORSE AND FIELD ARTILLERY.—TERRITORIAL FORCE.		
		Lieutenant-Colonels.		
		Talbot, G. R. Fitz. R. Capt. ret. pay (*Res. of Offr.*)	1 Apl. 08	
		Bedford-Pim, E. H. Maj. ret. pay (*Res. of Offr.*)	27 July 08	
		Cheke, E. G. Maj. ret. pay (*Res. of Offr.*)	20 Oct. 11	
		Hayward, F. G., T.D.	8 June 12	
		Powell, H. L. Maj. ret. pay (*Res. of Offr.*)	15 Jan. 13	
		Bruce-Kingsmill, J. C. de K., (*late Capt. R. A.*) (*o*) (207)	12 Dec. 14	
		Barton, P. H., T. D.	1 June 16	Comdg. 220th Bde.
		Majors.		
	1086	Flowers, E.	1 Jan. 12	
	1070	Carder, C. D.	1 July 13	
	1089	Thomas, C. J. H.	1 June 16	
22 Feb. 08	1088	House, P. J.	1 June 16	
	1072	Denne, R. T. (*actg. Major 27 May 18*)	1 June 16	
12 Mar. 08	—	House, M. H. N.	1 June 16	
	—	James, H. (57)	1 June 16	
	1104	Jobling, G. A.	1 June 16	
	1105	Williams, R. de V.	1 June 16	
		Captains.		
	1096	Bowdler, W. A.	13 Apl. 15	*Lv. ex Ind. 6 mos. 2 Apl. 18.*
5 May 08	1098	Blair, H. A.	1 June 16	
	1067	Brown, J. (*act'g. Major*)	1 June 16	
	102	Langley-Taylor, W. L.	1 June 16	
1 Jan.	1070	Wise, S. A.	1 June 16	
	—	Stevens, L. C.	1 June 16	Secy., Cen. Publicity Board, Simla.
	4	Denne, L. G. L. (*act'g. Maj. 27 May 18*)	1 June 16	
	77	Johnstone, C. H.	1 June 16	
	—	Schneider, L. W. (57)	1 June 16	
	79	Hoare, F. C. (*act'g. Maj., 18 Sep. 18*)	1 June 16	
7 Mar. 11	1066	Wade, H. R. (*act'g. Major*)	1 June 16	
1 Mar. 10	1081	Pybus, H. (*act'g. Maj. 26 May 18*)	1 June 10	
	90	Reid, R.	1 June 16	
30 Nov. 10	1094	Rickeard, C. R. (*act'g. Major*)	1 June 16	
9 Nov. 12	1091	Duke, A. B. C.	1 June 16	
	222 Bde.	*Claridge, J. W. L.*	*1 June 16*	*I. A. R. O.*
20 Apl. 11	74	Millett, P.	1 June 16	16 Bde., R. F. A.
11 May 12	1064	Edgar, L. A. (*act'g. Major*)	1 June 16	
23 July 13	—	*Chard, R. H.*	1 June 16	*Adjt., 222nd Bde.*
8 Nov. 12	1096	Manley, R. T. (57)	1 June 16	
	77	Squire, J. H.	1 June 16	
		Lieutenants.		
1 Mar. 13	1070	Stewart, H. L. G.	29 Aug. 14	
1 June 13	1103	Dyer, G. S.	29 Aug. 14	
1 Oct. 13	1070	Simmons, R. D. H.	1 Nov. 14	
13 Nov. 14	—	Hunt, A. E.	5 May 15	
1 Dec. 13	1098	Downton, A. M. (*act'g. Capt.*)	9 May 15	
6 Nov. 13	1064	Barker, J. W. (*act'g. Capt*)	1 June 16	
1 Mar. 14	1058	Colman, L. M. (*act'g. Major*)	1 June 16	
3 Aug. 14	215 Bde. a c.	Godwin, A. R. (*act'g. Capt.*)	1 June 16	
5 Aug. 14	1098	Jewell, C. J. S. (30, 59, 207*a*)	1 June 16	
12 Aug. 14	1086	Ealand, V. F.	1 June 16	
20 Aug. 14	1072	Budden, F. V. (*act'g. Capt.*)	1 June 16	
22 Aug. 14	—	Sidebottom, S. R.	1 June 16	217th Bde.
29 Aug. 14	—	Hughes, H. C.	1 June 16	
29 Aug. 14	1096	Hillman, A. J. H. S. (*temp. Capt. 23 June 18*).	1 June 16	Dy. Asst. Mily. Secy., Army Hdqrs.
	1089	Parke, C.	1 June 16	
	1089	Warner, A. H.	1 June 16	
11 Sep. 14	1093	Barker, R. R.	1 June 16	
11 Sep. 14	—	Burdon, W. B. C. (*act'g. Lt.-Col.*)	1 June 16	
11 Sep. 14	1066	Garle, H. A. (*act'g. Capt.*)	1 June 16	
11 Sep. 14	M	Hobgen, D. J. (*act'g. Capt, 16 Sep. 18*)	1 June 16	
11 Sep. 14	1105	Marrable, E. D.	1 June 16	
11 Sep. 14	1104	Rose, E. D. (*act'g. Capt.*)	1 June 16	

First Commn.	Batty. or Coy.	Names.	Regimental Rank.	Remarks.

ROYAL HORSE AND FIELD ARTILLERY.—TERRITORIAL FORCE—*contd.*

Lieutenants—*contd.*

First Commn.	Batty. or Coy.	Names.	Regimental Rank.	Remarks.
	W	Wilton, E. F.	1 June 16	
19 Sept. 14	1086	Barnes, S.	1 June 16	
19 Sept. 14	1086	Hall, P. M., M.C. (57)	1 June 16	
19 Sept. 14	1088	Lane, N. M. (*act'g. Capt.*) (57)	1 June 16	
19 Sept. 14	1098	Mac Ewan, A. K. (*act'g. Capt.*)	1 June 16	
19 Sept. 14	—	Nicholas, J. K. B.	1 June 16	
23 Sept. 14	1070	Hardie, W. B., M.C.	1 June 16	
		Brown, F. G.	1 June 16	74 Coy., R. G. A.
30 Sept. 14	25Mn.	Halden, K.	1 June 16	
1 Oct. 14	1103	Brailsford, R. W. (*act'g. Maj., 28 May 18*)	1 June 16	
	90	*Ironside, O. D.*	1 June 16	
1 Oct. 14	1098	Matthews, R. W. Y.	1 June 16	
5 Oct. 14	—	White, L.	1 June 16	Act'g. Adjt., 215 Bde. A.C.
8 Oct. 14	1072	Barton, H. E.	1 June 16	
8 Oct. 14	1067	*Cumberlege, R. M.* (*act'g. Capt.*)	1 June 16	
17 Oct. 14	—	Beverley, R.	1 June 16	*Adjt. 227th Bde.*
4 Nov. 14	—	*Nunn, R. L. H.*	1 June 16	*A. D. C., to G. O. C., R. A.*
9 Nov. 14	—	Peace, L. R.	1 June 16	Ord. Offr. 227th Bde.
1 Dec. 14	222 Bde. *a.c.*	Rand, T. A.	1 June 16	
11 Dec. 14	1068	Hughes, F. (*act'g. Capt.*)	1 June 16	
12 Dec. 14	1105	Pike, J. (*act'g. Capt.*)	1 June 16	
	6Mn.	Croil, T. M.	1 June 16	
30 Jan. 15	1064	Dobson, W. G.	1 June 16	
5 Feb. 15	4	*Beckingham, F.* (*act'g. Capt.*)	1 June 16	*Adjt. VII Bde.*
10 Feb. 15	1098	King, J. G.	1 June 16	
	90	Moore, W. E. C.	1 June 16	
12 Feb. 15	—	*Hill, P. E.* (57)	1 June 16	*Adjt. 217th Bde.*
11 Mar. 15	1096	Hole, H. E. M.	1 June 16	
14 Mar. 15	—	*Skinner, H. W.* (57)	1 June 16	*Adjt. 220th Bde.*
17 Mar. 15	12*a.c.*	Black, D. G.	1 June 16	
23 Mar. 15	1105	Wynn-Werninck, H.	1 June 16	
1 Apl. 15	77	Beausire, R. E.	1 June 16	
	215 Bde.	Farrow, A. R.	1 June 16	
12 Apl. 15	—	Eastwood, F. (57) (*act'g. Maj., 27 May 18*)	1 June 16	
24 Apl. 15	1066	Wreford, V. S. J.	1 June 16	
	74	Mouritz, C. E. A.	1 June 16	Staff Capt., R. A., 1st (Peshawar) Dn.
27 Apl. 15	79	Arneil, L. (*act'g. Lt.-Col., 3 Sep. 18*)	1 June 16	
27 Apl. 15	4	Phillips, A. A. H.	1 June 16	
1 May 15	38	Donne, B. G.	1 June 16	
5 May 15	1105	Waycott, A.	1 June 16	
5 May 15	79	Wilson, R. B.	1 June 16	
5 May 15	27Mtn	Ponsonby, H. P.	1 June 16	
7 May 15	222 Bde. *a.c.*	Kinnes, G. W.	1 June 16	
14 May 15	—	Depledge, O. F.	1 June 16	
	1094	Caink, A. S.	1 June 16	
	77	Gibb, H. M. (*act'g. Capt., 24 Sep. 18*)	1 June 16	
13 June 15	1064	Taylor, A. G. (57)	1 June 16	
16 June 15	1097	Smith, T. R.	1 June 16	
	38	Lynn, J. C.	1 June 16	
13 July 15	1089	Kemp, C. A.	1 June 16	
14 July 15	1089	Soutar, C. J. S.	1 June 16	
15 July 15	1093	Mac Intyre, W. F.	1 June 16	
16 July 15	1089	Butler, S. R., M.C.	1 June 16	
1 Aug. 15	1094	Edmunds, H. B. (*act'g. Capt.*)	1 June 16	
18 Aug. 15	—	Rance, C. J.	1 June 16	
25 Aug. 15	1089	Richardson, H. F.	1 June 16	
1 Sept. 15	1066	Annakin, A. H.	1 June 16	
3 Sept. 15	1089	Wheeler, F. O.	1 June 16	With 215 Bde. *a. c.*
4 Sept. 15	220 Bde. *a.c.*	Winton, M.	1 June 16	No. 5 Spl. Sec.
5 Sept. 15	1097	Baggar, F. A.	1 June 16	
13 Sept. 15	1093	Barford, G. E. (*act'g. Capt., 26 May 18*)	1 June 16	
20 Sept. 15	1103	Midwood, D. N.	1 June 16	
22 Sept. 15	1093	Shepton, W. S.	1 June 16	
2 Oct. 15	1067	Conway, F. J.	1 June 16	
2 Oct. 15	1098	Marsh, T.	1 June 16	
6 Oct. 15	74	Dick, N. R.	1 June 16	
6 Oct. 15	—	Nosworthy, J. L.	1 June 16	

ROYAL HORSE AND FIELD ARTILLERY.—TERRITORIAL FORCE—*concld.*

Lieutenants—*concld.*

First Commn.	Batty. or Coy.	Name.	Regimental rank.	Remarks.
8 Oct. 15	1098	Coverdale, G. L. B.	1 June 16	
18 Oct. 15	—	Spencer, R. D.	1 June 16	
22 Oct. 15	1098	Warlow, A. M.	1 June 16	
22 Oct. 15	1094	Dark, R. H.	1 June 16	
23 Oct. 15	12*a.c.*	Francis, W. J.	1 June 16	
25 Oct. 15	1097	King, A. J.	1 June 16	
26 Oct. 15	1096	Hawke, W.	1 June 16	
27 Oct. 15	1088	Beale, C. J.	1 June 16	
	1094	Campfield, O. G.	1 June 16	
28 Oct. 15	—	Musgrave, T. H.	1 June 16	
29 Oct. 15	215 Bde. *ac.*	Masterman, G. V.	1 June 16	
	101	Davis, M. D.	1 June 16	Lv., ex I.
15 Mar. 15	1097	Binney, A. L. (*act'g. Capt.*, *27 May 18*)	1 July 17	
10 Nov. 15	1098	Cargill, A.	1 July 17	1 June 16.
12 Nov. 15	1068	Vidler, C. H.	1 July 17	
23 Nov. 15	222 Bde. *a.c.*	Reeves, F. K.	1 July 17	
19 Dec. 15	—	May, A. H. (57)	1 July 17	
	—	Boyden, P.	1 July 17	
22 Nov. 15	222 Bde. *a.c.*	Chard, E. H.	1 July 17	
22 Nov. 15	1105	Smith, A. C.	1 July 17	
	1094	Backhouse, G. M.	1 July 17	*Lv., ex I.*
7 Apl. 16	—	Law, H. S.	7 Oct 17	

2nd-Lieutenants.

First Commn.	Batty. or Coy.	Name.	Regimental rank.	Remarks.
19 Mar. 15	1103	Luce, E. P. W.	19 Mar. 15	
15 Apl. 15	90	Hamer, R. A.	15 Apl. 15	
3 June 15	1097	Warry, C. F. G.	3 June 15	
27 Oct. 15	1097	Marwood, G.	27 Oct. 15	
3 Nov. 15	141	Watt, E. G.	3 Nov. 15	
3 Nov. 15	1067	Aitchison, H. M.	3 Nov. 15	
4 Nov. 15	1067	Hocking, E. A. E. (57)	4 Nov. 15	
6 Nov. 15	—	Webb, W. H.	6 Nov. 15	
6 Nov. 15	1070	Paine, W. H.	6 Nov. 15	
9 Nov. 15	1096	Side, A. D. B.	9 Nov. 15	
13 Nov. 15	1068	Geary, F.	13 Nov. 15	
15 Nov. 15	—	Spafford, D. N.	15 Nov. 15	
16 Nov. 15	1068	Barker, F., M.C.	16 Nov. 15	
26 Nov. 15	1098	DeBourcier, H. A.	26 Nov. 15	
27 Nov. 15	1070	Coombs, C. C.	27 Nov. 15	
2 Dec. 15	221 Bde. *a.c.*	Hirst, H.	2 Dec. 15	
7 Dec. 15	1094	Notley, R. A.	7 Dec. 15	
12 Dec. 15	1089	Edwards, J. C.	12 Dec. 15	
12 Dec. 15	—	Turner, B. C.	12 Dec. 15	Ord. Offr. 222nd Bde.
15 Dec. 15	—	Blanch, N. H.	15 Dec 15	Ord. Offr. 220th Bde.
19 Dec. 15	1097	May, A. H.	19 Dec. 15	
29 Dec. 15	1067	Carr, C. W.	29 Dec. 15	
1 Jan. 16	—	Danks, R. B.	1 Jan. 16	
14 Jan. 16	1104	Davies, D. A.	14 Jan. 16	
17 Jan. 16	—	DeWardt, T.	17 Jan. 16	
18 Jan. 16	—	Williams, F. T.	18 Jan. 16	Ord. Offr. 221st Bde.
24 Feb. 16	1097	Bagger, J. G.	24 Feb. 16	
3 Aug. 16	—	Gardner, K. A. R.	3 Aug. 16	
8 Aug. 16	74	Mugford, J.	8 Aug. 16	
10 Sep. 16	215 Bde. *a.c.*	Finch, R. E.	10 Sep. 16	
12 Sep. 16	1089	Hawley, N. W.	12 Sep. 16	
14 Oct. 16	1103	Terry, H. F.	14 Oct. 16	
2 Jan. 17	1097	Pillans, J. E.	2 Jan. 17	
23 Jan. 17	1094	Sutton, E. S.	23 Jan. 17	
3 Feb. 17	1089	Smith, W. F.	3 Feb. 17	
17 Feb. 17	1066	Howard, A.	17 Feb. 17	
7 May 17	8Mn.	Mackay, A. G. D.	7 May 17	
13 May 17	1091	Allison, D.	13 May 17	
30 July 17	1094	Brook, H. L.	30 July 17	
12 Aug. 17	*30Mn.*	*McIntyre, J.*	12 Aug. 17	
12 Aug. 17	1097	Simpson, D. T.	12 Aug. 17	
12 Aug. 17	1089	More, A. H.	12 Aug. 17	
5 Oct. 17	1094	Charles, R.	5 Oct. 17	
20 Oct. 17	90	Hutcheson, W. R. (57)	20 Oct. 17	
2 Dec. 17	28Mtn.	Murie, J. C. W.	2 Dec. 17	

ROYAL GARRISON ARTILLERY.

Lieutenant-Colonels.

First Commn.	Batty. or Coy.	Names.	Regimental rank.	Army rank.	Remarks.
*16 July 81	—	*Muspratt-Williams, C. A.*, C.I.E. (206)	25 May 09	. . . .	*Ch. Insptr. of Explo., G. of I.*
26 July 81	—	*Walker, M.* (65, 206)	1 Oct. 09	. . . .	*Ordce. Dept.*
22 Feb. 82	—	Dowell, G. C., C.M.G. (206)	1 July 10	*c.* 10 Oct. 10 *s.c.* 2 Oct. 15	
25 July 82	—	*Grier, H. D.*, C.B., D.S.O. (*temp. Brig.-Gen.*) (161, 170, 206)	11 Sep. 10	*s.c.* 16 Dec. 13	*Brig.-Gen., R. A.*
*1 Oct. 82	—	*Foote, H. B.* (*g*) (50, 206)	1 Apl. 11	. . . .	*Ordce. Dept.*
14 Feb. 83	—	*Williamson Oswald, O. C.*, C.B., C.M.G. (*temp. Brig.-Genl.*) (67, 107, 170, 206)	17 July 11	*s.c.* 15 Dec. 14	
14 Feb. 83	—	*Stanton, F. W. S.*, D.S.O. (*temp. Brig.-Gen.*) (207)	22 Nov. 11	. . . .	*Comdg. R.A., 2nd Dn.*
*28 July 83	—	Bertie-Clay, N. S. (206)	18 May 12	. . . .	*Resdg. out of India.*
5 July 84	—	Warren, W. L.	22 Nov. 12		
9 Dec. 84	—	Hardy, W. K. (169, 207)	29 Jan. 13		
16 Sep. 85	—	Davidson, F. M. (*g*) (207)	28 Jan. 14	. . . .	Comdg. 70 Coy., Coast Defence.
17 Feb. 86	—	Tancock, O. K., C.M.G. (156, 161, 169, 206)	30 Oct. 14	. . . .	*Lv., ex I., m. c., 6 m.*
18 Feb. 86	—	Brewster, R. F. (*g*) (162, 206)	30 Oct. 14	. . . .	Comdg. 60 Coy., R. G. A.
*24 July 86	—	*Smallwood, F. G.*, C.V.O. (*f*) (206)	30 Oct. 14	. . . .	*Ordce. Dept.*
16 Feb. 87	—	*Kenyon, L. R.*, C.B., *p.a.c.* (*h*) (*temp. Brig.-Gen.*)	30 Oct. 14	*c.* 29 Nov. 15	*Ordce. Dept.*
*16 Feb. 87	—	*Douglas, A. P.*, C.M.G. (*h*) (125*b*, 206)	30 Oct. 14	. . . .	*Ordce. Dept.*
16 Feb. 87	—	Stiffe, A. F. E. (57, 207)	30 Oct. 14	. . . .	
†23 July 87	—	Molesworth, R. P. (57, 206)	7 Nov. 14	*c.* 3 June 15	
23 July 87	—	*Patch, F. R.* (*temp. Brig.-Gen.*), C.M.G., D.S.O. (169, 206)	1 Feb. 15		
*17 Feb. 88	—	*Young, H. A.*, C.I.E. (*f*) (*temp. Br.-Genl., 8 Mar. 18*) (206)	20 May 15	21 Feb. 14	*Ordce. Dept.*
*17 Feb. 88	—	*Carter, B. F.* (*h*) (206)	7 Sep. 15	.	*Ordce. Dept.*
*17 Feb. 88	—	*Tomkins, E. L.* (*g*) (*f*) (206)	6 Jan. 16	. . .	*Ordce. Dept.*
*27 July 88	—	*Lethbridge, S.* (51, 206)	6 Jan. 16	. . . .	*Ordce. Dept.*
27 July 88	—	Mackintosh, J. B. (*g*) (Q 106, 206)	26 Feb. 16		
27 July 88	—	*Massie, R. H.* (*temp. Brig-Gen.*) (153, 162, 206)	7 Apl. 16	*c.* 1 Jan. 18	*Brig.-Gen., R. A.*
15 Feb. 89	104	Farmar, W. C. R. (206)	10 Apl. 16		
*15 Feb. 89	—	*Woods, G. G.* (*h*) (124, 206)	20 Apl. 16	. . . .	*Ordce. Dept.*
15 Feb. 89	—	Sherer, J. D., D.S.O. (*g*)	18 June 16	. . . .	Comdg. V Bde., R. F. A.
*14 Feb. 90	—	*Donovan, C. C.* (*h*) (170, 206)	1 May 17	. . . .	*Ordce. Dept.*
25 July 90	—	*Cotter, H. J.*, C.I.E., D.S.O. (170, 206)	1 May 17		
25 July 90	—	Easton, F. A. (57, 206)	1 May 17		
25 July 90	—	*Robinson, S. W.*, D.S.O. (*temp. Brig.-Gen.*) (162, 165, 169, 206)	1 May 17	*c.* 3 June 18	*D. A. & Q. M. G.*
*25 July 90	—	*Lawrence-Archer, J. H.* (*temp. Col.*) (57, 95, 206)	1 May 17	1 Jan. 17	*Ordce. Dept.*
25 July 90	—	Lowis, P. S., D.S.O., (162, 169, 206)	1 May 17		
*25 July 90	—	*Bowen, H. W.*, D.S.O. (*temp. Col.*) (206)	1 May 17	1 Jan. 17	*Dep. Dr. Ord. Services.*
13 Feb. 91	—	*Fergusson, A. C.*, C.M.G., D.S.O. (*g*) (64, 170, 206)	1 May 17	1 Jan. 16	*Comdg. VII Ind. Mtn. Arty. Bde.*
*13 Feb. 91	—	*Palmer, C. C.*, *p.a.c.* (206)	1 May 17	29 Nov. 15	*Mily. Asst. Ordce. Factories.*
* 4 Nov. 91	—	*Kennedy, A. C.*, D.S.O. (206)	1 May 17	. . . .	*Ordce. Dept.*
3 Nov. 91	—	Molesworth, H. E., D.S.O. (57, 207)	1 May 17	. . . .	Comdg. III Mtn. Bde.
1 Jan. 92	2 Mn. Bde.	*Wyatt, F. O.*, M.V.O. (Q. 57, 71, 94, 100, 162, 172*a*, 206)	1 May 17	. . . .	*Comdg. 2 Mtn. Arty. Bde.*
2 Feb. 92	—	*Freeth, C. J. D.* (*g*) (206)	1 May 17	. . . .	*Ordce. Dept.*
*3 Mar. 92	—	*Sturrock, G. C.* (*g*) (206)	1 May 17	. . . .	*Ordce. Dept.*
18 Mar. 92	4 Mn. Bde.	Corrie, G. G. W. (172*a*, 206)	1 May 17		
18 Mar. 92	—	Riddell, R. B.	1 May 17		
4 May 92	—	*Luke, T. M.* (*temp. Brig.-Gen.*), D.S.O., O.B.E. (206)	1 May 17	29 Nov. 15	*Dir. Admn. A. H. Q.*
*14 Sep. 92	—	*Carey, W. L. J.* [*l*] [*o*] (106, 161, 165, 169, 206)	13 July 17	. . . .	*Ordce. Dept.*
*13 Mar. 93	—	*Tyrrell, J. F.* (94, 160, 193, 206)	8 Sep. 17	. . . .	*Ordce. Dept.*
13 May 93	—	*Keogh, J. H.* [L] (94, 169, 206)	1 Mar. 18	. . . .	*Comdg. 1 Br. Mtn. Arty. Bde.*

(*h*) Passed Ordce. course (except A. O. duties).
† Acting rank.

ROYAL GARRISON ARTILLERY—*contd.*

Majors.

First Commn.	Batty. or Coy.	Names.	Regimental rank.	Army rank.	Remarks.
17 Nov. 94	82.	Beer, V. L.	30 Oct. 14		
*17 Nov. 94	—	*Walker, W. B.(h)* (12, 162, 169, 206)	30 Oct. 14	. . . .	*Ordce. Dept.*
15 June 95	—	*Gardiner, H. W.* (*o*), (12, 94, 162, 170, 206)	30 Oct. 14	. . . .	*Ordce. Dept.*
15 June 95	—	*Bayley, L. S.*, D.S.O. [L] (162, 170, 206)	30 Oct. 14	*l.c.* 8 Nov. 15	*Inspg. Offr., I. S. Troops.*
*26 Sep. 95	—	*Wilkes, G. S.* [*o*] (106, 206)	30 Oct. 14	. . . .	*Ordce. Dept.*
2 Nov. 95	—	MacFarlan, J. B., D.S.O. (12, 170, 206)	30 Oct. 14		
2 Nov. 95	—	*Greene, A. C. R., p.s.c.* (57, 94, 162, 169, 206)	30 Oct. 14	. . . .	*D. A. A. G.*
2 Nov. 95	—	*Climo, P. H.* (172*a*, 206)	30 Oct. 14		
*21 Mar. 96	—	*Thomas, C. S. W.* (12, 206)	30 Oct. 14	. . . .	*Ordce. Dept.*
*21 Sep. 96	—	*Spalding, W. B.* (12, 162, 206)	30 Oct. 14	. . . .	*Ordce. Dept.*
21 Sep. 96	26*Mn.*	*Campbell, K. G.*, D.S.O. (*g*) (206)	30 Oct. 14		
*21 Sep. 96	—	*Sitwell, N. S. H.* (206)	30 Oct. 14	. . . .	*Ordce. Dept.*
*16 Jan 97	—	*Riccard, J. S.* (*temp. Lt.-Col.*) (*h*), (12, 95, 169, 206)	30 Oct. 14	. . . .	*Ordce. Dept.*
23 Mar. 97	—	*Cruickshank, H. St. J., p.a.c.* (206)	30 Oct. 14	. . . .	*Ordce Dept.*
23 Mar. 97	—	*Hayes-Sadler, J.* (162, 206)	30 Oct. 14	. . . .	*Comdg. Ind. Mtn. Arty. Bde.*
* 1 Sep. 97	—	*Cowie, D. G.* (*temp. Lt.-Col., 25 Feb. 18*) (206)	30 Oct. 14	. . . .	*Ordce. Dept.*
23 Dec. 97	86	Alexander, R. C. J.	30 Oct. 14		
23 Dec. 97	74	Bennett, F. L. [*l*] (167, 172*a*, 207)	30 Oct. 14		
23 June 98	—	*Langhorne, J. A. D.*, D.S.O. *temp. Lt.-Col.* (*g*)	30 Oct. 14	. . . .	*A. Q. M. G.*
23 June 98	—	Snowdon, H. S. K. (*o*) (95, 206)	30 Oct. 14		
14 Sep. 98	—	Colbeck, B. B., D.S.O. (*temp. Lt.-Col.*) (57, 89, 207)	30 Oct. 14		
24 June 98	—	*Clarke, H.* (206)	30 Oct. 14	. . . .	*Ordce. Dept.*
23 Dec. 98	7 Mn.	Hall, B. A. M. (207)	30 Oct. 14		
23 Dec. 98	—	Loveday, F. W., *p.s.c.* (59, 169, 188*a*, 206)	30 Oct. 14		
23 Dec. 98	—	*Hassells-Yates, G. A.* (206)	30 Oct. 14	. . . .	*D. A. A. G., Army Hdqrs.*
25 June 99	—	*Brown, B. H., p.a.c.* (205*a*)	30 Oct. 14	. . . .	*Ordce. Dept.*
22 Dec. 98	—	Rynd, F. F., D.S.O. [L] (57)	30 Oct. 14		
14 Sep. 98	62	Holme, H. L. [L]	30 Oct. 14		
25 June 99	—	*Beamish, W. S.* (*o*)	30 Oct. 14	. . . .	*Ordce. Dept.*
25 June 99	—	Preston, P. H. H. (*g*)	30 Oct. 14		
25 June 99	—	Crowdy, C. R. (*g*)	30 Oct. 14		
9 Aug. 99	—	*Reed, C.*, D.S.O. (65, 169, 206)	30 Oct. 14	*l.c.* 3 June 18	*Ordce. Dept.*
22 Nov. 99	—	Barron, F. W., *p.sc.*, (*temp. Lt.-Col.*)	28 Nov. 14	*l.c.* 3 June 18	A. Q. M. G.
22 Nov. 99	21*Mn.*	Hoskyn, R. (*g*)	28 Nov. 14		
22 Nov. 99	—	Leah, T. C., D.S.O. (94. 170, 206)	18 Apl. 15		
6 Jan. 00	—	Hill, G. N. (12, 94, 105, 120, 125*a*, 162, 169, 206)	5 Sep. 15		
*10 Jan. 00	—	*Reeves, S. R.* (169, 206)	16 Oct. 15	. . . .	*Ordce. Dept.*
17 Mar. 00	—	*Morrison, J.* (12, 206)	16 Oct. 15	. . . .	*Ordce. Dept.*
17 Mar. 00	8 Mn.	Hunt, W. M., M.C. (95, 169, 206)	23 Nov. 15		
28 Mar. 00	—	*Fraser, T. C.* (206)	9 Dec. 15	. . . .	*D. A. A. G., Army Hdqrs.*
28 Mar. 00	31*Mn.*	*Perry, S.* (12, 206)	30 Dec. 15		
2 May 00	—	Hutton-Squire, R. H. E., D. S. O.	30 Dec. 15	. . . .	With 85 By., R. F. A.
2 May 00	—	*Finnis, F. A.* (*act'g. Lt.-Col.,*) [*O*] (12, 94, 162, 169, 200)	30 Dec. 15	. . . .	*Ordce. Dept.*
2 May 00	—	*Robertson-Glasgow, N.* (12, 153, 165, 169, 204)	30 Dec. 15	. . . .	*Ordce. Dept.*
22 Aug. 00	—	Lamb, B. (57)	30 Dec. 15		
5 May 00	—	*Griffith, G. M.* (*temp. Col.*)	30 Dec. 15	. . . .	
5 May 00	52	Dunne, T. F. K. (207)	30 Dec. 15		
12 May 00	—	*Jenkin, F. C.*, D.S.O. (12, 206)	30 Dec. 15	*l.c.* 10 Mar. [17.	*Ordce. Dept.*
23 May 00	2 Mn.	Morgan, F. A. S., D.S.O.	30 Dec. 15		
*26 May 00	—	*Fanshawe, L. A.*, D.S.O. (*actg. Lt.-Col*). [*o*] (206)	30 Dec. 15	. . . .	*Ordce. Dept.*
26 May 00	—	*Urquhart, A. M.* (206)	30 Dec. 15	. . . .	*Ordce. Dept.*
*26 May 00	—	*Lenfestey, L. d'E.* (12, 59, 206)	30 Dec. 15	. . . .	*Ordce. Dept.*

(*h*) Passed Ordce. course (except A. O. duties).
† Acting rank.

ROYAL GARRISON ARTILLERY—*contd.*

Majors—*concld.*

First Commn.	Batty. or Coy.	Names.	Date of rank.	Army rank.	Remarks.
26 May 00	—	*Vines, C. E., p. a. c.* (206)	30 Dec. 15	. . . .	*Ordce. Dept.*
*26 May 00	—	*Dickins, F.* (206)	30 Dec. 15	. . . .	*Ordce. Dept.*
18 July 00	—	*Waller, R. S. De W.* (*temp. Lt.-Col., 10 Apl. 18*) (59, 170, 206)	30 Dec. 15	. . . .	*Ordce. Dept.*
18 Aug. 00	—	Whitting, E. Le G., D.S.O., M.C. (59, 64, 165, 170, 206)	30 Dec. 15	. . . .	
18 Aug. 00	—	*Bryant, J. M.* (206)	30 Dec. 15	. . . .	*Ordce. Dept.*
18 Aug. 00	—	*Mascall, M. E.*, D.S.O., (94, 206)	30 Dec. 15	. . . .	*Staff Capt.*
18 Aug. 00	—	Lindsay, W. D. (172a, 206)	30 Dec. 15	. . . .	Comdg. 2/33. Ind. Mn. Batty.
18 Aug. 00	*37Mn.*	*Dreyer, G. V.* (12, 65, 162, 206)	30 Dec. 15		
28 Nov. 00	—	McGowan, W. H. (64, 206)	30 Dec. 15		
19 Dec. 00	—	Rothwell, R. S., D.S.O. (162, 169, 206)	30 Dec. 15		
19 Dec. 00	*28Mn.*	*Davies, L. M.* (59, 169, 206)	30 Dec. 15		
19 Dec. 00	—	*Johnston, C. E. L.* [L] (*temp. Lt.-Col., 25 Feb. 18*) (57, 106, 125a, 162, 169, 205a)	30 Dec. 15	. . . .	*Ordce. Dept.*
21 Dec. 00	—	Farfan, A. J. T., D.S.O. (42, 59, 64, 106, 125a, 165, 169, 206)	30 Dec. 15		
21 Dec. 00	—	*Little, E. M.* (162, 172a, 206)	30 Dec. 15		
21 Dec. 00	—	*Salvin, H. C. J.*	30 Dec. 15		
21 Dec. 00	—	*Routh, G. M.*, D.S.O. [L] (59, 88, 161, 172a, 204)	30 Dec. 15	. . . .	*Ordce. Dept.*
4 May 01	—	*D'Esterre, J. C. E.* (51, 207)	3 Apl. 16	. . . .	*Adjt., R. G. A.*
15 May 01	—	*Cole, A. F.* (12, 57, 206)	9 May 16	. . . .	*A. Inspg. Offr., Kashmir I. S. Arty.*
23 July 01	—	Medill, P. M. (206)	29 Sep. 16		
15 May 01	—	*Kirby, H. A.*, D.S.O., M.C. (206)	1 May 17		
23 July 01	—	*Kenyon, H. E.* (206)	1 May 17	. . . .	*Bde.-Maj.*
4 Dec. 01	—	*Forbes, R. M. N.*, M.C. (12, 162, 170, 206)	1 May 17	. . . .	*Bde.-Maj.*
4 Dec. 01	—	Thom, J. H., D.S.O. (172a, 206)	1 May 17		
21 Dec. 01	—	*Barron, A. F. N.* (12, 94, 162, 169, 206)	1 May 17	. . . .	*Ordce. Dept.*
21 Dec. 01	—	*Tute, C. S.* (o), (185, 205a)	1 May 17	. . . .	*Ordce. Dept.*
21 Dec. 01	—	*Campbell, D. D. H.* M.C. (65, 107, 206)	1 May 17	. . . .	*Burma Mily. Police.*
24 May 02	*34Mn.*	*Wilson, E. R. C.*, M.C. (12, 71, 206)	1 May 17		
31 July 02	*21Mn.*	*Garrett, R. C.* (205a)	1 May 17		
31 July 02	—	Healing, N. C., M.C. (12, 59, 207) (*act'g. Maj., 25 Sep. 16*)	1 May 17		
31 July 02	*35Mn.*	*Dutton, R. M. L.* (12, 146, 162, 172a, 206)	1 May 17		
31 July 02	—	*Larmour, L. O.* (o)	8 June 17	. . . .	*Ordce. Dept.*
24 Dec. 02	—	*Seys, R. C.*, D.S.O. [L]	14 June 17	. . . .	*Bde.-Maj.*
24 Dec. 02	*29Mn.*	*Somerville, J. A. H. B.* (12, 206)	13 July 17		
24 Dec. 02	—	*Rawson, C. D.*, D.S.O. (59, 105, 199, 206)	30 July 17	. . . .	*Adjt., VII Ind. Mtn. Bde.*
24 Dec. 02	73	Payne, D. W., M.C.	30 July 17		
24 Dec. 02	*30Mn.*	*Bell, M. D.* (172a, 205a)	20 Aug. 17		
24 Dec. 02	—	Gale, H. J. G. (59, 169, 206)	8 Sep. 17		
15 July 03	—	*Drake-Brockman, V., p.a.c.* [*l*]	1 Oct. 17	. . . .	*Ordce. Dept.*
15 July 03	—	Lewis, W. H., D.S.O., M.C. (65, 206) (*act'g. Lt.-Col., 16 Nov. 16*)	3 Oct. 17		
15 July 03	—	Trenchard, G. B. B. (42, 65, 127, 161, 205a)	3 Oct. 17		
15 July 03	*33Mn.*	*Wilson, R. E.*, D.S.O. (163, 169, 205a)	14 Nov. 17		
15 July 03	—	*Nixon, E. J.*, D.S.O., M.C. (57, 162, 172a, 206)	22 Nov. 17		
15 July 03	—	Ventris, E. F. V.	22 Nov. 17	. . . .	Q. M. G's. Bch., Army Hdqrs.
15 July 03	—	Kirkpatrick, C. D., M.C. (51)	3 Dec. 17		
15 July 03	*25Mn.*	*Jarrett, A. F. V.*, M.C. (94, 162, 172a, 206)	31 Dec. 17		
23 Dec. 03	—	Buckland, G. N., D.S.O. (162, 169, 188b, 205a)	13 Apl. 18	. . . .	D. A. & Q. M. G.
23 Dec. 03	—	Haskard, D. D., M.C. (66a, 95, 127, 139, 162, 169, 206)	31 July 18		

Captains.

First Commn.	Batty. or Coy.	Names.	Date of rank.	Army rank.	Remarks.
23 Dec. 03	—	*Long, W. E. L.* (o). (206)	30 Oct. 14	*Ordce. Dept.*	
4 June 04	—	*Knight, J. H.* (12, 57, 59, 165, 170, 206)	30 Oct. 14	*m* 3 June 17. With Sig. Serv.	
29 July 04	23Mn.	Landon, R. P., M.C. (*act'g. Maj., 30 June 18*) (12, 161, 169, 205a)	30 Oct. 14		
29 July 04	—	*Edwards, W. M.* (*act'g. Maj., 1 Jan. 17*) (12, 65a, 66a, 71, 95, 100, 161, 172a, 185, 205a)	30 Oct. 14	*S. & T. Corps.*	
29 July 04	2 Mn.	Crawford, A. E. (*act'g. Maj., 8 Oct. 16*) (206)	30 Oct. 14		
29 July 04	—	Price, J. (206)	30 Oct. 14		

(*h*) Passed Ordnance course (except A. O. duties).

ROYAL GARRISON ARTILLERY—*contd.*

Captains—*contd.*

First Commn.	Batty. or Coy.	Names.	Regimental rank.	Remarks.
29 July 04	—	Jones, C. G. P.	30 Oct. 14	
29 July 04	—	Newton, P. I., D.S.O. (71, 206)	30 Oct. 14	Comdg. No. 1 Mn. Batty.
20 Dec. 03	—	*Kinloch, R. I. H., p.a.c.* [*l*]	30 Oct. 14	*Ordce. Dept.*
21 Dec. 01	60	Gardiner, D. A. (*act'g. Maj., 1 Sept. 17*) (57)	30 Oct. 14	
3 Dec. 04	104	Guise, V. R.	30 Oct. 14	
20 Dec. 05	—	*Day, C. L.* (59, 206)	30 Oct. 14	With Sig. Serv.
25 July 06	60 Hy.	Russell, C. A. (*act'g. Maj. 16 Apl. 18*) (207)	30 Oct. 14	
25 July 06	—	*Stebbing, H. T.* (57, 94, 172*a*, 206)	30 Oct. 14	*Comdt., Sig. Schl*
18 Dec. 07	9	Fooks, H. E. (*act'g. Maj., 22 Feb. 17*) (95, 172*a*, 207)	30 Oct. 14	
29 July 08	3 Mn.	Montague, J. E. F. (*act'g. Maj., 26 Aug. 17*)	30 Oct. 14	
23 July 09	*26Mn.*	*Rossiter, F. N. C.*, M.C. (*act'g. Maj., 9 Sep. 18*)	23 July 15	
23 July 09	86	Lowndes, R. C. (57)	23 July 15	
23 July 09	4	Francis, H. F. (*act'g. Maj., 21 Aug. 17*)	23 July 15	
23 July 09	—	*Pargiter, R. B.* (42, 172*a*, 206)	23 July 15	
23 July 10	—	O'Leary, B. J.	23 July 16	
23 July 10	*35Mn.*	*Heaton-Ellis, J. S.* (206)	23 July 16	
23 July 10	—	Harker, A. W. A.	23 July 16	
23 July 10	—	Jenkins, W. M. (206)	23 July 16	
23 Dec. 10	*32Mn.*	*Troup, F. C. A* (*act'g. Maj., 5 Sep. 17*) (57, 206)	8 Aug. 16	
23 Dec. 10	5 Mn.	Rouquette, H. R. H. (*act'g. Maj.*)	8 Aug. 16	
23 Dec. 10	*22Mn.*	*Stevenson, J. H. M.*, M.C. (*act'g. Maj., 2 Feb. 18*) (206)	8 Aug. 16	
27 May 11	—	Underhill, N. (95, 206)	8 Aug. 16	
20 July 11	79	Hall, A. C. S.	8 Aug. 16	
20 July 11	64	Wilson, I. C.	8 Aug. 16	
20 July 11	—	Hay, A. [*l*]	8 Aug. 16	
9 Dec. 11	—	Skinner, R. O. (206)	8 Aug. 16	
9 Dec. 11	86	Williams, F. P. J. (207)	8 Aug. 16	
23 Dec. 11	104	Johnston, M. A. B.	8 Aug. 16	
23 Dec. 11	—	*Carew, R. L. O.* (206)	8 Aug. 16	
23 Dec. 11	—	Smith, G. L. S.	8 Aug. 16	
23 Dec. 11	86	Williams, E. J.	8 Aug. 16	
19 July 12	6 Mn.	Fardell, D. O. (*act'g. Maj., 19 Aug. 17*) (95, 172*a*).	8 Aug. 16	
19 July 12	28 Mn.	Collins, S. St. B. (57).	8 Aug. 16	
19 July 12	*31Mn.*	*Deane, J. A. L.* (172*a*), (206)	8 Aug. 16	
19 July 12	74	Wilson, R. R. (207)	8 Aug. 16	
19 July 12	—	English, J. W. (*act'g. Maj., 15 Oct 18*)	8 Aug. 16	
19 July 12	*21Mn.*	*Mackenzie, H. H.* (172*a*, 206)	8 Aug. 16	
20 Dec. 12	*32Mn.*	*Macgregor, A.* (207)	20 Dec. 16	
20 Dec. 12	—	Greene, H.	20 Dec. 16	
20 Dec. 12	*35Mn.*	*Hebbert, G. K. P.* (206)	20 Dec. 16	
20 Dec. 12	—	Allpress, H. V.	20 Dec. 16	Adjt., 60 (Heavy) Batty., R. G. A.
20 Dec. 12	*30Mn.*	*Armstrong, R. A.* (*act'g. Maj., 1 Oct. 18*)	20 Dec. 16	
20 Dec 12	—	Campbell, E. G. (*act'g. Maj., 3 Oct. 18*)	20 Dec. 16	
20 Dec 12	—	*McIntyre, H. M. J.*	20 Dec. 16	*Staff Capt. to C. R. A., 1st Dn.*
18 July 13	*25Mn.*	*Pemberton, A. L.*	18 July 17	
19 Dec. 13	—	Harrison, E. G. W. W., M.C.	3 Nov. 17	
17 July 14	—	de Fonblanque, E. B.	3 Nov. 17	
16 Sep. 14	7 Mn.	Tawney, A. E.	3 Nov. 17	
12 Jan. 15	—	Dean, F. W.	3 Nov. 17	
12 Jan. 15	—	Pirie, J. P. (95, 169)	3 Nov. 17	
12 Jan. 15	61	Carr, C.	3 Nov. 17	Instr. Gunnery.
12 Jan. 15	—	Ross, A. D.	3 Nov. 17	
12 Jan. 15	—	Moore, H. W. (*act'g. Maj., 23 July 18*) (95)	3 Nov. 17	Comdg. R. A. Depot., Roorkee.
12 Jan. 15	—	Glaysher, W. E.	3 Nov. 17	
12 Jan. 15	—	Napier, G.	3 Nov. 17	10-pr. Batty., Lucknow.
12 Jan. 15	—	King, C. T. W. (95)	3 Nov. 17	Mtn. Arty. Depôt, Ambala.
12 Jan. 15	—	*Taylor, J.*	3 Nov. 17	*Frontier Garr. Arty.*
12 Jan. 15	—	Devine, T.	3 Nov. 17	Adjt., 70 Coy.
12 Jan. 15	—	*Savage, T. A.* (*g*) (12)	3 Nov. 17	
12 Jan. 15	*27Mn.*	*Curry, M. J.* (*act'g. Maj. 16 Oct. 18*) (12)	3 Nov. 17	
12 Jan. 15	—	Steirn, F.	3 Nov. 17	
12 Jan. 15	—	Dorman, C. H.	3 Nov. 17	
12 Jan. 15	—	Smith, L. W.	3 Nov. 19	Asst. Instr. of Gunnery.
12 Jan. 15	—	Steele, F.	3 Nov. 17	
12 Jan. 15	51	Gardner, A. (59)	3 Nov. 17	
12 Jan. 15	—	Coombes, H. P.	3 Nov. 17	

First Commn.	Batty. or Coy.	Names.	Regimental rank.	Remarks.
		ROYAL GARRISON ARTILLERY—*contd.*		
		Captains—*contd.*		
12 Jan. 15	62	Whelan, J.	3 Nov. 17	
12 Jan. 15	—	Hill, B.	3 Nov. 17	
12 Jan. 15	—	Merrett, A. B.	3 Nov. 17	
12 Jan. 15	M	Willcox, G. F. (*act'g. Maj., 1 Oct. 18*)	3 Nov. 17	
12 Jan. 15	6 Mn.	Buchanan, A.	3 Nov. 17	
12 Jan. 15	74	Wood, R.	3 Nov. 17	
12 Jan. 15	101	Mead, P.	3 Nov. 17	
10 Feb. 15	*26 Mn.*	*Thuillier, H. S.*	3 Nov. 17	
10 Feb. 15	70	Lickman, H. S. (95)	3 Nov. 17	
		Lieutenants.		
22 Apl. 15	21 Mn.	*Armitage, H. M. A.*	1 July 17	
5 May 15	—	Marriott, F. C.	1 July 17	
15 May 15	*26 Mn.*	*Jones, P. H. (Act'g. Capt., 23 Jan. 18).*	*1 July 17*	
28 July 15	—	*Conville, L. H. G.*	1 July 17	*Comdt., Malay States Guides Batty.*
27 Oct. 15	—	Birmingham, W. R. (95)	1 July 17	
27 Oct. 15	—	Millinship, E. H.	1 July 17	
27 Oct. 15	62	Gilkes, G. H. (57)	1 July 17	
27 Oct. 15	21 Mn.	Gayer, W. M. (*Act'g. Capt.*) . . .	1 July 17	
27 Oct. 15	82	Wolfe-Murray, D. K.	1 July 17	Insp. to R. G. A.
19 Nov. 15	—	Shirley, E. A.	1 July 17	
19 Nov. 15	*33 Mn.*	*Beard, J. W. (act'g. Capt. 17 Oct. 18)*	1 July 17	
18 Dec. 15	5 Mn.	Sands, E. S.	1 July 17	
22 Dec. 15	76	Fisher, C. S. (*Act'g. Maj., 15 Apl. 18*)	1 July 17	
12 Apl. 16	101	Sims, R. J. (*Act'g. Capt., 24 Aug. 18*) (95).	12 Oct. 17	
4 May 16	—	Jones, H. R., M. C.	4 Nov. 17	
9 May 16	5 Mn.	Goodman, W. J.	9 Nov. 17	
11 May 16	—	*Froud, H. G.*, D.C.M. (*Act'g. Capt., 18 Nov. 17*).	11 Nov. 17	*Adjt., III Mn. Arty. Bde.*
13 June 16	—	Millar, P. P.	13 Dec. 17	
14 Oct. 16	—	Burton, T.	14 Apl. 18	
14 Oct. 16	*35 Mn.*	*Walker, F. McF.*	14 Apl. 18	
15 Oct. 16	—	Watts, A. T.	15 Apl. 18	
21 Nov. 16	—	Perigo, J.	21 May 18	
1 Dec. 16	*24 Mn.*	*Potter, S. (act'g. Capt., 1 Oct 18)* .	1 June 18	
1 Dec. 16	*33 Mn.*	*O'Day, T.*	1 June 18	
1 Dec. 16	*31 Mn.*	*Claxton, E.*	1 June 18	
1 Dec. 16	*25 Mn.*	*Riches, G. H.*	1 June 18	
9 Dec. 16	52	Rowney, F.	9 June 18	
9 Dec. 16	60	Cullen, P.	18 June 18	S. S. O., Roorkee.
26 Dec. 16	61	Dodd, A.	26 June 18	
27 Dec. 16	6 Mn.	Miller, E. C.	27 June 18	S. S. O., Khyra Gali.
29 Dec. 16	61	Donegan, M.	29 June 18	
30 Dec. 16	61	Donovan, J.	30 June 18	
1 Jan. 17	—	Hennessy, T. F.	1 July 18	
1 Jan. 17	—	Dickens, A. H. (*Act'g. Capt., 2 Sep. 17*)	1 July 18	Adjt., 61 Coy., R. G. A.
3 Jan. 17	73	Taylor, E.	3 July 18	
3 Jan. 17	—	Crudass, R. H.	3 July 18	
3 Jan. 17	—	Brown, W. C.	3 July 18	
4 Jan. 17	4 Mn.	Allen, J. (*Act'g. Capt., 28 June 18*) .	4 July 18	
4 Jan. 17	32 Mn.	Hawes, W.	4 July 18	
4 Jan. 17	—	Wheatley, W. A.	4 July 18	
6 Jan. 17	82	Faux, C. E.	6 July 18	
6 Jan. 17	61	Everton, R. F.	6 July 18	Asst. Provost Marshal, Bombay.
6 Jan. 17	76	Ratcliffe, W.	6 July 18	
11 Jan. 17	—	Lawn, A. J.	11 July 18	
11 Jan. 17	82	Lambert, V. G. (57)	11 July 18	
12 Jan. 17	*22 Mn.*	*Gulliver, J.*	11 July 18	
13 Jan. 17	68 Hy.	Mason, S. (57)	11 July 18	
27 Jan. 17	—	Packington, W. H. I.	27 July 18	10th I. M. A. Bde. Ammn. Column.
30 Jan. 17	32	Coxall, G.	30 July 18	
		2nd-Lieutenants.		
	—	Robertson, A.	1 Dec. 16	
	52	Perkins, H.	13 Jan. 17	
	35 Mn.	*Stewart, H.*	31 Jan. 17	
	76	Sleven, E. J., D.C.M. (58*a*)	6 Feb. 17	[Malakand.
	—	*Cooper, W.*	13 Mar. 17	*Frontier Garr. Artillery.* S. S. O.,
	31 Mn.	*Saxby, W.*	25 Mar. 17	
	8 Mn.	Wateridge, G. H.	4 May 17	
	—	*Delaney, M.* (207*a*)	29 May 17	
	27 Mn.	*Grant, C. D. (act'g. Capt., 23 Aug. 18)*	30 June 17	
	—	Miller, A. R.	2 Oct. 17	
	8 Mn.	Pound, F.	23 Oct. 17	
	—	Pitman, F.	25 Oct. 17	
	1	Peters, F.	29 Oct. 17	
	64	Yorke, W. H. W.	1 Nov. 17	
	75	Searls, A. G.	1 Nov. 17	

First Commn.	Batty. or Coy.	Names.	Regimental rank.	Remarks.
		ROYAL GARRISON ARTILLERY—*contd.*		
		2nd-Lieutenants—*contd.*		
	4	Juniper, A.	7 Nov. 17	
	70	Whiteside, J.	12 Nov. 17	
	61	Poor, W. A. J.	20 Nov. 17	
	68	Budge, G. S.	24 Nov. 17	*Lv., ex I.*
	68	Whiteman, W. M.	4 Dec. 17	
28 Dec. 17	—	Chimes, W. E. G.	28 Dec. 17	
9 Jan. 18	—	Alexander, S. W.	9 Jan. 18	
3 Feb. 18	—	Kirby, J. N.	3 Feb. 18	
1 June 18	—	*Bowers, H. (Act'g. Capt. 1 June 18)*	1 June 18	*Adjt., Luck. Gp. Garr. Arty., I. D. F.*
		Temporary Captains.		
	—	Slater, H.	31 Dec. 15	
		Temporary Lieutenants.		
	27Mn.	*Davies, F. L.*	3 Nov. 14	
	—	Elliott, F. A. B.	8 Oct. 15	
	52	Smallwood, E. S.	22 Feb. 16	Conv. Depot, Wellington.
	—	*Orr, G. B.*, M. C.	25 Feb. 16	
	7Mn.	Good, F. N.	25 Feb. 16	
	2Mn.	Church, B. H., M.C.	25 Feb. 16	
	—	Gee, T.	1 June 16	
	—	Kearton, A. N. M.	1 Oct. 16	
	—	*Jenkins, R. G. (temp. Capt.)*	1 Oct. 16	
	—	Balfour, W. M.	1 Oct. 16	
	15 *c.*	Gough, E. H.	1 Oct. 16	
	68 Hy.	Hardy, H. J.	1 Oct. 16	
	—	Ingoldby, E.	1 Feb. 17	
	73	Brooke, B. B.	1 July 17	
	—	Ellis, E. R.	1 July 17	
	69	Smith, E. W.	1 July 17	
	—	Thorpe, H. S.	1 July 17	*Lv., ex I., m.c., 6 m.*
	—	Organ, P. C.	1 July 17	
	69	Hawkins, A. S.	1 July 17	
	—	Tayler, H. F.	1 July 17	*Lv., ex I., m.c., 4 m.*
	—	*Walden, A. E.* (206)	1 July 17	
	—	Beveridge, A.	1 July 17	
	6Mn.	MacArthur, D. C.	1 July 17	
	—	Garnett, R. T.	1 July 17	*Lv., ex I., m. c., 6 m.*
	—	*Hodgins, J. N. M.*	*1 July 17*	*Guides Infy.*
	85	Davies, F. B.	1 July 17	
	64	Metcalfe, C.	1 July 17	
	—	Wilkinson, T. H. A.	1 July 17	
	22Mn.	Edwards, D.	1 July 17	
	—	Davies, N. D.	1 July 17	
	—	Townsend, F. C.	1 July 17	*Lv., m. c., 6 m.*
	—	Carr, J. F., M.C. (*act'g. Capt., 18 Sep. 18*).	1 July 17	
	22Mn.	Davies, O. G.	1 July 17	
		Temporary 2nd-Lieutenants.		
	—	Blake, L. F. M.	1 Oct. 14	
	104	Parsons, R. G.	4 Nov. 14	
	69	Collins, J. V.	10 Jan. 15	
	—	Draper, M.	27 Feb. 15	
	—	Hitchens, A. P.	27 Feb. 15	
	—	Hunter, W. M.	10 Mar. 15	
	85	Wishart, W. G. (57) (*temp. Capt.*)	26 Mar. 15	
		Special Reserve.		
		Captains.		
15 Aug. 14	*22Mn.*	Bowering, J. C.	3 Nov. 17	
	—	Pratt-Barlow, R. A.	3 Nov. 17	
30 Dec. 14	—	Clark, W. C.	3 Nov. 17	
30 Dec. 14	—	Lowson, J. McK. A.	3 Nov. 17	
		Lieutenants.		
	70	Carron, F. P. G. (*Act'g. Capt.*)	{ 3 Jun. 15 / 18 Jun. 16 }	
27 Jan. 15	75	McCammon, J. A.		
	—	Burchell-Crookes, W.	1 July 17	
	32Mn.	Lee, D. J. N. (206)	1 July 17	
	—	Jones, H. R.	1 July 17	
	—	Dickson, A. N.	1 July 17	
	74	Ellis, G. A.	1 July 17	
	75	Butcher, F. C.	1 July 17	
	2Mn.	Edge, A. B.	1 July 17	
	—	Mitchell, W. G.	1 July 17	
	64	Stephenson, G. T.	1 July 17	
	23Mn.	Tennant, W. R. (207*a*)	1 July 17	
	—	Howard, E.	1 July 17	
14 Dec. 15	—	Shaw, E.	1 July 17	
	—	Gissing, A. C.	1 July 17	
	51	Mann, J. R.	1 July 17	
	22Mn.	Kestin, C. A.	1 July 17	

ROYAL GARRISON ARTILLERY—*contd.*

Special Reserve.

2nd-Lieutenants.

First Commn.	Batty. or Coy.	Names.	Regimental Rank.	Remarks.
23 Feb. 15	—	Hilbourne, G. M.	23 Feb. 15	
26 May 15	—	Gray, A. H. (207a).	26 May 15	*Lv. ex I., m. c., 6 mos.*
8 June 15	—	Stevens, W. D.	8 June 15	
8 June 15	5Mn.	Peacock, A. D.	8 June 15	
31 Aug. 15	76	Bates, J. (Capt.)	31 Aug. 15	
3 Jan. 16	76	Fisher, G.	3 Jan. 16	
31 Aug. 16	68	Willan, H. C., M.C.	31 Aug. 16	
5 Oct. 16	74	Browne, F. G.	5 Oct. 16	
9 Nov. 16	76	Knight, A.	9 Nov. 16	
9 Nov. 16	73	Shallow, D. S.	9 Nov. 16	
11 Dec. 16	60	Entwistle, R.	11 Dec. 16	
20 Dec. 16	60	Theal, G.	20 Dec. 16	
20 Dec. 16	89	Smith, A. R.	20 Dec. 16	
3 Feb. 17	—	Weir, H. M., M.M.	3 Feb. 17	
26 Feb. 17	70	Ramsden, J. E.	26 Feb. 17	
1 Apl. 17	60	Murr, C.	1 Apl. 17	
9 Apl. 17	68	Douglas, W. G.	9 Apl. 17	
12 Mar. 17	60	Moore, A. H. W.	12 Mar. 17	
20 Apl. 17	60	Cockerton, G.	20 Apl. 17	
23 Apl. 17	70	Forsyth, J. A.	23 Apl. 17	
27 May 17	60	Munro, G. D.	27 May 17	
24 June 17	—	Shuttlewood, A. E.	24 June 17	
24 June 17	—	Meyer, S. S.	24 June 17	
24 June 17	60	Ridley, A. G.	24 June 17	
24 June 17	68	Bagg, H. A.	24 June 17	
24 June 17	—	Matthews, S. A.	24 June 17	
24 June 17	—	Carlill, C.	24 June 17	
30 June 17	—	Cowie, P. P.	30 June 17	
30 June 17	79	Barrett, L. J. S.	30 June 17	
30 June 17	60	Clark, G.	30 June 17	
22 July 17	76	Roe, F. F.	22 July 17	
31 July 17	—	Newell, E. L.	31 July 17	
1 Aug. 17	52	Fontain, P.	1 Aug. 17	
4 Aug. 17	60	Barber, W. H.	4 Aug. 17	
4 Aug. 17	—	Warwick, E. A.	4 Aug. 17	
16 Aug. 17	73	Withers, P. L.	16 Aug. 17	
19 Aug. 17	82	Scoones, E. H. C.	19 Aug. 17	
19 Aug. 17	64	Sykes, C.	19 Aug. 17	
19 Aug. 17	74	Gilbert, H.	19 Aug. 17	

ROYAL GARRISON ARTILLERY—*concld.*

Territorial Force.

Captains.

First Commn.	Batty. or Coy.	Names.	Regimental Rank.	Remarks.
. . . .	73	Hamilton, F. M.	. . . .	
. . . .	76	Mattey, A. G.	1 June 16	
. . . .	52	Macdonald, W. R. (*act'g. Maj., 1 Aug. 17*)	1 June 16	

Lieutenants.

First Commn.	Batty. or Coy.	Names.	Regimental Rank.	Remarks.
. . . .	70	Thornback, A. H.	29 Aug. 14	
. . . .	—	Oakley, G. C.	1 Jan. 16	
. . . .	—	Downs, W. S. (*act'g. Capt., 9 June 17*)	1 June 16	*Lv., ex I., m.c., 6 m.*
. . . .	101	Colbeck, H. K.	1 June 16	
. . . .	60	Walker, G. W. H.	1 June 16	
. . . .	75	Boyack, J. M.	1 June 16	
. . . .	—	Boyd, A.	1 June 16	No. 4 (spl.) 10 pr. Bullock Batty., R. G. A.
. . . .	75	Gill, H.	1 June 16[2]	
. . . .	—	MacIver, N. (*act'g. Capt., 11 Aug. 17*)	1 July 17	Adjt., I. Br. Mtn. Arty. Bde.
. . . .	6 Mn.	Gaskill, T.	8 July 17	

2nd-Lieutenants.

First Commn.	Batty. or Coy.	Names.	Regimental Rank.	Remarks.
2 July 16	61	Morgan, H. (*act'g. Capt., 4 July 18*) .	2 July 16	
20 Feb. 17	101	Beck, W. G.	20 Feb. 17	
23 Apl. 17	1 Mn.	Hunter, R. J. L.	23 Apl. 17	
28 May 17	1 Mn.	Jenkins, A. J.	28 May 17	

THE ROYAL (INDIAN) ARTILLERY.

NOTE.—The names shown in *italics* are those of Officers transferred to the Unemployed Supernumerary List.

Names.	Date of Birth.	2nd-Lieut.	Lieut.	Captain.	Major.	Lieut.-Col.	Colonel.	Major-Genl.	Lieut.-Genl.	General.	Colonel Comdt.
General.											
Chapman, Sir Edward Francis, K.C.B., *p.s.c.* (*B.*)®	14 Nov. 40	12 June 58	27 Aug. 58	21 Jan. 72	1 Dec. 72	2 Nov. 79	31 Dec. 81	20 Sep. 89	3 Dec. 92	15 Mar. 96	14 Jan. 03
Lieutenant-General.											
Major-Generals.											
Elliott, Edward Draper (*B.*)	30 Sep. 38	11 Dec. 57	27 July 58	29 May 66	16 Jan. 75	16 Jan. 82	16 Jan. 86	1 Apl. 95	. . .	. . .	4 Sep. 01
Ward, Francis William, C.B. (*B.*)	6 Apl. 40	11 Dec. 57	27 Aug. 58	5 Oct. 66	16 Jan. 75	16 Jan. 82	16 Jan. 86	25 May 95	. . .	. . .	1 May 02
Saward, Michael Henry (*B.*)	22 Dec. 40	. . .	9 Dec. 59	11 Dec. 72	26 May 80	28 July 85	28 July 89	23 June 96	. . .	. . .	22 Feb. 05

DISTRIBUTION OF THE ROYAL REGIMENT OF ARTILLERY IN INDIA.

Royal Horse Artillery.

Battery.	Brigade.	Station.	Date of arrival in India.	Remarks.
A	I	. . . .	4 Jan. 13	
M	VIII	. . . .	12 Dec. 13	
N	IX	. . . .	9 Feb. 06	
Q	VIII	. . . .	12 Dec. 13	
S	IX	. . . .	9 Feb. 06	
U	XI	. . . .	25 Oct. 10	
V	XII	. . . .	4 Nov. 07	
W	XII	. . . .	4 Nov. 07	
X	XIII	. . . .	9 Dec. 09	
"A"	Amn. Col.	. . . .	. . . . .	
"B"	Amn. Col.	. . . .	. . . . .	
"C"	Amn. Col.	. . . .	. . . . .	
"F"	Amn. Col.	. . . .	. . . . .	
"G"	Amn. Col.	. . . .	. . . . .	
"H"	Amn. Col.	. . . .	. . . . .	

Royal Field Artillery.

Battery.	Brigade.	Station.	Date of arrival in India.	Remarks.
2nd	XIII	. . . .	11 Nov. 06	
4th	VII	. . . .	8 Feb. 04	
7th	IV	. . . .	28 Oct. 02	
8th	XIII	. . . .	11 Nov. 06	
14th	IV	. . . .	28 Oct. 02	
19th	IX	. . . .	9 Feb. 05	
20th	IX	. . . .	9 Feb. 05	
28th	IX	. . . .	9 Feb. 05	
38th	VII	. . . .	8 Feb. 04	
44th	XIII	. . . .	11 Nov. 06	
59th	XVIII	. . . .	24 Nov. 10	
63rd	X	. . . .	29 Oct. 02	
64th	V	. . . .	29 Oct. 02	
66th	IV	. . . .	29 Oct. 02	
73rd	V	. . . .	29 Oct. 02	
74th	XVI	. . . .	4 Nov. 02	
76th	X	. . . .	9 Feb. 06	
77th	VII	. . . .	4 Nov. 02	
78th	VII	. . . .	8 Feb. 04	
79th	VI	. . . .	4 Nov. 02	
81st	V	. . . .	9 Feb. 06	
82nd	X	. . . .	9 Feb. 06	
83rd	XI	. . . .	10 Nov. 06	
84th	XI	. . . .	10 Nov. 06	
85th	XI	. . . .	10 Nov. 06	
89th	XVI	. . . .	Nov. 08	
90th	XVI	. . . .	Nov. 08	
91st	XVI	. . . .	Nov. 08	
93rd	XVIII	. . . .	24 Nov. 10	
94th	XVIII	. . . .	24 Nov. 10	
101st	XXI	. . . .	12 Dec. 13	
102nd	XXI	. . . .	12 Dec. 13	
*428th	. . . . .	. . . .	1 Nov. 94	
*429th	. . . . .	. . . .	14 Oct. 04	
1st	Amn. Col.	. . . .	. . . . .	
2nd	Amn. Col.	. . . .	. . . . .	
3rd	Amn. Col.	. . . .	. . . . .	
4th	Amn. Col.	. . . .	. . . . .	
6th	Amn. Col.	. . . .	. . . . .	
7th	Amn. Col.	. . . .	. . . . .	
8th	Amn. Col.	. . . .	. . . . .	
9th	Amn. Col.	. . . .	. . . . .	
10th	Amn. Col.	. . . .	. . . . .	
11th	Amn. Col.	. . . .	. . . . .	
12th	Amn. Col.	. . . .	. . . . .	

* Transferred from R. G. A.—I. A. O. 599 of 1918.

Battery.	Brigade.	Station.	Date of arrival in India.	Remarks.

DISTRIBUTION OF THE ROYAL REGIMENT OF ARTILLERY IN INDIA—*concld.*

Royal Garrison Artillery (Mountain Division).

Battery or Company.	Brigade.	*Equipment.	Station.	Date of arrival in India.	Remarks.
No. 1	II		. . . .	6 Nov. 83	
„ 2	III		. . . .	28 Jan. 86	
„ 3	IV		. . . .	8 Dec. 78	
„ 4	IV		. . . .	. . . .	
„ 5	III		. . . .	7 Dec. 78	
„ 6	I		. . . .	9 Nov. 85	
„ 7	III		. . . .	4 Mar. 86	
„ 8	I		. . . .	20 Oct. 87	
„ 9	II		. . . .	21 Oct. 84	

Royal Garrison Artillery.

Battery or Company.	Brigade.	*Equipment.	Station.	Date of arrival in India.	Remarks.
51			. . . .	30 Jan. 99	
52			. . . .	21 Jan. 02	
60		H.	. . . .	8 Mar. 02	
61			. . . .	8 Oct. 98	
62			. . . .	6 Jan. 05	
64			. . . .	6 Jan. 05	
68		H.	. . . .	5 Feb. 04	
69			. . . .	2 Oct. 94	
70			. . . .	1 Nov. 94	
73			. . . .	1 Nov. 94	
74			. . . .	Feb. 03	
75			. . . .	2 Dec. 96	
76			. . . .	29 Jan. 01	
79			. . . .	Nov. 01	
82			. . . .	12 Feb. 02	
85			. . . .	Mar. 02	
86		H.	. . . .	7 Nov. 02	
94			. . . .	7 Nov. 02	
101			. . . .	5 Feb. 04	
104		H.	. . . .	Nov. 01	

* *Equipment*—H.—Heavy.

1st (Calcutta Port Defence) Group, Garrison Artillery.

Formed 21st Sep. 1883. Called Calcutta Naval Volunteers, G. G. O. 964 of 15th Nov. 1889.

Reorganised and given present designation by Mily. Dept. Notfn. No. 1121, dated 13th October 1899. Reconstituted as an I. D. F. Unit with present designation, 1st Apl. 1917. Establishment—1 Active and 1 Res. Coy., 1 Coy. Elec. Engrs.

HONORARY COMMANDANT—*H. E. the Rt. Hon.* Lawrence John Lumley Dundas, *Earl of* Ronaldshay G.C.I.E. . . . . *(Governor of Bengal).* 1 Apl. 17.

First Commn.	Names.	Date of rank.	Date of appointment to corps.	Remarks.
	Lieutenant-Colonel.			
1 Apl. 17	Constable, Ernest Algernon, A.D.C.	1 Apl. 17	1 Apl. 17	Commdt. Hon. Col. late Ind. Volrs.
	Majors.			
1 Apl. 17	*Taylor, John*	*1 Apl. 17*	*1 Apl. 17*	*Lv., Ind., m.c.*
1 Apl. 17	*Dods, Alexander Waddell*	*1 Apl. 17*	*1 Apl. 17*	*Tempy Capt. in Army.*
1 Apl. 17	*Robertson, James Frank (D.I., 82)*	*1 Apl. 17*	*1 Apl. 17*	*Lv., ex Ind., m.c.*
1 Apl. 17	*MacGregor, Alexander*	*1 Apl. 17*	*1 Apl. 17*	*I. A. R. O.*
1 Apl. 17	Ross, Norman McLeod	1 Apl. 17	1 Apl. 17	
1 Apl. 17	*Morgan, Charles Carey*	*1 Apl. 17*	*1 Apl. 17*	*Tempy. Lt. R.F.A.*
1 Apl. 17	Child, Robert Harold	1 Apl. 17	1 Apl. 17	
	Captains.			
1 Apl. 17	*Vine, Horace*	*1 Apl. 17*	*1 Apl. 17*	*Tempy. Comr. R. N.*
1 Apl. 17	Carter, Alexander John (82)	1 Apl. 17	1 Apl. 17	
1 Apl. 17	Hamlyn, William Thomas (82)	1 Apl. 17	1 Apl. 17	
1 Apl. 17	*Cochran, Alexander,* O.B.E.	*1 Apl. 17*	*1 Apl. 17*	
1 Apl. 17	*Hosie, Alexander Lindsay* (82) (54)	*1 Apl. 17*	*1 Apl. 17*	*I. A. R. O.*
1 Apl. 17	*Batty, Archibald Douglas George Staunton*	*1 Apl. 17*	*1 Apl. 17*	*I. A. R. O.*
	Lieutenants.			
1 Apl. 17	*Munro, Thomas Mathew,* V.D.	*1 Apl. 17*	*1 Apl. 17*	*Tempy. Comr. R. N.*
1 Apl. 17	Chaitres, Charles Borthwick	1 Apl. 17	1 Apl. 17	
1 Apl. 17	Glover, William Morton (82)	1 Apl. 17	1 Apl. 17	
1 Apl. 17	Diggory, Samuel Luther	1 Apl. 17	1 Apl. 17	
1 Apl. 17	Towler, Reginald Alfred Gottlieb (82)	1 Apl. 17	1 Apl. 17	
	2nd-Lieutenants.			
1 Apl. 17	Bowrey, Ernest William Cantley	1 Apl. 17	1 Apl. 17	
1 Apl. 17	Langford-James, Reginald Hugh Lloyd	1 Apl. 17	1 Apl. 17	
1 Apl. 17	Coulthard, John Robert	1 Apl. 17	1 Apl. 17	
26 Apl. 18	Brierley, William Richard Cecil	26 Apl. 18	26 Apl. 18	
	Supernumerary.			
1 Apl. 17	Wyness, Maj. James Peter, V.D. (82)	1 Apl. 17	1 Apl. 17	
1 Apl. 17	Dickson, Maj. Raymond	1 Apl. 17	1 Apl. 17	
1 Apl. 17	MacIver, Maj. Robert Graves (82) (54)	1 Apl. 17	1 Apl. 17	
1 Apl. 17	Stansbury, 2nd-Lt. Vivian	1 Apl. 17	1 Apl. 17	

No. 1 Electrical Engineer Company.

First Commn.	Names.	Date of rank.	Date of appointment to corps.	Remarks.
	Major.			
1 Apl. 17	Stovold, Herbert William	1 Apl. 17	1 Apl. 17	
	Lieutenant.			
1 Apl. 17	Luke, Norman Ritchie	1 Apl. 17	1 Apl. 17	
	Supernumerary.			
1 Apl. 17	Crawford, Lt. Maurice Newenham	1 Apl. 17	1 Apl. 17	
	Staff.			
18 Oct. 15	Mooien, Lt. (Tempy. Capt.) L.G., D.C.M., R.F.A.	1 July 17	18 Oct. 15	Adjutant.
1 Apl. 17	Newbery, Revd. Charles Alfred	1 Apl. 17	1 Apl. 17	Hony. Chaplain.
1 Apl. 17	Allan, Capt. William Smith, M.B.	1 Apl. 17	1 Apl. 17	Med. Offr.

Uniform—Blue.
Working dress—Khaki drill.

2nd (Madras) Group Garrison Artillery. "The Duke's Own."

Raised, 11th March 1879, and reconstituted as an I. D. F. with present designation 1st Apl. 1917.

Establishment—*1 Active, 1 Res. and 1 Combined Active & Res. Coy., 1 Coy, Elec. Engrs.*

HONORARY COMMANDANT—*H. E. The Rt. Hon.* JOHN, *Baron* PENTLAND of LYTH, P.C., G.C.S.I., G.C.I.E. (*Governor of Madras.*)

1 Apl. 1917

First Commn.	Names.	Date of rank.	Date of appointment to corps.	Remarks.
	Lieutenant-Colonel.			
1 Apl. 17	*Mitchell, Hugh Henry Gordon*, A.D.C., (*54, 80, 82*)	*1 Apl. 17*	*1 Apl. 17*	Comdt. (Hon. Col. late Ind. Vol'rs.) (*Lv., ex* ***Ind.***, *12 mos.*)
	Majors.			
1 Apl. 17	Wright, William Owen (80)	1 Apl 17	1 Apl. 17	Comdt. (*Offg.*)
1 Apl. 17	*Cunliffe, John Bertram*, C.I.E., (54)	*1 Apl. 17*	*1 Apl. 17*	(*Temp. Maj., R. A.*)
1 Apl. 17	*Payne, Charles Augustus William* (80,82)	*1 Apl. 17*	*1 Apl. 17*	*Adjutant.*
1 Apl. 17	Eccles, Percy (80)	1 Apl. 17	1 Apl. 17	
1 Apl. 17	Reynolds, Bernard Oscar	18 Dec. 17	1 Apl. 17	
	Captains.			
1 Apl. 17	*Ayling, John Charles* (54, 80, 82)	*1 Apl. 17*	*1 Apl. 17*	*Army Signal Serv.*
1 Apl. 17	*Mellor, Robert Wood* (82)	*1 Apl. 17*	*1 Apl. 17*	*Employed under Munitions Board.*
1 Apl. 17	*Robertson, Henry Stewart* (80)	*1 Apl. 17*	*1 Apl. 17*	
1 Apl. 17	Hignett, Horace Arthur DuCane (80)	1 Apl. 17	1 Apl. 17	
1 Apl. 17	Bennett, Arthur Ernest	18 Dec. 17	21 Nov. 17	
1 Apl. 17	Buckney, Frank (82)	13 June 18	1 Oct. 18	
	Lieutenants.			
1 Apl. 17	McKean, Hailey (80)	1 Apl 17	1 Apl. 17	
1 Apl. 17	Macaulay, Frederic Gilbert Aulay (80)	1 Apl. 17	1 Apl. 17	
1 Apl. 17	*Sewell, Sydney Ewart*	*1 Apl. 17*	*1 Apl. 17*	(*Temp. Capt. in Army, 29 May 16*).
1 Apl. 17	*Todd, Alfred Sydney*	*1 Apl. 17*	*1 Apl. 17*	(*Temp. Lt. in Army, 20 Sep. 15*) *Ind. Ord. Dept.*
1 Apl. 17	*Birch, Alexander Russell* (80)	*1 Apl. 17*	*1 Apl. 17*	
1 Apl. 17	*Carrick, Robert Black*	*1 Apl. 17*	*1 Apl. 17*	(*Temp. Lt. in Army.*)
1 Apl. 17	*Guthrie, Alan*	*1 Apl. 17*	*1 Apl. 17*	(*Temp. Capt. in Army, 11 Aug. 15*).
1 Apl. 17	*Bracken, Geoffrey Thomas Hirst* (193)	*1 Apl. 17*	*1 Apl. 17*	(*Temp. 2nd-Lt. in Army.*)
1 Apl. 17	Colledge, John Edward (8.)	1 Apl. 17	1 Apl. 17	
1 Apl. 17	Jackson, Philip Sidney Edward (80)	1 Apl. 17	1 Apl. 17	O. C. Elec. Eng. Coy.
1 Apl. 17	*Huggins, Gerald Francis* (80)	*1 Apl. 17*	*1 Apl. 17*	(*Temp. Lt. in Army.*)
1 Apl. 17	*Rose, Ernest Reyment* (80)	*1 Apl. 17*	*1 Apl. 17*	(*Temp. Lt. in Army.*)
1 Apl. 17	Campbell, Duncan Donald	20 Jan. 18	20 Jan. 18	
1 Jan. 18	Todd, Stuart Ernest Rubie	20 Jan. 18	1 Jan. 18	
	2nd-Lieutenants.			
1 Apl. 17	Swan, James Stewart (80)	1 Apl. 17	1 Apl. 17	
1 Apl. 17	*Denniston, Robert Dale*	*1 Apl. 17*	*1 Apl. 17*	(*Temp. Lt. in Army.*)
1 Apl. 17	Wilkie, Harold Douglas (80)	1 Apl. 17	1 Apl. 17	
1 Apl. 17	Walker, George Keith (52)	1 Apl. 17	1 Apl. 7	
9 Apl. 18	Lavender, George Alexander	9 Apl. 18	9 Apl. 18	
1 Apl. 17	*Farquharson, John Robert*	*1 Apl. 17*	*1 Apl. 17*	
6 Aug. 18	Fraser, Ian Simon	6 Aug. 18	6 Aug. 18	
16 Oct. 18	Cotton, William Francis	16 Oct. 18	16 Oct. 18	
	Supernumerary.			
1 Apl. 17	Cole, Lt. Claud Willoughby	18 Dec. 17	20 Jan. 18	
1 Apl. 17	Pelly, 2nd-Lt Charles Brent Neville	1 Apl. 17	1 Apl. 17	
1 July 17	Reiner, 2nd-Lt. George Veitch	1 July 17	1 July 17	
1 July 17	Green, 2nd-Lt. Reginald Kellocksly	1 July 17	1 July 17	
11 Aug. 17	Steuart, 2nd-Lt. Alan	11 Aug. 17	11 Aug. 17	
	Staff.			
1 Apl. 17	Payne, Maj. C. A. W. (82)	1 Apl. 17	1 Apl. 17	Adjutant.
. . . .	Giles, Rev. C., M.A.	. . . .	. . . .	Hon. Chaplain (C. of E.).
. . . .	Aelen, Rt. Rev. Doctor John	. . . .	. . . .	Ditto (R. C.)

3rd (Rangoon Port Defence) Group, Garrison Artillery.

Raised, 30th September 1892. Reorganised by G. G. O. 442 of 19th May 1905 and reconstituted as an I. D. F. Unit with present designation, 1st Apl. 1917.

Establishment—1 Active Coy., 2 Combined Active & Reserve Coys., 1 Cadet Coy. 1 Coy. Elec. Engrs.

HONORARY COMMANDANT—

First Commn.	Names.	Date of rank.	Date of appointment to corps.	Remarks.
	Artillery Companies. *Raised 20th October 1879.* *Uniform*—White. Working dress—Khaki.			
	Lieutenant-Colonel.			
1 Apl. 17	Evans, George Henry, C.I.E., A.D.C., C.V.D. (82, 105, 206)	1 Apl. 17	1 Apl. 17	Comdt. (Hon. Col. Late, Ind. Volrs.)
	Majors.			
1 Apl. 17	*Anderson, Alexander James* (82)	*1 Apl. 17*	*1 Apl. 17*	*Temp. Maj. in Army, 11 Aug. 15.*
1 Apl. 17	*Rennie, Thomas* (82)	*1 Apl. 17*	*1 Apl. 17*	*With 75 Coy. R.G.A.*
1 Apl. 17	*Smith, Algernon Haslehurst* (82)	*1 Apl. 17*	*1 Apl. 17*	*Temp. Capt. R.A.*
1 Apl. 17	*Mossop, John Craig Christie*	*1 Apl. 17*	*1 Apl. 17*	*I.A.R.O.*
1 Apl. 17	*Pullar, Edgar John* (82)	*1 Apl. 17*	*1 Apl. 17*	*I.A.R.O.*
1 Apl. 17	Brigstocke, Arthur Montagu (82)	1 Apl. 17	1 Apl. 17	
1 Apl. 17	James, Douglas Hamlyn	1 Apl. 17	1 Apl. 17	
1 Apl. 17	Trench, Robert Hamilton	13 June 18	1 Apl. 17	
	Captains.			
1 Apl. 17	*Donald, Colin Dunlop*	*1 Apl. 17*	*1 Apl. 17*	*I.A.R.O.*
1 Apl. 17	*Shipster, George Cecil*	*1 Apl. 17*	*1 Apl. 17*	
1 Apl. 17	*Keeling, Edward Herbert*	*1 Apl. 17*	*1 Apl. 17*	*I.A.R.O.*
1 Apl. 17	*Hicks, Mathew Bede*	*1 Apl. 17*	*1 Apl. 17*	*Temp. Capt. R. F. A. (Vol. Arty. By.)*
1 Apl. 17	MacDonald, Stuart (82a)	1 Apl. 17	1 Apl. 17	
1 Apl. 17	Boyd, Archibald Edward	1 Apl. 17	14 Aug. 18	
1 Apl. 17	*Dunk, Edgar Frank*	*13 June 18*	*1 Apl. 17*	*With R. G. A.*
	Lieutenants.			
1 Apl. 17	*Casson, Randal*	*1 Apl. 17*	*1 Apl. 17*	*I.A.R.O.*
1 Apl. 17	*Doyle, John Patric*	*1 Apl. 17*	*1 Apl. 17*	*I.A.R.O. Attd., 22 Mt. By.*
1 Apl. 17	*Davern, Francis Beta*	*1 Apl. 17*	*1 Apl. 17*	*Temp. Lt. in Army, 1 July 17.*
1 Apl. 17	Goodwillie, David	13 June 18	1 Apl. 17	
1 Apl. 17	McGregor, Frederick Hugh	13 June 18	1 Apl. 17	
1 Apl. 17	*Hepburn, William Smith*	*13 June 18*	*1 Apl. 17*	*I. A. R. O.*
	2nd-Lieutenants.			
1 Apl. 17	*Rees, Owen Meredith*	*1 Apl. 17*	*1 Apl. 17*	*Acting Capt. in Army, 30 Sep. 16.*
1 Apl. 17	*Brakspear, Cecil Hayward*	*1 Apl. 17*	*1 Apl. 17*	*Temp. Lt. R. F. A.*
1 Apl. 17	Simpson, Raymond Muir	1 Apl. 17	1 Apl. 17	
1 Apl. 17	McCreath, Alexander Thomson	1 Apl. 17	1 Apl. 17	
1 Apl. 17	Tew, Gilbert Charles	1 Apl. 17	1 Apl. 17	
1 Sep. 18	Compton, George Gabriel	1 Sep. 18	1 Sep. 18	
1 Oct. 18	Dry, Roy	1 Oct. 18	1 Oct. 18	
1 Oct. 18	Goss, John Louis Leo	1 Oct. 18	1 Oct. 18	
1 Oct. 18	Fraser, Alexander	1 Oct. 18	1 Oct. 18	
31 Oct. 18	Robertson, Charles	31 Oct. 18	31 Oct. 18	
	No. 2 Electrical Engineer Company. *Raised, 1st April 1902.* *Uniform*—Khaki.			
	Major.			
1 Apl. 17	Cameron, Hugh Porteous	1 Apl. 17	1 Apl. 17	
	Lieutenants.			
	Dickie Robert Spiers			
1 Apl. 17	*Lynn, Guy*	*1 Apl. 17*	*1 Apl. 17*	*I.A.R.O.*
	2nd-Lieutenants.			
1 Apl. 17	*Bruce, Edwin*	*1 Apl. 17*	*1 Apl. 17*	*I.A.R.O.*
1 Apl. 17	Jones, Alfred Edward	1 Apl. 17	1 Apl. 17	
1 Apl. 17	Wilson, George Purdie	1 Apl. 17	1 Apl. 17	
1 Apl. 17	Rintoul, Joseph Philip	1 Apl. 17	1 Apl. 17	
1 July 18	Wigg, Ernest Charles	1 July 18	1 July 18	
3 July 18	McNeil, John	3 July 18	3 July 18	
	Staff.			
10 Oct. 11	Casson, Temp. Capt. R.	1 Apl. 15	16 July 15	I.A.R.O. Tempy. Adjutant.
10 Mar. 18	Bishop, Lt. G. J.	10 Mar. 18	10 Mar. 18	Qr.-Mr.

4th (Bombay) Group Garrison Artillery.

Uniform—Khaki drill.

Formed 6th June 1887, and reconstituted as an I. D. F. Unit with present designation 1st Apl. 17.

Establishment—2 Active Coys. 1 Res. Coy.

Cos. at Bombay.

HONORARY COMMANDANT—*H. E. the Rt. Hon.* FREEMAN FREEMAN-THOMAS, *Baron* WILLINGDON O RATTON, G.C.S.I., G.C.I.E., G.B.E. . . . . . 1 Apl. 17. (*Governor of Bombay.*)

First Comm".	Names.	Date of rank.	Date of appointment to corps.	Remarks.
	Lieutenant-Colonel.			
1 Apl. 17	Milne, John Sidney Wardlaw	1 Apl. 17	1 Apl. 17	Commdt.
	Majors.			
1 Apl. 17	Pennington, Robert William Rudall, V.D.	1 Apl. 17	1 Apl. 17	Offg. Comdt.
1 Apl. 17	Sowerby, Maurice Herbert (79)	1 Apl. 17	1 Apl. 17	
1 Apl. 17	*Retallack, Percy Hamilton* (12), (79)	*1 Apl. 17*	*1 Apl. 17*	Adjt.
1 Apl. 17	Starkey, Harold Augustus (79)	1 Apl. 17	1 Apl. 17	
	Captains.			
1 Apl. 17	Royal, Harry	1 Apl. 17	1 Apl. 17	
1 Apl. 17	Green, Allan Michael	1 Apl. 17	1 Apl. 17	*Lv., ex. I., 9 mos.*
1 Apl. 17	Murray, George William	1 Apl. 17	1 Apl. 17	
	Lieutenants.			
1 Apl. 17	Dunkerley, William	1 Apl. 17	1 Apl. 17	
1 Apl. 17	Cartland, Wilfred Neville	1 Apl. 17	1 Apl. 17	
31 July 18	Wilkinson, Noel	31 July 18	31 July 18	
31 July 18	Jones, William Ellis	31 July 18	31 July 18	
1 Apl. 17	Weber, Henry Guy	1 Apl. 17	1 Apl. 17	
	2nd-Lieutenants.			
31 July 18	Walker, John Charles	31 July 18	31 July 18	
31 July 18	Hamley, Herbert Russell	31 July 18	31 July 18	
31 July 18	Boyagis, George	31 July 18	31 July 18	
31 July 18	Grantham, Vincent Alpe	31 July 18	31 July 18	
	Supernumerary.			
29 Oct. 17	Hayward, Lt.-Col. Maurice Henry Weston	29 Oct. 17		
1 Apl. 17	Simcox, Maj. Arthur Henry Addenbrooke	1 Apl. 17	1 Apl. 17	
	Staff.			
9 Mar. 18	Davey, Lt., H. B.	9 Mar. 18	9 Mar. 18	Qr.-Mr.

No. 3 Company Electrical Engineers.

Constituted as a separate unit (of one Reserve Company) of the I. D. F. from 1st April 1918 under I. A. O. 376 of 1918.

First Comm".	Names.	Date of rank.	Date of appointment to corps.	Remarks.
	Major.			
1 Apl. 17	Taylor, Newman	1 Apl. 17	1 Apl. 17	Comdg.
	Lieutenants.			
3 Aug. 18	Lewis, Thomas William Lutwyche	3 Aug. 18	3 Aug. 18	
3 Aug. 18	Maughan, Nicholas	3 Aug. 18	3 Aug. 18	

F (Kirkee) Battery Mobile Artillery.

Establishment—1 Combined Active and Reserve Batty.

First Commn.	Names.	Date of rank.	Date of appointment to corps.	Remarks.
	Major.			
1 Apl. 17	Fox, Charles James John	1 Apl. 17	1 Apl. 17	Commdt.
	Captain.			
1 Apl. 17	Pratt, Hubert Evans Hills	1 Apl. 17	1 Apl. 17	
	Lieutenants.			
16 Apl. 18	Oddin Taylor, Harry Willoughby	16 Apl. 18	16 Apl. 18	
16 Apl. 18	Freke, Cecil George	16 Apl. 18	16 Apl. 18	
	Supernumerary.			
1 Apl. 17	Amor, Maj., Walter	1 Apl. 17	1 Apl. 17	
	Attached.			
17 Nov. 17	Coombs, Lt. Ernest Edwin	21 May 18	. . . .	35th Poona Bn.

Uniform—Full dress—Blue serge or cloth.
Working dress—Khaki drill.

1st (Cossipore) Brigade, Mobile Artillery.

Formed 1st February 1884. Reorganised 7th September 1888. Reconstituted, G. G. O. 437 of 1898. Reconstituted as an I. D. F. Unit with present designation, 1st Apl. 1917.

Establishment—2 Active and 2 Res. Battys.

HONORARY COMMANDANT—*His Excellency the Right Hon'ble* LAWRENCE JOHN LUMLEY DUNDAS, *Earl of* Ronaldshay, G.C.I.E. . . . . 1 May 17.

(*Governor of Bengal.*)

First Comm.	Names.	Date of rank.	Date of appointment to corps	Remarks.
	Lieutenant-Colonel.			
1 Apl. 17	Tyrie, David Alexander, V.D. (82)	1 Apl. 17	1 Apl. 17	Comdt., Hon. A.-D.-C. to Govr., Bengal. (Hon. Col., late Ind. Volrs.)
	Majors.			
1 Apl. 17	Ashworth, Sidney Herbert, V.D. (82)	1 Apl. 17	1 Apl. 17	
1 Apl. 17	Ferguson, William Alexander, V.D.	1 Apl. 17	1 Apl. 17	
1 Apl. 17	Mudie, David, V.D.	1 Apl. 17	1 Apl. 17	
1 Apl. 17	*Thoms, Robert Shand (54)*	*1 Apl. 17*	*1 Apl. 17*	*I. A. R. O.*
1 Apl. 17	*Kinloch, Graham*	*1 Apl. 17*	*1 Apl. 17*	*Temp. Maj. in Army, 20 Sep. 14.)*
1 Apl. 17	*Watt, William Douglas Bruce*	*1 Apl. 17*	*1 Apl. 17*	*I. A. R. O.*
1 Apl. 17	*Simson, Alaric*	*1 Apl. 17*	*1 Apl. 17*	*(Temp. Capt., R. G.A.)*
1 Apl. 17	Tyndall, William Wilson	1 Apl. 17	1 Apl. 17	
	Captains.			
1 Apl. 17	Abbott, Ernest George (54)	1 Apl. 17	1 Apl. 17	
1 Apl. 17	*Will, Robert Ross*	*1 Apl. 17*	*1 Apl. 17*	*2nd High. Bde., R.F.A., (T. F.)*
1 Apl. 17	Thoms, Alexander Keay	1 Apl. 17	1 Apl. 17	
1 Apl. 17	*Rose, Percy Thomas*	*1 Apl. 17*	*1 Apl. 17*	*I. A. R. O.*
1 Apl. 17	*Banks, Harry Kinnear* (82)	*1 Apl. 17*	*1 Apl. 17*	*(2nd-Lt., R.F.A., T. F.)*
1 Apl. 17	Crawford, John (54, 82)	1 Apl. 17	1 Apl. 17	
1 Apl. 17	Burns, James Charles Hamilton (54)	1 Apl. 17	1 Apl. 17	
	Lieutenants.			
1 Apl. 17	*Martin, Frank Reddaway*	*1 Apl. 17*	*1 Apl. 17*	*(Temp. Capt. in Army, 7 July 16.)*
1 Apl. 17	Whyte, David Edward (54)	1 Apl. 17	1 Apl. 17	
1 Apl. 17	*Darby, Bailey Gordon*	*1 Apl. 17*	*1 Apl. 17*	*(2nd-Lt., R.F.A., T.F.)*
1 Apl. 17	*Spence, Robert Maxwell*	*1 Apl. 17*	*1 Apl. 17*	*(Temp. Lt. in Army, 11 Aug. 15.)*
1 Apl. 17	Douglas, Thomas (82)	1 Apl. 17	1 Apl. 17	
1 Apl. 17	Scott, Sydney Keith	1 Apl. 17	1 Apl. 17	
1 Apl. 17	*Donald, Andrew*	*1 Apl. 17*	*1 Apl. 17*	*I. A. R. O.*
1 Apl. 17	Williams, Gomer Silvanus (82)	1 Apl. 17	1 Apl. 17	
1 Apl. 17	Phillips, Francis Edward	1 Apl. 17	1 Apl. 17	
1 Apl. 17	Lamb, Thomas	1 Apl. 17	1 Apl. 17	
1 Apl. 17	Lawson, David	1 Apl. 17	1 Apl. 17	
1 Apl. 17	Martin, John George	1 Apl. 17	1 Apl. 17	
1 Apl. 17	Paterson, James David	1 Apl. 17	1 Apl. 17	
1 Apl. 17	Prentice, James	1 Apl. 17	1 Apl. 17	*Lv. ex I., m.c.*
1 Apl. 17	Hall, David Kinnear	1 Apl. 17	1 Apl. 17	*Lv. ex. I., m.c.*
1 Apl. 17	Laird, Robert Balfour	1 Apl. 17	1 Apl. 17	
	2nd-Lieutenants.			
1 Apl. 17	Gray, Norris Rupert	1 Apl. 17	1 Apl. 17	
1 Apl. 17	Murray, John Alexander	1 Apl. 17	1 Apl. 17	
1 Mar. 18	Gray, Seymour Mowbray	1 Mar. 18	1 Mar. 18	
	Supernumerary.			
1 Apl. 17	Tosh, Lt.-Col. Charles William, V.D.	1 Apl. 17	1 Apl. 17	
	Staff.			
1 Jan. 16	Toohey, Capt. John	1 Jan. 16	1 Jan. 16	I. A. R. O., Adjt.
6 Mar. 18	Baker, Lt. Frank	16 Mar. 18	16 Mar. 18	Qr.-Mr.

2nd (Karachi) Brigade, Mobile Artillery.

Uniform—Khaki drill.

Formed 25th October 1892. Reorganised, G. O. C. C. 327 of 29th April 1901. Reconstituted as an I. D. F. Unit with present designation, 1st Apl. 1917.

Establishment—1 Combined Active & Reserve Batty. 1 Coy. Elec. Engrs.

HONORARY COMMANDANT—*Hon'ble* MR. HENRY STAVELEY LAWRENCE, C.S.I., I.C.S. 1st Apl. 1917 (*Commr. in Sind.*)

First Commn.	Names.	Date of rank.	Date of appointment to corps.	Remarks.
	Lieutenant-Colonel.			
1 Apl. 17	Sawyer, Henry Hollington (80) . .	1 Apl. 17	22 Dec. 17	
	Major.			
	Captain.			
1 Apl. 17	*Humphrey, John*	*1 Apl. 17*	*1 Apl. 17*	*I. A. R. O.*
1 Apl. 17	*Beaumont, Thomas Cuthbert*	*1 Apl. 17*	*1 Apl. 17*	*With No. 1 British Mn. Arty. Bde.*
1 Apl. 17	Pratt, Hubert Evans Hills	1 Apl. 17	10 July 18	Lv., 5 mos. fm. 31 Oct. 18 to 31 March 19.
1 Apl. 17	Pearson, Ernest Archbell	2 Oct. 18	1 Apl. 17	
	Lieutenant.			
	2nd-Lieutenant.			
1 Apl. 17	*Turton, Oliver*	*1 Apl. 17*	*1 Apl. 17*	*Tempy. 2nd-Lt., R. F. A.*

Staff.

No. 4 Electrical Engineer Company.

Formed 1st April 1902.

First Commn.	Names.	Date of rank.	Date of appointment to corps.	Remarks.
	Captain.			
1 Apl. 17	Backhouse, Robert Salmon	1 Apl. 17	1 Apl. 17	
	Lieutenant.			
1 Apl. 17	Shute, Arthur Fitzgerald	1 Apl. 17	1 Apl. 17	

Uniform—Officers—Khaki drill.

5th (Lucknow) Group Garrison Artillery.

Establishment—2 Combined Active & Reserve Coys. 1 Armoured Motor Batty.

(*Formed 10th Jan. 1872. Reorganised from Oudh Volunteer Rifles by M. D. Notfn. No. 779 of 24th July 1903. Reconstituted as an I. D. F. Unit with present designation, 1st Apl. 1917.*)

HONORARY COMMANDANT—

First Commn.	Names.	Date of rank.	Date of appointment to Corps.	Remarks.
	Lieutenant-Colonels.			
1 Apl. 17	*Warburton, Harris Grant*, A.D.C. (54)	*1 Apl. 17*	*1 Apl. 17*	(*I. A. R. O.*) *Hony. Col., late Ind. Volrs.*
1 Apl. 17	Stuart, Louis	1 Apl. 17	1 Apl. 17	Comdt.
	Majors.			
1 Apl. 17	Hellier, James (54, 169)	1 Apl. 17	1 Apl. 17	
1 Apl. 17	Crawford, James Charles (54)	1 Apl. 17	1 Apl. 17	
	Captains.			
1 Apl. 17	*Graham, Edward Charles*, V. D.	*1 Apl. 17*	*1 Apl. 17*	*Hony. Maj., late Ind. Volrs.*
1 Apl. 17	Garnett, Charles Leslie Stuart (54)	1 Apl. 17	1 Apl. 17	
1 Apl. 17	*Corlett, Malcolm James John*	*1 Apl. 17*	*1 Apl. 17*	*I. A. R. O.*
1 Apl. 17	Rowland-Hill, Thomas	1 Apl. 17	1 Apl. 17	
	Lieutenants.			
1 Apl. 17	English, Burleigh Arthur (54, 205*a*)	1 Apl. 17	1 Apl. 17	(*Capt., late Ind. Volrs.*)
1 Apl. 17	*Augier, Douglas Eric*	*1 Apl. 17*	*1 Apl. 17*	*I. A. R. O.*
1 Apl. 17	*Wood, Arthur Westmoreland*	*1 Apl. 17*	*1 Apl. 17*	*I. A. R. O.*
1 Apl. 17	*Davidson, George Stainforth*	*1 Apl. 17*	*1 Apl. 17*	*I. A. R. O.*
1 Apl. 17	Hammond, William Henry (54)	1 Apl. 17	1 Apl. 17	
1 Apl. 17	*Tarran, Frederick Percy*	*1 Apl. 17*	*1 Apl. 17*	*I. A. R. O.*
1 Apl. 17	*Thurston, Edward Thomas*	*1 Apl. 17*	*1 Apl. 17*	*I. A. R. O.*
13 Apl. 18	Davies, Lawrence Stanley Richard Gilbert	13 Apl. 18	13 Apl. 18	
	2nd-Lieutenants.			
1 Apl. 17	Lane-Brown, Henry	1 Apl. 17	1 Apl. 17	
4 Feb. 18	Martin, Aubrey Wilson	4 Feb. 18	4 Feb. 18	
13 Apl. 18	Warren, Alfred Aubrey Grenville	13 Apl. 18	13 Apl. 18	
13 Apl. 18	Swinchatt, Percival Hall	13 Apl. 18	13 Apl. 18	
13 Apl. 18	Coates, William Joseph	13 Apl. 18	13 Apl. 18	
13 Apl. 18	Rees, John Conway	13 Apl. 18	13 Apl. 18	
	Supernumerary.			
1 Apl. 17	DeGruyther, Capt. Thomas Donald Dunbar, V.D.	1 Apl. 17	1 Apl. 17	Hon. Maj., late Ind. Volrs.
1 Apl. 17	Kempster, Capt. James Charles	1 Apl. 17	1 Apl. 17	
1 Apl. 17	Burton, Capt. Frederick	1 Apl. 17	1 Apl. 17	
1 Apl. 17	Holmes, 2nd-Lt. Arthur Clement	1 Apl. 17	1 Apl. 17	
1 Apl. 17	Hobart, 2nd-Lt. Robert Charles	1 Apl. 17	1 Apl. 17	
	Staff.			
1 June 18	Bowers, 2nd-Lt. (Actg. Capt.) H., R. G. A.	1 June 18	1 June 18	Adjt.
1 Apl. 17	Menzies, Revd. Harry	1 Apl. 17	1 Apl. 17	Hon. Chaplain.
1 Apl. 17	Souza, Lt. Augustus, F.R.C.S.	1 Apl. 17	. . . .	Medl. Offr.

www.ingramcontent.com/pod-product-compliance
Ingram Content Group UK Ltd.
Pitfield, Milton Keynes, MK11 3LW, UK
UKHW010157180926
475500UK00003B/64

9 781843 422938

By Authority.

The Quarterly

INDIAN ARMY LIST

JANUARY 1919.

Volume 2

A DISTRIBUTION LIST OF OFFICERS ON THE ACTIVE LIST OF THE ARMY IN INDIA, INDIAN DEFENCE FORCE, ETC.

TO WHICH IS ADDED, HALF-YEARLY, ORDERS OF KNIGHTHOOD, ETC., INDIAN ORDERS, FOREIGN ORDERS, NON-EFFECTIVE LISTS, AND WAR SERVICES OF OFFICERS OF THE INDIAN ARMY.

ARMY HEADQUARTERS, INDIA.

CALCUTTA
SUPERINTENDENT GOVERNMENT PRINTING, INDIA
1919

NOTE.—The Indian Army List being an Official publication, and the Press having no power to make alterations in it, all communications regarding deaths, errors, or omissions should be addressed to "Military Secretary to the Commander-in-Chief" with the words "Army List" written on the upper left-hand corner of the cover.

All correspondence regarding the War Services of Officers to be addressed to the Adjutant-General in India ***through Brigade or Divisional Commanders.*** (***Vide*** India Army Order 31 of 1911.)

CORPS OF ROYAL ENGINEERS.

THE ROYAL ARMS AND SUPPORTERS.—"*Ubique*," and "*Quo Fas et Gloria ducunt*."
Uniform—Scarlet. *Facings*—Blue.

COLONEL-IN-CHIEF—THE KING.

OFFICERS OF THE ROYAL ENGINEERS BELONGING TO THE INDIAN ESTABLISHMENT.

NOTE.—An asterisk * prefixed to an officer's name indicates that he has elected for continuous service in India. The symbol † before an officer's name indicates that he has elected continuous M. W. employment to the age of 55 years, with final removal therefrom at that age; *vide* M. D. notification No. 5, dated 1st January 1904.

REMOVED FROM THE CORPS AND STILL ON THE ACTIVE LIST.

First Commn.	Names.	Army rank.	Remarks.
	General Officers.		
27 July 80	VC Aylmer, *Sir* F. J., K.C.B. (57, 206)	*l.g.* 11 June 15	
6 Apl. 79	Kelly, F. H., C.B. (57, 93, 161, 170, 185, 206)	*m. g.* 4 Dec. 09	
*18 Dec. 78	Mullaly, *Sir* H., K.C.M.G., C.B., C.S.I. (206)	*m. g.* 9 Oct. 10	Dn. Comdr., England.
30 June 85	Kirkpatrick, *Sir* G. M., K.C.B., K.C.S.I., *p. s. c.* (*temp. Lt.-Genl.*) (206)	*m.g.* 3 June 15	Chief of the Genl. Staff, Army Hdqrs., India.
†*15 Feb. 84	Rimington, J. C., C.B., C.S.I. (57, 161, 170, 206)	*m.g.* 3 June 16	
*31 Jan. 78	Williams, *Sir* G., K.C.I.E., C.B. (169, 207)	*m.g.* 1 Jan. 17	Dir.-Genl., Mily. Works, Army Hdqrs., India.
*6 Jan. 86	Twining, P. G., C.B., C.M.G., M.V.O. (162, 206)	*m.g.* 1 Jan. 17	D. F. W. War Office.
	Colonels.		
23 Feb. 81	Cowie, C. H., C.I.E. (207)	29 Nov. 06	Dy. Dir. of Ry. Trans.
*14 Feb. 83	Renny-Tailyour, T. F. B., C.S.I.	29 Nov. 06	Survey Dept.
*6 Apl. 79	Browne, C. A. R. (57, 207)	3 Jan. 07	*Resdg. ex India.*
*30 June 85	Nanton, H. C., C.B. (*temp. Brig.-Genl.*) (162, 170)	1 Feb. 07	Chief Engr., S. C.
*6 Apl. 79	Burrard, *Sir* S. G., K.C.S.I. (207)	15 Nov. 07	Surveyor-General of India.
30 July 79	Dallas, J. (206)	22 Apl. 08	
*18 Feb. 80	Maclagan, R. S., C.B., C.S.I. (170, 206)	12 Aug. 08	C. R. E. & Secy. (P. W. D.), Quetta.
*27 July 80	Baddeley, C. E., C.M.G. (50, 57, 162, 169, 206)	22 Oct. 08	Dn. Engr. Comdr.
26 July 81	Glanville, F., D.S.O. (*temp. Brig-Genl.*) (206)	1 Apl. 09	Br.-Genl., Admn., 8th (Lucknow) Dn.
*26 July 81	Ellis, W. M., C.I.E. (37, 207)	1 Apl. 09	
*22 Feb. 82	Walton, E. W. (207)	25 July 09	Ry. Dept.
*22 Feb. 82	Craster, S. L., C.B., C.I.E. (162, 206)	13 Aug. 09	*Resdg. ex India.*
†*14 Feb. 83	Houston, E. (207)	19 July 11	*Resdg. in India.*
*14 Feb. 83	Rose, C. S. (207)	6 Dec. 11	Ry. Dept.
*15 Feb. 84	Roe, C. H., C.I.E. (*temp. Brig.-Genl.*) (50, 53, 57, 105, 192, 206)	6 June 12	Dir. of Works.
†*5 July 84	Brewin, J. P.	30 Oct. 12	C. R. E.
9 Dec. 84	Powell, S. H., C.B. [*l*] (*temp. Brig.-Genl.*) (50, 162, 169, 206)	2 June 13	Chief Engr.
†*15 Feb. 84	Dealy, J. A., D.S.O. C.M.G., (*temp. Brig.-Gen.*) (207)	2 June 13	C. R. E.
9 Dec. 84	Ewbank, W., C.I.E. *p.s.c*, (206)	2 June 13	Asst. Engr.-in-Chief.
29 Apl. 85	Sorsbie, R. F., C.I.E. (*temp. Brig.-Genl.*) (50, 192, 206).	16 Dec. 13	Chief Engineer, N. C.
*16 Sep. 85	Lenox-Conyngham, G. P.	15 June 14	Survey Dept.
16 Sep. 85	Atkinson, E. H. de V., C.B., C.M.G., C.I.E. (*temp. Maj.-Genl.*) (169, 206)	15 June 14	Princpl., Thomason Civil Eng. Coll., Rurki.
*16 Sep. 85	Johnson, E. P., C. B. (*temp. Br.-Genl.*) (193, 206)	15 Dec. 14	Chief Engineer.
*16 Sep. 85	Scudamore, W. V. (112, 206)	15 Dec. 14	P. W Dept.
*16 Sep. 85	Swiney, A. J. H., C.B., C.M.G. (*temp. Brig.-Genl.*) (207)	15 Dec. 14	Chief Engineer.
*6 Jan. 86	Duff, G. M. (207)	25 July 15	C. R. E., 2nd Dn.
*6 Jan. 86	Joly deLotbinière, A. C.deL., C.B., C.S.I., C.I.E. (207)	6 Aug. 15	
*6 Jan. 86	Picton, R. E. (83, 207)	13 Aug. 15	Offg. C. R. E., 9th Dn.
6 Jan. 86	Kirby, N., C.I.E. (105, 207)	18 Aug. 15	C. R. E., 5th Dn.
*17 Feb. 86	Murray, V., C.B., C.M.G. (206)	22 June 16	Ry. Dept.
*17 Feb. 86	Weedon, F. F. (105, 159, 206) [L]	1 July 16	C. R. E., 8th Dn.
*17 Feb. 86	Blakeway, I. P., C.M.G. (105, 207)	1 Oct. 16	P. W. Dept.
*†8 Feb. 86	Fraser, T., (*temp. Maj Genl. 25 Sep. 18*) C.B., C.S.I., C.M.G. *p.s.c.* [L] (*n*) (50, 206)	19 Dec. 16	Divl. Comdr.
*8 Feb. 86	Fraser, H. A. D.	16 Apl. 17	C. R. E. & Secy. (P. W. D.), N. W. F. P., Peshawar.
12 May 86	Campbell, G. P. (172*a*, 207)	4 July 17	C. R. E., 16th (Ind.) Dn.
†*24 July 86	Ainslie, C. (106, 206)	10 July 17	A. C. R. E., Deolali.
24 July 86	Ryder, C. H. D., C.I.E., D.S.O. [*l*] (124*a*, 162, 170, 206).	14 July 17	Survey Dept.
16 Sep. 85	Dunsterville, E. L. (206)	20 May 18	Dy. Dir. of Works.

First Commn.	Names.	Regimental rank.	Army rank.	Remarks.
	ROYAL ENGINEERS.			
	Lieutenant-Colonels.			
*9 Dec. 84	Evans, U. W., C.B., C.M.G. (*temp. Brig.-Genl.*) [L] (170, 206)	22 Apl. 10	*s.c.* 16 Dec. 13	Bde. Comdr.
6 Jan. 86	*Wilson, F. A.*, C.B. (66a, 207)	*25 Sep. 11*	. . . .	*M. W. S.—*
*17 Feb. 86	Heycock, C. H. (206)	18 Jan. 12	. . . .	2 S. & M.
*17 Feb. 86	Murray, J. H. S. (162, 169, 206)	26 Aug. 12	. . . .	A. C. R. E., Pindi.
17 Feb. 86	Coffin, C., C.M.G., *p.s.c.* (206)	1 Oct. 12	*s.c.* 1 Oct. 16	D. D. G., M. W., Army Headqrs.
18 Feb. 86	Pilcher, A. J. (206)	25 Dec. 12	. . . .	A. C. R. E., Dehra Dun and Garhwal Dists.
17 Feb. 86	Burn, E. M. J. [L] (207)	1 Sep. 13	. . . .	A. C. R. E., Jullundur.
24 July 86	Robertson, C. L., C.M.G. (207)	25 Sep. 13	. . . .	Survey Dept.
24 July 86	Barton, H. J. [*l*] (206)	5 Dec. 13	. . . .	A. C. R. E., Ambala.
*†16 Feb. 87	Ogilvie, E. C., C.M.G. (30, 169, 207)	21 Dec. 13	*c.* 3 June 16	A. C. R. E., Quetta.
16 Feb. 87	Marshall, H. J. M., C.M.G. (*temp. Br. Genl. 22 Nov. 17*) (207)	5 Jan. 14	. . . .	M.W.S.—
†*16 Feb. 87	Waghorn, W. D., C.M.G. (*temp. Br.-Genl. 30 Nov. 15*) (207)	21 Mar. 14	*c.* 1 Jan. 17	Ry. Dept.
16 Feb. 87	*Austin, H. H.*, C.B., C.M.G., D.S.O., *p.s.c.* (*temp. Brig.-Gen.*) [*l*] (162) 170, 206)	*4 May 14*	*l. c. 18 Jan. 11* *c. 3 June 17*	*Offg. A. C. R. E., Peshawar.*
†*16 Feb. 87	Lathbury, H. O. (207)	16 May 14	. . . .	A. C. R. E., Poona.
16 Mar. 87	Leslie, G. A. J., C.B., C.M.G. (*temp. Maj.-Genl.*) (93, 169, 185, 206)	20 Sep. 14	*c.* 3 June 16	M. W. S.—Bde. Comdr.
*23 July 87	Bond, R. F. G. (169, 206)	27 Oct. 14	. . . .	A. C. R. E., Delhi.
†*23 July 87	Rivett-Carnac, S. G.	30 Oct. 14	. . . .	P. W. Dept.
23 July 87	Thuillier, H. F., C.B., C.M.G. (*temp. Maj. Genl.*) (169, 206)	30 Oct. 14	*c.* 1 Jan. 17	M. W. S.—
*23 July 87	Rushton, H. W. D.S.O. (*temp. Brig.-Genl.*) (207)	30 Oct. 14	. . . .	P. W. Dept.
23 July 87	Stockley, H. R., C.I.E. (50, 169, 205)	30 Oct. 14	. . . .	Inspr. of R.E. & Pioneers, Army Hdqrs.
†*17 Feb. 88	Winsloe, A. R., D.S.O. (169, 206)	30 Oct. 14	. . . .	M. W. S.—C. R. E., Aden.
*17 Feb. 88	Smyth, W. C. (207)	30 Oct. 14	. . . .	Ry. Dept.
17 Feb. 88	Hume, A. H. B. [*l*] (156, 207)	29 Mar. 15	. . . .	A. C. R. E., Loralai.
†*1 Apl. 88	Riach, A. H. D. (161, 169, 206)	12 Aug. 15	. . . .	Offg. A. C. R. E., Wellington.
1 Apl. 88	Close, L. H., C.M.G. (170, 207)	12 Aug. 15	. . . .	M. W. S.—
1 Apl. 88	Nugent, C. H. H., C.I.E. (174, 207)	4 Oct. 15	. . . .	Army Hdqrs., India.
27 July 88	Coldstream, W. M. (169, 207)	22 Oct. 15	. . . .	Survey Dept. *Lv., ex Ind.*
27 July 88	Saunders, F. W. (206)	22 Nov. 15	. . . .	Offg. A. C. R. E., Quetta.
*27 July 88	Wilkinson, C. W., C.M.G., D.S.O. (207)	1 Apl. 16	. . . .	Ry. Dept.
†*27 July 88	McCormick, A. L. C. (207)	1 Apl. 16	. . . .	Mint Master, Calcutta. *Lv. 6 mos., 2 Aug. 18.*
†*27 July 88	Sanders, G. A. F., *p.s.c.* (*temp. Brig.-Genl.*) (50, 106, 170)	1 Apl. 16	3 June 15	
27 July 88	Kelsall, H. W. (106, 125c)	30 Apl. 16	. . . .	A. C. R. E., Abbottabad.
28 July 88	Lesslie, W. B., C.M.G. (*temp. Brig.-Genl.*) (162, 169, 206)	18 July 16	*c.* 1 Jan. 18	Chief Engineer.
†*28 July 88	Panet, A. E., C.M.G., D.S.O. (*temp. Br.-Genl.*) (30, 207)	25 July 16	*c.* 3 June 18	M. W. S.—
*28 July 88	Bremner, A. G., C.M.G. (Q. 207)	30 July 16	*c.* 1 Jan. 18	3rd S. & M.
*15 Feb. 89	Westropp, F. M., D.S.O.	13 Aug. 16	. . . .	P. W. Dept.
*15 Feb. 89	Cameron, H. A., C.I.E. (207)	18 Aug. 16	. . . .	Ry. Dept.
*27 July 89	𝔄. 𝔆. Colvin, J. M. C., *p.s.c., q.s.* (Q. 65, 170, 206)	18 Jan. 17	.	3rd S. & M.
*27 July 89	Dickson, W. E. R., C.I.E., (*temp. Brig.-Genl.*), [*l*] (50, 169, 206)	2 Mar. 17	*c.* 17 Sep. 17	
*27 July 89	Cunningham, A. H. (50, 162, 170, 206)	1 June 17	. . . .	1st S. & M.
*27 July 89	Lubbock, G., C.M.G., D.S.O. (*temp. Brig.-Genl.*) (169, 206)	1 June 17	*c.* 1 Jan. 18	*Dir. of Rys.*
*27 July 89	Crookshank, S. D'A., C.B., C.I.E., D.S.O., M.V.O. (*temp. Brig.-Genl.*) (207).	22 June 17	*c.* 1 Jan. 18	P. W. Dept.
*27 July 89	Beach, W. H., C.M.G., D.S.O. (*temp. Br.-Genl.*) (162, 166, 169, 206)	1 July 17	*c.* 3 June 18	Br.-Genl., Genl. Staff.
*27 July 89	Cumberlege, A. F. (*temp. Col., 10 Apl. 18*) (206)	26 Aug. 17	1 Jan. 16	Dy. D. G., M. W., Army Hdqrs., India.
*23 Nov. 89	Turner, H. H.	1 Oct. 17	. . . .	Survey Dept.
*14 Feb. 90	Beazeley, G. A., D.S.O. (Q)	19 Dec. 17	. . . .	Survey Dept.
*14 Feb. 90	Sheppard, S. H., C.B., C.M.G., D.S.O., *p.s.c.* (Q. 206)	13 Mar. 18	*c.* 24 July 15	
*14 Feb. 90	Boileau, G. H., C.M.G., D.S.O. (66a, 206)	1 Apl. 18	3 June 16	M. W. S.—
14 Feb. 90	Rundle, F. P., D.S.O. (*temp. Col.*) (71, 207)	10 July 18	1 Jan. 18	M. W. S.—
14 Feb. 90	Rolland, A., D.S.O. (*temp. Br. Genl.*) (206)	14 July 18	3 June 17	M. W. S.—
*29 Mar. 90	Crosthwait, H. L. (Q. 172a)	1 Aug. 18	. . . .	Survey Dept.
4 July 90	Campbell, H. B. D.	27 Aug. 18	. . . .	*On 6 mos. lv. m.c., 25 Feb. 18.*
*25 July 90	Elliott, C. A., D.S.O. (*temp. Br.-Genl.*) (Q. 169, 207)	24 Sep. 18	1 Jan.	M. W. S.—

Majors.

First Commn.	Names.	Regimental rank.	Army rank.	Remarks.
17 Feb. 88	Greer, R. E. (30, 94, 169, 207)	1 June 07	. . . .	A. C. R. E., Allahabad.
*15 Feb. 89	Muter, R. S. (206)	1 Aug. 08	. . . .	Offg. A. C. R. E., Sialkot.
25 July 90	Hunter, C. G. W., D.S.O.	25 July 10	*l.c.* 11 Jan. 16	M. W. S.—
25 July 90	MacGeorge, J. B. (*act'g. Lt.-Col., 11 July 18*)	25 July 10	. . . .	P. W. D. Dept.—A. C. R. E., Bangalore.
*25 July 90	Hearn, G. R., D.S.O. (*act'g. Br.-Genl., 18 July 16*) (206)	25 July 10	*l.c.* 1 Jan. 18	Ry. Dept.
13 Feb. 91	Greenstreet, C. B. L. (*act'g. Lt.-Col., 11 July 18* (205)	13 Feb. 11	*l.c.* 3 June 18	*Offg.* A. C. R. E., Sec'bad.
*13 Feb. 91	Tandy, E. A. (207)	13 Feb. 11	. . . .	Survey Dept.
*13 Feb. 91	Barnardiston, E. (*act'g. Lt.-Col., 19 Jan. 17*) (207)	13 Feb. 11	. . . .	Ry. Dept.
*13 Feb. 91	Freeland, H. F. E., C.B., D.S.O., M.V.O., (*temp. Maj.-Genl.*) (Q. 121a, 162, 170, 206)	13 Feb. 11	*l.c.* 3 June 15	Ry. Dept.
**12 Feb. 92*	*Tylden-Pattenson, E. C.*, D.S.O. (31, 207)	*12 Feb. 12*	*l.c. 6 June 12*	*1st S. & M., C. R. E.*
12 Feb. 92	Garwood, F. S.	12 Feb. 12	. . . .	3rd S. & M.
12 Feb. 92	Wood, H. (206)	12 Feb. 12	*l.c.* 1 Jan. 18	Survey Dept.
*12 Feb. 92	Gillespie, R. St. J., C.I.E. (*act'g. Lt.-Col., 11 July 18* (200)	12 Feb. 12	. . . .	A. C. R. E., Ferozepore.
22 July 92	Winsloe, H. E., D.S.O. (105, 162, 169, 206)	22 July 12	*l.c.* 3 June 16	
22 July 92	Broughton, T. D.	22 July 12	. . . .	
*22 July 92	Hopkins, L. E., D.S.O. (*temp. Lt.-Col., 19 Nov. 17*) (Q. 207)	22 July 12	. . . .	Ry. Dept. Dy. Dir. of Works.
*22 July 92	Anderson, C. F., D.S.O. [L] (206)	22 July 12	*l.c.* 3 June 16	Ry. Dept.
22 July 92	LeMesurier, H. G., C.I.E., (*temp. Lt.-Col.*, 25 Feb. 18)	22 July 12	. . . .	Asst. D. G., M. W., Army Hdqrs., India.
24 July 91	King, R. G. (172a, 206)	22 July 12	. . . .	G. E., Quetta.
*10 Feb. 93	Loch, S. G. D.S.O. *p.s.c.* (*temp. Brig.-Genl.*), (Q. 30, 209)	10 Feb. 13	*c.* 17 Sep. 17	Brig.-Genl., Genl. Staff, N. C.
10 Feb. 93	Ogilvie, D. (207)	10 Feb. 13		
*10 Feb. 93	Cowie, H. McC.	10 Feb. 13	. . . .	Survey Dept.
10 Feb. 93	Gervers, F. R. S (51, 66a, 207)	10 Feb. 18	. . . .	Offg. A. C. R. E., Nowshera.
*10 Feb. 93	Russell, W. K., D.S.O. (*temp. Col.*) (206)	10 Feb. 13	. . . .	Ry. Dept.
10 Feb. 93	Gunter, C. P. (170, 206)	10 Feb. 13	*l.c.* 3 June 17	Survey Dept.
*10 Feb. 93	Cowie, H. E. C., D.S.O. (*temp. Lt.-Col.*) (207)	10 Feb. 13	. . . .	Ry. Dept.
5 July 93	Rich, E. T., C.I.E. (105, 125a, 140, 148b), 162, 169, 206)	25 July 13	16 July 08	Survey Dept.
July 93	Hodgson, P. E., D.S.O. (*act'g. Lt.-Col. 24 Dec. 16*) (94, 100, 104, 162, 169 206)	25 July 13	*l.c.* 3 June 18	C. R. E., M. W. S.—
July 93	McClintock, R. L., D.S.O. (*temp. Lt.-Col.*) (Q. 161, 206)	25 July 13	*l.c.* 1 Jan. 17	2nd S. and M.
*25 July 93	McHarg, A. A., D.S.O. [*l*] (162, 170, 206)	25 July 13	*l.c.* 1 Jan. 18	Survey Dept.
*25 July 93	Chaldecott, W. H. (207)	25 July 13	. . . .	I. S. Sappers.
25 July 93	Tandy, M. O'C.	25 July 13	. . .	Survey Dept.
27 Feb. 94	Clarke, C. J. (*act'g. Lt.-Col., 7 Apl. 17*) (172a, 206)	27 Feb. 14	. . . .	Ry. Dept.
27 Feb. 94	Smyth, G. J. W., D.S.O. (*temp. Lt.-Col.*) (206)	27 Feb. 14	. . . .	Ry. Dept. *Offg.* A. C. R. E., Jhansi.
*27 June 94	Osborne, G. F. F., D.S.O.	27 June 14	. . . .	Ry. Dept. *Lv., ex. Ind., m.c., 1 yr. 4 Jan. 18.*
25 July 93	*Manley, E. N.*, D.S.O. (207)	*5 Aug. 14*	. . .	*Ry. Dept.*
17 Aug. 94	Garrett, A. ff. (*act'g. Lt.-Col., 11 July 18* (207)	17 Aug. 14	. . . .	P. W. Dept.—A. C. R. E., Karachi.
*17 Aug. 94	Charles, J. R. E., C.B., D.S.O., *p.s.c.* (*temp. Maj.-Genl.*, 162, 206)	17 Aug. 14	*c.* 3 June 16	Divl. Comdr.
17 Aug. 94	Holme, A. S. [L] (200)	17 Aug. 14	. . . .	P. W. Dept.
17 Aug. 94	Hogg, C. C. H., C.M.G., (162, 172a, 206)	17 Aug. 14	*l.c.* 3 June 18	P. W. Dept.
24 Oct. 94	Earle, R. G., D.S.O., (42, 57, 206)	24 Oct. 14	*l.c.* 3 June 17	2nd S. & M.
*17 Nov. 94	Magniac, C. L., C.M.G. (*temp. Col.*) (206)	30 Oct. 14	*l.c.* 1 Jan. 18	Ry. Dept.
15 Mar. 95	Brough, A., (*temp. Brig.-Genl.*) D.S.O. (207)	30 Oct. 14	*l.c.* 3 June 16	Ry. Dept.
†*3 Aug. 95	Willis, G. H., C.I.E., M.V.O. (207)	30 Oct. 14	. . . .	*Offg.* Sr. Mint Master, Bombay.
*3 Aug. 95	Barron, W. E. (207)	30 Oct. 14	. . . .	Ry. Dept.—A. C. R. E., Mhow.
*8 Aug. 95	Eustace, F. R. H., D.S.O. (*temp. Lt.-Col.*) (207)	30 Oct. 14	. . . .	Ry. Dept.

First Commn.	Names.	Regimental rank.	Army rank.	Remarks.
	Majors—*contd.*			
21 Dec. 95	Pollard-Lowsley, H. deL., C.I.E., D.S.O. (*act'g. Lt.-Col., 15 Oct. 16*) (207)	30 Oct. 14	*l.c.* 3 June 17	P. W. Dept.
11 Mar. 96	Mathews, W. F. (207) . . .	30 Oct. 14	. . . .	Ry. Dept.
*21 June 96	Thomas, R. H.	30 Oct. 14	. . . .	Survey Dept.
21 Sep. 96	Evans, W. H., D.S.O. (*act'g. Lt.-Col., 11 Sept. 16*) (207)	30 Oct. 14	. . . .	M. W. S.—
25 Sep. 96	Evans, A. S.	30 Oct. 14	. . . .	Dy. Ch. Engr., N. C.
23 Dec. 96	Forster, D., D.S.O., *p.s.c.* (*temp. Lt.-Col.*) (104, 125*a*, 172*a*, 206)	30 Oct. 14	*l.c.* 3 June 17	G. S. O., 1st gr.
*31 Dec. 96	Izat, W. R., D.S.O., (*act'g. Lt.-Col.*) (207)	30 Oct. 14	. . . .	Ry. Dept.
†*31 Dec. 96	Malan, L. N. (207) . . .	30 Oct. 14	*l.c.* 1 Jan. 17	M. W. S.—
3 Feb. 97	Oldham, G. M., D.S.O. (*temp. Col.*) (50, 170, 207)	30 Oct. 14	*l.c.* 1 Jan. 18	M. W. S.—
†*22 Feb. 97	Giles, V., D.S.O. (*act'g.* Lt.-Col.) [L] (120, 128, 207)	30 Oct. 14	*l c.* 1 Jan. 18	P. W. Dept.
8 Mar. 97	Skinner, R. B. (102, 169, 206) .	30 Oct. 14	. . . .	M. W. S.—
23 June 97	Wagstaff, C. M., C.M.G., C.I.E., D.S.O., *p.s.c.* (51, 162, 169, 206)	30 Oct. 14	*l.c.* 1 Jan. 17	A. A. & Q. M. G.
*25 July 97	Walton, C., D.S.O., (*temp. Lt.-Col.*) (207)	30 Oct. 14	*l.c.* 1 Jan. 18	Ry. Dept.
27 July 97	Kealy, P. H. (207)	30 Oct. 14	. . . .	Offg. A. C. R. E., Fort William.
*27 Aug. 97	Hogg, P. G. H. D.S.O. (161, 206) .	30 Oct. 14	. . . .	Offg. A. C. R. E., Bareilly.
31 Aug. 97	Hepper, A. W. (30, 207) . . .	30 Oct. 14	. . . .	Asst. Inspr. of Machinery, S. C.
23 Sep. 97	Maxwell, W. F., D.S.O. (30, 170, 207)	30 Oct. 14	. . . .	With Sig. Serv.
*23 Dec. 97	Browne, C. M., C.M.G., D.S.O., (*act'g. Lt.-Col. 2 June 17*)	30 Oct. 14	*l.c.*, 1 Jan. 16	Survey Dept.
23 Mar. 98	Bell, A. H., D.S.O. [L] (206) . .	30 Oct. 14	. . . .	*Offg.* A. C. R. E., Bannu.
23 June 98	Evans, L., D.S.O. (207) . . .	30 Oct. 14	. . . .	M. W. S.—
23 June 98	Birney, C. F., D.S.O. (*temp. Col.*) (206)	30 Oct. 14	*l.c.* 3 June 18	Ry. Dept.
23 June 98	Noble, W. J. W. (207) . .	30 Oct. 14	. . . .	Offg. A. C. R. E., Mhow.
23 June 98	Stack, G. H. (*act'g. Lt.-Col., 19 July 16*) (206)	26 Jan. 15	*l.c.* 3 June 17	M. W. S.—
23 June 98	Phillimore, R. H. D.S.O. (207) .	29 Mar. 15	. . . .	Survey Dept.
23 June 98	Barker, J. S., M.V.O. (206) . . .	9 Apl. 15	. . . .	M. W. S.—
*23 June 98	Loch, G. C. R., *p.s.c.* (*act'g. Lt.-Col.*) (42, 51, 109, 206)	22 Apl. 15	. . . .	M. W. S.—
23 Dec. 98	Paris, A. L. (207)	28 May 15	. . . .	3rd S. & M.
*23 Dec. 98	Hill, E. F. J., M.C. (*act'g. Lt.-Col., 28 Mar. 17*) (206)	17 June 15	*l.c.* 3 June 17	1st S. & M. C. R. E.
†*23 Dec. 98	Walker, A. D. [L] (207) . . .	22 June 15	. . . .	Ry. Dept.
*9 Mar. 99	Thomson, D. A. (59, 207) . .	12 Aug. 15	. . . .	With Sig. Serv.
23 Mar. 99	*Burgess, B. (207)* . . .	*4 Oct. 15*	. . . .	*M. W. S.,*
†*25 June 99	Haswell, C. H. (*act'g. Lt.-Col. 11 July 18*) (169, 207)	5 Dec. 15	*l.c.* 1 Jan. 18	*Offg.* A. C. R. E., Kohat.
†*9 Aug. 99	Macrae, W. (207)	1 Apl. 16	. . . .	Ry. Dept., M. W. S.
20 Aug. 99	Tillard, E. D., D.S.O. (95, 207)	30 Apl. 16	. . . .	3rd S. & M.
25 Sep. 99	Hill, R. C. R., D.S.O. 45*a*, 207)	4 July 16	1 Jan. 16	1st S. & M.
25 Sep. 99	Cusins, A. R., M.C. (207)	6 July 16	. . . .	M. W. S.—
6 Jan. 00	Turner, J. F., D.S.O. (*act'g.* Lt.-Col., 26 Feb. 17) (169, 207)	6 Sep. 16	. . . .	C. R. E.
*6 Jan. 00	Molesworth, F. C. (42, 50, 162, 169, 206)	1 Oct. 16	. . . .	M. W. S.—
6 Jan. 00	Campbell, A., (*actg. Lt.-Col., 28 May 17*) D.S.O. [*l*] (162, 169, 206)	24 Oct. 16	. . . .	M. W. S.—
6 Jan. 00	Ogilvy, D., D.S.O. (*act'g. Lt.-Col., 19 May 17*) (50, 207)	2 Nov. 16	. . . .	M. W. S.—
2 May 00	MacLeod, M. N., D.S.O., M.C. (172, 207	2 Nov. 16	. . . .	Survey Dept.
2 May 00	Conran, W. D. B., D.S.O. (207)	2 Nov. 16	. . . .	2nd S. & M.
2 May 00	Watson, C. S. M. C., D.S.O. (207)	2 Nov. 16	. . . .	Ry. Dept.—G. E., Sialkot.
2 May 00	Smith, V. P. (167, 169, 207) . .	2 Nov. 16	. . . .	Instr., Schl. of Mily. Eng., Chatham.
2 May 00	Burn, R. N. (206) . .	2 Nov. 16	. . . .	Ry. Dept.
2 May 00	Cardew, A. M. [L]	2 Nov. 16	. . . .	
25 June 00	Archbold, F. H. W. (51, 65, 71, 128, 207)	2 Nov. 16	. . . .	M. W. S.—
18 Aug. 00	Battye, B. C., D.S.O. (207) .	2 Nov. 16	3 June 15	P. W. Dept.—G.S.O., 2nd gr.
18 Aug. 00	Couchman, H. J., M.C. (*act'g. Lt.-Col., 2 July 17*) (94, 170, 207)	2 Nov. 16	. . . .	Survey Dept.
18 Aug. 00	Day, J., D.S.O. (*temp. Lt.-Col.*)	2 Nov. 16	. . . .	Insp., Army Signals.
18 Aug. 00	Young, E. deL. (94, 162, 169, 185, 204)	2 Nov. 16	*l.c.* 3 June 17	M.W.S.—
18 Aug. 00	Broke-Smith, P. W. L., D.S.O. (42, 50, 162, 169, 206)	2 Nov. 16	. . . .	A. C. R. E., Peshawar (Offg.).
18 Aug. 00	Gaskell, H. S., D.S.O. (207) . .	2 Nov. 16	. . . .	M. W. S.—

First Commn.	Names.	Regimental rank.	Army rank.	Remarks.
	Majors—*concld.*			
18 Aug. 00	Boyd, S., D.S.O. (50, 53, 88, 206)	2 Nov. 16	. . . .	3rd S. & M.
18 Aug. 00	Prince, P. E., D.S.O. (162, 172a, 206)	2 Nov. 16	. . . .	2nd S. & M.
18 Aug. 00	Lynch, C. St. J., D.S.O. (95, 172a, 206)	2 Nov. 16	. . . .	Ry. Dept.
18 Aug. 00	Haig, D. de H. (207)	2 Nov 16	. . . .	A. C. R. E., Jubbulpore. *Lv., in I., 4 mos., 5 days.*
18 Aug. 00	Hawtrey, H. C., C.M.G., D.S.O. (*temp Lt.-Col.*) (57)	2 Nov. 16	*l.c.* 1 Jan. 17	A. C. R. E., Bombay Bed.
21 Dec. 00	Bird, A. J. G., D.S.O. (169, 206)	2 Nov. 16	*l.c.* 1 Jan. 18	1st S. & M.
21 Dec. 00	North, H. N., D.S.O. (94, 169, 205a)	2 Nov. 16	*l.c.* 3 June 18	M. W. S.—
21 Dec. 00	*Kent, L. M.* (207)	2 Nov. 16	. . . .	Ry. Dept.—G. E., Aden.
13 Jan. 01	Anderson, E. P., D.S.O. (*act'g. Lt.-Col., 22 Nov. 16*) (206)	2 Nov. 16	. . . .	Ry. Dept.
***4 Feb. 01**	Porter, M. T., M.C. (169, 207)	2 Nov. 16	. . . .	Ry. Dept.
4 Feb. 01	Loring, E. J., M.C. (51, 170, 206)	2 Nov. 16	. . . .	D. A. A. G., Army Hdqrs.
4 Feb. 01	Trenchard, O. H. B. (207)	2 Nov. 16	. . . .	Survey Dept.
23 July 01	Garforth, W., D.S.O., M.C. (207)	2 Nov. 16	. . . .	P. W. Dept.—G. S. O., 2nd gr.
23 July 01	Roberts, W. H., D.S.O., M.C. (94, 169, 206)	2 Nov. 16	. . . .	P. W. Dept.—*Lv., ex I., m.c., 6 mos.*
23 July 01	Fox, B. H. (169, 206)	2 Nov. 16	. . . .	M. W. S.—*Offg.* A. C. R. E., Derajat.
23 July 01	Hamilton, S. W. S., D.S.O. (207)	2 Nov. 16	. . . .	Survey Dept., Ch. Instr., R. M. Academy.
23 July 01	Borton, W. N. (207)	2 Nov. 16	. . . .	
23 July 01	Palmer, C. H. (172a, 207)	2 Nov. 16	. . . .	
23 July 01	LeBreton, E. P. (50, 89a, 162, 206)	2 Nov. 16	. . . .	1st S. & M. G. S. O., 3rd gde.
28 Aug. 01	Agnew, H. C. (51, 52, 162, 206)	2 Nov. 16	. . . .	2nd S. & M.
21 Dec. 01	Bushell, C. W. (51, 104, 206)	21 Dec. 16	. . . .	2nd S. & M.
21 Dec. 01	Sopwith, G. E., M.C. (207)	21 Dec. 16	. . . .	1st Peshawar Dn.—M. W. S. (temp.)
21 Dec. 01	Campbell, J. D. (170, 206)	21 Dec. 16	. . . .	Survey Dept.
21 Dec. 01	Chesney, C. H. R., D.S.O. (207)	21 Dec. 16	. . . .	P. W. Dept.
21 Dec. 01	Pye, K. W., D.S.O.	21 Dec. 16	. . . .	M. W. S., 5th Divn.
21 Dec. 01	Fenton, G. C. V., D.S.O. (*temp. Lt.-Col., 22 July 18*) (207)	21 Dec. 16	. . . .	Asst. Dir. Genl. of Mily. Works, Army Hdqrs.
21 Dec. 01	Græme, J. A., D.S.O. (*act'g Lt.-Col., 9 Aug. 17*) (207)	21 Dec. 16	. . . .	P. W. Dept.
21 Dec. 01	Morshead, H. T., D.S.O. (206)	21 Dec. 16	. . . .	Survey Dept.
21 Dec. 01	Pakenham-Walsh, W. P. (207)	21 Dec. 16	. . . .	3rd S. & M.
1 Jan. 01	Festing, J. E. G. (170, 207)	11 Jan. 17	. . . .	*Offg.* A. C. R. E., Ahmednagar.
***19 June 02**	Edgar, D. K., D.S.O. (*temp. Col.*) (207)	19 June 17	. . . .	Dy. Dir. of Works.
31 July 02	Kirke, E. St. G., D.S.O. (169, 207)	31 July 17	. . . .	M. W. S.—*Lv., m.c.*
31 July 02	Kelly, E. H., D.S.O., M.C. (50, 206)	31 July 17	. . . .	1st S. & M. G. S. O., 2nd gr.
31 July 02	Bocquet, R. N. (*temp. Lt.-Col., 25 Oct. 18*) (207)	31 July 17	. . . .	A. D. G., M. W. Army Hdqrs.
31 July 02	Bassett, T. P., D.S.O. (193, 207)	31 July 17	. . . .	2nd S. & M.
31 July 02	Walker, E. C. [*l*] (88, 101, 206)	31 July 17	. . . .	Deputy Chief Engineer, S. C.
31 July 02	Barnes, L. E., *p.s.c.* (163, 206)	31 July 17	. . . .	A. Hdqrs. Appt. G. S. O., 2nd gr.
31 July 02	Pemberton, S., D.S.O., M.C. (206)	31 July 17	. . . .	2nd S. & M.
31 July 02	Pollock, H. C. (172a, 207)	31 July 17	. . . .	P. W. Dept.—G. E., Ranikhet.—*Lv. ex Ind., m.c., 12 mos.*
1 Oct. 02	Mount, A. H. L. (*temp. Lt.-Col.*) (207)	1 Oct. 17	1 Jan. 16	Ry. Dept.
23 July 01	Bowen, J. A. B. P. (170, 207)	1 Oct. 17	. . . .	M. W. S.—
24 Dec. 02	Kidner, W. E., M.C. (51, 207)	24 Dec. 17	. . . .	2nd S. & M., G. S. O., 2nd gr., Army [Hdqrs.
17 Jan. 03	Squires, E. K., D.S.O., M.C. (50, 53, 101, 206)	17 Jan. 18	. . . .	M. W. S.—
24 Mar. 03	*Trench, A. H. C.* (50, 95, 207)	*24 Mar. 18*	. . . .	*A. D. W.*
24 Mar. 03	Morris, G. H. J. (51, 207)	24 Mar. 18		
15 July 03	Binney, E. V., D.S.O. (51, 94, 162, 169, 206)	15 July 18	. . . .	3rd S. & M.
15 July 03	Bond, L. V. (51, 59, 172a, 206)	15 July 18	1 Jan. 17	1st S. & M.—I. S. S.
15 July 03	Haig, W. de H., D.S.O. (51, 169, 206)	15 July 18	1 Jan. 17	P. W. Dept.—D. A. A. & Q.-M.-G.
15 July 03	Nottidge, G. (51, 207)	15 July 18	. . . .	M. T. employment.
	Captains.			
22 Nov. 99	Sandes, E. W. C., M.C. (51, 206)	22 Nov. 08	. . . .	M. W. S.—
2 May 00	Arbuthnot, A. D. S. (30, 57, 169, 207)	2 May 10	. . . .	3rd S. & M.
18 Aug. 00	Stace, R. E. (207)	18 Aug. 10	. . . .	M. W. S.—
21 Dec. 01	Tomlinson, H. W. (172a, 207)	21 Dec. 12	. . . .	M. W. S.—
10 Jan. 03	King, F. J. M. (51)	5 Aug. 14	. . . .	Survey Dept.
1 Sep. 03	Colbeck, C. E. (51, 207)	1 Sep. 14	. . . .	Asst. Inspg. Offr., I. S. Sappers.
1 Sep. 03	Pringle, J. C. (51, 207)	1 Sep. 14	. . . .	2nd S. & M.
3 Dec. 03	Stallard, R. H. (207)	30 Oct. 14	. . . .	Ry. Dept.
23 Dec. 03	Hunter, R. G. P. (*act'g. 28 Oct. 16*) (94, 207.)	30 Oct. 14	. . . .	M. W. S.—
26 Dec. 03	Harris, T. E., D.S.O. (51, 105, 207)	30 Oct. 14	. . . .	2nd S. & M.
13 Jan. 04	Deed, L. C. B. (51, 162, 165, 169, 206)	30 Oct. 14	. . . .	M. W. S.—G. E., Nowshera.
14 Feb. 04	Marryat, J. R. (*temp. Maj.*) (207)	30 Oct. 14	. . . .	Ry. Dept.
29 July 04	Stoehr, C. F. (51, 57, 101, 206)	30 Oct. 14	. . . .	3rd S. & M.
29 July 04	Wickham, J. C. (51, 57, 206)	30 Oct. 14	. . . .	2nd S. & M.

First Commn.	Names.	Regimental rank.	Army rank.	Remarks.
	Captains—*contd.*			
29 July 04	Hobart, P. C. S., D.S.O., M.C. (50, 162, 169, 206)	30 Oct. 14 *m.*	*m.* 1 Jan. 18	1st S. & M.
21 Dec. 04	Lewis, C. G. (206)	30 Oct. 14	. . . .	Survey Dept.
21 Dec. 04	Rawlence, M., D. S. O. (51, 65, 172*a*, 206)	30 Oct. 14	. . . .	3rd S. & M.
21 Dec. 04	Bird, C. A., D.S.O. (51, 162, 169, 206)	30 Oct. 14	*m.* 1 Jan. 18	1st S. & M.
14 Jan. 05	Scott, M. A. H., M.C. (*act'g. Maj. 11 July 18*) (162, 206)	30 Oct. 14	. . . .	M. W. S.—Offg. A. C. R. E.
16 Mar. 05	Perry, W. E., M.C. (206) . .	30 Oct. 14	. . . .	Survey Dept.
27 July 05	Stagg, M. (*temp. Maj., 25 Feb. 18*) (95, 206)	30 Oct. 14	. . . .	Dy. Asst. Dir. Genl., Mily. Works, [Army Hdqrs.
26 Aug. 05	Martin, A. D. De R. (207) . .	30 Oct. 14	. . . .	1st S. & M.
5 Sep. 05	Collin, F. S. (169, 207) . .	30 Oct. 14	*m.* 1 Jan. 18	1st S. & M.
17 Sep. 05	Cave-Browne, W. (*act'g.* Maj.) (51, 206)	30 Oct. 14	. . . .	1st S. & M.
1 Oct. 05	Lord, R. C., M.C. (207) . . .	30 Oct. 14	. . . .	M. W. S.—
20 Dec. 05	Hamilton, R. (46, 51, 206) . .	30 Oct. 14	. . . .	2nd S. & M.
25 July 06	Woodhouse, H. L., M.C. (95, 206)	30 Oct. 14	. . . .	Ry. Dept.
25 July 06	Townshend, F. H. E., M.C. (162)	30 Oct. 14	. . . .	P. W. Dept.
5 Aug. 06	Douie, F. Mc C., D.S.O. (*act'g. Maj., 22 Feb. 18*) (162, 205*a*)	30 Oct. 14	*m.* 3 June 17	1st S. & M.
6 Aug. 06	Johnston, D. S. (207)	30 Oct. 14	. . . .	P. W. Dept.
20 Dec. 06	Everett, M., D.S.O. (206) . .	30 Oct. 14	*m.* 3 June 18	P. W. Dept.—Attd. 3rd S. & M.
20 Dec. 06	Manton, L., D.S.O. (*temp. Lt.-Col.*) [*l*] (207)	30 Oct. 14	. . . .	Ry. Dept.
20 Dec. 06	Mason, K., M.C. (207) . .	30 Oct. 14	*m.* 3 June 18	Survey Dept. G. S. O., 2nd gr.
20 Dec. 06	Hughes, H B. W., D.S.O. (169, 207)	30 Oct. 14	. . . .	M. W. S.—
20 Dec. 06	Hogg, D. McA., M. C. [*l*] (65, 162, 205)	30 Oct. 14	. . . .	2nd S. & M.
27 June 07	Rhodes, G. D., D.S.O. (*temp. Lt.-Col., 25 Nov. 16*) (94, 207)	30 Oct. 14	*m.* 3 June 18	Ry. Dept.—Asst. Dir. of Rys.
23 July 07	Witts, F. V. B., D.S.O. (50, 206)	30 Oct. 14	. . . .	1st S. & M.
23 July 07	Kisch, F. H., D.S.O. (94, 172*a*) .	30 Oct. 14	. . . .	G.S.O., 2nd gr., War Office.
23 July 07	Clarke, E. H. (104, 193, 206) . .	30 Oct. 14	. . . .	2nd S. & M.
23 July 07	Preedy, C. (*temp. Major*) . .	30 Oct. 14	. . . .	Asst. Dir. of Works.
23 July 07	Chater, A. F. (*act'g.* Maj.) (162, 205*a*, 206)	30 Oct. 14	. . . .	1st S. & M.
23 July 07	Gill, A. G.	30 Oct. 14	. . . .	G. E., Rangoon.
24 July 07	Budden, F. H., M.C. (*act'g. Maj. 24 Jan. 17*)	30 Oct. 14	*m.* 1 Jan. 18	Ry. Dept.
1 Aug. 07	Fox, E. B., (207)	30 Oct. 14	. . .	3rd S. & M.
18 Dec. 07	Nosworthy, F. P., M.C. (207) .	30 Oct. 14	*m.* 1 Jan. 17	Survey Dept.—G. S. O. 2nd gr.
18 Dec. 07	Roome, H. E., M.C. (*act'g. Maj. 22 Sep. 18*) (206)	30 Oct. 14	. . . .	Survey Dept. 1st S. and M.
18 Dec. 07	Walton, G. (*temp. Lt.-Col., 16 Mar. 17*) (169, 207)	30 Oct. 14	. . . .	Ry. Dept.—Brig.-Maj.
26 June 08	Carson, C. F., M.C. (206) . . .	30 Oct. 14	. . . .	Ry. Dept.
29 July 08	Gompertz, A. V., M.C. (*temp. Maj., 27 Nov. 18*) (*l*) (71, 95, 207)	30 Oct. 14	. . . .	M. W. S.—G. S. O., 2nd gr., Army [Hdqrs.
29 July 08	Campbell, M. G. G., M.C. (57, 94, 207)	30 Oct. 14	. . . .	3rd S. & M.
29 July 08	Gidley-Kitchin, F. G. (205*a*, 206)	30 Oct. 14	. . . .	1st S. & M.
18 Dec. 08	Fitzpatrick, N. T., D.S.O., M.C. (172*a*, 207)	18 Dec. 14	*m.* 1 Jan. 17	P. W. Dept.
18 Dec. 08	Rohde, J. H. (169, 207) . . .	18 Dec. 14	. . . .	3rd S. & M.
18 Dec. 08	Geary, H. N. G. (95)	18 Dec. 14	. . . .	Inspg. Offr., I. S. Sappers.
3 July 09	Gosset, H. H. E., M.C. (207) .	23 July 15	. . . .	With M. W. S.
3 July 09	Buller, F. E., M.C. (207) . . .	23 July 15	. . . .	
31 July 09	Roberts, I. K., M.C. (172*a*, 207) .	31 July 15	. . . .	Ry. Dept.
20 Sep. 09	Bradney, E. (206)	20 Sep. 15	*m.* 3 June 16	2nd S. & M.
23 Dec. 09	Britten, W. E. (169, 207) . . .	23 Dec. 15	. . .	2nd S. & M.
23 Dec. 09	Farley, E. L., M.C. (207) . . .	23 Dec. 15	. . . .	1st S. & M.
23 Dec. 09	Gordon, R. E., M.C. (207) . .	23 Dec. 15	. . . .	Ry. Bn.
26 Jan. 10	Morse, A. H.	26 Jan. 16	. . . .	
9 Feb. 10	Phipps, C. C., M.C. (207) . .	9 Feb. 16	. . . .	Ry. Dept.
23 June 10	Wheeler, E. C., M.C., (*act'g. Maj.*) (207)	23 June 16	. . . .	1st S. & M.
23 July 10	King, C. J. S. (206) . . .	23 July 16	. . . .	2nd S. & M.
23 July 10	Sandford, H. R. (169, 205*a*, 206) .	23 July 16	. . . .	Offg. A. C. R. E., Derajat Dist.
23 July 10	Slater, O. (207) . .	23 July 16	. . . .	
23 July 10	Glennie, E. A., D.S.O. (*act'g. Maj 2 Dec. 16*) .	23 July 16	. . . .	G. E., Wellington.
23 July 10	Hounsell, R. H. S., (*temp. Maj. 25 Oct. 18*) (207)	23 July 16	. . . .	Dy. A. D. G., M. W., Army Hdqrs.
23 Dec. 10	Yearsley, K. D. (172*a*, 207) . .	23 Dec. 16	. . . .	M. W. S.
23 Dec. 10	Sykes, A. C. (207)	23 Dec. 16	. . . .	M. W. S.
23 Dec. 10	Douglas, J. K.	23 Dec. 16	. . . .	3rd S. & M.
23 Dec. 10	Rait Kerr, R. S., D.S.O., M.C. (207)	23 Dec. 16	. . . .	M. W. S.
23 Dec. 10	Cruickshank, A. J., D.S.O. (172*a*, 206)	23 Dec. 16	. . . .	2nd S. & M.—G. S. O., 2nd gr.

First Commn.	Names.	Regimental rank.	Army rank.	Remarks.
	Captains—*concld.*			
23 Dec. 10	Wagstaff, H.W., M.C. (172a, 207)	23 Dec. 16	. . . .	M. W. S.—G. S. O., 3rd gr.
23 Dec. 10	Guinness, V. E. G. (*act'g. Maj., 17 Sep. 18*) (207)	23 Dec. 16	. . . .	M. W. S.
21 June 11	Joly deLotbinière, H. A. (207)	21 June 17	. . . .	Attd. 1st S. & M.
20 July 11	Mason, A., M.C. (*temp. Capt., 23 Sep. 15*)	26 June 17	. . . .	1st S. & M.
20 July 11	Dewing, R. H., D.S.O.	26 June 17	. . . .	2nd S. & M.—G. S. O., 3rd gr.
20 July 11	Watson, G. D. (207)	26 June 17	. . . .	1st S. & M.
20 July 11	Johnston, E. F. (207)	26 June 17	. . . .	Attd. 3rd S. & M.
23 Dec. 11	Drew, F. G. (*act'g. Maj.*) (207)	26 June 17	. . . .	1st S. & M.
23 Dec. 11	Hamilton, H. W. R., M. C. (207)	26 June 17	. . . .	M. W. S.
23 Dec. 11	Papillon, A. H. B.	26 June 17	. . . .	Attd. 2nd S. & M.
23 Dec. 11	Greenwood, C. J. E.	26 June 17	. . . .	M. W. S.
23 Dec. 11	Matthews, A. B., D.S.O., M.C.	26 June 17	. . . .	Attd. 3rd S. & M.
23 Dec. 11	Crawford, K. B. S.	26 June 17	. . . .	Attd. 3rd S. & M.
19 July 12	Aitken, A. B., M.C., (*act'g. Maj., 26 Oct. 18*)	25 June 17	. . . .	Attd. 2nd S. & M.
16 Sep. 14	Morris, D.	3 Nov. 17	. . . .	3rd S. & M.
10 Feb. 15	Watson, R. A. D.	27 Mar. 18	. . . .	1st S. & M.
	Lieutenants.			
20 July 11	Drake-Brockman, R. A.	16 Aug. 13		M. W. S.
10 Feb. 15	Pim, G. R., M.C.	11 May 17		1st S. & M.
28 July 15	Lethbridge, J. S. (207a)	1 July 17		Attd. 1st S. & M.
27 Oct. 15	Dawes, L. F. S.	1 July 17		1st S. & M.
27 Oct. 15	Burrows, C. H. C.	1 July 17		1st S. & M.
20 Nov. 15	Farmer, G. C. (*act'g. Capt., 24 Sep. 18*)	1 July 17		Divl. Retg. Offr.
20 Nov. 15	Holmes, H. G.	1 July 17		Attd. Mily. Works Branch, Army Hdqrs.
20 Nov. 15	Parselle, J. (Capt.)	1 July 17		M. W. S.
20 Nov. 15	Battman, A. R. (*act'g. Capt. 24 Sep. 18*)	1 July 17		G. E., Peshawar.
20 Nov. 15	Wogan, H. J.	1 July 17		1st S. & M.
20 Nov. 15	Wilmer, T. G.	1 July 17		M. W. S.
20 Nov. 15	Pearson, A. F. (*act'g. Capt. 24 Sep. 18*)	1 July 17		G. E., Ambala.
20 Nov. 15	Abbott, T. W.	1 July 17		G. E., Wellington.
20 Nov. 15	Heenan, J.	1 July 17		G. E., Malakand.
20 Nov. 15	Brown, G.	1 July 17		G. E., Shabkadr.
20 Nov. 15	Lyall, W. J. (*act'g. Capt., 26 Oct. 17*) (206)	1 July 17		1st S. & M.
20 Nov. 15	Stevens, A. E. (104)	1 July 17		2nd S. & M.
19 Feb. 16	Obbard, H. N.	19 Aug. 17		1st S. & M.
19 Feb. 16	Crowther, C. T. B.	19 Aug. 17		2nd S. & M.
19 Feb. 16	Bird, C. F. A.	19 Aug. 17		1st S. & M.
19 Feb. 16	Le Sueur, J. H. R.	19 Aug. 17		2nd S. & M.
19 Feb. 16	Cameron, A. M., M.C.	19 Aug. 17		1st S. & M.
26 May 16	Costello, W. H. G.	26 Nov. 17		2nd S. & M.
26 May 16	Mitchell, A. C.	26 Nov. 17		2nd S. & M.
	2nd-Lieutenants.			
28 Feb. 17	Longland, R. H. B. (*act'g. Capt. 24 Sep. 18*)	28 Feb. 17		2nd S. & M.
28 Feb. 17	Bogs, W. W.	28 Feb. 17		3rd S. & M.
28 Feb. 17	Foley, P. F.	28 Feb. 17		3rd S. & M.
6 June 17	Hotine, M.	6 June 17		2nd S. & M.
6 June 17	Gueterbock, E. A. L.	6 June 17		3rd S. & M.
6 June 17	Porter, W. (46)	6 June 17		3rd S. & M.
6 June 17	Price, T. J. P.	6 June 17		1st S. & M.
6 June 17	Roberts, O. L. (207)	6 June 17		1st S. & M.
6 June 17	Vernon, G. F. D.	6 June 17		2nd S. & M.
6 June 17	Stillingfleet, R.	6 June 17		2nd S. & M.
6 June 17	Baker, S. J.	6 June 17		3rd S. & M.
6 June 17	Sydenham, P. S. B.	6 June 17		3rd S. & M.
6 June 17	Harrison, H. B.	6 June 17		3rd S. & M.
6 June 17	Capron, G. T. H.	6 June 17		3rd S. & M.
4 Sep. 18	Garcia-Henry, A.	4 Sep. 18		With Sig. Serv.
	SUPPLEMENTARY.			
	Captains.			
24 Jan. 15	Boal, K., D.S.O. (*act'g.* Maj.)	24 Mar. 18		1st S. & M.
24 Jan. 15	Dolan, M. J.	24 Mar. 18		3rd S. & M.
24 Jan. 15	Johnson, H. E. (193, 206)	24 Mar. 18		With Sig. Ser.
	Lieutenants.			
26 Apl. 15	Pocock, J. A.	1 July 17		
	2nd-Lieutenants.			
28 May 18	Kilcoin, C. L.	28 May 18		
28 May 18	Bradbury, W. E.	28 May 18		
28 May 18	Fitzpatrick, P. J.	28 May 18		
28 May 18	Macdonald, E. G.	28 May 18		
28 May 18	Mattocks, F.	28 May 18		
28 May 18	Beard, A. W.	28 May 18		
28 May 18	Bickley, G. W.	28 May 18		

First Commn.	Names.	Regimental rank.	Remarks.
	2nd-Lieutenant—*contd.*		
28 May 18	Woolner, P. C.	28 May 18	
28 May 18	Cam, G.	28 May 18	
28 May 18	Green, H. W.	28 May 18	
28 May 18	Daw, H. H.	28 May 18	
28 May 18	Hastings, F. J. B.	28 May 18	
28 May 18	Dawkins, H. (*actg. Capt. 15 Oct. 18*)	28 May 18	1st S. & M.
28 May 18	Maun, J. W.	28 May 18	
28 May 18	Dawson, F. T. G.	28 May 18	
28 May 18	Dower J. H.	28 May 18	1st S. & M.
28 May 18	Coltham, W. E.	28 May 18	1st S. & M.
28 May 18	Deverell, J. J. D. C. M.	28 May 18	1st S. & M.
28 May 18	Argent, J.	28 May 18	
28 May 18	Lloyd, H. S.	28 May 18	
	SPECIAL RESERVE.		
	Captains.		
	Gurney, C. R.	26 June 17	2nd S. & M.
	Morris J. H.	26 June 17	Ry. Bn.
	Roche, R. T. H.	26 June 17	Ry. Bn.
	Stevenson, C. Y.	26 June 17	
	Rhodes, G. E.	26 June 17	Ex. Engr., Works Dept., Ordce. Factories, Calcutta.
	Pennell, K. E. L., M.C.	26 June 17	3rd S. & M.
	Collier, A. E., M.C.	3 Nov. 17	2nd S. & M.
	TERRITORIAL FORCE.		
	Captain.		
	Wilson, J. L.	1 June 16	
	Lieutenant.		
	Symonds, P. R.	1 June 17	
	2nd-Lieutenants.		
	Wilkinson, F. S.	5 Oct. 15	
	Knight, E. H. (*act'g Capt.*)	4 Dec. 15	3rd S. & M.
	Vickers, T. B.	1 Apl. 17	2nd S. & M.

OFFICERS, NOT SHEWN IN REGIMENT OR CORPS EXTRA REGIMENTALLY EMPLOYED.

First Commn.	Names.	Regimental rank.	Remarks.
	2nd-Lieutenants.		
30 Apl. 16	Henry, G. C. (*actg. Capt.*)	30 Apl. 16	With Sig. Ser.
25 Nov. 16	Ryan, P. (*act'g. Lt., 25 Sep. 17*)	25 Nov. 16	With Sig. Ser.
9 Dec. 16	Leng, C.	9 Dec. 16	With Sig. Ser.
	Temporary Lieut.-Colonels.		
	Barton, G. B. B.	9 Aug. 16	Q. M. G.'s Branch, A. Hdqrs.
	Temporary Majors.		
	Craster, E. E.	18 May 15	A. C. R. E.. Lucknow Dist.
	Maxwell, B. C. D'A. (*act'g. Lt.-Col.*)	1 Sep. 16	Q. M. G.'s Branch, A. Hdqrs.
	Temporary Captain.		
26 June 15	Boucher, C. L. (172*a*)	18 Mar. 16	M. W. S.—
	Martin, J. Mc. E.	15 May 15	1st S. & M. *Lv., ex I., m. c., 12 mos.*
	Norrie, C. M., D.S.O.	. . .	Dy. Asst. Dir. of Works.
	Temporary Lieutenants.		
	Bennett, H. D.	16 Mar. 15 9 Oct. 14	3rd S. & M.
	Power, A.E. (*actg. Capt., 24 Sep. 18*)	30 Mar. 15	O. C. No. 1 Engr. Fd. Park and G. E., Peshawar.
	Murray, H. K. (*act'g. Capt., Feb. 17*)	2 June 15	
	Arnell, O. R.	11 July 15	Attd. 2nd S. & M.
	Singer, C. A. A. *temp. Maj., 11 Aug. 18.*	11 July 15	With Sig. Ser. Dy. Asst. Dir. of Army Signals.
	Mawson, G. H. (*act'g Capt.*)	11 July 15	With Sig. Ser.
	Woodrow, F. H.	11 July 15	With Sig. Ser.
	Adams, K. L. (*temp. Maj., 25 Feb. 18.*)	29 Sep. 15	D. A. A. G., Army Hdqrs.
	Byrne, H. S. (*act'g. Capt.*)	1 Jan. 16	G. E., Bannu.
	Shepherd, J. D., M.C. (207*a*)	9 Apl. 16	Attd. 1st S. & M.
	Gross, A. A.	12 July 16	Attd. 3rd S. & M.
	Tree, P. B. LeD. (*act'g. Capt., 16 June 17*)	23 July 16	
	Philp, R. E.	28 Aug. 16	Attd. 3rd S. & M.
	Mathews, C. W.	2 Sep. 16	2nd S. & M.
	Hegarty, T. C. V.	4 Oct. 16	G. E.—Ahmednagar.
	Paton, J.	28 Mar. 17	Attd. 1st S. & M. *Lv., ex I., m. c., 6 mos.*
	Prescot, C. P. (*atc'g. Capt., 23 Oct. 17*)	1 July 17	With Sig. Ser.
	Wallis, S. A.	1 July 17	Attd. 3rd S. & M.
	Hill, J. H.	1 July 17	Attd. 3rd S. & M.
	FitzGerald, W. R. (207*a*)	1 July 17	Attd. 3rd S. & M.
	Curtis, H.	1 July 17	3rd S. & M.
	Wrightson, N. P. E. (46)	1 July 17	Attd. 1st S. & M.
	Macdonald, Clark, A. F. (*act'g. Capt. 17 May 18.*)	8 Dec. 14	Attd. 1st S. & M.

First Commn.	Names.	Regimental rank.	Remarks.
	Temporary Lieutenants—*contd.*		
	Leedam, W. D.	1 July 17	With Sig. Serv.
	Allen, H.	1 July 17	
	Turnbull, W. G.	1 July 17	Ind. Mun. Board.
	Chapple, C. H.	15 July 17	
	Joyce, P. E.	30 July 17	Attd. 2nd S. & M.
	Waugh, A. D.	5 Aug. 17	Attd. 3rd S. & M.
	Stevenson, J. G.	13 Oct. 17	Attd. 3rd S. & M.
	Friedlander, O. A.	7 Nov. 17	3rd S & M.
	Brewer-Little, H.	1 Dec. 17	Attd. 3rd S. & M.
	Ingram, G.	3 Dec. 17	Ind. Mun. Board.
	Hemsworth, T. G. F.	30 July 17	Attd. 3rd S. & M.
	Kirby, S. J.	5 Aug. 17	3rd S. & M.
	Geoghegan, G.	2 Jan. 18	Attd. 3rd S. & M.
	McLay, D. B. (46)	2 Jan. 18	2nd S. & M.
	McAndrew, G. D.	23 Jan. 18	Attd. 3rd S. & M.
	Pearse, G. E.	23 Jan. 18	Attd. 3rd S. & M.
	Marshall, E. N. C. (207*a*)	19 Feb. 18	Attd. 2nd S. & M.
	Newman, B. L.	19 Feb. 18	Munitions Board.
	Otto, C. R.	19 Feb. 18	Attd. 3rd S. & M.
	McClelland, A. A. (*act'g. Capt., 14 Sep. 18*)	4 May 18	2nd S. & M.
	Walter, J. A.	4 May 18	Attd. 3rd S. & M.
	Riccomini, J. A.	4 May 18	G. E., Chitral.
	Spencer, W. N.	4 May 18	Attd. 3rd S. & M.
	Temporary 2nd-Lieutenants.		
	Minner, C. (12, 30, 51)	13 Feb. 16	
	Bowen, E. B. (*act'g.* Lt.)	19 Aug. 16	G. E.—Wana.
	Bennett, S. G.	19 Aug. 16	3rd S. & M.
	Lyon, W. A.	4 Nov. 16	Attd. 1st S. & M.
	Willis, V. J. (207*a*)	4 Nov. 16	Attd. 3rd S. & M.
	Ogborn, F. E.	4 Nov. 16	Attd. 2nd S. & M.
	Craig, M. T.	4 Nov. 16	Attd. 1st S. & M.
	Baxter, G. D.	4 Nov. 16	Attd. 1st S. & M.
	Pike, J. P.	4 Nov. 6	2nd S. & M.
	Turnbull, W. J.	4 Nov. 16	Attd. 3rd S. & M.
	Powell, W. G.	4 Nov. 16	2nd S. & M.
	Gardner, J. (*act'g. Capt., 24 Sep. 18*).	4 Nov. 16	Attd. 3rd S. & M.
	Taylor, J. H. C.	4 Nov. 16	Attd. 3rd S. & M.
	Pringle, J. E.	2 Dec. 16	Attd. 2nd S. & M.
	Barnes, L. F.	2 Dec. 16	3rd S & M.
	Wynne-Roberts, L. W.	2 Dec. 16	2nd S. & M.
	Jones, M. P.	27 Jan. 17 7 June 16	3rd S. & M.
	Read, M. A.	11 Jan. 17	Asst. G. E., Peshawar.
	Davis, R. W. J.	19 May 17	Attd. 1st S. & M.
	Pendlebury, R.	3 June 17	3rd S. & M.
	Bone, C. C. (*aac'g. Capt., 24 Sep. 18*).	28 July 17	M. W. S.—R. Pindi.
	Watt, G. T.	28 July 17	
	Pringle, J.	28 July 17	Attd. 1st S. & M.
	Mason, T. K.	28 July 17	
	Hall, H.	31 July 17	Attd. 1st S. & M.
	Anderson, H. C. (*act'g. Lieut., 24 Sep. 18*).	11 Aug. 17	M. W. S.
	Burt, S.	11 Aug. 17	M. W. S.
	Cropper, H.	11 Aug. 17	Ry. S. & M.
	Johnston, T. E.	11 Aug. 17	Attd. 1st S. & M.
	Postlethwaite, J. L.	11 Aug. 17	Bombay.
	Nichols, C. E.	11 Aug. 17	Asst. Gr., Engr. Peshawar.
	Routhledge, L. G. F.	11 Aug. 17	M. W. S.—Mhow.
	Crowley, B. E.	23 Sep. 17	3rd S. & M.
	Patterson, H. (207*a*)	17 Nov. 17	3rd S. & M.
	Forster, T.	11 Aug. 17	Attd. 1st S. & M.

H. Company of Royal Engineers.

(Formed G. G. O. No. 418, 22nd April 1868, and I. A. Cir., 1891, clause 76.)
Designation changed by India Army Circular, 1885, clause 97.
Head-Quarters—**Simla.**

Loring, Maj. E. J., M.C., R.E. . . . . . Commanding.

THE ROYAL (INDIAN) ENGINEERS.

NOTE.—All these officers are on the Unemployed Supernumerary List.

Names.	Date of birth.	2nd-Lieut.	Lieut.	Captain.	Major.	Lieut.-Col.	Colonel.	Major-General.	Lieut.-General.	General.	Colonel Comdt.
Generals.											
Walker, George Warren (M.)	6 Apl. 23	9 Dec. 42	10 Dec. 47	23 Jan. 56		30 Sep. 61	30 Sep. 66	6 Mar. 68	15 Aug. 78	28 Nov. 85	25 July 99
Tomkins, William Percival, C.I.E. (B)	11 June 41		8 June 60	3 Aug. 72	1 July 81	26 May 88	12 Apl. 93	21 May 96	1 Apl. 97	11 Apl. 98	12 June 98
Lieutenant-General.											
Strahan, Charles (B)	6 May 43		8 June 60	20 Aug. 72	2 Mar. 81	1 July 88	12 Apl. 93	27 Nov. 96	28 Apl. 01		23 Dec. 02

ALPHABETICAL LIST OF REGIMENTS.

The Regiments or Battalions shown in italics are not on the Indian Establishment.

Present title.		Late title.	Army List abbreviations.	Page.
Argyll and Sutherland Highlanders.	*1st Bn.*	*91st Foot*	*Arg. & Suth'd. Highrs.*	...
	2nd Bn.	*93rd Foot*		
Bedfordshire Regt.	*1st Bn.*	*16th Foot*	Bedf. R	708
	2nd Bn.			709
	1st Garr. Bn.			710
	2nd Garr. Bn.			
	3rd Garr. Bn.			
Berkshire Regt., Royal	*1st Bn.*	*49th Foot*	*R. Berks. R.*	...
	2nd Bn.	*66th Foot*		
Border Regt.	*1st Bn.*	*34th Foot*	Bord. R.	...
	2nd Bn.	*55th Foot*		...
	1-4th Bn.	. . . . .		724
	2-4th Bn.	. . . . .		725
Cameron Highlanders	*1st Bn.*	*79th Foot*	*Cam'n Highrs.*	...
	2nd Bn.	. . . . .		
Cheshire Regt.	*1st Bn.*	*22nd Foot*	*Ches. R.*	...
	2nd Bn.			
Connaught Rangers	1st Bn.	88th Foot	Conn. Rang.	769-70
	2nd Bn.	*94th Foot*		
Cornwall's Light Infantry, Duke of	*1st Bn.*	*32nd Foot*	D. of Corn. L.I.	...
	2nd Bn.	*46th Foot*		...
	2-4th Bn.	. . . . .		722
Devonshire Regt.	*1st Bn.*	*11th Foot*	Devon. R.	...
	2nd Bn.			
	1-4th Bn.	. . . . .		696
	2-4th Bn.	. . . . .		697
	1-6th Bn.	. . . . .		698-99
	2-6th Bn.	. . . . .		700
Dorsetshire Regt.	*1st Bn.*	*39th Foot*	Dorset. R.	...
	2nd Bn.	54th Foot		736-37
	1-4th Bn.	. . . . .		738-39
	2-4th Bn.	. . . . .		740
Dublin Fusiliers, Royal	*1st Bn.*	*102nd Foot*	*R. Dub. Fus.*	...
	2nd Bn.	*103rd Foot*		
Durham Light Infantry	1st Bn.	68th Foot	Durh. L. I.	761
	2nd Bn.	*106th Foot*		
Essex Regt.	*1st Bn.*	*44th Foot*	Essex R.	
	2nd Bn.	*56th Foot*		747
	2nd Garr. Bn.			
Gloucestershire Regt.	*1st Bn.*	*28th Foot*	*Glouc. R.*	...
	2nd Bn.	*61st Foot*		
Gordon Highlanders	*1st Bn.*	*75th Foot*	Gord. Highrs.	
	2nd Bn.	*92nd Foot*		766
	1st Garr. Bn.			
Hampshire Regt.	*1st Bn.*	*37th Foot*	Hamps. R.	...
	2nd Bn.	*67th Foot*		
	1-4th Bn.	. . . . .		729
	1-5th Bn.	. . . . .		730
	1-6th Bn.	. . . . .		731
	1-7th Bn.	. . . . .		732
	2-7th Bn.	. . . . .		733
	1-9th Bn.	. . . . .		734
Highland Light Infantry	1st Bn.	71st Foot	High. L. I.	762-63
	2nd Bn.	*67th Foot*		
Inniskilling Fusiliers, Royal	*1st Bn.*	*27th Foot*	*R. Innis. Fus.*	...
	2nd Bn.	*108th Foot*		
Irish Fusiliers, Royal	*1st Bn.*	*87th Foot*	R. Ir. Fus.	768
	2nd Bn.	*89th Foot*		
	1st Garr. Bn.			
Irish Regt., Royal	*1st Bn.*	*18th Foot*	*R. Ir. Regt.*	...
	2nd Bn.			
Irish Rifles, Royal	*1st Bn.*	*83rd Foot*	R. Ir. Rif.	767
	2nd Bn.	*86th Foot*		
	1st Garr. Bn.			
Kent Regt., East	*1st Bn.*	*3rd Foot*	E. Kent R.	...
	2nd Bn.			
	1-4th Bn.	. . . . .		683-84
	1-5th Bn.	. . . . .		685-86
Kent Regt., Royal West	*1st Bn.*	*50th Foot*	R. W. Kent R.	
	2nd Bn.	97th Foot		...
	1-4th Bn.	. . . . .		749-50
	1-5th Bn.	. . . . .		751
				752
King's Royal Rifles Corps	*1st Bn.*	*60th Foot*	*K. R. Rif. C.*	
	2nd Bn.			
	3rd Bn.			
	4th Bn.			

ALPHABETICAL LIST OF REGIMENTS.

The Regiments or Battalions shown in italics are not on the Indian Establishment.

Present title.		Late title.	Army List abbreviations.	Page.
Lancashire Fusiliers	*1st Bn.* *2nd Bn.*	*20th Foot*	*Lan. Fus.*	...
Lancashire Regt., East	*1st Bn.* *2nd Bn.*	*30th Foot* *59th Foot*	*E. Lan. R.*	...
Lancashire Regt., Loyal North	*1st Bn.* 2nd Bn.	*47th Foot* 81st Foot	N. Lan. R.	748
Lancashire Regt., South	1st Bn. *2nd Bn.*	40th Foot *82nd Foot*	S. Lan. R.	741-42
Lancaster Regt., Royal	*1st Bn.* *2nd Bn.*	*4th Foot*	*R. Lanc. R.*	...
Leicestershire Regt.	*1st Bn.* 2nd Bn.	17th Foot	Leic. R.	711
Leinster Regt.	*1st Bn.* *2nd Bn.*	*100th Foot* *109th Foot*	*Leins. R.*	...
Lincolnshire Regt.	*1st Bn.* *2nd Bn.* 1st Garr. Bn.	*10th Foot*	Linc. R.	695
Liverpool Regt.	*1st Bn.* 2nd Bn.	8th Foot	L'pool R.	689-90
Manchester Regt.	1st Bn. *2nd Bn.* 1st Garr. Bn.	63rd Foot *96th Foot*	Manch. R.	758 759
Middlesex Regt.	*1st Bn.* *2nd Bn.* *3rd Bn.* 1-9th Bn. 1-10th Bn.	*57th Foot* *77th Foot*	Midd'x R.	... ... 753 754
Munster Fusiliers, Royal	*1st Bn.* *2nd Bn.*	*101st Foot* *104th Foot*	*R. Muns. Fus.*	...
Norfolk Regt.	*1st Bn.* 2nd Bn. 1st Garr. Bn	9th Foot	Norf. R.	691-92 693-94
Northamptonshire Regt.	*1st Bn.* *2nd Bn.*	*48th Foot* *58th Foot*	*North'n R.*	...
Northumberland Fusiliers	*1st Bn.* *2nd Bn.* 2nd Garr. Bn.	*5th Foot*	North'd Fus.	687-88
Nottinghamshire and Derbyshire Regt.	*1st Bn.* *2nd Bn.*	*45th Foot* *95th Foot*	*Notts. & Derby R.*	...
Oxfordshire and Buckinghamshire Light Infantry.	1st Bn. *2nd Bn.* 1st Garr. Bn.	43rd Foot *52nd Foot*	Oxf. & Bucks. L. I.	745 746
Rifle Brigade	*1st Bn.* *2nd Bn.* *3rd Bn.* *4th Bn.* 18th Bn. 21st Bn. 23rd Bn. 24th Bn.	*Rifle Brigade*	Rif. Brig.	776 776a 777 778
Royal Fusiliers	*1st Bn.* *2nd Bn.* *3rd Bn.*	*7th Foot*	R. Fus.	...
Royal Highlanders	*1st Bn.* 2nd Bn.	*42nd Foot* 73rd Foot	R. Highrs.	743-44
Scots, Royal	*1st Bn.* *2nd Bn.*	*1st Foot*	*R. Scots.*	...
Scots Fusiliers, Royal	*1st Bn.* *2nd Bn.* 1st Garr. Bn.	*21st Foot*	R. Sc. Fus.	715-16
Scottish Borderers, King's Own	1st Bn. *2nd Bn.*	25th Foot	K. O. Sco. Bord.	718
Scottish Rifles	*1st Bn.* *2nd Bn.* 1st Garr. Bn.	*26th Foot* *90th Foot*	Sco. Rif.	719
Seaforth Highlanders	1st Bn. *2nd Bn.*	72nd Foot *78th Foot*	Sea. Highrs.	764-65
Shropshire Light Infantry	*1st Bn.* *2nd Bn.*	*53rd Foot* *85th Foot*	*Shrops. L. I.*	... ...
Somerset Light Infantry	*1st Bn.* 2nd Bn. 1-4th Bn. 2-4th Bn. 2-5th Bn. 1st Garr. Bn.	13th Foot	Som. L. I.	701 702-03 704 705 706

ALPHABETICAL LIST OF REGIMENTS.

The Regiments or Battalions shown in italics are not on the Indian Establishment.

Present title.		Late title.	Army List abbreviations.	Page.
Staffordshire Regt., North	*1st Bn.*	*64th Foot*	N. Staff. R.	760
	2nd Bn.	98th Foot		
Staffordshire Regt., South	*1st Bn.*	*38th Foot*	*S. Staff R.*	735
	2nd Bn.	*80th Foot*		
	1st Garr. Bn.	. . . . . .		
Suffolk Regt.	*1st Bn.*	*12th Foot*	*Suff. R.*	...
	2nd Bn.			
Surrey Regt., East	*1st Bn.*	*31st Foot*	E. Surr. R.	...
	2nd Bn.	*70th Foot*		...
	1-5th Bn.	. . . . .		720
	1-6th Bn.	. . . . .		721
Surrey Regt., Royal West	*1st Bn.*	*2nd Foot*	R. W. Surr. R.	...
	2nd Bn.			
	1-4th Bn.	. . . . .		679-80
	1-5th Bn.	. . . . .		681-82
Sussex Regt., Royal	1st Bn.	35th Foot	R. Suss. R.	726-27
	2nd Bn.	*107th Foot*		
	2-6th Bn.			728
Wales Borderers, South	*1st Bn.*	*24th Foot*	S. Wales Bord.	717
	2nd Bn.			
Warwickshire Regt., Royal	*1st Bn.*	*6th Foot*	*R. War R.*	...
	2nd Bn.			
Welsh Fusiliers, Royal	*1st Bn.*	*23rd Foot*	*R. W. Fus.*	...
	2nd Bn.			
Welsh Regt.	*1st Bn.*	*41st Foot*	*Welsh R.*	...
	2nd Bn.	*69th Foot*		
West Riding Regt.	1st Bn.	33rd Foot	W. Rid. R.	723
	2nd Bn.	*76th Foot*		
Wiltshire Regt.	*1st Bn.*	*62nd Foot*	Wilts. R.	...
	2nd Bn.	*99th Foot*		...
	1-4th Bn.	. . . . .		755
	2-4th Bn.	. . . . .		756-57
Worcestershire Regt.	*1st Bn.*	*29th Foot*	*Worc. R.*	...
	2nd Bn.	*36th Foot*		
	3rd Bn.	. . . . .		
	4th Bn.	. . . . .		
York and Lancaster Regt.	*1st Bn.*	*65th Foot*	*York. & Lanc. R.*	...
	2nd Bn.	*84th Foot*		
Yorkshire Light Infantry	*1st Bn.*	*51st Foot*	*York. L. I.*	...
	2nd Bn.	*105th Foot*		
Yorkshire Regt.	1st Bn.	*19th Foot*	York. R.	712-13
	2nd Bn.			
	1st Garr. Bn.			714
Yorkshire Regt., East	*1st Bn.*	*15th Foot*	E. York. R.	707
	2nd Bn.			
	1st Garr. Bn.			
Yorkshire Regt., West	*1st Bn.*	*14th Foot*	*W. York R.*	...
	2nd Bn.			
Kent. Bn.	1-1st Bn.	. . . . .	Kent. R.	780
London Regt.	1-25th Bn.	. . . . .	Lond. R.	779

Name of Corps.	Army List Abbreviations.	Page.
45th Aden Rifles	Aden Rif. . . .	860
19th Agra Company . . .	Agra Coy. . . .	829
8th Allahabad Rifles . . .	Alld. Rif. . . .	814
6th Assam Valley Light Horse	Assam V. L. H. .	606
42nd Assam-Bengal Railway Battalion.	A.-B. Ry. Bn. .	857
27th Baluchistan Company .	Bal. Coy. . . .	842
6th Bangalore, Coorg and Mysore Battalion.	Bang. C. & M. Bn.	808
22nd Bengal and North-Western Railway Battalion.	B. & N.-W. Ry. Bn.	834
36th Bengal-Nagpur Railway Battalion (1st & 2nd Bn.).	B.-N. Ry. Bn. .	850
1st Bihar Light Horse . .	Bihar L. H. . .	601
17th Bombay, Baroda and Central India Railway Battalion (1st & 2nd Bn.).	B. B. & C. I. Ry. Bn.	825
4th Bombay Light Horse. .	Bo. L. H. . . .	604
4th (Bombay) Group Garrison Artillery.	Bo. Gp. Garr. Arty.	654
15th Bombay Battalion . .	Bo. Bn. . . .	821
21st Burma Railways Battalion.	Bur. Ry. Bn. . .	832
3rd Calcutta Light Horse. .	Cal. L. H. . .	603
1st (Calcutta Port Defence) Group Garrison Artillery.	Cal. P. D. Gp. Garr. Arty.	651
44th Calcutta Scottish . .	Cal. Sco. . . .	859
5th Calcutta Battalion . .	Cal. Bn. . . .	807
37th Calcutta Presidency Battalion.	Cal. Pres. Bn. .	852
16th Cawnpore Rifles . .	Cawnpore Rif. .	823
39th Chota Nagpur Regt. .	Chota Nag. R. . .	855
1st-5th (Cossipore) Brigade, Mobile Artillery.	Coss. Bde. Mob. Arty.	656
38th East Coast Battalion .	E. Coast Bn. . .	854
th East Indian Railway Battalion.	E. I. Ry. Bn. . .	810
12th Eastern Bengal Railway Battalion.	E. B. Ry. Bn. .	817
41st Eastern Bengal Company.	E. B. Coy. . .	856
13th Great Indian Peninsula Railway Battalion (1st & 2nd Bn.).	G. I. P. Ry. Bn. .	818
26th Hyderabad Rifles . .	Hyd. Rif. . . .	841
2nd Kalachi Brigade, Mobile Artillery.	Kar. Bde., Mob. Arty.	657
"F" Kirkee Battery, Mobile Artillery.	Kirkee By. Mob. Arty.	655
43rd Kolar Gold Fields Battalion.	Kolar G. F. Bn. .	858
5th (Lucknow) Group Garrison Artillery.	Luck. Gp. Garr. Arty.	658
2nd (Madras) Group Garrison Artillery (*The Duke's Own*).	Mad. Gp. Garr. Arty.	652
32nd Madras and Southern Mahratta Railway Rifles (1st & 2nd Bn.).	M. & So. Mah. Ry. Rif.	844
1st Madras Guards . . .	Mad. Gds. . .	801
9th Mussoorie Battalion . .	Muss. Bn. . . .	815
2nd Nagpur Rifles . . .	Nag. Rif. . . .	803
10th Naini Tal Company .	N'-Tal. Coy. . .	816
24th North Western Railway Battalion (1st & 2nd Bn.).	N. W. Ry. Bn. .	836
9th Northern Bengal Mounted Rifles.	N. Beng. M. Rif.	610
25th Oudh and Rohilkhand Railway Battalion.	O. & R. Ry. Bn. .	840
35th Poona Battalion . .	Poona Bn. . .	849
5th Punjab Light Horse . .	Punj. L. H. . .	605
3rd Punjab Rifles . . .	Punj. Rif. . . .	804
3rd (Rangoon Port Defence) Group Garrison Artillery.	Rang'n P. D. Gp. Garr. Arty.	653
18th Rangoon Battalion .	Rang'n Bn. . .	827
4th Simla Rifles . . .	Simla Rif. . . .	806
23rd Sind Battalion. . .	Sind Bn. . . .	835
28th South Andaman Company.	S. And. Coy. . .	842
29th South Indian Railway Battalion.	S. I. Ry. Bn. . .	843
10th Southern Provinces Mounted Rifles.	S. Prov. M. Rif. .	611
2nd Surma Valley Light Horse.	Surma V. L. H. .	602
14th Tennaserim Battalion.	Tenn'm Bn. . .	820
7th(S) United Provinces Horse.	U. P. Horse . .	608
8th (N) United Provinces Horse.	U. P. Horse . .	609
34th Upper Burma Battalion	U. Bur. Bn. . .	848
20th Nilgiri Malabar Battalion.	Nilgiri Malabar Battalion.	830

Corps in order of precedence.	Date of formation.	Page.
Light Horse.		
1st Bihar Light Horse		601
2nd Surma Valley Light Horse		602
3rd Calcutta Light Horse		603
4th Bombay Light Horse		604
5th Punjab Light Horse		605
6th Assam Valley Light Horse		606
7th (S.) United Provinces Horse		608
8th (N.) United Provinces Horse		609
9th Northern Bengal Mounted Rifles		610
10th Southern Provinces Mounted Rifles		611
Artillery.		
1st (Calcutta Port Defence) Group, Garrison Artillery		651
2nd (Madras) Group Garrison Artillery, "The Duke's Own."		652
3rd (Rangoon Port Defence) Group, Garrison Artillery.		653
4th (Bombay) Group, Garrison Artillery		654
"F" Kirkee Battery, Mobile Artillery		655
1st-5th (Cossipore) Brigade, Mobile Artillery		656
2nd Karachi Brigade, Mobile Artillery		657
5th Lucknow Group Garrison Artillery		658
Infantry.		
1st Madras Guards		801
2nd Nagpur Rifles		803
3rd Punjab Rifles		804
4th Simla Rifles		806
5th Calcutta Battalion		807
6th Bangalore, Coorg and Mysore Battalion		808
7th East Indian Railway Battalion		810
8th Allahabad Rifles		814
9th Mussoorie Battalion		815
10th (Naini Tal) Company		816
11th Lucknow Rifles		
12th Eastern Bengal Railway Battalion		817
1-13th Great Indian Peninsula Railway Battalion		818
2-13th Great Indian Peninsula Railway Battalion		819
14th Tennaserim Battalion		820

Corps in order of precedence.	Date of formation.	Page.
Infantry—*contd.*		
15th Bombay Battalion		821
16th Cawnpore Rifles		823
1-17th Bombay, Baroda and Central India Railway Battalion		825
2-17th Bombay, Baroda and Central India Railway Battalion		826
18th Rangoon Battalion		827
19th Agra Company		829
20th West Coast Battalion		830
21st Burma Railways Battalion		832
22nd Bengal and North-Western Railway Battalion		834
23rd Sind Battalion		835
1-24th North-Western Railway Battalion		836
2-24th North-Western Railway Battalion		838
25th Oudh and Rohilkhand Railway Battalion		840
26th Hyderabad Rifles		841
27th Baluchistan Company		842
28th South Andaman Company		842
29th South Indian Railway Battalion		843
1-32nd Madras and Southern Mahratta Railway Rifles		844
2-32nd Madras and Southern Mahratta Railway Rifles		846
34th Upper Burma Battalion (I. D. F.)		848
35th Poona Battalion		849
1-36th Bengal-Nagpur Railway Battalion		850
2-36th Bengal-Nagpur Railway Battalion		851
37th Calcutta Presidency Battalion		852
38th East Coast Battalion		854
39th Chota Nagpur Regiment		855
41st Eastern Bengal Company		856
42nd Assam Bengal Railway Battalion		857
43rd Kolar Gold Fields Battalion		858
44th Calcutta Scottish		859
45th Aden Rifles		860

Regimental District, No. 2. [No. 10 District.]

Uniform—Scarlet. *Facings*—Blue.

The Queen's (Royal West Surrey Regiment).

1-4TH BATTALION.

Arrived in India, 3rd December 1914, from England.

"South Africa, 1900-02."

HONORARY COLONEL—*Sir* F. T. Edridge, *Kt.* . . . . . 7 Apl. 06.

First Commn.	Names.	Regimental rank.	Army rank.	Remarks.
	Lieutenant-Colonel.			
	Majors.			
24 Mar. 06	Atkins, H. R. (*actg. Lt.-Col., 30 Jan. 18*) (57, 71)	1 June 16	. . . .	Comdg.
15 Sep. 06	Potter, R. W. (51)	1 June 16		
6 Oct. 06	Gosney, H. J. (50)	1 June 16		
3 July 08	Dibdin, L. G. (51, 52, 59) . . .	1 June 16	. . . .	With 1-5th Bn.
	Captains.			
31 Aug. 08	*Helps, M. D.* (46, 48*a*, 51, 71) .	1 June 16	. . . .	*Platoon Comdr., Cadet Unit, Lower Topa.*
18 Nov. 08	*Evans, B. L.* (71, 162, 172*a*, 205*a*)	1 June 16	. . . .	*Bde. Maj., Allahabad Bde.*
8 Apl. 09	Featherstone, O. (71, 207) . .	1 June 16		
17 June 10	Fearon, J. G. (51)	1 June 16	. . . .	With 2-4th Bn.
24 June 10	*Turner, E. L.* (71, 172*a*) . . .	1 June 16	. . . .	*I. A.* (*on probn.*)
2 Dec. 10	Potter, A. G. (71)	1 June 16	. . . .	With 1-5th Bn.
23 Nov. 12	*Falcon, R. R. B.* (57)	1 June 16	. . . .	*I. A.* (*on probn.*)
12 Mar. 13	Brown, E. J. (51, 71) . . .	1 June 16	. . . .	With 1-5th Bn.
18 June 13	*Moss, R. L.* (51, 52, 71) . . .	1 June 16	. . . .	*M. G. Corps.*
22 Aug. 14	Dibdin, A. (48*a*, 50, 71, 207) . .	1 June 16		
29 Aug. 14	Caroe, O. K. (51, 52, 71, 205*a*, 206, 207)	1 June 16		
3 Oct. 14	Groves, G. L. (48*a*, 51, 52) . .	1 June 16		
21 Oct. 14	Reilly, H. D. R. (51)	1 June 16	. . . .	With 1-5th Bn.
27 Oct. 14	Gibsone, H. C. (57, 61, 71) . . .	1 June 16	. . . .	Attd. 1st G. Bn., S. Staff. R.
27 Oct. 14	*Maud, W. M.* (66*a*)	16 Aug. 16	. . . .	*Adjutant.*
27 Oct. 14	Bacon, S. F. (51, 66*a*, 71) . .	16 Aug. 16	. . . .	With 1-5th Bn.
29 Oct. 14	Frost, E. G. (57)	13 Feb. 17		
	Lieutenants.			
18 Feb. 15	*Charlesworth, E. F.* (46)	1 June 16	. . . .	*A. M. Batty.*
6 Mar. 15	Gosney, L. L. (48*a*, 51)	1 June 16	. . . .	With 1-5th Bn.
29 May 15	Edenborough, L. A. B.	11 Sep. 16		
9 June 15	Howell, L. B. (61, 66*a*)	11 Sep. 16	. . . .	With 1-5th Bn.
12 June 15	MacLennan, N. L. (46, 61) . . .	1 July 17	. . . .	Supvg. Offr., Phys. & Bayo. Trg., 2nd (R'pindi) Dn.
24 June 15	Cutler, P. H.	1 July 17		
6 Aug. 15	Stone, H. C. (51)	1 July 17		
12 Aug. 15	Johnson, D. G. K. (46)	1 July 17		
21 Sep. 15	Sharpe, A. J. (48*a*, 51)	1 July 17		
3 Oct. 15	Hughes, F. W. T. (12, 48*a*, 51) .	1 July 17		
22 Nov. 15	Menhinick, J. S. (46)	1 July 17		
18 Dec. 15	Barrenger, H. E. (46, 51, 66*a*) .	1 July 17	. . . .	Attd. M. A. Dept., Lahore.
24 Dec. 15	Brandt, A. C.	1 July 17		
11 Mar. 16	Bates, L. (12, 48*a*, 51)	11 Sep. 17		
7 July 16	Read, W. D. B.	7 Jan. 18		
26 Sep. 16	*Groombridge, K. C.*	26 Mar. 18	. . . .	*I. A. R. O.*
	2nd-Lieutenants.			
11 Aug. 17	Greener, N. B.	1 Aug. 17		
16 Mar. 18	Howell, R. E.	16 Mar. 18		
19 Aug. 18	Mason, H. (61)	19 Aug. 18		
	Attached.			
30 July 15	Barclay, Lt. T. M., 5th R. Scots. (46)	1 July 17		
6 Nov. 15	*Rodger, Lt. F. McG.*, 4th R. Scots.	1 July 17	. . . . .	Asst. Embn. Staff Offr., Karachi.
12 Nov. 15	Wood, Lt. A. R. B., 9th R. Scots. (46).	1 July 17		
16 Dec. 15	Christie, Lt. J. McI., 5th R. Scots.	1 July 17	. . . . .	I. A. (*on probn.*)
15 July 16	Benson, Lt. A. H., 5th R. Scots. (46).	15 Jan. 18		
26 Sep. 17	Enfiley, 2nd-Lt. W. G. (*temp.*), Notts. & Derby. R.	26 Sep. 17		
21 Oct. 17	Hurry, 2nd-Lt. W. F.	21 Oct. 17		
31 Oct. 17	Riley, 2nd-Lt. M. H. (*temp.*) Notts. & Derby. R.	31 Oct. 17		
27 Oct. 14	**Maud, Capt. W. M.**	**8 Nov. 16**	. . . .	**Adjt.**
11 July 94	**Greer, Lt.-Col. J.** (*ret. pay*) . .	**3 June 18**	. . . .	**Qr.-Mr.**

Regimental District No. 2.
[No. 10 District.]

Uniform—Green.
Facings—Scarlet.

The Queen's (Royal West Surrey Regiment).

1-5TH BATTALION.

Arrived in India, 3rd December 1914, from England.

"South Africa, 1900-02."

HONORARY COLONEL—

First Commn.	Names.	Regimental rank.	Army rank.	Remarks.
	Lieutenant-Colonel.			
7 Apl. 90	Brodrick, *Hon.* A. G., T.D.	1 July 12	. . . .	*Lv., ex I., m.c.*
	Majors.			
	Smallpeice, F. W.	1 July 12	. . . .	A. I. A. G., N. Army.
25 July 03	Hodges, W. L. (51, 52, 71)	1 June 16	. . . .	*Lv., ex I., m.c., 6 mos.*
5 Feb. 09	Bray, F. E. (51, 71)	1 June 16		
	Captains.			
19 Mar. 09	Whittington, P. R. (51, 52)	1 June 16		
9 Apl. 09	*Yetts, L. M.*, M.C. (51)	1 June 16	. . . .	*Asst. Prov. Marshal.*
	Whittington, L. (51, 71)	1 June 16	. . . .	Staff Capt., 58th Bde.
30 Aug. 11	Jardine, J. E. B. (51, 57, 71)	1 June 16	. . . .	Adjt.
2 Jan. 04	Morton, G. C.	1 June 16		
12 Sep. 06	*Young, J. H.*	1 June 16	. . . .	*42nd L. T. Mortar Battery.*
13 Aug. 14	*Mercer, J. L. C.*	1 June 16	. . . .	*Adjt., No. 2 Res. Bn.*
20 Jan. 12	Spens, W. P. (51)	1 June 16		
17 Feb. 12	Mountford, E. W. (51, 52)	6 Dec. 16	. . . .	No. 3 Res. Bn. (Ind.)
21 Sep. 14	Evetts, E. F.	3 Apl. 17		
11 Sep. 12	Cleverly, O. S. (66*a*)	3 Apl. 17		
16 Nov. 12	Jardine, L. W. (57, 71, 207)	3 Apl. 17		
25 Jan. 13	*Ardagh, F. D.* (52, 71)	26 June 17	. . . .	*M. G. Corps.*
5 Aug. 14	Campion, K. F.	25 June 17		
	Lieutenants.			
5 Aug. 14	Wigan, C. R., M.C. (12, 48*a*, 50, 71)	1 June 16		
5 Aug. 14	Cleverly, G. C. (71)	1 June 16		
5 Aug. 14	Stoop, A. D. (*act'g. Capt., 20 July 17*) (71)	1 June 16		
5 Sep. 14	Harrison, G. B. (*temp. Capt., 21 Sep. 18*) (57, 71)	1 June 16	. . . .	Staff Capt., 42nd Infy. Bde.
13 Oct. 14	Godlonton, B. G.	1 June 16		
15 Nov. 14	Angell, J. W.	1 June 16	. . . .	*Lv., ex I., m.c.*
14 Mar. 15	Crane, L. F.	1 June 16	. . . .	*Lv., ex I., m.c.*
6 Aug. 15	Cross, D. I. C. A.	1 July 17		
3 Sep. 15	Willis, G. P. (71)	1 July 17		
19 Sep. 15	Nice, C. G.	1 July 17	. . . .	No. 3 Res. Bn. (Ind.).
3 July 15	*Fison, C. H.* (71)	1 July 17	. . . .	*I. A. (on probn.)*
3 Oct. 15	Spencer, H.	1 July 17		
9 Oct. 15	Bell, W. A. (71)	1 July 17		
27 Oct. 15	Hayward, S. D.	1 July 17		
30 Oct. 15	Smart, G. R.	1 July 17	. . . .	No. 3 Res. Bn. (Ind.)
5 Dec. 15	Gillott, A.	1 July 17		
7 Dec. 15	Taylor, A. J.	1 July 17	. . . .	No. 3 Res. Bn. (Ind.)
31 Dec. 15	Brown, E. (71)	1 July 17		
17 Jan. 16	Hyslop, A. J.	17 July 17	. . . .	*Lv., ex I., m.c*
	2nd-Lieutenants.			
1 Mar. 17	Brown, H. W.	1 Mar. 17		
1 Mar. 17	Handscombe, H. T.	1 Mar. 17		
16 Mar. 18	Dillow, D'A. H.	16 Mar. 18	. . . .	Depôt, Ry. Bn. (S. & M.)
16 Mar. 18	McLean, A. D.	16 Mar. 18	. . . .	No. 3 Res. Bn. (Ind.)
19 Aug. 18	Paddison, S.	19 Aug. 18		
	Attached.			
3 July 08	Dibdin, Maj. L. G., 1-4th Bn.	1 June 16		
2 Dec. 10	Potter, Capt. A. G., 1-4th Bn. (71)	1 June 16		
27 Oct. 14	Bacon, Capt. S. F., 1-4th Bn. (51, 66*a*, 71)	16 Aug. 16		
	Morgan, Capt. T. H. (*temp.*), 1st Bn.	27 Aug. 17		
7 Nov. 14	Edwards, Capt. H. S., A.S.C.	25 Dec. 17	. . . .	No. 3 Res. Bn. (Ind.)
6 Mar. 15	Gosney, Lt. L. L., 1-4th Bn.	1 June 16		
9 June 15	Howell, Lt. L. B., 1-4th Bn.	11 Sep. 16		
25 Jan. 15	Mathews, Lt. W. A. (*temp.*)	1 July 17	. . . .	R. A. F., Egypt.
7 Oct. 15	Bickell, Lt. E. G. (*temp.*)	1 July 17		
24 Nov. 15	Bolton, Lt. E. F., 1st Bn.	1 July 17	. . . .	No. 3 Res. Bn. (Ind.)

Regimental District No. 2. [No. 10 District.]

Uniform—Green. *Facings*—Scarlet.

The Queen's (Royal West Surrey Regiment)—*contd.*

First Commn.	Names.	Regimental rank.	Army rank.	Remarks.
	Attached—*contd.*			
22 Jan. 16	Robson, Lt. A., 6th Bn.	22 July 17	. . . .	*Lv., ex I., m.c.*
20 Jun. 16	Savereux, Lt. E. (*temp.*)	20 Dec. 17		
8 Aug. 16	Boswall, 2nd-Lt. A. P. (*temp.*), 7th Bn.	8 Aug. 16	. . . .	No. 3 Res. Bn. (Ind.)
28 Mar. 17	Arnold, 2nd-Lt. G., 3rd Bn. . . .	28 Mar. 17	. . . .	With 1-5th E. Surr. R.
28 Mar. 17	Mitchell, 2nd-Lt. A. H., 3rd Bn. . .	28 Mar. 17		
28 Mar. 17	Osmond, 2nd-Lt. F. E., 3rd Bn. (46)	28 Mar. 17	. . . .	No. 3 Res. Bn. (Ind.)
30 May 17	Bishop, 2nd-Lt. H. C., M.M., 3rd Bn. (57)	30 May 17	. . . .	No. 3 Res. Bn. (Ind.)
29 Aug. 17	Clark, 2nd-Lt. V. T. G., 3rd Bn. .	29 Aug. 17	. . . .	No. 3 Res. Bn. (Ind.)
29 Aug. 17	*Davies, 2nd-Lt. A. 3rd Bn.* (52) . .	29 Aug. 17	. . . .	*M. G. Corps (on probn.)*
29 Aug. 17	Gunner, 2nd-Lt. H., 3rd Bn. . . .	29 Aug. 17		
26 Sep. 17	Cooke, 2nd-Lt. R. H., 3rd Bn. . .	26 Sep. 17	. . . .	No. 3 Res. Bn. (Ind.)
26 Sep. 17	Wraight, 2nd-Lt. G. J. R., 3rd Bn.	26 Sep. 17	. . . .	No. 3 Res. Bn. (Ind.)
	Bentley, 2nd-Lt. J., S. Lan. R. . .			
30 Aug. 11	Jardine, Capt. J. E. B.	1 Sep. 18	. . . .	Adjutant.
15 Sep. 14	Reader, Capt. E. J. W.	15 Sep. 14	. . . .	Qr.-Mr.
19 June 09	Bailey, Capt. L. D., M.C., R.A.M.C. (T.F.) (207*a*)	4 Aug. 13	. . . .	Medl. Offr. *Lv. ex I.*

Regimental District No. 3.
[No. 10 District.]

Uniform—Scarlet.
Facings—Buff.

The Buffs (East Kent Regiment).

1-4TH BATTALION.

Arrived in India, 3rd December 1914, from England.

"South Africa, 1900-02."

HONORARY COLONEL—Col. *Lord* G. R. C. Harris, G.C.S.I., G.C.I.E., C.B., T. D., A.-D.-C. . 10 Mar. 89.

First Commn.	Names.	Regimental rank.	Army rank.	Remarks.
	Lieutenant-Colonel.			
9 Oct. 95	Gosling, G., T. D. (*Hon. Capt. in Army, 10 July 01*)	8 Feb. 13		
	Majors.			
15 May 96	Dunstan, F. M., T.D.	26 June 15		
6 Oct. 06	Kingsford, G. M. (Q51)	1 June 16		
15 Apl. 07	Heggs, T. B. (51)	1 June 16	. . . .	Tempy. Maj., R. A. M. C.
	Captains.			
6 Feb. 09	Collard, B. S. (51)	26 Apl. 12	. . . .	
1 Apl. 09	*Grant, L. B. (temp. Maj.)* (51, 207)	6 May 14	. . . .	*D. A. A. G.*
1 June 10	Grigg, A. C. (51)	26 June 15		
28 Feb. 12	*Harbord, R. E.* (51)	26 June 15	. . . .	Bde. Maj., Ahmednagar Bde.
17 July 12	*Witts, C. S. F.* (51, 207*a*)	1 June 16	. . . .	*I. A. (on probn.)*
15 Jan. 13	*Cremer, H. L. H.* (12, 48*a*, 51)	1 June 16	. . . .	*Adjt.*
13 Nov. 13	Howell, C. J. (51)	1 June 16		
20 May 14	Harrison, W. A. (51, 60*a*)	1 June 16		
27 May 14	Harrison, E. G. (48*a*, 51, 57)	1 June 16		
19 Aug. 14	Green, E. C.	16 Jan. 17	. . . .	Bde. Maj., Garhwal Bde.
	Lieutenants.			
7 Feb. 14	Taunton, H. L. (51)	26 June 15	. . . .	M. G. Corps.
22 Aug. 14	*Mockett, V. (temp. Capt. 23 July 17)*, (51)	1 June 16	. . . .	*Staff. Capt.*
29 Aug. 14	Monins, J. E.	1 June 16	. . . .	Attd. Bareilly Bde.
29 Aug. 14	*Hubble, S. L.* (51, 52)	1 June 16	. . . .	*M. G. Corps.*
29 Aug. 14	Whitfield, H. F. (51, 52)	1 June 16		
31 Aug. 14	Bowden, E. M.	1 June 16		
19 Sep. 14	*Baker, S.* (52)	1 June 16	. . . .	*R. A. F.*
21 Oct. 14	*Hearn, E. R.*	1 June 16	. . . .	*I. A. R. O.*
4 Nov. 14	*Killick, A. S.* (57)	1 June 16	. . . .	*With Sig. Ser.*
4 Nov. 14	MacDonnell, R. G. (*act'g. Capt*)	1 June 16		
4 Nov. 14	Brockman, F. H. D.	1 June 16		
27 Nov. 14	Igglesden, G. H.	1 June 16		
16 Mar. 15	Brewer, G. L. (*temp. Capt. 22 Jan. 18*), (46, 51)	24 Feb. 17	. . . .	Instr., Bombing Schl., Mhow.
19 Mar. 15	*Saunders, G. N.* (51)	24 Feb. 17	. . . .	*I. A. (on probn.)*
19 Mar. 15	*Fulton, R.* (46)	1 July 17	. . . .	*I. A. on probn.)*
2 Apl. 15	Seale, J. H. (48*a*, 51)	1 July 17		
30 Aug. 15	*Russell, B. H.* (51)	1 July 17	. . . .	*I. A. R. O.*
6 Sep. 15	Kear, J. E. (57, 61)	1 July 17		
9 Sep. 15	Ruffer, R. (57, 60*a*)	1 July 17		
14 Sep. 15	Millar, E. G. (50)	1 July 17		
16 Aug. 15	Dolben, A. C. (48*a*, 51, 95, 207*a*.)	1 July 17	. . . .	M. G. Corps.
11 Oct. 15	Schulz, B. R. (57)	1 July 17		
5 Sep. 16	Goldring, H. (48*a*, 49, 51)	5 Mar. 18		
5 Sep. 16	Jackson, P. W. (46)	5 Mar. 18		
	2nd-Lieutenants.			
25 Oct. 16	Martin, E. L.	25 Oct. 16	. . . .	M. A. Dept.
1 Mar. 17	Merry, D. G.	1 Mar. 17		
28 Mar. 17	Bartter, A. S.	28 Mar. 17		
28 Mar. 17	Colven, A. F.	28 Mar. 17		
28 Mar. 17	Hale, T. H.	28 Mar. 17		
30 May 17	Roberts, V. L.	30 May 17	. . . .	R. A. F.
30 May 17	Colegrave, G. R. (46)	30 May 17		
30 May 17	Gale, S. T.	30 May 17		
30 May 17	Cotching, A.	30 May 17		
30 May 17	Lloyd-Jones, R. W. A. (*temp. Capt.*)	30 May 17	. . . .	Offr. i/c. A. C. Depot.
30 May 17	Shorter, L. V. H.	30 May 17		
31 Oct. 17	Harris, W. H., M.M.	31 Oct. 17		
31 Oct. 17	Middleton, C. W.	31 Oct. 17		
31 Oct. 17	Sherwood, J.	31 Oct. 17		

The Buffs (East Kent Regiment)—*contd.*

First Commn.	Names.	Regimental rank.	Army rank.	Remarks.
	Attached.			
19 Mar. 15	**Taylor, Lt. J. B.** (46) T. F. Res. (Gen. List)	1 July 17	. . . .	Asst. Postal Censor, Bombay.
28 Apl. 17	Southwell, 2nd-Lt. W. B. (48*a*, 51)	28 Apl. 17	. . . .	4th R. W. Surr. R.
31 Oct. 17	**Lee**, 2nd-Lt. J., 3 Bn., Notts. & Derby, R.	. . . .	31 Oct. 17	
15 Jan. 13	Cremer, Capt. H. L. H.	12 Mar. 17	. . . .	Adjutant.
25 Jan. 13	**Keeler**, Capt. F. H.	25 Jan. 13	. . . .	Qr.-Mr.
26 Feb. 96	**Paget**, Maj. P., T. D., R. A. M. C. (T. F.)	28 Dec. 07	. . . .	Medl. Offr.—*Lv., ex I., m.c. 6 mos., 20 Sep. 16.*

Regimental District No. 3.
[No. 10 District.]

Uniform—Scarlet.
Facings—Buff.

The Buffs (East Kent Regiment).

1-5TH (The Weald of Kent) BATTALION.

Arrived in India, 3rd December 1914, from England.

"South Africa, 1900-02."

HONORARY COLONEL—E. W. Hussey, V.D. . . 21 Oct. 11.

First Commn.	Names.	Regimental rank.	Army rank.	Remarks.
	Lieutenant-Colonel.			
	Majors.			
13 Jan. 94	Body, J., D.S.O. (*act'g. Lt.-Col.*) (51)	14 Jan. 16	. . . .	Comdg.
13 Dec. 05	Clarke, E.	1 June 16	. . . .	With 3-4th Bn., E. Kent R.
	Captains.			
10 Oct. 14	Maynard, P. A.	10 Oct. 14	. . . .	Bde. Maj., Nasirabad Bde.
29 Sep. 09	Collings, T. H. O.	29 Oct. 14	. . . .	With 25th G. B. Rif. Bde.
23 Nov. 09	Buss, B., (51)	14 Jan. 16	. . . .	*Lv., ex I., m.c. 6 mos.*
2 Oct. 09	Marchant, F. O., M.C. (51) . . .	1 June 16		
11 May 12	Wilkinson, D. S. (57, 207*a*) . .	1 June 16		
15 Dec. 12	Blomfield, H. M. (51, 206) . . .	1 June 16		
24 Jan. 14	Burns, E. B. (51)	1 June 16	. . . .	With 2nd R. W. Kent R.
	Lieutenants.			
22 Aug. 14	Walker, R. (*temp. Capt.*)	4 Nov. 14	. . . .	Adjt., No. 3 Res. Bn.
22 Aug. 14	Macfadyen, W. A., M.C.	8 Jan. 16	. . . .	*Lv., ex I., m. c.*
22 Aug. 14	Jessel, G., M.C. (51)	14 Jan. 16		
22 Aug. 14	Hardcastle, A. E. L. (*act'g. Capt.*) (48*a*, 51, 52)	21 Jan. 16		
12 Sep. 14	Fleuret, F. S. (*act'g. Capt.*) (51, 52)	1 June 16	. . . .	With 3-4th Bn., E. Kent R.
6 Sep. 14	*Thomas, C. E.*	1 June 16	. . . .	*I. A. R. O.*
18 Oct. 14	Rice, R. V.	1 June 16		
18 Oct. 14	*Leaf, C. S.* (57)	1 June 16	. . . .	*M. G. Corps.*
9 Dec. 14	Stabb, C. T.	1 June 16	. . . .	With 2-6 Bn., Devon. R.
10 Feb. 15	Weldon, S. W.	1 June 16		
4 Nov. 15	Pittock, J. A.	1 July 17		
9 Dec. 15	Liddle, F. A.	1 July 17		
9 Dec. 15	Liddle, A.	1 July 17		
10 June 16	Hook, C. W. T.	10 Dec. 17	. . . .	With Stokes Mortar Batty.
10 June 16	O'Brien, H.	10 Dec. 17		
10 June 16	Scarr, K. R.	10 Dec. 17		
8 July 16	Moore, E. A.	8 Jan. 18		
8 July 16	Ramsey, L. A.	8 Jan. 18		
5 Aug. 16	Couchman, C. C.	5 Feb. 18		
5 Aug. 16	Aylard, R. F. O.	5 Feb. 18		
	2nd-Lieutenants.			
29 Aug. 17	Boulding, E.	29 Aug. 17		
16 Mar. 18	Rabbitt, C.	16 Mar. 18		
16 Mar. 18	Body, G. A.	16 Mar. 18		
16 Mar. 18	Blaxall, M. L.	16 Mar. 18		
16 Mar. 18	Bull, C. A.	16 Mar. 18		
19 Aug. 18	Worthington, C.	19 Aug. 18		
19 Aug. 18	Kelway, R. T.	19 Aug. 18		
19 Aug. 18	Powter, H. F.	19 Aug. 18		
29 Oct. 14	Filmer, Lieut. (act'g. Capt.) W. H. G. 1-4th E. Kent. R.	1 June 16	. . . .	Adjt.
25 Aug. 16	Hawkins, Lt. A., M.C.	25 Aug. 16	25 Aug. 16	Qr.-Mr.
. . . .	Woolrich, Capt. W. G., R.A.M.C. .	. . . .	. . . .	Medical Officer.

The Buffs (East Kent Regiment)—*contd.*

1-5TH (The Weald of Kent) BATTALION.

First Commn.	Names.	Regimental rank.	Army rank.	REMARKS.
	Attached.			
	Majors.			
27 Aug. 14	Howard-Vyse, G. A., L'pool	1 Sep. 15		
	Piggott, J. T. (*temp. Maj.*), York. L. I.	1 Mar. 16		
	Captains.			
27 Feb. 09	Watney, R. D., 1-4th R. W. Kent R. (51, 52, 57)	31 Dec. 14		
1 June 10	Grigg, A. C. (51), 1-4th E. Kent R.	26 June 15		
5 Jan. 01	Laverton, W. R. C., M. C., Res. of Offrs. (*act'g. Maj., 15 Feb. 18*)	6 Dec. 15		
1 Oct. 12	Avison, J. L., 8th Midd'x R.	21 Jan. 16		
3 Mar. 09	Stone, F. L. (51), 1-4th R. W. Kent R.	1 June 16		
17 July 12	Harrison, G. K. (66), 1-5th E. Surr. R.	1 June 16		
20 May 14	Harrison, W. A. (51, 60*a*) 1-4th E. Kent R.	1 June 16		
	Lieutenants.			
22 Aug. 14	Mockett, V. (*temp. Capt.*) (51)	1 June 16		
31 Aug. 14	Bowden, E. M., 1-4th E. Kent. R.	1 June 16		A. D. C.
20 Nov. 14	Angus, W. C. S. (51), 2-4th Bord. R.	1 June 16		
22 May 15	Scott, T. J., 1-10th Midd'x R.	1 June 16		
7 Aug. 15	Hazzard, F., 2nd Bn., E. Kent R. (*temp.*)	1 June 16		
20 Oct. 15	Western, F. H.	1 June 16		
5 May 15	Linfield, W. A., 1st Garr. Bn., Bedf. R.	1 July 17		
12 May 15	Geach, P. (*act'g. Capt.*), 5th D. of Corn. L. I.	1 July 17		
14 Aug. 15	Searle, H., 2nd E. Kent R.	1 July 17		
15 Aug. 15	Mitchell, K. W. S., 5th Hamps. R.	1 July 17		
23 Sep. 15	Davies, J. W., 1-4th E. Kent R.	1 July 17		
12 Dec. 15	Hancox, G. R. (46), 2-6th Devon. R.	1 July 17		
18 Dec. 15	Pinhey, H. T. (46), 2-6th Devon. R.	1 July 17		
21 Jan. 15	Creswick, M. de J., 6th E. Kent R.	1 July 17		
22 Dec. 15	Knight, C. R. B., 1st E. Kent R. (51, 60*a*)	1 July 17		
23 Dec. 15	Gibson, A. D., 4th E. Kent R.	1 July 17		
10 Jan. 16	Goode, W. H. E. N., (*act'g. Capt.*) 4th E. Kent R.	10 July 17		
22 Jan. 16	Turpin, R. J., 2nd E. Kent R. (*temp.*)	22 July 17		
21 Sep. 15	Brooke, H. A., 7th Essex R.	11 Dec. 17		
7 July 16	Ellis, B. G., 2nd E. Kent R. (*temp.*)	7 Jan. 18		
8 July 16	Arnold, N., 6th Devon. R.	8 Jan. 18		
19 July 16	Turner, A. F. St. A., 2nd E. Kent R.	19 Jan. 18		
19 Dec. 16	Matheson, M., 8th R. Scots	19 June 18		
	2nd-Lieutenants.			
7 July 16	Bonner, E. H., 1st E. Kent R.	7 July 16		
5 Feb. 17	Kilbee, F. W., 2nd E. Kent R.	5 Feb. 17		
1 Mar. 17	Merry, D. G., 1-4th E. Kent R. (47*a*, 48*a*)	1 Mar. 17		
26 Apl. 17	Pert, J. E., 4th Black Watch	26 Apl. 17		
30 May 17	Swan, E., 4th Black Watch	30 May 17		
30 May 17	Grosvenor, V. L., 3rd E. Kent R.	30 May 17		

Regimental District No. 5.
[No. 5 District.]

Uniform—Scarlet.
Facings—Gosling Green.

The Northumberland Fusiliers.

2ND GARRISON BATTALION.

Arrived in India, 13th March 1916.

HONORARY COLONEL—Maj.-Genl. P. S. Wilkinson, C.B., C.M.G.

First Commn.	Names.	Regimental rank.	Army rank.	Remarks.
	Lieutenant-Colonel.			
94	Kent, H. H. (*temp.*) (51, 71) . .	12 Oct. 15		
	Major.			
29 Dec. 75	Synge, R. F. M. F. M. (*Lt.-Col., ret. pay*) (51, 67, 71)	13 Oct. 15	18 Oct. 02	2nd-in-Comd.
	Captains.			
28 Dec. 14	Tregoning, E. A. (*temp.*) . . .	2 Dec. 15	28 Dec. 14	Staff Capt., Nasirabad.
15 Dec. 14	Arnott, S. L. (*temp.*) . . .	2 Dec. 15		
	Lieutenants.			
22 Feb. 15	Maccoy, N. (*temp.*)	14 Dec. 15	4 May 15	Graves Registration Offr.
22 Sep. 14	Platt, W. P. (*temp.*) (95) . . .	2 Dec. 15	. . . .	M. A. Dept., Quetta.
27 Nov. 14	Strologo, R. C. (*temp.*) (51) . . .	2 Dec. 15	. . . .	Attd. Stokes & Mortar Batty.
9 Dec. 14	Geach, E. S. (*temp.*)	2 Dec. 15	. . . .	Adjt. & Qr.-Mr., Bazar Hill Camp, Wellington.
24 July 15	McDowall, J. P. K. (*temp.*) (*act'g. Capt.*, *3 Aug. 17*) (52).	3 July 16	. . . .	Adjt.
22 Sep. 14	Taylor, W. W.	1 July 17	22 Sep. 14	*Lv., ex I., m.c., 9 mos.*
16 Aug. 15	Court, R. H. (*temp.*)	1 July 17	. . . .	*Lv., ex I., m.c.*
	2nd-Lieutenants.			
11 Jan. 15	Yool, W. M. (*temp.*) (51) . . .	11 Jan. 15	. . . .	I. A. R. O.
28 May 15	Lee-Warner, J. (*temp.*)	25 Jan. 16	28 May 15	I. A. R. O.
11 Mar. 17	Johnson, H. (*temp.*) (51, 52). .	11 Mar. 17	. . . .	Supy. R. T. O., Bombay.
26 Apl. 17	Wilkins, L. J. (*temp.*)	26 Apl. 17		
13 May 17	Ravenscroft, G. E. (*temp.*) . . .	13 May 17	. . . .	Asst. Staff Capt., Ahmednagar Bde.
17 June 17	Halkett, F. G. (*temp.*)	17 June 17		
17 June 17	Thomson, J. R. (*temp.*)	17 June 17		
15 July 17	Barton, H. H. (*temp.*)	15 July 17		
	Attached.			
10 Aug. 14	Austin, Capt. H. N. (*temp.*), 8 Bn., R. War. R.	1 Nov. 14	. . . .	Supy. R. T. O., Poona.
14 Apl. 14	Macdonald, Capt. R. R. M., 4 Bn., Sea. Highrs.	2 June 17		
12 Jan. 15	Cowley, Capt. C., M.C. (*temp.*), 25 Bn., North'd. Fus.	20 Sep. 17		
15 Apl. 15	Murgatroyd, Lieut. J. O. (*temp.*), 17 Bn., R. Fus. (51)	10 July 16		
23 Aug. 15	Lingwood, Lieut. H. P. L. (*temp.*), 7 Bn., North'd Fus.	1 July 17		
19 Dec. 15	Ball, Lt. W. M. (*temp.*), 4 Bn., Border R.	1 July 17		
22 July 15	Greenwood, 2nd-Lt. J. (*temp.*), 11 Bn., S. Staff. R.	22 July 15		
28 Mar. 17	Colinge, 2nd-Lt. J. (*temp.*), 10 Bn., Lan. Fus.	28 Mar. 17	. . . .	Ind. Mun. Board.
26 Apl. 17	Hopton, 2nd-Lt. W. J. (*temp.*), 3 Res. Garr. Bn., R. W. Fus. (51).	26 Apl. 17		
29 May 17	Rawcliffe, 2nd-Lt. J. (*temp.*), 4 Bn., Lan. Fus. (51).	29 May 17		
26 Sep. 17	Sorrell, 2nd-Lt. F. G. (*temp.*), 5 Bn., Midd'x. R.	26 Sep. 17		
26 Sep. 17	Gilbert, 2nd-Lt. H. J. (*temp.*), 8 Bn., Lan. Fus. (95).	26 Sep. 17		

24 July 15	McDowall, Lt. (*actg. Capt.*) J. P. K. (*temp.*)	1 May 16	. . . .	Adjt.
4 Oct. 15	Rogers, Lt. T. (*temp.*)	4 Oct. 15	. . . .	Qr.-Mr.

Regimental District No. 8. [No. 3 District.]

Uniform—Scarlet. *Facings*—Blue.

The King's (Liverpool Regiment).

2ND BATTALION (*late 8th Foot*).

Arrived in India, 6th January 1909, from England.

THE WHITE HORSE WITHIN THE GARTER. "*Nec aspera terrent.*" IN EACH OF THE FOUR CORNERS THE ROYAL CYPHER ENSIGNED WITH THE IMPERIAL CROWN. THE SPHINX, SUPERSCRIBED "EGYPT." "Blenheim" "Ramillies" "Oudenarde" "Malplaquet" "Dettingen" "Martinique, 1809" "Niagara" "Delhi, 1857" "Lucknow" "Peiwar Kotal" "Afghanistan, 1878-80" "Burma, 1885-87" "Defence of Ladysmith" "South Africa, 1899-1902."

COLONEL—General *Sir* W. H. Mackinnon, G.C.B., K.C.V.O. . . . 2 Mar. 16.

First Commn.	Names.	Regimental rank.	Army rank.	Remarks.
	Lieutenant-Colonel.			
	Majors.			
17 Jan. 91	Hyslop, F. (*act'g. Lt.-Col.*) (Q. 51)	12 Dec. 08		
19 July 93	Fox, R. P. (50, 57, 71, 172*a*) . .	1 Sep. 15		
	Captains.			
4 May 07	Skinner, G. E. M. (*tem. Maj., 15 Oct. 18*) (51, 52, 57, 207) . . .	1 Oct. 14	. . . .	G. S. O., 2nd gr., Cadet College, Wellington.
8 Feb. 08	Harden, F. G. (46, 51, 52, 172*a*) .	21 Oct. 14		
3 Nov. 09	Briggs, F. C. C., D.S.O. (51, 57, 66*a*, 71, 207)	12 Jan. 15	. . . .	Bde. Maj., Jullundar Bde.
5 Oct. 10	Levick, A. U. (51, 172*a*)	9 Aug. 16		
	Lieutenants.			
21 Dec. 12	Tate, A. R. W. (*temp. Capt.*) (12, 48*a*, 50, 52, 207)	21 Oct. 14		
25 Dec. 12	*Cross, E. C.* (*temp. Capt., 21 Dec. 16*) (51, 52, 66, 95, 172*a*)	25 Oct. 14	. . . .	*Adjt.*
15 Aug. 14	Fulton, C. F. V., M.C. (60*a*) . .	10 June 15	. . . .	I. A. (*on probn.*)
7 Nov. 14	Kent, C. E. (*temp. Capt., 5 Dec. 18*) (48*a*, 51, 71, 95, 169) . . .	6 Sep. 15	. . . .	Staff Capt., Ahmednagar Bde.
7 Nov. 14	Holland, G. W. (51, 57, 71, 95, 169, 170, 206)	22 June 16	. . . .	I. A. (*on probn.*)
19 Apl. 15	Agar, C. S. (*temp.*) (12) . . .	22 June 16		
7 Nov. 14	Collins, J. (46, 48*a*, 51, 95, 172*a*) .	9 Aug. 16		
17 June 15	Rogers, A. G. (*act'g. Capt., 5 July 17*) (51, 52, 172*a*)	3 Sep. 16	. . . .	Comdt.—Attd. Section, Bombay.
	2nd-Lieutenants.			
5 May 17	Thompson, J. W. (48*a*, 51, 52, 57, 59, 60*a*)	5 May 17		
29 Aug. 17	Cooper, T. W. (*temp.*) (12, 46) .	29 Aug. 17		
29 Aug. 17	Staunton, J. W. (*temp.*) . . .	29 Aug. 17	. . . .	I. A. (*on probn.*)
16 Mar. 18	Treharne, T. H. (*temp.*) . . .	16 Mar. 18		
19 Aug. 18	Slynn, H. J., D.C.M. (*temp.*) . .	19 Aug. 18		
19 Aug. 18	Capell, G. C. (*temp.*)	19 Aug. 18		

The King's (Liverpool Regiment)—*contd.*

First Commn.	Names.	Regimental rank.	Army rank.	Remarks.
	Attached.			
20 June 15	Pierey, Lt. C. H. Z. (*temp.*) Bedf. R.	25 Aug. 16	....	Attd. M. G. Corps.
2 Feb. 16	Clark, Lt. C., Scottish Horse	2 Aug. 17	....	Attd. G. S. Bch., Army Hdqrs.
15 July 15	Wilson, 2nd-Lt. J., K. R. Rif. C. (46, 95).	15 July 15		
6 Aug. 15	Ancill, 2nd-Lt. G. C., R. Scots	6 Aug. 15	....	Attd. M. G. Corps.
12 Dec. 15	Goody, 2nd-Lt. R. (12, 48*a*, 51)	12 Dec. 15	....	Attd. Staff, 4th (Quetta) Dn.
1 Jan. 16	Kingsley, 2nd-Lt. H. H., 10 Bn., R. Lanc. R. (95)	1 Jan. 16	....	Attd. Staff, 4th (Quetta) Dn.
7 Apl. 16	Webb, 2nd-Lt. J. E. T., Dorset. R. (46, 57)	7 Apl. 16		
5 Aug. 16	Macnabb, Lt. H., Lovat Scouts. (12)	5 Feb. 18	....	Attd. G. S. Bch., Army Hdqrs.
28 Mar. 17	Clancy, 2nd-Lt. G. D. (46, 95), R. Dub. Fus.	28 Mar. 17		
25 Sep. 17	Vickers, 2nd-Lt. S., K. R. Rif. C.	25 Sep. 17		
26 Sep. 17	Lewis, 2nd-Lt. B. (*temp.*) 2nd Bn., Welsh R.	26 Sep. 17		
26 Sep. 17	Hourahane, 2nd-Lt. F. (*temp.*), 2nd-Bn., Welsh R.	26 Sep. 17		
31 Oct. 17	Crabtree, 2nd-Lt. C., W. York. R.	31 Oct. 17		
31 Oct. 17	Grassick, 2nd-Lt. W. J., Gord. Highrs.	....	31 Oct. 17	
26 Nov. 17	Padbury, 2nd-Lt. G. F., K. R. Rif. C.	26 Nov. 17		
27 Nov. 17	Irwin, 2nd-Lt. W. J., R. Ir. Rif.	....	27 Nov. 17	
5 Dec. 12	Cross, Lieut. (*temp. Capt.*) E. C.	25 Dec. 17	....	Adjt.
14 Jan. 15	Caddy, Capt. T., M.C.	14 Jan. 18	....	Qr.-Mr.

Agents :—Messrs. Cox & Co.

Regimental District No. 9. [No. 9 District.]

The Norfolk Regiment.

Uniform—Scarlet. *Facings*—Yellow.

2ND BATTALION (*late 9th Foot*).
Arrived in India, 3rd March 1911, from Gibraltar.

THE FIGURE OF BRITANNIA.
"Havannah" "Martinique, 1794" "Roliça" "Vimiera" "Corunna" "Busaco" "Salamanca" "Vittoria" "St. Sebastian" "Nive" "Peninsula" "Cabool, 1842" "Moodkee" "Ferozeshah" "Sobraon" "Sevastopol" "Kabul, 1879" "Afghanistan, 1879-80" "Paardeberg" "South Africa, 1900-02."

COLONEL-IN-CHIEF—THE KING.
COLONEL—Maj.-Genl. E. P. Strickland, C.B., C.M.G., D.S.O. . . 24 Dec. 17.

First Commn.	Names.	Regimental rank.	Army rank.	Remarks.
	Lieutenant-Colonel.			
	Majors.			
12 Mar. 92	Lodge, F. C., C.M.G., D.S.O. (*temp. Lt.-Col., 13 Sep. 15*) (Q. 50, 59)	15 Nov. 11		
14 Feb. 94	Cramer-Roberts, W. E. (50, 57, 72*b*)	11 Mar. 14	*l.c.* 3 June 16	
16 Feb. 98	Day, F. R.	1 Sep. 15	. . . .	*Lv., ex I., m.c.*
5 May 00	Willett, H. L. (51, 72*b*)	1 Sep. 15	. . . .	No. 3, Res. Bn.
	Captains.			
4 Dec. 01	Higson, F. (*actg. Lt.-Col.*) (50, 72)	13 Mar. 12		
6 Jan. 04	deGrey, G., D.S.O. (*actg. Maj.*) (50, 53, 59)	1 Sep. 12	*m.* 3 June 17	
29 Nov. 05	Horner, E. T., M.C. (51, 72)	1 June 14		
24 Jan. 06	*Shand, N. P.*, M.C. (50, 59)	20 June 14	. . . .	*Adjt.*
6 Oct. 06	Downs, R. A. (*temp. Maj.*) (51, 52)	16 Feb. 15	. . . .	D. A. Q. M. G., Army Hdqrs.
17 Nov. 06	Shakeshaft, A. J. [*l*] (50, 52, 190)	2 Apl. 15		
9 Sep. 08	Floyd, A. B. (51, 64)	25 Apl. 15		
6 Nov. 09	Frere, R. T. (51, 59)	6 Nov. 15	. . . .	S. S. O., Belgaum.
9 Jan. 15	Hallam, S. (*temp.*)	28 May 16		
	Lieutenants.			
3 Sep. 13	Read, J. F. W., M.C. (207)	12 Dec. 14		
3 Sep. 13	Peacocke, H. L.	12 Dec. 14		
17 Sep. 13	Bullock, J. H.	16 Feb. 15	. . . .	
7 Nov. 14	Campbell, T., M.C. (51)	6 Nov. 15		
7 Nov. 14	Monaghan, T. (51)	6 Nov. 15	. . . .	
7 Nov. 14	Portsmouth, F. V.	25 Nov. 15	. . . .	*Lv., ex I., m.c., 6 m.*
17 Mar. 15	Kirk, F. C. deL.	1 Feb. 17	. . . .	
27 Nov. 14	Dugdale, J. A.	1 July 17		
18 Dec. 14	Peake, S. C.	1 July 17	. . . .	Adjt. & Qr. Mr. Kurdish Labour Corps.
7 Jan. 15	Selwyn, H. G. (*temp.*)	1 July 17		
30 Oct. 14	Stocker, C. L. (*temp.*) (48*a*)	1 July 17		
20 Apl. 15	Shand, W. K. W.	1 July 17		
26 Jan. 15	Abbott, W. (*temp.*)	1 July 17		
27 Nov. 15	Porter, H. (*temp. Capt.*) (51)	1 July 17	. . . .	Asst. Prov. Marshal.
1 Apl. 15	Stead, J. (*temp.*)	1 July 17		
9 Apl. 15	Woodger, J. H. (*temp.*)	1 July 17		
10 Apl. 15	Child, P. (*temp.*)	1 July 17	. . . .	With No. 2 Res. Bn.
16 May 15	Denis, W. (*temp.*)	1 July 17	. . . .	M. T. Course, R.'Pindi,
29 Apl. 15	Moffat, J. (*temp. Capt.*)	1 July 17	. . . .	Staff Capt., Kurna,
17 June 15	Parsley, W., M.C. (*temp. Capt.*) (51, 60*a*)	1 July 17	. . . .	Emb. Staff Officer,
17 June 15	Clarke, A. J., M.C., D.C.M. (*temp. Capt.*) (51, 72*b*)	1 July 17	. . . .	D. A. Q. M. G.
6 July 15	Bell, G. F. St. J. (*temp.*)	1 July 17		
7 July 16	Boddy, G. W. (*temp.*)	7 Jan. 18		
7 July 16	Grummett, W. J. (*temp.*) (48*a*, 51)	7 Jan. 18		
7 July 16	Hirons, J. T. C. (*temp.*)	7 Jan. 18		
2 Aug. 16	Cullington, E.	2 Feb. 18		
5 Aug. 16	Jermy, W. F. (*temp.*)	5 Feb. 18		
14 Nov. 16	Brown, F. H., M.C. (51, 60*a*)	14 May 18		
4 Jan. 17	Bulleo, C.	4 July 18		
4 Jan. 17	Nicklin, S. G.	4 July 18		
	2nd-Lieutenants.			
15 Mar. 15	Little, R.	15 Mar. 15		
7 July 16	Sacre, L. G. B.	7 July 16		
9 Dec. 16	Connors, J. P. (*temp. Capt., 13 Mar. 18*)	9 Dec. 16	. . . .	Coy. Cmdr., N.-C. O.'s Schl. Belgaum'

Agents :—MESSRS. HOLT & CO.

The Norfolk Regiment—*contd.*

First Commn.	Names.	Regiment rank.	Army rank.	Remarks.
	2nd-Lieutenants—*contd.*			
19 Feb. 17	Cozens, S. J.	19 Feb. 17		
1 Mar. 17	Buckland, R. J.	1 Mar. 17		
1 Mar. 17	Lynes, F. C.	1 Mar. 17		
30 May 17	Browne J. C.	30 May 17	. . .	Kasauli.
30 May 17	Turner, H. L., M. M.	30 May 17		
29 Aug. 17	Baldock, W. H. (*temp. Capt.*)	29 Aug. 17		
13 Nov. 17	Hurrell, A. E.	13 Nov. 17		
16 Mar. 18	Alexander, L. F.	16 Mar. 18	. . .	M. W. S., Meerut.
16 Mar. 18	Sutton, C. B.	16 Mar. 18		
16 Mar. 18	Whitwell, H. E.	16 Mar. 18		
5 May 18	Wake, T. A.	5 May 18		
24 Jan. 06	Shand, Capt. N. P., M.C.	1 Feb. 17	. . .	Adjt.
. . .	Gamble, A.	. . . .		
. . .	Haggar, J., M.C.	. . . .	. . .	Qr.-Mr.
	Attached.			
	Captains.			
1 May 00	Taylor, G. C. R. (*temp. Maj., 24 Sep. 18*), 1-8th Essex Regt.	3 Nov. 14	. . .	D. A. A. G.
	Dallas, H. W., 8th Norf. R.	2 Feb. 15	. . .	I. A. (*on probn.*)
	McGregor, R. S., 3rd Norf. R.	10 Apl. 15		
	Holland, J. R., 1st Norf. R.	20 Apl. 15		
	Keighley, G. C., M.C., 3rd Norf. R.	21 Apl. 15		
	Montgomery, H. R. G., M. C., 7th Norf. R.	25 Sep. 16		
	Lieutenants.			
	Berners, C. H., 3rd Norf. R.	1 June 15		
13 Mar. 15	Marsham, R. H. P., 2-6th Norf. R.	3 Dec. 15		
23 Dec. 14	Kenny, H. T., 1st Norf. R. (*act'g. Capt., 7 Sep. 18.*)	5 Sep. 16		
3 Feb. 15	Daniel, A. C. E., 3rd Norf. R.	21 Apl. 16		
30 Sep. 16	Walsha, J. W., 9th Norf R. (*act'g. Capt., 9 Nov. 18*)	12 Nov. 16		
11 Apl. 15	Lyus, A. O., 3rd Norf. R.	1 Apl. 17	. . .	*Lv., ex I., 6 m.*
23 Apl. 15	Price, C. T., 3rd Norf. R.	1 July 17		
21 May 15	Lloyd, R. S., M. C., 3rd Norf. R.	1 July 17		
17 June 15	Webster, H. (*temp. Capt. 15 Jan. 18*), 3rd Norf. R.	1 July 17		
1 July 15	Farmer, F. S., 3rd Norf. R.	1 July 17		
6 July 15	Palmer, E. de S., 3rd Norf. R.	1 July 17	. . .	Adjt., 23rd Arab Corps.
4 Sep. 15	Plant, J. C. G., 3rd Norf. R.	1 July 17		
22 Sep. 14	Clark, P. G., 9th Norf. R.	1 July 17		
1 Oct. 15	Stanforth, L. H. (*temp.*) 2nd G. B. Essex Regt.	1 July 17		
5 Oct. 14	Fisher, B. O. F. de C., 7th Norf. R.	1 July 17		
1 Nov. 15	Turner, J. C., 3rd Norf. R.	1 July 17	. .	Adjt. and Qr.-Mr., 22nd Arab Labour Corps.
1 Nov. 15	Hornor, B. F., D.S.O., 3rd Norf. R.	1 July 17		
1 Nov. 15	Wilder, J. McM., 3rd Norf. R.	1 July 17		
29 Nov. 15	Clarke, J. W., 9th Norf. R. (46)	1 July 17		
18 Feb. 16	Heath, F. M., 3rd Norf. R.	18 Aug. 17		
7 Apl. 16	Hayes, E. C. (*temp. Capt., 15 Jan. 18*), 3rd Norf. R.	7 Oct. 17		
5 Aug. 16	Reeve, E. G., 3rd Norf. R.	5 Feb. 18		
5 Aug. 16	Webster, J. F., 3rd Norf. R.	5 Feb. 18		
5 Aug. 16	Fellowes, C. A., 3rd Norf. R.	5 Feb. 18		
7 Nov. 14	Richardson, A. M., 3rd Norf. R.	5 Feb. 18		
22 Nov. 10	Harrison, V. M., M. C., 3rd Norf. R.	22 May 18		
	2nd-Lieutenants			
27 Oct. 16	Mithchell, G. A., 3rd Norf. R.	27 Oct. 16		
25 Jan. 17	Cubitt, F. L., 3rd Norf. R.	25 Jan. 17		
25 Jan. 17	Winfield, L., 3rd Norf.	25 Jan. 17		
26 Apl. 17	Bunn, A. E., 4th Norf. R.	26 Apl. 17		
30 May 17	Archer, B., 4th Norf. R.	30 May 17		
28 Aug. 17	axter, W. E., 3rd Norf. R.	28 Aug. 17		
28 Aug. 17	Maiden, C. J. D., 3rd Norf.	28 Aug. 17		
28 Aug. 17	Tipping, F. B., 3rd Norf. R.	28 Aug. 17	. . .	With 1st Garr. Bn. Oxf. Bucks. L. I.
25 Jan. 17	Witthaus, G. T., 2nd Norf. R.	25 Aug. 17		
30 May 17	Brown, R. E., 5th Norf. R.	30 May 17		With 1-5th R. W. Kent. R.
30 May 17	Payne, C. C., 4th Norf. R.	30 May 17	. . .	With 1-5th R. W. Kent. R.
30 May 17	Swann, L. G., 5th Norf. R.	30 May 17		With 1-5th R. W. Kent. R.

Regimental District No. 9. [No. District.]

Uniform—Scarlet
Facings—Yellow.

The Norfolk Regiment.

1st GARRISON BATTALION.

Arrived in India, 23rd December 1915.

Honorary Colonel—

First Commn.	Names.	Regimental rank.	Army rank.	Remarks.
	Lieutenant-Colonel.			
Jan. 69	Martin, R. H., C.B., C.M.G., C.I.E. (ret. pay.)	26 Oct. 92	*c* 26 Oct. 96	Comdg. Mhow.
	Majors.			
Dec. 00	Norrie, Maj. (*act'g. Lt.-Col.*) J., 4 Bn., Linc. R.	24 May 16	. . . . .	Comdg.
	Captains.			
Mar. 89	Beeching, C. F. (*Hon. Maj.*)	20 Sep. 14	. . . . .	Comdg. Br. Attd. Section, Poona.
21 Oct. 14	Drysdale, J. R.	14 Dec. 14		
Dec. 10	Wall, C. E.	13 Jan. 15	. . . .	A. D. C. to G. O. C. 5th Divn.
15 Aug. 14	Goldingham, D. D. (95)	20 Mar. 15	. . . .	D. A. A. & Q. M. G., Karachi Bde.
28 Feb. 00	Burnand, H.	16 June 15		
2 Aug. c8	Hodgson, H. E. A. (*temp.*)	24 Nov. 15		
28 Feb. 14	Brown, S. D.	1 June 16		
3 Oct. 14	*Borland, G. D.*	*20 May 17*	*20 Oct. 15*	*Adjt.*
	Lieutenants.			
1 Sep. 15	Palin, D. V.	1 Sep. 15		
19 Oct. 15	Heath, F.	19 Oct. 15	. . . . .	Adjt., Br. Attd. Section, Poona.
	2nd-Lieutenants.			
5 Aug. 16	Twiddy, G. A. (95)	5 Aug. 16	. . . . .	Attd. Hdqrs., 4th (Quetta) Dn.
12 Mar. 17	Toplis, G. A.	12 Mar. 17		
28 Mar. 17	Bradley, W. C.	28 Mar. 17		
22 May 17	Denney, L. E. (*temp.*)	22 May 17		
18 Aug. 17	Golden, C. H. (*temp.*)	18 Aug. 17		
19 Aug. 17	Ross, W. E. T. (*temp.*)	19 Aug. 17		
28 Aug. 17	Birkett, W. G. N.	28 Aug. 17		
19 Aug. 18	Thear, H. T. (*temp.*)	19 Aug. 18		
	Attached.			
4 Mar. 15	Smithwick, Maj. D. J.	12 Jan. 16	. . . .	Comdt., Young Offrs.' Camp Poona.
Oct. 00	Burnett, Maj. S. J.	8 Aug. 16	. . . . .	Act'g. 2nd-in-Comd.
19 Nov. 14	Dewhurst, Capt. W.	2 Oct. 15		
7 Jan. 17	Race, 2nd-Lt. W. C.	7 Jan. 17		
26 Apl. 17	Leahy, 2nd-Lt. D.	26 Apl. 17		
16 Sep. 17	Newman, 2nd-Lt. F. C.	16 Sep. 17		
31 Oct. 17	Morley, 2nd-Lt. A. R.	31 Oct. 17		
18 Nov. 17	Sanderson, 2nd-Lt. G. M.	18 Nov. 17		
18 Apl. 18	Sommerlad, 2nd-Lt. H.	18 Apl. 18		

3 Oct. 14	Borland, Capt. G. D.	15 May 17	. . .	Adjt.
4 Oct. 15	Ballard, Lt. G. H. (*temp.*)	4 Oct. 15	. . .	Ch. Instr., Sch. of Army Cookery, Poona.

Regimental District No. 10.
[No. 6 District.]

Uniform—Scarlet.
Facings—White.

The Lincolnshire Regiment.

1st GARRISON BATTALION.

Arrived in India, 22nd December 1915.

Honorary Colonel—Maj.-Genl. C. R. Simpson, C.B., *p.s.c.* [L]. . . . 10 May 14.

First Commn.	Names.	Regimental Rank.	Army rank.	Remarks.
	Lieutenant-Colonel.			
2 May 65	Shepherd, C. H., C.B.E., D.S.O. (51), (*ret. pay*).	1 May 93	*c.* 28 Oct. 98	Hony. A.-D.-C. to H. E. the Govr. of Bengal—Comdg.
	Major.			
5 Feb. 02	Cooper, O., T.D. (*4th Bn.*) (*Act'g. Lt.-Col.*)	13 Oct. 16	27 Apl. 12	2nd-in-Comd.
24 Nov. 14	Clark, J. C. (*temp.*) (51, 52) . . .	24 Aug. 16	. . . .	With 2-6th Devon. R.
	Captains.			
23 Sept. 14	Vyner, R. J. (*temp.*)	26 Mar. 18	23 Sep. 14	
30 Sep. 14	Stonor, E. C. (*temp.*).	2 Oct. 15	30 Sep. 14	
11 Dec. 14	Day, R. E. (*temp.*) (46, 51) . .	28 Sep. 16	7 Jan. 15	Staff Capt., 5th Infy. Bde.
4 Oct. 15	Stephens, J. W. (*temp.*) (60*a*) .	1 Apl. 16	. . . .	Adjt.. Chota Nagpur Rifles, I.D.F.
3 Jan. 15	Gray, J. (*temp.*) (51)	24 Aug. 16		
	Lieutenants.			
1 Mar. 15	Maidlow, J. F., (*temp.*)	28 Sep. 16	1 May 15	Ind. Ord. Dept.
31 Mar. 15	Taylor, C. H. (*temp.*) (12, 51, 52) .	5 Nov. 15	24 Aug. 15	
14 Dec. 14	Machin, H. S. (*temp.*) (59) . . .	2 Oct. 15	. . . .	Sig. Service.
4 Jan. 15	Smith, F. N. (*temp.*)	13 Oct. 15	. . . .	I. Ord. Dept.
30 Oct. 15	Jackson, O. (*temp.*)	2 Feb. 16	. . . .	Attd. R.E., Calcutta.
12 June 15	Hobbs, C. F. G.	1 July 17		
29 Dec. 14	Lumsdaine, A. A. M. (*temp.*) (57) .	1 July 17	. . . .	Attd. 2-7th Hamps. R.
24 July 15	Joiner, R. H. T. (*temp.*)	3 Aug. 17	1 July 17	
4 May 15	Brodrick, H. (*temp.*)	1 July 17	. . . .	Attd S. & T. Corps.
11 Mar. 15	Roskilly, S. T. H. (*temp.*) . . .	1 July 17	. . . .	I. Ord. Dept.
27 Mar. 15	Marrian, A. G. (*temp.*) (60*a*). . .	1 July 17		
28 Mar. 16	Markwick, W. L. (*temp.*) . . .	28 Sep. 17	. . . .	Ind. Munitions Board.
	2nd-Lieutenants.			
15 Apl. 17	Chard, S. B. (*temp.*)	15 Apl. 17	. . . .	In charge of Wireless, Calcutta.
1 Aug. 17	Hughes, E. W. (*temp.*)	1 Aug. 17		
19 Aug. 17	Bond, G. H. (*temp.*)	19 Aug. 17		
19 Aug. 17	Esilman, A. (*temp.*)	19 Aug. 17		
19 Aug. 17	Hindmarsh, F. (*temp.*)	19 Aug. 17		
19 Aug. 17	Threlfall, C. H. (*temp.*)	19 Aug. 17		
	Attached.			
17 Aug. 14	Riggall, Capt. H. B., 5 Bn. Linc., R. (51)	1 June 16		
5 Sep. 15	Stevens, Capt. S. G. (46, 52) 5 Bn., Linc. R.)	1 June 16	. . . .	Adjt., 12th E. B. Ry. Bn. I. D. F.
12 July 15	Challis, Lt. C. L. (*temp.*) Dorset R.	18 Dec. 16		
21 May 15	Pollard, Lt. E. A. (*act'g. Capt.. 27 July 18*)	1 June 16	. . . .	Provost Marshal, Calcutta.
7 July 15	Hall, Lieut. C. B. (*temp.*) . . .	1 July 17		
5 Sep. 15	Dyer, Lieut C. J. F. (*temp.*) (46, 60*a*)	. . . .	5 Sep. 15	M. A. Dept.
21 Sep. 15	Martin, Lieut. G., 4 Bn., Linc. R. (46, 52, 60*a*)	1 July 17		
21 Aug. 15	Sandison, 2nd-Lt. J., 4 Bn., R. Scots (46)	21 Aug. 15	. . . .	With. M. A. Dept.
4 Oct. 15	Stephens, Capt. J. W. (60*a*) . .	21 June 16	. . . .	Adjt.
22 July 16	Brett, Lt. H.	22 July 16	. . . .	Qr.-Mr.

Regimental District No. 11.
[No. 8 District.]

Uniform—Green.
Facings—Black.

The Devonshire Regiment.

1-4TH BATTALION.

Arrived in India, 10th November 1914, from England.

"South Africa, 1900-01."

HONORARY COLONEL—*Rt. Hon. Sir* J. H. Kennaway, *Bt.* C.B., V.D. . . . . 31 Dec. 02

First Commn.	Names.	Regimental rank.	Army rank.	Remarks.
	Lieutenant-Colonel.			
31 Jan. 00	Tremlett, C. P. (51)	5 June 18		
	Majors.			
1 Mar. 00	Townsend, H.	1 June 16		
1 Mar. 00	Pollard, L.	1 June 16		
5 Jan. 98	Forward, W. G.	19 May 17		
	Captains.			
11 Nov. 04	Anderson-Morshead, R. Y. (51)	1 Aug. 09		
1 Apl. 06	Carter, F.	7 Mar. 13	. . . .	*Lv., ex I., m.c.*
5 May 03	*Stamp, W. A. E.* (57)	4 Aug. 14		
3 Dec. 10	Thoday, F. A. (59)	1 June 16	. . . .	*Lv., ex I., m.c.*
24 May 11	Logan, W.	1 June 16		
1 Nov. 12	Orchard, J. W. (52)	1 June 16	. . . .	*Lv., ex I., m.c.*
22 May 13	Snell, W.	10 Apl. 17	. . . .	With 2nd Dorset. R.
4 Aug. 14	Kennaway, J. (51)	12 May 17		
6 Feb. 13	Carpenter, H. R. (51)	18 May 17		
20 Sep. 14	Dodd, C. K. (57)	24 Sep. 17		
29 Aug. 14	*Thompson, D. C.* (52)	19 May 18	. . . .	*With 24th Labour Corps.*
	Lieutenants.			
24 Sep. 15	James, E.	1 July 17	. . . .	Mily. Accts. Dept.
10 Dec. 15	Simmons, C. D.	1 July 17		
12 Dec. 15	Smith, M. R., M.C.	1 July 17		
18 Dec. 15	Bird, R. (*temp. Capt.*)	1 July 17	. . . .	S. S. O., Secunderabad.
6 Jan. 16	Osborn, C. R.	6 July 17	. . . .	*Lv., ex I., m. c.*
22 Apl. 16	Cusse, C. A. (*temp. Capt.*)	22 Oct. 17		
	2nd-Lieutenants.			
16 Mar. 18	Hearn, L. A. (*temp.*)	16 Mar. 18	. . . .	With Sig. Serv.
19 Aug. 18	Hatton, F. H.	19 Aug. 18		
19 Aug. 18	Staddon, T. W.	19 Aug. 18		
	Attached.			
29 July 15	Trevor-Roper, Capt. A. M., 4 Bn.	4 Nov. 15		
	Brown, Capt. J. A., 5th Bn.	1 June 16	. . . .	Adjt., 35th Labour Corps.
7 Nov. 14	Withers, Capt. J. R., 2nd N. Staff R.	30 Mar. 17	1 May 18	Adjt.
29 Aug. 14	Ward, Capt. H. J., 4 Bn.	2 June 17		
9 Jan. 15	Harrison, Lt. C. W., 1 Bn., R. Suss. R.	20 Apl. 16		
1 Oct. 14	Gibson, Lt. R. J. T., 4 Bn.	1 June 16		
14 Oct. 14	Kingwell, Lt. G. D., 4 Bn.	1 June 16		
11 Dec. 15	Allen, Lt. W. G., 4 Bn.	1 July 17		
17 Dec. 15	Hunter, Lt. R. T., 4 Bn. (*Act'g. Capt.*)	1 July 17	. . . .	With 39th P. L. C.
13 June 15	Gilbert, Lt. S. W. (*Act'g. Capt.*)	1 July 17	. . . .	Comdt., Ry Camp.
. . . .	Frossard, Lt. C. B. S., 3rd Devon. R.	22 Mar. 16	. . . .	*Lv. ex I., m. c.*
13 Mar. 15	Wetherall, Lt. C. R.	1 July 17		
21 Apl. 15	Earl, Lt. E.A., 5th Devon R. (48*a*, 51)	1 July 17		
22 Sep. 15	Buck, Lt. H. (*temp. Capt.*) 5th Suff. R.	1 July 17	. . . .	Staff Capt., Nowshera Bde.
15 Oct. 15	Shapter, Lt. L. J., 4 Bn. (51)	1 July 17		
16 Nov. 15	Edwards, Lt. R. C., 4 Bn. (51)	1 July 17	. . . .	Comdt., P. of W. Camp.
26 Mar. 16	Rowe, Lt. M., 4 Bn.	26 Sep. 17		
. . . .	Pugh, Lt. J. H., 3 Bn., North'd Fus	1 July 17		
. . . .	Kent, Lt. D. G., (*temp.*), Devon R.	20 July 17		
7 Apl. 16	Elford, Lt., 3rd Devon R. (52)	7 Oct. 17		
9 June 16	Harding, Lt. H., 4 Bn.	9 Dec. 17	. . . .	Adjt. & Qr.-Mr.
11 July 16	Bourne, Lt. P.R. 4 Bn.	11 Jan. 18		
11 July 16	Chadwell, Lt. A. E., 4 Bn.	11 Jan. 18		
5 Aug. 16	Bond, Lt. S. J., 6 Res. Bn.	5 Feb. 18		
5 Aug. 16	Foyster, Lt. G., 6 Res. Bn.	5 Feb. 18		
29 Aug. 17	Scales, 2nd-Lt. A.M., 4 Bn.	29 Aug. 17		
1 Mar. 17	Hambly, 2nd-Lt. W.P., 3rd Devon R.	1 Mar. 17		
13 May 17	Ballamy, 2nd-Lt. G. R., 2nd Garr Bn. Devon R.	13 May 17		
30 May 17	Allard, 2nd-Lt. G. H., 3rd Devon R.	30 May 17		
27 June 17	Peterkin, 2nd-Lt. R. C., 4th Gord. Highrs.	27 June 17		
27 June 17	Farquhar, 2nd-Lt. W., 4th Gord. Highrs.	27 June 17		
30 July 17	Young, 2nd-Lt. J. H., Arg. and Suth'd Highrs	30 July 17		
1 Aug. 17	Dunbar, 2nd-Lt. W. C., Arg. and Suth'd Highrs	1 Aug. 17		
28 Mar. 17	Perry, 2nd-Lt. J. P., (*temp.*)	28 Mar. 17		
26 Sep. 17	Gough, 2nd-Lt. J. A., 10 Bn., Manch R.	26 Sep. 17		
16 Jan. 17	Robins, Lt. J. H.	16 Jan. 17	. . . .	Qr. Mr. *Lv., ex I., m. c.*
20 Jan. 13	Fisher, Capt. T. A., R. A.M.C. (T. F.)	1 Apl. 15	. . . .	Medl. Offr.

Regimental District No. 11. [No. 8 District.]

Uniform—Green. *Facings*—Black.

The Devonshire Regiment.

2-4TH BATTALION.

Arrived in India, 5th January 1915, from England.

"South Africa, 1900-01."

HONORARY COLONEL—*Rt. Hon. Sir* J. H. Kennaway, *Bt.* C.B., V.D. . . . 31 Dec. 02.

First Commn.	Names.	Regimental rank.	Army rank.	Remarks.
	Lieutenant-Colonel.			
Feb. 96	**Cosens,** F.R. S., T.D. (*temp. Maj., R. A. M. C.*) (51)	5 June 18		
	Majors.			
4 July 00	Harvey, F. J. (51)	19 May 17	. . . .	Hill Conv. Section, Wellington.
	Captain.			
7 Oct. 14	Cole, H. M. (*act'g. Maj.*) (51)	7 Oct. 14		
29 July 15	Trevor-Roper, A. M.	4 Nov. 15	. . . .	With 1-4th Bn.
29 Aug. 14	*Mitchell, C. A.* (59)	19 May 17	. . . .	*M. G. Corps.*
29 Aug. 14	Ward, H. J.	2 June 17	. . . .	With 1-4th Bn.
	Lieutenants.			
29 Sep. 14	Reed, H. S. (*act'g. Capt.*)	1 June 16		
30 Sep. 14	Halford, E. H. J. (51) (*act'g. Capt.*)	1 June 16		
1 Oct. 14	Birchall, J. R. (*act'g. Capt., 4 Nov. 17*) (51)	1 June 16	. . . .	Staff Capt.
1 Oct. 14	Gibson, R. J. T.	1 June 16	. . . .	With 1-4th Bn.
1 Oct. 14	Shield, G. E. R. (51, 52) (*Act g. Capt.*)	1 June 16	. . . .	
22 Oct. 14	*Tayleur, G.* (51, 57)	1 June 16	. . . .	*I. A.* (*On probn.*)
16 Nov. 14	Locke, D. W.	1 June 16	. . . .	
14 Oct. 14	Kingwell, G. L.	1 June 16	. . . .	With 1-4th Bn.
22 Oct. 14	*Howarth, R. A. S.* (51, 52)	1 June 16	. . . .	*M. G. Corps.*
10 Dec. 14	Pothury, H. W. (51, 52)	1 June 16		
8 Oct. 14	Greenwood, E. T. (*act'g. Capt.*) (51, 52)	18 May 17		
24 Mar. 15	Lefroy, A. L. M. (*temp. Capt.*)	1 July 17	. . . .	R. T. O., Quetta.
7 Oct. 15	Humphrys, D. H.	1 July 17		
15 Oct. 15	Shapter, L. J. (51)	1 July 17	. . . .	With 1-4th Bn.
16 Nov. 15	Edwards, R. C. (51)	1 July 17	. . . .	With 1-4th Bn.
11 Dec. 15	Allen, W. G.	1 July 17	. . . .	With 1-4th Bn.
17 Dec. 15	Hunter, R. T.	1 July 17	. . . .	With 1-4th Bn.
27 Jan. 15	Smith, E. C. (*act'g. Capt., 3 Aug. 17*)	1 July 17	. . . .	Adjt.
11 Jan. 16	Smerdon, H. M. (51)	11 July 17		
26 Mar. 16	Rowe, M.	26 Sep. 17	. . . .	With 1-4th Bn.
9 June 16	Harding, H.	9 Dec. 17	. . . .	With 1-4th Bn.
8 July 16	Robinson, W. W. (51)	8 Jan. 18		
8 July 16	Hill, E. R.	8 Jan. 18		
11 July 16	Bourne, P. R.	11 Jan. 18	. . . .	With 1-4th Bn.
11 July 16	Chadwell, A. E.	11 Jan. 18	. . . .	With 1-4th Bn.
11 July 16	Short, F. P.	11 Jan. 18		
20 July 16	Plowden, A. H. C.	20 Jan. 18		
5 Aug. 16	Bulleid, J. F. (51)	5 Feb. 18		
5 Aug. 16	Francis, J. E. (51)	5 Feb. 18		
5 Sep. 16	Dunstan, W. (51)	5 Mar. 18		
5 Sep. 16	*Jones, L. W.*	5 Mar. 18	. . . .	*I. A.* (*On probn.*)
5 Sep. 16	Palmer F. O. (46)	5 Mar. 18		
5 Sep. 16	Russell, P	5 Mar. 18		
26 Sep. 16	Huish, L.	26 Mar. 18		
	2nd-Lieutenants.			
23 July 16	Ellis, W. J.	23 July 16	. . . .	With 1-4th Bn.
	Attached.			
27 Jan. 15	Smith, Lt. (*act'g. Capt.*) E. C.	1 July 17	. . . .	Adjutant.
13 Apl. 15	Coombe, Lt. J. R.	13 Apl. 15	. . . .	Qr.-Mr.

Regimental District No. 11.
[No. 8 District.]

Uniform—Scarlet.
Facings—Lincoln Green.

The Devonshire Regiment.

1—6TH BATTALION.

Arrived in India, 10th November 1914, from England.

"South Africa, 1900-01."

HONORARY COLONEL—Colonel E. S. Walcott, C.B., T.D. (retd. I. A.) . . . 15 Apl. 07.

First Commn.	Names.	Regimental rank.	Army rank.	Remarks.
	Lieutenant-Colonel.			
	Flick, C. L., C.M.G., Can. Local Forces (*temp.*)	26 Apl. 16	29 July 15	
	Major.			
7 June 02	Bazeley, W. N.	2 Jan. 18		
2 Apl. 09	*Trevenen, W. B.*	2 Jan. 18	. . . .	*M. A. Dept.*
	Captains.			
2 Mar. 10	Smyth-Richards, F. G. (51) . . .	8 Mar. 15	. . . .	Asst. Postal Censor, Bombay.
28 May 10	*Mason, A. S.* (*temp. Maj.*) . . .	18 May 15	. . . .	*D. A. Q. M. G., 16th (Indian) Dn.*
15 Apl. 12	*Reavell, H. S.* (66a)	1 June 16	. . . .	*Ind. Ord. Dept.*
14 Sep. 12	Waldram, H. G (51)	1 June 16	. . . .	Attd., 2nd Dorset. R.
5 Aug. 14	Sidmouth, G. W., *Visct.* (51) . .	1 June 16		
21 Aug. 14	Wigan, T. K. (51)	1 June 16		
22 Aug. 14	Driver, C. E., M.C. (48a, 50) . .	1 June 16		
11 Sep. 14	Gray, D. P. G. (51)	12 May 17		
	Lieutenants.			
7 Oct. 14	Pennefather, E. M. (59)	1 June 16	. . . .	With Sig. Service.
27 Mar. 15	Gribble, H. J. C.	1 July 17		
27 July 15	Salmon, G. H.	1 July 17		
20 Nov. 15	*Thomas, A. R.* (temp. Capt., *1 Nov. 17*)	1 July 17	. . . .	Mily. Accts. Dept.
8 June 16	Bowen, G. M.	9 Dec. 17		
9 June 16	Mathews, L. D.	9 Dec. 17		
13 June 16	Hudson, W. G., D.C.M. (*act'g. Capt.*) .	13 Dec. 17		
	2nd-Lieutenant.			
4 June 16	Samuels. L.	4 June 16		
16 Mar. 18	Clarke, C. G. (*temp.*)	16 Mar. 18		
	Attached.			
21 Jan. 01	Roseveare, Capt. E., 1-5 Bn., Devon. R.	4 Aug. 06	. . . .	R. T. O., Poona.
1 Nov. 11	Abraham, Capt. A. C., Devon. R.	1 June 16	. . . .	Offrs.' Schl. of Instn., Nasik.
4 Feb. 15	Bowden, Lt. (*temp. Capt.*) W. J., (*temp.*) Devon. R.	13 Sep. 15	. . . .	Embkn. Staff.
1 Oct. 14	Young, Lieut. G. B., 6 Bn., Devon. R.	1 June 16		
21 Oct. 14	*Bennett, Lieut.* (*act'g. Capt.*) *A. H., 5 Bn. Devon. R.*	1 June 16	. . . .	*Combd. Depot.*
10 Dec. 14	Fursdon, Lt. A. A., 2-6 Devon. R.	1 June 16		
2 Nov. 14	Thompson, Lt. C. C., 9 Devon. R.	1 July 16		
5 Nov. 14	Butland, Lt. R., 9 Devon. R. . .	2 July 16		
9 Dec. 14	Hughes, Lt. F., 5 Bn., Devon. R.	1 July 17	. . . .	Ind. Ord. Dept.

The Devonshire Regiment—*concld.*

First Commn.	Names.	Regimental rank.	Army rank.	Remarks.
	Attached—*concld.*			
12 Dec. 14	*Orchard, Lt. M. F., 6 Bn., Devon. R.*	1 July 17	....	*M. G. Corps.*
28 Mar. 15	*Watts, Lt. R. B.*, 5 Bn., Devon. R.	1 July 17	....	*I. A. (on probn.)*
14 May 15	Allison, Lt. D. W., 5 Bn., Devon. R.	1 July 17		
30 May 15	Wilson, Capt. W. H., 5 Bn., Devon. R.	11 Apl. 18		
29 Oct. 15	Mason, Lt. C. M., 22 Bn., Lond. R.	1 July 17		
1 Nov. 15	Holmes, Lt. E. J., 5 Bn., Glouc. R.	1 July 17		
30 June 16	Vos, Lt. S., 3-6 Devon. R.	30 Dec. 17		
2 Sep. 15	Hill, Lt. R. F., 5 Bn., D. of Corn. L.I.	1 July 17		
2 Nov. 15	Cannell, Lt. J. H., 5 Suff. R.	1 July 17	....	*I.v. in & ex I.*
19 Mar. 16	Ormesher, Lt. C. I., 1 Devon. R.	19 Sep. 17		
8 July 16	Hodges, Lt. H. H., 2-6 Devon. R.	8 Jan. 18		
16 Aug. 16	Beer, Lt. G. H. B., Devon. R.	16 Feb. 18		
25 Oct. 16	Yeates, 2nd-Lt. S. P. J., 3-5 Devon. R.	25 Oct. 16	....	Attd. R. A. F.
	Kitson, Capt. W. K., 2-5th Devon. R.	1 June 16		
3 June 11	Robinson, Capt. H. I., 1-4th R. W. Kent R.	1 June 16		
29 June 15	Williams, Lt. (*temp. Capt.*) J. H., (46), 2-5th Devon R.	1 July 17	....	Instr., Bombing Schl.
5 Sep. 15	Archer, Lt. G, 3rd Devon R.	1 July 17		
3 Mar. 15	Brand, Lt. J. K., (*temp.*) R. W. Kent. R.	3 Mar. 15		
4 Jan. 16	Kelly, Lt. F. C., 4th Gord. Highrs.	4 July 17		
	Gemmill, Lt. A., 3rd Sco. Rif.	1 July 17		
3 Oct. 14	Wilson, Lt. W. H. (51, 52), 2-6th Devon. R.	1 June 16		
	Clover, Lt. (*act'g. Capt.*) P. T., 2-5th Devon R.	1 June 16	....	Adjt., Conv. Depot.
10 Dec. 14	Williams, Lt. N. T., 2-5th Devon. R.	1 July 17		
17 Dec. 15	Johnson, 2nd-Lt. R. H. K., 2-6th Devon R.	17 Dec. 15		
5 Aug. 16	Carswell, 2nd-Lt. J. S., 3rd Black Watch	5 Aug. 16		
5 Sep. 16	Wilkinson, 2nd-Lt. A. E., Sea. Highrs.	5 Sep. 16		
	Bowler, 2nd-Lt. D., Notts. &Derby. R.			
	Craze, Lt. F., 2-5th D. of Corn. L. I.	1 July 17		
4 Dec. 15	Elliott, 2nd-Lt. L., 8th Devon. R.	4 Dec. 15	....	*I.v. ex, I., m.c., 6 mos.*
3 May 15	Banner, Lt. J. W. (*temp.*), 1 Bn., York R.	30 June 15		
.....	Gilbee, Lt. S. J., R. Fus.	1 July 17		
18 Sep. 15	Lloyd, Lt. H. C. 1 Bn., York R.	1 July 17		
26 Sep. 17	Roe, 2nd-Lt. J. H., 4 Bn., Notts. & Derby. R.	26 Sep. 17		
26 Sep. 17	Blore, 2nd-Lt. A. V., 4 Bn., Notts. & Derby. R.	26 Sep. 17		
26 Sep. 17	Platts, 2nd-Lt. T. C., 6 Bn., Manch. R.	26 Sep. 17		

First Commn.	Names.	Regimental rank.	Army rank.	Remarks.
10 Sep. 15	Batt, Lt. (*act'g. Capt.*) R. H. (51), 3-18th Lond. R.	16 Apl. 18	....	Adjt. & Qr.-Mr.
19 Feb. 17	Yeo, Lt. W. H.	19 Feb. 17	....	Qr.-Mr.

Regimental District No. 11. [No. 8 District.]

The Devonshire Regiment.

Uniform—Scarlet. *Facings*—Lincoln Green.

2-6TH BATTALION.

Arrived in India, 6th January 1915, from England.

" South Africa. 1900-01."

HONORARY COLONEL—Colonel E. S. Walcott, C.B. T.D. (retd. I. A.) . . . 15 Apl. 07.

First Commn.	Names.	Regimental rank.	Army rank.	Remarks.
	Lieutenant-Colonel.			
	Major.			
	Captains.			
10 Feb. 14	Jewell, H. A. (51)	1 June 16	. . . .	With 1-6th Devon. R.
22 Mar. 14	*Crew, F. A. E.*	1 June 16	. . . .	*With R. A. M. C.*
10 Sep. 14	*Molesworth, G. F. N.* (51, 57)	12 May 17	. . . .	I. A. (*On probn.*)
1 Oct. 14	*Palairet, R. C. N.* (*temp. Maj., 7 Sep. 18*)	29 Aug. 17	. . . .	*Dy. Asst. Dir. of Equipment and Ordnance Stores, Army Hdqrs.*
1 Oct. 14	Cooper, C. A.	31 Aug. 17		
1 Oct. 14	*Pope, S. F.*	9 Mar. 18	. . . .	*D.A. Q. M.G.*
3 Oct. 14	*Watson, E. L.*	9 Mar. 18	. . . .	*Worc. R.*
	Lieutenants			
1 Oct. 14	Young, G. B.	1 June 16	. . . .	With 1-6th Devon. R.
3 Oct. 14	Wilson, W. R. (51, 52)	1 June 16	. . . .	With 1-6th Devon. R.
8 Oct. 14	Buckingham, T. N. (51)	1 June 16	. . . .	With 1-6th Devon. R.
15 Oct. 14	Prideaux, C. H.	1 June 16	. . . .	*Lv., ex I., m. c.*
6 Oct. 14	Jeffrey, E. J. B.	1 June 16	. . . .	*Lv., ex I., m.c., 6 m.*
17 Oct. 14	*Denny, T. H.* (51, 57, 207*a*)	1 June 16	. . . .	*I. A.* (*on probn.*)
10 Dec. 14	Fursdon, A. A. (51, 52)	1 June 16	. . . .	With 1-6th Devon. R.
11 Dec. 14	Fulford, G. L. (51)	5 Mar. 17		
11 Dec. 14	Bawden, P. R. (51, 52)	1 July 17	. . . .	Attd. Works Directcrate.
12 Dec. 14	Orchard, M. F. (51, 52)	1 July 17	. . . .	With 1-6th Devon. R.
12 Dec. 14	Minshall, C. H.	1 July 17	. . . .	*Lv., ex I., m.c.*
12 Dec. 14	*Date, W. A.* (51, 52)	1 July 17	. . . .	*With M. G. Corps.*
3 June 15	Giles, E. P.	1 July 17		
13 Mar. 15	Daimpre, M. R. W.	1 July 17	. . . .	With 1-6th Devon. R.
4 Nov. 15	Treeby, J. W. C. (46)	1 July 17		
10 Dec. 15	Mills, B. A. J.	1 July 17		
18 Dec. 15	Pinhey, H. T. (46)	1 July 17	. . . .	With 1-5th E. Kent R.
22 Dec. 15	Curry, R. C.	1 July 17		
16 Jan. 16	*Brede, H. G.* (*temp. Capt., 21 June 18*) (46)	16 July 17		
11 June 16	Bush, S. L. (66*a*)	11 Dec. 17		
21 June 16	Adams, A. K. (46)	21 Dec. 17		
8 July 16	Arnold, N.	8 Jan. 18	. . . .	With 1-5th E. Kent R.
8 July 16	*Winn, H.*	8 Jan. 18	. . . .	*R. A. F.*
8 July 16	Howells, P. I. (46)	8 Jan. 18		
8 July 16	Barber, A. L.	8 Jan. 18		
8 July 16	Grose, N. A.	8 Jan. 18		
8 July 16	Hodges, H. H.	8 Jan. 18	. . . .	With 1-6th Devon. R.
26 Sep. 16	*Prideaux, V. H.*	26 Mar. 18	. . . .	*M. G. Corps.*
	2nd-Lieutenants.			
12 Dec. 15	Hancock, G. R. (46)	12 Dec. 15	. . . .	With 1-5th E. Kent R.
7 Dec. 15	Johnson, R. H. K.	17 Dec 15	. . . .	With 1-6th Devon. R.
8 Apl. 17	Baker, J.	28 Apl. 17		
8 Apl. 17	Vittle, D. H.	28 Apl. 17	. . . .	*Lv., ex I., m. c., 6 mos., 13 June 18.*
8 Apl. 17	Smith, H. D.	28 Apl. 17		
8 Apl. 17	Baker, R. C. N.	28 Apl. 17		
29 Aug. 17	Kelsall, J. (60*a*)	29 Aug. 17	. . . .	N. C. O.'s Schl., Belgaum.
29 Aug. 17	Cole, E. J. (51)	29 Aug. 17		
29 Aug. 17	*Jeffery, H. E.*	29 Aug. 17	. . . .	I. A. R. O.
29 Aug. 17	Burridge, R. A.	29 Aug. 17		
16 Mar. 18	Selfe, G. V. H. (*temp.*)	16 Mar. 18		
16 Mar. 18	Jenkins, L. C. (*temp.*)	16 Mar. 18		

Regimental District No. 11. [No. 8 District.]

The Devonshire Regiment.

Uniform—Scarlet. *Facings*—Lincoln Green.

2·6TH BATTALION.

Arrived in India, 6th January 1915 from England.

"South Africa, 1900-01."

HONORARY COLONEL—Colonel E. S. Walcott, C.B. T D. (retd. I. A.) . . . 15 Apl. 07.

First Commn.	Names.	Regimental rank.	Army rank.	Remarks.
	Attached.			
15 Mar. 02	Woollcombe, Maj. L. D. (*act'g. Lt.-Col.*), Devon. R.	15 Mar. 17	. . . .	Comdg.
May 00	Clark, Maj. J. C. (51, 52) . . .	Aug. 16		
2 Sep. 05	Timmins, Maj. T. B., 1-4 Bn. Som. L. I.	22 Feb. 18	. . . .	
. . . .	Guy, Capt. G. R. (*temp.*), 1 Bn., Oxf. & Bucks. L. I.	23 June 16	. . . .	
28 Nov. 14	*Cade, 2nd-Lt. C. M. D.*, **T. F. Res. (Gen. List)** (169, 206)	28 Nov. 14	. . . .	*I. A. R. O.*
	Andrews, Capt. C. G., 2nd Bn., N. Staff. R.	13 June 17		
26 July 13	Fleming, Lt. G. T., W. Rid. R. .	25 Oct. 14	. . . .	Adjt.
14 Sep. 14	Dobbie, Lt. A. G., K. O. Sco. Bord.	19 Apl. 15		
9 Jan. 15	Parrott, Lt. A., York. R. . . .	10 June 15		
9 Dec. 14	Stabb, Lt. C. T., 1-5 Bn., E. Kent R.	1 June 16	. . . .	
15 Mar. 15	Scott, Lt. A. W., Durh. L. I. . .	12 June 16		
7 Nov. 14	Hazell, Lt. A. J., 1 Bn., W. Rid. R.	2 Mar. 17	. . . .	
. . . .	Sayers, Lt. W. W., 1st R. Suss. R.	1 July 17		
24 Jan. 15	Grey, 2nd-Lt. J. W. A., Durh. L. I.	24 Jan. 15		
8 May 15	Sheriff, 2nd-Lt. G., Sco. Rif. . . .	8 May 15		
11 Sep. 15	Woodhams, 2nd-Lt. D., R. Suss. R.	11 Sep. 15		
. . . .	Isbell, 2nd-Lt. H., 3rd Leic. R. .	. . . .		
16 Mar. 18	Simth, 2nd-Lt. F. H., 1 Bn., Durh. L. I.	16 Mar. 18	. . . .	
. . . .	Williamson, 2nd-Lt. 1 Bn., Oxf. & Bucks L. I.	. . . .	. . . .	
. . . .	Bicknell, 2nd-Lt. F., 1 Bn., Durh. L. I.	. . . .	. . . .	
. . . .	Cogell, 2nd-Lt. C. C., K. R. Rif. C.	. . . .	. . . .	
26 July 13	Fleming, Lt. (*act'g.* Capt.) G. T. (W. Rid. R.)	15 Dec. 17	. . . .	Adjt.
6 Dec. 14	Edgecombe, Capt. F.	6 Dec. 14	. . . .	Qr.-Mr. *Lv., ex I., m.c.*

Regimental District No. 13. [No. 8 District.]

Uniform—Scarlet *Facings*—Blue.

Prince Albert's (Somerset Light Infantry).

2ND BATTALION (*late 13th Foot*).

Arrived in India, 21st November 1913, from Tientsin.

THE SPHINX SUPERSCRIBED "EGYPT." A MURAL CROWN, SUPERSCRIBED "JELLALABAD." "Gibraltar, 1704-5" "Dettingen" "Martinique, 1809" "Ava" "Ghuznee, 1839" "Affghanistan, 1839" "Cabool, 1842" "Sevastopol" "South Africa, 1878-9" "Burma, 1885-87" "Relief of Ladysmith" "South Africa, 1899-1902."

COLONEL—Major-General R. L. Payne, C.B., D.S.O., *ret. pay*. 5 Apl. 14.

First Commn.	Names.	Regimental rank.	Army rank.	Remarks.
	Lieutenant-Colonel.			
5 Feb. 87	Platt, E. H. R. C. R. (51, 72*b*, 162, 206).	21 Apl. 14	*c.* 21 Apl. 18	
	Majors.			
20 May 99	Worrall, E. W. (*act'g. Lt.-Col., 24 Oct. 18*) ([L] (51)	1 Sep. 15		
5 Jan. 01	Stone, N. H. (51, 59, 71, 72) . .	1 Sep. 15		
	Captains.			
11 Apl. 08	Williams, C. A. (*act'g. Maj., 24 Oct. 88*) (51, 52, 94, 100) . . .	10 June 15		
5 Feb. 09	Roche, V. W. (51, 52)	10 June 15		
9 Mar. 10	*Molesworth, G. N.* (12, 50, 53, 94, 100, 127, 206)	9 Mar. 16	. . . .	Adjt.
4 Oct. 11	Howe, W. T. (51, 57, 95) . . .	25 Mar. 17	. . . .	Sig. Serv.
	Lieutenants.			
3 Sep. 13	Willis, N. O. (*act'g. Capt.*) (51, 52)	20 Dec. 14		
10 Dec. 13	Ludlow, C. W. P. (*act'g. Capt.*,) (51, 52, 95, 172*a*)	24 Feb. 15		
23 Jan. 14	McCririck, D. H. G. (*act'g. Capt., 24 Oct. 18*) (51, 52, 60*a*) . . .	26 Mar. 15		
24 Jan. 14	Romilly, H. A. (57)	26 Mar. 15	. . . .	With Sig. Serv.
29 Oct. 14	Browne, H. J. H. (*temp.*) (50, 52, 95)	10 June 15		
7 Nov. 14	Street, A. R. (*temp. Capt.*) . . (51, 52, 66, 95)	10 June 15	. . . .	Platoon Comdr. Officer Cadet Unit, Sialkot.
7 Nov. 14	Cunnington, J. (57, 66, 72, 95) . .	30 Sep. 15		
22 Feb. 15	Pavey, G. P. (51, 60*a*, 95) . . .	1 July 17		
16 June 15	Ranger, W. J. (*temp. Capt.*) (51, 60*a*, 172*a*)	1 July 17	. . . .	Comdt., Schl. of Physl. Trg., So. Comd.
17 June 15	Seamark, F. (51, 52, 72*a*) . . .	1 July 17		
17 June 15	Goode, T. C. (*act'g. Capt.*) . . (50, 53, 71)	1 July 17		With No. 1 Res. Bn.
11 Aug. 15	Mahony, N. E. B. C. (12, 51) . .	1 July 17		
15 Sep. 15	Snow, A. E. (46, 94)	1 July 17		
19 Nov. 15	Jones, J. (65*a*, 66, 72, 89*a*, 207) .	1 July 17	. . . .	Kurdish Labour Corps.
24 Nov. 15	Scott, S. (46)	1 July 17		
	2nd-Lieutenants.			
1 Jan. 17	Jimby, C. W. (*temp. Capt., 2 Mar. 18*)	1 Jan. 17	. . . .	Platoon Comdr., Offrs.' Schl. of Instn., Bangalore.
1 Jan. 17	Woodman, W. (*act'g. Capt.*) . .	1 Jan. 17	. . . .	Adjt., N. C. O.'s Schl.
1 Jan. 17	Pratt, G. H.	1 Jan. 17		
1 Jan. 17	Woods, J. (*act'g. Capt.*) . . .	1 Jan. 17	. . . .	Adjt., Offrs.' Schl. of Instn., Bangalore.
18 Feb. 18	Abbott, J.	18 Feb. 18		
18 Feb. 18	Emery, W. C.	18 Feb. 18		
16 Mar. 18	Clarke, H. B. J. (*temp.*).	16 Mar. 8		
16 Mar. 18	Gentle, F. H. (*temp.*)	16 Mar. 18		
16 Mar. 18	Vaux, R. (*temp.*)	16 Mar. 18		
17 Apl. 18	Lamb, R. J. (*temp. Capt., 15 Sept. 18.*)	17 Apl. 18	. . . .	R. T. O. Secunderabad.
	Attached.			
10 Apl. 15	Bailey, 2nd-Lt. H. G. (12, 46, 51, 94) (*temp.*) 8th Som. L. I.	10 Apl. 15		
10 Apl. 15	Giles, 2nd-Lt. L. R. (*temp.*) 8th Som. L. I. (12, 51)	10 Apl. 15		
9 Mar. 10	Molesworth, Capt. G. N. . .	16 Nov. 16	. . .	Adjt.
3 May 05	Owens, Capt. D. J. (50, 95, 170) .	3 May 15	. . .	Qr.-Mr.

Agents:—Messrs. Cox & Co.

Regimental District No. 13. [No. 8 District.]

Uniform—Scarlet. *Facings*—Blue.

Prince Albert's (Somerset Light Infantry).

1-4TH BATTALION.

Arrived in India, 11th November 1914, from England.

"South Africa, 1900-01."

HONORARY COLONEL—V. U. Langworthy, V.D. . . . 11 Aug. 00.

First Commn.	Names.	Regimental rank.	Army rank.	Remarks.
	Lieutenant-Colonel.			
11 Aug. 80	Cox, W. C. (*Maj. ret. pay*) (*Res. of offr.*) (51, 57, 72b, 170, 207)	7 Jan. 12		
	Majors.			
7 Mar. 00	Baily E. M.	1 June 16	. . . .	Ind. Mun. Board.
2 Feb. 02	*Strachey, Hon. E.* (51)	20 Jan. 18	. . . .	*Staff Capt., Karachi Bde.*
2 Sep. 05	Timmins, T. B.	22 Feb. 18	. . . .	With 2-6 Bn., Devon. R.
	Captains.			
12 Sep. 11	Phillips, W. E.	1 June 16		
11 Sep. 12	Bishop, G. W. R. (51, 67)	1 June 16	. . . .	With 2nd Dorset. R.
5 Sep. 13	Moger, W. (51)	1 June 16	. . . .	No. 8 Res. Bn.
2 Oct. 13	Tanner, H. R. (52)	1 June 16		
9 June 14	Wilson, S. (57)	1 June 16	. . . .	*Lv., ex I., m. c., 4 m.*
26 May 14	Coombs, A. P.	1 June 16		
29 June 14	Lewis, W. (51, 52)	1 June 16	. . . .	Adjt.
29 Aug. 14	Baines, F. J. T. (51)	23 July 17		
19 Sep. 14	*Openshaw, E. R.*	25 Oct. 17	. . . .	*R. A. F.*
19 Sep. 14	Fox, F. E. (57)	25 Oct. 17		
23 Sep. 14	Derrick, L.	27 Oct. 17	. . . .	With W. Rid. R.
7 Oct. 14	*Milne, E. O.*	22 Feb. 18		
	Lieutenants.			
29 Aug. 14	Stileman, R. F. (*act'g. Capt.*) (*Lt. Som. L. I.*) (51)	1 June 16		
7 Oct. 14	Burgess, F.	1 June 16	. . . .	F. T. C. O.
7 Oct. 14	Samler, W. H. G.	1 June 16		
9 Sep. 15	Nalder, F. W.	1 July 17	. . . .	*Lv., ex I., m. c., 6 mos.*
16 Oct. 15	*Hyatt, P. T.* (*temp. Major*)	1 July 17	. . . .	With Labour Corps.
7 Nov. 15	Helps, A. O.	13 July 17	. . . .	R. A. F.
	2nd-Lieutenant.			

Prince Albert's (Somerset Light Infantry)—*contd.*

First Commn.	Names.	Regimental rank.	Army rank.	Remarks.
	Attached.			
5 Jan. 01	Stone, Maj. N. H., 2nd Som. L. I. (*act'g. Lt.-Col.*)	1 Sep. 15		
21 Oct. 14	Proctor, Maj. J., K. R. Rif. C. (*attd.*)	1 Aug. 16		
6 Oct. 04	Gifford, Capt. R. E., 5 Som. L. I. (51)	1 July 11		
10 Apl. 09	Harris, Capt. F. J., 5 Som. L. I.	1 June 16		
29 Aug. 14	Morse, Capt. B. S., 4 Som. L. I.	8 Dec. 16		
29 Aug. 14	Stephenson, Capt. J. E., 4 Bn., Som. L. I.	14 Sep. 17		
29 Aug. 14	Elliot, Capt. G. O., 4 Bn., Som. L. I.	18 Oct. 17	. . . .	*Lv., ex I., m. c.*
5 May 15	Thomson, Lt. R. J., 1st. Garr. Bn., Som. L. I.	9 Feb. 17		
19 Dec. 14	Lee, Lt. A. K., 4 Bn., Som. L. I.	1 July 17		
3 Apl. 15	Morgan, Lt. H. G., 4th Bn.	1 July 17		
3 Apl. 15	Helson, 2nd-Lt. G. ((*temp.*) *Lt., 5 May 17*), E. Lanc. R.	3 Apl. 15		
20 Apl. 15	Bailey, 2nd-Lt. H. G., Som. L. I.	10 Apl. 15		
9 July 15	Hennessy, Lt. L. J., 5 Bn., Som. L. I.	1 July 17	. . . .	Adjt., 43rd Kurdish Labour Corps.
29 Oct. 15	Minns, Lt. A., 4 Bn., Som. L. I.	1 July 17		
1 Nov. 15	Page, Lt. L. H., 4th Bn.	1 July 17	. . . .	With I. W. T.
3 Nov. 15	Lewis, Lt. E. H., 4 Bn., Som. L. I.	1 July 17		
8 Nov. 15	Bird, Lt. D. R., 4 Bn., Som. L. I.	1 July 17		
26 Nov. 15	Whillier, 2nd-Lt. E. C., 4 Bn., Som. L. I.	26 Nov. 15	. . . .	*Lv., in I., m. c.*
14 Dec. 15	Dickinson, Lt. R., 4 Bn., Som. L. I.	1 July 17	. . . .	56th L. T. Mortar Batty.
16 Dec. 15	Williams, Lt. C., 4th Bn.	1 July 17	. . . .	*Lv., ex I., m.c.*
25 Dec. 15	Cuninghame, Lt. A. W., 4th Bn. (*temp. Capt.*)	1 July 17	. . . .	Asst. Prov. Marshal.
7 Jan. 16	Claxton, Lt. B., 4th Bn.	7 July 17	. . . .	R. A. F.
15 Jan. 16	Vallis, Lt. E. W. H., 4th Bn.	15 July 17		
15 Jan. 16	Thomas, Lt. B. S., 4th Bn.	15 July 17	. . . .	Poll. Employ.
24 Jan. 16	Parsons, Lt. L. C., 4th Bn.	24 July 17		
31 July 15	Jones, Lt. S. W. E., 4th Bn.	19 Aug. 17	31 July 15	
6 Mar. 16	Bradshaw, Lt. C. H., 4 Bn., Som. L. I.	6 Sep. 17	. . . .	56th L. T. Mortar Batty.
2 June 16	Humphrey, 2nd-Lt. G. P., D. of Corn. L. I.	2 June 16		
16 Mar. 18	Smith, 2nd-Lt. W. (*temp.*)	16 Mar. 18		
16 Mar. 18	Gilbert, 2nd-Lt. C. T. (*temp.*)	16 Mar. 18		
16 Mar. 18	Read, 2nd-Lt. E. S. (*temp.*)	16 Mar. 18		
9 Oct. 15	Watterson, Lt. J. S., 4th Bn., Som. L. I. (48*a*, 50)	9 Oct. 15	. . . .	No. 8 Res. Bn.
28 Apl. 17	Houghton, 2nd-Lt. A. T., 2-5th Bn., Som. L. I.	28 Apl. 17	. . . .	No. 8 Res. Bn.
15 Jan. 16	Parsons, 2nd-Lt. D. S., 1st Garr. Bn., Som. L. I.	15 Jan. 16		
9 Sep. 14	Wasborough, Lt. H. S., 1st Garr. Bn., Oxf. & Bucks. L. I.	20 Nov. 15		
19 Oct. 15	Crocker, Lt. E. G. M. (*temp. Lt.*), 3rd Bn., Som. L. I.	1 July 17	. . . .	No. 8 Res. Bn.
1 Mar. 17	Roberts, 2nd-Lt. C. E. L., 4th Bn., Som. L. I.	1 Mar. 17		
22 Feb. 16	Bradfield, 2nd-Lt. J. T., 4th Bn., Som. L. I.	22 Feb. 16		

29 June 14	Lewis, Capt. W.	1 Aug. 16	. . . .	Adjt.
. . . .	Wallace, Capt. A. D., 2-4 R. Sco. Fus.	18 May 15	. . . .	Qr.-Mr.

Regimental District No. 13. [No. 8 District.]

Uniform—Scarlet. *Facings*—Blue.

Prince Albert's (Somerset Light Infantry).

2-4TH BATTALION.

Arrived in India, 8th January 1915, from England.

"South Africa, 1900-01."

HONORARY COLONEL—V. U. Langworthy, V.D. 11 Aug. 00.

First Commn.	Names.	Regimental rank.	Army rank.	Remarks.
	Lieutenant-Colonel.			
	Captains.			
7 Oct. 14	Harford, E. B. (*Hon. Lt. in Army 7 Feb. 03*) (51, 72*f*)	7 Oct. 14		
9 Oct. 14	Farwell, E. W. (*act'g. Maj.*) (51)	9 Oct. 14		
27 Aug. 14	Miles, W. H.	8 Dec. 16	....	Adjt.
29 Aug. 14	Stephenson, J. E.	14 Sep. 17	....	With 1-4 Som. L. I.
29 Aug. 14	Elliott, G. O.	18 Oct. 17	....	With 1-4 Som. L. I.
23 Sep. 14	Mackie, J. R. (*act'g. Capt., 21 July 17*) (206)	22 Nov. 17		
2 Oct. 14	*Burt, W. H.* (51)	22 Feb. 18	....	*With 1st Garr. Bn., Oxf. & Bucks. L. I.*
7 Oct. 14	*Thatcher, A. F. B.*	22 Feb. 18	....	I. A. (*on probn.*)
	Lieutenants.			
31 Oct. 14	Ingram, A. I. (*act'g. Capt.*) (51, 52)	1 June 16		
1 Dec. 14	Bartlett, C. E. (*act'g. Capt.*)	1 June 16		
1 Dec. 14	Flower, H. G. S. (*act'g. Capt.*)	8 Dec. 16		
1 Dec. 14	Clutterbuck, E. L.	1 July 17	....	With 1-4 Som. L. I.
2 Dec. 14	*Mackie, J. B.*	1 July 17		
2 Dec. 14	*Davis, H.*	1 July 17	....	I. A. (*on probn.*).
14 Dec. 14	Campbell, R.C.	1 July 17		
19 Dec. 14	Lee, A. K.	1 July 17	....	With 1-4 Som. L. I.
21 Dec. 14	*White, R. S.* (95)	1 July 17		
12 Feb. 15	Mann, D. B. U.	1 July 17		
12 Feb. 15	Vallis, R. W. H. (46)	1 July 17		
19 Feb. 15	Watterson, J. W.	1 July 17		
9 Oct. 15	Watterson, J. S. (46)	1 July 17		
10 Oct. 15	Coney, G. B.	1 July 17		
28 Oct. 15	Morgan, F. T. (46)	1 July 17		
28 Nov. 15	Brown, G. B.	1 July 17	....	*Lv., ex I., m. c.*
12 Dec. 15	*Walker, J. C.*	1 July 17	....	*M. G. Corps.*
12 Dec. 15	Speed, B.	1 July 17		
12 Dec. 15	Speed, A. V.	1 July 17		
7 Jan. 16	Vaughan, E. N. L. (57)	7 July 17		
15 Jan. 16	Wynes, F. B. G.	15 July 17	....	M. G. Corps.
6 Mar. 16	Bradshaw, C. H.	6 Sep. 17	....	Attd. 1-4 Som. L. I.
14 Mar. 16	Williams, W. A. (52)	14 Sep. 17		
14 Mar. 16	Leggatt, A.F. (57)	14 Sep. 17		
14 Mar. 16	Fortt, N. H.	14 Sep. 17		
14 Mar. 16	Thomas, G. A.	14 Sep. 17		
14 Mar. 16	Richards, H. L.	14 Sep. 17		
23 Apl. 16	Fortt, W. G.	23 Oct. 17		
20 July 16	*Hook, T. V. B.*	20 Jan. 18		
20 July 16	Date, M.	20 Jan. 18		
	2nd-Lieutenants.			
27 Aug. 14	Miles, Capt. W. H.	25 Oct. 17		Adjt.

Regimental District No. 13.
[No. 8 District.]

Uniform—Scarlet.
Facings—Blue.

Prince Albert's (Somerset Light Infantry).

2-5TH BATTALION.

Arrived in India, 8th January 1915, from England.

"South Africa, 1900-1901."

First Commn.	Names.	Regimental rank.	Army rank.	Remarks.
	Lieutenant-Colonel.			
20 Feb. 89	Paull, J. R., T.D.	1 June 16		Comdg., Dinapore.
	Major.			
19 Nov. 14	Speke, W. H. (*Hon. Capt. in Army, 9 July 01.*) (*Lt.-Col., ret., T. F.*) (*act'g. Lt.-Col., 31 May 18*)	19 Nov. 14		
	Captains.			
1 Apl. 08	Goodland, C. H. (*temp. Maj.*)	7 May 12		D. A. A. G., Burma Dn.
3 Oct. 14	Goode, W. W. (51, 52)	3 Oct. 14		S. S. O., Lucknow.
3 Oct. 14	Jackson, C. W (51)	3 Oct. 14		Attd. A. Rem. Dept.
10 Apl. 09	Harris, F. J.	1 June 16		Attd. 1-4 Som. L. I.
	Lieutenants.			
22 Aug. 14	Mellor, J. S. P. (51)	1 June 16		
29 Sep. 14	Marsh, R. B. (57)	1 June 16		
3 Oct. 14	Blake, N. D. (*act'g. Capt.*) (51, 66*a*)	1 June 16		
3 Oct. 14	*Armstrong, A. H.* (52)	1 June 16		*M. G. Corps.*
3 Oct. 14	*Luscombe, A. M. P.* (51, 52)	1 June 16		*I. A.* (*on probn.*)
3 Oct. 14	Cook, J. (51, 105) (*act'g. Capt.*)	1 June 16		
3 Oct. 14	Spurway, F. E. (46, 51) (*act'g. Capt.*)	1 June 16		
3 Oct. 14	Pulman, T. M., (*act'g Capt., 29 Nov. 18*) (51).	1 June 16		
9 Oct. 14	*Broadmead, H. H.* (*temp. Capt.*) (51).	1 June 16		*Staff Capt., 43rd Bde.*
15 Oct. 14	Brooks-King, M. (57, 59)	1 June 16		Bde. Sig. Instr., Presy. Bde.
20 Oct. 14	*Hobhouse, R. O.*	1 June 16		*R. A. F.*
1 Dec. 14	*Pearce, C. F. B.* (48*a*, 51, 61)	6 Sep. 16		*I. A.* (*on probn.*)
1 Dec. 14	Vallance, G. A. G. (*temp. Capt., 18 Nov. 18*).	1 July 17		Bde. Maj., Sialkot Bde.
2 Dec. 14	Irving, D. F. (46, 51, 52)	1 July 17		
3 Dec. 14	Gawler, A. E. J. (51, 61)	1 July 17		
8 Dec. 14	Steele, J. C. (*temp. Maj., 17 Sep. 18*) (51).	1 July 17		Asst. Comdt., P. of War Camp, [Thayetmyo.
10 Dec. 14	*Price, C. N.* (46, 48*a*, 50, 61)	1 July 17		(*I. A. on probn.*)
11 Dec. 14	Riddell, R. C. H. (48*a*, 51)	1 July 17		
11 Dec. 14	*Bell, J.* (*act'g. Capt., 10 July 18*) (51).	1 July 17		*Adjt.*
11 Dec. 14	Whiting, H. N. (48*a*, 51, 105)	1 July 17		
7 May 15	Alford, B. T. (48*a*, 51)	1 July 17		
22 July 15	Beechey, O. P. (57)	1 July 17		
19 Sep. 15	Virgoe, C. G. B.	1 July 17		
18 Nov. 15	*Hunt, W. E.* (52)	1 July 17		*M. G. Corps.*
15 Jan. 16	Fowler, J. T. F. (61)	16 July 17		
	2nd-Lieutenants.			
19 Dec. 16	Smith, R. N.	19 Dec. 16		
28 Apl. 17	*Martin, G. N.* (51, 205*a*, 206)	28 Apl. 17		*I. A.* (*on probn.*)
28 Apl. 17	Dickinson, G. H. (48*a*, 51)	28 Apl. 17		
28 Apl. 17	Revell, F. (48*a*, 51)	28 Apl. 17		
29 Aug. 17	Sharland, W.	29 Aug. 17		
29 Aug. 17	Macey, C. T. (48*a*, 51, 52)	29 Aug. 17		
26 Sep. 17	Kerridge, T. X.	26 Sep. 17		S. S. O., Dinapore.
26 Sep. 17	Timmins, L. M.	26 Sep. 17		
26 Sep. 17	Vickers, S. G.	26 Sep. 17		
28 Nov. 17	Ives, R. B.	28 Nov. 17		
28 Nov. 17	Cousins, W. B., D.C.M.	28 Nov. 17		
16 Mar. 18	Isaac, A. J. (*temp.*)	16 Mar. 18		
19 Aug. 18	Stewart, D.	19 Aug. 18		
11 Dec. 14	Bell, Lieut. (*act'g. Capt.*) J.	10 July 18		Adjutant.
11 Dec. 14	Hood, Capt. T. H.	11 Dec. 17		Qr.-Mr.

Regimental District No. 13. [No. 8 District.]

Uniform—Scarlet. *Facings*—Blue.

Prince Albert's (Somerset Light Infantry).

1ST GARRISON BATTALION.

Arrived in India, 6th May 1917.

COLONEL—Major-General R. L. Payne, C.B., D.S.O. *ret. pay* . . . 5 Apl. 14.

First Commn.	Names.	Regimental rank.	Army rank.	Remarks.
	Lieutenant Colonel.			
May 81	North, *Hon.* W. F. J. (*temp.*)	10 Jan. 17	25 Dec. 15	
	Major.			
5 May 93	Cooper, J. H., 5 Bn., R. Scots	24 Jan. 17	22 Sep. 14	
......	Waddy, J. M. E. (*Lt.-Col., ret. pay*)	24 Aug. 17	29 Sep. 14	
Mar. 00	Campbell, N. A. H.	16 May 18	Oct. 15	
	Captains.			
Nov. 99	Campbell, A. (*temp.*)	8 Feb. 17	28 Oct. 14	M. A. Dept.
1 Apl. 12	Vincent, J. W. M. (12, 46, 51, 71, 82*a*, 128)	7 Feb. 17	1 Feb. 15	
27 Feb. 15	Blackwell, J. H. (*temp.*) (72*b*)	20 Jan. 17	27 Feb. 15	
3 Dec. 14	Timms, G. R. (*temp.*) (12, 51, 52)	18 Jan. 17	20 Sep. 15	
1 Nov. 99	Massey, J. W. (12, 43, 45*a*, 51, 65, 66, 68, 72*b*)	21 Jan. 17	16 Oct. 15	
5 Aug. 14	Castle, N. C. (*temp.*)	16 Aug. 18	13 Mar. 16	
	Lieutenants.			
6 May 15	Thomson, R. J. (*temp.*)	9 Feb. 17	....	Attd. 1-4 Bn., Som. L. I.
10 June 15	Jenkins, C. M. (*temp.*) (46)	12 Feb. 17		
30 Jan. 16	James, A. V. (46, 52, 54, 128)	12 Feb. 17		
20 Nov. 15	Hettler, E. W. C. (*temp.*)	10 Feb. 18	1 July 17	
30 July 15	Newhouse, A. W. (*temp.*)	25 Apl. 18	1 July 17	
22 Jan. 15	Marsh, G. F., M.C. (*temp.*)	22 Feb. 18	1 July 18	
1 Jan. 16	Austin, A. McN. (*temp.*)	9 Feb. 18	11 July 17	S. S. O., Subathu.
8 Dec. 15	Mc Carthy, A. W. (*temp.*)	2 May 18	8 June 16	Postal Censor, Bombay.
15 Dec. 16	Tribe, C. J. (*temp.*)	15 Jan. 18	15 Jan. 18	R. T.O., Lahore
22 Sep. 14	Chard, W. W. (*temp.*)	15 Oct. 18		
23 Jan. 16	Pyke, D. P. (*temp.*) (46)	15 Oct. 18		
16 Dec. 15	Walker, P. L. C (*temp.*)	8 Aug. 18	1 July 17	Attd. M. G. Centre, Mhow.
	2nd-Lieutenants.			
3 Apl. 15	Helson, G. (*act'g. Lt.*) (*temp.*) (46, 51, 66*a*, 128)	20 Feb. 17	3 Apl. 15	Attd. 1-4 Bn., Som. L. I.
5 May 15	Stout, J. C. (*act'g. Lt.*) (*temp.*) (66*a*, 128)	23 Feb. 17	5 May 15	*Lv., ex I., m.c., 6 mos.*
10 July 15	Carson, C. G. (*act'g. Lt.*) (*temp.*)	29 Jan. 17	10 July 15	
22 Jan. 16	Thomas, T. L. (*act'g. Lt.*) (*temp.*) (46, 172*a*)	21 Feb. 17	22 Jan. 16	
22 Nov. 16	Bridge, W. J. (*act'g. Lt.*) (*temp.*) (127, 129)	12 Feb. 17	22 Nov. 16	
17 June 17	Sarratt, N. D. (*temp.*)	10 Feb. 18	17 June 17	
28 Mar. 17	Joynson, S. D. (*temp.*)	10 Aug. 18	28 Mar. 17	
17 June 17	Good man, G. N. (*temp.*)	11 Feb. 18	17 June 17	
30 Oct. 17	Brown, A. J. (*temp.*)	2 Oct. 18	30 Oct. 17	

17 June 17	De la Haye, 2nd-Lt. (*act'g. Capt.*) P. G. E. G.	6 Aug. 18	....	Adjutant.
23 Jan. 18	Fuller, Lieut. W. (*temp.*)	4 Oct. 18	23 Jan. 18	Quartermaster.

Regimental District No. 15.
[No. 5 District.]

Uniform—Scarlet.
Facings—White.

The East Yorkshire Regiment.

1ST GARRISON BATTALION.

Arrived in India, 14th March 1916.

HONORARY COLONEL—Major-General *Sir* C. Grove, K.C.B., *p.s.c.*, ret. pay . . 26 Sep. 01

First Commn.	Names.	Regimental rank.	Army rank.	Remarks.
	Lieutenant-Colonel.			
8 Feb. 84	Stuart, R. P. Maj. (*temp. Lt.-Col.*) (*ret. pay*) (*Res. of Offr.*) (42,204)	1 Oct. 15		
	Majors.			
18 July 92	Benson, W. J. P. (*Hon. Capt. in Army, 1 Dec. 00*) (*temp.*) (*act'g. Lt.-Col.*)	1 Feb. 16	16 Nov. 14	Comdg., No. 4 Res. Bn., Bangalore.
23 Mar. 85	Capper, G. W. (*temp.*). . . .	1 Feb. 16	12 Jan. 15	2nd-in-Comd.
Jan. 10	Lawrence, G. E. (*4 Bn., E. York. R.*)	30 Aug. 18	7 Oct. 17	
	Captains.			
Nov. 02	Carroll, J. E. M. (*temp.*) . . .	8 Feb. 18	29 Aug. 14	
4 Aug. 14	Evans, E. J. (*temp.*)	28 Sep. 16	28 Nov. 14	*Lv., ex I, m. c., 6 mos., 28 March [18.*
Oct. 98	Stevens, I. M. (*Hon. Lt. in Army*) (*temp.*)	1 Dec. 15	7 Jan. 15	
30 Dec. 14	Bertin, H. (*temp.*)	1 Jan. 16	. . . .	Bde. Maj., Lucknow Bde.
1 June 96	Monckton, C. A. W. (*temp.*) . .	25 Jan. 16	5 Apl. 15	
7 Oct. 14	Cowan, E. S. (*temp.*)	1 Mar. 16		
6 Sep. 14	Burrell, E. A., 4 Bn., Hamps. R. .	18 June 18	1 June 16	
Mar. 10	Lees, A. S., 4 Bn., Cam'n Highrs.	10 June 16	. . . .	Cable Censor, Rangoon.
1 Dec. 14	Scobie, J., 5 Bn., N. Lan. R. . .	. . . .	7 Aug. 17	
30 Sep. 14	Hering, M. O. L., 5 Bn., W. York. R. (51)	. . .	9 Feb. 18	
	Lieutenants.			
5 Nov. 14	Towerton, H. J. (*temp.*) . . .	24 Jan. 16	1 Feb. 15	I. A. R. O.
30 July 15	Nicholson, W. H. (*temp.*) . . .	20 Nov. 15	. . . .	I. A. R. O.
2 Feb. 15	Fisher, J. W., 1 Bn., Mon. R. . .	30 Aug. 18	1 June 16	
7 May 15	Horsfield, G., 7 Bn., W. York. R.	9 Feb. 18	1 June 16	Attd. Chief Engr.'s Office, Murree.
26 Mar. 15	Brunton, F. H. (*temp. Capt., 24 July 18*) (12, 47*a*, 51, 61)	21 Nov. 16	. . . .	Instr., Sch. of Musky., Pachmarhi.
28 May 15	Petrie, C. A. B. (*temp.*) (51) . .	1 July 17	. . . .	*Lv., ex I., m.c., 6 mos., 12 Dec. 18.*
25 Sep. 15	Sheppard, S. V. (*temp.*) . . .	24 Apl. 18	1 July 17	Mily. Accts. Dept., Lahore.
23 Oct. 15	Bircumshaw, F. R. (*temp.*) . .	9 Feb. 18	1 July 17	
	2nd-Lieutenants.			
2 Apl. 15	Whiteman, W. H. (*temp.*) . . .	24 Jan. 16	2 Apl. 15	*Lv., ex. I.*
26 Apl. 15	Powell, T. H. G. (*temp.*) . . .	31 July 17	26 Apl. 15	
4 June 15	Webb, H. G. (*temp.*)	26 Jan. 16	4 June 15	Attd. Army Clothing Dept.
3 Nov. 15	Vagnolini, S. V. (*temp.*), R. W. Surr. R. (46, 51, 60*a*)	. . . . .	3 Nov. 15	
15 Dec. 16	Seed, C. J. (*temp.*)	13 Feb. 18	15 Dec. 16	Attd, Staff, Lucknow Bde.
12 Feb. 17	*Ives, V.G. Lav. N.* (*act'g. Capt., 18 Oct. 18.*) (*temp.*) (51, 52)	1 Aug. 17	12 Feb. 17	*Adjt.*
13 May 17	Gittings, F. S. (*temp.*)	10 Feb. 18	13 May 17	
15 May 17	Alcock, G. H. (*temp.*) (48*a*, 51) .	2 Feb. 18	15 May 17	
17 June 17	Newman, H. J. (*temp.*)	2 Feb. 18	17 June 17	
20 Aug. 17	McCann, J. S. (*temp.*)	9 Feb. 18	20 Aug. 17	Mily. Accts. Dept., Poona.
30 May 17	Biggin, J. V. (*temp.*)	13 Feb. 18	30 May 17	
29 Aug. 17	Barchard, L. B. (*temp.*)	11 May 18	29 Aug. 17	Mily. Accts. Dept., Peshawar.
19 Aug. 18	Mills, J. E., D.C.M. (*temp.*) . .	19 Aug. 18		

12 Feb. 17	Ives, 2nd-Lt. (*act'g. Capt.*) V. G. La V. N.	18 Oct. 18	. . . .	Adjutant.
4 Oct. 15	Dennett, Lt. G. (*temp.*)	4 Oct. 15	. . . .	Qr.-Mr.
	Cotterell, A. F. H. (*temp.*), 2 Bn., R. War. R.	5 Mar. 18	. . . .	Attd. No. 14 Armd. Motor Batty., Lucknow.

Regimental District No. 16.
[No. 9 District.]

Uniform—Scarlet.
Facings—White.

The Bedfordshire Regiment.

1st GARRISON BATTALION.

Arrived in India, 19th March 1916.

COLONEL—Maj.-General T. D. Pilcher, C.B., *p.s.c.*

First Commn.	Names.	Regimental Rank.	Army rank.	Remarks.
	Lieutenant-Colonel.			
10 May 82	Blakeney, Col. W. E. A., I.A. . .	10 May 08	*Col.* 23 Oct. 12	Comdg.
	Major (2nd-in-Comd.).			
13 Oct. 14	Sainsbury, H. (*temp.*)	16 Dec. 15	. . . .	2nd-in-Comd.
	Major.			
	Captains.			
30 Sep. 14	*Rushbrooke, R. B. W.* (3rd Suff. R.)	18 Dec. 15	24 Apl. 15	*Adjt.*
4 Jan. 02	Ellis, J. W. L. (E. Surr. R.) . .	25 May 16	17 Dec. 15	
15 Aug. 14	Curry, F. R. (*temp. Maj.*) . . .	2 Jan. 16	. . . .	D. A. A. & Q. M. G., Bangalore Bde.
19 July 12	Walker, J. D. (*temp.*)	2 Jan. 16		
11 Feb. 15	Thomson, G. C. (*temp.*)	25 Apl. 18	18 May 16	
	Lieutenants.			
20 Nov. 94	Callaway, H. W. (*temp.*) (*temp. Capt.*)	3 Jan. 16	8 Feb. 15	
5 Mar. 15	Bosanquet, H. P. (*temp.*) . . .	3 Jan. 16	5 Mar. 15	
25 Mar. 15	Chowns, W. P. (*temp.*) (95) . .	2 Jan. 16	1 Jan. 16	I. A. R. O.
29 May 15	Tanner, G. (*temp.*) (*temp. Capt.*, 28 *Oct.* 17).	3 Jan. 16	1 Jan. 16	S. S. O., Delhi.
16 Jan. 15	Walker, A C. (*temp.*)	6 Jan. 16	1 Jan. 16	R. A. F.
13 Dec. 14	Tod, A. K. (*temp.*)	1 May 18	11 May 16	
8 Oct. 15	Foster, R. C. (*temp.*)	10 Aug. 18	1 Jan. 17	
6 Dec. 14	Gwyther, E.N. (*temp.*) (*act'g. Capt.*) (51)	1 July 17	. . . .	Adjt., Chakrata Camp.
12 Dec. 14	Murray, K. R. (*temp.*)	7 Sep. 17	1 July 17	A. D. C. to G. O. C. 1st (Peshawar) Dn.
2 June 15	Brodie, K. (*temp.*)	10 Aug. 18	1 July 17	
6 Apl. 16	Mills, L. P. (*Bedf. R.*) (52) (51) .	2 Oct. 17	25 July 17	
5 Aug. 15	Spencer, J. S. (*temp.*)	1 July 17		
31 Aug. 14	Brown, D.A., 5 Bn., E. Surr. R. .	. . . .	1 June 16	M. W. S., Poona.
5 Jan. 15	Swift, G. A. (*temp.*) (57, 207*a*) .	1 July 17	5 Jan. 15	Zhob Militia.
	2nd-Lieutenants.			
12 May 15	Seymour-Higgins, A. C. P. (*temp.*)	13 Dec. 15	12 May 15	I. A. R. O.
13 May 17	Hill, C. I. H. (*temp.*)	30 Nov. 17	13 May 17	Asst. Censor. Telegraph Office, Simla.
30 May 17	Hartland, E. W. (*temp.*)	10 Feb. 18	30 May 17	
17 June 17	Forrest, C. (*temp.*) (52)	5 Feb. 18	17 June 17	
17 June 17	Grant, A. W. H. (*temp.*) . . .	7 Feb. 18	17 June 17	
17 June 17	Hall, F. R. (*temp.*)	7 Feb. 18	17 June 17	
4 Sep. 17	Levey, J. (*temp.*)	11 Feb. 18	4 Sep. 17	
16 Sep. 17	Kirwood, H. W. (*temp.*) . . .	10 Aug. 18	16 Sep. 17	
26 Sep. 7	Lloyd, C. H. (*temp.*)	17 July 18	26 Sep. 17	Attd. S. and T. Corps, Rawalpindi.
26 Sep. 17	Stone, J. E. (*temp.*)	17 July 18	26 Sep. 17	
	Attached.			
30 Dec. 15	Workman, Lt. D. J. G.	1 July 17	. . . .	Rif. Bde.—Ind. Ord. Dept.
31 Oct. 17	Shelford, 2nd-Lt. E. J. H., 4 Bn., R. Berks R.	. . . .	31 Oct. 17	

30 Sep. 14	Rushbrooke, Capt. R. B. W. . .	. . . .	. . . .	Adjt.
18 Dec. 15	Price, Lt. C. F.	18 Dec. 15	. . . .	Qr.-Mr.

Regimental District No. 16.
[No. 9 District.]

Uniform—Scarlet.
Facings—White.

The Bedfordshire Regiment.

2ND GARRISON BATTALION.

Arrived in India, 6th May 1917.

COLONEL.—Maj.-General T. D. Pilcher, C.B., *p.s.c.* . . . . . . 22 April 14.

First Commn.	Names.	Regimental Rank.	Army Rank.	Remarks.
	Lieutenant-Colonel.			
7 Mar. 83	Laidlaw, Lt.-Col. (*temp.*) D., V.D., Glasgow, T. F. A. (*Lt.-Col.*) *& Hon. Col. T. F. Res.*)	26 Dec. 16		
	Majors.			
Feb. 08	Danby, L. J., M.C.	3 Oct. 18	1 June 16	
3 Dec. 87	Ryan, H. W. (*temp.*) (*Maj., T. F. Res.*)	1 Jan. 17	. . . .	Asst. Prov. Marshal, Karachi.
29 Nov. 14	Felce, E., T. D. (*temp.*), *R. H. & R. F. A., T. F.*)	3 Jan. 17		
	Captains.			
Apl. 93	Milbourne, C. K. (*temp.*) . . .	6 Jan. 17	14 Oct. 14	Adjt., Rest Camp, Karachi.
14 Sep. 14	Hughes, S. A. (*temp.*)	23 Oct. 14	23 Oct. 14	
28 Dec. 14	Weigall, G. J. V. (*temp.*) . . .	27 Dec. 16	28 Dec. 14	
Jan. 15	Ludlow, L. (*temp.*)	17 Jan. 17	11 Oct. 15	
2 Dec. 14	Alford, E. N. (*temp.*) (*1 Bn., Essex R.*) (51)	2 Jan. 17	. . . .	Intell. Offr., Karachi Bde.
	Lieutenants.			
23 Dec. 14	Chesney, G. L. (*3 Bn., R. Lanc. R.*)	3 Oct. 18	5 May 16	
9 Feb. 15	Bray, S. A. (*7 Bn., Essex R.*) . .	27 Dec. 16	1 June 16	
18 Dec. 14	Symonds, E. T. (*temp.*) (57) . .	11 Jan. 17		
26 Mar. 16	Richards, W. V., M.C. (*temp.*) . .	13 Aug. 18	26 Mar. 17	
28 Dec. 14	Mills, R. G. (*3rd Bn., Hamps. R.*) (*temp.*)	12 Jan. 17	1 July 17	
20 Nov. 14	Micholls, R. M. B. (*temp.*) . .	31 Jan. 17		
29 Oct. 15	Liswell, J. A. (*temp*).	9 May 17	. . .	I. A. (*On probn.*)
20 Jan. 16	McAvoy, W. (*temp.*)	3 Oct. 18	20 July 17	
22 Nov. 16	*Hurn, A. S.* (*temp.*) (*Act'g. Capt.*)	22 May 18	. . . .	*Adjt.*
	2nd-Lieutenants.			
2 Sep. 15	Smith, H. M. (51, 95)	27 Dec. 16	2 Sep. 15	I. A. (*On probn.*)
15 Dec. 16	Spicer, N, (*temp.*)	25 Jan. 17	15 Dec. 16	Supvr., Phys. Trg., Poona Divn.
11 Mar. 17	Thompson, R. M. (*temp.*)	3 Feb. 18	11 Mar. 17	
28 Mar. 17	Willis, W. E. (*temp.*) . . .	3 Feb. 18	28 Mar. 17	
15 Apl. 17	Horne, A. E. M. (*temp.*) . . .	3 Feb. 18	15 Apl. 17	
26 Apl. 17	Butt, R. N. (*temp.*)	13 Aug. 18	26 Apl. 17	
13 May 17	Bevan, A. E. (*temp.*).	14 Feb. 18	13 May 17	
22 May 17	Frith, H. E. (*temp.*)	14 Feb. 18	22 May 17	
19 Aug. 17	Barker, K. C. (*temp.*)	14 Feb. 18	19 Aug. 17	
28 Mar. 17	Ginger, A. H. (*temp.*)	30 Mar. 18	28 Mar. 17	
15 Apl. 17	James, D (*temp.*)	30 Mar. 18	15 Apl. 17	
31 Oct. 17	Foote, B. A. M. (*temp.*	30 Mar. 18	31 Oct. 17	
18 Nov. 17	Garland, E. A. (*temp.*)	30 Mar. 18	18 Nov. 17	

22 Nov. 16	Hurn, Lt. (*Act'g. Capt.*) A. S. .	1 Aug. 18	. . . .	Adjt.
27 Nov. 16	Williams, Lt. C. G.	27 Dec. 16	27 Nov. 16	Qr.-Mr.

Regimental District No. 16.
[No. 9 District.]

Uniform—Scarlet.
Facings—White.

The Bedfordshire Regiment.

3RD GARRISON BATTALION.

Arrived in India, 12th June 1917.

HONORARY COLONEL—

First Commn.	Names.	Regimental rank.	Army rank.	Remarks.
	Lieutenant-Colonel.			
Sep. 87	Armstrong, Lt.-Col. W. Mc.G. (*temp.*)	6 Feb. 17		
	Major.			
19 Feb. 01	Crofton, G. H. S. (*temp.*) (*Capt. E. Surr. R.*)	6 Mar. 17	8 Dec. 14	*Lv., ex I., m.c.*
5 Nov. 14	Mumford, E. M. (*temp.*)	. . . .	1 Nov. 15	
	Captains.			
	Yorston, C. (*temp.*)	6 Mar. 17	27 Oct. 14	
29 Sep. 14	Terry, A. D. R. (*temp.*) (67) . .	6 Mar. 17	14 Oct. 15	
15 Dec. 14	Saunders, C. (*temp.*) (*Rif. Bde.*) (51, 52).	6 Mar. 17	7 Jan. 16	
28 Nov. 14	Anderson, D. C., M.C. (*temp.*) (52)	. . . .	28 Jan. 16	Asst Comdt., P. of W. Camp.
17 Oct. 14	Small, E., *Bedf. R.*	3 Feb. 17	. . . .	Asst. Comdt., P. of W. Camp.
	Lieutenants.			
10 Oct. 14	*Williams, S.* (*act'g. Capt., 8 Sep. 17*) (*R. W. Fus.*) (51, 52)	6 Mar. 17	31 Jan. 15	*Adjt.*
15 Aug. 14	Day, E. A. P. (*Lt., Midd'x. R., S. R.*) (52, 57, 128)	. . . .	1 Apl. 15	*Lv., ex I., m.c., 6 mos.*
7 Apl. 15	Barrett, R. O. (*temp.*)	. . . .	3 Mar. 16	
27 Feb. 15	Butler, G. W. (*temp.*) . . .	. . . .	21 Apl. 16	
15 July 15	Martin, R. Mc G. (*temp.*) (51) .	. . . .	5 Apl. 16	S. S. O. & Cant. Magte., Meiktila.
9 Sep. 15	Cox, R. A. (*temp.*) (67)	6 Mar. 17	20 June 16	
3 Apl. 15	Hawthorne, C. F. D. (*temp.*) (57).	1 July 17		
9 Feb. 15	Pennack, C. (*Linc. R.*)	. . . .	4 Apl. 18	
	2nd-Lieutenants.			
8 Apl. 15	Hayes, B. H. (*R. W. Surr. R.*) (128)	16 Mar. 17	8 Apl. 15	
1 Aug. 16	March, G. J. (*temp.*) (60*a*) . . .	8 Mar. 17	1 Aug. 16	
19 Dec. 16	Tyler, N. E. (*temp.*)	. . . .	19 Dec. 16	
11 Feb. 17	Bramwell, F. H. (*temp.*) . . .	11 Feb. 17	. . . .	No. 3 Res. Bn. (Ind.)
11 Feb. 17	deGray, R. (*temp.*)	11 Feb. 17		
11 Feb. 17	Gallichan, H. N. (*temp.*) (128) .	11 Feb. 17		
11 Feb. 17	George, A. C. (*temp.*) (51, 128) .	11 Feb. 17		
11 Feb. 17	Gould, W. G. (*temp.*)	11 Feb. 17	. . . .	No. 7 Res. Bn. (Ind.)
15 Apl. 17	Lisney, J.	. . . .	15 Apl. 17	
30 Aug. 17	Forge, F. W. (128)	. . . .	30 Aug. 17	
30 Aug. 17	Papps, F. J. (51)	. . . .	30 Aug. 17	
19 Aug. 18	Bradshaw, R. J. (*temp.*)	. . . .	19 Aug. 18	

10 Oct. 14	Williams, Lt. (*act'g. Capt.*) S. .	6 Mar. 17	. . . .	Adjt.
2 Jan. 17	Fryer, Lieut. F. W.	6 Feb. 17	2 Jan. 17	Qr.-Mr.

Regimental District No. 17. [No. 6 District.]

Uniform—Scarlet. *Facings*—White.

The Leicestershire Regiment.

2ND BATTALION (*late 17th Foot*).

Arrived in India, 11th October 1906, from England.

THE ROYAL TIGER SUPERSCRIBED "HINDOOSTAN."

"Namur, 1695" "Louisburg" "Martinique, 1762" "Havannah" "Ghuznee, 1839" "Khelat "Affghanistan, 1839" "Sevastopol" "Ali Masjid" "Afghanistan, 1878-79" "Defence of Ladysmith" "South Africa, 1899-1902."

COLONEL—Major-General *Sir* E. M. Woodward, K.C.M.G., C.B., *p.s.c.* 7 Dec. 16.

First Commn.	Names.	Regimental rank.	Army rank.	Remarks.
	Lieutenant-Colonel.			
	Majors.			
19 Feb. 96	Lewis, F., D.S.O. (*temp. Lt.-Col.*)	1 Sep. 15	*l. c.* 3 June 16	With 7 Bn., March. R.
21 Apl. 00	Brock, A. W. S., D.S.O. (*Act'g. Lt.-Col., 12 Aug. 17*).	1 Sep. 15	*l. c.* 3 June 17	
15 May 00	Colquhoun, A. F. R.	1 Sep. 15		
22 Oct. 02	Creagh, P. H., D.S.O.	22 Oct. 17		
	Captains.			
27 Jan. 04	Latham, F., D.S.O. (*Act'g.-Lt.-Col., 19 Sep. 16*)	8 Mar. 13	*m.* 1 Jan. 18	With Staff. R.
13 Oct. 09	Wardle, M. K., D.S.O., M. C.	13 Feb. 15		
14 Oct. 10	Chudleigh, C. A. E.	16 Aug. 15		
9 Dec. 11	Tunks, G. P. D'A. G.	9 Feb. 17		
4 Sep. 12	Lowther, R. A. N., M.C. (57)	1 Apl. 17		
	Lieutenants.			
4 Dec. 12	Grimble, H. N. H., M.C. (*temp. Capt.*).	22 Oct. 14		
22 Jan. 13	Grylls, T. R., M.C. (*temp. Capt.*)	22 Oct. 14	. . .	With E. York. R.
	Dakin, A. E. (*temp.*)	20 Feb. 15		
8 Aug. 14	Buxton, V. (*temp. Capt.*)	21 Feb. 15	. . . .	R. A. F.
16 Sep. 14	Sturdy, A. E.	16 Mar. 15		
30 Nov. 14	Redwood, J.	25 June 16		
30 Nov. 14	*Harris, J. E.*, (*Act'g. Capt.*)	18 Sep. 16	. . . .	Adjt.
13 Jan. 15	Copeman, C. N. C.	1 Apl. 17		
28 Mar. 15	Sutherland, D. W. (*Act'g. Capt.*)	16 May 17		
16 June 15	Wrixon, P. A. B.	1 July 17	. . . .	*Lv., ex I., m. c.*
26 June 15	Wilson, W.	1 July 17		
14 July 15	Hassall, T. C. D.	1 July 17		
1 Oct. 15	Belle, A.	1 July 17		
16 Jan. 16	Giband, A. B. (*temp.*)	16 July 17		
23 Jan. 16	Chudleigh, A. C. L.	23 July 17		
	2nd-Lieutenants.			
9 Feb. 17	McInerney, F.E.	9 Feb. 17		
9 Feb. 17	Matthews, C.	9 Feb. 17		
21 June 17	Bale, W., M.C.	21 June 17		
14 Sep. 17	Lyons, J.	14 Sep. 17		

30 Nov. 14	Harris, Lt. (*Act'g. Capt.*) J. E.	18 Sep. 16	. . . .	Adjutant.
4 Oct. 14	Brodie, Capt. H. C.	4 Oct. 17	. . . .	Qr.-Mr. *Lv., ex I., m. c.*

Agents:—MESSRS. COX & Co.

Regimental District No. 19.
[No. 5 District.]

Uniform—Scarlet.
Facings—Grass Green.

Alexandra, Princess of Wales's Own (Yorkshire Regiment).

1ST BATTALION (*late 19th Foot*).

Arrived in India, 6th February 1912, from Egypt.

THE CYPHER OF H. R. H. ALEXANDRA, PRINCESS OF WALES, IN GOLD (THEREON "ALEXANDRA") INTERLACED WITH THE DANNEBROG ENSCRIBED WITH THE DATE 1875, AND THE WHOLE SURMOUNTED BY THE CORONET OF THE PRINCESS.
"Malplaquet" "Alma" "Inkerman" "Sevastopol" "Tirah" "Relief of Kimberley" "Paardeberg" "South Africa, 1899-1902."

COLONEL-IN-CHIEF—QUEEN ALEXANDRA.

COLONEL—Major-General (*temp. Lieut.-Genl.*) *Sir* E. S. Bulfin, K.C.B., C.V.O., *q. s.* . . 28 Oct. 14.

First Commn.	Names.	Regimental rank.	Army rank.	Remarks.
	Lieutenant-Colonel.			
	Majors.			
20 Feb. 97	Carey, M. D. (*Act'g. Lt.-Col., 30 Aug. 16*) (50, 53, 59, 72*a*)	12 Feb. 14		
	Captains.			
16 Dec. 03	Ledgard, R. S. (51, 72, 207) . . .	28 Mar. 11		
23 Aug. 06	Hooton, A. C. (51, 65*a*, 72) . . .	31 Oct. 14		
4 Mar. 11	*Bagnall, H. S.* (51, 52, 59, 61, 207) (*Act'g. Maj.*)	27 Sep. 15	. . . .	*Adjt.*
4 Mar. 11	Bennedik, S. R. W. (50, 52, 66*a*) .	7 Oct. 15	. . . .	*Lv. ex I., m. c., 6 mos., 13 Dec. 18.*
17 May 11	Maude, G. W. E. (51, 52, 66*a*, 172*a*, 207)	2 Aug. 16		
16 Sep. 13	Errington, J. R. F. (52) . . .	17 June 17	. . . .	M. G. Corps.
	Lieutenants.			
24 Jan. 14	Beatson-Bell, A. (51, 59, 71, 207)	25 Jan. 15	. . .	With Sig. Serv.
7 Nov. 14	Barber, V. J. (*Act'g. Capt., 3 Aug. 17*) (52, 71)	15 Mar. 15	. . . .	Adjt., Br. Details Depôt, Deolali.
7 Nov. 14	Derham, W. (71)	18 May 15	. . . .	*Lv. ex. I., m. c., 6 mos., 13 Dec. 18.*
7 Nov. 14	McGovern, F. (57, 71) . . .	10 June 15	. . . .	With Sig. Serv.
22 Dec. 14	Hollom, G. H. S. (*temp.*) (51, 71)	10 June 15	. . . .	With 2-6 Bn., Devon. R.
9 Jan. 15	Parrott, A. (50, 52, 57, 71) (*Act'g. Capt., 16 Sep. 17*)	10 June 15	. . . .	With 2-6 Bn., Devon. R.
3 May 15	Banner, J. W. (*temp.*) (51) . . .	30 June 15	. . . .	With 1-6 Bn., Devon. R.
16 June 15	Clayton, W. D. (*Act'g. Capt., 31 Oct. 18.*) (51)	27 Sep. 15		
17 June 15	Hayes, T.	7 Oct. 15	. . . .	S. & T. Corps.
17 June 15	Clow, P. (71, 172*a*)	29 Mar. 16		
17 June 15	Crowsley, F. (46, 50, 61) (*Act'g. Capt., 25 Mar. 18*)	15 Apl. 16		Actg. Adjt.
18 Sep. 15	Lloyd, H. C. (51, 171)	1 July 17	. . . .	With 1-6 Bn., Devon. R.
21 Dec. 16	Hartley, F. (46, 59)	21 June 18	. . . .	
21 Dec. 16	Ball, G. (12, 48*a*, 51, 172*a*) . . .	21 June 18	. . . .	
	2nd-Lieutenants.			
13 May 15	Chouler, R. (*temp.*)	13 May 15	. . . .	I. A. R. O.
22 Nov. 16	Mitchell, W. H. (*temp.*)	22 Nov. 16		
29 Aug. 17	Turpin, C. F. (*temp.*) (61) . . .	29 Aug. 17		
29 Aug. 17	Paine, R. L. (*temp.*)	29 Aug. 17	. . . .	With R. A. F., Egypt.
29 Aug. 17	Lynn, T. P. (*temp.*) (57) . . .	29 Aug. 17		
29 Aug. 17	Heckle, E. W. (*temp.*) (46) . . .	29 Aug. 17		
27 Dec. 17	Mason, A.	27 Dec. 17	. . . .	Attd. 8 Bn., R. W. Fus.
16 Mar. 18	Crooks, B. D. (*temp.*) (12) . . .	16 Mar. 18	. . . .	Attd. M. W. S.
16 Mar. 18	Ure, H. O. (*temp.*)	16 Mar. 18	. . . .	Attd. Sig. Service.
16 Mar. 18	Hoare, L. J. (*temp.*)	16 Mar. 18	. . . .	With R. A. F., Lahore.
23 July 18	O'Callaghan, L. M.	23 July 18		
19 Aug. 18	Benson, F. W. (*temp.*)	19 Aug. 18		
19 Aug. 18	Lyle, A. (*temp.*)	19 Aug. 18		

AGENTS,—MESSRS. HOLT & CO.

Regimental District No. 19.
[No. 5 District.]

Uniform—Scarlet.
Facings—Grass Green

Alexandra, Princess of Wales's Own (Yorkshire Regiment)—*contd.*

First Commn.	Names.	Regimental rank.	Army rank.	Remarks.
	Attached.			
22 Jan. 13	Chandos-Pole, Lt. G. R., (*act'g Capt., 11 Aug. 18*) 7 Huss. (51, 52, 60a).	25 May 14		
14 Aug. 15	Kerr, Lt. W. B. (*temp.*) Gordon Highrs.	1 July 17		
7 July 16	Reed, Lt. E. W. (*temp.*) Linc. R. (12, 61).	7 Jan. 18		
26 Sep. 17	Draycott, 2nd-Lt. C. D., York. & Lanc. R.	26 Sep. 17		
26 Sep. 17	Watson, 2nd-Lt. A., Notts. & Derby. R.	26 Sep. 17		
26 Sep. 17	Newman, 2nd-Lt. E. B., York. & Lanc. R.	26 Sep. 17		
26 Sep. 17	Holland, 2nd-Lt. J. A., Midd'x. R.	26 Sep. 17		
31 Oct. 17	Barnett, 2nd-Lt. W., Notts. & Derby. R.	31 Oct. 17		
4 Mar. 11	Bagnall, Capt. H. S. (*act'g. Maj.) 26 Mar. 18*)	30 Aug. 16	. . . .	Adjt.
19 Mar. 13	Howes, Capt. W. T.	1 July 17	. . . .	Qr.-Mr.

Agents:—Messrs. Holt & Co.

Regimental District No. 19.
[No. 5 District.]

Uniform—Scarlet.
Facings—Grass Green.

Alexandra, Princess of Wales's Own (Yorkshire Regiment).

1st GARRISON BATTALION.

Arrived in India, 19th January 1916.

HONORARY COLONEL—

First Commn.	Names.	Regimental Rank.	Army rank.	Remarks.
	Lieutenant-Colonel.			
29 Nov. 90	Tucker, A. C., (*temp.*) (*Capt., Res. of Offr.*) (51, 67, 71, 72*a*).	2 Oct. 15		
	Major.			
8 Mar. 96	Chadwick, P. T.	6 Sep. 15	4 Oct. 14	
	Captains.			
1 June 01	MacLeod, R. C., T.D. (*6 Bn., R. Scots.*) (51, 53, 57, 65*a*, 66, 71)	28 Sep. 16	23 Oct. 09	*Lv., ex I., m. c., 6 mos.*
15 Sep. 00	Turner, A. P. (*Res. of Offr.*) (51, 72*b*)	5 Dec. 14	11 Mar. 02	
05	Cherry, J. W. (*5 Bn., Manch. R.*) .	. . . .	11 Sep. 14	
10 Oct. 00	Morier, W. E.	12 Oct. 15	5 Oct. 14	
22 Dec. 14	Whitehead, G. (*temp.*) (51, 60*a*, 170)	29 Oct 15		
23 Dec. 14	Mackay, J. M. (*temp.*) (51) . . .	7 Jan. 16		
28 Feb. 14	Wheatley, T. P. (*3 Bn., W. Rid. R.*) (51).	28 Sep. 16	9 June 16	
	Lieutenants.			
25 Jan. 15	Marshall, J. T. (*temp.*)(51, 60*a*, 170)	3 Nov. 15	. . . .	R. A. F.
6 May 15	*Roskin, H. H.* (*act'g Capt., 1 Nov. 18*) (*temp.*) (*3rd Bn.*) (51) . . .	*3 Nov. 15*	. . .	*Adjt.*
2 Feb. 15	Dutchman, R. H. (*temp.*)	23 Nov. 15	. . . .	*Lv., ex I., m. c., 6 mos.*
7 June 15	Grace, O. G.	13 Jan. 16	1 June 16	
28 Dec. 14	Noble, H. E.	15 May 16	21 July 16	
1 Jan. 16	Smith, J. L. L. (*temp.*) (*Manch. R.*)	22 May 16	24 Apl. 17	
14 Aug. 15	Illingworth, A. E. (*temp.*) (57) . .	1 July 17		
5 Nov. 15	Lee, H. (*temp.*) (*York. & Lanc. R.*)	. . . .	1 July 17	
28 Dec. 15	Annett, M. W. (*temp.*) (51) . .	1 July 17	28 Dec. 15	
	2nd-Lieutenants.			
16 Nov. 15	Kerslake, S. S.	3 Aug. 17	16 Nov. 15	
23 Jan. 16	Brown, A. E. (*temp.*) (51) . . .	23 Jan. 16		
6 July 16	McCrossan, A. D.	6 July 16	. . . .	S. S. O., Sialkot.
26 Sep. 16	Hancock, J.	. . . .	26 Sept. 16	
11 Feb. 17	Rayment, W. M.	11 Feb. 17	. . . .	Attd. M. W. S., Bombay.
11 Mar. 17	Warre, F. R.	11 Mar. 17	. . . .	Postal Censor, Bombay.
30 May. 17	Bryant, S. C. (57)	. . . .	30 May 17	
26 June 17	Rudd, C.	. . . .	26 June 17	
15 July 17	Noel, W. H. M.	. . . .	15 July 17	
29 Aug. 17	Love, S. H. (*temp.*) (*8 Bn., York. & Lanc. R.*)	. . . .	29 Aug. 17	
13 May 17	Leacock, E. E.	13 May 18	13 May 17	
1 May 18	Dowdeswell, G. A.	1 May 18		

6 May 15	Roskin, Lieut. (*act'g Capt.*) H. H.	1 Nov. 18	. . . .	Adjutant.
10 June 18	Leigh, Lt. J.	10 June 18	. . . .	Quarter-Master.

Regimental District No. 21. [No. 2 District.]

Uniform—Scarlet. *Facings*—Blue.

The Royal Scots Fusiliers.

1ST GARRISON BATTALION.

Arrived in India, 4th March 1916.

HONORARY COLONEL—Hon. Lieut.-General J. T. Dalyell, *ret. pay.*

First Commn.	Names.	Regimental rank.	Army rank.	Remarks,
	Lieutenant-Colonel.			
Dec. 99	Baird, R. E. W. (*temp.*) (51) (*3rd High. L. I.*)	2 May 18	29 Mar. 18	Comdg.
	Majors.			
May 00	Rankin, W. B. (*4th High. L. I.*)	3 Dec. 15	13 Mar. 12	2nd-in-Comd. *Lv., ex I., m. c., 6 mos., 24 Apl. 18.*
10 Oct. 14	Mitchell, T. G. (*temp.*)	2 May 18	. . . .	Act'g. 2nd-in-Comd.
	Captains.			
1 Nov. 83	Macaulay, J. (*temp.*) (*Hon. Maj. ret. Vols.*)	27 Nov. 15	1 Sep. 14	Comdt., Mt. Abu Sanitarium.
Aug. 11	Maclean, A. D. D. (51) (*3rd Cam'n Highrs.*)	18 Oct. 15	3 Feb. 15	*Lv., ex I., m. c., 6mos., 22 Nov. 17.*
Aug. 11	Colquhoun, A. R. (*3rd Arg. & Suth'd Highrs.*)	1 Jan. 16	14 Oct. 15	*Lv., ex I., m.c., 6 mos., 5 Sep. '8.*
2 Oct. 14	Thomson, J. (*3rd Gord. Highrs.*)	11 Mar. 16		
24 Nov. 14	Gray, O. E. (*3rd Gord. Highrs.*) (57)	4 May 17		
	Lieutenants.			
29 Oct. 15	Bethell, C. A. (*R. Sc. Fus.*) (46)	29 Oct. 15	27 Oct. 15	
15 Aug. 14	Munro, G. C. G. (*act'g. Capt., 7 July 18*) (*3rd Sea. Highrs.*) (57)	7 Mar. 16		
29 Mar. 15	Garson, I. G. M. (51, 52) (*temp.*)	19 Jan. 17	. . . .	Attd. Army Clothing Dept.
22 Apl. 15	Anderson, J. W. (*act'g. Capt.*) (*temp.*)	19 Jan. 17		
3 Oct. 15	Broad, G. C. W. (*temp.*) (*act'g. Capt., 23 Sep. 18*) (52)	15 Mar. 17		
23 Apl. 15	Hay, D. J. (51, 60a) (*3rd Sea. Highrs.*)	4 May 17	11 Jan. 16	
24 Mar. 15	Cleland, W. S. C. (*3rd High. L. I.*) (51)	1 July 17		
10 Apl. 15	Outerson, W. (*temp.*)	1 July 17	. . . .	Ind. Ord. Dept.
5 June 15	Bushell, T. O. (*4 Bn., Lan. Fus.*)	1 July 17		
8 Dec. 15	Todd, A. C. (*temp.*) (52, 60a)	1 July 17		
4 Jan. 16	Slack, H. B. (*4 Bn., Sea. Highrs.*)	4 July 17	. . . .	*Lv., ex I., m. c., 6 mos., 5 Sep. 18.*
28 Aug. 16	Ketley, E. W. (*temp.*)	5 Mar. 18		
	2nd-Lieutenants.			
30 Nov. 15	Grant, R. (*temp.*)	14 June 17	10 Nov. 15	I. A. R. O.
Sep. 16	Leighton, A. V. (51)	3 June 17	5 Sep. 16	Embn. Staff, Bombay.
5 Jan. 17	Allan, A. (*temp.*)	3 Nov. 17	25 Jan. 17	Attd. G. S. Bch., Army Hdqrs.
7 Feb. 17	Macleman, W. D. (*Sea. Highrs.*) (60a)	28 Jan. 18	7 Feb. 17	*Act'g.* Qr.-Mr.
11 Mar. 17	Askew, T. E. (*temp.*)	4 Aug. 17	11 Mar. 17	I. A. R. O.
23 Mar. 17	Allen, E. R. L. (*temp.*)	7 Feb. 18	23 Mar. 17	Bde. Sig. Offr., Jhansi Bde.
13 May 17	Wright, M. G. (*temp.*)	7 Feb. 18	13 May 17	Ind. Ord. Dept.
18 May 17	Kissock, T. (*temp.*)	7 Feb. 18	18 May 17	
16 June 17	Lochrie, C. (*temp.*)	7 Feb. 18	16 June 17	
15 July 17	Bell, T. W.	11 Feb. 18	15 July 17	Army Clothing Depôt, Jullundar.
5 Aug. 16	Wilson, A. T.	11 Feb. 18	5 Aug. 16	Army Clothing Depôt, Cawnpore.
28 Mar. 17	Grime, G. M. (*temp.*) (*4 Bn. Lan. Fus.*)	. . . .	28 Mar. 17	
30 Oct. 17	Christie, J. (*temp.*) (*R. Scots.*)	. . . .	30 Oct. 17	
18 Nov. 17	Reid, A. G. (*temp.*) (*6 Bn. High. L. I.*)	. . . .	18 Nov. 17	
26 Mar. 18	Bissett, P.	22 Nov. 18	26 Mar. 18	

First Commn.	Names.	Regimental rank.	Army rank.	Remarks,
15 Dec. 14	Murray, Capt. J., Cam'n. Highrs (60a)	1 Nov. 15	. . . .	Adjt.
16 Apl. 18	Hudson, Lt. J.	17 Oct. 18	16 Apl. 18	Quartermaster.

Regimental District No. 24.
[No. 4 District.]

Uniform—Scarlet.
Facings - Grass Green.

The South Wales Borderers.

1ST BRECKNOCKSHIRE BATTALION.

Arrived in India, 3rd December 1914, from England.

ON A MOUNT THE RED DRAGON PASSANT.
"South Africa. 1900-01."

HONORARY COLONEL—J. H. R. *Lord* Glanusk, C.B., D.S.O. . . . 1 June 07.

First Commn.	Names.	Regimental rank.	Army rank.	Remarks.
	Lieutenant-Colonel.			
85	Thomas, D. W. E., V.D. (51) .	15 Oct. 16	. . . . .	Comdt., Keti Camp.
	Major.			
	Careless, A. L. (51)	19 Aug. 14	. . . .	Comdt., Range Hill Camp.
11 June 00	Woodliffe, H. D.	1 June 16	. . . .	S. S. O. & Comdg., Dalhousie.
	Captains.			
2 May 08	Cockcroft, E. F. (*act'g. Maj., 17 Sep. 18*) (51)	19 Aug. 14		
20 Mar. 08	*Evans, H. W.*	2 Sep. 14	. . . . .	*Treasure Chest Offr., Karachi.*
5 June 08	Prichard, W. J.	1 June 16		
7 July 08	Thomas, G. D. E.	1 June 16		
23 Jan. 09	Woosnam, R. W. (51)	1 June 16	. . . .	With 4 Bn., S. Wales Bord.
23 Mar. 09	*Glynne-Jones, R. R.*	1 June 16	. . . .	*Staff Capt.*
26 Nov. 13	Butcher, A. C. S. (57)	1 June 16	. . . .	With Sig. Ser.
	Lieutenants.			
12 Nov. 13	*Adams, R. P.*	30 Mar. 15	. . . .	A. G.'s Office, Records Branch, Bombay.
14 Aug. 14	*Anson, H. A.*	1 June 16	. . . .	*R. A. F.*
26 Oct. 14	Hughes, H D. (*actg. Capt.*) (51) .	1 June 16		
26 Oct. 14	Williams, J. (*temp. Capt., 16 Mar. 18*).	1 June 16	. . . .	S. S. O.
29 Oct. 14	Williams, D. J. (*actg. Capt.*) (46) .	1 June 16		
9 June 15	Picton, A. G. C.	1 July 17	. . . .	4th Bn., S. Wales Bord.
16 June 15	Jones, T. J. (51)	1 July 17	. . . .	4th Bn., S. Wales Bord.
14 July 15	Williams, F. C., (46)	1 July 17	. . . .	Supvg. Offr. of Phys. & Bayonet Trg., 9th (Sec.'bad) Dn.
1 Sep. 15	Griffith, H. G. B. (*actg. Capt.*) (50)	1 July 17		
7 Oct. 15	Nicholls, C. W. (57)	1 July 17	. . . .	With Sig. Serv.
27 Nov. 15	Talbot, W. L.	1 July 17		
30 Dec. 15	Southron, A. H. (*temp. Capt.*) . .	1 July 17	. . .	Staff Capt., 5th (Mhow) Dn.
	2nd-Lieutenants			
28 Apl. 17	*Colsey, J. W. T.*	28 Apl. 17	. . . .	*I. A. (on probn.)*
28 Apl. 17	Graham, T. R. (61)	28 Apl. 17		
28 Apl. 17	Nicholson, A. T.	28 Apl. 17		
29 Aug. 17	Brown, F. L.	29 Aug. 17		
29 Aug. 17	Holmes, S. J. (46)	29 Aug. 17		
29 Aug. 17	Grisdale, W.	29 Aug. 17		
16 Mar. 18	Wilson, F. (*temp.*) . . .	16 Mar. 18	. . . .	Attd. M. G. Corps.
	Attached.			
30 Apl. 02	Wigley, Major A. B. (*act'g. Lt.-Col., 22 Sep. 18*) (51, 52, 66a, 72).	18 Jan. 17	. . . .	Comdg.
2 Dec. 14	French, Lt. J. L., 5 Bn., Essex R..	1 June 16		
29 Apl. 15	Edwards, Lieut. J. B., 3 Bn., Mon. R. (46)	1 June 16		
15 July 15	Edwards, Lieut. F. A., 3 Bn., Mon. R.	1 July 17		
30 May 17	Scourfield, 2nd-Lt. W. H., 4 Bn., Welsh R.	30 May 17		
30 May 17	Dodd, 2nd-Lt. W., 4 Bn., Welsh. R.	30 May 17		
30 May 17	Jones, 2nd-Lt. W. T. G., 5 Bn., Welsh. R.	30 May 17	. . . .	Attd. M. G. Corps.
29 Aug. 17	Jones, 2nd-Lt. H. G., 3 Bn., S. Wales. Bord.	29 Aug. 17		
29 Aug. 17	James, 2nd-Lt. J. A., S. Wales Bord	29 Aug. 17	. . . .	Attd. M. G. Corps.
26 Sep. 17	Lorange, 2nd-Lt. C. A., Welsh. R. .	26 Sep. 17		
26 Sep. 17	Davies, 2nd-Lt. S. H., Welsh. R. .	26 Sep. 17		
26 Sep. 17	Wootton, 2nd-Lt. H. E., Welsh. R.	26 Sep. 17		
26 Sep. 17	Walker, 2nd-Lt. P. J., 5 Bn., Notts. & Derby. R.	26 Sep. 17		
26 Sep. 17	Vinton, 2nd-Lt. J., 4 Bn., Welsh. R.	26 Sep. 17		
26 Sep. 17	Feldon, 2nd-Lt. L. T., 6 Bn., Manch. R.	26 Sep. 17		
26 Sep. 17	Henderson, 2nd-Lt. J., Rif. Brig. .	26 Sep. 17		
31 Oct. 17	Newman, 2nd-Lt. H., 3 Bn., Hamps. R.	26 Sep. 17		
1 Sep. 15	Griffith, Lt. (*actg. Capt.*) H. G. B.	12 Aug. 18	. . .	Adjt.
30 Sep. 14	Elvidge, Capt. R. W.	30 Sep. 14	. . . .	Qr.-Mr.
11 Nov. 14	Townley, Capt. J. S., M.C., R.A.M.C. (T.F.)	1 Apl. 15	. . . .	Medl. Offr. Comdg. Stn. Hosp., Mt. Abu.

Regimental District No. 25.
[No. 2 District.]

Uniform—Scarlet.
Facings—Blue.

The King's Own Scottish Borderers.

1st BATTALION (*late 25th Foot*).

Arrived in India, 13th February 1911, from Egypt.

THE CASTLE OF EDINBURGH, WITH THE MOTTO "*Nisi Dominus frustra.*" IN THE FIRST AND FOURTH CORNERS THE ROYAL CREST, WITH THE MOTTO "*In Veritate Religionis confido.*" IN THE SECOND AND THIRD CORNERS THE WHITE HORSE, WITH "*Nec aspera terrent.*"

THE SPHINX, SUPERSCRIBED "EGYPT."

"Namur, 1695" "Minden" "Egmont-op-zee" "Martinique, 1809" "Afghanistan, 1878-80" "Chitral" "Tirah" "Paardeberg" "South Africa, 1900-02".

COLONEL—Lt.-General *Sir* C. L. Woollcombe, K.C.B. . . . . . 31 Aug. 10.

First Commn.	Names.	Regimental rank.	Army rank.	Remarks.
	Lieutenant-Colonel.			
	Majors.			
9 Sep. 93	Welch, A. J. (Q. 51, 57, 71) . .	3 Dec. 10	*l. c.* 1 Jan. 16	
19 May 99	Whigham, R. D., D.S.O.	1 Sep. 15		
	Captains.			
28 Nov. 06	Ogilvy, G. M. H. (50, 52) . .	27 Nov. 14		
22 Feb. 08	Murray, C. A. G. O., D.S.O., (*act'g. Lt.-Col., 31 Aug. 17.*) 51, 52, 66a)	28 Apl. 15		
22 Feb. 08	Paterson, A. G., D.S.O., M.C. (53, 57, 89, 207)	28 Apl. 15		
5 Oct. 10	Renny, S. C. (71)	28 Sep. 15		
2 Jan. 13	Shorter, R. M.	3 Jan. 17		
	Lieutenants.			
23 Sep. 13	Agar, E. W. T. (*temp. Capt., 3 Sep. 17*)	27 Nov. 14		
17 Oct. 14	Lewis, A. N.	28 Apl. 15		
17 Mar. 15	Rowley, E. D. T.	26 Oct. 16	. . . .	A. D. C. to H. H. the L. G. of Bihar & Orissa.
17 Mar. 15	MacLeod, N., M.C.	15 Nov. 16		
17 Apl. 15	Moreton, C. A.	29 Feb. 17		
	2nd-Lieutenants.			
25 Jan. 08	Simpson, Capt. W. (12, 51, 60a, 67, 72, 72b, 207)	1 Jan. 17	. . . .	Qr. Mr

Agents:—MESSRS. COX & CO.

Regimental District No. 26.
[No. 2 District.]

Uniform—Green.
Facings—Dark Green.

The Cameronians (Scottish Rifles).

1ST GARRISON BATTALION.
Arrived in India, 13th March 1916.

HONORARY COLONEL—Maj.-Genl. P. R. Robertson, C.B., C.M.G. . . . . . . 14 June 18.

First Commn.	Names.	Regimental rank.	Army rank.	Remarks.
	Lieutenant-Colonel.			
15 Mar. 93	Haig, D. P. (*temp.*)	24 June 15		
	Majors.			
94	Valentine, A. (*temp.*), 4 Bn., R. Highrs (16, 51).	24 Sept. 14		
8 Sep. 08	Sharp, P. C. (*temp.*) (*Capt. W.I.R.*)	10 Aug. 15	. . . .	Attd. No. 1 Res. Bn. (India).
4 Dec. 84	Outram, J. D. (*3rd High. L. I.*) .	19 May 17	. . . .	Comdt., Kailana Camp.
	Captains.			
5 Dec. 03	Duncan, G. M. (*8 Bn., Arg. & Suth'd, Highrs.*) (51)	17 Dec. 10		
15 Aug. 14	Armour, J. S. D. M. (48*a*, 51, 52, 60*a*)	10 Aug. 15	. . . .	S. S. O., Bombay.
14 Feb. 06	McAllister, H. M. (*temp.*) . . .	2 Feb. 16		
30 Oct. 14	*Mumford, W. J. T.* (*High. L. I.*) .	1 Jan. 17	. . . .	*Adjt.*
	Lieutenants.			
15 Aug. 12	Robertson, S. R. (51)	6 Sep. 15	. . . .	Attd. Hdqrs., Meerut Divn.
15 May 01	Ralph, M. M. (*temp.*) (*temp. Capt.*)	27 Jan. 16	. . . .	S. S. O. & Cant. Magte., Pachmarhi.
June 01	Shaw, C. W. (*temp.*) (*act'g. Capt.*)	1 Feb. 16		
11 Nov. 14	Mackenzie, W. S., D.S.O., (*3 Bn., Sea. Highrs.*)	2 Feb. 16	. . . .	I. A. R. O.
15 June 15	Sladen, W. D. (Cam'n. Highrs.) .	23 Dec. 16		
18 Dec. 14	Duncan, T. (*temp.*) (48*a*, 51) . .	1 July 17	. . . .	Attd. Hdqrs. Delhi Bde.
19 Dec. 14	Munro, J. C. (51)	1 July 17	. . . .	With 1st Garr. Bn., Manch. R.
10 Apl. 15	Macintyre, J. (*temp.*) *act'g. Capt., 16 Sep. 18*) (61)	1 July 17.		
30 Apl. 15	Turner, W. (*temp.*) (46, 61, 207*a*) .	1 July 17	. . . .	Supvg. Offr., Phys. & Bay. Trg., 4th (Quetta) Dn.
8 May 15	Sheriff, G. (*temp.*) (51)	1 July 17	. . . .	Attd. 2-6 Bn., Devon. R.
12 July 15	*McBride, J.* (*3 Bn., K. O. Sco. Bord.*)	1 July 17	. . . .	*I. A. R. O.*
25 Aug. 15	Carr, S. G. (*5 Bn., R. Highrs.*) (46, 52).	1 July 17		
16 Dec. 15	Binnie, J. (*5 Bn., R. Scots.*) . .	1 July 17		
1 Jan. 16	Boynton-Lee, F. (*temp.*) (*High. L. I.*)	1 July 17	. . . .	I. A. R. O.
21 Jan. 16	Ritchie, T. K., 9th Arg. & Suth'd Highrds.	27 July 17	. . . .	*Lv, ex I., m.c., 6 mos*
17 Feb. 16	Page, A. J. (*temp.*) (51)	17 Aug. 17	. . . .	I. A. (T. C.)
	2nd-Lieutenants.			
28 Nov. 15	Henderson, J. T. (*temp. Capt., 22 Mar. 18*) (*8 Bn., R. Scots*) (46, 66*a*)	28 Nov. 15	. . . .	S. S. O., Landour.
17 July 16	Trotter, J. P., 9th Arg. & Suth'd. Highrs.	7 July 16		
11 Feb. 17	Waddell, W. (*temp.*)	11 Feb. 17		
11 Feb. 17	Semple, R. E. (*temp.*)	11 Feb. 17		
11 Feb. 17	Morrison, A. (*temp.*) (51, 52, 66*a*) .	11 Feb. 17	. . . .	With 1st Garr. Bn., Manch. R.
11 Mar. 17	Sharman, E. P. (*temp.*) (46) . .	11 Mar. 17	. . . .	Attd. Army Clothing Factory, Shahjahanpur.
28 Mar. 17	Macadam, A. McF. (*8 Bn., High. L. I.*)	28 Mar. 17		
15 Apl. 17	Keighley, K. (*temp.*)	15 Apl. 17		
13 May 17	Stewart, T. (*temp.*)	13 May 17		
13 May 17	Thomas, S. S. L. St. G. (*temp.*) . .	13 May 17		
26 May 17	Franklin, A. D. (*temp.*) (57) .	26 May 17		
17 June 17	Cockburn, T. S. (*temp.*)	17 June 17		
17 June 17	Nathan, M. (*temp.*) (61)	17 June 17		
29 Aug. 17	MacBain, J. R. G. (*8 Bn., Sco Rif.*) (46, 52)	26 Aug. 17		
26 Sep. 17	Curphey, R. L. (*4 Bn., E. Lan. R.*) (*temp. Capt.*)	26 Sep. 17		
18 Nov. 17	Laxton, A. A. (*5 Bn., High. L. I.*)	18 Nov. 17		

30 Oct. 14	Mumford, W. J. T., High. L. I. .	1 Jan. 17	. . . .	Adjt.
3 Oct. 18	Good-Gaunt, Lt. W. (*temp.*) . .	3 Oct. 18	. . . .	Quartermaster.

Regimental District No. 31.
[No. 10 District.]

Uniform—Scarlet.
Facings—White.

The East Surrey Regiment.

1-5TH BATTALION.

Arrived in India, 3rd December 1914, from England.

"South Africa, 1900-1902."

HONORARY COLONEL—E. Horsman Bailey, V.D. . . . 26 Mar. 09.

First Commn.	Names.	Regimental rank.	Army rank.	Remarks.
	Lieutenant-Colonel.			
8 Aug. 03	Roe, G. A. M.	27 Sep. 17	. . . . .	Comdg.
	Majors.			
28 May 04	Whiffen, G. G.	1 June 16	. . . .	Attd. A. G.'s Bch., Army Hdqrs.
23 June 06	Chetwynd-Stapylton, G. B. (206)	1 June 16	. . . .	G. S. O., 3rd gr.
	Captains.			
11 Apl. 06	Wheatley, A. R. (*act'g. Maj.*), 12 *Mar. 18.*	1 Oct. 09		
31 Dec. 08	Eggar, J. N.	6 Dec. 11		
11 June 08	Murray, C. H. (51)	24 Oct. 12		
24 Dec. 04	*Gray, R.*, M.C	25 May 14	24 Jan. 08	*Spec. Serv. Offr.*
2 Mar. 09	Garrett, J. E.	1 June 16		
1 Aug. 07	Hone, E. J.	1 June 16		
5 Oct. 09	Hardman, R.C.	1 June 16		
9 Feb. 10	*Ingram, C. D.* (52)	1 June 16	. . . .	*M. G. Corps.*
17 June 11	*Crawford, S. C. R.* (170)	1 June 16	. . . .	*D. J. A. G.*
17 July 12	*Harrison, G. K.* (66)	1 June 16	. . . .	*With 1-5th E. Kent R.*
27 Jan. 13	*Warren, M. C.* (51)	1 June 16	. . . .	*Spec. Serv. Offr.*
12 Mar. 13	Poston, G.	1 June 16	. . . .	No. 3 Res. Bn.
14 Aug. 14	Baker, A. V.	1 June 16	. . . .	*Lv., ex I., m. c., 6 mos.*
	Lieutenants.			
14 May 13	*Paterson, C. R. S.* (57)	1 June 16	. . . .	*I. A. (on probn.)*
11 Oct. 13	*Grant, J. F.*	1 June 16	. . . .	*I. A. R. O. (on probn.)*
29 Aug. 14	Chadwyck-Healey, H. P. (*act'g. Capt.*, 21 *Dec.* 17) (60a).	1 June 16		
29 Aug. 14	Orr, E. P., (*act'g. Capt.*, 28 *May* 18) (51)	1 June 16		
29 Aug. 14	*Churcher, W. D.*	1 June 16	. . . .	*I. A. (on probn.)*
29 Aug. 14	Hawes, E. L. (*act'g. Capt.*, 25 *Aug.* 18).	1 June 16		
29 Aug. 14	*Whiffen, N. H.* (51, 52)	1 June 16	. . . .	*M. G. Corps.*
31 Aug. 14	*Hatchard-Smith, W. H.* (*act'g. Capt.*, 1 *Mar.* 18) (66)	1 June 16	. . . .	*Adjt.*
31 Aug. 14	Brown, D. A. (51, 57)	1 June 16	. . . .	With 1st Garr. Bn., Bedf. R.
28 Sep. 14	Barber, F. H. (*act'g. Capt.*, 11 *Oct.* 18)	1 June 16		
24 Oct. 14	Blair, J. M. (59)	1 June 16	. . . .	With Sig. Serv.
24 Oct. 14	*Grahame, D.* (52)	1 June 16	. . . .	*M. G. Corps.*
9 Mar. 15	Dawson, C. W.	1 July 17		
5 Aug. 15	Mann, S. G. (51)	1 July 17		
27 Sep. 15	Hooper, H. G.	1 July 17		
1 Oct. 15	*Taylor, W. R.*	1 July 17	. . . .	*I. A. R. O.*
14 Oct. 15	Turketine, A. W. H. (51)	1 July 17		
14 Oct. 15	Turketine, T. L. (60a).	1 July 17		
14 Oct. 15	Austin, A. P. M.	1 July 17		
12 Nov. 15	Freeborn, F. V.	1 July 17		
24 Dec. 15	Gillham, B. C.	1 July 17	. . . .	Attd. L. T. Motor Batty.
25 Dec. 15	Barnby, J. E.	1 July 17		
15 Jan. 16	Green, A. L.	15 July 17	. . . .	With 1-6th Hamps. R.
16 Jan. 16	Fotheringham, J. A. R.	16 July 17		
3 Feb. 16	*Trethowan, R. F.*	3 Aug. 17	. . . .	*M. G. Corps.*
16 Feb. 16	Blackman, J. G.	16 Aug. 17		
16 Feb. 16	Fenwick, F. R.	16 Aug. 17	. . . .	No. 3 Res. Bn.
16 Feb. 16	Wells, E. H. (46, 51)	8 Jan. 18		
	2nd-Lieutenants.			
4 Sep. 16	Green, E. B. (57)	4 Sep. 16		
1 Mar. 17	Fitzsimons, G.	1 Mar. 17		
31 Oct. 17	Dear, G. W. V.	31 Otc. 17	. . . .	With No. 3 Res. Bn.
27 Nov. 17	Stevens, E. P.	27 Nov. 17	. . . .	Do.
28 Nov. 17	Lofthouse, J. D.	28 Nov. 17	. . . .	Do
28 Nov. 17	Tregenza, E. L.	28 Nov. 17	. . . .	Do.

Regimental District No. 31.
[No. 10 District.]

Uniform—Scarlet.
Facings—White.

The East Surrey Regiment—*contd.*

First Commn.	Names.	Regimental rank.	Army rank.	Remarks.
	2nd-Lieutenants—*contd.*			
16 Mar. 18	Handley, H. (60*a*)	16 Mar. 18	. . . .	With No. 3 Res. Bn.
16 Mar. 18	Marshall, F. S. (51)	16 Mar. 18		
16 Mar. 18	Courtenay, S. W.	16 Mar. 18		
16 Mar. 18	Thompson, A. G.	16 Mar. 18		M. G. Corps.
19 Aug. 18	Piper, N. L.	19 Aug. 18	. . . .	With No. 3 Res. Bn.
19 Aug. 18	Hoskin, F.	19 Aug. 18	. . . .	Do.
19 Aug. 18	Treleaven, W. J.	19 Aug. 18	. . . .	Do.
19 Aug. 18	Williams, J. P.	19 Aug. 18	. . . .	Do.
	Attached.			
22 May 15	Kingston, Lt. C. G., 3 Bn., E. Surr. R.	1 July 17	. . . .	With No. 3 Res. Bn.
17 June 15	James, Lt. J. H. (*temp.*), 9 Bn., E. Surr. R.	1 July 17		
3 Nov. 15	Stephenson, Lt. R. W. (*temp.*), E. Surr. R.	1 July 17		
22 Jan. 16	Robson, Lt. A., 6 Bn., R.W. Surr. R.	22 July 17	. . . .	*Lv., ex I., m.c., 6 mos.*
1 Mar. 17	Bertini, 2nd-Lt. M. A., 1-6th E. Surr. R.	1 Mar. 17		
28 Mar. 17	Arnold, 2nd-Lt. G., 3rd Bn., R. W. Surr. R.	28 Mar. 17		
30 May 17	Cameron, 2nd-Lt. K. K., Arg. & Suth'd. Highrs.	30 May 17		
17 June 17	Crutchley, 2nd-Lt. W. H., 1 Garr. Bn., Suff. R.	17 June 17		
31 Aug. 14	Hatchard-Smith, Lt. (*act'g. Capt.*) W. H.	1 June 16	. . . .	Adjutant.
19 Mar. 02	Tarran, Maj. W., T. D.	19 Mar. 02	. . . .	Qr.-Mr. —*Lv., ex I., m.c.*

Regimental District No. 31.
[No. 10 District.]

Uniform—Green.
Facings—Scarlet.

The East Surrey Regiment.

1-6TH BATTALION.

Arrived in India, 3rd December 1914, from England.

" Imperial Service."
" South Africa, 1900-02."

HONORARY COLONEL—J. L. G. Powell, V.D. . . . 12 Feb. 09.

First Commn.	Names.	Regimental rank.	Army rank.	Remarks.
	Lieutenant-Colonel.			
22 Dec. 97	Drayson, A. P., O.B.E., T.D. . .	4 June 13		
	Majors.			
27 Sep. 99	Hallett, P. (*Hon. Lt. in Army, 26 July, 02*) (Q) (51, 206)	9 Aug. 13	. . . .	G. S. O., 2nd gr., 16th Divn.
19 Sep. 03	Whitehouse, A. L. W. (Q) (51, 52, 170)	1 June 16		
24 Jan. 06	Sandover, W. L. (Q)	1 June 16	. . . .	*Lv., ex I., m. c.*
9 June 08	Buckingham, F.	7 Aug. 17	. . . .	Conv. Depot, Wellington.
	Captains.			
5 May 09	*Pullman, G. C.* (51)	29 Aug. 14	. . . .	*Staff Capt.*
6 Mar. 11	Chapple, G. L. (51, 52)	1 June 16	. . . .	Attd. 1-4th Dorset R.
8 Apl. 11	Plincke, J. F.	1 June 16	. . . .	Attd. 1-4th Dorset R.
28 Oct. 11	Baker, T. MacD. (51, 59) . . .	1 June 16		
4 Oct. 10	Elmslie, G. F. (51)	1 June 16		
5 Aug. 14	Lyle, P. (57)	1 June 16		
12 July 13	*Rickell, G. G.* (*temp. Maj., 25 June 18*) (51)	1 June 16	. . . .	*Dy. Asst. Dir. of Ry. Tpt., Karachi.*
12 July 13	*Woodfall, H. A. J.* (60*a*, 66) . .	1 June 16	. . . .	*Staff Capt.*
19 July 13	Webb-Johnson, S. (48*a*, 51, 52, 89*b*, 170)	1 June 16		
22 Apl. 14	Onyon, R. R. (48*a*, 51, 52) . . .	1 June 16		
29 Apl. 14	Field, C. D. (51)	1 June 16	. . . .	Attd. M. W. S.
23 July 14	Andras, J. B. (57)	1 June 16	. . . .	Adjt.
	Lieutenants.			
1 May 14	*Daniels, C. J. F. W.*	1 June 16	. . . .	*I. A. R. O.*
17 Aug. 14	Battershill, W. D. (57)	1 June 16	. . . .	Attd. 2nd R. W. Kent. R.
29 Aug. 14	de la Bere, R. (60*a*, 66)	1 June 16	. . . .	With R. A. F.
24 Mar. 15	Burgess, H. S. (48*a*, 51)	16 Dec. 16		
24 Mar. 15	Phillips, A. J. (51, 52)	16 Dec. 16		
25 Mar. 15	Nicholls, H. B. C.	1 July 17	. . . .	With R. A. F.
25 Mar. 15	Boulton, D. H. (57, 89*b*)	1 July 17		
29 Apl. 15	Swain, G. H.	1 July 17		
12 Sep. 15	Hicks, M. G.	1 July 17		
24 Dec. 15	Gillham, B. C.	1 July 17		
26 Dec. 15	Revell, P. J. (46)	1 July 17		
29 Dec. 15	Tompkins, R. H. (48*a*)	1 July 17		
3 Jan. 16	Aldous, A. M. (48*a*, 60*a*)	3 July 17	. . . .	*Lv., ex I., m. c., 6 mos., 3 Aug. 18.*
3 Jan. 16	Francis, H. C.	3 July 17	. . . .	With M. G. Corps.
8 Jan. 16	Sharpe, H. (46, 89*b*)	8 July 17		
9 Jan. 16	Weatherhead, E. S.	9 July 17	. . . .	*Lv., ex I., m. c.*
19 Jan. 16	Reeve, A. (48*a*)	19 July 17		
24 Jan. 16	Kirby, E. T. R. (57)	24 July 17		
11 Feb. 16	Farley, C. W. (46)	11 Aug. 17		
	2nd-Lieutenants.			
28 Apl. 17	Silvester, F. H. (48*a*)	28 Apl. 17		
16 Mar. 18	Simmonds, R.	16 Mar. 18		
16 Mar. 18	Smedley, E. A.	16 Mar. 18		
	Attached.			
19 June 11	Edsell, Capt. E. V. (Lt. E. Surr. R.)	1 June 16		
5 Aug. 16	Francis, Lt. P. E. C., E. Surr. R. .	5 May 18		
29 Aug. 17	Wells, 2nd-Lt. E. T. W., 8 Bn., Lan. Fus.	29 Aug. 17		
25 Sep. 17	Ames, 2nd-Lt. J. A. (*temp.*), York. L. I.	25 Sep. 17		
25 Sep. 17	White, 2nd-Lt. G. L., Notts. and Derby. R.	25 Sep. 17		
31 Oct. 17	Jolliffe, 2nd-Lt. G. W. V., R. Berks. R.	31 Oct. 17		
23 July 14	Andras, Capt. J. B.	1 June 16	. . . .	Adjt.
2 Nov. 95	Osborne, Maj. G. H. (*ret. pay*) . .	27 Jan. 12	. . . .	Qr.-Mr.

Regimental District No. 32. [No. 8 District.]

Uniform—Scarlet. *Facings*—White.

The Duke of Cornwall's Light Infantry.

2-4TH BATTALION.

Arrived in India, 9th January 1915, from England.

" South Africa, 1900-01."

HONORARY COLONEL—*Sir* G. J. Smith, *Kt.*, V.D. . . 10 Feb. 97.

First Commn.	Names.	Regimental rank.	Army rank.	Remarks.
	Lieutenant-Colonel.			
30 Jan. 84	Rogers, P. H., ret. pay (*Res. of Offr.*) (51, 67, 206)	11 Sep. 14	19 July 10	Comdg. Details Depot, Muttra.
	Majors.			
4 Jan. 00	Hamilton, W. H. (51)	23 Sep. 14	. . . .	With 1-4th Hants.
Sep. 96	Meadows, G. S. (*act'g. Lt.-Col.*)	28 Dec. 14		
1 Apl. 96	Ellis, J.	1 June 16		
	Captains.			
21 Apl. 14	Martin, H. S. (51)	1 June 16		
24 May 10	Pole Carew, G. A. (52)	1 June 16		
28 Aug. 14	*Huss, S. A.*	1 June 16	. . . .	*I. A. R. O.*
29 Aug. 14	Sandys, M. K. T. (48*a*, 51, 57)	1 June 16		
29 Aug. 14	*Seymour, A. H.* (51, 52)	1 June 16	. . . .	*I. A. (on probn.)*
29 Aug. 14	Thomas-Peter, G. F.	1 June 16		
	Lieutenants.			
12 Sep. 14	Bassett, R P. H. (51, 52, 60*a*)	1 June 16	. . . .	Attd. Sappers & Miners, Kirkee.
14 Sep. 14	Daunt, A. (51)	1 June 16		
17 Sep. 14	*Tresawna, H.* (66)	1 June 16	. . . .	*Staff Capt.*
1 Oct. 14	*Pocock, R. N. (temp. Capt.)* (51)	1 June 16	. . . .	*G. S. O., 3rd gr., Army Hdqrs.*
7 Oct. 14	Trusted, H. H. (*temp. Capt.*) (51)	1 June 16	. . . .	Staff Capt., Derajat Bde.
7 Oct. 14	*Read, N. J. (act'g Capt.)* (51, 52, 56)	1 June 16	. . . .	*Adjt.*
8 Oct. 14	Westlake, A. R. C. (51, 57, 95, 169, 170, 207)	1 June 16	. . . .	Attd. Kurdish Labour Corps.
10 Oct. 14	Williams, T. L. (51, 66)	1 July 17	. . . .	Staff Capt., Agra Bde.
10 Oct. 14	*Pascoe, F. J.* (57, 60*a*)	1 July 17	. . . .	*With Sig. Serv.*
1 Jan. 15	Moody, C. G. (60*a*) (*temp. Capt.*)	1 July 17	. . . .	Platoon, Comdr., Offrs.' Schl. of Instn., Nasik.
13 Apl. 15	Drake, F. H.	1 July 17	. . . .	M. A. Dept.
5 June 15	D'Ambrunenil, W. (46)	1 July 17	. . . .	S. S. O., Chakrata.
19 Aug. 15	*Gilbert, F.*	1 July 17	. . . .	*With R. A. F.*
19 Aug. 15	*Penberthy, W. F.* (52)	1 July 17	. . . .	*M. G. Corps.*
19 Aug. 15	*Lander, J. H.* (57)	1 July 17	. . . .	*I. A. (On probn.)*
19 Aug. 15	Rowse, H. C. (48*a*, 51, 60*a*)	1 July 17		
19 Aug. 15	*Pascoe, J. J.* (51)	1 July 17	. . . .	*M. G. Corps.*
10 Sep. 15	Trounson, W. (51)	1 July 17	. . . .	Attd. M. W. S.
4 Nov. 15	Andrew, B. H. (51)	1 July 17		
22 Nov. 15	Nivet, E. L. A. (48*a*, 51)	1 July 17		
9 June 16	Francis, R. G. (57)	9 Dec. 17		
14 June 16	Rule, G. C. S. (46)	14 Dec. 17		
23 June 16	Lawrence, R. R. S. (51)	23 Dec. 17		
9 July 16	*Strong, J. R.*	*9 Jan. 18*	. . . .	*M. G. Corps.*
21 July 16	Evans, R. E. (46, 49)	21 Jan. 18		
31 July 16	Hodge, W. E. (48*a*, 51, 60*a*)	31 Jan. 18	. . . .	Instr. of Lewis Gun, Dehra Dun [Bde.
5 Sep. 16	Stephens, E. J.	5 Mar. 18	. . . .	*Lv. ex I., m.c., 12 mos.*
26 Sep. 16	Worden, L. (51, 57, 66, 172*a*)	26 Mar. 18	. . . .	3 Divl. Supply Coy.
	2nd-Lieutenants.			
19 Aug. 15	Kearney, P. W.	19 Aug. 15	. . . .	With 2 Norf. R.
29 May 17	Stanton, P. E.	29 May 17		
30 May 17	Tucker, F. G.	30 May 17	. . . .	Ry. Labour Corps, Saharanpur.
26 July 17	Froggatt, W.	26 July 17		
29 Aug. 17	Shea, V. G. (88, 127)	29 Aug. 17	. . . .	Supply and Transport Corps.
29 Aug. 17	Webb, B. H.	29 Aug. 17		
26 Sep. 17	Dalgleish, A. (66*a*)	26 Sep. 17		
26 Sep. 17	Kneen, A. G.	26 Sep. 17		
26 Sep. 17	Oldham, J. B.	26 Sep. 17		
26 Sep. 17	Sharman, C. T.	26 Sep. 17		
26 Sep. 17	Chatterton, A. L.	26 Sep. 17		
26 Sep. 17	Collier, J.	26 Sep. 17		
28 Nov. 17	Mair, A.	28 Nov. 17		
16 Mar. 18	Stevens, K. G.	16 Mar. 18	. . . .	K. Supply Depôt Coy., Peshawar.
	Attached.			
19 Mar. 15	Ball, 2nd-Lt. H. F. (57, 206) (*temp. Lt.*) T. F. Res. (Gen. List)	19 Mar. 15	. . . .	With Sig. Serv.
7 Oct. 14	Read, Lt. (*act'g. Capt.*) N. J.	27 Nov. 17	. . . .	Adjt.
12 Sep. 14	Stear, Capt. B. T.	12 Sep. 14	. . . .	Qr.-Mr.

Regimental District No. 33.
[No. 5 District.]

Uniform—Scarlet.
Facings—Scarlet.

The Duke of Wellington's (West Riding Regiment).

1ST BATTALION (*late 33rd Foot*).
Arrived in India, 26th October 1905, from England.

THE LATE DUKE OF WELLINGTON'S CREST, WITH THE MOTTO IN AN ESCROLL ABOVE,
"*Virtutis fortuna comes.*"

THE ELEPHANT SUPERSCRIBED "HINDOOSTAN."
"Dettingen" "Mysore" "Seringapatam" "Ally Ghur" "Delhi, 1803" "Leswarree" "Deig" "Corunna"
"Nive" "Peninsula" "Waterloo" "Alma" "Inkerman" "Sevastopol" "Abyssinia"
"Relief of Kimberley" "Paardeberg" "South Africa, 1900-02."

COLONEL—Lt.-General *Sir* H. E. Belfield, K.C.B., K.C.M.G., D.S.O., *p.s.c.* ret. pay. 2 Aug. 09.

First Commn.	Names.	Regimental rank.	Army rank.	Remarks.
	Lieutenant-Colonel.			
21 Jan. 93	Maffett, R. E. (51, 52, 72*b*)	3 Feb. 16		
	Majors.			
	Captains.			
28 Jan. 03	Boutflower, E. C. (*act'g. Maj.*) (51, 52, 65, 66*a*, 67, 71, 72, 172*a*, 205*a*)	18 Dec. 14		
14 Oct. 08	Owen, R. H. W. (51, 52, 66*a*, 71)	10 June 15		
	Lieutenants.			
5 Sep. 11	Crane, M. E. (*temp. Capt., 5 Sep. 15*) (51, 52)	19 Sep. 14	5 Dec. 13	Comdt., Leave Details, Landour.
5 Feb. 13	*Davis, A. O. L.* (*temp. Capt., 5 Feb. 17*) (12, 51, 52)	9 Oct. 14	....	Adjt.
16 July 13	Fleming, G. T. (*temp. Capt., 5 Feb. 17*) (51, 52, 59)	25 Oct. 14	....	With 6 Bn., Devon. R.
3 Sep. 13	Faulkner, St. J. T. (*act'g. Capt., 17 Feb. 17*) (46, 48*a*, 51, 52)	8 Dec. 14		
17 Sep. 13	Tidmarsh, J. M. (51)	11 Dec. 14	....	With R. A. F.
7 Nov. 14	Baker, W. G. (*temp. Capt., 15 Aug. 18*) (51, 52, 61)	10 June 15		
7 Nov. 14	Wheeler, C. (*temp. Capt., 15 Aug. 18*) (57)	10 June 15	....	With Sig. Serv.
7 Nov. 14	Rowland, C. (*temp. Capt., 15 Aug. 18*) (60*a*)	10 June 15	....	With R. A. F.
7 Nov. 14	Sampson, H. S. (*temp. Capt., 15 Aug. 18*) (48*a*, 51)	10 June 15	....	
7 Nov. 14	Hazell, A. J. (*temp. Capt., 15 Aug. 18*)	2 Mar. 17	....	With 2-6 Bn., Devon. R.
9 Jan. 15	Looney, D. (*temp. Capt., 15 Apl. 17*)	1 July 17	....	Bde. Maj., Wellington Bde.
6 Feb. 15	Hoffman, J. A. (*temp.*) (*act'g. Capt., 24 Nov. 17*) (46, 48*a*, 51)	1 July 17		
10 Mar. 15	Fletcher, J. A. (*temp.*)	1 July 17	....	I. A. R. O.
10 Mar. 15	Smith, V. S. (*temp.*)	1 July 17	....	I. A. R. O.
2 Apl. 15	Marsh, G. P. (*temp.*) (48*a*, 51)	1 July 17		
29 June 15	Allan, J. N. (*temp.*) (46, 51)	1 July 17		
13 Jan. 16	Stilling, N. A. (57)	13 July 17	....	With Sig. Serv.
7 Apl. 16	Scott, J. W. (57, 61)	7 Oct. 17		
9 Dec. 16	Wilson, T. A. (48*a*, 51, 61)	9 June 18		
	2nd-Lieutenants.			
22 Oct. 14	Rogers, E. W. (*temp. Maj.*)	22 July 15	....	D. A. A. G., 2nd (Rawalpindi) Divn.
28 Jan. 17	Field, A. V. (*temp. Capt., 16 Jan. 18*)	28 Jan. 17	....	With Sig. Serv.
13 Apl. 17	Wood, R. (*act'g. Capt, 11 Sep 18*) (48*a*, 50)	13 Apl. 17	....	Adjt., N. C. O.s' Schl. of Instn., N. Comd.
29 Aug. 17	Gooding, W. P. (*temp.*)	29 Aug. 17		
29 Aug. 17	Sharpe, P. (*temp.*) (*temp. Capt., 19 Oct. 18*)	29 Aug 17	....	Platoon Comdr., Offrs'. Schl. of Instn., Nasik.

The Duke of Wellington's (West Riding Regiment)—*contd.*

First Commn.	Names.	Regimental rank.	Army rank.	Remarks.
	2nd-Lieutenants—*contd.*			
29 Aug. 17	Marriott, K. M. H. (*temp.*)	29 Aug. 17	....	With R. A. F.
29 Aug. 17	Owen, A. J. (*temp.*)	29 Aug. 17	....	M. G. Corps.
29 Aug. 17	Baker, R. C. N. (*temp.*)	29 Aug. 17	....	M. G. Corps.
16 Mar. 18	Ely, E. S. (*temp.*)	16 Mar. 18	....	*Lv., ex I., m.c., 6m.*
16 Mar. 18	Baker, V. H. E. (*temp.*)	16 Mar. 18	....	With R. A. F.
16 Mar. 18	Butler, C. C. F. (*temp.*)	16 Mar. 18	....	M. G. Corps.
16 Mar. 18	*Purchase, R. G.* (*temp.*)	16 Mar. 18	....	*I. A. R. O.*
23 Apl. 18	Woodfield, A. W.	23 Apl. 18		
19 Aug. 18	Allen, V. W. (*temp.*)	19 Aug. 18		
19 Aug. 18	Carman, C. R. (*temp.*)	19 Aug. 18		
5 Sep. 18	Turner, A.	5 Sep. 18	....	With Sig. Serv.
	Attached.			
30 Apl. 08	Charlesworth, M.	1 July 13		
7 July 16	Pringle, Lt. W. B. (*temp.*)	7 Jan. 18	....	S. S. O., Gharial.
14 Apl. 15	Wicks, G. (*temp.*) (12)	14 Apl. 15		
22 Nov. 16	Jennings, 2nd-Lt. W. G. (*temp.*)	22 Nov. 16		
28 Jan. 17	Burgess, 2nd-Lt. C. B. (*temp.*)	28 Jan. 17		
28 Mar. 17	Howarth, R. (*temp.*)	28 Mar. 17	....	With R. A. F.
28 Mar. 17	Moyle, S. (*temp.*)	28 Mar. 17	....	Do.
28 Mar. 17	Ross, A. T. (*temp.*)	28 Mar. 17	....	Do.
26 Apl. 17	Platt, J. (*temp.*)	26 Apl. 17	....	Do.
29 Aug. 17	Stewart, W. F. (*temp.*)	29 Aug. 17	....	Do.
25 Sep. 17	Macmillan, H. A. (*temp.*)	25 Sep. 17	....	Ind. Mun. Board.
26 Sep. 17	Bould, F. (*temp.*)	26 Sep. 17	....	With A. M. B.
26 Sep. 17	Foster, F. L. (*temp.*)	26 Sep. 17		
26 Sep. 17	Regan, W. A. V. (*temp.*)	26 Sep. 17	....	M. G. Corps.
26 Sep. 17	Reynolds, N. (*temp.*)	26 Sep. 17	....	Ind. Mun. Board.
26 Sep. 17	Tree, G. C. (*temp.*)	26 Sep. 17		
31 Oct. 17	Tomlinson, C. (*temp.*)	31 Oct. 17		
31 Oct. 17	Stocker, A. (*temp.*)	31 Oct. 17		
31 Oct. 17	Platt, P. (*temp.*)	31 Oct. 17		

Feb. 13	Davis, Lieut. (*temp. Capt.*) A. O. L.	1 Dec. 16	....	Adjt.
Mar. 14	Oliver, Capt. C.	4 Mar. 17	....	Qr.-Mr.

Agents :—Messrs. Cox & Co.

Regimental District No. 34.
[No. 3 District.]

Uniform—Scarlet.
Facings—Yellow.

The Border Regiment.

1-4TH (Cumberland and Westmorland) BATTALION.
Arrived in India, 3rd December 1914, from England.

"South Africa, 1900-02."

HONORARY COLONEL—*Lord* H. C. Cavendish-Bentinck, T.D. 8 July 93.

First Commn.	Names.	Regimental rank.	Army rank.	Remarks.
	Lieutenant-Colonel.			
	Majors.			
11 May 98	Bousfield, E. (51)	16 Oct. 14	....	Comdt., Advanced Base Depôt.
16 Feb. 98	Davidson, A. (51)	1 June 16	....	Comdg.
9 July 04	Lamonby, L. (51)	13 July 17	....	Attd. 1-4th Dorset R.
	Captains.			
31 Jan. 00	Crewdson, W. D. (51)	6 July 13		
16 July 04	Errington, J. P. (*temp. Maj.*) (51, 52, 57)	10 Oct. 14	....	Asst. Comdt., Officers' Schl. of Instn., Nasik.
16 July 04	Pratchitt, W. M. (51)	16 Oct. 14		
26 Nov. 04	Hele, J. W. (50)	4 Feb. 16		
24 Feb. 06	Chance, F. S. (48*a*, 51)	1 June 16		
20 Feb. 07	Story, G. A. (51, 52)	1 June 16		
3 May 09	*Banks, T. M.*	1 June 16	....	Staff Capt., N. Comd.
25 Nov. 09	*Hoggarth, L. S.* (*temp. Maj.*) (52)	1 June 16	....	*M. G. Corps.*
10 Jan. 11	*Hope, P. M.*	1 June 16	....	*Attd. R. E.*
15 July 12	Neville, R. B. (48*a*, 51, 57)	1 June 16		
15 Dec. 12	*Raikes, M. H.*	1 June 16	....	*I. A. R. O.*
10 Feb. 14	*Highet, H.*	1 June 16	....	*With R. A. M. C.*
29 Aug. 14	Dixon, A. F. W. (51, 58*a*, 60*a*)	1 June 16	....	Attd. 1-4th Dorset R.
27 Sep. 14	Wilson, A. P. (48*a*, 50)	12 May 17		
5 Oct. 14	Bendle, L. A.	12 May 17	....	Attd. 1-4th Dorset R.
24 Oct. 14	*Mashiter*, E. (51, 52)	15 Sep. 17	....	*M. G. Corps.*
29 Oct. 14	Whitehead, J. (46)	15 Sep. 17		
	Lieutenants.			
13 Aug. 15	*Ewers, L. F.* (52)	1 July 17	....	*M. G. Corps.*
20 Aug. 15	*Dickie, D. H.*	*1 July 17*	....	*M. A. Dept.*
15 Oct. 15	Heelis, G. H. (51)	1 July 17	....	Attd. 1-4th Dorset R.
21 Oct. 15	Hodgson, W. M. (60*a*)	1 July 17	....	Attd. 1-4th Dorset R.
9 Dec. 15	Alexander, W.	1 July 17	....	Attd. 1-4th Dorset R.
9 Dec. 15	Alexander, A. (46, 61)	1 July 17	....	Attd. M. W. S.
19 Dec. 15	Ball, W. M.	19 Dec. 15	....	Attd. 2nd Garr. Bn., North'd. Fus.
19 Dec. 15	Lewis, N. C. (57, 59, 205*a*)	1 July 17		
21 Dec. 15	Alexander, S.	1 July 17	....	Attd. 1-4th Dorset R.
2 Jan. 16	Lewthwaite, R. (12, 48*a*, 51)	2 July 17		
2 Jan. 16	Shepperd, C. B. (49)	2 July 17		
14 Mar. 16	Clark, W. S. (53)	14 Sep. 17		
4 Aug. 16	Webster, R.	4 Feb. 18		
5 Aug. 16	Gladstone, H. L. (12, 48*a*, 51)	5 Feb. 18	....	Attd. M. G. Corps.
	2nd-Lieutenants.			
29 Aug. 17	Langston, K. E. W. (59)	29 Aug. 17		
29 Aug. 17	Batten, A. E.	29 Aug. 17		
29 Aug. 17	Tyler, V. (50)	29 Aug. 17		
29 Aug. 17	Woodman, F. C.	29 Aug. 17		
16 Mar. 18	Dickens, G. W. W., (*temp.*)	16 Mar. 18	....	Attd. R A. F.
16 Mar. 18	Peters, H., (*temp.*)	16 Mar. 18	....	Attd. R. A. F.
16 Mar. 18	*Fraser, R. A. J.*, (*temp.*)	16 Mar. 18	....	*I. A. R. O.*
	Attached.			
7 Apl. 17	Davison, 2nd-Lt. W. M., E. Lanc. R.	7 Apl. 17		
28 Apl. 17	Harrison, 2nd-Lt. R. S., Border R. (*actg. Capt., 14 Dec. 17*) (48*a*, 50)	28 Apl. 17	....	Adjutant.
26 Sep. 17	Backhouse, 2nd-Lt. J., 5 Bn., R. Lanc. R.	26 Sep. 17		
26 Sep. 17	McKay, 2nd-Lt. J., Rif. Bde.	26 Sep. 17		

28 Apl. 17	Harrison, 2nd-Lt. (*actg. Capt.*) R. S.	14 Dec. 17	....	Adjt.
10 Oct 14	Howling, Capt. F. (51, 60*a*)	10 Oct. 17	....	Qr.-Mr. *Lv., m.c., 2 mos.*

Regimental District No. 34. [No. 3 District.]

Uniform—Scarlet. *Facings*—Yellow.

The Border Regiment.

2-4th (Cumberland and Westmorland) BATTALION.
Arrived in India, 31st March 1915, from England.

"South Africa, 1900-1902."

HONORARY COLONEL—*Lord* H. C. Cavendish-Bentinck, T.D. . . . 8 July 93.

First Commn.	Names.	Regimental rank.	Army rank.	Remarks.
	Lieutenant-Colonel.			
23 July 92	Haswell, J. F., V.D. (*Hon. Lt. in Army, 30 May 01*)	8 Oct. 14	. . . .	With R. A. M. C., Quetta.
11 Nov. 90	Halton, F. W., T.D. (51) . .	13 July 17		
	Major.			
26 Feb. 96	Heelis, G. H.	1 June 16	. . . . .	
	Captains.			
18 Mar. 96	Graham, J. E. C.	24 Oct. 14		
28 Jan. 00	Jones, V. S.	21 Dec. 14	. . . .	S. S. O., Peshawar.
22 Oct. 14	Hargreaves, R. (51)	20 July 17	. . . .	Mily. Accts. Dept.
22 Oct. 14	Jackson, J. (*act'g. Capt.*) (48*a*, 51, 172*a*).	17 Aug. 17		
	Lieutenants.			
14 Nov. 14	Smith, R. I. (48*a*, 51, 57) . . .	1 June 16	. . . .	R. A. F.
17 Nov. 14	Coates, N. E.	1 June 16	. . . .	R. A. F.
17 Nov. 14	Grierson-Jackson, H. C. (*act'g. Capt., 11 Nov. 17*) (51) . . .	1 June 16		
20 Nov. 14	*Strang, J. L.*, (48*a*, 51, 52, 66*a*)	1 June 16	. . . .	M. G. Corps.
20 Nov. 14	Hamilton, P. S. (*temp. Capt., 29 Nov. 17*) (51)	1 June 16	. . . .	G. S. O., 3rd gr., 1st (Peshawar) [Dn.
20 Nov. 14	Angus, W. C. S. (51)	1 June 16	. . . .	With 1-5th E. Kent R.
1 Dec. 14	*Hardy, E. P.* (51, 52, 172*a*, 207) .	1 June 16	. . . .	*I. A.* (*on probn.*).
7 Dec. 14	MacGlasson, L. (*act'g. Capt., 3 Aug. 17*) (51, 52).	1 June 16	. . . .	Adjt.
10 Dec. 14	Anderson, A. W. (*act'g. Capt., 11 Apl. 18*) (51, 66*a*) . . .	1 June 16		
12 Dec. 14	Thomson, H. (51, 72*b*, 172*a*)	1 June 16	. . . .	R. A. M. C.
3 Jan. 15	Bruckman, R. T. (48*a*, 51) . .	22 Oct. 16		
3 Jan. 15	Topham, G. H. (58*a*)	1 July 17	. . . .	With Sig. Serv.
3 Jan. 15	Fisher, M. G. (*temp. Capt., 5 Aug. 17*) (51, 52, 169, 172*a*)	1 July 17	. . . .	Staff Capt., Jubbulpore Bde.
3 Jan. 15	Walker, W. R. (*temp. Capt., 3 May 18*)	1 July 17	. . . .	Attd. A. G.'s Bch., Army Hdqrs.
5 Jan. 15	Glasson, J. (48*a*, 51, 57, 60*a*, 61) .	1 July 17		
15 Jan. 15	Kinghorn, E. C. (51)	1 July 17	. . . .	R. A. F.
3 Mar. 15	Ball, C. F. (51, 52, 58*a*, 172*a*, 207)	1 July 17	. . . .	With Sig. Serv.
3 Mar. 15	Rheam, P. M. (51).	1 July 17		
3 Mar. 15	Marshall, H. G. (*temp. Capt., 21 July 17*) (51, 172*a*) . . .	1 July 17	. . . .	Staff Capt., Peshawar.
5 Apl. 15	Richardson, L.	1 July 17	. . . .	With M. W. S.
27 Nov. 15	Thomson, J. S. (46, 61) . . .	1 July 17	. . . .	M. G. Corps.
11 Dec. 15	Brown, A. A.	1 July 17		
17 Dec. 15	*Cain, C. W.* (52)	1 July 17	. . . .	*M. G. Corps.*
27 Dec. 15	Barker, E. H. (46, 48*a*, 51) . .	1 July 17		
29 Dec. 15	Ashburner, E. H. (46, 48*a*, 51, 172*a*)	1 July 17		
2 Jan. 16	Harrison, A. F. (57)	2 July 17	. . . .	With Sig. Serv.
7 Jan. 16	Chester, B. F. (51)	7 July 17	. . . .	S. S. O., Cherat.
	2nd-Lieutenants.			
1 Sep. 17	Priestley, H. S.	1 Sep. 17	. . . .	With Mily. Accts. Dept.
1 Sep. 17	Dodkins, L. C.	1 Sep. 17	. . . .	With R. A. F.
16 Mar. 18	Snow, G. H.	16 Mar. 18	. . . .	
16 Mar. 18	Perry, D. G.	16 Mar. 18	. . . .	
16 Mar. 18	Anderson, D. A. T.	16 Mar. 18	. . . .	With R. T. O., Lahore.
16 Mar. 18	Deuchars, H.	16 Mar. 18	. . . .	
16 Mar. 18	Pepperett, W.	16 Mar. 18	. . . .	
	Attached.			
28 Apl. 17	Saint, 2nd-Lt. J. B., Border R. .	28 Apl. 17		
28 Apl. 17	Curtis, 2nd-Lt. P. V., Border R. .	28 Apl. 17		
28 Apl. 17	Gibbings, 2nd-Lt. O. D. Bordr. R. (57, 61)	28 Apl. 17		

7 Dec. 14	MacGlasson, Lt. (*act'g. Capt.*) L.	14 Feb. 17	. . . .	Adjt.
1 Dec. 14	Brooks, Capt. J. (51, 57) .	1 Dec. 17	. . . .	Qr-Mr.

Regimental District No. 35. [No. 10 District.]

Uniform—Scarlet. *Facings*—Blue.

The Royal Sussex Regiment.

1ST BATTALION (*late 35th Foot.*

Arrived in India, 4th November 1902, from South Africa.

THE WHITE (ROUSILLON) PLUME.

Gibraltar, 1704-5" "Louisburg" "Quebec, 1759" "Martinique, 1762" "Havannah" "St. Lucia, 1778" "Maida" "Egypt, 1882" "Abu Klea" "Nile, 1884-85" "South Africa, 1900-02."

COLONEL—Major-General J. C. Young, C.B., *p. s. c.* . . . 28 Mar. 14.

First Commn.	Names.	Regimental rank.	Army rank.	Remarks.
Lieutenant-Colonel.				
29 Oct. 90	Mackenzie, E. L., D.S.O. (50, 53, 65, 72*a*)	12 Aug. 15		
Majors.				
29 May 95	Gem, G. (51, 52, 59, 71)	1 Sep 15	*l.c.*, 3 June 17	Comdt., Bazaar Enteric Depôt, Wellington.
10 Jan. 00	Leachman, G. E, C.I.E. (50, 53, 59, 66*a*, 72*b*, 88, 172*a*, 206)	1 Sep. 15	*l.c.*, 3 June 16	On Spl. duty, Poll. Dept.
30 Apl. 02	Wigley, A. B. (*Act'g. Lt.-Col.*, 23 *Sep. 18*) (51, 52, 66*a*, 72)	18 Jan. 17	. . . .	Comdg. 1st Breck. Bn., S. Wales Bord.
Captains.				
4 Mar. 11	Malden, C. C. (*Actg. Maj.*, 17 *Feb.* 17) (51, 52, 172*a*)	4 Mar. 17		
21 June 11	Holderness, W. (51, 52, 66*a*, 172*a* 207)	4 Mar. 17		
20 Sep. 11	Hawkes, C. C. (48*a*, 51, 71, 169, 207)	13 Apl. 17		
Lieutenants.				
4 Sep. 12	Skinner, F. S. D. (*temp. Capt.*, 4 *Sep.* 16) (48*a*, 50, 52, 95, 172*a*, 207) . .	13 Feb. 15		
22 Oct. 14	Ramsden, G. C. F. (temp.) (*Actg. Capt.* 17 *Feb.* 17), (48*a*, 51, 71, 169)	10 June 15		
7 Nov. 14	Guthrie, W. (*temp. Capt.*, 24 *July* 18) (47*a*, 48*a*, 50, 52)	18 Aug. 15	. . . .	Asst. Instr., Schl. of Musk., Satara.
7 Nov. 14	Calver, J. E. (48*a*, 51, 61, 71, 172*a*) .	27 Sep. 15		
7 Nov. 14	Price, A. (*Actg. Capt.*)	6 Nov. 15	. . . .	Attd. Oxf. & Bucks. L. I.
9 Jan. 15	Harrison, C. W.	20 Apl. 16	. . . .	Attd. 1-4 Bn., Devon. R.
9 Jan. 15	Day, F. (59, 71)	10 June 16	. . . .	With Sig. Serv.
17 June 15	Cormody, E. K. (*temp. Capt.*, 16 *May* 17) (46, 48*a*, 50, 52, 71, 95, 172*a*) .	4 Mar. 17	. . . .	Platoon Comdr., Officer Cadet Unit, Lower Topa.
19 Nov. 15	Golden, H. (57, 71, 95, 170) . . .	1 July 17	. . . .	With Sig. Serv.
27 Jan. 16	*Haigh, F. H.* (*Act'g. Capt.*, 3 *Dec.* 18) (12, 48*a*, 51)	27 July 17	. . . .	*Adjt.*
2nd-Lieutenants.				
24 Dec. 16	Cook, A. (*Actg. Capt.*)	24 Dec. 16	. . . .	Adjt. & Qr.-Mr., Conv. Depôt.
20 Feb. 17	Ford, A. R. (59)	20 Feb. 17		
11 Aug. 17	Wise, H.	11 Aug. 17	. . . .	Extra Asst. Instr., Schl. of Musk., Pachmarhi.
29 Aug. 17	McDonald, H. G. (*temp.*)	29 Aug. 17		
29 Aug. 17	Watson, F. T. (*temp.*)	29 Aug. 17		
29 Aug. 17	Ashman, C. H. (*temp.*)	29 Aug. 17	. . . .	With M. G. Corps, Mhow.
29 Aug. 17	Walker, N. W. (*temp.*) (46, 95) . .	29 Aug. 17		
17 Dec. 17	Wilson, R. V. (59)	17 Dec. 17		
25 Feb. 18	Hook, W. J.	25 Feb. 18	. . . .	S. S. O., Barian.

Agents:—MESSRS. COX & CO.

The Royal Sussex Regiment—*contd.*

First Commn.	Names.	Regimental rank.	Army rank.	Remarks.

Attached.

First Commn.	Names.	Regimental rank.	Army rank.	Remarks.
13 Oct. 14	Tyler, Capt. A. T. (*temp.*), R. Suss. R. (48*a*, 50)	4 Jan. 16		
28 Jan. 15	Nixon, Lt. J. T. (*temp.*), R. Suss. R. (48*a*, 51)	4 Jan. 16		
23 Feb. 15	Tyler, F. R. (*temp.*), R. Suss. R. (46, 172*a*)	1 July 17	. . . .	With Sig. Serv.
9 Apl. 15	*Sayers, Lt. W. W.* (*temp. Capt.*, 3rd Bn. (46, 172*a*)	1 July 17	. . . .	*Staff Capt.*
22 Apl. 15	*Coulson, Lt. R. A.*, 3rd Bn. (169)	1 July 17	. . . .	*I. A. R. O.*
13 June 15	Pugh, Lt. J. H., 3 North'd. Fus.	1 July 17		
10 July 15	Rae, Lt. R. (61, 172*a*), Sco. H.	1 July 17	. . . .	With 21st (E. of I.) Lrs.
11 Aug. 15	Orme, Lt. R. F. P. (*temp.*), R. Suss. R. (48*a*, 51)	1 July 17		
2 Sep. 15	Burbidge, Lt. W. H. M. (*temp.*), (46, 49), R. Suss. R. (*temp. Capt.*)	1 July 17	. . . .	Inspr. of Bombing, N. Comd.
27 Oct. 15	Forbes, Lt. A. (*temp.*), R. Suss. R.	1 July 17	. . . .	With Armd. Motor Battery.
23 Dec. 15	Munro, Lt. R. C., 9 Arg. & Suth'd Highrs.	1 July 17		
1 Jan. 16	Andrews, Lt. H. (*temp.*), K.R.Rif.C. (48*a*, 51)	1 July 17		
21 Jan. 16	McIntyre, 2nd-Lt. J., (*temp.*) R. Sco. Fus. (60*a*)	21 July 17	. . . .	Attd. 1-6th Bn., Hamps. R.
13 Aug. 16	Gyll, Lt. G. W. F., 3 Bn., Essex R. (48*a*, 51)	23 Nov. 17		
27 July 16	Gordon, Lt. G. R., 2-1st Sco. H.	27 Jan. 18		
5 Aug. 16	Macdonald, Lt. J., Lovat's Scouts	5 Feb. 18		
29 Jan. 16	Dick, 2nd-Lt. R. (*temp.*), 11th Ryl. Hrs. (12)	29 Jan. 16		
5 Sep. 16	Ardron, 2nd-Lt. H. G., 3 Bn., Dorset R.	5 Sep. 16		
26 Sep. 17	Whitehead, 2nd-Lt. W. A. B. (*temp.*) 3 Bn., Glouc. R.	26 Sep. 17		
26 Sep. 17	Hallam, 2nd-Lt. H., D. C. M. (*temp.*) 3 Bn., Notts. & Derby. R.	26 Sep. 17		
26 Sep. 17	Campbell, 2nd-Lt. A. R. (*temp.*) 4 Bn., K. R. Rif. C.	26 Sep. 17		
26 Sep. 17	Taylor, 2nd-Lt. E. J. (*temp.*) 5 Bn., K. R. Rif. C.	26 Sep. 17		
26 Sep. 17	Scott, 2nd-Lt. H. G. (*temp.*) 4 Bn., Rif. Bde.	26 Sep. 17		
16 Mar. 18	Bryant, 2nd-Lt. B. (*temp.*) (46)	16 Mar. 18		
16 Mar. 18	Smith, 2nd-Lt. L. A. (*temp.*)	16 Mar. 18		
16 Mar. 18	Corbitt, 2nd-Lt. J. H. (*temp.*)	16 Mar. 18	. . . .	Sig. Depôt, Poona.
16 Mar. 18	Triebler, 2nd-Lt. F. C. (*temp.*)	16 Mar. 18	. . . .	M. G. Corps, Mhow.
16 Mar. 18	Griffiths, 2nd-Lt. J. Y. (*temp.*)	16 Mar. 18	. . . .	M. G. Corps, Mhow.
19 Aug. 18	Phillips, 2nd-Lt. W. G. (*temp.*)	19 Aug. 18		
19 Aug. 18	Lawrance, 2nd-Lt. E. R. (*temp.*)	19 Aug. 18		

First Commn.	Names.	Regimental rank.	Army rank.	Remarks.
27 Jan. 16	Haigh, Lieut. (*Actg. Capt.*) F. H.	3 Dec. 18	. . . .	Adjutant.
20 Oct. 01	Gilpin, Maj. G. (170, 207)	30 Oct. 16	. . . .	Qr.-Mr.

Agents :—MESSRS. COX & CO.

Regimental District No. 35.
[No. 10 District.]

Uniform—Scarlet.
Facings—Blue.

The Royal Sussex Regiment.

2-6th BATTALION.

Arrived in India, 26th February 1916.

HONORARY COLONEL—Maj.-Genl. C. J. Young, C.B., *p.s.c.*

First Commn.	Names.	Regimental rank.	Army rank.	Remarks.
	Lieutenant-Colonel.			
5 Aug. 14	Johnson, F. W. F., D.S.O.	1 June 16		
	Major.			
	Captains.			
26 July 02	Cockburn, J. G. H. (*Hon. Lt. in Army*)	5 Aug. 14	. . . . .	M. A. Dept., Poona.
1 Mar. 14	*Bennett, T. G.*	29 June 15	. . . .	*M. G. Corps.*
5 Aug. 14	Roberson, A. R. G. (51)	1 June 16		
5 Aug. 14	Wheler, *Sir* T. W., *Bt.* (46)	1 June 16	. . . . .	Mech. Transpt. Schl., R' Pindi.
13 Oct. 14	*Smithers, M. H. W.* (51, 52)	1 June 16	. . . . .	*M. G. Corps.*
30 Nov. 14	*Bonnerjee, K. K. E.*	1 June 16	. . . . .	*R. A. F.*
6 Dec 14	Holmes, H. R. J.	1 June 16		
12 Dec. 14	Johnson, F. B.	1 June 16		
	Lieutenants.			
22 Dec. 14	Mundy, H. A. G. (*temp. Capt.*)	1 June 16	. . . . .	Bde. Maj., Poona.
5 Jan. 15	*Tennent, A. K.* (*temp. Capt.*) (51, 60*a*)	1 June 16	. . . .	Base Comdt.
10 Feb. 15	*Mileham, C. A.*	1 June 16	. . . .	*I. A. R. O.*
17 Feb. 15	Vaughan, L. (*act'g. Maj.*) (51)	1 June 16		
19 Mar. 15	Boulton, R. A. (46)	1 June 16	. . . .	*Lv., ex I., 6 m.*
6 May 15	*Wright, E. R.*	1 June 16	. . . .	*R. A. F.*
24 June 15	Johnson, C. F. (*act'g. Capt.*) (51)	1 June 16	. . . .	Adjt.
6 Sep. 15	Brown, G. L. (12, 50)	1 June 16		
9 Sep. 15	Morton, R. C. (12)	1 July 17	. . . .	R. T. O.
9 Sep. 15	Lowe, P. W. C. S.	1 July 17		
11 Sep. 15	Woodhams, D. (51)	1 July 17		
28 Oct. 15	Burne, W. G. (66*a*)	1 July 17	. . . .	Ind. Mech. Transport, Nushki.
3 Dec. 15	Benson, A. J. (12, 51, 60*a*)	1 July 17		
6 Dec. 15	*Campbell, C.*	1 July 17	. . . .	*I. A. R. O.*
20 Dec. 15	Bannatyne, N. E. (*Act'g. Capt., 1 Aug. 18*) (66*a*)	1 July 17		
16 Mar. 18	Beckett, C. G.	16 Mar. 18		
16 Mar. 18	Berendt, H. (46)	16 Mar. 18	. . . .	M. G. Centre, Mhow.
16 Mar. 18	Lander, E. T. W.	16 Mar. 18	. . . .	R. A. F.
1 July 15	Symonds, F. H. (46)	19 July 17		
5 Sep. 16	Beringer, B.	5 Mar. 18		
	2nd-Lieutenants.			
26 Sep. 15	Norman, C. E. (46, 52, 60*a*)	26 Sep. 16		
25 Oct. 16	Bell, F. S. W. (51, 57)	25 Oct. 16		
29 Aug. 17	King, H. L. (51)	29 Aug. 17		
7 Jan. 16	Campbell, A. J. (*temp. Capt., 15 Sep. 18*)	7 July 17	. . . .	R. T. O., Lahore.
	Attached.			
Feb. 12	Hingston, Capt. C. V.	1 June 16		
5 Sep. 16	Jenkin, Lieut. A. J.	5 Mar. 18	. . . .	Mily. Accts. Dept., Poona.
1 Mar. 17	White, 2nd-Lt. L. G.	1 Mar. 17	. . . .	R. A. F.
29 May 17	Betts, 2nd-Lt. D. M.	29 May 17		
30 May 17	Curry, 2nd-Lt. J. H. (46)	30 May 17		
30 May 17	Venn, 2nd-Lt. W. G. W. (51)	30 May 17		
30 May 17	Credland, 2nd-Lt. H. W. (46)	30 May 17		
30 May 17	Cheshire, 2nd-Lt. T. A.	30 May 17	. . . . .	R. A. F.
26 Sep. 17	Millward, 2nd-Lt. L. J. W. (12)	26 Sep. 17		
26 Sep. 17	Billing, 2nd-Lt. W. E. H.	26 Sep. 17	. . . .	Attd. Army. Hdqrs.
26 Sep. 17	Boyle, 2nd-Lt. W. H. (51)	26 Sep. 17		
26 Sep. 17	Page, 2nd-Lt. H. F.	26 Sep. 17		
29 Aug. 17	Amphlett, 2nd-Lt. F. G.	29 Aug. 17		
30 Oct. 17	Poore, 2nd-Lt. L. L.	30 Oct. 17		
30 Oct. 17	Nesbitt, 2nd-Lt. J. D. W. (51)	30 Oct. 17		
24 June 15	Johnson, Lt. (*act'g. Capt.*) C. F.	1 June 16	. . . . .	Adjt.
13 Nov. 14	Gray, Capt. J. K.		. . . .	Qr.-Mr. S. S. O., Lahore.

Regimental District No. 37. (No. 8 District.)

The Hampshire Regiment.

Uniform—Scarlet. *Facings*—Yellow.

1-4TH BATTALION.

Arrived in India, 9th November 1914, from England.

" South Africa, 1900-1902."

HONORARY COLONEL—*Sir* T. S. Cave, K.C.B., C.B.E., V.D. 11 Apl. 09.

First Commn.	Names.	Regimental rank.	Army rank.	Remarks.
	Lieutenant-Colonel.			
	Majors			
4 Apl. 00	Footner, F. L. (57)	31 Jan. 14		
21 Feb. 03	Foster, H. M.	5 Sep. 16	....	D. A. A. G., 4th (Quetta) Divn.
	Captains.			
27 May 10	Reeks, N.	1 June 16		
17 July 13	Stilwell, J. G.	1 June 16		
8 Dec. 13	*Lester-Garland, A. G.*	1 June 16	....	I. A. (*on probn.*)
26 Feb. 14	Harris, J. H.	1 June 16		
11 Aug. 14	Forbes, A. G., M.C.	1 June 16		
6 Sep. 14	White, J. M.	1 June 16		
6 Sep. 14	Burrell, E. A.	1 June 16	....	With 1st Garr. Bn., E. York. R.
6 Sep. 14	Osborne, D. G. H. (52, 57)	8 Oct. 17		
19 Sep. 14	Elton, G. (66)	23 Jan. 18		
	Lieutenants.			
1 Oct. 14	Hamilton, A. F.	1 June 16		
24 Sep. 14	Chitty, C.	1 June 16		
1 Oct. 14	Andrews, G. D., M.C. (*Actg. Capt., 18 Sep. 17*) (57)	1 June 16	....	Adjt.
1 Oct. 14	*Butler, F. H. C.* (*temp. Capt.*)	1 June 16		
13 Oct. 14	Hamilton, J. A. de C., M.C. (52, 60*a*)	1 June 16		
26 Oct. 14	*Wyles, C. D.*	1 June 16		
16 Apl. 15	*Capes, G. A.* (57)	1 June 16	....	I. A. (*on probn.*)
27 Aug. 15	Hamley, A. E. J.	1 July 17	....	*Lv., ex I., m. c., 6 mos.*
27 Aug. 15	Hamley, D. W.	1 July 17		
28 Aug. 15	Fisher, L. E. R., M.C. (*2nd-Lt., 3 Hrs.*)	1 July 17		
7 Oct. 15	Wetherall, K. L.	1 July 17		
23 Oct. 15	Vawdrey, R. H.	1 July 17		
18 Nov. 15	Ellis, E. I. A. (48*a*, 51, 66*a*)	25 Nov. 17		
	2nd-Lieutenants.			
16 Mar. 18	Long, F. A.	16 Mar. 18		
19 Aug. 18	Parker, F. J. H.	19 Aug. 18		
19 Aug. 18	Hallaway, F. W.	19 Aug. 18		
19 Aug. 18	Knight, G. D.	19 Aug. 18		
19 Aug. 18	Minchin, V. R.	19 Aug. 18		
	Attached.			
	Matthews, Bt.-Lt.-Col. C. J. (*Act'g. Lt.-Col.*), Durh. L. I.	10 Mar. 17	....	Comdg.
4 Jan. 00	Hamilton, Maj. W. H.	23 Sep. 14	....	2-4th D. of Corn. L. I.
	Bowers, Capt. H. M., Hamps. R.	13 Dec. 12		
20 Apl. 09	Burnett, Capt. F. H., M.C., 2-5th Hamps. R.	2 Sep. 14		
...	Fardell, Capt. J. G., 8th Hamps. R.	27 Feb. 15		
9 Nov. 12	Walkinshaw, Capt. C. C., 1-7th Hamps. R.	1 June 16		
2 Sep. 12	Jensen, Capt. C. L. R., 6th Hamps. R	1 June 16		
	Parr, Capt. C. E. (*temp.*)	12 May 17		
	Mackenzie, Capt. D., 6 Bn., R. Highrs.	1 June 17		
22 Sep. 14	Bevis, Lt. E. C., 6th Hamps. R.	1 June 16		
10 Nov. 14	*Freeman, Lt. A. F. A.* (52), *6th Hamps. R.*	1 June 16		
3 Oct. 14	Mead, Lt. R. (*temp. Capt.*), 2-5th Hamps. R.	1 June 16	....	Staff Capt.
8 Sep. 15	Blackwell, Lt. P. W., (95) 4th North'n R.	1 July 17	....	*Lv., ex I., m. c., 6 mos.*

The Hampshire Regiment—*contd.*

First Commn.	Names.	Regimental rank.	Army rank.	Remarks.
	Attached—*contd.*			
23 Sep. 15	Hutchens, Lt. H. J., 7th Hamps. R.	1 July 17		
	Stokes, Lieut. H. F. S., M.C., 4 Bn. Hamps. R.	1 July 17		
	Wallace, Lieut. S. (*temp.*) 1 Garr. Bn., York. R.	1 July 17		
	Gow, Lieut. A. L., 7 Bn., Hamps. R.	1 July 17		
16 Jan. 16	Shaddock, Lt. J. C., 6th Hamps. R.	16 July 17		
	Hardie, Lieut. W. J., 9 Bn., Hamps. R.	11 Jan. 18		
	Greer, Lieut. A., 9 Bn., Hamps. R. .	11 Jan. 18		
	Stockdale, Lieut. S. R., 9 Bn., Hamps. R.	15 Jan. 18		
26 Sep. 17	Barrow, 2nd-Lt. H. G., 5 Bn., Manch. R.	26 Sep. 17		
26 Sep. 17	Waterfield, 2nd-Lt. C. H. (*temp.*), 3 Bn., Notts. & Derby. R	26 Sep. 17		
	Carew-Hunt, 2nd-Lt. A. L.	. . . . .		
1 Oct. 14	Andrews, Lieut. (*act'g. Capt.*) G. D., M.C.	18 Sep. 17	. . . .	Adjutant.
30 Jan. 14	Watson, Capt. A., D.S.O., M.B., R. A.M.C.	30 Mar. 15	9 July 16	Medl. Offr.

Regimental District No. 37.
[No. 8 District.]

Uniform—Scarlet.
Facings—Yellow.

The Hampshire Regiment.

1-5TH BATTALION.

Arrived in India, 9th November 1914, from England.

" South Africa, 1900-1902."

HONORARY COLONEL—*Hon. Sir* H. G. L. Crichton, K.C.B., T.D., A.-D.-C., ret. pay . . 8 Feb. 05.

First Commn.	Names.	Regimental rank.	Army rank.	Remarks.
	Lieutenant-Colonel.			
25 Nov. 85	Burford-Hancock, H. di S. B. (*Res. of Off.*) (51, 57, 67, 207).	13 Mar. 12		
	Majors			
19 Nov. 04	Gutteridge, R. F.	1 June 16		
20 Feb. 09	Dominy, J. C. (51)	1 June 16		
30 May 11	Hill, H. P. (51)	1 June 16		
20 Feb. 09	Hoare, C. E. (51)	1 June 16	. . . .	*Lv., ex I., m. c., 3 mos.*
	Captains.			
15 Mar. 09	Thomson, R. St. C.	12 June 14	. . . .	S. S. O., Bareilly.
25 May 09	Andrews, E. W. (48*a*, 61)	13 June 14		
25 Apl. 11	Fear, F. R. (51)	1 June 16	. . . .	Staff Capt., Rangoon Bde.
11 Oct. 12	Firth, D. G. (48*a*, 51)	1 June 16		
5 Dec. 12	*Pearce, W. K.*, (47*a*, 48*a*, 51, 57)	1 June 16	. . . .	*Asst. Instr., Musk. Schl., R'Pindi.*
2 Sep. 14	Wharton, A. F. (51)	1 June 16	. . . .	With 1-4th Hamps. R.
1 May 14	Kitchingman, H. (48*a*, 51, 60*a*, 66)	1 June 16		
16 Aug. 14	*Moseley, B.* (51, 52, 207*a*)	1 June 16	. . . .	*I. A.* (*on probn.*)
20 Aug. 14	*Hansard, J. H.* (57)	1 June 16	. . . .	*With. Sig. Serv.*
20 Aug. 14	*Baring, D. W.* (52)	20 July 17	. . . .	*M. G. Corps.*
2 Sep. 14	Harris, H. J. (51, 60*a*)	20 July 17	. . . .	Adjt.
2 Sep. 14	Drummond, R. H.	11 Nov. 17	. . . .	With 1-6th Hamps. R.
	Lieutenants.			
2 Sep. 14	Lucas, C. A. (51, 52)	1 June 16		
2 Sep. 14	Farquharson, S. (48*a*, 51)	1 June 16		
18 Sep. 14	MacLeod, A. (60*a*)	1 June 16		
3 Oct 14	Aldridge, N. C. (59)	1 June 16		
3 Oct. 14	Napier, M. (*temp. Capt.*) (46, 48*a*, 49, 51)	1 June 16	. . . .	Platoon Comdr., Officer Cadet [Unit.
3 Oct 14	Whitlock, D. K.	1 June 16	. . . .	With 1-4th Hamps. R.
11 Dec. 14	Palethorpe, C. N. deW. (48*a*, 50, 60*a*)	1 July 17		
13 Feb. 15	*Warren, H. F.*	1 July 17	. . . .	*Poll Employ.*
18 May 15	Stavert, G. H.	1 July 17	. . . .	With P. W. Dept., Maymyo.
28 May 15	Christmas, H. R. (46)	1 July 17		
22 June 15	McGovern, J. B. (51)	1 July 17	. . . .	Asst. Instr. Offrs.' Schl., Kirkee.
12 July 15	Malerbi, J. W. B.	1 July 17	. . . .	With 1-4th Hamps. R.
	2nd-Lieutenants.			
28 Apl. 17	Read, P. E. (46, 49)	28 Apl. 17		
29 Aug. 17	Hayward, C. W.	29 Aug. 17		
16 Mar. 18	Tatton-Winter, C. W.	16 Mar. 18		
19 Aug. 18	Fishburn, W. E.	19 Aug. 18		
19 Aug. 18	Norris, C. P.	19 Aug. 18		
	Attached.			
22 Jan. 16	Glover, Lt. H. R. ((*temp.*) 9 Bn., Norf. R. (46).	22 July 17		
28 Mar. 17	Halford, 2nd-Lt. H. W., 3 Bn., Bedf. R.	28 Mar. 17		
28 Mar. 17	Foster, 2nd-Lt. F. M., 5 Bn., Bedf. R.	28 Mar. 17		
30 May 17	Craven, 2nd-Lt. W. A. N. (*temp.*), Hamps. R.	30 May 17		
26 Sep. 17	Shrewsbury, 2nd-Lt. O. L., 6 Bn., Manch. R.	26 Sep. 17		
26 Sep. 17	Beard, 2nd-Lt. J. (*temp.*), Notts. & Derby. R.	26 Sep. 17		
26 Sep. 17	Stiles, 2nd-Lt. S., 6 Bn., Notts. & Derby. R.	26 Sep. 17		
26 Sep. 17	Hollingsworth, 2nd-Lt. H. J., (*temp.*) York. & Lanc. R.	26 Sep. 17		
26 Sep. 17	Chatterton, 2nd-Lt. C. E. (*temp.*) York. & Lanc. R.	26 Sep. 17		
31 Oct. 17	Petty, 2nd-Lt. J., 3 Bn., Hamps. R.	31 Oct. 17		
28 Nov. 17	Meiklejohn, 2nd-Lt. W. C. (*temp.*), Hamps. R.	28 Nov. 17		
28 Nov. 17	Menear, 2nd-Lt. S. J. (*temp.*), Hamps. R.	28 Nov. 17		
28 Nov. 17	Whittington, 2nd-Lt. E. F., 3 Bn., Hamps. R.	28 Nov. 17		
28 Nov. 17	Gallaway, 2nd-Lt. R. E. (*temp.*), Hamps. R.	28 Nov. 17		
2 Sep. 14	Harris, Capt. H. J.	11 Feb. 16	. . . .	Adjt.
26 Sep. 14	Baldwin, Lt. S. (51, 52)	26 Sep. 14	. . . .	Qr.-Mr.

Regimental District No. 37.
[No. 8 District.]

Uniform—Scarlet.
Facings—Yellow.

The Hampshire Regiment.

1-6TH (DUKE OF CONNAUGHT'S OWN) BATTALION.

Arrived in India, 10th November 1914, from England.

"South Africa, 1900—02."

HONORARY COLONEL—Field Marshal *H. R. H.* ARTHUR WILLIAM PATRICK ALBERT, *Duke of* CONNAUGHT AND STRATHEARN, K.G., K.T., K.P., G.C.B., G.C.S.I., G.C.M.G., G.C.I.E., G.C.V.O., G.B.E., V.D., A.-D.-C. . . 25 Mar. 92.

First Commn.	Names.	Regimental rank.	Army rank.	Remarks.
	Lieutenant-Colonel.			
13 May 96	Wyatt, A., T. D. (51)	6 Apl. 18		
	Major			
13 Dec. 05	Radwell, J. (51)	1 June 16		
	Captains.			
8 Dec. 08	Miller, M. R. (*act'g. Maj., 29 Sep 18*) (51)	21 Apl. 12		
22 Mar. 09	Webb, H. J. (51)	19 Apl. 14		
1 June 09	Curtis, G. R. (51)	1 June 14		
3 Sep. 09	*Addison, A.W.N.* (*temp. Maj.*) (51)	24 June 14		*Asst. Mily. Secy., Army Hdqrs.*
1 May 12	Blagg, F. O. (51)	13 Feb. 15		
	Lieutenants.			
8 Feb. 13	Bagg, R. P. J. (*act'g. Capt., 2 Mar. 18*).	1 Nov. 14		
4 July 14	*Wadeson, H. H.*	13 Feb. 15		*I. A. R. O.*
2 Sep. 14	Gillett, C. E. S. (*act'g. Capt.*) (52)	13 Feb. 15		
2 Sep. 14	Roper, C. R. (*act'g. Capt., 27 Jan. 18*) (*2nd-Lt., E. York. R.*)	13 Feb. 15		
2 Sep. 14	Breton, D. C. (51, 52)	13 Feb. 15		M. G. Corps.
2 Sep. 14	*Willison, W. L. P.*	1 June 16		*I. A. R. O.*
2 Sep. 14	Savill, E. M. (*temp. Capt. 15 Jan. 18*)	1 June 16		
6 Sep. 14	*Houston, J. F.*	1 June 16		*Dental Surgn., Meerut Divn.*
6 Sep. 14	*Palmer*, Hon. *W. J. L.* (*temp. Capt.*)	1 June 16		*Staff Capt.*
22 Sep. 14	Stallard, R. T.	1 June 16		
22 Sep. 14	*Read, J. E.*	1 June 16		*I. A.* (*on probn.*)
23 Sep. 14	Brown, N. A. S. (48*a*, 50, 60*a*, 61)	1 June 16		
24 Sep. 14	*Blake, R. A. E.* (57)	1 June 16		*With Sig. Serv.*
22 Mar. 15	Pulling, F. E. S. (60*a*)	1 June 16		
5 May 15	*Elliott, E. C.*	1 June 16		
22 May 15	*Craster, R. F.* (46)	1 July 17		I. A. (*on probn.*)
22 June 15	*Blackmore, L. W. S.* (60*a*)	1 July 17		*M. G. Corps.*
22 July 15	Beere, H. C. (46, 48*a*, 50, 60*a*, 61)	1 July 17		
20 Aug. 15	Gibson, F. A. S.	1 July 17		Stokes Mortar Batty.
23 Sep. 15	Thomas, *Sir* G. A., *Bart.*	1 July 17		
2 Sep. 16	Smith, J. H. (46)	5 Mar. 18		
	2nd-Lieutenants.			
20 May 15	Forbes Leith, H. L.	20 May 15		
16 Mar. 18	Mayers, N.	16 Mar. 18		
16 Mar. 18	Stone, E. W. R.	16 Mar. 18		
19 Aug. 18	Luton, F. G.	19 Aug. 18		
19 Aug. 18	Symes, R. W.	19 Aug. 18		
19 Aug. 18	Johnson, A. D.	19 Aug. 18		
19 Aug. 18	Randles, G. W.	19 Aug. 18		
	Attached.			
Feb. 12	Hingston, Capt. C. V., 5th Bn., Notts. & Derby. R.	1 June 16		
2 Sep. 14	Drummond, Capt. R. H., 1-5th Hamps. R.	11 Nov. 17		

The Hampshire Regiment—*continued.*

First Commn.	Names.	Regimental rank.	Army rank.	Remarks.
	Attached—*contd.*			
27 Nov. 14	Strologo, Lt. R. C. (*temp.*), 2nd Gn. Bn., North'd. Fus.	2 Dec. 15	. . . .	Stokes Mortar Batty.
22 Nov. 14	Blore, Lt. C. G., 1-9th Hamps. R. .	1 June 16		
15 Jan. 16	Green, Lt. A. L., 6th E. Surr. R. .	15 July 17		
21 Jan. 16	McIntyre, Lt. J. (*temp.*), R. Sc. Fus.	21 July 17		
5 May 15	May, Lt. H. R. (*temp.*), 1st G. B., Essex R.	1 Oct. 17		
19 July 16	Saunders, Lt. H. W. B., Gord. Highrs.	19 Jan. 18		
5 Aug. 16	Macdonald, Lt. J., 1st Lovats Scouts.	5 Feb. 18		
5 Aug. 16	Packer, Lt. E. V. (*temp.*), Essex R.	5 Feb. 18	. . . .	Stokes Mortar Batty.
5 Aug. 16	Low, Lt. F., 4th Bn., Gord. Highrs. .	5 Feb. 18		
15 Dec. 16	Joll, 2nd-Lt. A. E. (*temp.*), 2nd G. B., Bedf. R.	15 Dec. 16		
1 Mar. 17	Cowan, 2nd-Lt. N. M., 8th Bn., Arg. & Suth'd. Highrs.	1 Mar. 17		
28 Mar. 17	Mannering, 2nd-Lt. G. (*temp.*) Hamps. R.	28 Mar. 17		
28 Mar. 17	Erskine, 2nd-Lt. D. D. A., 5th Bn., Arg. & Suth'd. Highrs	28 Mar. 17		
30 May 17	Kendall, 2nd-Lt. S. (*temp.*), Hamps. R.	30 May 17		
30 May 17	Lakeman, 2nd-Lt. W. H. (*temp.*) Hamps. R.	30 May 17		
30 May 17	Ross, 2nd-Lt. R. D., 8 Bn., Sco. Rif. .	30 May 17		
29 Aug. 17	Gardner, 2nd-Lt. T. A. A., 8 Bn., Dorset R.	29 Aug. 17		
26 Sep. 17	Campbell, 2nd-Lt. C., 8 Bn., Manch. R.	26 Sep. 17		
26 Sep. 17	Jones, 2nd-Lt. D. V. (*temp.*) Notts. and Derby. R.	26 Sep. 17		
...	Wallace, 2nd-Lt. C.R.D., 11 Bn., York. L. I.	...		
26 Sep. 14	Pragnell, Capt. J. (51, 52) . . .	26 Sep. 17	. . . .	Qr.-Mr.

Regimental District No. 37.
[No. 8 District.]

Uniform—Scarlet.
Facings—Yellow.

The Hampshire Regiment.

1-7TH BATTALION.

Arrived in India, 8th November 1914, from England.

A STIRRUP. IN THE FIRST AND FOURTH CORNERS A WHITE ROSE, AND IN THE SECOND AND THIRD A RED ROSE.

"South Africa, 1900-1902."

HONORARY COLONEL—

First Commn.	Names.	Regimental rank.	Army rank.	Remarks.
	Lieutenant-Colonel.			
6 Apl. 96	Thomson, W. R., T. D. (*Hon. Lt. in Army*).	1 June 16		
	Majors.			
2 Dec. 05	Keene, H.	1 June 16	. . . .	
28 Mar. 06	Rebbeck, T. V.	1 June 16		
13 Feb. 08	Bacon, R. J. (51)	1 June 16	. . . .	S. S. O., Naini Tal.
	Captains.			
4 May 08	Davenport, G. C. (51, 60*a*)	10 Mar. 16	. . . .	Adjt.
7 June 09	Parke, L. S.	1 June 16	. . . .	S. S. O., Meerut.
12 Mar. 10	Jenkins, R. E. (51)	1 June 16		
8 Mar. 11	Druitt, G. T. (51, 57)	1 June 16		
18 Jan. 12	Stringer, F. W. D.C.M. (51)	1 June 16		
1 May 12	Elliott, C.E. (51)	1 June 16	. . . .	With 1st Bn., Oxf. & Bucks. L. I.
28 Oct. 12	Last, F. B. (51)	1 June 16	. . . .	
9 Nov. 12	Walkinshaw, C. C. (51, 52)	1 June 16	. . . .	With 1-4th Hamps. R.
3 Jan. 13	Warner, B. M. (51, 52, 60*a*)	1 June 16		
8 Jan. 13	Patmore, F. J. M. E.	1 June 16	. . . .	With 1-4th Hamps. R.
4 Mar. 13	Malim, A. W. (51)	1 June 16		
21 Jan. 14	*Lisby, V. C.* (51, 66*a*)	1 June 16	. . . .	*I. A (on probn.)*
19 Aug. 14	Streatfeild, H. G. C. (51, 60*a*, 206).	1 Feb. 17		
29 Aug. 14	*Coles, G. R.* (47*a*, 51, 60*a*)	19 Nov. 17	. . . .	*I. A. (on probn.)*
	Lieutenants.			
29 Aug. 14	*Kemp-Welch, H. A. (46, 51, 52)*	*1 June 16*	. . . .	*M. G. Corps.*
18 Sep. 14	Davis, D. (51, 60*a*)	1 June 16		
19 Sep. 14	*Peck, B. M.* (59)	1 June 16	. . . .	*With Sig. Serv.*
19 Sep. 14	*Jenkins, G. D.* (57)	1 June 16	. . . .	*With R. A. F.*
9 Oct. 14	Whittington-Ince, C.H. (46, 51, 60*a*)	1 June 16		
5 June 15	Black, J. G. McF. (51)	1 July 17		
11 June 15	*King, J. H. L.* (46, 51)	1 July 17	. . . .	*I. A. (on probn.)*
16 June 15	*Morse, W. S.*	1 July 17	. . . .	*With Sig. Serv.*
24 Aug. 15	Healls, C. F. (57, 60*a*)	1 July 17	. . . .	With Sig. Serv.
31 Dec. 15	Mortleman, C. C. (51)	1 July 17	. . . .	M. G. Corps.
	2nd-Lieutenants.			
28 Apl. 17	Palmer, H. R.	28 Apl. 17		
28 Apl. 17	Bateman, T. E. (51)	28 Apl. 17		
28 Apl. 17	Bush, W. P.	28 Apl. 17		
28 Apl. 17	Leeke-Roe, D. C.	28 Apl. 17	. . . .	With R. A. F.
29 Aug. 17	Carter, H. P.	29 Aug. 17	. . . .	With R. A. F.
28 Nov. 17	Keene, H.	28 Nov. 17	. . . .	Depot.
16 Mar. 18	Worth, J. G.	16 Mar. 18		
16 Mar. 18	Brown, I. T. J.	16 Mar. 18		
16 Mar. 18	Head, E. B.	16 Mar. 18		
16 Mar. 18	*Long, C. A.*	16 Mar. 18	. . . .	*I. A. R. O.*
	Attached.			
20 July 15	Morley, Lt. A. D., 2nd Bn., Hamps. R.	1 Jan. 17		
20 Oct. 15	Watts, Lt. J. W., 2nd Bn., Hamps. R.	1 Jan. 17	. . . .	Depot.
11 July 16	Reynolds, Lt. C. W. (57), 8th Bn., Hamps. R.	11 Jan. 18		
11 July 16	Wilcox, Lt. M. W. (52), 8th Bn., Hamps. R.	11 Jan. 18	. . . .	M. G. Corps.
11 July 16	Carter, Lt. A. B. (51), 8th Bn., Hamps. R.	11 Jan. 18		
15 July 16	Foyster, Lt. E. R. (51), 8th Bn., Hamps. R.	15 Jan. 18		
18 July 16	Cox, Lt. A. J., 8th Bn., Hamps R.	18 Jan. 18	. . . .	M. A. Dept., Poona.
25 Jan. 17	Marshall, 2nd-Lt. A. F., 5th Bn., Suff. R	25 Jan. 17		
4 May 08	Davenport, Capt. G. C., (51)	10 Mar. 16	. . . .	Adjt.
16 July 17	Williams, *Lt.* J., D.C.M. (51, 52, 60*a*, 72)	16 July 17	. . . .	Qr.-Mr.

Regimental District No. 37. [No. 8 District.]

Uniform—Scarlet. *Facings*—Yellow.

The Hampshire Regiment.

2-7TH BATTALION.

Arrived in India, 9th January 1915 from England.

A STIRRUP. IN THE FIRST AND FOURTH CORNERS A WHITE ROSE, AND IN THE SECOND AND THIRD A RED ROSE.

"South Africa, 1900-1902."

HONORARY COLONEL.—

First Commn.	Names.	Regimental rank.	Army rank.	Remarks.
	Lieutenant-Colonel.			
12 Aug. 02	Gribbon, H. H., D.S.O., Capt., Hamps. R.	*Act'g.* 12 Nov. 17	. . . . .	Comdg.
	Major.			
13 Feb. 08	Palmer, C. W. G. (71)	8 Oct. 17		
	Captains.			
1 Oct. 14	*Stanley, R. C. H. S.*	1 Oct. 14		
9 Dec. 14	Boucher, B. H., D.S.O. (ret. pay) (*temp. Lt.-Col., 1 Feb. 17*) (50, 52, 57, 71, 172*b*, 207).	9 Dec. 14	. . . .	Comdg. No. 6 Res. Bn. (India).
20 May 14	Meikle, R. (51, 162, 172*a*, 205*a*) . .	1 Feb. 17		
29 Aug. 14	Muspratt, C. K. (51, 71, 172*a*, 207)	4 Oct. 17		
	Lieutenants.			
1 Oct. 14	*Osborne, F. W.* (59)	1 June 16	. . . .	*I. A. R. O.*
1 Oct. 14	Duplessis, G. L. J. (*act'g. Capt.*) (51, 52)	1 June 16		
1 Oct. 14	deGeijer, W. A. W. G (51) . . .	1 June 16		
1 Oct. 14	*D'Angibau, W. E.* (12, 52, 60*a*). .	1 June 16	. . . .	*With M. G. Corps.*
1 Oct. 14	Mellor, J. E. P. (*act'g. Capt.*) (51) .	1 June 16		
12 Oct. 14	Pearce, R. E. S. (*temp. Capt.*) (51, 52, 71)	17 Nov. 16	. . . .	G. S. O., 3rd gr.
19 Oct. 14	Hake, H. D. (51, 52)	1 July 17		
26 Oct. 14	Parkes, R. G. (*act'g. Capt.*) (51, 71)	1 July 17	. . . .	Adjt.
3 Nov. 14	deSelincourt, L. (48*a*, 50, 57, 71) .	1 July 17		
21 Nov. 14	*Kite, P.* (51, 52)	1 July 17	. . . .	*M. A. Dept.*
23 Nov. 14	*Tanner, H.* (65*a*, 66)	1 July 17	. . . .	*I. A. R. O.*
12 Dec. 14	*Jessel, R. H.*	1 July 17		
29 May 15	*Hallum, S. S.*	1 July 17		
19 June 15	*Mead, B.* (59)	1 July 17	. . . .	*With Sig. Serv.*
22 July 15	*Long, P.* (57, 60*a*)	1 July 17	. . . .	*I. A.* (*on probn.*)
27 Aug. 15	Carpenter, E. H. O. (46, 51) . . .	1 July 17		
23 Sep. 15	Hutchens, H. J.	1 July 17	. . . .	Adjt. & Qr.-Mr., Cooperage Camp, Bombay.
21 Oct. 15	Dickinson, C. E. (*act'g. Capt., 15 Nov. 18*) (61).	1 July 17	. . . .	M. W. S., Calcutta.
23 Nov. 15	Taylor, H. A.	1 July 17	. . . .	B. A. Depôt.
30 Nov. 15	Talbot, C. C.	1 July 17		
30 Nov. 15	Raywood, W. (52)	1 July 17		
29 Dec. 15	Fitz, E. W. (51, 71)	1 July 17		
	2nd-Lieutenants.			
28 Nov. 14	*Pridgeon, R. C.* (51 52)	28 Nov. 14	. . . .	*With M. G. Corps.*
11 July 16	Hitchcock, L. (57)	11 July 16	. . . .	With Sig. Serv.
28 Apl. 17	Prince, A.	28 Apl. 17		
28 Apl. 17	Oakley, J.	28 Apl. 17		
28 Apl. 17	Cooper, T.	28 Apl. 17		
28 Aug. 17	*Dolman, H. G.*	28 Aug. 17	. . . .	*I. A.* (*on probn.*)
29 Aug. 17	*Preiss, R. B.* (50)	29 Aug. 17		
28 Nov. 17	Savage, R. J.	28 Nov. 17		
16 Mar. 18	Courtenay, C. E.	16 Mar. 18		
16 Mar. 18	Stephenson, P.	16 Mar. 18		
16 Mar. 18	Manser, A. R.	16 Mar. 18	. . . .	With No. 6 Res. Bn. (Ind.)
16 Mar. 18	Courtenay, H. G.	16 Mar. 18		
19 Aug. 18	Brown, C. E.	19 Aug. 18	. . . .	With No. 6 Res. Bn. (Ind.)
	Attached.			
30 Nov. 13	Campbell, Capt. I. P. F., 1-9th Hamps. R.	1 June 16		
5 Oct. 14	Smith, Lieut. A. L. F., M.V.O., 9 Bn., Hamps. Ry. (46, 50).	1 June 16		
7 Nov. 14	Cunnington, Lieut. J., 2nd Som. L. I.	30 Sep. 15		
15 Oct. 14	Dunning, Lt. G. K., 1-9th Hamps. R.	1 June 16		
17 Nov. 14	Norman, Lieut. A. L., 1st North'n. R.	14 May 15		
6 Dec. 15	Lewellin, Lt. D. J., High. L. I. (Attd.)	12 Jan. 17	. . . .	I. A. (*on probn.*)
29 Dec. 14	Lumsdaine, Lt. A. A. M., 1st Garr. Bn., Linc. R.	1 July 17		
28 Mar. 17	Hastings, 2nd-Lt. C., 3rd Bn., Hamps. R.	28 Mar. 17		
28 Mar. 17	Heaps, 2nd-Lt. W., 3rd Bn., Hamps. R.	28 Mar. 17		
26 Oct. 14	Parkes, Lt. (*act'g. Capt.*) R. G. .	26 Oct. 14	. . . .	Adjt.
8 Sep. 14	Marshall, Maj. R. (*ret. pay*) .	28 Sep. 14	. . . .	Qr.-Mr. *Lv, ex I., m.c.*

Regimental District No. 37.
[No. 8 District.]

Uniform—Scarlet.
Facings—Yellow.

The Hampshire Regiment.

1-9TH BATTALION.

Arrived in India, 25th February 1916.

HONORARY COLONEL—

First Commn.	Names.	Regimental rank.	Army rank.	Remarks.
	Lieutenant-Colonel.			
2 Apl. 92	Johnson, R. A., T.D. (*Hon. Lt. in Army, 4 Apl. 01*) (50, 53, 71, 72*b*)	14 Nov. 11		
	Majors.			
3 Aug. 03	Rose, D'A. L. (71, 206)	1 June 16		
1 May 08	Robson, A. (12)	27 July 17		
	Captains.			
18 Nov. 00	Low, S. P.	15 Sep. 14		
1 Feb. 13	Montefiore, L. N. [*l*] (48*a*, 51, 57)	26 May 15		
20 Feb. 08	Reader, F. E. M. (51)	1 June 16	....	Bde. Maj., Southern Bde.
1 Dec. 13	Price, W. D. (52, 60*a*, 61)	1 June 16		
30 Nov. 13	Campbell, I. P. F.	1 June 16	....	With 2-7th Hamps. R.
31 Mar. 14	*Boxer, F. P.* (66)	1 June 16	....	*I. A. R. O.*
2 Sep. 14	*Richards, A. R. T.* (53, 72*b*)	1 June 16	....	*M. G. Corps.*
31 Mar. 09	Barnes, K. R. (51)	1 June 16		
5 Sep. 14	Parker, *Sir* W. L., *Bt.* (48*a*, 50, 52)	1 June 16		
15 Sept. 14	*Dennison, G. D.*	1 June 16	....	*Staff Capt.*
31 Oct. 14	Howgrave-Graham, H. M. (50)	1 June 16		
	Lieutenants.			
5 Oct. 14	*Smith, A. L. F.*, M.V.O. (46, 50)	1 June 16	....	*With Light Trench Mortar By.*
4 Oct. 14	Buchanan-Riddell, W. R. (50)	1 June 16		
15 Oct. 14	Dunning, G. K. (60*a*, 61)	1 June 16	....	With 2-7th Hamps. R.
22 Oct. 14	Marshall, G. (50, 66)	1 June 16		
31 Oct. 14	*Earing, E. B.* (*act'g. Capt., 1 Aug. 18*) (46).	1 June 16	....	*Adjutant.*
12 Nov. 14	*Harvey, C. W. L.* (46, 60*a*, 61)	1 June 16	....	*I. A.* (*on probn.*)
22 Nov. 14	Blore, C. G. (57)	1 June 16	....	With 1-6th Hamps. R.
19 Jan. 15	Perkins, N. K. (51, 52)	1 June 16		
27 Feb. 15	*Howe, E. G.* (*temp. Capt.*) (51)	1 July 17	....	*Platoon Comdr., Officers' Schl. of Instn., Nasik.*
27 Feb. 15	Tyer, A. M. (48*a*)	1 July 17		
16 Apl. 15	Harrison, T. D. (50)	1 July 17		
24 July 15	Boardman, H. W. (51, 52)	1 July 17		
1 Nov. 15	Robinson, W. A. (48*a*, 49, 51)	1 July 17		
18 Nov. 15	Lordan, W. L. (45)	1 July 17		
18 Nov. 15	Adams, G. P. (58*a*, 59)	1 July 17	....	With Sig. Serv.
14 Jan. 16	Gosling, F. F., (48*a*, 51, 57, 60*a*)	14 July 17	....	Armoured Motor By.
27 Oct. 16	Whitaker, T. G. (48*a*, 51)	25 Apl. 18		
	2nd-Lieutenants.			
29 Aug. 17	Harris, A. F.	29 Aug. 17	....	With R. A. F.
16 Mar. 18	Childs, H. T. (61)	16 Mar. 18	....	Attd. M. G. Corps.
16 Mar. 18	Downs, G E.	16 Mar. 18		
19 Aug. 18	Cranton, D.	19 Aug. 18		
19 Aug. 18	Weatherley, J. E.	19 Aug. 18		
	Attached.			
11 July 16	Parish, Lt. W. E., 8th Bn., Hamps. R.	....		
28 Mar. 17	Philips, 2nd-Lt. T. E., 3rd Bn., Essex R. (46)	28 Mar. 17		
28 Mar. 17	Tarling, 2nd-Lt. T. D., 3rd Bn., Essex R. (12)	28 Mar. 17		
28 Apl. 17	Smith, 2nd-Lt. C. G., 10th Bn., Midd'x R. (48*a*, 59)	28 Apl. 17		
28 Apl. 17	Attwood, 2nd-Lt. W. W., 3rd Bn., (48*a*, 50)	28 Apl. 17	Attd. G. S. Bch., Army Hdqrs.	
28 Apl. 17	Ford, 2nd-Lt. R. S., 3rd Bn.	28 Apl. 17		
26 Sep. 17	Swindells, 2nd-Lt. T. A., M.M., 7th Bn., Manch. R.	26 Sep. 17		
26 Sep. 17	McCormick, 2nd-Lt. J., 5th Bn., Manch. R.	26 Sep. 17		

31 Oct. 14	Baring, Lieut. (*act'g. Capt.*) E. B.	1 June 16	....	Adjt.
1 Apl. 13	Wyard, Capt. F. J. (65*a*, 66)	1 Apl. 13	....	Qr.-Mr.
1 Jan. 14	Van Somren, Capt. A. G. V., V.B., F.R.C.S. (*Edin.*) R.A.M.C. (T.F.). (*attd.*).	1 Apl. 15	1 Jan. 14	Medl. Offr.—Surgical Specialist, 16th (Indian) Dn.

Agents.—Messrs. Grindlay & Co.

Regimental District No. 38.
[No. 6 District .]

Uniform—Scarlet.
Facings—White.

The South Staffordshire Regiment.

1ST GARRISON BATTALION.

Arrived in India, 6th May 1917.

First Commn.	Names.	Regimental rank.	Army rank.	REMARKS.
	Lieutenant-Colonel.			
	Petrie, Maj. (*temp. Lt.-Col.*) C. L. R., D.S.O., *ret. pay* (*Res. of Off.*)	9 Jan. 17	. . . .	*Lv., ex I., m. c., 6 m.*
	Major (2nd-in-Command).			
	Johnstone, B. A., *ret. Ind. Army* (*act'g. Lt.-Col.*)	1 Feb. 17	7 Oct. 14	Comdg. No. 5 Res. Bn., Kirkee.
	Major.			
4 Aug. 14	Humfrey, L. E. H., (*Capt. Res. of Off.*) (*act'g. Lt.-Col., 7 Mar. 18*)	4 Jan. 17	24 Feb. 16	Comdg.
23 Oct. 14	Weigall, L. A. F., 5 Bn., Linc. R.	31 Jan. 17	23 Oct. 14	2nd-in-Comd.
	Captains.			
Sep. 02	Trapman, Capt. A. H., 3rd Bn., Lond. R.	8 Nov. 18	15 Feb. 07	
23 Sep. 14	Hunt, G. E. D. (*temp.*)	2 Feb. 17	19 Feb. 15	Staff Capt., Q. M.G.'s Branch, Army Hdqrs.
15 Sep. 15	Bucknall, W. B. (*tempy.*) (*Hon. Capt. in Army, 10 July 01*).	27 Jan. 17	15 Sep. 15	Comdg. Lv. Details Camp, Deolali.
23 Sep. 14	Aulton, F. H., 6 Bn., S. Staff. R. .	23 June 17	1 June 16	Asst. Embn. Staff Offr., Bombay.
29 Oct. 14	Gibsone, H. C., 4 Bn., R. W. Surr. R.	30 Sep. 18	1 June 16	
12 May 14	Cranswick, H. S., 5 Bn., York. R. .	28 Jan. 17	15 Sep. 16	Bde. Signalling Offr., Bombay.
15 May 15	Gray, J. Mc. C. T., (*Lt., Bucks. Bn., Oxf. & Bucks. L. I.*)	17 June 17		
	Lieutenants.			
10 Oct. 14	Downing, P. H. (*actg. Capt., 1 Aug. 17*) 3rd Bn., W. York. R.	12 Jan. 17	14 June 15	R. T. O., Bombay.
25 July 14	Walpole-Simmons, C. R., 5 Bn., Worc. R.	27 Aug. 18	5 May 16	
5 Jan. 15	Richardson, F. (*temp.*) (*act'g. Capt., 7 May 18*	1 July 17	. . . .	Adjutant.
3 Feb. 15	Clark, R. J. C. (*temp.*), 5th Bn., Sco. Rif.	27 July 17	30 Jan. 17	
29 Mar. 15	Simpson, A. (*temp.*), 23 Bn., Manch R.	1 July 17	. . . .	Ind. Munitions Board.
11 Sep. 15	Rendell, E. A., 5 Bn., North'd Fus. (51).	1 July 17	. . . .	Armoured Motor Bde., Peshawar.
12 May 15	Scott, G. S., N. Lan. R. . .	1 July 17	. . . .	
26 Mar. 15	Kinloch, A. P. A. H., 3 Bn., R. War. R.	1 July 17	. . . .	
4 June 15	Tomlinson, A. E., S. Staff. R. .	1 July 17	. . . .	
7 Apl. 16	Firth, C. W., 4 Bn., York and Lanc. R., (*actg. Capt., 10 Nov. 18*).	7 Oct. 17		
	2nd-Lieutenants.			
15 June 16	Jupp, E. P.	29 Jan. 17	15 June 16	*Lv. ex I., m.c., 25 Sept. 17.*
19 Dec. 16	Cheetham, R. H. (temp.) 1 Garr. Bn., York. L. I.	31 Jan. 17	19 Dec. 16	
19 Dec. 16	Brown, V. H. (temp.)	31 Jan. 17	19 Dec. 16	
15 Apl. 17	Hirst, J. (*temp.*), 1 Garr. Bn., Suff. R.	8 Feb. 18	15 Apl. 17	Army Clothing Factory, Alipore.
15 Apl. 17	Woods, H. W., 1 Garr. Bn., Suff. R.	8 Feb. 18	15 Apl. 17	Army Clothing Factory, Jullundur.
4 May 17	Cooke, C. (*temp.*), 1 Garr. Bn., Suff. R.	8 Feb. 18	4 May 17	
13 May 17	Hollick, G. (*temp.*), 1 Garr. Bn., Suff. R.	8 Feb. 18	13 May 17	
22 May 17	Kent, N. G. (*temp.*), 1 Garr. Bn., Suff. R.	8 Feb. 18	22 May 17	
30 May 17	King, E. J. H. (*temp.*), 1 Garr. Bn., Suff. R.	8 Feb. 18	30 May 17	
30 May 17	Habgood, H. W. (*temp.*), 1 Garr. Bn., Worc. R.	10 Feb. 18	30 May 17	

The South Staffordshire Regiment—*contd.*

First Commn.	Names.	Regimental rank.	Army rank.	REMARKS.
	2nd-Lieutenants—*contd.*			
27 June 17	Hayward, E. J., M.M., (*temp.*) 1 Garr. Bn., Suff. R.	8 Feb. 18	27 June 17	Ry. Bn., Sappers & Miners, Sialkot.
2 Aug. 17	Tebbutt, H. J., (*temp.*) 1 Garr. Bn., Suff. R.	14 Jan. 18	2 Aug. 17	
2 Aug. 17	Gardiner, H. J., (*temp.*) 1 Garr. Bn., Suff. R.	14 Jan. 18	2 Aug. 17	
19 Aug. 17	Weaver, B. H., (*temp.*) 1 Garr. Bn., Suff. R.	14 Jan. 18	19 Aug. 17	No. 2 Mule Corps Depôt, Lahore.
28 Aug. 17	Tibbitt, W. T., (*temp.*) 1 Garr. Bn., Suff. R.	28 Jan. 18	28 Aug. 17	Mech. Tpt., A. S. Corps, Peshawar.
20 Nov. 16	Jamison, A. W., (*temp.*) 2nd Garr. Bn., Bedf. R.	13 Apl. 18	20 Nov. 16	Asst. Postal Censor, Dhanuskodi.
7 Sep. 17	Allen, E., (*temp.*) 3rd Bn., Suff. R.	1 May 18	7 Sep. 17	
19 Aug. 18	Guidon, G. J. R. (*temp.*) . . .	19 Aug. 18	. . . .	
16 Sep. 17	Chaplin, E. T. (*temp.*) 1 Res. Garr. Bn., Suff. R.	27 Aug. 18	16 Sep. 17	
	Attached.			
30 May 17	Wells, 2nd-Lt. C. M., (*temp.*) 4 Bn., E. Kent. R.	27 Dec. 17	30 May 17	Mily. Accts. Dept., Lahore.
5 Jan. 15	Richardson, Lt. (*act'g. Capt.*) F. (*temp.*)	1 July 17	. . . .	Adjt.
6 Jan. 17	Johnson, Lt. J. (*temp.*)	6 Jan. 17	. . . .	Qr.-Mr.

Regimental District No. 39.
[No. 8 District.]

Uniform—Scarlet.
Facings—Grass Green.

The Dorsetshire Regiment.

2ND BATTALION (*late 54th Foot*).

Arrived in India, 25th October 1906, from England.

"*Primus in Indis.*"

THE CASTLE AND KEY, SUPERSCRIBED "GIBRALTAR 1779-83," AND WITH THE MOTTO "*Montis Insignia Calpe*" UNDERNEATH. THE SPHINX, SUPERSCRIBED "EGYPT."

"Plassey" "Martinique, 1794" "Marabout" "Albuhera" "Vittoria" "Pyrenees" "Nivelle" "Nive" "Orthes" "Peninsula" "Ava" "Maharajpore" "Sevastopol" "Tirah" "Relief of Ladysmith" "South Africa, 1899-1902."

COLONEL—Maj.-General H. Cook, C.B. M., ret. pay . 17 June 10.

First Commn.	Names.	Regimental rank.	Army rank.	Remarks.
	Lieutenant-Colonel.			
3 Apl. 89	Clarkson, B. St. J., C.M.G., D.S.O. (51, 65, 66, 72, 72*a*, 206)	25 Oct. 14		
	Majors.			
1 Dec. 97	Herbert, G. M., D.S.O. (*temp. Lt.-Col.*) (51, 72, 72*b*).	1 Sep. 15	*l.c.* 1 Jan. 17	
17 Feb. 00	Case, H. A., D.S.O. (*act'g. Lt.-Col.*) (51)	1 Sep. 15	*l.c.* 3 June 16	
	Captains.			
10 Oct. 03	*Clemson, W.* (51, 52, 57, 72*a*) . .	26 May 10	. . . .	*Adjt.*
9 Oct. 07	Yeatman, G. D. (*act'g. Maj.*, 28 *May 18*) (51, 52, 206).	25 Oct. 14	. . . .	Attd. Suff. R.
18 Sep. 09	Miles, W. H. (51, 52)	23 Jan. 15		
20 Sep. 11	Hawkins, A. E.	1 Oct. 15	. . . .	With Sig. Serv.
27 May 12	Highett, C. T. (66*a*, 71)	1 Oct. 15		
4 Dec. 12	Stephenson, E. L., M.C.	1 Oct. 15		
24 Jan. 14	Thomson, W. B., M.C.	24 July 16		
24 Jan. 14	Simmons, D. A.	2 Oct. 16		
1 Oct. 14	Miller, S., M.C.	6 Feb. 17		
	Lieutenants.			
22 Oct. 14	Sigrist, A. E. (*temp. Capt.*, 23 *Dec.* 17).	12 Jan. 16	. . . .	*Lv. ex. I., m. c., 6 m*
7 Nov. 14	Mitchell, W. G. (*temp. Capt.*, 23 *Dec.* 17).	2 Mar. 16		
7 Nov. 14	Payne, J. H., M.C. (*temp. Capt.*, 23 *Dec.* 17).	3 Mar. 16		
7 Nov. 14	Barton, W., M.C.	21 Mar. 16		
7 Nov. 14	Bridgens, A. (*temp. Capt.*, 23 *Dec.* 17).	7 July 16	. . . .	*Lv. in I., m.c., 6 m.*
9 Jan. 15	Crawley, C. P.	30 July 16		
17 Feb. 15	Woodhouse, L. J. (*act'g. Capt.*, 18 *Sep.* 17).	30 July 16		
5 Apl. 15	Bishop, W. H. A. (*act'g. Capt.*) .	20 Oct. 16		
9 Nov. 15	Goode, F. V.	1 Jan. 17	. . . .	*Lv., m. c., 6 m.*
26 Jan. 16	Hewick, J. S. (*act'g. Capt.*) . .	26 Mar. 17		
7 Apl. 16	Oppenheimer, P. H. M.	7 Oct. 17		
	2nd-Lieutenants.			
18 Dec. 16	Newman, T., D.C.M.	18 Dec. 16		
7 Jan. 17	Race, W. C.	7 Jan. 17		
4 Apl. 17	Dowdell, R. A.	4 Apl. 17		
24 Oct. 17	Park, R. C. W.	24 Oct. 17		
24 Oct. 17	Everton, B.	24 Oct. 17		
30 Nov. 17	Passmore, M.	30 Nov. 17		
30 Nov. 17	Jackman, J.	30 Nov. 17		

The Dorsetshire Regiment—*contd.*

2ND BATTALION (*late 54th Foot*).

First Commn.	Names.	Regimental rank.	Army rank.	Remarks.

Attached.

First Commn.	Names.	Regimental rank.	Army rank.	Remarks.
	Boldero, Capt. A. J., 3rd Som. L. I.	4 Sep. 14		
	Ellis, Lt. F. R., 1-4th D. of Corn. L. I.	1 June 16		
	Bishop, Lt. G. W. R., 1-4th Som. L. I.	26 Feb. 14		
	Snell, Lt. W., 1-4th Devon. R.	1 Mar. 17		
	Waldram, Capt. H. G., 1-6th Devon. R.	1 June 16		
	Hall, Lt. R. S., 3rd Bn.	2 Feb. 15		
	Hall, Lt. F. G., 3rd Bn. (*temp. Capt.*)	2 Feb. 15		
10 Oct. 14	Watkins, Lt. H.	2 Feb. 15		
	Holland, Lt. A., 3rd Dorset R.	2 Feb. 15	. . . .	*Lv., ex I., m. c., 6 mos.*
31 Mar. 15	Taylor, Lt. C. H., 1st Garr. Bn., Linc. R.	24 Aug. 15		
	Forde, Lt. F. L., 3rd Dorset R.	27 Mar. 16		
1 Jan. 15	*Cole, Lt. J. T.,* Devon. R. (*temp.*).	24 July 16	. . . .	*I. A. R. O.*
14 July 15	Truscott, Lt. N. P., 3rd Bn., D. of Corn. L. I.	1 July 17		
	Tinley, Lt. R. G. W. N., 3rd Bn.	1 July 17	. . . .	*Lv., ex I., m. c., 4 ms.*
9 Jan. 15	Blucke, Lt. R. S., 3rd Dorset R.	1 July 17	. . . .	With R. A. F.
27 Feb. 15	Good, Lt. C. A., 3rd Bn.	1 July 17		
23 Mar. 15	Hands, Lt. A. S., M.C. (*act'g. Capt.*) 3rd Bn.	1 July 17		
8 Apl. 15	Dean, Lt. C. S. C., 3rd Bn.	1 July 17		
4 May 15	Jenkins, Lt. F. A. M., 3rd Bn.	1 July 17		
5 May 15	Paget, Lt. B. J., 3rd Bn.	1 July 17		
26 May 15	Hall, Lt. C. S., 3rd Bn.	1 July 17		
26 May 15	Mace, Lt. A. C., 3rd Bn.	1 July 17		
10 June 15	Evans, Lt. M. O., 3rd Bn.	1 July 17	. . . .	*Lv., in I., m. c., 4 ms. 28 ds.*
24 July 15	Brindley, Lt. W. W., 3rd Bn.	1 July 17		
4 Aug. 15	Matthews, Lt. F., 3rd Bn.	1 July 17		
26 Aug. 15	Bower, Lt. A. P., 3rd Dorset R.	1 July 17		
2 Sep. 15	Swallow, Lt. J. C., 3rd Bn.	1 July 17		
2 Sep. 15	Lighthall, Lt. W. S., 3rd Bn.	1 July 17		
9 Sep. 15	Jenks, Lt. A. L., 3rd Bn.	1 July 17		
7 July 16	Beauvais, Lt. L. (*temp.*), Bord. R.	7 Jan. 18		

First Commn.	Names.	Regimental rank.	Army rank.	Remarks.
10 Oct. 03	Clemson, Capt. W.	11 Dec. 16	. . . .	Adjt.
1 May 04	Brown, Capt. A., M.C.	11 May 14	. . . .	Qr.-Mr.

Agents :—MESSRS. COX & Co.

Regimental District No. 39.
[No. 8 District.]

Uniform—Scarlet.
Facings—Grass Green.

The Dorsetshire Regiment.

1-4TH BATTALION.

Arrived in India, 10th November 1914, from England.

"South Africa, 1900-01."

HONORARY COLONEL—*Sir* R. Williams, *Bt.*, V.D. . 1 Dec. 07.

First Commn.	Names.	Regimental rank.	Army rank.	Remarks.
	Lieutenant-Colonel.			
	Majors.			
7 Apl. 97	Kitson, H. L., T.D. (*Hony. Capt. in Army, 9 July 01*) (51)	18 May 17		
5 June 95	Hughes-D'Aeth, C. C., T.D. . .	1 June 16	. . . .	Censor.
25 May 06	Duke, H. E. (51)	18 May 17	. . . .	With No. 6 Res. Bn.
	Captains.			
28 Mar. 00	Fletcher, W. T. (51, 52)	15 June 12	. . . .	Poll. Employ.
5 Aug. 14	*Chelmsford,* Rt. Hon. *F. J. N.* Lord, G.M.S.I., G.C.M.G., G.M.I.E., G.B.E., P.C.	5 Aug. 14	. . . .	*Viceroy & Govr.-Genl. of India.*
30 Apl. 08	Symonds, D. G. (51)	3 Oct. 14	. . . .	
15 Jan. 09	*Roper, J.*	3 Oct. 14	. . . .	*Comdt., Purandhar.*
26 Nov. 10	Suttill, K. (51)	1 June 16	. . . .	*Lv., ex I., m.c., 17 Oct. 18.*
15 Mar. 11	Richards, C. E. (51)	1 June 16		
1 May 11	Tucker, A. C.	1 June 16	. . . .	*Lv., ex I., m.c., 15 July 18.*
1 May 11	*Burt, H. C.*	1 June 16	. . . .	*A. S. Corps.*
21 June 13	*Royds, G. F.*	1 June 16	. . . .	*Poll. Employ.*
23 June 13	Smith, F. J.	1 June 16		
22 Jan. 14	*Woodhouse, H. S.* (52)	1 June 16	. . . .	*M. G. Corps.*
26 Aug. 14	Carver, F. E. (52)	1 June 16		
29 Aug. 14	Chaning-Pearce, M. S. (51) . .	18 May 17		
29 Aug. 14	Symes, J. D.	12 July 17	. . . .	*Lv., in I., m. c., 6 mos.*
29 Aug. 14	Allen, D. J. (48*a*, 58*a*, 59) . . .	15 Sep. 17		
29 Aug. 14	*Dunkin, J. C.* (57)	8 Nov. 17		
12 Sep. 14	Ogle, J. E. (51)	8 Nov. 17		
	Lieutenants.			
3 Oct. 14	*Hay, W. R.* (51)	1 June 16	. . . .	*I. A.* (*on probn.*)
3 Oct. 14	Dibben, E. L.	1 June 16	. . . .	*Lv., ex I., m.c., 14 Oct. 18.*
3 Oct. 14	*Peele, C. R. De C.*	1 June 16	. . . .	*I. A. R. O.*
3 Oct. 14	Carus-Wilson, M. M.	1 June 16		
22 Oct. 14	Llewellin, W. W.	1 July 17	. . . .	Adjt., No. 6 Res. Bn. (India).
11 Dec. 14	Symonds, W. F.	1 July 17	. . . .	Local Audit Offr.
2 Feb. 15	Acheson-Gray, C. G. A.	1 July 17		
21 Feb. 15	Warnford-Davis, R. D.	1 July 17		
21 Feb. 15	Soper, V.	1 July 17	. . . .	Attd. R. A. F.
21 Feb. 15	Swan, L. M.	1 July 17	. . . .	L. P. O.
5 Mar. 15	Davies, E. E.	1 July 17	. . . .	*Lv., ex I., m.c., 17 Oct. 18.*
7 Mar. 15	Brown, D. H.	1 July 17		
14 Mar. 15	*Vizard, W. G.*	1 July 17		
3 Apl. 15	Molony, J. T.	1 July 17	. . . .	With 1st Garr. Bn., Oxf. & Bucks. L. I.
9 Apl. 15	Williams, G. G.	1 July 17		
29 Apl. 15	*Shaw, R. C.*	1 July 17	. . . .	*Empld. R. E.*
28 July 15	Claye, A. M.	1 July 17		
11 Nov. 15	Roe, A. E.	1 July 17		
19 Nov. 15	Holden, E. A. T. A.	1 July 17		
27 Nov. 15	Harrup, W. D.	1 July 17	. . . .	Adjt. & Qr.-Mr., Divl. Schl., Poona.
10 Dec. 15	Blount, C. B.	1 July 17		
4 June 16	*Base, E. H.*	4 Dec. 17		
4 June 16	*Outlaw, W. H.*	4 Dec. 17		
8 June 16	Garner, J. D.	4 Dec. 17		

The Dorsetshire Regiment—*contd.*

1-4TH BATTALION—*concld.*

First Commn.	Names.	Regtl. Rank.	Army Rank.	Remarks.
	2nd-Lieutenants.			
30 May 17	Barker, J. J.	30 May 17		
30 May 17	Beckworth, C.	30 May 17		
30 May 17	Cutts, E. J.	30 May 17		
30 May 17	Noyes, H.	30 May 17		
30 May 17	Missing, C. M.	30 May 17		
30 May 17	Vowles, A. C.	30 May 17		
30 May 17	Mercer, E. C.	30 May 17		
30 May 17	Tidd, G. M.	30 May 17		
30 May 17	Lawson, W. G.	30 May 17		
1 Aug. 17	Boldero, J. C.	1 Aug. 17		
29 Aug. 17	Troath, W. A.	29 Aug. 17		
29 Aug. 17	Gardner, T. A. A.	29 Aug. 17		
16 Mar. 18	Pinkerton, S. R.	16 Mar. 18		
16 Mar. 18	Eastwood, V.	16 Mar. 18		
19 Aug. 18	White, H. E. N. (*temp.*)	19 Aug. 18		
19 Aug. 18	Pursey, C. E. (*temp.*)	19 Aug. 18		
19 Aug. 18	Spurgen, J. C. (*temp.*)	19 Aug. 18		
	ATTACHED OFFICERS.			
	Majors.			
9 July 04	Lamonby, L. (51), 4th Bn., Bord. R.	13 July 17		
	Captains.			
	Barke, C. R. (*act'g. Lt.-Col.*), 5th Bn., N. Staff R.	19 Sep. 11		
6 Mar. 11	Chapple, G. L., 6th Bn., E. Surr R.	1 June 16		
8 Apl. 11	Plincke, J. F., 6th Bn., E. Surr R.	1 June 16		
29 Aug. 14	Dixon, A. F. W., 4th Bn., Bord. R. (51)	1 June 16		
5 Oct. 14	Bendle, L. A., 4th Bn., Bord. R.	20 July 17		
	Lieutenants.			
	Palmer, W. E., 2nd Bn., Dorset. R.	20 Oct. 15		
	Topley, A., 1st Bn., Bed R.	1 Nov. 15		
	Angus, H. F., 4th Bn., Wilts. R.	1 June 16		
15 Oct. 15	Heelis, G. H., 4th Bn., Bord. R.	1 July 17		
21 Oct. 15	Hodgson, W. M., 4th Bn., Bord. R.	1 July 17		
22 Nov. 15	Turner, W. F., 6th Bn., R. War R.	1 July 17	....	Act'g. Qr.-Mr.
9 Dec. 15	Alexander, W., 4th Bn., Bord. R. Att. L.T.M.B.	1 July 17		
21 Dec. 15	Alexander, S., 4th Bn., Bord. R.	1 July 17		
28 Dec. 15	Forrester, R. D., 3rd Bn., Dorset. R.	1 July 17		
16 Aug. 16	Dowling, J.R.H., 2nd Bn., Dorset. R.	16 Feb. 18		
	2nd-Lieutenants.			
8 Nov. 14	Campbell, R.J.T., 12th Bn., Arg. & Suth'd. Highrs.	28 Nov. 14		
12 July 15	Pell, L.R., 3rd Bn., Dorset.R.	12 July 15		
22 Nov. 16	Campbell, C.R., 3rd Arg. & Suth'd. Highrs.	22 Nov. 16		
1 Mar. 17	Powell, G.W., 3rd Bn., Dorset. R	1 Mar. 17		
1 Mar. 17	Carter, H.B., 3rd Bn., Dorset. R.	1 Mar. 17		
1 Mar. 17	Villar, A.W., 3rd Bn., Dorset. R.	1 Mar. 17		
28 Mar. 17	Field, L., 3rd Bn., Dorset. R.	28 Mar. 17	....	Adjt.
28 Mar. 17	Vaughan, A.A., 4th Gord. Highrs.	28 Mar. 17		
28 Mar. 17	Wilson, D.Mc.I., 4th Gord. Highrs.	28 Mar. 17		
29 Aug. 17	Read, E.H., 2nd Bn. Dorset R.	29 Aug. 17		
29 Aug. 17	Pugsley, W.H., 2nd Bn., Dorset R.	29 Aug. 17		

First Commn.	Names.	Regtl. Rank.	Army Rank.	Remarks.
28 Mar. 17	Field, 2nd-Lt. L., 3 Bn., Dorset. R.	28 Mar. 17	....	Adjutant.
22 Nov. 15	Turner, Lieut. W.F.	1 July 17	....	Qr. Mr. 8 Apl. 18.
2 Jan. 15	Molyneux, Capt. J.J., M.B., R.A.M.C., (S.R.)	2 July 15	....	Medl. Offr.
28 Feb. 08	Paterson, Capt. A.R., M.D., R.A.M.C.. (T.F.)	7 Dec. 11	....	Ditto.

Regimental District No. 39. [No. 8 District.]

Uniform—Scarlet. *Facings*—Grass Green.

The Dorsetshire Regiment.

2-4TH BATTALION.

Arrived in India, 9th January 1915, from England.

"South Africa, 1900-01."

HONORARY COLONEL—*Sir* R. Williams, *Bt.*, V.D. . . . . . . 1 Dec. 07.

First Commn.	Names.	Regimental rank.	Army rank.	Remarks.
	Lieutenant-Colonel.			
26 Nov. 14	Scott, F. W. A., D.S.O.	26 Nov. 14		
	Majors.			
2 Oct. 14	Woodall, H. W., C.I.E. (*temp. Lt.-Col.*)	3 Oct. 14	. . . .	A. Q. M. G., Army Hdqrs.
19 July 99	Lock, H. O. (57)	1 June 16		
	Captains.			
3 Oct. 14	Ramsay, B. F.	3 Oct. 14		
10 Jan. 13	Tucker, A. C.	1 June 16	. . . .	With 1-4th Dorset. R.
1 May 11	Burt, H. C.	1 June 16		
23 June 13	Smith, F. J. (57)	1 June 16	. . . .	With Sig. Serv.
11 Dec. 14	Arnold, A.	1 June 16	. . . .	Adjt.
12 Sep. 14	Davies-Bunton, A. E. K. (60*a*)	2 Feb. 18		
. . . .	Chamen, R. T. (*act'g. Capt.*)	31 May 18		
	Lieutenants.			
3 Oct. 14	Mauleverer, W. M. M. G. (59)	1 June 16	. . . .	With Sig. Serv.
3 Oct. 14	Carus-Wilson, M. M.	1 June 16	. . . .	With 1-4th Dorset. R
3 Oct. 14	*Wemyss, A. B.*	1 June 16	. . . .	*I. A. R. O.*
3 Oct. 14	Stable, H. H. (51, 52)	1 June 16	. . . .	I. A. (*Cn probn.*).
3 Oct. 14	Dammers, E. H. F. (*act'g. Capt.*) (*addl.*), (51)	1 June 16		
3 Oct. 14	Carter, C. C. (*act'g. Capt.*) (51)	18 May 17		
15 Oct. 14	Lester-Garland, G. H. (51, 52, 66*a*)	18 May 17		
22 Oct. 14	Llewellin, W. W. (51, 60*a*)	1 July 17	. . . .	With 1-4th Dorset. R.
29 Oct. 14	Sweet-Escott, W. W. (51)	1 July 17	. . . .	*Lv., ex I., m.c., 6 m.*
31 Oct. 14	Filleul, R. R.	1 July 17	. . . .	With 1-4th Dorset. R.
11 Dec. 14	Symonds, W. F. (51, 52)	1 July 17	. . . .	With 1-4th Dorset. R.
12 Dec. 14	Worsley, R. E. (51, 57)	1 July 17		
21 Feb. 15	Warnford-Davis, R. D. (51)	1 July 17	. . . .	With 1-4th Dorset. R.
21 Feb. 15	Soper, V. (51)	1 July 17		
21 Feb. 15	Swan, L. M.	1 July 17	. . . .	With 1-4th Dorset. R.
7 Mar. 15	Brown, D. H.	1 July 17	. . . .	With 1-4th Dorset. R.
14 Mar. 15	Vizard, W. G. (*temp. Capt.*) (57)	1 July 17	. . . .	Staff Capt.
3 Apl. 15	Moloney, J. T. (51)	1 July 17		
9 Apl. 15	Crick, W. H. R. (51)	1 July 17		
9 Apl. 15	Williams, G. G. (*temp. Capt.*)	1 July 17	. . . .	
22 Apl. 15	Cardwell, C. R. (46)	1 July 17		
29 Apl. 15	*Shaw, R. C.* (57)	1 July 17	. . . .	With Sig. Serv.
28 July 15	Powell, W. R.	1 July 17		
6 Aug. 15	Goody, G. R. (57)	1 July 17		
14 Oct. 15	Whittingdale, W. L.	1 July 17		
19 Nov. 15	Smith, R. S.	1 July 17		
23 Dec. 15	*Cheesman, E. J.*	1 July 17	. . . .	*M. G. Corps.*
16 Jan. 16	Wills, L. K.	16 July 17		
2 Feb. 16	Vincent, L. F.	2 Aug. 17		
9 May 16	Marshall, K. R. (57)	9 Nov. 17		
2 June 16	Sloot, J. N. C. (46)	2 Dec. 17		
8 June 16	Ferris, W. H.	8 Dec. 17		
12 June 16	Roper, E. W.	12 Dec. 17		
26 Sept. 16	Gaby, V. F.	26 Mar. 18		
26 Sept. 16	Garland, A. S. T.	26 Mar. 18		
	2nd-Lieutenants.			
11 Dec. 14	Arnold, Capt. A.	1 Feb. 15	. . . .	Adjt.
1 Oct. 14	Hoskins, Capt. J.	1 Oct. 17	. . . .	Qr.-Mr.—*Lv., ex I., m.c. 6 m.*

Regimental District, No. 40.
[No. 4 District.]

Uniform—Scarlet.
Facings—White.

The Prince of Wales's Volunteers.

(South Lancashire Regiment.)

1ST BATTALION (*late 40th Foot*).
Arrived in India, 7th February 1903, from Natal.

THE PLUME OF THE PRINCE OF WALES. THE SPHINX SUPERSCRIBED "EGYPT."
"Louisburg" "Martinique, 1762" "Havannah" "St. Lucia, 1778" "Monte Video" "Roliça" "Vimiera" "Corunna" "Talavera" "Badajoz" "Salamanca" "Vittoria" "Pyrenees" "Nivelle" "Orthes" "Toulouse" "Peninsula" "Niagara" "Waterloo" "Candahar, 1842" "Ghuznee, 1842" "Cabool, 1842" "Maharajpore" "Sevastopol" "Lucknow" "New Zealand" "Relief of Ladysmith" "South Africa, 1899-1902."

COLONEL—Major-General E. H. Sartorius, V. C., C.B., *p.s.c.*, ret. pay . . **26 Nov. 09.**

First Commn.	Names.	Regimental rank.	Army rank.	Remarks.
	Lieutenant-Colonel.			
24 Apl. 89	Willoughby-Osborne, A. deV. (51, 52, 57, 71, 95)	28 Feb. 16		
	Majors.			
16 Mar. 98	Woods, H. K. (50, 53, 60*a*, 72*b*, 207)	14 Sep. 14	*l. c.*, 1 Jan. 16	M. G. Corps.
	Captains.			
13 Dec. 07	Douglas-Withers, H. H., M.C. (51, 52, 71, 207)	4 Dec. 14	. . . .	Bde. Maj.
18 Sep. 09	Braide, R. W. (*act'g. Maj.* 50, 52, 58*a*, 66*a*, 95, 172*a*)	22 Jan. 15		
22 Oct. 02	St. John, H. B. (51, 52, 71) .	1 Oct. 15	. . . .	With 3rd (S. R.) Bn.
20 Dec. 11	Barker, R. E. (50, 57, 207) . . .	11 Oct. 15	. . . .	With Sig. Serv.
14 Feb. 12	Campbell, F. E. D. (51, 52) . . .	11 Nov. 15	. . . .	With A. I. F.
17 Sep. 13	Ward, W. D. (51, 52) . .	1 Apl. 16		
10 Dec. 13	Ransom, R. A. (48*a*, 51, 61, 95) . .	10 June 16	. . . .	Abbottabad.
20 Dec. 13	Macdonald, R. H. (52, 59, 95, 172*a*, 206)	2 July 16	. . . .	With Sig. Serv.
24 Jan. 14	Murray, K. M. (12, 52, 61, 95, 100, 206)	4 Oct. 16		
21 July 14	Hislop, J. F. G. (46, 52, 58*a*) . . .	1 Jan. 17		
7 Nov. 14	Duffy, L.	8 Aug. 17	. . . .	Staff Capt., Madras.
7 Nov. 14	Earle, A.	8 Aug. 17	. . . .	Adjt., Convalescent Depôt, Doolali.
7 Nov. 14	Pearce, A. (95)	8 Aug. 17	. . . .	Embrkn. Staff Offr., Karachi.
	Lieutenants.			
7 Nov. 14	Curry, L. H. (*temp. Capt.*) (48*a*, 51, 57)	1 Oct. 15		
7 Nov. 14	Shaw, N. (*act'g. Capt.*) (50, 58*a*) .	11 Oct. 15	. . . .	Adjt (*temp.*)
23 Oct. 14	Hobson, C. M. (*temp.*) (48*a*, 51) .	28 Feb. 16		
17 June 15	Smith, A. F. (52)	1 Apl. 16	. . . .	With M. G. Corps.
17 June 15	Earle, E. J. (51, 52)	10 June 16	. . . .	With M. G. Corps.
	2nd-Lieutenants			
7 Dec. 15	Smith, A. G. C.	7 Dec. 15		
4 Jan. 17	Howard, R. (*temp. Capt.*) (48*a*, 50)	4 Jan. 17	. . . .	Platoon Comdr., Officer Cadet Unit, Sialkot.
14 Mar. 17	Wright, W. S. (46)	14 Mar. 17		
14 Mar. 17	Dallas, P.	14 Mar. 17		
26 Mar. 17	O'Toole, A. H.	26 Mar. 17		
26 Apl. 17	Simon, W. M. (46, 61)	26 Apl. 17		

Agents:—MESSRS. COX & CO.

The Prince of Wales's Volunteers—*contd.*

(South Lancashire Regiment)—*contd.*

First Commn.	Names.	Regimental rank.	Army rank.	Remarks.
	2nd-Lieutenants—*contd.*			
29 Aug. 17	Breething, H. R. (temp.)	29 Aug. 17		
29 Aug. 17	Lunn, E. D. (temp.) (46, 61)	29 Aug. 17		
29 Aug. 17	Birkett, W. G. N. (temp.)	29 Aug. 17	. . . .	Attd. 1st G. B., Norf. R.
29 Aug. 17	Hawkins, J. (temp.)	29 Aug. 17	. . . .	With R. A. F.
16 Oct. 17	Mitchelson, G. (12)	16 Oct. 17		
8 Nov. 17	Jarman, A. J.	8 Nov. 17		
16 Nov. 17	Makin, R. (48*a*, 51)	16 Nov. 17		
16 Mar. 18	Bates, S. F. (*temp.*)	16 Mar. 18		
16 Mar. 18	McAlister, S G. (*temp*).	16 Mar. 18		
16 Mar. 18	Keywood, H. A. (*temp*).	16 Mar. 18		
16 Mar. 18	Baker, S. (*temp*).	16 Mar. 18		
21 Mar. 18	Jones, C. A. (95, 132, 153, 207)	21 Mar. 18		
11 June 18	Bentley, J. F.	11 June 18		
19 Aug. 18	Hyde, J. K. (*temp.*)	19 Aug. 18		
19 Aug. 18	Burghope, J. W. (*temp.*)	19 Aug. 18		
19 Aug. 18	Harker, G. A. (*temp.*)	19 Aug. 18		
	Attached.			
13 Jan. 15	Prynn, C. C., Lt., D. of Corn. L. I.	14 Apl. 16		
10 Sep. 15	Dodgson, E. C., temp. Lt., Seaf. Highrs.	1 July 17		
4 Jan. 16	Campbell, D., Temp. Lt., Sco. Rif.	4 July 17		
21 Jan. 16	Meikle, J. A. S., Temp. Lt., Sco. Rif.	21 July 17	. . . .	*Lv., ex I., m. c., 6 mos.*
22 Jan. 16	Malvern, J. F., Temp. Lt., Worc. Regt.	22 July 17		
30 May 17	Fairclough, R., 2nd-Lt. L. N. Lan. Regt.	30 May 17		
26 Apl. 17	Harty, E. P., Tem. 2nd-Lt., Muns. Fus.	26 Apl. 17		
29 Aug. 17	Amphlett, F. G., 2nd-Lt., 7th Bn., Lan. Fus	29 Aug. 17		
29 Aug. 17	Wells, E. T. W., 2nd-Lt., 8th Bn., Lan. Fus.	29 Aug. 17		
29 Aug. 17	Watkin, W., 2nd-Lt., 7th Bn., Lan. Fus.	29 Aug. 17		
29 Aug. 17	Blanche, R., 2nd-Lt., 8th Bn., Lan. Fus.	29 Aug. 17		
29 Aug. 17	Valentine, S. W., 2nd-Lt., 7th Bn., Lan. Fus.	29 Aug. 17		
29 Aug. 17	Lloyd, N. R., 2nd-Lt., 7th Bn., Lan. Fus.	29 Aug. 17		
29 Aug. 17	Williams, A. J., 2nd-Lt., 8th Bn., Lan. Fus.	29 Aug. 17		
29 Aug. 17	Gray, J. F., 2nd-Lt., 6th Bn., Lan. Fus.	29 Aug. 17		
26 Sep. 17	Woodcock, F. D., 2nd-Lt., 6th Bn., Lan. Fus.	26 Sep. 17		
26 Sep. 17	Savage, C. T., 2nd-Lt., 3rd Bn., S. Lan. R.	26 Sep. 17	. . . .	Army Clothing Factory, Alipore.
26 Sep. 17	Meyer, H. A., 2nd-Lt., 3rd Bn., S. Lan. R.	26 Sep. 17		
26 Sep. 17	Newton, 2nd-Lt. J. F., 4th Bn., R. Lanc. R.	26 Sep. 17		
26 Sep. 17	Ellis, 2nd-Lt. C. H., 5th Bn., Midd'x. R.	26 Sep. 17		
31 Oct. 17	Grassick, W. J., 2nd-Lt., 3rd Bn., Gord, Highrs.	31 Oct. 17		
31 Oct. 17	Watson, S. G., 2nd-Lt., 18th Bn., R. Ir. R.	31 Oct. 17		
31 Oct. 17	White, F. R., 2nd-Lt., 19th Bn., R. Ir. R.	31 Oct. 17		
28 Nov. 17	Irwin, W. J. A., 2nd-Lt., 19th Bn., N. R. Ir. R.	28 Nov. 17		
7 Nov. 14	Shaw, Capt. N. (50, 58*a*)	15 Jan. 18	. . . .	Temp. Adjutant.
20 July 10	Simon, Capt. T. A., D.C.M.	1 July 17	. . . .	Qr.-Mr.

***Agents*:—Messrs. Cox & Co.**

Regimental District No. 42.
[No. 1 District.]

Uniform—Scarlet.
Facings—Blue.

The Black Watch (Royal Highlanders).

2ND BATTALION (*late 73rd Foot*).
Arrived in India, 23rd October 1902, from South Africa.

THE ROYAL CYPHER WITHIN THE GARTER. THE BADGE AND MOTTO OF THE ORDER OF THE THISTLE. IN EACH OF THE FOUR CORNERS THE ROYAL CYPHER ENSIGNED WITH THE IMPERIAL CROWN. THE SPHINX, SUPERSCRIBED "EGYPT."
Guadaloupe, 1759" "Martinique, 1762" "Havannah" "North America, 1763-64" "Mangalore" "Mysore" "Seringapatam" "Corunna' "Busaco" "Fuentes d'Onor" "Pyrenees" "Nivelle" "Nive" "Orthes" "Toulouse" "Peninsula" "Waterloo" "South Africa, 1846-47, 1851-2-3" "Alma" "Sevastopol" "Lucknow" "Ashantee, 1873-4" "Tel-el-Kebir" "Egypt, 1882, 1884" "Kirbekan" "Nile, 1884-85" "Paardeberg" "South Africa, 1899-1902.".

COLONEL-IN-CHIEF—THE KING.

COLONEL—Lt.-General *the Rt. Hon'ble Sir* J. G. Maxwell, G.C.B., K.C.M.G., C.V.O., D.S.O. 11 Jan. 14.

First Commn.	Names.	Regimental rank.	Army rank.	Remarks.
	Lieutenant-Colonel.			
1 Mar. 90	Cameron, A. R., C.B., C.M.G., *p.s.c.* (*temp. Brig.-Genl.*)	25 Feb. 15	*c.* 1 Jan. 16	Brig.-Genl., Genl. Staff.
	Majors.			
22 Jan. 96	Wauchope, A. G., C.M.G., D.S.O. (51) (*temp. Br.-Genl.*)	15 Sep. 14	*c.* 23 Dec. 16	Bde. Comdr.
5 Jan. 00	Henderson, C. R. B. (*temp. Lt.-Col.*) (50).	5 Jan. 16	. . . .	
	Captains.			
21 Jan. 13	Gilmour, C. D., M.C.	2 Oct. 15	. . . .	*Lv., ex I.*
2 Sep. 14	Park, M. E., D.S.O.	2 Mar. 16		
. . . .	Kennedy, J., D.S.O., M.C.	27 Mar. 17	. . . .	With 13th Bn. Welsh R.
16 Dec. 14	*Ritchie, N. M.*, D.S.O. (*act'g. Capt.*)	19 Nov. 17	. . . .	Adjt.
	Lieutenants.			
20 Oct. 15	Gordon-Brown, D. S.	17 July 16		
4 Dec. 15	Barstow, J. A., M.C. (*act'g. Capt.*)	1 July 17		
18 Dec. 15	Grassie, J., D.S.O.	1 July 17		
	2nd-Lieutenants.			
15 Feb. 17	Houston, B., M.C., D.C.M.	15 Feb. 17		
16 Feb. 17	Kinnear, T.	16 Feb. 17		
16 Feb. 17	Gant, T.	16 Feb. 17		
29 May 17	Muir, A.	29 May 17		

Agents:—MESSRS. & Cox Co.

The Black Watch (Royal Highlanders)—*contd.*

First Commn.	Names.	Regimental rank.	Army rank.	Remarks.
	Attached.			
	Stewart, Maj. (*temp. Lt.-Col.*) J.	30 Oct. 14	. . . .	Comdg.
	Murray, Capt. (*temp. Maj.*) *Hon.* R. T. G., 3rd Bn.	19 Dec. 03		
	Cochrane, Capt. T. G. F., Spec. Res. (*temp. Maj.*)	18 Oct. 15		
	Grieve, Capt. A. McL., 3rd Bn.	18 Oct. 15		
29 July 08	Gibb, Capt. W. H., 3rd Bn.	18 Oct. 15	. . . .	Comdg. Details Depôt, Poona.
	Cocks, Capt. C. V. S., 3rd Bn.	19 Nov. 17	. . . .	*Lv., ex I.*
	Thorburn, Capt. M. M., M.C., 3rd Bn.	19 Nov. 17	. . . .	*Lv., ex I.*
	Dundas, Capt. R. H., 3rd Bn.	19 Nov. 17	. . . .	Asst. Censor, Army Hdqrs.
	MacEwen, Capt. D. W., M.C. (*temp. Capt.*), 3rd Bn.	19 Nov. 17		
	Stewart-Smith, Lt. D. C., 3rd Bn.	18 Oct. 15	. . . .	*Lv., ex I.*
	Macfarlane-Grieve, Lt. R. W., 3rd Bn.	8 Mar. 16	. . . .	*Lv., ex I.*
	Todd, Lt. J., 3rd Bn.	8 Mar. 16		
24 Dec. 14	Henderson, Lt. T. A., M.C. (*temp.*), 11th Bn.	1 July 17	. . . .	Bangalore.
23 Feb. 15	Buchanan, Lt. K., 11th Bn.	1 July 17		
23 Nov. 15	Cowie, Lt. T. McK. (*temp.*) (*temp. Capt.*)	1 July 17	. . . .	Asst. Embn. Staff Offr.
8 May 15	Woyka, Lt. A. G. (*temp.*), 11th Bn.	1 July 17		
20 May 15	Hunter, Lt. S. L. (*temp.*), 11th Bn.	1 July 17		
11 June 15	Jeff, Lt. J., 3rd Bn.	1 July 17		
12 July 15	Blair, Lt. A. E., 3rd Bn. (*temp. Capt.*)	1 July 17		
28 July 15	Stewart, Lt. G. V., 3rd Bn.	1 July 17	. . . .	With R. A. F.
6 Aug. 15	Hendry, Lt. C. V. (*temp.*), 11th Bn.	1 July 17		
14 Aug. 15	Crombie, Lt. A., 3rd Bn.			
14 Aug. 15	Stewart, Lt. T. W. (*act'g. Capt.*), 3rd Bn.	1 July 17		
10 Sep. 15	Smart, Lt. G. B., 3rd Bn.	1 July 17		
30 Oct. 15	Ryrie, Lt. G., 3rd Bn.	1 July 17		
. . . .	Edwards, Lt. A. L., C., M.C. (*temp. Capt., 1 May 18*), 4th Bn.	25 Sep. 15	. . . .	Instr., Bombg. Schl.
25 Sep. 15	Quine, 2nd-Lt. B. H. (*temp.*), 11th Bn.			
16 Dec. 14	Ritchie, Capt. N. M., D.S.O.	19 No4. 17	. . . .	Adjt.
28 Sep. 10	Anderson, Maj. J. (169)	3 June 18	. . . .	Qr.-Mr.

Agents:—MESSRS. COX & CO.

Regimental District, No. 43. [No. 7 District.]

Uniform—Scarlt. *Facings*—White.

The Oxfordshire and Buckinghamshire Light Infantry.

1ST BATTALION (*late 43rd Foot*).

Arrived in India, 15th October 1903, from England.

THE UNITED RED AND WHITE ROSE.

"Qutbec, 1759" "Martinique, 1762" "Havannah" "Mysore" "Hindoostan" "Martinique, 1794" "Vimiera" "Corunna" "Busaco" "Fuentes d'Onor" "Ciudad Rodrigo" "Badajoz" "Salamanca" "Vittoria" "Pyrenees" "Nivelle" "Nive" "Orthes" "Toulouse" "Peninsula" "Waterloo" "South Africa, 1851-2-3" "Delhi, 1857" "New Zealand" "Relief of Kimberley" "Paardeberg" "South Africa, 1900-02."

COLONEL—Major-General *Sir* J. Hanbury-Williams, K.C.B., K.C.V.O., C.M.G. . 6 Apl. 18.

First Commn.	Names.	Regimental rank.	Army rank.	Remarks.
	Lieutenant-Colonel.			
10 Oct. 85	Lethbridge, E. A. E., C.M.G., D.S.O. (51, 207)	25 Oct. 13		
	Majors.			
4 Jan. 93	Carter, L. J., D.S.O. (51, 57, 67) . .	23 June 13	. . . .	*Lv., ex I., m. c.*
3 Feb. 97	Stapleton, F. H., *p. s. c.* (51, 57) . .	1 Sep. 15	*l. c.* 3 June 16	A. A. & Q. M. G.
4 May 98	Henley, C. F. (51, 52, 57, 72*b*) . .	1 Sep. 15		
	Captains			
18 Jan. 02	Morland, W. E. T., D.S.O., M.C. (51, 66, 72)	22 Jan. 14		
6 June 03	Stephens, R. (51, 52, 207) (*act'g. Lt.-Col.*)	14 Mar. 14	. . . .	Comdg. Bn., Glouc. R.
24 Jan. 06	Whittall, G. F., M.C., (*act'g. Lt.-Col.*)	8 Jan. 15	*m* 3 June 17	Comdg.
30 Dec. 08	Powell, J. J. (50, 59)	10 June 15	. . . .	M. G. Corps.
9 June 09	Mundey, S. C. B. (207)	10 June 15	. . . .	
19 Jan. 12	Birch-Reynardson, H. T. . . .	25 Nov. 15	. . . .	*Lv., ex I., m. c., 6 m.*
2 Sep. 13	*Titherington, G. W.*	2 Mar. 16	*m* 3 June 18	*Adjt.*
1 Oct. 14	Meade, J. W.	1 Jan. 17	. . . .	*Lv., ex I., m. c., 6 m.*
7 Nov. 14	Loverock, R. C., D.S.O.	1 Jan. 17	. . . .	Staff Capt.
9 Nov. 14	Wooding, W. W.	1 Jan. 17		
	Lieutenants.			
7 Nov. 14	Mason, A. E.	10 June 15		
7 Nov. 14	Naylor, G.	27 Sep. 15		
7 Nov. 14	Murphy, D.	27 Oct. 15		
15 Sep. 15	Hardie, F. M.	11 Dec. 16	. . . .	*Lv., ex I., m. c.*
	2nd-Lieutenant.			
14 Feb. 17	Cowan, J. L.	14 Feb. 17		
8 Sep. 17	Murray, C. H.	8 Sep. 17		
20 Sep. 18	Cassidy, J. M.	20 Sep. 17		
2 Sep. 13	Titherington, Bt. Maj. G. W. . .	2 May 16	. . . .	Adjt.
22 Sep. 03	Ivey, Capt. T. (66*a*)	22 Sep. 13	. .	Qr.-Mr.

Agents:—MESSRS. Cox & Co.

Regimental District No. 43. [No. 7 District.]

Uniform—Scarlet. *Facings*—White.

The Oxfordshire and Buckinghamshire Light Infantry.

1ST GARRISON BATTALION.

Arrived in India, 4th March 1916.

First Commn.	Names.	Regimental rank.	Army rank.	Remarks.
	Lieutenant-Colonel.			
9 Dec. 14	Boucher, Lt.-Col. B.H., D.S.O., ret. pay (*temp.*) (50, 52, 57, 71, 172*b*, 207)	1 Feb. 17		
	Major (2nd-in-Command).			
	Majors.			
. . . .	Mitford, B. V. (3rd Devon R.) . .	. . . .	4 Nov. 14	Pension Paymaster, Madras
. . . .	Henderson, J. H. (K. R. R. C.) (51, 105, 106, 107, 137, 206) . .	. . . .	11 Jan. 15	
	Captains.			
. . . .	Maggs, C. W., V.D., (*Hon. Maj.*), *4 Wilts R.*	. . . .	11 Sep. 14	
. . . .	Collings, T. H., *5 E. Kent. R.* . .	. . . .	29 Oct. 14	
	Gidley, J., V.D.	17 Oct. 15	7 Dec. 14	S. S. O. & Cant. Magte., Bellary.
22 Dec. 14	Hallam, D. E. (*temp.*) (51) . .	2 Oct. 15	8 Jan. 15	
22 Feb. 15	Gilbert, B. J. (*temp.*)	2 Oct. 15	22 Feb. 15	Ind. Ord. Dept.
9 Apl. 15	Willis, R. G. (46) (*temp.*) . . .	10 Feb. 16		
14 Mar. 12	Pendlebury, J. W., M.C., E. Lan. R. (51)	28 Sep. 16	1 Apl. 16	
. . . .	Phelps, J. W., 4 Wilts R. . . .	. . . .	1 June 16	Asst. Cable Censor, Rangoon.
7 Oct. 14	Bone, C. B. (51)	2 Jan. 17	1 June 16	S. S. O. & Cant. Magte., Wellington.
	Lieutenants.			
. . .	*Gibbs, F. J. (temp.)*	4 Oct. 15	. . . .	*I. A. R. O.*
8 Aug. 14	Evans, E. A. (48*a*, 51)	. . . .	1 Sep. 15	I. A. (*on probn.*)
3 June 15	*Mattock, L. A. (act'g. Capt) 16 Oct. 18.* (48*a*, 51, 61)	. . . .	3 Jan. 16	*Adjt.*
2 Oct. 14	Burt, W. H. (51)	. . . .	1 June 16	Comdt., P. of W. Camp, Ramandrug.
10 June 15	Martin, A. J. (*temp.*)	10 Feb. 16		
28 May 15	Prior, E. G. (*temp.*)	16 June 16	. . . .	Ind. Ord. Dept.
27 Jan. 15	Cavies, R. T.	. . . .	1 June 16	S. S. O., Kasauli.
23 Nov. 15	Goodban, L., 10th Midd'x R. . .	. . . .	1 July 17	
. . . .	Molony, J. T., 4th Bn., Dorset R. .	. . . .	1 July 17	
	2nd-Lieutenants.			
7 Oct. 14	Orchard. G. F. (51)	. . . .	7 Oct. 14	
3 Dec. 14	Merton, J. (*temp.*) (51)	2 Oct. 15	3 Dec. 14	
29 Dec. 14	Burns, J. (*temp.*)	4 Oct. 15	29 Dec. 14	
16 Jan. 15	Jones, B. L. (*temp.*) (57) . . .	24 Nov. 15	16 Jan. 15	
10 June 15	Mare, E. J. (*temp.*)	2 Oct. 15	10 June 15	I. A. R.
10 Sep. 15	Lawson E., R. Highrs. (12, 52, 57)	. . . .	10 Sep. 15	
. . . .	Battishill, J. H., 3rd Bn., R. W. Kent R	. . . .	15 Sep. 15	
17 July 17	Lole, E. F.	. . . .	17 July 17	
17 July 17	Powell, C. A.	. . . .	17 July 17	
19 Aug. 17	Fowler, A. G. (51)	. . . .	19 Aug. 17	
19 Aug. 17	Mallett, G. W. L.	. . . .	19 Aug. 17	*Lv., ex I., m. c., 6 mos.*

3 June 15	Mattock, Lieut. (*act'g. Capt.*) L. A.	1 Oct. 18	. . . .	Adjutant.
22 Sep. 15	Hayes, Lt. G. F. L.	22 Sep. 15	22 Sep. 15	Qr.-Mr.

Regimental District No. 44
[No. 9 District.]

Uniform—Scarlet.
Facings—White.

The Essex Regiment.

2ND GARRISON BATTALION.

Arrived in India, 16th March 1916.

HONORARY COLONEL—

First Commn.	Names.	Regimental rank.	Army rank.	Remarks.
	Lieutenant-Colonel.			
	Majors.			
Feb. 80	Bagnall, C. E., (*act'g. Lt.-Col.*) (*Res. of Offr.*)	28 Jan. 16		
85	Christie, E. L. (*temp.*) (51) . . .	28 Jan. 16	4 Jan. 15	
	Sinton, C. McD. (5th Bn., E. York R.)	. . .	. . . .	2nd-in-Comd.
	Captains.			
15 Apl. 99	de St. Croix, C. (*act'g. Maj., 12 Apl. 18*), (3rd R. Suss. R.), (*Ret. pay*) (*51*)	28 Jan. 16	1 Aug. 13	
Feb. 01	Borissow, R. W. (*temp.*)	29 Jan. 16	29 Nov. 14	M. A. Dept., Lahore.
00	Herring, C. S. (*temp.*)	29 Jan. 16	19 Dec. 14	
18 Nov. 14	Gerard, M. F. M. J. (*temp.*) (51) .	28 Jan. 16	. . . .	M. A. Dept., Poona.
25 July 15	Healy, H. H. (*temp.*).	28 Jan. 16		
	Lieutenants.			
29 Jan. 15	Hubble, L. U. (*temp.*) (*act'g. Capt.*)	28 Jan. 16	29 Jan. 15	S. S. O., Jhansi.
Dec. 00	Levy, M. P. (*temp.*) (51)	28 Jan. 16	5 Feb. 15	
15 Aug. 14	Sinclair, W. V. S., (3rd E. Kent R.) .	28 Jan. 16	. . . .	S. S. O., Kamptee.
5 Nov. 14	Beney, G. E. (*temp.*)	28 Jan. 16	. . . .	I. A. R. O.
14 Jan. 15	Harwood, G. (*temp.*), (3rd Suff. R.)	28 Jan. 16	. . . .	With R. A. F.
18 Apl. 15	Newfield, M. (3rd Bedf. R.) (57, 207)	28 Jan. 16	15 Nov. 16	
2 Apl. 15	Knowles, J. B. (6th R. Fus.) . .	28 Jan. 16	22 Nov. 16	
5 May 15	Cox, R. N., (*act'g. Capt. 21 Feb. 18*) (6th Midd'x. R.)	1 July 17	5 May 15	Adjt.
19 May 15	Lester, H. N. (*3 Bn., R. Suss. R.*) .	1 July 17	19 May 15	I. A. (*On probn.*)
	2nd-Lieutenants.			
13 May 15	Down, G. O. (*temp.*)	28 Jan. 16	13 May 15	I. A. R. O.
13 May 15	Rae, A. D. (51)	28 Jan. 16	13 May 15	Supvg. Offr., Physical & Bay. Trg., 5th (Mhow) Dn.
22 May 15	Kingston, C. G. (*3rd Bn., E. Surr. R.*) (51)	28 Jan. 16	22 May 15	
17 June 15	Detmold, K. J. (*temp.*) (57) . . .	28 Jan. 16	17 June 15	With Sig. Serv.
1 Oct. 15	Stanforth, L. H. (*temp.*)	28 Jan. 16	1 Oct. 15	

5 May 15	Cox, Lieut. (*act'g. Capt.*) R. N. .	21 Feb. 18	. . . .	Adjutant.
13 Feb. 16	Harman, Lt. T. E.	13 Feb. 16	. . . .	Qr.-Mr.

Regimental District No. 47. [No. 3 District.]

Uniform—Scarlet. *Facings*—White.

The Loyal North Lancashire Regiment.

2ND BATTALION (*late 81st Foot*).

Arrived in India, 8th December 1909, from Mauritius.

THE RED ROSE.

"Louisburg" "Quebec, 1759" "Maida" "Corunna" "Tarifa" "Vittoria" "St. Sebastian" "Nive" "Peninsula" "Ava" "Alma" "Inkerman" "Sevastopol" "Ali Masjid" "Afghanistan, 1878-79" "Defence of Kimberley" "South Africa, 1899-1902."

COLONEL—General *Sir* J. Willcocks, G.C.M.G., K.C.B., K.C.S.I., D.S.O. . . . 12 Apl. 16.

First Commn.	Names.	Regimental rank.	Army rank.	Remarks.
	Lieutenant-Colonel.			
22 Aug. 88	Jourdain, C. E. A., D.S.O. (50, 53, 66*a*, 72*a*, 72*b*)	17 Dec. 13		
	Majors.			
7 Dec. 95	Berkeley, R. E., D.S.O. (51, 52, 57, 65, 71).	18 Aug. 15		
3 Sep. 02	Logan, R. H. (51, 57, 72, 72*b*) . .	9 Oct. 15		
	Captains.			
20 May 05	Knapp, A. F. P. (52, 72, 72*b*)	5 Nov. 14		
20 Apl. 10	Atkinson, G. P., M.C. (51, 66*a*, 72*b*) .	9 Feb. 15		
4 Dec. 12	*Withers, B. H.*	19 Nov. 15	. . . .	*Adjt.*
22 Jan. 13	Keays, R. L. C.	24 Nov. 15	. . . .	With 7th. Bn.
7 Nov. 14	Chambers, C. S.	20 Feb. 17		
7 Nov. 14	Parker, W. (50, 53, 71, 72*b*) . . .	31 July 17		
	Lieutenants.			
24 Jan. 14	Williams, G. G. R. (207)	15 Dec. 14		
7 Nov. 14	Crosby, C. (51, 58, 72, 72*b*)	1 Oct. 15		
7 Nov. 14	Chalk, P.	1 Oct. 15		
4 Sep. 15	Woodgate, H. P.	1 Jan. 17		
15 Sep. 15	Scholfield, C. G. (60*a*)	1 Jan. 17		
15 Sep. 15	Leonard, H. B. (51, 52)	1 Jan. 17		
1 Nov. 15	Wilkinson, H. (50, 53)	1 Jan. 17		
12 Mar. 16	Hooper, Lt. E. G. (71)	12 Sep. 17		
30 July 16	Spencer, J. (51, 60*a*, 66*a*, 72, 72*b*) .	30 Jan. 18		
	2nd-Lieutenant.			
16 Mar. 18	Waterworth, J. W. (*temp.*)	16 Mar. 18		
	Attached.			
	Seabroke, Maj. C., 7 R. War. R. .	29 Aug. 14		
	Walsh, Capt. H. P., 3 N. Lan. R. .	18 Aug. 15		
	Taylor, Capt. R. V., 8 R. War. R. .	19 Aug. 15		
	Morgan, Capt. F. L., 6 R. War. R. .	1 June 16		
	Brockbank, Capt. G. R., 5 R. War. R.	1 June 16		
	Jones, Capt. E. W., 6 R. War. R. .	1 June 16		
	McDonald, Lt. W. (temp.) 8th Cam'n Highrs.	24 Feb. 15		
	Walker, Lt. F. W., So. Af. Def. Forces.	28 Sep. 15		
	Cartwright, Lt. E. A., 3 Lan. Fus. .	8 Mar. 16		
	Martindale, Lt. J. B., 3 Lan. Fus. .	1 Apl. 16		
	Molyneux, Lt. E., 3 N. Lan. R. . .	26 Aug. 16		
	Martin, Lt. W. E., 3 N. Lan. R. . .	3 Oct. 16		
	Soppitt, temp. 2nd-Lt. H. J.. . .	11 Nov. 14		
	Baldwin, 2nd-Lt. N., 3 N. Lan. R. .	6 May 15		
	Marsh, temp. 2nd-Lt. F. H. . . .	6 May 15		
	Linaker, temp. 2nd-Lt. A. F. R., 12 S. Lan. R.	8 July 15		
	Evans, temp. 2nd-Lt. R. E., 9 L'pool R.	27 July 15		
	Thomson, temp. 2nd-Lt. P. R., E. Surr. R.	27 Oct. 15		
	Strang, 2nd-Lt. W., 7 R. War. R. .	25 June 16		
	Gisborne, 2nd-Lt. H. P., 3 W. Rid. R.	30 July 16		
	Dawson, 2nd-Lt. G. A., 4 R. Lanc. R.	19 Dec. 16		
	Mercer, 2nd-Lt. J. A., 4 R. Lanc. R.	19 Dec. 16		
	Hughes, 2nd-Lt. I., 3 N. Lan. R. . .	28 Mar. 17		
	Kirk, 2nd-Lt. G. B., 3 N. Lan. R. .	28 Mar. 17		
	Dutton, temp. 2nd-Lt. C. C. . . .	17 May 17		
	Nightingale, 2nd-Lt. C. G., D.C.M., 7 N. Lan. R.	17 Nov. 17		
4 Dec. 12	Withers, Capt. B. H.	28 June 17	. . . .	Adjt.
10 Apl. 12	Rowley, Lt. R.	10 Apl. 12	. . . .	Qr.-Mr.
23 Oct. 15	Lumb, temp. Capt. T. F., R.A.M.C.	23 Oct. 16	. . . .	Medl. Offir.

Agents:—MESSRS. COX & CO.

Regimental District, No. 50. [No. 10 District.]

Uniform—Scarlet. *Facings*—Blue.

The Queen's Own (Royal West Kent Regiment).

2ND BATTALION (*late 97th Foot*).

Arrived in India, 7th December 1908, from Singapore.

Quo Fas et Gloria ducunt.—THE SPHINX, SUPERSCRIBED "EGYPT."
"Vimiera" "Corunna" "Almaraz" "Vittoria" "Pyrenees" "Nive" "Orthes"
"Peninsula" "Punniar" "Moodkee" "Ferozeshah" "Aliwal" "Sobraon"
"Alma" "Inkerman" "Sevastopol" "Lucknow" "New Zealand"
"Egypt, 1882" "Nile, 1884-85" "South Africa, 1900-02."

COLONEL—Major-General *Sir* E. Leach, K.C.B., C. ret. pay . . 3 Apl. 04.

First Commn.	Names.	Regimental rank.	Army rank.	Remarks.
	Lieutenant-Colonel.			
26 June 89	Woulfe Flanagan, R. J. [*l*] (51, 71, 72*b*)	14 Mar. 16		
	Majors.			
23 Dec. 93	Kitson, C. E., D.S.O. (*temp. Lt.-Col.*) (Q. 12, 50, 57, 71)	21 Mar. 12	*l.c.* 1 Jan. 16	A. A. & Q.-M.-G., Poona Dn.
3 Jan. 00	Nelson, J. W., D.S.O. (50, 53, 65)	1 Sep. 15		
21 Apl. 00	Belgrave, H. D., D.S.O., *p. s. c.* (*temp. Lt.-Col.*) (50, 207)	1 Sep. 15	*l.c.* 3 June 16	A. Q. M. G.
9 Mar. 01	Ingram, C. R. (51, 59, 66*a*) . .	8 Jan. 16	. . . .	(Supy.) D. A. Q.-M.-G.
4 Dec. 01	Case-Morris, C. W. (*act'g. Lt.-Col.*) (51, 66*a*)	14 Sep. 16	. . . .	With No. 1 Res. Bn.
	Captains.			
14 Aug. 01	Dinwiddy, M. J. (51, 57, 60*a*) . .	28 Mar. 08		
18 Jan. 02	Hibbert, O. Y., D.S.O., *p. s. c.* (51) .	26 Mar. 13	. . . .	D. A. A.-G.
2 Mar. 04	Searight, A. K. [L] (51, 52, 72, 95, 169)	1 Sep. 14	. . . .	Spec. Serv. Offr.
24 Jan. 06	Whitty, N. I., D.S.O. (*temp. Maj.*, 27 *Sep.* 15) (51, 52)	19 Sep. 14	. . . .	With 6th Bn.
4 July 06	Hardy, A. E., M.C. (*act'g. Maj.*), (51)	28 Oct. 14		
27 May 08	Bredon, A. S. (60*a*)	24 Nov. 14	. . . .	*Lv., ex I.*
27 Aug. 10	Wilberforce-Bell, P. F., M.C. (51, 52, 60*a*, 172*a*)	23 July 15	. . . .	*Lv., ex I.*
4 Mar. 11	Clarke, V. S. (169, 207)	13 Sep. 15		
20 Dec. 11	Barnes, O. G. R. (172*a*, 207)	26 Nov. 15		
5 Oct. 10	Battye, C. F., M.C. (*temp. Maj.*, 16 *Feb.* 18 (52).	26 Mar. 16	. . . .	M. G. Coy.
10 Dec. 13	*Balbernie, A. G.*, M.C.	10 June 16	. . . .	*Adjt.*
7 Nov. 14	Howe, A.	21 Nov. 17	. . . .	With 26th R. Fus.
7 Nov. 14	Madgett, C.	21 Nov. 17	. . . .	*Lv., ex I., m. c.*
7 Nov. 14	Piggott, A. H.	1 Jan. 18		
	Lieutenants.			
25 Feb. 14	Chitty, A. A. E.	19 Sep. 14	. . . .	R. M. Coll.
7 Nov. 14	Harrison, R.	23 July 15	. . . .	*Lv., ex I., m. c.*
16 Dec. 14	Hilder, E. C.	3 May 16		
16 Dec. 14	Lees, R. M.	10 June 16		
23 Dec. 14	Fry, O. M.	11 Oct. 16		
8 Feb. 15	Hart, A. C. (*temp. Capt.*) . . .	10 Dec. 16	. . . .	Staff Capt.
20 Mar. 15	Bennett, H. B. (temp.)	27 Dec. 16		
June 15	Mills, J.	27 Dec. 16		
19 July 16	Butler, C. J.	1 Jan. 17		
16 Oct. 15	May, E. (*temp. Capt.*)	10 Apl. 17	. . . .	D. A. Q. M. G.
19 Nov. 15	Dix, W. (*actg. Capt.* 18 *Apl.* 18) .	1 July 17	. . . .	With S. & T. Corps.
5 Sep. 16	Hogbin, F. R. (*temp.*)	5 Mar. 18	. . . .	
24 Oct. 16	Capes, C. J.	27 Apl. 18		

The Queen's Own (Royal West Kent Regiment)—*contd.*

First Commn.	Names.	Regimental rank.	Army rank.	Remarks.
	2nd-Lieutenants.			
19 Nov. 15	Burns, H. J.	19 Nov. 15	....	R. A. F., Egypt.
13 Mar. 17	Johnson, P. F.	13 Mar. 17		
13 Mar. 17	Lander, F. J.	13 Mar. 17		
28 Mar. 17	Cowper, R. H. (*temp.*)	28 Mar. 17		
28 Mar. 17	Gains, A. T. (*temp.*)	28 Mar. 17		
30 May 17	Scott, J. (*temp.*)	30 May 17		
30 May 17	Hird, S. (*temp.*)	30 May 17	....	Army Clothing Factory.
30 May 17	White, W. F., M.M. (*temp.*)	30 May 17		
30 May 17	Woodward, D. L. J. (*temp.*)	30 May 17		
29 Aug. 17	Eastwick, A. J. (*temp.*)	29 Aug. 17		
29 Aug. 17	Lascelles, H. B. (*temp.*)	29 Aug. 17		
12 Nov. 17	Leigh, J. T.	12 Nov. 17		
13 Jan. 18	Craddock, C. W.	13 Jan. 18		
13 Jan. 18	Reeves, G. L.	13 Jan. 18		
16 Mar. 18	Swales, H. G., (*temp.*)	16 Mar. 18	....	Wireless Unit, Rawalpindi.
16 Mar. 18	Lawton, J., (*temp.*)	16 Mar. 18		
19 Aug. 18	Hewitt, B. E. (*temp.*)	19 Aug. 18		
	Attached.			
	Bonham-Carter, Capt. G., 3 Bn.	30 Oct. 14		
10 Feb. 10	Herman, Capt. G. L., 3rd Bn.	22 Aug. 15		
5 Nov. 13	Hooper, Capt. A. F., 1-5th R. W. Kent R.	2 Sep. 14		
	Aldworth, Capt. T. P., 3rd Bn.	26 Sep. 14		
	Robertson, Capt. M., 3 Bn.	20 Oct. 15		
	Boulter, Capt. F. H., 1st Kent. Bn.	1 June 16		
	Whicher, Capt. J. D'A., Notts. & Derby R.	1 Jan. 17		
	Pattisson, Capt. C. A. S., 3rd Bn.	21 Nov. 17		
4 Nov. 14	Huggan, Lt. G. H., 3rd Bn.	21 Nov. 17		
17 Aug. 14	Battershill, Lt. W. D., 6th E. Surr. R.	1 June 16		
	Booth, Lieut. W., 3 Bn.	1 July 17		
30 Dec. 14	Henn, Lt. J. F., 3rd Bn.	1 July 17		
7 Dec. 14	Elverson, Lt. C. G. L., 3rd Bn.	1 July 17		
21 Apl. 15	Chester, Lt. C. F. L., 3rd Bn.	1 July 17	....	R. A. F.
3 May 15	Cane, Lt. C. H., 3rd Bn.	1 July 17		
9 May 15	Chadwick, Lt. C. R., 3rd Bn.	1 July 17	....	Stokes Mortar Batty.
16 May 15	Vine-Stevens, Lt. L. T., 3rd Bn.	1 July 17		
16 May 15	Day, Lt. J. H., 3rd Bn.	1 July 17		
22 May 15	Tindall, Lt. A., 3rd Bn.	1 July 17		
17 June 15	Cockle, Lt. H. M., 7 Bn.	1 July 17		
22 July 15	Southan, Lt. J. H. P., 3rd Bn.	1 July 17	....	R. A. F.
2 Sep. 15	Bradley, Lt. G. G., 3rd Bn.	1 July 17	....	I. A. R. O.
23 Sep. 15	Morrish, Lt. J. H., 3rd Bn.	1 July 17		
	Lovell, Lieut. I. R., 1 Bn.	27 Apl. 18		
7 July 16	Layton, Lt. E. C.	7 Jan. 18		
28 July 16	Paulson, Lt. P. (*actg. Capt.*)	8 Feb. 18		
5 Aug. 16	Whitehead, Lt. H. E., 3rd Bn.	5 Feb. 18	....	Stokes Mortar Batty.
24 Oct. 16	Mansfield, 2nd-Lt. F. A.	24 Oct. 16	....	L. T. M. Batty.
22 Nov. 16	Pettigrew, 2nd-Lt. J. D.	22 Nov. 16	....	L. T. M. Batty.
28 Apl. 17	Crook, 2nd-Lt. C. J. (Genl. List)	28 Apl. 17	....	R. A. F., Egypt.
30 May 17	Highley, 2nd-Lt. H. F. (*temp.*) 3 Bn.	30 May 17		
30 May 17	Williams, 2nd-Lt. G. A. (*temp.*) 3 Bn	30 May 17		
27 June 17	Vercoe, 2nd-Lt. F. G. (*temp.*) 3 Bn.	27 June 17		
31 Oct. 17	Wells, 2nd-Lt. A. (*temp.*)	31 Oct. 17		
31 Oct. 17	Simpson, 2nd-Lt. J. (*temp.*)	31 Oct. 17		
31 Oct. 17	Barker, 2nd-Lt. B. M. (*temp.*)	31 Oct. 17		

10 Dec. 13	Balbernie, Capt. A. G., M.C.	18 Mar. 17	....	Adjt.
27 Nov. 12	Grey, Maj. F. (50, 52, 72*b*, 172*a*)	3 June 18	....	Qr.-Mr.
27 July 16	Colley, Capt. R., R.A.M.C.	2 Mar. 17	....	Medl. Offr.

Agents —MESSRS. COX & CO.

Regimental District No. 50. Uniform—Scarlet.
[No. 10 District.] *Facings*—Blue.

The Queen's Own (Royal West Kent Regiment).

1-4TH BATTALION.

Arrived in India, 2nd December 1914, from England.

"South Africa, 1900-02."

HONORARY COLONEL—C. S. Williams, V.D. (*Hon. Col., ret. vols*). 20 Dec. 06.

First Commn.	Names.	Regimental rank.	Army rank.	Remarks.
	Lieutenant-Colonel.			
6 Aug. 88	Watney, C. N., T. D. (*Hon. Lt. in Army, 30 May 01*)	4 Jan. 11		
	Major.			
2 Aug. 99	Cohen, A. M. [L] (51, 170) . . .	18 May 16	. . . .	P. A. to the A. G. G. in Baluchistan.
2 Dec. 05	Robb, C. B. (51, 57, 71, 170) (*temp. Lt.-Col., 30 Sep. 18*).	1 June 16	. . . .	A. A. & Q. M. G., Meerut Dn.
9 Feb. 06	Cheale, A. R. (51)	1 June 16		
6 Feb. 04	Henson, F. J.	1 June 16	. . . .	Bde. Maj. Ahmedabad Bde.
	Captains.			
1 Jan. 07	Kelsey, A. R. (51)	1 Sep. 13		
27 Feb. 09	*Watney, R. D.* (51, 52, 57) . .	31 Dec. 14		
3 Sep. 10	*Bourne, F. C.* (51)	1 June 16	. . . .	*Bde. Maj., 5th Infy. Bde., Sialkot.*
4 Dec. 10	Norman, C. L. (46, 51, 52, 172a) .	1 June 16		
3 Mar. 09	Stone, F. L. (51)	1 June 16		
5 June 11	Robinson, H. I.	1 June 16		
6 June 11	Locket, L. G. (51, 60, 60a, 63, 95, 100)	1 June 16		
3 May 12	Nash, W. (51, 66, 95)	1 June 16		
11 May 12	*Pardington, G. E. L.* (52) . . .	1 June 16	. . . .	Armoured Motor Bde., Peshawar.
17 Sep. 13	Peareth, W. F. E.	1 June 16		
3 Jan. 14	Styles, H. W. (51)	1 June 16		
5 Aug. 14	*Rooker, G.* (57)	1 June 16	. . . .	*I. A. R. O.*
21 Aug. 14	Wingfield-Stratford, E. C. (51) .	1 June 16	. . . .	Br. Infy. Details, Muttra.
21 Aug. 14	Golding, G. C. (51, 52)	1 June 16		
26 Aug. 14	Trought, T. (57)	1 June 16	. . . .	G. S. O., 3rd gr., N. Comd.
14 Oct. 14	Hartree, K. (57)	9 Nov. 16		
14 Oct. 14	Wilson, R. P. (57)	9 Nov. 16	. . . .	Aden Sig. Coy.
	Lieutenants.			
18 Oct. 14	Douglas, R. O. (46, 51, 60a, 94, 172a, 207).	1 June 16		
19 Oct. 14	*Swanston, J. F. A.* (51, 52). . .	1 June 16	. . . .	*M. G. Corps.*
11 Nov. 14	Stern, C. E. (51)	1 June 16		
12 Aug. 15	Macnicol, A. C.	1 July 17	. . . .	*Lv., ex I., m. c., 12 m.*
21 Aug. 15	Dunbar, W. R.	1 July 17	. . . .	*Lv., ex I., m. c., 6 m.*
9 Oct. 14	Tester, A. F.	1 July 17	. . . .	Attd. G. S. Bch., Army Hdqrs.
1 Sep. 15	Mitchell, A. J.	1 June 16	. . . .	
1 Nov. 15	*Taylor, L.* (*actg.-Capt*) (51, 52, 61, 71)	1 July 17	. . . .	Adjutant.
9 Dec. 15	Staniforth, J. R.	1 July 17		
	2nd-Lieutenants.			
25 Jan. 17	Quirke, J. P. F. (57)	25 Jan. 17	. . . .	Mily. Accts. Dept.
25 Jan. 17	Bowyer, F. D. (95).	25 Jan. 17		
25 Apl. 17	Etheridge, F. H. (12, 51) . . .	25 Apl. 17		
25 Apl. 17	Weedon, C. W. (51)	25 Apl. 17		
25 Apl. 17	Collins, G. H. (46)	25 Apl. 17	. . . .	With R. A. F.
25 Apl. 17	Bird, T. E. H.	25 Apl. 17	. . . .	Mily. Accts. Dept.
29 Aug. 17	Cave, A. (12)	29 Aug. 17		
16 Mar. 18	Gore, A., (*temp.*)	16 Mar. 18	. . . .	With R. A. F.
16 Mar. 18	Hammond, G. T., (*temp.*) . . .	16 Mar. 18	. . . .	With R. A. F.
16 Mar. 18	Brown, J. H., (*temp.*)	16 Mar. 18		
16 Mar. 18	Heath, E., (*temp.*)	16 Mar. 18	. . . .	Mech. Tpt., Rawalpindi.
19 Aug. 18	Pedgrift, A. E., D.C.M.	19 Aug. 18		
1 Nov. 15	Taylor, Lieut. (*actg. Capt.*) L. .	20 July 18	. . . .	Adjt.
24 Dec. 15	Carter, Lt., W. W.	24 Dec. 15	. . . .	Qr.-Mr.

Regimental District No. 50.
[No. 10 District.]

Uniform—Scarlet.
Facings—Blue.

The Queen's Own (Royal West Kent Regiment).

1-5TH BATTALION.

Arrived in India, 3rd December 1914, from England.

"South Africa, 1900-02."

First Commn.	Names.	Regimental rank.	Army rank.	Remarks.
	Lieutenant-Colonel.			
6 Jan. 00	Frazer, F. A.	9 Apl. 13		With No. 2 Res. Bn.
	Majors.			
21 Mar. 00	Clark, C. D.	9 Aug. 12		
19 July 06	Hills, A. E. (51)	1 June 16		
26 July 02	Neame, L. V. (51)	1 June 16		
1 Nov. 06	Hairston, J.	1 June 16		Attd. M. W. S.
	Captains.			
1 Apl. 08	*Satterthwaite, R. E.* (51)	7 Oct. 12		*Bde. Maj., 11th Infy. Bde.*
5 Mar. 10	Keyes, R. (51, 52)	9 Apl. 13		
1 July 10	Richardson, H. L. (51, 52)	27 Apl. 13		
23 July 10	*Hay, J. H.* (51)	19 May 14		
26 Nov. 00	*Brewis, R. D. (temp. Maj.)* (51)	19 Aug. 14	17 Feb. 12	*D. A. Q. M. G., Meerut Dn.*
15 Feb. 10	*Lepper, J. G.* (51)	16 Sep. 14		*G. S. O., 3rd gr.*
11 May 12	Marshall, T. D. (51)	1 June 16		
7 June 13	*Stokes, C. L. L.* (57)	1 June 16		*I. A. R. O.*
5 Nov. 13	Hooper, A. F.	1 June 16		With 2nd Bn.
24 Jan. 14	Pirie, G. B. (46)	1 Jnue 16		
6 Oct. 13	Brooks, L. G. (57)	1 June 16	6 Oct. 13	
22 Aug. 14	Pritchard, J. M. (51, 52)	1 June 16		
2 Sep. 14	*Engledow, F. L.* (51, 52)	1 June 16		*Adjt.*
2 Sep. 14	Carvosso, J. P. (51, 60*a*, 66)	1 June 16		
2 Sep. 14	Carvosso, E. W. (51)	1 June 16		
	Lieutenants.			
2 Sep. 14	Smithers, K. O. (51, 52)	1 June 16		S. S. O., Jubbulpore.
2 Sep. 14	Neame, E. M. (51)	1 June 16		With R. A. F.
2 Sep. 14	*Bakewell, W. B.* (51, 52)	1 June 16		*M. G. Corps.*
12 Sep. 14	Stokes, T. A. (57, 60*a*)	1 June 16		
29 Oct. 14	*Stokes, E. V.*	1 June 16		*With Labour Bn.*
6 Nov. 14	*Biggs, F. W. (temp. Capt.)* (46, 51, 60*a*)	1 June 16		
18 Mar. 15	Hawes, H. F. (46, 52)	9 May 17		
8 June 15	Jones, E. L.	9 May 17		*Lv., ex I., m.c., 6 mos.*
18 Sep. 15	*Sotham, R. C.*	1 July 17		*With R. A. F.*
18 Sep. 15	*Staddon, C. E.*	1 July 17		
19 Oct. 15	*Russell, V. H.* (46)	1 July 17		*With R. A. F.*
13 Dec. 15	*Taylor, H. P.* (51)	1 July 17		
1 Jan. 16	*Bennett, G.*	1 July 17		*M. G. Corps.*
15 Jan. 16	Millen, J. A. (60*a*)	15 July 17		
	2nd-Lieutenants.			
20 Jan. 16	Gripper, H. E.	20 Jan. 16		*Lv. ex I., m.c.*
25 Jan. 17	*Clark, A. N. W.*	25 Jan. 17		A.-D.-C.
29 Aug. 17	Proctor, R. H.	29 Aug. 17		
16 Mar. 18	Elphick, L. S.	16 Mar. 18		
16 Mar. 18	Schocktee, S. H.	16 Mar. 18		
16 Mar. 18	Baker, H. A. J.	16 Mar. 18		
	Attached.			
10 Mar. 15	Stockdale, Lt. T. C., 2nd Garr. Bn., North'd. Fus.	2 Dec. 15		
15 Jan. 15	Armstrong, Lt. G. G., 2nd-Garr. Bn., North'd. Fus.	1 July 17		
5 Sep. 16	Parker, Lt. T. C., 2nd Bn., R. W. Kent R.	5 Mar. 18		
19 Dec. 16	Key, 2nd-Lt. A. R., The Cambridgeshire R.	19 Dec. 16		
1 Mar. 17	Dunn, 2nd-Lt. J. Mc.L., Gord. Highrs.	1 Mar. 17		
1 Mar. 17	Waley, 2nd-Lt. R. P. S., 2nd Bn., R. W. Kent R.	1 Mar. 17		
26 Apl. 17	Powell, 2nd-Lt. G. M., 276 M. G. Coy.	26 Apl. 17		
28 Apl. 17	Ware, 2nd-Lt. W. J. K., 1-4th Bn., R. W. Kent R.	28 Apl. 17		
28 Apl. 17	Lane, 2nd-Lt. W. G., 1-1st Bn., Breck. Bn., S. Wales Bord.	28 Apl. 17		
28 Apl. 17	Drinkall, 2nd-Lt. R., 2nd Bn., R. W. Kent R.	28 Apl. 17		
30 May 17	Manning, 2nd-Lt. C.L., 2nd Bn., R.W. Kent R.	30 May 17		
1 Aug. 17	Watt, 2nd Lt. J. O., Gord. Highrs.	1 Aug. 17		
29 Aug. 17	Welsh, 2nd-Lt. G. J., 2nd Bn., R. W. Kent R.	29 Aug. 17		
29 Aug. 17	Gilham, 2nd-Lt. ., 2nd Bn., R. W. Kent R.	29 Aug. 17		
29 Aug. 17	Barrons, 2nd-Lt. J. E., 2nd Bn., R.	29 Aug. 17		

First Commn.	Names.	Regimental rank.	Army rank.	Remarks.
2 Sep. 14	Engledow, Capt. F. L.	1 June 16		Adjt.
1 July 08	Cooke, Lt. J. H., (50, 52, 72)	1 July 08		Qr.-Mr.
11 July 03	Jefferiss, Maj. F. B., F.R.C.S. (Edin). R.A.M.C. (T.F.)	11 July 15		Medl. Offr., i/c. No. 2 Sec. Hospital, R'Pindi.
1 Nov. 15	MacDougall, Capt. D.K., R.A.M.C.	18 Dec. 16		Medl. Offr.

Regimental District No. 57.
[No. 10 District.]

Uniform—Scarlet.
Facings—Lemon Yellow.

The Duke of Cambridge's Own (Middlesex Regiment).

1-9TH BATTALION.

Arrived in India, 3rd December 1914, from England.

"South Africa, 1900-02."

First Commn.	Names.	Regimental rank.	Army rank.	Remarks.
	Lieutenant-Colonel.			
	Majors.			
22 Feb. 99	Hewett, W. P. (*act'g. Lt.-Col., 17 Mar. 18.*)	5 Jan. 10		
4 Apl. 00	Montgomerie, H. S.	17 July 12	. . . .	
4 Apl. 00	Beach, G. (51, 52, 57, 95, 169, 207)	1 June 16		
4 Apl. 00	Bartholomew, A. G. (*temp. Lt.-Col.*) (172*a*)	1 June 16		
	Captains.			
8 Nov. 07	Stratton, F. C. G. (50, 57)	8 Mar. 10		
3 July 08	Cotton, P. V. (*temp. Maj.*) (51)	8 Mar. 10	. . . .	D. A. A. G., A. H. Qrs.
1 Apl. 09	*Tupper, G. W. H.* (51)	13 July 12		
26 Feb. 02	Hewlett, L. M.	22 Aug. 14	13 Nov. 05	
9 July 08	*Mellersh-Jackson, L.* (51)	26 Aug. 14	. . . .	*Staff Capt., Bushire.*
6 Jan. 12	Mellersh-Jackson, W. (51)	16 June 15		
28 Feb. 12	*Little, C. W. R.* (57)	1 June 16	. . . .	*With Spec. Corps.*
3 Nov. 08	Hewlett, J. H. (51)	1 June 16		
14 Aug. 12	*Somervell, D. B.* (170, 206)	1 June 16	. . . .	*Staff Capt.*
15 Oct. 12	Fisher, C. W. (46, 51)	1 June 16	. . . .	*Lv., ex I., m.c., 12 m.*
3 May 13	*Toovey, C. W.* (51, 206)	1 June 16	. . . .	*I. A. (on probn.)*
7 Mar. 14	Lamont, J. N., M.C. (51, 52)	1 June 16		
	Lieutenants.			
22 Apl. 14	*Holt, V.* (57, 59, 169, 207)	29 Oct. 14	. . . .	*With Sig. Serv.*
22 Apl. 14	Mieville, A. F. H. (*act'g. Capt.*) (51, 52).	29 Oct. 14		
29 Apl. 14	Sherwood, H. (46, 66)	16 June 15		
19 Aug. 14	*Frankau, G. N.* (66)	1 Nov. 15	. . . .	*I. A. R. O.*
26 Aug. 14	Higgins, G. L. (51)	1 June 16	. . . .	Attd. G. S. Bch., Army Hdqrs.
26 Aug. 14	Thompson, H. W. (51, 57)	1 June 16		
26 Aug. 14	deVesian, N. E. (66, 172*a*)	1 June 16		
26 Aug. 14	*Findlay, L. F.* (51, 52, 170, 207*a*)	1 June 16	. . . .	*M. G. Corps.*
24 Oct. 14	Noad, C. H. C. (*temp. Capt., 3 Aug. 17*) (51, 52)	1 June 16	. . . .	Adjt., 4th Simla Rif. I. D. F.
7 Oct. 14	Duirs, F. D.	1 July 17		
21 May 15	Harris, C. R. (170)	1 July 17		
21 May 15	Fox, G. L. (51, 57, 172*a*)	1 July 17		
19 June 15	*Ballantine, E. W.*	1 July 17	. . . .	*Asst. Cable Censor, Bombay.*
20 June 15	Cogswell, A.	1 July 17		
29 June 15	Scott, E. (*act'g. Capt.*) (51, 57)	1 July 17	. . . .	Actg. Adjt.
7 Sep. 15	Purry, P. S. (46, 51, 60*a*)	1 July 17		
27 Feb. 16	*Holloway, B. T.* (46, 51, 60*a*)	*27 Aug. 17*	. . . .	*M. G. Corps.*
	2nd-Lieutenants.			
7 Jan. 15	Lanfear, E. W.	7 Jan. 15		
28 Apl. 17	Wright, K. E.	28 Apl. 17		
28 Apl. 17	Truelove, C. W. (*temp. Capt.*)	28 Apl. 17		
28 Apl. 17	Martin, F. E.	28 Apl. 17		
28 Apl. 17	Blower, J. W.	28 Apl. 17		
28 Apl. 17	Grainger, R. L.	28 Apl. 17		
28 Apl. 17	Pannel, W. G. A.	28 Apl. 17		
29 Aug. 17	Spiller, G. C. (*temp. Capt.*)	29 Aug. 17		
29 Aug. 17	Griffiths, W. A. (*temp. Capt.*)	29 Aug. 17		
29 Aug. 17	Brothers, H. J.	29 Aug. 17		
26 Sep. 17	Price, S. H.	26 Sep. 17		
26 Sep. 17	Saunders, E. T.	26 Sep. 17		
26 Sep. 17	Kennedy, D. D.	26 Sep. 17		
16 Mar. 18	Wollstein, L. J. (*temp.*)	16 Mar. 18		
12 Apl. 18	Cross, H. E.	12 Apl. 18		
19 Aug. 18	Guest, R. E.	19 Aug. 18		
	Attached.			
21 Nov. 13	Williamson, Lt. F. (172*a*, 206), T.F. Res. (Gen. List)	29 Oct. 14	. . . .	*I. A. R. O.*
25 Feb. 15	Michie, Lt. G. F., 8th Bn. Midd'x R.	1 June 16		

1 May 12	Rogers, Capt. T. G.	11 May 12	. . . .	Qr.-Mr.

Regimental District No. 57. [No. 10 District.]

Uniform—Scarlet. *Facings*—Lemon Yellow.

The Duke of Cambridge's Own (Middlesex Regiment).

1-10TH BATTALION.

Arrived in India, 1st December 1914, from England.

HONORARY COLONEL—*Lord* F. Wolverton . . . . 29 Aug. 03.

First Commn.	Names.	Regimental rank.	Army rank.	Remarks.
	Lieutenant-Colonel.			
31 Dec. 97	Dimond, C. F. W. (51)	29 Oct. 14	. . . .	Comdg. No. 3 Res. Bn.
	Majors.			
1 Aug. 96	Hosken, C. C. F. (*act'g. Lt.-Col.*, 21 *Nov. 18.*) (51, 52)	8 Feb. 15		
1 Jan. 08	White, S. H. (50, 52)	1 June 16	. . . .	Bde. Maj., Ambala Bde.
9 Mar. 09	Overbury, J. G. (51)	10 May 17	. . . .	G. S. O., 3rd gr., So. Comd.
	Captains.			
1 Nov. 11	*Fisher, C. J.*, D.S.O. (51, 57)	6 May 14	. . . .	*Bde. Maj.*
6 Dec. 11	*Ocock, W. P.* (51)	5 Aug. 14	. . . .	*Adjt., No. 8 Res. Bn., Sec'bad.*
22 June 12	Wood, R. R. (*act'g. Maj.*, 2 *Nov. 18*), (51).	5 Aug. 14		
22 June 12	Mackenzie R. H. T. (51)	5 Aug. 14	. . . .	Bde.-Maj., R'pindi Addl. Bde.
5 Aug. 14	*Walker, T.*	5 Aug. 14	. . . .	*S. S. O., Darjeeling.*
25 Nov. 92	Hart, A. E.	18 Sep. 14	. . . .	S. S. O., Fyzabad.
15 Feb. 13	Munford, C. J. H. (51, 52)	8 Feb. 15		
22 Feb. 13	Bromage, J. A. R. (51, 52, 66)	18 Dec. 15		
18 June 13	*Coleman, R. E.* (59)	1 June 16	. . . .	*With Sig. Serv.*
11 Dec. 13	*Sturge, E.* (57)	1 June 16	. . . .	*With Sig. Serv.*
21 Mar. 14	*Templer, H. E.* (57)	1 June 16	. . . .	*I. A. (on probn.)*
27 Mar. 14	*Lisle, P. C.* (51, 52)	1 June 16	. . . .	*M. G. Corps.*
24 Nov. 14	Wicks, W. M.	1 June 16	. . . .	C. M. A.'s Office, Poona.
26 Aug. 14	*Elgood, C. L.* (57, 161, 204)	1 June 16	. . . .	*I. A. (on probn).*
5 Sep. 14	Loveday, S. G. (51, 57)	1 June 16		
16 Sep. 14	Bramall, B. (51, 60*a*)	1 June 16		
16 Sep. 14	Dardier, C. E. (46, 51, 57)	1 June 16	. . . .	Instr., Bomb. Schl., Mhow.
18 Sep. 14	*Griffin, J. B.* (*temp. Maj.*) (66)	1 June 16	. . . .	*D. A. A. G., 8th (Lucknow) Dn.*
18 Sep. 14	Watson, A. G. (51)	1 June 16		
	Lieutenants.			
9 June 15	Lang, H. S. (46)	1 July 17		
21 July 15	Snoxall, C. E. (51, 60*a*)	1 July 17		
30 July 15	Landragin, L. C. (*actg. Capt.*, 25 *Sep. 18*) (46).	1 July 17	. . . .	Adjt.
3 Sep. 15	Gardner, L. (51)	1 July 17		
28 Sep. 15	*Cressall, H. A.* (*temp. Capt.*)	1 July 17	. . . .	*Staff Capt., Lucknow Bde.*
19 Nov. 15	Viccars, L. J. (57)	1 July 17		
	2nd-Lieutenants.			
28 Mar. 17	Goodall, J. H. W.	28 Mar. 17	. . . .	With R. A. F., Egypt.
28 Mar. 17	Dee, J.	28 Mar. 17	. . . .	Ind. Ord. Dept.
28 Apl. 17	Norton, E. C.	28 Apl. 17	. . . .	Staff Capt., Dehra Dun Bde.
28 Apl. 17	Coomber, D. K. (48*a*, 51)	28 Apl. 17		
3 May 17	Simmonds, W. W.	3 May 17		
29 Aug. 17	Mills, H. (59)	29 Aug. 17	. . . .	Bde. Sig. Offr., Lucknow.
29 Aug. 17	Blakesley, W. P. (60*a*)	29 Aug. 17		
29 Aug. 17	Willis, T. A. (51)	29 Aug. 17		
29 Aug. 17	Murrell, P. A. (59)	29 Aug. 17		
26 Sep. 17	Boulter, A. J.	26 Sep. 17		
26 Sep. 17	Saunders, E. T.	26 Sep. 17	. . .	Mily. Accts. Dept.
26 Sep. 17	Geeson, I.	26 Sep. 17		
26 Sep. 17	Garside, H.	26 Sep. 17		
26 Sep. 17	Spurgeon, H. S.	26 Sep. 17		
26 Sep. 17	Saunders, M. C.	26 Sep. 17		
26 Sep. 17	Reeve, G. A.	26 Sep. 17		
28 Sep. 17	Clunie, T.	28 Sep. 17		
16 Mar. 18	Gaillard, P.	16 Mar. 18	. . . .	M. G. Corps.
16 Mar. 18	Smith, A. T. H.	16 Mar. 18		
16 Mar. 18	Fickling, W. A.	16 Mar. 18	. . . .	M. G. Corps.
16 Mar. 18	Grigg, C. A.	16 Mar. 18	. . . .	M. G. Corps.
16 Mar. 18	Holland, E. W.	16 Mar. 18		
19 Aug. 18	Ryder, F.	19 Aug. 18		
19 Aug. 18	Lloyd, W. H.	19 Aug. 18		
	Attached.			
30 July 15	Landragin, Lieut. (*act'g. Capt.*) L. C.	25 Sep. 18	. . . .	Adjt.
16 July 10	Carter, Capt. A. W.	16 July 10	. . . .	Qr.-Mr. *Lv. ex I, m. c.*
27 Dec. 15	Bowring, Lieut. E. H. W.	27 Dec. 15	. . . .	Qr.-Mr.
23 Mar. 09	Webb-Johnson, Capt. C., M.B., R. A. M. C. (T. F.)	4 Dec. 12	. . . .	Medl. Offr.

Regimental District No. 62.
[No. 8 District.]

Uniform.—Green.
Facings—Black.

The Duke of Edinburgh's (Wiltshire Regiment),

1-4TH BATTALION.

Arrived in India, 10th November 1914, from England.

" South Africa, 1900-02."

HONORARY COLONEL—Somerset, A., *Duke of*, T.D. . . . . 13 June 00.

First Commn.	Names.	Regimental rank.	Army rank.	Remarks.
	Lieutenant-Colonel.			
	Armstrong, A., D.S.O. (*ret. pay*) (*Res. of Offr.*) (170, 206)	6 May 17		
	Majors.			
	Willis, H. H., V. D. (51)	20 May 11	. . . .	*Lv., ex I., m.c., 6 ms.*
	Phillips, H. A.	1 June 16	. . . .	Mech. Tpt. Coy.
	Bennett, C. G. (51, 67)	6 May 17		
	Morrison, D. C. A.	25 Jan. 18	. . . .	Adjt.
3 Mar. 99	Mackay, E. A. S.	25 Jan. 18		
	Captains.			
3 Dec. 05	Herbert, *Hon.* G. S.	5 Feb. 14		
	Phelps, J. W.	1 June 16		
4 Oct. 09	Pye-Smith, T. E. B. (51, 52) . .	1 June 16		
28 Apl. 10	Arkell, J. G. (57)	1 June 16	. . . .	Platoon Comdr., Cadet Unit, Sialkot.
25 Nov. 11	*Stanley, G.* (57)	1 June 16	. . . .	*Sig. Serv.*
25 Oct. 12	Arkell, T. N. (46)	1 June 16		
14 Aug. 14	Elliot, H. W. A.	1 June 16		
14 Aug. 14	Folkestone, W.P.B., *Visct.*, (51,52)	2 Nov. 16		
20 Aug. 14	*Hiles, M.*	6 May 17	. . . .	*Poll. Dept.*
20 Aug. 14	Lockhart, J. G.	6 May 17		
26 Aug. 14	Homan, C. F. W. B. (59, 207) . .	6 May 17		
20 Aug. 14	Brown, W. A.	6 May 17		
7 Oct. 14	*Redman, H. G.*	6 May 17	. . . .	*M. G. Corps.*
	Lieutenants.			
8 July 14	*Bloxham, H. F.*	1 June 16	. . . .	*I. A. R. O.*
26 Aug. 14	Angus, H. F. (51) (207) . . .	1 June 16	. . . .	With 1-4th Dorset. R.
7 Oct. 14	Knight, R. H. (*act'g. Capt.*) . . .	1 June 16		
29 May 15	Hall, B. K. B. (*act'g. Capt.*) . .	1 July 17		
14 July 15	*Berry, W. J.*	1 July 17	. . . .	*I. A. R. O.*
24 July 15	*Joiner, R. H. T.*	1 July 17		
24 July 15	Thurgar, H. M. (51)	1 July 17		
30 July 15	*Evans, E. T.*	1 July 17	. . . .	*M. G. Corps.*
17 Aug. 15	*Teagle, C. H.*	1 July 17		
13 Mar. 15	Ware, J. R.	1 July 17		
12 Sep. 15	Tucker, G. E.	1 July 17		
17 Oct. 15	*Davies, D. A.* (51)	1 July 17		
29 Sep. 15	*Bottomley, J. E.*	1 July 17	. . . . .	*R. A. F.*
	2nd-Lieutenants.			
	Morrison, Capt. D. C. A. . . .	15 May 17	. . . .	Adjt.
1 Dec. 11	Taylor, Capt. A. A.	. . . .	. . . .	Qr.-Mr.
24 Apl. 12	Waylen, Capt. G. H. H., R. A. M. C. (T.F.) .	1 Apl. 15	. . . .	Medl. Offr.

Regimental District No. 62.
[No. 8 District.]

Uniform—Green.
Facings—Black.

The Duke of Edinburgh's (Wiltshire Regiment).

2-4TH BATTALION.

Arrived in India, 8th January 1915, from England.

"South Africa, 1900-02."

HONORARY COLONEL—Somerset, A., *Duke of*, T.D. . . 13 June 00.

First Commn.	Names.	Regimental rank.	Army rank.	Remarks.
	Lieutenant-Colonel.			
28 Sep. 14	Parsons, F. G., D.S.O. (50, 53, 71). (*Hon. Maj. in Army, 2 Apl. 02*)	28 Sep. 14	. . . .	Comdg., Jalapahar.
	Majors.			
	Captains.			
21 Mar. 00	Blake, G. R. (*act'g. Maj., 14 Dec. 17*) (57).	18 Aug. 10		
11 Sep. 14	Maggs, C. W. (*Hon. Maj., ret. T.F.*)	11 Sep. 14	. . . .	1st G. B., Oxf. & Bucks. L. I.
12 Sep. 14	*Hoare, A. S.*	12 Sep. 14		
7 Oct. 14	Callaway, J.	7 Oct. 14	. . . .	Attd. Staff of D. J. A. G., So. Comd.
5 Feb. 10	Phelps, J. W.	1 June 16	. . . .	1st G. B., Oxf. & Bucks. L. I.
8 Dec. 11	Awdry, N. J. (51)	1 June 16	. . . .	
20 Aug. 14	*Elliot, J. R. McD.* (51, 52)	6 May 17	. . . .	*M. G. Corps.*
	Lieutenants.			
7 Oct. 14	*Young, G. A.* (57)	1 June 16	. . . .	*With Sig. Serv.*
12 Oct. 14	*Hoare, G. G.* (51) (*temp. Capt.*)	1 June 16		
13 Oct. 14	*Laverton, G. A.*	6 May 17	. . . .	*A.-D.-C. to Govr., Bombay.*
29 Oct. 14	*Carr, C. T.* (*temp. Maj.*) (51, 207).	1 June 17	. . . .	*D. A. A. G., Army Hdqrs.*
27 Nov. 14	Wiles, J. J. (*act'g. Capt.*) (51, 52, 206)	1 June 17	. . . .	*Adjt.*
2 Dec. 14	Shippard, H. C. (51, 52)	1 July 17	. . . .	*Lv., ex I., m.c., 6 mos.*
11 Dec. 14	Warren, G. M. (*act'g. Capt.*) (51)	1 July 17		
11 Dec. 14	*Harris, E. R.* (66a)	1 July 17	. . . .	*M. G. Corps.*
12 Sep. 14	Crosbie, H. (*act'g. Capt.*)	1 July 17	12 Sep. 14	
. . .	Barnett, A. C.	1 July 17		
17 July 15	*Buchanan, R.* (51)	1 July 17	. . . .	*With R. A. F.*
19 July 15	*Tindle, T. A.* (51)	1 July 17	. . . .	*M. G. Corps.*
24 July 15	Robson, S. G. (46)	1 July 17	. . . .	R. A. F.
13 Aug. 15	*Harris, C. P.* (12)	1 July 17	. . . .	*I. A. R. O.*
26 Sep. 15	Evans, H. K.	1 July 17	. . . .	M. G. Corps.
16 Nov. 15	*Wilshin, H. L.*	1 July 17	. . . .	*Asst. Cable Censor.*
17 Dec. 15	Griffiths, E. G. (46, 51)	1 July 17		
26 Sep. 16	Tucker, W. L. (51)	26 Mar. 18		
	2nd-Lieutenants.			
28 Mar. 17	Warner, G. R.	28 Mar. 17		
29 Aug. 17	Ash, F. (46)	29 Aug. 17		
29 Aug. 17	Harrison, T. (51)	29 Aug. 17		
29 Aug. 17	*Billington, T. F.* (51, 52)	29 Aug. 17	. . . .	*M. G. Corps.*
16 Mar. 18	Bowers, S. J.	16 Mar. 18		
16 Mar. 18	Learner, J. H.	16 Mar. 18		
16 Mar. 18	Jeffery, R.	16 Mar. 18		

The Duke of Edinburgh's (Wiltshire Regiment)—*contd.*

First Commn.	Names.	Regimental rank.	Army rank.	Remarks.
	Attached.			
17 July 15	Cooper, Lt. W. P. (*temp. Capt.*), Suff. R. (46)	1 July 17	. . . .	Offr. i. c., Army Clothing Depôt, Rangoon.
1 Mar. 17	Dehn, 2nd-Lt. C. G., Wilts. R. (48*a*, 51)	1 Mar. 17	. . . .	S. S. O., Allahabad.
1 Mar. 17	Pyle, 2nd-Lt. S. T., Wilts. R. (57)	1 Mar. 17		
13 May 17	Ward, 2nd-Lt. W. N., Worc. R. (48*a*, 50)	13 May 17		
25 Jan. 17	Hurst, 2nd-Lt. C. (51) 7 Bn., Essex R.	25 Jan. 17		
28 Apl. 17	Davies, 2nd-Lt. W. V. (48*a*, 50, 59)	28 Apl. 17		
28 Apl. 17	Hayes, 2nd-Lt. M. E. (58*a*) . .	28 Apl. 17		
28 Apl. 17	Townsend, 2nd-Lt. F. J. . . .	28 Apl. 17		
28 Apl. 17	Smith, 2nd-Lt. J. P. (60*a*, 61) . .	28 Apl. 17		
28 Apl. 17	Rendell, 2nd-Lt. F. (48*a*, 50) . .	28 Apl. 17		
26 Sep. 17	McHale, 2nd-Lt. F. F., E. Lan. R.	26 Sep. 17		
26 Sep. 17	Fisher, 2nd-Lt A. H., W. Rid. R. .	26 Sep. 17	. . . .	*Lv., ex I., m. c., 6 mos.*
26 Sep. 17	Parry, 2nd-Lt. E. A. (*temp. Capt., 20 Aug. 18*), Notts. & Derby. R.	26 Sep. 17	. . . .	Platoon Comdr., Offr's. Schl. of Instn., Nasik.
19 Aug. 18	Webster, 2nd-Lt. R. O. (*temp.*), Wilts. R.	19 Aug. 18		

27 Nov. 14	Wiles, Lt. (*actg. Capt.*) J. J. .	1 June 17	. . . .	Adjt.

Regimental District No. 63. [No. 3 District.]

Uniform—Scarlet. *Facings*—White.

The Manchester Regiment.

1st BATTALION (*late 63rd Foot*).

Arrived in India, 19th December 1904, from Singapore.

THE SPHINX, SUPERSCRIBED "EGYPT."

"Guadaloupe, 1759" "Egmont-op-Zee" "Peninsula" "Martinique, 1809" "Guadaloupe, 1810" "New Zealand" "Alma" "Inkerman" "Sevastopol" "Afghanistan, 1879-80" "Egypt, 1882" "Defence of Ladysmith" "South Africa, 1899-1902."

COLONEL—Major-General W. O. Barnard, ret. pay . 8 Jan. 04.

First Commn.	Names.	Regimental rank.	Army rank.	Remarks.
	Lieutenant-Colonel.			
	Majors.			
1 Dec. 97	Hardcastle, R. N., D.S.O. (*act'g. Lt.-Col., 7 Sep. 17*)	27 Apl. 15	*l.c.* 3 June 16	
22 May 01	Irwin, C. D., M.C.	22 May 16	. . . .	*Lv., ex I., m.c., 1 yr. 22 days.*
19 Oct. 01	O'Meara, A. E.	24 Oct. 16		
19 Oct. 01	Browne, R. G., D.S.O. (50, 53, 172*a*)	24 Oct. 16		
	Captains.			
18 Sep. 09	Shipster, W. N., M.C. (51, 52, 66*a*, 172*a*)	18 Sep. 15		
24 Jan. 14	Henderson, G. S., D.S.O., M.C.	24 July 16		
8 Aug. 14	Moore, A. W. U.	1 Jan. 17	. . . .	R. T. O.
17 Sep. 14	Holmes, E. B., M.C.	1 Jan. 17		
	Lieutenants.			
2 Apl. 15	Smith, R. W. (*temp. Capt.*)	1 Oct. 15	. . . .	Staff Capt.
20 Aug. 15	*Adams, F. H.* (*act'g Capt.*)	8 Oct. 16	. . . .	*Adjt.*
4 Oct. 15	Pembroke, S. K.	1 Jan. 17		
25 Feb. 16	Clifford, W. (*act'g. Capt., 3 Aug. 17*)	16 Aug. 17		
	2nd Lieutenants.			
17 Dec. 16	Cummings, W. H. (*temp. Capt., 11 Aug. 17.*)	17 Dec. 16	. . . .	Asst. Prov. Marshal.
1 Oct. 14	Oxley, H. H.	12 Jan. 17	1 Oct. 14	
22 Mar. 17	Nicholson, J. R. (57)	22 Mar. 17		
28 Apl. 17	Staniland, D. L. (*temp.*)	28 Apl. 17		
6 July 17	Newbold, W.	6 July 17		
29 Aug. 17	Lund, J. H. (*temp.*)	29 Aug. 17		
6 Feb. 18	Harland, H., D.C.M.	6 Feb. 18		
16 Mar. 18	Jones, A. (temp.)	16 Mar. 18		
	Attached.			
	Hobkirk, Capt. R., 18 Bn.	18 Oct. 15		
15 Aug. 14	Cosgrove, Capt. E. T., 3rd Bn.	1 Feb. 16		
15 Aug. 14	Marley, Capt. V. D. K., 4th Bn. (52)	1 Jan. 17		
21 Sep. 14	Towers, Lt. F.	20 Jan. 15		
	Walker, Lt. M., 11 Essex R.	13 Sep. 15		
9 Jan. 15	Watson, Lt. B. S., 3rd Bn.	1 Feb. 16		
14 May 15	Thornton, Lt. A. M., 3rd Bn.	10 Mar. 16		
	Butterworth, Lt. B., 3rd Bn.	1 Feb. 16		
1 Mar. 15	Holden, Lt. H. D., 14th Bn.	1 Oct. 15		
12 Mar. 15	Sketchley, Lt. B.T., 1st Dur. L. I.	21 Jan. 17		
29 May 15	Cooper, Lt. E., 3rd Bn.	1 July 17		
25 Sep. 15	Best, Lt. W. R. P., 3 Bn., Conn. Rang.	1 July 17		
24 Dec. 15	Mann, Lt. G. E., 6 Bn., E. Surr. R.	1 July 17		
7 July 16	Hart, Lt. J. H. W., 3 Bn. Devon R.	7 July 16		
29 May 15	Marriott, Lt. S. A., 3rd Bn. (*act'g. Capt.*)	1 July 17		
2 Mar. 15	Brown, Lt. G. M., 13 Bn.	1 July 17		
10 Apl. 15	Rigby, Lt. L., 3rd Bn. (*act'g. Capt.*)	1 July 17		
14 May 15	Hughes, Lt. E. D. G., 4th Bn.	1 July 17		
2 Sep. 15	Symons, Lt. E. W., 1 Bn., E. Surr. R.	1 July 17		
5 Aug. 16	Clark, Lt. A., 3 Bn., Devon. R.	5 Feb. 18		
5 Aug. 16	Jack, Lt. R. H., 13 Essex R.	5 Feb. 18		
26 Sep. 16	Ingham, Lt. H.	26 Mar. 18		
26 Sep. 16	*Wayte, Lt. V. W.*	26 Mar. 18	. . . .	*I. A.* (*On probn.*)
26 Sep. 16	Farrington, Lt. R. L.	26 Mar. 18		
5 Aug. 16	Heath, 2nd-Lt. D. F., 20 Bn.	5 Aug. 16		
20 Aug. 15	Adams, Lieut. (*act'g. Capt.*) F. H.	4 Mar. 17	. . . .	Adjt.
4 Sep. 00	O'Brien, Maj. P.	3 June 18	. . . .	Qr.-Mr.

Agents:—MESSRS. COX & Co.

Regimental Distr ct No. 63.
[No. 3 District.]

Uniform—Scarlet.
Facings—White.

The Manchester Regiment.

1st GARRISON BATTALION.

Arrived in India, 17th March 1916.

First Commn.	Names.	Regimental rank.	Army rank.	Remarks.
	Lieutenant-Colonel.			
25 July 91	Bretherton, H. W. K. (*temp.*) (*Col. ret. T. F.*)	24 Dec. 15	14 Dec. 14	*Lv. ex I, m. c., 6 mos.*
	Majors.			
9 Dec. 99	Bonnyman, F. J. C. (*temp.*) (*act'g. Lt.-Col.*)	12 Feb. 16	8 Oct. 14	Comdg. No. 8. Res. Bn.
9 Feb. 89	Ashworth, R., T. D. (*temp.*) (*Col., ret. T. F.*) (*act'g. Lt.-Col., 2 May 18*)	23 Feb. 16	19 Mar. 15	
	Captains.			
12 Mar. 00	Adams, H. T. (*temp.*)	23 Feb. 16	27 Sep. 14	
Mar. 96	Jacques, S. (*temp.*)	17 Feb. 16	5 Oct. 14	
00	Sharples, J. L. (*temp.*)	18 Feb. 16	18 Oct. 14	
5 Dec. 14	Galloway, J. (*temp.*)	23 Feb. 16	15 Mar. 15	
30 Jan. 97	Palmer F. C. (*temp.*) (*Maj. ret. T. F.*)	23 Feb. 16	22 June 15	
12 Dec. 15	McKelvey, J. W. (*temp.*)	18 Feb. 16	18 Feb. 16	
	Lieutenants.			
17 Sep. 14	Smith, G. E. (*temp.*)	23 Feb. 16	1 Jan 15	
8 Feb. 15	Vincent, G. (*temp.*)	23 Feb. 16	8 Feb. 15	Adjt., 12th E. I. Ry. Bn., I.D.F.
22 Feb. 15	Turrell, J. W. (*temp.*)	12 Feb. 16	22 Feb. 15	Attd. M. A. Dept, Poona.
27 Feb. 15	Burrell, G. F. D.	1 Apl. 16	27 Feb. 15	
22 Dec. 14	Lewis, C. C. C. (*temp.*) (207)	23 Feb. 16	1 June 15	I. A. R. O.
10 Dec. 14	Frith, W. H. (*temp.*) (51)	30 Oct 15	2 June 15	Ind. Munitions Board.
10 Feb. 15	Williams, W. J. (*temp.*) (57)	23 Feb. 16	1 Aug. 15	
1 Jan. 15	Newman, C. J. (*temp.*)	23 Feb. 16	1 Aug. 15	
27 May 15	Carter, H. L. (*temp.*) (53)	11 Feb. 16	27 May 15	
19 July 15	*Goode, C. H.* (*temp.*) (*act'g. Capt.*)	11 Feb. 16	19 July 15	*Adjt.*
16 Dec. 14	Boorman, J. (*temp.*) *1st Garr. Bn. Norf. R.* (51, 60*a*)	22 Jan. 18	24 Nov. 15	With 2nd G. B., Bedf. R.
17 Feb. 15	Dawbarn, J. R., *4 North'n R.*(*temp.*)	28 Jan. 18	1 June 16	
19 Dec. 14	Munro, J. C., *4 Sco. Rif.*	18 Jan. 18	6 Jan. 17	
	2nd-Lieutenants.			
29 Dec. 14	Barrand, H. J. (*temp.*), 8 Ches. R.	4 Oct. 17	29 Dec. 14	
7 Apl. 15	Humfrey, B. (*temp.*) (57)	17 Feb. 16	7 Apl. 15	I. A. (*On probn.*)
7 Apl. 15	Rowechapple, W. C. (*temp.*)	17 Feb. 16	7 Apl. 15	
20 May 15	Pullen, A. P., 4th Lan. Fus.	18 Feb. 16	20 May 15	Inspr. of Munitions, Cawnpore.
1 June 15	Birks, O. M. M. (*temp.*) 8 R. W. Fus.	29 July 17	1 June 15	
3 June 15	Legge, G. E. (*temp.*)	17 Feb. 16	3 June 15	S. S. O., Benares.
8 July 15	Simpson, G. G. (*temp.*) (52)	29 Nov. 15	8 July 15	M. G. Corps.
9 Oct 15	Oliphant. I. M. B. (4th Lan. Fus.)	8 Feb. 16	9 Oct. 15	With Depot, 8th Rajputs.
7 May 16	Herbert, W. C. (*temp.*) 2nd Garr. Bn., Bedf R.	28 Feb. 18	7 May 16	
5 Sep. 16	Davey, H. W. (*temp.*) 2nd Garr. Bn., Norf. R. (172*a*)	22 Jan. 18	5 Sep. 16	
20 Oct. 16	Hubbard, E. A. (*7 Lond. R., T. F.*)	2 Aug. 17	20 Oct. 16	
11 Feb. 17	Herne, E. du P. (*temp.*), *1st Garr. Bn., Worc. R.*	3 Aug. 17	11 Feb. 17	
13 Feb. 17	Semple, R. E. (*temp.*), *1st Garr. Bn., High. L. I.*	18 Jan. 18	13 Feb. 17	
19 Aug. 17	Breed, F. G. (*temp.*), *1st Garr. Bn., Suff. R.*	11 Feb. 18	19 Aug. 17	
19 Aug. 17	Cavenaugh, P. H. C. (*temp.*) (51)	11 Feb. 18	19 Aug. 17	
19 July 15	Goode, Lt. (*act'g. Capt.*) C. H. (*temp.*)	17 Jan. 18	19 July 15	Adjt.
18 Feb. 16	Ward, Lt. W. W. (*temp.*)	18 Feb. 16	18 Feb. 15	Qr.-Mr.

Regimental District, No. 64. [No. 6 District.]

Uniform—Scarlet. *Facings*—White.

The Prince of Wales's (North Staffordshire Regiment).

2ND BATTALION (*late 98th Foot*).

Arrived in India, 26th October 1903, from England.

THE PRINCE OF WALES' PLUME. THE DRAGON SUPERSCRIBED "CHINA."

"Guadaloupe, 1759" "Martinique, 1794" "St. Lucia, 1803" "Surinam" "Punjaub" "Reshire" "Bushire" "Koosh-ab" "Persia" "Lucknow" "Hafir" "South Africa, 1900-02."

COLONEL—Major-General T. F. Lloyd, ret. pay . . . . 20 Aug. 11.

First Commn.	Names.	Regimental rank.	Army rank.	Remarks.
	Lieutenant-Colonel.			
7 Nov. 91	**Fox**, E. V. (Q, 51, 52, 57) . . .	1 Jan. 17		
	Majors.			
	Captains.			
11 Dec. 09	Tuck, S. A. (*act'g. Maj.* (50, 52) .	12 May 15		
22 Jan. 13	Morse, R. N. S. (51, 52) . . .	24 Apl. 16		
10 Dec. 13	Hatton, F. (48*a*, 51, 52) . . .	10 June 16		
24 Jan. 14	Stephenson, C. B. (48*a*, 51) . . .	10 Jan. 17		
7 Nov. 14	Withers, J. R. (57)	30 Mar. 17	. . . .	With 1-4th Bn., Devon. R.
5 Nov. 14	Planche-Hearn, C. R. (*temp.*) . .	13 June 17		
7 Nov. 14	Andrews, C. G.	13 June 17	. . . .	With 2-6th Bn., Devon. R.
22 Nov. 14	Thornewill, A. S. (*temp.*) (172) .	11 Oct. 17	. . . .	I. A. R. O.
9 Jan. 15	Squirrell, E. A. (48*a*, 51, 52) . . .	7 Dec. 17		
	Lieutenants.			
17 Apl. 15	Crickett, G. W. (*temp.*)	24 Mar. 16	. . . .	*Lv., ex I., m. c., 6 mos.*
7 Nov. 14	Bird, W. H. (46, 57, 58*a*) . . .	22 Aug. 16	. . . .	*Lv., ex I., m. c., 6 m.*
17 June 15	Mumford, J. J. (*act'g. Capt.*) (52) .	23 Nov. 16	. . . .	Adjt. and Qr.-Mr., Cadet College, Quetta.
17 June 15	Dodds, A.	14 Dec. 16		With R. A. F.
14 July 15	Beville, E. L. G.	15 Dec. 16		
16 Aug. 16	Thompson, B. E. (46)	16 Feb. 18		
13 Dec. 16	Munn, J.	13 June 18		
21 Dec. 16	Sims, J. H. (*act'g. Capt.*) . . .	21 June 18	. . . .	Attd. M. G. C.
21 Dec. 16	Stroud, T. J. (46)	22 June 18		
22 Dec. 16	Irwin, C. (1, 51) (*act'g. Capt.*) . .	22 June 18	. . . .	Instr., S. of M., Rawalpindi.
	2nd-Lieutenants.			
28 Apl. 17	Hayward, R. (*temp.*)	28 Apl. 17	. . . .	With R. A. F.
1 May 17	Lacey, R. (*temp.*)	1 May 17		
29 May 17	Keeling, E. J., M.C.	29 May 17		
29 Aug. 17	Horncastle, H. J. (*temp.*) (46) . .	29 Aug. 17		
29 Aug. 17	Scantlebury, L. A. (*temp.*) . . .	29 Aug. 17		
16 Mar. 18	Vining, C. (*temp.*)	16 Mar. 18		
16 Mar. 18	Goode, G. G. (*temp.*)	16 Mar. 18		
16 Mar. 18	Fuller, E. T. (*temp.*)	16 Mar. 18	. . . .	With R. A. F.
1 Aug. 18	Harrison, F. M. K.	1 Aug. 18	. . . .	Attd. M. G. C.
19 Aug. 18	Harris, S. (*temp.*)	19 Aug. 18		
19 Aug. 18	Savory, A. B. (*temp.*)	19 Aug. 18		
7 Oct. 18	Groom, E.	7 Oct. 18		
	Attached.			
28 Apl. 15	Ward, Lt. C. (46), 3rd Bn. . . .	31 Dec. 15		
12 Jan. 16	White, Lt. J. M.C., 5 Bn., Sea. Highrs. (170)	12 July 17	. . . .	With Sig. Depôt, Poona.
29 Jan. 16	Elder, Lt. E. W., M.C., (*temp.*), R. Highrs.	29 July 17	. . . .	With 7th Bn., N. Staff. R.
26 Sep. 16	Ford, Lieut. N. B.	26 Mar. 18		
26 Sep. 17	Frewin, 2nd-Lt. A. R.	26 Sep. 17		
26 Sep. 17	Horseman, 2nd-Lt. E. V. . . .	26 Sep. 17		
26 Sep. 17	Warner, 2nd-Lt. A. C.	26 Sep. 17	. . . .	Platoon Offr., N. C. O.'s Schl., Rawalpindi.
26 Sep. 17	Linton, 2nd-Lt. R.	26 Sep. 17	. . . .	Attd. Mech. Tpt., Meerut.
26 Sep. 17	Prior, 2nd-Lt. H. K.	26 Sep. 17		
26 Sep. 17	Hollinrake, 2nd-Lt. R.	26 Sep. 17	. . . .	Attd. Mech. Tpt., Peshawar.
9 Jan. 15	Squirrell, Capt. E. A.	1 Oct. 16	. . . .	Adjt.
22 June 07	Lowther, Capt. T. E.	22 June 17	. . . .	Qr.-Mr.

Agents:—MESSRS. COX & CO.

Regimental District No. 68. [No. 5 District.]

Uniform—Scarlet. *Facings*—Dark Green.

The Durham Light Infantry.

1ST BATTALION (*late 68th Foot*).

Arrived in India, 11th October 1902, from South Africa.

THE UNITED RED AND WHITE ROSE.

"Salamanca" "Vittoria" "Pyrenees" "Nivelle" "Orthes" "Peninsula" "Alma" "Inkerman" "Sevastopol" "Reshire" "Bushire" "Koosh-ab" "Persia" "New Zealand" "Relief of Ladysmith" "South Africa, 1899-1902."

COLONEL—Major-General R. Upcher, C.B., D.S.O. ret. pay . . 11 Sep. 08.

First Commn.	Names.	Regimental rank.	Army rank.	Remarks.
	Lieutenant-Colonel.			
	Major.			
2 June 94	Moore, E. DuP. H. (*act'g. Lt.-Col., 24 June 16*) (12, 50, 52, 61, 71, 72).	21 Sep. 14	. . . .	
	Captains.			
13 Oct. 09	Clifton, A. J., O.B.E. (12, 50, 52, 65*a*, 66, 67)	10 June 15	. . . .	Armd. Motor Unit, Peshawar.
20 Apl. 10	Hasted, F. O. C., D.S.O. (48*a*, 51, 57, 66*a*, 172)	22 June 16		
14 Feb. 12	Ferguson, J. F. (51, 52, 66, 172*a*) .	2 Dec. 17		
	Lieutenants.			
7 Nov. 14	Waiton, C. (*act'g. Capt.*, (12, 51, 172*a*).	15 May 15		
7 Nov. 14	Hartridge, P. (*act'g. Capt. 4 May 18*) (51, 172*a*).	10 June 15		
7 Nov. 14	Tilley, H. (*act'g. Capt., 3 Aug. 17*) (51).	10 June 15	. . . .	With R. A. F., Simla.
7 Nov. 14	Sibbald, C. W. (*act'g. Capt., 17 Feb. 17*) (12, 51).	10 June 15		
7 Nov. 14	Harwood, A. (57, 71)	10 June 15		
19 Nov. 14	Mercer, S. V. (*temp.*) (*act'g. Capt.*)	18 July 15		
8 Jan. 15	Macdonald, A. S.	13 May 16		
3 Mar. 15	Markham, R. W. (*temp.*) . . .	21 Jan. 17	. . . .	Dental Surgeon, Nowshera.
12 Mar. 15	Sketchley, B. T. (*temp.*) (172*a*) . .	21 Jan. 17		
17 June 15	Mahoney, J. (*temp. Capt., 11 Sep. 18*) (12, 172*a*),	1 July 17	. . . .	Coy. Comdr., N. C. O's. Schl. of Instn.
15 Sep. 15	Goldschmidt, G. T. (46, 53, 57) . .	1 July 17		
20 Oct. 15	Daintree, G. W. (46, 61) . . .	1 July 17	. . . .	Supvg. Offr., Phys. & Bayonet Trg., Meerut Dn.
27 Dec. 15	Taylor, F.	1 July 17		
29 Feb. 16	Hawke, A. W. (*temp. Capt., 15 Oct. 18*) (47*a*, 50, 53).	29 Aug. 17	. . . .	Asst. Instr. Schl. of Musk., Satara.
6 Jan. 17	Critchley, H. (51)	6 July 18		
	Attached.			
Dec. 09	Richardson, Capt. H., (*act'g. Maj.*) (50, 51, 59) (*ret.pay*) 3rd Bn.	26 Oct. 10	. . . .	2nd-in-Comd.
19 Feb. 02	Cockburn, Capt. J. L. K. (*temp.*) 69, 71, 172*a*).	8 Nov. 14	. . . .	Attd. S. & T. Corps.
4 Nov. 15	Callie, Lt. G. (*temp.*) E. Surr. R. .	1 July 17		
10 Jan. 16	Newey, Lt. R. H., 14 Bn., Worc. R. (48*a*, 51).	10 July 17	. . . .	Garr. Engr., Derajat.
23 Oct. 16	Mackintosh, J. D. (46)	23 Oct. 16		
19 Dec. 16	McMath, 2nd-Lt. R. F. (*temp.*) High. L. I. (46).	19 Dec. 16		
25 Jan. 17	Henshaw, 2nd-Lt. R. L., 3 Bn., R. Scots.	25 Jan. 17		
18 Dec. 17	Harley, 2nd-Lt. G. T.	18 Dec. 17	. . . .	With Sig. Serv.
16 Mar. 18	Smith, 2nd-Lt. F. H. (*temp.*) . .	16 Mar. 18	. . . .	With 2-6 Bn., Devon. R.
16 Mar. 18	Brown, 2nd-Lt. K. N. (*temp.*) .	16 Mar. 18		
16 Mar. 18	Goss, 2nd-Lt. E. C. (*temp.*) . . .	16 Mar. 18	. . . .	R. A. F., Egypt.
16 Mar. 18	Clark, 2nd-Lt. A., (*temp.*) (57) . .	16 Mar. 18		
21 Feb. 03	Freel, Maj. J. (51, 72)	10 Nov. 16	. . . .	Qr.-Mr.

Agents:—MESSRS. HOLT & Co.

Regimental District, No. 71. [No. 2 District.]

Uniform—Scarlet. *Facings*—Buff.

Highland Light Infantry.

1st BATTALION (*late 71st Foot*).

Arrived in India, 3rd November 1904, from Egypt.

THE CASTLE AND KEY SUPERSCRIBED "GIBRALTAR 1780-83" AND WITH THE MOTTO "Montis Insignia Calpe" UNDERNEATH
THE ELEPHANT, SUPERSCRIBED "ASSAYE."

"Carnatic" "Hindoostan" "Sholinghur" "Mysore" "Seringapatam" "Cape of Good Hope, 1806" "Roliça" "Vimiera" "Corunna" "Busaco" "Fuentes d'Onor" "Ciudad Rodrigo" "Badajoz" "Almaraz" "Salamanca" "Vittoria" "Pyrenees" "Nivelle" "Nive" "Orthes" "Toulouse" "Peninsula" "Waterloo" "South Africa, 1851-2-3" "Sevastopol" "Central India" "Tel-el-Kebir" "Egypt, 1882" "Modder River" "South Africa, 1899-1902."

COLONEL-IN-CHIEF—Field Marshal *H. R. H.* ARTHUR WILLIAM PATRICK ALBERT, *Duke of* CONNAUGHT AND STRATHEARN, K.G., K.T., K.P., G.C.B., G.C.S.I., G.C.M.G., G.C.I.E., G.C.V.O., G.B.E., A.-D.-C., 4 Sep. 01.

COLONEL—

First Commn.	Names.	Regimental rank.	Army rank.	Remarks
	Lieutenant-Colonel.			
	Major.			
21 Apl. 00	Leckie-Ewing, W. C.	1 Sep. 15		
	Captain.			
15 Dec. 14	Villiers, E. H. (*temp.*)	16 Mar. 17		
16 Dec. 14	McFarland, W. C. D.	16 Mar. 17		
12 May 15	Macdonald, I. M.	17 Nov. 17		
	Lieutenants.			
5 Dec. 15	Bartlett, A. O. (*temp.*)	12 Jan. 17		
1 Jan. 16	Lett, T.	31 Mar. 17		
	2nd-Lieutenants.			
7 Feb. 17	Warrender, G.	7 Feb. 17		
7 Feb. 17	Carney, J., D.C.M.	7 Feb. 17		
23 Mar. 17	Peden, W. (*temp. Capt.*)	23 Mar. 17	. . . .	Asst. Prov. Marshal.
21 Sep. 17	Morgan, E. H., D.C.M.	21 Sep. 17	. . . .	Ag. Qr.-Mr.
16 Mar. 18	Hill-Murray, N. J. G. (*temp.*)	16 Mar. 18		
16 Mar. 18	McMilian, P. K. T. (*temp.*)	16 Mar. 18		
16 Mar. 18	Oddie, E. (*temp.*)	16 Mar. 18		
11 Aug. 18	Stevenson, A.	11 Aug. 18		
	Attached.			
7 Nov. 91	Deacon, Bt.-Lt.-Col. (*temp.Lt.-Col.*) H. R. G., D.S.O., Conn. Rang.	4 Mar. 11	*l. c.*, 3 June 16	Comdg.
16 Aug. 15	McRoberts, Capt. B. A. K., 3 Gord. Highrs.	16 Aug. 15	. . . .	2nd-in-Comd.
25 Apl. 13	Mylles, Capt. C. C., Spl. Res.	8 Oct. 15		
8 Aug. 14	Leslie, Capt. F. S., 4th Bn.	8 Oct. 15		
15 Aug. 14	Peffers, Capt. A., 3rd Bn.	12 Dec. 15		
10 Oct. 14	Broadhurst, Capt. A. F. B., 3rd Bn.	25 May 16		

Agents :—Sir C. R. McGrigor, *Bt.*, & Co.

Highland Light Infantry—*contd.*

First Commn.	Names.	Regimental rank.	Army rank.	Remarks.
	Attached—*contd.*			
10 Oct. 14	Bendall, Capt. J., 3rd Bn.	25 May 16		
15 Aug. 14	Urquhart, Capt. F. M., 4th Bn.	12 Dec. 15	. . . .	No. 1 Res. Bn.
21 Oct. 14	Harley, Lt. J. R. H., 3rd Bn. (*act'g. Capt.*)	12 Dec. 15		
23 Oct. 15	Gray, Lt. G. C. (*temp.*) (51)	1 July 17	. . . .	No. 1 Res. Bn.
1 Sep. 14	Wilson, 2nd-Lt. (*temp.* Lt.) J. W., 4th Bn.	1 Sep. 14		
18 Dec. 14	Barnett, 2nd-Lt. S. H., 3rd Bn.	1 Dec. 14		
24 Feb. 15	Roger, Capt. W. W., 3rd Bn.	24 Feb. 15		
24 Feb. 15	McDonald, 2nd-Lt. W., 4th Bn.	24 Feb. 15		
20 Apl. 15	Wood, 2nd-Lt. J. H., 4th Bn.	20 Apl. 15	. . . .	
29 Apl. 15	Reid, Lt. A/ Capt. G. (*temp.*) 13th Bn	29 Apl. 15	. . . .	No. 1 Res. Bn.
8 May 15	Cumming, 2nd-Lt. A., 3rd Bn.	8 May 15		
10 May 15	Hendry, 2nd-Lt. J. M., (*temp. Capt.*) 3rd Bn.	10 May 15	. . . .	Spec. Serv. Offr.
13 May 15	Ross, 2nd-Lt. J. M., 4th Bn.	13 May 15	. . . .	*Lv. ex I., m. c., 6 mos.*
7 June 15	Phillips, 2nd-Lt. J. R. C., 3rd Bn.	7 June 15		
11 June 15	Tyler, 2nd-Lt. L. J., 3rd Bn.	11 June 15		
11 June 15	Cunningham, Lt. T. S., 3rd Bn.	1 July 17	. . . .	No. 1 Res. Bn.
26 June 15	Wilson, 2nd-Lt. P. F. H., 4th Bn.	26 June 15	. . . .	Actg. Adjt.
12 July 15	Glaister, Lt. J. D., 3rd Bn.	1 July 17		
30 July 15	Ross, 2nd-Lt. J. A., 4th Bn.	30 July 15		
30 July 15	Laughland, Lt. A. P., 3rd Bn.	1 July 17		
19 Oct. 15	Cameron, 2nd-Lt. H. A., 5th Bn.	19 Oct. 15		
30 Oct. 15	Needham, 2nd-Lt. J. V., 13th Bn.	30 Oct. 15		
1 Nov. 15	Dickson, 2nd-Lt. E. C. S., 5th Bn.	1 Nov. 15		
11 Nov. 15	Whitelaw, Lt. D., 3rd Bn.	1 July 17		
11 Nov. 15	Blair, 2nd-Lt. A., 12th Sco. Rif.	11 Nov. 15		
5 Dec. 15	Strachan, Lt. A. B., 3rd Bn.	5 Dec. 15	. . . .	No. 1 Res. Bn.
4 Jan. 16	James, Lt. J. L., 13th Bn.	4 Jan. 16		
5 Feb. 16	Barclay, 2nd-Lt. C. N., 1st Sco.	5 Feb. 16		
7 Apl. 16	Adams, Lt. H. A., 3rd Bn.	12 Aug. 17		
7 July 16	Woodford, 2nd-Lt. A. S., 12th Sco. Rif.	7 July 16		
5 Aug. 16	Coutts, 2nd-Lt. W. F., 3rd Bn.	5 Aug. 16		
5 Aug. 16	Baird, 2nd-Lt. J. W., 3rd Bn.	5 Aug. 16		
5 Aug. 16	Gilmour, 2nd-Lt. J. A., 3rd Bn.	5 Aug. 16		
29 June 15	Crawford, R. Mc G. C., Lt., 9th Bn.	1 July 17		No. 1 Res. Bn.
4 Nov. 15	Bonsor, R. B., Lt.	1 July 17		No. 1 Res. Bn.
25 Nov. 15	Conochie, N. C., Lt.	1 July 17		No. 1 Res. Bn.
?	Pitcairn, J. E., Lt.	1 Apl. 17		No. 1 Res. Bn.
19 Dec. 16	Tudhope, M.A., Lt.	19 June 18		No. 1 Res. Bn.
19 Dec. 16	Hamilton Kerr, D., Lt.	19 June 18		No. 1 Res. Bn.
25 Jan. 16	Blackley, J. Lt	25 July 18		No. 1 Res. Bn.
25 Jan. 16	Nisbet, J. D., Lt.	25 July 18		No. 1 Res. Bn.
25 Jan. 16	Paterson, G. M., Lt.	25 July 18		No. 1 Res. Bn.
25 Jan. 17	Pollock, C. A., 2nd-Lt.	25 Jan. 17		No. 1 Res. Bn.
1 Mar. 17	Bain, R. B., 2nd-Lt.	1 Mar. 17		No. 1 Res. Bn.
21 Mar. 17	Cameron, D. N., 2nd-Lt.	21 Mar. 17		No. 1 Res. Bn.
29 July 17	Quig, A., 2nd-Lt.	29 July 17		No. 1 Res. Bn.
1 Aug. 17	Robb, J. M., 2nd-Lt.	1 Aug. 17		No. 1 Res. Bn.
28 Aug. 17	McFarlane, W., 2nd-Lt.	28 Aug. 17		No. 1 Res. Bn.
28 Aug. 17	Hannah, W. F., 2nd-Lt.	28 Aug. 17		No. 1 Res. Bn.
28 Aug. 17	Corner, W.	28 Aug. 17		No. 1 Res. Bn.
28 Aug. 17	Raeside, D. A.	28 Aug. 17		No. 1 Res. Bn.
28 Aug. 17	McWalter, J. W.	28 Aug. 17		No. 1 Res. Bn.
26 Sept. 17	Campbell, C.	26 Sept. 17		No. 1 Res. Bn.
28 Sept. 17	Stevenson, J.	28 Sept. 17		No. 1 Res. Bn.
19 Aug. 18	Maclean, J. R.	19 Aug. 18		No. 1 Res. Bn.
19 Aug. 18	Prentice, J.	19 Aug. 18		No. 1 Res. Bn.
19 Aug. 18	McKay, G. S. M.	19 Aug. 18		No. 1 Res. Bn.
19 Aug. 18	Dowie, A. M.	19 Aug. 18		No. 1 Res. Bn.
19 Aug. 18	Macrae, M.	19 Aug. 18		No. 1 Res. Bn.
19 Aug. 18	Ross, W. B.	19 Aug. 18		No. 1 Res. Bn.
19 Aug. 18	Sugden, K. G. J.	19 Aug. 18		No. 1 Res. Bn.
19 Aug. 18	Knight, S. J.	19 Aug. 18		No. 1 Res. Bn.
26 Sept. 17	Kerr, J. C.	26 Sept. 17		No. 1 Res. Bn.
26 Sept. 17	Muir, W.	26 Sept. 17		No. 1 Res. Bn.
26 Sept. 17	Gossman, J.	26 Sept. 17		No. 1 Res. Bn.
5 Apl. 15	Dickie, J. C., Capt.	6 Mar. 17		No. 1 Res. Bn.

Regimental District, No. 72.
(No. 1 District.)

Uniform—Scarlet.
Facings—Buff.

Seaforth Highlanders.

(Ross-shire Buffs, The Duke of Albany's.)

1ST BATTALION (*late 72nd Foot.*)

Arrived in India, 12th February 1903, from Egypt.

IN EACH OF THE FOUR CORNERS THE LATE DUKE OF YORK'S CYPHER AND CORONET. THE MOTTO *"Cuidich'n Righ."* THE ELEPHANT, SUPERSCRIBED "ASSAY."
"Carnatic" "Hindoostan" "Mysore" Cape of Good Hope, 1806" "Maida" "Java" "South Africa, 1835" "Sevastopol" "Koosh-ab" "Persia" "Lucknow" "Central India" "Peiwar Kotal" "Charasiah" "Kabul, 1879" "Kandahar, 1880" "Afghanistan, 1878-80" "Tel-el-Kebir" "Egypt, 1882" "Chitral" "Atbara" "Khartoum" "Paardeberg" "South Africa, 1899-1902."

COLONEL—Hon. Maj.-General R. H. Murray, C.B., C.M.G. *q.s.* ret. pay 16 Apl. 14.

First Commn.	Names.	Regimental rank.	Army rank.	Remarks.
	Lieutenant-Colonel.			
	Majors.			
18 Sep. 97	Thomson, W. M., C.B., M.C. (51, 59, 72) (*temp. Maj.-Genl.*)	1 Sep. 15	*c.* 23 Dec. 16	
15 Nov. 99	Anstruther, P. G.	1 Sep. 15	*l. c.* 10 Mar. 17	
4 Dec. 01	Schomberg, R. C. F., D.S.O. (*act'g. Lt.-Col.*)	4 Dec. 16	10 Oct. 15	Comdg.
	Captains.			
13 Aug. 04	Scobie, I. H. McK.	10 May 15	25 Apl. 14	
18 Aug. 15	*Macrae, J. N.*, D.S.O.	18 Aug. 15	*m.* 3 June 16	Adjt.
20 Sep. 11	Muirhead, J.	1 Oct. 15		
3 Sep. 12	Murray, A. R.	7 Dec. 15		
2 Sep. 13	Barlow, R. G.	26 Feb. 16		
11 Nov. 14	Lynden-Bell, L. A., M.C.	18 Apl. 17	....	*Lv., ex I., m. c., 12 m.*
20 Jan. 15	Kenney, B., D.C.M.	19 Apl. 18	....	Asst. Prov. Marshal.
	Lieutenants.			
14 July 15	Wollaston, J. W. (46)	11 Sep. 16		
25 July 15	McNaughton, J.	3 Oct. 16		S.S.O., Madras.
2 Nov. 16	Bean, H.	2 Nov. 16		
4 Oct. 15	Miller, J. F.	1 Jan. 17		
. . .	Dunn, N. E. J. (*temp. Capt.*)	31 May 17	....	Platoon Comdr., Schl. of Instrn.
18 Nov. 15	Forbes, J., D.S.O.	1 July 17		
24 Mar. 16	Macleod, N., M.C., D.C.M.	24 Sep. 17	....	With R.A.F.
24 Mar. 16	Campbell, D., M.C.	6 Nov. 17		
4 Sep. 16	Denniston, J. G.	15 Nov. 17	....	*Lv., in I., m. c.*
16 May 16	Mackay, D. S., D.C.M. (*act'g. Capt.*)	16 Nov. 17	....	M.G. Corps.
	2nd-Lieutenants.			
7 Feb. 17	Macleman, W. D.	7 Feb. 17	....	With 1st G. B., R. Sco. Fus.
28 May 17	Fisher, J.	28 May 17		
29 June 17	Hennessy, T.	29 June 17		
1 Mar. 18	Smith, D.	1 Mar 18		

Agents:—MESSRS. COX & CO.

Seaforth Highlanders—*contd.*

First Commn.	Names.	Regimental rank.	Army rank.	Remarks.
	Attached.			
	Ranken, Capt. L. D.	23 June 15		
	Hadow, Capt. R. H., M.C., Arg. & Suth'd. Highrs.	8 July 16		
. . .	Ross, Capt. R. W., M.C., 3rd Bn., Sea. Highrs.	2 Feb. 16		
14 Aug. 14	McDermott, Capt. R. K., 3rd Bn. Sea. Highrs.	2 Feb. 16		
27 Mar. 15	Sherman, Capt. C. B., 4th Bn., Sea. High's.	28 Aug. 16		
	Dickson, Lt. (*actg. Capt.*) J. W. E., M.C.	2 Feb. 16		
	Elliott, Lt. P.	20 Mar. 16		
	Mackay, Lt. E. H.	4 May 17		
	Jackson, Lt. E. F.	1 July 17		
	Cameron, Lt. D. M.	1 July 17	. . . .	With No. 1 Res. Bn.
	Urquhart, Lt. D. S.	1 July 17		
	Bonar, Lt. A. W. T., M.C. . . .	1 July 17		
24 Mar. 15	Macroy, Lt. W. M., M.C., 3rd Bn., Sea. High's.	20 Mar. 16		
3 May 15	Gore, Lt. C. H., Attd. 1-2nd Bn., Sea. Highrs.	1 July 17		
14 Aug. 15	Oswald, Lt. J. A. R., Attd. 1-2nd K.O.S.B.	10 Nov. 17		
	Denniston, 2nd-Lt. J. G.	5 Aug. 15		
	Bonthrone, 2nd-Lt. A.	30 Jan. 16		
25 June 15	Wilson, 2nd Lt. N. J., 15th Bn., Arg. & Suth'd H ghrs	25 June 17		
16 Jan. 16	Allan, 2nd-Lt. J. R. N., 6th Bn., Sea. Highrs.	16 Jan. 16		
21 Jan. 16	Fortune, 2nd-Lt. F. R., 3rd Bn., Arg. &. Suth'd Highrs.	21 Jan. 16		
7 July 16	Hodgson, 2nd-Lt. C. H., 3-4th Bn., Sea Highrs.	7 July 16		
5 Sep. 16	Hoggart, 2nd-Lt. D., 3rd Bn., Sea. Highrs.	5 Sep. 16		
5 Sep. 16	Rodger, 2nd-Lt. T. T., Attd. 1-2nd Bn., Sea. Highrs.	5 Sep. 16		
5 Sep. 16	McClymont, 2nd-Lt. A. J., Attd. 1-2nd Bn., Sea. Highrs.	5 Sep. 16		
5 Sep. 16	Bell, 2nd-Lt. C. W. A., Attd. 1-2nd Bn., Sea. Highrs.	5 Sep. 16		
25 Jan. 17	Bower, 2nd-Lt. D. R., 3rd Cam'n. Highrs.	25 Jan. 17		
25 Jan. 17	Cavaye, 2nd-Lt. R. M., 1-2nd Cam'n. Highrs.	25 Jan. 17		
25 Jan. 17	Cullen, 2nd-Lt. W. R., Attd. 1-2nd Sea. Highrs.	25 Jan. 17		
25 Jan. 17	Fle't, 2nd-Lt. F., Attd. 1-2nd Sea. Highrs.	25 Jan. 17		
25 Jan. 17	McGregor, 2nd-Lt. G. N., 3rd Bn., Sea. Highrs.	25 Jan. 17		
1 Mar. 17	Marshall, 2nd-Lt. R. W., 7th Bn., Arg. & Suth'd Highrs.	1 Mar. 17		
1 Mar. 17	Dawsan, 2nd-Lt. J., 7th Bn., Arg. & Suth'd Highrs.	1 Mar. 17		
29 Aug. 17	Hill-Murray, 2nd-Lt. S.W. H., Attd. Sea. Highrs.	29 Aug. 17		

First Commn.	Names.	Regimental rank.	Army rank.	Remarks.
18 Aug. 15	Macrae, Bt.-Maj. J. N., D.S.O. .	18 Aug. 15	*m.* 3 June 16	Adjt.
24 Oct. 08	Reid, Capt. N.	3 June 16	. . . .	Qr.-Mr.

Regimental District No. 75. [No. 1 District.]

The Gordon Highlanders.

Uniform—Scarlet. *Facings*—Yellow.

1st GARRISON BATTALION.
Arrived in India 14th March 1917.

COLONEL—Genl. *Sir* I. S. M. Hamilton, G.C.B., D.S.O., *q.s.* A.-D.-C. . . 26 Oct. 14.

First Commn.	Names.	Regimental rank.	Army rank.	Remarks.
	Lieutenant-Colonel.			
19 Nov. 91	Gaunt, Maj. C. R., D.S.O. (*Bt.-Lt.-Col.*) (*temp. Lt.-Col. in Army*) (*4th Dragoon Guards, Res. of offrs.*) (43, 50, 64, 66*a*. 170).	16 Nov. 17	10 Dec. 04	Comdg.
	Majors.			
Feb. 89	Bunbury, W. C. H. (*Capt. Res. of Off.*). (51, 57, 66*a*, 127, 129, 169, 174, 206, 207).	1 Nov. 16	6 Nov. 16	
	Captains.			
22 Oct. 02	Gilmour, A. C., *4 Bn. Arg. & Suth'd. Highrs.* (*Lt. Ret. Pay*) (46, 51).	10 Oct. 16	4 Jan. 13	
June 01	Home, W. (*Res. of Off.*) (*Hon. Capt in Army*) (*act'g. Maj., 10 Nov. 18*) (51).	3 Oct. 16	5 Sep. 14	2nd-in-Comd.
21 Nov. 00	Cobb, H. S. (*temp.*) (48*a*, 51, 57, 72*b*)	19 Jan. 17	20 Oct. 14	
9 July 15	Sherwell, S. (*temp.*) *6 Bn.* (50, 52, 60*a*, 72, 170).	30 Sep. 16	9 July 15	
Feb. 05	McTavish, F. H. C., *R. Highrs.* (67)	Oct. 16	2 Oct. '15	Remount Depôt., Mona.
19 Nov. 14	Robertson, R. H. (*temp.*) *R. Highrs.* (46, 127).	8 Dec. 16	26 Oct. 15	M. A. Dept.
15 Aug. 14	Buchan-Hepburn, T. E. A., *3rd Bn., R. Scots* (46).	3 Oct. 16	16 Feb. 16	
June 98	Peareth, J. T. (*temp.*) (46, 51, 125*e*, 207).	13 Nov. 16	20 Jan. 17	
	Lieutenants.			
7 Jan. 15	Wood, A. D. (*temp.*) (*act'g. Capt.*) (57).	15 Jan. 17	1 Oct. 15	Adjt.
4 Nov. 14	Henderson, G. H. *3 Bn., Sco. Rif.* (46, 51).	5 Oct. 16	1 Feb. 16	Act'g. Qr.-Mr.
22 Sep. 14	Pitcairn, G. D. (temp.), *9 Bn., Gord. Highrs.* (207)	5 Feb. 18	1 July 17	
31 Dec. 14	Boothby, G. A. B. C. (*temp.*), *7 Bn., R. Highrs.* (46, 48*a*, 51, 146)	11 Aug. 18	1 July 17	
10 Feb. 15	Macdonald, J. K. (*temp.*) *7 Bn., Cam'n. Hrs.* (46, 51).	5 Jan. 17	1 July 17	Comdg. Attd. Section, Rawalpindi.
29 Mar. 15	Law, H. R. McK. (*temp.*) (127).	14 Nov. 16	1 July 17	
4 June 15	Burnham, G. K. (*temp.*), *4 Bn., R. Highrs.* (46, 61).	30 Aug. 18	1 July 17	
25 June 15	Chattanach, J. C. (*temp.*) *Cam'n. Highrs.* (46, 49)	17 Oct. 18	1 July 17	
30 July 15	Baxter, J., *3 Bn.* (48*a*, 51) . . .	17 Mar. 18	1 July 17	
25 Sep. 15	Astell, I. W. (temp.), *6 Bn., Sea. Highrs.*	5 Feb. 18	1 July 17	
6 Oct. 15	Reid, D. C. G. (*temp.*), *4 Bn., R. Highrs.* (46, 51)	11 Aug. 18	1 July 17	
12 Mar. 16	Swain, G. A. (*temp.*) *7 Bn., R. Sco. Fus.* (52, 61, 127).	4 Oct. 16	12 Sep. 17	
5 Sep. 16	Urquhart, D. I., *8 Bn., Sea. Highrs.* (46, 48*a*, 51).	2 Oct. 16	5 Mar. 18	
	2nd-Lieutenants.			
5 Sep. 16	Wilkinson, A. E. (*temp.*) *Sea. Highrs.*	17 July 17	5 Sep. 16	With 1-5th Devon R.
25 Oct. 16	McLean, A. (*temp.*) *R. Highrs.* (51, 52).	9 Dec. 16	25 Oct. 16	
30 May 17	Charlton, J. (*temp.*)	1 Aug. 18	30 May 17	
17 June 17	Hennessy, J. A. (*temp.*), *High. L.I.*	11 Feb. 18	17 June 17	
26 June 17	Alexander, A. P. (*temp.*), *High. L. I.* (46, 51)	11 Feb. 18	26 June 17	S. S. O., Burbhan.
8 Nov. 17	McLeod, R. D., *9 Bn., High. L. I.* (46)	1 Aug. 18	8 Nov. 17	
7 Jan. 15	Wood, Lt. (*act'g. Capt.*) A. D. .	16 Nov. 17	1 Oct. 15	Adjt.
5 May 05	Cooper, Maj. F. W. (*temp.*) (T. F., Genl. List.)	18 Oct. 16	1 Oct. 08	Qr.-Mr. *Lv., ex I., 6 mos.*

Regimental District No. 83.
[No. 11 District.]

Uniform—Green.
Facings—Dark Green.

The Royal Irish Rifles.

1ST GARRISON BATTALION.

Arrived in India, 13th March 1916.

First Commn.	Names.	Regimental rank.	Army rank.	Remarks.
	Lieutenant-Colonel.			
2 Dec. 76	Wood, Bt.-Col. A. H., C.B.E. (51, 206) (ret. pay.)	11 Nov. 15	8 Feb. 05	
	Majors.			
80	Jackson, C. (*temp.*) (71, 206) . .	5 Mar. 17	4 Sep. 14	
	Captains.			
Feb. 00	Webber, A. (*temp.*)	8 Dec. 16	26 Sep. 14	
05	Biggs-Atkinson, T. B. (*temp.*) (51) .	25 Jan. 16	7 Nov. 14	Asst. Cable Censor, Bhamo.
Sep. 14	*Ryall, W. D.* (*temp.*) (105, 162, 206).	2 Feb. 16	23 Nov. 14	
Oct. 06	Thompson, N. R. (*temp.*) (48*a*, 51)	29 July 17	23 Nov. 14	Platoon Comdr., Cadet Unit, [Sialkot.
Sep. 88	Allen, E. O. S. (temp.)	28 Dec. 15	23 Jan. 15	
17 Sep. 14	Johns, T. S. (temp.) (51)	5 Mar. 17	1 Apl. 15	
1 Mar. 15	Wood, R. T. (temp.) (52) . . .	14 Feb. 16	1 June 15	
8 May 15	*Toppin, A. J.* (*temp.*) (*Lt., 3 Bn.*) .	1 May 16	. . . .	*Adjt.*
	Lieutenants.			
10 Mar. 15	Dobbin, J. M. (*temp.*) . . .	20 Dec. 15	. . . . .	Asst. Censor, Simla.
19 Apl. 15	Dickson, J. W. (*temp.*) (57) . .	1 May 16		
3 May 15	McLellan, G. H. H. (*temp.*) . .	1 May 16	. . . . .	*Lv. ex. I, m.c., 6 mos.*
17 June 15	Monsell, H. H. (*temp.*) . . .	1 May 16		
1 Aug. 15	Bell, N. G. (*temp.*) (51) . . .	1 May 16	. . . . .	S. S. O., Cawnpore.
26 Apl. 15	Irvine, C. P. J. (*temp.*) (51) . .	1 July 17	. . . . .	Asst. Cable Censor, Rangoon.
12 Oct. 15	Morris, W. H. (*temp.*)	14 Feb. 18	1 July 17	
19 Apl. 15	McWhirter, R. J. (*temp.*) (66*a*) .	2 Aug. 18	1 July 17	
3 July 15	Waddell, C. M. (*3rd Bn., R. Ir. Rif.*)	1 July 17	. . . .	
1 Jan. 16	Huggard, E. H. (*temp.*) . . .	24 Mar. 18	1 July 17	
7 June 16	Barnes, T. L. (*7 Bn., L'pool R.*) .	10 Aug. 18	7 Dec. 17	
	2nd-Lieutenants.			
15 Dec. 16	McKerrow, H. B. (*temp.*) (*temp. Maj.*) (66*a*).	29 July 17	15 Dec. 16	Ind. Munitions Board.
30 May 17	Main, B. W. C. (*4 Bn., Leic. R.*) .	9 Aug. 18	30 May 17	
17 June 17	Naylor, T. (*temp.*)	14 Feb. 18	17 June 17	Mily. Accts. Dept.
17 June 17	Nield, H. M. (*temp.*)	14 Feb. 18	17 June 17	
17 June 17	Watson, H. T. (*temp.*)	14 Feb. 18	17 June 17	
17 June 17	Webb, R. (*temp.*)	14 Feb. 18	17 June 17	
17 June 17	Smith, C. (*temp.*) (51)	14 Feb. 18	17 June 17	Mily. Accts. Dept.
17 June 17	Botterill, P. W. (*temp.*) . . .	21 Feb. 18	17 June 17	Army Clothing Factory, Madras.
17 June 17	Davies, H. T. W. (*temp.*) . . .	14 Feb. 18	17 July 17	
17 June 17	Davis, A. S. (*temp.*)	14 Feb. 18	17 June 17	
26 Sep. 17	Hoyle, R. (3 Bn., S. Lan. R.) . .	2 Aug. 18	26 Sep. 17	
1 Sep. 13	Verschoyle, Capt. F. H. H. S., 1 Bn., R. Innis. Fus.	9 Nov. 18	17 Mar. 16	
30 May 17	Carr, 2nd-Lt. W., 3 Bn., R. Muns. Fus.	16 Mar. 16	30 May 17	

8 May 15	Toppin, Capt. A. J. (*temp.*) (*Lt., 3 Bn.*)	1 May 16	. . . .	Adjt.
4 Oct. 15	Rourke, Lt. D. (*temp.*) (58, 123, 129) (*g*)	4 Oct. 15	. . . .	Qr.-Mr.

Regimental District No. 87.
[No. 11 District.]

Uniform—Scarlet.
Facings—Blue.

Princess Victoria's (Royal Irish Fusiliers).

1ST GARRISON BATTALION.

Arrived in India, 4th March 1916.

First Commn.	Names.	Regimental rank.	Army rank.	Remarks.
	Lieutenant-Colonel.			
14 Sep. 87	Robinson, G. W., C.B. (*Lt.-Col., I. A.*) (50, 66*a*, 71, 170)	18 Feb. 18	14 Sep. 13	
	Majors.			
Apl. 92	Cross, P. W. (51) (*3rd Bn.*) . .	12 Sep. 15	10 Apl. 15	
21 Nov. 02	McLaughlin, W. (*temp. Capt., 3rd R. Ir. Rif.*) (51) .	2 Oct. 15	21 Jan. 08	Comdt., Conv. Sec., Deolali.
	Captains.			
20 Aug. 86	Ricketts, T. F. W., ret. pay (51) . .	12 Sep. 15	5 Aug. 14	
13 May 03	Algeo, L. W. A. (*4th R. Ir. Rif.*) .	22 Sep. 15	19 Dec. 14	
Nov. 01	Hazledine, H. St. J. C. (*3rd R. Dub. Fus.*)	24 Sep. 15	15 May 15	
Apl. 12	Kenny, M. J. (*4 Bn. Conn. Rang.*) (51)	18 Oct. 15	26 July 15	
June 08	Crean, G. J. (*4 Bn. R. Innis. Fus.*)	1 Nov. 15	27 Nov. 15	
June 12	Silver, B. J (*4 Bn. R. Innis. Fus.*)	1 Nov. 15	31 Mar. 16	S. S. O., Shwebo.
15 Aug. 14	Wilson, A. H. (*4th Bn.*)	22 Sep. 15	7 Feb. 16	M. A. Dept.
15 Aug. 14	Turner, H. (*3rd Bn.*) (51, 52) . .	28 Sep. 15	10 Mar. 16	Adjt.
	Lieutenants.			
7 Oct. 14	Pinkerton, A. T. H. (*4 Bn., R. Ir. Rif.*)	22 Nov. 15	15 Mar. 16	S. S. O. & Cant. Magte.-Thayetmyo.
6 Feb. 15	Byrne, G. L. (temp.) (*3rd Bn.*) . .	22 Nov. 15	8 Jan. 16	
24 Mar. 15	*McKeag, H. T. A.* (51) (*3rd Bn., R. Ir. Rif.*)	18 Sep. 15	29 Feb. 16	*Lv., ex I.*,
9 Apl. 15	Kenward, N. H. (*3rd Bn.*) (57) . .	28 Sep. 15	13 Dec. 16	*I. A. R. O.*
14 Oct. 14	Hill Dickson, J. E. (*2nd Garr. Bn., R. Ir. Fus.*)	18 Feb. 18	28 June 17	
19 Apl. 15	Wray, D. W. (*temp.*) (*17th Bn. R. Ir. Rif.*)	1 July 17		
	2nd-Lieutenants.			
19 Apl. 15	Watson, F. R. J. MacD. (temp.) .	11 Nov. 15	19 Apl. 15	
30 Apl. 15	Phillips, W.J.C. (*3rd R. Dub. Fus.*) (51, 60*a*)	3 Oct. 15	30 Apl. 15	I. A. (*on probn.*)
2 May 15	Norman, G. E. (*5th R. Ir. Rif.*) .	22 Sep. 15	2 May 15	With R. A. F.
18 May 15	Greene, J. (*4th R. Ir. Rif.*) . .	22 Sep. 15	18 May 15	With R. A. F.
6 Nov. 15	Stewart, H. J. L. (3rd R. Ir. Rif.)(51)	3 Feb. 16	16 Nov. 15	
1 Jan. 16	Potter, J. C. (*3rd Bn., R. Ir. Regt.*)	6 Mar. 17	1 Jan. 16	
29 Apl. 15	Suys, P. W. (*temp.*) (*act'g. Lt.*) .	8 May 17	29 Apl. 15	S. S. O., Mandalay.
22 Nov. 16	Jones, S. C. (*temp.*) (*3rd (Res.) Garr. Bn. R. W. Fus.*, (51) . . .	1 Sep. 17	22 Nov. 16	Ind. Ord. Dept.
	Attached.			
30 Apl. 15	Padley, 2nd-Lt. A. C., 4 Bn., R. Dub. Fus. (51)	30 Apl. 15		
2 Oct. 16	Cuming, 2nd-Lt. B. S. (*temp.*) R. Muns. Fus.	2 Oct. 16	. . . .	With No. 1 Res. Bn.
26 Sep. 17	Dowling, 2nd-Lt. J. C. (*temp.*), 11 Bn., R. Dub. Fus.	26 Sep. 17		
15 Aug. 14	Turner, Capt. H. (51, 52)	28 Sep. 15	10 Mar. 16	Adjt.
10 Dec. 15	Crowden, Lt. H. C.	1 Sep. 17	. . . .	Qr.-Mr.

Regimental District No. 88. [No. 12 District.]

The Connaught Rangers.

Uniform—Scarlet. Facings—Green.

1ST BATTALION (*late 88th Foot*).

Arrived in India, 11th March 1908, from Malta.

THE HARP AND CROWN, WITH THE MOTTO "*Quis Separabit.*" THE ELEPHANT. THE SPHINX, SUPERSCRIBED "EGYPT."
"Seringapatam" "Talavera" "Busaco" "Fuentes d'Onor" "Ciudad Rodrigo" "Badajoz" "Salamanca" "Vittoria" "Pyrenees" "Nivelle" "Orthes" "Toulouse" "Peninsula" "Alma" "Inkerman" "Sevastopol" "Central India" "South Africa, 1877-8-9" "Relief of Ladysmith" "South Africa, 1899-1902."

COLONEL—Maj.-General W. L. Dalrymple, C.B., *p.s.c.* ret. pay . 20 Jan. 12.

First Commn.	Name.	Regimental rank.	Army rank.	Remarks.
	Lieutenant-Colonels.			
8 Dec. 88	Ravenshaw, H. S. L., C.M.G. (*temp. Maj.-Genl.*)	16 July 14	*c.* 1 Jan. 16	Bde.-Comdr.
	Majors.			
7 Nov. 91	Deacon, H. R. G., D.S.O. (*temp. Lt.-Col.*) (50, 52, 59, 72*b*).	4 Mar. 11	*l.c.* 3 June 16	With High. L. I.
7 May 98	Truell, E. G. S. (51, 52, 72*b*)	1 Sep. 15	. . . .	With 12 Bn., Manch. R.
6 Dec. 99	Wratislaw, J. M. B. (51, 52, 72*b*) .	1 Sep. 15		
22 May 01	Underhill Faithorne, C. F. (51, 53)	22 May 16	. . . .	Comdt., Sch. of Musk.
11 Oct. 01	Nolan-Ferrall, H. J.	22 May 16		
5 Feb. 02	Eyre, R. G. (50, 52, 72, 72*b*) . .	18 Jan. 17	. . . .	With 12 Bn., R. Innis Fus.
7 Oct. 03	Callaghan, G. F., M.C. (51, 52, 66, 71)	22 Oct. 17		
	Captains.			
20 May 05	Moutray, A. G.	29 Nov. 14		
6 Nov. 09	Howard, S. W., D.S.O.	19 Mar. 15		
28 May 10	*Anderson, A. L. B.* (71) . .	29 Apl. 15	. . . .	*I. A.* (*On probn.*)
5 Feb. 13	*Hewitt, H. T.*, M.C.	1 Oct. 15	. . . .	*Adjt.*
24 Jan. 14	*Dennys, C. H. M.* (207) . . .	18 Aug. 16	. . . .	*I. A.* (*On probn.*)
14 Aug. 14	Beard, A. S.	1 Jan. 17		
17 Oct. 14	Booth, T., M.C.	4 Mar. 17	. . . .	With 4th Bn.
	Lieutenants.			
9 Nov. 14	Creery, W. F.	30 Apl. 15	. . . .	With R. A. F.
12 Nov. 14	O'Brien, W, M.C. (*act'g. Capt.*)	30 Aug. 15		
16 Feb. 15	Wallace, J. (*temp.*)	11 Nov. 15		
17 Feb. 15	Taaffe, W. A.	11 Nov. 15	. . . .	With Cadet Bn.
30 Mar. 15	Durrant, E. M., M.C., D.C.M. (temp.)	26 Nov. 15		
15 July 15	Reeve, F. (*temp.*)	19 Apl. 16		
11 Aug. 15	Davey, F.	10 Sep. 16	. . . .	With M. G. Corps.
4 Oct. 15	Beckett, E A.	1 July 17		
26 Oct. 15	Kelsey, T., M.C., D.C.M. . . .	1 July 17		
7 Nov. 15	Robey, A. E. (*temp.*)	1 July 17		
24 Nov. 15	Morrogh, V. M. (*temp.*)	1 July 17		
27 May 15	Gibbons, S.	27 Aug. 17		
30 Oct. 6	Whelan, J. P., D.C.M.	30 Apl. 18		
	2nd-Lieutenants.			
7 Aug. 16	Barry, E.	7 Aug. 16		
1 Mar. 17	Tolputt, W. L.	1 Mar. 17		
	Attached.			
11 Feb. 88	Murray, Lt.-Col. and Bt.-Col. S. J., 2nd Bn.	11 Dec. 14	*c.* 3 June 16	Comdg.
	Hamilton, Maj. & Bt.-Lt.-Col. (*act'g. Lt.-Col.*) E. G., D.S.O., M.C., 2nd Bn.	22 Oct. 17	*l.c.*, 23 Oct. 17	
	Dwyer, Capt. H. M. O'C., (*Res. of Offr.*) 4th Bn.	10 Nov. 14	. . . .	*Lv., ex I., m. c., 6 m.*
	Foott, Capt. F. D., 2nd Bn. . . .	7 Apl. 17		
	Roussel, Capt. P. B., 3rd Bn. . .	12 Sep. 16		
17 Feb. 15	Tittle, Lt. D. R., 2nd Bn., R. Dub. Fus.	30 Aug. 15		
	Minch, Lt. W., 3rd Bn.	10 Mar. 16		
22 Dec. 15	Yeates, Lt. R. A., 1st Bn. R. Dub. Fus.	1 July 17		
14 Aug. 15	Murray, Lt. H., 4th Bn., R. Dub. Fus.	1 July 17		
22 Apl. 15	Watkins, Lt. G. D., 5th Bn., R. Dub. Fus.	1 July 17		
24 Apl. 15	O'Rorke, Lt. M. D., 3rd Bn. . . .	1 July 17	. . . .	*Lv., ex I., m. c., 6 mos.*
26 June 15	McDowell, Lt. G. A., 3rd Bn. . .	1 July 17		
12 Aug. 15	Hearn, Lt. C. R. T., 3rd Fus. R. Dub. Bn.	1 July 17		
14 Aug. 15	Clarke, Lt. E. R., M.C., 3rd Bn. .	1 July 17		
24 Aug. 15	*Flynn, Lt. A. E., 3rd Bn.* . . .	1 July 17	. . . .	*I. A.* (*On probn.*)

Agents:—MESSRS. COX & CO.

The Connaught Rangers—*contd.*

First Commn.	Names.	Regimental Rank.	Army rank.	Remarks.
	Attached—*contd.*			
	Louw, Lt. D. L., 3rd Bn. . . .	1 July 17		
	Watson, Lt. F., 5th Bn. . . .	1 July 17		
	Foran, Lt. E., 3rd Bn. . . .	1 July 17		
	Gilfillan, Lt. N., 4th Bn., N.Staff. .	1 July 17		
	Heneghan, Lt. W. J.	1 July 17		
26 Sep. 16	Norman, Lt. C. G., 1st Bn., R. Dub. Fus.	26 Mar. 18		
30 Nov. 15	Hudson, 2nd-Lt. J., M.C., D.C.M. (temp. *Capt.*)(*temp.*), 5th Bn.	30 Nov. 15	. . . .	Asst. Prov. Marshal.
5 Feb. 13	Hewitt, Capt. H. T., M.C. . . .	12 July 16	. . . .	Adjt.
8 Mar. 11	Gorman, Capt. J. T.	3 June 17	. . . .	Qr.-Mr. Empld. at War Office.
15 June 17	Bridges, Lt. J. J.	15 June 17	. . . .	Qr.-Mr.

Uniform—Green.
Facings—Black.

The Rifle Brigade (The Prince Consort's Own).

18TH (LONDON) BATTALION.

Arrived in India, 5th January 1916.

HONORARY COLONEL—

First Commn.	Names.	Regimental rank.	Army rank.	Remarks.
	Lieutenant-Colonel.			
25 Feb. 80	Montgomery, C. A. S. (206) (Bt.-Col. ret. I. A.)	20 Sep. 15		
	Major.			
10 May 01	Thain, E. W. (*Hon. Capt. in Army, 29 Sep. 02.*)	1 June 16		
1 Jan. 98	Chambers, L. D.	1 June 16		
	Captains.			
94	Waters, S. (*Hon. Lt. in Army, 7 Feb. 03.*)	10 Oct. 14		
2 Nov. 14	Howie, J. (51)	4 Feb. 15		
31 May 00	Bridges, B. (51)	15 May 15		
11 Mar. 02	*McKean, M. C.*	1 June 16	....	*Cable Censor, Bombay.*
02	*Payne, C. H.* (206)	1 June 16		
June 07	Sinclair, W. C. B. (51)	1 June 16		
24 Jan. 15	*Tutt, A. R.*	12 May 17		
29 Sep. 14	Johnson, V. A.	12 May 17		
11 Nov. 14	Norman, C. (51)	10 Nov. 17		
28 Nov. 14	*Villiers, T. L.*	9 Apl. 18		Cable Censor, Rangoon.
28 Nov. 14	Morley, B. E.	9 Apl. 18		S. S. O., Maymyo.
	Lieutenants.			
24 May 15	Brook-Ascough, T. H. (51)	24 May 15		
1 Dec. 14	Wakefield, L. V. (*act'g. Capt.*) (51)	1 June 15	....	Adjt.
4 Dec. 14	Heanly, P. W.	4 June 15		
16 June 15	Geddes, T.	16 June 15	....	S. S. O , Port Blair.
10 Feb. 15	Graves, L. C. (51)	10 Aug. 15	....	A. D.C. to L. G. of Burma.
4 Mar. 15	*Martin, J.*	*4 Sep. 15*	....	*I. A. R. O.*
19 Mar. 15	Cull, A. C. B. (51)	19 Sep. 15		
11 May 15	Andrews, H. W. (51, 52)	27 Oct. 15	....	Attd. G. S. Bch., Army Hdqrs.
27 Oct. 15	*Parry, H. T.* (*temp. Capt.*)	1 July 17	...	*Comdt., P. of W. Camp, Shwebo.*
20 Nov. 15	Leader-Williams, B. C. (51)	1 July 17		
	2nd-Lieutenants.			
	Attached.			
6 May 15	Ellson, Lt. P., (*temp. Capt.*) 9 Bn., Midd'x R.	1 July 17	....	Platoon Comdr., Officers' Schl. of Instn., Bangalore.
19 Oct. 15	Herbert, Lt. W. C., 9 Bn., Midd'x R.	1 July 17		
24 Oct. 15	Duff, Lt. J. M., 7 Bn., Sco. Rif.	1 July 17		
25 Nov. 15	Luxton, Lt. F. C., 7 Bn., Lon. Regt.	1 July 17		
26 May 15	Churchill, Lt. W. F., 7th Bn., Lon. Regt.	30 Nov. 17		

1 Dec. 14	Wakefield, Lt. (*act'g.Capt.*) L.V. (51)	27 Oct. 15	....	Adjt.

Uniform—Green.
Facings—Black.

The Rifle Brigade (The Prince Consort's Own).

21ST (MIDLAND) BATTALION.

Arrived in India, 1st October 1918.

First Commn.	Names.	Regimental rank.	Army rank.	Remarks.
	Lieutenant-Colonel.			
13 Aug. 79	Sutton, H. G., *Bt.-Col., ret. Ind. Army.*	20 Sep. 15	*c.* 6 Aug. 08	
	Major.			
11 Dec. 95	Oakden, R.	1 June 16		
	Captains.			
7 Nov. 14	Garrad, W. E.	1 June 16		
—— 03	Williams, R.	1 June 16		
28 Jan. 15	Lloyd, C. F.	13 Dec. 17		
	Lieutenants.			
18 Dec. 15	Moody, W. P.	2 June 15		
12 May 15	*Stevens, E. T. D. (act'g. Capt.)*	23 Oct. 16	. . . .	*Adjutant.*
3 Dec. 15	Higgins, H.	1 July 17		
3 Dec. 15	Hawkins, E. N.	1 July 17		
	Attached.			
—— 00	Coode, Maj. A. P., 4 Bn. D. of Corn. L. I.	1 June 16		
1 July 16	Vyvyan, Capt. R. T., Oxf. and Bucks. L. I.	1 July 16		
11 Mar. 11	Coulston, Lieut. H. C., R. Lanc. R.	4 May 15		
20 Mar. 15	Lindley, Lieut. G. W., R. Berks. R.	2 Oct. 15		
5 May 15	Haynes, Lieut. G. W. (temp.), 1 Garr. Bn., Devon. R.	17 Mar. 17	26 Aug. 16	
12 Dec. 15	Orford, Lieut. C. J., 3rd County of London Yeo	1 July 17		
16 Nov. 15	Redshaw, Lieut. T. A., 4 Bn., N. Lan. R.	1 July 17		
5 Sep. 16	Gravestock, Lieut. G. T. (temp.), 9 Bn., Norf. R.	5 Mar. 18		
27 Feb. 17	Powell, 2nd-Lt. T. M. (temp.), 1st (Res.) Garr. Bn., Suff. R.	27 Feb. 17		
11 Mar. 17	Greville, 2nd-Lt. S. E. (temp.), 3rd (Res) Garr. Bn., R. W. Fus.	11 Mar. 17		
28 Mar. 17	Brewer, 2nd-Lt. G. H., 4 Bn., E. York. R.	28 Mar. 17		
28 Mar. 17	Crundwell, 2nd-Lt. E. F., 4 Bn., R. W. Surr. R.	28 Mar 17		
26 Apl. 17	Barnes, 2nd-Lt. B. L. (temp.), 17 Bn., R. Fus.	26 Apl. 17		
26 Apl. 17	Hutton, 2nd-Lt. H. W. (temp.), 32 Bn., R. Fus.	26 Apl. 17		
17 June 17	Baird, 2nd-Lt. J. (temp.), 1 (Res.) Garr. Bn., Suff. R.	17 June 17		
16 Sep. 17	Haynes, 2nd-Lt. F. J., (*temp.*) 1 (Res.) Garr. Bn., Suff. R.	16 Sep. 17		
26 Sep. 17	Gordon, 2nd-Lt. G. W., 4 Bn., N. Lan. R.	26 Sep. 17		
28 Nov. 17	Griffith, 2nd-Lt. H. G. (temp.), 1st (Res.) Garr. Bn., Suff. R.	28 Nov. 17		
18 Dec. 17	Whittaker, 2nd-Lt. C., 6 Bn., R. Fus.	18 Dec. 17		
14 Mar. 18	Steeds, 2nd-Lt. F. G. (*temp.*) 5, Bn., Som. L. I.	14 Mar. 18		
13 May 17	Cassell, 2nd-Lt. H. E. (temp.), 3 (Res.) Garr. Bn., R. W. Fus.	13 May 17		

12 May 15	Stevens, Lt. (*act'g. Capt.*) E. T. D.	23 Oct. 16	. . . .	Adjutant.
20 July 17	Ridge, Lt. J. H.	26 July 17	. . .	Quarter-Master.
. . . .	Cope, Capt. R. (temp.) R. A. M. C.	12 Oct. 17	. . . .	Medl. Offr.

Uniform—Green.
Facings—Black.

The Rifle Brigade (The Prince Consort's Own).

23RD (NORTH WESTERN) BATTALION.

Arrived in India, 8th February 1916.

HONORARY COLONEL—

First Commn.	Names.	Regimental rank.	Army rank.	Remarks.
	Lieutenant-Colonel.			
89	Turnbull, T. E., V.D.	8 Feb. 13	. . . . .	
	Majors.			
14 July 02	Stillwell, E. H.	20 Apl. 10	. . . . .	
Aug. 03	Greenslade, E. W., T.D.	11 Nov. 14	. . . . .	
	Captains.			
23 Jan. 95	Burns, A. J. (51)	12 Sep. 14	. . . . .	
00	Thomas, J. P. (51)	2 Dec. 14	. . . . .	
96	*Page-Turner, E. H. M.*	16 June 15	. . . . .	*Spec. Empld.*
9 Nov. 14	Murphy, J. P. (60a)	1 June 16	. . . . .	
2 Dec. 14	McCormick, A.	1 June 16	. . . . .	*Lv., m.c., 3 m. to 18 July 18.*
2 Dec. 14	Williams, P. D.	1 June 16	. . . . .	
3 Dec. 14	*Rimmer, H. A.*	1 June 16	. . . . .	*M. A. Dept.*
29 July 11	*Forrester, H. W.* (57)	1 June 16	. . . . .	*Ind. Ord. Dept.*
21 Jan. 15	*Haines, E.*	*12 May 17*	. . . . .	*Adjt.*
	Lieutenants.			
22 Jan. 15	Berey, C. W.	22 Jan. 15	. . . . .	Asst. Cable Censor, Karachi.
20 Feb. 15	Jones, A. G.	20 Feb. 15	. . . . .	
18 Apl. 15	Martin, W. A.	18 Apl. 15	. . . . .	
18 Apl. 15	*Sykes, A. W.* (60a, 61)	18 Apl. 15	. . . . .	*Spec. Empld.*
19 May 15	Haughton, C. S. (52)	19 May 15	. . . . .	
4 June 15	*Macfie, J.*	4 June 15	. . . . .	*I. A. R. O.*
8 June 15	Radford, E. J.	8 June 15	. . . . .	M. A. Dept.
3 May 15	Green, R. (*temp. Capt.*) (51). .	3 Nov. 15	. . . . .	Asst. Emb. Staff Offr., Bombay.
30 Sep. 15	Thompson, R.	1 July 17	. . . . .	S. S. O., Amritsar.
26 Sep. 16	Cozens, A. (*temp. Capt., 31 Dec. 17*)	26 Mar. 18	. . . . .	Coy. Comdr., N. C. O.'s Sch.
	2nd-Lieutenants.			
21 Jan. 15	Haines, Capt. E.	1 Nov. 17	. . . . .	Adjt.
23 Dec. 15	Russell, Lt. C. J.	23 Dec. 15	. . . .	Qr.-Mr.

Uniform—Green.
Facings—Black.

The Rifle Brigade (The Prince Consort's Own).

24TH (HOME COUNTIES) BATTALION.

Arrived in India, 11th February 1916.

HONORARY COLONEL—

First Commn.	Names.	Regimental rank.	Army rank.	Remarks.
	Lieutenant-Colonel.			
25 Aug. 86	Burbury, F. W., 3 Bn., R. W. Kent. R., Capt. ret. (*temp.*)	17 Nov. 15	2 Mar. 18	
	Major.			
3 Oct. 14	Gant, F. C. (*Maj., ret. Vols.*) .	3 Oct. 14	. . . .	Asst. Embn. Offr., Bombay.
	Captains.			
3 Oct. 14	Uridge, A. J. R. (51) . . .	3 Oct. 14	. . . .	
30 Sep. 14	Sturt, G. L.	1 June 16	. . . .	
11 Nov. 14	Tann, E. J.	1 June 16	. . . .	
4 Dec. 14	Carter, J. S.	1 June 16	. . . .	
8 Dec. 14	Neve, E.	13 Apl. 18	. . . .	
18 Feb. 15	*Auger, A. E.*	21 July 18	. . . .	*I. A. R. O.*
	Lieutenants.			
24 Mar. 15	Plant, J. A.	24 Mar. 15	. . . .	Army Clothing Depôt, Allahabad.
21 Sep. 14	Gadney, H. G. (51) . . .	28 Mar. 15	. . . .	
20 Apl. 15	McLellan, J. S. (206) . . .	20 Apl. 15	. . . .	M. A. Dept.
28 Apl. 15	*Lee, G. J.*	28 Apl. 15	. . . .	*I. A. R. O.*
3 Nov. 14	*Greenacre, E.* (*act'g. Capt.*) (12, 48*a*, 51)	3 May 15	. . . .	*Adjt.*
12 May 15	Gordon, C. F. (51) . . .	12 May 15	. . . .	A. D. C.
16 May 15	Cusack, J.	16 May 15	. . . .	
12 Nov. 14	Angier, C. T.	12 May 15	. . . .	Staff Capt., Bombay Bde.
11 Mar. 15	Mann, W. H.	11 Aug. 15	. . . .	Adjt., Divl. Schl. of Musk.
16 Jan. 15	Sturge, C. (48*a*, 51, 59, 61, 71)	1 July 17	. . . .	Mily. Accts. Dept.
14 Mar. 15	Ginger, L. S. (46, 51) . . .	1 July 17	. . . .	
2 Nov. 15	Stevens, C. (57)	1 July 17	. . . .	With Sig. Serv.
	2nd-Lieutenants.			
27 Apl. 17	Godley, J. L.	27 Apl. 17	. . . .	
27 Apl. 17	Crockwell, C. G. (52) . . .	27 Apl. 17	. . . .	
26 Sep. 17	Drummond, J.	26 Sep. 17	. . . .	
26 Sep. 17	Mead, A. W.	26 Sep. 17	. . . .	
26 Sep. 17	Cowieson, S.	26 Sep. 17	. . . .	
16 Mar. 18	Pinch, R. P.	16 Mar. 18	. . . .	Attd., R. A. F.
16 Mar. 18	Francis, A. K.	16 Mar. 18	. . . .	
16 Mar. 18	Grove, A.	16 Mar. 18	. . . .	
16 Mar. 18	Dowd, H.	16 Mar. 18	. . . .	
	Attached.			
02	Wills, Capt. W. A., 3-1st Essex Yeo. (*temp. Maj., 25 Mar. 15*).	10 Mar. 17	. . . .	
	Cann, Capt. H. D., 6 Welsh R. .	18 Jan. 17	2 Sep. 14	
20 June 15	Champion, Lt. P. H., West Kent Yeo. (46, 53).	7 Mar. 17	20 June 15	P. A. to Ch. Commr., Cen. P.
5 Sep. 16	Elliot, Lt. L. F., Manch. R. . .	26 Aug. 18	5 Feb. 18	
22 Mar. 15	Crump, Lt. D. J., 9 Midd'x. R. (*temp. Capt., 1 Mar. 18*).	1 July 17	. . . .	S. S. O., Jhelum.
23 Nov. 16	Murch, Lt. R. W., 7 Midd'x. R. .	23 May 18	. . . .	
23 Nov. 16	Lawrence, 2nd-Lt. W. H. A., 7 Bn., Essex R.	. . . .	23 Nov. 16	

3 Nov. 14	Greenacre, Lt. (*act'g. Capt.*) E. (12, 48*a*, 51)	31 Oct. 17	. . . .	Adjt.
1 Nov. 15	Etherington, Lt. J. (*temp.*) . .	1 Nov. 15	. . . .	Qr.-Mr.

Uniform—Grey.
Facings—Scarlet.

The London Regiment.

1-25TH (COUNTY OF LONDON) BATTALION.

Arrived in India, 26th February 1916.

HONORARY COLONEL—Col. G. Smith, T.D. . . . . . . 16 Apl. 13

First Commn.	Names.	Regimental rank.	Army rank.	Remarks.
	Lieutenant-Colonel.			
9 Aug. 99	Hynes, B. M. (*Maj., R. Suss. R.*) (temp.) (51, 52, 72)	21 May 17		
	Major.			
June 04	Easton, M. H.	4 Mar. 15	. . . .	*Lv., ex I., m. c.*
27 July 01	Swinnerton, H.	1 June 16		
6 June 00	Stafford, W. S. (48*a*, 50)	13 Mar. 17		
	Captains.			
1 Mar. 08	Paget, C. N. (51)	1 Apl. 13		
1 Nov. 11	*Long, E. W.*	1 Apl. 14	. . .	Comdg. No. 10 Armd. Motor Batty.
27 Nov. 12	Blunden, H. G.	4 Mar. 15		
22 Feb. 13	Gillett, N. C. (*Capt., R. Dub. Fus.*) .	10 June 15	. . .	S. S. O., Multan.
22 Aug 14	*Wildy, H. A.*	1 June 16	. . .	*Comdg. No. 1 Armd. Motor By.*
28 Jan. 16	Ambrose, J. (51)	1 June 16	. . .	A. G.'s Branch, Record Section, Bombay.
	Lieutenants.			
15 Aug. 14	*Dix, R. C. St. J.*, M.C.	4 Mar. 15	. . .	*With R. A. F.*
26 Sep. 14	*Kittoe, M. C. S. H.*	1 June 16	. . .	*I. A. R. O.*
26 Feb. 15	Downs, R. W. (*temp. Capt., 6 Aug. 18*).	1 June 16	. . .	Instr., M. T. Trg. Schl., R'Pindi.
27 Feb. 15	Morton, E. R.	1 June 16		
5 Mar. 15	Easton, A. C. A. (57)	1 June 16	. . .	
13 Mar. 15	*Wells, G. E. H.* (47*a*, 48*a*, 51) . .	1 June 16	. . .	
13 Mar. 15	Livermore, W. C. (46) . . .	1 June 16		
17 Apl. 15	*Bell, A. M.*	1 June 16	. . .	*Cable Censor, Madras.*
24 Apl. 15	*Sanders, H. R.* (*actg. Capt., 24 Sep. 18*) (46).	6 Aug. 16	. . .	*M. W. S.*
13 Mar. 15	*Hambly, R. L.*	1 July 17	. . .	*I. A. R. O.*
23 July 15	*Hillman G. E.* (46)	1 July 17	. . .	*With R. A. F.*
10 Aug. 15	*Hurburgh, L. H.*	1 July 17	. . .	*M. A. Dept.*
	2nd-Lieutenants.			
23 June 15	*Jones, T. J. W.*	23 June 15	. . .	*I. A. R. O.*
19 Sep. 15	*Wigfall, C. S.*	19 Sep. 15	. . .	*I. A. R. O.* Bde. S. & T. Offr., Dehra Dun.
12 Jan. 16	de la Coze, F. M.	12 Jan. 16	. . .	
28 Apl. 17	Francis, C. A. (59)	28 Apl. 17	. . .	Act'g. Adjt.
28 Apl. 17	Mosely, L. A.	28 Apl. 17	. . .	
16 Mar. 18	Lane, F. C. T. (59)	16 Mar. 18	. . .	Mily. Accts. Dept.
16 Mar. 18	Davis, V. C.	16 Mar. 18	. . .	With R. A. F.
16 Mar. 18	Macrostie, E. J.	16 Mar. 18	. . .	M. W. S., Bannu.
16 Mar. 18	*Beynon, G.*	16 Mar. 18	. . .	*I. A. R. O.*
16 Mar. 18	Snow, D. G. (46)	16 Mar. 18	. . .	
	Attached.			
30 June 04	Mitchell, Capt. H. C. B., 1st Bn., Hertford. R.	7 Jan. 12		
1 June 09	Hatch, Capt. W. E. J., (51) 17 Bn., Lond. R.	8 Dec. 14		
1 June 09	Austin, Capt. E. R., (51) 22 Bn., Lond. R.	17 July 16		
2 Dec. 15	Kennedy, Lt. J., (48*a*)	9 Aug. 17		
5 Aug. 16	Lamberton, Lt. N. C., 5 Bn., Sco. Rif. (46).	5 Feb. 18		
19 Dec. 16	Butt, 2nd-Lt. H., Midd'x R. (12) .	19 Dec. 16		
25 Jan. 17	Hayward, 2nd-Lt. L. G.(46) Suff. R.	25 Jan. 17		
30 May 17	Sheddon, 2nd-Lt. L. H. R. War. R. (12)	30 May 17		
30 May 17	Parrish, 2nd-Lt. A., R. War. R. .	30 May 17		
30 May 17	Lysaght, 2nd-Lt. D. R., Devon R. .	30 May 17		
30 May 17	Davis, 2nd-Lt. J. F., Worc. R. . .	30 May 17		
25 Sep. 17	Smith, 2nd-Lt. W., Notts. & Derby. R.	25 Sep. 17		
25 Sep. 17	Braithwaite, 2nd-Lt. C. V. C., Notts & Derby. R.	25 Sep. 17		
26 Sep. 17	Bailey, 2nd-Lt. W. G., Manch. R. .	26 Sep. 17		
26 Sep. 17	Bond, 2nd-Lt. C. E., Manch. R. .	26 Sep. 17		
26 Sep. 17	Cleere, 2nd-Lt. A. T. C., Manch. R.	26 Sep. 17		
26 Sep. 17	Frith, 2nd-Lt. N. R. C., Notts. & Derby. R.	26 Sep. 17	. . .	M. A. Depot.
. . . .	Woodcock, 2nd-Lt. F. B., Lan. Fus.	. . . .		
28 Apl. 17	Francis, 2nd-Lt. C. A.	28 Apl. 17	. . .	Adjt.
28 Jan. 16	Burt, Lt. C. A	28 Jan. 16	. . .	Qr.-Mr.

Agents:—MESSRS. COX & CO.

Uniform—Scarlet.
Facings—Black.

1-1st Kent Battalion.

Arrived in India, 2 Mar. 16.

HONORARY COLONEL—Bt.-Col. *Sir* H. Streatfeild, K.C.V.O.,C.B., ret. pay, *Eq.* . 25 Mar. 09.

First Commn.	Names.	Regimental rank.	Army rank.	Remarks.
	Lieutenant-Colonel.			
6 Mar. 09	Moore, W. G.	23 Nov. 16		
	Majors.			
Feb. 10	Oldendorff, A. J. A.	1 June 16		
28 Feb. 12	Pilditch, I.	28 July 17		
	Captains.			
5 June 12	Walmsley, E. S.	5 Aug. 14	. . . .	S. S. O., Ambala.
3 Jan. 14	Dunn, G. L.	1 June 16		
4 Feb. 14	*Sankey, G. B.* (*temp. Maj.*)	1 June 16	29 Aug. 14	D. A. A. G. A. H. Qrs.
4 Feb. 14	Boulter, F. H.	1 June 16	. . . .	Attd. 2 Bn. R. W. Kent R.
14 Mar. 14	*Phillips, C. S.* (206)	1 June 16	. . . .	*Staff Capt., Dehra Dun Bde.*
Jan. 09	Kendrick, T. L.	1 June 16		
13 Nov. 14	Smith, R. T. W. (12, 52)	1 June 16		
18 Nov. 14	Carlill, S. R. (51)	1 June 16		
	Lieutenants.			
24 Nov. 14	Harding, J. A.	1 June 16		
3 Dec. 14	Feild, R. (46)	1 June 16		
24 Dec. 14	LeFleming, J. N.	1 June 16		
7 Mar. 15	Bell, E. W. (57)	1 June 16	. . . .	With Sig. Serv.
17 Mar. 15	Raven-Hill, L. E.	1 June 16	. . . .	With 61st K. G. O. Prs.
3 Apl. 15	Fuller, T. A. (*act'g Capt.,*) (61)	1 July 17	. . . .	Ac'g. Adjt.
1 May 15	Fisher, S. M. (57)	1 July 17		
1 May 15	Hancock, F. B. (51)	1 July 17	. . . .	S. S. O., Ferozepore.
9 Oct. 15	Thomas, B. L. (12)	1 July 17		
11 Nov. 15	Pembrook, J. G. (51)	1 July 17		
28 Jan. 16	Baillie, E. A.	28 July 17		
	2nd-Lieutenants.			
28 Apl. 17	Bristow, G. M. (59)	28 Apl. 17		
29 Aug. 17	Charman, A. L. C. (46)	29 Aug. 17		
16 Mar. 18	Smith, A. F. (46)	16 Mar. 18	. . . .	M. G. Corps. (*On probn.*)
16 Mar. 18	Sheffield, D. E. (12)	16 Mar. 18		
16 Mar. 18	Wallington, G.	16 Mar. 18	. . . .	With R. A. F.
16 Mar. 18	Frisby, F.	16 Mar. 18	. . . .	M. G. Corps. (*On prcbn.*)
16 Mar. 18	Baseley, H. C. (12)	16 Mar. 18		
19 Aug. 18	Carlaw, N.	19 Aug. 8		
	Attached.			
15 Dec. 14	Canning, Capt. J. D., H. A. C.	1 Nov. 16	. . . .	Attd. Hdqrs.,Armd. Motor Bde.
28 Mar. 17	Cornwell, 2nd-Lt. W. S.	28 Mar. 17		
3 May 17	Lambert, 2nd-Lt. A.	3 May 17	. . . .	Attd. Hdqrs.,Armd. Motor Bde.
29 May 17	Swaine, 2nd-Lt. A. L. B. (4')	29 May 17		
29 May 17	Harvey, 2nd-Lt. E. L.	29 May 17		
30 May 17	Meadmore, 2nd-Lt. C. R. A.	30 May 17	. . . .	With 2- st G. R.
25 Aug. 17	Williams, 2nd-Lt. A. J.	25 Aug. 17		
25 Sep. 17	Wheeler, 2nd-Lt. T. W.	25 Sep. 17		
20 Oct. 17	Kirkwood, 2nd-Lt. P. W.	20 Oct. 17		

19 Mar. 14	Bain, Capt. K. D.	. . . .	. . . . .	Qr.-Mr.

Agents:—Sir C. R. McGRIGOR, *Bt.* & Co.

No. 1 Reserve Battalion (India).

First Commn.	Names.	Regimental rank.	Army rank.	Remarks.
	Majors.			
16 Mar. 01	Beaumont, *Sir* G. A. H., *Bart.*	3 Jan. 16	. . . .	K. R. Rif. C. (2nd-in-command).
4 Dec. 01	Case-Morris, C. W. (51, 66*a*) (*Act'g. Lt.-Col., 24 Dec. 17.*)	14 Sep. 15	. . . .	R. W. Kent. R.—Comdg.
	Captains.			
18 Sep. 08	Sharp, P. C. (51, 52) (*temp. Major*)	10 Aug. 15	. . . .	West India Regt.
5 Dec. 14	Galloway, J. (51)	5 Mar. 15	. . . .	Manch. R.
28 Feb. 13	Hewat, J. P. D. (46, 52, 67)	1 June 16	. . . .	Glasgow Highrs.—Attd. M. G. corps.
15 Aug. 14	Urquhart, F. M.	1 Jan. 17	. . . .	High. L. I.
5 Apl. 15	Dickie, J. C. (46)	16 Mar. 17	. . . .	Do.
	Lieutenants.			
2 Dec. 14	Craddock, S. (46, 71)	13 Mar. 16	. . . .	R. Dub. Fus.
20 May 15	Nuttall, B. A.	28 Sep. 16	. . . .	Yorks. R.—Employed D. D. O., Bolarum.
15 Sep. 15	Scott, K. G. (51, 52)(*Act'g. Capt.*)	15 Feb. 17	. . . .	R. Highrs.
29 Apl. 15	Reid, G. (46, 51) (*Act'g. Capt.*)	1 July 17	. . . .	High. L. I.
11 June 15	Cunningham, T. S. (46)	1 July 17	. . . .	Do.
17 June 15	Goode, T. C. (50, 53, 71) (*Act'g. Capt.*).	1 July 17	. . . .	Adjt., Som. L. I.
29 June 15	Crawford, R. McG. C.	1 July 17	. . . .	High. L. I.
22 Aug. 15	Godby, J. G.	1 July 17	. . . .	Ox. & Bucks. L. I.—Employed at Poonamallee.
23 Oct. 15	Gray, G. C. (temp.) (51)	1 July 17	. . . .	High. L. I.—*Leave ex India 1 month.*
25 Nov. 15	Conochie, N. C. (45)	1 July 17	. . . .	High. L. I.
3 Dec. 15	Serachan, A. B.	1 July 17	. . . .	Do.
4 Jan. 16	James, J. L. (52)	4 July 17	. . . .	High. L. I.—Employed at Wellington.
	Pitcairn, J. E. (51)	1 Apl. 17	. . . .	High. L. I.
16 Aug. 16	Isaac, A. H. (51)	16 Feb. 18	. . . .	Ox. & Bucks. L. I.
19 Dec. 16	Tudhope, M. A. (59)	19 June 18	. . . .	High. L. I.
19 Dec. 16	Hamilton-Kerr, D. (51, 52)	19 June 18	. . . .	Do.
25 Jan. 16	Blackley, J. (51, 52)	25 July 18	. . . .	Do.
25 Jan. 16	Nisbet, J. D.	25 July 18	. . . .	Do.
	2nd-Lieutenants.			
25 Feb. 15	Cave-Allan, W. N. (46, 52)	25 Feb. 15	. . . .	K. O. Sco. Bord.
2 Oct. 16	Cuming, B. S. (59)	2 Oct. 16	. . . .	R. Muns. Fus.
25 Jan. 17	Pollock, C. A. (60*a*)	25 Jan. 17	. . . .	High. L. I.
1 Mar. 17	Bain, R. B. (46)	1 Mar 17	. . . .	Do.
23 Mar. 17	Peden, W. (51, 66) (*Act'g. Capt.*)	23 Mar. 17	. . . .	A. P. M. Bangalore, High. L. I.

No. 1 Reserve Battalion (India)—*cont*

2nd-Lieutenants—*contd.*

1st Commn.	Names.	Regimental rank.	Army rank.	Remarks.
30 May 17	Byrne, J.	30 May 17	....	R. Dub. Fus.
29 July 17	Quig, A.	29 July 17	....	High. L. I.
1 Aug. 17	Robb, J. M. (46)	1 Aug. 17	....	Do.
28 Aug. 17	Hannah, W. F.	28 Aug. 17	....	Do.
28 Aug. 17	Corner, W.	28 Aug. 17	....	Do.
26 Sep. 17	Campbell, C.	26 Sep. 17	....	High. L. I.—Employed Arsenal, Allahabad.
28 Sep. 17	Stevenson, J. (46)	28 Sep. 17	....	High. L. I.—*Leave ex India 1 month.*
28 Nov. 17	Brasher, J. H. (60*a*, 5	28 Nov. 17	....	Ox. & Bucks. L. I.
16 Mar. 18	McMillan, P. K. T.	16 Mar. 18	....	High. L. I.
16 Mar. 18	Oddie, E.	16 Mar. 18	....	Do.
19 Aug. 18	Prentice, J. (*temp.*) (52)	19 Aug. 18	....	Do.—*Attd. M. G. Corps.*
19 Aug. 18	McKay, G. S. M. (*temp.*)	19 Aug. 18	....	Do.
19 Aug. 18	Dowie, A. M. (*temp.*) (52)	19 Aug. 18	....	Do.—*Attd. M. G. Corps.*
19 Aug. 18	Macrae, M. (*temp.*)	19 Aug. 18	....	Do.
19 Aug. 18	Ross, W. B. (*temp.*)	19 Aug. 18	....	Do.
19 Aug. 18	Sugden, K. G. J.	19 Aug. 18	....	Spec. Res.—High. L. I.—*Attd. M. G. Corps.*
20 Sep. 18	Cassidy, J. M.	20 Sep. 18	....	Ox. & Bucks. L. I.

17 June 18	Goode, (Acg. Captain) T. C. (50—53, 71)	1 July 17	.....	2nd Bn., Som. L. I.—Adjt.
...	Schofield, Lieut. F. W.	.....	....	Qr.-Mr., Genl. List.

No. 2 Reserve Battalion (India).

1st Commn.	Names.	Regimental rank.	Army rank.	Remarks.
	Lieut.-Colonels.			
Jan. 87	Mace, J. M., T.D. (Q) . . .	1 June 08	. . . .	T. F. Res.—Comdg.
23 Dec. 93	Kitson, C. E., D.S.O. (Q) (12, 50, 57, 71)	1 June 16	. . . .	2nd R. W. Kent R.—Staff, 6th Divn., Poona.
	Majors.			
19 July 06	Hills, A. E. (51)	1 June 16	. . . .	1-5th R. W. Kent R.—D. A. Q. M. G., Army. Hdqrs.
26 July 02	Neame, L. V. (51)	1 June 16	. . . .	1-5th R. W. Kent R.—2nd in Comman d.
1 Nov. 06	Harston, J.	1 June 16	. . . .	M. W. S. 1-5th R. W. Kent R.
	Captains.			
1 Aug. 08	Satterthwaite, R. E. (51) . . .	7 Nov. 12	. . . .	Bde. Major. Quetta 1-5th R. W. Kent R.
26 Nov. 00	Brewis, R. D. (*temp. Maj.*) (51) .	19 Aug. 14	17 Feb. 12	1-5th R. W. Kent R.—D. A. Q. M. G.—7th Divn.
10 Oct. 14	Maynard, P. A.	10 Oct. 14	. . . .	Bde. Major, Nasirabad Bde., 1-5th E. Kent R.
1 Jan. 89	Bonham Carter, G.	30 Oct. 14	. . . .	3rd R. W. Kent R.
1 May 00	Taylor, G. C. R. (*temp. Maj., 24 Sep. 18*)	3 Nov. 14	. . . .	8th Essex R. (attd. 2nd Norf. R.)—D. A. A. G., 16th Division.
6 Oct. 06	Downs, R. A. (*temp. Maj.*) (51) (52)	16 Feb. 15	. . . .	2nd Norf. R.—D. A. Q. M. G., Army Hdqrs.
6 Nov. 09	Frere, R. T. (51) (59)	6 Nov. 15	. . . .	S. S. O., Belgaum. 2nd Norf. R.
20 Dec. 11	Barnes, O. G. R.	26 Nov. 15	. . . .	2nd R. W. Kent R.
30 Dec. 08	Mercer, J. L. C. (52)	1 June 16	. . . .	Adjt. 1-5th R. W. Surr. R.
11 May 12	Wilkinson, D. S. (57) (207*a*) .	1 June 16	. . . .	1-5th E. Kent R.—Bde. Signalling Officer.
13 Nov. 13	Howell, C. J. (51)	1 June 16	. . . .	1-4th E. Kent R.—Actg. S. S. O., Deolali.
21 Mar. 14	Templer, H. E. (57)	1 June 16	. . . .	1-10th Middx. R.
	Lieutenants.			
. . . .	Berners, G. H.	1 June 15	. . . .	Armd. Motor Unit, Peshawar. 3rd Norf. R.
2 Sep. 14	Bakewell, W. B. (51) (52) . .	1 June 16	. . . .	M. G. Corps. 1-5th R. W. Kent R.
2 Sep. 14	Smithers, K. O. (51) (52) . .	1 June 16	. . . .	S. S. O., Jubbulpore. 1-5th R.W. Kent R.
8 Aug. 15	Park, T. W. U. (*act'g. Capt.*) .	1 June 16	. . . .	5th Arg. & Suth'd Highrs.
30 Sep. 16	Walsha, J. W. (57)	12 Nov. 16	. . . .	9th Norf. R.
5 Oct. 14	Fisher, B. O. F. de C. (12) . .	1 July 17	. . . .	7th Norf. R.—at Wellington.
30 Oct. 14	Stocker, C. L. (48*a*)	1 July 17	. . . .	2nd Norf. R.
19 Apl. 15	Wray, D. W.	1 July 17	. . . .	1st G. B. Royal Irish Fus.
20 Apl. 15	Shand, W. K. W. (52) (12) (46) (48*a*) (51)	1 July 17	. . . .	3rd Norf. R.
6 May 15	Linfield, W. A.	1 July 17	. . . .	1st G. B. Bedford R.
16 May 15	Denis, W.	1 July 17	. . . .	With Mech. T. Coy., Peshawar. 3rd Norf. R.
24 Sep. 15	Plant, J. C. G.	1 July 17	. . . .	3rd Norf. R.
1 Oct. 15	Stanforth, L. H. (51)	1 July 17	. . . .	2nd G. B. Essex R.—Arm. Motor Unit, Peshawar.
28 Oct. 15	Cameron, H.	1 July 17	. . . .	8th Arg. and Suth'd Highrs.
27 Nov. 15	Porter, H. (51) (*act'g. Capt.*) .	1 July 17	. . . .	A. P. M.. Wellington. 2nd Norf. R.
22 Dec. 15	Knight, C. R. B. (51) (60*a*) . .	1 July 17	. . . .	2nd E. Kent R.
24 Dec. 15	Clarke, L. P.	1 July 17	. . . .	9th Norf. R.
20 Oct. 15	Western, F. H.	1 June 16	. . . .	1st Cambridge R.
29 Nov. 15	Clarke, J. W.	1 July 17	. . . .	9th Norf. R.—Inspector of Munitions, Cawnpore.

No. 2 Reserve Battalion (India)—*contd.*

1st Commn.	Names.	Regimental rank.	Army rank.	Remarks.
	Lieutenants—*contd.*			
1 Jan. 16	Bennett, G.	1 July 17	. . . .	M. G. Corps. 1-5th R. W. Kent R.
10 Jan. 16	Goode, W. (46)	10 July 17	. . . .	4th E. Kent R.
7 July 16	Ellis, B. G.	7 Jan. 18	. . . .	1st East Kents.
8 July 16	Ramsey, L. A.	8 Jan. 18	. . . .	1-5th E. Kent R.
27 Oct. 16	Lovell, I. R. (60*a*) (61)	27 Apl. 18	. . . .	1st R. W. Kent R.
22 Nov. 16	Harrison, V. M., M. C. (46)	12 May 18	. . . .	3rd Norf. R.
19 Dec. 16	Matheson, M.	19 June 18	. . . .	*Lv. in I., m. c.* 8th R. Scots.
4 Jan. 17	Nicklin, S. G.	4 July 17	. . . .	2nd Norf. R.
1 Mar. 17	Buckland, R. J. (60*a*)	1 Sep. 18	. . . .	2nd Norf. R.
9 Dec. 16	Conners, J. P. (*temp. Capt.*)	9 Dec. 16	. . . .	Adjt. N. C. Os. School, Belgaum. 2nd Norf. R.
	2nd-Lieutenants.			
26 Apr. 17	Bunn, A. E. (51) (48*a*)	26 Apl. 17	. . . .	4th Norf. R.
30 May 17	Turner, H. L., M. M. (57)	30 May 17	. . . .	2nd Norf. R.—9th Divl Signal School.
30 May 17	Browne, J. C.	30 May 17	. . . .	2nd Norf. R.
30 May 17	Highley, H. F.	30 May 17	. . . .	2nd R. W. Kent R.
30 May 17	Hird, S.	30 May 17	. . . .	Army Clothing Dept. 2nd R. W. Kent R.
30 May 17	Vercoe, F. G. (71)	30 May 17	. . . .	3rd R. W. Kent R.—Asst. Adjt.
1 Aug. 17	Gavin, W. L.	4 Aug. 17	. . . .	4th Gord. Highrs.—attached G. S. Bch., Army Hqdrs.
28 Aug. 17	Baxter, W. C. (57)	28 Aug. 17	. . . .	3rd Norf. R.
28 Aug. 17	Tipping, F. B.	28 Aug. 17	. . . .	3rd Norf. R.
28 Aug. 17	Maiden, C. J. D.	28 Aug. 17	. . . .	3rd Norf. R.
29 Aug. 17	Lascelles, H.	29 Aug. 17	. . . .	P. T. Officer, 8th Division. 2nd R. W. Kent R.
29 Aug. 17	Eastwick, A. J.	29 Aug. 17	. . . .	Supernumerary R. T.O., Bombay.
28 Mar. 17	Gains, A. T.	28 Mar. 17	. . . .	2nd R. W. Kent R.
30 May 17	Payne, C. C.	30 May 17	. . . .	4th Norf. R.
31 Oct. 17	Bannochi, W. S.	31 Oct. 17	. . . .	5th R. Scots. Fus.
31 Oct. 17	Simpson, J.	31 Oct. 17	. . . .	2nd R. W. Kent R.
31 Oct. 17	Wells, A.	31 Oct. 17	. . . .	2nd R. W. Kent R.
31 Oct. 17	Barker, H. M.	31 Oct. 17	. . . .	2nd R. W. Kent R.
26 Sep. 17	Combe, H. W.	26 Sep. 17	. . . .	1-5th R. W. Kent R., Mily.-Accts. Dept., Poona.
16 Mar. 18	Bull, C. A.	16 Mar. 18	. . . .	1-5th E. Kent R.
16 Mar. 18	Body, G. A.	16 Mar. 18	. . . .	R. A. F. 1-5th E. Kent R.
16 Mar. 18	Blaxall, M. L.	16 Mar. 18	. . . .	1-5th E. Kent R.
16 Mar. 18	Whitwell, H. E.	16 Mar. 18	. . . .	2nd Norf. R.
16 Mar. 18	Swales, H. G.	16 Mar. 18	. . . .	With Sig. Serv., 2nd R. W. Kent R.
19 Aug. 18	Spare, H.	19 Aug. 18	. . . .	1-5th R. W. Kent R.
19 Aug. 18	Howlett, C. G.	19 Aug. 18	. . . .	1-5th R. W. Kent R.
19 Aug. 18	Kelway, R. T.	19 Aug. 18	. . . .	1-5th E. Kent R.
19 Aug. 18	Worthington, C.	19 Aug. 18	. . . .	1-5th E. Kent R.
19 Aug. 18	Williamson, G.	19 Aug. 18	. . . .	2nd Norfolks.
19 Aug. 18	Burnett, R. J.	19 Aug. 18	. . . .	2nd Norfolks.
19 Aug. 18	Hewitt, B. E.	19 Aug. 18	. . . .	2nd R. W. Kent R.

1 July 08	Cooke, J. H. (*Hony. Lt. & Qr. Mr.*)	1 July 08	. . . .	1-5th R. W. Kent R.

No. 3 Reserve Battalion (India).

Formed in India.

First Commn.	Names.	Regimental rank.	Army rank.	Remarks.
	Lt.-Colonel.			
1 Jan. 98	Dimond, C. F. W. (51)	29 Oct. 14		Comdg. 1-10th Midd'x. R.
	Majors.			
4 Apl. 00	Montgomerie, H. S.	17 July 12		1-9th Midd'x.
	Smallpeice, F. W.	1 Sep. 12		1-5th Queen's R. W. Surr. R. A. J. A. G., N. Comd.
5 May 00	Willett, H. L. (51, 52, 72*b*) (*temp.* Lt.-Col.)	1 Sep. 15	4 May 18	2nd Norf. R.
	Brough, W. P. (*temp.*)	18 Sep. 15		7th Glouc. R.
28 May 04	Whiffen, G. G.	1 June 16		1-5th E. Surr. R. A. G.'s Bch., A. H. Q., Delhi.
	Captains.			
26 Feb. 02	Hewlett, L. M.	13 Nov. 05		1-9th Midd'x. R.
Apl. 01	Knight-Eaton, H. D. H. U. M. (*temp.*) Hon. Lieut. in Army.	2 Jan. 15		1st G. B., Bedf. R.
1 June 12	Mellersh-Jackson, W. (51)	16 June 15		1-9th Midd'x. R.
28 Mar. 09	Whittington, L. (51, 71)	1 June 16		1-5th R. W. Surr. R. Staff Capt., Dhond, 58th Bde.
Dec. 10	Robertson, D. N., M.C.	1 June 16		6th Arg. & Suth'd. Highrs.
28 Oct. 12	Last F. B. (51)	1 June 16		1-7th Hamps. R.
12 Mar. 13	Poston, G.	1 June 16		1-5th E. Surr. R.
13 Aug. 14	Mercer, J. L.	1 June 16		1-5th R. W. Surr. R. Adjt. No. 2 R. B.
17 Feb. 12	Mountford, E. W. (51, 52)	6 Dec. 16		1-5th R. W. Surr. R.
12 Dec. 14	Finlinson, H. (12)	22 Feb. 17		3rd Essex. R.
7 Nov. 14	Edwards, H. S.	25 Nov. 17		A. S. C.
	Lieutenants.			
27 Mar. 15	Ward, H. A. (51)	27 Mar. 15		13th Glouc. R.
9 Apl. 15	Cartwright, H.	1 Aug. 15		2nd D. of Corn. L. I.
14 Feb. 15	Hawkins, R. I. (52)	2 Jan. 16		4th Glosters.
26 Aug. 14	Higgins, G. L. (51)	1 June 16		1-9th Midd'x. R.
15 June 15	Geddes, T. (51)	1 June 16		18th Rif. Bde.
9 Apl. 15	Kenward, N. H. (57, 61)	22 Nov. 16		3rd R. Ir. Fus.
26 Aug. 14	De Vesian, N. E. (66, 172*a*)	1 June 17		1-9th Midd'x. R.
25 Jan. 15	Sullivan, M. B. (51, 52, 95)	1 July 17		1st G. B. Norf. R.
30 Jan. 15	Hardie, J. F.	1 July 17		7th Glouc. R.
22 May 15	Kingston, C. G. (51)	1 July 17		3rd E. Surr. R.
13 June 15	Williamson, H. E. (46, 61)	1 July 17		Bedf. R.
15 June 15	Higman, J. W. (71)	1 July 17		3rd D. of Corn. L. I.
17 June 15	Leggat, W. K., M.C.	1 July 17		6th Gord. Highrs.
20 June 15	Cogswell, A.	1 July 17		1-9th Midd'x. R.
19 Sep 15	Nice, G. L.	1 July 17		1-5th R. W. Surr. R.
23 Sep 15	Pitt, W. A.	1 July 17		7th Glouc. R.
10 Oct 15	Morris, D. N. (46)	1 July 17		8th Arg. & Suth'd. Highrs.
30 Oct 15	Smart, G. R.	1 July 17		1-5th R. W. Surr. R.
24 Nov. 15	Bolton, E. F. (57)	1 July 17		1st R. W. Surr. R.
25 Nov. 15	Watt, W. B. F. (12, 52, 71)	1 July 17		7th Gord. Highrs.
3 Dec. 15	Ferrey, E. M. (61)	1 July 17		21st Midd'x. R.
3 Dec. 15	Taylor, F. H. (12)	1 July 17		7th Glouc. R.
7 Dec. 15	Taylor, A. J. (46, 51, 61)	1 July 17		1-5th R. W. Surr. R.
17 Jan. 16	Hyslop, A. J. (46, 51)	1 July 17		1-5th R. W. Surr. R.
24 Mar. 16	McArthur, J. (46, 61)	26 Nov. 17		3rd R. Scots.
27 June 16	Reeves, W. A.	27 Dec. 17		7th Glouc. R.
15 July 16	Middleton, J. (46, 71)	15 Jan. 18		5th Arg. & Suth'd. Highrs.
31 July 16	Currie, A. H.	31 Jan. 18		1st K. O. Sco. Bord.
5 Aug. 16	Joy, G. A.	5 Feb. 18		7th Glouc. R.
	Simmon, W. W. J.			

First Commn.	Names.	Regimental rank.	Army rank.	Remarks.
	2nd-Lieutenants.			
24 Nov. 14	Amies, E. L.	24 Nov. 14	. . . .	1-9th Midd'x. R.
19 May 15	Stevens, W. A.	19 May 15	. . . .	R. Ir. Rif.
10 Nov. 15	Clarke, F. E. (61, 71)	10 Nov. 15	. . . .	3rd Bedf. R.
5 Aug. 16	Crisp, J. H. (52, 61)	5 Aug. 16	. . . .	Suff. R.
8 Aug. 16	Boswell, A. P.	8 Aug. 16	. . . .	1-5th R. W. Surr. R.
15 Dec. 16	Moat, C. A.	15 Dec. 16	. . . .	1st G. B., Som. L. I.
11 Feb. 17	Bramwell, F. H.	11 Feb. 17	. . . .	3rd G. B., Bedf. R.
24 Mar. 17	Laughlin, G. E.	24 Mar. 17	. . . .	5th R. Berks. R.
28 Mar. 17	Cownie, J. C. (57)	28 Mar. 17	. . . .	4th Gord. Highrs.
28 Mar. 17	Inglis, F. A. G.	28 Mar. 17	. . . .	4th Gord. Highrs.
28 Mar. 17	Osmond, F. E. (46)	28 Mar. 17	. . . .	3rd R. W. Surr. R.
28 Mar. 17	Winckworth, A. R.	28 Mar. 17	. . . .	3rd Glouc. R.
28 Mar. 17	Gray, T. (46)	28 Mar. 17	. . . .	4th Gord. Highrs.
2 Apl. 17	Sabbatella, F.	2 Apl. 17	. . . .	7th Glouc. R.
25 Apl. 17	Grainger, R. L.	25 Apl. 17	. . . .	1-9th Midd'x. R.
26 Apl. 17	Magill, R. H. (46)	26 Apl. 17	. . . .	R. Dub. Fus.
26 Apl. 17	McKee, R.	26 Apl. 17	. . . .	K. O. Sco. Bord.
26 Apl. 17	Johnston, W. (46, 51)	26 Apl. 17	. . . .	3rd R. Scots.
28 Apl. 17	Pannell, W. G. A.	28 Apl. 17	. . . .	1-9th Midd'x. R.
May 17	Macpherson, A. (46, 57)	May 17	. . . .	K. O. Sco. Bord.
30 May 17	Mackenzie, D.	30 May 17	. . . .	3rd Arg. & Suth'd. Highrs.
30 May 17	Bishop, C. H., M. M. (57)	30 May 17	. . . .	3rd R. W. Surr. R.
27 June 17	Watson, J. (71)	27 June 17	. . . .	R. W. Fus.
31 July 17	Hare, P. V.	31 July 17	. . . .	8th Notts. & Derby. R.
26 Aug. 17	Jacobs, E. J. (51, 61, 71, 72*b*)	26 Aug. 17	. . . .	York L. I.
29 Aug. 17	Clarke, V. T. G.	29 Aug. 17	. . . .	3rd R. W. Surr. R.
26 Sep. 17	Wraight, G. J. R. (71)	26 Sep. 17	. . . .	3rd R. W. Surr. R.
26 Sep. 17	Cooke, R. H.	26 Sep. 17	. . . .	3rd R. W. Surr. R.
26 Sep. 17	Fourdrinier, C. D.	26 Sep. 17	. . . .	5th Midd'x. R.
26 Sep. 17	Chapman, R. S.	26 Sep. 17	. . . .	5th Midd'x. R.
26 Sep. 17	Wynn, W. A. (51)	26 Sep. 17	. . . .	8th Midd'x. R.
26 Sep. 17	Damerell, R. C.	26 Sep. 17	. . . .	5th Midd'x. R.
26 Sep. 17	Clay, N. C.	26 Sep. 17	. . . .	5th Midd'x. R. I. A. on probn.
26 Sep. 17	Hayward, E.	26 Sep. 17	. . . .	5th Midd'x. R.
26 Sep. 17	Price, S. H.	26 Sep. 17	. . . .	9th Midd'x. R.
26 Sep. 17	Kennedy, D. D.	26 Sep. 17	. . . .	9th Midd'x. R.
26 Sep. 17	Brittan, N. F.	26 Sep. 17	. . . .	7th Res. Midd'x. R.
18 Oct. 17	Brown, W. G.	18 Oct. 17	. . . .	9th Midd'x. R.
31 Oct. 17	Dear, G. W. V.	31 Oct. 17	. . . .	5th E. Surr. R.
31 Oct. 17	Rowbotham, A. S.	31 Oct. 17	. . . .	3rd Glouc. R.
17 Nov. 17	Roberts, R. J. (51, 52)	17 Nov. 17	. . . .	Glouc. R.
27 Nov. 17	Stevens, E. P.	27 Nov. 17	. . . .	5th E. Surr. R.
27 Nov. 17	Edeson, D. J. S.	27 Nov. 17	. . . .	3rd Hamps. R.
28 Nov. 17	Keene, H. (71)	28 Nov. 17	. . . .	1-7th Hamps. R.
28 Nov. 17	Lofthouse, J. D.	28 Nov. 17	. . . .	5th E. Surr. R.
28 Nov. 17	Tregenza, E. L.	28 Nov. 17	. . . .	5th E. Surr. R.
28 Nov. 17	Reed, W. A.	28 Nov. 17	. . . .	3rd Glouc. R.
28 Nov. 17	Bensley, L. F.	28 Nov. 17	. . . .	9th Midd'x. R.
1 Feb. 18	Blane, A. P. (46, 51, 52, 66)	1 Feb. 18	. . . .	Camm'n. Highrs.
16 Mar. 18	Dillow, D. H. (46)	16 Mar. 18	. . . .	Ry. Bn., Sappers & Miners Sialkot. 1-5th R. W. Surr. R.
16 Mar. 18	McLean, A. D.	16 Mar. 18	. . . .	1-5th R. W. Surr. R.
16 Mar. 18	Handley, H. (61)	16 Mar. 18	. . . .	1-5th E. Surr. R.
16 Mar. 18	Mackenzie, J. D. H. (71)	16 Mar. 18	. . . .	7th Glouc. R.
16 Mar. 18	Kennard, D. C.	16 Mar. 18	. . . .	7th Glouc. R. Attd. M. G. Corps, Mhow.
19 Aug. 18	Paddison, S.	19 Aug. 18	. . . .	1-5th R. W. Surr. R.
19 Aug. 18	Piper, N. L.	19 Aug. 18	. . . .	1-5th E. Surr. R.
19 Aug. 18	Williams, J. P.	19 Aug. 18	. . . .	1-5th E. Surr. R.
19 Aug. 18	Treleaven, W. J.	19 Aug. 18	. . . .	1-5th E. Surr. R.
19 Aug. 18	Hoskin, F.	19 Aug. 18	. . . .	1-5th E. Surr. R.
19 Aug. 18	Guest, R. E.	19 Aug. 18	. . . .	1-9th Midd'x. R.
19 Aug. 18	Gosney, C. G.	19 Aug. 18	. . . .	7th Glouc. R.

---

21 Aug. 14	Walker, Capt. R. (*temp.*)	14 Feb. 16	. . . .	Adjt., 1-5th E. Kent. R.
17 Nov. 15	Glazebrook, Lt. E. J. (51)	17 Nov. 15	. . . .	Qr.-Mr., 8th R. W. Fus.

No. 4 Reserve Battalion (India).

Formed in India, 1st December 1917.

1st Commn.	Names.	Regimental rank.	Army rank.	Remarks.
	Lieutenant-Colonel Commanding.			
	Majors.			
18 July 92	Benson, W. J. P. (50) (Hon. Capt. in Army, 1 Dec. 00) (*act'g. Lt.-Col., 24 Dect. 17*)	10 Nov. 14	. . . .	
	Captains.			
29 Dec. 14	Vasey, G. H. (*act'g. Maj.*)	23 Apl. 15	. . . .	Spl. Res. *Lv., mo., 22 dys., 10 Nov. 18.*
20 Jan. 15	Fagan, D. P. (72) (*temp.*)	20 Jan. 15	. . . .	Adjt.
11 Nov. 14	Lyons, J. A. de C. (12, 129)	21 Sep. 15	. . . .	Spl. Res.
Nov. 14	Richards, A. P. (*temp.*)	1 June 16	. . . .	
18 Dec. 14	Priestman, H. E. (51) (*temp.*)	29 June 16	. . . .	Dy. Asst. Mily. Secy. Army. Hdqrs.
Sep. 14	Varvill, J. K., M.C. (*temp.*)	2 Sep. 16	. . . .	
8 Sep. 15	Fradd, L. C. W. (*temp.*)	1 Mar 17	. . . .	
	Lieutenants.			
Jan. 15	Watson, J. H., M.C. (52) (*temp.*)	21 Sep. 15	. . . .	
17 Sep. 14	Shaw, L. H. (*temp.*) (51)	1 Jan. 16	. . . .	
10 Apl. 15	Chapman, V. V. D. (46) (*temp.*)	2 July 16	. . . .	
8 Feb. 15	Petter P. E. (52) (*temp.*)	1 July 17	. . . .	
8 Feb. 15	Ruglys, H. (51) (*temp.*)	1 July 17	. . . .	
10 Mar. 15	Ross, C. G. (51, 52)	1 July 17	. . . .	
28 July 15	Swindells, H. R. (51, 52)	1 July 17	. . . .	Spl. Res.
14 Aug. 15	White, C. M. (46) (*temp.*)	1 July 17	. . . .	
4 Oct. 15	Parry, M. J. (51)	1 July 17	. . . .	Spl. Res.
19 Oct. 15	Byrnell, J. G. (46) (*temp.*)	1 July 17	. . . .	
30 Jan. 16	Wood, J. (46) (*temp.*)	30 July 17	. . . .	
7 July 16	Bigham, R. A. (46)	7 Jan. 18	. . . .	Spl. Res.
22 Nov. 16	Turner, H. G. (46, 52)	22 May 18	. . . .	Spl. Res.
	2nd-Lieutenants.			
8 Feb. 15	Petter, P. E. (52) (*temp.*)	8 Feb. 15	. . . .	Asst. Postal Censor, Bombay.
8 Feb. 15	Ruglys, H. (51) (*temp.*)	8 Feb. 15	. . . .	
28 July 15	Swindells, H. R. (51, 52)	28 July 15	. . . .	Spl. Res.
4 Oct. 15	Parry, M. J. (51)	4 Oct. 15	. . . .	Spl. Res.
30 Jan. 16	Wood, J. (46) (*temp.*)	30 Jan. 16	. . . .	
22 Nov. 16	Turner, H. G. (46, 52)	22 Nov. 16	. . . .	Spl. Res.
25 Jan. 17	Belk, A. (*temp.*)	25 Jan. 17	. . . .	
25 Jan. 17	Hopkins, G. H. E.	25 Jan. 17	. . . .	T. F.
28 Mar. 17	Seller, J. R. (46)	28 Mar 17	. . . .	T. F.
28 Mar. 17	Knapman, A. A.	28 Mar. 17	. . . .	T. F.
28 Mar. 17	Brown, W. P. (51, 52)	28 Mar. 17	. . . .	T. F.
26 Apl. 17	Melville, E. C. (12, 46, 51) (*temp.*)	26 Apl. 17	. . . .	
30 May 17	Jacques, F. (51)	30 May 17	. . . .	Spl. Res. Ind. Mun. Board.

No. 4 Reserve Battalion (India)—*contd.*

1st Commn.	Names.	Regimental rank.	Army rank.	Remarks.
	2nd-Lieutenants—*contd.*			
30 May 17	Beaumont, L.	30 May 17	. . . .	Spl. Res.
30 May 17	Hone, H. L. (52)	30 May 17	. . . .	Spl. Res.
30 May 17	Birchall, S.	30 May 17	. . . .	Spl. Res.
30 May 17	Cossey, E. E. (52)	30 May 17	. . . .	T. F.
30 May 17	Ingham, B. (*temp.*)	30 May 17	. . . .	
30 May 17	Anderson, A. G (*temp.*)	30 May 17	. . . .	
30 May 17	Muir, T	30 May 17	. . . .	Spl. Res.
30 May 17	Turnbull, A. W. (*temp.*)	30 May 17	. . . .	
30 May 17	Truesdale, H. V. (*temp.*)	30 May 17	. . . .	
27 June 17	McLeod, D. McK.	27 June 17	. . . .	T. F.
29 Aug. 17	Blackhurst, J. J. (*temp.*)	29 Aug. 17	. . . .	
29 Aug. 17	Blanche R.	29 Aug. 17	. . . .	T. F.
29 Aug. 17	Lloyd, N. R.	29 Aug. 17	. . . .	T. F.
29 Aug. 17	Valentine, S. W. (57)	29 Aug. 17	. . . .	T. F.
29 Aug. 17	Watkin E. W.	29 Aug. 17	. . . .	T. F.
26 Sep. 17	Bilsborrow W. G.	26 Sept. 17	. . . .	T. F.
26 Sep. 17	Birkmyre A. (51,52)	26 Sep. 17	. . . .	T. F.
26 Sep. 17	Bullen, J P.	26 Sep. 17	. . . .	T. F.
26 Sep. 17	Connoy, E.P.	26 Sep. 17	. . . .	T. F.
26 Sep. 17	Elliot, T. R.	26 Sep. 17	. . . .	T. F.
26 Sep. 17	Garrett, C. S.	26 Sep. 17	. . . .	T. F.
26 Sep. 17	Wood, F.	26 Sep. 17	. . . .	T. F.
26 Sep. 17	Williamson, W. J.	26 Sep. 17	. . . .	T. F.
26 Sep. 17	Larcombe, J. N. ((*temp.*)	26 Sep. 17	. . . .	
26 Sep. 17	Appleyard, S. W.	26 Sep. 17	. . . .	T. F.
26 Sep. 17	Myer, H. A. (127, 129)	26 Sep. 17	. . . .	T. F.
26 Sep. 17	Periwee, J.W.	26 Sep. 17	. . . .	T. F.
26 Sep. 17	Vowles, R. D.	26 Sep. 17	. . . .	T. F. Duty. S. Bdge. Wellington
30 Oct. 17	Gardiner, D. (52)	30 Oct. 17	. . . .	T. F.
31 Oct. 17	Farrar, F. H.	31 Oct. 17	. . . .	Spl. Res.
31 Oct. 17	Jones, V.	31 Oct. 17	. . . .	Spl. Res.
31 Oct. 17	Paddock, E. A. (65)	31 Oct. 17	. . . .	Spl. Res.
10 Nov. 17	Rapson, R. A. (72)	10 Nov. 17	. . . .	
28 Nov. 17	Hall, R. V.	28 Nov. 17	. . . .	Spl. Res.
8 Dec. 17	Taylor. W. H.	8 Dec. 17	. . . .	
16 Mar. 18	Waterworth, J. W. (*temp.*)	16 Mar. 18	. . . .	
16 Mar. 18	Fairhurst, H. E. (*temp.*)	16 Mar. 18	. . . .	
16 Mar. 18	Phillips, G. W. (*temp.*)	16 Mar. 18	. . . .	
16 Mar. 18	Waite A. W. (*temp.*)	16 Mar. 18	. . . .	
27 Apl. 18	Ratcliffe, J.	27 Apl. 18	. . . .	
3 May 18	Riley, H. J. (51)	3 May 18	. . . .	
19 Aug. 18	Atkinson, V. H. (*temp.*)	19 Aug. 18	. . . .	
19 Aug. 18	Ayre. W. H. M. M. (*temp.*)	19 Aug. 18	. . . .	
19 Aug. 18	Bird. G. W. (46) (*temp.*)	19 Aug. 18	. . . .	
22 Aug. 18	Hinton, A. A.	22 Aug. 18	. . . .	

20 Jan. 15	Fagan, Capt. D. P. (*temp.*)	1 Dec. 17	. . . .	Adjt.
15 Apl. 18	Brice, Lt. A. (*temp.*)	15 Apl. 18	. . . .	Qr.-Mr.

No. 5 Reserve Battalion (India).

Formed in India.

First Commn.	Names.	Regimental rank.	Army rank.	REMARKS.
	Majors.			
23 Apl. 81	Johnstone, B.A., *ret. Ind. Army* (*Actg. Lt.-Col. 9, Nov. 17*)	7 Oct. 14	....	1st G. Bn., S. Staff. R.-Comdg.
9 Dec. 95	Yates, H. P., D.S.O., (*temp. Lt.-Col.*)	1 Sep. 15	L. C. 3 June 16	S. Wales. Bord.-Comdt., Cadet Sch., Lower Topa.
	Captains.			
Oct. 01	Evans, J. S. P.	3 Feb. 15	....	Attd. R. W. Fus.
14 Aug. 14	Kettlewell, L., D.S.O.	5 Mar. 15	....	5 Bn., Wilts.
Nov. 13	French, H. J.	1 June 16	....	Attd. Ches. R.-S. S. O. Ahmednagar.
15 Aug. 14	Bookless, G. T.	10 Aug. 15	....	Attd. Ches. R.
15 Aug. 14	MacSwiney, E. V. F.	2 Dec. 16	....	3 Bn., Ches. R.
21 Aug. 14	Braund, H. B. L.	12 Aug. 15	....	5 Bn., Wilts. R.-Comdg. Conv. Sec., Wellington.
2 Sep. 14	Bury, E. P.	26 Jan. 17	....	4 Bn., S. Wales Bord.Adjt., Cadet Unit, Sialkot.
8 Sep. 14	Bickford-Smith, W. N. V., M.C. (*act'g. (Maj., 6 Oct. 18*)	10 Aug. 15	....	4 Bn., S. Wales Bord. 2nd-in-comd.
Oct. 14	Moore, J. H.	23 Feb. 17	....	5 Bn., Wilts. R.
23 Nov. 14	Solomon, W. E. G.	9 Jan. 15	....	Welsh, R.
	Lieutenants.			
13 Oct. 14	Chandler, T. H.	14 Jan. 15	....	Wilts. R.
23 Nov. 14	Sutcliffe, L. B.	25 Oct. 16	....	8 Bn., R. W. Fus.-Ind. Mun. Board.
20 Nov. 14	Collins, E. G. W., M.C. (*act'g. Capt., 6 Oct. 18*)	23 Feb. 17	....	5 Bn., Wilts. R.
24 Dec. 14	Owen, L. W. G., M.C.	31 Mar. 17	....	13 Bn., Ches. R.
1 Apl. 15	Newman, C. J.	1 Aug. 15	....	16 Bn., R. W. Fus.
1 Mar. 15	Godfrey, R. L. (*actg. Capt.*)	11 Aug. 15	....	9 Bn., Welsh. R.
22 Mar. 15	White, C. T. F.	4 Feb. 18	....	3 Bn., Wilts. R.
7 Apl. 15	Peal, E. F.	22 Mar. 18	....	7 Bn., Wilts. R.
3 June 15	Dale, P. T.	22 Mar. 18	....	8 Bn., R. W. Fus.
3 June 15	Light, F., M.C.	1 July 17	....	4 Bn., S. Wales Bord.
3 June 15	*Jones, I. E. (actg. Capt.)*	*1 July. 17*	....	*R. W. Fus.-Adjutant.*
9 June 15	Webb, C. C. (*actg. Capt., 8 Oct. 18*)	2 Apl. 17	....	5 Bn., Wilts. R.
11 June 15	Lougheed, A. (*temp. Capt.*)	1 Jan. 16	....	Wilts. R.-Army Clothing Dept.
......	George, H. J.	15 May 16	....	R. W. Fus.-Adjt., Prisoner's Camp, Ahmednagar.
22 June 15	Kay, A. N.	22 Mar. 18	....	Attd. Ches. R.
15 July 15	Lewis, T.	22 Mar. 18	....	8 Bn., Welsh, R.
24 July 15	Jones, D. L.	1 July 17	....	4 Bn., S. Wales Bord.
23 Aug. 15	Shannon, W. T.	25 Mar. 17	....	3 Bn., Ches. R.
31 Oct. 15	Wood, J. C. (*actg. Capt.*)	7 Oct. 16	....	6 Bn., Wilts. R.
9 Sep. 14	Kelley, H. H.	1 July. 17	....	8 Bn., Ches. R.-Army Hdqrs., Simla.
2 Nov. 15	Proctor, A. L. (*actg. Capt., 6 Oct. 18*)	1 July 17	....	Attd. R. W. Fus.
5 Nov. 15	Spence, R. R. (*actg Capt., 4 Nov. 18*)	1 July 17	....	Attd. R. W. Fus.
18 Dec. 15	Russell, M.	1 July 17	....	Attd. S Wales Bord.
22 Jan. 16	Owen, T.	...	....	R. W. Fus.
1 Aug. 16	Ranson, H.	1 Feb 18	....	Attd. S. Wales Bord.
5 Aug. 16	Manners, W. McK.	5 Feb. 18	....	Attd. Welsh R.
	2nd-Lieutenants.			
29 Dec. 14	Barrand, H. J.	29 Dec. 14	....	8 Bn., Ches. R.-Attd. 1st G. B. Manch. R.

No. 5 Reserve Battalion (India)—*contd.*

Formed in India.

First Commn.	Names.	Regimental rank.	Army rank.	REMARKS.
	2nd-Lieutenants—*contd.*			
8 Feb. 15	Lewis, T. C.	8 Feb. 15	....	4 Bn., S. Wales Bord.-Attd. Conv. Section, Deolali.
13 May 15	Evans, D. M.	13 May 15	....	R. W. Fus.-Tel. Dept., Simla.
27 May 15	Miller, W. V.	27 May 15	....	8 Bn., Ches. R.
15 July 15	Evans, J.	15 July 15	....	12 Bn., Welsh. R.
24 July 15	Muller, C. J.	24 July 15	....	3 Bn., R. W. Fus.-Adjt. & Qr.-Mr., Conv. Section, Naini Tal.
5 Nov. 15	Ford, E. E.	5 Nov. 15	....	8 Bn., R. W. Fus.
4 Jan. 16	Coventry, G. St. J., M.C.	4 Jan. 16	....	4 Bn., S. Wales Bord.-Attd. Hdqrs., Poona Bde.
26 May 16	Palmer, P. R.	26 May 16	....	12 Bn., R. W. Fus.-Attd. Conv. Section, Deolali.
7 July 16	Barley, M. deM.	7 July 16	....	8 Bn., Ches. R.
7 July 16	Keys, F. H.	7 July 16	....	3 Bn., E. Surr. R.
7 July 16	Widdows, A.	7 July 16	....	11 Bn., D. of Corn. L. I.
11 July 16	Evans, C.	11 July 16	....	4 Bn., R. W. Fus.-Inspr. of Munitions, Agra.
10 July 16	Richards, J.	10 July 16	....	4 Bn., S. Wales Bord.
24 July 16	Homfray, N. E.	24 July 16	....	S. Wales Bord.
5 Aug. 16	Paterson, G. K., (*Actg. Capt.*)	5 Aug. 16	....	Attd. R. W. Fus.-Inspr., Bombg., S. Comd.
5 Aug. 16	Semple, A. G.	5 Aug. 16	....	Attd. S. Wales Bord.
24 Oct. 16	Purcell, C. H. C.	24 Oct. 16	....	4 Bn., Welsh R.
27 Oct. 16	Cook, E. N.	27 Oct. 16	....	S. Wales Bord.
22 Nov. 16	Treadgold, G. W. R.	22 Nov. 16	....	Attd. Wilts. R.
19 Dec. 16	Maxwell, W. R.	19 Dec. 16	....	Attd. Ches. R.
25 Jan. 17	Morrison, J.	25 Jan. 17	....	Attd Ches. R.
25 Jan. 17	Hilliard, F. H	25 Jan. 17	....	Attd. R. W. Fus.
25 Jan. 17	Blyth, C.	25 Jan. 17	....	Ches. R.
25 Jan. 17	Farquharson, J. C.	25 Jan. 17	....	Attd. Ches. R.
1 Feb. 17	Handley, F.	1 Feb. 17	....	3 Bn., Ches. R.
1 Mar. 17	Headrick, W. S.	1 Mar. 17	....	Attd. Ches. R.
27 Mar. 17	Connell, W. W.	27 Mar. 17	....	Attd. Welsh R.
28 Mar. 17	Roxburgh, W. A.	28 Mar. 17	....	Attd. Wilts. R.
28 Mar. 17	Bailey, B.	28 Mar. 17	....	Attd. S. Wales Bord.
29 May 17	Harley, J. E.	29 May 17	....	4 Bn., Ches. R.
29 May 17	Tonge, G. E.	29 May 17	....	7 Bn., R. W. Fus.
29 May 17	Sayee, F.	29 May 17	....	7 Bn., R. W. Fus.
30 May 17	Woods, A J.	30 May 17	....	4 Bn., Ches. R.
30 May 17	Watson, E Mcl.	30 May 17	....	Attd. Ches. R.
30 May 17	McKenna, W.	30 May 17	....	Attd. Welsh R.
30 May 17	Evans, T. F.	30 May 17	....	5 Bn., R. W. Fus.
30 May 17	Pugh, C. L.	30 May 17	....	4 Bn., R. W. Fus.
30 May 17	Sanders, C. F.	30 May 17	....	3 Bn., S. Wales Bord.
27 June 17	Stewart, J. R.	27 June 17	....	Attd. R. W. Fus.
27 June 17	Coutts, J.	27 June 17	....	Attd. S. Wales Bord.
28 Aug. 17	Tovey, R. C. D.	28 Aug. 17	....	3 Bn., S. Wales Bord.
28 Aug. 17	Williams, W. M.	28 Aug. 17	....	3 Bn., S Wales Bord.
29 Aug. 17	Farrell, R. A.	29 Aug. 17	....	8 Bn., Ches. R.
1 Sep. 17	Smith, J. C.	1 Sep. 17	....	8 Bn, Welsh R.
1 Nov. 17	Hadfield, W.	1 Nov. 17	....	3 Bn., Ches. R.
16 Mar. 18	Smith, C.	16 Mar. 18	....	8 Bn., R.W. Fus. Ind. Ord. Dept.
16 Mar. 18	Perry, F. H.	16 Mar. 18	....	5 Bn., Wilts. R.
16 Mar. 18	Lloyd, T. C.	16 Mar. 18	....	5 Bn., Wilts. R.

3 June 15	Jones, Lieut. (*Actg. Capt.*) I. E.	9 Nov. 17	....	Adjutant.
25 Aug. 14	Fricker, Capt. L.	28 Aug. 17	....	Quarter-master.

No. 6 Reserve Battalion (India.)

First Commn.	Names.	Regimental rank.	Army rank.	Remarks.
	Lieutenant-Colonel.			
	Majors.			
9 Dec. 14	Boucher, Capt. (*temp. Lt.-Col.*) B. H., D.S.O. (*ret. pay*) (50, 52, 57, 71, 172, 207) (*temp.*)	1 Feb. 17	. . . .	7 Bn., Hamps. R.—Comdg.
25 July 91	Martin, R. (51, 57, 66*a*, 88, 127, 206)	16 Feb. 15	. . . .	R. Ir. R.
5 July 99	Stilwell, W. B., D.S.O. . .	2 Oct. 14	. . . .	Comdg. Leave Camp, Kirkee.
26 May 06	Duke, H. E. (51, 66*a*) . . .	18 May 17	. . . .	2nd-in-Comd.
	Captains.			
3 Sep. 09	Addison, A. W. N. (*temp. Maj.*) (51)	24 June 14	. . . .	1-6 Bn., Hamps. R.—Asst. Mily. Secy., Army Hdqrs.
15 Jan. 09	Roper, J.	3 Oct. 14	. . . .	
1 May 12	Blagg, F. O. (51, 66*a*) . . .	1 Apl. 15	. . . .	1-6th Hamps. R.
16 Dec. 14	Shortt, G. A. (*temp.*) (190) . .	19 Oct. 15	. . . .	1st G. B., Yorks. R. with M. A. Dept.
2 Sep. 12	Jensen, C. L. R.	1 June 16	. . . .	1-6th Hamps. R.
1 May 11	Burt, H. C.	1 June 16	. . . .	1-4th Dorset. R.
10 May 15	Thomas, R. G. (127)	14 Oct. 16	. . . .	25th Midd'x. R.
	Lieutenants.			
Dec. 00	Levy, M. P. (51, 129)	28 Jan. 16	. . . .	2nd G. B. Essex R.
4 Jan. 15	Vick, R. I. (46, 51)	6 Apl. 16	. . . .	3rd Hamps. R., Adjt. Leave Camp, Kirkee.
23 Sep. 14	Brown, N. A. S. (*act'g. Capt.*) (50, 60*a*, 127)	1 June 16	. . . .	1-6th Hamps. R.
12 Sep. 14	Ogle, J. E. (*act'g. Capt., 7 Sep. 18*) (51)	1 June 16	. . . .	
3 Oct. 14	*Wemyss, A. B.*	1 June 16	. . . .	*I. A. R. O.*
29 Aug. 14	Symes, J. D.	1 June 16	. . . .	*Lv., ex I.*
24 Oct. 14	Smith, V. C. E. (*temp. Capt.*) .	4 June 16	. . . .	Hamps. R.—G. S. O., 3rd gr.
18 Apl. 15	Newfield, M. M. (57, 127, 129, 207)	15 Nov. 16	. . . .	3rd G. B. Bedf. R.
22 July 15	Beere, H. C. (46, 48*a*, 50, 60*a*) .	1 July 17	. . . .	1-6th Hamps. R. Instr., Divl. Musk. [Schl. (Br.) Poona Dn.
31 Oct. 14	Filleul, R. R.	1 July 17	. . . .	
21 Feb. 15	Warnford-Davis, R. D. (*act'g. Capt., 31 July 18*) (51, 127, 129)	1 July 17	. . . .	
21 Oct. 15	Dickinson, C. E. (*act'g. Capt., 15 Nov. 18*) (46, 60*a*)	1 July 17	. . . .	With M. W. S.
22 Oct. 14	Llewellin, W. W. (*act'g. Capt*) (51, 60*a*)	1 July 17	. . . .	Adjutant.
27 Nov. 15	Harrup, W. D.	1 July 17	. . . .	
10 Dec. 15	Blount, C. B.	1 July 17	. . . .	
23 Sep. 15	Hutchens, H. J.	1 July 17	. . . .	7th Hamps. R.
16 Jan. 16	Shaddock, J.	16 July 17	. . . .	6th Hamps. R. [Schl., Satara.
10 July 16	Woolgar, C. (*temp. Capt., 22 Nov. 18*)	14 Mar. 18	. . . .	5th W. Rid. R. Qr—Mr., Musk.
	2nd-Lieutenants.			
26 Oct. 14	Wyles, C. D.	26 Oct. 14	. . . .	
27 Aug. 15	Hamley, A.	27 Aug. 15	. . . .	
18 Nov. 15	Ellis, E. I. A. (66*a*, 51) . . .	25 May 16	18 Nov. 15	
19 Dec. 16	MacKenzie, T. C.	19 Dec. 16	. . . .	9th Arg. & Suth'd. Highrs.
19 Dec. 16	Smales, R. (51)	19 Dec. 16	. . . .	7th North'd Fus.
10 Mar. 17	Cake, S. F. (*temp. Capt.*) (51, 60*a*, 72)	10 Mar. 17	. . . .	Hamps. R.—Asst. Provost Marshal, [Poona.
30 May 17	Walker, J. B.	30 May 17	. . . .	5th Sco. Rif.
30 May 17	White, C. E. P.	30 May 17	. . . .	8th Sco. Rif.
30 May 17	Clarke, J. H. (46)	30 May 17	. . . .	8th Sco. Rif.
30 May 17	Godwin, A. H. V.	30 May 17	. . . .	
30 May 17	Lawson, W. G.	30 May 17	. . . .	Attd. M. W. Dept. K.
30 May 17	Mercer, E. C.	30 May 17	. . . .	
30 May 17	Missing, C. M. S.	30 May 17	. . . .	
30 May 17	Noyes, H.	30 May 17	. . . .	
30 May 17	Beckworth, C.	30 May 17	. . . .	Mily. Accts. Dept.
30 May 17	Vowles, A. C. (127)	30 May 17	. . . .	
27 June 17	Burns, M. M.	27 June 17	. . . .	4th Gord. Highrs.
27 June 17	Gibb, L. J. (46)	27 June 17	. . . .	4th Gord. Highrs.
27 June 17	Moodie, J.	27 June 17	. . . .	9th Arg. & Suth'd. Highrs.
1 Aug. 17	Wright, W.	1 Aug. 17	. . . .	4th Gord. Highrs.
26 Sep. 17	Cottle, S. G.	26 Sep. 17	. . . .	6 Bn., Manch. R.
16 Mar. 18	Manser, A. R.	16 Mar. 18	. . . .	
16 Mar. 18	Eastwood, V.	16 Mar. 18	. . . .	
16 Mar. 18	Pinkerton, S. R.	16 Mar. 18	. . . .	
16 Mar. 18	Mayers, N. (127)	16 Mar. 18	. . . .	
16 Mar. 18	Long, F. A. (127)	16 Mar. 18	. . . .	
28 Sep. 14	Marshall, Maj. R. (retd. pay)	28 Sep. 14	. . . .	Qr.-Mr.

No. 7 Reserve Battalion (India.)

First Commn.	Names.	Regimental rank.	Army rank.	Remarks.
	Lieutenant-Colonel.			
	Majors.			
Feb. 00	Wrangham, D. G. (*act'g. Lt.-Col.*), E. York R.	1 Sep. 15	. . . .	
Jan. 15	Sale, A. B., M.C. (12)	2 Feb. 17	. . . .	R. War. R.
	Captains.			
Jan. 04	Bankhead-Browne, A. E.	15 Apl. 08	. . . .	4 Bn., N. Staff. R.
20 Sep. 14	Siddall, T. (*act'g. Major, 9 Nov. 17*)	9 Dec. 14	. . . .	2nd-in-Comd.
2 Nov. 14	Howie, J. (*temp.*) (51)	4 Feb. 15	. . . .	Rif. Bde.
29 May 07	Graham, J. M.	15 Mar. 15	. . . .	Worc. R.
7 Oct. 15	Sybray, J. (*temp.*)	5 Feb. 16	. . . .	E. Surr. R.
14 Aug. 14	Gridley, R. F. (51)	31 Aug. 15	. . . .	R. W. Surr. R.
10 May 15	Hutton, T. A. (*temp.*)	26 Jan. 17	. . . .	Worc. R.
	Lieutenants.			
30 Nov. 14	Vance, H. R. (*temp.*) (*act'g. Capt., 1 Oct. 18*)	3 Feb. 16	. . . .	6th Worc. R.
17 Mar. 15	Hulbert, H. B.	3 Feb. 16	. . . .	4 Bn., Worc. R.
19 Sep. 15	Lucas, J. (*temp.*) (51)	17 May 16	. . . .	R. War. R.
1 Sep. 14	Starley, J. R., M.C. (*temp.*)	13 Dec. 16	. . . .	R. War. R.
11 June 15	Admans, H. G. (*act'g. Capt.*) (51)	28 Jan. 17	. . . .	R. War. R.
16 Aug. 15	Doyle, C. R. (*temp.*)	29 Jan. 17	. . . .	N. Staffs. R.
26 Apl. 15	Wilson, H. R. (*act'g. Capt.*) (*temp.*)	1 July 17	. . . .	R. War. R.
14 Aug. 15	Soanes, L. J. W. (*temp.*)	1 July 17	. . . .	Worc. R.
22 Apl. 15	Morris, R. G. (*temp.*) (52)	1 July 17	. . . .	Worc. R.
6 May 15	Roberts, C. S. (*temp.*)	1 July 17	. . . .	Worc. R.
16 Nov. 16	Tappin, W. H. (*temp.*)	1 July 17	. . . .	Northn. R.
8 Nov. 15	Shepherd, J. (*temp.*)	1 July 17	. . . .	Worc. R.
8 Dec. 15	Sim, A. J. A. (*temp.*)	1 July 17	. . . .	Worc. R.
15 Oct. 15	Smyth, S. W. (*temp.*)	1 July 17	. . . .	N. Staffs. R.
1 Apl. 15	Kirby, A. S. (*temp.*)	1 July 17	. . . .	N. Staffs R. Asst. Cable Censor, Rangoon.
24 Nov. 14	Davy, C. L. (*temp.*)	1 July 17	. . . .	R. War. R.
2 Dec. 15	Annandale, F. W. (*temp.*)	1 July 17	. . . .	S. Staff. R.—Spl. Res.
10 June 15	Crocker, G. (*temp.*) (51)	1 July 17	. . . .	Oxf. and Bucks. L. I.
8 Jan. 16	Daniell, C. I. (*temp.*) (46)	8 July 17	. . . .	Worc. R.
14 Jan. 15	Prodger, S. E. (*temp.*)	12 Aug. 17	. . . .	Worc. R.
7 July 16	Byrne, T. (*temp.*) (12, 51)	7 Jan. 18	. . . .	Worc. R.
7 July 16	McKillop, A. (*temp.*)	7 Jan. 18	. . . .	Worc. R.
	2nd-Lieutenants.			
5 Feb. 15	Newman, R. A. (*temp.*)	5 Feb. 15	. . . .	Worc. R.
10 Mar. 15	Bromage, W. A. (*temp.*) (12)	10 Mar. 15	. . . .	N. Staff. R.
10 Mar. 15	West, C. H., M.C. (*temp.*)	10 Mar. 15	. . . .	Linc R.
19 Mar. 15	Maver, A. M. (*temp.*)	19 Mar. 15	. . . .	N. Staff. R.
5 June 15	Ibbotson, H.	5 June 15	. . . .	3rd L'pool R.
5 Oct. 15	Meny-Gibert, M. A. (*temp.*)	5 Oct. 15	. . . .	Worc. R.
28 Dec. 15	Richardson, G. (*temp.*)	28 Dec. 15	. . . .	R. War. R.
5 Aug. 16	Fitz-Hugh, A. (*temp.*)	5 Aug. 16	. . . .	Worc. R.
5 Aug. 16	Calvert, A. P. (*temp.*)	5 Aug. 16	. . . .	R. Ir. Fus.
27 Oct. 16	Ransford, E. M.	27 Oct. 16	. . . .	Suff. R.
19 Nov. 16	Groom, C. R. N. (*temp.*)	19 Nov. 16	. . . .	E. Kent R.
22 Nov. 16	Saich, H. C. (*temp.*)	22 Nov. 16	. . . .	2nd G. B. Bedf. R.
19 Dec. 16	Miller, R. A.	19 Dec. 16	. . . .	Worc. R. (Spl. Res.)
11 Feb. 17	Gould, W. G. (*temp.*)	11 Feb. 17	. . . .	3rd G. B. Bedf. R.
1 Mar. 17	Noble, J. W. (51) (72)	1 Mar. 17	. . . .	Sco. Rif.
30 May 17	Sarginson, G. (*temp.*)	30 May 17	. . . .	R. War. R.
30 May 17	Harvey, E. G. (*temp.*)	30 May 17	. . . .	Worc. R.

No. 7 Reserve Battalion (India)—*contd.*

First Commn.	Names.	Regimental rank.	Army rank.	Remarks.
	2nd-Lieutenants—*contd.*			
30 May 17	Smith, H. C. (*temp.*)	30 May 17	....	Worc. R.
30 May 17	Williams, G. (*temp.*) (51)	30 May 17	....	R. War. R.
30 May 17	Harding, R. (*temp.*)	30 May 17	....	R. War. R.
30 May 17	Penn, G. G. (*temp.*)	30 May 17	....	R. War. R.
30 May 17	Randall, L. B. (*temp.*)	30 May 17	....	R. War. R.
30 May 17	Dainty, S. (*temp.*)	30 May 17	....	R. War. R.
30 May 17	Kilbourn, W. G.	30 May 17	....	R. War. R. (Spl. Res.).
30 May 17	Sabin, W. H. (*temp.*)	30 May 17	....	Worc. R.
30 May 17	Lewington, R. (*temp.*)	30 May 17	....	Worc. R.
30 May 17	Waggott, R. L. (*temp.*)	30 May 17	....	R. War. R.
30 May 17	Evans, J. H. (*temp.*)	30 May 17	....	Worc. R.
30 May 17	White, H. M. (*temp.*)	30 May 17	....	Worc. R.
29 Aug. 17	Burvill, H. A. (*temp.*) (51)	29 Aug. 17	....	Worc. R.
29 Aug. 17	Sutherland, D.	29 Aug. 17	....	K. O. Sco. Bord.
26 Sep. 17	Webb, N. D. (*temp.*)	26 Sep. 17	....	Worc. R.
26 Sep. 17	Sharpe, J. C. (*temp.*)	26 Sep. 17	....	Worc. R.
31 Oct. 17	Patterson, A. H.	31 Oct. 17	....	K. O. Sco. Bord.
31 Oct. 17	Jardine, E.	31 Oct. 17	....	K. G. Sco. Bord.
28 Nov. 17	Barnard, A. W. S. (*temp.*)	28 Nov. 17	....	R. War. R.
19 Aug. 18	Leonard, N. (*temp.*)	19 Aug. 18	....	R. War. R.
19 Aug. 18	Sims, A. W. (*temp.*)	19 Aug. 18	....	Worc. R.
19 Aug. 18	Saul, G. W. O. (*temp.*)	19 Aug. 18	....	Worc. R.
19 Aug. 18	Baldwin, C. (*temp.*)	19 Aug. 18	....	N. Staff. R.
18 Apl. 15	Sykes, Lt. (*act'g. Capt.*) A. W. (60*a*)	18 Apl. 15	....	Rif. Bde. Adjt.
11 Apl. 18	Thompson, Lt. H. E.	11 Apl. 18	....	Qr.-Mr.

No. 8 Reserve Battalion (India).

Formed in India.

First Commn.	Names.	Regimental rank.	Army rank.	Remarks.
	Majors.			
9 Dec. 99	Bonnyman, F. J. C. (*temp.*) (*act'g. Lt.-Col.*)	12 Feb. 16	8 Oct. 14	1st G. B., Manch. R.—Comdg.
......	Willis, H. H., V.D. (51)	20 May 11	....	4 Bn., Wilts. R.—*Lv., ex I., m. c., 6 mos.*
1 Mar. 00	Townsend, H.	1 June 16	....	4 Bn., Devon. R.
2 Apl. 09	*Trevenen, W. B.*	2 Jan. 18	....	*6 Bn., Devon. R.—M. A. Dept.*
2 Feb. 02	*Strachey*, Hon. *E.*	20 Jan. 18	....	*4 Bn., Som. L. I.—Staff Capt., Karachi Bde.*
12 Sep. 05	Timmins, T. B.	22 Feb. 18	....	4 Bn., Som. L. I.—2nd-in-Comd.
	Captains.			
11 Nov. 04	Anderson-Morshead, R. Y. (51)	1 Aug. 09	....	4 Bn., Devon. R.—*Lv., ex I., m. c.*
6 Dec. 11	Ocock, W. P. (51)	5 Aug. 14	....	10 Bn., Midd'x. R.
......	Hole, G. B. (*temp.*)	5 Nov. 14	....	Attd. Devon. R.
25 May 04	*Miles*, Sir *C. W.*, Bt. (*temp. Maj.*) (51, 66*a*).	20 Dec. 14	....	*Som. L. I. Asst. Mily. Secy., Army Hdqrs.*
......	Bowman, H. E. (*temp.*)	1 Jan. 15	....	Genl. List.—Dir. of Edn., Egyptian Admn.
2 Mar. 10	Smyth-Richards, F. G. (51)	8 Feb. 15	....	6 Bn., Devon. R.—O. C., Conv. Section, Trimulgherry.
28 May 10	*Mason, A. S.*	8 Feb. 15	....	*6 Bn., Devon R.—D. A. Q. M. G., 16th (Indian) Dn.*
......	Hill-Workman, J.	1 June 15	....	3 Bn., R. Scots.
15 Apl. 12	*Reavell, H. S.* (66*a*)	1 June 16	....	*6 Bn. Devon. R.—Ind. Ord. Dept.*
1 Nov. 12	Orchard, J. W. (52)	1 June 16	....	4 Bn., Devon. R.—Staff Capt., Madras.
5 Sep. 13	Moger, W. (51)	1 June 16	....	4 Bn., Som. L. I.
5 Aug. 14	Sidmouth, G. W., *Visct.* (51)	1 June 16	....	6 Bn., Devon. R.
......	Kitson, W. K.	1 June 16	....	5 Bn., Devon. R.—*Actg. Qr. Mr.*
......	Jervis, W. W.	10 Apl. 17	....	4 Bn., Devon. R.—Attd. Army Sig. Serv.
......	Abraham, A. C.	3 May 17	....	Devon. R.
10 Sep. 14	Molesworth, G. E. N. (51, 57)	12 May 17	....	6 Bn., Devon. R.—*I. A.* (*on probn.*)
11 Sep. 14	Gray, D. P. G. (51)	12 May 17	....	6 Bn., Devon. R.
1 Oct. 14	Palairet, R. C. N. (*temp. Maj., 7 Sep. 18*)	29 Aug. 17	....	6 Bn., Devon. R.—Dy. Asst. Dir.
	Lieutenants.			
......	Brand, J. K. (*act'g. Capt., 20 Apl. 18*)	3 Mar. 15	....	3 Bn., Devon. R.
......	Batt, R. H.	1 July 17	....	18 Bn., Lond. R.—Adjt., Keti Conv. Section, Wellington.
......	Holmes, J. (*temp.*)	3 Nov. 15	....	1 G. B., York. R.
......	Eden, P. R. (*temp.*)	1 Jan. 16	....	1 G. B., E. York. R.
......	Frossard, C. B. S.	22 Mar. 16	....	3 Bn., Devon. R.
9 Dec. 14	Stabb, C. T. (*act'g. Capt., 24 June 18*)	1 June 16	....	5 Bn., E. Kent R.
14 Oct. 14	Kingwell, G. L.	1 June 16	....	4 Bn., Devon. R.
1 Oct. 14	Young, G. B.	1 June 16	....	6 Bn., Devon. R.
3 Oct. 14	Wilson, W. R. (51, 52)	1 June 16	....	6 Bn., Devon. R.
6 Oct. 14	Jeffrey, E. J. B.	1 June 16	....	6 Bn., Devon. R.
......	Whiller, E. C.	1 June 16	....	5 Bn., Essex R.—Conv. Depôt, Talital.
5 Nov. 14	Butland, R. (*temp.*)	2 July 16	....	9 Bn., Devon. R.
......	Shalless, E. G. (*temp.*)	14 Oct. 16	....	Midd'x. R.
......	Braham, S. H. T. (*temp.*) (*temp. Capt.*)	14 Oct. 16	....	Midd'x. R.—Asst. Prov. Marshal, Secunderabad.
......	Tansley, L. R. H. (*temp.*)	27 Nov. 16	....	1st G. B., York. R.—*Lv., ex I., m. c.*

No. 8 Reserve Battalion (India)—*contd.*

First Commn.	Names.	Regimental Rank.	Army rank.	REMARKS.
	Lieutenants—*contd.*			
......	Fox, F. W. J.	1 July 17	....	3 Bn., North'n. R.
12 Dec. 14	Minshall, C. H.	1 July 17	....	6 Bn., Devon. R.
3 Apl. 15	Morgan, H. G.	1 July 17	....	4 Bn., Som. L. I.
......	Slessor, J. G.	1 July 17	....	4 Bn., Gord. Highrs.
......	Titchmarsh, F. L.	1 July 17	....	3 Bn., Suff. R.
......	Gilbee, S. J.	1 July 17	....	10 Bn., R. Fus.
......	Reeve, G. B. (*temp.*)	1 July 17	....	A. S. C.
......	Archer, G.	1 July 17	....	3 Bn., Devon. R.
......	Hughes, F.	1 July 17	....	5 Bn., Devon. R., Inspr. of Munitions, Ferozepore.
10 Dec. 14	Williams, N. T.	1 July 17	....	5 Bn., Devon. R.
......	Williams, J. H. (*tempy. Capt., 25 Nov. 18.*)	1 July 17	....	5 Bn., Devon. R.—Instr. Bombing Schl.
......	Crage, F.	1 July 17	....	5 Bn., D. of Corn. L. I.
21 Apl. 15	Earl, E. A.	1 July 17	....	5 Bn., Devon. R.
......	Crocker, E. G. M. (temp.)	1 July 17	....	Som. L. I.
9 Oct. 15	Watterson, J. S.	1 July 17	....	4 Bn., Som. L. I.
......	Hobson, P. W. (temp.)	1 July 17	....	1 G. B., E. York. R.
......	Cannell, J. H.	1 July 17	....	5 Bn., Suff. R.—*Lv., ex I., m. c.*
......	Goodman, R.	1 July 17	....	3 Bn., Suff. R. (T. F. Res.)
......	Kent, D. G. (temp.)	20 July 17	....	Devon. R.
7 Oct. 16	Elford, D. E.	7 Oct. 17	....	Devon. R.
30 June 16	Vos, S.	30 Dec. 17	....	6 Bn., Devon. R.
2 Sep. 15	Hill, R. F.	....	....	5 Bn., D. of Corn. L. I.—Attd. Hdqrs., 8th (Lucknow) Dn.
	2nd-Lieutenants.			
25 Aug. 15	Moor, J. A.	25 Aug. 15	....	5 Bn., Suff. R.—Con. Dept. Poonamallee.
22 Sep. 15	Buck, H.	22 Sep. 15	....	5 Bn., Suff. R.—Attd. Hdqrs., Nowshera Bde.
4 Dec. 15	Elliot, L.	4 Dec. 15	....	8 Bn., Devon. R.—*Lv., ex I., m. c.*
1 Feb. 15	Johnson, R. H. K.	1 Feb. 15	....	6 Bn., Devon. R.—*Lv., ex I., m. c.*
18 Dec. 15	Bird, R.	18 Dec. 15	....	4 Bn., Devon. R.—*Lv., ex I., m. c.*
3 June 16	Samuels, L.	3 June 16	....	6 Bn., Devon. R.—Qr.-Mr., Attd. Br. Section, Poona.
5 Sep. 16	Kinloch, J. (*temp.*)	5 Sep. 16	....	Sea. Highrs.
25 Jan. 17	Baillie, K. (*temp.*)	25 Jan. 17	....	3 Bn., Som. L. I.
28 Mar. 17	Dowling, T.	28 Mar. 17	....	3 Bn., Devon. R.
26 Apl. 17	Allan, R. M.	26 Apl. 17	....	4 Bn., R. Sco. Fus.
28 Apl. 17	Houghton, A. T.	28 Apl. 17	....	5 Bn., Som. L.I.—S. O. to Inspr. of Infy. (South).
28 Apl. 17	Vittle, D. H. (*temp.*)	28 Apl. 17	....	Devon. R.—*Lv., ex I., m. c.*
30 May 17	Goodhind, M. G. (*temp.*)	30 My 17	....	Som. L. I.
30 May 17	Robb, G.	30 May 17	....	8 Bn., Sco. Rif.-Offrs.' class, Nasik.
30 May 17	Andrews, C. W.	30 May 17	....	3 Bn., Som. L. I.
30 May 17	Denis, A. L., D. C. M.	30 May 17	....	3 Bn., Devon. R.
29 Aug. 17	Kellsall, J.	29 Aug. 17	....	6 Bn., Devon. R.—N. C. O's School, Belgaum.
29 Aug. 17	Cole, E. J.	29 Aug. 17	....	6 Bn., Devon. R.—*Lv., ex I., m. c.*
29 Aug. 17	Burridge, R. A.	29 Aug. 17	....	6 Bn., Devon. R.
29 Aug. 17	Moses, C. R.	29 Aug. 17	....	1 Camb. R.
29 Aug. 17	Diplock, A.	29 Aug. 17	....	1 Camb. R.
26 Sep. 17	Roe, J. H.	26 Sep. 17	....	4 Bn., Notts. & Derby. R.
26 Sep. 17	Blore, A. V.	26 Sep. 17	....	4 Bn., Notts. & Derby. R.
26 Sep. 17	Platts. T. C.	26 Sep. 17	....	6 Bn., Manch. R.
26 Sep. 17	Dawson, G. J. C.	26 Sep. 17	....	9 Bn., Manch. R.
26 Sep. 17	Flint, R. P.	26 Sep. 17	....	9 Bn., Manch. R.
26 Sep. 17	Gough, J. A.	26 Sep. 17	....	10 Bn., Manch. R.
26 Sep. 17	Strong, H. W. (*temp.*)	26 Sep. 17	....	11 Bn., R. Dub. Fus.
26 Sep. 17	Smyth, F. R.	26 Sep. 17	....	5 Bn., Rif. Bde.
26 Sep. 17	Scoggins, H. R.	26 Sep. 17	....	4 Bn., Suff. R.
......	Hearth, L. A.	....	....	4 Bn., Devon. R.—Sig. Ser. Depôt.
......	Selfe, G. V.	....	....	6 Bn., Devon. R.
......	Hack, T. W.	....	....	E. York. R.

Machine Gun Corps.

Arranged according to Army seniority.

TRAINING CENTRE.

Commandant	**Howson,** Maj. (*temp. Lt.-Col.*) G., M.C., 4th Cav., attd. M. G. Corps (*Cav.*)	21 Jan. 18
Instructor	**Slaughter,** Capt. (*temp. Major*) G. A., M.C., Worc. R. & M. G. Corps.	6 Aug. 15
Assistant Instructor	**Lecky,** Maj. J. G., 119th Inf.	21 Jan. 18
Adjutant	**Martin,** Lt. (*temp. Capt.*), W. H.	17 Sep. 17
Officer in charge of Indian Depôt.	**Green,** Lieut. (*temp. Capt.*) P. J., R. F.A.	1 July 17
Assistant to Officer in charge of Indian Depot.	**Avery,** Lt. A. H., I.A.R. O.	1 Sep. 17
Extra Assistant Instructor	**Humphries,** Lt. A., I.A.R.O.	5 Feb 18
Quartermaster	**Tanner,** Lt. G., Br. Ser. (Genl. List)	12 June 18

CAVALRY.

Captain.

Wright, A. . . . 26 Feb. 13
Sartorius, (*temp. Maj.*) G., 5th Cav. . . . 24 Jan. 15

Lieutenants.

Brierley, R. W. (*Lv., ex I., m. c., 6 mos.*) . . . 4 May 14
Myburgh, P. R. . . . 1 June 16
Guthrie, J. A. (*Act'g. Capt., 22 Jan. 18*) . . . 1 Mar. 17
Rennie, C. A. . . . 1 July 17
Lumby, L. J. . . . 1 July 17
Lavis, L. E. . . . 1 July 17
Austin, W. A. G. (61) . . . 1 July 17
Taylor, H. A. . . . 1 July 17
Vigne, F. . . . 29 Jan. 18
Hurndall, R. K. F. . . . 29 Jan. 18
Jackson, H. W. . . . 28 May 18

2nd-Lieutenants.

Sturges, F. G. . . . 26 Jan. 17
Jones, C. W. (12) . . . 26 Jan. 17
Chapman, J. J. (12) . . . 31 Jan. 17
Bartlett, G. . . . 25 Mar. 17
Thomas, C. V. . . . 25 Mar. 17
Cowell, R. O. . . . 25 Mar. 17
Doubleday, A. W. . . . 26 Apl. 17
Fitkin, S. A. . . . 26 Apl. 17
Stones, W. (12) . . . 28 May 17
Ransome, F. D. . . . 26 July 17
Nicholls, A. . . . 7 June 18
Jones, R. C. . . . 7 June 18

MOTORS.

Majors.

Molony, A. W., R. Dub. Fus., attd. M. G. Corps. . . . 13 Dec. 15

Captains.

Long, E. W., 1-25th Lond. R. . . . 1 Apl. 14
Clifton, A. J., O.B.E., (*temp. Maj.*) 1 Bn., Durh. L. I. . . . 12 Mar. 18
O'Neill, H. J. B., M.C., 127th Bal. Inf. . . . 29 Aug. 15
Wildy, H. A., 1-25th Lond. R. . . . 1 June 16
Pardington, G. E. L., 1-4th R. W. Kent R. . . . 1 June 16
Watkins, L. T., 3rd Punjab Rif. . . . 1 Apl. 17

Lieutenants.

Brown, H. G. (*Act'g. Capt. 7 Nov. 16*) M. G. Corps . . . 19 Nov. 15
Watson, E. L., 1st Gn. Bn., Norf. R. . . . 24 Nov. 15
Skelton, W., M. G. Corps (12) . . . 16 Dec. 15
Farmer, N. R. (*Act'g. Capt., 7 Nov. 16*), M. G. Corps . . . 9 Feb. 16
Windsor, E. P., M. G. Corps . . . 9 Feb. 16
Hargreaves, J., M. G. Corps . . . 6 Mar. 16
Marley, R. W., M. G. Corps . . . 1 June 16
Anderson, C. O. D. (*Act'g. Capt., 9 Dec. 16*) M. G. Corps . . . 5 July 16

Lieutenants—*contd.*

Charlesworth, E. F., 1-4th R. W. Surr. R.	1 Jun 16
Redfern, A. R., M. G. Corps	1 Dec. 16
Everard, G. H., M. G. Corps	1 Dec. 16
Graham, M. T., I.A.R.O.	1 Dec. 16
Farnell, F. B., M. G. Corps	1 Jan. 17
Wilson, D. G., M. G. Corps	1 Feb. 17
Knight, K. A., M. G. Corps	1 May 17
Crick, W., (*tempy. Capt.*) 26th Hyderabad Rif.	9 May 17
Farmer, C. R. H., 18th Huss.	1 July 17
Hobbs, C. F. G., 1st Leic. R.	1 July 17
Donnison, E. G. (12)	1 July 17
Forbes, A., 1st Sussex R.	1 July 17
Elmslie, C. S., I.A.R.O.	15 Sep. 17
Parr, A., M. G. Corps	5 Feb. 18
Gimson, A., M. G. Corps	5 Feb. 18
Tanner, J. T., I.A.R.O.	22 June 18

2nd-Lieutenants

Hirst, W. E. L., M. G. Corps	21 July 15
Corson, H., M. G. Corps (12)	5 Aug. 16
Farquhar, D., I.A.R.O.	5 Oct. 17
Martin, A. W., 5th (Lucknow) Group Gar. Artillery	4 Feb. 18

INFANTRY.

Majors.

Duckworth, G. S.	18 Feb. 18
Bleakley, J. F.	16 Feb. 18

Captains.

Fry, D. C.	13 Feb. 15
Bennet, T. G. (*actg. Maj., 19 Nov. 18*)	29 June 15
Hewat, J. P. D., (46, 67)	1 June 16
Moss, R. L.	1 June 16
Smithers, M. H. W.	1 June 16
Lisle, P. C.	1 June 16
Richards, A. R. T. (*Lv., ex I., m.c., 6 mos.*)	1 June 16
Scott, W. G. S.	2 Aug. 16
Leach, R. B., M.C. (*actg. Maj., 19 Nov. 18*)	2 Dec. 16
Saul, J. B.	1 Jan. 17
Errington, J. R. F.	1 July 17
Baring, D. W. (*Actg. Capt., 7 July 18*)	20 July 17

Lieutenants.

Breton, D. C. (*actg. Capt., 3 Nov. 18*)	13 Feb. 15
Whalley-Kelly, J. H.	28 May 15
Edwardes, H. W.	17 July 15
Munro, A. R.	1 Oct. 15
Murphy, J. E.	16 Nov. 15
Firth, J. R.	3 May 16
Findlay, L. F. (*actg. Capt., 26 Sep. 18*) (207*a*)	1 June 16
Grahame, D. (*actg. Capt. 27 Oct. 18*) (46, 51, 59)	1 June 16
Kemp-Welch, H. A.	1 June 16
Armstrong, A. H.	1 June 16
Bakewell, W. B. (*actg. Maj. 19 Nov. 18*)	1 June 16
D'Angibau, W. E. D.	1 June 16
Keller, R. C.	15 June 16
Dossetter, A. H.	6 July 16
Augus, G. W.	20 July 16
Bowerman, J. F. (*actg. Capt. 27 Oct. 18*) (207*a*)	1 Aug. 16
Piercy, C. H. Z.	25 Aug. 16
Hartle, J. C.	1 Oct. 16
Lea, C. M.	1 Nov. 16
Templeton, R. B. (S. S. O., Dehra Dun)	1 Feb. 17
Hollidge, S. G. (12)	1 Mar. 17
Donald, R. T. (207*a*)	1 Mar. 17
Shaw, J., M.C	18 Mar. 17
Lismore, T.	1 Apl. 17
Coombes, F. E.	1 May 17
Hill, R. A. C.	1 July 17
Simpson, G. G. (48*a*, 51)	1 July 17
Blackmore, L. W. S.	1 July 17
Fowler, A. C. O.	1 July 17
Spruce, P. A.	1 July 17

Lieutenants—*contd*

Atkins, A. T.	1 July 17
Penberthy, W. F.	1 July 17
Howell, W. E. (207*a*)	1 July 17
Beal, H. G.	10 Aug. 15
Dolben, A. C. (48*a*, 51, 95, 207*a*)	1 July 17
Parry, T. O.	25 Sep. 17
Hill, R. L.	4 Dec. 17
Frank, C. M., M. G. Corps	5 Feb. 18
Southon, H. M.	5 Feb. 18

2nd-Lieutenants.

Congreve-Pridgeon, R.	28 Nov. 14
Wilson, D. S. (57)	5 Aug. 15
Ainslie. C. J.	13 Apl. 16
Cole, J. (*Lv., ex I., m.c., 6 mos.*)	26 Sep. 16
Spark, F. E.	25 Mar. 17
Darnley-Smith, G.	28 Apl. 17
Appleton, H. T.	28 Apl. 17
Ashman, C. H. (*temp.*) (*Lv., ex I., m.c., 6 mos.*)	29 Aug. 17
Farr, C. A. C.	26 Sep. 17
Davies, A.	26 Sep. 17
Downes, G. F. (*temp.*)	16 Mar. 18
Fickling, W. A.	16 Mar. 18
Thompson, A. G.	16 Mar. 18
Berendt, H.	16 Mar. 18
Grigg, C. A.	16 Mar. 18
Gaillard, P.	16 Mar. 18
Frisby, F.	16 Mar. 18
Griffiths, J. V.	16 Mar. 18
Triebler, F. C.	16 Mar. 18
Smith, A. F.	16 Mar. 18

1st Madras Guards.

Uniform—Dark-blue.
Facings—Scarlet.
Lace—Gold.
Working dress—Khaki drill.

"Ready, aye Ready."

Raised, 2nd July 1857. Reconstituted as an I. D. F. Unit with present designation 1st Apl. 1917.
Establishment—3 Active Coys., 1 Res. Coy., 1 Cadet Coy., 1 Armoured Motor Battery.

HONORARY COMMANDANT—*H. E. the Rt. Hon.* JOHN, *Baron* PENTLAND of LYTH, P.C., G.C.S.I., G.C.I.E.
(*Governor, Madras.*) 1 Apl. 17.

First Commn.	Names.	Date of rank.	Date of appointment to corps.	Remarks.
	Lieutenant-Colonel.			
	Majors.			
1 Apl. 17	Smith, Walter Donald, V.D. (80)	1 Apl. 17	1 Apl. 17	
1 Apl. 17	Moresby, Charles (80)	1 Apl. 17	1 Apl. 17	
1 Apl. 17	*Lamb, William*	*1 Apl. 17*	*1 Apl. 17*	*(2nd-Lt. T. F.)*
1 Apl. 17	*Kirwan, Lionel Edward (72b)*	*1 Apl. 17*	*1 Apl. 17*	*I. A. R. O.*
1 Apl. 17	Ogden, Harold Greenwood (50, 54, 80)	1 Apl. 17	1 Apl. 17	*Lv., I., to 28 Feb. 19.*
	Captains.			
1 Apl. 17	Hooper, Frederick Edmed (80)	1 Apl. 17	1 Apl. 17	*Lv. Ind. 3 mos.*
1 Apl. 17	Wilson, Francis Henry (54, 80)	1 Apl. 17	1 Apl. 17	
1 Apl. 17	Necker, Harry Young (80)	1 Apl. 17	1 Apl. 17	
1 Apl. 17	*Smith, Wilfred Fraser*	*1 Apl. 17*	*1 Apl. 17*	
1 Apl. 17	*White, Gerald Sebastian (80)*	*1 Apl. 17*	*1 Apl. 17*	
1 Apl. 17	*Bowden, Frank Harold (80)*	*1 Apl. 17*	*1 Apl. 17*	*D. A. A. G., 9th (Sec'bad) Dn.*
1 Apl. 17	Madley, James Welby	1 Apl. 17	1 Apl. 17	
1 Apl. 17	Cartwright, Cecil Leonard (51, 80)	1 Apl. 17	1 Apl. 17	
1 Apl. 17	Coats, James Richardson (dl, 51, 54, 80)	1 Apl. 17	1 Apl. 17	
1 Apl. 17	Houston, Alexander Lindsay	1 Apl. 17	1 Apl. 17	
	Lieutenants.			
1 Apl. 17	*Johnston, William Hamilton Hall (54)*	*1 Apl. 17*	*1 Apl. 17*	
1 Apl. 17	*Barker, Howard William (80)*	*1 Apl. 17*	*1 Apl. 17*	
1 Apl 17	Molony, John Chartres (80)	1 Apl. 17	1 Apl. 17	
1 Apl. 17	*McGurk, Niall*	*1 Apl. 17*	*1 Apl. 17*	*I. A. R. O.*
1 Apl. 17	*Jacob, Edwin*	*1 Apl. 17*	*1 Apl. 17*	*I. A. R. O.*
1 Apl. 17	*Porrett, Cecil Robert (80)*	*1 Apl. 17*	*1 Apl. 17*	*I. A. R. O.*
1 Apl. 17	Warmington, Kenneth Payne (80)	1 Apl. 17	1 Apl. 17	
1 Apl. 17	*Adlard, Frank Addison,* M.C.	*1 Apl. 17*	*1 Apl. 17*	*I. A. R. O.*
1 Apl. 17	Wadsworth, Sidney (80)	1 Apl. 17	1 Oct. 18	
1 Apl. 17	Milne, James Leslie (80)	1 Apl. 17	1 Apl. 17	*Lv. ex Ind.*
	2nd-Lieutenants.			
1 Apl. 17	*Norfor, Frank Frederick*	*1 Apl. 17*	*1 Apl. 17*	
1 Apl. 17	Fyson, Phillip Furley (80)	1 Apl. 17	1 Apl. 17	
1 Apl. 17	*Hodges, Thomas Oliver*	*1 Apl. 17*	*1 Apl. 17*	
1 Apl. 17	Gill, James Archibald (80)	1 Apl. 17	1 Apl. 17	
1 Apl. 17	*Macfarlane, Jams Wallace*	*1 Apl. 17*	*1 Apl. 17*	
1 Apl. 17	MacBey, William David (80)	1 Apl. 17	1 Apl. 17	
1 Apl. 17	Dalgarno, Douglas James (80)	1 Apl. 17	1 Apl. 17	

1st Madras Guards—*contd.*

First Commn.	Names.	Date of rank.	Date of appointment to corps.	Remarks.
	2nd-Lieutenants—*contd.*			
1 Apl. 17	Bryant, John Forbes (80)	1 Apl. 17	1 Apl. 17	
1 Apl. 17	Steele, Allan Harmer (80)	1 Apl. 17	1 Apl. 17	*Lv. 1 to 31 Mar. 19.*
1 Apl. 17	Worke, Francis Edmund Lesslie (80)	1 Apl. 17	1 Apl. 17	
1 Apl. 17	Walker, William Austin (80)	1 Apl. 17	1 Apl. 17	*Lv., ex i., 1 yr.*
1 Apl. 17	*Leonard, Patrick Joseph*	*1 Apl. 17*	*1 Apl. 17*	
1 Apl. 17	Littlehailes, Richard (80)	1 Apl. 17	1 Apl. 17	
1 Apl. 17	Ross, Graham Fraser (80)	1 Apl. 17	1 Apl. 17	
1 Apl. 17	Rae, Hugh Paddon Meredydd (80)	1 Apl. 17	1 Apl. 17	
1 Sept. 17	Newman-Sanders, Cecil	1 Sep. 17	1 Sep. 17	
1 Apl. 18	Dodwell, Henry Herbert	1 Apl. 18	1 Apl. 18	
21 May 18	Freeman, Max Lee	21 May 18	21 May 18	
21 May 18	Paton, Eric Waldemar	21 May 18	21 May 18	
13 June 18	Sell, Wilfred Hazell	13 June 18	13 June 18	
1 Aug. 18	Kershaw, Henry Owen	1 Aug. 18	1 Aug. 18	
	Supernumerary.			
1 Apl. 17	Lacey, Lt.-Col. Joseph Melville V.D.	1 Apl. 17	1 Apl. 17	
1 Apl. 17	Adam, Capt. John Collyer	1 Apl. 17	1 Apl. 17	
1 Apl. 17	Henderson, Capt. Charles Alexander	1 Apl. 17	1 Apl. 17	
	Staff.			
10 Oct. 14	Arnold, W. W. Lt. (Tempy. Capt.) North'd Fus	29 May 15	3 Nov. 17	Adjt.
9 Mar. 18	Tomlinson, Lt. W. G.	9 Mar. 18	9 Mar. 18	Qr. Mr.

Uniform—Khaki drill.

2nd Nagpur Rifles.

Formed 29th Dec. 1860. Berar Volunteer Rifles amalgamated, 1st October 1904, G. G. O. 77 of 1905. Reconstituted as an I. D. F. Unit with present designation, 1st Apl. 1917. Companies at Hd.-Qrs., Amraoti and Jubbulpore.

Establishment—1 Res. Coy., 3 Combined Active & Reserve Coys., 2 Armoured Motor Battys. (B), Machine Gun Dett.

HONORARY COMMANDANT—*The Hon'ble Sir* BENJAMIN ROBERTSON, K.C.S.I., K.C.M.G., C.I.E., *Chief Commissioner, Central Provinces.* 1 Apl. 17.

First Commn.	Names.	Date of rank.	Date of appointment to corps.	Remarks.
	Lieutenant-Colonel.			
	Majors.			
1 Apl. 17	Gowan, Hyde Clarendon, V.D. (51)	1 Apl. 17	1 Apl. 17	Comdg.
1 Apl. 17	Green, Thomas George, M.B.E., V.D.	1 Apl. 17	1 Apl. 17	Nagpur.
1 Apl. 17	Danks, Eric	1 Apl. 17	1 Apl. 17	Amraoti.
1 Apl. 17	*Hallifax, Henry Francis,* V.D.	*1 Apl. 17*	1 *Apl. 17*	(*Temp. Lt. in Army, 23 Oct. 16.*)
	Captains.			
1 Apl. 17	*Phillips, George Alexander,* V.D.	*1 Apl. 17*	*1 Apl. 17*	*I. A. R. O.*
1 Apl. 17	Mitchell, David George (48*a*, 50, 80)	1 Apl. 17	1 Apl. 17	Jubbulpore.
1 Apl. 17	Allan, Robert George (80)	1 Apl. 17	1 Apl. 17	Nagpur.
1 Apl. 17	*Macnee, Eustace Alberic*	*1 Apl. 17*	*1 Apl. 17*	*I. A. R. O.*
1 Apl. 17	Dyer, James Ferguson (51, 80)	1 Apl. 17	1 Apl. 17	Nagpur.
1 Apl. 17	*Irwin, Cyril James (51, 52)*	*1 Apl. 17*	*1 Apl. 17*	*I. A. R. O.*
1 Apl. 17	Davies, Powys (48*a*, 50, 80)	1 Apl. 17	1 Apl. 17	Nagpur.
1 Apl. 17	*Plymen, Francis Joseph*	*1 Apl. 17*	*1 Apl. 17*	Nagpur.
	Lieutenants.			
1 Apl. 17	*Dewar, Francis*	*1 Apl. 17*	*1 Apl. 17*	*I. A. R. O.*
1 Apl. 17	Rudman, Francis Reginald Rider	1 Apl. 17	1 Apl. 17	Delhi.
1 Apl. 17	*Travers, Otho Algernon* (66*a*, 71, 105)	*1 Apl. 17*	*1 Apl. 17*	*I. A. R. O.*
1 Apl. 17	*Crosthwaite, Henry Robert,* C.I.E.	*1 Apl. 17*	*1 Apl. 17*	*I. A. R. O.*
1 Apl. 17	*Tostevin, Fred. Percy* (*80*)	*1 Apl. 17*	*1 Apl. 17*	*I. A. R. O.*
1 Apl. 17	Macnair, Robert Hill (80)	1 Apl. 17	1 Apl. 17	Nagpur.
1 Apl. 17	*Sams, Hurbert Arthur*	*1 Apl. 17*	*1 Apl. 17*	
1 Apl. 17	Cheshire, Francis Morton (80)	1 Apl. 17	1 Apl. 17	Nagpur.
	2nd-Lieutenants.			
1 Apl. 17	Marshall, Norman Edwyn (50, 80)	1 Apl. 17	1 Apl. 17	Nagpur.
1 Apl. 17	*Kelly, William James*	*1 Apl. 17*	*1 Apl. 17*	*I. A. R. O.*
1 Apl. 17	Durie, George Arthur (51, 80)	1 Apl. 17	1 Apl. 17	Jubbulpore. *Lv. Ind. to 31 Aug. 20.*
1 Apl. 17	*Evans, Geoffrey* (*80*)	*1 Apl. 17*	*1 Apl. 17*	*I. A. R. O.*
1 Apl. 17	*Rowlands, William Shaw* (*80*)	*1 Apl. 17*	*1 Apl. 17*	*I. A. R. O.*
1 Apl. 17	*Stow, Vincent Aubrey Stewart* (*80*)	*1 Apl. 17*	*1 Apl. 17*	*I. A. R. O.*
23 Mar. 18	Best, Hon'ble James William (51)	23 Mar. 18	23 Mar. 18	Hoshangabad.
23 Mar. 18	*Graham, Robert James Douglas* (*51*)	*23 Mar. 18*	*23 Mar. 18*	*I. A. R. O.*
23 Mar. 18	*Jackson, Robert John*	*23 Mar. 18*	*23 Mar. 18*	*I. A. R. O.*
23 Mar. 18	Waterfall Charles Francis	23 Mar. 18	23 Mar. 18	Akola.
23 Mar. 18	Seaman, Charles Kenningale	23 Mar. 18	23 Mar. 18	Kamptee.
23 Mar. 18	Kersey James	23 Mar. 18	23 Mar. 18	Kamptee.
23 Mar. 18	*Bailey Ronald Joseph*	*23 Mar. 18*	*28 Mar. 18*	*I. A. R. O.*
26 June 18	Davison, Arthur (80)	26 June 18	26 June 18	Jubbulpore.
12 Sep. 18	Bose, Lal Mohun Vivion	12 Sep. 18	12 Sep. 18	
	Staff.			
15 Sep. 17	Scott, 2nd-Lt. (*temp. Capt.*) W. A., Cam. Highdrs.	15 Sep. 17	15 Sep. 17	Adjt.
6 Mar. 18	Taylor, *Hony.* Lt. F.	6 Mar. 18	6 Mar. 18	Qr. Mr.
1 Apl. 18	Price, Ven'ble C. E., M.A.	1 Apl. 18	1 Apl. 18	Hony. Chaplain.

Uniform—Khaki drill.

3rd Punjab Rifles.

Formed by Punjab G. O. No. 664, dated 1st April 1861. Reorganised as the Administrative Battalion, 1st Punjab Volunteer Rifle Corps by G. G. O. No. 459, dated 12th May 1893, and reconstituted as an I. D. F. Unit with present designation, 1st Apl. 1917.

Establishment—4 Combined Active & Reserve Coys., 3 Armoured Motor Battys. (B), 1 Machine Gun Dett.

Platoons at Rawal Pindi, Murree, Sialkot, Delhi, Ferozepore, Peshawar, Amritsar, Srinagar and Ambala.

HONORARY COMMANDANT—*The Hon'ble Sir* MICHAEL FRANCIS O'DWYER, G.C.I.E., K.C.S.I., (*Lieutenant-Governor of the Punjab.*) 1 Apl. 17.

First Commn.	Names.	Date of rank.	Date of appointment to corps.	Remarks.
	Lieutenant-Colonel.			
1 Apl. 17	Wright, William Thomas, V.D., A.D.C. (54, 80).	1 Apl. 17	1 Apl. 17	Hon. Col., late Ind. Volrs. Comdt. Hony. A.D.C. to H. E. the Viceroy.
	Majors.			
1 Apl. 17	Hailey, *Hon.* William Malcolm, C.S.I., C.I.E. (54, 82).	1 Apl. 17	1 Apl. 17	(Hony. Lt.-Col., late Ind. Volrs.) Delhi.
1 Apl. 17	*Irving, Miles* (D. I.)	*1 Apl. 17*	*1 Apl. 17*	*I. A. R. O.*
1 Apl. 17	*Herdon Sydney Philip Norman*	*1 Apl. 17*	*1 Apl. 17*	*I. A. R. O.*
1 Apl. 17	*Longdin, Walter Gardner* (52)	*1 Apl. 17*	*1 Apl. 17*	*I. A. R. O.*
1 Apl. 17	*Egan, John Howard* (54)	*1 Apl. 17*	*1 Apl. 17*	
1 Apl. 17	*Boyd, Donald James* (80)	*1 Apl. 17*	*1 Apl. 17*	Lahore.
	Captains.			
1 Apl. 17	Ashford, John, O.B.E. (80)	1 Apl. 17	1 Apl. 17	Amritsar.
1 Apl. 17	*Woolner, Alfred Cooper* (54, 80)	*1 Apl. 17*	*1 Apl. 17*	Lahore.
1 Apl. 17	Harrison, Allan George, V.D.	1 Apl. 17	1 Apl. 17	(*Hony. Maj., late Ind. Volrs.*) Srinagar.
1 Apl. 17	Ferrier William (80)	1 Apl. 17	1 Apl. 17	
1 Apl. 17	*Watkins, Lawrence Theodore* (80)	*1 Apl. 17*	*1 Apl. 17*	*I. A. R. O.*
1 Apl. 17	Knowlton, Herbert Thomas (80)	1 Apl. 17	1 Apl. 17	Ambala.
1 Apl. 17	Heath, Hugh Lionel (54, 80)	1 Apl. 17	1 Apl. 17	Lahore.
1 Apl. 17	Carson, Charles William Charteris (54, 82)	1 Apl. 17	1 Apl. 17	Delhi.
1 Apl. 17	Harcourt, Henry (80)	1 Apl. 17	1 Apl. 17	Rohtak.
1 Apl. 17	Eyre Walker, William (51, 80)	1 Apl. 17	1 Apl. 17	Murree.
12 Mar. 18	Deane, Lewis Seymour (50, 72b, 80)	12 Mar. 18	12 Mar. 18	Lahore.
	Lieutenants.			
1 Apl. 17	*Ryan, Edward John.*	*1 Apl. 17*	*1 Apl. 17*	
1 Apl. 17	*Parker, Aubrey Hastings,* M.C.	*1 Apl. 17*	*1 Apl. 17*	*Temp. Maj. in Army.*
1 Apl. 17	*Hargreaves, Harold,* V.D. (54, 206)	*1 Apl. 11*	*1 Apl. 17*	*I. A. R. O.*
1 Apl. 17	*Swainson, Charles Anthony*	*1 Apl. 17*	*1 Apl. 17*	*I. A. R. O.*
1 Apl. 17	Smith, Ernest (51, 80)	1 Apl. 17	1 Apl. 17	Amritsar.
1 Apl. 17	*Williams, Albert*	*1 Apl. 17*	*1 Apl. 17*	*I. A. R. O.*

3rd Punjab Rifles—*concld.*

First Commn.	Names.	Date of rank.	Date of appointment to Corps.	Remarks.
	Lieutenants—*contd.*			
1 Apl. 17	*Garrett, Herbert Leonard Offley*	*1 Apl. 17*	*1 Apl. 17*	*I. A. R. O.*
1 Apl. 17	Brownlie, Thomas Archibald Millet (51, 80)	1 Apl. 17	1 Apl. 17	Lyallpur.
1 Apl. 17	Haward, Edwin (80)	1 Apl. 17	1 Apl. 17	Lahore.
1 Apl. 17	McNamara, Maurice Lionel (80)	1 Apl. 17	1 Apl. 17	Srinagar.
1 Apl. 17	Murray, William Henry (80)	1 Apl. 17	1 Apl. 17	Rawalpindi.
	2nd-Lieutenants.			
1 Apl. 17	Armstrong, John William, O.B.E.	1 Apl. 17	1 Apl. 17	Dhariwal.
1 Apl. 17	Burkitt, Francis Holy, O.B.E.	1 Apl. 17	1 Apl. 17	Peshawar.
1 Apl. 17	*Blacker, Harold Alfred Cecil*	1 Apl. 17	1 Apl. 17	*I. A. R. O.*
1 Apl. 17	*Farrell, Joseph (80)*	*1 Apl. 17*	*1 Apl. 17*	*I. A. R. O.*
1 Apl. 17	Eyre-Walker, Cecil (80)	1 Apl. 17	1 Apl. 17	Murree.
1 Apl. 17	Cockburn, Arthur Balcombe (80)	1 Apl. 17	1 Apl. 17	Lahore.
1 Apl. 17	Baylis, Roderick Llewellyn (80)	1 Apl. 17	1 Apl. 17	Delhi.
1 Apl. 17	Lincoln, Edward Henning (80)	1 Apl. 17	1 Apl. 17	Sialkot.
1 Apl. 17	Willison McQuillen, Douglas (51, 80)	1 Apl. 17	1 Apl. 17	Ferozepore.
1 Apl. 17	Anderson, Hedley Bertram (80)	1 Apl. 17	1 Apl. 17	Ferozepore.
12 Mar. 18	Rennie Francis Pepys	12 Mar. 18	12 Mar. 18	Peshawar.
1 Apl. 17	Whyte, Harold Adam	1 Apl. 17		
9 July 18	Grant Govan, Raymond Eustace	9 July 18	9 July 18	Delhi.
	Supernumerary.			
1 Apl. 17	Salkield, Maj. Tom	1 Apl. 17	1 Apl. 17	
1 Apl. 17	Stranack, Capt. Frederick William	1 Apl. 17	1 Apl. 17	Rawalpindi.
1 Apl. 17	Whitehouse, Capt. Richard Henry (54) (80)	1 Apl. 17	1 Apl. 17	Lahore.
1 Apl. 17	Ross, Lt. John, Archibald (54)	1 Apl. 17	1 Apl. 17	D. I. Khan.
1 Apl. 17	Howard, 2nd-Lt. Archibald	1 Apl. 17	1 Apl. 17	Peshawar.
	Staff.			
7 Nov. 15	Sumption, Lt. (*temp. Capt.*) F., Oxf. & Bucks L. I.	1 Jan. 15	7 Nov. 17	Adjutant.
8 Mar. 18	Bell, Lt. J.	8 Mar. 18	8 Mar. 18	Qr.-Mr.

Full dress—Khaki serge with green facings.
Working dress—Khaki drill.

4th Simla Rifles.

Formed by Punjab G. O.'s Nos. 101 and 171 of 2nd May and 21st October 1861. Designation changed. G. C. O. 367 of 1904 and reconstituted as an I. D. F. Unit with present designation, 1 Apl. 1917. Establishment—1 Active Coy., 1 Reserve Coy., 1 Combined Active & Reserve Coy., 2 Machine Gun Detts. Detachment at Sanawar.

HONORARY COMMANDANT—*H. E. the Rt. Hon.* FREDERIC JOHN NAPIER, *Baron* CHELMSFORD, P.C., G.M.S.I., G.C.M.G., G.M.I.E., G. B. E., 1 Apl. 17.
(Viceroy and Governor-General of India.)

First Commn.	Names.	Date of rank.	Date of appointment to corps.	Remarks.
	Lieutenant-Colonels.			
1 Apl. 17	*Sutherland, William*, V.D.	*1 Apl. 17*	*1 Apl. 17*	*(temp. Lt. in Army.)*
1 Apl. 17	*Dunn, Frederick Eustace*, V.D., (54)	*1 Apl. 17*	*1 Apl. 17*	*I. A. R. O.*
1 Apl. 17	Holland, Robert Erskine, S.I.E. (54) (80)	13 Apl. 18	1 Apl. 17	Comdt.
	Majors.			
1 Apl. 17	*Pepper, Henry*, V.D. (51, 52, 207)	*13 Apl. 18*	*1 Apl. 17*	
1 Apl. 17	*Stocks, Gerald Sumner* (D. I.)	*13 Apl. 18*	*1 Apl. 17*	*(Temp. 2nd-Lt. in Army, 25 Jan. 16.)*
1 Apl. 17	*Sanderson, Robert* (80)	*13 Apl. 18*	*1 Apl. 17*	
1 Apl. 17	Troup, Robert Scott (54) (80)	13 Apl. 18	1 Apl. 17	2nd-in-command.
1 Apl. 17	Marshall, George William, I.S.O., V.D., (D. I.) (54) (80)	13 Apl. 18	1 Apl. 17	
	Captains.			
1 Apl. 17	Sleigh, Alexander Irving (D.I.) (12, 48a, 51, 52, 54) (80)	1 Apl. 17	1 Apl. 17	
1 Apl. 17	Meikle, Henry George Watson (80)	1 Apl. 17	1 Apl. 17	
1 Apl. 17	Gaskell, Capt. William (80)	13 Apl. 18	1 Apl. 17	
1 Apl. 17	*Cooper, Philip Alan* (D. I.)	*1 Apl. 17*	*1 Apl. 17*	*I. A. R. O.*
1 Apl. 17	*Cox, Harold Ranyard*	*13 Apl. 18*	*1 Apl. 17*	*I. A. R. O.*
1 Apl. 17	*Duffy, Thomas Augustine* (54, (d.I)	*13 Apl. 18*	*1 Apl. 17*	*(temp. Capt. in Army).*
1 Apl. 17	*Tharle-Hughes, Reginald*, M.B.E., (54)	*13 Apl. 18*	*1 Apl. 17*	*I. A. R. O.*
1 Apl. 17	Bean, Basil Clarence Lahey (80)	13 Apl. 18	1 Apl. 17	
1 Apl. 17	*Bayley, Victor*, C.I.E.	*13 Apl. 18*	*1 Apl. 17*	*I. A. R. O.*
1 Apl. 17	Clarke, Geoffrey Rothe (80)	13 Apl. 18	24 Mar. 18	
	Lieutenants.			
1 Apl. 17	*Normand, Charles William Blyth*	*1 Apl. 17*	*1 Apl. 17*	*I. A. R. O.*
1 Apl. 17	Ford, Henry James (80)	1 Apl. 17	1 Apl. 17	
1 Apl. 17	*Firth, John Rupert* (52)	*1 Apl. 17*	*1 Apl. 17*	*(Temp. 2nd-Lt. in Army, 3 May 16.)*
1 Apl. 17	Prince, Thomas George (51) (80)	1 Apl. 17	1 Apl. 17	Sanawar.
1 Apl. 17	Harris, Douglas Gordon (80)	1 Apl. 17	1 Apl. 17	
1 Apl. 17	*Young, Patrick Charles*	*13 Apl. 18*	*1 Apl. 17*	*Actg. Lt.-Col. in Army.*
1 Apl. 17	Beynon, Frank Cecil (51) (80)	13 Apl. 18	1 Apl. 17	Sanawar.
1 Apl. 17	Grant, Thomas Eustace (80)	13 Apl. 18	1 Apl. 17	
1 Apl. 17	*Gordon-Walker, Alan Lachlan*	*13 Apl. 18*	*1 Apl. 17*	*Temp. Capt. in Army.*
1 Apl. 17	Meares, John Willoughby (80)	13 Apl. 18	1 Apl. 17	
	2nd-Lieutenants.			
11 Feb. 18	Woolmer, Frederick John	11 Feb. 18	11 Feb. 18	
11 Feb. 18	MacBean, Horatio Alistair	11 Feb. 18	11 Feb. 18	
	Supernumerary.			
1 Apl. 17	Henry, Lt.-Col. Sir William Daniel *Kt.* C.I.E., V.D., (54)	1 Apl. 17	1 Apl. 17	Hon. Col. late Ind. Volrs.
1 Apl. 17	DuBoulay, Lt.-Col. Sir James Houssemayne, K.C.I.E., C.S.I. (D. I) (54)	1 Apl. 17	1 Apl. 17	*Supy. List, 13 Apl. 18.*
1 Apl. 17	Barne, Major George Dunsford (80)	1 Apl. 17	1 Apl. 17	
1 Apl. 17	Fairley, James William	1 Apl. 17	1 Apl. 17	
1 Apl. 17	Warne, Lt. Francis Edward (54)	1 Apl. 17	5 June 18	
1 Apl. 17	Gould, Basil John	1 Apl. 17	1 Apl. 17	Persia 1 yr. leave, 6 June 18.
	Staff.			
24 Oct. 14	Noad, Lieut. (*temp. Capt.*) C. H. C., 1-9 Midd'x. R. (51. 52).	1 June 16	12 June 17	Adjutant.
4 Mar. 18	Ingram, Lt. A. G.	4 Mar. 18	4 Mar. 18	Qr.-Mr.
1 Apl. 17	Lewis, Revd. Hugh Mitchell, M.A.	1 Apl. 17	1 Apl. 17	Hon. Chaplain.

Uniform—Khaki drill.

5th Calcutta Battalion.

Formed by Military Department Notfn. No. 97 of 1863. Reconstituted, G.G.O. 437 of 1898 and reconstituted as an I. D. F. Unit with present Designation, 1st Apl. 1917.

Establishment—3 Active Coys. 1 Res. Coy. 1 Armoured Motor By. (*B*).

HONORARY COMMANDANT—*H. E. the Rt. Hon.* FREDERIC JOHN NAPIER, *Baron* CHELMSFORD, P.C., G.M.S.I., G.C.M.G., G.M.I.E., G.B.E. 1 Apl. 17.

(*Viceroy and Governor-General of India.*)

First Commn.	Names.	Date of rank.	Date of appointment to corps.	Remarks.
	Lieutenant-Colonel.			
1 Apl. 17	Grice, Walter Thomas, C.I.E., V.D., (D I) (54)	1 Apl. 17	1 Apl. 17	Comdt. (Hon. Col., late Ind. Volrs.) Hony. A.-D.-C. to C.-in-C.
	Majors.			
1 Apl. 17	Donald, James, C.I.E. (DI) (80)	1 Apl. 17	1 Apl. 17	
11 Apl. 17	Slater, Arthur William (54, 80)	1 Apl. 17	1 Apl. 17	
	Captains.			
1 Apl. 17	*Potter, Lawrence Robert* (54)	*1 Apl. 17*	*1 Apl. 17*	*I. A. R. O.*
1 Apl. 17	*Radice, William Archibald* (54, 80)	*1 Apl. 17*	*1 Apl. 17*	*Adjutant.*
1 Apl. 17	*Burroughs, Geoffrey Herbert Gore*	*1 Apl. 17*	*1 Apl. 17*	*I. A. R. O.*
1 Apl. 17	*Thurston, John Woodhouse* (D. I.) (54)	*1 Apl. 17*	*1 Apl. 17*	(Temp. Capt. in Army, 11 Nov. 16.)
1 Apl. 17	White, Bernard Alfred	1 Apl. 17	1 Apl. 17	
1 Apl. 17	Knox-Ord, Douglas (54, 80, 82)	1 Apl. 17	1 Apl. 17	
1 Apl. 17	*Campbell, Thomas Patrick*	*1 Apl. 17*	*1 Apl. 17*	*I. A. R. O.*
1 Apl. 17	Lloyd Alan Hubert (80, 104, 207)	1 Apl. 17	1 Apl. 17	
1 Apl. 17	*Marshall, James Douglas* (D. I.) (80)	*1 Apl. 17*	*1 Apl. 17*	*Temp. Capt. in Army.*
1 Apl. 17	Papworth, Walter Francis (DI) (80)	1 Apl. 17	1 Apl 17	
1 Apl. 17	*Biss, Evan Ebenezer* (54) (80)	*1 Apl. 17*	*1 Apl. 17*	*I. A. R. O.*
1 Apl. 17	Robins, Francis Joseph (54) (80)	1 Apl. 17	1 Apl. 17	
	Lieutenants.			
1 Apl. 17	Lomax, Charles Edward (54, 80)	1 Apl. 17	1 Apl. 17	
1 Apl. 17	*Smith, Basil St. John*	*1 Apl. 17*	*1 Apl. 17*	*I. A. R. O.*
1 Apl. 17	Gibson, Thomas Bainbridge (80)	1 Apl. 17	1 Apl. 17	
1 Apl. 17	Harper, Kenneth Brand (50, 80)	1 Apl. 17	1 Apl. 17	Lv., *ex* Ind., 6 mths.
1 Apl. 17	Eatwell, George Hunter Frankling (54, 80)	1 Apl. 17	1 Apl. 17	
	2nd-Lieutenants.			
1 Apl. 17	Andrews, William Edgar (51, 80)	1 Apl. 17	1 Apl. 17	
1 Apl. 17	Roberts, William Edward (80)	1 Apl. 17	1 Apl. 17	
1 Apl. 17	Abbott, Henry John Eastwood (80)	1 Apl. 17	1 Apl. 17	
2 Mar. 18	Chesterton, Percy Herbert (50, 80)	2 Mar. 18	2 Mar. 18	
2 Mar. 18	Murrell, Frederick Edmund (80)	2 Mar. 18	2 Mar. 18	
2 Mar. 18	Clarke, Francis Grenville (80)	2 Mar. 18	2 Mar. 18	
2 Mar. 18	Watson-Smyth, Robert Middleton (80)	2 Mar. 18	2 Mar. 18	
2 Mar. 18	Damiano, Nicholas Aristides (80)	2 Mar. 18	2 Mar. 18	
30 Sep. 18	Dixon. Edgar Geoffrey	30 Sep. 18	30 Sep. 18	

Attached.

....	Williams, Maj. G. B., R. E. (T. F.)	5 July 11	....	

Staff.

1 Apl. 17	Radice, Capt. W. A. (54, 80)	1 Apl. 17	1 Apl. 17	Adjutant.
14 Mar. 18	Mayes, Lt., A. T. (207)	14 Mar. 18	14 Mar. 18	Qr. Mr.

Uniform—Full dress—Red serge, *Facings*—Green.

6th Bangalore, Coorg and Mysore Battalion.

Raised as Bangalore Rifle Volunteers 21 Dec. 1868 (Mily. Dept. letter No. 998, dated 22 Mar. 1869; M. D. Notfn. No. 533, dated 10 May 1870), Coorg and Mysore Rifles amalgamated and reconstituted as an I. D. F. Unit with present designation 1st Apl. 1917.
Transferred to Coorg and Mysore. M. D. Notfn. No. 520, dated 15 May 1807
Establishment—1 Active Coy. 2 Combined Active & Reserve Coys. 1 Armoured Motor Batty. (B). Company at Mysore.

HONORARY COMMANDANT—*The Hon'ble* Mr. HENRY VENN COBB, C.S.I., C.I.E.
Resident in Mysore and Chief Commissioner, Coorg.

First Commn.	Names.	Date of rank.	Date of appointment to corps.	Remarks.
	Lieutenant-Colonels.			
1 Apl. 17	Barton, Alfred Percy, V. D.	1 Apl. 17	1 Apl. 17	Comdt.
	Majors.			
1 Apl. 17	*Scovell, George Vance Fielding, V.D. (80)*	*1 Apl. 17*	*1 Apl. 17*	*I. A. R. O.*
1 Apl. 17	Andrews, Alfred Stephenson (80)	1 Apl. 17	1 Apl. 17	2nd-in-Comd.
1 Apl. 17	*Oakley, Charles, Benjamin*	*1 Apl. 17*	*1 Apl. 17*	*I. A. R. O.*
1 Apl. 17	Cox, Talbot, V.D.	1 Apl. 17	1 Apl. 17	
	Captains.			
1 Apl. 17	*Drury, John (80)*	*1 Apl. 17*	*1 Apl. 17*	*I. A. R. O.*
1 Apl. 17	Campbell, James Robert (80)	1 Apl. 17	1 Apl. 17	
1 Apl. 17	Hannyngton, Frank	1 Apl. 17	1 Apl. 17	
1 Apl. 17	Mills, Edward Bridgeman (80)	1 Apl. 17	1 Apl. 17	
1 Apl. 17	*Collyer, Hugh Nataniel*	*1 Apl. 17*	*1 Apl. 17*	
1 Apl. 17	Anderson, Patrick James	1 Apl. 17	1 Apl. 17	
1 Apl. 17	Tacchella, Carl Frederick Hollinshead (80)	1 Apl. 17	1 Apl. 17	
	Lieutenants.			
1 Apl. 17	Wood, Herbert Cecil	1 Apl. 17	1 Apl. 17	
1 Apl. 17	Kann, John Hugo Conway	1 Apl. 17	1 Apl. 17	
1 Apl. 17	McAlpine, Edward George (80)	1 Apl. 17	1 Apl. 17	
1 Apl. 17	Paddock, William George (80)	1 Apl. 17	1 Apl. 17	
1 Apl. 17	*Stafford, Horace Christmas Cooke*	*1 Apl. 17*	*1 Apl. 17*	*I. A. R. O.*
1 Apl. 17	*Wailling, William*	*1 Apl. 17*	*1 Apl. 17*	*I. A. R. O.*
1 Apl. 17	Lake, Clement	1 Apl. 17	1 Apl. 17	
	2nd-Lieutenants.			
1 Apl. 17	Steele, Ronald Leslie (51)	1 Apl. 17	1 Apl. 17	
1 Apl. 17	*Benson, James Wilson Herbert*	*1 Apl. 17*	*1 Apl. 17*	*I. A. R. O.*
1 Apl. 17	*Kelly, Charance John*	*1 Apl. 17*	*1 Apl. 17*	*I. A. R. O.*
1 Apl. 17	*Pollett, John Charles St. Clair*	*1 Apl. 17*	*1 Apl. 17*	
1 Apl. 17	Scott, George Ernest Blaine	1 Apl. 17	1 Apl. 17	
1 Apl. 17	Watson, Herbert Edmeston	1 Apl. 17	1 Apl. 17	
1 Apl. 17	Rice, Harold Douglas	1 Apl. 17	1 Apl. 17	
1 Apl. 17	Tireman, Henry	1 Apl. 17	1 Apl. 17	
1 Apl. 17	*Stonor, Wilfred Thomas*	*1 Apl. 17*	*1 Apl. 17*	*I. A. R. O.*
1 Apl. 17	Oliver, Leonard Camrou	1 Apl. 17	1 Apl. 17	
	Supernumerary.			
1 Apl. 17	Cox, Major Alexander Ranken (80)	1 Apl. 17	1 Apl. 17	
1 Apl. 17	Stevenage, Capt. John James (80)	1 Apl. 17	1 Apl. 17	
1 Apl. 17	Fawcitt, Capt. Claude Scott	1 Apl. 17	1 Apl. 17	
1 Apl. 17	Foster, Lt. Reginald Guy	1 Apl. 17	1 Apl. 17	
1 Apl. 17	Hayward, 2nd-Lt. Cecil John	1 Apl. 17	1 Apl. 17	
7 Feb. 18	Vipan, 2nd-Lt. Alfred	7 Feb. 18	1 Apl. 18	
1 Apl. 17	Rollo, 2nd-Lt. John Carter	1 Apl. 17	1 July 18	
1 Apl. 18	Valentine, Capt. Charles Robret	1 Apl. 18	1 Apl. 18	
1 Apl. 18	Moore, Lieut. George	1 Apl. 18	1 Apl. 18	
	Staff.			
20 Nov. 17	Harrop, 2nd-Lt. (*Actg. Capt.*) E., L'pool. R.	20 Nov. 17	20 Nov. 17	Adjt.

Uniform—Khaki drill.

7th East Indian Railway Battalion.

(Formed by Mily. Dept. Notfn. No. 759, dated 17th July 1869.)

(Seebpore College Volunteer Rifle Corps amalgamated by Mily. Dept. Notfn. No. 505, dated 20th May 1890.) St. Michael's School Cadet Corps amalgamated and reconstituted as an I. D. F. Unit with present designation 1st Apl. 1917.

Establishment.

1st Bn.:—2 Reserve Coys., 2 Combined Active and Reserve Coys., 1 Armoured Motor Batty. (B).

2nd Bn.:—3 Reserve Coys., 1 Armoured Motor Batty. (B), 1 Armoured Train Dett.

3rd Bn.:—3 Reserve Coys., 1 Combined Active and Reserve Coy.

Companies at Sahebgunge, Howrah, Lillooah, Seebpore, Raneegunge, Burdwan, Sitarampore, Asansol, Dhanbaid, Madhupur, Dinapore, Bankipur, Gya, Jubbulpore, Allahabad, Cawnpore, Tundla, Ghaziabad, Delhi and Ambala.

"Strong without rage."

HONORARY COMMANDANT—*H. E. the Rt. Hon.* FREDERIC JOHN NAPIER, *Baron* CHELMSFORD, P.C., G.M.S.I., G.C.M.G., G.M.I.E., G.B.E., 1 APL. 17.

(Viceroy and Governor-General of India.)

First Commn.	Names.	Date of rank.	Date of appointment to corps.	Remarks.

RECOMMENDATION FOR I. D. F. COMMISSIONS NOT SUBMITTED.

Uniform—Khaki drill.

8th Allahabad Rifles.

Formed by Military Department Notfn. dated 11th Jan. 1871, and reconstituted as an I. D. F. Unit with present designation, 1st Apl. 1917.

Establishment—1 Active Coy., 1 Combined Active and Reserve Coy., 1 Armoured Motor Batty.

Company at Chunar and Mirzapur.

HONORARY COMMANDANT—*The Hon'ble Sir* HENRY GEORGE RICHARDS, *Kt.*, K.C., J.P., 1 Apl. 17.

First Commn.	Names.	Date of rank.	Date of appointment to corps.	Remarks.
	Lieutenant-Colonels.			
19 Sep. 18	Tudball, William, the Hon'ble	19 Sep. 18	19 Sep. 18	Comdt.
	Majors.			
1 Apl. 17	Abel, Walter Charles	1 Apl. 17	1 Apl. 17	2nd-in-Comd.
1 Apl. 17	*Mabbett, Percy Whereat . . . . .*	*1 Apl. 17*	*1 Apl. 17*	*I. A. R. O.*
1 Apl. 17	*Robertson, Frederick Arthur deVere (DI)*	*1 Apl. 17*	*1 Apl. 17*	*I. A. R. O.*
1 Apl. 17	*Crump, Ethelbert William Ernest (51) .*	*1 Apl. 17*	*1 Apl. 17*	*I. A. R. O.*
	Captains.			
1 Apl. 17	Bean, Salvador Michael Lahey (80) . .	1 Apl. 17	1 Apl. 17	
1 Apl. 17	*Key, Charles Taylor . . . .*	*1 Apl. 17*	*1 Apl. 17*	*I. A. R. O.*
1 Apl. 17	*Smyth, Gilbert Henry . . . .*	*1 Apl. 17*	*1 Apl. 17*	*I. A. R. O.*
1 Apl. 17	Watson, John Lawrence (80)	1 Apl. 17	1 Apl. 17	
	Lieutenants.			
1 Apl. 17	*Learmond, Alexander William . . . .*	*1 Apl. 17*	*1 Apl. 17*	*I. A. R. O.*
1 Apl. 17	*Cliff, Wilfred Pearse . . . .*	*1 Apl. 17*	*1 Apl. 17*	*I. A. R. O.*
1 Apl. 17	*Morris, William Gerald . . . . . .*	*1 Apl. 17*	*1 Apl. 17*	*I. A. R. O.*
1 Apl. 17	Crighton, David Wemyss (80)	1 Apl. 17	1 Apl. 17	
1 Apl. 17	Snow, Frederick William (80)	1 Apl. 17	1 Apl. 17	
1 Apl. 17	Weir, Ralph Somerville (80)	1 Apl. 17	1 Apl. 17	
	2nd-Lieutenants.			
1 Apl. 17	Harper, Claude Oswell (80)	1 Apl. 17	1 Apl. 17	
1 Apl. 17	Tinker, Harold (48a, 51, 80)	1 Apl. 17	1 Apl. 17	
1 Apl. 17	Stedman, Ralph Gager (51, 80)	1 Apl. 17	1 Apl. 17	
	Supernumerary.			
1 Apl. 17	MacKenzie, Capt. Arthur Henderson (80)	1 Apl. 17	1 Apl. 17	
1 Apl. 17	Clarke, Capt. Albert Ernest (80) . . .	1 Apl. 17	23 May 18	
1 Apl. 17	*Niblett, Lt. Robert Howard (80) . . .*	*1 Apl. 17*	*1 Apl. 17*	*I. A. R. O.*
1 Apl. 17	Howie, 2nd-Lt. James Reginald Stanhope (80)	1 Apl. 17	1 Apl. 17	
1 Apl. 17	*Nicoll, 2nd-Lt. William Atkins (51, 80) .*	*1 Apl. 17*	*1 Apl. 17*	*I. A. R. O.*
	Staff.			
18 May 18	Twigden, 2nd-Lt. (*a'ctg. Capt.*) C. R. R. Fus.	18 May 17	18 May 17	Adjt.

Uniform—**Khaki**

9th Mussoorie Battalion.

Formed by Mily. Dept. Notfn. No. 634, dated 27th July 1871. Thomason Coll. Vol. Rifles and Mussoorie Vol. Reserves incorporated by Mily. Dept. No. 867 B., dated 4th March 1901. Reconstituted as an I. D. F. Unit with present designation, 1st Apl. 1917. Establishment—2 Combined Active & Reserve Coys., 1 Armoured Motor Batty. B., 1 Machine Gun Dett. 1 Company at Mussoorie, Platoons at Meerut, Roorkee, Aligarh and Dehra Dun.

First Commn.	Names.	Date of rank.	Date of appointment to corps.	Remarks.
	Lieutenant-Colonel.			
	Majors.			
1 Apl. 17	McNair, Arthur Wyndham (80)	1 Apl. 17	1 Apl. 17	Lv., ex Ind., 11 mths, 20 Apl. 18.
1 Apl. 17	*Puech, Albert Gottelib*, O.B.E., V.D.	*1 Apl. 17*	*1 Apl. 17*	*Hon. Lt.-Col., late Ind. Volrs. (temp. Capt. in Ind. Army).*
	Captains.			
1 Apl. 17	*Sandes, Edward Warren Caulfield*, M.C.	*1 Apl. 17*	*1 Apl. 17*	
1 Apl. 17	Bell, William (80)	1 Apl. 17	1 Apl. 17	
1 Apl. 17	Slater, Alexander Frederick (80)	1 Apl. 17	1 Apl. 17	Aligarh.
	Lieutenants.			
1 Apl. 17	Jones, Frederick Theodore (80)	1 Apl. 17	1 Apl. 17	
1 Apl. 17	O'Callaghan, William Patrick (54) (80)	1 Apl. 17	1 Apl. 17	
1 Apl. 17	*Hearn, Edward Stephen*	*1 Apl. 17*	*1 Apl. 17*	*I. A. R. O.*
1 Apl. 17	McLaren, William David (80)	1 Apl. 17	1 Apl. 17	Roorkee.
1 Apl. 17	Angus, Malcolm Alexander	1 Apl. 17	1 Apl. 17	
	2nd-Lieutenants.			
1 Apl. 17	*Walsh, Hugh*	*1 Apl. 17*	*1 Apl. 17*	*I. A. R. O.*
1 Apl. 17	Gibbs, Harold Henry (80)	1 Apl. 17	1 Apl. 17	(*Lt. late Ind. Volr. Force.*)
1 Apl. 17	*Dubois, Harold Anthony* (54)	*1 Apl. 17*	*1 Apl. 17*	*I. A. R. O.*
1 Apl. 17	Pike, Francis Freeman (80)	1 Apl. 17	1 Apl. 17	Aligarh.
1 Apl. 17	*Drake, Arthur John Alexander* D.C.M.	*1 Apl. 17*	*1 Apl. 17*	*I. A. R. O.*
1 Apl. 17	Chunn, George Frederick (80)	1 Apl. 17	1 Apl. 17	
1 Apl. 17	*Corry, Percy Newton*	*1 Apl. 17*	*1 Apl. 17*	*I. A. R. O.*
26 Mar. 18	*Webster, Cyril Heathcock* (80)	*26 Mar. 18*	*26 Mar. 18*	*I. A. R. O.*
1 Apl. 17	Plomer, Edward Morris	1 Apl. 17	1 Apl. 17	
	Supernumerary.			
1 Apl. 17	Milne, Capt. Gerald Eustace (80)	1 Apl. 17	1 Apl. 17	
1 Apl. 17	Bond, Capt. Edgar Clinton James, V.D. (54).	1 Apl. 17	1 Apl. 17	
1 Apl. 17	Lawlor, Lt. Patrick Michael	1 Apl. 17	1 Apl. 17	
1 Apl. 17	*Morgans, 2nd-Lt. Clarence Gladwin* (80)	*1 Apl. 17*	*1 Apl. 17*	*I. A. R. O.*
1 Apl. 17	Perry, 2nd-Lt. Albert Edward Press (80)	1 Apl. 17	1 Apl. 17	
1 Apl. 17	Weldon, 2nd-Lt. William Edward	1 Apl. 17	1 Apl. 17	
	Staff.			
22 Dec. 15	Bowler, Lieut. (*temp. Capt.*) J., Arg. & Suth'd Highrs.	11 Apl. 17	22 Dec. 15	Adjutant.

Uniform—Khaki drill.

10th (Naini Tal) Company.

Formed as Naini Tal Volunteer Rifles by Mily. Dept. Notfn. No. 636, dated 26th July 1871.
Reconstituted as an I. D. F. Unit with present designation, 1st Apl. 1917.
Establishment—1 Combined Active & Reserve Coy.,
1 Armoured Motor Batty. (B).
Detachment at Bareilly.

HONORARY COMMANDANT.—*The Hon'ble Sir* Spencer Harcourt Butler, K.C.S.I., C.I.E., I.C.S. . 16 June 18.
(*Lieutenant-Governor of U. P.*)

First Commn.	Names.	Date of rank.	Date of appointment to corps.	Remarks.
	Majors.			
1 Apl. 17	Way, Theodore Alban Henry (54) . .	1 May 18	1 Apl. 17	Commdt.
	Captains.			
1 Apl. 17	Busher, Richard Charles, V.D. (54, 80) .	1 Apl. 17	1 Apl. 17	2nd-in-Comd.
	Lieutenants.			
1 Apl. 17	Fleming, Robert Patterson (54, 80) . .	1 Apl. 17	1 Apl. 17	
	2nd-Lieutenants.			
1 Apl. 17	Richardson, Frederick Charles (54, 80) .	1 Apl. 17	1 Apl. 17	Lt., late Ind. Volrs.
1 Apl. 17	Chambers, Walter Richard (48*a*, 51, 80) .	1 Apl. 17	1 Apl. 17	
1 Apl. 17	McIntosh, Kenneth Henry (80)	1 Apl. 17	1 Apl. 17	
	Supernumerary.			
1 Apl. 17	O'Donnell, Lt. Col. Samuel Perry . .	1 May 18	1 Apl. 17	To supy. list, 1st May 18.
1 Apl. 17	Teyen, Capt. Charles St. Leger (54, 80) .	1 Apl. 17	1 Apl. 17	
1 Apl. 17	Green, Capt. Charles Henry Vevers (80) .	1 Apl. 17	1 Apl. 17	
1 Apl. 17	Hearsey, Capt. Alfred John, V.D. (54) . .	1 Apl. 17	1 Apl. 17	
1 Apl. 17	Galloway, Capt. Frederick William . .	1 Apl. 17	1 Apl. 17	
1 Apl. 17	Chill, Lt. Waller Churchill (80) . . .	1 Apl. 17	1 Apl. 17	
1 Apl. 17	McGinn, Lt. Thomas Gordon	1 Apl. 17	1 Apl. 17	
1 Apl. 17	Kinns, Harold Edwin (48*a*, 80) . . .	1 Apl. 17	1 Apl. 17	
1 Apl. 17	Lilley, Lt. Urbane	1 Apl. 17	5 June 18	
1 Apl. 17	Stowell, 2nd-Lt. Robert Septimus (80) .	1 Apl. 17	1 Apl. 17	
	Staff.			
1 Nov. 15	Allison, Lieut. (*temp. Capt.*) T., Rif. Brig. (51, 52, 72*a*)	14 Mar. 16	1 Nov. 15	Adjutant.

Uniform—Khaki drill.

12th Eastern Bengal Railway Battalion.

Formed by Mily. Dept. Notfn. No. 223, dated 1st March 1873. Reconstituted G. G. O. 437 of 1898. Designation altered by A. D. Notfn. No. 146, dated 4th February 1916. Reconstituted as an I. D. F. Unit with present designation 1st Apl. 1917.

Establishment—3 Reserve Coys., 2 Machine Gun Detts. Companies at Sealdah, Kanchrapara and Saidpur.

HONORARY COMMANDANT—Mr. E. A. S. Bell.

First Commn.	Names.	Date of rank.	Date of appointment to corps.	Remarks.
	Lieutenant-Colonel.			
1 Apl. 17	Harnett, William Falkiner, *V.D.*	1 Apl. 17	1 Apl. 17	Comdt.
	Majors.			
1 Apl. 17	Coates, Joseph	1 Apl. 17	1 Apl. 17	2nd-in-Comd.
1 Apl. 17	Von Bibra, Frederick Hickie, V.D.	7 Sep. 18	1 Apl. 17	
	Captains.			
1 Apl. 17	Pearce, Walter Richard, V.D.	1 Apl. 17	1 Apl. 17	*Major, late Ind. Volrs.*
1 Apl. 17	*Kirkhope, Kenneth Macleay*	*1 Apl. 17*	*1 Apl. 17*	*I.A.R.O.*
1 Apl. 17	*McPherson, Duncan Louis*	*1 Apl. 17*	*1 Apl. 17*	*I.M.B.*
1 Apl. 17	Van Someren, Leonard William	1 Apl. 17	1 Apl. 17	
1 Apl. 17	*Cole, Frederick Eustace*	*1 Apl. 17*	*1 Apl. 17*	
1 Apl. 17	Robertson, Alan Cragie	1 Apl. 17	1 Apl. 17	
1 Apl. 17	Williams, Harry Edwin	26 June 18	1 Apl. 17	
1 Apl. 17	Goss, James Samuel	7 Sep. 18	1 Apl. 17	
	Lieutenants.			
1 Apl. 17	*Lilley, Abraham Ralph Grundy*	*1 Apl. 17*	*1 Apl. 17*	
1 Apl. 17	*Baxter, Robert Hugh Neville*	*1 Apl. 17*	*1 Apl. 17*	*I. A. R. O*
1 Apl. 17	*Smith, Bertram Gilbert*	*1 Apl. 17*	*1 Apl. 17*	
1 Apl. 17	*Newhouse, Cyril George*	*1 Apl. 17*	*1 Apl. 17*	*Temp. Lt. in Army.*
1 Apl. 17	Meakins, Harold Whayman	1 Apl. 17	1 Apl. 17	
1 Apl. 17	*Mitchell, William George*	*1 Apl. 17*	*1 Apl. 17*	
1 Apl. 17	Baylis, Herbert Theophilus	1 Apl. 17	1 Apl. 17	
1 Apl. 17	*Scruby, Joseph*	*1 Apl. 17*	*1 Apl. 17*	*I.A.R.O.*
1 Apl 17	Pringle, Ivan Terence St. Clair	1 Apl. 17	1 Apl. 17	
1 Apl. 17	*Guest-Williams, Warren Kirkham*	*1 Apl. 17*	*1 Apl. 17*	*Temp. Lt. in Army.*
1 Apl. 17	*Powis, Leonard John*	*1 Apl. 17*	*1 Apl. 17*	*I. A. R. O.*
1 Apl. 17	*Angell, Charles Henry Cooper*	*1 Apl. 17*	*1 Apl. 17*	
1 Apl. 17	*Moss, Richard Cecil*	......	......	
1 Apl. 17	Ritchie, Charles Stuart	26 June 18	1 Apl. 17	*Lt. late Ind. Vols.*
1 Apl. 17	Taylor, Robert William	26 June 18	1 Apl. 17	
1 Apl. 17	Gundry, Arthur Robert	7 Sep. 18	1 Apl. 17	
	2nd-Lieutenants.			
1 Apl. 17	*Muirhead, Alexander Kershaw*	*1 Apl. 17*	*1 Apl. 17*	
1 Apl. 17	*Cruttwell, Reginald Qwentin*	*1 Apl. 17*	*1 Apl. 17*	
1 Apl. 17	*Sims, Edward Alfred*	*1 Apl. 17*	*1 Apl. 17*	
1 Apl. 17	Parker, Harding Neville	1 Apl. 17	1 Apl. 17	
	Supernumerary.			
1 Apl. 17	Thorp, 2nd-Lieut. Frederick William	1 Apl. 17	1 Apl. 17	
1 Apl. 17	Gorton, 2nd-Lieut. William Hugh	1 Apl. 17	1 Apl. 17	
	Staff.			
5 Sep. 15	Stevens, Capt. S. G., (5th Linc. R.)	1 June 16	27 July 17	Adjt.
1 Apl. 17	Van Mulken, Rev. Fr. Peter, S. J.	1 Apl. 17	1 Apl. 18	Hony. Chaplain (R. C.).

Uniform—Khaki drill.

1—13th Great Indian Peninsula Railway Battalion.

Formed 29th Dec. 1875. Reconstituted as a corps of two Bns. and Midland Railway Volunteer Rifles amalgamated by Mily. Dept. Notfn. No. 624, dated 4th July 1902, and reconstituted as an I. D. F. Unit with present designation 1st Apl. 1917.

Establishment.—7 Reserve Coys., 2 Armoured Train Detts., 2 Machine Gun Detts.

Cos. at Igatpuri, Bhusawal, Budnera, Jubbulpore, Lanowli & Sholapur.

First Commn.	Names.	Date of rank.	Date of appointment to corps.	Remarks.
	Lieutenant-Colonels.			
1 Apl. 17	Bonner, Thomas William, O.B.E., V.D.	1 Apl. 17	1 Apl. 17	Comdt., Hon. Col., late Ind. Volrs.
	Majors.			
1 Apl. 17	*Webb, George Richard*, V.D.	*1 Apl. 17*	*1 Apl. 17*	*Hon. Lt.-Col., late Ind. Volrs. employed on Munitions.*
1 Apl. 17	Keily, Henry Gerald, V.D. (80)	1 Apl. 17	1 Apl. 17	Bhusawal.
1 Apl. 17	Scott, William Herbert Schroder, V.D. (80)	1 Apl. 17	1 Apl. 17	Bombay.
1 Apl. 17	Edwards, Percy Ashworth (80)	1 Apl. 17	1 Apl. 17	Bombay.
	Captains.			
1 Apl. 17	Stokes, Isaac Westley (51, 80)	1 Apl. 17	1 Apl. 17	Bombay.
1 Apl. 17	Goldney, Alfred Peel (80)	1 Apl. 17	1 Apl. 17	Bombay.
1 Apl. 17	*Tedman, Cyril*, M.C.	*1 Apl. 17*	*1 Apl. 17*	*Temp. Lt. I. A.*
1 Apl. 17	Barton, James Calvert (80)	1 Apl. 17	1 Apl. 17	Manmad.
1 Apl. 17	*Forbes, Alister Esme Buchan*	*1 Apl. 17*	*1 Apl. 17*	*I. A. R. O.*
1 Apl. 17	Luxton, Francis Willoughby (80)	1 Apl. 17	1 Apl. 17	Bombay.
1 Apl. 17	*Edwards, Osborne George*	*1 Apl. 17*	*1 Apl. 17*	*I. A. R. O.*
1 Apl. 17	*Fraser, Thomas (80)*	*1 Apl. 17*	*1 Apl. 17*	*Lv., ex Ind., 9 mths. from 1 Apl. 18.*
1 Apl. 17	*Tofts, Crosier Fullerton (207a)*	*1 Apl. 17*	*1 Apl. 17*	*I. A. R. O.* Adjutant.
1 Apl. 17	Scott, Joseph (51, 80)	1 Apl. 17	1 Apl. 17	Bombay.
1 Apl. 17	Costello, John Francis (80)	1 Apl. 17	1 Apl. 17	Betul.
1 Apl. 17	*Harrington, Frank*	*1 Apl. 17*	*1 Apl. 17*	
1 Apl. 17	Nicholl-Carne, Edward Dyffrig Stradling (80)	1 Apl. 17	1 Apl. 17	Jabalpur.
1 Apl. 17	*Mould, Arthur Ernest*	*1 Apl. 17*	*1 Apl. 17*	*I. A. R. O.*
1 Apl. 17	Remington, James Boileau (80)	1 Apl. 17	1 Apl. 17	Bhusawal.
1 Apl. 17	*Hitchcock, Ronald Victor*	*1 Apl. 17*	*1 Apl. 17*	*I. A. R. O.*
1 Apl. 17	McNeill, Kenneth John (80)	1 Apl. 17	1 Apl. 17	Nagpur.
	Lieutenants.			
1 Apl. 17	Jamieson, Harry William (80)	1 Apl. 17	1 Apl. 17	Bombay.
1 Apl. 17	*Anderson, Francis Edward (80)*	*1 Apl. 17*	*1 Apl. 17*	*Temp. Commn. in Army.*
1 Apl. 17	*Graham, Cyril George (79)*	*1 Apl. 17*	*1 Apl. 17*	*I. A. R. O.*
1 Apl. 17	Raper, John Hugh Francis (80)	1 Apl. 17	1 Apl. 17	Sholapur.
1 Apl. 17	Lincke, Sydney Howard Phipps (80)	1 Apl. 17	1 Apl. 17	Bombay.
1 Apl. 17	*Redmond, Lionel Godfrey*	*1 Apl. 17*	*1 Apl. 17*	Employed on Rlys.
1 Apl. 17	*Muir, James Logan*	*1 Apl. 17*	*1 Apl. 17*	*I. A. R. O.*
1 Apl. 17	*Carpenter, John Howard*	*1 Apl. 17*	*1 Apl. 17*	*I. A. R. O.*
1 Apl. 17	*Langdon, Frederick Guy*	*1 Apl. 17*	*1 Apl. 17*	
1 Apl. 17	MacBeth, Alexander (80)	1 Apl. 17	1 Apl. 17	Bombay.
1 Apl. 17	Brock, Henry Norman (80)	1 Apl. 17	1 Apl. 17	Lonavla.
1 Apl. 17	Wood, Percy (80)	1 Apl. 17	1 Apl. 17	Bombay.
1 Apl. 17	Berkley, Gerald Eustace (80)	1 Apl. 17	1 Apl. 17	Bhusaval.
1 Apl. 17	Henderson, William (80)	1 Apl. 17	1 Apl. 17	Bombay.
1 Apl. 17	Chilton, Alfred Henry (80)	1 Apl. 17	1 Apl. 17	Bombay.
1 Apl. 17	Hood, Thomas Christie (51, 80)	1 Apl. 17	1 Apl. 17	Manmad.
1 Apl. 17	Anderson, Cuthbert William	1 Apl. 17	1 Apl. 17	
1 Apl. 17	Bould, Reginald Perry	1 Apl. 17	1 Apl. 17	
	2nd-Lieutenants.			
14 Mar. 18	Firth, Harold (80)	14 Mar. 18	14 Mar. 18	Manmad.
19 July 18	Billington, Percy Robert	19 July 18	19 July 18	
19 July 18	Alexander, William, C.I.E.	19 July 18	19 July 18	
19 July 18	Flower, Ernest Hubert	19 July 18	19 July 18	
19 July 18	Dowling, Lionel Edward	19 July 18	19 July 18	
19 July 18	Russell, Thomas Guthrie	19 July 18	19 July 18	
19 July 18	Husband, Robert Henry	19 July 18	19 July 18	
19 July 18	Sharp, Samuel Henry	19 July 18	19 July 18	
19 July 18	Walker, Miles	19 July 18	19 July 18	
1 Apl. 17	Manson, Frederic Percy	1 Apl. 17	1 Apl. 17	
	Supernumerary.			
1 Apl. 17	Russell, Edward Wilson (80)	1 Apl. 17	1 Apl. 17	Bhusawal.
	Staff.			
7 Apl. 11	Tofts, Capt. Crosier Fullerton	1 Sep. 15	23 May 16	*I. A. R. O.* Adjutant.
9 Mar. 18	Richardson, *Hony.* Lt. A. E.	9 Mar. 18	9 Mar. 18	Qr. Mr.

2—13th Great Indian Peninsula Railway Battalion.

Establishment—3 Reserve Coys., 1 Armoured Motor Batty., 1 Armoured Train Dett.

First Commn.	Names.	Date of rank.	Date of appointment to corps.	Remarks.
	Lieutenant-Colonel.			
1 Apl. 18	McNeill, John Sm th, V.D.	1 Apl. 18	1 Apl. 18	Comdt.—*Hon. Col. late Ind. Volrs.*
	Majors.			
1 Apl. 17	Dangerfield Edward Rushton, V.D.	1 Apl. 17	1 Apl. 17	Jhansi.
1 Apl. 17	Cooke, Robert Francis Rickards, V.D.	1 Apl. 17	1 Apl. 17	Delhi.
	Captains.			
1 Apl. 17	Kennedy, David Waits	1 Apl. 17	1 Apl. 17	Cawnpore.
1 Apl. 17	*Cory, Edgar Myrie*	*1 Apl. 17*	*1 Apl. 17*	*I. A. R. O.*
1 Apl. 17	Williams, Alfred Edward	1 Apl. 17	1 Apl. 17	Jhansi.
1 Apl. 17	Richardson, Allan (80)	1 Apl. 17	1 Apl. 17	Jhansi.
1 Apl. 17	*Saunders, Howard Fauntleroy*	*1 Apl. 17*	*1 Apl. 17*	*I. A. R. O.*
1 Apl. 17	Everett, George Eliot	1 Apl. 17	1 Apl. 17	Jhansi.
1 Apl. 17	Cooke, Edward Bernard (80)	1 Apl. 17	1 Apl. 17	Bina.
	Lieutenants.			
1 Apl. 17	*Moberly, Ronald Stuart (54)*	*1 Apl. 17*	*1 Apl. 17*	*I. A. R. O.*
1 Apl. 17	*Rice, Eustace Meredith*	*1. Apl. 17*	*1 Apl. 17*	*I. A. R. O.*
	2nd-Lieutenants.			
1 Apl. 17	Bould, Hubert George (80)	1 Apl. 17	1 Apl. 17	Jhansi.
1 Apl. 18	Barton, John Everard	1 Apl. 18	1 Apl. 18	Mahoba.
29 July 18	Middleton, Percy Albert Hirst	29 July 18	29 July 18	Agra.
29 July 18	Barber, Lewis	29 July 18	29 July 18	Jhansi.
29 July 18	Tower, John Alexander	29 July 18	29 July 18	Jhansi.
29 July 18	Hussey-Freke, Percy Edward	29 July 18	29 July 18	Bina.
29 July 18	Darwin, Gerald Meredith	29 July 18	29 July 18	Orai.
29 July 18	Clarke, Robert Lucas Hyrapiet	29 July 18	29 July 18	Jhansi.
	Staff.			
26 Aug. 18	Munroe, 2nd-Lt. (*Actg. Capt.*) J., 3rd Middx. R.	26 Aug. 18	26 Aug. 18	Adjt.
6 Mar. 18	Cashmore, *Hony.* Lt. W.	6 Mar. 18	6 Mar. 18	Qr.-Mr.

Uniform—Khaki drill.

14th Tennaserim Battalion.

Raised as Moulmein Volunteer Rifles, 2nd July 1877, by Military Department Notfn. No. 852, dated 29th May 1877. Reserve Company raised, 24th January 1889. Moulmein Volunteer Artillery amalgamated and reconstituted as an I. D. F. Unit with present designation 1st Apl. 1917.

" Primus in Burma."

Establishment—*1 Active Coy.*, 1 Combined *Active and Reserve Coy.*, 2 *Machine Gun Detts.*

Companies at Tavoy and Mergui.

HONORARY COMMANDANT—The Hon'ble Sir Reginald Henry Craddock, K.C.S.I., I.C.S., 10th September 18 (Lieutenant-Governor of Burma.)

First Commn.	Names.	Date of rank.	Date of appointment to corps.	Remarks.
	Lieutenant-Colonels.			
1 Apl. 17	*Thane, Charles Stuart*	*1 Apl. 17*	*1 Apl. 17*	*Gen. List, Br. Serv.*
1 Apl. 17	Buchanan, Francis Henry Theodore (80, (82)	1 Apl. 17	1 Apl. 17	Commdt.
	Majors.			
1 Apl. 17	*Lakin, Colin Arthur*	*1 Apl. 17*	*1 Apl. 17*	*I. A. R. O.*
	Captains.			
1 Apl. 17	Beadon, William Robert Coleridge (80)	1 Apl. 17	1 Apl. 17	Tavoy.
1 Apl. 17	Darwood, Arthur John Sutherland (80)	1 Apl. 17	1 Apl. 17	Moulmein.
1 Apl. 17	Rhodes, Henry Seaton (80)	1 Apl. 17	1 Apl. 17	Mergui.
1 Apl. 17	*List, John Napier (207)*	*1 Apl. 17*	*1 Apl. 17*	*I. A. R. O.*
1 Apl. 17	Brander, William Browne (80)	1 Apl. 17	1 Apl. 17	Tavoy.
	Lieutenants.			
1 Apl. 17	*Anderson, Archibald Robert (105, 207)*	*1 Apl. 17*	*1 Apl. 17*	*I. A. R. O.*
1 Apl. 17	Chalmers David (80)	1 Apl. 17	1 Apl. 17	Thaton.
1 Apl. 17	*Couper, Thomas (80)*	*1 Apl. 17*	*1 Apl. 17*	*I. A. R. O.*
1 Apl. 17	*Law, Charles Ewan (temp. Capt.) (80, (82)*	*1 Apl. 17*	*1 Apl. 17*	*Staff Capt. Burma Divn.*
1 Apl. 17	Tennent, Jeffrey Bernard Hale (80)	1 Apl. 17	1 Apl. 17	Mergui.
1 Apl. 17	*Cuninghan, George Edmund (80), (82)*	*10 May 18*	*1 Apl. 17*	*I. A. R. O.*
1 Apl. 17	Mullins, Harold Lovel (80)	10 May 18	1 Apl. 17	Tavoy.
	2nd-Lieutenants.			
1 Apl. 17	Bruce, Courtenay Merceron Adair (80)	1 Apl. 17	1 Apl. 17	Moulmein.
26 Sep. 18	Perkins, Bryan Ward	26 Sep. 18	26 Sep. 18	
	Supernumerary.			
1 Apl. 17	Kinch, Capt. Edward Gerard Nöel	1 Apl. 17	1 Apl. 17	
1 Apl. 17	Clifford, Capt. Ronald Edward St. John	1 Apl. 17	1 Apl. 17	
	Staff.			
13 Jan. 8	Denney, 2nd-Lt. (Actg. Capt.) W. S. Lan. R. (12, 51, 52, 58).	13 Jan. 18	13 Jan. 18	Adjutant.

Uniform—Officers—Khaki drill.
N.-C. Officers and men—Khaki drill.

15th Bombay Battalion.

Formed 15th Aug. 1877 and reconstituted as an I. D. F. Unit with present designation, 1st Apl. 1917.
Establishment—3 Reserve Coys., 3 Combined, Active & Reserve Coys., 2 Armoured Train Detts.

HONORARY COMMANDANT—*H. E. The Rt. Hon.* FREEMAN FREEMAN-THOMAS, *Baron* WILLINGDON of Ratton, G.C.S.I., G.C.I.E., G.B.E. 1st Apl. 17.

(*Governor of Bombay.*)

First Commn.	Names.	Date of rank.	Date of appointment to corps.	Remarks.
	Lieutenant-Colonel.			
1 Apl. 17	Cadell, Patric Robert, C.I.E., V.D.	1 Apl. 17	1 Apl. 17	Comdt. A.-D.-C. to Govr., Bombay.
	Majors.			
1 Apl. 17	Bush, Harold Ferdinand	1 Apl. 17	1 Apl. 17	Lt.-Col., late Ind. Volrs.
1 Apl. 17	*Moberly, Charles Noel (54, 80)*	*1 Apl. 17*	*1 Apl. 17*	*Adjutant.*
1 Apl. 17	Gorrie, Harry Thomson (80)	1 Apl. 17	1 Apl. 17	
	Captains.			
1 Apl. 17	*Kemp, Kenneth McIntyre*	*1 Apl. 17*	*1 Apl. 17*	*I. A. R. O.*
1 Apl. 17	*Watson, John Francis*	*1 Apl. 17*	*1 Apl. 17*	*(Temp. Lt., R. E.)*
1 Apl. 17	*Cherry, John Arnold*	*1 Apl. 17*	*1 Apl. 17*	*I. A. R. O.*
1 Apl. 17	Dibbs, Henry James (80)	1 Apl. 17	1 Apl. 17	
1 Apl. 17	Kinnear, Norman Boyd	1 Apl. 17	1 Apl. 17	
1 Apl. 17	Ormerod, Harold Eric (51, 72*a*, 80)	1 Apl. 17	1 Apl. 17	
1 Apl. 17	Paton, Nigel Fairholt (80)	1 Apl. 17	1 Apl. 17	Hon. A.-D.-C. to Govr., Bombay.
1 Apl. 17	Hunt, Neville Greaves (80)	1 Apl. 17	1 Apl. 17	
1 Apl. 17	Mackenzie, Robert Theodore Hope (80)	1 Apl. 17	1 Apl. 17	
1 Apl. 17	Sharpe, William Rutton Searle (80)	1 Apl. 17	1 Apl. 17	
1 Apl. 17	Kemp, Norman Wright (80)	1 Apl. 17	1 Apl. 17	
1 Apl. 17	Begbie, Thomas Newton (51, 80)	1 Apl. 17	1 Apl. 17	
	Lieutenants.			
1 Apl. 17	*Ashton, Alexander*	*1 Apl. 17*	*1 Apl. 17*	*I. A. R. O.*
1 Apl. 17	*Puttick, Alec Walter,* M.C.	*1 Apl. 17*	*1 Apl. 17*	*(Temp. Capt. in Army.)*
1 Apl. 17	*Rich, Clive Noble*	*1 Apl. 17*	*1 Apl. 17*	*I. A. R. O.*
1 Apl. 17	Townend, Harry Douglas (80)	1 Apl. 17	1 Apl. 17	*Lv., ex India, 6 mos.*
1 Apl. 17	*Yates, Donald Russell Martin*	*1 Apl. 17*	*1 Apl. 17*	*I. A. R. O.*
1 Apl. 17	*Nicholson, Eustace Marmion Ferrer*	*1 Apl. 17*	*1 Apl. 17*	*I. A. R. O.*
1 Apl. 17	*Sheppard, Samuel Townsend, Temp. Capt.*	*1 Apl. 17*	*1 Apl. 17*	*Staff Capt., Bo. Bde.*
1 Apl. 17	Hoseason, William Sandford (80)	1 Apl. 17	1 Apl. 17	
1 Apl. 17	Brayne, Albert Frederic Lucas (51, 54, 80)	1 Apl. 17	1 Apl. 17	
1 Apl. 17	Wright, Guy Thomas (80)	1 Apl. 17	1 Apl. 17	
1 Apl. 17	Tyndale, Arthur Carrington (80)	1 Apl. 17	1 Apl. 17	
1 Apl. 17	Aldred, Thomas Robert (80)	1 Apl. 17	1 Apl. 17	
1 Apl. 17	Riley, Errol Samuel (80)	1 Apl. 17	1 Apl. 17	
1 Apl. 17	Soundy, Percy Reginald (80)	1 Apl. 17	1 Apl. 17	
1 Apl. 17	McGrath, James (51, 80)	1 Apl. 17	1 Apl. 17	
1 Apl. 17	McBain, James Anderson Dickson, C.I.E. (80)	1 Apl. 17	1 Apl. 17	
17 Nov. 17	Wiles, Gilbert (80)	21 May 18	1 July 18	
	2nd-Lieutenants.			
1 Apl. 17	Burnett-Hurst, Alexander Robert (80)	1 Apl. 17	1 Apl. 17	
1 Apl. 17	Botterill, Wilfred Reginald (51, 80)	1 Apl. 17	1 Apl. 17	
1 Apl. 17	Fox, Morley (80)	1 Apl. 17	1 Apl. 17	
1 Apl. 17	Barnes, Bertram Durell (51, 80)	1 Apl. 17	1 Apl. 17	
7 Feb. 18	Tait, John (80)	7 Feb. 18	7 Feb. 18	
7 Feb. 18	England, Richard Dubourdieu (80)	7 Feb. 18	7 Feb. 18	
23 June 17	Macnaughten, Henry Pelham Wentworth (80)	23 June 17	23 June 17	
7 Feb. 18	Robinson, Alfred Ernest (80)	7 Feb. 18	7 Feb. 18	
	Supernumerary.			
1 Apl. 17	Faith, Maj. Norman William, V.D. (80)	1 Apl. 17	1 Apl. 17	
1 Apl. 17	Forbes, Maj. Alister (80)	1 Apl. 17	1 Apl. 17	
1 Apl. 17	Bennett, Lt. Arthur Russell	1 Apl. 17	1 Apl. 17	

15th Bombay Battalion—*contd.*

First Commn.	Names.	Date of rank.	Date of appointment to corps.	Remarks.
	Attached.			
. . . . .	Mitchell, Capt. H. C. B., Herts. R. (51) .	7 Jan. 12		
24 Jan. 10	McNeill, Capt. H. L., Sco. Rif. (51) . .	21 June 15		
	Staff.			
1 Apl. 17	Moberly, Maj. Charles Noel	1 Apl. 17	1	Adjutant.
24 Mar. 18	Stone, Lt. A. S.	24 Mar. 18	24 Mar. 18	Qr.-Mr.
1 Apl. 17	Gordon, Revd. James Drummond . . .	1 Apl. 17	1 Apl. 17	Hony. Chaplain.

Uniform— haki drill.

16th Cawnpore Rifles.

(Formed 16th Aug. 1877 by Mily. Dept. Notfn. No 734 of 1877) Reconstituted as an I. D. F. Unit with present designation 1st Apl. 1917.

Establishment —1 Combined Active & Reserve Coy., 1 Armoured Motor Batty. (B), 1 Machine Gun Dett., 1 Artillery Coy.

Company at Jhansi.

HONORARY COMMANDANT—

First Commn.	Names.	Date of rank.	Date of appointment to corps.	Remarks.

Staff

Uniform—Khaki drill.

1-17th Bombay, Baroda and Central India Railway Battalion.

Formed 3rd Aug. 1877 and reconstituted as an I. D. F. unit with present designation 1st Apl. 1917.

Establishment—3 Reserve Coys., 1 Armoured Motor Batty. (B), 1 Armoured Train Dett., 1 Machine Gun Dett.
Companies at Rutlam, Ahmedabad and Gangapur.

HONORARY COMMANDANT—

First Commn.	Names.	Date of rank.	Date of appointment to corps.	Remarks.
	Lieutenant-Colonel.			
1 Apl. 17	Pigott, Robert Edward Pemberton, C.I.E., V.D.	1 Apl. 17	1 Apl. 17	Comdt. (*Hony. Col., late Ind. Volrs.*)
	Major.			
1 Apl. 17	Sidgwick, Charles Kater Dury, D.S.O., V.D.	1 Apl. 17	1 Apl. 17	2nd-in-Commd.
	Captains.			
1 Apl. 17	*Reynolds, John Richardson*, V.D. (207)	*1 Apl. 17*	*1 Apl. 17*	*I. A. R. O.*
1 Apl. 17	Page, Frederick James (54, 80)	1 Apl. 17	1 Apl. 17	
1 Apl. 17	Arnould, Francis Graham (80)	1 Apl. 7	1 Apl 17	
1 Apl. 17	McCanlis, William John	1 Apl. 17	1 Apl. 17	
1 Apl. 17	McDonald, Angus	1 Apl. 17	1 Apl. 17	
1 Apl. 17	Marryat, Nevil (80)	1 Apl. 17	1 Apl. 17	
1 Apl. 17	Jones, John Alexander (80)	1 Apl. 17	1 Apl. 17	
1 Apl. 17	*Jones, H. P.*	*1 Apl. 17*	*1 Apl. 17*	
	Lieutenants.			
1 Apl. 17	Ball, Hayward Percy	1 Apl. 17	1 Apl. 17	
1 Apl. 17	*Rollo, Andrew Duncan*	*1 Apl. 17*	*1 Apl. 17*	*Tempy. Capt. in Army, 17 Oct. 14.*
1 Apl. 17	Moore, Gerald Durie	1 Apl. 17	1 Apl. 17	
1 Apl. 17	*Lincoln, Norman* (80)	*1 Apl. 17*	*1 Apl. 17*	*I. A. R. O.*
1 Apl. 17	*Campion, A. H. M.*	*1 Apl. 17*	*1 Apl. 17*	
1 Apl. 17	Collins, Charles William Melville (54, 80)	1 Apl. 17	1 Apl. 7	
1 Apl. 17	*Thomson, Rowley Stuart*	*1 Apl. 17*	*1 Apl. 17*	*I. A. R. O.*
1 Apl. 17	Macmillan, Hugh Robert Munro	1 Apl. 17	1 Apl. 17	
1 Apl. 17	Shorrock, Arthur Gordon	1 Apl. 17	1 Apl. 17	
	2nd-Lieutenants.			
1 Apl. 17	May, Awdry John	1 Apl. 17	1 Apl. 17	
1 Apl. 17	Miles, Homer Henry (51, 80)	1 Apl. 17	1 Apl. 17	
1 Apl. 17	Slade, George Eric Rowland (80)	1 Apl. 17	1 Apl. 17	
1 Apl. 17	*Blake, William*	*1 Apl. 17*	*1 Apl. 17*	*I. A. R. O.*
1 Apl. 17	*Rowsell, Vere Norman*	*1 Apl. 17*	*1 Apl. 17*	*Spl. Res., C. Gds.*
1 Apl. 17	Moore, William James Alexander (80)	1 Apl. 17	1 Apl. 17	
1 Apl. 17	Paterson, John James Crosbie	1 Apl. 17	1 Apl. 17	
1 Apl. 17	Shaw, Stanley George Newell	1 Apl. 17	1 Apl. 17	
1 Apl. 17	Middleton, Robert Hugh	1 Apl. 17	1 Apl. 17	
	Supernumerary.			
1 Apl. 17	Thomason, Lt.-Col. Rowland Maxwell, V.D.	1 Apl. 17	17 July 18	
1 Apl. 17	Owen, Lt.-Col. Albert Cifford	1 Apl. 17	1 Apl. 17	
1 Apl. 17	Medley, Maj. Neville Rawlins	1 Apl. 17	1 Apl. 17	
1 Apl. 17	Master, Capt. Alfred	1 Apl. 17	1 Apl. 17	
1 Apl. 17	*Broomfield, 2nd-Lt. Robert Stonehouse*	*1 Apl. 17*	*Apl. 17*	
1 Apl. 17	Sievwright, Capt. Frederic James Hume, V.D.	1 Apl. 17	1 Dec. 18	*I. A. R. O.*
	Staff.			
24 Mar. 18	Brooks, 2nd-Lt. (*Actg. Capt.*) G. H., Cam'n. Highrs.	24 Mar. 18	24 Mar. 18	Adjutant.

2-17th Bombay, Baroda and Central India Railway Battalion.

Formed 3rd August 1877. Reconstituted as an I. D. F. Unit with present designation 1st Apl. 1917.
Establishment—3 Reserve Coys., 2 Combined Active & Reserve Coys., 2 Armoured Motor Battys. (B), 1 Armoured Train Dett., 2 Machine Gun Detts.

Companies at Abu Road, Mhow, and Fatehgarh.

Platoons at Phulera, Bandikui, Rewari, Sirsa, Jodhdpur, and Mount Abu.

HONORARY COMMANDANT—

First Commn.	Names.	Date of rank.	Date of appointment to corps.	REMARKS.
	Lieutenant-Colonel.			
	Majors.			
1 Apl. 17	*Gabriel, Edmund Vivian,* C.V.O., C.S.I.	*1 Apl. 17*	*1 Apl. 17*	*Temp. Lt.-Col. in Army.*
1 Apl. 17	Newton, Arthur Robert George, V.D.	1 Apl. 17	1 Apl. 17	*Hon. Lt.-Col. late Ind. Volrs.*
1 Apl. 17	*Waddington, Charles Willoughby* C.I.E., M.V.O.	*1 Apl. 17*	*1 Apl. 17*	*I. A. R. O.*
1 Apl. 17	Cooke, Conrad Allan, V.D. (80)	1 Apl. 17	1 Apl. 17	Comdt. (*Offg.*)
	Captains.			
1 Apl. 17	Skipwith Fulwar Estoteville (80)	1 Apl. 17	1 Apl. 17	*Lv. ex Ind. 1 year, m. c. from 26 Apl. 18.*
1 Apl. 17	Shipp, William Edgar, V.D. (80)	1 Apl. 17	1 Apl. 17	Hon. Maj. late Ind. Volrs.
1 Apl. 17	Fraser William Stuart, O.B.E. (80)	1 Apl. 17	1 Apl. 17	
1 Apl. 17	*Haig, Harry Graham*	*1 Apl. 17*	*1 Apl. 17*	*I. A. R. O*
1 Apl. 17	Clayton, James Adolphus (80)	1 Apl. 17	1 Apl. 17	
1 Apl. 17	Condon, Edmund Charles Hill (80)	1 Apl 17	1 Apl. 17	
1 Apl. 17	Cotesworth, Cuthbert Gilfillan (80)	1 Apl. 17	1 Apl. 17	
1 Apl. 17	Garrett, Reginald Sillifant (80)	1 Apl. 17	1 Apl. 17	
	Lieutenants.			
1 Apl. 17	Jamieson, Douglas James Topp (80)	1 Apl. 17	1 Apl. 17	
1 Apl. 17	*Covell, Beverley Carthew*	*1 Apl. 17*	*1 Apl. 17*	*I. A. R. O.*
1 Apl. 17	Martin, Leopold Julian (51, 69, 80)	1 Apl. 17	1 Apl. 17	
1 Apl. 17	Aldous, Wilfred Temple (69), (80)	1 Apl. 17	1 Apl. 17	
1 Apl. 17	Harrison, Rollo, Dallas (69, 80)	1 Apl. 17	1 Apl. 17	
1 Apl. 17	Donaldson, Roderick (80)	1 Apl. 17	1 Apl. 17	
1 Apl. 17	Stent, Dudley Hepburn	1 Apl. 17	1 Apl. 17	
1 Apl. 17	Remington George Henry Lewis (69, 80)	1 Apl. 17	1 Apl. 17	
1 Apl. 17	Sixby, Herbert Baber (80)	1 Apl. 17	1 Apl. 17	
1 Apl. 17	*Powell, Vivian Norman ffolliott*	*1 Apl. 17*	*1 Apl. 17*	*I. A. R. O.*
1 Apl. 17	*Cooper, Thomas*	*1 Apl. 17*	*1 Apl. 17*	
1 Apl. 17	*Rennick, Charles Moyle* (69)	*1 Apl. 17*	*1 Apl. 17*	
1 Apl. 17	Bird, Ernest Alfred	1 Apl. 17	6 June 18	
	2nd-Lieutenants.			
1 Apl. 17	Stirling, James Hay (69, 80)	1 Apl. 17	1 Apl. 17	
1 Apl. 17	*Clarke, Percy Scamander*	*1 Apl. 17*	*1 Apl. 17*	*I. A. R. O.*
1 Apl. 17	Lloyd, Llewellyn North (50, 80)	1 Apl. 17	1 Apl. 17	
1 Apl. 17	Madden, Samuel Fitzgerald	1 Apl. 17	1 Apl. 17	
1 Apl 17	Nash, Hubert Jephson (69, 80)	1 Apl. 17	1 Apl. 17	
1 Apl. 18	*Doran, Henry Francis*	*1 Apl. 7*	*1 Apl. 18*	*I. A. R. O.*
1 Apl. 17	*Lesmon, Cyril Aloysius*	*1 Apl. 17*	*1 Apl. 17*	*I. A. R. O.*
1 Apl. 17	Storrar, Adam Young	1 Apl. 17	1 Apl. 17	
1 Apl. 17	*Moseley, Frank Samuel, Hughes*	*1 Apl. 17*	*1 Apl. 17*	*I. A. R. O.*
17 Apl. 18	Fanshawe, Wilfred	17 Apl. 18	17 Apl. 18	
3 July 18	Landon, Cyril	3 July 18	3 July 18	
10 Aug. 18	Parry, Walter Leonard	10 Aug. 18	10 Aug. 18	
30 Sep. 18	Martelli, Charles Inglis Holroyd	30 Sep. 18	30 Sep. 18	
1 Apl. 17	Peyton, William Harry (80)	1 Apl. 17	1 Dec. 18	
1 Apl. 17	Baker, Percy Neville Humphry (80)	1 Apl. 17	1 Dec. 18	
	Staff.			
28 Aug. 18	Bell, 2nd-Lt. (*Actg. Capt.*) R. (50, 153)	28 Aug. 18	28 Aug. 18	Adjt.
6 Mar. 18	Rankin, Lt. E. M. (94, 17-*a*)	6 Mar. 18	6 Mar. 18	Qr. Mr.
1 Apl. 17	Morris, Revd. S. P.	1 Apl. 17	1 Apl. 17	Hony. Chaplain (C. of E.)
24 June 18	Ralph, Revd. Father	24 June 18	24 June 18	Hony. Chaplain (R. C.)
1 Apl. 17	Clayton, Capt. Robert Vickers, M.B., C.H., R D.P.H.	1 Apl. 17	1 Apl. 17	Medl. Offr.
1 Apl. 17	Wilkinson, Lieut. Sydney Arthur, M.R.C. S., L.R.C.P.	1 Apl. 17	1 Apl. 17	Medl. Offr.

Uniform—Khaki drill.

18th Rangoon Battalion.

Raised by Mily. Dept. Notfn. No 1062, dated 30th November 1877. Reserve Company raised, 10th July 1889. Reconstituted as an I.D.F. Unit with present designation 1 Apl. 17.

Establishment—4 Active Coys., 1 Reserve Coy., 2 Combined Active & Reserve Coys., 1 Platoon Mtd. Infy.

Detachments at Maubin, Myaungmya, Twante, Kanoungtoe, Syriam, and Akyab.

HONORARY COMMANDANT—*The Hon'ble Sir* Reginald Henry Craddock, K.C.S.I., I.C.S. 9 Mar. 18.

(*Lieutenant-Governor of Burma.*)

First Commn.	Names.	Date of rank.	Date of appointment to corps.	Remarks.
	Lieutenant-Colonels.			
1 Apl. 17	*Graham, Charles Townley* (82)	*1 Apl. 17*	*1 Apl. 17*	*I. A. R. O.*
1 Apl 17	*Cameron, Colin John Norman*	*1 Apl. 17*	*1 Apl 17*	*I. A. R. O.*
1 Apl. 17	Stevens, James Algernon (98, 206)	1 Apl. 17	1 Apl. 17	Comdt.
	Majors.			
1 Apl. 17	Lloyd, Idwal Geoffrey (104, 207)	1 Apl. 17	1 Apl. 17	2nd-in-Command.
1 Apl. 17	*Houldey, John Emeris (104, 207)*	*1 Apl. 17*	*1 Apl. 17*	*I. A. R. O.*
1 Apl. 17	Whigham, Gilbert Campbell	1 Apl. 17	1 Apl. 17	
	Captains.			
1 Apl. 17	*Henderson, Walter Lewis*	*1 Apl. 17*	*1 Apl. 17*	*I. A. R. O.*
1 Apl. 17	*Stevenson, William*	*1 Apl. 17*	*1 Apl. 17*	*I. A. R. O.*
1 Apl. 17	Hill, Cyril Patrick (107, 207*a*)	1 Apl. 17	1 Apl. 17	Adjutant (*tempy.*).
1 Apl. 17	Wells, Joseph Hamilton Primrose	1 Apl. 17	1 Apl. 17	
1 Apl. 17	Yeomans, Robert Roper (54)	1 Apl. 17	1 Apl. 17	
1 Apl. 17	Allan, Hugh Logie	1 Apl. 17	1 Apl. 17	Syriam.
1 Apl. 17	Oehme, Cecil Norman	1 Apl. 17	1 Apl. 17	
1 Apl. 17	McKeand, Andrew	1 Apl. 17	1 Apl. 17	Lv. ex Ind. 12 mths. from 24 May 18.
1 Apl. 17	*Roberts, Arthur Salisbury Browne,* (206)	*1 Apl. 17*	*1 Apl. 17*	(*Temp.* Maj., I. A.)
1 Apl. 17	*Glover, Richard Clarke*	*1 Apl. 17*	*1 Apl. 17*	(*Temp. 2nd-Lt., I. A.*)
1 Apl. 17	*Mann, Algernon Edward*	*1 Apl. 17*	*1 Apl. 17*	(*Temp. 2nd-Lt., R. E.*)
1 Apl. 17	*Faulkner, Alexander Keith*	*1 Apl. 17*	*1 Apl. 17*	(*Temp. 2nd-Lt. in Army*).
1 Apl. 17	Henstock, Philip	1 Apl. 17	1 Apl. 17	
1 Apl. 17	Somervaille, George	1 Apl. 17	1 Apl. 17	Lv. 7 mths. from 1 June 18.
1 Apl. 17	Henderson, Reginald Ernest	1 Apl. 17	1 Apl. 17	
1 Apl. 17	Grant, Clive Herbert	1 Apl. 17	1 Apl. 17	
1 Apl. 17	*Stephenson, Basil (207)*	*1 Apl. 17*	*1 Apl. 17*	(*Devon Yeomanry*).
1 Apl. 17	Moir, Ernest Templéton (82)	1 Apl. 17	1 Apl. 17	
1 Apl. 17	Turner, John Reginald	1 Apl. 17	1 Apl. 17	
1 Apl. 17	Annesley, Ian Douglas Campbell (51)	1 Apl. 17	1 Apl. 17	
	Lieutenants.			
1 Apl. 17	Anderson, John James	1 Apl. 17	1 Apl. 17	
1 Apl. 17	Dawson, Alexander Thomas	1 Apl. 17	1 Apl. 17	Maubin.
1 Apl. 17	*Jones, Frank Herbert*	*1 Apl. 17*	*1 Apl. 17*	*I. A. R. O.*
1 Apl. 17	*Ward, William Joseph*	*1 Apl. 17*	*1 Apl. 17*	*I. A. R. O.*
1 Apl. 17	*Tayler, Charles Hopewell*	*1 Apl. 17*	*1 Apl. 17*	*I. A. R. O.*
1 Apl. 17	*Henary, Frank Coutts,* M.C.	*1 Apl. 17*	*1 Apl. 17*	*I. A. R. O.*
1 Apl. 17	*Fogarty, Philip Christopher*	*1 Apl. 17*	*1 Apl 17*	*I. A. R. O.*
1 Apl. 17	*Stevenson, James Gemmell*	*1 Apl. 17*	*1 Apl. 17*	(*tempy. 2nd-Lt., R. E.*)
1 Apl. 17	Clover, Lt. William Bertram	1 Apl. 7	1 Apl. 17	
1 Apl 17	Hope, John Cunningham	1 Apl. 17	1 Apl. 17	Syriam.
1 Apl. 17	*Grant, Charles Frederick*	*1 Apl. 17*	*1 Apl. 17*	*I. A. R. O.*
1 Apl. 17	*Pearson, Henry Lawrence*	*1 Apl. 17*	*1 Apl. 17*	*Temp. Col., R. E.*
1 Apl. 17	Furnivall, John Sydenham	1 Apl 17	1 Apl. 17	Myaungmya.
1 Apl. 17	Bellars, Albert Ernest	1 Apl. 17	1 Apl. 17	
1 Apl. 17	*Barrington, Arthur Harry Manliffe*	*1 Apl. 17*	*1 Apl. 17*	*I. A. R. O.*

18th Rangoon Battalion—*contd.*

First Commn.	Names.	Date of rank.	Date of appointment to corps	Remarks.
	Lieutenants—*contd.*			
1 Apl. 17	Lely, William Gerald	1 Apl. 17	1 Apl. 17	
1 Apl. 17	Ferguson, James	1 Apl. 17	1 Apl. 17	
1 Apl. 17	*Campbell, Alexander*	*1 Apl. 17*	*1 Apl. 17*	*I. A. R. O.*
1 Apl. 17	*Tait, Arthur*	*1 Apl. 17*	*1 Apl. 17*	*I. A. R. O.*
1 Apl. 17	*Gill, Cyril Alfred Goddard*	*1 Apl. 17*	*1 Apl. 17*	*I. A. R. O.*
1 Apl. 17	*Watt William Wallace*	*1 Apl. 17*	*1 Apl. 17*	*I. A. R. O.*
1 Apl. 17	Bennett, William	1 Apl. 17	1 Apl. 17	
1 Apl. 17	Campbell, John	1 Apl. 17	1 Apl. 17	Mounted Platoon.
1 Apl. 17	Baum, Ernest Frederick	1 Apl. 17	1 Apl. 17	
	2nd-Lieutenants.			
1 Apl. 17	*Baum, Ernest Frederick*	*1 Apl. 17*	*1 Apl. 17*	*I. A. R. O.*
1 Apl. 17	*Rees, Thomas Owen*	*1 Apl. 17*	*1 Apl. 17*	*I. A. R. O.*
1 Apl. 17	Brown, Harold Arrowsmith	1 Apl. 17	1 Apl. 17	Myaungmya.
1 Apl. 17	Whyte, William	1 Apl. 17	1 Apl. 17	
1 Apl. 17	Marsh, Hubert Ronald (107)	1 Apl. 17	1 Apl. 17	
1 Apl. 17	*Nelson, John Ingis (105, 207)*	*1 Apl. 17*	*1 Apl. 17*	*I. A. R. O.*
1 Apl. 17	Rogerson, Arthur Marsden	1 Apl. 17	1 Apl. 17	
1 Apl. 17	Gibbon, Sam Harrison (105, 207*a*)	1 Apl. 17	1 Apl. 17	*Lv. ex. Ind., 6 mos. from 1-9-18.*
1 Apl. 17	Beamish Richard, Ronayne (207)	1 Apl. 17	1 Apl. 17	
1 Apl. 17	Maclean, Hugh Alexander (207*a*)	1 Apl. 17	1 Apl. 17	
1 Apl. 17	Buchanan, Walter	1 Apl. 17	1 Apl. 17	
1 Apl. 17	McDonnell, Thomas Francis Robert	1 Apl. 17	1 Apl. 17	
1 Apl. 17	Laidlaw, Hugh Alexander Lyon (207*a*)	1 Apl. 17	1 Apl. 17	
1 Apl. 17	Gray, William Athelstane (107, 207*a*)	1 Apl. 17	1 Apl. 17	
1 Apl. 18	McKay, Charles Hamilton	1 Apl. 18	1 Apl. 18	
1 Apl. 18	*Rawlings, Donald Arthur Russell*	*1 Apl. 18*	*1 Apl. 18*	*I. A. R. O.*
1 Apl. 18	*Baker, Robert Henry*	*1 Apl. 18*	*1 Apl. 18*	*I. A. R. O.*
1 Apl. 18	Britton, Charles John James	1 Apl. 18	1 Apl. 18	
24 May 18	Ritchie, Aden Beattie	24 May 18	24 May 18	
1 Oct. 18	McAllister, James	1 Oct. 18	1 Oct. 18	
1 Oct. 18	Harding, Edwin Albert Durtnall	1 Oct. 18	1 Oct. 18	
1 Oct. 18	Neale, Kenneth Geoffrey	1 Oct. 18	1 Oct. 18	

Supernumerary.

1 Apl. 17	Holdsworth, Capt. Edwin Richard	1 Apl. 17	1 Apl. 17	

Staff.

1 Apl. 17	Hill, Capt. Cyril Patrick (107, 207*a*)	1 Apl. 17	1 Apl. 17	Adjutant (tempy.)
. . . .	Magee, Lt. R.	. . . .	. . . .	Qr.-Mr.
1 Apl. 17	Tomlinson, Revd. Wilfred Harry Cowper	1 Apl. 17	1 Apl. 17	Hony. Chaplain.
1 Apl. 17	Spence, Capt. Arthur Dunbar, M.B.	1 Apl. 17	. . . .	Medl. Offr.
1 Apl. 17	Murray, Lieut. William Paterson, M.B.	1 Apl. 17	. . . .	Ditto.

Uniform—Full dress (Officers)—Drab
Facings—Grass Green.
Working dress.—Khaki drill.

19th Agra Company.

Formed 9th Aug. 1878, by G. G. O. No. 733 of 1878, and reconstituted as an I. D. F. Unit with present designation 1st Apl. 1917.

Establishment—Combined Active & Reserve Coy., 1 Armoured Motor Batty., 1 Machine Gun Dett.

HONORARY COMMANDANT—*The Hon'ble Sir* Spencer Harcourt Butler, K.C.S.I., C.I.E., I.C.S., ... 18 July 18.

First Commn.	Names.	Date of rank.	Date of appointment to corps.	Remarks.
	Major.			
1 Apl. 17	Jones, Thomas Cuthbertson, V.D. (54) (80)	1 Apl. 17	1 Apl. 17	Comdt. (Lt.-Col., late Ind. Volrs.)
	Captains.			
1 Apl. 17	*John, Anthony William Ulysses,* V.D. (54)	*1 Apl. 17*	*1 Apl. 17*	*I. A. R. O. (Major, late Ind. Volrs.)*
1 Apl. 17	Scott, Henry Lyon (80)	1 Apl. 17	1 Apl. 17	*Major, late Ind. Volrs.*
1 Apl. 17	*O'Donnell, Thomas Francis,* M.C.	*1 Apl. 17*	*1 Apl. 17*	*Tempy. Maj. in Army.*
1 Apl. 17	*Oliphant, Donald Aden*	*1 Apl. 17*	*1 Apl. 17*	*I. A. R. O.*
1 Apl. 17	*Ball, Cyril Francis (temp. Maj., 30 Sep. 18) (54)*	*1 Apl. 11*	*1 Apl. 17*	
1 Apl. 17	*Bliss, Cuthbert Vivian* C.M.G. *(54, 80)*	*1 Apl. 17*	*1 Apl. 17*	
1 Apl. 17	Field, James Hermann (54, 80)	1 Apl. 17	1 Apl. 17	*Lv., ex I., 2 yrs., 16th May 18.*
	Lieutenants.			
1 Apl. 17	*Collister, Harold James (80)*	*1 Apl. 17*	*1 Apl. 17*	*I. A. R. O.*
1 Apl. 17	*Brookes, Albert*	*1 Apl. 17*	*1 Apl. 17*	*I. A. R. O.*
Apl. 17	Plowden, Philip Peter Meredyth Chichele (54) (80)	1 Apl. 17	1 Apl. 17	
1 Apl. 17	Bradley, Terence Arthur Ernst (54) (80)	1 Apl. 17	1 Apl. 17	
	2nd-Lieutenants.			
1 Apl. 17	*Harwood, William Albert* (54)	1 Apl. 17	1 Apl. 17	*I. A. R. O.*
1 Apl. 17	Sully, Thomas Donald (80)	1 Apl. 17	1 Apl. 17	
6 Mar. 18	Lyle, David Ranken	6 Mar. 18	6 Mar. 18	
6 Mar. 18	Page, James Alfred	6 Mar. 18	6 Mar. 18	
	Supernumerary.			
1 Apl. 17	Key, Capt. Charles Taylor (54, 80)	1 Apl. 17	1 Apl. 17	
	Staff.			
1 Jan. 16	Catling, Lieut. (*temp. Capt.*) A., Dorset. R.	15 Jan. 17	1 Jan. 16	Adjutant.

20th Nilgiri Malabar Battalion.

Uniform—Khaki serge or drill according as stationed in the hills or plains.

Raised as Nilgiri Volunteer Rifles, 24th October 1878 (Madras G. O. 1032, 27th December 1878). Coimbatore Volunteer Rifle Corps, raised 7th August 1885. Incorporated, 4th January 1892. Malabar Volunteer Rifles amalgamated and reconstituted as an I. D. F. Unit with present designation, 1st Apl. 1917.

Establishment.—4 Combined Active & Reserve Coys., 1 Armoured Motor Batty., 1 Machine Gun Dett.

Detachments at Ootacamund, Lovedale, Coonoor, Coimbatore, Aruvankadu, Cannanore, Calicut, Tellicherry, Cochin, Alleppey, Quillon, and Trevandrum.

HONORARY COMMANDANT—

First Commn.	Names.	Date of rank.	Date of appointment to corps.	Remarks.
	Lieutenant-Colonels.			
1 Apl. 17	Chamier, William, V.D. (54)	1 Apl. 17	1 Apl. 17	Tempy. Lt.-Col. in Army.
25 May 18	Ward, Walter, V.D. (48a, 51, 71)	25 May 18	25 May 18	Commdt.
	Majors.			
1 Apl. 17	Porter, Leonard Lachlan (54)	1 Apl. 17	1 Apl. 17	I. A. R. O.
1 Apl. 17	Brock, Charles Henry (54)	1 Apl. 17	1 Apl. 17	I. A. R. O.
1 Apl. 17	Lockie, Hugh Reid, V.D (80)	1 Apl. 17	1 Apl. 17	
1 Apl. 17	Kirkpatrick, Herbert (80)	20 Sep. 18	1 Apl. 17	Staff Capt., Southern Bde.
	Captains.			
1 Apl. 17	Vaughan, Harry Russell	1 Apl. 17	1 Apl. 17	I. A. R. O.
1 Apl. 17	Hornsby, Norman	1 Apl. 17	1 Apl. 17	I. A. R. O.
1 Apl. 17	Campbell, William John (80)	1 Apl. 17	1 Apl. 17	Coimbatore.
1 Apl. 17	Jourdain, Percy Frederick Curry	1 Apl. 17	1 Apl. 17	Temp. Commn. in Army.
1 Apl. 17	Hodgson, Gerrald Hassell	1 Apl. 17	1 Apl. 17	Temp. Commn. in Army.
1 Apl. 17	Pigott, Thomas Francis (51)	1 Apl. 17	1 Apl. 17	I. A. R. O.
1 Apl. 17	Woosnam, Arthur George (54)	1 Apl. 17	1 Apl. 17	I. A. R. O.
1 Apl. 17	Barstow, John Anderson	1 Apl. 17	1 Apl. 17	Temp. Commn. in Army.
1 Apl. 17	Cobban, Henry Lee Norris	1 Apl. 17	1 Apl. 17	Temp. Commn. in Army.
1 Apl. 17	Walters, Thomas James Gray Stewart	1 Apl. 17	1 Apl. 17	I. A. R. O.
1 Apl. 17	Jones, Vaughan	1 Apl. 17	1 Apl. 17	Tempy. Commn. in Army.
1 Apl. 17	Evans, William Arthur Avenel	1 Apl. 17	1 Apl. 17	I. A. R. O.
1 Apl. 17	Henderson, John Lessels (80)	1 Apl. 17	1 Apl. 17	Quilon.
1 Apl. 17	Blackwood, John Richardson	1 Apl. 17	1 Apl. 17	I. A. R. O.
1 Apl. 17	Phillips, Charles Edward (52)	1 Apl. 17	1 Apl. 17	I. A. R. O.
1 Apl. 17	Courtenay, Reginald Herbert (80)	1 Apl. 17	1 Apl. 17	Madras.
1 Apl. 17	Reilly, Henry D'Arcy Cornelius (80)	1 Apl. 17	1 Apl. 17	
1 Apl. 17	Head, Harold Kenneth	1 Apl. 17	1 Apl. 17	I. A. R. O.
1 Apl. 17	Walker, Frederick Kennedy	1 Apl. 17	1 Apl. 17	I. A. R. O.
1 Apl. 17	Duke, Walter Stanley	1 Apl. 17	1 Apl. 17	I. A. R. O.
1 Apl. 17	Porteous, Ernest Roland (80)	1 Apl. 17	1 Apl. 17	Calicut.
1 Apl. 17	Richards, Frederick John (80)	1 Apl. 17	15 Oct. 18	
1 Apl. 17	Campbell, Alexander Farringdon (80)	1 Apl. 17	6 Nov. 18	
	Lieutenants.			
1 Apl. 17	Woodhouse, Albert Cyril (80)	1 Apl. 17	1 Apl. 18	
1 Apl. 17	Nugent, Thomas Dominic	1 Apl. 17	1 Apl. 17	I. A. R. O.
1 Apl. 17	Midgley, Ernest Arthur	1 Apl. 17	1 Apl. 17	I. A. R. O.
1 Apl. 17	Morris, Horace Stanley	1 Apl. 17	1 Apl. 17	I. A. R. O.
1 Apl. 17	Butcher, Frank Henry (80)	1 Apl. 17	1 Apl. 17	Ootacamund.
1 Apl. 17	Barker, Sydney George (80)	1 Apl. 17	1 Apl. 17	Trevandrum.
1 Apl. 17	Waller, David Grierson (80)	1 Apl. 17	1 Apl. 17	Coimbatore.
1 Apl. 17	Robb, Arthur Angus	1 Apl. 17	1 Apl. 17	I. A. R. O.
1 Apl. 17	Pitcairn, John Erskine (51, 80)	1 Apl. 17	1 Apl. 17	Temp. Commn. in Army.
1 Apl. 17	Anderson, William Torr (80)	1 Apl. 17	1 Apl. 17	Cochin.
1 Apl. 17	Buckner, Hugh Montmorency (80)	1 Apl. 17	1 Apl. 17	Ootacamund.
1 Apl. 17	Macbean, James Strong (80)	1 Apl. 17	1 Apl. 17	Cochin.

20th Nilgiri Malabar Battalion—*contd.*

First Commn.	Name.	Date of rank.	Date of appointment to corps.	Remarks.
	2nd-Lieutenants.			
1 Apl. 17	Deane, Stanley O'Meara (80)	1 Apl. 17	1 Apl. 17	Cochin.
1 Apl. 17	Hall John Frederick (80)	1 Apl. 17	1 Apl. 17	Calicut.
1 Apl. 17	Richardson, Albert Victor (80)	1 Apl. 17	1 Apl. 17	Lovedale.
28 Mar. 18	Sheldon, Norman Lindsay	28 Mar. 18	28 Mar. 18	Coonoor.
28 Mar. 18	Winter, Walter Ernest	28 Mar. 18	28 Mar. 18	Coimbatore.
	Supernumerary.			
1 Apl. 17	Baker, Capt., George Frederick (80) . .	1 Apl. 17	1 Apl. 17	Calicut.
1 Apl. 17	Hart, Capt., Vincent (80)	1 Apl. 17	1 Apl. 17	Coimbatore.
1 Apl. 17	Marsden, Capt. Arthur Peploe	1 Apl. 17	1 Apl. 17	
	Staff.			
1 Jan. 17	Sumner, 2nd-Lt. (*temp. Capt.*) W. S. Lan. R.	1 Jan. 17	1 Jan. 17	Adjutant.
19 Oct. 18	Baker, Hony. Lieut.	19 Oct. 18	19 Oct. 18	Qr.-Mr.

Uniform—Full dress (Officers only) Drab.
Facings—Maroon.
Working Dress—Khaki drill.

21st Burma Railways Battalion.

Raised, 18th April 1879, by M. D. Notfn. No. 370, dated 2nd May 1879. Designation altered by M. D. Notfn. No. 876 of 1897 and 720 of 1899. Reconstituted as an I. D. F. Unit with present designation, 1st Apl. 1917.

Establishment—4 Combined Active & Reserve Coys. 1 Armoured Train Dett. (Private). 1 Machine Gun Dett.

Companies at Prome, Toungoo, Pegu, Insein, Ywataung, Maymyo, Malagon, Bassein and Yamethin.

HONORARY COMMANDANT—

First Commn.	Names.	Date of rank.	Date of appointment to corps.	Remarks.
	Lieutenant-Colonels.			
1 Apl. 17	Bigg-Wither, Frederick Lovelace, V.D. (207).	1 Mar. 18	1 Apl. 17	Comdt. Rangoon.
	Majors.			
1 Apl. 17	Crosthwaite, Bertram Maitland	1 Apl. 17	1 Apl. 17	2nd-in-Comd. Rangoon
1 Apl. 17	Brereton, Randle Archange	1 Apl. 17	1 Apl. 17	Rangoon.
	Captains.			
1 Apl. 17	Jones, Thomas Martin	1 Apl. 17	1 Apl. 17	Toungoo.
1 Apl. 17	Cantor, Arthur Augustus	1 Apl. 17	1 Apl 17	Rangoon.
1 Apl. 17	Pratt, Walter Philip Raven	1 Apl. 17	1 Apl. 17	Rangoon.
1 Apl. 17	Hutton, Charles Inglis	1 Apl. 17	1 Apl. 17	Insein.
1 Apl 17	Moore, Arthur Esmond Du Pre	1 Apl. 17	1 Apl. 17	Martaban.
1 Apl. 17	Hicks, George Augustus	1 Apl. 17	1 Apl. 17	Rangoon.
1 Apl. 17	*Thomas, Humbert George Saxby*	*1 Apl. 17*	*1 Apl. 17*	
1 Apl. 17	*Darby, George Scott*	*1 Apl. 17*	*1 Apl. 17*	*I.A.R.O.*
1 Apl. 17	*Hardwick, Noel De Courcy*	*1 Apl. 17*	*1 Apl. 17*	*I. A. R.O.*
1 Apl. 17	Johnson, James (79)	1 Apl. 17	1 Apl. 17	Mandalay.
	Lieutenants.			
1 Apl. 17	Mackie, Ronald Hastings	1 Apl. 17	1 Apl. 17	Capt. late Ind. Volrs.
1 Apl. 17	McIver, Norman Alexander Harris	1 Apl. 17	1 Apl 17	Myitinge.
1 Apl. 17	Sparke, Walter	1 Apl. 17	1 Apl. 17	Reserve Co., Toungoo.
1 Apl. 17	*Power, Robin Turenne*	*1 Apl. 17*	*1 Apl. 17*	
1 Apl. 17	Hardy, Reginald Herbert	1 Apl. 17	1 Apl. 17	Rangoon.
1 Apl. 17	Bell, Charles Burnby	1 Apl. 17	1 Apl. 17	Ywataung.
1 Apl. 17	Glascott, John Richard Donovan	1 Apl. 17	1 Apl. 17	Rangoon.
1 Apl. 17	*Wharton, John*	*1 Apl. 17*	*1 Apl. 17*	Rangoon.
1 Apl. 17	*Beaman, Gerald William Devonish*	*1 Apl. 17*	*1 Apl. 17*	*I. A. R. O.*
1 Apl. 17	*Aikman, Edward Henry*	*1 Apl. 17*	*1 Apl 17*	*I. A. R. O.*
1 Apl. 17	Rickie, James Hepburn	1 Apl 17	1 Apl. 17	Henzada.
1 Apl. 17	*Hailes, Martin Hunter*	*1 Apl. 17*	*1 Apl. 17*	(*Temp. Capt. in Army*)
1 Apl. 17	Chance, William Henry	1 Apl. 17	1 Apl. 17	Rangoon.
1 Apl. 17	*Toller, George Gordon Tayle*	*1 Apl. 17*	*1 Apl. 17*	*I. A. R. O.*
1 Apl. 17	*Rowland, John Edward Maurice*	*1 Apl. 17*	*1 Apl. 17*	
	2nd-Lieutenants.			
1 Apl. 17	Murray, Norman	1 Apl. 17	1 Apl. 17	Lt. late Ind. Volrs.
1 Apl. 17	*Ferguson, Donald Harry,* M.C.	*1 Apl. 17*	*1 Apl. 17*	*I. A. R. O.*
1 Apl. 17	Harvey, Austin George	1 Apl. 17	1 Apl. 17	Lt. late Ind. Volrs.

21st Burma Railways Battalion—*contd.*

First Commn.	Names.	Date of rank.	Date of appointment to corps.	Remarks.
	2nd-Lieutenants.			
1 Apl. 17	*Doog, John Myles Gregory*	*1 Apl. 17*	*1 Apl. 17*	
1 Apl. 17	*Cardew, John Horace*	*1 Apl. 17*	*1 Apl. 17*	*I. A. R. O.*
1 Apl. 17	*Martin, Theobald Seymour Stewart*	*1 Apl. 17*	*1 Apl. 17*	*I. A. R. O.*
1 Apl. 17	Down, Edmund Charles	1 Apl. 17	1 Apl. 17	Bassein.
1 Apl. 17	*Bleeck, Erlysman*	*1 Apl. 17*	*1 Apl. 17*	*Lv., ex I., m.c., 1 yr.*
1 Apl. 17	*Greenway, Charles Mervyn*	*1 Apl. 17*	*1 Apl. 17*	
1 Apl. 17	Cobban, Stanley William	1 Apl. 17	1 Apl. 17	Myitnge.
1 Apl. 17	Van Der Beek, Aubrey Guilford	1 Apl. 17	1 Apl. 17	Rangoon.
1 Apl. 17	*Bevan, Gilbert John Beckford*	*1 Apl. 17*	*1 Apl. 17*	
1 Apl. 17	Walley, Arthur Frank	1 Apl. 17	1 Apl. 17	Rangoon.
1 Apl. 17	Pennell, Charles Stuart	1 Apl. 17	1 Apl. 17	*Lv., 6 mos., 4 Sep. 17.*
	Supernumerary.			
1 Apl. 17	Andrew, Maj. G. (P. 1207)	1 Apl. 17	7 Sep. 18	
1 Apl. 17	Colvin, Capt. Clement Preston	1 Apl. 17	1 Apl. 17	
1 Apl. 17	Cantrell, Capt. William Conyngham	1 Apl. 17	1 Apl. 17	
1 Apl. 17	Macdonell, Capt. Oliver Stephen	1 Apl. 17	1 Apl. 17	
1 Apl. 17	deRosario, Capt. Cyril Montague	1 Apl. 17	1 Apl. 17	
1 Apl. 17	Cohen, Lt. Alexander Joseph	1 Apl. 17	1 Apl. 17	
1 Apl. 17	Ward, 2nd-Lt. Frank Dudley	1 Apl. 17	1 Apl. 17	
1 Apl. 17	*Craig, Capt. Hugh Alexander*	*1 Apl. 17*	*1 Apl. 17*	
1 Apl. 17	Lowry, Lt. William Edward	1 Apl. 17	1 Apl. 17	
	Staff.			
7 May 17	Claydon, 2nd-Lt. (*temp. Capt.*) S., Leic. R.	7 May 17	7 May 17	Adjutant.
14 Mar. 18	Blythe, *Hony.* Lt. J.	14 Mar. 18	14 Mar. 18	Quarter-Master.

Uniform—Khaki drill.

22nd Bengal and North-Western Railway Battalion.

Formed 14 June 1879. Tirhut State Ry. Vol. Rifle Corps amalgamated 17 June 1892, by Mily. Dept. Notfn. No. 593, 17 June 1892.

(Designation altered by G. O. C. C. No 327, dated 20th Apl. 01.) Reconstituted as an I. D. F. Unit with present designation, 1st Apl. 1917.

Establishment—2 Reserve Coys., 1 Armoured Train Dett., 2 Machine Gun Detts.

Companies at Sonepore, Samastipore, Gonda, Benares and Bareilly.

HONORARY COMMANDANT—Lieutenant-Colonel (Hony. Colonel) James Walker, V.D. 1 September 1918.

First Commn.	Names.	Date of rank.	Date of appointment to corps.	Remarks.
	Lieutenant-Colonel.			
1 Apl. 17	Miller, William, V.D.	1 Sep. 18	1 Apl. 17	Comdt.
	Majors.			
1 Apl. 17	*Tainsh, Joseph Ramsay*, V.D.	*1 Apl. 17*	*1 Apl. 17*	*I. A. R. O.*
1 Apl. 17	Dundas, Robert Morison, V.D.	1 Apl. 17	1 Apl. 17	2nd-in-Comd., 1 Sep. 18.
1 Apl. 17	*Ferry, Edward Stanton Henry*, V.D.	*1 Apl. 17*	*1 Apl. 17*	
	Captains.			
1 Apl. 17	Paton, William	1 Apl. 17	1 Apl. 17	
1 Apl. 17	Westwood, John David	1 Apl. 17	1 Apl. 17	
1 Apl. 17	Smith, James Arthur	1 Apl. 17	1 Apl. 17	
1 Apl. 17	Jenkins, William	1 Apl. 17	1 Apl. 17	
	Lieutenants.			
1 Apl. 17	Anderson, David Young	1 Apl. 17	1 Apl. 17	
1 Apl. 17	*Clark, Robert Napier*	*1 Apl. 17*	*1 Apl. 17*	*I. A. R. O.*
1 Apl. 17	Hannay, Robert	1 Apl. 17	1 Apl. 17	
1 Apl. 17	Holmes, Geoffrey Meredith Kersley	1 Apl. 17	1 Apl. 17	
1 Apl. 17	*Lawler, John Rose*	*1 Apl. 17*	*1 Apl. 17*	
1 Apl. 17	Mann, George Kyd	1 Apl. 17	1 Apl. 17	
1 Apl. 17	Dyer, Alfred James Alexander	1 Apl. 17	1 Apl. 17	
	2nd-Lieutenants.			
1 Apl. 17	Watson, Robert	1 Apl. 17	1 Apl. 17	
1 Apl. 17	Mair, Andrew	1 Apl. 17	1 Apl. 17	
1 Apl. 17	Jolly, William	1 Apl. 17	1 Apl. 17	
28 June 18	Edwards, Lindsay Ernest	1 Apl. 17	1 Apl. 17	
	Supernumerary.			
1 Apl. 17	Blunt, Capt. Edward Arthur Henry	1 Apl. 17	1 Apl. 17	
1 Apl. 17	McArthy, Lt. Stanley Egerton Frank D'Arcy.	1 Apl. 17	1 Apl. 17	
	Staff.			
25 June 18	Ball, 2nd-Lt. (*actg. Capt.*) E., Leic. R.	25 June 18	25 June 18	Adjt.

Uniform—Khaki drill.

23rd Sind Battalion.

Formed 11 Nov. 1879. Reorganised, G. O. C. C 327 of 29 Apl. 1901, and reconstituted as an I.D.F. Unit with present designation 1 Apl. 1917.
Establishment—1 Active Coy., 1 Res. Coy., 1 Combined Active & Reserve Coy., 1 Armoured Motor Batty. (B).
Companies at Hyderabad and Sukkur.

HONORARY COMMANDANT—*The Hon'ble* MR. HENRY STAVELEY LAWRENCE, C.S.I., I.C.S. 1st Apl. 1917.
(*Commissioner in Sind.*)

First Commn.	Names.	Date of rank.	Date of appointment to corps.	Remarks.
	Lieutenant-Colonel.			
1 Apl 17	Anderson, Charles Stephen, V. D. (80)	1 Apl. 17	1 Apl. 17	Comdt., Hony. A.-D.-C. to C.-in-C.
	Majors.			
1 Apl. 17	*Mahon, Harry John*, C.I.E. (52, 71)	*1 Apl. 17*	*1 Apl. 17*	(*Tempy. Col. in Army.*) *Embkn. Comdt., Karachi.*
1 Apl. 17	*Hotson, John Ernest Buttery*, O.B.E.	*1 Apl. 17*	*1 Apl. 17*	*I. A. R. O.*
1 Apl. 17	*Arbuthnot, Clifford William Ernest* (46, 186)	*1 Apl. 17*	*1 Apl. 17*	*I. A. R. O.*
1 Apl. 17	Young, Ramsay Hannay (D, II, III) (80)	1 Apl. 17	1 Apl. 17	
	Captains.			
1 Apl. 17	Bolus, Edward John	1 Apl. 17	1 Apl. 17	Sukkur.
1 Apl. 17	Neilson, William Hardcastle (80)	1 Apl. 17	1 Apl. 17	
1 Apl. 17	Holland, Henry Tristram (69)	1 Apl. 17	1 Apl. 17	Hyderabad.
1 Apl 17	Gunter, Eustace Edward, O.B.E. (80)	1 Apl. 17	1 Apl. 17	
1 Apl. 17	*Nicholas, Richmond Horace* (54)	*1 Apl. 17*	*1 Apl. 17*	*I. A. R. O.*
1 Apl. 17	*Cooper, Joseph Albert William*, (71, 207)	*1 Apl. 17*	*1 Apl. 17*	(*Temp. Capt., Labour Corps.*)
1 Apl. 17	Demetriadi, Constantine Constantine (80, 82a)	1 Apl. 17	1 Apl. 17	
	Lieutenants.			
1 Apl. 17	*Anderson, Charles Edwin Lindsay*	*1 Apl. 17*	*1 Apl. 17*	*I. A. R. O.*
1 Apl. 17	*Walker, Arthur William*	*1 Apl. 17*	*1 Apl. 17*	*I. A. R. O.*
1 Apl. 17	*Culverweil, Joseph Stanley*	*1 Apl. 17*	*1 Apl. 17*	*I. A. R. O.*
1 Apl. 17	Flynn, Alfred Axen Leonhard (80)	1 Apl. 17	1 Apl. 17	
1 Apl. 17	Morgan, Godfrey Noel Richard, (80, 82a)	1 Apl. 17	1 Apl. 17	
1 Apl. 17	*Travers, Francis Guy* (*tempy. Capt.*) (80, 82a)	*1 Apl. 17*	*1 Apl. 17*	*Offg. Adjutant.*
1 Apl. 17	Petrie, Walter Macgregor (80, 82a)	*1 Apl. 17*	*1 Apl. 17*	
1 Apl. 17	Waite, Fabian Evans (51, 80)	1 Apl. 17	1 Apl. 17	
1 Apl. 17	Ravenscroft, Stanley (80)	1 Apl. 17	1 Apl. 17	
1 Apl. 17	Harris, Cyril James Watson (80)	1 Apl. 17	1 Apl. 17	
	2nd-Lieutenants.			
1 Apl. 17	*Menzies, Frederick Auber*, M.C.	*1 Apl. 17*	*1 Apl. 17*	*I. A. R. O.*
1 Apl. 17	*Furse, Gordon*	*1 Apl. 17*	*1 Apl. 17*	*Lt., Spec. Res., C. Gds.*
1 Apl. 17	Marshall, Alexander Walter (80)	1 Apl. 17	1 Apl. 17	
1 Apl. 17	Deane, William Stanley (80)	1 Apl. 17	1 Apl. 17	
1 Apl. 17	*Shuttleworth, John Thornton*	*1 Apl. 17*	*1 Apl. 17*	*I. A. R. O.*
1 Apl. 17	Marshal, Roger Davis (80)	1 Apl. 17	1 Apl. 17	
1 Apl. 17	Barker, Stanley (51, 80)	1 Apl. 17	1 Apl. 17	
	Supernumerary.			
1 Apl. 17	Anastasiadi, Maj. Stavro Eustace, V.D. (54)	1 Apl. 17	1 Apl. 17	
	Staff.			
16 Apl. 17	Travers, Lt. (*temp. Capt.*) F. G. (80, 82a)	1 Apl. 17	1 Apl. 17	Offg. Adjutant.
10 Mar. 18	McAndrew, Lt. Thomas (93)	10 Mar. 18	10 Mar. 18	Quarter-Master.

Uniform—Khaki drill.

1-24th North Western Railway Battalion.

Formed 5th Mar. 80. Reorganised 29th July 92. Reconstituted as an I. D. F. Unit with present designation, 1st Apl, 1917

Companies at

Establishment—5 Reserve Coys., 2 Armoured Train Detts.

HONORARY COMMANDANT—*The Hon'ble Sir* MICHAEL FRANCIS O'DWYER, G.C.I.E., K.C.S.I. Apl. 17.

(Lieutenant-Governor of the Punjab.)

First Commn.	Names.	Date of rank.	Date of appointment to corps.	Remarks.
	Lieutenant-Colonel.			
1 Apl. 17	White, Henry George Norman, V. D., (51, 71, 80, 207) (Q) q.	1 Apl. 17	16 Nov. 18	Commandant.
	Majors.			
1 Apl. 17	Calder, Norman Douglas	1 Apl. 17	3 July 18	2nd-in-Command.
1 Apl. 17	Wrench, William Thomas	1 Apl. 17	1 Apl. 17	
	Captains.			
1 Apl. 17	Chase, John Hartley	1 Apl. 17	1 Apl. 17	
1 Apl. 17	Flowerdew, Spencer Pelham (71)	1 Apl. 17	1 Apl. 17	
1 Apl. 17	Meares, George William (51, 59)	1 Apl. 17	1 Apl. 17	
1 Apl. 17	Scott, Ernest Lynch	1 Apl. 17	1 Apl. 17	
1 Apl. 17	Green, Harry Dale, V.D.	1 Apl. 17	1 Apl. 17	
1 Apl. 17	Joscelyne, Alan Hope (207)	1 Apl. 17	1 Apl. 17	
1 Apl. 17	Armitstead, Henry	1 Apl. 17	1 Apl. 17	
1 Apl. 17	Crosbie, Alfred (207)	1 Apl. 17	1 Apl. 17	
	Lieutenants.			
1 Apl. 17	Weller, Henry Owen (48A, 51)	1 Apl. 17	1 Apl. 17	
1 Apl. 17	Charlton, Henry Johnson	1 Apl. 17	1 Apl 17	
1 Apl. 17	Hunter, John Chalmers	1 Apl. 17	1 Apl. 17	
1 Apl. 17	Scott, Ernest Alexander	1 Apl. 17	1 Apl. 17	
1 Apl. 17	Phillips, Alan Andrew	1 Apl. 17	1 Apl. 17	
1 Apl. 17	Martin, Frederick Barclay (161, 169)	1 Apl. 17	1 Apl. 17	
1 Apl. 17	Mahoney, John Spencer (51)	1 Apl. 17	1 Apl. 17	
1 Apl. 17	Trevor, Dighy Bruce	1 Apl. 17	1 Apl. 17	
1 Apl. 17	Taylor, Ernest Bertram Neave	1 Apl. 17	1 Apl. 17	
1 Apl. 17	Ogle, John Wilfrid (51)	1 Apl. 17	1 Apl. 17	
1 Apl. 17	Dean, Vincent Lee (62*a*)	1 Apl. 17	1 Apl. 17	
	2nd-Lieutenants.			
1 Apl. 17	Griffith, Francis Gordon	1 Apl. 17	1 Apl. 17	
1 Apl. 17	Smith, Donald D'Arcy	1 Apl. 17	1 Apl. 17	
1 Apl. 17	Goulding, Herbert Percy Rayner	1 Apl. 17	1 Apl. 17	
	Supernumerary.			
1 Apl. 17	Royal-Dawson. Lt.-Col. Frederick George	1 Apl. 17	1 Apl. 17	
1 Apl. 17	Strong, Major Ernest Sidney, V.D.	1 Apl. 17	1 Apl. 17	
1 Apl. 17	Chase, Major Alfred James	1 Apl. 17	1 Apl. 17	
1 Apl. 17	Stowell, Captain Atrhur Terance	1 Apl. 17	1 Apl. 17	
1 Apl. 17	Hadow, Captain Frederick Austen	1 Apl. 17	1 Apl. 17	
1 Apl. 17	Sandys, Captain Ulric Capel	1 Apl. 17	1 Apl. 17	
1 Apl. 17	Davis, Lieut. Percy William	1 Apl. 17	1 Apl. 17	
1 Apl. 17	Proudlock, Lieut. Robin	1 Apl. 17	1 Apl. 17	
	Staff.			
14 Mar. 18	Annett, 2nd-Lt. (*Tempy. Capt.*) J. R. Ir. Fus.	14 Mar. 18	14 Mar. 18	Adjutant.
9 Mar. 18	Camm, Lt. John	9 Mar. 18	9 Mar. 18	Qr.-Mr.
1 Apl. 17	Younghusband, Revd. Oswald	1 Apl. 17	1 Apl. 17	Hony. Chap'ain.

2—24th North Western Railway Battalion.

Establishment.—3 Reserve Coys., 1 Armoured Train Dett.

HONORARY COMMANDANT—*The Hon'ble Sir* MICHAEL FRANCIS O'DWYER, G.C.I.E., K.C.S.I. 1 Apl. 17.

(*Lieutenant-Governor of the Punjab.*)

First Commn.	Names.	Date of rank.	Date of appointment to corps.	Remarks.
	Lieutenant-Colonel.			
1 Apl. 17	Sparke, Harry Charles, V.D. (206)	1 Apl. 17	1 Apl. 17	Comdt.
	Major.			
	Captains.			
1 Apl. 17	de Rhe-Philipe, George William Vitalli (80, 207)	1 Apl. 17	1 Apl. 17	Lahore.
1 Apl. 17	Hunt, Edward (207)	1 Apl. 17	1 Apl. 17	Sukkur.
1 Apl. 17	Hales, Thomas Christmas (207)	1 Apl. 17	1 Apl. 17	Lahore.
1 Apl. 17	Green, John Dal (80, 207)	1 Apl. 17	1 Apl. 17	Sukkur.
1 Apl. 17	Green, Alfred Dale, V.D. (207)	1 Apl. 17	1 Apl. 17	Lahore.
	Lieutenants.			
1 Apl. 17	Johnston, Augustus Frederick (167, 207)	1 Apl. 17	1 Apl. 17	Multan.
1 Apl. 17	Spurgeon, Christopher Edward (80, 207)	1 Apl. 17	1 Apl. 17	Karachi.
1 Apl. 17	White, Charles Fitzwilliam (80, 207)	1 Apl. 17	1 Apl. 17	Karachi.
1 Apl. 17	*Jackson, Harry*	*1 Apl. 17*	*1 Apl. 17*	
1 Apl. 17	August, Frederick Richard Bagley (207)	1 Apl. 17	1 Apl. 17	Layallpur.
1 Apl. 17	*Strange, Frank Horace Lumsdon*	*1 Apl. 17*	*1 Apl. 17*	
	2nd-Lieutenants.			
1 Apl. 17	Manico, Edward Leslie (80, 207)	1 Apl. 17	1 Apl. 17	Saharanpur.
1 Apl. 17	*Nagle, James Nicholas*	*1 Apl. 17*	*1 Apl. 17*	
1 Apl. 17	Avril, Sidney Marcel (80)	1 Apl. 17	1 Apl. 17	Karachi.
	Supernumerary.			
	Beatson, Capt. Ernest Bruce, V.D.			
	Staff.			
8 Nov. 15	Flanagan, Lt. (*Temp. Capt.*) G., Manch. R. (95)	19 Feb. 17	8 Nov. 15	Adjutant.
9 Mar. 18	Farman, Lieut. W. S., General List	9 Mar. 18	9 Mar. 18	Quarter-master.

Uniform—Full dress—Drab.
Working dress—Khaki drill.

25th Oudh and Rohilkhand Railway Battalion.

Formed from the Oudh Volunteer Rifles, and the Rohilkhand Volunteer Rifles, by Military Department Notfn. No. 779, dated 24th July 1903. Reconstituted as an I. D. F. Unit with present designation 1st Apl. 1917.

Establishment—2 Reserve Coys., 1 Armoured Train Dett.

Companies at Fyzabad, Bareilly and Moradabad.

HONORARY COMMANDANT—

First Commn.	Names.	Date of rank.	Date of appointment to corps.	Remarks.
	Lieutenant-Colonel.			
1 Apl. 17	Holmes, Henry Burvill, V. D. (207)	1 Apl. 17	1 Apl. 17	Comdt. Hon. Col., late Ind. Volrs.
	Major.			
	Captains.			
1 Apl. 17	Keelan, Douglas Hugh, V.D.	1 Apl. 17	1 Apl. 17	
1 Apl. 17	Cardew, David	1 Apl. 17	1 Apl. 17	
1 Apl. 17	*Jackson, Frank (51, 71)*	*1 Apl. 17*	*1 Apl. 17*	*Temp. Commn. in Army.*
1 Apl. 17	Vining, Leonard Edward (51, 79)	1 Apl. 17	1 Apl. 17	*Lv., ex I., 12 mos., 21-9-18.*
	Lieutenants.			
1 Apl. 17	Synnott, Neville William (79)	1 Apl. 17	1 Apl. 17	
1 Apl. 17	*Rose, George Wright Norton* (79)	*1 Apl. 17*	*1 Apl. 17*	*I.A.R.O.*
1 Apl. 17	Lacey, Onward Bayes (79)	1 Apl. 17	1 Apl. 17	
1 Apl. 17	*Marriot, Robert Eckstein*	*1 Apl. 17*	*1 Apl. 17*	*I.A.R.O.*
1 Apl. 17	Morgan, Frederick Robert (79)	1 Apl. 17	1 Apl. 17	
1 Apl. 17	McLeod, Norman Charles (80)	1 Apl. 17	22 June 18	
1 Apl. 17	Homan, Arthur Knox (79, 207)	1 Apl. 17	1 Apl. 17	
	2nd-Lieutenants.			
1 Apl. 17	Laughton, George Christian	1 Apl. 17	1 Apl. 17	
1 Apl. 17	Grindlay, Archibald Percival Reginald	1 Apl. 17	1 Apl. 17	
1 Apl. 17	Brown, Ivan Crisp	1 Apl. 17	1 Apl. 17	
1 Apl. 17	Hunter, Alexander Gardner Martin	1 Apl. 17	1 Apl. 17	
	Supernumerary.			
	Staff.			
20 Apl. 16	Powell, Lt. (temp. Capt.) G., K. O. Sco. Bord.	20 Apl. 16	20 Apl. 16	Adjutant.

Uniform—Khaki drill.

26th Hyderabad Rifles.

Formed 7th July 1882, and reconstituted as an I. D. F. Unit with present designation 1st Apl. 1917.
Detachments at Chudderghat, Hutti, Singareni and Gallagooda.

Establishment:—1 Active Coy., 2 Reserve Coys., 1 Combined Active and Reserve Coy.

HONORARY COMMANDANT—*The Hon'ble Sir* STUART MITFORD FRASER, K.C.S.I., C.I.E. 1 Apl. 17.

First Commn.	Names.	Date of rank.	Date of appointment to corps.	Remarks.
	Lieutenant-Colonel.			
1 Apl. 17	Carew Smyth, William, R. E.	1 Apl. 17	1 Apl. 17	Comdt.
	Majors.			
1 Apl. 17	*Hosken, Arthur Fayrer*	*1 Apl. 17*	*1 Apl. 17*	Temp. Capt., Notts. & Derby R.
1 Apl. 17	Sewell, Robert Arden Dalrymple, V.D.	1 Apl. 17	1 Apl. 17	Hon. Lt.-Col., late Ind. Volrs.
1 Apl 17	Peddie, Graham, V.D.	1 Apl. 17	1 Apl. 17	
	Captains.			
1 Apl. 17	Hudson, Ernsly Haskoll	1 Apl. 17	1 Apl. 17	
1 Apl. 17	Ruddle, Wilfrid Charles Hepburn	1 Apl. 17	1 Apl 17	
1 Apl 17	Burn, Eric Mackenzie	1 Apl. 17	1 Apl. 17	
1 Apl. 17	Wakefield, George Edward Campbell	1 Apl. 17	1 Apl. 17	
1 Apl. 17	Sewell, Louis Hilgrove	1 Apl. 17	1 Apl 17	
1 Apl. 17	*Munn, Leonard*	*1 Apl. 17*	*1 Apl. 17*	Temp. Lt., R. E.
1 Apl. 17	Hodgson, Alexander Ireland (50)	1 Apl. 17	1 Apl 17	
1 Apl. 17	Lynch-Blosse, Hely Richard	1 Apl. 17	1 Apl. 17	
	Lieutenants.			
1 Apl. 17	*Blackwood, Noel Pinkstan O'Reilly*	*1 Apl. 17*	*1 Apl. 17*	*I. A. R. O.*
1 Apl. 17	*Gregory, Reginald Victor*	*1 Apl. 17*	*1 Apl. 17*	*Tempy. 2nd-Lt., R. A.*
1 Apl. 17	Robertson Arthur	1 Apl. 17	1 Apl. 17	
1 Apl. 17	Soden, Ernest Edward	1 Apl. 17	1 Apl. 17	
1 Apl. 17	*Jervis, Archibald Elton Williams*	*1 Apl. 17*	*1 Apl. 17*	*I. A. R. O.*
1 Apl. 17	Parker, Llewellyn	1 Apl. 17	1 Apl. 17	
17 Apl. 17	Crer Alexander Norman Ley	17 Apl. 17	17 Apl. 17	
28 June 17	Cameron Roderick Harold	28 June 17	28 June 17	Indian Detail Camp, Deolali.
1 Feb. 18	Palairet Charles Roland	1 Feb 18	1 Feb. 18	
	2nd-Lieutenants.			
1 Feb. 18	Tripp, Harold Chamier	1 Feb. 18	1 Feb. 18	
	Supernumerary.			
1 Apl. 17	Dunlop, Maj. Henry Johnstone, V.D.	1 Apl. 17	1 Apl. 17	
1 Apl. 17	Burnett, Capt. Thomas Harding	1 Apl. 17	1 Apl. 17	
	Staff.			
25 Oct. 15	Crick, Lt. (*temp. Capt.*) W., R. Fus.	9 May 17	25 Oct. 15	Adjutant.
1 Mar. 18	Moreman, Lt. J.	1 Mar. 18	1 Mar. 18	Qr.-Mr.

Uniform—Khaki.

27th Baluchistan Company.

Formed 6th July 1883. Reconstituted as an I. D. F. Unit with present designation 1st Apl. 1917.

Establishment—1 Reserve Platoon. 1 Armoured Motor Battymen (B).

HONORARY COMMANDANT—

First Commn.	Names.	Date of rank.	Date of appointment to corps.	Remarks.
	Lieutenant.			
1 Apl. 17	Cornah, Joseph Ralph (51)	1 Apl. 17	1 Apl. 17	
	2nd-Lieutenant.			
23 Mar. 18	Lidierth, James Eugene	23 Mar. 18	23 Mar. 18	

Uniform—Khaki drill.

28th South Andaman Company.

Raised 8th February 1884. by Milv. Dept. letter No. 3087-K., dated 21st November 1883. Reconstituted as an I. D. F. Unit with present designation 1st Apl. 1917.

Establishment—1 Combined Active & Reserve Platoon.

	Captain.			
1 Apl. 17	Scott, George Edward (51, 79, 105, 207)	1 Apl. 17	1 Apl. 17	
	Lieutenant.			
1 Apl. 17	*Thirkell-White, Eric Howard*	1 Apl. 17	1 Apl. 17	*I. A. R. O.*
	2nd-Lieutenant.			
1 Apl. 17	*Tayler, Henry Holroyd*	1 Apl. 17	1 Apl. 17	*I. A. R. O.*

Uniform—Khaki drill

29th South Indian Railway Battalion.

Raised 5th August 1884, by Madras G. O. No. 24, dated 2nd August 1884, and reconstituted as an I. D. F. Unit with present designation 1st Apl. 1917.

Establishment—5 Reserve Coys., 1 Combined Active & Reserve Coy., 2 Machine Gun Detts.

Headquarters | | Negapatam |

Companies at Trichinopoly, Madura Madras, Podanur and Calicut.

HONORARY COMMANDANT—Mr. ALEXANDER MUIRHEAD, C.I.E. . . . 1 APL. 17

First Commn.	Names.	Date of rank.	Date of appointment to corps.	Remarks.
	Lieutenant-Colonels.			
1 Apl. 17	Crighton, Charles, V.D. (80)	1 Apl. 17	1 Apl. 17	Comdt. (*Hon. Col. late Ind. Volrs.*)
	Majors.			
1 Apl. 17	Scott, Beresford Charles (80)	1 Apl. 17	1 Apl. 17	Lv., ex. I., 9 mos.
1 Apl. 17	*Oliver, Robert Edward*, V.D.	*1 Apl. 17*	*1 Apl. 17*	*Tempy. Capt. in Army*
1 Apl. 17	Rowbotham, Charles Stuart, V.D. (80)	1 Apl. 17	1 Apl. 17	Trichinopoly.
1 Apl. 17	O'Shaughnessy, Hugh Patrick, V.D.	1 Apl. 17	1 Apl. 17	
	Captains.			
1 Apl. 17	***Tweedie, Frank Forbes***	*1 Apl. 17*	*1 Apl. 17*	
1 Apl. 17	Moore, Godfrey Wilson (80)	1 Apl. 17	1 Apl. 17	Trichinopoly.
1 Apl. 17	*McIntosh, Alexander*	*1 Apl. 17*	*1 Apl. 17*	Rlys. Mesopotamia. *I.A.R.O.*
1 Apl. 17	*Lambert, John Henry* (80)	*1 Apl. 17*	*1 Apl. 17*	Podanur
1 Apl. 17	Munro, Robert Philip (80)	1 Apl. 17	1 Apl 17	Cannanore.
1 Apl. 17	*Burns, William Hamilton*	*1 Apl. 17*	*1 Apl. 18*	*I. W. T.*
1 Apl. 17	Chabot, Vincent Howard (80)	1 Apl. 17	1 Apl. 17	Tanjore.
1 Apl 17	Newton, George Humphreys (80)	1 Apl. 17	1 Apl. 17	Podanur.
1 Apl. 17	*White, Colin Robert*	*1 Apl. 17*	*1 Apl. 17*	*I. A. R. O.*
1 Apl. 17	*Smythe, Archibald Hugh*	*1 Apl. 17*	*1 Apl. 17*	
1 Apl. 17	*Paul, Arthur Edward*	*1 Apl. 17*	*1 Apl. 17*	
1 Apl. 17	Stephens, Frederick Leonard (54, 80)	1 Apl. 17	1 Apl. 17	Madura.
1 Apl. 17	*Rothera, Percy* (80)	*1 Apl. 17*	*1 Apl. 17*	Rlys., Mesopotamia.
1 Apl. 17	*Williamson, Philip Ralph Clements*	*1 Apl. 17*	*1 Apl. 17*	*I. A. R. O.*
1 Apl. 17	Conway, Clement McCleester	1 Apl. 17	1 Apl. 17	Trichinopoly.
1 Apl. 17	*Flower, Arthur Coleman*	*1 Apl. 17*	*1 Apl. 17*	*I. A. R. O.*
1 Apl. 17	Heywood, William Henry Valentine (51, 80)	1 Apl. 17	1 Apl. 17	Madura.
1 Apl. 17	*Wigfall, Reginald Horace* (54)	*1 Apl. 17*	*1 Apl. 17*	*I. A. R. O.*
1 Apl. 17	*Cross, Alfred Butler*	*1 Apl. 17*	*1 Apl. 17*	*I. A. R. O.*
1 Apl. 17	*Chanter, Alfred Bernard*	*1 Apl. 17*	*1 Apl. 17*	Rlys., Mesopotamia.
1 Apl. 17	*Hickie, Allan Darrell*	*1 Apl. 17*	*1 Apl. 17*	*I. A. R. O.*
1 Apl. 17	*Millar, Alfred John Flude*	*1 Apl. 17*	*1 Apl. 17*	*I. A. R. O.*
1 Apl. 17	Gaby, Clive Osborne (80)	1 Apl. 17	1 Apl. 17	Trichinopoly.
	Lieutenants.			
1 Apl. 17	*Davies, Henry Noel* (80)	*1 Apl. 17*	*1 Apl. 17*	Rlys., E. Africa.
1 Apl 17	Rose, Bernard Alfred (80)	1 Apl. 17	1 Apl. 17	Negapatam.
1 Apl. 17	Stokes, Leonard Edward	1 Apl. 17	1 Apl. 17	Madras.
1 Apl. 17	Brown, Charles Mainwaring (80)	1 Apl. 17	1 Apl. 17	Madura.
1 Apl. 17	*McCloughin, Bertram Gordon* (80)	*1 Apl. 17*	*1 Apl. 17*	Trichinopoly.
1 Apl. 17	*Mainprice, Geoffrey Dennis*	*1 Apl. 17*	1 Apl. 17	Rlys., Mesopotamia.
1 Apl. 17	Carmichael, Wilfred George Hill (80)	1 Apl. 17	1 Apl. 17	Podanur.
1 Apl. 17	Ryle, Charles Danby, V. D.	1 Apl. 17	1 Apl. 17	Yercaud.
30 Mar. 18	Martin, Richard Harrington	30 Mar. 18	30 Mar. 18	Trichinopoly.
	2nd-Lieutenants.			
1 Feb. 18	Boyd, William Henry	1 Feb. 18	1 Feb. 18	Mesopotamia.
23 Apl. 18	Middleton, Cunliffe	23 Apl. 18	23 Apl. 18	Trichinopoly.
	Supernumerary.			
1 *Apl. 17*	*Lucas, Pierre Noel Arthur*	*1 Apl. 17*	*3 Jan. 18*	*I. A. R. O.*
1 *Apl. 17*	*Upshon, 2nd-Lt. Llewellyn Hector*	*1 Apl. 17*	*1 Apl. 17*	*I. A. R. O.*
1 Apl. 17	Robinson, 2nd-Lt. Augustus Lionel Clayton (80)	1 Apl. 17	15 June 18	Tuticorin.
	Staff.			
14 Nov. 17	Dobson, 2nd-Lt. (*Actg. Capt.*) J. H., W. Yorks R.	14 Nov. 17	14 Nov. 17	Adjutant.
12 Mar. 18	Macdonald, Lt. John	12 Mar. 18	12 Mar. 18	Qr. Mr.

Uniform.—Khaki drill.

1-32nd Madras and Southern Mahratta Railway Rifles.

"One and all."

Formed 21st Aug. 85, and reconstituted as an I. D. F. Unit with present designation 1st Apl. 1917.

Establishment—6 Reserve Coys.

Detachments at Arkonam, Gooty, Jalarpet, Pakala, Bitragunta and Rajahmundry.

HONORARY COMMANDANT.—LT.-GENERAL *Sir* JAMES WOLFE MURRAY, K.C.B. 1 Apl. 17.

First Commn.	Names.	Date of rank.	Date of appointment to corps.	Remarks.
	Lieutenant-Colonel.			
1 Apl. 17	Dwane, Herbert Milton, V.D. (80) . .	1 Apl. 17	1 Apl. 17	Comdt.
	Majors.			
1 Apl. 17	Fawcett, John Frederick (80)	1 Apl. 17	1 Apl. 17	Madras.
1 Apl. 17	*Browne, Charles Henry (54, 80.)* . . .	*1 Apl. 17*	1 *Apl. 17*	*Perambur.*
1 Apl. 17	Gibson, William Robert (80).	1 Apl. 17	1 Apl. 17	Madras.
	Captains.			
1 Apl. 17	Wathen, Frederick Blunt (80)	1 Apl. 17	1 Apl. 17	Madras.
1 Apl. 17	LeMarchand, Markham Maitland (80). .	1 Apl. 17	1 Apl. 17	Madras.
1 Apl. 17	Rendell, Alan Cleveland (80) . . .	1 Apl. 17	1 Apl. 17	Waltair.
1 Apl. 17	Waterworth, Thomas Laurence (80) . .	1 Apl. 17	1 Apl. 17	Nellore.
1 Apl. 17	Peters, Gerard Lovell (80) (48*a*, 50) . .	1 Apl. 17	1 Apl. 17	St. Thomas Mt.
1 Apl. 17	McLeod, Hugh Torquil	1 Apl. 17	24 Oct. 18	
1 Apl. 17	Thompson, Richard Dermott (80) . . .	1 Apl. 17	1 Apl. 17	Parambur.
1 Apl. 17	*Bowen, Charles Henry Croasdaile* .	*1 Apl. 17*	*1 Apl. 17*	*I. A. R. O.*
1 Apl. 17	Bisset, William Sinclair Smith (54, 80) .	1 Apl. 17	1 Apl. 17	Madras.
1 Apl. 17	Lingard, Harold (80)	1 Apl. 17	1 Apl. 17	Waltair.
1 Apl. 17	*Fraser, Hugh* (80)	*1 Apl. 17*	*1 Apl. 17*	*Lv., ex I., 1 yr., from 1 Dec. 18.*
	Lieutenants.			
1 Apl. 17	Walker, Maurice Ashley (80)	1 Apl. 17	1 Apl. 17	Perambur.
1 Apl. 17	Hanson, Frederick James Colin (80) . .	1 Apl. 17	1 Apl. 17	Godaveri.
1 Apl. 17	Woodd, Percy Charles (80).	1 Apl. 17	1 Apl. 17	Perambur.
1 Apl. 17	Turner, Arthur Cleveland	1 Apl. 17	24 Oct. 18	
1 Apl. 17	Sharpe, Arthur Rupert (80)	1 Apl. 17	1 Apl. 17	Perambur.
1 Apl. 17	Nunn, Richard Edward (80).	1 Apl. 17	1 Apl. 17	Waltair.
1 Apl. 17	Peebles, Archibald (80)	1 Apl. 17	1 Apl. 17	Madras.
1 Apl. 17	Dodd, Henry Black (80)	1 Apl. 17	1 Apl. 17	Arkonam.
1 Apl. 17	Lean, Richard	1 Apl. 17	24 Oct. 18	Parambur.
1 Apl. 17	Mann, Gerald Corry (50, 51, 80). . . .	1 Apl. 17	1 Apl. 17	Perambur.
	2nd-Lieutenants.			
1 Apl. 17	Plunkett, Arthur Lancelot Bonner (80) . .	1 Apl. 17	1 Apl. 17	St. Thomas Mt.
1 Apl. 17	Grayson, Norman (80)	1 Apl. 17	1 Apl. 17	Madras.
1 Apl. 17	Dinwoodie, John Frederick (80). . . .	1 Apl. 17	1 Apl. 17	Madras.
1 Apl. 17	Stuart, Arthur Reginald (80)	1 Apl. 17	1 Apl. 17	Madras.
1 Apl. 17	Wilson, Augustus Fritz (80)	1 Apl. 17	1 Apl. 17	Arkonam.
1 Apl. 17	Elliot, Stanley Steuart (50, 66)	1 Apl. 17	1 Apl. 17	Madras.
1 Apl. 17	Peebles, John Haughton	1 Apl. 17	24 Oct. 18	Gooty.

Supernumerary.

First Commn.	Names.	Date of rank.	Date of appointment to corps.
1 Apl. 17	Bamford, Lt.-Col. Robert George, V.D.	1 Apl. 17	1 Apl. 17
1 Apl. 17	Hobson, Lindsey Cecil	1 Apl. 17	24 Oct. 18
1 Apl. 17	Warburton, Capt. Wilfred (80)	1 Apl. 17	1 Apl. 17
1 Apl. 17	Crichton, 2nd-Lt. Thornton Elliot (80) . .	1 Apl. 17	18 Sep. 18

Staff.

First Commn.	Names.	Date of rank.	Date of appointment to corps.	Remarks.
2 Apl. 17	1 Pumfrey, 2nd-Lt. (*temp. Capt.*) R. J., The Rif. Brig.,	22 Apl. 17	22 Apl. 17	Adjutant.
8 Mar. 18	Morrow, Lt. A.	8 Mar. 18	8 Mar. 18	Qr.-Mr.

2-32nd Madras and Southern Mahratta Railway Rifles.

Formed 20th October 1886, and reconstituted as an I. D. F. Unit with present designation 1st Apl. 1917.
Establishment—4 Reserve Coys.

First Commn.	Names.	Date of rank.	Date of appointment to corps.	Remarks.
	Lieutenant-Colonels.			
1 Apl. 17	*Coen, William Whelan, V. D.*	*1 Apl. 17*	*1 Apl. 17*	*I. A. R. O.*
1 Apl. 17	Lyle, George Herbert	1 Apl. 17	1 Apl. 17	Comdt.
	Majors.			
1 Apl. 17	Sampson, Edward Dudley, V. D.	1 Apl. 17	1 Apl. 17	2nd-in-Comd.
1 Apl. 17	Sinclair, Harold David Blanchard, V. D.	1 Apl. 17	1 Apl. 17	
	Captains.			
1 Apl. 17	*Blomfield, Edward Valentine*	*1 Apl. 17*	*1 Apl. 17*	*I. A. R. O.*
1 Apl. 17	Williams, Arthur Vaughan	1 Apl. 17	1 Apl. 17	
1 Apl. 17	Saunders, Charles Henry Stanley (80)	1 Apl. 17	7 Nov. 18	Perambur.
	Lieutenants.			
1 Apl. 17	Wright, Herbert Lee	1 Apl. 17	1 Apl. 17	
1 Apl. 17	Jackson, Arthur Ivan	1 Apl. 17	1 Apl. 17	
1 Apl. 17	West, Leonard Clement	1 Apl. 17	1 Apl. 17	
1 Apl. 17	Fowle, William Thomas	1 Apl. 17	1 Apl. 17	
1 Apl. 17	Bazalgette, Edward	1 Apl. 17	1 Apl. 17	
1 Apl. 17	Atkinson, George Richard	1 Apl. 17	1 Apl. 17	
1 Apl. 17	*Cleary, Ralph Phillips*	*1 Apl. 17*	*1 Apl. 17*	
1 Apl. 17	Pipe, Thomas Sylvanus	1 Apl. 17	1 Apl. 17	
	2nd-Lieutenants.			
1 Apl. 17	*Cordon, Cecil Gilbert William*	*1 Apl. 17*	*1 Apl. 17*	
1 Apl. 17	Pearce, Arthur George	1 Apl. 17	1 Apl. 17	
1 Apl. 17	Creeke, William Russell	1 Apl. 17	1 Apl. 17	
	Supernumerary.			
1 Apl. 17	Rawlinson, Maj. Hugh George	1 Apl. 17	1 Apl. 17	
1 Apl. 17	Argyle, Capt. Reginald Edward Vere	1 Apl. 17	1 May 18	
1 Apl. 17	Molloy, Capt. Charles Patrick (80)	1 Apl. 17	19 July 18	
1 Apl. 17	Maynard, Lt. Richard de Kirklevington (80).	1 Apl. 17	1 May 18	
1 Apl. 17	Lane, Lt. Hugh Septimus (80)	1 Apl. 17	1 May 18	
1 Apl. 17	Tavernor, Capt. George Victor (80)	1 Apl. 17	5 Aug. 17	
1 Apl. 17	Sarkies, Lt. Otto St. Clayr.	1 Apl. 17	1 Apl. 17	
	Staff.			
....	Coen, Maj. W. W., I.A.R.O.	....	....	Adjutant.
....	Phillips, Lt. F.	....	....	Qr., Mr.

Uniform—Khaki drill.

34th Upper Burma Battalion.

Raised, 9th December 1886 and reconstituted as an I. D. F. Unit with present designation 1st Apl. 1917.
Establishment—4 Combined Active & Reserve Coys., 1 Mtd. Infy. Platoon.
Companies at Maymyo, Yenangyaung, Meiktila, Monywa, Mogok, Namtu and Myitkyina.

HONORARY COMMANDANT—

First Commn.	Names.	Date of rank.	Date of appointment to corps.	Remarks.
	Lieutenant-Colonel.			
1 Apl. 17	Aplin, *Hon'ble* Stephen Lushington, C.S.I.	1 Apl. 17	1 Apl. 17	Comdt.
	Majors.			
1 Apl. 17	Cuffe, *Sir* Otway Fortescue-Luke Wheeler, *Bart.* V.D.	1 Apl. 17	1 Apl. 17	(Hon. Lt.-Col., late Ind. Volrs.), Maymyo.
1 Apl. 17	Lee, Frederick Reginald, V.D. (Q.)	1 Apl. 17	1 Apl. 17	Taunggyi.
	Captains.			
1 Apl. 17	Allan, Douglas Hume, V.D.	1 Apl. 17	1 Apl. 17	Pakokku.
1 Apl. 17	Bacon, Arthur Lewis, V.D.	1 Apl. 17	1 Apl. 17	Mogôk.
1 Apl. 17	*Walton, Harry*	*1 Apl. 17*	*1 Apl. 17*	*I. A. R. O.*
1 Apl. 17	Heald, Benjamin Herbert	1 Apl. 17	1 Apl 17	Mandalay.
1 Apl. 17	Godber, Alexander Richard, V.D.	1 Apl. 17	1 Apl. 17	Myitkyina.
1 Apl. 17	*Stuart, John Matthew Blackwood*	*1 Apl. 17*	*1 Apl. 17*	*I. A. R O.*
1 Apl. 17	*Hill, Arther Halkett*	*1 Apl. 17*	*1 Apl. 17*	*I. A. R. O.*
1 Apl. 17	*Hughes, Peter*	*1 Apl. 17*	*1 Apl. 17*	*Tempy. Comm. in Army.*
1 Apl. 17	*Craw, Henry Hewart*	*1 Apl. 17*	*1 Apl. 17*	*I A. R. O.*
1 Apl. 17	Scott, Bernard Bodley	1 Apl. 17	1 Apl. 17	Mandalay.
1 Apl. 17	*Cooper, Charles Robert Plant*	*1 Apl. 17*	*1 Apl. 17*	*I. A. R. O.*
1 Apl. 17	Thorburn, John Milne	1 Apl. 17	1 Apl. 17	Yenangyaung.
	Lieutenants.			
1 Apl. 17	*Burne, Roy Douglas*	*1 Apl. 17*	*1 Apl. 17*	
1 Apl. 17	*Alderman, Robert Edward,* C.I.E.	*1 Apl. 17*	*1 Apl. 17*	*I. A. R. O.*
1 Apl. 17	*Lakeland, William John*	*1 Apl. 17*	*1 Apl. 17*	*I. A. R. O.*
1 Apl. 17	*Holme, Hugh Basil*	*1 Apl. 17*	*1 Apl. 17*	*I. A. R. O.*
1 Apl. 17	Tonkinson, Harry	1 Apl. 17	1 Apl. 17	Sagaing.
1 Apl. 17	*Farmer, Edward James*	*1 Apl. 16*	*1 Apl. 17*	*I. A. R. O.*
1 Apl. 17	Farr, Reginald Percy	1 Apl. 17	1 Apl. 17	Lashio.
1 Apl. 17	Rodger, Alexander	1 Apl. 17	1 Apl. 17	Maymyo.
1 Apl. 17	Smyth, William John	1 Apl. 17	1 Apl. 17	Mandalay.
1 Apl. 17	*Liddell, William Temple*	*1 Apl. 17*	*1 Apl. 17*	*I. A. R. O.*
1 Apl. 17	Watson, Joseph	1 Apl. 17	1 Apl. 17	Mandalay.
1 Apl. 17	Wood, Thomas Baird	1 Apl. 17	1 Apl. 17	Yenangyaung.
	2nd-Lieutenants.			
1 Apl. 17	*Lowis, Frank Currie,* C.I.E.	*1 Apl. 17*	*1 Apl. 17*	*I. A. R. O.*
1 Apl. 17	*Street, Walter*	*1 Apl. 17*	*1 Apl. 17*	*I. A. R. O.*
1 Apl. 17	*Dove, Ernest Percy*	*1 Apl. 17*	*1 Apl. 17*	*I. A. R. O.*
1 Apl. 17	*Nicholls, James Arthur*	*1 Apl. 17*	*1 Apl. 17*	*I. A. R. O.*
1 Apl. 17	Roberts, John William Basil Thornes	1 Apl. 17	1 Apl. 17	Maymyo.
1 Apl. 17	Layard, Cecil William Clayfield	1 Apl. 17	1 Apl. 17	Mandalay.
1 Apl. 17	Burbidge, Alfred Humphrey	1 Apl. 17	1 Apl. 17	Mandalay.
1 Apl. 17	*Baker, Richard Stanley*	*1 Apl. 17*	*1 Apl. 17*	*I. A. R. O.*
1 Apl. 17	Playne, Harold Francis	1 Apl. 17	1 Apl. 17	Lv. 1 month.
1 Apl. 17	Hampton, Septimus Herbert	1 Apl. 17	1 Apl. 17	Kyaukse.
1 Apl. 17	Stuart, James Douglas	1 Apl. 17	1 Apl. 17	Maymyo.
1 Apl. 17	Forrester, George Douglas	1 Apl. 17	1 Apl. 17	Yenangyaung.
1 Apl. 17	Middleton, John Edward	1 Apl. 17	1 Apl. 17	Yenangyaung.
	Supernumerary.			
1 Apl. 17	Wilkinson, Capt. Charles Robert	1 Apl. 17	1 Apl. 17	
1 Apl. 17	Stevenson, Lt. Henry Louis	1 Apl. 17	1 Apl. 17	
1 Apl. 17	Page, 2nd-Lt. Arthur John	1 Apl. 17	1 Apl. 17	
1 Apl. 17	Smales, Lieut., Charles Bertram	1 Apl. 17	1 Apl. 17	
	Staff.			
18 Feb. 18	Smith, 2nd-Lt. (*Actg. Capt.*) R., Cam'n Highrs.	18 Feb. 18	18 Feb. 18	Adjutant.
5 Apl. 18	Dye, Lt. Arthur	15 Apl. 18	15 Apl. 18	Qr.-Mr.
1 Apl. 17	Darne, Revd. Father August	1 Apl. 17	1 Apl. 17	Hony. Chaplain.
1 Apl. 17	Seeley, Revd. C. H.	1 Apl. 17	1 Apl. 17	Ditto.

Establishment—4 Composite Coys., 1 Mtd. Infy. Platoon (Yenangyaung).
Enrolled Strength,

35th Poona Battalion.

Uniform—Khaki drill.

Formed 4th Jan. 87, as a Vol. Force and re-constituted as an I. D. F. Unit with present designation, 1st Apl. 1917.

Establishment—1 Active Coy., 2 Reserve. Coys.

HONORARY COMMANDANT—

Governor of Bombay.

First Commn.	Names.	Date of rank.	Date of appointment to corps.	Remarks.
Lieutenant-Colonel.				
1 Apl. 17	Walker, George Kemp, C.I.E., O.B.E., F.R.C.V.S. (54, 170, 206)	1 Apl. 17	1 Apl. 17	Comdt. (Lt.-Col., I.C.V. D., 20 July 17. *Lv., ex Ind., 6 mths., 6 Apl. 18.*
Majors.				
1 Apl. 17	*Smith, Cyril Graham* (80)	*1 Apl. 17*	*1 Apl. 17*	*(Tempy. Capt. in Army) 10 March 17. Staff Capt., 6th (Poona) Dn.*
1 Apl. 17	*Inkson, Norman Leopold*, D.S.O. (51)	*1 Apl. 17*	*1 Apl. 17*	*I. A. R. O.*
1 Apl. 17	*Haigh, Percy Barnes* (51)	*1 Apl. 17*	*1 Apl. 17*	*I. A. R. O.*
1 Apl. 17	*Burns, William* (51, 54, 80, 82*a*, 151)	*1 Apl. 17*	*1 Apl. 17*	*I. A. R. O.*
1 Apl. 17	Brown, Robert Scott Cree (54, 80, 82*a*)	1 Apl. 17	1 Apl. 17	
Captains.				
1 Apl. 17	*Bourke, Dermot Richard Southwell*	*1 Apl. 17*	*1 Apl. 17*	*I. A. R. O.*
1 Apl. 17	Grier, John Arthur Bolton (80, 82*a*)	1 Apl. 17	1 Apl. 17	Sholapur.
1 Apl. 17	Eves, Graves William (80, 82*a*)	1 Apl. 17	1 Apl. 17	Kurduwadi.
1 Apl. 17	Duggan, Edwin Merle (80)	1 Apl. 17	1 Apl. 17	
1 Apl. 17	*Howie, Harry Campbell* (80, 82*a*, 207)(*g*)	*1 Apl. 17*	*1 Apl. 17*	*Asst. R. T. O.*
1 Apl. 17	Beyts, Clement Ayerst (80, 82)	21 May 18	1 Apl. 17	Ahmednagar.
1 Apl. 17	*Lory, Frederick Burton Pendarves* (80)	*15 July 18*	*1 Apl. 17*	*I. A. R. O.*
17 Nov. 17	Dobson, Tom (46, 80, 82*a*)	15 July 18	17 Nov. 17	
Lieutenants.				
17 Nov. 17	Coombs, Ernest Edwin (80)	21 May 18	17 Nov. 17	With "F" (Kirkee) Battery, Mobile Arty.
17 Nov. 17	Richards, Edwin Verney (80, 82*a*)	21 May 18	17 Nov. 17	Lanavla.
14 Dec. 17	*Marrs, Francis William* (80)	*21 May 18*	*4 Dec. 17*	*I. A. R. O., Asst. Recruiting Offr., Poona.*
4 Dec. 17	Ewbank, Robert Benson (80)	16 July 18	4 Dec. 17	
27 May 18	Thiselton-Dyer, George Henry	15 July 18	27 May 18	
27 May 18	Monteath, George	24 July 18	27 May 18	
2nd-Lieutenants.				
20 Sep. 18	Gebbie, Frederick St. John	20 Sep. 18	20 Sep. 18	
20 Sep. 18	Eric, Harper	20 Sep. 18	20 Sep. 18	

Supernumerary List.

1 Apl. 17	Rowan, Capt. Thomas	1 Apl. 17	1 Apl. 17

Staff.

6 Dec. 15	Kearsey, Lt. (*temp. Capt.*) Arthur Richard, R. Fus. (12, 51, 52, 72*b*)	1 July 17	6 Dec. 15	Adjutant.

Uniform—Khaki dril.

1-36th Bengal Nagpur Railway Battalion.

Formed 15th June 88. Orissa Vol. Rifles amalgamated by M. D. Notfn. No. 1275, dated 18th November 1898. Reconstituted as an I. D. F. Unit with present designation 1st Apl. 1917.

Establishment.—2 Reserve Coys., 1 Combined Active & Reserve Coy., 1 Machine Gun Dett, Companies at Calcutta, Khargpur and Khurda Road.

HONORARY COMMANDANT—

1st Commn.	Name.	Date of rank.	Date of appointment to corps.	REMARKS.
	Lieutenant-Colonel.			
1 Apl. 17	Godfrey, George Cochrane, V.D.	1 Apl. 17	1 Apl. 17	Simla.
1 Apl. 17	Jellett, Henry Holmes, V.D. (80)	1 Apl. 17	1 Apl. 17	Comdt., Calcutta.
	Majors.			
1 Apl. 17	Hughes, Robert Milward, V.D.	1 Apl. 17	1 Apl. 17	2nd-in-Comd.—Khargpur.
1 Apl. 17	*Greenham, Leonard*, V.D.	*1 Apl. 17*	*1 Apl. 17*	Ambulance—Overseas.
1 Apl. 17	*Duggan, Collingwood Ross*, V.D.	*1 Apl. 17*	*1 Apl. 17*	*Munitions.*
1 Apl. 17	*Tulloch, Sylvester Stephen Gregorie*, V.D. (*D* 1.), (12, 48a, 50, 54, 79)	*1 Apl. 17*	*1 Apl. 17*	*I. A. R. O.*
1 Apl. 17	*Hill, Alfred Rowley*, V.D.	*1 Apl. 17*	*1 Apl. 17*	*Khuragpur.*
	Captains.			
1 Apl. 17	Duncan, George Houston, V.D. (80)	1 Apl. 17	1 Apl. 17	Calcutta.
1 Apl. 17	Leslie David (80)	1 Apl. 17	1 Apl. 17	Khurda Road.
1 Apl. 17	*Eliot, Ernest Edward Marshall*, V.D.	*1 Apl. 17*	*1 Apl. 17*	*A. S. Corps.*
1 Apl. 17	Gayer. Ernest Conway (80)	1 Apl. 17	1 Apl. 17	Adra.
1 Apl. 17	Townshend, Robert Ponsonby Loftus (80)	1 Apl. 17	1 Apl. 17	Calcutta.
1 Apl. 17	*Neely, Thomas Robert*	*1 Apl. 17*	*1 Apl. 17*	*Temp. Lt. in Army.*
1 Apl. 17	*Watson, James Alexander Mytton*	*1 Apl. 17*	*1 Apl. 17*	*Lv., ex I., 12 mos., from 13 Sep. 18, m.c.*
1 Apl. 17	*Arnott, Harold Wright (54)*	*1 Apl. 17*	*1 Apl. 17*	Flying Corps—Overseas.
1 Apl. 17	*Grant, Rupert. Grenville*	*1 Apl. 17*	*1 Apl. 17*	*Temp. Lt. in Army.*
	Lieutenants.			
1 Apl. 17	Oldfield, Michael Humphrey (80)	1 Apl. 17	1 Apl. 17	Khargpur.
1 Apl. 17	Joyce, Henry William (80)	1 Apl. 17	1 Apl. 17	Calcutta.
1 Apl. 17	Mearns, Hector Grant (80)	1 Apl. 17	1 Apl. 17	Calcutta.
1 Apl. 17	*Gahan, Patrick Joseph* (80)	*1. Apl. 17*	*1 Apl. 17*	Munitions—Bombay.
1 Apl. 17	Gahan, Edward Charles Joseph (80)	1 Apl. 17	1 Apl. 17	Calcutta.
1 Apl. 17	Lambert, Henry (80)	1 Apl. 17	1 Apl. 18	Khurda Road.
	2nd-Lieutenants.			
1 Apl. 17	Hornett, William George (80)	1 Apl. 17	1 Apl. 17	Khargpur.
25 May 18	Pelly, George Sinclair (80)	25 May 18	25 May 18	Calcutta.
25 May 18	Farmer, Charles Rudolph Stahel (80)	25 May 18	25 May 18	Khargpur.
	Supernumerary.			
1 Apl. 17	Wyatt, 2nd-Lt., Archibald Frederic (80)	1 Apl. 17	1 Apl. 17	Calcutta.
	Staff.			
1 Oct. 14	Dallison, Capt. H. W., 1st G. B. Linc. R.	1 June 16	1 Nov. 18	Adjutant.

2-36th Bengal Nagpur Railway Battalion.

Establishment—4 Reserve Coys.

First Commn.	Names.	Date of rank.	Date of appointment to corps.	Remarks.
	Lieutenant-Colonel.			
1 Apl. 17	Clark, Aubrey Mathew, V. D., A. D. C. (54, 61, 80, 207).	1 Apl. 17	1 Apl. 17	Comdt. Calcutta.
	Major.			
1 Apl. 17	Shortland, Charles Vincent Nixon, V.D. (80).	1 Apl. 17	1 Apl. 17	2nd-in-Command, Nagpur.
	Captains.			
1 Apl. 17	Ridsdale, Cyril George, V. D. (80, 207*a*)	1 Apl. 17	1 Apl. 17	Adra.
1 Apl. 17	Neely, Alexander James (80)	1 Apl. 17	1 Apl. 17	Chakardharpur.
1 Apl. 17	Carroll, Charles William Alexander (137).	1 Apl. 17	1 Apl. 17	Calcutta.
1 Apl. 17	*Cunningham, Gordon*	*1 Apl. 17*	*1 Apl. 17*	
1 Apl. 17	Jarrad, Vivian Everard Donne (51)	1 Apl. 17	1 Apl. 17	Nagpur.
	Lieutenants.			
1 Apl. 17	Bayley, Archibald (80, 207)	1 Apl. 17	1 Apl. 17	Adra.
1 Apl. 17	Eaglesome, Arthur (80, 207*a*)	1 Apl. 17	1 Apl. 17	Kharagpur.
1 Apl. 17	Anderson, Charles William	1 Apl. 17	1 Apl. 17	
	2nd-Lieutenants.			
6 June 18	Lukis, Sydney	6 June 18	6 June 18	Kharagpur.
6 June 18	Remfry, Douglas Henry	6 June 18	6 June 18	Bhaga.
6 June 18	Carroll, Archibald Douglas	6 June 18	6 June 18	Panposh.
6 June 18	Blake, James Bryan (51)	6 June 18	6 June 18	Bilaspur.
6 June 18	Hunter, Ivon Guy Charles (80)	6 June 18	6 June 18	Chakardharpur.
6 June 18	Hughes, Frederic Selby	6 June 18	6 June 18	Adra.
	Supernumerary.			
1 Apl. 17	Beckett, Capt. Ernest Godfrey, V.D. (80).	1 Apl. 17	1 Apl. 17	
	Staff.			
27 Aug. 18	Matthews, 2nd-Lt. (*temp. Capt.*) George, R. Fus.	27 Aug. 18	27 Aug. 18	Adjt.

37th Calcutta Presidency Battalion.

Formed, 30th Mar. 88. Reorganised G. G. O. 437 of 22nd Apl. 1898. Designation changed by Mily. Dept. Notfn. No. 868 of 1901. Reconstituted as an I. D. F. Unit with present designation 1st Apl. 1917.

Establishment.—2 Active Coys., 2 Reserve Coys., 1 Cadet Coy.

HONORARY COMMANDANT—

First Commn.	Names.	Date of rank.	Date of appointment to corps.	Remarks.
	Lieutenant-Colonel.			
1 Apl. 17	Bullen, John Henry	1 Apl. 17	1 Apl. 17	Comdt.
	Majors.			
1 Apl. 17	*Newman, Henry*	*1 Apl. 17*	*1 Apl. 17*	*I. A. R. O.*
1 Apl. 17	*Pearse, George Gerald*	*1 Apl. 17*	*1 Apl. 17*	*Temp. Lt. I. A.*
1 Apl. 17	*Wootten, Eric Herbert Wootten* (48, 51)	*1 Apl. 17*	*1 Apl. 17*	*I. A. R. O.*
1 Apl. 17	*Collet, Arthur Lowe*	*1 Apl. 17*	*1 Apl. 17*	*I. A. R. O.*
1 Apl. 17	*Harvey, Godfrey Thomas Benedict*	*1 Apl. 17*	*1 Apl. 17*	*I. A. R. O.*
1 Apl. 17	Dixon, John	1 Apl. 17	1 Apl. 17	Temp. 2nd-in-Command.
1 Apl. 17	*Meade, George Alfred,* V.D.	*1 Apl. 17*	*1 Apl. 17*	
1 Apl. 17	*Hardinge, John George*	*1 Apl. 17*	*1 Apl. 17*	*Capt. R. W. Fus.*
1 Apl. 17	*Nichols, Harry*	*1 Apl. 17*	*1 Apl. 17*	
1 Apl. 17	Greaves, Edwin	7 Aug. 18	1 Apl. 17	
	Captains.			
1 Apl. 17	Wilson, James Deans	1 Apl. 17	1 Apl. 17	
1 Apl. 17	*Harvey, William Sidney*	*1 Apl. 17*	*1 Apl. 17*	
1 Apl. 17	*Cottle, Jack Lionel*	*1 Apl. 17*	*1 Apl. 17*	(*Tanks Corps.*)
1 Apl. 17	*Bagram, Charles Ernest*	*1 Apl. 17*	*1 Apl. 17*	
1 Apl. 17	Behrend, Ernest Stanley	1 Apl. 17	1 Apl. 17	
1 Apl. 17	Richards, Samuel Hope	1 Apl. 17	1 Apl. 17	
1 Apl. 17	*Leonard, Denis William*	*1 Apl. 17*	*1 Apl. 17*	*I. A. R. O.*
1 Apl. 17	*McMillan, William*	*1 Apl. 17*	*1 Apl. 17*	*I. A. R. O.*
1 Apl. 17	*Robertson, Guido George*	*1 Apl. 17*	*1 Apl. 17*	(*A. S. C. attd. R. F. C.*)
1 Apl. 17	*Hobbs, Charles Christie Arthur*	*1 Apl. 17*	*1 Apl. 17*	*I. A. R. O.*
1 Apl. 17	*Tredaway, George Herbert*	*1 Apl. 17*	*1 Apl. 17*	*I. A. R. O.*
1 Apl. 17	*Worster, Alexander Howett*	*1 Apl. 17*	*1 Apl. 17*	*I. A. R. O.*
1 Apl. 17	*Unger, Valentine Carl Whybrow*	*1 Apl. 17*	*1 Apl. 17*	*I. A. R. O.*
1 Apl. 17	Thompson, Edgar	7 Aug. 18	1 Apl. 17	
1 Apl. 17	Mackenzie, Alexander Edward	7 Aug. 18	1 Apl. 17	
	Lieutenants.			
1 Apl. 17	Remfry, Maurice	1 Apl. 17	1 Apl. 17	
1 Apl. 17	*Sanders, Henry Harold*	*1 Apl. 17*	*1 Apl. 17*	*I. A. R. O.*
1 Apl. 17	*Rivett, Harold Henry* (51)	*1 Apl. 17*	*1 Apl. 17*	*I. A. R. O.*
1 Apl. 17	Warren-Boulton, Clifford George	1 Apl. 17	1 Apl. 17	
1 Apl. 17	*Woolley, James*	*1 Apl. 17*	*1 Apl. 17*	*I. A. R. O.*
1 Apl. 17	Read, Richard Francis	1 Apl. 17	1 Apl. 17	
1 Apl. 17	Whitehorn, Roy Drummond	1 Apl. 17	1 Apl. 17	
1 Apl. 17	Chapman, Ernest Augustus John	15 Nov. 18	1 Apl. 17	
	2nd-Lieutenants.			
1 Apl. 17	Jenks, Robert Leonard	1 Apl. 17	1 Apl. 17	
7 Aug. 18	Gardiner, Edward John Rawson	7 Aug. 18	7 Aug. 18	
7 Aug. 18	Beecroft, Walter	7 Aug. 18	7 Aug. 18	
7 Aug. 18	Howe, Reginald	7 Aug. 18	7 Aug. 18	
	Supernumerary.			
1 Apl. 17	Leslie, Lt.-Col. Franklin Marston, V.D.	1 Apl. 17	1 Apl. 17	
1 Apl. 17	French, Maj. Vincent Cluge, V.D.	1 Apl. 17	1 Apl. 17	
1 Apl. 17	Oliver, Maj. Alfred James	1 Apl. 17	1 Apl. 17	
1 Apl. 17	Simpson, Capt. John Henry	1 Apl. 17	1 Apl. 17	
1 Apl. 17	Aldersmith, Lt. Claude Herbert	1 Apl. 17	1 Apl. 17	
	Staff.			
1 Apl. 17	Wootten, Maj. E. H. W.	1 Apl. 17	1 Apl. 17	Adjt.
1 Apl. 17	Grimes, Revd. C. J.	1 Apl. 17	1 Apl. 17	Chaplain (C. of E.)

Uniform—Khaki drill.

38th East Coast Battalion.

Formed 9th June 85. Godavari Rifle Volunteers and Vizagapatam Rifle Volunteers amalgamated, 14th October 1890. Reconstituted as an I. D. F. Unit with present designation 1st Apl. 1917. Establishment.—2 Combined Active & Reserve Coys.

Detachment at Vizagapatam, Coconada, Bezwada, Vizianagram, Berhampore,

First Commn.	Names.	Date of rank.	Date of appointment to corps.	Remarks.
	Lieutenant-Colonels.			
1 Apl. 17	***Arbuthnot,** Percy Bingham* (54), (79) . .	*1 Apl. 17*	*1 Apl. 17*	*Comdg. Depot, 1st Porter. Corps. Temp. Maj., I. A. Hony. Col. late Ind. Volrs. 1.*
1 Apl. 17	Mackay, Donald Stuart (51, 52, 79, 80) .	15 June 17	1 Apl. 17	Commdt.
	Majors.			
1 Apl. 17	*Foulkes, George Frederick Fischer D. I. (50, 54, 79.)*	*1 Apl. 17*	*1 Apl. 17*	
1 Apl. 17	Greg, Lionel Hyde (54, 79, 80) . . .	1 Apl. 17	1 Apl. 17	2nd-in-Comd.
	Captains.			
1 Apl. 17	*Lace, Earnest William*	*1 Apl. 17*	*1 Apl. 17*	*I. A. R. O.*
1 Apl. 17	*Dowley, Francis Michael*	*1 Apl. 17*	*1 Apl. 17*	*I. A. R. O.*
1 Apl. 17	*Nightingale, Arthur Winfield*	*1 Apl. 17*	*1 Apl. 17*	*I. A. R. O.*
1 Apl. 17	*Lecky, Reynolds Lamont*	*1 Apl. 17*	*1 Apl. 17*	*I. A. R. O.*
1 Apl. 17	*Shout, William Gordes*	*1 Apl. 17*	*1 Apl. 17*	*I. A. R. O.*
1 Apl. 17	Smith, James Ironside (79, 80)	1 Apl. 17	1 Apl. 17	
1 Apl. 17	Fowler, John Charles Hill (79, 80) . . .	1 Apl. 17	1 Apl. 17	
1 Apl. 17	Uzielli, Herbert Rex (79, 80)	1 Apl. 17	1 Apl. 17	
1 Apl. 17	Brown, Hilton (80)	1 Apl. 17	1 Apl. 17	
	Lieutenants.			
1 Apl. 17	*Marriott, Bertie Millar*	*1 Apl. 17*	*1 Apl. 17*	*I.A.R.C.*
1 Apl. 17	*Tate, James,* M. C.	*1 Apl. 17*	*1 Apl. 17*	*I.A.R.C.*
1 Apl. 17	Moodie, Alexander Lachlan Maclean (79, 80).	1 Apl. 17	1 Apl. 17	
1 Apl. 17	Lutman, John Blois (79, 80)	1 Apl. 17	1 Apl. 17	
26 Mar. 18	*Keer, William Horace*	*26 Mar. 18*	*26 Mar. 18*	*I.A.R.C.*
	2nd-Lieutenants.			
26 Mar. 18	Shores, Charles Dunbar Tatham . . .	26 Mar. 18	26 Mar. 18	
26 Mar. 18	Hunter, Reginald James	26 Mar. 18	26 Mar. 18	
26 Mar. 18	Connell, Cecil Bourke	26 Mar. 18	26 Mar. 18	
17 May 18	Robertson, Frederick William Rowntree .	17 May 18	17 May 18	
17 May 18	Leonard, George Russell	17 May 18	17 May 18	
1 Apl. 17	Flemying, Baldwyn Walter Buchanan Tull.	15 Oct. 18	1 Apl. 18	
	Staff.			
26 Nov. 17	Darby, 2nd-Lt. (*Actg. Capt.*) G. Ches. R. (5).	20 Nov. 17	20 Nov. 17	Adjutant.
10 Oct. 18	Julius, Revd. Leo, Cantab. B. A. . .	10 Oct. 18	10 Oct. 18	Hony. Chaplain (R. C.)

Uniform—Khaki drill.

39th Chota Nagpur Regiment.

Formed as Chota Nagpur Light Horse 7th Aug. 1891. Reorganised 2nd March 1910, and reconstituted as an I. D. F. Unit with present designation 1st Apl 1917.

Establishment—4 Combined Active & Reserve Coys.
2 Machine Gun Detts.
1 Combined Active & Reserve Sqdn.
1 Active Sqdn.

HONORARY COMMANDANT—*The Hon'ble Sir* Edward Albert Gait, K.C.S.I., C.I.E. . . 1 Apl. 17.

First Commn.	Names.	Date of rank.	Date of appointment to corps.	Remarks.
	Lieutenant-Colonel.			
1 Apl. 17	Foley, Blanchard (80)	1 Apl. 17	1 Apl. 17	Comdt.
	Majors.			
1 Apl. 17	Agabeg, Frank Joseph, V. D. (80, 54)	1 Apl. 17	1 Apl. 17	2nd-in-Comd. (*Hon. Col., late Ind. Volrs.*)
1 Apl. 17	Peppe, Arthur Tosco, V. D. (80)	1 Apl. 17	1 Apl. 17	
1 Apl. 17	Veasey, Harley Cyril, V. D. (80)	1 Apl. 17	1 Apl. 17	
	Captains.			
1 Apl. 17	*Douglas, Daniel*	*1 Apl. 17*	*1 Apl. 17*	*I. A. R. O.*
1 Apl. 17	Simpson, Robert Rowell (80)	1 Apl. 17	1 Apl. 17	
1 Apl. 17	Duke, William Victor	1 Apl. 17	1 Apl. 18	
1 Apl. 17	Fawcus, George Ernest (51, D. I, II, III, IV.)	1 Apl. 17	1 Apl. 17	
	Lieutenants.			
1 Apl. 17	*Heilgers, Frederick William*	*1 Apl. 17*	*1 Apl. 17*	*I. A. R. O.*
1 Apl. 17	*Chalmers, Hills Archie*	*1 Apl. 17*	*1 Apl. 17*	*I. A. R. O.*
1 Apl. 17	Miller, Nathan (80)	1 Apl. 17	1 Apl. 17	
1 Apl. 17	Mungall, William (80)	1 Apl. 17	1 Apl. 17	
1 Apl. 17	Dunbar, Noel Graham (51)	1 Apl. 17	1 Apl. 17	
1 Apl. 17	Luby, Thomas (80)	1 Apl. 17	1 Apl. 17	
	2nd-Lieutenants.			
1 Apl. 17	*Dixon, George*	*1 Apl. 17*	*1 Apl. 17*	*I. A. R. O.*
1 Apl. 17	Miller, Joseph (80)	1 Apl. 17	1 Apl. 17	
20 June 18	Turnbull, John James (80)			
	Supernumerary.			
1 Apl. 17	Fairhurst, George Henry	1 Apl. 17	1 Apl. 17	*Lv., ex I., 9 mos. fr.* 1-11-18
	Staff.			
16 Mar. 18	Foden, Lt. W. H.	16 Mar. 18	16 Mar. 18	Qr.-Mr. & Actg. Adjt.

Uniform—Khaki drill.
Officers—Drab serge.
Facings—Dark green.

41st Eastern Bengal Company.

Formed 23rd Jan. 1901. Reconstituted as an I. D. F. Unit with present designation 1st Apl. 1917.

Establishment—1 Combined Active & Reserve Coy., 2 Machine Gun Detts.

HONORARY COMMANDANT—*H. E. the Rt. Hon.* Lawrence John Lumley Dundas, *Earl of* Ronaldshay, G.C.I.E., (*Governor of Bengal.*) 1 Apl. 17.

First Commn.	Names.	Date of rank.	Date of appointment to corps.	Remarks.
	Lieutenant-Colonels.			
1 Apl. 17	Glen, Robert, V.D.	1 Apl. 17	1 Apl. 17	Comdt., Hon. A.-D.-C. to Govr., Bengal.
	Major.			
1 Apl. 17	Rankin, James Thomson	1 Apl. 17	1 Apl. 17	2nd-in-Comd.
	Captains.			
1 Apl. 17	Hely Hutchinson, John Walter	1 Apl. 17	1 Apl. 17	
1 Apl. 17	*Ramsbotham, Richard Bury*	*1 Apl. 17*	*1 Apl. 17*	*I.A.R.O.*
	Lieutenants.			
1 Apl. 17	Godden, Arthur Leigh	1 Apl. 17	1 Apl. 17	
1 Apl. 17	Leeson, George William	6 Nov. 18	1 Apl. 17	
	2nd-Lieutenants.			
1 Apl. 17	*Smith, Egerton*	*1 Apl. 17*	*1 Apl. 17*	*I.A.R.O.*
1 Apl. 17	Finlow, Robert Steel	1 Apl. 17	1 Apl. 17	
1 Apl. 17	Blackwood, John Ross	1 Apl. 17	1 Apl. 17	
	Supernumerary.			
1 Apl. 17	*McCormack, Robert Barker*	*1 Apl. 17*	*1 Apl. 17*	
	Staff.			
16 Jan. 16	Dillon, 2nd-Lt. (*Tempy. Capt.*) D., R. Ir. Regt.	16 July 17	16 Jan. 16	Adjutant.

Uniform—Summer—Khaki drill. *Winter*—Khaki serge with dark green facings.

42nd Assam-Bengal Railway Battalion.

(*Formed by M. D. letters Nos. 3323 B. and 1229 B., dated 16th Nov. 1900 and 10th April 1901.*)
Reconstituted as an I. D. F. Unit with present designation 1st Apl. 1917.
Head Quarters Chittagong.

Establishment—2 Combined Active & Reserve Coys., 2 Machine Gun Detts.

Detachments at Pahartali, Badarpur and Lumding.

HONORARY COMMANDANT—

First Commn.	Names.	Date of rank.	Date of appointment to corps.	Remarks.
	Lieutenant-Colonels.			
1 Apl. 17	*Berkeley, Arthur Mowbray*, C. I. E., V.D. (71)	1 Apl. 17	1 Apl. 17	*I. A. R. O.*
1 Apl. 17	Cooper, Arthur John, V.D.	1 Apl. 17	1 Apl. 17	Comdt.
	Major.			
1 Apl. 17	Nolan, Thomas Richard (80)	1 Apl. 17	1 Apl. 17	
	Captains.			
1 Apl. 17	Thomas, Ernest Whitney (80)	1 Apl. 17	1 Apl. 17	
1 Apl. 17	Baker, Charles Henry O'Neill (50, 54)(47*a*) (80)	1 Apl. 17	1 Apl. 17	
1 Apl. 17	*Kennedy Kenneth Hartley Heathcote*, M.C.	1 Apl. 17	1 Apl. 17	*I. A. R. O.*
1 Apl. 17	Bliss, Raymond Lewis (80)	1 Apl. 17	1 Apl. 17	
1 Apl. 17	Field, Wilfred Petrock (80)	1 Apl. 17	1 Apl. 17	
	Lieutenants.			
1 Apl. 17	*Salberg, Frank James*	1 Apl. 17	1 Apl. 17	*I. A. R. O.*
1 Apl. 17	Rochfort, Arthur Patrick Greenfield (51) (80)	1 Apl. 17	1 Apl. 17	
1 Apl. 17	*Marriott, Leslie John*	1 Apl. 17	1 Apl. 17	*I. A. R. O.*
1 Apl. 17	*Whitney, William Welsh (51)*	*1 Apl. 17*	*1 Apl. 17*	*Railways.*
1 Apl. 17	Bennett, George Ernest (80)	1 Apl. 17	1 Apl. 17	
1 Apl. 17	*Thomson, Herbert James*	1 Apl. 17	13 Dec. 18	*I. A. R. O.*
1 Apl. 17	*Biden Soden William*	1 Apl. 17	13 Dec. 18	*I. A. R. O.*
1 Apl. 17	Stapylton, Alan (80)	1 Apl. 17	13 Dec. 18	
	2nd-Lieutenants.			
1 Apl. 17	Manson, Robert George (80)	1 Apl. 17	1 Apl. 17	
1 Apl. 17	Vipan, Robert Stephen (80)	1 Apl. 17	1 Apl. 17	
1 Apl. 17	Swann, Albert Edward (80)	1 Apl. 17	1 Apl. 17	
	Supernumerary.			
1 Apl. 17	Bayley, Lt.-Col. George Anson	1 Apl. 17	1 Apl. 17	
1 Apl. 17	Hawkins, Lt.-Col. Rupert Skelton, V.D. (52)	1 Apl. 17	1 Apl. 17	Hony. Col. late Ind. Volrs.
1 Apl. 17	Venters, Capt. William Beattie (80)	1 Apl. 17	1 Apl. 17	*Lv., ex I., 12 mos.*
1 Apl. 17	Martin, Capt. Henry Thorpe (80)	1 Apl. 17	1 Apl. 17	
1 Apl. 17	Leishman, Capt. Alan Ross (80)	1 Apl. 17	1 Apl. 17	
	Staff.			
5 June 17	Walker, 2nd-Lt. (*temp. Capt.*) S. G., Notts. and Derby. R. (51, 58)	5 June 17	5 June 17	Adjutant.

Uniform—Khaki drill.

43rd Kolar Gold Fields Battalion.

(*Formed by Mily. Dept. Notfn. No. 659 of 26th June 1903 and reconstituted as an I. D. F. Unit with present designation 1st Apl. 1917.*)

Establishment—1 Reserve Coy., 1 Combined Active & Reserve Coy.

HONORARY COMMANDANT—

First. Commn.	Names.	Date of rank.	Date of appointment to corps.	Remarks.
	Lieutenant-Colonel.			
1 Apl. 17	Richards, Charles Henry, V.D.	1 Apl. 17	1 Apl. 17	Comdt., Hony A.-D.-C. to the C.-in-C.
	Majors.			
1 Apl. 17	*Ray, Henry Maurice* (51, 54)	*1 Apl. 17*	*1 Apl. 17*	*I. A. R. O.* Adjutant.
1 Apl. 17	*Stephenson, Frank Henry Albert*, V. D. (54)	*1 Apl. 17*	*1 Apl. 17*	*I. A. R. O.*
1 Apl. 17	*Weeks, Robert Foster Jeffrey* (54)	*1 Apl. 17*	*1 Apl. 17*	*Temp., Capt. R. E.*
1 Apl. 17	Keith, Charles Napier (51, 54)	1 Apl. 17	1 Apl. 17	2nd-in-Comd. (*offg.*)
	Captains.			
1 Apl. 17	Bogie, David (54)	1 Apl. 17	1 Apl. 17	
1 Apl. 17	Thomas, William (51)	1 Apl. 17	1 Apl. 17	
1 Apl. 17	*Daniell Percy Smith*	*1 Apl. 17*	*1 Apl. 17*	*Temp.* 2nd-Lt., R.E.
1 Apl. 17	*Davies, Percy Meyrick Morris* (46, 54)	*1 Apl. 17*	*1 Apl. 17*	*I. A. R. O.*
1 Apl. 17	Hammill, Albert Edward	1 Apl. 17	1 Apl. 17	
1 Apl. 17	Fairweather, John William	1 Apl. 17	1 Apl. 17	
	Lieutenants.			
1 Apl. 17	*Treloar, Richard Martin*	*1 Apl. 17*	*1 Apl. 17*	*I. A. R. O.*
1 Apl. 17	Baker, Arnold (51, 54)	1 Apl. 17	1 Apl. 17	
1 Apl. 17	Stapleton, Frederick	1 Apl. 17	1 Apl. 17	
1 Apl. 17	Jones, Gerald Routh	1 Apl. 17	1 Apl. 17	
1 Apl. 17	Southey, Archibald (54)	1 Apl. 17	1 Apl. 17	
1 Apl. 17	*Pryor, Thomas*, D.S.O.	*1 Apl. 17*	*1 Apl. 17*	*I. A. R. O.*
	2nd-Lieutenants.			
1 Apl. 17	Trythall, William John (54)	1 Apl. 17	1 Apl. 17	Lt. late Ind. Volrs.
1 Apl. 17	Griffith-Jones, Martin Leoline	1 Apl. 17	1 Apl. 17	Lt. late Ind. Volrs.
1 Apl. 17	*Harman, Oswald Aubrey*, M.C.	*1 Apl. 17*	*1 Apl. 17*	*I. A. R. O.*
1 Apl. 17	Pringle, John Archibald	1 Apl. 17	1 Apl. 17	
1 Apl. 17	Bennetts, Graham Henry	1 Apl. 17	1 Apl. 17	
1 Apl. 17	*Williams, Morley James*	*1 Apl. 17*	*1 Apl. 17*	*Temp. 2nd-Lt., R. E.*
1 Apl. 17	Line, Richard	1 Apl. 17	1 Apl. 17	
	Supernumerary.			
1 Apl. 17	Richards, Capt. Almond (54)	1 Apl. 17	1 Apl. 17	*Lv., ex I., 7 mos., from 7-11-18.*
1 Apl. 17	Osborn, Capt. Howard Harry	1 Apl. 17	1 Apl. 17	
1 Apl. 17	Stephens, Lt. John	1 Apl. 17	1 Apl. 17	
1 Apl. 17	Pool, 2nd-Lt. William Glasson	1 Apl. 17	1 Apl. 17	
1 Apl. 17	Taylor, 2nd-Lt Guy Llewellyn	1 Apl. 17	1 Apl. 17	
1 Apl. 17	Keene, 2nd-Lt. Alfred Horace (51)	1 Apl. 17	1 Apl. 17	
1 Apl. 17	*Champion, 2nd-Lt. Oscar Dennis*	*1 Apl. 17*	*1 Apl. 17*	*I. A. R. O.*
	Staff.			
4 Jan. 05	Ray, Maj. H. M. (Capt., I. A. Res.) (54)	1 Apl. 13	15 June 15	Adjutant.

44th Calcutta Scottish.

(*Formed 1st August 1914 by Army Department Notfn. No. 690 of 1914. Reconstituted as an I. D. F. Unit with present designation 1st Apl. 1917.*)

Establishment—2 Active Coys. 1 Reserve Coy. 1 Armoured Motor Batty. (B.) ½ Cyclist. Coy.

HONORARY COMMANDANT—
His Excellency the Right Hon'ble Lawrence John Lumley Dundas, *Earl of* Ronaldshay, G.C.I.E.

(*Governor of Bengal.*) **26 Mar. 1917.**

First Commn.	Names.	Date of rank.	Date of appointment to corps.	Remarks.
	Lieutenant-Colonel.			
1 Apl. 17	Richardson, Thomas William (79)	1 Apl. 17	1 Apl. 17	Comdt.
	Majors.			
1 Apl. 17	*McDonald, John Stuart* (D. I.) (54) (79)	*1 Apl. 17*	*1 Apl. 17*	*Temp. Capt. in Army.*
1 Apl. 17	*Bray, Hon.* Sir *Edward Hugh,* Kt. (54), (79)	*1 Apl. 17*	*1 Apl. 17*	*Temp. Brig.-Genl. in Army, Contr. of Contracts.*
1 Apl. 17	Stewart, Sir Francis Hugh, Kt., C.I.E.	1 Apl. 17	1 Apl. 17	
	Captains.			
1 Apl. 17	Brown, James Archibald (54, 79)	1 Apl. 17	1 Apl. 17	
1 Apl. 17	*Jarrad, Evelyn Digby Donne*	*1 Apl. 17*	*1 Apl. 17*	*I. A. R. O.*
1 Apl. 17	Mackenzie Gilbert Proby (79)	1 Apl. 17	1 Apl. 17	
1 Apl. 17	Tassie, James Alexander (d l) (54, 79)	1 Apl. 17	1 Apl. 17	
1 Apl. 17	*Macfadyen, Melville* (54, 79)	*1 Apl. 17*	*1 Apl 17*	
1 Apl. 17	Birrell, Norman (d l) (54, 79)	1 Apl. 17	1 Apl. 17	
1 Apl. 17	Carter, *Hon.* Frank Willington, C.I.E., C.B.E. (54)	1 Apl. 17	1 Apl. 17	
	Lieutenants.			
15 Mar. 18	McLean, Graham Campbell	15 Mar. 18	15 Mar. 18	
15 Mar. 18	Muir, Alexander Pinkerton	15 Mar. 18	15 Mar. 8	
15 Mar. 18	*Bryce, William* (51)	*15 Mar. 18*	*15 Mar. 18*	*I. A. R. O.*
15 Mar. 18	Lennox, James	15 Mar. 18	15 Mar. 18	
15 Mar. 18	Pomphrey, Charles (54)	15 Mar. 18	15 Mar. 18	
15 Mar. 18	Donaldson Roy Charles (54)	15 Mar 18	15 Mar. 18	
15 Mar. 18	*Wardley, Arthur Norton* (54)	*15 Mar. 18*	*15 Mar. 18*	*I. A. R. O.*
19 Nov. 18	Carey, Willoughby Langer	19 Nov. 18	19 Nov. 18	
	2nd-Lieutenants.			
1 Apl. 17	Mackintosh, William	1 Apl. 17	25 Sep. 18	
4 Oct. 18	Browne, Harold	4 Oct. 18	4 Oct 18	
19 Nov. 18	Wands, Robert Douglas	19 Nov. 18	19 Nov. 18	
19 Nov. 18	Chamberlain, Harwood	19 Nov. 18	19 Nov. 18	
19 Nov. 18	Wood, Harbert Dresser	19 Nov. 18	19 Nov. 18	
	Supernumerary.			
1 Apl. 17	Bell, Maj. The Hon. Mr. Nicholas Dodd Beatson, C.S.I., C.I.E.	1 Apl. 17	1 Apl. 17	
1 Apl. 17	Stuart, Capt. Murray (80)	1 Apl. 17	12 June 18	
4 Oct. 18	Barber, 2nd-Lt. Howard Charles Linney	4 Oct. 18	4 Oct. 18	
	Staff.			
5 Feb. 13	Horan, Capt. K., Oxf. & Buck. L.I. (50, 48a)	14 Dec. 15	5 June 16	Adjutant.
14 Mar. 18	Gibson, Lt. G.	14 Mar. 18	14 Mar. 18	Qr. Mr.

45th Aden Rifles.

(Formed 1st January 1917 by Army Department Notification No. 1588 of 1916, and reconstituted as an I. D. F. Unit with present designation 1st Apl. 1917.)

Establishment—1 Reserve Coy.

First Commn.	Names.	Date of rank.	Date of appointment to corps.	Remarks.
	Major.			
1 Apl. 17	*Berridge, Harold (88)*	*1 Apl. 17*	*1 Apl. 17*	*Garr. Engr., Aden.*
1 Apl. 17	Gray, James Bett	30 Apl. 18	1 Apl. 17	Comdt.
	Captain.			
1 Apl. 17	Rycroft, Henry Zoro	30 Apl. 18	1 Apl. 17	2nd-in-Comd.
	Lieutenants.			
1 Apl. 17	Leman, Cyril Eric	30 Apl. 18	1 Apl. 17	Lt., late Ind. Volrs.
5 Apl. 18	Day, Stanley Nicholas	30 Apl. 18	5 Apl. 18	
	2nd-Lieutenants.			
30 Apl. 18	Thubron, John Brown Sydney	30 Apl. 18	30 Apl. 18	
30 Apl. 18	Randall, Eric	30 Apl. 18	30 Apl. 18	

INDIAN DEFENCE FORCE MEDICAL CORPS.

Lieutenant-Colonels.

First Commn.	Names.	Date of Rank.	Remarks.
1 Apl. 17	Fuller, Charles Arthur, M.B.	1 Apl. 17	7th (S) United Provinces Horse.

Majors.

First Commn.	Names.	Date of Rank.	Remarks.
1 Apl. 17	Seal, Charles Edward Baldwin, V.D.	1 Apl. 17	9th Northern Bengal Mounted Rifles.
1 Apl. 17	Houseman, Edward Albert	1 Apl. 17	3rd Calcutta Light Horse.
1 Apl. 17	McCutcheon, Oliver Edward, M.B.	1 Apl. 17	9th Northern Bengal Mounted Rifles.
1 Apl. 17	Caddy, Adrian, M.D., F.R.C.S.	14 Apl. 17	3rd Calcutta Light Horse.
1 Apl. 17	Hewan, John	1 Apl. 17	......

Captains.

First Commn.	Names.	Date of Rank.	Remarks.
1 Apl. 17	Allan, William Smith, M.B.	1 Apl. 17	1st (Calcutta Port Defence) Group, Garr. Arty.
1 Apl. 17	Spence, Arthur Dunbar, M.B.	1 Apl. 17	18th Rangoon Bn.
1 Apl. 17	Clayton, Robert Vickers, M.B.	1 Apl. 17	2-17th B. B. & C. I. Ry. Bn.
1 Apl. 17	Scoresby-Jackson, Thomas	1 Apl. 17	......
1 Apl. 17	Macdonald, William Roy	1 Apl. 17	......
1 Apl. 17	Dunlop, John	1 Apl. 17	......

Lieutenants.

First Commn.	Names.	Date of Rank.	Remarks.
1 Apl. 17	McCombie, Frederick Champ	1 Apl. 17	6th Assam Valley Light Horse.
1 Apl. 17	Wilkinson, Sydney Arthur	1 Apl. 17	2-17th B., B. & C. I. Ry. Bn.
1 Apl. 17	Murray, William Paterson, M.B.	1 Apl. 17	18th Rangoon Bn.
1 Apl. 17	Forsyth, Charles Edwin Percy	1 Apl. 17	6th Assam Valley Light Horse.
26 Oct. 18	Souza, Augustus, F.R.C.S.	26 Oct. 18	5th Lucknow Group Garr. Arty.
6 Dec. 18	Stone, Charles Ainslie	6 Dec. 18	9th Northern Bengal Mounted Rifles.

INDIAN DEFENCE FORCE VETERINARY CORPS.

Date of First Commn.	Names.	Date of Rank.	Remarks.
	Captain.		
1 Apl. 17	Harris, William	1 Jan. 18	6th A. V. L. Horse.
	Lieutenant.		
1 Apl. 17	Hewlett, Kenelm (temp. Major)	1 Apl. 17	4th Bo. L. Horse.

LIST OF REGIMENTS AND BATTERIES, ETC., OF THE INDIAN ARMY IN ORDER OF PRECEDENCE.

Cavalry.

The Governor-General's Body-Guard.

The Governor's Body-Guard, Madras.

The Governor's Body-Guard, Bombay.

The Governor's Body-Guard, Bengal.

26th, 27th, 28th Light Cavalry.

1st Lancers (Skinner's Horse), 2nd Lancers (Gardner's Horse).

3rd Skinner's Horse.

31st, 32nd Lancers.

33rd Light Cavalry; 34th Poona Horse, 35th Scinde Horse.

4th, 5th, 6th Cavalry; 7th Hariana Lancers, 8th Cavalry.

36th Jacob's Horse.

Queen Victoria's Own Corps of Guides, Cavalry.

21st Cavalry (Daly's Horse), 22nd, 23rd, 25th Cavalry.

9th Hodson's Horse.

10th Lancers (Hodson's Horse), 11th Lancers (Probyn's Horse).

12th Cavalry.

13th Lancers (Watson's Horse), 14th Murray's Jat Lancers, 15th Lancers (Cureton's Multanis).

16th, 17th Cavalry.

18th Lancers, 19th Lancers (Fane's Horse).

Aden Troop.

37th Lancers (Baluch Horse).

20th Deccan Horse.

29th Lancers (Deccan Horse), 30th Lancers (Gordon's Horse).

38th, 39th Central India Horse.

Artillery.

21st Kohat, 22nd Deraját, 23rd Peshawar, 24th Hazara, 25th, 26th Jacob's, 27th, 28th, 29th, 30th Mountain Batteries.

31st and 32nd Mountain Batteries.

The Frontier Garrison Artillery.

Sappers and Miners.

2nd, 1st, 3rd Sappers and Miners.

Infantry.

61st Pioneers; 62nd Punjabis; 63rd Palamcottah L. I.; 84th Pioneers; 66th, 67th, 69th Punjabis; 72nd Punjabis; 73rd Carnatic Infantry; 74th Punjabis; 75th Carnatic Infantry; 76th Punjabis.

1st Brahmans.

79th, 80th Carnatic Infantry.

The 101st Grenadiers.

81st Pioneers; 82nd Punjabis; 83rd Wallajahbad L. I., 84th Punjabis; 86th Carnatic Infantry.

102nd Grenadiers; 103rd Mahratta L. I.; 104th Wellesley's Rifles; 105th Mahratta L. I.; 107th Pioneers; 108th, 109th Infantry; 110th Mahratta L. Infantry.

2nd Rajput Light Infantry.

87th Punjabis; 88th Carnatic Infantry; 89th Punjabis.

3rd Brahmans; 4th Rajputs.

112th Infantry.

90th, 91st, 92nd Punjabis; 93rd Burma Infantry.

113th Infantry; 114th, 116th, 117th Mahrattas.

5th Light Infantry; 6th Jat Light Infantry; 7th, 8th Rajputs.

19th Infantry (The Mooltan Regiment); 120th Rajputana I.; 121st Pioneers; 122nd Rajputana I.; 123rd Outram's Rifles; 124th Baluchistan Infantry; 125th Napier's Rifles.

9th Gurkha Rifles; 10th Jats.

1st Gurkha Rifles (The Malaun Regiment).

11th Rajputs.

126th Baluchistan Infantry.

12th Pioneers (The Kelat-i-Ghilzie Regiment); 13th Rajputs (The Shekhawati Regiment).

127th Baluch Light Infantry.

14th, 15th Sikhs.

2nd Gurkha Rifles (The Sirmoor Rifles), 3rd Gurkha Rifles.

Queen Victoria's Own Corps of Guides Infantry.

51st, 52nd, 53rd, 54th Sikhs.

128th Pioneers; 129th Baluchis.

55th Coke's Rifles; 56th Rifles; 57th Wilde's Rifles; 58th Vaughan's Rifles; 59th Scinde Rifles.

4th Gurkha Rifles.

16th Rajputs (The Lucknow Regiment), 17th Infantry (The Loyal Regiment); 18th Infantry; 19th Punjabis; 20th Infantry (Brownlow's Punjabis); 21st, 22nd Punjabis; 23rd Pioneers; 24th, 25th, 26th, 27th, 28th, 29th, 30th, 31st Punjabis; 32nd Pioneers.

5th Gurkha Rifles.

130th Baluchis.

33rd Punjabis; 34th Pioneers; 35th Sikhs; 36th Sikhs; 37th Dogras; 38th Dogras; 39th Garhwal Rifles; 40th Pathans.

6th Gurkha Rifles.

41st Dogras; 7th Gurkha Rifles; 8th Gurkha Rifles; 45th Rattray's Sikhs; 46th Punjabis; 47th Sikhs; 48th Pioneers.

94th Russell's Infantry; 95th Russell's Infantry; 96th Berar Infantry; 97th Deccan Infantry; 98th Infantry; 99th Deccan Infantry.

Resident's Escort (Nepal).

42nd Deoli Regiment.

9th Bhopal Infantry.

43rd Erinpura Regiment.

44th Merwara Infantry.

106th Hazara Pioneers.

10th Gurkha Rifles.

All Infantry Regiments of the late Punjab Frontier Force are considered Light Infantry.

OFFICERS IN COMMAND OF INDIAN REGIMENTS.

Regiments.	Rank and Names.	Date of Compltg. tenure.	Remarks.
1 Lancers	**Russell**, Lt.-Col. F. D., *p.s.c.*, [L]	6 July 1920	On compltg. 4 years' regtl. comd.
2 Lancers	**Gwyn-Thomas**, Lt.-Col. G., C.M.G., D.S.O., *p.s.c.* (*tmep. Br-Genl.*).	18 July 1921	On attaining 50 years of age.
3 Skinner's Horse	**Gaussen**, Lt.-Col. J. R., C.M.G., D.S.O.	19 Mar. 1921	On attaining 50 years of age.
4 Cavalry	**Macquold**, Lt.-Col. C. E. E. F. K., D.S.O.	1 Aug. 1919	On attaining 50 years of age.
5 Cavalry	**Hay**, Lt.-Col. L. T.	14 Apl. 1919	On attaining 50 years of age.
6 Cavalry	**Newnham**, Lt.-Col. C. C., D.SO., *p.s.c.*	3 Feb. 1922	On compltg. 4 years' regtl. comd.
7 Lancers	**Carnegy**, Lt.-Col. G. P. O., D.S.O.	14 Mar. 1920	On compltg. 4 years' regtl. comd.
8 Cavalry	**Chaplin**, Lt.-Col. R. E.	18 Feb. 1918	Tenure extended.
9 Hodson's Horse	**Beatty**, Lt.-Col. G. A. H., D.S.O.	21 June 1920	On attaining 50 years of age.
10 Lancers	**Ricketts**, Lt.-Col. R. L., *p.s.c.*	....	......
11 Lancers	**Norman**, Bt.-Col. W. H., D.S.O., *p.s.c.*	3 July 1920	On compltg. 4 years' regtl. comd.
12 Cavalry	**Humfrey**, Lt.-Col. F. G. C.	30 July 1919	On attaining 50 years of age.
13 Lancers	**Mackenzie**, Lt.-Col. C., D.S.O.	24 July 1919	On attaining 50 years of age.
14 Lancers	**Souter**, Lt.-Col. H. M. W., C.M.G., D.S.O.	13 Nov. 1920	On compltg. 4 years' regtl. comd.
15 Lancers	**Cheyne**, Lt.-Col. A. Y.	28 Nov. 1920	On compltg. 4 years of regtl. comd.
16 Cavalry	**Mears**, Lt.-Col. C. D., M.V.O.	25 July 1921	On attaining 50 years of age.
17 Cavalry	**Barry-Smith**, Lt.-Col. R. C.	18 June 1921	On compltg. 4 years' regtl. comd.
18 Lancers	**Keighley**, Lt.-Col. V. A. S., M.V.O.	1 Dec. 1921	On compltg. 4 years' regtl. comd.
19 Lancers	**Gregory**, Bt.-Col. C. L. (*temp. Br.-Genl.*), C.B.	27 Aug. 1919	On compltg. 4 years' regtl. comd.
20 Deccan Horse	**Adams**, Lt.-Col. F., *p.s.*	28 July 1920	On compltg. 4 years' regtl. comd.
21 Cavalry	**Johnson**, Lt.-Col. J. E. B.	30 June 1922	On compltg. 4 years' regtl. comd.
22 Cavalry	**Young**, Lt.-Col. H. G., D.S.O.	6 Mar. 1920	On attaining 50 years of age.
23 Cavalry	**Hawes**, Lt.-Col. C. H., M.V.O., D.S.O.	30 Aug. 1921	On attaining 50 years of age.
25 Cavalry	**Peck**, Bt.-Col. A. W., C.M.G.	22 Apl. 1919	On attaining 50 years of age.
26 Lt. Cavalry	**Henderson**, Lt.-Col. M. H.	16 Dec. 1919	On attaining 50 years of age.
27 Lt. Cavalry	**Birdwood**, Lt.-Col. H. B. (*temp. Br.-Genl.*).	28 May 1920	On attaining 50 years of age.
28 Lt. Cavalry	**Oakes**, Lt.-Col. H. C. [L]	2 May 1921	On attaining 50 years of age.
29 Lancers	**Sangster**, Lt.-Col. P. B., D.S.O.	6 Aug. 1920	On compltg. 4 years' regtl. comd.
30 Lancers	**Harbord**, Lt.-Col. C. R., D.S.O. (*temp. Br.-Genl.*).	31 Aug. 1922	On compltg. 4 years' regtl. comd.
31 Lancers	**Ellwood**, Lt.-Col. B. P.	28 July 1921	On compltg. 4 years' regtl. comd.
32 Lancers	**Keogh**, Lt.-Col. J. B., D.S.O.	24 Jan. 1921	On attaining 50 years of age.
33 Lt Cavalry	**Stack**, Lt.-Col. C. S., C.M.G.	27 Sep. 1918	On attaining 50 years of age.
34 Poona Horse	**Cooper**, Lt.-Col. W. G.	30 Dec. 1918	On attaining 50 years of age.
35 Scinde Horse	**Lukin**, Lt.-Col. R. C. W., D.S.O. (*temp. Brig.-Genl.*).	22 Dec. 1920	On attaining 50 years of age.
36 Jacob's Horse	**Fagan**, Lt.-Col. E. A., C.M.G., D.S.O. (*temp. Br.-Genl.*).	17 Feb. 1920	On compltg. 4 years' regtl. comd.
37 Lancers	**Sherer**, Lt.-Col. J. C.	25 Sep. 1918	On attaining 50 years of age.
38 C. I. H.	**Hewlett**, Lt.-Col. A.	14 Jan. 1922	On compltg. 4 years' regtl. comd.
39 C. I. H.	**Capper**, Lt.-Col. A. S., D.S.O., *p.s.c.*	14 Feb. 1921	On compltg. 4 years' regtl. comd.
40 Cavalry	**Gaussen**, Lt.-Col. J. R., C.M.G., D.S.O., 3rd Horse.	....	Temp. (Permt. Comdt. 3 Horse).
41 Cavalry	**Hallowes**, Maj. J. H. (*Actg. Lt-Col.*), 15th Lrs.	....	......
42 Cavalry	**Taylor**, Maj. P. H. M., 32nd Lrs.	....	......
43 Cavalry	**Gordon**, Bt. Lt.-Col. G. A. S., 13th Lrs.	....	......
44 Cavalry	**Battye**, Lt.-Col. A. B., 6th Cav.	....	......
45 Cavalry	**Taylor**, Maj. E. M., P.S.C., 22nd Cav.	....	......
Guides	**Buist**, Lt.-Col. A. H., M.V.O.	13 Nov. 1920	On compltg. 4 years' regtl. comd.
Guides, 2 Infy.	**Bogle**, Maj. J. S. (*Actg. Lt.-Col.*).	....	......
Guides, 3 Infy.	**Clementi**, Maj. J. (*Actg. Lt.-Col.*)	....	......
1st S. & M.			
2nd S. & M.	**Heycock**, Lt.-Col. C. H.	....	......
3rd S. & M.			

OFFICERS IN COMMAND OF INDIAN REGIMENTS—*contd.*

Regiments.	Rank and Names.	Date of Compltg. tenure.	Remarks.
1-1 Brahmans	**Douglas,** Lt.-Col. W. B	6 Oct. 1919	On compltg. 4 years' regtl. comd.
2-1 Brahmans	**Mactavish,** Maj. A. (*Actg. Lt.-Col.*), 3rd Brahmans.	. . . . .	......
1-2 Rajput Lt. Inf.	**Smith,** Lt.-Col. A. L. F.	22 June 1920	On attaining 52 years of age.
2-2 Rajput Lt. Infy.	**Strong,** Maj. R. G. (*Actg. Lt.-Col.*), 1-11th Rajputs.	. . . . .	......
3-2 Rajput Lt. Infy.	**Wood,** Maj. H. O. B., (*Actg. Lt.-Col.*) 8th Rajputs.	. . . . .	......
1-3 Brahmans	**Hawkes,** Lt.-Col. R.	9 May 1921	On compltg. 4 years' regtl. comd.
2-3 Gaur Brahmans.	**Lane,** Maj. H. V., 1-1st Brahmans	. . . .	......
1-4 Rajputs	**Ralph,** Lt.-Col. A. C., D.S.O.	24 Aug. 1919	On compltg. 4 years' regtl. comd.
2-4 Rajputs	**Cutler,** Maj. H. C. (*Actg. Lt.-Col.*), 1-8th Rajputs.	. . . . .	......
1-5 Lt. Inf.	**Cotton,** Lt.-Col. W. L.	7 July 1919	On compltg. 4 years' regtl. comd.
2-5 Lt. Infantry	**Henderson,** Maj. K. D. M., M.C., 1-17th Infy.	. . . .	......
1-6 Jat Lt. Inf.	**Bourne,** Lt.-Col. W. Fitz G.	30 Jan. 1920	On compltg. 4 years' regtl. comd.
2-6 Jat Lt. Infy.	**Hardie,** Maj. G. E. (*Actg. Lt.-Col.*), 123rd Rif.	. . . .	......
1-7 Rajputs	**Parr,** Bt.-Col. H. O.	22 Jan. 1919	On compltg. 4 years' regtl. comd.
2-7 Rajputs	**Meadows,** Maj. C. A. G. P. (*Actg. Lt.-Col.*), 1-7th Rajputs.	. . . .	......
1-8 Rajputs	**Abbott,** Lt.-Col. L. H., C.M.G., *p.s.c.* (*temp. Br.-Genl.*)	20 Nov. 1922	On compltg. 4 years' regtl. comd.
2-8 Rajputs	**McCleverty,** Maj. A. H. (*Actg. Lt.-Col.*), 13th Rajputs.	. . . .	......
1-9 Bhopal Infy.			
2-9 (Delhi) Infy.	**Lloyd,** Lt.-Col. W. E. E., 1-97th Infy.	. . . .	......
3-9 Bhopal Infy.	**Bally,** Maj. H. G., (*Actg. Lt.-Col.*), 1-10th Jats.	. . . .	......
4-9 Bhopal Infy.	**Rennick,** Maj. R. W. de C., (*Actg. Lt.-Col.*), 1-95th Infy.	. . .	......
1-10 Jats	**Robson,** Lt.-Col. C. G. [L]	4 Oct. 1919	On attaining 52 years of age.
2-10 Jats	**Marsh,** Maj. C. H., D.S.O. (*Actg. Lt.-Col.*), 18th Lrs.	. . . .	......
1-11 Rajputs	**Samborne-Palmer,** Lt.-Col. F. C. S.	3 Jan. 1920	On compltg. 4 years' regtl. comd.
2-11 Rajputs	**Sargent,** Maj. W. P. M. (*Actg. Lt.-Col.*), 104th Rif.	. . . .	
1-12 Pioneers	**Hawkes,** Lt.-Col. W. C. W., D.S.O.	19 Oct. 1922	On compltg. 4 years' regtl. comd.
2-12 Pioneers	**Hawkes,** Lt.-Col. W. C. W., D.S.O.	. . .	Temp. (Permt. Comdt. 1-12th Prs.)
13 Rajputs	**Stewart,** Lt.-Col. J. A.	5 June 1920	On attaining 52 years of age.
14 Sikhs	**Earle,** Lt.-Col. E.	7 Aug. 1920	On compltg. 4 years' regtl. comd.
1-15 Sikhs	**Gordon,** Lt.-Col. J. L. R., C.B.	16 Nov. 1919	On attaining 52 years of age.
2-15 Sikhs	**McRae,** Maj. H. St. G. M., D.S.O., *p.s.c.* (*Actg. Lt.-Col.*), 45th Sikhs.	. . . .	......
16 Rajputs	**Vanrenen,** Lt.-Col. G. R.	8 May 1920	On compltg. 4 years' regtl. comd.
1-17 Infantry	**Nicholson,** Lt.-Col. W. C.	15 June 1920	On attaining 52 years of age.
2-17 Infantry	**Smithers,** Maj. L. S. H. (*Actg. Lt.-Col.*), 1-17th Infy.	. . . .	......
1-18 Infantry	**Pryce,** Lt.-Col. H. E. ap. R., C.M.G., D.S.O., *p.s.c.*	16 Aug. 1922	On compltg. 4 years' regtl. comd.
2-18 Infantry	**Pratt,** Lt.-Col. E. G. W., 1-95th Infy.	. . . .	......
1-19 Punjabis	**Pennefather Evans,** Lt.-Col. G.	8 Dec. 1920	On attaining 52 years of age.
2-19 Punjabis	**Churchill,** Lt.-Col. G. R. D., (Q), 1-19th Punj.	. . . .	......
20 Infantry	**Fordham,** Lt.-Col. W. M., *p.s.c.*	21 Jan. 1922	On compltg. 4 years' regtl. comd.
1-21 Punjabis	**Murray,** Lt.-Col. W. G. P. *p.s.c.*	21 Jan. 1922	On compltg. 4 years' regtl. comd.
2-21 Punjabis	**Sherlock,** Maj. G. W. S., *p.s.c.* (*Actg. Lt.-Col.*), 1-6th G. R.	. . . .	......
1-22 Punjabis	**Costello,** Lt.-Col. E. W., C.M.G., D.S.O., *p.s.c.*	21 Sep. 1922	On compltg. 4 years' regtl. comd.
2-22 Punjabis	**Clifford,** Maj. R. (*Actg. Lt.-Col.*), 1-22nd Punj.	. . . .	......
1-23 Pioneers	**Ottley,** Lt.-Col. W. J.	20 Feb. 1922	On compltg. 4 years' regtl. comd.
2-23 Pioneers	**Ottley,** Lt.-Col. W. J., 1-23rd Prs.	. . . .	Temp. (Permt. Comdt., 1-23rd Sikh Prs).
3-23 Sikh Infy.	**Talbot,** Bt. Lt.-Col. F. E. G., *p.s.c.*, 14th Sikhs.	. . . .	......
24 Punjabis			
1-25 Punjabis	**Moberly,** Lt.-Col. F. J., D.S.O., *p.s.c.* (*temp. Br.-Genl.*)	24 Feb. 1919	On compltg. 4 years' regtl. comd.
2-25 Punjabis	**Pratt,** Maj. H. R. E., D.S.O. (*Actg. Lt.-Col.*), 1-35th Sikhs.	. . . .	......
1-26 Punjabis	**Cox,** Lt.-Col. A. D.	15 May 1921	On compltg. 4 years' regtl. comd.
2-26 Punjabis	**Magniac,** Maj. F. A. (*Actg. Lt.-Col.*), 1-22nd Punj.	. . . .	......
1-27 Punjabis			

Regiments.	Rank and Names.	Date of Compltg. tenure.	Remarks.

OFFICERS IN COMMAND OF INDIAN REGIMENTS—*contd.*

Regiments.	Rank and Names.	Date of Compltg. tenure.	Remarks.
2-27 Punjabis	**Tancred,** Maj. J. Y., 1-19th Punj.	....	......
1-28 Punjabis	**Chichester,** Lt.-Col. A. G. deV.	6 Nov. 1918	On attaining 52 years of age.
2-28 Punjabis	**Smith,** Maj. E. C. (*Actg. Lt.-Col.*), 1-22nd Punj.	....	......
29 Punjabis	**Ford,** Lt.-Col. H. B.	5 June 1920	On compltg. 52 years of age.
1-30 Punjabis	**James,** Lt.-Col. A. A., 31st Punj.	26 July 1921	On attaining 52 years of age.
2-30 Punjabis	**Keen,** Maj. P. H. (*Actg. Lt.-Col.*) 1-56th Punj.	....	......
3-30 Punjabis	**Isemonger,** 2nd-Lt. F. C. (*Actg. Lt.-Col.*)	....	(I. A. R. O.)
4-30 Punjabis	**Mercer,** 2nd-Lt. A. W., (*Actg. Lt.-Col.*)	....	(I. A. R. O.)
31 Punjabis	**Kemball,** Lt.-Col. A. G. (*temp. Br. Genl.*)	15 Oct. 1921	On attaining 52 years of age.
1-32 Pioneers	**Cooke,** Col. H. F., D.S.O. (*temp. Maj.-Genl.*)	....	On compltg. 4 years' regtl. comd.
2-32 Pioneers	**Clarke,** Maj. G. S. (*Actg. Lt.-Col.*), 1-34th Prs.	....	......
3-32 Pioneers	**Hodgson,** Maj. G. C., D.S.O. (*Actg. Lt.-Col.*) 1-23rd Prs.	....	......
1-33 Punjabis	**Harrison,** Lt.-Col. A. H. P. (*temp. Br.-Genl.*).	9 Nov. 1922	On compltg. 4 years' regtl. comd.
2-33 Punjabis	**Barrett,** Bt. Lt.-Col. A. L. (*Actg. Lt.-Col.*) 126th Bal. Infy.	....	......
1-34 Pioneers	**Cullen,** Lt.-Col. E. H. S., C.M.G., D.S.O., M.V.O.	23 Mar. 1919	On compltg. 4 years' regtl. comd.
2-34 Pioneers	**Turnbull,** Maj. B. (*Actg. Lt.-Col.*), 1-23rd Prs.	....	......
3-34 Pioneers	**Hutchinson,** Maj. A. G. C., *p.s.c.*, 1-23rd Prs.	....	......
1-35 Sikhs	**Rice,** Lt.-Col. S. K. B.	13 Mar. 1922	On compltg. 4 years' regtl. comd.
2-35 Sikhs	**Bourne,** Bt. Lt.-Col. W. K., *p.s.c.*, 2nd Lrs.	....	......
36 Sikhs	**Knox,** Lt.-Col. E. F., D.S.O.	14 Nov. 1921	On compltg. 4 years' regtl. comd.
37 Dogras	**Browne,** Lt.-Col. L. S.	14 June 1922	On compltg. 4 years' regtl. comd.
38 Dogras	**Longden,** Lt.-Col. A. B.	3 Jan. 1919	On compltg. 4 years' regtl. comd.
39 Garhwal Rif.	1 **Stewart,** Bt.-Col. J. H. K., D.S.O., *p.s.c.*	30 June 1920	On compltg. 4 years' regtl. comd.
	2 **Hogg,** Lt.-Col. J. McK. T.	19 May 1921	On attaining 52 years of age.
	3 **Stansfeld,** Maj. C. G., *p.s.c.* (*Actg. Lt.-Col.*), 8th G. R.	....	......
	4 **Lane,** Maj. J. T. H., M.C., 1-39th Rif.	....	......
40 Pathans	**Glasfurd,** Lt.-Col. A. I. R., C.M.G., D.S.O., *p.s.c.*	14 Nov. 1921	On compltg. 4 years' regtl. comd.
1-41 Dogras	**Macpherson,** Lt.-Col. R. G.	30 Mar. 1920	On compltg. 4 years' regtl. comd.
2-41 Dogras	**Bignell,** Maj. R. L., D.S.O., 1-41st Dogras.	....	......
1-42 Deoli Regt.	**Priestley,** Lt.-Col. C. E. N.	14 Jan. 1919	On compltg. 4 years' regtl. comd.
2-42 Deoli Regt.	**Brackenbury,** Lt.-Col. W. J. W., 1-43rd Regt.	....	Temp. (Permt. Comdt., 1-43rd R.)
1-43 Erinpura R.	**Brackenbury,** Lt.-Col. W. J. W.	31 Jan. 1922	On compltg. 4 years' regtl. comd.
2-43 Infantry	**Loudon,** Lt.-Col. J. A.	....	......
44 Merwara Inf.	**Alexander,** Lt.-Col. H. S., D.S.O.	15 May 1920	On compltg. 4 years' regtl. comd.
45 Sikhs	**McVean,** Lt.-Col. D. A. D.	28 July 1921	On compltg. 4 years' regtl. comd.
46 Punjabis	**Walton,** Lt.-Col. G.	19 Oct. 1921	On compltg. 4 years' regtl. comd.
47 Sikhs	**Davidson,** Lt.-Col. S. R., C.M.G. (*temp. Br.-Genl.*)	17 Feb. 1920	On compltg. 4 years' regtl. comd.
1-48 Pioneers	**Cuming,** Lt.-Col. R. J., D.S.O., *q.*	23 June 1922	On compltg. 4 years' regtl. comd.
2-48 Pioneers	**Hewett,** Maj. G., D.S.O., *q.* (*Actg. Lt.-Col.*), 1-48th Prs.	....	......
49 Bengalis	**Sandiford,** Maj. V. V. V., 62nd Punj.	....	......
1-50 Kumaon Rif.	**Lang,** Maj. E. M. (*Actg. Lt.-Col.*), Welsh R.	....	......
2-50 Kumaon Rif.	**Maxwell,** Maj. D W., *p.s.c.* (*Actg. Lt.-Col.*), 2-4th G. R.	....	......
51 Sikhs	**Lees,** Lt.-Col. C. H. B., C.M.G., *p.s.c.*	31 Jan. 1922	On compltg. 4 years' regtl. comd.
52 Sikhs	**Wynter,** Lt.-Col. C. P.	23 May 1920	On compltg. 4 years' regtl. comd.
53 Sikhs	**Gardiner,** Lt.-Col. R., *p.s.c.*	9 Mar. 1922	On compltg. 4 years' regtl. comd.
1-54 Sikhs	**Bainbridge,** Lt.-Col. W. F., C.M.G., D.S.O., *p.s.c.* (*temp. Br.-Genl.*)	23 May 1920	On compltg. 4 years' regtl. comd.
2-54 Sikhs	**Routh,** Maj. G. S. F. (*Actg. Lt.-Col.*), 1-56th Rif.	....	......
1-55 Rifles	**Herdon,** Lt.-Col. H. E., *p.s.c.*	1 Mar. 1922	On compltg. 4 years' regtl. comd.
2-55 Rifles	**Finnis,** Maj. B. H. (*Actg. Lt.-Col.*). 20th Infy.	....	......
1-56 Rifles	**Prissick,** Lt.-Col. C.	21 Sep. 1922	On compltg. 4 years' regtl. comd.
2-56 Rifles	**Ames,** Maj. C. G. (*Actg. Lt.-Col.*), 52nd Sikhs.	....	......

Regiments.	Rank and Names.	Date of Compltg. tenure.	Remarks.
	OFFICERS IN COMMAND OF INDIAN REGIMENTS—*contd.*		
57 Rifles	Willans, Lt.-Col. T. J., D.S.O.	7 Mar. 1922	On compltg. 4 years' regtl. comd.
58 Rifles	Knox, Bt.-Col. A. W. F., C.B., *p.s.c.*, [L] (*temp. Br.-Genl.*)	23 July 1920	On compltg. 4 years' regtl. comd.
59 Rifles	Leeds, Lt.-Col. T. L., C.M.G., D.S.O.	19 Feb. 1921	On compltg. 4 years' regtl. comd.
1-61 Pioneers	Scott Elliot, Lt.-Col. C. R., *q.*	4 May 1921	On compltg. 4 years' regtl. comd.
2-61 Pioneers	Bruce, Lt.-Col. G. D., D.S.O., *p.s.c.*	.....	Temp. (Permt. Comdt., 1-81st Prs.)
62 Punjabis	Morris, Bt.-Col. G. M., D.S.O. (*temp. Br.-Genl.*)	16 Mar. 1920	On attaining 52 years of age.
1-63 Lt. Infy.	Waring, Lt.-Col. E. H.	18 Aug. 1922	On compltg. 5 years' regtl. comd.
2-63 Lt. Infy.	Mouat, Maj. G. E. D., D.S.O., 1-63rd Infy.	.....	.....
64 Pioneers	Bliss, Lt.-Col. J. A., D.S.O., M.V.O.	19 Aug. 1920	On compltg. 4 years' regtl. comd.
1-66 Punjabis	Moberly, Lt.-Col. H. S., *p.s.c.*	31 May 1922	On compltg. 4 years' regtl. comd.
2-66 Punjabis	Stoney, Maj. P. S., *p.s.c.* (*Act'g. Lt.-Col.*) 1-26th Punj.	.....	.....
1-67 Punjabis	Colan, Lt.-Col. W. R. B., D.S.O.	26 Nov. 1921	On compltg. 4 years' regtl. comd.
2-67 Punjabis			
1-69 Punjabis	Houston, Lt.-Col. A. M.	9 Mar. 1922	On compltg. 4 years' regtl. comd.
2-69 Punjabis	Dunsford, Maj. E. H. (*Actg. Lt.-Col.*) 1-33rd Punj.	.....	.....
1-70 Burma Rif.	Stuart, Maj. D. N. (*Actg. Lt.-Col.*)	.....	.....
2-70 Burma Rif.	Rigg, Maj. H. M. (*Actg. Lt.-Col.*), 92nd Punj.	.....	.....
3-70 Burma Rif.	Morgan, Maj. B. E., D.S.O., (*Actg. Lt. Col.*) 93rd Infy.	.....	.....
4-70 Burma Rif.	Landon, Maj. C. R. H. P. (*Actg. Lt.-Col.*), 35th Horse.	.....	.....
71 Punjabis	Wright, Maj. C. H. B. (*Actg. Lt.-Col.*), 62nd Punj.	.....	.....
1-72 Punjabis	Prentis, Lt.-Col. W. S.	12 June 1920	On compltg. 4 years' regtl. comd.
2-72 Punjabis	Butler, Maj. H. M. (*Actg. Lt.-Col.*), 104th Rif.	.....	...
1-73 Infy.	Baker, Lt.-Col. H. R.	10 Nov. 1918	On compltg. 4 years' regtl. comd.
2-73 Infy.	Lawrenson, Maj. W. R., (*Actg. Lt.-Col.*), 86th Infy.	.....	.....
74 Punjabis	Alexander, Lt.-Col. A. V.	27 June 1921	On attaining 52 years of age.
1-75 Infy.	Field, Lt.-Col. C. D.	9 May 1921	On compltg. 4 years' regtl. comd.
2-75 Infy.	Watson, Maj. H. R., *q.* (*Act'g. Lt.-Col.*) 1-81st Prs.	.....	...
1-76 Punjabis			
2-76 Punjabis	Chamberlayne, Maj. A. (*Actg. Lt.-Col.*) 1st Lrs.	.....	.....
79 Infy.	Kauntze, Lt.-Col. B. C.	30 July 1922	On compltg. 4 years' regtl. comd.
80 Infy.	Hole, Bt.-Col. H. L.	29 Dec. 1920	On compltg. 4 years' regtl. comd.
1-81 Pioneers	Bruce, Lt.-Col. G. D., D.S.O., *p.s.c.*	12 Aug. 1922	On compltg. 4 years' regtl. comd.
2-81 Pioneers	Mackie, Maj. R., *q.* (*Actg. Lt.-Col.*)		.....
82 Punjabis	Milne, Lt.-Col. J. W.	24 Mar. 1922	On attaining 52 years of age.
83 Lt. Infy.	Keays, Lt.-Col. [illegible] W. C.	1 Dec. 1918	On compltg. 4 years' regtl. comd.
84 Punjabis	Crawford, Lt.-Col. A. G.	22 Aug. 1921	On attaining 52 years of age.
85 Burma Rif.	Burd, Maj. E. (*Temp. Lt.-Col.*), 93rd Infy.	.....	.....
86 Infy.	O'Reilly, Lt.-Col. C. M.	1 Jan. 1921	On attaining 52 years of age.
87 Punjabis	Prevost, Lt.-Col. G. H.	11 Apl. 1920	On attaining 52 years of age.
1-88 Infy.	McCaskill, Lt.-Col. J. C. H.	10 June 1921	On compltg. 4 years' regtl. comd.
2-88 Infy.	Malet, Maj. R. J. (*Actg. Lt.-Col.*), 64th Prs.	.....	.....
1-89 Punjabis	Jackson, Lt.-Col. H. W., D.S.O. (*temp. Br.-Genl.*)	9 Mar. 1922	On compltg. 4 years' regtl. comd.
2-89 Punjabis	Oldfield, Maj. G. E. B. (*Actg Lt.-Col.*), 74th Pun.	.....	.....
1-90 Punjabis	Murray, Lt.-Col. E. R. B. (*temp. Br.-Genl.*)	13 Sep. 1920	On compltg 4 years' regtl. comd.
2-90 Punjabis	Bayley, Maj. L. C. L. (*Actg. Lt.-Col.*), 6th Cav.	.....	.....
1-91 Punjabis	Basevi, Lt.-Col. W. H. F.	16 May 1920	On attaining 52 years of age.
2-91 Punjabis	Crawford, Lt.-Col. A. G., 84th Punj	.....	Temp. (Permt. Comdt., 84th Punj.)
92 Punjabis	Maclachlan, Bt.-Col. I. R., C.M.G., (*temp Br.-Genl.*)	2 July 1919	On compltg. 4 years' regtl. comd.
93 Bur. Infy.			
1-94 Infy.	Dobbs, Lt.-Col. C. F., D.S.O., *p.s.c.*	3 Feb. 1921	On compltg. 4 years' regtl. comd.
2-94 Infy.	Meiklejohn, Maj. W. L. S., *p.s.c.*, *q.* (*Actg. Lt.-Col.*), 106 Prs.	.....	.....
1-95 Infy.	Major, Lt.-Col. F. F.	10 Oct. 1918	On compltg. 4 years' regtl. comd.
2-95 Infy.	Savile, Maj. A. R. C. (*Actg. Lt.-Col.*), 31st Lrs.	.....	.....
1-96 Infy.	Nangle, Lt.-Col. K. E	2 May 1921	On compltg. 4 years' regtl. comd.
2-96 Infy.	Jardine, Bt. Lt.-Col. C H, 1-96th Infy.	.....	.....
1-97 Infy.	Preston, Lt. Col. W. J. P., D.S.O.	4 Nov. 1922	On compltg. 4 years' regtl. comd.
2-97 Infy.	Martin, Maj. J., M.C. (*Actg. Lt.-Col.*), 1-94th Infy.	.....	.....
1-98 Infy.	Adye, Bt.-Col. D. R. [L], (*temp. Br.-Genl.*)	29 Aug. 1919	On attaining 52 years of age.

Regiments.	Rank and Names.	Date of Compltg. tenure.	Remarks.
	OFFICERS IN COMMAND OF INDIAN REGIMENTS—*contd.*		
2-98 Infy.	**Grant,** Maj. W. O. (*Actg. Lt.-Col.*), 127th Bal. L. I.	....	......
1-99 Infy.	**Davidson,** Lt.-Col. F. D.	31 Mar. 1920	On compltg. 4 years' regtl. comd.
2-99 Infy.	**Nangle,** Maj. M. C., *p.s.c.*, (*l*), 92nd Punj.	....	......
1-101 Grs.	**Strong,** Lt.-Col. H. P.	29 Mar. 1922	On compltg. 4 years' regtl. comd.
2-101 Grenadiers	**Strong,** Lt.-Col. H. P., 101st Grs.	....	Temp. (Permt. Comd.)
1-102 Grs.	**Edwardes,** Lt.-Col. A. C.	11 Oct. 1921	On compltg. 4 years' regtl. comd.
2-102 Grenadiers	**Morris,** Maj. C. T. (*Actg. Lt.-Col.*), 82nd Punj.	....	......
1-103 Lt. Infy.	**Ross,** Lt.-Col. H.	16 Apl. 1921	On attaining 52 years of age
2-103 Lt. Infy.	**Bald,** Maj. J. A., C.M.G., 61st Prs.	....	......
3-103 Lt. Infy.	**Steen,** Maj. H. C.	....	......
104 Rifles	**Scott,** Lt.-Col. W.	5 Feb. 1922	On compltg. 4 years' regtl. comd.
1-105 Lt. Infy.	**Waller,** Lt.-Col. A. A. P.	27 Feb. 1922	On compltg. 4 years' regtl. comd.
106 Pioneers	**Carter,** Lt.-Col. G. L., C.I.E.	30 Sep. 1920	On compltg. 4 years' regtl. comd.
1-107 Pioneers	**Bickford,** Lt.-Col. W. W., C.I.E.	26 Feb. 1922	On compltg. 4 years' regtl. comd.
2-107 Pioneers	**Cunningham,** Maj. C. C., *p.s.c.*, (*Actg. Lt.-Col.*), 12th Prs.	....	......
108 Infy.	**Elderton,** Lt.-Col. M. F.	25 Dec. 1918	On attaining 52 years' of age.
1-109 Infy.	**Ommaney,** Lt.-Col. C. V.	29 May 1922	On compltg. 4 years' regtl. comd.
2-109 Infy.	**Smith,** Maj., C. A. J. (*Actg. Lt.-Col.*)	....	......
110 Lt. Infy			
111 Mahars.	**Dyke,** Lt.-Col. O. M., 21st Cav.	....	......
1-112 Infy.	**Winter,** Bt.-Col. C.B., C.M.G., D.S.O. [L]	22 July 1920	On compltg. 4 years' regtl. comd.
2-112 Infy.	**McCormick,** Maj. G. D., 1-72nd Punj.	....	......
1-113 Infy.	**Lloyd-Jones,** Lt.-Col. F. L.	11 Sept. 1919	On attaining 52 years of age.
2-113 Infy.	**Douglas,** Maj. G. S. (*Actg. Lt.-Col.*)	....	......
114 Mahrattas	**Wintle,** Lt.-Col. C. E. H.	28 Mar. 1922	On compltg. 4 years' regtl. comd.
1-116 Mahrattas	**Bridges,** Lt.-Col. A. H., D.S.O., *p.s.c.* [*l*]	8 Apl. 1922	On compltg. 4 years' regtl. comd.
2-116 Mahrattas	**Henderson,** Maj. C. G. H.	....	......
1-117 Mahrattas	**Hunter,** Lt.-Col. W. J. H.	18 Mar. 1922	On compltg. 4 years' regtl. comd.
2-117 Infy.	**Holmes,** Maj. C. W. A., P.S.C., (*Actg. Lt.-Col.*), 116th Mahs.	....	......
1-119 Infy.	**Darley,** Lt.-Col. J. R., D.S.O.	4 Jan. 1920	On attaining 52 years of age.
2-119 Infy.	**Carey,** Maj. T. deB. (*Actg. Lt.-Col.*) 119th Infy.	....	......
120 Infy.	**Ward,** Lt.-Col. C. H.	13 Apl. 1920	On attaining 52 years of age.
121 Pioneers	**Keelan,** Lt.-Col. H. P., Q.	13 Aug. 1920	On compltg. 4 years' regtl. comd.
122 Infy.	**Barclay,** Lt.-Col. P. C. R.	1 Apl. 1921	On compltg. 4 years' regtl. comd.
1-123 Rifles.	**Cassels,** Lt.-Col. G. R.	12 Jan. 1921	On compltg. 4 years' regtl. comd.
2-123 Rifles	**Byrne,** Maj. E. G. J. (*Actg. Lt.-Col.*), 104th Rif.	....	......
1-124 Baluchistan Infy.	**Tarver,** Bt.-Col. A. L., D.S.O., A.D.C., *p.s.c.*	24 June 1922	On compltg. 4 years' regtl. comd.
2-124 Baluchistan Infy.	**Mitchell,** Lt.-Col. W. J. (Q), 124th Infy.	....	......
3-124 Baluchistan Infy.	**Holbrooke,** Maj. B. F. R. (*Actg. Lt.-Col.*), 1-124th Infy.	....	......
1-125 Rifles	**Kelly,** Lt.-Col. E. P. C., C.M.G.	23 Sep. 1920	On compltg. 4 years' regtl. comd.
2-125 Rifles	**Hamer,** Maj. M. A., D.S.O., M.C., *p.s.c.* (*Actg. Lt.-Col.*)	....	......
126 Baluchistan Infy.	**Chitty,** Lt.-Col. A. W., *p.s.c.*	1 July 1922	On compltg. 4 years' regtl. comd.
1-127 Baluch. Lt. Infy.	**Hulseberg,** Lt.-Col. H., D.S.O.	25 Aug. 1922	On compltg. 4 years' regtl. comd.
2-127 Baluch. Lt. Infy.	**Chauncy,** Maj. C. H. K. (*Actg. Lt.-Col.*) 124th Infy.	....	......
1-128 Pioneers	**Andrew,** Lt.-Col. F. A., D.S.O. (*temp. Br.-Genl.*)	10 Sep. 1919	On compltg. 4 years' regtl. comd.
2-128 Pioneers	**Ashburner,** Maj. H. W., D.S.O., (*Actg. Lt.-Col*).	....	......
1-129 Baluchis	**Hay,** Lt.-Col. W. N., C.I.E., *p.s.c.*	22 Aug. 1922	On compltg. 4 years regtl. comd.
2-129 Baluchis	**Merriman,** Maj. A.B. (*Actg. Lt.-Col.*)	....	......
1-130 Baluchis	**Price,** Bt.-Col. C. U., C.M.G.	16 Apl. 1919	On compltg. 4 years' regtl. comd.
2-130 Baluchis			
1-131 Regt.	**Warner,** 2nd-Lt. H.R. (*Actg. Lt.-Col.*)	....	I. A. R. O.
2-131 Regt.	**Adams,** 2nd-Lt. R. J. (*Actg. Lt.-Col.*)	....	......
1-133 Regt.	**McDowell,** Maj. A.E. (*Actg. Lt.-Col.*)	....	......
1-140 Infy.	**MacMohan,** Maj. G. D. R. (*Actg. Lt.-Col.*)	....	......
1-141 Infy.	**Vander-Gucht,** Maj. G. T.	....	......
1-142 Infy.	**Money,** Maj. E. F. D., D.S.O.	....	......
1-143 Infy.	**Peters,** Maj. J. (*Actg. Lt.-Col.*), 10th Lrs.	....	......
1-144 Infy.	**Brown,** Maj. C. A.	....	......
1-145 Infy.	**Cruickshank,** Maj. J. E.	....	......
1-150 Infy.	**Orchard,** Maj. A. J. (*Actg. Lt.-Col.*), 8th Rajputs.	....	......

(*a*) Tempy. Br.-Genl.

Regiments.	Rank and Names.	Date of Compltg. tenure.	Remarks.
	OFFICERS IN COMMAND OF INDIAN REGIMENTS—*concld.*		
2-150 Infy. .	**Birch,** Maj. V. K. (*Actg. Lt.-Col.*) 29th Lrs.	. . . .	......
3-150 Infy. .	**Trueman,** Maj. H. J. H. (*Actg. Lt.-Col.*), 1-43rd R.	. . . .	......
1-151 Infy. .	**Strong,** Bt. Lt.-Col. A. D., D.S.O. (*Actg. Lt.-Col.*), 10th Lrs.	. . . .	......
2-151 Infy. .	**Smith,** Maj. A. A. (*Actg. Lt.-Col.*), 58th Rif.	. . . .	......
3-151 Infy. .	**Renny,** Maj. G. S. (*Actg. Lt.-Col.*), 25th Dogras.	. . . .	......
1-152 Infy. .	**Penton,** Maj. B. C. (*Actg. Lt.-Col.*), 25th Punj.	. . . .	......
2-152 Infy. .	**Wilkinson,** Maj. C. R. (*Actg. Lt.-Col.*), 52nd Sikhs.	. . . .	......
3-152 Infy. .	**Doveton,** Maj. H. J. (*Actg. Lt.-Col.*) 21st Punj.	. . . .	......
1-153 Infy. .	**Withers,** Maj. C. Mc G. (*Actg. Lt.-Col.* 87th Punj.	. . . .	......
2-153 Infy. .	**James,** Maj. C. A. (*Actg. Lt.-Col.*), 126th Infy.	. . . .	......
3-153 Infy. .	**Kidd,** Maj. B. G. B. (*Actg. Lt.-Col.*) 12th Rif.	. . . .	......
1-154 Infy. .	**Slingsby,** Maj. T. W. (*Actg. Lt.-Col.*), 22nd Cav.	. . . .	......
2-154 Infy. .	**Boxwell,** Maj. A., C.I.E. (*Actg. Lt.-Col.*), 119th Infy.	. . . .	......
3-154 Infy. .	**Davson,** Maj. H. J. H., (*Actg. Lt.-Col.*), 82nd Punj.	. . . .	......
1-155 Pioneers .	**Watson,** Bt.-Lt.-Col. S. B. Q., 64th Prs.	. . . .	......
2-155 Pioneers .	**Nicolas,** Maj. R., (*Actg. Lt.-Col.*), 23rd Prs.	. . . .	......
1-156 Infy. .	**Broadbent,** Maj. J. T. C. (*Actg. Lt.-Col.*), 88th Infy.	. . . .	......
1 G. R. .	1. **Anderson,** Bt.-Col. W. C., C.M.G. (*temp. Br.-Genl.*)	19 Feb. 1919	On attaining 52 years of age.
	2. **Dopping-Hepenstal,** Lt.-Col. M. E., D.S.O. [*l*].	10 Mar. 1922	On compltg. 4 years' regtl. comd
	3. **Ryder,** Lt.-Col. W. I.	. . . .	......
2 G. R. .	1. **Sweet,** Lt.-Col. E. H., D.S.O. . .	31 Dec. 1920	On compltg. 4 years' regtl. comd.
	2. **Boileau,** Lt.-Col. E. R. P. [L]	1 Apl. 1920	On compltg. 4 years' regtl. comd.
	3. **Arbuthnot,** Maj. A. H. (*Actg. Lt.-Col.*) 2-7th G. R.	. . . .	......
3 G. R. .	1. **Edwardes,** Lt.-Col. J. G. . . .	2 July 1921	On compltg. 4 years' regtl. comd.
	2. **Eastmead,** Lt.-Col. C. S . . .	27 Feb. 1920	On compltg. 4 years' regtl. comd.
	3. **Channer,** Maj. G. K., D.S.O. (*Actg. Lt.-Col.*), 3rd G. R.	. . . .	......
	4. **Nicolay,** Lt.-Col. B. U. . . .	. . . .	......
4 G. R. .	1 **Tillard,** Lt.-Col. A. B., D.S.O. . .	10 Nov. 1921	On compltg. 4 years' regtl. comd.
	2. **Nicolay,** Lt.-Col.. B. U. . . .	2 Aug. 1922	On compltg. 4 years' regtl. comd.
5 G. R. . . .	1. **Nightingale,** Lt.-Col. M. R. W., C.M.G., D.S.O. [L] (*temp. Br.-Genl.*)	28 Feb. 1921	On compltg. 4 years' regtl. comd.
	2. **Johnson,** Lt.-Col. C. R. .	21 Oct. 1920	On compltg. 4 years' regtl. comd
	3. **Barlow,** Maj. C. P. (*Actg. Lt.-Col.*,	. . . .	......
6 G. R. .	1. **Little,** Lt.-Col., W. C.	17 Mar. 1922	On compltg. 4 years' regtl. comd.
	2. **Preston,** Lt.-Col. G. A.	13 Jan. 1920	On attaining 52 years of age
	3. **Jewett,** Maj. A. E. (*Actg. Lt.-Col.*), 1-8th G. R.	. . . .	......
7 G. R. .	1 **Powell,** Lt.-Col. W. B., C.M.G.	27 Feb. 1920	On attaining 52 years of age.
	2. **Haldane,** Lt.-Col. C. L., C.M.G. .	23 Nov. 1918	On attaining 52 years of age.
	3. **Sealy,** Lt.-Col. A. E. . . .	. . . .	......
8 G. R. . . .	1. **Roosmale-Cocq,** Lt.-Col. C. A.	9 July 1920	On attaining 52 years of age.
	2. **Wall,** Lt.-Col. A. C. . . .	28 July 1922	On compltg. 4 years' regtl. comd.
	3. **Giffard,** Maj. D. H. R. (*Actg. Lt.-Col.*), 1-8th G. R. .	. . . .	......
9 G. R. . . .	1. **Bradshaw,** Lt.-Col. C. R., *p.s.c.*	22 June 1922	On compltg 4 years' regtl. comd.
	2 **Bateman-Champain,** Lt.-Col. H. F., C.M.G. (*temp. Br.-Genl.*)	4 Mar. 1921	On compltg. 4 years' regtl. comd.
	3. **Mackinnon,** Maj. A. C. B. (*Actg. Lt.-Col.*), 2-9th G. R.		......
10 G. R. . . .	1. **Coningham,** Lt.-Col. F. E., D.S.O.	2 Jan. 1922	On compltg. 4 years' regtl. comd.
	2. **Gale,** Lt.-Col. E. S.	2 July 1922	On compltg. 4 years' regtl. comd.
11 G. R. . .	1. **Johnson,** Maj. A. E. (*Actg. Lt.-Col.*)	. . . .	......
	2. **Gatherer,** Maj. R. G. T. (*Actg. Lt.-Col.*)	. . . .	......
	3. VC **Grant,** Maj. J. D. (*Actg. Lt.-Col.*), 2-8th G. R.	. . . .	......
	4. **Dundas,** Maj. W. L. (*Actg. Lt.-Col.*), 2-3rd G. R.	. . . .	......

INDIAN CAVALRY.

Governor-General's Body-Guard.

Dehra Dun.—*Arrived, 1st April 1912, from Ballygunge.*

Raised at Calcutta in 1774, by Captain S. Toone, under the official designation of 'The Governor General's Troop of Body-Guard,' but was at first generally known as the 'Governor's Troop of Moghals.' The present designation, which was never formally conferred upon the corps, was first used in 1781.

"Java" "Ava" "Maharajpore" "Moodkee" "Ferozeshah" "Aliwal" "Sobraon."

Uniform—*Scarlet*. Facings—*Blue*.

1st Commn. or date of entering service.	Names and Rank.	Army rank	Present appointment in Regiment.	Remarks.
	Commandant.			
5 Aug. 05	Brooke, Capt. A.	5 Aug. 14	1 June 17	Sqdn. Offr., 18th Lcrs.
	Adjutant.			
25 Jan. 08	Seymour, Capt. E. V. F.	1 Sep. 15	9 Sep. 18	Sqn. Offr., 9th Horse.

		Jemadar.	Ressaidar.	Risaldar.	Risaldar-Major.	
	Risaldar-Major.					
	Risaldar.					
8 Aug. 96	Jafar Husain	7 Feb. 09	1 Sep. 15	24 Nov. 15		
	Ressaidar.					
22 Nov. 97	Indar Singh	16 July 14	24 Nov. 15			
	Jemadars.					
1 June 95	Pal Singh	24 Nov. 15				

Governor's Body-Guard, Madras

Madras.

Formed in 1778 as a small body of European troopers. Expanded in 1781 to one European and one Native troop under Lt. P. Sullivan. Reduced to one Native troop, 1784. Reorganised in 1808, and again in 1820 as a body of details borne on the strength of the regular regiments of Madras Cavalry. Re-formed as a separate body, 1897.

"Seetabuldee."

Uniform—*Scarlet*. Facings—*Blue*.

1st Commn. or date of entering service.	Names and Rank.	Army rank.	Present appointment in Regiment.	Remarks.
	Commandant.			
23 July 15	Smith, Lt. D. E. (*Actg. Capt.*) .	23 July 16	14 Feb. 18	I. A. R. O.
	Adjutant.			

		Jemadar.	Ressaidar.	Risaldar.	Risaldar-Major.	
	Risaldar-Major.					
13 Jan. 86	*Malik Sher Bahadur Khan, Sardar Bahadur* (56) . . .	*2 Nov. 94*	*22 Sep. 97*	*1 Apl. 04*	*1 Feb. 05*	*Order of Br. I., 1st Cl. A.-D. C. to Govr., Madras.*
	Risaldar.					
25 Jan. 94	*Abdullah Sharif*	*15 May 16*	. . . .	15 Feb. 18		
	Jemadar.					
6 July 92	Narayan Singh	15 Feb. 18				

Governor's Body-Guard, Bombay.

Ganeshkhind.

Raised at Poona in 1865, from a portion of the Southern Mahratta Horse, retained for the prpuose when that corps was disbanded. Reorganised in 1895.

Uniform—*Scarlet*. Facings—*Blue*.

1st Commn. or date of entering service.	Names and Rank.	Army Rank.	Present appointment in Regiment.	Remarks.
	Commandant.			
25 Jan. 08	Greenway, Capt. C. K.	1 Sept. 15	14 Oct. 16	*Lv., ex I., m.c.*
	Adjutant.			
22 June 12	Savanur, Abdul Majid Khan, *Hony.* Capt., *Nawab* of	1 Jan. 18	24 July 17	(*temp.*)
		Jemadar.	Ressaidar.	Risaldar.
	Risaldar.			
2 July 97	Yasuf Ali Khan	1 Nov. 17	. . . . .	1 July 06
	Ressaidar.			
1 June 97	Albel Singh	. . . . .	1 Nov. 17	

Governor's Body-Guard, Bengal.

Calcutta.

Raised at Calcutta in 1912, by Captain R. B. Worgan, 20th Deccan Horse, from Volunteers from Indian Cavalry regiments.

Uniform—*Scarlet*. Facings—*Blue*.

	Commandant.			
27 Nov. 14	Douetil, Capt. P. V.	30 Aug. 18	25 Feb. 18	I. A. R. O.
	Adjutant.			
	Attached.			
		Jemadar.	Risaldar.	
	Risaldar.			
17 May 93	Malik Mehr Khan	1 Feb. 15	1 Apl. 15	
	Jemadar.			
22 Jan. 99	Samrath Singh	1 Apl. 15	With Remt. Trg. Depôt, Saugor.	

1st Duke of York's Own Lancers (Skinner's Horse).

Raised in 1803, by Capt. James Skinner, from a body of Perron's Horse (in Sindhia's service) who came over to the British after the battle of Delhi. Originally styled 'Captain Skinner's Corps of Irregular Horse,' a designation soon abbreviated into 'Skinner's Horse.' Became the 1st Regt. of Local Horse, 1823; the 1st Bengal irregular Cavalry, 1840; the 1st Regt. of Bengal Cavalry, 1861; the 1st Regt. of Bengal Lancers, 1896; the 1st the Duke of York's Own) Regt. of Bengal Lancers, 1899; the 1st (Duke of York's Own) Bengal Lancers (Skinner's Horse), 1901; present designation, 1903.

"Bhurtpore" "Candahar, 1842" "Afghanistan, 1879 80" "Pekin, 1900"

Composition—4 Squadrons of Hindustani Musalmans, and 1 Addl. Squadron.

Uniform—*Yellow*. Facings—*Black velvet*.

COLONEL-IN-CHIEF—THE KING.

HONORARY COLONEL—Lt.-General *His Highness Sir* MADHO RAO SCINDIA, *Bahadur, Maharaja* of Gwalior, G.C.S.I., G.C.V.O., G.B.E. A.-D.-C . . 1 January 1906.

Commn. or date of entering service.	Names and Rank.	Army rank.	Present appointment in Regiment.	Remarks.
	Commandant.			
30 Aug. 93	Russell, Lt.-Col. F. D., *p.s.c.* [L] .	3 June 16	7 July 16	
	Squadron Commanders. (4)			
10 Oct. 94	Macaulay, Maj. D. I. M. [L] (Q)	10 Oct. 12	7 July 16	2nd-in-Comd.
5 Aug. 96	Cheyne, Maj. H. B.	5 Aug. 14	1 Nov. 12	With 40th Cav.
28 Sep. 95	*Sexton, Maj. E. G.*	*19 Feb. 15*	*1 Apl. 12*	
16 Feb. 98	*Binny, Maj. A. J. M., p.s.c. (temp. Lt. Col.)*	*1 Sep. 15*	*7 July 16*	*A. Q. M. G. Derajat.*
	Squadron Officers. (9)			
27 Jan. 00	Chamberlayne, Maj. A. (*Actg. Lt.-Col.*)	30 Oct. 17	18 Jan. 03	With 2-76th Punj.
18 July 00	*Muirhead, Maj. J. A., p.s.c., (temp. Lt. Col.)* D.S.O.	*1 Sep. 15*	*9 Apl. 02*	*G. S. O., 1st gr.*
8 Jan. 01	*Pott, Maj. E. H.,* D.S.O. . .	*8 Jan. 16*	*24 Nov. 05*	*G. S. O., 2nd gr., Poona Divn.*
22 July 03	*Gillies, Maj. H.*	*12 Dec. 17*	*11 Mar. 06*	*D. A. A. G., 5th Mhow Div.*
22 July 03	Wetherall, Maj. G. A. C. . .	22 July 18	29 Sep. 06	
5 Aug. 05	Beaman, Capt. A. A. H., D.S.O.	5 Aug. 14	1 Apl. 07	*Lv., ex I. m. c.*
24 Jan. 06	*Wilson, Capt. H. M.*	*24 Jan. 15*	*19 Jan. 15*	*G. S. O., 2nd gr., 7th Dn.*
2 Sep. 10	Stevenson, Capt. M. M. . . .	1 Sep. 15	9 Oct. 12	Adjt., 29 Sep. 17.
24 Aug. 12	*Magnay, Capt. A. D.* . . .	*24 Aug. 16*	*4 Nov. 13*	*Bde.-Maj., Secunderabad.*
25 Aug. 17	*Zorawar Singh, Capt.,* M.C. .	*25 Aug. 17*	*25 Aug. 17*	*With I. S. Troops.*
14 Jan. 14	*Fulton, Capt. E. J.*	*14 Jan. 18*	*18 Oct. 14*	
15 Nov. 15	Wilson, Lt. A. D.	15 Nov. 16	19 Nov. 15	Attd. With 32nd Lrs.
15 Nov. 15	Stapylton, Lt. J. M.	15 Nov. 16	19 Nov. 15	Attd. With 23rd Cav.
5 July 17	Forrest, 2nd-Lt. A. E. G. . .	15 July 17	1 Dec. 17	Attd.
31 Jan. 18	Welsford, 2nd-Lt. E. P. . . .	31 Jan. 18	6 Feb. 18	Attd.
5 May 15	Turner, Lt. E. L. (207*a*) . . .	5 May 16	6 Aug. 17	I. A. With 14th Lrs.
	Attached.			
12 July 15	Dean, Lt. A. E. (*Actg. Capt.*) .	12 July 16	31 Mar. 18	New Armies, I. A. (*on probn.*)
21 July 15	Gelston, Lt. J. E. H. . . .	21 July 16	13 Feb. 18	A. S. C. I. A. (*on probn.*)
24 May 16	Mackenzie, Lt. I. D. M.C. . . .	24 May 17	27 Aug. 18	R. Dub. Fus. I. A. (*on probn.*)
27 Apl. 16	Rose, Lt. W. A. (*temp. Capt. 2 Feb. 18.*)	27 Apl. 17	. . . .	I. A. R. O.
29 July 16	Butterfield, Lt. J. (*Actg. Capt.*) .	29 July 17	29 Sep. 17	Essex. Yeo. I. A. R. O.
29 July 16	Meller, Lt. F. R.	29 July 17	2 Oct. 17	Essex. Yeo. I. A. R. O.
11 Oct. 16	Blyde, Lt. W. A., D.C.M. (51) .	11 Oct. 17	5 Aug. 17	Berks Yeo., I. A. (*on probn.*) Qr.-Mr. (*temp.*)
20 Nov. 16	Mckenzie-Muirson, Lt. K. B., M.C.	20 Nov. 17	12 Feb. 18	Aus. M. G. Coy. I. A. (*on probn.*)
27 Mar. 18	Pearson, Lt. C. V. B. . . .	27 Mar. 18	4 July 18	Suff. R., I. A. (*on probn.*)
1 June 17	Gruar, 2nd-Lt. R. F. (207*a*) .	1 June 17	27 Aug. 17	I. A. R. O. (*on probn.*)
25 July 17	Corley, 2nd-Lt. J. W. (207*a*), D.C.M.	25 July 17	7 Oct. 17	I. A. R. O. (*on probn.*)
8 Oct. 17	Kincaid, 2nd-Lt. J.	8 July 17	7 Feb. 18	I. A. R. O. (*on prabn.*)
4 June 18	Van Ingan, 2nd-Lt. C. T. . . .	4 June 18	27 June 18	I. A. R. O.
26 Sep. 18	Finch, 2nd-Lt. C. H.	26 Sep. 17	. . . .	I. A. R. O.
	Honorary Officer.			
10 Feb. 93	Skinner, *Hony. Capt.* S. E. .	12 Oct. 10	10 Feb. 93	Not attached for duty.

1st Duke of York's Own Lancers (Skinner's Horse)—*contd.*

Date of entering service.	Names.	Army rank. Jemadar.	Ressaidar.	Risaldar.	Risaldar-Major.	Remarks.
	Risaldar-Major.					
1 Apl. 86	Muhammad Akram Khan, *Bahadur* (56)	1 Nov. 96	16 Oct. 02	16 Sep. 08	16 Aug. 11.	*Order of Br. I., 2nd Cl.*
	Risaldars.					
1 Nov. 91	Ghulam Kadir Khan (56, 66*a*)	29 Aug. 01	10 Nov. 06	5 Nov. 12		
8 Nov. 89	Ghulam Muhammad Khan (56)	16 July 08	8 Jan. 09	1 Dec. 14		
16 Jan. 87	*Fais Muhammad Khan*, I.O.M.	*1 Sep. 02*	*16 Jan. 09*	*8 Dec. 14*		*A.-D.-C. to Govr. of* [*Bengal (temp.)*.
1 Jan. 94	Muhammad Ali Khan (I)	1 Mar. 06	4 Nov. 14	9 Feb. 16		
14 Jan. 89	Shaikh Abdul Majid (56)	3 Dec. 04	5 Nov. 12	1 Oct. 16		
16 Jan. 98	*Muhammad Ali Khan (II)* (45*a*, 56, 66*a*)	*1 Nov. 06*	*1 Dec. 14*	*6 Aug. 16*		*With 34th Horse.*
1 Aug. 94	Zabte Khan (56, 66*a*)	2 Jan. 11	11 Apl. 16	21 Nov. 17		With 19th Lrs.
	Muhammad Illias Khan	. . . . .	. . . . .	. . . . .		Re-employed 1 Nov. 16.
	Ressaidars.					
1 May 92	Faujdar Khan (66*a*)	10 Nov. 06	8 Dec. 14			
1 May 92	*Nathe Khan, Bahadur* (45*a*)	*1 Oct. 08*	*5 July 15*			*Order of Br. I., 2nd Cl.*
1 May 98	Muhammad Shafi Khan (66*a*)	16 Jan. 09	22 Nov. 15			
1 Oct. 01	Muhammad Akbar Khan (56)	16 June 14	1 Oct. 16			
13 Apl. 89	*Habibur Rahman Khan*, Khan Sahib.	*5 Nov 12*	*4 Apl. 17*			*Asst. Translation Officer., A. Hd.-qrs.*
1 Feb. 96	Barkat Khan (56, 66*a*)	16 Aug. 11	1 Jan. 18			
1 Jan. 98	Muhammad Ali Khan (III), (45*a*, 56, 66*a*)	29 Oct. 14	8 Jan. 18			
1 Jan. 96	Ruknuddin, I.D.S.M.	1 Dec. 14	8 Jan. 18			
1 July 03	Muhammad Umar Faruk Khan, I.O.M. (52, 56, 66*a*).	5 Jan. 15	1 Aug. 18			With 32nd Lrs.
	Jemadars.					
1 June 09	Muhammad Amir Khan	8 Dec. 14				With 21st Cav.
1 Oct. 05	Muhammad Taqi Khan	18 Feb. 15				With 32nd Lrs.
1 Jan. 97	Murtaza Khan (56)	5 Apl. 15				
1 Nov. 94	Sher Muhammad Khan (56)(I)	5 July 15				With 59th Camel Corps.
9 June 00	Muhammad Nawaz Khan	11 Apl. 16				
7 July 95	Abdul Hamid Khan (56)	13 May 16				*Remt. Depôt, Sehore.*
16 Jan. 04	Muhammad Sana Khan (56)	28 July 16				Woordie-Maj., 8 Jan. 18.
18 Mar. 01	Sher Muhammad Khan (II)	19 Sep. 16				
1 July 08	Abdul Razaq Khan	1 Oct. 16				With 59th Camel Corps.
4 July 04	Sayed Ahmad (52)	1 Oct. 16				
17 May 02	Ibrar Hasan Khan (56)	23 Mar. 17				
1 Jan. 97	Wali Muhammad Khan	1 Nov. 17				
1 July 92	Bahawal Khan	1 Jan. 18				
16 Jan. 06	Farzand Ali Khan (56)	8 Jan. 18				
1 Dec. 04	Abdul Ghani Khan	8 Jan. 18				With 59th Camel Corps.
1 Feb. 02	Abdul Rauf Khan	8 Jan. 18				
1 July 99	Abdullah	8 Jan. 18				
1 May 02	Wasil Muhd. Khan	9 May 18				
16 Sep. 98	Abdul Rahman	3 Aug. 18				
16 May 13	Ghulam Rasul Khan	10 May 18				With 19th Lrs.

2nd Lancers (Gardner's Horse).

Raised at Furruckabad and Mainpuri in 1809, by Lt.-Col. W. L. Gardner, and was originally styled 'Lieut.-Col. Gardner's Corps of Irregular Horse,' a term soon abbreviated into 'Gardner's Horse.' Became the 2nd Regt. of Local Horse, 1823; the 2nd Bengal Irregular Cavalry, 1840; the 2nd Regiment of Bengal Cavalry, 1861; the 2nd Regt. of Bengal Lancers, 1890; the 2nd Bengal Lancers, 1901; present designation 1903.
"Arracan" "Sobraon" "Punjaub" "Tel-el-Kebir" "Egypt, 1882."

Composition—1 Squadron of Sikhs, 1 of Rajputs, 1 of Jats, 1 of Hindustani Musalmans.

Uniform—*Blue.* Facings—*Light Blue.*

Date of entering service.	Names.	Army rank.	Present appointment in Regiment.	Remarks.
	Commandant.			
4 Mar. 91	*Gwyn-Thomas, Lt.-Col. G.*, C.M.G., D.S.O., *p.s.c.* (*temp. Br.-Genl.*)	*4 Mar. 17*	*2 Dec. 17*	
	Squadron Commanders. (4)			
21 Dec. 89	*Pritchard, Bt.-Col. A. G.*, C.M.G. (*Q*) (*temp. Br.-Genl.*)	*3 June 18*	*1 May 13*	
13 Aug. 92	Bennett, Maj. J F (Q)	13 Aug. 10	16 June 13	
19 July 93	Knowles, Maj. G., D.S.O.	19 July 11	2 Dec. 17	
	Squadron Officers. (9)			
27 July 98	*Ward, Bt.-Lt.-Col. H. C. S., p.s.c.* (*l*)	3 June 17	*13 Mar. 02*	*G. S. O., 1st. Gr., R. Pindi.*
17 Jan. 00	Scott, Maj. A. N. De V.	1 Sep. 15	1 Nov. 02	
1 Sep. 99	Salkeld, Maj. H. V., D.S.O.	1 Sep. 15	1 Sep. 06	2nd-in-Comd. (*temp.*).
8 May 01	Gould, Maj. G., D.S.O.	8 May 16	1 Nov. 06	
8 Jan. 01	Robertson, Maj. K.	4 Oct. 16	10 Feb. 06	
17 Aug. 07	Davison, Capt. D. S.	1 Sep. 15	20 Nov. 08	
29 Jan. 10	Whitworth, Capt. D. E., M.C.	1 Sep. 15	8 Mar. 11	Adjt., 19 Mar. 17.
22 Jan. 13	Vaughan, Capt. E. W. D.	22 Jan. 17	28 Mar. 14	
25 Aug. 17	Kanwar Amar Singh, Capt.	25 Aug. 17	25 Aug. 17	With 16th Cav.
14 Nov. 16	Harris, Lt. C. E. L.	14 Nov. 17	....	
23 July 15	Commeline, Lt. F. T.	23 Apl. 17	....	Attd.
14 Nov. 16	Mallam, Lt. G. L.	13 Sep. 16	....	Attd. With 44th Cav.
4 Oct. 17	Barnes, Lt. J. F. F.	4 Oct. 18	1 Jan. 18	Attd.
10 Apl. 16	Ambler, Lt. R.	10 Apl. 17	12 Sep. 17	Attd.
1 Sep. 18	Johnson, 2nd-Lt. S. H.	1 Sep. 18	4 Sep. 18	Attd.
	Attached.			
13 Aug. 04	Duke, Capt. O. A.	13 Aug. 13	23 July 15	22nd Cav.
24 Jan. 06	Smith, Capt. V. H. S.	24 Jan. 15	....	32nd Lrs.
23 Jan. 14	Banister, Capt. H.	25 Sep. 17	....	I. A. R. O. Qr.-Mr., 19 Nov. 17.
31 Dec. 14	Wilson, Capt. N. B.	30 Aug. 18	....	I. A. R. O.
2 Dec. 15	Hearsey, 2nd-Lt. G. C. M.	2 Dec. 15	2 Dec. 15	I. A. R. O.
16 June 16	Farrell, Lt. J. K. O'M.	16 June 16	23 Nov. 17	A. S. C., I. A. (*on probn.*)
13 July 15	Peck, Lt. L. J.	13 July 16	....	I. A. R. O.
9 Aug. 15	Bremner, Lt. S.	9 Aug. 16	....	L'pool R., I. A. (*on probn.*)
26 Aug. 16	Raymond, Lt. W. D.	26 Aug. 16	5 May 17	I. A. (*on probn.*)
15 Sep. 15	Turner, Lt. R. H. F.	15 Sep. 16		I. A. (*on probn.*)
10 Mar. 15	Duchesne, Lt. R. L.	10 Dec. 16	1 Mar. 18	Yorks. L. I., I. A. (*on probn.*)
12 Feb. 16	Macmillan, Lt. E. J. A. (*Actg. Capt.*)	12 Feb. 17	18 Oct. 16	I. A. R. O.
10 Mar. 16	Hearsey, Lt. L. D. W.	10 Mar. 17	4 May 16	I. A. R. O.
18 May 17	Bullock-Webster, Lt. A. P.	18 May 17	23 Nov. 17	R. F. A., I. A. (*on probn.*)
14 Nov. 16	Bloomfield, Lt. F. J. C.	14 Nov. 17	1 Oct. 17	
30 Jan. 17	Edgcome, Lt. G. E. R.	30 Jan. 18	17 July 18	
13 July 17	Lipscomb, Lt. L.	16 July 18	1 Oct. 17	I. A. R. O.
28 Oct. 17	Clapp, Lt. J. H.	28 Oct. 18	9 Dec. 17	I. A. (*on probn.*) Bedf. Yeo.
....	Sutherland-Hawes, 2nd-Lt. J. V. (207*a*)	....	25 Jan. 18	I. A. R. O. (*on probn.*)

2nd Lancers (Gardner's Horse)—*contd.*

Date of entering service.	Names.	Army rank. Jemadar.	Ressaidar.	Risaldar.	Risaldar-Major.	Remarks.
	Risaldar-Major.					
25 Oct. 85	Ganga Dat, *Bahadur* (56) . .	16 May 93	17 July 99	1 June 10	6 Feb. 12	*Order of Br. Ind., 2nd Cl.*
	Risaldars.					
4 Jan. 93	Mukand Singh, *Bahadur* (56, 58), M. C.	1 Nov. 95	1 Dec. 98	1 Apl. 07		*Order of Br. Ind., 2nd Cl.*
30 Mar. 90	Muhammad Raza Khan (56) .	11 Feb. 98	1 Apl. 10	1 Mar. 11		
16 Nov. 85	Bhaju Singh (56) . . .	3 Aug. 04	1 Oct. 07	6 Feb. 12		
7 Mar. 93	Suraj Singh, *Bahadur* I.O.M. (56, 66*a*).	1 Jan. 07	1 Apl. 10	26 Oct. 15		*Order of Br. Ind., 2nd Cl.*
16 July 88	Mukh Ram (56)	1 Apl. 07	1 Mar. 11	26 Oct. 15		
22 July 93	Diwan Chand (50, 56)	1 Nov. 12	10 Nov. 14	17 Dec. 17		
1 Nov. 00	Abdul Latif Khan, I.D.S.M. (45*a*, 66*a*)	1 Mar. 11	26 Oct. 15	17 Dec. 17		
1 Nov. 10	Krishna Chandra Singh, I.D.S.M. (45*a*).	1 Nov. 10	9 Feb. 17	9 Apl. 18		
1 May 00	Imdad Khan	16 Aug. 13	25 Aug. 17	3 June 18		
	Ressaidars.					
...	Sohan Singh (*Retd.*) *Bahadur*	...	1 June 10			Re-employed 16 Feb. 15, Depôt. *Order of Br. Ind., 2nd Cl.*
5 Oct. 92	Tek Singh (56)	1 Apl. 10	1 Aug. 13			
6 Mar. 02	Het Ram, I.D.S.M. (45*a*, 56). .	16 Oct. 09	19 May 16			
1 Feb. 89	Molar	16 Nov. 12	15 July 16			
1 Oct. 93	Kehri Singh (66*a*)	1 Apl. 10	4 Mar. 17			
7 Mar. 99	Jiwan Singh (56)	1 Aug. 10	25 Aug. 17			Woordie Major, 19 May 16.
24 Dec. 98	Bharpur Singh	1 Aug. 13	2 Dec. 17			
1 Feb. 01	Devi Chand	19 May 16	17 Dec. 17			
26 Nov. 09	Jang Bahadur Singh, I. D. S. M. .	27 June 16	29 Jan. 18			
9 Jan. 99	Lehri Singh	15 July 16	2 Feb. 18			
1 Oct. 97	Abdul Wahid Khan	8 Aug. 16	3 June 18			
	Jemadars.					
21 Dec. 92	Anokh Singh (60*a*)	26 Oct. 14				
16 Nov. 04	Muhammad Yusuf Khan . . .	19 May 16				
16 Oct. 04	Prag Singh	25 Aug. 17				
23 Dec. 03	Afzal Husain	25 Aug. 17				
19 Sep. 93	Ude Singh	15 Oct. 17				
23 Nov. 03	Brij Lall	2 Dec. 17				
16 Apl. 09	Teja Singh	3 Dec. 17				
1 Jan. 07	Rachha Ram	17 Dec. 17				
16 Aug. 09	Ata Muhammad Khan. . . .	30 Jan. 18				
16 July 02	Sarjit Singh	2 Feb. 18				
23 Dec. 03	Kanhiza	2 Feb. 18				
6 Dec. 02	Basti Singh	29 Mar. 18				
16 Oct. 12	Sanjam Singh	21 Apl. 18				

3rd Skinner's Horse.

Raised at Hansi in 1814, by Lt.-Col. James Skinner, and was originally the 'Second Corps of Lt.-Col. Skinner's Irregular Horse.' Became 'Baddeley's Frontier Horse,' 1821; the 4th Regt. of Local Horse, 1823; the 4th Bengal Irregular Cavalry, 1840; the 3rd Regt. of Bengal Cavalry, 1861; the 3rd Bengal Cavalry. **(Skinner's Horse), 1901; present designation, 1903.**

"Ghuznee, 1839" "Affghanistan, 1839" "Khelat" "Maharajpore."
"Moodkee" "Ferozeshah" "Aliwal" "Kandahar, 1880"
"Afghanistan, 1879-80" "Punjab Frontier."

Composition—1 Squadron of Sikhs, 1 of Jats, 1 of Rajputs, 1 of Musalman Rajputs.
Uniform—Blue. *Facings*—Yellow.

1st Commn. or date of entering service.	Names and Rank.	Army rank.	Present appointment in Regiment.	Remarks.
	Commandant.			
20 Feb. 92	Gaussen, Lt.-Col. J. R., C.M.G., D.S.O.	20 Feb. 18	1 Oct. 18	With 40th Cav.
	Squadron Commanders. (4)			
16 Jan. 95	*Conway-Gordon, Maj. E. C. W., p.s.c. (temp. Lt.-Col.).*	*16 Jan. 13*	*1 Oct. 18*	*2nd-in-Comd.—G. S. O., 1st gr. Poona Divn.*
28 Sep. 95	*Stokes, Bt.-Lt.-Col. C. B.,* C.I.E., *p.s.c.* [L]	*3 June 18*	*2 July 12*	*G. S. O., 1st gr.*
4 Aug. 97	*Webber, Maj. M. R. H.*	*4 Aug. 15*	*15 Dec. 12*	
22 Jan. 98	Coaker, Maj. V. A.	1 Sep. 15	1 Oct. 18	With Remt. Dept.
	Squadron Officers. (9)			
20 July 98	*Vander Gucht, Maj. G. T.*	*1 Sep. 15*	*15 Sep. 01*	*With I. S. Troops.*
24 Jan. 00	Bennett, Maj. J. H. [L] (*temp. Lt.-Col.*)	1 Sep. 15	2 Aug. 02	Comdt. (*temp.*)
17 Feb. 00	*Hulbert, Maj. T. E.*	*1 Sep. 15*	*26 Jan. 03*	*Comdt., Musky. Sch., Satara.*
2 May 00	*Medlicott, Maj. H. E. (temp. Lt.-Col.)* D.S.O.	*1 Sep. 15*	*1 July 07*	*A. A. & Q. M. G.*
22 Oct. 02	Polley, Maj. S. T.	22 Oct. 17	1 Nov. 06	With Remt. Dept.
9 Jan. 04	*Manderson, Capt. R. W. (temp. Maj.)*	*9 Jan. 13*	*1 Sep. 06*	With Remt. Dept.
18 Jan. 05	*Daniels, Bt. Maj. A. M.*	*10 Mar. 17*	*8 Jan. 08*	*Bde. Maj.*
24 Jan. 06	*Lorimer, Capt. J. M. (temp. Maj.)*	*24 Jan. 15*	*17 Sep. 09*	With Remt. Depot.
17 Aug. 07	*Metcalfe, Capt. E. D.,* M.C.	*1 Sep. 15*	*14 Nov. 09*	*With Sig. Serv.*
9 Sep. 08	*Gray, Capt. G. C. G. (temp. Maj.)*	*1 Sep. 15*	*26 Nov. 09*	*G. S. O., 2nd gr.*
25 Sep. 11	Randall, Capt. B. Fitz H.	5 Sep. 15	3 Mar. 14	With 40th Cav.
25 Aug. 17	Aga Cassim Shah, Capt.	25 Aug. 17	25 Aug. 17	With 40th Cav.
14 Jan. 14	Broadfoot, Capt. W. A. (*actg. Maj.*)	14 Jan. 18	10 Oct. 14	2nd-in-Comd. (*temp.*)
8 Aug. 14	Prinsep, Capt. D. W. M.	8 Aug. 18	24 Oct. 14	With Camel Corps.
	Attached.			
3 Sep. 15	Russell, Lt. A. M. H. (*actg. Capt.*)	3 Sep. 16	13 Apl. 18	I. A. R. O.
29 Aug. 14	Watson, Lt. H. B. (*tem. Capt.*)	29 Aug. 15	4 Aug. 17	I. A. R. O.
4 May 15	Jefferis, Lt. A. N. (*actg. Capt.*)	4 May 16	31 Dec. 17	I. A. (*on probn.*)
1 Sep. 14	Lyons, Lt. H. H., M.C. (*actg. Capt.*)	1 Sep. 15	27 Jan. 18	Conn. Rang. I. A. (*on probn.*)
31 Jan. 18	Francis, 2nd-Lt. W.	31 Jan. 18	1 Mar. 18	
31 Jan. 18	Ryper, 2nd-Lt. G H.. (207a)	31 Jan. 18	2 Mar. 18	
27 Apl. 17	Munckton, Lt. V. C. A. (66, 207a)	27 Apl. 18	9 May 17	I. A. R. O.
30 Jan. 17	Freeman-Thomas, Lt., *Hon.*, I. B.	24 June 17	4 Feb. 17	Attd. *Lv., ex Ind., m.c.*
17 Dec. 15	Hardie, Lt. J. D. (*actg. Capt.*)	1 July 17	1 Apl. 18	Sco. Horse, I. A. (*on probn.*) Adj. (*temp.*)
6 Dec. 16	Anderson, Lt. G.	6 Dec. 16	26 May 18	Life Gds., I. A. (*on probn.*)
21 Dec. 17	Fraser, 2nd-Lt. T. F. C.	. . . .	27 Mar. 18	I. A.
11 Sep. 17	Broom, Lt. J. T.	11 Sep. 18	27 Sep. 17	I. A. R. O.
28 Dec. 17	Brown, 2nd-Lt. S. B.	28 Dec. 17		S. Afr. Forces I. A. R. O. (*on probn.*)
. . . .	Chapman 2nd-Lt. A. R.	17 Nov. 17	. . . .	I. A. (*on probn.*)
. . . .	Turnbull, 2nd-Lt. W. R.	. . . .	. . . .	I. A. R. O. (*on probn.*)
. . . .	Macqueen, 2nd-Lt. R. B.	24 Apl. 18	. . . .	I. A. (U. L.)
31 Oct. 17	Bromham 2nd-Lt. A. W.	31 Oct. 17	. . . .	Attd.

3rd Skinner's Horse—*contd.*

Date of entering service.	Names.	Army rank. Jemadar.	Ressaidar.	Risaldar.	Risaldar-Major.	Present appointment in Regiment. / Remarks.
	Honorary Officer.					
	Risaldar-Major.					
4 Sep. 84	Balwant Singh, *Bahadur*	21 Dec. 00	16 Oct. 05	31 Aug. 12	22 Sep. 15	*Order of Br. I., 2nd Cl.*
	Risaldars.					
1 Oct. 94	Mir Shamshad Ali (56)	7 May 99	1 Feb. 01	16 Oct. 10		Remt. Depôt, Sehore.
4 Apl. 89	Sobha Singh (56, 62)	21 May 02	1 Jan. 06	24 Oct. 14		
16 Jan. 94	Nur Muhammad Khan (56, 66*a*, 71)	1 Jan. 05	16 Oct. 10	22 Sep. 15		
1 Jan. 06	Mukand Singh (56, 71)	1 Jan. 06	31 Aug. 12	3 Jan. 18		
	Ressaidars.					
3 Jan. 01	Raj Singh (45*a*, 56)	15 Feb. 11	24 Oct. 14			
1 Dec. 01	Amar Singh	31 Aug. 12	21 Jan. 15			
1 Feb. 79	Harnarayan Singh (Retd.)	. . . .	. . . .	. . . .		Re-employed, 5 Feb. 15.
	Jemadars.					
1 Aug. 99	Harbhaj Singh (56)	1 Feb. 13				
8 Aug. 00	Bhagwan Singh (56)	1 May 13				Woordie-Major, 6 Sep. 15.
1 May 95	Imam Ali Khan (56, 66*a*, 71)	1 May 14				
1 Nov. 93	Mahammad Shaffi Khan (71)	26 Aug. 14				
16 Nov. 00	Mangal Singh (56, 71)	24 Oct. 14				
1 June 98	Ashraf Khan (56)	22 Sep. 15				
3 Dec. 99	*Dalel Singh*	*3 Dec. 15*				*With Remt. Depôt.*
3 Dec. 99	Mamraj (58)	11 Apl. 16				
16 Nov. 03	Balwant Singh (51)	11 Apl. 16				
16 Apl. 01	Ratan Singh	. . . .				With 10th Camel Corps.
18 Mar. 07	Mukand Singh (*Ag. Risdr.*)	10 Nov. 16				With 3rd Camel Corps.
1 Feb. 09	*Ami Lal*	*28 Dec. 16*				*Supy.*
18 Dec. 03	Badshah Husain (*Ag. Ressdr.*)	1 Apl. 17				With Camel Corps.
19 Apl. 00	Wali Muhammad Khan	22 May 17				With Syce Corps.
1 Dec. 03	Bhagwan Singh (56)	3 Jan. 18				
3 July 99	Asghar Ali Khan	3 Jan. 18				
22 May 01	Waryam Singh	20 Feb. 18				With Syce Corps.
11 Oct. 94	Nur Mohomed	20 Feb. 18				With Syce Corps.
1 Apl. 02	Mohan Singh	20 Feb. 18				With Syce Corps.

4th Cavalry.

Raised at Sultanpore, Oudh, in 1838, by Captain C. Newbery, as the Cavalry Regt. of the Oudh Auxiliary Force. Became 6th Regiment of Bengal Irregular Cavalry, 1840; the 4th Regt. of Bengal Cavalry, 1861; th th Regt. of Bengal Lancers, 1900; the 4th Bengal Lancers, 1901; the 4th Lancers, 1903; presen esignation, 1904.

Granted an Honorary Standard for service in Sind in 1844, bearing the device of a lion passant regardant.—"Afghanistan, 1879-80."

Composition—1 Squadron of Rajput Musalmans, 1 of Hindustani Musalmans, 1 of Sikhs, 1 of Jats—1 Addl. Squadron.

Uniform—Scarlet. *Facings*—Blue.

COLONEL.—Major-General James Gibbon Turner, C.B. 2 Nov. 17.

1st Commn. or date of entering service.	Names and Rank.	Army rank.	Present appointment in Regiment.	Remarks.
	Commandant.			
01 Feb. 88	Macquoid, Lt.-Col. C. E. E. F. K., D.S.O.	11 Feb. 14	5 June 16	A. A. & Q. M.G., Meerut.
	Squadron Commanders. (4)			
28 June 93	Sproule, Maj. H. H. (Q) (*Actg. Lt.-Col.*)	28 June 11	5 June 16	2nd-in-Comd. *Offg.* Comdt.
27 July 98	*Johnston, Maj. R.*, D.S.O.	*1 Sep. 15*	*11 Aug. 11*	*D. A. A. G., A. Hdqrs.*
4 Feb. 00	*Farran, Maj. G. L.*, D.S.O., M.C. (*temp. Lt.-Col.*)	*1 Sep. 15*	*5 June 16*	*With South Persian Mtd. Rifles.*
4 Apl. 00	Pearse, Maj. J. C.	1 Sep. 15	21 July 18	2nd-in-Comd. (*Offg.*)
	Squadron Officers. (9)			
8 Jan. 01	*Williams, Maj. V. P. B. (temp. Lt.-Col.)*	*8 Jan. 16*	*21 Apl. 02*	*Burma Mily. Pol. Mounted Infy. Adj. & Q. M. G.*
8 May 01	*Wilson, Maj. T. A. A.*	*8 May 16*	*13 Nov. 02*	*D. A. A. G., Army Hdqrs.*
23 Apl. 02	*Bradley, Maj. C. R. S.*	*23 Apl. 17*	*21 Oct. 06*	*R. A. F.*
21 Jan. 03	*Howson, Maj. G.*, M.C. (*temp. Lt.-Col.*)	*21 Jan. 17*	*8 Oct. 05*	*Comdt., M. G. Trg. Centre.*
21 Jan. 03	Harenc, Maj. R. E.	21 Jan. 17	12 Nov. 05	Staff School, Saugor.
29 July 03	*Hotham, Capt. J. C.*	*29 July 12*	*5 Oct. 06*	*Army Remt. Dept.*
13 Aug. 04	Benton, Capt. H. G.	13 Aug. 13	31 July 06	
1 Jan. 07	Malik Mumtaz Muhammad Khan, Capt.	25 Aug. 17	25 Aug. 17	A. R. O., Campbellpore.
17 Mar. 15	Knight, Lt. A. B.	17 Mar. 16	....	With 9th Horse.
29 June 16	deSalis, Lt. R. A.	29 June 17	29 June 16	
22 June 15	Jennison, Lt. S. A. (*Actg. Capt.*)	22 June 16	16 May 18	Attd. Adjt. (*temp.*)
29 Apl. 17	Harland, Lt. F. W. (*Actg. Capt.*)	29 Apl. 18	8 Apl. 17	Attd.
31 Aug. 18	Blackwell, 2nd.-Lt. J. N.	31 Aug. 18	4 Sep. 18	Attd.
31 Aug. 18	Hotham, 2nd-Lt. G. I. M.	31 Aug. 18	7 Sep. 18	Attd.
26 June 17	Barlow, Lt. J. H.	26 June 18	2 Aug. 17	Attd.
31 Jan. 18	Phipps, 2nd-Lt. J. P.	....	6 Feb. 18	Attd.
21 Dec. 17	Moodie, 2nd-Lt. E. H.	21 Dec. 17	22 Mar. 18	Attd.
	Attached.			
18 Feb. 17	Hartmann, Lt. W. H.	18 Feb. 17	22 Jan. 18	Som. L. I. I. A. (*on probn.*)
6 Feb. 17	Innes, Lt. E. A. R.	6 Feb. 18	1 Mar. 17	I. A. R. O.
24 Mar. 17	Vanrenen, Lt. L. F. E. (*Actg. Capt.*)	24 Mar. 18	15 Apl. 17	I. A. R. O.
12 Sep. 17	Messum, 2nd-Lt. H. R.	12 Sep. 17	20 Sep. 17	I. A. R. O. Qr.-Mr. (*temp.*)
12 Aug. 16	Marshall, Lt. R. H.	12 Aug. 17	14 Feb. 18	High. Cyclist Bn., I. A. (*on probn.*)
28 Dec. 17	Brown, 2nd-Lt. J. B.	....	18 Jan. 18	I. A. R. O.
10 Nov. 15	Hussey, Lt. G. O. G.	10 Nov. 16	17 Mar. 18	I. A. (*on probn.*)
11 May 17	Herbison, 2nd-Lt. L. J.	....	1 Apl. 18	I. A. R. O. (*on probn.*)
19 Sep. 18	Scott, 2nd-Lt. W. S.	12 Sep. 16	14 Oct. 18	I. A. R. O.
21 Dec. 16	Owen, Lt. D. P.	21 Dec. 17	....	
12 Sep. 15	Sharpe, Lt. A. G.	19 Sep. 18	....	I. A. (*on probn.*)
...	Farrar, Lt. C. B. (*Actg. Capt.*)			

4th Cavalry—*contd.*

Date of entering service.	Names.	Army Rank. Jemadar.	Ressaidar.	Risaldar.	Risaldar-Major.	Remarks.
	Risaldar-Majors.					
18 May 89	Saddha Singh, *Bahadur* (56, 66*a*)	14 July 05	1 July 10	14 Nov. 14	*9 Feb. 16*	*Order of Br. I., 2nd Cl.*
	Risaldars.					
1 Dec. 88	Khem Singh (56)	1 June 99	1 Aug. 01	1 Sep. 09		
8 Feb. 08	Ram Singh (56)	16 Dec. 01	18 Nov. 09	1 July 10		
1 July 94	Udmi (56, 71) I.D.S.M.	1 Nov. 09	10 Feb. 13	15 Nov. 14		
6 May 96	Muhammad Yusuf Khan (45*a*, 56, 66*a*, 71)	20 Mar. 08	14 Nov. 14	9 Jan. 16		
23 May 06	Abdul Rauf Khan (56, 66*a*, 71) .	1 Jan. 12	8 Mar. 16	23 Sep. 18		
	Ressaidars.					
5 Feb. 86	*Muhammad Ali Khan (56, 66, 71)*	*1 Jan. 06*	*1 Jan. 12*			
20 Mar. 10	Surat Singh (60*a*, 66*a*) . . .	10 Feb. 13	15 Nov. 14			
1 Mar. 09	Musharraf Khan (56, 71) . . .	1 May 10	3 Oct. 15			
22 July 98	Daud Ali Khan (56)	27 Mar. 11	9 Jan. 16			
13 Sep. 00	Suleman Khan (56)	20 Aug. 14	27 June 16			
1 May 02	Abdul Shakur Beg	14 Jan. 16	23 Sep. 18			
	Jemadars.					
1 Jan. 03	*Harnam Singh* (45*a*, 56, 66*a*) .	*16 May 09*				(*Embarkation Duty, Bombay.*)
17 June 96	Moti Ram (56)	18 Nov. 96				Syce Jail Corps, Sehore.
1 Mar. 05	Kartar Singh	8 Nov. 14				
18 Jan. 06	Puran Singh	11 Dec. 14				
1 Nov. 11	Muhammad Usman Khan . .	9 Jan. 16				
16 Oct. 01	Amar Singh	14 Aug. 16				Remt. Training Depôt, Sehore.
1 Oct. 07	Indraj Singh	17 Aug. 16				
1 Apl. 01	Abdul Ghafar Khan	1 Feb. 17				
1 Nov. 99	Karam Singh, I.D.S.M.	22 Mar. 17				With 2nd Lrs.
14 Dec. 03	Ahmad Ali Khan	1 Oct. 17				
13 Nov. 11	Hira Singh	1 Oct. 17				
18 Aug. 98	Wajid Ali Khan	1 Oct. 17				
1 Jan. 88	Gafur Khan	3 Dec. 17				Qr.-Mr. Jemdr.
22 Oct. 02	Azam Beg	15 Feb. 18				Woordie-Major, 23 Sep. 18.
5 Nov. 90	Jiva Siingh	25 May 17				*Syce Jail Corps, Sitapur.*
1 Jan. 06	Allah Beli Khan	23 Sep. 18				
1 Jan. 05	Abdul Shakur Khan	23 Sep. 18				
27 Nov. 10	Dalip Singh	23 Sep. 18				

5th Cavalry.

Raised at Bareilly in 1841, by Captain F. Wheler, as the 7th Regt. of Bengal Irregular Cavalry. Became the 5th Regt. of Bengal Cavalry, 1861; the 5th Bengal Cavalry, 1901; present designation, 1903.
"Mooltan" "Punjaub" "Afghanistan, 1879-80."

Composition—2 Squadrons of Hindustani Musalmans, 1 of Rajputs, 1 of Jats.
Uniform—Scarlet. *Facings*—Blue.

1st Commn. or date of entering service.	Names and Rank.	Army rank.	Present appointment in Regiment.	Remarks.
	Commandant.			
9 Nov. 89	Hay, Lt.-Col. L. T.	9 Nov. 15	17 Aug. 16	
	Squadron Commanders. (4)			
18 June 92	*Hodgkinson, Lt-Col. J. (Q)*	*18 June 18*	*16 Mar. 12*	
25 Jan. 99	*Skinner, Maj. A. B.,* D.S.O. *(temp. Lt.-Col.)*	*1 Sep. 15*	*6 Feb. 13*	*Comdt., Offrs.' Sch. of Instn., Bangalore.*
17 Feb. 00	Norman, Maj. A. C. (*actg. Lt.-Col.*)	1 Sep. 15	6 Feb. 13	2nd-in-Comd. (*temp.*)
	Squadron Officers. (9)			
17 Oct. 00	*Williams, Bt. Lt.-Col. L. G.,* D.S.O.	*3 June 17*	*13 May 02*	
21 Dec. 00	*Mackintosh, Maj. G. H. T.*	*21 Dec. 15*	*21 Apl. 05*	
28 Jan. 03	de Gale, Maj. W. A. S.	17 Oct. 17	7 Aug. 06	
5 Aug. 05	*Clark, Capt. W. H.*	*5 Aug. 14*	*3 Nov. 06*	
16 Aug. 05	*Smythe, Capt. A. C. G.*	*16 Aug. 14*	*21 Nov. 08*	*With 2-70th Rifles.*
11 Dec. 07	*Hornsby, Capt. H. F. P.*	*1 Sep. 15*	*16 Sep. 10*	*With 1. S Troops.*
25 Jan. 08	*Copeland, Capt. A. G. J. (actg. Maj.)*	*1 Sep. 15*	*25 Mar. 09*	*With Sig. Serv.*
9 Sep. 08	*Walker, Capt. J. C.*	*1 Sep. 15*	*13 Nov. 0)*	*G. S. O., 3rd Gr. (temp.)*
22 Jan. 13	Harvey, Capt. G. B.	22 Jan. 17	16 Mar. 14	Adjt., 5 Jan. 17.
....	Kanwar Pirthi Singh, Capt.	25 Aug. 17	25 Aug. 17	
15 Nov. 15	Ranking, Lt. R. P. L.	15 Nov. 16	18 Nov. 15	Attd.
15 Nov. 15	Pratt, Lt. J.	15 Nov. 16	18 Nov. 15	Attd.
18 July 17	Crichton, Lt. T. C.	18 June 18	25 July 17	Attd.
31 Aug. 18	Day, 2nd-Lt H. M.	31 Aug. 18	5 Sep. 18	Attd.
27 Oct. 17	Whittaker, 2nd-Lt. E. H. (47*a*) (48*a*)	27 Oct. 17	21 Nov. 17	Attd.
	Attached.			
21 Dec. 00	*Weir, Maj. J. L. R.*	*21 Dec. 15*	*1 Nov. 14*	*Poll. Dept.*
5 May 16	Portal, Lt. G. E.	5 May 16	3 Aug. 17	I. A. (*on probn.*)
30 Mar. 16	Gove, Lt. R. V.	30 Mar. 16	8 Apl. 16	I. A. R. O.
9 Sep. 16	Tayler, Lt. E. R. N. a'W	9 Sep. 17	26 Sep. 16	I. A. R. O.
1 July 14	White, Lt. G. W. W. (207*a*) (*actg. Capt.*).	5 May 16	3 Aug. 17	I. A.
15 Sep. 16	Child, Lt. T. H.	15 Sep. 17	24 Sep. 16	I. A. R. O.
18 Apl. 17	Woolley, Lt. J.	18 Apl. 18	6 May 17	I. A. R. O.
15 Nov. 15	Clifford, Lt. H. H. (46,48*a* 51, 207*a*)	18 Nov. 16	3 Nov. 17	I. A. R. O. (*on probn.*)
5 May 17	McKerrow, Lt. J. A.	5 May 18	18 May 17	I. A. R. O.
24 Apl. 17	Bishop, Lt. E. C. G.	24 Apl. 18	3 July 18	I. A. (*on probn.*)
3 Jan. 15	Robinson, Lt. C. W.	3 Oct. 16	29 June 18	I. A. (*on probn.*)
24 July 17	Dove, 2nd-Lt. E. P. (47*a*, 57)	24 July 17	13 Aug. 17	I. A. R. O.
1 Sep. 17	Husband, Lt. W. F. (47*a*, 51, 207*a*)	1 Sep. 18	17 Sep. 17	I. A. (*T. C.*)
13 Sep. 17	Pereira, 2nd-Lt. J. E.	13 Sep. 17	24 Sep. 17	I. A. R. O.
10 July 15	Common, Lt. F. B. (*Actg. Capt.*). (207*a*)	10 July 16	29 Sep. 17	I. A. R. O.
27 June 16	Reddick, 2nd-Lt. G. P.	27 June 17	...	Linc. R. I. A. R. O.
	Honorary Officer.			
17 Sep. 13	Sardar Abdul Rahman Khan, Effendi.	17 Sep. 13	17 Sep. 13	Hony. Lieut. in Army.

		Jenadar.	Ressaidar.	Risaldar.	Risaldar-Major.	
	Risaldar-Major.					
1 Sep. 92	Chandar Singh (56)	1 Sep. 91	19 Jan 02	1 May 08	1 May 15	
	Risaldars.					
4 Aug. 93	Yakub Ali Khan (56)	1 Jan. 95	19 Apl. 04	20 Dec. 10		
17 May 02	Maya Ram (56)	19 Jan. 02	20 Dec. 10	16 Feb. 15		*Supy.*
25 Feb. 86	Nur Muhammad Khan (56)	1 Dec. 03	13 Sep. 08	1 May 15		Re-employed 8 Oct. 17.
16 Mar. 89	Luknan Khan	1 Oct. 11	8 Aug. 17	21 Apl. 18		

5th Cavalry—*contd.*

Date of entering service.	Names.	Army rank.		Remarks.
	Ressaidars.			
21 July 88	Siri Lal	13 Sep. 08	29 Oct. 14	
1 May 00	Nawab Singh (56, 58)	26 Nov 10	4 July 15	
1 Sep. 08	Randhir Singh (56)	20 Dec 10	21 Sep. 15	
1 July 00	Dost Muhammad Khan (45*a*, 56)	22 Sep. 17	22 Sep. 17	
16 Aug. 07	Hazur Ahmed (56)	29 Oct. 14	2 Feb. 18	With 2 Lrs.
1 Mar. 01	Amir Muhemmod Khan . . .	9 Nov. 12	21 Apl. 18	*Order of Br. I., 2 Class.*
	Jemadars.			
1 Mar. 99	Muhammad Husain Khan (45*a*, 56).	8 Aug. 14		
1 Jan. 11	Jot Ram	8 Sep. 14		With 2nd Lrs.
16 Aug. 89	Sundar Singh	1 Dec. 14		
1 Feb. 12	Bal Singh (56)	4 July 15		
1 Aug. 04	Imdad Ali Khan (58)	21 Sep. 15		
9 Apl. 93	Mumtaz Ali Khan (Ag. Ressdr.)	19 Oct. 15		With 62 Camel Corps.
22 Dec. 91	*Chandar Singh*	*14 Jan. 16*		
13 Mar. 13	Hari Singh (56)	14 Jan. 16		With 2nd Lrs.
26 Aug. 94	Chandar Singh	27 Jan. 16		
25 Sep. 96	Abdul Kaium Shah (56) . . .	27 Feb. 16		No. 1 Coolie Remt. Corps, Sehore.
1 Aug. 04	Lachhman Singh (56)	16 May 16		
1 Sep. 05	Zafar Muhammad Khan (56) . .	16 May 16		
1 Sep. 05	*Nabhu Singh*	*12 July 16*		
4 June 08	Zahir Uddin	1 Sep. 17		
11 Jan. 95	Muhammad Amin Khan . . .	22 Sep. 17		
19 Nov. 07	Abdul Rauf Khan	22 Sep. 17		Woordie-Maj., 22 Sep. 17.
22 July 11	Muhammad Ibrahim Khan (51, 52)	17 Nov. 17		
13 Feb. 89	Balwant Singh	3 Dec. 17		
12 July 09	Gokal Singh	26 Jan. 18		With 6 Cav.
15 Jan. 97	Man Singh (56)	26 Aug. 18		
	Mazhar Ali Shah	1 Aug. 18		
	Bisal Singh	1 Aug. 18		

6th King Edward's Own Cavalry.

Raised at Fatehgarh in 1842, by Lt. W. H. Ryves, as the 8th Regt. of Bengal Irregular Cavalry. Became the 6th Regt. of Bengal Cavalry, 1861; the 6th (the Prince of Wales') Regt. of Bengal Cavalry, 1883; the 6th (Prince of Wales') Bengal Cavalry, 1901; the 6th Prince of Wales' Cavalry, in 1903; present designation, 1st January 1906.

THE PLUME OF THE PRINCE OF WALES.
THE ROYAL AND IMPERIAL CYPHER OF KING EDWARD VII.
"Punniar" "Moodkee" "Ferozeshah" "Sobraon"
"Tel-el-Kebir" "Egypt, 1882" "Punjab Frontier."

Composition—1 Squadron of "Jat Sikhs," 1 of Sikhs other than Jat Sikhs, 1 of Jats, 1 of Hindustani Musalmans.
Uniform—Blue. *Facings*—Scarlet.

COLONEL-IN-CHIEF—THE KING.
COLONEL—General *Sir* Robert M. Jennings, K.C.B. . 13 May 1904.

1st Commn. or date of entering service.	Names and Rank.	Army rank.	Present appointment in Regiment.	Remarks.
	Commandant.			
25 Mar. 91	Newnham, Lt.-Col. C. C., D.S.O. *p.s.c.*	1 Jan. 16	4 Feb. 16	
	Squadron Commanders. (4)			
25 Dec. 91	Battye, Lt.-Col. A. B. (Q)	5 Dec. 17	4 Feb. 16	2nd-in-Comd.
18 June 92	Hodgkinson, Lt. Col. C.	18 June 18	3 Aug. 09	
4 Aug. 97	Craigie, Maj. P. N.	4 Aug. 15	1 May 14	
o July 98	Craster, Maj. G., D.S.O., *p.s.c.*	1 Sep. 15	4 Feb. 16	
	Squadron Officers. (9)			
25 Jan. 99	Bayley, Maj. L. C. L.	1 Sep. 15	5 Mar. 01	With 2-90th Punj.
27 Aug. 02	Oldham, Maj. J. S.	27 Aug. 17	8 Sep. 06	
16 Aug. 05	Porter, Capt. A. B.	16 Aug. 14	30 June 07	
24 Jan. 06	Wheatley, Capt. A. H. S.	24 Jan. 15	5 Aug. 07	Adjt., 10 May 13.
24 Jan. 06	*Sartorius, Capt. G. [l]*	*24 Jan. 15*	*30 Mar. 07*	*Qr.-Mr., 7 Oct. 13.*
20 Mar. 07	Ferguson, Capt. G. C. S.	1 Sep. 15	17 Sep. 09	
20 Dec. 07	Mackenzie, Capt. T. C.	1 Sep. 15	6 June 12	
9 Sep. 08	Smith, Capt. L.	1 Sep. 15	21 Nov. 09	
19 Jan. 12	Firth, Capt. F. J. W.	19 Jan. 16	30 Sep. 14	
4 Aug. 16	House, Lt. E. H.	4 Aug. 17	4 Aug. 16	
14 Nov. 16	Latter, Lt. F. A.	14 Nov. 17	24 Oct. 17	Attd.
14 Nov. 16	Wilson-Haffiden, Lt. E. L.	14 Nov. 17	18 Sep. 17	Attd.
30 Jan. 17	Griffin, Lt. G. W.	30 Jan. 18	13 Dec. 17	Attd.
19 Mar. 15	Newington, Lt. M. T. L.	19 Dec. 16	3 Aug. 16	Attd.
15 May 15	Abbott, Lt. A. A. (*actg. Capt.*)	15 May 15	10 Oct. 17	Attd. Adjt.
	Attached.			
27 Nov. 15	Macwhirter, Capt. D.	27 Nov. 15	9 Nov. 17	I. A. R. O.
8 July 17	Toogood, 2nd-Lt. L. G. C.	8 July 17	10 Dec. 17	I. A. R. O.
21 Nov. 17	Powell, Lt. W. S.	21 Nov. 18	17 Dec. 17	I. A. R. O.
. . . .	Murray, Capt. M. F.	. . . .	. . . .	6th Res. Regt. of Cav.
13 Nov. 14	Hills, Capt. A. C.	16 Feb. 18	16 Feb. 15	I. A. R. O.
8 Jan. 15	Leigh Clare, Lt. H. J. L., M.C.,	6 Jan. 16	4 May 15	I. A. R. O.
5 Mar. 15	Parmer, Lt. S. W. (*temp. Capt.*)	5 Mar. 15	30 July 15	I. A. R. O.
16 Dec. 14	Sykes, Lt. J. H.	4 Apl. 16	24 Sep. 16	3rd Dn. Gds.
13 Nov. 14	Nixon, Capt. C. N.	17 Aug. 18	. . . .	I. A. R. O.
18 Dec. 14	Bradshaw, Capt. J. W. H.	30 Aug. 18	. . . .	I. A. R. O.
24 Dec. 14	Haslam, Lt. R. E.	24 Oct. 15	. . . .	I. A. R. O.
5 Feb. 15	Streatfield, Lt. N.	5 Feb. 16	19 May 16	I. A. R. O.
12 Feb. 15	Burton, Lt. E. R.	12 Feb. 16	19 May 16	I. A. R. O.
21 Sep. 16	De la Fargue, Lt. D. G. H.	21 Sep. 17	11 June 16	I. A. (*on probn.*)
30 Oct. 14	Carpenter, Capt. J. H.	30 Oct. 18	3 Aug. 16	I. A.R. O.
22 Apl. 15	Eyton, Lt. J. S.	20 Apl. 16	13 Aug. 16	I. A. R. O.
10 June 16	Atkins, Lt. E. D.	10 June 17	13 Nov. 16	I. A. R. O.
14 June 16	Oliver, Lt. F. P.	14 June 17	22 Nov. 16	I. A. R. O.
16 Aug. 16	Parker, Lt. V. L.	16 Aug. 17	11 Nov. 17	5th Cav.
28 Oct. 15	O'Bryan, Lt. J. R.	28 Oct. 16	. . . .	I. A. (*on probn.*)
20 Dec. 16	Stepto, Lt. E.	20 Dec. 16	13 Aug. 17	I. A. (*on probn.*)
21 Dec. 14	Steward, Lt. R. O. B. (207*a*)	21 Dec. 14	15 Dec. 17	Asst. Imp. Force (on probn.)
18 Oct. 17	Mearns, Lt. J. P.	18 Oct. 18	14 Dec. 17	
. . . .	Macpherson, Lt. R. V.	. . . .	. . . .	Res. Regt. of Cav.
1 June 16	Johns, Lt. W. N.	1 June 16	. . . .	County of London Yeo.
. . . .	White, Lt. (*temp.*) G. H.	. . . .	6 Feb. 16	Res. Regt. of Cav.
. . . .	Maxted, 2nd-Lt. F. S.	. . . .	6 Feb. 16	Do.
28 Aug. 15	Turner, Lt. R. W.	28 Aug. 16	. . . .	I.A. (*on probn.*)

		Jemadar.	Ressaidar.	Risaldar.	Risaldar-Major.	
	Risaldar-Major.					
30 Oct. 85	Fateh Singh, *Bahadur* (68)	1 Nov. 04	1 Oct. 07	1 July 13	17 June 15	*Order of Br. I., 2nd Cl.*

6th King Edward's Own Cavalry—*contd.*

Date of entering service.	Names.	Army rank. Jemadar.	Ressaidar.	Risaldar.	Remarks.
	Risaldars.				
1 Mar. 93	Dayal Singh (56, 68) . . .	1 Feb. 08	1 Mar. 12	17 June 15	. . . . *Order of Br. I., 2nd Cl*
22 Mar. 90	Sher Singh (58, 68)	1 Oct. 07	16 Nov. 13	23 Mar. 16	
8 Mar. 90	Jai Singh, I.D.S.M. (45*a*, 56, 68)	1 Aug. 08	17 June 15	27 May 17	
19 Jan. 98	Jai Singh (68), M.C. . .	1 Feb. 13	14 Aug. 16	7 July 18	
13 Sep. 99	Dani Ram	31 Aug. 14	23 Mar. 16	4 Aug. 18	
	Ressaidars.				
	Ajaib Singh (*Retd.*)	. . .	. . . .		Re-employed, 20 Feb. 15. Depôt
20 Dec. 95	Ganda Singh	1 Oct. 11	27 Mar. 16		
1 Dec. 07	Bachittar Singh, I.D.S.M.	14 Sep. 15	19 Apl. 18		
1 May 98	Pahlwan Khan	11 Apl. 16	8 Sep. 18		
25 Mar. 12	Man Singh	16 Nov. 13	3 Sep. 18		
	Jemadars.				
16 June 98	Kirpa Ram (56)	1 Mar. 12			
16 Dec. 95	Bishan Singh	1 Jan. 15			Woordie-Major, 11 Apl. 16.
19 July 98	Anti Ram	25 Sep. 15			
23 May 00	Amir Singh, I.D.S.M.	11 Apl. 16			
14 Oct. 03	Shim Bhoo	11 Apl. 16			
13 Jan. 02	Bagga Singh	11 Apl. 16			
1 Mar. 97	Wali Muhammed Khan . . .	11 Apl. 16			
14 Apl. 03	Rala Singh	17 June 16			
19 Sep. 06	Amrik Singh	14 Aug. 16			
	Baldu Ram	8 Apl. 18			
	Fazal-Ali Khan	3 June 18			

7th Hariana Lancers.

Raised at Cawnpore and Meerut in 1846, by Capt. J. Liptrott, as the 16th Regt. of Bengal Irregular Cavalry. Became the 17th Regt., Bengal Irregular Cavalry, 1847; the 7th Regt. of Bengal Cavalry, 1861; the 7th Regt. of Bengal Lancers, 1900; the 7th Bengal Lancers, 1901; the 7th Lancers, 1903; present designation, 1904.

"Punjaub" "Burma, 1885-87."

Composition—1 Squadron of Jats, 1 of Sikhs, 1 of Dogras, 1 of Hindustani Musalmans.

Uniform—Scarlet. *Facings*—Blue.

COLONEL—Major-General James C. Stewart, ℳ., C.B. . 13 May 1904.

1st Commn. or date of entering service.	Names and Rank.	Army rank.	Present appointment in Regiment.	Remarks.
	Commandant.			
7 Dec. 95	Carnegy, Lt.-Col. G. P. O., D.S.O.	15 Mar. 18	15 Mar. 18	
	Squadron Commanders. (4)			
1 Mar. 90	Jones, Bt.-Col. L. C., C.M.G., M.V.O., *p.s.c.* (Q) [*l*]. (*temp. Maj.-Genl.*)	3 June 17	8 Sep. 14	2nd-in-Comd.
10 Oct. 91	*Harbottle, Maj. R. C.*	*10 Oct. 09*	*29 Apl. 05*	
24 June 98	Gretton, Maj. G. F.	1 Sep. 15	14 Apl. 15	
28 Mar. 00	*Egerton, Maj. S. W.*	*1 Sep. 15*	*15 Mar. 18*	*Adjt., I. C. Remt. Tg. Depôt, Bangalore.*
	Squadron Officers. (9)			
8 Jan. 01	Simonds, Maj. M. H.	8 Jan. 16	6 Feb. 03	2nd-in Comd. (temp.)
27 Aug. 02	*Hume, Maj. E. G.*	*27 Aug. 17*	*3 Nov. 04*	*With 42nd Cav.*
22 Oct. 02	*Williamson, Maj. A. G.*	*22 Oct. 17*	*31 Oct. 05*	*G. S. O., 2nd gr., Army Hdqrs.*
18 Jan. 05	*Trench, Capt. C. F.*, D.S.O. (*temp. Maj.*)	*18 Jan. 14*	*1 Sep. 06*	*G. S. O., 2nd gr., Derajat Bde.*
19 Sep. 08	*Harris, Capt. A. E. C.*, M.C.	*1 Sep. 15*	*4 Nov. 10*	*Adjt. & Qr.-Mr., Tg. Sch., Indore.*
3 Sep. 10	*Drake-Brockman, Capt. F. T.*	*1 Sep. 15*	*6 Nov. 11*	
11 Sep. 16	Champion, Lt. H. C.	11 Sep. 17	27 June 17	Attd.
1 July 15	Menzies, Lt. W. F.	1 July 16	9 May 17	Attd.
27 Sep. 16	Jones, Lt. G. F.	27 Sep. 17	9 May 17	Attd.
11 Aug. 15	Hilliard, Lt. L. M. (*Actg. Capt.*)	11 Aug. 16	1 Aug. 17	Attd. Adjt., 30 Mar. 18.
12 Sep. 17	Mcvean, 2nd-Lt. E. D.	12 Sep. 17	13 Jan. 18	Attd.
	Attached.			
13 Nov. 14	Forbes, Lt. C. A.	28 Feb. 15	29 Nov. 14	I. A. R. O.
19 Aug. 15	Butler, Lt. H. M.	9 Aug. 16	13 Feb. 18	I. A. (*on probn.*)
26 May 16	Lickman, Lt. E. W.	26 May 17	12 June 17	I. A. (*on probn.*)
25 Sep. 16	Barraclough, Lt. M. W. R.	25 Sep. 17	10 June 17	I. A. (*on probn.*)
5 Dec. 16	Balch, Lt. C. H.	5 Dec. 17	12 Jan. 18	I. A. (*on probn.*)
25 Jan. 17	Lyell, Lt. G. D. (*Actg. Capt.*)	25 Jan. 18	9 Feb. 17	I. A. R. O.
1 Mar. 17	Hill, Lt. F. R.	1 May 18	16 Jan. 18	I. A. (*on probn.*) *Lv. ex I. m.c.*
23 May 17	Pitt, Lt. W. H. (*temp. Capt.*)	23 May 17	7 June 17	I. A. R. O.
29 Mar. 18	Molyneux, 2nd-Lt. H. J.	29 Mar. 18	24 Apl. 18	I. A. (*on probn.*) Q. M. (*Offg.*)
29 July 17	Ogden, Lt. H. L. (*temp. Capt*)	25 July 18	22 Jan. 18	I. A. (*on probn.*) *N. C. Os.' School, Belgaum.*
9 Jan. 18	Hall, 2nd-Lt. R. G.	9 Jan 18	1 Feb. 18	I. A. R. O. (*on probn.*)
1 May 18	Watson, 2nd-Lt. E. L.	11 May 18	27 May 18	I. A. R. O.
6 May 18	Pratt, 2nd-Lt. C. T. E.	6 May 18	27 May 18	I. A. R. O.
22 July 18	Shattock, 2nd-Lt. E. K.	22 July 18	23 Aug. 18	I. A. R. O.
9 Aug. 18	Coleman, 2nd-Lt. H. C.	9 Aug. 18	4 Sep. 18	I. A. (*T.C.*)
	Wall, 2nd-Lt. H. J. G.	. . . .	26 Nov. 18	I. A. R. O. (*on probn.*)

Risaldar-Major.

		Jemadar.	Ressaidar.	Risaldar.	Risaldar-Major.	
9 Apl. 86	Muhammad Akbar Ali Khan, *Sardar Bahadur* (56, 58).	26 Aug. 92	13 Apl. 95	1 Dec. 99	15 Mar. 10	Order of Br. I., 1st Cl. *A. D. C. to H. E. the Viceroy.*
20 June 96	Khurshed Ali Khan (47*a*, 56)	20 June 96	16 Oct. 06	16 Dec. 10	1 Dec. 17	
. . .	Kalander Khan, *Bahadur*, I.O.M. (Retd.)	. . .	. . .	. . .	Re-employed.	*Order of Br. I. 2nd Cl.*

7th Hariana Lancers—*contd.*

Risaldars.

Date of entering service.	Names.	Army rank.	Present appointment in Regiment.		Remarks.
16 Sep. 83	Gopal Singh Bahadur (56) . .	16 June 93	15 July 10	1 Nov. 12	*Order of Br. Ind., 2nd Cl.*
23 Oct. 00	Balwant Singh (56)	23 Oct. 03	15 Mar. 10	1 Apl. 15	
1 May 99	Chet Ram (56), I.D.S.M. . . .	1 Nov. 12	1 Apl. 15	2 Jan. 16	
5 Oct. 92	*Mam Chand (56)* . . .	*10 Apl. 07*	*29 Oct. 14*	*14 Aug. 16*	*With 44th Cav.*
23 Mar. 01	*Ajaib Singh (53, 56, 66a, 170. 172a).*	*23 Mar. 01*	*19 Dec. 10*	*1 Dec. 17*	
16 Mar. 90	Suram Chand (56)	12 Mar. 01	16 Oct. 11	1 Dec. 17	
. . .	Suraj Bux (Retd.)	. . .	. . .	. . .	Re-employed, 12 Apl. 15 Depôt.

Ressaidars.

Date of entering service.	Names.	Army rank.	Present appointment in Regiment.	Remarks.
16 Apl. 02	*Muhammad Yusuf Khan (45a, 56).*	*1 May 06*	*29 Oct. 14*	
7 Jan. 96	Nanha Sing, I.D.SM.	16 Feb. 15	3 Jan. 16	
13 Aug. 03	Sheo Karan (56)	29 Oct. 14	18 July 16	
11 Jan. 96	Jai Singh (56)	16 Feb. 15	1 Dec. 17	
1 Nov. 98	Bishan Singh	30 Apl. 16	1 Feb. 18	
28 Feb. 01	Tara Singh (57)	7 Dec. 14	1 Feb. 18.	
1 Feb. 05	Bhagwan Singh (56)	2 Dec. 15	1 Feb. 18.	

Jemadars

Date of entering service.	Names.	Army rank.	Remarks.
26 Nov. 92	*Muhammad Umar Khan* . .	*29 Oct. 14*	
5 Apl. 92	*Dost Muhammad Khan* . . .	*1 Apl. 15*	
21 Feb. 95	*Mansa Singh*	*13 Apl. 15*	
1 Oct. 94	*Lajje Ram*	*14 Apl. 15*	*With Remt. Depôt, Bangalore.*
1 Apl. 11	Nihal Singh (47a)	23 Nov. 15	With 20th Horse.
3 Jan. 02	*Bela Singh*	*2 Dec. 15*	
16 Oct. 14	Abdul Rahim Khan (56) . . .	30 Apl. 16	
1 May 10	Ran Singh I.D.S.M. (47a, 56). .	15 June 17	
16 July 09	Harnam Singh	3 Sep. 17	
1 Apl. 17	Kapur Singh (56)	27 Sep. 17	
14 Aug. 99	Sri Chand (56)	27 Sep. 17	
6 Dec. 05	Mansa Ram (56)	27 Sep. 17	
1 Jan. 05	Khan Muhammad Khan . . .	1 Dec. 17	
1 Mar. 04	Mat Ram	3 Dec. 17	
2 Jan. 01	Sant Singh (47a, 56)	1 Feb. 18	
1 Mar. 08	Kehar Singh, I.D.S.M. (56) . .	1 Feb. 18	Woordie Maj.
21 July 00	Sant Singh	23 Feb. 18	
1 Apl. 00	Sucha Singh	1 June 18	

8th Cavalry.

Raised at Sultanpore in 1846, by Capt. W. H. Ryves, as the 17th Regt. of Bengal Irregular Cavalry. Became the 18th Regt. of Bengal Irregular Cavalry, 1847; the 8th Regt. of Bengal Cavalry, 1861; the 8th Regt. of Bengal Lancers, 1900; the 8th Bengal Lancers, 1901; the 8th Lancers, 1903; present designation, 1904.

"Afghanistan, 1878-80."

Composition—2 Squadrons of Hindustani Musalmans, 1 of Rajputs, 1 of Jats.

Uniform—Blue. *Facings*—Scarlet.

COLONEL—Major-General Hamilton Chapman, CB . . . 13 May 1904.

Commandant.

1st Commn. or date of entering service.	Names and Rank.	Army rank.	Present appointment in Regiment.	Remarks.
16 Nov. 87	Chaplin, Lt.-Col. R. E.	2 Nov. 13	2 Nov. 13	

Squadron Commanders. (4)

1st Commn. or date of entering service.	Names and Rank.	Army rank.	Present appointment in Regiment.	Remarks.
27 Apl. 89	*Becher, Lt.-Col. G. A.*, D.S.O.(*Q*)	*27 Apl. 15*	*20 Oct. 13*	*2nd-in-Comd.*
21 Sep. 89	*Cheyne, Lt.-Col. R. E.* (*Q*) (*l*)	*21 Sep. 15*	*31 Mar. 09*	
5 Aug. 96	*Fetherstonhaugh, Bt.-Lt.-Col. W. A.*, D.S.O., *p.s.c.*	*3 June 17*	*2 Nov. 13*	*A. A. & Q. M. G., 4th (Quetta) Divn.*
22 Jan. 98	Kay, Maj. H. C.	1 Sept. 15	22 Aug. 14	

Squadron Officers. (9)

1st Commn. or date of entering service.	Names and Rank.	Army rank.	Present appointment in Regiment.	Remarks.
18 May 98	*Craster, Maj. W. S.*	*1 Sep. 15*	*13 Feb. 04*	*2nd-in-Comd.* (*Offg.*)
12 Aug. 99	*Lane, Maj. F. B.*	*1 Sep. 15*	*5 Oct. 02*	*D. A. A. G.*
29 Dec. 00	Mansfield, Maj. C. B. H. . .	29 Dec. 15	1 Jan. 06	S. C. (*Temp.*) With 41st Cavy.
8 Jan. 01	Hamilton, Maj. F A. (*temp. Lt.-Col.*)	8 Jan. 16	26 Feb. 13	A. A. Q. M. G.
18 Jan. 05	*Carpendale, Capt. T. M.* . .	*18 Jan. 14*	*30 Apl. 07*	*Adjt., 1 Sep. 13.*
16 Sep. 05	Hudson, Capt. B. R. . . .	16 Sep. 14	15 Oct. 08	Adjt. Remt. Depot, Saugor.
17 Aug. 07	*Empson, Capt. A. H. A.* . .	*1 Sep. 15*	*2 Nov. 09*	*With Sig. Serv.*
18 Sep. 09	*Young, Capt. K. de L.*, M.C. . .	*1 Sep. 15*	*5 June 12*	*Bde. Maj., Sec'bad Cav. Bde.*
24 Aug. 12	De Wilton, Capt. J. G. B. . .	24 Aug. 16	21 Nov. 13	Ind. Contgt., Somaliland.
5 June 17	Coventon, Lt. J. A. (*Actg. Capt.*)	5 June 17	2 Aug. 17	*Adjt.* (*Offg.*)
10 July 15	Burch, Lt. G. W.	10 July 16	12 Feb. 18	E. York R. I. A. (*on probn.*)
29 Oct. 19	Hudson, Lt. A. R.	29 Oct. 16	29 Oct. 17	I. A. (*on probn.*)
24 May 17	Brigg, Lt. C. J.	6 Aug. 16	2 June 17	
18 June 17	Richards, Lt. C. D.	18 June 18	. . .	Attd. *Lv., ex I., m. c.*
21 Dec. 17	Vickers, 2nd-Lt. C. K.	21 Dec. 17	29 Mar. 18	Attd. *Lv., ex I., m. c.*

Attached.

1st Commn. or date of entering service.	Names and Rank.	Army rank.	Present appointment in Regiment.	Remarks.
27 Nov. 14	Fremlin, Capt. R.W. (*Actg. Capt.*)	27 Oct. 18	27 Nov. 14	I. A. R. O.
25 Feb. 18	Farrance, Lt. S. C. E. . . .	22 June 16	7 Mar. 18	I. A. (*on probn.*)
. . . .	Fearnside-Speed, Lt. K. A. . .	24 July 16	4 Sep. 16	A. S. C. I. A. (*on probn.*)
12 May 18	Taylor, Lt. J. S.	10 Apl. 17	21 May 18	Scottish Horse, I. A. (*on probn.*)
21 Dec. 16	Oecken, Lt. C.A.J. (*Actg. Capt.*)	21 Dec. 17	16 Jan. 18	North Irish Horse. I. A. R. O.
17 Mar. 17	Barrett, Lt. C. L. R.	17 Mar. 18	17 Mar. 17	I. A. R. O.
25 Sep. 17	Stephen, Lt. A. A.	25 Sep. 17	25 Sep. 17	I. A. (*T.C.*) *Lv., ex I., m.c.*
18 June 17	Moreland, 2nd-Lt. J. A., M. C. .	26 Jan. 18	28 Mar. 18	R. Ir. Rif. I. A. (*on probn.*)
12 May 18	English, 2nd-Lt. H. H. E. . .	25 May 18	21 May 18	Res. R. Cav. I. A. (*on probn.*)
. . . .	Philip, 2nd-Lt. C. L., I.C.S. . .	. . . .	28 Sep. 18	I. A. R. O.

Risaldar-Major.

1st Commn. or date of entering service.	Names and Rank.	Jemadar.	Ressaidar.	Risaldar.	Risaldar-Major.	
7 Dec. 97	Chattar Singh (56)	16 Oct. 98	1 Nov. 06	13 Apl. 10	23 May 16.	

Risaldars.

1st Commn. or date of entering service.	Names and Rank.	Jemadar.	Ressaidar.	Risaldar.		
7 Jan. 89	Faizullah Khan (56)	23 Oct. 00	1 Nov. 09	1 Nov. 10		
1 May 90	Hulas Singh (56)	15 Mar. 02	1 Nov. 09	29 Oct. 14		
12 Sep. 85	Chhannu Khan (56, 66*a*) . . .	1 Nov. 04	13 Apl. 10	3 May 16		
7 Feb. 07	Mirza Riyazul Hussain (56) . . .	26 Feb. 97	29 Oct. 14	4 Oct. 16		
21 July 94	*Ram Nath*	*18 May 15*	*1 Apl. 17*	*6 Nov. 17*		*14 Jat Lrs.*

8th Cavalry—*contd.*

Date of entering service.	Names and Rank.	Army rank	Present appointment in Regiment.	Remarks.
	Ressaidars.			
6 Jan. 94	Har-bakhsh Singh (66*a*)	1 Nov. 07	29 Oct. 14	
18 Aug. 05	*Mir Padshah (45a, 50, 66a)*	*9 May 08*	*4 Jan. 15*	*Rem't Sqdn.*
7 Oct. 89	*Sahib Singh (56, 66a)*	*1 Nov. 09*	*3 Dec. 15*	*I. C. R. Tg. Depot, Bangalore.*
2 Feb. 93	Shafi Mohammad Khan (56)	12 Apl. 10	5 Dec. 15	
5 Dec. 95	Ram Phal (45*a*, 51)	1 June 10	3 May 16	
2 Mar. 87	*Balwant Rao (66a)*	*1 Nov. 09*	*3 May 16*	*Br. Remt. Tg. Depot.*
11 Mar. 91	Gharam Singh (66*a*)	1 Nov. 10	3 May 16	
8 Sep. 96	*Niamat Ali Khan (56, 66a)*	*20 May 12*	*18 Nov. 17*	
31 Aug. 93	Jamaiyat Singh	29 Oct. 14	19 Nov. 17	
	Jemadars.			
2 Nov. 03	Malak Sher Khan (45*a*, 56)	29 Oct. 14	Woordie-Maj.	
20 Feb. 11	*Narain Singh (45a)*	*7 Dec. 14*	*22nd Cav.*	
18 Nov. 95	*Janak Singh*	*4 Jan. 15*	*Remt. Depôt, Deolali.*	
12 Sep. 90	*Kale Khan*	*1 July 15*	*With 2nd Lrs.*	
25 Jan. 92	*Taj Muhammad Khan*	*13 Oct. 15*	*Remt. Sqn.*	
5 Jan. 10	*Khazan Singh (56)*	*5 Dec. 15*	*Remt. Sqn.*	
26 Jan. 06	Muhammad Ismil Khan	19 Apl. 16		
1 Nov. 09	*Mohamed Mushtaq Khan*	*3 May 16*	*I. C. R. Training Depot, Bangalore.*	
12 Apl. 03	Khan Muhammad Khan (59)	10 May 16		
16 Nov. 96	*Sham Singh*	*11 May 16*		
1 Apl. 00	*Jan Muhammad Khan*	*9 July 16*	*Remt. Sqn.*	
27 Jan. 96	*Indar Singh*	*22 July 16*	*Remt. Sqn.*	
19 May 97	Darshan Singh	5 Oct. 16		
1 Jan. 04	*Nur Muhammad Khan*	*5 Oct. 16*	*Remt. Sqn.*	
10 Dec. 98	*Yusaf Khan*	*3 Oct. 16*	*With 22nd Cav.*	
29 Apl. 99	Sakhawal Khan (51)	3 Oct. 16		
1 Aug. 00	*Randhir Sing*	*3 Oct. 16*	*Remt. Sqn.*	
17 Dec. 01	Bharat Singh	13 Nov. 16		
10 Oct. 92	*Bakhshi Singh*	*1 Mar. 17*	*Syce Remt. Corps, Sehore.*	
1 May 09	Dhani Ram	7 July 17		
1 Dec. 00	*Suran Singh*	*1 Apl. 17*	*Remt. Sqn.*	
21 May 04	*Fazal Ali Khan,* I.D.S.M.	*14 Apl. 17*	*Remt. Sqn.*	
5 Dec. 95	*Raghbir Singh*	*16 Nov. 17*	*Syce Remt. Corps, Sehore.*	
22 May 02	*Nur Mahammad Khan*	*17 Nov. 17*	*Remt. Sqn.*	
17 Oct. 03	*Dhanpat Singh*	*17 Nov. 17*	*Remt. Sqn.*	
9 May 00	*Ibrahim Khan*	*17 Nov. 17*	*Remt. Sqn.*	
6 Jan. 04	Ahsan Ali Khan	18 Nov. 17		
1 Nov. 05	Ram Narain Singh	18 Nov. 17		
12 Mar. 02	Pyara Singh	29 Jan. 18		
8 Oct. 01	*Jagpal Singh*	*26 Feb. 18*	*Remt. Sqn.*	
21 July 00	*Puran Singh*	*4 Mar. 18*	*With Syce Corps.*	
12 Nov. 01	*Sheodan Singh*	*23 Mar. 18*	*Remt. Sqn.*	

9th Hodson's Horse.

Raised in separate risallahs in the Punjab in 1857, and embodied, under the designation of Hodson's Horse in the camp before Delhi, by Lieut. W. S. R. Hodson. Became the 1st Regt. of Hodson's Horse, 1858; the 9th Regt. of Bengal Cavalry, 1861; the 9th Regt. of Bengal Lancers, 1886; the 9th Bengal Lancers (Hodson's Horse), 1901; present designation, 1913.

"Delhi, 1857" "Lucknow" "Suakin, 1885" "Chitral" "Punjab Frontier."

Compositions—1½ Squadrons of Sikhs, ½ of Dogras, 1½ of Punjabi Musalmans, ½ of Pathans.

Uniform—Blue. *Facings*—White.

COLONEL—Major-General Frederick W. P. Angelo . . . 10 December 1912.

1st Commn. or date of entering service.	Names and Rank.	Army rank.	Present appointment in Regiment.	Remarks.
	Commandant.			
21 Dec. 89	Beatty, Lt.-Col. G. A. H., D.S.O. (*temp. Br. Genl.*)	21 Dec. 15	3 Oct. 16	
	Squadron Commanders. (4)			
19 Feb. 92	Rowcroft, Lt.-Col. C. H. (Q) . .	19 Feb. 18	3 Oct. 11	Comdt. (*temp.*),
22 Jan. 98	Dyce, Maj. H. L., M.C. . . .	1 Sep. 15	26 Nov. 11	
8 Jan. 01	*Smith, Maj. O. P.*	*8 Jan. 16*	*1 Dec. 17*	*G. S. Br., Army Hdqrs.*
	Squadron Officers. (9)			
28 Jan. 03	*deBurgh, Maj E.*, D.S.O. *p.s.c.* (*temp. Lt.-Col.*)	*21 June 17*	*17 Oct. 04*	
24 Jan. 06	Vigors, Capt. M.D., M.C. (*Actg. Maj.*)	24 Jan. 15	1 Dec. 07	2nd-in-Comd. (*temp.*)
19 Jan. 07	Corbett, Capt. T. W., M.C. . .	1 Sep. 15	8 Apl. 08	
25 Jan. 08	Seymour, Capt. E. V. F.	1 Sep. 15	2 Apl. 09	Adjt. Gov. Genl.'s Body Guard. With 43rd Cav.
3 Sept. 10	Graham, Capt. L. C. T., M. C. .	1 Sep. 15	23 Dec. 11	
20 Jan. 12	Luttman-Johnson, Capt. H. W.	20 Jan. 16	16 Mar. 13	
22 Jan. 13	Messervy, Capt. F. W. . . .	22 Jan. 17	16 Mar. 14	
14 Nov. 16	Murphy, Lt. J. R. K.	14 Nov. 17	. . . .	Attd.
30 Jan. 17	Clarke, 2nd-Lt. T. V.	30 Jan. 17	13 Dec. 17	Attd.
30 Jan. 17	Cumming, 2nd-Lt. R H. R. . .	30 Jan. 17	21 Mar. 18	Attd.
1 May 17	Croucher, Lt. F. E.	1 May 17	31 Dec. 17	Attd.
12 Sep. 17	Judd, Lt. E. L.	12 Sep. 18	1 Jan. 18	Attd.
31 Aug. 18	Watson, 2nd-Lt. N.	31 Aug. 18	4 Sep. 18	Attd.
31 Jan. 18	Townshend, 2nd-Lt. C. M. W. L.	31 Jan. 18	8 Feb. 18	
31 Aug. 18	Ryan, 2nd-Lt. S. E. L. . . .	31 Aug. 18	4 Sep. 18	Attd.
17 Mar. 15	Knight, Lt. A. B.	17 Mar. 16	. . . .	4th Cav.
	Attached.			
8 Sep. 11	Stevens, Capt. C. F. L., M.C.	6 Sep. 15	19 Jan. 16	10th Lrs.
22 Mar. 15	Morris, Capt. N. M.	16 Apl. 18	16 Apl. 15	I. A. R. O.
5 Feb. 15	Ewart, Lt. J. A., M.C. . . .	5 Feb. 16	7 Apl. 16	I. A. R. O.
4 Feb. 16	Brown, Lt. M. A.	4 Feb. 17	7 Nov. 17	I. A. (*on probn.*)
29 May 15	Ninis, 2nd-Lt. G. W.	29 May 15	. . . .	I. A. (*on probn.*)
23 July 15	Wilson, Lt. G.	23 July 16	14 June 16	I. A. R. O.
20 Aug. 15	Johnson, Lt. T. F.	20 Aug. 16	15 May 16	I. A. R. O.
13 Oct. 15	Wilson, Lt. J. M.	13 Oct. 16	13 Mar. 16	I. A. R. O.
1 Jan. 16	Graves, Lt. E.	1 Jan. 17	4 Oct. 16	I. A. R. O.
13 July 17	Morgan, Lt. H. S.	13 July 18	4 Oct. 17	I. A. R. O.
5 Nov. 17	Tate, Lt. J. W.	5 Nov. 18	11 Apl. 18	I. A. R. O.
4 July 18	Marsden, 2nd-Lt. P.	4 July 18	13 July 18	I. A. R. O.
	Honorary Officers.			
6 Feb. 95	Rampur, *His Highness Nawab Sir* Muhammad Hamid Ali Khan, *Bahadur*, G.C.I.E., G.C.V.O., A.-D.-C., *Nawab* of	3 June 10	28 Feb. 96	Hony. Col. in Army.
22 Oct. 02	Malik Muhammad Mubarak Khan, *Nawab*, O.B.E., Tiwana of Shahpur.	24 Apl. 12	16 Apl. 03	Hony. Capt. in Army.

		Jemadar.	Ressaidar.	Risaldar.	Risaldar-Major.	
	Risaldar-Major.					
24 Oct. 94	Malik Khan Muhammad, *Bahadur.*	5 Apl. 00	16 Oct. 07	1 Mar. 10	19 July 16	*Order of Br. I., 2nd cl.*
	Risaldars.					
10 May 87	Mir Dad Khan, *Bahadur* (56) .	12 Jan. 94	5 Apl. 00	20 June 07		*Order of Br. I., 2nd cl.*
1 Nov. 94	Muhammad Akram Khan *Bahadur* (56, 66*a*).	16 Jan. 06	11 Sep. 09	1 Feb. 15		*Order of Br. I., 2nd cl.*
25 Apl. 94	Tek Singh (56, 58) I.D.S.M. . .	12 Oct. 12	28 May 16	14 Nov. 16		
1 Apl. 02	Dost Muhammad (45*a*, 56, 66)	27 May 06	22 Dec. 14	27 July 17		
11 Dec. 01	Nur Ahmad Khan (45*a*, 50, 53) I.O.M.	16 Nov. 07	1 Feb. 15	18 Dec. 17		

9th Hodson's Horse—*contd.*

Date of entering service.	Names and Rank.	Army rank.		Remarks.
	Ressaidars.			
16 Oct. 09	Harband Singh (56, 66)	1 Mar. 10	1 Feb. 17	
15 Jan. 04	Malak Muhammad Hayat Khan (56, 66).	1 Jan. 11	1 Feb. 17	
1 Mar. 04	Bur Singh, M. C.	17 Mar. 15	25 Aug. 17	
	Sardar Khan (Retd.)	....	....	Re-employed, 9th Nov. 17.
4 Aug. 97	Fateh Muhammad Beg (56)	1 Apl. 14	25 Aug. 17	
17 Oct. 00	Hasan Shah (52, 56). M. C.	22 Dec. 14	13 Oct. 17	
24 Nov. 07	Sardar Khan (45a) I.O.M., I.D.S.M.	1 Feb. 15	14 Oct. 17	
1 Mar. 10	Ajaib Singh	11 Apl. 16	2 Dec. 17	
1 Sep. 01	Gandrabh Singh	22 Oct. 15	1 Dec. 17	
27 Nov. 15	Mewa Singh	12 Aug. 16	5 Dec. 17	
10 Jan. 01	Fateh Singh, I.D.S.M.	14 Nov. 16	18 Dec. 17	
9 Nov. 00	Habib Gul, I. D. S. M.	8 Mar. 17	18 Dec. 17	
	Jemadars.			
20 Nov. 93	Ahmad Shah	31 Aug. 14		
16 Jan. 01	Malik Nur Muhammad Khan	18 Oct. 15		*Supy.*
12 Aug. 95	Mir Alam Khan	19 May 16		
15 Mar. 12	Malik Muhammad Sher Khan	19 May 16		
18 Apl. 99	Muhammad Khan, I.D.S.M.	14 Nov. 16		
24 Apl. 01	Hira Singh, I.D.S.M.	1 Feb. 17		
1 July 94	Tohid Gul Khan, I.D.S.M.	21 Mar. 17		
13 Dec. 00	Fazal Khan	23 Mar. 17		
18 Apl. 99	Mangal Singh, I.D.S.M.	25 Aug. 17		
15 Aug. 05	Channan Singh	25 Aug. 17		
1 Oct. 01	Muhammad Khan	25 Aug. 17		
10 Dec. 01	Waryam Singh	25 Aug. 17		
1 Feb. 02	Bhagwan Singh	6 Oct. 17		
30 Aug. 06	Nauroz Khan	14 Oct. 17		
1 Jan. 10	Mir Alam Khan (56, 45a, 60a, 66)	1 Nov. 17		
5 Feb. 99	Jetha Singh	25 Nov. 17		
....	Deya Ram	1 Dec. 17		Woordie Maj. 8 Apl. 18.
10 Feb. 02	Abdul Satar Khan, I.D.S.M.	2 Dec. 17		
6 Feb. 05	Nawab Ali Khan, I.D.S.M.	2 Dec. 17		
	Sultan Muhammad Khan	3 Dec. 17		
8 Dec. 12	Indar Singh	5 Dec. 17		
1 Nov. 03	Indar Singh	18 Dec. 17		
18 Mar. 13	Makhan Singh	18 Dec. 17		
29 Nov. 08	Roop Singh	30 Jan. 18		
3 Feb. 99	Sultan Ahmad Khan	1 May 18		
21 Apl. 01	Mehta Gyan Chand	1 May 18		
16 Oct. 97	Indar Singh	13 July 18		
13 Aug. 07	Sher Khan	1 Oct. 18		
1 May 14	Khalil-ul-Rahman Khan (47(a), 56)	1 Oct. 18		
13 Oct. 08	Chanan Sing (56)	1 Oct. 18		
16 Dec. 13	Mohan Sing (56)	1 Oct. 13		

10th Duke of Cambridge's Own Lancers (Hodson's Horse).

Raised in separate risallahs in the Punjab in 1857, and under designation of 'Hodson's Horse,' embodied by Lt. W. S. R. Hodson in camp before Delhi. Became 2nd Regt. of Hodson's Horse, 1858; the 10th Regt. of Bengal Cavalry, 1861; the 10th Regt. of Bengal Cavalry (Lancers), 1864; the 10th Regt. of Bengal Lancers, 1874; the 10th Bengal (the Duke of Cambridge's Own) Lancers, 1878; the 10th (Duke of Cambridge's Own) Bengal Lancers (Hodson's Horse), 1901; present designation, 1903.

"Delhi, 1857" "Lucknow" "Abyssinia" "Afghanistan, 1878-80."

Composition—1½ Squadrons of Sikhs, 1 of Dogras, 1 of Punjabi Musalmans, ½ of Pathans.—
***Uniform*—Blue. *Facings*—Scarlet.**

COLONEL—Major-General Maitland Cowper, C.B., C.I.E.

1st Commn. or date of entering service.	Names and Rank.	Army rank.	Present appointment in Regiment.	Remarks.
	Commandant.			
3 Sep. 92	Ricketts, Lt.-Col. R. L., *p.s.c.* (Q)	8 Mar. 18	8 Mar. 18	
	Squadron Commanders. (4)			
17 Jan. 95	*Strong, Bt-Lt.-Col. A.D.*, D.S.O.	*3 June 18*	*8 Mar. 18*	*2nd-in-Comd. with 151st Ind. Inf.*
2 Aug. 99	Peters, Maj. J.	1 Sept. 15	24 Dec. 11	With 1-143rd. Infy.
17 Jan. 00	Chambers, Maj. P. R., D.S.O., *p.s.c.*	1 Sep. 15	13 Nov. 17	
20 Jan. 00	Kemmis, Maj. A. W. M., D.S.O.	1 Sep. 15	8 Mar. 18	
	Squadron Officers. (9)			
8 May 01	*Goldie, Maj. K. O.*	*8 May 16*	*1 Nov. 02*	
27 Aug. 02	Hodson, Maj. V. C. P.	27 Aug. 17	18 Feb. 04	
22 Oct. 02	*Foster, Maj. W. M. A.*, D.S.O.	*22 Oct. 17*	*5 Oct. 06*	*Bde-Maj., Rawalpindi Infy. Bde.*
9 Jan. 04	*Lamb, Capt. D.O.W. (temp. Maj.)*	*9 Jan. 13*	*25 May 06*	*D. A. A. G. A. Hdqrs. (temp.)*
24 Jan. 06	*Beresford, Capt. G. de la P.*, M.C.	*24 Jan. 15*	*24 Mar. 07*	*D. A. A. & Q. M. G.*
9 Oct. 07	Willoughby, Capt. M. G. P., M.C.	1 Sep. 15	11 Oct. 09	Adjt. 8 Jan. 15.
29 Jan. 10	Cahusac, Capt. S. D. N.	1 Sep. 15	3 May 12	
3 Sep. 10	Lawrence, Capt. R. T., M.C.	1 Sep. 15	28 Nov. 11	Adjt. (Offg.)
6 Sep. 11	*Stevens, Capt. C. F. L.*, M.C.	*6 Sep. 15*	*12 Dec. 12*	*With 9th Horse.*
14 Jan. 14	Bidie, Capt. A. G. C.	14 Jan. 18	14 Oct. 14	
8 Aug. 14	Whistler, Capt. A. R.	8 Aug. 18	12 Nov. 15	Attd.
3 July 15	Sheldon, Lt. H. G.	3 July 16	29 Dec. 17	Attd.
15 Nov. 15	Russell, Lt. O. B. P.	15 Nov. 16	17 Nov. 15	Attd.
20 May 16	Oates, Lt. A. T.	20 May 17	18 June 17	Attd.
....	Abercrombie, 2nd-Lt. D. E.	....	....	Attd.
....	Elkin, 2nd-Lt W. J.	....	....	Attd.
	Attached.			
24 Aug. 12	Welsh, Capt. R. H. W.	24 Aug. 16	....	16th Cav.
4 June 15	Wiles, Lt. B. H.	4 June 16	2 July 15	I. A. R. O.
13 Sep. 15	Lambert, Lt. D. H. C.	13 Sep. 16 25 Nov. 17	30 Nov. 17	I. A. R. O. (*on probn.*)
29 Oct. 15	Chater, Lt. C. W.	29 Oct. 16	8 Jan. 16	*I. A. R. O.* (*on probn.*)
22 Mar. 16	Wilks, Lt. E.	22 Mar. 17	11 Apl. 16	*I. A. R. O.* (*on probn.*)
29 Apl. 16	Moore, Lt. A. N. F., M.C.	29 Apl. 17	23 May 16	*I. A. R. O.* (*on probn.*)
14 Aug. 16	Drummond, Lt. J.G. (*actg. Capt.*)	14 Aug. 17	15 Mar. 17	I.A.R.O. Adjt.
17 Apl. 17	Morel, Lt. H. J. (*actg. Capt.*)	17 Apl. 18	1 Oct. 17	*I. A. R. O.* (*on probn.*) *Adjt. ½ Sq. Comdr. N. C. O. School, Belgaum.*
19 Apl. 17	Lowe, Lt. C. G.	19 Apl. 18	2 Feb. 18	*I. A. R. O.* (*on probn.*)
30 Apl. 18	Egan, 2nd-Lt. E. M.	30 Apl. 18	....	I. A. (*on probn.*)
28 Aug. 15	Sawyer, Lt. C. P. B.	28 Aug. 16	....	Dorset. R.-I. A. (*on probn.*)
12 Aug. 15	Charles, Lt. P. J.	12 Aug. 16	31 Aug. 18	Bucks Yeo. I. A. (*on probn.*)
24 May 15	Vernon, Lt. T. F.	....	7 Aug. 18	A. S. C. I. A. R. O. (*on probn.*)
8 Apl. 17	McCulloch, Lt. H. S.	8 Apl. 18	3 Aug. 18	I. A. (*on probn.*)
25 Apl. 17	Grant, 2nd-Lt. A. J.	25 Apl. 17	17 Aug. 18	Gord. Highrs—I. A. (*on probn.*)

		Jemadar.	Ressaidar.	Risaldar.	Risaldar Major.
	Risaldar-Major.				
9 July 87	Mahan Singh	1 July 02	7 July 14	1 July 15	16 Feb. 18
	Risaldars.				
18 July 90	Sant Singh (56)	1 Jan. 06	27 Apl. 10	29 Oct. 14	
3 Dec. 96	Labh Singh (56, 58, 66*a*)	16 Nov. 02	1 July 15	11 Apl. 16	A. D. C.
8 Sep. 05	Mal Singh (I) (45*a*, 50, 56, 63)	17 Dec. 09	14 Jan. 16	13 Aug. 16	
9 Mar. 91	Nur Khan. (45*a*)	28 Feb. 10	27 May 16	11 Sep. 16	
1 Apl. 94	Muhammad Ashruf Shah (56, 66*a*)	6 Jan. 10	29 Oct. 14	12 Sep. 16	
1 Apl. 90	Bishan Singh (56, 66)	4 Oct. 12	16 Feb. 18	......	
	Ganda Singh	......	......	......	Re-employed, 13 Aug. 16.
....	Abdul Ghafur Khan	....	....	....	Re-employed, 5 Feb. 18.
5 June 01	Muhammad Ali (56)	9 July 10	12 Sep. 16	13 Apl. 18.	

10th Duke of Cambridge's Own Lancers (Hodson's Horse)—*contd.*

Date of entering service.	Names and Rank.	Army rank.	Present appointment in Regiment.	Remarks.
	Ressaidars.			
1 Sep. 97	Muhammad Amin Khan (51, 66*a*)	27 May 10	9 July 14	
21 June 99	Sarup Singh (45*a*)	26 Apl 10	11 Apl. 16	
4 Feb. 00	Rai Singh (56, 66*a*)	12 Aug. 16		
2 Jan. 93	Jant Singh	1 July 15	13 Aug. 16	
1 Dec. 06	Amar Singh (56)	7 July 14	13 Apl. 18	
28 Dec. 00	Laurasib Khan	9 July 18		
	Jemadars.			
1 May 00	Hayat Khan (56)	27 May 13	Woordie-Maj.	
1 Aug. 97	Mal Singh (II)	7 Dec. 14		
16 Dec. 09	Sohan Singh	25 Jan. 15		
12 Aug. 03	Ishar Singh	14 Jan. 16		
28 Mar. 06	Matiullah Khan	14 Jan. 16		
21 June 03	Labh Singh (56)	12 Mar. 16		
10 Oct. 09	Jahandad Khan	11 Apl. 16		
16 Apl. 12	Chain Singh	16 May 16		
1 Nov. 96	Hari Singh	12 July 16		
22 Aug. 03	Khazan Singh	13 Aug. 16		
8 Sep. 11	Nawab Khan	13 Aug. 16		
16 May 11	Muhammad Ayub Khan	17 Sep. 16		
1 Jan. 03	Fauja Singh	1 Feb. 17		
8 Apl. 87	Lakhmi Das	1 Mar. 17		
15 Dec. 01	Mal Singh (III)	25 Aug. 17		
10 Dec. 01	Waryam Singh	25 Aug. 17	With 9th Horse.	
28 Dec. 01	Ram Singh	25 Aug. 17	With 19th Lrs.	
16 July 10	Saidan Shah	1 Sep. 17		
29 Dec. 00	Dheramal	3 Dec. 17		
21 July 10	Bhagwan Singh	21 Dec. 17		
3 Nov. 91	Kishorilal	22 Feb. 18		
4 May 02	Ghulam Jafir Shah	13 Apl. 18		
16 Mar. 04	Sultan Mohd.	3 Dec. 17	With 9th Horse.	
July 18	Sayyed Aftab Shah	9 July 18		

11th King Edward's Own Lancers (Probyn's Horse).

Raised at Lahore in 1857, by Capt. F. Wale, and was at first sometimes known as 'Wales's Horse.' Became the 1st Regiment of Sikh Irregular Cavalry, Sept. 1857; at a later period was known as 'Probyn's Horse'; became the 11th Regiment of Bengal Cavalry, 1861; the 11th Regt. of Bengal Cavalry (Lancers), 1864; the 11th Regt. of Bengal Lancers, 1874; the 11th (the Prince of Wales' Own) Regiment of Bengal Lancers, 1876; the 11th (the Prince of Wales's Own) Bengal Lancers, 1901; the 11th Prince of Wales's Own Lancers, 1903; the 11th Prince of Wales's Own Lancers (Probyn's Horse), 1904; present designation, 1st January 1906.

THE PLUME OF THE PRINCE OF WALES.

THE ROYAL AND IMPERIAL CYPHER OF KING EDWARD VII.

"Lucknow" "Taku Forts" "Pekin, 1860" "Ali Masjid" "Afghanistan, 1878-79" "Chitral" "Malakand" "Punjab Frontier."

Composition—2 Squadrons of Sikhs, 1 of Dogras, ½ of Punjabi Musalmans, ½ of Pathans.

Uniform—Blue. *Facings*—Scarlet.

COLONEL-IN-CHIEF—THE KING.

COLONEL—General *Rt. Hon. Sir* Dighton M. Probyn, VC, 𝔐., G.C.B., G.C.S.I., G.C.V.O., I.S.O., *Eq.* . . 13 May 1904.

1st Commn. or date of entering service.	Names and Rank.	Army rank.	Present appointment in Regiment.	Remarks.
	Commandant.			
8 Oct. 90	Norman, Bt.-Col. W. H., D.S.O., *p.s.c.*	3 June 18	4 July 16	G. S. O., 1st gr.
	Squadron Commanders. (4)			
29 Nov. 90	Sarel, Lt.-Col. G. B. M. (Q)	29 Nov. 16	4 July 16	2nd in-Comd., Comdt. (*temp.*)
22 Jan. 96	*Orr, Bt.-Col. G. M.,* D.S.O., *p.s.c.*	*3 June 18*	*29 Dec. 11*	*G. S. O., 1st gr. 5th Dn.*
4 Aug. 97	Anderson, Maj. R. H. . . .	4 Aug. 15	29 Dec. 11	2nd-in-Comd. (*temp.*)
4 May 98	*Rennick, Maj. A. de C.* . . .	*1 Sep. 15*	*29 Dec. 11*	
	Squadron Officers. (9)			
17 Jan. 00	*Robertson, Maj. D. E. (temp. Lt.-Col.)*	*1 Sep. 15*	*1 Aug. 02*	
8 Jan. 01	Hartley, Maj. A. F., D.S.O. . .	8 Jan. 16	10 May 05	
29 Aug. 06	*Macdonald, Bt.-Maj. H,* D.S.O.	*3 June 17*	*10 Nov. 07*	*Bde. Maj.*
19 Jan. 07	*Boyle, Capt. C. A.* [L] . . .	*1 Sep. 15*	*19 Mar. 08*	
17 Aug. 07	*Sherston, Capt. J. R. V.,* D.S.O., M.C.	*1 Sep. 15*	*21 Nov. 08*	*Bde Maj., 1st Cav. Bde.*
6 Sep. 11	Prinsep, Capt. E. S. MacL. . .	6 Sep. 15	19 Dec. 12	
6 Sep. 11	Campbell, Capt. I.	6 Sep. 15	8 Nov. 12	
24 Aug. 12	May-Somerville, Capt. J. A. C. .	24 Aug. 16	16 Oct. 13	
14 Jan. 14	Taylor, Capt. J. H.	19 Jan. 18	12 Oct. 14	
1 Oct. 14	*Hanwell, Capt. J. O.*	*1 Oct. 18*	*15 Dec. 14*	
15 Nov. 15	Leggatt, Lt. H. A.	15 Nov. 16	18 Nov. 15	Attd.
27 Oct. 16	*Baldwin, Lt. P. G. M.* . . .	*27 Oct. 17*	*11 Apl. 17*	*Attd.*
20 Dec. 17	Hog, 2nd-Lt. R. D.	20 Dec. 17	21 Mar. 18	Attd. A.D.C., G.O.C., 16th Div.
20 Dec. 17	ffolkes, 2nd-Lt. E. J. P. B. . .	20 Dec. 17	25 Mar. 18	Attd.
19 Jan. 17	Evetts, Lt. D. W.	19 Jan. 17	11 July 17	Attd.
20 June 16	Loyd, 2nd-Lt. E. B. K. . . .	20 June 16	5 Aug. 17	Attd.
3 Dec. 17	Hanna, Lt. A. L.	3 Dec. 17	9 Nov. 17	Aust. Horse, I. A.
23 Apl. 18	Stewart, 2nd-Lt. C. D. A. . .	23 Apl. 18	20 Oct. 18	Attd.
	Attached.			
22 Apl. 03	*Acland, Capt. H. E. P. D.* . .	. . .	*14 Apl. 16*	
13 Dec. 15	Bostock, Capt. J. A.	13 Dec. 16	17 May 18	Westminster Dns. I. A. (*on probn.*)
15 June 16	Elsworth, Lt. C. W. T. . . .	15 June 17	28 June 16	I. A. R. O.
15 Sep. 16	Oaten, Lt. E. F.	15 Sep. 17	23 Sep. 16	I. A. R. O.
3 Nov. 16	Pender, Lt. W. S. (*Actg. Capt.*)	3 Nov. 16	9 Nov. 17	Aust. L.H. I.A., (*on probn.*), Adjt.
15 Nov. 17	*Dickinson, Lt. W. G.* . . .	*15 Nov. 17*	*28 Nov. 17*	*I. A. R. O. Remt. Depôt.*
17 Apl. 17	Todd, Lt. H. J.	17 Apl. 18	15 Mar. 17	I. A. R. O. Qr. Mr.
1 Feb. 18	Ferguson, 2nd-Lt. J. M. . . .	1 Feb. 18	5 Dec. 17	Res. Cav. I. A. (*on probn.*)
15 Dec. 17	Neale, 2nd-Lt. F. C. L. . . .	15 Dec. 17	14 Dec. 17	M. G. Corps, I. A. (*on probn.*)
	Manders, Capt. H. C. . . .	. . . .	30 Apl. 18	I. A. R. O.
18 Dec. 17	Butt, 2nd-Lt. G. F.	18 Sep. 18	7 July 18	I. A. (*On probn.*)
31 Aug. 18	Wilkinson, 2nd-Lt. A. S. . .	31 Aug. 18	3 Sep. 18	I. A.
18 Sep. 18	Albert, 2nd-Lt. E. F.	18 Sep. 18	2 Oct. 18	I. A. R. O.

Risaldar-Major.

		Jemadar.	Ressaidar.	Risaldar.	Risaldar-Major.	
11 Feb. 89	Muhammad Aslam Khan, *Sardar Bahadur.*	2 Jan. 96	1 Nov. 03	1 Nov. 09	1 Mar. 17	*Order of Br. I. 1st Cl.*

11th King Edward's Own Lancers (Probyn's Horse)—*contd.*

Date of entering service.	Names and Rank.	Army rank.			Remarks.
	Risaldars.				
10 Mar. 01	Jaswant Singh	10 Mar. 01	1 Nov. 06	1 May 13	
1 Oct. 89	Bichitar Singh	1 Nov. 03	1 Nov. 09	10 Sep. 14	
1 Sep. 02	*Rab Nawaz Bahadur*	*1 Jan. 11*	*1 Mar. 14*	*1 Mar. 17*	*Order of Br. I., 2nd Cl.*
20 Jan. 94	Wassawah Singh (56, 58)	1 July 12	15 May 14	1 Mar. 17	
16 Mar. 00	Ram Singh (45a, 56)	9 Jan. 11	10 Sep. 14		
1 Dec. 94	Kahan Singh (58)	1 Nov. 09	7 Dec. 14	30 July 18	
	Ressaidars.				
1 Feb. 98	Nadir Ali Khan	9 Jan. 08	14 Nov. 16		
22 June 97	Nagina Singh	13 May 16	23 Dec. 16		
6 Nov. 93	Raundhi Singh (45a, 56, 58)	1 Nov. 12	1 Mar. 17		
8 Aug. 91	Prem Sing, I.O.M.	1 May 13	27 July 17		
1 Sep. 98	*Mir Jaman Khan*	*2 Mar. 14*	*23 Sep. 17*		*Consular Guard, Persia.*
1 Dec. 06	*Shah Nawaz Khan*	*15 May 14*	*27 Oct. 17*		*With 19th Lancers.*
1 Sep. 99	Bhagga Singh	15 Oct. 14	23 Sep. 17		
1 Oct. 00	Sayyid Khan (51)	16 May 16	23 Sep. 17		
. . . .	Muhammad Zaman	. . .	30 July 18		
2 Apl. 01	Dalip Singh	10 Sep. 14	25 Aug. 17		
	Jemadars.				
1 Nov. 06	*Waryam Singh*	*1 Jan. 14*			
15 Mar. 02	Bachan Singh	26 Mar. 14			
2 Apl. 01	Dhalip Singh	10 Sep. 14			
2 Jan. 94	Partab Singh	29 Oct. 14			
1 Nov. 99	Fauja Singh (45a, 55)	29 Oct. 14			Woordie-Major.
8 Jan. 97	Sawan Singh	27 Apl. 15			
1 Nov. 94	Mula Singh (58)	24 Dec. 15			
29 July 97	Rassia Ram (56)	21 Dec. 15			
2 Mar. 97	Bur Singh	10 Feb. 16			
1 Dec. 97	Kanshi Ram (56, 58)	2 Mar. 16			
7 Apl. 06	Bhagwant Singh (52)	12 Mar. 16			
4 Apl. 01	*Fateh Khan*	*18 June 16*			*With 7th Camel Corps.*
16 Oct. 93	*Arjan Singh*	*12 July 16*			
1 June 97	Chain Singh	16 Sep. 16			
1 Oct. 9[illegible]	Chattar Singh	6 Dec. 16			
1 Jan. 94	*Bhan Singh*	*20 Mar. 17*			
6 June 99	Mit Singh	23 Sep. 17			
4 Sep. 04	Tek Singh	23 Sep. 17			
1 Jan. 09	Lahrasib Khan	23 Sep. 17			
	Jagat Singh	1 June 18			
2 June 03	Balwant Singh	26 Apl. 18			
. . . .	Saidullah Khan	30 July 18			

12th Cavalry.

Raised at Lahore in 1857 by Capt. P. R. Hockin as the 2nd Regt. of Sikh Irregular Cavalry. Became the 12th Regiment of Bengal Cavalry in 1861; the 12th Bengal Cavalry in 1901; present designation in 1903.

"Abyssinia" "Peiwar Kotal" "Charasiah" "Kabul, 1879" "Afghanistan, 1878—80."

Composition—2 Squadrons of Sikhs, 1 of Dogras, 1 of Punjabi Musalmans.

Uniform—Blue. *Facings*—Blue.

1st Commn. or date of entering service.	Names and Rank.	Army rank.	Present appointment in Regiment.	Remarks.
	Commandant.			
22 Aug. 88	Humfrey, Lt.-Col. F. G. C.	22 Aug. 14	8 Feb. 16	*Lv., ex I. 6.*
	Squadron Commanders. (4)			
14 Sep. 93	Deane, Bt.-Lt.-Col. D., D.S.O., p.s.c. (Q) [*l*].	1 Jan. 17	8 Feb. 16	2nd-in-Comd. Comdt. *Offg.*
28 Nov. 94	Hesketh, Maj. W., D.S.O. (*Actg. Lt.-Col.*).	28 Nov. 12	1 Mar. 14	With 2-130th Baluchis.
20 Jan. 97	Baird, Bt.-Lt.-Col. H. B. D., C.M.G., D.S.O., *p. s. c.*	1 Jan. 17	8 Feb. 16	*G. S. O., 1st Gr., 4th Dn.*
22 Jan. 98	*Muspratt, Bt.-Lt.-Col. S. F.,* D.S.O., *p. s. c.*	*3 June 17*	*9 Feb. 16*	
	Squadron Officers. (9)			
4 May 01	*Fell, Maj. E. A. H.*	*22 Apl. 18*	*4 Oct. 05*	
4 Jan. 01	Shakespear, Maj. H. T.	8 Jan. 16	1 Aug. 02	
22 Apl. 03	Henry, Maj. W. R. P.	22 Apl. 18	14 Mar. 07	
29 Aug. 06	Watson, Bt.-Maj. T. N., M. C.	3 June 18	11 June 10	2nd-in-Comd.
15 Aug. 14	Davidson, Capt. J. W.	15 Aug. 18	8 Nov. 15	
19 Jan. 07	Sheepshanks, Capt. R. H., D.S.O.	1 Sep. 15	23 Dec. 09	*Lv., ex Ind., m. c.*
15 Nov. 15	Nadin, Lt. G.	15 Nov. 16	18 Nov. 15	
18 Apl. 16	Thomason, Lt. A. D. F.	18 Apl. 17	23 Apl. 16	With 13th Lrs.
29 June 16	Maconochie, Lt. A. E. F.	29 June 17	5 July 16	Attd.
27 Oct. 17	Winckley, Lt. S. F.	27 Oct. 18	5 Nov. 17	Attd.
31 Jan. 18	Davy, 2nd-Lt. H. N.	31 Jan. 18	5 Feb. 18	Attd.
31 Jan. 18	Noel, 2nd-Lt. E. H. F.	31 Jan. 18	5 Feb. 18	Attd.
31 Jan. 18	Welsh, 2nd-Lt. T. V.	31 Jan. 18	5 Feb. 18	Attd.
23 Apl. 18	Finch, 2nd-Lt. K. H.	23 Apl. 18	27 Oct. 18	Attd.
30 July 15	Evennett, Lt. H.C.	30 July 16	4 Feb. 17	Attd.
1 Feb. 18	Mooney, 2nd-Lt. H. F.	1 Feb. 18	30 Nov. 17	Attd.
	Attached.			
8 Aug. 14	Gordon, Capt. D. St. V.	8 Aug. 18	4 June 17	39th Horse.
14 Jan. 15	Cowell, Lt. H. C.	....	11 Feb. 18	I. A. R. O.
5 Feb. 15	Harrold, Lt. J. H.	5 Feb. 16	15 Nov. 16	I. A. R. O. Adjt.
28 May 15	Trousdell, Lt. J. P. (*Actg. Capt.*)	28 Nov. 15	27 Nov. 15	Adj., I. A. R. O.
8 Oct. 15	Elmes, Lt. J. P. J.	8 Oct. 16	16 Oct. 15	I. A. R. O. *Lv., ex I.*
4 Mar. 16	Smith, Lt. A. K.	4 Mar. 17	4 Feb. 17	I. A. R. O.
29 Apl. 16	Marriott, Capt. E. F.	29 Apl. 17	25 Oct. 16	I. A. R. O.
21 Aug. 16	Nicholson, Lt. J. W.	21 Aug. 17	4 June 17	I. A. R. O. Qr.-Mr.
7 Oct. 16	Wilson, Lt. C. C.	7 Oct. 17	10 Sep. 17	I. A. R. O.
23 May 17	Harper, Lt. K. J.	23 May 18	15 Sep. 17	I. A. R. O.
21 Sep. 17	Salt, Lt. A. C. J.	21 Sep. 18	7 July 18	I. A. (*on probn.*)
25 Sep. 16	Kennedy, Lt. K. M.	25 Sep. 17	1 Mar. 17	I. A. R. O.
19 Apl. 15	deFrieville, Lt. G. V.	19 Apl. 15	26 Oct. 13	I. A. R. O.
15 Apl. 17	Lowndes, Lt. R. C.	5 Apl. 18	29 Aug. 18	I. A. R. O.

12th Cavalry—*contd.*

Date of entering service.	Names and Rank.	Army rank. Jemadar.	Ressaidar.	Risaldar.	Risaldar-Major.	Remarks.
	Risaldar-Major.					
12 Nov. 92	Sukhdayal Singh	1 Jan. 96	22 Aug. 00	1 Feb. 11	20 Mar. 17	
	Risaldars.					
26 Sep. 01	Parmod Singh (56)	26 Sep. 01	15 Oct. 07	7 Nov. 14		
1 Jan. 02	Gonda Singh	1 Jan. 02	1 Jan. 08	27 Oct. 15		
22 July 91	Ghansar Singh Bahadur	23 Aug. 03	1 Feb. 13	27 Oct. 15		*Order of Br. I., 2nd Cl.*
17 July 92	Udham Singh (56)	17 July 04	1 Apl. 14	8 Mar. 16		
. . . .	Ganda Singh (*Retd.*)	. . . .	. . . .	1 June 16		Re-employed.
21 June 03	Ashak Ali Khan (45a, 56) I.D.S.M.	21 June 03	7 Nov. 14	5 Jan. 17		
20 Oct. 94	Bakhshish Singh (56)	21 Jan. 07	27 Oct. 15 8 Mar. 16	18 Aug. 17		
1 Dec. 07	*Khan Shirin Khan*	*1 Apl. 11*	*2 Aug. 18*			
26 Feb. 05	Jagat Singh (56)	26 Feb. 05	11 Apl. 16	11 Aug. 18		With 18th Lrs.
	Ressaidars.					
14 Oct. 93	Santa Singh (56, 66a)	15 Oct. 07	27 Oct. 15			
14 Aug. 94	Kishun Singh (45a, 56)	1 Jan. 08	27 Oct. 15			
16 Oct. 01	Khan Dilawar Khan (52, 56)	1 Feb. 11	5 Jan. 17			
8 Apl. 93	Gujar Singh	1 Feb. 13	5 Jan. 17			
	Jemadars.					
16 Oct. 07	Kartar Singh (50, 52)	29 Oct. 14				Woordie-Maj., 8 Feb. 16.
16 Apl. 06	Shamsher Ali Khan	29 Oct. 14				
28 Jan. 91	Partab Singh	7 Nov. 14				
1 Dec. 00	*Nadir Khan*	*22 Mar. 15*				*Supy.*
1 May 14	Muhammad Akbar Khan	1 July 15				
1 Jan. 04	Bishan Singh	18 Oct. 15				
5 Mar. 91	Karshasaf Khan	27 Oct. 15				
1 July 97	Ram Singh	27 Oct. 15				
8 Aug. 94	Karam Singh	16 Feb. 16				
16 Nov. 95	Rup Singh	8 Mar. 16				
1 Jan. 02	Partap Singh, I.O.M. (51, 56)	11 May 16				
7 Nov. 99	Muhammad Akbar Khan	6 Oct. 16				
11 Nov. 94	Labh Singh	5 Jan. 17				
26 Dec. 00	Bhan Singh, I.D.S.M.	5 Jan. 17				
16 Nov. 97	Nihal Singh	1 Aug. 17				
16 Nov. 11	Thakur Singh	18 Aug. 17				
8 Apl. 92	Wadhawa Singh	11 Sep. 17				
7 Feb. 01	Lachman Singh	23 Nov. 17				
. . . .	Mauladad Khan	11 Sep. 18				

13th Duke of Connaught's Lancers (Watson's Horse).

Raised at Lahore in 1858, by Lt. H. C. Cattley and Lt. J. Watson, as the 4th Regiment of Sikh Irregular Cavalry. Became the 13th Regt. of Bengal Cavalry, 1861; the 13th Regt. of Bengal Cavalry (Lancers), 1864; the 13th Regt. Bengal Lancers, 1874; the 13th (Duke of Connaught's) Regt. of Bengal Lancers, 1884; the 13th (Duke of Connaught's) Bengal Lancers, 1901; the 13th Duke of Connaught's Lancers, 1903; present designation 1904.
"Afghanistan, 1878-80" "Tel-el-Kebir" "Egypt, 1882" "Punjab Frontier."

Composition—2 Squadrons of Punjabi Musalmans, 1 of Sikhs, 1 of Dogras.

Uniform—Blue. *Facings*—Scarlet. *Lace*—Silver.

COLONEL-IN-CHIEF—Field Marshal *His Royal Highness* ARTHUR WILLIAM PATRICK ALBERT *Duke of* CONNAUGHT and STRATHEARN, K.G., K.T., K.P., G.C.B., G.C.S.I., G.C.M.G., G.C.I.E., G.C.V.O., G.B.E., A. D. C. . . . . 13 May 1904.
COLONEL—General *Sir* John Watson, V.C., K., G.C.B. . . 13 May 1904.

1st Commn. or date of entering service.	Names and Rank.	Army rank.	Present appointment in Regiment.	Remarks.
	Commandant.			
21 Dec. 89	Mackenzie, Lt.-Col. C., D.S.O. .	21 Dec. 15	20 May 16	
	Squadron Commanders. (4)			
27 July 88	*Cobbe, Bt.-Col. H. H.,* C.M.G., D.S.O. (Q).	*3 June 18*	*9 Apl 14*	*2nd-in-Comd.*
2 Jan. 93	Gordon, Bt Lt.-Col. G. A. S. (Q)	1 Jan. 18	13 Mar. 14	With 43rd Cav.
5 Aug. 96	Watson, Maj. J. H.	5 Aug. 14	16 Apl. 14	*Lv., ex I., m. c., 6 mos.*
7 May 98	Jennings, Maj. R. D.	1 Sept. 15	24 Mar. 18	
	Squadron Officers. (9)			
5 Jan. 01	Jameson, Maj. A. McC. .	1 Sept. 15	1 Mar. 06	S. C. (*temp.*)
27 Aug. 02	*Moody, Maj. F. H.,* M.C. (*temp. Lt.-Col.*)	*27 Aug. 16*	*16 Oct. 08*	*A. Q. M. G., A. Hdqrs.*
9 Jan. 04	*Haviland, Capt. L. P.* . . .	*9 Jan. 13*	*11 Jan. 08*	
29 Aug. 06	Connell, Capt. V. J. A. . . .	29 Aug. 15	17 Mar. 08	
6 Oct. 06	*Pott, Capt. D.,* D.S.O., M.C. .	*1 Sept. 15*	*26 Apl. 09*	*With Sig. Serv.*
9 Sep. 08	Mayne, Capt. A. G. O. M., D.S.O.	1 Sept 15	12 Jan. 10	Adjt., 8 May 15.
27 Jan. 09	Watson, Capt. G. N. .	1 Sept. 15	25 July 14	
8 Aug. 14	Harding, Capt. K. O'B. (*Actg. Capt.*)	8 Aug. 18	9 Nov. 15	Attd.
4 Feb. 11	Stubbs, Capt. J. B. (207*a*) . .	1 Sept. 15	1 Nov. 17	Attd.
6 Dec. 17	Bowe, Lt. R. W.	6 Dec. 18	7 Nov. 17	Attd.
	Attached.			
5 Mar. 15	Walsh, Lt. G. J. T. (*Actg. Capt.*)	5 Mar. 16	1 Aug. 16	I. A. R. O.
20 Aug. 15	Lovell, Lt. W. D.	20 Aug. 16	31 Aug. 15	I. A. R. O. Qr.-Mr. (*temp.,*
4 Oct. 15	Bartlett, Lt. P. J. A.	4 July 17	9 Jan. 18	Shrops. L. I., I. A. (*on probn.*)
20 Nov. 14	Weaver, Lt. J. Y.	20 Aug. 16	26 Jan. 18	A. S. C. I. A. (*on probn.*)
5 Feb. 18	Armstrong, 2nd.-Lt. C. F. . .	5 Feb. 18	12 Feb. 18	I. A. R. O.
26 Sep. 14	Stuart-William, Lt. W. L. . .	26 June 16	4 July 18	I. A. (*on probn.*)
1 Nov. 15	Schmidt, Lt. C. T.	1 Aug. 17	8 July 18	R. Highrs., I. A. (*on probn.*)
7 July 16	White, Lt. G. F., M.C.	7 Apl. 18	3 July 18	Notts. and Derby. R., I. A. (*on probn.*)
23 Apl. 18	Aitken, 2nd-Lt. R. G.	23 Apl. 18	28 Oct. 18	I. A. (U. L.).
30 Apl. 15	Carpenter, Lt. P. H. (Actg. Capt.)	30 Apl. 16	19 Sep. 18	I. A. R. O.
18 Apl. 16	Thomason, Lt. A. D. F. . . .	18 Apl. 17	. . . .	
3 Aug. 98	Hay, Maj. A. S.	1 Sep. 15	7 Apl. 04	

		Jemadar.	Ressaidar.	Risaldar.	Risaldar-Major.
	Risaldar-Major.				
20 Nov. 90	Fazil Husain, I.O.M. (56) . . .	23 Oct. 06	1 Dec. 14	17 Aug. 16	21 Nov. 17.
	Risaldars.				
26 Oct. 89	Pahlwan Khan	1 Nov. 03	1 Aug. 12	21 Dec. 15	
6 Dec. 00	Jit Singh (45*a*, 56, 58, 66*a*) . .	16 July 02	29 Oct. 14	27 July 17	
26 Oct. 03	Ajit Singh (45*a*, 56, 66) . . .	26 Oct. 03	10 June 15	5 Oct. 17	
19 Oct 98	Munsabdar Khan, M.C. (56, 58).	1 Aug. 12	21 Dec. 16	3 Nov. 17	
10 Apl. 00	Sohan Singh	10 June 15	6 Mar. 17	23 Nov. 17	

13th Duke of Connaught's Lancers (Watson's Horse)—*contd.*

Date of entering service.	Names.	Army rank.		Remarks.
	Ressaidars.			
24 May 96	Bhola Singh	24 Jan. 15	21 Dec. 15	
1 Dec. 97	Ganda Singh	17 Apl. 15	6 Sep. 17	
1 Jan. 06	Burhan Ali (56, 52) I.D.S.M. . .	17 Apl. 15	5 Oct. 17	
1 June 99	Rumal Singh	12 Feb. 16	5 Oct. 17	
12 Feb. 05	Narain Singh	1 Mar. 16	3 Nov. 17	
12 Mar. 16	Tikka Brij Mohan Pal . . .	12 Mar. 16	6 Nov. 17	On probn.
8 Nov. 92	Jiwan Singh	17 Dec. 16	6 Nov. 17	
12 Sep 00	Jahangir Khan (56, 66*a*) . . .	21 Dec. 16	23 Nov. 17	
15 Apl. 07	Teja Singh	11 Apl. 16	26 Apl. 18	With 19th Lrs.
	Jemadars.			
2 Mar. 99	Dost Muhammad (45*a*, 56) . .	1 Sep. 11		
14 Dec. 99	Harbhaj	17 Apl. 15		
1 Nov. 99	Lal Khan (*Ag. Ressdr.*) . . .	22 Oct. 15		With 59th Camel Corps.
28 May 98	Gul Muhammad	11 Apl. 16		
28 July 95	Hazara Singh	27 June 16		
29 Dec. 92	Nigah Ali Shah	27 Jan. 17		
1 Jan. 00	Jahan Khan	16 Feb. 17		
1 Apl. 15	Jagpal Singh, I.D.S.M. . . .	22 Mar. 17		
15 Mar. 07	Jhanda Singh	30 Apl. 17		
1 July 00	Firoz Khan (56)	7 July 17		
15 Apl. 02	Muhammad Khan	3 Nov. 17		
1 July 99	Mehr Singh	6 Nov. 17		
30 Apl. 14	Abdur Rahman	23 Nov. 17		
	Ganesha Singh	. . . .		Re-employed, 11 June 17.
	Shahamad Ali	28 Sep. 18		

14th Murray's Jat Lancers.

Raised at Aligarh in 1857, by Capt. J. I. Murray, as the Jat Horse Yeomanry. Became Murray's Jat Horse, 1859; the 14th Regiment of Bengal Cavalry, 1861; the 14th Regt. of Bengal Cavalry (Lancers), 1864; the 14th Regt. of Bengal Lancers, 1874; the 14th Bengal Lancers (Murray's Jat Horse), 1901; present designation, 1903.

"Charasiah" "Kabul, 1879" "Afghanistan, 1878--80."

Composition—4 Squadrons Jats.

Uniform—Blue. *Facings*—Scarlet. *Puggri*—Scarlet.

1st Commn. or date of entering service.	Names and Rank.	Army rank.	Present appointment in Regiment.	Remarks.
	Commandant.			
18 May 92	Souter, Lt.-Col. H. M. W., C.M.G., D.S.O.	14 Nov. 16	14 Nov. 16	
	Squadron Commanders. (4)			
16 Jan. 95	*Rivett-Carnac, Maj. J. S., p. s. c. (temp. Lt.-Col.) (Q)*	*16 Jan. 13*	*14 Nov. 16*	*2nd-in-Comd., G. S. O., 1st gr. 9th Divn.*
16 Feb. 98	Bally, Maj. F. K.	1 Sep. 15	18 Apl. 08	
25 Jan. 99	Hewlett, Bt. Lt.-Col. E. A.	3 June 17	30 Oct. 14	
21 Apl. 00	*Gell, Maj. P. F. (temp. Lt.-Col.)*	*1 Sept. 15*	*29 Nov. 16*	*Jodhpur I. S. Lrs.*
	Squadron Officers. (9)			
18 Aug. 00	*Maitland, Maj. G. R.*, D.S.O. *p. s. c.*	*1 Sept. 15*	*25 Apl. 05*	*Asst. Embarkation Comdt.*
23 Oct. 01	*Hammersley-Smith, Maj. R. H.*	*14 Feb. 16*	*11 Mar. 05*	
4 Dec. 01	Sutherland, Maj. R. O., D.S.O.	4 Dec. 16	1 Nov. 05	*s.c.* p.t.
21 Jan. 03	Bromilow, Maj. D. G., D.S.O.	21 Jan. 18	10 Dec. 04	*s.c.* p.t.
22 Mar. 03	Young, Maj. M. S.	22 Mar. 18	3 June 06	Army Remt. Depôt, Babugarh.
10 Oct. 03	*Malet, Maj. W. DeC.*	*10 Oct. 18*	*6 Oct. 06*	*Staff Capt., 5th Mhow Div.*
5 Aug. 05	Jopp, Capt. C. H. K. (*Actg. Maj.*)	5 Aug. 14	8 Dec. 06	With 2-94th Inf.
7 Aug. 07	Gibson, Capt. J. G., M. C.	1 Sept. 15	14 Nov. 08	s. c. *p.t.*
6 Sep. 11	Marriott, Capt. C. F.	6 Sept. 15	26 Jan. 17	
16 Dec. 14	Bradshaw, Lt. J. R. L. (*Actg. Capt.*)	16 Dec. 15	13 Mar. 15	Adjt. (*Offg.*).
12 Sep. 17	Dyer, Lt. G. M.	12 Sep. 18	10 Dec. 17	Attd.
21 Dec. 17	Huth, F. L.	21 Dec. 17	4 June 18	Attd.
30 Mar. 16	*Hargreaves, Lt. P. S.*	*30 Mar. 17*	4 Apl. 18	Attd. *A. P. M. (temp.)*
7 Sep. 17	Blackham, 2nd-Lt. J. C.	7 Sep. 17	8 Aug. 18	I. A. R. O.
5 May 15	Turner, Lt. E. L.	5 May 16	. . . .	Attd.
12 July 15	Gradidge, Lt. J. H.	12 July 16	24 Oct. 17	Attd.
4 May 17	Stanfield, Lt. E. A. F.	4 May 17	5 Nov. 17	Attd.
	Attached.			
7 Feb. 12	Maur, Capt. R. W. M. St.	7 Feb. 16	3 Aug. 18	14th Hrs.
26 Aug. 14	Elgood, Lt. C. L.	26 May 16	4 July 18	Midd'x R.
20 Nov. 14	Wylde, Lt. E. W.	18 Feb. 15	4 Mar. 17	I. A. R. O.
14 May 15	*Willis, Lt. R. D.*	*14 May 16*	*28 May 15*	*I. A. R. O.*
15 June 16	Nicholl, Lt. E. McK. (*Actg. Capt.*)	15 June 17	29 June 16	I. A. R. O. Qr.-Mr. (*temp.*)
14 Aug. 16	Martin, Lt. O. M.	14 Aug. 17	26 Sep. 17	I. A. R. O.
24 Jan. 18	Norris, 2nd-Lt. J. W.	. . .	1 Aug. 18	
23 July 15	*Case, Lt. J. B.*	*23 July 16*	*14 Jan. 17*	*I. A. R. O. R. A. F.*
20 Feb. 16	*O'Brien, Lt. D. P.*	*20 Feb. 17*	*15 Oct. 16*	*I. A. R. O.*
5 Nov. 17	Inman, Lt. W. F.	5 Nov. 18	14 Mar. 18	I. A. R. O.
5 Nov. 17	Kilburn, Lt. J. M.	5 Nov. 18	14 Mar. 18	I. A. R.O.
28 Oct. 17	Totham, 2nd-Lt. W. M.	28 Oct. 17	27 Aug. 18	

		Jemadar.	Ressaidar.	Risaldar.	Risaldar-Major.	
	Risaldar-Major.					
14 July 92	Mansa Ram, I.D.S.M.	1 Nov. 03	1 Nov. 10	6 July 16	1 May 18	
	Risaldars.					
. . .	Phul Singh	. . . .	. . . .	. . . .		Re-employed, 1 May 15.
7 July 90	Khazan Singh (56)	1 June 03	1 Nov. 09	16 Nov. 15		
1 Jan. 94	Hargyan Singh, I.O.M., I.D.S.M.	16 Mar. 09	16 Mar. 13	31 Dec. 16		
25 Nov. 92	Kure Singh (56)	1 May 09	17 Apl. 15	11 Apl. 17		
1 Mar. 04	Sardar Singh (56)	31 Oct. 10	31 Dec. 16	12 Feb. 17		
21 July 94	Ram Nath	18 May 15	1 Apl. 17	6 Nov. 17		
18 Oct. 94	Rati Ram, I.O.M.	17 Apl. 15	12 Feb. 17	29 Oct. 17		
19 Dec. 96	Nanhe Singh	16 Oct. 15	22 Oct. 17	15 July 18		

14th Murray's Jat Lancers—*contd.*

Date of entering service.	Names.	Army rank. Jemadar.	Army rank. Ressaidar.	Remarks.
	Ressaidars.			
. . . .	Surat Ram	......	......	Re-employed 1 Oct. 15.
16 Jan. 99	Ram Singh	14 Jan. 16	25 Apl. 17	
1 Nov. 07	Mansa Ram (56), I.D.S.M.	16 Sep. 15	1 May 17	
1 May 99	Bhani Singh	21 June 16	6 Nov. 17	
16 Aug. 00	Daya Kishan (56)	16 Mar. 13	15 July 18	
	Jemadars.			
1 Mar. 99	*Lakhpat Singh*	*17 Apl. 15*		
18 Feb. 99	Jagat Singh	14 May 16		
11 Dec. 01	Hukam Singh	16 May 16		
1 July 03	Not Ram	1 Jan. 17		
21 July 98	Rup Chand, I.O.M.	29 Jan. 17		
1 May 07	Bharat Singh	6 Feb. 17		
18 Oct. 99	Bujan Singh	25 Mar. 17		
4 Apl. 99	Mohar Singh	1 Apl. 17		
1 May 99	Sohan Singh	2 Apl. 17		
1 May 01	Kure Singh	25 Apl. 17		
15 June 05	Pahlad Singh, I.O.M.	6 Oct. 17		
12 June 07	Indar Singh	29 Oct. 17		
12 July 04	Jug Lal, I.D.S.M.	4 Nov. 17		
4 Feb. 90	Nawazish Ali	6 Nov. 17		
3 Dec. 17	Sheo Karan	3 Dec. 17		
. . . .	Laik Ram	20 July 18		

15th Lancers (Cureton's Multanis).

Formed at Lahore in 1858, by Capt. C. Cureton, from six risallahs of Multani Pathans, raised in 1857 by Ghulam Hasan Khan, and was originally designated 'The Multani Regt. of Cavalry.' Became Cureton's Multani Regt. of Cavalry, 1860; the 15th Regt. of Bengal Cavalry, 1861; the 15th (Cureton's Multani) Regt. of Bengal Cavalry, 1871; the 15th (Cureton's Multani) Regt. of Bengal Lancers, 1890; the 15th (Cureton's Multani) Bengal Lancers, 1901; present designation, 1903.

"Afghanistan, 1878—80."

Composition—2 Squadrons of Musalmans (Multani Pathans and Musalmans of the Derajat and Cis-Indus) 2 Squadrons of Jats—

Uniform—Blue. *Facings*—Scarlet.

COLONEL—Major-General Guy A. Prendergast, ℳ. . 13 May 1904.

1st Commn. or date of entering service.	Names and Rank.	Army rank.	Present appointment in Regiment.	Remarks.
	Commandant.			
29 Nov. 90	Cheyne, Lt.-Col. A. Y.	29 Nov. 16	29 Nov. 16	
	Squadron Commanders. (4)			
3 Sep. 92	Corlett, Lt.-Col. J. S. (Q) ГL	3 Sep. 18	5 Nov. 17	2nd-in-Comd.
23 Sep. 96	Hallowes, Maj. J. H. (*temp. Lt.-Col.*)	23 Sep. 14	1 Aug. 14	Comdt. 41st Cav. Regt.
4 Jan. 99	*Baddeley, Maj. R. J. H.*, M.C.	*1 Sept. 15*	*13 Sep. 16*	*With 29th Lrs.*
17 Jan. 00	*Howell, Maj. R. H.*	*1 Sept. 15*	*20 Feb. 02*	*With R. A. F.*
	Squadron Officers. (9)			
11 Aug. 00	Underwood, Maj. G. F.	1 Sept. 15	19 Nov. 02	
10 Oct. 03	Clarke, Maj. A. C. K. S., M.C.	10 Oct. 18	10 Jan. 07	
5 Aug. 05	Barbour, Capt. K. D.	5 Aug. 14	10 Nov. 06	
1st Jan. 11	*Geard, Capt. R. C.*	*1 Sept. 15*	*16 Mar. 12*	
20 Jan. 12	Debrath, Capt. N. S.	20 Jan. 16	18 Jan. 16	
24 Aug. 12	*Powle, Capt. J. Le C.*	*24 Aug. 16*	*3 Nov. 13*	*S. Persian Mtd. Rifles.*
	Attached.			
4 Dec. 14	Henderson, Capt. W. L.	4 Dec. 17	23 May 15	I. A. R. O.
20 Nov. 14	Fraser, Lt. A. D.	9 Apl. 15	9 Apl. 15	I. A. R. O.
7 Mar. 15	Hulbert, Lt. M. L.	7 Mar. 16	22 May 18	R. N. Devon Hrs.—I. A. R. O. (*on probn.*)
4 Aug. 14	Kirby, Lt. H.	4 Apl. 15	24 May 18	Notts & Derby R.—I. A. (*on probn*).
31 Dec. 14	Mackay, Lt. D. S.	30 Sep. 16	24 Jan. 18	Yorks Dns. I. A. (*on probn.*)
14 Oct. 15	Wright, Lt. G. A.	14 Oct. 16	3 June 18	Linc. R.—I. A. (*on probn.*)
29 Jan. 16	Bastin, Lt. H. E., M.C.	29 Jan. 17	1 Aug. 18	A. I. F.—I. A. (*on probn.*)
15 June 16	Smith, Lt. G. A.	15 June 17	9 Nov. 16	I. A. R. O.
18 June 16	Simmons, Lt. F.	18 June 17	19 Mar. 17	I. A. R. O. Qr.-Mr.
9 Dec. 16	Collinson, Lt. C. D. H.	9 Dec. 17	24 Apl. 18	I. A. R. O.
12 Sep. 17	Shannon, Lt. E. C. B.	12 Sep. 18	29 Apl. 18	York. R.—I. A. (*on probn.*)
26 Sep. 17	Beatty, Lt. H.	26 Sep. 18	13 Aug. 18	I. A. R. O.
1 Oct. 17	Newsam, Lt. W. O.	1 Oct. 18	12 Dec. 17	I. A. R. O.
6 Feb. 18	Garrett, 2nd-Lt. J. T.	6 Feb. 18	6 Feb. 18	U. L., I. A.
....	Dawson, 2nd-Lt. A. V.	......	4 Sep. 18	Attd.
....	Gosling, 2nd-Lt. E. A. C.	......	10 Sep. 18	Lond.R.—I.A.R.O. (*on probn.*)

		Jemadar.	Ressaidar.	Risaldar.	Risaldar-Major.	
	Risaldar-Major.					
1 Mar. 91	Ali Ahmad Khan (56)	1 Aug. 99	1 Nov. 05	1 July 10	16 May 17.	
	Risaldars.					
16 June 00	Ata Muhammad Khan (56)	29 Jan. 02	20 Feb. 09	31 Jan. 15		
14 Jan. 89	*Muhammad Afzal Khan (56)*	*19 May 08*	*17 Dec. 13*	*14 Apl. 15*		*Army Remt. Depôt.*
24 Dec. 88	Karim Nawaz Khan (56)	1 Sep. 09	31 Jan. 15	4 Jan. 18		
18 Feb. 02	Risal Singh	23 Nov. 15	1 June 17	1 Apl. 18		
	Ressaidars.					
1 Dec. 98	Muhammad Nawaz Khan	20 Feb. 09	14 Apl. 15			
11 Feb. 98	Didari	20 Aug. 14	1 June 17			
13 Jan. 94	Shiudan	18 Sep. 14	1 June 17			
15 July 00	Naubat Singh	13 Sep. 16	1 June 17			

15th Lancers (Cureton's Multanis)—*contd.*

Date of entering service.	Names.	Army rank.	Remarks.
	Jemadars.		
1 Apl. 11	Malik Mihr Khan, I.O.M. (45*a*)	1 Apl. 11	
3 Mar. 90	Muhammad Afzal Khan	29 June 11	
3 Dec. 98	Haqdad Khan	9 Aug. 14	
1 Sep. 99	*Fateh Sher Khan*	*28 Nov. 14*	
1 Mar. 12	Jan Mahammad Khan	14 Apl. 15	
16 Apl. 13	Amir Abdul Rahman Khan	25 Apl. 15	
18 Oct. 99	Randhir Singh (56)	1 June 17	
1 Aug. 98	Rupa Singh (56)	1 June 17	
1 June 03	Sarup Singh	1 June 17	
1 Dec. 04	Maya Ram	1 June 17	Woordie-Maj., 4 Jan. 18.
1 Apl. 01	Sar ar Khan	3 Dec. 17	
1 May 07	Ram Sarup	4 Jan. 18	
16 Dec. 96	Abdul Majid Shah	8 Apl. 18	
	Khan Muhammad Khan	24 Sep. 18	

16th Cavalry.

Raised at Haldwani in 1857 as the 'Rohilkhand Horse' by Capt. F. G. Crossman; became 16th Bengal Cavalry, 1862; was disbanded, 1882; re-raised at Ambala in 1885, by Lt.-Col. G. C. Ross, as the 16th Regiment of Bengal Cavalry; became the 16th Regiment of Bengal Lancers in 1900; the 16th Bengal Lancers in 1901; present designation, 1903.

"China, 1900."

Composition—2 Squadrons of Sikhs, 1 of Dogras, 1 of Jats.—1 Addl. Squadron.

Uniform—Blue. *Facings*—Blue.

COLONEL—Lieutenant-General George T. Halliday . . . 6 January 1905.

1st Commn. or date of entering service.	Names and Rank.	Army rank.	Present appointment in Regiment.	Remarks.
	Commandant.			
9 Sep. 91	Mears, Lt.-Col. C. D., M.V.O.	3 June 16	10 Nov. 17	
	Squadron Commanders. (4)			
21 Sep. 89	*Holman, Bt.-Col. H. C.*, C.B., C.M.G., D.S.O., *p.s.c.* [L]. (Q) (*temp. Maj.-Genl.*)	*1 Jan. 16*	. . . .	*2nd-in-Comd. D. A. & Q. M. G.*
9 Aug. 99	Hill, Maj. H. W. D.	1 Sept. 15	5 Mar. 15	2nd-in-Comd. (*temp.*)
21 July 02	Digges LaTouche, Maj. C. H.	31 July 17	20 Apl. 15	
21 Jan. 03	*Dick-Lauder, Maj. J. N. D.*	*21 Jan. 17*	*10 Nov. 17*	*G. S. O., 2nd gr., 8th Dn.*
	Squadron Officers. (9)			
9 Jan. 04	*Lang, Bt.-Maj. W. H.*	*3 June 17*	*26 June 05*	Bde. Maj.
27 Jan. 04	*Vicary, Bt.-Maj. E. S.*	*3 June 17*	*10 Nov. 05*	*D. A. Q. M. G.*
19 Jan. 10	Whitford-Hawkey, Capt. E. H.	1 Sept. 15	20 June 11	With 15th Lrs.
18 Jan. 11	*Landale, Capt. C. H.*	*1 Sept. 15*	*8 Mar. 12*	With 22nd Cav.
20 Sep. 11	Gordon, Capt. A. D.	20 Sep. 15	9 Aug. 17	With 42nd Cav.
20 Jan. 12	*Dickinson, Capt. T. M.*	*20 Jan. 16*	*10 Apl. 13*	*With R. A. F.*
24 Aug. 12	*Welsh, Capt. R. H. W.*	*24 Aug. 16*	*13 Oct. 13*	
28 Aug. 14	Wordsworth, Lt. R. H.	28 May 16	7 Nov. 17	
9 Apl. 15	Hobbs, Lt. C. E. (*actg. Capt.*)	9 Jan. 17	27 Aug. 17	Adjt.
28 Oct. 17	Tatham, 2nd-Lt. A. M.	28 Oct. 17	7 Nov. 17	With 14th Lrs.
28 Nov. 17	Read, Capt. J.	28 Nov. 18	21 July 17	Qr. Mr. (*temp.*)
20 Oct. 17	Brooke, 2nd-Lt. A. E.	28 Oct. 17	7 Nov. 17	
31 Aug. 18	Warren, 2nd-Lt. P.	31 Aug. 18	4 Sep. 18	Attd.
	Attached.			
19 Aug. 03	Slacke, Capt. G. C.	19 Aug. 18	27 Dec. 14	Spl. Ser. Offr. I. S. Troops.
25 Aug. 17	Kanwar Amar Singh, Capt.	25 Aug. 17	15 July 18	2nd Lrs.
23 July 16	Crittall, Lt. P. C.	23 July 16	1 Feb. 18	A. S. C. I. A. (*on probn.*)
1 Mar. 15	Brown, Lt. J. R. R.	1 Dec. 16	1 Mar. 18	I. A. (*on probn.*)
11 June 15	Cruickshank, Lt. F. C. M.	11 June 16	12 Mar. 16	I. A. R. O. P. of W. Camp, Sumerpur.
25 Oct. 15	Wing, Lt. R.	25 July 16	29 Mar. 18	Notts. S. R. Yeo. I.A.R.O. (*on probn.*)
19 Jan. 17	Bigwood, Lt. W., M.C.	19 Jan. 17	19 Jan. 18	Worc. R. I. A. (*on probn.*)
20 July 16	Scott, Lt. M.	20 Apl. 17	12 Feb. 18	Highl. Cyclist Bn. I. A. (*on probn.*)
29 June 17	Witchell, Lt. H.	29 June 18	28 Feb. 18	11th Hrs. I. A. R. O. (*on probn.*)
25 Aug. 17	Hamilton, Lt. A. S., M.M.	25 Aug. 18	12 Feb. 18	Res. R. Hrs. I. A. R. O. (*on probn.*)
5 Nov. 17	Wilks, 2nd-Lt. B. St. C.	5 Nov. 17	3 Apl. 18	I. A. R. O.
1 Feb. 18	Jasjit Singh, 2nd-Lt. Sirdar	1 Feb. 18	12 Aug. 18	
18 July 16	Whitaker, Lt. S. W.	18 July 17	. . . .	K. Afr. Rif. I. A. (*on probn.*)
1 Sep. 18	Drake, 2nd-Lt. J. C. B., I.C.S.	1 Sep. 18	21 Oct. 18	I. A. R. O.
17 Sep. 18	Bingemann, 2nd-Lt. L. S., I.C.S.	17 Sep. 18	26 Oct. 18	I. A. R. O.
8 Oct. 18	Munns, 2nd-Lt. P. G.	8 Oct. 18	4 Nov. 18	I. A. R. O.
28 Sep. 17	Franks, 2nd-Lt. H. S.	28 Sep. 17	. . . .	I. A. R. O.

16th Cavalry—*contd.*

Date of entering service.	Names.	Army rank. Jemadar.	Ressaidar.	Risaldar.	Risaldar-Major.	Remarks.
	Risaldar-Majors.					
18 Feb. 86	Agia Ram (56) *Bahadur*	13 Mar. 97	21 Oct. 02	1 July 06	1 Oct. 14	*Order of Br. I., 2nd cl.*
7 Sep. 86	Wariam Singh (56)	7 July 00	11 Jan. 07	1 Oct. 14	8 Feb. 16	
	Risaldars.					
8 Feb. 99	Hushyar Singh	16 Dec. 04	4 Oct. 12	10 Sep. 15		
12 Jan. 07	Bharat Singh (45*a*, 66)	12 Jan. 07	10 Sep. 15	8 Feb. 16		
1 Jan. 96	Raghunath Singh	23 Aug. 14	1 Oct. 17	1 Oct. 17		
	Ressaidars.					
1 July 04	Kesri Singh (52, 56, 66*a*)	1 July 04	4 Jan. 15			
12 Feb. 00	Sher Singh (45*a*, 51, 56, 66*a*)	11 Jan. 07	4 Mar. 15			
1 Dec. 00	Kala Singh (45*a*, 52, 56, 58)	10 July 11	10 Sep. 15			
6 Apl. 03	Hira Singh (66*a*)	18 May 12	8 Feb. 16			
14 Feb. 93	Munshi Ram (45*a*, 50)	1 Oct. 14	8 Feb. 16			
3 Apl. 98	*Ishar Singh* (50, 53)	*16 Mar. 12*	30 May 16			
27 July 99	Rattan Singh	10 Sep. 15	1 Oct. 17			
1 Mar. 04	Boor Singh, M.C.	17 Mar. 15	25 Aug. 17			With 9th Horse.
	Jemadars.					
28 Dec. 96	Ajaib Singh	16 Sep. 13				Woordie Maj. 26 July 16.
1 Mar. 97	Bachittar Singh (56)	4 Jan. 15				
10 July 00	*Sadhoo Singh* (56)	*4 Jan. 15*				*With Remt. Depôt, Sehore.*
11 Dec. 92	Lal Chand	30 Jan. 15				
7 July 96	Ran Singh	4 Mar. 15				
8 July 96	Gurdit Singh (56)	4 Jan. 16				
1 Apl. 04	Gourdhan	1 Oct. 17				With Remt. Depôt, Meerut.
2 Mar. 04	Amar Singh	1 Oct. 17				
1 June 11	Durga Parshad (56)	1 Oct. 17				
25 May 07	Kanwal Singh	1 Oct. 17				
27 July 99	Daya Ram	1 Dec. 17				With 9th Horse.
27 June 06	Jaswant Singh	22 Dec. 17				*With 44th Cav.*
25 Mar. 12	Risal Singh	5 Feb. 18				
1 Sep. 03	Partal Singh	1 Apl. 18				
13 Oct. 07	Ram Sing (56)	1 Apl. 18				
29 Dec. 96	Bishan Sing	5 May 18				

17th Cavalry.

Raised at Muttra in 1857, as the Muttra Horse. Became the Muttra Police Corps the same year; the Rohilkhand Auxiliary Police Levy, 1858; 'Robarts' Horse,' 1858; the 17th Regiment of Bengal Cavalry, 1861, was disbanded, 1882. Re-raised at Mian Mir, 1885, by Col. E. H. E. Kauntze, under its former designation. Became the 17th Regiment of Bengal Lancers, 1900; the 17th Bengal Lancers, 1901; present designation, 1903.

"Afghanistan, 1879-80."

Composition—2 Squadrons of Punjabi Musalmans, 2 of Pathans—1 Addl. Squadron.

Uniform—Blue. *Facings*—White.

COLONEL—Major-General William A. Lawrence 13 May 1904.

1st Commn. or date of entering service.	Names and Rank.	Army rank.	Present appointment in Regiment.	Remarks.
	Commandant.			
20 May 93	Barry-Smith, Lt.-Col. R. C. . .	19 June 17	19 June 17	
	Squadron Commanders. (4)			
5 Oct. 92	*Steel, Bt.-Col. R. A.,* C.I.E., *p.s.c.* (Q) [*l*].	*1 Jan. 18*	*5 Oct. 18*	
10 Oct. 94	Henderson, Maj R. W., D.S.O.	10 Oct. 12	19 June 17	*2nd-in-Comd.*
29 Nov. 99	*Wilson Maj. D.D.,* M.C., *p.s.c. (temp. Lt.-Col.)*	*1 Sept. 15*	*1 Oct. 13*	
23 May 00	Stewart, Maj. H. S.	1 Sept. 15	15 Dec. 13	*Lv., ex I., m. c., 6 mos.*
	Squadron Officers. (9)			
27 Aug. 02	*Dodd, Maj. A. H. R.*	*27 Aug. 17*	*19 Jan. 04*	
9 Sep. 03	*Barge, Maj. K. (temp. Lt-Col.),* D.S.O., M.C.	*18 Jan. 18*	*10 June 06*	*G. S. O., 1st gr.*
21 Jan. 03	Kirkwood, Maj. T. W. . .	21 Jan. 18	1 Mar. 06	G. S. O., 2nd gr.
5 Aug. 05	*Yeats-Brown, Capt F. C. C.* [*l*].	*5 Aug. 14*	*31 Dec. 06*	*Adjt., 15 Mar. 13.*
19 Jan. 07	*Atkinson, Capt. B. G.*	*1 Sept. 15*	*26 Mar. 08*	
11 Dec. 07	Foottit, Capt. J. A. W. . . .	1 Sept. 15	9 Oct. 09	With 2-26th Punj.
27 May 08	*Guthrie, Capt. I. D.,* M.C. . .	*1 Sept. 15*	*8 Oct. 11*	*With 2-25th Punj.*
8 Sep. 09	Skipwith, Capt C. G. Y. . .	1 Sept. 15	3 Jan. 11	Adjt. (*Temp.*)
16 June 15	Campbell-Harris, Lt. C. . . .	16 June 16	15 June 17	Attd. With Guides Cav.
21 Aug. 12	*Cochrane, Capt. R. P. I.* .	*24 Aug. 17*	*1 Nov. 13*	*R. A. F.*
16 Aug. 16	Ballentine, Lt. J. S. . . .	16 Aug. 17	19 Jan. 17	Attd. Qr.-Mr. (temp.)
18 Jan. 17	Pert, Lt. C. E.	18 June 18	23 Jan. 17	Attd. With 2-124th Baluchis.
21 Dec. 17	Freston, 2nd-Lt. T. A. . . .	21 Dec. 17	21 Mar. 18	Attd. With 46th Cav.
25 Sep. 14	*Platts, Lt. J. C.*	*25 June 16*	*9 May 17*	*Attd.*
31 Aug. 18	Dixon, 2nd Lt. V. H.	31 Aug. 18	3 Sep. 18	Attd.
23 Apl. 18	Wilson, 2nd-Lt. J. H. A. . . .	23 Apl. 18	21 Oct. 18	Attd.
	Attached.			
23 Jan. 14	Sleigh, Lt. G. P.	30 Oct. 15	1 Aug. 18	W. Rid. R.—I. A. (*on probn.*)
21 Sep. 14	Garle, Capt. J. A. B.	21 Sep. 18	29 Jan. 18	W. Kent Yeo. I. A. R. O.
23 Sep. 14	Persse, Capt. S. H.	23 Sep. 18	17 Apl. 18	Lond. R.- I. A. R. O.
16 Dec. 14	Williams, Lt. C. C. (*Actng Capt.*)	16 Dec. 15	3 Aug. 17	I. A. R. O.
26 June 14	MacCay, Lt. R. C., D.S.O. (*temp. Capt.*)	26 Mar. 16	30 Mar. 18	I. A. (*on probn.*) Staff Capt.
5 May 15	Hutchison, Lt. J. R. H. . . .	5 May 16	30 June 17	I. A.
2 Oct. 15	Cameron, Lt. W. W. L. . . .	2 Oct. 16	16 Sep. 17	I. A. R. O.
20 Mar. 15	Sydenham, Lt. A. F. . . .	20 Dec. 17	30 Mar. 18	13th Light Horse, A. I. F. I. A. (*on probn.*)
23 Dec. 15	Griffiths, Lt. W. S.	10 July 18	5 Mar. 18	I. A. (*on probn.*)
1 Mar. 16	Cazalet, Lt. R. G. (*Actg. Capt.*)	1 Mar. 17	10 Mar. 16	I. A. R. O.
13 May 18	Ritchie, 2nd-Lt. A. A. . . .	13 May 18	26 July 18	I. A. R. O.
7 June 18	Tait, 2nd-Lt. J.	7 June 18	29 June 18	I. A. R. O.
10 June 18	Barry, 2nd-Lt. C. B.	10 June 18	19 June 18	I. A. R. O.
3 July 18	Withinshaw, 2nd-Lt. L. . . .	3 July 18	17 Aug. 18	I. A. R. O.
. . . .	Dwyer, 2nd-Lt. J. C. O. . . .	. . . .	18 Dec. 18	I. A. R. O.

17th Cavalry—*contd.*

Date of entering service.	Names.	Army rank. Jemadar.	Army rank. Ressaidar.	Present appointment in Regiment. Risaldar.	Risaldar-Major.	Remarks.
	Risaldar-Major.					
1 Sep. 86	Zikriya Khan	26 Mar. 06	16 Mar. 08	1 Jan. 12	2 Feb. 18	
	Risaldars.					
20 May 86	Rukan Din (52, 56)	16 Nov. 01	15 Oct. 08	6 July 14		
18 May 94	Rahmat Sher (52, 56, 71)	1 Aug. 05	1 Jan. 12	11 Apl. 16		
19 Mar. 00	Bahadur Sher (53, 56, 66*a*)	12 Mar. 02	16 Oct. 09	20 Sept. 16		
6 May 02	Sultan Khan (52, 56, 71)	14 May 02	1 Dec. 10	26 Jan. 17		With 13th Lrs.
18 Jan. 94	Haji Gul (52, 56, 66*a*)	1 Aug. 05	6 July 14	1 Feb. 18		
10 May 85	Gauhar Ali	22 Dec. 06	8 Sep. 14	16 Apl. 18		
	Ressaidars.					
25 Feb. 80	Tikka Khan (56, 66*a*)	15 Oct. 08	22 Mar. 16			
16 Sep. 07	*Ata Muhammad Shah (45a, 53, 56, 71).*	*15 Mar. 08*	*14 Aug. 16*			
8 July 94	Wazir Khan, I.O.M. (66*a*)	1 Oct. 09	20 Sept. 16			
17 Feb. 98	Muhammad Raza (45*a*)	8 Sep. 14	1 Sep. 17			With 18th Lrs.
1 Oct. 00	Dost Muhammad Khan (71)	16 Oct. 04	22 Sep. 17			
1 Mar. 06	Malik Alam Sher Khan (66*a*) I.D.S.M.	1 Jan. 12	1 Feb. 18			
	Jemadars.					
24 Mar. 04	Muhammad Sarwar (56, 71)	6 July 14				
1 Jan. 04	Wazir Muhammad (56, 71)	8 Sep. 14				Remt. Depôt, Saugor.
19 Dec. 08	Malik Ghulam Husain Khan (50, 60*a*)	24 Dec. 15				Remt. Depôt, Bangalore
11 Mar. 13	Sultan Sikandar Khan (50, 71)	24 Dec. 15				
1 Aug. 95	Inayatullah Khan (50, 71)	22 Mar. 16				Woordie Maj. (*temp.*).
2 June 11	Arbab Muhammad Husain (45*a*, 56, 71)	26 Mar. 16				
4 May 96	Ata Muhammad Khan	1 Apl. 16				Remt. Depôt, Saugor.
18 Nov. 98	*Fateh Khan*	*12 Apl. 16*				No. 2nd Camel Corps.
12 July 16	Malik Ahmad Yar Khan	12 July 16				(*On probn.*)
1 Dec. 93	Muhammad Hasan Khan (52, 56)	13 July 16				
14 Aug. 16	Nizam-ud-din (71)	14 Aug. 16				(*On probn.*)
15 Dec. 93	Bahadur Ali Khan	27 Aug. 16				
8 Jan. 04	*Muhammad Akbar Khan*	*20 Sep. 16*				*Conv. Sec., Lucknow.*
2 Jan. 96	Sharif Khan	9 Nov. 16				
1 Oct. 09	Ahmad Ali Khan	22 Sep. 17				
1 Oct. 13	Gustasab Khan	22 Sep. 17				
1 Mar. 98	Pir Dad Khan	5 Oct. 17				With 13th Lrs.
6 Mar. 94	Sher Bahadur	3 Dec. 17				
15 Feb. 06	Said Raza	1 Jan. 18				
24 Mar. 01	Alam Khan	24 Jan. 18				
12 Apl. 18	Mahbub Khan	12 Apl. 18				(*On probn.*)
21 Oct. 02	Shah Gul	16 Apl. 18				
. . . .	Abdullah Khan	11 Sep. 18				

18th King George's Own Lancers.

Raised at Gwalior in 1858, by Capt. F. H. Smith, as the 2nd Regiment of Mahratta Horse. Became the 18th Regiment of Bengal Cavalry in 1861; the 18th Regiment of Bengal Lancers in 1886; the 18th Bengal Lancers in 1901; the 18th Tiwana Lancers in 1903; 18th Prince of Wales's Own Tiwana Lancers in 1906; present designation in 1910.

THE PLUME OF THE PRINCE OF WALES. THE ROYAL AND IMPERIAL CYPHER.
"Afghanistan, 1879-80" "Tirah" "Punjab Frontier."
Composition—3 Squadrons of Punjabi Musalmans, 1 of Sikhs.
Uniform—Scarlet. *Facings*—White.

COLONEL-IN-CHIEF—THE KING.

COLONEL—Lieut.-General *Sir* George L. R. Richardson, K.C.B., C.S.I., C.I.E.[R] . 1 January 1906.

1st Commn. or date of entering service.	Names and Rank.	Army rank.	Present appointment in Regiment.	Remarks.
	Commandant.			
28 Dec. 95	Keighley, Lt.-Col. V. A. S., M.V.O.	2 Dec. 17	2 Dec. 17	
	Squadron Commanders. (4)			
4 Oct. 93	*Wigram, Bt. Lt.-Col. C.,* C.B., C.V.O., C.S.I., *p.s.c., Eq.* (Q)	*3 June 15*	*11 Apl. 10*	*Asst. Pte. Secy. to the King.*
20 May 99	Marsh, Maj. C. H., D.S.O.	1 Sept. 15	5 Mar. 18	2nd-in-Comd. With 210th Jats
24 Mar. 00	Marsh, Maj. R. H.	1 Sept. 15	2 Dec. 17	
11 Aug. 00	Howell, Maj. C. H.	1 Sept. 15	5 Mar. 18	
	Squadron Officers. (9)			
4 May 01	Mills, Maj. A. M., D.S.O.	29 Mar. 16	10 Oct. 02	
13 July 01	*Risley, Maj. C. G.,* D.S.O.	*23 July 16*	*22 Dec. 05*	
19 Aug. 03	*Gwatkin, Maj., F.,* M.C.	*19 Aug. 18*	*12 May 06*	*Staff Capt.*
28 Nov. 03	Brooke, Capt. A. H. (*temp. Maj.*)	28 Nov. 12	12 May 06	
5 Aug. 05	Bray, Capt. N. N. E., M.C.	5 Aug. 14	25 Nov. 06	
5 Aug. 05	*Brooke, Capt. A.*	*5 Aug. 14*	*13 Feb. 07*	*Comdt., Govr. Genl.'s, Body-Guard, 1 June 17*
18 Dec. 07	Dening, Capt. R., M.C.	1 Sept. 15	24 Jan. 11	
Sep. 11	Royston, Capt. C. G.	6 Sept. 15	11 Dec. 12	
22 Jan. 13	Forbes, Capt. A. F. G.	22 Jan. 17	11 Mar. 14	(*Offg.*) Adjt.
14 Jan. 14	Frazer, Capt. D. S.	14 Jan. 18	12 Oct. 14	
6 Nov. 14	Collin, Lt. E. P. C.	6 Aug. 16	20 Jan. 16	Attd.
14 Nov. 16	Mackie, Lt. A. H.	1 July 16	. . . .	Attd.
12 July 16	Seager, Lt. R. L.	12 July 16	16 Jan. 18	Attd.
16 Aug. 16	Windham, Lt. J. B. R.	16 Aug. 17	14 Nov. 16	
14 Nov. 16	Monies, Lt. V. A.	14 Nov. 17	. . . .	Attd.
31 Aug. 18	Thompson, 2nd-Lt. W. G. M.	31 Aug. 18	4 Sep. 18	Attd.
1 May 17	Milne, 2nd-Lt. O. J. T.	1 May 17	4 Jan. 18	Attd.
1 May 17	Evans, 2nd-Lt. J. E.	1 May 17	14 Dec. 17	Attd.
22 Sep. 14	Comerford, Lt. T. J.	23 June 16	26 Nov. 17	Attd.
23 Apl. 18	Maxwell, 2nd-Lt. F. S.	23 Apl. 18	19 Oct. 18	Attd.
	Attached.			
20 Nov. 14	Danby, Capt. A. L.	30 Aug. 18	30 Aug. 15	I. A. R. O.
4 Dec. 14	Fulcher, Lt. G. W. (Actg. Capt.)	2 Dec. 15	2 Dec. 15	I. A. R. O.
8 Jan. 15	Watson, Capt. H. R. W.	30 Aug. 18	30 Aug. 15	I. A. R. O.
18 Feb. 15	Webster, Lt. F. G.	18 Feb. 16	11 Mar. 16	I. A. R. O.
18 Feb. 15	Abercrombie, Lt. J. R.	18 Feb. 16	23 June 16	I. A. R. O. Qr.-Mr. (*Offg.*)
. . . .	Flood, Lt. J. B. (*temp. Capt.*)	. . . .	15 Apl. 15	
30 July 15	Benfield, Capt. C. W. C.	30 July 18	30 Dec. 16	I. A. R. O.
14 Sep. 15	Williams, Lt. A. B.	14 Sep. 16	21 Sep. 16	I. A. R. O.
4 Mar. 16	Rothera, Lt. G. H.	4 Mar. 17	25 June 16	I. A. R. O.
8 June 15	Brayne, Tempy. Lt. F. L.	8 June 15	12 Sept. 16	
17 Sep. 16	Bletsoe, Lt. W. B.	17 Sep. 16	29 Jan. 17	I. A. (*on probn.*)
15 Sep. 16	Lloyd, Lt. S.	15 Sep. 17	. . . .	I. A. R. O. Qr.-Mr.
22 Apl. 16	Grosvenor, Lt. C. J. P.	22 Apl. 17	12 Feb. 18	I. A. (*on probn.*)
14 Sep. 15	Cruikshank, 2nd-Lt. J. R. H.	14 Sep. 15	. . . .	I. A. (*on probn.*)
31 Jan. 18	Strickland, 2nd-Lt. H. B.	31 Jan. 13	6 Feb. 18	
. . . .	Collings, 2nd-Lt. J. S.	1 Feb. 18	1 May 18	I. A. (*on probn.*)
30 Jan. 17	Hodgson, Lt. H. C.	30 Jan. 18		
5 Feb. 13	Watson, Capt. G. T. C.	5 Feb. 17	10 July 18	
	Honorary Officer.			
11 May 01	*Hon'ble Sir* Malik Umar Hayat Khan, K.C.I.E., M.V.O., Tiwana of the Shahpur Dist.	21 June 11	11 May 01	Hony. Maj. in Army
27 Jan. 18	Raja Sir Hari Singh	27 Jan. 18	27 Jan. 18	Hony. Capt. in Army.

18th King George's Own Lancers—*contd.*

Date of entering service.	Names.	Army Rank. Jemadar.	Ressaidar.	Risaldar.	Risaldar-Major.	Remarks.
	Risaldar-Majors.					
7 Dec. 92	Haji Gul Mawaz Khan (56, 71) *Bahadur.*	1 Mar. 94	1 Jan. 03	1 Nov. 07	21 July 12	*Order of Br. I., 1st cl.*
1 July 90	Malik Ahmadyar Khan, *Bahadur* (56, 71).	10 Apl. 02	1 July 07	21 July 12	14 Aug. 16	*Order of Br. I., 2nd cl.*
	Risaldars.					
5 Nov. 85	Muhammad Khan (56, 71) . .	22 Oct. 00	15 Apl. 04	19 Mar. 10		
9 Sep. 04	Malik Sultan Khan, I.D.S.M. . .	9 Sep. 04	25 Oct. 09	25 Oct. 14		
5 July 95	Lakha Singh *Bahadur* (56) . .	2 Dec. 00	19 Mar. 10	6 Aug. 16		*Order of Br. I., 2nd cl.*
24 Mar. 96	*Dost Muhammad Khan,* I.O.M., I.D.S.M., *Bahadur.*	*1 July 07*	*1 Jan. 11*	*14 Aug. 16*		*A. D. C. to H. E. the C.-in-C. Order of Br. I., 2nd cl. A. R. O. Sargodha.*
15 Aug. 02	Katar Singh (53, 56)	1 July 11	25 Aug. 17	12 May 18		
10 Sep. 09	Malik Sultan Ahmad Khan . .	10 Sep. 09	6 Aug. 16	10 May 18		
26 Feb. 05	Jagat Singh (56)	26 Feb. 05	11 Apl. 16	11 Aug. 18		12th Cav.
	Ressaidars.					
1 May 96	Muhammad Inayat Khan *Bahadur* (56, 71).	1 Nov. 07	1 Apl. 14			*Order of Br. I., 2nd cl.*
1 June 00	Wali Muhammad Khan (45*a*) .	10 Apl. 09	25 Oct. 14			
18 Feb. 04	Amar Singh (45*a*, 56) . . .	27 Apl. 07	11 Apl. 16			
1 May 90	Tawakli Khan *Bahadur* . . .	1 Jan. 11	14 Aug. 16			*Order of Br. I., 2nd cl.*
10 Jan. 91	Dhuman Khar	16 Sep. 14	25 Aug. 17			
17 Feb. 98	Muhammad Raza (45*a*)	8 Sep. 14	1 Sep. 17			
	Jemadars.					
16 Oct. 99	Muhammad Hayat Khan (45*a*) .	17 Dec. 13				
13 Feb. 05	Muhammad Khan, I.O.M. . . .	1 Apl. 14				
10 July 00	Sultan Khan	25 Oct. 14				
4 Mar. 02	Khuda Bakhsh	25 Oct. 14				
1 June 01	Adalat Khan, I.O.M., I.D.S.M. .	11 Apl. 16				Woordie-Maj.
15 June 00	Bakhshi Karam Chand	12 July 16				
	Bahadur Khan (*Retd.*)	. . . .				Re-employed, 8 Feb. 15, Depôt.
15 Apl. 95	Fateh Ali Khan	14 Aug. 16				
20 May 14	Malik Sultan Mahmud Khan . .	14 Aug. 16				
16 Sep. 99	Jagat Singh	10 Sep. 16				
10 Sep. 92	Khud Baksh Khan	30 Sep. 16				
16 Oct. 16	Azim Khan	. . . .				Re-employed, 16 Oct. 16.
18 Dec. 05	Shahwali Khan	25 Apl. 17				
1 June 98	Sher Ali Khan	2 Sep. 17				
16 July 01	Jwala Singh	18 Sep. 17				
20 Feb. 02	Muhammad Ashraf Khan . . .	7 Oct. 17				
16 May 05	Sher Muhammad Khan . . .	14 Oct. 17				17th Cav.
1 Apl. 96	Allahyar Khan	19 Jan. 18				
1 May 12	Muhammad Amir Khan . . .	3 Feb. 18				

19th Lancers (Fane's Horse).

Raised at Cawnpore in 1860, by Lt. Fane, in a great measure from volunteers from Hodson's Horse and was originally designated 'Fane's Horse.' Became the 19th Regt. of Bengal Cavalry, 1861; the 19th Regt. of Bengal Cavalry (Lancers), 1864; the 19th Regt. of Bengal Lancers, 1874; the 19th Bengal Lancers (Fane's Horse), 1901; present designation, 1903.

"Taku Forts" "Pekin, 1860" "Ahmad Khel" "Afghanistan, 1878-80."

Composition—1½ Squadrons of Sikhs, ½ of Dogras, 1 of Punjabi Musalmans, 1 of Pathans.

Uniform—Blue. *Facings*—French Grey. *Lace*—Silver.

COLONEL—Lieutenant-General William W. Biscoe, C.B. . . . 13 May 1904.

1st Commn. or date of entering service.	Names and Rank.	Army rank.	Present appointment in Regiment.	Remarks.
	Commandant.			
9 Nov. 89	Gregory, Bt.-Col. C. L., C.B., *p.s.c.* (*temp. Br.-Genl.*)	1 Jan. 17	28 Aug. 15	
	Squadron Commanders. (4)			
13 Feb. 91	McNeile, Lt.-Col. D. H. [L] (Q)	13 Feb. 17	28 Aug. 15	2nd-in-Comd.
13 July 92	*Taylor, Lt.-Col R. O'B.*, C.M.G., C.I.E., *p.s.c.* [L] (*temp. Br.-Genl*).	*13 July 18*	*2 Feb. 10*	
16 Jan. 95	Lance, Maj. F. FitzH. . . .	16 Jan. 13	28 Aug. 10	
20 Jan. 00	Whitby, Maj. H. F.	1 Sep. 15	28 Aug. 15	
	Squadron Officers. (9)			
24 July 01	MacLeod, Maj. R. D. C., D.S.O. .	4 May 16	11 Feb. 15	
18 Jan. 02	Curtis, Maj. F.	18 Jan. 17	8 Jan. 15	
27 Jan. 04	*Power, Maj. J. B.*	*28 Mar. 18*	*10 July 06*	
18 Jan. 05	Paterson, Capt. T. S. . . .	18 Jan. 14	1 Aug. 07	
18 Jan. 05	*Howell, Capt. G. B.*	*18 Jan. 14*	*31 Mar. 06*	*Bde.-Maj.*
19 Jan. 07	Baddeley, Capt. S. E. L. . . .	1 Sep 15	[illegible] Sep. [illegible]	Adjt., 8 Mar. 16.
8 Sep. 09	*Tweedy, Capt. R. J.*	*1 Sep. 15*	*8 Nov. 11*	*Staff Capt.*
8 Sep. 09	Scratton, Capt. G. W. B., M.C.	1 Sep. 15	Feb. 11	
29 Jan. 10	*FitzGerald, Capt. G. M.*, M.C. .	*1 Sep. 15*	*10 Mar. 11*	
3 Sep. 10	*Lane Capt. C. R. C. [l]*, M.C. .	*1 Sep. 15*	*16 Dec. 11*	*G. S. O., 3rd gr.*
8 Apl. 16	Davies, Lt. J. T.	8 Apl. 17	10 Dec. 17	Attd.
5 May 16	Howatson, Lt. C. N.	5 May 16	15 Jan. 18	Attd.
14 Nov. 16	Cartlidge, Lt. C.	14 Nov. 17	. . . .	
30 Jan. 17	Galbraith, 2nd-Lt. I. W., M.C. .	30 Jan. 17	9 Apl. 17	Attd.
30 Jan. 17	Benzie, 2nd-Lt J. H. S. . . .	30 Jan. 17	18 Sep. 17	Attd.
25 Mar. 17	Powell, Lt. L. K. M.	25 Mar. 17	26 Jan. 18	Attd.
1 May 17	Dulcken, 2nd-Lt. H. W. . . .	1 May 17	1 Dec. 17	Attd.
27 Oct. 17	Hilliard, Lt. P. J.	27 Oct. 18	4 Nov. 17	
27 Oct. 17	Reynolds, Lt. C. C. R. . . .	27 Oct. 18	3 Nov. 17	
31 Jan. 18	Collins, 2-Lt. M. R.	1 Feb. 18	6 Feb. 18	
	Burt, 2nd-Lt. P. J.	. .	. .	Attd.
	Attached.			
10 June 15	Carruthers, Lt. W. O. . . .	10 June 16	10 Sep. 17	
	Jacques, Capt. G. P. R. . . .	18 Dec. 14	17 Apl. 15	
20 Nov. 14	Lang, Capt. H. F.	20 Aug. 18	2[illegible] June 16	
30 Nov. 14	Walsh, Lt. C. L.	18 Jan. 15	4 Apl. 15	Res. Cav.
4 Dec. 14	Hindley, Capt. D. C. R. . . .	30 Aug. 18	30 Aug. 15	
11 Dec. 14	*Fremlin, Lt. H. J. L. (Actg. Capt.)*	*11 Dec. 15*	*30 July 15*	
5 Feb. 15	Maxwell, Lt. R M.	19 Jan. 16	9 Jan. 15	I. A. R. O.
12 Feb. 15	Denning, Lt. J. DuP.	12 Feb. 16	12 Mar. 15	I. A. R. O.
12 Feb. 15	Rust, Lt. L. D.	12 Feb. 16	2 Mar. 16	I. A. R. O. Qr.-Mr.
23 Apl. 15	Shee, Lt J. R. L.	23 Apl. 16	3 Aug. 16	I. A. R. O.
30 Apl. 15	Ashcroft, Lt. J. M.	30 Apl. 16	20 Jan. 16	I. A. R. O.
15 May 15	Holden, Lt. G. G.	5 May 16	. . . .	I. A. (*on probn.*)
6 June 15	Southgate, Lt. T. D.	16 June 16	25 Nov. 17	Essex R I. A. (*on probn.*)
4 Dec. 15	Pike, Lt. A.	4 Dec. 16	24 Jan. 17	I. A. R. O.
8 July 17	Chetwynd, 2nd-Lt. C. W. G. .	8 July 17	26 Jan. 18	I. A. R. O.
23 Nov. 17	Tye, 2nd-Lt. H. J. (*Temp. Capt.*)	23 Mar. 18	4 July 18	I. A. (*on probn.*) *Adj. & Qr. Mr., Sch. of Musky. R. Pindi.*
30 Nov. 17	Hadfield, 2nd-Lt. H. W. . . .	25 Sep. 18	6 Nov. 18	I. A. R. O. (*on probn.*)
25 Oct. 18	Blackburne, 2nd-Lt. H. H. . .	25 Oct. 18	26 Nov. 18	I. A. R. O.
6 Feb. 07	Ewings, Capt. J. L. S., M.C. . .	1 Sep. 15	7 Dec. 18	R. Highrs., I. A.

19th Lancers (Fane's Horse)—*contd.*

1st Commn. or date of entering service.	Names and Rank.	Army Rank. Jemadar.	Present appointment in Regiment. Ressaidar.	Risaldar.	Risaldar-Major.	Remarks.
	Risaldar-Major.					
9 Jan. 95	**Hira Singh**, *Bahadur*, I.D.S.M. (56).	19 Jan. 95	17 May 01	9 Mar. 12	31 Jan. 15	*Order of Br. I., 2nd cl.*
	Risaldars.					
2 July 94	Khushal Khan, *Bahadur* (56)	14 Mar. 00	3 Oct. 07	1 Feb. 13		*Order of Br. I., 2nd cl.*
27 Aug 87	Bakhshish Singh . . .	6 Feb. 01	6 June 08	25 Jan. 15		
2 Feb. 93	Ali Shan (56)	3 Oct. 07	31 Jan. 15	1 Aug. 16		
15 Feb. 01	Moti Singh, I.D.S.M. (56, 58, 60, 66a).	14 Jan. 07	1 Sep. 14	1 Mar. 17		
3 Feb. 96	Bhagwan Singh (51)	6 June 08	11 Apl. 16	27 July 17		
1 Apl. 96	Ghulam Husain, D.S.M. (45a, 51, 66a).	16 Oct. 16	11 Apl. 16	21 Nov. 17		
2 Mar. 93	Khwaja Muhammad *Bahadur* (66a)	9 Mar. 12	11 Dec. 16	9 Apl. 18		*Order of Br. I., 2nd cl.*
1 Sep. 94	Bishan Singh, I.O.M. (45a, 52, 56, 66a)	1 Feb. 13	1 Mar. 17	3 June 18		
	Ressaidars.					
. . . .	Muhammad Hanif Khan *Bahadur* (Retd)	. . . .	. . . .			*Order of Br. I. 2nd cl.*
13 Feb. 09	Mahendar Singh (66a) . . .	13 Feb. 09	1 Aug. 16			
13 July 07	Muhammad Khan	1 Sep. 14	27 July 1			
24 Dec. 09	Bhan Singh (45a)	1 Sep. 14	25 Aug. 17			
4 Feb. 01	Indar Singh	17 Mar. 16	21 Nov 17			
28 Mar. 96	Sawal Singh	11 Apl. 16	21 Nov 17			
24 Mar. 97	Sham Singh	12 Oct. 16	3 June 18			
30 Aug. 99	Didan Singh	19 May 16	8 May 18			
	Jemadars.					
15 Apl. 07	Teja Singh	11 Apl. 16	26 Apl. 18			
25 Apl. 98	Juma Khan	25 Jan. 15				
29 June 04	Hasham Khan, I.O.M.	11 Dec. 16	. . . .			Woordie Maj.
3 Apl. 08	Abdul Jabbar	11 Dec. 16				
16 Feb. 10	Sarbaland Khan, I.O.M., I.D.S.M.	25 Aug. 17				
12 June 02	Ram Singh	25 Aug. 17				
22 Nov. 00	Hari Singh	25 Aug. 17				
1 June 00	Narain Singh	25 Aug. 17				
1 May 04	Sardar Ali	13 Sep. 17				
20 Jan. 09	Kehar Singh	22 Oct. 17				
16 Nov. 08	Abdu Majid	21 Nov. 17				
1 Jan 95	Man Singh	3 Dec. 17	. . . .			Qr. Mr. Jem.
	Maula Dad	17 Feb. 8				
12 Sep. 02	Meher Singh	9 Apl. 18				
7 Dec. 15	Mubary Khan	8 May 18				
30 Mar. 06	Sant Singh	3 June 18				
5 Jan. 04	Udham Singh	16 June 18				
16 May 13	Ghulam Rasul Khan	10 May 18				
. . . .	Muhammad Islam	4 June 18				

20th Deccan Horse.

Formed in 1826, from Nawab Jalal-ud-daula's, Capt. Davis' and Capt. Clark's risallahs of the Nizam's Reformed Horse, and was originally styled the 1st Regiment, Nizam's Cavalry. Became the 1st Cavalry, Hyderabad Contingent, 1854; the 1st Lancers, Hyderabad Contingent, 1890; present designation, 1903.

"Central India."

Composition.—1 Squadron of Sikhs, 1 of Jats, 2 of Dekhani Musalmans.

Uniform—Rifle-green. *Facings*—White.

HONORARY COLONEL—*Hony. Lt.-Genl. His Exalted Highness* Asaf Jah Muzaffar-ul-Mamalik Nizam-ul-Mulk Nizam-ud-Daula *Nawab* Mir *Sir* Usman Ali Khan, *Bahadur* Fateh Jang, of Hyderabad, G.C.S.I., G.B.E. . . . 1 Jan. 18.

1st Commn. or date of entering service.	Names and Rank.	Army rank.	Present appointment in Regiment.	Remarks.
	Commandant.			
2 June 94	Adams, Lt.-Col. F., *p.s.*, D.S.O.	29 July 16	29 July 16	
	Squadron Commanders. (4)			
13 Nov. 95	Bourne, Bt.-Lt.-Col. W. K., *p.s.c.* (Q)	3 June 16	2 Sep. 17	2nd-in-comd. With 2-35th Sikhs.
20 May 99	*Wallis, Maj. H. J. K.*	*1 Sep. 15*	*22 Dec. 14*	*Assay Department. With 28th L. Cav.*
4 Mar. 99	*Ross, Maj. A. C.*, D.S.O.	*1 Sep. 15*	*11 Dec. 16*	*G. S. O., 2nd gr.*
21 Feb. 00	Worgan, Bt.-Lt.-Col. R. B., D.S.O.	1 Jan. 18	9 Feb. 17	
	Squadron Officers. (9)			
13 June 00	Ley, Maj. A. E. H.	1 Sep. 15	8 July 02	
18 Jan. 02	Jarvis, Maj. C., M.C.	18 Jan. 17	19 Sep. 05	
15 Feb. 02	Gregson, Maj. C. D.	15 Feb. 17	14 Apl. 04	
21 Jan. 03	Mackenzie, Capt. C. A. C.	21 Jan. 12	24 Oct. 07	
13 Aug. 04	Wilson, Capt. T. K.	13 Aug. 13	22 Mar. 17	With 45th Cav.
29 Aug. 06	Oswald, Capt. F.	29 Aug. 15	21 Nov. 07	S. O. to Inspr. of Cav.
6 Sep. 11	Tinley, Capt. F. B. N., M.C.	6 Sep. 15	3 Dec. 12	
14 Nov. 16	Wilson, Lt. J. F. S.	14 Nov. 17	24 Oct. 17	Attd.
14 Nov. 16	Humfrey, Lt. H. B.	14 Nov. 17	24 Oct. 17	
30 Jan. 17	Schofield, Lt. A. J. A.	30 Jan. 18	16 Sep. 17	Attd.
....	Burnad, 2nd-Lt.	....	....	Attd.
30 Jan. 17	Ryan, Lt. C. C. L.	30 Jan. 18	1 Nov. 17	With 45th Cav.
30 Jan. 17	Hopkins, Lt. W. J.	30 Jan. 18	16 Sep. 17	
30 Jan. 17	Pollitt, Lt. H. T.	30 Jan. 18	28 Nov. 17	Attd.
30 Jan. 17	Robinson, Lt. F. R. H.	30 Jan. 18	3 May 17	Attd.
31 Aug. 18	Story, 2nd-Lt. F. E.	31 Aug. 18	13 Sep. 18	Attd.
31 Aug. 18	Wardle, 2nd-Lt. J. L.	31 Aug. 18	13 Sep. 18	Attd.
	Attached.			
10 Oct. 94	Henderson, Maj. R. W., D.S.O.	10 Oct. 12	7 July 18	17 Cav. 2nd-in-Comd. (*Offg.*)
24 Feb. 13	Collins, Capt. J. R.	1 June 16	6 Dec. 16	Glasgow Yeo.
14 May 15	*Larkin, Lt. E. P.*	*14 May 16*	*12 Aug. 17*	*Do.*
7 Aug. 14	Hamley, Capt. W. S. C.	1 Sep. 15	....	I. A. R. O.
13 Nov. 14	Guthrie, Lt. F. C. (*Actg. Capt.*)	13 Nov. 15	24 Dec. 15	I. A. R. O. Adjt. (*Offg.*)
20 Nov. 14	Glasspool, Lt. L. A.	20 Nov. 15	2 Dec. 15	I. A. R. O.
27 Nov. 14	Lawford, Lt. E. E.	30 July 15	....	I. A. R. O.
19 Mar. 15	Maxwell, Lt. J. F.	19 Mar. 16	....	I. A. R. O.
23 Oct. 16	Hallifax, Ty. Lt. H. F.	23 Oct. 16	10 Dec. 16	
14 Nov. 14	Godfree, Ty. Lt. A. S., M.C.	28 Jan. 16	2 Feb. 16	10th Res. Cav.
8 Sep. 14	Armstrong, Lt. J. D. H.	8 Sep. 15	8 Sep. 15	I. A. R. O. Qr.-Mr. (*Offg.*) 1 May 18.
11 May 17	Carter, Lt. A. C. (57)	11 May 18	15 Nov. 17	I. A. R. O.
29 Aug. 17	Weitzel, Lt. R. H.	29 Aug. 18	25 Nov. 17	I. A. R. O.
17 Feb. 17	Henderson, Lt. C. H.	17 Feb. 18	28 Apl. 18	I. A. R. O. (*on probn.*)
8 July 17	Smith, 2nd-Lt. R. H.	8 Apl. 18	28 Apl. 18	I. A. (*on probn.*)
26 July 15	Simpson, Lt. J. N. (51, 61)	26 July 17	29 Aug. 18	I. A. (*on probn.*)
26 Oct. 15	Lowe, Lt. E. A. H.	26 Oct. 16	17 May 18	I. A. (*on probn.*)
	Honorary Officer.			
May 88	Muhammad Ali Beg, *Nawab, Sir* Afsar-i-Jang, Afsar-ud-daula, Afsar-ul-Mulk, *Bahadur*, K.C.I.E., M.V.O., A.D.C.	9 Aug. 02	22 May 03	Hony. Lt.-Col. in Army.
	Forester, 2nd-Lieut. G. D.	....	....	I. A. R. O.

20th Deccan Horse—*contd.*

Date of entering service.	Names.	Army rank. Jemadar.	Ressaidar.	Risaldar.	Risaldar-Major.	Remarks.
	Risaldar-Majors.					
	Hanooman Singh, *Sardar Bahadur* (Retd.)	. . . .	. . . .	. . . .	. . .	Re-employed 25 Feb. 15.
23 May 86	Jharmal Singh (43, 56) . . .	1 July 94	9 Oct. 98	1 Mar. 04	1 Apl. 12	
1 Apl. 90	Prem Singh, *Bahadur* (56) . .	1 July 01	1 July 04	21 Dec. 14	18 May 16	*Order of Br. I., 2nd cl.*
	Risaldars.					
10 June 92	Bachant Singh (52, 56, 66*a*) . .	1 July 04	3 May 10	{9 Nov. 15 18 May 16		
1 June 91	Konsal Singh Bahadur, I.D.S.M. (45*a*, 50, 53, 58, 66*a*)	1 Apl. 07	16 Dec. 14	10 Nov. 15		*Order of Br. I., 2nd cl.*
8 Mar. 90	Mir Sadik Ali (51, 52, 56, 65) . .	24 June 11	21 Dec. 14	12 July 16		
10 July 01	Shaikh Faiz-ud-din, I.D.S.M. (58)	1 Apl. 14	25 Aug. 17	4 Aug. 18		
	Ressaidars.					
1 July 93	Mahbub Khan (53, 56, 66*a*) . .	1 Apl. 06	10 Aug. 14			
17 Jan. 98	Mangal Singh (56)	17 Mar. 08	21 Dec. 14			
1 May 89	Hayat Mir Khan (50, 53, 71) . .	2 May 10	21 Dec. 14			
20 June 00	Rachpal Singh (56)	1 Apl. 13	17 May 16			
1 Jan. 97	Muhammad Khan	1 July 12	12 July 16			
6 Nov. 03	Gugan Singh (45*a*, 56) . . .	1 Oct. 13	13 July 16			
4 Nov. 03	Ganga Bishan (56)	16 Dec. 14	11 Oct. 17			
16 Dec. 86	Shaikh Mohi-ud-din	15 June 15	12 Oct. 17			
3 Nov. 94	Rahim Sher Khan	5 Nov. 15	4 Aug. 18			
	Jemadars.					
6 May 92	Ramji Lall (52, 56)	1 Apl. 12				
8 Feb. 10	Bharat Singh (56)	10 Aug. 14				
1 Feb. 02	Sher Singh	2 Dec. 14				
8 Mar. 87	Nabiyar Khan	21 Dec. 14				
21 Feb. 94	Dhulip Singh (I)	21 Dec. 14				
1 Jan. 01	Sirdar Singh, I.O.M. (56) . .	11 Apl. 16				
22 Sep. 03	Dalip Singh (II), I.D.S.M . . .	11 Apl. 16				
18 Aug. 93	Yusuf Ali Khan (I)	17 May 16				Woordie-Maj., 28 Dec. 17.
13 Aug. 91	Bhawani Singh	19 May 16				
6 Apl. 00	Sayyid Ahmed (56)	11 July 16				
9 Aug. 04	Yusuf Ali Khan (II)	14 July 16				
21 Dec. 02	Bharat Singh (66*a*)	24 July 17				
15 Aug. 03	Dalip Singh (III) (66*a*)	25 July 17				
9 Sep. 03	Sarwan Singh	25 Aug. 17				
24 Oct. 02	Jhanda Singh	25 Aug. 17				
21 Dec. 02	Lahri, I.D.S.M.	11 Oct. 17				
1 Mar. 02	Abdul Razak Khan	12 Oct. 17				

21st Prince Albert Victor's Own Cavalry (Frontier Force) (Daly's Horse).

Raised at Peshawar in 1849, by Lt. H. Daly, as the 1st Regt. of Punjab Cavalry. Became 1st Regt. of Cavalry, Punjab Irregular Force, 1851; the 1st Regt. of Cavalry, Punjab Frontier Force, 1865; the 1st (Prince Albert Victor's Own) Regt. of Punjab Cavalry, 1890; the 1st (Prince Albert Victor's Own) Punjab Cavalry, 1901; the 21st Prince Albert Victor's Own Cavalry (Frontier Force), 1903; present designation, 1904.

"Delhi, 1857" "Lucknow" "Ahmad Khel" "Afghanistan, 1878-80."

Composition—1½ Squadrons of Sikhs, ½ of Dogras, 1 of Hindustani Musalmans, 1 of Pathans.

Uniform—Blue. *Facings*—Scarlet.

COLONEL—Major-General Charles S. Maclean, 𝔐., C.B., C.I.E. 6 January 1905.

1st Commn. or date of entering service.	Names and Rank.	Army rank.	Present appointment in Regiment.	Remarks.
	Commandant.			
30 Aug. 93	Johnson, Lt.-Col. J. E. B.	1 July 18	1 July 18	
	Squadron Commanders. (4)			
23 July 90	Battine, Lt.-Col. R. St. C., D.S.O. p.s.c. [L]	23 July 16	19 Aug. 16	2nd-in-Comd.
17 Dec. 90	Souter, Lt.-Col. A. B. (Q) [*l*]	17 Dec. 16	10 June 07	
24 June 96	Jackson, Maj. F. A., D.S.O.	24 June 14	19 Feb. 11	*Lv., ex Ind., m. c.*
20 Jan. 97	Gunning, Maj. G. H., D.S.O.	20 Jan. 15	6 Oct. 12	
	Squadron Officers. (9)			
6 Apl. 98	Worsley, Maj. C. F. M.	1 Sep. 15	1 Apl. 00	*Lv, ex Ind., m. c.*
4 May 98	Dyke, Maj. O. M	1 Sep. 15	17 Nov. 02	With 111th Mahars.
8 Jan. 01	McCudden, Maj. J. H., D.S.O., M.C.	8 Jan. 16	4 Jan. 13	
5 Oct. 01	Smart, Maj. R. W. L. DeB.	5 Oct. 16	3 Mar. 04	
29 Jan. 02	Robertson, Maj. E. D. S.	29 Jan 17	15 Jan 05	With 2-21st Punj.
27 Jan. 04	Grace, Capt. H. G., M.C.	11 Apl. 13	1 Mar. 06	With 44 Cav.
5 Aug. 05	*Ismay, Bt.-Maj. H. L. (temp. Maj.)*	*5 Aug. 14*	*16 Sep. 07*	*G. S. O. 3rd gr.*
29 Aug. 06	Dane, Capt. R., M.C.	29 Aug. 15	18 Nov. 07	Adjt., 1 Aug. 14.
17 Aug. 07	Lewis, Capt. R. F.	1 Sep. 15	11 Nov. 08	
9 Sep. 08	Garstin, Capt. H. A., M.C.	1 Sep. 15	17 Nov. 09	
14 Jan. 14	Dening, Capt. J. P.	14 Jan. 18	10 Oct. 14	
15 Aug. 14	Logan, Capt. A. F., M.C.,	15 Aug. 18	25 May 16	Insptg. Officer, I. S. Troops.
29 June 16	Bevan-Petman, Lt. B. H.	2 July 16	29 June 16	
18 June 17	Hanmer, Lt. C. G.	18 June 18	18 Nov. 17	With 42nd Cav.
4 May 15	Brookes, 2nd-Lt. J. R.	4 May 15	4 Sep. 18	Attd.
31 Aug. 18	Meyer, 2nd-Lt. W. R. M.	31 Aug. 18	4 Sep. 18	Attd.
	Attached.			
20 Jan. 00	Connop, Maj. H. E.	1 Sep. 15	17 Oct. 1	35th Horse.
24 Aug. 12	Morrison, Capt. C. C.	24 Aug. 16	18 Dec. 17	35th Horse.
17 Nov. 16	Page, Lt. A. J.	17 Nov. 17	11 Feb. 18	I. A. (*on probn.*).
9 Nov. 14	Byrne, Lt. G. R.	9 Nov. 15	30 Nov. 17	I. A. R. O.
28 June 16	Latham, Lt. P. D.	28 June 17	4 July 16	I. A. (*on probn.*)
8 Oct. 16	Weaver, Lt. M. G.	8 Oct. 17	30 Apl. 18	Denbig. Yeo., I. A. (*on probn.*).
21 Aug. 16	Clough, Lt. T. (*actg. Capt.*)	21 Aug. 17	31 Aug. 16	I. A. R. O.
21 Aug. 16	Gude, Lt. N. C.	21 Aug. 17	28 Dec. 16	I. A. R. O.
25 Sep. 16	Maxwell, Lt. H. B.	25 Sep. 17	4 May 17	I. A. R. O. *Lv., ex. I., m. c., 6 mos.*
20 Oct. 16	Fitzpatrick, Lt. C.	20 Oct. 17	28 May 17	I. A. R. O.
19 May 15	Harris, Lt. H., M.C. (46, 48*a*, 51).	26 Aug. 16	13 July 18	13th K. L'pool Regt. I. A. (*on probn.*).
17 Mar. 17	Gerrard, Lt. H. J. F.	17 Mar. 18	5 Sep. 18	R. F. A. 77th Bty. I. A. (*on probn.*)
5 May 15	Stewart, Lt. I.	5 May 16	1 Aug. 18	I. A. (*on probation.*)
12 June 17	Thorniley, Lt. J. H.	12 June 18	23 Oct. 18	I. A. R. O.
10 Aug. 17	Cameron, Lt. E. G.	10 Aug. 18	9 Aug. 18	I. A. R. O.

Medical Officers.

		Jemadar.	Ressaidar.	Risaldar.	Risaldar Major.
	Risaldar-Major.				
19 July 98	Sundar Singh (53, 56, 71)	19 July 98	8 Apl. 02	17 Nov. 12	7 Mar. 17.
	Risaldars.				
1 Mar. 89	Kunda Singh, I.O.M., I.D.S.M. (52, 56)	1 Jan. 06	17 Nov. 12	6 Dec. 15	
29 Dec. 94	Latif Khan (53, 56)	1 Feb. 08	1 Apl. 13	16 Sep. 16	
25 Feb. 92	Mahbub Khan *Bahadur* (45*a*).	10 May 10	1 July 14	14 Feb. 17	*Order of Br. I., 2nd Class.*
21 Mar. 94	Narayan Singh *Bahadur* (53, 56, 66*a*).	1 July 10	11 Oct. 14	15 Sep. 17	*Order of Br. I, 2nd Class.*
....	Shad Muhammad Khan (Ret'd.)	....	....	26 Feb. 18	
2 Dec. 98	Farzand Ali	1 July 14	14 Feb. 17	28 June 18.	
18 Apl. 04	Jahangir Beg, I.O.M.	4 July 15	1 Apl. 17	5 Aug. 18.	
17 Apl. 94	Arbela Singh	11 Aug. 12	15 Sep. 17	3 Sep. 18.	

21st Prince Albert Victor's Own Cavalry (Frontier Force) (Daly's Horse).

Raised at Peshawar in 1849, by Lt. H. Daly, as the 1st Regt. of Punjab Cavalry. Became 1st Regt. of Cavalry, Punjab Irregular Force, 1851; the 1st Regt. of Cavalry, Punjab Frontier Force, 1865; the 1st (Prince Albert Victor's Own) Regt. of Punjab Cavalry, 1890; the 1st (Prince Albert Victor's Own) Punjab Cavalry, 1901; the 21st Prince Albert Victor's Own Cavalry (Frontier Force), 1903; present designation, 1904.

"Delhi, 1857" "Lucknow" "Ahmad Khel" "Afghanistan, 1878-80."

Composition—1½ Squadrons of Sikhs, ½ of Dogras, 1 of Hindustani Musalmans, 1 of Pathans.

Uniform—Blue. *Facings*—Scarlet.

COLONEL—Major-General Charles S. Maclean, M., C.B., C.I.E. 6 January 1905.

1st Commn. or date of entering service.	Names and Rank.	Army rank.	Present appointment in Regiment.	Remarks.
	Commandant.			
30 Aug. 93	Johnson, Lt.-Col. J. E. B.	1 July 18	1 July 18	
	Squadron Commanders. (4)			
23 July 90	Battine, Lt.-Col. R. St. C., D.S.O. p.s.c. [L]	23 July 16	19 Aug. 16	2nd-in-Comd.
17 Dec. 90	Souter, Lt.-Col. A. B. (Q) [*l*]	17 Dec. 16	10 June 07	
24 June 96	Jackson, Maj. F. A., D.S.O.	24 June 14	19 Feb. 11	*Lv., ex Ind., m. c.*
20 Jan. 97	Gunning, Maj. G. H., D.S.O.	20 Jan. 15	6 Oct. 12	
	Squadron Officers. (9)			
6 Apl. 98	Worsley, Maj. C. F. M.	1 Sep. 15	1 Apl. 00	*Lv, ex Ind., m. c.*
4 May 98	Dyke, Maj. O. M	1 Sep. 15	17 Nov. 02	With 111th Mahars.
8 Jan. 01	McCudden, Maj. J. H., D.S.O., M.C.	8 Jan. 16	4 Jan. 13	
5 Oct. 01	Smart, Maj. R. W. L. DeB.	5 Oct. 16	3 Mar. 04	
29 Jan. 02	Robertson, Maj. E. D. S.	29 Jan 17	15 Jan 05	With 2-21st Punj.
27 Jan. 04	Grace, Capt. H. G., M.C.	11 Apl. 13	1 Mar. 06	With 44 Cav.
5 Aug. 05	*Ismay, Bt.-Maj. H. L. (temp. Maj.)*	*5 Aug. 14*	*16 Sep. 07*	*G. S. O. 3rd gr.*
29 Aug. 06	Dane, Capt. R., M.C.	29 Aug. 15	18 Nov. 07	Adjt., 1 Aug. 14.
17 Aug. 07	Lewis, Capt. R. F.	1 Sep. 15	11 Nov. 08	
9 Sep. 08	Garstin, Capt. H. A., M.C.	1 Sep. 15	17 Nov. 09	
14 Jan. 14	Dening, Capt. J. P.	14 Jan. 18	10 Oct. 14	
15 Aug. 14	Logan, Capt. A. F., M.C.	15 Aug. 18	25 May 16	Insptg. Officer, I. S. Troops.
29 June 16	Bevan-Petman, Lt. B. H.	2 July 16	29 June 16	
18 June 17	Hanmer, Lt. C. G.	18 June 18	18 Nov. 17	With 42nd Cav.
4 May 15	Brookes, 2nd-Lt. J. R.	4 May 15	4 Sep. 18	Attd.
31 Aug. 18	Meyer, 2nd-Lt. W. R. M.	31 Aug. 18	4 Sep. 18	Attd.
	Attached.			
20 Jan. 00	Connop, Maj. H. E.	1 Sep. 15	17 Oct. 1	35th Horse.
24 Aug. 12	Morrison, Capt. C. C.	24 Aug. 16	18 Dec. 17	35th Horse.
17 Nov. 16	Page, Lt. A. J.	17 Nov. 17	11 Feb. 18	I. A. (*on probn.*).
9 Nov. 14	Byrne, Lt. G. R.	9 Nov. 15	30 Nov. 17	I. A. R. O.
28 June 16	Latham, Lt. P. D.	28 June 17	4 July 16	I. A. (*on probn.*)
8 Oct. 16	Weaver, Lt. M. G.	8 Oct. 17	30 Apl. 18	Denbig. Yeo., I. A. (*on probn.*).
21 Aug. 16	Clough, Lt. T. (*actg. Capt.*)	21 Aug. 17	31 Aug. 16	I. A. R. O.
21 Aug. 16	Gude, Lt. N. C.	21 Aug. 17	28 Dec. 16	I. A. R. O.
25 Sep. 16	Maxwell, Lt. H. B.	25 Sep. 17	4 May 17	I. A. R. O. *Lv., ex. I., m. c., 6 mos.*
20 Oct. 16	Fitzpatrick, Lt. C.	20 Oct. 17	28 May 17	I. A. R. O.
19 May 15	Harris, Lt. H., M.C. (46, 48*a*, 51)	26 Aug. 16	13 July 18	13th K. L'pool Regt. I. A. (*on probn.*).
17 Mar. 17	Gerrard, Lt. H. J. F.	17 Mar. 18	5 Sep. 18	R. F. A. 77th Bty. I. A. (*on probn.*)
5 May 15	Stewart, Lt. I.	5 May 16	1 Aug. 18	I. A. (*on probation.*)
12 June 17	Thorniley, Lt. J. H.	12 June 18	23 Oct. 18	I. A. R. O.
10 Aug. 17	Cameron, Lt. E. G.	10 Aug. 18	9 Aug. 18	I. A. R. O.

Medical Officers.

		Jemadar.	Ressaidar.	Risaldar.	Risaldar Major.
	Risaldar-Major.				
19 July 98	Sundar Singh (53, 56, 71)	19 July 98	8 Apl. 02	17 Nov. 12	7 Mar. 17.
	Risaldars.				
1 Mar. 89	Kunda Singh, I.O.M., I.D.S.M. (52, 56)	1 Jan. 06	17 Nov. 12	6 Dec. 15	
29 Dec. 94	Latif Khan (53, 56)	1 Feb. 08	1 Apl. 13	16 Sep. 16	
25 Feb. 92	Mahbub Khan *Bahadur* (45*a*)	10 May 10	1 July 14	14 Feb. 17	*Order of Br. I., 2nd Class.*
21 Mar. 94	Narayan Singh *Bahadur* (53, 56, 66*a*).	1 July 10	11 Oct. 14	15 Sep. 17	*Order of Br. I, 2nd Class.*
....	Shad Muhammad Khan (Ret'd.)	....	....	26 Feb. 18	
2 Dec. 98	Farzand Ali	1 July 14	14 Feb. 17	28 June 18.	
18 Apl. 04	Jahangir Beg, I.O.M.	4 July 15	1 Apl. 17	5 Aug. 18.	
17 Apl. 94	Arbela Singh	11 Aug. 12	15 Sep. 17	3 Sep. 18.	

20th Deccan Horse—*contd.*

Date of entering service.	Names.	Army rank.	Remarks.			
		Jemadar.	Ressaidar.	Risaldar.	Risaldar-Major.	
	Risaldar-Majors.					
	Hanooman Singh, *Sardar Bahadur* (*Retd.*)	. . . . .	. . . . .	. . . . .	. . .	Re-employed 25 Feb. 15.
23 May 86	Jharmal Singh (43, 56) . . .	1 July 94	9 Oct. 98	1 Mar. 04	1 Apl. 12	
1 Apl. 90	Prem Singh, *Bahadur* (56) . .	1 July 01	1 July 04	21 Dec. 14	18 May 16	*Order of Br. I., 2nd cl.*
	Risaldars.					
10 June 92	Bachant Singh (52, 56, 66*a*) . .	1 July 04	3 May 10	9 Nov. 15 18 May 16		
1 June 91	Konsal Singh Bahadur, I.D.S.M. (45*a*, 50, 53, 58, 66*a*)	1 Apl. 07	16 Dec. 14	10 Nov. 15		*Order of Br. I., 2nd cl.*
8 Mar. 90	Mir Sadik Ali (51, 52, 56, 65) . .	24 June 11	21 Dec. 14	12 July 16		
10 July 01	Shaikh Faiz-ud-din, I.D.S.M. (58)	1 Apl. 14	25 Aug. 17	4 Aug. 18		
	Ressaidars.					
1 July 93	Mahbub Khan (53, 56, 66*a*) . .	1 Apl. 06	10 Aug. 14			
17 Jan. 98	Mangal Singh (56)	17 Mar. 08	21 Dec. 14			
1 May 89	Hayat Mir Khan (50, 53, 71) . .	2 May 10	21 Dec. 14			
20 June 00	Rachpal Singh (56)	1 Apl. 13	17 May 16			
1 Jan. 97	Muhammad Khan	1 July 12	12 July 16			
6 Nov. 03	Gugan Singh (45*a*, 56) . . .	1 Oct. 13	13 July 16			
4 Nov. 03	Ganga Bishan (56)	16 Dec. 14	11 Oct. 17			
16 Dec. 86	Shaikh Mohi-ud-din	15 June 15	12 Oct. 17			
3 Nov. 94	Rahim Sher Khan	5 Nov. 15	4 Aug. 18			
	Jemadars.					
6 May 92	Ramji Lall (52, 56)	1 Apl. 12				
8 Feb. 10	Bharat Singh (56)	10 Aug. 14				
1 Feb. 02	Sher Singh	2 Dec. 14				
8 Mar. 87	Nabiyar Khan	21 Dec. 14				
21 Feb. 94	Dhulip Singh (I)	21 Dec. 14				
1 Jan. 01	Sirdar Singh, I.O.M. (56) . .	11 Apl. 16				
22 Sep. 03	Dalip Singh (II), I.D.S.M . . .	11 Apl. 16				
18 Aug. 93	Yusuf Ali Khan (I)	17 May 16	Woordie-Maj., 28 Dec. 17.			
13 Aug. 91	Bhawani Singh	19 May 16				
6 Apl. 00	Sayyid Ahmed (56)	11 July 16				
9 Aug. 04	Yusuf Ali Khan (II)	14 July 16				
21 Dec. 02	Bharat Singh (66*a*)	24 July 17				
15 Aug. 03	Dalip Singh (III) (66*a*)	25 July 17				
9 Sep. 03	Sarwan Singh	25 Aug. 17				
24 Oct. 02	Jhanda Singh	25 Aug. 17				
21 Dec. 02	Lahri, I.D.S.M.	11 Oct. 17				
1 Mar. 02	Abdul Razak Khan	12 Oct. 17				

21st Prince Albert Victor's Own Cavalry (Frontier Force) (Daly's Horse)—*contd.*

Date of entering service.	Names.	Army rank. Jemadar.	Ressaidar.	Remarks.
	Ressaidars.			
2 June 94	Atma Singh (45*a*, 56, 66*a*, 71).	17 Nov. 12	6 Dec. 15	
16 Mar. 04	Muhammad Hashum (56, 66*a*)	1 Apl. 10	6 May 16	
14 Sep. 10	Beant Singh (60*a*, 66*a*)	6 Dec. 15	13 Apl. 18	With 42nd Cav.
....	Indar Singh	....	28 June 18	
....	Muhammad Amir	....	3 Sep. 18	
	Jemadars.			
1 Oct. 94	Dewa Singh	21 Dec. 14	5 Aug. 18.	
9 June 01	Surayan Singh (66*a*)	11 Oct. 14		
10 Nov. 07	Sarab Singh (50)	29 Oct. 14	Woordie-Maj.	
3 Feb. 98	Santa Singh (56)	5 Apl. 15		
1 May 02	Saiyid Ahmad Shah (56)	12 Feb. 16	With 42 Cav.	
10 Oct. 03	Chain Singh	7 Apl. 16		
10 May 03	Shahbaz Khan	1 July 16		
1 Mar. 92	Harnam Singh	13 Aug. 16		
20 Aug 99	Mastan Singh (56, 66*a*)	16 Oct. 16		
5 Apl. 86	Haidar Khan	23 June 16		
17 Sep. 94	Hazara Singh	27 July 16		
1 July 91	Abdur Rahim	2 Apl. 17		
15 Mar. 98	Wakil Singh (66*a*)	5 Sep. 17		
30 July 01	Abdul Majid (56, 71)	16 July 17		
27 May 98	*Sher Jan*	*5 July 18*		
1 Aug. 00	Sahib Singh (58)	13 Apl. 18	With 42nd Cav.	
...	Suja Khan	5 Aug 18		
....	Bahadur Sher	11 Sep. 18		

22nd Sam Browne's Cavalry (Frontier Force).

Raised at Lahore in 1849, by Lieut. S. J. Browne, as the 2nd Regiment of Punjab Cavalry. Became the 2nd Regiment of Cavalry, Punjab Irregular Force, 1851; the 2nd Regiment of Cavalry, Punjab Frontier Force, 1865; the 2nd Punjab Cavalry, 1901; the 22nd Cavalry (Frontier Force), 1903; present designation 1904.

"Delhi 1857" "Lucknow" "Ahmad Khel" "Afghanistan, 1878-80."

Composition—1½ Squadrons of Sikhs, ½ of Hindustani Hindus, 1 of Punjabi Musalmans, ½ of Pathans, ½ of Hindustani Musalmans.

Uniform—Scarlet. *Facings*—Blue.

COLONEL—Lieut.-General Robert C. R. Clifford, C.B. . . 14 January 1913.

1st Commn. or date of entering service.	Names and Rank.	Army rank.	Present appointment in Regiment.	Remarks.
	Commandant.			
8 Oct. 90	Young, Lt.-Col. H. G., D.S.O.	8 Oct. 16	13 Nov. 17	
	Squadron Commanders. (4)			
8 Jan. 93	Taylor, Maj. E. M., *p.s.c.*(Q)	28 Jan. 11	21 Dec. 15	2nd-in-Comd. With 45th Cav.
2 Jan. 96	Coke, Maj. A. J. P.	22 Jan. 14	8 May 11	I. S. Troops.
9 Feb. 96	Slingsby, Maj. T. W.	1 Mar. 14	16 Feb. 12	With 1-154th Infy.
20 May 99	Stuart, Maj. W.	1 Sep. 15	29 Sep. 12	
	Squadron Officers. (9)			
21 Feb. 00	*Ramsay, Maj. A. Fitz G.*	*1 Sep. 15*	*13 Oct. 04*	
19 Oct. 01	Hastings, Maj. G. F.	17 Nov. 15	5 Mar. 07	
8 May 01	Yates, Maj. R. J. B., D.S.O.	8 May 16	1 Dec. 05	
13 Aug. 04	Duke, Cap. O. A. [L]	13 Aug. 13	29 Oct. 07	With 2nd Lrs.
18 Jan. 05	*Money, Capt. W. S. E.*	*18 Jan. 14*	*24 Apl. 06*	*Mily. Secy. to Govr., Madras.*
19 Jan. 07	Somerville, Capt. E. A.	1 Sep. 15	11 July 08	
25 Aug. 09	Hutton, Capt. G. M. (*temp. Maj.*)	1 Sep. 15	9 Nov. 10	Qr.-Mr., 9 Oct. 14.
3 Sep. 10	Morrogh Bernard, Capt. J. A. D.	1 Sep. 15	28 Jan. 12	With 42nd Cav.
18 Jan. 11	*Branfoot, Capt. D. C.*	*1 Sep. 15*	*10 Mar. 12*	Attd.
8 Aug. 14	Thwaytes, Capt. L. L.	8 Aug. 18	22 Mar. 16	Qr.-Mr. (*Offg.*)
15 Aug. 15	Melville, Lt. H. J. (*Actg. Capt.*)	15 Aug. 16	8 Aug. 17	Attd.
15 Nov. 15	Bullock, Lt. P. G. W.	15 Nov. 16	28 Nov. 15	Attd.
29 June 16	Manuk, Lt. T. M. (*Actg. Capt.*)	5 Aug. 16	11 July 16	Attd.
24 Apl. 18	Holder, 2nd-Lt. E. D.	24 Apl. 18	20 Oct. 18	Attd.
31 Aug. 18	Jones, 2nd-Lt. R. J. P.	31 Aug. 18	3 Sep. 18	Adgt. (*Offg.*)
31 Aug. 18	White, 2nd-Lt. E. W.	31 Aug. 18	3 Sep. 18	Attd.
	Attached.			
10 Oct. 03	Montagu-Smith, Capt. E.	15 Mar. 15	12 Nov. 17	35th Horse.
11 Dec. 15	Pegge, Lt. F. H.	11 Dec. 16	14 Jan. 16	I. A. R. O.
27 Jan. 16	Cliff, Lt. W. P.	27 Jan. 17	15 Mar. 16	I. A. R. O.
1 Oct. 14	Sheppard-Capurro, Lt. R. M. F. F. deS.	1 July 15	1 Dec. 17	I. A. (*on probn.*)
28 July 17	Ostler, Lt. C.	28 July 17	9 Feb. 17	I. A. R. O.
28 July 16	Brigg, Lt. W. L.	28 July 17	28 July 17	I. A. R. O.
7 Feb. 15	Porral, Lt. F. J. W.	7 Oct. 16	28 May 18	I. A. (*on probn.*)
18 June 11	Landale, Capt. C. H.	1 Sep. 15	10 Feb. 18	16 Cav.
26 July 15	Goldthorpe, Lt. H. D.	. . . . .	7 Aug. 18	I. A. (*on probn.*)
16 Apl. 15	Calder, Lt. J. T.	. . . . .	17 Aug. 18	I. A. R. O.

22nd Sam Browne's Cavalry (Frontier Force)—*contd.*

Date of entering service.	Names and Rank.	Army rank. Jemadar.	Ressaidar.	Risaldar.	Risaldar-Major.	Remarks.
	Risaldar-Major.					
13 Sep. 94	Sirdar Sant Singh (53, 56, 71)	13 Sep. 94	1 Jan. 06	24 Nov. 08	1 Nov. 14.	
	Risaldars.					
5 May 95	Kishan Singh (53, 56, 71)	1 Sep. 05	1 Jan. 08	6 May 11		
1 Apl. 96	*Asam Ali (53, 56)* I.D.S.M.	*1 Jan. 08*	*29 Dec. 08*	*1 July 14*		
18 Feb. 92	Mir Alam Khan* (53, 56, 71) *Bahadur.*	22 Jan. 03	6 May 17	27 Jan. 15		*Order of Br. I., 2nd* [*Cl.*
4 Jan. 94	Ganda Singh (53, 56, 65) *Bahadur*	1 Jan. 06	1 Oct. 12	1 Nov. 14		*Order of Br. I., 2nd* [*Cl.*
18 Apl. 05	Sirdar Autar Singh, I.O.M. (45*a*, 53, 56, 66*a*, 71).	18 Apl. 05	1 July 14	4 Jan. 15		
...	Bhagwan Singh (*Retd.*)	....	...	...		Re-employed. 25 June 15.
1 May 05	Sukhpal Singh (45*a*, 53,56, 66*a*,71)	1 Mar. 08	29 Oct. 14	28 Sep. 16		
1 May 06	Ram Kunwar Singh (53, 56, 66*a*, 71).	6 May 11	29 Aug. 16	20 June 17		
28 July 97	Ghulam Kadir	1 Nov. 14	28 Sep. 16	31 Mar. 18		With 34th Horse.
20 Nov. 00	Mahamad Yasin	21 Dec. 14	....	4 Aug. 18		
	Ressaidars.					
9 Sep. 03	Jamal Khan (45*a*, 71)	5 Apl. 15	9 Jan. 17			
1 Oct. 00	Idris Khan (45*a*, 53, 71)	7 Dec. 14	19 Jan. 17			
30 Mar. 94	Harnam Singh (I) (56)	1 Oct. 12	11 Feb. 17			
16 Oct. 04	Kasim Raza	11 Feb. 17	18 Feb. 17			
25 Dec. 06	Bostan Khan	19 Jan. 17	20 June 17			
5 Jan. 01	*Habib Khan (53, 56)*	*29 Dec. 08*	....			*With 2nd Lrs.*
1 Dec. 04	Tara Singh	15 Jan. 16	31 Mar. 18			
14 Aug. 99	Fateh Muhammad	24 May 16	31 Mar. 18			
6 Oct. 04	Abdul Qadir	28 Sep. 16	31 Mar. 18			
	Jemadars.					
11 Mar. 12	Gurbaksh Singh	1 Nov. 14				
30 Mar. 95	Harnam Singh (71)	7 Dec. 14				
20 Feb. 11	Narain Singh (45*a*)	7 Dec. 14				8th Cav.
13 May 95	Kesar Singh (II)	16 Jan. 15				
30 Sep. 95	Ali Asghar (*Ag. Ressldr.*)	7 Apl. 16				With 65th Camel Corps.
11 May 00	Rahim Bakhsh (*Ag. Ressldr.*)	9 Apl. 16				With 65th Camel Corps.
16 July 00	Khan Bahadur Khan	29 Aug. 16				
16 Sep. 10	Raghbir Singh	28 Sep. 16				
10 Dec. 98	Yusaf Khan	3 Oct. 16				8th Cav.
1 Nov. 98	*Sialu Singh*	*21 Dec. 16*				With Remt. Base.
1 Jan. 00	Chiragh Beg	9 Jan. 17				
8 Mar. 09	Lal Khan	19 Jan. 17				
1 Mar. 03	Baihram Khan	9 Feb. 17				
14 Aug. 08	Rauf Khan	18 Feb. 17				Woordie-Maj., 18th Feb. 17.
1 Jan. 10	Muhammad Usman	8 May 17				
1 Dec. 00	Qasim Ali	25 June 17				
3 Feb. 98	Rahim Dad Khan	18 Aug. 17				
5 Feb. 99	Jetha Singh	25 Nov. 17				With 9th Horse.
2 June 02	Dharam Singh	16 Dec. 17				
1 Jan. 95	Samund Sing.	13 Apl. 18				
1 Jan. 96	Bostan Khan	13 Apl. 18				With 42nd Cav.
7 May 06	Janmeje Singh	13 Apl. 18				With 42nd Cav.
5 Mar. 95	Ghulam Haider	5 May 18				
7 July 08	Amrik Sing	9 Aug. 18				

23rd Cavalry (Frontier Force).

Raised at Lahore in 1849, by Lieut. W. G. Prendergast, as the 3rd Regiment of Punjab Cavalry. Became the 3rd Regiment of Cavalry, Punjab Irregular Force, 1851; the 3rd Regiment of Cavalry, Punjab Frontier Force, 1865; the 3rd Punjab Cavalry, 1901; present designation, 1903.

"Kandahar, 1880" "Afghanistan, 1879-80."

Composition—1½ Squadrons of Sikhs, ½ of Dogras, 1 of Hindustani Musalmans, ½ of Punjabi Musalmans, ½ of Pathans.

***Uniform*—Blue. *Facings*—Scarlet.**

COLONEL—Field-Marshal *Sir* Charles C. Egerton, G.C.B., D.S.O. . . . . 13 May 1904.

1st Commn. or date of entering service.	Names and Rank.	Army rank.	Present appointment in Regiment.	Remarks.
	Commandant.			
7 Nov. 91	Hawes, Lt.-Col. C. H., M.V.O., D.S.O.	7 Nov. 17	1 May 18	
	Squadron Commanders. (3)			
16 Jan. 95	*Godwin, Bt.-Col (temp. Br.-Genl.) C. A. C., D.S.O., p.s.c. (Q)*	*3 June 18*	*1 May 18*	*2nd in-Comd.*
21 Jan. 98	*Crombie, Maj. D. C., p. s. c. (n)*	*1 Sep. 15*	*3 Sep. 16*	*G. S. O., 2nd gde., Bannu Bde.*
20 Jan. 00	Strettell, Maj. C. B. D. . . .	1 Sep. 15	1 May 18	
18 Apl. 00	Cameron, Maj. C. S.	1 Sep. 15	28 Oct. 18	
	Squadron Officers. (9)			
8 May 01	Kirkwood, Maj. C. H. K. (L) .	8 May 16	5 Mar. 04	
22 Oct. 02	Gannon, Maj. J. C. R. . . .	22 Oct. 17	4 Oct. 06	Comdt. R. T. Depot.
2 May 03	Allen, Maj. W. T. [*l*] . . .	2 May 18	1 Mar. 06	
28 Nov. 03	Paske, Maj. C. A. M. . . .	28 Nov. 18	18 July 06	
13 Aug. 04	*Lowry-Cory, Capt. W.* . . .	*13 Aug. 13*	*1 Mar. 06*	*P. A. to Ch. Commr., Assam.*
5 Aug. 05	*Cave Browne, Capt. H.* . .	*5 Aug. 14*	*9 Jan. 09*	*D. A. A. & Q. M. G.*
14 Jan. 14	Walker, Capt. B. W. G., M.C. .	14 Jan. 18	11 Oct. 14	
1 Oct. 14	Carr-White, Capt. G.	1 Oct. 18	15 Dec. 14	Qr.-Mr., 15 Dec. 16.
	Attached.			
1 Oct. 17	*Brierley, Capt. R.*	*1 Oct. 18*	*5 Mar. 15*	*I. A. R. O.*
21 Dec. 14	Chapman, Capt. H. R.	21 Dec. 15	17 May 18	Welsh R., I. A. (*on probn.*).
8 Jan. 15	Marsh, Lt. N. O. C.	8 Jan. 16	18 Jan. 15	I. A. R. O.
1 Apl. 15	Watkins, Lt. V. C.	1 June 16	14 Jan. 18	Derby Yeo. I. A. (*on probn.*)
23 Apl. 15	Hewett, Lt. H. P. (*Actg. Capt.*)	22 Mar. 16	22 Mar. 16	I. A. R. O.
19 Feb. 16	Onraët, Lt. P. T.	19 Feb. 17	31 Oct. 16	I. A. R. O.
28 Sep. 16	Warren, Lt. W. J.	28 Sep. 17	24 Oct. 16	I. A. R. O.
24 Jan. 17	Gardiner, 2nd-Lt. A. L. . . .	24 Jan. 17	27 Jan. 18	Norf. R. I. A. R. O. (*on probn.*).
26 Sep. 17	Gunson, 2nd-Lt. G. H. . . .	26 Sep. 17	26 Jan. 18	I. A. R. O. (*on probn.*).
31 Jan. 18	Titham, 2nd-Lt. P. R. . . .	31 Jan. 18	6 Feb. 18	I. A. (*on probn.*).
1 Sep. 18	Costello, 2nd-Lt. J. W. . . .	1 Sep. 18	4 Sep. 18	I. A. (*on probn.*).
15 Nov. 15	Stapylton, Lt. J. M.	15 Nov. 16	. . . .	
13 Aug. 15	Ewers, Lt. L. F.	11 Dec. 18	13 Oct. 18	M. G. C., I. A. (*on probn.*)
7 Aug. 14	Marriot, Lt. N. E.	1 Sep. 15	15 Feb. 15	25th Cav.

		Jemadar.	Ressaidar.	Risaldar.	Risaldar-Major.	
	Risaldar-Majors.					
6 Sep. 00	Husamuddin Khan, *Bahadur*, I.D.S.M. (53, 56, 66).	16 Sep. 00	25 Nov. 01	1 Oct. 08	1 July 13	*Order of Br. I., 2nd Class.*
	Arjun Singh, *Bahadur* (Retd.) .	. . . .	. . . .	. . .	. .	Re-empld. 13 Nov. 15, *Order of Br. I., 2nd cl.*

		Jemadar.	Ressaidar.	Risaldar.	
	Risaldars.				
1 May 90	Shibditt Singh (71)	11 Dec. 02	1 Oct. 08	3 Jan. 10	
28 Dec. 96	Muhammad Yákúb Khan (66*a*, 71)	May 06	3 Jan. 10	{30 Apl. 15 / 30 Apl. 16}	
2 Oct. 85	Dost Muhammad (71)	25 May 08	25 May 11	1 May 15	
1 Mar. 89	Sher Singh (56, 71)	1 May 08	1 July 13	11 Sep. 15	
	Mahmud Khan	. . . .	. . . .	. . . .	Re-employed 1 Aug. 17.
1 May 95	Sant Singh.	1 Apl. 09	10 Apl. 15	28 Nov. 17	

23rd Cavalry (Frontier Force)—*contd.*

Date of entering service.	Names.	Army rank.		Remarks.
		Jemadar.	Ressaidar.	
	Ressaidars.			
1 July 93	Phula Singh (45*a*, 50, 56, 66*a*, 71)	1 Oct. 08	1 May 15	
22 Nov. 98	Yakub Ali (66*a*)	1 May 13	30 Apl. 16	
26 Sep. 94	Bhagat Singh (56, 66*a*, 71) . .	27 Sep. 10	23 Nov. 17	
...	Jagat Sing	. . . . .	1 May 18	
1 Feb. 98	*Habibul Rahman Khan* (58, 66*a*)	*1 Apl. 11*	1 May 18	
	Jemadars.			
1 July 13	Balwant Singh	1 July 13		Woordie-Maj.
15 June 06	Sultan Muhammad (65, 66*a*) . .	29 Oct. 14		
5 Apl. 96	Sahib Dad	7 Dec. 14		
2 Mar. 92	Rur Singh (66*a*, 71)	4 Jan. 14		
29 May 99	Usman Khan (66, 71)	1 May 15		
5 Apl. 09	Muhammad Afzal (56)	2 July 15		
12 Dec. 94	Natha Singh (56, 71)	14 Feb. 16		
10 Nov. 14	Rajindar Singh	23 June 16		
4 Apl. 13	Musharaf Ali Khan	21 July 16		
7 Jan. 01	Jiwan Singh	1 Mar. 17		
5 Aug. 92	Mohkam Chand	1 May 17		
16 Oct. 94	Dittu Singh	28 Nov. 17		
22 Dec. 95	Chanda Singh	7 Jan. 18		
...	*Ismail Jan*	*5 July 18*		
27 Dec. 02	Sohbat Khan	15 Aug. 18		
. . . .	Adalat Khan	1 May 18		

25th Cavalry (Frontier Force).

Raised at Mooltan in 1849, by Capt. R. Fitzgerald, Bombay Army, as the 5th Regiment of Punjab Cavalry. Became the 5th Regiment of Cavalry, Punjab Irregular Force, 1851; the 5th Regiment of Cavalry, Punjab Frontier Force, 1865; the 5th Punjab Cavalry, 1901; present designation, 1903.

"Delhi, 1857" "Lucknow" "Charasiah" "Kabul, 1879" "Afghanistan, 1878-80."

Composition—1 Squadron of Sikhs, 1 of Dogras, 1 of Punjabi Musulmans, ½ of Hindustani Musalmans, ½ of Pathans.

Uniform—Dark-green. *Facings*—Scarlet. *Puggri*—Scarlet.

1st Commn. or date of entering service.	Names and Rank.	Army rank.	Present appointment in Regiment.	Remarks.
	Commandant.			
1 Mar. 90	Peck, Bt.-Col. A. W., C.M.G.	1 Jan. 17	19 June 17	
	Squadron Commanders. (4)			
3 Sep. 92	Watts, Lt.-Col. G. A. R.	3 Sep. 18	11 July 17	2nd-in-Comd. Comdt. (*temp.*)
7 May 98	*Tyrrell, Maj. A. C. L., p.s.c. (temp. Lt.-Col.)*	*1 Sep. 15*	*19 Jan. 14*	*A. A. & Q. M. G., S. Comd.*
1 Sep. 97	*Dening, Maj. H.*	*1 Sep. 15*	*19 Jan. 14*	*I. S. Troops.*
27 July 98	*McConaghy, Maj. J. G.*, D.S.O.	*1 Sep. 15*	*19 Aug. 14*	*D. A. Q. M. G., 2nd Dn.*
	Squadron Officers. (9)			
17 Jan. 00	Burrowes, Maj. P. W.	1 Sep. 15	1 July 08	2nd-in-Comd. (*temp.*)
20 Jan. 00	*Grant, Maj. E. P.*	*1 Sep. 15*	*29 Feb. 04*	*D.A.A.G., 1st Dn.*
27 Aug. 02	*Niven, Maj. D. S.*	*27 Aug. 17*	*17 Feb. 05*	*D.A.A.G.*
21 Jan. 03	Gall, Maj. L., M.C.	21 Jan. 18	31 Mar. 05	With Guides Cav.
24 Jan. 05	*O'Rorke, Capt. J. M. W.*, D.S.O.	*24 Jan. 14*	*23 Mar. 08*	
24 Jan. 06	Yule, Capt. H. V.	24 Jan. 15	25 Apl. 07	With 42nd Cav.
19 Jan. 07	Nethersole, Capt. J., M.C.	1 Sep. 15	13 July 08	With 7th Lrs.
9 Sep. 08	Buckland, Capt. L. L.	1 Sep. 15	4 Dec. 09	With 27th Cav.
8 Sep. 09	Fraser-Tytler, Capt. W. K., M.C.	1 Sep. 15	4 Nov. 11	Adjt., 3 Ocr. 18.
24 Aug. 12	Trehane, Capt. C. H.	24 Aug. 16	11 Nov. 13	With 5th Cav.
22 Jan. 13	*Hall, Capt. I. F. G.*	*22 Jan. 17*	*13 Mar. 14*	With Patiala Lrs.
7 Aug. 14	Marriott, Capt. N. E.	8 Aug. 18	15 Feb. 15	With 23rd Cav.
18 Apl. 16	Foster, Lt. J. W.	18 Apl. 17	27 Apl. 16	
14 Nov. 16	Pecker, Lt. C. B.	14 Nov. 17	29 Nov. 16	
14 Nov. 16	Jewell, Lt. J. W.	14 Nov. 17	30 Oct. 17	Attd.
12 Sep. 17	Burridge, Lt. R. A.	12 Sep. 18	30 Dec. 17	Attd. Adjt. (*Offg.*)
12 Sep. 17	Schokman, Lt. C. C.	12 Sep. 18	16 Jan. 18	Attd.
31 Aug. 18	Coleman, 2nd-Lt. C. M.	31 Aug. 18	4 Sep. 18	Attd.
31 Aug. 18	Turner, 2nd-Lt. M. R.	31 Aug. 18	4 Sep. 18	Attd.
	Attached.			
6 Nov. 09	James, Capt. (*temp. Maj.*) B. S.	1 Sep. 15	10 Jan. 17	R. Muns. Fus. I. A. (*on probn.*)
17 Sep. 13	Lynn, Capt. J. A. G.	7 Sep. 17	17 May 18	I. A. (*on probn.*)
20 Jan. 16	Leigh, Lt. A. P.	20 Jan. 17	8 July 18	Lan. Fus. I. A. (*on probn.*)
5 July 15	Ashbee, Lt. H. G. N.	5 Apl. 17	15 June 18	A. S. C. I. A. (*on probn.*)
17 Dec. 15	MacMahon-O'Byrne, Lt. D.	17 Dec. 16	2 July 17	I. A. (*on probn.*) *Lv., ex I., 6 mos., m. c.*
22 July 16	Ellis, Lt. R. (*Actg. Capt.*)	22 July 17	6 Aug. 16	I. A. R. O.
21 Aug. 16	Steven, Lt. T. M. (*Actg. Capt.*)	21 Aug. 17	31 Aug. 16	I. A. R. O.
29 Aug. 16	Steedman, Lt. L.	29 Aug. 17	23 May 18	North'd. Fus. I. A. (*on probn.*)
25 June 18	Dignam, 2nd-Lt. J. R.	25 June 18	2 June 18	Res. Cav. I. A. (*on probn.*)
26 Jan. 18	Clarke, 2nd-Lt. D. G.	26 Jan. 18	16 Aug. 18	Canadians. I. A. (*on probn.*)
21 Mar. 18	Gregson, 2nd-Lt. D. A.	21 Mar. 18	4 June 18	I. A. R. O. I. A. (*on probn.*)
27 May 18	Edmondson, 2nd-Lt. H. C.	27 May 18	....	I. A. R. O.
27 May 18	Ross, 2nd-Lt. H. C.	27 May 18	24 June 18	I. A. R. O.
1 June 18	Wilmer, 2nd-Lt. D. H.	1 June 18	....	I. A. R. O. Qr.-Mr. (*temp.*)

		Jemadar.	Ressaidar.	Risaldar.	Risaldar Major.
	Risaldar-Majors.				
3 Jan. 88	Kale Khan (71) I. D. S. M.	1 Dec. 03	29 Dec. 04	1 Feb. 11	1 Aug. 17
	Risaldars.				
5 Nov 01	Thakur Chand (66*a*)	15 Nov. 01	1 Oct. 05	15 Feb. 11	
1 Apl. 98	Surayan Singh (71)	5 Oct. 03	1 Nov. 09	16 Jan. 13	
1 Oct. 96	Mahmud Khan (45*a*, 56, 66*a*, 71)	1 Dec. 06	1 Feb. 11	1 Aug. 17	
13 July 99	Sripat (56)	1 Aug. 05	15 Feb. 11	1 Aug. 17	
1 Jan. 95	Amir Mahammad Khan (56, 71)	16 May 10	16 Jan. 13	18 Oct. 17	

25th Cavalry (Frontier Force)—*contd.*

Date of entering service.	Names.	Army rank. Jemadars.	Army rank. Ressaidar.	Remarks.
	Ressaidars.			
1 Mar. 98	*Hakim Singh (45a, 56, 71)*	*1 Nov. 09*	*1 June 14*	*Remt. Depôt, Sehore.*
1 Oct. 05	*Sher Ali Khan (50)*, I. D. S. M.	*1 Oct. 05*	*4 Apl. 16*	
1 Aug. 96	Sadhu Singh (56, 66a)	1 Apl. 10	1 Mar. 17	
5 Dec. 97	Harditt Singh	7 Dec. 14	1 Aug. 17	
24 Dec. 00	Hak-Nawaz (56, 66a)	29 Dec. 04	18 Oct. 17	
	Jemadars.			
1 Feb. 96	Rajwali Khan (45a, 71)	1 June 14		Woordie-Maj., 24 June 14.
1 Sep. 94	Narayan Singh (71)	15 Dec. 15		
1 Jan. 95	Kirath Singh	27 Dec. 15		
1 Jan. 96	Mumtaz Muhammad Khan	27 Jan. 16		
29 Jan. 16	Bidhi Chand	29 Jan. 16		*On probn.*
19 Oct. 00	Ram Singh (66a, 71)	8 Mar. 16		
1 Nov. 97	*Natha Singh*	*12 July 16*		
15 Oct. 02	Ashraf Khan (51)	1 Aug. 16		With 1st S. & M.
1 Mar. 98	Ali Shan, I. D. S. M.	1 Aug. 17		
16 Aug. 01	Jan Muhammad	1 Aug. 17		
1 Sep. 02	Abdul Latif	18 Oct. 17		
9 Jan. 00	Sobail Singh (71)	7 Nov. 17		
....	Lal Singh	....		Re-employed 25th Nov. 17.
16 July 02	Abdul Samad (45a)	22 Mar. 18		
11 June 02	Zaman Mehdi Khan	12 Apl. 18		
1 Oct. 02	Dina Nath	10 June 18		Qr.-Mr. Jem. 1st Sept. 18.

26th King George's Own Light Cavalry.

Raised in 1787, by Capt. H. Darley, as the 5th Regiment of Madras Native Cavalry. Became the 1st Regiment of Madras Native Cavalry, 1788; the 1st Regiment of Madras Light Cavalry, 1819; the 1st Regiment of Madras Lancers, 1886; the 1st Madras Lancers, 1901; the 26th Light Cavalry in 1903; 26th Prince of Wales's Own Light Cavalry in 1906; present designation in 1910.

THE PLUME OF THE PRINCE OF WALES. THE ROYAL AND IMPERIAL CYPHER.

"Mysore" "Seringapatam" "Ava"

"Afghanistan, 1879-80" "Burma, 1885-87."

Composition—1 Squadron of Madras and Dekhani Musalmans, 1 of Punjabi Musalmans, 1 of Rajputana Rajputs, 1 of Jats, 2 Addl. Squadrons.

Uniform—French Grey. *Facings*—Buff. *Lace*—Silver.

COLONEL-IN-CHIEF—THE KING.

HONORARY COLONEL—Hony. Colonel *His Highness* Maharaja *Sir* Sri Krishnaraja Wadia *Bahadur*, of MYSORE, G.C.S.I., G.B.E. . . . 24 June 1910.

1st Commn. or date of entering service.	Names and Rank.	Army rank.	Present appointment in Regiment.	Remarks.
	Commandant.			
4 Mar. 91	Henderson, Bt.-Col. M. H. . .	3 June 18	26 Apl. 17	
	Squadron Commanders. (4)			
5 Aug. 96	*Arthur, Maj. L. F.*, D.S.O., *p.s.c.* *[L]*.	*5 Aug. 14*	*25 Sep. 17*	*2nd-in-Comd., Asst. Comdt., Cadet College, Quetta.*
20 Jan. 17	Popham, Bt.-Lt.-Col. E. L., D.S.O.	3 June 18	18 June 12	
26 June 01	Mowat, Maj. J. S.	1 Mar. 16	26 Apl. 17	
8 May 01	Moir, Maj. J. S. S.	8 May 16	25 Sep. 17	
	Squadron Officers. (9)			
19 Jan. 04	*Summers, Bt. Maj. G. H. (temp. Lt.-Col.)*	*7 Jan. 14*	*25 May 07*	*O. C. Troops, Somaliland.*
13 Aug. 04	Collum, Capt. J. A. [L], D.S.O.	13 Aug. 13	29 May 07	
5 Aug. 05	*Johnson, Capt. H. A. B.* . .	*5 Aug. 14*	*27 Feb. 07*	*Instr. Young Offrs. Sch., Bangalore.*
4 May 07	Addington, Capt. Hon. R. A. .	1 Sep. 15	13 Nov. 12	
25 Jan. 08	*Greenway, Capt. C. K.* . . .	*1 Sep. 15*	*28 Mar. 09*	*Comdt. Bodyguard and Mily. Secy. to Govr., Bombay. Lv., ex I., m. c.*
9 Sep. 08	Armstrong, Capt. A. G. . . .	1 Sep. 15	7 Dec. 09	
3 Sep. 10	O'Donnell, Capt. B. H. . . .	1 Sep. 15	23 Dec. 11	
5 Oct 10	Handley, Capt. L. M., M.C. . .	1 Sep. 15	17 Oct. 13	With Aden Troop.
24 Aug. 12	*Worlledge, Capt. F. H* . . .	*24 Aug. 16*	*11 Nov. 13*	*With Som. Ind. Contg.*
28 May 15	Powell, Lt. W. N., M.C. (*Actg. Capt.*)	28 May 16	15 Apl. 17	
17 Aug. 15	Chalmers, Lt. T. C. (*Actg. Capt.*)	17 Aug. 16	20 Aug. 17	
31 Aug 18	Clarabut, 2nd-Lt. F. S. . . .	31 Aug. 18	14 Sep. 18	Attd.
31 Aug. 18	Brett, 2nd-Lt. W. H. . . .	31 Aug. 18	14 Sep. 18	Attd.
21 Sep. 17	Bruce-Smith, Lt. R.	21 Sep. 17	3 Oct. 17	Attd. *Lv., ex I., m. c., 6 mos.*
	Attached.			
4 Dec. 14	Taylor, Lt. H.	4 Dec. 15	24 Sep. 16	I. A. R. O.
3 Mar. 15	Noakes, Lt. E. J. W.	3 Mar. 16	22 May 18	Lond. R. I. A. (*on probn.*)
20 May 15	Mostyn-Owen, Lt. H. L. . . .	20 May 16	1 Sep. 17	I. A. (*on probn.*)
10 July 15	Williams, Lt. J. F. H. . . .	10 July 16	4 Nov. 17	I. A. R. O.
30 Dec. 15	Dunckley, Lt. P. J.	30 Dec. 16	24 Aug. 17	I. A. R. O.
12 Feb. 16	Wells, Ag. Capt. R. L. P. (*Actg. Capt.*)	16 Dec. 17	24 Sep. 17	I. A. R. O. Adjt.
31 Aug. 16	Keighley, Lt. F. W., M.C. . .	31 Aug. 16	13 June 18	I. A. R. O.
14 Mar. 17	Smith, Lt. L. W.	14 Mar. 18	21 Aug. 17	I. A. R. O.
24 Mar. 17	Hulatt, Lt. J. C.	24 Mar 18	8 Apl. 17	I. A. R. O. *Lv., ex I., m.c., 6 mos.*
6 Apl. 17	Williams, Lt. V. S.	6 Apl. 18	21 Apl. 17	I. A. R. O.
4 July 17	Turner, Lt. G. F.	4 July 18	24 July 17	I. A. R. O.
11 Aug. 17	Richardson, Lt. R. K. . . .	11 Aug. 18	16 Sep. 17	I. A. R. O. Qr.-Mr.
25 Sep. 17	Radcliffe, 2nd-Lt. C.	25 Sep. 17	3 July 18	I. A. (*on probn.*)
31 May 18	Kemp, 2nd-Lt. W. N. R. . .	31 May 18	29 June 18	I. A. R. O.
11 June 18	Gratton, 2nd-Lt. J. S. . . .	11 June 18	11 July 18	I. A. R. O.

26th King George's Own Light Cavalry—*contd.*

Date of entering service.	Names.	Army rank.				Remarks.
	Risaldar-Majors.					
		Jemadar.	**Ressaidar.**	**Risaldar.**	**Risaldar-Major.**	
	Risaldars.					
2 Feb. 09	Lakhpat Singh (56)	1 Sep. 09	29 Oct. 14	27 Oct. 16		
8 Mar. 96	Raghubir Singh (56)	10 Sep. 10	27 Oct. 16	1 Dec. 17		
11 Oct. 98	Feroze Khan (45*a*, 56), I.D.S.M.	1 May 10	19 Nov. 14	3 Jan. 18		
9 Mar. 04	Ahmad Khan (45*a*, 56)	29 Oct. 10	22 Sep. 17	31 July 18		
	Ressaidars.					
8 May 96	Ganpat Singh (56)	6 Mar. 09	4 Apl. 17			
13 July 04	Amrith Singh (56)	29 Oct. 14	22 Sep. 17			
4 Feb. 04	Alam Shah (56)	7 Oct. 15	3 Jan. 18			
13 Mar. 04	Ganga Pershad	13 Mar. 17	22 Mar. 18			
9 Mar. 04	Sher Bahadur (56, 66*a*)	10 Oct. 13	31 July 18			
	Jemadars.					
21 Nov. 90	*Abdul Latif (56)*	*1 Aug. 10*				*With Mysore I. S. Lrs.*
3 Oct. 00	*Saiyid Abdul Jalil*	*12 May 15*				With Syce Corps
2 Jan. 01	*Hargian Singh*	*1 Apl. 16*				*With K. T. Depôt.*
1 Dec. 99	Khazan Singh	17 July 16				
19 Mar. 04	*Ibrahim Khan*	*17 July 16*				*With Ind. Conv. Sec.*
1 Feb. 02	Guman Singh	3 Apl. 17				
1 Jan. 06	Paip Singh (56)	3 Apl. 17				
13 July 04	Tara Singh (45*a*, 52, 56, 58, 66*a*)	1 May 17				
6 Jan. 02	Khan Singh	7 July 17				
22 Apl. 04	Amichand	11 July 17				
1 Apl. 02	Chandar Singh	1 Dec. 17				
4 June 01	Nepal Singh	3 Dec. 17				Qr.-Mr. Jemdr.
9 Mar. 04	Muhammad Sher, I.D.S.M.	3 Jan. 18				Woordie Maj.
17 Apl. 04	Aidan Singh, I.D.S.M.	3 Jan. 18				
2 Mar. 04	Daliyaw Singh	3 Jan. 18				
12 Dec. 08	Toder Singh	2 Mar. 18				
14 May 04	Khazon Singh	22 Mar. 18				
9 May 04	Sher Mohammad	1 Aug. 18				

27th Light Cavalry.

Stevenson-ki-paltan.—Date of raising unknown; was in the service of the Nawab of Arcot prior to 1780. Received into the service of the East India Company, 1784, as the 3rd Regiment of Madras Native Cavalry. Became 1st Regiment of Madras Native Cavalry, 1784; the 4th Regiment, 1786; and the 2nd Regiment, 1788. Became the 2nd Regiment of Madras Light Cavalry, 1819; the 2nd Regiment of Madras Lancers, 1886; the 2nd Madras Lancers, 1901; present designation, 1903.

"Sholinghur" "Carnatic" "Mysore" "Seringapatam" "Burma, 1885-87."

Composition—1 Squadron of Rajput Musalmans, 1 of Punjabi Musalmans, 1 of Rathore Rajputs, 1 of Jats, 2 Addl. Squadrons.

Uniform—French Grey. *Facings*—Buff. *Lace*—Silver.

1st Commn. or date of entering service.	Names and Rank.	Army rank.	Present appointment in Regiment.	Remarks.
	Commandant.			
7 Nov. 91	Birdwood, Col. H. B. (temp. Br.-Genl.)	6 Apl. 18	20 June 16	Bde. Comdr.
	Squadron Commanders. (4)			
20 Feb. 95	*Wheatley, Maj. G. R. P. (Actg. Lt.-Col.)*	*20 Feb. 13*	*20 June 16*	
4 Aug. 97	*Casson, Bt. Lt.-Col. W. F. S., D.S.O., p.s.c. (Bt. Lt.-Col.)*	*4 Aug. 15*	*22 Nov. 13*	*G. S. O., 1st gr.*
21 Feb. 00	Abbay, Maj. B. N. (*temp. Lt.-Col.*)	1 Sept. 15	9 Mar. 18	Comdt. (*temp.*)
11 May 01	Jackson, Maj. V. L. F. P.	8 Nov. 16	13 Sep. 14	
	Squadron Officers. (9)			
24 Dec. 02	Durham, Maj. W. G. W. [L]	24 Dec. 17	1 Jan. 06	2nd-in-Comd. (*temp.*)
27 May 03	*Wise, Capt. A. D.*	*27 May 12*	*1 Mar. 06*	
4 June 04	*Moss-Blundell, Maj. R. S., M.C.*	*26 Aug. 18*	*1 Aug. 07*	*Inst. of Musky. Pachmari.*
6 Oct. 06	Phillips, Capt. H. S. (*temp. Maj.*)	1 Sept. 15	19 June 18	D. A. A. G.
19 Jan. 07	*Jobson, Capt. T. S.*	*1 Sept. 15*	*14 Mar. 08*	*Asst. Supdt., Remt. Dept.*
9 Sep. 08	*Hickson, Capt. B. E.*	*1 Sept. 15*	*31 Oct. 10*	*Asst. Supdt., Remt. Depôt*
10 June 15	Oldridge de la Hey, Lt. C. J.	16 June 16	27 Oct. 15	
14 July 15	Cooke, Lt. G. M. B.	14 July 16	18 Dec. 15	
29 June 16	Philpott, Lt. F. L.	23 Feb. 16	19 July 16	With 1-155th Prs.
18 June 17	Sproule, Lt. R. F.	18 June 18	18 June 18	Attd.
31 Aug. 18	Roberts, 2nd-Lt. V. P.	31 Aug. 18	4 Sep. 18	Attd.
31 Aug. 18	Russell, 2nd-Lt. H. L.	31 Aug. 18	4 Sep. 18	Attd.
1 Dec. 17	Lowson, 2nd-Lt. J.	1 Dec. 18	....	Attd.
	Attached.			
9 July 08	Buckland, Capt. L. L.	1 July 15	23 Dec. 17	25 Cav.
19 May 15	Dudley, Lt. S. (207*a*) (*Actg. Capt.*)	19 May 15	5 Dec. 17	I. A. (*on probn.*)
1 Oct. 16	Simons, Lt. F. H. (207*a*)	1 Oct. 17	12 Oct. 17	I. A. (*on probn.*)
28 May 16	O'Keeffe, Lt. A. R. W. M. (*Actg. Capt.*)	28 May 17	23 May 17	I. A. (*on probn.*)
16 Dec. 14	Fewtrell, Lt. A. H. H. (*Actg. Capt.*)	16 Dec. 15	25 Mar. 18	I. A. (*on probn.*) Adjt. 1 Aug. 18.
18 Jan. 16	Heath, Lt. S. F.	18 Jan. 17	5 Feb. 18	I. A. R. O.
7 Apl. 16	Dixon, Lt. W.	7 Apl. 17	16 Aug. 18	I. A. (*on probn.*)
5 May 15	Turner, Lt. J. W., M. C.	5 May 16	25 Mar. 18	I. A. (*on probn.*)
28 Aug. 17	Innocent, 2nd-Lt. A.	28 Aug. 17	23 Jan. 18	I. A. R. O.
28 Aug. 17	Bullmore, Lt. F. E.	28 Aug. 18	5 Feb. 18	I. A. (*on probn.*) Qr. Mr., 7 Mar. 18.
21 Dec. 16	Hopkins, 2nd-Lt. F. L.	21 Dec. 16	23 Jan. 18	I. A. R. O.
14 June 15	Claridge, Lt. D. P.	14 June 16	....	R. Ir. R. I. A. (*on probn.*)
7 Feb. 17	Newman, Lt. E. J.	7 Feb. 17	19 Feb. 17	I. A. R. O. (*on probn.*)
26 Feb. 17	Keston, Lt. T. H.	26 Feb. 18	31 Mar. 17	I. A. R. O. (*on probn.*)
15 Mar. 17	Gillespie, Lt. A. D.	15 Mar. 18	2 Apl. 17	I. A. R. O.
4 Apl. 17	Druitt, Lt. T. R. (207*a*)	4 Apl. 18	4 Apl. 18	I. A. R. O.
26 Apl. 17	Hawkeswood, Lt. H. W.	26 Apl. 18	18 May 17	I. A. R. O.
26 July 17	Alexander, Lt. H. B.	26 July 17	27 Aug. 17	I. A. R. O.
30 May 15	Young, Lt. G. M. K.	30 May 16	....	I. A. (*on probn.*)
21 Sep. 17	Peate, Lt. W. E.	21 Sep. 18	....	I. A. (*on probn.*) *Lv., ex I., m.c., 6m.*
29 May 18	Dyson, 2nd-Lt. F.	29 May 18	....	I. A. (*on probn.*)
19 Feb. 18	Wilson, 2nd-Lt. W., D.C.M.	19 Feb. 18	25 Mar. 18	High L. I.—I. A. R. O. (*on probn.*)
16 Mar. 18	McDougall, 2nd-Lt. F. A.	16 Mar. 18	16 Mar. 18	I. A. R. O. (*on probn.*)
....	Barr, 2nd-Lt. A.	....	16 Mar. 18	
1 Dec. 15	Houchen, Lt. J. B.	1 Dec. 16	1 June 18	I. A. R. O.

		Jemadar.	Ressaidar.	Risaldar.	Risaldar-Major.	
	Risaldar-Major.					
1 May 86	Dhulip Singh	18 Feb. 02	1 June 03	....	1 July 12	Order of Br. I. 2nd Cl.
	Risaldars.					
5 Apl. 83	Muhammad Khan	16 Nov. 02	1 June 03	16 June 08		
4 July 92	Mal Singh (45*a*)	1 Feb. 04	1 Apl. 10	16 Sep. 12		
26 Apl. 94	Ghulam Rasul Khan (52, 56)	1 July 07	5 Jan. 15	1 Jan. 17		
26 Nov. 85	Bhagwan Sahai (52, 56)	1 June 03	1 July 12	29 Sep. 17		
8 July 93	Yakub Khan (45*a*, 50)	1 Nov. 08	30 Aug. 13	1 Dec. 17		
1 Feb. 13	Gainda Singh (56, 57)	17 Sep. 15	29 Sep. 17	19 Feb. 18		
3 Aug. 92	Agar Singh (52, 56, 65)	1 Apl. 10	16 Sep. 12	29 Mar. 18		
14 Jan. 99	Balwant Singh	5 Dec. 14	22 Mar. 17	4 Aug. 18		34th Horse.

27th Light Cavalry—*contd.*

Date of entering service.	Names and rank.	Army rank. Jemadar.	Ressaidar.	Remarks.
	Ressaidars.			
9 June 98	*Najib Khan*	*18 Oct. 15*	*1 Jan. 17*	
28 Dec. 98	Ahmad Khan	10 Feb. 16	1 Mar. 17	
6 Mar. 95	*Raghubir Singh (52, 56, 66a)*	*4 Oct. 10*	*1 Dec. 17* *29 Sep. 17**	
18 Nov. 00	Nawal Singh	1 Jan. 17	29 Sep. 17	
10 June 97	*Indraj Singh (58, 66a)*	*1 July 12*	*29 Sep. 17*	
1 Oct. 01	Mustafa Khan (56)	15 May 16	14 Mar. 18	
5 May 00	Bharat Singh (66a)	16 Sep. 12	29 Mar. 18	
7 May 05	*Ahmed Khan (56)*	*9 July 16*	*22 Jan. 18*	
	Jemadars.			
1 July 00	*Ghulam Muhammad Khan (II)*	*5 Jan. 15*		
1 Apl. 03	Gig Singh (56)	15 May 16		
31 July 03	Lakpat Singh	19 Oct. 16		
24 Dec. 98	Sirdar Khan	1 Jan. 17		
1 Jan. 94	Birnarain	1 Mar. 17		
23 Apl. 09	Khairati Singh	1 Mar. 17		
26 Sep. 03	Bhur Singh	1 Mar. 17		
1 July 07	Fazal Khan	1 Mar. 17		
22 Sep. 03	Ranchor Singh	1 Mar. 17		
6 Jan. 04	Aulia Khan (52)	22 Mar. 17		
1 July 02	Dalpat Singh (52, 56) I.D.S.M.	20 Apl. 17		
11 Dec. 03	Ram Sarup	20 Apl. 17		
10 Oct. 95	Muhammad Abdul Qadir	1 May 17		Qr.-Mr. 3 Dec. 17.
16 July 02	Debi Singh	1 July 17		
1 May 08	Surat Ram (56)	29 Sep. 17		
17 Apl. 91	Rasul Khan	14 Mar. 18		
15 Mar. 12	Gulistan Khan (56)	29 Mar. 18		Woordie Maj. 29 Mar. 18.
1 Jan. 97	Muhammad Khan	1 July 12		1 July 17
10 Jan. 02	Ghulam Muhammad Khan (51)	18 Dec. 14		30 Dec. 17 34th Horse.
25 May 03	Muhammed Khan (51)	29 Mar. 18		
23 June 13	Dharam Singh (51)	29 Mar. 18		
2 Oct. 11	Muhammad Murtaza	29 Mar. 18		

* Without pay and Allowances.

28th Light Cavalry.

Campbell-ki-paltan.—Formed in 1784 from selected details of three regiments of Native Cavalry, which had been taken over that year from the Nawab of Arcot, but which had been disbanded for Mutiny, and was originally styled the 2nd Regiment of Madras Native Cavalry. Became the 1st Regiment of Madras Native Cavalry, 1786, and the 3rd Regiment, 1788; the 3rd Regiment of Madras Light Cavalry, 1819; the 3rd Regiment of Madras Lancers, 1891; the 3rd Madras Lancers, 1901; present designation, 1903.

"Mysore" "Seringapatam" "Maheidpoor."

Composition—1 Squadron of Madras and Dekhani Musalmans, 1 of Punjabi Musalmans, 1 of Rajputana Rajputs, 1 of Jats.

Uniform—French Grey. *Facings*—Buff. *Lace*—Silver.

COLONEL—General *Sir* Donald J. S. McLeod, K.C.B., K.C.I.E., D.S.O. . 7 May 1906.

1st Commn. or date of entering service.	Names and Rank.	Army rank.	Present appointment in Regiment.	Remarks.
	Commandant.			
12 Mar. 92	Oakes, Lt.-Col. H. C. [L]	17 Dec. 17	17 Dec. 17	
	Squadron Commanders. (4)			
4 Aug 97	Hawley, Maj. W. G. B.	4 Aug. 15	17 Dec. 17	2nd-in-Comd.
11 Aug. 00	*Dickinson, Maj. R. C.*	*1 Sep. 15*	*30 Mar. 15*	*With I. S. Troops.*
14 Sep. 01	*Nisbet, Bt.-Lt.-Col., T.,* D.S.O.	*3 June 18*	*17 Dec. 17*	*A. A. and Q. M. G.*
18 Jan. 02	*Marshall, Maj. A.,* D.S.O.	*18 Jan. 17*	*15 May 18*	*G. S. O., 2nd gr., Quetta Dn.*
	Squadron Officers. (9)			
7 May 02	Walker, Maj. C. C.	7 May 17	6 Nov. 05	*Lv. ex. I., m. c.*
22 Apl. 03	Kreyer, Capt. J. A. C.	22 Apl. 12	20 Sep. 05	
13 Aug. 04	James, Capt F.	13 Aug. 13	14 Nov. 05	
18 Jan. 11	Brownlow, Capt. F.	1 Sep. 15	7 Mar. 12	
22 Jan. 13	Uloth, Capt. G.	22 Jan. 17	13 Mar. 14	Adjt.
18 Aug. 15	Simkins, Lt. G. P.	18 May 17	24 June 18	Attd. With 1-150th Rif.
29 June 16	Gatehouse, Lt. R. O. A.	29 June 17	1 July 16	
	Attached.			
15 July 11	Muirhead, Lt. J., M.C.	5 May 16	1 Nov. 18	
5 Mar. 13	Lothair, Lt. J.	5 May 16	19 June 17	
18 Dec. 14	Fawkner, Lt. H. S.	. . . .	25 Apl. 18	I. A. (*on probn.*)
9 Apl. 15	*Turner, Lt. J. F. (Actg. Capt.)*	*16 Dec. 15*	*23 Apl. 15*	*I. A. R. O. With Zhob Militia.*
9 Jan. 17	Thomson, Lt. R. F.	9 Jan. 17	11 July 17	I. A. (*on probn.*)
14 Jan. 18	Turner, Lt. H. F. C.	14 Jan. 18	2 Feb. 18	I. A (*on probn.*)
14 May 15	Wood, Lt. A. E. F.	14 May 16	22 Jan. 18	I. A. R. O.
11 June 15	Forsyth, Lt. C. S.	11 June 16	5 Mar. 18	I. A. R. O.
12 Aug. 15	Cochrane, Lt. C. P.	. .	23 Oct. 18	I. A. (*on probn.*)
30 Nov. 15	Carter, Lt. H. L. M.	30 Aug. 17	9 Aug. 17	*Lv. ex. I., m. c.*
22 Dec. 15	Vines, Lt. R. G.	22 Sep. 17	3 July 17	
31 July 16	Hood, Lt. H. M. (*Actg. Capt.*)	31 July 17	22 Oct. 16	I. A. R. O.
5 Aug. 16	Askwith, Lt. A. V.	5 Aug. 17	20 Oct. 16	I. A. R. O.
28 Nov. 16	Falconer, Lt. G. A.	28 Aug. 18	25 Aug. 17	
30 Jan. 17	Hawkes, 2nd-Lt. T. Y.	30 Jan. 18	30 Jan. 17	
29 Apl. 17	Straker, Lt. W. R.	29 Apl. 18	25 Apl. 18	I. A. (*on probn.*)
11 May 17	Mortimer, Lt. W. G.	11 May 18	16 Nov. 17	I. A. R. O.
11 May 17	Wilson, Lt. J. P.	11 May 18	2 June 17	I. A. R. O.
12 May 17	Drinkwater, Lt. C. R.	12 May 18	12 May 17	I. A. R. O.
25 May 17	Cameron, Lt. W.	25 May 18	3 June 17	I. A. R. O.
1 July 17	Rhodes, Lt. B. B.	1 July 18	19 July 17	*Lv. ex. I., m.c.*
1 Sep. 17	Brown, Lt. A.	1 Sep. 18	16 Sep. 17	I. A. (*on probn.*) Qr. Mr. (*Offg.*)
8 Mar. 18	Laming, 2nd-Lt. S. A.	8 Mar. 18	2 Nov. 18	I. A. R. O. (*on probn.*)
25 Apl. 18	Preston, 2nd-Lt. A. K.	25 Apl. 18	12 May 18	I. A. R. O.
24 Aug. 18	Sturdy, 2nd-Lt. E. C.	24 May 18	16 Sep. 18	I. A. R. O.

		Jemadar.	Ressaidar.	Risaldar.	Risaldar-Major.
	Risaldar-Major.				
19 May 86	Miyan Din Khan (56)	21 Oct. 05	12 Sep. 10	9 Jan. 16	9 Jan. 1916.
	Risaldars.				
. . . .	Ismail Khan (Retd.)		. . . .	Re-employed 7-3-18.	
20 Dec. 95	Thakur Sawant Singh (51, 56)	18 Dec. 05	16 Oct. 11	1 May 14.	
14 Sep. 02	Sher Shah Khan (56)	1 June 08	16 Nov. 11	8 Apl. 16.	
5 Feb. 96	Harnam Singh (51, 56)	23 Nov. 06	1 Jan. 14	27 Jan. 18.	

27th Light Cavalry—*contd.*

Date of entering service.	Names and rank.	Army rank.		Remarks.
		Jemadar.	Ressaidar.	
	Ressaidars.			
9 June 98	*Najib Khan*	*18 Oct. 15*	*1 Jan. 17*	
28 Dec. 98	Ahmad Khan	10 Feb. 16	1 Mar. 17	
6 Mar. 95	*Raghubir Singh (52, 56, 66a)*	*4 Oct. 10*	*1 Dec. 17* / *29 Sep. 17**	
18 Nov. 00	Nawal Singh	1 Jan. 17	29 Sep. 17	
10 June 97	*Indraj Singh (58, 66a)*	*1 July 12*	*29 Sep. 17*	
1 Oct. 01	Mustafa Khan (56)	15 May 16	14 Mar. 18	
5 May 00	Bharat Singh (66a)	16 Sep. 12	29 Mar. 18	
7 May 05	*Ahmed Khan (56)*	*9 July 16*	*22 Jan. 18*	
	Jemadars.			
1 July 00	***Ghulam Muhammad Khan (II)***	*5 Jan. 15*		
1 Apl. 03	Gig Singh (56)	15 May 16		
31 July 03	Lakpat Singh	19 Oct. 16		
24 Dec. 98	Sirdar Khan	1 Jan. 17		
1 Jan. 94	Birnarain	1 Mar. 17		
23 Apl. 09	Khairati Singh	1 Mar. 17		
26 Sep. 03	Bhur Singh	1 Mar. 17		
1 July 07	Fazal Khan	1 Mar. 17		
22 Sep. 03	Ranchor Singh	1 Mar. 17		
6 Jan. 04	Aulia Khan (52)	22 Mar. 17		
1 July 02	Dalpat Singh (52, 56) I.D.S.M.	20 Apl. 17		
11 Dec. 03	Ram Sarup	20 Apl. 17		
10 Oct. 95	Muhammad Abdul Qadir	1 May 17		Qr.-Mr. 3 Dec. 17.
16 July 02	Debi Singh	1 July 17		
1 May 08	Surat Ram (56)	29 Sep. 17		
17 Apl. 91	Rasul Khan	14 Mar. 18		
15 Mar. 12	Gulistan Khan (56)	29 Mar. 18		Woordie Maj. 29 Mar. 18.
1 Jan. 97	Muhammad Khan	1 July 12		1 July 17
10 Jan. 02	Ghulam Muhammad Khan (51)	18 Dec. 14		30 Dec. 17 34th Horse.
25 May 03	Muhammed Khan (51)	29 Mar. 18		
23 June 13	Dharam Singh (51)	29 Mar. 18		
2 Oct. 11	Muhammad Murtaza	29 Mar. 18		

* Without pay and Allowances.

28th Light Cavalry.

Campbell-ki-paltan.—Formed in 1784 from selected details of three regiments of Native Cavalry, which had been taken over that year from the Nawab of Arcot, but which had been disbanded for Mutiny, and was originally styled the 2nd Regiment of Madras Native Cavalry. Became the 1st Regiment of Madras Native Cavalry, 1786, and the 3rd Regiment, 1788; the 3rd Regiment of Madras Light Cavalry, 1819; the 3rd Regiment of Madras Lancers, 1891; the 3rd Madras Lancers, 1901; present designation, 1903.

"Mysore" "Seringapatam" "Maheidpoor."

Composition—1 Squadron of Madras and Dekhani Musalmans, 1 of Punjabi Musalmans, 1 of Rajputana Rajputs, 1 of Jats.

Uniform—French Grey. *Facings*—Buff. *Lace*—Silver.

COLONEL—General *Sir* Donald J. S. McLeod, K.C.B., K.C.I.E., D.S.O. . 7 May 1906.

1st Commn. or date of entering service.	Names and Rank.	Army rank.	Present appointment in Regiment.	Remarks.
	Commandant.			
12 Mar. 92	Oakes, Lt.-Col. H. C. [L]	17 Dec. 17	17 Dec. 17	
	Squadron Commanders. (4)			
4 Aug 97	Hawley, Maj. W. G. B.	4 Aug. 15	17 Dec. 17	2nd-in-Comd.
11 Aug. 00	*Dickinson, Maj. R. C.*	*1 Sep. 15*	*30 Mar. 15*	*With I. S. Troops.*
14 Sep. 01	*Nisbet, Bt.-Lt.-Col., T.,* D.S.O.	*3 June 18*	*17 Dec. 17*	*A. A. and Q. M. G.*
18 Jan. 02	*Marshall, Maj. A.,* D.S.O.	*18 Jan. 17*	*15 May 18*	*G. S. O., 2nd gr., Quetta Dn.*
	Squadron Officers. (9)			
7 May 02	Walker, Maj. C. C.	7 May 17	6 Nov. 05	*Lv. ex. I., m. c.*
22 Apl. 03	Kreyer, Capt. J. A. C.	22 Apl. 12	20 Sep. 05	
13 Aug. 04	James, Capt F.	13 Aug. 13	14 Nov. 05	
18 Jan. 11	Brownlow, Capt. F.	1 Sep. 15	7 Mar. 12	
22 Jan. 13	Uloth, Capt. G.	22 Jan. 17	13 Mar. 14	Adjt.
18 Aug. 15	Simkins, Lt. G. P.	18 May 17	24 June 18	Attd. With 1-150th Rif.
29 June 16	Gatehouse, Lt. R. O. A.	29 June 17	1 July 16	
	Attached.			
15 July 11	Muirhead, Lt. J., M.C.	5 May 16	1 Nov. 18	
5 Mar. 13	Lothair, Lt. J.	5 May 16	19 June 17	
18 Dec. 14	Fawkner, Lt. H. S.	. . . .	25 Apl. 18	I. A. (*on probn.*)
9 Apl. 15	*Turner, Lt. J. F. (Actg. Capt.)*	*16 Dec. 15*	*23 Apl. 15*	*I. A. R. O. With Zhob Militia.*
9 Jan. 17	Thomson, Lt. R. F.	9 Jan. 17	11 July 17	I. A. (*on probn.*)
14 Jan. 18	Turner, Lt. H. F. C.	14 Jan. 18	2 Feb. 18	I. A (*on probn.*)
14 May 15	Wood, Lt. A. E. F.	14 May 16	22 Jan. 18	I. A. R. O.
11 June 15	Forsyth, Lt. C. S.	11 June 16	5 Mar. 18	I. A. R. O.
12 Aug. 15	Cochrane, Lt. C. P.	. .	23 Oct. 18	I. A. (*on probn.*)
30 Nov. 15	Carter, Lt. H. L. M.	30 Aug. 17	9 Aug. 17	*Lv. ex. I., m. c.*
22 Dec. 15	Vines, Lt. R. G.	22 Sep. 17	3 July 17	
31 July 16	Hood, Lt. H. M. (*Actg. Capt.*)	31 July 17	22 Oct. 16	I. A. R. O.
5 Aug. 16	Askwith, Lt. A. V.	5 Aug. 17	20 Oct. 16	I. A. R. O.
28 Nov. 16	Falconer, Lt. G. A.	28 Aug. 18	25 Aug. 17	
30 Jan. 17	Hawkes, 2nd-Lt. T. Y.	30 Jan. 18	30 Jan. 17	
29 Apl. 17	Straker, Lt. W. R.	29 Apl. 18	25 Apl. 18	I. A. (*on probn.*)
11 May 17	Mortimer, Lt. W. G.	11 May 18	16 Nov. 17	I. A. R. O.
11 May 17	Wilson, Lt. J. P.	11 May 18	2 June 17	I. A. R. O.
12 May 17	Drinkwater, Lt. C. R.	12 May 18	12 May 17	I. A. R. O.
25 May 17	Cameron, Lt. W.	25 May 18	3 June 17	I. A. R. O.
1 July 17	Rhodes, Lt. B. B.	1 July 18	19 July 17	*Lv. ex. I., m.c.*
1 Sep. 17	Brown, Lt. A.	1 Sep. 18	16 Sep. 17	I. A. (*on probn.*) Qr. Mr. (*Offg.*)
8 Mar. 18	Laming, 2nd-Lt. S. A.	8 Mar. 18	2 Nov. 18	I. A. R. O. (*on probn.*)
25 Apl. 18	Preston, 2nd-Lt. A. K.	25 Apl. 18	12 May 18	I. A. R. O.
24 Aug. 18	Sturdy, 2nd-Lt. E. C.	24 May 18	16 Sep. 18	I. A. R. O.

		Jemadar.	Ressaidar.	Risaldar.	Risaldar-Major.
	Risaldar-Major.				
29 May 86	Miyan Din Khan (56)	21 Oct. 05	12 Sep. 10	9 Jan. 16	9 Jan. 1916.
	Risaldars.				
. . . .	Ismail Khan (Retd.)	. . . .	. . . .	Re-employed 7-3-18.	
20 Dec. 95	Thakur Sawant Singh (51, 56)	18 Dec. 05	16 Oct. 11	1 May 14.	
14 Sep. 02	Sher Shah Khan (56)	1 June 08	16 Nov. 11	8 Apl. 16.	
5 Feb. 96	Harnam Singh (51, 56)	23 Nov. 06	1 Jan. 14	27 Jan. 18.	

28th Light Cavalry—*contd.*

Date of entering service.	Names.	Army rank.		Remarks.
		Jemadar.	Ressaidar.	
	Ressaidars.			
8 Aug. 95	Punnai Singh (51, 56, 58) . .	17 Apl. 12	1 May 14	
15 Aug. 99	Malik Muhammad Latif (45a, 51, 56).	1 Jan. 11	9 Jan. 16	
7 Nov. 93	Muhammad Akbar Khan (56, 52)	16 Nov. 11	24 Aug. 16	
7 Sep. 02	Sher Singh (I) (56)	13 Dec. 07	27 Jan. 18	
	Jemadars.			
1 Apl. 07	Muhammad Yakub Khan (56,56)	1 July 11		
2 Aug. 04	Sher Singh (II) (56, 66a) . . .	1 Apl. 13		
23 June 00	Rawat Singh (51, 56) . . .	1 May 14		
5 Dec. 03	Malik Sikandar Khan (51, 56) .	4 Jan. 15		
5 May 96	Ismail Sharif	16 Mar. 15		
10 Nov. 02	Harnam Singh	9 Jan. 16		
16 Oct. 97	Abdur Rahman Khan (50) . .	9 Jan. 16		
14 Sep. 02	Ranbaz Khan (50)	9 Jan. 16		
14 June 98	Fateh Singh (58)	16 Mar. 16		
25 Feb. 96	Shaikh Haidar	24 Aug. 16		
30 June 96	*Sayyid Qudratullah*	*26 Nov. 17*		
25 Nov. 03	Gobind Singh, V. C.	2 Dec. 17		
28 Sep. 03	Nur Khan	3 Dec. 17		
2 Aug. 03	Atma Ram	18 Jan. 18		
22 Jan. 99	Tej Singh (56)	18 Jan. 18		
22 Nov. 12	Kartar Chand	27 Jan. 18		
22 June 88	Muhammad Zahir-ud-din . .	20 May 18		

29th Lancers (Deccan Horse).

Formed in 1826, from Nawab Murtiza Yar Jung's, Captains Hallis' and Smith's risallahs of the Nizam's Reformed Horse, and was styled the 2nd Regiment Nizam's Cavalry. Became the 2nd Cavalry, Hyderabad Contingent, 1854; the 2nd Lancers, Hyderabad Contingent, 1890; present designation, 1903.

Composition—2 Squadrons of Jats, 1 of Sikhs, 1 of Dekhani Musalmans.

Uniform—Rifle-green. *Facings*—White.

COLONEL—Major-General George A. Cookson, C.B., C.M.G. 18 July 1913.

1st Commn. or date of entering service.	Names and Rank.	Army rank.	Present appointment in Regiment.	Remarks.
	Commandant.			
28 Nov. 94	Sangster, Lt.-Col. P. B., D.S.O.	7 Aug. 16	7 Aug. 16	
	Squadron Commanders. (4)			
7 Mar. 94	Lambert, Maj. W. J., D.S.O. (*temp. Lt.-Col.*)	7 Mar. 12	5 Aug. 18	2nd-in-Comd.
16 Jan. 95	Birch, Maj. V. K.	16 Jan. 13	10 May 11	With 2-150th Infy.
15 May 97	*Meynell, Maj. H.*	*15 May 15*	*8 July 15*	*With Remt. Dept.*
22 Jan. 98	Marchant, Maj. G., *p.s.c.*	1 Sept. 15	5 Aug. 18	With 43rd Cav.
	Squadron Officers. (9)			
6 Jan. 00	Henderson, Maj. C. G. H.	1 Sep. 15	1 Jan. 06	
26 June 01	Hunt, Maj. F. E., D.S.O.	18 Oct. 17	24 Oct. 06	With 2-116th Mah
28 Jan. 03	Dickson, Maj. H. R. P., C.I.E.	28 Jan. 18	20 June 07	
19 July 05	Jackson, Capt. M. H., M.C.	19 July 14	1 Mar. 07	
25 July 06	*Bacon, Capt. C. G.*	*25 July 15*	*24 Aug. 13*	*Staff Capt.*
9 Sep. 08	Wright, Capt. R. D.	1 Sep. 15	29 Dec. 10	
20 Jan. 09	Spurgin, Capt. E. W., M.C.	1 Sep. 15	17 Mar. 10	
20 Jan. 12	Bradley, Capt. R. O.	20 Jan. 16	10 Mar. 13	
3 Aug. 12	Boggs, Capt. A. B.	3 Aug. 16	20 Nov. 13	
14 Nov. 16	Harvey, Lt. C.	14 Nov. 17	....	Attd.
14 Nov. 16	Cashmore, Lt. A. E.	14 Nov. 17	....	Attd.
14 Nov. 16	Lewis, Lt. H. P.	14 Nov. 17	....	Attd.
28 Aug. 17	Robinson, Lt. F. W. O. (*temp. Capt.*)	28 Aug. 18	29 Sep. 17	Attd., *S. S. O., Poona Bde.*
	Attached.			
4 Jan. 99	Baddeley, Maj. R. J. H., M.C.	1 Sep. 15	5 Apl. 17	15th Lrs. 2nd-in-Comd. (*Offg.*)
20 Nov. 14	Rice, Capt. R. A., M.C.	8 July 18	8 July 15	I. A. R. O.
27 Nov. 14	Johnston, Lt. E. N. W.	20 Aug. 15	3 Apl. 16	I. A. R. O. Adjt. Cadet Coll., Wellington.
20 Nov. 14	Herrick, Capt. R. L. W., D.S.O.	30 Aug. 18	30 Aug. 15	I. A. R. O.
27 Nov. 14	Falconer, Capt. J. F.	30 Aug. 18	30 Aug. 15	I. A. R. O.
23 Apl. 15	Collett, Capt. C. M.	9 Sep. 18	7 Dec. 16	A. R. O., Secunderabad.
1 Sep. 15	King, Lt. F. H.	1 Sep. 16	18 Apl. 18	War. Yeo. I. A. (*on probn.*)
25 Mar. 15	Armstrong, Lt. A. H. H., M.C.	25 Mar 15	15 May 18	R. Innis. Fus. I. A. (*on probn.*)
21 May 15	Herbert, Capt. V. A., M.C.	30 Aug. 18	11 May 16	I. A. R. O. Adjt., 26 Feb. 17.
20 Sep. 15	Williams, Lt. J. H.	20 Sep. 16	28 Dec. 15	I. A. R. O.
23 Sep. 15	Bell, Lt. H. C.	23 Sep. 16	25 Oct. 16	I. A. R. O. Qr.-Mr., 8 Sep. 17.
31 Dec. 15	Gavaghan, Lt. E. L.	31 Dec. 16	7 Dec. 16	I. A. R. O.
30 Oct. 14	Clarke, 2nd-Lt. L. C.	30 Oct. 14	5 Feb. 16	Res. Regt. of Cav.
2 Apl. 15	O'Connor, Lt. J. C. J.	2 Apl. 16	19 May 17	I. A. R. O.
6 Sep. 16	Mercer, Lt. A. A.	8 Feb. 17	5 Apl. 15	Res. Regt. of Cav.
30 Jan. 17	Sawers, 2nd-Lt. C. M. O.	14 Feb. 17	19 Apl. 17	I. A. R. O.
17 July 16	Lory, Lt. F. B. P.	17 July 17	4 Jan. 17	I. A. R. O.
3 Sep. 17	Ashe, 2nd-Lt. H. D.	....	31 July 15	Volr. Officer.
....	Boggs, 2nd-Lt. G. B.	....	19 Apl. 16	
1 Jan. 17	Hallifax, 2nd-Lt. H. F.	....	8 July 17	Res. Regt. of Cav.
1 Jan. 17	Sayres, Lt. A. C.	1 Jan. 18	16 Jan. 17	I. A. R. O.
11 Apl. 17	Bois, Lt. H. E.	11 Apl. 18	15 Sep. 17	I. A. R. O.
25 Aug. 17	Hawkins, 2nd-Lt. M.	25 Aug. 17	20 May 18	Res. Cav. R. I. A. (*on probn.*)
14 Aug. 15	Davis, Lt. C. K.	14 Aug. 16	....	Res. Cav. R. I. A. (*on probn.*)
30 Jan. 17	King, Lt. R. S., M.C.	30 Jan. 18	11 Sep. 17	
24 Apl. 18	Howard, 2nd-Lt. C. W.	24 Apl. 18	....	I. A. (U. L.)

29th Lancers (Deccan Horse)—*contd.*

Date of entering service.	Names and Rank.	Army rank.	Present appointment in Regiment.	Remarks.

Honorary Officer.

Date of entering service.	Names and Rank.	Army rank.	Present appointment in Regiment.	Remarks.
22 June 13	Savanur, Abdul Majid Khan, *Nawab* of	1 Jan. 18	. . . .	Hony. Capt. in Army. Adjt. (temp.) Governor's Body-Guard, Bombay.

Risaldar-Major.

Date of entering service.	Names and Rank.	Jemadar.	Ressaidar.	Risaldar.	Risaldar-Major.	Remarks.
1 Nov. 92	Chanda Singh *Bahadur*, I.D.S.M. (43, 45*a*, 56, 66*a*)	1 June 08	1 Aug. 13	17 Apl. 15	9 Jan. 17	*Order of Br. I., 2nd Cl.*
. . . .	Ahmad Khan, *Bahadur* (*Retd.*)	. . . .	. . . .	. . . .	20 May 16	*Order of Br. I., 2nd Cl.*
. . . .	Ghulam Dastighir Khan, *Bahadur.*	. . . .	. . . .	. . . .	...	Re-employed 15 Jan. 18. *Order of Br. I., 2nd Cl.*

Risaldars.

Date of entering service.	Names and Rank.	Jemadar.	Ressaidar.	Risaldar.	Remarks.
18 June 92	Kabul Singh, *Bahadur*, I.D.S.M., I.O.M. (45*a*, 56)	26 Feb. 09	9 Sep. 13	22 Aug. 15	Attd. 31 Lrs.
1 Dec. 95	Hayat Ali Beg, *Bahadur*, I.O.M (56, 58, 66*a*, 71)	1 Sep. 12	16 Jan. 15	28 May 16	*Order of Br. I., 2nd Cl.*
5 Nov. 92	Subh Ram (56)	1 Sep. 10	24 Oct. 15	28 May 16	Attd. 14 Lrs.

Ressaidars.

Date of entering service.	Names and Rank.	Jemadar.	Ressaidar.	Remarks.
21 May 02	Harphul Singh (56)	1 Nov. 13	1 May 15	With 44th Cav.
11 Dec. 01	Bhulan Singh (56)	1 Aug. 13	11 Apl. 16	With 44th Cav.
4 Apl. 93	Daya Singh	24 Mar. 15	4 Feb. 17	
27 July 01	Abdul Rahim Khan, I.O.M. . .	15 May 15	25 Aug. 17	
16 Aug. 05	Khub Ram	22 Aug. 15	25 Dec. 17	
19 Aug. 08	Jaisi Ram	24 Oct. 15	19 Apl. 18	
. . . .	Abdul Sittar Khan (Retd.) . .	. . . .	16 May 16	Re-employed.
3 Jan. 01	Habib Khan	29 Dec. 08	11 Apl. 16	Attd. 22nd Cav.
1 July 97	Jaggan Singh	27 Mar. 15	25 Aug. 17	Attd. 31st Lns.

Jemadars.

Date of entering service.	Names and Rank.	Jemadar.	Remarks.
	Jagan Singh	27 Mar. 15	Attd. 14 Lrs.
26 July 02	Sis Ram Singh	23 Aug. 15	
11 Sep. 11	Gulab Singh	11 Apl. 16	
15 Oct. 00	Than Singh	11 Apl. 16	
16 July 07	Natha Singh, I.D.S.M. . . .	11 Apl. 16	Woordie Maj.
1 June 93	Ibrahim Khan	11 Apl. 16	
17 Aug. 07	Jag Singh	11 Apl. 16	
25 Apl. 05	Mahbob Ali Khan	17 May 17	
2 Oct. 95	Lal Singh, I.D.S.M.	25 Aug. 17	With 44th Cav
10 Apl. 02	Bakhtawar Singh	25 Aug. 17	
3 May 00	Abdul Ranman Khan	25 Aug. 17	
25 Jan. 04	Khan Singh	25 Aug. 17	
19 Oct. 04	Birkha Singh	27 Aug. 17	With 44th Cav.
. . . .	Sham Singh	1 Dec. 17	
15 Mar. 07	Nihal Singh	3 Dec. 17	
. . . .	Sitaram	14 Dec. 17	With 44th Cav.
20 Oct. 02	Mahan Singh, I.D.S.M. . . .	19 May 16	Attd. 31st Lrs.
19 Aug. 97	Rur Singh	18 Feb. 17	Attd. 33rd Cav.
27 Jan. 02	Kasim Ali, I.D.S.M.	25 June 17	Attd. 22nd Cav.

30th Lancers (Gordon's Horse).

Raised at Mominabad in 1826, by Sir John Gordon, Bart., as the 4th Regiment of Nizam's Cavalry. Became the 4th Cavalry, Hyderabad Contingent, 1854; the 4th Lancers, Hyderabad Contingent, 1890; present designation, 1903.

"Central India."

Composition—2 Squadrons of Sikhs, 1 of Jats, 1 of Hindustani Musalmans.

Uniform—Rifle-green. *Facings*—White.

COLONEL—Major-General Henry M. Mason . . 1 January 1906.

1st Commn. or date of entering service.	Names and Rank.	Army rank.	Present appointment in Regiment.	Remarks.
	Commandant.			
3 Sep 12	Harbord, Maj. C. R., D.S.O. (*temp. Br.-Genl.*)	1 Sep. 18	1 Sep. 18	
	Squadron Commanders. (4)			
28 Sep. 95	***Muspratt, Maj. V. R., p.s.c.*** [*l*] (*temp. Lt.-Col.*) (Q).	*28 Sep. 13*	*1 Sep. 14*	*2nd-in-Comd. A. Q. M. G., Army Hdqrs.*
5 Sep. 96	*Wilford, Maj. E. E.*, D.S.O. (*temp. Lt.-Col.*).	*5 Sep. 14*	*1 Sep. 14*	With R. D. Corps. T. F.
1 Dec. 97	Giedstanes, Maj. A. U. (*Actg. Lt.-Col.*)	1 Sep. 15	30 Jan. 17	Comdt. (*temp.*)
	Squadron Officers. (9)			
20 May 99	Walker, Maj. H. T.	1 Sep. 15	13 Aug. 03	2nd-in-Comd. (*temp.*)
18 Apl. 00	*Raymond, Maj. E. D.*, D.S.O., M.C.	*1 Sep. 15*	*12 Mar. 07*	*D. A. A. & Q. M. G., Presy. Bde.*
8 May 01	*Morris, Maj. G. P.*	*8 May 16*	*25 June 04*	*Bde. Maj.*
5 Aug. 05	***Butler, Bt.-Maj. R. B.***, M.C.	*1 June 18*	*15 Dec. 07*	*A. Q. M. G.*
5 Aug. 05	*Lucas, Capt. H. de N.* (*temp. Maj.*)	*5 Aug. 14*	*8 Sep. 07*	*D. A. A. & Q. M. G., N. C.*
3 Sep. 10	*Bird, Capt. W. T.*	*1 Sep. 15*	*16 Mar. 12*	*With I. S. Troops.*
24 Aug. 12	Spooner, Capt. F. V.	24 Aug. 16	27 Oct. 13	
24 Jan. 14	Law, Capt. C. H.	24 Jan. 18	8 Aug. 17	Attd.
12 Jan. 15	Hobson, Lt. E. M. C.	12 May 16	16 June 17	Attd.
29 May 15	Coles, Lt. G. R.	29 May 16	29 Apl. 17	Attd.
27 Oct. 17	Hay, Lt. C. E.	27 Oct. 18	6 Nov. 17	Attd. *Lv., ex I., m. c.*
17 Jan. 18	Bock, 2nd-Lt. C. C. (*Actg. Capt.*)	17 Jan. 18	29 Nov. 17	Attd. Adjt. 26 Oct. 18.
27 Oct. 17	Pearson, 2nd-Lt. R. F.	27 Oct. 17	6 Nov. 17	
	Attached.			
17 Oct. 14	Ridley, Lt. G. A.	17 July 17	16 May 17	E. York. R. I. A. (*on probn.*). *Lv., ex I., m. c.*
11 Dec. 14	Johnson, Lt. T. G.	11 Dec. 15	14 Oct. 18	Fus. I. A. R. O. (*on probn.*).
17 Mar. 16	Blair, Lt. J. R. (*Temp. Capt.*) .	17 Mar. 16	15 May 18	I. A. R. O. Staff Capt. 10 Cav. Bde.
3 May 15	Roth, Lt. E. J. H.	3 Feb. 17	16 May 18	R. Fus. I. A. (*on probn.*)
16 June 15	Ramsden, Lt. A. R.	16 June 16	30 Mar. 18	York. L. Infy. I. A. (*on probn.*)
7 Sep. 16	Standing, Lt. H.	7 Sep. 17	15 Feb. 18	I. A. (*on probn.*)
21 Dec. 16	Riddle, Lt. T. E. (*Actg. Capt.*) .	21 Dec. 17	25 Dec. 16	Shrops. Yeo. I. A. R. O.
24 Apl. 17	Hill, Lt. H. B. C.	24 Apl. 18	12 May 17	I. A. R. O. *A. D. C. to G. O. C., 1st (Pesh.) Dn*
15 May 17	Meyrick, Lt. R.	14 May 18	29 May 17	I. A. R. O.
25 July 17	McDougall, Lt. A.	25 July 18	17 Apl. 18	Sco. Horse I. A. (*on probn.*)
28 Aug. 17	Moorhouse, Lt. E. G.	28 Aug. 18	1 Mar. 18	Lan. Hrs. Yeo. I. A. (*on probn.*)
21 Sep. 18	Leonard, Lt. H. W.	19 Sep. 18	30 Mar. 18	Res. R. I. A. (*on probn.*)
10 Apl. 18	Bellairs, 2nd-Lt. S. S.	10 Apl. 18	22 Apl. 18	I. A. R. O.
13 June 18	Gasson, 2nd-Lt. L. B.	13 June 18	14 Aug. 18	I. A. R. O.

		Jemadar.	Ressaidar.	Risaldar.	Risaldar-Major.	
	Risaldar-Majors.					
27 May 89	Hira Singh (56, 66*a*), *Bahadur* .	20 Apl. 00	17 Aug 02	5 Sep. 10	4 Mar. 15	***Order of Br. I., 2nd Cl.***
	Risaldars.					
1 June 88	Akbar Ali Khan (56)	2 Nov. 00	16 July 12	1 July 14		
5 July 84	Partab Singh (56, 66*a*, 71) *Bahadur.*	22 Aug. 01	5 Sep. 10	21 Sep. 15		*Order of Br. I., 2nd* [*Class.*
	Mir Mumtaz Husain, *Bahadur* (*Retd.*)	. . . .	. . . .	. . . .		Re-employed 19 Jan. 15. *Order of Br. I., 2nd Cl.*
4 Jan. 91	Ratan Singh (56, 66*a*) . . .	29 Aug. 05	16 May 11	1 Sep. 17		

30th Lancers (Gordon's Horse)—*contd.*

Date of entering service.	Names.	Army rank.	Remarks.
	Ressaidars.		
13 Mar. 96	Rur Singh (56, 60, 66a) . . .	16 Mar. 08	1 July 11
22 Nov. 02	Todar Singh (45a, 66a, 72a) .	1 June 03	4 Sep. 14
23 Mar. 98	Mehar Singh (56, 66a) . .	16 Nov. 08	4 Mar. 15
9 June 93	Natha Singh (71)	5 Sep. 10	16 Sep. 15
22 Nov. 92	Mul Singh (66a)	1 July 11	21 Sep. 15
21 Nov. 93	Mahbub Khan	1 July 14	11 Apl. 16
	Jemadars.		
10 Mar. 98	Pohp Sing (56)	16 May 14	With Remt. Depôt, Sehore.
16 Nov. 93	Shaikh Jafar Ali (66a)	4 Sep. 14	With Remt. Depôt.
14 Nov. 98	Chet Ram (66a, 71, 45a) . . .	4 Sep. 14	
1 May 93	Bir Singh	22 Dec. 14	
5 Aug. 08	Muhammad Nazir Khan (45a, 66a, 52).	1 July 15	Woordie-Major, 9 Nov. 16.
5 June 05	Hazara Singh (52, 56) . . .	16 Sep. 15	
1 Aug. 91	Sadhu Singh	21 Sep. 15	With Remt. Depôt, Sehore.
27 July 01	Dost Muhammad Khan (56, 71) .	11 Apl. 16	With Camel Corps.
1 Oct. 11	Kapur Singh (52, 56)	17 Apl. 16	
24 Jan. 01	Makhan Lal	1 Feb. 17	
4 June 00	Sher Singh	23 June 17	
1 June 01	Wadhawa Singh (66a, 71) . .	14 Aug. 17	
16 Nov. 09	Muhammad Yasin Khan . . .	16 Aug. 17	
1 May 07	Fateh Singh	1 Sep. 17	
6 Dec. 99	Bodh Raj.	24 Dec. 17	Indian Qr.-Mr.
16 Feb. 12	Ajmer Singh	13 Mar. 18	
3 July 06	Mahfuz Ali Khan (47a, 52, 56) .	11 May 18	
15 Feb. 11	Bahadur Singh (47a, 56) . . .	25 May 18	
10 Aug. 94	Bachān Singh	19 June 18	
3 Jan. 00	Assa Singh	16 July 18	
23 Aug. 06	Mustafa Sharif	14 Sep. 18	

31st Duke of Connaught's Own Lancers.

Raised at Beerah in 1817, by Lt.-Col. R. Barclay, the first of two existing troops forming the nucleus, and was originally designated the 1st Regiment of Bombay Light Cavalry. Became the 1st Regiment of Bombay Light Cavalry (Lancers), 1842; the 1st Regiment of Bombay Silladar Light Cavalry, 1861; reverted to its previous designation, 1861; became the 1st Regiment of Bombay Light Cavalry, 1862; the 1st Bombay Lancers, 1880; the 1st (Duke of Connaught's Own) Bombay Lancers, 1890; present designation, 1903.

"Ghuznee," "Affghanistan, 1839."
"Mooltan" "Punjaub" "Central India" "Burma, 1885-87."
Composition—1 Squadron of Dekhani Mahrattas, 1 of Jats, 1 of Sikhs, 1 of Pathans.
***Uniform*—Blue. *Facings*—Scarlet.**

COLONEL-IN-CHIEF—Field Marshal *His Royal Highness* ARTHUR WILLIAM PATRICK ALBERT, *Duke of* CONNAUGHT and STRATHEARN, K.G., K.T., K.P., G.C.B., G.C.S.I., G.C.M.G., G.C.I.E., G.C.V.O., G.B.E., A. D. C. 13 May 1904.
COLONEL—Lieut.-General *Sir* Edward L. Elliot, K.C.B., D.S.O., 13 May 1904.

1st Commn. or date of entering service.	Names and Rank.	Army rank.	Present appointment in Regiment.	Remarks.
	Commandant.			
18 May 92	Ellwood, Lt.-Col. B.P.	29 July 17	29 July 17	
	Squadron Commanders. (4)			
14 Apl. 89	*Bannerman, Lt.-Col. W. P. (Q)*	*24 Apl. 15*	*6 Oct. 12*	*2nd-in-Comd.*
16 Nov. 92	*Savile, Lt.-Col. A. R. C. (Q)*	*16 Nov. 18*	*29 July 08*	*Comdt., 2-95th Infy.*
5 Aug. 96	*Williams, Maj. A. F. C.,* D.S.O., *p.s.c. (Q.) (temp. Lt.-Col.)*	*5 Aug. 14*	*1 Oct. 09*	*G. S. O., 1st gr.*
22 Jan. 98	*Macleod, Maj. W. M., p.s.c.*	*1 Sep. 15*	*29 July 17*	*G. S. O., 2nd gr., N. C.*
	Squadron Officers. (9)			
3 Aug. 98	Hay, Maj. A. S.	1 Sep. 15	7 Apl. 04	With 13th Lrs.
4 May 01	*Willis, Maj. G. C. W.,* D.S.O.	*4 May 16*	*17 Oct. 06*	*Comdt., Remt. Training Dept., Sehore.*
9 Jan. 04	*Dodd, Capt. P. C. R.*	*9 Jan. 13*	*31 Jan. 08*	*Kurram Militia.*
29 July 04	*Burmester, Capt. Z. G.*	*29 July 13*	*10 Nov. 08*	*With 2-15th Sikhs.*
9 Sep. 08	*Andrewes, Capt. C. L.,* M. C.	*1 Sep. 15*	*18 Nov. 09*	*With Sig. Ser.*
20 Jan. 09	Baker, Bt. Maj. W. H. G., D.S.O. (*temp. Maj.*)	1 Jan. 18	29 Mar. 10	2nd-in-Comd. (*temp.*)
29 Jan. 10	*Vickers, Capt. W. G. H.*	*1 Sep. 15*	*15 Mar. 11*	*Bde. Maj.*
6 Sep. 11	*Shearer, Capt. E. J.,* M.C.	*6 Sep. 15*	*1 Apl. 13*	
20 Jan. 12	Carr-White, Capt. R. A.	20 Jan. 16	2 Apl. 13	
15 Aug. 14	Bucher, Capt. F. R. R.	1 Sept. 15	1 Oct. 17	Adjt., 1 Apl. 18.
14 Nov. 16	Corner, Lt. R. J. (*Actg. Capt.*)	14 Nov. 17	21 Nov. 16	Attd.
29 Apl. 17	Hudson, Lt. T.	29 Apl. 18	11 May 17	Attd.
31 Aug. 18	Blackden, 2nd-Lt. J. F.	31 Aug. 18	5 Sep. 18	Attd.
	Attached.			
1 Aug. 15	Lochner, Lt. C. B., M.C. (*Actg. Capt.*)	1 Aug. 16	20 Aug. 17	S. Wales Bord. I. A. (*on probn.*)
28 Sep. 14	*Hastings, Lt. J. M.*	*28 June 16*	*29 Aug. 17*	*I. A. (on probn.) I. S. Troops.*
31 Oct. 14	Gilpin, Lt. E. L. P. P.	31 Oct. 15	23 Sep. 18	Derb. Yeo. I. A. (*on probn.*)
24 Dec. 14	Bennett, Lt. N. B.	24 Sep. 16	23 Jan. 18	I. A. (*on probn.*)
2 Mar. 15	Smith, Lt. A. A.	2 Dec. 16	3 Aug. 18	A. S. C. I. A. (*on probn.*)
20 Oct. 15	Burnley, Lt. W. D.	18 July 17	18 Jan. 18	Leic. Yeo. I. A. (*on probn.*)
22 Nov. 16	Muirhead, Lt. J. W.	22 Nov. 17	18 Feb. 18	Norf. Yeo. I. A. R. O.
17 Jan. 17	Kelso, 2nd-Lt. W. D.	17 Jan. 18	3 Feb. 18	S. I. Horse. I. A. (*on probn.*)
16 Apl. 17	Whinyates, Lt. S.	17 Apl. 18	18 Jan. 18	I. A. R. O. Qr.-Mr. (*temp.*)
7 Apl. 16	Boyle, Lt. H. McC., M.C.	17 Apl. 17	16 Feb. 18	I. A. (*on probn.*)
7 Apl. 16	O'Connor, Lt. T. J.	7 Apl. 17	3 Feb. 18	I. A. (*on probn.*)
27 Jan. 16	Campbell, Lt. G. R.	27 July 17	2 Mar. 18	I. A. R. O. Adjt. (*Actg.*)
19 Apl. 16	Parsons, Lt. J. H., M. C.	19 Apl. 17	25 Jan. 18	Worc. Yeo. I. A. (*on probn.*)
31 Oct. 17	Stevens, Lt. D. T.	31 Oct. 18	17 May 18	Tank Corps. I. A. (*on probn.*)
22 Nov. 16	Taylor, Lt. G. F.	22 Aug. 18	7 May 18	I. A. (*on probn.*)
22 Nov. 16	Fleming, Lt. F.	22 Aug. 18	7 May 18	Lond. Yeo. I. A. (*on probn.*)
29 Apl. 17	Williams, Lt. G. E. M.	29 Apl. 18	19 Apl. 18	S. R. Yeo. I. A. (*on probn.*)
2 July 17	Manley, Lt. W. B.	2 July 18	2 July 17	I. A. R. O. Staff Capt.
23 July 15	Paton, Lt. J. W. F.	23 July 16	....	I. A. (*on probn.*) Signal Service
28 Aug. 15	Henrotin, Lt. H. V.	28 Aug. 16	31 Aug. 18	A. S. C. I. A. (*on probn.*)
29 June 17	Hathaway, Lt. J. R.	29 June 18	....	

31st Duke of Connaught's Own Lancers—*contd.*

Date of entering service.	Names.	Army rank. Jemadar.	Ressaidar.	Risaldar.	Risaldar-Major.	Remarks.
	Risaldar-Majors.					
6 Dec. 85	*Muhi-ud-din Khan, Sardar Bahadur,* C.I.E., I.D.S.M. (56, 71).	*16 July 95*	*1 Feb. 06*	*26 Feb. 09*	*9 Aug. 12*	*Order of Br. I., 1st Cl. A. D. C. to H. E. the Viceroy.*
16 July 95	Asaram (50, 60)	1 Dec. 99	1 Apl. 05	1 Apl. 11	19 Apl. 14	
	Risaldars.					
25 Nov. 89	Ramcharan Singh (50, 66*a*)	1 Feb. 06	9 Aug. 12	19 Apl. 14		
1 June 07	Hakim Khan (45*a*, 50)	22 Jan. 9	8 Mar. 14	7 Sep. 16		
	Saidulla Khan (*Retd.*)	. . .	. . . .	. . . .	Re-employed.	
11 Sep. 05	Kehar Singh (50, 52)	11 Sep. 12	29 Oct. 14	17 Oct. 18		
1 July 03	Jaising Rao Mahadik (45*a*, 47*a*, 50)	9 Sep. 13	7 Sep. 16	17 Oct. 18		
	Abdulla Khan	...	...	...	Re-employed, 24 Jan. 17.	
	Ressaidars.					
15 Aug. 99	Ajun Khan (71) I.D.S.M.	8 Mar. 14	13 Sep. 16			
24 Aug. 05	Gauhar Rahman Khan (50, 53, 66*a*)	19 Apl. 14	13 Sep. 16			
7 July 08	Abaji Rao Mahadik (50)	16 Apl. 14	17 Oct. 18			
	Jemadars.					
6 Feb. 86	Muhammad Ishaq (64)	20 Sept. 14				
25 June 99	Indar Singh	29 Oct. 14	With 21st Cav.			
1 Jan. 98	Bakhshish Singh (52, 56, 66*a*, 71)	29 Nov. 14				
16 June 99	*Majid Gul (66a)*	*4 July 15*				
25 June 99	Badhawa Singh (45*a*, 50)	4 Oct. 15	Woordie-Maj. 7 Sep. 16.			
11 July 09	Ram Rao Powar	4 Oct. 15				
8 Dec. 99	Balwant Rao Sinde	23 Nov. 15	Remt. Training Depôt, Saugor.			
20 Oct. 02	Mahan Singh (66*a*)	. . .				
12 Dec. 15	Krishna Rao Khanvilkar (61)	14 Dec. 15				
1 Oct. 97	Hakikat Singh (66*a*)	17 June 16				
24 June 05	Muhammad Islam Khan (51)	12 July 16	With 19th Lrs.			
1 May 01	Imrat Singh (51, 58)	19 Aug. 16				
28 Jan. 98	Partap Singh (104)	3 May 17				
20 Oct. 02	Kehar Singh (56)	13 May 17				
1 Jan. 10	Todar Singh (47*a*)	6 June 17				
1 July 06	Mir Aslam Khan	30 June 17				
15 July 00	Bhanwar Singh (58)	18 Aug. 17				
12 June 07	Indar Singh	29 Oct. 17	14 Lrs.			
10 May 06	Baba Rao Ghorpude	9 Nov. 17				
10 Feb. 06	Jagdeo Singh (56, 47*a*)	3 Dec. 17				
1 Aug. 00	Maruti Rao Ghorse	2 Jan. 18				
	Baputi Rao Kadam	1 Oct. 18				
	Atar Singh	17 Oct. 18				

32nd Lancers.

Raised at Baroda in 1817, the second of two existing troops forming the nucleus, and was originally designated the 2nd Regiment of Bombay Light Cavalry. Became the 2nd Regiment of Bombay Silladar Light Cavalry, 1861, but reverted to its previous designation the same year; became the 2nd Bombay Lancers, 1883; present designation, 1903.

"Central India" "Afghanistan, 1879-80."

Composition—2 Squadrons of Musalman Rajputs, 1 of Rajputs, 1 of Sikhs other than Jat Sikhs.

***Uniform*—Blue. *Facings*—White.**

1st Commn. or date of entering service.	Names and Rank.	Army rank.	Present appointment in Regiment.	Remarks.
	Commandant.			
9 Sep. 91	Keogh, Lt.-Col. J. B., D.S.O.	23 Aug. 17	23 Aug. 17	
	Squadron Commanders. (4)			
20 Feb. 95	*Hennell, Maj. A. M.*	*20 Feb. 13*	*11 Oct. 14*	*Lv., ex I., m.c, 6 mos.*
22 Jan. 96	*Cassels, Bt.-Col. R. A.,* C.B. D.S.O., *p.s.c.* (*temp. Br.-Genl.*) (*Q*).	*10 Mar. 17*	*31 Mar. 15*	*Brig. Comdr.*
5 Aug. 96	*Taylor, Maj. P. H. M.*	*5 Aug. 14*	*1 Apl. 00*	*With 42nd Cav.*
	Squadron Officers. (9)			
27 Aug. 02	Reid, Maj. W. C.	27 Aug. 17	12 Apl. 04	2nd-in-Comd. (*temp.*)
18 Jan. 05	Mulloy, Capt. N. F. C., M.C. (*Actg. Maj.*)	18 Jan. 18	27 Apl. 06	
5 Aug. 05	Childers, Capt. H. F. E.	5 Aug. 14	17 Dec. 06	*Lv., ex Ind., m. c., 6 mos.*
24 Jan 06	*Smith, Capt. V. H. S.*	*24 Jan. 15*	*20 Aug. 07*	
29 Aug. 06	*Walker, Capt. E. J. P. T.* (*temp. Maj.*)	*29 Aug. 15*	*24 Nov. 07*	*G. S. O., 2nd gr., A. H. Qrs.*
7 Nov. 06	*Howard, Capt. C. A. L.*	*1 Sep. 15*	*24 Nov. 08*	With K. Afr. Rif.
15 Aug. 14	Fraser, Capt. J. D. (Capt.)	15 Aug. 18	8 Nov. 15	Attd.
15 Nov. 15	Wilson, Lt. A. D.	15 Nov. 16	....	Attd.
29 June 16	Denehy, Lt. F. P. G. (*Actg. Capt.*)	29 June 17	10 July 17	Attd. Adjt.
15 Feb. 15	Field, Lt. F. C. (207*a*)	15 Feb. 16	3 Dec. 17	Attd.
	Attached.			
20 Nov. 14	Mackenzie, Lt. J. M. (*Actg. Capt.*)	20 Nov. 15	11 Feb. 17	I. A. R. O.
30 Apl. 15	Lloyd-Williams, Lt. J. E. (*Actg. Capt.*)	30 Apl. 16	20 June 17	I. A. R. O.
3 Sept. 14	Greenhow, Lt. E. L.	3 June 16	3 July 18	I. A. (*on probn*) *Lv., ex. I., m. c.*
24 Aug. 15	Mathews, 2nd-Lt. S. L.	24 Aug. 16	27 Sep. 17	I. A. R. O.
11 Aug. 15	Danby, Lt. E. C.	25 Aug. 16	27 Dec. 16	I. A. R. O. *Lv., ex Ind., m.c., 6 mos.*
16 Sep. 15	Forbes, Lt. A. D. C.	16 Sep. 16	30 Sep. 16	I. A. R. O.
14 Jan. 16	Sheil, Lt. P. H. J. (207*a*)	14 Jan. 17	2 Mar. 18	3rd Res. R. Cav. I. A. (*on probn.*)
5 May 15	Cox, Lt. R. N.	1 July 17	27 Nov. 18	I. A. (*on probn.*)
11 Jan. 16	Black, Capt. J. S. C.	12 July 17	13 Oct. 18	I. A. R. O. (*on probn.*)
1 Jan. 16	Arrindell, Lt. G. D. S. M., M. C.	16 Aug. 16	14 Aug. 16	I. A. R. O.
24 Aug. 16	Quinlan, Lt. P. B.	24 Aug. 17	26 May 17	I. A. R. O.
1 Sept. 16	Dreyfus, Lt. B. E.	1 Sept. 16	11 Aug. 18	I. A. R. O.
10 Oct. 16	Boswell, Lt. R. B.	10 Oct. 17	26 May 17	I. A. R. O.
29 Nov. 16	Poole, Lt. E. R. L. (207*a*)	29 Nov. 17	29 Sep. 17	A. S. C. I. A. (*on probn.*)
26 Feb. 17	Clews, Lt. L. C.	26 Feb. 18	13 Apl. 18	York. L. I.—I. A. (*on probn.*)
23 Apl. 17	Tozer, Lt. D. C. E.	23 Apl. 18	25 Aug. 17	I. A. R. O.
5 Oct. 17	Bromley, Lt. R. E. (207*a*)	5 Oct. 18	13 Oct. 17	I. A. R. O.
23 Apl. 18	Love, 2nd-Lt. R. M.	23 Apl. 18	22 Oct. 18	I. A. (*on probn,*)

Risaldar-Majors.

1st Commn. or date of entering service.	Names and Rank.	Jemadar.	Ressaidar.	Risaldar.	Risaldar-Major.	Remarks.
....	Harnam Singh	....	....	....	Re-employed 15-12-17.	
13 Mar. 09	Thakur Jor Singh (56) *Bahadur*	7 Dec. 98	15 Nov. 04	11 May 13	14 Dec. 17	*Order of Br. I., 2nd cl.*

Risaldars.

1st Commn. or date of entering service.	Names and Rank.	Jemadar.	Ressaidar.	Risaldar.	Remarks.
....	Shaikh Mohi-ud-din	....	....	....	Re-employed 12-4-18.
9 June 86	Waryam Singh	1 Aug. 00	9 June 07	5 Dec. 13	
3 June 02	Wajid Ali Khan (56, 66*a*)	19 Aug. 06	16 Jan. 12	7 Dec. 14	
11 Dec. 07	Sundar Singh, I.O.M., I.D.S.M.	11 Dec. 07	24 Mar. 15	1 Dec. 16	
1 Feb. 00	Shaikh Hussain (45*a*, 56, 66*a*)	1 Mar. 13	9 Feb. 17	9 Oct. 17	
11 Mar. 93	*Yasin Khan* (66*a*)	*16 June 12*	*1 Aug. 18*	*2 Aug. 18*	*Base Remt. Depot.*

32nd Lancers—*contd.*

Date of entering service.	Names.	Army rank.		Remarks.
	Ressaidars.			
1 July 00	Hasan Ali Khan (56, 66*a*)	9 June 07	31 Oct. 14	
1 Oct. 87	Hari Singh (56) (Retd.)	1 Oct. 02	11 May 13	Re-employed 1 May 17.
2 Jan. 96	Sarup Singh (56)	16 June 09	1 Dec. 16	
1 Apl. 09	Muhammad Hayat Khan	29 Oct. 14	9 Oct. 17	
4 Mar. 97	Rewat Singh	5 Dec. 13	20 Nov. 16	
30 Mar. 02	Ratan Singh, I.D.S.M.	11 Apl. 16	19 Mar. 18	
	Jemadars.			
18 Feb. 96	Maqsud Ali Khan (66*a*)	11 May 13		
1 Apl. 96	Ali Bakhsh Khan	29 Oct. 14		
1 Aug. 94	*Zafuryab Khan*	*1 Nov. 14*		
1 Oct. 99	Abdul Hakim Khan	4 Jan 15		
17 Apl. 05	Adam Khan	1 May 15		
18 Dec. 95	Dhara Singh, I.D.S.M.	29 July 15		
. . . .	Magan Singh	. . . .		Re-employed 20 Feb. 18.
23 June 98	Bishan Singh	16 Dec. 15		
1 Aug. 95	Fazar Ali Khan, I.O.M., I.D.S.M. (56).	10 Feb. 16		
10 Nov. 91	Man Singh	16 May 16		
16 Sep. 99	Kalyan Singh	16 May 16		With 34th Horse.
1 Oct. 99	Sanowar Khan	1 July 16		
1 Apl. 01	Ashghar Ali Khan	1 July 16		
6 June 03	Nihal Singh (D)	2 Nov. 16		Wcordie-Maj., 1 Apl. 17.
18 Jan. 17	Harnam Singh	18 Jan. 17		Supy.
9 Mar. 96	Chamba Singh	29 Jan. 17		
1 May 06	Bichitar Singh	20 Mar. 17		
1 July 00	Ba'u Singh	9 Oct. 17		
15 June 05	Uttam Singh	22 Feb. 18		Qr. Mr. Jem.
24 May 02	Gurdat Singh	17 May 18		
10 Sept. 97	Bakhtawar Singh	10 June 18		
1 Dec. 06	Ijaz Ali Khan	1 July 18		

33rd Queen Victoria's Own Light Cavalry.

Raised at Sirur in 1820 by Major P. Delamotte, as the 3rd Regiment of Bombay Light Cvalry. Became the 3rd Regiment of Bombay Silladar Light Cavalry, 1861, but reverted to its previous designation, 1861; became the 3rd (the Queen's Own) Regiment of Bombay Light Cavalry, 1876; the 3rd (Queen's Own) Bombay Light Cavalry, 1901; 33rd Queen's Own Light Cavalry, 1903; present designation in 1911.

THE ROYAL AND IMPERIAL CYPHER OF QUEEN VICTORIA WITHIN THE GARTER.
"Ghuznee, 1842" "Cabool, 1842" "Hyderabad"
"Reshire" "Bushire" "Koosh-ab" "Persia"
"Central India" "Abyssinia" "Kandahar, 1880" "Afghanistan, 1879-80" "China, 1900"
Composition—1 Squadron of Jats, 1 of Sikhs, 1 of Kaimkhanis, 1 of Musalman Rajputs. 1 Addl. Squadron.

Uniform—Blue. *Facings*—Scarlet.

COLONEL—Major-General James F. Willoughby . . 5 November 1908.

1st Commn. or date of entering service.	Names and Rank.	Army rank.	Present appointment in Regiment.	Remarks.
	Commandant.			
11 Feb. 88	Stack, Lt.-Col. C. S., C.M.G.	11 Feb. 14	21 Sep. 15	
	Squadron Commanders. (4)			
1 Sep. 91	*Hopwood, Bt.-Col. H. R., p.s.c.* C.S.I., (*Q*).	*3 June 16*	. . . .	*2nd-in-Comd. G. S. O., 1st gr.*
5 Aug. 96	*Dening, Maj. L. E.*	*5 Aug. 14*	*17 May 11*	*With I. S. Troops.*
20 Jan. 00	*Gillies, Maj. F. G., p.s.c. (Lt.-Col.)*	*1 Sep. 15*	*22 Aug. 14*	*G. S. O., 1st gr., Meerut Divn.*
8 Jan. 01	Ferris, Maj. W. A. T.	22 Jan. 16	5 Oct. 16	2nd-in-Comd. (*Offg.*)
	Squadron Officers. (9)			
9 Jan. 04	Kenworthy, Capt. W. (*temp. Maj.*)	9 Jan. 13	12 Jan. 06	D. A. A. & Q. M. G., Ambala Bde.
16 Jan. 07	Edward-Collins, Capt. G., M.C.	1 Sep. 15	11 Mar. 09	Bde. Maj., 4th Cavy. Bde.
8 Sep. 09	Meiklejohn, Capt. J. F. (*temp. Maj.*)	1 Sep. 15	14 Nov. 10	D. A. Q. M. G., A. Hdqrs.
3 Sep. 10	Fulton, Capt. C. M. [*I*]	1 Sep. 15	5 Nov. 11	N.C. O's School, N. Comd.
20 Jan. 12	*Massy, Capt. D. F.*	*20 Jan. 16*	*13 Mar. 13*	*With R. A. F.*
22 Jan. 13	Baines, Capt. G. D.	22 Jan. 17	9 May 14	With 44th Cav.
14 Nov. 15	Newill, Lt. W. M. (*Actg. Capt.*)	14 Nov. 16	4 Aug. 17	Attd.
10 Sep. 15	Wallington, Lt. J. M. (*Actg. Capt.*)	10 Sep. 16	1 Apl. 17	Attd.
18 Feb. 16	*Tulloch, Lt. H. M.*	*18 Feb. 17*	*6 Nov. 16*	*Attd. R. A. F.*
27 July 16	*Vrooman, Lt. P. H. (Actg. Capt.)*	*27 July 17*	*15 Nov. 16*	*Attd.*
18 June 17	*Woodman, Lt. J. B.*	*18 June 18*	*24 June 17*	*Attd.*
12 Sept. 17	*Clarke, Lt. J. A.*	*12 Sept. 18*	*8 Jan. 18*	*Attd.*
. . . .	Ryley, 2nd-Lt. L. F. M.	. . . .	21 Oct. 18	Attd.
	Attached.			
10 Sep. 14	*Courthope-Munroe, Lt. C. H.*	*13 Aug. 15*	*7 Aug. 15*	*I. A. R. O.*
13 Dec. 15	Fetherstonhaugh, Lt. M. R.	13 Dec. 16	22 Jan. 18	I. A. (*on probn.*)
10 July 15	Cook, 2nd-Lt. H. C.	10 July 16	6 Oct. 17	Scot. Horse. I. A. R. O. (*on probn.*)
25 Jan. 16	Peel, Lt. W. R. B. (*temp. Capt.*)	25 Jan. 17	22 Nov. 17	Som. L. I. I. A. (*on probn.*) Offg. Adjt., 1 Aug. 18.
22 June 15	Cross, Lt. F. N., D.S.O.	22 June 16	11 Apl. 18	L'pool. R. I. A. (*on probn.*)
14 Nov. 16	Robertson, Lt. J.	14 Nov. 17	5 Jan. 18	I. A. R. O. (*on probn.*)
23 Aug. 17	Lane, Lt. V. C. T.	23 Aug. 18	26 Jan. 18	R. Highrs. I. A. (*on probn.*)
31 Oct. 17	Dasent, Lt. L. H.	31 Oct. 18	12 Feb. 18	Attd
23 Feb. 15	MacWilliam, Lt. T. A.	23 Feb. 16	12 Feb. 18	6th Innis. Dgs. I. A. R. O. (*on probn.*)
31 May 15	Cornell, Lt. W. E. L.	31 May 16	1 Mar. 18	A. S. C., I. A. (*on probn.*)
9 July 18	Codrington, 2nd-Lt. K. deB.	9 July 18	9 July 18	I. A.
19 Feb. 18	Reilly, 2nd-Lt. E. F. J.	19 Feb. 18	19 Mar. 18	I. A. R. O.
9 May 17	Noble, Lt. G.	9 May 18	15 Apl. 18	North'd. Fus. I. A. (*on probn.*)
1 Aug. 17	Hutton, Lt. E. D.	1 Aug. 17	31 Mar. 18	Aust. M. G. Corps. I. A. (*on probn.*)
10 Sep. 14	Booth, Lt. A. E. (*Actg. Capt.*)	10 Sept. 15	31 Mar. 18	W. York. R. I. A. R. O. (*on probn.*)
17 Feb. 16	King, Lt. E. R.	17 Feb. 18	. . . .	

33rd Queen Victoria's Own Light Cavalry—*contd.*

Date of entering service.	Names.	Army rank. Jemadar.	Ressaidar.	Risaldar.	Risaldar-Major.	Remarks.
	Risaldar-Major.					
1 Feb. 95	Santa Singh, I.O.M., *Bahadur* (52), (56)	11 Jan. 99	16 Oct. 04	1 Mar. 12	6 Mar. 16	*Order of Br. I., 2nd Cl.*
	Risaldars.					
2 Oct. 88	Aladad Khan (52, 56)	28 July 99	16 Feb. 11	16 Nov. 13		
1 Sep. 96	Udham Singh (56, 66*a*) . . .	16 Oct. 04	1 Mar. 12	18 Sep. 14		
23 Dec. 86	*Hidayat Ali Khan (43, 56, 66a),* I.D.S.M.	*21 Nov. 07*	*16 June 13*	*6 Mar. 16*		
8 Jan. 96	Muhammad Arshad Khan (56, 66*a*)	10 May 08	11 Sep. 13	29 Apl. 16		
21 Dec. 08	Abdul Sattar Khan (45*a*, 50, 53, 66*a*), I.D.S.M.	1 Apl. 09	16 Nov. 13	26 Sep. 17		
	Taj Muhammad Khan	. . . .	. . . .	. . . .		*Re-employed, 21 Jan. 17.*
	Ressaidars.					
12 Dec. 99	Alau-din Khan (45*a*, 52, 56, 66*a*), I.D.S.M.	1 Apl. 10	18 Sep. 14			
1 Jan. 98	Sundar Singh (50, 53, 60*a*, 66*a*) .	1 Mar. 12	6 Mar. 16			
1 May 93	Prem Singh (53, 56)	1 June 13	29 Apl. 16			
1 Apl. 94	Bhure Khan (52, 56) . . .	16 Nov. 13	26 Sep. 17			
1 Aug. 04	Gunga Man (71)	30 Jan. 15	26 Sep. 17			
	Jemadars.					
23 Nov. 94	*Bhag Mal (52, 56) (I)*	*1 Apl. 10*				*With Remt. Depôt, Saugor.*
13 Oct. 06	*Mahmud Khan (56, 66a)* . . .	*1 Sep. 13*				*With Remt. Depôt, Bangalore.*
11 Aug. 06	Muhammad Shafi Khan (I) (66*a*, 71)	6 Mar. 16				
21 Mar. 03	Muhammad Yusuf Khan, I.D.S.M. (45*a*, 52, 56, 66*a*).	29 Apl. 16				Woordie-Major, 26 Sep. 17.
	Hakim Singh (Re-employed) .	9 Aug. 16				
19 Aug. 01	Mazhar Ali Khan, I.D.S.M. (56, 66*a*)	11 Oct. 16				With 34 Horse.
1 May 99	Bhagmal (56, 66*a*) (II)	24 Oct. 16				
19 Aug. 97	*Rur Singh (56, 66a)* .	*18 Feb. 17*				*With 29 Lrs.*
15 May 97	*Muhammad Shafi Khan (66a) (II)*	*20 Mar. 17*				*With 62nd Grantee Camel Corps.*
1 Aug. 04	Harnarain (56, 60*a*) . .	1 June 17				
8 June 03	Rai Singh (47*a*, 56, 66*a*) . . .	1 June 17				
1 Jan. 94	Muhammad Ismail Khan (58) . .	1 July 17				Qr.-Mr., 3 Dec. 17.
1 Sep. 00	Siri Ram (47*a*, 56, 58, 66*a*) . .	1 Aug. 17				
1 Dec. 95	*Naneh Singh (66a)*	*26 Sep. 17*				*With Syce Corps, Sehore.*
1 Aug. 06	Digh Ram (47*a*, 56)	26 Sep. 17				
27 June 09	Alam Ali Khan, I.D.S.M. . . .	26 Sep. 17				
27 Apl. 02	*Hazara Singh (60a)*	*3 Dec. 17*				
1 Jan. 04	Dip Chand (58)	16 Jan. 18				
	Nur Muhammad Khan	. . . .				*Re-employed 27 May 17. With I. C. Sec.*
	Nand Singh	. . . .				*Re-employed 1 June 17. With I. C. Sec.*
4 Mar. 06	Tulsi Ram (47*a*, 56)	26 Apl. 18				
1 Nov. 04	Israil Khan (56)	1 Sep. 18				
20 June 12	Khan Muhammad Khan (56, 47*a*) .	1 July 18				
9 July 14	Daud Khan (56)	19 Aug. 18				

34th Prince Albert Victor's Own Poona Horse.

Raised in 1817, as the Auxiliary Horse. Became the Poona Auxiliary Horse, 1818; the Poona Irregular Horse, 1847; the Poona Horse, 1860; the 4th Regiment, Poona Silladar Cavalry, 1861; the 1st Regiment, Poona Horse, 1861; the Poona Horse, 1862; the 4th Bombay Cavalry (Poona Horse), 1885; the 4th (Prince Albert Victor's Own) Bombay Cavalry (Poona Horse), 1890; present designation, 1903.

Carries a **standard** surmounted by a silver hand and bearing a Persian inscription (captured at the battle of Kooshab.)

"Corygaum" "Ghuznee, 1839" "Affghanistan, 1839" "Candahar, 1842"
"Meeanee" "Hyderabad" "Reshire" "Bushire" "Kooshab" "Persia"
"Kandahar, 1880" "Afghanistan, 1879-80."

Composition—2 Squadrons of Rathore Rajputs, 1 of Kaimkhanis, 1 of Punjabi Musalmans.
Uniform—Blue. *Facings*—French Grey.

COLONEL—Major-General George C. Hogg, C.B. . 13 May 1904.
HONORARY COLONEL—*Hony.* Lieut.-General *His Highness* MAHARAJA *Sir* PRATAP SINGH, *Bahadur* G.C.B., G.C.S.I., G.C.V.O., A.D.C. . . . 19 November 1912.

1st Commn. or date of entering service.	Names and Rank.	Army rank.	Present appointment in Regiment.	Remarks.
	Commandant.			
22 Aug. 88	Cooper, Lt.-Col. W. G.	22 Aug. 14	3 Nov. 14	
	Squadron Commanders. (4)			
7 Nov. 91	*Davie, Lt.-Col. J. H. M. (Q)*	*7 Nov. 17*	*3 Nov. 14*	*2nd-in-Comd., D.A.A.G., Poona Dn.*
19 July 93	*Molloy, Maj. G. M., p.s.c. (Temp. Lt.-Col.) (Q)*	*19 July 11*	*19 Nov. 10*	*G. S. O., 1st gr., Army Hdqrs.*
20 July 98	Lucas, Maj. G. W. C.	1 Sep. 15	19 Nov. 10	
20 May 99	*Grimshaw, Maj. R. W. W.*	*1 Sep. 15*	*19 Nov. 10*	*With I. S. Troops.*
	Squadron Officers. (9)			
12 Aug. 99	Norbury, Maj. P. FitzG. [L]	1 Sep. 15	16 May 03	
22 Nov. 99	*Gray, Maj. E. St. C., M.C.*	*1 Sep. 15*	*15 Sep. 03*	
24 May 02	Paterson, Maj. R. H. O'D.	27 May 16	7 Apl. 06	With 2-8th Rajputs.
29 Jan. 02	Hildebrand, Maj. H. A.	29 Jan. 17	12 Oct. 06	
22 Mar. 04	*Raymond, Capt. M. C., M.C.*	*2 Mar. 13*	*1 Jan. 06*	*D. A. A. & Q. M. G.*
5 Aug. 05	Elphinston, Capt. W. G.	5 Aug. 14	17 Feb. 09	
29 Aug. 06	Campbell, Capt. W. E. D.	9 Aug. 15	12 Sep. 09	With 111th Mahars.
6 Sep. 08	MacGregor, Capt. R. G.	1 Sep. 15	14 Jan. 10	With 45th Cav.
6 Sep. 11	Simson, Capt. G. O.	6 Sep. 15	17 Nov. 12	Adjt., 1 July 16.
18 June 17	*Clark, Lt. C. D. LeG.*	*20 Apl. 16*	*18 June 17*	*With I. S. Troops.*
14 Nov. 16	Hatch, Lt. K.	30 Aug. 16	14 Nov. 16	Attd.
14 Nov. 16	Acworth, Lt. J. P.	14 Nov. 17	. . . .	Attd.
14 Nov. 16	Howland, Lt. C. C.	14 Nov. 17	. . . .	Attd.
14 Nov. 16	Beer, Lt. W. R.	15 Jan. 16	. . . .	Attd.
31 Jan. 18	Pearson, 2nd-Lt. H. S.	31 Jan. 18	22 Feb. 18	Attd.
31 Aug. 18	Ruttledge, 2nd-Lt. R. F.	31 Aug. 18	4 Sep. 18	Attd.
21 Dec. 17	Broughton, 2nd-Lt. C. E.	21 Dec. 17	14 Apl. 18	Attd.
1 Oct. 18	Spencer, 2nd-Lt. L. A.	1 Oct. 18	1 Oct. 18	Attd.
	Attached.			
1 Oct. 14	Scott, Capt. R. A.	1 Oct. 18	29 Mar. 18	Hrs. I. A. (*on probn.*)
7 Oct. 14	Roark, Lt. C. T. I.	19 Jan. 16	18 May 18	L. G. I. A. (*on probn.*)
27 Nov. 14	Nicholl, Capt. C. K.	30 July 18	30 July 15	I. A. R. O.
4 Dec. 14	Robertson, Capt. F. E.	30 Aug. 18	3 Aug. 15	I. A. R. O. Qr.-Mr., 1 July 16.
8 Jan. 15	Iles, 2nd-Lt. A. G.	18 Jan. 15	24 Oct. 15	I. A. R. O.
8 Jan. 15	Leigh-Clare, 2nd-Lt. H. J. L., M.C.	8 Jan. 15	6 Jan. 16	I. A. R. O.
12 Mar. 15	Pilcher, Capt. H. R.	30 Apl. 18	16 Apl. 15	I. A. R. O.
21 Apl. 15	Deacon, Lt. C. G.	9 Apl. 16	12 July 15	I. A. R. O. with Camel Corps, Agra.
30 Apl. 15	Durack, Lt. J. J.	30 Apl. 16	10 May 16	I. A. R. O.
2 June 15	Anson, Lt. E. E.	2 Mar. 17	24 Oct. 17	Duke of Lan. Yeo. I. A.
10 July 15	Hearn, Lt. W. S. G.	10 July 16	2 Mar. 18	
26 Jan. 16	Lunham, Lt. T. M.	26 Jan. 16	. . . .	
1 Sep. 17	Fishburn, Lt. J.	1 Sep. 18	25 Sep. 17	I. A. (T. C.)
18 Oct. 17	Butler, Lt. H. E.	18 Oct. 18	2 Nov. 17	I. A. R. O.
24 Oct. 17	Lynch-Bloss, 2nd-Lt. C. E. (*Actg. Capt.*)	24 Oct. 17	11 Sep. 17	I. A. R. O.
11 Sep. 18	Fanshawe, 2nd-Lt. W.	11 Sep. 18	11 Sep. 18	I. A. R. O.

		Jemadar.	Ressaidar.	Risaldar.	Risaldar-Major.
	Risaldar-Major.				
15 Aug. 93	Husain Bakhsh Khan *Bahadur* (52, 56).	17 Dec. 00	8 Dec. 04	10 Aug. 14	15 Jan. 15. *Order of Br. I., 2nd Cl.*
	Risaldars.				
27 Apl. 04	Hamir Singh *Bahadur* (52, 56)	27 Apl. 04	2 Apl. 08	1 Apl. 12	*Order of Br. I., 2nd Cl.*
21 Jan. 04	Rawat Singh (I), I.D.S.M. (45*a*, 52, 56, 71).	20 Apl. 08	2 Nov. 14	15 Jan. 15	
1 July 94	Sattar Shah, I.D.S.M. (53, 56, 62)	1 July 09	21 Dec. 14	15 Jan. 17	
16 Sep. 96	Firoz Khan	10 Aug. 14	25 Jan. 15	9 Mar. 18	
16 Nov. 01	Sultan Singh	2 Nov. 14	30 Jan. 17	19 Mar. 18	

34th Prince Albert Victor's Own Poona Horse—*contd.*

Date of entering service.	Names and Rank.	Army rank.		Remarks.
	Ressaidars.			
4 July 04	Balwant Singh	8 Dec. 04	3 Nov. 14	
4 Dec. 07	Amar Singh (56)	1 Apl. 12	21 Dec. 14	
14 Jan. 99	Bulwant Singh (51, 52) . . .	5 Dec. 14	22 Mar. 17	27th Cav.
1 July 97	Ganpat Singh	11 Apl. 16	25 Aug. 17	
2 Apl. 04	Amir Khan, I.D.S.M.	20 Sep. 16	25 Aug. 17	
10 Jan. 02	Ghulam Muhammad Khan (51) .	18 Dec. 14	29 Dec. 17	27 Cav.
26 Apl. 01	Shamsher Ali Khan	21 Dec. 14	4 Mar. 18	
30 Mar. 02	Jattan Singh, I. D. S. M. . . .	11 Apl. 16	19 Mar. 18	32 Lns.
21 Nov. 00	*Muhammad Yasin*	*21 Dec. 14*	. . . .	*22nd Cav.*
	Jemadars.			
11 Aug. 99	Shahzad Khan	9 Dec. 14		
1 July 97	Nathu Singh	15 Jan. 15		
7 Jan. 96	Mohan Singh	25 July 15		
17 Apl. 01	Jaswant Singh	16 Feb. 16		
13 Mar. 01	Ghulam Hasain Khan	11 Apl. 16		
17 Apl. 01	1 Unaol Singh	23 Feb. 17		
4 Sep. 08	1 Taj Muhammad Khan . .	25 Feb. 17		
28 May 13	1 Sher Bahadur Khan	22 Mar. 17		
5 Mar. 08	1 Jorawar Singh	5 Mar. 17		
6 Feb. 07	Anno Khan, I. D. S. M. . . .	12 Apl. 17		
1 Apl. 07	Sher Bahadur Khan	25 Aug. 17		
4 Mar. 06	Kale Khan	25 Aug. 17		
14 Mar. 10	Rawat Singh	25 Aug. 17		
15 Aug. 12	Umed Singh	16 Oct. 17		
1 Apl. 06	Ghulam Shah	3 Dec. 17		
6 Nov. 04	Ghulam Muhammad Khan . .	4 Mar. 18		
14 Sep. 09	Bahcrishal Singh	9 Mar. 18		
17 Feb. 13	Ugam Singh	19 Mar. 18		

35th Scinde Horse.

Raised at Hyderabad in 1839, by Capt. Ward, as the Scinde Irregular Horse. Became the 1st Regiment Scinde Irregular Horse, 1846; the 1st Regiment of Scinde Horse, 1860; the 8th Regiment, Scinde Silladar Cavalry, 1861, but reverted to its previous designation, 1861; became the 5th Bombay Cavalry (Jacob-ka-Risallah), 1885; the 5th Bombay Cavalry (Scinde Horse), 1888; present designation, 1903.

Granted an Honorary Standard for service in Sindh in 1843, bearing the device of a Native horseman with lance. Also carries a Standard surmounted by an open hand captured at Meeanee.

"Meeanee" "Hyderabad" "Cutchee" "Mooltan" "Goojerat" "Punjaub"
"Persia" "Central India" "Afghanistan, 1878-79."

Composition—2½ Squadrons of Derajat Musalmans, 1 of Pathans, 1½ of Sikhs—1 Addl. Squadron.

Uniform—Blue. *Facings*—White.

COLONEL—Major-General *Sir* George de S. Barrow, K.C.M.G., C.B. . . 20 March 1917.

1st Commn. or date of entering service.	Names and Rank.	Army rank.	Present appointment in Regiment.	Remarks.
	Commandant.			
3 May 90	Lu'tan, Lt.-Col. R. C. W., D.S.O. (*temp. Brig.-Genl.*)	3 May 16	20 Sep. 18	
	Squadron Commanders. (4)			
1 Mar. 90	*Hocken, Lt.-Col. C.A.F., p.s.c.* (*Q*) [*L*]	*1 Mar. 16*	*1 Jan. 17*	*2nd-in-Comd.—A. A. & Q. M. G.*
18 Sep. 97	Maunsell, Maj. E. B., *p.s.c.* . .	1 Sep. 15	2 Apl. 14	With 36th Horse.
25 Jan. 99	Landon, Maj. C. R. H. P. . . .	1 Sep. 15	1 Jan. 17	4-70th Burma Rifles.
12 Aug. 99	Gordon, Maj. G. S. (*Actg. Lt.-Col.*)	1 Sep. 15	28 Apl. 18	2nd-in-Comd. (*temp.*)
	Squadron Officers. (9)			
12 Aug. 99	*Giles, Maj. E. D.*, D.S.O., *p.s.c.* (*temp. Lt.-Col.*)	*1 Sep. 15*	*1 Jan. 02*	*G. S. O., 1st gr.*
10 Nov. 00	Dyer, Maj. H. R. D.S.O. (*temp. Lt.-Col.*)	1 Sep. 15	11 Mar. 04	Burma Md. Infy.
20 Jan. 00	Connop, Maj. H. E.	1 Sep. 15	29 June 11	With 21st Cav.
8 May 01	Thompson, Maj. J. P. . . .	8 May 16	16 Mar. 04	
28 Nov. 03	Simonds, Maj. J. N.	2 May 18	5 Apl. 09	
10 Oct. 03	Montagu-Smith, Capt. E. . .	15 Mar. 13	1 Feb. 09	With 22nd Cav.
8 Sep. 09	*Leslie, Capt. T. L.*	*1 Sep. 15*	*9 Nov. 11*	
18 Jan. 11	Booth, Capt. E. R. C. . . .	1 Sep. 15	16 Oct. 12	
20 Jan. 12	*Kennedy, Capt. E. E.*, M.C. . .	*20 Jan. 16*	*16 Mar. 13*	*With I. S. Troops.*
24 Aug. 12	Morrison, Capt. C. C. . . .	24 Aug. 16	22 Nov. 13	With 21st Cav.
	Attached.			
23 Oct. 14	Bird, Lt. R. J. (*Actg. Capt.*) .	23 Oct. 15	23 Jan. 18	I. A. R. O.
5 Oct. 14	Nokes, Lt. G. B. I.	5 July 16	16 Jan. 18	Bedf. R., I. A. (*on probn.*)
22 July 16	Ziman, Lt. S. N. (*Actg. Capt.*) .	22 July 16	6 Aug. 16	I. A. R. O. Staff Capt., 4th Cav Bde.
15 Dec. 15	Aucutt, Lt. W.	. . . .	22 Jan. 18	Midd'x Yeo., I. A. R. O. (*on probn.*)
11 Apl. 17	McGlashan, Lt. E. G.	11 Apl. 18	23 Apl. 17	I. A. R. O.
4 Sep. 17	Medwell, Lt. P. R.	4 Sep. 18	4 Sep. 17	I. A. R. O. (*on probn.*)
5 Dec. 15	Rorison, Lt. W. M. (*Actg. Capt.*)	5 Dec. 16	29 Sep. 17	I. A. R. O. (*on probn.*). Adjt. (*temp.*)
22 Dec. 17	Lloyd, 2nd-Lt. J. T.	22 Dec. 17	6 Jan. 18	I. A. R. O.
1 May 17	Bowles, 2nd-Lt. W. E.	. . . .	15 Jan. 18	Res. Dgs., I. A. R. O. (*on probn.*)
25 Aug. 17	Stewart, Lt. A. W.	25 Aug. 18	11 Feb. 18	Res. Hrs., I. A. R. O. (*on probn.*) Qr. Mr. (*Offg.*)
16 Jan. 18	Mackay, 2nd-Lt. D. H. . . .	16 Jan. 18	2 Feb. 18	I. A. R. O.
24 Dec. 14	Carmichael, Lt. T.	24 Dec. 15	19 May 18	I. A. R. O.
1 Mar. 17	Foreman, Lt. F. C.	. . . .	1 Mar. 17	I. A. R. O.
7 Aug. 15	Knox, Lt. J. H. G. (*Actg. Capt.*) .	7 Aug. 16	24 Apl. 18	I. A. R. O.
23 Apl. 18	Hartigan, 2nd-Lt. K. L. S. . .	...	21 Oct. 18	I. A. (U. L.)
31 Oct. 17	Harrison, 2nd-Lt. H. S. . . .	31 Oct. 17	...	
26 Sep. 17	Palmer, Lt. A. R. B.	26 Sep. 18	...	I. A. R. O.
. . . .	Mathews, 2nd-Lt. W. H. . . .	...	5 Oct. 18	
	Honorary Officer.			
7 Dec. 12	Ahmad Nawaz Khan-*Sadozai*, *Nawab*, of Dera Ismail Khan. .	7 Dec. 12	7 Dec. 12	Hony. Capt. in Army.

35th Scinde Horse—*concld.*

Date of entering service.	Names.	Army Rank. Jemadar.	Ressaidar.	Risaldar.	Risaldar-Major.	Remarks.
	Risaldar-Major.					
1 Mar 95	Muhammad Azim Khan . .	27 July 98	1 Dec. 04	16 Nov. 10	15 Nov. 16	
	Risaldars.					
2 Apl. 99	Kartar Singh (50, 56)	18 Aug. 03	6 Jan. 09	5 July 15		
16 Oct. 07	Muhammad Ayub Khan (45*a*, 56)	16 Oct. 07	18 Apl. 15	16 Nov. 16		
9 Feb. 08	Malik Muzaffar Khan (50, 56)	10 Nov. 10	17 Oct. 13	24 Aug. 17		
1 Jan. 00	Kartar Singh (45*a*)	6 Jan. 09	5 July 15	4 Oct. 17		
15 July 89	Fazl Haq	7 Dec. 12	16 Nov. 16	5 Jan. 18		
	Ressaidars.					
27 Oct. 94	Nur Muhammad	17 Oct. 13	24 Aug. 17			
1 Jan. 12	Faujdar Singh (50, 56, 66*a*, 71) .	1 Jan. 12	4 Oct. 17			
1 Nov. 96	Ahmad Baksh (56)	18 Apl. 15	4 Oct. 17			
22 Sep. 93	Ghulam Akbar Shah	29 Oct. 14	{*4 Oct. 17 30 Oct. 17}			
16 June 02	Sardar Khan	5 Jan. 17	5 Jan. 18			
	Jemadars.					
15 June 94	Jan Muhammad Khan	25 July 11				
1 Dec. 00	Mula Singh	18 Apl. 15				Woordie Maj. 4 Oct. 17.
11 Aug. 07	Akhtar Manir Khan	6 May 16				With 36th Horse.
24 June 99	Khan Beg Khan	15 May 16				
16 June 97	*Muhammad Khan*	*9 July 16*				
1 Oct. 01	Ghulam Hassan Shah (56) . .	4 Nov. 16				
1 Jan. 01	Sohan Singh	22 Nov. 16				
18 Aug. 02	Muzaffar Khan	24 Aug. 17				
11 May 97	Khuda Bakhash Khan . .	4 Oct. 17				
1 Sep. 10	Dalip Singh, I.D.S.M.	4 Oct. 17				
1 Aug. 01	Nijabat Husain (56)	30 Oct. 17				
16 Aug. 03	Rab Nawaz Khan	30 Oct. 17				
1 Oct. 05	Sangat Singh	3 Dec. 17				
1 July 11	Muhammad Nawaz	5 Jan. 18				
16 Nov. 01	Sultan Khan	21 Jan. 18				
1 Oct. 99	Bhagat Singh	22 Jan. 18				

* Without pay and allowances.

36th Jacob's Horse.

Raised at Hyderabad in 1846, by Capt. J. Jacob, as the 2nd Regiment of Scinde Irregular Horse. Became the 2nd Regiment of Scinde Horse, 1860; the 9th Regiment, Scinde Silladar Cavalry, 1861, but reverted to its previous designation, 1861; became the 6th Bombay Cavalry (Jacob-ka-Risallah), 1885; the 6th (Bombay Cavalry (Jacob's Horse), 1888; present designation, 1903.

Granted an Honorary Standard for service in Sindh in 1843, bearing the device of a Native horseman with lance

"Meeanee" "Hyderabad" "Cutchee" "Mooltan" "Goojerat"

"Punjaub" "Afghanistan, 1879-80."

Composition—2 Squadrons of Derajat Musalmans and Baluchis, 1 of Pathans, 1 of Sikhs.

Uniform—Blue. *Facings*—Primrose.

1st Commn. or date of entering service.	Names and Rank.	Army rank.	Present appointment in Regiment.	Remarks.
	Commandant.			
2 May 91	Fagan, Lt.-Col. E. A., C.M.G., D.S.O. (*temp. Br.-Genl.*)	18 Feb. 16	18 Feb. 16	
	Squadron Commanders. (4)			
7 Dec. 95	Green, Maj. W. G. K., D.S.O. (*temp. Bde. Genl.*)	7 Dec. 13	24 Sep. 17	2nd-in-Comd.
27 July 98	Nixon, Maj. E. M.	1 Sept. 15	12 Nov. 11	
14 Feb. 00	Davidson, Maj. H. M.	1 Sept. 15	1 Apl. 12	
4 May 01	Forteath, Maj. A. M.	4 May 16	24 Sep. 17	With 45th Cav.
	Squadron Officers. (9)			
4 May 01	Jones, Maj. A. N. G.[*l*], D.S.O.	4 May 16	9 Oct. 03	With 2-72nd Punj.
19 Aug. 03	Farquhar, Maj. F. R., M.C.	19 Aug. 18	1 Jan. 06	
14 Jan. 06	*Cahusac, Capt. C. F.*, D.S.O. (temp. Maj.)	*24 Jan. 15*	*19 Sep. 07*	
2 Mar. 07	Blacker, Capt. W. F.	1 Sept. 15	12 July 09	Adjt.
6 Sep. 11	Channer, Capt. K. F.	6 Sept. 15	8 Dec. 12	
22 Jan. 13	Dallas, Capt. D. A. G.	22 Jan. 17	14 Mar. 14	With R. A. F
14 Nov. 16	Stroud, Lt. G. M.	14 Nov. 17	....	Attd.
14 Nov. 16	Fry, Lt. G. C.	14 Nov. 17	....	Attd. With 45th Cav.
14 Nov. 16	Dormer, Lt. E. J.	14 Nov. 17	....	Attd.
14 Nov. 16	Watkins, Lt. F. W. S.	14 Nov. 17	....	Attd.
14 Nov. 16	Vosper, Lt. E. E.	14 Nov. 17	....	Attd.
	Attached.			
8 Sep. 97	Maunsell, Maj. E. B.	1 Sep. 15	9 Oct. 15	35th Horse. 2nd-in-Comd. (*Offg.*)
20 Nov. 14	Sheard, Capt. Lt. C.	30 Aug. 18	....	I. A. R. O.
8 Jan. 15	Chubb, Lt. W.	24 Oct. 15	....	I. A. R. O.
22 Feb. 15	Landale, Capt. H. R.	13 Nov. 18	....	I. A. R. O.
26 Feb. 15	Harrison, Lt. A. W.	16 Apl. 15	....	I. A. R. O.
7 May 15	Emerson, Lt. D. B.	11 Mar. 16	....	I. A. R. O.
8 Nov. 15	Caiger, Lt. A. C. E. (*Actg. Capt.*)	8 Aug. 17	4 Nov. 17	Middx. Yeo. I. A. Adjt. Depôt.
11 Oct. 17	Crofton, Lt. R. M.	11 Oct. 18	26 Oct. 17	I. A. R. O.
26 Nov. 17	Foster, 2nd-Lt. T. D.	26 Nov. 17	16 Dec. 17	I. A. (*temp.*).
25 Dec. 17	Huggins, 2nd-Lt. H. D.	25 Dec. 17	13 Jan. 18	I. A. R. O.
17 Jan. 17	Lightbody, 2nd-Lt. J. G.	17 Jan. 18	23 Jan. 18	Res. Cav. I. A. (*on probn.*).
1 Oct. 18	Gostling, 2nd-Lt. D. E.	1 Oct. 18	7 Oct. 18	I. A. (*on probn.*).
14 Sep. 18	Douglas, Lt. D.	14 Sep. 18	8 Oct. 18	I. A. R. O.
12 Aug. 16	Waller, Lt. B. C.	12 Aug. 17	....	Attd.

36th Jacob's Horse—*contd.*

Date of entering service.	Names.	Army rank.	Present appointment in Regiment.			Remarks.
		Jemadar.	Ressaidar.	Risaldar.	Risaldar-Major.	
	Risaldar-Major.					
18 May 88	Bagga Singh Bahadur (56)	20 Feb. 04	1 June 11	30 Aug. 14	15 May 16	
	Risaldars.					
1 Apl. 85	Sundar Singh	1 May 96	1 Dec. 03	1 Oct. 06		
16 Apl. 01	Sadik Muhammad Khan, *Bahadur*, I.O.M. (45*a*, 56).	17 Mar. 07	30 Aug. 14	1 Sep. 15		*Order of Br. I., 2nd Cl.*
24 Mar. 04	Muhammad Wazir Khan (56)	24 Mar. 04	1 Feb. 10	15 May 18		
24 Mar. 13	Malik Khan Muhammad Khan	24 Mar. 13	7 Mar. 15	16 Sep. 16		
	Ressaidars.					
5 Nov. 92	Ghulam Muhammad Khan	16 Sep. 05	23 Mar. 12			
16 Jan. 03	Madat Khan	23 Mar. 12	20 Sep. 15			
16 May 00	Bishan Singh (58)	10 Jan. 15	2 Oct. 15			
20 Sep. 00	Abu Khan	1 Feb. 10	11 Apl. 16			
1 Nov. 93	Bahadur Singh	10 May 15	15 Aug. 17			
15 Dec. 01	Muhammad Nur Khan, I.O.M.	1 Sep. 15	13 May 16			With 45th Cav.
1 Apl. 97	Wazir Singh, I.O.M.	11 Apl. 16	2 Dec. 17			With 45th Cav.
16 Jan. 05	Gul Jahan Khan	11 Apl. 16	2 Dec. 17			
24 Jan. 13	Iltaf Hussain Shah	15 May 16	3 Dec. 17			
	Jemadars.					
9 Nov. 93	Muhammad Khan	6 Mar. 15				
1 Feb. 06	Sucha Singh (56)	2 Oct. 15				
1 July 92	Man Singh	18 Dec. 15				
1 Jan. 01	Khazan Singh	11 Apl. 16				
16 Sep. 98	Khan Shirin Khan	15 May 16				
14 Nov. 07	Haji Ahmad Khan, I.O.M.	6 Nov. 16				
16 July 04	Amar Singh	25 Aug. 17				
1 Apl. 85	*Sahib Singh*	*10 Sep. 17*				*With Ind. Conv. Sec., Ambala.*
7 Feb. 12	Abdul Karim Khan	15 Oct. 17				
20 July 04	Hazara Singh	2 Dec. 17				
1 Dec. 04	Kartar Singh	3 Dec. 17				
	Sultan Khan (*Retd.*)	. . . . .				Depôt. Re-employed 1 Mar. 15.
1 Mar. 05	Ibrahim Khan	3 Dec. 17				

37th Lancers (Baluch Horse).

Raised at Shikarpur, Scinde, in 1885, by Major A. L. McNair, as the 7th Bombay Cavalry (Jacob-ka-Risallah). Became the 7th Bombay Cavalry (Baluch Horse), 1886; the 7th Bombay Lancers (Baluch Horse), 1890; present designation, 1903.

Composition—2 Squadrons of Derajat Musalmans and Baluchis, 1 of Pathans and 1 of Sikhs.

Uniform—Khaki Serge. *Facings*—Buff.

1st Commn. or date of entering service.	Names and Rank.	Army rank.	Present appointment in Regiment.	Remarks.
	Commandant.			
7 Mar. 94	*Orton, Lt.-Col. E. F., p.s.c. (Q) (temp. Col.)*	*26 Sep. 18*	*26 Sep. 18*	
	Squadron Commanders. (4)			
23 June 98	*Benn, Maj. J. P. (Actg. Lt.-Col.)*	*1 Sep. 15*	*26 Sep. 18*	*2nd-in-Comd.* Condt. (*temp.*)
17 Jan. 00	*Scott, Maj. A. E. S.*	*1 Sep. 15*	*1 July 14*	*D. A. Q. M. G.*
2 May 00	Stewart, Maj. J. St. C. D.	1 Sep. 15	26 May 18	2nd-in-Comd. (*temp.*)
19 May 00	*Reynolds, Maj. A. J.*	*1 Sep. 15*	*26 Sep. 18*	*With Jodhpur Lrs.*
	Squadron Officers. (9)			
8 May 01	Lorimer, Maj. E.	8 May 16	15 May 03	With 41st Cav.
14 Sep. 01	*Temple, Maj. P. De L.*	*18 Oct. 16*	*30 Mar. 07*	
8 Jan. 02	Lucas, Maj. M. H.	18 Jan. 17	22 May 03	With 2-17th Infy.
15 Oct. 02	*Foster, Maj. C. S. (temp. Lt.-Col.)*	*15 Oct. 17*	*19 June 05*	*D. A. Q. M. G. 9th Secunderabad Div.*
19 Aug. 03	Lee, Maj. W. C. N.	19 Aug. 12	1 Jan. 06	With 42nd Cav. Regt.
5 Aug. 05	Western, Capt. C. E. M.	5 Aug. 14	26 Nov. 06	
17 Aug. 07	*Burt, Capt. H. V.*	*1 Sep. 15*	*1 Jan. 11*	Inspr. Musketry & M. G., N. C.
25 Sep. 07	Mitchell, Capt. R. P. J.	1 Sep. 15	20 Jan. 13	*Ind. Conv. Sec., Lucknow.*
18 Jan. 11	Watkis, Capt. H. L., M.C.	Sep. 15	25 June 14	Adjt.
5 July 15	Collyns, Lt. G. G.	5 July 16	8 Aug. 17	
	Attached.			
28 May 10	Anderson, Capt. A. L. B.	1 Sep. 15	29 June 18	1st Conn. Rang.-I. A. (*on probn.*).
2 Nov. 15	Hughes Rowlands, Lt.	2 Nov. 16	4 Mar. 18	I. A. (*on probn.*)
29 Mar. 16	Webb, Lt. G. S. R., M.C.	29 Mar. 17	24 May 18	I. A. (*on probn.*)
3 Aug. 16	Orr, Lt. H. D.	3 Aug. 17	17 May 18	I. A. (*on probn.*)
8 Sep. 16	Boileau, Lt. C. M., M. C.	8 Sep. 17	16 Feb. 18	I. A. (*on probn.*)
18 June 17	Marshall, Lt. C.	18 June 18	24 June 17	I. A.
26 June 17	Steele, Lt. L.	26 June 18	7 Apl. 18	I. A. (*on probn.*)
24 Aug. 17	Bosworth, Lt. F. J.	25 Aug. 18	14 Feb. 18	I. A. R. O. (*on probn.*)
5 Nov. 17	Merrill, Lt. W. E.	5 Nov. 17	14 Mar. 18	I. A. (*T. C.*)
31 Jan. 18	Burkett, 2nd-Lt. R. N.	31 Jan. 18	7 Feb. 18	I. A.
5 Apl. 18	Holt, 2nd-Lt. E. M.	5 Apl. 18	2 Nov. 18	I. A. R. O. (*on probn.*)
4 May 18	Tilden, 2nd-Lt. E. D.	4 May 18	22 Oct. 18	I. A. (*on probn.*)
19 Oct. 18	Cockrell, 2nd-Lt. G. H.	19 Oct. 18	22 Nov. 18	I. A. R. O. (*on probn.*)

		Jemadar.	Ressaidar.	Risaldar.	Risaldar-Major.
	Risaldar-Major.				
7 Dec. 89	Saiyid Muhsin Shah (56)	7 Sep. 90	1 Apl. 96	1 July 08	1 Jan. 18.
	Risaldars.				
20 Apl. 86	***Kirpa Singh.***	*8 Dec. 99*	*1 May 05*	*1 Jan. 14*	
1 Sep 91	Nur Sher Khan (52, 56, 71)	1 Feb. 01	1 Jan. 14	15 Oct. 14	
20 Sep. 93	Karam Ilahi Khan (52, 56)	19 Nov. 01	16 July 07	5 Jan. 15	
1 Dec. 94	Khan Alam Khan (52, 56, 66)	16 July 07	5 Jan. 15	13 July 16	

	Ressaidars.			
1 June 92	*Dost Muhammad Khan*	*5 Nov. 02*	*16 July 07*	
7 Mar. 05	Ata Muhammad Khan (45a, 53, 56)	27 Mar. 05	15 Oct. 14	
1 Sep. 05	Muhammad Ibrahim Khan (45a, 52, 56)	1 Sep. 05	29 Oct. 14	With 41st Cav.
14 June 99	Yakub Khan (56)	15 Oct. 14	4 Dec. 15	
28 Sep. 01	Khaliqdad Khan (56)	30 Oct. 14	13 July 16	With 15th Lrs.
1 Dec. 95	Sabit Khan (56)	1 Apl. 13	1 Feb. 17	
1 Nov. 94	Indar Singh (52, 56)	4 July 15	20 Oct. 17	With 41st Cav.
11 May 90	Hukam Singh (56)	25 Mar. 16	1 Jan. 18	

37th Lancers (Baluch Horse)—*contd.*

Date of entering service.	Names.	Army rank.	Remarks.
	Jemadars.		
1 July 87	Lehna Singh	20 Sep. 05	Qr.-Mr. Jemr.
5 Jan. 93	Kutub Sher Shah	15 Oct. 14	
16 Jan. 09	*Alam Sher Khan*	*8 Dec. 14*	*With 36th Horse.*
10 Mar. 04	Muhammad Khan (52, 56, 66)	8 Dec. 14	Woordie Major, 25 Apl. 16.
5 Aug. 01	Mir Alam Khan (56)	5 Jan. 15	
15 Mar. 00	Muhammad Sharif Khan	4 July 15	
1 May 13	*Ghulam Muhammad Khan*	*17 Oct. 15*	*Remt. Depôt, Saugor.*
27 July 01	Mir Baz Khan	25 Apl. 16	
18 Feb. 02	Ali Ahmad Khan	13 May 16	
15 May 98	*Ghulam Haidar Khan*	*10 July 16*	*Ind. Conv. Sec., Lucknow.*
27 Apl. 97	Chattar Singh	13 July 16	
25 Dec. 03	Abdulla Khan	16 Nov. 16	
12 Feb. 01	*Qasi Khan (56)*	*24 Feb. 17*	*With 41st Cav. Regt.*
14 Nov. 14	Bahadur Sher Khan (47, 56)	7 Mar. 17	
1 Dec. 94	*Gul Hasan Khan*	*30 July 17*	*Ind. Conv. Sec., Ambala.*
23 Oct. 94	*Nawab Khan*	*30 July 17*	*Ind. Conv. Sec., Ambala.*
1 Jan. 06	*Abdul Hafiz*	*30 July 17*	*Ind. Conv. Sec., Ambala.*
22 Aug. 17	Muhammad Kazim Khan	22 Aug. 17	(*On probn.*)
20 Oct. 17	A'am Gir	20 Oct. 17	(*On probn.*)
1 Nov. 06	Bhagt Singh (45*a*)	3 Dec. 17	
16 Nov. 03	Fazal Karim	3 Dec. 17	
1 Aug. 94	*Hari Singh (52, 56)*	*1 Jan. 18*	*With 41st Cav.*
21 Mar. 96	Prem Singh	17 Mar. 18	Remt. Depôt, Marseilles.
	Sirbaland Khan	....	Re-employed, 17 May 17.

38th King George's Own Central India Horse.

Raised, 1858, by Capt. H. O. Mayne, on a nucleus of the faithful remains of the Cavalry of the Gwalior, Bhopal and Malwah Contingents, and originally styled "Mayne's Horse." Augmented, 1860, to a corps of three regiments (Beatson's Horse being incorporated), and the first of these became, the same year, the 1st Regiment of Central India Horse, the 38th Central India Horse in 1903; 38th Prince of Wales's Own Central India Horse, 1906; present designation, 1910.

THE PLUME OF THE PRINCE OF WALES. THE ROYAL AND IMPERIAL CYPHER.
"Kandahar, 1880" "Afghanistan, 1879-80" "Punjab Frontier."
Composition—2 Squadrons of Sikhs, 1 of Pathans, 1 of Musalman Rajputs. 1 Addl. Squadron.
Uniform—Drab. *Facings*—Maroon.

COLONEL-IN-CHIEF—THE KING.

1st Commn. or date of entering service.	Names and Rank.	Army rank.	Present appointment in Regiment.	Remarks.
	Commandant.			
28 Jan. 93	Hewlett, Lt.-Col. A.	15 Jan. 18	15 Jan. 18	
	Squadron Commanders. (4)			
16 Jan. 95	Goodfellow, Maj. R. C.	16 Jan. 13	14 Oct. 11	
21 Sep. 96	*Hogg, Bt. Lt.-Col., R. E. T.,* C.I.E. *(temp. Br.-Genl.)*	*1 Jan. 18*	*21 June 18*	*2nd-in-Comd. With R. A. F.*
26 July 99	Gourlie, Maj. J.	1 Sep. 15	27 Mar. 18	
19 May 00	Lewis, Maj. G.	1 Sept. 15	21 June 18	With 2-89th Punj.
	Squadron Officers. (9)			
15 May 07	Hutchison, Maj. J. R.	15 May 16	8 June 04	
27 Aug. 02	*Henderson, Maj. G.*	*27 Aug. 17*	*26 May 04*	*With R. A. F.*
4 June 04	Abbott, Bt.-Maj. R. S. *(temp. Maj.)*, M.C.	3 June 17	4 June 09	
25 July 06	Daunt, Capt. C. O'B., M.C.	25 July 15	25 Sep. 09	
9 Sep. 08	*Harvey, Capt. C. O.,* M.C.	*1 Sep. 15*	*1 Dec. 09*	
9 Sep. 08	*Leslie, Bt. Maj. Sir N. R. A. D., Bart.*	*1 Jan. 18*	*25 Nov. 09*	
20 Jan. 09	*Martin, Capt. N. M. (temp. Maj.)*	*1 Sep. 15*	*30 May 12*	*With R. A. F.*
8 Sep. 09	Williams, Capt. A. H.	1 Sep. 15	14 Dec. 10	Adjt., 2 Apl. 16.
24 Dec. 14	Roxburgh, Lt. G. A.	24 Sep. 16	19 May 15	Attd.
2 June 15	*Doudney, Lt. G. W.*	*2 June 16*	. . . .	*Attd. With I. S. Troops.*
14 Nov. 16	Cox, Lt. M.	14 Nov. 17	24 Oct. 17	Attd.
14 Nov. 16	Cousens, Lt. R. P.	14 Nov. 17	2 Nov. 17	Attd.
14 Nov. 16	George, Lt. R.	14 Nov. 17	2 Nov. 17	Attd.
14 Nov. 16	Williams, Lt. B. F.	14 Nov. 17	24 Oct. 17	Attd. With 44th Cav.
30 Jan. 17	Marshall, Lt. T. R.	30 Jan. 18	24 Oct. 17	Attd. With 44th Cav.
	Bates, 2nd-Lt. G. E. L.	. . . .	. . . .	Attd.
21 Dec. 16	Swann, 2nd-Lt. A. E.	. . . .	1 Dec. 17	Attd.
24 Apl. 18	Hastings, 2nd-Lt. P. G.	24 Apl. 18	24 Oct. 18	
	Alexander, 2nd-Lt. A. G. S.	23 Oct. 18		
	Marjoribanks, 2nd-Lt. S. D.	23 July 18		
21 Oct. 15	Martin, Lt. W. M.	21 July 17	26 Nov. 17	Attd.
	Attached.			
9 Jan. 04	*Colvin, Capt. E. J. D. (temp. Maj.)*	*9 July 14*	*7 Nov. 14*	*Poll. Dept. D. A. Q. M. G., A. Hdqrs.*
8 Aug. 14	Filose, Capt. A. A. E.	8 Aug. 18	22 Feb. 17	39th Horse.
24 Dec. 14	Allen, Lt. J. H. S.	24 Dec. 15	. . . .	I. A. R. O.
14 May 15	Milner, Lt. C. E.	14 May 16	28 Dec. 15	I. A. R. O.
8 Jan. 15	Walter, Capt. G. B.	30 Aug. 18	30 Aug. 15	I. A. R. O.
2 Oct. 15	Blakiston, Lt. J. F. *(Actg. Capt.)*	2 Oct. 16	7 Sep. 16	I. A. R. O.
	Keighley, Capt. H. D. S.	. . . .	25 Feb. 15	Res. Regt. of Cav.
	Page, Lt. W. K. *(Actg. Capt.)*, M.C.	. . . .	30 July 15	Indn. Vol. officer.
14 Nov. 16	Good, Lt. S. B.	14 Nov. 17	17 Sep. 17	
14 Nov. 16	Pigot, Lt. H. G.	14 Nov. 17	18 Sep. 17	
15 Oct. 17	Grigg, Lt. E. C.	15 Oct. 18	5 Nov. 17	I. A. R. O.
31 Aug. 17	Bowden, Lt. C. C. H.	31 Aug. 18	12 Nov. 17	With S. & T. Corps.
8 Nov. 17	Rutherford, Lt. T. G.	8 Nov. 18	18 Nov. 17	I. A. R. O.
	Westmacott, Capt. R.	. . . .	24 Oct. 17	
31 Jan. 18	Everett, 2nd-Lt. W. H.	31 Jan. 18	7 Feb. 18	
15 Aug. 14	Holmes, Lt. B., D.S.O. *(Actg. Capt.)*	15 May 16	13 Feb. 18	I. A. *(on probn.)*
15 Aug. 14	Chard, Lt. A. R.	15 May 16	21 May 18	I. A. *(on probn.)*
	Evers, 2nd-Lt. T. G.	. . . .	. . . .	Essex Yeo.
28 Jan. 17	Dalrymple, *Hony.* Lt. B. G. R.	28 Jan. 18	29 Apl. 18	
	Badham, 2nd-Lt. J. R. C.	. . . .	6 Sep. 18	I. A. R. O.

38th King George's Own Central India Horse—*contd.*

Date of entering service.	Names and Rank.	Army rank.	Remarks.

Honorary Officer.

7 Feb. 14	Baba Kartar Singh Bedi . .	7 Feb. 14	. . . .	Hony. Lieut. in Army.

Risaldar-Majors.

		Jemadar.	Ressaidar.	Risaldar.	Risaldar-Major.	
30 Apl. 90	Amar Singh (58), *Bahadur* .	16 Oct. 03	10 Jan. 09	15 Feb. 13	1 July 14	*Order of Br. I., 2nd cl.* [Depôt.
	Ali Haidar Khan, *Hony. Capt., Sardar Bahadur* (*Retd.*)	. . .	. . .	. . .	. . .	Re-employed 1 Mar. 15.

Risaldars.

		Jemadar.	Ressaidar.	Risaldar.
6 Feb. 04	Kamaluddin Khan (53, 56, 66*a*)	6 Feb. 04	20 Apl. 10	1 July 14
20 Apl. 98	Natha Singh, I.D.S.M. (45*a*, 56, 66*a*, 71).	10 Jan. 12	21 Sep. 15	1 Feb. 17
8 Oct. 99	Nihal Singh (45*a*)	1 Mar. 12	11 Apl. 16	21 Jan. 18
7 July 97	Jawand Singh (58), I.O.M. . .	1 Aug. 14	17 Aug. 16	22 Jan. 18

Ressaidars.

23 Nov. 96	Buta Singh (66*a*)	7 Jan. 13	14 Aug. 16	
16 Feb. 95	Lihaz Gul Khan, I.O.M. . .	24 Oct. 14	29 Aug. 16	
16 Jan. 99	Mal Singh (52, 56, 71) . . .	18 May 15	29 Aug. 16	
18 Sep. 94	Bostan Khan, I.D.S.M. (66*a*, 71)	1 July 14	31 Mar. 17	
22 Apl. 98	Dayal Singh	24 Oct. 14	25 Aug. 17	
1 Nov. 00	Kartar Singh (65)	11 Apl. 16	2 Feb. 18	
	Mohan Singh (*Retd.*) . . .	. . . .	26 Feb. 18	Re-employed, 1 Mar. 15. Depôt.
16 Oct. 02	Juma Khan, I.D.S.M. (58) . .	11 Apl. 16		
	Hayat Gul Khan	. . . .	. . . .	Re-employed, 15 July 18. Depôt.
2 Jan. 05	Gurbakhsh Singh (71) . . .	19 May 16	9 Apl. 18	
1 Apl. 98	Muhammad Akbar Khan . .	14 Aug. 16	9 Apl. 18	

Jemadars.

5 Apl. 98	Nadar Khan	11 Apl. 16	
10 Dec. 07	Thakur Singh	17 Aug. 16	Qr.-Mr. Jem., 3 Dec. 17.
20 Nov. 96	Mahmud Shah (51, 71) . .	29 Aug. 16	
16 Nov. 06	Mal Singh	9 July 17	Woordie-Maj., 22 Jan. 18.
23 Sep. 02	Chuhar Singh (51, 71) . . .	12 July 17	Woordie-Maj. Depôt, 9 July 18,
26 June 96	Karamdad Khan	10 Aug. 17	
23 Nov. 96	Mangal Singh (66*a*) . . .	25 Aug. 17	
27 Aug. 98	Kishan Singh (65, 66*a*) . . .	25 Aug. 17	
2 Nov. 99	Mirzaman Khan (71) . . .	25 Aug. 17	
16 Nov. 99	Shahwali Khan (71) . . .	5 Sep. 17	
27 May 12	Kartar Singh (71)	19 Sep. 17	
9 Apl. 02	Sherjan Khan, I.D.S.M. . . .	14 Nov. 17	
1 May 04	Balwant Singh (58)	2 Dec. 17	
1 Oct. 06	Lal Khan, I.D.S.M.	2 Dec. 17	
8 Apl. 01	Yakub Khan	3 Dec. 17	
18 Oct. 05	Mir Zaman Khan	8 Jan. 18	
11 Jan. 06	Ali Mardan Khan (65) . . .	22 Jan. 18	
16 July 01	Bhagat Singh (71)	2 Feb. 18	
1 Dec. 11	Balwant Singh	26 Mar. 18	
3 Nov. 01	Mir Aslam Khan	12 May 18	
6 Feb. 04	Labh Singh	9 Apl. 18	

39th King George's Own Central India Horse.

Raised at Hyderabad, in 1858, by Col. W. F. Beatson, as the 1st Regiment of Beatson's Horse. Became the 2nd Regiment of Mayne's Horse in 1860; the 2nd Regiment, Central India Horse, the same year. Two hundred men of Meade's Horse transferred to the Corps in 1861; the 39th Central India Horse in 1903; 39th Prince of Wales's Own Central India Horse, 1906; present designation, 1910.

THE PLUME OF THE PRINCE OF WALES. THE ROYAL AND IMPERIAL CYPHER.
"Kandahar, 1880" "Afghanistan, 1879-80" "Punjab Frontier, 1897-98."
Composition—2 Squadrons of Sikhs, 1 of Punjabi Musalmans, 1 of Musalman Rajputs, 1, Addl. Squadrons
Uniform—Drab. *Facings*—Maroon.

COLONEL-IN-CHIEF—THE KING.

Colonel—Maj.-Genl. W. A. Watson, C.B., C.M.G., C.I.E., M.V.O.

1st Commn. or date of entering service.	Names and Rank.	Army rank.	Present appointment in Regiment.	Remarks.
	Commandant.			
24 July 91	Capper, Lt.-Col. A. S., D.S.O., *p.s.c.*	15 Feb. 17	15 Feb. 17	
	Squadron Commanders. (4)			
1 Mar. 90	*Spence, Lt.-Col. A.H.O.*, C.I.E., *(temp. Col.)* [l]	*1 Mar. 16*	*12 Dec. 01*	*Dy. Secy., Army Dept.*
28 June 93	*Kettlewell, Maj. A. M., p.s.c.*	*28 June 11*	*15 Jan. 18*	*2nd-in-Comd.*
16 Feb. 98	*Pitcher, Bt.-Lt.-Col. D. LeG.,* C.M.G. *(temp. Br.-Genl.)*	*1 Jan. 17*	*15 Feb. 17*	*With R. A. F.*
8 May 01	*White, Maj. W. B. (temp. Lt.-Col.)*	*8 May 16*	*15 Jan. 18*	*Base Comdt.*
	Squadron Officers. (9)			
26 May 00	*Todd, Capt. J. F.*	*26 May 09*	*21 July 04*	
8 May 01	*Hoare, Maj. C G. (temp. Lt.-Col.)*	*8 May 16*	*13 Feb. 03*	*With R. A. F.*
28 Jan. 05	*Edwards, Capt. D. B.*	*28 Jan. 14*	*6 Mar. 07*	*Staff Capt.*
5 Aug. 05	*Fraser, Capt. W. A. K.,* M.C. (L) *(Temp. Lt.-Col.)*	*5 Aug. 14*	*12 Dec. 06*	*A. A. & Q. M. G.*
24 Jan. 06	Evans-Gordon, Capt. K. A. G. *(actg. Maj.)*	24 Jan. 15	28 June 09	2nd-in-Comd. *(temp.)*.
20 Jan. 09	Fraser, Capt. D. DeM. S.	1 Sep. 15	30 May 10	
29 Jan. 10	Wickham, Capt. E. T. R. [L]	1 Sep. 15	17 Mar. 10	
24 Aug. 12	Jones, Capt. M. S. H.	24 Aug. 16	7 Nov. 13	With I. C. M. G. Sqdn.
21 Jan. 13	Vansittart, Capt. G. N.	21 Jan. 17	4 Oct. 14	Offg. Adjt.
8 Aug. 14	*Filose, Capt. A. A. E.*	*8 Aug. 18*	*9 Nov. 15*	*Attd. with 38th Horse.*
8 Aug. 14	*Gordon, Capt. D. St. V.*	*8 Aug. 18*	*10 Nov. 15*	*Attd. with 12th Cav.*
18 June 17	Sansom, Lt. A. E.	18 June 18	24 June 17	Attd.
	Attached.			
20 Dec. 14	*Aitken, Lt. C. F.,* M.C.	*20 Sep. 16*	*12 Aug. 17*	Welsh Horse. *I. A. (on probn.).*
6 Oct. 16	Smith, Lt. S. B. *(Actg. Capt.)*	6 Oct. 17	6 Nov. 16	I. A. R. O.
6 Oct. 16	Connell, Lt. C. M.	6 Oct. 17	10 Nov. 16	I. A. R. O.
24 Feb. 15	Matthews, Lt. R. *(Actg. Capt.)*	24 Nov. 16	13 Jan. 18	Middx. Hrs. I. A. *(on probn.).*
13 Oct. 16	George, Lt. R. E. G.	13 Oct. 17	11 Nov. 16	*Lv., ex I., m. c., to 15 Feby. 19.*
15 May 17	Bagg, 2nd-Lt. W. F.	15 May 17	9 June 17	I. A. R. O.
23 June 17	Calvert, 2nd-Lt. G. R.	23 June 17	25 Aug. 17	I. A. R. O. Qr. Mr.
25 Dec. 17	Cochrane, 2nd-Lt. D. W.	25 Dec. 17	27 Jan. 18	I. A. R. O. *(on probn.)*
29 June 17	Bettles, Lt. A. L.	29 June 18	19 Jan. 18	I. A. R. O. *(on probn.)*
7 Mar. 16	Michell, Lt. J. H. *(Actg. Capt.)*	7 Mar. 17	18 Mar. 18	I. A. *(on probn.)*
19 Feb. 18	Pearson, 2nd-Lt. H. F.	19 Feb. 18	20 Mar. 18	No f. Yeo. I. A. R. O. *(on probn.)*
14 May 16	Lalor, Lt. D. C.	14 May 17	29 Mar. 18	R. Fus. I. A. *(on probn.)*
16 July 16	Skinner, Lt. L. S. C.	16 July 17	29 Mar. 18	A. S. C. I. A. *(on probn.)*
1 Feb. 18	Aylward, 2nd-Lt. T. B.	1 Feb. 18	16 Apl. 18	S. Ir. Horse. I. A. *(on probn.)*
3 July 15	Stable, Lt. H. H.	3 July 18	3 Sep. 18	Dorset. Yeo. I. A. *(on probn.)*
9 Oct. 15	Powell, Lt. H. N.	9 Oct. 16	28 Sept. 18	Ches. R. I. A. *(on probn.)*
15 July 18	Shirref, 2nd-Lt. A. G.	15 July 18	. . . .	

39th King George's Own Central India Horse—*contd.*

Date of entering service.	Names.	Army rank.	Present appointment in regiment.	Remarks.
	Honorary Officers.			
5 Feb. 08	Ratlam, *H. H. Raja Sir* Sajjan Singh, K.C.S.I., *Raja of* . .	1 Jan. 18	22 Oct. 10	Hony. Col. in Army.
5 Feb. 08	Jaora, *H. H. Nawab Sir* Muhammad Iftikar Ali Khan, *Bahadur*, K.C.I.E., *Nawab of* . . .	5 Feb. 08	22 Oct. 10	Hony. Major in Army.

Date of entering service.	Names.	Jemadar.	Ressaidar.	Risaldar.	Risaldar-Major.	Remarks.
	Risaldar-Major.					
1 July 91	Natha Singh I. D. S. M. (58, 68) .	16 Mar. 08	1 Mar. 13	1 Aug. 16		
	Risaldars.					
5 Oct. 89	Abdul Rahman Khan, I.D.S.M. . (56)	1 Oct. 04	10 May 13	3 Jan. 18		
21 Sep. 96	Bachan Singh 45*a*, 56) . . .	18 Nov. 14	1 Aug. 16	22 Sep. 17		
11 Feb. 04	Malik Muhammad Bahadur Khan (56, 162)	1 Feb. 04	19 Feb. 15	1 June 18		
1 Nov. 95	Surayan Singh	1 Feb. 11	10 Apl. 15	5 June 18		
	Ressaidars.					
8 July 93	Basant Singh	14 Oct. 11	11 Apl. 16			
1 Oct. 93	*Fauja Singh (66a)*	*1 Dec. 13*	*1 Aug. 16*			
1 Jan. 04	Bachittar Singh (56)	24 Aug. 14	22 Sep. 17			
15 May 11	Malikshe Muhammad Khan(52,56)	29 Oct. 14	22 Sep. 17			
20 Oct. 11	*Hasura Singh*	*1 Jan. 14*	*14 Nov. 17*			*With 38th Horse.*
1 May 97	Sulaiman Khan	29 Oct. 14	5 Jan. 18			
1 Jan. 01	*Malik Khuda Bakhsh Khan* . .	*15 Oct. 08*	*1 June 18*			
	Jemadars.					
1 Oct. 91	*Malik Khan Muhammad Khan*	*4 Jan. 15*				*With Remt. Tr. Depôt, Saugor.*
14 Apl. 96	*Moti Singh*	*18 Feb. 15*				*With General Base Depôt.*
1 Nov. 03	Labh Singh (56)	19 Feb. 15				With Remt. Depôt, Gujrat.
1 July 99	*Ghulam Muhammad Khan* (67).	*9 Oct. 15*				*With Remt. Depôt, Sehore.*
7 Jan. 99	Kehar Singh	9 Oct. 15				With 18th Lrs.
1 Jan. 07	Niranian Singh	16 Dec. 15				*With 38th Horse.*
1 July 03	*Muhammad Ajmeri Khan*. . .	*16 Mar. 16*				
5 Aug. 00	Anokh Singh	14 May 16				
6 July 00	Qasim Ali Khan (56)	1 Aug. 16				Woordie-Maj., 22 Sep. 17.
9 Nov. 04	Pal Singh (56)	14 Aug. 16				
1 July 03	Sohan Singh	11 Nov. 16				
14 June 99	*Suleman Khan*	*1 Feb. 17*				*With Remt. Depôt,*
22 Jan. 93	Abdul Ghani Khan	1 Mar. 17				With Remt. Tr. Depôt, Saugor.
1 Dec. 04	Makhan Singh (66*a*)	22 Sep. 17				
19 Oct. 01	Ratan Singh	22 Sep. 17				
30 May 94	Sanwal Khan	22 Sep. 17				
8 Oct. 07	Khan Muhammad Khan (52) . .	6 Nov. 17				
1 Nov. 99	Rustom Khan (52)	3 Jan. 18				
9 Mar. 03	Buta Singh	3 Jan. 18				
16 Nov. 02	Atma Singh	5 June 18				
1 Jan. 01	Muhammad Bashir Khan . . .	16 Aug. 18				
. . .	Gurdit Singh	4 Sep. 18				
. . .	Malik Mahmmad Khan . . .	17 Sep. 18				
. . .	Mehar Khan	20 Sep. 18				
1 Sep. 13	Muzaffar Khan	16 June 18				
17 Jan. 07	Man Singh	16 Nov. 17				

40th Cavalry Regiment.

Formed in April 1918, by Lieutenant-Colonel J. R. Gaussen, C.M.G., D.S.O., from 1 Squadron from 1st (D. Y. O. Lancers, 2 Squadrons from 3rd Skinner's Horse, 1 Squadron, 7th Hariana Lancers.

Composition—1 Squadron of Rajputs, ½ Squadron Jats, ½ Squadron Sikhs, ½ Squadron Dogras, 1½ Squadrons Musalmans.

1st Commn. or date of entering service.	Names and Rank.	Army rank.	Present appointment in Regiment.	Remarks.
	Commandant.			
20 Feb. 92	Gaussen, Lt.-Col. J. R., C.M.G., D.S.O.	20 Feb. 18	1 Apl. 18	3rd Horse.
	Squadron Commanders. (4)			
5 Aug. 96	Cheyne, Maj. H. B.	5 Aug. 14	19 June 18	1st Lrs. Ty. 2nd-in-Comd.
19 Sep. 08	*Harris, Capt. A. E. C.*, M.C.	*1 Sep. 15*	*1 Apl. 18*	*7th H. Lrs. Adjt. Qr. Mr. School Ind., Cadets. Indore.*
5 Sep. 11	Randall, Capt. B. Fitz H.	5 Sep. 15	1 Apl. 18	3rd Horse.
25 Aug. 17	Aga Cassim Shah, Capt.	25 Aug. 17	1 Apl. 18	3rd Horse.
	Squadron Officers.			
22 Mar. 15	Dickinson, Capt. H. B. H.	6 Apl. 18	21 Oct. 18	I. A. R. O.
29 Aug. 14	May, Capt. F. W.	29 Aug. 18	1 Apl. 18	I. A. R. O.
9 Sep. 14	Coles, Lt. N. R.	9 June 16	1 Apl. 18	E. York. R., I. A. (*on probn.*)
2 Feb. 16	Burn, Lt. R. F.	2 Feb. 17	13 Sep. 18	Worc. R., I. A., (*on probn.*)
7 Sep. 16	Heslop, Capt. G. D.	1 Apl. 18	1 Apl. 18	I. A. R. O., Adjt.
18 July 15	Hannah, Lt. G. K.	18 July 16	1 May 18	I. A. R. O. (*on probn*).
5 July 16	Tustin, H. C. V.	5 July 17	27 Sep. 18	E. Kent. R., I. A. (*on probn*).
6 Mar. 17	House, Lt. E. M.	6 Mar. 18	1 Apl. 18	I. A. R. O.
18 Aug. 17	Williams, Lt. W. D. C.	18 Aug. 18	23 May 18	I. A. R. O.
26 Sep. 17	Whitehead, Lt. T. A.	26 Sep. 18	1 Apl. 18	I. A. R. O.
30 Aug. 17	Bromham, 2nd-Lt. A. W.	30 Aug. 17	1 Apl. 18	I. A. R. O., Ty. Qr. Mr.
7 Dec. 17	Callow, 2nd-Lt. J. L.	8 Sep. 18	7 Sep. 18	R. W. Kent R., I. A.(*on probn.*)

Attached.

		Jemadar.	Ressaidar.	Risaldar.	Risaldar Major.
	Temporary Risaldar-Major.				
21 Dec. 91	Sheo Chand (52, 56, 58, 71)	1 Apl. 03	20 Jan. 10	16 Mar. 12	17 July 18
	Risaldars.				
26 June 94	Walidad Khan (45*a*, 66*a*)	1 Jan. 07	22 Sep. 15	7 Oct. 15	
16 Aug. 99	Sayed Raunak Ali Khan (66*a*)	18 June 04	16 Aug. 11	22 Nov. 15	
3 Jan. 95	Thakur Singh (56, 58, 66*a*), I.D.S.M.	16 Oct. 11	4 Apl. 16	1 Feb. 18	
	Ressaidars.				
11 Jan. 95	Milkhi Singh (66*a*)	1 May 09	13 Apl. 15		
1 Aug. 01	Chawand Singh	22 Sep. 15	20 Dec. 16		
11 Apl. 02	Mohan Singh (52, 56)	1 April 18			
1 Oct. 00	Fateh Muhammad Khan (45*a*, 56, 71)	28 July 15	15 Aug. 18		
	Jemadars.				
6 Dec. 98	Jan Muhammad Khan (60*a*, 66*a*)	11 Apl. 16			
1 Sep. 01	Khair-ud-din Khan (66*a*, 71)	1 Nov. 17			
1 Apl. 02	Bakhtawar Singh (45*a*)	3 Dec. 17			
1 Feb. 01	Sultan Singh (56, 71)	3 Jan. 18			
16 Dec. 03	Amar Singh	3 Jan. 18			
1 Mar. 01	Hem Singh (66*a*, 71)	1 Feb. 18			
8 Aug. 10	Bhajan Singh	1 Feb. 18	Woordie-Major (*temp.*)		
1 Aug. 07	Ghulam Mohiuddin Khan (56)	9 May 18			
. . .	Muhammad Husain	18 Aug. 18			

41st Cavalry Regiment.

Formed at Baleli.

Composition—½ Squadron Punjabi Musalmans, 1 Squadron Pathans, 1 Squadron Sikhs, 1 Squadron Derajat Musalmans, ½ Squadron Rajput Musalmans.

1st Commn. or date of entering service.	Names and Rank.	Army rank.	Present appointment in Regiment.	Remarks.
	Commandant.			
23 Sep. 96	Hallowes, Maj. J. H. (*Actg. Lt.-Col.*)	23 Sep. 14	1 Apl. 18	15 Lrs.
	Squadron Commanders.			
29 Dec. 00	Mansfield, Maj. C. B. H.	29 Dec. 15	6 Sep. 18	8th Cav. 2nd-in-Comd.
8 May 01	Lorimer, Maj. E.	8 May 16	17 July 18	2nd-in-comd. (*temp.*)
19 Jan. 01	Whitford-Hawkey, Capt. E. H.	1 Sep. 15	17 July 18	16th Cav.
	Squadron Officers.			
2 Oct. 14	Williams, Lt. A. H. K.	2 July 16	23 Aug. 18	Attd. Adjt.
31 Oct. 15	Thompson, Lt. D.	26 Aug. 17	17 July 18	
7 Jan. 16	Morris-Jones, Lt. J.	7 Oct. 17	17 July 18	Attd.
	Attached.			
20 Nov. 14	Monks, Lt. H. G. (*Actg. Capt., 19 Jan. 18*)	2 Oct. 15	17 July 18	I. A. R. O.
4 Dec. 14	Barrow, Lt. R. L.	4 Dec. 15	6 Aug. 18	I. A. R. O.
8 Sep. 15	Grant, Lt. R.	8 Sep. 16	17 July 18	I. A. R. O.
11 Apl. 17	Pratt, Lt. E. B.	11 Apl. 18	17 July 18	I. A. R. O.
11 May 17	Allen, Lt. W. B.	11 May 18	17 July 18	I. A. R. O.
21 May 17	Engeldow, Lt. G. R.	21 May 18	17 July 18	I. A. R. O. Qr, Mr. (*temp.*)
23 June 17	Mockler, Lt. G. S.	23 June 18	17 July 18	I. A. R. O.
11 Sep. 17	Napier-Ford, 2nd-Lt. G. S.	11 Sep. 17	17 July 18	I. A. R. O.
15 Feb. 18	Ball, 2nd-Lt. A. H.	15 Feb. 18	17 July 18	I. A. R. O.
23 Nov. 18	Brockliss, 2nd-Lt. A. A.	23 Nov. 18	7 Sep. 18	I. A. (*on probn.*)
23 Feb. 18	Lordan, 2nd-Lt. L. P.	23 Feb. 18	16 Sep. 18	I. A. R. O. (*on probn.*)
6 Jan. 16	McMullen, Lt. W. K.	6 May 18	....	I. A. (*on probn.*)

		Jemadar.	Ressaidar.		
	Risaldar-Major.				
1 Sep. 05	Muhammed Ibrahim Khan	1 Sep. 05	29 Oct. 14	17 July 18	(*temp.*)
	Risaldar.				
	Ressaidars.				
21 Feb. 94	Amar Ali Khan	14 Oct. 11	7 Dec. 14		
16 Nov. 05	Abdul Rahim Khan	16 Nov. 05	21 Dec. 14		
1 Nov. 94	Indar Singh	4 July 15	20 Oct. 17		
1 Oct. 91	Abdurrahman Khan	1 May 10	1 Jan. 18		
12 Apl. 01	Bhagat Singh	9 June 16	3 Jan. 18		
15 Aug. 97	Sadruddin Shah	7 Dec. 14	15 Jan. 18		
	Jemadars.				
1 Apl. 95	Muqarrab Khan	1 Aug. 13			
16 June 99	Yakub Khan	11 July 16			
13 Feb. 01	Qazi Khan	24 Feb. 17			
1 Feb. 05	Rahimullah Khan	23 Mar. 17			
16 May 05	Sher Muhammad Khan	14 Oct. 17			Woordi Maj.
16 Nov. 02	Harnam Singh	3 Dec. 17			
1 Aug. 94	Hari Singh	1 Jan. 18			
1 Jan. 97	Mouladad Khan	4 Jan. 18			
	Mohamed Newaz	5 Jan. 18			
10 June 00	Niaz Mohammed Khan	16 June 18			

42nd Cavalry Regiment.

Formed at Baleli.

Composition—1¾ Squadrons Sikhs, 1¼ Squadrons Punjabi Musalmans, ½ Squadron Pathans, ¼ Squadron Dogras, ¼ Squadron Hindustani Musalmans.

1st Commn. or date of entering service.	Names and Rank.	Army rank.	Present appointment in Regiment.	Remarks.
	Commandant.			
5 Aug. 96	Taylor, Maj. P. H. M. (*Actg. Lt.-Col.*)	5 Aug. 14	26 July 18	32nd Lns.
	Squadron Commanders.			
27 Aug. 02	Hume, Maj. E. G.	27 Aug. 17	3 Nov. 18	7th Lrs.
19 Aug. 03	Lee, Maj. W. C. N.	19 Aug. 18	10 Aug. 18	37th Horse.
24 Jan. 06	Yule, Capt. H. V.	24 Jan. 15	27 July 18	25th Cav.
3 Sep. 10	Morrogh-Bernard, Capt. J. A. D.	1 Sep. 15	27 July 18	22nd Cav.
	Squadron Officer.			
20 Sep. 11	Gordon, Capt A. D.	20 Sep. 15	4 Nov. 18	16th Cav.
18 June 17	Hanmer, Lt. G. C.	18 June 18	2 Aug. 18	21st Cav.
	Attached.			
13 Dec. 14	Figg, Lt. W. G. (*Actg. Capt.*)	13 Dec. 14	17 July 18	
3 Feb. 16	Gore, Lt. C. H.	3 Feb. 17	10 Sep. 18	Sea. Highrs. I. A. (*on probn.*)
7 Sep. 15	Simpson, Lt. S. R.	7 Sep. 16	20 Oct. 18	R. F. A. (T.), I. A. R. O. (*on probn.*)
12 Oct. 15	Gripper, Lt. H. B.	12 July 16	2 Aug. 18	Herts Yeo. I. A. (*on probn.*)
28 Oct. 16	Edge, Lt. F. L.	21 Oct. 18	21 Oct. 18	
19 Apl. 17	Lowe, Lt C G.	19 Apl. 18	30 Oct. 18	I. A R. O.
5 July 17	Tinkler, 2nd-Lt. D.	5 July 17	13 Sep. 18	I. A. R. O. (*on probn.*)
14 Aug. 16	Drummond, Lt. J. G.	14 Aug. 17	1 Aug. 18	I. A. R. O.
11 Oct. 17	Yorke, 2nd-Lt. R. L.	11 Oct. 17	1 Aug. 18	I. A. R. O.
25 Nov. 17	Brown, 2nd-Lt. E. H.	25 Nov. 17	2 Aug. 18	I. A. R. O. Adjt. & Qr.-Mr.
23 Nov. 18	Leader, 2nd-Lt. S.	23 Nov. 18	13 Sep. 18	I. A. R. O. (*on probn.*)

1st Commn. or date of entering service.	Names and Rank.	Jemadar.	Ressaidar.	Risaldar.	Risaldar-Major.	Remarks.
	Risaldar-Major.					
6 Jan. 99	Ali Sher Khan, I.D.S.M. (53, 56, 66*a*, 71).	16 Oct. 10	21 Oct. 14	13 July 18	7 Aug. 18	
	Risaldar.					
5 June 01	Mohammad Ali Khan (51, 66*a*)	9 July 10	12 Sep. 16	13 Apl. 18		
31 Oct. 96	Atta Mohammad Khan (56, 71)	25 May 11	23 June 16	13 Apl. 18		
	Ressaidars.					
14 Sep. 10	Baint Singh (60*a*, 66*a*)	6 Oct. 15	13 Apl. 18			
1 Dec. 06	Amar Singh (56)	7 July 14	13 Apl. 18			
	Jemadars.					
1 May 02	Said Ahmed Shah (56)	12 Feb. 16				
8 Sep. 11	Nawab Khan	13 Aug. 16				Woordie Maj.
26 Aug. 95	Sher Singh	15 Sep. 16				
15 Dec. 01	Mal Singh	25 Aug. 17				
1 Aug. 00	Sahib Singh (57)	13 Apl. 18				
1 Jan. 96	Bostan Khan	13 Apl. 18				22nd Cav.
7 May 06	Janemeja Sing	13 Apl. 18				22nd Cav.
28 Nov. 97	Adalat Khan	13 Apl. 18				

43rd Cavalry Regiment.

Formed at.—Kantara.

Composition.—1½ Squadrons Sikhs, ½ Dogras 1½ Punjabi Musa'mans, ½ Pathans.

1st Commn. or date of entering service.	Name and Rank.	Army Rank.	Present appointment in Regiment.	Remarks.
	Commandant.			
2 Jan. 93	Gordon, Bt. Lt.-Col. G. A. S.	1 Jan. 18	28 Aug. 18	Temp. Comdt. 13 Lrs.
	Squadron Commanders.			
22 Jan. 98	Marchant, Maj. G., *p.s.c.*	1 Sep. 15	3 Aug. 18	29th Lrs.
25 Jan. 08	Seymour, Capt. E. V. F.	1 Sep. 15	3 Aug. 18	9th Horse.
6 Sep. 11	Royston, Capt. C. G.	6 Sep. 15	3 Aug. 18	18th Lrs.
24 Aug. 12	May Somerville, Capt. J. A. C.	24 Aug. 16	3 Aug. 18	11th Lrs.
	Squadron Officers.			
8 Feb. 15	Webster, Lt. F. G.	8 Feb. 16	3 Aug. 18	I. A. R. O.
4 June 15	Fitzpatrick, Lt. J. O. C.	4 June 16	3 Aug. 18	I. A. R. O., Adjt. (*actg.*)
14 Nov. 16	Searle, Lt. W. A. A.	14 Nov. 17	3 Aug. 18	I. A., S. C.
14 Nov. 16	Gastrell, Lt. E. H.	14 Nov. 17	3 Aug. 18	I. A., Qr.-Mr. (*actg.*)
14 Nov. 16	Woolley, Lt. J. H. C.	14 Nov. 17	3 Aug. 18	I. A.
30 Jan. 17	Graig, Lt. G.	30 Jan. 18	3 Aug. 18	I. A.
23 Nov. 17	Pollard, 2nd-Lt. R. C.	23 Nov. 17	13 Sep. 18	I. A. R. O.
28 May 18	Wooldridge, 2nd-Lt. L. E.	28 May 18	13 Sep. 18	I. A. R. O.
1 Oct. 18	Pargiter, 2nd-Lt. J. G.	1 Oct. 18	12 Oct. 18	Attd.
24 Apl. 18	Gomer, 2nd-Lt. L. W.	24 Apl. 18	....	Attd.
10 July 18	Cartmel, 2nd-Lt. A. E.	10 July 18	2 Dec. 18	Attd.
	Attached.			
27 Mar. 18	Hardie, 2nd-Lt. J. J.	27 Mar. 18	6 Oct. 18	I. A. R. O. (*on probn.*)

		Jemadar.	Ressaidar.	Risaldar.	Risaldar-Major.
	Risaldar-Major.				
24 Oct. 94	Malik Khan Mohammed Khan Bahadur	5 Apl. 00	16 Oct. 07	1 Mar. 10	19 July 16
	Risaldars.				
	Ressaidars.				
24 June 05	Mohammed Aslam Khan	12 July 16	10 May 18		
24 Mar. 97	Sham Singh I	12 Sep. 16	3 June 18		
	Jemadars.				
6 June 05	Teja Singh	19 Mar. 10			
18 Apl. 14	Sher Bahadur Khan	11 Apl. 16			
10 Dec. 01	Waryam Singh	25 Aug. 17			
16 Oct. 06	Abdul Ghani Khan	25 Aug. 17			
14 June 02	Sham Singh II	13 Sep. 17			
19 Aug. 94	Mirza Khan	1 Dec. 17			
1 Feb. 01	Udham Singh	18 Feb. 18			

44th Cavalry Regiment.

Formed at—Kantara.
Composition.—6 Troops Sikhs, 6 Troops Jats, 2 Troops Punjabi Mussulmans, 2 Troops Hindustani Mussulmans.

1st Commn. or date of entering service.	Name and Rank.	Army Rank.	Present appointment in Regiment.	Remarks.
	Commandant.			
5 Dec. 91	Battye, Lt.-Col. A. B., (Q.)	5 Dec. 17	8 Aug. 18	6th Cav.
	Squadron Commanders.			
27 Jan. 04	Grace, Capt. H. G., M.C.	11 Apl. 13	19 Nov. 18	21st Cav. 2nd-in-Comd. (*temp.*)
22 Jan. 13	Baines, Capt. G. D.	22 Jan. 17	3 Aug. 18	33rd Cav.
24 Dec. 14	Murphy, Lt. E. L.	24 Dec. 15	3 Aug. 18	I. A. R. O.
5 Feb. 15	Streatfeild, Lt. N.	5 Feb. 16	3 Aug. 18	I. A. R. O.
	Squadron Officers.			
13 Sep. 15	Mallam, Lt. G. L.	13 Sep. 16	3 Aug. 18	Adjt., 3 Aug. 18.
14 Nov. 16	Latter, Lt. F. A.	14 Nov. 17	3 Aug. 18	
14 Nov. 16	Williams, Lt. B. F.	14 Nov. 17	3 Aug. 18	
14 Nov. 16	Cashmore, Lt. A. E.	14 Nov. 17	3 Aug. 18	Qr. Mr., 3 Aug. 18.
14 Nov. 16	Blakeney, Lt. E. C. W.	14 Nov. 17	3 Aug. 18	
	Attached.			
10 Dec. 15	Moffat, Lt. L. P.	1 Nov. 17	14 Sep. 18	I. A. (*on probn.*).
23 Nov. 17	Tully, 2nd-Lt. E. J. S.	23 Nov. 17	14 Sep. 18	I. A. (*on probn.*).
6 Apl. 18	Gordon, 2nd-Lt. W.	6 Apl. 18	6 Oct. 18	I. A. R. O. (*on probn.*).
14 Sep. 18	Hunt, 2nd-Lt. C. V.	14 Sep. 18	12 Oct. 18	I. A. R. O. (*on probn.*).
14 Nov. 16	Mallam, Lt. G. L.	13 Sep. 16	. . . .	
. . . .	Spencer, 2nd-Lt. W.	. . . .	5 Dec. 18	

		Jemadar.	Ressaidar.	Risaldar.	Risaldar-Major.	
	Risaldar-Major.					
1 Mar. 93	Dayal Singh	1 Feb. 08	1 Mar. 12	17 June 15	22 Jan. 18	6th Cav. (*actg.*)
	Risaldars.					
18 June 92	Kabul Singh	25 Feb. 09	9 Sep. 13	22 Aug. 15		
15 Oct. 92	Mam Chand	14 Aug. 07	29 Oct. 14	14 Aug. 16	7th Lrs.	
8 Mar. 90	Jai Singh	. . . .	17 June 15	13 Apl. 17	6th Cav.	
	Ressaidars.					
21 May 02	Harphul Singh	1 Nov. 13	1 May. 15	29th Lrs.		
11 Dec. 01	Bhulan Singh	1 Aug. 13	11 Apl. 16	29th Lrs.		
19 Jan. 98	Jai Singh, M.C.	1 Feb. 13	14 Aug. 16	6th Cav.		
7 May 07	Ahmed Khan	9 July 06	21 Jan. 18	27th Cav.		
16 Aug. 07	Hazur Ahmed	. . . .	2 Feb. 18	5th Cav.		
1 Jan. 11	Jot Ram	8 Sep. 14	22 Apl. 18	5th Cav.		
	Jemadars.					
23 Mar. 13	Hari Singh	14 Jan. 16	5th Cav.			
13 Jan. 02	Bagga Singh	11 Apl. 16	6th Cav.			
27 June 06	Jaswant Singh	18 Dec. 17	16th Cav. Act'g. Woordie Maj.—3 Aug. 18.			

45th Cavalry Regiment.

Formed at—Kantara.

Composition—1 Squadron Sikhs; 1 Squadron Punjabi Musulmans; 2 Troops Jats; 2 Troops Rajputs; 2 Troops Hindustani Musulmans; 2 Troops Kaimhanis.

1st Commn. or date of entering service.	Name and Rank.	Army Rank.	Present appointment in Regiment.	Remarks.
	Commandant.			
28 Jan. 93	Taylor, Maj. E. M., P.S.C., (*Actg. Lt.-Col.*) Q.	28 Jan. 11	29 Aug. 18	22nd Cav. (*temp.*)
	Squadron Commanders.			
4 May 01	Forteath, Maj. A. M. (*Actg. Lt.-Col.*)	4 May 16	3 Aug. 18	36th Horse, 2nd-in Comd. (*temp.*)
13 Aug. 04	Wilson, Capt. T. K. (*Actg. Maj.*)	13 Aug. 13	3 Aug. 18	20th Horse, 2nd-in-Comd. (*temp.*)
7 Sep. 08	MacGregor, Capt. R. G.	7 Sep. 15	3 Aug. 18	34th Horse.
8 Nov. 14	Hjrst, Lt. J. B. (*Actg. Capt.*)	15 Mar. 15	3 Aug. 18	3rd Res. Cav.
	Squadron Officers.			
22 Mar. 15	Phillips, Capt. G. C.	29 Apl. 18	3 Aug. 18	I. A. R. O.
28 May 15	Murray, Lt. V. W. (*Actg. Capt.*)	1 Mar. 17	3 Aug. 18	Attd. Adjt. 3 Aug. 18. (*Actg.*)
14 Nov. 16	Fry, Lt. G. C.	14 Nov. 17	3 Aug. 18	36th Horse. Qr.-Mr. 3 Aug. 18. (*Actg.*)
30 Jan. 17	Ryan, Lt. C. C. L.	30 Jan. 18	3 Aug. 18	20th Horse.
30 Jan. 17	McNeill, Lt. D. S. E.	30 Jan. 18	3 Aug. 18	I. A.
30 Jan. 17	Marshall, Lt. T. R.	30 Jan. 18	3 Aug. 18	38th Horse.
	Attached.			
25 Aug. 15	Oliver, Lt. J. S.	1 July 17	23 Oct. 18	I. A. R. O. (*on probn.*)
30 Apl. 16	Stenson, Lt. L. H. E.	30 Jan. 18	7 Sep. 18	Glouc. R., I. A. (*on probn.*) Qr.-Mr. (*Actg.*)
28 Aug. 18	Waddell, 2nd-Lt. J. W.	28 Aug. 18	24 Sep. 18	I. A. R. O.

		Jemadar.	Ressaidars.	Risaldar.	Risaldar Major.	
	Risaldar Major.					
1 Apl. 86	Dilawar Khan, *Bahadur*, I.O.M.	16 May 05	22 Oct. 11	24 Oct. 14	12 Oct. 18	38th Hrs., *Order of Br. Ind., 2nd cl.* (*Actg.*)
	Risaldars.					
1 Oct. 07	Kurshed Mohammed Khan	1 Nov. 09	1 Dec. 14	17 May 16		5th Cav.
	Ressaidars.					
6 Dec. 01	M. Mohammed Nur Khan, I.O.M.	1 Sep. 15	13 May 16	. . . .		36th Horse.
15 June 99	Pane Singh	3 Nov. 14	15 Jan. 17	. . . .		34th Horse.
11 Apl. 16	Wazir Singh, I.O.M.	2 Dec. 17	7 Sep. 18	. . . .		36th Horse.
	Jemadars.					
. . . .	Ram Lal Singh	3 Aug. 18	. . . .			
16 May 92	Ramji Lal	1 Apl. 12	. . . .			20th Horse.
11 Sep. 06	Ranjit Sing	19 May 16	. . . .			34th Horse.
25 Aug. 17	Mahfuz Ali Khan	25 Aug. 17	. . . .			21st Cav.—Woordie Maj., 3 Aug. 18.
30 Oct. 06	Durzan Singh	30 Dec. 17	. . . .			34th Horse.

Aden Troop.

Formed (under its present designation) at Jacobabad in 1867-68, by Lieut. C. B. Myers, of volunteers from the 1st and 2nd Scinde Horse and the Poona Horse.

Uniform—Khaki. *Loongi*—Khaki.

1st Commn. or date of entering service.	Names and Rank.	Army rank.	Date of appointment.	Remarks.
	Commandant.			
5 Oct 10	Handley, Capt. L. M., M.C. . .	1 Sep. 15	22 Nov. 17	Sqn. Offir., 26th Cav.
4 Dec. 15	Parker, Lt. G. T. M., I. A. R. O.	4 Dec. 16	29 May 17	I. A. R. O.
		Jemadar.	Risaldar.	
	Risaldar.			
1 Dec. 85	Malikdad Khan, *Khan Sahib Sardar Bahadur,* I.D.S.M. (71).	2 Aug. 03	1 Jan. 05	*Order of Br. I., 1st Cl.*
	Ressaidar.			
15 Aug. 93	Muhammad Isa Khan	25 Feb. 12	1 Sep. 17	
12 May 93	Bharat Singh	21 Sep. 16	15 July 18	
	Jemadars.			
10 July 99	Mahbub Ali Khan	1 Sep. 17		
1 June 05	Azimullah Khan	15 July 18		

Regt. Centre—Hoti Mardan.

Queen Victoria's Own Corps of Guides (Frontier Force) (Lumsden's).

(CAVALRY AND INFANTRY.)

Raised at Peshawar in 1846-47, by Lieut. H. B. Lumsden, as the Corps of Guides. Became the Corps of Guides, Punjab Frontier Force, in 1857; The Queen's Own Corps of Guides, Punjab Frontier Force, in 1876; Queen's Own Corps of Guides (Lumsden's), 1904, present designation in 1911.

THE CYPHER OF QUEEN VICTORIA WITHIN THE GARTER.
"Mooltan" "Goojerat" "Punjaub" "Delhi, 1857"
"Ali Masjid" "Kabul, 1879" "Afghanistan, 1878-80"
"Chitral" "Malakand" "Punjab Frontier"

Composition—1 Squadron of Sikhs, ½ of Dogras and Punjabi Hindus, 1 of Pathans, ½ of Punjabi Musalmans.
Composition—1 Company of Sikhs, ½ of Dogras, ½ of Gurkhas, 1 of Pathans, ½ of Punjabi Musalmans, 1 of mixed classes.

Uniform—Drab. *Facings*—Red velvet—Scarlet cloth for the ranks.

COLONEL—Major-General George Stewart, Kt., C.B. . 13 May 1904.

1st Commn. or date of entering service.	Names and Rank.	Army rank.	Present appointment in Regiment.	Remarks.
	Commandant.			
9 Sep. 91	Buist, Lt.-Col. A. H., M.V.O. . .	14 Nov. 16	14 Nov. 16	
	Squadron and Company Commanders. { Cav. 4 / Infy 4 }			
27 Jan. 92	Stewart, Lt.-Col. A. C. (Q.) . .	27 Jan. 18	23 Mar. 06	Comdg. Cav.
23 Nov. 92	Bogle, Maj. J. S. (Q) (*Actg. Lt. Col.*)	23 Nov. 10	5 Nov. 12	Comdg. Infy.
10 Oct. 94	Carey, Maj. C. W. . .	10 Oct. 12	13 July 13	Cav. With 35th Sikhs.
16 Jan. 95	Battye, Maj. I. U., D.S.O. (Q). . (*actg. Lt.-Col.*)	16 Jan. 13	12 Feb. 14	Infy. With 8th G. R.
22 Jan. 96	*Norman, Bt. Col. (temp. Br.-Genl.) C. L., D.S.O., M.V.O., p.s.c.* A.D.C.	*1 July 18*	*6 Nov. 99*	*Cav. Director, Staff duties, [A. Hqrs.*
20 Jan. 97	*Campbell, Maj. H., D.S.O. M.V.O. (actg. Lt.-Col.)*	*20 Jan. 15*	*12 Mar. 16*	*With I. S. Troops.*
4 Aug. 97	*Hay, Bt.-Lt.-Col. C. J. B., D.S.O., p.s.c.* [L].	*1 Jan. 17*	*14 Nov. 16*	*Infy. G. S. O., 1st gr.*
	Squadron and Company Officers. { Cav. 9 / Inf. 9 }			
8 Sep. 97	Clementi, Maj. J. (*actg. Lt.-Col.*)	1 Sept. 15	17 July 00	Cav. With 3rd Bn.
27 July 98	*Kirkpatrick, Bt. Lt.-Col. C., p.s.c.*	*1 Jan. 18*	*5 Dec. 00*	*Infy. G. S. O., 2nd gde.*
5 Dec. 00	*Kt., Wylly, Maj. G. G. E., D.S.O., p.s.c.*	*1 Sep. 15*	*11 Feb. 04*	*Cav. G.S.O., 2nd gr.*
8 Jan. 01	*Hensley, Maj. F. K.*	*8 Jan. 16*	*23 Feb. 04*	*Infy.* With 2nd Bn.
4 Dec. 01	*Morris, Maj. C. E.*	*4 Dec. 16*	*26 Apl. 04*	*Infy. Burma Mily. Police.*
27 Aug. 02	*Campbell, Maj. C. H.*	*27 Aug. 16*	*2 May 04*	*Infy. Khyber Rifles.*
21 Jan. 03	Sandeman, Maj. D. G. . . .	21 Jan. 18	9 Dec. 04	Infy. C. O. (*tempy.*)
19 Aug. 03	*McLeod, Maj. D. K.*, D.S.O. . .	*19 Aug. 18*	*1 Jan. 06*	*Cav. G. S. O., 2nd gr.*, A. Hdqrs.
18 Jan. 05	Ogilvie, Capt. J. F. W. . . .	18 Jan. 14	19 Mar. 06	Cav. S. C. (*tempy.*)
5 Aug. 05	Blood, Capt. W. H. (*temp. Maj.*) .	5 Aug. 14	20 May 07	D. A. Q. M. G.
24 Jan. 06	*Prendergast, Capt. N. H.* . .	*24 Jan. 15*	*4 Sep. 07*	*Infy. N. Waziristan Militia.*
29 Aug. 06	Erskine, Capt. C. E. T., M.C.	29 Aug. 15	9 Nov. 07	Infy. Adjt., 27 Jan. 16.
29 Aug. 06	*Murray, Capt. E. M.*, D.S.O., M.C. *(temp. Maj.)*	*29 Aug. 15*	*25 July 09*	*Cav. With R. A. F.*
19 Jan. 07	*Blacker, Capt. L. V. S. (temp. Maj.)*	*1 Sep. 15*	*27 Mar. 10*	*Infy. With For. & Pol. Dept.*
23 July 07	*Hankin, Capt. H. M.*	*1 Sep. 15*	*14 Oct. 11*	*Cav.*
6 Sep. 11	Hammond, Capt. A. V. . . .	6 Sep. 15	2 Dec. 12	Cav. Adjt., 1 Aug. 14.
20 Jan. 12	Grant, Capt. P.	20 Jan. 16	18 Mar. 13	
24 Aug. 12	*Dane, Capt. H.*	*24 Aug. 16*	*18 Nov. 13*	*Cav. With Sig. Serv.*
15 Aug. 14	Eales, Capt. C. H. H., M.C. . .	15 Aug. 18	10 Feb. 16	Attd. Cav.
15 Aug. 14	Money, Capt. H. D. K. . . .	15 Aug. 18	10 Dec. 15	Attd.
19 June 15	Edwards, Lt. C. G. N.	19 June 16	14 May 17	Attd. Infy.
5 Aug. 15	Wallis, Lt. C.	5 Aug. 16	10 May 17	Attd. Cav.
15 Nov. 15	Rich, Lt. E. P. (*Actg. Capt.*) . .	15 Nov. 16	19 Nov. 15	Attd.
15 Nov. 15	Barrett, Lt. M. L. (*Actg. Capt.*) .	15 Nov. 16	19 Nov. 15	Attd. Cav.
26 Jan. 16	Davies, Lt. F. A.	26 Jan. 17	19 Mar. 16	Attd. Cav. Infy.
29 June 16	Wainwright, Lt. J. G.	29 June 17	2 July 16	Attd. With 3-151 Infy.
30 Jan. 17	Spurway, Lt. W. R. P. . . .	30 Jan. 18	6 Feb. 17	Attd. Infy.
18 June 17	Prioleau Lt. C. P. J.	18 June 18	25 June 17	Attd. Cav. Infy.
18 June 17	Shipard, Lt. S. S. A.	18 June 18	25 June 17	Attd. Cav.
18 June 17	Bucknall, Lt. H. L.	18 June 18	25 June 17	Attd. Cav.
27 Oct. 17	Brooke-Hunt, Lt. R. C.	27 Oct. 18	4 Nov. 17	Attd. Infy. With 3-151 Infy.
. . . .	Bryant, 2nd-Lt. G. D.	16 Sep. 17	. . . . .	Attd.
. . . .	Knight, 2nd-Lt. R.	. . . .	. . . . .	Attd.
. . . .	Mawhood, 2nd-Lt. J. C. . . .	. . . .	. . . . .	Attd.
27 Aug. 15	Smelt, 2nd-Lt. M. V.	27 Aug. 15	. . . . .	Attd.

Queen Victoria's Own Corps of Guides—*continued.*

1st Commn. or date of entering service.	Names and Rank.	Army rank.	Present appointment in Regiment.	Remarks.
	Attached.			
20 Apl. 15	Blandy, Lt. R. D. (*Actg. Capt.*)	23 Apl. 16	29 Mar. 17	I. A. R. O. Infy.
16 June 15	Campbell-Harris, Lt. C.	16 June 16	. . . .	17 Cav.
15 Nov. 15	Coates, Lt. J. C., M.C.	15 Nov. 16	21 Feb. 17	5 G. R.
12 Feb. 16	Turner, Lt. E.	12 Feb. 17	30 May 18	I. A. (*on probn.*)
17 Feb. 16	Gordon, Lt. J. M.	17 Feb. 17	10 Dec. 17	Infy. I. A. R. O. (*on probn.*)
26 Feb. 17	Harvey, Lt. G. T. B.	26 Feb. 17	14 Mar. 17	I. A. R. O. Cav.
29 June 16	deCourcy, Lt. G. L.	29 June 17	. . . .	
29 July 16	Jaques, Lt. A. B.	29 July 17	. . . .	I. A. R. O. Cav.
1 Aug. 16	Tate, Lt. T. B. (*Actg. Capt.*)	1 Aug. 17	20 Aug. 16	Cav. (Pathans). I. A. R. O. A. R. O.
14 Aug. 16	Grose-Hodge, Lt. H.	14 Aug. 17	20 Aug. 16	I. A. R. O.
13 Sep. 16	Fagnani, Lt. H. H.	13 Sep. 17	5 Oct. 16	Cav. (Pathans.) Infy. with 3-151 Infy.
10 Nov. 16	Thorne, Lt. S. G.	10 Nov. 17	. . . .	Do. Infy. *Lv., ex Ind., m.c., 6 mos.*
15 Feb. 17	Eldon, Lt. A.	15 Feb. 18	. . . .	I. A. R. O.
25 May 17	Foster, Lt. H. B.	25 May 18	5 June 17	I. A. R. O.
26 June 17	Ker, Lt. J. S.	26 June 18	18 July 17	Cav. I. A. R. O.
8 Aug. 17	Mann, Lt. A.	8 Aug. 18	10 Nov. 17	I. A. R. O. Infy.
28 Nov. 16	Howe, 2nd-Lt. G. E.	. . . .	26 Aug. 17	Res. R. Cav. I. A. R. O. (*on probn.*)
19 Dec. 16	Reekie, Lt. J. (*Actg. Capt.*).	. . . .	26 Aug. 17	Cav. I. A. R. O (*on probn*). Adjt.
. . . .	Don, 2nd-Lt. P. J., M.M.	. . . .	. . . .	I. A. (*on probn.*)

CAVALRY.

		Jemadar.	Ressaidar.	Risaldar.	Risaldar Major.	
	Risaldar-Major.					
26 Sep. 80	*Bahadur Singh,* Sardar Bahadur, I.O.M. (*56, 58*).	*6 July 93*	*18 May 01*	*1 Sep. 04*	*1 Feb. 09*	*Order of Br. I., 1st cl. (Senior Indian A.-D.-C. to H. E. the C.-in-C., 1 Jan. 17.)*
	Risaldars.					
15 Jan. 92	***Khwaja Muhammad Khan, Sardar Bahadur*** (*56, 72a*), I.D.S.M.	*15 Jan. 92*	*1 Aug. 08*	*1 Feb. 09*		*Order of Br. I., 1st cl. A.-D.-C. to G. O. C., N. Army.*
22 Apl. 87	Abnashi Ram (56, 66*a*)	1 Sep. 04	1 Feb. 09	1 Feb. 12		
27 Oct. 91	Nur Khan (58)	1 Oct. 02	16 June 07	1 July 13		
13 Apl. 07	Muhammad Sarwar Khan (52, 56).	1 Feb. 08	10 Feb. 11	16 Apl. 15		
20 Nov. 01	Zardad Khan (45*a*, 56)	1 Feb. 09	17 Nov. 14	27 Oct. 17		
	Ressaidars.					
24 Feb. 09	Ratan Chand (66*a*)	1 Nov. 11	16 Apl. 15			
23 May 92	Jiwand Singh (56, 58)	1 Feb. 12	12 Feb. 16			
10 Feb. 94	Abdullah Khan Khuttak (56, 66*a*).	1 Feb. 12	6 July 16			
16 Dec. 99	Abdulla Khan	30 Oct. 14	27 Oct. 17			
16 Mar. 96	Bhagwan Singh.	17 Nov. 14	23 Sep. 17			

Queen Victoria's Own Corps of Guides—*contd.*

1st Commn. or date of entering service.	Names and Rank.	Army rank.	Remarks.
	Jemadars.		
14 Jan. 93	*Natha Khan (56)*	*1 July 13*	
27 Apl. 96	*Sharbat Ali Khan*, I.D.S.M.	*27 Nov. 13*	*Suty.*
16 Dec. 99	Raja Ram (56)	5 Apl. 15	Woordie-Maj., 25 May 1917.
1 Oct. 96	*Ganda Singh*	*16 Apl. 15*	
3 Jan. 10	Saadat Husain (56)	16 Apl. 15	
19 June 98	Beant Singh	12 Feb. 16	
9 Nov. 98	*Sher Muhammad*, I.D.S.M.	*1 June 16*	
1 Oct. 96	Aman Khan	6 July 16	
22 Sep. 97	*Anup Singh*	*12 July 16*	
26 Nov. 88	Zarif Khan	20 June 16	
17 July 03	Firdos Khan (45*a*, 60*a*)	26 July 16	
7 Jan. 94	Wasawa Singh	17 Nov. 16	
1 Aug. 01	Amal Baz (56)	1 July 13	
....	Khan Bahadur (Retd.)	....	Re-employed, 10 Jan. 16.
1 Sep. 00	Arjan Singh	23 Sep. 17	
20 May 05	Shah Pasand	23 Sep. 17	
13 Dec. 01	Buta Ram	27 Oct. 17	

INFANTRY.

1st Commn. or date of entering service.	Names and Rank.	Jemadar.	Subadar.	Subadar-Major.	Remarks.
	Subadar-Major.				
22 Dec. 94	Man Bir Thapa, I.O.M.	24 Jan. 10	1 May 14	1 Nov. 17	*Lv., m. c.*
	Subadars.				
19 June 05	Taj Muhammad (46, 56)	1 Oct. 06	24 Jan. 10		With 3rd Bn. (*Actg. Sub-Maj.*)
12 June 92	Afzal (66, 72*a*)	15 Sep. 09	5 Jan. 15		
1 Nov. 92	Bahadur Khan, *Bahadur* (56, 58, 66*a*).	24 Jan. 10	25 Feb. 15		*Order of Br. I., 2nd Cl. With 2nd Bn.* (*Actg. Sub-Maj.*)
18 Sep. 99	Alam Khan (56, 66*a*)	16 Sep. 08	6 Apl. 15		(*Actg. Sub-Maj.*)
21 Mar. 99	Khiyal Gul (56), (66*a*)	15 Apl. 13	17 Apl. 15		
28 May 01	Gulab Singh (56)	2 Apl. 14	16 May 15		
6 Aug. 99	Bishan Singh, I.D.S.M.	12 Oct. 15	14 Oct. 15		With 4th Bn. (*actg. Sub-Maj*).
6 Jan. 94	*Nidhan Singh (56)*	*5 Jan. 15*	*1 July 16*		*With 4th Bn.*
18 Oct. 97	Kehr Singh (56, 66*a*)	17 Apl. 15	16 Jan. 17		
18 Feb. 94	Chandra Bir	22 Aug. 16	....		
14 Dec. 97	Nur Muhammad	16 Jan. 17	26 Nov. 17		With 3-151 Infy.
3 Apl. 03	Gujar (56, 66*a*)	17 Mar. 15	1 Nov. 17		With 2nd Bn.
1 Jan. 02	Sher Singh	....	23 May 18		Re-employed, 31 Mar. 17.
1 Aug. 03	Hayat Khan	26 Nov. 17	23 May 18		
22 Feb. 98	Wazir	16 Jan. 17	9 June 18		3rd Bn.
22 Mar. 01	Khaista Khan (72*a*)	5 Feb. 15	1 Sep. 18		
11 May 98	*Ghafur Shah*, I.D.S.M.	*1 Sep. 15*	*5 July 18*		
....	Kajir Khan (retd.)	....	....		Re-employed, 25 Feb. 17.
	Jemadars.				
15 Mar. 89	Khushal (66*a*, 72*a*)	5 Jan. 15			
16 Aug. 04	Sardar Khan	22 Apl. 15			
18 Sep. 00	Mangtu (56, 66*a*)	16 May 15			
18 Mar. 99	Mula (56, 58)	10 Oct. 15			With 4th Bn.
1 May 00	*Samundar (66a)*	*1 Aug. 16*			*With 3-151st Rif.*
14 May 98	Narayan Singh	22 Aug. 16			
1 Apl. 04	Khan Zaman	5 June 17			
13 Mar. 06	Muhammad Khan	1 Nov. 17			
13 Mar. 06	Muhammad Khan	1 Jan. 18			With 3rd Bn.
3 Oct. 96	Samundar Khan	1 Jan. 18			With 4th Bn.
13 Mar. 06	Siraj-ud-Din	1 Jan. 18			
18 Nov. 08	Ram Singh	1 Mar. 18			
29 Mar. 98	Muhammad Khan	10 Mar. 18			Adjt., 9 July 18.
25 Mar. 12	Sher Jang	10 Mar. 18			
1 Apl. 99	Taj Mohammed	4 Apl. 18			
2 July 01	Lachho	9 Apl. 18			
18 Nov. 99	*Barna Bir*	*22 May 18*			*With 3-151 Infy.*
9 Apl. 08	Bir Ali	23 May 18			
29 June 09	Wazira	23 May 18			
1 Apl. 07	Khan Bahadur	8 June 18			
....	Mangal Singh	*5 July 18*			
1 Oct. 04	Chur Singh (retd.)	15 Aug. 18			
....	*Abdul Ghani*	....			Re-employed, 23 Feb. 17.
4 Jan. 95	Kaman Singh (retd.)	17 Apl. 15			Re-employed, 17 Apl. 18.

2nd Battalion, Queen Victoria's Own Corps of Guides (Frontier Force). (Lumsden's.)

(INFANTRY.)

Formed at Mardan on 12th January 1917.

Composition—½ Company of Sikhs, ½ of Dogras, 1 of Gurkhas, ½ of Pathans, 1½ of Punjabi Mussalmans.

1st Commn. or date of entering service.	Names and Rank.	Army rank.	Present appointment in regiment.	Remarks.
	Commandant.			
23 Nov. 92	Bogle, Maj. J. S. (*temp. Lt.-Col.*)	23 Nov. 10	18 Feb. 17	Guides.
	Company Commanders.			
8 Jan. 01	Hensley, Maj. F. K.	8 Jan. 16	10 June 17	Guides.
4 Dec. 01	Pollock, Maj. R. C. G.	4 Dec. 16	23 Jan. 17	51st Sikhs.
	Company Officers.			
29 June 16	*Petrie, Lt. J. F.*	*29 June 17*	*3 Feb. 17*	*Frontier Militia.*
16 Aug. 16	Barry, Lt. J. L.	16 Aug. 17	1 Feb. 17	Attd.
18 June 17	Fox-Strangways, Lt. V.	18 June 18	26 June 17	Attd.
12 Sep. 17	Archer, Lt. W. A.	12 Sep. 18	9 Dec. 17	Attd.
21 Dec. 17	Roberts, 2nd-Lt. G. E.	21 Dec. 17	21 Mar. 18	Attd.
31 Aug. 18	Malden, 2nd-Lt. W. J.	31 Aug. 18	4 Sep. 18	Attd.
26 Apl. 15	Lodge, Capt. A. B., D.S.O.	4 Aug. 15	1 Aug. 18	Aust. I. F., I. A. (*on probn.*)
10 Jan. 16	Fowler, Lt. G. H. A.	10 Oct. 17	31 Mar. 18	Midd'x. R. I. A. (*on probn.*)
22 Apl. 15	Dart, Lt. L. V.	1 Oct. 15	16 Feb. 18	Welsh R. I. A. (*on probn.*)
7 July 15	Hughes, Lt. C. R.	1 June 16	16 Feb. 18	R. W. Fus. I. A. (*on probn*)
28 Oct. 15	Wallace, Lt. A. A.	28 Oct. 16	11 May 17	Attd. With S. & T. Corps.
29 Oct. 15	Hodgins, Lt. J. N. M.	29 Oct. 16	6 Apl. 17	Attd.
14 Aug. 16	Kemm, Lt. A. H.	14 Aug. 17	4 Dec. 17	I. A. R. O.
10 Dec. 16	Chapman, Lt. R. V.	10 Dec. 16	10 Dec. 17	R. N. V. R. I. A. (*on probn.*)
5 May 17	Mitchell, Lt. C. H.	5 May 18	19 Aug. 17	I. A. R. O.
5 May 17	Leamon, Lt. H. W.	5 May 18	19 Aug. 17	I. A. R. O.
12 July 15	Collis, 2nd-Lt. F. S.	12 July 15	16 June 17	Attd.
20 June 17	Goulstone, 2nd-Lt. W. H.	20 June 17	14 Oct. 17	I. A. R. O.
1 Oct. 17	Rose, Lt. S. G. S.	1 Oct. 17	21 Jan. 18	I. A. (T.C.)
26 Oct. 17	Thew, 2nd-Lt. I. A.	26 Oct. 17	5 Nov. 17	Attd.

Subadar-Major.

		Jemadar.	Subadar.	Subadar-Major.
1 Nov. 92	Bahadur Khan, *Bahadur*, (56, 58, 66a)	24 Jan. 10	25 Feb. 15	1 Nov. 17 *Order of Br. I., 2nd. Cl.*

Subadars.

16 Jan. 93	Ran Bir	1 May 14	10 Oct. 15
19 Jan. 94	Shadi Khan	16 Apl. 15	12 Jan. 17
3 June 95	Fauja Singh	15 May 15	1 June 17
26 Sep. 98	Parshad Gurung	1 Jan. 17	1 Nov. 17
11 June 94	Ismail (72a)	16 May 15	1 Mar. 18
14 Nov. 93	Jaimal	1 Sep. 15	1 Nov. 17
5 May 00	Ahmed	29 Dec. 16	20 May 18

Jemadars.

10 Sep. 98	Mustafa	1 Jan. 17
15 Oct. 96	Mian Nur	24 July 17
6 Apl. 09	Fatteh Khan	29 Oct. 17
15 Nov. 05	Amar Singh (56)	1 Nov. 17
12 Dec. 97	Lahoru	1 Nov. 17
1 Apl. 07	Ram Dullah	1 Nov. 17
18 June 01	Tilok Singh	1 Nov. 17
18 Feb. 08	Ghulam Ali Khan	3 Dec. 17
25 July 99	Haidar Ali Shah	1 Jan. 18
17 Dec. 04	Santa Bir	1 Jan. 18
1 Apl. 04	Lal Khan	20 Jan. 18
9 Nov. 98	Nurbai Gurung	15 Feb. 18
15 Mar. 08	Ahmad Yar	1 Mar. 18
24 Aug. 95	Jaimal Singh	20 May 18
29 Oct. 08	Jodha Khan	20 May 18
21 Oct. 15	Abdul Rahman	1 Aug. 18
5 Mar. 18	Malik Khan	10 Aug. 18

3rd Battalion, Queen Victoria's Own Corps of Guides (Frontier Force) (Lumsden's).

(INFANTRY.)

Formed at Mardan on 22nd October 1917.

Composition—½ Company of Sikhs, ½ Company of Dogras, 1 Company of Gurkhas, 1 Company of Pathans, 1 Company of Punjabi Mussalmans.

1st Commn. or date of entering service.	Names and Rank.	Army rank.	Present appointment in regiment.	Remarks.
	Commandant.			
8 Sep. 97	Clementi, Maj. J. (*temp. Lt.-Col.*)	1 Sep. 15	15 Mar. 18	Guides.
	Company Commanders.			
9 Sep. 08	Bainbridge, Capt. D., M.C., (*Actg. Maj.*)	1 Sep. 15	7 Jan. 18	57th Rif. 2nd-in-Comd. (*temp.*)
19 June 15	Knight, Lt. L. R.	19 June 16	2 Mar. 18	L'pool R. I. A. (*on probn.*).
22 Sep. 14	Jameson, Lt. J. H., D.S.O. (*Actg. Capt.*)	21 June 16	24 Oct. 17	I. A. (*on probn.*). Adjt.
1 Aug. 15	Pritchard, Lt. A. J. L., M.C.	1 Aug. 16	2 Mar. 18	S. Staff R. I. A. (*on probn.*).
	Company Officers.			
5 Aug. 14	Free, Lt. C. W., M.C. (*Actg. Capt.*)	5 May 16	10 Dec. 17	I. A. (*on probn*). Adjt.
15 Aug. 14	Ferguson, Lt. F. E., M.C.	. . .	. . . .	
5 Oct. 14	Bourke, Lt. A. J. H.	5 July 16	24 Oct. 17	Attd.
6 Aug. 15	Mc Farlane, Lt. A. H.	1 July 17	19 Jan. 18	I. A. R. O. (*on probn.*).
21 June 16	Parker, Lt. E.	21 June 17	15 Feb. 18	York & Lanc. R. I. A. (*on probn.*).
4 Dec. 15	Griffiths, Lt. J. C.	. . . .	17 Feb. 18	S. Lan. R. I. A. (*on probn.*). *Lv. ex., I., m. c.*
25 Oct. 16	*Green, Lt. P. A.*	*25 July 18*	*29 Mar. 18*	*Monmouth R. I. A. (on probn.). With 3-30th Punj.*
5 May 17	Jeffries, Lt. H. R. M.	5 May 18	24 Oct. 17	I. A. R. O. Qr. Mr.
31 Oct. 17	Harrison, Lt. W. E. P.	31 Oct. 18	1 Feb. 18	I. A. R. O.
29 Aug. 17	Long, 2nd-Lt. H. A.	29 Aug. 17	28 Jan. 18	I. A. R. O.
12 Sep. 17	LeMesurier, Lt. T. H. F.	12 Sep. 18	24 Mar. 18	Attd.
19 Dec. 16	Robins, Lt. E., D.C.M.	19 Dec. 17	29 Mar. 18	I. A. R. O.
21 Dec. 17	*Devonshire, 2nd-Lt. N. G.*	*21 Dec. 17*	*22 Mar. 18*	*Attd.*
21 Dec. 17	*Evans, 2nd-Lt. S. F.*	*21 Dec. 17*	*22 Mar. 18*	*With 3-151st Rif.*
31 Jan. 18	*Wilkinson, 2nd-Lt. R. R.*	*31 Jan. 18*	*7 Feb. 18*	Attd.
. . . .	Meston, 2nd-Lt. A.	. . . .	. . .	I. A. R. O.
7 June 18	Yare, 2nd-Lt. F. R.	7 June 18	28 Oct. 18	I. A. (*T. C.*)
31 Aug. 18	*Ingram-Johnson, 2nd-Lt. R. E. S.*	*1 Aug. 18*	*5 Sep. 18*	*Attd. With 3-151st Rif.*
31 Aug. 18	*Price, 2nd-Lt. C. S.*	*1 Aug. 18*	*4 Sep. 18*	*Attd. With 3-151st Rif.*

		Jemadar.	Subadar.	Subadar-Major.	Remarks.
	Subadar-Major.				
9 June 05	Taj Mohammad	1 Oct. 06	24 Jan. 10	25 Dec. 17.	
	Subadars.				
1 Sep. 05	Nadir Khan	1 June 15	1 Jan. 17		
8 Jan. 94	Chuni Lal	29 Mar. 15	15 Jan. 17		
9 Jan. 91	Muhammad Akbar	5 Jan. 15	1 Nov. 17		
16 Jan. 04	Sukhbir Gurung	16 June 16	31 Aug. 18		
18 Nov. 08	Lalbahadur Thapa	1 Nov. 17	1 Sep. 18		
. . . .	Wadhawa Singh	1 Oct. 16	19 Oct. 18		
	Jemadars.				
1 Oct. 96	Muhammad Kazim	21 Sep. 14			
29 Feb. 04	*Taj Muhammad.*	*1 Nov. 17*			*With 2nd Porter Corps.*
20 Aug. 07	Shadi Khan	1 Nov. 17			
17 Feb. 03	Chatar Singh.	1 Nov. 17			
13 Mar. 06	Muhammad Khan	1 Feb. 18			Jem. Adjt., 1 Nov. 18.
12 Jan. 18	Sher Alam	6 Feb. 18			(*on probn.*)
. . . .	Karam Dad	14 Mar. 18			
18 Jan. 06	Lalbahadur Gurung	2 June 18			
8 Dec. 08	Maumesar Gharti	2 June 18			
19 Aug. 99	Sajawal Khan	28 June 18			
30 July 09	Dharam Singh	1 July 18			
22 Dec. 17	Fazl-i-Khaliq	1 Sep. 18			

INDIAN ARTILLERY.

Uniform—Blue. *Facings*—Scarlet.

Composition of each Battery :—Half Punjabi Musalmans and half Sikhs and other Hindus.

MOUNTAIN DIVISION. (*12 Batteries.*)

Ind. Mtn. Arty. Bde.	IV Ind. Mtn. Arty. Bde.	VII Ind. Mtn. Arty. Bde.	
Lt.-Col. Hayes-Sadler, Maj. (*Actg. Lt.-Col.*) J., R. A.	*Lt.-Col.*	*Lt.-Col.* Fergusson, Lt.-Col. A. C., C.M.G., D.S.O., R.A. (*g*) (*Tempy.*)	Chapman, L.-Col. L. R.A. (*Offg.*)
Adjutant. Grant, 2nd-Lt. (*Actg. Capt.*) C. D., R. A.,	*Adjutant.* Landon, Capt. R.P., R.A.	*Adjutant.* Buckland, Maj. G. N., D.S.O., R. A.	*Actg. Adjutant.*
23rd. Peshawar Battery	27th Battery.	21st Battery.	22nd Derajat Battery (F. F.)
30th Battery.	28th Battery.	26th Battery.	

(*Unallotted.*)

25th Battery. 29th Battery.	31st Battery. 32nd Battery.	24th Hazara Battery (F. F.)

21st, 22nd, 23rd & 24th Batts. are linked.

21st Kohat Mountain Battery (Frontier Force).

Raised at Bannu in 1851, by Lt. H. Hammond, from detachments of horse arty. formerly in the service of the Lahore Darbar. Originally styled No. 2 Horse Light Field Battery, Punjab Irregular Force, but more generally known as No. 2 Punjab Lt. Fld. Batty. Became No. 2 Horse Lt. Fld. Batty., Punjab Frontier Force, 1865; No. 1 Mtn. Batty., P. F. F., 1876; No. 1 (Kohat) Mtn. Batty., P. F. F., 1879; the Kohat Mtn. Batty., 1901; present designation, 1903.

"Peiwar Kotal" "Kabul 1879" "Afghanistan, 1878-80" "Tirah" "Punjab Frontier."

1st Commn. or date of entering service.	Names and Rank.	Army rank.	Present appointment in Batty.	Remarks.
	Commandant.			
31 July 02	Garrett, Maj. R. C.	1 May 17	29 Aug. 17	
	Captain. (1)			
19 July 12	Mackenzie, Capt. H. H. . . .	8 Aug. 16	22 Oct. 16	
	Subalterns. (3)			
7 Oct. 15	Ross, Lt. (*Act'g.* Capt.) R. E. (R. G. A., T.F.)	1 June 16	15 Dec. 17	
7 Nov. 15	Wells, Lt. (*Tempy.*) R. H., R. A..	1 Aug. 16	10 Feb. 16	
22 Apl. 15	Armitage, Lt. H. M. A., R. G. A.	1 July 17	11 Sep. 15	
27 Oct. 15	Gayer, Lt. W. M., R. A. . .	1 July 17	. . . .	
30 Nov. 15	Tennant, Lt. W. R. (Spl. Res.).	1 July 17	16 Nov. 17	

Attached.

		Jemadar.	Subadar.	Subadar-Major.
	Subadar.			
21 Jan. 93	Chanda, I.O.M.	16 Apl. 08	2 Nov. 15	
	Jemadars.			
26 Nov. 98	Khan Beg	16 May 12		
8 July 99	Puran Singh	30 Aug. 15		
26 Nov. 98	Raja	2 Nov. 15		

3rd Battalion, Queen Victoria's Own Corps of Guides (Frontier Force) (Lumsden's).

(INFANTRY.)

Formed at Mardan on 22nd October 1917.

Composition—½ Company of Sikhs, ½ Company of Dogras, 1 Company of Gurkhas, 1 Company of Pathans, 1 Company of Punjabi Mussalmans.

1st Commn. or date of entering service.	Names and Rank.	Army rank.	Present appointment in regiment.	Remarks.
	Commandant.			
8 Sep. 97	Clementi, Maj. J. (*temp. Lt.-Col.*)	1 Sep. 15	15 Mar. 18	Guides.
	Company Commanders.			
9 Sep. 08	Bainbridge, Capt. D., M.C., (*Actg. Maj.*)	1 Sep. 15	7 Jan. 18	57th Rif. 2nd-in-Comd. (*temp.*)
19 June 15	Knight, Lt. L. R.	19 June 16	2 Mar. 18	L'pool R. I. A. (*on probn.*).
22 Sep. 14	Jameson, Lt. J. H., D.S.O. (*Actg. Capt.*)	21 June 16	24 Oct. 17	I. A. (*on probn.*). Adjt.
1 Aug. 15	Pritchard, Lt. A. J. L., M.C. . .	1 Aug. 16	2 Mar. 18	S. Staff R. I. A. (*on probn.*).
	Company Officers.			
5 Aug. 14	Free, Lt. C. W., M.C. (*Actg. Capt.*)	5 May 16	10 Dec. 17	I. A. (*on probn*). Adjt.
15 Aug. 14	Ferguson, Lt. F. E., M.C. . .	. . .	. . . .	
5 Oct. 14	Bourke, Lt. A. J. H.	5 July 16	24 Oct. 17	Attd.
6 Aug. 15	Mc Farlane, Lt. A. H.	1 July 17	19 Jan. 18	I. A. R. O. (*on probn.*).
21 June 16	Parker, Lt. E.	21 June 17	15 Feb. 18	York & Lanc. R. I. A. (*on probn.*).
4 Dec. 15	Griffiths, Lt. J. C.	. . . .	17 Feb. 18	S. Lan. R. I. A. (*on probn.*). *Lv. ex., I., m. c.*
25 Oct. 16	*Green, Lt. P. A.*	*25 July 18*	*29 Mar. 18*	*Monmouth R. I. A.* (*on probn.*). *With 3-30th Punj.*
5 May 17	Jeffries, Lt. H. R. M.	5 May 18	24 Oct. 17	I. A. R. O. Qr. Mr.
31 Oct. 17	Harrison, Lt. W. E. P. . . .	31 Oct. 18	1 Feb. 18	I. A. R. O.
29 Aug. 17	Long, 2nd-Lt. H. A.	29 Aug. 17	28 Jan. 18	I. A. R. O.
12 Sep. 17	LeMesurier, Lt. T. H. F. . . .	12 Sep. 18	24 Mar. 18	Attd.
19 Dec. 16	Robins, Lt. E., D.C.M. . . .	19 Dec. 17	29 Mar. 18	I. A. R. O.
21 Dec. 17	*Devonshire, 2nd-Lt. N. G.* . .	*21 Dec. 17*	*22 Mar. 18*	*Attd.*
21 Dec. 17	*Evans, 2nd-Lt. S. F.*	*21 Dec. 17*	*22 Mar. 18*	*With 3-151st Rif.*
31 Jan. 18	*Wilkinson, 2nd-Lt. R. R.* . .	*31 Jan. 18*	*7 Feb. 18*	Attd.
. . . .	Meston, 2nd-Lt. A.	. . . . .	. . . .	I. A. R. O.
7 June 18	Yare, 2nd-Lt. F. R.	7 June 18	28 Oct. 18	I. A. (*T. C.*)
31 Aug. 18	*Ingram-Johnson, 2nd-Lt. R. E. S.*	*1 Aug. 18*	*5 Sep. 18*	*Attd. With 3-151st Rif.*
31 Aug. 18	*Price, 2nd-Lt. C. S.*	*1 Aug. 18*	*4 Sep. 18*	*Attd. With 3-151st Rif.*

		Jemadar.	Subadar.	Subadar-Major.
	Subadar-Major.			
9 June 05	Taj Mohammad	1 Oct. 06	24 Jan. 10	25 Dec. 17.
	Subadars.			
1 Sep. 05	Nadir Khan	1 June 15	1 Jan. 17	
8 Jan. 94	Chuni Lal	23 Mar. 15	15 Jan. 17	
9 Jan. 91	Muhammad Akbar	5 Jan. 15	1 Nov. 17	
16 Jan. 04	Sukhbir Gurung	16 June 16	31 Aug. 18	
18 Nov. 08	Lalbahadur Thapa	1 Nov. 17	1 Sep. 18	
. . . .	Wadhawa Singh	1 Oct. 16	19 Oct. 18	
	Jemadars.			
1 Oct. 96	Muhammad Kazim	21 Sep. 14	*With 2nd Porter Corps.*	
29 Feb. 04	*Taj Muhammad.*	*1 Nov. 17*		
20 Aug. 07	Shadi Khan	1 Nov. 17		
17 Feb. 03	Chatar Singh.	1 Nov. 17		
13 Mar. 06	Muhammad Khan	1 Feb. 18	Jem. Adjt., 1 Nov. 18. (*on probn.*)	
12 Jan. 18	Sher Alam	6 Feb. 18		
. . . .	Karam Dad	14 Mar. 18		
18 Jan. 06	Lalbahadur Gurung	2 June 18		
8 Dec. 08	Maumesar Gharti	2 June 18		
19 Aug. 99	Sajawal Khan	28 June 18		
30 July 09	Dharam Singh	1 July 18		
22 Dec. 17	Fazl-i-Khaliq	1 Sep. 18		

INDIAN ARTILLERY.

Uniform—Blue. *Facings*—Scarlet.

Composition of each Battery :—Half Punjabi Musalmans and half Sikhs and other Hindus.

MOUNTAIN DIVISION. (*12 Batteries.*)

Ind. Mtn. Arty. Bde.	IV Ind. Mtn. Arty. Bde.	VII Ind. Mtn. Arty. Bde.	
Lt.-Col. Hayes-Sadler, Maj. (*Actg. Lt.-Col.*) J., R. A.	*Lt.-Col.*	*Lt.-Col.* Fergusson, Lt.-Col. A. C., C.M.G., D.S.O., R.A. (*g*) (*Tempy.*)	Chapman, L.-Col. L. R.A. (*Offg.*)
Adjutant. Grant, 2nd-Lt. (*Actg. Capt.*) C. D., R. A.,	*Adjutant.* Landon, Capt. R.P., R.A.	*Adjutant.* Buckland, Maj. G. N., D.S.O., R. A.	*Actg. Adjutant.*
23rd Peshawar Battery	27th Battery.	21st Battery.	22nd Derajat Battery (F. F.)
30th Battery.	28th Battery.	26th Battery.	

(*Unallotted.*)

25th Battery. 29th Battery.	31st Battery. 32nd Battery.	24th Hazara Battery (F. F.)

21st, 22nd, 23rd & 24th Batts. are linked.

21st Kohat Mountain Battery (Frontier Force).

Raised at Bannu in 1851, by Lt. H. Hammond, from detachments of horse arty. formerly in the service of the Lahore Darbar. Originally styled No. 2 Horse Light Field Battery, Punjab Irregular Force, but more generally known as No. 2 Punjab Lt. Fld. Batty. Became No. 2 Horse Lt. Fld. Batty., Punjab Frontier Force, 1865; No. 1 Mtn. Batty., P. F. F., 1876; No. 1 (Kohat) Mtn. Batty., P. F. F., 1879; the Kohat Mtn. Batty., 1901; present designation, 1903.

"Peiwar Kotal" "Kabul 1879" "Afghanistan, 1878-80" "Tirah" "Punjab Frontier."

1st Commn. or date of entering service.	Names and Rank.	Army rank.	Present appointment in Batty.	Remarks.
	Commandant.			
31 July 02	Garrett, Maj. R. C.	1 May 17	29 Aug. 17	
	Captain. (1)			
19 July 12	Mackenzie, Capt. H. H.	8 Aug. 16	22 Oct. 16	
	Subalterns. (3)			
7 Oct. 15	Ross, Lt. (*Act'g.* Capt.) R. E. (R. G. A., T.F.)	1 June 16	15 Dec. 17	
7 Nov. 15	Wells, Lt. (*Tempy.*) R. H., R. A.	1 Aug. 16	10 Feb. 16	
22 Apl. 15	Armitage, Lt. H. M. A., R. G. A.	1 July 17	11 Sep. 15	
27 Oct. 15	Gayer, Lt. W. M., R. A.	1 July 17	. . . .	
30 Nov. 15	Tennant, Lt. W. R. (Spl. Res.).	1 July 17	16 Nov. 17	
	Attached.	Jemadar.	Subadar.	Subadar-Major.
	Subadar.			
21 Jan. 93	Chanda, I.O.M.	16 Apl. 08	2 Nov. 15	
	Jemadars.			
26 Nov. 98	Khan Beg	16 May 12		
8 July 99	Puran Singh	30 Aug. 15		
26 Nov. 98	Raja	2 Nov. 15		

22nd Derajat Mountain Battery (Frontier Force).

Raised at Dera Ghazi Khan in 1849, by Lt. D. McNeill, as a field battery, from a disbanded troop of horse arty. formerly in the service of the Lahore Darbar. Became No. 3 Horse Light Field Battery, Punjab Irregular Force, 1851, but was more generally known as No. 3 Punjab Light Field Battery; became No. 3 Horse Light Field Battery, Punjab Frontier Force, in 1865; No. 2 Mtn. Batty., P. F. F., 1876; No. 2 (Derajat) Mtn. Batty., P. F. F. in 1879; the Derajat Mtn. Batty., 1901; present designation, 1903.

"Charasiah" "Kabul, 1879" "Kandahar, 1880" "Afghanistan, 1878-80"

"Chitral" "Tirah" "Punjab Frontier."

1st Commn. or date of entering service.	Names and Rank.	Army rank.	Present appointment in Batty.	Remarks.
	Commandant.			
23 Dec. 10	Stevenson, Capt. (*act'g Maj.*) J. H. M., R. A.	8 Aug. 16	18 Jan. 18	
	Captain. (1)			
	Subalterns. (3)			
13 Nov. 14	Davies, Lt. O. G., R. A. . . .	1 July 17	13 Dec. 17	
30 Apl. 15	Edwards, Lt. D.	1 July 17	30 Nov. 17	
17 Apl. 17	Macgregor, Lt. A., I.A.R.O.. .	17 Apl. 18	21 Apl. 18	
20 June 17	Ogilvie, Lt. W. N., I.A.R.O.	20 June 18	22 Apl. 18	
12 Jan. 17	Gulliver, Lt. J., R.A.	11 July 18	20 May 18	
	ttached.			
		Jemadar.	Subadar.	
	Subadars.			
29 Nov. 95	Santa Singh, *Bahadur* . . .	18 Feb. 15	30 Oct. 15	Supy. *Order of Br. Ind., 2nd Cl.*
	Jemadars.			
26 Feb. 03	Nur Hasan	1 July 16		

23rd Peshawar Mountain Battery (Frontier Force).

Formed at Peshawar in 1853, by Capt. T. Brougham, as the Peshawar Mtn. Train Batty., manned by the 2nd Co., 2nd Battn., Bengal Arty. (now represented by No. 9 Mtn. Batty., Royal Garrn. Arty.). Reconstituted, 1854, as a Native Mtn. Batty. Became the Peshawar Mtn. Train Batty., Punjab Irregular Force, 1862; the Peshawar Mtn. Batty., Punjab Frontier Force, 1865; No. 3 (Peshawar) Mtn. Batty., P. F. F., 1876; the Peshawar Mtn. Batty, 1901; present designation, 1903.

"Afghanistan, 1878-79."

Date of entering service.	Names and Rank.	Army rank.	Present appointment in Battery.	Remarks.
	Commandant.			
29 July 04	Landon, Capt. (*act'g. Maj.*) R. P., M.C., R.A.	30 Oct. 14	5 June 18	Comdg.
	Captain. (1)			
	Subalterns. (3)			
5 Feb. 17	Bensly, Lt. (*act'g. Capt.*) H. F., I. A. R. O.	5 Feb. 18	6 Nov. 17	
14 Sep. 16	Stofford, Lt. *the Hon.* C. W., R.G.A.	14 Mar. 18	4 Dec. 18	
28 May 17	Jenkins, 2nd-Lt., A. J., R.G.A.	28 May 17	21 Oct. 18	
5 Nov. 17	Wilson, Lt. A. K., I.A.R.O.	5 Nov. 18	. . .	
2 Dec. 17	Frost, 2nd-Lt. A. W., R.G.A.	2 Dec. 17	16 Oct. 18	
6 Jan. 18	Plumer, 2nd-Lt. W. A., R.G.A.	6 Jan. 18	22 Nov. 18	
		Jemadar.	Subadar.	Subdr.-Major.
	Subadar-Major.			
	Subadar.			
22 Mar. 97	Kehr Singh	1 Dec. 07	16 Feb. 16	
20 Apl. 97	*Ghulam Jilani (58)*	*19 May 17*	1 Sep. 18	*Shiras.*
1 June 98	*Bagga Singh*	*1 Aug. 17*	1 Sep. 18	
	Jemadars.			
1 Sept. 99	Sajawal Khan	1 June 18		
27 Oct. 00	Ranjit Singh	1 June 18		
16 May 98	Muhammad Din	13 July 18		
. . . .	Bal Singh	19 Aug. 18		
11 June 99	Harnam Singh	1 Sep. 18		

24th Hazara Mountain Battery (Frontier Force).

Formed at Haripur Hazara, in 1851, by Lt. G. G. Pearse, Madras Arty., from a small body of Hazara gunners raised by Maj. Abbott, for local defence, 1848-49, and originally styled the Hazara Mountain Train. Became the Hazara Mtn. Train Batty., Punjab Irregular Force, 1856; the Hazara Mtn. Batty., Punjab Frontier Force, 1865, No. 4 (Hazara) Mtn. Batty., P. F. F., 1876; the Hazara Mtn. Batty., 1901; present designation, 1903.

"Ali Masjid" "Kabul, 1879" "Afghanistan, 1878-80" "Burma, 1885-87" "Chitral."

Date of entering service.	Names and Rank.	Army rank.	Present appointment Battery.	Remarks.
	Commandant.			
20 Dec. 12	Hebbert, Capt. (*act'g. Maj.*) G. K. P., R.A.	20 Dec. 16	1 Oct. 18	
	Captain. (1)			
8 Mar. 15	Horman, Lieut. (*act'g. capt.*) J. C. L., R.F.A. (T.F.).	1 June 16	1 Oct. 18	
	Subalterns. (3)			
23 Dec. 14	Walden, Lt. E. A. (*temp.*)	1 July 17	15 Nov. 17	
5 Oct. 17	Hume, Lt. R., I.A. (T. C.)	5 Oct. 18	20 Mar. 18	
10 Aug. 14	Moss, 2nd-Lt. D., R.G.A., (T.F.)	25 June 17	28 Sep. 18	
	Attached.			
22 Oct. 15	Cooper, 2nd-Lt. S. A., R.G.A. (S.R.).	2 Sep. 17	5 Oct. 18	
16 Oct. 14	Fiender, 2nd-Lt. C. M., R.G.A. (T).	24 Dec. 17	29 Sep. 18	
1 Feb. 17	Lightbody, 2nd-Lt. W. A.	18 May 17	1 Dec. 18	
		Jemadar.	Subadar.	Subadar-Major.
	Subadar-Major.			
6 Sep. 83	Gurditt Singh	1 Nov. 07	17 Apl. 10	1 Apl. 18
	Jemadars.			
24 Sep. 97	Fazlkhan	1 Apl. 15		
30 Sep. 98	Ghulam Muhammad	1 June 15		
5 Mar. 95	Jalal Singh	11 Aug. 16		

25th, 26th, 27th, 28th, 29th and 30th Batts. are linked.

25th Mountain Battery.

Raised prior to 1836 as the 8th Company of the Golandaz Battn., Bombay Foot Arty. Became the 1st Co., 4th Battn., Bomban Foot Arty., 1846; No. 1 Co., Bombay Golandaz Battn., 1862; No. 1 Co., Bombay Nat. Arty. 1863; No. 1 Bombay Mtn. Batty., 1876; No. 5 (Bombay) Mtn. Batty., 1889; the Quetta Mtn. Batty., 1901 present designation, 1903.

"Mooltan" "Punjaub" "Abyssinia" "Burma, 1885-87" "Tirah" "Punjab Frontier."

1st Commn. or date of entering service.	Names and Rank.	Army rank.	Present appointment in Battery.	Remarks.
	Commandant.			
15 July 03	Jarrett, Maj. A. F. V., M.C., R.A.	31 Dec. 17	6 Aug. 17	
	Captain. (1)			
21 Apl. 17	Watt, Capt. W. D. B., I. A. R. O.	. . .	1 Oct. 17	
	Subalterns. (3)			
19 Jan. 16	Hawes, Lt. E. J., R.F.A. (Spl. Res.)	19 July 17	2 May 18	
1 Dec. 16	Riches, 2nd-Lt. G. H., R.A.	1 Dec. 16	30 Nov. 16	Attd. 23rd Mtn. Batty.
30 Sept. 14	Halder, Lt. K., R. F. A. (T. F.)	1 June 16	1 Sep. 18	
18 Aug. 17	Daniel, 2nd-Lt. A. R., I. A. R. O.	18 Aug. 17	10 Aug. 18	
	Jemadar. Subadar.			
	Subadar.			
28 Mar. 97	Kehar Singh	1 Dec. 07	16 Feb. 16	Attd. 23rd Mtn. Batty.
23 Dec. 00	Kishen Singh	24 Mar. 10	22 Jan. 18	
	Jemadars.			
25 Oct. 08	Lahora Singh	16 Feb. 16		
19 Dec. 98	Nur Muhammad	9 Aug. 17		Attd. 37 Mtn. Batty.
. . .	*Pal Singh*	*4 Jan. 18*		Attd. 33 Mtn. Batty.
. . .	*Muhammad Khan*	*4 Jan. 18*		
. . .	Fateh Khan	23 Aug. 18		

26th Jacob's Mountain Battery.

Raised in 1843, as the 10th Co. of the Golandaz Battn., Bombay Foot Arty. Became the 3rd Co. 4th Battalion Bombay Foot Arty., 1846; No. 3 Co., Bombay Golandaz Battn., 1862; No. 2 Co., Bombay Nat. Arty., 1863, No. 2 Bombay Mtn. Batty., 1876; No. 6 (Bombay) Mtn. Batty., 1889; the Jullundur Mtn. Batty, 1901; present designation, 1903.

"Afghanistan, 1878-80."

	Commandant.			
21 July 01	Rossiter, Capt. (*act'g. Maj.*) F. N. C., M.C., R.A.	23 July 15	. . . .	
	Captain. (1)			
18 July 13	Pemberton, Capt. A. L., R.A.	18 July 17	19 Jan. 16	
10 Feb. 15	Thuillier, Capt. H. S., R.A.	3 Nov. 17	14 Jan. 16	
	Subalterns. (3)			
15 May 15	Jones, Lt. P. H., R.A.	1 July 17	6 Jan. 17	
1 Dec. 16	Claxton, 2nd-Lt. E.	1 Dec. 16	23 Nov. 17	
	Jemadar. Subadar.			
	Subadar.			
13 May 92	Hem Singh, I.D.S.M.	18 Feb. 15	1 May 15	
	Jemadars.			
. . .	Muhammad Bakhsh, I.O.M.	8 May 15		
. . .	Ghulam Hasan, I.D.S.M.	18 July 15		

27th Mountain Battery.

Raised at Rawal Pindi in 1886, by Capt. C. P. Triscott, as No. 1 Bengal Mtn. Batty. Became No. 7 (Bengal) Mtn. Batty. in 1889; the Gujrat Mountain Batty. in 1901; present designation in 1903.

"Burma, 1885—87."

1st Commn. or date of entering service.	Names and Rank.	Army rank.	Present appointment in Battery.	Remarks.
	Commandant.			
23 Dec. 11	Carew, Capt. R. L., R.A.	8 Aug. 16	1 Dec. 18	
	Captain. (1)			
12 Jan. 15	Curry, Capt. M. J., R.A.	3 Nov. 17	1 Oct. 18	
	Subalterns. (3)			
5 May 15	Ponsonby, Lt. H. P., R. A. (T.F.)	1 June 16	25 Aug. 18	
30 June 18	Grant, 2nd-Lt. C. D., R.A.	30 June 18	25 Feb. 18	
4 Nov. 16	Foddy, Lt. A. H., I.A.R.O.	4 Nov. 17	30 Sep. 18	
30 July 14	Elliott, 2nd-Lt. W., R.A. (T.F.)	30 July 17	14 Oct. 18	
1 Dec. 14	Butt, 2nd-Lt. A. W., R.A.	16 Sep. 17	13 Oct. 18	
15 Jan. 18	Craven, 2nd-Lt. L., I.A.	15 Jan. 18	8 Oct. 18	
30 Oct. 18	Mayhew, 2nd-Lt. A. G., I.A.R.O.	30 Oct. 18	30 Nov. 18	
31 Jan. 16	Henderson, 2nd-Lt. J. S. W., R.A	21 Jan. 16	30 Nov. 18	

Attached.

1st Commn.	Names and Rank.	Jemadar.	Subadar.	Sub-Maj.	Remarks.
	Subadar-Major.				
15 July 97	Sher Baz, I.O.M. *Bahadur*,	6 May 02	1 Dec. 10	1 Sep. 18.	*Order of Br. Ind., 2nd cl.*

Subadar.

Jemadars.

1st Commn.	Names and Rank.	Army rank.
14 Jan. 01	Bhan Singh, I.O.M.	1 Dec. 16
....	Shah Nawaz	18 July 17
....	Haidar Khan	19 Aug. 18
....	Natha Singh	19 Aug. 18
1 July 03	Kehar Singh	1 Oct. 18

28th Mountain Battery.

Raised at Meean Meer in 1886, by Capt. E. A. Smith, as No. 2 Bengal Mtn. Batty. Became No. 8 (Bengal) Mtn. Batty. in 1889; the Lahore Mountain Batty. in 1901; present designation in 1903.

"Burma, 1885—87" "Malakand" "Punjab Frontier."

1st Commn.	Names and Rank.	Army rank.	Present appointment in Battery.
	Commandant.		
19 Dec. 00	Davies, Maj. L. M., R.A.	30 Dec. 15	7 Mar. 16
	Captain.		
19 July 12	Collins, S. St. B., R. A.	18 Aug. 16	11 Apl. 18
	Subalterns. (3)		
6 July 15	Bowden, Lt. C. A., R.F.A. (S.R.)	1 July 17	21 Aug. 17
20 Dec. 16	Stonehouse, Lt. A. D., R.G.A. (T.F.).	20 June 18	12 Oct. 18
1 Nov. 17	Carter, Lt. L.H., I.A. (T.C.)	1 Nov. 18	15 Mar. 18
2 Dec. 17	Murie, 2nd-Lt. J. C. W., R. F. A. (T. F.)	2 Dec. 17	30 June 18
27 Nov. 16	Pilley, 2nd-Lt. F. C., R.G.A. (S.R.)	27 Nov. 16	17 Oct. 18
18 Sep. 18	Whelan, 2nd-Lt. J., I.A.R.O.	18 Sep. 18	15 Oct. 18

1st Commn.	Names and Rank.	Jemadar.	Subadar.
	Subadar-Major.		
	Subadars.		
11 Oct. 90	Lal Din	1 Apl. 08	15 July 17
	Jemadars.		
1 Feb. 97	Sant Singh, I.D.S.M.	29 Ma[illegible] 17	
6 June 98	Hayat Ali	19 Aug. 18	

29th Mountain Battery.

Raised at Abbottabad in 1898-99, by Capt. L. W. P. East, as No. 9 (Native) Mtn. Batty. Became the Murree Mountain Batty. in 1901; present designation in 1903.

1st Commn. or date of entering service.	Names and Rank.	Army rank.	Present appointment in Battery.	Remarks.
	Commandant.			
24 Dec. 02	Somerville, Maj. J. A. H. B.	13 July 17	28 Apl. 17	
	Captain. (1)			
10 Oct. 14	Rivers-Smith, Lt. P. E. (*temp.*) R. A.	3 Jan. 18	29 Sep. 16	
	Subalterns. (3)			
14 Dec. 15	Shaw, Lt. E. (Spl. Res.)	1 July 17	17 Apl. 18	
12 May 17	Sweeney, Lt. S., I. A. R. O.	12 May 18	1 June 17	
6 Sep. 17	Sargent, Lt. H. H., I. A. R. O.	6 Sep. 18	23 Sep. 17	
3 Feb. 18	Kirby, 2nd-Lt. J. N., R. A.	3 Feb. 18	3 Feb. 18	
		Jemadar.	Subadar.	
	Subadars.			
10 Apl. 92	Jiwan Khan	16 Sep. 09	14 Aug. 16	
	Jemadars.			
3 July 93	Ghunna Singh	18 Feb. 15		
6 Mar. 99	*Mehr Khan*	*23 July 16*		I. M. A. Depot.
4 Mar. 99	Bahadur Ali	14 Aug. 16		
24 Apl. 99	Indar Singh	23 Oct. 17		

30th Mountain Battery.

Raised at Abbottabad in 1900-01, by Capt. E. E. Edlmann, D.S.O., as No. 10 (Native) Mtn. Batty. Became the Abbottabad Mountain Batty. in 1901; present designation in 1903.

1st Commn. or date of entering service.	Names and Rank.	Army rank.	Present appointment in Battery.	Remarks.
	Commandant.			
21 Dec. 00	Little, Major E. M., R. A.	30 Dec. 15	11 Jan. 18	
	Captain. (1)			
12 Jan. 15	Buchanan, Capt. A. W., R.A.	3 Nov. 17	17 Oct. 18	
	Subalterns. (3)			
2 Feb. 17	Batty, Lt. A. D. G. S., I.A.R.O.	2 Feb. 17	16 Feb. 17	
12 Aug. 17	McIntyre, 2nd-Lt J., R.F.A. (T.F.)	12 Aug. 17	8 June 18	
1 Nov. 17	Martin, Lt. J. H. A., I. A. (*T. C.*)	1 Nov. 18	20 Sep. 18	
7 July 17	King, 2nd-Lt. A. H., R.G.A.	7 July 17	12 Oct. 18	
5 Jan. 18	Collett, 2nd-Lt. W. A., I.A. (*temp.*)	5 Jan. 18	4 Oct. 18	
12 May 18	Cameron, 2nd-Lt. A. P., R.G.A. (*temp.*)	12 Aug. 18	12 Oct. 18	
		Jemadar.	Subadar.	
	Subadar-Major.			
	Subadar.			
17 Feb. 01	Daya Singh, I.D.S.M.	16 Nov. 13	5 Aug. 17	
	Jemadars.			
....	Ali Akbar	19 Aug. 18		
....	Sham Singh	11 Oct. 18		

31st Mountain Battery.

Raised at Dehra Dun in 1907 by Major J. L. Parker, R.A.

1st Commn. or date of entering service.	Names and Rank.	Army rank.	Present appointment in Batty.	Remarks.
	Commandant.			
28 Mar. 00	Perry, Maj. S., R.A.	30 Dec. 15	25 July 17	
	Captain. (1)			
19 July 12	Deane, Capt. J. A. L., R.A. . .	8 Aug. 16	31 Jan. 17	
	Subalterns. (3)			
1 Mar. 17	Moncrieff, Lt. R. L., I.A.R.O. .	1 Mar. 18	23 Mar. 18	
27 Feb. 15	Wilkinson, Lt. T. H. A. (*temp.*), R. A.	1 July 17	14 Aug. 18	
3 Nov. 15	Fyffe, 2nd-Lt. F. C.	3 Nov. 15	29 Mar. 17	
25 Mar. 17	Saxby, 2nd-Lt. W.	25 Mar. 17	25 Mar. 17	
		Jemadar.	Subadar.	Subadar-Major.
	Subadar-Major.			
	Subadars.			
8 June 92	Khem Singh, *Bahadur*. . . .	15 Nov. 07	3 Sep. 14	*Order of Br. Ind., 2nd Cl.*
9 May 92	Mihar Khan	16 Oct. 14	4 Aug. 17	
16 Aug. 95	Prem Singh	10 Sep. 15	12 Aug. 18	
	Jemadars.			
14 Mar. 01	Sundar Singh	4 Aug. 17		
17 Oct. 07	Lal Khan	12 Aug. 18		

32nd Mountain Battery.

Raised at Dehra Dun in 1907 by Maj. O. C. Williamson, R.A.

	Commandant.			
23 Dec. 10	Troup, Capt. (*act'g. Maj.*) F. C. A., R.A.	8 Aug. 16	1 Sep. 17	
	Captain. (1)			
	Subalterns. (3)			
29 Apl. 15	Lee, Lt. (*act'g. Capt.*) D. J. N., R.A.	1 July 17	9 Sep. 16	
4 Jan. 17	Hawes, Lt. W., R. G. A. . . .	4 July 18	24 Apl. 18	
5 Nov. 17	Parsons, Lt. L. J., I.A. (*T.C.*). .	5 Nov. 18	. . . .	
. . . .	Whyte, 2nd-Lt. J.	. . . .	. . . .	
		Jemadar.	Subadar.	Subadar-Major.
	Subadar-Major.			
	Subadar.			
2 Nov. 99	Hari Singh †		19 Apl. 18	
	Jemadars.			
4 Apl. 94	Harnam Singh	31 Aug. 13		
1 Dec. 00	Gul Muaz	17 Jan. 15		With 26 M. B.
1 Oct. 07	Kifayat Allah	1 Aug. 17		
1 July 99	Bara Singh	19 Apl. 18		

† Acting rank.

The Frontier Garrison Artillery.

Raised at Bannu in 1851, as a Garrison Company, by Lt. S. W. Stokes, Bengal Horse Arty., from the supernumeraries of a Sikh detachment of artillery which had been broken up and drafted into the horse light field batteries. Became No. 4 or Garrison Co. of Artillery, Punjab Irregular Force, 1852; No. 4 or Garrison Battery, Punjab Frontier Force, 1876; No. 5 Garr. Batty., P. F. F., 1876; the Punjab Garr. Batty., 1889; present designation, 1901.

1st Commn. or date of entering service.	Names and Rank.	Army rank.	Present appointment in Battery.	Remarks.
	Commandant.			
12 Jan. 15	Taylor, Capt. J., R.A.	3 Nov. 17	21 Aug. 17	
	Subalterns. (2)			
20 Dec. 16	Warner, 2nd-Lt. J., R.A.	20 Dec. 16	16 Aug. 17	Malakand.
13 Mar. 17	Cooper, 2nd-Lt. W.	13 Mar. 17	13 Aug. 17	
		Jemadar.	Subadar.	
	Subadar.			
12 Nov. 96	Amir Singh	4 Sep. 09	29 Oct. 17	
	Jemadars.			
1 Aug. 88	Rahim Ali	1 Nov. 12		Supernumerary.
12 Dec. 00	Mani Ram	14 July 17		
19 Apl. 01	Atar Singh	24 Oct. 17		
1 Oct. 97	Lall Khan	29 Oct. 17		

33rd (Reserve) Mountain Battery.

Formed at Abbottabad in March 1917 from men at Indian Mountain Artillery Depôt by Major J. A. H. B. Somerville, R.A.

1st Commn. or date of entering service.	Names and Rank.	Army rank.	Present appointment in Battery.	Remarks.
	Commandant.			
15 July 03	Wilson, Maj. R.E., D.S.O., R.A.	14 Nov. 17	24 Apl. 17	
	Captain. (1)			
12 Jan. 15	Pirie, Capt. J. P., R.G.A.	3 Nov. 17	25 Oct. 18	
	Subalterns. (3)			
12 Feb. 15	Hollands, Lt. E. W., I. A. R. O.	12 Feb. 16	6 Aug. 18	
6 Oct. 15	Wilson, Lieut. J. C. J., R. G. A.	1 July 17	8 Oct. 18	
23 Feb. 15	Hilbourne, Lieut. G. M., R.G.A.	25 Jan. 18	26 Aug. 18	
1 Dec. 16	O'Day, 2nd-Lt. T., R. A.	1 Dec. 16	8 Mar. 17	
30 June 17	Riches, 2nd-Lt. R. J. B., R.G.A.	30 June 17	17 Oct. 18	
15 Jan. 18	Bugler, 2nd-Lt. J., I.A. (T.C.)	15 Jan. 18	14 Oct. 18	
29 Mar. 18	Adams, 2nd-Lt. F. A., I.A. (T.C.)	29 Mar. 18	2 Oct. 18	
		Jemadar.	Subadar.	
	Subadar.			
2 Nov. 94	Lal Khan	18 Feb. 15	27 Mar. 18	
	Jemadar.			
11 Feb. 02	Lal Beg	27 May 18		
	Madad Khan	19 Aug. 18		

34th (Reserve) Mountain Battery.

Raised by Major E. R. C. Wilson, M.C., R. A., at Abbottabad on 8th July 1917.

1st Commn. or date of entering service.	Names and Rank.	Army rank.	Present appointment in Battery.	Remarks.
	Commandant.			
24 May 02	Wilson, Maj. E. R. C., M.C., R.A.	1 May 17	23 July 17	
	Captain.			
12 Apl. 16	Sims, Lieut. R. J.	12 Oct. 17	9 Aug. 18	
	Subalterns.			
14 Oct. 16	Walker, 2nd-Lt. F. Mc F., R.A.	14 Oct. 16	30 July 17	
20 June 17	Dodds, Lt. B. W. P., I. A. R. O.	20 June 18	6 Oct. 17	
5 Nov. 17	Fullerton, Lt. L. W., I. A. (*T. C.*)	5 Nov. 18	15 Mar. 18	
		Jemadar.	Subadar.	
	Subadar.			
1 Feb. 98	Haidar Khan	7 Dec. 15	1 Apl. 18	
	Jemadars.			
17 Dec. 00	Hakum Singh	8 Sep. 16		
	Fateh Din	1 Apl. 18		

35th (Reserve) Mountain Battery.

Raised by Major R. M. L. Dutton, M. C., R.A., at Abbottabad in July 1917.

1st Commn. or date of entering service.	Names and Rank.	Army rank.	Present appointment in Battery.	Remarks.
	Commandant.			
31 July 02	Dutton, Maj. R. M. L., M.C., R.A.	1 May 17	23 July 17	
	Captain.			
23 July 10	Heaton-Ellis, J. S., R. A.	23 July 16	29 Mar. 18	
	Subalterns.			
2 May 16	Holliday, 2nd-Lt. J. M., R. A.	2 May 16	14 May 18	
23 June 17	Good, 2nd-Lt. O.N.H., R.A., I. A. R. O. (*on probn.*)	23 June 17	27 Aug. 17	
5 Nov. 17	Ford, Lt. R. A., I. A. (*T. C.*)	5 Nov. 18	16 Mar. 18	
5 Nov. 17	Sawyer, Lt. P. S., I. A. R. O.	5 Nov. 18	....	
		Jemadar.	Subadar.	
	Subadar.			
24 Dec. 01	Abas Khan	3 Sep. 14	10 Sep. 15	
	Jemadars.			
19 Feb. 96	Basant Singh	18 Feb. 15		
22 Apl. 01	Kehr Singh	5 Aug. 17		

39th (Reserve) Mountain Battery.

Raised at Maymyo, Burma in 1918 by Capt. F. C. A. Troup, R. A.

1st Commn. or date of entering service.	Names and Rank.	Army rank.	Present appointment in Battery.	Remarks.
	Commandant.			
18 Aug. 00	Lindsay, Major W. D., R. A. . .	30 Dec. 15	8 May 18	
	Captain.			
20 Dec. 12	Macgregor, Capt. A., R. A. . .	20 Dec. 16	22 Apl. 18	
	Subal terns.			
	Rouquette, Lt. L. P., (T. F.) . .	1 June 16	24 Apl. 18	
31 Aug. 16	Shaw, Lt. D. H., (T. F.)	1 Mar. 18	18 Apl. 18	
. . . .	Nicholson, 2nd-Lt. S., R. A. (S. R.)	. . . .	. . . . .	
		Jemadar.	Subadar.	Subadar-Major.
	Subadar-Major.			
9 Nov. 86	Muhammed Ismail, *Sardar Bahadur*, I. O. M.	22 Oct. 00	1 Sept. 07	1st Dec. 10 — *Order of Br. India, 1st class.*
	Subadar.			
	Jemadars.			
5 Jan. 03	Fateh Ali	19 Apl. 18		
27 Apl. 16	Pat Ram Singh	19 Apl. 18		

SAPPERS AND MINERS.

1st King George's Own Sappers and Miners.

Raised at Cawnpore in 1803, by Captain T. Wood, as the Bengal Pioneers, whereof 2 complete companies besides a large body of selected men from the other companies, were taken in 1819 to form the Corps of Bengal Sappers and Miners raised that year at Allahabad by Major Anburey, and into which the remainder of the Corps of Pioneers was absorbed in 1833. The Corps of Sappers and Miners thus formed became the Corps of Bengal Sappers and Pioneers in 1847; The Corps of Bengal Sappers and Miners, 1851; the 1st Sappers and Miners in 1903; 1st Prince of Wales's Own Sappers and Miners, 1906; present designation, 1910.

The Plume of the Prince of Wales. The Royal and Imperial cypher.

"Bhurtpore" "Cabool, 1842" "Ferozeshah" "Sobraon" "Mooltan" "Goojerat" "Punjaub" "Delhi, 1857" "Lucknow" "Ali Masjid" "Charasiah" "Kabul, 1879" "Ahmad Khel" "Afghanistan, 1878-80" "Burma, 1885-87" "Chitral" "Tirah" "Punjab Frontier" "China, 1900."

***Composition*—Sikhs, Hindustanis, Punjabi Musalmans, and Pathans.**

No. 1 Company, ½ Sikhs, ½ Punjabi Musalmans.
No. 2 Company, ½ Sikhs, ½ Punjabi Musalmans.
No. 3 Company, ½ Hindustani Hindus, ½ Punjabi Musalmans and Hindustani Musalmans.
No. 4 Company, ½ Sikhs, ½ Pathans.
No. 5 Company, ½ Sikhs, ½ Punjabi Musalmans.
No. 6 Company, ½ Hindustani Hindus, ½ *Punjab* Musalmans and Hindustani Musalmans.
B Company, mixed.
Field Troop—Mussulmans.
Karachi & Calcutta Defence Light Sections—Punjabi Musalmans.

***Uniform*—Scarlet. *Facings*—Blue.**

Colonel-in-Chief—THE KING.
Colonel—General Sir Bindon Blood, G.C.B., ret. pay 6 Mar. 18.

1st Commn. or date of entering service.	Names and Rank.	Army rank.	Present appointment in Corps.	Remarks.
	Commandant.			
	Staff Officers. (4)			
7 July 89	Cunningham, Lt.-Col. A. H., R. E.	1 June 17	20 Nov. 16	Comdt. (*tempy.*)
2 Feb. 92	*Tylden-Patterson, Bt.-Lt.-Col. (Act'g. Lt.-Col.), E.C.,* D.S.O. *R.E.*	*6 June 12*	*1 May 12*	*C. R. E.*
25 Sep. 99	Hill, Maj. R. C. R., D.S.O., R. E.	1 Jan. 16	1 Jan. 18	Supt. of Instn.
21 Dec. 04	Bird, Bt.-Maj. C. A., D.S.O., R. E.	1 Jan. 18	1 Dec. 17	Adjutant.
5 Aug. 06	Douie, Bt.-Maj. (*act'g. Maj.*) F. McC., D.S.O., M.C., R.E.	3 June 17	22 Feb. 18	Supdt. of Park.
	Company Commanders.			
23 Dec. 98	*Hill, Bt.-Lt.-Col. (act'g. Lt.-Col.) E. F. J.,* M.C., *R.E.*	*3 June 17*	*29 May 07*	*C. R. E.*
21 Dec. 00	*Bird, Bt.-Lt.-Col. (act'g. Lt.-Col.) A. J. G.,* D.S.O., *R. E.*	*1 Jan. 18*	*26 April 09*	*C. R. E.*
23 July 01	*LeBreton, Maj. (act'g. Lt.-Col.), E. P., R. E.*	*2 Nov. 16*	*17 May 10*	*C.R.E.*
31 July 02	*Kelly, Maj. (act'g Lt.-Col.) E. H.,* D.S.O., M.C., *R.E.*	*31 July 17*	*1 Nov. 12*	*G. S. O., 3rd gde.*
15 July 03	*Bond, Maj. L. V., R.E.*	*1 Jan. 17*	*27 Apl. 13*	*With Sig. Serv.*
29 July 04	*Hobart, Bt.-Maj. P. C. S.,* D.S.O., M.C., *R. E.*	*1 Jan. 18*	*1 Nov. 12*	*G. S. O., 2nd gde.*
5 Sep. 05	*Collin, Bt.-Maj. F. S., R. E.*	*1 Jan. 18*	*1 Nov. 13*	*R. M. Coll., Sandhurst.*
17 Sep. 05	Cave-Browne, Capt. (*act'g. Maj.*) W., M.C., R. E.	30 Oct. 14	. . .	No. 3 Coy.
23 July 07	Witts, Capt. F. V. B., D.S.O., M.C., R.E.	30 Oct. 14	22 May 13	No. 2 Bridging Tn.
23 July 07	Chater, Capt. (*act'g. Maj.*) A. F., R. E.	30 Oct. 14	2 Jan. 13	No. 5 Coy.
18 Dec. 07	Roome, Capt. (*act'g. Maj.*) H.E., M.C., R. E.	30 Oct. 14	16 Oct. 17	No. 52 Coy.
29 July 08	Gidley-Kitchin, Capt. E. G., R.E.	30 Oct. 14	30 Sep. 14	No. 5 Fd. Troop.
23 Jan. 10	Wheeler, Capt. (*act'g. Maj.*) E.O., M.C., R.E.	23 June 16	16 Aug. 14	No. 8 Coy.
23 Dec. 10	Guinness, Capt. (*act'g. Maj.*) V.E.G., R.E.	23 Dec. 16	6 Nov. 16	No. 6 Coy.
20 July 11	Watson, Capt. G. D., R. E.	26 June 17	1 July 16	No. 7 Coy.
20 July 11	Mason, Capt. A., M. C., R. E.	26 June 17	4 Nov. 17	No. 53 Coy.
23 Dec. 11	Drew, Capt. (*act'g Maj.*) F. G., R.E.	26 June 17	26 Feb. 16	No. 2 Coy.
23 Dec. 09	Farley, Capt. (*act'g. Maj.*) E.L., M. C., R. E.	23 Dec. 15	15 Aug. 14	No. 4 Coy.
21 June 14	Joly de Lotbiniere, Capt. (*act'g. Maj.*) H. A., R. E.	21 June 17	2 May 15	No. 1 Coy.
24 Jan. 14	Boal, Lieut. (*act'g. Maj.*) R., D.S.O. R.E.	23 July 16	24 Jan. 15	No. 51 Coy.
20 Nov. 1.	Lyall, Lieut. (*act'g. Maj.*) W. J., R. E.	1 July 17	20 Nov. 15	No. 54 Coy.

1st King George's Own Sappers and Miners—*contd.*

1st Commn. or date of entering service.	Names and Rank.	Army rank.	Present appointment in Corps.	Remarks.
	Company Officers.			
10 Feb. 15	Watson, Capt. R. A. D., R. E.	27 Mar. 18	2 Sep. 18	No. 56 Coy.
10 Feb. 15	Pim, Lieut. G. R., M. C., R. E.	11 May 17	16 Feb. 18	No 1 Fd. Troop.
22 Apl. 15	Wishart, Lieut. A. J. R., M.C., R. E.	1 July 17	6 Nov. 18	No. 58 Coy.
28 July 15	Lethbridge, Lt. J. S., R. E.	1 July 17	6 June 17	No. 60 Coy. (temp. Coy. Comdr.)
28 July 15	Crane, Lt. E. A., M.C., R. E.	1 July 17	8 Aug. 18	No. 53 Coy.
27 Oct. 15	Dawes, Lt. L. F. S., R. E.	1 July 17	16 Feb. 18	Leave to England M. C. (6 months).
27 Oct. 15	Burrows, Lt. C. H. C., R. E.	1 July 17	16 Feb. 18	No. 6 Coy.
19 Feb. 16	Obbard, Lieut. H. N., R. E.	19 Aug. 17	27 Feb. 18	No. 52 Coy.
19 Feb. 16	Bird, Lieut. C. F. A., R. E.	19 Aug. 17	16 Feb. 18	No. 56 Coy.
19 Feb. 16	Cameron, Lieut. A. M., M.C., R.E.	19 Aug. 17	16 Feb. 18	No. 54 Coy.
25 May 16	Lockhart, Lieut. K. H., R. E.	25 Nov. 17	2 Oct. 18	No. 60 Coy.
26 Sep. 16	Clutterbuck, Lieut. J. E., R. E.	26 Feb. 18	2 Oct. 18	No. 60 Coy.
6 June 17	Price, 2nd-Lt. T. J. P., R. E.	7 June 17	16 Feb. 18	No. 7 Coy.
6 June 17	Roberts, 2nd-Lt. O. L., R. E.	7 June 17	16 Feb. 18	No. 55 Coy.
23 Sep. 17	Williams, 2nd-Lt. H., R. E.	23 Sep. 17	3 Sep. 18	51 Coy.
24 Jan. 18	DeLinde, 2nd-Lieut. C. A., R. E.	24 Jan. 18	4 Oct. 18	No. 59 Coy.
28 May 18	Dawkins, 2nd-Lt. H., R. E.	28 May 18	27 May 18	6th (Art.) S. & M.
28 May 18	Dower, 2nd-Lt. J. H., D.C.M., R. E.	28 May 18	27 May 18	No. 56 Coy.
	Attached.			
4 Sep. 99	Sirmur, *Hony* Lt.-Col. *Kunwar* Bir Bikram Singh, C.S.I.,C.I.E. of-	30 Oct. 14	16 July 95	Comdt., Sirmur I. S. Sappers.
1 Mar. 15	Martin, Capt. J. M., (*temp.*) R. E.	13 Mar. 15	22 Jan. 18	B. Coy. Lv., ex In. 12 mos., m. c.
28 Aug. 15	Hobson, Capt. J. R., (*temp.*) R. E.	18 Sep. 17	6 Nov. 18	B. Coy.
23 Jan. 15	Harrington, Capt. F., (*temp.*) R. E.	9 Feb. 18	5 Oct. 18	B. Coy.
16 Nov. 14	Wrightson, Capt. N. P. E., (*temp.*) R. E.	1 July 18	26 Nov. 17	B. Coy.
30 Jan. 15	Troup, Capt. J. R., (*temp.*) R. E.	1 July 18	1 Nov. 18	B. Coy.
18 Dec. 14	Shepherd, Lt. J. D., (*temp.*) M.C., R. E.	1 Dec. 15	13 June 17	No. 8 Coy.
8 Dec. 14	Macdonald-Clark, Lt. (Act'g. Capt.) R. F., (*temp.*) R. E.	8 Dec. 14 7 May 18	17 May 8	B. Coy.
3 Oct. 15	Williams, Lt. F. S., D.S.O., (*terr.*) R. E.	1 June 16	6 Nov. 18	No. 59 Coy.
22 Apl. 15	Meikle, Lt. W. Q. R., (*temp.*) R. E.	30 Nov. 16	3 Oct. 18	B. Coy.
7 July 15	Paton, Lt. J., (*temp.*) R. E.	28 Mar. 17	25 June 17	Lv. *ex* India, 6 ms., m. c.
16 Dec. 15	Ford, Lt. G., (*temp.*) R. E.	1 July 17	3 Oct. 18	B. Coy.
26 Mar. 18	Smith, Lt. B. K., (*temp.*) R. E.	26 Mar. 18	1 Nov. 18	B. Coy.
9 Sep. 14	Baines, Lt. T. N., (*temp.*) R. E.	1 July 17 7 May 18	10 Nov. 18	B. Coy.
17 Oct. 14	Robinson, Lt. J. S., M.C., (*temp.*) R. E.	11 May 18	1 Nov. 18	No. 59 Coy.
21 May 15	Roberts, Lt. O. L., (*temp.*) R. E.	11 May 18	1 Nov. 18	B. Coy.
18 Jan. 16	Christie, Lt. J. I. T., (*temp.*) R.E.	11 May 18	6 Nov. 18	B. Coy.
16 Jan. 17	Best, Lt. A. J., (*temp.*) D.S.O., R. E.	14 July 18	5 Oct. 18	B. Coy.
4 Nov. 16	Lyon, 2nd-Lt. W. A., (*temp.*) R. E.	4 Nov. 16	20 June 17	
4 Nov. 16	Craig, 2nd-Lt. M. T., (*temp.*) R. E.	4 Nov. 16	20 June 17	No. 4 Coy.
4 Nov. 16	Baxter, 2nd-Lt. G. D., (*temp.*) R. E.	4 Nov. 16	7 July 17	No. 3 Coy.
27 Jan. 17	Augier, 2nd-Lt. E. H., (*temp.*) R. E.	27 Jan. 17	6 Jan. 18	52 Coy.
10 Mar. 17	Rowe, 2nd-Lt. G. C., (*temp.*) R. E.	10 Mar. 17	9 Nov. 18	B. Coy.
1 Apl. 17	Liddell, 2nd-Lt. W. S. (*terr.*) R. E.	1 Apl. 17	7 Nov. 18	B. Coy.
19 May 17	Davis, 2nd-Lt., R. W. J., (*temp.*) R.E.	19 May 17 3 Apl. 15	26 Nov. 17	E. E. F.
17 May 17	Williams, 2nd-Lt. H. T., (*S. Res.*) R. E.	17 May 17	6 Nov. 18	60 Coy.
16 June 17	Widdowfield, 2nd-Lt. C. S., (*terr.*) R. E.	16 June 17	7 Nov. 18	B. Coy.
1 July 17	Allan, 2nd-Lt. J. G., (*temp.*) R. E.	1 July 17	5 Oct. 18	B. Coy.
28 July 17	Pringle, 2nd-Lt. J. (*temp.*) R. E.	28 July 17	26 Nov. 17	58 Coy.
31 July 17	Hall, 2nd-Lt. H., (*temp.*) R. E.	31 July 17	16 Feb. 18	B. Coy.
11 Aug. 17	Johnston, 2nd-Lt. T. E., (*temp.*) R. E.	11 Aug. 17	29 Nov. 17	B. Coy.

1st King George's Own Sappers and Miners—*contd.*

Attached—*contd.*

1st Commn. or date of entering service.	Names and Rank.	Army rank.	Present appointment in Corps.	Remarks.
11 Aug. 17	Forster, 2nd-Lt. T., (*temp.*) R. E.	11 Aug. 17	21 Mar. 18	E. E. F.
25 Aug. 17	Kennedy, 2nd-Lt. J. S. (*temp.*) M.C., R. E.	25 Aug. 17	8 Nov. 18	B. Coy.
12 Jan. 18	Gibson, 2nd-Lt. W. G., (*temp.*) R. E.	12 Jan. 18	9 Nov. 18	B. Coy.
2 Feb. 18	Storrar, 2nd-Lt. J., (*terr.*) R. E.	2 Feb. 18	7 Nov. 18	57 Coy.
2 Feb. 18	Stewart, 2nd-Lt. N. T., (*temp.*) R. E.	2 Feb. 18	7 Nov. 18	B. Coy.
2 Feb. 18	Rodger, 2nd-Lt. R., (*temp.*) R. E.	2 Feb. 18	1 Nov. 18	1 Fd. Sqdn.
23 Feb. 18	Tatham, 2nd-Lt. W. H., (*temp.*) R. E.	23 Feb. 18	3 Nov. 18	1 Fd. Sqdn.
23 Feb. 18	Cook, 2nd-Lt. R. J., (*temp.*) R. E.	23 Feb. 18	7 Nov. 18	60 Coy.
30 Mar. 18	Beaumont, 2nd-Lt. W. R. L., (*temp.*) R. E.	30 Mar. 18	1 Nov. 18	6 Fd. Troop.
30 Mar. 18	Christian, 2nd-Lt. G. C., (*terr.*) R. E.	30 Mar. 18	7 Nov. 18	B. Coy.
27 Apl. 18	Matheson, 2nd-Lt. S. F., (*temp.*) R. E.	27 Apl. 18	7 Nov. 18	59 Coy.
1 June 18	Cohill, 2nd-Lt. J. L., (*temp.*) R. E.	1 June 18	10 Nov. 18	B. Coy.
18 Feb. 98	Proes, Maj. (*Ag. Lt.-Col.*) E. M.	1 Sep. 15	19 Oct. 14	I. A. R. O.
17 May 01	Hodges, Capt. R. W.	13 Sep. 14	19 Nov. 14	I. A. R. O.
26 Feb. 15	Berkeley, Capt. (*temp.Maj.*) A. M.	24 Oct. 16	13 Mar. 15	I. A. R. O.
4 Dec. 14	Cheshire, Capt. H. S.	4 Dec. 17	29 Dec. 14	I. A. R. O.
4 Sep. 14	Spink, Lt. H. H. M.	4 Sep. 15	19 Nov. 14	I. A. R. O.
30 Oct. 14	Menzies, Capt. F. A., M.C.	30 Oct. 18	6 Nov. 14	I. A. R. O. Staff Capt., A. Hd. Qrs.
13 Nov. 14	Hance, Capt. H. M., D.S.O., M.C.	13 Jan. 18	22 Nov. 14	I. A. R. O.
27 Nov. 14	Allan, Lt. (*Ag. Capt.*) J. C.	27 Nov. 15	9 Feb. 15	I. A. R. O. 59 Coy.
4 Dec. 14	Widdicombe, Lt. (*temp.Maj.*) E. P.	4 Dec. 15	3 Jan. 15	I. A. R. O. (Ag. Maj. R.E. B.E.F.)
11 Dec. 14	Mair, Lt. (*Ag. Capt.*) R., M.C.	11 Dec. 15	23 Jan. 17	I. A. R. O. Munitions Bd., Kirkee.
18 Feb. 15	Muir, Lt. (*Ag. Capt.*) J. L.	18 Feb. 16	26 Feb. 15	I. A. R. O. No. 1 Fd. Tp.
12 Mar. 15	Lyster, Lt. A. St. G.	12 Mar. 16	9 May 15	I. A. R. O.
12 Mar. 15	Farquharson, Lt. (*Ag. Capt.*) F. A., M.C.	12 Mar. 16	12 Mar. 15	I. A. R. O. No. 6 Fd. Tp.
21 May 15	Jones, Capt. R. T.	14 Mar. 16	25 July 15	I. A. R. O.
26 Mar. 15	Montagu, Lt. (*Ag. Capt.*) A. M. R.	26 Mar. 16	26 Mar. 15	I. A. R. O. No. 1 Bdg. Tn.
2 Apl. 15	Salusbury Trelawny, Lt. (*Ag. Capt.*) J. M., M.C.	2 Apl. 16	11 June 15	I. A. R. O. No. 1 Bdg. Tn.
2 Apl. 15	Sale, Lt. (*Ag. Capt.*) J. L.	2 Apl. 16	11 June 15	I. A. R. O. B Coy.
2 Apl. 15	Rose, Lt. (*Ag. Capt.*) G. W. N.	2 Apl. 16	11 June 15	I. A. R. O. Adjt. Dnl. Tps.
23 Apl. 15	Pollard, Lt. A. R.	23 Apl. 16	1 July 15	I. A. R. O.
30 Apl. 15	Courthorpe, E. A. Capt.	. . . .	23 Oct. 18	I. A. R. O. France, special duty.
7 May 15	Eastmond, Lt. (*Ag. Capt.*) A., M.C.	7 May 16	3 Aug. 15	I. A. R. O. No. 57 Coy.
2 June 15	Morgan Lt. P. R. J.	2 June 16	6 July 18	I. A. (*on probn.*) 6 Fd. Tp.
10 Dec. 14	Stafford, Lt. J. B.	1 June 16	19 May 18	I. A. R. O.
9 July 15	Austin, Lt. A. C.	9 July 16	25 Sep. 15	I. A. R. O. No. 5 Coy.
20 July 15	Fawcett, Lt. J. H.	20 July 16	19 Aug. 15	I. A. R. O. No. 2 Coy.
20 July 15	Crockwell, Lt. L. H. W.	20 July 16	4 Nov. 15	I. A. R. O. *Leave, ex India, 6 mos. m.c.*
25 July 15	Horton, Lt. K.	25 July 16	25 Aug. 16	I. A. R. O. No. 1 Coy.
30 July 15	Izat, Lt. J., C.I.E.	30 July 16	3 Sep. 15	I. A. R. O.
7 Sep. 15	Gunn, Lt. J. P.	7 Sep. 16	19 Sep. 16	I. A. R. O.
8 Sep. 15	Munsey, Lt. S. T. H.	8 Sep. 16	24 Nov. 15	I. A. R. O. No. 1 Coy.
1 Oct. 15	Bell, Lt. J. A.	1 Oct. 16	2 Dec. 15	I. A. R. O. Ry. Board.
29 Oct. 15	MacFadden, Lt. R. R.	29 Oct. 16	25 Aug. 16	I. A. R. O. Ry. Board.
1 Dec. 15	Izat, Lt. J. R.	1 Dec. 16	19 Jan. 16	I. A. R. O. Lab. Corps.
1 Dec. 15	Young, Lt. F.	1 Dec. 16	26 June 18	I. A. R. O. B. Coy.
9 Dec. 15	Miller, Lt. W. J. B.	9 Dec. 16	9 Apl. 17	I. A. R. O. No. 2 Coy.
7 Sep. 15	Fletcher, Lt. N.	7 Sep. 16	5 June 17	I. A. R. O. No. 1 Bdg. Tn.
23 Jan. 16	Roberts, Lt. H. C. W.	23 Jan. 17	12 Nov. 18	I. A. No. 7 Coy.
19 Aug. 16	Delamain, Lt. J. H.	19 Feb. 17	1 Jan. 18	I. A. (*on probn.*) 1 Fd. Sqdn.
27 Jan. 16	Fraser, Lt. E.	27 Jan. 17	14 Mar. 16	I. A. R. O. Ry. Board.
25 Mar. 16	Bender, Lt. W. E. G.	25 Mar. 17	29 May 16	I. A. R. O. Offg. Asst. Insp. Officer.
25 Mar. 16	Smith, Lt. (*Ag. Capt.*) G. M.	25 Mar. 17	26 Nov. 17	I. A. R. O. No. 55 Coy. Co. Cdr.
25 Mar. 16	Powis, Lt. (*Ag. Capt.*) L. T.	25 Mar. 17	29 May 16	I. A. R. O. B. Coy.
10 May 16	Hutcheson, Lt. J. H.	10 May 17	10 July 16	I. A. R. O. No. 6 Coy.
27 Jan. 16	Roberts, Lt. F. R.	27 Jan. 17	14 Mar. 16	I. A. R. O. M. W. S., Bannu.
10 Feb. 16	Harwood, Lt. W. A.	10 Feb. 17	15 Feb. 16	I. A. R. O.
12 July 16	Houghton, Lt. W. B.	12 July 17	17 Aug. 16	I. A. R. O. No. 3 Coy.
22 July 16	Bartholomew, Lt. G.	22 July 17	3 Aug. 16	I. A. R. O. No B. Coy.

1st King George's Own Sappers and Miners—*contd.*

Date of entering service.	Names.	Army rank.		Remarks.
	Attached—*contd.*			
29 July 16	Tinsley, Lt. W. S.	29 July 17	23 Sep. 17	I. A. R. O. No. 5 Coy.
12 Aug. 15	Murray, Lt. G.	12 Aug. 17	25 Aug. 16	I. A. R. O.
14 Aug. 16	Blomfield, Lt. D. J. (*act'g. Capt.*)	14 Aug. 17	21 Aug. 16	I. A. R. O. B Coy.
22 Aug. 16	Stein, Lt. J. A.	22 Aug. 17	19 Sep. 16	I. A. R. O. No. 3 Coy.
22 Aug. 16	Hogshaw, Lt. F. H.	22 Aug. 17	11 Sep. 16	I. A. R. O. No. 4 Coy.
22 Aug. 16	Melvin, Lt. (*Ag. Capt.*) W. G.	22 Aug. 17	11 Sep. 16	I. A. R. O. Engr. Field Park.
12 Sep. 16	Steedman, Lt. J. A.	12 Sep. 17	1 Oct. 17	I. A. R. O. Engr. Field Park.
21 Sep. 16	Cargill, Lt. A. K.	21 Sep. 17	10 Oct. 17	I. A. R. O. No. 2 Coy.
29 Aug. 17	Owen, 2nd-Lt. A. J. (*temp.*)	29 Aug. 17	4 Jan. 18	W. Rid. R.—B. Coy.
5 Oct. 17	Lacey, 2nd-Lt. G.	5 Oct. 18	20 Jan. 18	I. A. R. O. No. 57 Coy.
30 Sep. 16	Crowley, Lt. H. S.	30 Sep. 17	13 Oct. 16	I. A. R. O. 58 Coy.
10 Oct. 16	Roberts, Lt. S. A.	10 Oct. 1[illegible]	19 Oct. 16	I. A. R. O. No. 4 Coy.
28 Oct. 16	Trusler, Lt. C. H.	28 Oct. 17	13 Nov. 16	I. A. R. O No. 1 Coy.
5 Jan. 17	Saunders, Lt. W.	5 Jan. 18	19 Mar. 17	I. A. R. O. No. 5 Coy.
20 Jan. 17	Elton, Lt. A. C.	20 Jan. 18	15 Apl. 17	I. A. R. O. No. 8 Coy.
1 Mar. 17	Brown, Lt. G. L.	1 Mar. 18	15 Apl. 17	I. A. R. O.
1 Mar. 17	Haines, Lt. W. E.	1 Mar. 18	15 Apl. 17	I. A. R. O. No. 8 Coy.
5 Mar. 17	Robinson, Lt. J. G.	5 Mar. 18	15 Apl. 17	I. A. R. O.
5 Mar. 17	MacNiven, Lt. G. A.	5 Mar. 18	16 Apl. 17	I. A. R. O. No. 51 Coy.
5 May 17	Carter, Lt. T. F.	5 May 18	16 Aug. 17	I. A. R. O. No. 8 Coy.
5 May 17	Vickery, Lt. L. M.	5 May 18	10 Aug. 17	I. A. R. O.
5 May 17	Dewey, Lt. C. A. R.	5 May 18	10 Aug. 17	I. A. R. O. Ry. M. E. F.
5 May 17	Dixon, Lt. G.	5 May 18	20 Aug. 17	I. A. R. O. No. 2 Coy.
5 May 17	Stevenson, Lt. L. J.	5 May 18	20 Aug. 17	I. A. R. O.
5 May 17	Orr, Lt. J.	5 May 18	20 Aug. 17	I. A. R. O. B Coy.
5 May 17	Partridge, Lt. T. G.	5 May 18	20 Aug. 17	I. A. R. O. (Qr.-Mr.)
16 May 17	White, Lt. H. S.	16 May 18	18 July 18	I. A. (*on probn.*) M. E. F.
20 June 17	Cooper, Lt. F. A.	20 June 18	9 Oct. 17	I. A. R. O.
20 June 17	Batterbury, Lt. G. H.	20 June 18	5 Oct. 17	I. A. R. O.
20 June 17	Smart, Lt. J. H.	20 June 18	10 Oct. 17	I. A. R. O. Aden. D. L. S.
20 June 17	Batterbury, Lt. F. H.	20 June 18	5 Oct. 17	I. A. R. O.
8 Aug. 17	Cox, Lt. G. W.	8 Aug. 18	2 Nov. 17	I. A. R. O.
8 Aug. 17	George, Lt. H. S.	8 Aug. 18	2 Nov. 17	I. A. R. O.
8 Aug. 17	Stampe, Lt. W. L.	8 Aug. 18	2 Nov. 17	I. A. R. O.
8 Aug. 17	Park, Lt. E. R.	8 Aug. 18	2 Nov. 17	I. A. R. O.
8 Aug. 17	Club, Lt. A.	8 Aug. 18	2 Nov. 17	I. A. R. O.
12 Aug. 17	Butcher, Lt. C. R.	12 Aug. 18	20 Sep. 17	I. A. R. O. 6 Coy.
5 Oct. 17	Baker, Lt. G. M.	5 Oct. 18	24 Dec. 17	I. A. (*T. C.*) - B Coy.
29 Aug. 17	Jeffery, Lt. H. E.	29 Aug. 18	14 Apl. 19	I. A. R. O.-No. 54 Coy.
29 Sep. 17	Reilly, Lt. C. O. C.	29 Sep. 17	30 Nov. 17	I. A. R. O. No. 5 Fd. Troop.
5 Oct. 17	Hayward, Lt. R. D.	5 Oct. 17	14 July 18	I. A. R. O. B Coy.
5 Oct. 17	Lemon, Lt. R. E.	5 Oct. 18	6 Jan. 18	I. A. R. O. B Coy.
11 Aug. 17	Hays, 2nd-Lt. H.	11 Aug. 17	12 Feb. 18	I. A. R. O. (*on probn.*).
9 Oct. 17	Webber, 2nd-Lt. W. G.	9 Oct. 17	14 Mar. 18	I. A. R. O. (*on probn.*) B-7 Coy.
3 Nov. 17	Abbey, Lt. J. B.	3 Nov. 18	19 May 18	I. A. R. O.
5 Nov. 17	Henriques, 2nd-Lt. C. Q.	5 Nov. 17	13 Mar. 18	I. A. R. O. 52 Coy.
5 Nov. 17	Watson, 2nd-Lt. E. C.	5 Nov. 17	14 Mar. 18	I. A. R. O. 54 Coy.
5 Nov. 17	Barter, 2nd-Lt. E. G.	5 Nov. 17	13 Mar. 18	I. A. R. O. B Coy.
5 Nov. 17	Cook, 2nd-Lt. E. J. A.	5 Nov. 17	13 Mar. 18	I. A. R. O. Persia.
26 Nov. 17	Braimbridge, 2nd-Lt. R. G.	26 Nov. 17	6 June 18	I. A. R. O. No. 6 Fd. Troop.
15 Jan. 18	Betterton, 2nd-Lt. F. A.	15 Jan. 18	5 June 18	I. A. R. O.
15 Jan. 18	Cook, 2nd-Lt. H. W.	15 Jan. 18	6 June 18	I. A. R. O. 52 Coy.
15 Jan. 18	Determes, 2nd-Lt. A. C.	15 Jan. 18	5 June 18	I. A. R. O.
15 Jan. 18	Fox, 2nd-Lt. T. F.	15 Jan. 18	6 June 18	I. A. R. O.
15 Jan. 18	Squires, 2nd-Lt. A. W. T	15 Jan. 18	6 June 18	I. A. (*T. C.*)
15 Jan. 18	Turner, 2nd-Lt. J. H. R.	15 Jan. 18	6 June 18	I. A. R. O. 6 Coy.
15 Jan. 18	Watts, 2nd-Lt. F. H.	15 Jan. 18	6 June 18	I. A. (*T. C.*).
29 Mar. 18	Gomme, 2nd-Lt. D. E.	29 Mar. 18	14 July 18	I. A. R. O. No. 5 Fd. Troop.
29 Mar. 18	Hobson, 2nd-Lt. R. J.	29 Mar. 18	16 July 18	I. A. R. O.
29 Mar. 18	Short, 2nd-Lt. W. H. O.	29 Mar. 18	16 July 18	I. A. R. O.
29 Mar. 18	Gillie, 2nd-Lt. K. M.	29 Mar. 18	17 July 18	I. A. R. O. M. E. F.
29 Mar. 18	Bentham, 2nd-Lt. C. P.	29 Mar. 18	14 July 18	Do. 56 Coy.
20 June 18	Wolcroft, 2nd-Lt. T. C.	20 June 18	13 July 18	I. A. R. O. B Coy.
1 Aug 18	Mahony, 2nd-Lt. B. C.	1 Aug. 18	14 Aug. 18	I. A. R. O. B Coy.
7 June 18	Hardwicke, 2nd-Lt. S. W. B.	7 June 18	11 Oct. 18	I. A. R. O. B Coy.
14 June 18	Wontersz, 2nd-Lt. E. P. T.	14 June 18	1 July 18	Do.
14 June 18	Hall, 2nd-Lt. N. S.	14 June 18	25 July 18	Do.
17 June 18	Hodey, 2nd-Lt. W. R.	17 June 18	9 Oct. 18	Do.
19 June 18	Harrison, 2nd-Lt. A. S.	19 June 18	28 July 18	Do.
24 June 18	Mathewrs, 2nd-Lt. J. S.	24 June 18	22 July 18	Do.
24 Aug. 18	Lee, 2nd-Lt. J. S. S.	24 Aug. 18	15 Sep. 18	Do.
29 Aug. 18	Clay, 2nd-Lt. F. A.	29 Aug. 18	14 Sep. 18	Do.
18 Sep. 18	Dempster, 2nd-Lt. E. R.	18 Sep. 18	9 Oct. 18	Do.
28 Sep. 18	Jones, 2nd-Lt. F. L.	28 Sep. 18	14 Oct. 18	Do.
2 Oct. 18	Fitzherbert, 2nd-Lt. R. R.	2 Oct. 18	18 Oct. 18	Do.
10 Oct. 18	Slatter, 2nd-Lt. C. B.	10 Oct. 18	22 Oct. 18	Do.
10 Oct. 18	Corner, 2nd-Lt. E. S.	10 Oct. 18	2 Nov. 18	Do.

1st King George's Own Sappers and Miners—*contd.*

Date of entering service.	Names.	Army rank.			Remarks.
	Warrant Officer.				
	Eltham, Condr. F. J.	28 May 18	28 May 18		
	Williams, R. S. M., F. W.	24 Jan. 15	24 Jan. 15		
		Jemadar.	Subadar.	Subadar-Major.	
	Subadar-Major.				
12 June 86	Akbar Ali *Bahadur* (56)	1 Oct. 03	16 Nov. 07	16 Mar. 14	B Coy. *Order Br. Ind., 2nd Cl.*
26 Aug. 86	*Mihr Din,* I.D.S.M., *Sardar Bahadur*	*15 Mar. 05*	*1 June 10*	*13 Oct. 15*	*Labour Corps Coy. Order of Br. Ind., 1st Cl.*
	Subadars.				
5 May 84	Habibullah Khan, *Sardar Bahadur*	16 Mar. 06	20 Dec. 11		No. B Coy. *Order of Br. Ind., 1st Cl.*
12 Aug. 89	Ram Autar Patak, I.D.S.M.	26 June 08	1 June 12		No. 58 Coy.
17 June 86	Gauri Shankar Dube, *Bahadur,* M C (56)	16 Oct. 08	16 Mar. 14		No. B Coy. *Order of Br. Ind. 2nd Class.*
8 Dec. 90	Chanda Singh *Bahadur*	1 Jan. 11	16 Dec. 14		No. 1 Coy. *Order of Br. Ind. 2nd Cl.*
22 Feb. 90	Fazl Shah, I.D.S.M., *Bahadur* (56)	1 Jan. 11	12 Mar. 15		*Order of Br. Ind., 2nd Cl.* No. 1 Pontoon Park.
11 Feb. 98	Niamat Ullah, I.D.S.M., *Bahadur*	19 June 13	6 June 15		No. 7 Coy. *Order of Br. Ind., 2nd Cl.*
22 Sep. 91	Sultan, I.D.S.M.	15 Dec. 11	19 Oct. 15		No. 8 Coy.
23 Mar. 93	Abdul Aziz Khan, I.D.S.M., *Bahadur*	1 June 12	16 Mar. 16		No. 55 Coy. *Order of Br. Ind., 2nd Cl.*
24 Aug. 89	Shiu Shanker Misr	31 Aug. 09	1 July 16		No. 6 Coy.
28 Jan. 93	Bhan Singh (51)	20 Dec. 11	1 Dec. 16		No. B Coy.
13 Jan. 91	Nand Singh	6 June 13	1 Apl. 17		No. 51 Coy.
11 Dec. 95	Sada Singh, *Bahadur*	1 Jan. 13	1 June 17		No. 52 Coy. *Order of Br. Ind., 2nd Cl.*
17 Aug. 91	Nawab Khan, I.D.S.M., *Bahadur*	20 Feb. 13	1 June 17		*No. 5 Coy. Order of Br. Ind., 2nd Cl.*
28 Jan. 96	Ishar Singh	1 July 13	1 July 17		No. 53 Coy.
2 May 95	Yakhudin, I.D.S.M.	14 May 15	28 Oct. 17		No. 4 Coy.
1 Oct. 97	Saidan Shah (46)	15 Apl. 13	1 Nov. 17		No. 54 Coy.
1 Jan. 97	Diwan Ali, I.D.S.M. (56)	16 Apl. 13	1 Nov. 17		1 Field Troop.
30 June 98	Ram Rup Singh, I.O.M., *Bahadur*	16 Mar. 14	1 Nov. 17		No. 3 Coy. *Order of Br. Ind., 2nd Cl.*
18 Oct. 93	Bhola Singh	16 Mar. 14	1 Jan. 18		No. 56 Coy.
19 Apl. 00	Harnam Singh, I.D.S.M.	9 Aug. 14	13 Apl. 18		No. 57 Coy.
9 Oct. 96	Fazl Haqq	1 Oct. 14	1 May 18		No. 2 Coy.
20 July 97	Alam Khan, I.O.M. I.D.S.M. (56)	1 Oct. 14	1 May 18		Bridge Train.
25 Dec. 98	Chagatta, I.O.M., I.D.S.M.	11 Mar. 15	5 Sep. 18		
13 Apl. 97	Gujar Singh	6 June 15	5 Sep. 18		
25 Sep. 93	Kishan Singh	1 July 15	5 Sep. 18		No. 7 Coy.
15 Jan. 01	Narain Singh	23 Oct. 15	5 Sep. 18		
	Jemadars.				
12 Jan. 01	Sadar Din	16 Dec. 14			B. Coy. (Bridge Train).
16 Dec. 92	*Muhammad Din*	*1 July 15*			*Instr., Mily. Survey Cl., Roorkee.*
7 Feb. 00	Bholar	1 July 15			B Coy.
1 May 99	Nur Muhammad	1 July 15			No. 5 Coy.
30 July 98	Abdul Ghafur	1 July 15			No. 1 Coy.
19 July 98	Narayan Singh	28 Feb. 16			No. 2 Coy.
1 Mar. 00	Alla Rakha, I.D.S.M.	29 Feb. 16			No. 1 Bridging Train.
1 Feb. 00	Faqir Muhammad	29 Feb. 16			No. 6 Field Troop.
12 Jan. 01	Bishan Singh	23 Mar. 16 11 Apl. 16			No. 51 Coy.
4 June 01	Manowar Shah	30 Apl. 16			Base Depôt, Basra.
19 Oct. 98	Sitla Bakhsh Singh	7 June 16			No. 54 Coy.
19 Jan. 00	Muhammad Ali (56)	7 June 16			No. B Coy. (Jemdr. Adjt.)
16 Mar. 01	Rahmatullah Khan	23 June 16			B. Coy.
20 July 97	Ishar Din Pande	1 July 16			No. 55 Coy.
12 Jan. 01	Jiwa Singh, I.D.S.M. (I)	1 Dec. 16			No. 5 Coy.
5 Sep. 89	Hukam Singh	1 Mar. 17			B Coy.
17 Mar. 98	Harnam Singh, I.D.S.M.	1 Mar. 17			No. 5 Fd. Tr.
15 Jan. 98	Rambharose Misr, I. O. M., I.D.S.M.	1 Mar. 17			No. 52 Coy.
20 Oct. 00	Din Muhammad	1 Mar. 17			No. 8 Coy.
13 July 98	Ramayan Singh, I.D.S.M.	1 Mar. 17			No. 53 Coy.
5 June 01	Basant Singh, I.D.S.M.	1 June 17			No. 4 Coy.

1st King George's Own Sappers and Miners—*concld.*

Date of entering service.	Names.	Army rank.	Remarks.
	Jemadars—*contd.*		
5 June 01	Bag Khan	1 June 17	No. 51 Coy.
15 Sep. 00	Firoz Khan	1 June 17	No. 53 Coy.
6 Nov. 00	Fazal Shah (56)	1 June 17	No. 52 Coy.
26 Oct. 04	Ujagar Singh	1 June 17	No. 51 Coy.
20 Jan. 04	Bhawani Datt, I.D.S.M.	1 June 17	No. 7 Coy.
1 Dec. 99	Ishar Singh	14 July 17	No. 58 Coy.
8 Apl. 03	Thakur Singh	14 July 17	No. 8 Coy.
9 Jan. 00	Sahib Din	14 July 17	D. L. S., Karachi.
19 July 02	Khem Singh	14 July 17	No. 5 Coy.
20 Feb. 96	Fateh Mohd.	14 July 17	No. 2 Coy.
14 Apl. 02	Hashim Shah	14 July 17	No. 6 Coy.
8 Aug. 02	Chaudhri	14 July 17	No. 1 Coy.
1 Aug. 04	Samunder Khan	14 July 17	No. B. Coy.
3 Jan. 02	Jhande Khan	14 July 17	No. 54 Coy.
3 Nov. 03	Nausher Khan, I.D.S.M.	14 July 17	No. 4 Coy.
21 Nov. 04	Guman Singh	14 July 17	No. 8 Coy.
14 Apl. 02	Sawan Singh, I.O.M., I.D.S.M.	14 July 17	No. 3 Coy.
5 May 99	Kirpa Singh	1 Nov. 17	No. 52 Coy.
19 Sep. 04	Nawab	1 Nov. 17	No. 55 Coy.
1 Jan. 06	Hardit Singh, I.O.M. (56)	1 Nov. 17	No. 55 Coy.
31 July 99	Sucha Singh, I.D.S.M.	1 Nov. 17	B Coy.
8 Mar. 06	Fateh Muhmd. (56)	1 Nov. 17	No. 1 F. S.
1 Jan. 02	Firoz Khan	1 Nov. 17	No. 7 Coy.
6 June 04	Sayyid Akbar	1 Nov. 17	No. 6 Coy.
16 Oct. 98	Mahabir Pathak	1 Nov. 17	No. B Coy.
28 Sep. 02	Magh Singh	1 Jan. 18	No. 56 (Field) Coy.
30 Aug. 98	Sheoratan Dube	1 Jan. 18	No. 56 Coy.
11 Feb. 07	Abdul Rasul	1 Jan. 18	No. 56 Coy.
2 June 02	Bhulan Khan, I.O.M., I.D.S.M.	1 Jan. 18	No. 3 Coy.
17 May 98	Bhagga Singh	1 Jan. 18	No. 57 (Field) Coy.
27 Jan. 93	Dasaunda Singh	1 Feb. 18	B. Depot Coy.
5 Sep. 99	Fazal Nur	1 Feb. 18	B. Depot Coy.
1 Mar. 10	Guna Nand	26 Feb. 18	A. Depot Coy.
5 Nov. 00	Bishan Singh, I.D.S.M.	28 Feb. 18	No. 4 Coy.
23 Apl. 00	Man Singh	13 Apl. 18	No. 54 Coy.
10 Apl. 07	Nazir Muhammad	13 Apl. 18	No. 57 Coy.
26 Apl. 04	Shams Din	13 Apl. 18	No. 1 Fd Tr.
5 June 04	Amar Singh	13 Apl. 18	No. 6 (Field) Coy.
19 Jan. 01	Sujan Singh	15 May 18	No. 53 Coy.
15 Jan. 04	Rur Singh	27 May 18	No. 1 Coy.
26 Apl. 00	Nikka Singh	24 May 18	No. 2 B. T.
. . . .	Jowala Parshad	30 May 18	
19 Aug. 04	Kirpal Singh	11 June 18	No. S. F. T.
27 Apl. 06	Channan Shah	18 June 18	No. 58 Coy.
1 Feb. 08	Brij Bahadur	18 June 18	B Coy.
26 Mar. 99	Alla Ditta	18 June 18	D. L. S., Calcutta.
. . . .	Fazal Ahmad	18 June 18	
. . . .	Dasaunda Singh	1 Aug. 18	
21 Mar. 85	Dasaunda Singh	5 Sep. 18	B. Coy.
20 Aug. 92	Kuttra Singh	5 Sep. 18	B. Coy.
7 June 04	Kala Singh	5 Sep. 18	No. 1 F. S.
4 Nov. 08	Sakhi Muhammad	5 Sep. 18	No. 6 F. T.

2nd Queen Victoria's Own Sappers and Miners.

Raised at Madras, 1780, by Lt. Joseph Moorhouse, as two companies of Pioneers. Augmented, 1793, to 6 companies, and formed into a battalion styled the Madras Pioneer Battn. Augmented successively to 8, 10, 14, and finally in 1803 to 16 companies, organised in two battalions styled the 1st and 2nd Battns. of Madras Pioneers. The 1st Battn. became the Corps of Madras Sappers and Miners in 1831, into which the remaining battalion was absorbed, 1833. Became the Queen's Own Corps of Madras Sappers and Miners, 1876; 2nd Queen's Own Sappers and Miners, 1903; present designation, 1911.

THE CYPHER OF QUEEN VICTORIA WITHIN THE GARTER.

THE SPHINX.

A GOLDEN DRAGON WEARING AN IMPERIAL CROWN.

"Sholinghur" "Carnatic" "Mysore" "Seringapatam" "Egypt" "Assaye" "Java" "Nagpore" "Maheidpoor" "Ava" "China" "Meeanee" "Hyderabad" "Pegu" "Persia" "Lucknow" "Central India" "Taku Forts" "Pekin, 1860" "Abyssinia" "Afghanistan, 1878-80" "Tel-el-Kebir" "Egypt, 1882" "Tofrek" "Suakin, 1885" "Burma, 1885-87" "Chitral" "Malakand." "Tirah" "Punjab Frontier" "China, 1900."

Composition—Madrasis of all classes (mixed) and 2 companies of Burmans.

Uniform—Scarlet. *Facings*—Blue.

Colonel—Maj.-General Sir Arthur R. F. Dorward, K.C.B., D.S.O., ret. pay — 31 July 14.

1st Commn. or date of entering service.	Names and Rank.	Army rank.	Present appointment in Corps.	Remarks.
	Commandant.			
17 Feb. 86	Heycock, Lt.-Col. C. H., R. E.	18 Jan. 11	14 Dec. 16	
	Staff Officers. (3)			
25 July 93	*McClintock, Bt.-Lt.-Col. R. L., D.S.O., R.E. (Q.)*	*1 Jan. 17*	*1 Aug. 13*	
24 Oct. 94	*Earle, Bt.-Lt.-Col. R. G., D.S.O., R.E.*	*3 June 17*	*2 Aug. 13*	
2 May 00	Conran, Maj. W. D. B., D.S.O., R.E.	2 Nov. 16	7 Nov. 06	Supdt. of Instn.
18 Aug. 00	Prince, Maj. P. E., D.S.O., R.E.	2 Nov. 16	2 Dec. 07	Supdt. of Park.
23 Dec. 11	Papillon, Capt. A. H. B., R.E.	26 June 17	....	Adjt., 1 Sep. 18.
	Company Commanders (8)			
28 Aug. 01	Agnew, Maj. H. C., R.E.	2 Nov. 16	14 May 11	No. 12 Coy.
21 Dec. 01	Bushell, Maj. C. W., R.E.	21 Dec. 16	24 July 11	No. 15 (Burma) Coy.
31 July 02	Bassett, Maj. T. P., D.S.O., R.E.	31 July 17	1 Oct. 11	No. 10 Coy.
31 July 02	*Pemberton, Maj. S., D.S.O., M.C., R.E.*	*31 July 17*	*10 Mar. 13*	
24 Dec. 02	*Kidner, Major W.E., M.C., R.E.*	*24 Dec. 17*	*16 July 13*	*G. S. O., 2nd gr., Army Hdqrs.*
1 Sep. 03	Pringle, Capt. J. C., R. E.	1 Sep. 14	9 Feb. 06	No. 9 Coy.
26 Dec. 03	Harris, Capt. (*Act'g. Maj.*) T.B., D.S.O., R.E.	30 Oct. 14	6 Dec. 11	No. 16 Coy.
23 July 07	Clarke, Capt. E. H., M.C., R.E.	30 Oct. 14	13 Apl. 11	No. 62 (Burma) Coy.
	Company Officers. (9)			
29 July 04	Wickham, Capt. J. C., D.S.O., R.E.	30 Oct. 14	25 June 11	*Lv., ex I., m. c.*
20 Dec. 05	Hamilton, Capt. R., R.E.	30 Oct. 14	28 Nov. 09	No. 7 B. T.
20 Dec. 06	Hogg, Capt. D. McA., M.C., R.E. (*l*)	30 Oct. 14	19 Mar. 12	No. 2 F. T.
20 Sep. 09	Bradney, Bt.-Maj. E., R.E.	3 June 16	11 Sep. 13	No. D-4 Coy.
23 July 10	King, Capt. C. J. S., R.E.	23 July 16	2 Nov. 14	No. 11 Coy.
20 July 11	*Dewing, Capt. R. H., D.S.O., M.C., R.E.*	*26 June 17*	*17 May 14*	*G. S. O., 3rd gr.*
20 Nov. 15	Stevens, Lt. A. E., R.E.	1 July 17	4 Dec. 16	No. 15 (Burma) Co.
	Attached.			
26 Feb. 09	Britten, Capt. (*Act'g. Maj.*) W. E., R. E.	23 Dec. 15	....	No. 14 Coy.
23 Dec. 10	*Cruickshank, Capt. A. J., D.S.O., R.E.*	*23 Dec. 16*	*7 Dec. 14*	*G. S. O., 2nd. gr.*
19 July 12	Aitken, Capt. (*Act'g. Maj.*) A. B., M.C., R.E.	26 June 17	25 Mar. 15	D-6 Coy.
28 Jan. 09	Gurney, Capt (*Act'g. Maj.*) C. R., R.E. (Spl. Res.)	26 June 17	....	Comy. Comdr. 63 Coy.
2 June 15	Murray, Lt. (*temp. Capt.*) H. K.	2 June 15	....	No. 3 D-12.
6 June 13	Collier, Capt. (*Act'g. Maj.*) A. E., M.C., R.E. (Spl. Res.)	3 Nov. 17	....	No. 61 Coy.
9 Aug. 16	Marshall, 2nd-Lt. (*Act'g. Capt.*) E. N. C., (*temp.*) R.E.	9 Aug. 16	....	D. 5 Coy.

2nd Queen Victoria's Own Sappers and Miners—*contd.*

1st Commn. or date of entering service.	Names and Rank.	Army rank.	Present appointment in Corps.	Remarks.
	Attached—*contd.*			
2 July 16	Mc Lay, Lt. D.B., (*temp.*), R.E.	2 July 16	....	No. 67 Coy.
19 Feb. 16	Crowther, Lt. C. T. B., R. E.	19 Aug. 17	19 Feb. 18	I. A. R. O. 10 Coy.
19 Feb. 16	LeSueur, Lt. J. H. R., R.E.	19 Aug. 17	19 Feb. 18	No. 16 Coy.
26 May 16	Costelloe, Lt. W. H. G., R.E.	26 Nov. 17	19 Feb. 18	No. 3 F. T.
26 May 16	Mitchell, Lt. A. C., R.E.	26 Nov. 17	19 Feb. 18	No. 64 Coy.
28 Feb. 17	Longland, Lt. (*Act'g. Capt.*) R. H. B., R. E.	28 Feb. 17	19 Feb. 18	No. 3 F. T.
6 June 17	Hotine, 2nd-Lt. M., R. E.	6 June 17	19 Feb. 18	No. 63 Coy.
6 June 17	Vernon, 2nd-Lt. G. F. D., R.E.	6 June 17	4 Mar. 18	Burma Coys. Depôt.
6 June 17	Stillingfleet 2nd-Lt. R., R. E.	6 June 17	19 Feb. 18	No. 65 Coy.
4 Dec. 14	Bartholomew, Lt. H. L.	4 Dec. 15	19 Dec. 14	I. A. R. O.
27 Jan. 16	Johnston, Lt. J. H.	27 Jan. 17	....	Do. No. 13 Coy.
26 Feb. 15	Cullen, Lt. A. B., M.C.	26 Feb. 16	....	Do. No. 66 Coy.
23 Apl. 15	Grant, Lt. C. A.	23 Apl. 16	8 July 15	Do. D. 11 Coy.
7 May 15	Stewart, Lt. (*Actg. Capt.*) J. A., M.C.	7 May 16	8 May 15	Do. No. 68 (Burma) Coy.
11 June 15	Goepel, Lt. J. A.	11 June 16	24 Aug. 15	Do.
1 Aug. 16	Marshall, Lt. F.	1 Aug. 17	....	Do. Burma Coys. Depot.
28 Aug. 16	Barclay, Lt. A. G. G.	28 Aug. 17	....	Do. No. 9 Coy.
7 Oct. 16	Adlard, Lt. F. A., M.C.	7 Oct. 17	....	Do. *Lv., ex Ind., 6 mos., m.c.*
25 Nov. 16	Gompertz, Lt. G. H. M.	25 Nov. 17	3 Feb. 17	Do. No. 9 Coy.
12 Mar. 15	Woakes, Lt. (*Actg. Maj.*) R. B., M.C.	12 Mar. 16	....	Do. No. 65 F. Coy.
23 Apl. 15	Stuart, Lt. (*Actg. Capt.*) J. M. B.	23 Apl. 16	....	Do. Burma Coys. Depôt.
28 May 15	Tate, Lt. J., M.C.	28 May 16	....	Do. No. D-3 Coy.
16 Jan. 17	Fogarty, Lt. P. C.	16 Jan. 18	....	Do. No. 15 Coy.
4 Aug. 16	Vyall, Lt. F. S.	4 Aug. 17	1 Feb. 17	Do. No. 15 Coy.
5 May 15	Scott-Ruffle, Lieut. C. F.	....	....	Do. No. 2 F. T.
1 Feb. 17	White, Lt. T. H.	1 Feb. 18	....	Do. No. 13 Coy.
5 Oct. 15	Wilkinson, 2nd-Lt. F. S., R. E. (T. F.)	5 Oct. 15	....	C. C. D-7 Coy.
6 Oct. 15	Symonds, Lt. P. R., R. E. (T. F.)	1 June 17	....	No. 12 Coy.
30 Jan. 16	Joyce, 2nd-Lt. P. E. (*temp.*), R. E.	30 Jan. 16	....	D. 9 Coy.
4 Nov. 16	McClelland, 2nd-Lt. (*Act'g. Capt.*) A. A., R.E.	4 Nov. 16	16 June 17	No. D-10 Coy.
4 Nov. 16	Riccomini, 2nd-Lt. J. A., (*temp.*) R. E.	4 Nov. 16	16 June 17	Chitral Section.
2 Dec. 16	Pringle, 2nd-Lt. J. E. (*temp.*), R. E.	2 Dec. 16	....	No. 10 Coy.
4 Dec. 15	Mathews, Lt. C. W. (*temp.*), R. E.	2 Sep. 16	....	No. 10 Coy.
22 Jan. 15	Harvey-Hurst, Lt. A. J.	22 Jan. 16	....	I. A. R. O. D. 8 Coy. Qr.-Mr.
1 June 17	List, Lt. J. N.	1 June 17	....	Do. No. 15 Coy.
14 June 15	Broadbent, Lt. H. E.	14 June 16	....	Do. No. 7 B. T.
5 May 17	Medley, Lt. F. J.	5 May 18	....	Do. No. 12 Coy.
5 May 17	Penn, Lt. C. E.	5 May 18	....	Do. No. 7 B. T.
20 June 17	Clarke, Lt. G.	20 June 18	....	Do. No. 15 Coy.
8 Aug. 17	Chapman, Lt. A. F.	8 Aug. 18	....	Do. Burma Coys. Depôt.
8 Aug. 17	Brand, Lt. A. R.	8 Aug. 18	....	Do. No. 61 Coy.
8 Aug. 17	Littlejohn, Lt. A.	8 Aug. 18	....	Do. No. 14 Coy.
25 Mar. 16	Baikie, 2nd-Lt. H. Mc. A.	25 Mar. 16	....	Do. 9 Coy.
Aug. 17	Margerison, 2nd-Lt. L.A., R.E	Aug. 17	7 Dec. 17	Do. 11 Coy.
1 Apl. 17	Vickers, 2nd-Lt. T. B. (T. F.)	1 Apl. 17	19 Sep. 17	No. 14 Coy.
5 Oct. 17	Whitcombe, Lt. C.E.	5 Oct. 18	15 Jan. 18	I. A. R. O. No. 16 Coy.
5 Oct. 17	Hill, Lt. E. W.	5 Oct. 18	15 Jan. 18	Do. Burma Coys. Depôt.
5 Oct. 17	Lindsay, 2nd-Lt. J. G.	5 Oct. 17	15 Jan. 18	Do. No. 10 Coy.
5 Oct. 17	Drury, 2nd-Lt. J.	5 Oct. 17	15 Jan. 18	Do. No. 63 Coy.
5 Oct. 17	Josland, Lt. G. C.	5 Oct. 18	15 Jan. 11	I. A. (*T. C.*) D. 2 Coy.
5 Oct. 17	Bairsto, Lt. F. J.	5 Oct. 18	15 Jan. 18	I. A. R. O.
18 Dec. 14	Thorn, Lt. T. C., M.C.	4 May 17	26 Dec. 14	Do. No. 64 Coy.
12 Jan. 18	Bradbury, 2nd-Lt. C. H.	12 Jan. 18	3 June 18	I. A. R. O. No. 67 Coy.
15 Jan. 18	Butler, 2nd-Lt. S. G.	15 Jan. 18	3 June 18	I. A. R. O. No. C. Coy.
15 Jan. 18	Haffenden, 2nd-Lt. E. D.	15 Jan. 18	3 June 18	I. A. R. O. Burma Coys. Depôt.
29 Mar. 18	Casey, 2nd-Lt. J. W.	29 Mar. 18	3 June 18	I. A. (T. C.)
15 Jan. 18	Kirkby, 2nd-Lt. R. A.	15 Jan. 18	3 June 18	I. A. R. O. No. 1 Coy
15 Jan. 18	Bromley, 2nd-Lt. C. G.	15 Jan. 18	3 June 18	I. A. (T. C.) No. 65 Coy.

2nd Queen Victoria's Own Sappers and Miners—*concld.*

1st Commn. or date of entering service.	Names and Rank	Army rank.	Present appointment in Corps.	Remarks.
	Attached—*concld.*			
15 Jan. 18	Fairclough, 2nd-Lt. F. S.	15 Jan. 18	3 June 18	I. A. (T. C.) No. 61 Coy.
29 Mar. 18	Davis, 2nd-Lt. A. I.	29 Mar. 18	3 June 18	I. A. R. O. D. 5 Coy.
5 Jan. 18	Partridge, 2nd-Lt. (*Act'g. Capt.*) C.	5 Jan. 18	3 June 18	I. A. (T. C.) D. 2 Coy.
5 Nov. 17	Anderson, Lt. A. R.	5 Nov. 18	25 Mar. 18	I. A. R. O. D. 4 Coy.
6 June 17	Bogle, 2nd-Lt. H. C. L., R.E.	6 June 17	12 Mar. 18	No. 61 Coy.
6 June 17	Bourdillon, 2nd-Lt. P. A., R.E.	6 June 17	12 Mar. 18	No. 66.
2 Nov. 14	Tunnicliffe, 2nd-Lt. F. D.	2 Nov. 14	26 Jan. 18	I. A. R. O.
5 Nov. 17	Gasson, Lt. H. J.	5 Nov. 18	18 Mar. 18	I. A. R. O. No. 65 Coy.
19 Dec. 15	Hamilton, Lt. G. H.	19 June 17	19 Mar. 18	I. A. R. O. No. 63 Coy.
	Commissary.			
May 07	Gasson, Hony. Maj. H. T.	2 Oct. 11	26 Mar. 13	
	Warrant Officers.			
	King, Condr. W. H.	20 Aug. 14	1 Oct. 12	
	Dalgleish, Regtl. Sergt.-Maj. J.	30 Oct. 16	30 Oct. 16	
		Jemadar.	Subadar.	Subadar-Major.
	Subadar-Major.			
14 June 87	Michael, *Bahadur* (56)	28 Mar. 06	9 Apl. 13	1 Sep. 18, *Order of Br. Ind., 2nd Cl.* D. 2 Coy
23 Feb. 84	Maria Cruz, *Bahadur*	1 June 98	24 May 08	*Order of Br. Ind. 2nd Class.* Re-employed, 1 Sept. 17.
	Subadars.			
1 Sep. 85	Abdur Rahman Khan	2 Aug. 04	9 Jan. 12	D. Coy. 1 Re-employed, 1 Sep. 17.
17 Jan. 88	Bo Kin Po (56)	1 June 98	1 July 12	No. 15 (Burma) Coy.
9 Aug. 87	Sundaramurti (56)	16 June 06	25 June 13	D. 4 Coy.
25 Aug. 91	Ramasami (I) (46, 56)	17 Apl. 09	9 Oct. 13	D. 8 Coy.
1 May 89	Hanumantu, I.D.S.M., (56)	7 July 09	1 July 14	D. 9 Coy.
21 Oct. 89	Kuppusami (56)	16 Mar. 11	1 Feb. 16	No. 64 Coy.
31 May 93	Joseph, I.D.S.M., Bahadur (56)	15 Feb. 12	1 Feb. 16	No. 2 Field Troop. *Order of Br. Ind., 2nd Class.*
15 Aug. 94	Burland (56)	1 Feb. 13	27 Mar. 16	No. 12 Coy. I.O.M., 2nd Cl.
28 Nov. 93	Ramasami (II) (56)	7 June 12	16 Sep. 16	No. 10 Coy.
26 June 94	Michael (II) (56)	1 Aug. 12	1 Mar. 17	No. 14 Coy.
17 Mar. 96	Kothandapani I.D.S.M.	12 Dec. 13	1 Apl. 17	No. 16 Coy.
12 Mar. 95	Ponnusami, I.D.S.M.	8 Apl. 13	1 June 17	D. 1 Coy.
9 Aug. 95	Bo Chan, I.D.S.M.	1 July 12	12 June 17	No. 15 (Burma) Coy.
29 Oct. 98	Appasami (56, 66*a*)	5 July 13	1 Aug. 17	No. 61 Coy.
31 Mar. 98	Varadaraju	25 June 13	1 Mar. 17	No. 9 Coy.
12 July 98	Anthony, M.C., I.D.S.M.	1 Nov. 13	1 Oct. 17	No. 13 Coy.
18 Aug. 96	Nasiruddin	30 June 15 1 Feb. 16	1 Oct. 17	D. II Coy. Re-employed.
16 Apl. 98	Rangasami (I)	July 16	1 Feb. 18	No. 11 Coy.
7 Nov. 00	Kandasami	1 Aug. 16	1 Feb. 18	No. 63 Coy.
4 Sep. 03	Devaraju	1 Apl. 17	15 Apl. 18	D. Coy.
13 Nov. 16	Krishnasami (I)	9 Feb. 17	1 June 18	No. 12 Coy.
22 May 01	Subramoni (I)	1 Apl. 17	16 July 18	No. 7 B. T.
24 Aug. 03	Madurai	1 Aug. 17	16 July 18	No. 63 Coy.
8 Oct. 00	Murugesan (I)	1 July 15	7 Aug. 18	No. 14 Coy.
....	Abdul Rahim Khan	....	....	Re-employed, 1 Sep. 17.
....	Bo Thaw	....	19 June 18	
19 July 98	Govindasami (I)	9 Oct. 13	1 July 18	D. Coy.
18 June 87	Padaivittan	16 Sep. 16	1 Apl. 18	No. 65 Coy.
25 Oct. 00	Krishnasami (II) (46)	20 May 17	1 July 18	No. 10 Coy.
....	Moung Gyi	1 June 17	8 Aug. 18	Burma Coy.
....	Bo Sugyee	1 Sept. 17	22 Sept. 18	Burma Coy.
22 Jan. 90	Velanganni	9 Jan. 12	9 Oct. 18	D. 6 Coy.
16 Aug. 03	Appa Rao	1 Aug. 17	9 Oct. 18	D. 10 Coy.

2nd Queen Victoria's Own Sappers and Miners—*contd.*

1st Commn. or date of entering service.	Names and Rank.	Army rank.	Present appointment in Corps.	Remarks.
	Attached—*contd.*			
2 July 16	Mc Lay, Lt. D.B., (*temp.*), R.E.	2 July 16	. . . .	No. 67 Coy.
19 Feb. 16	Crowther, Lt. C. T. B., R. E.	19 Aug. 17	19 Feb. 18	I. A. R. O. 10 Coy.
19 Feb. 16	LeSueur, Lt. J. H. R., R.E.	19 Aug. 17	19 Feb. 18	No. 16 Coy.
26 May 16	Costelloe, Lt. W. H. G., R.E.	26 Nov. 17	19 Feb. 18	No. 3 F. T.
26 May 16	Mitchell, Lt. A. C., R.E.	26 Nov. 17	19 Feb. 18	No. 64 Coy.
28 Feb. 17	Longland, Lt. (*Act'g. Capt.*) R. H. B., R. E.	28 Feb. 17	19 Feb. 18	No. 3 F. T.
6 June 17	Hotine, 2nd-Lt. M., R. E.	6 June 17	19 Feb. 18	No. 63 Coy.
6 June 17	Vernon, 2nd-Lt. G. F. D., R.E.	6 June 17	4 Mar. 18	Burma Coys. Depôt.
6 June 17	Stillingfleet 2nd-Lt. R., R. E.	6 June 17	19 Feb. 18	No. 65 Coy.
4 Dec. 14	Bartholomew, Lt. H. L.	4 Dec. 15	19 Dec. 14	I. A. R. O.
27 Jan. 16	Johnston, Lt. J. H.	27 Jan. 17	. . . .	Do. No. 13 Coy.
26 Feb. 15	Cullen, Lt. A. B., M.C.	26 Feb. 16	. . . .	Do. No. 66 Coy.
23 Apl. 15	Grant, Lt. C. A.	23 Apl. 16	8 July 15	Do. D. 11 Coy.
7 May 15	Stewart, Lt. (*Actg. Capt.*) J. A., M.C.	7 May 16	8 May 15	Do. No. 68 (Burma) Coy.
11 June 15	Goepel, Lt. J. A.	11 June 16	24 Aug. 15	Do.
1 Aug. 16	Marshall, Lt. F.	1 Aug. 17	. . . .	Do. Burma Coys. Depot.
28 Aug. 16	Barclay, Lt. A. G. G.	28 Aug. 17	. . . .	Do. No. 9 Coy.
7 Oct. 16	Adlard, Lt. F. A., M.C.	7 Oct. 17	. . . .	Do. *Lv., ex Ind., 6 mos., m.c.*
25 Nov. 16	Gompertz, Lt. G. H. M.	25 Nov. 17	3 Feb. 17	Do. No. 9 Coy.
12 Mar. 15	Woakes, Lt. (*Actg. Maj.*) R. B., M.C.	12 Mar. 16	. . . .	Do. No. 65 F. Coy.
23 Apl. 15	Stuart, Lt. (*Actg. Capt.*) J. M. B.	23 Apl. 16	. . . .	Do. Burma Coys. Depôt.
28 May 15	Tate, Lt. J., M.C.	28 May 16	. . . .	Do. No. D-3 Coy.
16 Jan. 17	Fogarty, Lt. P. C.	16 Jan. 18	. . . .	Do. No. 15 Coy.
4 Aug. 16	Vyall, Lt. F. S.	4 Aug. 17	1 Feb. 17	Do. No. 15 Coy.
5 May 15	Scott-Ruffle, Lieut. C. F.	. . . .	. . . .	Do. No. 2 F. T.
1 Feb. 17	White, Lt. T. H.	1 Feb. 18	. . . .	Do. No. 13 Coy.
5 Oct. 15	Wilkinson, 2nd-Lt. F. S., R. E. (T. F.)	5 Oct. 15	. . . .	C. C. D-7 Coy.
6 Oct. 15	Symonds, Lt. P. R., R. E. (T. F.)	1 June 17	. . . .	No. 12 Coy.
30 Jan. 16	Joyce, 2nd-Lt. P. E. (*temp.*), R. E.	30 Jan. 16	. . . .	D. 9 Coy.
4 Nov. 16	McClelland, 2nd-Lt. (*Act'g. Capt.*) A. A., R.E.	4 Nov. 16	16 June 17	No. D-10 Coy.
4 Nov. 16	Riccomini, 2nd-Lt. J. A., (*temp.*) R. E.	4 Nov. 16	16 June 17	Chitral Section.
2 Dec. 16	Pringle, 2nd-Lt. J. E. (*temp.*), R. E.	2 Dec. 16	. . . .	No. 10 Coy.
4 Dec. 15	Mathews, Lt. C. W. (*temp.*), R. E.	2 Sep. 16	. . . .	No. 10 Coy.
22 Jan. 15	Harvey-Hurst, Lt. A. J.	22 Jan. 16	. . . .	I. A. R. O. D. 8 Coy. Qr.-Mr.
1 June 17	List, Lt. J. N.	1 June 17	. . . .	Do. No. 15 Coy.
14 June 15	Broadbent, Lt. H. E.	14 June 16	. . . .	Do. No. 7 B. T.
5 May 17	Medley, Lt. F. J.	5 May 18	. . . .	Do. No. 12 Coy.
5 May 17	Penn, Lt. C. E.	5 May 18	. . . .	Do. No. 7 B. T.
20 June 17	Clarke, Lt. G.	20 June 18	. . . .	Do. No. 15 Coy.
8 Aug. 17	Chapman, Lt. A. F.	8 Aug. 18	. . . .	Do. Burma Coys. Depôt.
8 Aug. 17	Brand, Lt. A. R.	8 Aug. 18	. . . .	Do. No. 61 Coy.
8 Aug. 17	Littlejohn, Lt. A.	8 Aug. 18	. . . .	Do. No. 14 Coy.
25 Mar. 16	Baikie, 2nd-Lt. H. Mc. A.	25 Mar. 16	. . . .	Do. 9 Coy.
Aug. 17	Margerison, 2nd-Lt. L.A., R.E	Aug. 17	7 Dec. 17	Do. 11 Coy.
1 Apl. 17	Vickers, 2nd-Lt. T. B. (T. F.)	1 Apl. 17	19 Sep. 17	No. 14 Coy.
5 Oct. 17	Whitcombe, Lt. C.E.	5 Oct. 18	15 Jan. 18	I. A. R. O. No. 16 Coy.
5 Oct. 17	Hill, Lt. E. W.	5 Oct. 18	15 Jan. 18	Do. Burma Coys. Depôt.
5 Oct. 17	Lindsay, 2nd-Lt. J. G.	5 Oct. 17	15 Jan. 18	Do. No. 10 Coy.
5 Oct. 17	Drury, 2nd-Lt. J.	5 Oct. 17	15 Jan. 18	Do. No. 63 Coy.
5 Oct. 17	Josland, Lt. G. C.	5 Oct. 18	15 Jan. 11	I. A. (*T. C.*) D. 2 Coy.
5 Oct. 17	Bairsto, Lt. F. J.	5 Oct. 18	15 Jan. 18	I. A. R. O.
18 Dec. 14	Thorn, Lt. T. C., M.C.	4 May 17	26 Dec. 14	Do. No. 64 Coy.
12 Jan. 18	Bradbury, 2nd-Lt. C. H.	12 Jan. 18	3 June 18	I. A. R. O. No. 67 Coy.
15 Jan. 18	Butler, 2nd-Lt. S. G.	15 Jan. 18	3 June 18	I. A. R. O. No. C. Coy.
15 Jan. 18	Haffenden, 2nd-Lt. E. D.	15 Jan. 18	3 June 18	I. A. R. O. Burma Coys. Depôt.
29 Mar. 18	Casey, 2nd-Lt. J. W.	29 Mar. 18	3 June 18	I. A. (T. C.)
15 Jan. 18	Kirkby, 2nd-Lt. R. A.	15 Jan. 18	3 June 18	I. A. R. O. No. 1 Coy
15 Jan. 18	Bromley, 2nd-Lt. C. G.	15 Jan. 18	3 June 18	I. A. (T. C.) No. 65 Coy.

2nd Queen Victoria's Own Sappers and Miners—*concld.*

1st Commn. or date of entering service.	Names and Rank	Army rank.	Present appointment in Corps.	Remarks.
	Attached—*concld.*			
15 Jan. 18	Fairclough, 2nd-Lt. F. S. . . .	15 Jan. 18	3 June 18	I. A. (T. C.) No. 61 Coy.
29 Mar. 18	Davis, 2nd-Lt. A. I. . . .	29 Mar. 18	3 June 18	I. A. R. O. D. 5 Coy.
5 Jan. 18	Partridge, 2nd-Lt.(*Act'g. Capt.*) C.	5 Jan. 18	3 June 18	I. A. (T. C.) D. 2 Coy.
5 Nov. 17	Anderson, Lt. A. R. . . .	5 Nov. 18	25 Mar. 18	I. A. R. O. D. 4 Coy.
6 June 17	Bogle, 2nd-Lt. H. C. L., R.E.	6 June 17	12 Mar. 18	No. 61 Coy.
6 June 17	Bourdillon, 2nd-Lt. P. A., R.E.	6 June 17	12 Mar. 18	No. 66.
2 Nov. 14	Tunnicliffe, 2nd-Lt. F. D. .	2 Nov. 14	26 Jan. 18	I. A. R. O.
5 Nov. 17	Gasson, Lt. H. J.	5 Nov. 18	18 Mar. 18	I. A. R. O. No. 65 Coy.
19 Dec. 15	Hamilton, Lt. G. H. . . .	19 June 17	19 Mar. 18	I. A. R. O. No. 63 Coy.
	Commissary.			
May 07	Gasson, Hony. Maj. H. T. .	2 Oct. 11	26 Mar. 13	
	Warrant Officers.			
	King, Condr. W. H. . . .	20 Aug. 14	1 Oct. 12	
	Dalgleish, Regtl. Sergt.-Maj. J.	30 Oct. 16	30 Oct. 16	
		Jemadar.	Subadar.	Subadar-Major.
	Subadar-Major.			
14 June 87	Michael, *Bahadur* (56) . .	28 Mar. 06	9 Apl. 13	1 Sep. 18, *Order of Br. Ind., 2nd Cl.* D. 2 Coy
23 Feb. 84	Maria Cruz, *Bahadur* . .	1 June 98	24 May 08	*Order of Br. Ind. 2nd Class.* Re-employed, 1 Sept. 17.
	Subadars.			
1 Sep. 85	Abdur Rahman Khan . . .	2 Aug. 04	9 Jan. 12	D. Coy. 1 Re-employed, 1 Sep. 17.
17 Jan. 88	Bo Kin Po (56)	1 June 98	1 July 12	No. 15 (Burma) Coy.
9 Aug. 87	Sundaramurti (56)	16 June 06	25 June 13	D. 4 Coy.
25 Aug. 91	Ramasami (I) (46, 56) . .	17 Apl. 09	9 Oct. 13	D. 8 Coy.
1 May 89	Hanumantu, I.D.S.M., (56)	7 July 09	1 July 14	D. 9 Coy.
21 Oct. 89	Kuppusami (56) . . .	16 Mar. 11	1 Feb. 16	No. 64 Coy.
31 May 93	Joseph, I.D.S.M., Bahadur(56)	15 Feb. 12	1 Feb. 16	No. 2 Field Troop. *Order of Br. Ind., 2nd Class.*
15 Aug. 94	Burland (56)	1 Feb. 13	27 Mar. 16	No. 12 Coy. I.O.M., 2nd Cl.
28 Nov. 93	Ramasami (II) (56) . . .	7 June 12	16 Sep. 16	No. 10 Coy.
26 June 94	Michael (II) (56)	1 Aug. 12	1 Mar. 17	No. 14 Coy.
17 Mar. 96	Kothandapani I.D.S.M. . .	12 Dec. 13	1 Apl. 17	No. 16 Coy.
12 Mar. 95	Ponnusami, I.D.S.M. . .	8 Apl. 13	1 June 17	D. 1 Coy.
9 Aug. 95	Bo Chan, I.D.S.M. . . .	1 July 12	12 June 17	No. 15 (Burma) Coy.
29 Oct. 98	Appasami (56, 66*a*) . . .	5 July 13	1 Aug. 17	No. 61 Coy.
31 Mar. 98	Varadaraju	25 June 13	1 Mar. 17	No. 9 Coy.
12 July 98	Anthony, M.C., I.D.S.M. .	1 Nov. 13	1 Oct. 17	No. 13 Coy.
18 Aug. 96	Nasiruddin	30 June 15 1 Feb. 16	1 Oct. 17	D. II Coy. Re-employed.
16 Apl. 98	Rangasami (I)	July 16	1 Feb. 18	No. 11 Coy.
7 Nov. 00	Kandasami	1 Aug. 16	1 Feb. 18	No. 63 Coy.
4 Sep. 03	Devaraju	1 Apl. 17	15 Apl. 18	D. Coy.
13 Nov. 16	Krishnasami (I)	9 Feb. 17	1 June 18	No. 12 Coy.
22 May 01	Subramoni (I)	1 Apl. 17	16 July 18	No. 7 B. T.
24 Aug. 03	Madurai	1 Aug. 17	16 July 18	No. 63 Coy.
8 Oct. 00	Murugesan (I)	1 July 15	7 Aug. 18	No. 14 Coy.
. . . .	Abdul Rahim Khan . . .	. . . .	. . . .	Re-employed, 1 Sep. 17.
. . . .	Bo Thaw	. . . .	19 June 18	
19 July 98	Govindasami (I)	9 Oct. 13	1 July 18	D. Coy.
18 June 87	Padaivittan	16 Sep. 16	1 Apl. 18	No. 65 Coy.
25 Oct. 00	Krishnasami (II) (46) . .	20 May 17	1 July 18	No. 10 Coy.
. . . .	Moung Gyi	1 June 17	8 Aug. 18	Burma Coy.
. . . .	Bo Sugyee	1 Sept. 17	22 Sept. 18	Burma Coy.
22 Jan. 90	Velanganni	9 Jan. 12	9 Oct. 18	D. 6 Coy.
16 Aug. 03	Appa Rao	1 Aug. 17	9 Oct. 18	D. 10 Coy.

2nd Queen Victoria's Own Sappers and Miners—*contd.*

Jemadars.

1st Commn. or date of entering service.	Name and rank.	Army rank.	Present appointment in Corps.	Remarks.
10 Dec. 95	Kesavelu (56)	9 Apl. 13		Chitral Section.
21 Aug. 97	Bo Pan Bu	6 May 13		Burma Coy.
31 July 00	Parasuraman	1 July 14		No. 11 Coy.
14 Apl. 97	Narayanasami (I)	14 Oct. 14		Rangoon D. L. Section.
2 July 98	Narasingasami	14 Oct. 14		D. 6 Coy.
15 Jan. 00	Venkatasami (I)	14 Oct. 14		D. 13 Coy.
11 Sep. 94	Gurusami (I)	1 July 15		D. 7 Coy.
24 Aug. 00	Murugesan (II)	1 July 15		D. Coy.
4 Aug. 99	Ramasami (III)	22 Apl 16		
20 Nov. 99	Parthasarathi	14 Aug. 16		No. 10 Coy.
....	Maung Po Thaw	22 Sep. 16		No. 15 (Burma) Coy.
12 Sept. 00	Munisami (II)	1 Oct. 16		No. 16 Coy.
30 Jan. 06	Sayyid Musa	26 Feb. 17		No. 12 Coy.
18 Jan. 88	Venkatasami (II)	....		Re-employed, 8 Feb. 17.
18 Aug. 88	Visvaningam	1 Mar. 17		No. 63 Coy.
16 July 00	Rajendram	1 Mar. 17		No. 9 Coy.
23 Oct. 00	Rangasami (II)	1 Apl. 17		D. 7 Coy.
18 July 93	Viraraghavelu	1 Apl. 17		D. 8 Coy.
23 Oct. 00	Govindasami (II)	1 Apl. 17		D. Coy.
3 Dec. 00	Abdul Rahim	1 Apl. 17		No. 16 Coy.
....	Maung Lugloy	12 June 17		Burma Coy.
....	Bo Sugalay	12 June 17		Burma Coy.
18 July 03	Subramoni (ii)	5 July 17		No. 65 Coy.
28 Aug. 00	Mahomed Hayat	14 July 17		No. 9 Coy.
....	Moung Hman	14 July 17		Burma Coy.
6 Aug. 02	Ramasami (IV)	13 July 17		No. 13 Coy.
17 Feb. 01	Gurusami (II)	14 July 17		No. D. 10 Coy.
27 Sept. 97	Dharmalingam	1 Aug. 17		D. 6 Coy.
23 Mar. 03	Michael (III)	1 Aug. 17		No. 14 Coy.
21 Feb. 05	Devanasan	1 Aug. 17		No. 61 Coy.
30 Oct. 06	Mariyan	1 Aug. 17		No. 65 Coy.
8 Dec. 03	Isac	1 Aug. 17		No. D. 13 Coy.
30 May 05	Murugesan (III)	1 Aug. 17		Chitral Section
26 June 04	Annamalai	1 Aug. 17		No. 13 Coy.
18 May 04	Sayyid Ahmad	1 Aug. 17		No. 10 Coy.
9 Jan. 06	Dominic	1 Aug. 17		D. 3 Coy.
10 June 06	Ramasami (V)	1 Aug. 17		No. 3 F. T.
26 June 06	Munisami (I)	1 Aug. 17		No. 11 Coy.
13 June 06	Narayansami (II)	1 Aug. 17		No. 64 Coy.
2 May 05	Shaikh Adbulla	1 Oct. 17		No. 61 Coy.
10 Feb. 08	Ayyam Perumal	1 Oct. 17		No. 63 Coy.
14 Feb. 06	Singaram	1 Oct. 17		No. 7 B. T.
24 Sept. 95	Narayanasami (III)	1 Nov. 17		No. 11 Coy.
19 June 07	Markasagayam	1 Nov. 17		No. D. 4 Coy.
12 Dec. 06	Rasundram	16 Nov. 17		No. 10 Coy.
11 July 08	Rajagopal	1 Dec. 17		No. 64 Coy.
24 Nov. 03	Muhammad Ali	1 Dec. 17		D. Coy.
28 Dec. 07	Kuppusami	22 Dec. 17		No. 1[illegible] Coy.
9 May 05	Raman	24 Jan. 18		No. 12 Coy.
26 Sep. 06	Chakrapani	1 Feb. 18		No. 64 Coy.
25 Aug. 00	Augustian	1 Feb. 18		No. 16 Coy.
18 Dec. 03	Ponnusami (II)	1 Feb. 18		No. 9 Coy.
8 Aug. 99	Appalasami	1 Feb. 18		No. 11 Coy.
13 Mar. 06	Sayyid Yusaf	1 Feb. 18		No. 65 Coy.
25 Apl. 08	Arogyasami	1 Feb. 18		No. D. 3 Coy.
1 Sep. 03	Raju	1 Feb. 18		No. 67 Coy.
2 May 05	Krishnasami (III)	1 Feb. 18		No. 2 F. T.
5 June 88	Sundaram	1 Feb. 18		No. D. 4 Coy.
6 Oct. 96	Mariyan	1 Feb. 18		No. D. 5 Coy.
28 Dec. 07	Kuppusami (I)	22 Dec. 17		No. 10 Coy.
5 Nov. 07	Kuppusami (II)	1 Mar. 18		No. 66 Coy.
24 Jan. 10	Abdul Razq	1 Mar. 18		No. D. 2 Coy.
....	Krishnasami (IV)	1 Apl. 18		
....	Munisami (III)	1 Apl. 18		
....	Perumal	1 Apl. 18		
....	Maduramuttu	1 Apl. 18		
....	Gopalsami	15 Apl. 18		
....	Loganadan	15 Apl. 18		

2nd Queen Victoria's Own Sappers and Miners—*concld.*

1st Commn. or date of entering service.	Name and rank.	Army rank.	Present appointment in Corps.	Remarks.
	Jemadars—*concld.*			
....	Innasimuttu	15 Apl. 18		
....	Jaganayakulu	15 Apl. 18		
....	Punnusami	9 May 18		
....	Mariasusai	1 June 18		
....	Krishnasami (V)	1 June 18		
....	Ramulu	1 June 18		
....	Anthony	1 June 18		
....	Rajaratnam	1 June 18		
....	Moung Ba Kin	19 June 18		
....	Maung Gyine	20 June 18		
....	Maung Nee	20 June 18		
....	Maung Zan Kin	20 June 18		
....	Maung Ba Gyan	20 June 18		
....	Sinduraju	1 July 18		
....	Saiyid Abdul Jafar	16 July 18		
....	Madurai	16 July 18		
....	Raju	16 July 18		
....	Kali	16 July 18		
....	Ponnusami	7 Aug. 18		
....	Venkatesape-Rumal	7 Aug. 18		
....	Abdul Qadir	7 Aug. 18		
....	Abdul Latif	7 Aug. 18		
....	Munisami (II)	8 Aug. 18		
....	Munisami (IV)	8 Aug. 18		
....	Muniratnam	8 Aug. 18		
....	Maung Tun Kah	8 Aug. 18		
....	Maung Tin	22 Sep. 18		
....	Rangasaml	9 Oct. 18		
....	Appalasaml	9 Oct. 18		
....	Shaikh Hussain	9 Oct. 18		
....	Qadir Velu	9 Oct. 18		
....	Ramasami	9 Oct. 18		
....	Lingayya	9 Oct. 18		
....	Maung Ba Thein	1 Nov. 18		
....	Maduranayakam	....	Re-employed, 1 Sep. 17.	

3rd Sappers and Miners.

Raised at Bombay in 1820 by Capt. T. Dickinson. Designated 'Engineers' in 1829, and the 'Engineer Corps' 1830, when a corps of Pioneers raised in the last decade of the 18th century was absorbed into it. Became the Corps of Bombay Sappers and Miners, 1840, present designation in 1903.

"Beni Boo Alli" "Ghuznee, 1839" "Affghanistan, 1839" "Khelat" "Mooltan" "Goojerat" "Punjaub" "Reshire" "Bushire" "Koosh-ab" "Persia" "Central India" "Abyssinia" "Kandahar, 1880" "Afghanistan, 1878-80" "Burma, 1885-87" "Tirah" Punjab Frontier" "China, 1900."

Composition – ⅓ Mahrattas, ⅓ Musalmans, ⅓ Sikhs.

Uniform—Scarlet. *Facings*—Blue.

1st Commn. or date of entering service.	Names and Rank.	Army rank.	Present appointment in Corps.	Remarks.
	Commandant.			
	Staff Officers. (3)			
28 July 88	*Bremner, Bt.-Col. (temp. Br.-Genl.) A. G., C.M.G., R.E. (Q)*	*30 July 16*	*31 Mar. 13*	
23 Dec. 10	Rait-Kerr, Capt. R. S., D.S.O., M.C., R.E.	23 Dec. 16	26 June 14	Adjt.
	Company Commanders. (8)			
23 Dec. 98	*Paris, Maj. A. L., R.E.*	*28 May 15*	*12 Sep. 06*	*No. 20 Coy.*
20 Aug. 99	*Tillard, Maj. E. D., D.S.O., R.E.*	*30 Apl. 16*	*1 Apl. 08*	
2 May 00	*Arbuthnot, Capt. A. D.S., R. E.*	*2 May 10*	*14 Feb. 09*	
18 Aug. 00	Boyd, Maj. S., D.S.O., R.E.	2 Nov. 16	1 July 13	No. 20 Coy.
4 Feb. 01	*Loring, Maj. E J., M.C., R.E.*	2 Nov. 16	21 Feb. 11	D. A. A. G., R. E.
21 Dec. 01	Pakenham-Walsh, Maj. W. P., R.E.	21 Dec. 16	....	24 F. Coy.
15 July 93	Binney, Maj. E. V., D.S.O., R. E.	15 July 18	1 July 13	Supdt. of Park (*temp.*)
29 July 04	Stoehr, Capt. C. F., R.E.	30 Oct. 14	9 Apl. 09	No. 19 Coy.
21 Dec. 04	Rawlence, Capt. M., R.E., D.S.O.	30 Oct. 14	16 Apl. 09	No. 18 Coy.
1 Aug. 07	Fox, Capt. E. B., R.E.	30 Oct. 14	14 Aug. 14	No. 3 Pontoon Park.
29 July 08	Campbell, Capt. (*act'g. Maj.*) M. G. G., M. C., R.E.	30 Oct. 14	6 Oct. 11	No. 72 Coy.
23 July 09	Buller, Capt. F. E., M.C., R.E.	23 July 15	23 Apl. 17	No. 71 Coy.
24 Jan. 15	Dolan, Lieut. (*act'g. Capt.*) M. J., R.E.	23 July 16	11 Feb. 15	No. 7 Coy.
	Company Officers. (9)			
18 Dec. 08	*Rohde, Capt. J. H., R E.*	18 Dec. 14	14 Feb. 14	No. 21 Coy.
8 Dec. 15	Cleaver, 2nd-Lt. L.R.E. (T.F.)	8 Dec. 15	20 Aug. 17	I A.R.O. (*on probn.*) No. 72 Coy.
8 Feb. 16	*Smeeth, 2nd-Lt. W. C., M.C., R.E.*	8 Feb. 16	....	Sick Lv. Extended to 26 Oct.
	Attached.			
27 July 89	VC Colvin, Lt-Col. J. M. C., R.E.	18 Jan. 17	24 Dec. 16	Comdt. *tempy.*
12 Feb. 92	Garwood, Maj. F. S., R.E.	12 Feb. 12	1 Dec. 17	Tempy. Supt. of Instn.
20 Dec. 06	Everett, Bt.-Maj. M., D.S.O., R.E.	3 June 18	25 Aug. 15	No. 21 Coy.
23 Dec. 10	Wagstaff, Capt. H. W., M.C., R.E.	23 Dec. 16	15 June 16	No. 73 Coy.
20 July 11	Johnston, Capt. E. F., R. E.	26 June 17	28 Aug. 17	No. 23 Coy.
23 Dec. 11	Hamilton, Capt. H. W. R., D.S.O., M.C., R.E.	26 June 17	2 Nov. 14	No. 20 Coy.
23 Dec. 11	Matthews, Capt. A. B., D.S.O., M.C., R.E.	26 June 17	10 Oct. 14	No. 22 Coy.
23 Dec. 11	Crawford, Capt. K. B. S., R.E.	26 June 17	10 Oct. 14	No. 22 Coy.
16 Sep. 14	Morris, Capt. D., R.E.	3 Nov. 17	17 Feb. 18	F. 3 Coy.
20 July 11	Drake-Brockman, Lt. R. A., R.E.	16 Aug. 13	6 June 16	No. 24 Coy.
27 Oct. 15	Craig, V. F. M.C., R.E.	1 July 17	17 Feb. 18	7th Field Troop.
26 May 16	Napier, Lt. C. S., R.E.	26 Nov. 17	27 Mar. 18	F. 3 Coy.
28 Feb. 17	Boggs, 2nd-Lt., W. W., R. E.	28 Feb. 17	17 Feb. 18	No. 74 Coy.
28 Feb. 17	Foley, 2nd-Lt., P. F., R. E.	28 Feb. 17	17 Feb. 18	F. 7 Coy.
6 June 17	Gueterbock, 2nd-Lt. E. L., R.E.	6 June 17	17 Feb. 18	F. 7 Coy.
6 June 17	Porter, 2nd-Lt. W., R.E.	6 June 17	17 Feb. 18	F. 5 Coy.
6 June 17	Neill, 2nd-Lt. F. H., R.E.	6 June 17	27 Mar. 18	F. 6 Coy.
6 June 17	Baker, 2nd-Lt. S. J., R.E.	6 June 17	17 Feb. 18	F. 6 Coy.
6 June 17	Sydenham, 2nd-Lt. P. S. B., R.E.	6 June 17	17 Feb. 18	F. 3 Coy.
6 June 17	Harrison, 2nd-Lt. H. B., R. E.	6 June 17	17 Feb. 18	No. 71 Coy.
6 June 17	Capron, 2nd-Lt. G. T. H., R.E.	6 June 17	17 Feb. 18	No. 73 Coy.
5 July 11	Pennell, Capt. K.E.L., M.C., R. E. (S. R.)	26 Jun. 17	29 Sep. 16	No. 21 Coy.
7 Sep. 14	Wilson, Capt. J. L., R.E., (T. F.)	1 June 16	12 June 16	
9 Oct. 14	Bennett Lt. (*act'g. Capt.*) H.D., (*temp.*) R. E.	9 Oct. 14	24 May 17	
23 Oct. 14	Gross, Lt. A. A., (*temp.*) R.E.	12 Aug. 16	6 Mar. 17	No. 2 Coy.
4 Feb. 15	Philp, Lt. (*act'g. Capt.*) R. E., (*temp.*) R.E.	1 July 15	28 Nov. 16	F. 4 Coy.
7 July 15	Wallis, Lt. S. A., (*temp.*) R.E.	1 July 17	28 Nov. 16	M. W. S., Bombay.
4 Aug. 15	Hill, Lt. J. H., (*temp.*) R.E.	1 July 17	13 June 17	Directorate of Works.
29 Aug. 15	FitzGerald, Lt. W. R., R. E.	1 July 17	2 Mar. 17	F. 6 Coy.
14 Nov. 15	Allen, 2nd-Lt. H., (*temp.*) R.E.	14 Nov. 15	18 June 16	No. 20 Coy.
5 Jan. 16	Kirby, 2nd-Lt. S. J., (*temp.*) R.E.	5 Jan. 16	30 Sep. 17	

3rd Sappers and Miners—*contd.*

1st Commn. or date of entering service.	Names and Rank.	Army rank.	Present appointment in Corps.	Remarks.
	Attached—*contd.*			
15 Jan. 16	Chapple, 2nd-Lt. C. H., (*temp.*) R.E.	15 Jan. 16	8 Feb. 17	*Lv., ex I., m.c.*
5 Feb. 16	Waugh, 2nd-Lt. A. D., (*temp.*) R. E.	5 Feb. 16	20 June 16	No. 21 Coy.
30 Jan. 16	Hemsworth, Lt. T. G. F., (*temp.*) R. E.	30 Jan. 16	2 Sep. 16	Irrigation Dept.
6 May 16	Friedlander, Lt. O. A., (*temp.*) R. E.	7 Nov. 17	24 Nov. 17	
12 Apl. 16	Stevenson, 2nd-Lt. J.G., R.E.	12 Apl. 16	18 June 16	No. 20 Coy.
1 June 16	Brewer-Little, 2nd-Lt. H., (*temp.*) R. E.	1 June 16	6 Mar. 17	S. & M. Sub-Depôt.
2 July 16	Geoghegan, 2nd-Lt. G., (*temp.*) R.E.	2 July 16	28 Nov. 16	No. 21 Coy.
23 July 16	McAndrew, 2nd-Lt. G. D., (*temp.*) R.E.	23 July 16	28 Nov. 16	Director of Works.
23 July 16	Pearse, 2nd-Lt. G. E., (*temp.*) R. E.	23 July 16	28 July 16	13 Divl. Engr.
19 Aug. 16	Bennett, 2nd-Lt. S. G., (*temp.*) R. E.	19 Aug. 16	30 Sep. 17	
19 Aug. 16	Otto, 2nd-Lt. C. R., (*temp.*) R.E.	19 Aug. 16	28 Nov. 16	No. 20 Coy.
	Curtis, Lt. H., (*temp.*) R.E.	1 July 17	....	No. 21 Coy.
4 Nov. 16	Willis, 2nd-Lt. V. J. (*temp.*) R.E.	4 Nov. 16	13 June 17	No. 71 Coy.
4 Nov. 16	Walter, 2nd-Lt. J.A., (*temp.*) R.E.	4 Nov. 16	13 June 17	
4 Nov. 16	Turnbull, 2nd-Lt. W. J., (*temp.*) R.E.	4 Nov. 16	13 June 17	
4 Nov. 16	Spencer, 2nd-Lt. W. N., (*temp.*) R.E.	4 Nov. 16	13 June 17	
4 Nov. 16	Gardner, 2nd-Lt. (*act'g. Lt.*) J., (*temp.*) R.E.	4 Nov. 16	13 June 17	M. W. S., Aden.
4 Nov. 16	Taylor, 2nd-Lt. J. H. C., (*temp.*) R.E.	4 Nov. 16	13 June 17	
15 Sep. 10	Corbould, Capt. P. S.	13 Sep. 14	1 Aug. 16	I. A. R. O. Cant. Magte.
2 Dec. 16	Barnes, 2nd-Lt. L. F., (*temp.*) R. E.	2 Dec. 16	15 Jan. 18	F. 3 Coy.
27 Jan. 17	Jones, 2nd-Lt. M. P., (*temp.*) R.E.	27 Jan. 17	30 Sep. 17	
3 June 17	Pendlebury, 2nd-Lt. R., (*temp.*) R. E.	3 June 17	26 Sep. 17	
25 July 17	Gilbert, Lt. W. H.	25 July 18	3 Oct. 17	I. A. R. O. No. 73 Coy.
11 Aug. 17	Evans, Lt. E. W., R. E.	11 Aug. 18	21 Sep. 17	I. A. R. O. No. 72 Coy.
23 Sep. 17	Crowley, 2nd-Lt. B. E., (*temp.*) R. E.	23 Sep. 17	4 Dec. 17	I. A. R. O. (*on probn.*) No. 4 F. Troop.
17 Nov. 17	Patterson, 2nd-Lt. H. E. (*T. F.*)	17 Nov. 17	17 Nov. 17	F. 4 Coy.
2 Aug. 12	Arbuthnot, Capt. C. W. E.	2 Aug. 16	25 Oct. 16	No. 4 Field Trp.
20 Nov. 14	Venables, Capt. A. V., M.C.	19 Sep. 16	11 May 15	I. A. R. O.
23 Apl. 15	Tedman, Lt. C., M.C.	23 Apl. 15	2 Dec. 15	I. A. R. O. No. 21 Coy.
13 Nov. 14	Gammon, Capt. J. C.	31 July 17	25 Nov. 14	I. A. R. O.
13 Nov. 14	Turner, Lt. D. M.	13 Nov. 15	18 Nov. 14	Do.
27 Nov. 14	Nicolle, Capt. A. V.	27 Nov. 18	18 Dec. 14	Do. No. 20 Coy.
2 Apl. 15	Yates, Lt. (*act'g. Maj.*) D. R. M.	16 Dec. 15	14 June 15	Do. Fd. Park, Basra.
5 Feb. 15	Buswell, Lt. G.	5 Feb. 16	23 Feb. 15	Do.
5 Feb. 15	Price, Lt. A. R. P.	5 Feb. 16	24 Apl. 18	Do. No. 20 Coy.
12 Feb. 15	Martyn, Lt. (*temp. Capt.*) J. K., M.C.	12 Feb. 16	20 Feb. 15	Do. No. 4 Labour Corps.
12 Feb. 15	Harvey, Lt. (*act'g. Capt.*) J. F. B.	12 Feb. 16	7 Dec. 17	Do. F. 5 Coy.
18 Feb. 15	Bennett, Lt. (*act'g. Capt.*) H. B.	18 Feb. 16	13 Sep. 15	Do. Recruiting, Bombay.
5 Mar. 15	Boyds, Lt. W. R.	5 Mar. 16	10 Sep. 15	Do. No. 17 Coy.
5 Mar. 15	Hobbs, Lt. C. C. A.	5 Mar. 16	23 Oct. 16	Do. *Lv., in I., m. c.*
19 Mar. 15	Sneyd, Lt. R. T. S., M.C.	19 Mar. 16	1 Apl. 15	Do. Faridkot I. S., S. & M.
19 Mar. 15	Watkins, Lt. H.	19 Mar. 16	23 May 18	Do. No. 7 Field Troop.
2 Apl. 15	Pearce, Lt. T.	2 Apl. 16	24 Apl. 16	Do. No. 18 Coy.
2 Apl. 15	Green, Lt. (*act'g. Capt.*) C. W. T.	2 Apl. 16	17 June 15	Do. E. Coy.
22 June 16	Bowers, Lt. (*act'g. Maj.*) P. L., M.C.	22 June 16	5 July 16	Do. Irrigation Dept.
8 July 15	Flewett, Lt. W. E.	8 July 16	5 May 16	Do. Maler Kotla I. S., S. & M.
18 Aug. 15	Nichols, Lt. H. J.	18 Aug. 16	1 Dec. 17	Do. Pontoon Park Advanced Base.
8 Sep. 15	Davidson, Lt. J. R., M.C.	8 Sep. 16	25 Nov. 15	Do. Tehri Garhwal I. S., S. & M.

3rd Sappers and Miners—*contd.*

1st Commn. or date of entering service.	Names and Rank.	Army rank.	Present appointment in Corps.	Remarks.
	Attached—*concld.*			
8 Sep. 15	Lister-Jackson, Lt. A. (*act'g. Capt.*)	8 Sep. 16	25 Nov. 15	I. A. R. O. No. 3 Bridging Train.
29 Oct. 15	Marshall, Lt. J. G., M.C.	29 Oct. 16	27 Dec. 15	Do. No. 18 Coy.
29 Oct. 15	Minnett, Lt. G. C.	29 Oct. 16	27 Dec. 15	Do. No. 18 Coy.
15 Dec. 15	Smart, Lt. C. N. B.	15 Dec. 16	16 Jan. 16	Do. No. 4 Labour Corps.
11 Feb. 17	Cooper, Lt. P. A.	12 Feb. 17	22 Apl. 17	Do. No. 19 Coy.
18 Feb. 17	Richard, Lt. T. J. R.	{ 18 Feb. 17 { 15 Mar. 18	15 Dec. 17	Do. F. 2 Coy.
20 Feb. 16	Preece, Lt. C. B.	20 Feb. 17	5 Apl. 16	Do. M. W. S.
20 Mar. 16	Myers, Lt. N. C.	20 Mar. 17	10 July 16	Do. Works Directorate.
14 Apl. 16	Stevenson, Lt. (*act'g. Capt.*) R. C.	14 Apl. 17	19 Apl. 16	Do.
18 May 16	Cory, Lt. E. M.	18 May 17	19 May 16	Do. No. 19 Coy.
13 June 16	Walker, Lt. F. H.	13 June 17	23 July 16	Do. No. 19 Coy.
17 July 16	Browne, Lt. (*act'g. Capt.*) D. R. H.	17 July 17	1 Aug. 16	Do. Works Directorate.
22 July 16	Williams, Lt. (*act'g. Maj.*) J. K.	22 July 17	1 Sep. 16	Do. Persia.
22 July 16	Bourke, Lt. D. R. S.	22 July 17	28 July 16	Do. 48 Pioneers.
22 July 16	Crossle, Lt. H. C.	22 July 17	10 Aug. 16	Do. Labour Corps.
22 July 16	Evershed, Lt. W. A.	22 July 17	10 Aug. 16	Do. Irrigation Dept.
10 Aug. 16	Mitchell, Lt. R. S. T. J.	10 Aug. 17	10 Oct. 16	Do. Railways.
22 Aug. 16	Bentley, Lt. T.	22 Aug. 17	2 Sep. 16	Do. E. and M. Section.
6 Sep. 16	Wardle, Lt. F. M.	6 Sep. 17	6 Oct. 16	Do. Tehrigarhwal I. S., S. & M.
1 Oct. 16	Milchem, Lt. H. G.	1 Oct. 17	20 Oct. 16	Do. No. 19 Coy.
6 Oct. 16	Cooper, Lt. L. K.	6 Oct. 17	21 Nov. 16	Do.
11 Oct. 16	Lawrence, Lt. E.	11 Oct. 17	2 May 17	Do.
10 Nov. 16	Scott, Lt. O. R.	10 Nov. 17	4 Feb. 17	Do. No. 24 Coy.
21 Nov. 16	Davies, Lt. H.	21 Nov. 17	31 Dec. 16	Do. F. 3 Coy.
5 Jan. 17	Howarth, Lt. H. S.	5 Jan. 18	18 Mar. 17	Do.
5 Jan. 17	Pirie, Lt. D. G. L.	5 Jan. 18	18 Mar. 17	Do. No. 18 Coy.
5 Jan. 17	Macdonald, Lt. A.	5 Jan. 18	. . .	Do.
5 Jan. 17	Wotherspoon, Lt. W. S.	5 Jan. 18	18 Mar. 17	Do. No. 2 Bridging Train.
1 Feb. 17	Bush, Lt. S. P.	1 Feb. 18	23 Apl. 17	I. A. R. O. F. No. 3 Coy.
1 Feb. 17	Morrison, Lt. A. O. H.	1 Feb. 18	23 Apl. 17	Do. No. 19 Coy.
5 Feb. 17	Wentworth-Hyde, Lt. E.	5 Feb. 18	23 Apl. 17	Do.
1 Mar. 17	Frattini, Lt. G. N.	1 Mar. 18	25 Apl. 17	Do. Tehrigarhwal, I. S., S. & M.
5 May 17	Bracy, 2nd-Lt. S. W.	5 May 17	21 Aug. 17	Do. No. 3 Bridging train.
5 May 17	Robertson, 2nd-Lt. W. V.	5 May 17	20 Aug. 17	Do.
5 May 17	Cambridge, 2nd-Lt. S. J.	5 May 17	20 Aug. 17	I. A. R. O. No. 3 Bridging Train.
5 May 17	Devonshire, 2nd-Lt. C. H.	5 May 17	25 Aug. 17	I. A. R. O. No. 19 Coy.
5 May 17	Phillips, 2nd-Lt. W. H.	5 May 17	20 Aug. 17	I. A. R. O. Sig. Ser.
25 Jan. 17	Russell, Lt. E. P., D.C.M.	25 Jan. 18	6 Nov. 17	I. A. R. O. Maler Kotla I. S., S. & M.
5 May 17	Fritchley, Lt. G. B.	5 May 18	10 Oct. 17	Do. Sirmoor S. & M.
12 May 17	Henderson, Lt. H. W.	12 May 18	. . .	Do. A. D. L. S.
20 June 17	DeBrunner, Lt. W. E.	20 June 18	8 Oct. 17	Do.
20 June 17	Swithenbank, Lt. R. S.	20 June 18	2 Oct. 17	Do. Tehrigarhwal I. S., S. & M.
8 Aug. 17	Chambers, Lt. T. B. W.	8 Aug. 18	5 Nov. 17	Do. Works Directorate.
5 Oct. 17	Morris, Lt. M. G.	5 Oct. 18	18 Jan. 18	Do. Bombay D.L.S.
1 Nov. 17	Brunton, Lt. T. S.	1 Nov. 18	16 Mar. 18	I. A. R. O. Bombay D. L. S.
5 Nov. 17	Innes, Lt. G. A.	5 Nov. 18	16 Mar. 18	I. A. (T. C.) F. No. 4 Coy.
8 Dec. 17	McAuliffe, 2nd-Lt. J. P. T.	8 Dec. 17	12 Jan. 18	I. A. R. O. F. No. 5 Coy.
20 Dec. 17	Marsh, 2nd-Lt. G. S.	20 Dec. 17	10 Jan. 18	Do. No. 72 Co.
5 Jan. 18	Flynn, 2nd-Lt. (*act'g. Capt.*) J. W.	5 Jan. 18	3 June 18	I. A. (T. C.)
15 Jan. 18	Bedford, 2nd-Lt. E. W.	15 Jan. 18	3 June 18	I. A. R. O.
15 Jan. 18	Clifford, 2nd-Lt. E. K.	15 Jan. 18	3 June 18	Do.
15 Jan. 18	Jackson, 2nd-Lt. A. W. J.	15 Jan. 18	3 June 18	Do.
15 Jan. 18	Kelly, 2nd-Lt. (*act'g. Capt.*) A. W. J.	15 Jan. 18	3 June 18	Do.
...	Miles, 2nd-Lt. E.	. . . .	. . . .	Do.
...	Shaw, 2nd-Lt. G. R.	. . . .	25 Aug. 18	Do.
7 Nov. 18	MacLean, 2nd-Lt. C. H. M.	7 Nov. 18	25 Nov. 18	Do.
	Honorary Officer.			
14 June 16	Maler Kotla, *His Highness Nawab Sir* Ahmad Ali Khan *Bahadur*, K.C.S.I., *Nawab of* ——	14 June 16	14 June 16	Hony. Major in Army.
	Commissary.			
28 Jan. 04	Hudson, *Hony.* Maj. J. R.	19 Sep. 05	10 June 97	Offg. Supdt. of Park.
	Warrant Officer.			
...	Miers, Sergt.-Major W. H.	3 Jan. 11	3 Jan. 11	

		Jemadar.	Subadar.	Subadar. Major.	
	Subadar-Majors.				
10 Sep. 89	Arjun Powar (56)	27 July 01	1 Oct. 07	1 July 14	F. No. 4 Coy.
20 Apl. 96	Firoz Ali, I.O.M., *Bahadur*	1 Nov. 12	18 Nov. 14	26 Oct. 15	F. No. 5 Coy.

3rd Sappers and Miners—*concld.*

1st Commn or date of entering service.	Names and Rank.	Army Rank.		Remarks.
	Subadars.			
1 Jan. 85	Poshati Mari, *Bahadur* (58)	26 Feb. 01	16 Oct. 06	F. No. 6 Co. *Order of Br. Ind., 2nd cl.*
1 Nov. 01	Bhawani Singh (50)	1 Nov. 01	16 Apl. 12	
6 May 91	Malla Singh, M.C., I.O.M. (56) (56).	2 Aug. 06	1 Sep. 13	F. No. 4 Co.
3 Nov. 93	Ganpat Mahadeo, *Bahadur*	1 July 09	1 July 14	No. 71 Coy. *Order of Br. Ind., 2nd cl.*
17 Oct. 94	Ismail Khan, I.D.S.M. *Bahadur*	1 Nov. 13	27 Dec. 14	F. No. 3 Coy. *Order of Br. Ind., 2nd cl.*
25 Aug. 96	Shah Sawar	1 Aug. 12	26 Apl. 16	No. 24 Coy.
20 Feb. 01	Muhammad Din, I.O.M. .	1 July 14	29 Sep. 15	No. 22 Coy.
12 Nov. 98	Khan Muhammad Khan .	15 Jan. 14	26 Oct. 15	No. 73 Coy.
30 June 94	Bir Singh I.O.M. (56), . .	1 Sep. 13	17 Apl. 16	No. 21 Coy.
15 Sep. 95	Karim Khan . . .	16 Apl. 12	22 June 16	No. 18 Coy. *Order of Br. Ind., 2nd cl.*
26 Sep. 98	Dhondibarao Barge, I. O. M. *Bahadur* (50)	1 Jan. 13	26 May 16	F. No. 24 Coy.
2 July 93	Gujar Singh (56) . . .	21 Nov. 11	1 Oct. 16	No. 19 Coy.
24 Mar. 07	Sardar Khan (*Bahadur*) . .	24 July 15	1 Apl. 17	No. 20 Coy. *Order of Br. Ind. 2nd cl.*
6 July 97	Baryam Singh	9 Sep. 14	16 July 17	No. 3 Pontoon Park.
10 Feb. 03	Liyakat Ali, I.O.M. . . .	23 July 17	24 July 17	No. 21 Coy.
1 Aug. 93	Lakshiman Powar, I D.S.M. (56)	1 Aug. 12	1 Sep. 17	No. 72 Coy.
1 July 06	Maroti Jadhao, I.D.S.M. . .	27 June 16	13 July 18	No. 18 Coy.
26 Dec. 02	Ghulam Nabi, I.O.M. . .	1 Jan. 16	1 Aug. 18	No. 71 Coy.
1 Apl. 94	Kailas Singh, I.D.S.M. . .	14 June 16	1 Sep. 18	F. 3 Coy.
	Jemadars.			
21 Mar. 94	Ganpatrao Jadhao . . .	1 Apl. 09	. . . . .	F. No. 1 Coy.
6 July 97	Sawan Singh	9 Sep. 14	. . . . .	No. 21 Coy.
11 Dec. 98	Sripat Barge	9 Sep. 14	. . . . .	F. 7 Coy.
20 Feb. 01	Datadin Badhai, I.D.S.M. .	30 Oct. 14	. . . . .	No. 23 Coy.
15 Nov. 93	Ashanna Poshati	5 Nov. 14		No. 24 Coy.
8 Nov. 99	Tek Singh	18 Nov. 14	. . . .	F. 2 Coy.
16 Dec. 02	Sitaramrao Vichare . . .	27 Dec. 14	. . . . .	F. 3 Coy.
17 Jan. 03	Nur Alam, I.D.S.M. (I) . .	16 Mar. 15	1 May 18	No. 74 Coy.
8 Dec. 99	*Fateh Khan, I.D.S.M.*	*15 Apl. 15*	. . . .	*No. 17 Coy.*
28 Jan. 03	Uttam Singh, I.O.M. . . .	20 Apl. 15	1 June 18	F. 3 Coy.
7 Nov. 02	Muhammad Shah	16 July 15	24 June 18	No. 19 Coy.
1 Jan. 96	Gangaji Rao, Khanvilkar, I.D.S.M.	25 July 15	. . . . .	F. 1 Coy.
1 May 86	Shankram Pille	29 Sep. 15	. . . . .	No. 22 Coy.
3 Sep. 98	Krishnaji Rao Bhonsle, I.D.S.M.	29 Sep. 15	. . . . .	No. 17 Coy.
7 Sep. 00	Bir Singh	12 June 16	. . . . .	No. 19 Coy.
5 June 01	Habibullah Khan, I.D.S.M. .	12 June 16	. . . . .	No. 3 Bridging Train.
1 Aug. 94	Farzand Khan I.D.S.M. . .	1 Oct. 16	. . . . .	Aden, D. L. S
8 Apl. 02	Nur Alam (II)	11 Nov. 16	. . . . .	No. 20 Coy.
17 Feb. 07	Gonda Singh	18 Jan. 17	. . . . .	No. 20 Coy.
15 Feb. 96	Shiubaran Misr	24 Apl. 17	. . . . .	No. 72 Coy.
1 July 02	Abdullah Shah	24 Apl. 17	. . . . .	No. 19 Coy.
21 Dec. 86	Sidram Pille	24 Apl. 17	. . . . .	F. 5 Coy.
10 Jan. 03	Ghulam Muhammad . . .	2 June 17	. . . . .	Attd. F. 4 Coy.
5 Jan. 86	Shaikh Hasan	2 June 17	. . . . .	F. 4 Coy.
15 Oct. 04	Muhammad Bakhsh, I.O.M., I.D.S.M.	2 June 17	. . . . .	No. 4 Fd. Trp.
20 Feb. 01	Ali Husain	2 June 17	. . . . .	No. 72 Coy.
24 May 03	Chanan Singh, I.D.S.M. . .	14 July 17	. . . . .	No. 21 Coy.
11 Aug. 03	Ali Bukhsh Khan	14 July 17	. . . . .	No. 20 Coy.
4 July 03	Sardar Khan	14 July 17	. . . . .	No. 18 Coy.
1 July 03	Khana	1 Sep. 17	. . . . .	No. 74 Coy.
16 May 94	Shiudatt Singh	1 Sep. 17	. . . . .	No. 73 Coy.
11 Jan. 87	Krishna Swami	1 Sep. 17	. . . . .	F. 6 Coy.
15 Jan. 99	Rama Limbalkar	1 Sep. 17	. . . . .	Bombay, D. L. S.
19 Mar. 04	Gurmukh Singh, I.O.M. . .	1 Sep. 17	. . . . .	F. 5 Coy.
18 Sep. 07	Sopana Thopta	1 Sep. 17	. . . . .	No. 71 Coy.
19 Nov. 00	Mirza Khan	1 Sep. 17	. . . . .	No. 73 Coy.
24 Apl. 08	Jalal Din	1 Sep. 17	. . . . .	No. 24 Coy.
29 Sep. 03	Fazal Din	1 Sep. 17	. . . . .	No. 72 Coy.

3rd Sappers and Miners—*concld.*

1st Commn. or date of entering service.	Names and Rank.	Army Rank.		Remarks.
	Jemadars—*contd.*			
4 Feb. 07	Ahmad Shah	1 Sep. 17	....	No. 7 Coy.
29 Jan. 09	Firoz Din	1 Sep. 17	....	No. 71 Coy.
6 July 06	Gulab Khan	1 Sep. 17	....	No. 4 Fd. Troop.
24 Nov. 09	Pahlwan Khan	1 Sep. 17	....	F. 2 Coy. Ind. Adjt.
3 Apl. 01	Ramji Bhonsle	1 Dec. 17	....	No. 73 Coy.
20 Feb. 01	Shankar Ashenna	1 Feb. 18	....	F. 4 Coy.
8 Apl. 07	Raghojirao Sinde	1 Feb. 18	....	F. 7 Coy.
	Sopana The Warkar	27 Apl. 18		
	Taj Din	27 Apl. 18		
	Said Alam	1 May 18		
	Muhammad Khan	1 May 18		
	Mahabir Singh	1 May 18		
	Chanda Singh	1 May 18		
	Ahmad Din	1 June 18		
	Muhammad Juma	1 June 18		
	Ganpatrao More	1 June 18		
	Sawariya Khan	1 June 18		
	Hari Singh	24 June 18		
	Ghulam Muhammad	24 June 18		
	Ghulam Haidar	24 June 18		
	Baryam Singh	28 June 18		
	Hayat Muhammad	13 July 18		
	Sheochand Singh	21 July 18		
	Lalu Rajena, I.O.M.	1 Aug. 18		
	Mihr Dad	1 Aug. 18		
	Ramji Sinde	1 Aug. 18		
	Chandika Parshad	1 Aug. 18		

Railway Battalion.

Formed in 1916 under Government of India, Army Department, letter No. H-2709, dated 11th April 1916.

Composition.— 25th and 27th Companies.—Punjabi Musalmans.
26th and 2$_9$th Punjabi Musalmans and Sikhs.
28th Company.—Punjabi Christians and Cis.-Sutlej Mahommedans.

Uniform – Khaki.

1st Commn. or date of entering service.	Names and Rank.	Army rank.	Present appointment in Corps.	Remarks.
	Commandant.			
31 Dec. 96	Malan, Bt.-Lt.-Col. L. N. R.E.	1 Jan. 17*	18 Dec. 16	
	Staff Officer.			
8 Jan. 10	Morris, Capt. J. H., R. E. (Spl. Res.)	26 June 17	18 Dec. 16	Adjt.
	Company Commanders.			
31 July 02	Kirke, Maj. E. St. G., D.S.O., R.E.	31 July 17	1 Oct. 14	*Lv., ex I., m. c.*
25 July 06	Woodhouse, Capt. H. L., M. C., R. E.	30 Oct. 14	7 May 11 / 6 June 15	Comdg. 25 Coy.
31 July 09	Roberts, Capt. J. R., M.C., R.E.	31 July 15	1 Oct. 12	No. 28th Coy. *Lv., ex I. m. c.*
23 Dec. 09	Gordon, Capt. R. E., M.C., R.E.	23 Dec. 15	7 Dec. 15	O. C. Depôt.
25 June 10	Roche, Capt. R. T. H., R.E. (Spl. Res.)	26 June 17	21 Apl. 17	Depôt.
21 Sep. 14	Watkins, Temp. Lieut. P. J., R.E.	21 Sep. 14	9 Apl. 17	26 Coy.
14 Dec. 14	Scrivener, Lt. (*act'g. Capt.*) R. F. S., R. E. (207*a*)	19 Feb. 15	30 Aug. 17	No. 29th Coy.
30 Apl. 15	Marriott, Lieut. R.E.	30 Apl. 16	27 July 18	I.A.R.O. Depôt.
	Company Officers.			
1 Dec. 15	Wigley, Lt. A. J., M.C.	1 Dec. 15	27 Jan. 18	I. A. R. O. (*on probn.*)
26 Mar. 15	Moylan, Lt. (*Actg. Capt.*) W. D	26 Mar. 16	8 Apl. 18	Do. No. 27th Coy.
30 Apl. 15	Bunting, Lieut. S. A. S., M.B.E.	30 Apl. 16	....	I. A. R. O. Depot.
8 Sep. 15	Anderson, Lt. A. T. D.	5 May 18	17 Oct. 17	I. A. R. O. No. 28 Coy.
7 May 16	Ellison, Lt. H. F. (*temp.*) R. E.	7 May 16	22 Aug 18	
14 May 15	White, Lieut. C. R., M.C.	14 May 16	1 May 16	Do. No. 25th Coy.
24 Apl. 16	Splane, Lieut. E. G.	24 Apl. 17	1 May 16	Do. Depot.
10 Nov. 16	Beveridge, Lt. R.	10 Nov. 17	1 Feb. 17	Do. No. 28 Coy.
....	Cardew, Lt. W. F. (*temp.*) R. E.	....	....	
....	Sim[illegible]s Lt. A. M. (*temp.*) R. E.	....	....	
19 Aug. 16	Delamain, 2nd-Lt. J. H.	19 Aug. 16	3 Jan. 18	106th Pioneers No. 25 Coy.
12 Aug. 17	Cropper, 2nd-Lt. H.	12 Aug. 17	21 Jan. 18	No. 26th Coy.
1 Oct. 17	Thomas, Lt. P. E.	1 Oct. 18	21 Jan. 18	I. A. R. O. No. 26 Coy.
5 Nov. 17	Julian, Lt. A. L.	5 Nov. 18	....	I. A. (*T. C.*) No. 29 Coy.
29 Dec. 17	Higby, 2nd-Lt. R. W. E. (*temp.*) R.E.	29 Dec. 17	21 Jan. 18	No. 28th Coy.
	Minniss, 2nd-Lt. A. O. (*temp.*) R.E.	....	21 Jan. 18	No. 28th Coy.
...	Wood, 2nd Lt. N. R. T. (48*a*, 51)	27 May 17	....	T. F. Depôt.
...	Browne, 2nd-Lt. V. S. T.	5 Jan. 18	....	I. A. R. O. 25 Coy.
...	Dillow, 2nd-Lt. D. H. (46)	...	....	T. F. Depôt.
5 Oct. 17	Bell, Lt. H. H.	5 Oct. 18	....	I.A.R.O.
....	Egan, 2nd-Lt. E.M. (*temp.* R.E.)	....	18 July 18	
....	Strange, 2nd-Lt. F.	....	....	I.A.R.O.
		Jemadar.	Subadar.	Subadar-Major.
	Subadar-Major.			
1 Feb. 03	Pande Khan, I.D.S.M., *Bahadur*	11 Feb. 03	1 May 14	15 Apl. 16 — *Order of Br. Ind., 2nd cl.*
	Subadars.			
22 May 88	Sher Baz, I.D.S.M., *Bahadur*	20 Feb. 06	1 Oct 13	No. 26 Coy. *Order of Br. Ind., 2nd cl.*
25 Jan. 97	Fazal Din, I.D.S.M. *Bahadur*	1 Oct. 13	1 Apl. 16	*Order of Br. Ind. 2nd cl.* No. 29th Coy.
14 Feb. 03	Alif Khan, I.D.S.M.	20 Feb. 06	1 Aug. 17	Depôt.
26 Feb. 03	Nathe Khan	1 Mar. 16	24 Sept. 7	No. 2 th Coy.
21 May 03	Fateh Khan, I.D.S.M.	1 Apl. 06	17 Nov. 17	No. 25 Coy.
	Jemadars.			
1 May 03	Mir Baz	1 Mar. 16	Jemdr. Adjt., No. 28 Coy.	
1 Jan. 06	Abdul Aziz	1 May 16	29 Coy.	
...	Mangal Singh, I.D.M.S.	1 May 16	Depôt.	
14 Feb. 03	Firoz Khan	15 May 16	25 Coy.	
1 Jan. 06	Muhammad Din	1 Aug. 16	28 Coy.	
1 July 96	Pala Singh, I.D.S.M.	14 July 17	26 Coy.	
15 Aug. 06	Amir Khan	1 Aug. 17	25 Coy.	
25 July 03	Shahzullah Khan	24 Sep. 17	27 Coy.	
16 Mar. 06	Ali Ahmad	17 Nov. 17	28 Coy.	
16 Feb. 03	Inayat Ali	29 Apl. 18	27 Coy.	
1 Mar. 03	Sajawal Khan	29 Apl. 18	29 Coy.	

SIGNAL UNITS.

NO. 31 (DIVISIONAL SIGNAL) COMPANY (Now known as the 1st Corps Signal Co.) Raised at Fatehgarh, 15th February 1911 by Capt. H. S. E. Franklin, 15th Sikhs.
NO. 32 (DIVISIONAL SIGNAL) COMPANY. Raised at Fatehgarh, 15th February 1911, by Capt. W. F. Maxwell, R.E.
NO. 33 (DIVISIONAL SIGNAL) COMPANY. Raised at Ahmednagar, 1st March 1911, by Capt L. H. Queripel, R.F.A.
NO. 34 (DIVISIONAL SIGNAL) COMPANY. Raised at Ahmednagar, 1st March 1911, by Capt. R. G. Earle, R.E.
NO. 35 (DIVISIONAL SIGNAL) COMPANY. Raised at Meerut, 18th August 1914, by Major R. G. Earle, R.E.
NO. 36 (DIVISIONAL SIGNAL) COMPANY. Formed 1st April 1916.
NO 37 (DIVISIONAL SIGNAL) COMPANY. Formed 5th March 1917. (Now 18th D. S. Coy.)
NO 38 (DIVISIONAL SIGNAL) COMPANY. Formed 8th January 1917.
NO. 39 (DIVISIONAL SIGNAL) COMPANY Formed 2nd June 1917.
NO. 40 (DIVISIONAL SIGNAL) COMPANY. Formed 2nd June 1917.
NO. 43 (DIVISIONAL SIGNAL) COMPANY Formed 21st April 1917.
NO. 44 (DIVISIONAL SIGNAL) COMPANY. Formed 27th May 1918.
NO. 45 (DIVISIONAL SIGNAL) COMPANY. Formed 19th June 1918.
NO. 46 (DIVISIONAL SIGNAL) COMPANY. Formed 19th July 1918.
NO. 47 (DIVISIONAL SIGNAL) COMPANY. Formed 1st Sept. 1918.
NO. 48 (DIVISIONAL SIGNAL) COMPANY. Formed 1st January 1918, as No. 2 L. of C. Company.
ADEN (DIVISIONAL SIGNAL) COMPANY. Formed 26th March 1917.
14TH (DIVISIONAL SIGNAL) COMPANY, Formed 21st September 1915.
15TH (DIVISIONAL SIGNAL) COMPANY. Formed 23rd March 1916.
17TH (DIVISIONAL SIGNAL) COMPANY. Formed 6th July 1917.
ARMY CORPS SIGNAL COMPANY. Formed 2nd June 1917.
NO 1 L. OF C. SIGNAL COMPANY.
PUNJABI BRIGADE SIGNAL SECTION. Formed 13th April 1917.
46TH BRIGADE SIGNAL SECTION.
47TH BRIGADE SIGNAL SECTION.
1ST CAVALRY BRIGADE SIGNAL TROOP. Formed 13th January 1917.
4TH CAVALRY BRIGADE SIGNAL TROOP. Formed 13th January 1917.
6TH CAVALRY BRIGADE SIGNAL TROOP. Formed 13th January 1917.
7TH CAVALRY BRIGADE SIGNAL TROOP. Formed 13th January 1917.
S'BAD CAVALRY BRIGADE SIGNAL TROOP. Formed 13th January 1917.
10TH CAVALRY BRIGADE SIGNAL TROOP. Formed 28th December 1917.
11TH CAVALRY BRIGADE SIGNAL TROOP. Formed 13th April 1918.
12TH CAVALRY BRIGADE SIGNAL TROOP. Formed 13th APRIL 1918.
13TH CAVALRY BRIGADE. Formed 13th Apl. 1918.
NO. 41 WIRELESS SIGNAL SQUADRON. (Now known as No. 2 (L. of C.) Wireless Signal Squadron.) Raised at Saugor 1st March 1911, by Capt. D. A. Thomson, R.E.
NO. 3 WIRELESS SIGNAL SQUADRON. Formed 5th May 1917.
NO. 5 WIRELESS SECTION.
SIGNAL SERVICE DEPOT.
WIRELESS SIGNAL SERVICE DEPOT

1st Commn. or date of entering service.	Names.	Corps.	Army Rank.	Date of appt. to signal service.	Remarks.
	Majors.				
23 Sep. 97	Maxwell, W. F., D.S.O.	R. E.	30 Oct. 14	15 Feb. 11	D. A. A. G., Army Hdqrs.
4 Aug. 97	*Franklin, H. S. E.*, D.S.O., *(Bt. Lt.-Col 3 June 16) (temp. Brig.-Genl.)*	*15th Sikhs*	*4 Aug. 15*	*15 Feb. 11*	Dir. Sigs.
11 Feb. 99	Torrie, C. J., D.S.O. *(temp. Lt.-Col.)*	30th Punj.	1 Sep. 15	14 Aug. 14	Comdt., S. S. D.
11 Aug. 00	*Johnston, P. D. C.*, D.S.O.	*8 Rajputs*	*1 Sep. 15*	....	Comdt., Army Sig. Coy.
12 Feb. 02	*Paulson, P. Z.*	*Manch. R.*	*24 Oct. 16*	....	Comdt., 17 D. S. C.
18 Aug. 00	Day, J., D.S.O. *(temp. Lt.-Col.)*	R. E.	2 Nov. 16	10 Aug. 11	I. A. S.
18 Aug. 00	Hawtrey, H. C., C.M.G., D.S.O. *(Bt. Lt.-Col., 1 Jan. 17.)*	R. E.	2 Nov. 16	....	Comdt., No. 3 W. S. S.
10 Oct. 03	*Cardew, H. S.*, M.C.	*39 Garh. R.*	*10 Oct. 18*		
	Captains.				
18 Jan. 05	Warden, W. R. *(actg. Maj.)*	75 C. I.	18 Jan. 14	10 May 11 / 2 Oct. 14	Comdt. 43 D.S.C.
5 Aug. 05	*Morse, L. H.* M.C.	*62 Punj.*	*5 Aug. 14*	*9 May 12*	*Lv., ex I., m.c., 6 mos.*
5 Aug. 05	*Morrell, H. G.*	*119th Inf.*	*5 Aug. 14*	*31 Feb. 11*	
20 Oct. 02	*Booth, F. (Bt.-Major, 3 June 15)*	*R. Lanc. R.*	*22 Oct. 14*	*16 Mar. 11*	
4 June 04	*Knight, J. H., (Bt.-Major, 3 June 17) (temp. Lt.-Col.)*	*R. G. A.*	*30 Oct. 14*	*15 Feb. 11*	*A. D. Sigs., 1st Corps.*

1st Commn. or date of entering service.	Names.	Corps.	Army Rank.	Date of apptt. to signal service.	Remarks.
	Captains—*contd.*				
25 July 06	Stebbing, H. T.	R. G. A.	30 Oct. 14	1 May 14	Comdt., S. C. Signal School.
20 Dec. 05	Day, C. L. (*temp. Maj.*)	R. G. A.	30 Oct. 14	9 May 12	Comdt., 33 D. S. Coy.
14 Oct. 08	Duke, V. W. H., M.C. (*temp. Maj.*)	Ches. R.	5 Dec. 14	20 Aug. 13	Comdt., 38 D. S. Coy.
26 Sep. 06	*Wersdell, G. B. (Bt. Maj., 1 Jan. 18.)*	*Yorks. R.*	*25 Jan. 15*	*15 Oct. 14*	*Comdt., Sig. Schl.*
22 Apl. 08	*Allen, H. I.,* D.S.O. *(Bt. Maj.) (temp. Lt.-Col.)*	*N. Staff. R.*	*1 Mar. 15*	*6 July 12*	*A. D. Sigs., G. H. Q.*
6 Oct. 06	Drayson, Fitz A., M.C.	Bord. R.	28 Apl. 15	16 Mar. 14	Ch. Instr., S. S. D.
9 Oct. 12	*St. John, F. O.,* D.S.O., M.C. *(temp. Maj.)*	*R. Scots.*	*21 Aug. 15*	*14 Aug. 14*	*Comdt., 13 D.S.C.*
29 Aug. 06	*LeCornu, C. J. S.,* M.C. *(temp. Maj.)*	*56 Rifs.*	*29 Aug. 15*	*6 Nov. 12*	*Comdt., 18 D.S.C.*
6 Oct. 06	*Pott, D.,* D.S.O., M.C. *(temp. Maj.)*	*13 Lrs.*	*1 Sep. 15*	*12 Mar. 15*	*Comdt., 32 D.S.C.*
20 Dec. 06	Taylor, A. T. W. (temp. Maj.)	43 Regt.	1 Sep. 15	25 Sep. 14	Comdt., 36 D. S. C.
17 Aug. 07	*Metcalfe, E. D.,* M.C.	*3 Horse*	*1 Sep. 15*	....	
17 Aug. 07	*Meredith, A. E.,* M.C. *(temp. Maj.).*	*114 Mah.*	*1 Sep. 15*	....	*Comdt., 15 D.S.C.*
17 Aug. 07	*De Butts, F. C.,* D.S.O., M.C. *(temp. Maj.)*	*55 Rifs.*	*1 Sep. 15*	*14 Aug. 14*	*Comdt., 35 D.S.C.*
17 Aug. 07	Empson, A. H. A. (*actg. Maj.*)	8 Cav.	1 Sep. 15	20 Feb. 13	Comdt., Depôt Coy.
25 Jan. 08	Copeland, A. G. J. (*actg. Maj.*), *4 Oct. 18*).	5 Cav.	1 Sep. 15	18 Feb. 14	Comdt., Depôt Coy.
9 Sep. 08	*Gatherer, A. H. N.,* M.C.	*46 Punj.*	*1 Sep. 15*	*14 Nov. 14*	
9 Sep. 08	*Pulverman, A. W. (temp. Maj.)*	*128 Prs.*	*1 Sep. 15*	....	
8 Sep. 09	*Osanne, G. D.*	*99 Inf.*	*1 Sep. 15*	....	*Lv. ex. I.*
29 Jan. 10	Deedes, R. B., M.C. (*actg. Maj.*, *17 Oct. 18*).	31 Punj.	1 Sep. 15	16 Oct. 18	2nd-in-Comd., S. S. Depôt.
18 Jan. 11	*Pollard, G. L. G.*	*106 Prs.*	*1 Sep. 15*	....	*Comdt., 48 D.S.C.*
18 Jan. 11	*Radley, H. P.,* M.C. *(temp. Maj.)*	*72 Punj.*	*1 Sep. 15*	....	*Comdt., 14 D.S.C.*
....	*Stoneham, H. F.*	*1 Bn. E. Surr. R.*	*6. Sep 15*	....	*1st Corps Sig. Coy.*
1 July 10	Ferguson, J. (*actg. Maj.*, *20 Nov. 97*).	R. E. (S. R.)	23 Sep. 15	1 Feb. 15	Comdt., 39 D.S.C.
20 Sep. 11	Hawkins, A. E. (*actg. Maj.*, *2 June 18*).	Dorset. R.	1 Oct. 15	23 Nov. 16	Comdt., 44 D.S.C.
20 Dec. 11	*Barker, R. R.*	*S. Lan. R.*	*11 Oct. 15*	....	
31 July 14	*Blake, A. L.*	*1-5 Som. L. I.*	*1 Jan. 16*	....	*2nd-in-Comd., 14 D. S. C.*
20 Jan. 12	Turner, C. J. R. (*temp. Maj.*).	15 Sikhs.	20 Jan. 16	1 Sep. 16	Comdt., 40 D.S.C.
8 Aug. 14	Drummond, J. L., M.C.	R. Sc. Fus.	17 Mar. 16	27 Oct. 16	W. S. S.
55 Nov. 11	*Stanley, G.*	*1-4 Wilts. R.*	*1 June 16*	*11 Dec. 15*	*2nd-in-Comd., 18 D. S.C.*
18 June 13	*Coleman, R. E.*	*1-10 Middx. R.*	*1 June 16*	*1 Aug. 16*	*Army Sig. Coy.*
7 June 13	*Stokes, C. L. L.*	*1-5 W. Kent. R.*	*1 June 16*	....	*2nd-in-Comd., 17 D. S.C.*
26 Nov. 13	Butcher, A. C. S.	S. Wales Bord.	1 June 16	....	33 D. S. C. (School).
11 Dec. 13	*Sturge, E.*	*1-10 Middx. R.*	*1 June 16*	....	
1 Aug. 14	Churchill, E. A. H. (*actg. Maj., 29 Oct. 18*).	1-5 S. L. I.	1 June 16	5 July 17	Comdt., 47 D. S. C.
20 Aug. 14	*Hansard, J. H.*	*1-5 Hamps. R.*	*1 June 16*	*17 Aug. 16*	*2nd-in-Comd., 13 D. S.C.*
20 Dec. 13	Macdonald R. H. (*temp. Maj.*)	S. Lan. R.	2 July 16	20 June 16	Comdt., 46 D. S. C.
24 Aug. 12	*Dane, H.*	*Guides*	*24 Aug. 16*	....	
1 Oct. 16	Foot, P. B. D.-H.	Midd'x. R.	1 Oct. 16	21 Dec. 15	Comdt., Signal School, Lucknow.
14 Oct. 14	*Wilson, R. P.*	*1-4 R. W. Kent R.*	*9 Nov. 16*	*17 Sep. 17*	*Comdt., Aden Sig. Coy.*
13 Dec. 10	*Sykes, A. C.,* D.S.O. *(temp. Maj.)*	*R. E.*	*23 Dec. 16*	....	*A. D. Sigs. (Wireless) G. H. Q.*
4 Oct. 11	*Howe, W. T.*	*2nd Bn., Som. L. I.*	*25 Mar. 17*	....	*2nd-in-Comd., 48 D. S.C.*
16 Sep. 14	*Whitestone, B. F.,* M.C.	*R. E.*	*3 Nov. 17*	....	*15 D. S. C.*
....	*Mitchell, C.A.*	*4 Bn. Devon. R.*	*19 May 17*	....	*17 D. S. C.*
....	*Harris, E.*	*R. E. (T. C).*	*9 Fb. 18*	....	*Attd. I. T. D.*
26 Mar. 15	Wishart, W. J.	R. G. A. (T. C.)	9 Feb. 18	5 Oct. 18	46 D. S. C.
8 Aug. 14	*Fausset-Baker, A. A.*	*21 Punj.*	*8 Aug. 18*	..	...
16 Sep. 14	*Marwood, S. L.*	*I. A. Res.*	*16 Sep. 18*	*25 May 15*	*Sig. School.*
10 Oct. 14	*Osborne, F. W.*	*I. A. R. O.*	*1 Oct. 18*	....	

1st Commn. or date of entering service.	Names.	Corps.	Army rank.	Date of appt. to Signal Service.	Remarks.
	Lieutenants.				
9 Dec. 11	Hutchison, B. O.	7 Huss.	9 Aug. 13	. . . .	
. . . .	Higgs, H. J.	R. E.	5 Aug. 14	. . . .	...
5 Aug. 14	Rooker, G.	I. A. Res.	1 Sep. 14	4 Sep. 16	Army Sig. Coy.
22 Apl. 14	Holt, V. (act'g. Capt.) (57)	1-9 Middx. R.	14 Oct. 14	17 Aug. 16	36 D. S. Coy. (School).
3 Sep. 13	Lysaght-Griffen, E. H. (act'g. Capt.)	Durh. L. I.	23 Oct. 14	17 Aug. 16	Comdt., No. 1 L. of C. Coy.
24 Jan. 14	Beatson-Bell, A.	Yorks. R.	25 Jan. 15	10 Nov. 16	14 D. S. C.
7 Dec. 14	McGovern, F.	Yorks. R.	10 June 15	. . . .	...
7 Dec. 14	Wheeler, C.	W. Rd. R.	10 June 15	. . . .	Lv., ex I., m. c., 6 mos.
5 Dec. 14	Wimberley, R. J.	R. E. (T.C.)	11 July 15	. . . .	...
6 Mar. 15	Singer, C. A. (temp. Maj., 11 Aug. 16).	R. E. (S R.)	11 July 15	6 Mar. 15	Dy. Asst. Dir. of Army Signals.
. . . .	Long, P.	2-7th Hamps. R.	20 July 15	. . . .	Sig. Park & Depôt.
29 Oct. 14	Shellick, H. D.	I. A. Res.	1 Sep. 15	3 Aug. 16	Sig.
28 May 15	Gurdon, W. N., D.C.M.	R. E.	12 Sep. 15	. . . .	...
14 Dec. 14	Machin, H. S.	1st G. B. Linc. R.	2 Oct. 15	6 Dec. 16	No. 1 L. of C. Coy.
5 Oct. 14	Paton, J. E. F.	31 Lancers	5 Oct. 15	5 Oct. 15	33 D. S. C.
30 Apl. 15	Holford, W. T.	3 R. W. Surr. R.	14 Nov. 15	29 May 17	6 Cav. Bde. Sig. Troops.
17 Nov. 14	Nott-Bower, E. E., M.C.	R. E.	12 Jan. 16	. . . .	...
14 Jan. 15	Adams, G.	I. A. R. O.	14 Jan. 16	20 July 18	S. S. D.
24 Jan. 14	Romilly, H. A.	Som. L. I.	26 Mar. 16	12 July 16	17 D. S. C.
26 Mar. 15	Reid, J. B. (act'g. Capt., 20 May 18)	I. A. Res.	26 Mar. 16	1 Nov. 16	Comdt., Depôt Coy.
2 Apl. 15	Cooper, C. H.	I. A. Res.	2 Apl. 16	. . . .	Attd. I. T. D.
14 Apl. 15	Wall, R. H. (act'g. Capt.)	I.A. (on probn.)	14 Apl. 16	1 June 16	N. C. Sig. Schl.
. . . .	Pearce, R. J. F.	3 Norf. R.	21 Apl. 16	. . . .	48 D. S. C.
30 Apl. 15	Macdonald, L. J. (temp. Capt.)	I. A. Res.	30 Apl. 16	. . . .	...
5 May 15	Ruck, C. F. L.	I. A.	5 May 16	23 Oct. 15	Lv., ex I., m.c.
5 May 15	Smith, H. S.	I. A. R. O.	5 May 16	21 Aug. 18	47 D. S. C.
15 May 15	Fitzpatrick, L. V.	I. A.	15 May 16	15 June 18	Corps Sig. Coy.
19 Mar. 15	Ball, H. F.	T. F. Res., Genl. List.	31 May 16	30 Sep. 16	...
19 Sep. 14	Peek, B. M. (temp. Capt.)	1-7 Hamps. R.	1 June 16	24 Nov. 16	W. S. S.
24 Sep. 14	Blake, R. A. E.	1-4 Hamp. R.	1 June 16	14 Oct. 16	...
3 Oct. 14	Mauleverer, W. G.	2-4 Dorset. R.	1 June 6	8 Nov. 16	33 D. S. Coy.
7 Oct. 14	Pennefather, E. M.	1-6 Devon. R.	1 June 16	. . . .	15 D. S. C.
7 Oct. 14	Young, G. A.	2-4 Wilts. R.	1 June 16	14 May 17	F. D. Cable Sec., A. C. S. Coy.
24 Oct. 14	Blair, J. M.	1-5 E. Surr. R.	1 June 16	. . . .	18 D. S. C.
4 Nov. 14	Killick, A. S. (57)	1-4 E. Kent R.	1 June 16	6 Dec. 16	S. S. D.
7 Mar. 15	Bell, E. W.	1-1 Kent Bn.	1 June 16	5 Nov. 17	38 D. S. Coy.
. . . .	Gardham, H. C. (act'g. Capt.)	R. F. A. (T)	1 June 16	. . . .	1st Corps. Sig. Coy.
. . . .	McHutcheon, E. G. (act'g. Capt.)	R. E. (T)	1 June 16	. . . .	No. 2 W. S. S.
. . . .	Singleton, C. (actg. Capt)	R. E. (T)	1 June 16	. . . .	No. 5 Wireless Section.
. . . .	Thomas, F.	R. E. (T)	1 June 16	. . . .	1 L. of C. Coy.
. . . .	Dalton, A. G.	R. E. (T)	1 June 16	. . . .	11 Cav. Bde. Sig. Troop.
9 Jan. 15	Day, F. (57)	R. Suss. R.	10 June 16	25 Mar. 15	18 D. S. C.
16 June 15	Tomlinson, F.	105 Mah. L. I.	16 June 16	. . . .	No. 2 W. S. S.
23 Apl. 15	Fitchew, H. P.	R. E.	10 July 16	. . . .	...
14 July 15	Rayner, R. H. (act'g. Capt. 17 June 18)	1-4 W. R. R.	14 July 16	20 Aug. 16	39 D. S. C. (School).
24 Jan. 15	Johnson, H. E.	R. E.	23 July 16	20 Mar. 15	Qr.-Mr., S. S. D.
25 July 15	Lawrence, H. S.	I. A. Res.	25 July 16	. . . .	...
30 July 15	McLean, A. M., M.C. (temp. Capt.)	I. A. Res.	30 July 16	. . . .	Comdt., Sig. Park & Depôt.
10 Sep. 15	Durnan, J. A. (temp. Capt.)	I. A. Res.	10 Sep. 16	10 Sep. 15	Adjutant, S. S. D.
12 Oct. 15	Mocatta, C. H. (act'g. Capt., 9 Oct. 18).	3rd G. R.	12 Oct. 16	12 Jan. 15	Comdt., Corps Sig. Coy.
2 Oct. 15	Shoulder, H. (act'g. Capt.)	L'pool. R.	14 Nov. 16	2 Oct. 15	Comdt., N. C. Sig. School.
18 Apl. 15	Newfield, M. M.	3 G. B., Bedf. R.	15 Nov. 16	. . . .	S. S. D.
10 Dec. 15	Hasluck, R. F. (act'g. Capt.)	I. A. R. O.	10 Dec. 16	10 Dec. 15	Sig. School, Lucknow.
29 July 15	Dawes, R. M.	6 Jat L. I.	13 Dec. 16	29 July 15	43 D. S. Coy.

1st Commn. or date of entering service.	Names.	Corps.	Army Rank.	Date of apptt. to Signal service.	REMARKS.
	Lieutenants—*contd.*				
15 Nov. 14	Tulloch, J. C.	R. E.	13 Dec. 16	10 Mar. 15	39 D. S. Coy.—(*On special duty*).
13 Dec. 15	Platts, C. S.	I. A. R. O.	13 Dec. 16	17 Jan. 18	S. S. D.
15 Mar. 15	Farrow, J. K.	2-8th G. R.	15 Dec. 16	28 May 18	47 D. S. C.
11 Nov. 14	Hinton, L.	5th Cheshire Regt.	1 June 16	17 May 18	47 D. S. C.
3 Jan. 16	Brennan, R., M.C.	5th G. R.	3 Jan. 17	21 Aug. 18	S. S. D.
5 May 15	Gowland, C. J.	7th G. R.	5 Nov. 16	31 Aug. 18	4[illegible] D. S. C.
7 Apl. 15	Humphrey, B.	I. A. (*on probn.*)	7 Jan. 17	28 Sep. 17	39 D. S. Coy.
11 Nov. 14	*Fraser, C. J. S.*	*54 Sikhs, F. F.*	*11 Feb. 17*	*17 Aug. 16*	*1st Corps. Sig. Coy.*
18 June 15	*Wilkinson, R. J.,* M.B.E. *(act'g. Maj.).*	*I. A. R. O.*	*18 June. 16*	. . . .	*Comdt., No. 2 W. S. S.*
30 Mar. 16	Small, G. A.	I. A. R. O.	30 Mar. 17	17 Jan. 18	No. 4 Cavy. Bde. S. Troop.
9 July 15	Simpson, A. R.	I. A. R. O.	9 Apl. 17	9 July 15	44 D. S. Coy.
28 Apl. 15	Farrar, C. B. (*act'g. Capt.*, 5 *Oct. 18*).	I.A. (*on probn.*)	18 Apl. 17	13 Sep. 14	Comdg. Depôt Coy.
29 July 15	Lewis, D. G., M.C.	I. A. (*on probn.*)	18 Apl. 17	27 Nov. 15	36 D. S. C.
18 Apl. 16	*Bell, H. G.*	*3rd Horse*	*18 Apl. 17*	*18 Jan. 18*	*18 D. S. Coy.*
18 Apl. 16	*Foster, J. W.*	*25 Cav.*	*18 Apl. 17*	*7 Aug. 17*	...
20 July 15	*Nichols, F. W.,* M.B.E.	*I. A.*	*20 Apl. 17*	. . . .	*Sig. Park & Depot.*
19 May 16	*Lander, J. H.*	*I. A.* (on probn.)	*19 May 17*	. . . .	*L. of C. Coy.*
22 Feb. 15	Pavey, G. P.	Som. L. I.	1 July 17	15 Oct. 17	33 D. S. C.
3 Mar. 15	Ball, C. F.	2-4 Bord. R.	1 July 17	24 Sep. 17	36 D. S. Coy.
19 Mar. 15	Maltby, G.	R. F. A.	1 July 17	13 May 18	36 D. S. C.
2 Apl. 15	Round, G. E. G.	I. A. (*on probn.*)	1 June 17	30 Apl. 18	10 Cav. Bde. Sig. Troop.
5 Aug. 14	*Dixson, H. F.*	*7 Huss.*	*1 July 17*	*10 Aug. 16*	*No. 1 L. of C. Coy.*
10 Oct. 14	*Pascoe, F. J.*	*2-4 D. C. L. I.*	*1 July 17*	*13 May 17*	*7 Cav. Bde. Sig. Troop.*
28 Nov. 14	Byers, J. A. (57)	8 Chestr. R.	1 July 17	. . . .	39 D. S. C.
29 Apl. 15	*Shaw, R. C.*	*2-4 Dorset. R.*	*1 July 17*	*11 July 16*	*17 D. S. C.*
15 May 15	Whitehead, R. B.	K. R. R. Rif. C.	1 July 17	27 Dec. 15	45 D. S. C.
9 June 15	*Mead, B.*	*2-7 Hamps. R.*	*1 July 17*	*26 Oct. 16*	*W. S. S.*
16 June 15	*Morse, W. S.*	*1-7 Hamps. R.*	*1 July 17*	*17 June 16*	*No. 2 W. S. S.*
17 June 15	Detmold, K. J.	2nd G. B. Essex. R.	1 July 17	2 Oct. 17	38 D. S. Coy.
23 Feb. 15	Tyler, F. R.	1 Bn., R. Suss. R.	1 July 17	2 Mar. 18	44 D. S. Coy.
23 Sep. 15	*Davies, J. W.*	*4 Bn., E. Kent. R.*	*1 July 17*	. . . .	*1 Corps. Sig. Coy.*
18 Nov. 15	*Adams, G. P.*	*1-9th Hants.*	*1 July 17*	*18 Aug. 18*	
19 Nov. 15	*Tudor, W. J. E.*	*R. F. A.*	*1 July 17*	*4 Feb. 18*	14 D. S. C.
11 June 16	Birtwistle, J. (*act'g Cap.*)	Ind. Army (*on probn.*)	11 June 17	16 May 18	Sig. Schod. S. C.
20 Apl. 15	Prescot, C. P. (*act'g. Maj.*, 29 *June 18.*)	R. E. (*T.C.*)	1 July 17	20 Apl. 18	Comdt., 45 D. S. C.
7 Oct. 15	Nicholls, C. W.	Brecknock Bn.	1 July 17	7 Oct. 17	40 D. S. Coy.
8 Oct. 15	Okell, C. F.	10 N. Staffs R.	1 July 17	28 Nov. 15	36 D. S. Coy.
11 Oct. 15	Leedam, W. D.	R. E.	1 July 17	11 Oct. 15	S. S. D.
27 Oct. 15	*Coghlan, E. M. B.*	*R. E.*	*1 July 17*	. . . .	*1st Corps Sig. Coy.*
27 Oct. 15	*Ferriday, T. B.*	*R. E.*	*1 July 17*	. . . .	
27 Oct. 15	Bull, W. J.	R. E.	1 July 17	27 Oct. 15	40 D. S. Coy.
2 Nov. 15	Stevens, C.	24th Rif. Brig.	1 July 17	7 Oct. 17	38 D. S. Coy.
5 Nov. 15	Bowden, J. H.	8 Bn. Arg. & Suth'd. Highrs.	1 July 15	21 June 18	46 D. S. C.
19 Nov. 15	Golden, H.	R. Sussex R.	1 July 17	22 Jan. 18	43 D. S. Coy.
1 Jan. 16	Harrison, A. F.	2-4th Border. R.	2 July 17	12 Jan. 18	36 D. S. Coy.
8 Feb. 16	*King, J. L.*	*R. E.*	*1 July 17*	. . . .	
20 June 17	King, E. H. (*act'g. Capt.*, 29 *Mar. 18*).	I. A. Res.	1 July 17	20 June 17	W. S. S.
. . . .	*Smith-Masters, E. L. (act'g. Capt.)*	*R. E. (T.C.)*	*1 July 17*	. . . .	*No. 2 W. S. S.*
. . . .	*Walker, R. (act'g. Capt.)*	*R. E. (T.C.)*	*1 July 17*	. . . .	*No. 2 W. S. S.*
. . . .	*Norris, C. E.*	*18 Bn. Lond. R.*	*1 July 17*	. . . .	*48 D. S. C.*
. . . .	*Greenwood, W.*	*R. F. A.*	*1 July 17*	. . . .	*18 D. S. C.*
. . . .	*Fulcher, E. W. P.*	*3 Norf. R.*	*1 July 17*	. . . .	*Adv. Sig. Park.*
. . . .	*Gyton, G. H.*	*R. E. (T)*	*1 July 17*	. . . .	*No. 2 L. of C. Coy.*
. . . .	*Sutcliffe, H.*	*R. E. (T)*	*1 July 17*	. . . .	*Attd. I. T. D.*
. . . .	*Wilmshurst, A. P.*	*R. E. (T.C.)*	*1 July 17*	. . . .	*No. 2 W. S. S.*
6 July 16	Frost, C. C. J.	10 G. R.	6 July 17	14 June 18	46 D. S. C.
14 July 16	Walters, R.	I. A. Res.	14 July 17	4 Nov. 17	43 D. S. Coy.
13 Jan. 16	*Stilling, N. A.*	*W. Rid. R.*	*13 July 17*	. . . .	15 D. S. C.
22 July 16	*Dalley, R. P.*	*I. A. Res.*	*22 July 17*	*2 Feb. 18*	*Civil Employment.*
	Jenkins, T. G.	R. E. (T. C.)	5 Aug. 17	5 Feb. 16	S. S. D.
20 Feb. 16	*Foxlee, F. J. (act'g. Capt., 27 Mar. 18.)*	*I. A. Res.*	*20 Aug. 17*	. . . .	*No. 2 W. S. S.*

Lieutenants—*concl d.*

1st Commn. or date of entering service.	Names.	Corps.	Army rank.	Date of apptt. to Signal Service.	Remarks.
25 Aug. 16	*Tysoe-Smith, C.*	*I. A. Res.*	*28 Aug. 17*	....	*Army Sig. Coy.*
7 July 16	*Friend, R. H.*	*R. E.*	*7 Jan. 18*	....	
25 May 16	Pratt, L. H.	*R. E. (T.C.)*	23 Nov. 17	....	48 D. S. C.
23 Dec. 16	Cherril, R. A.	I. A. Res.	23 Dec. 17	10 Jan. 17	Accts. Officer, S.S.D.
21 Nov. 16	*Lyons, C. P.*	*I. A. Res.*	*21 Nov. 17*	*27 Aug. 17*	*14 D. S. C.*
10 May 16	*Gordon, F. L.*	*12 Prs.*	*10 Feb. 18*	*3 Sep. 16*	*15 D. S. C.*
10 Mar. 17	*McGarry, J.*	*I. A. Res.*	*10 Mar. 18*	....	*15 D. S. C.*
5 May 17	*Thomas, E.*	*I. A. Res.*	*5 May 18*	*6 Nov. 17*	*Army Sig. Coy.*
5 May 17	Phillips, W. H.	I. A. R. O.	5 May 18	29 Aug. 17	33 D. S. Coy.
5 May 17	*Cooling, C. E.*	*I. A. R. O.*	*5 May 18*	*22 Aug. 17*	*Army Sig. Coy.*
5 May 17	*Lines, R. E.*	*I. A. R. O.*	*5 May 18*	*20 Aug. 17*	*No. 1 L. of C. Coy.*
29 June 17	Harry, W. T.	I. A. Res.	29 June 18	7 Feb. 18	38 D. S. Coy.
8 Aug. 17	*Gregory, H. G.*	*I. A. Res.*	*8 Aug. 18*	*15 Aug. 17*	*W. S. S.*
1 Sep. 17	Down, H. R.	8 Cavy.	1 Sep. 18	18 Mar. 18	33 D. S. C.
13 Sep. 17	McCuish A.	33 Cavy.	13 Sep. 18	25 Jan. 15	A. C. S. Coy.
1 Oct. 17	Lambden, C. B.	I. A. Res.	1 Oct. 18	18 Jan. 18	33 D. S. Coy.
5 Oct. 17	Williams, W. H.	I. A. Res.	5 Oct. 18	18 Jan. 18	33 D. S. Coy.
5 Oct. 17	*Armstrong, P. A.*	*I. A. Res.*	*5 Oct. 18*	*18 Jan. 18*	
1 Nov. 17	*Hiller, L. O.*	*I. A. (T. C.)*	*1 Nov. 18*	*1 Nov. 17*	*L. of C. Sig. Coy.*
1 Nov. 17	*Upchurch, L. C.,* D.C.M.	*I. A. (T. C.)*	*1 Nov. 18*	*18 Mar. 18*	*Army Sig. Coy.*
5 Nov. 17	Wood, C.	I. A. (T. C.)	5 Nov. 18	18 Mar. 18	45 D. S. C.
5 Nov. 17	Strudwick, W. A. J.	I. A. (T. C.)	5 Nov. 18	20 Mar. 18	47 M. B. S. S.
22 June 15	Kay, A. H.	1 Bn. Gord. Highrs.	1 July 18	30 Oct. 18	39 D. S. D.

1st Commn. or date of entering service.	Names.	Corps.	Army Rank.	Date of apptt. to Signal Service.	Remarks.
	2nd-Lieutenants.				
2 Apl. 15	Goulding, C. C.	I. A. Res.	2 Apl. 15	25 Apl. 15	W. S. S.
4 Aug. 15	*Russel, R. B.*	*R. E.*	*5 Aug. 15*	. . . .	
5 Aug. 15	Leigh, H.	10 Nor. R.	5 Aug. 15	23 Dec. 16	F. C. [Cable Sec., A. C. S. Coy.
12 Jan. 16	White, J., M.C.	N. Staff. R.	12 Jan. 16	6 Sep. 18	S. S. D.
30 Apl. 16	Henry, G. C. (*temp. Capt.*)	Gen. List	30 Apl. 16	30 Apl. 16	W. S. S.
1 May 16	*Scott, R. M.*	*Gen. List*	*1 May 16*	. . . .	
5 May 16	Fiddiman, J. E., M.C.	Gen. List	5 May 16	. . . .	S. S. D.
5 May 16	*McCulloch, E. A.*	*Gen. List*	*5 May 16*	. . . .	
5 June 16	*Collier, S.*	*Welsh R.*	*5 June 16*	. . . .	
7 July 16	*Pite, R. W.*	*R. E.*	*7 July 16*	. . . .	*17 D. S. C.*
5 Sep. 16	*Furner, A. S.*	*7 Glouc. R.*	*5 Sep. 16*	. . . .	*Sig. School.*
15 Sep. 16	Pyster, E. M.	A. I. F.	15 Sep. 16	10 Apl. 17	Sec'bad. Cav. Bde. Troop.
22 Sep. 10	*Forster, G. H.*	*R. E.*	*22 Sep. 16*	. . . .	*48 D. S. C.*
23 Sep. 16	Cosserat, E. W.	R. E.	23 Sep. 16	23 Sep. 16	S. S. D.
25 Nov. 16	Ryan, P.	Gen. List	25 Nov. 16	. . . .	W. S. S.
9 Dec. 16	*Leng, C.*	*Gen. List*	*9 Dec. 16*	*9 Dec. 16*	*No. 2 W. S. S.*
12 Dec. 16	*Reid, K. B.*	*R. E.*	*12 Dec. 16*	. . . .	*No. 1 L. of C. Sig. Coy.*
20 Dec. 16	*Pocock, H. S.*	*R. E.*	*20 Dec. 16*	. . . .	
28 Jan. 17	*Field, A. V.* (*temp. Capt., 16 Jan. 18*)	*W. Rid. R.*	*28 Jan. 17*	*28 Jan. 17*	
1 Feb. 17	*Roberts, J. O.*	*R. E.*	*1 Feb. 17*	. . . .	
6 Feb. 17	Pearson, H., M.M.	Gen. List	6 Feb. 17	6 Feb. 17	39 D. S. Coy.
7 Feb. 17	Sims, J. C.	Gen. List	7 Feb. 17	6 Feb. 17	A. C. S. Coy.
13 Feb. 17	*Russell, F.*	*Gen. List*	*13 Feb. 17*	. . . .	
14 Feb. 17	Veale, J. F., D.C.M.	Gen. List	14 Feb. 17	14 Feb. 17	S. S. D.
14 Feb. 17	*Butcher, J.*, D.C.M.	*Gen. List*	*14 Feb. 17*	*14 Feb. 17*	*1st Corps Sig. Coy.*
14 Feb. 17	Shrubsall, P. (58*a*)	Gen. List	14 Feb. 17	14 Feb. 17	Airline Sec., A. C. S. Coy.
14 Feb. 17	*Scheidweiler, F. C.*, D.C.M.	*Gen. List*	*14 Feb. 17*	*14 Feb. 17*	
14 Feb. 17	*Cowan, J. L.*	*Gen. List*	*14 Feb. 17*	*14 Feb. 17*	P. Bde. Sec.
19 Feb. 17	*Humphries, F.*, M.C., D.C.M.	*Gen. List*	*19 Feb. 17*	. . . .	*13 D. S. C.*
26 Feb. 17	Marshall, H.	Gen. List	26 Feb. 17	26 Feb. 17	44 D. S. Coy.
26 Feb. 17	*Yule, W.*, M.C.	*Gen. List*	*26 Feb. 17*	. . . .	*14 D. S. C.*
14 Mar. 17	Reynolds, C. E. J. (57)	Gen. List	14 Mar. 17	13 Feb. 17	33 D. S. Coy.
23 Mar. 17	*Younis, L. V.*	*Arg. & Suthd. Highrs.*	*23 Mar. 17*	. . . .	*1st Corps Sig. Coy.*
13 Apl. 17	Smitherman, H. S.	Gen. List	13 Apl. 17	13 Apl. 17	W. S. S.
8 July 17	Toogood, L. G. C.	6 Cavy.	8 July 17	18 Mar. 18	13 Cav. Bde. S. Troop.
17 Jan. 16	*Childs, E.*, M.B.E.	*R. G. A.*	. . . .	. . . .	*Sig. Park & Depôt.*
22 July 17	*Grant, L. J.*	*R. F. (T. C.)*	*22 July 17*	. . . .	*18 D. S. C.*
1 Aug. 17	*Bookless, A. C.*	*R. E. (T. C.)*	*1 Aug. 17*	. . . .	*48 D. S. C.*
15 Aug. 17	*Woods, W. L. J.*	*R. E. (T. C.)*	*15 Aug. 17*	. . . .	*5 Wireless Section.*
22 Aug. 17	*Shaw, W. H.*	*R. E. (T. C.)*	*22 Aug. 17*	. . . .	*5 Wireless Section.*
28 Aug. 17	Wrixon, A. G.	L'pool R.	28 Aug. 17	28 Aug. 17	40 D. S. Coy.
29 Aug. 17	McCarthy, A. J.	7 Glouc. R.	29 Aug. 17	29 Aug. 17	W. S. S.
14 Sep. 17	Pennell, G. E.	Gen. List	14 Sep. 17	14 Sep. 17	W. S. S.
1 Nov. 17	Carbould, J. H.	I. A. (*temp.*)	1 Nov. 17	19 Mar. 18	44 D. S. C.
18 Dec. 17	Harley, G. T.	Gen. List	18 Dec. 17	18 Dec. 17	39 D. S. C. *Lv., ex I., m.c., 6 mos.*
19 Dec. 17	Jones, E. H., D.C.M.	7 Hussars	19 Dec. 17	19 Dec. 17	38 D. S. Coy.
16 Mar. 18	Corbitt, J. H.	1 Royal Suss.	16 Mar. 18	12 Apl. 18	45 D. S. C.
15 June 18	Deacon, A. E.	Gen. List	15 June 18	15 June 18	W. S. S.
27 Nov. 17	Bound, C. G.	Glouc. R.	27 Nov. 17	31 Aug. 18	47 D. S. C.
5 Jan. 18	Hambly, W. E.	I. A. (T. C.)	5 Jan. 18	. . . .	W. S. S.
5 Jan. 18	Sands, E.	3 K. R. R.	5 Jan. 18	5 Jan. 18	36 D. S. Coy.
15 Jan. 18	*Cahill, J.*	*I. A. R. O.*	*15 Jan. 18*	*2 June 18*	*Army Sig. Coy.*
15 Jan. 18	Minhinnick, J. B.	I. A. R. O.	15 Jan. 18	. . . .	W. S. S.
15 Jan. 18	Short, R. L. B.	I. A. R. O.	15 Jan. 18	2 June 18	46 D. S. C.
15 Jan. 18	Spooner, J. D.	I. A. R. O.	15 Jan. 18	2 June 18	Corps Sig. Coy.
15 Jan. 18	Evans, A.	I. A. (T. C.)	15 Jan. 18	1 June 18	45 D. S. C.
15 Jan. 18	Morgan, R. W.	I. A. (T. C.)	15 Jan. 18	1 June 18	45 D. S. C.
15 Jan. 18	Osborne, H. M.	I. A. R. O.	15 Jan. 18	. . . .	S. S. D.
7 Mar. 18	Sirrell, E.	2 R. I. F.	7 Mar. 18	7 Mar. 18	S. S. D.
16 Mar. 18	*Ure, O. H.*	*1 Yorks*	*16 Mar. 18*	*18 Apl. 18*	*No. 1 L. of C. Coy.*
16 Mar. 18	Hearn, L. A.	1-4th Devons	16 Mar. 18	16 Apl. 18	S. S. D.
16 Mar. 18	Vaux, R.	2 S. L. I.	16 Mar. 18	10 Apl. 18	W. S. S.
16 Mar. 18	*Swales, H. G.*	*2 R. W. Kent*	*16 Mar. 18*	*20 Apl. 18*	*No. 3 W. S. S.*
19 Mar. 18	Bowen, J.	I. A. (T. C.)	19 Mar. 18	. . . .	47 D. S. C.
28 Mar. 18	Shaw, W. H.	I. A. (T. C.)	28 Mar. 18	23 Oct. 18	W. S. S.
29 Mar. 18	Small, J. H.	I. A. (T. C.)	29 Mar. 18	. . . .	S. S. D.
7 Apl. 18	McKechnie, A., D.C.M.	Genl. List	7 Apl. 18	21 May 18	39 D. S. C.
19 Apl. 18	*Key, F. H.*, D.C.M.	*Worc. R.*	*19 Apl. 18*	*19 Apl. 18*	Aden Sig. Coy.
28 Apl. 18	Bevan, J. D.	Genl. List	28 Apl. 18	27 Apl. 18	S. S. D.

1st Commn. or date of entering service.	Names.	Corps.	Army Rank.	Date of apptt. to Signal service.	REMARKS.
	2nd-Lieutenants—*concld.*				
30 Apl. 18	Cave, A. R.	Genl. List	30 Apl. 18	. . . .	12 Cav. Bde. Sig Troop.
13 May 18	Huffton, P. N.	I. A. R. O.	13 May 18	7 Oct. 18	45 D. S. C.
15 May 18	Cornish-Trestrail, A G.	I. A. R. O.	15 May 18	7 Oct. 18	43 D. S. C.
25 May 18	Church, F.	Genl. List	25 May 18	6 Sep. 18	S. S. D.
2 June 18	Newton, A.	Genl. List	2 June 18	2 June 18	S. S. D.
29 Mar. 18	McMurray, F. C.	I. A. R. O.	29 Mar. 18	17 July 18	40 D. S. C.
29 Mar. 18	Rushbrook, A. V.	I. A. R. O.	29 Mar. 18	17 July 18	46 D. S. C.
19 Mar. 18	Small, G. C.	I. A. R. O.	19 Mar. 18	19 Mar. 18	46 Bde. Sig. Section.
2 June 18	Lowes, G. W.	I. A. R. O.	2 June 18	7 Oct. 18	Corps. Sig. Coy.
2 June 18	Robinson, F. C.	I. A. (T. C.)	2 June 18	7 Oct. 18	W. S. S.
7 June 18	Mathews, F. C.	I. A. R. O.	7 June 18	7 Oct. 18	40 D. S. C.
7 June 18	Handley, P.	I. A. R. O.	7 June 18	7 Oct. 18	46 D. S. C.
7 June 18	Veneer, F. A.	I. A. R. O.	7 June 18	11 Oct. 18	46 D. S. C.
7 June 18	Randall, H. V.	I. A. R. O.	7 June 18	18 Oct. 18	W. S. S.
19 June 18	Toms, W. T.	Dorset R.	19 June 18	19 June 18	38 D. S. C.
7 Aug. 18	Roberts, E. C.	I. A. R. O.	7 Aug. 18	31 Aug. 18	S. S. D.
19 Aug. 18	Lloyd, W. H.	1-10 Bn., Middx. R.	19 Aug. 18	. . . .	S. S. D.
. . . .	Hackett, H. F.	I. A. R. O.	. . . .	. . . .	S. S. D.
. . . .	Harris, W. H.	1-4 Bn., E Kent R.	. . . .	. . . .	S. S. D.
. . . .	Randles, G. W.	1-6 Bn., Hamps. R.	. . . .	. . . .	S. S. D.
. . . .	Minchin, V. R.	1-4 Bn., Hamps. R.	. . . .	. . . .	S. S. D.
19 Aug. 18	Henderson, A. G.	1 Bn., Durh. L. I.	19 Aug. 18	. . . .	39 D. S. C.
19 Aug. 18	Wyatt, E. K.	1 Bn., Durh. L. I.	19 Aug. 18	. . . .	S. S. D.
5 Sep. 18	Turner, A.	W. Rid. R.	5 Sep. 18	5 Sep. 18	39 D. S. C.
4 Sep. 18	Garcia-Henry, A.	Genl. List	4 Sep. 18	. . . .	W. S. S.
16 Sep. 18	Lego, J. T.	Genl. List	16 Sep. 18	. . . .	W. S. S.
. . . .	Clarke, N. St. J. S.	I. A. R. O.	. . . .	. . . .	W. S. S.
	Attached,				
. . . .	*Newell, Major S. M. D. S. O. (actg. Lt.-Col.) (Bt.-Lt.-Col. 3 June 17).*	*R. E. (T),*	. . . .	. . . .	*A. D. Sigs. 1st Corps.*
. . . .	*Salmond, Major H. A. B. (Actg. Lt.-Col.)*	*Y. & L. Regt.*	. . . .	. . . .	*D. D. Sigs. Base.*
4 Jan. 04	Gould, Capt. J. C.	85th Burma R.	4 Jan. 13	. . . .	S. S. D.
2 Dec. 16	Packwood, *Hon.* Lt. I.	. . . .	2 Dec. 16	. . . .	S. S. D.
. . . .	*Moon, E. J.*, M.C.	*R. E. (T. C.)*	*11 July 15*	. . . .	*3 Corps Sig. Coy.*
. . . .	*Paige, J. F.*, M.C.	*R. E. (T. C.)*	*29 Sep. 15*	. . . .	*13 D. S. C.*
. . . .	*Turner, P. W.*	*R. E. (T. C.)*	*9 Nov. 15*	. . . .	*3rd Corps Sig. Coy.*
. . . .	*Saunderson, D. B.* M. C. *(actg. Capt).*	*R. E. (T.).*	*1 Jun. 16*	. . . .	*3rd Corps. Sig. Coy.*
7 Nov. 14	*Murphy, Lieut. E. J.*	*21st Lrs.*	*1 Jun. 17*	*17 Mar. 17*	*1 Cav. Bde. Sig. Troop.*
. . . .	*Anderson, J. W.*	*R. E. (T.)*	*1 July 17*	. . . .	*Army Sig. Coy.*
. . . .	*Vernon, R. C.*	*R. E. (T. C.)*	*1 July 17*	. . . .	*Army Sig Coy.*
. . . .	*Lucas, H. F.*	*R. E.*	*1 July 17*	. . . .	*3rd Corps Sig. Coy.*
. . . .	*Horner, R. S.*	*R. E.*	*1 July 17*	. . . .	*3rd Corps Sig. Coy.*
. . . .	*Campbell, R. S.*	*R. E. (T.)*	*1 July 17*	. . . .	*Attd I. T. D.*
. . . .	*Muddle, E. E.*	*R. E. (T.)*	*1 July 17*	. . . .	*13 D. S. C.*
. . . .	*Wilson, K. H.*, M.C.	*R. E. (T. C.)*	*27 Oct 17*	. . . .	*13 D. S. C.*
. . . .	*Ward, J. E.*	*R. E.*	*26 Nov. 17*	. . . .	*Army Sig. Coy.*
. . . .	*Hitchcock, L.*	*2-7 Hants*	*11 Jan. 18*	. . . .	*Army Sig. Coy.*
. . . .	*Swanston, E. R.*	*R. E. (T. C.)*	*29 Jan. 18*	. . . .	*Army Sig. Coy.*
. . . .	*Martin, H. E. L.*	*R. E. (T. C.)*	*1 Mar. 18*	. . . .	*Army Sig. Coy.*
. . . .	*Shaw, S.*	*R. E. (T. C.)*	*1 Mar. 18*	. . . .	*Army Sig. Coy.*
. . . .	*Gauld, H. A.*, M.C.	*R. E. (T. C.)*	*1 Mar. 18*	. . . .	*Attd. I. T. D.*
. . . .	*Garratt, R. B.*	*R. E. (T. C.)*	*1 Feb. 17*	. . . .	*3rd Corps Sig. Coy*
. . . .	*Fisher, A. W.*	*R. E. (T. C.)*	*1 Aug. 17*	. . . .	*3rd Corps Sig. Coy.*
. . . .	*Coxon, J. O.*	*R. E. (T. C.)*	*15 Sep. 17*	. . . .	*Army Sig. Coy.*

1st Commn. or date of entering service.	Names.	Corps.	Army Rank.	Date of apptt. to Signal service.	REMARKS.
	Subadar-Major.				
3 Nov. 86	*Kasim Ali, Sardar Bahadur*	*31 D. S. C.*	*16 Sep. 16*	*15 Feb. 11*	*Order of Br. Ind., 1st cl.*
	Subadars.				
1 June 91	*Lehru, Bahadur*	*58 R. F. F.*	*13 Oct. 08*	*15 Feb. 11*	*Order of Br. Ind., 2nd cl.*
26 July 87	Narasingaperumal	34 D. S. C.	25 Apl. 11	25 Apl. 11	S. S. D.
17 June 95	Diwan Singh, *Bahadur*	30 Punj	14 Aug. 14	14 Aug. 14	S. S. D.
13 Dec. 92	*Suchet Singh*	. . . . .	*15 Oct. 14*	*15 Feb. 11*	S. S. D.
29 Oct. 95	Srinivaslu	34 D. S. C.	18 Nov. 14	1 Mar. 11	S. S. D.
28 July 97	*Narainsami*	*33 D. S. C.*	*31 Aug. 15*		
. . . .	*Murugesan,* I.D.S.M.	*34 D. S. C.*	*2 Nov. 15*		
10 May 97	Abdur Raschid	75 C. I.	11 Feb. 16	18 Apl. 11	33 D. S. Coy.
2 Nov. 98	Dasaunda Singh	35 Sikhs	1 Aug. 16	1 Aug. 16	36 D. S. Coy.
19 Dec. 97	Allah Rakha Khan, I.O.M.	32 D. S. C.	8 Jan. 17	15 Feb. 11	38 D. S. Coy.
4 Mar. 93	*Mehar Baksh, Bahadur*	*31 D. S. C.*	*10 Jan. 17*	*15 Feb. 11*	*Order of Br. Ind., 2nd cl.*
7 Feb. 99	Narayan Singh	38 Dogras	12 Jan. 17	26 Dec 16	S. S. D.
5 Jan. 97	Firman Ali	54 S. F. F.	2 May 17	18 Feb. 11	S. S. D.
15 July 02	Jaimal Din	. . . . .	2 June 17	. . . .	39 D. S. Coy.
25 Oct. 02	Appalsami	33 D. S. C.	1 Oct. 17	1 Mar. 11	40 D. S. Coy.
5 June 16	Khan Zaman	Guides	1 Oct. 17	1 June 16	A. C. S. Coy.
. . . .	Asa Ram	. . . . .	22 Jan. 17	. . . .	
9 Sep. 02	Bagh Ali	31 D. S. C.	1 Mar. 18	15 Feb. 11	36 D. S. C.
1 Mar. 02	Amar Singh	92 Punj.	15 Apl. 18	15 Aug. 14	44 D. S. Coy.
16 Oct. 07	*Dharam Singh*	*55 R. F. F.*	23 May 18		
. . . .	Sawminathan	33 D. S. Coy.	15 Apl. 18	16 May 11	33 D. S. Coy.
22 July 05	Shunmagam	33 D. S. Coy.	31 July 18	2 Mar. 11	46 D. S. Coy.
. . . .	Dharam Singh	17 D. S. Coy.	15 Aug. 18	. . . .	17 D. S. Coy.

1st Commn. or date of entering service.	Names.	Corps.	Army rank.	Date of apptt. to Signal Service.	Remarks.
	Jemadars.				
21 Oct. 97	Hasan Ali Khan	. . . . .	1 Oct. 14	1 Oct. 14	W. S. S.
. . . .	*Sayyid Ibrahim*	*34 D. S. C.*	*24 Apl. 15*		
. . . .	Ratman	A. C. S. Coy.	2 Nov. 15	. . . .	A. C. S. Coy.
. . . .	Ghulam Ali Shah	. . . . .	9 Nov. 16	15 Feb. 11	18 D. S. Coy.
1 Nov. 01	Ahmed Nabi Khan	1 Lancers	10 Nov. 16	10 Nov. 16	45 D. S. C.
28 April 99	*Puran Singh*	. . . . .	*10 Jan. 17*		
8 June 03	*Mawas Din*	*31 D. S. C.*	*25 Jan. 17*		
. . . .	Subramoni	33 D. S. C.	2 May 17	1 Mar. 11	38 D. S. Coy.
. . . .	Mangal Singh	33 Cav.	2 June 17	. . . .	
. . . .	Simon	33 D. S. C.	2 June 17	. . . .	37 D. S. Coy.
1 Aug. 05	Mohammed Afzul	106 Prs.	1 Oct. 17	19 Aug. 15	44 D. S. Coy.
. . . .	Fateh Mohammed	. . . . .	. . . .	. . . .	36 D. S. Coy.
. . . .	Kala Singh	31 D. S. C.	1 Oct. 17	15 Nov. 14	A. C. S. Coy.
22 Mar. 99	Indar Singh	38 Dogras	1 Oct. 17	31 Oct. 16	40 D. S. Coy.
10 Oct. 99	Chinnasami Naidu	81 Prs.	1 Oct. 17	22 Oct. 14	39 D.S. Coy.
27 June 08	Panchanadan	81 Pio.	1 Oct. 17	22 Oct. 14	S. S. D.—Adjt.
1 Sep. 03	Ghulam Nabi	79th Infty.	1 Oct. 17	29 Sep. 15	S. S. D.
. . . .	Saifur Khan	. . . . .	22 Jan. 17	. . . .	
. . . .	Sardar Muhammad	39 D. S. Coy.	15 Apl. 18	15 Apl. 18	46 D. S. Coy.
. . . .	Ambarjit Rai	A. C. S. Coy.	15 Apl. 18	2 Oct. 16	36 D. S. C.
. . . .	Abraham Paul	33 D. S. C.	15 Apl. 18	. . . .	33 D. S. C.
. . . .	Bhao Rao Savant	S. S. D.	15 Apl. 18	15 Apl. 18	S. S. D.
. . . .	Bishan Singh	S. S. D.	15 Apl. 18	15 Apl. 18	S. S. D.
. . . .	G. Simon Peter	43 D. S. Coy.	1 May 18	4 July 17	43 D. S. Coy.
. . . .	R. Ganesaiyer	43 D. S. Coy.	1 May 18	21 July 17	43 D. S. Coy.
. . . .	L. D. Sandanasami	43 D. S. Coy.	1 May 18	21 July 17	43 D. S. Coy.
. . . .	Khan Muhammad	18 D. S. Coy.	13 May 18	. . . .	18 D. S. Coy.
. . . .	Sant Singh	31 D. S. Coy.	24 May 18	. . . .	31 D. S. Coy.
. . . .	Shanker Nath Dutt	43 D. S. Coy.	1 June 18	13 Aug. 17	43 D. S. Coy.
. . . .	Makhan Lall Banerjee	43 D. S. Coy.	1 June 18	26 Sep. 17	43 D. S. Coy.
. . . .	Ismnil Khan	17 D. S. Coy.	15 Aug. 18	. . . .	17 D. S. Coy.

LIST OF REGIMENTAL CENTRES

NORTHERN COMMAND.

Allahabad	1 Brahmans. 3 Brahmans.
Agra	2 Rajput L. Infantry. 4 Rajputs. 13 Rajputs. 16 Rajputs.
Benares	5 L. Infantry. 17 Infantry. 18 Infantry.
Meerut	6 Jat L. Infantry. 10 Jats.
Lucknow	7 Rajputs. 8 Rajputs. 11 Rajputs.
Bareilly	12 Pioneers. 48 Pioneers.
Multan	14 Sikhs. 15 Sikhs. 45 Sikhs.
Jhelum	19 Punjabis. 22 Punjabis. 24 Punjabis.
Lahore	20 Punjabis. 21 Punjabis. 26 Punjabis.
Ambala	23 S. Pioneers. 32 S. Pioneers. 34 S. Pioneers.
Ferozepore	25 Punjabis. 27 Punjabis. 28 Punjabis.
Peshawar	29 Punjabis. 30 Punjabis. 31 Punjabis.
Sialkot	33 Punjabis. 40 Pathans. 46 Punjabis.
Rawal Pindi	35 Sikhs. 36 Sikhs. 47 Sikhs.
Rawal Pindi	37 Dogras. 38 Dogras. 41 Dogras.
Lansdowne	39th Garhwal Rifles.
Bannu	51 Sikhs. 53 Sikhs. 54 Sikhs.
Kohat	52 Sikhs. 56 Rifles. 59 Rifles.
Dera Ismail Khan	55 Rifles. 57 Rifles. 58 Rifles.
Hoti Mardan	Guides Infy.
Dharmsala	1st Gurkha Rifles.
Dehra Dun	2nd Gurkha Rifles. 9th Gurkha Rifles.
Almora	3rd Gurkha Rifles. 2-10th Gurkha Rifles.
Bakloh	4th Gurkha Rifles.
Abbottabad	5th Gurkha Rifles 6th Gurkha Rifles.
Shillong	8th Gurkha Rifles.

SOUTHERN COMMAND.

Trichinopoly	61 Pioneers. 63 Lt. Infantry. 64 Pioneers. 81 Pioneers. 83 Lt. Infantry.
Secunderabad	62 Punjabis. 66 Punjabis. 84 Punjabis.
Bangalore	67 Punjabis. 69 Punjabis. 74 Punjabis.
Mandalay	72 Punjabis. 92 Punjabis. 93 Burma Infantry.
St. Thomas' Mount	73 Carnatic Infantry. 80 Carnatic Infantry. 88 Carnatic Infantry.
Secunderabad	75 Carnatic Infantry. 79 Carnatic Infantry. 86 Carnatic Infantry.
Secunderabad	76 Punjabis. 82 Punjabis. 87 Punjabis.
Meiktila	89 Punjabis. 90 Punjabis. 91 Punjabis.
Ahmednagar	101 Grenadiers 108 Infantry. 109 Infantry.
Deesa	102 Grenadiers. 112 Infantry. 113 Infantry.
Satara	103 Mahratta L. Infy. 105 Mahratta L Infy. 110 Mahratta L. Infy.
Nasirabad	104 Wellesley's Rifles. 123 Outram's Rifles. 125 Napier's Rifles.
Kirkee	107 Pioneers. 121 Pioneers. 128 Pioneers.
Poona	114 Mahrattas 116 Mahrattas. 117 Mahrattas.
Mhow	119 Infantry. 120 Rajputana Infantry. 122 Rajputana Infantry.
Quetta	106 Pioneers. 124 Baluchistan Infy. 126 Baluchistan Infy.
Karachi	127 Baluch L. Infy. 129 Baluchis. 130 Baluchis.
Quetta	7th Gurkha Rifles.
Maymyo	1-10 Gurkha Rifles.

Jemadars.

1st Commn. or date of entering service.	Names.	Corps.	Army rank.	Date of apptt. to Signal Service.	Remarks.
21 Oct. 97	Hasan Ali Khan	. . . . .	1 Oct. 14	1 Oct. 14	W. S. S.
. . . .	*Sayyid Ibrahim*	*34 D. S. C.*	*24 Apl. 15*		
. . . .	Ratman	A. C. S. Coy.	2 Nov. 15	. . . .	A. C. S. Coy.
. . . .	Ghulam Ali Shah	. . . . .	9 Nov. 16	15 Feb. 11	18 D. S. Coy.
1 Nov. 01	Ahmed Nabi Khan	1 Lancers	10 Nov. 16	10 Nov. 16	45 D. S. C.
28 April 99	*Puran Singh*	. . . . .	*10 Jan. 17*		
8 June 03	*Mawaz Din*	*31 D. S. C.*	*25 Jan. 17*		
. . . .	Subramoni	33 D. S. C.	2 May 17	1 Mar. 11	38 D. S. Coy.
. . . .	Mangal Singh	33 Cav.	2 June 17	. . . .	
. . . .	Simon	33 D. S. C.	2 June 17	. . . .	37 D. S. Coy.
1 Aug. 05	Mohammed Afzul	106 Prs.	1 Oct. 17	19 Aug. 15	44 D. S. Coy.
. . . .	Fateh Mohammed	. . . . .	. . . .	. . . .	36 D. S. Coy.
. . . .	Kala Singh	31 D. S. C.	1 Oct. 17	15 Nov. 14	A. C. S. Coy.
22 Mar. 99	Indar Singh	38 Dogras	1 Oct. 17	31 Oct. 16	40 D. S. Coy.
10 Oct. 99	Chinnasami Naidu	81 Prs.	1 Oct. 17	22 Oct. 14	39 D.S. Coy.
27 June 08	Panchanadan	81 Pio.	1 Oct. 17	22 Oct. 14	S. S. D.—Adjt.
1 Sep. 03	Ghulam Nabi	79th Infty.	1 Oct. 17	29 Sep. 15	S. S. D.
. . . .	Saifur Khan	. . . . .	22 Jan. 17	. . . .	
. . . .	Sardar Muhammad	39 D. S. Coy.	15 Apl. 18	15 Apl. 18	46 D. S. Coy.
. . . .	Ambarjit Rai	A. C. S. Coy.	15 Apl. 18	2 Oct. 16	36 D. S. C.
. . . .	Abraham Paul	33 D. S. C.	15 Apl. 18	. . . .	33 D. S. C.
. . . .	Bhao Rao Savant	S. S. D.	15 Apl. 18	15 Apl. 18	S. S. D.
. . . .	Bishan Singh	S. S. D.	15 Apl. 18	15 Apl. 18	S. S. D.
. . . .	G. Simon Peter	43 D. S. Coy.	1 May 18	4 July 17	43 D. S. Coy.
. . . .	R. Ganesaiyer	43 D. S. Coy.	1 May 18	21 July 17	43 D. S. Coy.
. . . .	L. D. Sandanasami	43 D. S. Coy.	1 May 18	21 July 17	43 D. S. Coy.
. . . .	Khan Muhammad	18 D. S. Coy.	13 May 18	. . . .	18 D. S. Coy.
. . . .	Sant Singh	31 D. S. Coy.	24 May 18	. . . .	31 D. S. Coy.
. . . .	Shanker Nath Dutt	43 D. S. Coy.	1 June 18	13 Aug. 17	43 D. S. Coy.
. . . .	Makhan Lall Banerjee	43 D. S. Coy.	1 June 18	26 Sep. 17	43 D. S. Coy.
. . . .	Ismail Khan	17 D. S. Coy.	15 Aug. 18	. . . .	17 D. S. Coy.

LIST OF REGIMENTAL CENTRES

NORTHERN COMMAND.

Allahabad	1 Brahmans. 3 Brahmans.
Agra	2 Rajput L. Infantry. 4 Rajputs. 13 Rajputs. 16 Rajputs.
Benares	5 L. Infantry. 17 Infantry. 18 Infantry.
Meerut	6 Jat L. Infantry. 10 Jats.
Lucknow	7 Rajputs. 8 Rajputs. 11 Rajputs.
Bareilly	12 Pioneers. 48 Pioneers.
Multan	14 Sikhs. 15 Sikhs. 45 Sikhs.
Jhelum	19 Punjabis. 22 Punjabis. 24 Punjabis.
Lahore	20 Punjabis. 21 Punjabis. 26 Punjabis.
Ambala	23 S. Pioneers. 32 S. Pioneers. 34 S. Pioneers.
Ferozepore	25 Punjabis. 27 Punjabis. 28 Punjabis.
Peshawar	29 Punjabis. 30 Punjabis. 31 Punjabis.
Sialkot	33 Punjabis. 40 Pathans. 46 Punjabis.
Rawal Pindi	35 Sikhs. 36 Sikhs. 47 Sikhs.
Rawal Pindi	37 Dogras. 38 Dogras. 41 Dogras.
Lansdowne	39th Garhwal Rifles.
Bannu	51 Sikhs. 53 Sikhs. 54 Sikhs.
Kohat	52 Sikhs. 56 Rifles. 59 Rifles.
Dera Ismail Khan	55 Rifles. 57 Rifles. 58 Rifles.
Hoti Mardan	Guides Infy.
Dharmsala	1st Gurkha Rifles.
Dehra Dun	2nd Gurkha Rifles. 9th Gurkha Rifles.
Almora	3rd Gurkha Rifles. 2-10th Gurkha Rifles.
Bakloh	4th Gurkha Rifles.
Abbottabad	5th Gurkha Rifles 6th Gurkha Rifles.
Shillong	8th Gurkha Rifles.

SOUTHERN COMMAND.

Trichinopoly	61 Pioneers. 63 Lt. Infantry. 64 Pioneers. 81 Pioneers. 83 Lt. Infantry.
Secunderabad	62 Punjabis. 66 Punjabis. 84 Punjabis.
Bangalore	67 Punjabis. 69 Punjabis. 74 Punjabis.
Mandalay	72 Punjabis. 92 Punjabis. 93 Burma Infantry.
St. Thomas' Mount	73 Carnatic Infantry. 80 Carnatic Infantry. 88 Carnatic Infantry.
Secunderabad	75 Carnatic Infantry. 79 Carnatic Infantry. 86 Carnatic Infantry.
Secunderabad	76 Punjabis. 82 Punjabis. 87 Punjabis.
Meiktila	89 Punjabis. 90 Punjabis. 91 Punjabis.
Ahmednagar	101 Grenadiers 108 Infantry. 109 Infantry.
Deesa	102 Grenadiers. 112 Infantry. 113 Infantry.
Satara	103 Mahratta L. Infy. 105 Mahratta L Infy. 110 Mahratta L. Infy.
Nasirabad	104 Wellesley's Rifles. 123 Outram's Rifles. 125 Napier's Rifles.
Kirkee	107 Pioneers. 121 Pioneers. 128 Pioneers.
Poona	114 Mahrattas 116 Mahrattas. 117 Mahrattas.
Mhow	119 Infantry. 120 Rajputana Infantry. 122 Rajputana Infantry.
Quetta	106 Pioneers. 124 Baluchistan Infy. 126 Baluchistan Infy.
Karachi	127 Baluch L. Infy. 129 Baluchis. 130 Baluchis.
Quetta	7th Gurkha Rifles.
Maymyo	1-10 Gurkha Rifles.

INDIAN INFANTRY.

Regt. Centre—Allahabad.
Linked with—3rd Brahmans.

Uniform—Scarlet.
Facings—White.

1st Battalion, 1st Brahmans.

Neelwar-ki-paltan.—Raised at Oudh in 1776 by Capt. T. Naylor, as part of a torce for the service of the Nawab Wazir. Transferred to the Company's Service, 1777, and styled the 30th Battn. of Bengal Sepoys. Became the 23rd Regt. of Bengal Sepoys, 1781; the 29th Regt., 1784; the 29th Bengal Sepoy Battn., 1786; the 2nd Bn., 9th Regt., Bengal Native Infantry, 1796; the 21st Regt. of Bengal N. I., 1824; the 1st Regt. of Bengal N. I. 1861; the 1st Regt. of Beng. Infy., 1885; the 1st Brahman Infy., 1901; present designation, 1903.

"Leswarree" "Bhurtpore" "Burma, 1885—87."

Composition—4 Companies Brahmans.

1st Commn. or date of entering service.	Names and Rank.	Army rank.	Present appointment in Regiment.	Remarks.
	Commandant.			
30 Jan. 89	Douglas, Lt.-Col. W. B.	30 Jan. 15	7 Oct. 15	
	Company Commanders. (4)			
12 Aug. 99	*Whitehead, Bt.-Lt.-Col. J.*, D.S.O., *p. s. c.*	*1 Jan. 17*	*10 May 17*	*2nd-in-Comd.*
5 May 00	Lane, Maj. H. V. (*Actg. Lt.-Col.*)	1 Sep. 15	1 Oct. 10	With 2-3rd Brahmans.
11 Aug. 00	*Willoughby, Maj. D. V.*, D.S.O.	*1 Sep. 15*	. . .	*Bde.-Maj.*
18 Aug. 00	Browning, Maj. E. R. L. [L]	1 Sep. 15	10 May 17	2nd-in-Comd.
	Company Officers. (9)			
8 Jan. 01	Kempson. Maj. E. C. R.	8 Jan. 16	1 Mar. 04	With 2-1st Brahmans.
24 Jan. 10	*Smyth, Maj. C. A. W.*	*24 July 16*	*19 Oct. 04*	*G. S. O. 2nd gr., Rawalpindi.*
23 Apl. 02	*Wilson, Maj. W. H. D.*	*6 Oct. 16*	*14 June 05*	*D. A. & Q. M. G., Army Hdqrs.*
29 July 03	Pope, Maj. P. F.	29 July 18	7 June 05	With 2-1st Brahmans.
29 Jan. 10	Sanders, Capt. C. W.	1 Sep. 15	13 Mar. 11	With 3rd Brahmans.
4 July 05	Khan Muhammad Akbar Khan, Capt, C.I.E.	25 Aug. 17	5 Oct. 17	
1 Oct. 14	Ackland, Capt. H. J.	1 Oct. 18	15 Dec. 14	
23 Dec. 14	Kellie, Lt. C. C. J.	23 Dec. 15	25 Feb. 15	With 5th L. Infy.
5 Nov. 15	Barton, Lt. F. V. G.	15 Nov. 16	18 Nov. 15	
23 Dec. 17	Wright, 2nd-Lt. C. C. V.	23 Dec. 17	21 Mar. 18	Attd.
	Attached.			
20 Feb. 95	Hirst, Maj. F. C.	20 Feb. 13	15 Jan. 18	I. A. (*Supy. List.*)
4 Nov. 13	Pinhey, Capt. R. A.	4 Nov. 17	15 Jan. 18	E. Kent R. I. A. (*on probn.*)
5 Mar. 15	Stowell, Lt. P. C.	5 Mar. 16	27 Feb. 18	I. A. R. O.
27 Jan. 16	Greenwood, Lt. J. A.	27 Jan. 17	21 Feb. 18	Do.
22 June 15	Cockayne, Lt. M. H. D.	22 June 16	21 Sep. 17	E. Surrey R. I. A. (*on probn.*)
30 Jan. 17	O'Sullivan, Lt. H. P.	30 Jan. 18	16 Mar. 17	3rd Brahmans.
21 May 15	Orr, Lt. C. H. H.	21 May 16	25 Nov. 17	R. Ir. Rif. I. A. (*on probn.*)
25 Aug. 16	Beaufort, Lt. P. M.	25 Aug. 17	18 June 16	I. A. R. O. Adjt. (*temp.*)
29 Aug. 17	Bell, 2nd-Lt. W. C.	29 Aug. 17	11 Dec. 17	I. A. R. O.
	Bennett, 2nd-Lt. W. A.	. . .	23 Jan. 18	I. A. R. O.
	Eaden, 2nd-Lt. C. L.	.	23 Jan. 18	I. A. R. O.
11 Dec. 15	Jay, Lt. C. E.	11 Dec. 16	16 Aug. 18	Suff. R., I. A. (*on probn.*)
20 July 15	Harrison, Lt. J. A.	20 Apl. 17	23 Jan. 18	Attd.
	Bramall, Lt. K. G.	. . .	23 Jan. 18	Royal Scots. I. A. (*on probn.*)
1 Dec. 17	Eccles, Lt. J. G. F.	1 Dec. 18	. . .	E. York R. I. A. (*on probn.*)
5 Nov. 17	Casling Lt. W	5 Nov. 18	31 Mar. 18	I. A. R. O.
26 Oct. 15	Bennett, Lt. W. H.	26 Oct. 16	. . .	Midd'x R., I. A. (*on probn.*)
7 June 18	Bishop, 2nd-Lt. W. E.	7 June 18	5 Oct. 18	I. A. (T. C.)
25 Aug. 16	Broon, Lt. C. W.	25 Aug. 17	. . . .	A. I. Forces, I. A. (*on probn.*)

Subadar-Major.

		Jemadar.	Subadar.	Subadar-Major.
7 Mar. 01	Bale Tewari (56, 72*a*)	1 Sep. 11	10 Oct. 14	1 June 16.

Subadars.

1 Nov. 06	Jalpaparshad Misir (56, 66*a*)	1 Nov. 06	1 Sep. 11	With 3rd Brahmans.
24 July 91	Mahesar Dube (56)	15 May 12	11 June 15	
14 July 97	Jawala Parshad Pande (56)	5 July 14	9 Mar. 16	
28 Dec. 97	Sita Ram Misr	10 Oct. 14	1 June 16	
20 Jan. 99	Shiubhajan Dube	1 Nov. 14	27 July 16	
25 May 01	Ramcharan Pande (52, 56)	5 Apl. 15	1 Jan. 17	
2 July 01	Ramharakh Misr (52, 56)	5 Apl. 15	20 Apl. 17	
26 June 01	Bansgopal Sukul	12 Ap. 15	8 Jan. 08	
1 Nov. 97	Ramdayal Misr	11 June 15	12 Jan. 18	
29 Mar. 05	Ramsudh Tewari	6 Jan. 16	27 May 18	

1st Battalion, 1st Brahmans—*contd.*

1st Commn. or date of entering service.	Names and Rank.	Army rank.	Present appointment in Regiment.	Remarks.
	Jemadars.			
1 June 97	*Ramesar Pathak* (56)	*6 Sep. 12*		
29 Sep. 12	Ramsuman Tewari	14 Jan. 16		
18 Oct. 01	Raghubardayal Tewari (58) . .	26 Feb. 16		
1 Aug. 02	Sri Ram Misr	9 Mar. 16		
2 July 01	Shiu Parshad Tewari	14 Mar. 16		
8 Sep. 98	Parbhudayal Sukul	1 June 16		
	Ramsumer Sukul	20 Dec. 16		
9 Aug. 04	Jwala Parshad Aginhotri . . .	20 Apl. 17		
13 Nov. 97	Shiudat Dube	18 Dec. 17		
20 May 02	Ramadhar Misr	27 Dec. 17		
18 June 03	Pragnarain Awasthi	27 Dec. 17		
12 July 01	Ram Sohan Tewari	27 Dec. 17		
1 Mar. 03	Shiuratan Dube	27 Dec. 17		
8 June 04	Sitladin Misr	27 Dec. 17		
1 Nov. 12	Ganpatdayal Tewari	8 Jan. 18		
14 Jan. 03	Raghunandanparshad Awasthi .	12 Jan. 18		
2 May 04	Bhaironparshad Pande (46) . .	12 Jan. 18		
2 July 03	Mahabir Tewari	27 May 18		
	Chaudhri Badan Singh	. . . .	(*temp. Hony.*)	
	Patandin Sukul	12 June 18		

2nd Battalion, 1st Brahmans.

Formed at Sitapur, 26th November 1917.

Composition.—1 Coy. Kanaujiya Brahmans, 1 Coy. Gaur Brahmans, 1 Coy. Sarwariya Brahmans, 1 Coy. "Sanadh" Brahmans.

1st Commn. or date of entering service.	Names and Rank.	Army rank.	Present appointment in Regiment.	Remarks.
	Commandant.			
20 Jan. 97	Mactavish, Maj. (Actg. Lt.-Col.)	20 Jan. 15	26 Nov. 17	3rd Brahmans.
	Company Commanders (4).			
8 Jan. 01	Kempson, Maj. E. C. R.	8 Jan. 16	2 Dec. 17	1st Brahmans: 2nd-in-Comd. (*temp.*)
29 July 03	Pope, Maj. P. F.	29 July 18	17 June 18	1-1st Brahmans.
19 Sep. 14	White, Lt. J. H. B. (*Actg. Capt.*)	. . . .	4 Feb. 18	R. Dub. Fus. Attd.
7 Oct. 14	Priestley-Evans, Capt. M.	7 Oct. 18	27 Jan. 18	I. A. R. O.
	Company Officers.			
	Amos, 2nd-Lt. T. E.	. . . .	23 Nov. 18	Lanc R., I. A R. O. (*on probn.*)
12 May 16	Stokes, Lt. H. A.	12 May 17	14 Feb. 18	Conn. Rang., I. A. (*on probn.*)
21 May 16	Smith, Lt. R. S., M. C.	21 May 17	19 Feb. 18	Midd'x R., I. A. (*on probn.*) Qr.-Mr (*temp.*)
19 July 16	Little, Lt. E. G., M. C.	19 Jul. 17	22 Jan. 18	R. Dub. Fus., I. A. (*on probn.*)
21 Oct. 16	Lane, Lt. A. D.	21 Oct. 17	16 Mar. 18	Lond. R., I. A. (*on probn.*)
31 Jan. 17	Whall, Lt. G. H. B.	31 Jan. 17	25 Feb. 18	
26 Jan. 18	Kirkhope, 2nd-Lt. J. L.	26 Jan. 18	26 Jan. 18	High L. I., I. A. (*on probn.*) Adjt.
1 Nov. 17	Hatton, Lt. W. A.	1 Nov. 18	14 Mar. 18	I. A. (*T. C.*)
4 Nov. 17	Bell, 2nd-Lt. W. A. J.	4 Nov. 17	14 Mar. 18	I. A. (*T. C.*)
5 Nov. 17	Hampton, Lt. C. W.	5 Nov. 18	14 Mar. 18	I. A. (*T. C.*)
26 Jan. 18	Lea, 2nd-Lt. W.	26 Jan. 18	1 Apl. 18	I. A. (*on probn.*)
15 Jan. 18	Finch, 2nd-Lt. H. C.	15 Jan. 18	7 June 18	I. A. (*T. C.*) *S. S. O. Sitapore.*
29 Aug. 17	Jones, 2nd-Lt. G. P.	29 Aug. 17	9 Sep. 18	Norf R., I. A. R. O. (*on probn.*)
26 Feb. 18	Whelton, 2nd-Lt. J. H.	26 Feb. 18	2 Sep. 18	M. G. Corps, I. A. (*on probn.*)
22 May 16	Brigg, Lt. G.	22 Mar. 17	3 Oct. 18	S. Staff. R., I. A. (*on probn.*)
9 July 16	Chamberlain, Lt. W. H.	9 July 17	9 Nov. 18	Dorset R., I. A. (*on probn.*)
26 June 18	Crowe, 2nd-Lt. H. W.	26 June 18	3 Oct. 18	Genl. List, I. A. R. O. (*on probn.*)
7 June 18	Watts, 2nd-Lt. W. W.	7 June 18	. . . .	I. A. R. O.
5 Nov. 18	Riley-Irving, 2nd-Lt. C. T.	5 Nov. 18	16 Nov. 18	I. A. R. O. (*on probn.*)
19 Apl. 18	Ram Prasad Naryan Sahi, 2nd-Lt.	19 Apl. 18	31 Oct. 18	I. L. Force.

Subadar Major.

		Jemadar.	Subadar.	Subadar Major.	
29 May 96	Ramesardat Pande (56, 58, 660)	1 Nov. 10	5 July 14	4 Jan. 18	(*temp.*)

Subadars.

21 Dec. 03	Indesri Parshad Awasthi	11 June 15	28 Dec. 17
4 May 97	Bholanath Pande	11 June 15	28 Dec. 17
6 Apl. 07	Ramdihal Pande	12 July 15	28 Dec. 17
11 Apl. 00	Chandikaparshad Tiwari	1 Sep. 15	28 Dec. 17
17 Nov. 97	Ramjiawan Sukul (56)	9 Dec. 15	28 Dec. 17
5 May 03	Bhagwalparshad Tewari	28 Dec. 17	1 July 18
15 July 02	Nagesar Tewari	28 Dec. 17	29 Oct. 18

Jemadars.

7 Jan. 02	Jagdama Sukul	28 Dec. 17	
30 Aug. 01	Shankardial Misir	28 Dec. 17	
11 May 05	Narsinghdat Pande	28 Dec. 17	
10 Aug. 10	Shiugopal Pande	1 July 18	
16 Feb. 06	Ramdulare Tewari	1 July 18	
28 Feb. 13	Hukum Chand	1 July 18	Indian Adjt.
21 July 15	Net Ram	1 July 18	
21 Mar. 04	Balakram Misr	1 July 18	
15 May 05	Bhazautidin Dube	1 July 18	
1 July 06	Lalta Pershad Upadhya	1 July 18	
17 June 13	Permeshwari Parshad Tewari	1 July 18	
28 Aug. 16	Jagdamma Parshad Misr	1 July 18	
21 Mar. 11	Bhupwarian Dube	1 July 18	
21 Apl. 10	Ramshankar Tewari	13 Sep. 18	
12 Dec. 02	Mahabir Pershad Misir	13 Sep. 18	
1 June 05	Bachhulal Aginhotri	13 Sep. 18	
23 Oct. 06	Bhagwant Tewari	1 Oct. 18	
16 July 07	Jhandulal Tewari	1 Oct. 18	
25 Nov. 05	Puttan Dube	1 Oct. 18	
4 Feb. 08	Ramnath Tewari	1 Oct. 18	
7 Aug. 03	Gangaparshad Tewari	29 Oct. 18	
1 Mar. 05	Brijlall Dichhit	29 Oct. 18	

Regt. Centre—Agra. *Uniform*—Scarlet.
Linked with—4th, 13th and 16th Rajputs. *Facings*—Blue.

1st Battalion, 2nd Queen Victoria's Own Rajput Light Infantry.

Broon-ki-paltan.—Raised at Sasseram in 1798, by Maj. G. S. Browne, as the 2nd Bn., 15th Regt., Bengal Native Infantry. Became the 31st Regt. of Bengal N. I., 1824; the 31st Regt. of Bengal Nat. (Light) Infy. 1858; the 2nd Regt. of Bengal Nat. (Light) Infy., 1861; the 2nd (The Queen's Own) Regt. of Bengal Nat (Light) Infy., 1876; the 2nd (The Queen's Own) Regt. of Bengal (Light) Infy., 1885; the 2nd (Queen's Own) Rajput Regt. of Bengal (Light) Infy., 1897; the 2nd (Queen's Own) Rajput Light Infy., 1901; 2nd (Queen's Own), Rajput Light Infantry, 1903; present designation in 1911.

Carries a third or honorary colour bearing the words "Lake and Victory," granted 1803 for distinguished service
THE ROYAL AND IMPERIAL CYPHER OF QUEEN VICTORIA WITHIN THE GARTER.
"Delhi, 1803" "Leswarree" "Deig" "Bhurtpore" "Affghanistan, 1839" "Khelat"
"Maharajpore" "Chillianwallah" "Goojerat" "Punjaub" "Central India"
"Afghanistan, 1879-80" "Burma, 1885-87" "China, 1900."

Composition—4 Companies Rajputs.

1st Commn. or date of entering service.	Names and Rank.	Army rank.	Present appointment in Regiment.	Remarks.
	Commandant.			
3 Dec. 90	Smith, Lt.-Col. A. LeF. . . .	3 Dec. 16	20 Oct. 18	With 1-7th Rajputs.
	Company Commanders(4).			
3 Aug. 98	McKenna, Maj. J. C., D.S.O., *p.s.c.*	1 Sep. 15	20 Oct. 18	2nd-in-Comd.
8 Jan. 01	Brander, Maj. E. H. C. . .	8 Jan. 16	22 Jan. 13	With 2-8th Rajputs.
18 Jan. 02	*Wagstaff, Maj. L. C.*, C.I.E. [L]	*18 Jan. 17*	*24 Feb. 16*	*G. S. O. 2nd gr.*
21 Jan. 03	Griffith, Maj. W. R. C., M.C. .	21 Jan. 18	1 Aug. 15	*Lv., ex I., m. c.*
	Company Officers. (9)			
21 Dec. 04	Noyes, Capt. C. D., M.C. (*Actg. Maj.*)	21 Dec. 13	23 Sep. 08	With 2-2nd Rajputs.
25 Aug. 17	Bala Sahib Daphle, Capt.. . .	25 Aug. 17	25 Aug. 17	With 2-2nd Rajputs.
29 June 16	Miller, Lt. G. R. B.	3 Feb. 17	. . .	With 16th Rajputs.
29 June 16	Mull, Lt. A.	3 Feb. 17	. . .	
18 Apl. 16	Mason, Lt. D. R. W. (*Actg. Capt.*)	18 Apl. 17	27 Apl. 16	
14 Nov. 16	Mearns, Lt. J. P.	18 Oct. 18	. . .	Attd.
30 Jan. 17	Short, Lt. V. C. G. O'B. . .	30 Jan. 18	. . .	With 13th Rajputs.
18 June 17	Herbert, Lt. H. A.	18 June 18		Attd.
31 May 15	Argles, Lt. E. V. (*temp. Capt.*) .	31 May 16	11 Dec. 17	Attd.
16 Aug. 16	Bussell, Lt. J. C.	16 Aug. 17	25 Nov. 17	Attd.
31 Aug. 18	A'Brook, 2nd-Lt. D.	31 Aug. 18	5 Sep. 18	Attd. With 1-150th Inf.
19 June 15	Pattinson, Lt. H. (*Actg. Capt.*)	19 June 16	28 July 18	Attd.
8 Jan. 01	Davis, Maj. T. A.	8 Jan 16	. . .	1-8th Rajputs.
	Attached.			
22 Jan. 96	Simpson, Bt.-Lt.-Col. J. C. . .	3 June 17	27 Aug. 16	16th Rajputs.
2 June 94	Barry, Maj. W. T.	2 June 12	4 Aug. 17	
14 Jan. 15	Bayly, Lt. P. F. B., M.C. . . .	14 Jan. 16	20 Jan. 15	I. A. R. O.
29 Jan. 15	Whitaker, Lt. C. F. (*temp. Capt.*)	29 Jan. 16	29 Aug. 15	I. A. R. O.
19 Apl. 15	Cook, Lt. T. P.	19 Apl. 16	28 July 15	I. A. R. O.
21 Oct. 15	Centeno, Lt. J. B.	21 Oct. 16	. . .	I. A. R. O.
5 Feb. 15	Adams, Lt. P. T.	5 Feb. 16	19 July 16	I. A. R. O.
20 July 15	Jack, Lt. A. J.	20 July 15	11 Dec. 17	I. A. (*on probn.*)
18 June 16	Thornton, Lt. W. W. . . .	18 June 17	15 May 18	I. A. (*on probn.*)
28 June 18	Williamson, 2nd-Lt. A. . . .	28 June 18	30 Aug. 18	I. A. R. O. (*on probn.*)
22 Sep. 15	Leys, Lt. J. F.	22 Sep. 16	25 May 18	
28 Nov. 17	Hind, 2nd-Lt. F. M. J. . . .	28 Nov. 17	19 May 18	I. A. R. O.
18 Mar. 16	Donovan, Lt. E.	18 Mar. 17	17 Nov. 16	I. A. R. O.
14 Nov. 16	Dagge, Lt. R. C.	14 Nov. 17	23 Feb. 18	
30 Jan. 17	Dawson, Lt. A. J. E.	30 Jan. 18	1 Mar. 18	1-117th Mah.
10 Mar. 16	Robson, Lt. E.	10 Mar. 17	24 July 18	North'd Fus., I. A. (*on probn.*)
10 Sep. 15	Wood, Lt. E., M.C.	10 Sep. 16	19 May 18	Manch. R., I. A. (*on probn.*).
4 June 18	Hollingbery, 2nd-Lt. G. R. V. .	4 June 18	1 July 18	I. A. R. O.
1 May 18	Cox, 2nd-Lt. H. G.	1 May 18	15 Aug. 18	I. A. R. O.
21 June 15	Thornton, Lt. E. A.	21 June 16	10 Sep. 18	I. A. (*on probn.*)
5 Nov. 17	Sergeant, Lt. A. H.	5 Nov. 18	8 July 18	I. A. (*T. C.*)
18 Sep. 15	Stobart, Lt. R. McI.	13 Sep. 16	15 Sep. 18	R. Dub. Fus., I. A. (*on probn.*)
27 Aug. 18	Brodie, 2nd-Lt. R..	27 Aug. 18	24 Sep. 18	I. A. R. O.
6 Sep. 18	Guthrie, 2nd-Lt. D. T. G.. . .	6 Sep. 18	25 Sep. 18	I. A. R. O.
2 Sep. 18	Inglis, 2nd-Lt. J. P. . . .	2 Sep. 18	9 Oct. 18	I. A. R. O.
26 June 18	Thompson, 2nd-Lt., A. B. . .	26 June 18	9 Oct. 18	I. A. R. O.
26 May 18	Ballantyne, 2nd-Lt. W. J. . . .	26 May 18	8 Nov. 18	M. G. Corps, I. A. (*on probn.*)

Subadar-Major.

		Jemadar.	Subadar.	Subadar-Major.	
2 Feb. 89	Indar Singh, I.D.S.M. (56) . .	1 Nov. 03	1 Mar. 09	24 Apl. 15	Depôt.
14 Dec. 91	Loknath Singh, I.O.M. . . .	29 June 07	16 Mar. 12	9 Mar. 16	

1st Battalion, 2nd Queen Victoria's Own Rajput Light Infantry—*contd.*

1st Commn. or date of entering service.	Names and rank.	Army rank.		Remarks.
	Subadars.			
14 Apl. 98	Assa Singh (56)	1 Nov. 11	9 Mar. 16	
28 Apl. 98	Pahlad Singh, I.D.S.M.	16 Mar. 12	9 Mar. 16	
17 Apl. 04	Partab Singh (53, 56)	1 Aug. 13	9 Mar. 16	
19 Apl. 97	Chhatar Singh	15 Oct. 14	19 Apl. 16	
4 May 03	Hari Singh	3 Feb. 16	6 Feb. 17	Indian Adjt.
6 Feb. 01	Pitambar Singh (56)	9 Feb. 17	6 Feb. 17	
8 Mar. 98	Khushiram Singh	3 Dec. 14	1 June 17	
23 Oct. 99	Hari Singh	3 Feb. 16	20 May 18	
	Mahabir Singh	. . . .	2 July 17	B. Mily. Police.
16 Apl. 04	Chiranji Singh	9 Mar. 16	31 Aug. 18	
	Lala Singh (*Retd.*)	. . . .	. . . .	Re-employed 26 May 15. Depôt.
	Puran Singh (*Retd.*)	. . . .	. . . .	Re-employed 1 June 15. Depôt.
	Dipnarayan Singh (Retd.)	. . . .	. . . .	Re-employed 26 Aug. 18.
	Jemadars.			
26 Dec. 00	Naurang Singh	1 Sep. 16		
1 Dec. 96	Jacob, Robinson Joseph	1 Oct. 16		Indian Qr.-Mr.
17 Jan. 99	Bhura Singh	1 Feb. 17		
7 June 99	Puran Singh	1 Feb. 17		
5 Jan. 01	Rustam Singh	1 Feb. 17		
23 May 99	Naurang Singh	1 Feb. 17		
16 Oct. 07	Shilu Singh	1 Feb. 17		
20 June 98	Bakhtawar Singh	1 June 17		
1 Mar. 02	Bhura Singh	2 July 17		
17 Jan. 99	Sarjit Singh	3 Dec. 17		
4 June 05	Rohtan Singh (56)	1 Jan. 18		
13 Dec. 03	Udai Singh	24 May 18		
6 Aug. 03	Rakhan Singh (56)	1 Aug. 18		
14 May 05	Shiama Singh (58)	1 Aug. 18		
10 June 05	Shiubakhsh Singh (56)	1 Aug. 18		
16 Feb. 02	Tikam Singh	31 Aug. 18		
13 Mar. 10	Sanwat Singh	31 Aug. 18		
11 Sep. 01	Hardwari Singh	31 Aug. 18		
16 Nov. 06	Raghubir Singh	31 Aug. 18		

The 2nd Rajput L. I. has an additional Jemadar for the third colour.

2nd Battalion, 2nd Queen Victoria's Own Rajput Light Infantry.

Formed at Agra on 15th June 1917.

Composition—4 companies Rajputs.

1st Commn. or date of entering service.	Names and Rank.	Army rank.	Present appointment in regiment.	Remarks.
	Commandant.			
18 Apl. 00	Strong, Maj. R. G. (*Actg. Lt.-Col.*).	1 Sep. 15	4 June 18	1-111th Rajputs.
	Company Commanders. (4)			
21 Dec. 04	Noyes, Capt. C. D., M.C. (*Actg. Maj.*)	21 Dec. 13	14 Dec. 17	1-2nd Rajputs. 2nd-in-Comd.
25 Aug. 17	Bala Sahib Daphle, Capt . . .	25 Aug. 17	11 Aug. 18	1-2nd Rajputs.
12 June 15	Stubbs, Lt. R. H. (*Actg. Capt.*) .	12 June 16	1 June 18	Durh. L. I., I. A. (*on probn.*)
30 Oct. 14	Butcher, Lt. F. G. (*Actg. Capt.*) .	30 July 16	26 Feb. 18	
	Company Officers.			
10 Aug. 15	Curry, Lt. A. V.	10 Aug. 16	16 Apl. 18	I. A. (*on probn.*)
23 Nov. 15	Carter, Lt. S. N.	23 Nov. 16	17 May 18	Worc. R., I. A. (*on probn.*)
5 Dec. 15	Craig, Lt. C. C. (*Actg. Capt.*) .	5 Dec. 16	25 Nov. 17	Attd.
19 Mar. 16	Rea, Lt. A. N.	19 Mar. 17	13 Apl. 18	Suff. R., I. A. (*on probn.*)
25 Oct. 17	Morgan, Lt. E. J.	25 Oct. 17	25 Nov. 17	I. A. R. O.
8 Jan. 17	Bedell, Lt. G. W. (*Actg. Capt.*) .	8 Jan. 18	13 Feb. 18	I. A. R. O. Adjt.
20 June 17	Clinton, Lt. A. G.	20 June 18	3 Oct. 17	I. A. R. O.
5 Oct. 17	Smith, Lt. S. A.	5 Oct. 18	19 Jan. 18	I. A. R. O.
5 Nov. 17	Lockstone, Lt. J.	5 Nov. 18	14 Mar. 18	I. A. (*T. C.*) Qr.-Mr.
28 Nov. 17	Dumbreck, 2nd-Lt. W. Y. . .	28 Nov. 17	3 Apl. 18	I. A. R. O. (*on probn.*)
15 Jan. 18	Taggart, 2nd-Lt. P. D. . . .	15 Jan. 18	8 June 18	I. A. (*T. C.*)
	Armstrong, 2nd-Lt. J. A. . .	. . . . .	. . . . .	Lan. Fus., I. A. (*on probn.*)
22 July 18	Moss, 2nd-Lt. A. G.	22 July 18	3 Dec. 18	I. A. (*T. C.*)
	Subadar-Major.			
		Jemadar.	Subadar.	Subadar-Major.
20 Sep. 95	Jaswant Singh	1 June 11	2 Jan. 16	21 July 17 I
	Subadars.			
23 July 00	Man Singh (58).	18 Sep. 11	31 Jan. 15	
20 Sep. 94	Shiuram Singh	9 Dec. 14	31 Jan. 17	
23 Mar. 96	Loknath Singh	3 Jan. 16	16 July 17	
17 July 08	Balwant Singh (66*a*)	17 July 08	2 Sep. 17	
16 Apl. 97	Arjun Singh	16 July 17	1 Dec. 17	
13 July 01	Khem Singh	16 July 17	1 Dec. 17	
20 Nov. 99	Shishpal Singh	31 July 17	1 Dec. 17	
	Jemadars.			
9 June 00	Phulander Singh	31 July 17		
4 Mar. 95	Surajpal Singh	2 Sep. 17		
21 Nov. 11	Sardar Singh	1 Dec. 17		
8 July 93	Ramdat Singh	1 Dec. 17		
31 May 04	Gulab Singh	1 Dec. 17		
1 Aug. 07	Chiranji Singh	1 Dec. 17		
18 Mar. 08	Arjan Singh (48*a*)	1 Feb. 18		
21 Oct. 07	Chandar Singh (56)	1 Apl. 18		
6 Jan. 11	Hira Singh	1 Aug. 18		
24 Apl. 07	Nihal Singh	1 Sep. 18		
5 May 10	Behari Singh	1 Oct. 18		

3rd Battalion, 2nd Queen Victoria's Own Rajput Light Infantry.

Formed at Allahabad on 3rd September 1918.

Composition—4 companies of Rajputs.

1st Commn. or date of entering service.	Names and Rank.	Army rank.	Present appointment in Regiment.	Remarks.
	Commandant.			
18 Apl. 00	Wood, Maj. H. O. B. (*Act'g. Lt.-Col.*)	3 June 15	8 Sep. 18	1-8th Rajputs.
	Company Commanders. (4)			
21 Dec. 04	Scott, Capt. F. B. (*Act'g. Maj.*)	21 Dec. 13	3 Sep. 18	Survey Dept.
29 May 15	Barham, Lt. W. C. (*Act'g. Capt.*)	29 May 16	3 Sep. 18	Lond. R., I. A. (*on probn.*)
18 June 15	Davies, Lt. F. C. (*Act'g. Capt.*)	18 June 16	3 Sep. 18	I. A. R. O.
22 June 15	Lake, Lt. F. T. M. (*Act'g. Capt.*)	22 June 16	9 Sep. 18	North'd Fus., I. A. (*on probn.*)
	Company Officers.			
7 Aug. 16	Taylor, Lt. A. C., D.S.O.	7 Aug. 17	10 Sep. 18	Midd'x R., I. A. (*on probn.*)
26 June 17	Bladworth, 2nd-Lt. H. M.	26 June 17	10 Sep. 18	I. A. R. O. (*on probn.*)
5 Oct. 17	Young, Lt. P. R.	5 Oct. 18	3 Sep. 18	I. A. R. O. Qr.-Mr. (*Actg.*)
9 May 18	Lowson, 2nd-Lt. J. M.	9 May 18	18 Sep. 18	I. A. (*on probn.*)
15 Jan. 18	Carey, 2nd-Lt. C. O'D.	15 Jan. 18	8 Sep. 18	I. A. R. O.
29 Jan. 18	Burns, 2nd-Lt. F.	29 Jan. 18	23 Sep. 18	Lan. R., I. A. R. O. (*on probn.*)
1 May 18	Norman, 2nd-Lt. F. E. B.	1 May 18	10 Nov. 18	R. Fus., I. A. (*on probn.*)
6 Oct. 16	Bramfitt, 2nd-Lt. A. E.	6 Oct. 17	23 Nov. 18	York & Lanc. R., I. A. (*on probn.*)
2 June 18	Willis, 2nd-Lt. H. A.	2 June 18	10 Oct. 18	I. A. R. O.
2 June 18	Wimpey, 2nd-Lt. S. G.	2 June 18	10 Oct. 18	I. A. R. O.
15 June 18	Barkway, 2nd-Lt. A. P. C.	15 June 18	16 Sep. 18	I. A. R. O.
10 July 18	Llewellyn, 2nd-Lt. R. G.	10 July 18	2 Dec. 18	I. A. (*T. C.*)
13 July 18	Barber, 2nd-Lt. J. F. A.	13 July 18	....	I. A. R. O.
22 July 18	Thomas, 2nd-Lt. F. G.	22 July 18	2 Dec. 18	I. A. R. O.
	Subadar-Major.	Jemadar.	Subadar.	Subadar-Major.
23 Feb. 98	Umrao Singh (56)	13 Feb. 10	3 Dec. 14	23 Sep. 18 (*Actg.*)
	Subadars.			
12 July 95	Hardwari Singh	6 Feb. 17		
24 Dec. 03	Tarif Singh	9 Feb. 16	23 Sep. 18	
20 July 00	Balwant Singh	9 Mar. 16	23 Sep. 18	
6 May 07	Dhal Singh	9 Mar. 16	23 Sep. 18	
	Jemadars.			
14 May 99	Shiucharan Singh	1 Sep. 16		
16 Oct. 06	Bharat Singh	1 Nov. 17		
22 Feb. 94	Raghunath Singh	1 Mar. 18		
8 July 01	Inder Singh (I)	3 Sep. 18		
10 Mar. 02	Sanwal Singh	23 Sep. 18		
20 Apl. 03	Mazal Singh	23 Sep. 18		
18 Feb. 04	Inder Singh (II)	23 Sep. 18		
5 July 07	Debi Singh	23 Sep. 18		
22 July 01	Ganpat Singh	23 Sep. 18		

Regt. Centre—Allahabad.
Linked with—1st Brahmans.

Uniform—Scarlet.
Facings—Black.

1st Battalion, 3rd Brahmans.

Guthrie-ki-paltan.—Raised at Jaunpur in 1798, by Lt.-Col. J. Guthrie, as the 1st Battn., 16th Regt. of Bengal Native Infantry. Became the 32nd Regt. of Bengal N. I., 1824; the 3rd Regt. of Bengal N. I., 1861; the 3rd Regt. of Beng. Infy., 1885; 3rd Brahman Infy., 1901; present designation, 1903.

"Bhurtpore" "Afghanistan, 1879-80."

Composition—4 companies Brahmans.

1st Commn. or date of entering service.	Names and Rank.	Army rank.	Present appointment in Regiment.	Remarks.
	Commandant.			
28 Jan. 93	Hawkes, Lt.-Col. R.	10 May 17	10 May 17	
	Company Commanders. (4)			
20 Jan. 97	Mactavish, Maj. A.	20 Jan. 15	19 Nov. 16	2nd-in-Comd. With 2-1st Brahmans.
18 Oct. 99	Hutcheson, Maj. S.	1 Sep. 15	12 June 11	With 1-41st Dogras.
7 Mar. 00	Kirkpatrick, Maj. K. E.	1 Sep. 15	26 Jan. 16	
11 Aug. 00	Hogg, Maj. W. L.	11 Aug. 15	19 Nov. 16	
	Company Officers. (9)			
4 Dec. 01	*Hutcheson, Maj. W. C.*	*4 Dec. 16*	*16 Nov. 06*	
18 Jan. 02	*Maxwell, Maj. P. A.*	*18 Jan. 17*	*19 Nov. 03*	
29 Nov. 05	Brickman, Capt. J. M.	29 Nov. 14	23 May 08	
19 Jan. 07	*Strover, Capt. E. J.*	*1 Sep. 15*	*22 Aug. 09*	
25 Jan. 08	Mackenzie, Capt. E. A. H.	1 Sep. 15	25 July 09	
19 Sep. 08	Brotherton, Capt. L. P. (*Actg. Maj.*)	1 Sep. 15	16 Mar. 14	With 2-3rd Brahmans.
25 Aug. 17	*Rana Jodha Jung Bahadur, Lieut.*, M.C., A.-D.-C.	*25 Aug. 17*	. . . .	*Comdt., Tehri State, Imp. Sap.*
30 Jan. 17	White, Lt. D. O.	30 Jan. 17	. . . .	Attd.
30 Jan. 17	O'Sullivan, Lt. H. P.	30 Jan. 18	. . . .	With 1-1st Brahmans.
5 May 15	Dowling, Lt. G.	5 May 16	8 Nov. 17	Attd. *S.S.O., Barrackpore.*
13 June 15	Furness, Lt. W. S., M.C. (*Actg. Capt.*)	13 June 16	14 Mar. 18	Attd.
27 Feb. 16	Porter, Lt. A. G.	27 Feb. 17	18 Jan. 16	Attd.
22 Feb. 16	Spencer, Lt. H. J. (*Actg. Capt.*)	22 Feb. 17	4 Nov. 17	Attd.
	Attached.			
15 May 15	Anderson, Lt. D. M.	15 May 16	10 Sep. 18	R. Ir. Rif., I. A. (*on probn.*)
29 Jan. 10	Sanders, Capt. C. W.	1 Sep. 15	. . . .	Adjt., 1st Brahmans.
27 Nov. 14	Tredway, Lt. G. H. (*Actg. Capt.*)	27 Nov. 15	3 Sep. 16	I. A. R. O.
14 Jan. 15	Barker, Lt. H. E. R. (*Actg. Capt.*)	14 Jan. 16	14 Jan. 15	I. A. R. O. Qr.-Mr. *Offg.*
14 Jan. 15	Reith, Lt. C. E. W. (*temp. Capt.*)	14 Jan. 16	14 Jan. 15	I. A. R. O.
6 Nov. 15	Topping, Lt. W. D.	6 Nov. 16	23 May 18	I. A. (*on probn.*)
27 Jan. 16	Fell, Lt. R. J. H. (*Actg. Capt.*)	27 Jan. 17	4 Nov. 16	I. A. R. O.
20 Feb. 16	Siedle, Lt. C. E.	20 Feb. 17	4 Aug. 16	I. A. R. O.
5 May 17	Pigott, Lt. T. F.	5 May 17	1 Mar. 18	I. A. R. O.
2 June 17	Gough, 2nd-Lt. F. B.	2 June 17	14 Mar. 18	I. A. R. O.
20 June 17	Forbes, Lt. E. R. C.	20 June 18	14 Mar. 18	I. A. R. O.
20 June 18	Burke, 2nd-Lt. W. G.	20 June 18	3 Dec. 18	I. A. R. O. [*Divn.*
14 Aug. 16	*Stork, Lt. H. C.* (*temp. Capt.*)	*14 Aug. 17*	*16 Nov. 16*	*I. A. R. O. G. S. O., 3rd Gr., 9th*
5 Nov. 17	Dirks, Lt. A. J.	5 Nov. 18	9 Mar. 18	I. A. R. O.
8 Jan. 17	Walker, Lt. L. W.	8 Jan. 18	1 Mar. 18	I. A. R. O.
8 Mar. 17	Rooke, Lt. I.	8 Mar. 18	16 Mar. 18	I. A. R. O.
20 June 17	Lund, Lt. R.	20 June 18	8 Oct. 17	I. A. R. O.
19 Sep. 16	Overton, Lt. H.	19 Sep. 17	. . . .	Notts. & Derby R., I.A. (*on probn.*)
7 Oct. 18	Fraser, 2nd-Lt. J. W.	7 Oct. 18	25 Oct. 18	I. A. (*T. C.*)
7 Oct. 18	Barrow, 2nd-Lt. A.	7 Oct. 18	30 Oct. 18	I. A. (*T. C.*)

Subadar-Major.

		Jemadar.	Subadar.	Subadar-Major.
1 May 98	Gangaparshad Sukul (56, 72*a*)	6 July 03	3 Jan. 11	8 June 15

Subadars.

9 June 98	Mahesh Narain Dube	9 June 98	1 Feb. 03	Depôt.
3 Dec. 97	Bachanu Dichit (56)	1 July 08	8 June 15	
6 Nov. 97	Shiunandan Misr	30 Dec. 10	9 Dec. 15	
21 Nov. 06	Girdharilal Pande (52, 56)	9 Aug. 14	9 Dec. 15	
6 Jan. 90	Udit Misr (56)	3 Jan. 11	22 Mar. 16	
16 Sep. 98	*Surajpal Tewari*	*15 Oct. 14*	12 June 16	
8 Aug. 01	Sattidin Sukul	8 June 15	24 Nov. 16	
5 Jan. 98	Ramphal Tiwari	9 Dec. 15	3 Dec. 16	
19 Oct. 05	Bholanath Awasthi	11 Mar. 16	3 June 17	
18 Jan. 00	Rampargat Pande	9 Dec. 15	18 June 17	
2 July 01	Shiu Parshad Tewari (56)	14 Mar. 16	7 July 17	
12 Oct. 00	Jagdeo Sukul	22 Mar. 16	7 July 17	
5 Jan. 99	Narayandat Dube	8 Oct. 15	18 July 17	
2 Dec. 97	Matadin Tewari	10 Apl. 12	6 Aug. 18	

1st Battalion, 3rd Brahmans—*contd.*

1st Commn. or date of entering service.	Names and Rank.	Army rank.	Remarks.
	Jemadars.		
5 Oct. 97	Thakurdin Misr (56)	19 Apl. 11	
24 Oct. 00	Ramjiawan Tiwari	18 May 16	Ind. Adjt.
16 Jan. 00	Nandlal Misr.	3 Dec. 16	
18 June 03	Shambhudayal Tiwari	3 Dec. 16	
1 Mar. 04	Kidarnath Tewari	18 June 17	
10 June 01	Rajaram Pande	7 July 17	
17 Dec. 00	Kithawardin Pande	7 July 17	
16 Nov. 11	Gokul Prasad Dube	1 Oct. 17	
7 Mar. 02	Bishan Ram	3 Dec. 17	
4 July 90	Becha Misr	1 Jan. 18	
28 Apl. 02	Datnarayan Dube	1 Jan. 18	
30 Aug. 02	Sukhpal Dube	1 Jan. 18	
1 Sep. 09	Shiunath Misr	1 Jan. 18	
27 Feb. 04	Ramsamujh Upadhya	1 Jan. 18	
6 Nov. 06	Shiumangal Tiwary (5a)	1 Jan. 18	
10 Jan. 00	Surajpal Dube	1 Mar. 18	With 2-131st R.
8 July 03	Paramsukh Misr	16 May 18	
18 July 01	Parmeshwardin Misr	12 June 18	
21 Dec. 10	Ramdas Pathak	12 June 18	
9 May 02	Mahadeo Tewari	24 June 18	
	Ramesardath Misr.	6 Aug. 18	

2nd Battalion, 3rd Gaur Brahmans.

Formed at Deesa.

Composition { 2 Companies Gaur Brahmans.
{ 2 Companies Oudh Brahmans.

1st Commn. or date of entering service.	Names and Rank.	Army rank.	Present appointment in regiment.	REMARKS.
	Commandant.			
5 May 00	Lane, Maj. H. V. (*Act'g. Lt.-Col.*)	15 Sep. 15	18 June 18	1st Brahmans. (*temp.*).
	Company Commanders. (4)			
19 Sep. 08	Brotherton, Capt. L. P. (*Actg. Maj.*)	15 Sep. 15	18 June 18	3rd Brahmans. 2nd-in-Comd. (*temp.*)
5 Nov. 14	Beney, Capt. G. E.	5 Nov. 18	21 Oct. 18	I. A. R. O.
3 Apl. 15	Steel, Lt. G. B. E.	1 July 17	24 June 18	I. A. R. O.
	Company Officers.			
7 Feb. 17	Melhuish, Lt. H.	7 Feb. 18	2 Oct. 18	W. York. R., I. A. (*on probn.*)
25 July 17	Austin, Lt. L. A., D.C.M.	25 July 18	23 Sep. 18	Norf. R., I. A. (*on probn.*) Adjt.
28 Dec. 17	Cranwill, 2nd-Lt. G. B.	23 Dec. 17	12 Sep. 18	R. Dub. Fus. I. A. (*on probn.*)
19 June 17	Bowen, Lt. G. L.	19 June 18	7 Sep. 18	Welsh R., I. A. (*on probn.*)
30 May 17	Kydd, 2nd-Lt. W. A.	30 May 17	23 Aug. 18	I. A. R. O. (*on probn.*)
26 Sep. 17	Barker, 2nd-Lt. W. D., M. M.	26 Sep. 17	26 June 18	I. A. R. O.
26 Sep. 17	Layfield. 2nd-Lt. F.	26 Sep. 17	3 Sep. 18	Innis. Fus., I. A. R. O (*on probn.*)
5 Oct. 17	Thurston, Lt. E. T.	5 Oct. 18	26 June 18	I A. R O.
31 Oct. 17	Praslow, 2nd-Lt. J. E.	31 Oct. 17	1 July 18	I. A. R. O. (*on probn.*)
5 Nov. 17	Clarke, 2nd-Lt. W. H.	5 Nov. 17	6 Sep. 18	I. A. R. O. (*on probn.*)
5 Dec. 17	Ross, 2nd Lt. R. P.	5 Dec. 17	1 July 18	I. A. R. O. (*on probn.*)
27 Feb. 18	Lomax, 2nd-Lt. G. W.	27 Feb. 18	3 Sep. 18	Lanc. R., I. A R. O. (*on probn.*)
29 Mar. 18	Hubball, 2nd-Lt. J. L.	29 Mar. 18	23 Aug. 18	I. A. R. O. (*on probn.*) Qr. Mr.
1 May 18	Gilliat, 2nd-Lt. L.	1 May 18	30 July 18	I. A. R. O. (*on probn.*)
29 May 18	Baldwin, 2nd-Lt. R. A. C.	29 May 18	4 Sep. 18	I. A. R. O. (*on probn.*)
1 Oct. 18	Carse, 2nd-Lt. W. M.	1 Oct. 18	....	Attd.

Subadar-Major.

		Jemadar.	Subadar.	Subadar-Major.
23 Nov. 02	Kedarnath Dube (66*a*)	3 Nov. 14	28 Dec. 17	16 Sep. 18

Subadar.

20 May 02	Baldeo Pershad Dube	28 Dec. 17	16 Sep. 18	
31 May 15	Neki Ram	1 Mar. 18	1 Oct. 18	

Jemadar.

	Jai Lal	1 Oct. 18		
	Sukhdeo	1 Oct. 8		
	Mulchand	1 Oct. 18		
	Harphul	1 Oct. 18		
	Manphul	1 Oct. 18		
	Kandhyalal	1 Oct. 18		
	Tirat Singh	1 Oct. 18		
	Hardwari	1 Oct. 18		
	Ramphal Sukul	1 Oct. 18		
	Reoti Parshad	1 Oct. 18		
	Sher Singh	1 Oct. 18		
	Kundan	1 Oct. 18		

Regt. Centre—Agra. *Uniform*—Scarlet.
Linked with—2nd, 13th and 16th Rajputs. *Facings*—Black.

1st Battalion, 4th Prince Albert Victor's Rajputs.

Hilliard-ki-paltan.—Raised at Baragaon in 1798, by Lt.-Col. John Hilliard, as the 2nd Battalion, 16th Regiment of Bengal Native Infantry. Became the 33rd Regiment of Bengal Native Infantry, 1824; the 4th Regiment of Bengal Native Infantry, 1861; the 4th Regiment of Bengal Infantry, 1885; the 4th (Prince Albert Victor's) Regiment of Bengal Infantry, 1890; the 4th (Prince Albert Victor's) Rajput Regiment of Bengal Infantry, 1897; the 4th (Prince Albert Victor's) Rajput Infantry, 1901; present designation, 1903.

"Leswarree" "Bhurrpore" "Cabool, 1842" "Ferozeshah" "Sobraon"
"Afghanistan, 1879-80" "Burma, 1885-87."
Composition—4 Companies Rajputs.

COLONEL—General Harry C. Anderson, M, C.B. . 13 May 1904.

1st Commn. or date of entering service.	Names and Rank.	Army rank.	Present appointment in Regiment.	Remarks.
	Commandant.			
22 Aug. 88	Ralph, Lt.-Col. A. C., D.S.O.	22 Aug. 14	25 Aug. 15	
	Company Commanders. (4)			
1 Dec. 97	Woodcock, Maj. E. E.	1 Sep. 15	1 May 14	With 2-11th Rajputs.
18 Jan. 02	Ames, Maj. W. R	18 Jan. 17	25 Aug. 15	
22 Apl. 03	*Ralston, Maj. W. J. H.*	*22 Apl. 18*	*9 May 16*	
	Company Officers. (9)			
10 Oct. 03	Cosens, Maj. F. R.	10 Oct. 18	14 July 07	*Lv., ex I., m.c.*
10 Oct. 03	Chesney, Maj. H. G.	10 Oct. 18	28 Apl. 06	
2 Aug. 05	Clarke, Capt. H. (*Actg. Maj.*)	2 Aug. 14	10 Jan. 08	
6 Oct. 06	Burroughs, Capt. S. V. G (*Actg. Maj.*)	1 Sept. 15	22 July 09	With 1-150th Infy.
23 Jan. 10	*O'Sullivan, Capt. J.*	*1 Sep. 15*	*9 Mar. 11*	
3 Sep. 10	*Penrose-Welsted, Capt. R.-H.*	*1 Sep. 15*	*15 Dec. 11*	*Staff Capt.*
15 Nov. 15	Gemmell, Lt. G. W. (*Actg. Capt.*)	15 Nov. 16	18 Nov. 15	Adjt. (*Actg.*)
15 Nov. 15	Snelling, Lt. A. H. J.	15 Nov. 16	18 Nov. 15	With 1-150th Infy.
26 Aug. 16	Smith, Lt. C. C. H.	26 Aug. 17	31 Aug. 17	Attd. With 1-131st Infy.
30 Jan. 17	Pocock, Lt. N. E. I.	30 Jan. 18	4 Dec. 17	
25 Aug. 17	Kunwar Savai Singhji, Lt.	25 Aug. 17	25 Nov. 17	*Lv., Ind., m. c.*
	Attached.			
27 Aug. 02	Gwyn, Capt. A. H.	27 Aug. 11	....	Survey Dept.
20 Nov. 14	Hendry, Lt. F. C., M.C.	20 Nov. 15	9 Mar. 15	I. A. R. O.
8 Jan. 15	Stevens, Lt. R. H. (*Actg. Capt.*)	8 Jan. 16	16 Dec. 16	I. A. R. O.
1 Mar. 17	Stacey, Lt. A. F.	1 Mar. 18	10 Mar. 17	I. A. R. O. Qr.-Mr. (*Offg.*)
15 Nov. 15	Bromham, Lt. A. C. (*Actg. Capt.*)	15 Nov. 16	14 Nov. 16	16th Rajputs.
16 Oct. 14	Cook, Capt. A. F.	16 Oct. 18	25 Nov. 17	I. A. R. O.
1 Oct. 17	Manthei, Lt. W. J.	1 Oct. 18	20 Jan. 18	I. A. (*T. C.*)
5 Nov. 17	Rowe, Lt. W. H.	5 Nov. 18	14 Mar. 18	I. A. (*T. C.*)
20 June 17	Ferguson, Lt. H. J.	20 June 18	3 Oct. 17	I. A. R. O.
30 Dec. 17	Bower, 2nd-Lt. F.	30 Dec. 17	29 Apl. 18	I. A. R. O.
29 Aug. 17	DuPre, 2nd-Lt. W. D.	29 Aug 17	13 Dec. 17	I. A. R. O.
29 Mar. 18	Hoskin, 2nd-Lt. J. G.	29 Mar. 18	14 July 18	I. A. (*T. C.*)
5 Aug. 16	Lilley, Lt. W. G.	5 May 18	3 July 18	Notts. & Derby. R., I. A. (*on probn.*)
29 Aug. 17	Hogbin, 2nd-Lt. C. J.	29 Aug. 17	12 July 18	I. A. R. O.
30 Aug. 18	Gidley, 2nd-Lt. D. J.	30 Aug. 18	30 Sep. 18	I. A. R. O. (*on probn.*)
3 Sep. 18	Maffat, 2nd-Lt. J.	3 Sep. 18	24 Sep. 18	I. A. R. O. (*on probn.*)
7 Oct. 18	Drew, 2nd-Lt. O. W.	7 Oct. 18	22 Oct. 18	I. A. (*T. C.*)
29 Jan. 18	Cruise, 2nd-Lt. W.	29 Jan. 18	7 Nov. 18	R. Lanc. R., I. A. R. O. (*on probn.*)

Subadar Major.

		Jemadar.	Subadar.	Subadar-Major.
25 Dec. 93	Hetram Singh (56, 72a)	16 Aug. 06	16 May 12	21 Feb. 16.

1st Battalion, 4th Prince Albert Victor's Rajputs—*concld.*

1st Commn. or date of entering service.	Names and Rank.	Army rank.	Present appointment in regiment.	REMARKS.
	Subadars.			
11 Dec. 02	Raghbir Singh (56, 66)	2 Mar. 06	15 Feb. 12	
5 Feb. 00	Behari Singh (56, 66)	16 Jan. 08	2 Apl. 14	
7 June 98	Gajraj Singh (56, 72*a*)	27 Apl. 10	1 July 14	
9 June 97	Hari Singh (66)	1 June 10	9 Dec. 14	
2 July 97	Jairam Singh (51)	24 Mar. 11	10 Dec. 14	
27 July 00	Faqira Singh, I.D.S.M.	18 Sep. 11	20 Jan. 15	
3 Mar. 09	*Simbhu Singh*	*13 Mar. 09*	*1 Feb. 17*	
16 Oct. 03	Bhura Singh	20 Jan. 15	1 Feb. 17	
11 May 00	Takhat Singh	10 Dec. 14	15 May 18	
10 Nov. 04	Munna Singh	15 Aug. 16	15 May 18	
12 Jan. 98	Maharam Singh	20 Jan. 15	30 June 18	
5 Apl. 04	Sispal Singh	19 Feb. 18	4 Sep. 18	
	Sundar Singh (*Retd.*)	. . . .	. . . .	Re-employed 6 Apl. 15.
	Pahar Singh (*Retd.*)	. . . .	. . . .	Re-employed 17 May 15.
	Jemadars.			
9 Nov. 98	Udmi Singh	20 Jan. 15		
15 Jan. 06	Mahtab Singh	19 Feb. 18		
1 Aug. 03	Rattu Singh	19 Feb. 18		
21 Aug. 08	Mohan Singh	19 Feb. 18		
6 Nov. 06	Hem Singh	19 Feb. 18	Indian Qr.-Mr.	
15 Apl. 06	Jwala Singh	1 Mar. 18		
10 June 05	Jhutha Singh	18 Apl. 18		
25 Dec. 00	Mukhtiyar Singh	1 May 18		
21 Nov. 02	Bhagwan Singh	1 May 18		
6 May 04	Genda Singh	15 May 18		
2 Jan. 05	Jagmal Singh	15 May 18		
5 June 06	Indar Singh	15 May 18		
4 June 04	Kalu Singh	5 July 18		
26 Feb. 02	Ganpat Singh	1 Aug. 18		
30 May 04	Mangtu Singh	1 Aug. 18		
10 Feb. 08	Birbal Singh	1 Aug. 18		

2nd Battalion, 4th Prince Albert Victor's Rajputs.

Formed at Agra.

Composition—4 Companies Rajputs.

1st Commn. or date of entering service.	Names and Rank.	Army rank.	Present appointment in Regiment.	Remarks.
	Commandant.			
20 Jan. 97	Cutler, Maj. H. C. (*Actg. Lt.-Col.*)	20 Jan. 15	16 Feb. 18	1-8th Rajputs.
	Company Commanders. (4)			
18 Jan. 02	Slater, Maj. D. McL. . . .	18 Jan. 17	21 Feb. 18	11th Rajputs, 2nd-in-Comd. (*temp.*)
28 Mar. 15	Folkard, Lt. F. T. (*Actg. Capt.*)	21 Feb. 16	16 Feb. 18	I. A. R. O.
5 Mar. 15	Salt, Lt. J. J.	5 Dec. 16	1 Oct. 18	Attd.
1 July 16	Ashton, Lt. A.	1 July 17	1 Oct. 18	I. A. R. O.
	Company Officers.			
23 July 17	Ellis, Lt. W. J. (*Actg. Capt.*) .	23 July 17	26 Feb. 18	I. A. R. O.
27 Oct. 16	Laughton, Lt. F. S.	27 Oct. 17	22 May 18	I. A. (*on probn.*)
27 Jan. 17	Joyce, Lt. B. J. M. R. . . .	27 Jan. 18	22 May 18	Yorks. L. I., I. A. R. O. (*on probn.*)
20 June 17	Wilcock, Lt. F. W. (*Actg. Capt.*)	20 June 18	16 Feb. 18	I. A. R. O. Adjt. (*temp.*)
25 Dec. 17	Higgs, 2nd-Lt. R. N. . . .	25 Dec. 17	. . . .	R. F. A., I. A. (*on probn.*)
1 Aug. 17	Maclean, 2nd-Lt. D., M. C. . .	1 Aug. 17	9 Sep. 18	R. Lanc. R., I. A. (*on probn.*)
28 Dec. 17	Lambie, 2nd-Lt. J.	28 Dec. 17	13 May 18	R. Sc. Fus., I. A. (*on probn.*)
15 Jan. 18	Lovering, 2nd-Lt. A. J. . . .	15 Jan. 18	5 June 18	I. A. (*T. C.*)
15 Jan. 18	Harvey, 2nd-Lt. C. A. F. . . .	15 Jan. 18	5 June 18	I. A. R. O. Qr.-Mr. (*Offg.*)
27 Feb. 18	Leonard, 2nd-Lt. R. E. . . .	27 Feb. 18	9 Sep. 18	Lan. R., I. A. R. O. (*on probn.*)
18 Mar. 18	Blanchford, 2nd-Lt. M. W. . .	18 Mar. 18	15 July 18	I. A. (*T. C.*)
29 Mar. 18	Bayley, 2nd-Lt. W. H.. . . .	29 Mar. 18	15 July 18	I. A. (*T. C.*)
29 Mar. 18	Burke, 2nd-Lt. P. H.	29 Mar. 18	15 July 18	I. A. R. O.
25 Oct. 18	Mitchell, 2nd-Lt. J.	25 Oct. 18	9 Sep. 18	I. A. R. O.
17 Feb. 18	Reeve, 2nd-Lt. H. L.	17 Feb. 18	3 Oct. 18	Norf. R., I. A. R. O. (*on probn.*)
2 June 18	Browning, 2nd-Lt. J. M. . . .	2 June 18	. . . .	I. A. R. O.
30 Oct. 18	Waterfall, 2nd-Lt. J.	30 Oct. 18	3 Oct. 18	Yorks. L. I., I. A. (*on probn.*)
5 July 18	Whiteside, 2nd-Lt. A. . . .	5 July 18	3 Oct. 18	Genl. List, I. A. R. O. (*on probn.*)
8 July 18	Blood, 2nd-Lt. C. D.	8 July 18	3 Dec. 18	I. A. R. O.
10 July 18	Amos, 2nd-Lt. C.	10 July 18	3 Dec. 18	I. A. (*T. C.*)

Subadar-Major.

		Jemadar.	Subadar.	Subadar-Major.	
15 July 02	Ramlal Singh, (56, 72a) . . .	15 July 02	16 July 11	15 Mar. 18	16th Rajputs (*temp.*)

Subadars.

3 Apl. 01	Pirbhu Singh	10 Dec. 14	15 Mar. 18	
28 Aug. 06	Ranjit Singh	19 May 15	15 Mar. 18	
24 Oct. 00	Sital Singh	10 July 16	15 Mar. 18	
15 Sep. 06	Sultan Singh	1 Feb. 17	15 Mar. 18	
14 Dec. 06	Durjan Singh	31 Mar. 17	15 Mar. 18	
15 Jan. 06	Jaswant Singh	16 Sep. 17	15 Mar. 18	
4 June 01	Puran Singh	15 Mar. 18	15 June 18	

Jemadars.

22 Aug. 04	Sagar Singh (51)	15 Mar. 18	Indian Adjt. (*Offg.*)
18 Mar. 08	Mohr Singh	15 Mar. 18	
14 Nov. 04	Mangal Singh	15 Mar. 18	
13 July 07	Ganeshi Singh	15 Mar. 18	
10 Jan. 08	Bijai Singh	15 Mar. 18	
5 Jan. 05	Hazari Singh	15 June 18	
16 Jan. 06	Girwar Singh	15 June 18	
8 May 05	Ramnath Singh	15 June 18	
18 May 07	Sughar Singh	15 June 18	
8 Sep. 08	Baldeo Singh	1 Sept. 18	
22 June 12	Ramjilal Singh	1 Sept. 18	
24 Feb. 12	Phul Singh	1 Sept. 18	

Regt. Centre—Benares.
Linked with—17th and 18th Infantry.

Uniform—Scarlet.
Facings—Yellow.

1st Battalion, 5th Light Infantry.

Jansin-ki-paltan.—Raised at Cawnpore in 1803, by Lieut. J. M. Johnson, as the 2nd Battalion, 21st Regiment Bengal Native Infantry. Became the 42nd Regiment of Bengal Native Infantry, 1824; the 42nd Regiment of Bengal Native (Light) Infantry, 1842; the 5th Regiment of Bengal Native (Light) Infantry, 1861; the 5th Regiment of Bengal (Light) Infantry, 1885; present designation, 1903.

"Arracan" "Affghanistan, 1839" "Candahar, 1842" "Ghuznee, 1842"
"Cabool, 1842" "Moodkee" "Ferozeshah" "Sobraon"
"Afghanistan, 1879-80" "Burma, 1885—87."

Composition—2 Companies Musalmans of the Eastern Punjab and Hindustan, and 2 Companies Ahirs.

1st Commn. or date of entering service.	Names and Rank.	Army rank.	Present appointment in Regiment.	Remarks.
	Commandant.			
27 Jan. 92	Cotton, Lt.-Col. W. L.	8 July 15	8 July 15	
	Company Commanders. (4)			
18 Oct. 99	Pott, Maj. H. P.	1 Sept. 15	31 Jan. 11	With 71st Punj.
26 June 01	Praed, Maj. G. A. M.	26 June 16	8 July 15	*Lv., ex I., m. c., 6 mos.*
3 Sep. 02	Tyndall, Maj. J. W. H. D.	27 June 16	2 July 17	
	Company Officers. (9)			
23 Apl. 02	Ball, Maj. L. P., D.S.O.	23 Apl. 17	2 Mar. 03	With 97th Infy.
16 Aug. 05	Hall, Capt. W. D., M.C.	16 Aug. 14	26 July 09	
29 Aug. 06	Strover, Capt. W. G. (*Actg. Maj.*)	29 Aug. 15	22 June 08	Adjt., 1 Jan. 13. 2nd-in comd. (*Offg.*)
4 May 07	*May, Capt. C. D.*	*1 Sept. 15*	*21 Sep. 11*	*S. & T. Corps.*
17 Aug 07	Boulton, Capt. C. E.	1 Sept. 15	31 Oct. 09	With 1-6th Infy.
17 Jan. 11	Malone, Capt. E. M.	1 Sept. 15	27 Mar. 13	
15 Sep. 15	Walmsley, Lt. J. B.	15 Sep. 16	....	With R. A. F.
30 Jan. 17	Farrar, Lt. W. B.	30 Jan. 18	....	With 95th Infy.
30 Jan. 17	McDonnell, Lt. F. A.	30 Jan. 18	....	With 2-95th Infy. *Lv., ex I., m. c.*
15 May 15	Evans, Lt. I. B.	15 May 16	23 Oct. 17	Attd. Adjt., 25 Feb. 18. (*Offg.*)
	Attached.			
23 Dec. 14	Kellie, Lt. C. C. J.	23 Dec. 15	....	1-1st Brahmans.
11 July 15	Rawbone, Lt. D. B. M. (*Actg. Capt.*)	11 July 16	9 June 18	Lond. R., I. A. (*on probn.*)
20 Jan. 17	Hainsworth, Lt. H. M.	20 Jan. 18	5 Feb. 17	I. A. R. O.
	Flower, Lt. H. G. S.	....	....	Som. L. I., I. A. R. O. (*on probn.*)
30 Jan. 17	White, 2nd-Lt. D. O.	30 Jan. 17	....	I. A. R. O.
8 Aug. 17	Brown, Lt. L. C.	8 Aug. 18	2 Nov. 17	I. A. R. O.
15 Jan. 18	Ovenden, 2nd-Lt. E. A.	15 Jan. 18	11 June 18	I. A. (*T. C.*)
15 Jan. 18	Slocombe, 2nd-Lt. W.	15 Jan. 18	11 June 18	I. A. (*T. C.*)
27 Dec. 17	Platt, 2nd-Lt. H. C.	27 Dec. 17	11 June 18	M. G. Corps., I. A. (*on probn.*)
10 Jan. 16	Ricketts, Lt. G. A. McG.	10 Jan. 17	11 June 18	Yorks. L. I., I. A. (*on probn.*)
29 Oct. 16	Edgson, Lt. T. J.	29 Oct. 17	17 May 18	Ches. R., I. A. (*on probn.*) Qr. Mr. *Offg.*
14 Mar. 17	Steel, 2nd-Lt. J.	14 Mar. 17	1 May 18	I. A. R. O. (*on probn.*) *Lv., ex I., m. c.*
27 June 17	Anderson, Lt. W. H.	27 June 18	1 May 18	I. A. R. O.
27 June 17	Reddick, Lt. G. P., M.M.	27 June 18	....	I. A. R. O.
15 June 18	Jones, 2nd-Lt. A.	15 June 18	....	I. A. R. O. (*on probn.*)
15 June 18	Knapten, 2nd-Lt. J. H.	15 June 18	....	I. A. R. O. (*on probn.*)
15 June 18	Mantle, 2nd-Lt. W. E.	15 June 18	....	I. A. R. O. (*on probn.*)
2 June 18	Dunne, 2nd-Lt. E. L.	2 June 18	19 Oct. 18	I. A. R. O.
27 Feb. 18	Walsh, 2nd-Lt V. P. T.	27 Feb. 18	13 Oct. 18	E. Kent. R., I. A. R. O. (*on probn.*)
31 July 18	Forster, 2nd-Lt. H. R.	31 July 18	21 Oct. 18	Genl. List, I. A. R. O. (*on probn.*)
22 July 18	McPherson, 2nd-Lt. W. H.	22 July 18	3 Dec. 18	I. A. (*T. C.*)
15 June 18	Smith, 2nd-Lt. E. H.	15 June 18	....	Genl. List, I. A. R. O. (*on probn.*)
21 July 16	Thomas, Lt. L. A. S.	21 July 17	....	North'd Fus., I. A. (*on probn.*)
5 Feb 18	Stone, 2nd-Lt. H. O' D.	5 Feb. 18	....	I. A. R. O.
1 Sep. 18	Etherington, 2nd-Lt. R. A.	1 Sep. 18	....	1-4th G. R.

Subadar-Major.

		Jemadar.	Subadar.	Subadar-Major.	
1 Apl. 89	Wahid Ali Khan, I.O.M.	29 June 03	17 Sep. 07	3 June 15	
7 Nov. 84	Amir Ali Khan (*Ret.*)	1 Mar. 99	....	....	Re-employed 7 Jan. 17

1st Battalion, 5th Light Infantry—*contd.*

1st Commn. or date of entering service.	Names and Rank.	Army rank.	Present. appointment.	Remarks.
	Subadars.			
20 June 95	Suleman Khan (46, 72*a*)	26 May 04	16 June 12	
1 July 00	Sharaf-ud-din (56)	1 Sep. 09	13 Dec. 13	
13 Apl. 04	Hans-Raj (56)	1 June 09	3 June 15	
6 May 97	Mubarak Ali Khan (56, 72*a*)	1 June 09	3 June 15	
15 Feb. 02	Aziz-ud-din Khan, I.D.S.M.	13 Dec. 13	3 June 15	
6 June 03	Ghulam Haidar Khan (46)	27 Mar. 14	9 Dec. 16	
21 Nov. 98	Rahmat Khan, I.D.S.M.	3 June 15	18 Apl. 17	
30 Aug. 98	Anwar	3 June 15	18 Apl. 17	
1 Jan. 04	Abdul Aziz Khan	3 June 15	30 June 17	With 2-5th Infy.
23 Nov. 94	Rahmat Khan	11 Jan. 10	3 June 15	With 2-5th Infy.
	Jemadars.			
1 Mar. 97	Ghafoor Khan		3 June 15	
22 Oct. 96	Muhammad Hassan Khan		3 June 15	
2 Oct. 99	Muhammad Ibrahim Khan		11 Dec. 16	
10 Mar. 04	Abdul Ghafur Khan		18 Apl. 17	
10 Jan. 00	Muhammad Umar Khan		18 Apl. 17	
3 Mar. 04	Sikhdar, I.D.S.M.		1 July 17	
18 May 03	Sazewar Khan		1 July 17	
8 Aug. 01	Haidar Ali Khan		21 Mar. 18	
7 Nov. 04	Suleman Khan		21 Mar. 18	
3 Dec. 03	Sazawar Khan		24 June 18	
1 Oct. 09	Ahmad Din		24 June 18	
3 Apl. 07	Ghulam Nabi Khan, I.D.S.M.		24 June 18	
4 Apl. 407	Habib-ur-Rahman Khan, I. O. M.		24 June 18	
1 July 06	Manzur Ahmad Khan		24 June 18	
16 Feb. 04	Subhan		24 June 18	
31 July 05	Suleman Khan		24 June 18	
	Bachan Singh			
	Mathura, I. D. S. M.			
	Har Narain Singh			
	Jhabbar			
	Bhim Singh			
	Lakh Ram			
	Hazari Singh			
	Suraj Mal			

2nd Battalion, 5th Light Infantry.

Formed at Quetta on 13th August 1918.

Composition:—2 Companies of Ahirs.
3 Companies of Musalmans of the Eastern Punjab and Hindustan.

1st Commn. or date of entering service.	Names and Rank.	Army Rank.	Present appointment in regiment.	REMARKS.
	Commandant.			
20 Jan. 00	Henderson, Maj. K. D. M., M.C.	1 Sep. 15	31 Aug. 18	1-17th Infy. (*temp.*)
	Company Commanders. (4)			
18 Jan. 02	Fox-Male, Maj. W. C. E. T., M.C.	18 Jan. 17	13 Aug. 18	93rd Infy. 2nd-in-Comd. (*temp.*)
6 Apl. 15	Lunnon, Lt. J. D. K. (*Actg. Capt.*)	6 Apl. 16	28 Aug. 18	Essex R., I. A. (*on probn.*)
13 May 15	Simms, Lt. F. H. (*Actg. Capt.*)	15 May 16	14 Sep. 18	R. War. R., I. A. R. O. (*on probn.*)
10 Oct. 15	Taitt, Lt. N. R. (*Actg. Capt*)	10 Oct. 15	6 Sep. 18	Bedf. R., I. A. (*on probn.*)
	Company Officers.			
13 June 16	James, Lt. J. R.	13 June 17	6 Sep. 18	York R., I. A. (*on probn.*)
11 Sep. 16	McKechnie, Lt. V. R.	11 Sep. 17	19 Nov. 18	C. R. Bl., I. A. (*on probn.*)
22 Nov. 16	Lawrie, Lt. T. M.	22 Nov. 17	20 Oct. 18	R. N. Vol. Reserve, I. A. R. O. (*on probn.*)
19 Sep. 17	Storey Cooper, Lt. E. S.	19 Sep. 18	24 Aug. 18	R. Welsh Fus., I. A. (*on probn.*), Adjt. 21st Aug. 18.
21 Dec. 16	Vaughan, Lt. S. G.	21 Dec. 16	28 Aug. 18	I. A. R. O. Qr. Mr.
1 May 17	Shelley, Lt. P.	1 May 18	24 Aug. 18	R. Lan. R., I. A. (*on probn.*)
31 Oct. 17	Bromby, 2nd-Lt. T. A.	31 Oct. 17	28 Aug. 18	K. R. Rif. C., I. A. R. O. (*on probn.*)
30 Jan. 18	Merry, 2nd-Lt. N. McI.	30 Jan. 18	14 Sep. 18	Midd'x R., I. A. R. O. (*on probn.*)
2 June 18	Lynch, 2nd-Lt. P. W.	2 June 18	20 Oct. 18	I. A. (*T. C.*)
7 June 18	[illegible], 2nd-Lt. H.	7 June 18	20 Oct. 18	I. A. (*T. C.*)
7 June 18	Goff, 2nd-Lt. H. H.	7 June 18	20 Oct. 18	I. A. R. O.
26 June 18	Allan, 2nd-Lt. H. K. S.	26 June 18	3 Oct. 18	Genl. List, I. A. R. O. (*on probn.*)
26 June 18	Dawson, 2nd-Lt. C.	26 June 18	3 Oct. 18	Genl. List, I. A. R. O. (*on probn.*)
26 June 18	Oldroyd, 2nd-Lt. O. E.	26 June 18	3 Oct. 18	Genl. List, I. A. R. O. (*on probn.*)

Subadar-Major.

		Jemadar	Subadar	Subadar-Major.	
	Amir Ali Khan (*Retd.*)	. . . .	. . . .	. . . .	Re-employed 7 Jan. 17.
3 Nov. 94	Rahmat Khan	11 Jan. 10	3 June 15	12 Sep. 18	1-5th Infy.

Subadar.

21 Jan. 04	Abdul Aziz Khan	3 Jan. 15	30 June 17	1-5th Infy.
9 Aug. 00	Muhmmad Yar Khan	24 June 16	1 July 17	

Jemadar

21 June 01	Ibrahim Khan	24 June 18
16 Nov. 02	Imtiyaz Muhammad Khan	24 June 18
1 Feb. 03	Rahim Sher Khan	24 June 18
15 Mar. 04	Imamuddin	24 June 18
7 Oct. 04	Muhammad Husain Khan	24 June 18
6 July 05	Keri	24 June 18
21 Oct. 05	Fazal Ahmad Khan	24 June 18
6 July 05	Faiz Muhammad Khan	24 June 18

Regt. Centre—Meerut. *Uniform*—Scarlet.
Linked with—10th Jats. *Facings*—White.

1st Battalion, 6th Jat Light Infantry.

***Kyne-ki-dahini paltan.*—Raised at Fatehgarh in 1803, by Major J. Malcolm and Lt. W. Ball, as the 1st Battalion, 22nd Regiment of Bengal Native Infantry. Became the 43rd Regiment of Bengal Native Infantry, 1824; the 43rd Regiment of Bengal Native (Light) Infantry, 1842; the 6th Regiment of Bengal Native (Light) Infantry, 1861; the 6th Regiment of Bengal (Light) Infantry, 1885; the 6th (Jat) Regiment of Bengal (Light) Infantry, 1897; the 6th Jat Light Infantry, 1901; present designation, 1903.**

"Nagpore" "Affghanistan, 1839" "Candahar, 1842" "Ghuznee, 1842"
"Cabool, 1842" "Maharajpore" "Sobraon" "Ali Masjid"
"Afghanistan, 1878-79" "China, 1900."
Composition—4 companies Jats.

COLONEL—Lieutenant-General Arthur G. Handcock, C.B. . 13 May 1904.

1st Commn. or date of entering service.	Names and Rank.	Army rank.	Present appointment in Regiment.	Remarks.
	Commandant.			
20 Oct. 90	Bourne, Lt.-Col. W. Fitz G.	31 Jan. 16	31 Jan. 16	
	Company Commanders. (4)			
3 Sep. 92	*Dundas, Lt.-Col. P. H.,* D.S.O., *p.s.c. (Q)*	*3 Sep. 18*	*8 May 16*	*2nd-in-Comd. A. Q. M. G., Embarkation Staff.*
20 Jan. 97	*Moore, Maj. F. G.*	*20 Jan. 15*	*29 Jan. 07*	*Asst. Secy., Army Dept.*
11 Aug. 00	*Ross, Bt. Lt.-Col. R. C.,* D.S.O. *(temp. Lt.-Col.)*	*20 Aug. 18*	*14 Mar. 11*	*A. Q. M. G., Army Hdqrs.*
8 May 01	*Wells, Maj. F. W. A.,* M.C.	*8 May 16*	*8 May 16*	*Lv., ex I., m. c.*
	Company Officers. (9)			
18 Jan. 02	*Thornhill, Bt.-Lt.-Col. C. J. M.,* D.S.O. [L] *(temp. Lt.-Col.)*	*3 June 18*	*19 Dec. 02*	
24 Jan. 06	*McPherson, Capt. A. B.,* M.C.	*24 Jan. 15*	*27 Mar. 07*	*Off. Sch. of Instn., Bangalore.*
29 Aug. 06	*Hailes, Capt. W. L.,* M.C.	*29 Aug. 15*	*1 Feb. 08*	*Bde. Maj.*
17 Aug. 07	*Peet, Capt. L. M.*	*1 Sep. 15*	*16 Apl. 09*	*A. D. C., Govr., Madras.*
15 Aug. 14	Young, Capt. G. A. G.	15 Aug. 18	7 Dec. 15	
29 June 16	*Dawes, Lt. R. M.*	*13 Dec. 16*	*4 Jan. 17*	*With Divl. Sig. Coy.*
18 Apl. 16	Glen, Lt. J. V. J.	18 Apl. 17	24 Apl. 16	
18 Apl. 16	Tester, Lt. C. K., M.C.	18 Apl. 17	25 Apl. 16	
29 June 16	McCarthy, Lt. T. C., M.C.	29 June 17	4 Jan. 17	
30 Jan. 17	O'Connor, Lt. M. J.	30 Jan. 18	. . .	
27 Oct. 17	Godley, Lt. B. R.	27 Oct. 18	5 Nov. 17	Attd.
27 Oct. 17	Allen, 2nd-Lt. R. F. W. K.	27 Oct. 17	5 Nov. 17	Attd. With 1-10th Jats.
31 Jan. 18	Dwane, 2nd-Lt. L. M. R.	31 Jan. 18	10 Feb. 18	Attd. With 2-150th Infy.
21 Dec. 17	Rootes, 2nd-Lt. A. H.	21 Dec. 17	21 Mar. 18	Attd. With 2-131st Infy.
8 Sep. 15	Thompson, Lt. C. G.	8 Sep. 16	16 Aug. 17	Attd.
	Attached.			
17 Aug. 07	Boulton, Capt. C. E.	1 Sep. 15	24 Apl. 18	1-5th Infy.
13 Nov. 14	Kelly, Lt. W. A. D.	13 Nov. 15	13 June 15	I. A. R. O.
11 Dec. 14	O'Connor, Lt. R. L.	11 Sep. 16	24 June 15	
17 Oct. 15	Fletcher, Capt. A. B.	5 Aug. 18	11 Sep. 15	1-10th Jats.
28 Apl. 16	Barckley, Lt. D. C.	28 Apl. 17	11 Sep. 16	I. A. R. O.
29 June 17	Colwill, Lt. R. F.	29 June 18	3 Oct. 17	I. A. R. O.
29 Aug. 17	Hancock, 2nd-Lt. A.	29 Aug. 17	12 Dec. 17	I. A. R. O.
30 July 16	Gilbert, Lt. E. C.	30 July 17	26 Jan. 18	I. A. (*on probn.*)
5 Oct. 17	Saunders, Lt. W. H.	5 Oct. 18	19 Jan. 18	I. A. (*T. C.*)
5 Nov. 17	Mooring, Lt. L. L.	5 Nov. 18	15 Mar. 18	I. A. (*T. C.*)
29 Mar. 18	Pillar, 2nd-Lt. E. J.	29 Mar. 18	15 July 18	I. A. (*T. C.*)
8 July 16	Longley, Lt. A. H.	8 July 17	8 July 18	Midd'x R., I. A. (*on probn.*)
	Paget, Capt. G. H.	. . . .	. . . .	N. Staff, R., I. A. (*on probn.*)
7 Oct. 18	Cooper, 2nd-Lt. F. L.	7 Oct. 18	21 Oct. 18	I. A. (*T. C.*)
5 Oct. 17	Higgs, 2nd-Lt. J. V.	5 Oct. 17	16 June 18	I. A. (*T. C.*)

Subadar-Major.

		Jemadar.	Subadar.	Subadar-Major.	
12 Feb. 02	Dalpat Singh, I.O.M. (56, 162)	12 Feb. 02	1 Sep. 05	21 May 15	*Order of Br. I., 2nd Cl.*

1st Battalion, 6th Jat Light Infantry—*contd.*

1st Commn. or date of entering service.	Names and Rank.	Army rank.	Present appointment in Regiment.	Remarks.
	Subadars.			
2 Mar. 97	Shib Lal, *Bahadur* (56, 72*a*)	2 Mar. 97	27 May 02	*Order of Br. I., 2nd Cl.* With 2-10th Jats.
17 July 91	Rup Chand (52, 56, 58)	21 May 06	5 Dec. 13	
23 May 92	Shiulal (I) (56, 72*a*)	13 Jan. 07	13 Aug. 14	
1 May 98	Lakhi Ram, M.C., I.D.S.M. (58)	13 Aug. 14	25 Nov. 14	
18 June 96	Sheolal (II) I.D.S.M.	24 Nov. 14	5 Jan. 16	
1 Aug. 99	Mula, I.D.S.M.	3 Jan. 15	14 May 16	
25 Dec. 03	Net Ram	24 Feb. 16	4 Dec. 17	
8 Sep. 96	Jug Lal (56)	1 Jan. 10	16 Nov. 14	Re-employed.
18 June 96	Ratna, I.D.S.M. (*Ret.*) (56)	3 Dec. 14	11 June 18	Re-employed.
15 Mar. 03	Shiu Dat	14 May 16	22 May 18	
29 Mar. 03	Rugha Singh	14 May 16	22 May 18	
28 May 98	Chandgi Ram (*Ret.*) (56)	23 Dec. 14	5 June 18	Re-employed.
	Jemadars.			
29 July 04	Hemraj	14 May 16		
18 Dec. 04	Sardara (I)	14 May 16		
2 Dec. 06	Ujala	27 June 17	With 2-150th Infy.	
1 Jan. 11	Sardara (II)	3 Oct. 16	Indian Adjt.	
8 Dec. 03	Ganga Ram	3 Dec. 17		
24 Dec. 08	Sheo Dat	4 Dec. 17		
21 Mar. 10	Daya Ram	4 Dec. 17		
12 Mar. 03	Sis Ram	1 Jan. 18		
29 Sep. 11	Ram Sarup	27 Mar. 18		
2 Dec. 06	Maru, I.D.S.M.	10 May 18		
12 June 07	Manbher	22 May 18		
16 Dec. 06	Sis Ram	22 May 18		
3 Dec. 06	Bhagmal	22 May 18		
15 Oct. 07	Phul Singh	9 May 18		
1 Sep. 18	Chiranji Lal	1 Sep. 18	(*On probn.*)	

2nd Battalion, 6th Jat Light Infantry.

Formed at Agra on 20th November 1917.

Composition—

1st Commn. or date of entering service.	Names and Rank.	Army Rank.	Present appointment in Regiment.	Remarks.
	Commandant.			
23 Sep. 96	Hardie, Maj. G. E. (*Actg. Lt.-Col.*).	23 Sep. 14	24 Nov. 17	123rd Rif.
	Company Commanders. (4)			
19 Aug. 03	Greig, Capt. R. M. (*Actg. Maj.*)	13 Feb. 13	1 Jan. 18	42nd R. 2nd-in-Comd. (*temp.*)
13 Oct. 14	Williams, Lt. G. B. (*Actg. Capt.*)	8 Apl. 16	20 Aug. 18	Attd.
9 Apl. 15	Drake, Lt. D. H. (*Actg. Capt.*) .	9 Jan. 17	29 July 18	Attd.
	Company Officers.			
5 June 17	Shaw, Lt. L. B.	5 June 18	31 Mar. 18	I. A. (*on probn.*)
26 Jan. 18	Arnold, 2nd-Lt. V. D.	26 Jan. 18	29 Mar. 18	I. A. (*on probn.*)
8 April 16	Heaton, Lt. A. D., M.C. (*Actg. Capt.*).	8 Apl. 17	13 Apl. 18	I. A. (*on probn.*)
15 Jan. 18	Roblin, 2nd-Lt. M. L. . . .	15 Jan. 18	4 June 18	I. A. R. O.
15 Jan. 18	Jeffreys, 2nd-Lt. W. J. . . .	15 Jan. 18	4 June 18	I. A. (*T. C.*)
5 Jan. 18	West, 2nd-Lt. W. G. . . .	5 Jan. 18	4 June 18	I. A. (*T. C.*)
9 Oct. 17	Robins, Lt. R. T.	9 Oct. 18	2 June 18	Suff. R., I. A. (*on probn.*)
1 Nov. 17	Davies, Lt. W. J. (*Actg. Capt.*).	1 Nov. 18	14 Mar. 18	I. A. (*T. C.*) Adjt.
5 Nov. 17	Brodie, Lt. J. W.	5 Nov. 18	14 Mar. 18	I. A. R. O. Qr. Mr.
30 Aug. 18	Fife, 2nd-Lt. D. McD. . . .	30 Aug. 18	14 Sep. 18	Attd.
28 Feb. 18	Pattullo, 2nd-Lt. H. A. . . .	28 Feb. 18	8 Nov. 18	R. Scots, I. A. (*on probn.*)
7 June 18	Stock, 2nd-Lt. C. A.	7 June 18	9 Oct. 18	I. A. (*T. C.*)
	Oldacre, 2nd-Lt. V. G. . .	. . . .	25 Nov. 18	Leic. R., I.A.R.O. (*on probn.*)
28 Aug. 17	Harrison, 2nd-Lt. P.	28 Aug. 17	18 Nov. 18	E. York. R.
1 Aug. 17	Wardell, 2nd-Lt. F. W. . . .	1 Aug. 17	18 Nov. 18	E. York. R.
31 Oct. 17	Vallentine, 2nd-Lt. W. C. . .	31 Oct. 17	18 Nov. 18	E. York. R.
13 Aug. 18	Kent, 2nd-Lt. L.	13 Aug. 18	20 Nov. 18	E. York. R.

Subadar-Major.

		Jemadar.	Subadar.	Subadar-Major.	
6 June 92	Bal Ram	1 Feb. 05	1 Aug. 02	1 Jan. 18	(*temp.*)

Subadars.

4 Jan. 02	Phul Chand (52, 56)	1 Nov. 15	1 Apl. 17	
11 July 02	Sukdhe	5 Jan. 16	20 Nov. 17	1 Mar. 18.
25 Nov. 03	Ramjilal	8 Jan. 16	20 Nov. 17	
16 June 02	Sultan	8 Jan. 16	20 Nov. 17	
25 Jan. 04	Chandgi, I.O.M.	3 Mar. 16	20 Nov. 17	
4 Nov. 03	Kurda	14 May 16	20 Nov. 17	
16 May 01	Ram Kala (I)	16 Sep. 17	1 Mar. 18	

Jemadars.

23 Aug. 98	Ram Kala (II)	16 Sep. 17
10 June 03	Maru (56)	20 Nov. 17
16 Mar. 05	Sukh Lal	20 Nov. 17
14 Apl. 05	Lal Chand (46)	20 Nov. 17
4 Jan. 06	Lajje Ram	20 Nov. 17
4 Aug. 08	Hari Singh	20 Nov. 17
15 Jan. 02	Sohbat	20 Nov. 17
1 Oct. 07	Manohar	1 Apl. 18
18 May 09	Sultan	26 June 18
2 Oct. 12	Kanaiya	1 Aug. 18
	Jai Ram	2 Sep. 18

Regt. Centre—Lucknow. *Uniform*—Scarlet.
Linked with—8th and 11th Rajputs. *Facings*—Yellow.

1st Battalion, 7th Duke of Connaught's Own Rajputs.

Craum-ki-paltan.—Raised in 1804 at Cawnpore, by Captain P. Crump, as the 1st Battalion, 24th Native Infantry. Became 47th Native Infantry, 1824. Disbanded same year, but raised immediately afterwards as 69th Native Infantry. Restored its former number of 47th Native Infantry, 1828. Became 7th Bengal Native Infantry, 1861; 7th (Duke of Connaught's Own) Bengal Native Infantry, 1883; 7th (Duke of Connaught's Own Rajput) Regiment of Bengal Infantry, 1893; present designation, 1903.

THE DUKE OF CONNAUGHT'S CREST AND CYPHER.

"Moodkee" "Ferozeshah" "Aliwal", "Sobraon" "China, 1858-59"

"Tel-el-Kebir" "Egypt, 1882" "Pekin, 1900."

Composition—4 companies Rajputs.

COLONEL-IN-CHIEF—FIELD MARSHAL *His Royal Highness* ARTHUR WILLIAM PATRICK ALBERT, *Duke of* CONNAUGHT AND STRATHEARN, K.G., K.T., K.P., G.C.B., G.C.S.I., G.C.M.G., G.C.I.E., G.C.V.O., G.B.E., A.-D.-C. . . . . 13 May 1904.

COLONEL—General *Sir* Edmund G. Barrow, G.C.B., G.C.S.I. [R] 13 May 1904.

1st Commn. or date of entering service.	Names and Rank.	Army rank.	Present appointment in Regiment.	Remarks.
	Commandant.			
16 Nov. 87	Parr, Bt.-Col. H. O.	3 June 16	23 Jan. 15	
	Company Commanders. (4)			
4 Mar. 91	*Tregear, Lt.-Col. F. C. (Q)* . .	*4 Mar. 17*	*23 Mar. 15*	*2nd-in-Comd.*
29 Nov. 90	Hume, Lt.-Col. S. G. W. (Q) .	6 July 18	8 Sep. 05	
7 May 98	Ogg, Maj. A. C., D.S.O., O.B.E., *p.s.c.*	1 Sep. 15	16 Jan. 08	*Lv., ex I., m.c.*
25 Jan. 99	Meadows, Maj. C. A. G. P. (*Actg. Lt.-Col.*).	1 Sep. 15	19 Feb. 13	With 2-7th Rajputs
	Company Officers. (9)			
4 May 01	Etheridge, Maj. F., D.S.O., *p.s.c.*	4 May 16	7 Apl. 03	
16 Aug. 05	Roberts, Capt. F. L.	16 Aug. 14	28 Feb. 08	With 1-94th Infy.
24 Jan. 06	*Rae, Capt. G. R.*	*24 Jan. 15*	*16 Aug. 08*	*S. & T. Corps.*
29 Aug. 06	*Thomson, Capt. A. R.*, M.C. .	*29 Aug. 15*	*9 Mar. 08*	
29 Jan. 10	Lilly, Capt. A. N. I. . . .	1 Sep. 15	18 Mar. 11	*Bde. Maj.*
22 Jan. 13	Norris, Capt. R. J. N. (*Actg. Maj.*).	22 Jan. 17	15 Mar. 14	
22 Jan. 13	Tindall, Capt. J. W. B. . . .	22 Jan. 17	20 Oct. 14	
15 Nov. 15	Titchmarsh, Lt. E. A. (*Actg Capt.*)	15 Nov. 16	15 Nov. 15	With 2-7th Rajputs.
7 Apl. 16	Wardle, Lt. F. M. (*Actg. Capt.*)	7 Apl. 17	5 Nov. 16	Adjt., 1 Oct. 17.
18 Apl. 16	Bridges, Lt. K. W.	18 Apl. 17	18 Apl. 16	Qr.-Mr. (*offg.*).
30 Jan. 17	Cleave, Lt. J. C. T.	30 Jan. 18	10 Feb. 17	Attd.
11 Dec. 14	*Halliley, Capt. W. S.*, M.C. . .	*11 Dec. 17*	*27 Mar. 15*	Attd.
	Attached.			
3 Dec. 90	Smith, Lt.-Col. A. LeF. (Q) . .	3 Dec. 16	18 Sep. 16	2nd Rajputs. Comdt.
29 Aug. 06	Hume-Wright, Bt. Maj. F. J. O. .	3 June 18	29 Sep. 16	16th Rajputs.
22 June 15	Taylor, Lt. H. D. (*temp. Capt.*) .	22 June 16	31 Dec. 17	R. W. Surr. R., I. A. (*on probn.*)
13 Nov. 14	Ronaldson, Lt. E. C. S'J. . . .	13 Nov. 15	27 Mar. 15	I. A. R. O. *Lv., ex I., m. c., 6 mos.*
8 Jan. 15	*Smith, Lt. M. L. C.*	*8 Jan. 16*	*24 Mar. 15*	*I. A. R. O.*
20 July 15	Thorne, Lt. H.	20 July 16	4 Oct. 15	I. A. R. O.
22 July 15	Yonge, Lt. G. E. H. (*temp. Capt.*)	22 July 16	*19 Jan. 16*	I. A. R. O.
30 July 15	Augier, Lt. G. D. E.	30 July 16	*19 Jan. 16*	*I. A. R. O.*
14 Aug. 15	*Furneaux, Lt. T. E.*	*14 Aug. 16*	... *Oct. 15*	*I. A. R. O.*
21 Oct. 15	Thompson, Lt. J. S.	21 Oct. 16	1 Oct. 18	York & Lanc. R., I. A. (*on probn.*)
26 Jan. 16	Rimmell, Lt. F. J.	26 Jan. 17	26 Nov. 17	13th Rajputs.
23 May 16	Bower, Lt. S. W.	23 May 17	10 Sep. 18	North'n R., I. A. (*on probn.*)
14 Aug. 16	Campbell, Lt. D. L.	14 Aug. 17	13 Oct. 16	I. A. R O.
5 Jan. 17	Alderson, Lt. R. C.	5 Jan. 18	25 Mar. 17	I. A. R. O.
8 Jan. 17	Haswell, Lt. P. L.	8 Jan. 18	16 Mar. 17	I. A R. O.
30 Jan. 17	Joyce, Lt. A. H. B., M.C. . .	30 Jan. 18	21 Sep. 17	62nd Punj.
3 Mar. 17	Butcher, Lt. R. E.	3 Mar. 18	21 Sep. 17	Midd'x R., I. A. R. O.
15 Aug. 16	Goss, Lt. J. E. S.	15 Aug. 17	26 Jan. 18	I. A. (*on probn.*)
8 Aug. 17	Wade, Lt. E. G.	8 Aug. 18	. . . .	I. A. R. O.
27 Oct. 17	Ropner, 2nd-Lt. G. V. . . .	27 Oct. 17	5 Nov. 17	I. A. R. O.
1 Nov. 17	Mills, Lt. E. G.	1 Nov. 18	14 Mar. 18	I. A. (*T. C.*)
1 Dec. 17	Bailly, Lt. O. H.	1 Dec. 18	25 June 18	R. Muns. Fus., I. A. (*on probn.*)
1 May 18	Jordan, 2nd-Lt. C.	1 May 18	3 June 18	R. W. Kent R., I. A. (*on probn.*)

1st Battalion, 7th Duke of Connaught's Own Rajputs—*contd.*

1st Commn. or date of entering service.	Names and Rank.	Army Rank.	Present appointment in Regiment.		Remarks.
	Subadar-Major.				
		Jemadar.	Subadar.	Subadar-Major.	
22 Sep. 94	Shiuambar Singh, *Bahadur*, I.O.M. (56).	22 Sep. 94	15 Dec. 01	1 Aug. 13	*Order of Br. Ind., 2nd Cl.*
4 May 94	Chandrapal Singh, I.D.S.M. (56, 162).	1 Nov. 10	28 Aug. 16	7 Apl. 17	
	Bakht Bali Singh (*Retd.*)	...	...	...	Re-employed 24 Aug. 18.
	Subadars.				
1 Oct. 90	*Ramdayal Singh (56)*	*1 Nov. 01*	*16 Nov. 05*		
15 May 03	*Rambaran Singh (56, 161)*	*15 May 03*	*1 May 10*		
2 Nov. 94	*Bhawani Parshad Singh (56, 162)*	*16 Nov. 05*	*15 Dec. 10*		
25 Nov. 93	*Brijmohan Singh,* I.D.S.M. *(56)*	*1 May 10*	*1 Aug. 14*		
4 Oct. 95	Gopal Singh	1 Aug. 13	28 Aug. 16		
4 July 02	Beni Singh	28 Aug. 16	1 Feb. 17		
15 Apl. 03	Ramautar Singh (56, 72*a*)	28 Aug. 16	19 Aug. 17		
8 Feb. 05	Benimadho Singh (I)	28 Aug. 16	19 Aug. 17		
1 Dec. 03	Benimadho Singh (II) (56)	28 Aug. 16	27 Oct. 17		
21 Feb. 01	Ahibaran Singh	15 Jan. 17	31 Dec. 17		
4 Oct. 96	Daljit Singh (56)	1 Feb. 17	13 Feb. 18		
5 July 04	Prem Singh (56)	9 Feb. 17	3 Mar. 18		
15 June 06	Raghunandan Singh	2 Apl. 17	29 Apl. 18		
	Basant Singh	...	...		(*tempy. Hony.*)
	Jemadars.				
23 Nov. 96	Baijnath Singh	2 Apl. 17			
12 Dec. 06	Rampal Singh	19 Aug. 17			
1 Sep. 04	Chandika Singh	29 Aug. 17			
21 Mar. 03	Narpat Singh	15 Oct. 17			
17 Nov. 11	Matan Singh	3 Dec. 17			
15 June 04	Sughar Singh (56)	3 Dec. 17			
1 Sep. 10	Ramraj Singh (56)	22 Dec. 17			
27 Jan. 02	Baijnath Singh II	13 Feb. 18			
9 July 04	Chandrapal Singh	3 Mar. 18			
4 Mar. 08	Jhandu Singh	29 Apl. 18			
3 Apl. 11	Matabakhsh Singh	29 Apl. 18			
15 June 12	Shiunarain Singh	29 Apl. 18			
15 Sep. 13	Mathura Singh	29 Apl. 18			
28 Sep. 14	Benimadho Singh	29 Apl. 18			
5 Aug. 08	Ahibaran Singh	11 June 18			
12 Sep. 04	Bikarmajit Singh	1 May 18			
16 Feb. 05	Hanuman Singh (51)	1 May 18			
17 Aug. 90	Shiudhari Singh	1 June 18			
24 May 09	Gangabaksh Singh	1 June 18			
15 Sep. 96	Shiuratan Singh	1 June 18			
5 Aug. 08	Ahibaran Singh	11 June 18			
	Kalka Singh	28 Sep. 15			

2nd Battalion, 7th Duke of Connaught's Own Rajputs.

Formed at Alipore.
Composition—3 companies Rajputs.
1 Company Kurmis.

1st Commn. or date of entering service.	Names and Rank.	Army rank.	Present appointment in Regiment.	Remarks.
	Commandant.			
25 Jan. 99	Meadows, Maj. C. A. G. P. (*Actg. Lt.-Col.*)	1 Sep. 15	16 June 18	1-7th Rajputs.
	Company Commanders. (4)			
18 Jan. 02	Breithaupt, Maj. E. A.	23 Feb. 17	....	63rd L. Infy. 2nd-in-Comd.
9 Aug. 15	Garrow, Lt. V. J. (*Actg. Capt.*)	9 Aug. 16	5 Sep. 18	Midd'x R., I. A. (*on probn.*)
11 Aug. 15	Mackenzie, Lt. H. C. (*Actg. Capt.*).	11 Aug. 16	26 July 18	S. Lan. R., I. A. (*on probn.*)
15 Nov. 15	Titchmarsh, Lt. E. A. (*Actg. Capt.*).	15 Nov. 16	31 Aug. 18	1-7th Rajputs.
	Company Officers.			
14 Jan. 15	Adams, Lt .G.	16 Jan. 16	....	I. A. R. O.
29 Aug. 17	Sheridan, 2nd-Lt. F. L. (*Actg. Capt.*)	29 Aug. 17	18 June 18	I. A. R. O. Adjt.
14 July 15	Raikes, Lt. A. F.	14 July 16	28 Nov. 18	Worc. R., I. A. (*on probn.*)
7 July 16	Smith, Lt. L. F.	7 July 17	11 Nov. 18	Midd'x. R., I. A. (*on probn.*)
11 Oct. 16	Porteous, Lt. H. A.	11 Oct. 17	27 Sep. 18	Devon. R., I. A. (*on probn.*)
10 Oct. 16	Carlile, Lt. T. A.	10 Oct. 17	3 Oct. 18	North'd Fus., I. A. (*on probn.*)
5 May 17	Asquith, Lt. E. J.	5 May 18	30 July 18	I. A. R. O.
25 July 17	Selway, Lt. H.	25 July 18	8 Sep. 18	W. Rid. R., I. A. (*on probn.*)
25 July 17	Colquhoun, 2nd-Lt. S. P.	25 July 18	22 Sep. 18	R. Ir. Fus., I. A. (*on probn.*)
19 Sep. 17	McCall, 2nd-Lt. R. G.	19 Sep. 17	2 Aug. 18	R. Ir. Fus., I. A. (*on probn.*)
15 Jan. 18	Shuell, 2nd-Lt. C. A. J.	15 Jan. 18	20 June 18	I. A. (*T. C.*) Qr.-Mr.
15 Jan. 18	Hoggett, 2nd-Lt. F. A. J.	15 Jan. 18	17 June 18	I. A. (*T. C.*)
15 Jan. 18	Grant, 2nd-Lt. A. E.	15 Jan. 18	17 June 18	I. A. R. O.
15 Jan. 18	Atkinson, 2nd-Lt. A. J. M.	15 Jan. 18	20 July 18	I. A. R. O.
18 Mar. 18	Lloyd, 2nd-Lt. E. T.	18 Mar. 18	30 July 18	I. A. (T. C.)
29 Mar. 18	Phillips, 2nd Lt. G. H.	29 Mar. 18	16 July 18	I. A. R O.
29 Mar. 18	Cabot, 2nd-Lt. C. E.	29 Mar. 18	16 July 18	I. A. R. O.
29 Mar. 18	Evenden, 2nd-Lt. J. M.	29 Mar. 18	16 July 18	I. A. (*T. C.*)
29 Mar. 18	Dear, 2nd-Lt. W. T.	29 Mar. 18	30 July 18	I. A. (*T. C.*)
29 Mar. 18	Waller, 2nd-Lt. L. G.	29 Mar. 18	17 Aug. 18	I. A. (*T. C.*)
7 June 18	Allum, 2nd-Lt. R. W.	7 June 18	8 Oct. 18	I. A. R. O.
26 Oct. 18	Moorhouse, 2nd-Lt. F. H.	26 Oct. 18	31 July 18	M. G. Corps, I. A. (*on probn.*)
22 July 18	Brewer, 2nd-Lt. A. A.	22 July 18	3 Dec. 18	I. A. (*T. C.*)
22 July 18	Bedford, 2nd-Lt. H.	22 July 18	3 Dec. 18	I. A. (*T. C.*)

Subadar-Major.

		Jemadar.	Subadar.	Subadar-Major.
28 Sep. 98	Rambali Singh	3 Mar. 15	15 Jan. 17	29 June 18

Subadars.

10 July 92	Debibakhsh Singh (56)	28 Aug. 16	16 June 18
8 May 03	Surajpal Singh		16 June 18
1 Oct. 06	Chandarpal Singh	9 Feb. 17	16 June 18
23 May 02	Surajbali Singh	1 Feb. 17	16 June 18
4 Jan. 04	Shiubodhan Singh	2 Apl. 17	16 June 18
1 Oct. 06	Basdeo Singh	28 July 17	16 June 18

Jemadars.

15 Feb. 18	Mahadeo Singh	15 Feb. 18	15 Feb. 18
29 Dec. 02	Rajwant Singh	14 Apl. 17	
1 Sep. 09	Ganesh Singh	16 June 18	
10 May 07	Sukhram Singh	16 June 18	
7 July 03	Shiudarshan Singh	16 June 18	
21 Aug. 09	Ganesh Singh	14 Apl. 17	
22 June 10	Bhim Singh	16 June 18	
1 Dec. 10	Sitlabakhsh Singh	16 June 18	
4 Oct. 10	Ujagar Singh	16 June 18	

Regt. Centre—Lucknow.
Linked with—7th and 11th Rajputs.

Uniform—Scarlet.
Facings—Yellow.

1st Battalion, 8th Rajputs.

Lauter-ki-paltan.—Raised at Dinapore in 1814, by Capt. F. Andrée, as the 1st Battalion, 30th Regiment of Bengal Native Infantry. Became the 59th Regiment of Bengal Native Infantry, 1824; the 8th Regiment of Bengal Native Infantry, 1861; the 8th Regiment of Bengal Infantry, 1885; 8th (Rajput) Regiment of Bengal Infantry, 1897; the 8th Rajput Infantry, 1901; present designation, 1903.

"Sobraon" "Afghanistan, 1879-80."
Composition—4 companies Rajputs.

1st Commn. or date of entering service.	Names and Rank.	Army rank.	Present appointment in Regiment.	Remarks.
	Commandant.			
29 Feb. 96	Abbott, Lt.-Col. L. H., C.M.G., *p.s.c.* (*temp. Br.-Genl.*)	21 Nov. 18	21 Nov. 18	
	Company Commanders. (4)			
20 Jan. 97	Cutler, Maj. H. C. (*Actg. Lt.-Col.*)	20 Jan. 15	4 Jan. 16	2nd-in-Comd. With 2-4th Rajputs.
5 Jan. 01	Orchard, Maj. A. J.	1 Sept. 15	27 Sep 13	With 1-150th Infy.
18 Apl. 00	Wood, Maj. H. O. B.	3 June 15	11 Mar. 15	With 3-2nd Rajputs.
11 Aug. 00	*Johnston, Maj. P. D. C.*, D.S.O.	*1 Sept. 15*	*4 Jan. 16*	*With Signal Coy.*
	Company Officers. (9)			
8 Jan. 01	Davis, Maj. T. A.	8 Jan. 16	1 Sep. 13	With 1-2nd Rajputs.
8 May 01	Ralph, Maj. S. L.	8 May 16	7 Nov. 07	
18 Jan. 02	*Black, Maj. G. C. S.*	*18 Jan. 17*	*19 Apl. 03*	*D. A. Q. M. G., 9th Dn.*
27 Jan. 04	Place, Capt. N. D.	27 Jan. 13	30 Jan. 08	*Lv., ex I., m. c.*
29 May 07	Durand, Capt. E. P. M.	1 Sept. 5	21 June 10	*Lv., Ind., m. c.*
25 Jan. 08	Spencer, Capt. L.	1 Sept. 15	30 Mar. 09	Adjt., 2 Dec. 14.
25 Jan. 12	Hindmarsh, Capt. J. H. L.	20 Sept. 16	10 July 14	Qr.-Mr., 26 Apl. 15.
1 Oct. 14	Ansell, Cpt. J. W.	1 Oct. 18	15 Dec. 14	
16 Dec. 14	Bagley, Lt. A. H.	16 Dec. 15	11 Feb. 15	
15 Nov. 15	Mizen, Lt. H. A.	15 Nov. 16	19 Nov. 15	Attd.
15 Nov. 15	Dunn, Lt. D. A. C.	15 Nov. 16	19 Nov. 15	Attd.
21 Aug. 18	Davies, 2nd-Lt. E. R. L.	21 Aug. 18	5 Sep. 18	Attd.
28 Sep. 15	Rea, Lt. A. (*Actg. Capt.*)	28 Sep. 16	....	Attd.
20 June 17	Croft, 2nd-Lt. F. R.	20 Mar. 18	...	Attd.
18 June 17	Seater, Lt. G. H.	18 June 18	....	Attd.
	Attached.			
23 Apl. 15	*Brown, Lt. L. N.*	*13 Apl. 16*	*5 July 15*	*I. A. R. O., A. R. O., Poona.*
15 Dec. 15	Moran, Lt. T.	15 Dec. 16	13 Jan. 16	I. A. R. O.
19 Jan. 17	Mann, 2nd-Lt. S. H.	19 Jan. 17	31 May 16	I. A. (*on probn.*)
1 Mar. 16	Russell, Lt. J. M.	1 Mar. 17	11 Mar. 16	I. A. R. O.
1 Feb. 17	Lawton, Lt. A. C.	1 Feb. 18	26 Aug. 17	I. A. R. O.
	Ewin, 2nd-Lt. H. L. H.	....	....	I. A. R. O. (*on probn.*)
29 Mar. 18	Kerley, 2nd-Lt. S.	29 Mar. 18	16 July 18	I. A. (*T.C.*)
18 Sep. 15	Beaton, Lt. D. J.	18 Sep. 16	27 Nov. 17	I. A. (*on probn.*)
9 June 16	Oliphant, Lt. I. M. B.	9 June 17	....	Lan. Fus. I. A. R. O.
12 May 15	Morton, Lt. A. F.	12 May 16	....	I. A. (*on probn.*)
22 Sep. 14	Bolland, Lt. L. G. S.	22 June 16	3 June 18	W. Red. R., I. A. (*on probn.*)
6 May 15	Chaundler, Lt. O. S. (*Actg. Capt.*)	6 May 16	....	I. A. (*on probn.*)
7 Nov. 14	Brewitt, Lt. A. C., M.C.	7 Aug. 16	18 Aug. 18	K. O. Sco. Bord., I. A. (*on probn.*)
5 Nov. 17	Brooke, Lt. R. W.	5 Nov. 18	26 Mar. 18	I. A. (*T. C.*)
27 Sep. 16	Baker, Lt. R. A. C.	27 June 18	4 Sep. 18	R. W. Fus., I. A. (*on probn.*)
25 June 15	Hildick-Smith, Lt. J. N	25 June 16	....	S. Staff R., I. A. (*on probn.*)
16 Sep. 15	Bellhouse, Lt. A. P.	16 Sep. 16	....	Welsh R., I. A. (*on probn.*)
	Hill, 2nd-Lt. W. A.	....	....	I. A. R. O. (*on probn.*)
	Hill, 2nd-Lt. C. F.	....	....	I. A. R. O. (*on probn.*)
27 Sep. 18	McCarthy, 2nd-Lt. M. P.	27 Sep. 18	....	I. A. R. O. (*on probn.*)

Subadar-Major.

		Jemadar.	Subadar.	Subadar-Major.
31 Jan. 87	Ram Ikbal Singh (56)	1 Dec. 97	1 Apl. 01	1 Jan. 18.

Subadars.

20 Oct. 84	Bije Bahadur Singh	1 Aug. 03	1 Apl. 10	
4 Nov. 90	*Daulat Singh*, I.D.S.M.	*1 Apl. 10*	....	
20 Sep. 92	Girwar Singh	1 Aug. 10	19 Apl. 15	
17 Dec. 92	Jang Bahadur Singh	7 Feb. 11	1 Sept. 15	
9 Feb. 93	Ramanand Singh (56)	7 Feb. 11	10 Jan. 16	
1 Oct. 93	Mathura Singh	16 Aug. 11	5 July 16	
19 Dec. 00	Shiunath Singh (I) (56)	5 Aug. 12	1 Jan. 18	With 1-131st Infy.
1 June 83	Chauharja Pershad Misr (56)	1 Aug. 17	20 Apl. 18	
5 Apl. 05	Sahib Bakhsh Singh	10 Apl. 16	20 May 18	
21 Jan. 03	Surajbali Singh	4 July 16	20 May 18	
3 Mar. 98	Indarjit Singh	14 July 16	6 July 18	
	Ajodhya, Singh	....	....	Re-employed 22 July 18.

1st Battalion, 8th Rajputs—*contd.*

1st Commn. or date of entering service.	Names and Rank.	Army rank.	Remarks.
	Jemadars.		
26 Mar. 98	Ramesar Singh	17 Apl. 15	
27 May 97	Ramsidh Singh	10 June 16	
13 Jan. 00	Nagesar Singh	4 July 16	
17 May 97	Ranjor Singh	20 Feb. 17	
1 Aug. 17	Rampal Singh	1 Aug. 17	
1 Aug. 17	Shiunath Singh	1 Aug. 17	
10 Feb. 03	Mahipal Singh	1 Dec. 17	
5 Oct. 95	Shiukaran Singh	1 Jan. 18	
14 Jan. 02	Ram Karan Singh	1 Apl. 18	
9 Feb. 97	Ramnawaz Singh	1 Apl. 18	
18 Aug. 06	Kidar Singh	1 Apl. 18	
9 Oct. 02	Arjun Singh	1 Apl. 18	
17 Apl. 06	Phul Singh	1 Apl. 18	
27 Mar. 07	Jodha Singh	20 Apl. 18	
16 Nov. 07	Shiupal Singh	20 Apl. 18	
23 Aug. 02	Brijlal Singh	20 Apl. 18	
28 Nov. 02	Chatrapal Singh	20 Apl. 18	
2 Oct. 02	Raja Ram Singh	20 May 18	
20 Feb. 01	Satnaram Singh	20 May 18	
15 Nov. 06	Shiu Pershad Singh	20 May 18	
15 Jan. 03	Gaya Bux Singh	20 May 18	
	Bajrang Singh	6 July 18	
	Sahdeo Singh	7 Oct. 18	
	Sarju Singh	....	Re-employed 22 July 18.

2nd Battalion, 8th Rajputs.

Formed at Lucknow on 19th March 1918.

Composition—4 Companies Rajputs.

1st Commn. or date of entering service.	Names and Rank.	Army Rank.	Present appointment in Regiment.	Remarks.
	Commandant.			
27 July 98	McCleverty, Maj. A. H. (*Act'g. Lt.-Col.*)	1 Sep. 15	13 Mar. 18	13th Rajputs.
	Company Commanders. (4)			
8 Jan. 01	Brander, Maj. E. H. C. . . .	8 Jan. 16	. . . .	2nd Rajputs. 2nd-in-Comd.
24 May 02	Paterson, Maj. R. H. O'D. . .	27 May 16	23 May 18	34th Horse.
30 Oct. 14	Rayneau, Capt. L. A. . . .	30 Aug. 18	7 Oct. 18	I. A. R. O.
4 Dec. 14	Hearn, Lt. E. S. (*Act'g. Capt.*)	4 Dec. 15	15 Mar. 18	I. A. R. O.
	Company Officers.			
9 Jan. 15	Saywer, Lt. C. P. B. (*Act'g. Capt.*)	9 Oct. 16	16 Nov. 18	Dorset R.
30 Jan. 15	Jones, Lt. P. J. D. (*Act'g. Capt.*)	30 Oct. 16	18 Mar. 18	L'pool R., I. A. (*on probn.*) Qr.-Mr. (*Offg.*)
8 July 16	Metcalfe, Lt. C. J. (*Actg. Capt.*) .	8 July 17	17 May 18	R Innis Fus., I. A. (*on probn.*)
18 June 17	Ommaney, Lt. F. R. . . .	18 June 18	18 Mar. 18	I. A. (*on probn.*)
18 June 17	Herbert, Lt. H. A.	18 June 18	26 Mar. 18	I. A. (*on probn.*)
1 Aug. 17	Edwards, 2nd-Lt. A. W. . . .	1 Aug. 17	5 Sep. 18	Mon R., I. A. R. O. (*on probn.*)
31 Oct. 17	Hicks, Lt. H. E.	31 Oct. 18	26 Mar. 18	I. A. R. O.
1 Nov. 17	McNab, Lt. A. (*Act'g. Capt.*) .	1 Nov. 18	15 Mar. 18	I. A. (*T. C.*) Adjt.
26 Jan. 18	Cranswick, 2nd-Lt. W. . . .	26 Jan. 18	6 Sep. 18	E. York R., I. A. (*on probn.*)
16 Mar. 18	Long, 2nd-Lt. C. A.	16 Mar. 18	13 Sep. 18	Hamps, R., I. A. R. O. (*on probn.*)
19 Mar. 18	De Coverley, 2nd-Lt. A. T. . .	19 Mar. 18	15 July 18	I. A. (*T. C.*)
21 Mar. 18	Hipsey, 2nd-Lt. A. W. L. . .	21 Mar. 18	23 July 18	I. A. R. O.
26 June 17	Marsh, 2nd-Lt. J. J., M.C. . .	26 Mar. 18	23 May 18	Bord, R., I. A. (*on probn.*)
27 May 18	Russell, 2nd-Lt. C. R.	27 Mar. 18	8 Sep. 18	Norf. R., I. A. R. O. (*on probn.*)
29 Mar. 18	Anscombe, 2nd-Lt. R. T. J. . .	29 Mar. 18	15 July 18	I. A. (*T. C.*)
29 May 18	Evans, 2nd-Lt. F. L.	29 May 18	6 Sep. 18	I. A. R. O. (*on probn.*)
29 May 18	Cook, 2nd-Lt. R.	29 May 18	4 Oct. 18	Lond R., I. A. R. O. (*on probn.*)
26 Feb. 18	Sniders, 2nd-Lt. A.	26 Feb. 18	23 Nov. 18	E. Kent R., I. A. R. O. (*on probn.*)
22 July 18	Cordwell, 2nd-Lt. E. T. . . .	22 July 18	2 Dec. 18	I. A. (*T. C.*)
	Hunter. 2nd-Lt. D. A. . . .	. . . .	3 Dec. 18	I. A. R. O. (*on probn.*)
12 July 18	Bridgnell, 2nd-Lt. E.	12 July 18	. . .	I. A. R. O.
	Tirwa, Raja of—2nd-Lt. . .	. . . .	1 Nov. 18	
	Partabner, Raja of—2nd-Lt. .	. . . .	1 Nov. 18	

Subadar-Major.

		Jemadar.	Subadar.	Subadar-Major.
3 Sep. 96	Jagesar Singh	24 Nov. 15	4 May 18	

Subadars.

20 Dec. 00	Ramradhe Singh	17 Apl. 15	25 Apl. 18		
17 Apl. 01	Atibal Singh	19 Apl. 15	25 Apl. 18		
28 Mar. 96	Ramnaryan Singh	19 Apl. 15	25 Apl. 18		
10 Feb. 97	Raibhan Singh	10 Jan. 16	4 May 18		
	Sukhraj Singh (*Retd.*)	. . . .	. . . .	. . . .	Re-employed 2 July 18.
23 Nov. 96	Ramswaarath Singh	6 Feb. 17	11 Sep. 18		
17 June 04	Frag Singh	25 Apl. 18	11 Sep. 18		
	Ajodhya Singh (*Retd.*) . . .	. . . .	. . . .	. . . .	Re-employed 22 July 18.

Jemadars.

10 Dec. 07	Jagratan Singh	25 Apl. 18
20 Mar. 00	Shiupal Singh	25 Apl. 18
11 Sept. 06	Brijpal Singh	4 May 18
13 Nov. 99	Ramesar Singh	4 May 18
18 Oct. 93	Ram Lal	18 July 18

Uniform—Drab.
Facings—Chocolate.

1st Battalion, 9th Bhopal Infantry.

Raised at Bhopal in 1859, by Capt. J. Peyton, Bombay Army, from the loyal remnants of the Bhopal, Gwalior and Malwa Contingents, and was originally designated the Bhopal Levy. Became the Bhopal Battalion in 1865; present designation in 1903.

"Afghanistan, 1878-79."

Composition—1 Companies of Sikhs, 1 of Rajputs, 1 of Brahmans, 1 of Musalmans.

Colonel—Hony. Col. Nawab Muhammad Nasrulla Khan, of Bhopal . . 1 Jan. 18.

1st Commn. or date of entering service.	Names and Rank.	Army rank.	Present appointment in Regiment.	Remarks.
	Commandant			
	Company Commanders (4)			
19 Oct. 92	Faunce, Maj. A. DeL.	19 Oct. 10	21 Aug. 16	2nd-in-comd.
7 May 98	Smith, Bt. Lt.-Col. H. H., D.S.O.	3 June 18	28 Dec. 07	Comdt. (*temp.*)
18 Apl. 00	*Gaskell, Maj. R. W.*	*1 Sep. 15*	*16 May 17*	
	Company Officers. (9)			
22 Oct. 03	Martin, Maj. G. D., M.C.	22 Oct. 17	10 Dec. 03	
4 June 04	*Wills, Capt. E. V.* (temp. Maj.)	*4 June 13*	*24 Aug. 09*	*R. O. for Hindustani Hindus.*
3 Dec. 04	Ralston, Capt. N. H. H. (*Act'g. Maj.*)	3 Dec. 13	10 Aug. 07	
3 Sep. 10	*Mullaly, Capt. J. C. D.*	*1 Sep. 15*	*11 Dec. 11*	
22 Jan. 13	Browning, Lieut. B. W.	22 Apl. 15	25 Mar. 14	*Lv., ex I., m. c.*
14 Jan. 14	*Moll, Capt. W. R.*	*14 Jan. 18*	*13 Oct. 14*	*With K. Afr. Rif.*
4 Dec. 14	Woledge, Lt. F. L.	4 Sep. 16	23 July 15	Attd.
14 Nov. 16	Stewart, Lt. C. E.	14 Nov. 17		Attd.
26 Feb. 16	Rodwell, Lt. L. H. (*Act'g. Capt.*)	26 Nov. 17	. . . .	Attd.
30 June 17	Bowers, Lt. T. I.	30 June 18	. . . .	With 11th Rajputs.
15 May 15	Holden, Lt. W. S. (*Actg. Capt.*)	15 May 16	24 Oct. 18	Attd.
5 May 15	Evans, Lt. A. L.	5 May 16	24 Nov. 17	Attd.
	Attached.			
20 Aug. 06	Ireland, Capt. G. (*Actg. Maj.*)	29 Aug. 15	. . . .	13th Rajput.
8 Jan. 15	Bousfield, Lt. H. T. W.	8 Jan. 16	. . . .	I. A. R. O.
18 Feb. 15	Weathrall, Lt. I. A. M.	18 Feb. 16	. . . .	Do.
9 Oct. 14	Flaxman, Lt. H. J. M.	9 Oct. 15	9 Feb. 16	I. A. R. O.
24 Dec. 14	Milne, Lt. E. M.	24 Dec. 15	5 July 16	I. A. R. O.
26 July 15	Roberts, Lt. F. C., M.C.	26 July 16	27 July 15	I. A. R. O.
1 Sep. 16	Malcolm, Lt. J. D.	1 Sep. 17	11 Oct. 16	I. A. R. O.
5 May 17	Owens, Lt. D. W. B.	5 May 18	1 Nov. 17	I. A. R. O.
12 Apl. 16	Grant, Lt. J. C.	12 Apl. 17	22 Jan. 18	Sea Highrs., I. A. (*on probn.*)
26 June 16	Cheverton, Lt. F. P.	26 June 17	1 May 18	I. A. (*on probn.*)
28 May 16	Woodliffe, Lt. L. F.	28 May 17	27 Apl. 18	Midd'x R., I. A. (*on probn.*)
5 May 15	Bell, Lt. J.	5 May 16	9 June 18	L'pool R., I. A. (*on probn.*)
20 June 17	Denness, Lt. R. B.	20 June 18	10 Oct. 17	I. A. R. O.
7 June 18	Corry-Smith, 2nd-Lt. L. B.	7 June 18	11 Oct. 18	I. A. (*T.C.*)
28 Aug. 18	Harper, 2nd-Lt. G. M.	28 Aug. 18	18 Sep. 18	I. A. R. O.
4 Sep. 18	Battersby, 2nd-Lt. R. S.	4 Sep. 18	29 Sep. 18	I. A. R. O.
5 Oct. 18	Stail, 2nd-Lt. C. J.	5 Oct. 18	30 Oct. 18	I. A. R. O.
4 Nov. 18	Wise, 2nd-Lt. H. H.	4 Nov. 18	20 Nov. 18	I. A. R. O.

Subadar-Major.

		Jemadar.	Subadar.	Subadar-Major.	
20 Jan. 90	Bhure Singh, I.D.S.M.	4 Dec. 01	1 May 07	1 July 14	
10 Feb. 93	Govind Singh, *Bahadur* (52, 56)	16 Oct. 11	28 Oct. 14	29 Feb. 15	***Order of Br. Ind., 2nd cl.***
11 Dec. 86	Sheikh Ali Muhammad (Retd.)	5 Dec. 93	15 Oct. 06	1 Nov. 07	Re-employed. 1 Feb. 17.

Subadars.

12 Apl. 05	Sheodarshan Tiwari	16 Oct. 13	28 Oct. 14	
15 Feb. 99	Shaikh Abdul Majid	23 Oct. 14	23 May 15	
6 Feb. 99	Thakur Singh	28 Apl. 15	22 Jan. 16	
3 Sep. 98	Abdul Aziz	28 Apl. 15	13 Apl. 16	
21 Sep. 04	Sher Muhammad Khan, I.D.S.M.	14 Jan. 16	15 Mar. 17	
12 Aug. 98	Ram Adin Tiwari	28 Oct. 14	1 Dec. 17	
16 July 04	Sheo Nath Singh	9 Aug. 14	15 Feb. 18	
7 Dec. 15	Hukam Chand	1 Aug. 16	15 Feb. 18	
25 Dec. 99	Amar Singh, I.D.S.M.	25 Feb. 12	6 Mar. 18	
	Dasaundha Singh (*Ret.*)	. . . .	. . . .	Re-employed, 15 July 17.
	Alam Ali Khan (*Ret.*)	. . . .	. . . .	Re-employed, 1 Aug. 17.
	Imam Bakhsh Khan (*Ret.*)	. . . .	. . . .	Re-employed, 1 Sep. 17.

1st Battalion, 9th Bhopal Infantry—*contd.*

1st Commn. or date of entering service.	Names and Rank.	Army rank.	Remarks.
	Jemadars.		
15 Feb. 94	Sawai Singh	25 Feb. 12	
13 Aug. 00	Husain Ali Khan	8 Mar. 15	
31 May 15	Jailal, I.O.M.	9 June 16	
20 Dec. 95	Rikkhi Singh	1 June 17	
10 Apl. 02	Deo Bakhsh	1 July 17	
11 Sep. 98	Basant Singh	1 Sep. 17	
6 June 11	Sayyid Muhammad Yaqub	1 Sep. 17	
21 Feb. 10	Shiu Balak Singh	1 Sep. 17	
1 Sep. 17	Shaikh Ishaq Muhammad	1 Sep. 17	
31 May 15	Harjiwan Lal	1 Oct. 17	
27 Sep. 02	Ismail Khan	26 Nov. 17	
4 May 15	Chandgi Ram (56, 48*a*)	26 Nov. 17	
22 Feb. 03	Sundar Singh	26 Nov. 17	
20 Oct. 00	Wazir Singh	26 Nov. 17	
5 May 10	Sukhdeo Pande	26 Nov. 17	
17 May 05	Suraj Prasad Misra	3 Dec. 17	
31 May 15	Tuhi Ram	15 Dec. 17	
22 Mar. 03	Rajmal Singh	19 Dec. 17	
31 May 15	Hira Lal	1 Mar. 18	
10 July 15	Gumani Singh	6 Mar. 18	
25 Nov. 03	Jan Muhammad Khan	1 May 18	
22 Oct. 07	Asghar Khan	1 May 18	
5 Feb. 11	Ram Din Dube	1 May 18	
17 Aug. 16	Roshan Lal	1 May 18	
	Achhaibar Singh	26 June 18	
	Hari Singh	2 Sep. 18	
23 June 03	Atar Singh	15 Aug. 18	
20 Feb. 04	Phera Singh	15 Aug. 18	
19 May 06	Abdul Haq Khan	15 Aug. 18	
25 Apl. 11	Bans Bahadur Singh	15 Aug. 18	
18 Sep. 16	Ram Nath Sharma	15 Aug. 18	
	Habib Khan (*Retd.*)	....	Re-employed, 1 Jan. 17.

2nd Battalion, 9th (Delhi) Infantry.

Formed at Raisina in February 1917.

Composition—1 Company Ahirs, 1 Company Gujars, 1 Company Hindustani Musalmans, 1 Company Jats.

1st Commn. or date of entering service.	Names and Rank.	Army rank.	Present appointment in Regiment.	Remarks.
	Commandant.			
1 Feb. 88	Lloyd, Lt.-Col. W. E. E.	11 Feb. 14	12 Jan. 17	1-97th Infy. *Lv., ex I., m. c., 6 mos.*
	Company Commanders. (4)			
10 Oct. 94	Trotter, Maj., E. G. S.	10 Oct. 12	4 Sep. 17	Cant. M'agte.'s Dept.
19 Dec. 00	Taylor, Maj. N. C., D.S.O. (*Actg. Lt.-Col.*)	19 Dec. 15	20 Sep. 17	92nd Punj.
19 Dec. 00	Bean, Maj. J. A.	23 Mar. 16	21 Nov. 17	101st Grs.
8 Aug. 14	Lambert, Capt. W. R.	8 Aug. 18	10 Apl. 17	97th Infy. Adjt. 10 Apl. 17.
	Company Officers.			
13 Nov. 14	Nightingale, Lt. G. O. (*Actg. Capt.*)	13 Aug. 16	29 Sep. 17	Attd.
26 Dec. 15	Loudoun, Lt. R. F.	26 Dec. 16	2 Feb. 18	Linc. R., I. A. (*on probn.*)
3 Jan. 16	Matthews, Lt. L. G. (*Actg. Capt.*)	3 Jan. 17	3 Aug. 17	North'd Fus., I. A. (*on probn.*)
13 Apl. 15	Axtell, 2nd-Lt. L. H. W.	26 Dec. 16	26 Nov. 17	Attd.
4 Jan. 16	Moore, Lt. H. L. (*Actg. Capt.*)	4 Oct. 17	3 Aug. 17	R. Fus., I. A. (*on probn.*)
9 Mar. 17	Pattinson, Lt. H. A., M.C.	9 Mar. 18	14 Aug. 17	I. A. (*on probn.*)
5 May 17	Beadon, Lt. W. C. (*Actg. Capt.*)	5 May 18	6 July 17	
27 June 16	Hennessey, Lt. D. M. L.	27 Mar. 18	12 Jan. 17	Attd.
9 Sep. 16	Bourne, 2nd-Lt. W. H.	9 June 17	17 Apl. 17	Attd.
29 June 16	*Twiddy, Lt. A. A. (temp. Capt.).*	*29 June 16*	*3 Aug. 17*	*I. A. R. O. Staff Capt. Nasirabad Bde.*
19 Dec. 16	Provis, Lt. F. (*Actg. Capt.*)	19 Dec. 17	3 Aug. 17	I. A. R. O.
23 Jan. 15	Brett, Lt. D. A., M.C.	23 Jan. 16	10 Aug. 18	I. A. R. O.
29 Aug. 17	Binns, 2nd-Lt. B. O.	29 Aug. 17	....	I. A. R. O.
7 June 18	O'Leary, 2nd-Lt. D. F. K.	7 June 18	8 Oct. 18	I. A. R. O.
29 Mar. 18	Sterl, 2nd-Lt. T. H.	....	....	I. A. (*T. C.*)
5 Nov. 17	Massey, Lt. A. P.	5 Nov. 18	27 May 18	I. A. (*T. C.*)
7 June 18	Hall, 2nd-Lt. F. R.	7 June 18	11 Oct. 18	I. A. (*T. C.*)
7 Oct. 18	Perrin, 2nd-Lt. F. E.	7 Oct. 18	24 Oct. 18	I. A. (*T. C.*)
7 Oct. 18	Fielding, 2nd-Lt. J.	7 Oct. 18	24 Oct. 18	I. A. (*T. C.*)
7 Oct. 18	Greenwood, 2nd-Lt. H.	7 Oct. 18	24 Oct. 18	I. A. (*T. C.*)
31 July 18	Mac Donald, 2nd-Lt. A. J.	31 July 18	9 Nov. 13	R. Scots, I. A. (*on probn.*)

Subadar-Major.

		Jemadar.	Subadar	Subadar-Major.
1 Aug. 16	Kehri Singh (48)	1 Aug. 16	12 Jan. 17	13 Feb. 17

Subadars.

16 Aug. 13	Behari Lal	16 Aug. 13	4 June 15
22 June 93	Mahbub Khan (56)	1 Feb. 09	27 Jan. 16
27 Dec. 97	Abdul Karim Khan, (56, 66*a*)	1 Nov. 05	13 May 15
4 Jan. 98	Richpal Singh	27 Aug. 15	7 Feb. 17
15 Dec. 00	Nizam-ud-din Khan (56, 66*a*)	1 July 10	1 Apl. 17
17 Apl. 15	Bhagat Ram	17 Apl. 15	1 Sep. 17
17 July 17	Jal Singh	17 July 17	14 Nov. 17
7 Feb. 17	Ramsarup	7 Feb. 17	14 Dec. 17
11 Mar. 17	Sultan Muhammad Khan	11 Mar. 17	13 Apl. 18
14 Nov. 17	Shaik Mahbub	14 Nov. 17	16 May 18
3 Dec. 17	Shiu Lal	17 Dec. 17	23 May 18
9 June 17	Puran Singh	17 Dec. 17	1 Aug. 18

Jemadars.

2 Jan. 17	Chiranji Singh	2 Jan. 17	
17 July 17	Sohan Lal (48, 56)	17 July 17	
17 Oct. 17	Bhoop Singh	17 Oct. 17	(*on probn.*)
14 Nov. 17	Durjan Singh	14 Nov. 17	
14 Nov. 17	Lekhi Singh	14 Nov. 17	
25 Nov. 15	Kishan Lal	21 May 18	
21 Aug. 08	Nathoo Khan	13 Apl. 18	
26 June 07	Ibrahim	13 Apl. 18	
29 July 10	Kanhaiya Singh	1 Sep. 17	
23 July 05	Jal Singh	1 Sep. 17	
15 July 09	Dalip Singh	1 May 18	
1 Feb. 17	Ram Sarup	23 May 18	
8 May 16	Nihal	23 May 18	
	Tirkha Ram (Retd.)		

3rd Battalion, 9th Bhopal Infantry.

Formed at Fyzabad in October 1917.

Composition—1 Company Ahirs, 1 Company Gaur Brahmins, 1 Company Gujars, 1 Company Rajputs.

1st Commn. or date of entering service.	Names and Rank.	Army rank.	Present appointment in Regiment.	Remarks.
	Commandant.			
25 Mar. 96	Bally, Maj. H. G. (*Actg. Lt.-Col.*)	25 Mar. 14	24 Oct. 17	1-10th Jats.
	Company Commanders. (4)			
19 May 00	McCleverty, Maj. P. H.	1 Sep. 15	1 Jan. 18	20th Infy. 2nd-in-Comd.
19 Jan. 07	Eales, Capt. A. C. H. A.	1 Sep. 15	6 Sep 18	1-119th Infy.
15 May 15	Robertson, Lt. J. (*Actg. Capt.*)	15 May 16	21 Aug. 18	Attd.
29 Jan. 16	Frankish, Lt. W. M.	29 Jan. 17	16 Aug. 18	Lond. R., I. A. (*on probn.*)
	Company Officers.			
20 May 16	Green, Lt. W.	20 May 17	26 Jan. 18	Bord. R., I. A. (*on probn.*)
8 Jan. 15	Harman, Lt. S. H.	8 Jan. 16	20 Oct. 18	I. A. R. O.
19 June 15	White, Lt. R. S. M.	19 June 16	17 Apl. 18	Linc. R., I. A. (*on probn.*)
11 Aug. 15	Diggory, Lt. P.	11 Aug. 16	26 Jan. 18	I. A. (*on probn.*)
11 Aug. 15	Moran, Capt. E. W.	11 Aug. 16	8 Sep. 18	Leins. R., I. A. R. O. (*on probn.*)
27 May 16	Saxton, Lt. A. C.	27 May 17	14 Feb. 18	I. A. (*on probn.*)
7 July 16	Trott, Lt. W. A., M.C.	7 July 17	5 Mar. 18	I. A. (*on probn.*) Adjt. 1 Aug. 18.
5 Nov. 17	Norman, Lt. S. H.	5 Nov. 18	15 Mar. 18	I. A. (*T. C.*)
5 May 17	Backhouse, Lt. J. R. C.	5 May 18	25 Oct. 17	I. A. R. O. Qr.-Mr. 25 Oct. 17.
5 Nov. 17	Rayner, Lt. T. A.	5 Nov. 18	15 Mar. 18	I. A. (*T. C.*)
29 Jan. 18	Robinson, 2nd-Lt. H.	29 Jan. 18	9 Sep. 18	Lanc. R., I. A. R. O. (*on probn.*)
27 Mar. 18	Nicholas, 2nd-Lt. J. R.	27 Mar. 18	4 Sep. 18	W. Fus., I. A. R. O. (*on probn.*)
7 June 18	Ryper, 2nd-Lt. H. J.	7 June 18	8 Oct. 18	I. A. R. O.
24 June 18	Vale, 2nd-Lt. H.	24 June 18	3 July 18	I. A. R. O.
	Brereton, 2nd-Lt. H.	. . . .	. . . .	Genl. List, I. A. R. O. (*on probn.*)
	Forbes, 2nd-Lt. A.	. . . .	. . . .	Genl. List, I. A. R. O. (*on probn.*)
22 July 18	Cockeram, 2nd-Lt. A. E.	22 July 18	3 Dec. 18	I. A. (*T. C.*)
28 June 18	Bryce, 2nd-Lt. W.	28 June 18	3 Dec. 18	I. A. R. O.
31 Jan. 18	Lowis, 2nd-Lt. H. R.	31 Jan. 18	6 Feb. 18	Attd.
19 Apl. 18	Hosain, 2nd-Lt. M. S. M.	19 Apl. 18	6 Aug. 18	I. L. F.
19 Apl. 18	Hosain, 2nd-Lt. M. N. A.	19 Apl. 18	6 Aug. 18	I. L. F.
7 June 18	Abbott, 2nd-Lt. P. D.	7 June 18	20 June 18	I. A. R. O.
24 June 18	Collins, 2nd-Lt. H. S.	24 June 18	2 July 18	I. A. R. O.
28 June 18	Shawe, 2nd-Lt. E. B	28 June 18	8 July 18	I. A. R. O.
26 July 18	Rundlett, 2nd-Lt. R. C.	26 July 18	10 Aug. 18	I. A. R. O.
7 Aug. 18	Barnett, 2nd-Lt. D. A	7 Aug. 18	31 Aug. 18	
7 Aug. 18	Fairlie, 2nd-Lt. W. S.	7 Aug. 18	28 Aug. 18	I. A. R. O.
9 Aug. 18	Anstiss, 2nd-Lt. L. A.	9 Aug. 18	31 Aug. 18	I. A. R. O. (*on probn*).
	Barua, 2nd-Lt G. N.		1 Nov. 18	
2 Mar. 18	Chaudhury, 2nd-Lt. B. N. S.			
	Subadar-Major.			
		Jamadar.	Subadar.	Subadar-Major.
21 Apl. 01	Banyad Ali	22 Apl. 15	1 Apl. 18	1 Apl. 18 (*temp.*)
	Subadars.			
14 Nov. 03	Digh Ram	1 July 15	1 Dec. 15	
4 Jan. 99	Abhaidatt Singh	. . . .	15 Mar. 17	*Indian Order of Merit, 2nd Class.*
11 June 99	Bhura Singh	7 Feb. 17	7 Nov. 17	
13 Aug. 10	Parbhu Singh	11 July 17	7 Nov. 17	
30 Oct. 16	Rameswar Singh	26 July 17	7 Nov. 17	
15 Jan. 00	Diwan Singh	28 Apl. 15	19 Dec. 17	
1 Nov. 04	Gurmukh Singh	29 Dec. 15	1 May 18	
31 May 15	Udmi Ram	1 Feb. 16	6 Nov. 18	
	Jemadars.			
22 May 17	Fateh Sher	22 May 17		
14 Aug. 13	Bhuri Singh	7 Nov. 17		
16 Sep. 13	Mohar Singh	7 Nov. 17		
25 Dec. 99	Balwant Singh	19 Dec. 17		
	Nadir Ali	1 May 18		
	Suraj Pal Misra	1 May 18		
	Udit Narain Singh	1 May 18		
3 Nov. 15	Pahlad Singh	1 Aug. 18		
31 July 15	Sher Singh	1 Aug. 18		
9 Aug. 09	Jaskeran Pande	1 Aug. 18		
26 Oct. 12	Prithipal Singh	1 Aug. 18		
3 July 09	Lal Singh	1 Aug. 18		
11 Jan. 18	Nand Lal	1 Aug. 18		
11 Dec. 17	Atma Singh	1 Oct. 18		
24 Dec. 05	Naubat Singh	1 Oct. 18		
31 May 15	Kashi Ram	1 Oct. 18		
29 Aug. 10	Sanman Singh	1 Oct. 18		
1 June 07	Jagmohan Singh	1 Oct. 18		
1 Jan. 04	Brijmohan Sukal	1 Oct. 18		
21 Sep. 10	Ram Lagan Singh	1 Oct. 18		
1 Dec. 10	Bhawari Din Pande	6 Nov. 18		

4th Battalion, 9th Bhopal Infantry.

Formed at Jhansi on the 16th February, 1918.

Composition—1 Company Lodhas, 1 Company Gaur Brahmans, 1 Company Hindustani Musalmans, 1 Company Rajputs.

1st Commn. or date of entering service.	Name and Rank.	Army Rank	Present appointment in Regiment.	REMARKS.
	Commandant.			
20 Jan. 97	Rennick, Maj. R. W. deC., (*Actg. Lt. Col.*)	20 Jan. 15	16 Feb. 18	1-95th Infy.
	Company Commanders. (4)			
29 Nov. 05	Cole, Capt. H. A. (*Actg. Maj.*)	29 Nov. 16	16 Feb. 18	44th Infy.
22 Dec. 14	Wilkinson, Lt. W. D., D.S.O., M.C. (*Actg. Capt.*)	22 Sep. 16	3 June 18	York R., I. A. (*on probn.*)
15 May 15	Henry, Lt. T. R. (*Actg. Capt.*)	15 May 16	22 Oct. 18	R. Ir. Fus., I. A. (*on probn.*)
15 Nov. 15	Amies, Lt. B. J. (*Actg. Capt.*)	15 Nov. 16	16 Feb. 18	1-21st Punj.
	Company Officers.			
23 Jan. 15	Baird, Lt. R. P. H.	23 Oct. 16	13 May 18	I. A. (*on probn.*)
16 Nov. 15	Harrison, Lt. L. H.	. . . .	19 May 18	I. A. (*on probn.*)
8 Mar. 17	Sheehan, Lt. T. P. (*Actg. Capt.*)	. . . .	13 May 18	I. A. R. O. Adjt.
	Robertson, 2nd-Lt. R.	. . . .	9 Oct. 18	Gord. Highrs., I. A. R. O. (*on probn.*)
26 Apl. 17	Marsland, Lt. K. D.	. . . .	13 May 18	I. A. R. O., Qr. Mr.
5 Jan. 18	Jennings, 2nd-Lt. H. W.	5 Jan. 18	5 June 18	I. A. (*T. C.*)
5 Jan. 18	King, 2nd-Lt. F. J. H.	5 Jan. 18	5 June 18	I. A. (*T. C.*)
15 Jan. 18	Longford, 2nd-Lt. G. W.	15 Jan. 18	5 June 18	I. A. R. O.
29 Mar. 18	Lake, 2nd-Lt. A. E.	29 Mar. 18	17 July 18	I. A. (*T.C.*)
29 Mar. 18	Teal, 2nd-Lt. F. H. D.	29 Mar. 18	23 July 18	I. A. R. O.
7 June 18	Willmore, 2nd-Lt. P. A.	7 June 18	9 Oct. 18	I. A. (*T C.*)
15 June 18	Sheffield, 2nd-Lt. J. C.	15 June 18	3 Oct. 18	Genl. List, I. A. R. O. (*on probn*).
30 Aug. 18	Graham, 2nd-Lt. D.	30 Aug. 18	8 Oct. 18	I. A. R. O.
25 June 18	Robertson, 2nd-Lt. R.	25 June 18	15 Oct. 18	I. A. R. O. (*on probn.*)
14 Sep. 18	Prosser, 2nd-Lt. P. H.	14 Sep. 18	23 Nov. 18	I. A. R. O.
24 Oct. 18	Arndt, 2nd-Lt. C. E.	24 Oct. 18	16 Nov. 18	I. A. R. O.
4 Nov. 18	Weinman, 2nd-Lt. A. N.	4 Nov. 18	23 Nov. 18	I. A. R. O.
24 Sep. 18	Priestley, 2nd-Lt. G. W.	24 Sep. 18	22 Oct. 18	I. A. R. O.
10 July 18	Jeans, 2nd-Lt. F.	10 July 18	2 Dec. 18	I. A. R. O.
17 July 18	Allsop, 2nd-Lt. J. J. W.	17 July 18	2 Dec. 18	I. A. R. O.
. . . .	Lewis, 2nd-Lt. R. E.	. . . .	26 Nov. 18	I. A. R. O.
1 Mar. 18	Watson, Armstrong, Capt., *The Hon'ble* W. J. M.	. . . .	17 Nov. 18	North'd Fus.
15 Jan. 15	Jamieson, Lt. W. H.	. . . .	2 Dec. 18	R. Highrs.
17 Apl. 15	Aiton, Lt. R. L.	. . . .	3 Dec. 18	R. Sc. Fus.
9 Oct. 15	Armstrong, Lt. G. B., M.C.	. . . .	17 Nov. 18	Northd's Fus.
16 Dec. 15	Paterson, Lt. H. E.	. . . .	3 Dec. 18	Lond. R.
5 Oct. 16	Rowe, Lt. V G.	. . . .	17 Nov. 18	North'd Fus.
29 Jan. 18	Warrior, 2nd-Lt. D.	29 Jan. 18	17 Nov. 18	North'd Fus.
29 Jan. 18	Hall, 2nd-Lt. J. W.	29 Jan. 18	17 Nov. 18	North'd Fus.
22 June 17	Briggs, 2nd-Lt. R. M.	22 June 18	3 Dec. 18	Lond. R.
	Johnson, Lt. A. G.	. . . .	. . . .	R. Innis. Fus., I. A. R. O. (*on probn.*)

Subadar-Major.

		Jemadar.	Subadar.	Subadar-Major.
22 Sep. 99	Nazir Khan	9 Aug. 14	24 Oct. 14	(*Actg.*)

Subadars.

22 Nov. 00	Mohammad Hasan	18 Feb. 15	1 Apl. 17
18 Nov. 01	Harnarain	4 June 15	13 Sep. 17
9 Nov. 03	Gopal Singh	1 Apl. 17	23 Mar. 18
11 July 04	Ram Sahai	7 Feb. 18	13 May 18
1 June 15	Lekh Ram	7 Feb. 18	13 May 18

Jemadars.

4 Feb. 99	Sukhdeo Singh	23 Mar. 18
31 July 15	Hardwari Lal	23 Mar. 18
29 Nov. 97	Desraj Singh	1 May 18
13 July 07	Shaikh Ali Muhammad	1 May 18
23 Sep. 03	Naiz Muhammad Khan	1 May 18
1 June 15	Chandan Singh	13 May 18
	Ntaha Ram	13 May 18
1 Nov. 04	Ram Lal	13 May 18

3rd Battalion, 9th Bhopal Infantry.

Formed at Fyzabad in October 1917.

Composition—1 Company Ahirs, 1 Company Gaur Brahmins, 1 Company Gujars, 1 Company Rajputs.

1st Commn. or date of entering service.	Names and Rank.	Army rank.	Present appointment in Regiment.	Remarks.
	Commandant.			
25 Mar. 96	Bally, Maj. H. G. (*Actg. Lt.-Col.*)	25 Mar. 14	24 Oct. 17	1-10th Jats.
	Company Commanders. (4)			
19 May 00	McCleverty, Maj. P. H.	1 Sep. 15	1 Jan. 18	20th Infy. 2nd-in-Comd.
19 Jan. 07	Eales, Capt. A. C. H. A.	1 Sep. 15	6 Sep 18	1-119th Infy.
15 May 15	Robertson, Lt. J. (*Actg. Capt.*)	15 May 16	21 Aug. 18	Attd.
29 Jan. 16	Frankish, Lt. W. M.	29 Jan. 17	16 Aug. 18	Lond. R., I. A. (*on probn.*)
	Company Officers.			
20 May 16	Green, Lt. W.	20 May 17	26 Jan. 18	Bord. R., I.A. (*on probn.*)
8 Jan. 15	Harman, Lt. S. H.	8 Jan. 16	20 Oct. 18	I. A. R. O.
19 June 15	White, Lt. R. S. M.	19 June 16	17 Apl. 18	Linc. R., I. A. (*on probn.*)
11 Aug. 15	Diggory, Lt. P.	11 Aug. 16	26 Jan. 18	I. A. (*on probn.*)
11 Aug. 15	Moran, Capt. E. W.	11 Aug. 16	8 Sep. 18	Leins. R., I. A. R. O. (*on probn.*)
27 May 16	Saxton, Lt. A. C.	27 May 17	14 Feb. 18	I. A. (*on probn.*)
7 July 16	Trott, Lt. W. A., M.C.	7 July 17	5 Mar. 18	I. A. (*on probn.*) Adjt. 1 Aug. 18.
5 Nov. 17	Norman, Lt. S. H.	5 Nov. 18	15 Mar. 18	I. A. (*T. C.*)
5 May 17	Backhouse, Lt. J. R. C.	5 May 18	25 Oct. 17	I. A. R. O. Qr.-Mr. 25 Oct. 17.
5 Nov. 17	Rayner, Lt. T. A.	5 Nov. 18	15 Mar. 18	I. A. (*T. C.*)
29 Jan. 18	Robinson, 2nd-Lt. H.	29 Jan. 18	9 Sep. 18	Lanc. R., I. A. R. O. (*on probn.*)
27 Mar. 18	Nicholas, 2nd-Lt. J. R.	27 Mar. 18	4 Sep. 18	W. Fus., I. A. R. O. (*on probn.*)
7 June 18	Ryper, 2nd-Lt. H. J.	7 June 18	8 Oct. 18	I. A. R. O.
24 June 18	Vale, 2nd-Lt. H.	24 June 18	3 July 18	I. A. R. O.
	Brereton, 2nd-Lt. H.	. . . .	. . . .	Genl. List, I. A. R. O. (*on probn.*)
	Forbes, 2nd-Lt. A.	. . . .	. . . .	Genl. List, I. A. R. O. (*on probn.*)
22 July 18	Cockeram, 2nd-Lt. A. E.	22 July 18	3 Dec. 18	I. A. (*T. C.*)
28 June 18	Bryce, 2nd-Lt. W.	28 June 18	3 Dec. 18	I. A. R. O.
31 Jan. 18	Lowis, 2nd-Lt. H. R.	31 Jan. 18	6 Feb. 18	Attd.
19 Apl. 18	Hosain, 2nd-Lt. M. S. M.	19 Apl. 18	6 Aug. 18	I. L. F.
19 Apl. 18	Hosain, 2nd-Lt. M. N. A.	19 Apl. 18	6 Aug. 18	I. L. F.
7 June 18	Abbott, 2nd-Lt. P. D.	7 June 18	20 June 18	I. A. R. O.
24 June 18	Collins, 2nd-Lt. H. S.	24 June 18	2 July 18	I. A. R. O.
28 June 18	Shawe, 2nd-Lt. E. B	28 June 18	8 July 18	I. A. R. O.
26 July 18	Rundlett, 2nd-Lt. R. C.	26 July 18	10 Aug. 18	I. A. R. O.
7 Aug. 18	Barnett, 2nd-Lt. D. A	7 Aug. 18	31 Aug. 18	
7 Aug. 18	Fairlie, 2nd-Lt. W. S.	7 Aug. 18	28 Aug. 18	I. A. R. O.
9 Aug. 18	Anstiss, 2nd-Lt. L. A.	9 Aug. 18	31 Aug. 18	I. A. R. O. (*on probn*).
	Barua, 2nd-Lt G. N.		1 Nov. 18	
2 Mar. 18	Chaudhury, 2nd-Lt. B. N. S.			

	Subadar-Major.	Jamadar.	Subadar.	Subadar-Major.
21 Apl. 01	Banyad Ali	22 Apl. 15	1 Apl. 18	1 Apl. 18 (*temp.*)
	Subadars.			
14 Nov. 03	Digh Ram	1 July 15	1 Dec. 15	
4 Jan. 99	Abhaidatt Singh	. . . .	15 Mar. 17	*Indian Order of Merit, 2nd Class.*
11 June 99	Bhura Singh	7 Feb. 17	7 Nov. 17	
13 Aug. 10	Parbhu Singh	11 July 17	7 Nov. 17	
30 Oct. 16	Rameswar Singh	26 July 17	7 Nov. 17	
15 Jan. 00	Diwan Singh	28 Apl. 15	19 Dec. 17	
1 Nov. 04	Gurmukh Singh	29 Dec. 15	1 May 18	
31 May 15	Udmi Ram	1 Feb. 16	6 Nov. 18	
	Jemadars.			
22 May 17	Fateh Sher	22 May 17		
14 Aug. 15	Bhuri Singh	7 Nov. 17		
16 Sep. 15	Mohar Singh	7 Nov. 17		
25 Dec. 99	Balwant Singh	19 Dec. 17		
	Nadir Ali	1 May 18		
	Suraj Pal Misra	1 May 18		
	Udit Narain Singh	1 May 18		
3 Nov. 15	Pahlad Singh	1 Aug. 18		
31 July 15	Sher Singh	1 Aug. 18		
9 Aug. 09	Jaskeran Pande	1 Aug. 18		
26 Oct. 12	Prithipal Singh	1 Aug. 18		
3 July 09	Lal Singh	1 Aug. 18		
11 Jan. 18	Nand Lal	1 Aug. 18		
11 Dec. 17	Atma Singh	1 Oct. 18		
24 Dec. 05	Naubat Singh	1 Oct. 18		
31 May 15	Kashi Ram	1 Oct. 18		
29 Aug. 10	Sanman Singh	1 Oct. 18		
1 June 07	Jagmohan Singh	1 Oct. 18		
1 Jan. 04	Brijmohan Sukal	1 Oct. 18		
21 Sep. 10	Ram Lagan Singh	1 Oct. 18		
1 Dec. 10	Bhawari Din Pande	6 Nov. 18		

4th Battalion, 9th Bhopal Infantry.

Formed at Jhansi on the 16th February, 1918.

Composition—1 Company Lodhas, 1 Company Gaur Brahmans, 1 Company Hindustani Musalmans, 1 Company Rajputs.

1st Commn. or date of entering service.	Name and Rank.	Army Rank	Present appointment in Regiment.	Remarks.
	Commandant.			
20 Jan. 97	Rennick, Maj. R. W. deC., (*Actg. Lt. Col.*)	20 Jan. 15	16 Feb. 18	1-95th Infy.
	Company Commanders. (4)			
29 Nov. 05	Cole, Capt. H. A. (*Actg. Maj.*)	29 Nov. 16	16 Feb. 18	44th Infy.
22 Dec. 14	Wilkinson, Lt. W. D., D.S.O., M.C. (*Actg. Capt.*)	22 Sep. 16	3 June 18	York R., I. A. (*on probn.*)
15 May 15	Henry, Lt. T. R. (*Actg. Capt.*)	15 May 16	22 Oct. 18	R. Ir. Fus., I. A. (*on probn.*)
15 Nov. 15	Amies, Lt. B. J. (*Actg. Capt.*)	15 Nov. 16	16 Feb. 18	1-21st Punj.
	Company Officers.			
23 Jan. 15	Baird, Lt. R. P. H.	23 Oct. 16	13 May 18	I. A. (*on probn.*)
16 Nov. 15	Harrison, Lt. L. H.	. . . .	19 May 18	I. A. (*on probn.*)
8 Mar. 17	Sheehan, Lt. T. P. (*Actg. Capt.*)	. . . .	13 May 18	I. A. R. O. Adjt.
	Robertson, 2nd-Lt. R.	. . . .	9 Oct. 18	Gord. Highrs., I. A. R. O. (*on probn.*)
26 Apl. 17	Marsland, Lt. K. D.	. . . .	13 May 18	I. A. R. O., Qr. Mr.
5 Jan. 18	Jennings, 2nd-Lt. H. W.	5 Jan. 18	5 June 18	I. A. (*T. C.*)
5 Jan. 18	King, 2nd-Lt. F. J. H.	5 Jan. 18	5 June 18	I. A. (*T. C.*)
15 Jan. 18	Longford, 2nd-Lt. G. W.	15 Jan. 18	5 June 18	I. A. R. O.
29 Mar. 18	Lake, 2nd-Lt. A. E.	29 Mar. 18	17 July 18	I. A. (*T.C.*)
29 Mar. 18	Teal, 2nd-Lt. F. H. D.	29 Mar. 18	23 July 18	I. A. R. O.
7 June 18	Willmore, 2nd-Lt. P. A.	7 June 18	9 Oct. 18	I. A. (*T C.*)
15 June 18	Sheffield, 2nd-Lt. J. C.	15 June 18	3 Oct. 18	Genl. List, I. A. R. O. (*on probn*).
30 Aug. 18	Graham, 2nd-Lt. D.	30 Aug. 18	8 Oct. 18	I. A. R. O.
25 June 18	Robertson, 2nd-Lt. R.	25 June 18	15 Oct. 18	I. A. R. O. (*on probn.*)
14 Sep. 18	Prosser, 2nd-Lt. P. H.	14 Sep. 18	23 Nov. 18	I. A. R. O.
24 Oct. 18	Arndt, 2nd-Lt. C. E.	24 Oct. 18	16 Nov. 18	I. A. R. O.
4 Nov. 18	Weinman, 2nd-Lt. A. N.	4 Nov. 18	23 Nov. 18	I. A. R. O.
24 Sep. 18	Priestley, 2nd-Lt. G. W.	24 Sep. 18	22 Oct. 18	I. A. R. O.
10 July 18	Jeans, 2nd-Lt. F.	10 July 18	2 Dec. 18	I. A. R. O.
17 July 18	Allsop, 2nd-Lt. J. J. W.	17 July 18	2 Dec. 18	I. A. R. O.
. . . .	Lewis, 2nd-Lt. R. E.	. . . .	26 Nov. 18	I. A. R. O.
1 Mar. 18	Watson, Armstrong, Capt., *The Hon'ble* W. J. M.	. . . .	17 Nov. 18	North'd Fus.
15 Jan. 15	Jamieson, Lt. W. H.	. . . .	2 Dec. 18	R. Highrs.
17 Apl. 15	Aiton, Lt. R. L.	. . . .	3 Dec. 18	R. Sc. Fus.
9 Oct. 15	Armstrong, Lt. G. B., M.C.	. . . .	17 Nov. 18	Northd's Fus.
16 Dec. 15	Paterson, Lt. H. E.	. . . .	3 Dec. 18	Lond. R.
5 Oct. 16	Rowe, Lt. V G.	. . . .	17 Nov. 18	North'd Fus.
29 Jan. 18	Warrior, 2nd-Lt. D.	29 Jan. 18	17 Nov. 18	North'd Fus.
29 Jan. 18	Hall, 2nd-Lt. J. W.	29 Jan. 18	17 Nov. 18	North'd Fus.
22 June 17	Briggs, 2nd-Lt. R. M.	22 June 18	3 Dec. 18	Lond. R.
	Johnson, Lt. A. G.	. . . .	. . . .	R. Innis. Fus., I. A. R. O. (*on probn.*)

Subadar-Major.

		Jemadar.	Subadar.	Subadar-Major.
22 Sep. 99	Nazir Khan	9 Aug. 14	24 Oct. 14	(*Actg.*)

Subadars.

22 Nov. 00	Mohammad Hasan	18 Feb. 15	1 Apl. 17
18 Nov. 01	Harnarain	4 June 15	13 Sep. 17
9 Nov. 03	Gopal Singh	1 Apl. 17	23 Mar. 18
11 July 04	Ram Sahai	7 Feb. 18	13 May 18
1 June 15	Lekh Ram	7 Feb. 18	13 May 18

Jemadars.

4 Feb. 99	Sukhdeo Singh	23 Mar. 18
31 July 15	Hardwari Lal	23 Mar. 18
29 Nov. 97	Desraj Singh	1 May 18
13 July 07	Shaikh Ali Muhammad	1 May 18
23 Sep. 03	Naiz Muhammad Khan	1 May 18
1 June 15	Chandan Singh	13 May 18
	Ntaha Ram	13 May 18
1 Nov. 04	Ram bal	13 May 18

Regt. Centre—Meerut. *Uniform*—Scarlet.
Linked with—6th Jat L. I. *Facings*—Yellow.

1st Battalion, 10th Jats.

Raised at Dinapore in 1823, as the 1st Battalion, 33rd Regiment of Bengal Native Infantry. Became the 63th Regiment of Bengal Native Infantry in 1824; the 10th Regiment of Bengal Native Infantry in 1861; the 10th Regiment of Bengal Infantry in 1885; the 10th (Jat) Regiment of Bengal Infantry in 1897; the 10th Jat Infantry in 1901; present designation in 1903.

"China, 1858-59" "Burma, 1885-87."
Composition—4 Companies Jats.

1st Commn. or date of entering service.	Names and Rank.	Army rank.	Present appointment in Regiment.	Remarks.
	Commandant.			
16 Nov. 87	Robson, Lt.-Col. C. G. [L] . .	16 Nov. 13	9 Sep. 16	
	Company Commanders. (4)			
25 Mar. 96	Bally, Maj. H.G. (*Actg. Lt.-Col.*)	25 Mar. 14	5 Dec. 16	2nd-in-Comd. With 3-9th Infy.
17 Jan. 00	*Gwyer, Bt.-Lt.-Col. V. E., p.s.c. (temp. Lt.-Col).*	*3 June 17*	*16 June 10*	*G. S. O., 1st Gr. Army Hdqrs.*
24 July 01	*Twiss, Maj. E. K.,* D.S.O. *(temp. Lt.-Col).*	*3 June 16*	*7 Sep. 13*	*Asst. Embn. Comdt., Karachi.*
22 Oct. 02	*Macpherson, Maj. D. I. . . .*	*22 Oct. 17*	*9 Sep. 16*	*Bengal Military Police.*
	Company Officers. (9)			
29 Aug. 06	*Haines, Capt. J. B. (L) . . .*	*29 Aug. 15*	*6 Feb. 08*	
19 Jan. 07	*Farquharson, Capt. C. W. (temp. Maj.)*	*1 Sept. 15*	*15 Nov. 09*	*D. A. Q. M. G., Embarkation, Bombay.*
20 Jan. 09	Crompton, Capt. J.	1 Sept. 15	10 Mar. 11	
29 Jan. 10	Standen, Capt. J. D.	1 Sept. 15	10 Mar. 11	Adjt., 2 Dec. 14.
24 Aug. 12	McCalmont, Capt. W. V. . .	24 Aug. 16	18 Oct. 13	With 2-10th Jats.
21 Jan. 13	Layng, Capt. T. M., M.C. . .	21 Jan. 17	9 Nov. 15	
14 Jan. 14	*Lewis, Capt. T. G. . . . .*	*14 Jan. 18*	*12 Oct. 14*	*With K. Afr. Rif.*
17 Oct. 15	Fletcher, Capt. A. B.	5 Aug. 18	24 Feb. 16	Attd. With 1—6th Jats.
1 Oct. 14	Henry, Capt. J. R. A. . . .	1 Oct. 18	18 Dec. 14	
22 Dec. 15	*Deakin, Lt. R. H. . . . .*	*28 Mar. 16*	*1 Mar. 16*	*Attd. With R. A. F.*
18 Apl. 16	Bragg, Lt. H. V.	18 Apl. 17	25 Apl. 16	Attd. With 1—99th Infy.
29 June 16	Boyton-Smith, Lt. E. S. . . .	29 June 17	3 July 16	Attd. *Lv., ex I., m.c.*
18 June 17	Robertson, Lt. D. G.	18 June 18	23 June 17	Attd. With 3—154th Infy.
27 Oct. 17	Buckton, 2nd-Lt. G. N. . .	27 Oct. 17	6 Nov. 17	Attd.
27 Oct. 17	Bodle, 2nd-Lt. R. T. . . .	27 Oct. 17	6 Nov. 17	Attd.
31 Aug 18	Brown, 2nd-Lt. W. F.. . . .	31 Aug. 18	14 Sept. 18	Attd.
2 Aug. 15	Peak, Lt. C. A. C. (*Actg. Capt.*)	2 Aug. 16	28 Nov. 17	Attd.
	Attached.			
8 May 01	Livesay, Maj. J.	8 May 16	30 Dec. 17	102nd Grs.
28 Jan. 05	Daly, Capt. C. F.	28 Jan. 14	14 Apl. 18	66th Punj. 2nd-in-Comd. (*temp.*).
25 Jan. 17	Turner, Lt. A. W..	25 Jan. 18	25 June 18	9th Lond. R. I. A. R. O. (*on probn.*)
8 Sep. 09	Jackson, Capt. L. H.	1 Sep. 15	. . . .	13th Rajputs.
30 July 15	Nicholson, Lt. W. H. (*c*) . .	30 July 16	27 Nov. 16	I. A. R. O.
21 Nov. 15	Thorogood, Lt. P. W. (*Actg. Capt*).	21 Nov. 16	31 Oct. 17	Lond. R., I. A. (*on probn.*)
26 Jan. 17	Sloane, Lt. W. H. St. J. . .	26 Jan. 18	24 Jan. 18	I. A. R. O. *Lv., ex I., m. c.*
28 Dec. 17	Walker, 2nd-Lt. L.	28 Dec. 17	28 Dec. 17	I. A. R. O.
5 May 17	Andrews, Lt. W. M.	5 May 18	17 Aug. 17	I. A. R. O.
20 June 17	Williams, Lt. C. W.	20 June 18	7 Oct. 17	I. A. R. O.
8 Aug. 17	Haslam, Lt. R. H. (*b*)	8 Aug. 18	3 Nov. 17	I. A. R. O. *A. R. O., Rajkote.*
27 Oct. 17	Allen, 2nd-Lt. R. F. W. K. . .	27 Oct. 17	2 Mar. 18	1—6th Jat. L. Infy.
25 July 17	Lee, Lt. W. M.	25 July 18	3 July 18	L'pool R., I. A. (*on probn.*)
29 July 17	Montgomerie, Lt. L. J. . . .	29 July 18	1 Aug. 17	
29 May 18	Beamount, 2nd-Lt. R.. . . .	29 May 18	29 Aug. 18	I. A. R. O. (*on probn.*)
31 Aug. 14	Larkin, Capt. A. S.	31 Aug. 18	19 Oct. 18	I. A. R. O. (*on probn.*).
19 Sep. 17	Purvis, 2nd-Lt. R. G. . . .	19 Sep. 18	. . . .	Sco. Rif., I. A. (*on probn.*)
21 July 17	Maunder, 2nd-Lt. G. H. O. .	21 July 17	6 Nov. 18	R. F. A., I. A. R. O. (*on probn.*)
24 Apl. 18	Walker, 2nd-Lt. W.	24 Apl. 18	20 Oct. 18	I. A. (U. L.)
20 Oct. 16	Hyland, Lt. J. P.	20 Oct. 17	. . . .	Conn. Rang., I. A. (*on probn.*)
31 July 18	Hill, 2nd-Lt. H. H.	31 July 18	20 Aug. 18	I. A. R. O.
15 Aug. 18	Hogshaw, 2nd-Lt. B. W. . .	15 Aug. 18	3 Sep. 18	I. A. R. O.
7 Oct. 18	Bennett, 2nd-Lt. S. B. . . .	7 Oct. 18	10 Oct. 18	I. A. (*T. C.*)

Subadar-Major.

		Jemadar.	Subadar.	Subadar-Major.	
12 Sep. 94	Nand Ram, *Bahadur* (52, 56)	1 Jan. 06	10 June 10	16 July 14	*Order of Br. Ind., 2nd Cl.*

1st Battalion, 10th Jats—*contd.*

1st Commn. or date of entering service.	Names and Rank.	Army rank.	Present appointment in Regiment.	Remarks.
	Subadars.			
2 Mar. 03	*Kunwar Chiranji Singh (52, 56)*	*2 Mar. 03*	*1 Nov. 08*	*47th Sikhs.*
6 Jan. 98	Raje Ram (52, 56)	2 Nov. 06	16 July 14	
19 Aug. 98	Umrao Singh (52, 56)	10 June 10	7 Dec. 14	
10 July 97	Molar (I) (52, 56, 66*a*)	5 Dec. 13	13 Apl. 16	
27 Mar. 00	Tirkha (56)	27 Oct. 14	14 Apl. 16	
19 Mar. 98	Shiuchand, (52, 56)	31 July 14	14 May 16	
27 Dec. 98	Molar (II) (52, 56)	27 Oct. 14	14 May 16	
5 Feb. 01	Bhola	16 May 15	9 Oct. 16	With 6 Jats.
5 Feb. 01	Kanhaiya	29 June 15	1 Jan. 18	
24 Jan. 02	Molar (52, 56)	8 Nov. 15	23 May 18	
24 Jan. 02	Digh Ram (56)	13 Apl. 16	23 May 18	
10 Sep. 98	Jotu	14 Apl. 16	31 May 18	
12 Apl. 99	Bhagwana	20 May 16	2 Oct. 18	
	Jemadars.			
15 June 03	*Gopal*	*13 Apl. 16*		*With 36 Sikhs.*
8 June 99	Gyani	14 Apl. 16		
18 Aug. 99	Mamraj	10 July 16		
26 Oct. 99	Hans Ram (66)	2 Oct. 16		
26 Jan. 02	Lok Ram	2 Oct. 16		
6 Jan. 06	Mansingh	15 Nov. 16		
24 June 01	Dani (60*a*)	7 Mar. 17		
18 Oct. 06	Sudhan	1 Apl. 17		
4 July 05	Shiuchand	1 Apl. 17		
23 May 08	Raje Ram (52, 56)	1 June 17		
1 Dec. 96	Ramdour Lal	1 July 17		
21 Sep. 09	Chunni Lal	1 Aug. 17		
5 June 07	Sheonath (52, 56)	16 Dec. 17		
11 Feb. 04	Harchand (56)	23 May 18		
18 Oct. 06	Giyani (56)	23 May 18		
6 May 03	Bhola	23 May 18		
23 May 08	Kalu	23 May 18		
6 June 05	Bhagwant (66)	16 Aug. 18		
26 Aug. 05	Dhig Ram (56)	27 Aug. 18		
30 Sep. 02	Sheochand	2 Oct. 18		

2nd Battalion, 10th Jats.

Formed at Jhansi on 12th June 1917.

Composition—4 companies Jats.

1st Commn. or date of entering service.	Names and rank.	Army rank.	Present appointment in Regiment.	Remarks.
	Commandant.			
20 May 99	Marsh, Maj. C. H., D.S.O. (Act'g. Lt.-Col.)	1 Sept. 15	23 May 18	18th Lrs.
	Company Commanders (4).			
24 Aug. 12	McCalmont, Capt. W. V.	24 Aug. 16	20 Feb. 18	1—10th Jats. 2nd-in-Comd. (*temp.*)
1 Oct. 15	O'Reilly, Capt. H. R.	1 Oct. 18	26 Apl. 18	Attd.
5 May 16	Greaves, Lt. J. (*Actg. Capt.*)	5 May 16	26 Apl. 18	Lan. Fus., I. A. (*on probn.*)
21 Jan. 16	Christie, Lt. G.	21 Jan. 17	1 Nov. 18	I. A. (*on probn.*)
	Company Officers.			
25 Dec. 15	Thompson, Lt. R. H. (*Actg. Capt.*)	25 Dec. 16	26 Mar. 18	Durh L. I., I. A. (*on probn.*)
18 May 16	Hill, Lt. E. D.	18 May 16	....	I. A. R. O.
26 June 15	Hanvey, Lt. M. P.	26 June 16	....	I. A. R. O. (*on probn.*)
27 Feb. 16	Porter, Lt. W. D.	27 Feb 17	....	I. A. (*on probn.*)
11 June 16	Birtwistle, Lt. J.	11 June 17	....	I. A. (*on probn.*)
8 Oct. 16	Kirkham, Lt. J. L.	8 Oct. 17	....	I. A. (*on probn.*)
5 Jan. 17	Farr, Lt. C. J.	5 Jan. 18	....	I. A. R. O.
5 May 17	Harrison, Lt. R. St. C.	5 May 18	....	I. A. R. O. Qr. Mr.
5 May 17	Dron, Lt. C. W.	5 May 18	....	Arg. & Suth'd. Highrs., I. A. (*on probn.*).
20 June 17	Perry, Lt. F. W. R.	20 June 18	....	I. A. R. O.
15 Oct. 17	McDonald, 2nd-Lt. M.	15 Oct. 17	....	I. A. R. O.
5 Jan. 18	Adkins, 2nd-Lt. E. C. A.	5 Jan. 18	....	I. A. (T. C.)
15 June 18	Gray 2nd-Lt. G. E.	15 June 18	....	I. A. R. O. (*on probn.*)
15 June 18	Horncastle, 2nd-Lt. J. A.	15 June 18	....	I. A. R. O. (*on probn.*)
27 Aug. 18	Hinton, 2nd Lt. F. E.	27 Aug. 18	....	I. A. R. O.
21 Aug. 18	Chapman, 2nd-Lt. J. J.	21 Aug. 18	....	I. A. R. O.
19 Oct. 18	Sherson, 2nd-Lt. D. B. N.	19 Oct. 18	....	I. A. R. O.
12 July 18	Heyter, 2nd-Lt. L. C. D.	12 July 18	....	I. A. R. O.
22 July 18	Jones, 2nd Lt. H. H.	22 July 18	....	I. A. (*T. C.*)
17 July 18	Bowen, 2nd- Lt. H. H.	17 July 18	....	I. A. R. O.
3 June 16	Prentice, Lt. J. F.	....	....	Leic. R., I. A. (*on probn.*)
5 July 15	Hyslop, Lt. A. G.	....	....	Leic. R., I. A. (*on probn.*)
18 July 17	Dixon, 2nd-Lt. C. R. B.	....	....	Leic. R., I. A. (*on probn.*)
25 Aug. 17	Norman 2nd-Lt. W., D.C.M., M.M.	....	....	Leic, R., I. A. (*on probn.*)
28 Nov. 17	Lee, 2nd Lt. T. E.	....	....	R. Dub. Fus., I. A. (*on probn.*)
31 Oct. 17	O'Donoghue, 2nd-Lt. E. J.	....	....	R. War. R., I. A. (*on probn.*).

Subadar-Major.

		Jemadar.	Subadar.	Subadar-Major.	
2 Mar. 97	Shib Lal, *Bahadur* (56, 72*a*)	2 Mar. 97	27 May 02	....	*Order of Br. Ind., 2nd cl. (temp.)*

Subadars.

2 Nov. 97	Harnand (56)	2 Dec. 14	12 June 17
30 May 10	Subhram	1 July 15	12 June 17
22 Dec. 94	Bakhtawar (52, 56, 66)	13 July 15	12 June 17
4 June 03	Bharat Singh (56)	5 Jan. 16	12 June 17

Jemadars.

5 Feb. 01	Moti (66)	12 June 17
5 Feb. 04	Nanda	12 June 17
21 Nov. 05	Gumani (48*a*)	12 June 17
22 Oct. 04	Ude Singh	18 Feb. 18
23 Nov. 06	Chandgi (48*a*. 56)	18 Mar. 18
8 Feb. 01	Sanwal (56)	18 Mar. 18
1 Sep. 05	Ramji Lal	15 Apl. 18
15 Dec. 07	Bujau	15 Apl. 18
25 Jan. 04	Chandgi Ram	28 Aug. 18
8 Feb. 01	Pohkar	28 Aug. 18

Regt. Centre—Lucknow.
Linked with—7th & 8th Rajputs.

1st Battalion, 11th Rajputs.

Uniform—Scarlet.
Facings—Yellow.

Raised at Cawnpore in 1825, as the 2nd Extra Regiment of Bengal Native Infantry. Became the 70th Regiment of Bengal Native Infantry, 1828; the 12th Regiment of Bengal Native Infantry, 1861; the 11th Regiment of Bengal Native Infantry, 1861; the 11th Regiment of Bengal Infantry, 1885; the 11th (Rajput) Regiment of Bengal Infantry, 1897; the 11th Rajput Infantry, 1901; present designation, 1903.

"Chillianwallah" "Goojerat" "Punjaub" "China, 1858-59"
"Afghanistan, 1878—80" "Burma, 1885-87."
Composition—4 Companies Rajputs.

1st Commn. or date of entering service.	Names and Rank.	Army Rank.	Present appointment in Regiment.	Remarks.

Commandant.

23 Oct. 89	Samborne-Palmer, Lt.-Col. F.C.S.	23 Oct. 15	4 Jan. 16	

Company Commanders. (4)

14 Sep. 87	*Watling, Lt.-Col. G. (Q)*	*14 Sept. 13*	*25 Aug. 15*	*2nd-in-Comd.*
10 Aug. 98	*Rogers, Maj. G. H., p.s.c. (temp. Lt.-Col.)*	*1 Sept. 15*	*1 June 11*	*G.S.O., 1st gr., Army Hdqrs.*
18 Apl. 00	Strong, Maj. R. G.	1 Sept. 15	1 Feb. 18	With 2-2nd Rajputs.
18 Jan. 02	Ovens, Maj. T. L.	18 Jan. 17	21 Nov. 18	2nd-in-Comd. (*Offg.*)

Company Officers. (9)

18 Jan. 02	Slater, Maj. D. McL.	18 Jan. 17	1 Aug. 07	With 2-4th Rajputs.
28 Jan. 03	Stoddart, Capt. H.	12 July 10	15 Jan. 04	With 76th Punj. *Lv., ex I., m. c.*
4 June 04	*Plinston, Capt. G. H.*	*4 June 13*	*15 May 08*	*Bde.-Maj., Bangalore Bde.*
13 Dec. 05	*Atkins, Capt. B. S.*	*13 Dec. 14*	*4 Nov. 08*	
9 Sept. 08	Staples, Capt. E. C., M.C.	1 Sept. 15	. . . .	*Grass Farm Dept.*
3 Sept. 10	Laville, Capt. T. T. (*Actg. Maj.*)	1 Sept. 15	21 Dec. 11	
18 Jan. 11	Scudamore, Capt. P. C.	1 Sept. 15	14 May 12	*S. S. O., Chitral,* & Adjt., 4 Oct. 15.
7 Apl. 16	Johnston, Lt. D. D. H.	7 Apl. 17	6 Nov. 16	Attd. With 1—143rd Infy.

Attached.

20 Jan. 12	Passy, Capt. W. De L. (*Actg. Maj.*)	20 Jan. 16	. . . .	1—113th Infy.
30 Oct. 15	Jones, Lt. P. J. D.	30 Oct. 16	. . . .	I. A. (*on probn.*)
30 Jan. 17	Bowers, Lt. C. I.	30 Jan. 18	3 Feb. 17	9th Infy.
29 May 15	Roper, Lt. T. P. (*Actg. Capt.*)	29 May 16	8 Aug. 17	I. A. (*on probn.*)
24 July 15	Gilmore, Lt. H. F. (*Actg. Capt.*)	24 July 16	3 Aug. 17	I. A. (*on probn.*)
7 July 17	Nicholls, Lt. R. L.	7 July 17	. . . .	I. A. R. O.
28 Jan. 17	Hale, Lt. E. C. (*Actg. Capt.*)	28 Jan. 18	. . . .	I. A. (*on probn.*)
5 May 17	Childs, Lt. F. D.	5 May 18	18 Aug. 17	I. A. R. O.
5 May 17	Lintott, Lt. P. J.	5 May 18	18 Aug. 17	I. A. R. O. Qr. Mr., 1 Feb. 18.
5 May 17	McDougall, Lt. P.	5 May 18	18 Aug. 17	I. A. R. O.
5 May 17	Rose, Lt. C. G.	5 May 18	18 Aug. 17	I. A. R. O.
1 Oct. 17	Leat, Lt. H. G.	1 Oct. 18	15 Jan. 18	I. A. (*T. C.*)
5 Jan. 18	Richardson, 2nd-Lt. J. H. S. (*Actg. Capt.*)	5 Jan. 18	5 June 18	I. A. (*T. C.*)
15 Jan. 18	Thompson, 2nd-Lt. W.	15 Jan. 18	5 June 18	I. A. (*T. C.*)
21 June 16	Haynes, Lt. A. E.	21 June 17	4 June 18	A. S. Corps. I. A. (*on probn.*)
19 Mar. 18	Knapman, 2nd-Lt. E. H.	19 Mar. 18	. . . .	I. A. R. O.

Subadar-Major.

		Jemadar.	Subadar.	Subadar-Major.
17 Jan. 94	**Jugraj Singh (56)**	**17 Jan. 94**	**1 Dec. 98**	**1 Mar. 10**

Subadars.

8 June 86	Matapalt Singh *Bahadur* (56)	16 Mar. 96	1 June 00	*Order of Br. Ind., 2nd Cl.*
17 Feb. 88	Durga Singh (56).	16 Sept. 99	15 May 02	
2 May 00	Narayan Singh, I.D.S.M. (56)	23 May 00	17 Nov. 03	
4 July 01	Mahabir Singh (56)	4 July 01	1 Oct. 09	
4 Nov. 93	Lal Bahadur Singh (56)	1 Jan. 11	6 Jan. 15	
7 Nov. 93	Bhagwan-bakhsh Singh	13 Jan. 11	6 Mar. 15	
5 Oct. 94	Manbhari Singh (56)	15 Feb. 12	12 Sept. 15 / 5 Nov. 15	With Labour Corps.
1 Feb. 96	Mahatam Singh (46, 56).	1 Oct. 13	13 Sept. 15	
9 Nov. 98	Jagatpal Singh	16 Apl. 11	19 Oct. 15	
11 Apl. 01	Harakh Bahadur Singh (56)	6 Jan. 15	6 Sept. 17	
	Jaipatra Singh (*Retd.*)	. . . .	Re-employed, 27 Apl. 15.	

1st Battalion, 11th Rajputs—*contd.*

1st Commn. or date of entering service.	Names and Rank.	Army Rank.	Present appointment in Regiment.	Remarks.
	Jemadars.			
7 Oct. 93	Bhagi Singh	6 Jan. 15		
11 Nov. 98	Ram Singh (56)	6 Jan. 15	Indian Adjt.	
9 July 00	Partab Singh (56)	6 Jan. 15		
29 Aug. 94	Jagesar Singh	6 Mar. 15	With Labour Corps.	
1 Feb. 99	Saltanat Singh	9 Sep. 15	With Labour Corps.	
22 Apl. 02	Sitla Bakhsh Singh (56)	14 Dec. 15 25 Mar. 16		
7 Sep. 07	Rampartab Singh (56)	15 Dec. 15		
1 Aug. 94	Parbhu Singh	6 June 16	With 2-11th Rajputs.	
17 Sep. 95	Jalesar Singh	6 June 16	With 2-11th Rajputs.	
22 Oct. 98	*Beni Singh*	*18 Oct. 16*		
1 Aug. 98	Rambiswas Singh	21 Oct. 16		
7 Apl. 05	Raghuraj Singh (56)	17 Apl. 17		
21 Jan. 96	Danbahadur Singh	9 May 17	With 1-150th Infy.	
20 Dec. 05	Pahlad Singh	6 Sep. 17		
20 Sep. 99	Surajbali Singh	6 Sep. 17		
15 Jan. 99	Bhawanibakhsh Singh	6 Sep. 17		
19 Sep. 07	Jagannath Singh	6 Sep. 17		
12 Jan. 02	Brijbhukan Singh	6 Sep. 17		
28 July 87	Beni Ram	3 Dec. 17		
26 Sep. 98	Daspati Singh	1 Mar. 18		
17 July 00	Mahabir Singh	1 Mar. 18		
16 Oct. 00	Dhenuk Singh	1 May 18		
17 Apl. 00	Indarpal Singh	1 May 18		
5 Feb. 00	Chattarpat Singh	1 Sep. 18		
1 Apl. 01	Dukhi Singh	1 Sep. 18		
23 Sep. 08	Sital Pershad Singh	14 Sep. 18		

2nd Battalion, 11th Rajputs.

Formed at Sitapore on 20th January 1918.

Composition—4 Companies Rajputs.

1st Commn. or date of entering service.	Names and Rank.	Army Rank.	Present appointment in Regiment.	REMARKS.
	Commandant.			
4 Aug. 97	Sargent, Maj. W. P. M. (*Actg. Lt.-Col.*)	4 Aug. 15	20 Jan. 18	104th Rif.
	Company Commanders. (4)			
1 Dec. 97	Woodcock, Maj. E. E.	1 Sep. 15	27 Sep. 18	1-4th Rajputs, 2nd-in-comd.
	Macqueen, Capt. E. T. L.	1 Sep. 15	4 Nov. 18	R. Ir. Rif. I. A. (*on probn.*)
2 June 15	Woodyer, Lt. J. M. (*Actg. Capt.*)	2 June 16	20 Jan. 18	I. A. (*on probn.*) Qr. Mr.
7 Apl. 16	Adams, Lt. H. V. (*Actg. Capt.*)	7 Apl. 17	4 Feb. 18	13th Rajputs.
	Company Officers.			
15 Feb. 15	Dempster, Lt. H. S.	15 Feb. 16	2 Feb. 18	I. A. R. O. Adjt.
22 Aug. 16	Wiseman, Lt. W.	22 Aug. 17	3 June 18	Gord. Highrs. I. A. (*on probn.*)
26 Oct. 16	Blackmore, Lt. E. W. H.	26 Oct. 17	2 May 18	I. A. (*on probn.*)
5 May 17	Burrows, Lt. A. C.	5 May 18	2 Feb. 18	I. A. R. O.
8 Nov. 17	Leake, 2nd-Lt. H.	8 Nov. 17	6 June 18	I. A. R. O.
5 Jan. 18	Denness, 2nd-Lt. S. W. G.	5 Jan. 18	6 June 18	I. A. (*T. C.*)
29 Mar. 18	Little, 2nd-Lt. W. J.	29 Mar. 18	16 July 18	I. A. (*T. C.*)
29 Mar. 18	Cooke, 2nd-Lt. C. A.	29 Mar. 18	4 Dec. 18	I. A. (*T. C.*)
29 Mar. 18	Reeves, 2nd-Lt. W. H.	29 Mar. 18	26 July 18	I. A. R. O.
7 June 18	Tee, 2nd-Lt. S. E.	7 June 18	23 Oct. 18	I. A. R. O.
7 June 18	Egginton, 2nd-Lt. J. F.	7 June 18	23 Oct. 18	I. A. (*T. C.*)
22 July 18	Coster, 2nd-Lt. G. W.	22 July 18	4 Dec. 18	I. A. (*T. C.*)
1 Oct. 18	Garrow, 2nd-Lt. R. K.	1 Oct. 18	12 Dec. 18	Attd.

Subadar-Major.

		Jemadar.	Subadar.	Subadar-Major.
10 June 92	Sardar Singh, I. D. S. M.	1 July 08	2 Nov. 13	1 Mar. 18

Subadar.

1 July 96	Rikkhi Singh	1 Aug. 13	1 Mar. 18	
1 Aug. 94	Perbhu Singh	6 June 16	1 Mar. 18	1-11th Rajputs.
17 Sep. 95	Jalesar Singh	6 June 16	1 Mar. 18	1-11th Rajputs.
18 Aug. 06	Harbaksh Singh	5 July 17	1 Mar. 18	

Jemadars.

1 Nov. 98	Rameswar Singh	23 July 15	
13 Oct. 94	Raghunath Singh	1 Mar. 18	
27 Oct. 03	Rampal Singh	1 Mar. 18	
4 Feb. 03	Basist Singh	1 Mar. 18	
17 Aug. 11	Shiunath Singh	1 Mar. 18	
13 June 06	Ram Lal Singh	1 Mar. 18	Indian Adjt., 1 Mar. 18.
21 Feb. 89	Ahmed Yar Khan	1 June 18	Indian Qr. Mr.
8 July 02	Sukhnandan Singh	26 Aug. 18	
2 Oct. 00	Shiubardan Singh	21 Oct. 18	
3 Dec. 00	Ratipal Singh	21 Oct. 18	
11 June 06	Ram Singh	21 Oct. 18	
7 July 06	Tirloki Singh	21 Oct. 18	
1 June 07	Kamta Singh	21 Oct. 18	
28 Oct. 08	Sughar Singh	21 Oct. 18	
4 Nov. 08	Ramtrich Singh	21 Oct. 18	

Regt. Centre—Bareilly. *Uniform*—Scarlet
Linked with—48th Pioneers. *Facings*—Black.

1st Battalion, 12th Pioneers (The Kelat-i-Ghilzie Regiment).

Raised at Ludhiana in 1838, by Capt. W. F. Beatson, as the 3rd Regiment of Infantry, Shah Shujah's Force. Became the Regiment of Kelat-i-Ghilzie, 1842; the 13th Regiment of Bengal Native Infantry, 1861; the 12th Regiment of Bengal Native Infantry, 1861, the 12th (the Kelat-i-Ghilzie) Regiment of Bengal Native Infantry, 1864; the 12th (the Kelat-i-Ghilzie) Regiment of Bengal Infantry, 1885; present designation, 1903.

Carries a special regimental colour (red, yellow and blue).

A MURAL CROWN SUPERSCRIBED *Invicta*.

"Kelat-i-Ghilzie" "Candahar, 1842" "Ghuznee, 1842" "Cabool, 1842"
"Maharajpore" "Afghanistan, 1878-79" "Burma, 1885-87" "Punjab Frontier."

Composition—2 Companies of Jats, 2 of Lobana Sikhs.

1st Commn. or date of entering service.		Names and Rank.	Army rank.	Present appointment in Regiment.	Remarks.
		Commandant.			
23 Dec.	93	Hawkes, Lt.-Col. W. C. W., D.S.O.	20 Oct. 18	20 Oct. 18	With 2-12th Prs.
		Company Commanders. (4)			
24 Jan.	00	Hooker, Maj. J. S., *q*. (*Actg. Lt.-Col.*)	1 Sep. 15	10 May 18	2nd-in-Comd. Comdt. (*temp.*)
19 Sep.	00	Marriott, Maj. J. H. G., *q*.	1 Sep. 15	19 Feb. 16	With 3-32nd Prs.
28 July	00	Cunningham, Maj. C.C., *p.s.c.*, *q*. (*Actg. Lt.-Col.*)	1 Sep. 15	10 May 18	With 2-107th Prs.
11 Aug.	*00*	*Pulley, Maj. H. C., q.[L]*	*1 Sep. 15*	*12 Oct. 18*	
		Company Officers. (9)			
18 Jan.	02	Clarke, Maj. G. A., *q*.	18 Jan. 17	15 Oct. 03	2nd-in-Comd. (*temp.*).
18 Jan.	*02*	*Massy, Maj. W. I. J., q.*	*18 Jan. 17*	*1 Feb. 06*	*Burma Mily. Police.—With 4-70th Rif.*
21 Jan.	03	Boyce, Maj. R. B., *q*.	21 Jan. 18	7 Nov. 04	With 1-34th Prs.
13 Aug.	04	Murray, Capt. S. G. C., C.I.E., q.	13 Aug. 13	1 Jan. 06	
9 Sep.	08	Macpherson, Capt. A. K., *q*.	1 Sep. 15	21 Feb. 10	With 3-34th Prs.
9 Sep.	08	Catterson-Smith, Capt. T.M.O., q.	1 Sep. 15	9 Sep. 09	*Lv., ex I., 6 mos., m. c.*
9 Sep.	*08*	*Yeates, Capt. E. P.*	*1 Sep. 15*	*16 Feb. 10*	*With 2-32nd Sikh Prs.*
20 Jan.	*09*	*Young, Capt. B. N., q.*	*1 Sep. 15*	*27 Mar. 10*	*M. T. Tg. Sch.*
23 Dec.	14	Hawkes, Lt. M. H.	23 Dec. 15	21 Feb. 15	With 48th Prs.
29 June	16	Hampton, Lt. A. V. F.	5 June 16	. . . .	Attd. With 1-34th Prs.
15 Nov.	15	Hirst, Lt. J. E.	15 Nov. 16	16 Nov. 15	Attd. With 2-23rd Prs.
15 Nov.	15	Teague, Lt. R. J. D.	15 Nov. 16	16 Nov. 15	Attd. *Bomb. Sch., Mhow.*
18 June	17	Parkin, Lt. J. H. D.	18 June 18	26 Apl. 17	Attd.
31 Jan.	18	Smith, 2nd-Lt. J. L. V.	31 Jan. 18	6 Feb. 18	Attd. Qr. Mr. (*temp.*)
31 Jan.	18	Wilson, 2nd-Lt. L. M.	31 Jan. 18	6 Feb. 18	Attd.
22 May	15	Virgin, Lt. A. H. (*Actg. Capt.*)	22 May 16	. . . .	Attd.
31 Apl.	18	Spurgin, 2nd-Lt. A. R.	31 Apl. 18	. . . .	Attd.
		Attached.			
5 Sep.	15	Bell, Lt. L. C., M.C. (*Actg. Capt.*)	5 June 17	. . . .	I. A. (*on probn.*) Adjt.
		Lee, Lt. D. H.	. . . .	. . . .	I. A. R. O.
8 Sep.	15	Mathew, Lt. W. C. (*Actg. Capt.*)	8 Sep. 16	. . . .	I. A. R. O.
7 July	17	Cameron, Lt. A. W.	7 July 17	1 Oct. 17	I. A. R. O.
18 Feb.	16	Cooke, Lt. C. W. (58)	18 Nov. 17	. . . .	A. Cyclist Corps, I. A. (*on probn.*).
1 Aug.	17	Seager, Lt. J. E. B.	1 Aug. 18	24 Apl. 18	
15 June	18	Peradon, 2nd-Lt. F. L.	15 June 18	. . . .	I. A. R. O. (*on probn.*).
15 June	18	Savage, 2nd-Lt. P. G.	15 June 18	. . . .	I. A. R. O. (*on probn.*).
		Bryant, Lt. R. J. C.	. . . .	. . . .	Wessex Bde., I. A. R. O. (*on probn.*).
7 June	18	Grover, 2nd-Lt. W.	7 June 18	. . . .	I. A. (*T. C.*)
26 Sep.	17	Crockett, Lt. N.	26 Sep. 18	. . . .	I. A. R. O.
		Subadar-Major.	Jemadar.	Subadar.	Subadar-Major.
10 Oct.	03	Hari Ram (56)	10 Oct. 03	16 Mar. 12	
		Phula Singh, Hony. Capt., Sardar Bahadur (Ret.)	. . . .	. . . .	Re-employed.
		Subadars.			
11 Nov.	88	Dewak Ram (56)	1 Apl. 03	7 May 09	
18 Apl.	89	Gyan Singh Bahadur (56)	1 July 10	1 July 13	Order of Br. Ind., 2nd cl. With 2-34th Prs.
2 May	97	Binda Singh (56)	24 Dec. 11	1 Mar. 14	
23 Mar.	95	Bhola Singh	24 May 11	9 Feb. 15	
18 May	93	Pala Singh	16 July 13	10 Feb. 15	
7 July	*00*	*Maingha Singh*, I.O.M. *(51)*	*4 Dec. 15*		
26 May	*98*	*Indar Singh (56)*	*22 Apl. 16*	. . . .	*With Labour Corps.*
Mar.	94	Shadi	5 Jan. 15	12 June 16	
May	03	Hira	5 Jan. 15	12 June 16	

1st Battalion, 12th Pioneers (The Kelat-i-Ghilzie Regiment) —*contd.*

1st Commn. or date of entering service.	Names and Rank.	Army rank.	Present appointment in Regiment.	Remarks.
	Jemadars.			
6 May 96	Balwant (56, 66*a*)	16 Mar. 12		
21 Sep. 94	Malla Singh	9 Feb. 15		
11 Dec. 98	Harnarain	29 Oct. 15		
24 Apl. 03	Nanak Singh (I)	30 Apl. 16		
19 Oct. 06	Mam Raj (56)	12 June 16		
16 Apl. 01	Ramji Lal	22 July 16		
23 June 17	Kanhaiya	23 June 17		
23 June 17	Bagat Singh			
16 July 17	Shamsuddin	16 July 17		
16 Apl. 01	Hira Singh	1 June 18		
19 May 03	Kanhaiya	1 July 18		
23 Apl. 03	Magh Singh	19 Aug. 18		
29 Aug. 05	Nihal Singh	19 May 18		
1 June 03	Indar Singh	1 June 18		

2nd Battalion, 12th Pioneers (The Kelat-i-Ghilzie Regiment).

Formed at Lahore Cant. on 6 June 1917.
Composition—2 Companies Jats, 1 of Sikhs, 1 of Hazaras.

1st Commn. or date of entering service.	Names and Rank.	Army rank.	Present appointment in Regiment.	Remarks.
	Commandant.			
23 Dec. 93	Hawkes, Lt.-Col. W. C. W., D.S.O.	20 Oct. 18	6 June 17	1-12th Prs.
	Company Commanders. (4)			
3 Aug. 04	Gover, Capt. A. C., M.C., (*Act'g. Maj.*)	13 Aug. 13	18 July 18	121st Prs. 2nd-in-Comd. (*temp.*)
22 Apl. 16	Girling, Lt. J. B. (*Actg. Capt.*)	22 Jan. 17	26 Feb. 18	I. A. (*on probn.*)
24 July 15	Comper, Lt. G. Q. (*Actg. Capt.*)	24 July 16	26 Feb. 18	I. A. R. O.
	Company Officers.			
1 Sep. 15	Silberbauer, Lt. H. L.	15 Sep. 16	28 July 18	Attd. With 2-5th G. R.
15 Sep. 16	Recknell, Lt. D. L.	15 Sep. 18	16 June 18	E. Surr. R., I. A. (*on probn.*)
9 May 17	Lundy, Lt. C. R. C.	9 May 18	15 Apl. 18	Austl. Impl. F., I. A. (*on probn.*)
5 June 17	Skinner, Lt. F. H. (*Actg. Capt.*)	5 June 18	3 Oct. 17	Attd. Adjt.
1 Oct. 17	Bowditch, Lt. C. E.	1 Oct. 18	20 Jan. 18	I. A. (*T. C.*)
11 Dec. 17	McCarthy, Lt. E. D.	11 Dec. 17	28 Nov. 17	Attd. Qr. Mr.
15 Jan. 18	Thomas, 2nd-Lt. G. S.	15 Jan. 18	6 Jan. 18	I. A. (*T. C.*)
22 Apl. 15	Eckel, Lt. P.	22 Apl. 16	11 Feb. 18	I. A. R. O. (*on probn.*)
18 Oct. 16	Webster, Lt. G. W.	18 Oct. 16	30 Oct. 17	I. A. R. O.
19 May 15	Atkins, Lt. E.R.N. (*Actg. Capt.*)	19 Feb. 18	4 Apl. 18	I. A. R. O.
13 Aug. 17	Wilkins, Lt. J. P. K.	13 Aug. 18	5 Nov. 17	I. A. R. O.
5 Nov. 17	Philp, Lt. W. H. S.	5 Nov. 18	14 Mar. 18	I. A. (*T. C.*)
26 June 18	Rahles-Rahbula, 2nd-Lt. E. H.	26 June 18	9 Nov. 18	M. G. Corps, I. A. (*on probn.*)
7 June 18	Johnston, 2nd-Lt. M.	7 June 18	14 Oct. 18	I. A. R. O.
	Murray, Lt. T.		19 Nov. 18	R. Scots.
	Kidsten, 2nd-Lt. A. H.		18 Nov. 18	Glouc. R.
	Mitchell, 2nd-Lt. N.		18 Nov. 18	Gord. Highrs.
	Ingram, 2nd-Lt. A. L.		18 Nov. 18	Gord. Highrs.
	Allan, 2nd-Lt. R. A. B.		18 Nov. 18	Gord. Highrs.

Subadar-Major.

		Jemadar.	Subadar.	Subadar-Major.
18 July 01	Man Singh (56)	1 Aug. 04	16 May 13	3 Sep. 17 (*temp.*)

Subadars.

Jan. 97	Mohammad Hassan, *Bahadur* I.D.S.M. (56,66).	1 Apl. 11	7 Apl. 15	*Order of Br. Ind. IInd Class.*
7 May 00	Hassan (56)	29 Jan. 14	9 Feb. 17	
16 May 03	Radha Singh	5 Jan. 15	6 June 17	
12 June 03	Labh Singh (56)	5 Jan. 15	6 June 17	
21 July 97	Nihala (56)	16 June 15	6 June 17	
21 Apl. 03	Maru (56)	12 June 16	6 June 17	
7 Sep. 05	Udai Singh (66)	6 June 17	1 July 18	

Jemadars.

4 Nov. 04	Nabi (56)	14 Dec. 15	
1 Dec. 04	Abdulla	9 Feb. 17	
2 July 06	Ram Ji Lal	6 June 17	
18 May 03	Jai Ram	6 June 17	
27 Nov. 07	Ram Sarup	6 June 17	
2 Jan 97	Kalbi Hussain	1 Jan. 18	
1 Mar. 18	Lachman Singh	1 Mar. 18	(*On probn.*)
4 Dec. 17	Kanhaiya	1 July 18	
7 May 03	Gugan (56)	1 Aug. 18	
10 May 03	Nand Singh	22 Aug. 18	

Regt. Centre—Agra. *Uniform*—Scarlet.
Linked with—2nd, 4th and 16th Rajputs. *Facings*—Blue.

13th Rajputs (The Shekhawati Regiment).

Shekhawati Paltan.—Raised in Shekhawati in 1835, by Major H. Forster, as the Infantry of the Shekhawat Brigade—Taken over from Jaipur, 1843. Became the Shekhawati Battalion, 1847; the 14th Regiment of Bengal Native Infantry, 1861; the 13th Regiment of Bengal Native Infantry, 1861; the 13th (The Shekhawati) Regiment of Bengal Native Infantry, 1884; the 13th (The Shekhawati) Regiment of Bengal Infantry, 1885; the 13th (The Shekhawati) Rajput Regiment of Bengal Infantry, 1897; the 13th (Shekhawati) Rajput Infantry, 1901; present designation, 1903.

"Aliwal" "Afghanistan, 1879-80" "Chitral."
Composition—4 companies Rajputana Rajputs.

HONORARY COLONEL—*Hony.* Major-General *His Highness Maharajadhiraja Sir* SAWAI MADHO SINGH *Bahadur*, of Jaipur, G.C.S.I., G.C.I.E., G.C.V.O., G.B.E. 13 May 1904.

1st Commn. or date of entering service.	Names and Rank.	Army rank.	Present appointment in Regiment.	Remarks.
	Commandant.			
12 Aug. 88	Stewart, Lt.-Col. J. A.	22 Aug. 14	17 Nov. 16	
	Company Commanders. (4)			
27 July 98	McCleverty, Maj. A. H., *p.s.c.* (*Actg. Lt.-Col.*)	1 Sep. 15	17 Nov. 16	2nd-in-Comd. With 2-8th Rajputs.
19 July 93	*Corbett, Maj. B. A.*	*19 July 11*	*12 Sep. 08*	
25 June 99	*Cole, Maj. B. L.*	*1 Sep. 15*	*23 Aug. 12*	*Rectg. Officer, Ajmer.*
17 Feb. 00	Seymour, Maj. C. R. F., *p.s.c.* (*temp. Lt.-Col.*) [L]	1 Sep. 15	21 Apl. 16	
	Company Officers. (9)			
13 Jan. 02	Bunbury, Maj. G. B. St. P.	18 Jan. 17	11 Apl. 03	
18 Jan. 02	Whitwell, Maj. H. W.	18 Jan. 17	26 Mar. 03	2nd-in-Comd. (*offg.*)
22 Oct. 02	Leapman, Maj. H. M.	22 Oct. 17	3 Aug. 07	
8 Nov. 05	*Falkland, Capt. H. L. F.*	*8 Nov. 14*	*4 Aug. 07*	*Assam Mily. Police.*
29 Aug. 06	Ireland, Capt. G (*Actg. Maj.*)	29 Aug. 15	1 Apl. 10	With 1-9th Infy.
7 Nov. 06	Hammick, Capt. St. V. F.	1 Sep. 15	29 Nov. 10	
8 Sep. 09	Jackson, Capt. L. H.	1 Sep. 15	29 Dec. 10	With 1-10th Jats.
2 Sep. 10	*Ferguson, Capt. J. A.*	*1 Sep. 15*	*27 June 16*	*Staff Capt.*
26 Jan. 16	Rimell, Lt. F. J.	26 Jan. 17	20 Mar 16	With 7th Rajputs.
24 Mar. 16	Gosnell, Lt. K. A. (*Actg. Capt.*)	24 Mar. 17	....	Attd.
7 Apl. 16	Adams, Lt. H. V. (*Actg. Capt.*)	7 Apl. 17	5 Nov. 16	Attd. With 2-11th Rajputs.
18 Apl. 16	Huelin, Lt. W. (*Actg. Capt.*)	18 Apl. 17	20 Apl. 16	Attd.
29 June 16	Skrine, Lt. P. R. H. (*Actg. Capt.*)	29 June 17	8 July 16	Attd. Adjt., 24 Apl. 18.
29 June 16	Davys, Lt. J. V. (*Actg. Capt.*)	29 June 17	8 July 16	Attd. With 1-150th Infy.
27 Oct. 16	Rogers, Lt. W. G.	27 Oct. 17	4 Aug. 17	Attd.
6 Oct. 16	Lindsley, 2nd-Lt. R. M.	6 July 17	3 Dec. 16	Attd. Qr. Mr.
30 Apl. 15	Andrews, Lt. J. S. T.	30 Jan. 17	29 Nov. 17	Attd.
15 Apl. 17	Beatson, 2nd-Lt. J.	15 Apl. 18	25 Nov. 17	Attd.
	Attached.			
23 Aug. 15	Baker, Lt. G. M.	23 Aug. 16	....	
24 May 16	Gosnell, Lt. K. A. (*Actg. Capt.*)	24 Mar. 17	11 May 17	I. A. (*on probn.*)
1 July 16	Mather, Lt. J. J. (*Actg. Capt.*)	1 July 17	20 June 18	I. A. R. O.
16 Aug. 16	Franklin, Lt. H. J.	16 Aug. 17	18 Feb. 18	1-17th Infy.
19 June 15	Sharp, Lt. C. J.	19 June 16	11 Sep. 17	I. A. (*On probn.*)
30 Jan. 17	Short, Lt. V.C.G. O'B.	30 Jan 18	19 Feb. 17	2nd Rajputs.
30 Jan. 17	Kilroy, 2nd-Lt. A. J. M.	30 Jan. 17	10 Oct. 17	122nd Infy.
8 Aug. 17	Sherard, 2nd-Lt. A. F. L.	8 Aug. 17	18 Feb. 18	I. A. R. O.
5 Oct. 17	Birthistle, 2nd-Lt. R. G. N.	5 Oct. 17	19 Jan. 18	I. A. R. O.
1 Nov. 17	Elis, Lt. S.	1 Nov. 18	14 Mar. 18	I. A. (*T. C.*).
27 July 18	Beard, 2nd-Lt. W. H. G.	27 July 18	3 June 18	M. G. Corps, I. A. (*On probn.*)
15 Oct. 18	Willmer, 2nd-Lt. E. J.	15 Oct. 18	21 Oct. 18	I. A. R. O.
21 Oct. 18	Haworth, 2nd-Lt. R.	21 Oct. 18	17 Nov. 18	I. A. R. O.
5 Nov. 18	Moir, 2nd-Lt. J. H.	5 Nov. 18	16 Nov. 18	I. A. R. O.
	Ellis, 2nd-Lt. A. R.			Glouc R., I. A. (*On probn.*)
	Wyer, 2nd-Lt. T. E.		26 Nov, 18	I. A. (*On probn.*)

Subadar-Major.

		Jemadar.	Subadar.	Subadar-Major.
16 June 94	Bakhtawar Singh, I.O.M. (56, 72a)	12 Feb. 08	16 Mar. 14	15 July 17.

Subadars.

15 Oct. 89	Arjan Singh (*Retd.*)	....	....	Re-employed, 13 July 15. Depôt.
11 Mar. 07	Dhan Singh	11 Mar. 07	1 Nov. 11	
16 Dec. 01	Mangal Singh	15 Feb. 13	1 May 15	
12 Apl. 00	Dal Singh (46)	1 May 15	18 May 16	
29 Mar. 98	Bhur Singh	4 Nov. 14	31 May 16	
16 Feb. 97	Fateh Din Khan (56, 66a)	13 May 15	6 Nov. 17	
25 Apl. 98	Indar Singh	14 Mar. 16	15 July 17	
1 July 04	Rukum Singh	14 Mar. 16	15 July 17	
25 Feb. 97	Gulsher Khan	1 Jan. 09	22 Nov. 17	

13th Rajputs (The Shekhawati Regiment)—*contd.*

1st Commn. or date of entering service.	Names and Rank.	Army rank.	Present appointment in Regiment.	Remarks.
	Jemadars.			
20 Dec. 98	Rur Singh	11 Sep. 14		
4 Dec. 95	Umrao Singh	9 Mar. 16		
9 Dec. 99	Maman Singh	31 May 16		
22 Aug. 09	Hukam Singh	31 May 16	Indian Adjt., 31 May 16.	
20 Mar. 00	Santa Singh	3 July 17		
29 May 04	Banji Singh (56)	3 July 17		
10 Nov. 00	Mohar Singh	3 July 17		
14 May 00	Shimbu Singh	3 July 17		
4 June 04	Hira Singh	15 July 17		
25 June 04	Balla Singh	15 July 17		
27 Oct. 06	Kallu Singh	22 Dec. 17		
23 Oct. 07	Jodh Singh	4 May 18		
20 Nov. 03	Panji Singh	6 Sep. 18		
1 Mar. 05	Harnarain Singh	6 Sep. 18		

Regt. Centre—Multan.
Linked with—15th and 45th Sikhs.

Uniform—Scarlet.
Facings—Yellow.

14th King George's Own Ferozepore Sikhs.

Raised at Ferozepore in 1846, as 'The Regiment of Ferozepore,' by Maj. Mackeson and Capts. Watt and Tebbs. Became the 15th Regiment of Bengal Native Infantry, 1861; the 14th Regiment of Bengal Native Infantry, 1861; the 14th (the Ferozepore) Regiment of Bengal Native Infantry, 1864; the 14th Regiment of Bengal Infantry (The Ferozepore Sikhs), 1885; the 14th (Ferozepore) Sikh Infantry, 1901; the 14th Ferozepore Sikhs in 1903; 14th Prince of Wales's Own Ferozepore Sikhs, 1906; present designation, 1910.

THE PLUME OF THE PRINCE OF WALES. THE ROYAL AND IMPERIAL CYPHER.

"Lucknow" "Ali Masjid" "Afghanistan, 1878-79"

"Defence of Chitral" "China, 1900."

Composition—4 Companies Sikhs.

COLONEL-IN-CHIEF—THE KING.

COLONEL—

1st Commn. or date of entering service.	Names and Rank.	Army rank.	Present appointment in Regiment.	Remarks.
	Commandant.			
5 Dec. 91	Earle, Lt.-Col. E. S.	8 Aug. 16	8 Aug. 16	
	Company Commanders. (4)			
4 Aug. 97	Franklin, Bt.-Lt.-Col. H. S. E., D.S.O. (*temp. Br.-Genl.*).	3 June 16	8 Aug. 16	2nd-in-Comd.
27 July 98	Talbot, Bt.-Lt.-Col. F. E. G., *p.s.c.*	1 Jan. 17	22 Dec. 09	With 3-23rd Sikh Infy.
8 May 01	Daniell, Maj. J. A. S., D.S.O.	8 May 16	14 May 15	
21 Jan. 03	Channer, Maj. G.	21 Jan. 18	5 June 15	2nd-in-Comd. (*temp.*)
	Company Officers. (9)			
10 Oct. 03	Wace, Maj. M.	10 Oct. 18	30 July 05	With 1-151st Infy.
14 Jan. 14	Savory, Capt. R. A., M.C.	14 Jan. 18	10 Oct. 14	Adjt.
15 Aug. 14	*Christie, Capt. D. A.*	*15 Aug. 18*	*13 Nov. 15*	*With I. S. Troops.*
15 Aug. 14	Bunbury, Capt. G. F.	15 Aug. 18	18 Nov. 15	
18 Apl. 16	Winthrop, Lt. H. E.	18 Apl. 17	25 Apl. 16	Qr. Mr. (*Offg.*)
18 Apl. 16	Pybus, Lt. G. D.	18 Apl. 17	25 Apl. 16	With 45th Sikhs.
29 June 16	O'Connor, Lt. K. K.	29 June 17	8 Mar. 17	
29 June 16	Cruickshank, Lt. A. S.	29 June 17	13 Feb. 17	With 53rd Sikhs.
29 June 16	Cameron-Webb, Lt. J. H.	29 June 17	26 Feb. 17	
30 Jan. 17	Church, Lt. C. J. B.	30 Jan. 18	3 Sep. 17	Attd.
19 Nov. 15	Spankie, Lt. H. V.	19 Nov. 16	12 June 18	Attd.
31 Aug. 18	Darnell, 2nd-Lt. W. R.	31 Aug. 18	3 Sep. 18	Attd.
	Attached.			
18 Jan. 05	Finnis, Capt. H. C.	18 Jan. 14	. . . .	*Poll. Dept.*
4 Dec. 14	Crocker, Lt. W. J. (*Actg. Capt.*)	4 Dec. 15	25 Mar. 16	I. A. (*on probn.*)
5 May 15	Milward, Lt. R. G.	5 May 16	1 May 18	Notts. & Derby. R. I. A. (*on probn.*)
11 Mar. 15	Booth, Lt. A. H. (*Ating. Capt.*)	11 Dec. 16	30 Nov. 17	I. A. (*on probn.*) Adjt. (*Offg.*)
9 Jan. 16	Palk, Lt. L. C.	9 Jan. 17	27 Jan. 18	N. Staff. R. I. A. (*on probn.*)
26 July 15	Maer, Lt. C. A.	10 Mar. 17	26 July 15	
27 Sep. 16	Woods, Lt. J. McB.	27 Sep. 17	24 May 18	I. A. (*on probn.*)
6 Oct. 16	Blair, Lt. F. R.	6 Oct. 17	13 Aug. 17	I. A. R. O.
5 May 17	Humphreys, Lt. D. W. G.	5 May 18	29 Sep. 17	I. A. R. O.
28 Feb. 18	Marigold, 2nd-Lt. A. J.	28 Feb. 18	24 Apl. 18	Oxf. & Bucks. L. I., I. A. (*on probn.*)
29 Mar. 18	Harman, 2nd-Lt. F. A.	29 Mar. 18	15 July 18	I. A. (*T. C.*)
7 June 18	Strip, 2nd-Lt. R. A.	7 June 18	11 Oct. 18	I. A. R. O.
7 Oct. 18	Hedger, 2nd-Lt. H.	7 Oct. 18	2 Nov. 18	I. A. (*T. C.*)
7 Oct. 18	Lord, 2nd-Lt. A. J.	7 Oct. 18	2 Nov. 18	I. A. (*T. C.*)

Subadar-Major.

		Jemadar.	Subadar.	Subadar-Major.	
20 Apl. 87	Sham Singh *Bahadur*, I.D.S.M. (56, 66*a*).	1 Nov. 07	24 Sep. 12	5 June 15	*Order of Br. Ind., 2nd Cl.*
	Bhagwan Singh, I. O. M. (*Hony. Capt.*) (56).	. . . .	. . . .	. . .	*Order of Br. Ind. 1st cl.* Re-employed 24 June 15.
	Subadars.				
27 Jan. 07	Narayan Singh, I.D.S.M. (56)	27 Jan. 07	17 Aug. 12		
15 Apl. 93	Jaimal Singh	1 Nov. 07	1 Nov. 12		
1 July 90	Wazir Singh	16 May 12	16 Oct. 14		
7 June 96	Sant Singh (58)	16 Oct. 14	15 Mar. 16		
7 July 94	Phuman Singh	16 Oct. 14	29 July 17		
1 Mar. 18	Jaimal Singh	1 Apl. 15	1 Mar. 18		
24 July 95	Ghamna Singh	5 June 15	16 May 18		
11 Jan. 96	Jaswant Singh	6 July 15	16 May 18		
14 Sep. 99	Jafar Shah	1 Sep. 17	10 June 18		
	Kehar Singh (*Retd.*)	. . . .	. . . .		*Order of Br. Ind., 2nd cl.* Re-employed, 14 Mar. 15 Depot.

13th Rajputs (The Shekhawati Regiment)—*contd.*

1st Commn. or date of entering service.	Names and Rank.	Army rank.	Present appointment in Regiment.	Remarks.
	Jemadars.			
20 Dec. 98	Rur Singh	11 Sep. 14		
4 Dec. 95	Umrao Singh	9 Mar. 16		
9 Dec. 99	Maman Singh	31 May 16		
22 Aug. 09	Hukam Singh	31 May 16	Indian Adjt., 31 May 16.	
20 Mar. 00	Santa Singh	3 July 17		
29 May 04	Banji Singh (56)	3 July 17		
10 Nov. 00	Mohar Singh	3 July 17		
14 May 00	Shimbu Singh	3 July 17		
4 June 04	Hira Singh	15 July 17		
25 June 04	Balla Singh	15 July 17		
27 Oct. 06	Kallu Singh	22 Dec. 17		
23 Oct. 07	Jodh Singh	4 May 18		
20 Nov. 03	Panji Singh	6 Sep. 18		
1 Mar. 05	Harnarain Singh	6 Sep. 18		

Regt. Centre—Multan.
Linked with—15th and 45th Sikhs.

Uniform—Scarlet.
Facings—Yellow.

14th King George's Own Ferozepore Sikhs.

Raised at Ferozepore in 1846, as 'The Regiment of Ferozepore,' by Maj. Mackeson and Capts. Watt and Tebbs. Became the 15th Regiment of Bengal Native Infantry, 1861; the 14th Regiment of Bengal Native Infantry, 1861; the 14th (the Ferozepore) Regiment of Bengal Native Infantry, 1864; the 14th Regiment of Bengal Infantry (The Ferozepore Sikhs), 1885; the 14th (Ferozepore) Sikh Infantry, 1901; the 14th Ferozepore Sikhs in 1903; 14th Prince of Wales's Own Ferozepore Sikhs, 1906; present designation, 1910.

THE PLUME OF THE PRINCE OF WALES. THE ROYAL AND IMPERIAL CYPHER.
"Lucknow" "Ali Masjid" "Afghanistan, 1878-79"
"Defence of Chitral" "China, 1900."
Composition—4 Companies Sikhs.

COLONEL-IN-CHIEF—THE KING.

COLONEL—

1st Commn. or date of entering service.	Names and Rank.	Army rank.	Present appointment in Regiment.	Remarks.
	Commandant.			
5 Dec. 91	Earle, Lt.-Col. E. S.	8 Aug. 16	8 Aug. 16	
	Company Commanders. (4)			
4 Aug. 97	Franklin, Bt.-Lt.-Col. H. S. E., D.S.O. (*temp. Br.-Genl.*).	3 June 16	8 Aug. 16	2nd-in-Comd.
27 July 98	Talbot, Bt.-Lt.-Col. F. E. G., *p.s.c.*	1 Jan. 17	22 Dec. 09	With 3-23rd Sikh Infy.
8 May 01	Daniell, Maj. J. A. S., D.S.O.	8 May 16	14 May 15	
21 Jan. 03	Channer, Maj. G.	21 Jan. 18	5 June 15	2nd-in-Comd. (*temp.*)
	Company Officers. (9)			
10 Oct. 03	Wace, Maj. M.	10 Oct. 18	30 July 05	With 1-151st Infy.
14 Jan. 14	Savory, Capt. R. A., M.C.	14 Jan. 18	10 Oct. 14	Adjt.
15 Aug. 14	*Christie, Capt. D. A.*	*15 Aug. 18*	*13 Nov. 15*	*With I. S. Troops.*
15 Aug. 14	Bunbury, Capt. G. F.	15 Aug. 18	18 Nov. 15	
18 Apl. 16	Winthrop, Lt. H. E.	18 Apl. 17	25 Apl. 16	Qr. Mr. (*Offg.*)
18 Apl. 16	Pybus, Lt. G. D.	18 Apl. 17	25 Apl. 16	With 45th Sikhs.
29 June 16	O'Connor, Lt. K. K.	29 June 17	8 Mar. 17	
29 June 16	Cruickshank, Lt. A. S.	29 June 17	13 Feb. 17	With 53rd Sikhs.
29 June 16	Cameron-Webb, Lt. J. H.	29 June 17	26 Feb. 17	
30 Jan. 17	Church, Lt. C. J. B.	30 Jan. 18	3 Sep. 17	Attd.
19 Nov. 15	Spankie, Lt. H. V.	19 Nov. 16	12 June 18	Attd.
31 Aug. 18	Darnell, 2nd-Lt. W. R.	31 Aug. 18	3 Sep. 18	Attd.
	Attached.			
18 Jan. 05	Finnis, Capt. H. C.	18 Jan. 14	. . . .	*Poll. Dept.*
4 Dec. 14	Crocker, Lt. W. J. (*Actg. Capt.*)	4 Dec. 15	26 Mar. 16	I. A. (*on probn.*)
5 May 15	Milward, Lt. R. G.	5 May 16	1 May 18	Notts. & Derby. R. I. A. (*on probn.*)
11 Mar. 15	Booth, Lt. A. H. (*Ating. Capt.*)	11 Dec. 16	30 Nov. 17	I. A. (*on probn.*) Adjt. (*Offg.*)
9 Jan. 16	Palk, Lt. L. C.	9 Jan. 17	27 Jan. 18	N. Staff. R. I. A. (*on probn.*)
26 July 15	Maer, Lt. C. A.	10 Mar. 17	26 July 15	
27 Sep. 16	Woods, Lt. J. McB.	27 Sep. 17	24 May 18	I. A. (*on probn.*)
6 Oct. 16	Blair, Lt. F. R.	6 Oct. 17	13 Aug. 17	I. A. R. O.
5 May 17	Humphreys, Lt. D. W. G.	5 May 18	29 Sep. 17	I. A. R. O.
28 Feb. 18	Marigold, 2nd-Lt. A. J.	28 Feb. 18	24 Apl. 18	Oxf. & Bucks. L. I., I. A. (*on probn.*)
29 Mar. 18	Harman, 2nd-Lt. F. A.	29 Mar. 18	15 July 18	I. A. (*T. C.*)
7 June 18	Strip, 2nd-Lt. R. A.	7 June 18	11 Oct. 18	I. A. R. O.
7 Oct. 18	Hedger, 2nd-Lt. H.	7 Oct. 18	2 Nov. 18	I. A. (*T. C.*)
7 Oct. 18	Lord, 2nd-Lt. A. J.	7 Oct. 18	2 Nov. 18	I. A. (*T. C.*)

Subadar-Major.

		Jemadar.	Subadar.	Subadar-Major.	
20 Apl. 87	Sham Singh *Bahadur*, I.D.S.M. (56, 66*a*).	1 Nov. 07	24 Sep. 12	5 June 15	*Order of Br. Ind., 2nd Cl.*
	Bhagwan Singh, I. O. M. (*Hony. Capt.*) (56).	. . . .	. . . .	. . .	*Order of Br. Ind. 1st cl.* Re-employed 24 June 15.

Subadars.

27 Jan. 07	Narayan Singh, I.D.S.M. (56)	27 Jan. 07	17 Aug. 12	
15 Apl. 93	Jaimal Singh	1 Nov. 07	1 Nov. 12	
1 July 90	Wazir Singh	16 May 12	16 Oct. 14	
7 June 96	Sant Singh (58)	16 Oct. 14	15 Mar. 16	
7 July 94	Phuman Singh	16 Oct. 14	29 July 17	
1 Mar. 18	Jaimal Singh	1 Apl. 15	1 Mar. 18	
24 July 95	Ghamna Singh	5 June 15	16 May 18	
11 Jan. 96	Jaswant Singh	6 July 15	16 May 18	
14 Sep. 99	Jafar Shah	1 Sep. 17	10 June 18	
	Kehar Singh (*Retd.*)	. . . .	. . . .	*Order of Br. Ind., 2nd cl.* Re-employed, 14 Mar. 15 Depot.

14th King George's Own Ferozepore Sikhs—*contd.*

1st Commn. or date of entering service.	Names and Rank.	Army rank.	Remarks.
	Jemadars.		
24 June 94	Partab Singh (66*a*)	16 Oct. 14	
13 Oct. 91	Mihan Singh (56)	16 Oct. 14	
4 July 94	Mewa Singh	28 Jan. 15	
4 July 89	Bhagwan Singh	1 July 16	*Burma Mily. Police.*
4 Dec. 99	Sarmukh Singh	4 Sep. 17	
15 June 97	Mastan Singh	1 Oct. 17	
4 Mar. 99	Amar Singh	1 Oct. 17	
10 May 02	Bogh Singh (56)	25 Nov. 17	
15 Nov. 98	Sawan Singh	11 Jan. 18	
18 Feb. 03	Magar Singh	11 Jan. 18	
22 Aug. 03	Nihal Singh	1 Mar. 18	*Burma Mily. Police,* Indian Adjt. (*Actg.*)
19 Mar. 05	Sapuran Singh	16 May 18	
13 May 05	Nand Singh	16 May 18	
1 Mar. 18	Sher Muhammad	1 Mar. 18	(*On probn.*)
28 Jan. 10	Nur Din	1 July 18	

Regt. Centre—Multan. *Uniform*—Scarlet.
Linked with—14th and 45th Sikhs. *Facings*—Emerald Green.

1st Battalion, 15th Ludhiana Sikhs.

Raised at Ludhiana in 1846, by Major P. Gordon, as 'The Regiment of Ludhiana.' Became the 16th Regiment of Bengal Native Infantry, 1861; the 15th Regiment of Bengal Native Infantry, 1861; the 15th (The Ludhiana) Regiment of Bengal Native Infantry, 1864; the 15th Regiment of Bengal Infantry (The Ludhiana Sikhs), 1885; the 15th (Ludhiana) Sikh Infantry, 1901; present designation, 1903.

"China, 1860-62" "Ahmad Khel" "Kandahar, 1880" "Afghanistan, 1878-80."
"Tofrek" "Suakin, 1885" "Chitral"
"Tirah" "Punjab Frontier."
Composition—4 Companies Sikhs.

COLONEL—Brigadier-General Henry A. Abbott, C.B. . . . 14 January 1913.

HONORARY COLONEL—*Hony.* Major-General *His Highness Maharajadhiraja* SRI *Sir* BHUPINDAR SINGH MAHINDAR *Bahadur*, G.C.I.E., G.B.E., OF PATIALA ....15 June 1918.

1st Commn. or date of entering service.	Names and Rank.	Army rank.	Present appointment in Regiment.	Remarks.
	Commandant.			
22 Aug. 88	Gordon, Lt.-Col. J. L. R., C.B. .	22 Aug. 14	14 Jan. 18	
	Company Commanders. (4)			
1 Aug. 97	*Macmullen, Bt.-Col. C. N.,* D.S.O., *p. s. c. (temp. Br.-Gen.)*	*1 Jan. 18*	*14 Jan. 18*	*2nd-in-Comd.*
16 Feb. 98	Hughes, Maj. C. F. W., M. C. (*Actg. Lt.-Col.*)	1 Sep. 15	25 Aug. 13	Comdt. (*temp.*)
3 Aug. 98	*Muir, Maj. W. W.*	*1 Sep. 15*	*8 Aug. 16*	*A.D.C. to H.E. the C.-in-C. in India.*
25 Jan. 99	*Macleod, Maj. N.,* D.S.O., *p.s.c.*	*1 Sep. 15*	*14 Jan. 18*	*Bde.-Maj., Rangoon.*
	Company Officers. (9)			
8 May 01	*MacBrayne, Maj. R. J.,* M.C. .	*8 May 16*	*2 Nov. 02*	*I. S. Troops.*
18 Jan. 05	*Muir, Capt. A. H. H. (temp. Maj.)*	*18 Jan. 14*	*9 Aug. 06*	*Coy. Comdr., Cadet Coll., Quetta.*
25 Jan. 08	*Barstow, Capt. A. E.,* M.C. . .	*1 Sep. 15*	*11 Mar. 09*	*Coy. Offr., Tg. Sch., Indore.*
20 Jan. 09	Betham, Capt. G. L.	1 Sep. 15	18 Mar. 10	2nd-in-Comd. (*temp.*)
29 Jan. 10	*Henderson, Capt. G. B.* . .	*1 Sep. 15*	*12 Mar. 11*	*With I. S. Troops.*
20 Jan. 12	*Turner, Capt. C. J. R. (Actg. Maj.)*	*20 Jan. 16*	*27 Mar. 13*	*With Sig. Serv.*
24 Aug. 12	VC Smyth, Capt. J. G. . . .	24 Aug. 16	5 Nov. 13	*Bde.-Maj., 43rd Bde.*
19 July 16	*Farwell, Lt. A. E.*	*19 July 17*	*11 Nov. 16*	*With Frontier Militia.*
16 Aug. 16	Carter, Lt. J. L.	16 Aug. 17	6 Nov. 16	Attd.
18 June 17	*Carter, Lt. R. F. F.*	*18 June 18*	*25 June 17*	*Attd. with Frontier Militia.*
18 June 17	Eastmead, Lt. C.	18 June 18	25 June 17	Attd. with 1-151st Infy. *Lv., ex I., [m.c.*
21 Dec. 17	Dalgety, 2nd-Lt. A. E. . . .	21 Dec. 17	22 Mar. 18	Attd.
31 Jan. 18	*Finlay, 2nd-Lt. J. R. P.* . .	*3 Jan. 17*	*7 Feb. 18*	*Attd. with Malay States Guides.*
31 Jan. 18	Guise, 2nd-Lt. C. P.	31 Jan. 18	7 Feb. 18	Attd.
24 Apl. 18	Dinwiddie, 2nd-Lt. H. W. . .	24 Apl. 18	21 Oct. 18	I. A. (U. L.)
31 Aug. 18	McDougall, 2nd-Lt. R. M. . .	31 Aug. 18	5 Sep. 18	Attd.
	Attached.			
18 Aug. 14	Nicholas, Capt. R. H.	18 Aug. 17	26 Sep. 18	I. A. R. O.
21 Oct. 14	Williams, Capt. V. S. M. . .	21 Oct. 18	29 Jan. 18	I. A. R. O.
6 Nov. 14	Keatinge, Lt. C. A. (*Actg. Capt.*)	6 Nov. 15	6 Feb. 18	I. A. R. O.
2 Dec. 15	Thacker, Lt. N., M.C.	2 Dec. 16	10 Apl. 18	Glouc. R., I. A. (*on probn.*). Adjt. (*temp.*)
25 June 16	Rutherford, Lt. J. S.	25 June 17	2 Nov. 18	Oxf. & Bucks. L. I., I. A. R. O. (*on probn.*)
1 Aug. 16	Hawken, Lt. G. H.	1 Aug. 17	19 Jan. 18	Bedf. R., I. A. (*on probn.*). Adjt.
20 Aug. 16	Morell, Lt. D. W.	20 Aug. 17	28 Jan. 18	Notts. & Derby. R., I. A. (*on probn.*) Qr.-Mr. (*temp.*)
26 Sep. 17	MacBeth, 2nd-Lt. C. J. S. . .	26 Sep. 17	28 Jan. 18	I. A. R. O.
29 Mar. 18	Robinson, 2nd-Lt. A. E. . . .	29 Mar. 18	15 July 18	I. A. (*T.C.*)
7 June 18	Swinburn, 2nd-Lt. H. R. . . .	7 June 18	22 Oct. 18	I. A. *T. C.*)
7 June 18	Fry, 2nd-Lt. E. N.	7 June 18	12 Oct. 18	I. A. R. O.
15 June 18	Bunt, 2nd-Lt. H.	15 June 18	15 Sep. 18	I. A. R. O. (*on probn.*)
7 Oct. 18	Spreadbrow, 2nd-Lt. F. J. . .	7 Oct. 18	3 Dec. 18	I. A. (*T. C.*)

Subadar-Major.

		Jemadar.	Subadar.	Subadar-Major.
18 Sep. 02	Lachhman Singh (66*a*, 72*a*, 162, 170)	18 Sep. 02	29 Jan. 07	1 Apl. 17

Subadars.

13 Dec. 90	Phuman Singh, I.O.M., I.D.S.M. (56, 66*a*) .	1 Apl. 09	1 July 14
1 Sep. 92	Bhagwan Singh (66*a*)	16 Apl. 12	13 Mar. 15
14 Sep. 95	Bishan Singh (56, 66*a*)	8 Aug. 14	3 Apl. 15
22 June 94	Muizullah Khan (56, 66*a*) . . .	13 Jan. 15	30 May 17
8 Jan. 96	Gurdit Singh (56)	8 Aug. 14	1 July 18
17 July 97	Dhanna Singh (46, 66*a*) . .	27 Oct 14	1 July 18

1st Battalion, 15th Ludhiana Sikhs—*contd.*

1st Commn. or date of entering service.	Names and Rank.	Army rank.	Remarks.
	Jemadars.		
21 July 93	Sundar Singh (56)	10 Mar. 15	
20 Jan. 98	Hira Singh (56)	3 Apl. 15	Indian Adjt., 20 May 15.
30 Mar. 98	Sucha Singh (56)	19 May 15	
10 Jan. 95	Mewa Singh (66*a*)	13 Mar. 16	
15 June 93	*Gopal Singh (66a)*	*1 July 16*	
1 May 98	Bishan Singh	2 July 17	
17 Oct. 98	Jaimal Singh	1 Aug. 17	
24 Dec. 98	Kishan Singh	1 Sep. 17	Indian Qr.-Mr., 3 Dec. 17.
28 Nov. 99	Bagh Singh (66*a*)	1 Sep. 17	
24 Dec. 97	Harnam Singh (72*a*)	1 May 18	
20 Apl. 97	Sardara Singh	1 July 18	
20 Aug. 02	Puran Singh	1 July 18	
1 Apl. 03	Bachan Singh	1 July 18	
23 Mar. 02	Mastan Singh	1 July 18	
30 May 17	Muhammad Husain	1 July 18	
29 Apl. 17	Mirza	1 July 18	
17 Nov. 14	Imam Din	24 July 18	

2nd Battalion, 15th Ludhiana Sikhs.

Formed at Hyderabad-Sind, on 6th June 1918.

Composition—2 Companies Sikhs, 2 Punjabi Musalmans.

1st Commn. or date of entering service.	Names and Rank.	Army rank.	Present appointment in Regiment.	Remarks.
	Commandant.			
4 Jan. 99	Mc Rae, Maj. H. St. G. M., D.S.O., *p.s.c.* (*Actg. Lt.-Col.*)	1 Sep. 15	18 Nov. 18	45th Sikhs.
	Company Commanders (4).			
29 July 04	Burmester, Capt. Z. G. (*Actg. Maj.*)	29 July 13	10 June 18	31st Lrs., 2nd-in-Comd. (*temp.*)
	Company Officers.			
21 June 15	Banks, Lt. R. E. (*Actg. Capt.*)	21 June 16	30 Aug. 18	I. A. (*on probn.*). Adjt. (*temp.*)
26 Nov. 15	Alteridge, Lt. L. F.	26 Nov. 16	24 Sep. 18	E. Kent R., I. A. (*on probn.*
3 Apl. 16	Walker, Lt. A. R. (*Actg. Capt*)	3 Apl. 17	29 July 18	Attd.
11 July 16	Edwards, Lt. M. F.	11 July 17	7 Aug. 17	Bedf. R., I. A. (*on probn.*)
26 Sep. 16	Osborne, Lt. C. A.	26 Sep. 17	. . .	I. A. (*on probn.*)
13 Aug. 17	Whiting, Lt. W. C.	13 Aug. 18	26 Aug. 18	R. Innis. Fus., I. A. (*on probn.*)
31 Oct. 17	Parry, Lt. R. E.	31 Oct. 18	24 June 18	I. A. R. O.
1 Nov. 17	Burgess, 2nd-Lt. B., M. M.	1 Nov. 17	24 Aug. 18	I. A. (*T. C.*)
5 Nov. 17	Edwards, Lt. E.	5 Nov. 18	26 Aug. 18	I. A. R. O. (*on probn.*)
15 Jan. 18	Otter, 2nd-Lt. R. B.	15 Jan. 18	26 Aug. 18	I. A. (*T. C.*)
27 Mar. 18	Young, 2nd-Lt. J. W. F.	27 Mar. 18	10 Sep. 18	R. W. Fus., I. A. (*on probn.*)
7 June 18	Nichols, 2nd-Lt. D.	7 June 18	18 Oct. 18	I. A. (*T. C.*)
15 Aug. 18	Steel, 2nd-Lt. E.	15 Aug. 18	11 Sep. 18	I. A. (*T. C.*)
31 Aug. 18	Gordon, 2nd-Lt. R.	31 Aug. 18	2 Sep. 18	Attd.
31 Aug. 18	Carles, 2nd-Lt. J. R.	31 Aug. 18	2 Sep. 18	Attd.
	Davidson, Lt. H. C.	. . . .	. . . .	I. A. (*on probn.*)
22 July 18	Longthorn, 2nd-Lt. J.	22 July 18	2 Dec. 18	I. A. (*T. C.*)

Subadar-Major.

		Jemadar.	Subadar.	Subadar-Major.
17 Mar. 85	Narayan Singh (56, 66*a*)	16 Dec. 06	21 July 11	3 July 18

Subadars.

25 Feb. 93	Maima Singh (66*a*)	21 July 11	10 Mar. 15
11 Dec. 96	Bir Singh, M.C., I.D.S.M. (56, 60, 66*a*)	8 Aug. 14	19 May 15
1 Feb. 01	Narayan Singh, I. O. M.	28 Oct. 14	6 Nov. 17
3 Aug. 14	Khan Bahadur	21 May 18	1 July 18
8 June 17	Ghulam Hussain	21 May 18	1 July 18

Jemadars.

22 July 97	Albel Singh	10 Mar. 15
13 Nov. 95	Basant Singh, I.O.M.	21 Mar. 15
28 Jan. 01	Waryam Singh	1 July 18
5 Mar. 02	Budh Singh	1 July 18

Regt. Centre—Agra. *Uniform*—Scarlet.
Linked with—2nd, 4th and 13th Rajputs. *Facings*—White.

16th Rajputs (The Lucknow Regiment).

Bailey Guard Paltan.—Formed at Cawnpore in 1857, by Col. H. Palmer, from the faithful remains of the 13th, 48th and 71st Bengal Native Infantry, and designated 'The Regiment of Lucknow.' Became the 20th Regiment of Bengal Native Infantry, 1861; the 16th Regiment of Bengal Native Infantry, 1861; the 16th (The Lucknow) Regiment of Bengal Native Infantry, 1864; the 16th (The Lucknow) Regiment of Bengal Infantry, 1885; the 16th (The Lucknow) Rajput Regiment of Bengal Infantry, 1897; the 16th (Lucknow) Rajput Infantry, 1901; present designation, 1903.

"Lucknow" (With a Turretted Gateway) "Afghanistan, 1879-80" "Burma, 1885-87."
Composition—4 Companies Rajputs.

1st Commn. or date of entering service.	Names and Rank.	Army rank.	Present appointment in Regiment.	Remarks.
	Commandant.			
11 Feb. 88	Vanrenen, Lt.-Col. G. R.	11 Feb. 14	9 May 16	
	Company Commanders. (4)			
22 Jan. 96	Simpson, Bt.-Lt.-Col. J. C.	3 June 17	9 May 16	2nd-in-Comd. With 2nd Rajputs.
4 Dec. 01	Greville, Maj. B. S. A. F.	4 Dec. 16	19 Nov. 10	
18 Jan. 02	Garratt, Maj. B. F.	18 Jan. 17	23 Mar. 16	2nd-in-Comd. (*Offg.*)
27 Aug. 02	Masters, Maj. J.	27 Aug. 17	20 Oct. 18	
	Company Officers. (9)			
9 Jan. 04	Grose, Capt. E.	9 Jan. 13	14 Oct. 06	With 73rd Infy. *Lv., ex I., m. c.*
13 Aug. 04	*Pulley, Capt. O. C. (temp. Maj.)*	*13 Aug. 13*	*27 Apl. 07*	*Asst. Comdt., Prisoners' War Camp.*
29 Aug. 06	Secondé, Capt. E. C.	29 Aug. 15	31 Oct. 08	
29 Aug. 06	Hume-Wright, Bt. Maj. F. J. O.	3 June 18	8 Sep. 11	With 7th Rajputs.
16 Feb. 07	*Herapath, Capt. C. B.*	*1 Sep. 15*	*6 Mar. 11*	*S. & T. Corps.*
20 Jan. 09	Hobson, Capt. J. W. S.	1 Sep. 15	25 Mar. 10	D. A. A. & Q. M. G.
21 Jan. 13	Blaxland, Capt. A. B.	21 Jan. 17	8 Nov. 15	Adjt. (*Offg.*)
13 Jan. 15	Laville, Lt. F. A. J.	13 Jan. 16	17 Mar. 15	
15 Nov. 15	Bromham, Lt. A. C.	15 Nov. 16	19 Nov. 15	Attd. with 4th Rajputs.
21 Dec. 17	Robinson, 2nd-Lt. J. G. K.	21 Dec. 17	21 Mar. 18	Attd.
30 May 16	Green, Lt. W. G.	30 May 17	26 Nov. 17	Attd.
6 Feb. 16	Dove, Lt. M.	6 Feb. 17	26 Nov. 17	Attd.
	Attached.			
12 May 16	Henderson, Lt. R. S.	12 May 17	16 Aug. 16	I. A. R. O.
29 June 16	Miller, Lt. G. R. B.	29 June 17	1 July 16	2nd Rajputs.
11 Aug. 16	Baker, Lt. F. C.	11 Aug. 17	25 Aug. 16	I. A. R. O.
18 Feb. 15	Rix, Lt. H. V.	18 Feb. 16	31 Oct. 16	I. A. R. O. *Lv., ex I., 6 mos., m. c.*
22 Mar. 15	Fagan, Lt. A. W.	22 Mar. 16	24 Nov. 16	I. A. R. O. Qr.-Mr. (*Offg.*)
29 July 16	Crocker, Lt. A.	29 July 17	16 Sep. 16	I. A. R. O.
22 Jan. 15	Woosnam, Lt. A. G. (*Actg. Capt.*)	22 Jan. 16	10 Sept. 17	I. A. R. O.
9 Apl. 15	Jack, Lt. M. M.	9 Apl. 16	18 May 18	I. A. R. O.
5 Nov. 17	Steele, Lt. H.	5 Nov. 18	14 Mar. 18	I. A. (*T. C.*)
10 Mar. 16	Carew-Hughes, Lt. C. F.	10 Mar. 17	10 July 18	Oxf. and Bucks. L. I., I. A. (*on probn.*)
15 June 18	Beale, 2nd-Lt. W. F. B.	15 June 18	8 July 18	I. A. R. O.
1 June 18	Davies, Lt. G. C.	1 June 16	2 Oct. 18	R. W. Fus., I. A. R. O. (*on probn.*)
6 Sep. 18	Andrews, 2nd-Lt. D. C.	6 Sep. 18	30 Sep. 18	I. A. R. O.
30 Aug. 18	Atkinson, 2nd-Lt. A. A.	30 Aug. 18	1 Oct. 18	I. A. R. O.
7 Oct. 18	Bowman, 2nd-Lt. W.	7 Oct. 18	22 Oct. 18	I. A. R. O.
26 June 18	Johnston, 2nd-Lt. A. J. S.	26 June 18	1 Nov. 18	R. Scots, I. A. (*on probn.*)

Subadar-Major.

		Jemadar.	Subadar.	Subadar-Major.
2 June 00	Lachhman Singh (56, 72*a*).	2 June 00	1 Mar. 08	16 Aug. 15.
	Subadars.			
15 May 98	Narayan Singh (56, 66*a*)	15 May 98	1 Jan. 07	
30 July 88	Sakaldip Singh (56)	23 Feb. 01	1 Apl. 08	
2 Nov. 98	Ram Singh (56)	2 Nov. 98	1 Sep. 09	
15 July 02	Ram Lal Singh (56, 72*a*)	15 July 02	16 July 11	With 2-4th Rajputs.
25 Aug. 88	Bulaki Singh (56, 66*a*)	1 Jan. 07	14 May 15	
11 July 90	Kan Singh (56)	1 Mar. 08	17 May 15	
15 Mar. 94	Surju Singh (56)	1 Oct. 10	2 June 15	
10 Apl. 95	Mohan Singh (56) I.	13 June 11	3 Aug. 15	
10 Mar. 96	Mahadeo Singh (I) (56, 58, 66*a*)	1 Apl. 13	15 Aug. 15	
19 Sep. 01	Hari Singh (56)	5 Apl. 15	16 Aug. 15	
31 Aug. 98	*Mohan Singh II*	*5 Apl. 15*	*10 Feb. 17*	
1 Oct. 98	*Puran Singh*	*5 Apl. 15*	*10 Feb. 17*	
9 Apl. 95	Mahadeo Singh (II)	14 May 15	24 July 17	
....	Saltanat Bahadur Singh (*Ret.*)	.....	....	Re-employed, 1 Oct. 17.

16th Rajputs (The Lucknow Regiment)—*contd.*

1st Commn. or date of entering service.	Names.	Army rank.	Remarks.
	Jemadars.		
8 July 96	Shankar Singh	14 May 15	
17 July 99	*Nawal Singh*	*17 May 15*	*With Bullock Corps.*
2 Dec. 96	Shiunath Singh	17 May 15	
7 Apl. 07	Merai Singh	2 June 15	
10 Jan. 99	*Jwana Singh*	*2 June 15*	
80 Oct. 01	Prag Singh (56)	2 June 15	
10 Nov. 98	Budha Singh	3 Aug. 15	
21 June 00	Chiragh Singh	3 Aug. 15	
23 Aug. 98	Mansha Singh	24 July 17	
20 Aug. 90	Sarabjit Singh	1 Sep. 17	
10 Dec. 99	*Dharam Singh*	*1 Sep. 17*	*With Sig. Ser.*
16 Sep. 01	Sohan Singh	1 Sep. 17	
4 July 04	Muhabbat Singh	31 Aug. 17	
23 Feb. 04	Jagmal Singh	3 Dec. 17	
	Thakur Surendra Bahadur Singh	....	*Temp. Hony.*

Regt. Centre—Benares. *Uniform*—Scarlet.
Linked with—5th and 18th Infy. *Facings*—White.

1st Battalion, 17th Infantry (The Loyal Regiment).

Wafadar Paltan.—Formed at Phillour in 1858, by Major J. C. Innes, from the faithful remains of the 3rd, 36th and 61st Bengal Native Infantry, and designated "The Loyal Purbiah Regiment." Became the 21st Regiment of Native Infantry, 1861; the 17th Regiment of Bengal Native Infantry, 1861; the 17th (The Loyal Purbiah) Regiment of Bengal Native Infantry, 1864 the 17th (The Loyal Purbiah) Regiment of Bengal Infantry, 1885; the 17th (The Loyal Regiment) of Beng l Infantry, 1898; the 17th Musalman Rajput Infantry (The Loyal Regiment), 1902; present designation, 19 3.

"Afghanistan, 1879-80" "Tofrek" "Suakin, 1885."

Composition—4 Compa ies Musalmans of the Eastern Punjab and Hindustan.

COLONEL—Major-Gener l Lawrence J. E. Bradshaw, C.B. . . 26 June 1908.

1st Commn. or date of entering service.	Names and Rank.	Army rank.	Present appointment in Regiment.	Remarks.
	Commandant.			
21 Dec. 89	Nicholson, Lt.-Col. W. C. . . .	21 Dec. 15	10 Oct. 16	*Lv., ex I., 1 yr., m. c.*
	Company Commanders. (4)			
6 Mar. 95	Fagan, Maj. B. J., D.S.O. (*Actg. Lt. Col.*)	6 Mar. 13	10 Oct. 16	2nd-in-Comd. Comdt. (*temp.*)
20 July 98	Smithers, Maj. L. S. H. (*Actg-Lt.-Col.*)	1 Sep. 15	15 Mar. 11	With 2-17th Infy.
12 Aug. 99	Porter, Maj. F. G.	1 Sep. 15	5 Sep. 13	*Conv. Sec., Weillington.*
20 Jan. 00	Henderson, Maj. K. D. M., M.C.	1 Sep. 15	24 Mar. 14	With 2-5th Infy.
	Company Officers. (9)			
2 May 03	Morris, Maj. A. L.	2 May 18	8 Feb. 07	2nd-in-Comd. (*temp.*)
24 Jan. 06	Fenton, Capt. J. N.	24 Jan. 15	25 Sep. 11	With 44 Infy.
20 Dec. 06	Condon, Capt. W. E. H. . .	1 Sep. 15	28 Sep. 11	With 76th Punj.
25 Jan. 08	Barron, Capt. A. H. D., M.C.	1 Sep. 15	21 Apl. 09	*Lv., ex I., m.c.*
6 Mar. 15	Powell, Lt. L. F.	26 Dec. 16	30 Apl. 15	Attd.
26 Aug. 16	Franklin, Lt. H. J.	16 Aug. 17	1 Dec. 17	Attd. With 13th Rajputs.
8 Aug. 17	Jelley, 2nd-Lt. C. A.	8 Aug. 17	2 Nov. 17	Attd.
	Attached.			
5 May 15	Applewhaite, Lt. G. R. . . .	5 May 16	1 Apl. 8	I. A. (*on probn.*)
4 May 15	Robertson, Lt. A. Mac. L., M.C. .	14 May 16	23 Feb. 16	I. A. R. O.
5 Mar. 16	McCarthy, Lt. W. O. . . .	5 Mar. 17	30 May 18	I. A. (*on probn.*)
5 Jan. 18	Sheppard, 2nd-Lt. E. A. . .	15 Jan. 18	5 June 18	I. A. (*T. C.*)
14 June 16	Clarke, Lt. J. R. R.	14 June 17	17 July 18	A. S. C., I. A. (*on probn.*)
26 Oct. 18	Barrett, 2nd-Lt. E. D. E. . .	26 Oct. 18	4 Sep. 18	M. G. Corps, I. A. (*on probn.*)
26 Jan. 18	Ballantine, 2nd-Lt. J.	26 Jan. 18	17 Aug. 18	Gord. Highrs. I. A. (*on probn.*)
6 Oct. 18	Carter, 2nd-Lt. A. J.	6 Oct. 18	29 Oct. 18	I. A. (*T. C.*)
21 June 18	Smith, 2nd-Lt. J. A.	21 June 18	4 July 18	I. A. R. O.
	Subadar-Major.			
		Jemadar.	Subadar.	Subadar-Major.
8 Dec 99	Muhammad Sulaiman Khan (66a, 72a), I.O.M.	8 Dec. 99	1 June 02	1 Dec. 17.
	Subadars.			
16 Apl. 95	Ashraf Ali Khan (66a) . . .	1 June 02	1 Nov. 05	
18 Mar. 04	Ali Bahadur Khan (56, 72a) . .	18 Mar. 04	17 Apl. 11	
18 July 89	Zahur Khan (56, 66a) . . .	8 Dec. 04	24 June 11	
12 Dec. 95	Amir Khan (56, 66a)	20 Dec. 09	28 Oct. 15	
1 June 93	Hayat Khan (65, 66)	5 Sep. 10	18 June 16	
7 June 99	Fajar Ali Khan	21 Aug. 14	20 May 17	
	Mussallam Khan (*Retd.*) . . .	. . . . .	. . . .	Re-employed. 11 Apl. 16.
28 July 99	Fiaz Muhammad Khan . . .	28 Oct. 15	23 Oct. 17	
8 Feb. 97	Imam Ali Khan	2 June 15	5 Nov. 17	
1 Feb. 94	Qudratullah Khan, (56) . . .	9 Nov. 15	6 May 18	
1 Sep. 00	Tangul Khan	7 Jan. 16	6 May 18	
15 Sep. 02	Chand Khan	1 Feb. 17	6 May 18	
6 Oct. 03	Rahmat Khan	20 May 17	6 May 18	
15 Aug. 05	Mumtaz Khan	4 Nov. 17	10 July 18	
	Muhammad Bakhsh (*Retd.*) .	. . . . .	. . . . .	Re-employed 21 Oct. 18.
	Fateh Muhammad Khan *Retd.*)	. . . . .	. . . . .	Re-employed 25 Jan. 18.

1st Battalion, 17th Infantry (The Loyal Regiment)—*contd.*

1st Commn. or date of entering service.	Names and Rank.	Army rank.	Present appointment in Regiment.	Remarks.
	Jemadars.			
2 Oct. 96	Muhammad Ismail Khan (66*a*)	11 Apl. 11		
10 Apl. 01	Yakub Khan	11 Mar. 17		
7 May 98	Karam Khan	4 July 17		
27 Apl. 03	Ali Muhammad Khan	4 July 17		
13 May 03	Ali Sher	11 July 17		
15 Aug. 91	Abdul Ghafur Khan	23 Oct. 17		
18 Apl. 09	Nur Din Khan	23 Oct. 17		
22 Apl. 98	Yusuf Khan	23 Oct. 17		
7 Oct. 03	Bhagu Khan	23 Oct. 17		
26 May 06	Sankar Khan	23 Oct. 17		
9 Oct. 02	Husain Khan	6 May 18		
24 May 00	Yasuf Khan	6 May 18		
13 Mar. 03	Nawab Ali Khan	6 May 18		
29 Nov. 06	Badar-ud-din Khan	6 May 18		
	Khawaj Bakhsh Khan	10 July 18		

2nd Battalion, 17th Infantry (The Loyal Regiment).

Formed at Dinapore on 20th April 1918.

Composition—4 Companies of Musalmans, of Eastern Punjab and Hindustan.

1st Commn. or date of entering service.	Names and Rank.	Army rank.	Present appointment in Regiment.	Remarks.
	Commandant.			
20 July 98	Smithers, Maj. L. S. H. (*Actg. Lt.-Col.*)	1 Sep. 15	20 Apl. 18	1-17th Infy.
	Company Commanders (4).			
8 Jan. 02	Lucas, Maj. M. H.	18 Jan. 17	30 Oct. 18	37 Lrs., 2nd-in-Ccmd. (*temp.*)
15 May 15	Jameson, Lt. C. E. E. (*Actg. Capt.*)	15 May 16	18 May 18	I. A. (*on probn.*)
3 July 15	Luscombe, Lt. A. M. P. (*Actg. Capt.*)	July 16	18 Sep. 18	Som. L. I., I. A. (*on probn.*)
1 Sep. 16	Pearce, Lt. C. F. B. (*Actg. Capt.*)	1 Sep. 17	5 Oct. 18	Som. L. I., I. A. (*on probn.*)
	Company Officers.			
23 June 16	Jordan, Lt. E. G.	23 June 17	17 May 18	I. A. (*on probn.*)
9 Aug. 16	Downie, Lt. A.	9 Aug. 17	20 Apl. 18	Attd.
7 Oct. 16	Carruthers, Lt. F. W. W.	7 Oct. 17	2 Oct. 18	Manch. R. I. A. (*on probn.*)
1 Nov. 17	Deane, Lt. F. J.	1 Nov. 18	20 Apl. 18	I. A. (*T. C.*)
5 Nov. 17	Taylor, Lt. W. (*Actg. Capt.*)	5 Nov. 18	20 Apl. 18	I. A. (*T. C.*) Adjt.
25 Nov. 17	Hanson, 2nd-Lt. H.	21 Nov. 17	6 June 18	I. A. R. O.
5 Jan. 18	Munro, 2nd-Lt. H.	5 Jan. 18	6 June 18	I. A. (*T. C.*) Qr. Mr.
15 Jan. 18	Richardson, 2nd-Lt. W.	15 Jan. 18	5 June 18	I. A. R. O.
25 Jan. 10	Grant, 2nd-Lt. A J.	25 Jan. 18	23 Sep. 18	Gord. Highrs. I. A. (*on probn.*)
28 Feb. 18	Prebble, 2nd-Lt. C. E.	28 Feb. 18	21 May 18	I. A. (*on probn.*)
19 Mar. 18	Harry, 2nd-Lt. H. T.	19 Mar. 18	18 July 18	I. A. (*T. C.*)
29 Mar. 18	Newman, 2nd-Lt. G. J.	29 Mar. 18	18 July 18	I. A. (*T. C.*)
3 May 18	Neale, 2nd-Lt. R. T.	3 May 18	10 Sep. 18	I. A. R. O. (*on probn.*)
28 May 18	Seth, 2nd-Lt. E. R.	28 May 18	28 Sep. 18	W. Surr. R., I. A. R. O. (*on probn.*)
28 May 18	Galland, 2nd-Lt. J. F.	28 May 18	4 Oct. 18	R. Fus., I. A. R. O. (*on probn.*)
29 May 18	North, 2nd-Lt. F. E.	29 May 18	4 Sep. 18	I. A. R. O. (*on probn.*)
2 June 18	Turner, 2nd-Lt. P. H.	2 June 18	11 Oct. 18	I. A. R. O.
7 June 18	Bisher, 2nd-Lt. D. R.	7 June 18	11 June 18	I. A. (*T. C.*)
10 July 18	Bullock, 2nd-Lt. J. W. H.	10 July 18	1 Dec. 18	I. A. R. O.
22 July 18	Barnes, 2nd-Lt. T.	22 July 18	3 Dec. 18	I. A. R. O.
22 July 18	Collins, 2nd-Lt. R. E.	22 July 18	1 Dec. 18	I. A. (*T. C.*)
14 Oct. 18	Booley, 2nd-Lt. G. H.	14 Oct. 18	25 Oct. 18	I. A. R. O.

Subadar-Major.

		Jemadar.	Subadar.	Subadar-Major.
1 Nov. 03	Yakub Ali Khan, I.D.S.M. (66*a*)	17 Apl. 11	27 Dec. 16	21 May 18

Subadars.

17 Jan. 99	Rahmat Khan	23 Nov. 14	22 Apl. 15	
15 July 11	Allah Baksh	15 July 11	1 Feb. 17	
22 Oct. 99	Wazir Khan	15 Sep. 17	14 Aug. 17	
18 Dec. 88	Mohammed Abdul Karim	27 Dec. 16	6 May 18	
2 May 02	Niaz Mohammed	14 Aug. 17	6 May 18	
10 Dec. 98	Umrao Ali Khan	23 Oct. 17	6 May 18	
	Mirza Wali Jan Beg (*Retd.*)	. . . .	. . . .	Re-employed.

Jemadars.

22 June 12	Mansab Khan	3 July 17
15 Feb. 01	Mubarak Ali Khan	23 Oct. 17
7 Apl. 00	Arshad Khan	23 Oct. 17
22 Dec. 05	Faiz Muhammed Khan	6 May 18
17 Sep. 04	Yesin Khan	6 May 18
6 July 03	Karim Ali Khan	11 May 18
1 Aug. 09	Nazar Ali Khan	17 May 18
20 Oct. 06	Wali Mohammad Khan	12 May 18
18 July 10	Nur Ahmad Khan	13 May 18
14 Aug. 05	Mahbub Khan	14 May 18
2 July 06	Mahmud Khan	20 July 18
9 Sep. 14	Nabi Bux	21 Aug. 18
20 Oct. 06	Barkat Khan	24 June 18

Regt. Centre—Benares. *Uniform*—Scarlet.
Linked with—5th and 17th Infantry. *Facings*—Black.

1st Battalion, 18th Infantry.

Castor-ki-paltan.—Raised at Calcutta in 1795, by Captain H. DeCastro, as The Calcutta Native Militia. Became the Alipore Regiment, 1859; the 22nd Regiment of Bengal Native Infantry, 1861; the 18th Regiment of Bengal Native Infantry, 1861; the 18th (The Alipore) Regiment of Bengal Native Infantry, 1864; the 18th Regiment of Bengal Infantry, 1885; the 18th Musalman Rajput Infantry, 1902; present designation, 1903.

"Burma, 1885-87."

Composition—4 Companies Musalmans of the Eastern Punjab and Hindustan.

COLONEL—Major-General *Sir* Alliston C. Toker, K.C.B. . . 16 June 1914.

1st Commn. or date of entering service.	Names and Rank.	Army rank.	Present appointment in Regiment.	Remarks.
	Commandant.			
16 Jan. 95	Pryce, Lt.-Col. H. E. ap. R., C.M.G., D.S.O., *p.s.c.* (*temp. Br. Genl.*).	3 June 16	17 Aug. 18	
	Company Commanders. (4)			
1 Jan. 98	Morgan, Maj. H. A.	1 Sep. 15	17 Aug. 18	2nd-in-Comd.
19 May 00	Douglas, Maj. G. S.	1 Sep. 15	30 Jan. 13	With 2/113th Infy.
11 Aug. 00	*Whitaker, Maj. C. L. D. H.*	*1 Sep. 15*	10 Oct. 16	*G. S. O., 2nd Gr., Bombay Bde.*
8 May 01	Hammond, Maj. W. P.	8 May 16	17 Aug. 18	
	Company Officers. (9)			
15 Jan. 02	Hankin, Maj. D.	19 Oct. 17	13 Apl. 06	With 2-18th Infy.
9 Jan. 04	Drake-Brockman, Capt. G. T.	9 Jan. 13	10 Apl. 05	
5 Aug. 05	Binny, Capt. A. C. M.	5 Aug. 14	4 Nov. 06	
9 Sep. 08	Brooke, Capt. A. S., M.C.	1 Sept. 15	12 Nov. 09	
28 May 10	Schofield, Capt. W. H. S.	1 Sept. 15	6 Dec. 12	*Lv., ex I., m. c.*
30 Aug. 11	Silver, Capt. H. E.	1 Sept. 15	11 Dec. 12	Qr.-Mr. (*Offg.*)
20 Jan. 12	Brett, Capt. E. W.	20 Jan. 16	9 Apl. 13	
1 Oct. 14	Kennedy, Capt. E. R.	1 Oct. 18	16 Dec. 14	
14 Nov. 16	Wingfield, Lt. R. W. D.	4 Apl. 17	8 Dec. 16	Attd.
	Attached.			
19 May 85	Wright, Col. E. G.	1 Jan. 17	17 Aug. 18	Comdt. (*temp.*)
7 May 02	Thomson, Maj. A. J. D.	24 Apl. 11	. . .	Civil.
20 Jan. 09	Grant, Capt. F. R. E.	1 Sep. 15	. . . .	122nd Infy.
3 Aug. 15	Stansfeld, Lt. C. A.	3 Aug. 16	. . . .	A. S. Corps, I. A. (*on probn.*)
20 Mar. 15	Greenop, Lt. M. E.	20 Dec. 16	. . . .	I. A. (*on probn.*)
12 May 15	Morton, Lt. A. F.	12 May 16	. . . .	I. A. (*on probn.*)
20 June 17	Howe, Lt. G.	20 June 18	. . . .	I. A. R. O.
11 Oct. 15	Nicholas, Lt. C. H.	11 Oct. 16	. . . .	I. A. (*on probn.*) Adjt.
15 Jan. 18	Watts, 2nd-Lt. B.	15 Jan. 18	. . . .	I. A. (*T. C.*)
29 Jan. 18	Frizzel, 2nd-Lt. W. G.	29 Jan. 18	7 Sep. 18	Innis. Fus., I. A. R. O. (*on probn.*)
15 Jan. 18	Libby, 2nd-Lt. W. H.	15 Jan. 18	9 Sep. 18	I. A. (*T. C.*)
1 Oct. 17	Dyson, Lt. H. P.	1 Oct. 18	. . . .	I. A. (*T. C.*)
7 Oct. 17	Welsman, 2nd-Lt. C. H.	7 Oct. 17	. . . .	I. A. R. O.
23 Apl. 15	Clayton-Jones, Lt. R.	23 Apl. 16	. . . .	I. A. R. O.
28 May 18	Preiss, 2nd-Lt. R. B.	28 May 18	. . . .	Hamps. R., I. A. (*on probn.*)
14 June 16	Kendrick, Lt. V. J. M.	14 June 17	. . . .	M. G. Corps, I. A. (*on probn.*)
20 Aug. 18	Stuart, 2nd-Lt. J.	20 Aug. 18	. . . .	I. A. R. O.
22 Aug. 18	King, 2nd-Lt. W. B.	22 Aug. 18	. . . .	I. A. R. O.

Subadar-Major.

		Jemadar.	Subadar.	Subadar-Major.	
23 June 83	Karam Sher Khan *Bahadur* (56, 60).	16 Jan. 96	1 Jan. 99	2 Nov. 13	*Order of Br. Ind., 2nd Cl.*

	Subadars.			
2 Feb. 03	Sher Muhammad Khan (56, 66*a*)	22 Sep. 12	16 Mar. 17	
19 Aug. 96	Sheikh Wilayat Ali	Nov. 14	17 Dec. 17	
20 May 96	Akbar Khan	27 Jan. 16	17 Dec. 17	
17 July 05	Niaz Muhammad Khan	16 Mar. 16	17 Dec. 17	
25 Sep. 98	Faiz Muhammad Khan	16 Mar. 16	17 Dec. 17	
	Isa Khan	24 June 17	1 July 18	
	Amir Khan (*Retd.*)	. . . .	. . . .	Re-employed.
	Jemadars.			
1 Oct. 00	Husain Bakhsh Khan	16 Dec. 15		
9 Aug. 00	Ismail Khan	16 Mar. 17		
24 Dec. 02	Zahur Ali Khan	16 Mar. 17		
	Nathu Khan	1 June 17		
4 Nov. 08	Ata Mohammed	1 Dec. 17		
16 Dec. 02	Muhammad Ali Khan	17 Dec. 17		
	Sakhawat Khan	17 Dec. 17		
5 Aug. 07	Fateh-ud-din Khan	17 Dec. 17		
9 Dec. 02	Qutbi Khan	17 Dec. 17		
30 July 07	Shah Muhammad Khan	17 Dec. 17		
11 Apl. 00	Karam Ali Khan	1 May 18		
	Mirdad Khan	1 July 18		
	Abdullah Khan	1 July 18		

2nd Battalion, 18th Infantry.

Formed at Shillong on 1st December 1917.

Composition :—4 Companies Musalmans of the Eastern Punjab and Hindustan.

1st Commn. or date of entering service.	Names and Rank.	Army rank.	Present appointment in Regiment.	Remarks.
	Commandant.			
3 Sep. 92	Pratt, Lt.-Col. E. G. W. (Q)	29 Nov. 17	29 Nov. 17	1-95th Infy. *Lv. ex. I., m. c.*
	Company Commanders (4).			
15 Jan. 02	Hankin, Maj. D. L. (*Actg. Lt.-Col.*)	19 Oct. 17	15 Feb. 18	1-18th Infy. 2nd-in-Comd. Comdt. (*Offg.*)
7 Nov. 06	Solano, Capt. B. A. (*Actg. Maj.*)	1 Sep. 15	1 May 18	46th Punj. 2nd-in-Comd. (*Offg.*)
5 Feb. 15	Thorburn, Capt. R., M.C.	10 Jan. 8	10 Jan. 18	I. A. R. O.
	Park, Lt. D. (*Actg. Capt.*)	....	M y 18	I. A. R. O.
	Company Officers.			
20 June 17	Neate, Lt. W. H. (*Actg. Capt.*)	1 Dec. 17	29 Nov. 17	I. A. R. O. Adjt.
4 Aug. 15	Peacock, Lt. L. V.	4 Aug. 16	2 Dec. 17	46th Punj.
7 Aug. 15	Coltman, Lt. A. C.	7 Aug. 16	12 Feb. 18	I. A. R. O.
19 Jan. 16	Allen, Lt. J. S.	19 Jan. 17	1 Dec. 17	I. A. (*on probn.*)
28 Feb. 16	Shaw, Lt. V. E.	28 Feb. 17	29 Mar. 18	M. G. Corps., I. A. (*on probn.*)
22 July 16	Cruickshank, Lt. J. G.	22 July 17	17 Mar. 18	R. Highrs. I. A. (*on probn.*)
30 July 16	Sommerville, Lt. R.	30 July 17	12 Feb. 18	R. Scots., I. A. (*on probn.*)
8 Sep. 16	Craven, Lt. A. G. W. S.	8 Sep. 17	5 Dec. 17	Hamps. R., I. A. (*on probn.*)
5 May 17	Cumming, Lt. J. F.	5 May 18	28 Jan. 18	Gord. Highrs., I. A. (*on probn.*)
28 Nov. 17	Richer, Lt. R. A.	28 Nov. 18	6 Feb. 18	I. A. R. O.
27 June 17	Barton, 2nd-Lt. W.	27 June 17	24 Apl. 18	I. A. R. O.
1 Nov. 17	Danagher, Lt. A. V. (*Actg. Capt.*)	1 Nov 18	17 Mar. 18	I. A. (*T. C.*)
1 Nov. 17	Fraser, Lt. R.	1 Nov. 18	31 May 18	I. A. (*T. C.*) Qr. Mr. 1 June 18.
7 July 16	Iles, Maj. F. W.	7 July 16	7 May 18	I. A. R. O.
16 Aug. 16	Darling, Lt. O. R.	16 Aug. 17	12 Apl. 18	R. Ir. Rif. I. A. (*on probn.*)
	Mackie, 2nd-Lt. J., M.C.	....	7 June 18	I. A. R. O.
15 Jan. 18	Minshall, 2nd-Lt. A. O.	15 Jan. 18	7 June 18	I. A. (*T. C.*)
13 Nov. 17	Mc Cluskey, Lt. F. J.	13 Nov. 18	2 July 18	North'd Fus., I. A. (*on probn.*)
25 Dec. 18	Davis, Lt. A. T.	25 Dec. 16	2 Oct. 18	R. Fus., I. A. (*on probn.*)
7 June 18	Barnes, 2nd-Lt. A. E. C.	7 June 18	10 Oct. 18	I. A. (*T. C.*)
23 Aug. 18	Morley, 2nd-Lt. E. A.	23 Aug. 18	13 Sep. 18	I. A. R. O.
30 Aug. 18	Danter, 2nd-Lt. A. W.	30 Aug. 18	23 Sep. 18	I. A. R. O.
27 Sep. 18	Congdon, 2nd-Lt J. C.	27 Sep. 18	27 Sep. 18	I. A. R. O.
7 Oct. 18	Jecks, 2nd-Lt. E. D.	7 Oct. 18	28 Oct. 18	I. A. (*T. C.*)
	Subadar-Major.			
		Jemadar.	Subadar.	Subadar-Major.
23 Sep. 87	Sardar Khan (56)	1 Nov. 02	1 Aug. 05	1 Mar. 18
	Subadars.			
11 Dec. 97	Kadir Nawaz Khan (56)	16 Nov. 02	1 July 10	
11 Dec. 00	Niyaz Ahmad Khan (56, (72*a*)	1 Aug. 08	16 Dec. 15	
16 June 93	Dost Muhammad Khan (66*a*), (71)	2 Nov. 13	16 Mar. 17	
8 Aug. 95	Niamuttullah Khan	5 Jan. 15	17 Dec. 17	
24 Oct. 00	Qutbuddin Khan	13 May 15	17 Dec. 17	
31 May 92	Shaikh Zaman Ali	16 Mar. 17	1 Apl. 18	
18 Feb. 05	Wali Mohammed	16 Mar. 17	1 Apl. 18	
4 Aug. 02	Nathu Khan	1 June 17	1 May 18	
12 Oct. 06	Isa Khan	26 June 17	1 July 18	
	Jemadars.			
2 Mar. 96	Mohammed Fazil	1 Dec. 17		
4 Nov. 08	Ata Mohammed	1 Dec. 17		
28 Feb. 99	Khan Mohammed Khan	17 Dec. 17		
16 Apl. 04	Abdullah Khan	17 Dec. 17		
18 Oct. 99	Khushi Mohammed	17 Dec. 17		
30 June 12	Nabi Bakhsh Khan	17 Dec. 17		
	Qasim Ali Khan	1 May 18		
10 May 96	Mir Aulad Ali	1 May 18		
28 Sep. 06	Muhammad Ali Khan	1 May 18		
	Badruddin	1 May 18		
	Rulya	1 May 18		
	Muhammad Shafi	18 Jan. 18		
	Sukawat Khan	17 Dec. 17		
	Foujdar Khan	1 Aug. 18	....	
	Alladia Khan	1 Oct. 18	....	
	Wali Muhammad	1 Oct. 18	....	

Regt. Centre—Jhelum. *Uniform*—Scarlet.
Linked with—22nd and 24th Punjabis. *Facings*—Blue.

1st Battalion, 19th Punjabis.

Sherdil-ki-paltan.—Formed at Phillour in 1857, by Lt. J. F. Stafford, of four companies transferred from the 2nd and four from the 7th Punjab Police Battalion, and designated the 7th Regiment of Punjab Infantry. Became the 23rd Regiment of Bengal Native Infantry, 1861; the 19th Regiment of Bengal Native Infantry, 1861; the 19th (Punjab) Regiment of Bengal Native Infantry, 1864; the 19th (Punjab) Regiment of Bengal Infantry, 1885; the 19th Punjab Infantry, 1901; present designation, 1903.

"Ahmad Khel" "Afghanistan, 1878-80."

Composition 2 Companies of Sikhs, 1 of Punjabi Musalmans, 1 of Pathans.

COLONEL—Major-General John Ruggles, M. 13 May 1904.

1st Commn. or date of entering service.	Names and Rank.	Army rank.	Present appointment in Regiment	Remarks.
Commandant.				
16 July 90	Pennefather Evans, Lt.-Col. G.	3 June 16	19 Nov. 18	With 1-22nd Punj.
Company Commanders. (4)				
25 July 91	Churchill, Lt.-Col. G. R. D. (Q.)	25 July 17	19 Nov. 18	2nd-in-Comd. With 2-19th Punj.
5 Aug. 96	Jarrett, Maj. H. C. D. . . .	5 Aug. 14	27 Apl. 15	
25 Jan. 99	Tancred, Maj. J. Y. . . .	1 Sep. 15	1 Nov. 16	With 2-27th Punj.
5 May 00	*Barnes, Maj. E. C.* . . .	*1 Sep. 15*	*19 Nov. 18*	*Divl. R.O., Lahore.*
Company Officers. (9)				
11 Aug. 00	Macrae, Maj. J. C., D.S.O., *p.s.c.* (*temp. Lt.-Col.*)	1 Sep. 15	16 Feb. 08	
18 Jan. 02	*Ross, Maj. D. B.*	*18 Jan. 17*	*12 Sep. 03*	*D. A. A. G.*
19 Aug. 03	Knollys, Maj. D. E. (*Actg. Lt.-Col.*)	19 Aug. 18	26 Oct. 05	
19 Aug. 03	Pemberton, Maj. E. H. . . .	19 Aug. 18	18 Nov. 04	
18 Jan. 05	*Heath, Capt. L. M.*, M.C. . .	*18 Jan. 14*	*25 Mar. 06*	*Instr., Cadet Coll., Quetta.*
16 Aug. 05	*Hornsby, Capt. R. W. (temp. Maj.)*	*16 Aug. 14*	*2 Aug. 07*	*G. S.O., 2nd gr.*
29 Aug. 06	Bennett, Capt. A. D., M. C. (*temp. Maj.*)	29 Aug. 15	6 May 08	
20 Jan. 09	Drummond, Capt. J. G. P. . .	1 Sep. 15	4 Apl. 10	
2 Sep. 13	Lindsay-Young, Capt. E. . . .	2 Sep. 17	11 Nov. 15	Attd. with 2-19th Punj.
1 Oct. 14	Shute, Capt. G. E. F.	1 Oct. 18	19 Dec. 14	
11 Nov. 14	Pigot, Lt. G.	11 Nov. 15	30 Jan. 15	
23 Dec. 14	Woolf, Lt. S. H.	23 Dec. 15	27 Feb. 15	With 2-128th Infy.
15 Nov. 15	Adams, Lt. R. F. G.	15 Nov. 16	16 Nov. 15	Attd. Adjt. (*Offg.*)
31 Jan. 18	Jones, 2nd-Lt. M. R.	31 Jan. 18	. . . .	Attd.
10 May 16	Hart, Lt. N. V.	10 Feb. 18	10 July 16	Attd.
10 June 16	Ingle, Lt. L. S.	10 Mar. 18	12 Aug. 16	Attd.
7 Nov. 14	Godsland, Capt. C. H., M.C. . .	7 Nov. 18	23 Nov. 17	Attd.
Attached.				
26 Aug. 15	Lind, Lt. E. N. A.	26 Aug. 16	. . . .	I. A. R. O.
1 Jan. 16	Stewart, Lt. F. W., M.C. . . .	1 Jan. 17	25 Feb. 16	I. A. R. O. Qr. Mr. (*Offg.*)
5 May 16	Dewar, Lt. H. S. L.	5 May 17	2 June 16	I. A. R. O.
6 Oct. 16	Millar, Lt. J. G.	6 Oct. 17	21 Nov. 16	I. A. R. O.
21 Nov. 16	Gipps, Lt. W. F.	21 Nov. 17	4 Jan. 17	I. A. R. O.
16 Mar. 18	Purchase, 2nd-Lt. R. G. . . .	16 Mar. 18	20 Oct. 18	I. A. R. O.
16 Aug. 18	Mackenzie, 2nd-Lt. C. A. . . .	16 Aug. 18	19 Sep. 18	I. A. R. O.
22 Aug. 18	Pollacco, 2nd-Lt. R. G. . . .	22 Aug. 18	16 Sep. 18	I. A. R. O.
26 Aug. 18	Collyer, 2nd-Lt. J.	26 Aug. 18	16 Sep. 18	I. A. R. O.
7 Oct. 18	Everett, 2nd-Lt. C. F. . . .	7 Oct. 18	1 Nov. 18	I. A. (*T. C.*)
31 Jan. 18	Bodenham, 2nd-Lt. K. L. . . .	31 Jan. 18	. . . .	1-4th G. R.
29 Mar. 18	Cullen, 2nd-Lt. S. E.	29 Mar. 18	15 July 18	I. A. R. O.
7 Apl. 17	Doherty, Lt. P. J.	7 Apl. 18	6 July 18	Dorset R., I. A. (*on probn.*)
15 June 18	Grabham, 2nd Lt. H. M. . . .	15 June 18	12 Sep. 18	I. A. R. O. (*on probn.*)
15 June 18	Jackson, 2nd-Lt. T.	15 June 18	12 Sep. 18	I. A. R. O. (*on probn.*)
31 Oct. 17	Smith, Lt. C. A. B.	31 Oct. 18	. . . .	I. A. R. O.
7 June 18	Jones, 2nd-Lt. S. V.	7 June 18	. . . .	I. A. (*T. C.*)
28 Nov. 14	Savage, Lt. E. G.	28 Nov. 15	. . . .	I. A. R. O.

Subadar-Major.

		Jemadar.	Subadar.	Subadar-Major.	
5 Aug. 85	Fateh Khan, *Sardar Bahadur* (56)	16 July 00	17 Oct. 03	1 June 13	*Order of Br. Ind., 1st Cl.*

1st Battalion, 19th Punjabis—*contd.*

1st Commn. or date of entering service.	Names and Rank.	Army rank.	Present appointment in Regiment.	Remarks.
	Subadars.			
5 Mar. 88	Isar Singh	1 Nov. 02	7 Nov. 04	
25 Dec. 88	Ralla Singh *Bahadur*, I.D.S.M. . (56).	1 Aug. 03	1 May 10	*Order of Br. Ind., 2nd-Cl.*
1 Aug 03	Maula Bakhsh (56) . . .	7 Jan. 11	1 Aug. 13	
18 Jan. 95	*Lal Khan (51, 72a)*	*1 Aug. 13*	*6 Jan. 15*	
12 July 94	Ghulam Muhammad, I. D.S.M. . (56, 66a)	6 Jan. 15	16 June 15	
19 Feb. 92	Hukam Singh (56)	6 Jan. 15	1 Aug. 15	
4 Nov. 95	Fazl Khan (56, 66a)	6 Jan. 15	1 Aug. 15	
29 Mar. 04	Nawab Khan (66a)	6 Jan. 15	1 Aug. 15	
23 July 98	Bal Singh (56)	6 Jan. 15	14 Jan. 17	
13 Mar. 97	Azizullah (66a)	1 Feb. 16		
	Jemadars.			
1 Aug. 00	Bak Khan 56)	1 Feb. 15		
26 Apl. 96	Sohel Sing (66a)	10 Feb. 15		
1 June 00	Nihal Singh (58)	1 Aug. 15		
5 Jan. 98	Harnam Singh	1 Aug. 15		
31 May 97	Fateh Muhammad (66a) . . .	1 Aug. 15		
14 Mar. 98	Fazal Khan (56, 66a)	1 Aug. 15		
1 Aug. 99	Saida Khan (66a)	14 Jan. 17		
9 Sep. 98	Rur Singh	14 Jan. 17		
31 July 99	Santa Singh	14 Jan. 17		
11 Jan. 02	Chaghta Khan (66a)	14 Jan. 17		
9 July 96	Faiz Talab (66a)	1 May 17		
28 Oct. 98	Rajwali	9 Sep. 17		
9 Feb. 18	Zarullah Khan	9 Feb. 18	Indian Adjt., 27 Apl. 18.	
13 Aug. 00	Kasim Din			

2nd Battalion, 19th Punjabis.

Formed at Hyderabad Sind on 12th January 1917.
Composition—2 Companies Sikhs, 1 Company Punjabi Mussalmans, 1 Company Pathans.

1st Commn. or date of entering service.	Names and Rank.	Army rank.	Present appointment in Regiment.	Remarks.
	Commandant.			
25 July 91	Churchill, Lt.-Col.. G. R. D.	25 July 17	2 Mar. 17	1-19th Punj.
	Company Commanders (4).			
18 Sep. 09	McLean, Capt. F. G. S. (*Actg. Maj.*)	1 Sep. 15	6 Dec. 17	67th Punj.
4 Mar. 11	Bremner, Capt. C. E. U., M.C. (*Actg. Maj.*)	1 Sep. 15	13 Mar. 18	29th Punj. 2nd-in-Comd. (*temp.*).
	Company Officers.			
2 Sep. 13	Lindsay Young, Capt. E.	2 Sep. 17	12 Jan. 17	19th Punj.
18 June 17	Reeve, Lt. M. B. P.	18 June 18	14 July 17	Attd.
5 May 15	Smith, Lt. W. S.	5 May 16	9 Aug. 18	Attd. Adjt.
22 June 15	Clarke, Lt. F. E. M.	22 June 16	7 July 17	Attd.
21 Oct. 15	Bates, Lt. H. S.	21 Oct. 15	11 Aug. 17	I. A. R. O.
27 Oct. 16	Spark, Lt. J. A.	27 Oct. 17	21 July 17	
1 Mar. 17	Roach, Lt. J. P.	1 Mar. 18	28 Apl. 17	I. A. R. O.
1 Feb. 17	Leonard, Lt. G. T. J.	1 Feb. 18	28 Apl. 17	I. A. R. O.
18 Sep. 15	Driver, Lt. R. A.	18 Sep. 16	19 May 17	I. A. (*on prob n.*) Qr.-Mr.
15 May 15	Fitzpatrick, Lt. L. V.	15 May 16	18 Aug. 17	Attd. *With Sig. Ser.*
5 May 17	Scott, Lt. R. E.	5 May 18	25 Aug. 17	I. A. R. O.
5 May 17	Cuthbert, Lt. L.A.	5 May 18	25 Aug. 17	I. A. R. O.
18 Feb. 15	Sorby, Lieut. W. J.	18 Feb. 16	13 Mar. 18	I. A. R. O. *Lv., ex. I., m.c.*
8 Sep. 15	Unsworth, Lieut. N.	8 Sep. 16	5 Jan. 18	I. A. R. O. *Lv., ex. I., m.c.*
9 Sep. 16	Latham, Lt. E. B., M.M.	9 Sep. 16	29 Sep. 17	I. A. R. O.
1 Dec. 17	Peffers, Lt. J.	1 Dec. 18	4 Feb. 18	Highl. Cyclist., I. A. (*on probn.*)
18 June 17	Tottenham, 2nd-Lt. J. L.	18 June 17	14 July 17	1-22nd Punj.
29 Mar. 18	Hyde, 2nd-Lt. H. S.	29 Mar. 18	18 July 18	I. A. R. O.
15 June 18	Libby, 2nd-Lt. H. J.	15 June 18	15 Sep. 18	R. Fus., I. A. R. O. (*on probn.*)
15 June 18	Peacocke, 2nd-Lt. A.	15 June 18	15 Sep. 18	I. A. R. O. (*on probn.*)
1 May 18	Giblett, 2nd-Lt. J. V.	1 May 18	4 Aug. 18	I. A. R. O. (*on probn.*)
7 June 18	Eden, 2nd-Lt. A. W. B.	7 June 18	11 Oct. 18	I. A. (*T. C.*)
	Hamilton, 2nd-Lt. E. W. H.		7 Nov. 18	Devon. R., I. A. (*on probn.*)
4 Oct. 18	Morgans, 2nd-Lt. C. G.	4 Oct. 18	23 Oct. 18	I. A. R. O.
8 Oct. 18	Scotson, 2nd-Lt. R.	8 Oct. 18	13 Nov. 18	I. A. R. O
	Suys, Lt. P. W.		27 Nov. 18	R. Ir. Rif., I. A. (*on probn.*)

Subadar-Major.

		Jemadar.	Subadar.	Subadar-Major.
4 Feb. 05	Teja Singh (56, 72*a*)	4 Feb. 05	2 May 10	19 Feb. 17 (*temp.*)

Subadars.

30 Sep. 16	Jalal Khan, *Bahadur*, I.D.S.M. (72*a*.)	15 Feb. 02	7 Jan. 11	*Order of Br. Ind., 2nd Class.*
10 May 90	Sohan Singh (56, 58)	2 May 10	1 June 13	
17 July 92	Surayan Singh (56)	1 June 13	6 Jan. 15	
13 Feb. 94	Azizullah	6 Apl. 15	16 Aug. 16	
21 May 97	Budha Singh (56)	5 Apl. 15	12 Jan. 17	
16 Sep. 98	Khuda Bakhsh	1 Oct. 16	12 Jan. 17	
11 Feb. 01	Ghafur Khan	15 Oct. 16	12 Jan. 17	
19 Mar. 99	Mardan Khan	16 Oct. 15	10 Feb. 18	
12 Feb. 96	Sundar Singh	7 June 16	7 May 18	
3 Jan. 01	Sahib Shah	15 Apl. 16	26 July 18	

Jemadars.

23 Apl. 00	Sundar Singh	12 Jan. 17	
1 Sep. 99	Talok Singh	12 Jan. 17	
16 Mar. 03	Nand Singh (56)	12 Jan. 17	Indian Adjt.
9 Sep. 00	Surayan Singh (56)	12 Jan. 17	
16 Feb. 99	Ghulam Haidar I	12 Jan. 17	
9 Nov. 01	Ismail Khan (56)	12 Jan. 17	
2 June 09	Makhmad Janab (56)	12 Jan. 17	
28 Aug. 15	Hukmat Khan (45)	12 Jan. 17	
8 Oct. 02	Lalkhan	10 Feb. 18	
8 Feb. 98	Waryam Singh	7 May 18	
19 Dec. 95	Sanwar Khan	15 May 18	
6 July 06	Quasim Khan	25 June 18	
25 May 98	Rahim Shah	26 June 18	
19 July 00	Ghulam Haidar II	26 July 18	
29 Dec. 03	Madat Ali	1 Aug. 18	
27 July 03	Sultan Khan	23 Aug. 18	
20 Nov. 05	Qasim Ali	29 Aug. 18	

Regt. Centre—Lahore.
Linked with—21st and 26th Punjabis.

Uniform—Drab.
Facings—Emerald Green.

20th Duke of Cambridge's Own Infantry (Brownlow's Punjabis).

Formed at Nowshera in 1857, by Lt. C. H. Brownlow, of 8 companies transferred from 4th and 5th Punjab Infy. and designated the 8th Regt. of Punjab Infy. Became the 24th Regt. of Bengal Native Infantry, 1861; the 20th Regt. of B. N. I., 1861; the 20th (Punjab) Regt. of B. N. I., 1864; the 20th (The Duke of Cambridge's Own Punjab) Regt. of B. N. I., 1883; the 20th (The Duke of Cambridge's Own Punjab) Regt. of Bengal Infy, 1885; the 20th (The Duke of Cambridge's Own) Punjab Infy.. 1901; present designation, 1904.

"Taku Forts" "Pekin, 1860" "Ali Masjid" "Afghanistan, 1878-80"
"Tel-el-Kebir" "Egypt, 1882"
"Punjab Frontier" "China, 1900."

Composition—2 companies of Pathans, 1 of Sikhs, 1 of Dogras.

COLONEL—

1st Commn. or date of entering service.	Names and Rank.	Army rank.	Present appointment in Regiment.	Remarks.
	Commandant.			
14 Aug. 95	Fordham, Lt.-Col. W. M., *p.s.c.*	22 Jan. 18	22 Jan. 18	
	Company Commanders. (4)			
22 Jan. 96	*St. John, Bt. Lt.-Col. R. S.,* D.S.O., *p.s.c.* (*b*).	*3 June 16*	*2 May 18*	*2nd-in-Comd.—Embarkation Comdt., Bombay.*
20 July 98	Finnis, Maj. B. H. (*Atg. Lt.-Col.*)	1 Sep. 15	15 Nov. 14	With 2-55th Rif.
19 May 00	McCleverty, Maj. P. H.	1 Sep. 1	22 Jan. 18	With 3-9th Infy.
28 Jan. 03	Saxton, Maj. P. D., M.C.	25 Dec.	27 May 18	*Lv., ex I., m. c.*
	Company Officers. (9)			
18 Jan. 02	Elliott, Maj. P. W.	18 Jan. 17	31 Mar. 03	
21 Jan. 03	Hawes, Maj. C. M.	21 Jan. 18	21 Apl. 04	
25 Jan. 08	*Dawson, Capt. A. A. F. C. H.*	*1 Sep. 15*	*21 Apl. 09*	*Adjt., Mewar Bhil Corps.*
10 Jan. 12	*Burn-Murdoch, Capt. C. T.*	*20 Jan. 16*	*30 Mar. 13*	*Frontier Militia.*
13 Jan. 14	Kidd, Capt. G. R.	13 Jan. 18	16 Feb. 16	
30 Jan. 17	Hadaway, Lt. J. C. S.	30 Jan. 18	30 Jan. 18	
31 Aug. 18	Raper, 2nd-Lt. C. F.	31 Aug. 18	4 Sep. 18	Attd.
	Attached.			
9 Jan. 04	Walker, Capt. H. R. O.	9 Jan. 13	8 Sep. 17	25th Punj.
27 Oct. 16	O'Sullevan, 2nd-Lt. E. T.	27 Oct. 16	19 May 17	66th Punj.
27 Nov. 14	Barclay, Lt. F.	27 Nov. 15	22 July 16	I. A. R. O.
2 Aug. 15	Fleming, Lt. T.	2 Aug. 16	22 July 16	I. A. R. O.
12 Feb. 15	Hayes, Lt. L., M. C.	12 Feb. 16	16 June 15	I. A. R. O.
22 Jan. 13	Bird, Capt. E. K.	22 Jan. 17	26 Apl. 17	29th Punj.
29 Jan. 15	Raschen, Lt. G. H.	29 Jan. 16	13 Mar. 17	I. A. R. O.
12 May 15	Harry, Lt. F. L.	12 May 16	30 Nov. 17	I. A. (*on probn.*)
9 July 15	Hillier, Lt. V. F. (*temp. Capt.*)	9 July 16	7 July 17	I. A. R. O.
22 Feb. 16	Webber, Lt. C. D.	22 Feb. 16	. . .	I. A. (*on probn.*)
26 June 17	Ballard, Lt. V.	26 June 18	8 Nov. 17	I. A. (*on probn.*)
27 Oct. 17	Billing, Lt. E. W.	27 Oct. 18	4 Nov. 17	
26 June 18	Langley, 2nd-Lt. L. C.	26 June 18	. . .	Midd'x. R., I. A. (*on probn.*)
5 Nov. 17	Dale, Lt. C. H.	5 Nov. 18	17 July 18	I. A. R. O.
20 June 17	Naismith, 2nd-Lt. W. N.	20 June 17	5 May 18	I. A. R. O.
	Tilfer, Lt. A. F.	. . .	9 May 18	I. A. R. O.
13 Sep. 15	Hunter, Lt. J.	. . .	9 May 18	I. A. (*on probn.*)
21 Oct. 16	Cowie, Lt. J. M.	21 Oct. 17	9 May 18	I. A. (*on probn.*)
2 Oct. 14	McKerron, Lt. P. A. B.	2 Oct. 15	8 Dec. 17	Gord. Highrs., I. A. R. O.
2 June 18	Pignon, 2nd-Lt. H. L.	2 June 18	. . .	I. A. R. O.
7 June 18	Robertson, 2nd-Lt. R. H.	7 June 18	. . .	I. A. R. O.
4 Oct. 18	Danter, 2nd-Lt. J. E.	4 Oct. 18	29 Oct. 18	I. A. R. O.
24 Sep. 18	Keats, 2nd-Lt. A. W.	24 Sep. 18	3 Nov. 18	I. A. R. O.
8 Nov. 18	Williams, 2nd-Lt. P. M.	8 Nov. 18	28 Nov. 18	I. A. R. O.

Subadar-Major.

		Jemadar.	Subadar.	Subadar-Major.
18 Feb. 88	Mota Singh (56, 58)	1 Sep. 06	15 Aug. 10	18 Feb. 17.

Subadars

21 May 87	Ganda (56)	16 Aug. 04	7 July 14
22 Jan. 92	Brahwan (56)	7 July 14	18 Nov. 14
4 Feb. 09	Amin Gul	14 Aug. 14	15 Aug. 15
15 May 94	Ghulam Rasul (56)	15 Aug. 15	1 Oct. 15
19 May 97	Kashmir Singh (58)	14 Aug. 14	18 Dec. 15
14 July 00	Said Gul	23 Mar. 16	1 Jan. 17
10 Apl. 97	Khushala (66*a*, 72)	18 Nov. 14	20 Oct. 16
4 June 94	Habib Khan	29 Sep. 15	18 Feb. 17
29 Nov. 02	Makhmad	1 Jan. 17	18 Feb. 17
28 Nov. 95	Ladha Singh (66*a*)	16 May 15	16 Feb. 17

20th Duke of Cambridge's Own Infantry (Brownlow's Punjabis) —*contd.*

1st Commn. or date of entering service.	Names and Rank.	Army rank.	Present appointment in Regiment.	Remarks.
	Jemadars.			
8 Nov. 97	Udham Singh (51,72)	14 Aug. 14		
19 Dec. 99	Niaz Gul (56)	12 Nov. 15		
15 Mar. 91	Bua Ditta (56, 72)	16 Feb. 16		
20 Apl. 99	Alam Khan	18 May 16		
28 May 04	Sansaru	20 Oct. 16		
2 Aug. 05	Mohammed Yakub Khan	3 Jan. 17		
9 Oct. 92	Sant Singh	18 Feb. 17		
18 Sep. 03	Sharif	18 Feb. 17		
4 Jan. 02	Lakha Singh	18 Feb. 17		

Regt. Centre—Lahore. *Uniform*—Drab.
Linked with —20th and 26th Punjabis. *Facings*—Scarlet.

1st Battalion, 21st Punjabis.

Formed at Kohat in 1857, by Lt. N. E. Boileau, of 8 companies transferred from the 3rd and 6th Punjab Infy. and designated the 9th Regt. of Punjab Infy. Became the 25th Regt. of Bengal Native Infantry, 1861; the 21st Regt. of Bengal N. I., 1861; the 21st (Punjab) Regt. of Bengal N. I., 1864; the 21st (Punjab) Regt. of Bengal Infy., 1885; the 21st Punjab Infy., 1901; present designation, 1903.

"Abyssinia" "Afghanistan, 1878-80."

Composition—1½ Companies of Pathans, ½ of Punjabi Musalmans, 1½ of Sikhs, ½ of Dogras.
Colonel.—Major-Genl. Sir V. B. Fane, K.C.S.I., C.B.

1st Commn. or date of entering service.	Names and Rank.	Army rank.	Present appointment in Regiment.	Remarks.
	Commandant.			
28 Jan. 93	Murray, Lt.-Col. W. G. P., *p.s.c.*	22 Jan. 18	22 Jan. 18	
	Company Commanders. (4)			
4 Jan. 99	Doveton, Maj. H. J.	1 Sep. 15	22 Jan. 18	2nd-in-Comd. With 3-152nd Infy.
8 Jan. 01	Wardell, Maj. C. G. V. M.	8 Jan. 16	26 Apl. 15	
19 Aug. 03	*Palmer, Maj. J. H. G.* [l]	*19 Aug. 18*	*16 May 17*	*Bde.-Maj.*
10 Oct. 03	Enriques, Maj. C. M. D. (*Actg. Maj.*)	10 Oct. 18	22 Jan. 18	With 85th Rif.
	Company Officers. (9)			
9 Jan. 04	Doveton, Capt. A. C.	19 Nov. 13	25 May 05	With 1-30th Punj.
16 Aug. 05	Haynes, Capt. W. F.	16 Aug. 14	23 Aug. 08	Adjt., 10 Nov. 13.
17 Aug. 07	*Jackson, Capt. C. H.*	*1 Sep. 15*	*16 Nov. 08*	*S. Waziristan Militia.*
17 Aug. 07	Messum, Capt. L. R.	1 Sep. 15	22 Nov. 08	With 1-152nd Infy.
11 Dec. 07	Malyon, Capt. F. H. [L]	1 Sep. 15	21 Sep. 09	
20 Jan. 09	Mackenzie, Capt. C. F.	1 Sep. 15	14 Mar. 10	
8 Aug. 14	*Fausset-Baker, Capt. A. A.*	*8 Aug. 18*	*26 Oct. 14*	*With 36 (Divl. Sig.) Coy.*
1 Oct. 14	Percy-Jones, Capt. C. M.	1 Oct. 18	17 Dec. 14	With 2-25th Punj.
17 Apl. 15	Wickham, Lt. J.	17 Apl. 16	28 Oct. 15	Attd. With 1-41st Infy.
15 Nov. 15	Amies, Lt. B. J.	15 Nov. 16	24 Dec. 15	Attd. With 4-9th Infy.
18 Apl. 16	Doyle, Lt. P. J.	18 Apl. 17	26 Apl. 16	Attd.
29 June 16	*Johnston, Lt. R. H. G.*	*29 June 17*	*2 July 16*	*Attd. Frontier Militia.*
29 June 16	*Beamish, Lt. R. P.*	*29 June 17*	*2 July 16*	*Attd. Frontier Militia.*
5 May 15	Maynard, Lt. C. I. F.	5 May 16	18 Aug. 17	Attd.
1 Sep. 15	Willoughby, Lt. B. D.	1 Sep. 16	21 Aug. 17	Attd.
7 Sep. 15	Barnes, Lt. R. M.	7 Sep. 16	21 Aug. 17	Attd.
18 Apl. 16	Hepworth, Lt. M. A.	18 Apl. 17	21 Aug. 17	Attd.
21 Dec. 17	Jones, 2nd-Lt. H. I. K.	21 Dec. 17	22 Mar. 18	Attd.
	Attached.			
18 Sep. 09	Bonham-Carter, Capt. B. H. (*Actg. Maj.*)	1 Sep. 15	....	40th Pathans.
28 July 15	Rowley, Lt. B. G.	28 July 16	20 Aug. 17	I. A. (*on probn.*)
18 June 17	Pouncey, Lt. G. R.	18 June 18	18 June 17	
28 Oct. 17	Hopper 2nd-Lt. F. C.	28 Oct. 17	4 Nov. 17	
28 Aug. 16	Saxton, Lt. W. T.	28 Aug. 17	13 Nov. 16	I. A. R. O.
5 May 17	Fry, Lt. H. B.	5 May 18	17 Aug. 17	I. A. R. O.
29 Aug. 17	Westgarth, 2nd-Lt. S. H.	29 Aug. 17	19 Jan. 18	I. A. R. O. (*on probn.*)
5 Nov. 17	Cunningham, Lt. W. J.	5 Nov. 18	14 Mar. 18	I. A. (*T. C.*)
5 May 15	Walsh, Lt. L. W.	5 May 16	....	Norf. R. I. A. (*on probn.*)
26 Oct. 15	Charter, Lt. W. F., M.C.	26 Oct. 16	....	Sco. Rif., I. A. (*on probn.*)
14 May 16	Lee, Lt. T. R.	14 May 17	4 July 18	Som. L. I., I. A. (*on probn.*)
19 Sep. 17	Clarke, Lt. J. W. A.	19 Sep. 18	....	Manch. R., I. A. (*on probn.*)
1 Nov. 18	Labanti, 2nd-Lt. V. W.	1 Nov. 18	23 Nov. 18	I. A. R. O.
....	Loftus, 2nd-Lt. G. M.	....	....	I. A. R. O.

Subadar-Major.

		Jemadar.	Subadar.	Subadar-Major.	
18 June 84	Waryam Singh, *Sardar Bahadur*	1 July 96	1 Sep. 99	1 Nov. 16	*Order of Br. Ind., 1st Cl*

Subadars.

20 May 04	Allah Dad Khan (46, 53, 56, 66*a*)	20 May 04	1 Oct. 13
13 Dec. 99	Habib Khan, I.D.S.M. (52, 56)	21 May 12	6 Oct. 00
31 Mar. 96	Nadir Ali	1 Nov. 96	17 Apl. 15
15 Aug. 93	Jawala Singh, (52, 56)	22 Apl. 15	28 Apl. 15
3 May 99	Achhar Singh (56)	2 Aug. 10	25 Nov. 17
18 Nov. 93	Lachman Singh	16 May 13	2 May 17
11 Oct. 98	Shad Muhammad (66*a*)	7 Jan. 15	29 June 18

1st Battalion, 21st Punjabis—*contd.*

1st Commn. or date of entering service.	Names and Rank.	Army rank.	Remarks.
	Jemadars.		
12 Nov. 08	*Gul Badshah*	*1 Jan. 16*	
30 Jan. 17	Munshi Ram	1 July 17	
24 Mar. 08	Hazara Singh	1 Oct. 17	
21 Aug. 03	Makhana Khan	1 Oct. 17	
4 July 00	Kirpa Singh (56)	1 Dec. 17	
1 Mar. 09	Dalel Singh	14 Dec. 17	
26 Aug. 07	Hakim Singh	1 Jan. 18	
17 June 01	Sohbat	1 Feb. 18	
27 July 12	Nizam Din	1 Feb. 18	Indian Adjt.
23 Feb. 17	Sapuran Singh	10 Feb. 18	
7 Feb. 17	Ram Kishan	2 Oct. 17	
18 Dec. 00	Fateh Ali	3 Mar. 18	
5 Feb. 18	Hidayatullah	17 Feb. 18	(*On probn.*)
5 Jan. 05	Sadhu Singh	1 July 18	
19 Aug. 00	Wariam Singh	1 Sep. 18	
	Shah Jahan	16 Aug. 18	
	Ghagga Singh	1 Sep. 18	

2nd Battalion, 21st Punjabis.

Formed at Jubbulpore on—16th October, 1917.

Composition—2 Companies Punjabi Musalmans, ¾ Company Sikhs, ¼ Company Dogras, 1 Company Pathans.

1st Commn. or date of entering service.	Names and Rank.	Army rank.	Present appointment in Regiment.	Remarks.
	Commandant.			
20 Jan. 97	Sherlock, Maj. G. W. S., *p. s. c.* (*Acting Lt-Col.*).	20 Jan. 15	23 Sep. 18	1-6th G. R. (*temp.*)
	Company Commanders. (4)			
29 Jan. 02	Robertson, Maj. E. D. S.	29 Jan. 17	16 May 18	21st Cav. 2nd-in-Comd. (*temp.*)
7 Sep. 14	Armitage, Capt. K. L. F.	7 Sep. 18	19 Oct. 18	I. A. R. O.
7 Nov. 14	Holland, Capt. G. W.	7 Nov. 18	12 Apl. 18	I. A. (*on probn.*)
19 July 15	Scott, Lt. W. (*Actg. Capt.*)	19 July 16	12 Feb. 18	A. S. C., I. A. (*on probn.*)
	Company Officers.			
15 Oct. 15	Latham, Lt. H. C. (*Actg. Capt.*)	15 July 17	10 Nov. 17	Adjt., 10 Nov. 17.
30 Nov. 14	Preece, Lt. D. E. G.	16 Sep. 16	25 Mar. 18	Shrops. L. I., I. A. (*on probn.*)
24 July 15	Rowe, Lt. S. G. (*Acting Capt.*)	24 July 16	5 June 18	I. A. R. O. Qr.-Mr. (*Offg.*) 5 June 18.
2 Sep. 15	Turner, Lt. F. R. (*Acting Capt.*)	2 Sep. 16	4 Dec. 17	I. A. R. O.
9 Oct. 15	Thorn, Lt. A. P.	9 Oct. 16	18 Feb. 18	Linc. R., I. A. (*on probn.*)
22 Dec. 15	Manley, Lt. W. J., M.C.	22 Dec. 16	31 May 18	S. Wales Bord., I. A. (*on probn.*)
3 Sep. 16	Fleming, Lt. J. G. G., D.S.O., M.C.	3 Sep. 17	9 Mar. 18	Gord. Highrs., I. A. (*on probn.*)
31 Oct. 17	de Gale Lt. H. O.	31 Oct. 18	25 Oct. 18	I. A. R. O.
20 Sep. 17	Warren, Lt E. T.	20 Sep. 18	20 Sep. 18	I. A. R. O.
30 Sep. 17	Tully 2nd-Lt. E. J.	30 Sep. 17	2 Oct. 18	Leins. R., I. A. (*on probn.*)
5 Nov. 17	D'Arcy, Lt. M. A. A.	5 Nov. 18	25 Oct. 18	I. A. (*T. C.*)
28 Nov. 17	Read, Lt. J.	28 Nov. 18	24 Jan. 18	I. A. R. O.
19 Dec. 17	Brewer, 2nd-Lt. C. E.	19 Dec. 17	11 Mar. 18	I. A. R. O. (*on probn.*)
20 Dec. 17	Evans, 2nd-Lt. T. L.	21 Dec. 17	20 Mar. 18	Attd.
31 Jan. 18	FitzMaurice, 2nd-Lt. W. H.	31 Jan. 18	16 Jan. 18	Attd.
27 Mar. 18	St. John, 2nd-Lt. C. R.	27 Mar. 18	31 Aug. 18	Suss. R. I. A. R. O. (*on probn.*)
10 May 18	Davis, 2nd-Lt. F. G. T.	10 May 18	3 Dec. 18	D. of Corn. L. I., I. A. R. O. (*on probn.*)
28 May 18	Manning-Press, 2nd-Lt. C. A.	28 May 18	31 Aug. 18	I. A. R. O. (*on probn.*)
29 May 18	Mills, 2nd-Lt. S. W.	29 May 18	31 Aug. 18	I. A. R. O. (*on probn.*)
2 June 18	Harrison, 2nd-Lt. H.	2 June 18	16 Oct. 18	I. A. (*T.C.*)
10 July 18	Bell, 2nd-Lt. W. S. B.	10 July 18	2 Dec. 18	I. A. (*T. C.*)
27 July 18	Corcoran, 2nd-Lt. B. J.	27 July 18	29 Aug. 18	I. A. R. O.
3 Aug. 18	Grose, 2nd-Lt. J. S.	3 Aug. 18	9 Sep. 18	I. A. R. O.
22 Aug. 18	Hereford, 2nd-Lt. R. J.	22 Aug. 18	6 Sep. 18	I. A. R. O.
29 Aug. 18	Harrison, 2nd-Lt. A.	29 Aug. 18	27 Nov. 18	Ches. R., I. A. (*on probn.*)
27 Nov. 18	Sawle, 2nd-Lt. J. F.	27 Nov. 18	2 Oct. 18	Welsh R., I. A. (*on probn.*)

Subadar-Major.

		Jemadar.	Subadar.	Subadar-Major.
5 May 96	Wazir Singh	1 June 99	1 Dec. 09	1 Dec. 00 1 (*temp.*)

Subadars.

28 Oct. 87	Hira Singh (52, 56)	12 Feb. 07	21 Nov. 11
21 Apl. 90	Piran Ditta (56, 66*a*)	1 Oct. 11	5 Feb. 15
24 Feb. 95	Nur Mast	1 Jan. 10	1 Jan. 16
24 June 96	Amir Khan (66*a*)	6 Oct. 13	25 Nov. 17
28 May 07	Isar Singh (I)	10 Dec. 14	25 Nov. 17
3 May 05	Muhammad Latif	9 July 16	25 Nov. 17
23 Aug. 01	Bhag Singh	6 Jan. 15	14 Dec. 17

Jemadars.

21 Feb. 96	Damodar	17 Apl. 15
24 Feb. 00	Ishar Singh, I.D.S.M.	1 Aug. 16
29 June 01	Sher Baz (56)	9 Dec. 16
24 Jan. 96	Hasan Shah	25 Nov. 17
23 Oct. 96	Dulip Singh	25 Nov. 17
24 Feb. 94	Hakim Singh	25 Nov. 17
1 May 18	Ghulam Jafer Shah	1 May 18
1 May 18	Assanullah Khan	1 May 18
15 May 18	Ghulam Kadir Khan	15 May 18

Regt. Centre—Jhelum. *Uniform*—Scarlet.
Linked with—19th and 24th Punjabis. *Facings*—Blue.

1st Battalion, 22nd Punjabis.

Formed at Mooltan in 1857, by Capt. E. S. Denniss, of 8 companies transferred from the 1st Sikh Infy. and the 3rd Punjab Police Battalion, and was designated the 11th Regt. of Punjab Infy. Became the 26th Regt. of Bengal Native Infantry, 1861; the 22nd Regt. of Bengal N. I., 1861; the 22nd (Punjab) Regt. of Bengal N. I., 1864; the 22nd (Punjab) Regt. of Bengal Infy., 1885; the 22nd Punjab Infy., 1901; present designation, 1903.

"China, 1860-62" "Afghanistan, 1879-80" "Punjab Frontier."

Composition—2 Companies of Sikhs, 1½ of Punjabi Musalmans, ½ of Pathans.

1st Commn. or date of entering service.	Names and Rank.	Army rank.	Present appointment in Regiment.	Remarks.
	Commandant.			
19 Nov. 92	𝔅𝔖 Costello, Lt.-Col. E.W., C. M. G., D. S. O., *p.s.c.* (*temp. Brig. Genl.*)	3 June 16	22 Sep. 18	
	Company Commanders. (4)			
1 Mar. 90	*Davis, Lt.-Col. G. J.*, D.S.O. (Q)	*1 Mar. 16*	*29 Sep. 16*	*2nd-in-Comd. With Kurram Militia.*
23 Dec. 93	Furney, Bt.-Lt.-Col. J. L. (Q)	3 June 16	22 Feb. 11	With 24th Punj.
14 Aug. 95	Magniac, Maj. F. A. (*Actg. Lt.-Col.*)	14 Aug. 13	1 Aug. 11	With 2-26th Punj.
22 Jan. 98	Clifford, Maj. R. (*Actg. Lt.-Col.*)	1 Sep. 15	22 Sep. 18	With 2-22nd Punj.
	Company Officers. (9)			
11 Feb. 99	Smith, Maj. E. C.	1 Sep. 15	1 Sep. 03	With 2-28th Punj.
20 Jan. 00	*Sutherland, Maj. A. O.*	*1 Sep. 15*	*10 June 03*	
8 May 01	Webb, Maj. W. F. R., D.S.O.	8 May 16	11 Oct. 02	2nd-in-Comd. (*temp.*)
9 Jan. 04	Wallace, Capt. C. W.	9 Jan. 13	14 June 06	
23 May 06	*Matthews, Capt. F. M. (temp. Maj.)*	*23 May 15*	*16 July 00*	*D. A. D. R. T., Bombay.*
29 Aug. 06	Leslie-Smith, Capt. C.	29 Aug. 15	21 Oct. 07	With 1-140th Infy.
20 Jan. 09	Johnson, Capt. G. T.	1 Sep. 15	16 Mar. 10	*Lv., ex I., m. c.*
6 Feb. 09	*Warner, Capt. C. T.*	*1 Sep. 15*	*19 Aug. 14*	
22 Jan. 13	Beanland, Capt. D.	22 Jan. 17	13 Oct. 14	With 2-56th Punj.
15 Nov. 15	Pine-Coffin, Lt. E. C.	15 Nov. 16	17 Nov. 15	Attd. With 2-27th Punj.
8 Sep. 15	Hathaway, Lt. C. J. (*Actg. Capt.*)	8 June 17	14 Aug. 16	Adjt. (*Offg.*)
30 Jan. 17	Moysey, Lt. R. A. L.	30 Jan. 18	14 Sep. 18	
18 June 17	Tottenham, Lt. J. L.	18 June 18		With 2-19th Punj.
24 Aug. 15	Joyce, Lt. W. D.	24 May 17	5 Sep. 16	Attd. Qr. Mr. (*Offg.*)
	Attached.			
16 July 90	Pennefather Evans, Lt.-Col. G. (Q)	3 June 16	13 June 16	19th Punj. Comdt. (*temp.*)
7 Apl. 11	*Huddleston, Capt. G.R. G.*	*1 Sep. 15*	*22 Sep. 14*	*I. A. R.O.*
21 May 15	*Potter, Capt. L. R.*	*21 May 18*	*24 Sep. 15*	*I. A. R. O.*
24 Dec. 14	Mears, Lt. H.	24 Dec. 15	25 May 15	I. A. R. O.
20 July 15	Drew, Lt. J. D. (*Actg. Capt.*)	20 July 16	7 June 16	I. A. R. O.
1 Oct. 15	Donald, Lt. A. (*Actg. Capt.*)	1 Oct. 16	27 Dec. 17	I. A. R. O.
25 Mar. 16	Delahunty, Lt. R. A. (*Actg. Capt.*)	25 Mar. 17	. . .	I. A. R. O.
31 Oct. 17	Stewart, Lt. R. R.	31 Oct. 18	21 Jan. 18	I. A. R. O.
5 Oct. 17	Legg, Lt. P. C.	5 Oct. 18	21 Jan. 18	I. A. (*T. C.*)
27 Oct. 17	Warren, Lt. A. G.	27 Oct. 18	2 Nov. 17	I. A. (*on probn.*)
15 Jan. 18	Williamson, 2nd-Lt. A.	15 Jan. 18	10 June 18	I. A. (*T. C.*)
19 Sep. 18	Anderson, 2nd-Lt. R. C.	19 Sep. 18	26 Aug. 18	R. Dub. Fus., I. A. (*on probn.*)
31 Oct. 17	Ginns, 2nd-Lt. R.	31 Oct. 17	26 Aug. 18	I. A. (*on probn.*)
	Smith, 2nd-Lt. R. J.	. . . .	. . . .	A Cyclist Corps. I. A. (*on probn.*).
29 July 15	Hind, Lt. H. W., D.S.O., M.C.	29 July 16	28 Oct. 18	Lond. R., I. A. (*on probn.*)
12 Apl. 16	Dunkerton, Lt. E. L. H.	12 Apl. 17	23 Nov. 18	York & Lanc. R., I. A. (*on probn.*).
27 Oct. 16	Morrison, Lt. B. H.	27 Oct. 17	11 Sep. 18	I. A. (U. L.)
. . . .	Ross, 2nd-Lt. J. C.	. . . .	8 Nov. 18	I. A. (*on probn.*)
7 June 18	Hopkinson, 2nd-Lt. G. G.	7 June 18	20 Oct. 18	I. A. (*T. C.*)
22 July 18	Towner, 2nd-Lt. C.	22 July 18	3 Dec. 18	I. A. (*T. C.*)
16 Sep. 18	Wright, 2nd-Lt. F. C.	16 Sep. 18	2 Oct. 18	I. A. R. O.
23 Oct. 16	Hitchcock, Lt. L. W.	23 Oct. 17	. . . .	M. G. Corps, I. A. (*on probn.*)
	Subadar-Major.	**Jemadar.**	**Subadar.**	**Subadar-Major.**
15 July 93	**Raja Khan (52, 56)**	**24 Apl. 05**	**1 Sep. 10**	**25 July 18.**
	Subadars.			
25 July 84	Dasaundha Singh, *Sardar Bahadur (72a)*	1 Jan. 00	16 July 06	***Order of Br. Ind., 1st Cl.*** **With 2-22nd Punj.**
12 Oct. 97	Rala Singh (53, 56, 58, 60, 66*a*)	16 July 06	1 Feb. 11	
4 July 91	Badan Singh (52, 56)	1 Feb. 11	11 Oct. 14	
17 Aug. 93	Mehar Singh	11 Oct. 14	29 Sept. 15	
8 Oct. 93	Gurditt Singh (52, 56, 66*a*)	1 July 13	24 Nov. 15	
5 June 93	Roda Singh (56)	1 June 14	24 Nov. 15	
17 Sep. 95	Ali Beg (50)	1 Apl. 11	29 Apl. 16	
2 July 00	Ahmad Din	11 Oct. 14	12 Oct. 16	
16 May 03	Labh Singh	1 July 00	1 May 17	With 2-67th Punj.
6 Nov. 99	Azim Khan	11 Oct. 14	5 Nov. 17	With 2-56th Rif.
6 July 94	Suhel Singh	11 Oct. 14	11 Nov. 17	
26 Jan. 06	Ali Akbar	9 June 17	19 Mar. 18	
17 Dec. 03	Bichitar Singh (56)	11 Oct. 14	10 Aug. 18	

1st Battalion, 22nd Punjabis—*contd.*

1st Commn. or date of entering service.	Names and Rank.	Army rank.	Remarks.
		Jemadars	
17 Feb. 97	Katha Singh	7 June 16	With 2-56th Rif.
2 Apl. 02	Qusim Khan	7 June 16	Indian Adjt.
19 Feb. 96	Harnam Singh	29 Sep. 15	
2 Feb. 92	Indar Singh	29 Sep. 15	
14 Oct 00	Lal Singh, I.O.M.	29 Sep. 15	
8 Jan. 00	Hakim Khan, I.D.S.M.	29 Sep. 15	
9 Nov. 00	Fatohi Khan	24 Nov. 15	
18 Nov. 99	Muhammad Sher	24 Nov. 15	
1 Oct. 95	Firoz Khan	24 Nov. 15	
4 Dec. 97	Udham Singh	24 Nov. 15	
18 July 95	Wazir Singh	24 Nov. 15	
20 Oct. 04	Sayyid Ahmad (58)	24 Nov. 15	
29 July 87	Devi Raj	15 Jan. 17	
18 July 95	Labh Singh	27 Mar. 17	
11 Feb. 97	Fateh Singh	16 May 17	
28 Nov. 00	Rahim Ali Khan (46)	5 Oct. 17	
29 July 00	Saida Khan	5 Nov. 17	
10 July 06	Muhabat Khan (58)	11 Nov. 17	With 2-56th Rif.
19 Oct. 00	Muhammad Amir	3 Dec. 17	Indian Qr. Mr.
25 Apl.	Ghulam Rasul	19 Mar. 18	
17 Feb. 97	Hardit Singh	10 Aug. 18	
23 Sep. 99	Dula Singh	10 Aug. 18	
29 Nov. 03	Sardar Khan	10 Aug. 18	
28 Jan. 04	Khan Muhammad	10 Aug. 18	
21 Sep. 09	Bakhsh Muhammad	1 Sep. 18	
19 Dec. 06	Abbas Ali	1 Sep. 18	

2nd Battalion, 22nd Punjabis.

Formed at Rawalpindi on 30th January 1918.

Composition—1 Company Sikhs, 2½ Companies Punjabi Musalmans, ½ Company Pathans, 1 Special Company of Hindu Aroras.

1st Commn. or date of entering service.	Names and Rank.	Army rank.	Present appointment in Regiment.	Remarks.
	Commandant.			
22 Jan. 98	Clifford, Maj. R. (*Actg. lt.-Col.*)	1 Sep. 15	19 Jan. 18	1-22nd Punj.
	Company Commanders. (4)			
13 Aug. 04	Middlemass, Capt. G deS. H. (*Actg. Maj.*)	1 Sep. 15	11 Jan. 18	122nd Infy. 2nd-in-Command.
6 Jan. 15	Cordon, Lt. C. I., M.C. (*Actg. Capt.*)	1 Oct. 16	2 Mar. 18	I. A. (*on probn.*)
25 Dec. 15	Carey, Lt. J. S. (*Actg. Capt.*) .	25 Dec. 16	13 Feb. 18	I. A. (*on probn.*)
14 July 15	Bourdillon, Lt. F. H. (*Actg. Capt.*	14 July 16	19 Aug. 18	36th Sikhs.
	Company Officers.			
14 Dec. 15	King, Lt. S. J.	14 Sep. 17	29 Jan. 18	I. A. (*on probn.*)
20 Jan. 16	Bygate, Lt. C. M.	20 Oct. 17	27 Jan. 18	I. A. (*on probn.*)
24 Jan. 15	Widdows, Lt. F. M.	24 Oct. 16	10 Sep. 18	Manch. R., I. A. (*on probn.*)
12 Sep. 17	Jones, Lt. J. L.	12 Sep. 18	12 Jan. 18	1-91st Punj.
24 Dec. 15	Hull, Lt. J. R.	24 Dec. 16	13 Feb. 18	I. A. R. O.
5 May 17	Noyle, Lt. H. W.	5 May 18	14 Mar. 18	I. A. R. O.
20 June 17	Bailey, Lt. D. C.	20 June 18	12 Jan. 18	I. A. R. O. Qr.-Mr. (*Actg.*)
31 Oct. 17	Cook, Lt. E. H.	31 Oct. 18	23 Jan. 18	I. A. R. O.
5 Nov. 17	Baker, Lt. V. A.	5 Nov. 18	14 Mar. 18	I. A. (*T.C.*)
23 Dec. 14	Hervey, Lt. E. G. (*Actg. Capt.*)	23 Dec. 15	1 Feb. 18	I. A. R. O. Adjt.
24 Apl. 16	Glover, Lt. M.	24 Apl. 17	23 Sep. 18	S. Lanc. R., I. A. (*on probn.*)
26 Jan. 18	Lees, 2nd-Lt. E. A.	26 Jan. 18	22 Nov. 18	Lan. Fus., I. A. (*on probn.*)
7 June 18	Clark, 2nd-Lt. L. A. F.	7 June 18	10 Oct. 18	I. A. R. O.
22 July 18	Wardle, 2nd-Lt. F.	22 July 18	2 Dec. 18	I. A. (*T. C.*)
8 Oct. 18	Banyard, 2nd-Lt. E. F.	8 Oct. 18	26 Oct. 18	I. A. R. O.
8 Oct. 18	Cockrane, 2nd-Lt. J. A.	8 Oct. 18	30 Oct. 18	I. A. R. O.
	Medical Officer.			
31 July 09	Burke, Capt. G. T.	31 July 12	6 Mar. 18	

		Jemadar.	Subadar.	Subadar-Maor.	
	Subadar-Major.				
25 July 84	Dasaundha Singh, *Sardar Bahadur* (72*a*).	1 Jan. 00	16 July 06	30 Jan. 18	*Order of Br. Ind. 1st Cl.* 1-22nd. Punj (*Actg.*)
	Subadars.				
23 Apl. 98	Ghulam Hussain (53, 56, 58) .	1 Sep. 10	24 Nov. 15		
25 May 99	Jiwan Singh (56)	25 Sep. 17	30 Jan. 18		
10 Oct. 94	Nur Khan	. . . . .	30 Jan. 18		
1 Mar. 05	Najabat Ali (56)	30 Jan. 18	1 July 18		
27 Nov. 03	Haji Khan	29 Mar. 18	1 July 18		
2 Jan. 12	Zardad Khan (48*a*, 56). . . .	29 Mar. 18	1 July 18		
. . .	Nathe Khan	. . . . .	. . . . .	Re-employed.	
	Jemadars.				
5 June 99	Chajja Singh (56)	30 Jan. 18			
30 Oct. 99	Khusal Singh	30 Jan. 18			
2 Mar. 00	Kajir Khan (56)	30 Jan. 18	Indian Adjt.		
. . .	Nur Muhammad	13 Apl. 18			
. . .	Abdullah Khan	27 Apl. 18			
. . .	Muhammad Khan	22 Apl. 18	*On probn*		

Regt. Centre—Ambala. *Uniform*—Drab.
Linked with—32nd and 34th S. Pioneers. *Facings*—Chocolate.

1st Battalion, 23rd Sikh Pioneers.

Raised at Lahore in 1857, by Lt. R. H. Shebbeare, as the 15th (Pioneer) Regt. of Punjab Infy. Became the 27th Regt. of Bengal Native Infantry, 1861; the 23rd Regt. of Bengal N. I., 1861; the 23rd (Punjab) Regt. of Bengal N. I. (Pioneers), 1864; the 23rd (Punjab) Regt. of Bengal Infy. (Pioneers), 1885; the 23rd Punjab Pioneers, 1901; present designation, 1903.

"Taku Forts" "Pekin, 1860" "Abyssinia"
"Peiwar Kotal" "Charasiah" "Kabul, 1879," "Kandahar, 1880" "Afghanistan, 1878-80" "Chitral."
Composition—4 Companies Mazbi and Ramdasia Sikhs.

1st Commn. or date of entering service.	Names and Rank.	Army rank.	Present appointment in Regiment.	Remarks.
	Commandant.			
9 Apl. 92	Ottley, Lt.-Col. W. J., *q.*	21 Feb. 18	21 Feb. 18	With 2-23rd Sikh Prs.
	Company Commanders. (4)			
20 Jan. 97	Hodgson, Maj. G. C., D.S.O. (*Actg. Lt.-Col.*)	20 Jan. 15	22 Feb. 18	2nd-in-Comd. With 3-32nd Prs.
4 Aug. 97	Nicolas, Maj. R., *q.*	4 Aug. 15	8 Oct. 09	With 2-155th Prs.
27 July 98	*Drew, Bt.-Lt.-Col. B. C. H., p.s.c.*	*1 Jan. 17*	*8 Oct. 12*	
3 Aug. 98	Hutchinson, Maj. A. G. C., *p.s.c.* (*Actg. Lt.-Col.*)	1 Sep. 15	21 Sep. 14	With 3-34th Prs.
	Company Officers. (9)			
28 July 00	Turnbull, Maj. B. (*Actg. Lt.-Col.*)	1 Sep. 15	19 Oct. 01	With 2-34th Prs.
18 Jan. 02	Renny, Maj. H. B., M.C., *q.*	18 Jan. 17	19 Dec. 05	
10 Oct. 03	Campbell, Maj. R. N. B.	10 Oct. 18	25 Mar. 05	With 2-23rd Prs.
9 Jan. 04	Bird, Capt. W. A. H., *q.*	9 Jan. 13	20 Mar. 05	With 2-32nd Prs.
9 Jan. 04	Ricketts, Capt. E. W. C. (*Actg. Maj.*)	9 Jan. 13	13 Mar. 06	With 1-34th Prs.
24 Jan. 06	Chesney, Capt. D. P. [L] *q.*	24 Jan. 15	23 Mar. 07	Qr.-Mr., 9 Nov. 13.
29 Aug. 06	Duberly, Capt. M. R. W., *q.*	29 Aug. 15	2 July 12	Adjt., 8 July 15.
18 Jan. 11	*Gabbett, Capt. K. J.*	*1 Sep. 15*	*7 Mar. 12*	*Asst. Supdt., Army Clothing Dept.*
20 Jan. 12	Hogge, Capt. A. H. F.	20 Jan. 16	9 Mar. 13	With 2-23rd Prs.
14 Jan. 14	Fletcher, Capt. R. C.	14 Jan. 18	11 Oct. 14	
29 June 16	Oates, Lt. W. J. S.	15 Feb. 17	10 Jan. 17	Attd. With 2-23rd Prs.
18 Apl. 16	Wallis, Lt. S. I.	18 Apl. 17	25 Apl. 16	Attd.
27 Oct. 17	Hobbs, Lt. A. J.	27 Oct. 18	4 Nov. 17	
27 Oct. 17	McDonald, Lt. J. B.	27 Oct. 18	4 Nov. 17	With 3-34th Prs.
4 Nov. 15	Christison, Lt. F. H. (*temp. Capt.*)	4 Nov. 16	23 Oct. 17	Attd.
31 Aug. 18	Hutchinson, 2nd-Lt. G. W.	31 Aug. 18	4 Sep. 18	Attd.
	Attached.			
16 Sep. 14	Longden, Lt. D. J.	16 June 16	1 Apl. 18	I. A. (*on probn.*)
4 Dec. 14	Treloar, Capt. R. M.	4 Dec. 18	22 Sep. 15	I. A. R. O.
20 Dec. 14	Stewart, Lt. R. H.	20 Feb. 16	12 Jan. 16	I. A. R. O.
18 Feb. 15	Hearn, Lt. J. (*Actg. Capt.*)	18 Feb. 16	3 Jan. 18	I. A. R. O.
14 May 15	Barrack, Lt. J. C.	14 May 16	22 Oct. 15	I. A. R. O.
7 Mar. 16	Wilkinson, Lt. F. R.	7 Mar. 17	25 Mar. 18	I. A. (*on probn.*)
10 Mar. 16	Warren, Lt. D. F. W.	10 Mar. 17	26 Mar. 18	I. A. (*on probn.*)
25 Sep. 16	Parkinson, Lt. H. L.	25 Sep. 17	7 July 18	I. A. R. O.
20 June 17	Cullis, Lt. L.	20 June 18	4 Oct. 17	I. A. R. O.
20 June 17	Ritson, Lt. F.	20 June 18	4 Oct. 17	I. A. R. O.
12 Sep. 17	McKenzie, 2nd-Lt. A. T., M.M.	12 Sep. 17	26 Mar. 18	I. A. R. O.
5 Oct. 17	Armstrong, Lt. A. J. E. (*Actg. Capt.*)	5 Oct. 18	21 Jan. 18	I. A. (*T. C.*)
30 Oct. 17	Cross, 2nd-Lt. S. G. N.	30 Oct. 17	25 Mar. 18	I. A. (*on probn.*)
5 Nov. 17	Philpott, Lt. G. R.	5 Oct. 18	14 Mar. 18	I. A. (*T. C.*)
5 Nov. 17	Wyatt, Lt. R. W.	5 Nov. 18	14 Mar. 18	I. A. (*T. C.*)
25 Nov. 17	Redfern, 2nd-Lt. T.	25 Nov. 17	24 July 18	I. A. R. O.
3 June 18	Rees, 2nd-Lt. H. V.	3 June 18	17 June 18	I. A. R. O.
7 June 18	Green, 2nd-Lt. J.	7 June 18	9 Oct. 18	I. A. R. O.
5 Sep. 18	Swales, 2nd-Lt. F. W.	5 Sep. 18	26 Sep. 18	I. A. R. O.
7 Sep. 18	Bickers, 2nd-Lt. A. A. D.	7 Sep. 18	18 Sep. 18	I. A. R. O.

Subadar-Major.

		Jemadar.	Subadar.	Subadar-Major.
27 Oct. 85	Man Singh *Bahadur*	16 Apl. 01	15 Apl. 06	29 May 15 *Order Br. Ind.*

1st Battalion, 23rd Sikh Pioneers—*contd.*

1st Commn. or date of entering service.	Names and Rank.	Army rank.		Remarks.
	Subadars.			
1 Mar. 95	Kala Singh (56)	15 Apl. 06	1 Mar. 12	
	Sant Singh (*Retd.*)	. . . .	. . . .	Re-employed, 4 Apl. 15.
	Ganesha Singh (*Retd.*)	. . . .	. . . .	Re-employed, 18 Oct. 16.
23 Nov. 91	Bhola Singh	1 Feb. 10	9 Sep. 16	
10 July 96	Sohan Singh (56)	1 Mar. 12	6 Nov. 16	
5 Feb. 95	Chanda Singh	11 Oct. 14	3 Feb. 17	
13 Feb. 01	Bawa Singh	6 Nov. 16	12 May 17	
20 Mar. 01	Basawa Singh	6 Nov. 16	7 Dec. 17	
28 Feb. 89	Jawahir Singh	6 Nov. 16	1 Jan. 18	
16 June 91	Harnam Singh (I)	6 Nov. 16	1 Jan. 18	
27 June 87	Bishan Singh	3 Feb. 17	1 Jan. 18	
1 June 05	Teja Singh, I.O.M.	3 Feb. 17	12 June 18	
8 Oct. 94	Hira Singh	6 Sep. 17	1 Aug. 18	
2 May 02	Sant Singh	16 Jan. 18	12 June 18	
	Jemadars.			
18 June 96	Harnam Singh (III)	3 Feb. 16		
2 July 01	Bhagat Singh	3 Feb. 17		
17 Nov. 92	Kishan Singh	1 Dec. 17		
12 Jan. 06	Dewa Singh	1 Dec. 17		
21 May 06	Kehar Singh	1 Dec. 17		
22 Feb. 10	Narayan Singh	3 Dec. 17		
26 Feb. 95	Sundar Singh	1 Jan. 18		
16 May 98	Binda Parshad	1 Jan. 18		
11 June 98	Kapur Singh	1 Jan. 18		
15 Oct. 98	Ishar Singh	1 Jan. 18		
16 May 09	Kartar Singh	1 Jan. 18		
25 May 97	Jhanda Singh	16 Jan. 18		
8 Apl. 01	Atma Sing	16 Jan. 18		
23 Mar. 03	Balwant Singh	16 Jan. 18		
20 June 08	Sher Singh	26 Apl. 18		
14 July 92	Sundar Singh	1 June 18		
21 Apl. 05	Katha Singh	12 June 18		
15 May 05	Bassawa Singh	12 June 18		
12 Nov. 06	Nihal Singh	12 June 18		
14 July 07	Sohan Singh	12 June 18		
16 Oct. 08	Puran Singh	4 July 18		
12 Apl. 04	Dhanna Singh	1 Aug. 18		
11 June 04	Pala Singh	1 Aug. 18		
15 May 05	Dalip Singh	1 Aug. 18		

2nd Battalion, 23rd Sikh Pioneers.

Formed at Shallufa in Egypt on 3rd, 1917.

Composition—4 Companies of Mazbi and Ramdasia Sikhs.

1st Commn. or date of entering service.	Names and Rank.	Army rank.	Present appointment in Regiment.	Remarks.
	Commandant.			
9 Apl. 92	Ottley, Lt.-Col. W. J.	21 Feb. 18	3 Feb. 17	1-23rd Prs.
	Company Commanders. (4)			
10 Oct. 03	Campbell, Maj. R. N. B.	10 Oct. 18	3 Feb. 17	1-23rd Prs.
20 Jan. 12	Hogge, Capt. A. H. F.	20 Jan. 16	3 Feb. 17	
4 Dec. 14	Homer, Lt. N. H. T.	4 Dec. 15	3 Feb. 17	I. A. R. O.
	Company Officers.			
29 June 16	Oates, Lt W. J. S.	29 June 16	15 Feb. 17	1-23rd Prs.
19 July 16	Lillies, Lt. G. L.	19 July 17	26 May 17	83rd Infy.
25 Aug. 16	Higginson, Lt. N., M.C.	25 Aug. 17	12 Jan. 17	I. A. R. O.
20 Oct. 14	Jenner, Lt. C. H.	20 July 16	11 Oct. 17	Hamps. R., I. A. (*on probn.*)
5 June 17	Dening, Lt. S. J. C.	5 June 18	. . . .	Devon R., I. A. (*on probn.*)
5 May 17	Kidson, Lt. R. A. B.	5 May 18	20 Oct. 17	I. A. R. O.
27 Mar. 18	Cole, 2nd-Lt. A. H.	27 Mar. 18	. . .	I. A. (*on probn.*)
20 Sep. 18	Grahamslaw, 2nd-Lt. J. G.	28 Sep. 18	12 Oct. 18	I. A. R. O.

Subadar-Major.

		Jemadar.	Subadar.	Subadar-Major.	
3 Feb. 91	Mota Singh, *Bahadur*, I. D. S. M. (56)	1 July 06	1 Feb. 10	3 Feb. 17	*Order of Br. Ind., 2nd Cl.*

Subadars.

26 Apl. 93	Sundar Singh	19 Nov. 13	6 Nov. 16
5 Feb. 95	Chanda Singh	11 Oct. 14	3 Feb. 17
14 Oct. 96	Hira Singh, I. D. S. M.	29 May 15	3 Feb. 17
1 Nov. 95	Jhanda Singh	6 Nov. 16	3 Feb. 17
27 Dec. 95	Sadda Singh	6 Nov. 16	8 Oct. 17
18 Mar. 96	Harnam Singh (II)	6 Nov. 16	22 Oct. 17
6 une 04	Mall Singh	15 Feb. 17	28 Mar. 18
26 July 07	Ala Singh	15 Feb. 17	26 May 18
5 Sep. 01	Bir Singh	18 Apl. 17	26 May 18
1 Dec. 90	Pala Singh	23 Aug. 17	26 May 18
15 Sep. 05	Ram Singh	17 Jan. 18	13 June 18

Jemadars.

27 Dec. 02	Hira Singh	18 Feb. 17
17 Apl. 06	Bir Singh	2 May 17
3 May 98	Gurdial Singh	8 Oct. 17
10 Sep. 07	Narayan Singh	3 Dec. 17
22 May 08	Sundar Singh	17 Jan. 18
10 Dec. 98	Harnam Singh	21 Feb. 18
24 Aug. 10	Tara Singh	26 May 18
23 Feb. 00	Mangal Singh	26 May 18
19 Jan. 06	Phaga Singh	26 May 18
1 Sep. 02	Harnam Singh	13 June 18
7 Aug. 09	Daula Singh	13 June 18

3rd Battalion, 23rd Sikh Infantry.

Formed at Jullundur on 20th January, 1918.

Composition—4 Companies Mazbi and Ramdasia Sikhs.

1st Commn. or date of entering service.	Names and Rank.	Army rank.	Present appointment in Regiment.	Remarks.
	Commandant.			
27 July 98	Talbot, Bt.-Lt.-Col. F. E. G., *p. s. c.*	1 Jan. 17	21 Jan. 18	14th Sikhs.
	Company Commanders. (4)			
18 Jan. 02	Carey, Maj. P. G.	18 Jan. 17	6 Feb. 18	31st Punj. 2nd-in-Comd. (*temp.*)
11 Nov. 14	Stead, Capt. E. A.	11 Nov. 18	30 Jan. 18	35th Sikhs.
1 June 15	Hamlyn, Lt. H. W. A. (*Actg. Capt.*)	1 June 16	13 Aug. 18	1-88th Infy.
19 Nov. 14	Martin, Lt. V. J. (*Actg. Capt.*)	6 July 16	16 Feb. 18	M. G. Corps, I. A. (*on probn.*)
	Company Officers.			
	Nesbitt, 2nd-Lt. H. U. S.		19 Nov. 18	W. Kent R., I. A. R. O. (*on probn.*)
28 Apl. 15	Morgan, Lt. K. S.	28 Apl. 16	28 Feb. 18	*Lv., ex I., 1 yr., m. c.*
29 Jan. 16	Miller, Lt. L. H.	29 Jan. 17	29 May 18	I. A. (*on probn.*)
6 Feb. 16	James, Lt. H. V.	6 Feb. 17	16 May 18	Welsh R., I. A. (*on probn.*)
27 June 16	Couldrey, Lt. O., M.C.	27 June 17	2 Oct. 18	M. G. Corps, I. A. (*on probn.*)
22 Nov. 16	Digby, 2nd-Lt. P. R. (*Actg. Capt.*)	22 Nov. 17	28 Feb. 18	I. A. R. O. Adjt. (*Offg.*)
5 May 17	Trinkle, Lt. P.	5 May 18	19 Aug. 18	I. A. R. O.
26 Nov. 17	Wright, 2nd-Lt. R. J.	26 Nov. 17	5 June 18	I. A. R. O.
15 Jan. 18	Joaquim, 2nd-Lt. B. J. P.	15 Jan. 18	1 June 18	I. A. R. O.
15 Jan. 18	Eldred, 2nd-Lt. H. S.	15 Jan. 18	5 June 18	I. A. (T. C.)
27 Mar. 18	Cole, 2nd-Lt. A. H.	27 Mar. 18	16 May 18	I. A. (*on probn.*) Qr.-Mr. (*Offg.*)
27 Mar. 18	Crane 2nd-Lt. A.	27 Mar. 18	18 Nov. 18	Notts. & Derby. R.
29 Mar. 18	Pleasance, 2nd-Lt. A. G.	29 Mar. 18	15 July 18	I. A. (T. C.)
29 Mar. 18	Richardson, 2nd-Lt. F. J.	29 Mar. 18	15 July 18	I. A. (T. C.)
30 Apl. 18	Allen, 2nd-Lt. H. K.	30 Apl. 18	22 Nov. 18	Notts. & Derby. R.
1 May 18	Cook, 2nd-Lt. R. D.	1 May 18	18 Nov. 18	Notts. & Derby. R.
15 June 18	Martin, 2nd-Lt. G. F.	15 June 18	5 Oct. 18	Genl. List, I. A. R. O. (*on probn.*)
25 June 11	Shaw, 2nd-Lt. P. R.	25 June 18	18 Nov. 18	Notts. & Derby. R.
22 July 18	Gerrey, 2nd-Lt. G. W.	22 July 18	1 Dec. 18	I. A. (*T. C.*)
31 July 18	Denyer, 2nd-Lt. P. H.	31 July 18	17 Oct. 18	Genl. List, I. A. R. O. (*on probn.*)
6 Sep. 10	Ross, 2nd-Lt. J. R.	6 Sep. 18	6 Sep. 18	Attd.
21 Sep. 18	Miller, 2nd-Lt. O. G.	21 Sep. 18	6 Oct. 18	I. A. R. O.
	Milligan, Lt. J. W.		...	R. Ir. Rif., I. A. (*on probn.*)

		Jemadar.	Subadar.	Subadar-Major.
	Subadar-Major.			
17 Mar. 91	Thakur Singh	1 Oct. 02	1 Mar. 07	1 Feb. 18
	Subadar.			
21 Jan. 99	Ladha Singh	3 Feb. 17		
14 July 99	Harnam Singh	3 Feb. 17		
2 Dec. 94	Tilok Singh	3 Feb. 17		
24 Apl. 00	Fateh Singh	3 Feb. 17		
8 Aug. 99	Kehr Singh	3 Feb. 17		
17 Apl. 03	Hari Singh	1 Feb. 18	1 Aug. 18	
30 Apl. 02	Sapuran Singh	1 Feb. 18	23 Oct. 18	
	Jemadars.			
25 June 00	Khushal Singh I (48*a*, 51)	1 Feb. 18		
7 Aug. 01	Narayan Singh	1 Feb. 18		
18 Apl. 02	Ghanaiya Singh	1 Feb. 18		
31 Jan. 04	Surain Singh	1 Aug. 18		
	Mangal Singh	23 Oct. 18		
	Khushal Singh II	23 Oct. 18		
	Mal Singh	23 Oct. 18		
	Bishan Singh	23 Oct. 18		
	Jaswant Singh	23 Oct. 18		
	Asa Singh	23 Oct. 18		
	Bhagat Singh	23 Oct. 18		
	Madho Singh	23 Oct. 18		

Regt. Centre—Jhelum. *Uniform*—Scarlet.
Linked with—19th and 22nd Punjabis. *Facings*—White.

24th Punjabis.

Raised at Peshawar in 1857, by Capt. G. N. Cave, as the 16th Regt. of Punjab Infy. Became the 28th Regt. o Bengal Native Infantry, 1861; the 24th Regt. of Bengal N. I., 1861; the 24th (Punjab) Regt. of B. N. I., 1864 the 24th (Punjab) Regt. of Bengal Infy., 1885; the 24th Punjab Infy., 1901; present designation, 1903.
"Kandahar, 1880" "Afghanistan, 1878-80" "Malakand" "Punjab Frontier" "Pekin, 1900."
Composition—2 Companies of Sikhs, ½ of Dogras, 1 of Afridis, ½ of Punjabi Musalmans.
COLONEL—Major-General *Sir* John G. Ramsay, K.C.B. . . 6 January 1911.

1st Commn. or date of entering service.	Names and Rank.	Army rank.	Present appointment in Regiment.	Remarks.
	Commandant.			
	Company Commanders. (4)			
21 Sep. 89	*Cummins, Bt.-Col. H. A. V., p.s.c. (Q).*	*3 June 16*	*13 Feb. 12*	*2nd-in-Comd.*
21 Dec. 89	*Rawlins, Lt.-Col. A. K., C. I. E., D.S.O.*	*21 Dec. 15*	*13 Feb. 05*	*Bikaner Camel Corps.*
20 July 98	*McWatters, Bt.-Lt.-Col. H. C., D.S.O., p.s.c.*	*3 June 17*	. . .	
	Company Officers. (9)			
9 Jan. 04	*Trevor, Capt. A. C. H.*	*9 Jan. 13*	*22 Mar. 05*	
18 Jan. 05	Haig, Capt. A. B.	18 Jan. 14	19 Mar. 06	
29 Aug. 06	Thornhill, Capt. C. M., D.S.O., M.C. (*Actg. Maj.*)	29 Aug. 15	19 Nov. 07	
20 Jan. 09	Scroope, Capt. C. F.	1 Sep. 15	4 Oct. 18	66th Punj.
29 Jan. 10	Langhorne, Capt. C. C.	1 Sept. 15	20 June 13	
24 Aug. 12	Pim, Capt. H. M., M.C.	24 Aug. 16	7 Oct. 14	
15 Nov. 15	Dawes, Lt. K. C. B.	15 Nov. 16	18 Nov. 15	With 2-125th Rif.
15 Nov. 15	Troughton-Dean, Lt. G. P.	15 Nov. 16	18 Nov. 15	
18 Apl. 16	Yarrow, Lt. T. B. (*Actg. Capt.*)	18 Apl. 17	28 Apl. 16	Attd.
18 Apl. 16	Sevenoaks, Lt. C. L.	18 Apl. 17	28 Apl. 16	Attd. with 1-152nd Infy.
19 July 16	*Attfield, Lt. C. J.*	*19 July 17*	*8 Nov. 16*	*Attd. Frontier Militia.*
1 May 17	Gault, Lt. C. (*actg. Capt.*)	1 May 18	29 Nov. 17	Attd.
25 Aug. 16	Stubbs, Lt. H. G. P.	25 May 18	10 Mar. 17	Attd. Adjt.
27 Oct. 17	Cotton, 2nd-Lt. J. C.	27 Oct. 17	2 Nov. 17	Attd.
	Attached.			
23 Dec. 93	Furney, Bt.-Lt.-Col. J. L. (Q)	3 June 16	26 Feb. 17	22nd Punj., Comdt. (*temp.*)
18 Jan. 02	Atkins, Maj. G. W., M.C.	18 Jan. 17	1 Feb. 17	25th Punj.
22 Jan. 13	Lockhart, Capt. R. Mac. G. M., M.C.	22 Jan. 17	3 Nov. 18	51st Sikhs.
30 Apl. 15	*Stapleton, Capt. H. E.*	*30 Apl. 18*	. . .	*I. A. R. O.*
3 July 15	Hay, Lt. W. R.	3 July 16	. . .	*I. A. (on probn.)* Qr.-Mr. 30 Sept. 16.
27 Nov. 14	*Duxbury, Lt. E. L.*	*27 Nov. 15*	*24 Apl. 15*	*I. A. R. O.*
14 Jan. 15	Phillips, Lt. C. H. K., M.C. (*Actg. Capt.*)	14 Jan. 16	15 Jan. 15	I. A. R. O.
14 Jan. 15	*Phillips, Lt. W. W.*	*14 Jan. 16*	*15 Jan. 15*	*I. A. R. O.*
12 Feb. 15	*Browne, Lt. H.*	*12 Feb. 16*	*19 Aug. 15*	*I. A. R. O.*
14 May 15	Kemp, Capt. K. McI.	16 Dec. 17	15 Feb. 17	I. A. R. O.
29 May 17	Penny, 2nd-Lt. C. E.	29 May 17	10 Dec. 17	I. A. R. O.
14 Nov. 16	Colbourne, Lt. W. A.	14 Nov. 17	28 Dec. 17	25th Punj.
3 Apl. 15	Fison, Lt. C. H.	3 Apl. 16	16 May 18	I. A. (*on probn.*)
15 Jan. 18	Miller, 2nd-Lt. R. G.	15 Jan. 18	28 June 18	I. A. (*T. C.*)
8 Aug. 16	Cunningham, Lt. A. P.	8 Aug. 17	16 June 18	S. Lan. R. I. A. (*on probn.*)
31 July 18	Dobney, 2nd-Lt. T. A.	31 July 18	30 Apl. 18	I. A. (*on probn.*)
5 Oct. 17	Page, Lt. C.	5 Oct. 18	16 May 18	I. A. (*T. C.*)
14 Nov. 16	Barry, Lt. F. C.	3 May 17	13 Feb. 17	52nd Sikhs.
27 Oct. 17	Gwyn, 2nd-Lt. P. E. C. J.	27 Oct. 17	20 May 17	I. A. (*on probn.*)
29 Mar. 18	Carey, 2nd-Lt. T. W.	29 Mar. 18	. . .	I. A. (*T. C.*)
23 Apl. 18	De Silva, 2nd-Lt., W. R. I.	23 Apl. 18	23 Oct. 18	I. A. (U. L.)
7 June 18	Shannon, 2nd-Lt. A. G. L.	7 June 18	. . .	I. A. R. O.
7 Oct. 18	Sampson, 2nd-Lt. C. A.	7 Oct. 18	1 Nov. 18	I. A. (*T. C.*)
7 Oct. 18	Shackell, 2nd-Lt. A.	7 Oct. 18	1 Nov. 18	I. A. (*T. C.*)

Subadar-Majors.

		Jemadar.	Subadar.	Subadar-Major.	
15 Aug. 87	Sakt Chand, *Sardar Bahadur* [(56, 66*a*)	15 Aug. 87	1 Feb. 91	1 Nov. 14	*Order of Br. Ind., 1st Cl.*
14 Aug. 82	*Diwan Singh, Sardar Bahadur*	*9 Apl. 99*	*6 May 07*	*15 July 15*	*Order of Br. Ind., 1st Cl.*
1 Nov. 01	Labh Singh (56) *Bahadur*	21 Nov. 01	16 Mar. 10	7 Apl. 17	*Order of Br. Ind., 2nd Cl.*

24th Punjabis—*contd.*

1st Commn. or date of entering service.	Names and Rank.	Army rank.	Present appointment in Regiment.	Remarks.
	Subadars.			
5 Feb. 92	Gul Akbar *Bahadur*, M.C., I.D.S.M. (72*a*)	12 Dec. 01	1 Feb. 07	*Order of Br. Ind., 2nd Cl.*
1 Aug. 89	*Umar Khan*, I.O.M. (66*a*) . . .	*1 Dec. 06*	*24 May 14*	
25 July 93	*Kishen Singh* (56, 66*a*) . . .	*14 Apl. 12*	*1 Nov. 14*	
23 Feb. 97	Amir Khan (56)	24 May 14	15 Apl. 15	
21 Mar. 89	*Jhanda Singh* (56)	*1 Aug. 12*	*14 May 15*	
28 Nov. 94	Lal Mir	15 Apl. 15	15 July 15	
12 May 13	Narain Singh	12 May 13	23 Nov. 15	
[illegible] June 98	Ilam Din	5 July 15	23 Nov. 15	
1 Nov. 10	*Prithi Chand* (56)	*16 Oct. 14*	*2 Dec. 15*	
9 June 93	Fateh Singh, I. O. M. (56) *Bahadur*	16 Oct. 14	25 Sep. 16	*Order of Br. Ind., 2nd Cl.*
9 Mar. 94	Ujagar Singh	1 Nov. 14	25 Sep. 16	
18 May 03	*Sarfaraz Khan*	*13 Oct. 15*	*7 Apl. 17*	
4 Dec. 02	Sudama	25 Sep. 16	20 July 17	
7 June 01	Shadam Khan	16 June 15	16 Apl. 18	
23 July 03	Saba Khan (56)	14 Feb. 17	13 May 18	
13 Jan. 96	Hakam Singh (56)	14 Feb. 17	3 Oct. 18	
	Jemadars.			
12 July 93	*Narain Singh*	*4 May 15*		
2 Nov. 93	*Sher Nur*	*2 June 15*		
12 Oct. 91	*Sawan Singh*	*23 Nov. 15*		
26 Oct. 95	*Sher Baz*	*23 Nov. 15*		
9 Apl. 96	*Nathu*	*2 Dec. 17*		
9 Mar. 04	Kehr Sing (II)	25 Sep. 16		
21 Jan. 96	Firoz Khan	14 Feb. 17		
6 Mar. 04	Lal Singh, I.O.M.	14 Feb. 17		
23 Jan. 87	Ala Singh	23 Mar. 17		
21 Mar. 96	Suba	1 July 17		
16 June 05	Karam Ilahi	28 May 17		
13 Oct. 08	Shamsher Singh	20 July 17		
11 Oct. 98	Saiyid Ahmad Khan	19 Sep. 17		
10 Nov. 83	Lachman Singh	3 Dec. 17		
23 Oct. 99	Wadhawa Singh	16 Apl. 18		
15 Jan. 88	Kehr Singh (*Retd.*)	. . . .	Re-employed 12 June 15. Depôt.	
20 Oct. 05	Aksar Khan	25 May 18		
20 Dec. 17	Faiz Ahmad Khan	12 Apl. 18		
18 Jan. 18	Zafar Husain	12 Apl. 18		
28 Jan. 04	Garbhu	25 May 18		
29 July 03	Niamat Khan	25 May 18		
3 Sep. 06	Ali Muhammad	19 June 18		
4 July 08	Khadam Khan (56)	14 July 18		
2 July 06	Ram Saran (56)	1 Sep. 18		
10 Sep. 06	Gurdit Singh	3 Oct. 18		
2 July 03	Ganda Singh, I.D.S.M.	3 Oct. 18		

Regt. Centre—Ferozepore.
Linked with—27th and 28th Punjabis.

1st Battalion, 25th Punjabis.

Uniform—Scarlet
Facings—White

Raised at Mian Mir in 1857, by Capt. R. Larkins, as the Lahore Punjab Battn., and was designated the 17th Regt. of Punjab Infy., 1857. Became the 29th Regt. of Bengal Native Infantry, 1861; the 25th Regt. of Bengal N. I., 1861; the 25th (Punjab) Regt. of Bengal N. I., 1864; the 25th (Punjab) Regt. of Bengal Infy., 1885; the 25th Punjab Infy., 1901; present designation, 1903.

"Ahmad Khel" "Kandahar, 1880" "Afghanistan, 1878-80" "Chitral."

Composition—1½ Companies of Sikhs, 1 of Dogras, 1 of Punjabi Musalmans, ½ of Pathans.

1st Commn. or date of entering service.	Names and Rank.	Army rank.	Present appointment in Regiment.	Remarks.
	Commandant.			
22 Aug. 88	Moberly, Lt.-Col. F. J., D.S.O., *p.s.c.* (*temp. Br.-Genl.*)	13 Feb. 13	25 Feb. 15	*Dir. of Mily. Operations, G. S. Br.*
	Company Commanders. (4)			
25 Jan. 99	Hunt, Maj. H. R. A., D.S.O., *p.s.c.* [*l*]	1 Sep. 15	24 July 18	
25 Jan. 99	Penton, Maj. B. C., *p.s.c.*	1 Sep. 15	7 Nov. 14	With 1-152nd Infy.
28 July 00	Tyrrell, Maj. C. H	1 Sep. 15	2 Aug. 16	
18 Jan. 02	Atkins, Maj. G. W., M. C.	18 Jan. 17	24 July 18	With 24th Punj.
	Company Officers. (9)			
10 Oct. 03	Steel, Maj. C. N.	10 Oct. 18	1 Oct. 06	With I. S. Troops.
9 Jan. 04	Walker, Capt. H. R. O.	9 Jan. 13	12 Oct. 05	With 20th Infy.
18 Jan. 05	Finnis, Capt. S. W. (*Actg. Maj.*)	18 Jan. 14	19 June 06	2nd-in-Comd. (*temp.*)
19 Jan. 07	Gardner, Capt. R. G.	1 Sep. 15	18 May 10	With 28th Punj.
9 Sep. 08	Coats, Capt. N. G. R.	1 Sep. 15	14 Nov. 09	With 1-152nd Infy.
17 Mar. 15	Berkeley, Lieut. E. L. F.	17 Mar. 17	2 May 15	
22 Apl. 15	Carleton, Lt. K. O.	22 Apl. 16	5 Nov. 15	Attd. With 1-143rd Infy.
15 Nov. 15	Kilpin, Lt. J. G.	15 Nov. 16	18 Nov. 15	Attd.
15 Nov. 15	Henson, Lt. R. K.	15 Nov. 16	18 Nov. 15	Attd. With 45th Sikhs.
18 Apl. 16	Neale, Lt. R. K.	18 Apl. 17	26 Apl. 16	Attd.
14 Nov. 16	Colbourne, Lt. W. A.	14 Nov. 17	22 Nov. 16	Attd. With 24th Punj.
18 June 17	Roxburg, Lt. W.	18 June 18	26 June 17	Attd. With 1-152nd Infy.
18 June 17	Needham, Lt. J. O.	18 June 18	26 June 17	Attd. With 1-152nd Infy.
27 Oct. 17	Penney, Lt. N. A.	27 Oct. 18	7 Nov. 17	Attd.
9 Sep. 16	Rowson, Lt. J. C. (*Actg. Capt.*)	9 Sep. 17	12 Mar. 18	I. A. Attd.
4 July 16	Duggan, Lt. R. D.	4 July 17	. . .	Attd. Adjt. (*Offg.*)
25 Oct. 15	Harris, Lt. A. A. F. (*Actg. Capt.*)	25 Oct. 16	17 June 17	Attd.
29 May 15	White, Lt. F. (*Actg Capt.*)	29 May 16	9 Aug. 18	Attd.
31 Aug. 18	Auton, 2nd-Lt. J. H.	31 Aug. 18	4 Sep. 18	Attd.
	Attached.			
16 May 01	Mottram, Capt. P. J. MacQ.	1 Sep. 15	26 Sep. 17	I. A. R. O.
29 Dec. 15	McNeil, Lt. W. A.	29 Dec. 16	30 Apl. 18	R. Innis. Fus., I. A. (*on probn.*)
15 Mar. 16	Best, L. W. J.	15 Mar. 17	. . . .	I. A. (*on probn.*)
14 Apl. 15	Rowland Lt. J. W. (*Actg. Capt.*)	14 Apl. 16	20 Mar. 18	I. A. R. O.
5 May 17	Jackson, Lt. W. E.	5 May 18	28 Aug. 17	I. A. R. O.
5 May 17	Pitts, Lt. G.	5 May 18	28 Aug. 17	I. A. R. O. *Lv., ex I., m. c.*
5 May 17	Frend, Lt. G. V. R.	5 May 18	28 Aug. 17	I. A. R. O.
19 Sep. 18	O'Collins, 2nd-Lt. P. F.	19 Sep. 18	. . . .	Aust'n I. F., I. A. (*on probn.*)
2 Aug. 16	Rains-Bath, Lt. J. G.	2 Aug. 17	28 Sep. 18	Devon R., I. A. (*on probn.*)
3 Sep. 18	Harrison, 2nd-Lt. A. A.	3 Sep. 18	23 Sep. 18	I. A. R. O.
7 June 18	Gould, 2nd-Lt. P. R.	7 June 18	28 Oct. 18	I. A. (*T. C.*)
8 Oct. 18	Allen, 2nd-Lt. D. S.	8 Oct. 18	28 Oct. 18	I. A. R. O.
23 Aug. 17	Murray, Lt. P.	23 Aug. 18	6 Nov. 18	R. Berks R. I. A. (*on probn.*)
9 Nov. 18	Stallard, 2nd-Lt. G. B.	9 Nov. 18	28 Nov. 18	I. A. R. O.
	Subadar-Major.	Jemadar.	Subadar.	Subadar-Major.
24 May 87	Sudama *Bahadur* (56)	1 Dec. 96	1 Nov. 05	1 May 17. *Order of Br. Ind., 2nd cl.*
9 Dec. 90	Mawaz Khan (56, 72*a*)	1 July 07	1 Apl. 12	21 July 18.
	Pir Muhammad Khan (*Ret.*)	. . . .	. . . .	. . . Re-employed. 8 Dec. 17.
	Subadars.			
6 Feb. 06	*Lochan Singh (72a)*	*6 Feb. 06*	*16 Jan. 13*	*With 28th Punjabis.*
1 July 88	Indar Singh (56, 66*a*) I.D.S.M.	1 Apl. 07	1 May 13	
11 Oct. 96	Sher Muhammad Khan (II)	1 Jan. 15	17 May 15	
21 May 93	Ghulam Muhammad (52, 56)	1 Nov. 14	8 Feb. 17	
4 Oct. 93	Harnam Singh	1 July 14	21 May 17	
8 May 05	Muhammad Khan	21 Aug. 17	1 Nov. 17	
	Zaid Gul (*Ret.*)	. . . .	. . . .	Re-employed. 8 Mar. 18.
4 June 92	Sadhu Singh, I.D.S.M. (50)	16 Jan. 13	26 May 18	
10 Apl. 00	Basant Singh	1 May 17	26 May 18	
1 Jan. 98	Maghar Singh	22 May 17	21 July 18	

1st Battalion, 25th Punjabis—*contd.*

1st Commn. or date of entering service.	Names and Rank.	Army rank.	Present appointment in Regiment.	Remarks.
	Jemadars.			
4 Apl. 97	Lal Khan (I)	1 Aug. 16		
1 Feb. 98	Shah Sowar	8 Feb. 17		
20 Aug. 04	*Lal Khan* (II)	*8 Feb. 17*	*With 28th Punj.*	
25 June 02	Gurdas Singh	21 May 17		
15 Feb. 02	Sahib Gul	28 May 17		
5 Feb. 98	Harnam Singh (I)	2 July 17		
5 Aug. 05	Adalat Khan	10 Sep. 17		
16 Jan. 97	Ghulam Rasul	10 Sep. 17		
23 July 00	Maula Dad	29 Sep. 17		
22 Jan. 98	Johli	1 Nov. 17		
27 Jan. 00	Hira	1 Nov. 17		
29 Dec. 02	Kabul Shah (56)	1 Nov. 17		
5 Feb. 98	Harnam Singh (II)	1 Nov. 17		
7 Dec. 01	Mansha Singh	9 Nov. 17		
24 June 06	Mir Afzal	16 Apl. 18		
25 May 05	Zaman Khan	26 May 18		
30 Sep. 98	Bhikho	27 May 18		
20 Nov. 02	Chet Ram	22 July 18		
7 Jan. 99	Kapur Singh	15 Aug. 18		
12 Aug. 08	Ahmad Khan	15 Aug. 18		
11 Feb. 04	Hari Singh	15 Aug. 18		
6 July 08	Nagina Singh	2 Sep. 18		

2nd Battalion, 25th Punjabis.

Formed at Jhansi on 16th October 1917.

Composition— 2 Companies Punjabi Musalmans.
½ Company Bannuchis and Marwats.
½ Company Sikhs.
½ Company Dogras.
½ Company Pathans.

1st Commn. or date of entering service.	Names and Rank.	Army rank.	Present appointment in Regiment.	Remarks.
	Commandant.			
14 Aug. 95	Pratt, Maj. H. R. E., D.S.O. (*Actg. Lt.-Col.*)	14 Aug. 13	11 June 18	1—35th Sikhs.
	Company Commanders (4).			
23 May 00	Swayne, Maj. F. G.	1 Sep. 15	2 Sep. 18	1—35th Sikhs, 2nd-in-Comd.
27 May 08	Guthrie, Capt. I. D., M.C.	1 Sep. 15	23 Oct. 17	17th Cav.
8 Aug. 14	Mahon, Capt. B. MacM., D.S.O., M.C.	20 Feb. 18	29 Mar. 18	Lond R., I. A. (*on probn.*)
1 Oct. 14	Percy-Jones, Capt. C. M.	1 Oct. 18	9 Dec. 17	1—21st Punj.
	Company Officers.			
18 Sep. 15	Hibberd, Lt. A. S. (*Actg. Capt.*)	18 Sep. 16	8 Feb. 18	I. A. (*on probn.*) Adjt.
20 Dec. 15	Priday, Lt. A. K. (*Actg. Capt.*)	20 Dec. 16	14 Apl. 18	I. A. (*on probn.*)
29 July 16	Garrod, Lt. W. F.	29 July 17	15 Apl. 18	Worc. R., I. A. (*on probn.*)
10 Mar. 16	Hobley, Lt. A. J.	10 Mar. 17	. . .	North'd. Fus., I. A. (*on probn.*)
24 Oct. 16	Harwood, Lt. G. E.	24 Oct. 17	3 Apl. 18	I. A. R. O. (*on probn.*)
30 Jan. 17	O'Donnell, Lt. F. G.	30 Jan. 18	25 Jan. 18	Norf. R., I. A. (*on probn.*)
5 May 17	Dand, Lt. J.	5 May 18	24 Oct. 17	I. A. R. O. Qr.-Mr.
29 Aug. 17	Dixon, 2nd-Lt. R.	29 Aug. 17	15 Jan. 18	I. A. R. O.
29 Aug. 17	Moriarty, 2nd-Lt. P. P.	29 Aug. 17	26 Jan. 18	I. A. R. O.
26 Sep. 17	Evans, 2nd-Lt. J.	26 Sep. 17	26 Jan. 18	I. A. R. O.
5 Oct. 17	Cowie, Lt. A.	5 Oct. 18	19 Jan. 18	I. A. R. O.
5 Nov. 17	Muschamp, 2nd-Lt. W. A. H.	5 Nov. 17	21 Apl. 18	I. A. (*T. C.*)
21 Dec. 17	Nolan, 2nd-Lt. M. C.	21 Dec. 17	. . . .	Attd.
31 Jan. 18	Tittle, 2nd-Lt. D. E.	31 Jan. 18	8 Feb. 18	Attd.
29 Mar. 18	Gurney, 2nd-Lt. A.	29 Mar. 18	15 June 18	I. A. R. O.
29 Jan. 18	Courley, 2nd-Lt. J. Mc C.	29 Jan. 18	2 Oct. 18	Innis. Fus., I. A. R. O. (*on probn.*)
. . . .	Harrold, 2nd-Lt. P. H.	. . . .	2 Oct. 18	Bord R., I. A. R. O. (*on probn.*)
26 June 18	Shallcroft, 2nd-Lt. L. E.	26 June 18	2 Oct. 18	Genl. List, I. A. R. O. (*on probn.*)
5 Oct. 17	Inglis, Lt. H. A.	5 Oct. 18	18 Oct. 18	I. A. R. O.
6 July 18	Cox, 2nd-Lt. G. W. S.	6 July 18	1 Dec. 18	I. A. R. O.
10 July 18	Gunn, 2nd-Lt. A.	10 July 18	1 Dec. 18	I. A. (*T. C.*)
	Subadar-Major.			
		Jemadar.	Subadar.	Subadar-Major
20 Jan. 94	Jalal Din (56)	17 May 15	1 Nov. 17	
	Subadars.			
17 Aug. 91	Dhani Ram	18 Jan. 12	1 Nov. 17	14 Oct. 18.
13 Jan. 93	Prem Singh (58)	1 Aug. 15	1 Nov. 17	
8 June 91	Jiwan Singh (58)	1 Aug. 15	1 Nov. 17	
17 Aug. 96	Miran Baksh	13 July 16	1 Nov. 17	
2 Jan. 04	Makhane Khan (56)	1 Aug. 16	1 Nov. 17	
1 Oct. 02	Nawazish Ali	16 Aug. 17	9 Oct. 18	
18 Apl. 07	Nadir Khan	1 Nov. 17	15 Oct. 18	
1 July 10	Barbari Khan	1 Jan. 18	5 Oct. 18	
15 Nov. 11	Firoz Khan	1 Sep. 18	1 Nov. 18	
	Jemadars.			
27 Apl. 17	Ghulam Hussain	13 July 17		
10 June 96	Bhagwan Singh I	1 Nov. 17		
21 Dec. 96	Chatar Singh	1 Nov. 17		
11 Nov. 02	Bhagwan Singh II (58)	1 Nov. 17		
18 Feb. 02	Shah Sawar	1 Nov. 17		
10 Apl. 17	Nazar Muhammad	1 Mar. 18		
14 Jan. 18	Fazal Ilahi	1 Apl. 18		
27 Sep. 02	Indar Singh	1 Sep. 18		
9 May 04	Mani Ram	1 Sep. 18		
15 Jan. 13	Cheba Khan	1 Sep. 18		
13 May 04	Amar Singh	1 Oct. 18		
20 Dec. 08	Gulistan	5 Oct. 18		
28 Mar. 18	Khaja Mohamed Khan	. . . .	(*on probn.*)	
8 May 18	Sher Khan	29 Nov. 18		

Regt. Centre—Lahore.
Linked with—20th and 21st Punjabis.

Uniform—Drab.
Facings—Scarlet.

1st Battalion, 26th Punjabis.

Raised at Peshawar in 1857, by Capt. H. T. Bartlett, as the 18th Regt. of Punjab Infy. Became the 30th Regt. of Bengal Native Infantry, 1861; the 26th Regt. of B. N. I., 1861; the 46th (Punjab) Regt. of B. N. I., 1864; the 26th (Punjab) Regt. of Bengal Infy., 1885; the 26th Punjab Infy., 1901; present designation, 1903.

"Afghanistan, 1878-79" "Burma, 1885-87."

Composition—2 Companies of Sikhs, 1 of Afridis, 1 of Punjabi Musalmans.

1st Commn. or date of entering service.	Names and Rank.	Army rank.	Present appointment in Regiment.	Remarks.
	Commandant.			
29 Oct. 90	Cox, Lt.-Col. A. D.	29 Oct. 08	16 May 17	
	Company Commanders (4)			
22 Jan. 98	*Turnbull, Maj. G. O.,* D.S.O. *(temp. Lt.-Col.)*	*1 Sep. 15*	*16 Apl. 18*	*2nd-in-Comd. A. A. G.*
25 Jan. 99	Stoney, Maj. P. S., *p.s.c.*	1 Sep. 15	22 Aug. 14	With 2-66th Punj.
14 Sep. 01	*Salusbury, Maj. R. T. G.*	*1 Sep. 15*	*27 Jan. 17*	*S. S. O., Bombay.*
7 Mar. 00	Ivens, Maj. H. T. C.	1 Sep. 15	8 Mar. 18	
	Company Officers. (9)			
8 Feb. 92	*Maude, Maj. E. A.,* D.S.O.	*1 Sep. 15*	*17 Jan. 04*	*Bde.-Maj., 12th Bde.*
11 Aug. 00	*Cargill, Maj. R. J.*	*1 Sep. 15*	*24 Aug. 03*	*Asst. Comdt., Pris. of War Camp.*
19 Aug. 03	*Bennett, Maj. O. D.*	*19 Aug. 18*	*1 Sep. 06*	*Khyber Rif.*
9 Jan. 04	Anderson, Capt. G. W.	9 Jan. 13	13 Oct. 05	
12 Oct. 07	Torrie, Capt. L. J.	1 Sep. 15	20 Sep. 09	*Lv., ex I., m. c.*
9 Sep. 08	*Fulton, Capt. J. D.,* M.C.	*1 Sep. 15*	*17 Nov. 09*	*Staff Capt., 36th Bde.*
3 Sep. 10	Shearer, Capt. J. E., M.C.	1 Sep. 15	1 Nov. 11	
1 Oct. 14	Farwell, Capt. G. A. L.	1 Oct. 18	16 Apl. 15	
30 Jan. 17	Robertson, Lt. H. L. C.	30 Jan. 18	18 Aug. 17	
31 Aug. 18	Lane, 2nd-Lt. C. M.	31 Aug. 18	5 Sep. 18	Attd.
	Attached.			
19 Oct. 15	Graham, Capt. C. T.	19 Oct. 15	12 Nov. 15	I. A. R. O.
25 July 15	Thornhill, Lt. M. B., M.C.	25 July 15	1 Aug. 18	I. A. (*on probn.*
20 July 15	Penniger, Lt. R. C.	20 July 16	25 Aug. 15	I. A. R. O.
30 July 15	Cunninghame, Lt. T. B.	30 July 16	5 Sep. 15	I. A. R. O.
7 Aug. 15	Warren, Lt. D. S.	7 Aug. 16	7 Sep. 15	I. A. R. O.
27 Oct. 15	Copleston, Lt. R. T.	27 Oct. 16	23 Sep. 18	R Lanc R., I. A. R. O. (*on probn.*)
29 Jan. 16	Thomas, Lt. J. E. (*Actg. Capt.*)	29 Oct. 17	26 Jan. 18	I. A. (*on probn.*) Adjt.
1 July 16	Jessop, Lt. F. T.	1 July 17	27 Apl. 17	I. A. R. O.
10 Aug. 16	Perry, Lt. A. S. H.	10 Aug. 17	27 Sep. 17	I. A. R. O.
30 Mar. 17	Brommage, Lt. J. C.	30 Mar. 18	3 Sep. 18	I. A. R. O. (*on probn.*)
27 Mar. 18	Hughes, 2nd-Lt. T.	27 Mar. 18	3 June 18	Welsh R., I. A. (*on probn.*)
1 May 18	Sevenoaks, 2nd Lt. P. L.	1 May 18	25 Aug. 18	I. A. (*on probn.*)
16 Aug. 18	Alexander, 2nd-Lt. W. L.	16 Aug. 18	16 Sep. 18	I. A. R. O.
7 Oct. 18	Bonniface, 2nd-Lt. B. H.	7 Oct 18	21 Oct. 18	I. A. R O.
7 Oct. 18	Croghan, 2nd-Lt. S. C.	7 Oct. 18	21 Oct. 18	I. A. R. O.
7 Oct. 18	Brown, 2nd-Lt. F. H. J.	7 Oct. 18	27 Oct. 18	I A. (*T. C.*)
7 Nov. 18	Almond, 2nd-Lt. H. R.	7 Nov. 18	18 Nov. 18	I. A. R. O.
	Subadar-Major.	Jemadar.	Subadar.	Subadar-Major.
7 June 87	Harnam Singh.	25 June 03	2 Dec. 04	6 Jan. 16.
5 May 88	Ishar Singh (58)	16 Mar. 04	11 Nov. 06	9 Mar. 16.
	Subadars.			
	Jan Gul, I. O. M. (56)	. . . .	. . . .	Re-employed. 13 Jan. 18.
19 Jan. 01	Mawaz Khan (56)	1 Jan. 12	1 May 15	
6 June 01	Suraj Din (51)	1 Feb. 13	5 Apl. 16	
4 Dec. 95	Sher Akhmad (56) I.O.M.	5 Jan. 10	28 Nov. 16	
20 Dec. 01	Ata Muhammad (52, 56, 66*a*)	1 Jan. 16	27 Jan. 17	
20 Mar. 96	Buta Singh (56, 58, 72*a*)	21 Oct. 10	9 Mar. 17	
18 Aug. 99	Pal Singh (56)	27 Jan. 12	9 Mar. 17	
20 Aug. 03	Dasanda Singh	12 Feb. 16	1 July 17	
21 Jan. 98	Jind Singh	9 Mar. 16	1 July 17	
18 Apl. 03	Sardar Khan	10 Oct. 16	17 Dec. 17	
	Jemadars.			
4 Mar. 02	Miran Shah	1 Jan. 16	Indian Adjt.	
26 Nov. 92	Jummah Khan	3 May 16		
25 Aug. 99	Mir Dast, I.D.S.M.	13 Aug. 16		
20 Sep. 93	Dalip Singh	27 Jan. 17		
4 Feb. 96	Gonda Singh	27 Jan. 17		
	Fazal Din	27 Jan. 17		
	Hayat Muhammad (48*a*, 51)	27 Jan. 17		
6 Nov. 98	Narayan Singh	27 Jan. 17		
	Ghulam Hussain	25 Feb. 17		
20 Aug. 17	Narayan Singh	1 Nov. 17		
	Sultan Mahmud	17 Dec. 17		
9 Nov. 99	Hukam Singh	17 Dec. 17		
9 Feb. 99	Narayan Singh	25 May 18		
. . . .	Ali Bakhsh	25 May 18		

2nd Battalion, 26th Punjabis.

Formed at Bareilly.

Composition.—2 Companies Punjabi Hindus, 1 Company Sainis, 1 Company South Western Punjab Musalmans.

1st Commn. or date of entering service.	Names and Rank.	Army rank.	Present appointment in Regiment.	Remarks.
	Commandant.			
14 Aug. 95	Magniac, Maj. F. A. (*Actg. Lt.-Col.*)	14 Aug. 13	5 June 18	1-22nd Punj.
	Company Commanders. (4)			
20 Jan. 00	Ransford, Maj. F. M.	1 Sep. 15	. . . .	6th G. R., 2nd-in-Comd. (*temp.*)
11 Dec. 07	Foottit, Capt. J. A. W.	1 Sep. 15	. . . .	17th Cav.
	Company Officers.			
	Saywell, Capt. J. R.	. . . .	. . . .	Dorset. R., I. A. R. O. (*on probn.*)
30 Aug. 15	Howell, Lt. V.	30 Aug. 16	. . . .	Welsh R., I. A. (*on probn.*)
10 Sep. 15	Pallock, Lt. A. H., M.C.	10 Sep. 16	. . . .	R. Lan. R., I. A. (*on probn.*)
23 Jan. 15	Raven-Hill, Lt. A. L.	23 Oct. 16	. . . .	I. A. (*on probn.*)
13 June 16	Maclaren, Lt. W. de B. (*Actg. Capt.*)	13 June 17	. . . .	High. L. I., I. A. (*on probn.*)
13 Feb. 16	Hughes, Lt. H. T. (*Actg. Capt.*)	13 Feb. 17	. . . .	I. A. (*on probn.*)
20 Oct. 16	Kerridge, Lt. S. C., M.C.	20 Oct. 17	. . . .	North'd Fus., I. A. (*on probn.*)
5 May 17	Furminger, Lt. H. J., M.C. (*Actg. Capt.*)	5 May 18	. . . .	Bord. R., I. A. (*on probn*)
30 May 17	Green, 2nd-Lt. T. W.	30 May 18	. . . .	Manch. R., I. A. R. O. (*on probn.*)
12 Feb. 16	Parmenter, Lt. F.	12 Feb. 18	24 Oct. 18	Attd.
12 Sep. 17	Rudolph, Lt. W.	12 Sep. 18	. . . .	I. A. (*on probn.*)
5 Oct. 17	Newnham, Lt. H. J.	5 Oct. 18	. . . .	I. A. (*T. C.*)
21 Dec. 17	Daly, 2nd-Lt. F. O. W. A.	21 Dec. 17	. . . .	Attd.
15 Jan. 18	Wragg, 2nd-Lt. W. F.	15 Jan. 18	. . . .	I. A. (*T. C.*)
19 Mar. 18	Hicks, 2nd-Lt. E. S.	19 Mar. 18	. . . .	I. A. R. O.
19 Mar. 18	Bibby, 2nd-Lt. E.	19 Mar. 18	. . . .	I. A. (*T. C.*)
22 Mar. 18	MacQueen, 2nd-Lt. S. W.	22 Mar. 18	. . . .	Lan. R., I. A. R. O. (*on probn.*)
29 Mar. 18	Woods, 2nd-Lt. G. H.	29 Mar. 18	. . . .	I. A. R. O. (*on probn.*)
22 May 18	Fitzpatrick, 2nd-Lt. R. H. J.	22 May 18	. . . .	I. A. R. O.
	Davis, 2nd-Lt. N. E.	. . . .	. . . .	Genl. List I. A. R. O. (*on probn.*)
	Holmes, 2nd-Lt. C.	. . . .	. . . .	Notts & Derby. R., I. A. R. O. (*on probn.*)
22 July 18	Craig, 2nd-Lt. S. F.	22 July 18	. . . .	I. A. R. O.
22 July 18	King, 2nd-Lt. J. H.	22 July 18	. . . .	I. A. (*T. C.*)

Subadar-Major.

		Jemadar.	Subadar.	Subadar-Major.
6 Oct. 94	Parab Dayal (56)	21 Feb. 12	25 Sep. 15	15 Sep. 18

Subadars.

7 May 03	Bodh Raj	28 Nov. 15	1 Oct. 16
15 Jan. 96	Imam Ali	5 June 17	12 June 18
14 Jan. 03	Samundar Khan	5 June 17	12 June 18

Jemadars.

4 Jan. 01	Muhammad Ramzan Khan	16 June 17	. . . .
5 Aug. 02	Asghar Ali (60a)	1 Aug. 17	. . . .
21 Feb. 17	Mohamad Yasin Khan	20 Aug. 17	. . . .
24 Feb. 17	Khuda Bakhash	20 Aug. 17	. . . .

Regt. Centre—Ferozepore.
Linked with—25th and 28th Punjabis.

1st Battalion, 27th Punjabis.

Uniform—Drab.
Facings—Scarlet.

Doran Sahib-ki-Paltan.—Raised at Rawal Pindi in 1857, by Captain J. Doran, and was originally designated the Regt. of Rawal Pindi. Became the 19th Regt. of Punjab Infy., 1857; the 31st Regt. of Bengal Native Infantry, 1861; the 27th Regt. of B. N. I., 1861; the 27th (Punjab) Regt. of B. N. I., 1864; the 27th (Punjab) Regt. of Bengal Infy., 1885; the 27th Punjab Infy., 1901; present designation, 1903.

"China, 1860-62" "Ali Masjid" "Afghanistan, 1878-80" "Burma, 1885-87."

Composition—1½ Companies of Sikhs, ½ of Dogras, 1 of Punjabi Musalmans, 1 of Pathans.

COLONEL—Major-General Alexander Wallace, C.B. . . . 15 December 1911.

1st Commn. or date of entering service.	Names and Rank.	Army rank.	Present appointment in Regiment.	Remarks.
	Commandant.			
	Company Commanders. (4)			
22 Jan. 96	*Scott, Maj. G. B.*, D.S.O. . . .	*22 Jan. 14*	*1 Nov. 10*	*N. Waziristan Militia.*
20 Jan. 00	Reinhold, Maj. H. E., M.C. [L] (*Actg. Lt.-Col.*)	1 Sep. 15	30 June 11	
	Company Officers. (9)			
22 Oct. 02	Vernon, Maj. H. B., M.C. . .	22 Oct. 17	1 July 07	
9 Jan. 04	Quinan, Bt. Maj. E. P. . . .	1 Jan. 18	30 Mar. 05	With 2-117th Mah.
9 Jan. 04	Smith, Capt. E. D. D.	9 Jan. 13	12 Mar. 05	With 1-54th Sikhs.
5 Aug. 05	*Herz-Smith, Capt. G. V.*, M.C. .	*5 Aug. 14*	*3 Nov. 06*	*Bde.-Maj.*
29 Aug. 06	Todd, Capt. R.	29 Aug. 15	23 Nov. 07	With 2-105th Infy.
8 Sep. 09	Pitman, Capt. C. R. S., D.S.O., M.C.	1 Sep. 15	31 Dec. 10	
18 Jan. 11	Shewen, Capt. D. G. P. M. .	1 Sep. 15	3 Apl. 12	With 120th Infy.
6 Sep. 11	Rule, Capt. D. G.	6 Sep. 15	2 Dec. 12	With 3-152nd Infy.
29 June 16	Nicholas, Lt. A. A.	13 Dec. 16	12 Feb. 17	
29 June 16	Roper, Lt. R. B.	29 June 17	12 Feb. 17	With 3-152nd Infy.
30 Jan. 17	Chamberlain, Lt. L. L. . . .	30 Jan. 18	18 Mar. 18	
27 Oct. 17	Edge, Lt. J.	27 Oct. 18	5 Nov. 17	Attd.
31 Aug. 18	Robinson, 2nd-Lt. J. G. . .	31 Aug. 18	4 Sep. 18	Attd.
1 Oct. 18	Mallaby, 2nd-Lt. A. W. S. . .	1 Oct 18	8 Oct. 18	Attd.
	Attached.			
20 Dec. 06	Goldman-Monk, Capt. M. E. deB. (L) (*Actg. Maj.*) . . .	1 Sep. 15	29 Oct. 18	1-91st Punj.
18 July 15	Wilson, Lt. L. E.	18 July 16	17 Aug. 17	I. A. (*on probn.*)
25 July 16	Bates, Lt. H. V. R.	25 July 17	29 Apl. 18	E. Surr. R., I. A. (*on probn.*)
15 July 15	Rodgers, Lt. J.	15 July 16	23 Apl. 18	I. A. (*on probn.*)
26 June 17	King, Lt. L. C., M.C. (*Actg. Capt.*)	26 June 18	1 May 18	Sea. Highrs., I. A. (*on probn.*) Adjt. 22 June 18.
9 Oct. 14	Mold, Lt. G. L.	1 Mar. 16	16 Apl. 17	40th Pathans.
8 Jan. 15	Moriarty, Lt. J., M.C. . . .	8 Jan. 16	3 May 15	I. A. R. O.
14 Jan. 15	Rolt, Lt. P. J.	14 Jan. 16	19 Jan. 15	I. A. R. O.
30 Oct. 15	Marks, Lt. R. M.	30 Oct. 16	. . . .	I. A. R. O.
29 Oct. 15	Byron, Lt. G.	29 Oct. 16	29 Sep. 16	I. A. R. O.
28 Apl. 16	Proctor, Lt. D. P.	28 Apl. 17	27 Oct. 16	I. A. R. O.
20 Feb. 16	Murphy, Lt. R. S.	20 Feb. 17	17 Feb. 17	I. A. R. O.
3 June 18	Crowhurst, 2nd Lt. J. D. J. . .	3 June 18	22 June 18	I. A. R. O. (*on probn.*)
8 Oct. 16	Tyres, Lt. F. W.	8 Oct 17	4 Sep. 18	R. War. R., I. A. (*on probn.*)
3 Apl. 16	Burnett, Lt. R. R.	3 Apl. 17	1 Oct. 18	R. Scotts, I. A. (*on probn.*)
7 June 18	Eden, 2nd-Lt. H. G. A. . . .	7 June 18	11 Oct. 18	I. A. (*T. C.*)

Subadar-Majors.

		Jemadar.	Subadar.	Subadar-Major.	
31 Oct. 90	Mir Akbar, *Bahadur*, I.D.S.M. .	23 Apl. 03	1 Apl. 12	1 Apl. 16	*Order of Br. Ind., 2nd Cl.*
11 Sep. 88	Khushal Singh, *Sardar Bahadur* (56).	1 Sep. 05	1 Nov. 12	27 Sep. 16	*Order of Br. Ind., 1st Cl.*

Subadars.

9 June 99	Arjun Singh (56)	1 Nov. 10	28 July 14	
7 Dec. 95	Bhagat Singh, I.O.M. (72a) . .	2 July 11	11 Oct. 14	
22 Oct. 94	Aman Gul, I.D.S.M. (56) . . .	11 Oct. 14	16 Apl. 16	
13 Dec. 92	Sher Shah *Bahadur*	26 Nov. 15	18 Apl. 16	*Order of Br. Ind., 2nd Cl.*
14 Dec. 01	Rama, I.O.M.	. . . .	18 Apl. 16	
11 Jan. 92	Naim Gul (51, 56)	1 Apl. 12	20 Apl. 16	
18 Aug. 02	Fateh Khan, I.D.S.M.	29 Apl 16	22 May 16	
3 Oct. 99	Sher Ali	16 Apl. 16	21 May 17	
14 Mar. 01	Ishar Singh	23 Nov. 15	16 Feb. 18	
25 June 04	Muhammad Zaman (56) . . .	18 Apl. 16	16 Feb. 18	
13 June 00	Labh Singh	20 Apl. 16	24 May 18	
23 Mar. 02	Zain Din, I.D.S.M.	11 Sep. 17	24 May 18	
17 Dec. 00	Tor Khan	11 Oct. 14	20 Sep. 18	
9 Apl. 02	Mustafa	30 Jan. 16	20 Sep. 18	

1st Battalion, 27th Punjabis—*contd.*

1st Commn. or date of entering service.	Names and Rank.	Army rank.	Remarks.
	Jemadars.		
23 Apl. 97	Badri Das	1 May 15	
20 Nov. 98	Chanda Singh	18 Dec. 15	
21 Feb. 96	Hayat Khan	27 Nov. 15	
31 Mar. 04	Pala, I.O.M.	18 Apl. 16	
13 July 01	Muhammad Azim	22 May 16	
26 Aug. 01	Nand Singh	27 Sep. 16	
18 May 02	Surjan	6 Nov. 16	
20 Dec. 02	Hikmat	21 May 17	
7 Dec. 03	Karam Bakhsh	16 Feb. 18	
27 June 03	Indar Singh	24 May 18	
10 July 07	Fateh Khan	24 May 18	

2nd Battalion, 27th Punjabis.

Formed at Agra on 25th July 1918.

Composition—1 Company of Punjabi Musalmans, 1 Company of Pathans, 1½ Companies of Sikhs, ½ Company of Dogras.

1st Commn. or date of entering service.	Names and Rank.	Army rank.	Present appointment in Regiment.	Remarks.
	Commandant.			
25 Jan. 99	Tancred, Maj. J. Y.	1 Sep. 15	29 July 18	1-19th Punj. (*temp.*)
	Company Commanders. (4)			
5 Aug. 05	Grylls, Capt. W. M.	5 Aug. 14	29 July 18	52nd Sikhs. 2nd-in-Comd.
	Company Officers.			
26 June 15	Smith, Lt. D. G. S.	26 June 16	28 Nov. 18	A. S. C., I. A. (*on probn.*)
15 Nov. 15	Pine-Coffin, Lt. E. C. (*Actg. Capt.*)	15 Nov. 16	29 July 18	1-22nd Punj. Adjt.
7 Apl. 16	Bradshaw, Lt. A. R. (*Actg. Capt.*)	7 Apl. 17	14 Aug. 18	E. Lan. R., I. A. (*on probn.*)
13 July 15	Lancaster, Lt. A. S. (*Actg. Capt.*)	13 Apl. 17	26 Aug. 18	Lond. R., I. A. (*on probn.*)
18 Apl. 16	Leese, Lt. F. W.	18 Apl. 17	29 July 18	29th Punj.
25 July 15	Hoare, Lt. H. P. B. (*Actg. Capt.*)	25 Apl. 17	26 Aug. 18	Hamps. R., I. A. (*on probn.*)
4 May 16	Powell, Lt. F. N. H.	4 May 17	30 Aug. 18	I. A. (*on probn.*)
	Barnes, Lt. C. S.	. . . .	18 Nov. 18	War. Yeom., I. A. (*on probn.*)
	Baldwin, Lt. L. W.	. . . .	29 July 18	I. A. (*on probn.*)
4 July 16	Toomey, Lt. W. G.	4 July 17	29 July 18	Leins. R., I. A. (*on probn.*)
	Olivier, Lt. R. T.	. . . .	20 Nov. 18	Notts. & Derby. R., I. A. (*on probn.*)
1 Aug. 16	Willoughby, Lt. J. J.	1 Aug. 17	3 Sep. 18	Essex R., I. A. R. O. (*on probn.*)
10 Apl. 17	Brewer, Lt. V.	10 Apl. 18	. . . .	I. A. R. O.
	Whiteside, 2nd Lt. H. H. . .	. . . .	18 Nov. 18	Ches. R., I. A. (*on probn.*)
18 Dec. 17	Steward, 2nd-Lt. D. H. . . .	18 Dec. 17	28 Aug. 18	I. A. R. O. (*on probn.*)
21 Dec. 17	Kitchen, 2nd-Lt. A. C. . . .	21 Dec. 17	14 Aug. 18	
	Allport, 2nd-Lt. D. H. . . .	. . . .	18 Nov. 18	Notts. & Derby. R., I. A. (*on probn.*)
21 Dec. 17	Stuart, 2nd-Lt. A.	. . . .	29 July 18	I. A. (*T. C.*)
10 July 18	Hubbard, 2nd Lt. H.	10 July 18	2 Dec. 18	I. A. (*T. C.*)
31 Aug. 18	Eyre, 2nd-Lt. J. S. G. . . .	31 Aug. 18	16 Sep. 18	Attd.
12 Sep. 18	Johnstone, 2nd-Lt. R. B. . . .	12 Sep. 18	3 Oct. 18	I. A. R. O.
18 Sep. 18	Hollingbery, 2nd-Lt. N. H. . .	18 Sep. 18	3 Oct. 18	I. A. R. O.
18 Sep. 18	Moncur, 2nd-Lt. J. O. . . .	18 Sep. 18	3 Oct. 18	I. A. R. O.
11 Nov. 18	Franks, 2nd-Lt. S. W. . . .	11 Nov. 18	24 Nov. 18	I. A. R. O.
17 June 18	Millin, 2nd-Lt. D. J.	7 June 18	. . . .	I. A. (*T. C.*)

Subadar-Major.

		Jemadar.	Subadar.	Subadar-Major.

Subadar.

20 Nov. 93	Dilawer Khan (56)	29 Aug. 15	22 Feb. 16	
9 May 04	Sher Bahadur, I.D.S.M.	8 Feb. 17	15 Aug. 18	
19 Feb. 96	Khan Gul	11 Oct. 14	15 Aug. 18	
11 Apl. 92	Tara Singh, M.C.	27 Nov. 15	15 Aug. 18	

Jemadars.

22 July 01	Ram Singh	1 June 16		
25 Nov. 93	Firoz Khan	16 Oct. 17		
26 Mar. 01	Bela Singh, I.D.S.M.	3 Dec. 17		
5 June 09	Murtaza Khan, ...	1 Aug. 18		
	Harnam Singh	15 Aug. 18		
	Khadim Shah	15 Aug. 18		
	Khoda Bakhsh	15 Aug. 18		
	Munshi	15 Aug. 18		
	Mendi Khan	15 Aug. 18		

Regt. Centre—Ferozepore.
Linked with—25th and 27th Punjabis.

Uniform—Scarlet.
Facings—Emerald Green.

1st Battalion, 28th Punjabis.

Raised, at Ferozepore, in 1857, by Capt. C. S. Salmon, and was at first designated the Ferozepore Punjab Battn. Became the 20th Regt. of Punjab Infy., 1857; the 32nd Regt. of Bengal Native Infantry, 1861; the 28th Regt. of Bengal N. I., 1861; the 28th (Punjab) Regt. of Bengal N. I., 1864; the 28th (Punjab) Regt. of Bengal Infy., 1885; the 28th Punjab Infy., 1901; present designation, 1903.

"Charasiah" "Kabul, 1879" "Afghanistan, 1878-80."

Composition—1¼ Companies of Sikhs, ¾ of Dogras, 1¼ of Pathans, ¾ of Punjabi Musalmans.

COLONEL—Major-General Charles L. Prendergast . 13 May 1904.

1st Commn. or date of entering service.	Names and Rank.	Army rank.	Present appointment in Regiment.	Remarks.
	Commandant.			
16 Nov. 87	Chichester, Lt.-Col. A. G. deV. . .	16 Nov. 13	20 Nov. 14	
	Company Commanders. (4)			
29 Oct. 90	Tomkins, Lt.-Col. H. L., C.M.G., D.S.O. (Q)	3 June 15	20 Nov. 14	2nd-in-Comd.
28 Sep. 92	Keene, Maj. C. W.	28 Sep. 10	1 Mar. 10	*Lv., ex. I., m. c.*
27 July 01	*Philips, Maj. J. C.*	*1 Sep. 15*	*17 Feb. 12*	With 2-56th Rif.
18 Apl. 00	*Harcourt, Maj. C. B.*	*1 Sep. 15*	*1 Nov. 12*	
	Company Officers. (9)			
8 May 01	*Carey, Maj. J. E.*	*8 May 16*	*3 Jan. 14*	
5 Aug. 05	*Yearsley, Capt. R. A. (temp. Maj)*	*5 Aug. 14*	*30 Jan. 08*	*Cadet Coll., Wellington. G. S. O., 2nd gr.*
29 Aug. 06	Coffin, Capt. A. S.	29 Aug. 15	30 Jan. 08	*Lv., ex I., m. c.*
16 Jan. 07	Boswell, Capt. W. R.	1 Sep. 15	3 Nov. 08	
20 Jan. 09	Scott, Capt. R. N. G. . .	1 Sep. 15	14 May 10	
20 Jan. 09	Nugent, Capt. J. F. H., D.S.O.	1 Sep. 15	29 Mar. 10	*Bde.-Maj., 57th Infy. Bde.*
15 Nov. 15	Dann, Lt. C., M.C.	15 Nov. 16	18 Nov. 15	
15 Nov. 15	Tinney, Lt. G. E.	15 Nov. 16	18 Nov. 15	With 2-23th Punj.
30 Jan. 17	Boxall, Lt. M. W. J.	30 Jan. 18	. . . .	With 1-152nd Infy.
	Attached.			
9 July 15	Robinson, Capt. A. E. W. . .	3 June 18	6 June 16	I. A. R. O.
16 July 15	Norrish, Lt. E. H. B. (*temp. Capt.*)	16 July 16	19 July 15	I. A. R. O.
7 Aug. 15	Oddie, Lt. W. P., M.C. . . .	7 Aug. 16	. . . .	I. A. R. O.
23 Oct. 15	Bartlett, Lt. G. D. N. . . .	23 Oct. 16	. . . .	I. A. R. O.
15 Nov. 15	Barlow, Lt. W. N. C., M.C. . .	15 Nov. 16	. . . .	30th Punj.
28 Apl. 16	Allen, Lt. F.	28 Apl. 17	. . . .	I. A. R. O.
30 Jan. 16	Walters, Lt. E. C.	30 Jan. 17	. . . .	I. A. (*on probn.*)
6 Sep. 16	Curtis, Lt. A. G. C.	6 Sep. 17	. . . .	I. A. (*on probn.*)
17 Mar. 16	Baldwin, Lt. J. W.	17 Mar. 17	4 June 18	R. Muns. Fus., I. A. (*on probn.*)
14 Feb. 16	Wilson, Lt. J. W., M.C. . . .	14 Feb. 17	27 Jan. 18	I. A. (*on probn.*)
20 June 17	Fraser, Lt. W. C.	20 June 18	11 Oct. 17	I. A. R. O.
29 Aug. 17	Swan, Lt. S. R.	29 Aug. 18	27 Jan. 18	I. A. R. O.
7 Apl 17	Pemberton, Lt. J. D.	7 Apl 18	4 Sep. 18	R. Ir. Rif., I. A. (*on probn.*)
1 Aug. 17	Cooper, 2nd-Lt. N. E. . . .	1 Aug. 17	7 Sep. 18	W. Kent R., I. A. R. O. (*on probn.*)
7 June 18	Milner, 2nd Lt. F. D. . . .	7 June 18	11 Oct. 18	I. A. R. O.
9 Sep. 18	Munn, 2nd-Lt. A. E. . . .	9 Sep. 18	4 Oct. 18	I. A. R. O.
19 Oct. 18	Brooks, 2nd-Lt. C. W. . . .	19 Oct. 18	31 Oct. 18	I. A. R. O.
19 Oct. 18	Jackson, 2nd-Lt. R. J. . . .	19 Oct. 18	7 Nov. 18	I. A. R. O.

Subadar-Major.

		Jemadar.	Subadar.	Subadar-Major.	
1 Oct. 90	Sundar *Bahadur*	1 Sep. 08	1 June 10	14 Jan. 16	*Order of Br. Ind., 2nd Cl.*

Subadars.

1 Apl. 99	Muhammad Nawaz Khan (56) .	1 Apl. 99	16 Mar. 08
19 Oct. 96	Amar Singh (56, 66)	7 Mar. 08	1 Aug. 12
16 May 88	Hassan Muhammad (56) . . .	21 Jan. 06	1 July 15
3 Jan. 97	Mangal Singh	1 Aug. 12	27 Nov. 15
27 May 00	Kaka Singh (56)	26 Mar. 14	20 Jan. 16
5 June 99	Mirza Khan	8 Jan. 16	7 Apl. 16
6 July 95	Thakur Singh	13 Jan. 16	10 Aug. 16
27 Oct. 00	Ghulam Sarwar	1 June 16	13 Oct. 16
5 Mar. 99	Gulab Din, I.O.M.	7 Apl. 16	22 May 18
29 Sep. 99	Dalip Singh	10 Aug. 16	24 May 18
14 Nov. 99	Dewan Ali	22 Apl. 17	20 Sep. 18
22 July 01	Ram Singh	1 June 16	26 Sep. 18

1st Battalion, 28th Punjabis—*contd.*

1st Commn. or date of entering service.	Names and Rank.	Army rank.	Remarks.
		Jemadars.	
5 Aug. 97	Azim Khan	1 July 15	
5 Mar. 98	Muhammad	27 Nov. 15	
10 June 97	Mehtab Din	7 Apl. 16	
7 Mar. 00	Wasawa Singh	8 Apl. 16	
6 July 04	Sultan Sarkhru (56)	10 Apl. 16	
25 Mar. 00	Sundar Singh	24 Apl. 16	
11 Sep. 06	Shad Muhammad	13 July 16	
12 Mar. 98	Ranya	11 Mar. 17	
8 Sep. 00	Niamat Khan	22 Apl. 17	
17 Aug. 02	Azmat Khan	22 Aug. 17	
22 Jan. 01	Allah Lok (I)	3 Dec. 17	
....	Adalat Zar Khan	2 Jan. 18	
19 May 97	Teja Singh (46)	2 Mar. 18	
	Ghafur Khan	22 May 18	Indian Adjt.
	Fateh Sher	24 May 18	
	Chaudri Khan	5 July 18	

2nd Battalion, 28th Punjabis.

Formed at Jhelum on 13th August 1918.

Composition.—1 Company Pathans, 1 Company Punjabi Mussalmans, 1 Company Dogras, 1 Company Sikhs.

1st Comm. or date of entering service.	Names and Rank.	Army Rank.	Present appointment in Regiment.	Remarks.
	Commandant.			
11 Feb. 99	Smith, Maj. E. C., (*Actg. Lt.-Col.*)	1 Sep. 15	13 Aug. 18	1-22nd Punj. (*temp.*)
	Company Commanders. (4)			
20 June 12	Scott, Capt. J. B., M. C.	20 Jan. 16	23 Aug. 18	1-33rd Punj. 2nd-in-Comd. (*temp.*)
	Company Officers.			
5 May 15	Watkin, Lt. H. G. (*Actg. Capt.*)	5 May 16	3 Oct. 18	Ches. R., I. A. (*on probn.*)
18 June 15	Berkeley, Lt. H. F.	18 June 16	18 Aug. 18	I. A. R. O.
24 Oct. 15	Aldworth, Lt. E. E. (*Actg. Capt.*)	24 Oct. 16	12 Oct. 18	S. Lan. R., I. A. (*on probn.*)
15 Nov. 15	Tinney, Lt. G. E. (*Actg. Capt.*)	15 Nov. 16	27 Aug. 18	1-28th Punj. Adjt. 18 Oct. 18.
16 Dec. 15	Williams, Lt. E. W.	16 Dec. 16	3 Oct. 18	Suss. R., I. A. (*on probn.*)
20 Dec. 15	Stevens, Lt. C. W.	20 Dec. 16	20 Aug. 18	Sco. Rif. I. A. (*on probn.*)
23 Dec. 15	Tucker, Lt. V.	23 Dec. 16	14 Oct. 18	Lond. R., I. A. R. O. (*on probn.*)
2 Jan. 16	Batchelor, Lt. A. J.	2 Jan. 17	27 Aug. 18	S. Lan. R., I. A. (*on probn.*)
25 July 17	Moriarty, Lt. D.	25 July 18	17 Aug. 18	Lond. R., I. A. O. (*on probn.*)
25 Dec. 17	Kelly, 2nd-Lt. A.	25 Dec. 17	25 Aug. 18	I. A. R. O.
7 June 18	Bristow, 2nd-Lt. A. C.	7 June 18	10 Sep. 18	I. A. R. O.
15 June 18	Brown, 2nd-Lt. R. S.	15 June 18	11 Oct. 18	Genl. List, I. A. R. O. (*on probn.*)
15 June 18	Lloyd, 2nd-Lt. W.	15 June 18	11 Oct. 18	Genl. List, I. A. R. O. (*on probn.*)
13 Sep. 18	Macgregor, 2nd-Lt. J. S. I.	13 Sep. 18	16 Oct. 18	Genl. List, I. A. R. O. (*on probn.*)
9 Oct. 18	Bredin, 2nd-Lt. J. A.	9 Oct. 18	2 Nov. 18	I. A. R. O.
8 Oct. 18	Gleeson, 2nd-Lt. J.	8 Oct. 18	30 Oct. 18	I. A. R. O.
22 July 18	Warr, 2nd-Lt. C. H.	22 July 18	1 Dec. 18	I. A. (*T. C.*)
16 June 16	Dixon, Lt. T.	16 June 17	....	R. Lanc. R., I. A (*on probn.*)
1 Oct. 18	Stacey, 2nd-Lt. L. B.	1 Oct. 18	....	Attd.

Subadar Major.

		Jemadar.	Subadar.	Subadar-Major.
1 Aug. 93	Allah Baksh	1 July 15	14 June 16	10 Aug. 16

Subadars.

		Jemadar.	Subadar.
16 Apl. 95	Sundar	30 Nov. 14	...
18 Mar. 99	Wazir Singh	21 Jan. 15	19 Oct. 17
24 Dec. 95	Alam Sher	1 June 16	19 Sep. 18

Jemadars.

19 Mar. 94	Hazara Singh	....	26 Sep. 15	
27 May 01	Nawab Khan	....	6 Sep. 18	Indian Qr. Mr.
19 Oct. 02	Jagat Singh	....	19 Sep. 18	Indian Adjt.
8 Nov. 97	Sardar Shah	....	25 Sep. 18	
2 Feb. 03	Lal Khan	....	25 Sep. 18	
23 Jan. 07	Tora Baz	....	25 Sep. 18	
18 Jan. 98	Basant Singh	....	1 Oct. 18	

Regt. Centre—Peshawar. *Uniform*—Scarlet.
Linked with—30th and 31st Punjabis. *Facings*—Blue.

29th Punjabis.

Raised, Jullundur, 1857, by Capt. T. Tulloh, as the 21st Regt. of Punjab Infy. Became the 33rd Regt. of Bengal Native Infantry, 1861; the 29th Regt. of Bengal N. I., 1861; the 29th (Punjab) Regt. of Bengal N. I., 1864, the 29th (Punjab) Regt. of Bengal Infy., 1885; the 20th Punjab Infy., 1901; present designation, 1903.

"Peiwar Kotal" "Afghanistan 1878-80" "Chitral."

Composition—2 Companies of Sikhs, 1 of Dogras, 1 of Punjabi Musalmans.

1st Commn. or date of entering service.	Names and Rank.	Army rank.	Present appointment in Regiment.	Remarks.
	Commandant.			
14 Mar. 88	Ford, Lt.-Col. H. B.	14 Mar. 14	23 June 16	
	Company Commanders. (4)			
19 Nov. 92	*Haslehust, Maj. G.W., p.s.c. (Q) (temp. Lt-Col.)*	*19 Nov. 10*	*6 Feb. 18*	*2nd-in-Comd., G. S. O. 1st Gr., Meerut.*
23 Nov. 98	Skinner, Maj. H. T., D.S.O.	1 Sep. 15	15 July 15	
28 July 00	Smith, Maj. C. W. J., D.S.O.	1 Sep. 15	22 Jan. 18	With 2-54th Sikhs.
8 Jan. 01	Massy, Maj. S. D., D.S.O.	8 Jan. 16	6 Feb. 18	2nd-in-Comd. (*temp.*)
	Company Officers. (9)			
7 May 02	Edwards. Maj. E.	7 May 17	24 June 04	
4 July 03	*Campbell, Maj. J., M.C.*	*6 Oct. 17*	*19 Sep. 05*	
24 Jan. 06	Bignell, Capt. G. N., M.C.	24 Jan. 15	22 May 07	Adjt.
24 Jan. 06	Stranack, Capt. H. R.	24 Jan. 15	24 Mar. 07	
4 Mar. 11	Bremner, Capt. C. E. U., M.C.	1 Sep. 15	11 Feb. 14	With 2-19th Punj.
22 Jan. 13	Bird, Capt. E. K.	22 Jan. 17	12 Mar. 14	With 20th Infy.
18 Apl. 16	Leese, Lt. F. W.	18 Apl. 17	23 Apl. 16	With 2-27th Punj.
18 Apl. 16	Wernicke, Lt. A. E.	18 Apl. 17	23 Apl. 16	
30 Jan. 17	Butler, Lt. G. R.	30 Jan. 18	3 Feb. 17	Attd.
18 June 17	Hayes-Newington, Lt. E. A.	18 June 18	. . .	Attd. with 102nd Grs.
26 Sep. 15	Evans, Lt. T. R.	26 Sep. 16	30 June 17	Attd.
1 July 15	Hill, Lt. H. V. J.	1 July 16	30 June 17	Attd.
24 Aug. 15	O'Morchoe, Lt. N. C.	24 Aug. 16	2 Aug. 17	Attd.
	Attached.			
25 Mar. 15	Bradley, Lt. S. M. (*Actg. Capt.*)	25 Mar. 16	25 June 17	I. A. R. O.
19 Apl. 15	Gibson, Lt. D. V.	19 Apl. 16	13 Dec. 17	I. A. R. O.
15 May 15	Colhoun, Lt. W. A., M.C.	15 May 16	4 June 18	R. Ir. Fus. I. A. (*on probn.*)
14 May 15	Clifford, Lt. C. H. W. (*Actg. Capt.*)	14 May 16	23 June 15	I. A. R. O.
17 June 15	Underwood, Lt. H. J. (*Actg. Capt.*)	17 June 16	10 Apl. 17	I. A. (*on probn.*)
10 July 17	Rayner, Lt. G. H.	. . . .	8 Dec. 17	I. A. R. O. (*on probn.*)
20 Sep. 16	Dunt, Lt. R. B.	20 Sep. 16	28 Aug. 17	I. A. R. O.
28 Aug. 15	Hide, Lt. T. G. S. (*temp. Capt.*)	28 Aug. 16	2 June 17	I. A. (*on probn.*) *Asst. Instr., Bomb. Sch., Mhow.*
20 Jan. 15	Sykes, Lt. F. C.	20 Oct. 16	19 Aug. 18	Manch. R., I. A. (*on probn.*)
17 Oct. 15	Hemsworth, Lt. N. E. C.	17 Oct. 16	24 June 17	I. A. (*U. L.*)
14 Nov. 16	Maffett, Lt. D. B.	14 Nov. 17	9 Dec. 16	31st Punj.
29 May 18	Ross, Lt. C.	. . . .	30 Aug. 18	I. A. R. O. (*on probn.*)
29 May 18	Walton, Lt. G.	. . . .	30 Aug. 18	I. A. R. O. (*on probn.*)
17 Apl. 16	Murray, Lt. N. T. G.	7 Apl. 17	. . . .	Dorset R., I. A. (*on probn.*)

Subadar-Major.

		Jemadar.	Subadar.	Subadar Major.	
	Mehr Khan, *Hony.* Capt. (*Ret.*)	. . . .	. . . .	. . . .	Re-employed, 15 Feb. 18.
7 May 90	Man Singh (56, 58)	16 Jan. 06	1 Dec. 08	13 Feb. 17	

Subadars.

25 Nov. 96	Rai Singh	1 Jan. 01	15 Dec. 01	
20 Nov. 86	Fazl Muhammad, *Sardar Bahadur* (56, 58)	15 Dec. 95	1 Aug. 07	*Order of Br. Ind., 1st Cl.*
4 July 02	Kanwar Udham Singh, I.D.S.M. (56, 66*a*)	14 July 02	1 Jan. 12	
1 June 06	Sirdar Piar Singh (56)	1 July 06	9 Aug. 14	
10 Nov. 91	Roshan Khan (56, 66)	11 Mar. 12	7 Sep. 14	
24 May 88	Gurditt Singh	1 Dec. 98	22 Feb. 17	
20 Mar. 93	Ali Bahadur (56)	26 Feb. 16	13 June 17	
17 Apl. 94	Kesar Singh	9 Aug. 14	16 June 17	
1 Mar. 92	Indar Singh	18 Dec. 15	4 Aug. 17	
18 Mar. 99	Wazir Singh	21 Jan. 15	19 Oct. 17	
21 May 96	Ganda Singh (56, 58, 66*a*)	12 Mar. 11	26 Mar. 18	

29th Punjabis—*contd.*

1st Commn. or date of entering service.	Names and Rank.	Army rank.	Remarks.
	Jemadars.		
18 May 96	Lachman Singh (56)	1 Aug. 11	
23 Dec. 94	Basawa Singh (46)	9 Aug. 14	
1 Sep. 00	Kaldip Singh (56)	11 Sep. 14	
6 Mar. 00	Gokal (56)	16 Jan. 16	
1 Jan. 95	Labh Singh	3 June 16	
7 Nov. 99	Muhammad Khan	1 Mar. 17	
1 July 00	Kamal Khan	13 June 17	
31 Dec. 94	Isar Singh	13 June 17	
1 Jan. 94	Bhagwan Singh	16 June 17	
1 May 02	Saida Khan	16 Aug. 17	
24 Oct. 90	Baz Khan, I.D.S.M.	14 Oct. 17	
18 Dec. 97	Balla Singh	19 Oct. 17	
19 May 00	Mangtu	19 Oct. 17	
1 Dec. 99	Lakhu	24 Dec. 17	
	Udham Singh	16 Apl. 18	

Regt. Centre—Peshawar.
Linked with—29th and 31st Punjabis.

Uniform—Scarlet.
Facings—White.

1st Battalion, 30th Punjabis.

Raised at Ludhiana in 1857, by Capt. R. O. T. Nicolls, Madras Army, and designated the 22nd Regiment of Punjab Infy. Became the 34th Regt. of Bengal Native Infantry, 1861; the 30th Regt. of Bengal N. I., 1861; the 30th (Punjab) Regt. of Bengal N. I., 1864; the 30th (Punjab) Regt. of Bengal Infy., 1885; the 30th Punjab Infy., 1901; present designation, 1903.

"Afghanistan, 1879-80" "Chitral" "Tirah" "Punjab Frontier."

Composition—2 Companies of Sikhs, 1 of Dogras, 1 of Punjabi Musalmans.

COLONEL—

1st Commn. or date of entering service.	Names and Rank.	Army rank.	Present appointment in Regiment.	Remarks.
	Commandant.			
12 Mar. 92	James, Lt.-Col. A. A.	12 Mar. 18	11 Sep. 18	
	Company Commanders. (4)			
23 Dec. 93	*Murphy, Bt. Lt.-Col. C.C.R.* [L]	*3 June 15*	*9 Aug. 09*	*G. S. O. 2nd Gr.*
14 Aug. 95	Wallace, Maj. E. C. L., D.S.O., (Q.) (l).	14 Aug. 13	22 Mar. 11	*Lv., ex I., m. c.*
5 Aug. 96	*Keene, Maj. G. N. S.*	*5 Aug. 14*	*12 Dec. 13*	*Lv., ex I., m. c.*
	Company Officers. (9)			
11 Feb. 99	*Torrie, Maj. C. J.*, D.S.O., *(actg. Lt.-Col.)*	*1 Sep. 15*	*4 Apl. 02*	
14 Feb 00	Webster, Maj. C. D.	1 Sep. 15	1 Sep. 01	
28 July 00	*Robertson, Maj. R. A. H. (temp. Lt.-Col.)*	*1 Sep. 15*	*26 Dec. 01*	*Army Clothing Dept.*
4 Aug. 00	Murray, Maj. H. A.	1 Sep. 15	21 July 05	With 2-35th Sikhs.
8 Jan. 01	Keppel, Maj. T. R. H.	8 Jan. 16	1 Sep. 06	2nd-in-Comd. (*temp.*)
22 Oct. 02	Wilson, Maj. W. K. P.	22 Oct. 17	30 Jan. 07	*Lv. ex I., m. c.*
10 Oct. 03	Pottinger, Maj. J. A., M.C.	10 Oct. 18	24 Nov. 07	With 36th Sikhs.
18 Jan. 05	*Moss, Capt. T. (temp. Maj.)*	*18 Jan. 14*	*1 Mar. 07*	*G. S. O. 2nd Gr., Army Hdqrs.*
29 Aug. 06	Digby, Capt. T. A.	29 Aug. 15	18 Nov. 08	With 2-153rd Infy.
9 Sep. 08	Mitchell-Carruthers, Capt. N. L.	1 Sep. 15	20 Nov. 09	Adjt., 12 May 16.—
8 Sep. 09	*Barton, Capt. L. E.*	*1 Sep. 15*	*13 Dec. 10*	*Khyber Rif.*
11 Dec. 09	Kellie, Capt., G. J. D., M.C.	1 Sep. 15	6 Sep. 14	*Bde.-Maj., Jhelum Bde.*
9 Jan. 04	Doveton, Capt. A. C.	19 Nov. 13	....	1-21st Punj.
11 Nov. 14	*Powell, Capt. D.*	*11 Nov. 18*	*25 Jan. 15*	
23 Dec. 14	Clarke, Lt. C. P.	Dec. 10	24 Feb. 15	
15 Nov. 15	Barlow, Lt. W. N. C., M.C.	15 Nov. 16	19 Nov. 15	Attd. with 28th Punj.
	Attached.			
6 Sep. 14	Thomas, Capt. C. E.	6 Sep. 18	27 Nov. 16	I. A. R. O. *Asst. Embarkation Officer, Bombay.*
5 May 15	Stubbings, Lt. R. A. (*Actg. Capt.*)	5 May 16	17 Aug. 17	Essex R., I. A. (*on probn.*)
3 Aug. 15	Smith, L. R.	3 Aug. 16	12 Feb. 18	I. A. R. O. (*on probn.*)
19 Dec. 15	Saunders, Lt. G. N. (*Actg. Capt.*)	19 Dec. 16	21 Feb. 18	E. Kent R. I. A. (*on probn.*)
8 Apl. 16	Dilks, Lt. A. C.	8 Apl. 17	1 Nov. 17	Lond. R., I. A. (*on probn.*)
30 April 16	Otto, 2nd-Lt. L. A., M.C.	30 Apl. 16	10 Sep. 17	
8 June 16	Spencer, Lt. G. W.	8 June 17	13 Feb. 18	R. W. Fus., I. A. (*on probn.*)
24 July 16	Jones, Lt. C. Y.	24 July 17	23 Mar. 18	Lond R., I. A. (*on probn.*)
2 Oct. 16	Reeve, Lt. A. J.	2 Oct. 17	30 Mar. 18	M. G. Corps, I. A. (*on probn.*)
1 May 17	Corke, Lt. F. S.	1 May 18	26 Nov. 17	Shrops, L. I., I. A. (*on probn.*)
29 Aug. 17	Pearson, Lt. T. F.	29 Aug. 18	13 Dec. 17	I. A. R. O.
1 Apl. 18	Jack, 2nd-Lt. H. A.	1 Apl. 18	1 Mar. 18	I. A. (*on probn.*)
26 Apl. 17	Couche, 2nd-Lt. T. J.	26 Apl. 17	18 Nov. 18	Notts. and Derby R.
15 July 17	Gemmell, 2nd-Lt. J. C.	15 July 17	15 Sep. 18	I. A. R. O.
15 Sep. 17	Sterwin, 2nd-Lt. R. J.	15 Sep. 17	18 Feb. 18	I. A. R. O. (*on probn.*)
27 Sept. 17	Bowron, 2nd-Lt. R. A.	29 Sep. 17	18 Nov. 18	Notts. and Derby R.
5 Oct. 17	Woods, Lt. H. (*Actg. Capt.*)	5 Oct. 18	21 Jan. 18	I. A. (*T. C.*) Adjt. 6 Sep. 18.
5 Nov. 17	Rice, Lt. H. G.	5 Nov. 18	15 Mar. 18	I. A. (*T. C.*)
18 Dec. 17	Hancock, 2nd-Lt. G. A.	18 Dec. 17	18 Nov. 18	Notts and Derby R.
10 Jan. 18	Seward, 2nd-Lt. J. B.	10 Jan. 18	24 Oct. 18	I. A. R. O.
7 June 18	Green, 2nd-Lt. V. N.	7 June 18	10 Oct. 18	I. A. (*T. C.*)
26 June 18	Sheriston, 2nd-Lt. J. A.	26 June 18	4 Dec. 18	Notts. and Derby R.
22 July 18	Brown, 2nd Lt. H.	22 July 18	1 Dec. 18	I. A. R. O.
22 July 18	Moody, 2nd-Lt. F. W.	22 July 18	1 Dec. 18	I. A. (*T. C.*)
9 Oct. 18	Grant, 2nd Lt. J.	9 Oct. 18	24 Oct. 18	I. A. R. O.
7 June 00	Arthur, 2nd-Lt. J. R. H.	7 June 18	...	I. A. R. O.
26 Oct. 17	Makenzie, Lt. F.	26 Oct. 18	...	......

1st Battalion, 30th Punjabis—*contd.*

1st Commn. or date of entering service.	Names and Rank.	Army rank.		Remarks.
	Subadar-Major.			
		Jemadar.	Subadar.	Subadar-Major.
24 Nov. 01	Thakur Singh, I.D.S.M. (56, 66, 72*b*).	1 May 07	17 Sep. 14	5 Nov. 17.
	Hakim Khan (Retd.)	. . . . .	. . . . .	Re-employed 26 May 15.
	Subadars.			
17 June 95	*Diwan Singh Bahadur (56, 72a)* .	*1 Dec. 07*	*14 Aug. 14*	*Order of Br. Ind., 2nd Cl. With [Sig. Ser.*
30 July 94	Chhajja Singh (56, 60*a*) . . .	17 Aug. 14	19 Aug. 16	
14 Feb. 00	Lal Singh *Bahadur* (56, 66*a*) .	17 Aug. 14	1 Dec. 16	*Order of Br. Ind., 2nd Cl.*
20 Mar. 00	Haidar Ali	1 Feb. 16	7 Jan. 17	
24 Feb. 88	Allah Ditta	1 July 15	14 Oct. 17	
18 May 02	Jamit Singh (72*a*)	21 Jan. 15	13 Nov. 17	
25 Oct. 98	Achhar Singh	1 Jan. 16	19 Oct. 17	
18 Jan. 96	Gurdit Singh	19 Aug. 16	19 Oct. 17	
31 July 93	Hakim Singh	22 Apl. 17	19 Oct. 17	
19 Dec. 99	Sham Singh	1 Sep. 17	19 Oct. 17	
12 Feb. 94	Karam Dad	11 May 15	2 Nov. 17	
	Jemadars.			
24 Feb. 95	Lakha	1 Aug. 17		
7 July 01	Prabh Dayal, I.D.S.M.	29 Sep. 17		
24 May 00	Gurdit Singh	19 Oct. 17		
20 Jan. 99	Kesar Singh	19 Oct. 17		
13 Feb. 00	Ram Singh	19 Oct. 17		
15 Oct. 00	Sahab-ud-din	1 Nov. 17		
4 Jan. 99	Muhammad Khan	2 Nov. 17		
18 Mar. 01	Fazal Husain	3 Nov. 17	. . . .	With 3-30th Punj.
6 Apl. 08	Ram Saran	1 May 18		
	Haidar Khan	1 Aug. 18		

2nd Battalion, 30th Punjabis.

Formed at Lahore on 28th December 1917.

Composition—2 Companies Punjabi Musalmans, 1 Company Sikhs, 1 Company Dogras.

1st Commn., or date of entering service.	Names and Rank.	Army rank.	Present appointment in Regiment.	Remarks.
	Commandant.			
16 Feb. 98	Keen, Maj. P. H. (*Act'g. Lt. Col.*)	1 Sep. 15	16 Sep. 18	1-56th Rif.
	Company Commanders. (4)			
30 Oct. 01	Maynard, Maj. F. H., M.C.	30 Oct. 16	6 Jan. 18	125th Rif., 2nd-in-Comd.
23 Dec. 14	Clarke, Lt. C. P. (*Actg. Capt.*)	23 Dec. 15	1 Mar. 18	
13 July 15	Kearey, Lt. H. P. (*Actg. Capt.*)	13 July 16	1 Mar. 18	N. Staff R., I. A. (*on probn.*)
3 Sep. 15	Hunter, Lt. T. C. L. (*Actg. Capt.*)	3 Sep. 16	30 Mar. 18	I. A. R. O. (*on probn.*)
	Company Officers.			
8 May 01	Foster, Maj. R.	8 May 16	8 May 18	Survey Dept.
5 May 15	Horner, Lt. B. S. (*Actg. Capt.*)	5 May 16	....	Surr. R., I. A. (*on probn.*) Adjt.
16 Nov. 15	Crothers, Lt. A. V.	16 May 17	13 Feb. 18	I. A. R. O.
13 Mar. 15	Omealy-Piggot, Lt. F. M.	1 Nov. 16	13 Feb. 18	
11 Mar. 16	Banon, Lt. H. L.	11 Mar. 17	8 Jan. 18	I. A. R. O.
24 Aug. 15	Gilligan, Lt. C. H.	24 May 17	28 Feb. 18	R. Ir. Rif., I. A. (*on probn.*)
9 Sep. 15	Edwards, Lt. F. O.	1 Apl. 17	6 Jan. 18	Qr.-Mr. (*offg.*)
19 July 16	Jameson, 2nd-Lt. G. G.	19 July 17	5 Jan. 18	Leins. R.
20 Dec. 15	Tyrell, Lt. H. E.	1 July 17	21 Jan. 18	
22 Mar. 17	Crawley, Lt. N. L.	22 Mar. 18	30 Mar. 18	I. A. (*on probn.*)
27 June 17	Jones, 2nd-Lt. J. H.	27 June 17	17 Apl. 18	I. A. R. O.
5 Nov. 17	Whittington, Lt. H.	5 Nov. 18	11 Mar. 18	I. A. (*T. C.*)
8 Nov. 17	Padfield, 2nd-Lt. C. A.	23 Nov. 17	31 Jan. 18	I. A. R. O. (*on probn.*)
26 Jan. 18	Hoad, 2nd-Lt. H. R.	26 Jan. 18	....	I. A. (*on probn.*)
1 Jan. 18	Mallinson, 2nd-Lt. E. H. P.	31 Jan. 18	6 Feb. 18	Attd.
3 Jan. 18	Dowse, 2nd-Lt. B. T. J.	31 Jan. 18	6 Feb. 18	Attd.
7 June 18	Brick, 2nd-Lt. P. J.	7 June 18	10 Oct. 18	I. A. R. O.
7 June 18	Ireland, 2nd-Lt. B. J.	7 June 18	10 Oct. 18	I. A. R. O.
	Subadar-Major.	Jemadar.	Subadar.	Subadar-Major.
16 July 06	Warriam Singh	1 Jan. 14	1 Feb. 18	(*temp.*)
	Ghulam Ali Khan, *Bahadur*, (*Hony. Capt.*) (*Retd.*)	....	....	Re-employed.
	Subadars.			
20 Jan. 96	Hussain Khan	15 Jan. 15	30 Apl. 18	
4 Apl. 91	Tota Ram	29 Sep. 16	1 Feb. 18	
12 July 93	Chet Singh	5 Oct. 16	1 Feb. 18	
10 Dec. 95	Sultan Alam	1 Feb. 16	1 Feb. 18	
19 July 96	Bagh Ali	10 Aug. 17	1 Feb. 18	
26 Apl. 98	Abbas Ali	12 June 17	1 Feb. 18	
23 Feb. 94	Labha	1 Feb. 18	1 Feb. 18	
3 May 95	Moghul Khan	1 Nov. 17	1 Feb. 18	
	Jemadars.			
13 June 91	Uttam Singh	1 Dec. 17		
22 Dec. 96	Buta Khan	1 Jan. 18		
2 Apl. 96	Bahadur Singh	1 Feb. 18		
13 Oct. 99	Munshi Khan	1 Feb. 18		
3 June 05	Labhu Ram	1 Feb. 18		
29 May 98	Brijlal	1 Feb. 18		
20 June 06	Diwan Ali	1 Feb. 18		
5 Mar. 06	Sawar Khan	1 Feb. 18		
6 June 03	Maula Bakhsh	6 May 18		
17 July 96	Mula Singh	22 May 18		
30 June 98	Surayan Singh	3 Apl. 18		
10 Jan. 95	Kahan Singh	1 Oct. 18		
1 July 01	Atta Mohammad	1 Oct. 18		
26 Nov. 02	Mehar Singh	1 Oct. 18		
21 Aug. 07	Walayat Khan	1 Oct. 18		
3 Apl. 18	Indar Singh	5 Nov. 18		

3rd Battalion, 30th Punjabis.

Formed at Phillaur on 22nd June 1918.

Composition—2 Companies Punjabi Musalmans.
1 Company, Punjabi Hindus.
½ Company, Garhwalis.
¼ Company, Dogras.
¼ Company, Sikhs.

1st Commn. or date of entering service.	Names and Rank.	Army rank.	Present appointment in Regiment.	Remarks.
	Commandant.			
6 June 17	Isemonger, 2nd-Lt. F. C. (*Actg. Lt.-Col.*)	22 June 18	25 June 18	I. A. R. O.
	Company Commanders. (4)			
22 June 05	Jukes, Bt.-Maj. A. H., D.S.O.	1 Jan. 17	29 Aug. 18	2-9th G. R., 2nd-in-Comd.
20 Aug. 15	Goodall, Lt. L. M.	20 Aug. 16	1 Nov. 18	A. S. C., I. A. (*on probn.*)
18 Feb. 15	Hoade, Lt. J. R., (*Actg. Capt.*)	18 Feb. 16	13 Sep. 18	I. A. R. O.
26 Mar. 15	Crawford, Lt. C. G., (*Actg. Capt.*)	26 Mar. 16	30 Sep. 18	I. A. R. O.
	Company Officers.			
28 Mar. 17	Logan, Lt. E. W.	28 Mar. 18	14 Sep. 18	I. A. R. O.
5 Sep. 16	Meadmore, Lt. T. B. R.	5 June 18	14 Sep. 18	M. G. Corps, I. A. (*on probn.*)
25 Oct. 16	Green, Lt. P. A., (*Actg. Capt.*)	25 July 18	13 Sep. 18	I. A. (*on probn.*) Adjt.
20 June 17	Bush, Lt. A. J.	20 June 18	10 Oct. 18	I. A. R. O.
29 Mar. 18	Holt, 2nd-Lt. D. R.	29 Mar. 18	19 Sep. 18	I. A. R. O.
27 June 18	Price, 2nd-Lt. G. W., (*Actg. Capt.*)	27 June 18	1 July 18	I. A. R. O.
28 Mar. 17	Parkinson, 2nd-Lt. A. G.	28 Mar. 17	26 Sep. 18	I. A. R. O.
27 Aug. 18	Hinge, 2nd-Lt. F. G. H.	27 Aug. 18	16 Sep. 18	I. A. R. O.
5 Jan. 18	Cleator, 2nd-Lt. A. E.	5 Jan. 18	10 Oct. 18	I. A. (*T. C.*)
29 Sep. 15	Johnson, Lt. R. A.	29 Sep. 16	22 Oct. 18	Sud. R., I. A. (*on probn.*)
1 May 17	Percy Smith, Lt. H. K.	1 May 18	1 Nov. 18	40th Pathans, Gr. Mr. 1 Nov. 18.
20 June 17	Jones, Lt. D. J.	20 June 18	12 Oct. 18	I. A. R. O.
21 Dec. 17	Orgill, 2nd-Lt. A. W.	21 Dec. 17	20 Sep. 18	Attd.
31 Jan. 18	Coldstream, 2nd-Lt. G. A. P.	31 Jan. 18	26 Sep. 18	37th Rif.
21 Sep. 18	Collins, 2nd-Lt. T. C.	21 Sep. 18	15 Oct. 18	I. A. R. O.
15 July 18	Crowhurst, 2nd-Lt. A. D.	15 July 18	1 Dec. 18	I. A. R. O.
22 July 18	Hutchings, 2nd-Lt. F. S.	22 July 18	1 Dec. 18	I. A. (*T. C.*)
24 Oct. 18	Jones, 2nd-Lt. H. R.	24 Oct. 18	5 Dec. 18	I. A. R. O.

Subadar-Major.

		Jemadar.	Subadar.	Subadar-Major.	
24 Mar. 91	Abdulla Khan	11 Sep. 14	29 July 16	13 Sep. 18	(*Temp.*)

Subadars.

1 May 02	Pahlwan Khan	5 July 15	29 Dec. 15
25 June 05	Harnam Singh	24 Mar. 15	5 June 17
16 June 07	Kapur Singh	26 June 18	19 July 18
1 Sep. 99	Yar Ghulam	1 June 14	20 July 18
23 Nov. 98	Parja	14 Jan. 16	29 July 18

Jemadars.

5 Oct. 02	Fazl Dad Khan	9 June 15	
16 Feb. 95	Labh Singh	1 Nov. 15	
18 Aug. 03	Nathwa Ram	1 Jan. 17	
15 Feb. 01	Shamsher Singh	12 July 17	
5 Mar. 06	Akbar Khan	22 May 18	
1 July 18	Fateh Mohammad	26 June 18	
5 July 18	Sapuran Singh	5 July 18	
1 July 18	Muzaffar Khan	1 July 18	
1 Aug. 18	Abdul Haq	1 Aug. 18	
23 Feb. 91	Sher Mohammad	25 Aug. 18	
1 July 18	Mata Din Singh	7 Sep. 18	
12 Sep. 18	Balmokand	12 Sep. 18	
27 May 18	Hadi Hussain Shah	1 Oct. 18	
4 July 18	Lachman Singh	17 Oct. 18	Indian Qr. Mr.
12 Feb. 98	Fateh Mohammad	25 Oct. 18	
5 Feb. 97	Tek Singh	25 Oct. 18	
1 Dec. 05	Lall Khan	25 Oct. 13	
12 Dec. 08	Puran Singh	1 Nov. 18	Indian Adjt.
18 Mar. 01	Fazal Hussain	3 Nov. 17	1.30th Punj.

4th Battalion, 30th Punjabis.

Formed at

Composition—

1st Commn. or date of entering service.	Names and Rank.	Army rank.	Present appointment in Regiment.	Remarks.

Commandant.

21 June 18	Mercer, 2nd-Lt. A. W. (*Actg. Lt.-Col.*)	21 June 18	. . . .	

Company Commanders. (4)

18 Apl. 16	Gompertz, Lt. M. C. T. (*Actg. Capt.*)	18 Apl. 17	. . . . .	1-8th Rajputs.
21 June 18	Hadow, Capt. D. S. (*Actg. Maj.*)	21 June 18		
29 June 16	Davis, Lt. J.	29 June 17		

Company Officers.

5 Mar. 15	Bright, Lt. F. C.	5 Mar. 16	. . . .	I. A. R. O.
21 Aug. 16	Nevill, Lt. J. B. (*temp. Capt.*) .	21 Aug. 17	. . . .	
17 Apl. 15	Mass, Lt. L. (*Actg. Capt.*) .	17 Apl. 16	. . . .	
30 June 17	Skead, Lt. B. (*Actg. Capt.*) . .	30 June 18	. . . .	Lond. R., I. A. (*on probn.*)
7 June 18	Quinton, 2nd-Lt. A. W. . . .	7 June 18	. . . .	I. A. (*T. C.*)

Subadar-Major.

		Jemadar.	Subadar.	Subadar-Major.

Subadar.

Jemadar.

Regt. Centre—Peshawar.
Linked with—29th and 30th Punjabis.

Uniform—Scarlet.
Facings—White.

31st Punjabis.

Raised at Ferozepore in 1857, by Capt. G. C. Bloomfield, and originally styled 'Van Cortlandt's Levy. Became the 23rd Regt. of Punjab Infantry, 1857; the 35th Regt. of Bengal Native Infantry, 1861; the 31st Regt. of Bengal Nat. Infy., 1861; the 31st (Punjab) Regt. of Bengal Nat. Infy., 1864; the 31st (Punjab) Regt. of Bengal Infy., 1885; the 31st Punjab Infy., 1901; present designation, 1903.

"Afghanistan, 1879-80" "Malakand" "Punjab Frontier."

Composition—2 Companies of Sikhs, 1 of Dogras, 1 of Punjabi Musalmans.

COLONEL—General *Sir* Henry B. B. Watkis, K.C.B. . . . 26 June 1908.

1st Commn. or date of entering service.	Names and Rank.	Army rank.	Present appointment in Regiment.	Remarks
	Commandant.			
1 Mar. 90	Kemball, Lt.-Col. A. G. (*temp. Br.-Genl.*)	1 Mar. 16	22 Jan. 18	
	Company Commanders. (4)			
29 Feb. 96	*Leslie, Bt.-Col. W. S., D.S.O., p.s.c.*	*3 June 18*	*11 Sep. 18*	*2nd in-Comd. G. S. O., 1st gr.*
3 Aug. 98	*Duff-Sutherland-Dunbar, Maj. Sir G., Bart.*	*1 Sep. 15*	*22 Jan. 14*	*Bde.-Maj., Jhansi.*
12 Aug. 99	*Tancock, Maj. A. C.*	1 Sep. 15	23 June 10	*Mohmand Militia.*
8 Jan. 01	McLeod, Maj. K. (*Actg Lt.-Col.*)	8 Jan. 16	11 Sep. 18	
	Company Officers. (9)			
8 May 01	Venning, Maj. F. E. W., D.S.O. .	8 May 16	20 Nov. 02	Adjt., 15 July 15.
18 Jan. 02	Macdonald, Maj. W. .	18 Jan. 17	2 Oct. 03	
18 Jan. 02	Carey, Maj. P. G.	18 Jan. 17	7 Feb. 04	With 3-23rd Sikh Infy.
10 Oct. 03	Dennys, Maj. G. T.	10 Oct. 18	1 Aug. 05	
29 Jan. 10	*Deedes, Capt. R. B., M.C.* . .	*1 Sep. 15*	*18 Mar. 11*	
18 Jan. 11	Gaisford, Capt. P.	1 Sep. 15	5 Apl. 12	Qr.-Mr., 1 July 15.
6 Sep. 11	Graham, Capt. R. A.	6 Sep. 15	19 Nov. 12	
20 Sep. 11	Scott, Capt. K. R., M.C. . . .	20 Sep. 15	26 Feb. 14	
8 Aug. 14	Dymott, Capt. B. K.	8 Aug. 18	11 Nov. 15	Attd. Staff Capt., 57th Bde.
15 Aug. 14	Briggs, Capt. H. R.	15 Aug. 18	. . . .	Attd. With 1-152nd Infy.
12 May 15	Brennan, Lt. C. G. O.	12 May 16	6 Nov. 15	Attd.
15 Nov. 15	Walker, Lt. F. R.	15 Nov. 16	18 Nov. 15	Attd.
20 Apl. 16	Cox, Lt. G. W.	20 Apl. 17	3 Sep. 15	
8 June 16	Corrin, Lt. T. S.	8 June 17	. . . .	
29 June 16	Moorhead, Lt. H. D.	29 June 17	24 Jan. 17	Attd.
14 Nov. 16	Maffett, Lt. D. B.	14 Nov. 17	. . . .	Attd.
21 May 15	Ingles, Lt. H. N. G.	21 Feb. 17	. . . .	Attd.
	Attached.			
29 Aug. 14	Westmacott, Capt. P. G. . .	29 Aug. 18	15 Nov. 15	I. A. R. O.
31 July 15	Hayson, Lt. R. A.	31 July 16	3 Sep. 15	I. A. R. O.
20 Feb. 16	Dawson, Lt. T. E.	20 Feb. 17	23 Sep. 16	I. A. R. O.
18 Nov. 16	Booth, Lt. A. R.	18 Nov. 17	22 Oct. 17	I. A. (*on probn.*)
3 Aug. 15	Hunt, Lt. R. E.	3 Aug. 16	26 Nov. 17	I. A. (*on probn.*)
10 June 16	Hoare, Lt. J. G.	10 June 17	26 Nov. 18	I. A. (*on probn.*)
26 Jan. 17	Jennings, Lt. C. K.	26 Jan. 18	24 Apl. 18	I. A. (*on probn.*)
26 June 17	Nunn, 2nd-Lt. H.	26 June 17	29 Apl. 18	I. A. (*on probn.*)
29 Mar. 18	Webber, 2nd-Lt. C. J. . . .	29 Mar. 18	22 July 18	I. A. (*T. C.*)
22 Sep. 14	Kinns, Capt. P. C.	22 Sep. 18	7 Dec. 17	I. A. R. O.
17 June 17	Inrig, Lt. L.	17 June 18	5 Oct. 17	I. A. R. O.
29 Mar. 18	Venton, Lt. W. F.	29 Mar. 18	19 July 18	I. A. R. O.
12 Apl. 16	Anstey, Lt. D.	12 Apl. 17	. . . .	Worc. R., I. A. (*on probn.*)
5 Dec. 16	Figgins, Lt. A. P.	5 Dec. 17	3 July 18	Welsh R., I. A. (*on probn.*)
23 Dec. 15	Bisley, Lt. F. de B.	23 Dec. 16	. . . .	I. A. R. O.
22 Feb. 16	Reid, Lt. W. F.	22 Feb. 17	. . . .	R. Ir. Fus., I. A. (*on probn.*)
	Cornock, Lt. W. B.	. . . .	. . . .	Glouc. R. I. A. (*on probn.*)
5 Oct. 18	Willie, 2nd-Lt. E. E.	5 Oct. 18	10 Oct. 18	I. A. (*T. C.*)
8 Oct. 18	Finch, 2nd-Lt. C.	8 Oct. 18	23 Oct. 18	I. A. R. O.
19 Oct. 18	Gilrey, 2nd-Lt. A.	19 Oct. 18	30 Oct. 18	I. A. R. O.
21 Oct. 18	Donald, 2nd-Lt. J.	21 Oct. 18	7 Nov. 18	I. A. R. O.

31st Punjabis—*contd.*

1st Commn. or date of entering service.	Names and Rank.	Army rank.	Present appointment in Regiment.	Remarks.
	Subadar-Major			
		Jemadar.	Subadar.	Subadar Major
1 Apl. 00	Sahib Khan (52, 56, 66a, 72a)	1 Dec. 02	1 Oct. 07	11 May 16
3 Oct. 00	Muhammad Feroz Khan	13 Aug. 14	5 July 16	6 July 18 (*Supy*).
	Subadars.			
14 Oct. 86	Mangal Singh (56)	9 June 02	1 July 09	
7 Apl. 89	Kor Singh (72a)	1 Mar. 03	23 Apl. 10	
26 Apl. 87	Dewa Singh (56, 66a)	16 Apl. 09	1 Apl. 12	
8 Jan 98	*Harnam Singh (66a)*	*22 Feb. 08*	7 Apl. 15	*With 24th Punj.*
26 July 94	*Prem Singh (52, 56, 60a)*	*1 Apl. 12*	*28 Apl. 15*	
15 Mar. 95	Beli Ram (72a)	1 Jan. 13	17 May 15	
1 Jan. 89	Bishan Singh (66a)	1 Feb. 11	4 Sep. 15	
8 Mar. 98	Punjab Singh	9 Feb. 14	4 Sep. 15	
10 Feb. 95	Kishn Singh (66a)	1 Sep. 13	4 July 16	
1 Jan. 94	*Karam Khan*	*7 Apl. 15*	*1 Mar. 16*	
4 July 99	Arjan	5 Apl. 15	2 May 17	
20 Oct. 90	Partab Singh	5 Feb. 15	13 May 17	
1 Nov. 99	Narayan Singh	17 Mar. 15	13 June 17	
10 Sep. 98	Phnman Singh	4 Sep. 15	16 Feb. 18	
	Jemadars.			
3 Mar. 95	Mihan Singh	28 Apl. 15		
2 Oct. 03	Tulsi Ram	17 May 15		Indian Adjt.
19 Dec. 95	Muhammad Khan	4 Sep. 15		
1 Feb. 97	Shahbaz Khan	28 Oct. 15		
29 June 04	Sherbaz Khan	28 Oct. 15		
26 May 95	Sadda Singh	12 Feb. 17		
16 Mar. 00	Munshi Ram	2 May 17		
28 Jan. 99	Kishan Singh	13 May 17		
1 Feb. 01	Kirpa Ram	3 Dec. 17		
1 Apl. 01	Qasim Ali	12 Jan. 18		
17 Oct. 99	Lal Singh	16 Feb. 18		
	Wakila Dhian Singh	1 Sep. 18		
	Mainga Singh	1 Sep. 18		

1st Battalion, 32nd Sikh Pioneers.

Regt. Centre—Ambala. *Uniform*—Scarlet.
Linked with—23rd and 34th Pioneers. *Facings*—Blue.

Raised at Madhopore in 1857, by Lt. Dyas, and was known at the time both as the 'Punjab Sappers' and 'Punjab Pioneers.' Became the 24th (Pioneer) Regt. of Punjab Infy., 1858; the 36th Regt. of Bengal Native Infantry, 1861; the 32nd Regt. of Bengal N. I., 1861; the 32nd (Punjab) Regt. of Bengal Nat. Infy. (Pioneers), 1864; the 32nd (Punjab) Regiment of Bengal Infy. (Pioneers), 1885; the 32nd Punjab Pioneers, 1901; present designation, 1903.

"*Aut viam inveniam aut faciam.*"
"Delhi, 1857" "Lucknow" "Afghanistan, 1878-80" "Chitral."
Composition—4 Companies Mazbi and Ramdasia Sikhs.

COLONEL—Lieut.-General *Sir* Ernest DeBrath, K.C.B., C.I.E. [R] 6 January 1905.

1st Commn. or date of entering service.	Names and Rank.	Army rank.	Present appointment in Regiment.	Remarks.
	Commandant			
27 Jan. 92	*Cooke, Col. H. F.,* D.S.O. *(temp. Maj.-Genl.)*	*8 June 17*	*3 June 16*	*D. A. G., Army Hdqrs.*
	Company Commanders. (4)			
28 Jan. 93	Hope, Maj. A. V. W. *q* (Q) (*Act'g. Lt.-Col.*)	28 Jan. 11	23 Oct. 15	2nd-in-Comd. Comdt. (*temp.*)
19 May 00	Mitchell, Maj. H. S., *q.*	1 Sep. 15	2 Oct. 14	2nd-in-Comd. (*temp.*)
19 Oct. 01	Croslegh, Maj. E. L., *q.*	1 Sep. 15	21 Feb. 18	With 2—32nd Prs.
22 Oct. 02	deCourcy, Maj. Hon. M. W. R. D.S.O., *q.*	22 Oct. 17	24 Mar. 15	
	Company Officers. (9)			
18 Jan. 05	Burn-Murdoch, Capt. I., *q.*	18 Jan. 14	1 Feb. 07	
29 Aug. 06	Ommanney, Capt. W. M., *q*	29 Aug. 15	3 Feb. 09	With 2-155th. Infy.
9 Sep. 08	Andrewes, Capt. H. W., *q.* (*Actg. Maj.*)	1 Sep. 15	12 Dec. 09	
20 Oct. 08	Christie, Capt. B.	1 Sep. 15	6 Jan. 11	With 1—155th Infy.
20 Jan. 09	Taylor, Capt. N. S., *q.*	1 Sep. 15	23 Mar. 10	Adjt., 16 Mar. 16.
18 Jan. 11	North, Capt. K. A., M.C.	1 Sep. 15	4 May 13	With 34th Prs.
22 Jan. 13	Cambier, Capt. V.	22 Jan. 17	15 July 15	
15 Aug. 14	Teversham, Capt. M. S.	15 Aug. 18	9 Nov. 15	
23 Dec. 14	Critchley, Lt. D. R.	23 Dec. 15	1 Mar. 15	With 34th Prs.
13 Jan. 15	Paterson, Lt. L. M., M.C.	13 Jan. 16	27 Feb. 15	
15 Nov. 15	Gazzard, Lt. T.	15 Nov. 16	18 Nov. 15	With 34th Prs.
15 Nov. 15	Hayward, Lt. T. E.	15 Nov. 16	18 Nov. 15	With 2—81st Prs.
29 June 16	Combe, Lt. L. M.	20 Mar. 16	. . .	With 2—32nd Sikh Prs.
18 Apl. 16	Humby, Lt. L. J.	28 Aug. 16	25 Apl. 16	
1 May 17	Leake, Lt. J. I.	1 May 18	1 Dec. 17	Attd. with 1-155th Prs.
27 Oct. 17	Grieves, 2nd-Lt. R. B.	27 Oct. 17	4 Nov. 17	With 2-155th Infy.
11 Aug. 15	Litton, Lt. G. E.	11 Aug. 16	25 Nov. 17	Attd. with 3-32nd Prs.
31 Aug. 18	Alston, 2nd-Lt. W. L.	31 Aug. 18	31 Aug. 18	Attd.
1 Oct. 18	Coole, 2nd-Lt. A. E.	1 Oct. 18	12 Oct. 18	Attd.
	Attached.			
18 Dec. 14	Duncan, Capt. L. F.	3 June 18	. . . .	I. A. R. O.
14 Jan. 15	Morris, Lt. C. M.	14 Jan. 16	. . .	I. A. R. O.
26 Mar. 15	Ranger, Lt. C. A. (*Act'g. Capt.*)	26 Mar. 16	1 Dec. 17	I. A. R. O.
2 Sep. 16	Moberly, Lt. R. S.	2 Sep. 17	2 Oct. 16	I. A. R. O. Qr.-Mr.
8 Sep. 16	Harlow, Lt. C. M. (*Act'g. Capt.*)	8 Sep. 17	8 Oct. 16	I. A. R. O.
9 June 16	Ellis, Lt. A. B.	9 June 17	5 Oct. 18	Attd.
15 Jan. 18	Dickson, 2nd-Lt. W. A.	15 Jan. 18	31 May 18	I. A. R. O.
10 Sep. 17	Pritchard, Lt. T. A.	10 Sep. 18	20 July 18	Tank Corps, I. A. (*on probn.*)
1 Dec. 17	Bowden, Lt. S. A.	1 Dec. 18	2 Oct. 18	Suff. R., I. A. (*on probn.*)
24 Apl. 18	Brampton, 2nd-Lt. C.	24 Apl. 18	17 Oct. 18	I. A. (U. L.)
24 Apl. 18	Hughes, 2nd-Lt. A. H.	24 Apl. 18	23 Oct. 18	I. A. (U. L.)

Subadar-Major.

		Jemadar.	Subadar.	Subadar-Major.	
20 Dec. 82	Bishn Singh, *Sardar Bahadur* [(56).	1 July 95	1 Apl. 02	1 July 12	*Order of Br. Ind., 1st Cl.*

Subadars

16 May 86	Wassawa Singh, Sardar Bahadur, I. O. M., I. D. S. M.	1 Aug. 95	1 Apl. 03	Order of Br. Ind., 1st Cl.
9 Dec. 86	Lachman Singh *Bahadur* (56)	1 Aug. 02	1 Mar. 10	*Order of Br. Ind., 2nd Cl.*
12 Jan. 99	Sundar Singh	1 June 13	15 Sep. 15	
4 May 94	Gopal Singh	2 Nov. 14	12 May 16	
9 Apl. 01	Arjan Singh (56)	2 Nov. 14	12 June 16	
15 July 01	Sant Singh (56)	30 Nov 14	4 Oct. 16	
27 Feb. 93	Sham Singh	16 July 11	9 Sep. 16	
11 Apl. 93	Budha Singh (56)	30 Nov. 14	20 Sep. 16	
10 Mar. 97	Kesar Singh	30 Nov. 14	20 Sep. 16	
12 Jan. 97	Jhanda Singh	30 Nov. 14	2 May 17	
19 Mar. 01	Sawan Singh	1 Nov. 15	13 Sep. 18	
	Sher Singh (*Retd.*)	. . . .	. . . .	Re-employed 20 Sep. 18.

1st Battalion, 32nd Sikh Pioneers—*contd.*

1st Commn. or date of entering service.	Names and Rank.	Army rank.	Present appointment in Regiment.	Remarks.
	Jemadars.			
30 Oct. 96	Sher Singh	15 Sep. 15		
1 Apl. 08	Rup Singh	1 Oct. 15		
26 July 02	Natha Singh	12 May 16		
1 Aug 99	Sundar Singh	12 June 16		
28 June 99	Ishar Singh	12 June 16		
14 July 95	Sant Singh	2 July 16		
2 Aug. 16	Kirpa Singh	2 July 16		
4 Sep. 16	Mit Singh	4 Sep. 16		
31 Aug. 16	Arjan Singh	31 Aug. 16		
20 Sep 16	Ganesha Singh	20 Sep. 16		
20 Sep. 16	Hira Singh	20 Sep. 16		
4 Oct. 16	Laihra Singh	4 Oct. 16		
18 June 17	Lachhman Singh	2 May 17		
	Buta Singh	24 July 18		

2nd Battalion, 32nd Sikh Pioneers.

Formed at Sialkot on 11th December 1916.

Composition—4 companies of Mazbi and Ramdasia Sikhs.

1st Commn. or date of entering service.	Names and Rank.	Army Rank.	Present appointment in Regiment.	Remarks.
	Commandant.			
14 Aug. 95	Clarke, Maj. G. S. (*Act'g. Lt.-Col.*)	14 Aug. 13	16 Jan. 17	1-34th Sikh Prs.
	Company Commanders. (4)			
19 Oct. 01	Croslegh, Maj. E. L.	1 Sep. 15	13 Feb. 18	1-32nd Prs., 2nd-in-Comd. (*temp.*)
27 Aug. 02	Bailey, Capt. P. H. H., M. C.	9 Oct. 12	6 Jan. 18	34th Sikh Prs.
9 Jan. 04	Bird, Capt. W. A. H. (*q*)	9 Jan. 13	6 Jan. 18	23rd Prs.
29 Aug. 06	Chapman, Capt. E. H.	9 Aug. 15	13 Feb. 18	107th Prs.
9 Sep. 08	Yeates, Capt. E. P.	1 Sep. 15	15 Feb. 18	12th Prs.
	Company Officers.			
3 Sep. 13	Dawes, Capt. H. D.	22 Oct. 14	4 June 17	
29 June 16	Combe, Lt. L. M.	20 Mar. 16	11 Dec. 16	1-32nd Sikh Prs. Adjt., 11 Dec. 16.
5 May 15	Paterson, Lt. C. R. S.	5 May 16	29 Aug. 17	E. Surr. R., I. A. (*on probn.*)
16 Aug. 17	Sivary, Lt. H. S.	16 Aug. 17	27 Nov. 17	I. A. (*on probn.*)
1 Sep. 16	Weston, Lt. F. H.	1 Sep. 17	1 Mar. 17	I. A. R. O. Qr.-Mr.
3 Feb. 17	King, 2nd-Lt. J.	3 Feb. 17	16 Apl. 17	I. A. R. O.
9 Apl. 17	Flitcroft, 2nd-Lt. E. A. S.	9 Apl. 17	18 Jan. 18	I. A. (*on probn.*)
5 June 17	Webster, Lt. H. P. (*Act'g. Capt.*)	5 Jun. 18	13 Feb. 18	I. A. (*on probn.*)
	Coldham, 2nd-Lt. P. C.	....	....	I. A. (*T. C.*)
	Chamings, 2nd-Lt. R. G.	....	....	I. A. R. O. (*on probn.*)
	Windsor-Aubrey, 2nd-Lt. W. A.	....	....	Attd.
12 Sep. 17	Lambert, Lt. W. G. A.	12 Sep. 18	....	Attd.
29 May 18	Little, 2nd-Lt. D. D.	29 May 18	....	Sco. Bord., I. A. (*on probn.*)
24 Apl. 18	Woodman, 2nd-Lt. W. H. L.	24 Apl. 18	....	I. A. (U. L.)

Subadar-Major.

		Jemadar.	Subadar.	Subadar-Major.
18 Nov. 86	Hukam Singh *Bahadur*	1 Oct. 02	1 May 07	1 Feb. 17 (*Actg.*) *Order of Br. Ind., 2nd Cl.*

Subadars.

3 Dec. 83	Ram Singh, Sirdar Bahadur. (56)	28 Oct. 03	1 June 12	*Order of Br. Ind., 1st Class.*
5 Apl. 90	Prem Singh (56)	4 Apl. 08	29 May 15	
14 July 92	Indar Singh	23 May 09	29 May 15	
3 Sep. 97	Mal Singh	1 Sep. 12	30 Aug. 16	
15 May 93	Garbha Singh	30 Nov. 14	31 Aug. 16	
12 Apl. 99	Lehna Singh (56)	11 Oct. 14	1 Feb. 18	
20 Apl. 98	Natha Singh (56)	3 Aug. 16	13 June 18	
12 July 05	Santokh Singh	4 Oct. 16	13 June 18	

Jemadars.

9 Sep. 01	Bishan Singh	1 Sep. 15
16 Jan. 00	Mehnga Singh (56)	12 May 16
14 July 96	Sham Singh	1 July 17
17 Apl. 06	Daulat Singh	1 Aug 17
17 Nov. 02	Jwala Singh	1 Aug. 17
13 Mar. 99	Sundar Singh	1 Aug. 17
6 Jan. 01	Thakur Singh	1 Aug. 17
7 Sep. 96	Gurdit Singh	1 Aug. 17
7 Sep. 96	Gonda Singh	19 Sep. 17
5 Oct. 07	Saman Singh	3 Dec. 17
	Ralla Singh	13 June 18

3rd Battalion, 32nd Sikh Pioneers.

Formed at Sialkot on 13th October 1918.

Composition.—4 Companies Mazbi and Ramdasias Sikhs.

1st Commn. or date of entering service.	Names and Rank.	Army Rank.	Present appointment in Regiment.	Remarks.
	Commandant.			
20 Jan. 97	Hodgson, Maj. G. C., D.S.O., (*Actg. Lt.-Col.*)	20 Jan. 15	22 Nov. 18	1-23rd Prs.
	Company Commanders. (4)			
19 Sep. 00	Marriott, Maj. J. H. G.	1 Sep. 15	22 Nov. 18	1-12th Prs., 2nd-in-Comd.
	Savage, Lt. E. G.	. . .	18 Oct. 18	I. A. R. O.
30 May 17	Coles, Lt. T. H.	13 May 18	22 Nov. 18	I. A. R. O.
30 Jan. 17	Fordyce, 2nd-Lt. J. F. D.	30 Jan. 17	20 Oct. 18	1-34th Prs.
	Company Officers.			
11 Aug. 15	Litton, Lt. G. E.	11 Aug. 16	15 Oct 18	1-32nd Prs.
21 Dec. 17	Newland, 2nd-Lt. J. H. F.	21 Dec. 17	19 Oct. 18	Attd.
5 Jan. 18	Wreford, 2nd-Lt. A. P.	5 Jan. 18	14 Oct. 18	I. A. (*T. C.*)
29 Mar. 18	Townsend, 2nd-Lt. T.	29 Mar. 18	16 Oct. 18	I. A. (*T. C.*)
7 June 18	Tidmarsh, 2nd-Lt. C. F.	7 June 18	14 Oct. 18	I. A. (*T. C.*)
	Subadar Major.			
		Jemadar.	Subadar.	Subadar-Major.
27 June 92	Hardit Singh	26 May 07	11 Oct. 14	
	Subadars.			
21 Jan. 99	Sundar Singh	1 June 13	15 Sep. 15	
12 July 91	Gurdit Singh	29 May 15	3 Feb. 17	
25 June 02	Assa Singh	10 July 15	27 Sep. 17	
19 May 96	Atma Singh	3 Feb. 17	2 Feb. 18	
	Jemadars.			
9 Sep. 01	Bishan Singh	1 Sep. 15		
30 Oct. 96	Sher Singh	15 Sep. 15		
1 Apl. 08	Rup Singh	1 Oct. 15		
16 Jan. 00	Manga Singh	12 May 16		
26 July 02	Natha Singh	12 May 16		
28 June 99	Ishar Singh	12 June 16		
2 Feb. 05	Sadhu Singh	3 Feb. 17		
17 Nov. 02	Jawala Singh I	1 Aug. 17		
8 Oct. 94	Hira Singh	6 Sep. 17		
18 Nov. 01	Jawala Singh II	16 Jan. 18		
16 May 01	Bisakha Singh	16 Jan. 18		
27 Nov. 00	Jagat Singh	1 Feb. 18		
14 May 04	Chanda Singh	25 May 18		

Regt. Centre—Sialkot.
Linked with—40th Pathans and 46th Punjabis.

Uniform—Drab.
Facings—Emerald Green.

1st Battalion, 33rd Punjabis.

Raised at Allahabad in 1857, by Lt. E. H. Langmore, as the 'Allahabad Levy.' Became the 37th Regt. of Bengal Native Infantry, 1861; the 33rd Regt. of Bengal N. I., 1861; the 33rd (Allahabad) Regt. of Bengal Nat. Infy., 1864; the 33rd Regt. of Bengal Infy., 1885; the 33rd (Punjabi Mahomedan) Regt. of Bengal Infy., 1890; the 33rd Punjab Infy., 1901; present designation, 1903.

"Burma, 1885-87."

Composition—2 companies of Punjabi Musalmans, 1 of Pathans, 1 of Sikhs.

1st Commn. or date of entering service.	Names and Rank.	Army rank.	Present appointment in Regiment.	Remarks.
	Commandant.			
28 June 93	*Harrison, Lt.-Col. A. H. P. (temp. Br. Genl.)*	*10 Nov. 18*	*10 Nov. 18*	*Dir. of Rectg.*
	Company Commanders (4)			
5 Aug. 96	Dunsford, Maj. E. H.	5 Aug. 14	1 Nov. 12	With 2-69th Punj.
27 Aug. 02	*Chamier, Maj. J. A.*, D.S.O.	*27 Aug. 17*	*15 Nov. 17*	*With R. A. F.*
5 Aug. 05	Broughton, Capt. G (*Act'g. Maj.*)	5 Aug. 14	22 Mar. 18	2nd-in-Comd. (*temp.*)
	Company Officers. (9)			
29 Aug. 06	*Wells, Capt. T. R*, M.C.	*29 Aug. 15*	*8 Dec. 07*	*With R. A. F.*
20 Jan. 12	Scott, Capt. J. B., M. C.	20 Jan. 16	23 Mar. 13	With 2-28th Punj.
15 Nov. 15	Connor, Lt. G. H.	15 Nov. 16	18 Nov. 15	With 90th Punj.
18 Apl. 16	Keegan, Lt. D.	18 Apl. 17	25 Apl. 16	Attd. with 1-151st Infy.
18 Apl. 16	Jenkins, Lt. N. L.	18 Apl. 17	25 Apl. 16	Attd. with 1-140th Infy.
29 June 16	Murray-Shirreff, Lt. B. G.	29 June 17	....	Attd. with 1-151st Infy.
29 June 16	Eccles-Holmes, Lt. J. F.	29 June 17	....	Attd.
30 Jan. 17	Gerrard, Lt. R. J.	30 Jan. 18	....	Attd. with 2-69th Punj.
18 June 17	Yiend, Lt. E. K.	18 June 18	....	With 46th Punj.
1 Feb. 18	Price, 2nd-Lt. C. J.	1 Feb. 18	6 Feb. 18	Attd.
31 Aug. 18	West, 2nd-Lt. P. S.	31 Aug. 18	....	Attd.
	Attached.			
4 Dec. 14	Pender, Lt. G. T., M. C. (*Act'g. Capt.*)	4 Dec. 15	11 Sep. 15	I. A. R. O.
24 Dec. 14	Field, Lt. W. H.	24 Dec. 15	29 Jan. 16	I. A. R. O.
15 May 15	Dempster, Lt. H. F.	15 May 16	1 May 15	I. A. R. O.
3 Aug. 15	Blomeley, Lt. C. (*Act'g Capt.*)	3 Aug. 16	29 Mar. 18	I. A. (*on probn.*)
23 Sep. 15	Stranger, Lt. F. R.	23 Sep. 16	29 Mar. 18	I. A. (*on probn.*)
26 Oct. 15	Jackson, Lt. N. C (*Act'g. Capt.*)	26 Oct. 16	15 Dec. 17	
12 Nov. 15	Richmond, Lt. J.P.	12 Nov. 16	18 Apl. 18	I. A. (*on probn.*)
1 May 16	Stevenson, Lt. W. H.	1 May 17	18 Apl. 18	I. A. (*on probn.*)
30 Aug. 16	Beall, Lt. J. S.	30 Aug. 17	17 May 18	I. A. (*on probn.*)
21 Oct. 16	Cowie, Lt. J. M.	21 Oct. 17	29 Mar. 18	I. A. (*on probn.*)
26 Apl. 17	Chapman, Lt. G. P.	26 Apl. 18	9 Dec. 17	I. A. R. O.
5 June 17	Jones, Lt. L. W.	5 June 18	17 May 18	I. A. (*on probn.*)
23 Aug. 17	Graham, Lt. F. (*Act'g. Capt.*)	23 Aug. 18	29 Mar. 18	I. A. (*on probn.*) Adjt. 8 Sep. 18.
1 Nov. 17	Frost, Lt. H. R.	1 Nov. 18	12 Mar. 18	
15 Jan. 18	Tuffin, 2nd-Lt. R.	15 Jan. 18	7 June 18	I. A. (*T. C.*)
15 Jan. 18	Barker, 2nd-Lt. F. R.	15 Jan. 18	7 June 18	I. A. R. O.
7 June 18	Kelly, 2nd-Lt. R. G.	7 June 18	....	I. A. (*T.C.*)

Medical Officer.

		Jemadar.	Subadar.	Subadar-Major.	
	Subadar-Major.				
18 June 84	Atar Khan, *Bahadur*, I.O.M. (56)	1 Jan. 00	16 July 03	22 Jan. 18	*Order of Br. Ind., 2nd Cl.*
	Subadars.				
15 Oct. 02	Karam Dad Khan (56, 72a)	15 Oct. 02	16 Nov. 03		
10 Mar. 04	Amar Singh	10 Mar. 04	22 May 10		
5 Mar. 04	Kallandar Khan (66)	11 Sep. 14	11 Oct. 16		
2 Sep. 96	Waris Khan	9 Aug. 15	24 June 17		
10 Jan. 01	Kirpal Singh (56)	11 Sep. 14	22 Jan. 18		
	Jemadars				
1 Feb. 00	Malook	9 Aug. 15			
19 Mar. 94	Hazara Singh	26 Sep. 15			
27 June 99	Indar Singh	26 Sep. 15			
14 Apl. 99	Gakhar	26 Sep. 15			
14 Mar. 08	Shah Baz	26 Sep. 15			
17 Oct. 96	Allah Ditta	29 Sep. 15			
27 Dec. 00	Bela Singh	20 July 17			
19 Aug. 03	Tora Baz Khan (58)	7 Aug. 17	Indian Adjt., 11th June 17.		

2nd Battalion, 33rd Punjabis.

Formed at Rawalpindi on January 1918.

Composition—2 companies Punjabi Musalmans, 1 company Sikhs, ½ company Dogras, ½ company Pathans.

1st Comm. or date of entering service.	Names and rank.	Army Rank.	Present appointment in Regiment.	REMARKS.
	Commandant.			
30 Aug. 93	Barrett, Bt. Lt.-Col. A. L. (*Act'g. Lt.-Col.*)	30 Aug 11	29 Oct. 18	126th Bal. Infy.
	Company Commanders.			
5 Aug. 05	Segar, Capt. E. (*Actg. Maj.*)	5 Aug. 14	15 Jan. 18	40th Pathans. 2nd-in-Comd. (*temp.*)
18 Feb. 15	Fitton, Lt. C. V.	18 Feb. 16	12 Mar. 18	I. A. R. O.
22 June 15	Dodd, Lt. A. C.	22 Jun. 16	17 Apl. 18	I. A. (*on probn.*)
23 May 16	Winnall, E. I. (*Actg. Capt.*)	23 May 17	4 June 18	I. A. (*on probn.*)
	Company Officers.			
22 Jan. 15	Hart, Lt. J. H. B. B.	22 Jan. 16	7 Jan. 18	I. A. R. O.
22 July 15	Tayleur, Lt. G.	22 July 16	10 June 18	
15 Jan. 16	Helme, Lt. F. S.	15 Jan. 17	18 Apl. 18	I. A. R. O.
25 Mar. 16	Duffus, Lt. A. H. H.	25 Mar. 17	4 Jan. 18	I. A. R. O. Qr. Mr. (*Offg.*)
28 Feb. 17	Roberts, 2nd-Lt. J. R.	28 Feb. 18	30 May 17	I. A. (*on probn.*)
8 Aug. 17	MacDougal, 2nd-Lt. J.	8 Aug. 17	7 Jan. 18	I. A. R. O.
31 Oct. 17	Sharpe, Lt. G. F.	31 Oct. 18	24 Jan. 18	I. A. R. O.
28 Nov. 17	MacIntyre, 2nd-Lt. W.	28 Nov. 17	1 Feb. 18	I. A. R. O.
31 Jan. 18	MacGuckin, 2nd-Lt. C. E.	31 Jan. 18	7 Feb. 18	I. A. (*on probn.*)
28 Nov. 17	Farley, 2nd-Lt. F. A. (*Actg. Capt.*)	28 Nov. 17	1 Mar. 18	I. A. R. O. Adjt.
5 Nov. 17	Girling, Lt. G. F. H.	5 Nov. 18	19 Mar. 18	I. A. (T. C.)
5 Nov. 17	Cunningham, Lt. T. C.	5 Nov. 18	14 Mar. 18	I. A. (T. C.)
29 May 18	Thompson, 2nd-Lt. N. L.	29 May 18	6 Nov. 18	Midd'x R., I. A. R. O. (*on probn.*)
7 June 18	Irving, 2nd-Lt. M. C.	7 June 18	11 Oct. 18	I. A. (*T. C.*)

Subadar-Major.

		Jemadar.	Subadar.	Subadar-Major.	
11 Feb. 93	Khan Gul (56)	1 July 12	26 Sep. 15	13 Nov. 18	(*Actg.*)

Subadars.

10 Apl. 99	Gheba Khan	11 Oct. 16	20 Sep. 18
9 Apl. 98	Lal Khan	5 Dec. 16	1 Oct. 18

Jemadars.

6 Apl. 87	Jwala Singh	16 Feb. 18	
4 June 03	Bhagat	1 Mar. 18	
22 June 01	Said Husain	1 Mar. 18	
1 Oct. 00	Chand Pir Shah	1 Mar. 18	
6 May 01	Hukam Dad	29 Apl. 18	
9 Jan. 01	Bachan Singh	29 Apl. 18	
16 Mar. 10	Muhammad Aslam Khan	20 May 18	
1 June 18	Muhammad Akbar Khan	1 June 18	*On probn.*
	Amir Baz	1 Oct. 18	
	Sultan Khan	1 Oct. 18	
	Hazara Singh	1 Oct. 18	
	Faiz Ahmad	1 Oct. 18	

Regt. Centre—Ambala.
Linked with—23rd and 32nd S. Pioneers.

Uniform—Scarlet.
Facings—Blue.

1st Battalion, 34th Sikh Pioneers.

Raised at Mian Mir in 1887, by Col. A. C. W. Crookshank, as the 34th (Punjab) Regiment of Bengal Infantry (Pioneers). Became the 34th Punjab Pioneers, 1901; present designation, 1903.

"Chitral" "Punjab Frontier" "China, 1900."

Composition—4 companies Mazbi and Ramdasia Sikhs.

1st Commn. or date of entering service.	Names and Rank.	Army rank.	Present appointment in Regiment.	Remarks.
	Commandant.			
1 Mar. 90	Cullen, Lt.-Col. E. H. S., C.M.G., D.S.O., M.V.O. *q.*	24 Mar. 15	24 Mar. 15	
	Company Commanders. (4)			
14 Aug. 95	Clarke, Maj. G. S.(*Actg. Lt.-Col.*)	14 Aug. 13	16 Jan. 17	2nd-in-Comd. With 2-32nd Prs.
27 July 98	*Wilson, Maj. G. E. H., q.*	*1 Sep. 15*	*3 Nov. 13*	
27 Aug. 02	Bailey, Maj. P. H. H., M. C., *q.*	27 Aug. 17	30 Apl. 17	With 2-32nd Prs.
9 Jan. 04	Paterson, Capt. G. F. J., M.C., *q.*	9 Jan. 13	8 Apl. 17	
	Company Officers. (9)			
28 July 00	*Hunt, Capt. C. R., M.C., p.s.c.*	*28 July 09*	*6 Nov. 01*	
16 Aug. 05	Ponsonby, Capt. D. G., D.S.O.	16 Aug. 14	29 Jan. 11	
29 Aug. 06	*Hickman, Capt. H. T. D., M.C., q.*	*29 Aug. 15*	*28 Oct. 07*	
20 Jan. 09	Donaldson, Capt. A.	1 Sep. 15	10 Aug. 12	
20 Jan. 12	Browne, Capt. R. B.	20 Jan. 16	9 Mar. 13	*Off. Sch. of Instn., Sabathu.*
14 Jan. 14	Gordon, Capt. D. H., D.S.O.	14 Jan. 18	12 Oct. 14	*Off. Sch. of Instn., Bangalore.*
18 Apl. 16	Hurly, Lt. M. F. F.	18 Apl. 17	. . .	Attd.
18 June 17	Macmillan, Lt. D. MacL.	18 June 18	. . . .	With 2-34th Prs.
30 Jan. 17	Fordyce, 2nd-Lt. J. F. D.	30 Jan. 17	. . . .	Attd. With 3-32nd Prs.
29 Oct. 17	Alfieri, 2nd-Lt. F. J.	29 Oct. 17	. . . .	Attd.
	Attached.			
21 Jan. 03	Boyce, Maj. R. B. G.	21 Jan. 18	7 Nov. 04	1-12th Prs.
9 Jan. 04	Ricketts, Capt. E. W. C. (*Actg. Maj.*)	9 Jan. 13	. . . .	1-23rd Prs.
18 Jan. 11	North, Capt. K. A., M. C.	1 Sep. 15	27 Jan. 17	32nd Prs.
23 Dec. 14	Critchley, Lt. D. R.	23 Dec. 15	1 Mar. 15	32nd Prs.
24 Dec. 14	Bracher, Lt. H. C.	24 Dec. 15	. . . .	I. A. R. O.
13 Jan. 15	Paterson, Lt. L. M., M.C.	13 Jan. 16	27 Feb. 15	
9 July 15	Fitzgibbon, Lt. A. F.	9 July 16	. . . .	I. A. R. O.
4 Mar. 16	Smith, Lt. D. R.	4 Mar. 17	. . . .	I. A. (*on probn.*)
29 June 16	Hampton, Lt. A. V. F.	29 June 17	. . . .	1-12th Prs.
	Barton, Lt. A. G. A.	. . . .	. . . .	S. Staff R. I. A. R. O. (*on probn.*)
5 Oct. 17	White, Lt. P. R.	5 Oct. 18	15 Jan. 18	I. A. (*T. C.*)
5 May 17	Dawson, Lt. T.	5 May 18	26 Mar. 18	Ches. R., I. A. (*on probn.*).
1 Oct. 17	Rendell, Lt. C.	1 Oct. 18	26 Mar. 18	I. A. (*T. C.*)
28 Nov. 17	Bray, 2nd-Lt. N. D.	28 Nov. 17	31 May 18	I. A. R. O.
15 Nov. 15	Gazzard, 2nd-Lt. T.	15 Nov. 15	18 Nov. 15	32nd Prs.
24 Apl. 18	Bayley, 2nd-Lt. J. P.	24 Apl. 18	. . . .	I. A. (U. L.)
18 Jan. 18	Smith, 2nd-Lt. A. R.	18 Jan. 18	. . . .	I. A. R. O.
1 July 17	Worssam, Lt. C.	1 July 18	. . . .	A. S. C., I. A. (*on probn.*)
1 July 17	Edwards, Lt. S. W.	1 July 18	. . . .	S. Staff R., I. A. (*on probn.*)

Subadar-Major.

		Jemadar	Subadar.	Subadar-Major.	
15 July 87	Jwala Singh, I.O.M., *Sardar Bahadur* (56)	19 Oct. 93	7 July 97	1 Mar. 11	*Order of Br. Ind., 1st Cl.*
7 Sep. 87	Natha Singh, *Sardar Bahadur*, I.O.M.	1 May 04	1 Sep. 08	18 Oct. 16	*Order of Br. Ind., 1st Cl.*

1st Battalion, 34th Sikh Pioneers—*contd.*

1st Commn. or date of entering service.	Names and Rank.	Army rank.		Remarks.
	Subadars.			
23 Jan. 93	Sant Singh, M. C. (56)	11 Apl. 06	16 Mar. 11	
23 Aug. 89	Mundar Singh (56)	16 Oct. 06	1 June 12	
16 Oct. 93	Kharak Singh, I. O. M. (56)	1 June 12	16 Dec. 15	
23 Sep. 99	Binda Singh	11 Oct. 14	26 Mar. 17	
14 Jan. 95	Malla Singh	24 Nov. 14	28 June 17	
4 June 94	Sohawa Singh	30 Nov. 14	16 Sep. 17	
25 June 02	Asa Singh	16 July 15	27 Sep. 17	
23 Jan. 99	Bhanga Singh	8 Aug. 14	24 Mar. 18	
2 Apl. 94	Dewa Singh	8 Aug. 14	17 July 18	
23 Aug. 97	Ganesha Singh	24 Nov. 14	22 May 18	
11 Apl. 99	Mangal Singh, I.D.S.M.	23 Jan. 17	20 June 18	
1 June 97	Karam Singh (46)	23 June 17	15 Sep. 18	
12 Jan. 01	Bishan Singh	23 Jan. 17	19 Aug. 18	
	Natha Singh (*Retd.*)	. . . .	. . . .	Re-employed 2 May 1918.
	Harnam Singh (*Retd.*)	. . . .	. . . .	Re-employed 1 Jan. 1918.
	Jemadars.			
26 June 98	Tara Singh	23 Mar. 16		
10 Mar. 02	Dhanna Singh, M. M.	25 Mar. 17		
5 Sep. 01	Fateh Singh	28 June 17		
16 Feb. 01	Sundar Singh	18 Sep. 17		
27 Mar. 04	Phumman Singh, I.D.S.M.	27 Sep. 17		
21 Apl. 00	Gyan Singh, M. M.	3 Dec. 17		
14 Mar. 93	Waryam Singh	26 Mar. 18		
10 Apl. 03	Sohan Singh	26 Mar. 18		
5 May 97	Mihan Singh	25 May 18		
18 Nov. 99	Bisakha Singh	25 May 18		
17 July 05	Mool Singh	25 May 18		
19 May 04	Chanda Singh	25 May 18		
1 July 04	Charan Singh	29 June 18		
	Gurditt Singh	17 July 18		
	Sudha Singh	1 Aug. 18		
	Harnam Singh	13 June 18		
	Khem Singh	20 June 18		
	Mehar Singh	1 Oct. 18		

2nd Battalion, 34th Sikh Pioneers.

Formed at Sialkot on 7th June 1917.

Composition—4 companies Ramdasia and Mazbi Sikhs.

1st Commn. or date of entering service.	Names and Rank.	Army rank.	Present appointment in Regiment.	Remarks.
	Commandant.			
28 July 00	Turnbull, Maj. B. (*Actg. Lt.-Col.*)	1 Sep. 15	2 July 17	1-23rd Sikh Prs.
	Company Commanders. (4)			
4 May 01	Paterson, Maj. H. F. W.	8 Sep. 17	2 July 17	Cantt. Magte.'s Dept. 2nd-in-Comd.
20 Sep. 11	Tinney, Capt. L. H.	20 Sep. 15	26 Feb. 18	Attd.
1 Jan. 16	Jervis, Lt. C. W. J. (*Actg. Capt.*)	1 Jan. 17	23 Jan. 18	Glouc. R., I. A. R. O.
8 Aug. 14	Campbell, Capt. A. D. P.	8 Aug. 18	1 Oct. 18	106th Prs.
	Company Officers.			
22 Sep. 14	Ede, Lt. H. S.	22 Sep. 15	24 Jan. 18	I. A. R. O.
5 Feb. 15	Jellicoe, Lt. J. A.	15 Feb. 16	28 Aug. 17	I. A. R. O.
17 Feb. 15	Whittome, Lt. J. E.	15 Feb. 16	. . . .	I. A. R. O.
27 Mar. 16	Morgan, Lt. W. L.	27 Mar. 17	26 Feb. 18	I. A. R. O.
19 July 16	Smith, Lt. V. T.	19 July 17	17 Oct. 17	Attd.
12 Dec. 16	McLean, Lt. D. A.	12 Dec. 17	13 Feb. 18	I. A. F., I. A. (*on probn*)
3 Feb. 17	London, Lt. H. R.	3 Feb. 18	12 Oct. 18	I. A. R. O.
1 Mar. 17	Fowler, Lt. G. V.	1 Mar. 18	13 June 17	I. A. R. O. Qr.-Mr., 13 June 17.
20 Apl. 17	Griffith, Lt. E. S.	20 Apl. 18	6 Oct. 18	I. A. R. O.
18 June 17	Macmillan, Lt. D. MacL.	18 June 18	25 June 17	1-34th Sikh Prs.
18 July 17	Philbrick, Lt. H. A.	18 July 18	20 Sep. 18	I. A. R. O.
10 Sep. 17	Gordon, Lt. S. V. L. deM.	10 Sep. 18	7 Oct. 18	I. A. R. O.
26 Sep. 17	Anderson, Lt. F. S.	26 Sep. 18	23 Jan. 18	I. A. R. O.
5 Nov. 17	Keelan, Lt. A. P. (*Actg. Capt.*)	5 Nov. 18	11 May 18	I. A. R. O. Adjt., 1 Oct. 18.
20 Jan. 18	Davies, 2nd-Lt. W. V.	28 Jan. 18	27 Sep. 18	Wilts R., I. A. (*on probn.*)
31 Jan. 18	Ker, 2nd-Lt. T. M.	31 Jan. 18	6 Feb. 18	Attd.
31 Jan. 18	Curtoys, 2nd-Lt. O. F.	31 Jan. 18	6 Feb. 18	Attd.
28 Feb. 18	Gourlie, 2nd-Lt. N. E. J.	28 Feb. 18	28 Sep. 16	Cam'n. Highrs. I. A. (*on probn.*)
29 Mar. 18	Lambert, 2nd-Lt. H.	29 Mar. 18	14 July 18	I. A. (*T. C.*)
22 July 18	Crandell, 2nd-Lt. W.	28 July 18	1 Dec. 18	I. A. (*T. C.*)
31 July 18	Durrant, 2nd-Lt. G.	31 July 18	7 Sep. 18	Suff. R., I. A. (*on probn.*)
31 July 18	Wallers, 2nd-Lt. R. W.	31 July 18	1 Sep. 18	Lond. R., I. A. (*on probn.*)
26 Aug. 18	Thomas, 2nd-Lt. D. V.	26 Aug. 18	5 Sep. 18	I. A. R. O. (*on probn.*)
31 Aug. 18	Whitley, 2nd-Lt. A. P.	31 Aug. 18	3 Sep. 18	I. A. Attd.
1 Oct. 18	Akehurst, 2nd-Lt. C. H. I.	1 Oct. 18	22 Oct. 18	Attd.
1 Oct. 18	Price, 2nd-Lt. C. H.	1 Oct. 18	6 Oct. 18	Attd.

Subadar-Major.

		Jemadar.	Subadar.	Subadar-Major.	
15 Sep. 96	Fateh Singh (56)	5 Jan. 11	2 Nov. 14	1 July 18	(*temp.*)

Subadars.

16 July 91	Kishen Singh (56, 72*a*)	27 Mar. 07	18 Nov. 13
8 Dec. 93	Ala Singh (56, 66*a*)	7 Aug. 12	18 Oct. 14
2 June 91	Teja Singh (58)	11 Oct. 14	1 July 17
26 Jan. 98	Atma Singh (56)	9 Dec. 14	1 July 17
21 Nov. 99	Jagat Singh (I) (46, 56)	11 Oct. 14	1 Sep. 17
4 Sep. 93	Gurditt Singh	1 July 17	1 Apl. 18
17 Apl. 06	Bir Singh (48, 56)	2 May 17	1 May 18
1 June 03	Jagat Singh, (II) (56, 61, 66*a*)	1 Jan. 18	2 Sep. 18

Jemadars.

3 Sep. 06	Thakur Singh	16 Apl. 17	Indian Qr. Mr.
20 Sep. 98	Uttam Singh (46, 48, 56)	1 July 17	
1 Feb. 04	Daya Singh	1 July 17	
20 Jan. 04	Kishen Singh (56)	1 July 17	
23 May 01	Partab Singh (56, 6*a*)	1 Dec. 17	
2 Nov. 05	Mundar Singh, I.D.S.M.	1 Jan. 18	
1 June 06	Sohan Singh (56)	1 Jan. 18	
5 July 07	Ganga Singh (56)	1 Jan. 18	
1 Sep. 07	Isar Singh	1 Jan. 18	
10 May 07	Warriam Singh	1 Jan. 18	
1 Jan. 09	Sapuran Singh (46, 48, 56)	1 Apl. 18	
27 Apl. 08	Teja Singh	1 May 18	
1 July 12	Lachhman Singh (48, 50)	1 May 18	Indian Adjt.
17 Feb. 09	Paritam Singh (56, 61)	1 May 18	
21 Jan. 98	Mehr Singh (56)	1 Oct. 18	

3rd Battalion, 34th Sikh Pioneers.

Formed at Ambala on 13th June 1918.

Composition—4 companies Mazbis and Ramdasia Sikhs.

1st Commn. or date of entering service.	Names and Rank.	Army rank.	Present appointment in Regiment.	Remarks.
	Commandant.			
3 Aug. 98	Hutchinson, Maj. A. G. C., *p.s.c.* (*Actg. Lt.-Col.*)	1 Sep. 15	....	1-23rd Prs. (*temp.*)
	Company Commanders. (4)			
9 Sep. 08	Macpherson, Capt. A. K. (*Actg. Maj.*)	1 Sep. 15	....	1-12th Prs. 2nd-in-Comd. (*temp.*)
20 Nov. 14	Eiloart, Lt. B. H., M.C. (*Actg. Capt.*)	20 Nov. 15	26 June 18	I. A. R. O.
24 Oct. 15	Carr, Lt. H. J. (*Actg. Capt.*)	24 Oct. 16	2 Oct. 18	Attd.
22 June 15	Smith, Lt. N. V.	22 June 16	2 Oct. 18	Durh. L. I., I. A. (*on probn.*)
	Company Officers.			
18 Jan. 17	Forth, Lt. C. O. (*Actg. Capt.*)	18 Jan. 18	....	I. A. (*on probn.*) Adjt. 15 July 18.
7 Oct. 15	Peel, Lt. W.	7 Oct. 16	....	I. A. R. O.
1 July 17	Martyn, Lt. R. O. F.	1 July 18	18 Oct. 18	N. Lan R., I. A. (*on probn.*)
24 Sep. 16	Preddle, Lt. A. R.	24 Sep. 17	16 Sep. 18	I. A. (*on probn.*)
28 Aug. 16	Lee, Lt. E. O.	28 Aug. 17	6 Oct. 18	I. A. R. O.
27 Oct. 17	McDonald, 2nd-Lt. J. B.	27 Oct. 17	....	1-23rd Sikh Prs. Qr. Mr.
17 Dec. 17	Veale, 2nd-Lt. R. de S.	17 Dec. 17	14 Oct. 18	Labour Corps, I. A. R. O. (*on probn.*)
5 Jan. 18	Johnstone, 2nd-Lt. A. M.	5 Jan. 18	....	I. A. R. O.
1 May 18	Rogers, 2nd-Lt. R. L.	1 May 18	17 Oct. 18	W. Surr. R., I. A. R. O. (*on probn.*)
2 June 18	Richards, 2nd-Lt. R. W.	2 June 18	9 Oct. 18	I. A. (*T. C.*)
7 June 18	Loder, 2nd-Lt. L. N.	7 June 18	9 Oct. 18	I. A. (*T. C.*)
26 June 18	Wycherley, 2nd-Lt. R. H.	26 June 18	4 Oct. 18	Genl. List, I. A. R. O. (*on probn.*)
22 July 18	Whitaker, 2nd-Lt. W. A.	22 July 18	30 Nov. 18	I. A. (*T.C.*)
31 Aug. 18	Pirouet, 2nd-Lt. H. N.	31 Aug. 18	4 Sep. 18	Attd.
1 June 16	James, Lt. A. V.	1 June 17	19 Nov. 18	Lond. R.
9 Jan. 18	Brimelow, Lt. E. (*Actg. Capt.*)	9 Jan. 18	20 Nov. 18	Lond. R.
26 Apl. 17	Laithwaite, Lt. A., D. S. O.	26 Apl. 18	19 Nov. 18	Lond. R.
26 Apl. 17	Phelps, Lt. H. H. V.	26 Apl. 18	20 Nov. 18	Lond. R.
27 June 17	Dabbs, 2nd-Lt. S. E.	27 June 17	20 Nov. 18	Lond. R.

Subadar-Major.

		Jemadar.	Subadar.	Subadar-Major.	
5 June 88	Maghar Singh, I.O.M., I.D.S.M.	20 Oct. 04	1 Oct. 09	4 July 18	(*temp.*)

Subadars.

16 Apl. 97	Indar Singh (56)	13 Sep. 10	24 Nov. 14
5 Feb. 95	Chanda Singh	11 Oct. 14	3 Feb. 17
4 June 94	Sohawa Singh	30 Nov. 14	16 Sep. 17
4 Aug. 00	Udham Singh	3 Feb. 17	1 Jan. 18
1 June 97	Karm Singh	23 June 17	15 Sep. 18
25 Sep. 97	Jagat Singh	1 Jan. 17	15 Sep. 18

Jemadars.

3 Apl. 03	Atma Singh	9 Mar. 17
10 Oct. 93	Kishan Singh	1 July 17
3 Mar. 01	Gurditt Singh	16 Sep. 17
1 Nov. 95	Nand Singh	16 Jan. 18
11 Mar. 02	Bhagat Singh	26 Mar. 18
10 Jan. 01	Maghar Singh	1 July 18

Regt. Centre—Rawal Pindi.
Linked with—36th and 47th Sikhs.

Uniform—Scarlet.
Facings—Yellow.

1st Battalion, 35th Sikhs.

Raised at Ferozepore in 1887, by Lt.-Col. D. W. Inglis, as the 35th (Sikh) Regiment of Bengal Infantry. Became the 35th Sikh Infantry, 1901; present designation, 1903.

"Malakand" "Punjab Frontier."

Composition—4 companies Sikhs.

1st Commn. or date of entering service.	Names and Rank.	Army rank.	Present appointment in Regiment.	Remarks.
	Commandant.			
3 Sep. 92	Rice, Lt.-Col. S. K. B.	14 Mar. 18	14 Mar. 18	
	Company Commanders. (4)			
14 Aug. 95	Pratt, Maj. H. R. E., D.S.O. (Q) (*Actg. Lt.-Col.*)	14 Aug. 13	14 Mar. 18	2nd-in-Comd.—With 2-25th Punj.
5 Aug. 96	*Mackensie, Maj. J.*, C.I.E.	*5 Aug. 14*	*8 Nov. 12*	*Comptroller, Viceroy's Household.*
4 Aug. 97	*Freeland, Bt. Lt.-Col. J. C.*, p.s.c.	*3 June 18*	*8 Nov. 12*	*G. S. O., 2nd gr.*
23 May 00	Swayne, Maj. F. G.	1 Sep. 15	14 Mar. 18	With 2-25th Punj.
	Company Officers. (9)			
18 Jan. 02	*Williams, Maj. R. C. B.*	*18 Jan. 17*	*8 Apl. 03*	*Bde.-Maj., 44th Infy. Bde.*
27 Jan. 04	*Story, Capt. H. F. (temp. Maj.)*	*27 Jan. 13*	*1 Sep. 06*	*D. A. A. & Q. M. G., Bannu Bde.*
3 Dec. 04	*Marshal, Bt. Maj. J. S.*, D. S. C. (*temp. Maj.*)	*3 June 18*	*25 Sep. 07*	*G. S. O., 2nd gr., Presy. Bde.*
13 Aug. 04	Allan, Capt. K. D. (*Actg. Maj.*)	13 Aug. 13	4 Apl. 07	With 1-97th Infy.
24 Jan. 06	Burridge, Capt. T. F. S.	24 Jan. 15	13 Dec. 07	With 2-72nd Punj.
29 Aug. 06	*Thomson-Glover, Capt. J. W.*	*29 Aug. 15*	*17 Feb. 09*	*Poll. employ.*
16 Feb. 07	Gardner, Capt. C. F.	1 Sep. 15	1 Feb. 09	*Lv. ex I., m.c.*
12 June 07	Collings, Capt. E. G. (*Actg. Maj.*)	1 Sep. 15	15 July 09	With 36th Sikhs.
11 Nov. 14	Stead, Capt. E. A., M.C.	11 Nov. 18	26 Jan. 15	With 3-23rd Sikhs.
16 Dec. 14	Ford, Lt. C. W. W.	16 Dec. 15	11 Feb. 15	With Nabha State Infy.
15 Sep. 15	Conolly, Lt. J. J. P.	15 Sep. 16	27 Oct. 15	Attd. With 47th Sikhs.
15 Nov. 15	Purves, Lt. J. J.	15 Nov. 16	18 Nov. 15	Attd.
18 Apl. 16	*Voysey, Lt. C. C. A.*	*18 Apl. 17*	*26 Apl. 16*	*Attd. With Malay States Guides.*
14 Nov. 16	Good, Lt. S. B.	14 Nov. 17	23 Nov. 16	Attd. With 38th Horse.
18 June 17	Bampfield, Lt. R.	18 June 18	25 June 17	Attd. With 36th Sikhs.
18 June 17	*Poulton, Lt. H. M.*	*18 June 18*	*8 Sep. 17*	*Attd. With Frontier Militia.*
31 Jan. 18	Putnam, 2nd-Lt. W. A.	31 Jan. 18	6 Feb. 18	Attd. With 1-144th Infy.
31 Jan. 18	*Wells, 2nd-Lt. H. V.*	*31 Jan. 18*	*6 Feb. 18*	Attd. *With S. & T. Corps, Rawal Pindi.*
21 Dec. 17	Voit, 2nd-Lt. R. A.	21 Dec. 17	20 Mar. 18	Attd. Qr. Mr. (*temp.*)
31 Aug. 18	Williams, 2nd-Lt. D. H. J.	31 Aug. 18	4 Sep. 18	Attd.
1 Oct. 18	Pollock, 2nd-Lt. V. D.	1 Oct. 18	6 Oct. 18	Attd.
	Attached.			
17 Oct. 00	Sealy, Maj. H. G.	17 Oct. 15	21 June 18	130th Baluchis.
6 May 16	Andrew, Lt. H. W. (*Actg. Capt.*)	6 May 17	18 Apl. 18	I. A. (*on probn.*)
21 Aug. 17	Parsons, Lt. F. S.	21 Aug. 18	13 Feb. 18	I. A. R. O.
1 Oct. 17	Bugler, Lt. E. F. (*Actg. Capt.*)	1 Oct. 18	21 Jan. 18	I. A. (*T. C.*) Qr.-Mr. (*Temp.*)
5 Oct. 17	Mugford, Lt. T. G. (*Actg. Capt.*)	5 Oct. 18	21 Jan. 18	I. A. (*T. C.*)
5 Nov. 17	Matthews, Lt. S. F. W.	5 Nov. 18	15 July 18	I. A. (*T. C.*)
28 Feb. 18	White, 2nd-Lt. T. N.	28 Feb. 18	15 Feb. 18	Suff. R., I. A. (*on probn.*)
29 Mar. 18	Phillips, 2nd-Lt. E. C.	29 Mar. 18	15 July 18	I. A. (*T. C.*)
31 Aug. 18	Bunn, 2nd-Lt. C. W.	31 Aug. 18	. . . .	I. A. (U. L.) *Lv., ex I., m.c.*
7 June 18	Haye, 2nd-Lt. N.	7 June 18	11 Oct. 18	I. A. R. O.

Subadar-Major.

		Jemadar.	Subadar.	Subadar-Major.	
1 Aug. 87	Mangal Singh, *Bahadur* I.D.S.M. (53, 56).	16 Nov. 99	16 Dec. 03	1 July 14	*Order of Br. Ind., and Cl.*

1st Battalion, 35th Sikhs—*contd.*

1st Commn. or date of entering service.	Names and Rank.	Army rank.		Remarks.
		Jemadar.	Subadar.	

Subadars.

1st Commn. or date of entering service.	Names and Rank.	Jemadar.	Subadar.	Remarks.
23 Dec. 91	*Sundar Singh (53, 56, 58)*	*20 Apl. 08*	*1 Feb. 14*	*With 47th Sikhs.*
10 Apl. 89	Dula Singh (66*a*)	1 May 10	13 Jan. 15	
29 Apl. 97	Partab Singh	18 Nov. 14	9 July 15	
1 Apl. 96	Buta Singh (56)	1 Feb. 14	1 Sep. 15	
25 Dec. 95	Harnam Singh (56)	18 Nov. 14	1 Sep. 15	With 2-30th Punj.
13 Jan. 99	Atar Singh	9 Dec. 14	1 Sep. 15	
23 Oct. 93	Harditt Singh (56, 66)	16 July 13	20 Feb. 16	
17 Apl. 00	Arjan Singh	15 Apl. 15	10 Dec. 17	
1 Aug. 16	Gurditt Singh	1 Aug. 16	2 Nov. 18	

Jemadars.

1st Commn. or date of entering service.	Names and Rank.	Jemadar.	Remarks.
5 May 90	*Sahib Singh*	*5 July 16*	
10 July 02	Narindar Singh	1 Aug. 17	
5 Apl. 97	Gainda Singh	10 Dec. 17	
24 Sep. 02	Arjan Singh (56)	10 Dec. 17	Indian Adjt.
10 July 94	Sundar Singh (46)	10 Dec. 17	With 2-30th Punj.
5 Nov. 04	Punjab Singh (58)	10 Dec. 17	
24 Jan. 04	Mansha Singh	10 Dec. 17	
1 Aug. 05	Surain Singh II	1 Mar. 18	With 2-30th Punj.
19 May 06	Makhan Singh	1 Oct. 18	
12 July 05	Lal Singh	1 Oct 18	
19 July 04	Narinjan Singh	1 Oct. 18	
15 Jan. 04	Fateh Singh	1 Oct 18	
17 Aug. 01	Bhagat Singh	1 Oct 18	
26 Jan. 07	Chattar Singh	1 Oct 18	
17 July 05	Sohan Singh	15 Oct. 18	
16 May 05	Bela Singh	15 Oct. 18	

2nd Battalion, 35th Sikhs.

Formed at Lahore Cantonment on 20th November 1917.

Composition—2 companies of Sikhs and 2 of Jats.

1st Commn. or date of entering service.	Names and Rank.	Army rank.	Present appointment in Regiment.	Remarks.
	Commandant.			
13 Nov. 95	Bourne, Bt. Lt.-Col. W. K., *p.s.c.*	3 June 16	29 Nov. 17	20 Horse.
	Company Commander. (4)			
4 Aug. 00	Murray, Maj. H. A.	1 Sep. 15	27 Nov. 17	30th Punj. 2nd-in-Comd. (*temp.*)
8 Sep. 09	Ogilvy, Capt. A. G.	1 Sep. 15	26 Aug. 18	38th Dogras.
23 Dec. 14	Barry, Lt. E. H. (*Actg. Capt.*).	23 Dec. 15	14 Feb. 18	I. A. R. O.
5 May 15	Adams, Lt. D. R. (*Actg. Capt.*).	5 May 16	26 Jan. 18	I. A. (*on probn.*)
	Company Officers.			
1 Feb. 16	Lloyd, Lt. J. E., M.C. (*Actg. Capt.*)	1 Feb. 17	29 Mar. 18	I. A. (*on probn.*) Adjt.
5 Aug. 14	Lee, Lt. A. E.	5 Aug. 15	6 Aug. 18	I. A. R. O.
29 May 15	Kirby, Lt. N. C.	29 May 16	30 Mar. 18	I. A. (*on probn.*)
23 July 16	Adderley, Lt. A. H.	23 July 16	5 Mar. 18	I. A. (*on probn.*)
25 Sep. 15	Fischer, Lt. N. W. (*Actg. Capt.*)	25 Sep. 16	16 Dec. 17	I. A. R. O.
30 Jan. 15	Brothers, Lt. J. A.	30 Oct. 16	6 Aug. 18	I. A. (*on probn.*)
27 Nov. 15	Gault, Lt. J.	27 Nov. 16	2 Sep. 18	R. Ir. Rif. I. A. (*on probn.*)
28 Mar. 17	Castle, Lt. J. P., D.S.O.	28 Mar. 18	1 Mar. 18	I. A. R. O.
23 Feb. 16	Atkins, Lt. W. V.	28 Feb. 17	23 Feb. 18	I. A. (*on probn.*)
12 Apl. 16	Pickersgill, Lt. R.	12 Apl. 17	26 Jan. 18	I. A. (*on probn.*) Qr. Mr.
22 Apl. 16	Bayliss, Lt. C. L.	22 Apl. 17	5 Oct. 18	York. R., I. A. (*on probn.*)
22 July 16	Scorggie, Lt. J. A.	22 July 17	....	I. A. R. O.
15 Sep. 16	Anderson, Lt. G. A.	15 Sep. 17	10 Oct. 18	I. A. R. O.
1 Dec. 17	Magenty, Lt. E. R.	1 Dec. 18	1 Mar. 18	I. A. (*on probn.*)
21 Dec. 17	Hope, 2nd-Lt. A. A.	21 Dec. 17	21 Mar. 18	Attd.
15 Jan. 18	Rodd, 2nd-Lt. N. J., M.M.	15 Jan. 18	6 Aug. 18	I. A. (*T. C.*)
26 Jan. 18	Collett, 2nd-Lt. N. T.	26 Jan. 18	17 Apl. 18	I. A. (*on probn.*)
31 Jan. 18	Stephenson, 2nd-Lt. G. E. B.	31 Jan. 18	....	Attd.
31 Jan. 18	George, 2nd-Lt. S. P.	3 Jan. 18	....	Attd.
19 Mar. 18	Burrough, 2nd-Lt. I. J.	19 Mar. 18	15 July 18	I. A. (*T. C.*)
29 Mar. 18	Askew, 2nd-Lt. J.	29 Mar. 18	15 July 18	I. A. (*T. C.*)
29 Mar. 18	De Beer, 2nd-Lt. E. S.	29 Mar. 18	1 Dec. 18	I. A. (*T. C.*)
29 Apl. 18	Hayes, 2nd-Lt. A. W.	29 Apl. 18	24 Aug. 18	I. A. (*T. C.*)
7 June 18	Foweraker, 2nd-Lt. H. A.	7 June 18	10 Oct. 18	I. A. R. O.
7 June 18	Lockhart, 2nd-Lt. W.	7 June 18	10 Oct. 18	I. A. R. O.
10 July 18	Drever, 2nd-Lt. G. D.	10 July 18	1 Dec. 18	I. A. (*T. C.*)
26 Aug. 18	McDonald, 2nd-Lt. C. K.	26 Aug. 18	5 Sep. 18	I. A. R. O.
26 Aug. 18	Thomas, 2nd-Lt. A. H.	26 Aug. 18	5 Sep. 18	I. A. R. O. (*on probn.*)
31 Aug. 18	Hindle, 2nd-Lt. M. D.	31 Aug. 18	3 Sep. 18	Attd.
1 Oct. 18	Arratoon, 2nd-Lt. J.	1 Oct. 18	19 Oct. 18	I. A. R. O.

2nd Battalion, 35th Sikhs—*contd.*

1st Commn. or date of entering service.	Names and Rank.	Army.			Remarks.
		Jemadar.	Subadar.	Subadar-Major.	
	Subadar-Major.				
15 June 87	Attar Singh	16 Apl. 09	18 Nov. 14	20 Nov. 17	(*temp.*)
	Subadars.				
27 Feb. 97	Hatti (52, 56)	16 Oct. 07	27 Oct. 14		
1 Jan. 97	Bhola Singh (52, 56)	21 Oct. 14	30 Oct. 16		
23 Sep. 97	Katha Singh (56)	15 Apl. 15	20 Nov. 17		
19 Oct. 99	Kaur Singh	1 Sep. 15	20 Nov. 17		
13 Feb. 97	Thakur Singh	1 Sep. 15	20 Nov. 17		
5 Feb. 01	Ji Ram	30 July 15	21 Jan. 18		
11 July 02	Sukh Dev (46)	5 Jan. 16	21 Jan. 18		
20 June 03	Jaimal Singh (56)	20 Nov. 17	20 Aug. 18		
2 Feb. 05	Sheo Chand I (56)	20 Nov. 17	1 Aug. 18		
14 June 99	Sheo Chand II	21 Jan. 18	7 Oct. 18		
	Jemadars.				
	Ram Sarup Singh (*Ret.*)	. . . . .			Re-employed.
6 Mar. 98	Kirpal Singh	20 Nov. 17			
15 Feb. 99	Partab Singh	20 Nov. 17			
23 Aug. 02	Badlu	21 Jan. 18			Indian Adjt., 7 Oct. 18.
7 Mar. 18	Kashmir Singh	7 Mar. 18			
20 Jan. 97	Pritam Singh	8 June 18			Indian Qr. Mr., 7 Oct. 18.
21 Jan. 06	Narain Singh	21 June 18			
14 Aug. 07	Sundar Singh	5 July 18			
9 Feb. 04	Sant Singh	27 July 18			
23 Dec. 14	Ami Chand	14 Aug. 18			
20 May 01	Bakshi Singh	26 Aug. 18			
22 Sep. 01	Harbant Singh	26 Aug. 18			
18 Sep. 08	Nand Singh	26 Aug. 18			
17 Jan. 04	Achchra Singh	26 Aug. 18			
15 Nov. 89	Abbi Ram	26 Aug. 18			
26 June 14	Dip Chand	26 Aug. 18			
9 July 12	Siri Ram (56)	26 Aug. 18			
26 Jan. 15	Choti Lal	26 Aug. 18			
12 Sep. 18	Bhup Singh	12 Sep. 18			

Regt. Centre—Rawal Pindi. *Linked with*—35th and 47th Sikhs.

36th Sikhs.

Uniform—Scarlet. *Facings*—Yellow.

Raised at Jullundur in 1887, by Lt.-Col. J. Cook, as the 36th (Sikh) Regiment of Bengal Infantry. Became the 36th Sikh Infantry in 1901; present designation in 1903.

"Samana" "Tirah" "Punjab Frontier."

Composition—3 Companies Sikhs,
1 Company Hindu Jats.

COLONEL—Major-General James Cook, C.B. . 13 May 1904.

1st Commn. or date of entering service.	Names and Rank.	Army rank.	Present appointment in Regiment.	Remarks.
	Commandant.			
29 Nov. 90	Knox, Lt.-Col. E. F., D.S.O.	29 Nov. 16	15 Nov. 17	
	Company Commanders. (4)			
25 Aug. 86	*Moore, Lt.-Col. H. A. (Q)*	*23 Aug. 12*	*5 May 10*	*2nd-in-Comd.*
23 Mar. 89	Munn, Lt.-Col. R.G. (Q)	23 Mar. 15	26 Feb. 01	*Lv., ex I., m. c.*
27 July 98	*Wilson-Johnston, Bt. Lt.-Col. W. E.*, D.S.O., *p.s.c.*	*3 June 18*	*28 Jan. 15*	
20 Jan. 00	Lyall, Maj. C. R.	1 Sep. 15	15 Nov. 17	With 1-151st Infy.
	Company Officers (9)			
28 July 00	Blair, Maj. G. L. [L]	1 Sep. 15	26 Nov. 01	*Lv., ex I., m. c.*
27 Aug. 02	*Haughton, Maj. H. L.*	*27 Aug. 17*	*18 Jan. 04*	*Sr. Intsr., Tg. Sch., Indore.*
23 Dec. 03	*Saunders, Capt. M.*, D.S.O. *[L] (temp. Maj.)*	*23 Dec. 12*	*12 Nov. 07*	*G. S. O., 2nd gr.*
13 Aug. 04	Hay, Capt. J. B. W.	13 Aug. 13	26 Dec. 06	With 47th Sikhs.
29 Jan. 10	Robinson, Capt. D. C., M. C.	1 Sep. 15	19 May 11	*Lv, Ind., m c.*
3 Sep. 10	Elkington, Capt. E. H. O., M.C.	1 Sep. 15	27 Nov. 11	
12 June 07	Collings, Capt. E. G. (*Actg. Maj.*)	1 Sep. 15	. . . .	1-35th Sikhs.
14 July 15	Bourdillon, Lt. F. H.	14 July 16	28 Oct. 15	Attd. with 2-22nd Punj.
15 Nov. 15	Birdwood, Lt. F. T. (*Actg. Capt.*)	15 Nov. 16	19 Nov. 15	Attd. Adjt.
15 Nov 15	Mackintosh, Lt. G. N.	15 Nov. 16	19 Nov. 15	Attd.
15 Nov. 15	Mitchell, 2nd-Lt. J. H.	15 Nov. 15	19 Nov. 15	Attd.
29 June 16	Scotland, Lt. A. G.	29 June 17	. . . .	Attd.
	Attached.			
10 Oct. 03	Pottinger, Maj. J. A., M.C.	10 Oct. 18	5 Dec. 17	30th Punj.
22 Sep. 15	Fox, Lt. E B.	22 Sep. 16	13 Oct. 15	I. A. R. O.
1 Oct. 15	Milburne, Lt. W.	1 Oct. 16	31 Mar. 16	I. A. R. O.
25 Sep. 15	Webb, Lt. J. E. N.	25 Sep. 16	22 Sep. 16	I. A. R. O.
28 May 15	Keddie, Lt. C. M.	28 May 16	15 May 17	I. A. R. O.
21 Aug. 16	Dow, Lt. T. M.	21 Aug. 17	31 July 17	I. A. R. O.
5 May 17	Humphreys, 2nd-Lt. D. W. G.	5 May 17	28 Aug. 17	I. A. R. O.
19 Mar. 18	Wellings, 2nd-Lt. D.	19 Mar. 18	16 July 18	I. A. (*T. C.*)
29 May 18	Whitwam, 2nd-Lt. R.	29 May 18	30 Aug. 18	I. A. R. O. (*on probn.*)
28 Feb. 18	McGowan, 2nd-Lt. T. E.	28 Feb. 18	20 June 18	R. Ir. Fus., I. A. (*on probn.*)
26 June 17	Westoby, Lt. C. B. W.	26 June 18	10 July 18	Lan. Fus., I. A. (*on probn.*)
7 July 16	Hill, Lt. D. N.	7 July 17	1 Aug. 18	R. Ir. Fus., I. A. (*on probn.*)
1 Nov. 15	Tickell, Lt. G. W.	1 Nov. 16	3 Oct. 18	Arg. and Suth'd Highrs., I. A. (*on probn.*)
24 Apl. 18	Murray, 2nd-Lt. J. R.	24 Apl. 18	19 Oct. 18	I. A. (U. L.)
29 Aug. 17	Ryall, Lt. H. H. (*Actg. Capt.*)	29 Aug. 18	23 Jan. 18	I. A. R. O., Adjt.
18 July 18	Davies, 2nd-Lt. W. A.	18 July 18	7 Aug. 18	I. A. R. O.
7 Sep. 18	Laws, 2nd-Lt. W.	7 Sep. 18	30 Sep. 18	I. A. R. O.
7 Oct. 18	John, 2nd-Lt. W. H.	7 Oct. 18	21 Oct. 18	I. A. R. O.
7 Oct. 18	Hawkins, 2nd-Lt. F. L.	7 Oct. 18	28 Oct. 18	I. A. R. O.

Subadar-Major.

		Jemadar.	Subadar.	Subadar-Major.	
1 July 87	Kishn Singh, *Sardar Bahadur* (56)	13 Sep. 95	9 Feb. 02	1 Nov. 05	*Order of Br. Ind., 1st Cl.*

Subadar.

20 June 87	**Khem Singh *Bahadur* (56)**	1 Feb. 99	19 Apl. 05	*Order of Br. Ind., 2nd Cl.*
1 June 92	**Pertab Singh (56)**	1 June 08	6 June 14	
28 Oct. 92	**Sher Singh (66a)**	1 Mar. 10	19 Nov. 15	
22 Jan. 96	**Narang Singh**	1 Oct. 15	30 Jan. 17	
1 May 04	**Sewa Singh**, I.O.M.	9 Apl. 16	30 Jan. 17	
11 Aug. 08	**Harbans**	1 Jan. 17	12 Apl. 18	
1 May 98	**Bishan Singh**	1 Oct. 15	15 May 18	
2 Feb. 00	**Narayan Singh**	12 Apl. 16	15 May 18	
17 May 00	**Sant Singh**	12 Apl. 16	17 Sep. 18	

36th Sikhs—*contd.*

1st commn. or date of entering service.	Names and Rank.	Army rank.	Present appointment in Regiment.	Remarks.
	Jemadars.			
27 Nov. 94	Chajja Singh, I.O.M.	2 May 14		
4 Jan. 00	Makhan Singh	1 Oct. 15		
28 Apl. 98	Mastan Singh	1 Feb. 16		
17 June 00	Dal Singh	11 Apl. 16		
28 July 04	Kartar Singh (48*a*)	17 May 16		
24 July 00	Fateh Sing	2 Feb. 17		
4 Aug. 03	Jodh Singh	2 Feb. 17		
1 Apl. 91	Nand Lal	1 Mar. 17		
17 May 00	Dal Singh	1 July 17		
29 June 04	Bhagat Singh	1 July 17		
1 Jan. 00	Harnam Singh	16 Nov. 17		
12 Aug. 93	Ram Lal	3 Dec. 17		
30 Sep. 02	Tek Ram	19 Dec. 17		
	Chob Singh	12 Apl. 18		
	Uttam Singh	15 May 18		

Regt. Centre—Rawal Pindi. *Uniform*—Scarlet.
Linked with—38th and 41st Dogras. *Facings*—Yellow.

37th Dogras.

Raised at Sialkot in 1887, by Lt.-Col. V. Rivaz, as the 37th (Dogra) Regiment of Bengal Infantry. Became the 37th Dogra Infantry in 1901; present designation in 1903.

"Chitral" "Punjab Frontier."

Composition—4 Companies Dogras.

HONORARY COLONEL *Hony.* Lieut.-General *His Highness Maharaja Sir* PARTAB SINGH, *Bahadur, Maharaja* of JAMMU AND KASHMIR, G.C.S.I., G.C.I.E., G.B.E. 4 May 1888.

1st Commn. or date of entering service.	Names and Rank.	Army rank.	Present appointment in Regiment.	Remarks.
	Commandant.			
19 Nov. 92	Browne, Lt.-Col. L. S.	15 June 18	15 June 18	
	Company Commanders. (4)			
10 Oct. 94	Cox, Maj. T. S. (*Actg. Lt.-Col.*)	10 Oct. 12	15 June 18	2nd-in-Command.—With 2-41st [Dogras. *Lv., ex I., m.c.*
14 Aug. 97	*Cloete, Maj. L. B., p.s.c.*	*4 Aug. 15*	*19 Oct. 12*	*Lv., ex I., m.c.*
12 Aug. 99	*Dunlop, Maj. C. A. M.*, M.C.	*1 Sep. 15*	*8 June 14*	*G.S.O. 2nd gr., Army Hdqrs.*
8 Jan. 01	Moirce, Maj. R. C. L., M.C.	8 Jan. 16	15 June 18	
	Company Officers. (9)			
27 Aug. 02	*Walker, Maj. C. W. G.*, D.S.O., *p.s.c.*	*27 Aug. 17*	*26 Dec. 03*	*D.A.Q.M.G.*
18 Nov. 03	Nicholson, Maj. F. L., D.S.O., M.C.	18 Nov. 18	23 June 05	
3 Sep. 10	Andrews, Capt. H. A., M.C.	1 Sep. 15	28 Nov. 11	With 2-41 Dogras.
22 Jan. 13	*Lennox, Capt. G., (temp. Maj.)*	*22 Jan. 17*	*11 Mar. 14*	*Asst. Instr. Sch. of Musky., Pach-[mari.*
8 Aug. 14	Churchill, Capt. A. P. ff.	8 Aug. 18	9 Nov. 15	
15 Nov. 15	Weatherley, Lt. E. F. C. F. (*Actg. Capt.*)	15 Nov. 16	19 Nov. 15	With 2-152nd Infy.
15 Nov. 15	Hayne, Lt. M. L. (*Actg. Capt.*)	15 Nov. 16	19 Nov. 15	Adjt.
18 Apl. 16	Warren, Lt. V. A.	18 Apl. 17	25 Apl. 16	
18 Apl. 16	Simpson, Lt. F. C.	18 Apl. 17	25 Apl. 16	
18 Apl. 16	Power, Lt. H. R.	18 Apl. 17	25 Apl. 16	Attd.
18 June 17	McIntyre, Lt. H. F.	18 June 18	....	Attd.
27 Oct. 17	Hewett, 2nd-Lt. J. P. O. C.	...	4 Nov. 17	Attd.
27 Oct. 17	Standish, Lt. N.	27 Oct. 18	4 Nov. 17	Attd. With 2-152nd Infy.
31 Jan. 18	Powell, 2nd-Lt. R. E.	31 Jan. 18	6 Feb. 18	Attd.
21 Dec. 17	Seller, 2nd-Lt. C.	21 Dec. 17	22 Mar. 18	Attd.
31 Aug. 18	Hill, 2nd-Lt. F. O. R.	31 Aug. 18	7 Sep. 18	Attd.
31 Aug. 18	Moreton, 2nd-Lt. B. M.	31 Aug. 18	4 Sep. 18	Attd.
	Attached.			
8 Jan. 15	Galvin, Lt. A. (*Actg. Capt.*)	8 Jan. 16	23 June 15	I. A. R. O.
14 May 15	Walmesley, Lt. H. J. (*Actg. Capt.*)	14 May 16	31 July 15	I. A. R. O.
5 May 17	Heath, Lt. C. J.	5 May 18	18 Aug. 17	I. A. R. O., *Lv., ex I., m. c.*
9 July 16	Hunt, Lt. J. L. (*Actg. Capt.*)	9 July 17	13 Nov. 17	I. A. (*on probn.*)
15 Nov. 15	Moore, Lt. W. S.	15 Nov. 16	....	
18 Apl. 16	Townley, Lt. N. P.	18 Apl. 17	....	41st Dogras.
22 June 17	Smellie, Lt. C. S.	22 June 18	12 June 18	I. A. R. O.
13 Mar. 17	Jackson, Lt. C.	13 Mar. 18	....	I. A. R. O.
27 Oct. 17	Ries, Lt. J. L.	27 Oct. 18	12 June 18	
24 Apl. 16	Fairgrieve, Lt. A. O.	24 Apl. 17	1 June 18	M. G. Corps. I. A. (*on probn.*)
29 Mar. 18	Roblin, 2nd-Lt. C. H.	29 Mar. 18	12 Aug. 18	I. A. R. O.
...	Clark, 2nd-Lt. A. H.	...	30 Aug. 18	I. A. R. O. (*on probn.*)
11 Apl. 17	Evans, Lt. D. W.	11 Apl. 18	....	Lond. R., I. A. (*on probn.*)
1 Oct. 18	Kerr, 2nd-Lt. J. N.	1 Oct. 18	....	Attd.
	Honorary Officer.			
1 Jan. 88	Lambagraon, *The Hon'ble* Raja *Sir* Jai Chand, K.C.I.E., C.S.I., Raja of—	25 Jan. 11	4 May 88	Hony. Lt.-Col. in Army.

37th Dogras—*contd.*

1st Commn. or date of entering service.	Names and Rank.	Army rank.	Present appointment in Regiment.	Remarks.

Subadar-Major.

1st Commn. or date of entering service.	Names and Rank.	Jemadar.	Subadar.	Subadar-Major.	Remarks.
23 July 95	Bhag Singh, I.O.M., *Bahadur* (66*a*, 72*a*)	1 Feb. 00	1 Apl. 08	1 Nov. 14	*Order of Br. Ind., 2nd Cl.*
29 Dec. 96	Chagar Singh, *Bahadur* (56, 66*a*, 72*a*, 72*b*).	1 Jan. 02	16 Jan. 12	21 Jan. 16	*Order of Br. Ind., 2nd Cl.*

Subadars.

1st Commn. or date of entering service.	Names and Rank.	Army rank.	Present appointment in Regiment.	Remarks.
23 July 84	Attar Singh	1 Dec. 91	1 Aug. 09	*Burma Mily. Police.*
1 Oct. 87	Tega, *Bahadur* (56)	1 Apl. 07	2 June 12	*Order of Br. Ind., 2nd Cl.*
26 Apl. 93	Udho (56, 66*a*, 72*a*)	1 July 08	1 Nov. 14	
19 June 90	Janka (66*a*)	16 Jan. 12	1 Nov. 14	
17 Nov. 90	Ran Singh (51, 56)	16 Jan. 12	30 Nov. 14	
7 Jan. 02	Ranjodh Singh (50)	1 June 13	1 Nov. 15	
17 Dec. 95	Mussadi (52, 56)	1 Nov. 14	24 Dec. 15	
14 Jan. 93	Rai Singh (56)	1 June 14	6 Jan. 16	
6 Nov. 96	Hakim Singh (56, 58)	30 Nov. 14	6 Jan. 16	
16 Apl. 95	Sundar (52, 56)	30 Nov. 14	8 Jan. 16	
15 Sep. 00	Nanak Chand	. . . .	18 Apl. 16	Attd.
16 June 92	Kahna (56)	1 Nov. 14	13 Feb. 17	
23 Dec. 95	Phandi	1 Aug. 15	1 May 17	
21 Dec. 95	Chand	14 Sep. 15	6 Dec. 17	
26 Nov. 98	Jamit Singh	1 Nov. 15	18 May 18.	

Jemadars.

1st Commn. or date of entering service.	Names and Rank.	Army rank.	Remarks.
2[?] June 04	Janak Singh, M.C. (56, 61)	1 Nov. 15	
[?]1 Nov. 99	Nag (66*a*)	1 Nov. 15	
22 Dec. 96	Sundar	30 Dec. 15	
1 July 99	Tek Singh (58)	6 Jan. 16	
5 Mar. 99	Hardayal (66*a*)	7 Apl. 16	
16 Mar. 00	Shib Dyal (56, 61)	8 June 16	
24 Dec. 98	Hira	10 June 16	
18 Oct. 05	Anant Ram	1 Oct. 16	Attd.
	Jiwan Singh	13 Dec. 16	38th Dogras.
24 Mar. 00	Govinda	21 Dec. 16	
7 Nov. 98	Lachhman (66*a*)	13 Feb. 17	
24 Jan. 99	Barru (66*a*)	13 Feb. 17	
	Rao, I.O.M.	13 Feb. 17	
23 Mar. 15	Bishen Dass	25 Feb. 17	Attd.
17 Dec. 14	Devi Das	1 Aug. 17	Attd.
5 July 00	Sadhu, I.O.M. (56)	12 Aug. 17	
24 Jan. 00	Teku	18 May 18	
19 Jan. 98	Sohanu	1 July 18	

Regt. Centre—Rawal Pindi.
Linked with—37th and 41st Dogras.

Uniform—Scarlet.
Facings—Yellow.

38th Dogras.

Raised at Agra in 1858, by Lt. G. L. Fraser, as the Agra Levy. Became the 42nd Regt. of Bengal Native Infantry, 1861; the 38th Regt. of Bengal Nat. Infy., 1861; the 38th (Agra) Regt. of Bengal Nat. Infy., 1864; the 8th Regt. of Bengal Infy., 1885; the 38th (Dogra) Regt. of Bengal Infy., 1890; the 38th Dogra Infy., 1891; present designation, 1903.

"Malakand" "Punjab Frontier."
Composition—4 Companies Dogras.

1st Commn. or date of entering service.	Names and Rank.	Army rank.	Present appointment in Regiment.	Remarks.
	Commandant.			
11 Feb. 88	Longden, Lt.-Col. A. B.	11 Feb. 14	14 Jan. 15	
	Company Commanders. (4)			
22 Jan. 96	*Barstow, Bt.-Lt.-Col. H. (temp. Br.-Genl.)*	*1 Jan. 17*	*6 July 16*	*2nd-in-Comd.—Dir. of Persl. Ser. A. H. Qrs.*
22 Jan. 98	***Rowlandson**, Bt.-Lt.-Col. M. G. D.*, D.S.O., *[L] p.s.c.*	*3 June 16*	*5 Oct. 12*	*G. S. O., 2nd Gr.*
4 Jan. 99	Renny, Maj. G. S.	1 Sep. 15	2 Nov. 14	With 3-151st Infy.
14 Feb. 00	Young, Maj. G. H.	1 Sep. 15	2 Nov. 14	
	Company Officers. (9)			
18 Jan. 02	*Kerr, Maj. A. N.*, M.C.	*18 Jan. 17*	*8 Apl. 03*	*I.S. Troops.*
19 Aug. 03	Barrow, Maj. R. E.	19 Aug. 18	1 Jan. 05	
18 Jan. 05	Cunningham, Capt. W. B.	18 Jan. 14	30 Apl. 96	Adjt., 23 Dec. 14.
5 Aug. 05	Malet, Capt. A. W. (*Actg. Maj.*)	5 Aug. 14	1 Nov. 96	With 1-70th Burma Rif.
21 Nov. 06	*Bickford, Capt. M. H.*	*1 Sep. 15*	*30 Oct. 08*	*I.S. Troops.*
9 Sep. 08	Trail, Capt. R. M.	1 Sep. 15	4 June 10	
8 Sep. 09	Ogilvy, Capt. A. G.	1 Sep. 15	12 Jan. 11	With 2-35th Sikhs.
25 Dec. 09	Jones, Capt. B. I., M.C.	1 Sep. 15	14 Oct. 14	With 2-152nd Punj.
1 Oct. 14	Reiche, Capt. A. D. E.	1 Oct. 18	19 Dec. 14	2-41st Dogras.
13 Jan. 15	Sanders, Lt. R. G. (*Actg. Capt.*)	13 Jan. 16	25 Feb. 15	With 3-154th Infy.
17 Feb. 15	Hall, Lt. E. R. M.	17 Feb. 16	5 Jan. 17	Attd.
18 June 17	Morton, Lt. H. K.	18 June 18	25 June 17	Attd. With 1-151st Infy.
31 Jan. 18	Johnson, 2nd Lt. A. H. G.	31 Jan. 18	6 Feb. 18	Attd. With 3-151st Infy.
22 Oct. 15	Black, Lt. R. (*Actg. Capt.*)	22 Oct. 16	29 July 18	Attd. With 1-41st Dogras.
21 Dec. 17	Dreffen, 2nd-Lt. J. E.	21 Dec. 17	22 Mar. 18	Attd. With 2-50th Rif.
15 Oct. 15	Schneider, Lt. A.	15 Oct. 16	26 Nov. 17	Attd.
30 Aug. 18	Phelps, 2nd-Lt. J. N.	30 Aug. 18	4 Sep. 18	Attd.
21 Dec. 17	Connolly, 2nd-Lt. D. R. V.	21 Dec. 17	....	Attd. With 1-131st Reg.
	Attached.			
8 Sep. 09	Drysdale, Capt. A. E. (*Actg. Maj.*), D.S.O., M.C.	1 Sep. 15	....	47th Sikhs.
20 June 17	Audley, 2nd-Lt. W. D.	20 June 17	5 Oct. 17	I. A. R. O.
10 Sep. 15	Charlesworth, Lt J. H.	10 Sep. 16	30 Aug. 18	W. Rid. R., I. A. (*on probn.*)
9 Apl. 16	Cook, Lt. S.	25 Apl. 17	30 Sep. 17	I. A. R. O. (*on probn.*)
14 May 17	Finlayson, Lt. R. E. M.	14 May 18	30 Aug. 18	North'd Fus., I. A. (*on probn.*)
29 May 18	Murray, 2nd-Lt. J. W.	29 May 18	30 Aug. 18	I. A. R. O. (*on probn.*)
28 Aug. 18	Beverage, 2nd-Lt. J. W. P.	28 Aug. 18	17 Aug. 18	High. L. I., I. A. (*on probn.*)
7 June 18	Young, 2nd-Lt. L. E.	7 June 18	....	I. A. R. O.

Subadar-Major.

		Jemadar.	Subadar.	Subadar-Major.
23 Mar. 91	Unkar Chand (56)	18 July 06	1 Jan. 13	1 June 16

Subadars.

14 Feb. 91	Kapur Singh (56, 66*a*)	19 Mar. 11	5 Jan. 15
8 Feb. 91	Ushnaki (56, 66*a*)	1 Nov. 11	5 Jan. 15
7 Mar. 91	Narsinghu (52, 56)	17 Apl. 12	5 Jan. 15
11 Jan. 96	Hako (56)	5 Jan. 15	15 Apl. 16
1 Dec. 93	Jewan Singh	5 Jan. 15	1 June 16
25 May 15	Sil Singh	1 Jan. 16	20 July 17
14 Sep. 95	Kanhaiya (53, 56, 66*a*)	5 Jan. 15	21 Sep. 17
6 Oct. 00	Teg Singh (46, 56)	1 Apl. 15	9 June 18
15 Feb. 95	Gujar Singh, I.O.M., I.D.S.M.	7 Apl. 15	9 June 18

38th Dogras—*contd.*

1st Commn. or date of entering service.	Names and Rank.	Army rank.	Remarks.
	Jemadars.		
4 Apl. 01	Sujada	15 Apl. 16	
1 Feb. 01	Gurdas (46)	1 June 16	
11 Mar. 03	Bhagta	22 June 16	
25 May 01	Pars Ram	1 Mar. 17	
14 Jan. 15	Lekhu Ram	20 July 17	
12 May 02	Suchet Singh (56)	15 Aug. 17	
8 Mar. 91	Nathu	1 Dec. 17	
18 Mar. 07	Mohar Singh	1 Dec. 17	
11 Mar. 00	Lachhman	1 Dec. 17	
1 May 00	Mehar Singh	1 Dec. 17	
21 Dec. 10	Basant Singh	1 Dec. 17	
	Brahm Dat	6 Mar. 18	
	Narayan Dat	6 Mar. 18	
7 May 03	Mali	9 June 18	
28 June 04	Dina Nath	9 June 18	
1 Feb. 02	Jai Singh	9 June 18	

Regt. Centre—Lansdowne.

39th Garhwal Rifles.

Uniform—Dark Green. *Facings*—Black.

1st Battalion . . . . . 2nd Battalion . . . . .

Raised at Almora in 1887, by Lt.-Col. E. P. Mainwaring, as the 2nd Bn. of the 3rd (the Kamaon) Goorkha Regiment. Became the 39th (The Garhwali) Regt. of Bengal Infantry in 1890; the 39th (The Garhwal Rifles) Regt. of Bengal Infy. in 1892; present designation in 1901.

The 2nd Battalion was raised in 1901 by Lt.-Col. J. T. Evatt, D.S.O., as the 49th (Garhwal Rifle) Regiment of Bengal Infantry. Received its present designation in November of the same year.

"Punjab Frontier."

COLONEL—Brig.-General John T. Evatt, D.S.O. . . . . 3 March 1914.

1st Commn. or date of entering service.	Names and Rank.	Army rank.	Present appointment in Battalion.	Remarks.
	Commandants.			
29 Mar. 90	2 Hogg, Lt.-Col. J. McK. T. .	29 Mar. 16	12 Dec. 17	
3 Sep. 92	1 Stewart, Bt.-Col. J. H. K., D.S.O.	10 Mar. 17	1 July 16	*Lv., ex I., m. c.*
	Company Commanders (8)			
25 Mar. 91	*1 Home, Maj. P. M. (Q)* . . .	*25 Mar. 09*	*30 July 06*	*Censor, Quetta*
14 Aug. 95	2 Henderson, Maj. K., D.S.O. *p.s.c. (Q)*	14 Aug. 13	1 July 16	2nd-in-Comd. With 3-39th Rif.
4 Aug. 97	1 Lumb, Maj F. G. E., M.C. (*Actg. Lt.-Col.*)	4 Aug. 15	1 July 16	Comdt., 1st Bn. (*temp.*)
18 Oct. 99	2 Blair, Maj. D. A. . . .	1 Sep. 15	23 Mar. 16	
17 Jan. 00	1 Lane, Maj J. T. H., M.C. . .	1 Sep. 15	23 Mar. 16	With 4-39th Rif.
17 Jan 00	*2 Reed, Maj. H. R. B., M.C.*	*1 Sep. 15*	*1 July 16*	*With 3-39th Garh. Rif.*
28 July 00	*1 Carruthers, Maj. J. H. DeW., M.C., p.s.c.*	*1 Sep. 15*	*1 July 16*	*G. S. O. 2nd gr., 2nd Dn.*
11 Aug. 00	2 Orton, Bt.-Lt.-Col. S. B., *p.s.c.*	3 June 18	12 Apl. 17	
	Company Officers. (18)			
8 Jan. 01	2 Lvell, Maj. A. G.	8 Jan. 16	12 Feb. 04	2nd-in-Comd. (*offg.*)
4 July 03	*1 Etherton, Maj. P. T. [I]* . .	*4 July 18*	*4 Aug. 06*	*With For. & Poll. Dept.*
10 Oct. 03	*1 Cardew, Maj. H. S., M.C.*	*10 Oct. 18*	*28 Oct. 06*	*With Sig. Service.*
13 Aug. 04	2 Berryman, *Bt.-Maj.* E. R. P. .	13 Aug. 13	7 July 06	
5 Aug. 05	1 Mainwaring, Capt. G. R. (Actg. Maj.)	5 Aug. 14	2 Nov. 06	
22 Feb. 08	2 Fox, Capt. F N., M.C. . . .	1 Sep. 15	19 Feb. 14	Adjt., 29 Mar. 17.
9 Sep. 08	1 Rogers, Capt G. S. .	1 Sep. 15	18 Oct. 10	With 2-50th Rif.
6 Sep. 11	2 Rodgerson, Capt. A. P., D.S.O.	6 Sep. 15	14 Jan. 15	*Lv. ex I., m.c.*
20 Jan. 12	2 Clarke, Capt. A. E (*Actg. Maj.*)	20 Jan. 16	11 Mar. 13	
14 Jan. 14	2 Gore, Capt. C. H. C. . . .	14 Jan. 18	19 Oct. 14	With 3-39th Garh. Rif.
8 Aug. 14	2 Powell, Capt. F., M.C. .	8 Aug. 18	14 Nov. 15	
8 Aug 14	1 Eustace, Capt. R. . . .	8 Aug. 18	10 Nov. 15	Adjt., 7 May 17.
29 June 16	1 Phillips, Lt. R. S.	16 Jan. 16	. . . .	
15 Nov. 15	1 Robinson, Lt. E. C. . . .	15 Nov. 16	18 Nov. 15	With 3-9th G. R.
26 Jan. 16	2 Davies, Lt. H. L. . . .	26 Jan. 17	18 Mar. 16	With 3-39th Garh. Rif.
18 Apl. 16	2 Chappell, Lt. H	18 Apl. 17	27 Apl. 16	Qr.-Mr. (*offg.*)
29 June 16	2 Tayler, Lt. S. E. . . .	29 June 17	. . . .	
16 Aug. 16	1 Prendergast, Lt. L. C. .	16 Aug. 17	12 Nov. 17	Attd.
30 Jan. 17	2 Meyer, Lt. B. (*temp. Capt.*) .	30 Jan. 18	31 Jan. 17	Attd.
30 Jan. 17	1 Holmes, Lt. R. E. J. . . .	30 Jan. 18	31 Jan. 17	Attd.
18 June 17	1 Prophit, Lt. G. C. . . .	18 June 18	. . . .	Attd.
1 May 17	2 Orr, Lt. J. H. R.	1 May 18	1 Dec. 17	Attd.
12 Sep. 17	2 Plowden, Lt. J. M. B. C. . .	12 Sep. 18	9 Dec. 17	Attd.
27 Oct. 17	1 Fox, Lt. E. L. W.	27 Oct. 18	. . . .	Attd.
27 Oct. 17	2 Aspinall, Lt. A. W. . . .	27 Oct. 18	6 Nov. 17	Attd.
11 Oct. 15	1 Molyneux, Lt. P. L. (*Actg. Capt.*)	11 Oct. 16	5 Aug. 17	Attd.
1 Nov. 15	1 Marshall, Lt. J. D. C. . .	1 Nov. 16	5 Aug. 17	Attd.
3 Sep. 16	1 Wilkinson, Lt. G. N. . .	3 Sep. 17	6 July 17	Attd.
31 Aug. 18	2 Bulgin, 2nd-Lt. J.	31 Aug. 18	31 Aug. 18	Attd.
31 Apl. 18	1 McOstrich, 2nd-Lt. A. H. M.	31 Apl. 18	31 Aug. 18	Attd.
31 Aug. 18	2 Thornton, 2nd-Lt. C. A. . .	31 Aug. 18	31 Aug. 18	Attd.
1 Oct. 18	1 Lowndes, 2nd-Lt. D. G. . .	1 Oct. 18	21 Oct. 13	Attd.
	Attached.			
30 Nov. 14	2 Collins, R. T. (temp. Lt.) .	30 Nov. 14	22 May 15	Sher. Foresters.
24 Dec. 14	2 Hayne, Lt. C. W.	24 Dec. 15	28 Sep. 15	I. A. R. O.
22 Jan. 15	2 France, Lt. E.	22 Jan. 16	28 Jan. 15	I. A. R. O.,
18 June 15	1 Tresham, Lt. B. F.	18 June 16	24 Aug. 15	I. A. R. O. Qr.-Mr. (*temp.*)
25 June 15	2 MacLear, Lt. G. D'O., M.C. .	25 June 16	16 Jan. 17	I. A. R. O.
10 Nov. 16	2 Hunter, 2nd-Lt. H. C. A. .	10 Nov. 16	28 Jan. 17	I. A. R. O.
1 Aug. 16	2 Fitzpatrick, Lt. P. W. .	1 Aug. 17	22 Apl. 17	I. A. R. O.
15 Feb. 16	1 Brown, Lt. O. S.	15 Feb. 17	. . . .	I. A. (*on probn.*)
18 May 15	1 Haines, Lt. M. A.	18 May 16	28 Aug. 17	I. A. (*on probn.*)
11 Oct. 15	1 Shepherd, Lt. H. V. . . .	11 Oct 16	5 Aug. 17	I. A. R. O.
12 May 15	1 Hall, Lt. G. H.	12 May 16	5 Aug. 17	I. A. R. O.
20 June 17	2 Martin, 2nd-Lt. A. S. W. . .	20 June 17	5 Oct. 17	I. A. R. O.
5 May 17	2 Walsh, Lt. L. W.	5 May 18	5 Oct. 17	I. A. R. O.
8 Aug. 17	2 Elliot, 2nd-Lt. E. T. . . .	8 Aug. 17	2 Nov. 17	I. A. R. O.
10 Nov. 15	1 Salt, Lt. R. F.	10 Nov. 16	24 Jan. 18	I. A. (*on probn.*)
29 Mar. 18	2 Burt, 2nd-Lt. W. G. . . .	29 Mar. 18	15 July 18	I. A. R. O.
15 Aug. 14	2 Upton, Lt. R. B. E. (Actg. Capt.)	1 Sep 15	. . . .	1-1st G. R.
11 Mar. 16	2 King, Lt. J. H. L.	11 Mar. 17	2 July 18	Hamps. R., I. A. (*on probn.*)

39th Garhwal Rifles—*contd.*

1st Commn. or date of entering service.	Names and Rank.	Army rank.	Present appointment in Battalion.	Remarks.
	Attached—*contd.*			
17 Nov. 16	2 Preston, Lt. N.	. . . .	31 Aug. 18	Ches. R., I. A. R. O. (*on probn.*)
4 Dec. 14	1 Banon, Lt. H. M.	4 Dec. 15	5 Nov. 17	I. A. R. O.
14 May 15	1 Paterson, Lt. G. S. V.	14 May 16	13 Apl. 18	I. A. R O.
2 June 15	1 Park, Lt. L.	2 June 16	16 May 18	I. A. (*on probn.*)
24 Apl. 18	1 Desborough, 2nd-Lt. J. N. C.	24 Apl. 18	21 Oct. 18	I. A. (U. L.)
24 Apl. 18	2 Banks, 2nd-Lt. A. T.	24 Apl. 18	22 Oct. 18	I. A. (U. L.)
7 June 18	1 Glover, 2nd-Lt. C. A. W.	7 June 18	12 Oct. 18	I. A. R. O.
12 Feb. 15	2 Browne, Lt. C. E. M.	12 Feb. 16	19 Nov. 18	I. R. R. O.
7 July 16	2 Winters, Lt. L. P.	7 July 17	29 Sep. 1	Bedf. R., I. A. (*on probn.*)
	Honorary Officer.			
14 June 16	Tehri, *His Highness Raja* Narendra Sah, of—	14 June 16	14 June 16	Hony. Lieut. in Army.

Subadar-Majors.

		Jemadar.	Subadar.	Subadar-Major.	
	2 Nathu Sing Sunar, *Hony. Capt., Sardar Bahadur (Retd.)*	. . .	. . . .	. . .	Re-employed.
6 May 88	2 Nain Sing Chinwarh, M.C., *Bahadur.*	1 Apl. 01	16 Feb. 04	*1 Mar. 16*	*Order of Br. Ind., 2nd Cl.*
23 Feb. 88	1 Kedar Sing Rawat, I.D.S.M. (72*a*)	1 May 01	5 Sep. 12	11 July 16	

Subadars.

24 Sep. 87	2 Lal Sing Gusain	1 Apl. 01	1 Oct. 01	Re-employed.
4 Nov. 94	1 Madho Sing Rawat	9 May 13	24 Jan. 15	
4 May 92	2 Gopi Sing Rawat (56)	20 May 12	11 Mar. 15	
7 May 01	2 Indar Sing Negi	9 Aug. 14	1 Jan. 16	
23 Dec. 99	1 Kamli Sing Rawat (66*a*)	11 Mar. 15	1 May 16	
4 May 01	2 Amar Sing Negi	9 Aug. 14	20 Aug. 16	
4 May 02	1 ✠ Darwan Sing Negi (56)	9 Aug. 15	1 Oct. 16	
13 Oct. 04	1 Gobar Sing Butola	17 Mar. 16	1 Dec. 16	
17 May 01	2 Tilok Sing Sauntiyal, I.O.M. (56)	10 Mar. 15	1 Mar. 17	
11 Feb. 03	2 Meharwan Sing Butola	10 Mar. 15	1 Mar. 17	
18 May 01	2 Ratan Sing Butola (46)	20 Aug. 16	30 Sep. 17	
28 Oct. 01	2 Bir Sing Danu, I.D.S.M., I.O.M. (56)	20 Aug. 16	10 Oct. 17	
17 Aug. 08	1 Autar Sing Negi	20 Aug. 16	27 May 18	
11 Apl. 03	1 Pan Sing Ghariya	20 Aug. 16	27 May 18	
7 Jan. 07	1 Bhawan Sing Rawat	20 Aug. 16	12 Sep. 18	
29 Jan. 02	2 Ranjit Sing Rawat	20 Aug. 16	15 Oct. 18	
23 Sep. 04	1 Jit Sing Pundir	1 Oct. 16	2 Sep. 18	

Jemadars.

8 Feb. 04	2 Dulat Sing Sunar	20 Aug. 16
25 June 01	2 Bahadur Sing Mahar	20 Aug. 16
1 Apl. 04	1 Rupchand Bisht	1 Dec. 16
14 July 03	1 Dewan Sing Negi (58)	1 Jan. 17
27 Apl. 01	1 Umrao Sing Chaudhari	1 Jan. 17
10 Apl. 02	2 Harsing Gusain	30 July 17
29 Oct. 01	2 Rupsing Rawat	20 Sep. 17
12 Apl. 05	2 Jamansing Bisht, I.O.M.	23 Sep. 17
15 Mar. 00	1 Dalip Sing Rana	27 Sep. 17
22 Nov. 99	2 Indar Sing Kathait	1 Mar. 18
20 Aug. 07	1 Dewan Sing Takuli	27 May 18
11 Nov. 02	1 Netar Sing Negi	27 May 18
15 Mar. 07	1 Kundan Sing Negi	17 June 18
25 Mar. 03	2 Rupchand Sing Bisht	12 Sep. 18
31 May 01	2 Madan Sing Gusain	12 Sep. 18
3 Apl. 04	2 Autar Sing Butola	21 Sep. 18
25 Mar. 02	2 Keshar Sing Danu	21 Sep. 18
12 Apl. 13	1 Dhan Sing Negi	2 Sep. 18
29 Apl. 10	1 Nether Sing Bisht	12 Sep. 18
1 May 06	1 Jot Sing Rawat	12 Sep. 18
10 Oct. 08	1 Nain Sing Bisht	12 Sep. 18
28 Oct. 03	1 Bhim Sing Rawat	1 Apl. 17
22 July 12	1 Jot Sing Gusain	1 Aug. 17

3rd Battalion, 39th Garhwal Rifles.

Formed, 20th August 1916.

1st Commn. or date of entering service.	Names and Rank.	Army rank.	Present appointment in Regiment.	Remarks.
	Commandant.			
13 Aug. 92	Stansfeld, Maj. C. G., *p.s.c.* (Q) (*Actg. Lt.-Col.*)	13 Aug. 10	1 Nov. 17	8th G. R.
	Company Commanders. (4)			
17 Jan. 00	Reed, Maj. H. R. B., M.C.	1 Sep. 15	20 Aug. 16	2-39th Garh. Rif.
14 Jan. 14	Gore, Capt. C. H. C.	14 Jan. 18	20 Aug. 16	2-39th Garh. Rif.
2 June 15	Challen, Lt. P. (*temp. Capt.*)	2 June 16	. . . .	
3 June 16	Renwick, Lt. H. (*Actg. Capt.*)	3 June 17	11 May 17	
	Company Officers.			
5 May 17	Honnor, Lt. A. J.	5 May 18	21 Sep. 17	I. A. R. O. Qr.-Mr.
18 June 17	Foster, Lt. E. L. P.	18 June 18	. . .	
6 Feb. 16	Clark, Lt. J. (*Actg. Capt.*)	6 Feb. 17	20 Aug. 18	Attd. Adjt.
1 May 17	Orpen, Lt. G. E.	1 May 18	26 Nov. 17	Attd.
5 Oct. 17	Buley, Lt. S. H.	5 Oct. 18	23 Jan. 18	I. A. (T. C.) *Lv., ex I., m.c.*
1 Nov. 17	Gronow, 2nd-Lt. I. A.	1 Nov. 17	27 Jan. 18	I. A. R. O. (*on probn.*)
9 Sep. 16	Fanning, Lt. R. J. (*Actg. Capt.*)	9 Sep. 17	14 Feb. 18	I. A. (*on probn*).
30 May 17	Cooper, 2nd-Lt. A. C.	30 May 17	14 Feb. 18	I. A. R. O.
29 Aug. 17	Shearer, 2nd-Lt. F. C.	29 Aug. 17	5 Dec. 17	I. A. R. O. (*on probn.*)
19 May 16	Prince, Lt. H.	19 May 17	1 Apl. 18	I. A. (*on probn.*)
5 May 17	Neale, Lt. E. A. C.	5 May 18	17 Apl. 18	I. A. (*on probn.*)
15 Jan. 18	Cattermole, 2nd-Lt. F. C.	15 Jan. 18	8 Apl. 18	I. A. R. O.
15 June 18	Burns, 2nd-Lt. J.	15 June 18	. . .	I. A. R. O. (*on probn.*)
15 June 17	Hawkins, 2nd-Lt. F. H.	15 June 18	. . .	I. A. R. O. (*on probn.*)
19 Apl. 18	Lester, Lt. L.	19 Apl. 17	. . .	Midd'x R., I. A. (*on probn.*)
28 Feb. 18	Toms, 2nd-Lt. R. G.	28 Feb 18	. . .	R. Fus., I. A. (*on probn.*)
24 Apl. 18	Collard, 2nd-Lt. B. H. A.	24 Apl. 18	. . .	I. A. (U. L.)
8 July 18	Plumb, 2nd-Lt. W. C.	8 July 18	. . .	I. A. R. O.

Subadar-Major.

		Jemadar.	Subadar.	Subadar-Major.	
21 Sep. 87	Makhar Sing Kawar, *Bahadur*	1 Oct. 01	17 Apl. 04	20 Aug. 16	*Order of Br. I, 2nd Cl.*

Subadars.

3 June 01	Gaur Sing Negi	9 Aug. 14	1 Jan. 16
24 Apl. 01	Fateh Sing Gusain	21 Apl. 15	11 July 16
15 Sep. 93	Ram Sing Bhandari	21 Jan. 15	20 Aug. 16
28 May 01	Bakhtwar Sing Rawat	11 July 16	30 Nov. 17
14 May 01	Ratan Sing Tulsara (56)	20 Aug. 16	31 Dec. 17
20 Feb. 03	Dhum Sing Chauhan	20 Aug. 16	31 Dec. 17
27 July 01	Mangal Sing Bisht	8 Aug. 17	12 Oct. 18
7 Sep. 96	Jwala Sing Mahar	30 Nov. 17	12 Oct. 18
1 Nov. 05	Har Sing Rawat	30 Nov. 17	12 Oct. 18

Jemadars.

8 May 01	Gaur Sing Negi	17 Mar. 16
20 Aug. 06	Narain Sing Pawar	31 Dec. 17
27 Feb. 04	Chillu Singh Rawat	9 Apl. 18
14 Dec. 01	Tilok Sing	9 Apl. 18
25 Feb. 10	Girdhari Sing Negi	9 Apl. 18
31 July 07	Khyat Sing Negi	9 Apl. 18
4 Feb. 10	Ram Sing Rawat	9 Apl. 18
19 July 02	Sankar Sing Kathait	21 June 18
10 Oct. 02	Suraj Sing Gusain	21 June 18
10 Oct. 02	Mohan Sing Rawat	21 June 18
30 Oct. 02	Buddhi Sing Negi	21 June 18
	Guman Sing Dangwal	
14 July 03	Bijai Sing Rawat	14 Sep. 18

4th Battalion, 39th Garhwal Rifles.

Formed at Dehra Dun on 28th October 1918.

Composition:—4 Companies Garhwalis.

1st Commn. or date of entering service.	Names and Rank.	Army rank.	Present appointment in Battalion.	Remarks.
	Commandant.			
17 Jan. 00	Lane, Maj. J. T. H., M.C.	1 Sep. 15	23 Oct. 18	1-39th Rif.
	Company Commanders. (4)			
29 Jan. 02	Orchard, Capt. D. S.	29 Jan. 11	24 Nov. 18	1-8th G. R.
	Company Officers.			
22 Jan. 13	Lemarchand, Capt. F. F.	22 Jan. 17	13 Nov. 18	2-5th G. R.
15 Sep. 15	Valentine, Lt. W. B.	15 Sep. 16	5 Nov. 18	I. A. (*on probn.*)
12 May 15	Jones, Lt. B. M.	12 May 16	28 Oct. 18	I. A. (*on probn.*)
12 May 15	Johnson, Lt. H. B.	12 May 16	23 Nov. 18	Ches. R., I. A. (*on probn.*)
16 Aug. 16	Hughes, Lt. F. E. C.	16 Aug. 17	....	1-4th G. R.
30 Jan. 18	Smith, 2nd-Lt. J.	30 Jan. 18	23 Nov. 18	I. A. (*on probn.*)
31 Aug. 18	Kenny, 2nd-Lt. W. D.	31 Aug. 18	5 Nov. 18	I. A. (*on probn.*)

Subadar-Major.

		Jemadar.	Subadar.	Subadar-Major.
	Subadars.			
19 Nov. 94	Gudal Sing Gusain	17 Jan. 15	17 Mar. 16	
17 Nov. 00	Padam Sing Rawat	11 Feb. 15	17 Mar. 16	
10 Sep. 00	Gulab Sing Negi	24 Jan. 15	20 Aug. 16	
28 May 01	Teg Sing Kufola	10 Mar. 15	14 Nov. 16	
30 Apl. 01	Indar Sing Bisht (56)	17 Mar. 16	6 Dec. 16	
1 Aug. 97	Bhageo Sing Rawat	7 Feb. 12	1 Aug. 17	
9 Dec. 01	Butha Sing Negi, I.O.M. (56)	1 Jan. 16	23 Sep. 17	
1 Dec. 98	Jhagar Sing Bisht (60*a*)	20 Aug. 16	27 Sep. 17	
20 May 89	Gundar Sing Bhandari	1 Sep. 07	1 Mar. 18	
6 June 01	Jatni Sing Negi	8 Aug. 17	15 Sep. 18	
	Kishan Sing Bisht (*Retd.*)	....	....	Re-employed.
	Jemadars.			
13 Nov. 96	Man Sing Bisht	1 May 16		
	Diwan Sing Padhiyar, I.D.S.M.	20 Aug. 16		
16 Apl. 01	Mastu Sing Gussain	20 Aug. 16		
12 Nov. 00	Bahadur Sing Bartwal (56)	20 Aug. 16		
4 Nov. 04	Gaur Sing Rawat	1 Sep. 16		
1 Nov. 01	Mohan Sing Negi, I.D.S.M.	1 Sep. 16		
18 Apl. 01	Pancham Sing Negi	1 Oct. 16		
21 May 01	Nain Sing Bhandari	1 Oct. 16		
3 Nov. 03	Shankar Sing Rawat	1 Apl. 17		
26 Sep. 93	Prem Sing Kunwar	18 Sep. 17		
22 Mar. 02	Mastu Sing Rawat	3 Nov. 17		
3 May 02	Bhup Sing Chauhan	30 Nov. 17		
3 May 07	Makar Sing Chauhan	31 Dec. 17		
4 Nov. 04				

Regt. Centre—Sialkot.
Linked with—33rd and 46th Punjabis.

Uniform—Drab.
Facings—Emerald Green.

40th Pathans.

Raised at Shahjehanpur in 1858, by Lieut. E. Dandridge, as the Shahjehanpur Levy. Became the 44th Regt. of Bengal Native Infantry, 1861; the 40th Regt. of Bengal Nat. Infy., 1861; the 40th (Shahjehanpur) Regt. of Bengal N. I., 1864; the 40th Regt. of Bengal Infy., 1885; the 40th (Baluch) Regt. of Bengal Infy., 1890; the 40th (Pathan) Regt. of Bengal Infy., 1892; the 40th Punjab Infy., 1901; present designation, 1903.

Composition—1 Company of Orakzais, ½ of Afridis, ½ of Yusufzais, 1 of Punjabi Musalmans, 1 of Dogras.

COLONEL—Lieut.-General *Sir* Frederick Campbell, K.C.B., D.S.O. . 15 December 1911.

1st Commn. or date of entering service.	Names and Rank.	Army rank.	Present appointment in Regiment.	Remarks.
	Commandant.			
29 Mar. 90	Glasfurd, Lt.-Col. A. I. R., C.M.G., D.S.O., *p.s.c.*	29 Mar. 16	15 Nov. 17	
	Company Commanders. (4)			
16 Jan. 95	Tyndall, Maj. H. S., D.S.O.	16 Jan. 13	6 June 16	2nd-in-Comd.
17 Jan. 00	Cochran, Maj. A. C. P.	1 Sep. 15	5 June 16	
5 May 00	Waters, Maj. R. S.	1 Sep. 15	6 June 16	
30 Apl. 02	*Lee, Maj. M. G.*	*25 Feb. 16*	*19 Apl. 17*	*With R. A. F.*
	Company Officers. (9)			
5 Aug. 05	Segar, Capt. E.	5 Aug. 14	15 June 07	With 2-33rd Punj.
29 Nov. 05	*Fowle, Capt. T. C. W. [L]*	*29 Nov. 14*	*14 Sep. 07*	*Poll. employ.*
19 Jan. 07	*Campbell, Capt. F. C. G.*, M.C.	*1 Sep. 15*	*17 Mar. 08*	*Mohmand Militia.*
18 Sep. 09	Bonham-Carter, Capt. B. H. (*Actg. Maj.*)	1 Sep. 15	25 Aug. 13	With 1-21st Punj.
15 Nov. 15	Jepson, Lt. S.	15 Nov 15	19 Nov. 15	Qr.-Mr. (*offg.*)
29 June 16	Mold, Lt. G. L.	1 Mar. 16	. . . .	With 27th Punj.
14 Nov. 16	Davis, Lt., G. G.	14 Nov. 17	. . . .	
30 Jan. 17	Caunter, Lt. H. E.	30 Jan. 18	. . . .	
1 May 17	Percy-Smith, Lt. H. K.	1 May 18	30 Nov. 17	Attd. With 3-30th Punj.
1 Oct. 18	Mitchell, 2nd-Lt. S. A.	1 Oct. 18	6 Oct. 18	Attd.
	Attached.			
19 Oct. 01	Hill, Maj. H. A.	28 Feb. 16	26 Sep. 18	Midd'x. R.
22 Aug. 15	Shelswell, Capt. O. H. C.	22 Aug. 18	2 Sep. 18	I. A. R. O.
14 Sep. 15	Ellis, Lt. H. C.	14 Sep. 16	. . . .	I. A. (*on probn.*)
26 Nov. 15	Davenport, Lt. T.	26 Nov. 16	30 Apl. 18	I. A. (*on probn.*)
8 Feb. 16	Charles, Lt. F. D.	8 Feb. 17	30 Mar. 18	I. A. (*on probn.*) Adjt.
5 June 17	Wilson, Lt. W. L.	5 June 18	30 Mar. 18	I. A. (*on probn.*)
1 Dec. 17	Wilton, Lt. C. E. P.	1 Dec. 18	29 Mar. 18	I. A. (*on probn.*)
14 Jan. 15	Thornton, Lt. R. T., M.C.	14 Jan. 16	25 Mar. 15	I. A. R. O.
5 Mar. 15	Whigham-Teasdale, Lt. D. G., M.C.	5 Mar. 16	15 Nov. 15	I. A. R. O. *G. S. O., 16th Dn.*
14 Jan. 15	Johnson, Lt. E.	14 Jan. 16	25 Mar. 15	I. A. R. O.
28 May 15	Evans, Lt. E. A.	20 May 16	3 July 18	Oxf. & Bucks. L. I., I. A. (*on probn.*)
30 May 17	Twelftree, 2nd-Lt. A. E.	30 May 17	9 Feb. 18	I. A. R. O.
15 Jan. 18	Waite, 2nd-Lt. W. H.	15 Jan. 18	10 June 18	I. A. (*T. C.*)
15 June 18	Howfield, 2nd-Lt. J. A. J.	15 June 18	15 Sep. 18	I. A. R. O. (*on probn.*)
1 Nov. 17	Nicholson, Lt. R. L.	1 Nov. 18	15 Mar. 18	I. A. (*T. C*)
28 Nov. 17	Vaughan, 2nd-Lt. F. B.	28 Nov. 17	30 Mar. 18	I. A. R. O.
19 June 18	Rogers, 2nd-Lt. G. D.	19 June 18	3 July 18	I. A. R. O.
11 Oct. 18	Toussaint, 2nd-Lt. J. B.	11 Oct. 18	3 Nov. 18	I. A. R. O.
11 Oct. 18	Speldewinde, 2nd-Lt. C. G. O.	11 Oct. 18	3 Nov. 18	I. A. R. O.
25 July 16	Brown, Lt. J.	25 July 17	. . . .	R. Highrs., I. A. (*on probn.*)
24 Apl. 18	Bryant, 2nd-Lt. E. J. A.	24 Apl. 18	17 Oct. 18	I. A. (U. L.)

Subadar-Major.

		Jemadar.	Subadar.	Subadar-Major.	
1 Jan. 86	Malapa, *Sardar Bahadur* (63)	7 Apl. 01	1 July 01	1 Apl. 16	*Order of Br. Ind., 1st Cl. Depôt.*

Subadars.

4 Jan. 03	Gul Zaman, I.O.M. (56, 66*a*, 72*a*)	4 Jan. 03	1 June 13	
19 Apl. 04	Ata Muhammad (56)	11 Apl. 06	26 Feb. 15	Depôt.
18 Jan. 96	Ghulam Ali, M.C.	26 Feb. 15	27 Apl. 15	
19 Dec. 94	Abdullah Jai	27 Apl. 15	2 May 16	
3 Jan. 07	Sher Afzal (66*a*, 72*a*)	10 May 15	13 July 16	
17 Apl. 01	Ghulam Muhi-ud-din	26 Mar. 15	23 Jan. 18	
25 Aug. 01	Maluk Singh	13 July 16	23 Jan. 18	

40th Pathans—*contd.*

1st Commn. or date of entering service.	Names and Rank.	Army rank.	Remarks.
	Jemadars.		
6 May 01	Fazal Dad	26 Feb. 16	With 4-30th Punj.
10 Aug. 91	Multani Singh	11 Dec. 16	
18 Apl. 02	Surjan	1 Jan. 17	
17 Sep. 01	Nikku	13 Jan. 17	
30 Apl. 02	Ali Akbar (56)	26 Feb. 17	
3 July 06	Tor	26 Feb. 17	
20 Jan. 03	Sultan Khan (56)	11 June 17	
. . .	Mukhtiara, I.D.S.M.	19 Nov. 17	*With Branch Sch. of Musky. R. Pindi.*
27 Aug. 02	Gandaki	15 Sep. 17	
	Durani	6 June 18	

Regt. Centre—Rawal Pindi. *Uniform*—Scarlet.
Linked with—37th and 38th Dogras. *Facings*—Yellow.

1st Battalion, 41st Dogras.

Raised at Jullundur in 1900, by Major E. T. Gastrell as the 41st (Dogra) Regiment of Bengal Infantry Became the 41st Dogra Infantry in 1901; present designation in 1903.

Composition—4 Companies Dogras.

1st Commn. or date of entering service.	Names and Rank.	Army rank.	Present appointment in Regiment.	Remarks.
	Commandant.			
12 Sep. 88	Macpherson, Lt.-Col. R. G.	12 Sep. 14	31 Mar. 16	*Lv., ex I., m. c.*
	Company Commanders (4)			
15 Feb. 93	Quin, Lt.-Col. R. P. W. [L]	12 Sep. 18	20 Sep. 16	2nd-in-Comd., Comdt. *Offg.*
7 Nov. 91	*Hutchinson, Col. C. A. R.,* D.S.O., *p. s. c.* (Q).	*31 Aug. 17*	*20 May 02*	*Asst. Comdt., Cadet College, Wellington.*
20 Jan. 00	Bignell, Maj. R. L., D.S.O.	1 Sep. 15	22 Jan. 16	With 2-41st Dogras.
4 Dec. 01	*Fleming, Maj. W. E.,* M.C.	*4 Dec. 16*	*21 Jan. 16*	
	Company Officers. (9)			
19 Aug. 03	Hobday, Maj. E. M.	19 Aug. 18	2 Nov. 04	With 82nd Punj.
23 May 06	Nixon, Capt. B. G. M. F.	23 May 15	29 Oct. 08	
6 Feb. 09	*Mackenzie, Capt. J. D. (temp. Maj.)*	*1 Sep. 15*	*14 Aug. 14*	*D. A. Q. M. G.*
20 Apl. 10	*Mardall, Capt. F. H.*	*1 Sep. 15*	*26 Feb. 14*	*With R. A. F.*
14 Jan. 14	Fullerton, Capt. J. P.	14 Jan. 18	12 Oct. 14	*Coy. Comdr., N. C. O's. Sch., R'pindi*
15 Nov. 15	Devenish-Meares, Lt. B. J.	15 Nov. 16	20 Nov. 15	With 2-101st Grs.
18 Apl. 16	Townley, Lt. N. P.	18 Apl. 17	23 Apl. 16	With 37th Dogras.
29 June 16	Oatts, Lt. H. A.	29 June 17		
30 Jan. 17	Macdonald. Lt. N.	30 Jan. 18	. . . .	Attd.
18 June 17	Preston, Lt. G. A.	18 June 18	. . . .	Attd.
18 June 17	Platt, Lt. E. P.	18 June 18	. . . .	Attd.
18 June 17	Hendrick, Lt. C. A.	18 June 18	. . . .	Attd.
31 Jan. 18	Hourihane, 2nd-Lt. T. C.	31 Jan. 18	. . . .	Attd. With 1-131st Infy.
3 Oct. 15	Bodenham, Lt. A.	3 Oct. 16	. . . .	Attd. Qr. Mr. (*Offg.*)
1 May 17	Piper, Lt. W. J. F.	1 May 18	2 Dec. 17	Attd.
10 Nov. 15	Macintyre, Lt. Ian H.	10 Nov. 16	29 July 18	Attd.
8 Mar. 16	Heaver, Lt. L. G. H.	8 Mar. 17	29 July 18	Attd. With M. G. Corps.
27 Nov. 14	Vaughan, Capt. H. R.	6 Nov. 17	. . . .	Attd. With Burma Mily. Police.
31 Aug. 18	Palmer, 2nd-Lt. G. W.	31 Aug. 18	5 Sep. 18	Attd.
31 Aug. 18	Power, 2nd-Lt. F. W.	31 Aug. 18	5 Sep. 18	Attd.
	Attached.			
18 Oct. 99	Hutcheson, Maj. S.	1 Sep. 15	20 Oct. 18	1-3rd Brahmans.
26 Aug. 14	Clement, Lt. J. M.C. (*Actg. Capt.*)	26 Aug. 14	. . . .	I. A. R. O.
15 Sep. 15	Evanson, Lt. E. A. (*Actg. Capt.*)	15 Sep. 16	18 Feb. 18	I. A. (*on probn.*)
22 Oct. 15	Black, Lt. R. *Actg. Capt.*)	22 Oct. 16	3 Sep. 17	38th Dogras.
12 Nov. 15	Gain, Lt. J. C., M.C.	12 Nov. 16	19 Apl. 18	
26 Nov. 15	Rudd, Lt. C. H. C.	26 Nov. 16	14 Dec. 17	Lan. Fus., I. A. (*on probn.*)
11 Dec. 15	Kane, Lt. G. B. *ctg. Capt.*)	11 Dec. 16	. . . .	I. A. (*on probn.*) Adjt. (*Offg.*)
25 July 17	Hartley, Lt. C. R. B.	25 July 18	. . . .	I. A. (*on probn.*)
5 May 17	Doubt, 2nd-Lt. H. E.	5 May 17	. . . .	I. A. R. O. *Lv. ex I., 6 mos.*
26 Sep. 17	Burton, 2nd-Lt. C. G.	26 Sep. 17	26 Jan. 18	I. A. R. O.
1 Oct. 17	Partridge, 2nd-Lt. E.	1 Oct. 17	25 Jan. 18	I. A. R. O.
28 Oct. 17	Dudley, Lt. F.	28 Oct. 18	4 Feb. 18	I. A. R. O.
15 June 18	Bishop, 2nd-Lt. H. G. S.	15 June 18	10 Sep. 18	I. A. R. O. (*on probn.*)
15 June 18	Laird, 2nd-Lt. A. B.	15 June 18	10 Sep. 18	I. A. R. O. (*on probn.*)
28 Nov. 17	Crapp, Lt. W. V.	28 Nov. 18	4 Feb. 18	I. A. R. O.
24 Apl. 18	Holmes, 2nd-Lt. L. V.	24 Apl. 18	. . . .	I. A. (U. L.)
31 Oct. 17	Bapty, 2nd-Lt. P. A.	31 Oct. 17	. . . .	I. A. R. O.
6 Jan. 16	McMorran, Lt. J.	6 Jan. 17	. . . .	I. A. R. O.
	Honorary Officer.			
7 Aug. 18	Bilaspur, *His Highness Raja Sir* Bije Chand, K.C.I.E., C.S.I., *Raja of*	7 Aug. 18	7 Aug. 18	Hony. Capt. (*temp.*)

Subadar-Major.

		Jemadar.	Subadar.	Subadar-Major.	
4 Feb. 01	Mehar Singh *Bahadur* (56, 72a).	4 Feb. 01	1 Dec. 06	13 July 15	*Order of Br. Ind., 2nd Cl.*

1st Battalion, 41st Dogras—*contd.*

1st Commn. or date of entering service.	Names and Rank.	Army rank.	Present appointment in Regiment.	Remarks.
	Subadars.			
17 Sep. 00	Sundar Singh *Bahadur* (56, 68)	5 July 02	21 Sep. 07	*Order of Br. Ind., 2nd Cl.*
1 Feb. 01	Suba Singh, I.D.S.M. (56, 66*a*)	1 Feb. 11	1 July 14	
11 May 98	Chur Singh	26 July 14	4 May 15	
12 Nov. 00	Ganga Ram	10 Aug. 14	10 May 15	
1 Feb. 97	Dulo	10 Aug. 14	20 Aug. 15	
12 Nov. 00	Tara Chand, I.O.M.	20 Dec. 14	14 Jan. 16	
3 Apl. 97	Labh Singh (56)	11 Nov. 14	15 May 18	
12 Nov. 00	Lachhi Ram (46)	10 May 15	18 Sep. 18	
20 Feb. 01	Gobinda (56)	5 May 17	18 Sep. 18	
	Jemadars.			
9 Feb. 01	Nand Lal	10 May 15		
15 Feb. 01	Gilja	27 Mar. 17		
20 Feb. 01	Rao	13 Feb. 17	With 37th Dogras.	
12 Nov. 00	Pohlo	6 June 17		
19 Feb. 01	Chartu	6 June 17		
12 Nov. 00	Baichand	6 June 17		
24 Aug. 17	Kanwar Lachmi Singh	24 Aug. 17	(*On probn.*)	
12 Nov. 00	Jodha	6 June 17		
27 Feb. 01	Chanda Singh	1 Sep. 17		
13 Mar. 91	Jai Singh	24 Dec. 17		
13 July 17	Mian Shamsher Singh	13 July 17	(*On probn.*)	
9 Feb. 01	Kirpa Ram	18 Dec. 17		
16 Oct. 09	Jhanda Singh	26 Apl. 18		
21 Feb. 01	Sunka (56)	18 Sep. 18		

2nd Battalion, 41st Dogras.

Formed at Jubbulpore on 16th October 1917.

Composition—4 Companies Dogras.

1st Commn. or date of entering service.	Names and Rank.	Army rank.	Present appointment in Regiment.	Remarks.
	Commandant.			
20 Jan. 00	Bignell, Maj. R. L., D.S.O.	1 Sep. 15	24 Sep. 18	1-41st Dogras.
	Company Commanders. (4)			
24 Jan. 06	Conder, Capt. H. (*Actg. Maj.*)	24 Jan. 15	16 Oct. 17	112th Infy. 2nd-in-Comd. (*temp.*)
3 Sep. 10	Andrews, Capt. H. A., M.C.	1 Sep. 15	18 Sep. 18	37th Dogras.
1 Oct. 14	Reiche, Capt. A. D. E.	1 Oct. 18	14 Feb. 18	38th Dogras.
4 Aug. 14	Wilton, Lt. F. (*Actg. Capt.*)	4 Aug. 15	6 Jan. 18	Devon. R., I. A. R. O.
	Company Officers.			
7 Nov. 15	Webb, Lt. S. R. (*Actg. Capt.*)	May 17	2 June 18	I. A. R. O. Adjt. (*temp.*)
29 Apl. 16	White, Lt. S. E. C.	29 Apl. 17	29 Jan. 18	Hereford R., I. A. (*on probn.*)
29 July 16	Liswell, Lt. J. A.	29 July 17	18 Dec. 17	R. Fus., I. A. (*on probn.*)
5 Aug. 16	Moreton, 2nd-Lt. E. A.	5 May 17	14 Jan. 18	W. Rid. R., I. A. (*on probn.*)
29 Aug. 17	Williams, Lt. A.	29 Aug. 18	20 Apl. 18	I. A. R. O.
26 Sep. 17	Forrester, Lt. H. E.	26 Sep. 18	2 Jan. 18	I. A. R. O.
5 Oct. 17	Curley, Lt. W. L.	5 Oct. 18	2 Jan. 18	I. A. (T. C.)
5 Oct. 17	Ainsworth, Lt. R.	5 Oct. 18	19 Jan. 18	I. A. R. O.
28 Nov. 17	Birnie, Lt. A.	28 Nov. 18	29 Jan. 18	I. A. R. O.
28 Nov. 17	Linton, Lt. J. A.	28 Nov. 18	29 Jan. 18	I. A. R. O.
25 Dec. 17	Cunningham, 2nd-Lt. E. C. T.	25 Dec. 17	5 Feb. 18	I. A. R. O.
31 Jan. 18	Saunders, 2nd-Lt. W. E. P.	31 Jan. 18	31 Jan. 18	Attd.
24 Apl. 16	Jones, Lt. C. D.	24 Apl. 17	6 Sep. 18	A. C. Corps, I. A. (*on probn.*)
	Kellaway, 2nd-Lt. G. D.	. . . .	29 Aug. 18	Dorset R., I. A. R. O. (*on probn.*)
27 Sep. 16	Norris, Lt. W. J.	27 June 18	4 Sep. 18	R. W. Fus., I. A. (*on probn.*)
26 June 18	Chanin, 2nd-Lt. P. R.	26 June 18	30 Sep. 18	Genl. List, I. A. R. O. (*on probn.*)
26 June 18	Severn, 2nd-Lt. F.	26 June 18	30 Sep. 18	Genl. List, I. A. R. O. (*on probn.*)
26 June 18	Thornley, 2nd Lt. R.	26 June 18	30 Sep. 18	Genl. List, I. A. R. O. (*on probn.*)
4 Mar. 16	Riley-Lloyd, Lt. C.	4 Mar. 17	12 Jan. 18	Durh. L. I., I. A. (*on probn.*)
19 Dec. 17	Cornwell, 2nd-Lt. V. J. S.	19 Dec. 17	26 Mar. 18	I. A. R. O.
6 Jan. 16	McMorran, Lt. J.	6 Jan. 17	16 Mar. 18	I. A. R. O.
25 Aug. 15	Moor, Lt. J. A.	25 Aug. 16	13 Aug. 18	I. A. R. O.

Subadar-Major.

		Jemadar.	Subadar.	Subadar-Major.
4 Apl. 95	Dhani Ram (56, 72*a*, 72*b*)	26 Jan. 06	9 May 12	13 Jan. 16

Subadars.

11 Mar. 00	Kara (56, 66)	5 Jan. 15	12 Nov. 17
16 Jan. 96	Kanshi Ram (56, 66)	13 Feb. 15	12 Nov. 17
6 Apl 96	Charat Singh	11 Mar. 15	12 Nov. 17
19 Mar. 03	Dalip Singh (56)	10 May 15	12 Nov. 17
	Ram Das (56)	24 Dec. 15	12 Nov. 17
12 Apl. 01	Hira Singh (66*a*)	24 Dec. 15	12 Nov. 17
19 July 99	Kesar Singh	12 Nov. 17	12 Nov. 17
12 Feb. 01	Sarju (50, 61)	12 Nov. 17	26 Aug. 18

Jemadars.

11 Nov. 00	Khusi Ram (72*a*)	12 Nov. 17	
6 July 00	Makhan	12 Nov 17	
18 Nov. 95	Lehnu	12 Nov. 17	
2 Aug. 12	Rai Singh	12 Nov. 17	Indian Adjt.
19 Jan. 03	Gauri Saran	3 Dec. 17	
14 Apl. 02	Sundar I	3 Dec. 17	
16 Dec. 08	Tara	29 July 18	
28 June 04	Udham Singh	23 Aug. 18	
21 Feb. 01	Sundar II	23 Aug. 18	
2 Jan. 03	Lachhman	26 Aug. 18	
6 Dec. 00	Rohlu	5 Sep. 18	
	Dharam Singh	15 May 18	

Uniform—Dark Green.
Facings—Scarlet.

1st Battalion, 42nd Deoli Regiment.

Raised at Deoli in 1857 as "the Meena Battalion." Became the Infantry of the Deoli Irregular Force in 1860; present designation in 1903.

"Central India" "Afghanistan, 1879-80."

Composition—4 Companies of Rajputana Hindus and Musalmans.

COLONEL—Major-General *Sir* James A. Bell, K.C.V.O. . . . . 23 April 1912.

1st Commn. or date of entering service.	Names and Rank.	Army rank.	Present appointment in Regiment.	Remarks.
	Commandant.			
5 Feb. 87	Priestley, Lt.-Col. C. E. N. . .	5 Feb. 13	15 Jan. 15	
	Company Commanders. (4)			
10 Oct. 94	Hogg, Maj. A. G. M. (Q) (*Actg. Lt.-Col.*)	10 Oct. 12	10 Sep. 15	2nd-in-Comd. With 1-43rd Regt.
9 Dec. 96	George, Maj. D. S.	9 Dec. 14	20 Mar. 16	*Comdg. Deoli.*
20 Jan. 00	Lyon, Maj. C. W.	1 Sep. 17	11 Dec. 10	2nd-in-Comd. (*offg.*)
8 May 01	Connor, Maj. A. D.	8 May 16	16 May 16	
	Company Officers. (8)			
23 July 01	Burnett, Maj. R. F. D., M.C. .	23 July 16	3 Oct. 06	With 2-42nd Regt.
22 Oct. 02	*Tayler, Maj. F. L.*	*22 Oct. 17*	*16 July 04*	With 2-42nd Regt.
22 Apl. 03	Sutton, Maj. H. G.	22 Apl. 18	28 Feb. 06	With 3-154th Infy.
16 Dec. 03	*Patrick, Capt. R. M. F.*, M.C. [*L*]	*16 Dec. 12*	*3 May 05*	*Burma Mily. Police.*
9 Jan. 04	*Mallock, Capt. A. R. O.*	*9 Jan. 13*	*6 Nov. 09*	*Bde.-Maj., 58th Infy. Bde.*
19 Aug. 03	Greig, Capt. R. M. (*Actg. Maj.*)	13 Feb. 13	12 Nov. 04	With 2-6th Jats.
5 Aug. 05	Spankie, Capt. C. G.	5 Aug. 14	15 Mar. 07	
29 Aug. 06	*Capendale, Bt.-Maj. F. M.* . .	*1 Jan. 18*	*18 Feb. 09*	
20 Jan. 12	Guise, Capt. F. E. B. . .	20 Jan. 16	10 Mar. 13	*Lv., ex I., m. c.*
13 Jan 15	Smith, Lt. A. B. (*Actg. Capt.*)	13 Jan. 16	6 Mar 15	With 2-124th Infy.
12 May 15	Salvesen, Lt. H. K. (*Actg. Capt.*)	12 May 16	1 Nov. 15	Attd. Adjt (*offg.*)
19 July 16	Allen, Lt. C. A. (*Actg. Capt.*) . .	19 July 17	12 Nov. 16	Attd.
19 July 16	Abram, Lt. G. F.	19 July 17	12 Nov. 16	Attd.
	Attached.			
17 Mar. 15	Cline, Lt. A. H.	17 Mar. 16	. . . .	Midd'x. R., I. A. (*on probn.*)
11 May 15	Ingman, Lt. G. E.	11 May 16	11 Dec. 17	K. R. Rif. C., I. A. R. O. (*on probn.*) *Lv., ex. I., m. c.*
21 July 15	Swan, Lt. J. H.	21 July 16	2 May 18	I. A. (*on probn.*)
11 Aug. 15	Cole, Lt. R. St. G.	11 Aug. 16	. . . .	I. A. (*on probn.*)
10 May 16	Graham, Lt. P. H.	10 May 17	26 July 16	I. A. R. O.
14 Aug. 16	McDermott, Lt. F.	14 Aug. 17	2 May 18	I. A. (*on probn.*)
10 Nov. 16	Cornish, Lt. H.	10 Nov. 17	30 Jan. 17	1-43rd Regt.
14 Nov. 16	Dodds, Lt. H. C.	14 Nov. 17	27 Nov. 16	Do.
20 June 17	Anson, 2nd-Lt. C.	20 June 17	6 Oct. 17	I. A. R. O.
15 Jan. 18	Bowden, 2nd-Lt. C. G. . . .	15 Jan. 18	6 June 18	I. A. R. O.
15 Jan. 18	Howells, 2nd-Lt. T. R. R. . .	15 Jan. 18	6 June 18	I. A. (*T. C.*)
30 Oct. 18	Welch, 2nd-Lt. J. F.	30 Oct. 18	16 Aug. 18	Shrops. L. I., I. A. (*on probn.*)
29 May 18	Rogers, 2nd-Lt. J. W. . . .	29 May 18	30 Aug. 18	I. A. R. O. (*on probn.*)
24 Dec. 15	Laurie, Lt. J.	24 Dec. 16	12 Nov. 18	R. Sc. Fus., I. A. (*on probn.*)
21 Aug. 18	Darwin, 2nd-Lt., J. H. . . .	21 Aug. 18	19 Sep. 18	I. A. R. O.
24 Aug. 18	Saubelle, 2nd-Lt. C. E. . . .	24 Aug. 18	15 Sep. 18	I. A. R. O.
7 Oct. 18	Kniveton, 2nd-Lt. F.	7 Oct. 18	22 Oct. 18	I. A. R. O.
7 Oct. 18	Harden, 2nd-Lt. H. A. . . .	7 Oct. 18	28 Oct. 18	I. A. R. O.
	Honorary Officer.			
1 Jan. 03	Kotah, *His Highness Sir* Umed Singh, *Bahadur*, G.C.S.I., G.C.I.E., G.B.E., *Maharao* of—	1 Jan. 15	1 Jan. 03	Hony. Lt.-Col. in Army.

Subadar-Major.

		Jemadar.	Subadar.	Subadar-Major.	
28 Mar. 00	Hira Singh (45*a*, 52, 56, 71) . .	1 May 08	1 Nov. 13		
	Girdhari (*Hon. Capt.*) *Sardar Bahadur, Rai Bahadur, Retd.* Re-empld.	20 June 15	. . . .	. . .	*Order of Br. Ind. 1st Cl.*
	Subadars.				
14 July 94	Sukha (52, 56)	1 June 10	17 May 15		
5 Feb. 91	Jagnatha (56)	15 June 10	1 Feb. 17		
25 July 01	Ram-dayal Singh (56, 66*a*) . .	15 June 10	1 Feb. 17		
19 Apl. 01	Bhaguta (66*a*)	1 May 13	8 Feb. 17		
1 Sep. 06	Zorawar Singh (53, 507) . . .	17 May 15	1 Apl. 17		
19 Apl. 01	Kaliyana (56)	17 May 15	24 May 18		
12 Sep. 04	Kalu	18 Aug. 16	24 May 18		
15 July 01	Sonpal	24 May 16	16 July 18		
23 Apl. 01	Srikishna	5 May 17	25 July 18		
17 Aug. 01	Paip Singh	8 Feb. 17	20 Sep. 18		

1st Battalion, 42nd Deoli Regiment—*contd.*

1st Commn. or date of entering service.	Names and Rank.	Army rank.	Present appointment in Regiment.	Remarks.
	Jemadars.			
1 Feb. 98	*Sheodan Singh*	*15 July 16*		
1 Dec. 04	Gulab Singh	28 Sep. 16		
23 Jan. 97	Surjan Singh	12 Dec. 16		
21 Nov. 09	Sunwat Singh	8 Feb. 17		
22 Feb. 08	Gega	8 Feb 17		
1 Apl. 08	Rup Singh	1 Apl. 17		
1 Feb. 98	Jaggu Singh	10 July 17		
9 Mar. 02	Jagannath Singh	28 June 17		
3 Jan. 01	Ganga Ram (56)	29 July 17		
10 July 01	Pirthi Singh	3 Dec. 17		
17 Jan. 03	Chhagna	17 Jan. 18		
19 Apl. 01	Nanga	5 Feb. 18		
17 Sep. 01	Balwant Singh	22 Feb. 18		
14 Apl. 02	Bholu	24 May 18		
10 Nov. 05	Ladu	24 May 18		
28 Mar. 05	Maha Dewa	24 May 18		
1 Mar. 04	Dewa Singh	24 May 18		
3 Apl. 03	Kesra	16 June 18		
15 Mar. 97	Niaz Muhammad Khan	25 July 18		
23 May 04	Baldewa (I)	2 Sep. 18		
18 Sep. 06	Manohar Singh	20 Sep. 18		
1 June 10	Baldewa (II)	20 Sep. 18		

2nd Battalion, 42nd Deoli Regiment.

Formed at Deoli on 12th January 1917.

Composition—4 Companies of Rajputana Hindus and Musalmans.

1st Commn. or date of entering service.	Names and Rank.	Army rank.	Present appointment in Regiment.	Remarks.
	Commandant.			
4 Mar. 91	Brackenbury, Lt.-Col. W. J. W.	4 Mar. 17	21 Feb. 17	1-43rd Regt.
	Company Commanders (4).			
23 July 01	Burnett, Maj. R. F. D., M.C.	23 July 16	20 Mar. 18	1-42nd Regt..
22 Oct. 02	Tayler, Maj. F. L.	22 Oct. 17	23 Aug. 17	1-42nd Regt.
18 Jan. 05	Eden, Capt. J. (*Actg. Maj.*)	18 Jan. 14	23 Feb. 18	1-43rd Regt.
	Company Officers.			
28 Feb. 13	Simpson, Capt. V. E. W., M.C.	28 Feb. 17	31 Oct. 17	Attd. Adjt., 24 Dec. 17.
2 Nov. 14	Stewart, Lt. M. J.	2 Jan. 16	13 Dec. 17	I. A. (*on probn.*)
29 Dec. 14	Brownlie, Lt. J. K. C.	29 Sep. 16	6 Oct. 17	Attd.
22 May 15	Ganniclift, Lt. A. L.	22 Feb. 17	5 Aug. 17	Attd.
27 July 15	Meehan, Lt. C. F.	29 Apl. 17	1 Aug. 17	Attd.
24 Nov. 15	Torrance, Lt. I. S.	24 Nov. 16	6 Oct. 17	Attd.
24 Apl. 15	Brodie, Lt. R. P.	24 Apl. 16	2 Nov. 17	I. A. R. O. (*on probn.*) Qr. Mr., 22 Oct. 18.
31 Dec. 15	Seaton, Lt. O. C.	25 Sep. 17	1 Dec. 17	I. A. (*on probn.*)
23 Jan. 16	Morgan, Lt. W. M., M.C.	23 Jan. 17	24 May 18	I. A. (*on probn.*)
6 Oct. 16	Lloyd, Lt. G. T. (*Actg. Capt.*)	6 Oct. 17	12 May 17	I. A. R. O. *Staff Capt., Karachi Bde.*
11 Oct. 17	Jacks, 2nd-Lt. W. R.	11 Oct. 17	9 Dec. 17	I. A. R. O. (*on probn.*)
26 Aug. 17	Tancred, 2nd-Lt. N. G. W.	26 Aug. 17	13 Dec. 17	I. A. R. O.
5 Oct. 17	Ashton, Lt. C. T.	5 Oct. 18	22 Jan. 18	I. A. (*T. C.*)
29 Aug. 17	Bacon, 2nd-Lt. W. D.	29 Aug. 17	30 Jan. 18	I. A. R. O.
31 Oct. 17	Atkins, 2nd-Lt. C. R. (*Actg. Capt.*)	31 Oct. 17	14 Feb. 18	I. A. R. O.
31 Oct. 17	Willcock, 2nd-Lt. H. A.	31 Oct. 17	14 Feb. 18	I. A. R. O. (*on probn.*)
29 May 18	Craigie, 2nd-Lt. J. H.	29 May 18	29 Aug. 18	I. A. R. O. (*on probn.*)
31 Aug. 18	Florence, 2nd-Lt. G. D.	31 Aug. 18	14 Sep. 18	Attd.
18 June 17	Metcalfe, Lt. M. R.	18 June 18	26 June 17	Attd.
	Groom, 2nd-Lt. C. O.	. . . .	. . . .	Linc. R., I. A. R. O. (*on probn.*)

Subadar-Major.

		Jemadar.	Subadar.	Subadar-Major.
5 Dec. 94	Sheo Singh (56, 58, 52)	1 June 06	1 July 10	1 Feb. 17
	Subadars.			
29 Aug. 88	Nanda (52, 56, 58)	3 Apl. 06	1 May 13	
26 May 97	Goru Ram (52, 56, 66*a*, 72*a*)	1 June 06	1 Aug. 13	
15 Oct. 00	Jag Singh (45*a*, 52, 56, 60*a*, 71)	8 Jan. 13	26 Aug. 15	
7 Dec. 12	Bhur Singh (56, 66*a*)	28 July 15	1 Feb. 17	
18 Jan. 06	Ganga Ram (52, 56)	20 Sep. 15	1 Feb. 17	
17 Jan. 99	Birdha (66*a*, 72*a*)	20 Nov. 15	1 Feb. 17	
1 Apl. 08	Kesar Lal	18 Mar. 16	1 May 17	
1 May 02	Ranjit Singh (56)	13 July 16	11 May 18	
15 July 12	Bhollu (46, 56)	1 Feb. 17	11 May 18	
	Jemadars.			
7 Nov. 03	Mahadewa (I) (66*a*)	1 Feb. 17		
30 Jan. 04	Dhaunkal Singh (53)	1 Feb. 17		
14 Mar. 01	Ghasi	1 Feb. 17		
28 Apl. 04	Mahadewa (II) (56)	1 Feb. 17		
12 Dec. 01	Sukha	1 Feb. 17		
11 Apl. 02	Bajranga (5)	1 Feb. 17		
3 Apl. 03	Gopi	1 July 17		
13 Apl. 04	Sirilala	1 Aug. 17		
19 Feb. 02	Jagnatha	4 Jan. 18		
25 Oct. 04	Sedu	1 Feb. 18		
1 Apl. 08	Umrao Singh	1 Mar. 18		
19 Apl. 98	*Pertab Singh*	*22 Feb. 18*		
3 Nov. 04	Umda	11 May 18		
18 Apl. 03	Unkara	12 May 18		
1 June 06	Mahadewa (III)	12 May 18		

Uniform—Dark Green.
Facings—Scarlet.

1st Batttalion, 43rd Erinpura Regiment.

Raised in 1860 as the Infantry of the Erinpura Irregular Force. Present designation in 1903.
Composition—4 Companies of Rajputana Hindus and Musalmans.

1st Commn. or date of entering service.	Names and Rank.	Army rank.	Present appointment in Regiment.	Remarks.
	Commandant.			
4 Mar. 91	Brackenbury, Lt.-Col. W. J. W.	4 Mar. 17	1 Feb. 18	With 2-42nd Regt.
	Company Commanders. (4)			
11 Feb. 99	Trueman, Maj. H. J. H. (*Actg. Lt.-Col.*)	1 Sep. 15	1 Feb. 18	2nd-in-Comd. With 3-150th Infy.
30 Aug. 93	Dalrymple-Hay, Maj. S. F. B. (Q)	30 Aug. 11	7 Nov. 08	
26 May 00	Cobban, Maj. C. L. [L]	1 Sep. 15	10 Aug. 16	
8 May 01	*Hughes, Maj. G. R.*	*8 May 16*	*1 Feb. 18*	*I. O. Rajputana & C. I., I. S. Infy.*
	Company Officers. (8)			
8 May 01	Thompson, Maj. G. N.	8 May 16	29 Feb. 08	
2 Mar. 04	*Bovill, Capt. W. J.*	*2 Mar. 13*	*29 Jan. 11*	*G. S. O., 3rd gr.*
18 Jan. 05	Eden, Capt. J. L.	18 Jan. 14	8 Mar. 06	
20 Dec. 06	*Taylor, Capt. A. T. W. (Actg. Maj.)*	*1 Sep. 15*	*21 Feb. 14*	*With Sig. Ser.*
18 Jan. 11	Craig, Capt. H. T.	1 Sep. 15	27 Mar. 12	With 2-105th Mah.
13 Jan. 15	Palin, Lt. C. E. E.	13 Jan. 16	28 Feb. 15	
13 Mar. 15	*Seccombe, Lt. H. S.*	*13 Dec. 16*	*2 Nov. 17*	*With I. S. Troops.*
19 Sep. 16	McIntyre, Lt. J. M.	19 Sep. 17	21 Sep. 17	
14 Nov. 16	Williamson, Lt. G. A.	14 Nov. 17	20 Sep. 17	
14 Nov. 16	Dodds, Lt. H. C.	14 Nov. 17	....	Attd.—With 1-42nd R.
30 Jan. 17	Russell, Lt. A.	30 Jan. 18	21 Sep. 17	
9 Apl. 15	Denny, Lt. M. P. R.	9 Jan. 17	21 Jan. 17	Attd.
	Attached.			
10 Oct. 94	Hogg, Maj. A. G. M. (Q) (*Actg. Lt.-Col.*)	10 Oct 12	....	1-42nd Regt.
4 Dec. 14	McCollum, Lt. D. V.	4 Dec. 15	....	I. A. R. O.
18 Dec. 14	Dunning, Capt. A. G. A., M.C.	18 Dec. 18	....	I. A. R. O.
18 Feb. 15	Wilkinson, Lt. R.	18 Feb. 16	3 Aug. 15	I. A. R. O.
18 Feb. 15	Gordon, Lt. J. W. S.	18 Feb. 16	21 Apl. 15	I. A. R. O.
2 Apl. 15	Crisp, Lt. R. H. (*Actg. Capt.*)	2 Apl. 16	16 May 15	I. A. R. O.
14 June 16	Barrow, Lt. G. E.	14 June 17	....	I. A. R. O.
15 Nov. 15	Newton-King, Lt. R. M.	15 Nov. 16		
5 Jan. 18	Watts, 2nd-Lt. H. E. (*Actg. Capt.*)	5 Jan. 18	....	I. A. (*T. C.*) Adjt. 15 Oct. 18.
5 June 18	Brohier, 2nd-Lt. C. P.	5 June 18	28 June 18	I. A. R. O.
29 May 18	Smallman, 2nd-Lt. W. F.	29 May 18	28 Aug. 18	I. A. R. O. (*on probn.*)
15 Sep. 15	Barrow, Lt. R.	15 Sep. 16	....	86th Infy.
1 July 17	Dade, Lt. R. T. (*Actg. Capt.*)	1 July 18	....	R. Dub. Fus., I. A. (*on probn.*)
11 Apl. 17	Borwick, Lt. R. H.	11 Apl. 18	....	Bord. R., I. A. (*on probn.*)
5 June 17	Wilton, Lt. V. R.	5 June 18	11 Sep. 18	W. York R., I. A. (*on probn.*)
21 July 16	Barfield, Lt. F. H.	21 July 17	22 Oct. 18	A. S. Corps, I. A. (*on probn.*)
29 Aug. 18	Smith, 2nd-Lt. W. H.	29 Aug. 18	15 Sep. 18	I. A. R. O.
30 Aug. 18	Blewitt, 2nd-Lt. A. E.	30 Aug. 18	21 Sep. 18	I. A. R. O.
7 Oct. 18	Down, 2nd-Lt. E. A.	7 Oct. 18	27 Oct. 18	I. A. R. O.

Subadar-Major.

		Jemadar.	Subadar.	Subadar-Major.
15 Jan. 02	Goma (52, 56, 66*a*)	1 June 09	1 Nov. 11	1 Aug. 17

Subadars.

16 Nov. 95	Wazira (52, 56, 66*a*)	13 Dec. 10	1 Mar. 13	
27 Oct. 00	Jalal Singh (66*a*, 72*a*)	18 Aug. 08	1 Feb. 15	
14 Mar. 02	Raghunath Singh (52, 56, 66*a*)	20 June 11	1 Mar. 15	
11 Mar. 01	Udham Singh (56)	1 Nov. 11	27 Sep. 15	
4 Jan. 03	Bagh Singh	1 Feb. 15	22 Oct. 15	
20 Nov. 99	Alladad Khan	13 Oct. 15	1 Aug. 17	
9 Oct. 07	Sadhu Singh	18 May 16	3 Aug. 17.	
...	Goda (*Retd.*)	....	....	Re-employed, 1 Oct. 15.
...	Muhammad Yusuf (*Retd.*)	....	....	Re-employed, 13 Oct. 15.

1st Battalion, 43rd Erinpura Regiment—*contd.*

1st Commn or date of entering service.	Names and Rank.	Army rank.	Present appointment in Regiment.	Remarks.
	Jemadars.			
7 Feb. 93	Mota	1 Jan. 12		
1 Dec. 96	Harnam Singh (45*a*, 56, 71)	1 Mar. 13		
28 May 98	Panna Singh	1 Mar. 15		
18 Aug. 97	Umar Khan	28 July 15		
30 Sep. 00	Nanne Khan	28 July 15		
7 Dec. 03	Bhur Singh	28 July 15		
27 Dec. 04	Gulaba	27 Sep. 15		
9 Dec. 00	Nishan Ali	13 Oct. 15		
10 Apl. 08	Shamrath Singh	1 Aug. 17		
13 Sep. 01	Karim Khan (51)	1 Aug. 17		
29 Apl. 00	Gopal Singh	3 Aug. 17		
9 Jan. 04	Salab Singh	18 May 18		
15 Dec. 03	Sanwat Singh	18 May 18		
4 Mar. 98	Banna	20 Aug. 18		
14 May 04	Kajju Khan	20 Aug. 18		
19 Nov. 09	Prem Singh	20 Aug. 18		

2nd Battalion, 43rd Erinpura Regiment.

Formed at Ahmedabad on 23rd June 1917.

Composition—2 Companies, Rajputana Hindus and Musalmans. 2 Companies Gaur Brahmans.

1st Commn. or date of entering service.	Names and rank.	Army rank.	Present appointment in Regiment.	Remarks.
	Commandant.			
10 Mar. 83	Loudon, Lt.-Col. J. A.	10 Mar. 09	1 July 17	
	Company Commanders (4).			
12 Aug. 02	Ross, Maj. E. F.	12 Aug. 17	6 July 17	88th Infy. 2nd-in-Comd. (*temp.*)
5 May 15	Robertson, Lt. E. (*Actg. Capt.*)	5 May 16	26 Mar. 18	K. O. Sco. Bord., I. A. (*on probn.*)
2 June 15	McClenaghan, Lt. H. C. (*Actg. Capt.*)	2 June 16	24 July 17	Attd.
9 Oct. 15	Laird, Lt. W. R. (*Actg. Capt.*)	9 Oct. 16	26 Feb. 18	Attd.
	Company Officers.			
14 Feb. 12	Mills, Capt. S. D., M.C.	14 Feb. 16	17 Oct. 17	Bedf. R., I. A. (*on probn.*)
23 Aug. 15	Addinsell, Lt. T. A.	23 Aug. 16	28 Nov. 18	York. R., I. A. (*on probn.*)
19 Mar. 15	Evans, Lt. W. A. A.	19 Mar. 16	9 July 17	I. A. R. O. (*on probn.*) Adjt.
8 June 15	Randall, Lt. L. E.	8 June 16	22 Oct. 17	Attd.
2 Apl. 17	Breadon, Lt. G. W. D.	2 Apl. 18	6 Dec. 18	I. A. R. O.
15 Apl. 16	Hey, Lt. B. S.	15 Apl. 17	31 Dec. 17	I. A. (*on probn.*)
26 Sept. 17	Barker, Lt. G. M.	26 Sep. 18	25 Jan. 18	I. A. R. O.
1 Oct. 17	Jones, 2nd-Lt. T. P.	1 Oct. 17	19 Jan. 18	I. A. (*on probn.*)
5 Oct. 17	Bell, Lt. W. F.	5 Oct. 18	19 Jan. 18	I. A. (*on probn.*)
5 Oct. 17	Woods, Lt. H. A.	5 Oct. 18	19 Jan. 18	I. A. (*on probn.*)
5 Nov. 17	Wyatt, Lt. C. H.	5 Nov. 18	15 Mar. 18	I. A. (*T. C.*)
5 Nov. 17	Musto, Lt. C. R.	5 Nov. 18	15 Mar. 18	I. A. (T. C.)
1 Dec. 17	Holmes, Lt. J. A.	1 Dec. 18	30 Sep. 18	Durh. L. I., I. A. (*on probn.*)
31 Oct. 17	Harris, 2nd-Lt. E. J.	31 Oct. 17	11 Feb. 18	I. A. R. O. (*on probn.*)
26 Jan. 18	Snook, 2nd-Lt. C. G.	26 Jan. 18	26 Nov. 18	Lan. Fus., I. A. (*on probn.*)
28 Feb. 18	Bruce, 2nd-Lt. W.	28 Feb. 18	25 Apl. 18	I. A. R. O. (*on probn.*)
26 Mar. 18	Allequen, 2nd-Lt. E. J.	26 Mar. 18	30 Sep. 18	Bedf. R., I. A. R. O. (*on probn.*)
29 Mar. 18	Oke, 2nd-Lt. H.	29 Mar. 18	17 July 18	I. A. (*T. C.*)
29 May 18	Ellis, 2nd-Lt. A. W.	29 May 18	1 Oct. 18	R. Fus., I. A. R. O. (*on probn.*)
22 July 18	Reyner, 2nd-Lt. A. G.	22 July 18	3 Dec. 18	I. A. (*T. C.*)
12 Aug. 18	Wright, 2nd-Lt. F. N.	12 Aug. 18	5 Sep. 18	I. A. R. O.
14 Aug. 18	Douglas, 2nd-Lt. V. E.	14 Aug. 18	5 Sep. 18	I. A. R. O.
14 Aug. 18	Pugh, 2nd-Lt. H. V.	14 Aug. 18	5 Sep. 18	I. A. R. O.

Subadar-Major.

		Jemadar.	Subadar.	Subadar-Major.
22 June 95	Abdul Wahab Khan (58)	1 Jan. 09	1 Feb. 12	1 Feb. 15
	Subadars.			
4 June 15	Sobha Ram	4 June 15	28 Oct. 15	
26 Feb. 98	Mendu	15 Feb. 15	23 June 17	
8 Jan. 03	Ram Singh	1 Mar. 15	23 June 17	
	Jemadars.			
5 May 97	Teja	27 Jan. 15		
26 Mar. 01	Bhura	23 June 17		
7 Mar. 10	Rupa (46,66)	23 June 17		
9 Apl. 02	Choga	23 June 17		
31 Mar. 97	Kana	23 June 17		
5 Aug. 01	Duja	23 June 17		
3 June 15	Chandulal	1 Oct. 17		
15 June 04	Kehar Singh	13 Dec. 17		
20 Aug. 15	Shimbu Lal	7 Feb. 18		
9 Oct. 16	Badri Parshad Sharma	7 Feb. 18		
31 July 15	Ude Ram	1 May 18		
21 Feb. 16	Parbhu Dyal	1 May 18		
1 May 18	Kantiprasad B. Pathak	1 May 18	(*on probn.*)	

Uniform—Scarlet.
Facings—Gosling Green.

44th Merwara Infantry.

Raised at Beawar in 1822, by Capt. H. Hall, as the Mhairwara Local Battn. Became the 14th (or Mhairwara Local Battalion in 1823; the 9th (or Mhairwara) Local Battalion in 1826; the Mhairwara Battalion in 1843; the Ajmer and Mhairwara Police Corps in 1861; the Mhairwara Battalion in 1871; present designation in 1903.

"Central India" "Afghanistan, 1878-79."

Composition—2 Companies of Mers, Musalman Merats, 2 Companies of other classes of Rajputana.

1st Commn. or date of entering service.	Names and Rank.	Army rank.	Present appointment in Regiment.	Remarks.
	Commandant.			
23 Mar. 89	Alexander, Lt.-Col. H. S., D.S.O.	23 Mar. 15	16 May 16	
	Company Commanders. (4)			
4 Feb. 91	Butler, Lt.-Col. A. H. (Q.) . . .	4 Feb. 17	2 Feb. 15	2nd-in-Comd.
16 Jan. 95	*Hoskyn, Maj. J. C. M.*, D.S.O. *[L] (temp. Lt.-Col.)*	*16 Jan. 13*	*24 Aug. 11*	*A. A. & Q. M. G.*
9 Dec. 99	Browne, Maj. W. S. W. . .	1 Sep. 15	8 Feb. 16	
8 May 01	*Radcliffe Smith, Maj. N. M. R.* M.C.	*8 May 16*	*8 Feb. 16*	
	Company Officers. (8)			
18 Jan. 05	*Deane, Capt. C. H. S.*	*18 Jan. 14*	*20 Mar. 06*	
5 Aug. 05	*Harbord, Capt. L. B.*, M.C. *(temp. Maj.)*	*5 Aug. 14*	*8 Mar. 06*	*G. S. O., 2nd gr., A. Hdqrs.*
29 Nov. 05	Cole, Capt. H. A. . . .	29 Nov. 14	31 Aug. 08	With 4-9th Infy.
29 Aug. 06	*Waller, Capt. A. G. (temp. Maj.)*	*29 Aug. 15*	*25 July 09*	
3 Sep. 10	*Hebbert, Capt. P. B.* . . .	*1 Sep. 15*	*2 Dec. 11*	*Assam Mily. Police.*
14 Nov. 16	Walter, Lt. E. F.	14 Nov. 17	17 Oct. 17	
14 Nov. 16	Hemnon, Lt. C. R. *(Actg. Capt.)*	14 Nov. 17	21 Aug. 17	Adjt.
30 Jan. 17	Curtin, Lt. A. F. S.	30 Jan. 18	1 Sep. 17	
20 Feb. 16	Hawksworth, Lt. P. *(Actg. Capt.)*	20 Nov. 17	6 June 16	Attd.
	Attached.			
24 Jan. 06	Fenton, Capt. J. N. *(Actg. Maj.)*	24 Jan. 15	7 July 18	17 Infy.
16 Aug. 14	Steuart, Capt. F. G., M.C. . .	16 Aug. 18	26 June 15	I. A. R. O.
18 Dec. 14	Thomson, Lt. R. S.	18 Dec. 15	1 Jan. 15	I. A. R. O.
14 Jan. 16	Mumford, Lt. S. E. A. . . .	14 Jan. 17	24 Feb. 16	I. A. R. O.
9 July 15	Thomas, Lt. M. P. *(Actg. Capt.)*	9 July 16	26 Mar. 17	I. A. R. O.
15 Feb. 15	Morbey, Lt. E. H. *(Actg. Capt.)*	15 Feb. 16	16 Jan. 17	I. A. R. O. *A. R. O., Raj. & C. I. Musalmans.*
18 Apl. 17	Souter, Lt. W. D.	18 Apl. 18	2 Oct. 18	Cam'n Highrs., I. A. *(on probn.)*
20 Feb. 16	Stevenson, Lt. J. F. B., M.C. .	20 Feb. 17	27 Nov. 17	I. A. *(on probn.)*
5 May 17	Walker, Lt. C. A. P.	5 May 18	15 May 18	I. A. *(on probn.)*
19 Sep. 17	McCarthy, Lt. J. W.	19 Sep. 18	29 Apl. 18	I. A. *(on probn.)*
5 Jan. 18	Wearing, 2nd-Lt. W.	5 Jan. 18	5 Jan. 18	I. A. *(T. C.)*
5 Nov. 17	Mannes-Abbott, Lt. W. S. E.	5 Nov. 18	17 July 18	I. A. R. O.
19 Mar. 18	Obourn, 2nd-Lt. N. A. . . .	19 Mar. 18	17 July 18	I. A. *(T. C.) S. S. O., Ajmere.*
26 Sep. 17	Belfield, 2nd-Lt. G. H. . . .	26 Sep. 17	22 Jan. 18	I. A. R. O.
8 Aug. 18	Grant, 2nd-Lt. P. W. . . .	8 Aug. 18	8 Aug. 18	I. A. R. O.
24 Apl. 18	Benson, 2nd-Lt. C. R. . . .	24 Apl. 18	19 Oct. 18	I. A. (U. L.)
30 Aug. 18	Patton, 2nd-Lt. S. L. S. . . .	30 Aug. 18	25 Sep. 18	I. A. R. O.
27 Oct. 18	Pentland, 2nd-Lt. A. . .	27 Oct. 18	27 Oct. 18	I. A. *(T. C.)*
27 Oct. 18	Brown, 2nd-Lt. A.	27 Oct. 18	1 Nov. 18	I. A. *(T. C.)*

Subadar-Major.

		Jemadar.	Subadar.	Subadar-Major.
14 Nov. 96	Bhura (52, 56, 66*a*)	1 July 07	1 Jan. 11	29 Oct. 17
	Subadars.			
26 Feb. 02	Hamira (56)	7 June 14	1 July 16	
22 July 03	Lala (56)	23 Jan. 15	23 Feb. 17	
17 Dec. 00	Moti	23 Jan. 15	7 Feb. 17	
15 Nov. 07	Padma	17 Aug. 15	24 May 18	
15 Feb. 04	Dewa	28 Jan. 16	24 May 18	
10 July 00	Bhura	18 Feb. 16	24 May 18	
26 Jan. 01	Kundan Lal	7 Feb. 17	24 May 18	
	Dhira *(Retd.)*	. . . .	. . . .	Re-employed, 28 June 15.

44th Merwara Infantry—*contd.*

1st Commn. or date of entering service.	Names and Rank.	Army rank.	Remarks.
	Jemadars.		
27 Mar. 05	Jora	1 June 16	
21 Jan. 05	Wazira	1 July 16	
2 Apl. 02	Uda (56)	26 Aug. 16	
28 Jan. 02	Anna (I)	7 Feb. 17	
20 Mar. 03	Anna (II)	7 Feb. 17	
6 Aug. 00	Bhaira	22 May 17	
8 May 01	Lala	1 July 17	
4 July 07	Hira	1 Oct. 17	
12 Jan. 07	Persa	24 May. 18	
5 Nov. 04	Dhanna	24 May 18	
10 Oct. 07	Anna	1 Aug. 18	
5 Nov. 04	Himta	1 Aug. 18	

Regt. Centre—Multan.
Linked with—14th and 15th Sikhs.

Uniform—Scarlet.
Facings—White.

45th Rattray's Sikhs.

Raised at Lahore in 1856 by Capt. T. Rattray, as the Bengal Military Police Battalion. Became the 1st Bengal Mily. Police Battn. in 1858; the 45th (Rattray's Sikh) Regt. of Bengal Native Infantry in 1864; the 45th (Rattray's Sikh) Regt. of Bengal Infy. in 1885; the 45th (Rattray's) Sikh Infy. in 1901; present designation in 1903.

"Defence of Arrah" "Behar" "Ali Masjid" "Afghanistan, 1878-80."
"Malakand" "Punjab Frontier."
Composition—4 Companies Sikhs.

1st Commn. or date of entering service.	Names and Rank.	Army rank.	Present appointment in Regiment.	Remarks.
	Commandant.			
7 Nov. 91	McVean, Lt.-Col. D. A. D., D.S.O.	29 July 17	29 July 17	
	Company Commanders (4)			
31 Jan. 94	Keen, Maj. F. S., D.S.O., *p.s.c.* (Q)	31 Jan. 12	29 July 17	2nd-in-Comd.
5 Aug. 96	*Van Someren, Maj. W. W., D. S. O.* [*L*]	*5 Aug. 14*	*5 Apl. 12*	
27 July 98	Cadell, Maj. J. G., D.S.O., p.s.c.	1 Sep. 15	2 Feb. 17	
4 Jan. 99	McRae, Maj. H. St. G. M., D.S.O., p.s.c.	1 Sep. 15	29 July 17	With 2-15th Sikhs.
	Company Officers. (9)			
6 Dec. 99	Shuttleworth, Maj. B. W., *p.s.c.*	1 Sep. 15	5 Oct. 03	C. C. *s.p.t.*, 29 July 17.
8 May 01	Anderson, Maj. R. H.	8 May 16	2 Sep. 03	
18 Jan. 05	Waller, Capt. J. E.	18 Jan. 14	1 Feb. 07	
24 Jan. 06	*Cates, Capt. K. G. H. (Actg. Lt.-Col.)*	*24 Jan. 15*	*5 Mar. 09*	*I. S. Troops.*
17 Aug. 07	*Macausland, Capt. R. A.*	*1 Sep. 15*	*9 Aug. 12*	*Staff Capt., Abbottabad Bde.*
8 Aug. 14	Curtis, Capt. A. C., M.C.	8 Aug. 18	13 Nov. 15	
8 Aug. 14	Guise, Capt. J. W.	8 Aug. 18	15 Oct. 14	With 2-55th Rif.
1 Oct. 14	Key, Capt. B. W.	1 Oct. 18	18 Dec. 14	
20 Mar. 15	Preston, Lt. K. H. (*Actg. Capt.*)	20 Dec. 16	13 Sep. 16	Attd. With 1-151st Infy.
3 Jan. 16	Finlay, Lt. J. A.	3 Jan. 17	13 Mar. 18	Attd. With I. S. Troops.
18 June 17	Fox, Lt. R. de V. R.	18 June 18	24 Oct. 17	Attd.
15 Aug. 14	Minchin, Capt. R. H. L.	15 Aug. 18	....	With 1-69th Punj.
1 Sep. 18	Clapp, 2nd-Lt. C. D.	1 Sep. 18	3 Sep. 18	Attd.
1 Sep. 18	Will, 2nd-Lt. H. St. C. O.	1 Sep. 18	3 Sep. 18	Attd.
	Attached.			
29 Jan. 16	Watson, Lt. G. L.	29 Jan. 17	....	I. A. R. O.
18 Apl. 16	Pybus, 2nd-Lt. G. D.	18 Apl. 16	....	14th Sikhs.
15 Nov. 15	Henson, Lt. R. K.	15 Nov. 16	....	25th Punj.
17 Aug. 16	Ramsbotham, Lt. R. B.	17 Aug. 17	17 Oct. 17	I. A. R. O.
8 Aug. 17	Sime, Lt. E. A.	8 Aug. 18	2 Nov. 17	I. A. R. O.
5 Oct. 17	Benton, Lt. G. W.	5 Oct. 18	24 Jan. 18	I. A. R. O.
22 Aug. 17	Worden, Lt. G. M.	22 Aug. 18	15 May 18	I. A. (*on probn.*)
25 Oct. 15	Butcher, Lt. A. L. (*Actg. Capt.*)	25 Oct. 16	21 May 18	I. A. (*on probn.*) Adjt.
5 Nov. 17	Kelly, Lt. R. W.	5 Nov. 17	....	I. A. R. O.
21 Sept. 14	Fyldes, Lt. G. B.	22 June 16	3 Aug. 18	Notts. & Derby. R., I. A. (*on probn.*).
3 Jan. 16	Buckley, Lt. F.	3 Jan. 17	10 July 18	E. Lan. R., I. A. (*on probn.*).
1 June 15	Hewett, Lt. H. S.	1 June 16	31 Aug. 18	I. A. R. O. (*on probn.*).
28 Sep. 15	Bowtell-Harris, Lt. J. F., M.C.	28 June 17	1 Aug. 18	Aus. Imp. Forces, I. A. (*on probn.*).
24 Apl. 18	Angus, 2nd-Lt. T. H.	24 Apl. 18	23 Oct. 18	I. A. (U. L.)
7 Oct. 18	Vanston-Rumney, 2nd-Lt. W. F.	7 Oct. 18	25 Oct. 18	I. A. (*T. C.*)
7 Oct. 18	Jeffery, 2nd-Lt. F. B.	7 Oct. 18	2 Nov. 18	I. A. (*T. C.*)
7 Oct. 18	Tumber, 2nd-Lt. W. A.	7 Oct. 18	2 Nov. 18	I. A. (*T. C.*)
8 Oct. 18	Grant, 2nd-Lt. N. M	8 Oct. 18	26 Oct. 18	I. A. R. O.

Subadar-Major.

		Jemadar.	Subadar.	Subadar-Major.	
6 Jan. 90	Sundar Singh *Bahadur*, I.O.M. (56)	1 Sep. 98	25 July 03	1 July 15	*Order of Br. Ind., 2nd cl.*
....	Bhagat Singh (Retd)., *Sardar Bahadur*	....	....	....	Re-employed, 1 Feb. 18.

Subadars.

10 Oct. 90	Nidhán Singh (56)	25 July 03	28 Oct. 12	
31 Jan. 03	Wariam Singh (56)	1 Feb. 05	1 Nov. 12	With 1—151st Infy.
7 Mar. 93	Narain Singh, I.O.M. (56, 66*a*)	1 Feb. 10	1 Aug. 14	
5 Apl. 91	Kehar Singh (56, 60, 66*a*)	5 June 13	10 June 16	
17 Dec. 95	Labh Singh	17 Apl. 15	2 Feb. 17	
3 Mar. 95	Lal Singh	16 Sep. 15	2 Feb. 17	
25 Mar. 97	Partab Singh	2 Feb. 17	15 June 18	
27 Oct. 99	Kesar Singh	2 Feb. 17	8 July 18	
7 July 91	Abdul Hamid	1 Jan. 17	26 Apl. 18	
	Mall Singh *Bahadur* (Retd.)	....	....	*Order of Br. Ind., 2nd cl.* Re-employed, 14 May 15. Depôt.
	Sundar Singh (Retd.)	....	....	Re-employed, 14 May 15. Depôt.

45th Rattray's Sikhs—*contd.*

1st Commn. or date of entering service.	Names and Rank.	Army rank.	Remarks.
	Jemadars.		
17 Mar. 97	Buta Singh	16 May 15	
25 Dec. 98	Bishan Singh	2 Feb. 17	Indian Adjt., 1 May 17.
2 Aug. 01	Attar Singh	2 Feb. 17	
19 Jan. 01	Harnam Singh (I)	2 Feb. 17	
1 Jan. 99	Harnam Singh (II)	2 Feb. 17	
23 Aug. 99	Wadhawa Singh	2 Feb. 17	
22 Aug. 00	Abdul Aziz	1 June 17	
5 Jan. 00	Sewa Singh	14 Mar. 18	
6 June 98	Bishan Singh	15 June 18	
15 Feb. 00	Sundar Singh	15 June 18	
8 June 92	Kishan Singh	21 May 18	
Feb. 01	Sant Singh	8 July 18	
18 Feb. 18	Bahawal Sher	10 Sep. 18	

Regt. Centre—Sialkot.
Linked with—33rd Punjabis and 40th Pathans.

Uniform—Drab.
Facings—Emerald Green.

46th Punjabis.

Raised at Sialkot in 1900, by Major G. P. Ranken, as the 46th (Punjab) Regiment of Bengal Infantry. Became the 46th Punjab Infantry in 1901; present designation in 1903.

Composition—2 companies of Punjabi Musalmans, ½ of Afridis, ½ of Orakzais, 1 of Labana Sikhs.

1st Commn. or date of entering service.	Names and Rank.	Army rank.	Present appointment in Regiment.	Remarks
	Commandant.			
20 Feb. 92	Walton, Lt.-Col. G.	3 June 16	20 Oct. 17	
	Company Commanders. (4)			
7 Mar. 94	Graham, Bt.-Lt.-Col. R. B. (Q.)	3 June 18	15 Nov. 17	2nd-in-Comd.
7 May 98	Robinson, Maj. D.G., D.S.O., *p.s.c.*	1 Sep. 15	13 Apl. 15	*G.S.O., 2nd gr., A. Hdqrs.*
18 Jan. 02	Beadle, Maj. R. D.	18 Jan. 17	13 Apl. 15	With 2-151st Infy.
18 Jan. 02	Hardcastle, Maj. J.	18 Jan. 17	20 Oct. 17	
	Company Officers. (9)			
9 Jan. 04	Ring, Capt. J. S. H.	9 Jan. 13	5 Apl. 05	
5 Aug. 05	Poole, Capt. E. J. E.	5 Aug. 14	21 Oct. 06	
7 Nov. 06	Solano, Capt. B. A.	1 Sep. 15	25 Mar. 09	With 2-18th Infy.
17 Aug. 07	Paget, Capt. H. E. G.	1 Sep. 15	27 Dec. 09	Adjt., 13 Dec. 17.
9 Sep. 08	*Gatherer, Capt. A. H. N.*, M.C.	*1 Sep. 15*	*13 Dec. 09*	*With Sig. Ser.*
20 Jan. 09	Munn, Capt. F. L.R., M.C.	1 Sep. 15	14 Mar. 10	
20 Jan. 12	Clarke, Capt. F. S.	24 Sep. 16	10 Mar. 13	
18 June 17	Jackman, Lt. H. R.	18 June 18	25 June 17	Attd. Qr.-Mr. (*Offg.*)
4 Aug. 15	Peacock, Lt. L. V.	4 Aug. 16	30 July 18	Attd. With 2-18th Infy.
26 Feb. 16	Hoskins, Lt. C. J.	26 Feb. 17	4 Sep. 17	Attd.
	Attached.			
8 May 01	*Batten, Maj. H. H.*	*8 May 16*	. . .	*Civil, Divl. R. O. Rawalpindi.*
5 May 17	Robinson, Lt. D. B.	5 May 18	19 Aug. 17	I. A. R. O.
5 May 17	Amor, Lt. L. E.	5 May 18	19 Aug. 17	I. A. R. O.
5 May 17	Bramwell, Lt. W. H.	5 May 18	19 Aug. 17	I. A. R. O.
22 Apl. 16	Elliott, Lt. C. D.	22 Apl. 17	27 Aug. 17	I. A. (*on probn.*)
7 Sep. 16	Dunn, Lt. C. L.	7 Sep. 16	30 Nov. 17	I. A. R. O.
26 Dec. 15	Lavender, Lt. S. S.	26 Dec. 16	20 Jan. 18	M. G. Corps, I. A. (*on probn.*).
29 Aug. 17	Blyth, Lt. J. C.	29 Aug. 18	15 Dec. 17	I. A. R. O.
26 Sep. 17	Rowlinson, 2nd-Lt. P. J.	26 Sep. 17	29 Jan. 18	I. A. R. O. (*on probn.*).
26 Sep. 17	Shaw, 2nd-Lt. W.	26 Sep. 17	29 Jan. 18	I. A. R. O., (*on probn.*).
13 Oct. 16	Thomas, Lt. A. F. F.	13 Oct. 17	30 Nov. 17	I. A. (*on probn.*)
18 June 17	Yiend, 2nd-Lt. E. K.	18 June 17	25 June 17	1-33rd Punj.
1 Mar. 17	Scott, 2nd-Lt. K. D.	1 Mar. 17	17 May 18	I. A. R. O. (*on probn.*).
26 Nov. 17	Putnam, 2nd-Lt. T. C.	26 Nov. 17	6 June 18	I. A. R. O. (*on probn.*).
29 Mar. 18	Pope, 2nd-Lt. H. W.	29 Mar. 18	18 July 18	I. A. (*T. C.*).
5 June 17	Jewell, Lt. E. C.	5 June 18	19 Sep. 18	E. Lan. R., I. A. (*on probn.*).
. . .	Von Hogt, 2nd-Lt. A. L.	. . . .	12 Nov. 18	I. A. R. O.

Subadar-Major.

		Jemadar.	Subadar.	Subadar-Major.	
31 Mar. 91	Habibullah Khan, *Bahadur*, I.D.S.M. (56).	31 Mar. 91	1 Dec. 96	20 Sep. 07	*Order of Br. Ind., 2nd Cl.*

Subadars.

		Jemadar.	Subadar.
4 Dec. 88	Jota Singh	1 Oct. 99	16 Apl. 01
18 Dec. 00	Abdul Ghafur Khan, I.D.S.M. (52, 56, 72a)	18 Dec. 00	1 Aug. 04
26 Sep. 98	Badi-ul-Zaman, I.D.S.M. (66a)	25 Jan. 01	20 Sep. 07
17 June 01	*Najibullah Khan (66a, 72a)*	*17 June 01*	*16 Mar. 08*
21 Apl. 96	Amir Khan (53, 56, 66a)	6 Nov. 06	16 July 12
1 July 90	Khushal Singh (52, 56, 72a)	1 May 03	21 Feb. 14
3 June 93	*Firoz Khan*	*8 Nov. 10*	*25 Jan. 15*
18 July 94	Ghulam Muhammad (I)	23 Sep. 12	1 July 15
14 Nov. 90	*Budha Singh (58)*	*1 July 13*	*1 July 15*
6 Aug. 94	Ghulam Muhammad (II)	1 July. 14	1 July 15
10 Apl. 96	Karam Din (56)	25 Jan 15	1 July 15
12 Mar. 98	Juma Gul	1 Aug. 14	13 Aug. 15
1 Oct. 95	*Ahmad Khan*	*25 Jan. 15*	*1 Nov. 15*
10 Dec. 95	*Muhammad Bakhsh*	*25 Jan. 15*	*1 Nov. 15*
19 Jan. 95	Nur Alam	26 Apl. 15	22 Dec. 15
8 Aug. 92	Ghulam Muhiyuddin	1 May 12	1 Nov. 16
12 May 96	Punjab Singh	21 Feb. 14	1 July 17
14 Aug. 00	Allahdad Khan	26 Apl. 15	8 Aug. 17
10 May 02	Zaman Ali	1 Nov. 15	19 Feb. 18
16 Sep. 01	Ditta Khan	1 Nov. 15	19 Feb. 18
7 Apl. 96	Habib Khan (56)	13 Jan. 16	19 Feb. 18
17 July 97	Prem Singh	26 Apl. 15	19 May 18

46th Punjabis—*contd.*

1st Commn. or date of entering service.	Names and Rank.	Army rank.	Remarks.
	Jemadars.		
26 June 96	Nanak Singh	1 July 15	
23 July 05	Hassan Shah	1 Apl. 16	
20 Feb. 01	*Budha Singh*	*8 July 16*	
23 Jan. 00	Fazal Ahmad	1 Sep. 16	
31 Dec. 95	Sher Muhammad	1 Nov. 16	
18 Oct. 00	Ghazi Khan	1 Nov. 16	
6 Sep. 08	Ayub Khan	1 Nov. 16	
8 June 01	Punjab Singh	1 July 17	
1 Nov. 99	Ilahi Baksh	18 Aug. 17	
20 Dec. 01	Sultan Mahmud	17 Dec. 17	
2 Mar. 08	Shukul Chand	24 Dec. 17	Indian Qr. Mr.
4 Apl. 10	Muhammad Sarwar Khan	19 Jan. 18	
8 Aug. 03	Fateh Ali	8 Feb. 18	
26 Mar. 04	Ghulam Muhi-ud-din	3 May 18	
20 Jan. 06	Amir Ali	8 May 18	
1 July 04	Nur Alam	19 May 18	

Regt. Centre—Rawal Pindi. *Uniform*—Scarlet.
Linked with—35th and 36th Sikhs. *Facings*—Yellow.

47th Sikhs.

Raised at Sialkot in 1901, by Major P. G. Walker, as the 47th (Sikh) Regiment of Bengal Infantry. Became the 47th Sikh Infantry in 1901; present designation in 1903.

Composition—4 companies Sikhs.

1st Comm n or date of entering service.	Names and Rank.	Army rank.	Present appointment in Regiment.	Remarks.
	Commandant.			
16 Apl. 90	Davidson, Bt.-Col. S. R., C.M.G. (*temp. Br.-Genl.*)	3 June 16	18 Feb. 16	
	Company Commanders. (4)			
22 Jan. 98	Combe, Bt.-Lt.-Col. S. B., D.S.O., M.C.[*l*]	. . . .	18 Feb. 16	2nd-in-Comd. Comdt. (*temp.*)
27 July 98	Hogge, Maj. L. R. [*l*] . . .	1 Sep. 15	1 Mar. 15	
18 Jan. 02	Ralston, Maj. W. H., D.S.O., M.C.	18 Jan. 17	18 Feb. 16	2nd-in-Comd. (*temp.*)
27 Sep. 05	Francis, Capt. R. F., M.C. . .	27 Sep. 14	27 Oct. 16	*G. S. O. 3rd gr.*
	Company Officers. (9)			
8 Sep. 09	Drysdale, Capt. A. E., D.S.O., M.C. (*Actg. Maj.*)	1 Sep. 15	6 Nov. 10	With 38th Dogras.
18 Jan. 11	Brunskill, Capt. G. S., M.C. . .	1 Sep. 15	13 Mar. 12	
15 Nov. 15	*Forrest, Lt. L. H.*	*15 Nov. 16*	*19 Nov. 15*	*With R. A. F.*
15 Nov. 15	Walker, Lt. S. E.	15 Nov. 16	20 Nov. 15	
18 Apl. 16	Bennett, Lt. W. G.	18 Apl. 17	24 Apl. 16	
29 June 16	Tahourdin, Lt. R. J. L.	29 June 17	. . . .	
29 June 16	Short, Lt. J. McL.	29 June 17	4 Apl. 17	
30 Jan. 17	Evans, Lt. N. W.	30 Jan. 18	23 July 17	
1 May 17	Rayne, Lt. P. C.	1 May 18	4 Dec. 17	Attd.
24 Nov. 15	Watkins, Lt. A. N. M. . . .	24 Nov. 16	29 Nov. 17	Attd.
31 Aug. 18	Byers, 2nd-Lt. E.	31 Aug. 18	5 Sep. 18	Attd.
1 Oct. 18	O'Ferral, 2nd-Lt. J. L. F. . .	1 Oct. 18	. . . .	Attd. With 1-2nd G. R.
	Attached.			
13 Aug. 04	Hay, Capt. J. B. W.	13 Aug. 13	. . . .	36th Sikhs.
9 Jan. 15	May, 2nd-Lt. L. H.	9 Jan. 15	. . . .	D. L. Infy. *Lv., ex I., 6 mos., m.c.*
12 Mar. 15	Field, Lt. F. M. S.	12 Mar. 16	. . . .	I. A. R. O.
9 July 15	Fitzpatrick, Lt. L. J.	9 July 16	. . . .	I. A. R. O.
23 Aug. 15	Down, Lt. F. D.	23 Aug. 16	. . . .	I. A. (*on probn.*)
15 Sep. 15	Conolly, Lt. J. J. P.	15 Sep. 16	. . . .	35th Sikhs.
18 Feb. 15	Phelps, Capt. A. A.	19 Sep. 16	. . . .	I. A. R. O.
22 Dec. 14	Marsh, Lt. C. C. A.	22 Sep. 16	15 Dec. 17	Worc. R., I. A. (*on probn.*)
11 Jan. 16	Tyte, Lt. A. N.	11 Jan. 17	7 Mar. 18	I. A. R. O. (*on probn.*)
10 Apl. 16	Miller, Lt. A. B. (*Actg. Capt.*) .	10 Apl. 17	16 Aug. 18	I. A. (*on probn.*) Adjt., 27 Sep. 18.
22 Aug. 17	West, Lt. E.	21 Aug. 18	29 Dec. 17	I. A. R. O. (*on probn.*)
25 Aug. 16	Tiffen, Lt. W. L.	25 Aug. 17	25 Aug. 17	I. A. R. O.
29 Mar. 18	Wicks, 2nd-Lt. E. L.	29 Mar. 18	16 July 18	I. A. (*T. C.*)
28 Aug. 18	Steedman, 2nd-Lt., W. McQ. .	28 Aug. 18	17 Oct. 18	High L. I., I. A. (*on probn.*) . .
27 Mar. 18	Fisher, 2nd-Lt. A. R. G. . . .	27 Mar. 18	7 Nov. 18	Hamps. R., I. A. (*on probn.*) . .
24 Apl. 18	James, 2nd-Lt. L.	24 Apl. 18	17 Oct. 18	I. A. (U. L.).
18 July 18	Doherty, 2nd-Lt. J. A. . . .	18 July 18	20 Aug. 18	
27 July 18	Jefford, 2nd-Lt. P. E. . . .	27 July 18	18 Aug. 18	
7 Oct. 18	Morgan, 2nd-Lt. I. M. . . .	7 Oct. 18	28 Oct. 18	
7 Oct. 18	Nicholas, 2nd-Lt. T.	7 Oct. 18	21 Oct. 18	

Subadar-Major.

		Jemadar.	Subadar.	Subadar-Major.	
17 Mar. 01	Thakur Singh, *Bahadur* M.C. (56, 72*a*, 72*b*)	17 Mar. 01	1 Mar. 04	1 Apl. 17	*Order of Br. Ind., 2nd cl.*

Subadars.

18 May 87	Ram Singh (56)	18 Feb. 04	1 Mar. 10
24 June 87	Bakhshi Singh, I.O.M. . . .	18 Apl. 05	16 Sep. 13
8 Feb. 06	Mit Singh, M.C. (56, 66*a*) I.D.S.M.	8 Apl. 10	8 Aug. 14
3 Dec. 95	Uttam Singh (56, 66*a*) . . .	3 Oct. 11	27 Oct. 14
22 Oct. 00	Hira Singh	8 Aug. 14	16 Mar. 15
16 Mar. 01	Mehar Singh	27 Oct. 14	6 Nov. 17
	Fauja Singh, I.D.S.M.	19 Apl. 16	6 Nov. 17
19 Mar. 01	Dewa Singh	20 Jan. 17	6 Nov. 17
31 July 01	Bishan Singh	27 Apl. 15	19 Sep. 18
	Ram Lal	. . . .	16 Apl. 18
	Uttam Singh	6 Nov. 17	19 Apl. 18

47th Sikhs—*contd.*

1st Commn. or date of entering service.	Names and Rank.	Army rank.	Remarks.
	Jemadars.		
25 May 03	Girdhari	5 Jan. 16	
1 May 01	Gurmukh Singh	26 Aug. 16	
19 Jan. 01	Kesar Singh	17 Jan. 17	
. . . .	Vir Singh	20 Jan. 17	
19 Mar. 01	Harnam Singh	20 June 17	
	Mangal Singh	24 Oct. 17	
	Dewa Singh	6 Nov. 17	
2 Apl. 01	Bishan Singh	6 Nov. 17	
19 Mar. 01	Jagat Singh	3 Dec. 17	
18 June 03	Harphul	1 Feb. 18	
9 Apl. 18	Sucha Singh	9 Apl. 18	
. . . .	Lal Singh (*Ret.*)	. . .	(*On probn.*)
26 Mar. 01	Sundar Singh	16 Apl. 18	Re-employed.
29 July 00	Rura	28 Aug. 18	
20 Nov. 17	Tikam Singh	1 Oct. 18	
24 June 01	Bhag Singh	13 Oct. 18	
	Hari Singh	23 Sep. 18	
	Channan Singh	23 Sep. 18	

Regt. Centre—Bareilly.
Linked with—12th Pioneers.

Uniform—Scarlet.
Facings—Black.

1st Battalion, 48th Pioneers.

Raised at Lucknow in 1901, by Maj. C. LeG. Justice, as the 48th Regiment of Bengal Infantry (Pioneers). Became the 48th Bengal Pioneers the same year; present designation in 1903.
Composition—2 companies of Jats, 2 of Labana Sikhs.

1st Commn. or date of entering service.	Names and Rank.	Army rank.	Present appointment in Regiment.	Remarks.
	Commandant.			
7 Mar. 94	Cuming, Bt.-Lt.-Col. R. J., D.S.O., q.	3 June 17	24 June 18	
	Company Commanders. (4)			
20 July 98	*Neumann, Bt. Lt.-Col. C. W., q.*	*1 Jan. 17*	*24 June 18*	2nd-in-Comd.
27 July 98	Johnson, Maj. M. E. S., *p.s.c., q.*	1 Sep. 15	17 Nov. 09	
20 May 99	*Vickers, Maj. A., q.*	*1 Sep. 15*	*19 Feb. 14*	*Assam Mily. Police.*
20 Jan. 00	Hewett, Maj. G., D.S.O., q. (*Actg. Lt.-Col.*)	1 Sep. 15	24 June 18	With 2-48th Prs.
	Company Officers. (9)			
5 Jan. 01	Drinan, Maj. G. C., *q.*	5 Jan. 16	6 Jan. 06	
23 Dec. 03	*Gray, Capt. D. B., q.*	*23 Dec. 12*	*13 Feb. 06*	*R. A. F.*
24 Jan 06	*Davies, Capt. G. B., q.*	*24 Jan. 15*	*23 Dec. 07*	*Assam Mily. Police.*
17 Aug. 07	Burdett, Capt. E. W., M.C.	1 Sep. 15	12 Nov. 08	
25 Jan. 08	Leslie, Capt. F. D.	1 Sep. 15	8 Apl. 09	*With 1-155th Infy.*
20 Jan. 09	*Ferrier, Capt. I., q.*	*1 Sep. 15*	*15 Mar. 10*	*With 2-128th Prs.*
8 Sep. 09	Raynor, Capt. C. A., M.C.	1 Sep. 15	24 Oct. 10	
6 Sep. 11	Marshall, Capt. H. H., M.C.	6 Sep. 15	30 Oct. 12	With 1-107th Prs.
15 Nov. 15	Shaw, Lt. R. L.	15 Nov. 16	19 Nov. 15	Attd. With 2-81st Prs.
18 Apl. 16	Reed, Lt. R. R.	18 Apl. 17	24 Apl. 16	Attd.
29 June 16	Hewett, Lt. W. G. O'C.	29 June 17	2 Mar. 17	Attd.
20 Nov. 14	Jonas, Lt. R. M.	20 Aug. 16	. . . .	Attd.
5 Feb. 15	Lovat e-Fraser, Lt. W. A.	5 Nov. 17	. . . .	92nd Punj.
1 Oct. 18	Gordon, 2nd-Lt. R. D. L.	1 Oct. 18	31 Oct. 18	
27 Aug. 15	Rayner, Lt. C. S. W. (Actg. Capt.)	27 Aug. 16	12 Feb. 18	1-95th Infy.
	Attached.			
13 Aug. 04	Maturin, Capt. H. G. (*q*) [L]	13 Aug. 13	. . . .	61st Prs.
23 Dec. 14	Hawkes, Lt. M. H.	23 Dec. 15	. . . .	12th Prs.
2 Apl. 15	Pearce, Lt. T.	2 Apl. 16	. . . .	I. A. R. O.
23 Apl. 15	Rackstraw, Lt. G., M.C.	23 Apl. 16		
18 Dec. 14	*Biden, Lt. S. W.*	*18 Dec. 15*	*26 May 15*	*I. A. R. O.*
2 Apl. 15	*Souter, Lt. L. F.*	*2 Apl. 16*	*20 Sep. 15*	*I. A. R. O.*
8 Feb. 16	Warburton, Lt. J. F.	8 Feb. 17	9 Nov. 17	I. A. (*on probn.*).
27 Oct. 17	Gee, 2nd-Lt. W. H. S.	27 Oct. 17	4 Nov. 17	I. A. (*on probn.*)
12 Feb. 15	Hodson, Lt. J. S. F.	12 Feb. 16	. . . .	I. A. R. O.
22 Feb. 16	Rudd, Lt. N. B.	22 Feb. 16	7 Dec. 17	I. A. R. O.
3 Nov. 16	Jennison, Lt. F. G.	3 Nov. 16	7 Dec. 17	I. A. R. O.
5 May 17	Fitzpatrick, Lt. M. J. J.	5 May 18	20 Nov. 17	I. A. R. O.
15 Jan. 18	Shaw, 2nd-Lt. M. H.	. . . .	4 June 18	I. A. (*T. C.*)
29 Mar. 18	Creasy, 2nd-Lt. R. L.	29 Mar. 18	. . . .	I. A. R. O.
	Butcher, 2nd-Lt. E. A.	. . . .	. . . .	I. A. (*T. C.*)
15 June 17	Howlett, Lt. A. H. J.	15 June 18	. . . .	I. A. (*on probn.*)

Subadar-Major

		Jemadar.	Subadar.	Subadar-Major.	
30 Nov. 86	Dula Singh (56, 60) *Bahadur*	4 June 01	31 July 04	3 Jan. 12	*Order of Br. Ind., 2nd Cl.*
18 Apl. 89	Gyan Singh, *Bahadur* (56)	1 July 10	1 July 13	30 Apl. 16	*Order of Br. Ind., 2nd Cl.*
	Surja, *Sardar Bahadur* (*Retd.*)	. . . .	. . . .	. . . .	Re-employed 26 May 15, Depôt.

Subadars.

29 Dec. 85	Lehna Singh *Bahadur* (56)	31 July 04	1 Sep. 11	*Order of Br. Ind., 2nd cl.*
12 Dec 01	Hari Singh (56)	12 Dec. 01	3 Jan. 12	
1 July 93	Lehri (56)	3 Jan. 12	17 Aug. 14	
13 Mar. 96	Shib Lal (56)	16 May 13	8 July 15	
17 Feb. 93	Pala Singh	1 Feb. 12	24 July 15	
15 Apl. 97	Shiu Singh	17 Aug. 14	31 July 15	
8 Jan. 97	Hari Ram	16 May 15	30 Apl. 16	
10 May 97	Canda Singh	. . . .	30 Apl. 16	
24 Apl. 03	Het Ram	15 Dec. 15	3 July 16	
13 May 99	Gurdit Singh (56)	13 Aug. 15	12 June 17	
25 Nov. 95	Narayan Singh	30 Apl. 16	19 July 17	
6 June 01	Hukmi	30 Apl. 16	6 Aug. 17	
16 May 01	Tokha (56)	30 Apl. 16	13 Nov. 17	
	Mansa Singh (Retd.)	. . . .	. . . .	Re-employed, 1 Feb. 18.

1st Battalion, 48th Pioneers—*contd.*

1st Commn. or date of entering service.	Names and Rank.	Army rank.	Remarks.
	Jemadars.		
19 Feb. 98	Ramji Lal	17 Aug. 14	
12 Mar. 01	Achhar Singh	31 July 15	
16 Apl. 01	Ramji Lal	22 July 16	
24 Apl. 03	Nanak Singh	30 Apl. 16	
12 Mar. 97	Sohan Singh	30 Apl. 16	
1 July 01	Fakir Singh (48*a*, 51)	30 Apl. 16	
. . . .	Sultan	30 Apl. 16	
11 Apl. 97	Natha Singh (56)	22 May 17	
8 Nov. 02	Dale Ram (56)	1 June 17	
22 Nov. 01	Bagh Singh	19 July 17	
29 Mar. 01	Het Ram	6 Aug. 17	
12 June 01	Teja Singh (56)	14 Mar. 18	
20 Aug. 01	Hardhan (56	14 Mar. 18	
27 Feb. 05	Arjan Singh (56)	1 May 18	
	Nihal Singh	19 May 18	
	Indar Singh	1 June 18	
9 Sep. 01	Gulab Singh (56)	1 July 18	
3 July 08	Umar Singh (6)	1 July 18	

2nd Battalion, 48th Pioneers.

Formed at Arangaon on 26th July 1918.

Composition—2 Companies Jats.
1 Company Lobana and Saini Sikhs.
1 Company Heris.

1st Commn. or date of entering service.	Names and Rank.	Army rank.	Present appointment in Regiment.	Remarks.
	Commandant.			
20 Jan. 00	Hewett, Maj. G., D.S.O., *q.* (*Actg. Lt.-Col.*)	1 Sep. 15	16 Aug. 18	1-48th Prs.
	Company Commanders. (4)			
5 Aug. 05	Story, Capt. J. A., M.C., *q.* (*Actg. Maj.*)	5 Aug. 14	14 Aug. 18	1-61st Prs. 2nd-in-Comd.
	Hawkes, Capt. H., M.C.	....	....	S. Staff. R., I.A.R.O. (*on probn.*)
5 June 15	Gordon, Lt. T. J., M.C. (*Actg. Capt.*)	5 June 16	....	Attd.
18 Feb. 16	Tollit, Lt. W. E. (*Actg. Capt.*)	18 Feb. 17		
	Company Officers.			
5 May 17	Seymour, Lt. C. L. (*Actg. Capt.*)	5 May 18	....	I. A. R. O. Adjt.
20 June 17	Chalk, Lt. J.	20 June 18	....	I. A. R. O. Qr.-Mr.
1 Mar. 17	Kent, 2nd-Lt. H. T.	1 Mar. 17	....	Attd.
30 May 17	Hood, 2nd-Lt. D. A.	30 May 17	....	I. A. R. O.
5 Jan. 18	Hewitt, 2nd-Lt. D. J.	5 Jan. 18	....	I. A. R. O.
15 Jan. 18	Jacob, 2nd-Lt. L. W. R.	15 Jan. 18	....	I. A. (*T. C.*)
15 Jan. 18	Nightingale, 2nd-Lt. V. R. J.	15 Jan. 18	....	I. A. (*T. C.*)
31 Jan. 18	Marsh, 2nd-Lt. F. C.	31 Jan. 18	....	
28 Feb. 18	Weeks, 2nd-Lt. H. A.	28 Feb. 18	....	Lond. R., I.A. (*on probn.*)
31 Aug. 18	Jones, 2nd-Lt. V. W.	31 Aug. 18	....	Attd.
27 Sep. 18	Christie, 2nd-Lt. J. S.	27 Sep. 18	....	I. A. R. O.
10 July 18	Thorn, 2nd-Lt. G.	10 July 18	....	I. A. (*T. C.*)
15 July 18	Marklew, 2nd-Lt. E. G.	15 July 18	....	I. A. R. O.

Subadar Major.

		Jemadar.	Subadar.	Subadar-Major.
8 Jan. 97	Hari Ram		16 May 15	

Subadars.

11 May 93	Kehir Singh	30 Oct. 15	19 Mar. 18
20 May 03	Gopal Singh	29 Oct. 15	24 Aug. 18
7 May 03	Nanak Singh	25 May 17	24 Aug. 18
19 Sep. 00	Thakur Singh I	23 June 17	24 Aug. 18
13 Aug. 01	Thakur Singh II	6 June 17	24 Aug. 18
29 Mar. 01	Udmi	6 June 17	24 Aug. 18
	Kanhaiya	19 Mar. 18	24 Aug. 18

Jemadars.

16 July 98	Labh Singh	24 Aug. 18
18 Feb. 99	Dewan Singh	24 Aug. 18
23 Apl. 03	Mela Singh	24 Aug. 18
6 May 63	Menga Singh	24 Aug. 18
3 May 03	Dip Chand	24 Aug. 18
11 Oct. 06	Shiv Lal	24 Aug. 18

49th Bengalis.

Formed at Karachi on 1st July 1917.

Composition— Bengalis.

1st Commn. or date of entering service.	Names and Rank.	Army rank.	Present appointment in Regiment.	Remarks.
	Commandant.			
28 July 00	Sandiford, Maj. V. V. V.	1 Sep. 15	7 July 17	62nd Punj. 2nd-in-Comd.
	Company Commanders.			
25 Jan. 08	Darell, Capt. R. D. E. (*Actg. Maj.*)	1 Sep. 15	1 July 17	76th Punj.
	Webster, Capt. C. G.	25 Sep. 15	18 July 17	Ches. R.
27 Feb. 15	Pearse, Capt. G. G.	27 Feb. 18	15 July 17	I. A. (*T. C.*)
7 May 16	Jones, Lt. H. (*Actg. Capt.*)	7 May 17	1 Dec. 17	Essex R., I. A. (*on probn.*)
	Company Officers.			
11 Aug. 15	Harley, Lt. N.	11 Aug. 16	15 July 18	I. A. R. O. (*on probn.*)
7 Apl. 16	Fitzherbert, Lt. R. F. (*Actg. Capt.*)	7 Apl. 17	26 Dec. 17	Attd.
20 Apl. 16	Watson, Lt. J. H. L.	20 Apl. 17	5 Sep. 18	Leins. R., I. A. (*on probn.*)
10 May 16	Long, Lt. W. H. (*Actg. Capt.*)	10 Feb. 18	20 July 17	Attd. Qr.-Mr.
14 Aug. 16	Douglas, Lt. R.	14 Aug. 17	30 Oct. 18	I. A. R. O.
15 Sep. 16	Taylor, Lt. S. G.	15 Sep. 17	28 Feb. 17	I. A. R. O.
6 Oct. 16	Monies, Lt. S. O.	6 Oct. 17	1 June 17	I. A. R. O.
14 Nov. 16	Gordon, Lt. S. H.	14 Nov. 17	18 July 17	1-99th Infy.
30 Jan. 17	Gibbs, Lt. E. W. B. (*Actg. Capt.*)	30 Jan. 18	8 Apl. 17	Adjt.
1 Feb. 17	Hussey, Lt. N. G. E.	1 Feb. 18	10 June 18	I. A. R. O.
1 Feb. 17	Atkinson, Lt. A. W.	1 Feb. 18	29 July 17	I. A. R. O.
5 Mar. 17	Fookes, Lt. T. G.	5 Mar. 18	27 Aug. 17	I. A. R. O.
5 May 17	Wilks, 2nd-Lt. D. J.	5 May 18	20 Oct. 17	I. A. R. O.
5 May 17	Griffen, 2nd-Lt. F. W. B.	5 May 18	24 Oct. 17	I. A. R. O.
5 May 17	McGavin, Lt. D. G. K.	5 May 18	10 Oct. 18	I. A. R. O.
26 June 17	Battie, Lt. J. W.	26 June 18	1 July 18	Som. L. I., I. A. (*on probn.*)
8 Aug. 17	Faulkner, 2nd-Lt. W. G.	8 Aug. 17	10 Nov. 17	I. A. R. O.
5 Oct. 17	Johnson, Lt. J. R.	5 Oct. 18	5 Feb. 18	I. A. R. O.
15 Jan. 18	Ledingham, 2nd-Lt. J. J.	15 Jan. 18	19 July 18	I. A. (*T. C.*)
28 Jan. 18	Colsey, 2nd-Lt. J. W. T.	28 Jan. 18	13 July 18	S. Wales Bord., I. A. (*on probn.*)
29 Mar. 18	Driver, 2nd-Lt. G. L. S.	29 Mar. 18	15 July 18	I. A. R. O.
16 May 18	Staig, 2nd-Lt. B. M.	16 May 18	2 Dec. 18	I. A. R. O.
7 June 18	Fraser, 2nd-Lt. C. D.	7 June 18	17 Oct. 18	I. A. R. O.
18 July 18	Hogg, 2nd-Lt. G. P.	18 July 18	19 Aug. 18	I. A. R. O.
23 July 18	Mills, 2nd-Lt. S. H. H.	23 July 18	2 Oct. 18	I. A. R. O.
30 July 18	West, 2nd-Lt. M. P.	30 July 18	4 Sep. 18	I. A. R. O.
23 July 18	Gasper, 2nd-Lt. B. C.	23 July 18	24 Aug. 18	I. A. R. O.

Subadar-Major.

		Jemadar.	Subadar.	Subadar-Major.
14 Oct. 17	S. N. Basu, I.D.S.M.		1 Feb. 17	25 July 18
	Subadars.			
1 Feb. 17	Kumar Adhikram Mazumdar	. . . . .	1 June 17	
30 June 17	Hirendra Nath Sarkar	1 June 17	5 Mar. 18	
30 June 17	Durgapoda Banerji	30 June 17	5 Mar. 18	
30 June 17	Mor bahadur Singh	30 June 17	5 Mar. 18	
30 June 17	Dhirendra Kumar Sen	30 June 17	1 May 18	
29 Aug. 17	Phani Bhusan Dutt	29 Aug. 17	1 Mar. 18	
	Jemadars.			
30 June 17	Sarat Kumar Roy	30 June 17		
30 June 17	Bhupendra Nath Ghosal	30 June 17		
25 Aug. 17	Ranoda Prosad Saha	25 Aug. 17		
25 Aug. 17	Bireswar Mukherji	25 Aug. 17		
29 Aug. 17	Rajendra Lal Mukherji	29 Aug. 17		
29 Aug. 17	Pramatha Krishna Santra	29 Aug. 17		
21 Nov. 17	Gopal Chandra Ghose	21 Nov. 17		
21 Nov. 17	Joshada Kinkar Ghosh	21 Nov. 17		
23 Feb. 18	Nawab Khajeh Habibullah	23 Feb. 18		
5 Mar. 18	Bhumendra Mohan Roy	5 Mar. 18		
1 July 18	Kiron Kumar Mitra	1 July 18		
1 July 18	Prikity Kumar Ghose	1 July 18		
	Sachindra Kumar Roy	15 July 18		
	Nagendra Vehari Mukerjee	1 Aug. 18		

1st Battalion, 50th Kumaon Rifles.

Formed at Ranikhet as "4-39th Garhwal Rifles" 23rd October 1917.
Name changed to "4-39th Kumaon Rifles" 15th November 1917.
Name changed to "1-50th Kumaon Rifles" 30th April 1918.
Composition—4 companies "Kumaonis."

1st Commn. or date of entering service.	Names and Rank.	Army rank.	Present appointment in Regiment.	REMARKS.
	Commandant.			
27 July 89	Lang, Maj. E. M. (*Act. Lt. Col.*)	11 Nov. 14	23 Oct. 17	Welsh R.
	Company Commanders (4).			
11 Apl. 30	Holderness, Maj. H.	1 Sep 15	23 Nov. 17	1st G. R.
28 July 00	Latham, Maj. A.	1 Sep. 15	31 Oct. 17	1st G. R.
29 Jan. 10	Meade, Capt. H.R. C. (*Act'g. Maj.*)	29 Jan. 14	29 Jan. 18	1-8th G. R.
1 Oct. 14	Smith, Capt. M. H. H. . . .	1 Oct. 18	20 Feb 18	2-6th G. R.
	Company Officers			
18 Dec. 14	Duncan, Lt. J. G. (*Actg. Capt.*)	18 Dec. 15	28 Oct. 17	I. A. R. O., Adjt.
24 May 15	Catron, Lt. J. E. T., M.C. (*Actg. Capt.*)	24 May 16	9 Feb. 18	Aus. Imp. Force, I. A. (*on probn.*)
18 June 15	Harrop, Lt. H.R. (*Act'g. Capt.*) .	18 June 16	9 Dec. 17	I. A. R. O., Qr.-Mr.
26 Feb. 15	Hayes, Lt. E. J. T.	26 Nov. 16	9 Feb. 18	M. G. Corps, I. A. (*on probn.*)
20 Mar. 15	Milward, Lt. H. H.	20 Mar. 16	16 Feb. 18	I. A. (*on probn.*)
26 Mar. 15	Jenkins, Lt. A. G.	26 Dec. 16	9 Feb. 18	I. A. (*on probn*).
14 Aug. 15	Cotton Swanston, Lt. W. L. . .	14 May 17	18 Feb. 18	I. A. (*on probn.*)
18 Aug. 15	Oglethorpe, Lt. R. S.	19 May 17	9 Feb. 18	I. A. (*on probn*).
21 July 16	Quin, Lt. R. B.	21 July 17	23 Oct. 17	I. A. (*on probn.*)
9 June 16	Wagstaff, Lt. H.	9 Mar. 18	9 Feb. 18	Aug. Imp. Force, I. A. (*on probn.*)
18 June 17	Chilcott, Lt. J. S.	18 June 18	29 Oct. 17	Attd.
29 Aug. 17	Smith, 2nd-Lt. R. C. K. . . .	29 Aug. 17	29 Jan. 18	I. A. R. O. (*on probn.*)
5 Oct. 17	Bonner, Lt. F. G.	5 Oct. 18	21 Jan. 18	I. A. (*T. C.*)
5 Oct. 17	Calder, Lt. O. V.	5 Oct. 18	25 Feb. 18	I. A. R. O.
31 Jan. 18	Wallace, 2nd-Lt. R. W. A. . .	31 Jan. 18	14 Feb. 18	Attd.
31 Jan. 18	Adames, 2nd-Lt. F. H. . . .	31 Jan. 18	14 Feb. 18	Attd.
6 Oct. 17	Hussey, Lt. D.	6 Oct. 17	2 July 18	I. A. R. O.
8 Aug. 15	Giffen, Lt. L. E.	8 Aug. 18	18 Feb. 18	I. A. R. O. (*on probn.*)
	Mc Lean, Lt. A. E. J.	. . . .	. . . .	
18 Aug. 15	Simkins, Lt. G. P.	18 May 17	. . . .	28th Cav.
1 May 17	Brewester, Lt. F.	1 May 18	. . . .	I. A. R. O. (*on probn.*)
11 July 18	Willmott, 2nd-Lt. M. J. . . .	11 July 18	. . . .	I. A. R. O.
13 Sep. 18	Newson, 2nd-Lt. W.	13 Sep. 18	. . . .	I. A. R. O.
1 Oct. 18	Prichard, 2nd-Lt. H. M. . . .	1 Oct. 18	. . . .	I. A. R. O.
26 June 18	Barker, 2nd-Lt. P.	26 June 18	. . . .	I. A. R. O.
18 May 18	Gardner, 2nd-Lt. O. J.	18 May 18	. . . .	I. A. R. O.

Subadar-Major.

		Jemadar.	Subadar.	Subadar-Major.	
7 Aug. 91	Nain Singh Thakur	. . . .	. . . .	1 May 18	*Sub Maj., Chin Hill Bn.*
14 Aug. 93	Lachham Singh Rawat., I.D.S.M. .	16 Jan. 14	11 Mar. 15	19 May 18	

Subadars.

27 Mar. 89	Madan Singh Panwar.	1 Nov. 08	17 June 14
3 Nov. 98	Nain Singh Bhandari	. . . .	19 June 17
7 Apl. 01	Man Singh Rawat.	1 Jan. 16	23 Nov. 17
5 Oct. 00	Bala Singh Chauhan.	1 Apl. 17	1 Jan. 18

Jemadars.

3 Aug. 96	Par Singh Rawat.	14 Nov. 16
8 Nov. 07	Bhawan Singh Nasarkoti . .	1 Mar. 17
5 Sep. 95	Umapati Pande, I.D.S.M. . . .	1 May 17
5 May 01	Shib Singh Danu	1 Jan. 18
3 Dec. 01	Rup Singh Negi.	1 Jan. 18
17 Apl. 01	Mohan Singh Rawat, I.D.S.M. .	1 Jan. 18
20 May 01	Bagh Singh Negi	1 Jan. 18
10 Apl. 07	Fateh Singh Bisht	1 Jan. 18
26 Nov. 06	Padam Singh Karki	14 May 18
23 Mar. 05	Dan Singh Rawat	14 May 18
19 May 17	Hoshiar Singh Mahar Bisht . .	14 May 18
1 July 10	Hira Singh Bisht	15 June 18
4 July 13	Thepar Singh	15 June 18
31 Mar. 04	Deb Singh Negi	15 June 18

2nd Battalion, 50th Kumaon Rifles.

Formed at Bareilly on 6th June 1918.

Composition—2 companies Kumaonis, 2 companies Brahmans.

1st Commn. or date of entering service.	Names and Rank.	Army rank.	Present appt. in Regiment.	Remarks.
Commandant.				
29 Feb. 95	Maxwell, Maj. D. W., *p.s.c.* (*Act'g. Lt.-Col.*)	20 Feb. 13	18 June 18	2-4th G. R.
Company Commanders (4).				
9 Sep. 08	Rogers, Capt. G. S.	1 Sep. 15	15 July 18	1-39th Garhwal Rif. 2nd-in-Comd.
3 June 15	Martin, Lt. W. J. R., M.C. (*Act'g. Capt.*)	3 June 16	5 Sep. 18	R. Fus., I. A. (*on probn.*)
20 June 15	Lloyd, Lt. W. L., M. C. (*Act'g. Capt.*)	20 June 16	2 July 18	Shrops. L. I., I. A. (*on probn.*)
17 Mar. 16	Carling, Lt. T. E. (*Act'g. Capt.*)	17 Mar. 17	17 June 18	I. A. (*on probn.*)
Company Officers.				
18 Jan. 15	Northy, Capt. W. B.	18 Jan. 14	3 Oct. 18	1-1st G. R.
4 July 18	Ambler, Lt. C. T. M.	4 July 18	21 July 18	R. Fus., I. A. R. O.
31 July 16	Craig, Lt. J. H.	30 Apl. 18	2 Oct. 18	Oxf. & Bucks. L. I., I. A. (*on probn.*)
5 May 17	Papworth, Lt. F. A. G.	5 May 18	18 Sep. 18	I. A. R. O.
25 July 17	Tomkins, Lt. C. J. D. (*Act'g. Capt.*)	25 July 18	23 June 18	Lond. R., I. A. (*on probn.*) Adjt. 6 Aug. 18.
25 Oct. 16	Oswell, Lt. G. C.	25 July 18	9 Sep. 18	N. Staff. R., I. A. (*on probn.*)
8 Aug. 17	McKenzie, Lt. H. N.	8 Aug. 18	17 June 18	I. A. R. O.
29 Aug. 17	Trowsdale, Lt. H. J.	29 Aug. 18	5 Oct. 18	I. A. R. O.
25 Oct. 17	Smith, Lt. D. F.	25 Oct. 18	19 Sep. 18	Mon. R., I. A. (*on probn.*)
5 Nov. 17	Mills, Lt. C. F.	5 Nov. 18	19 Sep. 18	I. A. (*T. C.*)
	Mehew, 2nd-Lt. R. S.	. . . . .	17 June 18	I. A. R. O.
20 Dec. 17	Williams, 2nd-Lt. S. M.	20 Dec. 17	18 June 18	I. A. (U. L.)
21 Dec. 17	Draffen, 2nd-Lt. J. E.	21 Dec. 17	30 June 18	28th Dogras.
31 Dec. 17	Morrison, 2nd-Lt. J. W.	31 Dec. 17	3 Oct. 18	Yorks. L. I., I. A. R. O. (*on probn.*)
7 June 18	Livesay, 2nd-Lt. E.	7 June 18	10 Oct. 18	I. A. (*T. C.*)
25 June 18	Roberts, 2nd-Lt. W. C.	25 June 18	30 Oct. 18	Genl. List. I. A. R. O. (*on probn.*)
26 June 18	Cowper-Smith, 2nd-Lt. S. C.	26 June 18	3 Oct. 18	Genl. List. I. A. R. O. (*on probn.*)
31 Aug. 18	Tomlinson, 2nd-Lt. R. A. H.	31 Aug. 18	5 Sep. 18	Attd.
31 Aug. 18	Tennant, 2nd-Lt. J. H.	31 Aug. 18	5 Sep. 18	Attd.
20 Sep. 18	Cooke, 2nd-Lt. H. L.	20 Sep. 18	7 Oct. 18	I. A. R. O.

Subadar-Major.

		Jemadar.	Subadar.	Subadar-Major.
21 July 86	Ganga Singh Rawat	1 May 01	1 Nov. 08	28 Nov. 17

Subadar.

2 June 96	Chandri Chand, I.O.M.	15 Aug 11	1 Jan. 16
15 Sep. 93	Ram Singh Bhandari	21 Jan. 15	20 Aug. 16

Jemadar.

5 Sep. 95	Dharam Datt	1 Apl. 15
20 Feb. 95	Dalip Chand	1 Jan. 16
27 Aug. 92	Nain Singh	. . . .
5 Apl. 04	Joga Chand	8 July 16
1 Nov. 04	Tara Datt	1 May 17
19 Aug. 98	Partab Singh	1 Oct. 17
20 June 01	Lal Singh Rawat	1 Jan. 18
9 Oct. 01	Deb Singh Chupal	14 May 18
3 Nov. 14	Jai Singh Bhandari	14 May 18
1 Aug. 17	Naraln Datt	1 Aug. 17

Regt. Centre—Bannu.
Linked with—53rd and 54th Sikhs.

Uniform—Drab.
Facings—Yellow.

51st Sikhs (Frontier Force).

Raised at Hoshiarpur in 1846-47, by Capt. J. S. Hodgson, as the 1st Regiment of Infantry of the Frontier Brigade. Became the 1st Regt. of Sikh Local Infy., 1847; the 1st Regt. of Sikh Infy., 1857; the 1st Regt. of Sikh Infy., Punjab Irregular Force, 1857; the 1st Regt. of Sikh Infy., Punjab Frontier Force, 1865; the 1st Sikh Infy., 1901; present designation, 1903.

"Punjaub" "Ali Masjid" "Afghanistan, 1878-79" "Pekin, 1900."

Composition—2 companies of Sikhs, ½ of Dogras, 1 of Pathans, ½ of Punjabi Musalmans.

COLONEL—Major-General John A. H. Pollock, C.B. . . 11 April 1911.

1st Commn. or date of entering service.	Names and Rank.	Army rank.	Present appointment in Regiment.	Remarks.
	Commandant.			
27 Jan. 92	Lees, Lt.-Col. C. H. B., C.M.G., *p.s.c.*	27 Jan. 10	1 Feb. 18	*With Malay States Guides.*
	Company Commanders (4)			
20 Jan. 97	Beddy, Bt.-Lt.-Col. P. L., D.S.O.(Q)	3 June 16	24 May 16	2nd-in-Comd.
25 Jan. 99	*Shea, Maj. A.G.,* D.S.O. *(temp. Lt. Col.)*	*1 Sep. 15*	*9 Sep. 14*	*A. A. G., A. Hdqrs.*
17 Mar. 00	Koebel, Maj. F. E., D.S.O.[L] . .	1 Sep. 15	9 Sep. 14	*Lv., ex I., m. c.*
26 June 01	Mackenzie, Maj. H. J., D.S.O. . .	26 June 16	24 May 16	
	Company Officers. (9)			
4 Dec. 01	Pollock, Maj. R. C. G. .	4 Dec. 16	10 Oct. 04	With 2nd Bn. Guides.
18 Jan. 17	*More, Maj. J. C.* D.S.O. [L]	*18 Jan. 17*	*12 Feb. 06*	
9 Jan. 04	Burrows, Capt. H. M (*Actg. Maj.*)	9 Jan. 13	2 Dec. 17	With 1-72nd Punj.
16 Aug. 05	Williams, Capt. S. F. (*Actg. Maj.*)	16 Aug. 14	4 Aug 07	
23 July 09	Buist, Capt. C. N.	1 Sep. 15	8 July 14	
8 Sep. 09	O'Neill, Capt. W. H. L., M.C.	1 Sep. 15	9 Jan. 11	
22 Jan. 13	Lockhart, Capt. R. Mac G. M., M.C.	22 Jan. 17	14 Mar. 14	Adjt. With 24th Punj.
14 Jan. 14	Garrett, Capt. K. A., M.C. . .	14 Jan. 18	11 Oct. 14	
18 Apl. 16	Cristall, Lt. B. P.	13 Apl. 17	15 May 16	
18 Apl. 16	MacLachlan, Lt. A.	18 Apl. 17	25 Apl. 16	With 2-151st Punj.
29 June 16	Perry, Lt. E. G.	29 June 17	. . . .	
21 Dec. 17	Sangster, 2nd-Lt. D. B.	21 Dec. 17	. . . .	Attd.
22 May 15	Peters, Lt. J. H. G.	22 May 16	. . .	Attd.
31 Aug. 18	Barnes, 2nd-Lt. E. C.	31 Aug. 18	10 Sep. 18	
	Attached.			
9 Jan. 15	Macgregor, Lt. L. E.	20 Apl. 16	20 Aug. 17	I. A. R. O.
9 Apl. 15	Rice, Lt. P. H.	9 Apl. 16	28 Oct. 15	I. A. R. O. *Lv., ex I., 6 mos., m. c.*
30 Apl. 15	Learmond, Lt. A. W.	30 Apl. 16	24 Jan. 16	I. A. R. O. *Lv., ex I., 6 mos., m. c. 22 May 18.*
5 May 17	Brent Smith, Lt. C.	5 May 18	25 Oct. 17	I. A. R. O.
5 Oct. 17	Bain, Lt. G. W.	5 Oct. 18	19 Jan. 18	I. A. R. O.
1 June 18	Farrell, 2nd-Lt. J.	1 June 18	17 June 18	I. A. R. O.
19 Mar. 18	Walker, 2nd-Lt. J.	19 Mar. 18	22 July 18	I. A. (*T. C.*)
8 Nov. 18	Parsons, 2nd-Lt. J.	8 Nov. 18	11 Aug. 18	I. A. R. O.
1 May 18	Sales, 2nd-Lt. H. D.	1 May 18	2 Oct. 18	Norf. R., I. A. R. O. (*on probn.*)
24 Apl. 18	Mackenzie, 2nd-Lt. D. A. L. .	24 Apl. 18	19 Oct. 18	I. A. (U. L.)
24 Apl. 18	Dunley-Owen, 2nd-Lt. A. H. U.	24 Apl. 18	15 Oct. 18	I. A. (U. L.)
13 July 18	Ebert, 2nd-Lt. W. E. R. B. . .	13 July 18	3 Dec. 18	I. A. R. O.
8 Nov. 18	Dawson, 2nd-Lt. R. M. . . .	8 Nov. 18	27 Nov. 18	I. A. R. O.

Subadar-Major.

		Jemadar.	Subadar.	Subadar-Major.
29 July 85	Mir Baz Khan *Bahadur* (56) .	12 Oct. 04	22 Sep. 11	1 May 18 *Order of Br. Ind., 2nd Cl.*
	Turra Baz Khan (*Ret.*) . . .	. . . .	. . . .	Re-employed, 15 Aug. 17.

Subadars.

1 Nov. 91	Kesar Singh (56)	16 Jan. 11	3 Nov. 15	
17 Aug. 94	Sher Gul, *Bahadur*, I.D.S.M. .	1 Jan. 12	8 Jan. 16	*Order of Br. Ind., 2nd Cl.*
22 June 95	Akbar Khan, I.D.S.M.	12 Oct. 14	8 Jan. 16	
18 Oct. 97	Chanda Singh, I.O.M.	12 Oct. 14	8 Jan. 16	
1 Feb. 94	Labh Singh, *Bahadur*, I.D.S.M. .	12 Oct. 14	22 Jan. 16	*Order of Br. Ind., 2nd Cl.*
15 Mar. 99	Arjan Singh, I.O.M., I.D.S.M. .	21 Feb. 16	23 Jan. 16	
15 May 95	Tika Khan, I.O.M. (56) . .	20 Aug. 04	7 Apl. 16	
8 Feb. 92	Gokal (56)	26 Feb. 09	8 Oct. 16	
3 Sep. 94	Makhmud Din, I.O.M. . . .	8 Jan. 16	10 Mar. 17	
17 Oct. 94	Burhan Ali	4 Jan. 16	22 Mar. 17	
17 Feb. 92	Jagat Singh	3 June 16	23 Apl. 17	
16 Mar. 95	Surayan Singh (56)	7 Jan. 16	16 June 18	
	Ahmad Ali Khan (*Retd.*) . . .	. . . .	. . . .	Re-employed, 7 June 18.

51st Sikhs (Frontier Force)—*contd.*

1st Commn. or date of entering service.	Names and Rank.	Army rank.	Remarks.
	Jemadars.		
11 Jan. 97	Ghulam Shah	8 Jan. 16	
1 Nov. 98	Niazbat Khan	8 Jan. 16	
11 June 00	Harnam Singh	9 Jan. 16	
11 Feb. 94	Gurmukh Singh	22 Jan. 16	
1 Jan. 95	Prem Singh	22 Jan. 16	
31 Aug. 93	Sundar Singh, I.D.S.M.	21 Feb. 16	
14 Dec. 01	Hakim Singh	28 Mar. 16	
2 July 99	Mangal Singh, I.O.M.	7 Apl. 16	
23 Sep. 03	Prem Singh, I.D.S.M.	7 Apl. 16	
1 Aug. 02	Shiraz Gul	17 May 16	
7 Nov. 00	Basawa Singh	23 Feb. 17	
23 Mar. 03	Muhammad Khan	23 Feb. 17	
24 Apl. 95	Zaman Ali	1 Apl. 17	
8 Jan. 02	Kartara Singh	15 Apl. 17	
4 June 07	Amar Din	15 Apl. 17	
	Wadhawa Singh (*Ret.*)	. . . .	Re-employed, 10 Mar. 18.
22 June 00	Sham Singh	16 June 18	
1 Nov. 01	Hira Singh, I.D.S.M.	16 June 18	
9 Jan. 05	Muzaffar Khan	1 Oct. 18	
	Karam Din	25 Aug. 18	

Regt. Centre—Kohat.
Linked with—56th Rifles and 59th Rifles.

Uniform—Drab.
Facings—Scarlet.

52nd Sikhs (Frontier Force).

Dogra paltan.—Raised at Kangra in 1846-47, Capt. J. W. V. Stephen, as the 2nd Regt. of Infantry of the Frontier Brigade. Became the 2nd (or Hill) Regt. of Sikh Local Infantry, 1847; the 2nd (or Hill) Regt. of Sikh Infy., 1857; the 2nd (or Hill) Regt. of Sikh Infy., Punjab Irregular Force, 1857; the 2nd (or Hill) Regt. of Sikh Infy., Punjab Frontier Force, 1865; the 2nd (or Hill) Sikh Infy., 1901; present designation), 1903.

"Punjaub" "Ahmad Khel" "Kandahar, 1880" "Afghanistan, 1878-80."

Composition—1½ companies of Dogras, 1 of Sikhs, 1 of Punjabi Musalmans, ½ of Pathans.

COLONEL—Major-General Charles G. M. Fasken, C.B. . . . . . 1 May 1909.

1st Commn. or date of entering service.	Names and Rank.	Army rank.	Present appointment in Regiment.	Remarks.
	Commandant.			
17 Jan. 91	Wynter, Lt.-Col. C. P.	24 May 16	24 May 16	
	Company Commanders. (4)			
30 Aug. 93	*O'Grady, Bt.-Col. H. DeC., p. s. c. (Q)*	*3 June 18*	*24 May 16*	*2nd-in-Comd. Bde. Comdr. 10th Infy. Bde.*
5 Aug. 96	Ames, Maj. C. G. (*Actg. Lt.-Col.*)	5 Aug. 14	16 Feb. 13	With 2-56th Rifles.
20 Jan. 97	Wilkinson, Maj. C. K.	20 Jan. 15	16 Feb. 13	With 2-152nd Infy.
27 July 98	*Gilchrist, Maj. W. F. C.*, C.I.E., *p.s.c.*	*1 Sep. 15*	*9 May 16*	
	Company Officers. (9)			
27 July 98	*Moens, Bt.-Lt.-Col. A. W. H. M.*, D.S.O., *p. s.c. (temp. Br. Genl.)*	*1 Jan. 17*	*31 July 01*	*G. S. O., 1st Gr.*
4 May 01	Gordon, Maj. J. G. B., M.C.	1 Sep. 15	27 May 05	*Lv., ex I., m.c.*
5 Aug. 05	Grylls, Capt. W. M.	5 Aug. 14	17 June 07	With 2-27th Punj.
5 Aug. 05	Wynter, Capt. J. R., D.S.O., (*temp. Maj.*)	5 Aug. 14	9 Dec. 06	
4 Apl. 06	*Parsons, Capt. A. B. B.*	*4 Apl. 15*	*30 Sep. 08*	*N. Waziristan Militia.*
29 Jan. 10	Crew, Capt. R. D.	1 Sep. 15	8 Mar. 11	With 53rd Sikhs.
17 Jan. 11	Wallace, Capt. A. J. L. (*Actg. Maj.*)	1 Sep. 15	30 Mar. 13	With 82 Punj.
12 Jan. 13	Minchin, Capt. H. C. S.	22 Jan. 17	23 Mar. 14	Adjt., 15 Nov. 16.
14 Jan. 14	Campbell, Capt. W. F.	14 Jan. 18	11 Oct. 14	Qr.-Mr., 15 Nov. 16.
1 Oct. 14	Feltham, Capt. W. I. P., M.C.	1 Oct. 18	18 Dec. 14	
13 Jan. 15	Bookey, Lt. J. T. B.	13 Jan. 16	. . . .	Attd. with 1-6th G. R.
12 May 15	Barlow, Lt. C. B.	12 May 16	29 Oct. 15	Attd.
15 Nov. 15	Hansen, Lt. G. E.	15 Nov. 16	19 Nov. 15	Attd. With 59th Rif.
29 June 16	Woodward, Lt. R. G., M.C.	29 June 17	. . . .	Attd.
29 June 16	Bannerman, Lt. H. D'A.	29 June 17	. . . .	Attd.
14 Nov. 16	Barry, Lt. F. C.	3 May 17	24 Nov. 16	Attd. With 24th Punj.
14 Nov. 06	Elliott, Lt. J. G.	14 Nov. 17	. . . .	Attd.
18 June 17	Graham, Lt. J. P.	18 June 18	. . . .	Attd. *N. Waziristan Militia.*
31 Jan. 18	Wilder, 2nd-Lt. B. L.	31 Jan. 18	6 Feb. 18	Attd.
21 Dec. 17	Cunningham, 2nd-Lt. W. E.	21 Dec. 17	. . . .	Attd.
7 Sep. 15	Lewis, Lt. A. M. (*Actg. Capt.*)	7 Sep. 16	. . . .	Attd.
21 Jan. 16	Moore, Lt. F. M.	21 Jan. 17	. . . .	Attd.
12 June 16	Thorpe, Lt. C. W. G.	12 June 17	. . . .	Attd. With 1-131st Infy.
31 Aug. 18	Power, 2nd-Lt. W. N.	31 Aug. 18	. . . .	Attd.
	Attached.			
4 Aug. 15	*Dash, Lt. A. J.*	*4 Aug 16*	*14 Sep. 15*	*I. A. R. O. Addl. A. R. O. for Punjabi Musalmans.*
18 Feb. 16	Heath, Lt. H. C. S. (*Actg. Capt.*)	18 Feb. 17	. . . .	I. A. (*on probn.*)
5 May 17	Shaw, Lt. R.	5 May 18	6 Aug. 17	I. A. R. O.
23 June 17	Watson, 2nd-Lt. J.	23 June 17	20 Aug. 17	I. A. R. O.
7 July 16	Hay, Lt. E. B.	7 July 17	20 Aug. 17	I. A. R. O.
19 June 16	Johnson, Lt. H. H. (*Actg. Capt.*)	19 June 17	. . . .	I. A. (*on probn.*) Adjt. (*offg.*)
17 May 15	Thorne, Lt. M.	17 Mar. 16	. . . .	Middx. R., I. A. (*on probn.*).
15 Dec. 15	Scott, Lt. W.	15 Dec. 16	. . . .	M. G. Corps, I. A. (*on probn.*)
	Liddell, 2nd-Lt. R. E. A.	. . .	5 Jan. 18	I. A. R. O.
14 Aug. 16	Scott, Lt. T. T.	14 Aug. 17	10 Oct. 16	I. A. R. O.

Subadar-Major.

		Jemadar.	Subadar.	Subadar-Major.
11 Apl. 90	Dayal Singh	16 July 97	10 Nov. 03	10 July 14

Subadars.

14 July 84	Bhagat Singh (72*a*)(*Ret.*)	15 Nov. 02	30 Dec. 05	Re-employed.
11 Nov. 87	Chattar Singh (56)	30 Dec. 05	1 June 11	
1 Jan. 01	Bachitter Singh (56, 66*a*)	16 July 10	16 Mar. 13	
9 June 10	Sansar Chand	9 June 10	5 Feb. 15	With 59th Rifles.
20 Jan. 95	Rasam Sing (56)	16 Mar. 13	28 July 15	
31 Dec. 02	Sahib Gul	15 May 15	17 June 17	
27 Feb. 94	Natha Singh (56)	2 Nov. 14	10 Dec. 17	
	Jodhbir Singh	1 July 17	11 Dec. 17	
	Sunit Chand, I. O. M.	. . . .	3 Nov. 17	With 1-56th Rif.
26 Dec. 94	Nihal Sing (46)	9 Dec. 14	14 Sep. 18	
1 Dec. 11	Gul Mawaz Khan	15 May 15	14 Sep. 18	
	Sawal Singh (*Ret.*)	. . . .	. . . .	Re-employed 11 Oct. 17.

52nd Sikhs (Frontier Force)—*contd.*

1st Commn. or date of entering service.	Names and rank.	Army rank.	Remarks.
	Jemadars.		
18 Feb. 98	Bir Singh (II)	26 Apl. 15	With 59th Rifles.
10 Apl. 16	Rai Singh	10 Apl. 16	(*On probn.*)
7 July 05	Ghulam Nabi	29 June 16	
3 Sep. 98	Nandu	1 Oct. 16	
	Indar Singh	26 Mar. 17	
2 July 02	Akbar Khan (48*a*)	17 June 17	
10 June 01	Khuda Baksh (46)	1 July 17	
1 Aug. 05	Abdulla	28 July 17	
8 Apl. 01	Sohan Singh	1 Aug. 17	
9 Jan. 85	Mehr Muhammad	1 Sep. 17	
1 May. 05	Muhammad Shah (48*a*)	19 Oct. 17	
14 Aug. 01	Waryan Singh	19 Oct. 17	
	Burhan Ali (*Ret.*)	. . . .	Re-employed.
	Lal Khan	13 May 18	
	Niaz Ali Khan	6 July 18	
	Ali Afsar Khan	14 Sep. 18	
	Faqir Chand	14 Sep. 18	
	Rasila	14 Sep. 18	
	Asa Singh	14 Oct. 18	
	Bhup Singh	14 Oct. 18	

Regt. Centre—Bannu.
Linked with—51st and 54th Sikhs.

Uniform—Drab.
Facings—Black.

53rd Sikhs (Frontier Force).

Renny-ki-paltan.—**Raised at Ferozepore in 1846-47, by Capt. F. Winter, as the 3rd Regiment of Infantry or the Frontier Brigade. Became the 3rd Regiment of Sikh Local Infantry, 1847; the 3rd Regiment of Sikh Infantry, 1857; the 3rd Regiment of Sikh Infantry, Punjab Irregular Force, 1857; the 3rd Regiment of Sikh Infantry, Punjab Frontier Force, 1865; the 3rd Sikh Infantry, 1901; present designation, 1903.**

"Kabul, 1879" "Kandahar, 1880" "Afghanistan, 1879-80."

"Tirah" "Punjab Frontier."

Composition—2 companies of Sikhs, ½ of Dogras, 1 of Khattaks, ½ of Punjabi Musalmans.

1st Commn. or date of entering service.	Names and Rank.	Army rank.	Present appointment in Regiment.	Remarks.
	Commandant.			
6 Mar. 95	Gardiner, Lt.-Col. R., *p.s.c.* [*l*]	10 Mar. 18	10 Mar. 18	
	Company Commanders. (4)			
4 Aug. 97	*Milward, Bt. Lt. Col. C. A.*, D.S.O., *p.s.c., Maj.*	*3 June 18*	*10 Mar. 18*	*2nd-in-Comd.*
27 Aug. 02	Shepherd, Maj. C. I., D.S.O.	27 Aug. 17	30 Oct. 14	
21 Jan. 03	Bruce, Maj. G. E., M.C.	21 Jan. 18	1 Feb. 18	
10 Oct. 03	*Macartney, Maj. F. A.*, M.C.	*10 Oct. 18*	*10 Mar. 18*	*A. D. C. to H. E. the C.-in-C.*
	Company Officers. (9)			
18 Jan. 05	*Rainsford-Hannay, Capt. D.*	*18 Jan. 14*	*11 Oct. 06*	
18 Jan. 05	Browne, Capt. St. J. A.	18 Jan. 14	15 Mar. 06	
29 Aug. 06	Waller, Capt. E. de W. (*Actg. Maj.*)	29 Aug. 15	....	1-72nd Punj.
9 Sep. 08	Watts, Capt. E. P.	1 Sep. 15	9 Dec. 09	
9 Sep. 08	Pearson, Capt. H. G. A. (*Actg. Maj.*)	1 Sep. 15	18 Nov. 09	
8 Sep. 09	*Finnis, Capt. H.*, M.C.	*1 Sep. 15*	*22 Feb. 11*	*Adjt., 18 Feb. 16. Coy. Comdr., Cadet Coll., Wellington.*
18 Apl. 16	Dean, Lt. W. E.	18 Apl. 17	25 Apl. 15	
18 Apl. 16	Rodwell, Lt. E. H.	18 Apl. 17	25 Apl. 16	*Lv., ex I., m.c.*
13 Mar. 15	Hugh-Jones, Lt. N. I.	13 Mar. 16	....	With 2-151st Infy.
30 Jan. 17	Moss, Lt. C. H. deV.	30 Jan. 18	....	Attd.
18 June 17	Shuttleworth, Lt. E. H. S.	18 June 18	22 Sep. 17	
31 Jan. 18	Clery, 2nd-Lt. P. A. A.	31 Jan. 18	6 Feb. 18	Attd.
21 Dec. 17	Clease, 2nd-Lt. E. K. J. (*Actg. Capt.*)	21 Dec. 17	21 Mar. 18	Attd. Adjt. (*Offg.*)
21 Dec. 17	Danson, 2nd-Lt. E. T.	21 Dec. 17	20 Mar. 18	Attd. With 2-151st Infy.
	Attached.			
30 Apl. 15	Fieldsend, Lt. C. E., M.C.	30 Apl. 16	28 Oct. 15	I. A. R. O.
7 Aug. 15	Newitt, Capt. D. M.	3 June 18	....	I. A. R. O.
11 Nov. 15	Carter, Lt. A. B.	11 Nov. 16	....	I. A. R. O.
13 Nov. 14	Culverwell, Capt. T. S.	3 June 18	27 Aug. 16	I. A. R. O.
19 Dec. 15	Fulton, Lt. R.	19 Dec. 16	....	The Buffs. I. A. (*on probn.*)
5 Jan. 18	Borthwick, 2nd-Lt. C. J.	5 Jan. 18	....	I. A. (*T. C.*)
30 May 17	Shepherd, 2nd-Lt. F. A.	30 May 17	14 Jan. 18	I. A. R. O.
5 Oct. 17	Maconachie, Lt. R. R.	5 Oct. 18	20 Jan. 18	I. A. R. O.
27 June 17	Goode, Lt. L. C. H., M.C.	27 June 18	31 Jan. 18	I. A. R. O.
23 Mar. 18	Terry, 2nd-Lt. J. H. S.	23 Mar. 18	14 July 18	I. A. (T. C.)
22 Aug. 17	Beaty, 2nd-Lt. A. P.	22 Aug. 18	14 Oct. 18	R. Ir. Fus., I. A. (*on probn.*)
24 Apl. 18	Graham, 2nd-Lt. R. M.	24 Apl. 18	15 Oct. 18	I. A. (U. L).
25 June 18	Arnold, 2nd-Lt. H.	25 June 18	2 Oct. 18	Genl. List, I. A. R. O. (*on probn.*)
4 Aug. 15	Dyer, Lt. I. R.	4 Aug. 16	15 Oct. 18	Dorset R., I. A. (*on probn.*)
4 Oct. 18	Whitworth, 2nd-Lt. H. J.	4 Oct. 18	22 Oct. 18	I. A. R. O.

Subadar-Major.

		Jemadar.	Subadar.	Subadar-Major.	
25 Feb. 93	Mangal Singh, *Bahadur* (56)	6 Aug. 00	1 Aug. 06	1 July 14.	*Order of Br. Ind., 2nd Cl.*

Subadars.

15 Apl. 87	*Bostan Khan, Bahadur (56)*	*24 June 05*	*16 June 11*	*Order of Br. Ind., 2nd Cl.*
16 June 88	Arsla Khan, *Bahadur*, I.O.M. (56, 72*a*)	1 Feb. 05	16 June 13	*Order of Br. Ind., 2nd Cl.*
18 Feb. 99	Mula Singh (66*a*)	1 Nov. 10	12 Oct. 14	
8 Jan. 92	Chanan Singh, I.O.M.	12 Oct. 14	14 Jan. 16	
11 Mar. 97	Ghulam Muhammad	14 Jan. 16	8 Apl. 16	
23 Jan. 00	Dost Muhammad (52, 56)	15 Feb. 15	17 June 16	
26 Mar. 00	Hazara Singh	9 Mar. 16	23 Feb. 17	
1 Apl. 14	Indra Singh	1 Apl. 14	10 Mar. 17	
2 Oct. 95	Gagan Singh	12 Oct. 14	15 Apl. 17	
27 July 98	Hira Singh	14 Jan. 16	31 May 18	
26 Sep. 87	Latif Khan I.D.S.M. (52, 56, 106)	2 Dec. 99	....	1-72nd Punj.
18 Apl. 00	Lal Khan	1 Nov. 15	20 Sep. 18	

53rd Sikhs (Frontier Force)—*contd.*

1st Commn. or date of entering service.	Names and Rank.	Army rank.	Remarks.
	Jemadars.		
29 June 93	Shah Baz	1 July 14	
2 July 95	Sapuran Singh	14 Jan. 16	
9 Nov. 97	Shamir Singh	14 Jan. 16	
20 Oct. 98	Samandar Khan	14 Jan. 16	
1 Nov. 02	Mir Awas	14 Jan. 16	
. . . .	Sharhab Khan	8 Apl. 16	
1 Dec. 05	Sajan Singh	1 June 16	
20 Mar. 05	Kehr Singh, I.O.M.	4 June 16	
16 June 16	Gurdayal Singh	16 June 16	
4 May 88	Santokh Singh	17 June 16	
. . . .	Muhammad Sher	11 July 16	
. . . .	Hukam Dad	22 Sep. 16	
1 Jan. 00	Fateh Muhammad, I.O.M. (56) .	9 May 17	
	Ghulam Mehdi	31 May 18	
21 Mar. 06	Muhammad Sayyid (56) . . .	1 Aug. 17	1-72nd Punj.
	Bishun Singh	1 July 18	
23 July 98	Muhammad Khan	1 Aug. 18	

Regt. Centre—Bannu.
Linked with—51st and 53rd Sikhs.

Uniform—Drab.
Facings—Emerald Green.

1st Battalion, 54th Sikhs (Frontier Force).

Burmah paltan.—Raised at Ludhiana in 1846-47 by Capt. C. Mackenzie, Madras Army, as the 4th Regiment of Infantry of the Frontier Brigade. Became the 4th Regiment of Sikh Local Infantry, 1847; the 4th Regiment of Sikh Infantry, 1857; the 4th Regiment of Sikh Infantry, Punjab Irregular Force, 1857; the 4th Regiment of Sikh Infantry, Punjab Frontier Force, 1865; the 4th Sikh Infantry, 1901; present designation, 1903.

"Pegu" "Delhi, 1857" "Chitral."

Composition—2 companies of Sikhs, ½ of Dogras, 1 of Punjabi Musalmans, ½ of Pathans.

COLONEL—

1st Commn. or date of entering service.	Names and Rank.	Army rank.	Present appointment in regiment.	Remarks.
	Commandant.			
7 Nov. 91	Bainbridge, Lt.-Col. W. F., C.M.G., D.S.O., *p.s.c.*	7 Jan. 14	24 May 16	*G. S. O., 1st Gr.*
	Company Commanders. (4)			
30 Aug. 93	*Storr, Maj. C. L., p.s.c. (Q)*	*30 Aug. 11*	*21 Feb. 16*	*2nd-in-Comd. G. S. O., 2nd Gr.*
6 June 96	Woodward, Maj. C. F. (*Actg. Lt.-Col.*)	6 June 14	6 Apl. 08	Comdt. (*temp.*)
8 Jan. 01	*Channer, Maj. B. G.*	*8 Jan. 10*	*21 Dec. 14*	*D. A. Q. M. G., S. Comd.*
2 Apl. 02	Ruck, Maj. O. L.	23 Apl. 17	22 Aug. 15	2nd-in-Comd. (*temp.*)
	Company Officers. (9)			
12 Dec. 03	*Lethbridge, Maj. A.*	*12 Dec. 18*	*20 Mar. 07*	*Burma Mily. Police.*
13 Aug. 04	Nicoll, Capt. N. F.	13 Aug. 13	1 Sep. 06	With 2-75th Infy.
5 Aug. 05	Shirley, Capt. S. R., M.C. (*Actg. Maj.*)	5 Aug. 14	19 Nov. 06	
5 Aug. 05	*Beazley, Bt.-Maj. W. E.*, M.C.	*1 Jan. 18*	*12 Dec. 06*	
29 Jan. 10	Dennys, Capt. L. E.	1 Sep. 15	7 Mar. 11	Adjt., 1 Aug. 15.
29 Jan. 10	Matheson, Capt. B. H.	1 Sep. 15	9 Mar. 11	
7 Aug. 14	Barwis, Capt. C. W. A.	7 Aug. 18	9 Nov. 15	With 2-70th Rif.
11 Nov. 14	*Fraser, Lt. C. J. S.*	*11 Nov. 15*	*26 Jan. 15*	*With Sig. Ser.*
15 Nov. 15	Hawkes, Lt. G. E., M.C.	15 Nov. 16	18 Nov. 15	Attd.
18 June 17	Featherstone, Lt. B. K.	18 June 18	25 June 17	
31 Jan. 18	Fitch, 2nd-Lt. K. S.	31 Jan. 18	6 Feb. 18	Attd.
21 Dec. 17	Bryceson, 2nd-Lt. D. J.	21 Dec. 17	23 Aug. 18	Attd.
31 Aug. 18	Bates, 2nd-Lt. F. P. E.	31 Aug. 18	4 Sep. 18	Attd.
	Attached.			
8 Mar. 15	Prestwick, Lt. E.	8 Mar. 16	26 Jan. 18	Manch. R., I. A. R. O. (*on probn.*)
15 Aug. 15	Meeks, Lt. F. S.	15 Aug. 16	20 Sep. 17	I. A. R. O.
2 Aug. 16	Beard, Lt. F.	2 Aug. 17	2 Aug. 18	E. Surr. R., I. A. (*on probn.*)
18 Feb. 16	Mackay, Lt. A. J., M.C.	18 Feb. 17	18 Jan. 18	Sea. Highrs., I. A. (*on probn.*)
7 July 16	Macdonald, Lt. S.	7 July 17	17 Aug. 17	
15 Nov. 16	Hyde, Lt. J. A.	15 Nov. 17	17 Nov. 17	I. A. R. O.
13 Dec. 16	Uff, Lt. G. T. R.	13 Dec. 17	27 Nov 17	M. G. Corps, I. A. (*on probn.*)
5 May 17	Monk, Lt. F. F.	5 May 18	1 Dec. 18	I. A. R. O.
8 Aug. 17	Beauchamp, Lt. W. G.	8 Aug. 18	3 Nov. 17	I. A. R. O.
29 Aug. 17	Ellery, Lt. A. C.	29 Aug. 18	18 Jan. 18	I. A. R. O.
27 Oct. 17	Nicholl, Lt. M. C.	27 Oct. 8	4 Nov. 17	
28 Nov. 17	Moore, 2nd-Lt. R. N.	28 Nov. 17		I. A. R. O.
5 Jan. 18	Evans, 2nd-Lt. A.	5 Jan. 18	7 Oct. 17	I. A. (*T.C.*)
	Medical Officer.			
1 Sep. 06	Millar, Capt. G. Mc. G.	1 Sep. 09	25 May 13	

Subadar-Major.

		Jemadar.	Subadar.	Subadar-Major.
12 Feb. 91	Sher Ali	21 Sep. 06	16 Feb. 07	16 Mar. 16

Subadars.

1 Aug. 00	Janus Khan, I.D.S.M. (56, 72a)	16 July 11	1 Jan. 14
23 Feb. 92	Hukam Dad (56, 72a)	16 Feb 07	15 Feb. 15
5 Apl. 90	Kesar Singh	16 Apl. 09	1 May 15
27 May 99	Bhikham Singh, M.C., I. D. S. M. (52, 56)	5 Feb. 15	1 May 15
21 Oct. 95	Hira Singh (56)	1 May 15	26 Sep. 15
11 Aug. 04	Teja Singh	1 Aug. 15	23 Mar. 16
1 Jan. 0[illegible]	Indar Singh *M. C., I. D. S. M.* (50, 53, 66a)	16 Mar. 13	9 Apl. 16
28 Mar. 96	Ragha (58)	1 May 15	5 June 17
29 Jan. 03	Qasim Ali, I.O.M.(46, 56, 66a)	20 Sep. 15	22 June 17
6 June 98	Ibrahim Khan (56)	2 May 15	31 May 18
8 Sep. 96	Rur Singh (50)	1 Dec. 15	27 June 18
13 Mar. 08	Khwaja Muhammad(50)	5 Feb. 15	1 Oct. 18

1st Battalion, 54th Sikhs (Frontier Force)—*contd.*

1st Commn. or date of entering service.	Names and Rank.	Army rank.	Remarks.
	Jemadars.		
13 Nov. 01	Dan Singh (50, 53, 66*a*)	16 Mar. 13	
10 May 90	Attar Singh	1 June 15	
31 Aug. 95	Chanda Singh	1 June 15	
31 Oct. 96	Bir Singh	9 Apl. 16	Indian Adjt.
1 Apl. 03	Sher Baz	9 Apl. 16	
3 Feb. 00	Makhan Singh	14 Oct. 16	
15 Mar. 04	Mukarrab Khan	28 Jan. 17	
5 June 05	Gurnam Singh	1 Apl. 17	With 53rd Sikhs.
10 June 07	Isa Khan	5 June 17	
1 Nov. 03	Partab Singh	5 June. 17	
22 Apl. 05	Dayal Singh (56)	1 May 18	
25 Nov. 99	Sher Dil	1 May 18	
21 Jan. 00	Bir Singh (56)	1 May 18	
26 Mar. 06	Harnam Singh	1 May 18	
10 June 08	Bishan Singh (61)	1 May 18	
25 Apl. 95	Amir Ali Khan	1 July 18	
4 Dec. 08	Ahmad Khan	31 May 18	
11 Sep. 07	Nur Muhammad	31 May 18	
	Lal Khan	27 June 18	
15 Apl. 07	Bachint Singh	1 Sep. 18	
22 May 12	Mohammad Said	21 Sep. 18	

2nd Battalion, 54th Sikhs (Frontier Force).

Formed at Jullundur on 5th June 1917.
Composition—1½ companies Sikhs, ½ Kumaonis, ½ company Yusufzais, ½ company Punjabi Musalmans, 1 company Chhachi Pathans.

1st Commn. or date of entering service.	Names and Rank.	Army rank.	Present appointment in regiment.	Remarks.
	Commandant.			
16 Dec. 96	Routh, Maj. G.S.F. (*Act'g. Lt-Col.*)	16 Dec. 14	15 July 17	1-56th Rif.
	Company Commanders. (4)			
28 July 00	Smith, Maj. C. W. J., D.S.O.	1 Sep. 16	22 Aug. 17	29th Punj. 2nd-in-Comd.
30 June 15	Blood, Lt. J. A. (*Act'g. Capt.*)	30 June 16	18 Oct. 17	R. Ir. Fus., I. A. (*on probn.*).
15 Mar. 15	Anderson, Lt. C.B. (*Act'g. Capt*)	11 July 16	24 Sep. 17	I. A. R. O. (*on probn.*)
10 Feb. 16	Bailey, Lt. F. R. (*Act'g. Capt.*)	10 Feb. 17	16 July 17	I. A. R. O. Adjt., 16 July 17.
	Company Officers.			
22 Dec. 14	Dempster, Lt. D. E. (*Act'g. Capt.*)	22 Sep. 16	14 Sep. 17	I. A. (*on probn.*) Adjt. (*Offg.*) 24 May 18.
3 Sep. 15	Furney, Lt. S. K.	3 Sep. 16	1 Nov. 18	R. Muns. Fus., I. A. (*on probn.*)
4 Nov. 15	Beckerson, Lt. R. E., M.C.	4 Nov. 16	4 June 18	Norf. R., I. A. (*on probn.*)
4 Aug. 15	Price, Lt. H.	4 Aug. 16	16 Jan. 18	I. A. R. O. Qr.-Mr. (*offg.*)
15 Nov. 15	Gilmore-Ellis, Lt. P. D.	15 Nov. 16	18 Oct. 17	Som. L.I., I. A. (*on probn.*)
25 Jan. 16	Michie, 2nd-Lt. J. B.	25 Jan. 16	2 Aug. 17	I. A. R. O. (*on probn.*)
5 Oct. 17	Gibson, Lt. P. G. H.	5 Oct. 18	11 Jan. 18	I. A. T. C. (*on probn.*)
19 Dec. 17	Pickard, Lt. L. A.	19 Dec. 18	30 Mar. 18	I. A. R. O.
20 Dec. 17	Robinson, 2nd-Lt. C. P. W.	20 Dec. 17	21 Mar. 18	Attd.
5 Jan. 18	Bird, 2nd-Lt. A. G.	5 Jan. 18	6 June 18	I. A. R. O.
15 Jan. 18	Ryde, 2nd-Lt. A. A.	15 Jan. 18	6 June 18	I. A. (*T. C.*)
7 June 18	Nicholas, 2nd-Lt. O. R.	7 June 18	11 Oct. 18	I. A. (*T. C.*)
1 Oct. 18	Watkins, 2nd-Lt. B. S.	1 Oct. 18	6 Oct. 18	Attd.
22 July 18	Tutton, 2nd-Lt. H. K.	22 July 18	2 Dec. 18	I. A. (*T. C.*)
31 Aug. 18	Officer, 2nd-Lt. H. R.	31 Aug. 18	4 Sep. 18	Attd.

Subadar-Major.

		Jemadar.	Subadar.	Subadar-Major.	
17 Mar. 91	Kehar Singh (72*a*)	1 Jan. 00	16 Dec. 08	5 June 17	(*temp.*)

Subadars.

5 Mar. 92	Allah Dad	1 June 15	5 June 17	
9 June 99	Sewa Singh	5 June 17	5 June 17	
22 Oct. 96	Partab Singh	1 Jan. 16	1 Feb. 18	
18 Nov. 03	Muhammad Alam	1 June 17	1 Feb. 18	
20 May 05	Ali Bahadur, I.O.M	5 June 17	1 May 18	

Jemadars.

22 Jan. 17	Nawab Khan	22 Jan. 17	
1 Nov. 02	Chumel Singh	5 June 17	Indian Adjt.
13 July 06	Mian Khan	5 June 17	
23 Feb. 03	Mangal Singh	5 June 17	
11 Oct. 10	Sardar Khan	5 June 17	
11 Jan. 03	Ilam Din	5 June 17	
8 Oct. 17	Abdul Qadus	8 Oct. 17	
1 Jan. 03	Kartar Singh	1 Feb. 18	
11 Feb. 97	Bhola Dut	1 Feb. 18	
18 Feb. 12	Khan Bahadur	1 Feb. 18	
15 Mar. 04	Muhammad Khan	1 May. 18	Indian Adjt.
10 Mar. 05	Sundar Singh	1 May 18	
14 Aug. 17	Nand Singh	1 May 18	
13 Sep. 17	Amanat Khan	1 May 18	

Regt. Centre—Dera Ismail Khan.
Linked with—57th and 58th Rifles.

Uniform—Dark Green.
Facings—Scarlet (piping).

1st Battalion, 55th Coke's Rifles (Frontier Force).

Coke-paltan.—Raised at Peshawar in 1849, by Capt. John Coke, as the 1st Regiment of Punjab Infantry. Became the 1st Regiment of Punjab Infantry, Punjab Irregular Force, 1851; the 1st Regiment of Punjab Infantry, Punjab Frontier Force, 1865; the 1st Punjab Infantry, Punjab Frontier Force, 1901; present designation, 1903.

"Delhi, 1857" "Afghanistan, 1878-79."

Composition—½ of Yusufzais, ¼ of Khattaks, 1 of Punjabi Musalmans, 1 of Sikhs, 1 of Dogras.

COLONEL—Major-General William du G. Gray, C.B. . . 26 May 1908.

1st Commn. or date of entering service.	Names and Rank	Army rank.	Present appointment in Regiment.	Remarks.
	Commandant.			
13 July 92	Herdon, Lt.-Col. H. E., *p.s.c.* .	2 Mar. 18	2 Mar. 18	
	Company Commanders. (4)			
28 Sep. 95	*Alexander, Bt.-Lt.-Col. E. C.,* D.S.O., *p.s.c.* (Q).	*3 June 17*	*2 Mar. 18*	*2nd-in-Comd.*
25 Jan. 99	*Villiers-Stuart, Bt. Lt.-Col. J. P.,* D.S.O., *p.s.c.* (*temp. Lt.-Col.*)	*1 Jan. 18*	*1 Jan. 17*	*G. S. O., 1st gr., A. Hdqrs.*
7 Mar. 00	Mahon, Maj. A. E., D.S.O. . . .	1 Sep. 15	27 Apl. 17	2nd-in-Comd. (*temp.*)
8 May 01	*Milne, Maj. T.*	*8 May 16*	*2 Mar. 18*	
	Company Officers. (9)			
9 June 01	Galbraith, Maj. E. D., D.S.O. . .	19 June 16	29 Aug. 05	With 3-151st Infy.
9 Jan. 04	*Leith-Ross, Capt. W.,* M. C. . .	*9 Jan. 13*	*26 July 07*	*G. S. O. 2nd gr.*
5 Aug. 05	*Acworth, Capt. D. H.,* M.C. (*temp. Maj.*).	*5 Aug. 14*	*14 Mar. 07*	
29 Aug. 06	Evans, Capt. D. M., M.C. .	29 Aug. 15	11 Nov. 07	With 56th Rif.
17 Aug. 07	*De Butts, Capt. F. C.,* D.S.O., M.C.	*1 Sep. 15*	*6 Nov. 08*	*With Sig. Ser.*
9 Sep. 08	Tate, Capt. W. M. (*Actg. Maj.*).	1 Sep. 15	14 Dec. 09	With 1-125th Rif.
8 Sep. 09	Mein, Capt. D. B., M.C. (*Actg. Maj.*)	1 Sep. 15	11 Nov. 10	
1 Oct. 14	Ekin, Capt. R. G.	1 Oct. 18	16 Dec. 14	With 58th Rif.
11 Nov. 14	Richardson, Lt. R., M. C. (*Act'g. Capt.*	11 Nov. 15	26 Jan. 15	
15 Nov. 15	Woodhouse, Lt. F. V. R. . . .	15 Nov. 16	9 May 16	
29 June 16	Raynor, Lt. E. V.	29 June 17	2 July 16	Attd. With 2-55th Rif.
1 Jan. 18	Ross-Hurst, 2nd-Lt. B. P. . .	1 Jan. 18	7 Feb. 18	Attd. Qr. Mr. (*Offg.*)
31 Jan. 18	Stewart, 2nd-Lt. F. G. . . .	31 Jan. 18	7 Feb. 18	Attd.
31 Aug. 18	Wallace, 2nd-Lt. D. J. C. . .	31 Aug. 18	1 Sep. 18	Attd.
31 Aug. 18	Wilkins, 2nd-Lt. R.	31 Aug. 18	1 Sep. 18	
	Attached.			
14 Nov. 14	Imrie, Lt. W.	14 Aug. 16	7 June 18	Sco. Rif., I. A. (*on probn.*)
1 Sep. 14	Kynaston, Capt. N. E. V. . .	14 Aug. 18	1 May 18	I. A. (*on probn.*)
25 Nov. 17	Lyte, 2nd-Lt. E. F. N. . . .	25 Nov. 17	9 July 18	I. A. R. O.
5 Jan. 18	Thompson, 2nd-Lt. G. L. . .	5 Jan. 18	31 May 18	I. A. (*T. C.*)
15 Apl. 19	Clappen, Lt. S. G.	15 Jan. 18	7 June 18	Lon. R., I. A. (*on probn.*)
26 Jan. 18	Ellen, 2nd-Lt. W.	26 Jan. 18	7 June 18	R. W. Surr. R., I. A. (*on probn.*)
5 Jan. 17	Perrott, Lt. L. C.	5 Jan. 18	17 Mar. 17	I. A. R. O.
5 Jan. 17	Hutchens, Lt. F. K.	5 Jan. 18	17 Mar. 17	I. A. R. O.
	Boyle, Lt. J.	. . . .	13 Dec. 17	I. A. R. O.
1 Mar. 17	Steele, Lt. F. R.	1 Mar. 18	13 Dec. 17	I. A. R. O.
14 May 17	Matthews, 2nd-Lt. G. L. . . .	14 May 17	4 Mar. 18	I. A. (*on probn.*)
29 Mar. 18	Smith, 2nd-Lt. W. S.	29 Mar. 18	19 July 18	I. A. R. O.
5 Nov. 17	Stark, Lt. C. W.	5 Nov. 18	18 July 18	I. A. (*T. C.*)
	Gray, 2nd-Lt. F.	. . . .	. . .	R. Lanc. R., I. A. R. O. (*on probn.*)
26 Jan. 15	Smyth, Lt. J. (*Actg. Capt.*) . .	26 Jan. 16	29 Sep. 17	I. A. R. O.
16 May 15	Smyth, Lt. S.	16 May 16	17 Jan. 18	I. A. R. O.
7 Apl. 17	Felix-Williams, Lt. A. . . .	7 Apl. 18	11 Oct. 18	R. W. Fus., I. A. (*on probn.*)
	Smerdon, 2nd-Lt. H. E. . . .	. . .	11 Oct. 18	K. R. Rif. C., I. A. R. O. (*on probn.*)
	Dutton, 2nd-Lt. W. H. R. . .	. . . .	14 Nov. 18	R. Suss. R., I. A. R. O. (*on probn.*)
28 Dec. 17	Meikle, 2nd-Lt. W. A. . . .	28 Dec. 17	. . . .	High. L. I., I. A. (*on probn.*)
	Subadar-Major.	Jemadar.	Subadar	Subadar-Major.
14 Nov. 90	Gheba Khan (52, 56)	. . . .	17 July 07	1 Sep. 17.
	Subadars.			
16 Apl. 92	Dugla (56)	1 Sep. 12	1 June 14	
13 Apl. 95	Rupa Singh	19 Jan. 15	1 Aug. 15	
2 May 94	Dheru Khan	22 Aug. 14	1 Apl. 16	
21 Oct. 96	Anar Gul	21 Nov. 14	18 Sep. 16	
6 Dec. 07	Ibrahim Khan, I.D.S.M. (50, 53) .	16 July 09	18 Nov. 16	
1 Nov. 93	Tikhu	1 June 14	1 Dec. 16	
26 May 94	Gul Din (*66a*)	26 Apl. 15	21 Mar. 17	
1 Aug. 96	Hardit Singh, I.D.S.M. (51) . .	27 Nov. 14	15 Apl. 17	
13 May 95	Isa Khan.	8 April 17	1 Sep. 17	
10 Jan. 96	Karam Khan	1 Sep. 15	16 Sep. 18	

1st Battalion, 55th Coke's Rifles (Frontier Force)—*contd.*

1st Commn. or date of entering service.	Names and Rank.	Army rank.	Remarks.
	Jemadars.		
30 Oct. 94	Sheraban	1 Aug. 12	
16 May 98	Husain	16 Oct. 15	
12 Dec. 96	Fazal Dad	1 Aug. 16	
5 Oct. 02	Haidar Khan	1 Aug. 16	
27 June 02	Abdullah Khan	18 Nov. 16	
18 Jan. 96	Teja Singh, I.O.M. (47*a*)	29 Jan. 17	
15 Feb. 97	Muhammad Khan	29 Jan. 17	
10 Jan. 03	Jahan Khan	1 July 17	
24 Mar. 99	Gobinda	30 Aug. 17	
9 Feb. 97	Shah Baz Khan	1 Sep. 17	
7 Sep. 02	Dalel Singh	22 Oct. 17	
14 Feb. 96	Mota Singh	23 Nov. 17	
1 Feb. 02	Ayub Khan	25 Nov. 17	
28 Apl. 91	Giyan Chand	5 Dec. 17	Indian Qr.-Mr.
25 June 98	Kishan Singh	7 Apl. 18	
10 Apl. 04	Saifur	22 Jan. 17	

2nd Battalion, 55th Coke's Rifles (Frontier Force).

Formed at Ambala.

Composition—8 platoons Cis-Frontier Pathans, 4 platoons Sikhs, 2 platoons Dogras, 2 platoons Punjabi Hindus.

1st Commn. or date of entering service.	Names and Rank.	Army rank.	Present appointment in Regiment.	Remarks.
	Commandant.			
20 July 98	Finnis, Maj. B. H. (*Actg. Lt.-Col.*)	1 Sep. 15	21 Feb. 18	20th Infy.
	Company Commanders. (4)			
18 Jan. 02	Champion de Crespigny, Maj. H., M.C.	18 Jan. 17	18 Feb. 18	56th Rif. 2nd-in-Comd. (*temp.*)
8 Aug. 14	Guise, Lt. J. W.	1 Sep. 15	1 May 18	45th Sikhs.
14 Oct. 14	Wilcox, Lt. F. J. (*Actg. Capt.*)	14 Oct. 15	15 Feb. 18	I. A. R. O.
29 July 15	Stileman, Lt. C. E. C. (*Actg. Capt.*)	29 July 15	16 May 18	Genl. List, I. A. (*on probn.*)
	Company Officers.			
27 Jan. 15	McCoy Hill, Lt. L. M. C. C.	27 Oct. 16	15 Feb. 18	I. A. (*on probn.*)
24 May 16	Flynn, Lt. A. E.	24 May 17	12 Nov. 18	Conn. Rang. I. A. (*on probn.*)
14 Nov. 15	Brilliant, Lt. L. (*Actg. Capt.*)	15 Nov. 16	4 July 18	129th Baluchis. Adjt.
27 Oct. 17	Cooper-Chadwick, Lt. I. L.	27 Oct. 18	13 Apl. 18	
27 Oct. 17	Milliken, Lt. E. K.	27 Oct. 18	23 Feb. 18	
1 Jan. 18	Ross-Hurst, 2nd-Lt. B. P.	1 Jan. 18	1 May 18	1-55th Rif.
31 Jan. 18	Bateman, 2nd-Lt. H. E. R.	31 Jan. 18	16 Feb. 18	Attd.
1 Oct. 18	Coleman, 2nd-Lt. G. V. L.	1 Oct. 18	10 Oct. 18	Attd.
28 Nov. 17	Gow, 2nd-Lt. H.	28 Nov. 17	13 Feb. 18	I. A. R. O.
15 Jan. 18	Davenport, 2nd-Lt. C. V.	15 Jan. 18	5 June 18	I. A. (*T. C.*)
15 Jan. 18	Jones, 2nd-Lt. J. S.	15 Jan. 18	5 June 18	I. A. (*T. C.*)
7 June 18	Coleman, 2nd-Lt. A. J.	7 June 18	10 Oct. 18	I. A. (*T. C.*)
31 Aug. 18	Wilkins, 2nd-Lt. G. A. W.	31 Aug. 18	20 Sep. 18	I. A. R. O.
5 Sep. 18	Keegan, 2nd-Lt. R. J.	5 Sep. 18	23 Sep. 18	I. A. R. O.
26 Sep. 18	O'Brien, 2nd-Lt. T.	26 Sept. 18	14 Oct. 18	I. A. R. O.
7 Nov. 18	Smith, 2nd-Lt. H. V.	7 Nov. 18	25 Nov. 18	I. A. R. O.
19 Sep. 15	Deuber, Capt. A. G. C.	26 Sep. 15	19 Nov. 18	Yorks. L. I.
22 Oct. 15	Moodie, Lt. A. E. S.	1 July 17	18 Nov. 18	Arg. and Suth'd. Highrs.
	Laskie, Lt. A. L.	12 Sep. 17	19 Nov. 18	Genl. List.
13 Mar. 18	Craig, 2nd-Lt. J. A.	13 Mar. 18	18 Nov. 18	Sea. Highrs.
22 July 18	Nicholas, 2nd-Lt. C. F.	22 July 18	2 Dec. 18	I. A. (*T. C.*)
22 July 18	Wall, 2nd-Lt. W. H.	22 July 18	2 Dec. 18	I. A. (*T. C.*)
22 July 18	Burscough, 2nd-Lt. G. E.	22 July 18	2 Dec. 18	I. A. (*T. C.*)

Subadar-Major.

		Jemadar.	Subadar.	Subadar-Major.	
12 Oct. 86	Bela Singh, *Bahadur*, I.O.M.	16 Oct. 12	16 Apl. 09	15 Apl. 18	*Order of Br. Ind., 2nd Cl.*

Subadars.

10 Feb. 00	Hamid Khan, I.D.S.M. (66)	16 Apl. 13	. . . .
31 Oct. 03	Hukmat Khan (66*a*)	18 May 15	2 June 16
20 Nov. 96	Hazrat Gul	8 Mar. 15	12 May 18
21 Jan. 96	Murafat Shah (66*a*)	20 Aug. 14	6 Apl. 18
20 Oct. 93	Gur Charan	13 Nov. 14	5 Apl. 18
16 June 96	Pala Singh	16 May 15	15 Apl. 18

Jemadars.

10 Oct. 98	Sundar	29 Jan. 17
1 May 98	Wazir Singh	15 Apl. 17
19 Sep. 95	Muhammad Sharif	1 Sep. 17
23 Sep. 99	Afzal Khan	15 Sep. 17
1 July 00	Juma Khan	30 Oct. 17
31 May 02	Nawdar Khan	6 May 18
3 June 99	Ghazi Khan	6 May 18
17 Sep. 03	Fatteh Singh	1 Oct. 18
16 Jan. 14	Ghulam Nabi	18 Oct. 18

Regt. Centre—Kohat.
Linked with—52nd Sikhs and 59th Rifles.

Uniform—Drab.
Facings—Black.

1st Battalion, 56th Punjabi Rifles (Frontier Force).

Johnston-ki-paltan.—Raised at Mianwali in 1849, by Lt. J. C. Johnston, from disbanded Sikh Darbar regiments, and originally designated the 2nd Regiment of Punjab Infantry. Became the 2nd Regiment of Infantry, Punjab Irregular Force, 1851; the 2nd Regiment of Infantry, Punjab Frontier Force, 1865; the 2nd Punjab Infantry, 1901; 56th Infantry Frontier Force, 1903; present designation, 1906.

"Delhi, 1857" "Lucknow" "Peiwar Kotal" "Afghanistan, 1878-79" "Tirah" "Punjab Frontier."

Composition—1 Company of Sikhs, 1 of Dogras, 1 of Punjabi Musalmans, 1 of Khattaks.

1st Commn. or date of entering service.	Names and Rank.	Army rank.	Present appointment in Regiment.	Remarks.
	Commandant.			
23 Dec. 93	Prissick, Lt.-Col. C.	3 June 17	22 Sep. 18	
	Company Commanders. (4)			
16 Dec. 96	*Routh, Maj. G. S. F. (Actg. Lt.-Col.)*	*16 Dec. 14*	*22 Sep. 18*	2nd-in-Comd. With 2-54th Sikhs.
4 Aug. 97	*Moberly, Bt.-Lt.-Col. B. R.,* D.S.O., *p.s.c. (temp. Lt.-Col.)*	*3 June 18*	*27 Nov. 13*	*G.S.O., 1st gr., Army Hdqrs.*
16 Feb. 98	Keen, Maj. P. H. (*Actg. Lt.-Col.*)	1 Sep. 15	18 Jan. 16	With 2-30th Punj.
4 May 01	Skinner, Maj. G. S.	4 May 16	22 Sep. 18	
	Company Officers. (9)			
30 Apl. 02	Wolley, Maj. H. S. L.	4 June 16	13 Dec. 06	Divl. R. O., Jullundur.
18 Jan. 02	Champion de Crespigny, Maj. H., M.C.	18 Jan. 17	11 Aug. 07	With 2-55th Rif.
29 Aug. 06	Le Cornu, Capt. C. J. S., M.C.	29 Aug. 15	16 Sep. 09	*With Sig. Service.*
9 Sep. 08	Curtis, Capt. O. O.	1 Sep. 15	13 Nov. 09	
24 Aug. 12	Weld, Capt. C. J.	24 Nov. 16	8 Nov. 13	
15 Nov. 15	Wilkinson, Lt. H. A.	15 Nov. 16	18 Nov. 15	*Lv., ex I., m. c.*
18 Apl. 16	Millar, Lt. A.B.	18 Apl. 17	26 Apl. 16	Attd.
8 Sep. 15	Tapsell, Lt. N. B.	8 June 17	25 Apl. 17	Attd.
29 June 16	Morris, Lt. H. G.	29 June 17	29 June 16	
31 Jan. 18	Gilbert, 2nd-Lt. C.	31 Jan. 18	6 Feb. 18	Attd.
19 Dec. 15	Owen, Lt. A. T. J.	19 Dec. 16	26 Apl. 17	Attd. With Patiala Infy.
1 May 17	Stephen, Lt. A. T.	1 May 18	30 Nov. 17	Attd. With 2-131st Infy.
21 Dec. 17	Hughes, 2nd-Lt. H.	21 Dec. 17	22 Mar. 18	Attd. With 2-151st Infy.
1 Oct. 18	Broadway, 2nd-Lt. A. C.	1 Oct. 18	1 Oct. 18	Attd.
	Attached.			
29 Aug. 06	Evans, Capt. D. M.	29 Aug. 15	2 Sep. 16	55th Rif.
29 Nov. 14	Nathan, Capt. S. L.	29 Nov. 18	24 Apl. 18	I. A. R. O.
11 Nov. 15	Shallow, Lt. G.	11 Nov. 16	11 Nov 15	I. A. R. O.
19 Oct. 18	*Cox, Lt. H. R.*	*19 Oct. 16*	. . .	*I. A. R. O.*
1 Jan. 16	Wilson, Lt. J. McL.	1 Jan. 17	28 Sep. 16	I. A. R. O.
5 May 15	Huthwaite, Lt. H. Y., M.C.	5 May 16	26 Nov. 17	I. A. (*on probn.*)
6 Oct. 16	Grieve, Lt. W.	6 Oct. 17	15 Aug. 17	I. A. R. O.
27 Oct. 17	Goode, 2nd-Lt. R. L.	27 Oct. 17	3 Nov. 17	
15 Jan. 18	Francis, 2nd-Lt. T. V.	15 Jan. 18	5 June 18	I. A. (*T. C.*)
	Jones, 2nd-Lt. J. L.	. . . .	8 Nov. 18	Genl. List, I. A. R. O. (*on probn.*)
25 Oct. 15	Boyd, Lt. N. R.	26 Oct. 16	6 Nov. 18	R. N. Vol. Reserve, I. A. R. O. (*on probn.*)

Subadar-Major.

		Jemadar.	Subadar.	Subadar-Major.	
11 July 87	Harnam Singh, *Bahadur* (56) I.O.M.	1 Oct. 07	28 July 15	7 Jan. 16	*Order of Br. Ind., 2nd Cl.*
1 Sep. 03	Bakhshi Chand (56) I.D.S.M.	1 Sep. 03	5 Dec. 12	11 July 16	

Subadars.

2 Oct. 82	Sohel Singh, *Sardar Bahadur* (56)	10 Feb. 00	20 Oct. 03	*Order of Br. Ind., 1st Cl.*
5 May 89	Mardan Ali (56)	1 Aug. 04	8 Oct. 05	
11 Mar. 85	Nanak Singh Bahadur (56, 72)	16 Jan. 05	13 Apl. 08	
16 Feb. 01	Jagat Singh (56)	24 Sep. 07	26 Oct. 14	
15 Mar. 95	Sher Singh	24 Oct. 14	1 Dec. 15	
11 June 94	Khwaja Muhammad, I.O.M.	4 Feb. 12	3 Jan. 16	
27 Mar. 95	Sangara	31 Oct. 15	7 Jan. 16	
	Allah Ditta Khan (*Retd.*)	. . . .	. . . .	Re-employed, 5 July 15. Depôt.
12 Nov. 05	Jodh Khan, I.O.M.	24 Oct. 14	27 June 16	With 2-151st Infy.
21 June 00	Bakhshish Singh	8 Mar. 16	11 July 16	
1 July 03	Kan Beg	13 Jan. 16	23 Feb. 17	
1 Dec. 98	Mitt Singh	24 Oct. 14	23 Feb. 17	
	Sunit Chand, I.O.M.	. . . .	3 Nov. 17	52nd Sikhs.
17 Dec. 95	Lachhman	1 Dec. 15	1 Dec. 17	
30 Nov. 98	Spin Gul	10 Mar. 17	1 June 18	
	Shah Zaman (*Ret.*)	. . . .	. . . .	Re-employed.
	Sarfraz Khan (*Ret.*)	. . . .	. . . .	Do.

1st Battalion, 56th Punjabi Rifles (Frontier Force)—*contd.*

1st Commn. or date of entering service.	Names and Rank.	Army rank.	Remarks.
	Jemadars.		
30 Dec. 99	Karam Dad	1 Dec. 15	With 2-151st Infy.
	Sher Ahmad Khan	15 Dec. 15	
20 May 96	Lachhman Singh	7 Jan. 16	
5 Jan. 95	Wazir Singh	13 Jan. 16	
21 June 00	Dost Muhammad	8 Mar. 16	With 2-151st Infy.
31 May 02	Nadir Shah	9 Mar. 16	
26 Feb. 01	Sapuran Singh, I.O.M.	17 May 16	
22 June 00	Karam Sher	24 Oct. 16	With 2-151st Infy.
19 Mar. 02	Ishar Singh	23 Feb. 17	
11 June 94	Nadir Khan	10 Mar. 17	
21 Mar. 02	Sham Nang	10 Mar. 17	
4 Apl. 06	Hoshan Khan	10 Mar. 17	With 2-151st Infy.
28 Sep. 04	Gulab Din	23 Apl. 17	
27 July 12	Muhammad Azam	23 Apl. 17	
13 Aug. 07	Chaudhri	10 Aug. 17	
25 Oct. 98	Gul Khan	24 Jan. 18	
23 Mar. 06	Muhammad Khan	1 June 18	
6 Dec. 04	Ram Singh	5 Aug. 18	
1 July 99	Gopal Singh	3 Dec. 17	
17 Feb. 06	Madar Khan	31 May 18	
	Rakhim Gul	....	Re-employed.

2nd Battalion, 56th Punjabi Rifles (Frontier Force).

Formed at Multan on 5 June 1917.

Composition—

1st Commn. or date of entering service.	Names and Rank.	Army rank.	Present appointment in Regiment.	Remarks.
	Commandant.			
5 Aug. 96	Ames, Maj. C. G. (*Actg. Lt.-Col.*)	5 Aug. 14	8 Aug. 17	52nd Sikhs.
	Company Commanders. (4)			
27 July 01	Philips, Maj. J. C.	1 Sep. 15	28 Aug. 17	28th Punj. 2nd-in-Comd. (*temp.*)
19 May 00	Moreton-Marshall, Maj. F. W.	1 Sep. 15	17 Apl. 18	130th Baluchis.
22 Jan. 13	Beanland, Capt. D.	22 Jan. 17	1 Jan. 18	1-22nd Punj.
1 June 15	Maude, Lt. E. W., M.C. (*Actg. Capt.*)	1 June 16	21 Aug. 17	Oxf. and Bucks. L. I., I. A. (*on probn.*).
	Company Officers.			
5 May 15	Wade, Lt. H. (*Actg. Capt.*)	5 May 16	21 Aug. 18	Attd. Adjt. (*Offg.*).
15 May 17	Barton, Lt. R. F.	15 May 18	29 Sep. 18	R. Scots., I. A. (*on probn.*).
13 Nov. 14	Phelps, Lt. A. E.	15 Aug. 15	12 Dec. 17	Attd. Qr. Mr. (*Offg.*)
18 Nov. 16	McCutcheon, Lt. G. H.	18 Nov. 17	30 July 18	Attd.
30 May 17	Perraton, Lt. C. W.	30 May 18	1 Apl. 18	I. A. R. O.
31 Oct. 17	Wright, Lt. R. A.	31 Oct. 18	13 Feb. 18	I. A. R. O.
29 Mar. 18	Swynnerton, 2nd-Lt. F. R.	29 Mar. 18	. . . .	I. A. R. O.
31 Aug. 18	May, 2nd-Lt. C. R. A.	31 Aug. 18	31 Aug. 18	Attd.
31 Aug. 18	Richardson-Gibson, 2nd-Lt. F. J.	31 Aug. 18	31 Aug. 18	Attd.
31 Aug. 18	Dolphin, 2nd-Lt. H. M. B.	31 Aug. 18	31 Aug. 18	Attd.
21 Dec. 17	Radcliff, 2nd-Lt. H. G.	21 Dec. 17	29 July 18	I. A. (U. L.)
7 June 18	Finelli, 2nd-Lt. C. W. F.	7 June 18	19 Oct. 18	I. A. R. O.
7 June 18	Smith, 2nd-Lt. F.	7 June 18	19 Oct. 18	I. A. R. O.
10 July 18	Badhams, 2nd-Lt. O. G.	10 July 18	2 Dec. 18	I. A. (*T. C.*)
22 July 18	Jodrell, 2nd-Lt. F. J.	22 July 18	2 Dec. 18	I. A. R. O.

Subadar-Major.

		Jemadar.	Subadar.	Subadar-Major.
5 Dec. 98	Atta Khan, I.D.S.M. (56, 72*a*)	16 Oct. 11	5 June 17	

Subadars.

27 Mar. 04	Bishan Singh, I.D.S.M. (46, 56, 66*a*)	20 Nov. 07	5 June 17	
26 Apl. 92	Lorinda (66*a*)	5 Jan. 15	5 June 17	
23 Sep. 98	Ahmad Shah (56, 58, 66*a*)	. . . .	5 June 17	
12 Oct. 03	Fakir Muhammad	10 Jan. 17	6 Nov. 17	59th Rif.
17 Feb. 97	Raj Mal (56)	5 June 17	21 Jan. 18	
21 Jan. 18	Amin Khan	. . . .	21 Jan. 18	
6 Nov. 99	Azim Khan	11 Oct. 14	12 Oct. 16	1-22nd Punj.

Jemadars.

23 Dec. 97	Bhagel Singh, I.D.S.M.	5 June 17	
20 Aug. 17	Wali Mohomed Khan (46)	20 Aug. 17	
27 May 10	Alam Sher	15 June 18	
10 Apl. 03	Muhammad Saraj	15 June 18	
30 May 06	Ata Muhammad	25 June 18	
19 Mar. 07	Jalal Khan	25 June 18	
1 Feb. 97	Katha Singh	7 June 16	1-22nd Punj.
10 July 06	Muhabat Khan (58)	11 Nov. 17	1-22nd Punj.

Regt. Centre—Dera Ismail Khan. *Uniform*—Drab.
Linked with—55th and 58th Rifles. *Facings*—Prussian Blue.

57th Wilde's Rifles (Frontier Force).

Wilde-ki-paltan.—Raised at Lahore in 1849 by Captain G. G. Denniss, from drafts from disbanded Sikh Darbar regiments and originally styled the 4th Regiment of Punjab Infantry. Became the 4th Regiment of Punjab Irregular Force, 1851; the 4th Regiment of Infantry, Punjab Frontier Force, 1865; the 4th Punjab Infantry, 1901; present designation, 1903.

"Delhi, 1857" "Lucknow" "Afghanistan, 1879-80" "China, 1900."

Composition—1 Companies of Sikhs, 1 of Dogras, 1 of Punjabi Musalmans, 1 of Pathans.

1st Commn. or date of entering service.	Names and Rank.	Army rank.	Present appointment in Regiment.	Remarks.
	Commandant.			
1 Oct. 93	Willans, Lt.-Col. T. J., D.S.O.	8 Mar. 18	8 Mar. 18	
	Company Commanders. (4)			
17 Jan. 91	*Swifte, Lt.-Col. E. L. (Q)*	*17 Jan. 17*	*16 Aug. 16*	*2nd-in-Comd. Lv. ex. I. m. c.*
27 July 98	*Pepys, Maj. G. L.*, D.S.O., *p.s.c.* [L] *(temp. Lt.-Col.)*	*1 Sep. 15*	*3 Nov. 14*	*G. S. O., 1st gr., A. Hdqrs.*
24 July 01	Forbes, Maj. L.	24 July 16	8 Mar. 18	*Lv., ex I., m.c.*
21 Jan. 03	*Radford, Maj. O. H. (temp. Lt.-Col.)*	*21 Jan. 17*	*10 Aug. 18*	*Comdt., N. C. O's. Schl., Secunderabad.*
	Company Officers. (9)			
5 Aug. 05	*Trail, Capt. W. S.*, M.C. *(temp. Maj.)*	*5 Aug. 14*	*22 Apl. 07*	*A. M. S., A. Hdqrs.*
29 Aug. 06	Fowler, Bt.-Maj. E. K., M.C. *(Actg. Maj.)*	1 Jan. 18	19 Mar. 09	2nd-in-Comd. *(temp.).*
25 Jan. 08	Molony, Capt. C. W.	1 Sep. 15	2 Aug. 09	With 2nd Guides.
9 Sep. 08	Bainbridge, Capt. D., M.C.	1 Sep. 15	9 Dec. 09	With 3rd Bn. Guides.
14 Jan. 14	Glegg, Capt. J. A.	14 Jan. 18	11 Oct. 14	
14 May 15	Leonard, Lt. D. J. *(Actg. Capt.)*	14 May 16	8 Aug. 17	Attd. Adjt.
18 Apl. 16	Guy, Lt. N. G.	18 Apl. 17	26 Apl. 16	Attd.
18 Apl. 16	Edward, Lt. W. D.	18 Apl. 17	26 Apl. 16	Attd. With 58th Rif.
1 Oct. 18	Tower, 2nd-Lt. H. G. E., M.C.	1 Oct. 18	1 Oct. 18	Attd.
18 June 17	Iliff, Lt. W. A. B.	18 June 18	17 Nov. 17	Attd. With 36th Sikhs.
18 June 17	Maclaren, Lt. C. S.	18 June 18	17 Nov. 17	Attd. With 58th Rif.
12 Sep. 17	Amsden, Lt. B. H.	12 Sep. 18	. . . .	Attd.
21 Dec. 17	Kitchen, 2nd-Lt. A. C.	21 Dec. 17	. . . .	Attd. With 2-27th Punj.
31 Jan. 18	Coldstream, 2nd-Lt. G. A. P.	31 Jan. 18	7 Feb. 18	Attd. With 3-30th Punj.
31 Jan. 18	Fogerty, 2nd-Lt. W. P.	31 Jan. 18	7 Feb. 18	Attd. With 58th Rif.
1 Oct. 18	Dent, 2nd-Lt. R. C.	1 Oct. 18	1 Oct. 18	Attd.
21 Dec. 17	Jefcoate, 2nd-Lt. H. M.	21 Dec. 17	. . . .	Attd.
	Attached.			
22 Dec. 14	Pedley, Lt. J. E., M.C.	22 Dec. 15	22 Jan. 18	I. A. R. O.
5 May 15	Clarke, Lt. F. A. S., D.S.O. *(Actg. Capt.)*	5 May 16	. . . .	I. A. *(on probn.)*.
5 June 15	Moorcroft, Lt. R.	5 June 18	8 Feb. 18	I. A. *(on probn.)*.
7 Apl. 16	Bradshaw, Lt. A. T. R.	7 Apl. 17	17 Apl. 18	E. Lan. R., I. A. *(on probn.)*.
15 Feb. 16	Ellis, Lt. A. H. J. *(Actg. Capt.)*	15 Feb. 17	. . . .	I. A. *(on probn.)*
23 Feb. 16	Rainier, Lt. W. H. C.	23 Feb. 17	31 Mar. 18	*I. A. (on probn.)*
29 June 16	Bion, Lt. R. H.	29 June 17	. . . .	13th Lrs.
25 July 17	Butler, Lt. F. H.	25 July 18	30 Mar. 18	I. A. *(on probn.)*.
22 Aug. 17	Connel, Lt. J. H.	22 Aug. 18	25 Mar. 18	I. A. *(on probn.)*.
26 Sep. 17	Francis, Lt. W. D.	26 Sep. 18	30 Mar. 18	I. A. R. O.
7 June 18	Mount, 2nd-Lt. T.	. . . .	12 Oct. 18	I. A. *(T. C.)*.
. . . .	Allen, 2nd-Lt. W. A.	. . . .	. . . .	I. A. (U. L.)

Subadar-Major.

		Jemadar.	Subadar.	Subadar-Major.	
9 Aug. 96	Arsla Khan, *Sardar Bahadur*, M.C., I.O.M. (66*a*, 72*a*)	16 Dec. 99	16 Jan. 03	1 June 15	*[Cl. Order of Br. Ind., 1st*

57th Wilde's Rifles (Frontier Force)—*contd.*

1st Commn. or date of entering service.	Names and Rank.	Army rank.		Remarks.
	Subadars.			
12 Apl. 84	Madaman, *Sardar Bahadur* (56)	1 Apl. 95	16 Dec. 02	*Order of Br. Ind., 1st Cl.*
1 Jan. 03	Bhagat Singh, I.D.S.M. (56)	17 Sep. 04	16 Mar. 12	
19 Jan. 04	Muhammad Azim Khan (56)	19 Jan. 04	1 July 12	
23 Dec. 90	Chur Singh (56)	1 Nov. 07	8 Nov. 13	
15 Feb. 93	Jagat Singh, I.O.M., I.D.S.M. (56)	21 May 08	15 Aug. 14	
6 Aug. 90	Amar Singh	15 Aug. 14	31 Oct. 14	
10 June 90	Imam Ali, *Bahadur* (56, 58)	1 Nov. 09	31 Mar. 15	*Order of Br. Ind., 2nd Cl*
28 Mar. 02	Rabel Singh (56)	15 Aug. 14	5 Dec. 15	
26 Mar. 00	Ram Ditta	12 Oct. 15	6 Dec. 15	With 58th Rif.
12 Aug. 98	Shah Baz	1 Nov. 12	18 Mar. 16	
27 June 92	Ahmed Khan (56)	1 July 12	14 Oct. 18	
1 Mar. 00	Ganga Ram (46)	31 Oct. 14	15 Oct. 18	
	Jemadars.			
8 Nov. 13	Dewa Singh	8 Nov. 13		
18 Mar. 98	Bur Singh, I.D.S.M.	31 Oct. 14		
19 Apl. 02	Zargir	13 Mar. 15		Indian Adjt.
8 Aug. 00	Fazal Ahmad	31 Mar. 15		With 58th Rif.
18 June 98	Mangal Singh, I.O.M.	8 Apl. 15		*Instructor N. C. O's. Sch.*
21 Feb. 03	Badhawa Sing	26 Apl. 15		
31 Jan. 04	Paras Ram	26 Apl. 15		
12 July 07	Hira Sing, I.D.S.M.	5 June 15		
11 Aug. 03	Sultan Khan	17 Apl. 17		
2 Jan. 00	Allah Ditta	22 May 18		
23 Dec. 00	Surayan Singh	20 July 18		
2 Jan. 03	Rahmat Khan	17 Sep. 18		
19 Dec. 00	Firoz Ali	3 Oct. 18		
15 Sep. 03	Qasim Khan	4 Oct. 18		
....	Sadhu Singh	14 Oct. 18		
....	Hira Singh	14 Oct. 18		
....	Gajindar Singh	14 Oct. 18		
....	Prabh Dayal	14 Oct. 18		

Regt. Centre—Dera Ismail Khan.
Linked with—55th and 57th Rifles.

Uniform—Drab.
Facings—Emerald Green.

58th Vaughan's Rifles (Frontier Force).

Vaughan-ki-paltan.—Raised at Leiah in 1849, by Capt. J. E. Gastrell, as the 5th Regiment of Punjab Infantry. Became the 5th Regiment of Infantry, Punjab Irregular Force, in 1851; the 5th Regiment of Infantry, Punjab Frontier Force, in 1865; the 5th Punjab Infantry in 1901; present designation in 1903.

"Peiwar Kotal" "Charasiah" "Kabul, 1879" "Afghanistan, 1878-80."

Composition—1½ Companies of Sikhs, ½ of Dogras, 1½ of Pathans, ½ of Punjabi Musalmans.

1st Commn. or date of entering service.	Names and Rank.	Army rank.	Present appointment in Regiment.	Remarks.
	Commandant.			
2 May 91	Knox, Bt.-Col. A. W. F., C.B., *p.s.c.* [L]. (*temp. Br.-Genl.*)	1 Jan. 17	24 July 16	
	Company Commanders. (4)			
21 Oct. 93	Thomson Bt.-Lt. Col. A.G., D.S.O., *p.s.c.* (Q).	3 June 18	24 July 16	2nd-in-Comd.
22 Oct. 97	*Finlay, Maj. R. F.*, D.S.O.	*1 Sep. 15*	*3 Jan. 08*	*G. S. O., 2nd gr., Army Hdqrs.*
1 Dec. 97	Lind, Maj. A. G., D. S. O. (*Actg. Lt.-Col.*)	1 Sep 15	17 Sep. 09	Comdt. (*temp.*).
1 Dec. 97	Smith, Maj. A.A. (*Actg. Lt.-Col.*)	1 Sep. 15	5 Apl. 14	With 2-151st Infy.
	Company Officers. (9)			
11 Feb. 99	Willis, Maj. E. S. C., D.S.O.	1 Sep. 15	25 May 04	
26 May 00	Flood, Maj. J. D. M.	1 Sep. 15	27 May 04	
4 May 01	*Pope, Maj. S. B.*, D.S.O., *p.s.c. (Actg. Lt.-Col.)*	*4 May 16*	*25 Feb. 04*	*G. S. O., 1st gr.*
21 Jan. 13	Mackenzie, Capt. D. B.	21 Jan. 17	12 Nov. 15	Adjt.
13 Jan. 14	McConkey, Capt. A. I. G.	14 Jan. 18	7 Nov. 15	Qr.-Mr. (*offg.*)
14 Jan. 14	Dowland, Capt. G. R., M.C.	14 Jan. 18	15 Oct. 14	
8 Jan. 15	*Jameison, Lt. J. W. E.*	*8 Oct. 16*	*16 Aug. 15*	*Attd. With R. A. F.*
4 Nov 15	Gillespie, Lt. D. S.	4 Nov. 16	11 May 17	Attd.
29 June 16	Cumming, Lt. A. E., M.C.	28 Dec. 16	12 Feb. 17	With 53rd Sikhs.
18 Apl. 16	Spear, Lt. C. R.	18 Apl. 17	25 Apl. 16	
18 Apl. 16	Jackson, Lt. H. J. R.	18 Apl. 17	25 Apl. 16	
29 June 16	Ross-Hurst, Lt. W. J. O'B.	29 June 17	29 Mar. 17	
30 Jan. 17	Aberrethy, Lt. P. P.	30 Jan. 18	13 June 17	
29 July 15	Ashton, Lt. P. W.B. (*Actg. Capt.*)	29 July 16	5 Nov. 17	Attd.
14 May 16	Babtie, Lt. J. W. A.	14 May 17	5 Dec. 17	Attd.
	Attached.			
23 July 01	Waller, Maj. R. de W.	23 July 16	16 Nov. 15	108th Infy.
1 Oct. 14	Ekin, Capt. R. G.	1 Oct. 18	. . . .	55th Rif.
20 Nov. 14	Gray, Capt. S. T.	20 Nov. 18	17 May 15	I. A. R. O.
12 Feb. 15	Farley, Lt. G. P. (*temp. Capt.*)	12 Feb. 16	4 May 16	I. A. R. O.
5 May 15	Snell, Lt. W. J. B.	5 May 16	28 Feb. 18	I. A. (*on probn.*)
30 Jan. 17	Frost, Lt. R. L.	30 Jan. 18	21 Dec. 17	1-69th Punj.
29 Aug. 17	Harris, Lt. D. B. B.	29 Aug. 18	. . . .	I. A. R. O.
27 Oct. 17	Pearson, Lt. G. T.	27 Oct. 18	3 Nov. 17	
26 Sep. 17	Breton, Lt. B. O.	26 Sep. 18	29 Jan. 18	I. A. R. O.
5 May 17	Green, Lt. D. C.	5 May 18	27 June 18	I. A. R. O.
17 June 15	Manlove, Lt. J. E. D.	17 June 16	31 July 18	I. A. (*on probn.*)
20 June 17	Dunn, Lt. O. H. B.	20 June 18	10 July 18	I. A. R. O.
20 June 17	Lyon, Lt. H. B.	20 June 18	18 July 18	I. A. R. O.
16 Nov. 14	Locke, Lt. D. W.	1 June 16	17 Aug. 18	Devon. R.,
10 Dec. 14	Potbury, Lt. H. W.	1 June 16	11 Sep. 18	Devon. R.,
5 Aug. 16	Budleid, Lt. J. F.	5 Feb. 18	17 Aug. 18	Devon. R.
31 Oct. 17	Allen, Lt. A. H. C., M.C.	31 Oct. 18	17 Aug. 18	Devon. R.,
18 June 17	Maclaren, Lt. C. S.	18 June 18	7 Sep. 18	57th Rif.
7 June 18	Peace, 2nd-Lt. J. N.	7 June 18	11 Oct. 18	I. A. (*I. C.*)
31 Jan. 18	Fogerty, 2nd-Lt. W. P.	31 Jan. 18	1 Nov. 18	I. A. (U. L.)
31 Jan. 18	Roach, 2nd-Lt. J. O.	31 Jan. 18	2 July 18	I. A. R. O.
7 Oct. 18	Hazzard, 2nd-Lt. P. H.	7 Oct. 18	2 Nov. 18	I. A. R. O. [*Offrs. Campbellpur.*
7 Oct. 18	Pouncy, 2nd-Lt. T. C.	7 Oct. 18	2 Nov. 18	57th Rif.
12 Oct. 18	Hastings, 2nd-Lt. J. E.	12 Oct. 18	31 Oct. 18	I. A. R. O. *Sch. of Instn. for*

Subadar-Major.

		Jemadar.	Subadar.	Subadar-Major.
6 May 93	Tika Khan (56, 58, 66, 72*a*, 72*b*)	1 Mar. 01	6 May 04	1 Dec. 17
20 Apl. 94	Indar Singh, M.C., I.D.S.M.	3 Nov. 14	14 Feb. 15	7 July 18

58th Vaughan's Rifles (Frontier Force)—*contd.*

Subadars.

1st Commn. or date of entering service.	Names and Rank.	Army rank. Jemadar.	Present appointment in Regiment. Subadar.	Remarks.
8 Aug. 91	Gujar Singh (56, 66)	25 Oct. 01	1 Aug. 07	
6 Sep. 05	Ghulam Haidar Khan (56)	16 Sept. 05	1 July 09	
7 Jan. 97	Raj Talab, *Bahadur*, I.D.S.M.	13 Aug. 14	1 Nov. 14	*Order of Br. Ind., 2nd Cl.*
8 Sep. 96	Karm Singh, I.O.M.	1 Nov. 14	24 Nov. 14	
1 May 00	Harchand Singh, I.O.M.	24 Nov. 14	7 Mar. 15	
23 Aug. 00	Muhammad Arbi, I.O.M.	8 Feb. 15	26 Sep. 15	
1 May 01	*Hamesh Gul (72a, 72b)*	*4 Mar. 15*	*26 Jan. 16*	
1 July 94	Suba (58)	1 Apl. 12	2 June 16	
24 July 06	Fateh Roz Khan	26 Sep. 15	31 Mar. 18	
31 Oct. 14	Ajun Khan, I.D.S.M.	18 Sep. 16	1 Jan. 18	
	Surjan (*Ret.*)	. . . .	. . . .	Re-employed.
	Khan Bahadur (*Ret.*)	. . . .	. . . .	Do.
17 Feb. 12	Kehr Singh	7 Sep. 15	1 Dec. 17	
11 July 07	Nur Akhmad	1 Sep. 15	6 July 18	(*Supy.*)
16 Mar. 96	Madat	1 Dec. 15	19 July 18	

Jemadars.

1st Commn. or date of entering service.	Names and Rank.	Army rank.	Remarks.
12 Apl. 92	Rasila	25 Oct. 14	
18 Oct. 99	Santu (72*a*)	1 Aug. 15	
11 Oct. 06	*Chaman*	*1 Aug. 15*	
24 Feb. 07	Bishnu Singh	26 Sep. 15	
18 July 99	Diwany, I.O.M., (56, 72)	26 Jan. 16	
15 July 93	Rang Baz Khan	1 Apl. 17	
9 Feb. 00	Kehr Singh	1 May 17	
5 Nov. 06	Sowar Khan	1 May 17	
	Juma Khan	30 Oct. 17	55th Rif.
30 Apl. 97	Harnam Snigh	3 Dec. 17	
18 Mar. 01	Thakur Sing	31 Mar. 18	
14 Nov. 11	Muhammad Sher	31 Mar. 18	
27 Dec. 08	Hira Singh	24 May 18	
14 June 01	Budh Singh	24 May 18	
14 Feb. 08	Faiz Muhammad Khan	1 June 18	
13 May 05	Sher Khan, I.O.M.	13 June 18	
3 Mar. 13	Misri Khan	19 July 18	
25 Sep. 09	Baidullah, I.D.S.M.	11 Aug. 18	
3 Dec. 91	Mirali	1 Oct. 18	
5 July 01	Channan Singh	21 July 18	
28 May 04	Wali Dad	21 Aug. 18	

Regt. Centre—Kohat. *Uniform*—Drab.
Linked with—52nd Sikhs & 56th Rifles. *Facings*—Scarlet.

59th Scinde Rifles (Frontier Force).

Raised at Karachi in 1843 as the Scinde Camel Corps. Became the 6th Regiment of Infantry, Punjab Irregular Force, or the Scinde Rifle Corps, 1853; the 'Scinde Rifle Corps' the same year; the 6th Regiment of Punjab Infantry, 1856; the 6th Regiment of Infantry, Punjab Frontier Force, 1865; the 6th Punjab Infantry, 1901; present designation, 1903.

"Ready, Aye Ready!"

Composition—1½ companies of Pathans, ½ of Punjabi Musalmans, 1 of Sikhs, 1 of Dogras.

1st Commn. or date of entering service.	Names and Rank.	Army rank.	Present appointment in Regiment.	Remarks.
	Commandant.			
9 Sep. 91	Leeds, Lt.-Col. T. L., C.M.G., D.S.O.	3 June 15	20 Feb. 17	
	Company Commanders. (4)			
1 Mar. 96	Kirkpatrick, Lt.-Col. E. (Q)	1 Mar. 16	27 Oct. 13	2nd-in-Comd.
11 Feb. 99	*Murray, Maj. K. D. B.,* D.S.O., *p.s.c. (temp. Lt.-Col.)*	*1 Sep. 15*	*25 Aug. 13*	*G. S. O., 1st gr., Meerut Divn.*
8 May 01	Anderson, Maj. B. E., D.S.O., *p s.c.*[L].	18 Feb. 15	20 Feb. 17	
9 Jan. 04	Burn, Capt. A. H.	9 Jan. 13	3 June 18	
	Company Officers. (9)			
5 Aug. 05	Inskip, Capt. R. D., D.S.O., M.C.	5 Aug. 14	8 May 08	*Bde. Maj. 28th Bde.*
22 Jan. 13	Stuart-Prince, Capt. C. E.	22 Jan. 17	11 Mar. 14	With 2-151st Infy.
14 Jan. 14	Hamilton, Capt. H. M.	14 Jan. 18	14 Oct. 14	With 3-151st. Infy.
15 Nov. 15	Twinberrow, Lt. J. D.	15 Nov. 16	18 Nov. 15	
18 Apl. 16	Malden, Lt. C. M.	18 Apl. 17	. . . .	Attd.
16 Aug. 16	Simms, Lt. G. H.	16 Aug. 17	25 Dec. 17	
2 July 17	Stirling, Lt. J. M.	2 July 17	10 Dec. 18	
30 Jan. 17	Lindsay, Lt. D. M.	30 Jan. 18	25 Dec. 17	
18 June 17	Clarke, Lt. P. T.	18 June 18	18 July 18	Attd.
12 Sep. 17	Arbuthnot, Lt. J. St. C.	12 Sep. 18	12 Sept. 17	Attd. With 3-151st Infy.
31 Jan. 18	Seccombe, 2nd-Lt. J. B. P.	31 Jan. 18	6 Feb. 18	Attd.
	Attached.			
27 Nov. 14	Burne, Capt. R. H., M.C.	27 Nov. 17	. . . .	I. A. R. O.
23 July 15	Cooper, Capt. H. L. (*Actg. Maj.*)	18 Aug. 17	1 Apl. 17	I. A. R. O.
18 Dec. 14	Jarrad, Lt. E. D. D.	18 Dec. 15	15 Aug. 15	I. A. R. O. *A. R. O., Dogras.*
15 Feb. 16	Waite, Lt. J. J.	18 Aug. 17	21 Sep. 18	R. Ir. Rif.
29 Mar. 16	Dudley, Lt. R. S.	29 Mar. 17	. . . .	I. A. R. O.
6 May 16	Harris, Lt. C. E.	16 May 17	29 May 16	I. A. R. O.
15 Nov. 15	Hansen, Lt. G. E.	15 Nov. 16	19 Nov. 15	I. A. R. O.
4 May 17	Lindsay, Lt. H. N.	4 May 18	17 Oct. 18	R. Innis. Fus. I. A. (*on prob'n.*).
12 Sep. 17	Wilson, Lt. R. E.	12 Sep. 18	14 Oct. 18	Midd'x. R., I. A. (*on probn.*)
29 Mar. 18	Wilkins, 2nd-Lt. H. W. J.	29 Mar. 18	14 July 18	I. A. (*T. C.*)
7 June 18	Cotton, 2nd-Lt. C. W.	7 June 18	20 June 18	I. A. R. O.

Subadar-Majors.

		Jemadar.	Subadar.	Subadar-Major.	
22 May 83	Nasir Khan, I.O.M., *Sardar Bahadur*	16 May 03	1 Nov. 06	19 Dec. 14	*Order of Br. Ind., 1st Cl.*
20 Mar. 07	Parbat Chand, M.C. (53, 56, 66*a*)	20 Mar. 07	16 June 14	10 Mar. 15	

Subadars.

1 Sep. 83	Mukhmad Jan, *Sardar Bahadur*, I.D.S.M. (56, 60)	1 July 02	1 Dec. 07	*Order of Br. Ind., 1st Cl.*
28 Nov. 98	Khan Gul, I. O. M.	26 Nov. 14	26 Dec. 14	
27 Oct. 94	Zarin Khan (66*a*)	9 Aug. 14	10 Mar. 15	
1 Jan. 03	Gauri Charan, *Bahadur*, I. O. M.	9 Aug. 14	12 Mar. 15	*Order of Br. Ind., 2nd Cl.*
4 Apl. 03	Ali Akbar	20 Dec. 14	8 Mar. 16	
28 Mar. 87	Makhmad Ali (56)	1 Dec. 07	17 June 16	
12 Dec. 98	Jowahir Singh	. . . .	. . . .	
1 Sep. 03	Sahib Haq	8 Mar. 16	10 Jan. 17	
18 Dec. 00	Muhammad Zaman	1 Dec. 16	10 Jan. 17	With 3-151st Infy.
30 June 94	Hardit Singh	22 Dec. 14	9 Feb. 17	
6 Mar. 00	Santa Singh	. . . .	8 Mar. 17	
29 Apl. 02	Partab Singh	10 Jan. 17	6 Nov. 17	
12 Oct. 03	Fakir Muhammad	10 Jan. 17	6 Nov. 17	With 2-56th Rif.
16 Feb. 01	Malang Khan	10 Jan. 17	15 Mar. 18	
17 Dec. 97	Dost Muhammad	6 Nov. 17	22 June 18	With 3-151st Infy.
23 Mar. 04	Ruqam Din	26 Mar. 17	15 July 18	
	Shah Wali (Retd.)	. . . .	. . . .	Re-employed, 29th Mar. 15. Depôt.

59th Scinde Rifles (Frontier Force)—*contd.*

1st Commn. or date of entering service.	Names of Rank.	Army rank.	Remarks.
	Jemadars.		
5 Mar. 97	Chunak Singh (**56, 66*a***)	1 July 11	
4 Mar. 97	Gulab Singh	29 Oct. 14	
10 Mar. 97	Jabar Khan	10 Mar. 15	
26 July 00	Karam Din	9 June 15	
8 Aug. 03	Ram Singh	13 Oct. 15	
9 Apl. 00	Har Dayal	8 Mar. 16	
0 Dec. 02	Akhmad Khan	4 June 16	
4 Feb. 09	Bhan Singh	10 Jan. 17	With 3-151st Infy.
23 Aug. 04	Hazarat Khan	24 Mar. 17	
	Indar Singh	26 Mar. 17	
2 Aug. 07	Nadir Khan	6 Nov. 17	With 3-151st Infy.
25 Mar. 03	Qalandar Khan, I.O.M.	6 Nov. 17	
1 Feb. 18	Brij Mohan Lal	1 Feb. 18	(*On probn.*)
13 Dec. 07	Paru	15 Mar. 18	
27 Jan. 04	Nathu	16 Mar. 18	
18 Mar. 06	Niyaz Gul	16 Mar. 18	
	Amir Ali	17 June 18	With 3-151st Infy.
13 Oct. 08	Raja Khan	22 June 18	With 3-151st Infy.
26 Aug. 10	Hukum Dad	22 June 18	
25 Aug. 95	Nadhan Singh	15 July 18	
12 Jan. 03	Sadar Din	15 July 18	
30 Mar. 07	Karam Singh	15 July 18	
	Natha Singh	. . .	Re-employed 16 Apl. 17.

Regt. Centre—Trichinopoly. *Uniform*—Scarlet.
Linked with—64th & 81st Pioneers. *Facings*—White.

1st Battalion, 61st King George's Own Pioneers.

Shaikh Mudin-ki-purani paltan.—Formed at Fort St. George in December, 1758, from independent companies that had been in existence some years, and originally styled the 1st Battalion of Coast Sepoys. Became the 1st Carnatic Battalion, 1769; the 1st Madras Battalion, 1784; the 1st Battalion, 1st Regiment, Madras Native Infantry, 1796; disbanded, 1806; reformed as the 1st Battalion, 24th Regiment, of M. N. I., 1807; restored to its original position as the 1st Battalion, 1st Regiment of M. N. I., 1818; became the 1st Regiment of M. N. I., 1824; the 1st Regiment of M. N. I. (Pioneers), 1883; the 1st Regiment of Madras Infantry (Pioneers), 1885; the 1st Madras Pioneers, 1901; the 61st Pioneers, 1903; the 61st Prince of Wales's Own Pioneers, 1906; present designation, 1910.

THE PLUME OF THE PRINCE OF WALES. THE ROYAL AND IMPERIAL CYPHER.
"Carnatic" "Mysore" "Seringapatam" "Seetabuldee" "Nagpore" "Av " "Pegu"
"Central India" "Afghanistan, 1879-80" "Burma, 1885-87" "China, 1900."
Composition—2 Companies of Tamils. 1 of Madrasi Musalmans, 1 of Paraiyans and Christians.

COLONEL-IN-CHIEF—THE KING.

COLONEL—Major-General Edward C. W. Mackenzie-Kennedy, C.B. . . 23 April 191

1st Commn. or date of entering service.	Names and Rank.	Army rank.	Present appointment in Regiment.	Remarks.
	Commandant.			
28 Jan. 93	Scott-Elliot, Lt.-Col. C. R., *q.*	3 June 16	5 May 17	
	Company Commanders. (4)			
5 Aug. 96	Bald, Maj. J. A., C.M.G., *q.*	5 Aug. 14	13 Aug. 18	2nd-in-Comd. With 2-103rd Mah. L. I.
28 Aug. 01	Robertson, Maj. C. A., *q.*	28 Aug. 16	3 Nov. 14	With 2-81st Prs.
12 Nov. 02	Plumer, Maj. C. G. M.	11 Jan. 17	1 June 16	2nd-in-Comd. (*temp.*).
27 Jan. 04	Hamilton, Capt. A. F., M.C. *q.*	27 Jan. 13	13 Aug. 18	With 2-63rd Infy.
	Company Officers. (9)			
13 Aug. 04	Maturin, Capt. H. G., *q.* [L]	13 Aug. 13	1 Jan. 05	With 48th Prs.
5 Aug. 05	Story, Capt. J. A., M. C. *q.* (*Actg. Maj.*)	5 Aug. 14	16 Feb. 08	With 2-48th Prs.
16 June 06	*Nairne, Capt. R. G., q.*	*16 June 15*	*16 Aug. 08*	*Burma Mily. Police.*
7 Nov. 06	Ford, Capt. J. M. R., M.C., *q.* (*Actg. Maj.*).	1 Sep. 15	20 Sep. 08	With 1-73rd Infy.
9 Sep. 08	*Dawes, Capt. A. M., q*	*1 Sep. 15*	*8 Dec. 09*	*Coy. Comdr., Sch. of Instr., Subathu.*
5 Sep. 11	*Ewing, Capt. R. R.*	*5 Sep 15*	*16 Oct. 13*	*Burma Mily. Police.*
24 Aug. 12	Johnson, Capt. G. S., M.C.	24 Aug. 16	14 Jan. 14	
15 Nov. 15	Blackburn, Lt. B. M.	15 Nov. 16	23 Nov. 15	With 2-61st Prs.
29 June 16	Revell, Lt. A. J.	29 June 17	23 Feb. 16	
14 Nov. 16	Robinson, Lt. N. P.	14 Nov. 17	. .	Attd.
30 Jan. 17	Daldy, Lt. A. J.	30 Jan. 18	10 Apl. 18	With 64th Prs.
18 June 17	Kirk, Lt. J. C. M.	18 June 18	19 May 18	
20 Oct. 15	Missy, Lt. I. P.	20 Oct. 16	27 Apl. 17	81st Prs.
12 Sep. 17	Badham, Lt. R. A. A.	12 Sep. 18	9 Dec. 17	Attd.
11 Aug. 16	Bigger, Lt. T. A. L. (*Actg. Capt.*)	11 Aug. 17	29 Nov. 17	Attd. Adjt. (*Offg.*)
	Attached.			
27 Nov. 14	Venables, Lt. E. P. A.	27 Nov. 15	29 Oct. 15	I. A. R. O.
28 Feb. 16	Herbert, Lt. R. H. (*Actg. Capt.*)	28 Feb. 17	28 Feb. 16	I. A. R. O.
9 May 16	Lethbridge, Lt. W. B.	9 May 17	9 May 16	I. A. R. O.
20 June 17	Dawson, Lt. C.	20 June 18	29 Sep. 17	I. A. R. O. Qr. Mr. (*Offg.*)
15 Jan. 18	Baily, 2nd-Lt. P. E. C.	15 Jan. 18	1 June 18	I. A. (*T. C.*)
12 Apl. 16	Doig, Lt. E. W.	12 Apl. 17	3 June 18	M. G. Corps, I. A. (*on probn.*)
26 Jan. 18	Evans, 2nd-Lt. R. M.	26 Jan. 18	3 June 18	R. W. Fus., I. A. (*on probn.*)
26 June 18	Garner, 2nd-Lt. E. M.	26 June 18	3 June 18	Lan. Fus., I. A. (*on probn.*)
15 Nov. 15	Jones, Lt. G. W. (*Actg. Capt.*)	15 Nov. 16	18 Apl. 18	I. A. (*on probn.*) *S. S. O., Cawnpore.*
15 Feb. 15	Monypenny, Lt. R. d'A.	15 Feb. 16	1 Aug. 18	Essex R., I. A. (*on probn.*)
23 July 18	Holland, 2nd-Lt. G. M.	23 July 18	5 Sep. 18	I. A. R. O.
28 Mar. 17	Maycock, 2nd-Lt. H. W.	28 Mar. 17	23 Aug. 18	I. A. R. O. (*on probn.*)
24 July 15	Baily, Lt. B. W. S. S.	24 July 16	2 Nov. 18	Welsh R., I. A. (*on probn.*).
26 Oct. 15	Kenworthy, Lt. L. A. (*Actg. Capt.*)	26 Oct. 16	21 Sep. 18	I. A. (U. L.).
22 July 15	Belder, Lt. H. B.	1 July 17	18 Nov. 18	H. A. C.
25 Oct. 16	Newland, Lt. C. J.	1 Apl. 18	13 Nov. 18	H. A. C.
25 July 17	Le May, Lt. A. G.	25 July 18	28 Nov. 18	M. G. Corps.
5 Oct. 17	Brookes, Lt. T. L.	5 Oct. 18	12 Sep. 18	I. A. (*T. C.*)
17 June 17	Thorburn, 2nd-Lt. R.	17 June 17	18 Nov. 18	H. A. C.
27 Mar. 18	Rowe, 2nd-Lt. H. C.	27 Mar. 18	18 Nov. 18	H. A. C.
26 Apl. 18	Pillar 2nd-Lt. F. V.	26 Apl. 18	18 Nov. 18	H. A. C.
1 May 18	Scott, 2nd-Lt. R. L.	1 May 18	20 Nov. 18	R. A. M. C.
7 June 18	Blaylock, 2nd-Lt. J.	7 June 18	11 Oct. 18	I. A. R. O.
10 Aug. 18	Whalley, 2nd-Lt. C. A.	10 Aug. 18	26 Oct. 18	I. A. R. O.
	Subadar-Major.	Jemadar.	Subadar	Subadar-Major.
12 Nov. 86	Rangayya (53 56)	16 Nov. 03	9 June 11	7 Sep. 15
	Subadars.			
11 Mar. 89	Muhammad Abdul Hafiz *Bahadur*, I.D.S.M. (51, 56, 60).	1 Feb. 11	19 Apl. 13	*Order of Br. Ind., 2nd Cl.*
6 Sep. 90	Saiyid Abdul Ghafur (53, 56, 60).	9 June 11	1 Nov. 13	
6 Sep. 91	Kaliaperumal *Bahadur* (53, 56)	6 Oct. 08	29 Aug. 14	*Order of Br. Ind., 2nd Cl.*
9 Dec. 98	Veerappen *Bahadur* (56)	18 Feb. 12	3 Nov. 14	*Order of Br. Ind., 2nd Cl.*
5 May 93	Laul Khan, I.D.S.M. (56)	19 Apl. 13	1 Sep. 16	
6 Aug. 91	Paparayadu *Bahadur* (56)	5 Apl. 14	13 Sep. 16	*Order of Br. Ind., 2nd Cl.*
12 Aug. 93	Shaikh Muhiuddin (56)	1 Nov. 13	30 Dec. 16	
	Abdul Wahab Khan (*Retd.*)	. . . .	. . . .	Re-employed, 9 May 18.

1st Battalion, 61st King George's Own Pioneers—*contd.*

1st Commn. or date of entering service.	Names and Rank.	Army rank.	Remarks.
	Jemadars.		
29 Mar. 99	Ratnasabapathy (21)	11 Aug. 14	
26 Apl. 97	Katha Nadan (21, 52, 56)	15 Sep. 14	
20 Sep. 98	Ayyakannu, I.D.S.M. (21)	13 Sep. 16	
8 May. 97	Perumal Reddy	13 Mar. 17	
1 Apl. 03	Muhammad Abdul Azim (21)	13 Mar. 17	
6 Oct. 04	Murtaza Sharif	26 Apl. 17	
10 June 02	Tambusami I.D.S.M.	18 May 17	
12 May 02	Duraisami (21)	11 June 17	
30 Jan. 91	William Cotter	15 June 17	
2 Nov. 08	Sangili	27 June 17	
8 June 87	Narainsami (21, 56)	1 Nov. 17	
30 Aug. 04	Muttusami Nayudu (51, 56)	1 July 18	
23 Sep. 05	Joseph (21)	1 July 18	
8 Nov. 07	Kannayram, I.D.S.M. (58)	1 July 18	
12 Aug. 08	Jagannathan, I.D.S.M.	1 July 18	
24 Jan. 05	Muhammad Husain (21, 56)	21 July 18	
19 June 06	Abdul Subhan (48*a*, 53, 56)	21 July 18	
2 Dec. 08	Chinnayya (21)	21 July 18	
20 Aug. 07	Abdul Aziz (58)	21 July 18	

2nd Battalion, 61st (King George's Own) Pioneers.

Formed at Dhond on 20th July 1918.

Composition:—2 Companies Mussalis.
2 Companies Madrassis.

1st Commn. or date of entering service.		Names and Rank.	Army rank.	Present appointment in Regiment.	Remarks.
		Commandant.			
20 May	93	Bruce, Lt.-Col. G. D., D.S.O., *p.s.c.*	1 Jan. 17	9 Aug. 18	1-31st Prs.
		Company Commanders. (4)			
27 Aug.	02	Alexander, Maj. C., M.C.	27 Aug. 17	9 Aug. 18	1-81st Prs. 2nd-in-Comd.
		Company Officers.			
15 Nov.	15	Blackburn, Lt. B. M.	15 Nov. 16	2 Aug. 18	1-61st Prs. Adjt.
30 Nov.	15	Williams, Lt. L. J.	30 Nov. 16	28 Aug. 18	I. A. R. O.
5 Nov.	17	Milward, Lt. A. G.	5 Nov. 18	30 July 18	I. A. R. O.
5 Jan.	18	Kent, 2nd-Lt. J. B.	5 Jan. 18	28 July 18	I. A. (*T. C.*)
15 Jan.	18	Ward, 2nd-Lt. F. H. T.	15 Jan. 18	3 Aug. 18	I. A. R. O. Qr. Mr.
28 Sep.	16	Hanson, Lt. F. A.	28 Sep. 17	2 Sep. 18	W. York R., I. A. (*on probn.*)
15 Jan.	18	Beard, 2nd-Lt. N. T.	15 Jan. 18	28 July 18	I. A. R. O.
19 Mar.	18	Beer, 2nd-Lt. F. E.	19 Mar. 18	25 July 18	I. A. R. O.
7 June	18	Culpeck, 2nd-Lt. T. C.	7 June 18	7 Oct. 18	I. A. R. O.
15 June	18	Rose, 2nd-Lt. W.	15 June 18	16 Sep. 18	I. A. R. O.
15 June	18	Woolner, 2nd-Lt. R. J.	15 June 18	16 Sep. 18	I. A. R. O.
12 Sep.	18	Smerdon, 2nd-Lt. C. P.	12 Sep. 18	25 Sep. 18	I. A. R. O.
14 Sep.	18	Meredith, 2nd-Lt. F. B. S.	14 Sep. 18	30 Sep. 18	I. A. R. O.
4 Sep.	14	Wardle, Lt. H. M.	30 June 16	17 Nov. 18	Durh. L. I.
5 Aug.	16	Braidford, Lt. F. G.	5 Dec. 18	17 Nov. 18	Durh. L. I.
30 Oct.	17	Blayney, 2nd-Lt. A., M. M.	30 Oct. 17	17 Nov. 18	Durh. L. I.
26 Sep.	17	Broadley, 2nd-Lt. J. F.	26 Sep. 17	17 Nov. 18	Durh. L. I.
1 July	16	Stick, Lt. L. T.	1 July 17	15 Oct. 18	I. A. R. O.
8 July	17	Thorne, 2nd-Lt. A. M.	8 July 17	14 Oct. 18	I. A. R. O.
8 Oct.	15	Davidson, Lt. G. F.	1 July 16	21 Nov. 18	Durh. L. I.
10 July	18	Magee, 2nd-Lt. J. H.	10 July 18	2 Dec. 18	I. A. (*T. C.*)
29 Mar.	18	Downing, 2nd-Lt. L. D.	29 Mar. 18	2 Dec. 18	I. A. R. O.
22 July	18	Petersen, 2nd-Lt. R. G.	22 July 18	2 Dec. 18	I. A. R. O.
29 May	18	Ottley, 2nd-Lt. A. R. C.	29 May 18	2 Dec. 18	I. A. R. O.
25 July	18	Ryves, 2nd-Lt. C. H.	25 July 18	....	I. A. R. O.

Subadar-Major.

			Jemadar.	Subadar.	Subadar-Major.
4 May	92	Gangayya	5 May 05	18 Feb. 12	21 July 18

Subadar.

5 June	96	Sivasamy	4 Nov. 14	21 July 18
4 Sep.	94	Abdur Rahim, I.D.S.M.	0 Dec. 16	21 July 18
5 June	97	Raja Gopal	30 Dec. 16	21 July 18

Jemadar.

14 Sep.	03	Muhammad Abdul Haq	16 Jan. 17
4 Sep.	04	Karam Elahi	6 Nov. 17
3 Feb.	96	Sankara Pondian	21 July 18
1 May	00	Vaidalingam	21 July 18
11 June	06	Sodalamuthu	21 July 18
25 Sep.	96	Arpasamy	1 Sep. 18
19 Apl.	01	Michael	23 Mar. 17
28 July	98	Abdul Rahman	1 Sep. 18

Regt. Centre—Secunderabad. *Uniform*—Scarlet.
Linked with—66th & 84th Punjabis. *Facings*—Emerald Green.

62nd Punjabis.

Mootoo Naik-ki-paltan.—Formed at Madras in 1759, from independent companies already in existence, and was originally designated the 3rd Battalion of Coast Sepoys. Became the 2nd Carnatic Battalion, 1769; the 2nd Madras Battalion, 1784; the 1st Battalion, 2nd Regiment of Madras Native Infantry, 1796; the 2nd Regiment of Madras N. I., 1824; the 2nd Regiment of Madras Infantry, 1885; the 2nd Madras Infantry, 1901; present designation, 1903.

THE ELEPHANT—A GOLDEN DRAGON WEARING AN IMPERIAL CROWN.
"Carnatic" "Mysore" "Assaye" "Nagpore" "China"

Composition—2 companies of Punjabi Musalmans, 1 of Sikhs, 1 of Rajputs from Western Rajputana and the Eastern Punjab.

1st Commn. or date of entering service.	Names and Rank.	Army rank.	Present appointment in Regiment.	Remarks.
	Commandant.			
28 June 90	Morris, Bt.-Col. G. M., D.S.O. [L] (*temp. Br.-Genl.*)	3 June 18	20 Dec. 16	
	Company Commanders. (4)			
15 Dec. 97	Wright, Maj C. H. B. (*Actg. Lt.-Col.*)	1 Sep. 15	8 Dec. 11	With 71st Punj.
28 July 00	Sandiford, Maj. V. V. V.	1 Sep. 15	22 Jan. 16	With 49th Bengalis.
28 July 00	*Morgan, Maj. M. H. L.*, D.S.C. [*L*]	*1 Sep. 15*	*17 Nov. 16*	*G. S. O., 2nd gr., Cadet Coll., Quetta. Lv., ex I., m. c.*
	Company Officers. (9)			
18 Aug. 00	Underhill, Maj. G. E. C., *p.s.c.* (*Actg. Lt.-Col.*)	1 Sep. 15	18 Oct. 03	Comdt. (*temp.*)
8 Jan 01	*King, Maj. J. St. A.*	*8 Jan. 16*	*20 Sep. 06*	
21 Jan. 03	Auchinleck, Maj. C. J. E., D.S.O.	21 Jan. 18	16 Apl. 04	
5 Aug. 05	*Morse, Capt. L. H.*, M.C.	*5 Aug. 14*	*25 Dec. 05*	*With Sig. Ser.*
24 Jan. 06	Young, Capt. R.	24 Jan. 15	25 Mar. 07	
16 June 09	Anderson, Capt. G. D.	1 Sep. 15	2 July 12	With 3-152nd Infy.
6 Sep. 11	Watson, Capt. E. E.	6 Sep. 15	7 Dec. 12	With 2-152nd Infy.
5 Mar. 15	Hudson-Heaven, Capt. L. G.	14 Oct. 16	. . . .	Attd.
14 Oct. 15	Wemyss, Lt. A. F., M.C.	14 Oct. 16	14 Jan. 15	Offg. Qr.-Mr.
14 Oct. 15	Reeves, Lt. A. J. M.	14 Oct. 16	. . .	
18 Apl 16	Ronnfeldt, Lt. W. E.	18 Apl. 17	24 Apl. 16	With 2-152nd Infy.
30 Jan. 17	Ellison, Lt. E. C.	30 Jan. 18	. . . .	
30 Jan. 17	Joyce, Lt. A. H. B., M.C.	30 Jan. 18	. . . .	With 7th Rajputs.
28 Feb. 18	John, 2nd-Lt. D. C.	28 Feb. 18	26 Nov. 17	Attd.
	Attached.			
17 Aug. 07	Griffith-Griffin, Bt. Maj. F. M., M. C.	1 Jan. 17	. . . .	1-129th Baluchis.
20 Nov. 14	Fisk-Aldis, Lt. A. (*Actg. Capt.*)	20 Nov. 15	26 Aug. 15	I. A. R. O.
14 Jan. 15	Thompson, Lt. R. D.	14 Jan. 16	14 Jan. 15	I. A. R. O.
30 July 15	Bell, Lt. L.	30 July 16	24 May 18	I. A. R. O.
16 May 16	Wilson, Lt. W. M. C.	16 May 17	26 Nov. 17	North'd Fus., I. A. (*on probn.*).
21 June 16	Desborough, Lt. B.	21 June 17	2 Nov. 18	E. York. R., I. A. (*on probn.*).
20 July 16	Spencer, Lt. J. C.	20 July 17	28 Oct. 18	Linc. R., I. A. (*on probn.*).
25 Oct. 16	Rollinson, 2nd-Lt. R. G. (58)	25 Oct. 16	30 Nov. 17	I. A. R. O.
5 Jan. 17	Stoddart, Lt. J. S. P. (*Actg. Capt.*)	5 Jan. 18	19 Mar. 17	I. A. R. O. Adjt.
26 Jan. 18	Pearcey, 2nd-Lt. R. A.	26 Jan. 18	16 Aug. 18	L'pool R., I. A. (*on probn.*).
27 Mar. 18	Berrisford, 2nd-Lt. E. A.	27 Mar. 18	16 Aug. 18	S. Staff. R., I. A. (*on probn.*).
8 Aug. 17	Rowley, Lt. W. J. D.	8 Aug. 18	2 Nov. 17	I. A. R. O.
5 Nov. 17	Kent, Lt. W. H.	5 Nov. 18	15 Mar. 18	I. A. (*T. C.*).
5 Nov. 17	Hanney, Lt. R. J.	5 Nov. 18	15 Mar. 18	I. A. R. O. (*T. C.*).
26 May 18	Crisp, 2nd-Lt. M. F.	26 May 18	20 Aug. 18	I. A. R. O. (*on probn.*).
29 May 18	Edmiston, 2nd-Lt. A. W.	29 May 18	30 Aug. 18	I. A. R. O. (*on probn.*).
7 Oct. 18	Miles, 2nd-Lt. J. W.	7 Oct. 18	21 Oct. 18	I. A. (*T. C.*).
15 Oct. 18	Oakeshott, 2nd-Lt. D. B.	15 Oct. 18	1 Nov. 18	I. A. R. O.
18 Oct. 18	Court, 2nd-Lt E. T.	18 Oct. 18	18 Nov. 18	I. A. R. O.

Subadar-Major.

		Jemadar.	Subadar.	Subadar Major.	
10 Apl. 86	Muhammad Zaman, *Bahadur*	8 May 05	24 June 07	3 Jan. 14	*Order of Br. Ind., 2nd Cl.*

62nd Punjabis—*contd.*

1st Commn. or date of entering service.	Names and Rank.	Army rank.		Remarks.
	Subadars.			
24 Aug. 02	Muhammad Mir Afzal Khan (56)	. . . . .	24 Aug. 02	
9 June 87	Kalu Khan, *Bahadur* (56, 68)	2 June 02	11 Sep. 14	*Order of Br. Ind., 2nd Cl.*
1 Sep. 02	Muhammad Hashim Khan (56, 68)	1 Sep. 02	17 Sep. 14	
14 Oct. 98	Jaman Singh	1 Nov. 13	21 Jan. 16	With 2-152nd Infy.
25 Mar. 92	Sundar Sing, *Bahadur* (68)	3 Jan. 14	22 Jan. 16	*Order of Br. Ind., 2nd cl.*
5 Aug. 02	Bir Singh	11 Sep. 14	22 Jan. 16	
30 May 98	Wazir Khan (68)	11 Sep. 14	29 Jan. 16	
23 Dec. 13	Sher Afzal Khan	23 Dec. 13	29 Jan. 16	
27 May 97	Sher Zaman	11 Sep. 14	29 Jan. 16	
11 Oct. 99	Anokh Singh	11 Sept. 14	4 Feb. 16	
10 July 00	Dasaundha Singh	27 Mar. 15	9 Mar. 16	84th Punj.
9 Nov. 96	Jahandad Khan	14 Oct. 14	19 Jan. 17	
21 Jan. 16	Abdulla Khan	21 Jan. 16	27 Jan. 17	
5 Dec. 02	Sham Lal Singh	9 July 08	8 Feb. 17	(*temp*).
1 April 82	Sohna Singh	1 Oct. 15	15 June 17	
3 Feb. 03	Maghan Singh (68)	22 Jan. 16	23 Feb. 18	
	Jemadars.			
14 June 02	Samandar Khan (56)	11 Sep. 14		
10 May 98	Ghulam Muhammad Khan . .	17 Sep. 14		
12 June 02	Badar Bakhsh (48*a*, 50) . . .	10 Jan. 16		
25 July 02	Lal Singh	4 Feb. 16		
26 July 02	Roda Singh (68)	9 Mar. 16		
5 July 02	Sher Gul	17 May 16		
16 Jan. 98	Man Singh	19 Jan. 17		With 2-152nd Infy.
24 Dec. 05	Manir Shah	27 Jan. 17		
28 Aug. 87	Jung Bahadur Singh	22 Apl. 17		(*temp.*).
	Nur Muhammad Khan (*Retd.*) .	. . . . .		Re-employed 24 Feb. 15. Depôt.
19 Jan. 02	Bhim Singh	27 June 18		
9 Feb. 03	Sohan Singh	27 June 18		
1 Dec. 02	Pir Muhammad Khan	24 June 18		
	Abdul Ghafur Khan	16 Feb. 18		
15 June 87	Sheodath Singh	1 Sep. 18		
	Abdulla Khan	2 Sep. 18		
	Hukam Dad	5 Oct. 18		
	Jhabu Singh	5 Oct. 18		
	Rakhman Shah	5 Oct. 18		

Regt. Centre—Trichinopoly.
Linked with—83rd W. L. Infy.

Uniform—Scarlet.
Facings—Emerald Green.

1st Battalion, 63rd Palamcottah Light Infantry.

Turing-ki-paltan.—Formed at Madras in 1759, from independent companies already in existence, and was originally the 4th Battalion of Coast Sepoys. Became the 4th Carnatic Battalion, 1769; the 3rd Carnatic Battalion, 1770; the 3rd Madras Battalion, 1784; the 1st Battalion, 3rd Regiment of Madras Native Infantry, 1796; the 1st Battalion, 3rd Regiment of Madras Native Infantry, or Palamcottah Light Infantry, 1812; the 3rd Regiment of Madras N. I., or Palamcottah Light Infantry, 1824; the 3rd (or Palamcottah) Regiment of Madras (Light) Infantry, 1885; the 3rd (Palamcottah) Madras Light Infantry, 1901; present designation, 1903.

" Now or Never."
"Sholinghur" "Carnatic" "Mysore" "Maheidpoor"
"Ava" "Burma, 1885-87" "China, 1900."
Composition—2 Companies of Madrasi Musalmans, 1 of Tamils, 1 of Paraiyans and Christians.

COLONEL—Major-General Arthur W. L. Anderson 13 May 1904.

1st Commn. or date of entering service.	Names and Rank.	Army rank.	Present appointment in Regiment.	Remarks.
	Commandant.			
14 Aug. 95	Waring, Lt.-Col. E. H.	19 Aug. 18	19 Aug. 18	With 1-90th Punj.
	Company Commanders. (4)			
12 Aug. 88	Ainslie, Lt.-Col. H. P. (Q)	22 Aug. 14	12 Oct. 12	2nd-in-Comd.
25 Jan. 99	Mouat, Maj. G. E. D., D.S.O.	1 Sep. 15	6 Oct. 12	With 2-63rd Infy.
21 Feb. 00	Phillips, Maj. L. *Actg. Lt.-Col.*)	1 Sep. 15	12 Oct. 12	Comdt. (*temp.*).
18 Jan. 02	Breithaupt, Maj. E. A.	18 Jan. 17	24 May 17	With 7th Rajputs.
	Company Officers. (5)			
20 Jan. 09	*Windeler, Capt. W. (T. Actg. Maj.)*	*1 Sep. 15*	*24 May 13*	*Staff Capt., S. Comd.*
15 Aug. 14	Hurford Jones, Lt. M. (*Actg. Maj.*)	15 May 16	31 July 17	Attd. 2nd-in-comd. (*temp.*)
29 Sep. 14	Digges La Touche, Lt. H. N.	22 June 16	20 Mar. 17	
29 Sep. 15	Urquhart, Lt. D. D. M. (*Actg. Capt.*)	29 Sep. 16	9 May 17	Attd.
1 Oct. 18	Brown, 2nd-Lt. R. T.	1 Oct. 18	11 Oct. 18	Attd.
	Attached.			
5 Jan. 15	Colchester, Lt. M. T. (*Actg. Capt.*)	5 Jan. 16	27 Jan. 18	I. A. R. O.
21 Jan. 16	Glendening, Lt. C. J.	21 Jan. 17	13 Nov. 18	Midd'x R. I. A. (*on probn.*).
3 Mar. 16	Hamill, Lt. G. E.	3 Mar. 17	17 Sep. 18	Suff. R., I. A. (*on probn.*).
26 Mar. 16	Morphy, Lt. E. C. O. (*Actg. Capt.*)	26 Mar. 17	27 Jan. 18	I. A. (*on probn.*).
3 Dec. 15	*Conyers-Brown, Lt. H. P. (temp. Capt.)*	*3 Sep. 17*	*7 Dec. 15*	*Bombing Insptr., Mhow.*
18 Jan. 16	Determes, Lt. M. J.	18 Jan. 17	18 Jan. 16	I. A. R. O. *Lv., ex I., m.c.*
5 May 17	Bishop, Lt. H. C.	5 May 18	19 Aug. 17	I. A. R. O. Qr.-Mr. (*Offg.*) [19 Aug. 18.
27 June 17	Jeynes, Lt. A. G. H. (*Actg. Capt.*)	27 June 18	18 Feb. 18	I. A. R. O. Adjt. (*Offg.*)
23 June 17	Turton, 2nd-Lt. W. H.	23 June 17	25 Aug. 17	I. A. R. O.
5 Oct. 17	Gillon, Lt. J. B.	5 Oct. 18	1 Feb. 18	I. A. R. O.
5 Oct. 17	Parsons, Lt. W. G. J.	5 Oct. 18	1 Feb. 18	I. A. R. O.
11 Feb. 17	Coy, Lt. F. C.	11 Nov. 18	17 Dec. 17	R. W. Fus., I. A. (*on probn.*).
19 Dec. 17	Young, 2nd-Lt. J. E.	19 Dec. 17	7 Apl. 18	I. A. R. O. (*on probn.*)
15 Jan. 18	Towell, 2nd-Lt. A. F. H.	15 Jan. 18	18 June 18	I. A. R. O.
25 Jan. 17	Garrett, 2nd-Lt. J. N.	25 Jan. 17	7 Apl. 18	I. A. (*T. C.*)
29 May 18	Limb, 2nd-Lt. H. E.	29 May 18	22 Oct. 18	Genl. List. I. A. R. O. (*on probn.*)
7 June 18	Chirgwin, 2nd-Lt. S. H.	7 June 18	5 Oct. 18	I. A. R. O.
15 June 18	Jones, 2nd-Lt. G. E.	15 June 18	23 Sep. 18	Genl. List. I. A. R. O. (*on probn.*)
28 June 18	Morton, 2nd-Lt. J. I.	28 June 18	5 July 18	I. A. R. O.
25 July 18	Macpherson, 2nd-Lt. A. I.	25 July 18	3 Aug. 18	I. A. R. O.
31 July 18	Webster, 2nd-Lt. J.	31 July 18	22 Oct. 18	Genl. List. I. A. R. O. (*on probn.*)
23 Aug. 18	Stewart, 2nd-Lt. W.	23 Aug. 18	13 Sep. 18	I. A. R. O.
27 Aug. 18	Beale, 2nd-Lt. E. J.	27 Aug. 18	27 Sep. 18	I. A. R. O.
8 Nov. 18	Whittaker, 2nd-Lt. A. P.	8 Nov. 18	19 Nov. 18	I. A. R. O.
	Burrows, 2nd-Lt. E.			L'pool R., I. A. R. O. (*on probn.*)

Subadar-Major.

		Jemadar.	Subadar.	Subadar-Major.
27 Mar. 05	Mir Asghar Ali (53, 56, 72*a*)	27 Mar. 05	3 Aug. 11	16 Oct. 17

Subadars.

9 Nov. 92	Maduramuttu (52, 56, 66*a*)	1 Dec. 12	5 Dec. 14
17 Oct. 92	Devasagayam (56)	15 Feb. 13	25 Mar. 16
2 Sep. 92	Shaikh Ali	5 Nov. 14	25 June 16
1 July 97	Hari Singh (53, 56)	15 July 14	16 Oct. 17
20 July 97	Murugayya (56)	4 Aug. 14	8 Aug. 18

1st Battalion, 63rd Palamcottah Light Infantry—*contd.*

1st Commn. or date of entering service.	Names and Rank.	Army rank.	Remarks.
	Jemadars.		
9 June 97	Daud Khan (52, 56, 60*a*, 66*a*)	5 Dec. 14	
24 Feb. 00	Sivalingam	25 Mar. 16	
17 Aug. 98	Kuppusami (56)	27 Oct. 16	
7 May 97	Sayyid Ali (56)	21 Nov. 16	
	Ganesh Singh	22 Mar. 17	
16 Jan. 00	Narainsawmy	3 Dec. 17	Indian Qr.-Mr.
17 Nov. 99	Partab Singh	11 Apl. 18	
24 Mar. 98	Tiruvangadam	1 June 18	Indian Adjt.
12 Mar. 98	Murugason	1 June 18	
	Rajogopal	1 July 18	
19 Feb. 00	Abdul Karim	1 Aug. 18	
2 July 99	Abdul Wahid	8 Aug. 18	

2nd Battalion, 63rd Palamcottah Light Infantry.

Formed at Cawnpore on 30th October 1918.

Composition:—1 Company of Tamils, 1 company of Paraiyans and Christains, 1 Company of Telugues, 1 Company of Madrasi Mohamadans and Maharattas.

1st Commn. or date of entering service.	Names and Rank.	Army Rank.	Present appointment in Regiment.	Remarks.
	Commandant.			
25 Jan. 99	Mouat, Maj. G.E.D., D.S.O.	1 Sep. 15	6 Nov. 18	1-63rd Infy. *Comdg. 9th Dn.*
	Company Commanders. (4)			
27 Jan. 04	Hamilton, Capt. A.F., M.C.	27 Jan. 13	6 Nov. 18	1-61st Prs. 2nd-in-Comd. (*temp.*)
5 Nov. 15	Clay, Lt. R.D.	1 July 17	8 Nov. 18	Notts. & Derby R., I. A. R. O. (*on probn.*)
5 Aug. 16	Chamberlain, Lt. W.G.	5 Aug. 17	30 Oct. 18	I. A. R. O.
22 Nov. 16	Corin, Lt. T.	22 May 18	22 Nov. 18	Yorks. L. I.
	Companny Officers.			
5 Jan. 17	Clark, Lt. G. A.	5 Jan. 18	10 Nov. 18	I. A. R. O. (*on probn.*)
10 Jan. 17	Dowland, 2nd-Lt. W.	10 Jan. 17	20 Nov. 18	Yorks L. I. Qr.-Mr. (*Offg.*)
5 Nov. 17	Godley, Lt. E. L.	5 Nov. 18	30 Oct. 18	I. A. R. O.
7 June 18	Simmons, 2nd-Lt. F. S.	7 June 18	10 Nov. 18	I. A. (*T. C.*)
25 June 18	Kilner, 2nd-Lt. G. A.	25 June 18	20 Nov. 18	Yorks. L. I.
25 June 18	Marr, 2nd-Lt. H. R.	25 June 18	20 Nov. 18	Yorks. L. I.
25 June 18	Spooner, 2nd-Lt. A. F.	25 June 18	6 Dec. 18	I. A. R. O.
12 July 18	Minto, 2nd-Lt. A. H.	12 July 18	12 July 18	I. A. R. O.
14 Aug. 18	Abbott, 2nd-Lt. G.	14 Aug. 18	20 Nov. 18	Yorks. L. I.
1 Oct. 18	Nicholls, 2nd-Lt. A. W.	1 Oct. 18	1 Nov. 18	I. A. (*on probn.*) Adjt. (*Offg.*)

Subadar Major.

		Jemadar.	Subadar.	Subadar-Major.
13 May 89	Muhammad Yassin (56)	2 May 08	1 July 11	8 Nov. 18

Subadars.

4 May 88	Khadir Sharief (53, 56, 66*a*)	19 Dec. 07	12 Nov. 11
1 Dec. 94	Sankarapandian Pillai (53, 56, 66*a*).	18 Mar. 11	4 Aug. 14
25 May 94	Saiyed Ahmed	5 Nov. 14	30 July 18

Jemadars.

24 Feb. 99	Arokiasami (46, 52, 56)	5 Dec. 14
3 Mar. 01	John Francis (56)	6 July 16
2 Nov. 04	V. Jacob (56)	26 July 16
20 Nov. 01	Basith Khan (56)	1 Nov. 16
13 Dec. 98	Saiyed Abdulla	1 June 18

Regt. Centre—Trichinopoly.
Linked with—61st & 81st Pioneers.

Uniform—Scarlet.
Facings—White.

64th Pioneers.

Baillie-ki-paltan.—Formed at Madras in 1759, from independent companies already existing, and was originally the 5th Battalion of Coast Sepoys. Became the 5th Carnatic Battalion, 1769; the 4th Carnatic Battalion, 1770; the 4th Madras Battalion, 1784; the 1st Battalion, 4th Regiment of Madras Native Infantry, 1796; the 4th Regiment of Madras N. I., 1824; the 4th Regiment of Madras N. I. (Pioneers), 1883; the 4th Regiment of Madras Infantry (Pioneers), 1885; the 4th Madras Pioneers, 1901; present designation, 1903.

THE ELEPHANT

"Sholinghur" "Carnatic" "Mysore" "Assaye" "Afghanistan, 1879-80."

Composition—2 Companies of Tamils, 1 of Madrasi Musalmans, 1 of Paraiyans and Christians.

1st Commn. or date of entering service.	Names and Rank.	Army rank.	Present appointment in Regiment.	Remarks.
	Commandant.			
3 May 90	Bliss, Lt.-Col. J. A., D.S.O., M.V.O., q.	3 May 16	20 Aug. 16	
	Company Commanders. (4)			
28 Jan. 93	*Rice, Maj. S. M.*, C.I.E., *p.s.c., q.* (*temp. Lt.-Col.*)	*28 Jan. 11*	*30 Sep. 02*	*A. A. & Q.M.G., N. Comd.*
10 Oct. 94	Watson, Bt.-Lt.-Col. S. B., q. (Q)	3 June 16	22 Sep. 16	2nd-in-Comd. With 155 Prs.
4 Aug. 97	Mackie, Maj. R., q. (*Actg. Lt.-Col.*)	4 Aug. 15	1 June 12	With 2-81st Prs.
22 Jan. 98	Malet, Maj. R. J., q. (*Actg. Lt.-Col.*)	1 Sep. 15	16 Aug. 14	With 2-88th Infy.
	Company Officers. (9)			
17 Jan. 00	Harvey-Kelly, Maj. H.St.G.H., q.	1 Sep. 15	1 Sep. 01	
4 July 03	Skinner, Maj. T. B., q.	26 Sep. 16	12 Dec. 06	With 155th Prs.
27 Aug. 02	*Murland, Maj. H. F., q.*	*27 Aug. 17*	*23 Feb. 04*	*D. A. A. G., 9th Dn.*
16 Dec. 03	*Smith, Maj. M. C., q.*	*3 July 18*	*26 May 06*	*Instr., Sch. of Musky., Pachmarhi.*
19 Aug. 03	*Hemsley, Maj. C.*, D.S.O., *q.*	*19 Aug. 18*	*18 July 05*	*Bde. Maj., 50th Bde.*
17 Aug. 07	Burne, Capt. F. O. N., M.C., *Actg. Maj.*	17 Aug. 16	19 Nov. 08	
25 Jan. 08	Hare, Capt. I. L. O'H.	1 Sep. 15	30 Mar. 09	With 2-107th Prs.
8 Sep. 09	Deane, Capt. H. H. R.	1 Sep. 15	7 Mar. 11	
11 Dec. 09	*Urmson, Capt. D. G. S.*	*1 Sep. 15*	*20 Feb. 14*	*Staff Capt., 12th Bde.*
11 Aug. 15	Rossel, Lt. H. G. T. (*Actg. Capt.*)	11 Aug. 16	7 Nov. 15	Attd.
15 Nov. 15	Slade, Lt. M. (*Actg. Capt.*)	15 Nov. 16	28 Nov. 15	Adjt., 27 May 18.
1 Dec. 15	Sydney-Smith, Lt. D. C.	1 Sep. 17	26 Jan. 16	Attd.
1 Sep. 18	Jennings, 2nd-Lt. L. W. B.	1 Sep. 18	24 Sep. 18	Attd.
1 Sep. 18	Worthington, 2nd-Lt. R. F.	1 Sep. 18	14 Sep. 18	Attd.
	Attached.			
28 Jan. 18	Smith, 2nd-Lt. J. C.	28 Jan. 18	. . . .	I. A. (*on probn.*)
18 Dec. 14	Aylward, Lt. D., M.C. (*Actg. Capt.*)	18 Dec. 15	12 Aug. 15	I. A. R. O.
24 Dec. 14	Stonehewer, Lt. E. H. (*Actg. Capt.*)	24 Dec. 15	18 Sep. 16	I. A. R. O.
4 Apl. 16	Barnes, Lt. T. J.	4 Apl. 17	30 Nov. 17	I. A.
11 Aug. 15	Curle, Lt. V. (*Actg. Capt.*)	11 Aug. 16	28 Oct. 15	81st Prs.
1 Jan. 16	Banister, Lt. F. (*Actg. Capt.*)	1 Jan. 17	9 Mar. 16	I. A. R. O. Adjt. (*temp.*) 14 May 18.
26 May 16	Deed, Lt. L. H.	26 May 17	21 Oct. 17	I. A. R. O. Qr.-Mr. (*temp.*) 14 May 18.
8 Aug. 17	Johnstone, 2nd-Lt. C. W. H.	8 Aug. 17	30 Jan. 18	Do.
29 June 16	Hamilton, Lt. L. S.	29 June 17	20 Apl. 18	1-81st Prs.
30 Jan. 17	Daldy, Lt. A. J.	30 Jan. 18	21 May 18	1-61st Prs.
5 Jan. 17	Cornwell, Lt. R. B.	5 Jan. 18	. . . .	I. A. R. O.
1 Mar. 17	Scott, Lt. E. H.	1 Mar. 18	18 Aug. 18	I. A. R. O.
29 Aug. 17	Mills, Lt. F. C. R.	29 Aug. 18	8 Dec. 17	I. A. R. O.
1 Mar. 18	Carey, 2nd-Lt. A. B.	1 Mar. 18	19 July 18	I. A. R. O. Qr.-Mr., 1 Oct. 18.
29 May 18	Fairclough, 2nd-Lt. J. E. B.	29 May 18	28 Aug. 18	I. A. R. O. (*on probn.*)
29 Mar. 18	Wilson, 2nd-Lt. J.	29 Mar. 18	19 Aug. 18	I. A. (*T. C.*)
15 Jan. 18	Batchelor, 2nd-Lt. G. F. F.	15 Jan. 18	28 Aug. 18	I. A. R. O.
7 June 18	Bryant, 2nd-Lt. L.	7 June 18	8 Oct. 18	I. A. (*T. C.*)
13 May 18	Taylor, 2nd-Lt. J. F. S.	13 May 18	7 Oct. 18	I. A. R. O.

Subadar-Major.

		Jemadar.	Subadar.	Subadar-Major.	
1 Oct. 86	Krishnasami *Bahadur* (30)	8 Oct. 02	25 July 07	15 May 14	*Order of Br. Ind., 2nd Cl.*

64th Pioneers—*contd.*

1st. Commn. or date of entering service.	Names and Rank.	Army rank.	Present appointment in Regiment.	Remarks.
	Subadars.			
28 Feb. 91	Anantadri Nayudu (56)	25 July 07	2 Mar. 12	
1 Apl. 94	Nanjappa, *Bahadur* (56)	26 May 08	27 Mar. 12	*Order of Br. Ind., 2nd Cl.*
14 July 92	Narayanasami (I)(56)	19 Nov. 08	13 Dec. 13	
17 Feb. 91	Durgayya (56)	28 May 10	4 May 14	
25 Mar. 95	Narayanasami (II) (56, 58)	5 Apl. 11	15 May 14	
20 Mar. 01	Viraraghavulu, *Bahadur* (56, 66*a*)	1 Oct. 12	11 Feb. 16	*Order of Br. Ind., 2nd Cl.*
11 Apl. 95	Krishnamurti (56)	2 Mar. 12	8 Aug. 16	
29 Apl. 94	Muttuveli (56, 66*a*)	27 Mar. 12	8 Apl. 17	
9 Aug. 98	Gopalsami, I.D.M.S. (56)	13 Dec. 13	8 Apl. 17	
15 June 94	Anthony David, I.D.S.M. (51)	13 July 11	17 Feb. 18	
26 June 93	Shamsuddin (56)	4 May 14	19 May 18	
6 May 03	Punyakoti	28 Sep. 17	19 May 18	
	Jemadars.			
15 Jan. 02	Gnanaprakasam (56, 58)	20 July 15		
22 June 00	Perumal Pillai	10 Oct. 15		
12 Aug. 95	Muhammad Abdul Karim (48, 52, 56)	11 Feb. 16		
18 Dec. 04	Dharmalingam	19 May 16		Indian Adjt., 19 May 16.
1 Nov. 02	P. Narayanasami, I.D.S.M.	9 Aug. 16		Indian Qr. Mr.
27 Oct. 02	Narayanasami IV.	9 Aug. 16		
16 Oct. 98	Vadamalai	9 Aug. 16		
30 May 04	Balaraman	3 Dec. 17		
6 May 03	Puniakoti Mudali	1 Jan. 18		
31 Aug. 02	Babu	1 Feb. 18		
25 Nov. 97	Sauriyappan	1 Apl. 18		
26 Feb. 02	Muhammad Burhan	1 Apl. 18		
17 Jan. 98	Shaikh Jamal-ud-din	19 May 18		
19 Nov. 04	Saurinayagam	19 May 18		
7 July 07	Narayanasami V.	19 May 18		
9 Mar. 11	Arogyasami	19 May 18		
12 Aug. 04	Munisami	1 June 18		
12 Aug. 04	Chinnappan	1 Aug. 18		
18 May 94	Tirumalay.	1 Sep. 18		
22 June 12	Balasundaram	5 Sep. 18		

Regt. Centre—Secunderabad. *Uniform*—Scarlet.
Linked with—62nd & 84th Punjabis. *Facings*—Emerald Green.

1st Battalion, 66th Punjabis.

Mackenzie-ki-paltan.—Raised at Trichinopoly in 1761, and was originally the 7th Battalion of Coast Sepoy. Became the 7th Carnatic Battalion, 1796; the 6th Carnatic Battalion, 1770; the 6th Madras Battalion, 1784; the 1st Battalion, 6th Regiment of Madras Native Infantry, 1796; the 6th Regiment of Madras Native Infantry, 1824; the 6th Regiment of Madras Infantry, 1885; the 6th Madras Infantry, 1901; present designation, 1903.

A GOLDEN DRAGON WEARING AN IMPERIAL CROWN.

"Sholinghur" "Carnatic" "Mysore" "Seringapatam" "Bourbon" "China."

Composition—2 Companies of Punjabi Musalmans, 1 of Sikhs, 1 of Rajputs from Western Rajputana and the Eastern Punjab.

1st Commn or date of entering service.	Names and Rank.	Army rank.	Present appointment in Regiment.	Remarks.
	Commandant.			
30 Aug. 98	Moberly, Lt.-Col. H. S., *p.s.c.*	3 June 16	1 June 18	*G. S. O., 1st gr., 8th Dn.*
	Company Commanders. (4)			
6 July 89	*Moore, Lt.-Col. A. (Q)*	*6 July 15*	*18 Aug. 14*	*2nd-in-Comd.*
27 July 98	Burton, Maj. H. F. (*Actg. Lt.-Col.*)	1 Sep. 15	11 June 11	
17 Jan. 00	*Bennett, Maj. E. E. S., p.s.c.*	*1 Sep. 15*	*1 June 14*	
8 May 01	*Stockley, Maj. C. H.,* M.C.	*8 May 16*	*1 June 18*	
	Company Officers. (9)			
18 Jan 02	Smith, Maj. Y. E.	18 Jan. 17	31 Oct. 03	With 2-66th Punj.
22 Oct. 02	Paterson, Maj. F. W. J., M.C.	22 Oct. 17	23 Jan. 05	
2 Mar. 04	Allan, Capt. P. A.	2 Mar. 13	15 Sep. 06	With 1-153rd Infy.
28 Jan. 05	Daly, Capt. C. E.	28 Jan. 14	29 July 07	With 1-10th Jats.
20 Jan. 09	Scroope, Capt. C. F.	1 Sep. 15	27 Feb. 11	With 24th Punj.
18 Jan. 11	Steward, Capt. C. W. d'A.	1 Sep. 15	23 Sep. 13	With 1-112th Infy.
27 Oct. 16	O'Sullivan, Lt. E. T.	27 Oct. 17	....	With 20th Infy.
18 June 17	Sandham, Lt. W. F. N.	18 June 18	28 June 17	Attd. With 2-72nd Punj.
4 Dec. 14	Waite, Lt. E. W.	4 Sep. 16	12 Mar. 18	Attd.
1 May 17	Stiven, Lt. R.	1 May 18	....	Attd.
18 Aug. 15	Nunn, Lt. R. N., M.C.	18 Aug. 16	29 July 18	Attd.
24 Aug. 15	Brear, Lt. B. H.	24 Aug. 16	6 Dec. 17	Attd.
5 Jan. 16	Dagg, Lt. A. V.	5 Jan. 17	6 Dec. 17	Attd.
7 May 16	Sainter, Lt. J. D., M.C.	7 May 17	16 Sep. 17	Attd. (*Offg. Qr.*) M..
22 Aug. 17	Barden, Lt. E. A.	22 Aug. 18	6 Dec. 17	Attd.
28 Feb. 18	Bennett, 2nd-Lt. M. C.	28 Feb. 18	6 Dec. 17	Attd.
	Attached.			
8 Sep. 09	Perram, Capt. H. C.	1 Sep. 15	13 Mar. 18	84th Punj.
29 May 15	Wodehouse, Lt. C. G.	29 May 16	12 Aug. 18	I. A. R. O.
6 Jan 16	Openshaw, Lt. S. N.	6 Jan. 17	7 Nov. 18	Lan. Fus., I. A. (*on probn.*)
3 Aug. 14	Atkins, Lt. E. A.	28 Jan. 17	16 Feb. 18	I. A. R. O.
24 Sep. 16	George, Lt. A.	24 Sep. 17	18 Jan. 18	I. A. (*on probn*).
8 Aug. 17	Mackay, Lt. J. (*Actg. Capt.*)	8 Aug. 18	5 Nov. 17	I. A. R. O. Adjt.
	Williams, Lt. R. F.	....	7 Nov. 18	R. N. Vol. Reserve, I. A. R. O. (*on probn.*)
28 Dec. 17	Bree, 2nd-Lt. P. G.	28 Dec. 17	4 Nov. 18	Suff. R., I. A. (*on probn.*)
5 Oct. 17	Stephen, Lt. P. W. R.	5 Oct. 18	22 Jan. 18	I. A. (*on probn.*)
29 Aug. 17	Corbould-Warren, Lt. W.	29 Aug. 18	18 Jan. 17	I. A. R. O.
2 Nov. 15	Billson, Lt. J. V.	31 Oct. 18	16 Feb. 18	I. A. R. O.
27 Mar. 18	Brawn, 2nd-Lt. J. A.	27 Mar. 18	15 Oct. 18	Norf. R., I. A. (*on probn.*)

Subadar-Major.

		Jemadar.	Subadar.	Subadar-Major.
1 Oct. 02	Girdhari Singh (46)	....	1 Oct. 02	5 Jan. 17

1st Battalion, 66th Punjabis—*contd.*

1st Commn. or date of entering service.	Names and Rank.	Army rank.	Present appointment in Regiment.	Remarks.
	Subadars.			
		Jemadar.	Subadar.	
1 June 02	*Bostan Khan (56)*	*1 June 02*	*1 May 08*	
3 Aug. 92	Jamal Khan (56)	21 Mar. 08	29 May 09	
23 Oct. 99	Akbar Khan (56, 72*a*)	1 May 08	1 Dec. 10	
15 July 98	Muhammad Khan	1 Dec. 10	2 July 15	
14 July 02	Said Muhammad (56)	2 Dec. 14	31 Dec. 16	
9 July 02	Abdul Rahaman (46)	7 Aug. 15	1 May 17	
1 July 02	Dharm Singh	7 Feb. 16	1 May 17	
....	Labh Din Khan	....	1 May 17	
8 May 94	Sher Zaman	31 Dec. 16	3 Nov. 18	
3 July 02	Moti Singh	31 Dec. 16	15 Sep. 18	
24 June 02	Ran Singh (52, 56)	31 Dec. 16	3 Nov. 18	
	Hazara Singh (*Retd.*)	....	....	Re-employed.
	Jemadars.			
29 Aug. 97	Nur Husain	1 Nov. 08		
14 Jan. 95	Bishn Singh (56, 66*a*)	4 Feb. 10		
1 Oct. 02	Arjun Singh (56)	7 Feb. 10		
3 Jan. 96	Muhammad Kazim	1 Dec. 10		
10 June 93	Gulzara Singh	2 Dec. 14		
1 June 02	Aki Khan	8 Mar. 15		
24 May 04	Jawahir Singh	6 Aug. 15		
30 Nov. 95	Ahmed Khan	31 Dec. 16		
1 Dec. 02	Kamal Khan	31 Dec. 16		
29 Apl. 01	Faqir Muhammad	1 May 17		
16 Jan. 09	Afsar Khan	1 Oct. 17		
27 June 02	Mit Singh	15 Dec. 17		
10 Sep. 03	Jahandad	16 Dec. 17		
16 Jan. 09	Jahandad Khan	16 July 18		
1 Feb. 06	Simrat Singh	15 Sep. 18		
4 July 07	Jodh Singh	15 Sep. 18		

2nd Battalion, 66th Punjabis.

Formed at Sitapur on 5th October 1918.

Composition:—2 Companies Punjabi Musalmans.
2 Companies Oudh Rajputs.

1st Commn. or date of entering service.	Names and Rank.	Army rank.	Present appointment in Regiment.	Remarks.
	Commandant.			
25 Jan. 99	Stoney, Maj. P. S., *p. s. c.* (*Act'g. Lt.-Col.*)	1 Sep. 15	27 Oct. 18	1-26th Punj.
	Company Commanders. (4)			
18 Jan. 02	Smith, Maj. Y. E.	18 Jan. 17	1 Dec. 18	1-66th Punj.
8 Mar. 15	Horner, Lt. W. J. (*Act'g. Capt.*)	8 Mar. 16	5 Nov. 18	I. A. R. O.
15 Mar. 15	Tatham, Lt. C. W. (*Act'g. Capt.*)	15 Mar. 16	20 Oct. 18	I. A. R. O. Adjt. (*Offg.*)
9 June 16	Witts, Lt. C. A. (*Act'g. Capt.*)	9 June 17	5 Nov. 18	E. Kent R., I. A. (*on probn.*)
	Company Officers.			
15 Jan. 18	Cameron, 2nd-Lt. H. H. (*Act'g Capt.*)	15 Jan. 18	14 Oct. 18	I. A. R. O. Qr. Mr. (*Offg.*)
2 June 18	Goldsmith, 2nd-Lt. M. L.	2 June 18	5 Nov. 18	I. A. R. O.
20 June 18	Martin, 2nd-Lt. H. L.	20 June 18	5 Nov. 18	Genl. List I. A. R. O. (*on probn.*)
7 Oct. 18	Williams, 2nd-Lt. E. P.	7 Oct. 18	1 Nov. 18	I. A. R. O.
5 Sep. 18	Falloon, 2nd-Lt. P. H.	5 Sep. 18	14 Oct. 18	I. A. R. O.
22 July 18	Osborne, 2nd-Lt. S.	22 July 18	3 Dec. 18	I. A. R. O.

Subadar Major.

		Jemadar.	Subadar.	Subadar-Major.
13 July 02	Fazel Dad Khan	17 Jan. 15	29 Aug. 15	22 Nov. 18

Subadars.

5 Nov. 03	Sultan Ali	2 July 15	1 May 17
9 June 02	Sucha Singh	31 Dec. 16	2 May 17
	Durga Singh	1 Sep. 15	25 Apl. 18
22 Oct. 04	Pudan Singh	25 Apl. 18	11 Sep. 18

Jemadars.

	Ghulam Muhammad	17 Sep. 14
	Said Ghulam	12 Apl. 15
	Arjan Singh	23 Apl. 15
	Lal Singh	4 Feb. 16
	Sarju Singh	
4 July 02	Ali Akbar	1 May 17
1 Dec. 02	Dhonkal Singh	29 Apl. 18
	Sheo Prasad Singh	1 Apl. 18
	Fatehbahadur Singh	1 Sep. 18
	Surajbaksh Singh	1 Sep. 18

Regt. Centre—Bangalore.
Linked with—69th & 74th Punjabis.
Uniform—Scarlet.
Facings—Emerald Green.

1st Battalion, 67th Punjabis.

Cooke-ki-paltan.—Raised at Trichinopoly in 1761 as the 8th Battalion of Coast Sepoys. Became the 8th Carnatic Battalion, 1769; the 7th Carnatic Battalion, 1770; the 7th Madras Battalion, 1784; the 1st Battalion, 7th Regiment of Madras Native Infantry, 1796; the 7th Regiment of Madras Native Infantry, 1824; the 7th Regiment of Madras Infantry, 1885; the 7th Madras Infantry, 1901; present designation, 1903. Reformed 30th April, 1916.

"Carnatic" "Mysore" "Ava."

Composition —2 Companies of Punjabi Musalmans, 1 of Sikhs (other than Jats and Mazbis), 1 of Punjabi Hindus.

1st Commn. or date of entering service.	Names and Rank.	Army rank.	Present appointment in Regiment.	Remarks.
	Commandant.			
3 May 90	Colan, Lt.-Col. W. R. B., D.S.O.	3 May 16	27 Nov. 17	
	Company Commanders. (4)			
25 Mar. 96	*Cox, Bt.-Lt.-Col. C. E. S.*	*1 Jan. 16*	*27 Nov. 17*	*2nd-in-Comd.*
11 Feb. 99	Dickson, Maj. E.	1 Sep. 15	13 Nov. 13	
1 Jan. 99	Luck, Maj. T.	1 Sep. 15	1 Jan. 13	With 2-67th Punj.
8 May 01	*Ogle, Bt.-Lt.-Col. N.*, D. S. O., *p. s. c.*	*3 June 18*	*13 Nov. 13*	*G. S. O. 2nd gr. Cadet College, Quetta.*
	Company Officers. (9)			
27 July 01	Colan, Maj. H. N., D.S.O.	27 July 16	3 Mar. 03	
24 Jan. 06	Atkins, Capt. R. F.	24 Jan. 15	4 Aug. 08	
17 Aug. 07	Manisty, Capt. G. J. E.	1 Sep. 15	2 Nov. 08	With 1-98th Infy.
18 Sep. 09	McLean, Capt. F. G. S.	1 Sep. 15	11 May 14	With 2-19th Punj.
5 Sep. 11	Armstrong, Capt. F. H. C.	5 Sep. 15	19 Mar. 14	
20 Jan. 12	Crowther, Capt. A. H.	20 Jan. 16	11 Mar. 13	Adjt. 28 Dec. 16.
15 Nov. 15	Longbottom, Lt. J. L.	15 Nov. 16	2 Aug. 16	With 2-152nd Infy.
18 Apl. 16	Dalison, Lt. J. B.	18 Apl. 17	30 Apl. 16	With 1-143rd Infy.
18 Apl. 16	Guiton, Lt. L. W.	18 Apl. 17	30 Apl. 16	With 129th Baluchis.
12 Sep. 17	Brown, Lt. L. C. H.	12 Sep. 18	10 Dec. 17	Attd. With 1-131st Reg.
27 Oct. 17	Bannon, Lt. J. D.	27 Oct. 18	4 Nov. 17	
9 July 15	Nash, Lt. A. V.	9 Apl. 17	7 Nov 16	Attd.
1 June 15	Harrison, Lt. E. L.	1 June 16	24 Sep. 17	Attd.
	Geere, 2nd-Lt. A. J. K.	....	....	Attd.
	Attached.			
21 Jan. 03	Thompson, Maj. C. M.	21 Jan. 18	15 Mar. 17	Civil.
20 Nov. 14	Curnow, Lt. E. (*Actg. Capt.*)	20 Aug. 16	....	I. A. R. O.
20 July 15	Hannah, Lt. A. J.	20 July 16	28 Mar. 17	I. A. R. O.
10 June 16	Cable, Lt. R. W.	10 June 17	28 Mar. 17	I. A. R. O.
30 Jan. 17	McCaughey, Lt. F. G.	30 Jan. 18	29 Mar. 17	74th Punj.
21 Jan. 03	Bagnall, Maj. R.	21 Jan. 18	....	74th Punj.
28 May 15	Hayne, Lt. A. R. S. (*Actg. Capt.*)	28 May 16	....	I. A. R. O. Qr. Mr.
9 July 15	Sexton, Lt. C. B.	9 July 16		I. A. R. O.
29 Aug. 14	Churcher, Lt. W. D. (*Actg. Capt.*)	29 May 16	24 Nov. 16	E. Surr. R., I. A. (*on probn.*)
20 June 17	Crilly, Lt. A. J.	20 June 18	4 Oct. 17	I. A. R. O.
8 Aug. 17	Yorke, Lt. H. B.	8 Aug. 18	2 Nov. 17	I. A. R. O.
4 July 16	Crawford, Lt. C. O., M.C.	4 July 17	6 Sep. 18	R. Ir. Rif., I. A. (*on probn.*)
2 Jan. 17	Pepperall, Lt. R. A.	2 Jan. 18	2 July 18	M. G. Corps, I. A. (*on probn.*)
5 May 17	Henwood, Lt. F. H.	5 May 18	2 Sep. 18	Lond. R., I. A. (*on probn.*)
1 May 18	Fuller, 2nd-Lt. B. H., M.M.	1 May 18	14 Aug. 18	I. A. R. O. (*on probn.*)
26 Apl. 17	Rostron, Lt. H.	26 Apl. 18	....	I. A. R. O.
1 Aug. 18	Gildea, 2nd-Lt. J. P.	1 Aug. 18	19 Aug. 18	I. A. R. O.
1 Aug. 18	Madeley, 2nd-Lt. W. Y.	1 Aug. 18	23 Aug. 18	I. A. R. O.
28 Aug. 18	White, 2nd-Lt. H. R.	28 Aug. 18	8 Oct. 18	I. A. R. O.
7 Oct. 18	Holmes, 2nd-Lt. W. G.	7 Oct. 18	26 Oct. 18	I. A. R. O.

Subadar-Major.

		Jemadar.	Subadar.	Subadar-Major.	
3 Dec. 89	Lachman Singh (56)	3 Jan. 11	15 Mar. 15	30 Apl. 16	*Ind. Distgd. Ser. Medal.*

Subadars.

30 Nov. 02	Bakhshi Autar Singh (56, 66*a*, 72*a*)	30 Nov. 02	26 Sep. 08
1 Apl. 87	Shihab Din (66*a*)	23 Feb. 07	1 Nov. 10
12 Oct. 88	Muhammad Din (56)	23 June 08	7 Mar. 15
17 Jan. 08	Hayat Khan (56, 66)	16 May 10	7 May 15
20 Mar. 91	Sher Khan (56, 66*a*)	1 Nov. 10	15 July 15
4 Jan. 92	Budh Singh (56, 66*a*)	5 Feb. 10	30 Apl. 16
12 Dec. 92	Gopal Singh	7 Mar. 15	30 Apl. 16
18 Nov. 97	Shakar Khan (56, 72*a*)	7 Mar. 15	30 Apl. 16
7 Mar. 98	Ahmed Khan, I.D.S.M.	7 May 15	30 Apl. 16
1 Jan. 11	Karamdad Khan	13 July 15	30 Apl. 16
17 Jan. 03	Indar Singh, I.D.S.M.	15 July 15	30 Apl. 16
3 Mar. 00	Sharif Khan, I.D.S.M.	30 Apl. 16	25 Feb. 17
17 Jan. 03	Ghulam Khan	30 Apl. 16	29 Oct. 17

1st Battalion, 67th Punjabis—*contd.*

1st Commn. or date of entering service.	Names and Rank.	Army rank.	Remarks.
	Jemadars.		
19 Mar. 96	Kalu (56)	7 Mar. 15	
17 Sep. 99	Rahmat Ali	3 Mar. 15	
16 May 98	Rur Singh	17 Mar. 15	
....	Fazal Hassan	17 Mar. 15	
15 Aug. 99	*Raunaq Singh*	*24 July 15*	
1 Dec. 02	Mahammad Alam	15 July 15	
18 Nov. 02	Khushia	25 Dec. 15	
1 May 08	Muhammad Afzal Khan	30 Apl. 16	Indian Adjt., 1 Aug. 16.
17 Jan. 03	Atta Muhammad	30 Apl. 16	
3 Dec. 02	Mangal Singh	30 Apl. 16	
26 Nov. 02	Wilayat Khan (50, 48*a*)	23 Oct. 17	
11 Apl. 04	Shamas Din	23 Oct. 17	
16 Nov. 02	Piaru	29 Oct. 17	

2nd Battalion, 67th Punjabis.

Formed at Loralai on 17th March 1915.

Composition—2 companies of Punjabi Musalmans, 1 of Sikhs, 1 of Punjabi Hindus.

1st Commn. or date of entering service.	Names and Rank.	Army rank.	Present appointment in Regiment.	Remarks.
	Commandant.			
	Company Commanders. (4)			
1 Jan. 99	Luck, Maj. T.	1 Sep. 15	25 Sep. 17	1—67th Punj. 2nd-in-Comd. (*temp.*)
	Company Officers.			
27 Nov. 14	Wilkinson, Lt. H. A. (*Actg. Capt.*)	27 Nov. 15	18 Feb. 18	I. A. R. O.
17 Feb. 16	Trechmann, Lt. A. D.	17 Feb. 17	18 Jan. 18	Durh. L. I., I. A. (*on probn.*)
20 Dec. 15	Bassett, Lt. F. S.	20 Dec. 16	4 Dec. 17	I. A. (*on probn.*)
7 Oct. 17	Barnard, Lt. L. G.	7 Oct. 18	3 Dec. 17	I. A. R. O.
29 Aug. 17	Flower, Lt. L. T.	29 Aug. 18	16 Dec. 17	I. A. R. O.
5 Oct. 17	Waight, Lt. G.	5 Oct. 18	23 Jan. 18	I. A. (T. C.)
31 Oct. 17	Deakin, Lt. C. C.	31 Oct. 18	15 Feb. 18	I. A. R. O.
29 Jan. 16	Oates, Lt. A., D.S.O.	29 Jan. 17	1 Apl. 18	I. A. (*on probn.*)
26 May 18	Stephens, 2nd-Lt. K. T., M.C.	26 May 18	. .	Worc. R., A. A. (*on probn.*)
7 June 18	Morgan, 2nd-Lt.	7 June 18	. . . .	I. A. (*T. C.*)
7 June 18	Mackey, 2nd-Lt. D. D.	7 June 18	. . . .	I. A. (*T. C.*)

Subadar-Major.

		Jemadar.	Subadar.	Subadar-Major.	
22 Feb. 03	Sultan Asghar Ali Khan (52, 56, 72*a*)	22 Feb. 03	17 Mar. 15	1 June 17	(*temp.*)

Subadars.

14 Sep. 96	Bal Mukand	24 Feb. 08	17 Mar. 15	
1 Apl. 98	Jahandad Khan (72*a*)	1 Apl. 11	17 Mar. 15	
1 June 95	Sham Singh (72*a*, 72*b*)	1 Mar. 14	17 Mar. 15	
12 Apl. 92	Ghulam Hassan	17 Mar. 15	9 June 15	
16 May 83	Labh Singh	1 July 00	1 May 17	1-22 Punj.
9 June 99	Dasaundi Ram	17 Mar. 15	1 June 17	
27 Nov. 98	Dalip Singh (72*a*)	15 Mar. 17	11 Jan. 18	
14 Sep. 96	Alla Dad Khan	7 Mar. 15	22 Jan. 18	

Jemadars.

11 Jan. 97	Jiwan Singh (56)	17 Mar. 15	
21 Apl. 04	Abdulah Khan (56)	27 Sep. 15	Indian Adjt., 15 Mar. 17.
15 Jan. 03	Ahmad Khan	22 Nov. 15	
17 May 06	Barkat Shah	25 July 16	
27 Mar. 98	Kala Singh	11 Nov. 16	
30 Apl. 02	Bahadur Singh (56)	2 May 17	
28 Dec. 96	Tabiyat Singh	1 June 17	
17 Jan. 03	Muhammad Khan (I)	1 Sep. 17	
14 Dec. 02	Muhammad Khan (II)	1 Nov. 17	
9 Dec. 02	Allah Ditta Khan	3 Dec. 17	Indian Qr.-Mr.
14 Dec. 02	Gopal Singh	11 Jan. 18	
1 Aug. 05	Husain Khan (56)	22 Jan. 18	

Regt. Centre—Bangalore. *Uniform*—Scarlet.
Linked with—67th & 74th Punjabis. *Facings*—Emerald Green.

1st Battalion, 69th Punjabis.

Kelly-ki-paltan.—Raised at Madura, between the years 1762 and 1765, as the 10th Battalion of Coast Sepoys. Became the 10th Carnatic Battalion, 1769; the 9th Carnatic Battalion, 1770; the 9th Madras Battalion, 1784; the 1st Battalion, 9th Regiment of Madras Native Infantry, 1796; the 9th Regiment of Madras Native Infantry, 1824; the 9th Regiment of Madras Infantry, 1885; the 9th Madras Infantry, 1901; present designation, 1903.

A Galley with the Motto—خشکی و تری

"Sholinghur" "Carnatic" "Mysore" "Ava" "Pegu."

Composition—2 companies of Punjabi Musalmans, 1 of Sikhs (other than Jats and Mazbis), and 1 of Punjabi Hindus

COLONEL—Lieut.-Genl. *Sir* Herbert Vaughan Cox, K.C.B., K.C.M.G., C.S.I. . . . 4th July 16

1st Commn. or date of entering service.	Names and Rank.	Army rank.	Present appointment in Regiment.	Remarks.
	Commandant.			
5 Dec. 91	Houston, Lt.-Col. A. M.	5 Dec. 17	10 Mar. 18	
	Company Commanders. (4)			
22 Jan. 96	Bingham, Maj. W. H. [L] . .	22 Jan. 14	10 Mar. 18	2nd-in-Comd.
3 Aug. 98	Torkington, Maj. G. S. .	1 Sep. 15	10 Mar. 13	*Lv., ex I., m.c.*
12 Aug. 99	*Catty, Bt. Lt. Col. T. C.,* D.S.O., *p.s.c. (temp. Lt.-Col.)*	*3 June 18*	*20 Dec. 13*	*A. Q. M. G.*
17 Jan. 00	Greenaway, Maj. H.	1 Sep. 15	20 Dec. 13	
	Company Officers. (9)			
5 May 00	*Compton, Maj. C. W. McG.*	*1 Sep. 15*	*24 Sep. 03*	*G. S. O., 2nd gr., 9th Divn.*
18 May 01	*Hanson, Maj. G. J. G* .	*8 May 16*	*18 Apl. 04*	
18 Jan. 02	*McGann, Maj. H. H.* . . .	*18 Jan. 17*	*1 Jan. 06*	*Burma Mily. Police.*
20 Jan. 09	*Gulland, Capt. J. P.* . .	*1 Sep. 15*	*16 Mar. 10*	
3 Sep. 10	Gell, Capt. H. V. .	1 Sep. 15	10 Dec. 11	
5 Sep. 11	*Lumby, Capt. A. F. R. (temp. Maj.)*	*5 Sep. 15*	*15 Oct. 13*	*G. S. O., 2nd gr., A. Hdqrs.*
30 Jan. 17	Frost, Lt. R. L.	30 Jan. 18	. . . .	Attd. With 8th Rif.
30 Jan. 18	Burch, Lt. A. P.	30 Jan. 18	17 Dec. 17	Attd.
1 May 17	Stanley, Lt. D. R.	1 May 18	. . . .	Attd.
9 June 15	Marsh, Lt. B. C. (*Actg. Capt.*) .	9 June 16	. . . .	Attd.
27 July 16	Walsh, Lt. M. C.	27 July 17	. . . .	Attd.
11 Mar. 17	Breton, Lt. C. G.	11 Mar. 17	. . . .	Attd.
	Attached.			
1 May 01	Parkin, Maj. J. F., D.S.O. . .	1 May 16	. . . .	113th Infy.
15 Aug. 14	Mirchin, Capt. R. H. L. . . .	15 Aug. 18	3 Nov. 17	45 Sikhs.
20 Nov. 14	*Brookes, Lt. A.*	*20 Nov. 15*	*9 July 15*	*I. A. R. O., Provost Marshal.*
14 Mar. 15	Bousfield, Lt. H. N.	14 Mar. 16	. . . .	I. A. (*on probn.*)
24 July 16	Smalley, Lt. S.	24 July 16	. . . .	I. A. R. O.
15 Nov. 15	Williams, Lt. J. W. (*Actg. Capt.*)	11 Apl. 16	24 Sep. 16	87th Punj.
29 June 16	Proudlock, Lt. R. V.	29 June 17	24 Sep. 16	92nd Punj. Adjt. (*Actg.*)
21 Aug. 16	Anderson, Lt. E. D.	21 Aug. 17	. . . .	
30 Jan. 17	Telford, Lt. B. D.	30 Jan. 18	10 Apl. 18	93rd Infy. *Lv., ex I., m.c.*
18 June 17	Humphreys, 2nd-Lt. C. E. . .	18 June 17	25 Dec. 17	74th Punj.
19 Sep. 17	Reilly, Lt. T. J.	19 Sep. 18	. . . .	I. A. (*on probn.*)
27 Sep. 17	Lange, Lt. B. D.	27 Sep. 17	9 Feb. 18	I. A. R. O.
29 Mar. 18	Logan, 2nd-Lt. J. S.	29 Mar. 18	. . . .	I. A. (*T. C.*)
24 Apl. 18	Fitz Gerald, 2nd-Lt. K. P. . .	24 Apl. 18	. . . .	I. A. (U. L.)
7 June 18	Simpson, 2nd-Lt. H. B. . . .	7 June 18	. . . .	I. A. (*T. C.*)
31 July 18	Berry, 2nd-Lt. A.	31 July 18	. . . .	Genl. List, I. A. R. O. (*on probn.*)
27 Aug. 18	Keeling, 2nd-Lt. F. G. . . .	27 Aug. 18	. . . .	I. A. R. O.
28 Aug. 18	Mac Bey, 2nd-Lt. S. G. . . .	28 Aug. 18	. . . .	I. A. R. O.
29 Aug. 18	Edwards, 2nd-Lt. F.	29 Aug. 18	. . . .	I. A. R. O.
7 Oct. 18	Price, 2nd-Lt. E. J.	7 Oct. 18	. . . .	I. A. R. O.
28 Oct. 18	Kettleband, 2nd-Lt. H. W. B. .	28 Oct. 18	. . . .	I. A. R. O.
10 June 15	Molesworth, Lt. G. E. N. . .	10 June 16	. . . .	Devon R., I. A. (*on probn.*)
27 Mar. 18	Brannon, 2nd-Lt. A. C. . . .	27 Mar. 18	. . . .	Hamps. R., I. A. (*on probn.*)

Subadar-Major.

		Jemadar.	Subadar.	Subadar-Major.
6 Feb. 92	Muhammad Khan, I.O.M. (52, 56)	6 Nov. 07	3 July. 11	13 Apl. 17

1st Battalion, 69th Punjabis—*contd.*

1st Commn. or date of entering service.	Names and Rank.	Army rank.		Remarks.
	Subadars.			
20 Jan. 96	Rahmat Khan (66*a*)	15 Dec. 07	26 May 13	
1 July 02	Bulram Singh (66*a*)	22 Nov. 04	3 July 15	
2 Mar. 97	Rajwali (58)	3 July 11	25 Sep. 15	
4 Dec. 02	Indar Singh	Sep. 14	25 Sep. 15	
25 May 93	Fazal Dad	1 Oct. 14	25 Sep. 15	
10 Jan 03	Ahmad Khan	11 Oct. 14	27 June 16	
. . . .	Nishan Ali Bahadur	. . . .	23 July 16	*Order of Br. Ind., 2nd Cl.*
1 Jan. 94	Mansha Singh	2 Sep. 14	13 Apl. 17	
14 Dec. 02	Daulat Singh	11 Oct. 14	5 Oct. 17	
15 July 01	Sunkha Singh	25 Sep. 15	31 Jan. 18	
10 Sep. 03	Sardar Khan	11 Oct. 14	7 May 18	
15 Sep. 03	Sardar Khan, II, I.O.M.	11 Oct. 14	8 Aug. 18	
11 Jan. 03	Nabi Bakhsh	25 Sep. 15	8 Aug. 18	
14 Dec. 02	Labh Singh, I. D. S. M.	23 July 16	8 Aug. 18	
. . . .	Abdulla Khan (*Retd.*)	. . . .	. . . .	Re-employed, 1 May 15.
. . . .	Fazal Khan (*Retd.*)	. . . .	. . . .	Re-employed, 19 Jan. 16.
	Jemadars.			
4 July 03	Gheba Khan, I. D. S. M.	25 Sep. 15		
2 Dec. 02	Gurdit Singh	13 Apl. 17		
24 Nov. 03	Aulya Khan	16 May 17		
16 June 16	Ghulam Mohammad	15 Oct. 17		
16 Mar. 01	Gainda Singh	23 Oct. 17		
1 Nov. 00	Muhammad Wali	3 Dec. 17		
27 Nov. 03	Suba Ram	31 Jan. 18		
1 Sep. 06	Gopal Dass	7 Feb. 18		
14 Dec. 02	Waryam Singh	3 May 18		
. . . .	Pala Ram	1 June 18		
	Abdulla Khan	8 June 18		
11 June 04	Lal Khan	8 Aug. 18		
14 Jan. 03	Muhammad Bakhsh	21 Aug. 18		
14 Jan. 03	Kapura Ram	21 Aug. 18		
25 July 03	Nur Khan	21 Aug. 18		

2nd Battalion, 69th Punjabis.

Formed at Ambala.

Composition—2 companies Bishnois Jats, 1 company South-Western Punjabi Musalmans, 1 company Gadarias.

1st Commn. or date of entering service.	Names and Rank.	Army rank.	Present appointment in Regiment.	Remarks.
	Commandant.			
5 Aug. 96	Dunsford, Maj. E. H. (*Actg. Lt.-Col.*)	5 Aug. 14	1 June 18	1-33rd Punj.
	Company Commanders. (4)			
1 Sep. 99	Salkeld, Maj. H. Y., D.S.O.	1 Sep. 15	24 May 18	2nd Lrs. 2nd-in-Comd. (*temp.*).
16 Nov. 14	Marple, Lt. R. G. (*Actg. Capt.*)	15 July 17	28 June 18	I. A. R. O. (*on probn.*)
2 Apl. 15	Walkinshaw, Lt. I. C., M.C. (*Actg. Capt.*)	2 Jan. 17	27 June 18	I. A. (*on probn.*)
1 Aug. 15	Donaldson, Lt. S. (*Actg. Capt.*)	1 Aug. 16	12 July 18	I. A. R. O.
	Company Officers.			
7 May 16	Meers, Lt. P. J.	7 May 17	26 June 18	I. A. (*on probn.*) Qr. Mr.
30 Jan. 17	Gerrard, Lt. R. J. (*Actg. Capt.*)	30 Jan. 18	12 July 18	1-33rd Punj. Adjt.
26 Apl. 17	Williams, 2nd-Lt. G. H.	26 Apl. 17	5 July 18	I. A. R. O.
30 May 17	Meggett, 2nd-Lt. H. A.	30 May 17	25 Aug. 18	I. A. R. O.
29 May 18	Edmeades, 2nd-Lt. F. D.	29 May 18	. . . .	Suff R., I. A. (*on probn.*)
31 Oct. 17	Pocock, Lt. L. J.	30 Oct. 17	31 Oct. 18	I. A. R. O. (*on probn.*)
1 Nov. 17	Michael, Lt. T. J.	1 Nov. 18	15 Sep. 18	I. A. (*T. C.*)
5 Nov. 17	Flye, Lt. T. H.	5 Nov. 18	15 July 18	I. A. (*T. C.*)
25 Dec. 17	Corbett, 2nd-Lt. H. R.	25 Dec. 17	26 Aug. 18	I. A. R. O.
29 Mar. 18	Burton, 2nd-Lt. V. M.	29 Mar. 18	15 July 18	I. A. (*T. C.*)
22 July 18	Irwin-Caster, 2nd-Lt. E.	22 July 18	12 Dec. 18	I. A. (*T. C.*)
22 July 18	Walker, 2nd-Lt. A.	22 July 18	1 Dec. 18	I. A. R. O.
31 July 18	Will, 2nd-Lt. G. D.	31 July 18	17 Oct. 18	Genl. List, I. A. R. O. (*on probn.*)
30 Aug. 18	St. Pourcain, 2nd-Lt. E. A.	30 Aug. 18	21 Sep. 18	I. A. R. O.
3 Sep. 18	O'Connor, 2nd-Lt. T. G.	3 Sep. 18	30 Oct. 18	I. A. R. O.
4 Sep. 18	Goutiers, 2nd-Lt. G. F. C. de M.	4 Sep. 18	21 Sep. 18	I. A. R. O.
13 Sep. 18	Goutiers, 2nd-Lt. E. M. de M.	13 Sep. 18	9 Oct. 18	I. A. R. O.

Subadar-Major.

Jemadar.	Subadar.	Subadar-Major.

Subadars.

9 Jan. 18	Ghulam Sadiq	. . . .	21 Jan. 18
24 Dec. 04	Khiali Ram	1 Dec. 16	26 June 18
23 Oct. 17	Sohan Ram	23 Oct. 17	26 June 18
26 Feb. 03	Raja Khan	21 Jan. 18	27 June 18

Jemadars.

27 Feb. 17	Ghulam Hussain Khan	20 Aug. 17	
21 Feb. 17	Noorullah Khan	21 Jan. 18	
30 Aug. 01	Jan Muhammad	24 June 18	Indian Qr. Mr.
3 Jan. 11	Pat Ram	26 June 18	
30 Sep. 16	Tulsi Ram	26 June 18	

1st Battalion, 70th Burma Rifles.

Formed at Maymyo on the 19th September 1917.

Composition—3 Companies Burmans. 1 Company Karens.

1st Commn. or date of entering service.	Names and Rank.	Army rank.	Present appointment in Regiment.	Remarks.
	Commandant.			
19 Dec. 00	Stuart, Maj. D. N. (*actg. Lt.-Col.*)	19 Dec. 15	19 Sep. 17	
	Company Commanders. (4)			
	Company Officers.			
5 Aug. 05	Malet, Capt. A. W. (*actg. Maj.*)	5 Aug. 14	24 Nov. 17	38th Dogras, 2nd-in-Comd.
16 Aug. 05	Kenny, Capt. E. C. (*Actg. Maj.*)	16 Aug. 14	8 Oct. 17	1-98th Infy.
9 July 17	Grigson, Lt. P. St. J. B. (*actg. Capt.*)	9 July 18	27 Oct. 17	I. A. R. O.
6 Oct. 17	Hay, 2nd-Lt. E. F. A. (*Actg. Capt.*)	6 Oct. 17	20 Dec. 17	Welsh R., I. A. (*on probn.*)
16 Feb. 16	Clarke, Lt. M. J. (*Actg. Capt.*)	16 Feb. 17	16 Jan. 18	I. A. (*on probn.*) Adjt.
5 Oct. 17	Hartnoll, Lt. E. S.	5 Oct. 18	18 Jan. 18	I. A. R. O. (*on probn.*)
31 Jan. 18	Higginbotham, 2nd-Lt. P. P.	31 Jan. 18	4 Feb. 18	I. A. (*on probn.*)
31 Jan. 18	Blaber, 2nd-Lt. H. L.	31 Jan. 18	4 Feb. 12	Attd.
26 June 17	Davis, Lt. D. A.	26 June 18	8 Mar. 18	I. A. (*on probn.*)
5 Jan. 17	Cole, Lt. R. D.	5 Jan. 18	14 Mar. 18	I. A. R. O. Qr.-Mr.
20 Nov. 17	Collis, Lt. M. S.	20 Nov. 18	16 Ma . 18	I. A. R. O. (*on probn.*)
5 Nov. 17	Spargo, Lt. F. W.	5 Nov. 18	16 Mar. 18	I. A. R. O.
5 Nov. 17	Smith, Lt. H. C.	5 Nov. 18	16 Mar. 18	I. A. R. O. (*on probn.*)
8 Aug. 17	Dawn, Lt. A.	8 Aug. 18	17 Mar. 18	I. A. R. O.
27 Sep. 16	Foster, Lt. R. F.	17 Sep. 17	5 Apl. 18	I. A. (*on probn.*)
21 Aug. 17	Aue, Lt. W. C. R.	22 Aug. 18	28 Apl. 18	R. Dub. Fus., I. A. (*on probn.*)
15 Jan. 18	Cane, 2nd-Lt. S. P.	15 Jan. 18	29 Apl. 18	Suff. R., I. A. (*on probn.*)
21 Aug. 15	Prideaux, Lt. G. V. A. (*Actg. Capt.*)	21 Aug. 16	6 May 18	I. A. R. O.
29 Mar. 18	Young, 2nd-Lt. S.	29 Mar. 18	28 July 18	I. A. R. O.
7 June 18	Purser, 2nd-Lt. C. R.	7 June 18	. . . .	I. A. R. O.
30 July 18	Pusey, 2nd-Lt. A. G.	30 July 18	. . . .	I. A. R. O.

		Jemadar.	Subadar.	Subadar-Major.	
	Subadar-Major.				
	Mg. On Kyaw	23 Apl. 17			
	Subadars.				
	Mg. Po San	24 Apl. 17			
	Mg. Pa Pon	1 Apl. 17			
	Maung U	14 Nov. 17	30 Apl. 1.		
	Jemadars.				
1 June 95	Mg Maung	1 Mar. 17			
	Mg Nyan Gyi	15 June 17			
	Mg Nyo	1 Nov. 17			
	Mg Kyaw Nyun	1 Nov. 17			
	Mg Thein Maung	1 Nov. 17			
	Mg Aye	1 Dec. 17			
	Maung Kin Maung	1 Dec. 17			
	Maung Ba Sein	1 Dec. 17			
	Maung Ba Than	1 Dec. 17			
	Maung Ba Zan	1 Jan. 18			
	Maung Tin	14 Jan. 18			
	Maung Ba Gyan	1 Feb. 18			
	Maung San Mya	1 Feb. 18			
	Maung Tun Gyaw	30 Apl. 18			
	Mg. Ba Han	18 Apl. 18			
	Mg. Htang Law	30 Sep. 18			
	Silas San Wa	3 Sep. 18			
	Mg. Ba Ket	10 Sep. 18			

2nd Battalion, 70th Burma Rifles.

Formed at

Composition. 2 Companies Burmans. 1 Company Karens. 1 Company Arakanese.

1st Commn. or date of entering service.	Names and Rank.	Army rank.	Present appointment in Regiment.	Remarks.
	Commandant.			
4 Jan. 99	Rigg, Maj. H. M. (*actg. Lt.-Col.*)	1 Sep. 15	. . .	92nd Punj.
	Company Commanders. (4)			
28 July 00	Macdonald, Maj. R. W., D.S.O.	1 Sep. 15	24 May 18	2nd-in-Comd. (*temp.*)
30 Dec. 14	Griffin, Lt. S. F. H. (*actg. Capt.*)	30 Dec. 15	20 Aug. 18	I. A. R. O.
1 May 15	Macpherson, Lt. H. A. (*actg. Capt.*)	1 May 16	23 Jan. 18	I. A. (*on probn.*)
5 Oct. 17	Webster, Lt. W. H. A. (*actg. Capt.*)	5 Oct. 18	23 Jan. 18	I. A R. O.
	Company Officers.			
1 Sep. 16	Palmes, Lt. W. T.	1 Sep. 17	30 Dec. 17	I. A. R. O.
7 Sep. 16	Underwood, Lt. W. R.	7 Sep. 17	29 Dec. 17	I. A. R. O.
15 Feb. 16	Holme, Lt. H. B.	15 Feb. 17	9 Apl. 18	I. A. R. O. Adjt.
26 June 17	Bleathman, Lt. W. E. P.	26 June 18	20 Jan. 18	I. A. (*on probn.*) Qr. Mr. 16 May 18.
7 June 18	Shead, 2nd-Lt. C. G.	7 June 18	7 Oct. 18	I. A. R. O.
18 July 17	Paddon-Row, Lt. D. A. T.	18 July 18	14 Sep. 18	I. A. R. O.
1 Oct. 17	Nichols, 2nd-Lt H. L.	1 Oct. 17	26 Mar. 18	I. A. R. O.
5 Nov. 17	Gordon, Lt. H. V.	5 Nov. 18	. . .	I. A. R. O.
28 Nov. 17	Mayers, Lt. E. W.	28 Nov. 18	13 Feb. 18	I. A. R. O.
15 Jan. 18	Woledge, 2nd-Lt. C. G. C.	15 Jan. 18	19 June 18	I. A. R. O.
31 Jan. 18	Bullard, 2nd-Lt. D. A.	31 Jan. 18	13 Feb. 18	Attd.
31 Jan. 18	Smythe, 2nd-Lt. L. G. F.	31 Jan. 18	13 Feb. 18	Attd.
31 Jan. 18	Barnes, 2nd-Lt. J. H.	31 Jan. 18	24 Oct. 18	1—129th Baluchis.
29 Mar. 18	Thompson, 2nd-Lt. A. K.	29 Mar. 18	. . .	I. A. R. O.
10 May 18	Davsi, 2nd-Lt. A. P.	10 May 18	3 Dec. 18	I. A. R. O.
29 May 18	Barrington, 2nd-Lt. A. H. M.	29 May 18	13 Aug. 18	I. A. R. O.
30 July 18	Harding, 2nd-Lt. L. O'B.	30 July 18	30 Aug. 18	I. A. R. O.
30 July 18	Nelson, 2nd-Lt. J. I.	30 July 18	30 Aug. 18	I. A. R. O.
29 Aug. 18	Nichols, 2nd-Lt. J. A.	29 Aug. 18	29 Aug. 18	I. A. R. O.

Subadar Major.

	Jemadar.	Subadar.	Subadar-Major.
Mg. Nyun U.	. . . .	8 Feb. 18	5 May 18

Subadars.

Yaba, A.	7 Feb. 18	6 Mar. 18
Mg Aung Din	. . . .	12 Mar. 18
Mg. Kyaw Zan	20 Apl. 18	20 May 18
Mg. Gyi	10 Feb. 18	6 Sep 18

Jemadars.

Mg. Shwe Gyaw	17 Jan. 18
Mg. Pe Thein (60*a*)	17 Jan. 18
Mg. Ba Shwe	5 Mar. 18
Mg. Ba Than	5 Mar. 18
Mg. Po Byu	21 Apl. 18
Mg. Ba Chein	5 June 18
Mg. Mg. Gale	5 June 18
San U Kaing	15 June 18
Mg. Tun Aung Gyaw	15 June 18
Mg. Ba Saw	17 June 18
Mg. Saw	17 June 18
Mg. Tun Maung	31 Aug. 18
Mg. Myit	31 Aug. 18
Mg. Maung	23 Nov. 18
San Tun Tha	23 Nov. 18

3rd Battalion, 70th Burma Rifles.

Formed at Meiktila on 23rd April 18.
Composition—3 Companies Burmans. 1 Company Arakanese.

1st Commn. or date of entering service.	Names and Rank.	Army rank.	Present appointment in Regiment.	Remarks.
	Commandant.			
20 May 99	Morgan, Maj. B. E., D.S.O. (*Actg. Lt.-Col.*)	1 Sep. 15	17 Apl. 18	93rd Infy.
	Company Commanders. (4).			
19 Aug. 03	Beadon, Maj. V., M.C.	19 Aug. 18	26 May 18	2nd-in-Comd. (*temp.*)
7 Aug. 14	Barwis, Capt. C. W. A.	7 Aug. 18	17 Apl. 18	
4 July 17	Hill, Lt. A. H. (*Actg. Capt.*)	4 July 17	24 May 18	I. A. R. O.
16 May 17	Tucker, Lt. J. D.	16 May 18	17 Apl. 18	Glouc. R., I. A. (*on probn.*)
	Company Officers.			
5 Dec. 14	Beatty, Lt. H. P.	5 Dec. 15	20 Nov. 18	I. A. R. O.
5 Aug. 15	Holmes, Lt. E.	5 Aug. 16	17 Apl. 18	I. A. (*on probn.*)
20 Sep. 15	Smith, Lt. F. M.	20 Sep. 16	17 Apl. 18	North'n R., I. A. (*on probn.*)
1 Apl. 17	Cooper, Lt. C. R. P. (*Actg. Capt.*)	29 May 18	17 June 18	I. A. R. O. Adjt. (*Offg.*)
5 Oct. 17	Zaiman, Lt. B. A.	5 Oct. 18	17 Apl. 18	I. A. R. O.
5 Oct. 17	Sheehy, Lt. J. F.	5 Oct. 18	17 Apl. 18	I. A. R. O.
1 Nov. 17	Hewett, Lt. D. P.	1 Nov. 18	22 May 18	I. A. R. O.
1 Nov. 17	Bailey, Lt. A. H.	1 Nov. 18	17 Apl. 18	I. A. R. O.
5 Nov. 17	Jennings, Lt. S. P.	5 Nov. 18	17 Apl. 18	I. A. R. O.
26 Nov. 17	Bisset, 2nd-Lt. J. D.	26 Nov. 17	12 June 18	I. A. R. O.
15 Jan. 18	Clark, 2nd-Lt. H.	15 Jan. 18	12 June 18	I. A. R. O.
15 Jan. 18	McGregar, 2nd-Lt. A.	15 Jan. 18	12 June 18	I. A. R. O.
26 Nov. 17	Shearer, 2nd-Lt. W.	26 Nov. 17	12 June 18	I. A. R. O.
15 Jan. 18	Webster, 2nd-Lt. T. G. C.	15 Jan. 18	12 June 18	I. A. R. O. Qr. Mr. (*Offg.*)
15 Jan. 18	Williamson, 2nd-Lt. A.	15 Jan. 18	12 June 18	I. A. R. O.
15 Jan. 18	Holt, 2nd-Lt. G. H.	15 Jan. 18	. . . .	I. A. R. O.
10 May 18	Killick, 2nd-Lt. A. E. H.	10 May 18	2 Dec. 18	I. A. R. O.
7 June 18	Hutton, 2nd-Lt. E. A.	7 June 18	7 Oct. 18	I. A. (*T. C.*)
17 July 18	Chalk, 2nd-Lt. W. G.	17 July 18	3 Dec. 18	I. A. R. O.
22 July 18	Searl, 2nd-Lt. A. E. G.	22 July 18	3 Dec. 18	I. A. (*T. C.*)
27 July 18	Gill, 2nd-Lt. C. A. G.	27 July 18	16 Aug. 18	I. A. R. O.
30 July 18	Fleming, 2nd-Lt. J. H.	30 July 18	20 Aug. 18	I. A. R. O.
8 Aug. 18	Johnston, 2nd-Lt. J. A.	8 Aug. 18	13 Sept. 18	I. A. R. O.
9 Aug. 18	Lowry, 2nd-Lt. J.	9 Aug. 18	13 Sept. 18	I. A. R. O.
12 Aug. 18	Buchanan, 2nd-Lt. J. P.	12 Aug. 18	17 Sept. 18	I. A. R. O.
2 Oct. 18	Hunneybun, 2nd-Lt. B.	2 Oct. 18	30 Oct. 18	I. A. R. O.

Subadar-Major.

		Jemadar.	Subadar.	Subadar-Major.
28 May 18	Po Than			28 May 18.

Subadar.

21 Feb. 18	Ba Pe	. . . .	11 Nov. 18	
24 June 18	Ba Thaw	. . . .	24 June 18	
18 July 16	Thein Kin	. . . .	20 Aug. 18	

Jemadars.

13 Feb. 18	Win Pe	13 Feb. 18	
22 Feb. 18	Maung Nyein	22 Feb. 18	Indian Adjt.
22 Apl. 18	Ba Shin	22 Apl. 18	
29 Apl. 18	Ba Nyun	29 Apl. 18	
7 Apl. 18	San Pe	6 May 18	
7 May 18	Sein Po	7 May 18	
27 May 18	Ba Min	27 May 18	
7 June 18	Ba Gyan	7 June 18	
13 June 18	Ba Oh	13 June 18	
10 July 18	Po Mya	10 July 18	
16 May 18	Tun Sein	25 July 18	
25 May 18	Le Maung	25 July 18	
30 July 18	Hla Bu	16 Sep. 18	Indian Qr. Mr.

4th Battalion, 70th Burma Rifles.

Formed at Meiktila on 10th June 1918.

Composition—
2 Companies Burmans.
1 Company Karens.
1 Company Shans.

1st Commn. or date of entering service.	Names and Rank.	Army rank.	Present appointment in Regiment.	Remarks.
	Commandant.			
25 Jan. 99	Landon, Maj. C. R. H. P. (*Actg. Lt.-Col.*)	1 Sep. 15	10 June 18	35th Horse.
	Company Commanders. (4)			
18 Jan. 02	Massy, Maj. W. I. J. *q*	18 Jan. 17	18 June 18	12th Prs. 2nd-in-Comd. (*temp.*)
	Company Officers.			
2 July 18	Houldey, Capt. J. E.	2 July 18	25 July 18	I. A. R. O. Adjt.
29 Sep. 17	Lanktree, Lt. H. R. (*Actg. Capt.*)	29 Sep. 18	10 June 18	I. A. R. O.
5 Oct. 17	L'Estrange, Lt. T. W.	5 Oct. 18	10 June 18	I. A. R. O.
27 June 17	Glenie, 2nd-Lt. J. W.	27 June 17	10 June 18	I. A. (*on probn.*) Qr. Mr. (*Offg.*)
1 Dec. 17	Leach, 2nd-Lt. F. B. (*Actg. Capt.*)	1 Dec. 17	24 July 18	I. A. R. O.
15 Jan. 18	McManus, 2nd-Lt. A. A.	15 Jan. 18	12 June 18	I. A. (*T. C.*)
15 Jan. 18	Tunstall, 2nd-Lt. J. E.	15 Jan. 18	12 June 18	I. A. R. O.
15 Jan. 18	Wilson, 2nd-Lt. E. H. E.	15 Jan. 18	12 June 18	I. A. R. O.
15 Jan. 18	Corrie, 2nd-Lt. A. B.	15 Jan. 18	12 June 18	I. A. R. O.
15 Jan. 18	Gawthorne, 2nd-Lt. C. H.	15 Jan. 18	12 June 18	I. A. R. O.
15 Jan. 18	Sullivan, 2nd-Lt. C. S.	15 Jan. 18	12 June 18	I. A. R. O.
15 Jan. 18	Corban-Lucas, 2nd-Lt. A. J. L.	15 Jan. 18	12 June 18	I. A. R. O.
15 Jan. 18	Searle, 2nd-Lt. H. F. (*Actg. Capt.*)	15 Jan. 18	12 June 18	I. A. R. O.
29 Mar. 18	Fairweather, 2nd-Lt. A.	29 Mar. 18	27 July 18	I. A. R. O.
10 May 18	Clague, 2nd-Lt. J.	10 May 18	21 July 18	I. A. R. O.
28 Mar. 18	Guinee, 2nd-Lt. J. A.	28 Mar. 18	2 July 18	I. A. (*on probn.*)

Subadar-Major.

		Jemadar.	Subadar.	Subadar-Major.	
30 May 18	Mg Po Chu		30 May 18	26 Oct. 18	

Subadars.

17 Mar. 18	Mg Po Myit	15 July 18	15 July 18	
1 July 18	Mg San Tun	21 Aug. 18	21 Aug. 18	
16 Oct. 18	Mg Ba On	16 Oct. 18	16 Oct. 18	

Jemadars.

11 July 18	Mg So Maung	11 July 18		
16 July 18	Mg So Hlaing	16 July 18		
23 July 18	Maung Pu	23 July 18		
26 July 18	Mg Ba Than	26 July 18		
17 June 18	Maung Shwe	1 Aug. 18		
5 June 18	Mg Tun Win	10 Aug. 18		
16 Mar. 18	Mg Pe Gyi	10 Aug. 18		
18 Sep. 18	Maung Po Shwe	18 Sep. 18		
15 Aug. 18	Maung Tun Aung	20 Sep. 18		
19 July 18	Maung Khin	20 Sep. 18		
16 July 18	Maung Sao Ohn	1 Oct. 18		
6 Nov. 18	Maung Gyee	6 Nov. 18		

71st Punjabis.

Formed at Montgomery on 9th June, 1917.
Composition—4 Companies of Punjabi Christians.

1st Commn. or date of entering service.	Names and Rank.	Army Rank.	Present appointment in Regiment.	Remarks.
	Commandant.			
15 Dec. 97	Wright, Maj. C. H. B. (*Actg. Lt.-Col.*)	1 Sep. 15	15 June 18	62nd Punj.
	Company Commanders. (4)			
1 Jan. 98	Beyts, Maj. R. F. S. . .	1 Sep. 15	1 May 18	*Cant. Magte.'s Dept.*
25 Jan. 08	Bacon, Capt. R. G.	1 Sep. 15	5 May 18	124th Bal. Inf.
17 Nov. 14	Skey, Lt. L. G. (*Actg. Capt.*) .	17 Nov. 15	16 Jan. 18	I. A. R. O.
11 Feb. 15	Greenwood, Lt. R. S. (*Actg. Capt.*)	11 Feb. 16	12 Dec. 17	I. A. R. O.
	Company Officers.			
18 Oct. 99	Pott, Maj. H. P.	1 Sep. 15	. . . .	1-5th Infy.
30 Sep. 15	Moritz, Lt. F. E. B.	30 Sep. 16	16 Feb. 18	I.A. (*on probn.*).
31 July 15	Hood, Lt. F. W. (*Actg. Capt.*) .	31 July 16	12 Dec. 17	I. A. R. O.
1 May 16	Atkins, Lt. P. C. (*Actg. Capt.*) .	1 May 17	15 Dec. 17	Attd. Adjt.
22 July 16	Norman, Lt. D.	22 July 17	17 June 17	*I. A. R. O., A. R. O., Poona.*
5 May 17	Berry, Lt. O. D. W.	5 May 18	18 Aug. 17	I. A. R. O.
5 May 17	Joblin, Lt. G. W.	5 May 18	7 Oct. 17	I. A. R. O.
5 May 17	Wilken, Lt. W.	5 May 18	18 Aug. 17	I. A. R. O.
5 June 17	Chapman, Lt. C. A.	5 June 18	5 Nov. 17	Attd. Qr.-Mr.
20 June 17	Paterson, Lt. A. D.	20 June 18	5 Oct. 17	I. A. R. O.
20 June 17	Fisk, Lt. L. E.	20 June 18	5 Oct. 17	I. A. R. O. *Lv., ex I.*
1 Oct. 17	Moore, Lt. A. J.	1 Oct. 18	21 Jan. 18	I.A. (*T. C.*).
31 Oct. 17	Buckley, 2nd-Lt. P. W. . . .	31 Oct. 17	14 Feb. 18	I. A. R. O.
1 July 16	Godfrey, Lt. W. E. (*Actg. Capt.*)	1 July 17	6 May 18	I. A. R. O.
	Taylor, 2nd-Lt. S.	. . . .	1 Apl. 18	I. A. R. O.
5 Jan. 18	Halstead, 2nd-Lt. J.	5 Jan. 18	6 June 18	I. A. (*T. C.*)
15 Jan. 18	Pibworth, 2nd-Lt. H. A. . . .	15 Jan. 18	9 June 18	I. A. (*T. C.*)
31 July 18	Hourihane, 2nd-Lt. G. O. B. .	31 July 18	31 Oct. 18	R. Ir. Fus., I. A. (*on probn.*)

Subadar-Major.

		Jemadar.	Subadar.	Subadar-Major.	
14 July 85	Abdai Razzak, *Bahadur* . .	1 July 01	23 Jan. 05	4 Aug. 14	*Order of Br. Ind., 2nd Cl.*

Subadars.

1 May 87	Jiwan Singh	. . . .	29 Apl. 18	
	Arjan Sing (*Retd.*)	. . . .	. . . .	Re-employed. 19 Aug. 17.
	Natha Singh (*Retd.*), I.O.M. . .	. . . .	. . . .	Re-employed. 1 Sep. 17.
22 Oct. 15	Gulam Khan	22 Oct. 15	1 Oct. 17	
17 Nov. 15	Shahbaz Khan	17 Nov. 15	1 Oct. 17	
15 July 16	Wadhawa Mall	15 July 15	1 Oct. 17	
9 Oct. 17	Khazan Singh	9 Oct. 17	2 July 18	
10 Oct. 17	Jamal Din	10 Oct. 17	3 July 18	
15 Mar. 17	Vivian Sheriff	15 Mar. 17	1 July 18	

Jemadars.

1 Sep. 17	B. A. Joseph	1 Sep. 17
15 Oct. 17	Mathew	15 Oct. 17
21 May 17	Paul Singh	27 Oct. 17
15 Oct. 15	Inayat Masih (46)	15 Feb. 18
30 Sep. 03	Dhana Singh	1 Mar. 18
16 Apl. 97	Bhola Singh	1 Mar 18
21 July 00	Labh Singh	1 Mar. 18
5 Nov. 15	Andrew	26 Apl. 18
22 Oct. 15	Alexander Hari Singh . . .	26 Apl. 18
15 July 16	Dina Nath Wadhawa Mal . .	26 Apl. 18
29 Nov. 00	Kehar Singh	26 Apl. 18
11 June 06	Nawab Singh	26 Apl. 18
16 June 96	Labh Singh	26 Apl. 18
. . . .	Khem Singh	26 Apl. 18
	Arjan Dass	3 June 18
	Ahmad Shah Gilani	3 June 18

Regt. Centre—Mandalay. *Uniform*—Drab Serge.
Linked with—92nd Punjabis & 93rd Burma Infy. *Facings*—White.

1st Battalion, 72nd Punjabis.

Barim-ki-pultan.—Raised at Cuddalore in 1767, by Capt. R. Mathews, as the 16th Battalion of Coast Sepoys. Became the 13th Carnatic Bn., 1769; the 12th Carnatic Bn., 1770; the 12th Madras Bn., 1784; the 2nd Bn. 8th Regt. of Madras Native Infantry, 1796; the 12th Regt. of Madras N. I., 1824; the 12th Regt. of Madras N. I., 1885; the 2nd Regt. of Burma Infy., 1890; the 12th Regt. (2nd Burma Bn.) of Madras Infy., 1891; the 12th Burma Infy., 1901; present designation, 1903.

"Sholinghur" "Carnatic" "Ava" "Burma, 1885–87."

Composition—2 Companies of Sikhs, 1 of Punjabi Musalmans, 1 of Pathans.

1st Commn. or date of entering service.	Names and Rank.	Army rank.	Present appointment in Regiment.	Remarks.
	Commandant.			
21 Sep. 89	Prentis, Lt.-Col. W. S. . . .	21 Sep. 15	13 June 16	
	Company Commanders. (4)			
23 Feb. 98	McCormick, Maj. G. D. . .	1 Sep. 16	28 Aug. 17	2nd-in-Comd. With 2-112th Infy.
8 May 01	*Davrell, Maj. W. S.*, D.S.O. . .	*8 May 16*	*11 Feb. 04*	*Bde.-Major., 42nd Infy. Bde.*
18 Jan. 02	*Whitworth, Maj. T. S.* . . .	*18 Jan. 17*	*4 Feb. 15*	*Bde.-Maj. Nowshera.*
28 Jan. 03	*Burnett, Maj. A. H.*, D.S.O. . .	*28 Jan. 18*	*28 Aug. 17*	*Bde.-Maj., 36th Infy. Bde.*
	Company Officers. (8)			
29 July 04	Howlett, Capt. E. G.	29 July 13	19 Mar. 09	*Bde. Maj. 19th Infy. Bde.*
24 Jan. 06	Pughe, Capt. O. L.	24 Jan. 15	3 Apl. 07	With 85th Rif.
29 Aug. 06	Waller, Capt. E. de W. (*Actg. Maj.*)	29 Aug. 15	22 July 09	With 53rd Sikhs.
28 May 10	Longbotham, Capt. C. M. . .	1 Sep. 15	5 Oct. 12	*Lv., ex I., m.c.*
18 Jan. 11	*Radley, Capt. H. P.*, M.C. (*Actg. Maj.*)	*1 Sep. 15*	*6 Mar. 12*	*With 14th Signal Company.*
1 Oct. 14	Clarke, Capt. G. R.	1 Oct. 18	17 Dec. 14	*Lv. ex I., 6 mos.*
15 Nov. 15	Challen, Lt B. S. (*Actg. Capt.*) .	15 Nov. 16	19 Nov. 15	With 2-125th Rif.
9 June 16	Barker, Lt. D. H.	29 June 17	3 July 16	
19 July 16	Smithett, Lt. W. H.	19 July 17	11 Nov. 16	Attd. With 28th Punj.
14 Nov. 16	*Blakeney, 2nd-Lt. E. C. W.* . .	*14 Nov. 16*	*22 Nov. 16*	*Attd.*
7 Apl. 17	Keenan, Lt. F. W.	7 Apl. 18	2 Aug. 17	Attd.
5 July 17	Barr, Lt. M. C.	5 July 17	12 Dec. 17	Attd.
	Attached.			
29 July	Finnie, Maj. A. McL. . . .	29 July 18	4 Feb. 15	Civil. 2nd-in-Comd. (*temp.*)
9 Jan. 04	Burrows, Capt. H. M. (*Actg. Maj.*)	9 Jan. 13	9 Jan. 18	51st Sikhs.
18 Jan. 11	Taylor, Capt. V. C. L. . . .	1 Sep. 15	12 Aug. 16	93rd Infy.
26 Feb. 16	Jenkins, Lt. F. A. M. B. (*Actg. Capt.*)	26 Feb. 17	15 May 17	York. R., I. A. (*on probn.*)
21 June 16	Brewis, Lt. R. S., M.C. (*Actg. Capt.*)	21 June 16	14 Aug. 17	E. Surr. R., I. A. (*on probn.*)
20 June 17	Johnston, Lt. W. C.	20 June 18	4 Oct. 17	I. A. R. O.
19 Oct. 17	Oliphant, Lt. R. W.	19 Oct. 18	9 Dec. 17	I. A. R. O.
1 Dec. 14	Hardy, Lieut. E. P.	1 Sep. 16	1 Jan. 18	Bord. R., I. A. (*on probn.*)
19 Sep. 14	Burrows, Lt. G. T.	19 Sep. 15	16 Jan. 18	I. A. R. O.
5 Oct. 17	Turner, Lt. G. K.	5 Oct. 18	20 Jan. 18	I. A. R. O.
5 Oct. 17	Waitt, Lt. W. E. (*Actg. Capt.*) .	5 Oct. 18	20 Jan. 18	I. A. (*T. C.*) Adjt. 17 Aug. 18.
25 Oct. 16	Thompson, Lt. F. C. A. . . .	25 Oct. 17	23 Jan. 18	I. A. R. O.
29 Mar. 18	Ford, 2nd-Lt. C. W.	29 May 18	15 July 18	I. A. (*T. C.*)
27 Nov. 15	Rowe, Lt. W. C.	27 Nov. 16	30 June 18	Oxf. & Bucks. L. I., I. A. (*on probn.*)
21 June 13	Hartless, 2nd-Lt. L. C. . . .	21 June 18	7 July 18	I. A. R. O.
29 June 18	Honeywell, 2nd-Lt. B. . . .	29 June 18	14 July 18	I. A. R. O.
24 Apl. 18	Plewman, 2nd-Lt. T. L. . . .	24 Apl. 18	14 Oct. 18	I. A. (U. L.)
13 Sep. 18	Kenderdine, 2nd-Lt. N. W. . .	13 Sep. 18	5 Oct. 18	I. A. R. O.
13 Sep. 18	Sinclair, 2nd-Lt. A. V. . . .	13 Sep. 18	5 Oct. 18	I. A. R. O.
7 Oct. 18	Nicholson, 2nd-Lt. J.	7 Oct. 18	27 Oct. 18	I. A. R. O.
7 Oct. 18	Phillips, 2nd-Lt. H. T.	7 Oct. 18	27 Oct. 18	I. A. R. O.
	Subadar-Major.	**Jemadar.**	**Subadar.**	**Subadar-Major.**
	Ghulam Muhammad, Hony. Capt. *Sardar Bahadur* (*Retd.*)	. . . .	. . . .	Re-employed 20 Apl. 18 Depôt.
21 Oct. 92	Kishun Singh (52, 56, 66*a*) .	16 July 09	16 July 10	25 Jan. 18.
	Subadars.			
. . . .	Hakam Khan, Hony. Lt. *Sardar Bahadur* (*Retd.*)	. . . .	. . . .	Re-employed 30 Sep. 18.
1 Jan. 89	Chand Khan (52, 56, 66*a*) . .	11 Feb. 05	16 May 11	
26 Sep. 87	Latif Khan, I.D.S.M. (52, 56, 106)	2 Dec. 99	1 July 14	With 53rd Sikhs.
Feb. 93	Fazil Din (52, 56)	7 Apl. 10	13 Jan. 15	
17 June 88	Lal Singh (52, 56, 66*a*, 105) . .	1 June 10	29 Aug. 15	
1 Aug. 91	Murad Bakhsh (52, 56, 66*a*) .	16 May 11	12 Jan. 16	
1 June 98	Jhanda Singh (53, 50, 104) . .	6 Feb. 14	16 Oct. 17	
17 Feb. 95	Lal Singh (52, 56, 66*a*) . . .	10 June 14	1 Nov. 17	
7 Nov. 94	Moula Dad	8 Mar. 15	22 Feb. 18	
5 May 07	Atma Singh (56)	1 Nov. 17	10 July 18	
	Sopan Singh, *Bahadur* (*Retd.*) .	. . . .	. . . .	Re-employed 3 Aug. 18.
	Ahmed Din (*Retd.*)	. . . .	. . . .	Re-employed 17 July 18.
.	Sohan Singh, *Bahadur* (*Retd.*) .	. . . .	. . . .	Re-employed 3 Aug. 18.

1st Battalion, 72nd Punjabis—*contd.*

1st Commn. or date of entering service.	Names and Rank.	Army rank.	Remarks.
	Jemadars.		
13 July 93	Santokh Singh	29 Aug. 15	
19 Sep. 92	Jhanda Singh	29 Aug. 15	
27 June 98	Phuman Singh (56)	29 Aug. 15	
1 July 97	Nizam-ud-din (56, 61, 66*a*)	29 Aug. 15	Indian Adjt.
15 Dec. 96	Shams-ud-din (56, 105)	12 Jan. 16	
1 Apl. 94	Imam Shah	1 Nov. 16	
21 Mar. 06	Muhammad Sayyid (56)	1 Aug. 17	With 53rd Sikhs.
26 Nov. 00	Mal Singh (56)	16 Oct. 17	
16 Nov. 99	Hira Singh (56)	16 Oct. 17	
1 Mar. 97	Fateh Muhammad	22 Feb. 18	
23 Sep. 02	Ahmad Khan	23 May 18	
3 Dec. 02	Mardan Ali	23 June 18	

2nd Battalion, 72nd Punjabis.

Formed at Kamptee on 16 October 1917.

Composition—2 Companies Sikhs, 1 of Punjabi Mussalmans, 1 of Pathans.

1st Com[illegible] or date of entering service.	[illegible]es and Rank.	Army rank.	Present appointment in Regiment.	Remarks.
	Commandant.			
20 Jan. 97	Butler, Maj. H. M. (*Actg. Lt.-Col.*).	20 Jan. 15	16 Oct. 17	104th Rif.
	Company Commanders. (4)			
4 May 01	Jones, Maj. A. N. G., D.S.O.	4 May 16	16 Oct. 17	36th Horse. 2nd-in-Comd.
24 Jan. 06	Burridge, Capt. T. F. S.	24 Jan. 15	30 Sep. 18	1-35th Sikhs.
	Company Officers.			
18 June 17	Sandham, Lt. W. F. N.	18 June 18	16 Oct. 17	66th Punj.
18 June 17	Davies, Lt. R. M. (*Actg Capt.*)	18 June 18	16 Oct. 17	104th Rif.
1 Mar. 17	Hutchison, Lt. E. (*Actg. Capt.*)	1 Mar. 18	25 Jan. 18	I. A. R. O.
25 July 17	Adams, Lt. H. R.	25 July 18	12 Feb. 18	I. A. (*on probn.*) Adjt. (*offg.*)
28 Mar. 17	Brown, Lt. D. H.	28 Mar. 18	28 Mar. 18	I. A. R. O.
29 Aug. 17	Druett, Lt. H. C.	29 Aug. 18	12 Feb. 18	I. A. R. O.
5 Oct. 17	Stillman, Lt. A. C. H.	5 Oct. 18	19 Jan. 18	I. A. (*T. C.*)
5 Oct. 17	Shepard, Lt. A. J.	5 Oct. 18	19 Jan. 18	I. A. (*T. C.*)
5 Nov. 17	Goldsack, Lt. A. G. R.	5 Nov. 18	17 Mar. 18	I. A. (*T. C.*) Qr. Mr. (*offg.*)
5 Nov. 17	Hoare, Lt. A. W.	5 Nov. 18	23 Mar. 18	I. A. (*T. C.*)
13 Jan. 18	Moore, 2nd-Lt. G.	13 Jan. 18	31 Mar. 18	E. York. R., I. A. (*on probn.*)
31 Jan. 18	Peckover, 2nd-Lt. R. G.	31 Jan. 18	5 Feb. 18	Attd.
31 Jan. 18	Henshaw, 2nd-Lt. A. G.	31 Jan. 18	5 Feb. 18	Attd.
29 Mar. 18	Heselton, 2nd-Lt. R. T.	29 Mar. 18	14 July 18	Midd'x R., I. A. R. O. (*on probn.*)
10 Sep. 17	Pitt, Lt. G.	10 Sep. 18	14 Sep. 18	T. Corps, I. A. (*on probn.*)
	Pumfrey, Lt. A. I.	....	31 Oct. 18	L'Pool R., I. A. R. O. (*on probn.*)
	Cave, 2nd-Lt. H.	....	4 Oct. 18	Genl. List I. A. R. O. (*on probn.*)
28 Feb. 16	Hay, Capt. T. W. D.	28 Feb. 17	4 Oct. 18	R. Scots, I. A. (*on probn.*)
	Powell, 2nd-Lt. C. T.	....	4 Oct. 18	Genl List, I. A. R. O. (*on probn.*)
28 Aug. 16	Wace, Lt. F. B.	28 Aug 17	16 Nov. 18	I. A. R. O.
22 July 18	Pearce, 2nd-Lt. G. A.	22 July 18	...	I. A R. O.
2 July 18	Giddens, 2nd-Lt. H. L.	2 July 18	...	I. A. R. O.

Subadar-Major.

		Jemadar.	Subadar.	Subadar-Major.	
25 Jun. 83	Gyan Singh, *Bahadur* (52, 56, 66*a*)	1 Jan. 05	16 Oct. 09	7 Nov. 17	*Order of Br. Ind. 2nd Cl*

Subadars.

26 June 00	Sher Singh, I.O.M.	15 Sep. 14	15 Oct. 16
8 June 99	Bahadur Sher Khan (56)	12 Feb. 16	12 Apl. 17
24 May 93	Thaman Singh (52, 56, 66*a*)	26 Mar. 13	16 Oct. 17
8 Apl. 04	Said Ghulam	1 Aug. 16	16 Oct. 17
20 July 02	Sher Khan	8 Mar. 16	16 Oct. 17
2 Nov. 98	Taihal Singh	2 Apl. 16	16 Oct. 17
17 Aug. 02	Khushi Muhammed	2 Feb. 17	16 Oct. 17
19 June 03	Ambar Khan	1 Jan. 18	1 Oct. 18

Jemadars.

1 May 93	Daud Khan	18 May 16	
2 Oct. 00	Kala Singh	26 Mar. 17	
1 Feb. 01	Muhammad Khan	12 Apl. 17	
21 Mar. 06	Musharraf Khan	12 Apl. 17	
29 Oct. 94	Wadhawa Singh	16 Oct. 17	
13 July 03	Hussain Shah	16 Oct. 17	
17 Oct. 01	Balwant Singh	16 Oct. 17	
11 Sep. 01	Uttam Singh	16 Oct. 17	
13 Feb. 03	Buta Singh	16 Oct. 17	
13 Feb. 03	Nizam Din	16 Oct. 17	
11 Jan. 18	Muhammad Nasar Nazir	11 Jan. 18	(*On probn.*)
16 July 18	Atta Ullah Khan	16 July 18	(*On probn.*)
17 Nov. 02	Walait Shah	1 Sep. 18	
	Lal Khan	1 Sep. 18	
28 May 06	Abdul Khan	1 Sep. 18	
15 Sep. 09	Gulab Din	1 Sep. 18	

Regt. Centre—St. Thomas' Mount. *Uniform*—Scarlet.
Linked with—80th & 88th Carnatic Infy. *Facings*—White.

1st Battalion, 73rd Carnatic Infantry.

Alcock-ki-paltan.—Formed at Madras in 1776, by Capt. H. R. Alcock, of drafts from the 4th, 7th and 11th Carnatic Battalions and was originally styled the 13th Carnatic Battn. Became the 13th Madras Bn., 1784; the 2nd Bn., 3rd Regt. of Madras Native Infantry, 1796; the 13th Regt. of Madras N. I., 1824; the 13th Regt. of Madras Infy., 1885; the 13th Madras Infy., 1901; present designation, 1903.

"Sholinghur" "Carnatic" "Mysore" "Seringapatam" "Burma, 1885—87."

Composition—2 Companies of Madrasi Musalmans, 1 of Tamils, 1 of Paraiyans and Christians.

1st Commn. or date of entering service.	Names and Rank.	Army rank.	Present appointment in Regiment.	Remarks.
	Commandant.			
15 Feb. 88	Baker, Lt.-Col. H. R.	15 Feb. 14	11 Nov. 14	
	Company Commanders. (4)			
27 July 98	*Jeffery, Maj. W. H.,* C.I.E. *(temp. Lt.-Col.)*	*1 Sep. 15*	*11 June 17*	*2nd-in-Comd. G. S. O., 1st gr., A. Hdqrs.*
28 July 00	Ebden, Maj. F. T. P.	1 Sep. 15	26 Oct. 12	2nd-in-Comd. (*temp.*)
18 Jan. 02	Bell, Maj. H. C. F.	18 Jan. 17	1 July 13	*Lv. ex I., m.c.*
3 Dec. 04	*Gardner, Capt. C. P.*	*3 Dec. 13*	*11 June 17*	*Coy. Comdr., Officers Sch. of Instr. Bangalore.*
	Company Officers. (5)			
29 Jan. 10	Harward, Capt. R. B.	1 Sep. 15	10 Feb. 14	With 2-113th Infy.
20 Jan. 12	*Candy, Capt. H. R.*	*20 Jan. 16*	*25 Mar. 13*	*Burma Mily. Pol.*
22 Jan. 15	Roy, Lt. R. M.	22 Jan. 16	22 Mar. 15	With 83rd Infy.
15 Nov. 15	Rees, Lt. T. W.	15 Nov. 16	24 Nov. 15	
27 Dec. 15	Salt, Lt. G. (*Actg. Capt.*)	27 Dec. 16	. . . .	Attd.
	Attached.			
7 Nov. 06	Ford, Capt. J. M. R., M.C. (*Actg. Maj.*)	1 Sep. 15	14 Aug. 18	1-61st Prs.
9 Jan. 04	Grose, Capt. E.	9 Jan. 13	14 Sep. 17	16 Rajputs, *Lv. ex. I.m.c.*
8 Sep. 15	Grant, Lt. R.	8 Sep. 16	30 Sep. 16	I. A. R.O.
1 Oct. 15	*Breithaupt, Lt. G. W.*	*1 Oct. 16*	*27 Nov. 15*	*I. A. R. O., A.R. O., Coorg.*
17 Nov. 16	Bull, Lt. L. C., M.C.	17 Nov. 17	28 June 18	S. Staff R., I. A. (*on probn.*)
29 June 16	Bird, Lt. G. G.	14 Dec. 16	. . . .	
16 June 16	Auston, Lt. G. T.	16 June 17	19 Aug. 16	I. A. R. O.
29 June 16	Connor, Lt. T. S.	29 June 17		
5 May 17	Collett, Lt. H. S.	5 May 18	. . . .	I. A. R. O.
27 June 17	Gidden, Lt. J. A.	27 Jan. 18	. . . .	I. A. R. O., Adjt.
29 Mar. 18	Shortland, 2nd-Lt. R. B.	29 Mar. 18	20 July 18	I. A. R. O.
23 Aug. 17	Seed, Lt. R. B.	23 Aug. 18	. . . .	N. Lan. R., I.A. (*on probn.*)
19 Aug. 16	Coates, Lt. H. L.	19 Aug. 17	18 Oct. 18	Linc. Yeomanry, I. A. (*On probn.*)
	Subadar-Major.	**Jemadar.**	**Subadar.**	**Subadar-Major.**
31 May 88	Innasimuttu	13 Mar. 09	16 Dec. 11	16 May 14.
14 Sep. 89	Duraisami (56)	1 Apl. 08	12 Jan. 12	
	Abdur Razack *Bahadur*	. . . .	. . . .	*Order of Br. Ind., 2nd cl.* Re-employed.
	Subadars.			
9 Apl. 90	Abdul Ghafur (56)	1 May 06	23 Apl. 12	
5 May 96	Bhagavati (56)	12 Jan. 12	1 Apl. 13	
12 Nov. 00	Alexander (56, 58, 66*a*)	29 Apl. 11	1 Sep. 16	
8 Dec. 94	Abdul Aziz Khan (56, 66*a*)	16 Mar. 12	16 May 17	
12 Mar. 98	Veerappa	2 Dec. 13	3 Aug. 17	
7 Jan. 15	Muhi-ud-din	7 Jan. 15	1 Nov. 17	
3 Sep. 15	Jati Ram	3 Sep. 15	1 Nov. 17	
6 Feb. 16	Golakh Nath Chatterji	6 Feb. 16	9 Apl. 18	
10 Nov. 93	Muhammad Abdus-Subhan (56, 66*a*)	1 Mar. 14	14 Feb. 18	
	Venkatachallum (*Retd.*)	. . . .	. . . .	Re-employed.
	Kumarasami (*Retd.*)			Re-employed.
	Madhava Menon	. . . .	21 May 18	With 2-73rd Infy.
10 Jan. 00	Arokiasami	3 Aug. 17	21 May 18	
12 Apl. 99	Abdul Azim Khan	4 Apl. 18	21 May 18	
13 Mar. 95	Mirasadulla (56)	16 May 14	12 Sep. 18	

1st Battalion, 73rd Carnatic Infantry—*contd.*

1st Commn. or date of entering service.	Names and Rank.	Army rank.	Remarks.
	Jemadars.		
1 Nov. 93	Shaikh Inayatullah (56, 66*a*) .	23 Apl. 12	
1 Aug. 94	*Gopalsami* (66*a*)	*1 Apl. 13*	
7 Jan. 15	Elachola Muhi-ud-din . . .	7 Jan. 15	
20 Dec. 15	Manavazi, Madavhavan Menon .	20 Dec. 15	
4 Oct. 95	*Muhammad Ibrahim*	*1 July 16*	
19 Apl. 98	*Kuppusami*	*1 July 16*	
7 July 97	Azhagirisami	1 Oct. 16	
17 May 98	Abdur Rahaman	1 Oct. 16	
14 May 97	Ramasami	1 Feb. 17	
21 July 97	Murray, C.	1 June 17	
	Ersu Nukayya	14 Feb. 18	
25 Apl. 04	Venketachellam	5 Apl. 18	
5 June 01	Madurai Muthu.	5 Apl. 18	
	Samuel David	5 Apl. 18	
	Samuel Peter	21 May 18	
	Ahmad Kutti	21 May 18	
	Ananda Nambiar	21 May 18	
	Muhammad Khwaja Hussain. .	1 June 18	
	Pachiappan	1 June 18	
	Hussain Muhammad Khan .	1 June 18	
	Muzhur Husain	22 June 18	
	George G. Aseervatham . . .	1 Aug. 18	
	C. Narayanan Nambiyar . . .	1 Aug. 18	With 2-73rd Infy.
	Thamburan, ...	2 Sep. 18	

2nd Battalion, 73rd Malabar Infantry.

Formed at Cannanore, on 6th June 1918.

Composition—1 Company Mapillas, 1 Company Tiyyans & Christians, 2 Companies Nayars.

1st Commn. or date of entering service.	Names and Rank.	Army rank.	Present appointment in Regiment.	Remarks.
	Commandant.			
20 July 98	Lawrenson, Maj. W. R. (*Actg. Lt.-Col.*)	1 Sep. 15	15 July 18	86th Infy.
	Company Commanders (4).			
29 Jan. 10	Dawe, Capt. T. R. (*Actg. Maj.*)	1 Sep. 15	. . .	79th Infy. 2nd-in-Comd. (*temp.*)
27 Feb. 15	Maxwell, Lt. R. (*Actg. Capt.*)	27 Feb. 16	24 Aug. 18	I. A. R. O.
11 June 15	Butler, Lt. M.	25 Aug. 16	15 Nov. 18	Wilts R., I. A. R. O. (*on probn.*)
29 July 16	Theak, Lt. L. F. (*Actg. Capt.*)	29 July 17	28 June 18	Linc R., I. A. (*on probn.*)
	Company Officers.			
1 Aug. 16	Rowcliffe, Lt. N. H. B.	1 Aug. 17	. . . .	I. A. R. O.
29 Oct. 16	Foskett, Lt. W., M.C.	29 Oct. 17	6 July 18	Manch. R., I.A. (*on probn.*)
14 Mar. 17	Field, Lt. R. T. (*Actg. Capt.*)	14 Mar. 18	1 July 18	Lord. R., I.A. (*on probn.*) Adjt.
8 Aug. 17	Foster, 2nd-Lt. R.	8 Aug. 17	30 Aug. 17	I.A.R.O.
12 July 17	Thurgood, Lt. W. J.	12 July 18	2 July 18	Durh. L. I., I.A. (*on probn.*) Qr. Mr.
5 Oct. 17	Hitchcock, Lt. R. H.	5 Oct. 18	15 Nov. 18	I. A. R. O.
11 Sep. 17	Simpson, 2nd-Lt. E. S.	11 Sep. 17	23 Sep. 18	Attd.
15 Jan. 18	McDermott, 2nd-Lt. N. W. E.	15 Jan. 18	21 Jan. 18	I. A. R. O.
29 Mar. 18	Hewett, 2nd-Lt. H.	29 Mar. 18	24 Aug. 18	I. A. R. O.
7 June 18	Wibby, 2nd-Lt. E. J.	7 June 18	6 Oct. 18	I. A. (*T. C.*)
7 June 18	Pilbeam, 2nd-Lt. F. W.	7 June 18	6 Oct. 18	I. A. (*T. C.*)
15 June 18	Gibson, 2nd-Lt. P. R.	15 June 18	19 Sep. 18	Sco. Rif., I. A. R. O. (*on probn.*)

Subadar-Major.

	Jemadar.	Subadar.	Subadar-Major.	
P. Syed Ali	. . . . .	. . . .	. . . .	Re-employed.

Subadar.

	Jemadar.	Subadar.	
Madhava Menon	. . . . .	21 May 18	1-73rd Infy.

Jemadar.

		Jemadar.		Remarks.
14 Jun. 16	V. Sunkunni Menon	14 June 16		
3 Mar. 17	Kunhi Raman Menon	3 Mar. 17		
1 Jan. 18	Velu Kutty Nair	1 Jan. 18		
	K. Gopalan	. . . .	. . . .	(*on probn.*)
1 Aug. 18	C. Narayanan Nambier	1 Aug. 18	. . . .	
	Kunhi	13 Nov. 18	. . . .	(*on probn.*)
	Ahamed Baig	2 Sep. 18	. . . .	(*on probn.*)
	Sanjeeva Menon	11 Nov. 18	. . . .	(*on probn.*)

By Authority.

The Quarterly

INDIAN ARMY LIST

JANUARY 1919.

Volume 4

A DISTRIBUTION LIST OF OFFICERS ON THE ACTIVE LIST OF THE ARMY IN INDIA, INDIAN DEFENCE FORCE, ETC.

TO WHICH IS ADDED, HALF-YEARLY, ORDERS OF KNIGHTHOOD, ETC., INDIAN ORDERS, FOREIGN ORDERS, NON-EFFECTIVE LISTS, AND WAR SERVICES OF OFFICERS OF THE INDIAN ARMY.

ARMY HEADQUARTERS, INDIA.

CALCUTTA
SUPERINTENDENT GOVERNMENT PRINTING, INDIA
1919

NOTE.—The Indian Army List being an Official publication, and the Press having no power to make alterations in it, all communications regarding deaths, errors, or omissions should be addressed to "Military Secretary to the Commander-in-Chief" with the words "Army List" written on the upper left-hand corner of the cover.

All correspondence regarding the War Services of Officers to be addressed to the Adjutant-General in India ***through Brigade or Divisional Commanders.*** (***Vide*** India Army Order 31 of 1911.)

NOTE.

It is requested that any errors or omissions that may be observed in the War Services of Officers may be brought to the notice of the Adjutant General in India through Divisional Commanders.

WAR SERVICES OF BRITISH OFFICERS OF THE INDIAN SERVICE, THE INDIAN VOLUNTEER FORCES, THE UNATTACHED LIST, AND THE SUBORDINATE MEDICAL SERVICE.

A

Abbay, B. N.—
South African War, 1901-02.—Operations in the Transvaal and Orange River Colony, between 6th Dec. 1901 and 31st May 1902.—Queen's Medal with 4 clasps.

Abbey, W. B. T.—
The War of 1914—16, Despatches.
London Gazette, 4th July 1916.

Abbot F. B.—
The War of 1914-18.—Despatches, *Lond. Gaz.*, 25 Aug. and 5 Nov. '15.—D.S.O.

Abbott, G. M. (*Lt.-Col. retd., Ind. S.C.*)—
Abyssinia, 1868.—Action of Arogee and capture of Magdala.—Medal.

Afghanistan, 1879-80.—Action of Ahmed Khel and Patkao Shana. Despatches. G. G. Os. 326 and 493 of '80.—Medal with clasp; Brev. of Maj.

Abbott, H. A. (*Hon. Brig.-Genl. u.s.l., I. A.*)—
Afghanistan, 1878-79-80.—Battle of Ahmed Khel. Action of Urzu. March from Kabul to the relief of Kandahar and battle of Kandahar. Despatches, G.G.O. 582 of '80.—Medal with 2 clasps. Bronze star.

Soudan, 1885.—Suakin. Actions of Hasheen, Tofrek, and Tamai. Despatches. *Lond. Gaz.*, 25 Aug. 85.—Medal with 2 clasps. Bronze star. Brev. of Maj.

N.-W. Frontier of India, (2nd) Miranzai, 1891.—Despatches. G. G. O. 632 of '91.—Medal with clasp.

N.-W. Frontier of India, 1897-98.—Operations on the Samana and in the Kurram Valley during Aug. and Sep. '97. Operations of the Flying Column in the Kurram Valley under Colonel Richardson, 20 Aug. to 1 Oct. '97.—Medal with clasps.

2 *Tirah, 1897-98.*—Actions of Chagru Kotal, Dargai and of the Sampagha and Arhanga Passes. Reconnaissance of the Saran Sar and action of 9 Nov. '97. Operations in the Waran Valley and action of 16 Nov. '97. (Twice severely wounded.) Despatches. G. G. Os. 58, 244 and 304 of '98.—Clasp. C.B.

Abbott, J. F.—
Operations in the Abor country, 1911-12.—Medal with clasp. *The War of 1914-18.*—Despaches, *Lond. Gaz.*, 15 Aug. '17. M.C.

Abbot, L. H.—
The War of 1914-18.—Despatches, *Lond. Gaz.*, 21 June '16, 13 July '16, 4 Jan. '17 and 15 May '17, Brev. of Lt.-Col. C.M.G.

Abbot, R. S.—
The War of 1914-18.—Despatches, *Lond. Gaz.*, 15 June '16 and 15 May '17. Brev. of Maj. M.C.

Abbott, S. H. L.—
Tibet, 1903-04.—Medal.

Abergh-Mackay, J. L. (*Lt.-Col. u.s.l., I.S.C.*)—
Afghanistan, 1880.—Medal.

Acton, H. L. B.—(*Bt. Col. retd., I. A.*)
Soudan, 1885.—Suakin as Transport Officer.—Medal with clasp; bronze star.

Acworth, D. H.—
N.-W. Frontier of India, 1908.—Operations in the Mohmand Country. Engagement of Kargha, 24th May 1908. Medal with clasp. *The War of 1914-18.*—M.C.

Adam, F. J. S. (*Hon. Brig.-Gen. u.s.l., Ind. S.C.*)—
Crimea, 1855-56.
India, 1858-59.
Afghanistan, 1879-80.—Defence of Kandahar. Despatches, G. G. O. 566 of '80.—Medal.

Adams, C. (*Lt.-Col. retd., I.M.S.*)—
Burma, 1885—87.—In Medical charge, 2nd M. L.—Medal with clasp.

Adams, Sir R. B. (Maj.-Genl., retd., I. A.)—
Afghanistan, 1879-80.—Operations at and around Kabul. Action of Chihildukhteran.—Medal with clasp.

Chitral, 1895.—Relief of Chitral.—Action of the Malakand Pass and at Khar. Despatches. G. G. O. 998 of '95.—Medal with clasp. Brev. of Lt.-Col.—Medal with clasp.

N.-W. Frontier of India, 1897-98.—Defence of Malakand. Relief of Chakdara, Malakand Action at Landakai. Operations in the Mohmand country, Buner. Despatches. G. G. Os. 1089 and 1317 of '97, and 178 and 217 of '98.—2 clasps. VC., C.B.

Adamson, C. H. E. (*Lt.-Col., retd., Ind. S.C.*)—
Burma, 1885-89.—Employed on Civil Duties Mentioned by Govt. of India. Despatches. *Lond. Gaz.*, 2 Sep. 87.—Medal with 2 clasps.

Adamson, D. (*Hon. Col., retd., Ben. S.C.*)—
Afghanistan, 1879-80.—Served on the Khyber Line—Medal.

Adey, A. W. G. (*Surg.-Maj., retd., Ind. Med. Serv.*)—
China, 1856.—As a Volunteer with expedition in pursuit of Chinese pirates.

Adye, A. (*Lieut.-Col., retd., I. A.*)—
Afghanistan, 1878-79-80.—With Kabul and Kurram Field Forces.—Medal.

Soudan, 1885.—Suakin. Medal. Bronze star.

Adye, D. R.—
China, 1900.—Relief of Pekin. Action of Peitsang. Action at Yangtsun—Special Service Officer. Despatches. *Lond. Gaz.*, 14 May '01.—Medal with clasp. Brev. of Maj. *The War of 1914-18.* Brev. of Col.

Ahmed, Z. A. (*Lt.-Col., retd., I.M.S.*)—
N.-W. Frontier of India, Hazara, 1891.—Medal with clasp.
N.-W. Frontier of India, 1897-98. Malakand Operations in Bajaur and Mohmand country. Despatches. G. G. O. 178 of '98.—Medal with clasp. *Tirah, 1897-98.*—Operations in the Bara Valley, 7 to 14 Dec. '97.—Clasp.
N.-W. Frontier of India, Waziristan, 1901-2.

Ainslie, H. P.—
China, 1900.—Medal.

Ainsworth, H.—
Tirah, 1897-98.—Actions of Chagra Kotal and Dargai.—Actions of Sampagha and Arhanga Passes.—Operations at and around Dwatoi and action of 24 Nov. '97.—Operations in the Bara Valley, 7 to 14 Dec. '97.—Medal with 2 clasps.

Airy, G.—
N.-W. Frontier of India, 1902.—Operations against the Darwesh Khel Waziris. (Slightly wounded.)

Aitchison, C. T. (*Gen. u.s.l. Ind., S.C.*)—
South Mahratta Campaign, 1844-45.—As Field Engineer. Storm and capture of Sumunghur, Boodurghur, Punella (wounded), Pounghur, Munohur, Munsuntosh. Despatches. G. G. O. 29 Dec. '44.
Persia, 1856-57.—As Brig.-Maj. Assault and capture of the Fort of Reshire, surrender of Bushire, expedition to Borazjoon, and action at Khooshab. Despatches. G. G. O., 6 Feb. '59. Thanked by the Government of India.—Medal with clasp. Brev. of Maj.
India, 1857-58.—March on, surrender, and occupation of Shorapore.—Medal.

Aitken, A. E. (*Colonel, I. A., retired*)—
Soudan, 1885.—Suakin. Actions of Hasheen and Tofrek. Various attacks on convoys. Advance on Tamai. Affair of Takul.—Medal with 2 clasps; bronze star.

Alban, C. F. S. (*Maj., retd., Ind. S.C.*)—
Afghanistan, 1879-80.—Defence of Kandahar and battle of 1 Sep.—Medal with clasp.
Burma, 1885-89.—Despatches. G. G. O. 864 of '87.—Medal with 2 clasps.

Alexander, A. de V. (*Col., retd., I. A.*)—
Afghanistan, 1879-80.—With Khyber Field Force.—Medal.
China, 1900.—Despatches. *Lond. Gaz.*, 14 May 01—D.S.O.

Alexander, C.—
The War of 1914-18.—M.C.

Alexander, C. H.—
China, 1900.—Relief of Pekin.—Medal with clasp.
East Africa, 1902-04.—Operations in Somaliland.—Medal with clasp.

Alexander, E. C.—
Tirah, 1897-98.—Actions of Chagru Kotal, Dargai, and of the Sampagha and Arhanga Passes.—Reconnaissance of the Saran Sar and action of 9 Nov. '97.—Operations in the Waran Valley and action of 16 Nov. '97.—Operations in the Bara Valley, 7 to 14 Dec. '97.—Medal with 2 clasps.
N.-W. Frontier of India, Waziristan, 1901-02.—Despatches. G. G. O. 611 of '02.—Clasp. D.S.O.
N.-W. Frontier of India, 1908.—Operations in the Mohmand country. Engagement of Kargha, 24th May 1908.—Medal with Clasp. *The War of 1914-18.*—Despatches, *Lond. Gaz.*, 31 Oct. '17. Brev. of Lt.-Col.

Alexander, F. G. (*Maj., I. A., retd.*)—
Afghanistan, 1878-79.—Medal.

Alexander, H. M.—
The War of 1914-18.—Despatches, *Lond. Gaz.*, 5 Nov. '15.—D.S.O.

Alexander, H. S.—
The War of 1914-18.—Despatches, *Lond. Gaz.*, 15 Aug. '17.

Alexander, R. (*Maj.-Gen., retd.*)—
South Africa, 1879.—Zulu Campaign.—Battle of Ulundi. Despatches. *Lon. Gaz.*, 21 Aug. '79.—Medal with clasp.
Afghanistan, 1880.—Medal.

Alexander, R. D.—
N.-W. Frontier of India, Waziristan, 1901-02.—Medal with clasp. *The War of 1914-18.*—Despatches, *Lond. Gaz.*, 17 Feb. '15.

Alexander, R. G.—
South African War, 1899-1900.—Relief of Ladysmith, including action at Colenso; operations (17th to 24th Jan. 1900) and action at Spion Kop.—Queen's medal with clasp. *The War of 1914-18.*—Despatches, *Lond. Gaz.*, 22 June '15, 1 Jan. and 19 Oct. '16.—M.C.

Allanson, C. J. L.—
The War of 1914-18.—Despatches, *Lond. Gaz.*, 4 Nov. '15 and 28 Jan. '16.—D.S.O.

Allen, B. M. (*Lt.-Col., retd., Ind. S.C.*)—
Afghanistan, 1878-79.—Medal.
N.-W. Frontier of India, Mahsud-Waziri, 1881.—Attached to the Signalling Department. Mentioned in Despatches.
N.-W. Frontier of India, Zhob Valley, 1884.

Allen, G. F. S.—
South African War, 1900-02.—Operations in the Orange River Colony, May to 29th November 1900, including actions at Biddulphsberg (29th May), Lindley (1st June), Bethlehem (6th and 7th July), and Wittebergen (1st to 29th July). Operations in the Transvaal, East of Pretoria, July to 29th November 1900, including action at Rhenoster Kop (29th November).—Queen's medal with three clasps, King's medal with 2 clasps.

Allen, H. M. (*Maj., retd., I. A.*)—
N.-W. Frontier of India, 1897-98.—*Malakand* Mohmand.—Medal with clasp.
N.-W. Frontier of India, Waziristan, 1901-02.—Clasp.
N.-W. Frontier of India, 1908.—Operations in the Zakka Khel country. D. A. A. G. 2nd Bde. Operaions in the Mohmand country. D. A. A. G. 2nd Bde. Despatches, G. G. O. No. 521 A of 1908.—Medal with clasp. D.S.O.
The War of 1914-18.—Despatches, *Lond. Gaz.*, 1 Jan. and 15 June '16, C.M.G.

Alphonso, A. G.—
N.-W. Frontier of India, 1897-98.—Operations on the Samana.—Medal with 2 clasps.
Tirah, 1897-98.—Actions of the Sampagha and Arhanga Passes. Reconnaissance of the Saran Sar and action of 9th November, 1897.
Operations at and around Dwatoi and action of 24th November 97. Operations in the Bara Valley, 7th to 14th December 97.—Clasp.

Alpin, W. G. P. (*Lt.-Col., retd., I.M.S.*)—
Soudan, 1885.—Suakin. Action of Tofrek.—Medal with 2 clasps. Bronze star.
Chitral, 1895.—Relief of Chitral.—Medal with clasp.

Alves, W. (*Hony. Capt., Dy. Comy. retd., S. and T. Corps*)—
N.-W. Frontier of India, Hazara, 1888.—Despatches. G. G. O. 978 of '88.—Medal with clasp. Promoted Sub-Conductor.

Ames, C. G.—
East Africa, 1903-04.—Operations in Somaliland.—Medal with clasp.
N.-W. Frontier of India, 1908.—Operations in the Mohmand country. Engagement of Kargha, 24th May 1908. Despatches. G. G. O. No. 521 A of 1908.—Medal with clasp.

Amesbury, W. L. R.—
N.-W. Frontier of India, 1908.—Operations in the Mohmand country. Engagement of Matta, 24th April 1908.—Medal with clasp.

Anderson, A. M.—
Burma, 1890.—Tonhon Expedition.
Burma, 1891-92.—North-Eastern Column.—Medal and clasp.

Burma, 1893-94.

N.-W. Frontier of India, 1897-98.—Operations on the Samana and in the Kurram Valley during Aug. and Sep. '97.—Medal with 2 clasps.

Tirah, 1897-98.—Actions of Chagru Kotal and Dargai. Operations in the Bazar Valley, 25 to 30 Dec. '97.—Clasp.

Anderson, A. V. (*Lt.-Col., retd., I.M.S.*)—
N.-W. Frontier of India, (2nd) *Miranzai, 1891.*—Medal with clasp.

Anderson, A. W. L. (*Maj.-Gen., retd., Ind. Army*)—
Burma, 1885—87.—(Severely wounded.)—Medal with clasp.

Anderson, B. E.—
N.-W. Frontier of India, 1908.—Operations in the Zakka Khel country. Operations in the Mohmand country. Engagement of Matta, 24th April 1908. Engagement of Kargha, 24th May 1908. Despatches. G. G. O. No. 521 A of 1908.—Medal with clasp.
The War of 1914-18.—Despatches, *Lond. Gaz.*, 17 Feb. '15 and 19 Oct. '16. Brev. of Maj. D.S.O.

Anderson, C. C. (*Lt.-Col., retd., I. A.*)—
Burma, 1886 and 1888—90.—Operations of the 2nd Brigade.—Medal with 2 clasps.
Soudan, 1896.—Dongola expedition.—2 Medals.
N.-W. Frontier of India, 1908.—Operations in the Zakka Khel country.—Medal with clasp.

Anderson, C. J. (*Col. u.s.l. Ind. S.C.*)—
India, 1857-58.—As A.-D.-C. to G. O. C., Gwalior Divn. Mutiny at Nusseerabad, first attack on Awah, siege of Kotah and subsequent pursuit, action at Kotah-ke-serai, capture of Gwalior and action of Ranode. Despatches. *Lond. Gaz.*, 18 April '59.—Medal with clasp.

Anderson, D. N. (*Maj. retd., I.M.S.*)—
China, 1900.—Medal.

Anderson, E. B. (*Col., retd., I. A.*)—
Burma, 1885-86.—Brigade Transport Officer. Present at the occupation of Mandalay and subsequent operations.—Medal with clasp.

Anderson, E. S. J.—
N.-W. Frontier of India, Waziristan, 1901-02.—Medal with clasp.
The War of 1914-18.—Despatches, *Lond. Gaz.*, 15 Aug. '17.

Anderson, F. J.—
The War of 1914-18.—Despatches, *Lond. Gaz.*, 22 Jun. '15. M.C.

Anderson, H. C. (*Gen. u.s.l. Ind. S.C.*)—
Sutlej Campaign, 1846.—Battle of Ferozeshah.—Medal with clasp.
India, 1857-58.—Brig. Maj. Rohilcund Field Force. Actions of Narnoul and Nagal, occupation of Nujeebabad, action of Nugeenah, occupation of Moradabad, actions of Bareilly and at Ford of Burnai, capture of Fort Mahomdee, and affairs with Jhansi mutineers. Despatches. *Lond. Gaz.*, 28 July '58.—Medal; Brev. of Maj.

Anderson, J. (*Lt.-Col., retd., I.M.S.*)—
Afghanistan, 1879-80.—Actions with the Ghilzais and at Jagdallak.—Medal.

Anderson, J. F. H.—
The War of 1914-18.—Despatches, *Lond. Gaz.*, 19 Oct. '16, 14 Nov. '16 and 15 Aug. '17. Brev. of Maj. D.S.O.

Anderson, J. McK.—
The War of 1914-18.—Despatches, *Lond. Gaz.*, 31 Oct. '17. M.C.

Anderson, J. W. M. (*Col., retd., Bo. S.C.*)—
Persia, 1856-57.—March to Borazjoon, action of Kooshab, and capture of Mohumra.—Medal with clasp.
India, 1858-59.—Served in the Deccan, Khandeish, and Central India, and at action at Shamshapoor.—Medal.

Anderson, J. W. T. (*Lt.-Col., retd., I.M.S.*)—
Burma, 1886-87

Anderson, L. A. P.—
The War of 1914-18.—Despatches, *Lond. Gaz.*, 5 Apl. '16.

Anderson, R. H.—
Tirah, 1897-98.—Medal with 2 clasps.
N.-W. Frontier of India, Waziristan, 1901-02.—Clasp.

Anderson, R. H.—
N.-W. Frontier of India, 1908.—Operations in the Zakka Khel country.—Medal with clasp.
The War of 1914-18.—Despatches, *Lond. Gaz.*, 15 Aug. '17.

Anderson, R. H. B.—
East Africa, 1903.—Operations in Somaliland.—Medal with clasp.

Anderson, S.—
China, 1900.—Medal.
N.-W. Frontier, Waziristan, 1901-02.—Medal with clasp.
The War of 1914-18.—Despatches, *Lond. Gaz.*, 5 Apl. '16, 19 Oct. '16, and 15 Aug. '17, Brev. of Lt.-Col.

Anderson, W. C.—
N.-W. Frontier of India, 1897-98.—Mohmand.—Medal with clasp.
Tirah, 1897-98.—Actions of the Sampagha and Arhanga Passes. Operations in the Bazar Valley, 25 to 30 Dec. '97.—Clasp.
South African War, 1901-02.—Operations in the Transvaal between 31st October 1901 and 31st May 1902. Operations in Cape Colony between 27th and 30th September 1901. Operations in the Orange River Colony between 1st and 2nd October 1901.—Queen's Medal with 5 clasps.
East Africa, 1902—04.—Operations in Somaliland.—Medal with clasp.
The War of 1914-18.—Despatches, *Lond. Gaz.*, 1 Jan. and 19 Oct. Dangerously wounded. C.M.G. Order of St. Anne's (3rd class) with Swords. Brev. of Col.

Anderson, W. P. (*Maj., retd., I.A.*)—
N.-E. Frontier of India, Chin-Lushai, 1889-90.—Medal with clasp.

Andrew, A. W.—
South African War, 1901—02.—Operations in the Transvaal, March to July 1901, and October to May 1902. Operations in the Orange River Colony, July 1901, and August to October 1901. Operations in Cape Colony, July and August 1901. Commanded 6th New Zealand Mounted Rifles from 6th July 1901 to 9th April 1902. Commanded 10th New Zealand Contingent (two regiments) from 17th May 1902 to 18th August 1902. Despatches, *Lond. Gaz.*, 3rd December 1901.—Queen's medal with 5 clasps. Brev. of Major.
The War of 1914-18.—Despatches, *Lond. Gaz.*, 15 Aug. '17.

Andrew, D. C. A.—
Burma, 1891-92.—Momeik Column. Various actions against Kachins.—Medal with clasp.
N.-E. Frontier of India, Chin Hills, 1892-93.—Clasp.
N.-W. Frontier of India, 1897-98.—Defence of Malakand. Action at Landakai, Utman Khel, Buner. Action of the Tanga Pass.—Medal with 2 clasps.
N.-W. Frontier of India, 1908.—Operations in the Mohmand country.—Medal with clasp.

Andrew, F. A.—
N.-W. Frontier of India, Waziristan, 1901-02.—Medal with clasp.
The War of 1914-18.—Despatches, *Lond. Gaz.*, 15 Aug. 17. D.S.O.

Andrews, L. J.—
Burma, 1887—80.—Post Transpt. Offr. Zagabiu post.—Medal with 2 clasps.

Andrews, H. A.—
The War of 1914-18.—Despatches, *Lond. Gaz.*, 19 Oct. '16. M.C.

Andrewes, H. W.—
Operations in the Abor country, 1911-12.—Medal with clasp.

Angelo, F. W. P.—(*Maj.-Genl., retd., I. A.*)—
Soudan, 1885.—Suakin. Actions of Hasheen and Tamai.—Medal with clasp. Bronze star.
Chitral, 1895.—Relief of Chitral.—Medal with clasp.
China, 1900.—Deputy Assistant Adjutant General, Cavalry Brigade. Despatches, 14 May '01.—Medal; Brev. of Lt.-Col.

Angelo, J. C. C. (*Lt.-Col., retd., I. A.*)—
N.-W. Frontier of India, Waziristan, 1894-95.—Action at Wana.—Medal with clasp.

Angelo, R. F. (*Hon. Col., retd., Ben. S.C.*)—
India, 1858.—Actions at Terai Forest, Nujeebabad, Nugeenah, Moradabad, Dojora Nudee, Bareilly, Fort Bunuai and Mahomdee. Despatches. *Lond. Gaz.*, 28 July '58.—Medal.

Annesley, A. S. R.—
N.-E. Frontier of India, Chin-Lushai. 1889-90.—Burma Column.—Medal with clasp.
N.-W. Frontier of India, 1897-98.—Tochi.—Medal with clasp.

The War of 1914-18.—Despatches, *Lond. Gaz.*, 5 Apl. and 19 Oct. '16. Brev. of Col.

Anson, G. F. W.—
N.-W. Frontier of India, Waziristan, 1901-02.—Medal with clasp.

Anthony, R. W.—
British East Africa, 1901.—Operations against the Ogaden Somalis in Jubaland.—Medal with clasp.

Anwyl-Passingham, R. T. (*Maj., retd., I. A.*)—
Burma, 1886—89.—Operations of the 6th and 2nd Brigades.—Medal with 2 clasps.
Burma, 1889—92.—Clasp.

Aplin, P. J. H.—
Afghanistan, 1879-80.—Medal.
Burma, 1885—89.—Medal with 2 clasps.
East Africa, 1902—04.—Operations in Somaliland. Despatches. *Lond. Gaz.*, 2 Sep. '04.—Medal with clasp. D.S.O.

Aplin, S. L.—
Burma, 1887-88.—Medal with clasp.
N.-E. Frontier of India, Chin-Lushai, 1889-90.—Transport Officer, Northern Chin Field Force.—Clasp.

Apthorp, S. E.—
South African War, 1900-02.—Operations in the Transvaal and Orange River Colony from April 1901 to 31st May 1902.—Queen's medal with 5 clasps.

The War of 1914-18.—Despatches, *Lond. Gaz.*, 15 Mar. and 27 July '16. D.S.O.

Apthorpe, E. H. F.—
The War of 1914-18.—Despatches, *Lond. Gaz.*, 19 Oct. '16.

Arbery, J.—
N.-E. Frontier of India, Manipur, 1891.—Medal with clasp.

Arbuthnot, G. H. (*Lt.-Col., retd., I. A.*)—
Burma, 1886-87.—As orderly officer to Brig.-Genl. R. C. Stewart during the operations of the 5th Brigade, Ruby Mines Expedition, advance from Khannyath; served subsequently as Addl. A.-D.-C. to Commander-in-Chief, Madras Army.—Medal with clasp.

Archer, A. J.—
Chitral, 1895.—Relief of Chitral.—Medal with clasp.
China, 1900.—Medal.

Archer, C.—(*Lt.-Col., retd., I. A.*)
N.-W. Frontier of India, Zhob Valley, 1890.

Archer, E. J.—
The War of 1914-18.—Despatches, *Lond. Gaz.*, 15 Aug. '17.

Argles, O. C. (*Major, retd., I. A.*)—
West Africa, 1887-88, Youneis Expedition.—Medal with clasp.

N.-W. Frontier of India, Waziristan, 1901-02.—Medal with clasp.

The War of 1914-18.—Despatches, *Lond. Gaz.*, 11 Dec. 17.

Armstrong, H. (*Lt.-Col., retd., I. M. S.*)—
Rumpa Rebellion, 1879.
Afghanistan, 1879-80.—Medal.
China, 1900.—Medal.

Armstrong, O. C. (*Col., retd., I. A.*)—
Burma, 1888.—Medal with clasp.
South African War, 1900—02.—Organising Indian details, South Africa. Financial Secretary to Military Governor of Pretoria. Financial Adviser to General Officer Commanding-in-Chief. Specially thanked by S. of S. for War for services during and subsequent to the war.—Queen's Medal with 3 clasps. King's Medal with 2 clasps. D.S.O.

Arnold, A. S.—
South African War, 1899-00.—Operations in the Orange Free State, Feb. '00, including operations at Paardberg (17 and 18 Feb.). Operations in the Transvaal, East of Pretoria, July to 9 Nov. '00, including action at Belfast (26 and 27 Aug. '00).—Queen's medal with 3 clasps.

The War of 1914-18.—Despatches, *Lond. Gaz.*, 31 Oct. '17.

Arnott, J. (*Brig.-Surg.-Lt.-Col., retd., I. M. S.*)—
Abyssinia, 1867-68.—Served with the Field Hospital.—Medal.

Afghanistan, 1879-80.—As Principal Medical Officer of Reserve Brigade in Bolan Pass, and in Medical charge of Hd.-Qrs. Staff, and Sanitary Officer at defence of Kandahar. Present at sortie of Deh Khoja and battle of 1 Sep. Despatches. *Lond. Gaz.*, 3 Dec. '80.—Medal with clasp.

Arthur, J. R.—(*Commsy. and Hon. Capt., retd., S. & T.*)

N.-W. Frontier of India, 1891.—2nd Miranzai Expedition.—Medal with clasp.

N.-W. Frontier of India, Waziristan, 1894-95.—Despatches. G. G. O. No. 473 of 1895. Promoted Sub-Conductor.—Medal with clasp.

Tirah, 1897-98.—Medal with 2 clasps. Promoted Conductor.

Aden, 1903-04.—Operations in the interior.

Arthur, L. F.—
The War of 1914-18.—Despatches, *Lond. Gaz.*, 21 Jun. '16 and 4 Jan. '17. D.S.O.

Ashburner, H. W.—

China, 1900.—Relief of Pekin.—Medal with clasp.
East Africa, 1903-04.—Operations in Somaliland.—Medal with clasp.

The War of 1914-18.—Despatches, *Lond. Gaz.*, 19 Oct. 16. D.S.O.

Ashby, J. S. (*Lt.-Col., retd., I. A.*)—

Afghanistan, 1880.—Medal.

Aslett, W. C.—

Afghanistan, 1879-80.—Affair at Girishk. Action of Maiwand. Defence of Kandahar, and battle of 1 Sep.—Medal with clasp.

Burma, 1885—87.—Medal with clasp.

Aspinall, H. H. H. (*Lt.-Col., u. s. l., I.A.*)—

Burma, 1886—89.—Present at the operations round Pyinmana, Yamethin, and in the operations of the 3rd Brigade.—Medal with 2 clasps.

Atkins, B. S.—

The War of 1914-18.—Despatches, *Lond. Gaz.*, 5 Apl. '16.

Atkins, G. W.—

South African War, 1902.—Operations in Orange River Colony and Cape Colony in May 1902.—Queen's medal with 4 clasps.

N.-W. Frontier of India, 1908.—Operations in the Zakka Khel country.—Medal with clasp.

The War of 1914-18.—Despatches, *Lond. Gaz.*, 28 Jan. '16. M.C.

Atkins, R. F.—

The War of 1914-18.—Despatches, *Lond. Gaz.*, 5 Apl. '16.

Atkinson, E. G.—

The War of 1914-18.—Despatckes, *Lond. Gaz.*, 28 Dec. 17.

Atkinson, F. G.—

Egypt, 1882.—Actions of Kassassin and Tel-el, Kebir.—Medal with clasp. Bronze star.
N.-W. Frontier of India, 1897-98.—Engagement near Shabkadr, 9 Aug. '97. Despatches. G. G. O. 1188 of '97.—Mohmand.— Medal with clasp. Brev. of Lt.-Col.

Atkinson, J. R. B. (*Maj.-Gen., retd., Ben. Inf.*)—

Afghanistan, 1878-80.—Actions of Saif-u-deen Takht-i-pul, Ahmed Khel, Urzoo and Patka Shana. Despatches. G. G. Os. 610 of '79 and 326 and 493 of '80.—Medal with clasp; Brev. Maj. and Lt.-Col.

Auchinlek, C. J. E.—

The War of 1914-18.—Despatches, *Lond. Gaz.*, 15 Aug. '17. D.S.O.

Audain, G. M. (*Maj., retd., I. A.*)—

China, 1900.—Medal.

Avetoom, S. T. (*Lt.-Col., retd., I. M. S.*)—

Soudan, 1885.—Suakin.—Medal with clasp.
Burma, 1888-89.—Medal with clasp.
Uganda, 1897-8.—Medal with clasp.

Ayerst, W. (*Lt.-Col., retd., I.A.*)—

Burma, 1885—89.—Affair near Mingoon, 29 June '86. (Slightly wounded. Despatches, 18 Aug. '86, and G. G. O. 339 of '87).—Medal with 2 clasps.
Chitral, 1895.—Relief of Chitral.—Medal with clasp.

Aylmer, G.—

South African War 1900-01.—Served in Impl. Yeo. Operations in Orange River Colony (May to Nov. 00) including actions at Rhenoster River, Bethlehem, and Wittenberg. Operations in Cape Colony, 00. Operations in the Transvaal 30 Nov. 00 to Feb. 01. Operations in Orange River Colony, Feb. to May 01. Operations in Cape Colony, May to June 01. Queen's Medal with 4 clasps.

The War of 1914—17.—Despatches, *Lond. Gaz.* 13th July '16. M. C., Croix de Guerre.

Ayscough, G. M.—

South African War, 1902.—Operations in Cape Colony, (Apr. and May 02). Queen's Medal with 2 clasps.

B

Babington, A. H.—

N.-W. Frontier of India, Waziristan, 1901-02.—Medal with clasp.
N.-W. Frontier of India, 1908.—Operations in the Mohmand country. Supply Officer, 2nd Bde., from 3rd May 1908 to 2nd June 1908.—Medal with clasp.

Babington, C. W. (*Col. retd., Ben. Inf.*)—

Afghanistan, 1878—80.—Medal.

Babonau, R. G.—

Tirah, 1897-98.—Medal with 2 clasps.

Bacon, R. G.—

N.-W. Frontier of India, 1908.—Operations in the Mohmand Country.—Medal with clasp.

Badcock, F. F.—

N.-W. Frontier of India, Hazara, 1888.—Medal with clasp.
N.-W. Frontier of India, Hazara, 1891.—Despatches. G.G. O. 816 of 1891.—Clasp.
N.-W. Frontier of India, (2nd) Miranzai, 1891.—Clasps.
N.-W. Frontier of India, Hunza-Nagar, 1891-92.—Capture of Nilt Fort. (Wounded.) Despatches G. G. O. 397 of 1892.—Clasp, D.S.O.
N.-W. Frontier of India, Waziristan, 1894-95.—Clasp.
Tirah, 1897-98.—Field Intelligence Officer, Army Staff. (Severely wounded).—Medal with 2 clasps.
N.-W. Frontier of India, 1908.—Operations in the Zakka Khel country.—Medal with clasp.
The War of 1914—18.—Despatches, *Lond. Gaz.*, 15 August 1917.

Badcock, G. H. (*Maj., retd., I. A.*)—

N.-W. Frontier of India, 1897-98.—Operations or the Flying Column in the Kurram Valley undef Colonel Richardson during Sep. '97.—Medal with clasp.

Tirah, 1897-98.—Clasp.

Baddeley, R. J. H.—

The War of 1914—18.—Despatches, *Lond. Gaz.*, 13 July '16 and 6th December '16. M.C.

Badgley, W. F. (*Col. retd., Ben. S.C.*)—

India, 1857-58-59.—Siege of Lucknow. Minor affairs in Central India, in Oude, and on the Nepal Frontier.—Medal.
N.-E. Frontier of India, Lushai, 1871-72.—Medal with clasp.
Survey Expeditions to Lushai and Naga Hills, 1872—78.—(Four times wounded.)

Bagot-Chester, W. G.—

The War of 1914-18.—Despatches, *Lond. Gaz.*, 15 Jun. 16.

Bagshawe, F. W.—

South African War, 1899—1901.—Relief of Ladysmith, including action at Colenso (15th December 1899). Operations on Tugela Heights (14th to 27th February 1900). Operations in Cape Colony, South of Orange River, 1899-1900.—Queen's Medal with 3 clasps.

Bailey, C.—

Burma, 1885—87.—Operations of the 2nd Brigade. Operations of the Lower Burma Field Force in the Shwegyn District.—Medal with clasp.

South African War, 1900.—Operations in the Orange Free State, February to May '00, including operations at Paardeberg, 17 to 26 Feb.—Action at Poplar Grove (9 Mar.), Driefontein (10 Mar.) and Karee Siding (29 Mar.) Queen's Medal with 3 clasps.

The War of 1914-18.—Despatches, *Lond. Gaz.*, 30 Jun. '16.

Bailey, F. M.—

Tibet, 1903-04.—Action at Niani, 26th June 1904. Operations at and around Gyantse, between 5th May and 6th July 1904. March to Lhassa, 14th July to 3rd August 1904.—Medal with clasp.

Bailey, P. H. H.—

N.-W. Frontier of India, 1908.—Operations in the Mohmand country.—Medal with clasp.

The War of 1914-18.—Despatches, *Lond. Gaz.*, 13 Jun. '16 and 19 Oct. '16. M.C.

Bailey, T. J. (*Lt.-Col. retd., Ind. S.C.*)—

Afghanistan, 1878—80.—Medal.

N.-W. Frontier of India, 1884, Zhob Valley Expedition.

Burma, 1886-87.—Medal with clasp.

Bailey, W. B.—

N.-W. Frontier of India, Waziristan, 1901-02.—Medal with clasp.

Operations in the Abor country, 1911-12.—Medal with clasp.

Baille, S. M. H.—

South African War, 1899—01.—Relief of Ladysmith, including operations on Tugela Heights (4th to 27th February 1900). Operations in the Transvaal from May 1900 to 4th July 1901.—Queen's Medal with 4 clasps.

Baillie, J. C. P. (*Gen., u.s.l., Ben. Inf.*)—

India, 1857—59.—Affairs of Trimmo Ghat, actions of Gungaree, Puttialee, Mynpoorie and Baugaon. Oude Campaign, 1858-59. Capture of Forts Dehagu and Turowl, operations in the Terai at Ruttonpore and in the Nepaul hills. (Twice wounded, once severely).—Medal.

Baillie, R.—

China, 1900.—Medal.

Bainbridge, D.—

The War of 1914-18.—Despatches, *Lond. Gaz.*, 1 Jan. '16. M.C.

Bainbridge, W. F.—

N.-W. Frontier of India, 1897-98.—Tochi.—Medal with clasp.

China, 1900.—Relief of Pekin.—Action of Peitsang. Action of Yangtsun. Despatches, *Lond. Gaz.* of 6 Nov. '00 and 14 May '01.—Medal with clasp. D.S.O.

The War of 1914-18.—Despatches, *Lond. Gaz.*, 21 Jun. '16. C.M.G.

Baines, E. F. E.—

East Africa, 1902-04.—Operations in Somaliland.—Medal with clasp.

The War of 1914-18.—Despatches, *Lond. Gaz.*, 13 July and 19 Oct. '16.

Baines, R. H.—

The War of 1914-18.—Despatches, *Lond. Gaz.*, 15 Aug. '17. M.C.

Baird, H. B. D.—

Tirah, 1897-98.—Operations in the Bara Valley 7 to Dec. 14 '97.—Medal with 2 clasps. D.S.O.

The War of 1914-18.—Despatches, *Lond. Gaz.*, 19 Oct. '14, 22 Jun. '15, 4 Jan. and 11 Dec. '17, Brev. of Lt.-Col. C.M.G., D.S.O.

Baird, R. F.—

China, 1900—Medal.

Tibet, 1903-04.—Action at Niani, 26th June 1904. Operations at and around Gyantse, between 5th May and 6th July 1904.—Medal with clasp.

Bairnsfather, P. R. (*Lt.-Col. retd., I. A.*)—

N.-W. Frontier of India, Jowaki, 1877-78.—Medal with clasp.

Afghanistan, 1878-79-80.—Action of Charasiah. Operations in Kohistan and at and around Kabul.—Medal with 2 clasps.

Baker, C. H. N.—

The War of 1914-18.—Despatches, *Lond. Gaz.*, 15 Aug. '17. M.C.

Baker, C. N.—

China, 1900.—Field Paymaster. Relief of Pekin.—Medal with clasp.

Baker, G. H. (*Lt.-Col., retd., I. M. S.*)—

Burma, 1886-87.—Operations of the 6th Brigade.—Medal with clasp.

Tirah, 1897-98.—Medal with 2 clasps.

Baker, L. S. (*Lt.-Col., retd., I. A.*)—

Afghanistan, 1879-80.—Action of Charasiah and battle of Khandahar (slightly wounded) and Mari Expedition. Despatches, Lond. Gaz., 30 July 80. Medal with clasp; bronze star.

N.-W. Frontier of India, Zhob Valley, 1890.

N.-W. Frontier of India, Waziristan, 1894-95.—Medal with clasp.

N.-W. Frontier of India, 1897-98.—Tochi.—Medal with clasp.

N.-W. Frontier of India, 1902.—Operations against the Darwesh Khel Waziris.

Baker, R. J. (*Lieut.-Col., retd., I.M.S.*)—

N.-W. Frontier of India, Zhob Valley, 1884.

Baker, W. B.—

China, 1900.—Medal.

Baker, W. H. G.—

The War of 1914-18.—Despatches, *Lond. Gaz.*, 4 Jan. 22 May, 17 and 24 Dec. '17. Brev. of Maj. D.S.O.

Bakhle, C. R.—

China, 1900.—Medal.

Bal, N. K.—

The War of 1914-18.—Despatches, *Lond. Gaz.*, 5 Apl. '16. M.C.

Bald, J. A.—

China, 1900.—Relief of Pekin.—Medal with clasp.

The War of 1914-18.—Despatches, *Lond. Gaz.* 28 Jan. '16. C.M.G.

Baldry, T.—

Burma, 1889-90.—Medal with clasp.

N.-W. Frontier of India, 1897-98.—Malakand. Operations in the Mamund country. Utman Khel, Buner. Action of the Tanga Pass.—Medal with clasp.

Baldwin, F. E. W.—
South African War, 1902.—Queen's Medal with clasps.
Aden, 1903-04.—Operations in the Interior.

Baldwin, G. M.—
N.-W. Frontier of India, Hazara, 1888.—Action of Kotkai.—Medal with clasp.
Chitral, 1895.—Relief of Chitral.—Medal with clasp, D.S.O.
N.-W. Frontier of India, 1897-98.—Defence of Malakand.—Relief of Chakdara.—(Severely wounded). Malakand.—Operations in Bajaur and in the Mamund country. Buner.—Despatches. G. G. O. 1089 of '97.—2 clasps. Brev. of Maj.
The War of 1914—1918.—Despatches, *Lond. Gaz.*, 20th March 1916. Brev. of Col.

Baldwin, M. C.—
The War of 1914—18.—Despatches, *Lond. Gaz.*, 19th October '16. M. C.

Balfe, E. (*Col. retd., I. A.*)—
N.-W. Frontier of India, Hazara, 1891.—Despatches. G. G. O. 816 of '91.—Medal with clasp.
Tirah, 1897-98.—Principal Provost Marshal, Army Staff.—Actions of the Sampagha and Arhanga Passes.—Despatches. G. G. O. 244 of '98.—Medal with 2 clasps.

Balfour, J. H.—
Afghanistan, 1878-79-80.—Action of the Peiwar Kotal.—Medal with clasp.
Egypt, 1882.—Actions of Kassassin and Tel-el-Kebir.—Medal with clasp. Bronze star.
Chitral, 1895.—Relief of Chitral.—Medal with clasp.
N.-W. Frontier of India, 1897-98.—Mohmand.—Clasp.

Ball, E. P.—
The War of 1914-18.—Despatches, *Lond. Gaz.*, 5 Apl. '16.

Ball, L. P.—
South African War, 1900-01.—Operations in Orange Free State, April to July 1900, including action at Bethlehem (6th and 7th July). Operations in the Transvaal from August 1900 to March 1901. Operations in Cape Colony from March to June 1901.—Queen's Medal with 4 clasps.

Bally, H. G.—
South African War, 1902.—Operations in the Transvaal between 30 Nov. '00 and 31 May '02. Operations in the Orange River Colony between 30 Nov. '00 and 31 May '02. Queen's Medal with 3 clasps.

Bamber, C. J. (*Col., retd., I.M.S.*)—
N.-W. Frontier of India, Mahsud-Waziri, 1881.
Burma, 1886-87.—Operations of the 1st Brigade.—Medal with clasp.
N.-W. Frontier of India, 1897-98, Malakand.—Medal with clasp.

Bamfield, H. J. K.—
N.-W. Frontier of India, 1897-98.—Action of the Ublan Pass, 27 Aug. '97.—Despatches. G. G. O. 1141 of '97.—Operations on the Samana and in the Kurram Valley during Aug. and Sep. '97.—Medal with 2 clasps.
Tirah, 1897-98.—Clasp.
N.-W. Frontier of India, 1902.—Operations against the Darwesh Khel Waziris.
N.-W. Frontier of India, 1908.—Operations in the Zakka Khel country.—Medal with clasp.
The War of 1914-18.—Despatches, *Lond. Gaz.*, 22 Jun. '15 and 1 Jun. '17. D.S.O.

Bampfied, G. C.—
The War of 1914-18.—Despatches, *Lond. Gaz.*, 15 May '17.

Banatvala, H. E.—
Burma, 1886-89.—Operations of the 1st Brigade. Pursuit of Hlaoo.—Medal with 2 clasps.
N.-E. Frontier of India, Lushai, 1892.

Banbury, W. E.—
Burma, 1886-87.—Operations of the 1st Brigade.—Medal and clasp.
N.-W. Frontier of India, 1897-98.—Malakand.—Operations in Bajaur and in the Mamund country. Despatches. G. G. O. 1317 of '97. Buner. Action of the Tanga Pass.—Medal with clasp.
The War of 1914-18.—Despatches, *Lond. Gaz.*, 1 Jan. '16, 15 June '16, 4 Jan. '17, 15 May '17 and 11 Dec. '17. Brev. of Col. C. M. G.

Banerjee, U. N.—
The War of 1914-18.—Despatches, *Lond. Gaz.*, 19 Oct. '16.

Banks, S. O'B. (*Colonel retd., I.M.S.*)—
Abyssinia, 1867-68.—Storming of Magdala.—Medal.

Bannatyne, F. C.—
South African War, 1899—02.—Operations in Cape Colony, South of Orange River, 1899-1900.—Operations in the Transvaal in May and June 1900.—Operations in Natal, March to June 1900, including action at Laing's Nek (6th to 9th June).—Operations in Orange River Colony, May to 29 Nov. '00.—Operations in the Transvaal, May 1901 to January 1902.—Queen's Medal with 3 clasps. King's Medal with 2 clasps.
East Africa, 1902-04.—Operations in Somaliland. Action at Jidballi.—Medal with 2 clasps.
The War of 1914-18.—Despatches, *Lond. Gaz.*, 31 Oct. '17. Brev. of Maj.

Bannatyne, N. C.—
East Africa, 1902—04.—Operations in Somaliland. Action at Jidballi.—Medal with 2 clasps.
The War of 1914-18.—Despatches, *Lond. Gaz.*, 1 Jan. '16 and 6 July '17.

Bannerman, W. B.—
Burma, 1885—89.—Medal with 2 clasps.

Bannerman, W. P.—
Soudan, 1896.—*Dongola Expedition.*—Medal. Also Egyptian medal.

Barber, C. H.—
N.-W. Frontier of India, 1908.—Operations in the Zakka Khel country. Operations in the Mohmund country. Engagement of Matta, 24th April 1908. Engagement of Kargha, 24th May 1908.—Medal with clasp.
The War of 1914-18.—Despatches, *Lond. Gaz.*, 5 Apl., 13 July and 19 Oct. '16.

Barclay, E. A. (*Col., retd., I. A.*)—
Afghanistan, 1879-80.—Medal.
Burma, 1887—89.—Medal with 2 clasps.

Barclay, P. C. R.—
British Central Africa, 1899.—Expedition against Kwamba and Awemba.—Medal with clasp.
British East Africa, 1901.—Operations against the Ogaden Somalis in Jubaland.—Clasp.
East Africa, 1902—04.—Operations in Somaliland.—Clasp.

Barge, K.—
South African War, 1902.—Operations in Cape Colony, Feb. to 31 May 02.—Queen's medal with 2 clasps.
The War of 1914-18.—Despatches, *Lond. Gaz.*, 15 May and 11 Dec. 17.—D.S.O., M.C.

Barlow, C. P.—
N.-W. Frontier of India, 1897-98.—Operations on the Samana and in the Kurram Valley during Aug. and Sep. 1897.—Medal with 2 clasps.
Tirah, 1897-98.—Operations against the Khani Khel Chamkannis.—Clasp.

Barlow, W. N. C.—
The War of 1914-18.—M. C.

Barnard, A. E.—
N.-W. Frontier of India, (1st) Miranzai, 1891.
East Africa, 1902-04.—Operations in Somaliland.—Medal with clasp.
N.-W. Frontier of India, 1908.—Operations in the Zakka Khel country. Operations in the Mohmand country.—Medal with clasp.
The War of 1914-18.—Despatches, *Lond. Gaz.*, 1 Jun. '17.

Barnardo, F. A. F.—
South African War, 1901.—Served as Civil Surgeon.—Queen's medal with 3 clasps.
East Africa, 1904—Operations in Somaliland. Action at Jidballi.—Medal with 2 clasps.

Barnes, E. (*Maj., retd., I. A.*)—
South African War, 1899-1900.—(Severely wounded.)

Barnes, E. C.—
South African War, 1899—1901.—Relief of Ladysmith, including action at Colenso (15th December 1899). Operations of 17th to 24th January 1900, and action at Spion Kop; operations of 5th to 7th February 1900, and action at Vaal Krans; operations on Tugela Heights (14th to 27th February 1900) and action at Pieter's Hill (27th February 1900). Operations in Natal, March to June 1900, including action at Laing's Nek (6th to 9th June). Operations in the Transvaal, east of Pretoria, July to September 1900. Operations in Orange River Colony, October 1900 to January 1901, and March to September 1901. Operations in Cape Colony, February 1901.—Queen's Medal with 6 clasps.

Barnes, J. P.—
Tirah, 1897-98.—Medal with 2 clasps.

Barnes, O. (*Col., u.s.l., Ind. S.C.*)—
India, 1857-58-59.—In Central India, siege of Kotah actions of Sanganeer, Kotahia, Jowra, and Purawas.—Medal with clasp.
Abyssinia, 1868.—Medal.
Afghanistan, 1878-79.—Action at Kam Dakka.—Mentioned in Despatches.—Medal.
N.-W. Frontier of India, Zhob Valley, 1884.—Mentioned in Despatches.

Barnett, J. W.—
East Africa (Somaliland), 1908—10.—Medal with clasp.

Barnett, R. P. S. (*Capt. retd., Ind. S.C.*)—
Burma, 1887—89.—Medal with 2 clasps.

Barr, J. H. V.—
N.-W. Frontier of India, Waziristan, 1901-02.—Medal with clasp.
The War of 1914—18—Despatches, *Lond. Gaz.*, 4 July 1916.

Barratt, W. C.—
Soudan, 1885.—Suakim. Reconnaissance to Hasheen. Actions at Hasheen and Tofrek. Operations at, and destruction of, Tamai.—Medal with 2 clasps. Bronze star.
N.-W. Frontier of India, Zhob Valley, 1890.
N.-W. Frontier of India, Waziristan, 1894-95.—Medal with clasp.
East Africa (British), 1896.—Commandant, Indian Contingent. Operations against Mazrui Rebels. Mentioned in despatches.—Medal. D.S.O. Brilliant Star of Zanzibar, 1st grade, 2nd class.
Uganda, 1897-98.—Commandant, Indian Contingent. Actions of Kakanguru Kabegambe, Jass Camp and Mruli. Mentioned in despatches.—Medal with 2 clasps. Brev. of Maj.
China, 1900.—Relief of Pekin. Action of Peitsang. Action of Yangtsun.—Despatches, *Lond. Gaz.*, dated 14 May 01.—Medal with clasp.
N.-W. Frontier of India, 1902.—Operation against the Darwesh Khel Waziris.

Barrett, Sir A. A.—
Afghanistan, 1879-80.—Affairs at and near the Shutur-gardan Pass. Operations at and around Kabul. March from Kabul to the relief of Kandahar, and battle of Kandahar. Operations against the Maris.—Medal with 2 clasps; bronze star.
N.-W. Frontier of India, Hazara, 1888.—Medal with clasp.
N.-W. Frontier of India, (2nd) Miranzai, 1891.—Clasp.
N.-W. Frontier of India—Hunza-Nagar, 1891.—Clasp.
Tirah, 1897-98.—D. A. Q. M. G., 2nd Brigade and subsequently A. Q. M. G., 1st Division. Actions of the Sampagha and Arhanga Passes. Operations against the Khani Khel Chamkannis. Operations in the Bara Valley, 25 to 30 Dec. '97. Despatches. G. G. O. 244 of '98.—Medal with 2 clasps. Brev. of Lt.-Col.
N.-W. Frontier of India, 1908.—Operations in the Zakka Khel country. G. O. C. 2nd Bde. Despatches, G. G. O. No. 188 of 1908. Operations in the Mohmand country. G. O. C. 2nd Bde. Despatches, G. G. O. No. 521 A of 1908.—Medal with clasp, K.C.B.
The War of 1914—18.—K.C.S.I.

Barrett, A. L.—
China, 1900.—Medal.
The War of 1914-18.—Despatches, *Lond. Gaz.*, 15 Aug. '17.

Barrett, C. C. J.—
The War of 1914-18.—Despatches, *Lond. Gaz.*, 15 Aug. '17.

Barron, A. H. D.—
N.-W. Frontier of India, 1908.—Operations in the Mohmand country.
The War of 1914-18.—M. C.

Barron, R. M.—
The War of 1914-18.—Despatches, *Lond. Gaz.*, 22 Jun. '15.

Barrow, Sir E. G.—
Afghanistan, 1878.—Action of Ali Musjid.—Medal with clasp.
Egypt, 1882.—D. A. Q. M. G., Indian Contingent Action of Tel-el-Kebir.—Medal with clasp. Bronze star, 5th class of the Medjidie.
Tirah, 1897-98.—A. A. G., Army Staff. Actions of the Sampagha and Arhanga Passes. Reconnaissance of the Saran Sar and action of 9 Nov. '97. Operations against the Khani Khel Chamkannis. Operations in the Bara Valley, 7 to 14 Dec. '97.—Despatches. G. G. O. 244 of '98.—Medal with 2 clasps, C. B.
China, 1900.—Chief of Staff.—Relief of Pekin. Action of Peitsang. Action at Yangtsun. Despatches, *Lond. Gaz.*, 14 May '01.—Medal with clasp, K.C.B.

Barrow, G. de S.—

N.-W. Frontier of India, 1894-95.—Waziristan.—Orderly Officer to G. O. C. the Force.

China, 1900.—Special Service Officer.—Despatches *Lond. Gaz.,* 14 May '01.—Medal with clasp.

The War of 1914-18.—Despatches, *Lond. Gaz.,* 19 Oct. '14, 17 Feb. '15, 22 Jun. '15, 15 Jun. '16 and 4 Jan. 17. Promd. Maj.-Genl. C. B.

Barrow, R. E.—

N.-W. Frontier of India, 1908.—Operations in the Zakka Khel country. Signalling officer, attached Line of Communication Staff.—Medal with clasp.

Barry, C. C. S.—

Chitral, 1895.—Relief of Chitral.—Medal with clasp.

Barry, J. F.—

South Africa, 1881.—Transvaal Campaign.
Burma, 1885—88.—Medal and 2 clasps.
Burma, 1891-92.—Operations against the Kachins.—Clasp.

China, 1900.—Despatches, *Lond. Gaz.,* 14 May '01.—Medal.

Barry, W. T.—

South African War, 1900—02.—Operations in Cape Colony, South of Orange River, '99-00, including action at Kheis (29 May). Operations in Orange River Colony (May to 29 Nov. 00), including actions at Lindley (1 June), Rhenoster River (11 June), Lindley (26 June). Operations in the Transvaal, west of Pretoria, July to 29 Nov. '00, including action at Ventes kroon (7 and 9 Aug.). Operations in the Transvaal, between 30 Nov. '00 and 31 May '02. Queen's medal with 3 clasps. King's medal with 2 clasps.

Barry-Smith, R. C.—

The War of 1914-18.—Despatches, *Lond. Gaz.,* 8 Feb. '17.

Barstow, H.—

N.-W. Frontier of India, 1897-98.—Mohmand.—Medal with clasp.

N.-W. Frontier of India, Waziristan, 1901-02.—Clasp.

Bartleman, W. F. (*M.-Gen., u.s.l., Ben. Inf.*)—

India, 1858.—With Sarun Field Force.—Medal.

N.-E. Frontier of India, Cossyah and Jyntiah Hills, 1862-63.

Bartlett, E. L.—

N.-W. Frontier of India, (2nd) Miranzai, 1891.—Medal with clasp.

China, 1900.—Medal with clasp. Promoted Conductor.

The War of 1914-18.—Despatches, *Lond. Gaz.,* 21 Jun. and 25 Sept. '16.

Bartley, W. D.—

Burma, 1885-1889.—Medal with 2 clasps.
N.-W. Frontier of India, Manipur, 1891.—Clasp.

Barton, A. E.—

Burma, 1887.—Medal with 2 clasps.

Barton, H. J. (*Hon. M.-Gen., retd., Ben. S.C.*)—

Afghanistan, 1878—80.—At Kandahar, Ghirisk, Jelalabad and Kabul. Mentioned in Despatches.—Medal; Brev. of Lt.-Col.

Basevi, C. E. (*Hon. Col. retd.*)—

Persia, 1857.—Medal.
India, 1858.—Action of Amherpanee (wounded).—Medal.

Basevi, W. H. F.—

Burma, 1894-96.

The War of 1914-18.—Despatches, *Lond. Gaz.,* 1 Jan. '16.

Basu, B. D. (*Maj. retd., I. M. S.*)—

Soudan, 1896.—Dongola Expedition.—Medal.
N.-W. Frontier of India, 1897-98.—Malakand—Buner.—Action of the Tanga Pass.—Medal with clasp.

Bate, T. E. L. (*Col., retd., I. M. S.*)—

Afghanistan, 1879-80.—Medal.

Bateman-Champain, C. E.—

Chitral, 1895.—Relief of Chitral.—Medal with clasp.

N.-W. Frontier of India, 1902.—Operations against the Darwesh Khel Waziris.

N.-W. Frontier of India, 1908.—Operations in the Zakka Khel country.—Medal with clasp.

Bateman-Champain, H. F.—

N.-W. Frontier of India, Waziristan, 1894-95.—Action at Wano, 3 Nov. '94.—Medal with clasp.

N.-W. Frontier of India, 1897-98.—Mohmand.—Medal with clasp.

Tirah, 1897-98.—Provost Marshal, 1st Brigade, 1st Division. Actions of the Sampagha and Arhanga Passes. Operations in the Bazar Valley 25 to 30 Dec. '97.—Clasp.

The War of 1914-18.—Despatches, *Lond. Gaz.,* 16 July '16.

Batten, F. G.—

China, 1900.—Relief of Pekin.—Despatches, *Lond. Gaz.,* 14 May '01.—Medal with clasp.

The War of 1914-18.—Despatches, *Lond. Gaz.,* Jun. '16. D.S.O.

Battine, R. St. C.—

N.-W. Frontier of India, Waziristan, 1894-95.—Medal with clasp.

Chitral, 1895.—Relief of Chitral.—Medal with clasp.

N.-W. Frontier of India, 1897-98.—Tochi.—Clasp.

Batty, R. H. (*Brig. Surg., retd., Ind. Med. Serv.*)—

Abyssinia, 1867-68.—Medal.
Afghanistan, 1879-80.—Medal.
Burma, 1886-87.—Medal with clasp.

Battye, A. B.—

Chitral, 1895.—Relief of Chitral.—Medal with clasp.
Tirah, 1897-98.—2 clasps.
The War of 1914-18.—Despatches, *Lond. Gaz.,* 4th Jan. 1917.

Battye, A. H.—

Burma, 1885-87.—Operations of the 1st Brigade, Shan States Column, including advance to, and occupation of, Fort Stedman.—Medal with clasp.

N.-E. Frontier of India, Chin-Lushai, 1889-90.—Clasp.

N.-E. Frontier of India, Manipur, 1891.—Clasp.

Chitral, 1895.—Relief of Chitral.—Medal with clasp.

Tirah, 1897-98.—Reconnaissance of the Saran Sar and actions of 9 and 11 Nov. 97.—Operations against the Khani Khel Chamkannis. Operations in the Bazar Valley, 25 to 30 Dec. '97.—2 clasps.

China, 1900.—Medal.

attye, C. F. (*Captain, retd., Ben. S. C.*)—

Served with Field Force under Brigadier-General Douglas, C. B. in Behar and Ghazeepore districts, 1858.—Medal.—With China Expeditionary force 1860-61, and in various operations against Taipings, 1862.—Medal.

Battye, I. U.—

N.-W. Frontier of India, 1897-98.—Relief of Malakand and Chakdara.—Malakand. Operations in Bajaur.—Utmankhel.—Buner.—Medal with 2 clasps.

The War of 1914-18.—Despatches, *Lond. Gaz.*, 4 July 16. D.S.O.

Battye, W. R.—

China, 1900.—Medal.

The War of 1914-18.—Despatches, *Lond. Gaz.*, 28 Jan., 21 Jun. and 13 July 16. Legion of Honour (5th class) D.S.O.

Baugh, G. S. (*Maj., retd., Ind. S.C.*)—

Burma, 1885—89.—Medal with 2 clasps.

Burma, 1892-93.—Operations in the Northern Chin Hills. Despatches G. G. O. 733 of '93.—Clasp.

Baugh, M. W. (*Maj., retd., I. A.*)—

Burma, 1886-87-88.—Medal with 2 clasps.

N.-W. Frontier of India, 1897-98.—Malakand—Buner.—Action of the Tanga Pass.—Medal with 2 clasps.

Bax, W. I. (*Col., retd., Ben. Cavy.*)—

N.-W. Frontier of India, Jowaki, 1877-78.—Mentioned in Despatches.—Medal with clasp.

Afghanistan, 1878-79.—Capture of Ali Musjid.—Medal with clasp.

Bayley, E. C.—

East Africa, 1903-04.—Operations in Somaliland.—Medal with clasp.

Bayly, A. W. L. (*Maj. Gen., retd., I. A.*)—

Afghanistan, 1879-80.—Defence of Kandahar and battle of 1 Sep.—Medal with clasp.

Soudan, 1885.—Suakin. Various attacks on convoys. Action of Tamai.—Medal with clasp; bronze star.

Burma, 1886-87.—As D. A. A. and Q. M. G. Despatches. G.G. O. 339 of '87.—Medal with 2 clasps. D.S.O.

South African War, 1899-1900.—Staff Officer, 4th Brigade and D. A. A. G., 4th Infantry Division. Relief of Ladysmith, including action at Colenso (15 Dec. '99); operations of 17 to 24 Jan. '00, and action at Spion Kop (slightly wounded); operations of 5 to 7 Feb. '00 and action at Vaal Kranz; operations on Tugela Heights (14th to 27 Feb. '00) and action at Pieter's Hill (27 Feb. '00) (as Staff Officer, 4th Brigade). Operations in Natal, March to June '00, including action at Laing's Nek (6 to 9 June) (as D. A. A. G., 4th Division) operations in the Transvaal east of Pretoria Aug. to Oct. '00, including actions at Belfast (26 and 27 Aug. '00), Lydenburg (5 to 8 Sep.), as D. A. A. G., 4th Division.—Despatches (*twice*), *Lond. Gaz.*, dated 8 Feb. '01.—Queen's medal with 5 clasps, C.B.

Baynes, C. E.—

South African War, 1899-1900.—Relief of Kimberley.—Operations in the Orange Free State, Feb. to May '00, including operations at Paardeberg, 17 to 26 Feb. Actions at Poplar Grove, 7 Mar. Actions at Houtnek, 1 May, Vet river, 5 and 6 May, and Zand river, 10 May.—Operations in the Transvaal in May and June '00, including actions near Johannesburg, Pretoria, and Diamond Hill, 11 and 12 June.—Severely wounded.—Queen's Medal with 4 clasps.

Bazett, S.—

Tirah, 1897-98.—Medal with 2 clasps.

Beadon, V.—

The War of 1914-18.—Despatches, *Lond. Gaz.*, 15 Aug. 17.

Beale, A. (*Bt.-Col. retd I. A.*)—

Burma, 1885-89.—Medal with 2 clasps.

Beaman, A. A. H.—

The War of 1914-18.—Despatches, *Lond. Gaz.*, 4th July 1916. D.S.O.

Bean, J. A.—

South African War, 1900-01.—Operations in Cape Colony, from March to November 1900. Operations in the Orange River Colony, from November 1900 to August 1901.—Queen's Medal with 3 clasps.

Beatson, C. H. (*Col. retd., I. M. S.*)—

Afghanistan, 1878-79.—Medal.

N.-E. Frontier of India, Manipur, 1891.—Despatches. G. G. O. 585 of '91.

Beatson Bell, J.—

N.-W. Frontier of India, Hazara, 1891.—Medal with clasp.

Tibet, 1903-04.—Action at Niani, 26th June 1904. Operations at and around Gyantse, between 5th May and 6th July 1904. March to Lhassa, 14th July to 3rd August 1904.—Medal with clasp.

The War of 1914-18.—Despatches, *Lond. Gaz.*, 22 Jun. 15.

Beatty, G. A. H.—

N.-W. Frontier of India, 1897-98.—Tochi.—Provost Marshal, 2nd Brigade.—Medal with clasp.

China, 1900.—Despatches, *Lond. Gaz.*, 14 May '01.—Medal.

The War of 1914-18.—Despatches, *Lond. Gaz.*, 1st June 1917. D.S.O.

Beatty, L. N.—

Soudan, 1896, Dongola Expedition.—Medal. Also Egyptian medal.

N.-W. Frontier of India, 1897-98.—Malakand, Buner.—Medal with clasp.

Tirah, 1897-98.—Mentioned in despatches. G. G. O. 483 of '98. Operations in the Bazaar Valley, 25 to 30 Dec. '97.—Clasp.

East Africa, 1903-04.—Operations in Somaliland. Medal with clasp.

Beaumont, S. J. G.—

South African War, 1902.—Operations in the Transvaal, May—02.—Queen's medal with 4 clasps.

Beavan, R. (*Col., retd., Ind. S.C.*)—

Afghanistan, 1878—80.—With Kandahar, Mari and Bozdar Field Force. Mentioned in Despatches. —Medal.

Beazley, W. E.—

N.-W. Frontier of India, 1908.—Operations in the Zakka Khel country.—Medal with clasp.

The War of 1914-18.—Despatches, *Lond. Gaz.*, 30 Jun. 16. Brev. of Maj. M.C.

Becher, A. W. R. (*Col., retd., Ben. Cav.*)—

N.-E. Frontier of India, Bhootan, 1864-65.—Storming of Chamoochee. Mentioned in despatches.—Medal with clasp.

N.-W. Frontier of India, Hazara, 1868.—Clasp.

Afghanistan, 1878—80.—Mentioned in despatches. —Medal.

Becher, G. A.—

N.-W. Frontier of India, 1897-98.—Operations on the Samana and in the Kurram Valley during Aug. and Sep. '97. Relief of Gulistan.—Medal with 3 clasps.

East Africa, 1903-04.—Operations in the Somaliland. Special service officer.—Medal with clasp.

Beck, R. C.—

N.-W. Frontier of India, 1908.—Operations in the Zakka Khel country. Operations in the Mohmand country. Engagement of Matta, 24th April 1908. Engagement of Kargha, 24th May 1908.—Medal with clasp.

Beckett, S. (*Col., retd., Ben. S.C.*)—

India, 1858-59.—Campaign in Oude, passage of the Gogra at Fyzabad, skirmish near Bunkussia fort, action near Toolsepore (dangerously wounded). Despatches, *Lond. Gaz.*, 29 Jan. '59.—Medal.

N. W. Frontier of India, 1863.—2nd Eusofzai Expedition. Capture of Umbeyla Pass.—Despatches, *Lond. Gaz.*, 19 Mar. '64.—Medal with clasp.

Afghanistan. 1870.

Soudan Expedition, 1885.—Suakin. In charge of transport from India.—Despatches, *Lond. Gaz.*, 25 Aug. '85. — Medal with clasp; bronze star. C.B.

Beckham, L. G. (*Lt.-Col., u. s. l., I. A.*)—

Afghanistan, 1880.—March from Quetta to the relief of Kandahar.—Medal.

Burma, 1886-87.—Medal with 2 clasps.

Beddy, E. (*Col., u.s.l., Ind. S.C.*)—

N.-E. Frontier of India, Bhootan, 1865.—In command of storming party at capture of Dewangiri.—Medal with clasp.

Afghanistan, 1880.—Medal.

N.-W. Frontier of India, Hazara, 1888.—Mentioned in Despatches.—Clasp.

Beddy, P. L.—

N.-W. Frontier of India, 1897-98.—Malakand.—Operations in Bajaur.—Mohmand.—Medal with 2 clasps.

Tirah, 1897-98.—Actions of the Sampagha and Arhanga Passes.—Operations in the Bazar Valley 25 to 30 Dec. '97.—Clasp.

China, 1900.—Relief of Pekin—Medal with clasps.

N.-W. Frontier of India, 1902.—Operation against the Darwesh Khel Waziris.

The War of 1914—18.—Despatches, *Lond. Gaz.*, 19 Oct. '16. Brev. of Lt.-Col.

Bedell, E. A.—

Burma, 1885-86.—Medal with clasp.

Chitral, 1895.—Relief of Chitral.—Medal with clasp.

Beech, L. (*Surg. Lt.-Col., retd., Ind. Med. Ser.*)—

Afghanistan, 1879-80.— In Southern Afghanistan.—Medal.

Begbie, A. S. (*Lt.-Col., retd., I. A.*)—

N.-W. Frontier of India, 1897-98.—Malakand—Utman Khel—Buner—Action of the Tanga Pass.—Medal with clasp.

Begbie, E. W. (*Maj.-Genl., u.s.l.*)—

Abyssinia, 1867-68.—Served as Sub-Assistant Commissary General.—Mentioned in Despatches, *Lond. Gaz.*, 30 June '68.—Medal.

N.-E. Frontier of India, Daphla, 1874-75.—Served as Superintendent of Army Signalling.—Mentioned in Despatches, *Gaz. of India*, 13 April '75.

Burma, 1885-86.—Served as Superintendent of Army Signalling. Present at the capture of Minhla and occupation of Mandalay and Bhamo.—Mentioned in Despatches. *Lond. Gaz.*, 22 June '86.—Medal with clasp.

Begbie, F. R. (*Bt. Col., retd., I. A.*)—

N.-W. Frontier of India, Jowaki, 1877.—Medal with clasp.

Afghanistan, 1878-79.—Action of Ali Musjid.—Medal with clasp.

N.-W. Frontier of India, Mahsud-Waziri, 1881.—Despatches. Brigadier-General T. G. Kennedy's Despatch, dated 20 June '81.

N.-E. Frontier of India, Lushai, 1889.—Clasp.

N.-E. Frontier of India, Chin-Lushai, 1889-90.—Despatches, G. G. O. 677 of '90.—Clasp.

N.-W. Frontier of India, 1897-98.—Karram-Kohat Force.—Medal with 2 clasps.

Tirah, 1897—99.—Line of communications.—Clasp.

Belt, F. V. O.—

N.-W. Frontier of India, Waziristan, 1901-02.—Medal with clasp.

Bell, A. W. C. (*Col. retd. I.A.*)—

Soudan, 1885.

Bell, G. J. H.—

Burma. 1887—89.—Medal with clasp.

N.-E. Frontier of India, Lushai, 1889.—Clasp.

Bell, H. C. F.—

South African War, 1902.—Operations in Cape Colony from 7 to 9 April 1902. Operations in the Orange River Colony from 10 to 12 April 1902. Operations in the Transvaal from 13 April to 31 May 1902.—Queen's medal with 4 clasps.

Bell, H. P. (*Lt.-Col. retd. I. A.*)—

Burma, 1891.—Wunthoo Field Force.—Medal with clasp.

Chin Hills.—Clasp.

Bell, Sir J. A.—

Afghanistan, 1878-79.—Medal.

Bell, L. H. I.—

South African War, 1900-01.—Operations in Cape Colony, between 6th November 1900 and 5th January 1901, and from 27th January to 4th June 1901. Operations in the Transvaal between 6th and 26th January 1901.—Queen's medal with 3 clasps.

Bell, L. M.—

Burma, 1887—89.—Operations against the Chins, Upper Burma.—Medal with clasp.

N.-W. Frontier of India, 1st Miranzai, 1891.

Do. do. Isazai, 1892.

Do. do. 1897-98.—Malakand-Utman Khel.

Buner.—Action of the Tanga Pass.—Medal with clasp.

Bell, R. C.—

N.-W. Frontier of India, 1897-98.—Operations in the Kurram Valley during Aug. and Sep. '97.—Medal with 2 clasps.

Tirah, 1897-98.—Reconnaissance of the Kharmana Defile and action of 7 Nov. '97. Operations against the Khani Khel Chamkannis.—Clasp.

The War of 1914—18.—Despatches, *Lond. Gaz.*, 4 January '17. D.S.O.

Bell, R. M.—

Burma, 1887-88.—Medal with clasp.

N.-E. Frontier of India, Chin Hills, 1892-93.—Clasp.

Bell, Revd. W. L., M.A.—

South African War, 1900-01.—Served in Natal and South-Eastern Transvaal as acting Chaplain to the Forces.—Queen's Medal with 2 clasps.

Bell-Kingsley, H. E. W.—

The War of 1914—18.—Despatches, *Lond. Gaz.*, 15 August '17. D.S.O.

Belli-Bivar, H. F.—

East Africa (Somaliland), 1908—10.—Medal with clasp.

Benn, J. P.—
N.-W. Frontier of India, 1908.—Operations in the Mohmand country. Engagement of Kargha, 24th May '08.—Medal with clasp.

Benn, R. A. E. (*Maj., Supy. List, I. A.*)—
Burma, 1885—87 and 1887—89.—Medal with 2 clasps.

Bennett, A. D.—
N.-W. Frontier of India, 1908.—Operations in the Mohmand country.—Medal with clasp.

Bennett, C. H. (*Lt.-Col., retd., I. M. S.*)—
Afghanistan, 1879.

Bennett, E. E. S.—
N.-W. Frontier of India, Waziristan, 1901-02.—Medal with clasp.
The War of 1914—18.—Despatches, *Lond. Gaz.*, 4 July '16, and 5 Aug. 17.

Bennett, Hugh—
N.-W. Frontier of India, 1897-98.—Operations on the Samana. Action of the Ublan Pass.—Medal with 2 clasps.
Tirah, 1897-98.—Clasp.
South African War, 1902.—Operations in Cape Colony, January to April, May '02. Operations in the Transvaal, April '02. Operations in Orange River Colony, April 1902.—Queen's medal with 4 clasps.

Bennett, J. (*Asst. Comsy. and Hony. Lieut., retd., Ord. Dept.*)—
Tirah, 1897-98.—Medal with 2 clasps.

Bennett, J. H.—
South African War, 1900-01.—Operations in the Orange Free State, Apl. and May '00. Operations in the Orange River Colony, Sept. to 29 Nov. '00. Operations in Cape Colony, South of Orange River, '00. Operations in Orange River Colony, 30th Nov. '00 to May' 01.—Queen's Medal with 3 clasps.

Bennett, V. B.—
Tirah, 1897-98.—Medal with 2 clasps.

Benson-Cooke, P. G.—
South African War, 1901-02.—Operations in the Transvaal, between 1st and 21st May 1902. Operations in Orange River Colony, between 27th October 1901 and 30th April 1902, and between 21st and 31st May 1902. Operations in Cape Colony, between 21st and 26th October 1901.—Queen's Medal with 5 clasps.

Bensley, C. H.—
N.-W. Frontier of India, 1897-98.—Malakand, Utmen Khel, Buner, action of the Tanga Pass.—Medal with clasp.

Bentinck, R. J. (*Capt., retd., Ind. S. C.*)—
South African War, 1899-1901.—Employed with Thorneycroft's Mounted Infantry.—Queen's medal with 3 clasps.

Benton, W. B.—
The War of 1914—18.—Despatches, *Lond. Gaz.*, 5 April '16, 13 July 16 and 15 August '17. M.C.

Beresford, G. W. (*Hon. Lt.-Col., retd., Ben. S.C.*)—
N.-E. Frontier of India, Garo Hills, 1871-72.
N.-E. Frontier of India, Naga Hills, 1877-78.
N.-E. Frontier of India, Akha, 1883-84.—Mentioned in Despatches.

Beresford, G. de la P.—
The War of 1914—18.—M.C.

Beresford, J. H. B. dela P. (*Maj. retd. I. A.*)—
Burma, 1886-87.—Medal with clasp.
N.-W. Frontier of India, Hazara, 1888.—Clasp.
N.-W. Frontier of India, Hazara, 1891.—Clasp.
N.-W. Frontier of India, (1st) Miranzai, 1891.
N.-W. Frontier of India, Waziristan, 1894-95.—Action at Wana, 3rd Nov. '94.—Clasp.
N.-W. Frontier of India, 1897-98.—Operations on the Samana and in the Kurram Valley during Aug. and Sep. '97.—Medal with 2 clasps.
Tirah, 1897-98.—Actions of Dargai and of the Sampagha Pass. Reconnaissance of the Saran Sar and action of 11 Nov. '97. Operations against the Khani Khel Chamkannis. Operations in the Bazar Valley, 25 to 30 Dec '97.—Clasp.
South African War, 1899—01.—Operations in the Orange Free State, March to May '00, including actions at Houtnek (1 May), Vet River (5 and 6 May), Zand River (10 May). Operations in the Transvaal in May and June '00, including actions near Johannesburg (29 May) and Pretoria (4 June). Operations in the Transvaal, east of Pretoria, July to 29 Nov. '00. Operations in the Transvaal, west of Pretoria, July to 29 Nov. '00, including action at Zilikat's Nek (2 Aug.). Operations in Cape Colony between 30 Nov. and 5 Dec. '00. Despatches. *London Gaz.*, 10 Sep. '01.—Queen's Medal with 3 clasps.—Brevet of Major.
N.-W. Frontier of India, 1902.—Operations against the Darwesh-Khel Waziris.

Berkeley, R. E. G.—
South Africa, 1900-01.—Operations in Cape Colony, north of Orange River May to July 1900; operations in the Transvaal, west of Pretoria, July to 29th November 1900; operations in the Transvaal, 30th November 1900 to May 1901.—Queen's medal with 3 clasps.

Berkeley, R. B.—
Burma, 1886-87.—Medal with clasp.

Berkeley-Hill, O. A. R.—
The War of 1914—18.—Despatches, *Lond. Gaz.*, 8 Feb. '17.

Bernard, E. H. (*Lieut.-Col., retd., I. A.*)—
Soudan, 1884-85.—Nile.—Medal with clasp. Bronze star.
Burma, 1886-87.—Medal with 2 clasps.
N.-W. Frontier of India, (2nd) Miranzai, 1891.—Clasp.

Bernard, R. P. St. V.—
The War of 1914—17.—Despatches, *Lon. Gaz.*, 15 August '17. M.C.

Berry, A. E.—
N.-W Frontier of India, 1908.—Medal with clasp.
The War of 1914—18.—Despatches, *Lond. Gaz.*, 1 Jan. '16.

Berry, E. (*Depy. Comsy. and Hony. Capt., retd.*)—
Afghanistan, 1878—80—Medal with clasp.

Berryman, E. R. P.—
The War of 1914—18.—Brev. of Maj.

Berthon, H. W. (*Maj. Supy. List, I. A.*)—
Tirah, 1897-98.—Medal with 2 clasps.

Betham, G. L.—
The War of 1914—18.—Despatches, *Lond. Gaz.*, 17 Feb. '15.

Betham, R. M.—
N.-W. Frontier of India, Zhob Valley, 1884.

Bethell, L. A.—
South African War, 1899-'01.—Relief of Ladysmith, including operations of 17th to 24th Jan. 1900 and action at Spion Kop. Operations of 5th to 7th Feb. '00 and action at Vaal Krantz. Operations on Tugela Heights (14th to 27th Feb. 1900) and action at Pieter's Hill (27th Feb. '00). Operations in Natal, March to June '00, including action at Laing's Nek (6th to 9th June). Operations in Transvaal, July '00 to April '01. Operations on the Zululand frontier of Natal, April to July '01.—Queen's Medal with 7 clasps.

Tibet, 1903-04.—Action at Niani, 26th June '04 Operations at and around Gyantse, between 5th May and 6th July '04. March to Lhassa, 14th July to 3rd Aug. '04.—Medal with clasp.

Operations in the Abor country, 1911-12.—Medal with clasp.

The War of 1914—18.—Despatches, *Lond. Gaz.*, 22 June 15.

Beville, C. H.—
Burma, 1886-87.—Medal with clasp.

N.-W. Frontier of India, Hazara, 1891.—Despatches. G. G. O. 816 of '91.—Clasp.

Chitral, 1895.—Relief of Chitral.—Medal with clasp.

N.-W. Frontier of India, 1897-98.—Malakand—Relief of Malakand. Action at Landakai.—Despatches. G. G. O. 1089 of '97. Operations in Bajaur and in the Mamund country—Buner—Action of the Tanga Pass. Despatches. G. G. O. 217 of '98.—Clasp.

The War of 1914—18.—Despatches, *Lond. Gaz.*, 13 July '16. C.B.

Bevington, L. F.—
N.-W. Frontier of India, 1908.—Operations in the Mohmand country.—Medal with clasp.

Beynon, W. G. L.—
N.-W. Frontier of India, Hazara, 1888.—Medal with clasp.

Chitral, 1895.—Relief of Chitral.—Despatches. G. G. O. 530 of '95.—Medal with clasp, D.S.O.

N.-W. Frontier of India, 1897-98.—Operations on the Samana and in the Kurram Valley during Aug. and Sep. '97.—Relief of Gulistan.—2 clasps.

Tirah, 1897-98.—Actions of Chagru Kotal, Dargai, and of the Sampagha and Arhanga Passes. Reconnaissance of the Saran Sar and action of 9 Nov. '97. Operations at and around Dwatci and action of 24 Nov. '97. Operations in the Bara Valley, 7 to 14 Dec. '97. Affair at Shinkamar on 29 Jan. '98.—Despatches, G. G. O. 58 of '98.—Clasp. Brev. of Maj.

Somaliland, 1901.—Despatches, *Lond. Gaz.*, 18th April '02.—Medal with clasp.

Tibet, 1903-04.—Action at Niani, 26th June 1904. Operations at and around Gyantse, between 5th May and 6th July 1904. March to Lhassa, 14th July to 3rd August 1904. D. A. Q. M. G. Despatches, *Gazette of India*, No. 1065 of 1904.—Medal with clasp. Brevet of Lieut.-Colonel.

Operations in the Abor Country, 1911-12.—Medal with clasp.

The War of 1914—18.—Despatches, *Lond. Gaz.*, 4 July 16. Promd. Maj.-Gen. C.I.E., K.C.I.E.

Bharucha, P. B.—
The War of 1914—18.—Despatches, *Lond. Gaz.*, 5 Apr. '16 and 13 July '16. D.S.O.

Bickford, W. W.—
British East Africa, 1898.—Operations against the Ogaden Somalis—Medal with clasp.
The War of 1914—18.—Brev. of Lt.-Col.

Bickley, G. W.—
Afghanistan, 1878-80.—Action of Charasia, 6th Oct' 1879. Operations at and around Kabul, 10th to 23rd Dec. 1879.—Medal with 2 clasps.
Burma, 1886-87.—Medal with clasp.

Biddulph, John (*Col. u.s.l. Ind. S.C.*)—
India, 1858-59.—Action of Doondia Kaira, capture of forts Chundra and Musjidia, and action on the bank of the Raptee.—Medal.

Biddulph, S. F. (*Bt.-Col., retd., I. A.*)—
Afghanistan, 1878-79.—Despatches. Major-General *Sir* M. A. S. Biddulph's Despatch No. 194, dated 16th May 79.—Medal.

N.-W. Frontier of India, (2nd) Miranzai, 1891.—Medal with clasp.

N.-W. Frontier of India, 1897-98.—Buner. Divisional Provost Marshal, Action of the Tanga Pass.—Medal with clasp.

N.-W. Frontier of India, 1908.—Operations in the Zakka Khel country. Commanding Line of Communication. Despatches, G. G. O. No. 188 of '08.—Medal with clasp.

Bigger, S. F. (*Lt.-Col. retd., I. M. S.*)—
Afghanistan, 1879-80.—Zaimukht Expedition. Action of Zawa.—Medal.

N.-E. Frontier of India, Chin-Lushai, 1889-90.—Despatches, G. G. O. 677 of 90.—Medal with clasp.

N.-W. Frontier of India, Waziristan, 1894-95.—Clasp.

Tirah, 1897-98.—Action of Dargai. Despatches, G. G. O. 244 of '98.—Medal with 2 clasps.

Bignell, B. H.—
Tibet, 1903-04.—Action at Niani, 26th June 1904. Operations at and around Gyantse, between 5th May and 6th July 1904. March to Lhassa, 14th July to 3rd August 1904. Despatches, *Gazette of India*, No. 1065 of 1904.—Medal with clasp.

Bignell, G. N.—
The War of 1914—18.—Despatches, *Lond. Gaz.* 8 and 13 Feb. '17 M.C.

Bignell, R. L.—
Operations in the Abor Country, 1911-12.—Medal with clasp.

The War of 1914—18.—Despatches, *Lond. Gaz.* 19 Oct. '16. Brev. of Maj.

Bikaner, H. H. *Sir* G. S., Bahadur, Maharaja of—
China, 1900.—Despatches, *Lond. Gaz.*, 14 May 01 K.C.I.E.

The War of 1914-18.—Despatches, *Lond. Gaz.*, 17 Feb. '15 and 21 Jun. '16.

Bellimoria, S. D. (temp. Lt. Ind. Med. Serv.)—
The War of 1914-18.—Despatches *Lond. Gaz.*, 15 Aug. '17.

Bingham, W. H.—
China, 1900.—Medal.

The War of 1914-18.—Despatches, *Lond. Gaz.*, 15 June '16.

Bingley, A. H.—
Burma, 1891-02.—Medal with clasp.

China, 1900.—Action at Peitsang (5 Aug. '00) Action at Yangtsun (6 Aug. '00), Relief of Pekin (14 Aug. '00).—Despatches, *Lond. Gaz.*, 14 May '01.—Medal with clasp. Brev. of Maj.

War, 1914-18.—Despatches, Lond. Gaz., 21 June '17.—Promd. Maj.-Genl.

Binny, A. J. M.—
The War of 1914-18.—Despatches, *Lond. Gaz.*, 20 Mar. '16.

Bir Bikram Singh of Sirmur, Raj Kumar—
N.-W. Frontier of India, 1897-98.—Operations on the Samana.—Medal with 2 clasps.

Tirah, 1897-98.—Mentioned in Despatches, G. G. O. No. 244 of '98.—Clasp.

Birch, W. J. A. (*Col. retd. Ind. S.C.*)—
Afghanistan, 1878-79.—Capture of Ali Masjid.—Medal with clasp.

Bird, W. A. H.—
N.-W. Frontier of India, 1908.—Operations in the Zakka Khel Country.—Medal with clasp.
East Africa (Somaliland).—*1914-15.*—Despatches, *Lond. Gaz.* 2 Aug. '16.—Medal with clasp.

Birdwood, G. C. McD. (*Lieut.-Col., retd., I. A.*)—
Burma, 1885—89.—Medal with 2 clasps.

Birdwood, G. T.—
N.-E. Frontier of India, 1894.—Abor.
N.-W. Frontier of India, Waziristan, 1894-95.—Medal with clasp.
N.-W. Frontier of India, 1897-98.—Operations on the Samana and in the Kurram Valley during Aug. and Sep. 97.—Medal with 3 clasps.

Birdwood, H. B.—
N.-W. Frontier of India, 1897-98.—Operations on the Samana and in the Kurram Valley during Aug. and Sep. 97.—Medal with 2 clasps.
Tirah, 1897-98.—Reconnaissance of the Kharmana Defile and action of 7 Nov. 97.—Clasp.
South African War, 1901-02.—Operations in Transvaal between 16 July '01 and 31 May '02. Operations in Orange River Colony between 14 July '01 and 16 July '01. Operations in Cape Colony between 10 July 01 and 13 July '01.—Queen's Medal and 5 clasps. Despatches, *Lond. Gaz.*, 29th July '02.
Persia, 1911.—Attacks on detachment of 39th Horse. Brev. of Lt.-Col.

Birdwood, W. R.—
N.-W. Frontier of India, Hazara, 1891.—Medal with clasp.
N.-W. Frontier of India, Isazai, 1892.
N.-W. Frontier of India, 1897-98.—Operations on the Samana—Medal with 2 clasps.
Tirah, 1897-98.—Orderly Officer to G. O. C. 2nd division.—Actions of Chagru Kotal, Dargai, and of the Sampagha and Arhanga Passes.—Reconnaissance of the Saran Sar and action of 9 Nov. '97.—Operations in the Waran Valley and action of 16 Nov. 97.—Operations at and around Dwatoi and action of 24 Nov. 97.—Operations in the Bara Valley, 7 to 14 Dec. 97.—Despatches. G. G. O. 483 of 98.—Clasp.
South African War, 1899—1902.—As Staff Officer, Mounted Troops, Southern Natal Field Force, Brigade Major, 2nd Cavalry and 3rd Mounted Brigades, Natal Field Force and Chief Staff Officer, Natal Volunteer Brigade. Deputy Assistant Adjutant-General, Head-Quarters. S. A. Field Force. Operations in Natal, 99. Relief of Ladysmith, including action at Colenso (15 Dec. 99), operations of 17 to 24 Jan. 00 and action at Spion Kop; operations of 5 to 7 Feb. 00 and action at Vaal Krantz; operations on Tugela Heights (14 to 27 Feb. 00) and action at Pieter's Hill (27 Feb. '00). Operations in Natal, March to June 00, including action at Laing's Nek (6 to 9 June). Operations in the Transvaal in June 00. Operations in Transvaal, east of Pretoria, July to 29 Nov. 00, including actions at Belfast (26 and 27 Aug. 00.), Lydenberg (5 to 8 Sep.) Severely wounded. Horse killed. Operations in Orange River Colony and Cape Colony, Nov. 00 Operations in the Transvaal, between 30 Nov. 00 and 31 May 02. Operations in the Orange River Colony, between 30 Nov. 00 and 31 May 02. Operations in Cape Colony, between 30 Nov 00 and 31 May 02. Despatches, *Lond. Gaz.*, 8 Feb. 01 (*Sir* R. H. Buller, 30 Mar. 19 June and 9 Nov. 00) and *Lond. Gaz.*, 29 July 02.—Queen's Medal with 6 clasps. King's South African Medal, with 2 clasps. Brev. of Maj. and Lt.-Col. Placed on list of officers considered qualified for Staff employment in consequence of service on the Staff in the Field.
N.-W. Frontier of India, 1908.—Operations in the Mohmand country. Chief Staff Officer. Engagement of Kargha, 24th May 1908. Despatches, G. G. O. No. 521A of 1908.—Medal with clasp. D.S.O.
The War of 1914-18.—Despatches, *Lond. Gaz.*, 6 July '15, 20 Sept. '15, 28 Jan. '16, 10 Apl. '16, 13 July '16, 25, Sept. '16, 4, Jan. '17, 15 May '17 and 11 Dec. '17, Legion of Honour (2nd class) French War Cross. K.C.B., K.C.M.G.

Birdwood, W. S. (*Col. retd., I. A.*)—
Afghanistan, 1879-80.—Medal.
N.-W. Frontier of India, Zhob Valley, 1884.

Biscoe, J. V. M.—
South African War, 1902.—Queen's Medal with 2 clasps.

Biscoe, W. T.—
The War of 194-18.—Despatches, *Lond. Gaz.*, 30 Jun. '16.

Biscoe, Wm. W. (*Lt.-Gen., retd., I.S.C.*)—
Afghanistan, 1878-79-80.—Medal. Brev. of Lt.-Col.
N.-W. Frontier of India, (2nd) Miranzai, 1891.—Despatches, G. G. O. 632 of 1891.—Medal with 2 clasps.

Bishop, E. B. (*Col., u. s. l., I. A.*)—
Afghanistan, 1878-79-80.—Action of Saif-u-din. Battle of Ahmed Khel. Action of Urzu. Despatches. Maj.-Genl. *Sir* M. A. S. Biddulph's Despatch No. 117, dated 5 March 79, and Lt.-Genl. *Sir* D. M. Stewart's Despatch No. 640K., dated 24 June '79, and G. G. O. 326 of '80.—Medal with clasp. Brev. of Maj.
Burma, 1886-87.—Shan States column. Advance to, and occupation of, Fort Stedman. Despatches. G. G. O. 434 of '87.—Medal with clasp. Brev of Lt.-Col.

Bishop, J. (*Asst. Comy. and Hony. Lt., S. and T. Corps, retd.*)—
Afghanistan, 1879-80.—Affair of Khush-ki-Nakhud. Battle of Maiwand, and covering retreat from Maiwand. Defence and battle of Kandahar.—Medal with clasp, and medal for distinguished conduct in the field.

Bishop, L. T. (*Lt.-Col. retd. I. A.*)—
N.-W. Frontier of India, Dour Valley, 1872.
N.-W. Frontier of India, Jowaki, 1877.—Medal with clasp.
Afghanistan, 1879-80.—D.-A.-Q.-M.-G. to the Ghazni Field Force; action of Takht-i-pul, occupation of Kandahar, second occupation of Khelat-i-Ghilzai, actions of Ahmed Khel, Urzoo, and Patkao Shana (horse wounded). Despatches, G. G. O. 326 and 493 of '80.—Medal with clasp; Brev. of Maj.

Bisset, E.—
The War of 1914-18.—Despatches, *Lond. Gaz.*, 21 July '17.

Black, G. C. S.—
The War of 1914-18.—Despatches, *Lond. Gaz.*, 19 Oct. '16.

Black, W. C. (*Maj.-Genl., retd., I. A.*)—
Burma, 1886.—Medal with clasp.
Chitral, 1895.—Relief of Chitral.—Medal with clasp.
N.-W. Frontier of India, 1897-98.—Operations on the Samana and in the Kurram Valley during Aug. and Sep. 97.—2 clasps.
Tirah, 1897-98.—Reconnaissance of the Kharmana Defile and action of 7 Nov. '97. Operations against the Khani Khel Chamkanis.—Clasp.

Black, W. C.—
Operations in Chitral 1895.—With the Relief Force. Medal with clasp.
N.-W. Frontier of India.—1897-98.—Operations on the Samana and in the Kurram Valley during August and September 1897. 2 clasps.
Tirah, 1897-98.—Reconnaissance of the Kharmana Defile and action of 7 Nov. 1897. Operations against the Khana Khel Chamkanis. Clasp.

Blair, D. A.—
South African War, 1899—1902.—Advance on Kimberley, including actions at Modder River and Magersfontein. Operations in the Orange Free State to May 1900. Operations in Orange River Colony (May to 29th November 1900) including actions at Wittebergen (1st to 29th July) and Witpoort. Operations in Orange River Colony, 30th November 1900 to February 1901. Operations in Cape Colony, February 1901 to 31st May 1902. Despatches, *London Gazette*, 7th May and 10th September 1901.—Queen's medal with 2 clasps. King's medal with 2 clasps.
The War of 1914-18.—Despatches, *Lond. Gaz.*, 22 Jun. 15.

Blair, R. W. C.—
N.-W. Frontier of India, Waziristan, 1901-02.—Medal with clasp.
East Africa, 1902—04.—Operations in Somaliland. On Staff; afterwards Special Service Officer.—Medal with clasp.

Blakeney, W. E. A.—
Egypt, 1882.—Medal. Bronze star.
N.-W. Frontier of India, 1897-98.—Operations on the Samana and in the Kurram Valley during Aug. and Sep. 97—Relief of Gulistan.—Medal with 2 clasps.
Tirah, 1897-98.—Clasp.

Blakeway, D. B.—
N.-W. Frontier of India, Isazai, 1892.
N.-W. Frontier of India, 1897-98.—Operations on the Samana and in the Kurram Valley during August and September 1897.—Medal and two clasps.
Tirah, 1897-98.—Actions of Chagru Kotal, Sampagha and Arhanga Passes, Reconnaissance of the Saran Sar and action of 9th November 1897. Operations at and around Dwatoi and action of 24th November 1897. Operations against the Khani Khel Chamkanis. Operations in Bara Valley, 7th to 14th December 1897. Operations in the Bazar Valley, 25th to 30th December 1897.—Clasp.
N.-W. Frontier of India, 1908.—Chief Political Officer. Operations in the Mohmand country. Engagement of Kharga, 24th May 1908. Despatches, *Lon. Gaz.*, 14 Aug. 08.—Medal with clasp, C.I.E.

Bland, G. H.—
The War of 1914-18.—M. C.

Blandy, R.—
N.-W. Frontier of India, 1908.—Operations in the Mohmand country—Medal with clasp.
The War of 1914-18.—Despatches, *Lond. Gaz.*, 17 Aug. 17. M. C.

Blandford, A. W.—
The War of 1914-17.—Despatches, *Lon. Gaz.*, 30 Jun. '16.

Blanford, P. T.—
The War of 1914—18.—Despatches, *Lond. Gaz.*, 4 July 16. M. C.

Blanshard, F. H. (*Lt.-Col. retd., Ind. S. C.*)—
Afghanistan, 1880.—Medal.

Blathwayt, L. (*Hon. Lt.-Col. retd., Ben. S. C.*)—
India, 1858-59.—Siege and capture of Lucknow, attack on Fort Rooyah, actions of Allygunge, Bareilly and Shahjehanpore, passage of the Gogra at Fyzabad, capture of Rampore Kussia and subsequent operations in Oudh.—Medal with clasp.
China, 1860.—Medal. Also operations against the Taepings in 1861.
N.-E. Frontier of India, Bhootan, 1864-65.—Medal.

Blenkinsop, F. L.—
Tirah, 1897-98.—Medal with 2 clasps.
China, 1900.—Relief of Pekin.—Medal with clasp.
East Africa, 1903-04.—Operations in Somaliland Action at Jidballi. Despatches, *London Gazette*, 2nd September 1904.—Medal with 2 clasps.

Bliss, J. A.—
N.-E. Frontier of India, Chin Hills, 1892-93.—Medal with clasp.
N.-W. Frontier of India, 1897-98.—Operations on the Samana and in the Kurram Valley during Aug. and Sep. 97.—Medal with 2 clasps.
Tirah, 1897-98.—Actions of Chagru Kotal and Dargai. Operations in the Bazar Valley, 25 to 30 Dec. 97.—Clasp.
South African War, 1900—1902.—Adjutant, 22 Bn., Imperial Yeomanry, from 24 Mar. 01 to 23 Jan. 02. Operations in Cape Colony between 25 and 30 April 01. Operations in the Orange River Colony, between 1 May 01 and 20 Nov. 01. Operations in the Transvaal, between 21 Nov. 01 and 8 Jan. 02.—Queen's Medal with 5 clasps.
Tibet, 1903-04.—Medal.
The War of 1914-18.—Despatches, *Lond. Gaz.*, 4 July 16 and 15 Aug. '17.

Blood, J. (*Surg. Maj., retd., Ind. Med. Serv.*)—
Afghanistan, 1878-79.—With the Kandahar and Thal Chotiali Field Forces.—Medal.

Blyth, F. A.—
N.-W. Frontier of India, Zhob Valley, 1884.
N.-W. Frontier of India, 1897-98.—Buner.—Despatches, G. G. O. 217 of 98.

Boddam, E. B. C.—
N.-W. Frontier of India, Zhob Valley, 1890.
N.-W. Frontier of India, Isazai, 1892.
Chitral, 1895.—Relief of Chitral.—Orderly Officer to G. O. C., 4th Brigade. Despatches, G. G. O. 998 of 95.—Medal with clasp.
N.-W. Frontier of India, 1897-98.—Tochi.—Clasp.
China, 1900—Relief of Pekin.—Medal with clasp.
The War of 1914-18.—Despatches, *Lond. Gaz.*, 21 June '16.

Boermel, R. A.—
N. W. Frontier of India, Hazara, 1891.—Medal with clasp.
Frontier of India, 1897-98.—Tochi. Medal with clasp.

Bogle, J. S.—
Chitral, 1895.—Relief of Chitral. Action of the Malakand Pass.—Medal with clasp.
N.-W. Frontier of India, 1897-98.—Malakand. Operations in Bajaur and in the Mamund country. Utman Khel.—Buner.—Clasp.
China, 1900.—Medal.
N.-W. Frontier of India, 1908.—Operations in the Mohmand country.—Medal with clasp.
The War of 1914-18.—Despatches, *Lond. Gaz.*, 21 June '16 and 6 July '17.

Boileau, E. R. P.—
Chitral, 1895.—Relief of Chitral.—Medal with clasp.
N.-W. Frontier of India, 1897-98.—Operations on the Samana during Aug. and Sep. 97. Relief of Gulistan.—2 clasps.

Tirah, 1897-98.—Actions of Dargai and of the Sampagha and Arbanga Passes. Operations in the Waran Valley and action of 16 Nov. 97. Operations at and around Dwatoi and action of 24 Nov. 97. Operations against the Khani Khel Chamkannis. Operations in the Bara Valley, 7 to 14 Dec. 97.—Clasp.
Tibet, 1903-04.—Medal.

The War of 1914-18.—Despatches, *Lond. Gaz.*, 22 June 15.—Brev. of Lt.-Col.

Boileau, L. M. (*Maj., retd., Ind. S.C.*)—
Afghanistan, 1878-79.—Medal.

Boisragon, G. H.—
N.-W. Frontier of India, Hazara, 1888.—Medal with clasp.
N.-W. Frontier of India, Hazara, 1891.—Clasp.
N.-W. Frontier of India, (1st) Miranzai, 1891.—Orderly Officer to G. O. C.
N.-W. Frontier of India, (2nd) Miranzai, 1891.—Clasp.
N.-W. Frontier of India, Hunza Nagar, 1891.—Capture of Nilt Fort. Despatches, G. G. O. 397 of 1892.—Clasp. V. C.
N.-W. Frontier of India, Waziristan, 1894-95.—Clasp.
N.-W. Frontier of India, 1897-98.—Operations on the Samana and in the Kurram Valley during Aug. and Sep. 97. Operations of the Flying Column in the Kurram Valley under Colonel Richardson, 20 Aug. to 1 Oct. 97.—Medal with 2 clasps.
Tirah, 1897-98.—Reconnaissance of the Kharmana Defile and action of 7 Nov. 97.—Operations against the Khani Khel Chamkannis.—Clasp.
The War of 1914-18.—Despatches, *Lond. Gaz.*, 21 June 16. Brev. of Col.

Bolton, H. J. (*Maj., retd., Ind. S. C.*)—
Afghanistan, 1878-80.—Capture of Ali Musjid, operations at and around Kabul in Dec. 79, march from Kabul to relief of Kandahar, and battle of 1 Sep. Mari expedition. Despatches, G. G. O. 137 of 80.—Medal with 3 clasps; bronze star.
N.-E. Frontier of India, Chin-Lushai Expedition, 1889-90.

Bond, G. W.—
Operations in the Abor Country, 1911-12.—Despatches, *Gazette of India*, Notifn. No. 480 of 1912.—Medal with clasp.
The War of 1914-18.—Despatches, *Lond. Gaz.*, 22 June 15, 19 Oct. 16. D. S. O.

Bond, C. R. A. (*Maj. retd., I. S. C.*)—
N.-W. Frontier of India, Isazai, 1892.
Soudan, 1896.—Dongola Expedition.—Medal.
N.-W. Frontier of India, 1897-98.—Malakand.—Operations in Bajaur and in the Mamund country.—Utman Khel.—Medal with clasp.

Bone, A. F.—
South African War, 1900-02.—King's medal with 2 clasps.

Bonham, J. (*Hon. Col., retd., Ben. Art.*)—
India, 1857.—(Three times wounded.) Action at Chinhut. Defence of Lucknow. Despatches, *Lond. Gaz.*, 16 Jan. 58.—Medal with clasp. Brev. of Major. Granted one year's service for Lucknow.

Bonnar, T. H.—
N.-W. Frontier of India, Hazara, 1888.—Medal with clasp.
Tirah, 1897-98.—Medal with clasp.

Bookey, J. T. B. (*Col. retd., I. M. S.*)—
N.-W. Frontier of India, Jowaki, 1877-78.—Meda with clasp.
N.-W. Frontier of India, Mahsud Waziri, 1881.
Burma, 1886-87.—Operations of the 2nd and 5th Brigades, including the Wuntho Expedition. Despatches, G. G. O. 434 of 1887.—2 clasps.
N.-W. Frontier of India, Hazara, 1888.—Despatches. G. G. O. 978 of '88.—Clasp.
N.-W. Frontier of India, (2nd) Miranzai, 1891.—Despatches, G. G. O. 632 of 91.—Clasp.
N.-W. Frontier of India, Waziristan, 1894-95.—Despatches, G. G. O. 473 of '95.—Clasp.
N.-W. Frontier of India, 1897.—Malakand, Despatches, *Lond. Gaz.*, 5 Nov. 97.
China, 1900.—Despatches, *Lond. Gaz.*, 14th May 01. - Medal with clasp. C. B.
N.-W. Frontier of India, 1902.—Operations against the Darwesh Khel Waziris, Despatches, *Lond. Gaz.*, June 03.

Boome, E. H.—
N.-E. Frontier of India, Chin-Lushai, 1889-90.—Medal with clasp.
China, 1900.—Medal.
Tibet, 1903-04.—Medal.
The War of 1914-18.—Despatches, *Lond. Gaz.*, 19 Oct. 16 and 15 Aug. 17. Brev. of Col. C. M. G.

Booth, H. C. (*Asst. Comsy. and Hon. Lt. retd.*)—
N.-W. Frontier of India, Hazara, 1891.—Medal with clasp.
N.-W. Frontier of India, Isazai, 1892.
N.-W. Frontier of India, Waziristan, 1894.—Action at Wana, 3 Nov. '94. Clasp.
Chitral, 1895.—Medal with clasp.
N.-W. Frontier of India, 1897-98.—Malakhand. Action at Landakai. Operations in Bajaur and in the Mamund country. Buner. Action of the Tanga Pass.—2 clasps.
Promoted Sub-Conductor.
Tibet, 1903-04.—Action at Naini, 26th June 1904. Operations at and around Gyantse, between 5th May and 6th July 1904. March to Lhassa, 14th July to 3rd August 1904. Despatches, *Gazette of India*, No. 1065 of 1904.—Medal with clasp.
N.-W. Frontier of India, 1908.—Operations in the Mohmand country. Engagement of Matta, 24th April 1908. Engagement of Kargha, 24th May 1908.—Despatches.
Gazette of India, No. 521 of 1908.—Medal with clasp. Promoted Assistant Commissary.

Borradaile, H. B. (*Col., retd., I. A.*)—
Burma, 1885—89.—Medal with 2 clasps.
Chitral, 1895.—Relief of Chitral. Despatches, G. G. O. 530 of 95.—Medal with clasp. Brev. of Maj. D.S.O.

Boswell, W. L. (*Maj., retd., I. S. C.*)—
Burma, 1886—88.—Operations of the 4th Brigade. —Medal with clasp.
N.-W. Frontier of India, Waziristan, 1894-95.—Clasp.

Bott, R. H.—
N.-W. Frontier of India, 1908.—Operations in the Zakka Khel country.—Medal with clasp.

Boudier, E. W.—
East Africa, 1902—04.—Operations in Somaliland —Medal with clasp.

Boulderson, H. A. B. (*Lt.-Col., retd., I. A.*)—
Afghanistan, 1878-79-80.—March on Kandahar and Kelat-i-Ghilzai. March from Kandahar to Ghuzni and Kabul. Actions of Ahmed Khel and Urzu. March from Kabul to Kandahar. Battle of Kandahar.—Medal with 2 clasps; bronze star.
Burma, 1886-87.—Operations of the 3rd Brigade.—Medal with clasp.

Boulton, H.—
East Africa, 1901.—Operations in Somaliland. Despatches, *London Gazette*, 18th April 1902. —Medal with clasp.
The War of 1914-18.—Despatches, *Lond. Gaz.*, 22 Jun. 15, 1 Jan. and 19 Oct. 16. Brev. of Lt.-Col.

Bourke, J. J.—
N.-W. Frontier of India, Waziristan, 1894-95. —Medal with clasp.
China, 1900.—Medal.

Bourne, W. FitzG.—
Chitral, 1895.—Relief of Chitral.—Medal with clasp.

Bourne, W. K.—
The War of 1914-18.—Despatches, *Lond. Gaz.*, 22 Jun. 15 and 15 Jun. 16. Brev. of Lt.-Col.

Boustead, R. (*Surg.-Maj., retd., Ind. Med. Serv.*)—
Crimea, 1855-56.—Served with Turkish Contingent and in medical charge of Osmani Irregular Cavalry 5th class Medjidie.
India, 1857-59.—Served with the Field Force against the Rohillas, and in the Central India Campaign, including pursuit of Tantia Topee, Rao Sahib, Feroze Shah, Chattur Saol, Adil Mahomed Khan, Nawab of Banda, etc. Also served against the Bheels in the Ahmednagar District, and the mutineers on the Bheema-Krishna Frontier.—Medal with clasp.
China, 1860.—In medical charge of troops at Hong-Kong, Canton, and Wamphao.
Abyssinia, 1867-68.—In medical charge of 1st Balooch Regt. and subsequently as Chief Sanitary Officer at Zoula.—Medal.
Soudan, 1885.—Suakim. Action at Hasheen, outside Zariba at Tofrek, and action at and burning of Tamai.—Medal with clasp; bronze star.

Bowder, W. E.—
South African War, 1899-1900.—Defence of Ladysmith.—Medal with clasp.

Bower, Sir H. (*Maj. Genl., retd., I. A.*)—
Soudan, 1896.—Dongola Expedition. D. A. Q. M. G., Intelligence Branch, Suakin.—British and Egyptian medals. Brev. of Maj.
China, 1900.—Relief of Tientsin. Relief of Pekin. Commanding Chinese Regiment. Despatches, *Lond. Gaz.*, 14 May 01.—Medal with clasp. Brev. of Lieut.-Col.
Operations in the Abor Country, 1911-12.—In command. Despatches, *Lond. Gaz.*, 16 July 12. K. C. B.

Bowle-Evans, C. H.—
N.-W. Frontier of India, Waziristan, 1894-95.—Medal with clasp.
Chitral, 1895.—Relief of Chitral. Action of the Malakand Pass.—Medal with clasp.
N.-W. Frontier of India, 1897-98.—Tochi.—Clasp.
China, 1900.—Medal.
The War of 1914-18.—Despatches, *Lond. Gaz.*, 22 Jun. 15. C. M. G.

Bown, A. T. (*Lieut-Col., retd., I. M. S.*)—
N.-W. Frontier of India, Hazara, 1888.—Medal with clasp.
N.-W. Frontier of India, Hazara, 1891.—Clasp.
Chitral, 1895.—Relief of Chitral. Action of the Malakand Pass. Action at Khar.—Medal with clasp.
N.-W. Frontier of India, 1897-98.—Present at Fort Jamrud.—Clasp.

Bowring, G. (*Lt.-Col., retd., I. A.*)—
Burma, 1887—89—92.—Medal with 2 clasps.

Boxwell, A.—
South African War, 1901-02.—Operations in the Transvaal, Mar. 01 to 13 May 02.—Queen's medal with 3 clasps.
The War of 1914-18.—Despatches, *Lond. Gaz.*, 19 Dec. 16. C. I. E.

Boyce, C. J.—
South African War, 1901-02.—Operations in the Transvaal, October 1901 to 31 May 1902.—Queen's medal with 3 clasps.
The War of 1914-18.—Despatches, *Lond. Gaz.*, 15 Aug. 17.

Boyd, H. (*Surg. Lt.-Col., retd., Ind. Med. Serv.*)—
N.-W. Frontier of India, Jowaki, 1878-78.—Medal with clasp.
Afghanistan, 1878-79.—Capture of Ali Musjid, and operations in the Khyber.—Medal with clasp.
N.-W. Frontier of India, Hazara, 1888.—Medal with clasp.

Boyd, I. M. (*M.-Gen., u.s.l., Ind. S.C.*)—
India, 1858.—Operations in Khandesh.
Abyssinia, 1868.—Despatches, *Lond. Gaz.*, 30 June '68.—Medal.
Afghanistan, 1879-80.—March from Quetta to the relief of Kandahar. Despatches. G. G. O. 693 of '80.—Medal.

Boyd, O. T.—
The War of 1914-18.—Despatches, *Lond. Gaz.*, 19 Aug. 16. M. C.

Boyle, Hon. R. E. (*Col., u. s. l., I.A.*)—
India, 1857—59.—Operations in Delhi District and Bundelcund. Thanked by Govts. of India and N.-W. Provinces.—Medal.
Afghanistan, 1878-79.—Capture of Ali Musjid. Mentioned in Despatches.—Medal with clasp.

Brackenbury, W. J. W.—
South African War, 1902.—Queen's medal with 3 clasps.

Bradfield, E. W. C.—
N.-W. Frontier of India, 1908.—Operations in the Mohmand country—Medal with clasp.

Bradley, C. R. S.—
South African War, 1901-02.—Operations in Cape Colony, Jan. 01 to 31 May 02.—Queen's medal with 2 clasps.
The War of 1914-18.—Despatches, *Lond. Gaze.*, 1 Jan. and 19 Oct. 16.

Bradley, F. G. (*Lt.-Col., retd., I. A.*)—
Burma, 1885—87.—Medal with clasp.

Bradley, R. J.—
N.-W. Frontier of India, 1908.—Medal with clasp.

Bradshaw, C. R.—
N.-W. Frontier of India, 1897-98.—Mohmand.—Medal with clasp.
Tirah, 1897-98.—Operations in the Bara Valley, 7 to 14 Dec. 97.—Clasp.
The War of 1914-18.—Despatches, *Lond. Gaz.*, 4 July 16.

Bradshaw, L. J. E. (*Maj.-Genl., retd., I. A.*)—
Afghanistan, 1878-79-80.—(2nd) Bazar Valley Expedition. Affairs with the Ghilzais at and near Jugdalak. March from Kabul to the relief of Kandahar and battle of Kandahar. Despatches, G. G. Os. 177 and 327 of 80.—Medal with clasp. Bronze star.
N.-W. Frontier of India, Hazara, 1891.—Despatches, G. G. O. 816 of 91.—Medal with clasp.
N.-W. Frontier of India, Hunza-Nagar, 1891.—Despatches. G. G. O. 397 of 92.—Clasp.
British Central Africa.—Operations against slave traders, 1895. Mentioned in Despatches (India Office letter, 29 Sep. 96). Brev. Lt.-Col.—Medal with clasp.

Soudan, 1896.—Dongola Expedition.—Queen's medal and Khedive's medal.
N.-W. Frontier of India, 1897-98.—Relief of Chakdara. Malakand. Operations in Bajaur and Mamund country Utman Khel. Despatches, G. G. O. 89 of 97.—Medal with 2 clasps.

Brander, H. R. (*Col., retd., I. A.*)—
N.-W. Frontier of India, Hazara, 1891.—Medal with clasp.
Chitral, 1895.—Relief of Chitral.—Medal with clasp.
N.-W. Frontier of India, Waziristan, 1901-02.—Clasp.
Tibet, 1903-04.—Action at Niani, 26th June 1904. Operations at and around Gyantse, between 5th May and 6th July 1904. March to Lhassa, 14th July to 3rd August 1904. Despatches. *Gazette of India*, No. 1065 of 1904.—Medal with clasp. C.B.

Brandon, J. (*Sr. Asst. Surgn. Hon. Capt., retd. I. S. M. D.*)—
Afghanistan, 1880.—Medal.
China, 1900.—Medal.

Braneon, C. E. D. (*Col., retd., Ben. Inf.*)—
N.-W. Frontier of India, 1863.—Umbeyla Campaign. —Medal with clasp.
N.-E. Frontier of India, Lushai, 1871 72.—Clasp.
Afghanistan, 1879-80.—Attack on Ali Khel, Zaimusht expedition, capture of Zawa, Hissarak expedition. Mentioned in Despatches.—Medal.
Soudan, 1885.—Suakin.—Medal with clasp; bronze star.

Branson, L. H.—
N.-W. Frontier of India, Waziristan, 1901-02.—Despatches. G. G. O. No. 611 of 1902.—Medal with clasp.
The War of 1914-18.—Despatches, *Lond. Gaz.*, 5 Apl. and 10 May 16.

Brassey, L. P.—
The War of 1914-18.—Despatches, *Lond. Gaz.*, 28 Jan. 16.

Bray, E. M.—
Soudan, 1885-86.—Action of Giniss.—Medal; Khedive's Bronze star.
N.-W. Frontier of India, Miranzai, 1891.—Medal with clasp.
N.-E. Frontier of India, Chin Hills, 1893-94.
N.-W. Frontier of India, 1897-98.—Malakand.—Medal with clasp.
China, 1900.—Relief of Pekin.—Medal with clasp. Promoted Conductor.
Operations in the Abor country 1911-12.—Medal with clasp.

Bray, N. N. E.—
The War of 1914-18.—Despatches, *Lond. Gaz.*, 17 Sept. 17.

Breakey, W. (*Commy. and Hony. Major retd., I.O.D.*)—
Afghanistan, 1880.—Medal.

Bredin, A. N.—
South African War, 1902.—Operations in the Transvaal, Mar. to 31 May 02. Operations in Orange River Colony, Feb. to Mar. 02. Despatches, *Lond. Gaz.*, 29 July 02.—Queen's medal with 5 clasps.

Bremner, C. E. V.—
The War of 1914-18.—Despatches, *Lond. Gaz.*, 8 and 13 Feb. 17. M.C.

Bremner, H. J. (*Bt.-Col., retd., I.A.*)—
Burma, 1885—88.—Medal with 2 clasps.

Brett, J. A.—
N. W. Frontier of India, 1915.—Operations in the Mohmand Swat and Buner countries. Served as Political Officer Despatches, *Lond. Gaz.*, 4 July 16.

Brewer, J. C.—
The War of 1914-18.—M.C.

Brickmann, F. I. O.—
The War of 1914-18.—Despatches, *Lond. Gaz.*, 5 Apl. and 13 July 16. M.C.

Bridge, J. H.—
The War of 1914-18.—Despatches, *Lond. Gaz.*, 22 Jun. 15.

Bridge, R. E. A.—
The War of 1914-18.—Despatches, *Lond. Gaz.*, 15 Aug. 17. D.S.O.

Bridges, A. H.—
South African War, 1899-1900.—Relief of Ladysmith, including operations of 17 to 24 Jan. 00 and 5 to 7 Feb. 00 and action at Vaal Krans; operations on Tugela Heights (14 to 27 Feb. 00 and action at Pieter's Hill (27 Feb. 00). Operations in Natal, Mar. to June 00, including action at Laing's Nek (6 to 9 June). Despatches, *Lond. Gaz.*, 10 Sep. 01. Brev. of Maj.—Queen's medal with 4 clasps.
British East Africa, 1901.—Operations against the Ogaden Somalis in Jubaland.—Medal with clasp.
The War of 1914-18.—Despatches, *Lond. Gaz.*, 4 Jan. 15 May and 11 Dec. 17. Brev. of Lt.-Col. D.S.O.

Brierley, C. I.—
South African War, 1901-02.—Operations in Cape Colony, Dec. 1901 to January 1902. Operations in Orange River Colony, Feb. 1902. Operations in the Transvaal, March to May 1902.—Queen's medal with 5 clasps.

Britton, J. A. H.—
South African War, 1901-02.—Operations in Cape Colony, 25th to 30th Oct. 1901. Operations in the Transvaal, 1st Nov. 1901 to 31st May 1902.—Queen's medal with 4 clasps.

Brock, B. de L.—
South African War, 1901-02.—Operations in Cape Colony, 9th August 1901 to 31st May 1902. —Queen's medal with 3 clasps.
The War of 1914-18.—Despatches, *Lond. Gaz.*, 22 Jun. 15. M.C.

Brockman, P. W. D. (*Lt.-Col., retd., I.A.*)—
N.-W. Frontier of India, 1897-98.—Tochi Brigade Transport Officer to 2nd Brigade, Tochi Field Force.—Medal with clasp.
China, 1900.—D. A. Q. M. G. Lines of Communication. China Field Force. Despatches, *Lond. Gaz.*, 13 Sept. 01.—Medal.

Bromhead, Sir B. P., Bt. (*Col., u.s.l., Ind. S.C.*)—
Afghanistan, 1878—80.—Both the Bazar Valley Expeditions. Mentioned in Despatches.—Medal.
Soudan, 1885.—Suakim.—Medal with clasp; bronze star.
N.-E. Frontier of India, Sikkim, 1888.—(Severely wounded, right hand amputated, horse wounded.) Mentioned in Despatches.—Medal with clasp; C.B.
N.-W. Frontier of India, Hazara, 1891.—Despatches, G. G. O. 816 of 91.—Clasp.

Bromilow, D. G.—
The War of 1914-18.—Despatches, *Lond. Gaz.*, 4 July 16.

Brook, T. F.—
South African War 1900-02.—Operations in Cape Colony between 2nd Jan. 1902 and 31st May 1902. Queen's Medal with 2 clasps.

Brooke, A. S.—
The War of 1914-18.—Despatches, *Lond. Gaz.*, 15 Aug. 17.

Brooke, H. M. M.—
Operations in the Abor country, 1911-12—Despatches, *Gaz. of India*, Notn. 480, dated 23rd May 12.—Medal with clasp.

Brooking, H. T.—
Burma, 1885—89.—Present at the taking of Minhla and occupation of Mandalay. Ruby Mines Expedition. Operations of the 2nd Brigade. Despatches. G. G. O. 295 of 86.—Medal with 2 clasps.
Chin Hills, 1892-93.—Operations against the Siyin and Nowengal Chin tribes.—Clasp.
N.-W. Frontier of India, 1897-98.—Operations on the Samana and in the Kurram Valley during Aug. and Sep. 97.—Medal with 2 clasps.
Tirah, 1897-98.—Actions of Chagru Kotal and Dargai. Operations in the Bazar Valley, 25 to 30 Dec. 97.—Clasp.
China, 1900.—Relief of Pekin. Served as D. A. Q. M. G., 1st Brigade. Despatches, *Lond. Gaz.*, 14 May 01.—Medal with clasp. Brev of Maj.
The War of 1914-18.—Despatches, *Lond. Gaz.*, 27 July 16, 14 Nov. 16, 19 Oct. 16, and 15 Aug. 17, Promd. Maj.-Genl. K.C.B., K.C.M.G.

Broome, A. P. (*Col., retd., Ben. Inf.*)—
N.-W. Frontier of India, 1869.—Expedition against Bizoti Urakzais and Thull Billam Khel Waziris.
Afghanistan, 1878-79.—Capture of Ali Musjid, action of Charasiah, operations at and around Kabul. Mentioned in Despatches.—Medal with 3 clasps.

Broome, G. S. (*Bt. Col., retd., I. A.*)—
Egypt, 1882.—Defence of Alexandria, occupation of Kafr Dowar, surrender of Damietta.—Medal; bronze star.

Broome, W. A. (*Col., retd., I. A.*)—
Afghanistan, 1878—80.—Actions at Takht-i-pulai u subsequent occupation of Kandahar and Girishk.—Medal.
Egypt, 1882.—Battle of Tel-el-Kebir. Forced march on, and seizure of, Zagazig, and occupation of Cairo.—Medal with clasp; bronze star.
Burma, 1886—88.—Operations against the Karen Tribes, 1887-88.—Medal with clasp. (*Karen Field Force, 1888-89.*)—Engagement at Nga-Kyaing. Forcing of the Nanko defile and capture of Sawlon.—Clasp.
Uganda, 1897-98.—Medal with clasp.
British East Africa Protectorate, 1898.—Operations in the Teita District.—In command.

Brown, E. H. (*Lt.-Col., retd., I. M. S.*)—
Tirah, 1897-98.—Medal with 2 clasps.

Brown, H. R.—
N.-W. Frontier of India, 1897-98.—Malakand. Operations in Bajaur and in the Mamund country, Utman Khel. Buner. Action of the Tanga Pass.—Medal with clasp.
China, 1900.—Medal.
The War of 1914-18.—Despatches, *Lond. Gaz.*, 5 Apl. 16.

Brown, H. H.—
The War of 1914-18.—Despatches, *Lond. Gaz.*, 15 Aug. 17.

Brown, H. T. (*Lt.-Col., retd., I. A.*)—
N.-E. Frontier of India, Lushai, 1889.—Medal with clasp.
Chitral.—Relief of Chitral.—Medal with clasp.
Tirah, 1897-98.—2 clasps.

Brown, R. J. R.—
Burma, 1885—89.—Operations of the 2nd and 5th Brigades.—Medal with 2 clasps.

Brown, S.—
Burma, 1885-89.—Served with Northern Frontier Flying Column.—Medal with 2 clasps.
N.-E. Frontier of India, Chin Lushai, 1889-90.—Served with Northern Chin Field Force.—Clasp.
N.-W. Frontier of India, 1897-98.—Operations on the Samana and in the Kurram Valley during August and September 1897.—Medal with 2 clasps
Tirah, 1897-98.—Clasp.
East Africa, 1903-04.—Operations in Somaliland.—Medal with clasp.

Brown, W.—
South African War, 1901-02.—Operations in Orange River Colony, Apr. 01 to 31 May 02. Queen's medal with 4 clasps.
The War of 1914-18.—Despatches, *Lond. Gaz.*, 19 Oct. 16, and 15 Aug. 17.

Brown, W. H.—
The War of 1914-18.—Despatches, *Lond. Gaz.*, 5 Apl. 13 July and 13 Oct. 16. C. B.

Browne, Sir A. G. F. (*Lieut.-Genl., retd., I. A.*)—
Afghanistan, 1878-79.—Medal.
Burma, 1886.—Operations of the 1st Brigade.—Medal with clasp.
N.-E. Frontier of India, Lushai, 1889.—Despatches. G.G.O. 592 of 89.—Clasp.
N.-E. Frontier of India, Chin-Lushai, 1889-90.—Despatches, G.G.O. 677 of 90.—Clasp. D.S.O.
Burma, 1892.—Clasp. Commanded the Tlantlang Column.
N.-E. Frontier of India, Chin Hills, 1892-93.—Clasp.
Chitral, 1895.—Relief of Chitral.—Medal with clasp.
Tirah, 1897-98.—Actions of the Sampagha and Arhanga Passes.—Reconnaissance of the Saran Sar and actions of the 9 and 11 Nov. 97 Operations against the Khani Khel Chamkannis. Operations in the Bazar Valley, 25 to 30 Dec. 97. Despatches G.G.O. 244 of 1898.—2 clasps. Brev. of Col.

Browne, A. J. W.—
China, 1900.—Medal.
The War of 1914-18.—Despatches, *Lond. Gaz.*, 5 Apl. 16.

Browne, A. P.—
N.-W. Frontier of India, 1897-98.—Operations in the Kurram Valley during Aug. and Sep. 97.—Medal with 2 clasps.
Tirah, 1897-98.—Operations against the Khani Khel Chamkannis.—Clasps.
China, 1900.—Relief of Pekin. Action of Peitsang. Action at Yangtsun. Despatches, *Lond. Gaz.*, 14 May 01.—Medal with clasp.
The War of 1914-18.—Despatches, *Lond. Gaz.*, 15 Jun. 16.

Browne, E. W.—
N.-W. Frontier of India, Waziristan, 1901-02.—Medal with clasp.
Aden, 1903-04.—Operations in the Interior.

Browne, H. E. (*Major, retd., I. A.*)—
East Africa, 1903-04.—Operations in Somaliland Action at Jidballi.—Medal with 2 clasps.

Browne, H. F. M.—
N.-W. Frontier of India, 1897-98.—Medal with clasp.
Tirah, 1897-98.—Operations in the Bara Valley, 7 to 14 Dec. 97.—Clasp.
N.-W. Frontier of India, 1908.—Operations in the Mohmand country.—Medal with clasp.

Browne, H. J. P.—
N.-W. Frontier of India, Waziristan, 1894-95.—Medal with clasp.
N.-W. Frontier of India, 1897-98.—Tochi.—Medal with clasp.—Operations on the Samana and in the Kurram Valley during Aug. and Sep. 97. Operations of the Flying Column in the Kurram Valley under Colonel Richardson, 20 Aug. to 1 Oct. 97.—Clasp.
Tirah, 1897-98.—Reconnaissance of the Kharmana Defile and action of 7th November 1897. Operations against the Khani Khel Chamkannis.—Clasp.

N.-W. Frontier of India, 1908.—Operations in the Mohmand country. Provost Marshal, 2nd Brigade.—Medal with clasp.
Burma, 1887-88.—Medal with clasp.
South African War, 1899-1900.—Acted as Senior Press Censor, Lines of Comm. from 28 Mar. 00. Afterwards as Comdt. of Prisoners of War (graded as an A. A. G.) from 13 Aug. 00. Subsequently Asst. Press Censor (graded as a D. A. A. G.), Lines of Comm., from 18 Oct. 00.—Queen's medal with 3 clasps.
The War of 1914-18.—Despatches, *Lond. Gaz.*, 5 Apl. 16, 19 Oct. 16 and 15 Aug. 17. Brevs. of Lt.-Col. and Col.

Browne, L. S.—
N.-W. Frontier of India, 1897-98.—Relief and defence of Malakand. Relief of Chakdara. Malakand.—Operations in Bajaur and in the Mamund country.—Medal with 2 clasps.
N.-W. Frontier of India, Waziristan, 1901-02.—Clasp.

Browne, S. H. (*Col., retd., I. M. S.*)—
Afghanistan, 1878-79.—Mentioned in Despatches. Thanks of the Govt. of India. (G. O. C. C., 14 October 1879.)—Medal.

Browne, St. J. A.—
N.-W. Frontier of India, 1908.—Operations in the Zakka Khel country.—Medal with clasp.

Browne, W. (*Capt., retd., Ind. S. C.*)—
S. Africa, 1881.—Transvaal Campaign. Served with the Natal Field Force.
Burma, 1888-89.—Medal with clasp.
N.-W. Frontier of India, 1897-98.—Tochi.—Medal with clasp.

Browne, W. H. (*Lt.-Col., retd., Ind. S. C.*)—
N.-W. Frontier of India, 1863.—2nd Eusufzal Expedition. Umbeyla Campaign.—Medal with clasp.
Afghanistan, 1878-79.—Medal.

Browne, W. H. (*Lt.-Col., retd., I. A.*)—
Afghanistan, 1879-80.—With the Kurram Field Force.—Medal.
Burma, 1886-89.—Medal with 2 clasps.
N.-E. Frontier of India, Manipur, 1891.—Clasp.

Browne, W. R. (*Surg. Genl., retd., I.M.S.*)—
Rumpa Rebellion, 1879-80.

Browning, H. A. (*Lt.-Col., retd., I. A.*)—
Burma, 1886—89.—Medal with 2 clasps.

Browning, T. C.—
Tirah, 1897-98.—Medal with 2 clasps.

Brownlow, C. B. (*Col., retd., I. A.*)—
Afghanistan, 1878-79-80.—Medal.
N.-W. Frontier of India, Mahsud-Waziri, 1881.
N.-W. Frontier of India, (1st) Miranzai, 1891.
N.-W. Frontier of India, (2nd) Miranzai, 1891.—Medal with clasp.
N.-W. Frontier of India, Waziristan, 1901-02.—Medal with clasp.

Brownlow, C. C. (*Col., u. s. l., I. A.*)—
N.-W. Frontier of India, (2nd) Eusufzai, 1863.—Umbeyla.—Medal with clasp.
N.-W. Frontier of India, Utmanzai, 1866.
N.-W. Frontier of India, Dour Valley, 1892.
N.-W. Frontier of India, Jowaki, 1877-78.—Clasp.
Afghanistan, 1879.—Medal.
N.-W. Frontier of India, (1st) Miranzai, 1891.
N.-W. Frontier of India, (2nd) Miranzai, 1891.—Despatches, G. G. O. 632 of 91.—Clasp. Brev. of Col.

Brownlow, d'A. C.—
N.-W. Frontier of India, 1897-98.—Malakand, Utman Khel, Buner.—Action of the Tanga Pass.—Medal with clasp.
N.-W. Frontier of India, 1908.—Operations in the Mohmand country.—Medal with clasp.
The War of 1914-18.—Despatches, *Lond. Gaz.*, 5 Apl. 16, 19 Oct. 16 and 15 Aug. 17. C. M. G., C. I. E.

Brounlow, J.—
The War of 1914-18.—Despatches, *Lond. Gaz.*, 31 Oct. 17.

Browse, G.—
The War of 1914-18.—Despatches, *Lond. Gaz.*, 1 Jan. 16 and 15 Aug. 17. Brev. of Lt.-Col. D. S. O.

Bruce, A. McC. (*Col., u. s. l., Ben. Inf.*)—
N.-E. Frontier of India, Bhutan, 1865-66.—Medal with clasp.
N.-W. Frontier of India.—Besoti-Afridi Affair, 1869.—Capture of Garoh village. Special thanks of Government of India.
N.-W. Frontier of India, Miranzai, 1869.
N.-W. Frontier of India, Jowaki, 1877-78.—Clasp.
Afghanistan, 1879-80.—Chukmani and Zaimusht Expeditions and capture of Zawa.—Medal.
N.-W. Frontier of India, Mahsood Wuzeeree, 1881.—Mentioned in despatches.
N.-W. Frontier of India, Zhoo Valley, 1884.
N.-W. Frontier of India, Hazara, 1888.—Clasp.
N.-W. Frontier of India, (1st) Miranzai, 1891.—Commanded 1st column. Despatches, G. G. O. 632 of 1891.

Bruce, C. E.—
China, 1900.—Medal.

Bruce, *Hon'ble* C. G.—
Burma, 1889.—Medal with clasp.
N.-W. Frontier of India, Hazara, 1891.—Clasp.
N.-W. Frontier of India, (2nd) Miranzai, 1891.—Despatches.—Clasp.
N.-W. Frontier of India, Waziristan, 1894-95—Clasp.
Tirah, 1897-98.—Actions of the Chagru Kotal and of the Sampagha and Arhanga Passes. Reconnaissance of the Saran Sar and action of 11 Nov. 97. Operations at and around Dwatoi and action of 24 Nov. 97. Operations against the Khani Khel Chamkannis. Operations in the Bara Valley, 7 to 14 Dec. 97. Operations in the Bazar Valley, 25 to 30 Dec. 97. Despatches G. G. Os. 58 and 244 of 98.—Medal with 2 clasps. Brev. of Maj.
The War of 1914-18.—Despatches, *Lond. Gaz.*, 20 Sep. 15, 5 Nov. 15 and 21 Jun. 16 Brev. of Col.

Bruce, E. A. (*Col., u.s.l., Ind. S. C.*)—
Afghanistan, 1878-79-80.—As Executive Commissariat Officer with Vitakri and Lower Kurram Field Forces.—Medal.
Burma, 1885—88.—As Executive Commissariat Officer. Despatches, G. G. O. 339 of 87.—Medal with clasp; Brev. of Col.

Bruce, G. D.—
N.-W. Frontier of India, 1897-98.—Operations on the Samana and in the Kurram Valley during Aug. and Sep. 97.—Medal with 2 clasps.
Tirah, 1897-98.—Actions of the Chagru Kotal and Dargai.—Operations in the Bazar Valley, 25 to 30 Dec. 97.—Clasp.
China, 1900.—Medal with clasp.
The War of 1914-18.—Despatches, *Lond. Gaz.*, 1 Jan. 16, 4 Jan. 17 and 11 Dec. 17. Brev. of Lt.-Col. D. S. O.

Bruce, G. E.—
N.-W. Frontier of India, 1908.—Operations in the Zakka Khel country. Operations in the Mohmand country. Engagement of Matta, 24th April 1908. (Slightly wounded.)—Medal with clasp.
The War of 1914-18.—Despatches, *Lond. Gaz*, 19 Oct. 16. M.C.

Bruce, M. E. L.—
N.-W. Frontier of India, Waziristan, 1901-02. —Medal with clasp.
The War of 1914-18.—Despatches, *Lond. Gaz.*, 21 Jun 16.

Bruce, W. A. M.—(*Lt.-Col., retd., I.A.*)—
China, 1900.—Medal.

Brunskill, G. S.—
The War of 1914-18.—Despatches, *Lond. Gaz.*, 17 Feb. 15. M.C.

Bryson, R.—
Tirah, 1897-98.—Medal with 2 clasps.
China, 1900.—Medal.

Buchanan, A.—
N.-E. Frontier of India, Chin-Lushai, 1889-90.—Medal with clasp.

Buchanan, K. J.—
Burma, 1887.—Operations of the 2nd Brigade.—Medal with clasp.
N.-W. Frontier of India, 1892, Isazai Expedition.
Chitral, 1895.—Relief of Chitral. (Slightly wounded.)—Medal with clasp.
N.-W. Frontier of India, Waziristan, 1901-02.—Clasp.
N.-W. Frontier of India, 1908.—Operations in the Zakka Khel country. Despatches, G. G. O. No. 188 of 1908. Operations in the Mohmand country.—Medal with clasp; C.B.

Buchanan, W. J.—
N.-E. Frontier of India, Chin-Lushai, 1889-90.—Medal with clasp.

Buck, C. H.—
Chitral, 1895.—Relief of Chitral.—Medal with clasp'

Buckland, G. C. B.—
Operations in the Abor country 1911-12—Medal with clasp.
The War of 1914-18.—Despatches, *Lond. Gaz.*, 17 Feb. 15, 19 Oct. 16, D. S. O., M.C. Croix de Guerre.

Buckland, P. A. (*Lt.-Col., retd., I. A.*)—
Afghanistan, 1878-79.—Medal with clasp.

Budd, N. A. H.—
N.-E. Frontier of India, Chin-Lushai, 1889-90.—Medal with clasp.
East Africa, 1890.—Medal with clasp.
Tirah, 1897-98.—Medal with clasp.

Buist, A. H.—
N.-W. Frontier of India, 1897-98.—Tochi.—Medal with clasp.
The War of 1914-18.—Despatches, *Lond. Gaz.* 4 July 16.

Buist, D. S. (Lt.-Col., retd., *I. A.*)—
Burma, 1885-89.—Medal with 2 clasps.

Bull, G. H. (*Lieut.-Col., retd., I.M.S.*)—
Burma, 1885-89.—Medal with 2 clasps.
Soudan, 1896.—Dongola Expedition.—Medal. Also Egyptian Medal.

Buller, F. E.—
The War of 1914-18.—Despatches, *Lond., Gaz.*, 17 Feb. 15. M. C.

Buller, J. H.—
South African War, 1901-02.—Operations in Orange River Colony, May to June 01, and Jan. to Feb. 02. Operations in Cape Colony, June 01 to Jan. 02, and Feb. to 31 May 02.—Queen's medal with clasps.
The War of 1914-18.—Despatches, *Lond. Gaz.*, 22 Sep. 17. D. S. O.

Bunbury, W. E.—
Afganistan, 1880.—Medal.
N.-W. Frontier of India, Mahsud-Waziri, 1881.
N.-W. Frontier of India, Isazai, 1892.
Chitral, 1895.—Relief of Chitral, D. A. A. and Q. M. G., 1st Brigade.—Action of the Malakand Pass. Action at Khar.—Despatches, G. G. O. 998 of 95.—Medal with clasp.
N.-W. Frontier of India, Waziristan, 1901-02.—Despatches, G. G. O. 611 of 1902.—Clasp.

Burd, E.—
The War of 1914—1916.—Despatches, *Lond. Gaz.*, 4th July 1916.

Burden, H.—
Chitral, 1895.—Relief of Chitral.—Medal with clasp.
N.-W. Frontier of India, 1897-98.—Clasp.
Tirah, 1897-98.—Clasp.

The War of 1914-18.—Despatches, *Lond. Gaz.*, 21 Jun 16.

Burdett, E. W.—
The War of 1914-18.—Despatches, *London Gazette*, 13 July 16. M. C.

Burgess, A. J. (*Asst. Comy. and Hony. Lt., retd., I. O. D.*)—
N.-W. Frontier of India, Hazára, 1888.—Medal with clasp.
N.-W. Frontier of India, 1897-98.—Tochi.—Medal with clasp.

Burhe, J. T.—
The War of 1914-18.—Despatches, *Lond. Gaz.*, 30 Jun 16.

Burke. W. H. (*Lt.-Col., retd., I. M. S.*)—
Burma, 1886-87.—Despatches, G. G. O. 561 of 87.—Medal with clasp.

Burkitt, F. H.—
The War of 1914—18.—Despatches, *Lond. Gaz.*, 4 July 16.

Burlton, A. R.—
N.-W. Frontier of India, Waziristan, 1894-95.—Medal with clasp.
N.-W. Frontier of India, 1897-98.—Malakand. Operations in Bajour and in the Mamund country.—Utman Khel—Buner. Action of the Tanga Pass. Despatches. G. G. Os. 178 and 217 of 98.—Medal with clasp.
East Africa, 1902—04.—Operations in Somaliland. Despatches, *London Gazette*, 2nd September 1904.—Medal with clasp.

Burlton, P. S. M.—
N.-W. Frontier of India, Hazara, 1891.—Meda with clasp.

Burlton, W. E. F.—
Burma, 1885—89.—Operations of the 2nd and 5th Brigades.—Medal with 2 clasps.
N.-W. Frontier of India, 1897-98.—Malakand.—Medal with clasp.
Tirah, 1897-98.—Clasp.
N.-W. Frontier of India, 1908.—Operations in the Mohmand country. As Supply Officer, 1st Advanced Depôt, Shabkadar.—Medal with clasp.

Burn, A. G. (*Lieut.-Col., retd., I. A.*)—
Burma, 1887.—Medal with 2 clasps.

Burn, A. H.—
N.-W. Frontier of India, 1908.—Operations in the Zakka Khel country. Operations in the Mohmand country. Engagement of Kargha, 24th May 1908.—Medal with clasp.

Burn-Murdoch, I.—
Operations in the Abor country, 1911-12.—Medal with clasp.

Burne, F. O. N.—
The War of 1914-18.—Despatches, *Lond. Gaz.*, 15 Aug. 17.

Burne, K. P. (*Col., retd., I.A.*)—
Egypt, 1882.—Battle of Tel-el-Kebir.—Medal with clasp and bronze star.
Burma, 1886-87.—Despatches. G. G. O. 434 of 87.—Medal with clasp.
N.-W. Frontier of India, 1897-98.—Malakand.—Medal with clasp.
N.-W. Frontier of India, Waziristan, 1901-02.—Clasp.

Burne, N. A. K. (*Lt.-Col., retd., I. A.*)—
Burma, 1887-89.—Medal with clasp.
Burma, 1893-94.
Chitral, 1895.—Relief of Chitral.—Medal with clasp.
Tibet, 1903-04.—Action at Niani, 26th June 1904. Operations at and around Gyantse, between 5th May and 6th July 1904.—Medal with clasp.
The War of 1914-18.—Despatches, *Lond. Gaz.*, 4 Jan. 17.

Burnett, A. H.—
South African War, 1902.—Served in Mediterranean.—Medal.
The War of 1914-18.—Despatches, *Lond. Gaz.*, 19 Oct. 16 and 15 Aug. 17. D. S. O.

Burnett, R. F. D.—
The War of 1914-18.—Despatches, *Lond. Gaz.*, 2 Oct. 15, 1 Jan. 16 and 19 Oct. 16. M. C.

Burnett, S. H.—
N.-W. Frontier of India, 1897-98.—Medal with clasp.
Tirah, 1897-98.—Clasp.

Burrard, C. (*Major, retd., I. A.*)—
Tirah, 1897-98.—Medal with 2 clasps.

Burrows, H. M.—
N.-W. Frontier of India, 1908.—Operations in the Zakka Khel country. Operations in the Mohmand country.—Medal with clasp.

Burton, C. W. W. (*Lt.-Col., retd., I.A.*)—
Burma, 1888-89.—Expedition against the Karennees. As Staff Officer with the Lower Burma column Mentioned in Despatches.—Medal with clasp.
N.-E. Frontier of India, Chin-Lushai, 1889-90.—As Staff Officer to Brigadier-General Symons. Despatches. G. G. O. 677 of 90.—Clasp.

Burton, E. B.—
N.-W. Frontier of India, Mahsud-Waziri, 1881.
Egypt, 1882.—Action of Tel-el-Kebir.—Medal with clasp. Bronze star.
Burma, 1886-87.—Served as Transport Officer in the Shan States column, including advance to, and occupation of, Fort Stedman. Despatches. G. G. O. 434 of 87.—Medal with clasp.
Chitral, 1895.—Relief of Chitral. Served as Field Intelligence Officer.—Medal with clasp.

Burton, F. C. (*Col., retd., I. A.*)—
Afghanistan, 1878-79-80.—Services acknowledged by Government. Despatches. Lt.-Col. A. R. D. Mackenzie's Despatch, dated 1 Jan. 80, and G. G. O. 315 of 81.—Medal with clasp. Brev. of Maj.

Burton, F. N. (*Maj., retd., I. A.*)—
Burma, 1887—89.—Medal with clasp.

Burton, J. A. (*Lt-Col., retd., I.M.S.*)—
China, 1900.—Medal.

Burton, R. G.—
The War of 1914-18.—Despatches, *Lond. Gaz.*, 22 Sep. 17.

Butler, C. W.—
South African War, 1899-1902.—Operations in Cape Colony, South of Orange River, 99-00, including action at Colesberg (1 to 6 Jan.). Operations in the Orange Free State, Feb. to April 00. (Severely wounded.)—Queen's medal with 2 clasps.

Butler, H. M.—
Aden. 1903-04.—Operations in the Interior.
The War of 1914-18.—Despatches, *Lond. Gaz.*, 5 Apl. and 19 Oct. 16.

Butler, R. B.—
The War of 1914-18.—Despatches, *Lond. Gaz.*, 1 Jan. and 25 Sep. 16. Brev. of Maj. M. C.

Butterfield, E.—
The War of 1914-18.—Despatches, *Lond. Gaz.*, 5 Apl. 16. D. S. O.

Byers, C. B.—
N.-W. Frontier of India, Waziristan, 1901-02.—Medal and clasp.

Byrne, E. G. J.—
The War of 1914-18.—Despatches, *Lond. Gaz.*, 5 Apl. 16. Brev. of Maj.

C

Cadell, A.—
N.-E. Frontier of India, Chin-Lushai, 1889-90.—Medal with clasp.
N.-W. Frontier of India, Waziristan, 1894-95.—Clasp.
Soudan, 1896.—Dongola Expedition.—Medal. Also Khedive's medal. Brev. of Maj.
N.-W. Frontier of India, 1897-98.—Malakand. Staff Officer at the Base. Operations in Bajaur and in the Mamund country. Buner. Staff Officer at the Base. Despatches. G. G. O. 217 of '98.—Medal with clasp.

Cadell, J. G.—
N.-W. Frontier of India, Waziristan, 1901-02.—Medal with clasp.
N.-W. Frontier of India, 1908.—Operations in the Zakka Khel country.—Medal with clasp.
The War of 1914-18.—Despatches, *Lond. Gaz.*, 30 Jun 16, 8 Feb. 17 and 22 Sept. 17, D.S.O.

Cadell, J. M. (*Lt.-Col., retd., I. M. S.*)—
Burma, 1886—89.—Operations of the 4th Brigade, Mogoung column.—Medal with 2 clasps.
N.-W. Frontier of India, Hazara, 1891.—Clasp.
Tirah, 1897-98.—Medal with 2 clasps.

Cadell, Thos. (*Col. u.s.l. Ind. S.C.*)—
India, 1857.—Action of Budlee-ke-Serai, siege, storm, and capture of Delhi, operations in Delhi district. Oude campaign. Operations in Bundelcund. Thanks of Government. Despatches, *Lond Gaz.*, 22 Feb. 59.—Medal with clasp; *V. C.*
Operations against the Bheels, 1860-61.—Thanks of Government.

Cadge, W. H. (*Surg.-Lt.-Col., retd., I. M. S.*)—
Afghanistan, 1879.—With the Peshawar Valley Field Force. Mentioned in despatches.

Cahusac, A. N.—
The War of 1914-18.—Despatches, *Lond. Gaz.*, 19 Oct. 16, and 15 Aug. 17, D.S.O.

Cahusac, S. D. N.—
The War of 1914-18.—Despatches, *Lond. Gaz.*, 15 Aug. 17.

Caldeira, M. S. A. (*Sr. Asst. Surg. and Hony. Capt., retd., I. S. M. D.*)—
Tirah, 1897-98.—With Tirah Exdy. Force.—Medal with 2 clasps.

Calvert, J. T.—
N.-E. Frontier of India, Manipur, 1891.—Medal with clasp.
Tirah, 1897-98.—Medal with 2 clasps.

Cama, R. H. (*Lt.-Col., retd., I. M. S.*)—
Burma, 1885.—Present at the occupation of Mandalay and Bhamo.—Medal with clasp.
N.-E. Frontier of India, Chin-Lushai, 1889-90.—Medal.

Cameron, A.—
The War of 1914-18.—Despatches, *Lond. Gaz.*, 8 Feb. 17, 7 Mar. 18.

Cameron, C. (*Surg.-Lt.-Col., retd., Ind. Med. Serv.*)—
N.-E. Frontier of India, Bhootan, 1865.—Medal.

Cameron, C. S.—
South African War, 1900.—Served in Mediterranean.—Medal.
The War of 1914-18.—Despatches, *Lond. Gaz.*, 15 Aug. 17.

Cameron, J. P.—
East Africa, 1903-04.—Operations in Somaliland. Action at Jidbali.—Medal with 2 clasps.

Camilleri, J. M. (*Lt.-Col., retd., I.A.*)—
N.-E. Frontier of India, Manipur, 1891.—Medal with clasp.
N.-E. Frontier of India, Chin-Lushai, 1892-93.—Despatches, G. G. O. 733 of 1893.
N.-W. Frontier of India, Abor, 1894.
N.-W. Frontier of India, 1897-98.—Malakand. Action of Landakai.—Despatches G. G. O. 1089 of 97. Operations in Bajour and in the Mamund country. *Buner.* Action of the Tanga Pass. Despatches G. G. O. 217 of 98.—Medal with clasp.
Tirah, 1897-98.—Clasp.
South African War, 1899—1902.—Relief of Kimberley. Operations in the Orange Free State, Feb. to May 00, including operations at Paardeberg (17 to 26 Feb.). Actions at Poplar Grove (17 March), Driefontein (10 March), Vet River (5 and 6 May) and Zand River (May 10). Operations in the Transvaal in May and June 00, including actions near Johannesberg (29 May), Pretoria (4 June), and Diamond Hill (11 and 12 June). Operations in the Transvaal and Orange River Colony, June to Oct. 01, including action at Wittebergen (1 to 29 July 00). Despatches, *Lond. Gaz.*, 3 Dec. 01. Brev. of Maj.—Queen's medal with 7 clasps.

Campbell, A. A. E.—
N.-W. Frontier of India, Isazai, 1892.
N.-W. Frontier of India, Waziristan, 1894-95.—Medal with clasp.

Campbell, A. E. (*M.-Gen. u.s.l. Ind. S.C.*)—
India, 1855-56.—Operations against Sonthals.
India, 1857-58.—Numerous affairs in Saugor district (severely wounded).—Medal with clasp.

Campbell, A. J.—
Burma, 1889.—Medal with clasp.
Burma. 1891-92.—North-Eastern Column.
N.-E. Frontier of India, Chin Hills, 1893-94.—Expedition against the Pimpi Rebels.
The War of 1914-18.—Despatches, *Lond. Gaz.*, 31 Oct. 17, D.S.O.

Campbell, A. W. D. (*Lt.-Col., retd., Ind. S.C.*)—
Afghanistan, 1878—80.—Occupation of Kandahar and Khelat-i-Ghilzai, and operations in Yarkistan.—Medal.

Campbell, C. C. (*Col., retd., Ind. S.C.*)—
Burma, 1885-86-87.—Commanded Expedition to Kandat. Despatches, G. G. O. 295 of 86. D.S.O.

Campbell, C. F. (*Lieut.-Col., retd., I.A.*)—
N.-W. Frontier of India, 1897-98.—Operations on the Samana in the Kurram Valley during Aug. and Sep. 97.—Medal with 2 clasps.
Tirah, 1897-98.—Operations against the Kihan Khel Chamkannis.—Clasp.

Campbell, C. H.—
N.-W. Frontier of India, 1908.—Operations in the Mohmand country.—Medal with clasp.
The War of 1914-18.—Despatches, *Lond. Gaz.*, 4 July 16.

Campbell, C. P. (*Lt.-Col., retd., I. A.*)—
Bikanir Field Force, 1883, for the suppression of rebellion of Thakurs in Bikanir in Dec. 83.
Chitral, 1895.—Defence of Chitral. Severely wounded, 3 March 95. Despatches, G. G. O. 531 of 95.—Medal with clasp. Brev. of Maj.
N.-W. Frontier of India, 1897-98.—Operations in the Kurram Valley during Aug. and Sep. 97.—2 clasps.
Tirah, 1897-98.—Reconnaissance of the Kharmana Defile and action of 7 Nov. 97.—Operations against the Khani Khel Chamkannis.—Clasp.

Campbell, Chas. Walter (*Hon. Lt.-Col. retd., Ben. Cav.*)—
Afghanistan, 1879-80.—With Kurram Valley Field Force.—Medal.

Campbell, F.—
N.-W. Frontier of India, Hazara, 1888.—Assistant Superintendent, Signalling, 3rd Column.—Medal with clasp.
Chitral, 1895.—Relief of Chitral. Action of the Malakand Pass. Despatches twice. G. G. O. 998 of 95.—Medal with clasp. Brev. of Maj.
N.-W. Frontier of India, 1897-98.—Malakand. Operations in Bajaur and in the Mamund country, Utman Khel. Despatches, G. G. O.'s Nos. 1317 of 1897 and 178 of 1898. *Buner.* Despatches. G. G. O. No. 217 of 1898.—Clasp; D.S.O.
Tibet, 1903-04.—Operations at and around Gyantse, between 5th May and 6th July 1904. (Slightly wounded.) March to Lhassa, 14th July to 3rd August 1904. Despatches, *Gazette of India*, No. 1065 of 1904.—Medal with clasp. Brevet of Colonel.
The War of 1914-18.—Despatches, *Lond. Gaz.*, 4 July 16. K. C. B. Promd. Lt.-Gen.

Campbell, F. C. G.—
N.-W. Frontier of India, 1908.—Operations in the Zakka Khel country. Operations in the Mohmand country.—Medal with clasp.
The War of 1914-18.—Despatches, *Lond. Gaz.*, 17 Feb. 16 and 1 Jan. 16, M.C.

Campbell, H.—
Tirah, 1897-98.—Actions of Chagru Kotal, Dargai, and of the Sampagha and Arhanga Passes. Operations in the Waran Valley and action of 16 November 97. Operations in the Bara Valley, 7 to 14 Dec. 97.—Medal with 2 clasps.
China, 1900.—Medal.
The War of 1914-18.—D.S.O.

Campbell, John—
South African War, 1901-02.—Operations in Cape Colony, Dec. 01 to 31 May 02.
The War of 1914-18.—Despatches, *Lond. Gaz.*, 19 Oct. 16, M.C.

Campbell, L. W. Y.—
The War of 1914-18.—Despatches, *Lond. Gaz.*, 1 Jan. 16, 19 Oct. 16 and 15 Aug. 17, Brev. of Col. C.M.G.

Campbell, M. H. A.—
The War of 1914-18.—Despatches, *Lond. Gaz.*, 4 July 16.

Campbell, R. N. (*Col., retd., I. M.S.*)—
N.-E. Frontier of India, Naga Hills, 1879-80.—Despatches. G.G.O. 123 of 80.—Medal with clasp.

Campbell, R. N. B.—
N.-W. Frontier of India, 1908.—Operations in the Zakka Khel country.—Medal with clasp.
The War of 1914-18.—Despatches, *Lond. Gaz.*, 6 July 17.

Campbell, W.—
South African War, 1899—01.—Operations in Cape Colony from 30 Nov. 00 to 1 March 01. Operations in the Transvaal from 4 March 01 to 10 Nov. 01. (Severely wounded).—Queen's medal with 4 clasps.
Tibet, 1903-04.—Medal.
The War of 1914-18.—Despatches, *Lond. Gaz.*, 15 Jun. and 19 Oct. 16, D.S.O.

Campbell, W. N. (*Maj., retd., I. A.*)—
Burma, 1886-87.—Despatches. G. G. O. 434 of 87. —Medal with clasp.
The War of 1914-18.—Despatches, *Lond. Gaz.*, 19 Oct. 16, 14 Nov. 16 and 15 Aug. 17, C.M.G., D.S.O.

Cantor, C. H. (*Col. u.s.l. Ind. S.C.*)—
India, 1857.—Siege, assault, and capture of Delhi and subsequent operations.—Medal with clasp.
N.-E. Frontier of India, Bhootan, 1864-65.—Medal with clasp.

Capper, A. S.—
Tirah, 1897-98.—Medal with clasp.
The War of 1914-18.—Despatches, *Lond. Gaz.*, 15 Aug. 17, D.S.O.

Carden, H. S.—
The War of 1914-18.—Despatches, *Lond. Gaz.*, 5 Apl. 16.

Cardew, F. G. (*Major, retd., I. A.*)—
Burma, 1886-87.—Served as Brigade Intelligence Officer.—Medal with clasp.

Cardew, H. S.—
The War of 1914-18.—Despatches, *Lond. Gaz.*, 13 July and 19 Oct. 16, M.C.

Carey, C. W.—
N.-W. Frontier of India, 1897-98.—Operations on the Samana and in the Kurram Valley during Aug. and Sep. 97. Relief of Gulistan.—Operations of the Flying Column in the Kurram Valley under Colonel Richardson, 20 Aug. to 1 Oct. 97.—Medal with 2 clasps.
Tirah, 1897-98.—Clasp.
South African War, 1902.—Operations in Cape Colony from 1 Jan. to 31 May 02, Adjutant, 26th Battalion, Imperial Yeomanry. —Queen's medal with 2 clasps.

Carey, H. R. LeM. (*Lt.-Col. u.s.l. Ind. S.C.*)—
N.-W. Frontier of India, Hazara, 1868.—Medal with clasp.

Carey, J. E.—
South African War, 1901-02.—Operations in the Transvaal, Oct. 01 to 31 May '02.—Queen's medal with 5 clasps.

Carey, O. W.—
Burma, 1885—89.—Medal with 2 clasps.
N.-W. Frontier of India, 1897-98.—Operations on the Samana and in the Kurram Valley during Aug. and Sep. 97. Operations of the Flying Column in the Kurram Valley under Colonel Richardson, 20 Aug. to 1 Oct. 97.—Medal with 2 clasps.
Tirah, 1897-98.—Actions of Chagru Kotal, Dargai, and of the Sampagha and Arhanga Passes. Reconnaissance of the Saran Sar and action of 9 Nov. 97. Operations in the Waran Valley and action of the 16 Nov. 97.—Clasp.
N.-W. Frontier of India, Waziristan, 1901-02.—Clasp.
N.-W. Frontier of India, 1908.—Operations in the Zakka Khel country. Operations in the Mohmand country.—Medal with clasp.
The War of 1914-18.—Despatches, *Lond. Gaz.*, 19 Oct 16 and 15 Aug. 17. Brev. of Col., C.M.G.

Carey, T. de B.—
Tibet, 1903-04.—Action at Niani. Operations at and around Gyantse, March to Lhassa. Medal with clasp.
The War of 1914-18.—Despatches, *Lond. Gaz.*, 5 Apl. 16.

Carey, W. R. (*Captain Lt.-Col. tempy.*) *retd. I.A.*—
South African War, 1900.—Operations in Cape Colony, south of the Orange River, from 1 March to 9 July 00.—Queen's medal with clasp.
N.-W. Frontier of India, 1902.—Operations against the Darwesh-Khel Waziris.

Cargill, R. J.—
South African War, 1901.—Operations in the Transvaal between March 01 and June 01. Operations in Cape Colony between 17 Jan 01 and 27 Jan. 01.—Queen's medal with 3 clasps.

Carleton, H. A. (*Col., retd., I. A.*)—
Afghanistan, 1878-79-80.—Battle of Ahmed Khel. March from Kabul to the relief of Kandahar and battle of Kandahar.—Medal with 2 clasps. Bronze star.

Carleton, Henry Anthony—
N.-E. Frontier of India, Lushai, 1889-92.—Served with Kanhow Column.—Medal with clasp.
N.-E. Frontier of India, Chin Hills, 1892-93.—Clasp.
The War of 1914-18.—Despatches, *Lond. Gaz.*, 1 Jan. 16 and 15 Aug. 17, D.S.O.

Carlos, N. H. (*Sr. Asst. Surgn. and Hon. Lt., retd.*)—
East Africa, 1902-04.—Medal.

Carnegy, C. G.—
Burma, 1886—88.—Medal with 2 clasps.

Carnegy, G. P. O.—
The War of 1914-18.—Despatches, *Lond. Gaz.*, 5 Apl. 16, 19 Oct. 16 and 12 Jun. 17, D.S.O.

Carnegy, P. M.—
Afghanistan, 1878-79-80.—Action of Charasiah, affair of Doaba. Operations at and around Kabul. Despatches. Brigadier-General H. Macpherson's Despatch, dated 21 Nov. 79.—Medal with 2 clasps.
Burma, 1885-86.—Taking of Minhla, occupation of Mandalay, and subsequent operations of 2nd Brigade. (Severely wounded near Shwebo.) Despatches. G. G. O. 295 of 86.—Medal with clasp.
N.-E. Frontier of India, Chin-Lushai, 1889-90.—(Burma column.)—Clasp.
Burma, 1891.—Thetta column.—Clasp.
N.-E. Frontier of India, Manipur, 1891.—(Severely wounded at Bapam.) Despatches. G. G. O. 585 of 91.—Clasp.

Chitral, 1895.—Relief of Chitral.—Medal with clasp.
Tirah, 1897-98.—Actions of the Sampagha and Arhanga Passes.—Reconnaissance of the Saran Sar and actions of the 9 and 11 November 97.—Operations against the Khani Khel Chamkannis.—Operations in the Bazar Valley, 25 to 30 Dec. 97. Despatches G. G. O. 244 of 98.—2 clasps.
China, 1900.—Medal.
The War of 1914-18.—Despatches, *Lond. Gaz.*, 17 Feb. 15.

Carpendale, F M.—
The War of 1914-18.—Despatches, *Lond. Gaz.*, 15 Aug. 17.

Carpendale, J. M. (*Lieut.-Col., retd., I. A.*)—
Afghanistan, 1878-79-80.—Operations in the Lughman Valley.—Medal.

Carpendale, P. M.—(*Bt.-Col., retd., I. A.*)—
Afghanistan, 1880.—With Kurram Field Force.—Medal.
N.-W. Frontier of India, Mahsud-Waziri, 1881.
N.-W. Frontier of India, 1897-98.—Malakand, Utman Khel. Buner.—Action of the Tanga Pass.—Medal with clasp.
N.-W. Frontier of India, 1908.—Operations in the Mohmand country. Post Comdt. Dand.—Medal with clasp.

Carpendale, W. M. (*Lt.-Col., retd., I. A.*)—
Burma, 1885-87.—Medal with clasp.
The War of 1914-18.—Despatches, *Lond. Gaz.*, 22 Jun 15.

Carr-Calthrop, C. W. (*Col., retd., I. M. S.*)—
Afghanistan, 1879-80.—Affair of Ali Khel. Operations at and around Kabul.—Medal with clasp.
Soudan, 1896.—Dongola Expedition. Principal Medical Officer.—Medal; also Khedive's medal.

Carr-Hall, R. E.—
Tirah, 1897-98.—Medal with 2 clasps.
The War of 1914-18.—Despatches, *Lond. Gaz.*, 1 Jan. 16.

Carroll, B. M.—
South African War, 1901-02.—Operations in Orange River Colony, July 01 to 31 May 02. Queen's medal with 3 clasps.

Carroll, E. R. W. C. (*Lt.-Col., retd., I. M. S.*)—
Burma, 1886-87.—Operations of the 4th Brigade and on the Western Frontiers, Salin column.—Medal with clasp.

Carruthers, H. St. C. (*Col., retd., I. M. S.*)—
Afghanistan, 1879-80.—Medal.

Carruthers, J. G. T. (*Col. u.s.l. Ben. Inf.*)—
Abyssinia, 1867-68.—Medal.
Afghanistan, 1878-79.—Action of Matun.—Despatches, G. G. O. 2-A of 79.—Medal.

Carruthers, J. H.deW.—
The War of 1914-18. Despatches, *Gaz. of India*, Notfn. 757 of 15, *Lon. Gaz.*, 5 Apr. 16. M.C.

Carruthers, R. A.—
Chitral, 1895.—Relief of Chitral.—Medal with clasp.
N.-W. Frontier of India, 1908.—Operations in the Zakka Khel country. Operations in the Mohmand country. Engagement of Matta, 24 April 1908. Engagement of Kargha, 24 May 08. Despatches G. G. O. No. 521A of 1908.—C. B. Medal with clasp.
The War of 1914-18.—Despatches, *Lond. Gaz.*, 5 Aug. 15, 5 May 16, 13 July 16, 4 Jan. 17, 15 May 17, and 11 Dec. 17. C.M.G.

Carr-White, P.—
Burma, 1891-92.—Irrawaddy column.—Medal with clasp.
N.-W. Frontier of India, 1897-98.—Tochi—Medal with clasp.

Carson, W. P. (*Lt.-Col., retd., I. M. S.*)—
Afghanistan, 1880.—Medal.
Burma, 1885—87.—Medal with clasp.

Carter, E. J.—
Tirah, 1897-98.—Medal with 2 clasps.

Carter, G. L.—
Egypt, 1889.—Nile Expedition, against Wall-el-Ndgume.
N.-W. Frontier of India, 1897-98.—Operations on the Samana and in the Kurram Valley during Aug. and Sept. 97. Operations of the Flying Column in the Kurram Valley under Colonel Richardson, 20 Aug. to 1 Oct. 97.—Medal with 2 clasps.
Tirah, 1897-98.—Reconnaissance of the Kharmana Defile and action of 7 Nov. 97. Operations against the Khani Khel Chamkannis.—Clasp.
N.-W. Frontier of India, 1898-99.—Operations against the Para Chamkannis.
N.-W. Frontier of India, 1902.—Operations against the Darwesh-Khel Waziris.
The War of 1914-18.—Despatches, *Lond. Gaz.*, 4 July 16, and 31 Oct. 17.

Carter, J. F. C.—
N.-W. Frontier of India, Waziristan, 1901-02.—Medal with clasp.

Carter, R. M.—
Aden, 1903-04.—Operations in the interior.
N.-W. Frontier of India, 1908.—Operations in the Zakka Khel country. (Severely wounded).—Medal with clasp.
The War of 1914-18.—Brev. of Lt.-Col.

Carthew-Yorstoun, M. E. (*Lieut.-Col., retd., I. A.*)—
South Africa, Zulu Campaign, 1879.—Battle of Ulundi. Medal with clasp.
Soudan, 1896.—Dongola Expedition.—Medal, also Khedive's Medal.
Tirah, 1897-98.—Actions of Dargai and of Sanpagha Pass. Reconnaissance of the Saran-Sar. Operations in the Bara Valley, 7 to 14 Dec. 97. Affair of Shinkamar on 29 Jan. 98.—Medal with 2 clasps.
South African War, 1900.—In command of the New Zealand Hotchkiss Battery. Operations in the Orange Free State, May 00, including actions at Vet River (5 and 6 May) and Zand River, operations in the Transvaal in May and June 1900, including actions near Johannesburg and Pretoria.—Queen's Medal with 4 clasps.

Cartman, S. A.—
Burma, 1885—87.—Medal with clasp.
N.-E. Frontier of India, Chin-Lushai, 1889-90.—Clasp.

Caruana, A. J.—
N.-W. Frontier of India, Zhob Valley, 1890.
N.-W. Frontier of India, (1st) Miranzai, 1891.
N.-W. Frontier of India, (2nd) Miranzai, 1891.—Medal with clasp.
N.-W. Frontier of India, 1897-98.—Tochi.—Medal with clasp.—Operations on the Samana and in the Kurram Valley during August and Sep. 97.—Clasp.
Tirah, 1897-98.—Operations in the Bara Valley, 7 to 14 Dec. 97.—Clasp.

Carwithen, E. T.—
Tirah, 1897-98.—Medal with 2 clasps.

Cartwright, C. M.—
The War of 1914-18.—Despatches, *Lond. Gaz.*, 1 Jan. 16, 15 Jun 16, and 11 Dec. 17, C.B., C.M.G.

Cassels, G. R.—
Soudan, 1896.—Dongola Expedition.—Medal, also Khedive's Medal.
N.-W. Frontier of India, 1897-98.—Relief of Malakand. — Malakand. Operations in Bajaur and in the Mohmand country. — Dangerously wounded.—Medal with 2 clasps.
N.-W. Frontier of India, Waziristan, 1901-02.—Severely wounded.—Clasp.
N.-W. Frontier of India, 1902.—Operations against the Darwesh Khel Waziris.

Cassels, K. S.—
N.-W. Frontier of India, Waziristan, 1901-02.—Medal with clasp.

Cassels, R. A.—
The War of 1914-18.—Despatches. *Lond. Gaz.*, 5 Apr. 16, 10 May 16, 13 July 16, 19 Oct. 16, and 15 Aug. 17. Brevs. of Lt.-Col. and Col.

Casserly, J. H. (Major, retd., I. A.)—
China, 1900.—Medal.

Casson, W. F. S.—
The War of 1914-18.—Despatches, *Lond. Gaz.*, 17 Feb. 15, and 1 Jan. 16, D S.O.

Castor, R. H.—
Burma, 1887-89.—Medal with clasp.

Cathcart, G. E.—
The War of 1914-1918. Despatches, *Lond. Gaz.*, 4 July 16.

Catty, T. C.—
The War of 1914-18.—Despaches, *Lond. Gaz.*, 4 July 16.

Caulfeild, G. N. (*Bt. Col., retd., I. A.*)—
Burma, 1885-89.—Present at the taking of Minhla and occupation of Mandalay. Operations of the 1st Brigade.—Wuntho Field Force. In command of advance party, Northern column, in action at Kyauk-Ton-Lon.—Medal with 3 clasps.
N.-E. Frontier of India, Chin Hills, 1892-93.—Despatches, G. G. O. 733 of 93. Clasp. D. S. O.

Cawood, G. C. (*Lieut.-Col., retd., I. A.*)—
Burma, 1887-88.—Medal with clasp.

Chaldecott, A. W. (*Lt.-Col. retd., I. A.*)—
N.-W. Frontier of India, Wazirisian, 1894-95.—Action at Wana, 3 Nov. 94. Despatches, G. G. O. 268 of 95.—Medal with clasp.
N.-W. Frontier of India, 1897-98.—Tochi.—Medal with clasp.
N.-W. Frontier of India, Waziristan, 1901-02.—Despatches, G. G. O. 611 of 02.—Clasp.

Chalmers, R. C. H. (*Capt., retd., I. A.*)—
N.-W. Frontier of India, Hazara, 1888.—Medal with clasp.
China, 1900.—Medal.

Chamberlain, Sir N. F. FitzG. (*Col., retd., I.S.C.*)—
Afghanistan, 1878-79 80.—Action of the Peiwar Kotal and in the Sapari Pass. Operations in the Khost Valley. Action of Charasia. Operations at and around Kabul. March from Kabul to the relief of Kandahar and battle of Kandahar. Wounded. Despatches, G.G.O.'s. 1226 of 78, 1136 of 79 and 137 of 80.—Medal with 4 clasps. Bronze star.
Burma, 1886-87.—As D. A. A. and Q. M. Genl., 5th Brigade. Ruby Mines Expedition, including advance from Khannyat. Despatches. G. G. O. 434 of 1887.—Medal with clasp. Brev. of Lt.-Col.
South African War, 1899-1900.—On Staff (as Priv. Sec. to F. M. Comg.-in-Chf. the Forces, 23 Dec. 99 to Aug. 00). Operations in the Orange Free State Feb. to May 00, including operations at Paardeberg (17 to 26 Feb.); actions at Poplar Grove, Dreifontein, Vet River (5 and 6 May) and Zand River. Operations in the Transvaal in May and June 00, including actions near Johannesburg, Pretoria and Diamond Hill (1 to 12 Jun.). Operations in the Transvaal, east of Pretoria July 00, Despatches, *London Gazette*, 8 Feb. 01. Queen's medal with 5 clasps. C. B.

Chamberlayne, A.—
South African War, 1899-1901.—Operations in the Orange Free State, April and May 00. Operations in the Transvaal. East of Pretoria, Sept. to Nov. 00. Operations in the Transvaal, West of Pretoria, July and Aug. 00, including action at Zilihats Nek (2 Aug.) Operations in Orange River Colony (May to July 00). Operations in Cape Colony, South of Orange River, 00. Operations in Transvaal, Nov. 00 to June 01.—Queen's medal with 4 clasps.

Chambers, G. H.—
The War of 1914-18.—Despatches, *Lond. Gaz.*, 24 Dec. 17.

Chambers, P. R.—
The War of 1914-18.—Despatches, *Lond. Gaz.*, 13 July 16, Brev. of Maj. D. S. O.

Chambers, R. M. (*M.-Gen. u.s.l. Bo. Inf.*)—
India, 1860.—Pursuit of Wagheers in Kattywar.
China.—Campaign against the Taepings, attack on the Camp at Wang-Kaza, capture of Tseerpoo and Nazeau, storming of Tsingpoo, defence of Kahding, and minor affairs.—Medal.
Afghanistan, 1880.—Medal.

Chamier, F. E. A. (*M.-Gen. u.s.l. Ind. S.C.*)—
India, 1857-58-59.—Defence, siege and capture of Lucknow. Despatches, *Lond. Gaz.*, 17 Feb. and 25 May 58, and 31 Jan. 59.—Medal with 2 clasps; Brev. of Maj.

Chamier, J. A.—
The War of 1914-18.—Despatches, *Lond. Gaz.*, 21 Jun. 16 and 4 Jan. 17, D. S. O.

Chamier, R. O.—
The War of 1914-18.—Despatches, *Lond. Gaz.*, 13 July and 19 Oct. 16.

Charnier, W. St. G.—
China 1900.—Relief of Pekin.—Medal with clasp.

Champion, de Crespigny. H.—
The War of 1914-18.—Despatches, *Lond. Gaz.*, 21 Jun. 16 and 4 Jan. 17. M. C.

Chandler, H. G. W.—
Tirah, 1897-98.—Medal with 2 clasps.

Channer, B. G.—
N.-W. Frontier of India, 1908.—Operations against the Darwesh Khel Waziris.
N.-W. Frontier of India, 1908.—Operations in the Zakka Khel country. Operations in the Mohmand country.—Medal with clasp.

Channer, G. K.—
N.-W. Frontier of India, 1897-98.—Operations on the Samana and in the Kurram Valley during Aug. and Sep. 97. Operations of the Flying Column in the Kurram Valley under Colonel Richardson, 20 Aug. to 1 Oct. 97.—Medal with 2 clasps.
Tirah, 1897-98.—Clasp.
N.-W. Frontier of India, Waziristan, 1901-02.—Despatches, G. G. O. 611 of 02. Clasp.

Channer, G. O. De. R.—
The War of 1914-18.—Despatches, *Lond. Gaz.*, 5 Apr. and 19 Oct. 16.

Channer, O. H. (*Lieut.-Col., retd., I.M.S.*)—
Afghanistan, 1879-80.—Despatches. G. G. O. 395 of 79.—Medal.

Chapman, A. R. (*Col. u.s.l., Ind. S.C.*)—
India, 1857-58.—Action of Budlee-ke-Serai, siege, assault and capture of Delhi, actions of Narnoul, Gungaree, Puttialee and Mynpoorie, storm and capture of Lucknow, and operations in Oude.—Medal with 2 clasps.
Afghanistan, 1878-80.—Medal.

Chapman, F. R. H. (*Maj., retd., Ind. S.C.*)—
Afghanistan, 1880.—Medal.

Chapman, H. (*M.-Gen., u.s.l. Ind. S.C.*)—
India, 1857-58-59.—Outbreak at Meean Meer, affairs of Narnoul, Gungaree, Puttialee and Mynpoorie. In Etawah District, at affairs of Oreyah and Deweah, operations in Rajpootana and Bundelkund (horse twice wounded). Despatches, *Lond. Gaz.*, 5 May 59.—Medal.
Afghanistan, 1878-80.—Action of Baghao and expedition against the Achakazai Tribes (horse wounded). Despatches, G. G. O. 610 of 79.—Medal. Brev. of Lt.-Col.

Chatterton, B. R.—
N.-W. Frontier of India, 1897-98.—Operations on the Samana. Medal and 2 clasps.
Tirah, 1897-98. Clasp.
The War of 1914—18.—Despatches, *Lond. Gaz.*, 4 July 16.

Chatterton, F. W. (*Col. u.s.l. Ben. Inf.*)—
N.-W. Frontier of India, 1863.—2nd Eusofzai Expedition. Forcing of the Umbeyla Pass.—Medal with clasp.
N.-E. Frontier of India, Bhutan, 1865-66.—(Slightly wounded.)—Clasp.

Chatterton, G. D. L.—
N.-W. Frontier of India, Waziristan, 1894-95.—Medal with clasp.
Soudan, 1896.—Dongola Expedition.—Medal; Khedive's medal.
N.-W. Frontier of India, 1897-98—Tochi.—Medal with clasp.
China, 1900.—Relief of Pekin.—Medal with clasp
The War of 1914-18.—Despatches, *Lond. Gaz.*, 5 Apr. and 13 July 16. D.S.O.

Chodhuri, M. N.—
The War of 1914-18.—Despatches, *Lond. Gaz.*, 15 Aug. 17.

Chahncy, C. H. K.—
The War of 1914-8.—Despatches, *Lond. Gaz.*, 19 Oct. 16 and 15 Aug. 17.

Chaytor-White, J. (*Lieut.-Col., retd. M. S.*)—
Chitral, 1895.—Relief of Chitral.—Medal with clasp.

Cheeke, A. M.—
The War of 1914-18.—Despatches, *Lond. Gaz.*, 5 Apr. 16.

Chenevix-Trench, R. H.—
China, 1900.—Relief of Pekin.—Medal with clasp.

Cherry, F. S. (*Col., retd., Mad. Cav.*)—
India, 1858-59.—Actions of Shahabad, Burhampore, Kheree, Saugor, Jugdespore. See Kreta and Koatach Khaz.—Medal with clasp.
Afghanistan, 1878-79.—Asst. Supdt. of Transport, Peshawar Valley Field Force.

Chesney, K. (*Lt.-Col., retd., I.A.*)—
Burma, 1885-87.—Operations of the 3rd Brigade. Despatches, G. G. O. 434 of 87.—Medal with clasp.
N.-W. Frontier of India, Zhob Valley, 1890.
N.-W. Frontier of India, 1897-98.—Operations of the Samana and in the Kurram Valley during Aug. and Sep. 97. Operations of the Flying Column under Colonel Richardson, 20 Aug. to 1 Oct. 97.—Medal with 2 clasps.
Tirah, 1897-98.—Clasp.
South African War, 1900-02.—Operations in Natal, April to June 1900, including action at Laing's Nek (6th to 9th June). Operations in the Transvaal, east of Pretoria, August to 29th November 1900, including actions at Belfast (26th and 27th August 1900), and Lydenberg (5th to 8th September 1900). Operations in Cape Colony, Orange River, and the Transvaal, 1901-02.—Queen's medal with 3 clasps, King's medal with 2 clasps.

Cheyne, A. Y.—
N.-W. Frontier of India, 1897.—Operations against Mohmands; action near Shabkadr (slightly wounded—horse shot). Despatches. G. G. O. 1188 of 97. Medal with clasp.
South African War, 1902.—Operations in the Transvaal, from 28 to 29 May 02. Operations in the Orange River Colony, from 29 to 31 May 02.—Queen's Medal with 3 clasps.

Cheyne, H. B.—
N.-W. Frontier of India, 1897-98.—Malakand. Utman Khel, Buner. Action of the Tanga Pass.—Medal with clasp.
S. African War, 1899-1900.—Employed with Kitchener's Horse.
China, 1900.—Relief of Pekin. Action of Peitsang. Action at Yangtsun.—Medal with clasp.

Cheyne, R. E.—
N.-W. Frontier of India, 1897-98.—Tochi.—Medal with clasp.
South African War, 1899-1900.—Operations in the Orange Free State, Feb. to May 00, including operations at Paardeberg (17 to 26 Feb.), actions at Poplar Grove (17 March), Dreefontein (10 March), Karee Siding (29 March), Houtnek (1 May). Operations in Orange River Colony (May to 29 Nov. 00), including action at Wittebergen (1 to 29 July). Operations in the Transvaal in May and June 00, including actions near Pretoria (4 June), Diamond Hill (11 and 12 June). Operations in the Transvaal, west of Pretoria, July to 29 Nov. 00, including action at Elands River (4 to 16 August). Wounded. Despatches, *London Gazette*, 16 April 01. Brev. of Maj.—Queen's medal with 6 clasps.
The War of 1914-18.—Despatches, *Lond. Gaz.*, 25 May 17.

Chichester, A. G. de V.—
N.-W. Frontier of India, Waziristan, 1901-02.—Medal with clasp.
N.-W. Frontier of India, 1908.—Operations in the Zakka Khel country. Operations in the Mohmand country.—Medal with clasp.
The War of 1914-18.—Despatches, *Lond. Gaz.*, 19 Oct. 16.

Chitty, A. W.—
China, 1900.—Relief of Pekin.—Despatches, *Lond. Gaz.*, 14 May 01.—Medal with clasp.
The War of 1914—18. Despatches, *Lond. Gaz.*, 5 Apr. 16.

Chitty, E. R. I.—
The War of 1914-18.—Despatches, *Lond. Gaz.*, 15 Aug. 17.

Chitty, W. W.—
British East Africa, 1898.—Operations against Ogaden Somalis.—Medal with clasp.
Uganda, 1898-99.—Operations against Kabarega.—Clasp.
The War of 1914-18.—Despatches, *Lond. Gaz.*, 5 Apr., 13 July and 19 Oct. 16. Brev. of Col.

Chope, A. J. H.—
Operations in the Abor country, 1911-12.—Medal with clasp.
The War of 1914-18.—Despatches. *Lond. Gaz.*, 21 June 16. D.S.O.

Christian, C.—
The War of 1914—18.—Despatches, *Lond. Gaz.*, 4 July 16.

Christison, Sir A. Bt. (*Surgn.-Genl. retd., I.M.S.*). Served with expedition to Burma, present at operations in vicinity and capture of Rangoon. April, 1852.—Medal with clasp.
India, 1857-58.—Capture of Gwalior.—Medal with clasp.

Christopher, C. deL.—
The War of 1914-18.—Despatches, *Lond. Gaz.*, 15 Aug. 17.

Christopher, L. W. (*Maj.-Gen., retd., I. A.*) —
Afghanistan, 1878-80. Mentioned in despatches. Medal.
Egypt, 1882.—Medal; bronze star.
Tirah, 1897-98.—Tirah Exped. Force. On Staff. Despatches, *Lond. Gaz.*, 5 Apl. 98. Medal with 2 clasps. C. B.

Christophers, S. R.—
The War of 1914-18.—Despatches, *Lond. Gaz.*, 15 Aug. 17.

Christie, B.—
Operations in the Abor country, 1911-12.—Medal with clasp.

Christie, R. C.—
The War of 1914-16.—Despatches, *Lond. Gaz*, 4 July 16 and 15 Aug. 17. M.C.

Chrystie, G. (*Col. u.s.l., I.A.*)—
India, 1858-59.—Operations in Saugor and Nerbudda Districts and in Bundelcund.—Medal.

Churchill, A. P. H.—
The War of 1914-18.—Despatches, *Lond. Gaz.*, 15 Aug. 17.

Churchill, G. R. D.—
Chitral, 1895.—Relief of Chitral.—Medal with clasp.
Tibet, 1903-04.—Medal.

Churchill, W. M.—
The War of 1914-18.—Despatches, *Lond. Gaz.*, 28 Jan. 16.

Clarke, A. C. K. S.—
The War of 1914—18.—Despatches, *Lond. Gaz.*, 15 Aug. 17.

Clarke, B. G. S.—
N.-W. Frontier of India, 1908.—Operations in the Zakka Khel country. Operations in the Mohmand country.—Medal with clasp.

Clarke, G. C. C.—
Tibet, 1903-04.—Medal.

Clarke, G. S.—
Tirah, 1897-98.—Operations in the Bara Valley, to 16 Dec. 97.—Medal with 2 clasps.
China, 1900.—Medal.
N.-W. Frontier of India, 1908.—Operations in the Mohmand country.—Medal with clasp.

Clarke, H. S. (*Hon. Col., retd.*)—
India, 1857.—Present at Cawnpore, and final capture of Lucknow.
N.-W. Frontier of India, 1863.—Operations in the Umbeyla Pass.—Medal with clasp.
N.-E. Frontier of India, Bhootan, 1865.—Clasp.

Clarke, R. C.—
The War of 1914-18.—Despatches, *Lond. Gaz.*, 15 Aug. 17. M.C.

Clark-Kennedy, A. H. (*Col. retd., I. A.*)—
Afghanistan, 1879-80.—Action of Kam Dakka.—Medal.
Burma, 1887-89.—Karennee Expedition.—Supdt., Comst.-Transport arrangements, Lower Column.—Medal and clasp.
N.-E. Frontier of India, Chin-Lushai, 1889-90.—Despatches. G. G. O. 449 of 90. Brev. of Maj.—Clasp.

Clarkson, F. C. (*Lieut.-Col., retd., I.M.S.*)—
N.-E. Frontier of India.—Chin-Lushai, 1889-90 (Lushai and Burma Column.)—Medal with clasp.
Burma, 1891.—(Thetta Column.)—Clasp.
N.-E. Frontier of India, Manipur, 1891.—Clasp.

Clay, C. H.—
Burma, 1886-87.—Operations of the 1st Brigade, Ruby Mines Expedition, including advance from Khanyat.—Medal with clasp.
N.-E. Frontier of India, Manipur, 1891.—Clasp.

Cleghorn, J. (*Surg.-Gen. ret., Ind. Med. Serv.*)—
N.-E. Frontier of India, Bhootan, 1865-66.—Medal with clasp.

Clementi, M. (*Col. retd., Ben. S.C.*)—
N.-W. Frontier of India, 1863.—Forcing of Umbeyla Pass.—Medal with clasp.
Egypt, 1882.—Dept. Judl. Adv. Ind. Cont. Battle of Tel-el-Kebir.—Medal with clasp; bronze star; 4th class, Osmanieh.

Clements, J. E.—
China, 1900.—Medal.

Clery, C. B. L.—
East Africa, 1902—04.—Operations in Somaliland. Action at Jidballi. Despatches, *Lond. Gaz.*, 2 Sep. 04.—Medal with 2 clasps.
The War of 1914-18.—Despatches, *Lond. Gaz.*, 5 Apr. 16 and 13 July 16, Brev. of Col. C.B.

Cleveland, H. F.—
• *N.-W. Frontier of India, 1897-98.*—Malakand. Mohmand.—Medal with clasp.
Tirah, 1897-98.—Clasp.
China, 1900.—Medal.

Cliff, W. (*Hony. Capt. Dy. Comy., retd., S. and T. Corps*)—
N.-E. Frontier of India, Chin-Lushai, 1889-90.—Medal with clasp.
N.-W. Frontier of India, (2nd) Miranzai, 1891.—Clasp.

Clifford, R.—
N.-W. Frontier of India, 1908.—Operations in the Mohmand country. Engagement of Kargha, 24th May 1908.—Medal with clasp.

Clifford, R. C.—
The War of 1914-18.—Despatches, *Lond. Gaz.*, 5 Apr. and 19 Oct. 16. D.S.O. M.C.

Clifford, R. C. R. (*Lt.-Genl. u.s.l., I.A.*)—
India, 1857-58-59.—Re-occupation of Oudh and Jugdispore.—Medal.
N.-W. Frontier of India, Jowaki, 1877.—Despatches, G. G. O. 758 of 78.—Medal with clasp.
Afghanistan, 1878-79-80.—Occupation of Kandahar. Battle of Ahmed Khel. Despatches. Generals *Sir* M. A. S. Biddulph and *Sir* D. M. Stewart, No. 117, 5 March 79, and No. 640K., 24 June 79, respectively, and G. G. O. 326 of 80.—Medal with clasp. Brev. of Lieut.-Col.
N.-W. Frontier of India, Mahsud-Waziri, 1881.—Despatches. Brigadier-General T. G. Kennedy's Despatch, 20 June 81.

Climo, S. H.—
N.-W. Frontier of India, 1897-98.—Defence of Malakand. Relief of Chakdara. Malakand. Operations in Bajaur and in the Mamund country. Despatches, G. G. O. 89 of 97.—Medal with 2 clasps. Brev. of Maj., D. S. O.
China, 1900.—Relief of Pekin. Action of Peitsang. Action at Yangtsun. Despatches, *Lond. Gaz.*, 14 May 01.—Medal with clasp.
Tibet, 1903-04.—Medal.
N.-W. Frontier of India, 1908.—Operations in the Mohmand country. Engagement of Kargha, 24th May 1908. Despatches, G. G. No. 521A of 1908.—Medal with clasp.
The War of 1914-18.—Despatches, *Lond. Gaz.*, 5 Apr., 21 June and 13 July 16. Brev. of Col. C.B.

Climo, V. C.—
Ashanti Expedition, 1895-6.—Star.
S. African War, 1901.—Spec. Serv. Offr. (for Mtd. Inf.) operations in Cape Colony, Apr. and May 01.
The War of 1914-18.—Despatches, *Lond. Gaz.*, 4 Jan. 17.

Clive, H. W. F.—
South African War, 1902.—Operations in the Transvaal, April and May 1902. Queen's medal with 3 clasps.
The War of 1914-18—Despatches, *Lon. Gaz.*, 4 July 16.

Cloete, H. D.—
South African War, 1902.—St. Helena. Queen's medal.
The War of 1914-18.—Despatches, *Lond. Gaz.*, 19 Oct. 16. M.C.

Cloete, I B.—
N.-W. Frontier of India.—Buner.—Action of the Tanga Pass.—Medal with clasp.

Close, J. K.—
N.-E. Frontier of India. Sikkim, 1888.—Attack of Jalapla.—Medal with clasp.
N.-E. Frontier of India, Manipur, 1891.—Clasp.

Coaker, V. A.—
The War of 1914-18.—Despatches, *Lond. Gaz.*, 1 Jun 17.

Coape-Smith, Henry (*M.-Gen. u.s.l. Ind. S.C.*)—
N.-W. Frontier of India, 1854-55.—Expedition to Shah Moosah Khel against the Momunds, 1854. Expedition against the Akka Khel and Bassee Khel Tribes, 1855. Mentioned in Despatches.—Medal with clasp.

Coates, S. B.—
The War of 1914-18.—Despatches, *Lond. Gaz.*, 1 July 16 and 19 Oct. 16. M.C.

Coates, V.—
N.-W. Frontier of India, Waziristan, 1901-02.—Medal with clasp.

Coates, W. (*Lieut.-Col., retd., I.M.S.*)—
Afghanistan, 1878-79-80.—Actions of Ali Musjid and Charasia. Operations at and around Kabul. March from Kabul to the relief of Kandahar, and battle of Kandahar. Despatches, G. G. O. 137 of '80.—Medal with 4 clasps. Bronze star.
N.-W. Frontier of India, Mahsud-Waziri, 1881.

Coats, G. H. B. (*Bt.-Col., retd., I. A.*)—
Afghanistan, 1878-79-80.—Action of Saif-u-din. Battle of Ahmed Khel. Action of Urzu. March from Kabul to the relief of Kandahar and battle of Kandahar.—Medal with 2 clasps. Bronze star.
N.-W. Frontier of India, Isazai, 1892.
Chitral, 1895.—Relief of Chitral.—Medal with clasp.
N.-W. Frontier of India, 1897-98.—Tochi.—Despatches, G. G. O. 1432 of '97.—Clasp.

Cobb, C.—
N.-W. Frontier of India, 1908.—Operations in the Mohmand country. Engagement of Kargha, 24th May 1908.—Medal with clasp.
The War of 1914-18.—Despatches, *Lond. Gaz.*, 21 Jun 16.

Cobbe, A. S.—
Chitral, 1895.—Relief of Chitral. Despatches. Lieut.-Col. J. G. Kelly's Despatch No. 8 C., 26 July '95.—Medal with clasp.
British Central Africa, 1898.—Southern Angoniland expedition.—Medal with clasp.
British Central Africa, 1899.—Expedition against Nkwamba. Secretary of State's Despatch No. 51, para. 19, 19th April 1900.—Medal with clasp.
West Africa, 1900.—Operations in Ashanti. Severely wounded. Despatches, *London Gazette*, 4th December 1900, and 8th March 1901.—Medal with clasp, D. S. O.
East Africa, 1902-04.—Operations in Somaliland, action at Jidballi. On Staff. Despatches, *London Gazette*, 7th August 1903. Despatches (Lieut.-Col. Swayne, 20th Feb. 1904, Brigadier-Genl. Manning, 17th August 1903, and Sir C. C. Egerton, 30th May 1904), *London Gazette*, 3rd Sep. 1904. Brevet of Lieut.-Col. 2 clasps. V. C.
The War of 1914-18.—Despatches, *Lond. Gaz.*, 17 Feb. 15, 22 Jun 15, 1 Jan. 16, 15 Jun 16 and 10 July 17. Promd. Maj. Gen. C.B. K.C.B.

Cobbe, H. H.—
Burma, 1895-96.
South African War, 1900-02.—Special Service Officer, afterwards on Staff, also as Comdt., Durbanville, and employed with Peninsula Horse. Relief of Mafeking. Operations in the Transvaal, May and June 1900. Operations in the Transvaal, east and west of Pretoria, July to 29th November 1900, including action at Zilikat's Nek (2nd August). Operations in Orange River Colony (May to 29th Nov. 1900). Operations in Cape Colony, south and north of Orange River, 1900. Operations in Cape Colony, 30th Nov. 1900 to 31st May 1902. Despatches, *London Gazette*, 16th April 1901. Queen's Medal with 3 clasps. King's medal with 2 clasps, D. S. O.
The War of 1914-18.—Despatches, *Lon. Gaz.*, 1 Jan. and 4 Jan. 17. C.M.G.

Cobham, H. W. (*Major, retd., I.A.*)—
South Africa, 1884-85.—Bechuanaland Expedition.
Tirah, 1897-98.—Action of the Sampagha Pass. Operations in the Waran Valley and action of 16 Nov. 97. Operations in the Bara Valley, 7 to 14 Dec. 97.—Medal with 2 clasps.
The War of 1914-18.—Despatches, *Lond. Gaz.*, 15 June 16, 4 Jan. 17, 15 May 17 and 11 Dec. 17 Brev. of Lt.-Col. D.S.O.

Cochran, G. W.—
The War of 1914-18.—Despatches, *Lond. Gaz.*, 5 Apr. 16, 13 July 16 and 15 Aug. 17. D.S.O.

Cochrane, A. W. R.—
N.-W. Frontier of India, 1897-98.—Malakand.—Medal with clasp.

Cockburn, H. A. (*Gen. u.s.l. Ind. Army*)—
India, 1857-59.—Affairs near Shahgunge and Agra and several minor engagements in the Doab. (Twice wounded; once severely; one horse wounded and one killed). Commanded Sipri Field Force. Services acknowledged by Government of North-Western Provinces on four occasions.—Medal.

Cockerill, R. C. (*Capt., retd., Ind. S.C.*)—
Burma, 1888-89.—Medal with clasp.

Codrington, H. de B.—
Tirah, 1897-98.—Medal with 2 clasps.
East Africa, 1903-04.—Operations in Somaliland. Despatches, *Lond. Gaz.*, 2 Sep. 04. Brev. of Maj.—Medal with clasp.
The War of 1914-18.—Despatches, *Lond. Gaz.*, 22 Jun 15.

Codrington, H. W.—
N.-W. Frontier of India, (1st) Miranzai, 1891.—
N.-W. Frontier of India, Hazara, 1891.—Despatches, G. G. O. 816 of '91. —Medal with clasp.
Chitral, 1895.—Relief of Chitral. Action of the Malakand Pass.—Medal with clasp.
N.-W. Frontier of India, 1897-98.—Malakand. Operations in Bajaur and in the Mamund country. Despatches, G. G. O.'s 1317 of 97 and 178 of 98. Utman Khel. Buner.—Clasp.
N.-W. Frontier of India, 1908.—Operations in the Mohmand country.—Medal with clasp.

Coffin, A. S.—
N.-W. Frontier of India, 1908.—Operations in the Zakka Khel country. Operations in the Mohmand country.—Medal with clasp.

Coke, A. J. P.—
N.-W. Frontier of India, Waziristan, 1901-02.—Medal with clasp.

Colan, H. N.—
South African War, 1899-1902.—Operations in the Transvaal, west of Pretoria, July to 29 Nov. 00. Operations in Orange River Colony (May to 2 Nov. 00.) Operations in Cape Colony, south of Orange River, 99-00. Operations in the Orange River Colony, between 30 Nov. 00 and 31st May 02. Operations in Cape Colony, between 30 Nov. 00 and 31 May 02.—Queen's Medal with 3 clasps. King's medal with 2 clasps.
The War of 1914-18.—Despatches, *Lond. Gaz.*, 5 Apr. 16. DSO.

Colan, W. R. B.—
The War of 1914-18.—Despatches, *Lond. Gaz.*, 19 Oct. 16 and 15 Aug. 17. D.S.O.

Coldstream, J. C.—
N.-W. Frontier of India, Waziristan, 1901-02.—Medal with clasp.
The War of 1914-18.—Despatches, *Lond. Gaz.*, 1 Jan. 16.

Cole, B. L.—
N.-W. Frontier of India, Waziristan, 1901-02.—Medal with clasp.

Cole, E. H.—
Burma, 1886-87.—Operations of the 1st Brigade.—Medal with 2 clasps.
N.-W. Frontier of India, Hazara, 1891.—Clasp.
Chitral, 1895.—Relief of Chitral.—Medal with clasp.
N.-W. Frontier of India, 1897-98.—Relief of Malakand. Relief of Chakdara, Malakand.—Action at Landakai. Operations in Bajaur and in the Mamund country. Mohmand. Despatches, G. G. O. 1317 of 97.—2 clasps.
China, 1900.—Special Service Officer. Despatches, *Lond. Gaz.*, 13 Sep. 01.—Medal.
The War of 1914-18.—Despatches, *Lond. Gaz.*, 28 Jan. 16 and 15 Aug. 17.

Cole, H. W. G.—
N.-W. Frontier of India, Hazara, 1888.—Medal with clasp.
N.-E. Frontier of India, Lushai, 1889.—Clasp.—*Chin-Lushai, 1889-90.*—Despatches, G. G. O. 677 of 90.—Clasp.
N.-E. Frontier of India, 1890-91.—Defence of Changsil.—Services acknowledged by Govt.
N.-E. Frontier of India, Manipur, 1891.—Despatches, G. G. O. 585 of 91.—Clasp.

Coleman, A.—
Burma, 1888-89.—Medal with clasp.
N.-E. Frontier of India, Chin-Lushai, 1889-90.—Clasp.
N.-E. Frontier of India, Manipur, 1891.—Clasp.

Coleridge, J. F. S. D.—
Tibet, 1903-04.—Action at Niani, 26th June 1904. Operations in and around Gyantse, between 5th May and 6th July 1904. March to Lhassa, 14th July to 3rd August 1904.—Medal with clasp.
Operations in the Abor country 1911-12.—Despatches, *Gazette of India*, Notification 480, dated 23rd May 1912.—Medal with clasp.
The War of 1914-18.—Despatches, *Lond. Gaz.*, 27 Mar. 16, 13 July 16, 4 Jan. 17, 15 May 17 and 11 Dec. 17. Brev. of Lt.-Col. D.S.O.

Coles, C. H. R. (*Maj., retd., I. A.*)—
Burma, 1887.—Medal with 2 clasps.
N.-W. Frontier of India, Hazara, 1891.—Clasp.
Chitral, 1895.—Relief of Chitral.—Action of the Malakand Pass. Action at Khar.—Medal with clasp.

Colligridge, H. F.—
South African War, 1899-1900.—Operations in Cape Colony, south of Orange River, 1899-1900.—Queen's medal with clasp.
Tibet, 1903-04.—Action at Niani, 26th June 1904. Operations at and around Gyantse, between 5th May and 6th July 1904. March to Lhassa, 14th July to 3rd August 1904.—Medal with clasp.

Collins, G. A. (*Lt.-Col. retd., I. A.*)—
Afghanistan, 1878.—Zaimukht Expedition.—Medal.

Collins, L. P.—
The War of 1914-18.—Despatches, *Lond. Gaz.*, 8 May, 22 June 15, and 13 July 16. Brev. of Maj. D.S.O.

Collins, R. (*Sr. Asst. Surgn. Hony. Lt. retd. I.S.M.D.*)—
Burma, 1886-87.—Promoted 2nd grade Apothy—Medal with clasp.

Collum, J. A.—
The War of 1914-18.—Despatches, *Lond. Gaz.*, 31 Oct. 17. D.S.O.

Colomb, F. C. (*Col., retd., I. A.*)—
Burma, 1885—89.—Operations against the Chins—Medal with 2 clasps.
N.-E. Frontier of India, Chin-Lushai, 1889-90.—Clasp.
China, 1900.—As D. A. A. G., 4th Brigade, O. Despatches, *Lond. Gaz.*, 14 May 01.—Medal

Colomb, G. H. C.—
Burma, 1885-87.—Medal with clasp.
N.-W. Frontier of India.—*Waziristan, 1894-95.*—Clasp.
China, 1900.—As D. A. A. G., Base, Wei-hai-wei—Despatches, *Lond. Gaz.*, 14 May 01. Medal
The War of 1914-18.—Despatches, *Lond. Gaz.*, 27 July 17. C. M. G.

Colomb, R. P. (*Lt.-Col., retd., I.A.*)—
Burma, 1885—90.—Medal with 2 clasps

Colquhoun, H. W. C. (*Major, retd., I. A.*)—
N.-W. Frontier of India, 1897-98.—Operations on the Samana and in the Kurram Valley during Aug. and Sep. 97.—Medal with 2 clasps.
Tirah, 1897-98.—Reconnaissance of the Kharmana Defile and action of 7 Nov. 97. Operations against the Khani Khel Chamkannis.—Clasp.

Colvin, J. R. C. (*Lt.-Col. retd., I. A.*)—
Soudan, 1885.—Suakin.—Medal with clasp.

Combe, S. B.—
China, 1900.—Medal.
N.-W. Frontier of India, 1902.—Operations against the Darwesh Khel Waziris.
The War of 1914-18.—Despatches, *Lond. Gaz.*, 1 Jan. 16 and 15 Aug. 17. Brev. of Lt.-Col. M.C.

Compelline, F. H. B.—
South African War, 1899-1900.—Employed with Remounts during the operations in the Transvaal, Orange River Colony and Cape Colony in 1900.—Queen's medal with 3 clasps.

Comyn, G. V.—
The War of 1914-18. Despatches, *Lond. Gaz.*, 13 July 16 and 15 Aug. 17.

Condon, de V.—
China, 1900.—Relief of Pekin.—Medal with clasp.

Congreve, S. M. (*Maj., retd., Ind. S. C.*)—
N.-W. Frontier of India, Zhob Valley, 1884.

Coningham, F. E.—
N.-W. Frontier of India, Waziristan, 1894-95.—Medal with clasp.
N.-W. Frontier of India, 1897-98.—Operations on the Samana and in the Kurram Valley during Aug. and Sep. 97.—Medal with 2 clasps.

Tirah, 1897-98.—Reconnaissance of the Kharmana Defile and action of 9 Nov. 97. Operations against the Kham Khel Chamkannis.—Clasp.

Tibet, 1903-04.—Action at Niani, 26th June 1904. Operations at and around Gyantse, between 5 May and 6 July 04. March to Lhassa, 14 July to 3 Aug. 04.—Medal with clasp.

The War of 1914-18.—Despatches, *Lond. Gaz.*, 15 Aug. 17. D.S.O.

Coningham, R. E.—
The War of 1914-18.—Despatches, *Lond. Gaz.*, 5 May and 13 July 16. Brev. of Maj.

Connell-McDowell, A. E. J.—
South African War, 1901-02.—Operations in the Transvaal, Oct. 01 to 31 May 02. Queen's medal with 5 clasps.

Connop, H. E.—
South African War, 1899-1902.—Operations in Orange River Colony, June to Nov. 00. Operations in the Transvaal, west of Pretoria, July to 29 Nov. 00. Operations in the Transvaal, 30. Nov. 00 to July 01. Operations in Cape Colony, July 01 to 31 May 02. Queen's medal with 3 clasps, King's medal with 2 clasps.

Connor, F. P.—
The War of 1914-18.—Despatches, *Lond. Gaz.*, 19 Oct., 1 and 15 Aug. 17.

Connor, T. A. S. (*Hony. Capt. and Sr. Asst. Surg., retd., I. S. M. D.*)—
Khond Rebellion, 1882.
Burma, 1886.—Medal with clasp.

Conolly, J. J. P.—
The War of 1914-18.—Despatches, *Lond. Gaz.*, 15 Aug. 17.

Conway-Gordon, E. C. W.—
N.-W. Frontier of India, 1897-98.—Operations on the Samana and in the Kurram Valley during Aug. and Sep. 97.—Medal with 2 clasps.

Tirah, 1897-98.—Clasp.

East Africa, 1903-04.—Operations in Somaliland. Action at Jidballi.—Medal with 2 clasps.

The War of 1914-18.—Despatches, *Lond. Gaz.*, 1 Jan. 16 and 11 Dec. 17.

Cook, C. C.—
The War of 1914-18.—Despatches, *Lond. Gaz.*, 5 Apr., 13 July, and 19 Oct. 16. D.S.O.

Cook, H. (*Surg.-Gen., retd., Ind. Med. Serv.*)—
Persia, 1857.—Medal.
Abyssinia, 1867-68.—Medal.

Cook, J. (*Maj.-Genl., retd.*)—
N.-W. Frontier of India (2nd) Eusufzai, 1863.—Umbeyla.—Medal with clasp.

Afghanistan, 1878-79-80.—Action of Ali Musjid, 1st Bazar Valley Expedition. Operations in the Laghman Valley. Served as Bde.-Maj. to Brig.-Genl. C. G. Arbuthnot, C.B., Comdg. 2nd Bde., Khyber Division, Kabul Field Force. Despatches, G. G. O. 610 of 79.—Medal with clasp. Brev. of Maj.

N.-W. Frontier of India, Mahsud-Waziri, 1881.

Soudan, 1885.—Suakin. Brigade-Maj., Indian Contingent, Actions of Hasheen and Tofrek.—Despatches, *Lond. Gaz.*, 25 Aug. 85.—Medal with 2 clasps. Bronze star. Brev. of Lt.-Col.

Cook, J. H. E.—
The War of 1914-18.—Despatches, *Lond. Gaz.*, 15 Jun 16.

Cook, L.—
The War of 1914-18.—Despatches, *Lond. Gaz.*, 5 Apr. 16.

Cook, W. (*Lt.-Col. u.s. l., I.A.*)—
Afghanistan, 1878-79-80.—Action of Ali Musjid Bazar Valley Expedition. Affairs at and near the Shuturgardan Pass. Operations at and around Kabul.—(Severely wounded). Despatches. G.G.O. 1136 of 79. March from Kabul to the relief of Kandahar and battle of Kandahar. Operations against the Maris.—Medal with 3 clasps. Bronze star.
Burma, 1885—89.—Medal with 2 clasps.

Cook, W. K.—
The War of 1914-18.—Despatches, *Lond. Gaz.*, 5 Apr. 16.

Cooke, C. B. (*Lt.-Col. u.s.l. Ind. S.C.*)—
Burma, 1885-87.—Medal with clasp.

Cooke, H. F.—
Chitral, 1895.—Relief of Chitral.—Medal with clasp.
Tirah, 1897-98.—2 clasps.
N.-W. Frontier of India, Waziristan, 1901-02.—Clasp.
Tibet, 1903-04.—Action at Niani, 26th June 1904. Operations at and around Gyantse, between 5 May and 6 July 04. March to Lhassa, 14 July to 3 Aug. 04.—Medal with clasp.
The War of 1914-18.—Despatches, *Lond. Gaz.*, 1 Jan. 16, 13 July 16, 6 Dec. 16, 1 Jan. 17, D.S.O., 4 Jun 17, Brev. of Col., 21 July 17.

Cooke, W. (*Brev.-Col., retd., Ind. S.C.*)—
Afghanistan, 1879.—Kurram Valley Field Force in charge of Commissariat; action at Ali Kheyl. Advance force to Kabul. Operations at Shuturgardan and Khushi, battle of Charasia, and operations at Sherpur.—Medal with 2 clasps.
Burma, 1885-87.—With the Commissariat Department at Rangoon. Thanked by Government of India. Despatches, G. G. O. 864 of 87, Brev. of Lt.-Col.

Cooke, W. N. M. (*Capt., retd., Ind. S.C.*)—
Egypt, 1882.—Medal; bronze star.

Cookson, G. A.—
N.-W. Frontier of India, 1897-98.—Operations on the Samana during Aug. and Sep. 97.—Medal with 2 clasps.
Tirah, 1897-98.—Actions of Dargai and of the Sampagha and Arhanga Passes. Reconnaisance of the Saran Sar and action of 9 Nov. 97. Operations in the Waran Valley and action of the 16 Nov. 97. Despatches, G. G. O. 244 of 98.—Clasp.
South African War, 1899—1902.—Commanded Kitchener's Horse, VI Corps Mounted Infy., and Mobile Column. Relief of Kimberley. Operation in the Orange Free State, Feb. to May 00, including operations at Paardeberg (17 to 26 Feb.); actions at Poplar Grove, Driefontain (10 March), Karree Siding, Houtnek, Vet River (5 and 6 May), Zand River. Operations in the Transvaal, May and June 00, including actions near Johannesburg, Pretoria and Diamond Hill (11 and 12 June). Operations in the Transvaal, West of Pretoria, July to 29 Nov. 00, including action at Eland's River, 4 to 6 August. Operations in Orange River Colony (May to 29 Nov. 00), including action at Wittebergen (1 to 29 July). Operations in the Transvaal between 30 Nov. 00 and 31 May 02. Despatches, *Lond. Gaz.*, 23 April; 7 May and 10 Sep. 01. Queen's medal with 6 clasps. King's medal with 2 clasps. Brev. of Lt.-Col.
The War of 1914-18.—Despatches, *Lond. Gaz.*, 15 June 16. C.M.G.

Cookson, H. (*Surg.-Major, retd., I. M. S.*)—
N.-W. Frontier of India, Jowaki, 1877-78.—Mentioned in Despatches, G. G. O. 738 of 78.—Medal and clasp.
Afghanistan, 1878-79.—Mentioned in Despatches. G. O. C. C. of 14 Oct. 79.—Medal with clasp, Ali Musjid.

Cookson, S. M.—
N.-W. Frontier of India, 1908.—Operations in the Mohmand country.—Medal with clasp.
The War of 1914-18.—Despatches, *Lond. Gaz.*, 15 Jun 16.

Cooper, C. H.—
N.-W. Frontier of India, 1897-98.—Medal with clasp.
Tirah, 1897-98.—Clasp.
South African War, 1900-02.—Operations in the Transvaal in May and June 00 including action near Johannesburg (29 May) and Diamond Hill, (11 and 12 June). Operations in the Transvaal east of Pretoria, July to 29 November 00, including action at Belfast (26 and 27 August, 1900). Operations in Cape Colony, North of Orange River, 99-00. Operations in Orange Free State, February to May 1900. Queen's medal with 5 clasps.

Cooper, E. S. (*Maj., retd. I. A.*)—
Soudan, 1885-86.—Frontier Field Force. Action of Ginnis.—Medal. Bronze star.
N.-W. Frontier of India, Hazara, 1888.—Action of Kotkai.—Medal with clasp.
N.-W. Frontier of India, 1897-98.—Defence of Malakand. Relief of Malakand. Action at Landakai. Operations in the Mohmand country.—Medal with 2 clasps.

Cooper, G. C.—
The War of 1914-18.—Despatches, *Lond. Gaz.*, 30 Jun 16, M.C., Croix de Guerre.

Cooper, H. G.—
The War of 1914-18.—Despatches, *Lond. Gaz.*, 8 Feb. 17.

Cooper, K. E.—
The War of 1914-18.—Despatches, *Lond. Gaz.*, 5 Apr. 16. M.C.

Corbett, B. A.—
Tibet, 1903-04.—Medal.

Cooper, W. G.—
The War of 1914-18.—Despatches, *Lond. Gaz.*, 1 Jun 17.

Copeland, A. G. J.—
The War of 1914-18.—Despatches, *Lond. Gaz.*, 22 Jun 15.

Corbett, T. W.—
The War of 1914-18.—Despatches, *Lond. Gaz.*, 1 Jan. 16, and 15 May 17.

Corbould, P. S.—
South African War, 1900.—Operations in the Orange Free State, February to May 00. Operations in the Transvaal in May and June 00, including actions near Johannesburg (29 May), Pretoria (4 June), and Diamond Hill (11 and 12 June). Operations in the Transvaal, east of Pretoria (July to September 00). Operations in Orange River Colony (September to October 00).—Queen's Medal with 3 clasps.

Corbyn, M. McL.—
South African War, 1900-01.—Operations in Natal, March to June 00, including action at Laing's Nek (6 to 9 June). Operations in the Transvaal, May to June 01.—Queen's Medal with 5 clasps.

Cordeaux, H. E. S. (*Major, retd., I.A.*)—
East Africa, 1901—04.—Operations in Somaliland. Action at Jidballi. Vice-Consul at Berbera. Despatches, *Lond. Gaz.*, 18 Apr. 02, and Despatches (Lt.-Col. Swayne, 20 Feb. 04, Brig.-Gen. Manning, 17 Aug. 03, and Sir C. C. Egerton, 30 May 04). *Lond. Gaz.*, 2 Sept. 04. Medal with 3 clasps.

Corker, T. M.—
N.-W. Frontier of India, 1897-98.—Malakand; Operations in the Mohmand country.—Medal with clasp.
Tirah, 1897-98.—Action of the Sampagha Pass. Action of the Arhanga Pass. Operations against the Khani Khel Chamkanis. Operations in the Bazar Valley, 25th to 30th December 1897. Despatches, *Lond. Gaz.*, 5 Apr. 98. Clasp.

Corlett, J. S.—
South African War, 1902.—Operations in the Transvaal, Feb. 02. Operations in the Orange River Colony, February—March 02. Operations in Cape Colony, March to 31 May 02.—Queen's Medal with 4 clasps.

Cormack, H. S.—
The War of 1914-17.—Despatches, *Lond. Gaz.*, 22 June 15, and 1 Jan. 16. M.C.

Cornish, A. W. D.—
The War of 1914-18.—Despatches, *Lond. Gaz.*, 28 Jan. 16. M.C.

Corridon, R.—
Chitral, 1895.—Relief of Chitral.—Medal with clasp.
Tibet, 1903-04.—Action at Niani, 26 June 04. Operations at and around Gyantse, between 5 May and 6 July 04. March to Lhassa 14 July to 3 Aug. 04. Despatches, G. G. O. No. 1065 of 04. Medal with clasp. Promoted Senior Asst. Surgeon.

Corse-Scott, E. J.—
The War of 1914-18.—Despatches, *Lond. Gaz.*, 17 Feb. 15, 22 Jun 15. M.C.

Costello, E. W.—
N.-W. Frontier of India, 1897-98.—Defence of Malakand. (Twice severely wounded.)—Despatches, G. G. O. 1089 of 97.—Medal with 2 clasps. V. C.
N.-W. Frontier of India, 1908.—Operations in the Mohmand country. Medal with clasp.
The War of 1914-18.—Despatches, *Lond. Gaz.*, 5 Apr 16 and 19 Oct. 16 and 15 Aug. 17. Brev. of Lt.-Col. C.M.G. D.S.O.

Cotgrave, E. C. B. (*Lt.-Col., retd., I. A.*)—
Tirah, 1897-98.—Actions of the Sampagha and Arhanga Passes. Operations at and around Dwatoi and action of 24 Nov. 97. Operations in the Bara Valley, 7 to 14 Dec. 97. Despatches, G. G. O. 244 of 98.—Medal with 3 clasps.

Cotton, W. L.—
The War of 1914-18.—Despatches, *Lond. Gaz.*, 31 May 16.

Cotter, V. R.—
N.-W. Frontier of India, 1902.—Operations against the Darwesh Khel Waziris.

Cotterill, H. E. (*Maj., retd., I. A.*)—
N.-W. Frontier of India, 1897-98.—Defence of Malakand. Relief of Malakand, Relief of Chakdar Utman Khel.—Medal with 2 clasps.
N.-W. Frontier of India, Waziristan, 1901-02.—Clasp.

Cotton, A. A.—
Burma, 1885—87.—Medal with 2 clasps.
China, 1900.—Medal.

Couper, E. E. (*Lt.-Col., retd., I.A.*)—
N.-W. Frontier of India.—Waziristan, 1894-95.—Medal with clasp.
China, 1900.—Medal.

Courtney, W. M. (*Surgn. Lt.-Col. retd., Ind. Med. Serv.*)—
N.-E. Frontier of India, Duffla, 1873-74.
Burma, 1886-87.—Medal with clasp.
N.-E. Frontier of India, Chin-Lushai, 1899-90.—Clasp.

Coussmaker, G. (*Hon. Lt.-Col., retd., Bom. S. C.*).—*India, 1858-59.*

Cowle, W. A. L.—
Burma, 1887—89.—Medal with clasp.
N.-W. Frontier of India, 1897-98.—Operations on the Samana and in the Kurram Valley during Aug. and Sep. 97.—Operations of the Flying Column in the Kurram Valley under Colonel Richardson, 20 Aug. to 1 Oct. 97.—Medal with 2 clasps.
Tirah, 1897-98.—Clasp.

Cowley, J. W. (*Bt. Col. retd., I.A.*)—
N.-E. Frontier of India, Akha, 1883—84.
Burma, 1886—88.—Operations of the 1st Brigade. Ruby Mines Expedition, including advance from Khannyat. Exploration of the Mettaung and Nanka Passes.—Medal with 2 clasps.
N.-E. Frontier of India, Manipur, 1891.—Despatches, G. G. O. 585 of '91.—Clasp.

Cowper, M.—
Chitral, 1895.—Relief of Chitral.—Medal with clasp.
The War of 1914-18.—Despatches, *Lond. Gaz.*, 10 May 16, 19 Oct. 16 and 14 Nov. 16. C.B.

Cox, C. E. S.—
The War of 1914-18.—Despatches, *Lond. Gaz.*, 5 Apr. 16. Brev. of Lt.-Col.

Cox, F. W. H.—
Burma, 1886—89.—Operations of the 1st Brigade.—Despatches.—Major-Genl. Sir G. White's despatch No. 56 B. C., dated 1 April 88.—Medal with 2 clasps.
N.-E. Frontier of India, Manipur, 1891.—With Mounted Infantry. Action at Langkatel. (Severely wounded.) Despatches, *Lond. Gaz.*, 14 Aug. '91.—Clasp.
Burma, 1894.—Baungshe column.

Cox, G. (*Hony. Capt. Commy. retd., Ordnce. Deptt.*)—
Afghanistan, 1879-80.—Medal.
South African War, 1889—1901.—Defence of Ladysmith.—Despatches, *Lond. Gaz.*, 10 Sep. 01. Queen's medal with clasp.—Promoted Assistant Commissary.

Cox, Sir H. V.—
Afghanistan, 1879-80.—Khyber Line Force.—Medal.
Burma, 1885—89.—Present at the taking of Minhla and occupation of Mandalay. Operations of the 5th Brigade.—Despatches, *Lond. Gaz.*, 22 June '86.—Medal with 2 clasps.
N.-W. Frontier of India, 1897-98.—Mohmand. Despatches, G. G. O. 145 of 98. Buner.—Medal with clasp.
Tirah, 1897-98.—Clasp.
China, 1900.—Medal.
The War of 1914-18.—Despatches, *Lond. Gaz.*, 20 Sep. 15, 6 and 21 Jan. 16, 21 Jun. 16, 25 Sep. 16, and 4 Jan. 17. Promd. Lt.-Gen. K.C.M.G.

Cox, T.—
South African War, 1900.—Operations in Cape Colony. North of Orange River, 28th March to 4th April 1900. Operations in the Orange Free State, 5th April to 26th May 1900, including actions at Houtnek (1st May), Vet River (5th and 6th May), and Zand River (10th May). Operations in the Transvaal 27th May to 26th June, including actions at and near Johannesburg (29th May) and Pretoria (4th June). Operations in the Transvaal, west of Pretoria, 22nd to 28th August 1900. Operations in the Transvaal, east of Pretoria, 29th August to 3rd November 1900.—Queen's medal with 3 clasps.

Cox, Sir P. Z.—
The War of 1914-18.—Despatches, *Lond. Gaz.*, 5 Apr. 16, 10 May 16, 19 Oct. 16, 14 Nov. 16, and 10 July 17. G.C.I.E., K.C.S.I.

Cox, T. S.—
N.-W. Frontier of India, 1897-98.—Tochi.—Medal with clasp.
China, 1900.—Medal.

Craig, J. MacA.—
N.-W. Frontier of India, 1908.—Operations in the Mohmand country. Medal with clasp.
Chitral, 1895.—Relief of Chitral.—Medal with clasp. Promoted Deputy Assistant Commissary.
N.-W. Frontier of India, 1897-98.—Operations on the Samana and in the Kurram Valley during Aug. and Sep. 97.—2 clasps.
Tirah, 1897-98.—Clasp. Promoted Asstt. Comsy.
China, 1900.—As Personal Assistant to Chief Transport Officer. Despatches, *Lond. Gaz.*, 13 Sep. 01. Medal. Honorary rank of Major.

Craigie, P. N.—
N.-W. Frontier of India, 1897-98.—Buner.—Action of the Tanga Pass.—Medal with clasp.

Craik, J.—
The War of 1914-18.—Despatches, *Lond. Gaz.*, 11 Dec. 17. D.S.O.

Cramer-Roberts, M. T.—
South African War, 1900-2.—Operations in the Transvaal, Feb. 02. Operations in Orange River Colony and Cape Colony, Nov. 01 to Apr. 02. Queen's medal with 3 clasps. D. S. O.
The War of 1914-18.—Despatches, *Lond. Gaz.*, 22 Jun. 15. D.S.O.

Craster, G.—
The War of 1914-18.—Despatches, *Lond. Gaz.*, 11 Dec. 17. D.S.O.

Craster, J. C. B. (*Lt.-Col., retd., I. A.*)—
Afghanistan, 1879-80.—With Kurram Field Force. Operations in the Khost Valley. Advance on Kabul. Action of Charasia.—Medal with clasp.
Burma, 1887—89.—Salay Expedition.—Medal with clasp.

Crawford, A. G.—
The War of 1914-18.—Despatches, *Lond. Gaz.*, 15 Jun. 16.

Crawford, J. D.—
The War of 1914-18.—Despatches, *Lond. Gaz.*, 1 Jan. 16, 2 Jun. 16, 19 Oct. 16, and 15 Aug. 17. Brev. of Maj. D.S.O., M.C.

Crawford, J. H.—
The War of 1914-18.—Despatches, *Lond. Gaz.*, 4 Jan. 17.

Crawford, J. M.—
N.-E. Frontier of India, Lushai, 1892-93.
N.-W. Frontier of India, 1897-98.—Malakand.—Buner.—Action of the Tanga Pass.—Medal with clasp.
China, 1900.—Medal.

Crawfurd, H. P. R. F. (*Col. u.s.l. Ind. S.C.*)—
Burma, 1885-86.—Medal with clasp.

Crawley, G. B. (*Lieut.-Col. retd., I. A.*)—
Afghanistan, 1880.—Medal.

Creagh, A. H. D.—
Mekran, 1898.
The War of 1914-18.—Despatches, *Lond. Gaz.*, 21 June and 19 Oct. 16. C.M.G.

Creagh, E. C. (*Lt.-Col., retd., I. A.*)—
N.-W. Frontier of India, Waziristan, 1894-95.—Medal and clasp.
China, 1900.—Medal.
N.-W. Frontier of India, 1908.—Operations in the Mohmand country. Engagement of Matta, 24th April 1908. Engagement of Kargha, 24th May 1908.—Medal with clasp.

Creagh, Sir O'M.—
Afghanistan, 1878—80.—In the Peshawar Valley F. F., operations in the Bazar Valley, and action of Kam Dakka with the Kurram F. F.; as D. A. Q. M. G. and subsequently as A. Q. M. G.; action of Ali Khel. (Despatches, G. G. O. 1222 of '79. A. G. in I.'s Lr. 2761A of '81.) V. C. Medal, and Brev. of Maj.

N.-W. Frontier of India, Zhob Valley, 1890.—(Despatches, G. G. O. 148 of '91.)
China, 1900.—Medal and K. C. B. Mentioned in Despatches. *Lond. Gaz.*, 26 June 03.

Cresswell, C. A. (*Lt.-Col., retd., Ind. S. C.*)—
Burma, 1886-87.—As Political Officer.—Medal with clasp.

Crighton, McV.—
The War of 1914-18.—Despatches, *Lond. Gaz.*, 5 Apr. and 19 Oct. 16.

Crick, C. C.—
N.-W. Frontier of India, 1908.—Operations in the Mohmand country. Engagement of Matta, 24th Apr. 1908.—Medal with clasp.

Crimmin, J.—
Burma, 1886—89.—Karen Field Force, 1888-89. As S. M. O. Engagement at Nga Kyaing; forcing of the Nanko defile and capture of Sawlon. Despatches, special mention, G. O. G. 274 of '89.—Medál with 2 clasps. V.C.

Cripps, A. W.—
N.-W. Frontier of India, (1st) Miranzai, 1891.
Tirah, 1897-98.—Operations in the Bazar Valley, 25 to 30 Dec. 97.—Medal with 2 clasps.
China, 1900.—Relief of Pekin. Despatches, *Lond. Gaz.*, 14 May 01. Medal with clasp.
The War of 1914-18.—Despatches, *Lond. Gaz.*, 1 Jan. 16.

Crocker, S. F.—
Burma, 1886-87.—Operations of 4th Bridge.—Medal with clasp.
Chitral, 1895.—Relief of Chitral.—Medal with clasp.
Tirah, 1897-98.—Operations in the Bara Valley, 7 to 14 Dec. 07.—2 clasps.
The War of 1914-18.—Despatches, *Lond. Gaz.*, 4 July 16. C.B.

Crombie, D. C.—
N.-W. Frontier of India, 1902.—Operations against the Darwesh Khel Waziris.
The War of 1914-18.—Despatches, *Lond. Gaz.*, 20 Mar. 16.

Crommelin, C. Y. (*Bt. Col., retd., I. A.*)—
N.-W. Frontier of India, Waziristan, 1894-95.—Medal with clasp.

Cronin, J. J. (*Lieut.-Col., I. A., Retired*)—
Burma, 1887—89.—Medal with clasp.
China, 1900.—Medal.

Crosbie, C. S.—
China, 1900.—Relief of Pekin. Actions of Peitsang and Yangtsun.—Medal with clasp. Promoted Conductor.
The War of 1914-18.—Despatches, *Lond. Gaz.*, 21 Jun. 16.

Croslegh, E. L.—
South African War, 1900-01.—Operations in Cape Colony, South of Orange River, 2 to 9 June 00. Operations in Orange River Colony, 10 June to 30 July 00. Operations in Cape Colony, North of Orange River, 31 June to 29 Nov. 00. Operations in Cape Colony, between 30 Nov. 00 and 25 Jan. 01, between 7 Feb. and 6 May 01 and between 27 Sep. and 13 Nov. 01. Operations in Orange River Colony, between 26 Jan. and 6 Feb. 01. Operations in the Transvaal, between 7 May and 26 Sep. 01—Queen's medal with 4 clasps.
Tibet, 1903-04.—Action at Niani, 26th June 1904. Operations at and around Gyantse, between 5th May and 6th July 1904.—Medal with clasp.

Crosthwaite, C. G.—
N.-W. Frontier of India, Waziristan, 1901-02.—Medal with clasp.

Crosthwaite, J. S.—
East Africa, 1903-04—Operations in Somaliland—Medal with clasp.
N.-W. Frontier of India, 1908.—Medal with clasp.

Crowe, T. C. (*Hon. M. Gen. ret. R. (Bo. Art.)*)—
Persia, 1856-57—Action of Bushire.—Medal with clasp.
India, 1857-59.—Siege and capture of Rhatghur, action of Barodia, relief of Saugor, capture of Gharakota, pursuit to Behra; forcing of the Maddenpore Pass, siege and storming of Jhansi, battle of Betwa, battles of Koonch and Gallowlie, capture of Calpee, pursuit to Surceela, capture of Gwalior, battle of Morar, pursuit and defeat of the rebels at action of Jawra Alipore; operations in Bundelcund; capture of Garotha and pursuit of Shah Zada Ferozeshah.—Medal with clasp.

Crowdy, J. D.—
The War of 1914-18.—Despatches, *Lond. Gaz.*, 15 Aug. 17.

Crowther, A. H.—
The War of 1914-18.—Despatches, *Lond. Gaz.*, 5 Apl. 16.

Crowther, R. T. (*Bt. Col. retd. I. A.*)—
N.-W. Frontier of India, 1st Miranzai, 1891.
N.-W. Frontier of India, Waziristan, 1901-02.—Medal with clasp.

Cruddas, H. M.—
China, 1900.—Medal.
N.-W. Frontier of India, 1902.—Operations against the Darwesh Khel Waziris.
N.-W. Frontier of India, 1908.—Operations in the Mohmand country.—Medal with clasp.
The War of 1914-18.—Despatches, *Lond. Gaz.*, 1 Jan. 16. C.M.G.

Cruikshank, J. A.—
The War of 1914-18.—Despatches, *Lond. Gaz.*, 13 July 16. M.C.

Cruickshank, J. E.—
South African War, 1901-02.—Operations in the Orange River Colony between 15th March 1901 and 30th September 1901. Operations in the Transvaal between 1st October 1901 and 31st May 1902.—Queen's medal with 5 clasps.
The War of 1914-18.—Despatches, *Lond. Gaz.*, 1 Jan. 16.

Cubbon, F. R.—
The War of 1914-18.—M.C. Clasp to M.C.

Cubbon, R. (*Dy. Comsy. and Hony. Capt., retd.*)—
Egypt, 1882. El Magfar. Capture of Mahsameh. Actions at Kassassin and Tel-el-Kebir, and occupation of Cairo.—Medal with clasp. Khedive's Star.
N.-W. Frontier of India, Waziristan, 1894-95.—Medal with clasp.

Cullen, E. H. S.—
Chitral, 1895.—Relief of Chitral.—Medal with clasp.
N.-W. Frontier of India, 1897-98.—Mohmand.—Clasp.
Tirah, 1897-98.—Operations in the Bara Valley, 7 to 14 Dec. 97.—Clasp.
N.-W. Frontier of India, Waziristan, 1901-02.—Clasp.
Tibet, 1903-04.—Action at Niani, 26th June 1904. Operations at and around Gyantse, between 5th May and 6th July 1904. Despatches, *Gazette of India*, No. 1065 of 1904.—Medal with clasp.
Operations in the Abor country, 1911-12.—Despatches, *Gazette of India*, Notn. 480, dated 23 May 12. Medal with clasp. D. S. O.
The War of 1914-18.—Despatches, *Lond. Gaz.*, 1 Jan. and 19 Oct. 16. C.M.G.

Cullen, W. B. A. K.—
The War of 1914-18.—Despatches, *Lond. Gaz.*, 15 Jun. 16.

Cumberlege, C. J.—
N.-W. Frontier of India, Waziristan, 1901-02.—Medal with clasp.

Cuming, R. J.—
China, 1900.—Medal.
The War of 1914-18.—Despatches, *Lond. Gaz.*, 15 Jun 16 and 15 Aug. 17. Brev. of Lt.-Col.

Cumming, G. P. (*Capt., Ben. S. C., retd.*)—
India, 1857-58.—With the 10th Foot in the advance to Lucknow. With the Jounpore Field Force under Br. Genl. Franks, C.B., at the actions of Chanda, Umeerpore and Sultanpore. Siege and capture of Lucknow in March 1858 under Sir Colin Campbell, G.C.B., including the storming of the Emambara and the Kaiserbagh. With Major General *Sir* B. Lugard's column at the passage of the Tonse, relief of Azimghur, capture of the town and stronghold of Jugdeespore, action of Chitowrah, and operations in its vicinity.—Medal with clasp.
N.-E. Frontier of India, Bhutan, 1865-66.—With the 44th Bengal Native Infantry.—Medal with clasp.

Cummins, H. A. V.—
China, 1900.—Medal with clasp.
The War of 1914-18.—Despatches, *Lond. Gaz.*, 13 July and 19 Oct. 16. Brev. of Col.

Cuningham, C. A. (*Maj.-Genl. u.s.l. I. S. C.*)—
Afghanistan, 1880.—As A. A. G. of Division, and subsequently as A. A. G., Line of Communication. —Medal.

Cuninghame, D. S.—
Afghanistan, 1878-79-80.—Action of Saif-u-din, Battle of Ahmed Khel. Actions of Patkao Shana and Urzu. Despatches, G. G. O.'s 326 and 493 of '80—Medal with clasp. Brev. of Maj.
N.-W. Frontier of India, Mahsud-Waziri, 1881.
N.-W. Frontier of India, (2nd) Miranzai, 1891. —Medal with clasp.
N.-W. Frontier of India, Waziristan, 1894-95.—Clasp.
N.-W. Frontier of India, 1897.—Tochi.—Section Road Commdt., Lines of Commn. (7 July to Nov. 97). Despatches, *Lond. Gaz.*, 11 Feb. 98.

Cunliffe, J. B.—
The War of 1914-18.—Despatches, *Lond. Gaz.*, 4 July 16.

Cunningham, C. C.—
The War of 1914-18.—Despatches, *Lond. Gaz.*, 15 Jun 16. Brev. of Major.

Cunningham, J.—
The War of 1914-18.—Despatches, *Lond. Gaz.*, 4 July 16.

Cunningham, P. H.—
Soudan, 1885.—Suakin. Actions at Hasheen and Tofrek and operations at Tamai.—Medal with 2 clasps; bronze star.
Soudan, 1885-86.—Frontier Field Force. Action at Ginnis.
N.-E. Frontier of India, Manipur, 1891.—As Transport Officer.
N.-W. Frontier of India, 1897-98.—Tochi. Commissariat Transport Officer.—Medal with clasp.
China, 1900.—Transport Officer, 3rd Brigade.—Medal.
East Africa, 1902-04.—Operations in Somaliland.—Medal with clasp.
The War of 1914-18.—Despatches, *Lond. Gaz.*, 30 Jun 16.

Curran, J. F.—
Burma, 1885-87.—Medal with clasp.
N.-W. Frontier of India, Manipur, 1891.—Clasp. Promoted 2nd Grade Apothecary.

Currey, H. P.—
N.-W. Frontier of India, 1908.—Operations in the Mohmand country. Engagement of Kargha, 24th May 1908.—Medal with clasp.

Cursetjee, H. J. M.—
The War of 1914-18.—Despatches, *Lond. Gaz.*, 5 Nov. 15.

Currie, J. W. (*Lt.-Col., retd., I. S. C.*)—
Burma, 1886-87.—Medal with 2 clasps.

Curtis, A. C.—
The War of 1914-18.—Despatches, *Lond. Gaz.*, 3 July 15, 1 Jan. 16 and 15 Aug. 17. M.C.

Curtis, F.—
N.-W. Frontier of India, 1908.—Asst. Sig. Offr. Operations in the Mohmand country. Medal with clasp.

Curtis, O. O.—
The War of 1914-18.—Despatches, *Lond. Gaz.*, 19 Oct. 16.

Custance, H. L. (*Major retd., I. A.*)—
Soudan, 1885.—Suakin. Actions of Hasheen, Tofrek, and Tamai.—Medal with 2 clasps. Bronze star.
Burma, 1886-89.—Medal with 2 clasps.
N.-W. Frontier of India, 1897-98.—Operations on the Samana and in the Kurram Valley during Aug. and Sep. 97.—Medal with 2 clasps.
Tirah, 1897-98.—Actions of Changru Kotal and of the Sampagha and Arhanga Passes. Reconnaissance of the Saran Sar and action of 9 Nov. 97. Operations in the Waran Valley and action of 16 Nov. 97. (Severely wounded.) Despatches, G. G. O. 244 of 98.—Clasp. Brev. of Maj.

Cutler, H. C.—
Tirah, 1897-98.—Operations in the Bara Valley, 7 to 14 Dec. 97.—Medal with 2 clasps.
China, 1900.—Relief of Pekin. Action of Peitsang. Action at Yangtsun.—Medal with clasp.

D

Dabholkar, A. Y.—
The War of 1914-18.—M.C.

Dale, G. A.—
N.-W. Frontier of India, Waziristan, 1894-95.—Medal.
The War of 1914-18.—Despatches, *Lond. Gaz.*, 31 Oct. 17. C.M.G.

Dale, H. M. (*Lt.-Col., retd., Mad. Inf.*)—
Burma, 1885-86.—Medal with clasp.

Dallas, A. E.—
Burma, 1889-92.—Irrawaddy Column.—Medal with clasp.
The War of 1914-18.—Despatches, *Lond. Gaz.*, 5 Apl. and 19 Oct. 16. C.M.G.

Dallas, C. M. (*Lieut.-Col., retd., I. A.*)—
N.-W. Frontier of India, 1st and 2nd Miranzai, 1891.—Despatches, G. G. O. 632 of 91.—Medal with clasp.

Dallas, J. S.—
Burma, 1891-92.—North Eastern Column.—Medal with clasp.

Dalmahoy, P. C. (*M.-Genl. u. s. l. Ben. Inf.*)—
India, 1857-58-59.—Siege and capture of Delhi, subsequent operations, and operations in Bundelcund.—Medal with clasp.

Dalrymple-Hay, S. F. B.—
Chitral, 1895.—Relief of Chitral.—Medal with clasp.

Daly, C. E.—
The War of 1914-18.—Despatches, *Lond. Gaz.*, 4 July 16.

Daly, E. D. H. (*Lt., retd., Ben. S. C.*)—
South African War, 1899-1902.

Daly, Hon. Sir H.—
Burma, 1886-7.—Political Asst., Upper Chindwin. Despatches, *Lond. Gaz.*, 2 Sept. 87.

Daly, J. T.—
Waziristan, 1894-5.—Medal with clasp.
China, 1900.—Medal.
N.-W. Frontier of India, 1908.—Operations in the Zakka Khel country. Operations in the Mohmand country.—Medal with clasp.

Damania, P. J. (*Lt.-Col., retd., Ind. Med. Serv.*)—
Afghan War, 1880-81.—Attached to Base Hospital, Kandahar, and in Medical Charge of Depôt Hospital. Gulistan.—Medal.
Burma, 1885-87.—Medal with clasp.

Dane, R.—
N.-W. Frontier of India, 1908.—Operations in the Mohmand country. Engagement of Kargha, 24th May 1908.—Medal with clasp.
The War of 1914-18.—Despatches, *Lond. Gaz.*, 15 Aug. 17. MC.

Daniell, F. W. (*Major, retd., I. A.*)—
South African War, 1900-02.—Operations in the Transvaal, March and April 02. Operations in Orange River Colony, Jan. to Mar. 02. Operations in Cape Colony, April and May 1902—Queen's Medal with 3 clasps.
The War of 1914-18.—Despatches, *Lond. Gaz.*, 4 Jan. and 18 Dec. 17. DSO.

Daniell, H. S. (*Lt.-Col., retd., I. A.*)—
Persia, 1857.—Medal with clasp.
India, 1857-9.—With the Central India Field Force at siege and capture of Ratghur.
Action of Burodea (wounded), relief of Saugor. Capture of Gurrakhota, forcing of the Muddenpore Pass, and siege and storm of Jhansi.
Action at Koonah. Affairs on the Jumna, capture of Calpee; battle and capture of Gwalior; action of Morar. Battle of Jowra Alipore; action of Mhow Mohoni. Medal with clasp.

Daniell, J. A. S.—
The War of 1914-18.—Despatches, *Lond. Gaz.*, 17 Feb. and 22 Jun. 15. D.S.O.

Daniell, R. H. (*Maj. retd., Ind. S. C.*)—
Afghanistan, 1878-80.—Medal.

Daniels, A. M.—
The War of 1914-18.—Despatches, *Lond. Gaz.*, 15 May and 15 Aug. 17. Brev. of Maj.

Darby, W. A.—
The War of 1914-18.—Meritorious Service Medal.

Daunt, W. D.—
Chitral, 1895.—Relief of Chitral.—Medal with clasp.

Darell, R. D. E.—
The War of 1914-18.—Despatches, *Lond. Gaz.*, 5 Apl. 16.

Darley, J. R.—
The War of 1914-18.—Despatches, *Lond. Gaz.*, 5 Apr. 16, 13 July 16 and 15 Aug. 17. D.S.O.

Darley, W. H. La T.—
The War of 1914-18.—Despatches, *Lond. Gaz.*, 11 Dec. 17.

Davidson, C. (*Lieut.-Col., retd., I. A.*)—
N.-W. Frontier of India, (1st and 2nd) Miranzai, 1891.—Medal with clasp.
Tirah, 1897-98.—Medal with 2 clasps.
East Africa, 1903.—Operations in Somaliland.—Medal with clasp.

Davidson, H. B.—
N.-W. Frontier of India, 1908.—Operations in the Zakka Khel country. Operations in the Mohmand country. Engagement of Matta, 24th April 1908. Medal with clasp.
East Africa (Somaliland), 1914-15.—Medal with clasp.

Davidson, Jas. (*Brig. Surgn.-Lt.-Col., retd., I. M. S.*)—
Abyssinia, 1867-68.—Asst. Surgn. attached to General Hospital, Zoula.—Medal.
Afghanistan, 1880.—Served in Southern Afghanistan in medical charge of Base Hospital at Pir Chowky.—Medal.

Davidson, J.—
Waziristan, 1894-95.—Medal with clasp.
Chitral, 1895.—Relief of Chitral.—Medal with clasp.
Soudan, 1896.—Dongola Expedition.—British medal and Khedive's medal.
Tirah, 1897-98.—2 clasps.
Tibet, 1903-04.—Action at Niani, 26th June 1904. Operations at and around Gyantse between 27th June and 6th July 1904. March to Lhassa, 14th July to 25th July 1904.—Medal with clasp.
Operations in the Abor country, 1911-12.—Medal with clasp.

Davidson, S. R.—
Chitral, 1895.—Relief of Chitral.—Medal with clasp.
N.-W. Frontier of India, 1897-98.—Mohmand.—Clasp.
Tirah, 1897-98.—Reconnaissance of the Kharmana Defile and action of 7 Nov. 97.—Clasp.
East Africa, 1903-04.—Operations in Somaliland. As special service officer. Despatches, *Lond. Gaz.*, 2 Sept. 04.—Medal with clasp.
The War of 1914-18.—Despatches, *Lond Gaz.*, 17 Feb. 15, 19 Oct. 16 and 15 Aug. 17. Brevs. of Lt.-Col. and Col. C.M.G.

Davie, J. H. M.—
South African War, 1902.—Operations in the Transvaal, March 1902.—Queen's medal with 3 clasps.
The War of 1914-18.—Despatches, *Lond. Gaz.*, 5 Apl. 16, 13 July 16, and 19 Oct. 16. Brev. of Lt.-Col.

Davies, C. H.—
N.-W. Frontier of India, 1897-98.—Operations on the Samana and in the Kurram Valley during Aug. and Sep. 97.—Medal with 2 clasps.
Tirah, 1897-98.—Action of Dargai and of the Samrpagha and Arhanga Passes. Reconnaissance of the Saran Sar and action of the 9 Nov. 97. Operations against the Khani Khel Chamkhanis. Operations in the Bazar Valley, 25 to 30 Dec. 97.—Clasp.
China, 1900.—Medal.
N.-W. Frontier of India, 1902.—Operations against the Darwesh-Khel Waziris. (Slightly wounded). Despatches, G. G. O. 392 of 03.—D. S. O.
N.-W. Frontier of India, 1908.—Operations in the Zakka Khel country. Operations in the Mohmand country. Engagement of Kargha, 24th May 1908.—Medal with clasp.
The War of 1914-18.—Despatches, *Lond. Gaz.*, 4 July 16, 21 Jun. 16, 19 Oct. 16 and 15 Aug. 17. Brev. of Col. C.M.G.

Davies, D. R.—
Burma, 1885-87.—Medal with clasp.
N.-W. Frontier of India, 1897-98.—Medal with clasp.

Davies, F. G. H.—
Chitral, 1895.—Relief of Chitral.—Medal with clasp.
N.-W. Frontier of India, 1897-98.—Malakand. Operations in Bajaur and the Mamund country. Buner.—Clasp.
South African War, 1902.—Operations in the Transvaal, from 25 April to 31 May 02.—Queen's medal with 2 clasps.
The War of 1914-18.—Despatches, *Lond. Gaz.*, 4 July 16.

Davies, H. W.—
Aden, 1903-04.—Operations in the Interior.

Davies, R. D.—
East Africa (Somaliland), 1908-10.—Medal with clasp.

Davis, A. T. (*Lt.-Col. retd., Ben. Inf.*) –
N.-W. Frontier of India, 1863.—Umbeyla Campaign.—Medal with clasp.
N.-E. Frontier of India, Bhootan Expedition, 1865-66.—Clasp.

Davis, C. (*Col., retd., I. A.*) –
Tirah, 1897-98.—Brigade Transport Officer, 3rd Brigade.—Action of the Sampagha Pass.—Operations in the Bara Valley, 7 to 14 Dec. 97.—Affair at Shinkamar on 29 Jan. 98. Despatches, G. G. O. 244 of 98.—Medal with 2 clasps.
South African War, 1899-1900.—Operations in the Orange Free State, May to 29 Nov. 00. Operations in Cape Colony, south of Orange River, 99-00. Operations in Cape Colony, north of Orange River.—Queen's medal with 3 clasps.
China, 1900.—Medal.

Davis, C. T.—
East Africa, 1903-04.—Operations in Somaliland.—Medal with clasp.
The War of 1914-18.—Despatches, *Lond. Gaz.* 15 Aug. 17.

Davis, G. J.—
N.-W. Frontier of India, 1897-98.—Relief of Malakand. Malakand. Operations in Bajaur—Mohmand.—Medal with 2 clasps.
Tirah, 1897-98.—Clasp.
N.-W. Frontier of India, 1898-99.—Operations against the Para-Chamkanis.
East Africa, 1902-04.—Operations in Somaliland. Action at Jidballi.—Medal with 2 clasps.
The War of 1914-18.—Despatches, *Lond. Gaz.* 1 Jan. and 15 Jun. 16. D.S.O.

Davis, L. M.—
The War of 1914-18.—Despatches, *Lond. Gaz.* 8 Feb 17.

Davis, R. E. S. (*Lt.-Col., retd., I.M.S.*)—
Burma, 1885-87.—Despatches, G. G. O. 34 of '87. —Medal with clasp.

Davis, T. (*Hony. Capt. Sr. Assist. Surgeon retd., I. S. M. D.*)—
N.-E. Frontier of India, Chin-Lushai, 1889-90.—Medal with clasp.

Davis, W. E.—
N.-W. Frontier of India, Waziristan, 1894-95.—Despatches, G. G. O. 473 of '95.—Medal with clasp.—Promoted Sub-Conductor.

Davis, W.—
N.-W. Frontier of India, Hazara, 1891.—Medal with clasp.
Chitral, 1895.—Relief of Chitral.—Medal with clasp.

Davison, K. S.—
N.-W. Frontier of India, Waziristan, 1894-95.—Despatches, C. 473 of '95.—Medal with clasp.
The War of 1914-18.—Despatches, *Lond. Gaz.* 5 Apl. 16 and 10 May 16.

Davson, H. J. H.—
The War of 1914-18.—Despatches, *Lond. Gaz.*, 15 Aug. 17.

Davys, G. I.—
Tibet, 1903-04.—Action at Niani, 26th June 1904. Operations at and around Gyantse, between 27th June and 6th July 1904. March to Lhassa, 14th July to 28th August 1904.—Medal with clasp.

Dawes, A. M.—
The War of 1914-18.—Despatches, *Lond. Gaz.*, 8 Feb. 17.

Dawes, C. D.—
N.-W. Frontier of India, 1897-98.—Mohmand. —Medal with clasp.
Tirah, 1897-98.—Clasp.
China, 1900.—Medal.
Operations in the Abor Country, 1911-12.—Medal with clasp.

Dawes, W. M. (*Bt. Col. retd., I.A.*)—
Burma, 1885-86.—Medal with clasp.
Burma, 1891-92.
China, 1900.—Medal.

Dawson, A. G.—
South African War, 1901-02.—Operations in the Orange River Colony, 25th June to 9th July 1901. Operations in Cape Colony, 10th July 1901 to 31st May 1902. Queen's medal with 4 clasps.

Dawson, A. W. (*Lt.-Col., retd., I.M.S.*)—
N.-E. Frontier of India, Chin-Lushai, 1889-90.—Medal with clasp.
N.-W. Frontier of India, Hazara, 1891.—Clasp.
Chitral, 1895.—Relief of Chitral.—Medal with clasp.
Tirah, 1897-98.—Actions of Dargai and of the Sampagha and Arhanga Passes. Operations at and around Dwatoi and action of 24 Nov. 97. Operations in the Bara Valley from 7th to 14th Dec. 97.—2 clasps.
China, 1900.—Medal.
Tibet, 1903-04.—Medal.
N.-W. Frontier of India, 1908.—Operations in the Mohmand country.

Dawson, E. (*Moulmein Vol. Rifles*)—
South African War, 1900.—Operations in Orange Free State, Mar. to May 00. Queen's medal with 3 clasps.

Dawson, F. D.—
N.-W. Frontier of India, 1897-98.—Malakand.—Medal with clasp.
Tirah, 1897-8.—Actions of the Sampagha and Arhanga Passes.—Promoted Sub-Conductor. Clasp.
The War of 1914-18.—Despatches, *Lond. Gaz.*, 19 Oct. 16.

Dawson, H. L. (*Col., retd., I. A.*)—
Soudan, 1885.—Suakin. Actions of Hasheen and Tamai.—Medal with clasp. Bronze star.
Chitral, 1895.—Relief of Chitral.—Medal with clasp.
Tirah, 1897-98.—2 clasps.
South African War, 1900.—Relief of Kimberley. Operations in the Orange Free State, Feb. to May 00, including operations at Paardeberg (17 to 26 Feb.). Actions at Poplar Grove (7 March). Driefontein (10 March). Actions at Houtre (1 May), Vet River (5 and 6 May), and Zand River (10 May). Operations in the Transvaal in May and June 00, including actions near Johannesburg (29 May), Pretoria (4 June), and Diamond Hill (11 and 12 June). Operations in the Transvaal, west of Pretoria, July to 10 Oct. 00, including actions at Venterskron (7 and 9 Aug.). Operations in Orange River Colony (May to 10 Oct. 00), including actions at Wittebergen 1 to 29 July. Commanded Roberts's Horse, 12 Feb. to 15 March, 5th Corps Mounted Infy., 15 March to 10 Oct. 00. Despatches, *Lond. Gaz.*, 16 April 01.—Queen's medal with 6 clasps. C. B.

Day, A. D.—
China, 1900.—Medal.

Day, H. (*Sr. Asst. Hony. Capt. retd., I.S.M.D.*)—
Afghanistan, 1878-79-80.—March from Kabul to the Relief of Kandahar and battle of Kandahar, 1st Sep. 1880.—Medal with clasp. Bronze star.

Dayrell, W. S.—
South African War, 1900-02.—Operations in the Transvaal between 30th Nov. 1900 and 31st May 1902.—Queen's medal with 2 clasps.
The War of 1914-18.—Despatches, *Lond. Gaz.*, 15 Aug. 17.

Deane, D.—
Tirah, 1897-98.—Orderly Officer to G. O. C. Line of Communications and subsequently to G. O. C., 2nd Division.—Action of Chagru Kotal.—Medal with 2 clasps.
South African War, 1901-02.—Operations in the Transvaal, 7 to 18 Jan. 02. Operations in the Orange River Colony, 19 to 21 Jan. 02, and during Feb. 02. Operations in Cape Colony, 22 Jan. to 31 May 02. Special service officer with Imperial Yeomanry from 20 Dec. 01 to 24 Jan. 1902. Adjutant, 15th Battalion, Imperial Yeomanry, from 2 Jan. to 31 May 02. Horse killed.—Queen's medal with 5 clasps.

Deane, E. S.—
South African War, 1900-02.—Operations in the Orange Free State, February to May 1900. Operations in the Transvaal, May and June 1900. Operations in Cape Colony, South of Orange River Colony, 1899-1900. Operations in Cape Colony, North of Orange River, May 1900. Queen's medal with 3 clasps.

Deane, G. W. (*Col., retd., I.A.*)—
Afghanistan, 1878-79.—With Khyber and Kurram Field Forces.—Medal.
Egypt, 1882.—Action of Tel-el-Kebir. Despatches, *Lond. Gaz.*, 2 Nov. 82.—Medal with clasp Bronze star.
Burma, 1887-88.—Medal with clasp.
N.-W. Frontier of India, 1897-98.—Mohmand. Despatches, G. G. O. 1818 of 97.—Medal with clasp.

Deane, L. S.—
South African War 1901-02.—Queen's Medal with 4 clasps.

Deare, B. H.—
N.-W. Frontier of India, 1897-98.—Operations on the Samana and in the Kurram Valley during Aug. and Sep. 97. Relief of Gulistan.—Medal with 2 clasps.
Tirah, 1897-98.—Actions Chagru Kotal—Dargai and of the Sampagha and Arhanga Passes. Reconnaissance of the Saran Sar and action of the 11 Nov. 97.—Clasp.

Deas, L. J. M.—
N.-W. Frontier of India, Waziristan, 1901-02.—Medal with clasp.
The War of 1914-18.—Despatches, *Lon. Gaz.* 15 Aug. 17. Brev. of Lt.-Col.

Dease, C. E.—
South African War, 1900-02.—Operations in the Orange Free State, Feb. to May 1900. Operations in the Transvaal, east of Pretoria, July to 29th Nov. 1900, including actions at Belfast (26th and 27th Aug. 1900), Lydenberg (5th to 8th Sep.). Operations in Orange River Colony (May to July 1900), including actions at Biddulphsberg (29th May), Lindley (26th June), Bethlehem (6th and 7th July), and Wittebergen (1st to 29th July) Operations in the Transvaal between 30th Nov. 1900 and 31st May 1902.—Queen's Medal with 3 clasps. King's medal with 2 clasps.

DeBrath, E. (*Lieut.-General, retd., I.A.*)—
Afghanistan, 1879-80.—Affair of Mazina.—Medal.
N.-W. Frontier of India, Mahsud-Waziri, 1881.—Despatches, Brigadier-General T. G. Kennedy's Despatch, 20 June '81. *Hazara, 1891.*—Medal with clasp.
Soudan, 1896.—Dongola Expedition. Brigade Maj. Suakin Force.—Medal. Brev. of Lt.-Col.

deBurgh, E.—
South African War, 1902.—Operations in Cape Colony, 7th to 28th February 1902. Operations in the Transvaal, 1st to 7th March 1902. Operations in the Orange River Colony, 8th March to 31st May 1902.—Queen's medal with 4 clasps.
The War of 1914-18.—Despatches, *Lond. Gaz.*, 15 Jun. 16, Brev. of Maj. D.S.O.

De Butts, F. C.—
The War of 1914-18.—Despatches, *Lond. Gaz.*, 22 Jun. 15, 15 Aug. 17 and 22 Sept. 17. D.S.O. M.C.

deCourcy, Hon. M. W. R.—
Tibet, 1903-04.—Medal.
Operations in the Abor country, 1911-12.—Despatches, *Gazette of India*, Notn. 480, dated 23 May 12.—Medal with clasp.
The War of 1914-18.—Despatches, *Lond. Gaz.*, 19 Oct. 16, 14 Nov. 16 and 15 Aug. 17. D.S.O.

Dee, P.—
N.-W. Frontier of India, 1897-98.—Tochi.—Medal with clasp.
Tirah, 1897-98.—Clasp.

Deedes, R. B.—
The War of 1914-18.—Despatches, *Lond. Gaz.*, 1 Jan. 16. M. C.

deGale, W. A. S.—
South African War, 1902.—Operations in Cape Colony, May 02. Queen's Medal with 3 clasps.

DeLabilliere, E. G. D.—
N.-W. Frontier of India, Waziristan, 1894-95.—Medal with clasp.
N.-W. Frontier of India, 1897-98.—Malakand.—Medal with clasp.
China, 1900.—Medal.

Delamain, F. G. (*Bt.-Col., retd. I. A.*)—
Afghanistan, 1878-79-80.—With Kandahar and Khyber Field Forces.—Medal.
N.-W. Frontier of India, Hazara, 1891.—Medal with clasp. *Isazai, 1892.*
Chitral, 1895.—Relief of Chitral.—Medal with clasp.
N.-W. Frontier of India, 1897-98.—Defence of Malakand—Relief of Chakdara.—Malakand.—Action at Landakai. Operations in Bajaur and in the Mamund Country.—Mohmand.—2 clasps.

Delamain, W. S.—
Egypt, 1882.—Surrender of Kafr Dowar.—Medal. Bronze star.
Burma, 1885—88.—Medal with 2 clasps.
Arabia, Zaila Field Force, 1890.—Affairs at Wad, Damel and Husain Zareeba.
China, 1900.—Medal. Mentioned in Despatches, *London Gazette*, 14 May 01.
N.-W. Frontier of India, Waziristan, 1901-02.—Medal with clasp.
Aden, 1903-04.—Operations in the Interior.—Despatches, M. D. Notfn. No. 43 of 1905. D.S.O.
The War of 1914-18.—Despatches, *Lond. Gaz.*, 5 Apl. 16, 13 July 16, and 19 Oct. 16. Promd. Maj.-Genl.

Dela Motte, R. B.—
South African War, 1900-02.—Served with Yeomanry and Army Service Corps.—Operations in the Orange Free State and Transvaal, April to July 1901, operations in Cape Colony, Aug. 1901 to 31st May 1902.—Queen's medal with 5 clasps.
The War of 1914-18.—Despatches, *Lond. Gaz.*, 19 Oct. 16. D.S.O.

Dela Motte, H. T.—
The War of 1914-18.—Despatches, *Lond. Gaz.*, 18 and 13 Feb. 18. M. C.

DeLanty, H. W.—
N.-W. Frontier of India, Hazara, 1888.—Medal with clasp.
N.-W. Frontier of India, Isazai, 1892.
N.-W. Frontier of India, 1897-1898.—Utman Khel-Buner. Action of the Tanga Pass.—Medal with clasp.

deLautour, E. J. (*Col. retd.*, B. Arty.).—
N.-W. Frontier of India, 1863.—Engagements in the Umbeyla Pass Despatch, G. G. O. 76 of 64.—Medal with clasp. *Hazara Campaign, 1868.*—Expedition against the tribes of the Black Mountain.—Clasp.
N.-E. Frontier of India, Duffla Expedition, 1874-75.—Mentioned in Despatches.
N.-W. Frontier of India, Jowaki Expedition, 1877-78.—Mentioned in Despatches.—Clasp.
Afghanistan, 1878-79.—Capture of Ali Musjid and expedition to Lughman Valley. Despatches. G. G. O. 2-A of 79.—Medal with clasp. Brev.-Maj.
N.-W. Frontier of India, Miranzai (2nd) Expedition, 1891.—Commanded the Artillery of the Force. Despatches, G. G. O. 1163 of 91.—Brev. of Col.

Delany, T. H.—
China, 1900.—Medal.
Major Retd. I. M. S.

DeLisle, G. de S.—
N.-W. Frontier of India, 1897-98.—Malakand Buner.—Action of the Tanga Pass.—Medal with clasp.

Delme-Radcliffe, S. A.—
The War of 1914-18.—Despatches, *Lond. Gaz.*, 6 July 17.

Dempster, C. (*Col., retd., I. A.*)—
N.-W. Frontier of India.—Mahsud-Waziri Expedition, 1881.
Hazara Expedition, 1891.—Despatches, G. G. O. 819 of 91.—Medal with clasp.
Isazai Expedition, 1892.
Chitral, 1895.—Relief of Chitral. Action of the Malakand Pass.—Despatches, G. G. O. 998 of 95—Brev. of Col.

Dempster, J.—
China, 1900.—Medal.

Dening, H.—
N.-W. Frontier of India, Waziristan, 1901-02.—Despatches, G. G. O. 611 of 02.—Medal with clasp.

Dening, J. P.—
The War of 1914-18.—Despatches, *Lond. Gaz.*, 15 Aug. 17.

Dening, L. E.—
China, 1900.—Medal.
The War of 1914-18.—Despatches, *Lond. Gaz.*, 6 July 17.

Dening, R.—
The War of 1914-18.—Despatches, *Lond. Gaz.*, 11 Jan. 17.

Denne, R. W. A.—
N.-E. Frontier of India, Manipur, 1891.—Tamu column.—Medal with clasp.

Dennys, A. H.—
The War of 1914-18.—Despatches, *Lond. Gaz.*, 1 Jan. 16.

Dennys, C. J. (*Lieut.-Col. retd., I.A.*)—
Afghanistan, 1878-79-80.—Affair of Matun.—Action of Charasiah and Lataband. Operations at and around Kabul.—Medal with 2 clasps.

Dennys, G. W. P. (*Col. Retd. I.M.S.*)—
Afghanistan, 1879-80.—Medal.

Dennys, W. A. B.—
Burma, 1886-88.—Operations of the 2nd Brigade. Wuntho Expedition, including advance from Katha and Manle to Wuntho, and occupation of Wuntho.—Medal with clasp.

Dent, W.—
The War of 1914-18.—Despatches, *Lond. Gaz.*, 5 Apl. 16, 16 Nov. 16, and 15 Aug. 17. D.S.O.

Desvœux, H.—
Burma, 1891.

Devine, A.—
N.-W. Frontier of India, 1897-98.—Buner.—Clasp.
Tirah, 1897-98.—Medal with clasp.

Dew, A. B. (*Major, Supy. List, I. A.*)—
N.-W. Frontier of India, Hazara, 1891.—Medal with clasp.
The War of 1914-18.—Despatches, *Lond. Gaz.*, 31 Oct. 17.

Dewes, F. J.—
Burma, 1888-89.—Medal with clasp.
N.-E. Frontier of India, Chin-Lushai, 1889-90.—Clasp.
N.-W. Frontier of India, 1897-98.—Mohmand.—Medal with clasp.
Tirah, 1897-98.—Clasp.

deWilton, A. W. (*Maj. retd., I. A.*)—
Burma, 1885—87.—Operations of the 2nd and 5th Brigades.—Medal with clasp.

Dewey, H. V.—
The War of 1914-18.—Despatches, *Lond. Gaz.*, 5 Aug. 17.

Dewing, R. H. (*Lt.-Col., retd., I. A.*)—
Burma, 1877-89.—Operations of the 3rd Brigade.—Medal with 2 clasps.

Dhondy, B. S.—
The War of 1914-18.—Despatches, *Lond., Gaz.*, 31 Oct. 17.

Dick, A. R.—
N.-W. Frontier of India, Waziristan, 1894-95.—Medal with clasp.
N.-W. Frontier of India, 1897-98.—Malakand. Orderly Officer to G. O. C. Force. Action at Landakai. Despatches, G. G. O. 1039 of 97. Operations in Bajaur and in the Mamund country. Mohmand. Despatches, G. G. O. 1317 of 97.—Medal with clasp.
China, 1900.—Special Service Officer.—Medal Brevet of Lieut.-Col.
N.-W. Frontier of India, 1908.—Operations in the Mohmand country. Base Comdt. Despatches, G. G. O. No. 521A of 1908. Brevet of Colonel.
The War of 1914-18.—Despatches, *Lond. Gaz.*, 4 Jan. 17.

Dick, M. (*Maj., retd., I.M.S.*)—
Tirah, 1897-98.—Affair at Shinkamar on 29th January 1898.—(Slightly wounded). Despatches, G. G. O. No. 483 of 1898.—Medal with 2 clasps.

Dick-Lauder, T. N. D.
The War of 1914-18.—Despatches, *Lond. Gaz.*, 19 Oct. 16.

Dickinson, T. M.—
The War of 1914-18.—Despatches, *Lond. Gaz.*, 19 Oct. 16.

Dickson, A. N.—
Balkan Campaign, 1912-13.—British Red Cross Commission with Bulgarean Forces.
The War of 1914-18.—M. C.

Dickson, H. R. P.—
The War of 1914-18.—Despatches, *Lond. Gaz.*, 15 Aug. 17. C. I. E.

Dickson, J. H.—
Chitral, 1895.—Medal with clasp.
Soudan, 1896.—Dongola Expedition.—Medal.
Tirah, 1897-98. Operations against the Khani Khel Chamkannis.—Clasp.

Dimmock, H. P. (*Lt.-Col., retd., I. M. S.*)—
N.-W. Frontier of India. Expedition against the Marri Tribe. Dec. 80, Jan. 81. Medical charge, 2nd Belooch Regt.

Ditmas, F. R. (*Lt.-Col., retd., I.A.*)—
Afghanistan, 1879.—With Vitakri Field Force.—Medal.

Divecha, F. R. (*Surg.-Maj., retd., Ind. Med. Serv.*)—
Soudan, 1885-86.—Suakin.

Dobbie, H. H.—
N.-W. Frontier of India, 1897-98.—Engagement near Shabkadr, 9 Aug. 97. Operations on the Samana and in the Kurram Valley during Aug. and Sep. 97.—Medal with 2 clasps.
Tirah, 1897-98.—Operations in the Bazar Valley, 25 to 30 Dec. 97.—Clasp.

Dobbie, W. H.—
Burma, 1885-87.—Operations in the Eastern and Western movable columns. Operations of the 6th Brigade. Despatches, G. G. O.'s 295 of 86 and 434 of 87 and Despatches by Major-General B. L. Gordon. G. G. O. 864 of 87.—Medal with clasp.
N.-E. Frontier of India, Manipur, 1891.—Clasp.
South African War, 1899-1900.—Employed with Remounts.

Dobbin, W. J. K. (*Lt.-Col., retd., I. A.*)—
N.-W. Frontier of India, Mahsud-Waziri, 1881. Zhob Valley, 1890. Waziristan, 1894-95.—Medal with clasp.
N.-W. Frontier of India, 1897-98.—Tochi.—Medal with clasp.
China, 1900.—Relief of Pekin. Action of Peitsang. Action of Yangtsun.—Medal with clasp.

Dobbs, A. H. (*Lieut.-Col., retd. I. A.*)—
Burma, 1888-89.—Mogoung column.—Medal with clasp.

Dobbs, C. F.—
Aden, 1903-04.—Operations in the Interior.
The War of 1914-18.—Despatches, *Lond. Gaz.*, 30 Jun. 16 and 1 Feb. 17. Brev. of Lt.-Col., D.S.O.

Dobbs, G. C. (*Col. retd. I. A.*)—
Afghanistan, 1879-80.—Battle of Maiwand. Defence of Kandahar.—(Mentioned in Despatches.)—Medal.

Dods, E. R. S.—
The War of 1914-18.—M.C.

Domenichetti, F. H.—
Burma, 1888-89.—Medal with clasp.
N.-E. Frontier of India, Chin-Lushai, 1891—Chinbok column.—Clasp.
China, 1900.—Medal.

Donald, C. H.—
South African War 1900.—Operations in the Transvaal, Cape Colony and Orange Free State. Queen's medal with three clasps.

Donaldson, C. E. (*Hony. Capt. Dy. Commy. retd. India Miscellaneous List.*)—
China, 1900.—Relief of Pekin. Action of Peitsang. Action at Yangtsun. Despatches, *London Gazette*, Sep. 01. Medal with clasp. Promoted Deputy Assistant Commissary.

Donnan, W. (*Lt.-Col. retd. I. A.*)—
East Africa, 1904.—Operations in Somaliland. Medal with clasp.
The War of 1914-18.—Despatches, *Lond. Gaz.*, 1 Jan. 16.

Donovan, C.—
Tirah, 1897-98.—Actions of Dargai and of the Sampagha and Arhanga Passes.—Operations in the Bara Valley, 7 to 14 Dec. 97.—Medal with 2 clasps.

Dopping-Hepenstal, M. E.—
Tirah, 1897-98.—Medal with clasp.
N.-W. Frontier of India, Waziristan, 1901-02.—Clasp.
The War of 1914-18.—Despatches, *Lond. Gaz.*, 15 Jun. 16, 27 July 16, and 15 Aug. 17. D.S.O.

Dorkins, A. J.—
Egypt, 1882.—Medal; bronze star.
Soudan, 1884-85.—Nile.—Clasp.
China, 1900.—Medal. Promoted Deputy Assistant Commissary.

Douglas, G. S.—
South African War, 1899—1901.—Served with Impl. Lt. Inft. Operations in the Transvaal, Dec. 00 to June 01. Queen's medal with 2 clasps.

Douglas, H. A.—
N.-W. Frontier of India, 1908.—Operations in the Mohmand country.—Medal with clasp.
The War of 1914-18.—Despatches, *Lond. Gaz.*, 1 January 16.

Douglas, J. A.—
N.-W. Frontier of India, Waziristan, 1894-95.—Field Intelligence Officer, 2nd Brigade.—Medal with clasp.
China, 1900.—Relief of Pekin. Served as Special Service Officer. Despatches, *London Gazette*, 14 May 01.—Medal with clasp.
The War of 1914-18.—Despatches, *Lond. Gaz.*, 5 Apl. 16, 19 Oct. 16, and 15 Aug. 17. C.I.E.

Douglas, W. B.—
Tirah, 1897-98.—Assistant Brigade Transport Officer with Peshawar column and afterwards as Orderly Officer to G. O. C. 5th Brigade.—Operations in the Bara Valley, 7 to 14 Dec. 97.—Medal with 2 clasps.

Doveton, A. C.—
N.-W. Frontier of India, 1908.—Operations in the Mohmand country.—Medal with clasp.

Doveton, A. M. (*Capt., retd., I. A.*)—
Aden, 1903-04.—Operations in the interior.

Dowden, T. F. (*Col., retd., Royal (Bo.) Engineers*)—
Afghanistan, 1879-80.—March from Quetta to the relief of Kandahar. Despatches, G. G. O. 693 of 81.—Medal.

Downing, J. G. (*Lt.-Col., retd., I. A.*)—
Afghanistan, 1878-79.—Affair of Matun.—Medal

Doyle, B. (*Lt.-Col., retd., Ind. Med. Serv.*)—
Afghanistan, 1879-80.—Medal.

Doyle, I. P. (*Maj., retd., I. M. S.*)—
Burma, 1888-89.—Despatches, *London Gazette*, 15 Nov., 89. D. S. O.
Burma, 1891-92.—Operations on N.-E. Frontier; Chin Hills.

Doyle, M. J.—
Egypt, 1882.—Action at Tel-el-Kebir, wounded. Medal with clasp. Bronze star.
Burma, 1885-89.—Medal with 2 clasps.
N.-W. Frontier of India, Waziristan, 1894-95.—Clasp.
N.-W. Frontier of India, Waziristan, 1901-02.—Medal with clasp.

Drage, G. (*Capt., retd., I. A.*)—
Burma, 1892-93.—Operations in the Northern Chin Hills. Medal with clasp.
The War of 1914-18.—Despatches, *Lond. Gaz.*, 28 Jan. 16 and 6 Dec. 16. D.S.O.

Drake, H. B.—
The War of 1914-18.—Despatches, *Lond. Gaz.*, 1 Dec. 16.

Drake-Brockman, D. H.—
N.-W. Frontier of India, 1897-98.—Mohmand.—Medal with clasp.
Tirah, 1897-98.—Actions of Dargai and of the Sampagha and Arhanga Passes. Operations in the Waran Valley and action of 16 Nov. 97. Operations in the Bara Valley, 7 to 14 Dec. 97.—Clasp.
China, 1900.—Medal.
Tibet, 1903-04.—Medal.
The War of 1914-18.—Despatches, *Lond. Gaz.*, 17 Feb. and 22 Jun. 15. C.M.G.

Drake-Brockman, H. E.—
N.-W. Frontier of India (1st) Miranzai, 1891.

Drake-Brockman, V. G. (*Lt.-Col. I.M.S., retd.*)—
N.-W. Frontier of India, 1897-98.—Tochi.—Medal with clasp.

D'Ravara, A. J. (*Sr. Asst. Surgeon and Hony. Lieut. retd. I. S. M. D.*)—
N.-W. Frontier of India, Hazara, 1891.—Medal with clasp.

Drew, A. B. H.—
N.-W. Frontier of India, Hazara, 1888.—Action of Kotkai.—Medal with clasp. (*1st and 2nd*) *Miranzai, 1891.*—Clasp.
Chitral, 1895.—Relief of Chitral.—Medal with clasp.
N.-W. Frontier of India, 1897-98.—Malakand—D. A. A. and Q. M. G. at Malakand.—Despatches, G. G. O. 178 of 88.—Clasp.
N.-W. Frontier of India, Waziristan, 1901-02.—Clasp.
The War of 1914-18.—Despatches, *Lond. Gaz.*, 30 June 16 and 1 Feb. 17. Brev. of Col.

Drew, B. C. H.—
China, 1900.—Medal.
N.-W. Frontier of India, Waziristan, 1901-02.—Medal with clasp.
Tibet, 1903-04.—Medal.
The War of 1914-18.—Despatches, *Lond. Gaz.*, 15 Jun. 16.

Drew, G. (*Hony. Capt. Commissary retd. Barrack Deptt.*)—
Afghanistan, 1878-79-80.—Affair of Doaba operations in Kohistan. Operations at and around Kabul.—Medal with clasp.

Drinan, G. C.—
South African War, 1901-02.—Served with Mtd. Inf. Operations in the Transvaal, in Orange River Colony and in Cape Colony, June 01 to 31 May 02.—Queen's medal with 5 clasps.

Drummond, A. B. (*Capt. Supy. List I. A.*)—
N.-E. Frontier of India, Chin-Hills, 1892-93.—Medal with clasp.

Drummond, F. H. R.—
Afghanistan, 1878-79-80.—Affairs with the Ghilzais at and near Jagdalak. Operation at and round Kabul, despatches, Lieut.-Col. F. B. Norenian's despatch, 27 April 79, and G. G. O. 177 of 80.—Medal with clasp.

Drury, F. M. (*Lt.-Col. retd., I.S.C.*)—
Afghanistan, 1879-80.—Medal.
Soudan, 1885.—Suakin. Actions of Hasheen and Tofrek. (Severely wounded; charger killed.) Despatches, *Lond. Gaz.*, 25 Aug. 85.—Medal with 2 clasps. Bronze star.
N.-E. Frontier of India, Chin-Lushai, 1889-90.—(Lushai and Burma Columns.)—Medal with clasp.
Manipur, 1891.—(Severely wounded.) Despatches, G. G. O. 585 of 91.—Clasp.
Chitral, 1895.—Relief of Chitral.—Medal with clasp.
N.-W. Frontier of India, 1897-98.—Buner—Clasp.

Dripdale, A. E.—
The War of 1914-18.—Despatches, *Lond. Gaz.*, 3 July 15, 1 Jan. 16 and 19 Oct. 16. D.S.O., M.C.

D'Souza, A. J.—
The War of 1914-18.—Despatches, *Lond. Gaz.*, 15 Aug. 17.

Duberly, M. R. W.—
The War of 1914-18.—Despatches, *Lond. Gaz.*, 6 July 17.

Ducat, C. M. (*Col. retd., I.A.*)—
Burma, 1885—89.—Medal with 2 clasps.
China, 1900.—Special service officer. Despatches, *London Gazette*, 13th September 1891.—Medal. Brevet of Lieut.-Col.

Ducat Hamersley, H. C. C. (*Maj. retd. I. A.*)—
N.-E. Frontier of India, Sikkim, 1888.—Attack of Jalapla.—Medal with clasp.
N.-E. Frontier of India, Chin-Lushai, 1889-90 (Lushai and Burma Columns.)—Clasp. *Manipur, 1891.*—Clasp.
Burma, 1891.—Thetta Column.—Clasp.
Chitral, 1895.—Relief of Chitral. Action of the Malakand Pass.—Medal with clasp.
Tirah, 1897-98.—Reconnaissance of the Saran Sar and action of the 11 Nov. 97.—Operations against the Khani Khel Chamkannis.—Operations in the Bazar Valley, 25 to 30 Dec. 97.—2 clasps.

Duff-Sutherland-Dunbar, Sir G., Bt.—
Operations in the Abor Country, 1911-12.—Despatches, *Lond. Gaz.*, 16 July 12.

Duffy, T. A.—
Chitral, 1895.—Relief of Chitral.—Medal with clasp.
China, 1900.—Medal.
The War of 1914-18.—Despatches, *Lond. Gaz.*, 5 Apl. 16, 10 May 16, 19 Oct. 16 and 15 Aug. 17.

Duke, A. A. G.—
South African War, 1900-02.—Served at St. Helena.
The War of 1914-18.—Despatches, *Lond. Gaz.*, 15 Jun. 16.

Duke, J. (*Lt.-Col. retd. I. M. S.*)—
Afghanistan, 1879-80.—Action of Charasia. Operations at and around Kabul. March from Kabul to the relief of Kandahar and battle of Kandahar. Operations against the Maris. Despatches, G.G.O. 137 of 80.—Medal with 3 clasps. Bronze star.

Duncan, H. C.—
The War of 1914-18.—Despatches, *Lond. Gaz.*, 21 Jun. 16. D. S. O.

Duncan, R. C.—
N.-W. Frontier of India, 1908.—Operations in the Zakka Khel country.—Medal with clasp.

Duncan, W. (*Surg.-Maj. retd. Ind. Med. Serv.*)—
Afghanistan, 1878-79.—Medal.
China, 1900.—Medal.

Dundas, P. H.—
N.-W. Frontier of India, 1897-98.—Tochi.—Medal with clasp.
China, 1900.—Medal.
The War of 1914-18.—Despatches, *Lond. Gaz.*, 10 Feb. 15 and 15 Aug. 17, Brev. of Lt.-Col. D.S.O.

Dundas, W. L.—
N.-W. Frontier of India, 1897-98.—Tochi.—Medal with clasp.

Dunlop, C. A. M.—
The War of 1914-18.—Despatches, *Lond. Gaz.*, 4 Jan. 17.

Dunlop, H. H.—
Tibet 1903-04.—Medal.
The War of 1914-18.—Despatches, *Lond. Gaz.*, 5 Apl. 16.

Dunlop, W. B.—
Tibet, 1903-04.—Despatches, *Gazette of India*, No. 1065 of 1904.—Medal.
Operations in the Abor country, 1911-12. Despatches, *Gazette of India*, Notification 480, dated 23rd May 12.—Medal with clasp.
The War of 1914-18.—Despatches, *Lond. Gaz.*, 30 Jan. 16 and 8 Feb. 17, Brev. of Lt.-Col. D.S.O.

Dunn, C. L.—
South African War, 1900-01.—Operations in Cape Colony, March and April 1900. Operations in Orange River Colony (May to Nov. 1900). Operations in the Transvaal, Nov. 1900 to Jan. 1901, and again from June to Sept. 1901. Operations in Cape Colony, Feb. to May 1901.—Queen's medal with 4 clasps.
Tibet, 1903-04.—Action at Niani, 26th June 1904. Operations at and around Gyantse, June and July 1904.—Medal with clasp.

Dunscombe, N. B. (*Major retd. I. A.*)—
East Africa, 1902-04.—Operations in Somaliland.—Medal with clasp.

Dunsford, E. H.—
N.-W. Frontier of India, 1897-98.—Tochi.—Medal with clasp.

Dunsford, F. P. S.—
N.-E. Frontier of India, Manipur, 1891.—Medal with clasp.
China, 1900.—Medal.
The War of 1914-18.—Despatches, *Lond. Gaz.* 21 Jan. 16 and 15 Aug. 17, D. S. O.

Dunsterville, L. C.—
N.-W. Frontier of India.—Action at Wana, 3rd Nov. 1894.—Medal with clasp.
N.-W. Frontier of India, 1897-98.—Buner. Action of the Tanga Pass.—Medal with clasp.
China, 1900.—Despatches, *Lond. Gaz.* of 13 Sep. 01.—Medal.
The War of 1914-18.—Despatches, *Lond. Gaz.*, 4 July 16.

Durand, A. G. A.—(*Col. retd. B. S. C.*)—
Afghanistan, 1878-79-80.—With Khyber Field Force. March from Kabul to the relief of Kandahar and battle of Kandahar. Despatches. Brigadier-General J. Doran's despatch, 18 June 80.—Medal with clasp. Bronze star.
N.-W. Frontier of India, Hunza-Nagar, 1891-92.—In command of the force (severely wounded). Services acknowledged by the Government of India. Despatches, G. G. O. 397 of 92.—Medal with clasp.—Brev. of Lt.-Col., C. B.

Durand, C. J. (*Hon. Col. retd. Ben. S.C.*)—
India, 1857-58-59.—Actions of Cawnpore, Serai Ghat, Kalee Nuddee, Shumsabad, siege of Lucknow, storming of the Martinière, action of Aligunge and Rooyah, capture of Bareilly.—Medal with clasp.
N.-E. Frontier of India, Cossyah and Jyntiah Hills, 1862-63.—Storming of Surtung stockades. Thanks of Government. Mentioned in Despatches.
N.-E. Frontier of India, Bhootan, 1864-65.—Recapture of stockades at Bala Pass and Nadoo.—Medal with clasp.

Duthie, T.
The War of 1914-18.—Despatches, *Lond. Gaz.*, 30 Jun. 16.

Dutton, H. R.—
Tibet, 1903-04.—Medal.

Duxbury, C. D.—
China, 1900.—Relief of Pekin.—Medal with clasp.
Aden, 1903-04.—Operations in the interior. Served as S. and T. Officer, Nobat Dakin.
The War of 1914-18.—Despatches, *Lond. Gaz*, 21 Jun. 16.

Dyce, G. H. C. (*Col. retd. I. A.*)—
N.-W. Frontier of India, Hazara, 1868.—Medal with clasp.
Afghanistan, 1878-79-80.—Affair of Matun. Affairs at and near the Shuturgardan Pass. Despatches G. G. O. 1136 of 79.—Medal.
N.-W. Frontier of India, Mahsud Waziri, 1881.
Burma, 1886-87.—Despatches, G.G.O. 434 of 87.—Clasp.
N.-W. Frontier of India.—Waziristan, 1894-95.—Clasp.
Chitral, 1895.—Relief of Chitral.—Served as Asst. Quartermaster-General of the Forces. Despatches, G. G. O. 998 of 95.—Medal with clasp. C. B.

Dyce, H. L.—
The War of 1914-18.—Despatches, *Lond. Gaz.*, 22 Jun. 15 and 1 Jan. 16. M.C.

Dyer, H. R.—
South African War, 1899-1900.—Operations in the Orange Free State, Feb. to May 00, including actions at Karee Siding, Vet River (5 and 6 May), and Zand River. Operations in the Transvaal in May and June 00, including action near Johannesburg.—Queen's Medal with 3 clasps.

Dyer, R. E. H.—
Burma, 1886-87.—Operations of the 3rd Brigade.—Medal with 2 clasps.
N.-W. Frontier of India, Hazara, 1888.—Action of Kotkai.—Clasp.
Chitral, 1895.—Relief of Chitral.—Medal with Clasp.
N.-W. Frontier of India, Waziristan, 1901-02—Clasp.
N.-W. Frontier of India, 1908.—Operations in the Zakka Khel country.—Medal with clasp.
The War of 1914-18. Despatches, *Lond. Gaz.* 31 Oct. 17. C. B.

Dyke, O. M.—
Nile Expedition, 1898.—Battle of Khartoum.—Khedive's medal with clasp. Also Queen's medal.
N.-W. Frontier of India, 1902.—Operations against the Darwesh Khel Waziris.
N.-W. Frontier of India, 1908.—Operations in the Mohmand country. Engagement of Kargha, 24th May 1908. Despatches, G. G. O. No. 521 A of 1908.—Medal with clasp.

Dyke, P. H.—
Uganda, 1897-98.—Action of Jeruba. Staff Officer to O. C. Troops, Unyoro—Mentioned in despatches—Medal with 2 clasps.
N.-W. Frontier of India, 1908.—Operations in the Zakka Khel country. Comdt., 6th Mule Corps. Despatches, G. G. O. No. 188 of 1908.—Medal with clasp.
East Africa (Somaliland) 1908-10.—Medal with class
The War of 1914-18.—Brev. of Lt.-Col. D. S. O.

Dymott, D. F. (*Surg.-Maj. retd. Ind. Med. Serv.*)—
Burma, 1885-86.—Medal with clasp.

Dyson, J. B.—
N.-W. Frontier of India, Hazara, 1888.—Medal with clasp.
N.-W. Frontier of India, Manipur, 1891.—Clasp.

E

Eadie, J. I.—
The War of 1914-18.—Despatches, *Lond. Gaz.*, 19 Oct. 16 and 15 Aug. 17. D.S.O.

Eales, A. C. H. A.—
East Africa (Somaliland) 1914-15.—Medal with clasp.

Eales, C. H. H.—
The War of 1914-18.—Despatches, *Lond. Gaz.*, 1 Jan. 16. M.C.

Eales, C. M. (*Lt.-Col. retd., I. A.*)—
N.-W. Frontier of India, 2nd Miranzai, 1891.—Medal with clasp, *Waziristan, 1894-95.*—Clasp.
N.-W. Frontier of India, 1897-98.—Action of the Ublan Pass, 27 Aug. 97. Operations on the Samana and in the Kurram Valley during Aug. and Sep. 97. Relief of Gulistan.—Medal with 2 clasps.
Tirah, 1897-98.—Operations in the Bara Valley 7 to 14 Dec. 97.—Despatches, G. G. O. 244 of 98.—Clasp.

Eardley-Willmot, H. (*Maj. retd. I.S.C.*)—
Afghanistan, 1879-80.—Medal.
Burma, 1886.—Medal with clasp.

Eardley-Willmot, I. (*Col. retd. I. A.*)—
N.-W. Frontier of India, 1897-98.—Operations on the Samana end in the Kurram Valley during Aug. and Sep. 97.—Operations of the Flying Column under Colonel Richardson, 20 Aug. to 1 Oct. 97. Despatches, G. G. O. 304 of 98.—Medal with 2 clasps.
Tirah, 1897-98.—Clasp.

Eardley-Wilmot, R. (*M.-Gen. retd. Ben. Inf.*)—
N.-W. Frontier of India, 1864-65.—Bhutan Expedition. (Slightly wounded—horse wounded)—Medal with clasp.
N.-W. Frontier of India, Jowaki Expedition, 1877-78.—Mentioned in despatches.—Clasp.
Afghanistan, 1878-80.—Capture of Ali Musjid, battle of Charasiah, and operations at and around Kabul in Dec. 79. Despatches, G. G. O. 137 of 80.—Medal with 3 clasps; Brev. of Lt.-Col.

Earle, E. S.—
N.-W. Frontier of India, Waziristan, 1894-95.—Medal with clasp.
N.-W. Frontier of India, 1897-98.—Tochi.—Medal with clasp.
China, 1900.—Medal.
The War of 1914-18.—Despatches, *Lond. Gaz.*, 13 July 16.

Earle, E. J.—
N.-W. Frontier of India, Waziristan Expedition, 1894-95.—Medal with clasp.
N.-W. Frontier of India, 1897-98.—Tochi.—Medal with clasp.
China, 1900.—Medal.

Earle, H. M. (*Lt.-Col. retd., I.M.S.*)—
Egypt, 1896.—Medal.
N.-W. Frontier of India, 1908.—Operations in the Mohmand country.

Eastmead, C. S.—
Chitral, 1895.—Relief of Chitral.—Medal with clasp.
N.-W. Frontier of India, Waziristan, 1901-02—Clasp

Eate, G. W. (*Sr. Asst. Sgn., Hy. Capt., M.*)—
Burma, 1885—87.—Medal with clasp.

Ebden, F. T. P.—
N.-W. Frontier of India, Waziristan, 1901-02.—Medal with clasp.
Aden, 1903-04.—Operations in the Interior.

Eberhardie, R. C.—
The War of 1914-18—Despatches, *Lond. Gaz.*, 20 Mar. 16.

Eckford, J. J. H. B. (*Major retd. I. A.*)—
Burma, 1885—89.—Operations of the 2, 3, 4 and 5 Brigades.—Medal with 2 clasps.

Eccles, M.—
The War of 1914-18.—Despatches, *Lond. Gaz.*, 5 Apr. 16.

Edgerley, W.—
The War of 1914-18.—Despatches. *Lond. Gaz.*, 19 Oct. 16.

Edward-Collins, C. E.—
South African War, 1900—02.—Operations in the Transvaal, east of Pretoria, July to 29th Nov. 1900, including actions at Belfast (26th and 27th Aug.) and Lydenberg (5th to 8th Sep.) Operations in the Transvaal, 30th Nov. 1900 to Dec. 1901.—Queen's medal with clasp. King's medal with 2 clasps.
Operations in the Abor country, 1911-12.—Despatches, *Gazette of India*, Notification 480, dated 23rd May 12.—Medal with clasp.—Brev. of Major.

Edward, Collins G.—
The War of 1914-18.—Despatches, *Lond. Gaz.*, 5 Apr. 16, 19 Oct. 16 and 12 Jun. 17 M.C.

Edwardes, A. C.—
Aden, 1903-04.—Operations in the Interior.
The War of 1914-18.—Despatches, *Lond. Gaze.*, 4 July 16. 19 Oct. 16, 14 Nov. 16 and 15 Aug. 17, Brev. Lt.-Col.

Edwardes, J. G.—
Chitral, 1895.—Relief of Chitral.—Medal with clasp.
N.-W. Frontier of India, 1897-98.—Operation on the Samana and in the Kurram Valley during Aug. and Sep. 97.—2 clasps.
Tirah, 1897-98.—Actions of Chagru Kotal, Dargai and of the Sampagha and Arhanga Passes Reconnaissance of the Saran Sar and action of 11 Nov. 97.—Clasp.
N.-W. Frontier of India, Waziristan, 1901-02.—Clasp.

Edwardes, S. M.—
Chitral, 1895.—Relief of Chitral.—Despatches. G. G. O. 530 of 95.—Medal with clasp. D. S. O.
N.-W. Frontier of India, 1897-98.—Clasp.
Tirah, 1897-98.—Clasp.
Aden, 1903-04.—Operations in the Interior.
The War of 1914-18.—Despatches. *Lond. Gaz.*, 15 Aug. 17. C.M.G.

Edwards, A. (*Hony. Capt. Commy. retd. I.M.L.*)—
N.-W. Frontier of India, 1897-98.—Malakand. Operations in Bajaur and the Mohmand country. Buner.—Action of the Tanga Pass. Despatches, G. G. O. 217 of 98. Promoted Deputy Assistant Commissary.—Medal with clasp.

Edwards, F. J. M.—
N.-W. Frontier of India, 1897-98.—Mohmand. Despatches, G. G. O. 244 of 98.—Medal with clasp.
Tirah, 1897-98.—Actions of Dargai and of the Sampagha and Arhanga Passes. Reconnaissance of the Saran Sar and action of 9 Nov. 97. Operations at and around Dwatoi and action of 24 Nov. 1897. Operations in the Bara Valley, 7 to 14 Dec. 97.—Clasp. D. S. O.
China, 1900.—Medal.
The War of 1914-18.—Despatches, *Lond. Gaz.*, 1 Jan. 16, 15 June 16, and 4 Jan. 17. C.M.G.

Edwards, J. B. (*Col. retd., I.A.*)—
Afghanistan, 1878-79-80.—Action of the Peiwar Kotal.—Medal with clasp.
Chitral, 1895.—Relief of Chitral.—Despatches. G. G. O. 998 of 95.—Medal with clasp.—D.S.O.

Edwards, R. M. (*Lt.-Col., retd., I. A.*)—
Burma, 1886—89.—Operations of the 3rd Brigade. Expedition up the Chindwin River.—Medal with 2 clasps.
N.-E. Frontier of India, Manipur, 1891.—Clasp.
N.-E. Frontier of India, Lushai, 1892.—With Eastern Sonai Column.
South African War, 1901-02.—Operations in the Transvaal, between 10th September 1901 and 31st May 1902. Despatches, *Lond. Gaz.*, 29th July 1902.—Queen's medal with 3 clasps.
The War of 1914-18.—Despatches, *Lond. Gaz.*, 15 June 16. C.M.G.

Edwards, W. R.—
South African War, 1899-1900.—Operations in Cape Colony in March 1900. Operations in the Orange Free State in April 1900. Operations in the Transvaal in May and June 1900, including actions near Johannesburg (29th May) and Diamond Hill (11th and 12th June). Operations in the Transvaal east of Pretoria, July to 29th November 1900, including action at Belfast (26th and 27th August, 1900). Despatches, *Lond. Gaz.*, 2nd April 1901. C. M. G. Queen's medal with 5 clasps.

Egerton, *Sir* C. C.—
Afghanistan, 1880.—March from Kabul to the relief of Kandahar and battle of Kandahar. Operations against the Maris. Despatches, G. G. O. 582 of 80.—Medal with clasp. Bronze star.
N.-W. Frontier of India, Hazara, 1888.—Served as Assistant Adjutant-General of the Force.—Despatches, G.G.O. 978 of 88.—Medal with clasps, *Hazara, 1891*—Clasp. (*1st and 2nd*) *Miranzai, 1891.*—Served as Assistant Adjutant General. Severely wounded. Despatches, G. G. O. 632 of 91.—Clasp, Brev. of Lt.-Col., D.S.O.
Waziristan, 1894-95.—Commanded Bannu Column. Despatches, G. G. O. 473 of 95.—Clasp, C. B.
Soudan, 1896.—Dongola Expedition. Commanded Suakin Force.—Despatches, *Lond. Gaz.*, 3 Nov. 96.—Medal.—Appointed A.-D.-C. to the Queen.
N.-W. Frontier of India, 1897-98.—Tochi.—Commanded 1st Brigade. Despatches, G. G. O. 1432 of 97.—Medal with clasp.
N.-W. Frontier of India, Waziristan, 1901-02.—In command. Thanks of Government of India.—Clasp.
N.-W. Frontier of India, 1902.—Operations against the Darwesh Khel Waziris. In command.
East Africa, 1903-04.—Operations in Somaliland. Action at Jidballi.—Medal with 2 clasps. G.C.B.

Egerton, R. G.—
N.-W. Frontier of India, Hazara, 1888.—Served with Kashmir Contingent.—Medal with clasp. *Isazai, 1892. Waziristan, 1894-95.*—Clasp.
Chitral, 1895.—Relief of Chitral.—Medal with clasp.
Soudan, 1896.—Dongola Expedition.—Medal, also Khedive's medal. Brev. of Maj.
N.-W. Frontier of India, Waziristan, 1901-02.—Despatches, G. G. O. No. 611 of 02.—Clasp.
The War of 1914-18.—Despatches, *Lond. Gaz.*, 17 Feb. 15, 1 Jan. 16, 19 Oct. 16 and 15 Aug. 17. K.C.B., K.C.I.E.

Elcum, D. (*Surg.-Maj. retd., Ind. Med. Serv.*)—
Afghanistan, 1879-80.—Medal.
Burma, 1886-87.—Despatches, G. G. O. 434 of 87.

Elias, A. H. W.—
East Africa, 1903.—Operations in Somaliland. Medal with clasp.
Nandi, 1905-06.—Clasp.

Elliot, G. L. (*Col. Ind. S. C.*)—
N.-E. Frontier of India, Duffla, 1874-75.
N.-W. Frontier of India, Jowaki, 1877-78.—Medal with clasp.
Afghanistan, 1878-79.—Medal.

Elkington, E. H. O.—
The War of 1914-18.—Despatches, *Lond. Gaz.*, 19 Oct. 16. M.C.

Elliott, F. H. (*Lt.-Col. retd., I. A.*)—
Burma, 1888-89.—Medal with clasp.

Elliot, C. R. (*Lt.-Col., I. A., retd.*)—
Burma, 1886-87.—Medal with 2 clasps.

Elliot, Sir E. L. (*Lt.-Genl. retd., I.A.*)—
Afghanistan, 1878-79.—Medal.
Burma, 1885-89.—Despatches, G. G. O. 339 of '87. D. S.O. —Medal with clasp.
Soudan, 1896.—Dongola Expedition.—Medal, Egyptian Medal. C. B.
South African War, 1901-02.—As Major-General on Staff in Command Mobile Cavalry Division, Orange River Colony, March 01 to June 02. Despatches, *Lond. Gaz.*, 17 June and 29 July 02.—Queen's Medal with 4 clasps. K.C.B.
The War of 1914-18.—Despatches, *Lond. Gaz.*, 1 Jan. 16. Legion of Honour (3rd class).

Elliot, H. R. (*M.-Gen. u.s.l. Mad. Inf.*)—
India, 1858.
Abyssinia, 1867-68.—Battle of Arogee and capture of Magdala.—Medal.

Elliot, R. H. (*Lieut.-Col. retd., I. M. S.*)—
N.-E. Frontier of India, Chin Hills, 1892-93.—Medal with clasp.

Elliot, W. H. W. (*Lieut.-Col. retd., I. M.S.*)—
N.-W. Frontier of India, Hazara, 1888.—Medal with clasp.
N.-W. Frontier of India, (1st) Miranzai, 1891.
N.-W. Frontier of India, 1897-98.—Tochi.—Medal with clasp.
South African War, 1899-1901.—Operations in Natal, 1899, including action at Lombard's Kop, 30th October 1899. Defence of Ladysmith, including action of 6th January 1900. Operations in Natal, March to June 1900. Operations in the Transvaal, east of Pretoria, 21st October 1900. Operations in the Transvaal, 23rd to 27th February 1901, 27th to 30th June 1901 and 2nd to 10th October 1901. Despatches, twice, *London Gazettes*, 8th February 1901 and 10th September 1901.—Queen's medal with 3 clasps. D.S.O.

Elliott, G. H. (*Brev.-Col. retd., Ind. S. C.*)—
Afghanistan, 1879-80.—Action of Ali Boghan; Hissarak Expedition; march from Kabul to the relief of Kandahar, and battle of 1 Sep. Despatches, G. G. O. 566 of 80.—Medal with clasp; bronze star. Brev. of Maj.
Soudan Expedition, 1885.—Suakin. With the Transport Department, Indian Contingent Despatches, *Lond. Gaz.*, 25 Aug. 85.—Medal with clasp; bronze star.

Elliott, H. H.—
N.-W. Frontier of India, 1897-98.—Medal with clasp.
Tirah, 1897-98.—Operations in the Bara Valley, 7 to 14 Dec. 97.—Operations in the Bazar Valley, 25 to 30 Dec. 97.—Clasp.

Ellis, S. G. V.—
West Africa, 1900.—Operations in Ashanti. Relie of Kumasi. (Wounded.)—Medal with clasp.
The War of 1914-18.—Despatches, *Lond. Gaz.*, 1 Jan. 16 and 1 June 17. D.S.O.

Elliston, E. C. (*Lt.-Col. retd., Ind. S. C.*)—
Afghanistan, 1879-80.—In the Kurram Valley.—Medal.
N.-E. Frontier of India, Akha Expedition, 1883-84.—Mentioned in Despatches. *Chin-Lushai Expedition, 1889-90.*—Medal with clasp.

Ellwood, B. P.—
Soudan, 1896.—Dongola Expedition.—Medal. Egyptian Medal.

Elsmie, A. M. S.—
N.-W. Frontier of India, (1st and 2nd) Miranzai, 1891.—Medal with clasp. *Waziristan, 1894-95.*—Clasp.
N.-W. Frontier of India, 1897-98.—Action of the Ublan Pass, 27 Aug. 97. Despatches, G. G. O. 1141 of 97. Operations on the Samana and in the Kurram Valley during Aug. and Sep. 97. Relief of Gulistan. Despatches. G. G. O. 1417 of 97.—Medal with 2 clasps.
Tirah, 1897-98.—Operations in the Bara Valley, 7 to 14 Dec. 97.—Clasp.
The War of 1914-18.—Despatches, *Lond. Gaz.*, 21 Jun. 16, 4 July 16, 19 October 16, C. M. G. Brev. of Col.

Elton, H. S. (*Maj.-Genl. u.s.l.*)—
Burma, 1886-88.—Operations round Nyingyan. Operations of the 3rd Brigade. Despatches, G. G. O. 434 of 87. Brev. of Col.—Medal with 2 clasps.

Elwes, F. F.—
China, 1900.—Relief of Pekin.—Medal with clasp.

Emerson, G. A. (*Lt.-Col. retd. I.M.S.*)—
Afghanistan, 1878-80.—Medal with clasp.
Soudan, 1885.—Suakin —Medal with 2 clasps. Bronze star.

Empson, A. H. A.—
The War of 1914-18.—Despatches *Lond. Gaz.*, 4 Jan. 17.

Engledue, R. S.—
The War of 1914-18.—Despatches, *Lond. Gaz.*, 5 Nov. 15 and 21 June. 16. Brev. of Maj.

Enriquez, A. D. (*Lieut.-Col. u. s. l. I. A.*)—
Afghanistan, 1878-9-80.—Medal.

Enriquez, C. M. D.—
N.-W. Frontier of India, 1908.—Operations in the Mohmand country.—Medal with clasp.

Entrican, J.—
Burma, 1889-92.—Medal with clasp.
China, 1900.—Medal.

Erck, J. C. W.—
Burma, 1887.—Operations of the 1st Brigade.—Medal with clasp.
China, 1900.—Medal.

Erskine, C. E. T.—
N.-W. Frontier of India, 1908.—Operations in the Mohmand country.—Medal with clasp.

Ensine, K. C. S.—
The War of 1914-18.—Despatches, *Lond. Gaz.*, 5 Nov. 15. 28 Jan. 16 and 21 Jun. 16.—M.C.

Etheridge, F.—
The War of 1914-18.—Despatches, *Lond. Gaz* 19 Oct. 16. and 15 Aug. 17.

Etherton, P. T.—
South African War, 1901.—Queen's medal with 4 clasps.
The War of 1914-18.—Despatches, *Lond. Gaz.*, 1 Jan. 16.

Eustace, A. H.—
N.-W. Frontier of India, Hazara, 1888 and 91.—Field Intelligence Officer, 1st and 2nd Brigades, Hazara Field Force, 1891.—Medal with 2 clasps.
East Africa, 1903-04.—Operations in Somaliland. Despatches, *Lond. Gaz.*, 2 Sep. 04.—Medal with clasp. D. S. O.

Evans, A. O. (*Col. Retd. I. M. S.*)—
Burma, 1885-87.—Medal.

Evans, D. MacN.—
The War of 1914-18.—M. C.

Evans, Sir H. M. (*Genl. u.s.l., I.A.*)—
N.-E. Frontier of India, Daffla, 1874-75.—Despatches, G. G. O. 375 of 75. *Naga Hills*, 1879-80.—Despatches, G. G. O. 405 of 80.—Medal with clasp. Brev. of Lt.-Col.
N.-E. Frontier of India, Manipur, 1891.—Despatches, G. G. O. 585 of 91.—Clasp.

Evans, J. W. (*Lt.-Col. retd. L.M.S.*)—
Burma, 1885-6.—Medal with clasp.

Evans, R. J.—
The War of 1914-18. Despatches, *Lond. Gaz.*, 5 April 16, *Gaz. of Ind.* Notfn. 1241 of 16.

Evans, T. J. C.—
Operations in the Abhor Country, 1911-12.—Medal with clasp.
The War of 1914-18.—Despatches, *London Gazette* 5 Nov. 15 and 13 July 16.—M.C.

Evans, W. J.—
The War of 1914-18.—Despatches, *London Gazette*, 22 Jun. 15.

Evatt, F. W. (*Maj. retd. I. A.*)—
N.-W. Frontier of India, Hazara, 1888 and *1891.*—Medal with 2 clasps. *(1st) Miranzai, 1891.*—Superintendent, Army Signalling. *Isazai, 1892.*
N.-W. Frontier of India.—Assistant Superintendent. Army Signalling, Waziristan Delimitation Escort—Action at Wana, 3 Nov. 94. Despatches, G. G. O. 268 of 95. *Waziristan*, 1894-95. Assistant Superintendent, Army Signalling 1st Brigade.—Clasp.

Evatt, J. T. (*Col. retd., I.A.*)—
Burma, 1892.—Tlangtlang and Tashon Columns.—Medal with 2 clasps.
N.-E. Frontier of India, Chin-Hills, 1892-93.—Clasp.
N.-W. Frontier of India, 1897-98.—Mohmand Malakand. Operations in Bajour.—Medal with clasp.
Tirah, 1897-98.—Clasp.
British East Africa, 1898.—Expedition against the Ogaden Somalis.—Medal with clasp.
Uganda, 1899.—Operations against Kaberega - in Command.—Despatches, *Lond. Gaz.*, 2 Jan. 00.—Clasp. D. S. O.
Uganda, 1900.—Operations against the Nandi tribe. In command—Despatches, *Lond. Gaz.*, 10 Sep. 01—Medal with clasp. Brev. of Lt.-Col.

Even, G. E. (*Br. Col. retd., I. A.*)—
Afghanistan, 1879-80.—As Transport officer.—Medal.
N.-E. Frontier of India, Chin-Lushai, 1889-90.—Medal with clasp.
Mekran, 1898.—Staff Officer of the Field Force—Action of Gok Parosh. Despatches, G. G. O. 648 of 98.

Ewart, J. A.—
The War of 1914-18.—Despatches, *Lond. Gaz.*, 4 July 16.

Ewart, R. F.—
South African War, 1899-1902.—Operations in the Orange Free State, Feb. to May 00. Operations in the Transvaal, east of Pretoria, July to 29 Nov. 00. Operations in Orange River Colony, May to 29 Nov. 00, including action at Wittebergen (1 to 29 July). Operations in Cape Colony, south of Orange River, 99-00. Queen's medal with 2 clasps. King's medal with 2 clasps.
The War of 1914-18.—Despatches, *London Gazette*, 15 Aug. 17.

Ewart, R. H.—
N.-W. Frontier of India, Hazara, 1891.—Medal with clasp. *Isazai, 1892. Waziristan, 1894-95.*—Brigade Commissariat Officer, 3rd Brigade. Despatches, G. G. O. 473 of 95.—Clasp. D.S.O.
The War of 1914-18.—Despatches, *Lond. Gaz.*, 17 Feb. 15, 20 and 30 Jun. 16 and 17 Jan. 17. Promd. Maj. Gen., K. C. M. G.

Exham, H.—
The War of 1914-18.—Despatches, *Lond. Gaze.*, 5 Apl. 16.

Eyre, E. H. (*Col. u. s. l. Ind. S.C.*)—
N.-W. Frontier of India, Zhob Valley Expedition 1884.
Burma, 1885-87.—Mentioned in Despatches.—Medal with clasp. Brev. of Col.

Eyre, G. S. (*Lt.-Col. retd. I. S. C.*)—
Afghanistan, 1878-79.—Action of the Peiwar Kotal.—Medal with clasp.
Burma, 1885-87.—Taking of Minhla. Operations round Hlinnedet and Yemethin. Cavalry affair at Yindaw (horse shot). Ponk Column during the operations in Yuw. Yuw Column. Ponk Column during the operations on the Western Frontier Despatches, G. G. O. 434 of 87.—Medal with clasp.
Burma, 1887-89.—Expedition against the Chins.—Clasps.

Eyre, V. G. L. (*Lt.-Col. retd. Ind. S.C.*)—
Afghanistan, 1878-80.—Assault and capture of the Piewar Kotal, passage of Mungiar defile, expedition into the Khost Valley, action of Charasia, operations at and around Kabul in Dec. 79, affair of Sidabad. March from Kabul to relief of Kandahar, and battle of 1 Sep.—Medal with 4 clasps; bronze star.

Ezechiel, J. A. (*Hony. Maj. retd. Comst. Dept.*)—
Afghanistan, 1880.—Medal. Promoted from Sub-Conductor to Dy. Asst. Commy., with the Hony. rank of Lieut.
Burma, 1886-88.—Despatches, G.G.O. 434 of 1887—Medal with 2 clasps. Promoted Asst. Commy.
N.-W. Frontier of India, Tirah, 1897-98.—Actions of Sampagha and Arhanga Passes. Operation in the Bara Valley, 7 to 14 December 1897. Despatches, G. G. O. 244 of 1898.—Medal with 2 clasps Promoted Dy. Commy. with the Hony. rank of Capt.

F

Fagan, B. T.—
The War of 1914-18.—D. S. O.

Fagan, C. G. F. (*Lt.-Col. retd. I.A.*)—
Afghanistan, 1879-80.—Medal.

Fagan, E. A.—
Tirah, 1897-98.—Operations in the Bara Valley, 7th to 14th December 1897.—Medal with 2 clasps.
Tibet, 1903-04.—Medal.
The War of 1914-18.—Despatches, *London Gaz.*, 15 Jun. 16, 4 Jan. 17 and 11 Dec. 17—C. M. G., D. S. O.

Fagan, H. H. F. (*Maj. retd. I. A.*)—
Afghanistan, 1878-79-80.—Action of Charasiah, 6 Oct. 79, and operations at and around Kabul in Dec. 79.—Medal with 2 clasps.

N.-W. Frontier of India, 1897-98.—*Buner.*—Medal with clasp.

Fagan, H. R. (*Maj. retd. I. A.*)—
N.-W. Frontier of India, (1st) Miranzai, 1891.
Soudan, 1896.—Dongola Expedition—Medal, also Khedive's medal.

N.-W. Frontier of India, 1897-98.—*Tochi.*—Medal with clasp.

China, 1900.—Medal.
N.-W. Frontier of India, 1902.—Operations against the Darwesh Khel Waziris.

Fagan, J. L. (*Maj.-Genl. u.s.l.*)—
Afghanistan, 1879-80.—March from Quetta to relief of Kandahar. Despatches, G. G. O 975 of 80 Medal.

Fairbrother, W. T.—
Afghanistan, 1878-79-80.—Bazar Valley Expedition Action on the Gara heights near Kam Dakka Zaimukhit Expedition. Action of Zawa.—Medal.

N.-E. Frontier of India, Sikkim, 1888.—Medal with clasp.

N.-E. Frontier of India, Abor, 1894.
Chitral, 1895.—Relief of Chitral.—Medal with clasp.
N.-W. Frontier of India, Waziristan, 1901-02.—Clasp.

Fairweather, J. (*Dep. Surg.-Gen. retd. Ind. Med. Serv.*)—

N.-W. Frontier of India, 1857.—Expedition to the Bozdar Hills—Medal with clasp.

India, 1858.—Siege and capture of Delhi, action at Bulundshuhr, Alighar and Agra, relief of Lucknow, defeat of Gwalior Contingent at Cawnpore, action at Shamshabad, siege and capture of Lucknow, action of Aligunge and capture of Bareilly—Medal with 3 clasps.

Faithful, G. F. H.—
The War of 1914-18.—Despatches, *Lond. Gaz.*, 1 Jan. 16 and 15 May 17.

Faithfull, H. T. (*Lt.-Col. retd. I.S.C.*)—
Afghanistan, 1878-79-80.—Advance on the Helmund. With Ghuzni Field Force. Battle of Ahmed Khel. Action of Urzu—Medal with clasp.

Faithfull, W. C. (*Lt.-Col. retd. I. A.*)—
Afghanistan, 1878-79-80.—Action of Ali Musjid. Advance on Gandamak. With Kurram Field Force.—Medal with clasp.
N.-W. Frontier of India, Mahsud Waziri, 1881.
N.-W. Frontier of India, 1897-98.—Malakand.—Utman Khel—Buner. Action of the Tanga Pass.—Medal with clasp.

Falcon, R. W.—
Burma, 1885-89.—Medal with 2 clasps.
N.-W. Frontier of India, 1892, Isazai Expedition.
Chitral, 1895.—Relief of Chitral.—Medal with clasp.

N.-W. Frontier of India, Waziristan, 1901-02.—Clasp.
N.-W. Frontier of India, 1908.—Operations in the Zakka Khel country. Operations in the Mohmand country. Despatches, G. G. O. No. 521 A of 1908.—Medal with clasp.

Falk, H.—
The War of 1914-18.—Despatches, *Lond. Gaz.*, 15 Aug. 17.

Fane, V. B.—
N.-W. Frontier of India, Waziristan, 1894-95.—Served as Orderly Officer to G. O. C., 2nd Brigade.—Medal with clasp.
N.-W. Frontier of India, 1897-98, Tochi.—D. A. A. G., 1st Brigade.—Medal with clasp.
China, 1900.—D. A. Q. M. G., Cavalry Brigade. Despatches, *Lond. Gaz.*, 14 May 01.—Medal. Brevet of Lt.-Col.

N.-W. Frontier of India, 1908.—Operations in the Mohmand country. Engagement of Matta, 24th April 1908. Engagement of Kargha, 24th May 1908. Despatches, G. G. O. No. 521 A of 1908.—Medal with clasp.
The War of 1914-18.—Despatches, *Lond. Gaz.*, 20 March 16, 15 Aug. 17. Promd. Maj.-Gen. —C. I. E.

Fanshawe, L. A.—
The War of 1914-18.—Despatches, *Lond. Gaz.*, 4 July 16.

Farebrother, F. H.—
The War of 1914-18.—Despatches, *Lond. Gaz.*, 21 Jun. and 16 Oct. 16.

Farquhar, F. R.—
The War of 1914-18.—M. C.

Farran, G. L.—
N.-W. Frontier of India, Waziristan, 1901-02.—Medal with clasp.

The War of 1914-18.—Despatches, *Lond. Gaz.*, 19 Oct. 16 and 31 Oct. 17.—D. S. O., M. C.

Farrant, H. Fitz L. (*Lt.-Col., retd., I.A.*)—
Burma, 1885-87.—Present at the taking of Minhla and occupation of Mandalay and Bhamo.—Medal with clasp.

Farrington, G.—
Burma, 1887-89.—Medal with clasp.

Farwell, G. A. L.—
The War of 1914-18.—Despatches, *Lond. Gaz.*, 15th Aug. 17.

Fasken, C. G. M. (*Major-Genl., retd., I.A.*)—

Afghanistan, 1878-79-80.—Battle of Ahmed Khel. Action of Urzu. March from Kabul to the relief of Kandahar and battle of Kandahar. Aitchakzi Expedition. Operations against the Maris.—Medal with 2 clasps. Bronze star.

N.-W. Frontier of India, Hazara, 1888.—Served as Brigade-Major, Reserve column. Despatches, Brigr.-Genl. G. N. Channer's despatch, 11 Nov. 88.—Medal with clasp.

East Africa, 1902-04.—Operations in Somaliland. Action at Jidballi. Commanded 2nd Brigade with local rank of Brigadier-General. Despatches, (Brigr.-Genl. Manning, 17th Aug. 1903, and Sir C. C. Egerton, 13th May 1904) *Lond. Gaz.*, 2 Sep. 04.—Medal with 2 clasps. C.B.

Fasken, W. H.—
The War of 1914-18.—Despatches, *Lond. Gaz.*, 1 Jan. 16.—C.B.

Faunce, R. de L.—
Burma, Chin Field Force, 1888-89.—Present at the first actions against the Chin Tribes as Orderly Officer to Brigadier-General E. Faunce, C.S.B.—Despatches, G. G. O. 489 of 90.—Medal with clasp.

Fawcett, E. (*Lt.-Col. retd. Ind. Med. Serv.*)—
Burma, 1886-87.—Despatches, G. G. O. 434 of 87.—Medal with clasp.

Fayrer, J. O. S. (*Lieut.-Col. retd. I. A.*)—
N.-W. Frontier of India, Hazara, 1888 and 1891.—Medal with 2 clasps (*2nd*) *Miranzai, 1891.*—Clasp. *Waziristan, 1894-95.*—Clasp.

Fearnside, C. F. (*Lt.-Col. retd. I. M S.*)—
N.-E. Frontier of India, Chin Hills, 1892-93.—Medal with clasp.

Fell, E. A. H.—
South African War, 1900-02.—Operations in the Transvaal in June 1900. Operations in Natal, Mar. to June 1900, including action at Laing's Nek (6 to 9 June). Operations in Orange River Colony in June 1900. Operations in the Transvaal, 30th Nov. 1900 to June 1901.—Queen's medal with 5 clasps.

The War of 1914-18.—Despatches, *Lond. Gaz.*, 15 Aug. 17.

Fellowes, H. Le M. (*Maj., retd., I.A.*)—
Soudan, 1896.—Dongola Expedition.—Medal. Also Egyptian medal.

N.-W. Frontier of India, 1897-98.—Relief of Malakand. Relief of Chakdara.—Malakand. Action at Landakai. Operations in Bajaur and in the Mamund Country—Utman Khel.—Medal with 2 clasps.

Fellows, C. S.—
East Africa, 1902-04.—Operations in Somaliland.—Medal with clasp.

Fennell, A. E. S.—
Tibet, 1903-4.—Medal.

Fenton, A.—
China, 1900.—Medal.

Fenton, A. B.—(*Hony. Brig.-Genl., retd. I. A.*)—
Burma, 1886-89.—As Brigade-Major, 5th Brigade, Staff Officer to Col. Farrington, Comdg. Shwebo column in advance upon and occupation of Wuntho. Despatches, G. G. O. 434 of 87. Served also as Brigade-Major, 1st Brigade, and subsequently as Depy. Asst. Qr.-Mr.-Gen. for Intelligence. Brev. of Maj.—Medal with 2 clasps.

N.-E. Frontier of India. Chin-Lushai, 1889-90.—As Depy. Asst. Qr.-Mr.-Gen. for Intelligence and Intelligence Officer, Fort White column. Present during operations round Fort White.—Clasp.

Fenwick, G. C. (*Brev.-Col., retd. Ind. S. C.*)—
Afghanistan, 1878-80.—Operations in the Besud Valley.—Mentioned in Despatches.—Medal.
Burma, 1885-87.—Medal with clasp.

Fenwick, J. (*Sr. Asst. Surgn. Hony. Lt., retd. I.S.M.D.*)—
Afghanistan, 1878-1880.—Medal.
Burma, 1889.—Medal with clasp.

Fenwicke-Clennell, E. C.—
The War, 1914-18.—Despatches, *London Gaz.*, 14 Nov. 16—M. C.

Ferrar, M. L.—
N.-W. Frontier of India, 1897-98.—Operations on the Samana and in the Kurram Valley during Aug. and Sep. 1897.—Medal with clasps.
Tirah, 1897-98.—Operations against the Khani Khel Chamkannis.—Clasp.

Ferris, J. E. C. (*Surg. Lt.-Col. retd., I. M. S.*)—
Afghanistan, 1879-80.—Medal.

Ferris, J. L. (*Col. retd. Ind. S.C.*)—
N.-E. Frontier of India, Cossyah and Jyntiah Hills Campaign, 1862-63.
Bhutan Expedition, 1864-65.—Capture of Dewangiri.—Medal with clasp.
Burma, 1886-87.—Despatches, G. G. O. 434 of 87.

Ferris, W. A. T.—
South African War, 1901-1902.—Operations on the Transvaal, May to September 1901 and October 1901 to 31st May 1902. Operations on the Zululand Frontier of Natal, September and October 1901.—Queen's medal with 3 clasps.

Fetherstonhaugh, W. A.-
Tirah, 1897-98.—Actions at Chagru Kota Dargai. Arhanga Pass. Reconnaissance of the Saran Sar. Operations in the Waran Valley and action of 16 Nov. 97. Operations in the Bara Valley, 7 to 14 Dec. 97.—Medal with 2 clasps.

The War of 1914-18.—Despatches, *Lond. Gaz.*, 22 Jun. 15, 1 Jan. 16, 4 Jan. 17 and 15 May 17. Brev. of Lt.-Col. D.S.O.

French, T. (*Lt.-Col., retd. I. A.*)—
Burma, 1899-92.—Tonhon Expedition and Momeik Column.—Medal with clasp.

French-Mullen, J. L. W.—
Burma, 1892-3.—Operations in the Kachin Hills.—Medal with clasp.

Field, C. D.—
Burma,—1891-92—Momeik column.—Medal with clasp.
Burma, 1892-93.—Chin Hills.—Clasp.
China, 1900.—Medal.

Field, C. W.—
Burma, 1886-88.—Operations of the 2nd Brigade.—Medal with clasp.

Field, J. H.—
South African War, 1901-02—Queen's Medal with 3 clasps.

Field, W. C. F. (*Lt.-Col., retd., I. A.*)—
N.-W. Frontier of India, Zhob Valley, 1884. Burma, 1887.—Medal with clasp.

Fielding, C. H.—
The War of 1914-18.—Despatches, *Lond. Gaz.*, 21 Jun. 16.

Finch, C.—
Burma, 1885-87.—Brigade Transport Officer, 1st Brigade.—Medal with 2 clasps.
China, 1900.—Relief of Pekin. Action of Peitsang. Action at Yangtsun.—Medal with clasp.

Fink, G. H.—(*Maj. retd., I. M. S.*)—
N.-W. Frontier of India, 1897-98.—Medal with clasp.
Tirah, 1897-98.—Clasp.

Finlay, R. F.—
N.-W. Frontier of India, Waziristan, 1901-02.—Medal with clasp.
N.-W. Frontier of India, 1902.—Operations against the Darwesh Khel Waziris.
Tibet, 1903-04.—Medal.

N.-W. Frontier of India, 1908.—Operations in the Mohmand country. Engagement of Kargha, 24th May 1908. (Slightly wounded.)—Medal with clasp.
The War of 1914-18.—Despatches, *Lond. Gaz.*, 4 July 16.—D.S.O.

Finlayson, W. T.—
The War of 1914-18.—D.S.O.

Finnie, A. McL.—
South African War, 1899-1901.

Finnimore, J. P. (*Hony. Capt. Dy. Comm. retd. I.M.S.*)—
South Africa, 1879.—Zulu Campaign.—Medal with clasp.

Finnis, B. H.—
China, 1900.—Medal.

Finnis, H.—
The War of 1914-18.—Despatches, *Lond. Gaz.*, 31 May 16. M.C.

Finnis, S. W.—
N.-W. Frontier of India, 1908.—Operations in the Zakka Khel country.—Medal with clasp.

Firth, R. A.—
N.-E. Frontier of India, Chin-Hills, 1892-93.—Medal with clasp.
The War of 1914-18.—Despatches, *Lond. Gaz.*, 28 Jan. 16. D.S.O.

Fisher, J.—
N.-E. Frontier of India, Manipur, 1891.—Medal with clasp.
Tirah, 1897-98.—Medal with clasp.
N.-W. Frontier of India, Waziristan, 1901-02.—Clasp.
Operations in the Abor Country, 1911-12.—Despatches, *Lond. Gaz.*, 16 July 12.—C.B.

Fisher, J.—
N.-W. Frontier of India, 1897-98.—Utman Khel. Buner. Operations in Rajaur. Operations in the Mamund country. Despatches, *London Gazette*, 11th January and 18th March 1898. D. S. O.—Medal with clasp.

FitzGerald, A.—
South African War, 1900-2.—(Severely wounded.)—Relief of Kimberley. Operations in Cape Colony and Orange River Colony, Feb. 01 to 31 May 02.—Queen's medal with 3 clasps.

FitzGerald, C. M. (*Lt.-Col. retd. I. A.*)—
Afghanistan, 1878-79-80.—Affair of Matun. Action of Chihildakhteran. March from Kabul to the relief of Kandahar and battle of Kandahar.—Medal with clasp. Bronze star.
N.-E. Frontier of India, Akka, 1883-84.—Despatches, Brigadier-General, R. S. Hill's despatch, 31 Jan. 84. *Manipur, 1891.*—Chief Commissariat Officer, Kohima Column, and afterwards of the entire force. Despatches, G. G. O. 585 of 91.—Medal.
Chitral, 1895.—Relief of Chitral.—Served as Divisional Transport Officer. Despatches, G. G. O. No. 968 of 95.—Medal with clasp.

FitzGerald, C. R. L.—
South African War, 1901-02—Operations in Cape Colony from 1st May 1901 to 29th April 1902.—Queen's Medal with 3 clasps.
The War of 1914-18.—Despatches, *Lond. Gaz.*, 13 July 16.—D.S.O.

FitzGerald, D. C. V.—
N.-W. Frontier of India, 1908.—Operations in the Mohmand country.—Medal with clasp.
The War of 1914-18.—Despatches *London Gaz.*, 19 Oct. 16.—M. C.

Fitz-Gerald, G. M.—
The War of 1914-18.—Despatches, *Lond. Gaz.*, 14 Nov. 16.—M. C.

FitzHugh, A. (*Hon. Maj.-Gen. retd. Ben. S. C.*)—
N.-W. Frontier of India, 1863-64.—Affair with the Momunds near Shubkudder.—Medal with clasp. *Dour Valley Expedition, 1872.*—Mentioned in Despatches. *Jowaki Expedition, 1877.* Mentioned in Despatches.—Clasp.
Afghanistan, 1878-79-80.—Actions of Surkh (wounded), Kotal, and Charasiah, forcing of the Peiwar Kotal action at Sapari Pass, action at Charasiah and operations at and around Kabul in December 1879, march from Kabul to relief of Kandahar and battle of 1 Sep. Expedition against the Atchakzais. Despatches, G. G. Os. 1136 of 70 and 137 and 566 of 80.—Medal with 4 clasps; bronze star. Brev. of Lt.-Col. C. B.

Flagg, A.—
N.-W. Frontier of India, 1908.—Operations in the Zakka Khel country.—Operations in the Mohmand country.—Medal with clasp.

Fleming, A. N.—
N.-W. Frontier of India, 1897-98.—Medal with clasp.
Tirah, 1897-98.—Clasp.
The War of 1914-18—Despatches, *Lond. Gaz.*, 1 Jan. 16 and 28 Dec. 17.—D.S.O.

Fleming, J. K. S.—
The War of 1914-18.—Despatches, *Lond. Gaz.*, 4 July 16.

Fleming, J. M. (*Lt.-Col. retd. I. A.*)—
Afghanistan, 1879-80.—Medal.

Fleming, W. E.—
The War of 1914-18.—Despatches, *London Gazette*, 27 Mar. 15, 22 Jun. 15 and 15 Aug. 17.—M.C.

Fletcher, H. A. (*Lt.-Col. retd. Ben. Cav. Genl.-at-Arms.*)—
N.-W. Frontier of India, 1863-64.—Affair with Momands near Shubkudder.—Medal with clasp.
Egypt, 1882.—Medal; bronze star.

Flowerdew, R. E.—
The War of 1914-18.—Despatches, *Lond. Gaz.*, 5 Apl. 16.

Flood, J. D. M.—
South African War, 1901-02.—Operations in Orange River Colony, 25th June to 10th Sep. 1900. Operations in the Transvaal, 31st Nov. 1900 to Aug. 1901. Operations on the Zululand Frontier of Natal in Sep. and Oct. 1901. Operations in the Transvaal, Nov. 1901 to 31st May 1902.—Queen's medal with 3 clasps. King's medal with 2 clasps.

Fooks, G. E. (*Lt.-Col. retd. I. M. S.*)—
N.-W. Frontier of India, Zhob Valley, 1884.
East Africa (British), 1896.—Operations against the Mazrui Rebels.—Medal.
China, 1900.—Served as Senior Medical Officer, Tientsin, during relief of Pekin. Despatches, *London Gazette*, 14 May 01.—Medal with clasp.

Fooks, H.—
N.-E. Frontier of India, Chin-Lushai, 1889-90.—Medal with clasp.
N.-W. Frontier of India, Zhob Valley, 1890.
N.-E. Frontier of India, Manipur, 1891.—Clasp.
N.-W. Frontier of India, Waziristan, 1894-95.—Clasp.
China, 1900.—Medal.

Foord, W. O. (*Hon. Maj.-Genl. retd. Mad. S.C.*)—
India, 1858.—Southern Mahratta and North Canara Field Forces.—Medal.

Foote, F. B. (*Hon. Col. retd. Ben. S.C.*)—
Expeditions against Hill Tribes near Peshawar, 1851-52.
India, 1857-58.—With the Goorkha Force at the capture and re-occupation of Goruckpore, actions of Shahgunge and Pandoonudee; siege and capture of Lucknow. Despatches, *Lond. Gaz.*, April and 7 May 58.—Medal with clasp.

Foottit, J. A. W.—
N.-W. Frontier of India, 1908.—Operations in the Mohmand Country. Medal with clasp.

Forbes, A. W.—
Afghanistan, 1879-80.—Medal.

Forbes, E. E. (*Capt., retd., Ind. S. C.*)—
Burma, 1886-9.—Medal with 2 clasps.

Forbes, E. E.—
N.-W. Frontier of India, 1897-98.—Operations on the Samana and in the Kurram Valley during Aug. and Sep. 97.—Medal with 2 clasps.
Tirah, 1897-98.—Operations against the Khani Khel Chamkannis.—Clasp.

The War of 1914-18.—Despatches, *Lond. Gaz.*, 24 Jun. 16.—M.C.

Forbes, L.—
South African War, 1900.—Operations in the Orange Free State, March to May 1900, including action at Zand River (10th May). Operations in the Transval, west of Pretoria, July and August 1900. Operations in Cape Colony, South of Orange River, February and May 1900.—Queen's Medal with 3 clasps.
N.-W. Frontier of India, 1908.—Operations in the Zakka Khel country. Orderly Officer to G. O. C., 1st Bde., 14th Feb. to 1st March 1908.—Medal with clasp.
The War of 1914-18.—Despatches, *Lond. Gaz.*, 4 Jan. 17.

Forbes, L. A. (*Maj., retd., I. A.*)—
Burma, 1891.—Tlangtlang Column. (Severely wounded.)—Medal with clasp.

Ford, C. A. W. (*Major, retd., I. A.*)—
British East Africa, 1898.—Operations against the Ogaden Somalis. Staff Officer to the O. C. the Troops.—Medal with clasp.
N.-W. Frontier of India, Waziristan, 1901-02.—Medal with clasp.

Ford, H. B.—
N.-W. Frontier of India, (2nd) Miranzai, 1891.—Medal with clasp.
Waziristan, 1894-95.—Clasp.
N.-W. Frontier of India, 1897-98.—Defence and Relief of Malakand.—(Severely wounded.)—Malakand.—Operations in the Mamund country.—Buner. Despatches, G. G. O. 1889 of 97.—Medal with 2 clasps.
East Africa, 1903-04.—Operations in Somaliland. Action at Jidballi.—Medal with 2 clasps.

Ford, J. M. R.—
The War of 1914-18.—Despatches, *Lond. Gaz.*, 30 Jun. 16.

Fordham, W. M.—
N.-W. Frontier of India, 1897-98.—Engagement near Shabkadr, 9 Aug. 97. *Mohmand, Buner.*—Action of the Tanga Pass.—Medal with clasp.
China, 1900.—Medal.

Fordyce, A. D. (*Maj. retd. Ind. S.C.*)—
Burma, 1886-87.—Medal with clasp.

Fordyce, H. L. D.—
Burma, 1892-93.—Medal with clasp.
Chitral, 1895.—Relief of Chitral.—Medal with clasp.
Tirah, 1897-98.—2 clasps.
N.-W. Frontier of India, 1908.—Operations in the Zakka Khel country. Base Supply Officer.
The War of 1914-18, Despatchest. *Lond. Gaz.*, 5 Apr. 16, Brev. of Col.

Fordyce, J. F. D. (*Col. retd. I.S.C.*)—
Afghanistan, 1878-79.—Medal.

Forster, P. B. B.—
N.-W. Frontier of India, Hazara, 1888.—Medal with clasp. *Zhob Valley, 1890.*
N.-W. Frontier of India, Waziristan, 1901-02.—(Wounded).—Medal with clasp.
East Africa, 1903.—Operations in Somaliland.—Medal with clasp.

Forster, W. C. H.—
N.-W. Frontier of India, Waziristan, 1901-02.—Medal with clasp.

Forteath, A. M.—
South African War, 1901.—Queen's medal with 4 clasps.

Forteath, F. W. H.—
Tirah, 1897-98.—Medal with 2 clasps.
China, 1900.—Medal.

Forth, C. T. W. (*Major retd. I. A.*)—
N.-E. Frontier of India, Lushai, 1890-91.—Medal with clasp.
Chitral, 1895.—Relief of Chitral.—Medal with clasp.

Foss, K. M. (*Lieut.-Col. retd. I. A.*)—
Rumpa Rebellion, 1879-80.—Thanked by Government, P. G. No. 7657, 7 Dec. 80.
Burma, 1886-87.—Despatches, G. G. O. 434 of 87.—Medal with clasp.

Foster, C. S.—
South African War, 1899-1901.—Operations in Cape Colony, March to June 1900. Operations in the Transvaal, July 1900 to July 1901.—Medal with 2 clasps.
N.-W. Frontier of India, 1908.—Operations in the Zakka Khel country.—Medal with clasp.

Foster, L. M. (*Lt.-Col., retd., I.A.*)—
N.-E. Frontier of India, Chin-Lushai, 1889-90.—Medal with clasp.
N.-W. Frontier of India, 1897-98.—Mohmand. Despatches, G. G. O. 1318 of 97.—Medal with clasp.
Tirah, 1897-98.—Action of the Sampagha Pass. Operations at and around Dwatoi and action of 24 Nov. 97. Operations in the Bara Valley, 7 to 14 Dec. 97.—Clasp.

Foster, R. B. B.—
East Africa, 1903-04.—Operations in Somaliland.—Medal with clasp.

Foster, W. M. A.—
The War of 1914-18.—Despatches, *Lond. Gaz.* 15 Jun. 16 and 15 Aug. 17. D. S. O.

Foulkes, T. H.—
Burma, 1894-96.
Tirah, 1897-98.—Operations in the Bara Valley.—Medal with clasp.
China, 1900.—Medal.

Fowle, T. C. W.—
N.-W. Frontier of India, 1908.—Operations in the Mohmand country.—Medal with clasp.

Fowler, C. A.—
N.-W. Frontier of India, (1st) Miranzai, 1891.
N.-W. Frontier of India, 1902.—Operations against the Darwish Khel Waziris.
N.-W. Frontier of India, 1908.—Operations in the Mohmand country. Engagement of Kargha, 24th May 1908. Despatches, G. G. O. No. 521 A of 1908.—Medal with clasp. D.S.O.
The War of 1914-18.—Despatches, *Lond. Gaz.*, 1 Jan. 16, C. B

Fowler, F. J.—
N.-W. Frontier of India, Zhob Valley, 1884.
Burma, 1887-88.—With Maukme Expeditionary Column, 1888. Capture of Moni and engagement at Kedugle. *Karen Field Force, 1888-89.*—Engagement at Nga Kyaing. Forcing of the Nanko defile and capture of Sawlon.—Despatches, Medal with clasp. D. S. O.
Uganda, 1898-99.—Commanded Column in pursuit of mutineers and rebels.—Actions at Jaruba, Kuveri and Kyanginti.—Mentioned in Despatches.—Medal with 2 clasps. Brev. of Maj.
China, 1900.—Medal.
East Africa (Somaliland, 1908 10).—Commanding Troops.—Medal with clasp.

Fox, C. J.—
Chitral, 1895.—Relief of Chitral.—Medal with clasp.

Fox-Male, W. C. T.—
The War of 1914-18.—Despatches, *Lon. Gaz.*, 19 Oct. 16. M. C.

Francis, G. F. (*Col. retd. I. A.*)—
Afghanistan, 1878-79.—Medal.
N.-W. Frontier of India, Zhob Valley, 1884.—Affair of Dowlutzai.
Soudan, 1885.—Suakin. Action of Tamai.—Medal with clasp; bronze star.

Francis, R. F.—
The War of 1914-18.—Despatches, *Lon. Gaz.*, 1 Jan. 16. M.C.

Franklin, G. D.—
Tibet, 1903-04.—Operations at and around Gyantse, between 5th May and 6th July 1904. March to Lhassa, 14th July to 3rd August 1904.—Medal with clasp.

Franklin, H. S. E.—
Tirah, 1897-98.—Medal with clasp.
China, 1900.—Medal.

Franks, K. F.—
N.-W. Frontier of India, 1908.—Operations in the Zakka Khel country.—Medal with clasp.

Franks, W. A. (*Maj.-Gen. u.s.l. Ben. Inf.*)—
India, 1857-58.—Mutiny at Nowgong and Nagode, Pursuit of Feroze Shah.—Medal.
Afghanistan, 1878-79.—With the Peshawar Valley Field Force, 2nd Bazar Valley Expedition, and action at Halwai Heights.—Medal.

Fraser, A. C.—
N.-W. Frontier of India, Waziristan, 1901-02.—Medal with clasp.

Fraser, E. W.—
Burma, 1885-87.—Medal with clasp.
China, 1900.—Medal.

Fraser, F. J.—
Chitral, 1895.—Relief of Chitral.—Medal with clasp.
N.-W. Frontier of India, 1897-98.—Tochi clasp.

Fraser, H. St. J. (*Maj. retd. I.M.S.*)—
N.-W. Frontier of India, 1897-98.—Malakand.—Medal with clasp.
Tirah 1897-98.—Actions of the Sampagha and Arhanga Passes. Operations in the Bazar Valley, 25 to 30 Dec. 97.—Clasp.

Fraser, W. A. K.—
The War of 1914-18—Despatches, *Lond. Gaz.*, 1 Jan. 16 and 28 Dec. 17. Brev. of Major M.C.

Fraser, W. F. M. I. (*Capt. retd. Ind. S.C.*)—
Burma, 1886-89.—Medal with 2 clasps.

Fraser, W. S. (*Maj., retd., I.A.*)—
N.-W. Frontier of India, (2nd) Miranzai, 1891.—Medal with clasp.
N.-W. Frontier of India, 1897-98.—Utman Khel.—Buner.—Orderly Officer to G. O. C. Force.—Action of the Tanga Pass. Despatches, G. G. O. 217 of 98.—Medal with clasp.
South African War, 1901-02.—Operations in the Transvaal, 11 June 01 to 1 March 02. Operations in the Orange River Colony, 2 March to 8 March 02. Operations in the Zululand Frontier of Natal, 24 Sept. to 14 Oct 01. Operations in Cape Colony, 15 April to 31 May 02.—Queen's medal with 4 clasps.

Frazer, G. S.—
Burma, 1887–88. Medal with 2 clasps.

The War of 1914-18.—Despatches, *Lond. Gaz.*, 5 Apr. and 19 Oct. 16. C. M. C.

Freeman, T.—
South Africa, 1879.—Zulu Campaign.—Medal.
N.-W. Frontier of India, 1897-98.—Malakand.—Medal with clasp.

Freeland, J. C.—
The War of 1914-18.—Despatches, *Lond. Gaz.*, 22 Jan. 15, 1 Jan 16 and 4 Jan. 17.

Frere, A. G.—
South Africa, 1899-1902.—Relief of Kimberley. Operations in the Orange Free State, February to May 1900 including operations at Paardeburg (17th to 26th February) action at Driefontien. Operations in Cape Colony. South of Orange River Colony, 1899-00 including actions at Cole berg (1st January to 12th February). Operations in Transvaal, 1901. Operations in Orange River Colony and Transvaal, June 1901 to 31st May 1912.—Queen's medal with 4 clasps. King's medal with 2 clasps.

Frost, F. D.—
South African War 1900-1902.—Served in ranks, Imperial Yeomanry (to June 19 1). Operations in Cape Colony, March to April 1900. Operations in Orange River Colony, May to July 1909 including actions at Bechlehem (6th and 7th July), and Witbenberg (1st to 29th July). Operations in Transvaal, August to 29th November 19 0. Operations in Transvaal, 30th November 1900 to February 1901. Operations in Cape Colony, February and March 1901. Operations in Orange River Colony March to May 1901 and February to 31st May 1902. Queen's Medal with 3 clasps. King's Medal with 2 clasps.
The War of 1914-18.—Despatches, *Lond. Gaz.*, 19 Oct. 16 and 15 Augt 17. M.C.

Frost, G. H.—
South African War, 1899-1901.—Operations in Natal, 1899, including action of Lombard's Kop, 30th October 1899. Defence of Ladysmith, including action of 6th January 1900. Operations in Natal, March to June 00. Operations in the Transvaal, east of Pretoria, 2 and 3 Aug. 00. Operations in the Transvaal from 11 Dec. 00 to 9 July 01. Operations in the Orange River Colony from 10 July to 13 Nov. 01.—Queen's medal with 5 clasps.

Fry, A. B.—
N.-W. Frontier of India, Waziristan, 1901-02.—Despatches, G. G. O. No. 611 of 02.
N.-W. Frontier of India, 1902.—Operations against the Darwesh Khel Waziris.
Tibet, 1903-04.—Medal.

The War of 1914-18—Despatches, *Lond. Gaz.*, 15 Aug 17. D. S. O.

Fry, C. I.—
N.-W. Frontier of India, Zhob Valley, 1884.
Soudan, 1885.—Suakin.—Medal with clasp; bronze star.
British East Africa, 1898.—Operations against the Ogaden Somalis.—Medal with clasp; Brilliant Star of Zanzibar, 4th class.
East Africa, 1902.—Operations in Somaliland.—Medal with clasp.
The War of 1917-18.—Despatches, *Lond. Gaz.*, 5 Apr. 16.

Fryer, G. W. S.—
Afghanistan, 1880.—Medal.

Fryer, L. C. (*Col. retd. I.A.*)—
Burma, 1886-87.—(Severely wounded.) Despatches, G. G. O. 434 of 87.—Medal with clasp.
N.-W. Frontier of India, 1897-98.—Tochi.—Medal with clasp.
N.-W. Frontier of India, Waziristan, 1901-02.—clasp.
N.-W. Frontier of India, 1908.—Operations in the Zakka Khel country.—Medal with clasp.

Fullerton, J. C. (*Lt.-Col., retd., I. M. S.*)—
Afghanistan, 1878-79-80.—Action of Saif-u-din. Advance from Quetta to the relief of Kandahar.—Medal.
N.-W. Frontier of India, Zhob Valley, 1890.

Fulton, E. J.—
The War of 1914-18.—Despatches, *Lond. Gaz.*, 5 Apr. and 13 July 16.

Fulton, F. C. R.—
China, 1900.—Despatches, *Lond. Gaz.*, 14 May 1901.—Medal.
The War of 1914-18.—Despatches, *Lond. Gaz.* 13 Oct. and 14 Nov. 16. Brev. of Col.

Fulton, J. D.—
The War of 1914-18.—Despatches, *Lond. Gaz.*, 19 Oct. 16 and 15 Aug. 17. M.C.

Fulton, R. (*Col. retd. I. A.*)—
N.-E. Frontier of India, Sikkim, 1888.—Attack of Jalapla.—Medal with clasp.

Furlong, M. (*Col., u. s. l., Ind. S. C.*)—
India, 1857-58.—Capture of Copul.

Furney, J. L.—
N.-W. Frontier of India, 1897-98.—Malakand. Operations in Bajaur—Mohmand.—Medal with clasp.
Tirah, 1897-98.—Clasp.
China, 1900.—Medal with clasp.
The War of 1914-18.—Despatches *Lond. Gaz.*, 5 Apr. and 13 July 16. Brev. of Lt.-Col.

G

Gabbett, A. C. (*Major, retd., I.A.*)—
Tirah, 1897-98.—Medal with 2 clasps.

Gaisford, J. (*Capt., retd., I. A.*)—
N.-W. Frontier of India, Isazai, 1892.
Chitral, 1895.—Relief of Chitral.—Medal with clasp.
N.-W. Frontier of India, 1897-98.—Tochi.—Clasp.

Galbraith, E. D.—
N.-W. Frontier of India, 1908.—Operations in the Zakka Khel country. Operations in the Mohmand country.—Medal with clasp.
The War of 1914-18.—Despatches, *Lond. Gaz.*, 4 Jan. and 11 Dec. 17. Brev. of Major. D. S. O.

Galbraith, I. W.—
The War of 1914-18.—M.C.

Gale, A. (*Dy. Comy. and Hon. Capt., retd., I.-A. Depts.*)—
Soudan, 1885.—Suakin.—Medal with clasp; Khedive's star.

Gale, E. S.—
The War of 1914-18.—Despatches, *Lond. Gaz.* 4 July 16.

Gall, L.—
The War of 1914-18.—Despatches, *Lond, Gaz.*, 13 July 16. M.C.

Galvin, M.—
Chitral, 1895.—Relief of Chitral.—Medal with clasp.

Gamlen, R. L.—
Operations in the Abor country, 1911-12.—Medal with clasp.

Garbett, C. H. V. (*Col., u. s. l., I. A.*)—
Afghanistan, 1879-80.—D. A. A. G., 2nd Brigade, Kyber Field Force. March from Kabul to the relief of Kandahar and battle of 1st September. Despatches. Lieut.-Colonel A. R. D. Mackenzie's despatch, 1st January 1880.—Medal with clasp. Bronze Star.

Garden, A. R. H. (*Major, retd., I.A.*)—
N.-W. Frontier of India, Hazara, 1891.—Medal with clasp.
N.-W. Frontier of India, 1897-98.—Malakhand. Operations in Bajaur and in the Mamund country. Buner.—Medal with clasp.
N.-W. Frontier of India, 1902.—Operations against the Darwesh Khel Waziris.
N.-W. Frontier of India, 1908.—Operations in the Mohmand country.—Medal with clasp.

Gardiner, A. B.—
China,—1900.—Medal.

Gardiner, W. A.—
The War of 1914-18.—Despatches, *Lond. Gaz.*, 6 July 17. M.C.

Garnett, C. L. S. (*Lucknow Volr. Rifles.*)—
South African War, 1901-02.—Operation in Cape Colony, 7th to 14th April 1901. Operations in the Orange River Colony, 15th to 30th April 1901. Operations in the Transvaal, 1st May 1901 to 31st May 1902.—Queen's medal with 5 clasps.

Garratt, H. S.—
N.-W. Frontier of India, 1897-98.—*Malakand, Buner.*—Action of the Tanga Pass.—Medal with clasp.
China, 1900.—Medal.

Garrett, K. A.—
The War of 1914-18.—M. C.

Garratt, R.—
Tirah, 1897-98.—Medal with clasp.

Garrett, A. J. (*Lt.-Col., retd., Ind. S. C.*)—
Afghanistan, 1879-80.—Defence of Kandahar and battle of 1 Sep. (wounded).—Medal with clasp.

Garrett, N. D. (*Capt. h.p. Ben. Art.*)—
India, 1858-59.—Medal.
N.-W. Frontier of India, 1863.—Umbeyla Expdn.—Medal with clasp.

Garstin, G. L. (*Col., retd., I.A.*)—
Afghanistan, 1880.—Medal.
Soudan, 1885.—Suakin. Actions of Hasheen and Tamai.—Medal with clasp. Bronze star.
Chitral, 1895.—Relief of Chitral.—Medal with clasp.
Tirah, 1897-98.—Operations in the Bara Valley, 7 to 14 Dec. 97.—2 clasps.

Garstin, H. A.—
The War of 1914-18.—M. C.

Gartside-Tipping, R. F. (*Lt.-Col. u. s. l., I. A.*)—
Afghanistan, 1879-80.—With Kurram Field Force.—Medal.
N.-W. Frontier of India, 1st Miranzai, 1891.—Isazai, 1892.
China, 1900.—Despatches, *Lond. Gaz.*, 14th May 1901.—Medal with clasp. C. B.

Garvie, J.—
N.-W. Frontier of India, Zhob Valley, 1890.

Gaskell, R. W.—
South African War, 1901-02.—Operations in the Transvaal, 18 Dec. 01 to 31 May 02.—Queen's medal with 3 clasps.

Gastrell, E. T.—
Burma, 1885-87.—Raised and commanded the "Punjabi Levy" (Mandalay Military Police Battalion). Thanks of Local Government No. 236 of 27th December 1888.—Medal with clasp.
Burma, 1887-89.—Commanded Special Military Police operations against the Setkya Mintha Pretender. Thanks of Local Government No. 5671 of 5th August 1889.—Clasp.
Burma, 1889-92.—In command of Mandalay Military Police Battalion. Thanks of Loca Government No. 381 of 4th June 1891, and by telegram in 18—2.—Clasp.

Gatacre, *Sir* J. (*Maj.-Genl. u.s.l.*)—
India, 1858.—Operations in Khandeish.
China, 1860.—Medal.
Okhamundel, 1865-68.
Afghanistan 1879-80.—Medal.
Burma, 1886-88.—(Despatches, twice. Special mention. G. G. O. 339 of 87.)—Medal with 2 clasps, C. B.

Gatherer, A. H. N.—
The War of 1914-18.—Despatches, *Lond. Gaz.*, 13 July and 19 Oct. 16. M. C.

Gaudoin, G. R. (*Senior Asst. Surgeon and Hony. Capt. retd. I. S. M. D.*)—
Burma, 1885-86.—Promoted 1st Grade Assistant Apothecary.—Medal with clasp.

Gaussen, J. R.—
China, 1900.—Relief of Pekin. Wounded. Despatches, *Lond. Gaz.*, 14 May 01.—Medal with clasp. D. S. O.
The War of 1914-18.—Despatches, *Lond. Gaz.*, 15 Jun. 16, and 4 Jan. 17 C. M. G.

Gayer, E. A.—
Burma, 1885-88.—Present at the taking of Minhla and occupation of Mandalay. Operations round Hlinedet, Yemethen, and Myingyan. Operations of the 2nd and 5th Brigade.—Medal with 2 clasps.
Tirah, 1897-98.—Medal with 2 clasps.

Gee, F. W.—
Chitral, 1895.—Relief of Chitral.—Medal with clasp.
N.-W. Frontier of India, 1897-98.—Tochi.—Clasp.
China, 1900.—Medal.
East Africa, 1903-04.—Operations in Somaliland Despatches, *London Gazette*, 2nd September 1904—Medal with clasp.
The War of 1914-18—Despatches, *Lond. Gaz.*, 15 Aug. 17.

Gell, H. V.—
The War of 1914-18.—Despatches, *Lond. Gaz.*, 13 July 16.

Gell, P. F.—
South African War, 1900.—Served in Mediterranean.—Medal.

Gell, W. T.—
China, 1900.—Medal.

Geoghegan, F. E.—
N.-W. Frontier of India, 1897-98.—Tochi Brigade. Commissariat Officer, 2nd Brigade.—Medal with clasp.
China, 1900.—Base Transport Officer. Despatches *London Gaz.*, 14 May 01.—Medal with clasp.

Geoghegan, N. M.—
The War of 1914-18.—Despatches, *Lond. Gaz.*, 15 Jan. and 19 Oct 16 Brev. of Lt.-Col.

Geoghegan, O. A.—
Operations in the Abor country, 1911-12.—Medal with clasp.

Geoghegan, S.—
Burma, 1889-92.—Momeik Column.—Medal with clasp.
Burma, 1892-93.—(Severely wounded in skirmish near Pimpi.)
Burma, 1894-95-96.
China, 1900.—Medal.
The War of 1914-18.—Despatches, *Lond. Gaz.*, 21 Jun. 16 C.B.

Georges, H. W. E.—
Rumpa Rebellion, 1879-80.—Operations in the Rumpa and Golconda Hills.
Burma, 1886-87.—Operations of the 6th Brigade.—Medal with 2 clasps.

Gerrard, H. D. (*Maj. retd Ind. S. C.*)—
Burma, 1885-86.—Medal with clasp.

Gharpurey, K. G.—
East Africa (Somaliland), 1908-10. Medal with clasp.

Gib, H. A.—
N.-W. Frontier of India, 1897-98.—Defence of Malakand. Operations on the Samana and in the Kurram Valley during Aug. and Sep. 97.—Medal with 3 clasps.
Tirah, 1897-98.—Actions of the Sampagha and Arhanga Passes.—Clasp.
N.-W. Frontier of India, Waziristan, 1901-02.—Clasp.
N.-W. Frontier of India, 1908.—Operations in the Mohmand country.—Medal with clasp.

Gibb, J. (*Sr. Asst. Surgn. Hony. Capt., retd., I.S.M.D.*)—
Afghanistan, 1878—80.—Defence and battle of Kandahar.—Medal with clasp.

Gibbs, A. A.—
East Africa, 1903-04.—Operations in Somaliland.—Medal with clasp.
East Africa (Somaliland), 1908-10.—Clasp.

Gibbs, M. I. (*Lt.-Col., retd, Ind. S. C.*)—
Afghanistan, 1879-80.—Action at Kam Dakka—Medal.
N.-W. Frontier of India, 1897.—Defence of Malakand.—Night attacks on camp of 26 and 27 July.—Despatches, G. G. O. 1089 of 98.—Medal with clasp.

Gibson, H. R. B.—
The War of 1914—18, Despatches, Lon. Gaz., 15 Aug. 17.

Gidney, H. A. J.—
China, 1900.—Relief of Pekin. Action of Peitsang.—Medal with clasp.

Giffard, D. H. R.—
Tibet, 1903-04.—Medal. Operations in the Abor country, 1911-12.—Medal with clasp.

Giffard, G. G.—
N.-E. Frontier of India. Manipur, 1891. Medal and clasp.

Gilbert, C. E. L. (*Lt.-Col., retd., I.M.S.*)—

N.-E. Frontier of India, Sikkim, 1888.—Medal with clasp.
N.-E. Frontier of India, Manipur, 1891.—Clasp.
N.-W. Frontier of India, Isazai, 1892. Waziristan, 1894-95.—Clasp.
Chitral, 1895.—Relief of Chitral.—Medal with clasp.
N.-W. Frontier of India, 1897-98.—Engagement near Shabkadr on 9 Aug. 97. Operations on the Samana and in the Kurram Valley during Aug. and Sep. 97.—2 clasps.
Tirah, 1897-98.—Operations in the Bazar Valley, 25th to 30th December 1897.—Clasps.
East Africa, 1902-04.—Operations in Somaliland.—Medal with clasp.

Gilbert, G. E. L. (*Lt.-Col., retd., I.A.*)—

N.-W. Frontier of India, Hazara, 1888.—Action of Kotkai.—Medal with clasp. *Isazai, 1892.*
Chitral, 1895.—Relief of Chitral.—Medal with clasp.
Tirah, 1897-98.—Operations in the Bara Valley, 7 to 14 Dec. 97.—2 clasps.
China, 1900.—Medal.
N.-W. Frontier of India, 1908.—Operations in the Mohmand country. Despatches, G. G. O. No. 521 A of 1908.—Medal with clasp. D.S.O.
The War of 1914-18.—Despatches, *Lond. Gaz.*, 4 Jan. 16.

Gilbertson, G.—
The War of 1914-18.—Despatches, *Lond. Gaz.*, 19 Oct. 16 and 15 Aug. 17.

Gilchrist, W. F. C.—

East Africa, 1903-04.—Operations in Somaliland. Action at Jidballi. Served as Asstt. Provost Marshal, 2nd Brigade.—Medal with 2 clasps.
The War of 1914-18.—Despatches, *Lond. Gaz.*, 5 Apr., 13 July and 19 Oct. 16.

Giles, A.—
N.-E. Frontier of India, Sikkim, 1888.—Medal with clasp.
N.-E. Frontier of India, Manipur, 1891.—Despatches, G. G. O. 585 of 91.—Clasp.
Chitral, 1895.—Relief of Chitral.—Medal with clasp.
N.-W. Frontier of India, Waziristan, 1901-02.—Clasp.

Giles, E. D.—
The War of 1914-18.—Despatches, *Lond. Gaz.*, 15 Jun. 16, 4 Jan. 17, and 11 Dec. 17, D. S. O.

Giles, G. M. J. (*Lt.-Col., retd., I.M.S.B.*)—
South Africa.—Operations against the Galekas, 78. Zulu Campaign, 1879.—Medal with clasp.

Giles, W. (*Lieut.-Col., retd. I. A.*)—
Afghanistan, 1879-80.—Medal.
Burma, 1887-89.—Medal with 2 clasps.
N.-W. Frontier of India, 1897-98.—Buner. Action of the Tanga Pass.—Medal with clasp.
China, 1900.—Medal.

Gill, C. A.—
South African War, 1900.—Operations in the Orange Free State, May and June 1900. Operations in the Transvaal, west of Pretoria, July and Aug. 1900. Operations in the Transvaal, east of Pretoria, Aug. to Nov. 1900.—Queen's medal with 3 clasps.

Gillan, G. V. B.—
The War of 1914-18.—Despatches, *Lond. Gaz.*, 15 Aug. 17.

Gillies, F. G.—
The War of 1914-18.—Despatches, *Lond. Gaz.*, 5 Apr. 16 and 12 Jun. 17.

Gillies, H.—
South Africa War, 1900-1902.—Operations in the Orange river Colony and Cape Colony, March to the 31st May 1902.—Queen's medal with 3 clasps.

Gellet, W.—
The War of 1914-18—Despatches, *Lond. Gaz.* 19 Oct. 16.—C.I.E.

Gillmon, J. C.—
N.-E. Frontier of India, Sikkim, 1888.—Medal with clasp. Promoted 1st class Asst. Surgeon.

Gimlette, G. H. D. (*Lt.-Col., retd., I.M.S.*)—
Egypt, 1882.—Action of Tel-el-Kebir, and subsequent pursuit to Zagazig.—Medal with clasp. Bronze star.

Glasfurd, A. I. R.—
N.-W. Frontier of India, 1908.—Operations in the Mohmand country. D. A. A. G. 3rd Bde.—Medal with clasp.
The War of 1914-18—Despatches, *Lond. Gaz.*, 15 Jun. 16 and 15 May 17. C.M.G., D.S.O.

Glegg, J. A.—
The War of 1914-18—Despatches, *Lond. Gaz.*, 8 Feb. 17.

Glynton, G. M.—
The War of 1914-18.—Despatches, *Lond. Gaz.*, 5 Apr. 16.

Goad, H. (*Col. retd. I.A.*)—
Afghanistan, 1879-80.—Medal.

Goad, C. E. G. B.—
The War of 1914-18—Despatches, *Lond. Gaz.*, 15 Aug. 17. M.C.

Godfrey, S. H. (*Lieut.-Col., retd., I. A.*)—
N.-W. Frontier of India, Waziristan, 1901-02.—Medal with clasp.
N.-W. Frontier of India, 1908.—C.I.E.

Godkin, S. R.—
The War of 1914-18—Despatches, *Lond. Gaz.*, 15 Aug. 17. D. S. O.

Godson, C. A.—
N.-W. Frontier of India, 1908.—Operations in the Zakka Khel country.—Medal with clasp.
The War of 1914-18.—Despaches, *Lond. Gaz.*, 15 Aug. 17.

Godwin, C. A. C.—
N.-W. Frontier of India, Waziristan, 1901-02.—Medal with clasp.
N.-W. Frontier of India, 1902.—Operations against the Darwesh Khel Waziris.
The War of 1914-18.—Despatches, *Lond. Gaz.*, 1 Jan. 16 and 11 Dec. 17, Brev. of Lt.-Col.

Goldfrap, H. W.—
The War of 1914-18.—Despatches, *Lond. Gaz.*, 15 Mar., 5 Apr., 13 July and 19 Oct. 16. M.C.

Goldie, J. O. (*Col. u. s. l.*)—
Soudan, 1885-86.—Suakin.

Goldie, K. O.—
The War of 1914-18.—Despatches, *Lond. Gaz.*, 15 Aug. 17.

Goldman-Monk, M. E. de B.—
The War of 1914-18.—Despatches, *Lond. Gaz.*, 15 Aug. 17.

Goldney, F. C. N. (*Maj., retd., Ben. S. C.*)—
N.-E. Frontier of India, Duffla Expdition, 1874-75.
Afghanistan, 1879.—Medal.

Goldsmid, G. S. (*Maj., retd., Ind. S. C.*)—
N.-W. Frontier of India, Hazara Expedition, 1888.—Medal with clasp.
N.-E. Frontier of India. Suhai, 1892.

Goldsmith, H. A.—
The War of 1914-18.—Despatches, *Lond. Gaz.*, 5 Apr. 17. 13 July and 19 Oct. 16. M.C.

Goldsmith, J.—
N.-W. Frontier of India, Jowaki, 1877-78.—Medal with clasp.
N.-W. Frontier of India, Waziristan, 1894-95.—Clasp.

Goldthorp, F. H.—
N.-W. Frontier of India, Waziristan, 1894-95.—Medal with clasp.
N.-W. Frontier of India, 1897-98.—Operations in the Samana and in the Kurram Valley during Aug. and Sep. 97.—Medal with 2 clasps.
Tirah, 1897-98.—Clasp.
China, 1900—Medal with clasp.

Good, I.—
The War of 1914-18.—Despatches, *Lond. Gaz.*, 1 Jan. 16.

Goodbody, C. M.—
N.-W. Frontier of India, Waziristan, 1901-02.—Medal with clasp.
The War of 1914-18.—Despatches, *Lond. Gaz.*, 17 Feb. 15, 19 Oct. 16 and 15 Aug. 17. Brev. of Lt.-Col. D.S.O.

Goodfellow, N. G. B.—
The War of 1914-18.—Despatches, *Lond. Gaz.*, 15 Aug. 17. Brev. of Lt.-Col.

Goodhall, E. J.—
Burma, 1887-89.—Medal with clasp.
Burma, 1891.—Wuntho Field Force.—Clasp.

Gordon, D. H.—
The War of 1914-18.—Despatches, *Lond. Gaz.*, 15 Aug. 17. D.S.O.

Gordon, G. A. S.—
The War of 1914-18.—Despatches, *Lond. Gaz.*, 4 July 16.

Gordon, H. W. (*Col. u.s.l.*)—
India, 1855-56.—Operations against Sonthals.
China, 1860.—Action of Sinho, capture of the Taku Forts, and occupation of Pekin.—Medal with 2 clasps.
N.-W. Frontier of India, 1863.—Umbeyla Campaign.—Medal with clasp. *Hazara Campaign, 1868.*—Clasp. *Jowaki Expedition, 1877-78.*—Clasp.
Afghanistan, 1878-80.—Capture of Ali Musjid, advance to Gundamuk, operations in Kurram, and Zaimukht Expedition. Despatches, G. G. Os. 2 A and 610 of 79.—Medal with clasp.
N.-W. Frontier of India, Mahsud-Waziri Expedition, 1881.—Mentioned in Despatches.

Gordon, I. H.—
N.-W. Frontier of India, 1902.—Operations against the Darwesh-Khel Waziris.

Gordon, J. A. D. (*Lt.-Col., retd., Ben. Inf.*)—
N.-E. Frontier of India, Bhootan Expedition, 1865-66.—Medal.
Afghanistan, 1879.—Medal.
Burma, 1885-87.—Despatches, G. G. O. 434 of 87.—Clasp.

Gordon, J. C. F. (*Col. retd. I. A.*)—
N.-W. Frontier of India, Mahsud Waziri, 1881.—Despatches: Brigadier-General J. J. H. Gordon's despatch, 26 May 81.
Egypt, 1882.—Action of Tel-el-Kebir. Despatches, *Lond. Gaz.*, 2nd Nov. 82.—Medal with clasp. Bronze star.
N.-W. Frontier of India, 1897-98.—Operations on the Samana and in the Kurram Valley during Aug. and Sep. 97.—Medal with 2 clasps.
Tirah, 1897-98.—Operations against the Khani Khel Chamkannis. Despatches, G. G. O. 244 of 98.—Clasp.
South African War, 1902.—Queen's medal with 2 clasps.

Gordon, J. de la H.—
The War of 1914-18.—Despatches, *Lond. Gaz.*, 1 Jan. 16. M. C.

Gordon, J. G. B.—
South African War, 1900—02.—(Slightly wounded.) Operations in Orange River Colony (May to 29th Nov. 1900). Operations in Cape Colony, south of Orange River, Feb. to April 1900. Operations in the Transvaal, July 1901 to 31st May 1902. Operations in the Orange River Colony, 30th Nov. 1900 to July 1901. Despatches, *London Gazette*, 25th April 1902.—Queen's medal with 3 clasps. King's medal with 2 clasps.
The War of 1914-18.—Despatches, *Lond. Gaz.*, 19 Oct. 16. M. C.

Gordon, J. H. (*Col., u. s. l., Ind. S. C.*)—
Burma, 1885-87.—Commanded 23rd Madras Light Infantry. Despatches, G. G. O. 295 of 86.—Medal with clasp. *D. S. O.*

Gordon, J. L. R.—
Chitral, 1895.—Relief of Chitral.—Medal with clasp.
N.-W. Frontier of India, 1897-98.—Operations on the Samana and in the Kurram Valley during Aug. and Sep. 97. Operations of the Flying Column in the Kurram Valley under Colonel Richardson. 20 Aug. to 1 Oct. 97.—2 clasps.
Tirah, 1897-98.—Actions of Chagra Kotal and Dargai and of the Sampagha and Arhanga Passes. Reconnaissance of the Saran Sar and action of 9 Nov. 97. Operations in the Waran Valley and action of 16 Nov. 97. Despatches, G. G. O. 244 of 98.—Clasp.
The War of 1914-18.—Despatches, *Lond. Gaz.*, 21 Jun. 16. C. B.

Gordon, L. A.—
Afghanistan, 1880.—As Transport Officer and Orderly Officer to Brig.-Genl. Burrows.—Medal.
South African War, 1900.—Despatches, *Lond. Gaz.*, 10 Sept. 01.—Queen's medal with 3 clasps.

Gordon, P. J. (*Lt.-Col., retd., I. A.*)—
N.-W. Frontier of India, Hazara, 1888.—Medal with clasp.

Gordon, Robert (*Col., u. s. l., I. A.*)—
N.-W. Frontier of India, Jowaki, 1877-78.—Medal with clasp. (*1st*) *Miranzai, 1891.*
N.-W. Frontier of India, 1897-98.—Relief of Malakand. Malakand Operations in Bajaur Mohmand. Despatches, G. G. O. No. 178 of 98. —Medal with 2 clasps.
Tirah, 1897-98.—Despatches, G. G. O. No. 244 of 98.—Clasp.

Gordon, S.—
The War of 1914-18.—Despatches, *Lond. Gaz.*, 17 Feb. 15 and 15 Jun. 16 M. C.

Gordon, S. D. (*Bt. Col. retd., I.A.*)—
Afghanistan, 1878-79.—Battle of Ahmed Khel. Actions of Patkao Shana and Urzu. Despatches, G. G. O. 493 of 80.—Medal with clasp.
Egypt, 1882.—Action of Tel-el-Kebir.—Medal with clasp. Bronze star. 5th class Medjidie.
South African War, 1901-02.—Chief Staff Officer to Colonel Remington's Force and subsequently commanding Mobile column during driving operations in Transvaal, and Orange River Colony from Jan. to June 02. Despatches, *London Gazette*, 29 July 02.—Queen's medal with 4 clasps.

Gordon, S. V.—(*Bt. Col. retd. Ind. Army*)—
Afghanistan, 1878-79-80.—Action in the Sapari Pass. Operations in the Khost Valley. Action of Charasiah. Subsequent pursuit under General T. D. Baker. Action of Latabund. Affair at Shekhabad. March from Kabul to the relief of Kandahar and battle of Kandahar. Despatches, G. G. Os. 137 and 582 of 80.—Medal with 3 clasps. Bronze star. Brev. of Maj.
N.-W. Frontier of India, (*1st*) *Miranzai, 1891.*
Chitral, 1895.—Relief of Chitral. Despatches G. G. O. 998 of 95.—Medal with clasp. D. S. O.

Gordon, William (*Gen., u. s. l., Ind. S.C.*)—
India, 1848-49.—*Punjaub Campaign.*—Siege of Mooltan and battle of Goojerat.—Medal with 2 clasps.
India, 1857-58.—Medal.

Gore, F. L.—
East Africa (Somaliland), 1908-10.—Medal with clasp.

Goss, E. S.—
The War of 1914-18.—Despatches, *Lond. Gaz.*, 19 Oct. 16. M. C.

Gough, H. A. K.—
Chitral, 1895.—Relief of Chitral. Mentioned in Despatches.—Medal with clasp.

Gough, W. G. H.—
The War of 1914-18.—Despatches, *Lond. Gaz.*, 27 July 16. M. C.

Gould, G.—
South African War, 1901-02.—Operations in Cape Colony, 17th to 25th October 1901. Operations in the Transvaal, 1st November 1901 to 31st May 1902.—Queen's medal with 3 clasps.

Gould, J.—
N.-W. Frontier of India.—Waziristan, 1894-95. —Medal with clasp.
Chitral, 1895.—Relief of Chitral.—Medal with clasp.
N.-W. Frontier of India, 1897-98.—Malakand. Operations in Bajaur and in the Mamund country.—Utman Khel.—Buner. Action of the Tanga Pass.—Clasp.
China, 1900.—Medal.
Tibet, 1903-04.—Medal.
The War of 1914-18.—Despatches, *Lond. Gaz.*, 15 Aug. 17.

Gould, S. C.—
Afghanistan, 1879-80.—Medal.

Gourlie, J.
The War of 1914-18.—Despatches, *Lond. Gaz.*, 22 Jun. 15.

Gout, P. J.—
South African War, 1901-02.—Operations in Cape Colony from 27th March to 7th April 1901. Operations in the Transvaal and Orange River Colony from 7th April 1901 to 31st May 1902. - Medal with 5 clasps.
The War of 1914-18.—Despatches, *Lond. Gaz.*, 13 July 16 and 21 July 17.

Gover, A. C.—
The War of 1914-18. M. C.

Gow, P. F.—
The War of 1914-18.—Despatches, *Lond. Gaz.* 22 Jun. 15, 27th July 16 and 15 Aug. '17.—D S.O.

Gowan, W. E. (*Lt.-Col. retd., Ben. Inf.*)—
Afghanistan, 1879-80.—Medal.

Grace, H. G.—
N.-W. Frontier of India, 1908.—Operations in the Mohmand country.—Medal with clasp.
The War of 1914-18. M. C.

Gracey, D. D.—
The War of 1914-18. M.C.

Graham, B. C.—
Tirah, 1897-98.—Medal with 2 clasps.

Graham, G. F.—
The War of 1914-18.—Despatches, *Lon. Gaz.*, 5 Apr. 16.

Graham, G. F. (*Hon. Maj.-Gen. retd., Ben. S. C.*)—
N.-E. Frontier of India, Cossiah and Jyntiah Hills Campaign, 1862-63.

Graham, G. F. I. (*Maj.-Genl. retd., Bl. S. C.*)—
India, 1857-58-59.—Served Adjutant of the Etawah Yeomanry Levies and Military Police Battalion during the rebellion in that District. Actions at Ajeetmul, 6th and 8th July 1858 (commanded a detachment of cavalry on the former occasion); capture of Fort Bhurree, 28th Aug. 1858 (commanded the infantry, cavalry and artillery); action at Chuchernugger in Sep. 1858 (commanded the skirmishers), and other minor affairs. Mentioned by A. O. Hume, Esq., C.S., Magistrate of Etawah, in a letter to the Govt. of the N.-W. P., dated 18th Nov. 1858. Also served with the Force under the command of Brigadier St. G. D. Showers, C.B., in the action at Dowsa, 14th January 1859, against Tantia Topee and other Chiefs. Despatches, G. G. O. No. 315 of 1859, and Supplement to *London Gazette* of 3rd May 1859.—Medal.

Graham, J. D.—
N.-W. Frontier of India, Waziristan, 1901-02.—Medal with clasp.
The War of 1914-18.—Despatches, *Lond. Gaz.*, 21 Jun '16 and 15 Aug '17. Brev. of Lt.-Col.

Graham, J. S.—
West Africa, 1901.—Gambia expedition.—Medal with clasp.

Graham, L. C. T.—
The War of 1914-18.—Despatches, *Lond. Gaz.*, 28 Dec. '17.

Graham, R.—
Soudan, 1896.—Dongola Expedition-Medal; Egyptian Medal. Promoted Sub-Conductor.
China 1900—Medal.

Graham, R. B.—
N.-W. Frontier of India, 1897-98. Tochi.—Medal with clasp.
South African War, 1899-1900.—Advance on Kimberley, including action at Magersfontein. Severely wounded.—Queen's medal with clasp.
The War of 1914-18.—Despatches, *Lond. Gaz.*, 15 Jun. '16.

Graham, S. B.—
Soudan, 1884-85.—Nile. With Desert Column.—Medal with clasp; bronze star.
Burma, 1887.—Medal with 2 clasps.

Grainger, T.—
N.-E. Frontier of India, Sikkim, 1888.—Attack of Jalapla.—Medal with clasp.
N.-W. Frontier of India, Hazara, 1891.—Clasp.
Tirah, 1897-98.—Actions of Dargai, and of the Sampagha and Arhanga Passes. Operations in the Bara Valley, 7th to 14th December 1897. Despatches, G. G. O. No. 244 of 1898. Promoted to Surgeon Lieutenant-Colonel.—Medal with 2 clasps.

Granger, T. A.—
Mekran, 1898.—
N.-W. Frontier of India, 1898-99.—Operations against the Para Chamkannis.
The War of 1914-18.—Despatches, *Lond. Gaz.*, 11 Dec. 17.

Grant, A.—
N.-E. Frontier of India, Chin-Lushai, 1889-90.—(Lushai and Burma Columns.)—Medal with clasp.
Burma, 1891.—Thetta Column.—Clasp.
N.-E. Frontier of India, Manipur, 1891.—Clasp.
Chitral, 1895.—Relief of Chitral.—Medal with clasp.
Soudan, 1896.—Dongola Expedition.—Medal and Khedive's medal.
N.-W. Frontier of India, 1897-98.—Tochi—Orderly Officer to G. O. C. 1st Brigade.—Clasp.
Tirah, 1897-98.—Actions of the Sampagha and Arhanga Passes.—Clasp.
China, 1900.—Medal.

Grant, C. J. P. MacA.—
South African War, 1899—1901.—Operations in Cape Colony and Orange Free State, Feb. to May 00. Operations in Orange River Colony May to June 00.—Queen's medal with 2 clasps.

Grant, C. J. W.—
Burma, 1885—87.—Medal with clasp.
N.-E. Frontier of India, Manipur, 1891.—Capture and defence of Thobal. (Severely wounded, horse shot in the attack and occupation of Imphal.) Despatches, G. G. O. 585 of 91. V.C.—Promoted Captain and Brev. of Maj.—Clasp.

Grant, D. St. J. D. (*Col. retd. I.M.S.*)—
Akha Expedition, 1883-84.

Grant, E. P.—
South African War, 1900-02.—Operations in Natal, March to June 00, including action at Laing's Nek (6 to 9 June). Operations in Transvaal between Jan. and Sep. 01 and Jan. to March 02. Operations on the Zululand Frontier of Natal in Sep. and Oct. 01. Operations in the Orange River Colony, April to May 02.—Queen's medal with 3 clasps. King's medal with 2 clasps.

Grant, F. C. (*Bt. Col., retd. I.A.*)—
N.-W. Frontier of India, 1897-98.—Operations on the Samana and in the Kurram Valley during Aug. and Sep. 97.—Medal with 2 clasps.
Tirah, 1897-98.—Reconnaissance of the Kharmana Defile and action of 7 Nov. 97. Operations against the Khani Khel Chamkannis.—Clasp.

Grant, F. W. (*Col., retd., Ben. S. C.*)—
India, 1857-58.—Actions of Cawnpore, Serai Ghat, Kallee Nuddee, Futteghur and Shumsabad; siege and capture of Lucknow.—Medal with Clasp.

Grant, G. P.—
Mekran, 1901.—Attack and capture of Nodiz Fort 20 December 01. (Severely wounded.) Despatches G. G. O. 415 of 02. D. S. O.

Grant, I. H. (*Maj. retd., I. A.*)—
Burma, 1892.—Medal with clasp.
Chitral, 1895.—Relief of Chitral.—Medal with clasp.

Grant, J. (*Col. u. s. l., I. A.*—)
Afghanistan, 1879-80.—Affair on the Helmund. Battle of Maiwand (severely wounded) and defence of Kandahar.—(Despatches, G. G. O. 399 of 79).—Medal.

Grant, John (*Maj. retd. Bo. S. C.*)—
Afghanistan, 1880.—Medal.
Egypt, 1882.—Medal; bronze star.

Grant, J. D.—
Tibet, 1903-04.—Action at Niani, 26th June 1904. Operations at and around Gyantse, between 5th May and 6th July 1904. March to Lhassa, 14th July to 3rd Aug. 1904. (Twice wounded.) Despatches, *Gaz. of India*, No. 1065 of 1904.—Medal with clasp. V.C.
The War of 1918-19.—Despatches, *Lond. Gaz.*, 19 Oct. 16.

Grant, J. P. (*Col., retd., Bo. S.C.*)—
India, 1858.—Suppression of rebellion in Sawant Wari State, and affair of Dir Sing.—Medal.

Grant, R. A. P.—
The War of 1914-18—Despatches *Lond. Gaz.*, 19 Oct., 16.

Grant, J. W.—
Chitral, 1895.—Relief of Chitral.—Medal with clasp.

Grant, W. O.—

Uganda, 1897-98.—Unyoro.—Medal with clasp.

Grantham, C. F.—(*Col. retd., I. A.*)—

China, 1900.—Medal.

Grant-Smith, H. G. H.—

China, 1900.—Defence of Tientsin. Relief of Tientsin. Relief of Pekin. Action of Peitsang. Action at Yangtsun.—Medal with clasp.

South African War, 1901-02.—Operations in the Orange River Colony, May 1901 to 31st May 1902.—Queen's medal with 4 clasps.

Graves, S. H. P. (*Col. retd. I. A.*)—

Afghanistan, 1878-79.—Medal.

Graveson, H.—

N.-W. Frontier of India, Hazara, 1891.—Medal with clasp.—Promoted Sub-Conductor.
Tirah, 1897-98.—Medal with clasp.
China, 1900.—Medal.
Tibet, 1903-04.—Medal.

Gray, E. St. C.—

South African War, 1900-01.—Operations in Cape Colony, 27th February to March 1900. Operations in the Orange Free State, April and May 1900. Operations in Orange River Colony, May to 29th November 1900, including action at Wittenbergen. Operations in Orange River Colony, 30th Nov. 1900 to Feb. 1901. Operations in Cape Colony, south of Orange River, March to August 1901.—Queen's medal with 3 clasps.

The War of 1914-18.—Despatches *Lond. Gaz.* 1 Jan. '16 and 15 May 17. M. C.

Gray, F. W. B.—

N.-W. Frontier of India, Waziristan, 1894-95.—Medal with clasps.

China, 1900.—Medal.
N.-W. Frontier of India, 1908.—Operations in the Mohmand country. Engagement of Matta, 24th April 1908. Engagement of Kargha, 24th May 1908. Despatches, G. G. O. No. 521 A of 1908. —Medal with clasp. D. S. O.

The War of 1914-18—Despatches, *Lond. Gaz.*, 17 Feb. '15, 21 July and 28 Nov. '17 Brev. of Col. C.M.G.

Gray, W. duG. (*Maj. Genl. retd., I. A.*)—

Afghanistan, 1879-80.—Medal.

N.-W. Frontier of India, Hazara, 1888.—Action of Kotkai. Despatches, G. G. O. 1022 of 88.—Medal with clasp.

(*1st and 2nd*) *Miranzai, 1891.*—Despatches, G. G. O. 632 of 91.—Clasp. Brev. of Maj.

Waziristan, 1894-95.—Clasp.
N.-W. Frontier of India, 1897-98.—Tochi.—Medal with clasp.

N.-W. Frontier of India, Waziristan, 1901-02.—Despatches, G. G. O. 611 of 02.—Clasp. Brev. of Col.

Greany, H. (*Lt.-Col., retd., I. M. S.*)—
China, 1900.—Medal with clasp.

Green, C. R. M.—
N.-W. Frontier of India, Hazara, 1888.—Medal with clasp.

Tirah, 1897-98.—Medal with clasp.

Green, D. R. (*Lieut.-Col. retd., I.M.S.*)—

Tirah, 1897-98.—Actions of Chagru Kotal, Dargai and of the Sampagha and Arhanga Passes. Operations at and around Dwatoi and action of 24 Nov. 97. Operations in the Bara Valley, 7 to 14 Dec. 97.—Medal with 2 clasps.

Green, G. E. T. (*Lt.-Col., retd. I.A.*)

Burma, 1885-89.—Medal with clasp.

Green, W. G. K.—

The War of 1914-18.—Despatches, *Lond. Gaz.* 4 Jan. and 28 Dec. 17. D.S.O.

Green, Armytage V. B.—

The War of 1914-18.—Despatches, *Lond. Gaz.*, 1 Jan. and 19 Oct. 16.

Greenway, C. K.—

The War of 1914-18.—Despatches, *Lond. Gaz.*, 5 Apr. 16.

Greenwood, A.—

China, 1900.—Relief of Pekin.—Medal with clasp. Promoted 1st class Asst. Surgeon.

Gregg, C. N.—

Burma, 1885-87.—Medal with clasp.

Burma, 1887-88.—Clasp.

China, 1900.—Medal.

Gregory, C. L.—

N.-W. Frontier of India, 1908.—Operations in the Zakka Khel country. D. A. Q. M. G. Field Force Staff. Despatches, G. G. O. No. 188 of 1908. —Medal with clasp.

The War of 1914-18.—Despatches, *Lond. Gaz.*, 15 June 16 and 4 Jan. 17. Brev. of Col.

Greig, J. G.—

N.-W. Frontier of India, 1897-98.—Mohmand. —Medal with clasp.

Tirah, 1897-98.—Action of the Sampagha Pass. Operations at and around Dwatoi and action of 24 Nov. 97. Operations in the Bara Valley, 7 to 14 Dec. 97.—Clasp.

East Africa, 1902-04.—Operations in Somaliland —Medal with clasp.

Gretton, J. C.—

South African War, 1899—1902.—Operations in Natal, 1899. Relief of Ladysmith, including action at Colenso. Operations of 17th to 24th Jan. 1900 and action at Spion Kop. Operations of 5th to 7th Feb. 1900, and action at Vaal Krantz Operations on Tugela Heights (14th to 27th Feb 1900); wounded. Operations in Natal, March to June 1900, including action at Laing's Nek (6th to 9th June). Operations in the Transvaal, east and west of Pretoria, July to 29th Nov. 1900. Operations in the Transvaal and Orange River Colony, 30th Nov. 1900 to Mar. 1902. Operations on the Zululand Frontier of Natal, Sept. and Oct. 1901.—Queen's medal with 5 clasps. King's medal with 2 clasps.

Grey, A. J. H.—

(Major u. s. l. Political Employ)—
China, 1900.—Medal.

The War of 1914-18.—Despatches, *London Gazette*, 5 Apr. 16.

Grey, E. (*Col. retd., I.A.*)—
N.-E. Frontier of India, Sikkim, 1888.—Medal with clasp.
N.-E. Frontier of India, Manipur, 1891.—Clasp.
South African War, 1901-02.—Attd. to A. S. Corps.—Queen's medal with 3 clasps. King's medal with 2 clasps.

Grey, L. J. H. (*Col. u. s. l.—Ind. S. C.*)—
India, 1857-58-59.—Medal.
N.-E. Frontier of India, Bhootan Expedition, 1865-66.—Medal.

Grey, P—
The War of 1914-18.—Despatches, *Lond. Gaz.*, 15 Aug. 17.

Griffin, C. P. G. (*Lt.-Col. retd., I.A.*)—
Tirah, 1897-98.—Medal with 2 clasps.
China, 1900.—Relief of Pekin. Action of Peitsang. Action at Yangtsun. Despatches, *Lond. Gaz.*, 14 May 01.—Medal with clasp. D. S. O.

Griffin, L.—
The War of 1914—18. Despatches, *Lond. Gaz.*, 22 Jun. 15.

Griffith, R. (*Col., u. s. l., I. A.*)—
Egypt, 1882.—Commandant at Suez.
Burma, 1886.—Medal with clasp.

Griffich, R. E. H.—
The War of 1914-18.—Despatches, *Lond. Gaz.*, 4 July 16.

Griffich, W, R, C.—
The War of 1914-18.—Despatches, *Lond. Gaz.*, 19 Oct. 16, M.C.

Griffith-Griffin, F. M.—
N.-W. Frontier of India, 1908.—Operations in the Zakka Khel country.—Medal with clasp.

The War of 1914-18.—Despatches, *Lond. Gaz.*, 22 June 15 M.C.

Griffiths, C. (*Lt.-Col. retd., I. A.*)—
Egypt, 1882.—Action of Tel-el-Mahuta and both actions at Kassassin; action of Tel-el-Kebir.—Medal with Clasp. Bronze star.
China, 1900.—Medal.

Griffiths, G. S. (*Lt.-Col. retd., I. M. S.*)—
Afghanistan, 1878-79-80.—Arambi Karez Expedition. Action of Baghao.—Medal.
N.-W. Frontier of India, Zhob Valley, 1884.
N.-E. Frontier of India, Sikkim, 1888.—Medal with clasp.

Grimston, S. B.—
Burma, 1885-87.—Occupation of Mandalay. Operations of the 6th Brigade.—Medal with clasp.
N.-W. Frontier of India, 1897-98.—Operations on the Samana and in the Kurram Valley during Aug. and Sep. 97. Operations of the Flying Column in the Kurram Valley under Colonel Richardson, 20 Aug. to 1 Oct. 97.—Medal with 2 clasps.
Tirah, 1897-98.—Clasp.
China, 1900.—Medal.

The War of 1914-18.—Despatches, *Lond. Gaz.* 15 Jun. 16. Legion of Honour 4th class. C. M. G.

Grindley, W. H.—
N.-E. Frontier of India, Sikkim, 1888.—Medal with clasp.
N.-E. Frontier of India, Lushai, 1888-89.—Clasp.
Chitral, 1895.—Relief of Chitral.—Medal with clasp.

Grisewood, A. E.—
N.-W. Frontier of India, 1908.—Operations in the Mohmand country. Engagement of Kargha, 24th May 1908.—Medal with clasp.
The War of 1914-18.—Despatches, *Lond. Gaz.*, 8 Feb. 17.

Groube, G. P. T.—
China, 1900.—Medal.

Grover, M. H. S.—
Afghanistan, 1879-80.—With Thal-Chotiali Field Force.—Medal.
Soudan, 1885.—Suakin. Action of Tamai. Affair of Thakool.—Medal with clasp. Bronze star.
N.-W. Frontier of India, Waziristan, 1894-95.—Served as D. A. A. G. Divisional Headquarters. Despatches, G. G. O. 473 of 95.—Medal with clasp.
N.-W. Frontier of India, 1897-98.—Tochi.—D. A. Q. M. G., 2nd Brigade. Despatches, G. G. O. 1432 of 97.—Medal with clasp. Brev. of Lt.-Col.
South African War, 1901-02.—Operations in the Transvaal.—Queen's medal with 3 clasps.

Guise, J. W.—
The War of 1914-18.—Despatches *Lond. Gaz.*, 15 Aug. 17.

Gunn, A. D.—
The War of 1914-18.—Despatches, *Lond. Gaz.*, 19 Oct. 16.

Gunning, G. H.—
Tirah, 1897-98.—Medal with 2 clasps.
N.-W. Frontier of India, Waziristan, 1901-02.—Clasp.
N.-W. Frontier of India, 1902.—Operations against the Darwesh Khel Waziris.

Gurdon, B. E. M.—
Chitral, 1895.—Defence of the Fort of Chitral. Despatches, G. G. O. 531 of 1895.—Medal with clasp. D. S. O.

Gurdon, E. F. (*Maj.-Gen. u. s. l. Ind. S. C.*)—
India, 1857.—Actions at Nimbhera and Jeerun, and siege of Neemuch. Mentioned in Despatches.—Medal.

Gurr, G. W.—(*Comry. & Hony. Maj. retd.*)
N.-W. Frontier of India, 1897-98.—Tochi. Despatches, G. G. O. No. 1432 of 1897.—Medal with clasp.—Promoted Deputy Assistant Commissary.

Guthrie, J. C.—
The War of 1914-18.—Despatches, *Lond. Gaz.*, 15 May 17.

Gwalior, H. H. Maharaj Adhiraj, *Sir* M. R. S. Maharaja of—
China, 1900.—Despatches, *Lond. Gaz.*, 1 May 01, A. D. C. to the King. Medal.

Gwatkin, F. S. (*Col. retd., I. A.*)—
Afghanistan, 1878-79-80.—Both the Bazar Valley Expeditions. Action of Deh-Sarak. Zaimukht Expedition.—Medal.
Egypt, 1882.—Actions of Kassassin and Tel-el-Kebir.—Medal with clasp. Bronze star.
Soudan, 1885.—Suakin.—Clasp.
N.-E. Frontier of India, Lushai, 1889.—As Staff Officer. Despatches, G. G. O. 592 of 89.—Clasp.
Chin-Lushai, 1889-90.—As Staff Officer. Despatches, G. G. O. 677 of 90.—Medal with clasp. Brev. of Maj. *Manipur, 1891.*—As A. A. Genl. Despatches, G. G. O. 585 of 91.—Clasp.
Chitral, 1895.—Relief of Chitral.—Despatches, G. G. O. 998 of 95.—Medal with clasp. Brev. of Lt.-Col.
Tirah, 1897-98.—A. A. and Q. M. G., Peshawar Column. Despatches, G. G. O. 224 of 98.—2 clasps.

Gwyer, V. E.—
The War of 1914-18.—Despatches, *Lond. Gaz.*, 4 July 16 and 31 Oct. 17 Brev. of Lt.-Col.

Gwyn-Thomas, G.—
The War of 1914-18.—Despatches, *Lond. Gaz.*, 14 Nov. 16, 4 Jan. 17 and 11 Dec. 17. D.S.O.

Gwyther, A.—
N.-W. Frontier of India, 1897-98.—Medal with clasp.

H

Hackett, H. M. M.—
N.-W. Frontier of India, 1908.—Operations in the Zakka Khel country.—Medal with clasp.

The War of 1914-18.—Despatches, *Lond. Gaz.*, 15 Aug. 17.

Hackett, T. D. H.—
South African War, 1902.—Served with Impl. Yeo. Operations in Cape Colony.—Queen's medal with 2 clasps.

Hadow, R. C. (*Bt.-Col., retd., I.S.C.*)—
Afghanistan, 1878-79-80.—Advance on Ghuzni Battle of Ahmed Khel. Action of Urzu. March from Kabul to the relief of Kandahar and battle of Kandahar.—Medal with 2 clasps. Bronze star.
Soudan, 1885.—Suakin. Actions of Tofrek and Tamai.—Medal with 2 clasps. Bronze star.
N.-W. Frontier of India, (2nd) Miranzai, 1891.—Medal with clasp.

Chitral, 1895.—Relief of Chitral. Despatches G. G. O. 998 of 95.—Medal with clasp.—D. S. O.
N.-W. Frontier of India, 1897-98.—Operations of the Samana and in the Kurram Valley during Aug. and Sep. 97. Operations of the Flying Column in the Kurram Valley under Colonel Richardson 20 Aug. to 1 Oct. 97.—2 clasps.
Tirah. 1897-98.—Actions of the Chagru Kotal and of Dargai, (severely wounded).—Clasp.

Haig, A. B.—
The War of 1914-18.—Despatches, *Lond. Gaz.*, 13 July and 19 Oct. 16.

Haig, P. B.—
N.-W. Frontier of India, Waziristan, 1894-95.—Medal with clasp.
N.-W. Frontier of India, 1897-98.—Malakand. Utman Khel. Buner. Action of the Tanga Pass.—Medal with clasp.
Uganda, 1899.—Operations against Kabrega. Despatches, *Lond. Gaz.*, 2 Jan. 00. Medal with clasp.
Uganda, 1900.—Operations against the Nandi Tribe. Despatches, *Lond. Gaz.*, 10 Sep. 01.—Medal with clasp.

Haig, P. deH. (*Lieut.-Col., retd., I. M. S.B.*)—
Afghanistan, 1878-79-80.—With Peshawar Valley Field Force. Expedition to the Tonk Frontier.—Medal.
N.-W. Frontier of India.—Action at Wana, 3 Nov. 94, (severely wounded) *Waziristan, 1894-95.*—Medal with clasp.

Haig, T. W.—
Burma, 1887-88.—Medal with clasp.

Haigh, F. F.—
The War of 1914-17.—Despatches, *Lond. Gaz.*, 15 Aug. 17.

Halles, W. L.—
The War of 1914-18.—Despatches. *Lond. Gaz.*, 15 Aug. 17.

Hall, H. S. G.—
The War of 1914-18.—M.C.

Haldau, C. L.—
The War of 1914-18.—Despatches, *Lond. Gaz.*, 21 Jun 16, C. M. G.

Hall, E. G.—
The War of 1914-18.—Despatches, *Lond. Gaz.*, 5 Apr. and 13 July 16.

Hall, E. S.—
N.-W. Frontier of India, Isazai, 1892.
Chitral, 1895.—Relief of Chitral.—Medal with clasp.
Tirah, 1897-98.—Clasp.

Hall, G. J. C. (*Major, retd., I. A.*)—
Chitral, 1895.—Relief of Chitral.—Medal with clasp.
Tirah, 1897-98.—Clasp.

Hall, H. C.—
South African War, 1901-02.—Operations in Orange River Colony, February to March 1902. Operations in Cape Colony, July 1901 to Feb. 1901, and March to 31st May 1902.—Queen's medal with 4 clasps.
Operations in the Abor country, 1911-12.—Medal with clasp.

The War of 1914-18.—Despatches, *Lond. Gaz.* 13 July 16.

Hall, H. S. G., (*Col., retd., I. A.*)—
Tirah, 1897-98.—As Asst. to Commy. Gen. (*Oct. 97 to Apr. 98*).—Despatches, *Lond. Gaz.*, 5 Apr. 98.—Medal with 2 clasps.

Hall, L. M. M.—
N.-E. Frontier of India, Chin-Lushai, 1889-90.—Despatches, G. G. O. 677 of 90.—Medal with clasp.
N.-W. Frontier of India, 1897-98.—With Kohat-Kurram Force.—Medal with clasp.

Hall, W. D.—
The War of 1914-18.—M.C.

Halliday, G. T. (*Lt-Gen. u. s. l. Ben. Cav.*)—
N.-W. Frontier of India, Hazara Expedition, 1888.—Medal with clasp.

Hallowes, F. W.—
N.-W. Frontier of India, Hazara, 1888.—Medal with clasp.
Chitral, 1895—Relief of Chitral.—Medal with clasp.
East Africa, 1903-04.—Operations in Somaliland. Despatches, *Lond. Gaz.*, 2 Sep. 04.—Medal with clasp.

Hallowes, J. H.—
N. W. Frontier of India, 1897-98.—Medal with clasp.

Hamer, M. A.—
The War of 1914-17—Despatches, *Lond. Gaz.*, 22 June 15, 1 Jan. 16, 19 Oct. 16, D. S. O, M. C.

Hamilton, A. (*Bt.-Col., retd., I.A.*)—
N.-W. Frontier of India, Zhob Valley, 1884.
N.-W. Frontier of India, 1897-98.—Tochi.—Medal with clasp.
N.-W. Frontier of India, 1908.—Operations in the Zakka Khel country.—Medal with clasp

Hamilton, A. F.—
The War of 1914-18.—Despatches, *Lond. Gaz.*, 30 Jan. 16 and 8 Feb. 17 M.C.

Hamilton, C. (*Major, retd., I. A.*)—
Burma, 1886-87.—Expedition up the Chindwin River. Operations of the 6th Brigade.—Medal with clasp.
N.-E. Frontier of India, Lushai, 1889.—Clasp.
China, 1900.—Medal.

Hamilton, C. A. B.—
South African War, 1899—1902.—Operations in Cape Colony, South of Orange River, 99-00. Operations in the Orange Free State (Feb. to May 00). Operations in the Orange River Colony (May to 29 Nov. 00). Operations in the Transvaal, west of Pretoria, July to 29 Nov. 00. Operations in the Transvaal between 30 Nov. 00 and 31 May 02.—Queen's medal with 4 clasps.
Tibet, 1903-04.—Medal.

Hamilton, C. L.—
N.-W. Frontier of India, 1897-98.—Operations on the Samana and in the Kurram Valley during Aug. and Sep. 97.—Relief of Gulistan—Medal with 2 clasps.
Tirah, 1897-98.—Clasp.

Hamilton, F. A.—
S. African War of 1901-02.—Operations in Orange, River Colony, Dec. 01 to 31 May 02. Queen's Medal with 4 clasps.

Hamilton, G. F. D. (*Lt.-Col., retd., I. A.*)—
N.-W. Frontier of India, Hazara, 1888.—Medal with clasp.

Hamilton, H.—
Afghanistan, 1878-79-80.—Action of Charasiah. Operations at and around Kabul. Affair of Shekhabad. March from Kabul to the relief of Kandahar, and battle of Kandahar. Despatches, G. G. O. 582 of 80.—Medal with 3 clasps. Bronze star.
Chitral, 1895.—Relief of Chitral—Medal with clasp.
N.-W. Frontier of India, 1897-98.—Operations on the Samana and in the Kurram Valley during Aug. and Sep. 97. Operations of the Flying column under Colonel Richardson, 20 Aug. to Oct. 97. Despatches, G. G. O. 304 of 98.—2 clasps.
Tirah, 1897-98.—Despatches, G. G. O. 244 of 1898. Promoted Brigade Surgeon Lt.-Colonel.—Clasp.
China, 1900.—Medal.

Hamilton, H. W.—
South African War, 1900—02.—Operations in Orange River Colony, 25th June to 10th September 1900. Operations in the Transvaal, 30th November 1900 to August 1901. Operations on Zululand Frontier of Natal in September and October 1901. Operations in Transvaal, November 1901 to 31st May 1902.—Queen's medal with 3 clasps. King's medal with 2 clasps.

Hamilton, J. A.—
China, 1900.—Relief of Pekin.—Medal with clasp.
The War of 1914.18.—Despatches, *Lond. Gaz.*, 1 Jan. 16. C. M. G.

Hamilton, R. E. A.—
Chitral, 1895.—Relief of Chitral.—Medal with clasp.
East Africa (British), 1896.—Operations against the Mazrui Rebels.—Medal.
Tirah, 1897-98.—Actions of the Sampagha and Arhanga Passes. Reconnaissance of the Saran Sar and actions of the 9th and 11th November 1897. Operations against the Khani Khel Chamkannis. Operations in the Bazar Valley, 25th to 30th December 1897.—2 clasps.

Hamilton, W. H.—
Operations in the Abor country, 1911-12.—Medal with clasp.
The War of 1914-18.—Despatches, *Lond. Gaz.*, 10 May 16, and 15 Aug. 17. D. S. O.

Hammersley-Smith, R. H.—
South African War, 1900-02.—Operations in Cape Colony, 14th April to 20th August 1900, 14th February to 8th March 1901, and 7th May 1901 to 31st May 1902. Operations in Orange River Colony, 20th August 1900 to 13th February 1901, and 9th March to 6th May 1901.—Queen's medal with 2 clasps. King's medal with 2 clasps.

Hammond, *Sir* A. G. (*Col., u. s. l., I.S.C.*)—
N.-W. Frontier of India, Jowaki, 1877-78.—Despatches, G.G.O. 738 of 1878.—Medal with clasp.
Afghanistan, 1878-79-80.—Action of Ali Musjid. Despatches. Lieutenant-General *Sir* S. J. Browne's despatch, 8 July 79. Operations at and around Kabul. Action of Chihildakhteran. Despatches, G. G. O. 137 of 80.—Medal with 2 clasps. V.C.
N.-W. Frontier of India, Hazara, 1888 and 1891.—Commanded 2nd Brigade, Hazara Field Force of 1891. Despatches, G. G. O. 978 of 88, and Brigadier-General G. N. Channer's despatch 11 Nov. 88, and G.G.O. 816 of 91.—2 clasps. D. S. O., C. B.
Isazai, 1892.—Commanding a Brigade.
Chitral, 1895.—Relief of Chitral. Commanding Line of Communications and 4th Brigade.—Despatches, G. G. O. 908 of 95.—Medal with clasp.
Tirah, 1897-98.—Commanding Peshawar Column. Despatches, G. G. O. 244 of 98.—2 clasps.

Hammond, F. A. L.—
China, 1900.—Medal.

Hammond, W. P.—
N.-W. Frontier of India, 1908.—Operations in the Mohmand country.—Medal with clasp.

Hanafin, J. B.—
The War of 1914-18,—Despatches, *Lond. Gaz.*, 31 Oct. 17.

Hancock, F. H.—
N.-W. Frontier of India, Mashud Waziri, 1881.
Burma, 1886-88.—Operations of he 2nd Brigade. —Medal with clasp.
Soudan, 1896.—Dongola Expedition.—Medal; Egyptian medal.
N.-W. Frontier of India, 1897-98.—Mohmand.— —Medal with clasp.

Handcock, A. G. (*Lt.-Genl. u.s.l.*)—
N.-E. Frontier of India, Bhutan, 1865-66.—Medal with clasp.
Afghanistan, 1878-79-80.—Battle of Ahmed Khel. Action of Urzu. Despatches, G. G. O. 326 of 80.—Medal with clasp. Brev. of Lt.-Col.

Handley, L. M.—
The War of 1914-18.—Despatches, *Lond. Gaz.*, 31 Oct. 17.

Hankin, D. L.—
South African War, 1899-1902.—Operations in Orange River Colony. Operations in Cape Colony, north of Orange River. Operations in the Transvaal, Feb. to 31 May, 02. Queen's medal with 3 clasps. King's medal with 2 clasps.

Hankin, H. M.—
The War of 1914-18.—Despatches, *Lond. Gaz.*, 22 June 15.

Hankins, R. F. H.—
Burma, 1887-92. Medal with 2 clasps.

Hannyngton, J. A.—
East Africa (British), 1896.—Operations against the Mazrui Rebels.—Medal.
Uganda, 1897-98.—Severely wounded.—Medal with clasp. D.S.O., C.M.G.
East Africa (Somaliland), 1908-10.—In command of troops from 19th August 1908 to 20th January 1909. In command L. of C. from 21st January to 28th February 1909. In command of Field Force from 3rd June to 29th December 1909. Despatches, *London Gazette*, 17 June 1910. C. M. G.—Medal with clasp.
The War of 1914-8.—Despatches, *Lond. Gaz.*, 22 June 15, 20 June 16 and 17 Jan. 17, Brev. of Col. D.S.O.

Hanson, G. J. G.—
South African War, 1901-02.—Operations in the Transvaal and Orange River Colony, between 6th December 1901 and 31st May 1902.—Queen's meda with 4 clasps.

Harbord, C. R.—
South African War, 1899-02.—As Special Service Officer. Operations in the Transvaal, west of Pretoria, July to 29 Nov. 00, including actions at Venterskroon (7 and 9 Aug.). Operations in the Transvaal between 30 Nov. 00 and 31 May 02. Operations in Cape Colony between 30 Nov. 00 and 31 May 02.—Queen's medal with 3 clasps. King's medal with 2 clasps.
East Africa, 1902.—Operations in Somaliland. —Medal with clasp.
The War of 1914-18.—Despatches, *Lond. Gaz.*, 6 July 17.

Harbord, L. B.—
The War of 1914-18.—M. C.

Harcourt, C. B.—
South African War, 1900-01.—Railway Staff Officer. Operations in the Transvaal, 30th Nov. 1900 to Nov. 1901.—Queen's medal with 4 clasps.
N.-W. Frontier of India, 1908.—Operations in the Zakka Khel country. Despatches, G. G. O. No. 188 of 1908. Operations in the Mohmand country. —Medal with clasp.

Harcourt, E. S.—
The War of 1914-18.—Despatches, *Lond. Gaz.*, 5 Apr. 16. M. C.

Harcourt, J. S. M.—
Operations in the Abor country, 1911-12.—Medal with clasp.

Harcourt, P. H. (*Hon. Maj.-Gen., retd.*)—
India, 1857-58.—Capture of Jhansi. Mentioned in Despatches.—Medal with clasp.

Hardaker, G. B.—
South African War, 1899-1900.—Defence of Ladysmith.—Despatches, *London Gazette*, 10th September 1901. Queen's medal with clasp.

Hardcastle, J.—
South African War, 1902.—Operations in Orange River Colony, April to 31 May 1902. Queen's medal with 3 clasps.
The War of 1914-18.—Despatches, *Lond. Gaz.*, 4 July 16.

Hardie, G. E.—
Aden, 1903-04.—Operations in the interior.

Harsington, H. R.—
Operations in the Abor country, 1911-12.—Medal with clasp.

Harding, E. F. (*Capt., retd., I. A.*)—
China, 1900—Medal.

Harding, M. F.—
The War of 1914-18.—Despatches, *Lond. Gaz.*, 21 Jun. 16 and 31 Oct. 17.

Hardy, T. H.—
Tirah, 1897-98.—Medal with 2 clasps.

Hare, E. C. (*Lt.-Col., retd., I. M. S.*)—
N.-W. Frontier of India, 1897-98.—Operations on the Samana and in the Kurram Valley.—Medal with 2 clasps.
Tirah, 1897-98.—Action of Chagru Kotal.—Clasp.

Hareno, R. E.—
The War of 1914-18.—Despatches, *Lond. Gaz.*, 15 Jun. and 19 Oct. 16.

Harington, H. R.—
Operations in the Abor Country, 1911-12.—Medal with clasp.
The War of 1914-18.—Despatches, *Lond. Gaz.*, 15 Jun. 16.

Harley, A. B.—
Aden, 1903-04.—Operations in the interior.
The War of 1914-18.—Despatches, *Lond. Gaz.*, 15 Aug. 17. D. S. O.

Harper, A. F.—
The War of 1914-18.—Despatches, *Lond. Gaz.*, 15 Aug. 17.

Harrington, Sir J. L. (*Maj., retd., I.A.*)—
East Africa, 1901.—Operations in Somaliland. Agent and Consul-Genl. at Addis Abbaba. Despatches, *Lond. Gaz.*, 18 Apr. 02.
The War of 1914-18.—Despatches, *Lond. Gaz.* 15 Jun. 16.

Harris, A. E. C.—
The War of 1914-18.—Despatches, *Lond. Gaz.*, 12 Jun. 17, M. C.

Harris, A. P. D. (*Bt. Col., retd., I.A.*).—
Burma, 1885-87.—Taking of Minhla. Operations in Yaw, including Pauk and Yaw columns. Operations of the 6th Brigade. Despatches, G. G. O. 295 of 86.—Medal with clasp.
N.-W. Frontier of India, 1897-98.—Operations on the Samana and in the Kurram Valley during Aug. and Sep. 97.—Medal with 2 clasps.
Tirah, 1897-98.—Actions of Dargai and of the Sampagha Pass.—Clasp.
N.-W. Frontier of India, Waziristan, 1901-02.—Clasp.

Harris, E. T.—
The War of 1914-18.—Despatches, *Lond. Gaz.*, 30 Jun. 16, 8 Feb. 17 and 22 Sept. 17, D.S.O.

Harris, G. F. A.—
Afghanistan, 1880.—Medal.

Harris, T. M. (*Col. retd.*)—
Persia, 1856-57.—Storming and capture of Reshire, surrender of Bushire, bombardment of Mohumrah.—Medal with clasp.
India, 1857-59.—Siege and capture of Kotah, taking of Chandaree, action of Kotahkeserai, taking of Gwalior and Powree, actions of Sundwaho and Kurai, pursuit at Koondrye (severely wounded). Despatches, *Lond. Gaz.*, 31 Jan. and 18 April 59.—Medal with clasp.

Harrison, A. H. P.—
N.-W. Frontier of India, Waziristan, 1894-95.—Medal with clasp.
N.-W. Frontier of India, 1897-98.—Tochi.—Medal with clasp.

Harrison, A. P.—
South African War, 1901-02.—Operations in Cape Colony from 22nd January to 28th February 1901. Operations in the Transvaal from 1st March to 30th June 1901. Operations in the Orange River Colony, from 1st July 1901 to 31st May 1902. —Queen's medal with 3 clasps.

Harrison, C. B. (*Lt.-Col., I. M. S. retd.*)—
N.-W. Frontier of India, 1897-98.—Malakand. —Medal with clasp.
Tirah, 1897-98.—Action of the Sampagha and Arhanga Passes.—Clasp.

Harrison, D. C. W. (*Capt., retd., Ind. S.C.*)—
Afghanistan, 1880.—Defence of Kandahar and battle of 1 Sep.—Medal with clasp.

Harrison, L. G. B.—
South African War, 1899-1901.—Operations in Natal, 1899. Relief of Ladysmith including operations of 17th to 24th January 1900, and action at Spion Kop. Operations of 5th to 7th February 1900, and action at Vaal Kranz. Operations on Tugela Heights (14th to 27th Feb. 1900). Operations in the Transvaal, May and June 1900. Operations in the Transvaal, west of Pretoria, July to 29th November 1900. Operations in Orange River Colony, May to 29th November 1900), and Cape Colony, South of Orange River, 1899-1900.—Queen's medal with 5 clasps.

Harrison, W. C. W. (*Maj., retd., I. A.*)—
N.-W. Frontier of India, 1908.—Operations in the Zakka Khel country. Supply Officer, attached, 2nd Brigade.—Medal with clasp.

Harriss, S. A.—
China, 1900.—Medal with clasp.

Hart, E. G.—
China, 1900.—Medal.
Mekran, 1901.—Attack and capture of Nodiz Fort. Despatches, G. G. O. No. 415 of 1902.
Aden, 1903-04.—Operations in the interior.
The War of 1914-18.—Despatches, *Lond. Gaz.*, 15 Aug. 17.

Hart, W. P. B.—
Burma, 1886.—Promoted, 1st grade Assist. Apothy. —Medal with clasp.
N.-E. Frontier of India, Chin-Lushai, 1889-90 —With the Gangaw Column.—Clasp.
N.-W. Frontier of India, Chin Hills, 1892-93.—Clasp.

Hartigan, E. R. (*Bt. Col., retd., I. A.*)—
Soudan, 1885.
Burma, 1887-89.—Medal with 2 clasps.

Hartley, A. F.—
South African War, 1901-02.—Operations in the Transvaal, May 01 to 31 May 02. Operations on the Zululand Frontier of Natal in Sept. and Oct. 01.—Queen's medal with 3 clasps.

The War of 1914-18.—Despatches, *Lond. Gaz.*, 25 Sept. 16; 21 and 23 July 17. D.S.O.

Hartell, J. R.—
The War of 1914-8.—Despatches *London Gazette*, 1 Nov. 16. D.S.O.

Harvey, A. B.—
Burma, 1886-89.—Medal with 2 clasps.
N.-W. Frontier of India, 1897-98.—Malakand Utman Khel.—Buner—Action of the Tanga Pass.—Medal with clasp.

Harvey, A. W. M.—
The War of 1914-18.—Despatches *Lond. Gaz.*, 17 Feb. 15 and 15 Aug. 17. Brev. of Lt. Col.

Harvey, C. O.—
The War of 1914-18.—Despatches, *Lond. Gaz.*, 25 Sept. 16.

Harvey, W. F.—
The War of 1914-18.—Despatches, *Lond. Gaz.*, 15 Aug. 17.

Harvey-Kelly, C. H. G. H.—
East Africa (Somaliland), 1908-10.—Medal with clasp.

The War of 1914-18.—Despatches, *Lond. Gaz.*, 31 Oct. 17. Brev. of Maj.

Harvey Kelly, H. St. G. H.—
Tibet, 1903-04.—Action at Niani, 26th June 1904. Operations at and around Gyantse, between 5th May and 6th July 1904. March to Lhassa, 14th July to 3rd August 1904.—Medal with clasp.

Harward, A. J. N.—
N.-W. Frontier of India, 1897-98.—Medal with clasp.

The War of 1914-18.—Despatches, *Lond. Gaz.*, 5 Apr., 13 July and 19 Oct. 16, Brev. of Col.

Haslehust, G. W.—
South African War, 1899-02.—(Slightly wounded). Served with Mtd. Inf. Operations in Natal, 1899. Relief of Ladysmith, including operations on Tugela Heights (14th to 27th Feb. 1900), and action at Pieter's Hill (27th Feb. 1900). Operations in the Transvaal and Orange River Colony between April 1901 and 31st May 1902.—Queen's medal with 5 clasps. King's medal with 2 clasps.
The War of 1914-18.—Despatches, *Lond. Gaz.*, 30 Jun. 16.

Hassan, S. (*Maj. retd. I. M. S.*)—
Burma, 1886-87.—Medal with clasp.
N.-E. Frontier of India, Chin Lushai Expedition, 1889-90.—Clasp.
N.-W. Frontier of India, Miranzai (1st) Expedition, 1891.
N.-W. Frontier of India, 1897.—Defence of the Malakand. Despatches, G. G. O. 1089 of 97.

Hastings, E. S.—
Afghanistan, 1879-80.—Operations against the Shinwaris and in the Kama District.—Medal.
Basuto War, 1881.—Operations on the Orange River and in the Transkei.—Medal with 2 clasps.
Burma, 1885-87.—Taking of Minhla and occupation of Mandalay and Bhamo Expedition to Mogaung. Operations of the 2nd Brigade. Despatches (thrice), *Lond. Gaz.*, 22 Sep. 87—G. G. O. 434 of 87.—Medal with clasp. D S O.
Burma, 1887-89.—Operations of the 6th Brigade. Slightly wounded in night affair with dacoits, 2 April 87. Commanded special Military Police operations in Popa District, Feb. to April 88. Thanks of Local Government, Nos. 858, 28 April, and 82, dated 4 May 88.—Clasp.
N.-E. Frontier of India, Manipur, 1891.—Tamu column.—Clasp.
Burma, 1889-92.—Commanded Chinbôk Punitive column, 91. Thanks of Local Government, No. 1250, 3 April 91.—Clasp.

Hastings, G. F.—
South African War, 1900-02.—Operations in Cape Colony, South of Orange River. Operations in Orange River Colony, May to 29th Nov. 1900. Operations in the Orange River Colony, 6th December 1900 to 12th August 1901. Operations in the Transvaal, 12th August 1901 to 31st May 1902. Supply Officer to Col. Young's column.—Queen's medal with 5 clasps.

Hastings. W. H.—
The War of 1914-18.—Despatches, *Lond. Gaz.*, 28 Jan. 16. 21 Jun. 16, 13 July 16, and 4 Jan. 17. Brev. of Maj. D.S.O.

Hatch, A. E. (*Lt.-Col., retd., I.A.*)—
N.-W. Frontier of India, Zhob Valley, 1884.
N.-W. Frontier of India, Zhob Valley, 1890.

Hatch, A. V.—(*Col, retd., I. A.*)
N.E Frontier of India, Sikkim, 1888.—Attack on Jalapla.—Medal with clasp.
N.-W. Frontier of India, 1897-98.—Mohmand.—Medal with clasp.
Tirah, 1897-98.—Action of the Sampagha and Arhanga Passes. Operations in the Bazar Valley, 25 to 30 Dec. 97.—Clasp.

Hatch, W. K. (*Lt.-Col., retd., I.M.S.*).—
Afghanistan, 1880.—Medal.

Hatchell, C. F. W. (*Revd.*)—
Aden, 1903-04.—Operations in the Interior.

Hatherell, W. G. (*Lt.-Col., retd., I.A.*).—
Burma, 1885-89.—Medal with 2 clasps.
N.-E. Frontier of India, Chin-Lushai, 1889-90.—Clasp.
China, 1900.—Medal.

Hathornthwaite, J C —
Aden, 1903-04.—Operations in the interior.

Haughton, E. T H.—
The War of 1914-18.—Despatches, *Lond. Gaz.*, 15 Aug. 17.

Haughton, T. H (*Col., retd., I. A.*)—
Burma, 1889.—Medal with clasp.

Haviland, L. P.—
The War of 1914-18.—Despatches, *Lond. Gaz.*, 15 Jun. 16.

Hawes, C. H.—
N.-W. Frontier of India, 1897-98.—Operations on the Samana and in the Kurram Valley during Aug. and Sep. 97. Relief of Gulistan. Operations of the Flying column in the Kurram Valley under Col. Richardson, 20 Aug. to 1 Oct. 97.—Medal with 2 clasps.
Tirah, 1897-98.—Clasp.
N.-W. Frontier of India, 1902.—Operations against the Darwesh Khel Waziris.
The War of 1914-18.—Despatches, *Lond. Gaz.*, 19 Oct. 16. D.S.O.

Hawkes, F. G.—
N.-W. Frontier of India, Hazara, 1888.—Medal with clasp.

Hawkes, G. (*Bt.-Col., u.s.l., I. A.*)—
Rumpa Rebellion, 1880.—Served with the Rumpa Field Force under Maj.-Genl. L. W. Buck.
N.-W. Frontier of India, Hazara, 1888 and 1891.—Medal with clasps.
N.-W. Frontier of India, Isazai, 1892.

Hawkes, Sir H. M. P. (*Maj. Genl., retd. I. A.*)—
Afghanistan, 1879-80.—March from Kabul to the relief of Kandahar and battle of Kandahar. Despatches, G. G. O. 582 of 80.—Medal with clasp. Bronze star.
N.-W. Frontier of India.—Isazai, 1892.
Tirah, 1897-98.—Medal with 2 clasps.

Hawkes, R.—
Tirah, 1897-98.—Medal with 2 clasps.

Hawkes, W. C. W.—
The War of 1914-15.—Despatches, *Lond. Gaz.*, 4 Jan. 17. D. S. O.

Hawkins, F.—
Afghanistan, 1879-80.—Action on the Gara Heights near Kam Dakka.—Medal.
N.-W. Frontier of India, Zhob Valley, 1884.
Burma, 1885-87.—Operations of the 5th Brigade—Medal with clasp.

Hawkins, G. A.—
Chitral, 1895.—Relief of Chitral.—Medal with clasp.
N.-W. Frontier of India, 1897-98.—Relief of Malakand. *Malakand.*—Operations in Bajour and in the Mamand country. *Buner.*—Action of the Tanga Pass. Despatches, G. G. O. 178 and 217 of 98.—2 clasps.

Hawks, F. W. (*Maj., retd., I. A.*)—
China, 1900.—Despatches, *Lond. Gaz.*, 14 May 01—Medal.

Hawks, G. A.—
N.-W. Frontier of India, Waziristan, 1901-02.—Medal with clasp.

Hawks, T. B. (*Lt.-Col., retd., I. A.*)—
Burma, 1886.—Operations round Shwebo.—Medal with clasp.

Haworth, L. B.—
Tirah, 1897-98.—Affair at Shinkamar on 29 Jan. 98.—Medal with 2 clasps.

Hay, C. J. B.—
N.-W. Frontier of India, 1897-98.—Medal with clasp.
Tirah, 1897-98.—Clasp.
N.-W. Frontier of India, Waziristan, 1901-02.—Served as Assistant Superintendent of Army Signalling, Jandola and Datta Khel columns. Despatches, G. G. O. 611 of 02.—Clasp.
The War of 1914-18.—Despatches, *Lond. Gaz.*, 1 Jan. '15, 4 Jan. and 11 Dec. 17. Brev. of Lt.-Col. D. S. O.

Hay, W. N.—
China, 1900.—Medal.

Haycraft, W. C. S.—
The War of 1914-18.—Despatches, *Lond. Gaz.*, 19 Oct. 16. M.C.

Haydon, W. P.—
The War of 1914-18.—Despatches, *Lond. Gaz.*, 15 Aug. 17. D.S.O.

Hayes, C. H. (*Bt. Col., retd., I. A.*)—
China, 1900.—Relief of Pekin. Action of Peitsang. Action at Yangtsun. Despatches, *Lond. Gaz.*, 1 May 01.—Medal with clasp.

Hayes-Sadler, Sir J. (*Lt.-Col., retd., I. A.*)—
East Africa, 1901.—Operations in Somaliland. Medal with clasp.
East Africa, Nandi, 1905-06.

Hayward, W. D.—
Chitral, 1895.—Relief of Chitral.—Medal with clasp.
N.-W. Frontier of India, 1897-98.—Operations in Bajaur and in the Mamund Country.—Clasp.

Hazelgrove, H. S. (*Bt. Col., retd., I. A.*)—
Aden, 1903-04.—Operations in the Interior.

Heard, R.—
N.-W. Frontier of India, Waziristan, 1894-95.—Medal with clasp.
Chitral, 1895.—Relief of Chitral. Action of the Malakand Pass.—Medal with clasp.

Heath, H. H. R. (*Col., retd., I.A.*)—
N.-W. Frontier of India, Jowaki, 1877-78.—Despatches, G.G.O. 738 of 78.—Medal with clasp.
Afghanistan, 1878-79.—Capture of Ali Musjid. Despatches, G. G. O. 610 of 79.—Medal with clasp.
N.-W. Frontier of India, Hazara, 1891.—Clasp.
Chitral, 1895.—Relief of Chitral.—Despatches, G. G. O. 998 of 95.—Medal with clasp. Brev. of Lt.-Col.

Heath, L. M.—
The War of 1914-18.—Despatches, *Lond. Gaz.*, 19 Oct. 16. M.C.

Heathcock, W. (*Sr. Asst. Surgn. Hony. Capt. retd. I. S. M. D.*)—
Afghanistan, 1878-79.—Medal.
Chitral, 1895.—Relief of Chitral.—Medal with clasp.

Heathcote, O. D. (*Commy. and Hony. Capt. retd. I. O. D.*)—
Afghanistan, 1878-79.—Medal.

Hehir, P.—
Burma, 1886-87.—Medal with clasp.
Tirah, 1897-98.—Medal with 2 clasps.
The War of 1914-18.—Despatches, *Lond. Gaz.*, 5 Apl., 13 July and 19 Oct. 16. C. B., C.M.G.

Hemsley, C.—
The War of 1914-18.—Despatches, *Lond Gaz.*, 15 Aug. 17.

Henderson, J. H. (*Capt., retd., I. A.*)—
South African War, 1899-1901.—Empld. with Mtd. Inf. Operations in the Orange Free State, Feb. to May 00. Operations in the Orange River Colony, May to Nov. 00.—Queen's medal with 4 clasps.

Henderson, K.—
Uganda, 1900.—Operations against the Nandi tribe. Acted as Staff Officer to Officer Commanding. (Severely wounded). Despatches, *Lond. Gaz.*, 10 Sep. 01.—Medal with clasp.
South African War, 1900-02.—Operations in the Transvaal between 30 Nov. 00 and 31 May 02. Operations in the Orange River Colony between 30 Nov. 00 and 31 May 02. Operations in Cape Colony between 30 Nov. 00 and 31 May 02.—Queen's medal with 4 clasps.
The War of 1914-18.—Despatches, *Lond. Gaz.*, 1 Jan. 16, 15 Jun. 16, and 4 Jan. 17. D.S.O.

Henderson, K. D. M.—
South African War, 1900-01.—Operations in the Orange Free State, April and May 1900. Operations in the Orange River Colony, May to 29th November 1900, including action at Wittebergen (1st to 29th July), Operations in Cape Colony, south of Orange River, 1900. Operations in Orange River Colony, 30th November 1900 to May 1901 —Queen's medal with 3 clasps.
The War of 1914-18.—C. M.

Henderson, M. H.—
The War of 1914—18.—Despatches, *Lon. Gaz.*, 21 Jun. 16, 25 Sept. 16 and 6 July 17.

Henderson, R. W.—
Chitral, 1895.—Relief of Chitral.—Medal with clasp.
N.-W. Frontier of India, 1897-98.—Operations on the Samana and in the Kurram Valley during Aug. and Sep. 97. *Buner.*—2 clasps.
Tirah, 1897-98.—Clasps.
South African War, 1902.—Operations in the Transvaal between April and 31st May 1902. Queen's medal with 2 clasps.
Tibet, 1903-04.—Medal.

Henderson, T. H.—
Burma, 1885-87.—Medal with clasp.
East Africa, 1904.—Operations in Somaliland Field Controller. Despatches, *Lond. Gaz.*, 2 Sept. 04.—Medal with clasp.

Hendley, A. G. (*Lieut.-Col., retd. I. M. S.*)—
N.-W. Frontier of India, 1897-98 —Medal with clasp.

Hendley, C. E.—
Burma, 1885-89.—Operations of the 2nd and 5th Brigades.—Medal with 2 clasps.
N.-E. Frontier of India, Chin Hills, 1892-93.—Despatches, G. G. O. 733 of 93.—Clasp.
N.-W. Frontier of India, 1897-98.—Operations on the Samana and in the Kurram Valley during Aug. and Sep. 97.—Medal with 2 clasps.
Tirah, 1897-98.—Actions of Chagru Kotal and Dargai. Operations in the Bazar Valley, 25 to 30 Dec. 97. —Clasp.
The War of 1914-18.—Despatches, *Lond. Gaz.*, 5 Apl. 16.

Hendley, H.—
Soudan, 1885.—Suakin.—Medal with clasp. Bronze star.
Burma, 1886-88.—Operations of the 1st Brigade.—Medal with 2 clasps.

Henegan, J.—
Burma, 1885-89.—Operations of the 4th and 2nd Brigades.—Medal with 2 clasps.
N.-E. Frontier of India, Lushai, 1892.—Clasp.
N.-E. Frontier of India, Chin Hills, 1892-93.—Despatches, G. G. O. 733 of 93.—Clasp—D. S. O.
Burma, 1895-96.

Hennell, Sir R. (*Lt.-Col. retd. Bo. Infy.*)—
Abyssinia, 1867-68.—Medal.
Afghanistan, 1879-80.—Medal.
Burma, 1885-87.—Despatches, G. G. O. 434 of 87. —Medal with clasp; *D.S.O.*

Hennessy, J. P. C.—
South African War, 1899-1901.—Relief of Ladysmith, including action at Colenso (15 Dec. 99). Operations of 17 to 24 Jan. 00 and action at Spion Kop; operations of 5 to 7 Feb. 00 and action at Vaal Krantz; operations on Tugela Heights (14 to 27 Feb. 00) and action at Pieter's Hill (27 Feb. 00). Operations in Natal, March to June 00 Operations in the Transvaal from November 1900 to June 1901 and again from August 1901 to December 1901. Operations in the Orange River Colony and Cape Colony in July 1901. Despatches, *Lond. Gaz.*, 8 Feb. and 10 Sep. or Brevt. of Maj.—Queen's medal with 7 clasps.
The War of 1914-18.—Despatches, *Lond. Gaz.*, 17 Feb. and 23 Jun. 15. C.M.G.

Hennessey, T.—
Afghanistan, 1880.—Medal.
Egypt, 1882.—Battle of Tel-el-Kebir. Forced march to, and occupation of Zagazig.—Medal with clasp. Bronze star.

Henry, J. R. A.—
The War of 1914-18.—Despatches, *Lond. Gaz.*, 15 Aug. 17.

Henry, W. R. P.—
The War of 1914-18.—Despatches, *Lond. Gaz.*, 15 Jun. 16.

Hepper, E. C.—
N.-W. Frontier of India, 1908.—Operations in the Zakka Khel country.

Herbert, Charles (*Lieut.-Col., retd., I. A.*)—
Afghanistan, 1879-80.—Zaimukht Expedition. Despatches. Br.-Genl. J. A. Tytler's despatch, dated 20 Dec. 79.—Medal.

Herbert, Claude (*Maj., Supy. List, I. A.*)—
Burma, 1886-89.—Medal with 2 clasps.
N.-E. Frontier of India, Sikkim, 1888-89.—Clasp.
N.-W. Frontier of India.—Action at Wano, 3 Nov. 94. (Dangerously wounded). Despatches, G. G. O. 268 of 95.—Clasp. D. S. O.

Herbert, D.—
N. E. Frontier of India, Manipur, 1891.—Medal with clasp.

Herbert, H. (*Lieut.-Col., retd., I. M. S.*)—
Somali Coast, 1890.—(Zaila Field Force).—Affair at Husain Zareeba.

Herbert, L. *Maj.-Genl., retd., I. A.*)—
Chitral, 1895.—Relief of Chitral. Served as D. A. A. and Q. M. G., 3rd Brigade. Action of the Malakand Pass. Despatches, G. G. O. 998 of 95.—Medal with clasp. Brev. of Maj.
N.-W. Frontier of India, 1897-98.—Defence of Malakand as D. A. A. and Q. M. G., Malakand Brigade, (Severely wounded). Despatches, G. G. O. 1089 of 97. *Utman Khel.*—As D. A. Q. M. G. *Buner.*—As Staff Officer, Lines of Communication. Despatches, G. G. O. 217 of 98.—2 clasps. Brev. of Lt.-Col.

Herbert, L. N.—
N.-W. Frontier of India, (1st) Miransai, 1891.
N.-W. Frontier of India, 1897-98.—Relief of Malakand—*Malakand.*—Operations in Bajaur—Mohmand.—Medal with 2 clasps.
Tirah, 1897-98.—Clasp.
Tibet, 1903-04.—Depatches, *Gazette of India*, No. 1065 of 1904.—Medal.
N.-W. Frontier of India, 1908.—Operations in the Mohmand country.—Medal with clasp.
The War of 1914-18.—Detpatches, *Lond. Gaz.*, 6 Dec. 16. C.B.

Herdon, H. E.—
Tirah, 1897-98.—Medal with 2 clasps.
N.-W. Frontier of India, Waziristan, 1901-02.—Clasp.

Herlz-Smith, G. V.—
The War of 1914-18.—Despatches, *Lond. Gaz.*, 21 Jun. 6 and 15 Aug. 17. M.C.

Heron, D.—
The War of 1914-18.—Despatches, *Lond. Gaz.*, 3 Oct. 17.

Herrick, R. de S. B.—
The War of 1914-18.—Despatches, *Lond. Gaz.*, 19 Oct. 16. D.S.O.

Hesketh, W.—
Chitral, 1895.—Relief of Chitral. Action of the Malakand Pass. (Severely wounded).—Medal with clasp.
N.-W. Frontier of India, 1897-98.—Malakand operations in the Mamund country.—Clasp.
The War of 1914-18.—Despatches, *Lond. Gaz.*, 15 Aug. 17.

Hesterlow, J. N. (*Sr. Asstt. Surgn., Hony. Capt. retd., M.*)—
Afghanistan, 1878-80.—Action of Charasiah, engagements around Kabul and defence of Sherpur. —Medal with 2 clasps.

Hewes, E. E.—
N.-W. Frontier of India, 1897-98.—Mohmand. Medal with clasp.
Tirah, 1897-98.—Clasp.

Hewett, G.—
N.-W. Frontier of India, Waziristan, 1901-02.—Medal with clasp.
The War of 1914-18.—Despatches, *Lond. Gaz.*, 5 Apr. 16. D.S.O.

Hewett, W. S. (*Lt.-Col., u. s. l., I.A.*)—
Afghanistan, 1878-79-80.—Battle of Ahmed Khele.—Medal with clasp.

Hewitt, D. R.—
South African War, 1900-01.—Operations in the Transvaal, West of Pretoria, 15th August 1900 to 29th November 1900. Operations in the Transvaal, 30th November 1900 to 16th January 1901. —Queen's medal with 3 clasps.

Hewlett, A.—
N.-W. Frontier of India, 1897-98.—Operations on the Samana and in the Kurram Valley during Aug. and Sep. 97.—Medal with 2 clasps.
Tirah, 1897-98.—Reconnaissance of the Kharmana Defile and action of 7 Nov. 97.—Operations against the Khani Khel Chamkannis.—Clasp.

Hewlett, E. A.—
The War of 1914-18.—Despatches, *Lond. Gaz.*, 15 Aug. 17. Brev. of Lt-Col.

Hext, C. W.—
South African War, 1900-1902.—Operations in the Transvaal, east of Pretoria, July to 29th November 1900, including actions at Belfast (26th and 27th August 1900) and Lydenburg (5th to 8th September). Operations in the Transvaal and Orange River Colony, 30th November 1900 to December 1901. Operations on the Zululand Frontier of Natal in September and October 1901.—Queen's medal with 3 clasps, King's medal with 2 clasps. Operations in the Abor Country, 1911-12.—Despatches, *Gazette of India*, No. 480 of 23rd May 1912—Medal.

Hext, G. T. B.—
The War of 1914-18.—Despatches, *Lond. Gaz.*, 30 Jun. 16. D.S.O.

Hext, J. E.—
South African War, 1900-02.—Operations in Orange River Colony, 3rd May to 16th July 1900. Operations in the Transvaal in 1901. Operations in Cape Colony 7th July 1901 to 30th January 1902. —Queen's medal with 3 clasps. King's medal with 2 clasps.

Heyland, A. K.—
British East Africa, 1901.—Operations against the Ogaden Somalis in Jubaland.—Medal with clasp.
Aden, 1903-04.—Operations in the Interior.
The War of 1914-18.—Despatches, *Lond. Gaz.*, 22 Jun. 15. D.S.O.

Heyland, G. D.—
The War of 1914-18.—Despatches, *Lond. Gaz.*, 15 Aug. 17.

Hickey, P. F. B.—
The War of 1914-18.—Despatches, *Lond. Gaz.*, 15 Aug. 17.

Hickie, C. C.—
South African War, 1901-02.—Served in Mediterranean.—Medal.

Hickley, A. C.—
N.-W. Frontier of India, Hazara, 1888.—Medal with clasp.
N.-E. Frontier of India, Manipur, 1891.—Clasp.
N.-W. Frontier of India, 1897-98.—Operations on the Samana and in the Kurram Valley during Aug. and Sep. 97. Relief of Gulistan.—Medal with 2 clasps.
Tirah, 1897-98.—Actions of Chagru Kotal, Dargai and of the Sampagha and Arhanga Passes.—Reconnaissance of the Saran Sar and action of 11 Nov. 97. Operations in the Bara Valley, 7 to 14 Dec. 97.—Clasp.
N.-W. Frontier of India, Waziristan, 1901-02.—Clasp.

Hickley, H. D.—
The War of 1914-18.—Despatches, *Lond. Gaz.*, 5 Apr. 16.

Hickman, H. T. D.—
N.-W. Frontier of India, 1908.—Operations in the Mohmand country.—Medal with clasp.
The War of 1914-18.—Despatches, *Lond. Gaz.*, 19 Oct. 16 and 15 Aug 17. M.C.

Higham, B.—
N.-W. Frontier of India, 1908.—Operations in the Mohmand country.—Medal with clasp.

Hight, E. L. (*Maj., retd., Ind., S. C.*)—
Afghanistan, 1878—80.—Action of Charasiah and operations at and around Kabul in Dec. 79. —Medal with 2 clasps.
N.-W. Frontier of India, Miranzai (1st and 2nd) Expeditions, 1891.—Medal with clasp.

Hildebrand, H. A.—
South African War, 1902.—Operations in the Transvaal, from 10th April to 31st May 1902. —Queen's medal with 2 clasps.

Hildebrand, W. H. (*Lt.-Col., retd., I. A.*)—
Burma, 1887-88.—Kalewa column in the operations on the Western Frontier.—Medal with clasp.
N.-E. Frontier of India, Chin-Lushai, 1889-90.—(Burma column.) Despatches, G. G. O. 677 of 90.—Clasp.
N.-W. Frontier of India, 1897-98.—Operations during Aug. and Sep. 97.—Operations of the Flying Column in the Kurram Valley under Colonel Richardson, 20 Aug. to 1 Oct. 97.—Medal with 2 clasps.

Hill, F. B. (*Lt.-Col., retd., I. A.*)—
N.-W. Frontier of India, Waziristan, 1894-95.—Medal with clasp.
Chitral, 1895.—Relief of Chitral.—Medal with clasp.

Hill, H. A.—
South African War, 1901.—Operations in Cape Colony between 25th March and 30th September 1901. Queen's medal with 2 clasps.

Hill, H. C.—
The War of 1914-18.—Despatches, *Lond. Gaz.*, 5 Apr. and 13 Jul. 16.—D. S. O.

Hill, H. W. D.—
The War of 1914-18.—Despatches, *Lond. Gaz.*, 19th Oct. and 15 Aug. 17.

Hill, J.—
N.-E. Frontier of India. Chin-Lushai, 1889.—Medal with clasp.
N.-W. Frontier of India (2nd) Miranzai, 1891.—Clasp.
Chitral, 1895.—Relief of Chitral.—Medal with clasp.
N.-W. Frontier of India, 1897-98.—Operations on the Samana and in the Kurram Valley during Aug. and Sep. 97. Operations of the Flying Column in the Kurram Valley under Colonel Richardson, 20 Aug. to 1 Oct. 97.—2 clasps.
Tirah, 1897-98.—Actions of the Chagru Kotal and Dargai, and of the Sampagha and Arhanga Passes.—Operations in the Waran Valley and action of 16 Nov. 97.—Despatches, G. G. O. 483 of 98.—Clasp—D. S. O.
The War of 1914-18.—Despatches, *Lond. Gaz.*, 22 Jan. 15, 13 July 16, 4 Jan 17 and 15 May 17. Brev. of Col.

Hill, R. D. O.—
Operations in the Abor country, 1911-12.—Medal with clasp.

Hill, R. H. T. (*Col., retd., Mad. Inf.*)—
Burma, 1885-86.—Director of Transport to the force. Mentioned by Government of India. Despatches, G. G. O. 295 of 86. Brev. of Lt.-Col.

Hill, W. A.—
N.-W. Frontier of India, Hazara, 1891.—Medal with clasp.
N.-W. Frontier of India, 1897-98.—With Kohat-Kurram Force.—Medal with 2 clasps.
Tirah, 1897-98.—Clasp.
N.-W. Frontier of India, Waziristan, 1901-02.—Clasp.

Hilliard, H. N.—
Tirah, 1897-98.—Assistant to Base Commissariat Officer and subsequently Divisional Transport Officer, 2nd Division.— Operations in the Bara Valley, 7 to 14 Dec. 97.—Despatches, G. G. O. 244 of 98.—Medal with 2 clasps.

Hills-Johnes, *Sir* James (*Lt.-Gen., retd., Ben. Art.*)—
India, 1857-58.—Actions on the Hindun, battles of Budleekeserai and Nujjufghur, siege and capture of Delhi (severely wounded), siege and capture of Lucknow, Rohilcund campaign, taking of Bareilly, action of Mohumdee. Despatches, *Lond Gaz.*, 16 Jan. 58. Specially mentioned in General Order of Governor-General of India in Council.—Medal with 2 clasps, VC Br. Maj.
Eusofzai Expedition, 1863-64.
Abyssinia, 1868.—In command of mortars at capture of Magdala. Despatches, *Lond. Gaz.*, 16 June 68. —Medal; Brev. of Lt.-Col.
N.-E. Frontier of India, Lushai Expedition, 1871-72. —Mentioned in Despatches.—Medal with clasp; *C. B.*
Afghanistan, 1878-79-80.—Operations at Sherpur. Action of Patkao Shana and defence of Kandahar. Despatches, G. G. Os. 1136 of 79 and 137. 493 and 566 of 80. Thanked by both Houses of Parliament, 5 May 81. *K. C. B.*—Medal with 2 clasps.

Hilson, R. J.—
Chitral, 1895.—Relief of Chitral.—Medal with clasp.
China, 1900.—Medal.

Hinde, H. T. L.—
The War of 1914-18.—Despatches, *Lond. Gaz.*, 8 Feb. 17.

Hinds, W. A. G.—
The War of 1914-18.—Despatches, *Lond. Gaz.*, 15 Mar. and 19 Oct. 16.—M.C.

Hinston, R. W. G.—
The War of 1914-18.—Despatches, *Lond. Gaz.*, 25 Nov. 16.—M. C.

Hislop, J. H.—
The War of 1914-18.—Despatches, *Lond. Gaz.*, 5 Ap. and 19 Oct. 16.—M. C.

Hitchens, C. H. M.—
The War of 1914-18.—Despatches, *Lond. Gaz.*, 28 Nov. 17.

Hobday, E. M.—
The War of 1914-18.—Despatches, *Lond. Gaz.*, 4 July 16.

Hobday, J. R. (*Col., retd., I. A.*)—
Afghanistan, 1879.—With Kandahar Field Force.—Medal.
Burma, 1886-90.—Ponkhan and Tonhon Expeditions. Despatches, G. G. O. 434 of 87.—Medal with 3 clasps. Brev. of Maj.

Hobday, T. F. (*Hony. Maj.-Genl., retd., I.A.*)—
Afghanistan, 1878-79.—With Kandahar Field Force. Despatches. General *Sir* D. M. Stewart's despatch No. 640 K., 24 June 79.—Medal. Brev. of Maj.

Hobson, A. C.—
China, 1900.—Medal.

Hocken, C. A. F.—
Chitral, 1895.—Relief of Chitral.—Medal with clasp
China, 1900.—Medal.

Hodding, J. S. (*Major, retd., I. A.*)—
Burma, 1889-91.—Wuntho Expedition —Medal with clasp.

Hodgkinson, C.—
Tirah, 1897-98.—Medal with 2 clasps.

Hodgkinson, J.—
Tirah, 1897-98.—Operations in the Bazar Valley from 25 to 30 Dec. 97.—Medal with 2 clasps.

Hodgson, C. N. (*Col., retd., Ben. Inf.*)—
N.-E. Frontier of India, Bhutan Expedition, 1865. Medal with clasp.
Afghanistan, 1879-80.—With the Kurram Column —Medal.
N.-W. Frontier of India, 1881.—*Mashud-Waziri Expedition.*—Mentioned in Despatches.

Hodgson, G. B.—
Afghanistan, 1880.—Medal.

Hodgson, G. C.—
N.-W. Frontier of India, Waziristan, 1901-02. Medal with clasp.
Tibet, 1903-04.—Action at Niani, 26th June 1904. Operations at and around Gyantse, between 5th May and 6th July 1904. (Severely wounded) March to Lhassa, 14th July to 3rd August 1904 Despatches, *Gazette of India*, No. 1065 of 1904 —Medal with clasp. D. S. O.
Operations in the Abor country, 1911-12. Medal with clasp.

Hodgson, W. H.—
South African War, 1899—1901.—Advance on Kimberley, including action at Magersfontein. Operations in the Transvaal east of Pretoria, July to 29 Nov. 00, including actions at Lydenberg (5 to 8 Sept.) Despatches, *Lond. Gaz.*, 16 Mar. 00.—Queen's medal with 3 clasps. Medal for distinguished conduct in the field.

Hogan, W J. A.—
N.-E. Frontier of India, Chin-Lushai, 1889-90.—Medal with clasp. Promoted, G.G.O. No. 71 of 1891.

Hogan, E. M. A. J.—
South African War, 1900-02.—Severely wounded. Operations Orange River Colony Aug. 01 to 31 May 02.—Queen's medal with 5 clasps, King's medal with 2 clasps.

Hogg, C. M. T.—
The War of 1914-18.—Despatches, *Lond. Gaz.*, 22 Jun. 15 and 4 Jan. 17.—D. S. O.

Hogg, G. C. (*Maj.-Gen. u. s. l. Bo. Cav.*)—
Afghanistan, 1879-80.—Action of Girishk, cavalry affair of Khush-ki-Nakud, battle of Maiwand, affair at Kairabad, defence of and sortie from Kandahar (horse shot). Despatches, G. G. Os. 480, 551 and 566 of 80.—Medal.

Hogg, G. E. M.—
The War of 1914-18.—Despatches, *Lond. Gaz.*, 30 Jun '16.

Hogg, H. C. (*Maj., retd., Bo. S. C.*)—
Afghanistan, 1878-79.—Action of Kush-ki-Nakud, Despatches, G. G. O. 610 of 1879.—Medal.

Hogg, J. MacK. T. (*Lt.-Col., retd., I. A.*)—
Tirah, 1897-98.—Operations in the Bara Valley, 7 to 14 Dec. 97.—Medal with 2 clasps.

Hogg, R. E. T.—
The War of 1914-18.—Despatches, *Lond. Gaz.*, 5 May 16.

Hogg, T. C. McK. T.—
N.-W. Frontier of India, Hazara, 1888.—Action of Kotkai.—Medal with clasp.
N.-E. Frontier of India, Kachin Hills, 1892-93.—Clasp.
South African War, 1901-02.—Queen's medal with 4 clasps.

Holbrooke, B. F. R.—
East Africa (British), 1896.—Operations against the Mazrui Rebels —Medal.
China, 1900.—Relief of Pekin. Medal with clasp.
South African War, 1901-02.—Operations in the Transvaal, between 30th November 1901 and 15th March 1902. Operations in the Orange River Colony, between 16th March and 31st May 1902. —Queen's medal with 4 clasps.

Holderness, H.—
N.-W. Frontier of India, Waziristan, 1901-02.—Medal with clasp.

Holdich, H. A.—
Tirah, 1897-98.—Assistant Survey Officer.—Actions of the Sampagha and Arhanga Passes.—Reconnaissance of the Saran Sar and action of 9 Nov. 97.—Operations in the Waran Valley and action of the 16 Nov. 97.—Operations in the Bazar Valley, 25 to 30 Dec. 07.—Medal with 2 clasps.
The War of 1914-18.—Despatchs, *Lond. Gaz.*, 5 Apl. 13 July and 19 Oct. 16. Brev. of Lt-Col. D.S.O.

Holland. E. F. (*Major, retd., I.A.*)—
Tirah, 1897-98.—Action of the Sampagha and Arhanga Passes.—Medal with 2 clasps.

Holland, G. L.—
N.-E. Frontier of India Sikkim 1888.—Attack of Jalapla. Despatches, G. G. O. 1889 of 88.—Medal with clasp.
Chitral, 1895.—Relief of Chitral.—Medal with clasp.
China, 1900.—Medal.
N.-W. Frontier of India, Waziristan, 1901-02.—Clasp.

Holland, J. A.—
Afghanistan, 1878-79-80.—Medal.
N.-W. Frontier of India, 1897-98.—Mohmand.—Medal with clasp.

Holland, P. (*Col., retd., I. A.*)—
Egypt, 1882.—Actions of El-Magfar, Tel-el-Mahuta, and Kassassin, 28 Aug. and 9 Sep. 82, and battle of Tel-el-Kebir.—Medal with clasp and bronze star.
Burma, 1885-87.—Despatches, G. G. O. 339 of 87.—Medal with 2 clasps.
N.-W. Frontier of India, Miranzai (1st), 1891.
N.-W. Frontier of India, 1897-98.—Operations on the Samana and in the Kurram Valley during Aug. and Sep. 97. Operations of the Flying Column in the Kurram Valley under Colonel Richardson, 20 Aug. to 1 Oct. 97.—Medal with 2 clasps.
China, 1900.—Medal.

Holland-Pryor, P.—
N.-W. Frontier of India, Waziristan, 1894-95.—Served as Brigade Transport Officer.—Medal with clasp.
N.-W. Frontier of India, 1897-98.—Mohmand.—Medal with clasp.
Tirah, 1897-98.—Served as Brigade Transport Officer, 5th Brigade. Operations in the Bara Valley, 7 to 14 Dec. Despatches, G. G. O. 244 of 98. —Clasp.
South African War, 1899-1902.—Served as A. A. G. Bloemfontein and as D. A. A. G. Kimberley District. Operations in Cape Colony, south of Orange River, Feb. 00. Operations in the Orange Free State, March to May 00. Operations in Orange River Colony, May to 29 Nov. 00. Operations in the Transvaal, west of Pretoria, July to 29 Nov. 00. Operations in the Orange River Colony between 30 Nov. and 31 May 02. Operations in Cape Colony between 30 Nov. 00 and 31 May 02. Despatches, *Lond. Gaz.*, 16 April 01. Services acknowledged by High Commissioner for South Africa; Military Despatch No. 165 of 28 Nov. 02 from the Right Hon'ble the Secretary of State for India.—Queen's Medal with 3 clasps. King's medal with 2 clasps. Brev. of Maj.
The War of 1914-18.—Despatches, *Lon. Gaz.*, 4 July 16 and 15 Aug. 17—C.M.G., D.S.O.

Holloway, B.—
Burma, 1885-86.—Despatches, G. G. O. 295 of 86. —Medal with clasp.

Holloway, E. L. (*Bt.-Col., retd., I. A.*)—
Burma, 1891-92.—Baungshe column.—Medal with clasp.

Holman, H. C.—
Burma, 1891-92.—Wuntho Expedition.—Acted as Intelligence Officer. (Slightly wounded.) Despatches, G. G. O. 985 of 91.—Staff Officer, Irrawaddy column, in 1891-92.—Medal with clasp.
China, 1900.—Interpreter to the China Expeditionary Force. Special Service Officer, graded as D. A. A. G.; and Railway Staff Officer, graded as 1st class Post Staff Officer. Despatches, *Lond. Gaz.* 13 Sep. 01.—Medal. D.S.O.
Russo-Japanese War, 1904-05.—Attached to the Russian Army in Manchuria. Order of St. Stanislas, 2nd class, with swords, and Russian medal.
The War of 1914-18.—Despatches, *Lond. Gaz.*, 17 Feb. 15, 22 Jan. 15, 1 Jan. 16, 15 Jun. 16, 15 May 17 and 11 Dec. 17. Brev. of Col. Legion of Honour (4th class).—C.B., C.M.G.

Holmes, B. W.—
South African War, 1900.—Operations in the Orange Free State, 3rd April to 25th May 1900, including actions at Houtnek (1st May), Vet River (5th and 6th May), and Zand River (10th May). Operations in the Transvaal in May and June 1900, including actions near Johannesburg (29th May) and Pretoria (4th June). Despatches, *Lond. Gaz.*, 10 Sep. 01.—Queen's medal with 3 clasps.

Holmes, C. W. A.—
British E. Africa, 1901.—Operations against the Ogaden Somalis in Jubaland.—Medal with clasp.
The War of 1914-18.—Despatches, *Lond. Gaz.*, 15 Aug. 17.

Holmes, G. V.—
N.-E. Frontier of India, Lushai, 1889.—Medal with clasp.
Burma, 1890-92.—Staff Officer, Kanhow column.

Holroyd, G.—
N.-W. Frontier of India, 1908.—Operations in the Mohmand country.—Medal with clasp.

Home, D. C.—
Tibet, 1903-04.—Medal.

Home, J. E.—
The War of 1914-18.—Despatches, *Lond. Gaz.*, 15 Aug. 17

Home, J. M.—
N.-E. Frontier of India, Chin-Lushai, 1889-90.—Medal with clasp. *Manipur, 1891.*—Clasp.
N.-W. Frontier of India, Waziristan, 1901-02.—Medal with clasp.
Russo-Japanese War, 1904—Attached to the Russian Army in Manchuria.—Order of St. Anne, 2nd class with swords and Russian medal.

Home, P. M.—
N.-W. Frontier of India, 1897-98.—Mohmand.—Malakand. Operations in Bajaur.—Medal with clasp.
Tirah, 1897-98.—Clasp.

Home, W. E. (*Capt., retd., I. A.*)—
N.-W. Frontier of India, Waziristan, 1901-02.—Medal with clasp.

Hood, E. F. (*Lt.-Col., retd., I. A.*)—
Burma, 1888.—Medal with clasp.
China, 1900.—Relief of Pekin. Action of Peitsang. Action at Yangtsun. Despatches, *Lond. Gaz.*, 14 May 01.—Medal with clasp.

Hooker, J. S.—
South African War, 1900-01.—Served with Mounted Infantry under Colonel Pilcher, during operation in the Orange River Colony and Cape Colony from 29 Dec. 00 to 30 Aug. 01.—Queen's medal with 4 clasps.

Hooton, A.—
N.-W. Frontier of India, 1897-98.—Mohmand.—Medal with clasp.
Tirah, 1897-98.—Action of the Sampagha and Arhanga Passes. Reconnaissance of the Kharmana Defile and action of the 7th Nov. 97. Operations against the Khani Khel Chamkanis. Operations in the Waran Valley and action of the 6th Nov. 97. Operations in the Bazar Valley, 25th to 30th Dec. 97.—clasp.

Hopkinson, G. H.—
N.-W. Frontier of India, 1908.—Operations in the Zakka Khel Country. Medal with clasp.

Hopwood, H. R.—
China, 1900.—Medal.
The War of 1914-18.—Despatches, *Lond. Gaz.*, 5 Apr. 16, 13 July 16, 10 May 16, 19 Oct. 18, 14 Nov. 16 and 15 Aug. 17 Brevs. of Lt. Col. and Col.

Hornby, M. L. (*Maj., retd., I. A.*)—
N.-W. Frontier of India.—Orderly Officer to G. O. C. Waziristan Delimitation Escort, 1894. Action at Wana, 3 Nov. 94. (Severely wounded). Despatches, G. G. O. 268 of 95.
Waziristan, 1894-95.—Orderly Officer to G. O. C. 1st Brigade.—Medal with clasp. D. S. O.
Tirah, 1897-98.—Operations in the Bara Valley, 7th to 14th Dec. 97.—Medal with 2 clasps.
British East Africa, 1898.—Operations against the Ogaden Somalis. Served as Adjutant and Quarter-master, 1st Bn., Uganda Rifles. Medal with clasp.
Uganda, 1899.—Operations against mutineers and rebels, Unyoro 99, including the defeat of Kabarega. Served as Adjutant and Quarter-master, 1st Bn., Uganda Rifles, and Transport and Signalling Officer to the Force. Clasp.
Uganda, 1900.—Operations against the Nandi Tribe. Commanding No. I column. Despatches *Lond. Gaz.*, 10 Sep. 01.—Medal with clasp.
East Africa, 1902-03.—Operations in Somaliland. As Chief Transport Officer. Despatches, *Lond. Gaz.*, 2 Sep. 04. Brev. of Maj.—Clasp.
The War of 1914-18.—Despatches, *Lond. Gaz.*, 15 Jun. 16, 4 Jan. 17, 15 May 17 and 11 Dec. 17. Brev. of Lt.-Col. C. M. G.

Hornen F. H. F.—
The War of 1914-18.—Despatches, *Lond. Gaz.*, 1 Jan. 16.—M. C.

Hoskyr, J. C. M.—
The War of 1914-18.—Despatches, *Lond. Gaz.*, 19 Oct. and 14 Nov. 16.

Houlson-Craufurd, J. A. (*Maj., retd., I. A.*)—
N.-W. Frontier of India, Zhob Valley, 1884.
Burma, 1885-89.—Medal with 2 clasps.
China, 1900.—D. A. Q. M. G. 2nd Brigade, Despatches, *Lond. Gaz.*, 13 Sep. 01—Medal.
East Africa, 1902-04.—Operations in Somaliland. —Medal with clasp.

Houston, A. M.—
N.-W. Frontier of India, 1897-98.—Tochi.—Medal with clasp.
N.-W. Frontier of India, Waziristan, 1901-02.—Clasp.
N.-W. Frontier of India, 1908.—Operations in the Mohmand country. Engagement of Kargha, 24th May 1908. Despatches, G. G. O. No. 521 A of 1908.—Medal with clasp.

Houston, W. M.—
East Africa, 1902-04.—Operations in Somaliland. —Medal with clasp.

Howard, C. A. L.—
East African (Somaliland) 1914-15.—Slightly wounded. Despatches, *Lond. Gaz.*, 2 Aug. 16. —Medal with clasp.

Howe, R. B. B. (*Major, retd., I. A.*)—
Burma, 1890-91.—Medal with clasp.
N.-W. Frontier of India, 1897-98. Tochi.—Medal with clasp.

Howell, A. A. (*Lt.-Col., retd., I. A.*)—
N.-W. Frontier of India, Manipur, 1891. Medal with clasp.

Howell, G. B.—
N.-W. Frontier of India, 1908.—Medal with clasp.
The War of 1914-18.—Despatches, *Lond. Gaz.*, 18 May 17.

Howell, C. H.—
The War of 1914-18.—Despatches, *Lond. Gaz.*, 21 Jan. 16 and 15 Aug. 17.

Howell, P.—
N.-W. Frontier of India, 1908.—Operations in the Zakka Khel country. Provost Marshal and Treasure Chest Officer, attached F. F. Staff.—Medal with clasp.

Howell, R.—
N.-W. Frontier of India, Hazara, 1891—Medal with clasp.
N.-W. Frontier of India, 1897-98.—Mohmand. Operations in the Mohmand country.—Medal with clasp.
Tirah, 1897-98.—Operations in the Bazaar Valley, 25th to 30th December 1897. Actions at Dargai and of the Sampagha Pass. Despatches, G. G. O. 244 of 1898. Promoted Sub-Conductor. Clasp.

Howey, W. (*M. Gen. u.s.l. Ben. Inf.*)—
India, 1858-59.—Capture of Fort Turowl, action of Jugdespore and other affairs.—Medal.

Howlett, A. (*Col., retd., I. A.*)—
Perak, 1875-76.—Asst. Field Engr.—Medal with clasp.
N.-E. Frontier of India, Manipur, 1891.—Despatches, *Lond. Gaz.*, 14 Aug. 91.—Clasp.
Burma, 1892.—Commanded the Tashon Column.

Howlett, C. G.—
The War of 1914-18.—Despatches, *Lond. Gaz.*, 15 Aug. 17.

Howson, G.—
The War of 1914-18.—Despatches, *Lond. Gaz.*, 22 Jun. 15.—M. C.

Hudson, C.—
China, 1900.—Medal.
East Africa, 1903-04.—Operations in Somaliland, Action at Jidballi. Despatches, *Lond. Gaz.*, 2 Sep. 04.—Medal with 2 clasps. D.S.O.
Balkan Campaign, 1912-13.—British Red Cross Commissioner with Bulgarian Forces.
The War of 1914-18.—Despatches, *Lond. Gaz.*, 1 Jan. 16.

Hudson, C. T.—
N.-W. Frontier of India, 1897-98.—Operations on the Samana and in the Kurram Valley—Medal with clasp.
Tirah, 1897-98.—Clasp.
The War of 1914-18.—Despatches, *Lond. Gaz.*, 31 Oct. 17.—C. M. G.

Hudson, H.—
N.-W. Frontier of India, (2nd) Miranzai, 1891.—Medal with clasp.
N.-W. Frontier of India, 1897-98.—Mohmand.—D. A. Q.-M. G., 2nd Brigade.—Medal with clasp.
China, 1900.—D. A. Q.-M. G., 3rd Brigade, China Expeditionary Force. Despatches. *Lond. Gaz.*, 14 May 01.—Medal. Brev. of Maj.
The War of 1914-18.—Despatches, *Lond. Gaz.* 17 Feb. 15, 1 Jan. 16, 15 Jun. 16, 4 Jan. 17. Promd. Maj.-Gen.

Hudson, J. R.—
The War of 1914-18.—Despatches, *Lond. Gaz.*, 19 Oct. 16.

Huggins, P. G. (*Lt.-Col., retd., I. A.*)—
Afghanistan, 1878-79.—Medal.
Burma, 1885-89.—Present at the taking of Minhla and occupation of Mandalay. Operations of the 5th Brigade—Despatches, *Lond. Gaz.*, 2 Sep. '87. D. S. O.—Medal with 2 clasps.
N.-E. Frontier of India, Chin Hills, 1892-93.—Clasp.
Tirah, 1897-98.—Actions of Chagru Kotal and Dargai.—Operations in the Bazar Valley, 25 to 30 Dec. 97. Despatches, G. G. O. 244 of 98.—Medal with 2 clasps.

Hughes, A. J.—
The War of 1914-18.—Despatches, *Lond. Gaz.*, 1 Dec. 16.

Hughes, C. F. (*Major-Genl., u. s. l. I. A.*)—
Arabia.—Expedition against Foodhlee Arabs, 1865-66.—(Despatches, G. G. O. 98 of 66).
Afghanistan, 1878-80.—Action at Ahmed Khel.—Despatches, G. G. O. 74 and G. G. O. I. 326 of 80. Brev. of Maj.—Medal with clasp.
Burma, 1886-87.—(Despatches, G. G. O. 339 of 87. Brev. of Col.—Medal with 2 clasps.)

Hughes, C. F. W.—
The War of 1914-18.—Despatches, *Lond. Gaz.*, 21 June 16.

Hughes, G. R.—
The War of 1914-18.—Despatches, *Lond. Gaz.*, 6 July 17.

Hughes, G. V.—
The War of 1914-18.—Despatches, *Lond. Gaz.*, 15 Aug. 17.

Hughes, L. W.—
The War of 1914-18.—Despatches, *Lond. Gaz.*, 19 Oct. 16.

Hugo, E. V.—
N.-W. Frontier of India, Waziristan, 1894-95.—Medal with clasp.
Chitral, 1895.—Relief of Chitral.—Medal with clasp.
N.-W. Frontier of India, 1897-98.—Defence and Relief of Chakdara. Buner.—Despatches, G. G. O. 1317 of 97.—2 clasps.
The War of 1914-18.—Despatches, *Lond. Gaz.*, 15 Aug. 17. C. M. G.

Hugo, J. H.—
N.-W. Frontier of India, 1897-98.—Defence of Malakand—*Malakand.* Action at Landakai.—Operations in Bajaur and in the Mamund country.—Despatches, G. G. O. 1089 of 97.—Medal with 2 clasps.—D. S. O.

Hulbert, J. G. (*Lt.-Col. retd., I. M. S.*)—
N.-E. Frontier of India, Kachin Hills, 1892-93.—Medal with clasp.
N.-W. Frontier of India, 1897-98.—Mohmand.—Medal with clasp.
Tirah, 1897-98.—Operations in the Bara Valley, 7 to 14 Dec. 97.—Clasp.

Hulbert, T. E.—
South African War, 1900-01.—Operations in the Orange Free State. April and May 1900. Operations in the Orange River Colony May to 29th November 1900, including action at Wittebergen, 1st to 20th July. Operations in Cape Colony, South of Orange River. 12th to 14th April 1900. Operations in the Orange River Colony, 30th November 1900 to May 1901. Queen's medal with 3 clasps.
The War of 1914-18.—Despatches, *Lond. Gaz.*, 1 Jun. 17.

Hulseberg, H.—
East Africa Protectorate, 1898.—Operations in Teita District, 11th May to 31st December.
Uganda, 1898-99.—Actions of Kuoeri and Kijanguti. S. of S. despatch No. 49, dated 18 May 1899.—Medal with clasp.
Mekran, 1901.—Attack and capture of Nodiz Fort. Mentioned in despatches. G. G. O. 415 of 02.
East Africa (Somaliland), 1908-10.—Medal with clasp.
The War of 1914-18.—Despatches, *Lond. Gaz.*, 1 Jan. 16 and 22 Sep. 17.—D. S. O. Clasp to D. S. O.

Hume, S. G. W.—
N.-W. Frontier of India, Waziristan, 1894-95.—Medal with clasp.
South African War, 1901-02.—Operations in the Transvaal from 26th January to 31st May 1902. Operations in the Orange River Colony from 20th January to 25th January 1902. Operations in Cape Colony from 5th January to 20th January 1902.—Queen's medal with 5 clasps.

Hume-Spry, W. E.—
The War of 1914-18.—Despatches, *Lond. Gaz.*, 19 Oct. 16.—D. S. O.

Humfrey, F. G. C.—
Chitral, 1895.—Relief of Chitral.—Medal with clasp.

Humfrey, F. T. (*Col. u. s. l. Ind. S.C.*)—
N.-W. Frontier of India, Hazara, 1863-64.
Afghanistan, 1878-80.—Affair at Synd Boot. Despatches, G. G. O. 610 of 79.—Medal Brev. of Lt.-Col.

Hungerford, S. A. H.—
The War of 1914-18.—Despatches, *Lond. Gaz.*, 19 Oct. 16.

Humphrys, F. H.—
South African War, 1900-01.—Operations in the Transvaal and Orange River Colony between 30th Nov. 00 and 31st Oct. 01.—Queen's medal with 3 clasps.

Hunt, C. E.—
Tibet, 1903-1904.—Action at Niani, 26th June 1904.—Operations at and around Gyantse, between 5th May and 6th July 1904.—Medal with clasp.
N.-W. Frontier of India, 1908.—Operations in the Mohmand country.—Medal with clasp.
The War of 1914-18.—Despatches, *Lond. Gaz.*, 22 June 15.—M. C.

Hunt, F. E.—
South African War, 1900-01.—Operations in Cape Colony between 30th November 1900 and 31st May 1902. (Wounded). Despatches, *Lond. Gaz.*, 9th July 1901.—Queen's medal with 2 clasps. D. S. O.

Hunt, H. R. A.—
North-West Frontier of India, 1908.—Operations in the Zakka Khel Country.—Medal with clasp.
The War of 1914-18.—Despatches, *Lond. Gaz.*, 1 Jan. 16, 27 July 16, 4 Jan. 17, 15 May 17 and 11 Dec. 17.—Brev. of Maj. Legion of Honour (5th Class) D. S. O.

Hunt, R. A. C. (*M.-Gen. u. s. l. Bo. Inf.*)—
China, 1860-63.—Storm and capture of the camp at Wang-Kaza, and town of Tseerpo. Action at Nazain, storming of Kahding and Tsingoo, and storming and re-capture of Kahding.—Medal.
Abyssinia, 1867-68.—
Afghanistan, 1879-80.—March from Quetta to the relief of Kandahar. Despatches, G. G. O. 693 of 80.—Medal.

Hunt, R. V.—
South African War, 1900.—Operations in the Transvaal, west of Pretoria. Operations in Orange River Colony.

Hunt, S.—
China, 1900.—Medal.

Hunter, C. B. (*Lt.-Col., retd., Ind. Med. Serv.*)—
Afghanistan, 1879-80.—Medal.
N.-W. Frontier of India, Zhob Valley Expedition, 1890.—Miranzai (1st and 2nd) Expeditions, 1891. —Medal with clasp.

Hunter, F. F.—
China, 1900.—Medal.
The War of 1914-18.—Despatches, *Lond. Gaz.*, 31 Oct. 17.—D. S. O.

Hunter, J. (*Surg.-Maj., retd., Ind. Med. Serv.*)—
Afghanistan, 1880.—With Khyber Line Force —Medal with clasp.

Hunter, J. G.—
Afghanistan, 1878-79.—Medal.
Burma, 1885-89.—Operations of the 1st Brigade and in Pagyi.—Despatches. Maj.-General *Sir* G. S. White's despatch No. 56 B. C., 1 Apl. 88.—Medal with 2 clasps.
N.-E. Frontier of India, Chin-Lushai, 1889-90.—(Burma Column.) Despatches, G. G. O. 677 of 90.—Clasp.
The War of 1914-18.—Despatches, *Lond. Gaz.*, 4 Jan. 17.

Hunter, T.—
China, 1900—Medal.

Hunter, W. J. H.—
British East Africa, 1901.—Operations against the Ogaden Somalis in Jubaland.—Medal with clasp.

Hurly, M. R.—
Burma, 1891-92.

Husband, J.—
N.-W. Frontier of India, 1908.—Operations in the Zakka Khel country. Operations in the Mohmand country. Engagement of Kargha, 24th May 1908. —Medal with clasp.

Hutcheson, G.—
China, 1900.—Medal.

Hutcheson, S.—
South African War, 1899-1901.—Relief of Ladysmith, including action at Colenso (15th Dec. 1899). Operations of 17th to 24th January 1900, and action at Spion Kop. Operations of 5th to 7th Feby. 1900, and action at Vaal Krantz. Operations on Tugela Heights (14th to 27th Feby. 1900) and action at Pieter's Hill (27th Feby. 1900). Operations in the Orange Free State, May 1900. Operations in the Transvaal in May and June 1900. Operations in Natal, March and April 1900. Operations in the Transvaal, east of Pretoria, July to 29th November 1900, including action at Reit Vlei (16 July). Operations in Cape Colony, south of Orange River, April 1900. Operations in Cape Colony, north of Orange River, April and May 1900. Operations in the Transvaal, 30th November to December 1900. Operations in Orange River Colony, December 1900 to June 1901. Operations in Cape Colony, December 1900 to February 1901 and June to September 1901. Despatches, *London Gazette*, 10th September 1901.—Queen's medal with 6 clasps.

Hutchesson, W. C.—
The War of 1914-18.—Despatches, *Lond. Gaz.*, 15 Aug. 17.

Hutchins, H. L. (*Col., retd., I. A.*)—
Burma, 1885-87.—Present at the occupation of Mandalay. Despatches, G. G. O. 434 of 87.—Medal with 2 clasps.
Tirah, 1897-98.—Action of Dargai.—Medal with 2 clasps.

Hutchinson, A. G. C.—
N.-W. Frontier of India, Waziristan, 1901-02.—Medal with clasp.
East Africa, 1903-04.—Operations in Somaliland. Actions at Jidballi.—Medal with 2 clasps.
N.-W. Frontier of India, 1908.—Operations in the Zakha Khel country.—Medal with clasp.
The War of 1914-18.—Despatches, *Lond. Gaz.*, 31 Oct. 17.

Hutchinson, C. A. R.—
N.-W. Frontier of India, 1897-98.—Tochi.—Medal with clasp. Operations on the Samana and in the Kurram Valley during Aug. and Sep. 97.—Clasp.
Tirah, 1897-98.—Actions of Dargai and of the Sampagha and Arhanga Passes. Operations against the Khani Khel Chamkhanis. Operations in the Bazar Valley, 25 to 30 Dec. 97.—Clasp.
Operations in the Abor country, 1911-12.—Medal with clasp. (G. S. O. 2nd Grade.)
The War of 1914-18.—Despatches, *Lond. Gaz.*, 1 Jan. 16,—D. S. O.

Hutchinson, C. R. M.—
Chitral, 1895.—Relief of Chitral.—Medal with clasp.
China, 1900.—Medal.

Hutchinson, F. H. G.—
China, 1900.—Medal.

Hutchinson, F. P.—
Afghanistan, 1879-80.—Medal.
N.-E. Frontier of India, Chin-Lushai, 1889-90.—Medal with clasp. *Lushai, 1890-91.*—Clasp.
N.-W. Frontier of India, 1897-98.—Operations on the Samana and in the Kurram Valley during Aug. and Sep. 97.—Medal with clasp.
Tirah, 1897-98.—Clasp.
N.-W. Frontier of India, Waziristan, 1901-02.—Clasp. Despatches, G. G. O. 611 of 02.

Hutchinson, G. S. M.—
The War of 1914-18.—Despatches, *Lond. Gaz.*, 22 Jun. 15.

Hutchinson, J. R.—
Operations in the Abor country, 1911-12.—(Severely wounded). Medal with clasp.
The War of 1914-18.—Despatches, *Lond. Gaz.*, 15 May 17.

Hutchinson, J. W. C. (*Lt.-Col., retd., I. A.*)—
Afghanistan, 1878-80.—Peshawar Valley Field Force and Khyber Line Force.—Medal.
N.-W. Frontier of India, Miranzai (2nd), 1891.—Medal with clasp. *Waziristan, 1894-95.*—Clasp.

Hutchinson, W. G.—
Tirah, 1897-98.—Operations in the Bara Valley, 7 to 14 Dec. 97.—Medal with 2 clasps.

Hutchison, G.—
China, 1900.—Medal.

Hutchison, H. D. (*Lieut.-Genl., retd., I. A.*)—
Chitral, 1895.—Relief of Chitral.—Medal with clasp.
Tirah, 1897-98.—2 clasps.

Huxford, H. J.—
The War of 1914-18.—Despatches, *Lond. Gaz.*, 21 July 17.

Hynds, J. (*Asst. Comsy., Hon. Lt., retd.*)—
Chitral, 1895.—Relief of Chitral.—Medal with clasp.

I

Idar, Maharaj Dhiraj, *Sir* Pertab Singh, Bahadur of (*Hon. Maj.-Genl.*)—
N.-W. Frontier of India, 1897.—Operations of the Mohmand Field Force as extra A.-D.-C. to G. O. C. Despatches, G. G. O. 1318 of 97.
Tirah, 1897-98.—As extra A.-D.-C. to G. O. C. (wounded). Services acknowledged by H. M.'s Government G. G. O. 671 of 98. Despatches, *London Gazette*, 5th April 1898. *C. B.* Granted hony. rank of Col.
China, 1900.—Despatches, *London Gazette*, 14th May 1901. K. C. B.

Iles, F. W. (*Capt., retd., I. A.*)—
China, 1900.—Medal.

Ilington, E. M.—
N.-W. Frontier of India, 1897-98.—Malakand.—Medal with clasp.

Illus, H. W.—
N.-W. Frontier of India, Waziristan, 1901-02.—Medal with clasp.
Tibet, 1903-04.—Action at Niani, 26th June 1904. Operations at and around Gyantse between 5th May and 6th July 1904.—Medal with clasp.

Inman, H. M.—
East Africa (Somaliland), 1908-10.—Medal with clasp.

Innes, G. V. d'A.—
South African War, 1902.—Operations in Cape Colony. Queen's medal with 2 clasps.

Innes, H.—
N.-W. Frontier of India, Waziristan, 1901-02.—Medal with clasp.

Innes-Lillingston, F. F.—
Tirah, 1897-98.—Medal with clasp.
N.-W. Frontier of India, 1908.—Operations in the Zakka Khel country. Operations in the Mohmand country. Chief Transport Officer, Line of Commn. Staff from 29th April 1908 to end of operations in addition to duties as Bde. Transport Officer, 3rd Bde.—Medal with clasp.
The War of 1914-18.—Despatches. *Lond. Gaz.*, 5 Apl 16.

Inskip, R. D.—
N.-W. Frontier of India, 1908.—Operations in the Zakka Khel country.—Medal with clasp.
The War of 1914-18.—Despatches, *Lond. Gaz.*, 1st Jan. 16, D.S.O., M.C.

Iredell, F. S. (*Lt.-Gen. u. s. l. Bo. Inf.*)—
India, 1860-61.—Against the Wagheers in Kattywar.
Afghanistan, 1879-80.—March from Quetta to the relief of Kandahar, affair at the Kojak Pass, in command of Chaman Garrison, commanded a Field Force in the Argandab Valley. Despatches, G. G. O. 693 of 80.—Medal.

Iredell, J. S. (*Hon. Col., retd., Bo. S.C.*)—
India, 1857-58.—Served at Shorapore and Nurgoond, operations against the Dessaies in the Sawunt Waree Districts.
Afghanistan, 1879-80.—Action of Girishk, battle of Maiwand (severely wounded). Mentioned in Despatches.—Medal.

Ireland, G.—
N.-W. Frontier of India, 1908.—Operations of the Mohmand country.—Medal with clasp.

Irvine, A. A.—
Chitral, 1895.—With the Relief Force as Orderly Officer to the G. O. C. (24th March to August 95). Despatches, *London Gazette*, 15th November 1895.

Irvine, G. B.—
N.-W. Frontier of India, Waziristan, 1894-95.—Medal with clasp.
N.-W. Frontier of India, 1897-98.—Tochi—Medal with clasp.
East Africa, 1903-04.—Operations in Somaliland.—Medal with clasp.
The War of 1914-18.—*Lond. Gaz.*, 5 Apl. 16, 10 May 16, 19 October 16 and 15 Aug. 17. C.B.

Irvine, T. W.—
China, 1900.—Medal.

Ismay, H. L.—
N.-W. Frontier of India, 1908.—Operations in the Mohmand country.—Medal with clasp.
East Africa, 1914-15.—Operations in Somaliland. Action at Shimber-Berris. Despatches, *Lond. Gaz.*, 2nd August 1916.—Medal with clasp.

Ivens, H. T. C.—
N.-W. Frontier of India, 1908.—Operations in the Mohmand country. (Dangerously wounded).—Medal with clasp.

J

Jackson, C. (*Major, retd., Ind. S.C.*)—
N.-W. Frontier of India, 1897-1898.—Operations on the Samana and in the Kurram Valley during Aug. and Sep. 97. Relief of Gulistan. Medal with 2 clasps.
Tirah, 1897-1898.—Clasp.

Jackson, D. R. H.—
South African War, 1901-02.—Operations in the Transvaal between 20th February 1902 and 31st May 1902. Operations in the Orange River Colony, between 11th February 1902 and 19th February 1902. Operations in Cape Colony between 28th February 1901 and 10th February 1902. (Severely wounded).—Queen's medal with 5 clasps.

Jackson, E. C. S. (*Lt.-Col., retd., Ben. Inf.*)—
Afghanistan, 1878-79.—Medal.
N.-W. Frontier of India, 1881.—Mahsud-Waziri Expedition.

Jackson, F. A.—
Soudan.—Nile Expedition, 1898.—Battle of Khartoum.—Medal with clasp; also Egyptian medal.
N.-W. Frontier of India, 1897-98.—Malakand, action at Landakai.—Operations in Bajaur and in the Mamund country. (Slightly wounded.) Despatches, G. G. O. 178 of 98.—Medal with clasp.
N.-W. Frontier of India, Waziristan, 1901-02.—Clasp.
N.-W. Frontier of India, 1908.—Operations in the Mohmand country. Engagement of Matta, 24th April 1908. Despatches, G. G. O. No. 521 A of 1908.—Medal with clasp.

Jackson, G. D'A. (*M.-Gen., retd., Ind. Army*)—
Afghanistan, 1878-79.—Medal.

Jackson, H. W.—
Burma, 1895-96.
The War of 1914-18.—Despatches, *Lond., Gaz.* 4 Jan. and 11 Dec. 17. Brev. of Lt. Col., D.S.O.

Jackson, J.—(*Lt.-Col., retd., I. A.*)
Soudan, 1896.—Dongola Expedition.—Medal, Egyptian medal.
N.-W. Frontier of India, 1897-98.- Tirah. Samana —Medal with 2 clasps.

Jackson, M. H.—
The War of 1914-18—M. C.

Jackson, S. G.—
N.-E. Frontier of India, Chin-Lushai, 1889-9 With the Gangaw Column.—Medal with 2 clasps.
Burma, 1891-1892.—Wuntho Field Force.—Clasp.

Jackson, T.—
Soudan, 1896.—Dongola Expedition. Egyptian medal.—Medal.
N.-W. Frontier of India, 1897-98.—Medal with 2 clasps.

Jackson, W. (*Surg. Maj., retd., Ben. M. S.*)—
Afganistan, 1878-1879.—Capture of the Peiwar Kotal.—Medal with clasp.
N.-W. Frontier of India, Mahsud-Waziri, 1881.

Jackson-Jellie, R.(*Dy. Asst. Comsy. Hony. Capt., retd.*)—
Afghanistan, 1878-80.—Present during advance to cover retreat from Maiwand. Siege of Kandahar. Sortie of Dehkhoja. Battle of Kandahar.—Medal with clasp.

Jacob, A. LeG.—
N.-W. Frontier of India, Zhob Valley, 1890.
Mekran, 1898.—Action of Gok Parosh. Despatches, G. G. O. 648 of 98. D. S.O.
N.-W. Frontier of India, Waziristan, 1901-02.—Medal with clasp.

Jacob, C. W.—
N.-W. Frontier of India, Zhob Valley, 1890.
N.-W. Frontier of India, Waziristan, 1901-02.-Medal with clasp.
The War of 1914-18.—Despatches, Lond. Gaz., 17 Feb. 15. 22 Jun 15, 1 Jan. 16, 15 Jun 16, 4 Jan. 17, 15 May 17, and 11 Dec. 17, Promd. Maj.-Gen. and Lt.-Gen. Order of St. Vladimir (4th class with Swords). C.B.,K. C.B.

Jacob, H. B. (*Col. u. s. l. Ind. S.C.*)—
India, 1858.—Campaign in Central India and Bundelcund.

Jacob, H. F.—
The War of 1914-18.—Despatches, *Lond. Gaz.*, 31 Oct. 17.

Jacob, R. M.—
The War of 1914-18.—M. C.

Jacob, W. (*M.-Gen. u. s. l. Ind. S.C.*)—
India, 1858.—Siege and capture of Kotah.—Medal with clasp.
Afghanistan, 1880.—Mentioned in Despatches.—Medal.

James, A. A.—
Chitral, 1895.—Relief of Chitral.—Medal with clasp.
N.-W. Frontier of India, 1897-98.—Present at Hari Singh Ka Burj.—Clasp.
N.-W. Frontier of India, Waziristan, 1901-02.—Clasp.

James, C. A.—
China, 1900.—Relief of Pekin.—Medal with clasp.

James, E. H. S.—
N.-W. Frontier of India, Waziristan, 1901-02.—Medal with clasp.

James, F. H.—
The War of 1914-18.—Despatches, *Lond. Gaz.*, 31 May 16. M. C.

James, F. P (*Capt., retd. I. A.*)—
N.-W. Frontier of India, 1897-98.—Operations on the Samana and in the Kurram Valley during Aug. and Sep. 97. Operations of the Flying Column in the Kurram Valley under Colonel Richardson, 20 Aug. to 1 Oct. 97.—Medal with 2 clasps.

James, G. G.—
The War of 1914-18.—Despatches, *Lond. Gaz.*, 15 Aug. 17.

James, H.—
South African, Zulu Campaign, 1879.—Action of Ginghinlovo and relief of Etshowe.—Medal with clasp.
Afghanistan, 1879-80.—Medal.
N.-W. Frontier of India, Hazara, 1888.—Action of Kotkai. Despatches, G. G. Os. 978 and 1022 of 88.—Medal with clasp.
N.-W. Frontier of India, 1897-98.—Tochi—Divisional Transport Officer. Despatches, G. G. O. 1432 of 97.—Medal with clasp.
N.-W. Frontier of India, Waziristan, 1901-02.—Clasp.

James, J. F.—
N.-W. Frontier of India, 1908.—Operations in the Zakka Khel country.—Medal with clasp.

James, M. (*Lt.-Col., retd., I. A.*)—
Afghanistan, 1879-80.—Action of Hurnai Mentioned in Despatches.—Medal.

James, M. G. (*Major, retd., I. A.*)—
Burma, 1887-89.—Medal with clasp.
N.-E. Frontier of India, Manipur, 1891.—Clasp.
Tirah, 1897-98.—Medal with 2 clasps.

James, S. P.—
N.-W. Frontier of India, 1897-98.—Tochi.—Medal with clasp.
China, 1900.—Medal.
The War of 1914-18.—Despatches, *Lond. Gaz.*, 14 Nov. '16.

James, W. B.—
South African War, 1902.—Operations in Cape Colony between 20 and 31 May 02.—Queen's medal with 2 clasps.
The War of 1914-18.—Despatches, *Lond. Gaz.*, 22 Jun 15. Brev. of Col.

James, W. R.—
The War of 1914-18.—Despatches, *Lond. Gaz.*, 1 Jan. '16 and 27 July 16. M. C.

Jamson, A. McC.—
South African War, 1900-02.—Operations in Orange River Colony (May to Nov. 00). Operations in Cape Colony, north of Orange River, Mar. to Sept. 00. Operations in the Transvaal, Apl. 01 to 31 May 02. Operations in Orange River Colony, 30 Nov. 00 to Mar. 01.—Queen's medal with clasp King's medal with 2 clasps.

Jameson, J. B. (*Lt.-Col., retd., I. M. S.*)—
N.-W. Frontier of India, (2nd) Miranzai, 1891.—Medal with clasp.
N.-W. Frontier of India, 1897-98.—Mohmand Despatches, G. G. O. 1318 of 97.—Medal with clasp.
Tirah, 1897-98.—Action of the Sampagha Pass. Operations at and around Dwatoi and action of 24 Nov. 97. Operations in the Bara Valley, 7 to 14 Dec. 97.—Clasp.

Jamieson, A. J.—
N.-W. Frontier of India, 1897-98.—Tochi.—Medal with clasp.
China, 1900.—Medal.

Jamieson, A. W. (*Col., retd., I. A.*)—
Burma, 1891-92.—Medal with clasp.

Jardine, C. H.—
South Africa, 1896-97.—Served in ranks.—Medal with clasp.
South African War, 1899-1902.—Advance on Kimberley, including action at Belmont. Operations in the Transvaal, east and west of Pretoria, July to 29 Nov. 00. Operations in Orange River Colony, May to 29 Nov 00 including actions at Lindley (26 June), Bethlehem (6 and 7 July), and Wittebergen (1 to 29 July). Operations in Cape Colony, south of Orange River, 99-00. Operations in Cape Colony, north of Orange River. Operations in the Transvaal, 30 Nov. 00 to July 01. Operations in Orange River Colony, Feb. to Mar. 02. Operations in Cape Colony, July 01 to 31 May 02. Despatches, *Lond. Gaz.*, 10 Sept. 01.—Queen's medal with 3 clasps. King's medal with 2 clasps.
The War of 1914-18.—Despatches, *Lond. Gaz.*, 22 June '15, and 15 Aug. '17. Brevs. of Maj. and Lt.-Col.

Jarret, H. C. D.—
N.-W. Frontier of India, 1897-98.—Medal with clasp.
Tibet, 1903-04.—Medal.

Jarret Kerr, H. T.—
S. African War, 1901-02.—Operations in the Transvaal and Orange River Colony. Operations in Cape Colony, March 01 to Jan. 12.—Queen's medal with 5 clasps.

Jarrett, H. S. (*Col. u. s. l. Ind. S.C.*)—
India, 1858.—Outbreak at Mooltan. Mentioned in Despatches.—Medal.
N.-W. Frontier of India, 1860 and 1863.—Mahsud Waziri Expedition, 1860.—Forcing the Berara Pass. (*2nd*) *Eusofzai Expedition.*—Operations in the Umbeyla Pass.—Medal with 2 clasps.

Jarvis, C.—
The War of 1914-18.—M. C.

Jatar, N. S.—
The War of 1914-18.—Despatches, *Lond. Gaz.*, 12 June 17. D. S. O.

Jennings, *Sir* R. M. (*Genl. u. s. l. I. A.*)—
N.-W. Frontier of India, Mohmunds, 1863-64.—Medal with clasp.
Egypt, 1882.—Action of Tel-el-Kebir. Despatches, *Lond. Gaz.*, 2 Nov. 82.—Medal with clasp.—Bronze star, 4th class, Osmanieh. Brev. of Lt.-Col.

Jermyn, T. (*Col., retd., I. A.*)—
N.-W. Frontier of India, Hazara, 1888.—Medal with clasp.
Waziristan, 1894-95.—Clasp.
Tirah, 1897-98.—Section Road Commandant, Line of Communications.—Medal with 2 clasps.
China, 1900.—Relief of Pekin. Action of Peitsang. Action at Yangtsun. D. A. A. G., 1st Infantry Brigade. Despatches, *Lond. Gaz.*, 14 May 01. —Medal with clasp. Brev. of Major.

Jervis, H. P. (*Lt.-Col., retd., I. M. S.*)—
Burma, 1895-98.—Medal with 2 clasps.

Jeudwine, W. W.—
N.-W. Frontier of India, 1908.—Operations in the Zakka Khel country.—Medal with clasp.
The War of 1914-18.—Despatches, *Lond. Gaz.*, 1 Jan. '16. C. M. G.

Jewett, A. E.—
Tirah, 1897-98.—Actions of the Sampagha and Arhanga Passes.—Medal with 2 clasps.

Johnson, C. A. K. (*Major., retd., I. A.*)—
N.-W. Frontier of India, Hazara, 1891.—Medal with clasp.
China, 1900.—Medal.

Johnson, C. E.—
Burma, 1886-87.—Medal with 2 clasps.
British Central Africa, 1892-93.—Medal. Brev. of Maj.
N.-W. Frontier of India, Waziristan, 1901-02.—Medal with clasp.

Johnson, C. R.—
N.-E. Frontier of India,—Lushai, 1892. Services acknowledged by Govt., G. G. O. 977 of 1892. —Medal with clasp.
N.-W. Frontier of India.—Waziristan, 1894-95.—Medal with clasp.
N.-W. Frontier of India, 1908.—Operations in the Zakka Khel country.—Medal with clasp.

Johnson, E. R. (*Lt.-Col., retd., I.M.S.*)—
Looshai Expedition, 1871-72.—Mentioned in despatches.—Medal with clasp.
Akha Expedition, 1883-84.—Mentioned in despatches.

Johnson, G. W. (*Lt.-Col., retd., I. A.*)—
Burma, 1889.—Medal with clasp.
N.-W. Frontier of India, 1890.—Zhob Valley.
China, 1900.—Relief of Tientsin.—Relief of Pekin. Provost Marshall, Lines of Communication. Despatches, *Lond. Gaz.*, 13 Sep. 01.—Medal with clasp.

Johnson, J. E. B.—
Chitral, 1895.—Relief of Chitral.—Action of the Malakand Pass.—Medal with clasp.
N.-W. Frontier of India, 1897-98.—Relief of Malakand, Malakand. Action at Landakai.—Operations in Bajaur and in the Mamund country, Buner.—2 clasps.
N.-W. Frontier of India, 1908.—Operations in the Mohmand country.—Medal with clasp.
The War of 1914-18.—Despatches, *Lond Gaz*, 4 July 16.

Johnson, M. E. S.—
N.-W. Frontier of India, Waziristan, 1901-02.—Medal with clasp.
The War of 1914-18.—Despatches, *Lond. Gaz.*, 5 Apl., 13 July and 19 Oct. 16.

Johnston, A.L.—
The War of 1914-18.—Despatches, *Lond. Gaz.*, 15 Aug. 16.

Johnston, C.—
China, 1900.—Relief of Pekin. Medal with clasp.

Johnston, C. A.—
N.-E. Frontier of India, Manipur, 1891.—Medal with clasp.
Burma, 1891-92.—Clasp.
Tirah, 1897-98.—Medal with clasp.
China, 1900.—Medal.
The War of 1914-18.—Despatches, *Lond. Gaz.*, 30 June 16. D. S. O.

Johnston, H. M.—
N.-W. Frontier of India, Hazara, 1888.—Medal with clasp.
East Africa, 1902-04.—Operations in Somaliland.—Medal with clasp.

Johnston, P. D. C.—
The War of 1914-18.—Despatches, *Lond. Gaz.*, 4 July 16 and 15 Aug. 17.

Johnston, R.—
The War of 1914-18.—Despatches, *Lond. Gaz.*, 1 Jan. and 19 Oct. 16. D. S. O.

Johnston, W.—
The War of 1914-18.—Despatches, *Lond. Gaz.*, 5 Apl. and 13 July 16. M. C.

Johnstone, A. A. J.—
N.-W. Frontier of India, (1st) Miranzai, 1891.
N.-W. Frontier of India, Waziristan, 1901-02.—Medal with clasp.

Johnstone, B. A. (*Maj., retd., I.A.*)—
Burma, 1892-93.—Chin Hills.—Medal with clasp.
China, 1900.—Relief of Pekin.—Medal with clasp.

Johnstone, J. W. H. (*Col., retd., Ben. S.C.*)—
India, 1857-58.—Attack on Thanna Bhawan (wounded), actions of Gungaree, Puttialee, Mynpoorie and Bungaon (wounded).—Medal.
N.-W. Frontier of India, 1872.—Dawar Valley Expedition.

Johnstone, R. FitzR. M. (*Major, retd., Ind. S. C.*)—
Burma, 1888-89.—Medal with clasp.
South African War, 1899-1902.—District Commdt., from 20 June 01. Operations in Cape Colony, May 01 to 31 May 02.—Queen's medal with 3 clasps.

Jollie, F.—
Soudan, 1888-89.—Suakin. Battles of Gamaizah and Toski.—Medal with clasp. Bronze star. Toski. 4th class Order of Medjidie.

Jones, A. E.—
Afghanistan, 1878-79-80.—Actions of the Peiwar Kotal and in the Sapari Pass. Operations in the Khost Valley. Action of Charasiah. Pursuit under General T. D. Baker. Operations at and around Kabul. Affair of Shekhabad. March from Kabul to the relief of Kandahar, and battle of Kandahar.—Medal with 4 clasps. Bronze star.
N.-W. Frontier of India, Mahsud Waziri, 1881. (*1st*) *Miranzai, 1891.*
N.-E. Frontier of India, Manipur, 1891 —Medal with clasp.
N.-W. Frontier of India, Isazai, 1892.
Chitral, 1895.—Relief of Chitral.—Medal with clasp.

Jones, A. N. G.—
The War of 1914-18.—Despatches, *Lond. Gaz.*, 28 Dec. 17. D. S. O.

Jones, B. I.—
The War of 1914-18.—Despatches, *Lond. Gaz.*, 15 Aug. 17. M. C.

Jones, F. S.—
N.-W. Frontier of India, 1897-98.—Tochi.—Medal with clasp.—Promoted Sub-Condr.

Jones, G. T. (*Lt.-Gen. u.s.l., Ben. Inf.*)—
Egypt, 1882.—Medal; bronze star.

Jones, H. H.—
Tirah, 1897-98.—Operations in the Bara Valley, 7 to 14 Dec. 97. Operations in the Bazar Valley, 25 to 30 Dec. 97.—Medal with 2 clasps.

Jones, H. J.—
Soudan, 1884-85.—Nile. With Desert Column. Actions at Abu Klea.—Medal with clasp. Bronze star.
N.-W. Frontier of India, Hazara, 1888.—Medal with clasp.
Burma, 1891-92.
Chitral, 1895.—Relief of Chitral. (Twice wounded). Despatch, G. G. O. 530 of 95.—Medal with clasp. D. S. O.
N.-W. Frontier of India, 1897-98.—Tochi.—Clasp.
China, 1900.—Medal.

Jones, J. W.—
The War of 1914-8.—Despatches, *Lond. Gaz.*, 22 Sept. and 19 Oct. 16. M. S. O.

Jones, L. A. M.—
South African War, 1900-1901.—Attached to the 1st Battn. Royal Dublin Fusiliers and employed with Scottish Horse. Operations in the Transvaal, June to 29th November 1900 and from 30th November 1900 to February 1901.—Queen's medal with clasp.
N.-W. Frontier of India, 1908.—Operations in the Zakka Khel country. Operations in the Mohmand country. Engagement of Kargha, 24th May 1908. —Medal with clasp.
The War of 1914-18.—Despatches, *Lond. Gaz.*, 12 Jun. 17.

Jones, L. C.—
N.-W. Frontier of India, Isazai, 1892.
South African War, 1900-02.—Operations in Orange River Colony between May and 29 Nov. 00. Operations in Cape Colony, south of Orange River, 00. Operations in the Transvaal between 30 Nov. 00 and 31 May 02. Operations in the Orange River Colony between 30 Nov. 00 and 31 May 02. Operations in Cape Colony between 30 Nov. 00 and 31 May 02. Despatches, *Lond. Gaz.*, 29 July 02. Brev. of Maj. G. G. O. 1051 of 02. —Queen's medal with 3 clasps. King's medal with 2 clasps.
The War of 1914-18.—Despatches, *Lond. Gaz.*, 15 Jun. 16 and 15 Aug. 17. Brev. of Col. C. M. G.

Jones, L. J.—
West Africa (S. Nigeria), 1901-02.—Aro Expedition. Despatches, *Lond. Gaz.*, 12 Sep. 02.—Medal with clasp.

Jones, R. G. (*Lt.-Col., u. s. l., I. A.*)—
Afghanistan, 1879-80.—Medal.
Burma, 1886-89.—Operations of the 4th and 6th Brigades. (Severely wounded).—Medal with 2 clasps.
N.-E. Frontier of India, Chin-Lushai, 1888-89.—Present at the first action against the Chins.—Clasp.

Jones, S. W.—
N.-W. Frontier of India, 1908.—Operations in the Mohmand country.—Medal with clasp.
The War of 1914-18—Despatches, *Lond. Gaz.*, 21 Jun. 16.

Jones, W. H. D. (*Lt.-Col. u.s.l. Ind. S.C.*)—
Afghanistan, 1879-80.—Medal.

Jopp, John *Hon.* (*Br.-Gen., u.s.l. Ind. S.C.*)—
Arabia, 1858.—Destruction of Fort of Sheikh Othman.
Abyssinia, 1868.—Medal.
Afghanistan, 1879-80.—As A.-Q.-M.-G. March from Quetta to the relief of Kandahar. Despatches, G. G. O. 693 of 80.—Medal.

Joyner, H.—
N.-W. Frontier of India, 1908.—Operations in the Zakka Khel Country. Operations in the Mohmand country. Engagement of Kargha, 24th May 1908. Medal with clasp. Promoted Conductor.
The War of 1914-18.—Despatches, *Lond Gaz.*, 19 Oct 16. Promd. Assistant Commissary.

Jukes, A. H.—
The War of 1914-18—Despatches, *Lond. Gaz.*, 15 Mar. 16, 15 Jun. 16, 4 Jan. 17 and 15 May 17. Brev. of Maj. D. S. O.

Justice, C. Le G.—
South Africa, Zulu Campaign, 1879.—Battle of Ginghinlovo.—Medal with clasp.
N.-E. Frontier of India, Sikkim, 1888.—Attack of Jalapla.—Medal with clasp.
Chitral, 1895.—Relief of Chitral.—Medal with clasp.

Juxon-Jones, J. H.—
Afghanistan, 1879-80.—Medal.
Burma, 1886-87.—Operations of the 6th Brigade.—Medal with clasp.
N.-E. Frontier of India, Chin-Lushai Expedition, 1889-90.

K

Kanga, J. K. (*Lt.-Col., retd., I.M.S.*)—
Soudan, 1896.—Dongola Expedition.—Medal. Egyptian Medal.
China, 1900.—Medal.

Kanwar Amar Singh—
China, 1900.—Medal.
The War of 1914-18.—Despatches, *Lond Gaz.*, 22 Jun. 15.

Kanwar Pirthi Singh—
N.-W. Frontier of India, 1908.—Operations in the Zakka Khel country. Operations in the Mohmand country. Despatches, G. G. O. No. 521 A of 1908.

Kauntze, B. C.—
The War of 1914-18.—Despatches, *Lond. Gaz.*, 15 Aug. 17.

Kaye, C.—
N.-W. Frontier of India, 1897-98.—Malakand. Provost Marshal, 3rd Brigade. Operations in Bajaur—Mohmand—Utman Khel—Buner. Action of the Tanga Pass.—Medal with clasp.

Keary, H. D'U.—
Afghanistan, 1880.—Action of Kam Dakka.—Medal.
Burma, 1885-87-89.—Operations of the 1st and 5th Brigades.—Despatches, G. G. O. 434 of 87. —Medal with 2 clasps.

Burma, Wuntho Expedition, 1891.—As Comdt., Shwebo Police Battalion.—Mentioned in Despatches.—Clasp. D. S. O.
N.-E. Frontier of India, Chin Hills, 1892-93.—Mentioned in Despatches.—Clasp. Brev. of Maj.
Burma, 1895-96.
China, 1900.—Despatches, *Lond. Gaz.*, 14 May 01.—Medal.
The War of 1914-18.—Despatches, *Lond. Gaz.*, 17 Feb. 15, 22 Jun. 15, 1 Jan. 16, 19 Oct. 16 and 15 Au. 17. K.C.B., K.C.I.E. Karageorge, 2nd class with swords.

Keays, R. W. C.—
Tibet, 1903-04.—Medal.

Kee, W. J.—
The War of 1914-18.—Despatches, *Lon. Gaz.*, 5 Apr. 16, *Gaz. of India* Notfn. 1242 of 16. Promd. Assistant Commissary.

Keefer, W. N. (*Surg.-Maj., retd., Ind. Med. Serv.*)—
N.-E. Frontier of India, Lushai Expedition, 1871-72.—Medal with clasp.
Duffla Expedition, 1874-75.—Mentioned in Despatches.
N.-W. Frontier of India, Jowaki Expedition, 1877-78.—Clasp.
Afghanistan, 1878-80.—Capture of Ali Musjid, Lughman Valley expedition, Zaimusht expedition, capture of Zawa. Mentioned in Despatches. —Medal with clasp.
Egypt, 1882.—Action of Kassassin, battle of Tel-el-Kebir. Medal with clasp; bronze star; 4th class, Osmanieh.

Keegan, D. F. (*Brig.-Surg.-Lt.-Col., retd., Ind. Med. Serv.*)—
Afghanistan, 1879-80.—March from Kabul to relief of Kandahar and battle of 1 Sep.—Medal with clasp; bronze star.

Keelan, H. P.—
East Africa, 1902-04.—Operations in Somaliland. —Medal with clasp.
The War of 1914-18.—Despatches, *Lond. Gaz.*, 15 Jun. 16 and 15 Aug. 17. Brev. of Lt.-Col.

Keely, I. E.—
South Africa, 1900-02.—Operations in the Orange River Clony between 30th November 1900 and 31st May 1902. Operations in Cape Colony between 3th November 1900 and 31st May 1902. Queen's medal with 4 clasps.

Keen, F. S.—
Tirah, 1897-98.—Operations in the Bara Valley, 9 to 14 Dec. 97.—Medal with 2 clasps.
British East Africa, 1898.—Operations against Ogaden Somalis.—Medal with clasp.
Uganda, 1899.—Operations against Kabarega.—Clasp.
N.-W. Frontier of India, 1902.—Operations against the Darwesh Khel Waziris.
N.-W. Frontier of India, 1908.—Political Offr. Operations in the Mohmand country. Despatches *Lond. Gaz.*, 14 Aug. 08.
The War of 1914-8.—Despatches, *Lond. Gaz.*, 30th June 1916 and 1st February 1917. D. S. O.

Keen, W. J. (*Capt., Supy. List, I. A.*)—
Chitral, 1895.—Relief of Chitral.—Action of the Malakand Pass. Action at Khar.—Medal with clasp.
N.-W. Frontier of India, 1908.—Operations in the Mohmand country. Engagement of Matta, 24th April 1908. Despatches G. G. O. No. 521 A of 1908.—Medal with clasp.
The War of 1914-18.—Despatches, *Lond. Gaz.*, 4 July 16.

Keene, C. W.—
East Africa, 1903-04.—Operations in Somaliland. Action at Jidballi.—Medal with 2 clasps.

Keene, G. N. S.—
N.-W. Frontier of India, 1897-98.—Malakand. Operations in Bajaur and the Mamund country.—Dangerously wounded.—Medal with clasp.

Keighley, V. A. S.—
N.-W. Frontier of India, 1897-98.—With Tochi and Tirah Expdy. Forces.—Medal with clasp.

Kelly, F. P. C.—
East Africa (British), 1896.—Operations against the Mazrui rebels.—Medal.
East Africa, 1902-04.—Operations in Somaliland.—Medal with clasp.
The War of 1914-18.—Despatches, *Lond. Gaz.*, 15 Apr. 16 and 15 Aug. 17. C.M.G., D.S.O.

Keith, J. F. (*Brig.-Surgn.-Lt.-Col., retd., Ind. Med. Serv.*)—
Abyssinia, 1868.—Attached to General Hospital, Zoula.—Medal.
Afghanistan, 1880.—Defence of Kandahar, sortie of Deh Khoja, and battle of 1 Sep. Despatches, G. G. O. 582 of 80.—Medal with clasp.
Burma, 1886-88.—Medal with 2 clasps.

Kelawala, M. J. (*Lt.-Col., retd., I. M. S.*)—
Burma, 1886-88.—Operations of the 1st Brigade under Brigadiers-Genl. C. J. East and G. B. Wolseley.—Medal with 2 clasps.
Burma, 1890-92.—Tlang-Tlang column.—Clasp.

Kellie, A. H. (*Major, retd., I.S.C.*)—
Egypt, 1882.—Action of Mabala Junction.—Battle of Tel-el-Kebir.—Medal with clasp and bronze star.
Burma, 1886.—Medal with clasp.
N.-E. Frontier of India, Chin-Lushai, 1889.—Clasp.

Kellie, E. C. (*Lieut.-Col., retd., I. A.*)—
Afghanistan, 1880.—Medal.
Burma, 1885-87.—As D. A. A. and Q. M. G.—Despatches, G. G. O. 339 and 434 of 87.—Medal with 2 clasps.

Kellie, G. J. (*Col., retd., I. M. S.*)—
Burma, 1888-89.—Medal with clasp.
Tirah, 1897-98.—Action of the Arhanga Pass. Operations in the Bara Valley, 7 to 14 Dec. 97.—Medal with 2 clasps.

Kelly, J. G. (*Col., u. s. l., I.A.*)—
N.-W. Frontier of India, Hazara, 1891.—Medal with clasp. (*2nd*) *Miranzai, 1891.*—Clasp.
Chitral, 1895.—Relief of Chitral.—Commanded the force from Gilgit which relieved Chitral.—Thanked by the Government of India. Despatches, G. G. O. 530 of 95.—Medal with clasp. Appointed A.-D.-C. to the Queen. Brev. of Col.—C. B.

Kelly, T. B.—
Tibet, 1903-04.—Action at Niani, 26th June 1904. Operations at and around Gyantse, between 5th May and 6th July 1904. March to Lhassa, 14th July to 3rd August 1904. (Slightly wounded.) Despatches, *Gazette of India*, No. 46 of 1904.—Medal with clasp.
The War of 1914-18.—Despatches, *Lond. Gaz.*, 4 July 16 and 15 Aug. 17.

Kelsall, R.—
South African War, 1900-02.—Operations in Natal April to June 1900, including action at Laing's Nek (6th to 9th June). Operations in the Transvaal east of Pretoria, July to 29th November 1900. Operations in the Transvaal, December 1900 to February 1901. Wounded at Zuickerbosch February 1901. Operations in the Transvaal April and May 1902. Queen's medal with 4 clasps.
The War of 1914-18.—Despatches, *Lond. Gaz.*, 19 Oct. 16. D. S. O.

Kemball, A. G.—
Chitral, 1895.—Relief of Chitral.—Medal with clasp.
N.-W. Frontier of India, 1897-98.—Malakand.—Operations in the Mamund Country—Buner.—Clasp.
The War of 1914-8.—Despatches, *Lond. Gaz.* 15th August 1917.

Kemball, J. S.—
N.-W. Frontier of India.—Action at Wana, 3rd November 1894. *Waziristan, 1894-95.*—Medal with clasp.
Chitral, 1895.—Relief of Chitral.—Medal with clasp.
China, 1900.—Orderly Officer to G. O. C., 3rd Brigade, China Field Force.—Medal.
N.-W. Frontier of India, Waziristan, 1901-02.—Clasp.

Kemble, P. B.—
N.-W. Frontier of India, Waziristan, 1901-02.—Medal with clasp.

Kemmis, A. W. M.—
The War of 1914-18.—Despatches, *Lond. Gaz.*, 15 Aug. 17.

Kemp, T. N. C.—
War of 1914-18.—

Kemp, W. I. J.—
N.-W. Frontier of India, 1897-98.—
Tirah.—Medal with 2 clasps.
South African War, 1900-1902.—Operations in Cape Colony, Actions at Johannesburg, Diamond Hill and Wiltebergen.—Queen's medal with 4 clasps. King's medal with 2 clasps.

Kempson, E. C. R.—
South African War, 1901-02.—Operations in Cape Colony, June 1901. Operations in the Orange River Colony, July 1901. Operations in the Transvaal, August 1901. Operations on the Zululand Frontier of Natal, September and October 1901 including the defence of Forts Itala and Prospect—Queen's medal with 5 clasps.

Kendall, J. (*Major, retd., I. A.*)—
Burma, 1887-88.—Medal with 2 clasps.

Kennedy, A. M.—
Operations in the Abor country, 1911-12.—Medal with clasp.

Kennedy, E. E.—
The War of 1914-18.—Despatches, *Lond. Gaz.*, 28 Nov. 17. M. C.

Kennedy, H.—
Burma, 1885-89.—Operations of the 2nd and 5th Brigades.—Medal with 2 clasps.
N.-W. Frontier of India, Zhob Valley, 1890.
N.-W. Frontier of India, 1897-98.—Operations on the Samana and in the Kurram Valley during Aug. and Sep. 97. Operations of the Flying Column in the Kurram Valley under Colonel Richardson, 20 Aug. to 1 Oct. 97. Staff Officer to O. C. Flying Column.—Despatches, G. G. O. 304 of 98.—Medal with 2 clasps.
Tirah, 1897-98.—Clasp.
N.-W. Frontier of India, 1902.—Operations against the Darwesh Khel Waziris.

Kennedy, J. P.—
South African War, 1900.—Operations in the Orange Free State, February to May, 1900, including actions at Houtnek (1st May,) Vet River (5th and 6th May), and Zand River (10th May). Operations in the Transvaal in May and June 1900, including actions near Johannesburg (29th May), Pretoria (4th June). Operations in the Transvaal east of Pretoria, July to 29th November, 1900.—Queen's Medal with 3 clasps.

Kennedy, R. S.—
Operations in the Abor Country, 1911-12.—Medal with clasp.
The War oy 1914-18.—M. C.

Kennedy, W. P. (*Lt.-Col., retd., I. A.*)—
Afghanistan, 1880.—In the Pol. Dept.—Medal

Kennedy-Craufurd-Stuart, C.—
South African War, 1900-02.—Operations in Cape Colony, south of Orange River, February to April 1900. Operations in Orange Free State, April to May 1900. Operations in Natal, June 1900. Operations in the Transvaal, from 30th November 1900 to 31st May 1902.—Queen's medals with 3 clasps. King's medal with 2 clasps.
The War of 1914-18.—Despatches, *Lond. Gaz.*, 5 Nov. 15 D.S.O.

Kenny, H. T. (*Col., retd., I. A.*)—
Tirah, 1897-98.—Actions of the Sampagha and Arhanga Passes. Reconnaissance of the Saran Sar and action of 9 Nov. 97. Operations against the Khani Khel Chamkannis. Operations in the Bazar Valley, 25 to 30 Dec. 97.—Medal with 2 clasps.
N.-W. Frontier of India, 1908.—Operations in the Zakka Khel country.—Medal with clasp.

Kenrick, W. H.—
N.-W. Frontier of India, 1897-98.—Tochi.—Medal with clasp. Operations on the Samana and in the Kurram Valley during Aug. and Sep. 97.—Clasp.
Tirah, 1897-98.—Actions of Dargai and of the Sampagha and Arhanga Passes.—Clasp.
China, 1900.—Relief of Pekin. Action of Pietsang. Action at Yangtsun. Despatches, *Lond. Gaz.*, 14 May 01.—Medal with clasp.

Kensington, E. C.—
Aden, 1903-04.—Operations in the interior.
The War of 1914-18.—Despatches, *Lond. Gaz.*, 1 Jan. 16, 19 Oct. 16, and 15 Aug. 17. Brev. of Lt.-Col., M.C.

Kenworthy W.—
The War of 1914-18.—Despatches, *Lond. Gaz.*, 19th October 1916.

Keogh, J. B.—
The War of 1918.—Despatches, *Lond. Gaz.*, 19 Oct. 16, 3 Mar. 17 and 15 Aug. 17, D. S. O.

Keppel, T. R. H.—
South African War, 1901-2.—Operations in the Transvaal, May 01 to 31 May 02.—Queen's medal with 5 clasps.

Kerans, G. C. L.—
N.-W. Frontier of India, Waziristan, 1901-02.—Medal with clasp.
The War of 1914-18.—Despatches *Lond. Gaz.*, 1 Jan. 16 D. S. O.

Kerr, A. N.—
N.-W. Frontier of India, 1908.—Operations in the Mohmand country.—Medal with clasp.
The War of 1914-18.—Despatches, *Lond. Gaz.*, 30 Jun. 16 and 8 and 13 Feb. 17. M. C.

Kerr, M. A. (*Bt.-Col., retd., I. A.*)—
Tibet, 1903-04.—Action at Niani, 26th June 1904. Operations at and around Gyantse, between 5th May and 6th July 1904. Despatches, *Gazette of India*, No. 1065 of 1904.—Medal with clasp. C.B.

Kerrich, L. W. C. (*Col., retd., I. A.*)—
South African War, 1900-02.—Relief of Kimberley. Operations in the Orange Free State, Feb. to May 00, including operations at Paardeberg (17 to 26 Feb.); action at Driefontain (10 March). Operations in the Transvaal in May and June 00, including actions near Johannesburg (29 May) and Diamond Hill (11 and 12 June). Operations in the Transvaal, east of Pretoria, July to Nov. 00, including at Belfast (26 and 27 Aug).—Queen's medal with 6 clasps. King's medal with 2 clasps.

Kerwick, R. (*Hony. Capt., Commy., retd.*)—
Egypt, 1882.—Present at second action of Kassassin and battle of Tel-el-Kebir.—Medal with clasp. Khedive's Bronze star.
China, 1900.—Relief of Pekin.—Medal with clasp.
The War of 1914-18.—Despatches, *Lond. Gaz.*, 24 Dec. 17.

Kettle, H. D. (*Asst. Commy. Hony. Lieut. S. and T. C. Bo.*)—
N.-E. Frontier, of India, Chin Hills, 1892-93.—Medal with clasp.

Kettlewell, E. A. (*Bt.-Col., retd., I. A.*)—
Burma, 1885-87.—Medal with clasp.
N.-W. Frontier of India, 1897-98.—Operations on the Samana and in the Kurram Valley during Aug. and Sep. 87.—Medal with 2 clasps.
Tirah, 1897-98.—Reconnaissance of the Kharmana Defile and action of 7 Nov. 97.—Operations against the Khani Khel Chamkannis.—Despatches, G. G. O. 244 of 08.—Clasp.
China, 1900.—Medal.

Key, J. A. (*Hony. Lieut., Assist. Commy., retd.*)—
South Africa, 1877-78-79.—*Kaffir Campaign.*—Operations against the Galekas and Gaikas. Affair at Perie Bush. *Zulu Campaign.*—Medal with clasp.

Keyes, T. H.—
Tirah, 1897-98.—Action of Chagru Kotal. Wounded. Despatches, G. G. O. 58 of '98.—Medal with 2 clasps.
The War of 1914-18.—Despatches, *Lond. Gaz.*, 31 Oct. 17.

Keys, W. D. A.—
N.-W. Frontier of India, Waziristan, 1901-02.—Medal with clasp.
Aden 1903-04.—Operations in the interior.
The War of 1914-18.—Despatches, *Lond. Gaz.*, 15 Aug. 17.

Khambatta, D. D. (*Poona Volunteer Rifle Corps*)—
Afghanistan, 1878-80.—March from Kandahar to Ghazni and Kabul. Action of Ahmed Khel. —Medal with clasp.

Kahn Muhammed Akbar Khan.—
The War of 1914-18.—Despatches, *Lond. Gaz.*, 21 Jun. 16.

Kharegat, M. P. (*Lt.-Col., retd., I. M. S.*)—
Burma, 1886-87.—Medal with clasp.

Kidby W. T.—
The War of 1914.18.—Meritorious Service Medal.

Kidd, B. G. B.—
N.-W. Frontier of India, Waziristan, 1901-02.—Medal with clasp.
Aden, 1903-04.—Operations in the interior.

Kiddle, T. (*Hony. Capt. Sr. Asst. Surgn., retd., I.S.M.D.*)—
Perak, 1875-76.—Medal with clasp.

Kilkelly, P. P.—
Soudan, 1896.—Dongola Expedition.—Medal; bronze star.
British East Africa, 1898.—Expedition against the Ogaden Somalis.—Medal with clasp.—Brilliant star of Zanzibar.
China, 1900.—Medal with clasp.

King, A. F. W.—
Tirah, 1897-98.—Medal with 2 clasps.
British East Africa, 1901.—Operations against the Ogaden Somalis in Jubaland.—Medal with clasp.
South African War, 1902.—Queen's Medal with 3 clasps.

King, H. W. G. (*Sr. Asst. Surgn. Hony. Capt., retd., I. S. M. D.*)—
Burma, 1886-1889.—Medal with 2 clasps.
N.-E. Frontier of India, Chin Lushai, 1889-90.—Served under Brig.-Genl. Symons in the Southern column. Specially promoted.—Clasp.

King, J. A.—
China, 1900.—Medal.

Kingdon, J.—
East Africa, 1914-15.—Operations in Somaliland, Action at Shimber-Berris, Medal with clasp.

King-Harman, M. J. (*Col. u. s. l. I.S.C.*)—
N.-W. Frontier of India, Hazara, 1868.—Medal with clasp.
N.-E. Frontier of India, Lushai, 1871-72.—Clasp.
Chin-Lushai, 1889-90.—Despatches.—Clasp.

Kingston, F. A. (*Major, retd., I. A.*)—
Chitral, 1895.—Relief of Chitral.—Medal with clasp.

Kirby, W. T.—
The War of 1914-18.—Despatches. *Lond. Gaz.*, 5 Apr. 16 and 15 Aug. 17. D.S.O.

Kirkpatrick, C.—
The War of 1914-18.—Despatches, *Lond. Gaz.*, 28th Nov. 17. M. C.

Kirkpatrick, C. H.—
N.-W. Frontier of India, 1897-98.—Buner. Action of the Tanga Pass.—Medal with clasp.
Tirah, 1897-98.—Action of the Sampagha Pass.—Clasp.

Kirkpatrick, E.—
N.-W. Frontier of India, Waziristan, 1894-95.—Medal with clasp.
Tibet, 1903-04.—Medal.
N.-W. Frontier of India, 1908.—Operations in the Zakka Khel country. Operations in the Mohmand country. Engagement of Karga, 24th May 1908. —Medal with clasp.

Kirkpatrick, K. E.—
South African War, 1900-01.—Operations in Orange Free State, April 1900. Operations in Transvaal, east of Pretoria, August to September 1900 and November 1900. Operations in Orange River Colony, May to 28th August 1900 and 2nd September to 29th November 1900, including action at Wittebergen. Operations in Cape Colony, south of Orange River, April 1900.—Queen's medal with 4 clasps.

Kirkpatrick, W. (*Maj., retd., I. A.*)—
Burma, 1885-87.—Medal with clasp.
N.-W. Frontier of India, (1st) Miranzai, 1891.
N.-W. Frontier of India, 1897-98.—Tochi.—Medal with clasp.
N.-W. Frontier of India, Waziristan, 1901-02.—Despatches, twice, G. G. O. No. 611 of 1902. —D. S. O.; clasp.

Kirkwood, C. K. H.—
The War of 1914-18.—Despatches, *Lond. Gaz.*, 19th Oct. 16.

Kirkwood, J. N. S. (*Col. u. s. l. I.S.C.*)—
Afghanistan, 1880.—Medal.

Kirkwood, T. M.—
Tibet, 1903-04.—Medal.

Kirkwood, W. C.—
South African War, 1899-1901.—Operations in Rhodesia (11th October 1899 to 25th May 1900) Operations in the Transvaal, west of Pretoria (July to 29th November 1900). Operations in Cape Colony, South of Orange River, 1899-1900. (Severely wounded.) Queen's medal with 4 clasps.
The War of 1914-18.—Despatches, *Lond. Gaz.* 15 Aug. 17,

Kirtikar, K. R. (*Lt.-Col., retd., I. M. S.*)—
Afghanistan, 1879-80.—Sungsilla and Sibi outposts. Girishk Column. Battle of Maiwand and siege of Kandahar.—Medal.

Kirwan, E. W. O'G.—
The War of 1914-18,—Despatches, *Lond. Gaz.* 19 Oct. 16.

Kirwan, L. E. (*Madras Vol. Guards*)—
South African War, 1900.—Operations in Cape Colony, north of Orange River, 28 Mar. to 4 Apl. 00. Operations in Orange Free State, 5th Apl. to 26 May 00, including actions at Houtnek (1 May), Vet River (5 and 6 May) and Zand River (10 May), (wounded). Operations in the Transvaal, May to June, including actions near Johannesberg (29 May), Pretoria (4 June). Operations in the Transvaal west of Pretoria (22nd to 28th Aug. 1900). Operations in the Transvaal, east of Pretoria (29 Aug. to 3 Nov. 00). Queen's medal with 3 clasps.

Kishangarh, H. H. *Umdae Rajahae Makam Maharajadhiraga* Sir Madan Singh.
The War of 1914-18.—Despatches, *Lond. Gaz.*, 17 Feb. 15.

Kitson, E. (*Capt., retd., Ind. Army*)—
India, 1858.—Capture of Lucknow.—Medal with clasp.
China, 1860-61.—Medal.

Knaggs, C. F.—
South African War, 1902.—Operations in the Transvaal and Orange Free State, 25th January to 31st May 1902.—Queen's medal with 3 clasps.
The War of 1914-18.—Despatches, *Lond. Gaz.*, 30 May 16.

Knapp, H. G. G.—
South African War, 1900-02.—Operations in the Transvaal, April to July 1901. Operations in Cape Colony, August 1901.—Queen's medal with 4 clasps.

Knapton, H. A. F.—
Tirah, 1897-98.—Actions of Sampagha and Arhanga Passes.—Operations in the Waran Valley, and action of 16th November 1897. Operations against the Khani Khels Chamkanis. Operations in the Bara Valley 7th to 14th December 1897. Medal with 2 clasps.
N.-W. Frontier of India, 1897-98.—Operations on the Samana and in the Kurman Valley during August and Septemper 1897.—Clasp.
China, 1900.—Medal.

Knatchbull, G. W. C.—
Burma, 1885—87.—Despatches, G. G. O. 561 of '87.—Medal with clasp.
China, 1900.—Medal.
The War of 1914-18.—Despatches, *Lond. Gaz.*, 28 Jan. 16. C.M.G.

Knight, W. C.—
N.-W. Frontier of India, 1897-98.—Mohmand. As Road Commandant, Despatches, G. G. O. 1318 of 97.—Medal with clasp.
Tirah, 1897-98.—Provost Marshal, 2nd Division Action at Dargai; action of the Sampagha and Arhanga Passes, operations at and around Datoi and action of 24 Nov. 97.—Despatches, G. G. O. 244 of 98.—Clasp.
South African War, 1900-02.—As special service officer and afterwards employed on staff with Transport (graded as D. A. A. G.) and as D. A. A. G., South Africa, and D. A. A. G. for Imperial Yeomanry. Relief of Kimberly. Operations in the Orange Free State, Feb. to May 00, including operations at Paardeberg (17 to 26 Feb.) Actions at Poplar Grove (7 March). Driefontien (10 March). Operations in Orange River Colony (May to 29 Nov. 00), including actions at Wittebergen (1 to 29 July). Operations in the Transvaal between 30 Nov. 00 and 31 May 02. Operations in the Orange River Colony between 30 Nov. 00 and 13 May 02.—Despatches, *Lond. Gaz.*, 16 April 01 and 29 July 02.—Queen's medal with 5 clasps, King's medal with 2 clasps. Brev. of Maj. D. S. O.

Knollys, R. W. E.—
N.-W. Frontier of India, 1897-98.—Operations on the Samana and in the Kurram Valley during Aug. and Sep. 97. Operations of the Flying Column in the Kurram Valley under Colonel Richardson, 20 Aug. to 1 Oct. 97.—Medal with clasp.
Tirah, 1897-98.—Assistant Transport Officer, 1st Division.—Operations in the Bazar Valley, 25 to 30 Dec. 97.—Clasp.

Knowles, G.—
West Africa (South Nigeria), 1901.—Ishan-Ulia expedition.—Medal with clasp.
West Africa (South Nigeria), 1901-02.—Aro expedition. (Severely wounded.) Despatches, *London Gazette*, 12th September 1902.—Clasp.
East Africa, 1903-04.—Operations in Somaliland.—Clasp.
The War of 1914-18.—Despatches, *Lond. Gaz.*, 28 Dec. 17.

Knowles, J. K.—
N.-W. Frontier of India; Waziristan, 1901-02.—Medal with clasp.

Knowles, R.—
The War of 1914-18.—Despatches, *Lond. Gaz.*, 5 Apr. 16.

Knox, A. W. F.—
N.-W. Frontier of India, Waziristan, 1901-02.—Despatches, G. G. O. No. 702 of 1902.—Medal with clasp.
The War of 1914-18.—Brev. of Col. C. B.

Knox, E. F.—
Tirah, 1897-98.—Medal with 2 clasps.

The War of 1914-18.—Despatches, *Lond. Gaz.*, 30 May and 19 Oct. 16, D.S.O.

Knox, R. W.—
The War of 1914-18.—Despatches, *Lond. Gaz.*, 21 Jun. 16, and 6 July 17. D.S.O.

Knox, S. G. (*Capt. Supy. List. I. A.*)—
Mekran, 1898.—Political Officer.—Action of Gok Parosh.—Despatches. G. G. O. 648 of 98.
The War of 1914-18.—Despatches, *Lond. Gaz.*, 19 Oct. 16, 14 Nov. 16 and 15 Aug. 17. C.S.I.

Knox-Niven, H. W. (*Major Retd. I. A.*)—
N.-W. Frontier of India, Waziristan, 1901-02.—Medal with clasp.
East Africa, 1902—04.—Operations in Somaliland. Action at Jidballi. Despatches, *Lond. Gaz.*, 2nd September 1904.—Medal with 2 clasps.

Kœbel, F. E.—
South African War, 1900.—Operations in Cape Colony, May and June 00.—Queen's Medal with clasp.

The War of 1914-18.—Despatches, *Lond. Gaz.*, 15 Aug. 17, D.S.O.

Kreyer, F. A. C. (*Lt.-Col., retd., I. A.*)—
Afghanistan, 1878-79.—Served in the two Bazar Valley Expeditions.—Medal.

Kukday, K. V.—
British East Africa, 1898.—Expedition against the Ogaden Somalis.—Medal with clasp.

Kunhardt, F. G.—
South African War, 1901-02.—Operations in the Transvaal, between 30th November 1900 and 23rd October 1901. Operations in the Orange River Colony, 24th October to 5th December 1901. Operations in Cape Colony, 5th to 15th December 1901.—Queen's medal with 5 clasps.

Kunhardt, J. C. G.—
East Africa, 1903-04.—Operations in Somaliland.—Medal with clasp.

L

Lack, L. A. H.—
The War of 1914-18.—Despatches, *Lond. Gaz.*, 19 Oct. 16.

Lafond, H. A. (*Hony. Captain Retired*)—
Burma, 1886-87.—Medal.

Lake, E. A. W.—
East Africa (Somaliland) 1908-10.—Medal with clasp.

Laing, F. C.—
Chitral, 1895.—Relief of Chitral.—Medal with clasp.
Tirah, 1897-98.—Reconnaissance of the Kharmana. Defile and action of 7 Nov. 97. Operations against the Khani Khel Chamkannis.—2 clasps.

Laing, S Van B.—
The War of 1914-18.—Despatches, *Lond. Gaz.*, 5 Apl. and 19 Oct. 16. M.C.

Lalor, N. P. O'G.—
N.-W. Frontier of India, 1897-98.—Mohmand.—Medal with clasp.
China, 1900.—Medal.

Lamb, H. C. (*Col., retd., I.A.*)—
Afghanistan, 1879-80.—Medal.

Lambert, W. J.—
The War of 1914-18.—Despatches, *Lond. Gaz.*, 25 Sep. and 6 Dec. 16. D.S.O.

Lambragon, Jai Chand, *Raja of*—(*Hony. Maj.*)—
N.-W. Frontier of India, Hazara, 1891.—Medal with clasp.

Lamont, J. C. (*Lt.-Col., retd., I.M.S.*)—
N.-E. Frontier of India, Chin-Lushai, 1889-90.—Medal with clasp. *Manipur, 1891.*—Clasp.

Lancaster, E. H.—
N.-W. Frontier of India, 1908.—Operations in the Zakka Khel country. Operations in the Mohmand country. Engagement of Kargha, 24th May 1908. Despatches, G. G. O. No. 521 A of 1908.—Medal with clasp.

Landale, C. H.—
The War of 1914-18.—Despatches, *Lond. Gaz.*, 5 Apl. 16.

Lance, F. F. H.—
South African War, 1900-02.—Operations in Cape Colony, south of Orange River, 1899-1900. Operations in the Orange Free State, including defence of Wepener. Operations in the Orange River Colony (May to 29th November 1900) including action at Wittebergen (1st to 29th July). Operations in the Transvaal between 30th November 1900 and 31st May 1902. Severely wounded. Queen's medal with 4 clasps.—King's medal with 2 clasps.
N.-W. Frontier of India, 1908.—Operations in the Zakka Khel country. Operations in the Mohmand country. Provost Marshal to Division. Engagement of Matta, 24th April 1908. Engagement of Kargha, 24th May 1908.—Medal with clasp.

Land, J. J. (*Hony. Capt. Commissary, retd.—Ordnance Dept.*)—
Afghanistan, 1878-80.—Medal.
Tirah, 1897-98.—Actions of Sampagha and Arhanga Passes. Despatches 990244 of 98.—Medal with 2 clasps. Promoted Deputy Assistant Commissary.

Landon, A. (*Col., retd., Ben. Inf.*)—
N.-W. Frontier of India, 1863-64.—Umbeyla expedition.—Medal with clasp.
Afghanistan, 1878-79.—With Kuram Valley Field Force.—Medal.

Lane, C. A.—
N.-W. Frontier of India, 1897-98.—Operations on the Samana and in the Kurram Valley during Aug. and Sep. 97 (G. G. O. 1417 of 97). Operations of the Flying Column, in the Kurram Valley, under Col. Richardson, 20 Aug. to 1 Oct. 97 (G. G. O. 304 of 98).—Medal with 2 clasps.
Tirah, 1897-98.—Clasp.

Lane, C. R. C.—
The War of 1914-18.—Despatches, *Lond. Gaz.*, 15 May 17.

Lane, D. T.—
N.-W. Frontier of India (1st and 2nd), Miranzai, 1891.—Medal with clasp. *Hazara, 1891.*—Clasp.
Chitral, 1895.—Relief of Chitral.—Medal with clasp.

Lane, F. B.—
South African War, 1899—01.—Relief of Ladysmith, including action at Colenso (15 Dec. 99). Operations of 17 to 24 Jan. 00 and action at Spion Kop operations of 5 to 7 Feb. 00 and action at Vaal Krantz; operations on Tugela Heights (14 to 23 Feb. 00) and action at Pieter's Hill. Dangerously wounded. Operations in the Transvaal, between 30 Nov. 00 and Aug. 01.—Queen's medal with 6 clasps.

Lane, F. M. (*Maj., retd., I.A.*)—
N.-W. Frontier of India, 1897-98.—Mohmand.—Medal with clasp.
Tirah, 1897-98.—Action of the Sampagha Pass. Operations at and around Dwatoi and action of 24 Nov. 97.—Operations in the Bara Valley, 7 to 14 Dec. 97.—Clasp.

Lane, H. A.—
East Africa, 1903-04.—Operations in Somaliland. Despatches, *London Gazette,* 2nd September 1904—Medal with clasp.
The War of 1914-18.—Despatches, *Lond. Gaz.,* 15 Jun. 16.

Lane, H. P.—
East Africa, 1902-04.—Operations in Somaliland.—Medal with clasp.
The War of 1914-8.—Despatches, *Lond. Gaz.,* 27 July and 19 Oct. 16.

Lane, H. R. C.—
N.-W. Frontier of India, 1908.—Operations in the Zakka Khel country.—Medal with clasp.

Lane, J. T. H.—
The War of 1914-18.—Despatches, *Lond. Gaz,* 17 Feb. 15. M. C.

Lane, W. B.—
N.-W. Frontier of India, Hazara, 1891.—Medal with clasp. *Waziristan, 1894-95.*—Clasp.
Chitral, 1895.—Relief of Chitral.—Medal with clasp.

Lang, A. G. B. (*Lt.-Col., retd., I. A.*)—
N.-W. Frontier of India.—Action at Wano, 3 Nov. 94. (Severely wounded.) *Waziristan, 1894-95.*—Medal with clasp.

Lang, C. F. G.—
China, 1900.—Medal.

Lang, E. B.—
Burma, 1885-86.—Medal with clasp.
The War of 1914-18.—Despatches, *Lond. Gaz.,* 19 Oct. 16.

Lang, L. E.—
The War of 1914-18.—Despatches, *Lond. Gaz.,* 31 Oct. 17. M. C.

Lang, W. F.—
N.-W. Frontier of India, 1897-98.—Mohmand—Medal with clasp.

Lang, W. H.—
The War of 1914-18.—Despatches, *Lond. Gaz.,* 1 Jan. 16, and 15 Aug. 17. Brev. of Maj.

Langston, T. A. O. (*Major retd. I. M. S.*)—
N.-W. Frontier of India, Waziristan, 1894-95.—Medal with clasp.
N.-W. Frontier of India, 1897-98.—Malakand, Defence of Malakand.—Relief of Chakdara. Action at Landakai.—Despatches, G. G. O. 1089 of 97.—Medal with 2 clasps.

Lapsley, W.—
N.-W. Frontier of India, Waziristan, 1901-02—Medal with clasp.

Latham, A.—
N.-W. Frontier of India, Waziristan, 1901-02.—Medal with clasp.

Lawford, E. E. M.—
Afghanistan, 1879-80.—Action of Girishk. Battle of Maiwand. Defence and battle of Kandahar.—Medal with clasp.
Burma, 1886-88.—Operations of the 4th and 6th Brigades. Operations in the Yaw Country with the Pauk Column.—Medal with 2 clasps.
Burma, 1888-89.—Present at the first action against the Chins.

Lawrence, J. (*Comy. & Hony. Capt., retd. S. & T.*)—*Burma, 1887-89.*—Medal with clasp.
N.-E. Frontier of India, Chin Hills, 1892-93.—Clasp.
N.-W. Frontier of India, 1897-98.—Malakand.—Medal with clasp.

Lawrence, T. C. (*Sr. Asst. Surgn. and Hv. Capt. retd.*)—
Afghanistan, 1878-80.—Medal.
Burma, 1886-87.—Medal with 2 clasps.
N.-E. Frontier of India, Chin Lushai, 1889-90.—Clasp.
Burma, 1892.—Clasp.

Lawrence, W. A. (*Maj.-Gen., u.s.l., I.A.*)—
Afghanistan, 1878-79-80.—Battle of Ahmed Khel.—Medal with clasp. Brev. of Lt.-Col.

Lawrenson, T. G. P.—
N.-W. Frontier of India, 1897-98.—Malakand.—Medal.
Tirah, 1897-98.—Clasp.

Lawson, H. (*Lt.-Col. u.s.l., I.A.*)—
Burma, 1887-88.—Medal with clasp.

Layng, T. M.—
The War of 1914-18.—Despatches, *Lond. Gaz.,* 1 Jan. 16. M. C.

Lean, W. W. (*Lieut.-Col. u.s.l., I. A.*)—
Afghanistan, 1878-79-80.—Actions of the Peiwar Kotal and Charasiah. Operations at and around Kabul. Affair of Mazina.—Medal with 3 clasps.

LeBailly, A. C. (*Capt. retd., Ind. S. C.*)—
Egypt, 1882.—Battle of Tel-el-Kebir. Medal with clasp; bronze star.

Leckie, F. W. V.—
Afghanistan, 1879-80.—Staff Officer of Transport and served as Brig.-Maj. of the Force; sent out to cover the retreat from Maiwand, and during the defence of Kandahar, including the sorties on 12 and 16 Aug. Present in the same capacity at the battle of 1 Sep.—Medal with clasp. Despatches, G. G. O. 868 of 80.

Le Cornu, C. J. S.—
N.-W. Frontier of India, 1908.—Operations in the Mohmand country.—Medal with clasp.
The War of 1914-18.—Despatches, *Lond. Gaz.,* 1 Jan. and 19 Oct. 15. M. C.

Lee, A. W. H.—
East Africa, 1903-04.—Operations in Somaliland. Action at Jidballi.—Medal with 2 clasps.
The War of 1914-18.—Despatches, *Lond. Gaz.,* 4 July 16.

Lee, J.—
N.-W. Frontier of India, 1897-98.—Malakand.—Action at Landakai. Operations in the Mamund country. Buner. Action of the Tanga Pass.—Medal with clasp.
N.-W. Frontier of India, 1908.—Operations in the Mohmand country Despatches, G. G. O. No 521 A of 1908. Promoted Senior Asst. Surgeon and Hony. Lieut.—Medal with clasp.

Lee, M. G.—
South African War, 1901-02.—Operations in the Transvaal, March 1901 to January 1902. Operations in the Orange River Colony, January 1902 to 31st May 1902.—Queen's medal with 4 clasps.
N.-W. Frontier of India, 1908.—Operations in the Mohmand country.—Medal with clasp.

Lee, W. C. N.—
N.-W. Frontier of India, 1908.—Operations in the Zakka Khel country.—Medal with clasp.

Leeds, T. L.—
China, 1900.—Medal.
N.-W. Frontier of India, 1908.—Operations in the Zakka Khel country. Operations in the Mohmand country.—Medal with clasp.
The War of 1914-18.—Despatches, *Lond. Gaz.*, 22 Jun 15 and 19 Oct. 16. Brev. of Lt.-Col. D. S. O.

Lees, C. H. B.—
N.-W. Frontier of India, 1897-98.—Tochi.—Medal with clasp.
The War of 1914-18.—Despatches, *Lond. Gaz.*, 4 July 16 and 31 Oct. 17 C.M.G.

Legat, T. (*Hy. Lt., Dy. Asstt. Comsy., retd.*)—
Soudan, 1896.—Dongola Expedition.—In charge Ordnance Field Park. Promoted Deputy Assistant Commissary.—Medal, Egyptian medal.

Leggett, J. B. (*Col., retd., Mad. S. C.*)—
India, 1858.
Rumpa Rebellion, 1880.
Burma, 1887-88.—Medal with 2 clasps.

Le Geyt, P. H. (*M.-Gen. u.s.l. Ind. S.C.*)—
Persia, 1856-57.—Assault and capture of the fort of Reshire, surrender of Bushire expedition to Borazjoon, and action at Khooshab.—Medal with clasp.

Leicester, J. C. H.—
China, 1900.—Medal.
The War of 1914 18.—Despatches, *Lond. Gaz.*, 15 Aug. 17.

Leigh, H. P. P. (*Lt.-Col., retd., I. A.*)—
N.-W. Frontier of India (1st and 2nd), Miranzai, 1891.—Served as Political Officer. Despatches, G. G. O. 632 of 91.—Medal with clasp.

Leitch, G.—
The War of 1914-18.—Despatches, *Lond. Gaz.*, 15 Aug. 17.

Leith-Ross, W.—
N.-W. Frontier of India, 1908.—Operations in the Mohmand country. Engagement of Kargha, 24th May 1908.—Medal with clasp.
The War of 1914-18.—Despatches, *Lond. Gaz.*, 15 Aug. 17.

Leith, T. (*Maj., retd., I. A.*)—
Punjab Campaign, 1848-49.—Siege and surrender of Mooltan. Medal with clasp.
Persia, 1857.—With the Sind Irregular Horse. Bombardment and capture of Mohumra. Medal.
India, 1858-59.—With the Delhi Field Force in pursuit of Rebels in the Rewaree, Ulwar and Jeypur districts. Medal.

LeMessurier, G. D. S.—
N.-W. Frontier of India, 1908.—Operations in the Mohmand country.—Medal with clasp.

Lemon, W. S.—
South African War, 1900.—Operations in the Transvaal, Cape Colony, and Orange Free State. Queen's medal with 3 clasps.
Operations in the Abor country, 1911-12.—Medal with clasp.

Lentaigne, E. C.—
The War of 1914-18.—Despatches, *Lond. Gaz.*, 15 Aug. 17. D. S. O.

Leonard, A. W.—
Soudan, 1884-85.—Nile.—Medal with clasp bronze star.
Burma, 1885-89.—Medal with 2 clasps.
China, 1900.—Orderly Officer to G. O. C. 1st Brigade. Action at Yangtsun. Relief of Pekin. Horse shot.—Medal with clasp.

Leonard, W. H.—
Tibet, 1903-04.—Action at Niani, 26th June 1904. Operations at and around Gyanste, between 5th May and 6th July 1904. March to Lhassa, 14th July to 3rd August 1904.—Medal with clasp.

Leslie, W. C. C. (*Lt.-Col., Supy. List, I. A.*)—
Egypt, 1882.—Actions of Kassassin and Tel-el-Kebir. Medal with clasp. Bronze star.
South African War, 1902.—Empld. with Remounts Operations in Cape Colony. Queen's medal with 2 clasps.

Leslie, W. S.—
N.-W. Frontier of India, 1897-98.—Malakand. Action at Landakai. Operations in Bajaur and in the Mamund country. Buner. Action of the Tanga Pass.—Medal with clasp.
The War of 1914-18.—Despatches, *Lond. Gaz.*, 1 Jan. 16, 19 Oct. 16 and 15 Aug. 17. Brev. of Lt.-Col. D. S. O.

Leslie-Smith, C.—
N. W. Frontier of India, 1908—Operations in the Mohmand country. Engagement of Kargha Medal with clasp.

L'Estrange, E. B.—
South African War, 1900-02.—Operations in Natal. 00. Operations in the Transvaal, east of Pretoria, July to 29 Nov. 00, including actions at Belfast (26 and 27 Aug.) and Lyndenberg (5 to 8 Sept.). Operations in the Transvaal, Mar. to 31 May 02. Operations in Orange River Colony, Feb. to Mar. 02. Queen's medal with 4 clasps.

Lewin, W. H. (*Lt.-Col. retd. I. A.*)—
Ashanti, 1881.
Burma, 1886-88.—Served in the operations of the 4th and 6th Brigades.—Medal with 2 clasps.

Lewis, D. S. (*Bt.-Col. retd. I. A.*)—
Burma, 1889-92.—Poukan Expedition. Mentioned in Despatches.—Medal with clasp.

Lewis, G.—
South African War, 1900-1902.—Operations in Orange River Colony, June and July 1900, including actions at Lindley, Bethlehem and Wittebergen. Operations in the Transvaal July 1900 to March 1901, including actions at Frederickstad and Rhenoster Kop. Operations in Orange River Colony, March 1901 to May 1902, with mounted infy.—Queen's medal with 3 clasps. King's medal with 2 clasps.

Lewis, H. Y.—
The War of 1914-18.—Despatches, *Lond. Gaz.*, 22 June 15, 8 and 13 Feb. and 22 Sep. 17. D.S.O. M.C. clasp to M.C.

Lewis, R. F.—
The War of 1914-18.—Despatches, *Lond. Gaz.*, 19 Oct. and 14 Nov. 16.

Lewtas, J. (*Lt.-Col. retd. I. M.S.*)—
Afghanistan, 1878-79-80.—Action of Ali Musjid Both the Bazar Valley Expeditions. Action of Futehabad. Operations at and around Kabul. Despatches (twice), G. G. O. 328 of 80.—Medal with 2 clasps.

Lewty, W. H. (*Sr. Asstt. Sgn., Hony. Capt. retd. I. S. M. D.*)—
N.-W. Frontier of India, (2nd) Eusufzai, 1863.—Umbeyla.—Medal with clasp. *Hazara, 1868*—Clasp.
Afghanistan, 1878-79.—Medal with clasp.
N.-W. Frontier of India, Mahsud-Waziri, 1881.

Light, R. H. (*Lt.-Col. retd. I. A.*)—
Afghanistan, 1879-80.—Medal.

Light, W. A.—
South African War, 1900.—Operations in Cape Colony, 00. Operations in the Orange Free State, Feb. to May 00, including operations at Paardeberg (17 to 26 Feb.); actions at Poplar Grove (17 March), Driefontein (10 March). Actions at Houtnek (1st May), Vet River (5 and 6 May) and Zand River (10 May). Operations in the Transvaal in May and June 00, including actions near Johannesburg (29 May), Pretoria (4 June) and Diamond Hill (11 and 12 June). Operations in the Transvaal, east of Pretoria, up to 29 Nov. 00, including action at Belfast (26 and 27 Aug. 00).—Queen's medal with 6 clasps.

Lightfoot, T. W. (*Maj. retd. I. A.*)—
N.-W. Frontier of India, 1908.—Operations in the Mohmand country.—Medal with clasp.

Lilly, G. W. (*Lt.-Col. retd. I. A.*)—
Burma, 1885-88.—Operations round Hlinedet, Yemethen, and Myingyan, and operations of the 4th Brigade.—Medal with 2 clasps.
China, 1900.—Medal.

Lincoln, S. W. (*Lt.-Col., retd., I.A.*)—
Burma, 1886-87.—Operations of the 3rd and 4th Brigades under Brig.-Genl. H. S. Anderson and R. C. Low, C.B., respectively; repulse of the night attack on Ling-adaw.—Medal with 2 clasps.
N.-W. Frontier of India, Miranzai (2nd), 1891.—Operations on the Samana Range. Affair at Mastaon.—Clasp.
Chitral, 1895.—Relief of Chitral.—Medal with clasp.
N.-W. Frontier of India, 1897-98.—Malakand, Base Commissariat Officer.—Despatches, G.G.O. 178 of 98.—Buner, Base Commissariat Officer.—Despatches, G. G. O. 217 of 98.—Clasp.

Lend, A. G.—
The War of 1914-18.—Despatches, *Lond. Gaz.*, 6 July 17.

Lindesay, A. L.—
N.-W. Frontier of India, Hazara, 1888.—Medal with clasp.

Lindesay, G. W. G. (*Maj. retd., I. A.*)—
N.-W. Frontier of India—Isazai, 1892.
Soudan, 1896.—Dongola Expedition—Medal, Khedive's medal.
N.-W. Frontier of India, 1897-98.—Tochi—Malakand.—Medal with clasp.
East Africa, 1903.—Operations in Somaliland. Spec. Serv. Offr.—Medal with clasp.

Lindesay, V. E. H.—
China, 1900.—Medal.

Lindsay, H. A. P.—
N.-W. Frontier of India, 1897-98.—Tochi.—Medal with clasp.
The War of 1914-18—Despatches, *Lond. Gaz.* 22 June 15, 19 Oct. 16, and 14 Nov. 16, Brev. of Col. C.M.G.

Lindsell, P. B. (*Maj. retd. Ind. S.C.*)—
Afghanistan, 1878—80.—With Transport Department Kabul Field Force in expedition to Sherpur. Dec. 79.—Medal with clasp.

Lister, A. E. J.—
East Africa, 1903-04.—Operations in Somaliland. Action at Jidballi.—Medal with 2 clasps.

Liston, R. C. C.—
The War of 1914-18.—Despatches, *Lon. Gaz.* 4 July 16.

Little, S. (*Lieut.-Col. retd. I.M.S.*)—
Afghanistan, 1879-80.—Operations in the Hissarak District.—Medal.
N.-W. Frontier of India, Mahsud-Waziri, 1881.

Little, W. C.—
N.-W. Frontier of India, Waziristan, 1901-02.—Medal with clasp.

Llewhellin, N. G. R.—
South Africa, 1900-02.—Relief of Ladysmith including operations of 17th to 24th January 1900 and action at Spion Kop.—Operations of 5th to 7th February 1900 and action at Vaal Kranz.—Operations on Tugela Heights (14th to 27th February 1900) and action at Pieters Hill (27th February 1900.)—Relief of Mafeking.—Operations in the Transvaal between 30th November 1900 and 31st May 1902.—Queen's Medal with 4 clasps.

Lloyd, A. M. (*Maj. retd. I. S. C.*)—
Afghanistan, 1879-80.—Medal.

Lloyd, C. E. (*Maj. retd. I. A.*)—
Soudan, 1884-85.—Suakin. As Staff Officer to H. E. the Governor-General, Red Sea Littoral, during the defence of Suakin. Present at the action of Hasheem, battle of Tofrek, and succeeding convoy actions.—Medal with 2 clasps. Bronze star.
Burma, 1885-86.—As Orderly Officer to Brigadier-Genl. H. H. Foord. Present at the taking of Minhla and occupation of Mandalay.—Medal with clasp.
Tirah, 1897-98.—Despatches, G. G. O. 483 of 98.—Medal with 2 clasps.

Lloyd, C. G.—
The War of 1914-18.—Despatches, *Lond. Gaz.*, 15 Aug. 17.

Lloyd, C. R.—
The War of 1914-18.—Despatches, *Lond. Gaz.*, 15 Aug. 17. D.S.O.

Lloyd, J. H.—
N.-W. Frontier of India, 1897-98.—Mohmand. Medal with clasp.
Tirah, 1897-98.—Action of the Sampagha Pass. Operations at and around Dwatoi and action of 24 Nov. 97. Operations in the Bara Valley, 7 to 14 Dec. 97.—Clasp.
East Africa, 1903-04.—Operations in Somaliland. Action at Jidballi. As Staff Officer to Chief Transport Officer and as D. A. A. and Q. M. G. 1st Brigade. Despatches, Brigadier-Genl. Manning's 17th August 1903 and Sir C. C. Egerton's, 30th May 1904. *Lond. Gaz.*, 2 Sep. 04.—Medal with 2 clasps.
The War of 1914-18.—Despatches, *Lond. Gaz.*, 5 Apl. 16.

Lloyd, R. A.—
N.-W. Frontier of India, 1908.—Operations in the Mohmand country.—Medal with clasp.
The War of 1914-18.—Despatches, *Lond. Gaz.*, 19 Oct. 16, and 15 Aug. 17, D.S.O.

Lloyd, R. E.—
Tibet, 1903-04.—Medal.

Lloyd, R. M. (*Col. retd. Bo. S. C.*)—
India, 1858.—Against the Naikra Bheels in the Rewa Kanta, and in pursuit of Tantia Topee (wounded).—Medal.
Burma, 1886-87.—Despatches, G. G. O. 864 of 87.

Lloyd, W. E. E.—
Tirah, 1897-98.—Operations in the Bara Valley, 7 to 14 Dec. 97.—Medal with 2 clasps.
South African War, 1900-01.—Queen's Medal with 3 clasps.

Lloyd-Jones, F. L.—
East Africa, 1902-04.—Operations in Somaliland.—Medal with clasp.
Aden, 1903-04.—Operations in the Interior. (Severely wounded.) Despatches, G. G. O. No. 43 of 1905.
East Africa, Somaliland, 1908-10.—Medal with clasp.

Loch, E. C.—
Tirah, 1897-98.—Medal with 2 clasps.

Loch, G. H. (*Lt.-Col., Supy. List, I. A.*)—
Afghanistan, 1880.—Medal.
N.-W. Frontier of India, Zhob Valley, 1884.
Burma, 1886—89.—Shan States column. Advance to and occupation of, Fort Stedman.—Medal with 2 clasps.

Loch, H. F. (*Br.-Gen. retd., I.A.*)—
N.-W. Frontier of India, Zhob Valley, 1884.
Burma, 1886-89.—Raised and commanded Chindwin Battalion Military Police.—Medal with 2 clasps.

Loch, J. C.—
China, 1900.—Relief of Pekin. Action of Peitsang. Action at Yangtsun.—Medal with clasp.

Loch, J. H. (*Dep. Surg.-Gen. retd. Ind. Med. Ser.*)—
India, 1858.—Relief of Azimghur.—Medal.

Lock, F. R. E.—
Aden, 1903-04.—Operations in the Interior.
The War of 1914-18.—Despatches, *Lond. Gaz.*, 15 Aug. 17. D.S.O.

Lockhart, R. Mac. G. M.—
The War of 1914-18.—Despatches, *Lond. Gaz.*, 19 Oct. 16.

Logan, A. F.—
The War of 1914-18.—Despatches, *Lond. Gaz.*, 4 Nov. 15 and 1 Jan. 16. M. C.

Lonergan, J. (*Comsy. and Hony. Capt. retd.*)—
Afghanistan, 1878-79-1880.—Medal.

Long, W. C. (*Major, retd. I.M.S.*)—
East Africa, 1903-04.—Operations in Somaliland.—Medal with clasp.

Longden, A. B.—
Chitral, 1895.—Relief of Chitral.—Medal with clasp.
N.-W. Frontier of India, 1897-98.—Mohmand. Clasp.
The War of 1914-18.—Despatches, *Lond. Gaz.*, 4 July 16.

Longhurst, A. L.—
The War of 1914-18.—Despatches, *Lond. Gaz.*, 4 July 16 and 19 Oct. 16. Brev. of Lt.-Col.

Lorimer, D. L. R.—
N.-W. Frontier of India, 1897-98.—Present at Forts Bara and Jamrud.—Medal with clasps.

Lorimer, E.—
N.-W. Frontier of India, 1908.—Operations in the Zakka Khel country.—Medal with clasp.

Lorimer, J. M.—
N.-W. Frontier of India, 1908.—Operations in the Zakka Khel country. Operations in the Mohmand country. Medal with clasp.

Loudon, F. A. (*Maj. retd. I. A.*)—
Burma, 1885—87.—Present at the taking of Minhla and occupation of Mandalay and Bhamo. Operations of the 1st Brigade.—Medal with clasp.

Loudon, J. A.—
Burma, 1886—89.—Medal with 2 clasps.

Lovel, C. L.—
Operations in the Abor country, 1911-12.—Medal with clasp.

Lovett, B. (*M.-Gen. retd.*)—
N.-W. Frontier of India, 1878.—Jowaki Expedition.—Medal with clasp.
Afghanistan, 1878—79.—Capture of Ali Musjid. Mentioned in Despatches.—Medal with clasp; Brev. of Lt.-Col.
N.-W. Frontier of India, Hazara, 1888.—As Commanding Royal Engineer. Mentioned in Despatches.—Clasp; C.B.

Low, J. A. (*Capt. retd. Art.*)—
India, 1858.—Siege and capture of Lucknow.—Medal with clasp.

Low, R. B.—
N.-E. Frontier of India, Lushai, 1889.—Clasp.
N.-W. Frontier of India, Hazara, 1891.—Medal with clasp.
Chitral, 1895.—Relief of Chitral.—Served as A.-D.-C. to the Lieut.-General Commanding Chitral Relief Force.—Despatches; G. G. O. 998 of 1895.—Medal with clasp. D. S. O.
Tirah, 1897-98.—Operations in the Bara Valley, 7 to 14 Dec. 97.—2 clasps.
China, 1900.—Relief of Pekin. Action of Peitsang. Action at Yangtsun. Provost Marshal and Head Quarter Camp Commandant, China Field Force. Despatches, *Lond. Gaz.* 14 May 01. Medal with clasp. Brev. of Maj.

Lowis, H. E.—
The War of 1914-18.—Despatches, *Lond. Gaz.* 20 Mar. 16. Brev. of Col.

Lowis, R. M. (*Lt.-Col. retd. I. A.*)—
N.-W. Frontier of India, Hazara, 1888.—Medal with clasp. (*1st*) *Miranzai, 1891.*

Lowry, F. J. S. (*Maj., retd., Ind. S. C.*)—
N.-W. Frontier of India, Zhob Valley, 1890.
Tirah 1897-98.—Asst. Judge Adv.-Gen. Capture of the Sampagha and Arhanga Passes.—Medal with 2 clasps.
The War of 1914-18—Despatches, *Lond. Gaz.*, 15 Jun. 16 and 28 Dec. 17.

Lowry, H. W. (*Lt.-Col. retd., I. A.*)—
Burma, 1887—89.—Ruby Mines column.—Medal with clasp.

Lowry, W. H. (*Bt.-Col. retd., I. A.*)—
Burma, 1885—89.—Operations of the 1st Brigade. Present during the operations in the Chin-Lushai country, 1889.—Medal with 2 clasps.
N.-E. Frontier of India, Chin-Lushai, 1891.—Chinbok column.
Burma, 1891-1892.
China, 1900.—Despatches, *London Gazette*, 4th May 1901. Medal.

Lowther, H.—
Tirah, 1897-98.—Medal with 2 clasps.

Lowry-Corry, W.—
East Africa (Somaliland) 1914-15—Severely wounded. Medal with clasp.

Luard, H. B. (*Capt. retd. I.M.S.*)—
N.-W. Frontier of India, (*2nd*) *Miranzai, 1891.*—Medal with clasp. *Hunza-Nagar, 1891.* Despatches, G. G. O. 397 of '92.—Clasp.
Chitral, 1895.—Relief of Chitral.—Principal Medical Officer, Gilgit Force.—Medal with clasp.

Lucas, C. A.—
East Africa, Somaliland 1908-10.—Medal with clasp.

Lucas, F. G.—
N.-W. Frontier of India, Hazara, 1891.—Medal with clasp. *Isazai, 1892.*
Tirah, 1897-98.—Commanded the Gurkha Scouts.—Actions of the Chagru Kotal and of the Sampagha and Arhanga Passes.—Reconnaissance of the Saran Sar and action of 11 Nov. 1897.—Operations at and around Dwatoi and action of 24 Nov. 1897.—Operations against the Khani Khel Chamkannis.—Operations in the Bara Valley, 7 to 14 Dec. 97.—Operations in the Bazar Valley, 25 to 30 Dec. 97.—Despatches, G. G. O. 58 and 244 of 98—Medal with 2 clasps. Brev. of Maj.—D. S. O.
N.-W. Frontier of India, Waziristan, 1901-02.—Despatches, G. G. O. 611 of 1902.—Clasp.

N.-W. Frontier of India, 1908.—Operations in the Zakka Khel country. Despatches, G. G. O. No. 188 of 1908.—Medal with clasp.
The War of 1914-18.—Despatches, *Lond. Gaz.*, 15 Aug. 17, Brev. of Col. C. B.

Lucas, G. W. C.—
The War of 1914-18.—Despatches, *Lond. Gaz.*, 1 Jun. 17.

Lucas, H. C. E. (*Col. retd. I. A.*)—
Afghanistan, 1878-79.—Occupation of Kandahar and Girishk. Despatch of Maj.-Genl. M. A. S. Biddulph, C.B., No. 344, 6 Mar. '79.—Medal.
Soudan, 1885.—Suakin.—Medal with clasp, bronze star.

Lucas, H. de N.—
The War of 1914-18.—Despatches, *Lond. Gaz.*, 15 Jun. 16.

Luck, C. A.—
N.-W. Frontier of India, Waziristan, 1894-95.—Medal with clasp.
The War of 1914-18.—Despatches, *Lond. Gaz.*, 4 Jan. 17.

Luck, T.—
N.-W. Frontier of India, 1908.—Operations in the Mohmand country.—Medal with clasp.

Lugard, E. J. (*Major retd. I.A.*)—
Burma, 1888-89.—Expedition against the Chins.—Medal with clasp.
N.-E. Frontier of India, Chin-Lushai, 1889-90.—(Burma Column) Despatches, G. G. O. 677 of '90. Clasp. D. S. O.—*Manipur, 1891.*—(Slightly wounded.)—Clasp.
South African War, 1899-1900.—D. A. A. G. Line of Communications and on Staff (graded as D. A. A. G.) of Military Governor of Johannesburg.—Queen's Medal with 3 clasps.

Lukin, R. C. W.—
Chitral, 1895.—Relief of Chitral.—Medal with clasp.
Tirah, 1897-98.—Operations in the Bara Valley, 7 to 14 Dec. 97.—2 clasps.
South African War, 1900-1902.—Operations in the Transvaal, between 30th November 1901 and 31st May 1902. Operations in the Orange River Colony, between 30th November 1901 and 31st May 1902.—Queen's Medal with 5 clasps.
The War of 1914-18.—Despatches, *Lond. Gaz.*, 22 Jun. 15. D. S. O.

Lumb, F. G. E.—
China, 1900.—Medal.
The War of 1914-18.—Despatches, *Lond. Gaz.*, 17 Feb. 15. M. C.

Lumsden, H. R. W.—
Afghanistan, 1878-79-80.—Occupation of Kandahar and Girishk. With Thal-Chotiali, Khyber and Kabul Field Forces. Operations in Kohistan and affair at Shekhabad.—Medal.
N.-E. Frontier of India, Sikkim, 1888.—Taking of Jeluk. (Severely wounded). Horse wounded. Despatches.—Medal with clasp.
N.-W. Frontier of India, 1897-98.—Operations on the Samana and in the Kurram Valley during Aug. and Sep. 97.—Medal with 2 clasps.
Tirah, 1897-98.—Actions of the Chagru Kotal and Dargai.—Operations in the Bara Valley, 7 to 14 Dec. 97.

Lumsden, P. J.—
N.-W. Frontier of India, (1st) Miranzai, 1891.
The War of 1914-18.—Despatches, *Lond. Gaz.*, 21 Jun. 16.

Lumsden, *Sir* P.S. (*Gen. u. s. l. Ind. S. C.*)—
N.-W. Frontier of India.—Operations against the Ootman Khel and Ranezai tribes, 1852. Mentioned in Despatches.
Operations against Momunds, 1852. Mentioned in Despatches.
Operations against the Bori Afridis, 1853-54. Mentioned in Despatches.
Expedition to Shah Moosah Khel against the Momunds, 1854. Mentioned in Despatches.
Expedition into Miranzai and Upper Kurram, 1856-57. Mentioned in Despatches.—Medal with clasp.
India, 1858.—With Central India Field Force as Asst. Qr.-Mr. Genl. Pursuit and capture of Tantia Topee; action at Ranode. Despatches, *Lond. Gaz.*, 18 Apl. '59.—Medal.
China, 1860-61.—Action of Sinho, capture of Tonghoo and Taku Forts; occupation of Pekin. Mentioned in Despatches.—Medal with 2 clasps; Brevs. of Maj. and Lt.-Col.
N.-E. Frontier of India Bhutan, 1865.—Clasp.

Lunham, J. L.—
N.-W. Frontier of India, 1908.—Operations in the Mohmand country. Engagement of Kargha, 24th May 1908.—Medal with clasp.

Lunn, A. J.—
The War of 1914-18.—Despatches, *Lond. Gaz.*, 15 Aug. 17. M.C.

Lushington, E. C. M. (*Lieut.-Col. retd. I.A.*)—
Burma, 1886-88.—Medal with 2 clasps.

Lushington, W. N.—
N.-W. Frontier of India, 1908.—Operations in the Zakka Khel country. Despatches, G. G. O. No. 188 of 1908. Operations in the Mohmand country.—Medal with clasp.
The War of 1914-18.—Despatches, *Lond. Gaz.*, 1 Jan. 16. Brev. of Lt.-Col.

Luxmoore, C. de J.—
N.-W. Frontier of India, 1897-98.—Operations on the Samana.—Medal with 2 clasps.
China, 1900.—Medal.

Luxmoore, C. T. P. (*Col. u.s.l. Ind. S.C.*)—
India, 1857-58.—Medal with clasp.

Lyall, R. A.—
N.-W. Frontier of India, 1897-98.—Medal with clasp.
The War of 1914-18.—Despatches, *Lond. Gaz.*, 30 Jun. 16. D. S. O.

Lynch, E. H.—
South African War, 1900—02.—Operations in Cape Colony in May 1900. Operations in the Orange River Colony, May to July 1900. Operations in the Transvaal, July 1900 to May 1902. Queen's medal with 3 clasps. King's medal with 2 clasps.
Tibet, 1903-04.—Operations at and around Gyantse, between 5th May and 6th July 1904. March to Lhassa, 14th July to 3rd August 1904.—Medal with clasp.

Lynch, J. B. (*Brev. Col. retd. Ind. S. C.*)—
Afghanistan, 1878-80.—Actions of the Peiwar Kotal and Charasiah and operations at and around Kabul in Dec. '79.—Queen's Medal with 3 clasps.

Lyne, C. V. N. (*Maj., retd. I. A.*)—
Burma, 1891-92.—As Transport Officer with the Irrawaddy column.—Medal with clasp.
East Africa, 1903-04.—Operations in Somaliland. Special Service Officer.—Medal with clasp.

Lynsdale, A. W. J.—
Burma, 1885-86.—Promoted 1st Grade Assistant Apothecary.—Medal with clasp.
N.-E. Frontier of India, Chin Lushai, 1889-90.—Clasp.

Lynn, G. R.—
The War of 1914-18.—Despatches, *Lond. Gaz.*, 17 Feb. 15.

Lyon, C. W.—
South African War, 1899—1901.—Operations in the Transvaal west of Pretoria, July to 29th November 1900. Operations in Orange River Colony, May to 29th November 1900.—Queen's medal with 4 clasps.

Lyons-Montgomery, H. F. (*Col. retd. I. A.*)—
Afghanistan, 1878-79-80.—Affair of Matun. Action of Charasiah. Operations at and around Kabul. March from Kabul to the relief of Kandahar and battle of Kandahar. Despatches, G. G. O. 582 of '80.—Medal with 3 clasps. Bronze star.
N.-W. Frontier of India, Hazara, 1891.—Served as Chief Commissariat Officer of the Force. Despatches, G. G. O. 810 of '91.—Medal with clasp Brev. of Maj.
Tirah, 1897-98.—Divisional Commissariat Officer, 2nd Division.—Operations in the Bara Valley, 7 to 14 Dec. 97. Despatches, G. G. O. 244 of 98.—Medal with 2 clasps. Brev. of Lt.-Col.

Lys, A. M. (*Col. retd. Mad. S. C.*)—
Parlah Kimedy Insurrection, 1856-57.

Lyster, C. J. H.—
South African War, 1899—1902.—Operations in Natal, 1899. Relief of Ladysmith, including action at Colenso (15th December 1899). Operations of 17th to 24th January 1900, and action at Spionkop; Operations of 5th to 7th February 1900 and action at Vaal Kranz. Operations on Tugela Heights (14th to 27th February 1900), and action at Pieter's Hill (27th February). Operations in Natal (March to June 1900), including action at Laings Nek (6th to 9th June). Operations in the Transvaal, east and west of Pretoria (July to 29th November 1900). Operations in the Transvaal (30th November 1900 to April 1901 and April to 31st May 1902). Operations in Orange River Colony (November 1901 to April 1902). Despatches, *London Gazette*, 8th February and 10th September 1901. Queen's medal with 6 clasps. King's medal with 2 clasps.
N.-W. Frontier of India, 1908.—Operations in the Mohmand country.—Medal with clasp.
The War of 1914-18.—Despatches, *Lond. Gaz.*, 19 Oct. 16. M.C.

Lyster, H. H. (*Lt.-Gen. u.s.l. Ind. S. C.*)—
India, 1857-58. A.-D.-C. to Maj.-Gen. Commanding Central India Field Force; capture of Fort Rathghur, action of Baroda, forcing Pass of Muddenpore, siege and storm of Jhansi, actions of the Betwa and Koonch, attacks on camp at village of Muttra and at Golowlee, capture of town and fort of Calpee, and capture of Morar. Despatches, *Lond. Gaz.*, 17 July, 10 Aug. and 17 Sep. '58. 18 Apl. and 29 Nov. 59.—Medal with clasp; V.C.; Brev. of Maj.
Afghanistan, 1878-80.—Actions of Ahmed Khel and Urzoo. Despatches, G. G. O. 326 of 80.—Medal with clasp; C. B.

Lyster, W. H. (*Col. retd. I. A.*)—
Afghanistan, 1880.—Medal.

Lyttle, W. J.—
N.-W. Frontier of India, Isahai, 1892.
Tibet, 1903 04.—Action at Naini, 26th 1904.—Operations at and around Gyantse, between 5th May and 6th July 1904. March to Lhassa, 14th July to 3rd August 1904.—Medal with clasp.
N.-W. Frontier of India, 1908.—Operations in the Zakka Khel Country.—Operations in the Mohmand Country. Engagement of Kargha 24th May 1908. Despatches G. G. O., No. 188 of 1908.—Medal with clasp.
The War of 1914-18.—Despatches, *Lond. Gaz.*, 5 Apl. 16, 19 Oct. 16, and 15 Aug. 17.

M

McBarnet, A. E.—(*Major, Retd., I. A.*)—
N.-E. Frontier of India, Manipur, 1891.—Medal with clasp.

N.-W. Frontier of India, Waziristan, 1901-02.—Despatches, G. G. O. No. 611 of 4 July 02.—Medal with clasp. D. S. O.

McCarrison, R.—
The War of 1914-18.—Brev. of Lt.-Col.

McCartie, C. J. (*Lt.-Col. retd. I. M. S.*)—
Afghanistan, 1878-79-80.—Actions of Saif-u-din and Shahjui. Despatches. Colonel T. G. Kennedy's despatch, 24 Oct. '79. (Severely wounded near Kandahar).—Medal.

Burma, 1891-92.—Medal with clasp.
Chitral, 1895.—Relief of Chitral.—Medal with clasp.
N.-W. Frontier of India, 1897-98.—Mohmand-Malakand. Operations in Bajaur.—Clasp.
Tirah, 1897-98.—Clasp.

McCaskill, J. C. H.—
N.-W. Frontier of India, Waziristan, 1894-95.—Medal with clasp.

Chitral, 1895.—Relief of Chitral.—Medal with clasp.
N.-W. Frontier of India, 1897-98.—Defence and relief of Malakand. *Malakand*—Operations in Bajaur and in the Mamund Country.—Utman Khel—Buner—2 clasps.

East Africa, 1901.—Operations against the Ogaden Somalis in Jubaland. Medal with clasp.
N.-W. Frontier of India, 1908.—Operations in the Mohmand country.—Medal with clasp.

McCay, D.—
China, 1900.—Relief of Pekin.—Medal with clasp.

McConaghy, J. G.
The War of 1914-18.—Despatches, *Lond. Gaz.*, 5 Nov. 15, 13 July 16, 4 Jan. 17, D. S. O. Orders of the White Eagle, 4 class (with swords).

McCleverty, A. H.—
China, 1900.—Medal
The War of 1914-18.—Despatches, *Lond. Gaz.*, 27 Mar. 16.

McCleverty, G. M.—
Operations in the Abor country, 1911-12.—Medal with clasp.

The War of 1914-18.—Despatches, *Lond. Gaz.*, 17 Feb. 15 and 15 Aug. 17. M. C.

McCloghry, J. (*Col. retd. I.M.S.*)—
Afghanistan, 1878—80.—Affair of Bhagao.—Medal.

McCormick, G. D.—
The War of 1914-18.—Despatches, *Lond. Gaz.*, 31 Oct. 17.

McCowan, W. T.—
Persia, 1911.—Attacks on detachment of 39th Horse. Brev. Maj.

McCrea, A. C.—
N.-W. Frontier of India, Hazara, 1891.—Medal with clasp.
N.-E. Frontier of India, Chin Hills, 1892-1893.—Clasp.
Chitral, 1895.—Relief of Chitral. Action of the Malakand Pass. Action at Khar.—Medal with clasp.
The War of 1914-18.—Despatches, *Lond. Gaz.*, 19 Oct. 16. C.M.G.

McCudden, J. H.—
Aden, 1903-04.—Operations in the interior.
East Africa (Somaliland) 1908-10.—Staff Officer to O. C. Troops, Somaliland.—Medal with clasp.
The War of 1914-18.—Despatches, *Lond. Gaz.*, 16 May 16. M.C.

McDermott, J. (*Hony. Maj. Commy. retd. India Misls. List*)—
Tirah, 1897-98.—Actions of the Sampagha and Arhanga Passes.—Despatches, G. G. O. 244 of '98.—Medal with 2 clasps.—Promoted Assistant Commissary.

McElwaine, E. J D.—
The War of 1914-18.—Despatches, *Lond. Gaz.*, 5 Apl. 16.

McEnery, R. T.—
N.-W. Frontier of India, 1908.—Operations in the Mohmand Country. Engagement of Kargha, 24th May 1908. Medal with clasp.

McEwen, W. L.—
The War of 1914-18.—Despatches, *Lond. Gaz.*, 22 Jun. 15 and 1 Jan. 16. M. C.

McGaun, H. H.—
The War of 1914-18—.Despatches, *Lond. Gaz.*, 22 Jun. 15.

McInroy, C. (*Col. u.s.l. Ind. S.C.*)—
India, 1857-59.—Actions of Sahao, Maharajpore, Girwasa, Jakoli, and Mowmahoni. Despatches, *Lond. Gaz.*, 31 Jan. '59.—Medal with clasp.
Abyssinia, 1867-68.—Commanded a division of the Transport Train. Despatches, *Lond. Gaz.*, 10 July '68.—Medal.
Egypt, 1882.—Battle of Tel-el-Kebir.—Medal with clasp: bronze star.
Burma, 1885-86.—Medal with clasp.

McIntyre, H. D.— (*Bt.-Col. retd., I. A.*)—
Burma, 1886—89.—Transport Officer, 3rd Brigade, and with Yaw Expedition, 1887.—Medal with 2 clasps.
Tirah, 1897-98.—Provost Marshal, 2nd Division.—Medal with 2 clasps.
China, 1900.—Relief of Pekin. Despatches, *Lond. Gaz.*, 14 May 01.—Medal with clasp.

McKay, H. K.—
N.-E. Frontier of India, Naga Hills, 1875.—Despatches, G. G. O. 574 of '75.
Afghanistan, 1878-79-80.—Medal.
N.-W. Frontier of India, Mahsud-Waziri, 1881.

McKay, T. (*Lieut.-Col. retd. I. A.*)—
N.-W. Frontier of India, Hazara, 1888.—Medal with clasp.

Mckechnie, W. E.—
China, 1900—Medal.
N.-W. Frontier of India, 1908.—Operations in the Mohmand country. Medal with clasp.

McKendrick, A. G.—
East Africa, 1903-04.—Operations in Somaliland.—Action at Jidballi.—Medal with clasp.

McKenna, J. C.—
The War of 1914-18.—Despatches, *Lond. Gaz.*, 5th Apl. 1 Jun. and 13 July 16. D.S.O.

McLaughlin, W. P. M. D.—
East Africa, 1902—04.—Operations in Somaliland.—Medal with clasp.
The War of 1914-18.—Despatches, *Lond. Gaz.*, 17 Feb. 15.

McLeod, Sir D. J. S. (*General retd. I. A.*)—
Burma, 1886-87.—Despatches, G. G. O. 434 of 87.—Medal with clasp. D S.O.

McLeod, D. K.—
N.-W. Frontier of India, 1908.—Operations in the Zakha Khel country. Brigade Signalling officer, 2nd Brigade.—Medal with clasp.
The War of 1914-18.—Despatches *Lond. Gaz.*, 1 Jan. 16 15 Jun. 16 and 4 Jan. 17. D. S. O.

McLeod, R. D. C.—
South African War, 1899—1902.—Relief of Ladysmith, including action at Colenso (15th Dec. 1899). Operations of 5th to 7th February 1900 and action at Vaal Kranz; operations on Tugela Heights (14th to 27th February 1900) and action at Pieter's Hill (27th Feb. 1900). Operations in Natal, March to June 1900, including action at Laing's Nek (6th to 9th June). Operations in the Transvaal, East of Pretoria, July to November 1900.—Queen's medal with 6 clasps. King's medal with 2 clasps.
Tibet, 1903-1904.—Medal.
N.-W. Frontier of India, 1908.—Operations in the Zakka Khel country. Staff Officer, Lines of Communications. Operations in the Mohmand country. Engagement of Matta, 24th April 1908.—Medal with clasp.

McLeod, T. J.—
N.-W. Frontier of India (2nd) Miranzai, 1891.—Medal with clasp.—*Waziristan, 1894-95.*—Clasp.
N.-W. Frontier of India, 1897-98.—Mohmand.—Medal with clasp.
Tirah, 1897-98.—Actions of the Sampagha and Arhanga Passes.—Operations in the Bazar Valley, 25 to 30 Dec. 97.—Clasp.
The War of 1914-18.—Despatches, *Lond. Gaz.*, 5 Apr. 16. D. S. O.

McLean F. G. S.—
The War of 1914-18.—Despatches, *Lond. Gaz.*, 5 Apl. 10.

McMahon, Sir A. H. (*Lt.-Col., retd. I. A.*)—
N.-W. Frontier of India, Zhob Valley, 1890.
The War of 1914-18.—Despatches, *Lond. Gaz.*, 25 Sep. 16.

McMullin, J. (*Col. u. s. l. Ind. S. C.*)—
China, 1857-58.—Operations before and capture of Canton.—Medal with clasp.

McMurray, B. S. J.—
South African War, 1902.—Operations in the Transvaal, May 02. Queen's medal with 4 clasps.

McNair, A. L. (*Col. u.s.l.*)—
Abyssinia, 1867-68.—Medal.
Afghanistan, 1878.—Medal.

McNair, E. J. (*Lt.-Gen. u. s. l. Bengal Inf.*)—
India, 1857.—Action of Budlee-ki-Serai, siege and capture of Delhi and subsequent operations in the Delhi district.—Medal with clasp.
N.-W. Frontier of India, 1868, Hazara.—The occupation of Black Mountain.—Medal with clasp.
Afghanistan, 1879-80.—Mentioned in Despatches.—Medal.

McNelle, D. H.—
N.-W Frontier of India, 1908.—Medal with clasp.
The War of 1914-18.—Despatches, *Lond. Gaz.* 14 Jan. 17.

McNeill, D. (*Col. u. s. l. Ind. S.C.*)—
India, 1857—59.—Pursuit of Tantia Topee, expedition to Puchmari Hills and capture of Hurraketi —Medal.

McPherson, A. B.—
The War of 1914-18.—Despatches, *Lond. Gaz.*, 1 Jan. and 19 Oct. 16. M. C.

McPherson, G.—
China, 1900.—Medal.

McPherson, J.—
Aden, 1903-1904.—Operations in the interior. Despatches, G. G. O. No. 43 of 1905.

McRae, H. St. G. M.—
N.-W. Frontier of India, 1902.—Operations against the Darwesh-Khel Waziris.
The War of 1914-18.—Despatches, *Lond. Gaz.*, 21 Jun. 16, 4 Jan. 17 and 15 May 17. D. S. O.

McSwiney, H. F. C.—
The War of 1914-18,—Despatches, *Lond. Gaz.*, 17 Feb. 15. M. C.

McVean, D. A. D.—
N.-W. Frontier of India, 1897-98.—Malakand.—Utman Khel.—Medal with clasp.
N.-W. Frontier of India, Waziristan, 1901-02.—(Severely wounded.) Despatches, G. G. O. No. 61 of 1902.—Clasp. D. S. O.
N.-W. Frontier of India, 1908.—Operations in the Zakka Khel country. Operations in the Mohmand country.—Medal with clasp.

Macbay, W. G. W. (*Lt.-Col. u.s.l.*)—
Egypt. 1882.—Battle of Tel-el-Kebir, and subsequent advance on Zagazig and occupation of Cairo, (Despatches. *Lond. Gaz.*, 2 Nov. '82.) Brev. of Maj.—Medal with clasp : bronze star.

McWalters, W.—
Egypt, 1882.—Medal & Bronze Star.
N.-E. Frontier of India, Sikkim, 1888.—Medal with clasp.
N.-W. Frontier of India, 1897-98.—Tochi.—Medal with clasp.
China, 1900.—Medal.
N.-W. Frontier of India, Waziristan, 1901-02.—Clasp.
Tibet, 1903-04.—Mentioned in Despatches, *Gazette of India*, No. 1065 of 1904, promoted Hony. Lieutenant.—Medal.

MacWatters, G. A.—
N.-W. Frontier of India, 1897-98.—Operations on the Samana and in the Kurram Valley during Aug. and Sep. 97.—Medal with 2 clasps.
Tirah, 1897-98.—Actions of Chagru Kotal and Dargai.—Clasp.

MacWatters, M. R. C.—
N.-W. Frontier of India, 1908.—Operations in the Mohmand country.

MacBrayne, J. O—
The War of 1914-18.—Despatches, *Lond. Gaz.*, 1 Feb. 17. M. C.

McWatters, H. C.—
N.-W. Frontier of India, 1902.—Operations against the Darwesh Khel Waziris.
N.-W. Frontier of India, 1908.—Operations in the Mohmand country. Engagement of Kargha, 24th May 1908.—Medal with clasp.
The War of 1914-18.—Despatches, *Lond. Gaz*, 19 Oct. 16 and 15 Aug. 17. Brev. of Lt.-Col. D. S. O.

Mac Watters, M. R. C—
N.-W. Frontier of India, 1908.—Operations in the Mohmand country. Medal with clasp.

MacAndrew, H. J. M.—
Tirah, 1897-98.—Actions of the Sampagha and Arhanga Passes. Operations against the Khani Khel Chamkannis. Operations in the Bazar Valley, 25 to 30 Dec. 97. Despatches, G. O. 483 of 98.—Medal with 2 clasps.
South African War, 1900—02.—Operations in Orange Free State, Feb. to May 00, including operations at Paardeberg (17 to 26 Feb.); actions at Poplar Grove (7 March) Driefointein (10 March) and Karee Siding (29 March). Action at Houtuck (1 May). Operations in the Transvaal in May and June 00 including actions near Johannesburg (29 May), Pretoria (4 June). Despatches, *Lond. Gaz.*, 2 April 01. Operations in east of Pretoria, July to Sep. 00. Operations in Transvaal, west of Pretoria, Sep. to 29 Nov. 00. Operations in the Transvaal between 30 Nov. 00 and 31 May 02. Operations in Orange River Colony between 30 Nov. 00 and 31 May 02. Operations in Cape Colony between 30 Nov. and 31 May 02. Despatches, *Lond. Gaz.*, 8 March 01.—Queen's Medal with 4 clasps. King's Medal with 2 clasps.—D. S. O.—Placed on list of officers considered qualified for Staff Employment in consequence of service on the Staff in the Field.
The War of 1914-18.—Despatches, *Lond. Gaz.*, 22 Jun. 15, 4 Jan. 17 and 11 Dec. 17. Brev. of Col. C. B.

Macartney, F. A.—
N.-W. Frontier of India, 1908.—Operations in the Zakka Khel country. Operations in the Mohmand country. Engagement of Matta, 24th April 1908.—Medal with clasp.
The War of 1914-18.—M. C.

Macartney, H. F. T. (*Lieut.-Col. retd. I. A.*)—
N.-W. Frontier of India, Hazara, 1891.—Medal with clasp.
China, 1900.—Medal.

Macaulay, C. E. (*Hon. Col. retd. Ben. S. C.*)—
India, 1858-59.—Capture of Fort Amethi ; action of Shunkerpore ; action of Daundhiakhera ; capture of Fort Umeria (Oudh) ; and Trans-Gogra operations.—Medal with 2 clasps.
China, 1860.—Action of Singho (severely wounded); capture of Taku Forts ; actions of Chankiawan and Palichao and occupation of Pekin. Despatches, G. G. O. No. 167 of 1861.—Medal with 2 clasps.
N.-W. Frontier of India, 1863.—Umbeyla campaign.—Medal with clasp.
Afghanistan, 1879-80.—Action of Jhandola. Mentioned in Despatches.
N.-W. Frontier of India, Mahsud-Waziri Expedition, 1881.—Political Officer with the Expedition. Mentioned in Despatches.

Macaulay, D. I. M.—
China, 1900.—Action of Peitsang. Action at Yangtsun. Relief of Pekin.—Medal with clasp.

Macausland, R. A.—
The War of 1914-18.—Despatches, *Lond. Gaz.*, 15 Aug. 17.

Macausland, R. C. S. (*Lt.-Col. retd. I. A.*)—
Afghanistan, 1878-79.—Action of Ali Musjid. Despatches. Lt.-General *Sir* S. J. Browne's despatch, 8 July '79. (1) Bazar Valley Expedition. —Medal with clasp.
Burma, 1888.—Medal with clasp.

Macdonald, C. E. W. (*Lieut.-Col. retd. I. A.*)—
Afghanistan, 1878-79-80.—Medal.
Egypt, 1882.—Actions of Kassassin and Tel-el-Kebir. —Medal with clasp. Bronze star.
South Africa.—Bechuanaland Expedition, 1884-85.
N.-W. Frontier of India, 1897-98.—Operations on the Samana and in the Kurram Valley during Aug. and Sep. 97.—Medal with 2 clasps.
Tirah, 1897-98.—Clasp.
South African War, 1900-02.—Commanded the 23rd Battalion Imperial Yeomanry, 4 April 01 to 15 Feb. 02.—Operations in the Transvaal between 30 Nov. 00 and 31 May 02.—Operations in the Orange River Colony between 30 Nov. 00 and 31 May 02. —Operations in Cape Colony between 30 Nov. 00 and 31 May 02.—Queen's Medal with 5 clasps.

Macdonald, C. H. (*Col. retd. I. A.*)—
Burma, 1885—87.—Medal with clasp.

Macdonald, D. T.—
The War of 1914-18.—Despatches, *Lond. Gaz.*, 15 Aug. 17. M. C.

Macdonald, H. (*Maj. retd. I. A.*)—
Tirah, 1897-98.—Medal with 2 clasps.

Macdonald, H.—
The War of 1914-18.—Despatches, *Lond. Gaz.*, 19 Oct. 16 and 15 Aug. 17. Brev. of Major.

Macdonald, J. V.—
The War of 1914-18.—Despatches, *Lond. Gaz.*, 8 and 13 Feb. 17. M. C.

Macdonald, R. W.—
The War of 1914-18.—Despatches, *Lond. Gaz.*, 19 Oct. 16. D. S. O.

Macdonald, T R. (*Maj. retd. Ind. Med. Serv.*)—
Egypt, 1882.—Battle of Tel-el-Kebir.—Medal with clasp; bronze star.

MacDonnell, H. N. F.—
China, 1900.—Relief of Pekin.—Medal with clasp.
The War of 1914-18.—Brev. of Lt.-Col.

Macfarlav, J. B.—
The War of 1914-16.—Despatches, *Lond. Gaz.*, 20 Mar. 16.

MacGilchrist, A. C.—
China, 1900.—Medal.

Macgregor, J. (*Surg.-Lt.-Col. retd. Ind. Med. Serv.*) —*Burma, 1885-87.*—Despatches, G. G. O. 434 of 87. Medal with 2 clasps.

Mcgregor, R. F. D.—
The War of 1914-18.—Despatches, *Lond. Gaz.*, 19 Oct. 16. M. C.

MacIntyre, D. C. F.—
Afghanistan, 1880.—Medal.
N.-E. Frontier of India, Manipur, 1891.—Despatches, G. G. O. 585 of '91.—Medal with clasp.
N.-W. Frontier of India, 1897-98.—Operations on the Samana during Aug. and Sep. 97. Relief of Gulistan.—Medal with 2 clasps.
Tirah, 1897-98.—Actions of Chagru Kotal, Darga and of the Sampagha and Arhanga Passes. Operations in the Waran Valley and action of 16 Nov. 97.—Operations at and around Dwato and action of 24 Nov. 97. Operations against the Khani Khel Chamkannis. Operations in the Bara Valley, 7 to 14 Dec. 97. Despatches, G. G. Os. 58 and 244 of '98.—Clasp. Brev. of Maj.
Operations in the Abor country, 1911-12.—Despatches, *Gazette of India*, Notification 480, 23 May 12.—Medal.—C. B.

MacKelvie, M.—
N.-W. Frontier of India, Waziristan, 1901-02.—Medal with clasp.
N.-W. Frontier of India, 1902.—Operations against the Darwesh Khel Waziris.
East Africa, 1903-04.—Operations in Somaliland. —Medal with clasp.

Mackenzie, Sir A. R. D. (*Col. u. s. l. Ben. Cav.*)—
India, 1857-58.—Siege and capture of Delhi, siege and capture of Lucknow (slightly wounded). —Medal with 2 clasps.

Afghanistan, 1879-80.—Expedition to the villages of Banda, Kuddi, and Roghani, march from Kabul to the relief of Kandahar, reconnaissance of the 31st Aug. and battle of the 1 Sep.; expedition against the Aitchakzai tribes. Despatches, G. G. O. 566 of 80.—Medal with clasp; bronze star.

Mackenzie, C.—
N.-W. Frontier of India, 1897-98.—Mohmand.—Medal with clasp.
The War of 1914-18.—Despatches, *Lond. Gaz.*, 4 July 16, and 15 Aug. 17. D. S. O.

Mackenzie, E. A. H.—
N.-W. Frontier of India, 1908.—Operations in the Mohmand country.—Medal with clasp.

Mackenzie, H. J.—
South African War, 1900-1901.—Operations in Orange Free State, February to May 1900. Operations in Orange River Colony (May to 29th November 1900) including actions at Biddulphsburg (29th May), Lindley (1st June and 26th June), Bethlehem (6th and 7th July) and Wittebergen (1st to 29th July). Operations in Transvaal, between 30th November 1900 and March 1901. Operations in Orange River Colony, March and April 1901. Operations in Cape Colony, between April and June 1901 —Queen's Medal, with 4 clasps.
N.-W. Frontier of India, 1908.—Operations in the Zakka Khel country.—Medal with clasp.
The War of 1914-18.—Despatches, *Lond. Gaz.*, 19 Oct. 16. D. S. O.

Mackenzie, J.—
N.-W. Frontier of India, 1897-98.—Malakand.—Utman Khel.—Medal with clasp.

Mackenzie, K. W. (*Capt., retd., I. M. S.*)—
N.-W. Frontier of India, 1908.—Operations in the Mohmand Country. Medal with clasp.

MacKenzie, L. H. L.—
The War of 1914-18.—Despatches, *Lond. Gaz.*, 15 Aug. 17. Brev. of Maj.

Mackenzie, T. H. (*Lt.-Col. retd. I.S.C.*)—
Afghanistan, 1880.—Medal.

Mackenzie-Kennedy, E. C. W.—
Burma, 1885-86-87.—Present during advance on, and occupation of, Mandalay. Operations around Hlinedet, Yemethin, and Nyingyan.—Operations against Boo Shwe and Ootama.—Medal with clasp.
N.-W. Frontier of India, Hazara, 1891.—Present at the night attack on Ghazikot.—Clasp.
China, 1900.—Relief of Pekin. Despatches, *Lond. Gaz.*, 14 May 01—Medal with clasp. Brev. of Lt.-Col.

MacKinnon, A. C. B.—
The War of 1914-18.—Despatches, *Lond. Gaz.*, 15 Jun. and 19 Oct. 16.

Mackrell, A.—
South African War, 1899-1901.—Empld. with Impl. Yeo. Operations in Orange River Colony, May to Nov. 00, including actions at Bethlehem (6 and 7 July), and Wittebergen (1 to 29 July).

Mackworth, N. W.—
The War of 1914-18.—Despatches, *Lond. Gaz.*, 19 Oct. 16.

MacLachlan, T. R.—
Tibet, 1903-04.—Operations at and around Gyantse, between 5th May and 6th July 1904. March to Lhassa, 14th July to 3rd Aug. 1904. Despatches, *Gazette of India*, No. 1065 of 1904.—Medal with clasp.
N.-W. Frontier of India, 1908.—Operations in the Mohmand country.—Medal with clasp.
The War of 1914-18.—Despatches, *Lond. Gaz.*, 21 Jun. 16 and 15 Aug. 17. Brev. of Col. C.M.G.

MacLaren, J. F.—
Soudan, 1885.—Suakin. Action of Hasheen and Tamai.—Medal with clasp. Bronze star.

Maclean, C. S. (*M.-Gen. u.s.l. Ind. S. C.*)—
India, 1857.—Siege and capture of Delhi, actions of Bolundshuhur, Alighur, Akrabad, and Agra (twice severely wounded—horse severely wounded).—Medal with clasp.
China, 1860.—Actions of Sinho, Chankiawan and Tangchow, and occupation of Pekin.—Medal with clasps.
Afghanistan, 1878-80.—Actions of Saif-u-deen, Ahmed-Khel, Urzoo, and Patkao Shana. Despatches, G. G. O. 610 of 79 and 327 and 493 of 80.—Medal with clasp; Brev. of Lt.-Col.; C.B.
N.-W. Frontier of India, 1881—Mahsud-Waziri Expedition.—Mentioned in Despatches.

Maclean, F. A.—
South African War, 1900—02.—Operations in the Orange Free State, April to May 1900, including action at Houtnek (1st May). Operations in Orange River Colony, May to Nov. 1900 including actions at Biddulphsberg (29th May) and Wittebergen (1st to 29th July). Operations in the Transvaal, between Dec. 1900 and April 1901, also between Oct. 01 and 31st May 02. Operations in the Orange River Colony, Sept. and Oct. 01. Operations on the Zululand Frontier of Natal in Sept. and Oct. 01. Served as Adjutant of 14th M. I. Battn. from 1 Nov. 1900 to July 1901. Queen's medal with 3 clasps. King's medal with 2 clasps.

Maclean, G. G. C.—
South African War, 1900-02.—Operations in the Transvaal between 5th July 1901 and 10th Feb. 1902, and 20th February and 20th May 1902. Operations in Orange River Colony between 11th and 19th February, 22nd and 24th May 1902. Operations in Cape Colony between 25th and 31st May 1902.—Queen's medal with 5 clasps.
The War of 1914-18.—Despatches, *Lond. Gaz.*, 19 Oct. 16. Brev. of Maj.

Maclean, L. L.—
The War of 1914-18.—Despatches, *Lond. Gaz.*, 16 Jan. 18.

Macleod, N.—
Nandi, 1905-06.—Medal with clasp.
The War of 1914-18.—Despatches, *Lond. Gaz.*, 15 Aug 17. D.S.O.

MacLeod, J. N. (Lt.-Col., I.M.S., retd.)—
Chitral, 1895.—Relief of Chitral.—Medal with clasp.
N.-W. Frontier of India, 1897-98.—Tochi.—Clasp.

MacLeod, R. W. (*Col. retd. I. A.*)—
N.-W. Frontier of India, Jowaki Expedition, 1877-78.—Medal with clasp.
Afghanistan, 1878-79-80.—Action of the Peiwar Kotal. Zaimukht Expedition. Action of Zawa.—Medal with clasp.
N.-W. Frontier of India, Hazara, 1888.—Clasp. *(1st and 2nd) Miranzai, 1891.*—Clasp.
Chitral, 1895.—Relief of Chitral.—Medal with clasp.
N.-W. Frontier of India, Waziristan, 1901-02.—Despatches, *Lond. Gaz.*, 8 Aug. 02. Clasp.

MacMahon, H. F. E.—
N.-W. Frontier of India, Waziristan, 1901-02—Medal with clasp.
The War of 1914-18.—Despatches, *Lond. Gaz.*, 22 Jun. 15, 19 Oct. 16, and 15 Aug. 17. M.C.

Macmullen, C. N.—
N.-W. Frontier of India, 1897-98—Buner.—Action of the Tanga Pass.—Medal with clasp.
Tibet, 1903-04.—Medal.
The War of 1914-18.—Despatches, *Lond. Gaz.*, 5 Nov. 15, 5 May 16, 21 Jun. 16, 13 July 16, 25 Sept 16, 4 Jan. 17, 15 May 17 and 11 Dec. 17. Brevs. of Lt. Col. and Col. D.S.O.

Macmullen, W. H. F. (*Lt.-Col. u. s. l. I.S.C.*)—
Afghanistan, 1879-80.—Medal.

Macnab, A. J.—
N.-W. Frontier of India, Hazara, 1891.—Present at the night attack on Ghazikot. Medal with clasp.
Chitral, 1895.—Relief of Chitral.—Medal with clasp.
N.-W. Frontier of India, 1897-98.—Defence and relief of Malakand—Relief of Chakdara—Malakand, action at Landakai—Operations in the Mamund Country—Utman Khel—Buner.—2 clasps.
East Africa, 1903-04.—Operations in Somaliland.—Medal with clasp.
The War of 1914-18—Despatches, *Lond. Gaz.*, 1 Jan. 16, and 11 Dec. 17. Brev. of Col.

MacNabb, R. J.—
The War of 1914-18.—Despatches, *Lond. Gaz.*, 1 Jan. 17.

Macnaghten, A. W. (*Col. u.s.l. Bo. Cav.*)—
India, 1858-59.—Served in Rajpootana and Central India in pursuit of Tantia Topee.
Abyssinia, 1868.—Storm and capture of Magdala. Despatches, *Lond. Gaz.*, 16 June 68.—Medal.
Afghanistan, 1880.—In command of 2nd Bombay Lt. Cav.—Medal.

Macnamara, R. J.—
Burma, 1887-89.—Medal with clasp.

Maconchy, E. W. S. K. (*Col., retd., I.A.*)—
N.-W. Frontier of India, Hazara, 1888.—Served as Transport Officer, No. 1 Column.—Medal with clasp.

N.-W. Frontier of India, Hazara, 1891.—Present at the affair of Ghazikot. (Slightly wounded.) Mentioned in Despatches, G. G. O. 816 of 91.—Clasp. D.S O.

N.-W. Frontier of India, 1892.—Isazai.

Chitral, 1895.—Relief of Chitral—Field Intelligence Officer.—Medal with clasp.

N.-W. Frontier of India, 1897-98.—Operations on the Samana and in the Kurram Valley during Aug. and Sep. 97.—Operations of the Flying Column in the Kurram Valley under Colonel Richardson, 20 Aug. to 1 Oct. 97.—2 clasps.

Tirah, 1897-98.—D. A. Q. M. G. for Intelligence Army Staff.—Actions of the Sampagha and Arhanga Passes.—Reconnaissance of the Saran Sar and action of 9 Nov. 97.—Operations in the Waran Valley and action of 16 Nov. 97.—Operations at and around Dwatoi and action of 24 Nov. 97.—Operations against the Khani Khel Chamkannis—Operations in the Bazar Valley, 25 to 30 Dec. 97. Despatches, G. G. O. 244 of 98.—Clasp. Brev. of Maj.

N.-W. Frontier of India, Waziristan, 1901-02.—Clasp.

Macpherson, A. D.—

N.-W. Frontier of India, 1897-98.—With Tirah Expd. Force. Asst. to Div. Commt. Offr., 2nd Divn. (12 Oct. 97 to Apr. 98). Medal with 2 clasps.

Macpherson, D. A. A. (*Lt.-Col. retd., B. S. C.*)—

Burma, 1885-86.—Field Paymaster—Despatches, *Lond. Gaz.*, 22 June 86.—Medal with clasp.

Macpherson, J.—

China, 1900.—Despatches, *Lond. Gaz.*, 14 May 01.—Medal.—D. S. O.

Macpherson, J. D. (*Col. u.s.l.* I. S. C.)—

India, 1858-59.—Capture of Tiroul.—Medal.

N.-W. Frontier of India, 1868. Hazara.—Mentioned in Despatches.—Medal.

Afghanistan, 1879-80.—March from Kabul to the relief of Kandahar and battle of 1 Sep. Mari expedition. Despatches, G. G. O. 566 of 80.—Medal with clasp; bronze star.

Macpherson, R. G.—

N.-W. Frontier of India, Hazara, 1891.—Medal with clasp. (*2nd*) *Miranzai, 1891.*—Clasp.

N.-W. Frontier of India, 1897-98.—Malakand. Action at Landakai.—Medal with clasp.

Macqueen, K. G. S.—

China, 1900.—Medal.

The War of 1914-18.—Despatches, *Lond. Gaz.*, 22 Jun. 15.

Mcquoid, C E E. F. K.—

N.-W. Frontier of India, 1897-98.—Mohmand Field Intelligence Officer. Despatches, G. G. O. 1318 of '97. Operations on the Samana and in the Kurram Valley during Aug. and Sep. 97.—Medal with 2 clasps.

Tirah, 1897-98.—Field Intelligence Officer, 1st Division.—Actions of the Sampagha and Arhanga Passes.—Reconnaissance of the Saran Sar and action of the 9 Nov. 97.—Operations against the Khani Khel Chamkannis.—Operations in the Bazar Valley, 25 to 30 Dec. 97. Despatches, G. G. O. 244 of 98.—Clasp. D. S. O.

South African War, 1899—01.—Queen's medal with clasp.

The War of 1914-18.—Despatches, *Lond. Gaz.*, 19 Oct. 16.

Macrae, J. C.—

Aden, 1903-04.—Operations in the Interior.

The War of 1914-18.—Despatches, *Lond. Gaz.*, 3 Apr. 16. 10 May 16, 19 Oct. 10, and 15 Aug. 17. D. S. O.

Macrae, J. L.—

N.-W. Frontier of India, 1897-98.—Buner.—Medal with clasp.

China, 1900.—Medal.

Mactavish, A.—

Tirah, 1897-98.—Actions of the Sampagha and Arhanga Passes.—Operations at and around Dwatoi.—Medal with 2 clasps.

South African War, 1900—02.—Queen's Medal with clasp.

MacWatt, R. C.—

N.-W. Frontier of India, Hazara, 1888.—Medal with clasp.

N.-E. Frontier of India, Lushai, 1889.—Clasp.

N.-W. Frontier of India, Hazara, 1891.—Clasp.

N.-W. Frontier of India, Miranzai (2nd) Expedition, 1891.—Clasp.

Maddox, R. H.—

N.-W. Frontier of India.—Waziristan, 1894-95.—Medal with clasp.

Soudan, 1896.—Dongola Expedition.—Medal.

Tirah, 1897-98.—Action of the Sampagha Pass.—Medal with clasp.

The War of 1914-18.—Despatches, *Lond. Gaz.*, 15 Aug. 17.

Magniac, F. A.—

N.-W. Frontier of India, 1897-98.—Tochi.—Medal with clasp.

N.-W. Frontier of India, 1908.—Operations in the Mohmand country. Engagement of Kargha, 24th May 1908.—Medal with clasp.

Magrath, H. A. F.—

N.-W. Frontier of India.—Waziristan, 1894-95.—Medal with clasp.

N.-W. Frontier of India, 1897-98.—Tochi.—Medal with clasp.

Mahon, A. E.—

South African War, 1900-1901.—Operations in the Transvaal, east of Pretoria, July to 29th November 1900. Operations in the Orange River Colony, July to November 1901. Queen's Medal with 4 clasps.

The War of 1914-18.—Despatches, *Lond. Gaz.*, 1 Jan. 16.

Mainwaring, C. V.—

Burma, 1887-88.—Medal with clasp.

Mainwaring, E. P. (*Col. u.s.l., Ben. Inf.*)—

N.-W. Frontier of India, 1868, Hazara Campaign.—Medal with clasp.

N.-E. Frontier of India, 1871-72, Lushai Expedition.—Clasp.

Afghanistan, 1878—80.—Capture of Ali Musjid, in Jugdulluck Pass and in the Maidan and Wardak country. Affair of Saidabad, operations at and around Kabul in Dec. 79, march from Kabul to the relief of Kandahar and battle of 1 Sep.—Mari Expedition. Despatches, G. G. Os. 327 and 582 of 80.—Medal with 3 clasps; bronze star.

Burma, 1891.—Clasp.

Mainwaring, F. G. L. (*Lt.-Col., retd., Ind. S. C.*)—
Afghanistan, 1878—80.—Medal. Action on the Gara Heights near Kam-Dakka. Operations in the Lughman Valley.—Medal.
Chitral, 1895.—Relief of Chitral. With the Relief Force.—Medal with clasp.

Mainwaring-White, E. P.—
South African War, 1900.—Served in Mediterranean. Medal.

Maisey, F. C. (*Lieut.-Col. u.s.l. I. A.*)—
Afghanistan. 1878-79-80.—Action of the Peiwar Kotal. Operations in the Kama District.—Despatches. Brigadier-General J. Doran's despatch, 18 June '80.—Medal with clasp.
N.-W. Frontier of India, Mahsud-Waziri, 1881. Miranzai, (1st and 2nd) 1891.—Despatches, G. G. O. 632 of '91.—Medal with clasp.
Chitral, 1895.—Relief of Chitral.— Medal with clasp.
N.-W. Frontier of India, 1897-98.—Operations on the Samana and in the Kurram Valley during Aug. and Sep. 97.—2 clasps.—Tirah, 1897-98. Operations in the Bazar Valley, 25 to 30 Dec. 97. Despatches, G. G. O. 244 of 98.—Clasp.

Maitland, G. R.—
South African War, 1901-02.—Operations in the Transvaal, Orange River Colony and Cape Colony from December 1901 to May 1902. Queen's medal with 5 clasps.
The War of 1914-18.—Despatches, *Lond. Gaz.*, 1 Jan. 16.

Maitland, P. J. (*Maj-Genl. retd. I. A.*)—
Afghanistan, 1878-79.—Action of Khush-ki-Nakhud. Despatches, G. G. O. 610 of 79.—Medal.

Major, F. F.—
Aden, 1903-04.—Operations in the Interior.
The War of 1914-18.—Despatches, *Lond. Gaz.*, 4 July 16.

Malcolm, P. (*Maj. retd. I. A.*)—
N.-E. Frontier of India, Chin-Lushai, 1889-90.—Burma column.—Medal with clasp.
Chitral, 1895.—Relief of Chitral.—Action of the Malakand Pass.—Medal with clasp.
N.-W. Frontier of India, 1897-98.—Tochi—Provost Marshal to the Force and subsequently D. A. A. G., 1st Brigade. Despatches, G. G. O. 1432 of 97.—Clasp.
N.-W. Frontier of India, Waziristan, 1901-02.—Officiating A. A. G., Derajat District, and Chief Staff Officer to Brigadier-General L. Dening, C.B., D.S.O., Commanding District and Column. Despatches, G. G. O. 61? of 02.—Clasp. D.S.O.
The War of 1914-18.—Despatches, *Lond. Gaz.*, 4 Jan. 17.

Malcolm, W. L.—
Chitral, 1895.—Relief of Chitral.—Medal with clasp.
N.-W. Frontier of India, 1897-98.—Malakand.—Action at Landakai. Operations in the Mamund country.—Clasp.

Malcolmson, J. H. P. (*Hon. M.-Gen. retd. Bo. S.C.*)—
India, 1857—59.—Siege and capture of Kotah, battle of Bunass, action at Kooshana (horse wounded). Despatches, *Lond. Gaz.*, 5 May 59.—Medal with clasp.
Afghanistan, 1878-79-80.—Action near Atta Karez, actions of Kushki Nakhud (wounded) and Girishk, battle of Maiwand, defence of, and sortie from, Kandahar (wounded). Despatches, G. G. Os. 610 of 79, and 480 and 451 of 80.

Malhotra, R. C.—
The War of 1914-18.—Despatches, *Lond. Gaz.*, 15 Aug. 17.

Malik Mumtaz Muhammad Khan—
The War of 1914-18.—Despatches, *Lond. Gaz.*, 15 Aug. 17.

Malik Umar Hyat Khan, (*Hon. Sir*)—
East Africa, 1903 04.—Operations in Somaliland. A. D. C. to G. O. C. Field Force. Action at Jidballi. Mentioned in despatches. Medal with clasp.
Tibet, 1903-04.—Medal.
The War of 1914-18.—Despatches, *Lond. Gaz.*, 17 Feb 15, 1 Jan. 16, 19 Oct. 16, 14 Nov. 16, and 15 Aug. 17. Granted Hon. rank of Maj.

Malleson, W.—
The War of 1914-18.—Despatches, *Lond. Gaz.*, 30th Jun. 16. C. B.

Mallins, C. (*Lt.-Col. retd. Ind. Med. Serv.*)—
N.-W. Frontier of India, 1897-98.—Tochi.—Medal with clasp.

Mallock A. R. O.—
The War of 1914-18.—Despatches, *Lond. Gaz.*, 19 Oct. 16.

Maltby, C. M.—
The War of 1914-18.—Despatches, *Lond. Gaz.*, 4 July 16. M. C.

Maltby, F. C. (*Lt.-Col., retd., I.A.*)—
Burma, 1887-88.—Medal with clasp.
Chin-Lushai Expedition, 1889-90.—Clasp.

Malyon, F. H.—
N.-W. Frontier of India, 1908.—Operations in the Mohmand country.—Medal with Clasp.
The War of 1914-18.—Despatches, *Lond. Gaz.*, 15 Jan. 16.

Mallock, H. A. (*Col. u. s. l. Ind. S.C.*)—
India, 1857-58.—Central India; affair near Mundesore and siege of Ratghur Fort.—Medal with clasp.
N.-E. Frontier of India, 1865-66.—*Bhutan Expedition.* In charge of field telegraph.—Medal with clasp.

Mander, F. D. (*M.-Gen. retd. Ind. Army*)—
Afghanistan, 1879-80.—As General Transport Officer, Line of Communications in Southern Afghanistan. —Medal.

Mangin, E. B.—
The War of 1914-18.—Despatches, *Lond. Gaz.*, 22 June 15. M. C.

Manifold, C. C.—
Tirah, 1897-98.—Operations in the Waran Valley. Operations in the Bazar Valley.—Medal with clasp.
China, 1900.—Relief of Pekin. Action of Peitsang. Action at Yangtsun. Despatches, *London Gazette*, 14th May 1901.—Medal with clasp. Promoted Lieut.-Colonel.
The War of 1914-18.—Despatches, *Lond. Gaz.*, 4 Jan. and 1 Jun. 17. C. M. G.

Manning, Sir W. H. (*Col., retd. I. A.*)—
Burma, 1887—89.—Intelligence Officer, Mogoung Field Force (wounded).—Medal with clasp.
N.-W. Frontier of India, (1st) Miranzai, 1891. Hazara, 1891.—Clasp.
Central Africa, 1893.—Commanded Mlanje (Nyassera and Mranda) and Chirad-Zulu Expeditions. —Medal.
British Central Africa and Rhodesia, 1895--98.—Expeditions against Kawinga, Odeti, and Mkoma, Chilwa, and in command of operations against the Chief Mpezeni. Mentioned in despatches. Brevet of Major and Lieut.-Col.—Clasp.
East Africa, 1902-04.—Operations in Somaliland. Action at Jidballi. On Staff (in command of field force, afterwards in comnd. of a Brigade). Despatches, *London Gazette*, 2nd September 1904. Medal with 2 clasps. K.C.M.G., C.B.

Mansel, C. G. (*Lt.-Col. retd. Ind. S. C.*)—
N.-W. Frontier of India, 1881, Mahsud Waziri Expedition.
Waziristan Expedition, 1894-95.—Medal with clasp.
N.-W. Frontier of India, 1897-98.—Operations on the Samana and in the Kurram Valley during Aug. and Sep. 97. Relief of Gulistan. Operations of the Flying Column in the Kurram Valley under Col Richardson, 20 Aug. to 1 Oct. 97.—Medal with 2 clasps.

Mansel, W. G.—(*Col. retd. I. A.*)—
Afghanistan, 1879-80.—Hissarak expedition, attack on Fort Battye.—Medal.
N.-E. Frontier of India, 1889—92.—Lushai.—Relief of Aijal and Changsil.
Burma, 1892-93.

Mansfield, C. B. H.—
South African War, 1899—1902.—Operations in the Orange Free State, Feb. to May 00, including actions at Houtnek (1 May), Vet River (5 and 6 May).—Operations in the Transvaal in May and June 00, including actions near Johannesburg (29 May), Pretoria (4 June).—Queen's Medal with 4 clasps.

Mansfield, H. (*Maj.-Gen. retd. I. A.*)—
Afghanistan, 1878-79-80.—Action of Ali Musjid. Despatches. Lieut.-General *Sir* S. J. Browne's despatch, 8 July 79. Zaimukht Expedition. Despatches, Br.-General J. A. Tytler's despatch, 20 Dec. 79.—Medal with clasp.
N.-E. Frontier of India, Sikkim, 1888.—Attacks of Jeluk, Gnathong, and Jelapla. Despatches, G. G. O. 889 of 88.—Medal with clasp; Brev. of Maj.
Tirah, 1897-98.—Chief Transport Officer, Actions of the Sampagha and Arhanga Passes. Despatches, G. G. O. 244 of 98.—Medal with 3 clasps. Brev. of Lt.-Col.

Manson, J. A. P.—
Tibet, 1903-04.—Action at Niani, 26th June 1904. Operations at and around Gyantse, between 5th May and 6th July 1904. Medal with clasp.

Marett, J. R. (*Col. u.s.l., Ind. S. C.*)—
N.-E. Frontier of India, 1862-63. Cossiah and Jynteah Hills Expedition.—Capture of several stockades. Mentioned in Despatches. *Bhutan Expedition, 1865.*—Assault and capture of Chamoochee stockades.—Medal with clasp.

Marjoribanks, A.—
East Africa, 1903-04.—Operations in Somaliland. Action at Jidballi.—Medal with 2 clasps.
N.-W. Frontier of India, 1908.—Operations in the Mohmand country. Engagement of Kargha, 24th May 1908.—Medal with clasp.

Marjoribanks, R. D.—
East Africa, 1902-04.—Operations in Somaliland. Action at Jidballi. Despatches, *London Gazette*, 2nd September 1904. Medal with 2 clasps.

Marks, R. J.—
N.-E. Frontier of India, 1891.—Lushai.
N.-W. Frontier of India, 1897-98.—Tochi.—Medal with clasp.

Marlow, B. W.—
Burma, 1886.—Medal with clasp.
Soudan, 1896.—Dongola Expedition.—Medal.

Marr, C. F.—
N.-W. Frontier of India, Waziristan, 1901-02.—Medal with clasp.

Marriott, E. M. L. (*Col. u.s.l., Ind. Army*).—
N.-E. Frontier of India, 1871-72.—Lushai Expedition.—Mentioned in Despatches.—Medal with clasp.
Afghanistan, 1878-79.—Medal.

Marriott, J. H. G.—
South African War, 1900.—Served on Lines of Communication in Cape Colony and Orange River Colony from March to Oct. 00.—Queen's Medal with 2 clasps.

Marsh, C. H.—
South African War, 1899—1901.—Relief of Ladysmith, including operations of 17 to 24 Jan. 00 and action at Spion Kop; operations of 5 to 7 Feb. 00 and action at Vaal Krans; operations on Tugela Heights (14 to 27 Feb. 00) and action at Pieter's Hill (27 Feb. 00). Wounded. Operations in Natal, March to June 00, including action at Laing's Nek (6 to 9 June 00). Operations in the Orange River Colony in June 00. Operations in the Transvaal (3 July 00 to 20 Aug. 01). Operations in Cape Colony (28 Aug. 01 to 5 Sep. 01).—Despatches, *Lond. Gaz.*, 8 Feb. 01.—Queen's Medal with 7 clasps.—D. S. O.
The War of 1914-18.—Despatches, *Lond. Gaz.*, 4 Jan. 17.

Marsh, C. H.—
South African War, 1900.—Operations in the Transvaal, Orange River Colony and Cape Colony, from Apr. 1900 to Dec. 1900. Queen's Medal with 3 clasps.

Marsh, F. G.—
N.-W. Frontier of India, 1897-98.—Operations on the Samana and in the Kurram Valley during Aug. and Sep. 97.—Medal with 2 clasps.
Tirah, 1897-98.—Actions of Chagru Kotal and Dargai.—Operations in the Bazar Valley, 25th to 30 Dec. 97.—Clasp.
The War of 1914-18.—Order of St. Vladimir (4th Class with swords and Bow), C. M. G.

Marsh, F. H. B. (*M.-Gen. u.s.l., Ind. Army*)—
N.-W. Frontier of India, 1863.—Umbeyla Expedition Despatches, *G. G. O. 76 of 64.*—Medal with clasp.
Abyssinia, 1867-68.—Action of Arogee.—Medal.

Marsh, H. C. (*Hon. Col. retd. Ben. S. C.*)—
Afghanistan, 1879-80.—Medal.
Egypt, 1882.—Medal; bronze star; 4th class, Osmanieh.

Marsh, R. H.—
South African War, 1900-01.—Operation in the Transvaal, west of Pretoria, July to 29th November 1900 (including action at Frederickstadt 17th to 25th October). Operations in the Transvaal, 30th November 1900 to October 1900.—Queen's medal with 4 clasps.

Marshall, A.—
South African War, 1901-02.—Operations in Cape Colony from 30th November to 11th December 1901. Operations in Orange River Colony from 12th December 1901 to 31st May 1902.—Queen's medal with 3 clasps. D. S. O.
The War of 1914-18.—Despatches, *Lond. Gaz.*, 3 July 15 and 1 Jan. 16. D. S. O.

Marshall, D. G. (*Maj. retd. I. M. S.*)—
Burma, 1889—92.—Medal with clasp.
N.-W. Frontier of India, Isazai, 1892.
N.-W. Frontier of India, 1897-98.—Tochi.—Medal with clasp.
China, 1900.—Relief of Pekin.—Medal with clasp.

Marshall, J. S.—
South African War, 1902.—Operation in Cape Colony, Apr. to 31 May 02. Queen's medal with 2 clasps.
The War of 1914-18.—Despatches, *Lond. Gaz.*, 30 Jun. 16 and 13 July 16. D. S. O.

Marshall, W. S. (*Bt. Col. retd. I. A.*)—
Afghanistan, 1878-79-80.—With Ghuzni Field Force Battle of Ahmed Khel.—Medal with clasp.
N.-W. Frontier of India, Hazara, 1891.—Medal with clasp.
Chitral, 1895.—Relief of Chitral. Action of the Malakand Pass. Action at Khar. Despatches, G. G. O. 998 of 95.—Medal with clasp.

Martelli, N. C. (*Lt.-Col. u.s.l. Ind. S. C.*)—
Afghanistan, 1879-80.—Medal.

Martin, Sir A. R. (*Lt.-Gen., retd., I. A.*)—
N.-W. Frontier of India, Jowaki, 1877-78.—Medal with clasp.
Afghanistan, 1878-79-80.—Actions of the Peiwar Kotal and in the Sapari Pass. Despatches, G. G. O. 610 of 79. Action of Charasiah. Despatches, G. G. O. 1136 of 79. Operations at and around Kabul. March from Kabul to the relief of Kandahar and battle of Kandahar. Despatches, G.G. O. 582 of 80. Aitchakzai Expedition. Operations against the Maris.—Medal with 4 clasps. Bronze star.
N.-W. Frontier of India, Hasara, 1888.—Clasp. (*2nd*) *Miransai, 1891.*—Despatches, G. G. O. 633 of 91.—Clasp; Brev. of Maj. *Isasai, 1892.*—Served as A. A. G. of the Force.
Wasiristan, 1894-95.—Served as A. A. G. of the Force. Despatches, G. G. O. 473 of 95.—Clasp, Brev. of Lt.-Col.
Tirah, 1897-98.—A. A. G., 2nd Division. Operations in the Bara Valley, 7 to 14 Dec. 97. Despatches, G. G. O. 244 of 98.—Medal with 2 clasps.

Martin, C. (*Col. u. s. l. Ben. Cav.*)—
India, 1857.—Capture of Dhar and action of Mundisore (severely wounded). Mentioned in Despatches.—Medal with clasp.
Abyssinia, 1867-68.—Capture of Magdala. Mentioned in Despatches.—Medal.
Afghanistan, 1880.—Operations in the Kama district, and action at Besud, march from Kabul to the relief of Kandahar, and battle of 1 Sep. Despatches, G. G. O. 582 of 80.—Medal with clasp; bronze star; C. B.

Martin, D. N. (*Lt.-Col., retd., Ind. Med. Serv.*)—
Afghanistan, 1879-80.—Medal.
N.-W. Frontier of India, 1881, Mahsud-Waziri Expedition.

Martin, E. H.—
N.-W. Frontier of India, 1908.—Operations in the Mohmand country.—Medal with clasp. Noted for special promotion to the rank of Asst. Comy. with the hony. rank of Lieut. (su py.) on attaining the rank of Conductor.

Martin, F.—
N.-W. Frontier of India, 1897-98.—Tochi.—Medal with clasp.
N.-W. Frontier of India, 1908.—Operations in the Zakka Khel Country. Medal with clasp.

Martin, G. D.—
The War of 1914-18.—Despatches, *Lond. Gaz.*, 17 Feb. 15 and 15 Aug. 17. M. C.

Martin, G. W. (*Lt.-Col., retd., Ind. S. C.*)—
Afghanistan, 1899.—Actions of Killa Kazi, Ali Khel, Karej-i-Mir, and in the Chardeh Valley, attack on Asmai Heights, operations at and around Kabul in Dec. 79. Despatches, G. G. O. 137 of 80.—Medal with clasp.
N.-W Frontier of India, 1881. Mahsud-Waziri Expedition.—Mentioned in Despatches.

Martin, J.—
The War of 1914-18.—Despatches, *Lond. Gaz.*, 19 Oct. 16. M. C.

Martin, J. S. S.—
The War of 1914-18.—Despatches, *Lond. Gaz.*, 19 Oct. 16.

Martin, R. V.—
The War of 1914-18.—Despatches, *Lond. Gaz.*, 5 Apr. 16.

Martindale, C. S. deB. (*Maj., retd., I. S. C.*)—
Burma, 1886—89.—Medal with 2 clasps.

Masani, H. D. (*Surg. Lt.-Col., retd., I.M.S.*)—
Afghanistan, 1879.—Served in Southern Afghanistan as Sanitary Officer in the Bolan Pass. Medal.
N.-W. Frontier of India, Zhob Valley, 1890,
East Africa, 1896.—Operations against the Mazrui rebels. Medal.

Mason, H. M. (*Maj. Genl., retd., I. A.*)—
Burma, 1886-87.—Medal with 2 clasps.

Massy, F. S.—
South African War, 1901-02.—Operations in the Transvaal, May 1901 to January 1902. Queen's Medal with 3 clasps.

Massy, H. S. (*Col., retd., I. A.*)—
N.-W. Frontier of India, Jowaki, 1877-78.—Medal with clasp.
Afghanistan, 1878-79-80.—Battle of Ahmed Khel (wounded). Actions of Urzu and Patkao Shana. Despatches, G. G. O. 493 of 80.—Medal with clasp.
Burma, 1886—89.—Served as Superintendent of Signalling and Brigade-Major, 3rd Brigade. Advance to, and occupation of, Fort Stedman. Operations of the 3rd Brigade. Despatches, G. G. O. 434 of 87.—2 clasps.

N.-E. Frontier of India, Chin-Lushai, 1889-90.—Served as Staff Officer to Major-General B. L. Gordon.—Clasp.
N.-W. Frontier of India (2nd), Miransai, 1891—Clasp.
N.-W. Frontier of India, 1897-98.—*Mohmand.*—Provost Marshal to Force.—Medal with clasp.
Tirah, 1897-98.—D. A. Q. M. G., 3rd Brigade.—Actions of Chagru Kotal, Dargai and of the Sampagha and Arhanga Passes.—Operations in the Waran Valley and action of (16 Nov. 97.)—Operations in the Bara Valley, 7 to 14 Dec. 97.—Despatches, G. G. O. 244 of 98.—Clasp.

Massy, S. D.—
South African War, 1901-02.—Queen's medal with 5 clasps.
The War of 1914-18.—Despatches (Sir J. Maxwell, 19 Aug. 15 and 16 Apr. 16), *Lond. Gaz.*, 21 June 16 and 19 Oct. 16. Brev. of Maj., D. S. O.

Masters, A. (*Col., retd., I. A.*)—
Afghanistan, 1879-80.—Affair of Kama. March from Kabul to the relief of Kandahar, and battle of Kandahar. Despatches. Br.-Genl. J. Doran's despatch, 18 June 80.—Medal with clasp; Bronze star.
N.-W. Frontier of India, 1897-98.—Malakand.—Action at Landakai.—Operations in Bajour and in the Mamund Country.—*Mohmand Utman Khel.*—*Buner*, Action of the Tanga Pass.—Despatches, G. G. Os. Nos. 1089 and 1317 of 97, and 217 of 98.—Medal with clasp.

Masters, J.—
Operations in the Abor country, 1911-12.—Despatches, *Gazette of India*, Notification No. 480, dated 23rd May 1912.—Medal with Clasp.

Masters, O.—
The War of 1914-18.—Despatches, *Lond. Gaz.*, 8 Feb. 17.

Mathewes, J. R.—
Burma, 1886.—Medal with clasp.

Mathias, L. W.—
The War of 1914-18.—Despatches, *Lond. Gaz.*, 19 Oct. 16.

Matthews, C. R. (*Col., u.s.l., I.A.*)—
China, 1858.—Capture of Canton. Medal.

Matthews, E. A. C.—
N.-W. Frontier of India, 1908.—Operations in the Mohmand country. Medical with clasp.
The War of 1914-18.—Despatches, *Lond. Gaz.*, 15 Jun. 16.

Maude, E. A.—
South African War, 1900-02.—Operations in Cape Colony, July 1900 to July 1901. Operations in Orange River Colony, July 1901 to January 1902. Queen's medal with 2 clasps. King's medal with 2 clasps.
The War of 1914-18.—Despatches, *Lond. Gaz.*, 15 Aug. 17. D.S.O.

Maunsell, E. B.—
Mekran, 1901.—Attack and capture of Nodiz Fort.

Maunsell, E. C. C.—
Aden, 1903-04.—Operations in the interior.

Maunsell, I. O'G.—
South African War, 1902.—Operations in the Transvaal, 13th April to 31st May 1902. Queen's Medal with 2 clasps.

Mawhinny, R. J. W.—
The War of 1914-18.—Despatches, *London Gazette*, 4 July 16.

Mawson, W. A. (*Lieut.-Col., retd., I. M. S.*)—
Afghanistan, 1878-79-80.—Actions of Ali Musjid and Charasiah. Operations at and around Kabul. —Medal with 3 clasps.
N.-W. Frontier of India, 1897-98.—Tochi.—Medal with clasp.
N.-W. Frontier of India, Waziristan, 1901-02.—Despatches, G. G. O. No. 611 of 02.—Clasp.

Maxwell, A. G. (*Maj., retd., I. A.*)—
Tirah, 1897-98.—Medal with clasp.
South African War, 1899-1900.—Operations in the Orange Free State, Feb. to May 00, including operations at Paardeberg (17 to 26 Feb.). Actions at Poplar Grove (7 March), Driefontein (10 March). —Operations in the Transvaal in May and June 00, including actions near Johannesburg (29 May), Pretoria (4 June).—Operations in the Transvaal, east of Pretoria, July to 29 Nov. 00, including actions at Belfast (26 and 27 Aug. 00). Despatches, *Lond. Gaz.*, 16 April 01. Queen's Medal with 5 clasps. Brev. of Maj.
East Africa, 1903-04.—Operations in Somaliland Medal with clasp.

Maxwell, D. W.—
N.-W. Frontier of India, 1897-98.—Malakand. Mohmand.—Operations in Bajaur and in the Mamund Country, Buner.—Action of the Tanga Pass.—Despatches, G. G. O. 178 of 98.—Medal with clasp.
China, 1900.—Medal.

Maxwell, G. W.—
Burma, 1887—89.—Medal with clasp.

Maxwell, H. G.—
N.-W. Frontier of India (*1st and 2nd*) *Miranzai, 1891.*—Served as Orderly Officer to Br.-Genl. Sir W. S. A. Lockhart, K.C.B., in the 1st Expedition.—Medal with clasp.
Burma, 1892-93.—Clasp.
N.-W. Frontier of India, 1897-98.—Tochi.—Medal with clasp.
China, 1900.—Medal.

Maxwell, H. St. P. (*Lt.-Col. M.S.L., I.A.*)—
N.-E. Frontier of India, Daphla, 1874-75.—Despatches, G. G. O. 375 of 75.—*Naga Hills, 1879-80.* —Despatches, G. G. Os. 123 and 405 of 80.—Medal with clasp.—*Akha, 1883-84.*—Despatches. Brigadier-General R. S. Hill's despatch, 31 Jan. 84.
N.-E. Frontier of India, Manipur, 1891.—Thanks of the Government of India. Despatches, G. G. O. 585 of '91.—Clasp.

Maxwell, L. L.—
N.-W. Frontier of India, Waziristan, 1894-95.—Medal with clasp.
Tirah, 1897-98.—Operations in the Bazar Valley, 25th to 30th December 1897.—Medal with 2 clasps.
South African War, 1900.—Relief of Mafeking—(Severely wounded). Despatches, *Lond. Gaz.*, 16 April 01.—Queen's Medal with 3 clasps. Brev. of Maj.
The War of 1914-18.—Despatches, *Lond. Gaz.*, 4 Jan. 17. C. M. G.

Maxwell, P. A.—
The War of 1914-18.—Despatches (Sir J. Maxwell, 19 Aug. 15 and 9 Apr. 16), *Lond. Gaz.*, 21 Jun. 16.

Maydwell, H. S. L.—
South African War, 1899-01.—Station Staff Officer from 4 July 1901. Operations in the Transvaal, Feb. to July 1901. Operations in Cape Colony, July to Oct. 1901. Queen's medal with 3 clasps.

Maynard, C. W. (*Capt. retd. Ben. Art.*)—
India, 1857—59.—Medal.

Maynard, F. H.—
The War of 1914-18.—Despatches, *Lond. Gaz.*, 19 Oct. 16 M. C.

Maynard, F. P.—
N.-W. Frontier of India, Hazara, 1891.—Medal with clasp.

Mayne, A. G. O. M.—
The War of 1914-18.—Despatches, *Lond. Gaz.*, 15 Aug. 17. D. S. O.

Mayne, C. E. M.—
China, 1900.—Medal.

Mayne, R. C. G. (*Col., retd., I. A.*)—
Afghanistan, 1879-80.—Skirmish of Sir-i-asp, march from Khelat-i-Ghilzai to relief of Kandahar and battle of 1 Sep.—Medal with clasp; bronze star.
Egypt, 1882.—Battle of Tel-el-Kebir. Forced march to, and occupation of, Zagazig and occupation of Cairo.—Medal with clasp; bronze star.
N.-W. Frontier of India, Zhob Valley, 1890.
Mekran, 1898.—Commanded the Field Force. Action of Gok Parosh. Despatches, G. G. O. 648 of 98. *C.B.*
China, 1900.—Medal.

Mayo, G. L. L.—
N.-W. Frontier of India, 1908.—Operations in the Mohmand Country. Engagement of Kargha. 24th May 1908. Medal with clasp.
The War of 1914-18.—Despatches, *Lond. Gaz.*, 21 Jun. 16.

Mead, H. R.—
Soudan, 1885-86.—Frontier Field Force. Action of Giniss.—Medal; bronze star.
British East Africa, 1901.—Operations against the Ogaden Somalis in Jubaland. Medal with clasp.
The War of 1914-18.—Despatches, *Lond. Gaz.*, 6 July 17.

Meade, H. R. C.—
Operations in the Abor country, 1910-11.—Medal with clasp.

Meade, J. de C. D. (*Lieut.-Col., retd., I. A.*)—
Afghanistan, 1880.—Medal.

Meade, J. W. B.—
Afghanistan, 1878-79-80.—Battle of Ahmed Khel. Action of Urzu.—Medal with clasp.
Burma, 1886—88.—Medal with 2 clasps.

Meade, W. L.—
Aden, 1903-04.—Operations in the Interior.

Meadows, C. A. G. P.—
China, 1900.—Relief of Pekin, Action of Peitsang. Medal with clasp.

Mears, C. D.—
Chitral, 1895.—Relief of Chitral.—Medal with clasp.
N.-W. Frontier of India, 1897-98.—Mohmand-Buner.—Clasp.
Tirah, 1897-98.—Clasp.
The War of 1914-18.—Despatches, *Lond. Gaz.*, 19 Oct. 16. Brev. of Lt.-Col.

Medley, A. G.—
N.-W. Frontier of India, (2nd) Miranzai, 1891.—Medal with clasp.
N.-W. Frontier of India, 1897-98.—Malakand.—Medal with clasp.

Medley, E. J. (*Lieut.-Col., retd., I. A.*)—
Afghanistan, 1878-79-80.—Medal.

Medlicott, H. E.—
South African War, 1902.—Operations in Cape Colony, May 1902.—Queen's Medal with 2 clasps.
The War of 1914-18.—Despatches, *Lond. Gaz.*, 15 Jun. 16 and 11 Dec. 17. Brev. of Maj.

Megaw, J. W. D.—
China, 1900.—Medal.

Meiklejohn, J. F.—
The War of 1914-18.—Despatches, *Lond. Gaz.*, 5 Apl. and 19 Oct. 16.

Meiklejohn, W. L. S.—
Tibet, 1903-04.—Action at Niani, 26 June 04. Operations at and around Gyantse, between 5 May and 6 July 04. March to Lhassa, 14 July to 13 Aug. 04.—Medal with clasp.
The War of 1914-18.—Despatches, *Lond. Gaz.*, 15 Aug. 17.

Men, A. B. (*Lt.-Col., retd. Ind. S. C.*)—
Afghanistan, 1879-80.—Medal.

Mein, D. B.—
The War of 1914-18.—Despatches, *Lond. Gaz.*, 3 July 15 and 1 Jan. 16. M.C.

Mein, J. E. (*Col., retd. I. A.*)—
N.-W. Frontier of India, Jowaki, 1877-78.—Medal with clasp.
Afghanistan, 1878-79-80.—Operations at and around Kabul.—Medal with clasp.
N.-W. Frontier of India, Mahsud-Waziri, 1881.
N.-W. Frontier of India, (2nd) Miranzai, 1891.—Clasp.
N.-W. Frontier of India, Waziristan, 1894-95.—Clasp.

Melhuish, H. M. H.—
The War of 1914-18.—Despatches, *Lond. Gaz.*, 22 Jun. 15 and 15 Aug. 17. Brev. of Lt.-Col.

Mell, F. O. N.—
The War of 1914-18.—Despatches, *Lond. Gaz.*, 15 Aug. 17. C.I.E.

Melliss, C. J.—
East Africa (British), 1896.—Operations against the Mazrui Rebels.—Medal. Brilliant Star of Zanzibar, 2nd class.
N.-W. Frontier of India, 1897-98.—Operations in the Kurram Valley during Sep. 97.
Tirah, 1897-98.—Operations in the Bara Valley, 7 to 14 Dec. 97.—Medal with 3 clasps.
West Africa, 1900.—Operations in Ashanti. Relief of Kumassi (four times wounded, once severely). Despatches, *Lond. Gaz.*, 4 Dec 00 and 8 March 01. Brev. of Lt.-Col. *V. C.*—Medal with clasp.
East Africa, 1902-04.—Operations in Somaliland. Despatches, *London Gazette*, 2nd September 1904.—Medal with clasp.
N.-W. Frontier of India, 1908.—Operations in the Zakka Khel country.—Medal with clasp.
The War of 1914-18.—Despatches, *Lond. Gaz.*, 5 Apl., 21 Jun., 13 July and 19 Oct. 16. K.C.B.

Melliss, Sir H. (*Col. u. s. l., I.S.C.*)—
Abyssinia, 1867-68.—Storm and capture of Magdala.—Medal.
Afghanistan, 1879.—As D. A. Q. M. G., Reserve Division, Sind.
Egypt, 1882.—As A. Q. M. G., Indian Contgt. Battle of Tel-el-Kebir.—Despatches, *London Gaz.*, 2 Nov. 82. Brev. of Maj., 4th class, Medjidie.—Medal with clasp; bronze star.
Burma, 1885.—As Military Attaché to the Naval Commander-in-Chief. Despatches.—Medal with clasp.

Mellor, G. F.—
South African War, 1900-02.—Served with Imperial Yeomanry. Operations in the Transvaal, Nov. 01 to 31st May 02. Operations in Orange River Colony.—Queen's medal with 3 clasps. King's medal with 2 clasps.

Melville, C. W. F.—
Operations in the Abor Country, 1911-12.—Despatches, *Gazette of India*, Notification No. 480, dated 23rd May 1912.—Medal with clasp.

Melville, E. P. A.—
N.-W. Frontier of India, Waziristan, 1901-02.—Medal with clasp.
N.-W. Frontier of India, 1908.—Operations in the Zakka Khel country.—Medal with clasp.

Melville, H. B.—
N.-E. Frontier of India, Lushai, 1890.

Melville, J. S. (*Bt.-Col. retd. I. A.*)—
Burma, 1886-87.—Medal with clasp.
N.-W. Frontier of India, 1897-98.—Malakand.—Medal with clasp.
China, 1900.—Medal.

Mennie, J. O.—
Uganda, 1897-98.—Medal with clasp.
China, 1900.—Medal.
East Africa (Somaliland), 1908-10.—Medal with clasp.

Menzies, O. (*Col. u. s. l. Ind. S. C.*)—
India, 1858-59.—Capture of Fort Rampore Kussia and operations in Oudh.—Medal.

Mercer, C. A. (*Col. retd. I. A.*)—
N.-W. Frontier of India, Hazara, 1868.—Medal with clasp.
N.-E. Frontier of India, Lushai, 1871-72.—Clasp.
Afghanistan, 1878-79-80.—Action of Ali Musjid. Operations at and around Kabul. Operations in Maidan, Wardak and Kohistan. Affair of Shekhabad. March from Kabul to the relief of Kandahar and battle of Kandahar. Operations against the Maris. Despatches, G. G. O. 582 of 80.—Medal with 3 clasps; bronze star.
N.-W. Frontier of India, Waziristan, 1894-95—Clasp.

Mercer, L. F.—
The War of 1914-18.—Despatches, *Lond. Gaz.*, 15 Aug. 17.

Mercer, S. M. (*Hony. Capt. Commy., retd. S. and T. Corps*)—
N.-W. Frontier of India, Waziristan, 1894-95.—Despatches, G. G. O. 473 of 95.—Medal with clasp. Promoted Deputy Assistant Commissary.
N.-W. Frontier of India, 1897-98.—Operations on the Samana.—Medal with clasp.
Tirah, 1897-98.—2 clasps.

Merewether, J. W. B.—
East Africa, 1901.—Operations in Somaliland.—Medal with clasp.
Aden, 1903-04.—Operations in the Interior.

Messenger, H. T. K.—
South African War, 1900-02.—Operations in the Transvaal and Orange River Colony,—30th November 1900 to December 1901.—Queen's medal with 3 clasps; King's medal with 2 clasps.
The War of 1914-18.—Despatches, *Lond. Gaz.*, 5 Apl. 16.

Messum, L. R.—
N. W. Frontier of India, 1908.—Operations in the Zakka Khel Country.—Medal with clasp.

Metcalfe, E. D.—
The War of 1914-18.—Despatches, *Lond. Gaz.*, 15 Aug. 17. M.C.

Michell, J. W. A. (*Col. u.s.l. Ind. S. C.*)—
N.-E. Frontier of India, 1865-66, Bhutan Expedition.—Recapture of Dewangiri.—Medal with clasp. *Duffla Expedition, 1874-75. Sikkim Expedition, 1888.*—Second-in-command of the expedition.—Clasp.

Middlemass, R. W. H.—
China, 1900.—Medal.

Middleton, H. J. J.—
South Africa, 1879, Zulu Campaign.—Action of Inyezane and occupation of Etshowe.—Medal with clasp.
N.-W. Frontier of India, 1897-98.—Operations on the Samana and in the Kurram Valley during Aug. and Sep. 97.—Medal with 2 clasps.
Tirah, 1789-98.—Clasp.

Middleton-West, S. H.—
The War of 1914-18.—Despatches, *Lond. Gaz.*, 17 Feb. 15, 1 Jan. 16 and 15 Aug. 17. M.C.

Miles, P. J.—
N.-W. Frontier of India, Hazara, 1888.—Action of Kotkai.—Medal with clasp.—(*1st*) *Miranzai, 1891.*
China, 1900.—Medal.
N.-W. Frontier of India, 1908.—Operations in the Zakka Khel country. Operations in the Mohmand country. Engagement of Matta, 24th April 1908. Engagement of Kargha, 24th May 1908. Despatches, G. G. O. No. 521A of 1908.—Medal with clasp. Brevet of Lieut.-Col.

Milford, E.—
The War of 1914-18.—Despatches, *Lond. Gaz.*, 19 Oct. 16.

Millar, G. McG.—
N.-W. Frontier of India, 1908.—Operations in the Zakka Khel country. Operations in the Mohmand country.—Medal with clasp.
Operations in the Abor country, 1911-12.—Clasp.

Millar, W. H.—
Burma, 1886—88.—Operations of the 3rd Brigade, Southern Shan Column.—Medal with 2 clasps.
N.-W. Frontier of India, Hazara, 1891.—Clasp.
(*2nd*) *Miranzai, 1891.*—Clasp.

Miller, A.—
N.-W. Frontier of India, 1897-98.—Malakand. Operations in Bajaur and in the Mamund Country, Utman Khel-Buner. Action of the Tanga Pass. Medal with clasp.
China, 1900.—Medal with clasp.

Miller, G. N.—
The War of 1914-18.—Despatches, *Lond. Gaz.*, 15 Aug. 17.

Miller, H. (*Captain, retd. I. A.*)—
N.-W. Frontier of India, 1897-98.—Tochi.—Medal with clasp.

Miller, J. (*Hon. M.-Gen. retd. Ben. S. C.*)—
India, 1857—59.—At Peshawar and in Bundelcund.—Medal.

Miller, T. (*Hony. Capt. Dy. Commy., retd. Misc. List*)—
China, 1900.—Medal.

Mills, A. M.—
South African War, 1901.—Operations in the Transvaal, 20 July to 8 Oct. 01.—Operations in the Orange River Colony from 1 May to 19 July 01 and from 9 Oct. to 20 Dec. 01.—Queen's medal with 5 clasps.
The War of 1914-18.—Despatches, *Lond. Gaz.*, 4 Jun. 17. D.S.O.

Mills, H. G. C.—
The War of 1914-18.—Despatches, *Lond. Gaz.*, 15 Aug. 17.

Mills, P. S.—
N.-W. Frontier of India, 1908.—Medal with clasp.
The War of 1914-18.—Despatches, *Lond. Gaz.*, 15 Aug. 17.

Milne, C.—
N.-W. Frontier of India, Waziristan, 1894-95.—Medal with clasp.
Soudan, 1896.—Dongola Expedition.—Medal; Egyptian Medal.

Milne, T.—
N.-W. Frontier of India, 1908.—Operations in the Mohmand country. Engagement of Kargha, 24th May 1908.—Medal with clasp.

Milward, C. A.—
N.-W. Frontier of India, Waziristan, 1901-02.—Medal with clasp.
N.-W. Frontier of India, 1902.—Operations against the Darwesh Khel Waziris.
N.-W. Frontier of India, 1908.—Operations in the Zakka Khel country. Despatches, G. G. O. No. 188 of 1908. Operations in the Mohmand country. Engagement of Matta, 24th April 1908. Engagement of Kargha, 24th May 1908.—Medal with clasp.

The War of 1914-18.—Despatches, *Lond. Gaz.*, 1 Jun. 17. D. S. O.

Minchin, A. B. (*Capt., Supy. List, I.A.*)—
N.-W. Frontier of India, Isazai, 1892
Chitral, 1895.—Relief of Chitral.—Medal with clasp.
N.-W. Frontier of India, 1897. Defence with Chakdara Fort. Assistant Political Agent. Despatches, *Lon. Gaz.*, 5th November 1897.

Minchin, C. F. (*Maj., Sup. List, I. A.*)—
N.-W. Frontier of India, Hazara, 1888.—Medal with clasp.
Chitral, 1895.—Relief of Chitral. Medal with clasp.
South African War, 1899—1902.—Railway Staff Officer, Durban, from 1st Dec. 1899 to 30th Jan. 1900. Employed with Thorneycroft's Mounted Infantry as Captain from 4th Feb. 1900 to 20th April 1901. Commanded T. M. I. from 2nd November to 7th Dec. 1900. Granted local rank of Major, 21st April 1901 and Lieutenant-Colonel, 1st May 1901. Commanded 4 squadrons, 18th Battalion, and 2 squadrons, 22nd Battalion, Imperial Yeomanry, with Colonel Thorneycroft's column from 21st April 1901 to 6th Feb. 1902. In command of a Mobile Column from 17th July to 22nd Dec. 1901. Relief of Ladysmith, including operations of 5 to 7 Feb. 00 and action at Vaal Kranz; operations on Tugela heights (14 to 27 Feb. 00) and action at Pieter's Hill (27 Feb. 00); operations in Natal, March to June 00, including action at Laing's Nek (6 to 9 June); operations in the Transvaal, east of Pretoria, July to November 00; operations in the Transvaal, November and Dec. 00; operations in Cape Colony, Dec. 00 to March 1901; operations in Orange River Colony, Dec. 00 and March 01 to Feb. 02. Despatches, twice, *London Gazette*, 8th Feb. 01 (Sir R. H. Buller, 19 June and 9 Nov. 00) and *London Gazette*, 3rd Dec. 01.—Queen's Medal with 6 clasps; King's Medal with 2 clasps. D. S. O.

Minty, T. W.—
Burma, 1886-87.—Medal with 2 clasps.
Chitral, 1895.—Relief of Chitral.—Medal with clasp.

Miskin, W. L.—
The War of 1914-18.—Despatches, *Lond. Gaz.*, 5 Apl. and 19 Oct. 16.

Mistri, K. H. (*Lieut.-Col. retd. I. M. S.*)—
Burma, 1886-88.—Medal with clasp.

Mitchell, G. W. (*Col., retd. I.A.*)—
Afghanistan, 1880.—Medal.
Aden, 1903-04.—Operations in the Interior.

Mitchell, H. S.—
South African War, 1899-1900.—Operations in the Orange Free State, Feb. to May 00, including operations at Paardeberg (17 to 26 Feb.); action at Driefontein (10 Mar.); operations in the Transvaal in May and June 00, including actions near Johannesburg (29 May). Operations in Cape Colony from 1 Dec. 99 to 8 Feb. 00.—Queen's medal with 4 clasps.
Tibet, 1903-04.—Operations at and around Gyantse between 5th May and 6th July 1904. (Severely wounded.)—Medal with clasp.
N.-W. Frontier of India, 1908.—Operations in the Mohmand country.—Medal with clasp.
Operations in the Abor country, 1911-12.—Clasp.

Mitchell, W. (*1st Punjab Vol. Rifles*)—
N.-W. Frontier of India, Hunza Nagar, 1891-92.—Medal with clasp.
Chitral, 1895.—Relief of Chitral.

Mitchell, W. J.—
Chitral, 1895.—Relief of Chitral.—Medal with clasp.
East Africa (British), 1896.—Operations against the Mazrui Rebels.—Medal.
East Africa, 1903-04.—Operations in Somaliland. Action at Jidbaili. Despatches, *London Gazette*, 2 Sep. 1904.—Medal with 2 clasps.

Mitford, J. P.—
The War of 1914-18.—D. S. O.

Mitford, R. C. W. (*Hon. M.-Gen., retd. Ben. S.C.*)—
India, 1857-58.—Actions of Agra and Futtehpore Sikri. Capture of Lucknow and minor affairs (wounded). Despatches, *Lond. Gaz.*, 31 Jan. 58.—Medal with clasp.
Afghanistan, 1879-80.—Action of Charasiah, occupation of Kabul, defence of Sherpur.—Medal with 2 clasps.

Mitter, R. K.—
N.-W. Frontier of India, 1897-98.—Mohmand—Medal with clasp.
Tirah, 1897-98.—Actions of the Sampagha and Arhanga Passes. Operations in the Waran Valley 14th to 16th Nov. 1897. Operations in the Bara Valley, 10th to 14th Dec. 1897.—Clasp.

Moberly, B. R.—
N.-W. Frontier of India, Waziristan, 1901-02.—Medal with clasp.
East Africa, 1903-04—Operations in Somaliland. Action at Jidballi. Despatches, *London Gazette*, 2nd September 1904.—Medal with 2 clasps. D.S.O.
The War of 1914-8—Despatches, *Lond. Gaz.*, 28 Jan., 21 Jun. and 13 July 16. D. S. O.

Moberly, F. J.—
N.-E. Frontier of India, Manipur, 1891.—Asstt. Transport Officer, Tammu Column.—Medal with clasp.
Burma, 1891.—Wuntho Field Force.—Assistant Transport Officer.—Clasp.
N.-W. Frontier of India, Chilas, 1893.—(Severely wounded.) Despatches, G. G. O. 466 of 93.—D.S.O.
Chitral, 1895.—Relief of Chitral. Despatches, G. G. O. 530 of 95.—Medal with clasp. Promoted Captain.
N.-W. Frontier of India, 1897-98.—Present at Fort Jamrud—Clasp.
South African War, 1899-1900.—Operations in the Orange Free State, Feb. to May 00, including operations at Paardeberg (17 to 26 Feb.) actions at Poplar Grove (7 March); Driefontein (10 March), Vet River (5 and 6 May) and Zand River (10 May). Operations in the Transvaal in May and June 00, including actions near Johannesberg (29 May), Pretoria (4 June) and Diamond Hill (11 and 12 June). Operations in the Transvaal, east of Pretoria, July to 29 Nov 00, including action at Belfast (26 and 27 Aug. 00). Despatches, *Lond. Gaz.*, 10 Sep. 01.—Queen's medal with 6 clasps. Brev. of Maj.
N.-W. Frontier of India, 1908.—Operations in the Zakka Khel Country.—Medal with clasp.

Moberly, H. S.—
China, 1900.—Medal.
The War of 1914-18.—Despatches, *Lond. Gaz.*, 5 Apl. and 19 Oct. 16. Brev. of Lt.-Col.

Mocatta, D. E. (*Lt.-Col., retd. I.A.*)—
Burma, 1891-92—With Tlang-tlang Column. Medal with clasp. Services acknowledged by Government.
Tirah, 1897-98.—Medal with 2 clasps.

Mockler, E. (*Major.-Gen, retd. Ind. Army*)—
Abyssinia, 1867-68.—Political Officer at Zoulla. Despatch, *Lond. Gaz.*, 30 Jun. 68.

Mockler, G. H. G.—
Burma, 1889-92.—With the Wuntho Field Force.—Medal with clasp.
Burma, 1893-94.—
China, 1900.—Relief of Pekin. Special Service Officer. Despatches, *London Gazette*, 14th May 1901.—Medal with clasp. Brevet of Major.

East Africa, 1902-1904.—Operations in Somaliland. Special Service Officer.—Medal with clasp.
Russo-Japanese War, 1904-05.—Attached to the Russian Army in Manchuria. Order of St. Anne, 2nd Class, with swords, and Russian Medal.
The War of 1914–1918.—Despatches. *Lond. Gaz.*, 4 July 16.

Moens, A. W. H. M.—
East Africa, 1903-04.—Operations in Somaliland. Action at Jidballi—Medal with 2 clasps.
The War of 1914-18.—Despatches, *Lond. Gaz.*, 15 Aug. 17. D. S. O.

Molesworth, A. L. M.—
Operations in the Abor country, 1911-12.—Despatches, *Gazette of India*, Notification No. 480, dated 23rd May 1912.—Medal with clasp.

Molesworth, E. H. (*Hony. Br.-Genl., retd. I.A.*)—
Afghanistan, 1879-80.—Zaimukht Expedition.—Medal.
N.-E. Frontier of India, Akha, 1883-84.—Despatches. Brigadier-General R. S. Hill's despatch, 31st Jan. 84.
N.-E. Frontier of India, Abor, 1894.

Molesworth, W.—
Burma, 1891.—Wuntho Field Force.—Medal with clasp.

Molloy, G. M.—
East Africa, 1903-04. Operations in Somaliland. D. A. A. G. Lines of Commn. Special Service Officer.—Medal with clasp.
The War of 1914-18.—Despatches, *Lond. Gaz.*, 17 Feb. 15.

Molloy, H. T.—
N.-W. Frontier of India, 1908.—Operations in the Zakka Khel country.—Medal with clasp.

Molony, O. H. C.—
South African War, 1900-02.—Operations in the Transvaal, west of Pretoria, July to 29 Nov. 00, including actions at Frederickstad (17 to 25 Oct.). Operations in Orange River Colony, May to 29 Nov. 00. Operations in the Transvaal from Dec. 00 to 19 May 02.—Queen's medal with 3 clasps. King's medal with 2 clasps.

Monck-Mason, G. N. G.—
South African War, 1901-02—Queen's medal with 5 clasps.

Monck-Mason, H. E.—
Burma, 1886-89.—Operations against Ootarra in Minbu. With the Chin Field Force.—Medal with 2 clasps.

Money, E. A. (*Col., u. s. l. I. S. C.*)—
N.-W. Frontier of India, Mahsud Waziri, 1881.—Despatches. Brigadier-General T. G. Kennedy's despatch, 20 June 81.

Money, E. D.—
N.-W. Frontier of India, Isazai, 1892.
Waziristan, 1894-95.—Medal with clasp.
Tirah, 1897-98.—Medal with clasp.
The War of 1914–18.—Despatches, *Lond. Gaz.*, 4 July 16.

Money, E. F. D.—
The War of 1914-18.—Despatches, *Lond. Gaz.*, 30 Jun. 16. D. S. O.

Money, G. A.—
Afghanistan, 1879-80.—Medal.
N.-W. Frontier of India, Mahsud-Waziri, 1881.
Zhob Valley, 1890.
N.-W. Frontier of India, 1897-98.—Operations on the Samana and in the Kurram Valley during Aug. and Sep. 97. Operations of the Flying Column in the Kurram Valley under Colonel Richardson, 20 Aug. and 1 Oct. '97. (Wounded.)—Medal with 2 clasps.
Tirah, 1897-98.—Clasp.

Money, W. S. E.—
The War of 1914-18.—Despatches, *Lond. Gaz.*, 15 Aug. 17.

Money-Simons, J. J. (*Lt.-Col., u.s.l. Ind. S.C.*)—
Afghanistan, 1878-80.—With the Thal Chotiali Field Force, advance to, and defence of, Jugdulluck, action of Saidabad, march from Kabul to relief of Kandahar, and battle of 1 Sep. Despatches, G. G. O. 177 of 80.—Medal with clasp; bronze star.
N.-W. Frontier of India, Hazara, 1888.—Medal with clasp.

Monisse, C. B.—
Burma, 1880-88.—Medal with 2 clasps.
Burma, 1891-92.—Wuntho expedition. Maingwan column.—Clasp
Chitral, 1895.—Relief of Chitral.—Medal with clasp.
N.-W. Frontier of India, 1897-98.—Malakand. Buner. Action of the Tanga Pass.—Clasp.

Montanaro, A. (*Colonel, retd. I. A.*)—
Afghanistan, 1879-80.—With Khyber Field Force.—Medal.
Burma, 1886-89.—Medal with 2 clasps.
N.-W. Frontier of India, 1897-98.—Malakhand, Utman Khel. Despatches, G. G. O. 57 of 98.—Buner.—Action of the Tanga Pass.—Despatches, G. G. O. 217 of 98.—Medal with clasp.

Monteath, A. M. (*Maj., retd. Ind. S.C.*)—
Afghanistan, 1878-80.—Action at Takht-i-pul, battle of Maiwand (wounded), defence of Kandahar and battle of 1 Sep.—Despatches, G. G. O. 551 of 80.—Medal with clasp.

Monteith, J. (*Col., retd. I. A.*)—
Afghanistan, 1879-80.—Action at Girishk, battle of Maiwand, defence of, and sortie from, Kandahar, and battle of 1 Sep.—Despatches, G. G. Os. 743 and 860 of '80.—Medal with clasp.
Soudan, 1885.—Suakin.—Medal with clasp, bronze star.
N.-W. Frontier of India, Zhob Valley, 1890.

Montgomery, C. A. S. (*Bt.-Col., retd. I. A.*)—
Burma, 1886-87.—Medal with clasp.

Montford, D. R.—
The War of 1914-18.—Despatches, *Lond. Gaz.*, 8 Feb. 17.

Moody, F. H.—
Tibet, 1903-04.—Action at Niani, 26 June 04. Operations at and around Gyantse, between 5 May and 6 July 04. March to Lhassa, 14 July to 31 Aug. 1904.—Medal with clasp.
N.-W. Frontier of India, 1908.—Operations in the Mohmand country.—Medal with clasp.
The War of 1914-18.—Despatches, *Lond. Gaz.*, 1 Jan. 16. M. C.

Moore, A.—
N.-W. Frontier of India, 1897-98—Tochi.—Medal with clasp.
The War of 1914-18—Despatches, *Lond. Gaz.*, 13 July and 19 Oct. 16.

Moore, C. A. (*M.-Gen., u. s. l. Bo. Cav.*)—
India, 1858-60.—With the Rajpootana Field Force, attacking of Rowah, siege of Awah. With the Central India Field Force, at the operations in the Mohi Kanta, night attack at Byrannah, Operations in Kathiawar and the taking of the Burda Hills.—Medal with clasp.
Abyssinia, 1867-68.—Commanded 3 Lt. Cav. at capture of Magdala.—Despatches, *Lond. Gaz.*, 16 June and 10 July 68.—Medal.

Moore, C. F. F.—
The War of 1914-18.—Despatches, *Lond. Gaz.*, 22 Jun. 15. M. C.

Moore, C. H. G.—
N.-W. Frontier of India, 1897-98.—Malakand. Operations in Bajaur-Mohmand. Despatches G. G. O. 178 of 98.—Medal with clasp.
Tibet, 1903-04.—Action at Niani, 26 June 04. Operations at and around Gyantse, between 5 May and 6 July 04. March to Lhassa, 14 July to 3 Aug. 04. Despatches, *Gazette of India*, No. 1065 of 1904.—Medal with clasp. D. S. O.
N.-W. Frontier of India, 1908.—Operations in the Mohmand country. Engagement of Kargha, 24th May 1908. Despatches, G. G. O. No. 521A of 08.—Medal with clasp.
The War of 1914-18.—Despatches, *Lond. Gaz.*, 22 Jun. 15 and 15 Jun. 16. C. M. G.

Moore, F. G.—
Tirah, 1897-98.—Actions of the Sampagha and Arhanga Passes. Reconnaissance of the Saran Sar and action of 9 Nov. 97. Operations at and around Dwatoi. Operations in the Bara Valley, 7 to 14 Dec. 97. Operations in the Bazar Valley, 25 to 30 Dec. 97.—Medal with 2 clasps.
China, 1900.—Medal.

Moore, F. L.—
Tirah, 1897-98.—Medal with 2 clasps.

Moore, F. T. T. (*Major, retd. I. A.*)—
South African War, 1902.—Adjutant, 28th Battalion, Imperial Yeomanry, from 11 Jan. 02 to 21 Oct. 02.—Queen's medal with 2 clasps.
Tibet, 1903-04.—Action at Niani, 26th June 1904. Operations at and around Gyantse, between 5th May and 6th July 1904. March to Lhassa, 14th July to 3rd August 1904.—Medal with clasp.
The War of 1914-18.—Despatches, *Lond. Gaz.*, 4 Jan. and 21 Dec. 17.

Moore, G. D. M.—
N.-W. Frontier of India, 1897-98.—Mohmand.—Medal with clasp.
Tirah, 1897-98.—Action of the Sampagha Pass. Operations at and around Dwatoi and action of 24 Nov. 97. Operations in the Bara Valley, 7 to 14 Dec. 97.—Despatches, G. G. O. 244 of 98.—Clasp.
South African War, 1902.—Operations in the Transvaal between 15 April and 31 May 02.—Queen's medal.

Moore, H. A. (*Lt.-Col., retd. I. A.*)—
N.-W. Frontier of India, Hazara, 1888.—Medal with clasp. (*1st and 2nd*) *Miranzai, 1891.*—Clasp.
N.-W. Frontier of India, 1897-98, Malakand.—Operations in Bajaur and in the Mahmund country.
Tirah, 1897-98.—Operations in the Bara Valley, 7 to 14 Dec. 97.—Medal with 2 clasps.
N.-W. Frontier of India, Waziristan, 1901-02.—Clasp.
The War of 1914-18.—Despatches, *Lond. Gaz.*, 4 Jan. 17.

Moore, H. D.—
The War of 1914-18.—Despatches, *Lon. Gaz.*, 8 Feb. 17 and 24 Sept. 17. D. S. O.

Moore, J. (*Hon. Capt. Sen. Asst. Surg., retd. I.S.M.D.*)—
South African War, 1899-1900.—Operations in Natal, 1899, including action at Talana. Defence of Ladysmith, including sortie of 10th Dec. 1899, and action of 6th Jan. 1900. Operations in Natal (March to June 1900), including action at Laing's Nek. Operations in the Transvaal, east of Pretoria, including actions at Reit Vlei, Belfast (26th and 27th August) and Lyndenberg (5th to 8th September).

Moore, T. C. (*Major, retd. I.M.S.*)—
China, 1900.—Medal.

More, J. C.—
The War of 1914-18.—Despatches, *Lond. Gaz.*, 10 May 16 and 15 Aug. 17.

More, P. St. C.—
Chitral, 1895.—Relief of Chitral.—Medal with clasp.

Morgan, B. E.—
The War of 1914-18.—Despatches, *Lond. Gaz.*, 19 Oct. 16 and 15 Aug. 17. D. S. O.

Morgan, E. J. (*Major, I. A. retired*)—
N.-W. Frontier of India, 1897-98.—Tochi.—Medal with clasp.

Morgan, H. A.—
N.-W. Frontier of India, 1908.—Operations in the Mohmand Country.—Medal with clasp.

Morgan, J. H. W.—
Tirah, 1897-98.—Operations in the Bara Valley, 7 to 14 Dec 97.—Medal with 2 clasps.

Morgan, M. H. L.—
The War of 1914-18.—Despatches, *Lond. Gaz.*, 21 Jun. 16. D. S. O.

Morice, R. C. L.—
The War of 1914-18.—Despatches, *Lond. Gaz.*, 15 Apl. 16. M. C.

Morrell, H. G.—
The War of 1914-18.—Despatches, *Lond. Gaz.*, 5 Apl. 16.

Morris, C. E.—
South African War, 1902.—Queen's Medal with 4 clasps.
N.-W. Frontier of India, 1908.—Operations in the Mohmand country.—Medal with clasp.

Morris, C. O. —
The War of 1914-18.—Despatches, *Gaz. of India* Notifn. 1241 of 16. D. S. O.

Morris, C. T.—
Tirah, 1897-98.—Medal with 2 clasps.
N.-W. Frontier of India, 1908.—Operations in the Momand country. Engagement of Kargha, 24th May 1908.—Medal with clasp.
The War of 1914-18.—Despatches, *Lond. Gaz.*, 19 Oct. 16.

Morris, G. M.—
The War of 1914-18.—Despatches, *Lond. Gaz.*, 15 Aug. 17. D. S. O.

Moorls, J. A.—
Burma 1889-93.—Medal with clasp.
Chin Hills, 1892-93.—Clasp.
The War of 1914-18.—Despatches, *Lond. Gaz.*, 22 Jun. 15. Promd. Assistant Commissary.

Morris, R. L.—
South African War, 1901.—Operations in the Transvaal, Aug. to Sep. 01.—Queen's medal with 2 clasps.

Morrison, H. (*Hony. Capt., retd. Ind. Misc. List*)—
Tirah, 1897-98.—Actions of the Sampagha and Arhanga Passes.—Despatches, G. G. O. 244 of 98.—Medal with 2 clasps.—Promoted Assistant Commissary.

Morrison, J.—
Egypt, 1882.—Action at Kassasin, 9 Sep. 82. Battle of Tel-el-Kebir, 13 Sep. 82.—Medal with clasp, bronze star.

Morse, L. H.—
The War of 1914-18.—Despatches, *Lond. Gaz.*, 1 Jan. 16, 19 Oct. 16 and 15 Aug. 17. M. C.

Morton, A. W. (*Asst. Comsy. and hony. Lieut., retired S. and T. Corps*)—
N.-W. Frontier of India, 1897-98.—Malakand. Operations in Bajaur and in the Mahmund Country. *Buner*—Action of the Tanga Pass.—Medal with clasp.
East Africa, 1903-04.—Operations in Somaliland. Despatches (Br.-Genl. Mannings, 17 Aug. 03) *London Gazette*, 2 Sept. 04. Medal with clasp. Promoted Assist. Commy.
Tibet, 1903-04.—Medal.

Morton, B. W. D. (*Lt.-Gen., u. s. l. Ind. S. C.*)—
N.-E. Frontier of India, 1863.—Cossiah and Jynteah Hills.—Capture of Stockades.

Morton, E. R. (*Lt.-Col., retd., I.A.*)—
N.-W. Frontier of India, Hazara, 1891.—Medal with clasp.
N.-W. Frontier of India, 1897-98.—Defence and relief of Malakand.—Malakand.—Action at Landakai.—Medal with 2 clasps.

Morton-Marshall, F. W.—
South African War, 1899-1900.—Relief of Ladysmith, November 1899 to February 1900. Action at Laing's Nek in June 1900.—Queen's medal with 2 clasps.
East Africa, 1902-04.—Operations in Somaliland.—Medal with clasp.

Morwood, J.—
N.-W. Frontier of India, Hazara, 1888.—Medal with clasp.
Tirah, 1897-98.—Medal with 2 clasps.

Moss-Blundell, R. S.—
South African War, 1902.—Served with Imperial Yeomanry. Operations in Cape Colony.—Queen's medal with 2 clasps.
The War of 1914-18.—Despatches, *Lond. Gaz.*, 12 Jun. 17. M. C.

Mosse, C. O. R.—
The War of 1914-18.—Despatches, *Lond. Gaz.*, 5 Apl. 16. M. C.

Mosse, W. O. M. (*Col., retd. I. A.*)—
Soudan, 1896.—Dongola Expedition.—Medal.

Mouat G. E. D.—
The War of 1914-18.—Despatches, *Lond. Gaz.*, 30 Jun 16, D. S. O.

Mowat, I. S.—
South African War, 1900-01.—Operations in the Transvaal, West of Pretoria, from 1st July to 29th November 1900. Operations in Orange River Colony, from 2nd March to 1st September 1901. Operations in Cape Colony, from 2nd September to 14th December 1901.—Queen's medal with 4 clasps.
The War of 1914-18.—Despatches, *Lond. Gaz.*, 5 Apl. and 10 May 16.

Muir, C. W. (*Col. retd. I.A.*)—
Afghanistan, 1880.—March from Quetta to the relief of Kandahar.—Medal.
Soudan, 1885.—Actions of Hasheen, Tofrek and Tamai.—Medal with 2 clasps; bronze star.
Burma, 1885-87.—Medal with clasp.
N.-W. Frontier of India, 1897-98.—Operations on the Samana and in the Kurrum Valley during Aug. and Sep. 97.—Medal with 2 clasps.
Tirah, 1897-98.—A. A. C., 1st Division.—Actions of the Sampagha and Arhanga Passes. Operations in the Bazar Valley, 25 to 30 Dec. 97. Despatches G. G. O. 244 of 98.—Clasp.

Muir, J. H.—
N.-W. Frontier of India, 1908.—Medal with clasp.

Muirhead, J. A.—
The War of 1914-18.—Despatches, *Lond. Gaz.*, 1 Jan. 10 and 15 May 17. D S.O.

Mukerji, U. N. (*Lt.-Col., retd. I. M. S.*)—
N.-W. Frontier of India, Zhob Valley, 1890.
N.-E. Frontier of India, Manipur, 1891.—Medal with clasp.

Mullane, P. (*Lt.-Col., retd. I. M. S.*)—
Burma, 1886-87.—Operations of the 1st and 6th Brigades.—Medal with 2 clasps.

N.-W. Frontier of India, 1st Miranzai, 1891.—*Chitral, 1895.* Relief of Chitral.—Medal with clasp.
Tirah, 1897-98.—Actions of Dargai and of the Sampagha and Arhanga Passes. Operations in the Bara Valley, 7 to 14 Dec. 97.—2 clasps.

Mullins, W. B. (*Bt.-Col., retd. I. A.*)—
Burma, 1887-89.—Operations of the 4th Brigade.—Medal with 2 clasps.
N.-W. Frontier of India, Waziristan, 1894-95.—Clasp.
N.-W. Frontier of India, Waziristan, 1901-02.—Medal with clasp.
East Africa, 1902-04.—Operations in Somaliland. Action at Jidballi. Despatches, *London Gazette*, 2 Sep. 04.—Medal with 2 clasps.
The War of 1914-18.—Despatches, *Lond. Gaz.*, 13 July 16.

Mulloy, N. F. C.—
The War of 1914-18.—M. C.

Mulvany, J.—
Chitral, 1895.—Relief of Chitral.—Medal with clasp.
N.-W. Frontier of India, 1897-98.—Tochi.—Clasp.
China, 1900.—Medal.

Mumford, R. (*Comsy. Hony. Capt., retd.*)—
Afghanistan, 1878-79-80.—Occupation of Kandahar. Advance on Khilat-i-Ghilzai and Ghazni. Battle of Ahmed Khel. Action of Urzu.—Medal with clasp.

Munn, F. L. R.—
The War of 1914-18.—Despatches, *Lond. Gaz.*, 1 Jan. 16. M. C.

Munn, R. G.—
N.-W. Frontier of India, Hazara, 1891.—Served as Brigade Transport Officer.—Medal with clasp.
Chitral, 1895.—Relief of Chitral.—Medal with clasp.
N.-W. Frontier of India, 1897-98.—Operations on the Samana and in the Kurrum Valley during Aug. and Sep. 97. Relief of Gulistan. Despatches, G. G. O. 1417 of 97.—2 clasps.
Tirah, 1897-98.—Actions of the Sampagha and Arhanga Passes. Reconnaissance of the Saran Sar and action of 9 Nov. 97. Operations in the Waran Valley and action of 16 Nov. 97. (Severely wounded.) Despatches, G. G. O. 244 of 1898.—Clasp.
N.-W. Frontier of India, Waziristan, 1901-02.—Despatches, G. G. O. 611 of 02.—Clasp.
N.-W. Frontier of India, 1902.—Operations against the Darwesh Khel Waziris.
East Africa, 1903-04.—Operations in Somaliland Served as A. D. C. to G. O. C., also as D. A. A. G. Head Qrs. Staff, Somaliland Field Force. Went on Special Mission to Ogaden Somalis. Despatches, *London Gazette*, 2nd Sep. 1904.—Medal with clasp.

Munro, D.—
The War of 1914-18.—Despatches, *Lond. Gaz.*, 15 Aug. 17. C. I. E.

Murison, C. C.—
East Africa, 1902-04.—Operations in Somaliland.—Medal with clasp.

Murphy, C. C. R.—
South African War, 1900—1902.—On Staff from 30th Sep. 1900 to 23rd Jan 1901 and again from 6th Aug. 1901 to 9th Feb. 1902. Operations in Cape Colony and Orange River Colony, July 1900. Operations in the Transvaal, east of Pretoria, July to 29th Nov. 1900. Operations in the Transvaal, 30th Nov. 1900 to March 1902.—Queen's medal with 3 clasps. King's medal with 2 clasps.

Operations in the Persian Gulf, 1913-14, including Tangistan June 1913; Naval Intelligence Officer. —Medal with clasp.

Murphy, P. (*Surg. Lt.-Col., retd. I.M.S.*)—
Egypt, 1882.—Battle of Tel-el-Kebir.—Medal with clasp; bronze star.

Murphy, R. T.—
N.-W. Frontier of India, Hazara, 1888.—Medal with clasp.
Chitral, 1895. Relief of Chitral.—Medal with clasp.

Murphy, W. R. (*Lieut.-Col., retd. I. M. S.*)—
Afghanistan, 1878-80.—Action of Saif-u-Deen and occupation of Khandahar. Battle of Ahmed Khel, actions of Urzoo and Patkao Shana. Despatches *Lond. Gaz.*, 22 Oct. 88.—Medal with clasp.
N.-W. Frontier of India, Hazara, 1888.—Despatches, G. G. O., 78 of 88.—Medal with clasp.
N.-E. Frontier of India, Lushai, 1889.—Clasp.
N.-E. Frontier of India, Chin-Lushai, 1889-90.—Mentioned by Govt. of India. Despatches, *Lond. Gaz.*, 12 Sept. 90.—Clasp. D. S. O.
Chitral, 1895.—With the Relief Force.—Medal with clasp.
N.-W. Frontier of India, 1897.—Served with the Kurram-Kohat Force; action in the Ublan Pass. Despatches, *Lond. Gaz.*, 11 Feb. 98. Also with Tirah Exped. Force as P. M. O., Kurram Movable Column—Despatches, *Lond. Gaz.*, 5 Apr. 98.

Murray, Alan (*Hon., M.-Gen., retd. Ben. S. C.*)—
India, 1857-58-59.—Siege and capture of Delhi, actions of Bolundshuhur and Agra, second relief of Lucknow and pursuit of Tantia Topee.—Medal with 2 clasps.

Murray, A. B.—
N.-E. Frontier of India, Chin Hills, 1892-93.—Medal with clasp.
Burma, 1895-96.

Murray, A. C.—
China, 1900.—Medal.

Murray, C. F. T.—
N.-W. Frontier of India, Miranzai, 1891.—Medal with clasp.
N.-W. Frontier of India, 1897-98.—Action of the Ublan Pass, 27 Aug. 97.—Operations on the Samana and in the Kurrum Valley during Aug. and Sep. 97. Relief of Gulistan. Despatches, G. G. O. 1417 of 97.—Medal with 2 clasps.
Tirah, 1897-98.—Reconnaissance of the Kharmana Defile and action of 7 Nov. 97. Operations against the Khani Khel Chamkannis.—Despatches, G. G. O. 244 of 98.—Clasp.
The War of 1914—18.—Despatches, *Lond. Gaz.*, 4 July 16. C.B.

Murray, E. M.—
The War of 1914-18.—Despatches, *Lond. Gaz.*, 15 Apl. 16. M.C.

Murray, F. S. J.—
N.-W. Frontier of India, Waziristan, 1901-02.—Medal with clasp.

Murray, G. (*Lieut.-Col., retd. I. A.*)—
N.-W. Frontier of India, Waziristan, 1894-95.—Medal with clasp.
N.-W. Frontier of India, 1897-98.—Operations on the Samana and in the Kurrum Valley during Aug. and Sep. 97. Relief of Gulistan. Operations of the Flying Column in the Kurrum Valley under Colonel Richardson, 20 Aug. to 1 Oct.—Medal with 2 clasps.

Murray, H. A.—
South African War, 1900-1901.—Queen's medal with 3 clasps.
West Africa (South Nigeria), 1901-1902.—Aro Expedition.—Medal with clasp.

Murray, H. B. (*Maj., retd. I. A.*)—
Burma, 1886-87.—Operations of the 4th and 6th Brigades. Minhla Flying Column. Staff Officer to No. 6 Sidottia Column.—Medal with clasp.
N.-W. Frontier of India, Hazara, 1888.—Clasp.
Zhob Valley, 1890.

Murray, J. G. P.—
China, 1900.—Medal.

Murray, J. H.—
Tibet, 1903-4.—Operations at and around Gyantse.—Medal with clasp.

Murray, K. D. B.—
South African War, 1902.—Operations in the Orange River Colony between January and May 1902.—Queen's Medal with 2 clasps.
N.-W. Frontier of India, 1908.—Operations in the Zakka Khel country. Operations in the Mohmand country. Engagement of Kargha, 24th May 1908.—Medal with clasp.
The War of 1914—18.—Despatches, *Lond. Gaz.*, 4 Jan. 17, 11 Dec. 17. D.S.O.

Murray, R. D. (*Col., retd. I.M.S.*)—
Burma, 1886-87.—Operations of the 1st Brigade. Despatches, G. G. O. 434 of 87.—Medal with clasp.

Murray, R. G. H.—
The War of 1914-18.—Despatches, *Lond. Gaz.*, 8 May 18, 22 Jan. 15 and 15 May 17. M.C.

Murray, W. G. P.—
Chitral, 1895.—Relief of Chitral.—Medal with clasp.

Muscroft, W. St. C.—
China, 1900.—Medal.

Muspratt, F. C.—
N.-W. Frontier of India, Isazai, 1892.
Chitral, 1895.—Relief of Chitral.—Medal with clasp.
Tirah, 1897-98.—2 clasps.
The War of 1914-18.—Despatches, *Lond. Gaz.*, 1 Jan. 16. C.M.G.

Muspratt, S. F.—
N.-W. Frontier of India, 1908.—Intell. Offr. Operations in the Zakka Khel country. Operations in the Mohmand country.—Medal with clasp.
The War of 1914-18.—Despatches, *Lond. Gaz.*, 1 Jan. 16, 15 Jun. 16 and 13 May 17. Brev. of Lt.-Col. D.S.O.

Muspratt, V. E.—
N.-W. Frontier of India, 1897-98.—Malakhand. Action at Landakai. Operations in Bajaur. Buner. Action of the Tanga Pass.—Medal with clasp.
The War of 1914-18.—Despatches, *Lond. Gaz.*, 1 Jan. 16.

N

Nandi, S. C. (*Lt.-Col., retd. I. M. S.*)—
Burma, 1886-88.—Operations of the 4th Brigade and in the Western Frontier.—Medal with 2 clasps.
N.-W. Frontier of India, (1st) Miranzai, 1891.
N.-W. Frontier of India, 1897-98.—With Tirah Exped. Force.—Medal with 2 clasps.
China, 1900.—Medal.
N.-W. Frontier of India, Waziristan, 1901-2.—Clasp.

Nand Singh, Sardar Bahadur—
The War of 1914-18.—India Order of Merit (2nd class)

Nangle, K. E.—
N.-W. Frontier of India, 1897-98.—Buner.—Transport Department.
South African War, 1900.—Special Service Officer under Brigadier-General Wood from Feb. to July 00. Operations in the Orange Free State, Feb. to May 00, including action at Karee Siding (29 March). Operations in Cape Colony, North and South of Orange River.—Queen's medal with 2 clasps.

Napier, A. H.—
N.-W. Frontier of India, 1908.—Operations in the Zakka Khel country. Operations in the Mohmand country.—Medal with clasp.
The War of 1914-18.—Despatches, *Lond. Gaz.*, 15 Aug. 17.

Napier, Hon'ble H. D.— (*Lt.-Col., retd. I. A.*)—
China, 1900.—Special Service Officer. Despatches, *Lond. Gaz.*, 14th May 1901. Brevet of Major.—Medal.

Neal, W. D.—
Burma, 1885-87.—Medal with clasp.
N.-W. Frontier of India, 1897-98.—Relief of Gulistan. Operations in the Kurram Valley under Col. Richardson.—Medal with clasp.

N.-W. Frontier of India, Mahsud-Waziri, 1881.
Hazara, 1888.—Medal with clasp, (*2nd*) *Miranzai, 1891.*—Clasp.—*Isazai, 1892.*
Tirah, 1897-98.—Operations against the Khani Khel Chamkannis. Operations in the Bazar Valley, 25 to 30 Dec. 97.—Medal with 2 clasps.

Needham, R. A.—
The War of 1914-18.—Despatches, *Lond. Gaz.*, 17 Feb. 15, 22 Jun. 15 and 1 Jan. 16. D.S.O.

Nelson, G. H. (*Hony. Maj.*) *Commissary, retd. S. and T. Corps* —
Afghanistan, 1879-80.—Defence and battle of Kandahar.—Medal.
Soudan, 1885.—Suakin.—Medal with clasp, bronze star.

Nekon, J. J. H.—
The War of 1914-18.—Despatches, *Lond. Gaz.*, 5 Apl. 16 and 15 Aug. 17.

Nepean, H. A. I. (*Col. retd. Mad. S. C.*)—
India, 1857-58.—Medal.
Burma, 1886-87.—Despatches, G. G. O. 864 of 1887.

Nepean, H. E. C. B.—
Soudan, 1885.—Frontier Field Force. Action of Giniss.—Medal, bronze star.
Burma, 1887-1889.—Medal with clasp.
Burma, 1889-1892.—Operations of the Touhon Column and in the Kachin Hills. Clasp.
The War of 1914-18.—Despatches, *Lond. Gaz.*, 15 Aug. 17. C.M.G.

Nesfield, V. B.—
Tibet, 1903-04—Medal.

Nethersole, F. R.—
Burma, 1892-1893.—Operations in the Kachin Hills.—Clasp.

Nethersole, J.—
The War of 1914-18.—Despatches, *Lond. Gaz.*, 22 Jan. 15.

Neumann, C. W.—
The War of 1914-18.—Despatches, *Lond. Gaz.*, 19 Oct. 16.

Neville, J. P. C. (*Hony. Br.-Genl., retd. I. A.*)—
Afghanistan, 1878-79-80.—Action of Charasiah. Operations at and around Kabul. Despatches G. G. O. 137 of '80.—Medal with 2 clasps; Brev. of Maj.

Newall, C. L. N.—
N.-W. Frontier of India, 1908.—Operations in the Zakka Khel country. Medal with clasp.
The War of 1914-18.—Despatches, *Lond. Gaz.*, 15 Jun. 16 and 11 Dec. 17. Brev. of Maj.

Newbold, A. W. (*Col., retd. I. A.*)—
Soudan, 1884-85.—Nile. With Desert column.—Medal with clasp; bronze star.
Burma, Wuntho Field Force, 1891-92.—Medal with clasp.
Kachin Hills, 1892-93.—Clasp.

Newell, H. A.—
N.-E. Frontier of India, Chin Hills, 1892-1893.—Medal with clasp.
Chitral, 1895.—With the Relief Force.—Medal with clasp.
Tirah, 1897-98.—Medal with 2 clasps.

Newington, C.—
South African War, 1901-02.—Operations in Orange River Colony, Sept. 1901. Operations in Cape Colony, Oct. 1901 to Jan. 1902. Operations in the Transvaal, Jan. to 31st May 1902.—Queen's medal with 5 clasps.

Newland, A. G. E. (*Maj., retd. Ind. Med. Serv.*)—
Soudan Expedition, 1885.—Suakin.
Burma, 1886-89.—Medal with 2 clasps.
N.-E. Frontier of India, 1891.—Manipur Expedition, 1891.—Clasp.
Burma, 1891-2-5-6.—Operations on N.-E. Frontier and in the Chin Hills; with Baungshe, Tlang-tlang and Tashon columns.—Clasp. Chin Hills, 1895—9.

Newman, E. A. R.—
N.-W. Frontier of India, 1897-98.—Defence of Malakand. Relief of Chakdara. Action of Landakai. Operations in Bajaur and in the Mamund country.—Medal with 2 clasps.
Tirah, 1897-98.—Operations in the Bara Valley, 7th to 14th Dec. 1897—Clasp.

Newman, J. H. (*Col., retd. Ind. Med. Serv.*)—
Abyssinia, 1868.—Medal.
Afghanistan, 1878-79.—Both the Bazar Valley expeditions.—Medal.

Newmarch, Sir O. R. (*Hon. M.-Gen., retd. Ben. S. C.*)—
India, 1857.—Actions of Agra, Gunga, and Pattialee.—Medal.

Newnham, C. C.—
N.-W. Frontier of India, 1897-98.—Operations in the Samana and in the Kurram Valley during Aug. and Sep. 97.—Medal with 2 clasps.

Tirah, 1897-98.—Operations against the Khani Khel Chamkannis.—Clasp.
South African War, 1899-1900.—Operations in the Orange Free State, Feb. to May 00. Operations in Orange River Colony (May to 29 Nov. 00), including action at Biddulphsberg (29 May), Wittebergen (1 to 29 July).—Queen's medal with 2 clasps.
The War of 1914-18.—Despatches, *Lond. Gaz.*, 17 Feb. 15 and 1 Jan. 16. Brev. of Lt.-Col.

Newnham-Smith, E. D. (*Lt.-Col., retd. I. S. C.*)—
Afghanistan, 1878-80.—Action at Khush-ki-Nakhud. Battle of Maiwand. Defence of Kandahar, sortie of Deh Kojah, and battle of Kandahar. Despatches, G. G. O. 610 of 79.—Medal with clasp.

Newton, G.—
N.-W. Frontier of India, 1897-1898.—Operations in the Samana and in the Kurram Valley, Aug. and Sept. 97.—Medal with 2 clasps.
Tirah, 1897-1898.—Clasp.
Tibet, 1903-1904—Medal.
Operations in the Abor Country, 1911-12.—Medal.

Nicholas, S. H. E.—
Tirah, 1897-98.—Medal with clasp.

Nicholls, A. (*Col., retd., I. A.*).—
Egypt, 1882.—Surrender of Kafr Dowar.—Medal; bronze star.
N.-W. Frontier of India, Zhob Valley, 1890.—Gomal column.
N.-W. Frontier of India, 1st and 2nd Miranzai, 1891.—Medal with clasp.

Tirah, 1897-98.—D. A. Q. M. General for Intelligence, 1st Division.—Actions of the Sampagha and Arhanga Passes. Reconnaissance of the Saran Sar and action of the 9 Nov. 97. Operations in the Bazar Valley, 25 to 30 Dec. 97—Despatches G. G. O. 483 of 98.—Medal with 2 clasps.
China, 1900.—D. A. A. G. 3rd Brigade. Despatches, *Lond. Gaz.*, 13 Sep. 01.—Medal.
N.-W. Frontier of India, 1908.—Operations in the Mohmand country. Engagement of Kargha, 24th May 1908.—Medal with clasp.

Nicholls, Thos. (*Lt.-Col., retd. Ben. Inf.*)—
N.-W. Frontier of India, 1863.—Action at Laloo forcing of Umbeyla Pass.—Medal with clasp.
N.-E. Frontier of India, 1865-66.—Bhutan Expedition.—Clasp.
Afghanistan, 1878-79.—Occupation of Kandahar, advance on the Helmund and Girishk ; also in Khyber Pass.—Medal.

Nicholson, F. L.—
The War of 1914-18.—Despatches, *Lond. Gaz.*, 19 Oct., 16 and 15 Aug. 17. M.C.

Nicholson, R. B.—
The War of 1914-18.—Despatches, *Lond. Gaz.*, 1 Jan. 16. M.C.

Nicholson, R. W.—
Soudan, 1884.—Battle of El-Teb. Relief of Tokar. Action of Tamai and at Tamanieb.—Medal with clasp, bronze star.
Chitral, 1895.—Medal with clasp.

Nicolas, C. P.—
South African War, 1899-1902.—Relief of Ladysmith, including operations of 17th to 24th January 1900. Operations on Tugela heights (14th to 27th February 1900) and action at Pieter's Hill (27th Feb. 1900). Operations in Natal, March to June 1900, including action at Laing's Nek (6th to 9th June). Operations in the Transvaal east of Pretoria, July to 29th Nov. 1900, including actions at Belfast (26th and 27th Aug. 1900), Lydenberg (5th to 8th Sep. 1900). Operations in the Transvaal between 30th Nov. 1900 and 30th Nov. 1901. Operations in Cape Colony between 1st Dec. 1901 and 31st May 1902. (Wounded at Ganvlei near Botha's Pass, 7th June 1900.)—Queen's medal with 6 clasps. King's medal with 2 clasps.

Nicolas, F. C.—
N.-W. Frontier of India, 1897-98—Operations on the Samana and in the Kurram Valley during Aug. and Sep. 97. Relief of Gulistan. Operations of the Flying Column in the Kurram Valley under Colonel Richardson, 20 Aug. to 1 Oct. 97.—Medal with 2 clasps.
Tirah, 1897-98.—Clasp.

Nicolas, R.—
China, 1900.—Medal.
Tibet, 1903-04.—Action at Niani, 26 June 04. Operations at and around Gyantse, between 5 May and 6 July 04. (Wounded.)—Medal with clasp.
N.-W. Frontier of India, 1908.—Operations in the Zakka Khel country.—Medal with clasp.

Nicolay, B. U.—
Tirah, 1897-98.—Medal with 2 clasps.
China, 1900.—Medal.
The War of 1914-18.—Despatches, *Lond. Gaz.*, 22 Jun. 15.

Nightingale, M. R. W.—
Tirah, 1897-98.—Served with Gurkha Scouts.—Medal with clasp.
China, 1900.—Medal.
The War of 1914-18.—Despatches, *Lond. Gaz.*, 25 Aug. and 5 Nov. 15. C.M.G., D.S.O.

Nisbet, T.—
South African War, 1901-02.—Operations in Cape Colony, Orange River Colony and the Transvaal, 1901-02.—Queen's Medal with 5 clasps.
The War of 1914-18.—Despatches, *Lond. Gaz.*, 15 Aug. 17. D.S.O.

Nisbett, W. G.—
Burma, 1889-92.—Wuntho Expedition, 1891.—Medal with clasp.

Nixon, C. H. F.—
The War of 1914-18.—Despatches, *Lond. Gaz.*, 19 Oct. 16. D.S.O.

Nixon, J. E.—
Afghanistan, 1879-80.—Zaimukht Expedition. Action of Zawa. Despatches. Brigadier-General J. A. Tytler's despatch, 20 Dec. 79.—Medal.
N.-W. Frontier of India, Mahsud Waziri, 1881.
Chitral, 1895.—Relief of Chitral.—Served as D. A. Q. M. G., Intelligence Branch. Action of the Malakand Pass. Action at Khar.—Despatches, G. G. O. 998 of 95.—Medal with clasp. Brevet of Lt.-Col.
N.-W. Frontier of India, 1897-98.—Tochi. Assistant Quarter Master General. Despatches, G. G. O. 1432 of 97.—Clasp.
South African War, 1901-02.—Commanded Cavalry Brigade in operations in Orange River Colony from 1 Jan. to 30 April and from 19 to 31 May 02 and in the Transvaal from 24 to 31 Dec. 01 and from 1 to 18 May 02. Despatches, *Lond. Gaz.*, 17 June 02.—Queen's medal with 4 clasps.—C.B.

Nolan, A. H. (*Sr. Asst. Surgn., Hony. Maj., retd. I. S. M. D.*)—
Afghanistan, 1878-80.—Medal.
N.-E. Frontier of India, Chin Lushai, 1889-90.—Medal with clasp.
Burma, 1891.—Wuntho Field Force.—Clasp.
N.-E. Frontier of India, Kachin Hills, 1892-93.—Clasp.

Norbury, P. Fitz G.—
The War of 1914-18.—Despatches, *London Gazette*, 4th July 1916.

Norie, C. E. deM.—
N.-E. Frontier of India, Manipur, 1891.—Medal with clasp.
N.-W. Frontier of India, 1897-98.—Operations on the Samana during Aug. and Sep. 97. Relief of Gulistan.—Medal with 2 clasps.
Tirah, 1897-98.—Actions of Chagri Kotal, Dargai and of the Sampagha and Arhanga Passes. Operations in the Waran Valley and action of 16 Nov. 97. Operations at and around Dwatoi and action of 24 Nov. 97. Operations against the Khani Khel Chamkannis. Operations in the Bara Valley, 7 to 14 Dec. 97. (Severely wounded.) Despatches, G. G. Os. 58 and 244 of 98. Clasp. Brev. of Maj.
South African War, 1901-02.—On Special duty, lines of communication. Despatches, *Lond. Gaz.*, 11 Sep. 01.—Medal with 2 clasps. D. S. O.
The War of 1914-18.—Despatches, *Lond. Gaz.*, 30 Nov. 14, 17 Feb. 15, 22 Jun. 15, 1 Jan. 16, 19 Oct. 16 and 15 Aug. 17. Brev. of Col. C.B., C.M.G.

Norman, C. L.—
N.-W. Frontier of India, 1897-98.—Operations of the Samana and in the Kurram Valley during Aug. and Sep. 97. Operations of the Flying Column in the Kurram Valley under Colonel Richardson, 20 Aug. to 1 Oct. 97.—Medal with 2 clasps.
Tirah, 1897-98.—Clasp.
East Africa, 1903-04.—Operations in Somaliland. Despatches, *London Gazette*, 2 Sep. 1904.—Medal with clasp.
The War of 1914-18.—Despatches, *Lond. Gaz.*, 15 Aug. 17. D.S.O.

Norman, W. H.—
N.-W. Frontier of India, Isazai, 1892.
Chitral, 1895.—Relief of Chitral.—Medal with clasp.
N.-W. Frontier of India, 1897-98.—Defence of Malakand. Relief of Chakdara-Malakand. Action at Landakai. Operations in Bajour and in the Mamund Country.—Mohmand.—2 clasps.
China, 1900.—Medal.
The War of 1914-18.—Despatches, *Lond. Gaz.*, 15 Aug. 17. D.S.O.

Norman, W. W. (*Bt. Col., retd. I. A.*)—
Tirah, 1897-98.—Operations against the Khan Khel Chamkannis.—Operations in the Bara Valley, 7th to 14th December 1897, and Bazar Valley, 25 to 30 Dec. 97.—Medal with clasp.
The War of 1914-18.—Despatches, *Lond. Gaz.*, 15 Jun. 16 and 4 Jan. 17. D.S.O.

Norris, H.—
South African War, 1899-1902.—Operations in Orange River Colony, 30 Nov. 00 to May 01. —Queen's medal with 3 clasps.

Norris, R. J. N.—
The War of 1914-18.—Despatches, *Lond. Gaz.*, 5 Apl. 16.

Norton, W. R. T. (*Lt., retd. I.A.*).—
South African War, 1902.—Operations in Orange River Colony, Apr. to 31 May 02.—Queen's medal with 3 clasps.

Nott, A. H.—
N.-W. Frontier of India, (*1st and 2nd*) *Miranzai, 1891.*—Medal with clasp.—*Zhob Valley, 1890.*
The War of 1914-18.—Despatches, *Lond. Gaz.*, 19 Oct. 16.

Noverre, W. L. (*Hon. Col., retd. Ben. S.C.*)—
India, 1858.—Medal.

Noyes, C. D.—
The War of 1914-18.—Despatches, *Lond. Gaz.*, 19 Oct. 16. M.C.

Nurse, C. G. (*Lieut.-Col, retd. I. A.*)—
Soudan, 1884.—Occupation of Fort Baker, battle of El-Teb, relief of Tokar. Battle of Tamai.—Medal with clasp, bronze star.
Somali Coast, 1890.—Zaila Field Force.

Nugent, J. F. H.—
The War of 1914-18.—Despatches, *Lond. Gaz.*, 19 Oct. 16. D.S.O.

Nurse, H. H. (*Major, retd. I. A.*)—
China. 1900.—Medal.
East Africa, 1902-04.—Operations in Somaliland.—Medal with clasp.
Aden, 1903-04.—Operations in the interior.

Nutt, H. R. -
N.-W. Frontier of India, 1902.—Operations against the Darwesh Khel Waziris.

Nuttall, J. R. (*Maj., retd., I. A.*)—
N.-W. Frontier of India, 1897-98.—Mohmand.-Medal with clasp.
Tirah, 1897-98.—Operations in the Bara Valley, 7 to 14 Dec. 97.—Clasp.

Nutting, H. B.—
The War of 1914-18.—Despatches, *Lond. Gaz.*, 4 Jan. 17.

Oakes, G. R.—
N.-E. Frontier of India, Chin Hills, 1892-93.—Medal with clasp.
China, 1900.—Medal.

Oakes, T. T.—
Aden, 1903-04.—Operations in the interior.

Obbard, E. N.—
Burma, 1887-89.—Medal with clasp.

Obbard, O. J. (*Maj., Sup. List, I. A.*)—
Burma, 1887.—Operations of the 1st Brigade.—Medal with clasp.

Obbard, T. (*Hon. Maj.-Gen., retd., Mad. S. C.*).—
India, 1858-59.—Capture of Nerowlee, action of Ghurra, re-occupation of Hunnepore, and engagement and destruction of Talpore. Despatches, *Lond. Gaz.*, 10 Aug. 58.—Medal.
Abyssinia, 1867-68.—Medal.

O'Brien, B. P. T.—
The War of 1914-18.—M.C.

O'Brien, Jos. (*Lt.-Col., retd. Ind. Med. Serv.*)—
N.-E. Frontier of India, 1879-80, Naga Hills Expedition.—Action of Konama. Despatches, G. G. Os. 123 and 405 of 80.—Medal with clasp.

O'Bryen, C. W.—
Burma, 1886-88.—Operations of the 3rd and 4th Brigades, Shan States column. Southern Shan column. Despatches, G. G. O. 434 of '87.—Medal with 2 clasps.
N.-W. Frontier of India, Hazara, 1891.—Clasp. (*2nd*) *Miranzai, 1891.*—Clasp.
East Africa, 1903-04.—Operations in Somaliland.—Medal with clasp.

O'Connor, P. F. (*Lieut.-Col., retd. I. M. S.*)—
Afghanistan, 1878-79-80.—Battle of Ahmed Khel. Actions of Urzu and Patkao Shana. Despatches, G. G. O. 493 of '80.—Medal with clasp.
Egypt, 1882.—Action of Tel-el-Kebir.—Medal with clasp; bronze star.
Burma, 1886-87.—Operations of the 2nd Brigade, Wuntho Expedition, including advance from Katha and Manle to Wuntho. Despatches, G. G. O. 434 of '87.—Medal with 2 clasps.
Chitral, 1895.—Relief to Chitral. Action of the Malakand Pass. Despatches, G. G. O. 998 of '95. —Medal with clasp.
N.-W. Frontier of India, 1897-98.—Malakand. Operations in Bajaur and in the Mamund country. Utman Khel, Buner. Action of the Tanga Pass. Despatches, G. G. O. 178 of 98. —Clasp.
China, 1900.—Medal.

O'Connor, W. F. (*Sr. Asst. Surg. Hony. Capt., retd., I. S. M. D.*)—
Burma, 1886-87-88.—Medal with 2 clasps.
N.-W. Frontier of India, 1897-98.—Operations on the Samana and in the Kurrum Valley during Aug. and Sep. 97.—Medal with 2 clasps.
Tirah, 1897-98.—Actions of the Chagru Kotal and of the Sampagha and Arhanga Passes. Operations at and around Dwatoi and action of the 24 Nov. 97. Operations against the Khani Khel Chamkannis. Operations in the Bazar Valley from 25 to 30 Dec. 97 Despatches, G. G. O. 244 of 98.—Clasp. Promoted Senior Assistant Surgeon.

O'Connor, W. F. T.—
N.-W. Frontier of India, 1897-98.—Malakand. Action at Landakai. Operations on the Samana and in the Kurram Valley during August and September 1897.—Medal with 2 clasps.
Tirah, 1897-98.—Action of Dargai. Action of the Sampagha Pass. Action of the Arhanga Pass. Operations at and around Dwatoi, and action of 24th November 1897.—Clasp.
Tibet, 1903-04.—As Secretary and Interpreter to British Commissioner. Action at Niani, 26th June 1904. Operations at and around Gyantse, between 5th May and 6th July 1904. (Severely wounded.) March to Lhassa, 14th July to 3rd August 1904. —Medal with clasp. C. I. E.

Odlum, W. H.—
South African War, 1900-02.—Operations in Orange River Colony (May to 29th November 1900), including actions at Biddulphsberg (29th May), Bethlehem (6th and 7th July), Wittebergen (1st to 29th July), Witpoort (15th August). Ladybrand (2nd to 5th September), Caledon River (27th to 29th November 1900). Operations in Orange River Colony, between 30th November 1900 and 31st May 1902 Despatches, *London Gazette*, 15th November 1901.—Queen's medal with 3 clasps. King's medal with 2 clasps.
The War of 1914-18.—Despatches, *Lond. Gaz.*, 22 June 15.

O'Donald, E.—
Tirah, 1897-98.—Medal with 2 clasps.

O'Donnell, C.—
Egypt, 1882.—Actions of Tel-el-Mahuta and Kassassin. Battle of Tel-el-Kebir, and forced march to Zagazig.—Medal with clasp Bronze Star.

O'Donnell, G. B. (*Lt.-Col., retd. I. A.*)—
N.-W. Frontier of India, Zhob Valley, 1884.—Intell. Branch, Q.M. G.'s Department.—Affair at Petao (Daulatzai).

O'Donoghue, M. E. (*Bt.-Col., retd. I. A.*)—
Afghanistan, 1880.—Medal.
Burma, 1886-89.—Expedition against the Karennees.—Medal with 2 clasps.

O'Dwyer, M. (*Lt.-Col., retd. Ind. Med. Serv.*)—
Afghanistan, 1879.—Medal.

O'Gorman, P. W.—
N.-W. Frontier of India, (1st and 2nd) Miranzai, 1891.—Medal with clasp.

O'Grady, H. de. C.—
N.-W. Frontier of India, 1902.—Operations against the Darwesh-Khel Waziris.
N.-W. Frontier of India, 1908.—Operations in the Mohmand Country. Engagement of Kargha, 24th May 1908. (Severely wounded.) Despatches, G. G. O. No. 521 A of 1908.—Medal with clasp.

O'Keefe, D. S. A.—
The War of 1914-18.—Despatches, *Lond. Gaz.*, 5 Apl. 16.

O'Kinealy, F.—
N.-W. Frontier of India, 1897-98.—Medal with clasp.
Tirah, 1897-98.—Clasp.

O'Malley, W. A. D'O. (*Col., retd. I. A.*)—
N.-W. Frontier of India, Mahsud Waziri, 1881.—Despatches, Brigadier-General T. G. Kennedy's, dated 20 June '81. *Zhob Valley, 1890.*
N.-W. Frontier of India, Waziristan, 1894-95.—Action at Wano, 3 Nov. 94. Despatches, G. G. O. 268 of 95.—Medal with clasp. Brev. of Lt.-Col.
N.-W. Frontier of India, 1897-98.—Tochi.—Medal with clasp.
N.-W. Frontier of India, 1902.—Operations against the Darwesh Khel Waziris.

O'Meara, C. A. E.—
N.-W. Frontier of India, 1897-98.—Mohmand.—Medal with clasp.
Tibet, 1903-04.—Medal with clasp.

Ogilvie, W. H.—
Soudan, 1896.—Dongola Expedition.—Medal; Khedive's medal.
China, 1900.—Relief of Pekin.—Medal with clasp.
Tibet, 1903-04.—March to Lhassa, 14th July to 3rd August 1904.—Medal.
The War of 1914-18.—Despatches, *Lond. Gaz.*, 15 Aug. 17. C. M. G.

Ogle, N.—
The War of 1914-18.—Despatches, *Lond. Gaz.*, 5 Apl. 16, 19 Oct. 16 and 15 Aug. 17. D. S. O.

Ogg, A. C.—
The War of 1914-18.—Despatches, *Lond. Gaz.*, 5 Apl. and 13 July 16. D.S.O.

Ogley, L. (*Hony. Lt., D. A.C., retd. S. and T. Corps*)—
N.-W. Frontier of India, Jowaki, 1877.—Medal with clasp.
Afghanistan, 1878-79-80.—Action of Ali Musjid.—Medal with clasp.
Egypt, 1882.—Action of Tel-el-Kebir.—Medal with clasp. Bronze star.
N.-W. Frontier of India, Hazara, 1888.—Clasp.

Ommanney, A. H. (*Capt., retd. I. A.*)—
Chitral, 1895.—Relief of Chitral.—Action of the Malakand Pass. (Wounded.)—Medal with clasp.
N.-W. Frontier of India, 1897-98.—Malakand. Operations in Bajaur and in the Mamund Country.—Clasp.

Ommanney, M. W. (*Col., retd. Ben. Art.*)—
North-West Frontier of India, 1863.—Umbeyla Campaign and expedition into the Eusofzai.
Afghanistan, 1879-80.—Expedition into the Hissarak Valley, and other operations.—Medal.

Ommanney, W. M.—
Operations in the Abor country, 1911-12.—Medal with clasp.

O'Neill, H. J. D.—
East Africa (Somaliland), 1908-10—Medal with clasp.
The War of 1914-18.—Despatches, *Lond. Gaz.*, 30 Jun. 16.—M. C.

O'Neill, J. S.—
Operations in the Abor country, 1911-12.—Despatches, *Gazette of India*, Notification No. 480, dated 23rd May 1912.—Medal with clasp.
The War of 1914-18.—Despatches, *Lond. Gaz.*, 22 Jun. 15.—M.C.

Onslow, F. M. (*Hon. Lt.-Col., retd. Mad. S.C.*)—
Abyssinia, 1867-68.—Attached to Land Transport Train. Despatches, *Lond. Gaz.*, 10 and 14 July '68.—Medal.
Afghanistan, 1878-79.—Capture of Ali Musjid.—Medal with clasp.

Onslow, R. C.—
Afghanistan, 1878-79-80.—Actions with Ghilzais, and at Jagdalak.—Medal.

Orchard, A. J.—
South African War, 1900-02.—Operations in the Orange Free State, Feb. to May 00. Operations in Orange River Colony (May to 29 Nov. 00). Operations in Cape Colony, South of Orange River, 00. Operations in the Transvaal, Mar. 01 to May 02.—Queen's medal with 4 clasps. King's medal with 2 clasps.

Orchard, D. S.—
Operations in the Abor country, 1911-12.—Medal with clasp.

O'Reilly, C. M.—
Tirah, 1897-98.—Medal with 2 clasps.
China, 1900.—Medal.
Burma, 1889.—Medal with clasp.

O'Reilly, L. J. (*Capt., retd. I. S. M. D.*)—
Afghanistan, 1878-79.—Medal.

O'Rorke, J. M. W.—
The War of 1914-18.—Despatches, *Lond. Gaz.*, 11 Dec. 17.—Legion of Honour (5th class).

Orman, F. L.—
N.-E. Frontier of India, Kachin Hills, 1892-93.—Medal with clasp.
Burma, 1893-94.

Orr, G. M.—
N.-W. Frontier of India, 1897-98.—Malakand.—Operations in Bajaur and in the Mamund Country. Mohmand.—Medal with 2 clasps.
The War of 1914-18.—Despatches, *Lond. Gaz.*, 30 Jun. 16 and 8 Feb. 17.—Brev. of Lt.-Col. D.S.O.

Orr, R.—
Chitral, 1895.—Relief of Chitral.—Promoted to Deputy Assistant Commissary.—Medal with clasp.
Soudan, 1896.—Medal; Egyptian Medal.
The War of 1914-18.—Despatches, *Lond. Gaz.*, Jan. and 15 Jun. 16. Granted hon. rank of Lt.-Col.

Orton, E. F.—
China, 1900.—Relief of Pekin. Railway Staff Officer, graded as Staff Captain. Mentioned in despatches. *Lond. Gaz.*, 13 Sep. 01.—Medal with clasp.
Mekran, 1901.—Attack and capture of Nodiz Fort. Employed as Special Service Officer (Intelligence). Mentioned in despatches. G. G. O. No. 1113 of 02.

Orton S. B.—
The War of 1914-18.—Despatches, *Lond. Gaz.*, 15 Jun. 16 and 15 Aug. 17.

Ostrehan, F. G. R.—
Afghanistan, 1880.—Medal.
N.-W. Frontier of India, 1897-98.—Mohmand.—Medal with clasp.
Tirah, 1897-98.—Operations in the Bara Valley, 7 to 14 Dec. 97.—Clasp.

Oswald, W. A. (*Col., retd. I. A.*)—
Burma, 1885—87.—Operations of the 1st Brigade—Medal with 2 clasps.
China, 1900.—Medal.

Ottley, W. J.—
N.-W. Frontier of India, 1897-98.—Utman Khel.—Buner.—Action of the Tanga Pass.—Medal with clasp.
Tirah, 1897-98.—Clasp.
N.-W. Frontier of India, Waziristan, 1901-02.—Clasp.
Tibet, 1903-04.—Action at Niani, 26th June 1904. Operations at and around Gyantse, between 5th May and 6th July 1904. March to Lhassa, 14th July to 3rd August 1904. Despatches, *Gazette of India*, No. 1065 of 1904.—Medal with clasp. Brevet of Major.
N.-W. Frontier of India, 1908.—Operations in the Mohmand country.—Medal with clasp.
The War of 1914-18.—Despatches, *Lond. Gaz.*, 4 July 1916 and 1 Dec. 16.

Overbeck-Wright, A. W.—
East Africa, 1903-04.—Operations in Somaliland.—Medal with clasp.

Owen, C. A. (*Lt.-Col., retd. I. A.*)—
Abyssinia, 1867-68.—Action at Arogee and storm and capture of Magdala. Despatches, *Lond. Gaz.*, 10 July 86.—Medal.
Afghanistan, 1880.—Medal.

Owen, C. W. (*Lt.-Col., retd. I.M.S.*)—
Afghanistan, 1879-80.—Action of Charasiah. Operations at and around Kabul (wounded slightly). Despatches, G. G. O. 137 of '80.—Medal with 2 clasps.
Egypt, 1882.—Battle of Tel-el-Kebir. Medal with clasp; bronze star.

Owen, R. J.—
N.-W. Frontier of India (2nd Miranzai), 1891.—Medal with clasp.
Chitral, 1895.—Relief of Chitral. Action of the Malakand. Action at Khar.—Medal with clasp.

Ozanne, G. D.—
The War of 1914-18.—Despatches, *Lond. Gaz.*, 30 Jun. 16.

Ozzard, F. R.—
N.-W. Frontier of India, (1st and 2nd) Miranzai, 1891.—Medal with clasp.
N.-W. Frontier of India, 1897-98.—Tochi.—Medal with clasp.
China, 1900.—Medal.
The War of 1914-18.—Despatches, *Lond. Gaz.*, 22 Jun. 15.

Ozzard, H. H. (*Maj., retd. Ind. S.C.*)—
Perak, 1875-76.—Medal with clasp.
Afghanistan, 1878-79.—Medal.

P

Paddon, S. S. W. (*Capt., retd. I. A.*)—
South African War, 1901-02.—Operations in Cape Colony, February and March 1901. Operations in Orange Free State and Transvaal, from March 1901 to 31st May 1902.—Queen's medal with 5 clasps.

Paige, C. P.—
The War of 1914-18.—Despatches, *Lond. Gaz.*, 31 Oct. 17—C. I. E.

Pais, S. J. (*Sr. Asst. Surgn. Hony. Capt., retd. I. S.M. D.*)—
Afghanistan, 1878-79-80.—Medal.
China, 1900.—Medal.

Palaeologus, W. C. (*Major, retd. I. A.*)—
N.-W. Frontier of India, 1908.—Operations in the Zakka Khel Country.—Medal with clasp.

Palin, C. W.—
The War of 1914-18.—M. C.

Palin, G. W.—
N.-W. Frontier of India, Zhob Valley, 1884.
N.-W. Frontier of India, Isazai, 1892.
Soudan, 1896.—Dongola Expedition.—Medal. Egyptian Medal.
Tirah, 1897-98.—Actions of the Sampagha and Arhanga Passes. Operations in the Bazar Valley, 25 to 30 Dec. 97.—Despatches, G. G. O. 483 of 98.—Medal with 2 clasps.

Palin, P. C.—
Burma, 1887—89.—Medal.
N.-W. Frontier of India, Hazara, 1888.—Medal.
Waziristan, 1894-95.—Clasp.
N.-W. Frontier of India, 1897-98.—Tochi.—Medal with clasp.
The War of 1914-18.—Despatches, *Lond. Gaz.*, 5 Nov. 15, 1 Jan. 16, 21 Jun. 16, 13 July 16 and 6 July 17.—Brev. of Col. C.B.

Palin, R. H.—
China, 1900.—Medal.

Palmer, C. H. (*Col., retd. Ind. S.C.*)—
N.-W. Frontier of India, 1868. Hazara Expedition.—Occupation of the Black Mountain.—Medal with clasp.
N.-E. Frontier of India, 1871-72.—*Looshai Expedition.*—Mentioned in despatches.—Clasp.
Burma, 1887.—Clasp.

Palmer, Gerald S.—
China, 1900.—Medal with clasp.

Palmer, H. I. E.—
N.-W. Frontier of India, (2nd) Miranzai, 1891.—Medal with clasp.
Chitral, 1895.—Relief of Chitral.—Medal with clasp.
N.-W. Frontier of India, 1897-98—Relief of Gulistan.—Malakand—Action at Landakai—(Severely wounded.)—2 clasps.
N.-W. Frontier of India, Waziristan, 1901-02.—Clasp.
The War of 1914-18.—Despatches, *Lond. Gaz.*, 51 Jun. 16.—C.M.G.

Palmer, T. H. G.—
The War of 1914-18.—Despatches, *Lond. Gaz.*, 15 Aug. 17.

Palmer, S. V.—
The War of 1914-18.—Despatches, *Lond. Gaz.* 15 Aug. 17.

Pank, P. D. (*Lt.-Col., retd. I. M. S.*)—
N.-W. Frontier of India, Mahsud Waziri, 1887.

Park, J.—
China, 1900.—Medal.

Parker, G. M. G.—
N.-W. Frontier of India, 1897-8.—With Tirah Exped. Force.—Medal with 2 clasps.
N.-W. Frontier of India, Waziristan, 1901-02.—Clasp.
The War of 1914—18.—Despatches, *London Gazette*, 4 July 1916.

Parker, H. P. E. (*Major, retd. I. A.*)—
East Africa (British), 1896.—Operations against the Mazrui Rebels.—Medal.

Parker, J. (*Lt.-Col., retd. I.M.S.*)—
Burma, 1886-9.—Medal with 2 clasps.

Parker, J (*Major, retd. I. A.*)—
Burma, 1886-87.—Medal with 2 clasps.
China, 1900.—Relief of Pekin.—Medal with clasp.

Parker, N. T.—
N.-W. Frontier of India, 1897-98.—Tochi.—Medal with clasp.
China, 1900.—Medal.

Parkin, H. (*Lt.-Col., retd. I. A.*)—
N.-W. Frontier of India, Zhob Valley, 1884.
Burma, 1886—89.—Despatches G. G. O. 434 of 87.—Medal with 2 clasps.
Burma, 1892-93.—Operations in the Kachin Hills—Received the thanks of the Government of India.—Foreign Department letter No. 60 C. 21 Nov. 93—Clasp.

Parkin, J. F—
The War of 1914-18.—Despatches, *Lond. Gaz.*, 22 Jun. 15.

Parkinson, J. T.—
N.-W. Frontier of India, Hazara, 1888.—Promoted 1st Grade Assist. Apothecary.—Medal with clasp.

Parr, H. O.—
Burma, 1892-93.
Chin Hills, 1892-93.—Medal with clasp. Despatches, G. G. O. 733 of 1893.
Tirah, 1897-98.—Action of Chagru Kotal.—Medal with 2 clasps. Despatches, G. G. O. 483 of 98.
China, 1900.—Relief of Pekin. Action of Peitsang. Action of Yangtsun. Despatches, *Lond. Gaz.*, 14 May 01.—Medal with clasp.
Tibet, 1903-04.—Medal.
The War of 1914-18.—Despatches, *Lond. Gaz.*, 5 Apl., 13 July and 19 Oct. 16. Brev. of Col.

Parrott, B. A. N. (*Lt.-Col., retd. Ind. S.C.*)—
Operations, Frontier Hills, Arakan, 1871-72.
Burma, 1886—89.—Employed on Civil duties. Mentioned by Govt. of India. Despatches, G. G. O. 434 of 87.—Medal with clasp.

Parry, E. R.—
Chitral, 1895.—Relief of Chitral.—Medal with clasp.
East Africa, 1903-4.—Operations in Somaliland.—Medal with clasp.

Parsons, A. E. B.—
The War of 1914-18.—Despatches, *Lond. Gaz.*, 20 Mar. 16.

Parsons, A. D. (*Col., retd. Mad. Cav.*)—
Afghanistan, 1879-80.—Medal.

Patch, F. R.—
The War of 1914—18.—Despatches, *Lond. Gaz.*, 20 Mar. 16.

Patch, R. (*Col., retd. I. A.*)—
N.-E. Frontier of India, 1865. Bhutan Expedition.—Medal with clasp.
Afghanistan, 1879-80.—With Peshawar Valley Field Force. Mentioned in Despatches.—Medal.
N.-W. Frontier of India, 1888. Hazara Expedition.—Mentioned in Despatches.—Brev. of Lt.-Col.
Miranzai Expedition, 1891.—Mentioned by Government of India.—Despatches, G. G. O. 709 of 91.

Paterson, A. R.—
N.-W. Frontier of India, Hazara, 1891.—Medal with clasp.
N.-W. Frontier of India, (2nd) Miranzai, 1891.—Clasp.

Paterson, F. W. J.—
The War of 1914-18.—Despatches, *Lond. Gaz.*, 5 Apl. 16.—M.C.

Paterson, G. F. J.—
The War of 1914-18.—Despatches, *Lond. Gaz.*, 1 Jan. 16 and 15 Aug. 17.—M.C.

Paterson, H. F. W.—
Tibet, 1903-04.—Action at Niani, 26th June 1904. Operation at and around Gyantse, between 5th May and 6th July 1904. March to Lhassa, 14th July to 3rd August 1904.—Medal with clasp.

Paterson, R. H. O'D.—
South African War, 1900-01.—Operations in Cape Colony, 30 Nov. 1900 to July 01.—Queen's medal with 2 clasps.

Paterson, T. G. F.—
The War of 1914-18.—Despatches, *Lond. Gaz.*, 15 Aug. 17.—D. S. O.

Paterson, T. S.—
N.-W. Frontier of India, 1908.—Operations in the Mohmand country.—Medal with clasp.

Paton, W. C.—
The War of 1914-18.—Despatches, *Lond. Gaz.*, 1 Jan. 16:—M. C.

Patrick, R. M. F.—
The War of 1914-18.—Despatches, *Lond. Gaz.*, 1 Jan. 16.—M. C

Patterson, S. B. A.—
N.-W. Frontier of India, Waziristan, 1894-95.—Medal with clasp.
N.-W. Frontier of India, 1897-98.—Defence and relief of Malakand. Relief of Chakdara—Malakand. Operations in Bajaur and Mamund country.—Medal with 2 clasps.

Patton, W. S.—
Aden, 1903-04.—Operations in the interior.
The War of 1914-18.—Despatches, *Lond. Gaz.*, 19 Oct. 16.

Paul, E. T.—
Egypt, 1882.—Action of Tel-el-Kebir and subsequent pursuit to Zagazig.—Medal with clasp. Bronze star.

Paul, R. S.—
China, 1900—Relief of Pekin.—Medal with clasp.
The War of 1914-18.—Despatches, *Lond. Gaz.*, 4 July 16.

Peacock, E. B.—
N.-W. Frontier of India, 1897-98.—Defence and Relief of Malakand.—Malakand.—Action of Landakai. Operations in the Mamund country. (Severely wounded.)—Medal with 2 clasps.

Peacocke, T. G.—
Chitral, 1895.—With the Relief Force.—Medal with clasp.

Pears, T. C. (*Lt.-Col., u. s. l. I.A.*)—
Afghanistan, 1878-79.—Thal-Chotiali Field Force. Despatches, G. G. O. 610 of '79.—Medal.

Pearse, J. C.—
South African War, 1900-1901.—Operations in Natal, April to June 1900. Operations in the Transvaal, and Orange River Colony, July to November 1900. Operations in the Transvaal, Feb. to May 1901. Served with Mounted Infantry.—Queen's Medal with 4 clasps.

Pearse, S. A. (*Maj., retd. I.A.*)—
Burma, 1885-89.—Operations of the 1st Brigade. Port Transport Officer, Zagabin.—Medal with 2 clasps.
The War of 1914-18.—Despatches *Lond. Gaz.*, 6th Dec. 16.—D.S.O.

Pearson, Sir A. A. (*Lieut.-Genl., retd. I. A.*)—
Afghanistan, 1880.—Medal.
Burma, 1886-87.—As D. A. A. and Q. M. G., 3rd Brig., Upper Burma Field Force.—Despatches, Bombay G. G. O. 339 of 87.—Brev. of Maj. Medal with clasp.
East Africa (British), 1896.—Commanded the Field Force against the Mazrui rebels. Received the thanks of Her Majesty's Government.—Medal and Order of Brilliant Star of Zanzibar, 1st class, 1st grade.

Pearson, H. F. A.—
N.-W. Frontier of India, (1st and 2nd) Miranzai, 1891.—Medal with clasp.

Chitral, 1895.—Relief of Chitral.—Medal with clasp.

Tibet, 1903-04.—Operations at and around Gyantse between 5th May and 6th July 1904. Despatches, *Gazette of India*, No. 1065 of 1904.—Medal with clasp.

N.-W. Frontier of India, 1908.—Operations in the Zakka Khel country.—Medal with clasp.

Pearson, W. Mac M.—
The War of 1914-18.—Despatches, *Lond. Gaz.*, 13 July 16.

Peart, C. L.—
China, 1900.—Relief of Pekin. Despatches, *Lond. Gaz.*, 14 May 01.—Medal with clasp.
N.-W. Frontier of India, Waziristan, 1901-02.—Medal with clasp.

Peart, G. R. (*Col., u.s.l. Ind. S.C.*)—
Afghanistan, 1878-79.—Medal.

Peck, A. W.—
N.-W. Frontier of India, Waziristan, 1894-95.—Medal with clasp.

South African War, 1900.—Operations in the Transvaal between 30 June and 31 Dec. 00. Operations in the Orange River Colony between 1 June and 12 June 00. Operations in Cape Colony, south of the Orange River, 99-00. Remount Depôt, Port Elizabeth, 11 to 24 June 00.—Queen's Medal with 3 clasps.
N.-W. Frontier of India, 1908.—Operations in the Zakha Khel country. D. A. A. G. Field Force Staff.—Medal with clasp.
The War of 1914-18.—Despatches, *Lond. Gaz.*, 7 Feb. 15, 4 Jan. 17, and 11 Dec. 17, Brevs. of Lt.-Col. and Col.—C.M.G.

Peck, E. S. (*Major, retd. I.M.S.*)—
Chitral, 1895.—Relief of Chitral.—Medal with clasp.

Peck, J. H.—
N.-W. Frontier of India, 1897-98.—Tochi.—Medal with clasp.

Pedder, D. C. (*Hon. Lt.-Col., retd. Bo. S.C.*)—
Crimea, 1855.—Siege and assault of Sevastopol.—Medal with clasp.

Pedrora, F. H. (*Surg. Maj., retd. I.M.S.*)—
Burma, 1886-9.—Medal with 2 clasps.

Peebles, A. S. M.—
N.-W. Frontier of India, 1908.—Operations in the Mohmand country.—Medal with clasp.

Peel, B. G.—
South African War, 1900.—Operations in Natal, 31 Mar. to 10 June, including action at Laing's Nek (6 to 9 June). Operations in the Transvaal east of Pretoria, 16 July to 15 Sep. 00. Operations in Orange River Colony, 11 to 15 Jun. 00. Operations in Cape Colony, south of Orange River, 29th Mar. 00.—Queen's medal with 4 clasps.
The War of 1914-18.—Despatches, *Lond. Gaz.*, 5 Apl., 13 July and 19 Oct. 16.—D.S.O.

Pelle, H. D.—
N.-W. Frontier of India, Waziristan, 1901-02.—Slightly wounded. Despatches, G. G. O. No. 611 of 1902.—Medal with clasp.

Pelle, S. C. F.—
Afghanistan, 1878-79.—Medal.
Burma, 1885-87.—Served as Commissariat Officer in the 3rd Brigade. Taking of Minhla. Occupation of Mandalay and Bhamo. Wuntho Expedition. Operations of the 5th Brigade. Thanks of the Governor-General in Council. Despatches, G. G. O. 434 and 864 of '87.—Medal with clasp.
Burma, 1887-89.—Clasp.
Burma, 1889-96.—Directed and in chief command of the operations against the Kachins at Sima. Thanks of Government of India—Foreign Department letter No. 60-C., 21 Nov. 93.—Brev. of Lt.-Col.

Pelly, H. R.—
The War of 1914-18.—Despatches, *Lond. Gaz.*, 19 Oct. 16.

Peirse, C. E.—
Afghanistan, 1879-80.—Was Orderly Officer of Brig.-Genl. Henderson.—Medal.

Pemberton, E. H.—
N.-W. Frontier of India, 1908.—Operations in the Mohmand country.—Medal with clasp.

Pemberton, R. (*Lieut.-Col., retd. I. M. S.*)—
Tirah, 1897-98.—Medal with 2 clasps.

Pengelly, E. A.—
The War of 1914-18.—Despatches, *Lond. Gaz.* 18 May 17—M.C.

Pennefather Evan, G.—
The War of 1914-18.—Despatches, *Lond. Gaz.*, 21 Jun 16 and 15 Aug. 17.—Brev. of Lt.-Col.

Pennington, A. W.—
Chitral, 1895.—Relief of Chitral.—Medal with clasp.
Tirah, 1897-98.—Operations in the Baraj Valley, 7 to 14 Dec. 1897.—2 clasps.
The War of 1914-18.—Despatches, *Lond. Gaz.*, 5 Jun 16.

Penny, J.—
N.-E. Frontier of India, Chin Hills, 1892-93.—Medal with clasp.

Penrose-Thackwell, E. R. (*Col., retd. I. A.*)—
S. Africa, Zulu Campaign, 1879.—Medal with clasp.
Burma, 1886—88.—Medal with clasp.

Penton, B. C.—
N.-W. Frontier of India, 1908.—Operations in the Zakka Khel country.—Medal with clasp.

Penton, H. E. (*Maj. Genl., retd. I. A.*)—
Burma, 1885—89.—Medal with 2 clasps.

Pepper, H.—
Tirah, 1897-98.—Actions of the Sampagha with Arhanga Passes. Operations in the Bara Valley, 7 to 14 Dec. 97. Operations in the Bazar Valley, 25 to 30 Dec. 97. Despatches, G. G. O. No. 244 of 98.—Medal with 2 clasps. Promoted Conductor.
East Africa, 1903-04.—Operations in Somaliland. Action at Jidballi. Despatches, *Lond. Gaz.*, 2 Sep. 04.—Medal with 2 clasps. Promoted Assistant Commissary.

Pepys, G. L.—
N.-W. Frontier of India, 1908.—Operations in the Mohmand country. Engagement of Matta, 24 Apl. 08.—Medal with clasp. D. S. O.
The War of 1914-18.—Despatches, *Lond. Gaz.*, 5 Nov. and 28 Jan. 16 and 21 Jun 16.—D.S.O.

Pereira, E. J. C.—
N.-W. Frontier of India, 1908.—Operations in the Mohmand country.—Medal.

Perkins, N. C. (*Major, retd. Ind. S. C.*)—
Burma 1886-89.—Medal with clasp.

Perry, E. L.—
N-W. Frontier of India, 1897-98.—Operations in Bajaur and in the Mohmand country.—Despatches, G. G. O. No. 1317 of 1897.—Medal with clasp.
N.-W. Frontier of India, Waziristan, 1901-02.—Clasp.
The War of 1914-18.—Despatches, *Lond. Gaz.*, 19 Oct. 16 and 15 Aug. 17.—D.S.O.

Perry, G. E. J. (*Major, retd. I. A.*)—
East Africa, 1904.—Operations in Somaliland. Medal with clasp.

Peters, C. T. (*Lt.-Col., retd. I. M. S.*)—
N.-W. Frontier of India, Zhob Valley Expedition, 1884.
Burma, 1885-87.—Medal with 2 clasps.

Peterson, C. H. (*Major, retd. I. A.*)—
Burma, 1887-88.—Medal with clasp.
N.-E. Frontier of India, Manipur, 1891.—Clasp.
N.-W. Frontier of India, Tochi, 1897-98.—Medal with clasp.
Tibet, 1903-04.—Action at Niani, 26 Jun. 04. Operations at and around Gyantse between 5th May and 6 July 1904. March to Lhassa, 14 July to 3 Aug. 04. Despatches, *Gaz., of India*, No. 1065 of 04.—Medal with clasp.

Peterson, F. H.—
N.-E. Frontier of India, Sikkim, 1888.—Attack of Jalapla.—Medal with clasp.
N.-W. Frontier of India, Hazara, 1891.—Clasp.
Chitral, 1895.—Relief of Chitral. Despatches, G. G. O. No. 718 of 95.—Medal with clasp. D.S.O.
Tibet, 1903-04.—Action at Niani, 26 Jun. 04. Operations at and around Gyantse between 5 May and 6 July 04. March to Lhassa, 14 July to 3 Aug. 04. Despatches, *Gaz. of India*, No. 1065 of 04.—Medal with clasp.
Operations in the Abor country, 1911-12.—Despatches, *Gaz. of India*, Notn. 480, dated 23 May 12.—Medal with clasp.

Peyton, A. G. (*Col., retd. I. A.*)—
Soudan, 1885.—Suakin. Actions of Hasheen and Tofrek.—Medal with 2 clasps. Bronze star.
Chitral, 1895.—Relief of Chitral.—Medal with clasp.

Peyton, A. H.—
The War of 1914-18.—Despatches, *Lond. Gaz.*, 15 Jun. 16 and 4 Jan. 17.

Peyton, L. S. (*Col., retd. I. A.*)—
Afghanistan, 1879-80.—Medal.
Tirah, 1897-98.—Provost Marshal. Line of Communications.—Medal with 2 clasps.

Phayre, A.—
Afghanistan, 1878-80.—Medal.
China, 1900.—Despatches, *Lond. Gaz.*, 14 May 01. Brev. of Lt.-Col.—Medal.
South African War, 1902.—Operations in the Transvaal between 14 and 31 May 02.—Queen's Medal.

Phayre, R. B.—
The War of 1914-18.—Despatches, *Lond. Gaz.*, 19 Oct., 16 and 15 Aug. 17.—M.C.

Phelps, A. (*Lt.-Gen., u. s. l. Bo. Inf.*)—
India, 1858-60.—Pursuit of rebels in the Nerbudda Valley and against the Wagheers.

Phillott, D. C. (*Lt.-Col., retd. I. A.*)—
N.-W. Frontier of India, Zhob Valley, 1890.—*Hazara, 1891*.—Field Intelligence Officer, Right Column, and Deputy Assistant Quarter-Master General for Intelligence.—Medal with clasp.
N.-W. Frontier of India, 1897-98.—Action of the Ublan Pass, 27 Aug. 97. Operations on the Samana and in the Kurram Valley during Aug. and Sep. 97.—Medal with 2 clasps.

Phillpps, I. (*Major, retd. I. A.*)—
Burma, 1885—89.—Operations of the 4th Brigade and in the Yaw Valley round Gungaw in expedition against the Chins.—Medal with 2 clasps.
N.-W. Frontier of India, (2nd) *Miranzai, 1891*.—Clasp.—*Isazai, 1892.*
N.-W. Frontier of India, 1897-98.—Operations on the Samana and in the Kurram Valley during Aug. and Sep. 97.—Medal with 2 clasps.
Tirah, 1897-98.—D. A. A. and Q. M. G. Lines of Communication. Action of the Chagru Kotal. Despatches, G. G. O. 58 and 244 of 98.—Clasp.
China, 1900.—Despatches, *Lond. Gaz.*, 14 May 01.—Medal with clasp. B. S. O.

Phillps, J. C.—
South African War, 1900-01.—Queen's medal with clasp.
N.-W. Frontier of India, 1908.—Operations in the Zakka Khel country.—Medal with clasp.

Phillps, J. W.—
The War of 1914-18.—Despatches, *Lond. Gaz.*, 5 Apl. 15.

Phillipps, C. R. (*Lt.-Col., retd. I. A.*)—
Afghanistan, 1879-80.—Defence and battle of Kandahar.—Medal with clasp.

Phillips, A. L. (*Col., retd. I. A.*)—
N.-E. Frontier of India, Chin-Lushai, 1888-90.—Lushai Column. Served as Assistant Commissariat Officer.—Medal with clasp.

Phillips, J. A. S.—
The War of 1914-18.—Despatches, *Lond. Gaz.*, 15 Aug. 17.

Phillips, P. (*Dy. Comsy. Hony. Capt., retd.*)—
Soudan, 1885.—Suakin. Medal with 2 clasps. Khedive's Bronze star.
N.-E. Frontier of India, Chin Hills, 1892-93.—Medal with clasp.
N.-W. Frontier of India, 1897-98.—Tochi. Despatches, G. G. O. No. 1432 of 1897. Promoted Deputy Assist. Comy. G. G. O. No. 961 of 1898.—Medal with clasp.

Phipsen, E. S.—
The War of 1914-18.—Despatches, *Lond. Gaz.*, 28 Jan. 16.—D.S.O.

Picot, H. P. (*Lt.-Col., retd. Indian Army*)—
Afghanistan, 1878-79-80.—Action of the Peiwar Kotal. Despatches, G. G. O. 610 of 79. Affair of Ali Khel. Despatches. Brigadier-General T. E. Gordon's despatch, 14 Oct. 79. Zaimukht Expedition. Action of Zawa.—Medal with clasp.
N.-W. Frontier of India, Hazara, 1891.—Medal with clasp.—(2nd) *Miranzai, 1891*.—Despatches, 15 Sep. 91.—Clasp.

Pim, H. M.—
The War of 1914-18.—Despatches, *Lond. Gaz.*, 5 Apl. 16—M.C.

Pinney, J. C. D.—
The War of 1914-18.—Despatches, *Lond. Gaz.*, 13 July 16.

Pinto, J. O. (*Major, retd. I. M. S.*)—
N.-W. Frontier of India, 1897-98.—Operations on the Samana and in the Kurram Valley during Aug. and Sep. 97.—Medal with 2 clasps.
Tirah, 1897-98.—Actions of Chagru Kotal and Dargai. Operations in the Bazar Valley, 25 to 30 Dec. 97.—Clasp.

Pirie, C. P. W.—
Afghanistan, 1880.—Medal.
The War of 1914-18.—Despatches *Lond. Gaz.*, 1 Jan. 16. C. B.

Pirrie, F. W. (*Maj., Supy. List, I. A.*)—
Chitral, 1895.—Relief of Chitral. Assistant Commissariat Officer, 1st Brigade.—Medal with clasp.
N.-W. Frontier of India, 1897-98—Tochi.—As Survey Officer.—Clasp.
The War of 1914-18.—Despatches, *Lond. Gaz.*, 5 Apl. 16, 19 Oct. 16, 14 Nov. 16 and 15 Aug. 17. Brev. of Col. C.I.E.

Pisani, L. J. (*Lt.-Col., retd. I. M. S.*)—
N.-W. Frontier of India, Hazara, 1888.—Action of Kotkai.—Medal with clasp.

Pitkeathly, J. S.—
The War of 1914-18.—Despatches *Lond. Gaz.*, 15 Aug. 17.

Pitman, C. R. S.—
The War of 1914 18.—Despatches, *Lond. Gaz.*, 19 Oct. 16 and 15 Aug. 17. D.S.O. M.C.

Plinston, C. H.—
The War of 1914-18.—Despatches, *Lond. Gaz.*, 27 July and 19 Oct. 16.

Plowden, C. H. (*Col., u.s.l. Ind. S.C.*)—
India, 1857-59.—Capture of the rebel Nasrain Ras at Kerwel, and engagement at Garuassah.—Medal with clasp.

Plowden, W. F. C. C. (*Lt.-Col., retd. I. A.*)—
Afghanistan, 1878-79-80.—Battle of Ahmed Khel. Advance to Kabul.—Medal with clasp.

Pluckrose, **E.** (*Comsy. Hony. Capt., retd.*)—
Burma, 1887—89.—Medal with clasp.

Plumer, C. G. M.—
South African War, 1901-02.—Operations in the Transvaal, Dec. 01 to Feb. 02. Operations in the Orange River Colony, Oct. 01 to Dec. 01. Operations in Cape Colony, Feb. to May 1902.—Queen's medal with 4 clasps.

Plumer, T. H.—
Burma, 1887-88.—Medal with clasp.

Pocock, P. F.—
Chitral, 1895.—Relief of Chitral.—Medal with clasp.
The War of 1914-18.—Despatches, *Lond. Gaz.*, 5 Apl. and 19 Oct. 16.

Pogson, C. A.—
The War of 1914-18.—Despatches, *Lond. Gaz*, 5 Apl. 16. M.C.

Pollock, **J. A. H.** (*Maj.-Genl., retd. I. A.*)—
N.-W. Frontier of India, Jowaki, 1877-78.—Medal with clasp.
Afghanistan, 1878-79.—Action of Ali Musjid.—Medal with clasp. Zaimukht Expedition. Despatches. Lieut.-Genl. *Sir* J. S. Browne's despatch, 8 July 79.
N.-W. Frontier of India, Mahsud Waziri, 1881.—Despatches. Brigadier-General T. G. Kennedy's despatch, dated 20 June 81—(*1st*) *Miranzai, 1891 Waziristan, 1894-95.*—Clasp.
N.-W. Frontier of India, 1897-98.—Tochi.—Medal with clasp. Despatches, G. G. O. No. 1432 of 97.
China, 1900.—Relief of Pekin. Action of Peitsang. Action of Yangtsun. Despatches, *Lond. Gaz.*, 14 May 01.—Medal with clasp. Brev. of Col.
N.-W. Frontier of India, 1902.—Operations against the Darwesh Khel Waziris.

Pond, A. D. C. (*Lieut.-Col., retd. I. A.*)—
Burma, 1886—89 and '91.—Operations of the 2nd Brigade, Wuntho Field Force, 91. Despatches, G. G. O. 985 of 91.—Medal with 3 clasps.
Tirah, 1897-98.—Operations in the Bazar Valley, 25 to 30 Dec. 97.—Medal with 2 clasps.

Ponsonby, D. G.
The War of 1914 18.—Despaches, *Lond. Gaz.*, 15 Aug. 17. D.S.O.

Ponting, T. J.—
N.-W. Frontier of India, 1908.—Operations in the Zakka Khel country.—Medal with clasp.
The War of 1914-18.—Despatches, *Lond. Gaz.*, 15 Aug. 17. M.C.

Poole, A. (*Hon. Col., retd. Bo. S.C.*)—
Abyssinia, 1867-68.—Medal.
Afghanistan, 1880.—Medal.
Burma, 1886-87.—Despatches, G. G. O. 434 of 1887.

Poole, I.M.C.—
The War of 1914-18.—Despatches, *Lond. Gaz.*, 19 Oct. 16. D.S.O.

Pope, S. B.—
N.-W. Frontier of India, 1908.—Operations in the Mohmand county. Engagement of Kargha, 24 May 08.—Medal with clasp.
The War of 1914-18.—Despatches *Lond. Gaz.*, 15 Jun. 16, 4 Jan. 17, 15 May 17 and 11 Dec. 17. D.S.O.

Pope, T. H. (*Lt.-Col., retd. I.M.S.*)—
Afghanistan, 1878-80.—Medal.

Popham, E. L.—
The War of 1914-18—Despatches, *Lond. Gaz.*, 3 Oct. 17. D.S.O.

Porch, E. A.—
South African War, 1899—01.—Operations in Natal 1899. Relief of Ladysmith, including action at Colenso (15 Dec. 99). Operations of 17 to 24 Jan. 00, and action at Spion Kop; operations of 5 to 7 Feb. 00, and action at Veal Kranz; operations on Tugela Heights (14 to 27 Feb. 00). (Wounded.) Operations in Natal, Mar. to Jun. 00, including action at Laing's Nek (6 to 9 June). Operations in the Transvaal, east and west of Pretoria, July to 29 Nov. 00. Operations in the Transvaal, 30 Nov. 00 to Apl. 01.—Queen's medal with 5 clasps.
North-West Frontier of India, 1908.—Operations in the Mohmand Country.—Medal with clasp.
The War of 1914-18.—Despatches, *Lond. Gaz.*, 13 July 16, and 15 Aug. 17.

Porteous, C. A. (*Col., u.s.l. Ind. S.C.*)—
India, 1857-59.—Capture of Lucknow, operations in Bundelcund.—Medal with 2 clasps.

Porteous, C. McL.—
N.-E. Frontier of India, Chin-Lushai, 1889-90.—Medal with clasp.
N.-W. Frontier of India, Isazai, 1892.
N.-W. Frontier of India, 1897-98.—Mohmand.—Medal with clasp.
Tirah, 1897-98.—Operations in the Bara Valley, 7 to 14 Dec. 97.—Clasp.

Porteous, D. G.—
The War of 1914-18.—Despatches, *Lond. Gaz.*, 5 Apr. 16.

Porteous, **J. E.** (*Col., retd. Ind. S.C.*)—
Abyssinia, 1867-68.—Medal.
Afghanistan, 1879-80.—Operations in the Elachipore, Waziri, and the Kama Districts, actions of Kam Dakka, and operations in Besud. Mentioned in despatches.—Medal.
Burma, 1891-92.—Operations in the Chin Hills.

Porter, A. R. (*Col., retd. I. A.*)—
Afghanistan, 1879.—With Kurram Field Force.—Medal.
N.-W. Frontier of India, Hazara, 1891.—Medal with clasp.

Porter, H. E. (*Maj., retd. I.S.C.*)—
Burma, 1885-86.—Despatches, G. G. O. 295 of 86.—Medal with clasp.

Pott, D.—
N.-W. Frontier of India, 1908.—Operations in the Mohmand country.—Medal with clasp.

The War of 1914-18.—Despatches, *Lond. Gaz.*, 1 Jan. 16, 19 Oct. 16 and 13 Aug. 17.—D.S.O. M. C.

Pott, E. H.—
South African War, 1900-1902.—Operations in the Transvaal and Orange River Colony from May 1901 to May 1902. Operations on the Zulland Frontier of Natal in September and October 1901, including the defence of Forts Itala and Project.—Queen's medal with 5 clasps.

The War of 1914-18.—Despatches, *Lond. Gaz.*, 4 Jan. and 15 May 17 D.S.O.

Pott, H. P.—
South African War, 1899-00.—Relief of Ladysmith including action at Colenso (15 Dec. 99), operations of 17 to 24 Jan. 00 and action at Spion Kop; operations of 5 to 7 Feb. 00 and action at Vaal Krans; operations on Tugela Heights (14 to 27 Feb. 00); action at Pieter's Hill (27 Feb. 00). (Severely wounded.)—Queen's medal with 3 clasps.

Potter, R. J. K.—
South African War, 1902.—Operations in Cape Colony, 26th March to 25th May 1902. Operations in Orange River Colony, May 1902.—Queen's medal with 3 clasps.

The War of 1914-18.—Despatches, *Lond. Gaz.*, 19 Oct. 16.

Pottinger, J. A.—
The War of 1914-18.—Despatches, *Lond. Gaz.*, 1 Feb. 17.—M.C., Clasp to M. C.

Powell, C. H.—
N.-W. Frontier of India, Waziristan, 1894-95.—Action at Wana, 3 Nov. 94, D. A. A. G., Waziristan Field Force. Despatches, G. G. Os. 473 of 95.—Medal with clasp.

N.-W. Frontier of India, 1897-98.—Malakand. Operations in Bajaur and in the Mamund country. Buner, action of the Tanga Pass. D. A. Q. M. G., 2nd Brigade in both Forces. Despatches, G. G. Os. 178 and 217 of 98.—Medal with clasp. Brev. of Lt.-Col.

China, 1900.—British representative with rank of D. A. G. on Field Marshal Count Von Waldersee's Staff, C.-in-C. International Forces, Chi-li Province.—Medal.

Powell, D. H.—
South African War, 1901-02.—Operations in Cape Colony from 27th March to 14th June 1901 and again from 6th September to 13th November 1901. Operations in the Transvaal and Orange River Colony from 15th June to 5th September 1901.—Queen's medal with 5 clasps.

The War of 1914-18.—M.C.

Powell, F.—
The War of 1914-18.—Despatches, *Lond. Gaz.*, 1 Jan. 16.—M. C.

Powell, W. B.—
The War of 1914-18.—Despatches, *Lond. Gaz.*, 13 July and 19 Oct. 16.—C.M.G.

Power, J. E.—
N.-W. Frontier of India, 1908.—Medal with clasp.

Power, R. (*Brig.-Surg., retd. I. M. S.*)—
N.-W. Frontier of India, 1868. Hazara Expedition.—Occupation of the Black Mountain.—Medal with clasp.

Jowaki Expedition, 1877-78.—Clasp.
Afghanistan, 1878-79.—Capture of Ali Musjid.—Medal with clasp.

Poynder, F. S.—
The War of 1914-18.—Despatches, *Lond. Gaz.*, 15 Jun. 16.

Praed, G. A. M.—
South African War, 1900-02.—Relief of Kimberley. Operations in the Orange Free State, Feb. to May 00, including operations at Paardeberg (17 to 26 Feb.); actions at Poplar Grove (7 March), Driefontein (10 March) (wounded). Operations in the Transvaal between 1st January 1900 and 30th November 1900 and between 16th June 1901 and 31st May 1902. Operations in the Orange River Colony, between 1st December 1900 and 15th June 1901. Despatches, *Lond. Gaz.*, 10 Sept. 01.—Medal for distinguished conduct in the field.—Queen's medal with 4 clasps. King's medal with 2 clasps.

Prasad, K. (*Lt.-Col., retd. I. M. S.*)—
N.-E. Frontier of India, Manipur, 1891.—Medal with clasp.

N.-W. Frontier of India, Isazai, 1892.

N.-W. Frontier of India, 1897-98.—Malakand; operations in Bajaur.—Mohmand.—Medal with clasp.

Tirah, 1897-98.—Clasp.

Pratt, E. G. W.—
Tirah, 1897-98.—Medal with 2 clasps.
East Africa, 1903-04.—Operations in Somaliland.—Medal with clasp.

Pratt, H. M. (*Col., u. s. l. Ind. S. C.*)—
India, 1857-58.—Medal.
China 1860—Action of Sinho, capture of Taku Forts and occupation of Pekin.—Medal with 2 clasps.

Afghanistan, 1878-80.—Assault and capture of the Peiwar Kotal, action of Charasiah, operations in and around Kabul in Dec. '79, march from Kabul to the relief of Kandahar and battle of 1 Sep. Aitchakzai and Mari expeditions. Despatches, G. G. O. 137 of 80.—Medal with 4 clasps. Bronze star. Brev. of Lt.-Col.

N-W. Frontier of India, Hazara, 1888.—Mentioned in Despatches.—Medal with clasp; C.B.

Pratt, H. R. E.—
N.-W. Frontier of India, 1897-98.—Operations on the Samana and in the Kurram Valley during Aug. and Sep. 97.—Defence of Gulistan.—Despatches, G. G. O. 1417 of 97.—Medal with 2 clasps.

Tirah, 1897-98.—Actions of the Sampagha and Arhanga Passes.—Reconnaissance of the Saran Sar and action of 9 Nov. 97.—Operations in the Waran Valley and action of 16 Nov. 97.—Operations at and around Dwatoi and action of 24 Nov. 97.—Operations in the Bara Valley, 7 to 14 Dec. 97.—Clasp. D. S. O.

Pratt, J. J. (*Lt.-Col., retd. I. M. S.*)—
N.-W. Frontier of India, Zhob Valley, 1884.

Prendergast, C. G.—
Burma, 1885—87—89.—Medal with 2 clasps.
N.-W. Frontier of India, (1st) Miranzai, 1891.
N.-W. Frontier of India, Waziristan, 1894-95.—Clasp.
China, 1900.—Despatches, *Lond. Gaz.*, 13 Sep. 01.—Medal.

Prendergast, C. L. (*M.-Gen., retd. Ben. Inf.*)—
N.-E. Frontier of India, Cossyah and Jyntia Hills, 1862-63.

Afghanistan, 1880.—Medal.
N.-E. Frontier of India, Chin-Lushai, 1889-90.—Medal with clasp.
N.-W. Frontier of India, Hazara, 1891.—Clasp.

Prendergast, G. A. (*M.-Gen., u.s.l. Ind. S.C.*)—
India, 1858.—Defence of Powaen, actions of Pusgaon. Russoolpore and Mitowlee. Mentioned in Despatches.—Medal.

Afghanistan, 1879.—Commanded Thal Chotiali and Vitakri Field Force.—Medal.

Prendergast, N. H.—
N.-W. Frontier of India, 1908.—Operations in the Mohmand country.—Medal with clasp.
The War of 1914-18.—Despatches, *Lond. Gaz.*, 20 Mar 16.

Prentle, J. (*Capt., retd. I. S. M. D.*)—
Afghanistan, 1878-79-80.—1st Bazar Valley expedition. Action of Fatehabad, 2nd April 1879. Operations at and around Kabul. March from Kabul to the relief of Kandahar, and battle of Kandahar 1st September 1880.—Medal with 2 clasps. Bronze star.

Prentis, W. S.—
China, 1900.—Medal.
The War of 1914-18.—Despatches, *Lond., Gaz.*, 21 Jun. 16.

Prescott, C. W.—
N.-W. Frontier of India, 1902.—Operations against the Darwesh Khel Waziris.

Prescott, H. C.—
South African War, 1901-02.—Operations in the Transvaal, May 02. Operations in Cape Colony and Orange River Colony, 01-02.—Queen's medal with 5 clasps.

Presgrave, E. R. J.—
Afghanistan, 1879-80.—Medal.
Burma, 1886—89.—Operations of the 1st Brigade. —Medal with 2 clasps.
N.-E. Frontier of India, Manipur, 1891.—Clasp.
Burma, 1892—94.—Tashon Column and operations in Northern Chin Hills. Despatches, G. G. O. 733 of 93. - Clasp. Brev. of Maj. and D. S. O.

Pressey, A. (*Lt.-Col., retd. I. A.*)—
Egypt, 1882.—Action of Kassassin.—Medal. Bronze star.
N.-W. Frontier of India, Hazara, 1891.—Medal with clasp.

Preston, E. E.—
Tibet, 1903-04.—Chief Transport Officer, Line of Communication.—Medal.
The War of 1914—18.—Despatches, *Lond. Gaz.* 19 Oct. 16.

Preston, F. G.—
Burma, 1885—87.—Operations of the 5th and 2nd Brigades. Present with the column in pursuit of Hlaoo.—Medal with 2 clasps.

Preston, G. A.—
Tibet, 1903-04.—Actions at Niani, 26th June 1904. Operations at and around Gyantse, between 5th May and 6th July 1904. March to Lhassa, 14th July to 3rd August 1904. (Slightly wounded.) Despatches, *Gazette of India*, No. 1065 of 1904. —Medal with clasp.

Preston, J. E.—
N.-W. Frontier of India, Jowaki, 1877.—Medal with clasp.
Afghanistan, 1880.—Medal.
Burma, 1885—89.—Attack on Lamaing Post, 27 June '86. (Severely wounded.) Despatches, *Lond. Gaz.*, 22 June '86 and 2 Sep. '87.—Medal with 2 clasps. D. S. O.
Burma, 1892.

Preston, W. J. P.—
The War of 1914-18.—Despatches, *Lond. Gaz.*, 19 Oct. 16. D.S.O.

Prevost, G. H.—
N.-E. Frontier of India, Manipur, 1891.—Medal with clasp.

Price, C. H. U.—
Burma, 1886-88.—(Severely wounded.)—Medal with 2 clasps.
Uganda, 1898-99.—Commanded troops operating against rebel Waganda in Singo and Ankole and against rebel Wanyoro in Unyoro. Commanded Column against Soudanese mutineers and rebels, capture of stockade at Jeruba. Mentioned in despatches.—Medal with 2 clasps. Brev. of Maj. D. S. O.

Price, C. U.—
N.-W. Frontier of India, Zhob Valley, 1890.
Uganda, 1898-99.—Operations against the Soudanese mutineers including actions of Jeruba and Kijangute.—Operations in Budda and Ankoli. Mentioned in Despatches.—Medal with clasp.
China, 1900.—Medal.
The War of 1914-18.—Despatches, *Lond. Gaz.* 30 Jun. 16. Brev. of Col. C.M.G.

Price, H. W.—
South African War, 1900-01.—Operations in Cape Colony, Mar. to 29 Nov. 00. Operations in Cape Colony, 30 Nov. 00 to May 1901.—Queen's Medal with 2 clasps.
Operations in the Abor Country, 1911-12.—Medal with clasp.

Price, R. H.—
N.-W. Frontier of India, 1897-98.—Operations on the Samana and in the Kurram Valley during Aug. and Sep. 97.—Medal with 2 clasps.
Tirah, 1897-98.—Clasp.
N.-W. Frontier of India, 1898-99.—Operations against the Para Chamkannis.
Nandi, 1905-1906.—Medal with clasp.

Price, T. R. C.—
N.-W. Frontier of India, 1897-98.—Malakand. Operations in Bajaur and in the Mamund country. *Buner.*—Action of the Tanga Pass.—Medal with clasp.
Tibet, 1903-04.—Medal.

Prideaux, F. B.—
The War of 1914-18.—Despatches, *Lond. Gaz.*, 31 Oct. 17.

Pridmore, W. G.—
Burma, 1892.—Tashon Column.
The War of 1914-18.—Despatches, *Lond. Gaz.*, 21 Jun. 16. C.M.G.

Pringle, A. (*Lt.-Col., retd. I.S.C.*)—
Afghanistan, 1880.—Medal.

Prins, E. G. A.—
Chitral, 1895.—Relief of Chitral.—Medal with clasp.
Tirah, 1897-98.—Operations in the Bara Valley, 7th to 14th December 1897—2 clasps.
Tibet, 1903-04.—Action at Niani, 26th June 1904. Operations at and around Gyantse, between 5th May and 6th July 1904.—March to Lhassa 14th July to 3rd August 1904. Medal with clasp.

Prinsep, E. S. Moch—
The War of 1914-18.—Despatches, *Lond. Gaz.*, 1 Jan. 16.

Prissick, C.—
East Africa, 1903-04.—Operations in Somaliland. Action at Jidballi. - Medal with 2 clasps.
The War of 1914-18.—Despatches, *Lond. Gaz.*, 15 Aug. 17.—Brev. of Lt.-Col.

Prior, W. (*Col., retd. I. A.*)—
N.-E. Frontier of India, Sikkim, 1888.—Attack of Jalapla.—Medal with clasp.
N.-E. Frontier of India, Manipur, 1891.—Clasp.
N.-E. Frontier of India, Abor, 1894.
Chitral, 1895.—Relief of Chitral.—Medal with clasp.
The War of 1914-18.—Despatches, *Lond. Gaz.*, 4 Jan. 17.

Pritchard, A. G.—
The War of 1914-18.—Despatches, *Lond. Gaz.*, 15 Jun. 16, 1 Jun. 17, 23 July 17, and 11 Dec. 17 Brev. of Col. C.M.G.

Pritchard, H. E.—
The War of 1914-18.—Despatches, *Lond. Gaz.*, 15 Jun. 16.

Probyn, Rt. Hon'ble Sir D. M. (*Genl., G. C. B., G. C. S. I., G. C. V. O., K. C. I. E., V. C., I. A.*)—
N.-W. Frontier of India, 1857.—Bozdar Expedition. —Medal with clasp.

India, 1857-58.—Battle of Najafgarh, siege and capture of Delhi, actions of Bulandshahr, Aligarh, Akbarpur, Agra and Kanouj, relief of Lucknow (November 1857), battle of Cawnpore, action of the Kali Naddi and siege and capture of Lucknow. Mentioned in despatches, G. G. O.'s Nos. 1383, 1546, 1627 and 1649 of 1857, and No. 315 of 1858; *London Gazettes*, 15 Dec. 1857, and 15 and 29 Jan., 19 Feb. and 23 Apl. 1858. Medal with 3 clasps; Brevet-Major; C. B. and V. C.
China, 1860.—Action of Sinho, capture of the Taku Forts, actions of Chang-tsin-wan and Tang-Chan, and occupation of Pekin. Mentioned in despatches, *London Gazette*, 27 Nov. 1860.—Medal with 2 clasps; Brevet Lt.-Col.
N.-W. Frontier of India, 1863.—Ambeyla Expedition, in command of the cavalry, skirmish in the Chamla Valley (22 Oct.) and action at the mouth of the Buner Pass (16 Dec.) Mentioned in despatches; G. G. O. No. 76 of 1864, *London Gazette*, 18 Mar. 1864.—Clasp.

Proctor, A. H.—
The War of 1914-18.—Despatches, *Lond. Gaz.*, 15 Aug. 17.

Proctor, C. N.—
N.-W. Frontier of India, 1908.—Operations in the Mohmand country. Engagement of Kargha, 24th May 1908.—Medal with clasp.
The War of 1914-18.—Despatches, *Lond. Gaz.*, 8 Feb. 17.

Proudfoot, A. W. (*Col. retd. I.A.*)—
Afghanistan, 1879-80.—Various affairs in the Kojak Pass.—Medal.

Pruce, C. A.—
South African War, 1901-1902.—Queen's medal.

Pryce, H. E. ap. R.—
Tibet, 1903-04.—Medal.
The War of 1914-18.—Despatches, *Lond. Gaz.*, 22 Jun. 15, 15 Jun. 16, 4 Jan. 17, 15 May 17 and 11 Dec. 17. Brev. of Lt.-Col. C.M.G., D.S.O.

Pughe, O. L.—
The War of 1914-18.—Despatches, *Lond. Gaz.*, 4 July 16.

Pulley, C. (*Col., retd. I.A.*)—
Afghanistan, 1878-79-80.—Battle of Ahmed Khel. Action of Urzu.—Medal with clasp.
Burma, 1886-87.—Shan States column. Advance to, and occupation of, Fort Stedman. (Wounded.)—Despatches, G. G. O. 434 of 87.—Medal with clasp. Brev. of Maj.
N.-E. Frontier of India, Lushai, 1892-93.
N.-W. Frontier of India, 1897-98.—Operations on the Samana and in the Kurram Valley during Aug. and Sep. 97. Relief of Gulistan. Despatches, G. G. O. 1417 of '97.—Medal with 2 clasps.
Tirah, 1897-98.—Actions of Chagru Kotal, Dargai, and of the Sampagha and Arhanga Passes. Reconnaissance of the Saran Sar and action of 11 Nov. 97. Operations at and around Dwatoi, and action of 24 Nov. 97. Operations in the Bara Valley, 7 to 14 Dec. 97. Despatches, G. G. O. 244 of '98.—Clasp. Brev. of Col.

Pulleyn, C. H.—
N.-W. Frontier of India, Hazara, 1888.—Medal with clasp.
N.-W. Frontier of India, Waziristan, 1894-95.—Despatches G. G. O. 473, 10 May 1895.—Clasp.
N.-W. Frontier of India, 1897-98—Malakand.—Medal with 2 clasps.
The War of 1914-18.—Despatches, *Lond. Gaz.*, 13 July 16.

Pulvermann, A. W.—
The War of 1914-18.—Despatches, *Lond. Gaz.*, 21 Jun 16.

Purdon, D. W.—
Burma, 1887-88.—With the Kalewa column during the operations on the Western Frontier.—Medal with clasp.

Puri, M. L.—
The War of 1914-18.—Despatches, *Lond. Gaz.*, 13 July and 19 Oct. 16.

Pye, W. E.—
Tirah, 1897-98.—Operations in the Bazar Valley, 25th to 30th December 1897.—Medal with 2 clasps.
China, 1900.—Medal.

Q

Qadir Buksh Khan.—
The War of 1914-18.—Despatches, *Lond. Gaz.*, 1 Jan. 16.

Quadros, L. G.—
Afghanistan, 1880.—Medal.

Quicke, W. H. (*Lt.-Col., retd. I. M. S.*)—
N.-W. Frontier of India, Zhob Valley, 1884.—Affair at Dowlatzai.
Burma, 1885-86.—Medal with clasp.

Quin, E. (*Lt.-Col., retd. Ind. S.C.*)—
Abyssinia, 1867-68.—Siege and storming of Magdala.—Medal.
Afghanistan, 1880.—Medal.
Burma, 1885–87.—Medal with clasp.

Quin, R. P. W.—
Central Africa, 1899.—Expedition against Kwamba. Medal with clasp.

Quin, T. (*Maj., retd. I. A.*)—
Egypt, 1882.—Medal Bronze star.
N.-W. Frontier of India, Hazara, 1888.—Medal with clasp.
(*1st and 2nd*) *Miranzai, 1891.*—Clasp.
Action at Wana, 3 Nov. 94.—*Waziristan, 1894-95.*—Clasp.
Chitral, 1895.—Relief of Chitral.—Action of the Malakand Pass.—Medal with clasp.
Tirah, 1897-98.—Actions of Dargai and of the Sampagha and Arhanga Passes.—Operations against the Khani Khel Chamkannis.—Operations in the Bazar Valley, 25 to 30 Dec. 97.—Despatches, G. G. O. 244 of '98.—2 clasps. Brev. of Maj.
N.-W. Frontier of India, Waziristan, 1901-02.—Despatches, *Lond. Gaz.*, 8 Aug. 02, clasp.

Quin, T. J. (*M.-Gen. u.s.l. Ind. Army*)—
N.-E. Frontier of India, Bhootan, 1865.—Capture of Dewangiri.—Medal.

Quinan, E. P.—
The War of 1914-18.—Despatches, *Lond. Gaz.*, 21 Jun. 16.

Quirke, M. J.—
East Africa (Somaliland), 1908—10.—Despatches, *London Gazette*, dated 17th June 1910.—Medal with clasp.

R

Radcliff, S. G. (*Lieut.-Col., and Bt.-Col., retd. I. A.*)—
Burma, 1887—89.—Medal with clasp.
Burma, 1889-92.—Clasp.

N.-E. Frontier of India, Kachin Hills, 1892-93.—Clasp.

China, 1900.—Relief of Pekin. Senior Transport Officer for Chinese Transport. Despatches, *Lond. Gaz.*, 14th May 1901.—Medal with clasp.

Radcliffe, A. W. T. (*Lt.-Col., retd. I. S. C.*)—
N.-W. Frontier of India, Jowaki, 1877-78.—Medal with clasp.—*Waziristan, 1894-95.*—Clasp.
N.-W. Frontier of India, 1897-98. -Tochi.—Medal with clasp.
Tirah, 1897-98 -Clasp.

Radford, J. T. (*Asst. Commy. and Hony. Lieut, retd.*)—
Afghanistan, 1878—80.—Action of Charasiah. Affair at Doaba. Expedition into Kohistan, and operations in Kabul and Serpur. Medal with 2 clasps.

Soudan, 1885.- Suakin.—Medal with clasp; Khedive's Bronze Star.

Burma, 1885—87.—Present at the occupation of Mandalay. Expedition to Bhamo.—Medal with clasp.

Radford, O. C. (*Bt. Col., retd. I. A.*)—
N.-W. Frontier of India, Jowaki, 1877-78.—Despatches, G. G. O. 738 of '78.—Medal with clasp.
Afghanistan, 1879.—Chakmani and Zaimukht Expeditions. Action of Zawa.—Medal.

N.-W. Frontier of India, Zhob Valley, 1884. Hazara, 1888.—Action of Kotkai. Wounded. Despatches, G. G. O. 1022 of '88.—Clasp. Brev. of Maj. (*1st*) *Miranzai, 1891*). *Waziristan 1894-95.*—Clasp.

N.-E. Frontier of India, Manipur, 1891.—Clasp
Burma, 1889—92.—(Chinbok column, 1890.) Baungshe column, 1891-92.—Clasp.
Burma, 1894 —Baungshe column.
China, 1900.—Despatches, *London Gazette*, dated 14 May and 13 Sep. 01.—C. I. E.—Medal.

N.-W. Frontier of India, 1902.—Operations against the Darwesh Khel Waziris.

Radford, O. H.—
N.-W. Frontier of India, 1908.—Operations in the Zakka Khel country. Operations in the Mohmand country. Engagement of Matta, 24th April 1908.—Medal with clasp.

Radley, H. P.—
The War of 1914-18.—Despatches, *Lond. Gaz.*, 5 Apl. 16, 27 July 16, 19 Oct. 16 and 15 Aug. 17. M.C.

Rae, G. R.—
The War of 1914-18.—Despatches, *Lond. Gaz.*, 19 Oct. 16.

Rahman, M. A.—
The War of 1914-18.—Despatches, *Lond. Gaz.*, 31 Oct. 17.

Rai, D. H.—
The War of 1914-18.—Despatches, *Lond. Gaz.*, 22 Jun. 15 and 19 Oct. 16. M.C.

Rainey-Robinson, R. M. (*Col., retd. I. A.*)—
N.-W. Frontier of India, Zhob Valley, 1884.

Burma, 1885—89.—Present at the taking of Minhla and occupation of Mandalay. Pursuit of Hlaoo. Yaw operations, Alon Kanle column. Operations of 1st, 3rd, and 6th Brigades. Affair near Popa. Wounded. Despatches, G. G. O. 434 of 87.—Medal with 2 clasps.

N.-E. Frontier of India, Chin-Lushai, 1889-90.—Clasp.

N.-E. Frontier of India, Manipur, 1891.—Clasp.

Burma, 1889-92.—Chinbok column, 1890. Baungshe column, 1891-92.—Clasp.

Burma, 1894.—Baungshe column.
The War of 1914-18.—Despatches, *Lond. Gaz.*, 6 Mar. 16. C.M.G.

Rainsford, J. A. G.—
Burma, 1885—87.—Medal with clasp.

Rainsford-Hannay, D.—
N.-W. Frontier of India, 1908.—Operations in the Zakka Khel country. Operations in the Mohmand country. Engagement of Matta, 24th April 1908. Engagement of Kargha, 24th May 1908.—Medal with clasp.

Rait, J. W. F.—
China, 1900.—Medal.
Raitt, H. G. B. (*Lt.-Col., retd. I. A.*)—
Burma, 1891-92.

Ralph, A. C.—
The War of 1914-18.—Despatches, *Lond. Gaz.*, 15 Aug. 17.

Ralston, W. H.—
The War of 1914-18.—Despatches *Lond. Gaz.*, 22 Jun. 15, 1 Jan. 16 and 15 Aug. 17. D.S.O., M.C.

Ramsay, A. Fitz G.—
South African War, 1900—02.—Employed with Mounted Infantry. Operations in the Transvaal and Orange River Colony, Nov. 00. Operations in Orange River Colony, 30 Nov. 00 to Dec. 00, and Mar. 01 to 31 May 02. Operations in Cape Colony, Dec. 00 to Mar. 01. Queen's medal with 3 clasps. King's medal with 2 clasps.

Ramsay, H. L. (*Lt.-Col., retd. I.S.C.*)—
Afghanistan, 1879-80.—Operations at and around Kabul. Despatches, G. G. O. 137 of '80.—Medal with clasp.

Ramsay, Sir J. G. (*Maj.-Gen., retd. I. A.*)—
Afghanistan, 1878-79-80.—(2nd) Bazar Valley Expedition. March from Kabul to the relief of Kandahar and battle of Kandahar.—Medal with clasp. Bronze star.

N.-W. Frontier of India, Hazara, 1888.—Medal with clasp.

N.-W. Frontier of India, 1897-98.—Defence of Malakand.—Relief of Chakdara.—Malakand. Action at Landakai. Operations in Bajaur and in the Mamund country. Despatches, G. G. O. 1089 of '97.—Medal with 2 clasps.

China, 1900.—Relief of Pekin. Action of Peitsang. Action at Yangtsun. Despatches, *Lond. Gaz.*, 14 May 01.—Medal with clasp. C. B.

N.-W. Frontier of India, 1908.—Operations in the Mohmand country. G. O. C. 3rd Brigade. Despatches, G. G. O. No. 521A of 1908.—Medal with clasp.

Ramsden, H. F. S. (*Col., retd. I.A.*)—
Afghanistan, 1879-80.—With the Kurram Valley Field Force.—Medal.
Egypt, 1885.—Suakin.—Medal with clasp. Bronze star.
Burma, 1886-87.—Thanked by Government of India. Despatches, G. G. O. 433 of 87.—Brev. of Maj.—Medal with 2 clasps.

Ramsden, Wm. C. (*Lt.-Col., retd. Ben. Inf.*)—
N.-E. Frontier of India, 1865-66.—Bhootan Expedition.—Capture of Dilingkote and re-capture of Bala.—Medal with clasp.
N.-W. Frontier of India, 1877-78.—Jowaki Expedition.—Clasp.
Afghanistan, 1879-80.—Medal.

Rana Jodha Jang Bahadur—
The War of 1914-18.—Despatches *Lond. Gaz.*, 4 Nov. 15 and 1 Jan. 16. M. C.

Ranken, G. P. (*Bt.-Col., retd. I. A.*)—
Burma, 1886-87.—Served as Staff Officer to the Director of Transport. Despatches, G. G. O. 434 of '87.—Medal with 2 clasps.
N.-W. Frontier of India, Hazara, 1888.—Clasp.
Chitral, 1895.—Relief of Chitral.—Medal with clasp.
N.-W. Frontier of India, 1897-98.—Malakand.—Operations in Bajaur and in the Mamund country.—Clasp.

Ransford, F. M.—
South African War, 1899-1900.—Operations in the Orange Free State, February to May 1900. Operations in the Transvaal, east of Pretoria, July to 29th November 1900. Operations in Orange River Colony, May to 29th November 1900. Operations in Cape Colony, south of Orange River, 1899-1900.—Queen's medal with 4 clasps.
East Africa, 1903-04.—Operations in Somaliland.—Medal with clasp.
East Africa (Somaliland), 1908-09.—Clasp.

Ratlam, H. H. *Rajah Sir* Sajjan Singh Bahadur—
The War of 1914-18.—Despatches *Lond. Gaz.*, 15 Jun. 16.

Ratton, J. J. L. (*Surg.-Maj., retd. Ind. Med. Serv.*)—
Abyssinia, 1868.—Medal.

Rattray, C.—
Soudan, 1896, Dongola Expedition.—Medal, also Egyptian medal.
N.-W. Frontier of India, 1897-98.—Mohmand.—Medal with clasp.
The War of 1914-18.—Despatches *Lond. Gaz.*, 5 Apr. 16, 19 Oct. 16, and 15 Aug. 17 Brev. of Col. C. B.

Rattray, R. H. (*Bt. Col., retd. I. A.*)—
Afghanistan, 1880.—Medal.
N.-W. Frontier of India, (1st) Miranzai, 1891.
N.-W. Frontier of India, 1897-98.—Relief of Malakand—Malakand. Operations in Bajaur—Mohmand. Despatches, G. G. O. 1780 of '98.—Medal with 2 clasps.
Tirah, 1897-98.—Action of Chagru Kotal.—Clasp
N.-W. Frontier of India, 1902.—Operations against the Darwesh Khel Waziris.

Ravenshaw, C. W. (*Lt.-Col., retd. I. A.*)—
Afghanistan, 1879-80.—Medal.
N.-W. Frontier of India, (1st) Miranzai, 1891.

Ravenshaw, H. A. (*Maj., retd. Ind. S.C.*)—
Afghanistan, 1878-79.—With Kandahar column—Medal.
Burma, 1886—88.—Medal with clasp.

Rawlins, A. K.—
N.-W. Frontier of India, 1897-98.—Defence of Malakand.—Relief of Chakdara.—*Malakand*, action at Landakai.—Operations in Bajaur and in the Mamund country.—Despatches, G. G. O. 1089 of '97.—Medal with 2 clasps. D. S. O.
China, 1900.—Medal with clasp.
East Africa, 1903-04.—Operations in Somaliland. Action at Jidballi. Despatches, *Lond. Gaz.*, 2 Sep. 1904.—Medal with 2 clasps.
The War of 1914–18.—Despatches *Lond. Gaz.*, 21 Jun. 16 and 6 July 17. C. I. E.

Raye, D. O'C. (*Col., retd. Ind. Med. Serv.*)—
N.-E. Frontier of India, 1874-75.—Daphla Expedition.

Raymond, E. D.—
South African War, 1900—1902.—Operations in Transvaal, east of Pretoria, including the action at Belfast (26th and 27th August 1900). Operations in Transvaal, west of Pretoria, including action at Frederickstad (17th to 25th October). Operations in the Transvaal and Cape Colony, 30th November 1900 to January 1902.—Queen's medal with 3 clasps; King's medal with 2 clasps.
The War of 1914-18.—Despatches, *Lond. Gaz.*, 1 Jan. 16 and 15 Aug. 17. D.S.O., M.C.

Raymond, M. C.—
The War of 1914-18.—Despatches *Lond. Gaz.*, 4 Jan. 17.

Raynor, C. A.—
The War of 1914-18.—Despatches *Lond. Gaz*, 5 Apr., 13 July and 19 Oct. 16. M. C.

Read, H.—
Afghanistan, 1879-80.—Medal.
Burma, 1887.—Medal with clasp.
Tibet, 1903-04.—Despatches, *Gazette of India*, No. 1065 of 1904.—Medal. C. B.

Rebsch, W. K.—
The War of 1914-18.—Despatches *Lond. Gaz.*, 13 July 16. D. S. O.

Redl, E. A. F.—
Nile Expedition, 1899.—Egyptian medal with clasp.
Aden, 1903-04.—Operations in the interior. Despatches, *Gazette of India*, No. 43 of 1905.

Redman, H. E.—
South African War, 1900—1902.—Operations in Cape Colony, Orange River Colony and Transvaal from January 1900 to March 1901.—Queen's medal with 4 clasps.

Reed, H. R. B.—
N.-W. Frontier of India, Waziristan, 1901-02.—Medal with clasp
The War of 1914-18.—Despatches *Lond. Gaz.*, 1 Jan. 16. M. C.

Reeves, W.—
Afghanistan, 1879-80.—Medal.
N.-W. Frontier of India, Hazara, 1888.—Medal with clasp.
N.-W. Frontier of India, Zhob Valley, 1890.
Chitral, 1895.—Relief of Chitral. Medal with clasp. Promoted Sub-Conductor. Despatches, G. G. O. No. 530 of 1895.
N.-W. Frontier of India, 1897-98.—Tochi.—Clasp.
Tibet, 1903-04.—Medal.

Reid, C. C. (*Maj., retd. I. A.*)—
N.-W. Frontier of India, Waziristan, 1894-95.—Medal with clasp.
Chitral, 1895.—Relief of Chitral. Action of the Malakand Pass.—Medal with clasp.
N.-W. Frontier of India, 1897-98.—Tochi.—Clasp.
N.-W. Frontier of India, Waziristan, 1901-02.—Clasp.

Reid, L. H.—
Afghanistan, 1879-80.—Medal.
Burma, 1886-87.—Operations of the 4th Brigade. Severely wounded.—Medal with clasp.
N.-W. Frontier of India, Hazara, 1891.—Clasp. (2nd) *Miranzai, 1891.*—Clasp.
China, 1900.—Despatches, *Lond. Gaz.*, 13 Sept. 01.—Medal.

Reid, W. D.—
Tirah, 1897-98.—Actions of Chagru Kotal and Dargai.—Operations in the Bazar Valley, 25th to 30th Dec. 97.—Medal with 2 clasps.

Reilly, E. W. (*Lieut.-Col., retd. I.M.S.*)—
Burma, 1886—88.—Operations of the 2nd Brigade. Actions at Kypoung and Sebenetba.—Medal with 2 clasps.

Reilly, H. L.—
N.-W. Frontier of India, 1908.—Operations in the Mohmand country. Engagement of Kargha, 24th May 1908.—Medal with clasp.
The War of 1914-18.—Despatches *Lond. Gaz.*, 5 Apr., 10 May, 21 Jun. 16. Brev. of Maj. D. S. O.

Reilly, R. E. D. (*Lieut.-Col., retd. I. A.*)—
Afghanistan, 1880.—Affair at Khairabad. Defence of Kandahar. Sortie of Deh Khojah and battle of Kandahar.—Medal with clasp.
Soudan, 1885.—Actions of Hasheem and Tofrek. Various attacks on convoys. Advance on Tamai Despatches, *Lond. Gaz.*, 25 Aug. '85 Medal with 2 clasps ; bronze star. Brev. of Maj.
N.-W. Frontier of India, 1897-98.—Mohmand. Despatches, G. G. O. 1318 of 97.—Medal with clasp.

Tirah, 1897-98.—Action of the Sampagha Pass.—Clasp. Brev. of Lt.-Col.

Reinhold, H. E.—
East Africa, 1903-04.—Operations in Somaliland. Action at Jidballi.—(Severely wounded.) Medal with 2 clasps.
The War of 1914-18.—Despatches *Lond. Gaz.*, 19 Oct. 16. M.C.

Rennick, R. H. F. (*Col., retd. I. A.*)—
Abyssinia, 1867-68—Medal.
Afghanistan, 1878-79.—Affairs of Ali Khel. Mentioned in Despatches.—Medal.
Burma, 1886-87.—Despatches, G. G. O. 434, 87.—Medal with clasp. Brev. of Lt.-Col.
N.-E. Frontier of India, 1891.—Manipur Expedition. Commanded the Silchar Column. Services acknowledged by Government of India. Despatches, G. G. O. 585 of 91.—Clasp.
Burma, 1891-92.—Operations in the Chin Hills. With the Eastern Sonai Column. Brev. of Col.

Rennick, R. W. de C.—
Tirah, 1897-98.—Medal with 2 clasps.

Rennie, P. M.—
The War of 1914-18.—Despatches *Lond. Gaz.*, 28 Jan. 16.

Rennison, A. J.—
South African War, 1901.—Operations in the Transvaal, Mar. to Aug. 1901.—Queen's Medal with 4 clasps.
The War of 1914-18.—Despatches *Lond. Gaz.*, 13 July 16, Brev. of Maj. D. S. O.

Renny. A. MacW. (*Bt.-Col., retd. I.A.*)—
Afghanistan, 1880.—Zaimukht Expedition.—Medal.
Burma, 1886-87.—Operations of the 1st Brigade.—Medal with 2 clasps.

Renny, G. B. (*Bt.-Col., retd. I. A.*)—
Afghanistan, 1879-80.—Medal.

Renny, H. B.—
South African War, 1902.—Operations in Cape Colony, between 11th and 13th April 1902. Operations in the Orange River Colony, between 14th and 15th April 1902. Operations in the Transvaal, between 16th April and 31st May 1902. Queen's medal with 4 clasps.
The War of 1914-18.—Despatches *Lond. Gaz.*, 19 Oct. 16. M.C.

Renton, R. R. (*Lieut.-Col., retd. I. A.*)—
Burma, 1886—89.—Operations round Hlindet, Yemethen and Nyingyan. Operations of the 1st and 3rd Brigades.—Medal with 2 clasps.
N.-E. Frontier of India, Manipur, 1891.—Clasp.
N.-W. Frontier of India, 1897-98.—Malakand.—Medal with clasp.

Repton. F. W. (*Maj. retd. I.A.*)—
N.-E. Frontier of India, Sikkim, 1888.—Attack on Jeluk.—Medal with clasp.

Retallick, J. M. A. (*Lt.-Col., retd. I. A.*)—
Afghanistan, 1878-79-80.— Bazar Valley Expedition. Affairs with the Ghilzai at and near Jagdala-Kand. Action of Chihildakhteran.—Medal.
N.-W. Frontier of India, 1884.—Zhob Valley Expedition.
China, 1900.—Despatches, *Lond. Gaz.*, 14 May 01, —Medal.

Reyne, G. Van R.—
The War of 1914-18.—Despatches *Lond. Gaz.*, 5 Apr., 13 July and 19 Oct. 16. D. S. O.

Reynolds, A. J.—
South African War, 1899—1901.—Advance on Kimberley, December 1899. Operations in Cape Colony, south of Orange River, including action at Colesburg, February 1900. Operations in Orange Free State, February to May 1900, including actions at Vet River (5th and 6th May), and Zand River (10th May). Operations in the Transvaal in May and June 1900, including actions near Johannesburg, (29th May), Pretoria (4th June) and Diamond Hill (11th and 12th June). Operations in the Transvaal, east of Pretoria, July to 29th Novr. 1900, including action at Belfast (26th and 27th August 1900). Operations in the Orange River Colony between April and November 1901. —Queen's medal with 6 clasps.
The War of 1914-18.—Despatches, *Lond. Gaz.*, 2 Oct. 16.

Reynolds, H. L.—
Chitral, 1895.—Relief of Chitral.—Action of the Malakand Pass. Action at Khar.—Medal with clasp.—Promoted Sub-Condr.
The War of 1914-18.—Despatches *Lond. Gaz.*, 19 Oct. 16.

Rice, B. A. McH.—
N.-W. Frontier of India, 1897-98.—Mohmand. —Medal with clasp.

Rice, G. D.—
South African War, 1901.—Served in Mediterranean.—Medal.

The War of 1914-18.—Despatches Lond. Gaz., 15 Aug. 17.

Rice, H. A. H. (*Major, retd. I. A.*)—
N.-W. Frontier of India, Waziristan, 1894-95.—Medal with clasp.
Chitral, 1895.—Relief of Chitral.—Medal with clasp.
N.-W. Frontier of India, Waziristan, 1901-02.—Clasp.
N.-W. Frontier of India, 1902.—Operations against the Darwesh Khel Waziris.
N.-W. Frontier of India, 1908.—Operations in the Zakka Khel country. Despatches, G. G. O. No. 188 of 1908.—Medal with clasp. Brevet of Major.

Rice, H. C. P. (*Hon. M.-Gen., retd. Ben. S.C.*)—
N.-E. Frontier of India, 1861.—Sikkim Expedition.
N.-W. Frontier of India, 1872.—Dour Valley Expedition.

Jowaki Expedition, 1877-78.—(Dangerously wounded.) Mentioned in Despatches.—Medal with clasp. Brev. of Lt.-Col.

Afghanistan, 1879.—Zaimukht Expedition.—Medal.

N.-W. Frontier of India, 1881.—Mahsud Waziri Expedition. Mentioned in Despatches.

Rice, S. K. B.—
Soudan, 1896.—Dongola Expedition.—Medal, also Khedive's medal.
N.-W. Frontier of India, 1897-98.—Relief of Malakand.—Relief of Chakdara—Malakand.—Operations in Bajaur and in the Mamund country—Utman Khel.—Medal with 2 clasps.
N.-W. Frontier of India, Waziristan, 1901-02.—Despatches, G. G. O. 611 of 02.—Clasp.
Tibet, 1903-04.—Medal.

Richards, E. W. (*Lt.-Col., retd. I. A.*)—
Chitral, 1895.—Relief of Chitral.—Medal with clasp.
N.-W. Frontier of India, 1897-98.—Clasp.

Rich, H. H.—
The War of 1914-18.—Despatches *Lond. Gaz.*, 5 Apr. and 19 Oct 16.

Richards, W. G. (*Major, retd. I.M.S.*)—
China, 1900.—Medal.

Richardson, C. W. G.—
Burma, 1888-89.—Medal with clasp.
N.-W. Frontier of India, Waziristan, 1894-95.—Clasp.

South African War, 1900—1902.—Employed with Imperial Yeomanry. On Staff; afterwards Adjutant, Maitland Camp.—Queen's Medal with 2 clasps.

Richardson, F. B. W.—
N.-E. Frontier of India, Chin-Lushai, 1889-90.—Medal with clasp. *Manipur, 1891.*—Clasp.

Richardson, *Sir* G. L. R. (*Lieut.-Genl., retd. I.A.*)—
N.-W. Frontier of India, Hazara, 1868.
Afghanistan, 1879-80.—Medal. Brev. of Maj.
N.-W. Frontier of India, Mahsud Waziri, 1881.—*Zhob Valley, 1890.*
N.-W. Frontier of India, 1897-98.—Operations on the Samana and in the Kurram Valley during Aug. and Sep. 97.—Commanded the Flying Column in the Kurram Valley, 20 Aug. to 1 Oct. 97. Despatches, G. G. O. 1417 of 97.—Medal with 2 clasps.
Tirah, 1897-98.—Action of the Sampagha Pass.—Clasp. C.B.
China, 1900.—In command of Cavalry Brigade. Despatches, *Lond. Gaz.*, 14 May 01.—Medal, C.I.E.

Richardson, H. L.—
N.-W. Frontier of India, 1897-98.—Tochi.—Medal with clasp.

China, 1900.—Medal.

The War of 1914-18.—Despatches *Lond. Gaz.*, 22 Jun. 15 Brev. of Col.

Richardson, R.—
The War of 1914-18.—M. C.

Ricketts, E. W. C.—
N.-W. Frontier of India, 1908.—Operations in the Zakka Khel country.—Medal with clasp.

Ricketts, F. R. R.—
N.-W. Frontier of India, Hazara, 1888.—Medal with clasp.

N.-W. Frontier of India, Hazara, 1891.—Clasp.

N.-W. Frontier of India, Miranzai, 1891.—Clasp.

N.-W. Frontier of India, Isazai, 1892.

Tirah, 1897-98.—Medal with clasp.

South African War, 1900—1902.—Operations in Orange Free State, May 1900. Operations in the Transvaal, October 1900. Operations in Orange River Colony, May to 29th November 1900. Operations in Cape Colony, south of Orange River 1899-1900 (January to May). Operations in the Transvaal, June 1901. Operations in the Orange River Colony, between 30th November 1900 and 31st May 1902. Operations in Cape Colony, between 30th November 1900 and 31st May 1902.—Queen's medal with 3 clasps. King's medal with 2 clasps.

The War of 1914-18.—Despatches, *Lond. Gaz.*, 15 Aug. 17.

Ricketts, H. C.—
N.-W. Frontier of India, Hazara, 1888.—Medal with clasp.

N.-W. Frontier of India, 1897-98.—Medal with clasp.

Tirah, 1897-98.—Clasp.

South African War, 1900.—Queen's medal with clasp

The War of 1914-18.—Despatches, Lond. Gaz., 22

Ricketts, L. H. (*Major, retd. I. A.*)—
Burma, 1892.—Tashon column.

Ricketts, P. E.—
Chitral, 1895.—Relief of Chitral. Action of the Malakand Pass. Despatches, G. G. O. 998 of '95.—Medal with clasp.

N.-W. Frontier of India, 1897-98.—Operations on the Samana and in the Kurram Valley during Aug. and Sep. 97. Operations of the Flying Column in the Kurram Valley under Colonel Richardson, 20 Aug. to 1 Oct. 97.—2 clasps.
Tirah, 1897-98.—Clasp.

The War of 1914-18.—Despatches *Lond. Gaz.*, 4 Jan. 17. D. S. O.

Ricketts, R. L.—
Chitral, 1895.—Relief of Chitral.—Medal with clasp.

N.-W. Frontier of India, 1897-98.—Buner.—Clasp.

The War of 1914-18.—Despatches *Lond. Gaz.*, 12 Jun. 17.

Ricketts, W. S. P. (*Lt.-Col., retd. I.M.S.*)—
East Africa, 1902-04.—Operations in Somaliland. Medal with clasp.

Riddell, A.—
South African War, 1902.—Operations in Cape Colony, 10th to 17th April 1902. Operations in the Orange River Colony, 17th to 25th April 1902. Operations in the Transvaal, 26th April to 31st May 1902. Queen's medal with 4 clasps
The War of 1914-18.—Despatches, *Lond. Gaz.*, 15 August 17. D. S. O.

Riddell, W. H.—
The War of 1914—18.—Despatches, *Lond. Gaz.*, 27 July 16. M. C.

Rideout, F. C. W. (*Col., retd I. A.*)—
Burma, 1887-88.—Medal with clasp.
N.-W. Frontier of India, 1897-98.—Mohmand Divisional Transport Officer. Despatches, G. G. O. 1318 of 97.—Medal with clasp.
Tirah, 1897-98.—Divisional Transport Officer, 1st Division. Despatches, G. G. O. 244 of 98.—Clasp. Brev. of Maj.
China, 1900.—Relief of Pekin.—Divisional Transport Officer.—Medal with clasp.

Ridge, K. J. T.—
The War of 1914—18.—Despatches, *Lond. Gaz.*, 31 Oct. 17. M. C.

Ridgeway, D. G.—
The War of 1914—18.—Despatches, *Lond. Gaz.*, 1 Jan. 16.

Ridgeway, *Sir* J. W. (*Col., retd. Ben. Inf.*)—
Afghanistan, 1879-80.—Operations in Shekabad, march from Kabul to the relief of Kandahar, and battle of 1st Sept. Despatches, G. G. O. 326 and 582 of 80. Brev. of Lt.-Col.

Ridgeway, R. K. (*Col., retd. I. A.*)—
N.-E. Frontier of India, Naga Hills, 1875 and 1879-80.—(Wounded.) Despatches, G. G. O. 574 of '75 and 123 of '80.—Medal with clasp. V.C.
N.-E. Frontier of India, Manipur, 1891.—Despatches. G. G. O. 585 of '91.—Clasp.

Ridgway, R. T. I.—
N.-W. Frontier of India, 1897-98.—Buner. Medal with clasp.
Tirah, 1897-98.—Clasp.
N.-W. Frontier of India, 1908.—Operations in the Mohmand country.—Medal with clasp.
The War of 1914—18.—Despatches, *Lond. Gaz.*, 15 and 21 Jun. 16, and 31 Oct. 17. C. B.

Riley, C. B. (*Major, retd. I. A.*)—
N.-W. Frontier of India, 1908.—Operations in the Zakka Khel country. Medal with clasp.

Rimington, M. B—
South African War, 1900—02.—Served with Rimmington's Guides. Spec. Serv. Offr., 18 May to 6 Dec. 00. Operations in the Transvaal, including action near Diamond Hill. Operations in the Orange River Colony, including action at Wittebergen. Operations in the Transvaal, 30 Nov. 00 to May 01. (Severely wounded.)—Queen's medal with 4 clasps.
N.-W. Frontier of India, 1908.—Operations in the Zakka Khel country.—Medal with clasp.

Rind, A. T. S. A. (*Brev. Col., retd. Ind. S.C.*)—
Afghanistan, 1879-80.—March from Kabul to the relief of Kandahar, and battle of 1 Sep. Operations against the Maris. Despatches, G. G. O. 585 of 91.—Medal with clasp; bronze star; Brev. of Maj.
Egypt, 1882.—Medal; bronze star.

Rind, G. B. A.—
Aden, 1903-04.—Operations in the interior.

Rippon, G. (*Lt.-Col., retd. I. A.*)—
Burma, 1887-89.—Medal with clasp.
Burma, Wuntho Field Force, 1891.—With Southern column, fight at Kyeik Wintoung, Military Police.—Clasp.

N.-E. Frontier of India, Manipur, 1891.—As D. A. Q. M. G., Tammu column, M. F. F. Mentioned in Despatches, G. G. O. 585 of 91.—Clasp.
Burma, 1891-92.
N.-E. Frontier of India, 1892.—Lushai.—Clasp.

Risley, C. G.—
The War of 1914—18.—Despatches, *Lond. Gaz.*, 4 Jan. 17. D. S. O.

Ritchie, John (*Hon. M.-Gen., retd. Bo. Art.*)—
India, 1857-58.—Actions of Nimbhera and Jeerun, defence of Neemuch, and capture of Kotah.—Medal with clasp.

Ritchie, W. D.—
The War of 1914—18.—Despatches, *Lond. Gaz.*, 21 Jun. 16.

Rivaz, V. (*Col. u.s.l. Ind. S. C.*)—
N.-W. Frontier of India, 1868.—Hazara Expedition—Occupation of the Black Mountain.—Medal with clasp. *Dour Valley Expedition, 1872.*
Afghanistan, 1878-79-80.—D.-A.-A. and Q.-M.-G., Kurram Valley Field Force; action of Jhondola. Mentioned in Despatches.—Medal.
N.-W. Frontier of India, 1881.—Mahsud Waziri Expedition.—Mentioned in Despatches. *Hazara Expedition, 1891.*—Despatches, G. G. O. 816 of 91.—Clasp.

Rivett-Carnac, E. H. (*Col., retd. I. A.*)—
Afghanistan, 1879-80.—Zaimukht Expedition. Action of Zawa.—Medal.
Egypt, 1882.—Actions of Kassassin and Tel-el-Kebir.—Medal with clasp. Bronze star; 5th class of the Medjidie.
N.-W. Frontier of India (2nd), Miranzai, 1891—Medal with clasp.

Rivett-Carnac, J. S.—
China, 1900.—Medal.

Roach, H. R. (*Sr. Asst. Surgn. and Hony. Capt., retd. I. S. M. D.*)—
Tirah, 1897-98.—Action of Chagru Kotal, action of Dargai, action of the Sampagha Pass, action of the Arhanga Pass, reconnaissance of the Kharmana Defile and action of 7th November 1897. Reconnaissance of the Saran Sar and action of the 9th November 1897. Operations in the Waran Valley and action of 16th November 1897. Operations at and around Dwatoi and action of 24th November 1897. Operations against the Khan Khel Chamkanis. Operations in the Bara Valley, 7th to 14th December 1897.—Medal with 3 clasps.

Roberts, C. S. H.—
South African War, 1902.—Queen's medal with 4 clasps.

Roberts, A. E. (*Lieut.-Col., retd. I. M.S.*)—
N.-W. Frontier of India, Hazara, 1888.—Medal with clasp.

Roberts, H. L.—
Tirah, 1897-98.—Medal with clasp.
The War of 1914—18.—Despatches, *Lond. Gaz.*, 13 July 16. C. B.

Roberts, J. (*Hon. Maj.-Gen., retd. Ben. S. C.*)—
Burma, 1852-53.—Medal.
Sonthal Rebellion, 1855-6.

Roberts, J. R.—
N.-E. Frontier of India, Chin-Lushai, 1889-90.—Medal with clasp.
N.-W. Frontier of India, Hunza-Nagar, 1891.—Principal Medical Officer. Capture of Nilt Fort. Despatches, G. G. O. 397 of '92.—Clasp.

Roberts, M. B. (*Lt.-Col., retd. I. A.*)—
N.-E. Frontier of India, Chin Hills, 1892-93.—Medal with clasp.
N.-W. Frontier of India, 1897-98.—Mohmand—Malakand. Operations in Bajaur.—Medal with clasp.

Roberts, V. H.—
China, 1900.—Medal.

Roberts, W. B.—
East Africa, 1902-04.—Operations in Somaliland. Despatches (twice), *Lond. Gaz.*, 2 Sep. 04.—Medal with clasp.
The War of 1914–18.—D. S. O.

Roberts, W. H. (*Sur.-Col., retd. Ind. Med. Serv.*)—
Burma, 1886-87.—Despatches, G. G. O. 434 of 87.

Robertson, C. R.—
The War of 1914–18.—Despatches, *Lond. Gaz.*, 8 Feb. 17.

Robertson, D. E.—
The War of 1914–18.—Despatches, *Lond. Gaz.*, 1 Jan. and 15 Jan. 16.

Robertson, J. C.—
Egypt, 1896.—Medal.
South African War, 1902.—Operations in the Transvaal, 02.—Queen's medal with 2 clasps.

Robertson, R.—
Burma, 1888-89.—Medal with clasp.

Robertson, R. A. H.—
Operations in the Abor country, 1911-12.—Medal with clasp.

Robin, N. E.—
N.-E. Frontier of India, Sikkim, 1888.—Medal with clasp.
Tirah, 1897-98.—Medal with clasp.
China, 1900.—Relief of Pekin. Action of Peitsang. Action at Yangtsun —Medal with clasp.

Robinson, D. C.—
The War of 1914–18.—M. C.

Robinson, F. A.—
The War of 1914–18.—Despatches, *Lond. Gaz.*, 5. Apl. 16. M. C.

Robinson, G. W.—
N.-W. Frontier of India, Hazara, 1891.—Medal with clasp. (*2nd*) *Miranzai, 1891.*—Clasp.
The War of 1914–18.—Despatches, *Lond. Gaz.*, 13 July 16. C. B.

Robinson, H. (*Hon. Capt. Dy. Commy., retd. I. A. Depts.*)—
Afghanistan, 1878-80.—Medal.
Tirah, 1897-98.—Medal with 2 clasps. Promoted Dy. Commy.

Robinson, W. H. B.—
Burma, 1885-89.—Despatches, G. G. O. 782 of '89 —Medal with 2 clasps.
N.-W. Frontier of India, Waziristan, 1894-95.—Clasp.
Chitral, 1895.—Relief of Chitral.—Medal with clasp.
Tirah, 1897-98.—Operations in the Bara Valley, 7 to 14 Dec. 97.—2 clasps.
The War of 1914–18.—Despatches, *Lond. Gaz.*, 21 June and 1 Dec. 16. C. B.

Robson, C. G.—
Burmese Expedition, 1887–0.—Medal with clasp.
Chin Lushai Expedition, 1889—90.—Clasp.

Roche, H. J. (*Colonel, retd. I.A.*)—
N.-W. Frontier of India, 1897-98.—Tochi.—Medal with clasp..
China, 1900 —Medal.
The War of 1914–18.—Despatches, *Lond. Gaz.*, 17 Feb. 15. C. B.

Roche, M. J.—
The War of 1914—18.—M. C.

Rocke, C. E. A. S.—
South African War, 1899—1902.—On Staff Relief of Ladysmith, including action at Colenso (15th Dec. 1899). Operations of 17th and 24th Jan. 1900, and action at Spion Kop. Operations of 5th and 7th Feb. 1900, and action at Vaal Kranz. Operations on Tugela Heights (14th to 27th Feb. 1900) and action at Pieter's Hill (27th Feb. 1900). Queen's medal with 5 clasps. King's medal with 2 clasps.

Roddy, H. H.—
N.-E. Frontier of India, Chin-Lushai, 1889-90.—Medal with clasp.
N.-W. Frontier of India, Lushai, 1890—92.—Services acknowledged by Govt., G. G. O. 977 of '92.—Clasp.
Tirah, 1897-98.—Medal with 2 clasps.
Tibet, 1903-04.—Despatches, *Gazette of India.* No. 1065 of 1904.—Medal.
The War of 1914-18.—Despatches, *Lond. Gaz.*, 30 June 16. C. B.

Rodrigues, A. R. G. (*Sr. Asst. Surgn. and Hony. Capt., retd. I. S. M. D.*)—
Egypt, 1882.—With No. 1 Company, Madras Sappers and Miners. Battle of Tel-el-Kebir and forced march on Zagazig.—Medal with clasp; Bronze star.
Burma, 1885-86.—Medal with clasp.
N.-E. Frontier of India, Chin-Lushai, 1889-90.—With the Fort White Column. Specially promoted.—Clasp.
East Africa, 1902—04.—Operations in Somaliland. —Medal with clasp.

Rodrigues, P. H.—
Tirah, 1897-98.—Medal and two clasps.
Aden, 1903-04.—Operations in the interior.

Rodwell, E. H.—
Afghanistan, 1878-79.—Medal.
N.-W. Frontier of India, Mahsud Waziri, 1881.—Despatches, Brigadier-General T. G. Kennedy's despatch, dated 20 June '81.
N.-W. Frontier of India, Waziristan, 1894-95.—Medal with clasp.
N.-W. Frontier of India, Waziristan, 1901-02.—Despatches, G. G. O. No. 611 of 02.—Medal with clasp.

Roe, C. D.—
The War of 1914–18.—Despatches, *Lond. Gaz.*, 24 July 15 and 1 Jan. 16. D. S. O.

Roffey, H. R.—
N-W. Frontier of India. 1897-98.—Operations on the Samana. Medal with 2 clasps.
Tirah, 1897-98.—Actions of Chagru, Kotal, Dargai, Sampagha and Arhanga Passes. Reconnaissance of the Saran Sar and around Dwatoi and action of 24th November 1897. Operations in the Bara Valley, 7th 14th December 1897. Clasp.
The War of 1914–18.—D. C. M.

Rogers, G. H.—
N.-W. Frontier of India, Waziristan, 1901-02.—Medal with clasp.
N.-W. Frontier of India, 1908.—Operations in the Mohmand country.—Medal with clasp.

Rolland, S. E. (*Hony. Br.-Genl., u.s.l. I. A.*)—
Burma, 1885—89.—In command Eastern Frontier from Dec. '85 to Jan. '86. In command Mappah and movable column, June to Sep. '86. Action at Segu, 1 Sep. '86. (Severely wounded.) Despatches, G.G.O.'s 376 and 434 of 87.—Medal with 2 clasps. Brev. of Lt.-Col.

Rolls, T. D.—
The War of 1914—18.—Despatches, *Lond. Gaz.*, 27 July 16.

Rombulow-Pearse, A. B.—
South African War, 1901-02.—Operations in Orange River Colony, May 01 to May 02. Operations in Cape Colony, May 01 to May 02.—Queen's medal with 4 clasps.

Rooke, B. P. S. (*Maj., retd. I. A.*)—
Burma, 1886-87.—Medal with clasp.

Roome, H. N.—
N.-W. Frontier of India, 1908.—Operations in the Zakka Khel country.—Medal with clasp.

Roome, R. E.—
N.-W. Frontier of India, Waziristan, 1901-02.—Medal with clasp.

Roos-Keppel, *Sir* G. O.—
Burma, 1887-89.—Medal with 2 clasps.
N.-W. Frontier of India, 1897-98.—Medal with 3 clasps.
N.-W. Frontier of India, 1899.—Operations against the Para Chamkannis. In Commd. Brev. of Major.
N.-W. Frontier of India, 1908.—Operations in the Zakka Khel country. Despatches, *Lond. Gaz.*, 22 May 08. Operations in the Mohmand country.—Medal with clasp. K. C. I. E.
The War of 1914—18.—Despatches, *Lond. Gaz.*, 4 July 16, and 11 Oct. 17.

Roosmale Cocq, C. A.—
N.-W. Frontier of India, 1897-98.—Malakand—Buner. Action of the Tanna Pass.—Medal with clasp.

Rose, H. (*Col. retd. I. A.*)—
Burma, 1886-87.—Shan States column. Advance to, and occupation of, Fort Stedman.—Medal with clasp.
N.-E. Frontier of India.—Chin-Lushai, 1892-93.—Clasp.
N.-W. Frontier of India, 1897-98.—Operations on the Samana and in the Kurram Valley during Aug. and Sept. 97. Relief of Gulistan.—Medal with 2 clasps.
Tirah, 1897-98.—Actions of Chagru Kotal, Dargai, and of the Sampagha and Arhanga Passes. Reconnaissance of the Saran Sar and action of 9 Nov. 97. Operations at and around Dwatoi and action of 24 Nov 97. Operations in the Bara Valley, 7 to 14 Dec. 97. Affair at Shinkamar on 29 Jan. 98. Despatches, G. G. Os. 58 and 244 of '98.—Clasp.
N.-W. Frontier of India, Waziristan, 1901-02.—Despatches, G. G. O. 611 of 02.—Clasp.

Rose, J. L.—
N.-W. Frontier of India, 1897-98.—Tochi-Mohmand.—Medal with clasp.
Tirah, 1897-98.—Actions of the Sampagha and Arhanga Passes. Operations in Bazar Valley.—Clasp.
China, 1900.—Relief of Pekin. Despatches, *Lond. Gaz.*, 13 Sep. 01.—Medal with clasp.
East Africa, 1902-04.—Operations in Somaliland. Special service officer, employed as S. and T. officer—Medal with clasp.

Ross, A. C.—
The War of 1914—18.—Despatches, *Lond. Gaz.*, 18 Feb. and 22 June 15. D. S. O.

Ross, C. E. (*Capt., retd. Ind. S. C.*).—
Burma, 1886-89.—Medal with 2 clasps.

Ross, D. B.—
Tibet, 1903-04.—Medal.

Ross, E. J.—
Operations in the Abor country, 1911-12.—Medal with clasp.
The War of 1914-18, Despatches, *Lond. Gaz.*, 15 Aug. 17.

Ross, G. W.—
China, 1900.—Medal.
China, 1900.—Medal.
The War of 1914—18.—Despatches, *Lond. Gaz.*, 19 Oct. 16. D. S. O.

Ross, H.—
Tirah, 1897-98.—Transport officer.—Medal with clasps.
East Africa, 1903-04.—Operations in Somaliland.—Medal with clasp.

Ross, H. (*Capt. I. M. S.*)—
N.-W. Frontier of India, 1902.—Operations against the Darwesh Khel Waziris.

Ross, J. W. (*Lt., retd. Ind. S. C.*)—
Miranzai (1st and 2nd) Expeditions, 1891.—Medal with clasp.

Ross, *Sir* R. (*Maj., retd. Ind. Med. Serv.*)—
Burma, 1886.—Medal with clasp.

Ross, R.—
The War of 1914—18.—Despatches, *Lond. Gaz.*, 15 Aug. 17.

Ross, R. C.—
South African, War, 1901-02.—Operations in the Transvaal, 17th February 1901 to 17th February 1902. Operations in Orange River Colony, 9th June 1901 to 3rd July 1901. Operations in Cape Colony, between 27th January 1901 and 14th February 1901.—Queen's medal with 5 clasps.
The War of 1914—18.—Despatches, *Lond. Gaz.*, 22 Jun. 15, 1 Jan. 16 and 19 Oct. 16. D. S. O. Clasp to D. S. O.

Ross, T. S.—
China, 1900.—Medal.

Ross, W. H. (*Col. u. s. l. Ind. S. C.*)—
Persia, 1857.—Capture of the Mohumra Forts—Medal with clasp.
India, 1858-59.—Operations in Kandeish.

Rouse, A. H. T. (*Major, retd. I. A.*)—
Burma, 1895.
N.-W. Frontier of India, 1897-98.—Medal with clasp.
Tirah, 1897-98.—Clasp.
China, 1900.—Relief of Pekin.—Medal with clasp.

Routh, C.—
Egypt, 1882.—Medal with clasp. Bronze Star, 4th class Medjidie.

Routh, G. S. F.—
N.-W. Frontier of India, Waziristan, 1901-02.—Medal with clasp.
The War of 1904—18.—Despatches, *Lond. Gaz.*, 21 Jun. 16.

Row, G. R.—
N.-E. Frontier of India, Akha, 1883-84.—Despatches. Brigadier-General R.S. Hill's despatch, 31 Jan. '84.
Burma, 1886—89.—Medal with 2 clasps.
N.-E. Frontier of India, Manipur, 1891.—Clasp
Abor Expedition, 1894.
Tibet, 1903-04.—Operations at and around Gyantse between 5th May and 6th July 1900. March to Lhassa, 14th July to 3rd August 1904.—Medal with clasp.

Rowan-Robinson, J. R.—
South African War, 1900-1902.—Employed with Mounted Infantry and attached to A. S. Corps. Operations in the Transvaal, Dec. 1900 to 31st May 1902. Operations in Orange River Colony, February 1902. Operations on the Zululand Frontier of Natal in October 1901.—Queen's medal with 4 clasps. King's medal with 2 clasps.
The War of 1914—18.—Despatches, *Lond. Gaz.*, 19 Oct. 16. D. S. O.

Ruck, J.—
South African War, 1877-8-9.—Kaffir Campaign, Zulu Campaign.—Medal with clasp.

Ruck, O. L.—
South African War, 1900-01.—Operations in Cape Colony south of Orange River (1899-1900). Operations in Orange Free State, February to May 1900, including action at Vet River (5th and 6th May) and Zand River (10th May). Operations in the Transvaal in May and June 1900, including actions near Johannesburg (29th May), Pretoria (4th June) and Diamond Hill (11th and 12th June). Operations in the Transvaal, 30th November 1900 to 31st January 1901. Queen's medal with 4 clasps.
N.-W. Frontier of India, 1908.—Operations in the Mohmand country.—Medal with clasp.

Ruegg, A. J.—
Chitral, 1895.—Relief of Chitral. Action of the Malakand Pass. Action at Khar.—Medal with clasp. Promoted Sub-Conductor.

Rowcroft, G. C. (*Maj.-Gen. u. s. l., I. A.*)—
Scind Frontier Expedition, 1852.
N.-W. Frontier of India, 1852-53.—Expedition against the Hussanzais.—Medal with clasp.
India, 1857.—At Peshawar.—Medal.
N.-W. Frontier of India, 1858.—Sittana Expedition. Mentioned in despatches.

Rowcroft, G. F. (*Bt.-Col., retd. I. A.*)—
Soudan, 1885.—Suakin. Actions of Hasheen and Tamai.—Medal with clasp. Bronze star.
Chitral, 1895.—Relief of Chitral.—Medal with clasp.
Tirah, 1897-98.—Reconnaissance of the Saran Sar and action of 9 Nov. 97. Operations in the Warran Valley and action of 16 Nov. 97. Despatches, G. G. O. 244 of 98.—2 clasps—D. S. O.

Rowlandson, H. W.—
N.-W. Frontier of India, 1908—Operations in the Mohmand Country. Engagement of Kargha, 24th May 1908.—Medal with clasp.

Rowlandson, M. A. (*Col. u.s.l. Ind. S. C.*)—
Abyssinia, 1867-68.—Sub.-Asst. Com.-Gen.—Medal.
Egypt, 1882.—Battle of Tel-el-Kebir.—Medal with clasp; bronze star; 4th class Osmanieh.

Rowlandson, M. G. D.—
N.-W. Frontier of India, Waziristan, 1901-02.—Medal with clasp.
N.-W. Frontier of India, 1908.—Operations in the Zakka Khel country.—Medal with clasp.
The War of 1914—18.—Despatches, *Lond. Gaz.*, 19 Oct. 16. Brev. of Lt.-Col.

Rowntree, H. K.—
The War of 1914—18.—Despatches, *Lond. Gaz.*, 19 Oct. 16. M. C.

Ruel, W. G.—
South African War, 1899-1902.—Operations in Natal, Mar. to June 00, including action at Laing's Nek (6 to 9 June). Queen's medal with 4 clasps.—King's medal with 2 clasps.
The War of 1914—18.—D. S. O.

Rugg, G. F.—
The War of 1914-18.—Despatches, *Lond. Gaz.*, 4 July 16.

Ruggles, J. (*Maj.-Genl., retd. I. A.*)—
India, 1846.—Expedition against, and surrender of Fort Kote Kangra under Brigadier H. M. Wheeler. C. B.
India, 1857.—Defence of Lucknow.—Medal with clasp.
China, 1860-61.—Served with the Regiment of Ludhiana (15th Ludhiana Sikhs) as Volunteer.—Medal.
N.-E. Frontier of India, Bhutan, 1865.—Commanded column of attack on the Bala Stockades which resulted in their recapture. Received the thanks of the C.-in-C., Sir Hugh Rose.—Medal with clasp.

Rumbold, W.—
N.-W. Frontier of India, Waziristan, 1894-95.—Medal with clasp.
Aden 1904.—Operations in the interior.
N.-W. Frontier of India, 1908.—Operations in Zakka Khel country.—Medal with clasp.

Rundall, F. M. (*Col. u.s.l. I. A.*)—
Burma, 1886-87.—Operations of the 3rd Brigade.—Medal with clasp.
N.-E. Frontier of India, Chin-Lushai, 1889-90.—Clasp.
Burma, 1890-91.—D. S. O.
N.-E. Frontier of India, Manipur, 1891.—Despatches, G. G. O. 585 of '91.—Clasp.
N.-W. Frontier of India, Waziristan, 1894-95.—Clasp.
China, 1900.—Medal.

Rundle, C. A. G.—
The War of 1914—18.—Despatches, *Lond. Gaz.*, 15 Aug. 17. M. C.

Rush, J. S.—
Burma, 1885-86.—Present at the occupation of Mandalay.—Medal with clasp.

Russell, A. R. P. (*Lieut.-Col., retd. I.M. S.*)—
Burma, 1887-89.—Medal with clasp.

Russell, F. D.—
N.-W. Frontier of India, 1897-98.—Buner.—Medal with clasp.
Tirah, 1897-98.—Actions of Dargai and of the Sampagha Pass.—Operations in the Bazar Valley, 25 to 30 Dec. 97.—Clasp.
South African War, 1900.—Operations in Orange Free State, Feb. to May 00. Operations in Orange River Colony, May to 29 Nov. 00, including action at Wittebergen (1 to 29 July).—Queen's medal with 2 clasps.
China, 1900.—Medal.
The War of 1914—18.—Despatches, *Lond. Gaz.*, 1 Jan. and 15 June 16. Brev. of Lt.-Col.

Ruzzak, S. A.—
East Africa, 1902-04.—Operations in Somaliland—Medal with clasp.

Ryall, E. C. (*Lieut.-Col., retd. I.A.*)—
N.-W. Frontier of India, Waziristan, 1894-95.—Medal with clasp.
China, 1900.—Medal.

Ryan, D. G. J.—
The War of 1914—18.—Despatches, *Lond. Gaz.*, 25 Aug. and 5 Nov. 15. D. S. O.

Rybot, N. V. L.—
Tibet, 1903-04.—Action at Niani, 26th June 1904. Operations at and around Gyantse, from 27th June to 6th July 1904. March to Lhassa, 14th July to 3rd August 1904.—Medal with clasp.
The War of 1914—8.—Despatches, *Lond. Gaz.*, 5 Apl. and 19 Oct. 16. D. S. O.

Ryder, W. I.—
N.-E. Frontier of India, Sikkim, 1888.—Medal with clasp.
N.-W. Frontier of India, 1897-98.—Malakand Relief of Malakand. Operations in Bajaur and in the Mamund country. (Wounded.) Despatches, G. G. O. No. 1317 of 1897.—Medal with 2 clasps. Brevet of Major.
Tirah, 1897-98.—Actions of the Sampagha and Arhanga Passes. Operations in the Bazar Valley, 25 to 30 Dec. 97.—Clasp.
South African War, 1900.—Operations in Cape Colony, south of Orange River.—Queen's medal with clasp.

Rynd, F. C.—
Burma, 1885-86.—Action of 16 Aug. '86 at Thayetgone. (Severely wounded.)—Medal with clasp.

Ryves, B. H.—
N.-W. Frontier of India, 1897-98.—Tochi.—Medal with clasp.

S

Sage, C. A. R. (*Brev.-Col., retd. Ind. S. C.*)—
India, 1881.—Operations against Sonthals.
Burma, 1886-87.—Despatches, G. G. O. 434 of 87. —Medal with clasp; Brev. of Maj.
N.-W. Frontier of India, Sikkim, 1888.—Clasp.
N.-W. Frontier of India, 1897.—Operations of the Mohmand Field Force. Despatches, G. G. O. 1418 of 87. Tirah, 1897-98 (severely wounded). Despatches, G. G. O. 244 of 98.

Saigol, R. D.—
Tibet, 1903-04.—Medal.

Salkeld, C. E. (*Col.. retd. Ben. S. C.*)—
Egypt, 1882.—Action of Kassassin, battle of Tel-el-Kebir.—Medal with clasp; bronze star.

Salmon, W. H. (*Col. u.s.l. I. A.*)—
Burma, 1885-89.—Medal with 2 clasps.

Salmon, W. H. B.—
N.-W. Frontier of India, 1908.—Operations in the Mohmand Country. Brigade Supply Officer, 3rd Brigade. Medal with clasp.

Salusbury, R. T. G.—
South African War, 1900-01.—Operations in the Orange Free State, 12th March to 30th April 1900. Operations in Orange River Colony, 1st May to 29th November 1900. Operations in Cape Colony, south of Orange River, 7th to 12th March 1900. Operations in the Transvaal, 17th to 21st September 1901. Operations in Cape Colony, 30th November to 31st December 1900, 1st January to 16th September 1901 and 22nd September to 23rd October 1901.—Queen's medal with 4 clasps.
N.-W. Frontier of India, 1908.—Operations in the Mohmand country. Engagement of Kargha, 24th May 1908.—Medal with clasp.

Samborne Palmer, F. C. S.—
Chitral, 1895.—Relief of Chitral—Medal with clasp.

Sandeman, D. G.—
N.-W. Frontier of India, 1908.—Operations in the Zakka Khel country. As signalling officer attached Line of Communication Staff. Operations in the Mohmand country.—Medal with clasp.
The War of 1914—18.—Despatches, *Lond. Gaz.*, 31 Oct 17.

Sandeman, J. E. (*Col. u.s.l. I. S. C.*)—
N.-W. Frontier of India, Hazara, 1868.—Medal.

Sanders, G. L'H.—
N.-W. Frontier of India, Waziristan, 1894-95.—Served as Bde. Comst. Officer, 2nd Bde.—Medal with clasp.
South African War, 1899—1901.—Operations in Natal, 1899, including actions at Talana and Lombard's Kop. Defence of Ladysmith. Operations in the Transvaal in May and June 00. Operations in Natal, March to June 00.—Queen's medal with 4 clasps.
N.-W. Frontier of India, 1908.—Operations in the Zakka Khel country. Chief Supply Officer, attached Line of Communication Staff. Despatches, G. G. O. No. 188 of 1908.—Medal with clasp.

Sanders, G. P.—
Tibet, 1903-04.—Medal.
N.-W. Frontier of India, 1908.—Operations in the Zakka Khel country.—Medal with clasp.

Sanderson, F. G. O.—
South African War, 1900.—Operations in the Transvaal in May and June 1900, including actions near Johannesburg (29th May), Pretoria (4th June) and Diamond Hill (11th and 12th June). Operations in the Transvaal, east of Pretoria, July to 29th November 1900, including action at Belfast (26th and 27th August 1900).—Queen's medal with 4 clasps.

Sandes, J. D.—
N.-W. Frontier of India, 1908.—Operations in the Zakka Khel country. Operations in the Mohmand country.—Medal with clasp.

Sandwith, J. R. (*Lieut.-Col., retd. I. A.*)—
Afghanistan, 1880.—March from Quetta to relief of Kandahar.—Medal.

Sanford, G. B.—
N.-W. Frontier of India, Isazai, 1892.
Chitral, 1895.—Relief of Chitral.—Medal with clasp.

Sangster, P. B.—
Tirah, 1897-98.—Reconnaissance of the Kharmana. Defile and action of 7 Nov. 97. Operations against the Khani Khel Chamkannis.—Medal with 2 clasps.
The War of 1914—18.—Despatches, *Lond. Gaz.*, 22 June 5, and 15 June 16. D. S. O.

Sanjana, K. C. (*Lieut.-Col., retd. I. M. S.*)—
Burma, 1886-87.—Medal with clasp.
China, 1900.—Medal.

Sarel, G. B. M.—
Chitral, 1895.—Relief of Chitral.—Medal with clasp.

Sargent, A. E. E.—
East Africa, 1902-04.—Operations in Somaliland.—Medal with clasp.
The War of 1914—18.—Despatches, *Lond. Gaz.*, 1 Jan. 16, and 28 Nov. 17. D. S. O., M. C.

Sargent, W. P. M.—
Aden, 1903-04.—Operations in the Interior.
East Africa, 1903-04.—Operations in Somaliland.—Medal with clasp.

Sarkies, C. J. (*Lieut.-Col., retd. I.M.S.*)—
Burma, 1887-88.—Medal with clasp.
Tirah, 1897-98.—Medal with 2 clasps.
East Africa, 1903.—Operations in Somaliland.—Medal with clasp.

Sarkies, S. C. (*Lieut.-Col., retd. I.M.S.*)—
Afghanistan, 1879-80.—Medal.

Sartorius, G.—
The War of 1914—18.—Despatches, *Lond. Gaz.*, 17 Feb. 15.

Saulez, E. (*Maj., retd. I.A.*)—
Chitral, 1895.—Relief of Chitral.—Medal with clasp.
N.-W. Frontier of India, 1897-98.—Tochi.—Clasp.

Saunders, A. R.—
Tirah, 1897-98.—Operations against the Khani Khel Chamkannis.—Medal with 2 clasps.
Tibet. 1903-04.—Medal.
The War of 1914—18.—Despatches, *Lond. Gaz.*, 15 Jun. 16.

Saunders, E. H.—
South African War, 1899—1902.—Employed with Mtd. Inf. Operations in Orange Free State March to April 1900. Operations in the Transvaal, January to November 1900. Operations in Cape Colony, 1899 to 1900. Despatches, *Lond. Gaz.*, 10th September 1901. Queen's medal with 3 clasps. King's medal with 2 clasps. D. S. O.
The War of 1914-18.—Despatches, *Lond. Gaz.*, 15 Jun. 16.

Saunders, M.—Despatches, *Lond. Gaz.*, 5 Nov. 15, 13 July 17, and 31 Oct. 17. D. S. O.

Saunders, P. C.—
South African War, 1900-01.—Operations i Orange River Colony, 1900-01. Operations in the Transvaal in May and June 1900, including actions near Johannesburg (29th May), Pretoria (4th June) and Diamond Hill (11th and 12th June). Operations in the Transvaal, east of Pretoria, July to September 1900, including action at Belfast (26th and 27th August 1900). Operations in Transvaal, west of Pretoria, October 1900, including action at Frederickstad (17th to 25th October 1900). Operations in Cape Colony, south of Orange River, January and February 1901. —Queen's medal with 6 clasps.
The War of 1914—18.—Despatches, *Lond. Gaz.*, 13 July 16.

Savage, W. H.—
N.-W. Frontier of India. Zhob Valley. 1884.
The War of 1914—18.—Despatches, *Lond. Gaz.*, 4 Jan. 17. C. M. G.

Savi, R. J. T.—
Burma, 1888-89.—Medal with clasp.

Savile, A. R. C.—
Soudan, 1896.—Dongola Expedition.—Medal. Egyptian medal.

Sawyer, H. A. (*Brev. Col., retd Ind. S.C.*)—
Burma, 1887-89.—Medal with clasp.
N.-W. Frontier of India, 1897.—Defence of the Malakand and relief of Chakdarra Fort. Despatches, G. G. O. 1089 of 97. Tirah, 1897-98. Despatches, G. G. O. 244 of '98.

Saxton, P. D.—
South African War, 1901-02.—Operations in Cape Colony, July 1901 to 31st May 1902.—Queen's medal with 3 clasps.

The War of 1914—18.—Despatches, *Lond. Gaz.*, 15 Aug. 17. M. C.

Scale, J D.—
The War of 1914—18.—Despatches, *Lond. Gaz.*, 12 Mar. and 27 July 17. D. S. O.

Scallon, R. I.—
Afghanistan, 1879-80.—Engagements of Khelat-i-Ghilzai and Sir-i-asp. March from Khelat-i-Ghilzai to the relief of Kandahar and battle of Kandahar. Despatches.—Medal with clasp, bronze star.

Burma, 1885-88.—Staff Officer, Kanle Column, 1886-87, and Comdg. Myaing Column, 1887-88. Despatches.—Medal with 2 clasps, D.S.O.

N.-W. Frontier of India, 1897-98.—Operations on the Samana and in the Kurram Valley during Aug. and Sep. 97.—Medal with 2 clasps.

Tirah, 1897-98.—Actions of the Chagru Kotal and of the Sampagha Pass.—Despatches, G. G. O. 244 of 98.—Clasp.

N.-W. Frontier of India, Waziristan, 1901-02.—Commanded Zhob Section of Blockading Force. Twice commended by the Government of India. Brev. of Col.—Clasp.

Aden, 1903-04.—Operations in the Interior. Despatches, G. G.O. No. 43 of 05.

Schalch, V. A. (*Col. retd. I. A.*)—
Afghanistan, 1878-79-80.—Action of Ali Khel.—Medal.

Burma, 1885-87.—Served as Brigade Major, 1st and 2nd Brigades. Occupation of Mandalay and Bhamo. Wuntho Expedition, including advance from Katha and Manle to Wuntho, and occupation of Wuntho. Despatches, G. G. Os. 295 of '86 and 434 of '87.—Medal with clasp. Brev. of Maj.

N.-W. Frontier of India, Hazara, 1891.—Clasp.

N.-W. Frontier of India, 1897-98.—Malakand.—Despatches, G. G. O. 1317 of '97 and G. G. O. 217 of '98.—Medal with clasp.

Sconce, J. (*Col. u.s.l. Ind. S.C.*)—
India, 1857-58.—Siege and capture of Delhi, actions of Bulundshahr, Alighur and Agra, relief of Lucknow, actions of Cawnpore, Shumshabad and Futeghur, and siege of Lucknow.—Medal with 3 clasps.

Scoones, G. A. P.—
The War of 1914—18.—Despatches, *Lond. Gaz.*, 1 Jan. '16, 4 Jan. 17 and 15 May 17. D. S. O., M. C.

Scott, A. E. S.—
N.-W. Frontier of India, 1908.—Operations in the Zakka Khel country.—Medal with clasp.

Scott, G. B.—
N.-W. Frontier of India, Waziristan, 1901-02.—Medal with clasp.
East Africa, 1903-04.—Operations in Somaliland.—Action at Jidballi.—Medal with 2 clasps.

The War of 1914-18.—Despatches, *Lond. Gaz.*, 20 Mar. 16. D. S. O.

Scott, H. L—
The War of 1914—18.—Despatches, *Lond Gaz*, 1 Jan. 16, 1 Oct. 16 and 15 Aug. 17. D. S. O., M. C.

Scott, H. St. G. S.—
South African War, 1901-02.—Operations in Cape Colony, 6th to 14th February 1901. Operations in the Orange River Colony, 14th to 18th February 1901 Operations in the Transvaal, 20th February 1901 to 30th January 1902.—Queen's Medal with 5 clasps. D. S. O.

The War of 1914-18. Despatches, *Lond. Gaz.*, 1 Jun. 16 and 15 Aug. 17. D. S. O., clasp, to D. S. O.

Scott, J. (*Lt.-Col., retd. I. M.S.*)—
N.-W. Frontier of India, 1897-98.—Tochi.—Medal with clasp.

Scott, J.—
The War of 1914—18.—Despatches, *Lond. Gaz.*, 1 Jan. 16.

Scott, J. B.—
The War of 1914—18.—M. C.

Scott, R.—
N.-W. Frontier of India, Waziristan, 1901-02.—Medal with clasp.

Scott, T. E.—
N.-W. Frontier of India, (1st) Miranzai, 1891.
Waziristan, 1894-95.—Medal with clasp.
East-Africa (British), 1896.—2nd-in-Command, Indian Contingent. Operations against the Mazarini rebels. Mentioned in Despatches.—Medal, D.S.O. Brilliant Star of Zanzibar, 1st grade, 2nd class.
Uganda, 1897-98.—2nd-in-Command, Indian Contingent. Commanded Flying Column for disarmament of the mutinous Nubian garrisons in Unyoro. Mentioned in Despatches.—Medal with 2 clasps Brev. of Maj.
China, 1900.—Action of Peitsang. Action at Yangtsun. Relief of Pekin. Road Commandant. Line of Communications. Despatches, *Lond. Gaz.*, 6 Nov. 00 and 14 May 01. C.I.E.—Medal with clasp.

Scott, Walter (*Col. u. s. l. Ind. S.C.*)—
Operations in Kathiawar, 1865-66.—Action at Bange Hill.

Scott, W. R. A.—
The War of 1914—18.—Despatches, *Lond. Gaz.*, 15 Aug. 17.

Scott-Elliot, C R.—
Tirah, 1897-98.—Assistant Superintendent of Army Signalling, Kurram Movable Column.—Reconnaissance of the Kharmana Defile and action of 7 Nov. 97.—Operations against the Khani Khel Chamkannis.—Medal with 2 clasps.
China, 1900.—Relief of Pekin. Action of Peitsang. Action at Yangtsun (horse shot).—Medal with clasp.

The War of 1914—18.—Despatches, *Lond. Gaz.*, 22 Jun. 15 and 15 Jun. 16. Brev. of Lt.-Col.

Scott-Moncrieff, W. E. (*Major, retd. I. M. S.*)
Chitral, 1895.—Relief of Chitral. Medal with clasp.
N.-W. Frontier of India, 1897-98.—Tochi.—Clasp.
China, 1900.—Medal.

Scratton, G. W. B.—
The War of 1914—18.—M. C.

Scroggie, W. R. J.—
East Africa, 1902-04.—Operations in Somaliland.—Medal with clasp.

Scroope, C. F.—
The War of 1914—18.—Despatches, *Lond. Gaz.*, 15 Aug. 17.

Sealy, A. E.—
N.-W. Frontier of India, (2nd) Miranzai, 1891.—Medal with clasp.
Tirah, 1897-98.—Actions of the Sampagha and Arhanga Passes.—Reconnaissance of the Saran Sar and actions of the 9 and 11 Nov. 97. Operations against the Khani Khel Chamkannis.—Operations in the Bazar Valley, 25 to 30 Dec. 97.—Medal with 2 clasps.

Sealy, H. G.—
South African War, 1901-02.—Operations in the Transvaal, between 16th February 1901 and 31st May 1902. Operations on the Zululand Frontier of Natal in September and October 1901.—Queen's medal with 3 clasps.

Seaman, A. B. (*Brig.-Surg.-Lt.-Col., retd. Ind. Med. Serv.*)—
Burma, 1885-86.—Medal with clasp.
Afghanistan, 1879-80.—Medal.
N.-E. Frontier of India, Lushai, 1889.—Honourably mentioned G. G. O. 592 of '89.
N.-W. Frontier of India, Waziristan, 1894.—Action at Wano.—Clasp.

Searle, A. E. S. (*Bt.-Col., retd. I. A.*)—
Burma, 1885-86.—Medal with clasp.
East Africa, 1902-04.—Operations in Somaliland.—Medal with clasp.

Searle, C. S.—
The War of 1914—18.—Despatches, *Lond. Gaz.*, 24 July 15 and 1 Jan. 16. M. C.

Seaton, F. D. R.—
South African War, 1901-2.—Operations, in the Transvaal, Jan. to May 01 and Aug. 01 to 31st May 02.—Queen's medal with 5 clasps.

Seddon, T. Y.—
N.-W. Frontier of India, Zhob Valley, 1890.
Chitral, 1895.—Relief of Chitral.—Medal with clasp.
Tirah, 1897-98.—Operations in the Bara Valley, 7 to 14 Dec. 97.—2 clasps.
China, 1900.—Medal.

Selby, W.—
Chitral, 1895.—Relief of Chitral.—Medal with clasp.
N.-W. Frontier of India, 1897-98.—Operations on the Samana during Aug. and Sep. 97.—Relief of Gulistan.—2 clasps.
Tirah, 1897-98.—Actions of Chagru Kotal, Dargai and of the Sampagha and Arhanga Passes.—Operations in the Waran Valley and action of 16 Nov. 97. Operations at and around Dwatoi and action of 24 Nov. 97. Operations against the Khani Khel Chamkannis. Operations in the Bara Valley, 7 to 14 Dec. 97.—Despatches, G. G. O. 244 of 98.—Clasp. D. S. O.

Sellens, S. H. (*Comsy. and Hony. Capt., retd.*)—
Chitral, 1895.—Relief of Chitral (severely wounded).—Medal with clasp.

Selwyn, C. H.—
Burma, 1886-87.—Medal with 2 clasps.
Chitral, 1895.—Relief of Chitral.—Medal with clasp.
China, 1900.—Special Service Officer. Relief of Pekin.—Despatches, *Lond. Gaz.*, 14 May 01.—Medal with clasp.

Sen, J. L.—
The War of 1914—18.—Despatches, *Lond. Gaz.*, 19 Oct. 16. M. C.

Senior, H. W. R.—
Burma, 1887-89.—Medal with clasp.
N.-E. Frontier of India, Manipur, 1891.—Clasp.
N.-W. Frontier of India, Waziristan, 1894-95.—Clasp.

N.-W. Frontier of India, 1897-98.—Tochi.—Mohmand.—Medal with clasp.
China, 1900.—Medal.

Seton, Sir B. G., Bart. (*Bt.-Col., retd. I. M. S.*)—
N.-W. Frontier of India, Waziristan, 1894-95.—(Severely wounded.)—Medal with clasp.
N.-W. Frontier of India, 1897-98.—Tochi.—Medal with clasp.

Seton, H. W.—
N.-W. Frontier of India, 1908.—Operations in the Mohmand country. Engagement of Matta, 24th April 1908.—Medal with clasp.

Seton Browne, C. L. S.—
N.-W. Frontier of India, 1897-98.—Action at Maizar, 10 June 97. (Severely wounded.) Despatches, G. G. O. 773 of 97.—Medal with clasp. D. S. O.
N.-W. Frontier of India, Waziristan, 1901-02.—Clasp.

Sewell, J. H.—
Afghanistan, 1878-80.—Occupation of Kandahar and Khelat-i-Ghilzai.—Operations in Yarkistan.—Affair of the Ghlo Pass with the Thal Chotiali Field Force and at relief of Kandahar.—Medal.
South Africa, 1881.—Transvaal Campaign.
Soudan, 1884-85.—Nile.—With Mounted Infantry.—Action of Abu Klea.—Medal with 2 clasps.—Bronze Star.
Burma, 1889-90.—Tonhon Expedition (severely wounded).—Medal with clasp.

Sewell, R. B. S.—
The War of 1914—18.—Despatches, *Lond. Gaz.*, 6 July 17.

Sexton, E. G.—
China, 1900.—Relief of Pekin. Action of Peitsang. Action at Yangtsun.—Medal with clasp.

Seymour, C. G.—
Operations in the Abor Country, 1911-12.—Medal with clasp.
The War of 1914—18.—Despatches, *Lond. Gaz.*, 30 Jun. 16.

Seymour, C. R. F.—
South African War, 1899—1902.—Operations in the Orange Free State, Feb. to May 00. Operations in the Transvaal in May and June 00. Operations in the Orange River Colony, June to Oct. 00.—Queen's medal with 4 clasps.

Shairp, A.—
East Africa, 1902-04.—Operations in Somaliland.—Medal with clasp.
The War of 1914—18.—Despatches, *Lond. Gaz.*, 21 Jun. 16. C. M. G.

Shairp, H. F.—
China, 1900.—Medal.
The War of 1914—18.—Despatches, *Lond. Gaz.*, 5 Apr. 16.

Shakespear, G. F. C.—
The War of 1914—18.—Despatches, *Lond. Gaz.*, 30 Jun. 16. D. S. O., M. C.

Shakespear, G. R. J. (*Maj.-Genl., u.s.l., I. A.*)—
Soudan, 1885.—Suakin. Despatches, *Lond. Gaz.*, 25 Aug. 85.—Medal with clasp. Bronze star. Brev. of Lt.-Col.
N.-W. Frontier of India, Hazara, 1891.—As A. A. Genl. Despatches, G. G. O. 816 of 91.—Medal with clasp.

Shakespear, J. (*Lt.-Col., retd. I. A.*)—
N.-E. Frontier of India, Lushai, 1889.—Honourably mentioned, G. G. O., 592 of 1889.
N.-E. Frontier of India, Chin-Lushai, 1889-90.—Despatches, *Lond. Gaz.*, 12 Sept. 90. D. S. O.
N.-E. Frontier of India, Lushai, 1892.—Operations in the South Lushai Hills. In command of the force.
The War of 1914—18.—Despatches, *Lond. Gaz.*, 4 Jan. and 22 May 17. C. M. G.

Shakespear, L. W.—
N.-E. Frontier of India, Lushai, 1889.—Medal with clasp.—*Chin-Lushai, 1889-90.*—Clasp.
The War of 1914-18.—Despatches, *Lond. Gaz.*, 5 Apl. 16 and 15 Aug. 17.

Shakespear, W. F. (*Lt.-Col., retd. I. A.*)—
Chitral, 1895.—Relief of Chitral. Despatches, G.G.O. 998 of 95.—Medal with clasp.
N.-W. Frontier of India, 1897-98.—Operations on the Samana and in the Kurram Valley during Aug. and Sep. 97.—2 clasps.
Tirah, 1897-98.—Operations against the Khani Khel Chamkannis.—Despatches, G. G. O. 244 of 98.—Clasp.

Shand, J. G. B.—
The War of 1914—18.—Despatches, *Lond. Gaz.*, 15 Jun. 16.

Sharman, E. H.(*Lt.-Col., retd. I.M.S.*)—
China, 1900.—Medal.

Shaw, C. C. C.—
The War of 1914—18.—Despatches, *Lond. Gaz.*, 5 Apl., 13 July and 19 Oct. 15.

Shaw, D. G. L.—
Afghanistan, 1879-80.—Operations round Kabul.—Medal with clasp.
N.-W. Frontier of India, Zhob Valley, 1890.
N.-W. Frontier of India, Waziristan, 1894-95.—Medal with clasp.
N.-W. Frontier of India, 1897-98.—Tochi.—Medal with clasp.

Shaw, E. (*Col., u.s.l., I. A.*)—
Afghanistan, 1879-80.—Medal.

Shaw, G. J. (*Col., retd, I. A.*)—
Burma, 1888-89.—Medal with clasp.
N.-W. Frontier of India, 1897-98.—Tochi.—Medal with clasp.

Shaw, W. L. P. (*Revd.*)—
N.-W. Frontier of India, 1897-98.—Medal with clasp.
Tirah, 1897-98.—Clasp.

Shea, A. G.—
China, 1900.—Relief of Pekin.—Medal with clasp.
N.-W. Frontier of India, 1902.—Operations against the Darwesh Khel Waziris.
The War of 1914—18.—Despatches, *Lond. Gaz.*, 15 Aug. 17. D. S. O.

Shea, J. S. M.—
Chitral, 1895.—Relief of Chitral.—Medal with clasp.
South African War, 1900-02.—With Thorneycroft's Mounted Infantry. Operations in Natal, June 00, including action at Lang's Nek (6 to 9 June); Operations in the Transvaal, east of Pretoria, July 00 to 29 Nov. 00. Operations in Orange River Colony, Dec. 00 and March 01 to March 02. Operations in Cape Colony, Dec. 00 to March 01. Operations in the Transvaal, March 02 to 31 May 02. Commanded 5th and 6th Australian Contingents from March 01 to March 02. Despatches, *Lond. Gaz.*, 10 Sep. and 15 Nov. 01.—Queen's medal with 4 clasps, King's medal with 2 clasps. Brev. of Maj. D.S.O. Placed on the list of officers considered qualified for Staff employment in consequence of service on the Staff in the field.
The War of 1914—18.—Despatches, *Lond. Gaz.*, 19 Oct. 14, 1 Jan. 16, 15 Jun. 16 and 4 Jan. 17. Brev. of Col. Promd. Maj.-Gen. C. B.

Shearer, E. J.—
The War of 1914—18.—M. C.

Shearer, J. E.—
The War of 1914—18.—Despatches, *Lond. Gaz.*, 15 Aug. 17. M. C.

Sheepshanks, R. H.—
The War of 1914—18.—Despatches, *Lond. Gaz.*, 5 Apl. 16. D. S. O.

Shepherd, C. I.—
N.-W. Frontier of India, 1908.—Operations in the Mohmand country. Engagement of Kargha, 24th May 1908.—Medal with clasp.
The War of 1914-18.—Despatches, *Lond. Gaz.*, 15 Aug. 17. D. S. O.

Sheppard, C. H. (*Lt.-Col., retd. Ind. S.C.*)—
Afghanistan, 1879-80.—Transport Officer, Khyber Line, at Bosawal and Jellalabad.—Medal.
Burma, 1885—88.—Officiated as Dep. Judge Advocate. Despatches, G. G. Os. 295 of 86 and 434 of 87.—Medal with 2 clasps.

Sheppard, G. S.—
The War of 1914-18.—Despatches, *Lond. Gaz.*, 22 Jun. 15, 19th Oct. 16, 15 Aug. 17, Brev. of Col. C. M. G.

Sherard, R. W. (*Col., retd. I. A.*)—
Afghanistan, 1879-80.—Medal.

Sherer, J. C.—
N.-W. Frontier of India, Waziristan, 1894-95.—Medal with clasp.
N.-W. Frontier of India, 1897-98.—Operations on the Samana and in the Kurram Valley during Aug. and Sep. 97.—Relief of Gulistan.—Operations of the Flying Column in the Kurram Valley under Colonel Richardson, 20 Aug. to 1 Oct. 97.—Medal with 2 clasps.
The War of 1914—18.—Despatches, *Lond. Gaz.*, 19 Oct. 16.

Sheridan, R. J.—
China, 1900.—Relief of Pekin. Actions of Peitsang and Yantsun.—Medal with clasp.
N.-W. Frontier of India, 1908.—Operations in the Zakka Khel country.—Medal with clasp.

Sheringham, A. T.—
South African War, 1900—02.—Queen's medal with 3 clasps. King's medal with 2 clasps. D. S. O.
The War of 1914—18.—Despatches, *Lond. Gaz.*, 22 Jun. 15. D. S. O.

Sherlock, G. W. S.—
N.-W. Frontier of India, 1897-98.—Mohmand.—Medal with clasp.
Tirah, 1897-98.—Actions of the Sampagha and Arhanga Passes.—Operations against the Khani Khel Chamkannis.—Operations in the Bazar Valley, 5 to 30 Dec. 97.—Clasp.
The War of 1914-18.—Despatches, *Lond. Gaz.*, 15 Jun. 16.

Sherrard, J. W.—
The War of 1914-16.—Despatches, *Lond. Gaz.*, 4 July 16 and 15 Aug. 17. Brev. of Lt.-Col.

Sherston, J. R. V.—
The War of 1914-18.—Despatches, *Lond. Gaz.*, 19 Oct. 14, 17 Feb. 15 and 11 Dec. 17. D. S. O., M. C.

Shettle, F. B.—
N.-W. Frontier of India, 1908.—Operations in the Mohmand country. Engagement of Matta, 24th April 1908. Engagement of Kargha, 24th May 1908.—Medal with clasp.
The War of 1914—18.—Despatches, *Lond. Gaz.*, 1 Jan. 16.

Shewen, M. T. (*Col., retd. I.A.*)—
Egypt, 1882.—Occupation of Alexandria. Operations at Mex. Occupation of Port Said and surrender of Fort Ghemil.—Medal; bronze star.
Burma, 1885—87.—Medal with clasp.
The War of 1914—18.—Despatches, *Lon. Gaz.*, 15 Jun. 16.

Shirley, S. R.—
N.-W. Frontier of India, 1908.—Operations in the Zakka Khel country. Despatches, G. G. O. No. 188 of 1908. Operations in the Mohmand country.—Medal with clasp.
The War of 1914—18.—Despatches, *Lond Gaz.*, 1 Jan. 16 and 4 Jan. 17. M. C.

Shore, O. B. S. F.—
N.-W. Frontier of India, 1897-98.—Malakand. Section Commandant, line of communications. Action at Landakai. Operations in Bajaur.—Medal with clasp.
Tirah, 1897-98.—Section Commandant (subsequently D. A. A. and Q. M. G.) line of communications. Action at Chagru Kotal. Despatches, G. G. O. 244 of 1898. Brev. of Maj.—Clasp.
South African War, 1900-02.—Operations in the Transvaal from 1 July 00 to 1 Feb. 01, and from 12 April to 31 May 02. Operations in the Orange River Colony from 3 March to 15 May 00, and from 2 to 13 May 02. Operations in Cape Colony from 6 Feb. to 2 March 00, 16 May to 30 June 00; 2 Feb. 01 to 15 March 02. Despatches, *Lond. Gaz.*, 10 Sep. 01.- Queen's medal with 3 clasps.—King's medal with 2 clasps. D. S. O.
The War of 1914—18.—Despatches. *Lond. Gaz*, 19 Oct. and 14 Nov. 16.

Short, H. E.—
The War of 1914—18.—Despatches, *Lond. Gaz.*, 5 Apl. 16.

Shoubridge, C. A. G.—
N.-W. Frontier of India, 1897-98.—Tochi.—Medal with clasp.

Shute, C. H.—
The War of 1914—18.—Despatches, *Lond. Gaz.*, 6 July 17.

Shuttleworth, A. R. B.—
Tibet, 1903-04.—March to Lhassa, 14th July to 3rd August 1904.—Medal.
N.-W. Frontier of India, 1908.—Operations in the Zakka Khel Country. Brigade Supply Officer, 3rd Brigade, and subsequently Asst. to S. and T. Officer, Advanced Depôt, Jamrud.—Medal with clasp.
The War of 1914—18.—Despatches, *Lond. Gaz.*, 5 Apl. 16.

Shuttleworth, B. W.—
South African War, 1899—1901.—Operations in Natal, March to June 00, including action at Laing's Nek (6 to 9 June). Operations in the Transvaal, east of Pretoria, July to 29 Nov. 00. Operations in the Transvaal, west of Pretoria, July to 29 Nov. 00. Operations in the Transvaal, 30th November 1900 to 20th July 1901.—Queen's medal with 4 clasps.
N.-W. Frontier of India, 1908.—Operations in the Zakka Khel country.—Medal with clasp.
The War of 1914—18.—Despatches, *Lond. Gaz.*, 27 Mar. 16.

Shuttleworth, D. I.—
The War of 1914—8.—Despatches, *Lond. Gaz.*, 1 Aug. 17. Brev. of Lt.-Col.

Sibthorpe, L. H. (*Hon. M.-Gen., retd. Bo. S. C.*)—
India, 1857.—With the Satpoora Field Force at the action of Ambapawnee.—Medal.
Afghanistan, 1879-80.—Skirmishes in the Kojak Pass, march from Quetta to the relief of Kandahar. Despatches, G. G. O. 693 of 81.—Medal.

Simmonds, W. A. (*Lt.-Col., retd. I.M.S.*)—
Afghanistan, 1878-79-80.—Action of Charasiah. Occupation of Kabul. Operations at and around Kabul.—Medal with 2 clasps.
Burma, 1885-86.—Occupation of Mandalay. Operations round Shwebo.—Medal with clasp.
N.-W. Frontier of India, Isazai, 1892.
Chitral, 1895.—Relief of Chitral. Despatches, G. G. O. 998 of '95.—Medal with clasp.
N.-W. Frontier of India, 1897-98.—Tochi.—Despatches, G. G. O. 1432 of 97.—Clasp.
Tirah, 1897-98.—Operations in the Bazar Valley, 25 to 30 Dec. 97.—Clasp.

Simon, J. R. (*Hony. Capt., Sr. Assist. Surgn. retd. I. S. M. D.*)—
Afghanistan, 1879-80.—Medal.

Simonds, J. N.—
South African War, 1902.—Operations in Cape Colony, January to February 1902. Operations in Orange River Colony, February to May 1902. —Queen's Medal with 3 clasps.

Simpson, *Sir* B. (*Sur.-Gen., retd. Ind. Med. Serv.*)—
N.-E. Frontier of India, Bhutan, 1865-66.—Medal with clasp.

Simpson, Benjamin (*Commy. and Hony. Capt., retd. I. M. L.*)—
Afghanistan, 1878-79-80.—Siege and battle of Kandahar.—Medal with clasp.

Simpson, J. C.—
Tirah, 1897-98.—Medal with 2 clasps.
The War of 1914—18.—Despatches, *Lond. Gaz.*, 15 Aug. 17. Brev. of Lt.-Col.

Simpson, R. J. B. (*Col., retd. Mad. S. C.*)—
Burma, 1885-86.—(Dangerously wounded.) Despatches, G. G. O. 295 of 86.

Sinha, N. P. (*Maj., retd. I. M. S.*)—
Burma, 1885—89.—Medal with 2 clasps.

Sinton, J. A.—
The War of 1914—18.—Despatches, *Lond. Gaz.*, 21 Jun. and 19 Oct. 16. V. C.

Skeen, A.—
N.-W. Frontier of India, 1897-98.—Defence of Malakand.—Relief of Chakdara. Malakand. Action at Landakai.—Operations in Bajaur and in the Mamund country.—Medal with 2 clasps.
China, 1900.—Medal.
East Africa, 1902-03-04.—Operations in Somaliland. —Action at Jidballi.—Medal with 2 clasps. Despatches, *Lond. Gaz.*, 2 Sep. 04. Brev. of Maj.
The War of 1914—18.—Despatches, *Lond. Gaz.*, 5 Aug. 15 and 23 Jan. 16. Brev. of Col. French War Cross. C. M. G.

Skinner, A. B.—
The War of 1914—18.—Despatches, *Lond. Gaz.*, 4 Jan. and 28 Dec. 17. D. S. O.

Skinner, F. J.—
N.-E. Frontier of India, Chin-Lushai, 1889-90.—Medal with clasp. Promoted Sub-Conductor.
East Africa, 1902-1904.—Operations in Somaliland. Despatches, *London Gazette*, 2nd September 1904. —Medal with clasp. Promoted Assist. Commissary

Skinner, G. J. (*Col., u.s.l., Ind. S. C.*)—
Afghanistan, 1880.—Expedition against the Marris at Mal. Mentioned in Despatches.—Medal.
N.-E. Frontier of India, Chin-Lushai, 1889-90.—Commanded Northern Column of the Lushai Force. Mentioned by Govt. of India. Despatches, G. G. O. 677 of 1890.—Medal with clasp; D.S.O.

Skinner, H. T.—
N.-W. Frontier of India, Waziristan, 1901-02.—Medal with clasp.
The War of 1914—18.—Despatches, *Lond. Gaz.*, 30 Jun. 16. D. S. O.

Skinner, R. H.—
South African War, 1902.—Operations in Cape Colony, 1st April to 31st May 1902.—Queen's medal with 2 clasps.

Skinner, T. B.—
South African War, 1901-02.—Attached to Army Service Corps. Operations in Cape Colony and Orange River Colony from September 1901 to April 1902. Operations in the Transvaal from April 1902 to 31st May 1902.—Queen's medal with 5 clasps.

Skipwith, C. G. Y.—
The War of 1914—18.—Despatches, *Lond. Gaz.*, 1 Jan. 16 and 4 Jan. 17.

Skipwith, F.—
China, 1900.—Medal with clasp.
Tibet, 1903-04.—Medal.
N.-W. Frontier of India, 1908.—Operations in the Zakka Khel country.—Medal with clasp.

Slacke, G. C.—
The War of 1914—18.—Despatches, *Lond. Gaz.*, 12 Jun. 17.

Slater, A. F. M.—
South African War, 1901-02.—Operations in the Transvaal between 31 Oct. 01 and 31 May 02. —Queen's medal with 5 clasps.

Slingsby, T. W.—
N.-W. Frontier of India, Waziristan, 1901-02.—Medal with clasp.

Smalley, J.—
The War of 1914—18.—Despatches, *Lond. Gaz.*, 1 Jan. 16.

Smallwood, St. Clair H.—
South African War, 1901-2.—Operations in the Orange River Colony between 30th November 1900 and 31st May 1902. Operations in Cape Colony between 30th November 1900 and 31st May 1902.

Smart, E. deS. (*Maj., retd. I. S. C.*)—
N.-W. Frontier of India, Hazara, 1888.—Medal with clasp. (*2nd*) *Miranzai, 1891.*—Clasp. *Isazai, 1892.*

Smart, R. W. L.deB.—
South African War, 1900-01.—Served with 16th Worcestershire Yeomanry, 1st Yeomanry Bde. Operations in Cape Colony, south of Orange River, 1900. Operations in the Orange Free State, Feb. to May 1900. Operations in Orange River Colony, May and June 1900, including actions at Lindley (1 June) and Rhenoster River (11 June). Operations in the Transvaal, west of Pretoria, July to 29 Nov. 1900, and between 30 Nov. 1900 and 30 June 1901 —Queen's medal with 4 clasps.
N.-W. Frontier of India, 1908.—Operations in the Mohmand country. Engagement of Kargha, 24th May 1908.—Medal with clasp.

Smith, A. A.—
N.-W. Frontier of India, Waziristan, 1901-02.—Medal with clasp.
N.-W. Frontier of India, 1908.—Operations in the Mohmand country. Engagement of Kargha, 24th May 1908.—Medal with clasp.

Smith, A. D.—
Operations in the Abor country, 1911-12.—Medal with clasp.
The War of 1914-18.—Despatches, *Lond. Gaz.*, 22 Jun. 15. M. C.

Smith, A. Le F.—
Tirah, 1897-98.—Action of the Sampagha Pass.—Medal with 2 clasps.
China, 1900.—Medal.

Smith, C. A. J.—
South African War, 1900.—Queen's medal with clasp.
The War of 1914—18.—Despatches, *Lond. Gaz.*, 31 Oct. 17.

Smith, C. W. J.—
N.-W. Frontier of India, Waziristan, 1901-02.—Medal with clasp.
The War of 1914—18.—D. S. O.

Smith, C. H.—
The War of 1914—18.—Despatches, *Lond. Gaz.*, 1 Jan. 16.

Smith, E. C.—
N.-W. Frontier of India, 1902.—Operations against the Darwesh Khel Waziris.
The War of 1914-18.—Despatches, *Lond. Gaz.*, 20 Mar. 16.

Smith, F. A.—
N.-E. Frontier of India, Chin-Lushai, 1889-90.—Medal with clasp.
N.-W. Frontier of India, 1897-98.—With Kohat-Kurrum Force.—Medal with 2 clasps.
Tirah, 1897-98.—Actions of Chagru Kotal and Dargai.—Clasp.
China, 1900.—Medal.
The War of 1914—18—Despatches, *Lond. Gaz.*, 21 Jun. 16. Brev. of Col.

Smith, F. F. S.—
The War of 1914—18.—Despatches, *Lond. Gaz.*, 6 Dec. 16.

Smith, F. M.—
Chitral, 1895.—Relief of Chitral.—Medal with clasp.

Smith, G. McI. C.—
Chitral, 1895.—Relief of Chitral.—Medal with clasp.
N.-W. Frontier of India, 1897-98.—Mohmand.—Clasp.
Tirah, 1897-98.—Action of Chagru Kotal. Action of Dargai. Action of the Sampagha Pass. Operations in the Waran and Bara valleys.—Clasp.
N.-W. Frontier of India, Waziristan, 1901-02.—Clasp.

Smith, G. R. de H. (*Maj., retd. I.A.*)—
N.-W. Frontier of India, 1897-98.—Malakand. Defence of Malakand. Relief of Chakdara. Action of Landakai. Despatches, G. G. O. 1089 of 97. —Medal with 2 clasps.
Tirah, 1897-98.—Actions of Chagru Kotal, Dargai, and of the Sampagha and Arhanga Passes. Reconnaissance of the Saran Sar and action of 9 Nov. 97. Operations at and around Dwatoi and action of 24 Nov. 97. Operations against the Khani Khel Chamkannis. Operations in the Bazar Valley, 25 to 30 Dec. 97.—2 clasps.
South African War, 1899-1900.—Operations in Natal, 99, including actions at Rietfontein (24Oct.) and Lombard's Kop (30 Oct.). Defence of Ladysmith and action of 6 Jan. 00. Operations in the Orange Free State, Feb. to May 00. Actions at Houtnek (1 May), Vet River (5 and 6 May) and Zand River (10 May). Operations in the Transvaal in May and June 00, including actions near Johannesburg (29 May), Pretoria (14 June) and Diamond Hill (11 and 12 June). Operations in the Transvaal, east of Pretoria, July to 29 Nov. 00, including action at Lydenberg (5 to 8 Sep.). Despatches, *Lond. Gaz.*, 10 Sept. 01.—Queen's medal with 5 clasps.—Brev. of Maj.

Smith, H. A.—
N.-E. Frontier of India, 1892-93.—Lushai.
Soudan, 1896.—Dongola Expedition.—Medal, also Khedive's medal.
N.-W. Frontier of India, 1897-98.—Tochi.—Medal with clasp.

Smith, H. H.—
South African War, 1899-1900.—Operations in Natal, 99, including actions at Rietfontein (24 Oct.) and Lombard's Kop (30 Oct.). Operations in Orange River Colony (May to 29th November 1900). (Severely wounded.) Despatches, *Lond. Gaz.*, 10 Sep. 01.—Queen's medal with 3 clasps.
The War of 1914—18.—Despatches, *Lond. Gaz.*, 19 Oct. 16. D. S. O.

Smith, J. (*Brig.-Surg.-Lt.-Col., retd. Ind. Med. Serv.*)—
N.-E. Frontier of India, Chin-Lushai, 1889-90.—Medal with clasp.

Smith, J. B. (*M.-Gen. u.s.l. Ind. S.C.*)—
N.-W. Frontier of India, 1863.—2nd Eusofzai expedition; forcing of the Umbeyla Pass.—Medal with clasp.
N.-E. Frontier of India, Lushai, 1871-72.—Clasp.
Afghanistan, 1878—80.—Medal; Brev. of Lt.-Col.

Smith, J. B.—
The War of 1914—18—Despatches, *Lond. Gaz*, 21 Jun. 16.

Smith, J. C. C. (*Lieut.-Col., retd. I.M.S.*)—
Afghanistan, 1879-80.—Medal.
Burma, 1886—88.—Operations of the 1st Brigade.—Medal with 2 clasps.

Smith, J. G. (*Col., retd. I. A.*)—
Aghanistan, 1879-80.—Operations in the Lughman Valley. Expedition against the Wazir Khugianis.—Medal.

Smith, J. M. (*Maj. Supy. List. I. A.*)—
N.-W. Frontier of India, Hunza Nagar, 1891.—Despatches, G. G. O. No. 397 of 1892.—Medal with clasp. V. C.
N.-W. Frontier of India, Isazai, 1892.

Smith, J. M.—
Chitral, 1895.—Relief of Chitral. Promoted Deputy Commissary.—Medal with clasp.

Smith, J. MacD. (*Col. u.s.l. Ind. S.C.*)—
China, 1857—59.

Smith, M. C.—
South African War, 1902.—Operations in Cape Colony, April and May 1902.—Queen's medal with 2 clasps.

Smith, O. L. (*Lt.-Gen. u.s.l. Ind. S.C.*)—
India, 1857-58-59.—At Lucknow, May and June 1857, action of Chinhut, defence of the Lucknow Residency, actions of Cawnpore and Serai Ghat. Occupation of Bithoor, action of Kalee Nuddee, occupation of Futehghur, action of Mow Shumshabad, siege of Lucknow, action of Sissaindee, occupation of Pertabghur, Shunkerpore and Rai Bareilly. (Twice severely wounded.) Thanked by Govt. of India.—Medal with 2 clasps.

Smith, S. B.—
Chitral, 1895—Relief of Chitral. Despatches. Lieut.-Col. J. G. Kelly's despatch No. 8-C., 26 July 95.—Medal with clasp.
N.-W. Frontier of India, Waziristan, 1901-02.—Clasp.
The War of 1914-18.—Despatches, *Lon. Gaz.*, 27 July 17. C. M. G.

Smith, S. H. P.—
South African War, 1899-1901.—Operations in the Transvaal, November 1900 to June 1901.—Queen's medal with 3 clasps.

Smith, T. H. (*Col. retd. I. A.*)—
N.-W. Frontier of India, 1897-98.—Buner.—Medal with clasp.
Tirah, 1897-98.—Assistant to Chief Transport Officer.—Actions of the Sampagha and Arhanga Passes.—Operations in the Bazar Valley.—25 to 30 Dec. 97.—Clasp.

Smith, W. (*Comsy. and Hony. Capt., retd.*)—
Afghanistan, 1880.—Defence of Kandahar.—Medal.
Egypt, 1882.—Medal; Bronze Star.

Smith, W. F.—
N.-W. Frontier of India, Hazara, 1891.—Medal with clasp.
The War of 1914-18.—Despatches, *Lond. Gaz.*, 22 Jun. 15. 18 May 18. Brev. of Lt.-Col.

Smith, W. G. (*Col., retd. Ben. Inf.*)—
Abyssinia 1868.—Mentioned in despatches.—Medal.

Smith, W. H. (*Hony. Capt. and Comy., retd. I. O. D.*)—
East Africa, 1902-04.—Operations in Somaliland. In charge of Ordnance Base Depôt.—Medal with clasp.

Smithers, L. S. H.—
Operations in the Abor Country, 1911-12—Despatches, *Lond. Gaz* 16 July 12.
The War of 1914—18.—Despatches, *Lond. Gaz.*, 5 Sept 16.

Smith-Wynne, W. A. (*Dep. Surg. Gen., retd. Ind. Med. Serv.*)—
India, 1857—59.—With a Movable Column and with the Saugor and Nerbudda Field Force. Pursuit of rebels in the Bundelcund jungles.—Medal with clasp.

Smyth, C. A. W.—
South African War, 1900-01.—With Imperial Yeomanry. Operations in the Transvaal, July to Sept. 1900, and Oct. 1900 to June 1901. Operations in Cape Colony, March to July 1900.—Queen's medal with 3 clasps.

Smyth, E. W. (*Col., retd. Ben. Inf.*)—
N.-E. Frontier of India, 1865-66.—Bhutan Expedition—Medal with clasp.
Afghanistan, 1879.—Medal.
N.-W. Frontier of India, 1892.—Isazai Expedition.
Chitral, 1895.—With the Relief Force. Despatches, G. G. O. 998 of 95—C.B.

Smyth, F. A. (*Brig.-Surg.-Lt.-Col., retd. Ind. Med. Serv.*)—
Burma, 1886-87.—Operations in and around Hlinedet and Yemethen.—Medal with 2 clasps.
N.-W. Frontier of India, 1897-98.—Was P. M. O. of the Malakand Brigade. Relief and Defence of the Malakand. Despatches, G. G. O. 1089 of 97.—Medal with 2 clasps.

Smyth, J.—
Soudan, 1885.—Suakin. Present at the battles of Hasheem and Tofrek. Siege and burning of Tamai. (Wounded.)—Medal with 2 clasps; bronze star.

Smyth, J. G.—
The War of 1914—18.—Despatches, *Lond. Gaz.*, 22 and 29 Jun. 15. V. C. Order of St. George (4th Class)

Soady, G. J. FitzM.—
N.-W. Frontier of India, Hazara, 1891.—Medal with clasp.
N.-W. Frontier of India, (2nd) Miranzai, 1891.—Clasp.
China, 1900.—Relief of Pekin. Action of Peitsang. Action at Yangtsun. Despatches, *Lond. Gaz.*, 14 May 01.—Medal with clasp.

Sodhi, N. S.—
The War of 1914—18.—Despatches, *Lond. Gaz.*, 15 Aug. 17. M. C.

Solbe, C. E. DeL. (*Lieut.-Col., retd., I. A.*)—
Burma, 1888-89-90.—Medal with 2 clasps.
N.-W. Frontier of India, 1897-98.—Operations on the Samana and in the Kurrum Valley during Aug. and Sep. 97.—Medal with 2 clasps.
Tirah, 1897-98.—Actions of Chagru Kotal, Dargai, and of the Sampagha and Arhanga Passes. Operations against the Khani Khel Chamkannis. Operations in the Bazar Valley, 20 to 30 Dec. 97.—Clasp.
N.-W. Frontier of India, Waziristan, 1901-02.—Clasp.
N.-W. Frontier of India, 1902.—Operations against the Darwesh-Khel Waziris.
N.-W. Frontier of India, 1908.—Operations in the Zakka Khel country. Operations in the Mohmand country. Engagement of Matta, 24th April 1908. Engagement of Kargha, 24th May 1908.—Medal with clasp.

Somerset, C. W.—
Burma, 1886—89.—Medal with 2 clasps.
Chitral, 1895.—Relief of Chitral. Served as D. A. Q. M. G. at the Base.—Medal with clasp.
N.-W. Frontier of India, 1897-98.—Operations on the Samana and in the Kurram Valley during Aug. and Sep. 97.—2 clasps.
Tirah, 1897-98.—Reconnaissance of the Kharmana Defile and action of 7 Nov. 97.—Operations against the Khani Khel Chamkannis.—Clasp.
The War of 1914—18.—Despatches, *Lond. Gaz.*, 15 Jun. 10 and 4 Jan. 17. C. M. G.

Somerville, E. A—
The War of 1914—18.—Despatches, *Lond. Gaz.*, 22 Jun. 15.

Sondhi, S. D.—
The War of 1914—18.—Despatches, *Lond. Gaz.*, 19 Oct. 16. M. C.

Souter, A. B.—
N.-W. Frontier of India, 1897-98.—Tochi.—Medal with clasp.
N.-W. Frontier of India, Waziristan, 1901-02.—Clasp.

Souter, H. M. W.—
Tibet, 1903-1904.—Action at Niani, 26th June 1904. Operations at and around Gyantse, between 5th May and 6th July 1904. March to Lhassa, 14th July to 3rd August 1904. Despatches, *Gazette of India*, No. 1065 of 1904.—Medal with clasp.
The War of 1914—18.—Despatches, *Lond. Gaz.*, 21 Jun. 16 and 15 Aug. 17. C. M. G., D. S. O.

Southam, T. (*Hy. Capt., Dy. Comsy., retd. Ord.*)—
Burma, 1885-86.—Present at the occupation of Mandalay.—Medal with clasp.
N.-W. Frontier of India, 1897-98.—Tochi.—Medal with clasp.

Southey, W. M.—
Uganda, 1898-99.—Actions of Jeruba in April and Aug. 98, Kuoeri and Kijanguti. Staff Officer to O. C. Troops. Mentioned in Despatches.—Medal with 2 clasps.
East Africa (Somaliland), 1908-10.—Medal with clasp.
The War of 1914—18.—Despatches, *Lond. Gaz.*, 17 Feb. and 22 Jun. 15. Brev. of Col. C. M. G.

Spackman, W. C.—
The War of 1914—18.—Despatches, *Lond. Gaz.*, 13 July and 19 Oct. 16.

Spankie, C. G.—
The War of 1914—18.—Despatehes, *Lond. Gaz.*, 19 Oct. 16.

Sparling, J. P. (*Lieut.-Col., retd. I. A.*)—
Afghanistan, 1878-79-80.—Actions of the Peiwar Kotal and Charasiah. Operations at and around Kabul.—Medal with 3 clasps.
N.-W. Frontier of India, Hazara, 1888.—Medal with clasp.

Spence, A. H. O.—
N.-W. Frontier of India, 1897-98.—Operations on the Samana in the Kurram Valley during Aug. and Sep. 97.—Medal with 2 clasps.
Tirah, 1897-98.—Reconnaissance of the Kharmana Defile and action of 7 Nov. 97. Operations against the Khani Khel Chamkannis.—Clasp.
China, 1900.—Medal.

Spencer, D. B. (*Lieut.-Col., retd. I.M.S.*)—
Soudan, 1885.—Suakin. Actions of Hasheen, Tofrek, and Tamai.—Medal with 2 clasps. Bronze star.
Burma, 1886—88.—Operations of the 2nd Brigade.—Medal with 2 clasps.
N.-E. Frontier of India, Lushai, 1892.
China, 1900.—Senior Medical Officer, Tientsin Despatches, *Lond. Gaz.*, 14 May 01.—Medal.

Spencer, T. C. H. (*Lt.-Col., retd. Ind. Med. Serv.*)—
Rumpa Rebellion, 1879-80.

Spencer, T. E. (*Col., retd. I. A.*)—
Soudan, 1885.—Despatches, *Lond. Gaz.*, 25 Aug. '85.—Medal with clasp. Bronze star. Brev. of Maj.
Burma, 1889.—Despatches, G. G. O. 782 of '80.—Medal with clasp.

Spitteler, A.—
East Africa, 1902-04.—Operations in Somaliland.—Medal with clasp.
The War of 1914—18.—Despatches, *Lond. Gaz.*, 5 Apl. 16.

Splane, G. G. (*Hon. Capt., Commy., retd. I. A. Depts.*)—
China, 1900.—Relief of Pekin.—Medal with clasp.

Sprawson, C. A.—
N.-W. Frontier of India, Waziristan, 1901-02.—Medal with clasp.

Sproule, H. H.—
China, 1900.—Relief of Pekin. Action of Peitsang. Action at Yangtsun.—Medal with clasp.
West Africa (S. Nigeria), 1902-03.—Operations in the Afikpo country.—Medal with clasp.
West Africa (S. Nigeria), 1903.—Operations in the Ebegga country.—In command.—Clasp.
West Africa (S. Nigeria), 1904.—Operations at Obokum. In command (wounded). Despatches, *London Gazette*, 25th August 1905.—Clasp.

Spurgin, E. W.—
The War of 1914—18.—M. C.

Stable, R. H.—
The War of 1914—18.—Despatches, *Lond. Gaz.*, 15 Aug. 17. D. S. O.

Stack, C. E. (*Col., u.s.l. Bo. Cav.*)—
India, 1857—59.—Served with the Kotah Contingent against rebels in Agra and Muttra Districts; battle of Agra, siege and capture of Ahwa, campaign in Rajputana including pursuit of Tantia Topee, and battle of Bunass, and other minor affairs. Mentioned in Despatches, 31 Jan. 59.—Medal with clasp.

Stack, C. S.—
N.-W. Frontier of India, Miranzai (2nd), 1891.—As Transport Officer.—Medal with clasp.
China, 1900.—Medal.
The War of 1914—18.—Despatches, *Lond. Gaz.*, 5 Apl., 13 July and 19 Oct. 16. C. M. G.

Stainforth, L. C. H. (*Bt.-Col., retd., I. A.*)—
N.-W. Frontier of India, Waziristan, 1894-95.—Medal with clasp.
N.-W. Frontier of India, 1897-98.—Relief and defence of Malakand.—Relief of Chakdara. Malakand.—Operations in Bajaur and in the Mamund Country.—Despatches, G. G. O. 1089 of 97.—Medal with 2 clasps.
N.-W. Frontier of India, Waziristan, 1901-02.—Clasp.

Stainforth, W. (*Col., retd. Mad. Inf.*)—
Abyssinia, 1867-68.—Medal.
Afghanistan, 1879.—Served with the Thal Chotial and Vitakri Field Forces.—Medal.

Stainton, T. H.—
The War of 1914—18.—Despatches, *Lond. Gaz.*, 15 Aug. 17.

Standage, R. F.—
Uganda, 1897—99.—Actions at Jeruba in April and August 1898.—Operations against Soudanese mutineers and rebels. Wuganda in Unyoro, June and July 98.—Medal with clasp.

Standbridge, C. H.—
South African War, 1902.—Operations in the Transvaal, April to 31 May 1902.—Queen's medal with 2 clasps.

Stanger, Leathes, H. E.—
The War of 1914—18.—Despatches, *Lond. Gaz.*, 5 Apl. 16.

Stanley, E. H. B.—
East Africa, 1902-04.—Operations in the Somaliland.—Medal with clasp.

Stansfeld, C. G.—
The War of 1914—18.—Despatches, *Lond. Gaz.*, 14 Dec. 17.

Staples, E. C.—
The War of 1914—18.—Despatches, *Lond. Gaz.*, 15 Mar. and 27 July 16. M. C.

Staples, E. H.—
N.-W. Frontier of India, 1908.—Operations in the Mohmand country. Engagement of Matta, 24th April 1908.—Medal with clasp.

Stedman, H.—
The War of 1914—18.—M. C.

Steel, C. N.—
N.-W. Frontier of India, 1908.—Operations in the Zakka Khel Country.—Medal with clasp.
The War of 1914—18.—Despatches, *Lond. Gaz.*, 15 Aug. 17. Brev. of Maj.

Steel, R. A.—
N.-W. Frontier of India, 1897-98.—Mohmand.—Medal with clasp.
China, 1900.—Action of Peitsang. Action at Yangtsun. Despatches, *Lond. Gaz.*, 14 May 01.—Medal with clasp
The War of 1914—18.—Despatches, *Lond. Gaz.*, 17 Feb. 15.

Steel, R. F.—
N.-W. Frontier of India, 1908.—Operations in the Mohmand country.—Medal with clasp.

Steele, St. G. L.—
Egypt, 1882.—Actions of Kassassin and Tel-el-Kebir.—Medal with clasp. Bronze star.
N.-E. Frontier of India, Manipur, 1891.—Served as Brigade Major, Silchar column. Despatches, G. G. O. 585 of '91.—Medal with clasp.
N.-W. Frontier of India, Waziristan, 1894-95.—Served as Assistant Superintendent, Army Signalling, 1st Brigade.—Clasp.
Soudan, 1896.—Dongola Expedition.—Brigade Transport Officer. Egyptian medal.—Medal.
Tirah, 1897-98.—Section Road Commandant. Despatches, G. G. O. 244 of 98.—Medal with 2 clasps.
China, 1900.—Served with the Transport Department and as Special Service Officer in Manchuria. Mentioned in Despatches, *Lond. Gaz.*, 13 Sep. 01.—Medal.

Steen, R.—
N.-W. Frontier of India, Waziristan, 1901-02.—Medal with clasp.
Tibet, 1903-04.—Medal with clasp.

Stephen, A. (*Col., retd. Ind. Med. Serv.*)—
Abyssinia, 1868.—Medal.

Stephens, A. E. R. (*Surg.-Lt.-Col., retd. Ind. Med. Serv.*)—
Afghanistan, 1878-80.—Action of Ahmed Khel.—Medal with clasp.
N.-W. Frontier of India, Hazara, 1888.—Medal with clasp.

Sterndale Bennet, J.—
East Africa, 1903-04.—Operations in Somaliland.—Medal with clasp.

Stevens, C. (*Maj., retd. I. A.*)—
Burma, 1892-93.—Operations in the Northern Chin Hills.—Medal with clasp.

Stevens, C. F. L.—
The War of 1914—18.—M. C.

Stevens, C. R.—
Chitral, 1895.—With the Relief Force.—Medal with clasp.
N.-W. Frontier of India, 1897-98.—With Tirah Expdy. Force.—2 clasps.

Stevens, G. B.—
Burma, 1886—88.—Served on special duty and as Staff Officer and Assistant Inspector, Mounted Infantry. Operations of the 1st Brigade.—Despatches, G. G. O. 434 of '87.—Medal with 2 clasps.
N.-E. Frontier of India, Lushai, 1889.—Despatches, G. G. O. 592 of '89.—Clasp.
Burma, 1890—92.—Comdg. Kanhow Column, 1891-92.—Brev. of Maj.

Stevens, M. W. (*Lt.-Col. Retd. I. A.*)—
Afghanistan, 1879-80.—As Transp. Offr., Cav. Brig. March from Kandahar to Kabul; battle of Ahmed Khel and occupation of Guzhni. Medal with clasp.

Stevens, N. M. C.—
Burma, 1891.—Wuntho Field Force.—Medal with clasp.

Chin Hills, 1892-93.—Clasp.
Tirah, 1897-98.—Operations in the Bazar Valley, 25 to 30 Dec. 97.—Medal with 2 clasps.
The War of 1914—18.—Despatches, *Lond. Gaz.*, 23 Jun. 15. C. M. G.

Stevenson, F. (*Col., retd. I. A.*)—
Afghanistan, 1879-80.—Actions of Ahmed Khel and Urzoo, march from Kabul to the relief of Kandahar, and battle of 1 Sep.—Despatches, G. G. O. 326 of 80.—Medal with 2 clasps; bronze star.
Soudan, 1885.—Suakin. Served as D. A. A. and Q. M. G.

Stevenson, H. D. M.—
The War of 1914—18.—Despatches, *Lond. Gaz.*, 19 Oct. 16 and 15 Aug. 17. M. C.

Stewart, A. C.—
Burma, 1889—92.—Medal with clasp.
Chitral, 1895.—Relief of Chitral.—Action of the Malakand Pass.—Medal with clasp.
N.-W. Frontier of India, 1897-98.—Malakand. Action at Landakai.—Operations in Bajaur and in the Mamund country.—Buner.—Clasp.
South African War, 1902.—Operations in the Orange River Colony, 20 Feb. to 15 March 02. Operations in the Transvaal, 15 March to 18 April 02.—Queen's medal with 3 clasps.
The War of 1914—18.—Despatches, *Lond. Gaz.*, 13 July 16.

Stewart, A. E.—
South African War, 1899-1900.—Operations in the Transvaal, east of Pretoria, July to 29th November 1900. Operations in the Transvaal west of Pretoria, July to 29th Nov. 1900, including action at Zilikat's Nek, 2nd August 1900. Operations in Orange River Colony, May to 29th November 1900. Operations in Cape Colony, south of Orange River, 1899-1900.—Queen's medal with 4 clasps.
The War of 1914—18.—M. C.

Stewart, Alexander F.—
The War of 1914—18.—Despatches, *Lond. Gaz.*, 5 Apl. 16.

Stewart, Archibald, F.—
The War of 1914—18.—Despatches, *Lond. Gaz.*, 19 Oct. 16. and 15 Aug. 17.

Stewart, A. K. (*Surn. Lt.-Col., retd. Ind. Med. Serv.*)—
Afghanistan, 1880.—Defence of Kandahar, affair at Khairabad, sortie against Deh Khoja (twice slightly wounded) and battle of 1 Sep.—Medal with clasp.

Stewart, Chas. (*Revd.*)—
N.-W. Frontier of India, 1908.—Operations in the Mohmand Country.—Medal with clasp.

Stewart, F. H.—
Operations in the Abor country, 1911-12.—Medal with clasp.

Stewart, F. T.—
British Central Africa, 1895-96.—Commanded Tambala and Chikusi Angoni Expeditions.—Medal with clasp.
N.-W. Frontier of India, 1897-98.—Malakand. Operations in Bajaur and in the Mamund Country.—Medal with clasp.
Tirah, 1897-98.—Operations in the Bara Valley, 7 to 13 Dec. 97.—Clasp.
N.-W. Frontier of India, Waziristan, 1901-02.—Despatches, G. G. O. 611 of 02.—Clasp.
N.-W. Frontier of India, 1908.—Operations in the Zakka Khel country.—Medal with clasp.

Stewart, G. (*Hon. M.-Gen., retd. Ben. S.C.*)— 2081
India, 1857-58.—Actions of Futehpor Aoung, Pandoo Nuddee, Cawnpore, Bithoor, Busseeputgunge, Mungawar, 1st relief, occupation, defence and capture of Lucknow, operations in Oude, occupation of Fyzabad (twice wounded), mentioned in Despatches. Medal with 2 clasps.

China, 1860.—Actions of Sinho and Palichow and occupation of Pekin. Mentioned in Despatches. Medal with 2 clasps.
N.-W. Frontier of India, 1863.—2nd Eusofzai Expedition.—Capture of Laloo and Umbeyla.
N.-W. Frontier of India, Jowaki, 1877-78.—Medal with clasp. Mentioned in Despatches. Clasp.
Surprise of Skakot and affair with Uthman Khels, 1878.
Afghanistan, 1878—80.—Capture of Ali Musjid, expeditions to the Kama and Lughman Valley, operations at and around Kabul in Dec. 79, and action of Chhildukhteran. Despatches, *Lond. Gaz.*, 28 March 79, and G. G. O. 137 and 328 of 80. Medal with 2 clasps; Brev. of Lt.-Col.

Stewart, G. E.—
China, 1900.—Medal.
Mekran, 1901.—Attack and capture of Nodiz Fort.
East Africa (Somaliland), 1908-10.—Medal with clasp.

Stewart, G. H.—
China, 1900.—Relief of Pekin.—Medal with clasp.

Stewart, J. A.—
Chitral, 1895.—Relief of Chitral.—Medal with clasp.
China, 1900.—Medal with clasp.
N.-W. Frontier of India, 1908.—Operations in the Mohmand country. Medal with clasp.
The War of 1914—18.—Despatches, *Lond. Gaz.*, 30 Jun. 16.

Stewart, J. C. (*M.-Gen. u.s.l. Ind. S.C.*)—
India, 1857-58.—Action of Meeangung, siege and capture of Lucknow, actions of Barree and Nuggur.—Medal with clasp.
N.-W. Frontier of India, Jowaki, 1877-78.—Medal with clasp.
Afghanistan, 1878-80.—Action of Matun, battle of Charasiah, operations at and around Kabul in Dec. 79. Despatches, G. G. Os. 610 of 79 and 137 of 80. Medal with 2 clasps. Brev. of Lt.-Col.
Burma, 1887-88.—Clasp.

Stewart, J. H. K.—
N.-W. Frontier of India, 1897-98.—Mohmand. Malakand.—Operations in Bajaur.—Medal with clasp.
Tirah. 1897-98.—Clasp.—D.S.O.
The War of 1914—18.—Despatches, *Lond. Gaz.*, 22 Jun. 15, 19 Oct. 16 a d 15 Aug. 17. Brevs. of Lt.-Col. and Col. D. S. O.

Stewart, J. M. (*Lt.-Gen. u. s. l. Ben. Inf.*)—
N.-E. Frontier of India, 1862-63.—Cossiah and Jynteah Hills Expedition. Capture of several stockades.
Abyssinia, 1867-68.—Medal.

Stewart, J. M.—
N.-W. Frontier of India, Hazara, 1888.—Medal with clasp.
N.-E. Frontier of India, Chin-Lushai, 1889-90.—(Despatches, G. G. O. 677 of 90).—Clasp.
N.-W. Frontier of India, Hazara, 1891.—Commanding Khyber Rifles.—Clasp. *Miranzai (2nd), 1891.*—Clasp. *Isazai, 1892.*
N.-W. Frontier of India.—As D. A. Q. M. G. Waziristan Delimitation Escort, 1894. Action at Wana, 3 Nov. 94. Despatches, G. G. O. No. 268 of 95. *Waziristan, 1894-95.*—As D. A. Q. M. G., 1st Brigade.—Clasp.
China, 1900.—As D. A. A. G., 2nd Brigade. Despatches, *Lond. Gaz.*, 14 May 01. Medal. Brev. of Maj.
Tibet, 1903-04.—Despatches. *Gazette of India*, No. 1065 of 1904.—Medal.
N.-W. Frontier of India, 1908.—Operations in the Zakka Khel country. Despatches, G. G. O. No. 188 of 1908. A. D. C. to the King. Medal with clasp.
The War of 1914—18.—Despatches, *Lond. Gaz.*, 31 Oct. 17. Promd. Maj.-Genl.

Stewart, Sir N. R., Bart., C. B. (*Maj.-Genl., retd. I. A.*)—
Afghanistan, 1878-79-80.—Battle of Ahmed Khel, Action of Urzu. Despatches. Lt.-Genl. *Sir* D. M. Stewart's despatch No. 640K., 24 June '79, and G. G. O. 326 of '80.—Medal with clasp. Brev. of Maj.
Soudan, 1885.—Suakin. D. A. A. and Q. M. Genl. and Offg. A. A. and Q. M. G., Indian Contingent. Actions of Hasheen, Tofrek, and operations at Tamai. Despatches, *Lond. Gaz.* 25 Aug. '85.—Medal with 2 clasps. Bronze star. Brev. of Lt.-Col.
China, 1900.—Relief of Pekin. Action at Peitsang and action at Yangtsun. Commanded 1st Brigade. Despatches, *Lond. Gaz.*, 14 May 01.—Medal with clasp.—C.B.

Stewart, J. M. Y.—
The War of 1914—18.—Despatches, *Lond. Gaz.*, 21 Jun, 16.

Stewart, R. J. T.—
N.-W. Frontier of India, 1897-98.—Malakand.—Operations in Bajaur.—Mohmand.—Medal with clasp.
Tirah, 1897-98.—Clasp.
South African War, 1899-1902.—Operations in the Orange Free State, Feb. to May 00, including operations at Paardeberg (17 to 26 Feb.)—Operations in the Transvaal between 30 Nov. 00 and 31 May 02.—Queen's medal with 3 clasps; King's medal with 2 clasps.
N.-W. Frontier of India, 1902.—Operations against the Darwesh Khel Waziris.

Stewart, T. W.—
Burma, 1887—89.—Medal with clasp. *Tonhon Expedition, 1889-90.*—Clasp.

Stewart, W. (*Maj., retd. Ind. S. C.*)—
N.-W. Frontier of India, 1898.—With Buner Field Force. Attack and capture of the Tanga pass.

Stirling, W. G. H.—
Burma, 1886—89.—Medal with 2 clasps.

St. John, O.—
South African War, 1901.—Served with Impl. Yeo. Operations in Cape Colony, and the Transvaal, Apr. to Nov. 01. (Slightly wounded.) Queen's medal with 3 clasps.

St. John, R. S.—
N.-W. Frontier of India, 1897-98.—Medal with clasp.
China, 1900.—Medal.

St. Maur-Wynch, H. St. M. (*Col. u.s.l. I.A.*)—
India, 1857—59.—Affair at Shahpore and operations in Bundlecund. Medal.

Stock, H. J. (*Hon. Col., retd. Bo. S. C.*)—
Afghanistan, 1884.—As Brig. Maj. Medal.
N.-W. Frontier of India, Zhob Valley Expedition, 1884.

Stockley, C. H.—
East Africa (Somaliland) 1908-10.—Medal with clasp.

Stockley, H. W. (*Col. retd. Bom. Art.*)—
Persia, 1856-57.—Bombardment of Mohumrah. Medal with clasp.

Stockley, V. M. (*Col., retd. I. A.*)
Egypt, 1882.—Actions of Kassassin and Tel-el-Kebir. Occupation of Cairo.—Medal with clasp. Bronze star.
China, 1900.—Medal.

Stodart, T.—
Soudan, 1896.—Dongola Expedition. Medal, Khedive's medal.
Tirah, 1897-98.—With Kohat Kurram Force. Operations against the Khani Khel Chamkannis. Medal with 3 clasps.

Stoddart, C. H. (*Col., retd. I.S.C.*)—
Afghanistan, 1879-80.—With Jelalabad Field Force.—Medal.
Burma, 1886—89.—Operations of the 1st Brigade.—Medal with 2 clasps.

Stoddart, G.—
The War of 1914—18.—Despatches, *Lond. Gaz.*, 5 Apl. 16.

Stoddart, H.—
South African War, 1902.—Operations in Cape Colony, Jan. to 31 May 02. Queen's medal with 2 clasps.

Stoker, R. N. (*Lt.-Col., retd. Ind. Med. Serv.*)—
Afghanistan, 1879.—Medal.
N.-E. Frontier of India, 1888.—Sikkim Expedition. Medal with clasp.

Stokes, C. B.—
N.-W. Frontier of India, 1897-98.—Malakand.—Operations in Bajaur and in the Mamund country.—Medal with clasp.

Stokes, G. J. (*Comry. and Hony. Capt., retd.*)—
Afghanistan, 1879-80.—Medal.

Stone, A. G.—
South African War, 1902.—Operations in Orange River Colony, and Cape Colony, from 16th January to 31st May 1902. Queen's medal with 3 clasps.
The War of 1914—18.—Despatches, *Lond. Gaz.*, 1 Dec. 16. M. C.

Stoney, P. S.—
N.-W. Frontier of India, 1908.—Operations in the Mohmand country. Medal with clasp.

Stooks, C. S.—
The War of 1914—18.—Despatches, *Lond. Gaz.*, 8 and 13 Feb. 17. D. S. O.

Stopford, W. H. J. (*Col., retd. Bo. Cav.*)—
Afghanistan, 1879-80.—March from Quetta to the relief of Kandahar. Despatches, G. G. O. 693 of 81. Medal.

Storr, C. L.—
Tirah, 1897-98.—Operations in the Bazar Valley, 25 to 30 Dec. '97.—Medal with 2 clasps.
China, 1900.—Medal.

Story, J. A.—
The War of 1914—18.—M. C.

Stotherd, E. A. W. (*Lt.-Col., retd. I.A.*)—
Burma, 1888-89.—Medal with clasp.
N.-W. Frontier of India, 1897-98.—Operations on the Samana, Sep. '97.—Medal with 2 clasps.
Tirah, 1897-98.—Actions of Dargai and of the Sampagha and Arhanga Passes. Operations in the Waran Valley and against the Khani Khel Chamkannis.—Operations in the Bara Valley—7 to 14 Dec. '97.—Clasp.
China, 1900.—Medal with clasp.

Strachey, B. (*Capt., retd. Indian Army*)—
Burma, 1887-88.—Operations of the 3rd Brigade.—Medal with clasp.
N.-E. Frontier of India, Manipur, 1891.—Clasp.

Strachey, J. (*Lieut.-Col., retd. I. A.*)—
Afghanistan, 1879-80.—Medal.
Burma, 1885—87.—Taking of Minhla. Operations of the 6th Brigade. Despatches, G. G. O. 434 of '87.—Medal with clasp.
N.-W. Frontier of India, Hazara, 1891.—Clasp.

Stranack, H. R.—
The War of 1914—18.—Despatches, *Lond. Gaz.*, 8 Feb. 17.

Street, A.—
N.-W. Frontier of India, Miranzai (2nd), 1891.—Medal with clasp.
N.-W. Frontier of India, 1902.—Operations against the Darwesh Khel Waziris.
N.-W. Frontier of India, 1908.—Operations in the Mohmand country. Engagement of Kargha, 24th May 1908.—Medal with clasp.

Strettell, A. D. (*Lt.-Col., retd. Ind. S.C.*)—
Abyssinia, 1867-68.—Despatches, *Lond. Gaz.*, 10 July '68.—Medal.
Afghanistan, 1878-79.—Capture of Peiwar Kotal. Medal with clasp.
N.-W. Frontier of India, Mahsud-Waziri, 1881.
N.-W. Frontier of India, (1st) Miranzai, 1891.

Strettell, C. B. D.—
N.-W. Frontier of India, Waziristan, 1901-02.—Medal with clasp.
The War of 1914—18.—Despatches, *Lond. Gaz.*, 15 Aug. 17.

Strickland, P. C. H.—
Burma, 1888—91.—Wunthu Field Force, 1891. Mogoung column.—Medal with clasp.

St. Romaine, E. A.—
N.-W. Frontier of India, 1897-98.—Malakand. Operations in Bajaur, and in the Mamund country. Action at Landakai—Buner. Action of the Tanga Pass.—Medal with clasp. Specially commended and promoted to 2nd class Assistant Surgeon.
South-African War, 1899-1902.—Operations in Natal, 1899, including actions at Elandslaagte (21st October), Rietfontein (24th October), and Lombard's Kop (30th October). Defence of Ladysmith, including sorties of 7th and 10th December 1899 and action of 6th January 1900. Operations in Natal, March to June 1900, including action at Laing's Nek (6th to 9th June). Operations in the Transvaal east of Pretoria, July to 29th November 1900, including actions at Reit Vlei (16th July). Belfast (26th and 27th August, 1900). Lydenburg, (5th to 8th September) and Rhenoster Kop (29th November). Operations in the Transvaal between 30th November, 1900 and 2nd January 1902, and from 22nd March 1902 to 31st May 1902. Operations in the Orange River Colony, between 3rd January 1902 and 21st March 1902. Despatches, *Lond. Gaz.*, 8th February 1901. Queen's medal with 5 clasps, King's medal with 2 clasps. Promoted Lieutenant and Sr. Assistant Surgeon, G. G. O. 323 of 1902.

Strong, A. D.—
N.-W. Frontier of India, 1897-98.—*Buner.*—Action of the Tanga Pass.—Medal with clasp.
The War of 1914—18.—Despatches, *Lond. Gaz.*, 1 Jan. 16 and 28 Dec. 17. D. S. O.

Strong, J. D.—
South African War, 1901.—Queen's medal with 3 clasps.

Strong, R. G.—
N.-W. Frontier of India, Waziristan, 1901-02.—Medal with clasp.

Stuart, A. P. D. C. (*Major, retd., I. A.*)—
Burma, 1888-89.—Medal with clasp.
Tibet, 1903-04.—Action at Niani, 26th June 1904. Operations at and around Gyantse, between 5th May and 6th July 1904. March to Lhassa, 14th July to 3rd August 1904.—Medal with clasp.

Stuart, C. J. L. (*Lt.-Col., retd. I.S.C.*)—
N.-W. Frontier of India, Jowaki, 1877.—Medal with clasp.
Afghanistan, 1879-80.—Action of Saif-u-din. Battle of Ahmed Khel (wounded).—Medal with clasp.

Stuart, D. F.—
Burma, 1885—87.—Taking of Minhla. Operation of the 6th Brigade.—Medal with clasp.
N.-W. Frontier of India, Hazara, 1891.—Clasp.
N.-W. Frontier of India, Waziristan, 1901-02.—Medal with clasp.

Stuart, W.—
South African War.—1899-1900.

Sturmer, A. J. (*Lt.-Col., retd. I.M.S.*)—
Afghanistan, 1878-79.—Medal.
N.-E. Frontier of India, Chin Hills, 1892-93.—Medal with clasp.

Sullivan, E. L.—
Burma, 1888-89.—Medal with clasp.
Soudan, 1896.—Dongola Expedition.—Medal; Khedive's medal.
Tirah, 1897-98.—(Severely wounded.)—Medal with 2 clasps.
N.-W. Frontier of India, Waziristan, 1901-02.—D. A. A. G. Brigade Staff. Despatches, G. G. O. 611 of 02.—Clasp.
The War of 1914-18.—Despatches, *Lon. Gaz.*, 30 May 16, 4 July 16, and 15 Aug. 17. C. M. G.

Sullivan, R. E.—
Burma, 1885-86.—Present at the occupation of Mandalay.—Medal with clasp.
Chitral, 1895.—Relief of Chitral.—Medal with clasp.
N.-W. Frontier of India, 1897-98.—Tochi.—Clasp.—Operations on the Samana during Sep. '97.—Clasp.
Tirah, 1897-98.—Clasp.

Summers, G. H.—
East Africa (Somaliland) 1914—15.—Despatches, *Lond. Gaz.*, 2 Aug. 16.—Medal. with clasp.

Sumner, F. W.—
N.-W. Frontier of India, Waziristan, 1901-02.—Medal with clasp.
The War of 1914—18.—Despatches, *Lond. Gaz.*, 8 Feb. 17.

Sunder, C. E.—
N.-E. Frontier of India, Sikkim, 1888.—Medal with clasp.
N.-W. Frontier of India, (1st) Miranzai, 1891.

Sutherland, A. O.—
South African War, 1899-1901.—Operations in the Transvaal, west of Pretoria, July to 29th November 1900.—Queen's medal with 4 clasps.
The War of 1914—18.—Despatches, *Lond. Gaz.*, 19 Oct. 16.

Sutherland, D. W.—
Chitral, 1895.—Relief of Chitral. Action of the Malakand Pass. Despatches, G. G. O. No. 998 of 1895.—Medal with clasp.

Sutherland, R. O.—
The War of 1914—18.—Despatches, *Lond. Gaz.*, 15 Aug. 17. D. S. O.

Sutherland, W. D.—
Burma, 1891-92.—Operations on the N.-E. Frontier.

Sutton, F. G. H.—
N.-E. Frontier of India, Chin Hills, 1892-93.—Medal with clasp.
Burma, 1893-94.
The War of 1914—18.—Despatches, *Lond. Gaz.*, 21 Jun and 13 July 16.

Sutton, H. G. (*Bt.-Col., retd. I. A.*)—
Burma, 1885—89.—Medal with 2 clasps.

Swaine, C. L. (*Lieut.-Col., retd. I. M. S.*)—
Burma, 1886—89.—Medal with 2 clasps.

Swan, C. T.—
N.-E. Frontier of India, Lushai, 1889.—Medal with clasp.
Burma, 1891-92.—Tlang Tlang and Tashon cols.
China, 1900.—Orderly Officer, 4th Brigade.—Medal.
The War of 1914-18.—Despatches, *Lond Gaz.*, 4 July 16.

Swan, J. S.—
N.-W. Frontier of India, 1897-98.—Mohmand.—Medal with clasp.
Tirah, 1897-98.—Clasp.
N.-W. Frontier of India, Waziristan, 1901-02.—Clasp.

Swann, J. C. (*Maj.-Genl., retd. I.A.*)—
Afghanistan, 1880.—Medal.
Soudan, 1885.—Suakin.—Medal with clasp; bronze star.
Burma, 1885—87.—Medal with clasp.
China, 1900.—Relief of Pekin. Assistant Adjt. and Q. M. G., Lines of Communication.—Medal with clasp. Despatches, *London Gazette*, 14 May '01. Brev. of Lt.-Col.
East Africa, 1902—04.—Operations in Somaliland. Base Commandant and afterwards Colonel on the Staff Commanding Lines of Communication. Despatches (twice), *London Gazette*, 2 Sep. '04.—Medal with clasp. C. B.

Swayne, Sir E. J. E. (*Brev. Col., retd. I.A.*)—
Burma, 1885-86.—Medal with clasp.
East Africa, 1898.—Operations against Ogaden Somalis. Medal with clasp.
East Africa, 1901-02.—Operations in Somaliland In commd. Despatches, *London Gazette*, 18 Apl. 02. Brev. of Lt.-Col. Medal with 2 clasps. C. B.

Sweeny, P. (*Hy. Lt. A. C., retd. Ord.*)—
Tirah, 1897-98.—Medal with 2 clasps.—Promoted Deputy Assistant Commissary.

Sweeny, R. L. C.—
The War of 1914—18.—Despatches, *Lond. Gaz.*, 30 Jun. 16. D. S. O., M. C.

Sweeny, T. H. (*Lieut.-Col., retd. I.M.S.*)—
Afghanistan, 1879-80.—Operations at and around Kabul.—Medal with clasp.

Sweet, E. H.—
N.-W. Frontier of India, 1897-98.—Operations on the Samana during Aug. and Sep. '97.—Medal with 2 clasps.
Tirah, 1897-98.—Clasp.
Operations in the Abor country, 1911-12.—Medal with clasp.
The War of 1914—18.—Despatches, *Lond. Gaz.* 1 Jan. 16 and 15 Aug. 17. D. S. O.

Sweet, R.—
The War of 1914—18.—Despatches, *Lond. Gaz.* 19 Oct. 16. D. S. O.

Swete, C. D. (*Lt.-Col., retd. Ind. S.C.*)—
Burma, 1887.—Medal with clasp.

Swifte, E. L.—
N.-W. Frontier of India, Waziristan, 1894-95.—Medal with clasp.
China. 1900.—Medal.
The War of 1914—18.—Despatches, *Lond. Gaz.*, 22 Jun. 15 and 19 Oct. 16.

Swiney, G. W. B. (*Lt.-Col., retd. I. A.*)—
Afghanistan, 1878-79-80.—Action of Charasiah. Operations at and around Kabul.—Medal with 2 clasps.
N.-W. Frontier of India, Zhob Valley, 1899. Waziristan, 1894-95.—Medal with clasp.
N.-W. Frontier of India, 1897-98.—Operations on the Samana and in the Kurram Valley during Aug. and Sep. '97.—Relief of Gulistan.—Medal with 2 clasps.
Tirah, 1897-98.—Clasp.
N.-W. Frontier of India, Waziristan, 1901-02.—Clasp.

Swinhoe, C. (*Col. u. s. l. Ind. S. C.*)—
Afghanistan, 1880.—Medal.

Swinley, G. (*M.-Gen., retd. B. Arty.*)—
N.-E. Frontier of India, 1865-66.—Bhutan Expedition.—Medal with clasp.
N.-W. Frontier of India, 1868.—Hazara Expedition—Clasp. *Jowaki Expedition, 1877.*—Clasp.

Afghanistan, 1878-79-80.—Action of Charasiah, Beni Shahr, Bala Hissar, operations near Kabul in Sept. and Oct., occupation of Sherpur, action of Charasiah, and march from Kabul to relief of Kandahar and battle of 1 Sep. Expedition against the Atchakzai tribes. Despatches, *Lond. Gaz.*, 31 March '79, and 16 Jan. '80, and G. G. Os. 137, 328 and 566 of '80. Brevs. of Maj. and Lt.-Col. —Medal with 3 clasps; bronze star.

Syer, H. H.—
South African War, 1901-02.—Operations in the Transvaal, October and November 1901. Operations in the Orange River Colony, 1st December 1901 to 31st May 1902. Queen's medal with 4 clasps.
East Africa, 1902-04.—Oparations in Somaliland. Special Service Officer. Medal with clasp.
The War of 1914—18.—Despatches, *Lond. Gaz.*, 19 Oct. 16.

Sykes, J.—
N.-W. Frontier of India, Mahsud-Waziri, 1881.
Egypt, 1882.—Actions of Kassassin and Tel-el-Kebir. —Medal with clasp. Bronze star.

Sykes, P. M.—
South African War, 1899—1901.—On Staff and employed with Imperial Yeomanry. Slightly wounded. Despatches, *Lond. Gaz.*, 3 Dec. '01. —Queen's medal with 3 clasps.
The War of 1914—18.—Despatches, *Lond. Gaz.*, 31 Oct. 17.

Sykes, W. A. (*Lt.-Col., I.M.S., retd.*)—
Soudan, 1885.—Suakin.—Medal with 2 clasps. Bronze star.
Burma, 1885—89.—Ruby Mines Expeditions. Despatches, G. G. O. 434 of '87.—Medal with 2 clasps. D.S.O.
N.-W. Frontier of India, Zhob Valley, 1890.—Waziristan, 1894-95.—Clasp.
N.-W. Frontier of India, 1897-98.—Malakand. Operations in Bajaur and in the Mamund country. *Buner.* Action of the Tanga Pass. Despatches, G. G. O. 217 of '98.—Medal with clasp.
China, 1900.—Medal.

Sym, *Sir* J. M. (*Maj.-Genl., u.s.l., I.S.C.*)—
N.-W. Frontier of India, (2nd) Eusufzai, 1863.—Umbeyla.—Medal with clasp. *Hazara, 1868.*—Clasp.
Afghanistan, 1879-80.—Action of Charasiah. Operations at and around Kabul. Despatches, G. G. O. 137 of '80. March from Kabul to the relief of Kandahar, and battle of Kandahar. Despatches, G. G. O. 552 of '80. Aitchakzai Expedition. Operations against the Maris. Despatches, Br. Genl. C. M. Macgregor's despatch, dated 16 Nov. '80.—Medal with 3 clasps. Bronze star. Brev. of Lt.-Col.
N.-W. Frontier of India, Hazara, 1888.—Commanded 1st column, 1st Brigade. Despatches, G. G. O. 978 of '88.—Clasp. C. B. *(2nd) Miranzai, 1891.*—Commanded 1st Brigade. Despatches, G. G. O. 632 of '91.—Clasp.

Synge, M.—
N.-W. Frontier of India, 1897-98.—Medal with clasp.
Tibet, 1903-04.—Action at Niani, 26th June 1904. Operations at and around Gyantse, between 5th May and 6th July 1904. March to Lhassa, 14th July to 3rd August 1904.—Medal with clasp.
Operations in the Abor country, 1911-12.—Medal with clasp.
The War of 1914—18.—Despatches, *Lon. Gaz.*, 15 Jun. '16, 4 Jan. '17. D.S.O.

Szczepanski, H. C.—
N.-W. Frontier of India, Waziristan, 1901-02.—Medal with clasp.

Szczepanski, H. C. A. (*M.-Gen., u.s.l., Ind. S.C.*)—
India, 1857-58.—Actions of Cawnpore, Kallee Nuddee, and Fatehghur, and pursuit of Tantia Topee.—Medal.

T

Tabuteau, T. R. (*Col., u.s.l., Ind. S.C.*)—
India, 1858.—Oude Campaign. Taking of Fort Koti.—Medal.
Abyssinia 1867-68.—Medal.

Talbot, F. E. G.—
China, 1900.—Medal.

Talbot, J.—
Burma, 1887-89.—Medal with clasp.

Tancock, A. C.—
N.-W. Frontier of India, 1908.—Operations in the Zakka Khel country. Operations in the Mohmand country.—Medal with clasp.

Tancred, J. Y.—
Tibet, 1903-04.—Medal.

Tancred, *Sir* T. S. (Major, retd. I. A.)
N.-W. Frontier of India, (1st and 2nd) Miranzai, 1891.—Medal with clasp.

Tandy, H. S. (*Col. u.s.l. Ind. S.C.*)—
Afghanistan, 1880.—March from Quetta to the relief of Kandahar.—Medal.
Bikanir Field Force, 1883-84.
Burma, 1886-87.—Medal with 2 clasps.

Tanner, C. O. O.—
British East Africa, 1898.—Operations against the Ogaden Somalis.—Medal with clasp; Brilliant star of Zanzibar, 4th class.
The War of 1914—18.—Despatches, *Lond. Gaz.*, 31 Oct. 17.

Tannyane, T.—
N.-W. Frontier of India, 1897-98.—Tochi.—Medal with clasp
China, 1900.—Medal.

Tarr, W.—
South African War, 1900-02.—Operations in Orange River Colony, between 30 Nov. 1900 to 27 Feb. 1901. Operations in Cape Colony, between 28 Feb. to 5 March 1901.—Queen's medal with 3 clasps.

Tarver, A. L.—
East Africa (British), 1896.—Operations against the Mazrui Rebels.—Medal.
Tirah, 1897-98.—Medal with 2 clasps.
N.-W. Frontier of India, 1908.—Operations in the Zakka Khel country. D. A. A. G., 1st Brigade. Despatches, G. G. O. No. 188 of 1908. Operations in the Mohmand country. D. A. A. G., 1st Bde. Engagement of Matta, 24th April 1908. Engagement of Kargha, 24th May 1908. Despatches, G. G. O. No. 521 A of 1908.—Medal with clasp. D. S. O.
The War of 1914-18.—Despatches, *Lond. Gaz.*, 4 July 16. Brev. of Lt.-Col.

Tate, G.—
China, 1900.—Medal.
N.-W. Frontier of India, 1908.—Operations in the Zakka Khel country. Operations in the Mohmand country—Medal with clasp.

Tayler, W. F. C. (*Maj., retd. I. A.*).—
N.-W. Frontier of India, 1897-98.—Tochi.—Medal with clasp.

Taylor, A. F. (*Lt.-Col., retd. Ind. S.C.*)—
Afghanistan, 1880.—Medal.

Taylor, A. W. N.—
Egypt, 1884-85.—Nile. With the Mounted Infantry Desert Column; Battles of Abu Klea, and El Gubat, action at the Metemmeh.—Medal with 2 clasps and bronze star.

Burma, 1891, Wuntho Field Force.—Medal with clasp.
N.-E. Frontier of India, Chin Hills, 1892-93.
China, 1900.—Brev. of Major.
The War of 1914-18.—Despatches, *Lond. Gaz.*, 19 Oct. 16.

Taylor, D. J. O. (*Lt.-Col., retd. I. A.*)—
Afghanistan, 1878-80.—With the Peshawar Valley Field Force and Khyber Line Force.—Medal.
N.-W. Frontier of India (Miranzai 2nd), 1891.—Medal with clasp. *Waziristan, 1894-95.*—Clasp.

Taylor, E. C.—
South African War, 1900-01.—Operations in the Orange River Colony (May to 29th November 1900), including actions at Middulphsberg (29th May). Wittebergen (1st to 29th July). Operations in the Orange River Colony, between 30th November 1900 and 31st May 1902.—Queen's medal with 3 clasps.

Taylor, F. H. (*Major, retd. I. A.*)—
N.-W. Frontier of India, Hazara, 1888 and '91.—Despatches, G. G. O. 816 of '91.—Medal with 2 clasps. (*1st and 2nd*) *Miranzai, 1891.*—Clasp.—*Hunza-Nagar, 1891.*—Despatches, G. G. O. 397 of '92.—Clasp. Action at Wana, 3 Nov. '94. *Waziristan, 1894-95.*—Clasp.
N.-W. Frontier of India, 1897-98.—Tochi.—Medal with clasp. Operations on the Samana and in the Kurram Valley during Aug. and Sep. '97.—Clasp.
Tirah, 1897-98.—Actions of Dargai and of the Sampagha and Arhanga Passes.—Operations against the Khani Khel Chamkannis.—Operations in the Bazar Valley, 25 to 30 Dec. '97.—Despatches, G. G. O. 244 of '98.—Clasp; Brev. of Maj.
N.-W. Frontier of India, Waziristan, 1901-02.—Clasp.

Taylor, G. (*Comm. and Hony. Captain, retd. I.A.*)—
N.-E. Frontier of India, Chin Lushai Expedition, 1889-90.—Medal with clasp.
N.-W. Frontier of India, 1897-98.—Medal with 2 clasps.
Tirah, 1897-98.—Clasp.

Taylor, H. N.—
Burma, 1887-89, 1889-92.—Meda with 2 clasps.
Burma, 1892-93.
N.-E. Frontier of India, Chin Hills, 1892-93.—Mentioned in Despatches, G. G. O. 733 of '93. D. S. O.—Clasp.

Taylor, J.—
N.-W. Frontier of India, 1908.—Operations in the Zakka Khel country. Medal with clasp.
The War of 1914-18.—Despatches, *Lond. Gaz.*, 22 Jun. 15. D.S.O.

Taylor, J. B. (*Col., retd. Mad. S.C.*)—
India, 1858-59.

Taylor, J. McL. G.—
The War of 1914-18.—Despatches, *Lond. Gaz.*, 5 Apl. 16.

Taylor, N. C.—
The War of 1914-18.—Despatches, *Lond. Gaz.*, 19 Oct 16. D.S.O.

Taylor, N. S.—
Operations in the Abor country, 1911-12.—Medal with clasp.

Taylor, P. H. M.—
South African War, 1900-01.—Attd. to Transport to 27 Oct. 01. Queen's medal with 5 clasps.

Taylor, R. O'B.—
South African War, 1900.—Operations in the Orange Free State, Feb. to May '00, including actions at Houtnek (1 May) and Zand River (10 May). Operations in the Transvaal in May and Jun '00, including actions near Johannesburg (29 May), Pretoria (4 June) and Diamond Hill (11 and 12 June). Operations in the Transvaal, west of Pretoria, action at Zillikats Nek (2 Aug.).—Queen's medal with 4 clasps.

The War of 1914-18.—Despatches, *Lond. Gaz.*, 11 Dec. 17. Brev. of Lt.-Col.

Taylor, T.—
N.-W. Frontier of India, 1897-98.—Tochi, Despatches, G. G. O. 1432 of 1898.—Medal with clasp. Promoted Conductor.

Teed, A. S. H.—
China, 1900.—Despatches, *London Gazette*, 14th May 1901.—Medal.

Temple, P. De L.—
South African War, 1899-01.—Operations in Cape Colony, Jan. to July '01. Queen's medal with 2 clasps.

Temple, Sir R. C., Bart. (*Lt.-Col., retd. I. A.*)—
Afghanistan, 1878-79.—Advance on Khelat-i-Ghilzai. Despatches, Maj. Genl. *Sir* M. A. S. Biddulph's Despatch No. 194, dated 16 May '79.—Medal.
Burma, 1885-87.—Medal with clasp.

Templer, C. B. (*Maj., retd. I. A.*)—
N.-W. Frontier of India, (2nd) Miranzai, 1891.—Medal with clasp.

Templer, C. F.—
Soudan, 1896.—Dongola Expedition.—Medal. Egyptian medal.
The War of 1914-18.—Despatches, *Lond. Gaz.*, 1 Jan. 16. C.M.G.

Templer, H. (*Col., retd. I.A.*)—
N.-W. Frontier of India.—(1st) Miranzai Expedition, 1891.
The War of 1914-18.—Despatches, *Lond. Ga*, 28 Dec. 17.

Tennant, C. A. M.—
The War of 1914-18.—Despatches, *Lond. Gaz.*, 4 July 16 and 15 Aug. 17.

Tennant, C. C.—
N.-E. Frontier of India, Chin-Lushai Expedition, 1889-90.—Medal with clasp.

Teversham, R. K.—
Burma, 1885-87.—As Staff Officer in Nyingyan column during the operations round Yemethin and Nyingyan, and operations of the 3rd Brigade.—Despatches, G. G. O. 434 of '87. D. S. O.—Medal with clasp.
China, 1900.—Medal.

Thackeray, Sir E. T. (*Col., retd. B. Engs.*)—
India, 1857-59.—Battles on the Hindun and Budlee-ke-serai, siege of Delhi, actions of Gungeree, Puttialee and Mynpoory, siege and capture of Lucknow, attack on Fort Rooyah, action of Alleygunge, capture of Bareilly. Despatches, *London Gazette*, 15 Dec. '57.—Medal with 2 clasps; VC.
Afghanistan, 1879-80.—Defeat of Ghilzais at Jugdulluck Kotal (severely wounded). Despatches, G. G. O. 137 of '80.—Medal.

Thackwell, A. C. A.—
South African War, 1902.—Operations in Cape Colony and Orange River Colony, Feb. to 31 May '02. Queen's medal with 3 clasps.
The War of 1914-18.—Despatches, *Lond. Gaz.*, 30 Jun. 16. M.C.

Thackwell, C. G. R.—
Afghanistan, 1878-79-80.—Operations in the Hissarak District and Lughman Valley. Affairs with the Ghilzais at and near Jugduluck. Severely wounded. Despatches. Lt.-Genl. R. O. Bright's Despatch, 8 July '80.—Medal.
N.-W. Frontier of India, Mahsud-Waziri, 1881.
Egypt, 1882.—Action of Tel-el-Kebir.—Medal with clasp. Bronze star.
Chitral, 1895.—Relief of Chitral. Served as Brigade Commissariat Officer, 2nd Brigade. Action of the Malakand Pass.—Medal with clasp.

N.-W. Frontier of India, 1897-98.—Malakand.—Divisional Transport Officer.—Operations in Bajaur and in the Mamund Country.—*Mohmand—Utman Khel—Buner.*—Action of the Tanga Pass. Despatches, G. G. O. 1317 of '97 and 217 of '98.—Clasp.—D. S. O.

Tharle-Hughes, R.—
N.-W. Frontier of India, 1897-98.—Medal with clasp.
Tirah, 1897-98.—Clasp.

Thomas, A. N.—
The War of 1914-18.—Despatches, *Lond. Gaz.*, 12 Jun. 17. D.S.O.

Thomas, F. H. S.—
Burma, 1886-08.—Medal with clasp.
N.-E. Frontier of India, Chin-Lushai, 1889-90.—As Commissariat Officer, Burma Northern Column. Despatches, G. G. O. 449 of '90.—Clasp.
N.-E. Frontier of India, Manipur, 1891.—As Chief Commissariat Officer, Burma-Tammu Column. Despatches, G. G. O. 372 of '91.—Clasps.

Thomas, H. St. G.—
Soudan, 1884-85.—Nile. With Desert Column.—Medal with clasp and Khedive's star.
Burma, 1885-89.—Medal with 2 clasps.
Burma, 1891-92.—With Irrawaddy Column. Clasp.
Tibet, 1903-04.—Special Service officer. Action at Niani, 26th June 1904. Operations at and around Gyantse, between 5th May and 6th July 1904.—Medal with clasp.
The War of 1914-18.—Despatches, *Lond. Gaz.*, 1 Jan. 16, 4 Jan. and 15 May 17. D.S.O.

Thomas, W.—
Afghanistan, 1879-80.—Medal.

Thomas. W. F. (*Lieut.-Col., retd. I.M.S.*)—
Rumpa Rebellion, 1879-80.
Burma, 1886—89.—Operations of the 1st Brigade.—Medal with 2 clasps.

Thomason, A. F.—
Chitral, 1895.—Relief of Chitral.—Medal with clasp.
N.-W. Frontier of India, Waziristan, 1901-1902.—Brigade S. and T. Officer, Jandola Brigade.—Clasp.

Thompson, D. M. (*Bt.-Col., retd. I. A.*)—
Afghanistan, 1879-80.—Medal.
N.-W. Frontier of India, Sikkim, 1888.—Medal with clasp.
South African War, 1899-1900.—On Staff. Defence of Ladysmith. Despatches, *Lond. Gaz.*, 8 Feb. '01. Queen's medal with clasp.

Thompson, F. S. C.—
China, 1900.—Medal.
East Africa, 1903-04.—Operations in Somaliland. S. M. O., Hospital Ship "*Hardinge.*"—Medal with clasp.
Aden, 1903-04.—Operations in the Interior.

Thompson, F. T.—
N.-W. Frontier of India, Waziristan, 1901-02.—Medal with clasp.

Thompson, G. N.—
Tibet, 1903-04.—Medal.

Thompson, H. A. H.—
N.-E. Frontier of India, Manipur, 1891.—Medal with clasp.
N.-W. Frontier of India, 1897-98.—Mohmand.—Medal with clasp.
Tirah, 1897-98.—Actions of the Sampagha and Arhanga Passes.—Operations in the Bazar Valley, 15 to 30 Dec. '97.—Clasp.

Thompson, P. T. L.—
The War of 1914-18.—Despatches, *Lond. Gaz.*, 28 Jan. 16.

Thompson, W. (*Hon. Capt. Sr. Asst. Surg., retd. I.S.M.D.*)—
N.-W. Frontier of India, Waziristan, 1894-95.—Medal with clasp.
China, 1900.—Relief of Pekin. Medal with clasp.

N.-W. Frontier of India, Waziristan, 1901-02.—Medal with clasp.

Thompson, W. A. (*Col., retd. I. A.*)—
Burma, 1885-89.—(Medal with clasps.)
The War of 1914-18.—Despatches, *Lond. Gaz.*, 17 Feb. 15.

Thompson, W. O. (*Col., retd. I. A.*)—
Afghanistan, 1878-79.—Medal.
N.-W. Frontier of India, Mahsud-Waziri, 1881.—*Hazara, 1888.*—Medal with clasp.—(*1st and 2nd*) *Miranzai, 1891.*—Despatches, G. G. O. 632 of '91.—Clasp. *Waziristan, 1894-95.*—Despatches, G. G. O.'s 268 and 473 of '95.—Clasp.—D. S. O.

Thompson, W. T. F. (*Captain, retd. I. A.*)—
China, 1900.—Medal.

Thomson, A. G.—
N.-W. Frontier of India, 1902.—Operations against the Darwesh Khel Waziris.—D. S. O.
The War of 1914-18.—Despatches, *Lond. Gaz.*, 22 Jun. 15 and 4 Jan. 17. D.S.O.

Thomson, A. R.—
The War of 1914—18.—Despatches, *Lond, Gaz.*, 5 Apl. and 19 Oct. 16. M.C.

Thomson, C.—
N.-W. Frontier of India, 1908.—Medal with clasp.

Thomson, D. B. (*Maj., retd. I. A.*)—
Soudan, 1885.—Suakin. Actions of Hasheen and Tofrek. (Wounded).—Medal with 2 clasps; bronze star.
N.-E. Frontier of India, Chin-Lushai, 1889-90.—Medal with clasp.

Thomson, G. S. (*Lieut.-Col. I. A., retd.*)—
Chitral, 1895.—With Reserve Brigade.
China, 1900.—Medal.

Thomson, H.—
Burma, 1886-87.—Operations of the 4th and 6th Brigades.—Medal with 2 clasps.

Thomson, M. T. (*Lt.-Col., retd. I. A.*)—
Afghanistan, 1879-80.—Zaimukht Expedition Action of Zawa.—Medal.
N.-W. Frontier of India.—*Waziristan, 1894-95.*—Medal with clasp.

Thomson, S. J. (*Lt.-Col., retd. I. M. S.*)—
Afghanistan, 1880.—Medal.
South African War, 1902.—Director of Burgher Camps, Transvaal.—Queen's Medal with 2 clasps.

Thomson, W. D.—
N.-E. Frontier of India, Manipur, 1891.—Medal with clasp.

Thomson Glover, J. W.—
The War of 1914-18.—Despatches, *Lond, Gaz.*, 21 Jun. and 19 Oct. 16.

Thorne, A. C. D.—
The War of 1914—18.—Despatches, *Lond, Gaz.*, 19 Oct. 16.

Thorne, T.—
Tirah, 1897-98.—Despatches, G. G. O. No. 244 of '98.—Medal with 2 clasps.—Promoted Deputy Assistant Commissary.

Thornley, M. H.—
North-West Frontier of India, Waziristan, 1901-02.—Medal with clasp.

Thornhill, C. J. M.—
The War of 1914-18.—D.S.O.

Thornhill, C. M.—
The War of 1914-18.—Despatches, *Lond, Gaz.*, 1 Jan. 16, 15 Aug. 17.—D.S.O., M.C.

Thornhill, *Sir* H. B. (*Lt.-Col., retd. I. A.*)—
Afghanistan, 1878-79.—Medal.

Thornhill, W. H. (*Lt.-Col., retd. Ind. Med. Serv.*)—
Afghanistan, 1879-80.—Medal.
Burma, 1887-88.—Medal with clasp.

Thornton, C. E.—
China, 1900.—Medal.
Tibet, 1903-04.—Medal.
The War of 1914-18.—Despatches, *Lond, Gaz.*, 5 Apl. and 13 July 16. C.M.G.

Thornton, *Sir* J. H. (*Dep. Surg.-Gen., retd. Ind. Med. Serv.*)—
India, 1857-58-59.—Action of Benares, relief of Arrah, capture of Jugdespore, actions of Surujpore, defence of the Alumbagh, capture of Lucknow, and operations in Oude and Gorukpore. Despatches, *Lond. Gaz.*, 13 Oct. '57.—Medal with 2 clasps; one year's service for Lucknow.
China, 1860.—Action of Sinho, capture of the Taku Forts, and occupation of Pekin.—Medal with 2 clasps.
N.-E. Frontier of India, 1862-63, Cossyah and Jyntiah Hills Expedition.—Capture of several stockades (wounded). Mentioned in Despatches. *Bhutan Expedition, 1865-66.*—Forcing Durunga Pass, capture and re-capture of Dewangiri.—Medal with clasp.

Thoy, G.—
N.-W. Frontier of India, 1897-98.—Tochi, Medal with clasp.
Egypt, 1882.—Medal; bronze star.
Soudan Expedition, 1885.—Suakin. As Prin. Med. Officer of the Indian Contingent. Despatches, *Lond. Gaz.*, 25 Aug. '85.—Clasp; C.B.
N.-W. Frontier of India, Hazara Expedition, 1888.—As Principal Medical Officer. Mentioned in Despatches.—Clasp.

Thring, R. H. D.—
Burma, 1886-89.—Operations of the 4th and 6th Brigades.—Medal with 2 clasps.

Thuillier, L. C.—
Tirah, 1897-98.—Actions of the Sampagha and Arhanga Passes.—Medal with 2 clasps.

Thuillier, W. (*Col., retd. I. A.*)—
N.-W. Frontier of India, 1897-98.—Malakand.—Utmankhel.—Buner.—Action of the Tanga Pass.—Medal with clasp.
Tirah, 1897-98.—Clasp.

Thurston, E. O.—
China, 1900—Medal.

Thyne, W.—
The War of 1914-18.—Despatches, *Lond. Gaz.*, 4 July 16 and 15 Aug. 17.

Tibbs, J.—
The War of 1914-18.—Despatches, *Lond. Gaz.*, 15 Aug. 17.

Tighe, M. A. (*Lt.-Col., retd. I.A.*)—
Mekran, 1898.—Despatches, G. G. O. 648 of '98.

Tighe, M. J.—
Burma, 1886-88.—Served with Mounted Infantry.—Medal with clasp.
Karen Field Force, 1888-89.—Served with Mounted Infantry.—Engagement at Nga Kyaing. Forcing of the Nanko defile and capture of Sawlon. (Slightly wounded).—Despatches.—G. G. O. 417 of '89.—D. S. O.—Clasp.
N.-E. Frontier of India, 1889-90.—Chin-Lusha clasps.
Burma, 1890-91.—Chin Hills—(Chinbock columns.) As Political Officer in '92.—Clasp.
ast Africa (British), 1896.—Operations against the Mazrui Rebels—Medal and Brilliant Star of Zanzibar, 2nd class.—*British East Africa Protectorate, 1898.*—Operations in the Teita District.
Uganda, 1898-99.—Commanded the operations in Aukole, operations against the Soudanese mutineers, action of Kiganguti. Mentioned in Despatches.—Medal with clasp.—Brev. of Maj.
Mekran, 1901.—Attack on and capture of Nodiz Fort. In command. Despatches, G. G. O. 415 of '02.—Brev. of Lt.-Col.
The War of 1914-18.—Despatches, *Lond, Gaz.*, 20 and 30 Jun. 16. Promd. Maj. Gen.

Tighe, S. M.—
Burma, 1890-91.—Medal with clasp.
N.-W. Frontier of India, 1897-98.—Engagement near Shabkadr, 9 Aug. '97.
Mohmand.—Despatches, G. G. O. 1318 of '97.
Buner.—Action of the Tanga Pass.—Medal with clasp.
China, 1900.—Medal.

Tillard, A. B.—
N.-E. Frontier of India, Lushai, 1892-93.
N.-W. Frontier of India, 1897-98.—Operations on the Samana and in the Kurram Valley during Aug. and Sept. '97. Relief of Gulistan.—Medal with 2 clasps.
Tirah, 1897-98.—Actions of Chagru Kotal, Dargai, and of the Sampagha and Arhanga Passes.—Reconnaissance of the Saran Sar and action of 11 Nov. '97.—Operations in the Waran Valley and action of 16 Nov. '97. Operations at and around Dwatoi and action of 24 Nov. '97.—Despatches, G. G. O.'s 58 and 244 of '98.—Clasp. D. S. O.
N.-W. Frontier of India, Waziristan, 1901-02.—Clasp.
Tibet, 1903-04.—Medal.
The War of 1914—18, Despatches, *Lon. Gaz.*, 22 Jun. 15, 13 July 16 and 15 Aug. 17.

Timbrell, T.—
China, 1900.—Medal.
Operations in the Abor country, 1911-12.—Medal with clasp.
The War of 1914-18.—Despatches, *Lond. Gaz.*, 15 Aug. 17. D.S.O.

Tinby, F. B. N.—
The War of 1914-18.—Despatches, *Lond, Gaz.*, 22 Jun. 15. M.C.

Tobin, H. W.—
East Africa, 1902-04.—Operations in Somaliland.—Medal with clasp.
The War of 1914-18.—Despatches, *Lond. Gaz.*, (Sir J. Maxwell, 16 Mar. and 19 Apl. 16), *Lond. Gaz.*, 21 Jun. 16.—Brev. of Maj. D.S.O.

Tod, J. K.—
Burma, 1886—88.—Served as A.-D. C. to Br.-Genl. C. J. East, Commanding 1st Brigade; Maj. Genl. *Sir* G. S. White, Commanding Upper Burma Force; and Brigadier-General *Sir* R. C. Low. Despatches, G. G. O. 434 of '87.—Medal with 2 clasps.
N.-W. Frontier of India, 1897-98.—Malakand, Field Intelligence Officer.—Operations in Bajaur. *Mohmand.—Utman Khel.—Buner.*—Medal with
Aden, 1903-04.—Operations in the interior.

Toker, *Sir* A. C. (*Maj.-Genl., u. s. l., I.A.*)—
N.-E. Frontier of India, Bhutan, 1864-65.—Medal with clasp.
Egypt, 1882.—D. A. A. Genl., Indian Contingent Action of Tel-el-Kebir and subsequent pursuit to Zagazig. Despatches, *Lond. Gaz.*, 2 Nov. '82.—Medal with clasp. Bronze star. 4th class of the Osmanieh.—Brev. of Lt.-Col.
Burma, 1886-87.—Commanded Chindwin District Destruction of rebel stronghold at Balet and occupation of Kindat. Despatches. G. G. O. 434 o '87.—2 clasps. C. B.

Tomkins, H. L.—
N.-W. Frontier of India.—Waziristan, 1894-95.—Medal with clasp.
Tirah, 1897-98.—Despatches, G. G. O. 244 of '98.—Medal with clasp. D. S. O.

N.-W. Frontier of India, Waziristan, 1901-02.—Clasp.
East Africa, 1903-04.—Operations in Somaliland.—Medal with clasp.
N.-W. Frontier of India, 1908.—Operations in the Zakka Khel country.—Medal with clasp.
The War of 1914-18.—Despatches, *Lond. Gaz.*, 22 Jun. 15 and 19 Oct. 16. Brev. of Lt.-Col. C.M.G.

Torrie, C. J.—
The War of 1914-18.—Despatches, *Lond. Gaz.*, 8 May 15, 22 Jun. 15 and 4 Jan. 17. D.S.O.

Torrie, T. G. J.—
China, 1900.—Relief of Pekin.—Medal with clasp.

Tottenham, R. L. (*Major, retd. I. A.*)—
Chitral, 1895.—Relief of Chitral.—Medal with clasp.

Townsend, R. S.—
The War of 1914-18.—Despatches, *Lond. Gaz.*, 30 Jun. 16. M.C.

Trafford, E. A.—
South African War, 1902.—Served in St. Helena, Mar. to 31 May 02.

Trafford, W. L.—
N.-W. Frontier of India, 1908.—Operations in the Mohmand country.—Medal with clasp.
The War of 1914-18.—Despatches, *Lond. Gaz.*, 22 Jun. 15.

Trail, W. S.—
N.-W. Frontier of India, 1908.—Operations in the Mohmand country. Engagement of Kargha, 24th May 1908. Despatches, G. G. O. No. 521A of 1908.—Medal with clasp.
The War of 1914-18.—Despatches, *Lond. Gaz.*, 22 Jun. 15 and 15 Jun. 16. M.C.

Tranchell, H. G.—
N.-W. Frontier of India, 1908.—Operations in the the Mohmand country. Medal with clasp.

Travers, R. E.—
Chitral, 1895.—Relief of Chitral.—Medal with clasp.
Tirah, 1897-98.—Action of the Sampagha and Arhanga Passes. Reconnaissance of the Saran Sar and actions of the 9 and 11 Nov. '97.—Operations against the Khani Khel Chamkannis.—Operations in the Bazar Valley, 25 to 30 Dec. '97.—2 clasps.

Tregear, F. C.—
West Africa, 1892.—Expedition to Tambaku Country.—Capture of Tambi.—Medal with clasp.—*West Coast of Africa, 1894.*—Expedition to the Gambia against Fodey Silah.—Clasp.
Tirah, 1897-98.—Medal with clasp.
The War of 1914-18.—Despatches, *Lond. Gaz.*, 5 Apl. and 13 July 16.

Tregear, V. F. W.—
China, 1900.—Medal.

Tregear, Sir V. W. (*Maj.-Genl., u. s. l., I.A.*)—
Afghanistan, 1879-80.—With Khyber Field Force.—Medal.
N.-E. Frontier of India, Lushai, 1889.—Commanded the Force. Thanks of Government. Despatches. G. G. O. 592 of '89.—*Chin-Lushai, 1889-90.*—Commanded the Chittagong Column. Mentioned by Government of India. Despatches, G. G. O. 677 of '90.—Medal with clasp. C. B.

Trelawny, L. C.—
The War of 1914-18.—Despatches, *Lond. Gaz.*, 15 Aug. 17. M.C.

Trench, C. F.—
The War of 1914-18.—Despatches, *Lond. Gaz.*, 13 July, 25 Aug. and 19 Oct. 16. D.S.O.

Trevelyan, W. R. F.—
The War of 1914-18.—Despatches, *Lond. Gaz.*, 15 Aug. 17.

Trevor, A. C. H.—
N.-W. Frontier of India, 1908.—Operations in the Mohmand country. Engagement of Kargha, 24th May 1908.—Medal with clasp.
The War of 1914-18.—Despatches, *Lond. Gaz.*, 5 Apl. and 13 July 16.

Trevor, H. (*Maj., retd. I. A.*).—
Soudan, 1885.—Suakin. Actions of Hasheen, Tofrek and Tamai.—Medal with 2 clasps. Bronze star.
Burma, 1887.—Medal with clasp.
N.-W. Frontier of India, 1897-98.—Tochi.—Medal with clasp.
Tirah, 1897-98.—Clasp.

Trelawny, L. C.—
The War of 1914-18.—Despatches, *Lond. Gaz.*, 15 Aug. 17. M.C.

Trotter, E. G. S.—
Aden, 1903-04.—Operations in the interior.

Trotter, J. M. (*Hon. Lt.-Col., retd. Ben. S.C.*)—
N.-W. Frontier of India, Eusafzai (2nd) Expedition, 1863-64.
N.-W. Frontier of India, Hazara, 1868.—Medal with clasp.
N.-W. Frontier of India, Jowaki, 1877-78.—As D. A. Q. M. G.—Mentioned in despatches—G. G. O. 738 of '78.—Clasp.
N.-W. Frontier of India, Mahsud-Waziri, 1881.—Mentioned in Despatches.

Trotter, R. F. (*Lieut.-Col., retd. I. A.*)—
N.-W. Frontier of India, Jowaki, 1877-78.—Medal.
Afghanistan, 1879-80.—Action of Ali Musjid.—Medal with clasp.

Trueman, H. J. H.—
The War of 1914-18.—Despatches, *Lond. Gaz.*, 15 Aug. 17.

Trydell, B. (*Lieut.-Col., retd., I. A.*)—
Burma, 1888.—Medal with clasp.
Burma, 1891.—Wuntho Field Force.—Clasp.
N.-E. Frontier of India, Chin Hills, 1892-93.—Clasp.

Tuck, C. H. A.—
South African War, 1900-01.—Operations in Orange River Colony, 7th November 1900 to 6th May 1901. Operations in Cape Colony, 7th May to 17th November 1901.—Queen's medal with 3 clasps.

Tucker, E. F. G.—
China, 1900.—Medal.

Tucker, H. B.—
The War of 1914-18.—Despatches, *Lond. Gaz.*, 8 Feb. 17.

Tucker, L. H. E. (*Maj.-Gen., retd. I A.*).—
N.-W. Frontier of India, 1863.—Forcing of the Umbeyla Pass. Despatches, *Lond. Gaz.*, 19 Mar. 64.—Medal with clasp.
N.-W. Frontier of India. Jowaki, 1877-78.—Mentioned in despatches.—Clasp.
Afghanistan, 1878-80.—Capture of Ali Musjid; operations at and around Kabul. Despatches, *Lond. Gaz.*, 7 Nov. 79, and 4 May 80. G. G. O. 315 of 81.—Medal with 2 clasps.

Tugwell, W. B. P.—
South African War, 1901-02.—Operations in the Transvaal, Sep. 01, and Oct. to Nov. 01. Operations in Orange River Colony, Mar. to Sep. 01 and Nov 01 to April 02. Operations on the Zululand Frontier of Natal in Sep. and Oct. 01. Operations in Cape Colony, Mar. 01.—Queen's medal with 5 clasps.
The War of 1914-18.—Despatches, *Lond. Gaz.* 19 Oct. 16.

Tuite, M. A. (*Bt. Col., retd., I. A.*)—
Afghanistan, 1879-80.—Operations around Kabul.—Medal with clasp.
China, 1900.—Medal.

Tulte-Dalton, E. G. T.—
The War of 1914-18.—Despatches, *Lond. Gaz.*, 6 July 17.

Tulloch, J. W. G.—
N.-W. Frontier of India, Zhob Valley, 1890.—Despatches, G. G. O. 214 of '91.—*Hazara, 1891.*—Despatches, G. G. O. 816 of '91.—Medal with clasp.
Chitral, 1895.—Relief of Chitral.—Despatches, G. G. O. 998 of 95.—Medal with clasp.
East Africa (British), 1896.—Operations against the Mazrui Rebels.—Medal. Brilliant Star of Zanzibar, 1st Grade, 2nd class.
Tirah, 1897-98.—A. A. and Q. M. G. Lines of Communication.—Action of Chagru Kotal.—Despatches G. G. O. 244 of '98.—2 clasps.—Brev. of Maj.
China, 1900.—Relief of Pekin. Despatches, *Lond. Gaz.*, 14 May '01.—Medal with clasp.—Brev. of Lt.-Col.
Russo-Japanese War, 1904-1905.—Attached to the 2nd Japanese Army in Manchuria. Order of the Rising Sun, 2nd class.—Japanese Medal. C. B.

Turkington, W. H. (*Hony. Lieut. Asst. Commissary, retd., Ordnance Deptt.*)—
N.-W. Frontier of India, 1897-98.—Tochi.—Medal with clasp.—Promoted Deputy Assistant Commissary.

Turnbull, B.—
N.-W. Frontier of India, Waziristan, 1901-02.—Medal with clasp.
Tibet, 1903-04.—Action at Niani, 26th June 1904. Operations at and around Gyantse, between 5th May and 6th July 1904. March to Lhassa, 14th July to 3rd August 1904.—Medal with clasp.
N.-W. Frontier of India, 1908.—Operations in the Zakka Khel country.—Medal with clasp.
The War of 1914-18. Despatches, *Lon. Gaz.*, 6 July '17.

Turnbull, G. O.—
Tirah, 1897-98.—Medal with clasp.
The War of 1914-18.—Despatches, *Lond. Gaz.*, 1 Jan. 16. D.S.O.

Turnbull, P. S. (*Surg.-Maj.-Gen., retd., I. M. S.*)—
Abyssinia, 1867-68.—Action of Arogee, assault and capture of Magdala.—Medal.

Turner, A. G. B.—
N.-W. Frontier of India, 1897-98.—Engagement near Shabkadr, 9 Aug. '97.—Mohmand.—Medal with clasp.
China, 1900.—Medal.
The War of 1914-18.—Despatches, *Lond. Gaz.*, 15 Aug. 17. D.S.O.

Turner, A. H. (*Col. u.s.l. Ind. S.C.*)—
N.-W. Frontier of India, Hazara Campaign, 1868.—Medal with clasp.
Bezoti Afridi Affair, 1869.
Afghanistan, 1878-80.—Capture of the Peiwar Kotal, action in Shuturgardan Pass and action of Chihildakhteran. Despatches, G. G. O. 632 of '81.—Medal with clasp.
N.-W. Frontier of India, 1881.—Mahsud-Waziri Expedition.—Mentioned in Despatches. *Zhob Valley Expedition, 1890.—Miranzai (1st and 2nd) Expeditions, 1891.*—Commanded 2nd Column. Despatches, G. G. O. 632 of '91.—Clasp; Brev. of Col.
Waziristan Expedition, 1894-95.—Commanded Delimitation Escort, including attack on camp at Wano, and afterwards commanded 1st Brigade. Despatches, G. G. O. 473 of '95.

Turner, G. H. (*Bt.-Col, retd., I. A.*)—
China, 1900.—Despatches, *Lond. Gaz.*, 14 May '01.—Medal with clasp.

Turner, J. G.—
Chitral, 1895.—Relief of Chitral.—Despatches, G. G.O. 998 of '95.—Medal with clasp.—Brev. of Maj.
Tirah, 1897-98.—Orderly Officer to G. O. C., 3rd Brigade.—Clasp.
China, 1900.—Despatches, *Lond. Gaz.*, 13 Sep. 01.—Medal.
South African War, 1902.—As temporary Lt.-Col. Commanding 28th Battalion, Imperial Yeomanry, operations in Cape Colony between 21 and 31 May '02.—Queen's Medal with 2 clasps.

Turner, R. G.—
N.-W. Frontier of India, Waziristan, 1894-95.—Medal with clasp.
Uganda, 1897-98.—Attack at Kymbo; capture of Kabagambiz and of Soudanese Forts near Mruli. Mentioned in Despatches.—Medal with 2 clasps.
China, 1900.—Medal.
The War of 1914-18.—Despatches, *Lond. Gaz.*, 30 Jun. 16. D.S.O.

Turton, H. M.—
N.-W. Frontier of India, 1897-98.—Tochi.—Medal with clasp.
China, 1900.—Medal.

Tweddell, F. (*Col., retd. Ben. Inf.*)—
India, 1858.—Action of Toolsepore.—Medal.
N.-E. Frontier of India, Bhutan Expedition, 1865.—Medal with clasp.
Afghanistan, 1879-80.—Action of Kam Dakkai operations in Lughman against Wazir Khugiar, and in Hisarak District. Mentioned in Despatches.—Medal; Brev. of Lt.-Col.

Tweddell, F.—(*Maj., retd. I. A.*)—
Soudan, 1885-86.—Present in the engagement at Giniss.—Medal.—Bronze star.
China, 1900.—Medal.
N.-W. Frontier of India, Waziristan, 1901-02.—Medal with clasp.

Tweddell, H.—
China, 1900.—Medal.
N.-W. Frontier of India, Waziristan, 1901-02.—Medal with clasp.
The War of 1914-18.—Despatches, *Lond. Gaz.*, 4 July and 19 Oct. 16.

Tweedie, M. (*Col. u.s.l. Ind. S.C.*)—
India, 1858-59.—Assault and capture of Kotah, pursuit of Tantia Topee, occupation of Sherkarpur, Trans-Gogra Campaign, and operations on Nepaul frontier.—Medal with clasp.

Twemlow, E. D'O. (*M.-Gen., retd.*)—
Afghanistan, 1879-80.—March from Quetta to the relief of Kandahar. Despatches, G. G. O 693 of '81.—Medal.

Twigg, E. F.—
East Africa (British), 1896.—Operations against the Mazrui Rebels.—Medal.

Twigg, R. H.—
N.-E. Frontier of India, Akha, 1883-84.
Burma, 1886-87.—Medal with clasp.
N.-W. Frontier of India, 1891-92.—Hunza Nagar.—Served as D. A. A. G. to the Force. Despatches.—G. G. O. 397 of '92.—Clasp.—Brev. of Maj.

Tirah, 1897-98.—Reconnaissance of the Kharmana Defile and action of 7 Nov. '97.—Operations against the Khani Khel Chamkannis.—Clasp.
N.-W. Frontier of India, 1897-98.—Operations on the Samana and in the Kurram Valley during Aug. and Sep. '97.—Medal with 2 clasps.

Twiss, E. K.—
South African War, 1900-1901.—Served in ranks Impl. Yeo.—Queen's Medal with 4 clasps.
The War of 1914-18.—Despatches, *Lond. Gaz.*, 17 Feb. 15, 15 Jun. 16, and 4 Jan. 17. Brev. of Maj. D.S.O.

Twiss, H. W. F.—
South African War, 1901-1902.—Operations in Natal, 1899, including actions at Elandsgate. (21st October), and Lombard's Kop (30th October.) Defence of Ladysmith, including sortie of 7th December 1899. (Wounded.) Operations in Natal, March to June 1900. Operations in the Transvaal, between 30th November 1900, and December 1901. Despatches, *London Gazette* 10th September 1901.—Queen's medal with 3 clasps. King's medal with 2 clasps.
The War of 1914—18.—Despatches, *Lond. Gaz.*, 15 Jun. 16 and 4 Jan. 17.

Twiss, W. L. O.—
China, 1900.- Relief of Pekin. Despatches, *Lond. Gaz.*, 14 May '01.—Medal with clasp.
Tibet, 1903-04.—Medal.

The war of 1914-18. Despatches, *Lon. Gaz.*, 22 Jun. 15, 1 Jan. and 15 Jun. 16, 15 May 17, and 11 Dec. 17, M.C. Legion of Honour (Chavalier) Brev. of Lt.-Col.

Tyndall, H. S.—
N.-W. Frontier of India, 1897-98.—Defence and Relief of Malakand.—Relief of Chakdara.—Malakand.—Operations in Bajaur and in the Mamund country.—Medal with 2 clasps.
The War of 1914—18.—D.S.O.

Tyndall, J. W. H. D.—
South African War, 1901.—Queen's medal with 4 clasps. King's medal with 2 clasps.

Tyrrell, A. C. L.—
South African War, 1899—1901.—Advance on Kimberley, Oct. '99. Operations in the Transvaal, west of Pretoria in Sep. '00. Operations in Cape Colony, north of the Orange River, April to Aug. '00. Operations in the Orange River Colony Feb. to Nov. '01.—Queen's medal with 4 clasps.
The War of 1914—18.—Despatches, *Lond. Gaz.*, 1 Jan. 16.

Tyrrell, F. H. (*Lt.-Gen. u.s.l. Mad. Inf.*)—
Abyssinia, 1867-68.—Medal.

Tytler, H. C.—
N.-E. Frontier of India, Sikkim, 1888.—Attacks of Jeluk, Gnathong, and Jalapla.—Despatches.—Medal with clasp.
N.-E. Frontier of India, Lushai, 1890.—Relief of Chargsil.—Services acknowledged by Government (M. D. letter No. 574 L. C., 1 Dec. '90).—Clasp.
N.-W. Frontier of India, Waziristan, 1901-02.—Medal with clasp.
The War of 1914—18.—Despatches, *Lond. Gaz.*, 30 Jun. 16. Brev of Col.

U

Underhill, G. E. C.—
Aden, 1903-04.—Operations in the interior.
The War of 1914—18.—Despatches, *Lond. Gaz.*, 21 Jun. 16.

Underwood, A. W.—
Egypt, 1882.—Medal; Bronze Star.

Underwood, G. F.—
The War of 1914-18. Despatches, *Lon. Gaz.*, 15 Aug. '17.

Unwin, G. B.—
N.-W. Frontier of India, Zhob Valley, 1890.
N.-W. Frontier of India, Waziristan, 1894-95.—Medal with clasp.
N.-W. Frontier of India, 1897-98.—Tochi.—Medal with clasp.
N.-W. Frontier of India, 1908.—Operations in the Mohmand country. Engagement of Kargha, 24th May 1908. Despatches, G. G. O. No. 521A of 1908.—Medal with clasp. D. S. O.

Unwin, W. H. (*Col., retd. Ben. S. C.*)—
India, 1857-58.—Siege of Lucknow.—Medal with clasp.

Urmson, D.G.S.—
The War of 1914—18.—Despatches, *Lond. Gaz.*, 15 Aug. 17.

Upperton, J. (*M.-Gen. u.s.l. Ind. S. C.*)—
India, 1857-58-59.—Siege of Delhi, action of Dound eakeira, capture of Amethee and Shunkergarh.—Medal with clasp.
China, 1860.—Capture of the Taku Forts, actions of Sinho, Chankiawan and Palikao; capture of Pekin.—Medal with 2 clasps.
N.-W. Frontier of India, 1869.—Affair with Bezoti, Afridis.
Egypt, 1882.—Action of Kassassin, battle of Tel-el-Kebir and occupation of Cairo. Despatches, *Lond. Gaz.*, 2 Nov. '82. Medal with clasp; bronze star. 3rd class Medjidil, C.B

V

Vaid, C. C. (*Lt.-Col., retd., I. M. S.*)—
N.-W. Frontier of India, Mahsud-Waziri, 1881.
Burma, 1885—87.—Taking of Minhla.—Medal with clasp.

Vander-Gucht, G. T.—
China, 1900.—Relief of Pekin.—Medal with clasp.

Vanrenen, G. R.—
N.-W. Frontier of India, 1897-98.—Malakand—Utman Khel—Buner.—Action of the Tanga Pass—Medal with clasp.

Vans Agnew, J. (*Lieut.-Col., retd., I. A.*)—
Burma, 1888.—As Unpaid Attaché, Intelligence Branch, Q. M. G.'s Department.—Medal with clasp.
South African War, 1900—02.—Relief of Kimberley. Operations in the Orange Free State, Feb. to May '00, including operations at Paardeberg (17 to 26 Feb.). Actions at Poplar Grove (7 March and Dreifontein (10 March).—Queen's medal with 4 clasps, King's medal with 2 clasps.

Vansittart, E.—
N.-W. Frontier of India, Mahsud-Waziri Expedition, 1881.
N. W. Frontier of India, Hazara Expedition, 1888.
N.-W. Frontier of India, Hazara Campaign, 1891.—Medal with clasp.
N.-W. Frontier of India, 1897-98.—Operations on the Samana and in the Kurram Valley during Aug. and Sep. '97. Operations of the Flying Column in the Kurram Valley under Colonel Richardson, 20 Aug. to 1 Oct. '97.—Medal with 2 clasps.
Tirah, 1897-98.—Reconnaissance of the Kharmana Defile and action of 7 Nov. '97.—Operations against the Khani Khel Chamkannis. (Slightly wounded.)—Clasp.

The War of 1914—18.—Despatches, *Lond. Gaz.*, 15 Jun. 16.

Van Someren, W. W.—
Tirah, 1897-98.—Actions of Chagru Kotal and of the Sampagha and Arhanga Passes.—Reconnaissance of the Saran Sar, and action of 9 Nov. 97.—Operations in the Waran Valley and action of the 16 Nov. '97. Operations at and around Dwatoi and action of the 24 Nov. '97. Operations in the Bara Valley, 7 to 14 Dec. '97. Affair at Shinkamar on 29 Jan. '98. Despatches G. G. O., No. 244 of '98. D. S. O.—Medal with 2 clasps.

Vaughan, E. G.—
N.-E. Frontier of India, Chin-Lushai, 1889-90.—Medal with clasp.
Burma, 1894.
N.-W. Frontier of India, 1897-98.—Malakand, Mohmand, Buner.—Medal with clasp.
Uganda, 1898-99.—Medal with clasp.
British East Africa, 1901.—Operations against the Ogaden Somalis in Jubaland. Medal with clasp.
N.-W. Frontier of India, 1908.—Operations in the Zakka Khel country. Operations in the Mohmand country. Medal with clasp.
Operations in the Abor country, 1911-12.—Despatches, *Gaz. of India*, Notn. No. 480, dated 23rd May 1912.—Medal with clasp. D.S.O.

Vaughan, H. B. (*Bt.-Col. retd. I. A.*)—
Egypt, 1882.—Medal. Bronze star.
Burma, 1891-92.—Served as Intelligence Officer, Kanhow Column.
China, 1900.—Relief of Pekin. Action of Peitsang. Action at Yangtsun. Despatches, *Lond. Gaz.*, 14 May 01. Medal with clasp.—Brev. of Lt.-Col.

Vaughan, J. C. S.—
N.-W. Frontier of India, (1st and 2nd) Miranzai, 1891.—Medal with clasp.
N.-W. Frontier of India, 1897-98.—Malakand. Action at Landakai. Buner. Action of the Tanga Pass.—Medal with clasp.

Vaughan, L. R.—
The War of 1914—18.—Despatches, *Lond. Gaz.*, 17 Feb. and 22 Jun. 15, 1 Jan. and 15 Jun. 16, 4 Jan. and 11 Dec. 17, Brevs. of Lt.-Col., and Col.—D.S.O.

Vaughan, P. B. (*Maj. retd. Ind. S. C.*)—
N.-E. Frontier of India, 1883-84.—Akka Expedition.
Burma, 1887-89.—Medal with clasp.
Soudan, 1896.—Dongola Expedition.—Egyptian medal.

Vaughan, R. E.—
China, 1900.—Relief of Pekin. As Brigade Commissariat-Transport Officer, 1st Infantry Brigade. Despatches, *Lond. Gaz.*, 14 May 01.—Medal with clasp. Brev. of Maj.
The War of 1914—18.—Despatches, *Lond. Gaz.*, 22 Jun. 15. Brev. of Col.

Venning, F. E. W.—
The War of 1914—18.—Despatches, *Lond. Gaz.*, 15 Aug. 17. D. S. O.

Vernon, H. B.—
The War of 1914—18.—Despatches, *Lond. Gaz.*, 19 Oct. 16. M.C.

Vibart, E. D. H. (*Col. u.s.l. Ind. S. C.*)—
India, 1857-8-9.—Outbreak at Delhi, siege, storm, and capture of Delhi, actions of Nujufghur and Narnoul, affairs of Gungaree, Puttialee, and Mynpoorie, siege and capture of Lucknow, capture of Bareilly, and operations in Central India.—Medal with 2 clasps.
Afghanistan, 1879-80.—Medal.
N.-W. Frontier of India, Hazara, 1888.—Medal with clasp.

Vicary, E. S.—
The War of 1914—18.—Despatches, *Lond. Gaz.*, 1 Jan. 16 and 15 Aug. 17. Brev. of Maj.

Vickers, A.—
Tibet, 1903-04.—Medal.

Vickers, C. (*Capt., retd. I. A.*)—
Chitral, 1895.—Relief of Chitral.—Medal with clasp.
N.-W. Frontier of India, Waziristan, 1901-02.—Clasp.

Vickers, H.—
N.-E. Frontier of India, Chin-Lushai, 1889-90.—Medal with clasp.
The War of 1914—18.—Despatches, *Lond. Gaz.*, 8 Feb. 17.

Vickers, W. G. H.—
The War of 1914—18.—Despatches, *Lond. Gaz.*, 31 Oct. 17.

Vidal, L. H. (*Maj., retd. Ind. S.C.*)—
Egypt, 1882.—Medal, bronze star.

Vigors, M. D.—
The War of 1914—18.—M.C.

Villiers-Stuart, J. P.—
China, 1900.—Medal.
N.-W. Frontier of India, Waziristan, 1901-02.—Medal with clasp.
East Africa, 1903-04.—Operations in Somaliland. Action at Jidballi.—Medal with 2 clasps.
N.-W. Frontier of India, 1908.—Operations in the Zakka Khel country. Orderly Officer to G. O. C. 2nd Brigade Bazar Valley Field Force. Despatches, G. G. O. No. 188 of 1908.—Medal with clasp.
The War of 1914—18.—Despatches, *Lond. Gaz.*, 22 Jun. 15, 4 Jan. 15th May and 11 Dec. 17. Brev. of *Lt.-Col.* Legion of Honour Chavalier.—D.S.O.

Villiers-Stuart, W. D.—
N.-W. Frontier of India, 1897-98.—Operations on the Samana and in the Kurram Valley during Aug. and Sep. 97. Operations of the Flying Column in the Kurram Valley under Colonel Richardson, 20 Aug. to 1 Oct. 97.—Medal with 2 clasps.
Tirah, 1897-98.—Reconnaissance of the Kharmana Defile and action of 7 Nov. 97. Operations against the Khani Khel Chamkannis. (Dangerously wounded).—Clasp.
N.-W. Frontier of India, Waziristan, 1901-02.—Clasp. D.S.O.

Vivian, F. G. (*Bt.-Col., retd. I. A.*)—
N.-W. Frontier of India, Waziristan, 1894-95.—Medal with clasp.
N.-W. Frontier of India, 1897-98.—Relief and defence of Malakand.—Relief of Chakdara.—*Malakand.*—Operations in Bajaur and in the Mamund Country.—Despatches, G. G. O. 178 of 98.—Medal with 2 clasps.

Vost, W.—
N.-W. Frontier of India, Hazara, 1888.—Medal with clasp. (*2nd*) *Miranzai, 1891.*—Clasp.
China, 1900.—Medal.

Voyle, F. R. C. (*Hon. Lt.-Col., retd. Ben. S. C.*)—
N.-W. Frontier of India, Jowaki Expedition, 1877-78.—Medal with clasp.
Afghanistan, 1878-79.—Capture of the Peiwar Kotal.—Medal with clasp.

Vyvyan, B. G. (*Hon. Col., retd. Ben. Inf.*)—
Egypt, 1882.—Commanded outpost at Serapeum Medal; bronze star.

W

Wace, S. L.—
N.-W. Frontier of India, 1908.—Medal with clasp.
The War of 1914—18.—Despatches, *Lond. Gaz.*, 15 Aug. 17.

Waddell, L. A. (*Lt.-Col., retd. I.M.S.*)—
Burma, 1886-87.—Operations of the 2nd Brigade. Wuntho Expedition. Advance from Manie to Wuntho, and occupation of Wuntho.—Medal with clasp.
Chitral, 1895.—Relief of Chitral.—Medal with clasp.
China, 1900.—Senior Medical Officer, Base, afterwards Pekin. Despatches, *Lond. Gaz., 14 May 01.*—Medal with clasp. C.I.E.
N.-W. Frontier of India, Waziristan, 1901-02.—Clasp.

Wadeson, F. W. G.—
The War of 1914—18.—Despatches, *Lond. Gaz.*, 17 Feb. and 22 Jun. 15. C.B.

Wagstaff, L. C.—
South African War, 1902.—Operations in the Transvaal between 19th April and 31st May 1902. —Queen's medal with 4 clasps.
The War of 1914—18.—Despatches, *Lond. Gaz.*, 21 Jun. 16 and 31 Oct. 17.—C. I. E.

Wake, E. St. A. (*Lt.-Col., retd. I.A.*)—
Burma, 1885-86.—Asst. Provost Marshal.— Medal with clasp.
N.-W. Frontier of India, 1897-98.—Action of the Ublan Pass, 27th Aug. '97. Operations on the Samana and in the Kurram Valley during Aug. and Sep. '97. Relief of Gulistan.—Despatches, G. G. O. 1417 of 97.—Medal with 2 clasps.
Tirah, 1897-98.—Actions of Dargai and of the Sampagha and Arhanga Passes.—Reconnaissance of the Kharmana Defile and action of 7 Nov. 97. Reconnaissance of the Saran Sar and action of 9th Nov. '97.—Operations at and around Dwatoi and action of 24 Nov. 97. Operations in the Bara Valley, 7 to 14 Dec. '97.—Despatches, G. G. O. 244 of 98.—Clasp.

Waldron, H. F. K. (*Bt.-Col., retd. I. A.*)—
Afghanistan, 1878-79-80.—With Southern Afghanistan and Kurram Field Forces.—Medal.
N.-W. Frontier of India, Zhob Valley, 1884.
China, 1900.—Medal.

Walker, A. M. E. H. T.—
South African War, 1899-1901.—Queen's medal with 4 clasps.

Walker, B. W. G.—
The War of 1914—18.—Despatches, *Lond. Gaz.*, 12 Jun. 17. M.C.

Walker, G.B.—
The War of 1914—18.—Despatches, *Lond. Gaz.*, 5 Apl. 16 and 20 May 16. M.C.

Walker, H. R. O.—
N.-W. Frontier of India, 1908.—Operations in the Zakka Khel country. Medal with clasp.

Walker, J. G. D. (*M.-Gen. u. s. l. Ind. Army.*)—
India, 1859.
Burma, 1887-89.—Medal with clasp.

Walker, J. N. (*Col. u.s.l., I.A.*)—
Afghanistan, 1880.—As Brig.-Maj., Kandahar Line of Communication.—Medal.
N.-W. Frontier of India, Zhob Valley, 1884.—As D. A. A. Genl. of the Force. Affair of Dowlatzai.

Walker, J. N. (*Major, I. M.S.*)—
China, 1900.—Medal.
N.-W. Frontier of India, 1902.—Operations against the Dharwesh Khel Waziris.
The War of 1914—18.—Despatches, *Lond. Gaz.*, 21 Jun. 16.

Walker, P. G. (*Bt.-Col., retd. I. A.*)—
N.-W. Frontier of India.—Action at Wana, 3rd November 1894.—Medal with clasp.
N.-W. Frontier of India, 1897-98.—Engagement near Shabkadr, 9 Aug. 97.—Despatches G. G. O. 1188 of 97.—*Mohmand.*—Despatches, G. G. O. 1318 of 97.—*Buner.*—Action of the Tanga Pass. —Medal with clasp.—Brev. of Maj.
China, 1900.—Medal.

Walker, W. G.—
N.-W. Frontier of India (2nd Miranzai), 1891.—Medal with clasp.
N.-W. Frontier of India, Waziristan, 1894-95.—Clasp.
East Africa, 1903-04.—Operations in Somaliland Action at Jidballi. Despatches, (Brig.-Genl. Manning, 17th Aug. 03. Maj.-Genl. C. C. Egerton, 30th May 04). *Lond. Gaz.*, 2 Sep. '04. Brev. of Lt.-Col., V. C. (*Gaz. of India* Notification No. 883 of 20 Aug. 03). Medal with 2 clasps.
The War of 1914—18.—Despatches, *Lond. Gaz.*, 22 Jun. 15, 1 Jan. 16, and 4 Jan. 17. Promd. Maj. Gen.

Walker, W. R.—
Chitral, 1895.—Relief of Chitral.—Medal with clasp.
China, 1900.—Medal.
The War of 1914-18.—Despatches, *Lond Gaz.*, 4 July 16.

Wall, E. W.—
The War of 1914—18.—Despatches, *Lond. Gaz.*, 15 Jun. 16, and 1 Jun. 17. C.M.G.

Wall, F.—
China, 1900.—Medal.
The War of 1914—18.—Despatches, *Lond. Gaz.*, 17 Feb. and 22 Jun. 15. C.M.G.

Wallace, A.—
Afghanistan, 1879-80.—Medal.
Burma, 1886-88.—Medal with 2 clasps.
N.-W. Frontier of India, Hazara, 1891.—Clasp.
N.-W. Frontier of India, (2nd) Miranzai, 1891.—Clasp.
N.-W. Frontier of India, 1897-98.—Tochi.—Buner Medal with clasps.
N.-W. Frontier of India, Waziristan, 1901-02.—Clasp.
East Africa, 1903-04.—Operations in Somaliland. Action at Jidballi.—Despatches, *Lond. Gaz.*, 2nd September 1904.—Medal with 2 clasps. Brevet-Colonel.
The War of 1914—18.—Despatches, (Sir J. Maxwell 19 Aug. 15, and 16. Mar. 16), *Lond. Gaz.*, 21 Jun. 16.

Wallace, C. W.—
N.-W. Frontier of India, 1908.—Operations in the Mohmand country. Engagement of Kargha, 24th May 1908.—Medal with clasp.
The War of 1914—18.—Despatches, *Lond. Gaz.*, 5 Apl. and 19 Oct. 16.

Wallace, E. C. L.—
China, 1900.—Medal with clasp.

Wallace-Dunlop, A. (*Maj., retd., I. A.*)—
Chitral, 1895.—Relief of Chitral.—Medal with clasp.
Tibet, 1903-04.—Action at Niani, 26th June 1904. Operations at and around Gyantse, between 5th May and 6th July 1904. March to Lhassa, 14th July to 3rd August 1904. (Severely wounded.) —Medal with clasp.
N.-W. Frontier of India, 1908.—Operations in the Zakka Khel Country. Medal with clasp.

Waller, A. A. P.—
Mekran, 1898.
The War of 1914—18.—Despatches, *Lond. Gaz.*, 15 Aug. 17.

Waller, A. G.—
The War of 1914—18.—Despatches, *Lond. Gaz.*, 15 Aug. 17.

Waller, E. (*Maj., Supy. List I.A.*)—
N.-W. Frontier of India, Hazara, 1891.—Medal with clasp.
(*2nd*) *Miranzai, 1891.*—Clasp.

Waller, J. E. (*M.-Gen., retd. Ben. Inf.*)—
N.-E. Frontier of India, Bhutan, 1865-66.—Capture of Bala stockades. Medal with clasp.
Afghanistan, 1878-80.—Action of Ahmed Khel and advance to Kabul. Mentioned in Despatches. Medal with clasp, Brev. of Lt.-Col.
N.-W. Frontier of India, Hazara, 1891.
N.-W. Frontier of India, Miranzai (*2nd*), *1891* — Clasp.

Waller, J. E.—
N.-W. Frontier of India, 1908.—Operations in the Zakka Khel country—Medal with clasp.

Waller, R. J. (*Maj., retd. Ben. S.C.*)—
Agfhanistan, 1878-80.—Actions at Futtehabad and Jugdulluk, and 2nd action of Charasiah.—Meda, with clasp.

Wallis, B. H.—
The War of 1914—18.—Despatches, *Lond. Gaz.*, 1 Jan. 16 M. C.

Wallis, H. R. (*Major, retd., I. A.*)—
N.-W. Frontier of India, Hazara, 1888.—Medal with clasp.
Tirah, 1897-98.—Operations in the Bara Valley, 7 to 14 Dec. 97.—Medal with 2 clasps.

Walmsley, R.—
Afghanistan, 1880.– March from Kabul to the relief of Kandahar. Battle of Kandahar.—Medal with clasp; Bronze Star.
South African War, 1899-1900.—Operations in Natal, 99. Defence of Ladysmith, Despatches, *London Gazette*, 10th September 1901. Promoted Lieut. G. G. O.'s 194 and 451 of 02.—Queen's Medal with clasp.

Walsh, J. H. T. (*Lt.-Col., retd. I. M. S.*)—
Burma, 1885—87.—Expedition up the Chindwin River.—Medal with clasp.

Walter, A. E.—
China, 1900.—Relief of Pekin—Medal with clasp.

Walters, L. (*Comsy. and Hony. Capt., retd.*)—
Burma, 1885-88.—Medal with 2 clasps.
N.-W. Frontier of India, Hazara, 1888.—Despatches. G. G. O. 978 of 1888. Clasp. Promoted Sub-Conductor.
N.-W. Frontier of India, Hazara, 1891.—Clasp.
N.-W. Frontier of India, (1st) Miranzai, 1891.
N.-W. Frontier of India, (2nd) Miranzai, 1891.—Despatches. G. G. O. 632 of 1891.—Clasp. Promoted Conductor.
China, 1900.—Relief of Pekin.—Medal with clasp. Promoted Dy. Asst. Commissary.

Walton, G.—
Soudan, 1896.—Dongola Expedition.—Medal; Egyptian medal.
N.-W. Frontier of India, 1897-98.—Mohmand.—Medal with clasp.
The War of 1914—18.—Despatches, *Lond. Gaz.*, 19 Oct. 16. Brev. of Lt.-Col.

Walton, H. J.—
N.-W. Frontier of India, 1897-98.—Medal with clasp.
China, 1900.—Relief of Pekin. Action of Peitsang. Action at Yangtsun.—Medal with clasp.
Tibet, 1903-04.- Operations at and around Gyantse between 5th May and 6th July 1904. March to Lhassa, 14th July to 3rd Aug. 1904.—Medal with clasp.

Walton, W. C.—
Burma, 1885-86.—Medal with clasp.
The War, 1914—16.—Despatches, *London Gazette*, 15 Jun. 16, 4 July 16, 31, Oct. 17.—C. B. C. M. C.

Wapshare, A. (*Lt.-Col., retd. I. A.*)—
Soudan, 1884-85.—Medal; bronze star.

Wapshare, R.—
Burma, 1886-88.—Medal with 2 clasps.
The War of 1914—18.—C. B.

Ward, A.—
Burma, 1887-89.—Medal with clasp.
N.-W. Frontier of India, Hazara, 1891.—Clasp.
N.-W. Frontier of India, (2nd) Miranzai, 1891.—Clasp.
Tirah. 1897-98.—Medal with clasp.
The War of 1914—18.—Despatches, *Lond. Gaz.*, 22 Sept. 17. D. S. O.

Ward, C. H.—
The War of 1914—18.—Despatches, *Lond. Gaz.*, 19 Oct. 16.

Ward, F. W. (*M.-Gen. u.s.l. R. Art.*)—
India, 1857-58.—Medal.
N.-W. Frontier of India, 1863-64.—Medal with clasp.
Afghanistan, 1879-80.—March from Quetta to the relief of Kandahar. Despatches, G. G. O. 694 of 81. Medal.

Ward, G. A.—
Burma, 1886-89.—Medal with 2 clasps.
Burma, 1894.

Ward, G. L. S. (*Major, retd. I. A.*)—
N.-E. Frontier of India, Lushai, 1892.
N.-E. Frontier of India, Abor Expedition, 1894.
Tibet, 1903-04.—Action at Niani, 26th June 1904. Operations at and around Gyantse between 5th May and 6th July 1904. March to Lhassa, 14th July to 3rd August 1904. Medal with clasp.

Ward, H. C. S.—
The War of 1914—17, Despatches, *Lon. Gaz.*, 21st June '16, 25th September 16th, 6th, July '17. Order of the White Eagle, 4th Class (with swords) Brev. of Lt.-Col.

Wardell, G. F. E.—
Tibet, 1903-04.—Medal.
The War of 1914—18.—Despatches, *Lon. Gaz.*, 15 Aug. 17.

Warden, A. W. (*Lt.-Col., retd. I. A.*)—
Burma, 1885-88.—Medal with 2 clasps.

Warden, W. R.—
The War of 1914—18.—Despatches, *Lond. Gaz.*, 19 Oct. 16.

Warliker, D. P. (*Lieut.-Col., retd., I. M. S.*)—
Rumpa Rebellion, 1879-80.
Khond Rebellion in Central Provinces, 1882.
Burma, 1886-88.—Medal with 2 clasps.
China, 1900.—Medal.

Warren, R. P. (*Major, retd. I. A.*)—
Tirah, 1897-98.—Operations in the Bara Valley, 7 to 14 Dec. 97.—Medal with 2 clasps.

Warton, C. P. F.—
Aden, 1903-04.—Operations in the interior.

Waterfield, A. C. M.—
N.-W. Frontier of India, Hazara, 1888 and *1891* - Medal with 2 clasps.
Chitral, 1895.—Relief of Chitral.—Medal with clasp.
N.-W. Frontier of India, 1897-98.—Relief and defence of Malakand. Relief of Chakdara, Malakand. Action at Landakai. Operations in Bajaur and in the Mamund Country, Mohmand, Utman Khel, Buner. Action of the Tanga Pass.—2 clasps.
South African War, 1900-01.—Operations in the Orange Free State, Feb. to May 00, including operations at Paardeberg (18 to 26 Feb.). Actions at Poplar Grove (7 March), Driefontein (10 March), Vet River (5 and 6 May) and Zand River (10 May). Operations in the Transvaal in May and June 00, including actions near Johannesburg (29 May), Pretoria (4 June) and Diamond Hill (11 and 12 June). Operations in the Transvaal, east of Pretoria, July to Sep. 00, including action at Belfast (26 and 27 Aug. 00). Operations in the Transvaal from Nov. 00 to Aug. 01. Operations in Cape Colony, 15 Aug. to 21 Sep. 01. Despatches, *Lond. Gaz.*, 16 April 01—Queen's medal with 6 clasps. Brev. of Maj.

Waterfield, B. C. (*Maj., retd. I. A.*)—
N.-W. Frontier of India, (2nd) Miranzai, 1891.—Medal with clasp.
N.-W. Frontier of India, 1897-98.—Mohmand.—Despatches, G. G. O. of 145 of 98.—Medal with clasp.
The War of 1914—18.—Despatches, *Lond. Gaz.*, 1 Jan. 16.

Waters, E. E.—
N.-W. Frontier of India, 1897-98.—Malakand.—Medal with clasp.
Tirah, 1897-98.—Actions of Dargai and of the Sampagha and Arhanga Passes. Operations in the Bazar Valley, 25 to 30 Dec. '97. Clasp.

Waters, R. S.—
South African War, 1900-01.—Operations in Orange River Colony, July to Dec. 00; Operations in the Transvaal, Jan. to July 01, Despatches, *Lond. Gaz.* 10 Sep. 01 —Queen's medal with 4 clasps.
Tibet, 1903-04.—Medal.

Watkis, H. B. B.—
N.-W. Frontier of India, Tochi.—D. A. A. G., 1st Brigade. Despatches, G. G. O. 1432 of 97.—Medal with clasp.—Brev. of Lieut.-Col.

Watkis, H. L.—
The War of 1914—18.—Despatches, *Lond. Gaz.*, 22 Jan. 15.—M. C.

Watling, F. H.—
Chitral, 1895.—Relief of Chitral.—Medal with clasp
China, 1900.—Medal.
The War of 1914—18.—Despatches, *Lond. Gaz.*, 15 Aug. 17.

Watling, G.—
N.-W. Frontier of India, 1897-98.—Malakand—Utman Khel—Buner.—Action of the Tanga Pass. —Medal with clasp.

Watling, L. D.—
China, 1900.—Medal with clasp.
The War of 1914—18.—Despatches, *Lond. Gaz.*, 31 Oct. 17.

Watney, C. W.—
South African War, 1901-02.—Operations in Cape Colony between 30th Novr. 1900 and 31st May 1902. Operations in the Orange River Colony between 30th Novr. 1900 and 31st May 1902. Operations in the Transvaal between 30th Novr. 1900 and 31st May 1902. Queen's medal with 5 clasps.

Watson, A. C.—
N.-W. Frontier of India, Hazara, 1891.—Medal with clasp.

Watson, C. H.—
N.-W. Frontier of India, 1897-98.—Tochi.—Operations on the Samana and in the Kurram Valley during August and Sept. 1897. Operations of the Flying column in the Kurram Valley under Col. Richardson, 20 Aug. to 1 Oct. 97.—Medal with 2 clasps.
N.-W. Frontier of India, 1908.—Operations in the Zakka Khel country.—Medal with clasp.
The War of 1914—18.—Despatches, *Lond. Gaz.*, 13 July 16.—D.S.O.

Watson, G. F. (*Major, retd. I. A.*)—
Burma, 1886-87.—Despatches, Lond. Gaz., 2 Sept. 87 and G. G. O. 864 of 87.
Burma, 1892-93.—Operations in the Northern Chin Hills. Despatches, G. G. O. 733 of 93.

Watson, G. H. (*Lt.-Col., retd. I. A.*)—
Afghanistan, 1880.—With Khyber Field Force.—Medal.

Watson, H. D.—
The War of 1914—18.—Despatches, *Lond. Gaz.*, 15 Jun. 16.

Watson, H. D.—
N.-E. Frontier of India, Sikkim, 1888.—Attack of Jalapla.—Medal with clasp.
N.-E. Frontier of India, Chin-Lushai, 1889-92.—Clasp.
China, 1900.—D. A. A. G., I. S. Troops.—Medal.
The War of 1914—18.—Despatches, *Lond. Gaz.*, 21 Jun. 16 and 6 July 17.—C. M. G.

Watson, Sir John (*Gen. u.s.l. Ind. S.C.*)—
India, 1848-49.—Punjab Campaign.—Siege of Mooltan, battle of Goojerat, and pursuit to Peshawar.—Medal with 2 clasps.
Bozdar Expedition, 1857.—Medal.
India, 1857-58.—Siege of Delhi, second siege and relief of Lucknow (wounded on three occasions). Medal with 3 clasps. Brev. of Maj.; V.C. and C.B.

N.-W. Frontier of India, 1863.—Umbeyla Expedition.—Clasp.
Afghanistan, 1879-80.—Commanded Kurram Field Force. Mentioned in Despatches. Received thanks of both Houses of Parliament, 5 May 81.—Medal.

Watson, J. H.—
N.-W. Frontier of India, 1897-98.—Mohmand.—Medal with clasp.
South African War, 1902.—Operations in the Transvaal between 7th January 1902 and 30th March 1902. Operations in the Orange River Colony between 1st April and 31 May 02.—Adjutant XI Battalion Imperial Yeomanry, 7 Jan. 02 to 31 May 02.—Queen's medal with 3 clasps.
The War of 1914—18.—Despatches, *Lond. Gaz.*, 13 July 16.

Watson, L. A.—
Burma, 1887-89.—Medal with clasp.
N.-W. Frontier of India, Isazai, 1892.
N.-W. Frontier of India, 1897-98.—Malakand.—Operations in the Mamund Country.—Buner.—Medal with clasp.

Watson, P. H.—
The War of 1914—18.—M.S.M.

Watson, S. B.—
The War of 1914—18.—Despatches, *Lond. Gaz.*, 19 Oct. 16.—Brev. of Lt.-Col.

Watson, T. E.—
Tirah, 1897-98.—Operations against the Khani Khel Chamkannis. Operations in the Bazar Valley, 25 to 30th December 1897. Medal with 2 clasps.
China, 1900.—Medal.

Watson, T. N.—
The War of 1914—18.—Despatches, *Lond. Gaz.*, 15 Aug. 17.—M. C.

Watson, W. A.—
Soudan, 1885.—Suakin—Action of Takul. Despatches, *Lond. Gaz.*, 25 Aug. 85.—Medal with clasp. Bronze star.
China, 1900.—Despatches, *Lond. Gaz.*, 13th Sep. 1901. Medal.
The War of 1914—18.—Despatches, *Lond. Gaz.*, 21 Jun 16, 25 Sept. 16, and 6 July 17.—C. M. G.

Watt, A. (*Hon. Capt. Commy., retd., I. A. Depts.*)—
Chitral, 1895.—With the Relief Force. Medal with clasp.
N.-W. Frontier of India, 1897-98.—Clasp.
China 1900.—Despatches, *Lond. Gaz.*, 13 Sept. 01. Medal.

Watt, D. M.—
Chitral, 1895.—Relief of Chitral.—Medal with clasp.
Tirah, 1897-98.—Actions of the Sampagha and Arhanga Passes. Operations in the Waran Valley and action of 16 Nov. 97. Operations at and around Dwatoi and action of 24 Nov. 97. Operations against the Khani Khel Chamkannis. Operations in the Bara Valley, 7 to 14 Dec. 97.—2 clasps.
East Africa, 1902-04.—Operations in Somalil and.—Medal with clasp.
The War of 1914—18, Despatches, *Lon. Gaz.*, 22 Jun. '15, 15 Jun. '16. D. S. O., Clasp. to D. S. O. Brev. Lt.-Col.

Watts, A. H.—
The War of 1914—18.—Despatches, *Lond. Gaz.*, 15 Aug. 17.

Watts, G. A. R.—
N.-W. Frontier of India, 1897-98.—Action of the Ublan Pass, 27 Aug. 97. Operations on the Samana and in the Kurram Valley during Aug. and Sep. 97. Operations of the Flying Column in the Kurram Valley under Colonel Richardson, 20 Aug. to 1 Oct. 97.—Medal with 2 clasps
The War of 1914—18.—Despatches, *Lond. Gaz.*, 15 Aug. 17.—Brev. of Lt.-Col.

Watts, H.—
China, 1900.—Relief of Pekin. Action of Peitsang. Action at Yangtsun. Despatches, *Lond. Gaz.*, 13 Sep. 01.—Medal with clasp. Promoted Deputy Assistant Commissary.

Watts, J. B. (*Brev. Col., retd. Ind. S.C.*)—
N.-W. Frontier, 1877-78, Jowaki Expedition.—Medal with clasp.
Afghanistan, 1879-80.—Action of Charasiah.—Medal with clasp.

Webb, A. E.—
N.-W. Frontier of India, Hazara, 1888.—Medal with clasp.
Chitral, 1895.—Relief of Chitral.—Medal with 2 clasps.
Tirah, 1897-98.—Medal with 2 clasps.

Webb, H. G. S.—
South African War, 1901-02.—Served as Civil Surgeon. Operations in the Transvaal and Orange River Colony, and Cape Colony, June of to 31 May 02.

N.-W. Frontier of India, 1908.—Operations in the Mohmand country—Medal with clasp.

Webb, H. N. (*Maj., retd., Ind. S.C.*)—
Afghanistan, 1879-80.—Operations in Kohistan, affairs of Saidabad, march from Kabul to relief of Kandahar and battle of 1st Sep.—Medal with clasp, bronze star.

Webb, W. F. R.—
N.-W. Frontier of India, 1902.—Operations against the Darwesh Khel Waziris.
N.-W. Frontier of India, 1908.—Operations in the Mohmand country. Engagement of Kargha, 24th May 1908. Despatches, G. G. O. No. 521 A of 1908—Medal with clasp.
The War of 1914—18.—Despatches *Lond. Gaz.*, 1 Jan. 16, 15 Jun. 16, 25 Nov. 16, and 4 Jan. 17.—Brev. of Maj.—D. S. O.

Webster, C. G. (*Lt.-Col., retd. I.M.S.*)—
China, 1900.—Medal.

Webster, T. (*Lt.-Col., retd., I.A.*)—
Burma, 1886-88.—Mogoung column.—Medal with 2 clasps.

Weekes, H. E.—
South African War, 1900.—Operations in the Orange Free State, February to May 1900, including action at Vet River (5th and 6th May), and Zand River (10th May). Operations in the Transvaal, west of Pretoria, July to October 1900, including action at Frederickstad (17th to 25th October).—Queen's Medal with 3 clasps.
The War of 1914—18.—Despatches, *Lond. Gaz.*, 21 Jun. 16.

Weeks, R. F. J.—
South African War, 1900-02.—Operations in Cape Colony, south of Orange River, 1899-1900. Operations in Orange Free State, February to October 1900. Operations in the Transvaal west of Pretoria, October to March 1902. Queen's medal with 3 clasps. King's medal with 2 clasps.

Weir, P. A. (*Col, retd.* I. M. S.)—
Afghanistan, 1879-80.—Medal.

Welchman, A. J. T. (*Col., retd., Ben. Inf.*)—
N.-E. Frontier of India, 1864-66, Bhootan Expedition.—Capture of Dewangiri.—Medal with clasp.
Afghanistan, 1878-79.—Capture of Ali Musjid.—Medal with clasp.

Welchman, E. W. St. G. (*Bt.-Col., retd., I. A.*)—
N.-W. Frontier of India, Jowaki, 1878.—Medal with clasp.
Afghanistan, 1879-80.—Action against the Mohmands.—Medal.
Burma, 1886-88.—Operations of the 1st Brigade, and in Pagyi.—Medal with 2 clasps.
Tirah, 1897-98.—Medal with 2 clasps.
China, 1900.—Medal.

Welchman, F. D. (*Maj., retd., Ind. S. C.*)—
Afghanistan, 1880.—Operations at and around Kabul in Dec. 79.—Medal with clasp.

Wellborne, C. de M.—
The War of 1914—18.—Despatches, *Lond. Gaz.*, 15 Aug. 17.

Weller, A. T. (*Maj., retd., Ind. S. C.*)—
Afghanistan, 1880.—On the Khyber line.—Medal.
Burma, 1886-87.—Medal with clasp.
N.-E. Frontier of India, 1889, Lushai Expedition.—Honourably mentioned. G. G. O. 592 of 89.
Chin-Lushai Expedition, 1889-90.—Despatches, G. G. O. 677 of 90.—Clasp.

Weller, G. H. (*Col., retd., I. A.*)—
N.-W. Frontier of India, 1897-98.—Operations on the Samana and in the Kurram Valley during Aug. and Sep. 97.—Medal with 2 clasps.
Tirah, 1897-98.—Actions of the Sampagha and Arhanga Passes.—Despatches, G. G. O. 244 of 98.—Clasp.

Wellesley, H. A.—
The War of 1914—18.—Despatches, *Lond. Gaz.*, 15 Aug. 17.—M. C.

Wells, F. W. A.—
South African War, 1902.—Operations in the Transvaal, Mar. to 31 May 02. Queen's medal with 2 clasps.
The War of 1914—18—Despatches, *Lond. Gaz.*, 19 Oct. 16.—M. C.

Wells, N. F.—
The War of 1914—18—Despatches, *Lond. Gaz.*, 13 July 16.

Wells, R. T.—
The War of 1914—18—Despatches, *Lond. Gaz.*, 19 Oct. 16.

Wells, T. F.—
The War of 1914—18.—Despatches, *Lond. Gaz.*, 5 Apl. 16.

Western, J. S. E. (*Col., retd., I. A.*)—
N.-W. Frontier of India, 1881.—Mahsud Waziri Expedition.
N.-W. Frontier of India, 1888.—Hazara Expedition. Served as A.-D.-C. to the G. O. C. Despatches. G. G. O. No. 978 of 1888. Medal with clasp.
N.-W. Frontier of India, Waziristan, 1894-95.—Clasp.
N.-W. Frontier of India, 1902.—Operations against the Darwesh Khel Waziris. A. A.-G., Derajat District and Staff Officer, Idak Column.

Westlake, A. P.—
Afghanistan, 1880-81.—Medal.
Burma, 1886-89.—Operations of the 4th and 6th Brigades. Despatches, G. G. O. 434 of 1887. D. S. O.—Medal with 2 clasps.
The War of 1914—18.—Despatches, *Lond. Gaz.*, 28 Dec. 17.

Westmacott, *Sir* R. (*Maj.-Genl. u. s. l., I.A.*)—
India, 1859.—Siege of Forts Beyt and Dwarka. *Operations against rebel Bheels, 1860. Operations against rebel Niakras Mahals, 1868.*—Action of Waruk.—(Received the commendation of Government, G. G. O. 161 of 68.)
Afghanistan, 1879-80.—March from Quetta to the relief of Kandahar. Engagements in Kojak Pass.—Despatches, G. G. O.'s 645 and 975 of '80.—Medal.
Soudan, 1885.—Suakin. Actions of Hasheen and Tofrek. Various attacks on convoys. Advance on Tamai.—Despatches, *Lond. Gaz.*, 25 Aug. '85. —Medal with 2 clasps; bronze star.
N.-E. Frontier of India, Chin-Lushai, 1889-90.—In command of the 28th Pioneers, commanded the advance column on Haka.—Despatches, G. G. O. 413 of '90. D. S. O.—Medal with clasp.
N.-W. Frontier of India, 1897-98.—Mohmand. Commanded 1st Brigade. Despatches. G. G. O. 1318 of 97.—Medal with clasp.
Tirah, 1897-98.—Commanded 4th Brigade.—Actions of Dargai and of the Sampagha and Arhanga Passes. Reconnaissance of Saran Sar and action of 9 Nov. 97. Operations at and around Dwatoi and action of 24 Nov. 97. Operations in the Bara Valley, 7 to 14 Dec. 97. Affair at Shinkamar on 29 Jan. 98. Despatches, G. G. O.'s 58 and 244 of 98.—Clasp.—K. C. B.

Weston, E. (*Sr. Asst. Surgeon and Hony. Capt., retd., I. S. M. D.*)
Burma, 1885-1888.—Medal with 2 clasps.

Weston, G. E. (*Col. u.s.l., Mad. Inf.*)—
Burma, 1887-90.—Medal with clasp.

Weston, G. O.—
Burma, 1887-89.—Medal with clasp.
N.-E. Frontier of India, Chin-Lushai, 1889-9 Clasp.
N.-W. Frontier of India, Isazai, 1892.
Chitral, 1895.—Relief of Chitral.—Medal with Clasp.
N.-W. Frontier of India, 1897-98. Buner.—Action of the Tanga Pass.—Clasp.
South African War, 1899-1902.—Operations in Natal in Oct. 1899, including action at Lombard's Kop (30th Oct.). Defence of Ladysmith, including action of 6th Jan. 1900. Operations in Natal, Mar. to June 1900, including action at Laing's Nek (6th to 9th June). Operations in the Transvaal 30th Nov. 1900 to 31st May 1901, and in the Orange River Colony to 31st May 1902.—Queen's medal with 4 clasps. King's medal with 2 clasps.
The War of 1914—18.—Despatches, *Lond. Gaz.*, 19 Oct. 16.

Westropp, G. R. C. (*Col., retd., I. A.*)—
Afghanistan, 1879-80.—As Asst. Supt. of Transport. Mentioned in Despatches. Medal.
N.-W. Frontier of India, Zhob Valley, 1884.—As an Exe. Commt. Officer.
N.-E. Frontier of India, Chin-Lushai, 1889-90.—As Commt. Transport Officer. Despatches, *Lond. Gaz.*, 12 Sept. 90. Medal with clasp.
N.-W. Frontier of India, 1897-98.—Mohmand. Chief Commt. Officer (Sept. to 7 Oct. 97). Despatches, *Lond. Gaz.*, 11 June 98. Medal with clasp.
Tirah, 1897-98.—Base Commt. Officer, Kohat, afterwards Chief Commt. Officer, Base and Lines of Communication. Clasp.
N.-W. Frontier of India, Waziristan, 1901-02.—Clasp.

Westropp White, W.—
The War of 1914—18.—Despatches, *Lond. Gaz.*, 22 Jun. 15 and 19 Oct. 16,—Brev. of Col. C. B.

Wetherall, F. G. B.—
East Africa (Somaliland), 1908-10.—Medal with clasp.

Wheatley, H. S. (*Col. u.s.l. I. A.*)—
Afghanistan, 1878-79-80.—Accompanied *Sir* C. Gough's Column to Sherpur. March from Kabul to the relief of Kandahar, and battle of Kandahar. Despatches, G. G. O. 582 of '80.—Medal with 2 clasps. Bronze star.

Wheatley, W. P. R.—
East Africa, 1903.—Operations in Somaliland. Action at Jidballi. 2nd class Post Staff Offr. 16 July to 8 Aug. 03. Medal with 2 clasps.
The War of 1914—18—Despatches, *Lond. Gaz.*, 15 Aug. 17.—D. S. O.

Wheeler, E. S.—
South African War, 1902.—Operations in Cape Colony, January to May 1902. Queen's Medal with 2 clasps.

Wheeler, G. (*Comsy. and Hony. Capt., retd.*)—
Afghanistan, 1878-79-80.—Action of Charasiah. Affair of Doaba. Mentioned in despatches. Expedition into Kohistan. Action of Mir Karez. Action in the Hardeh Valley and operations, etc., in Kabul and Sherpur. Medal with 2 clasps and medal for distinguished conduct in the field.—Promoted Sub-Condr.
Egypt, 1882.—Medal; Khedive's bronze star.
Burma, 1887.—Honourably mentioned in G. O. No. 764 of 31st Jan. '89.—Medal with clasp.

Wheeler G. C.—
The War of 1914—18—Despatches, *Lond. Gaz.*, 15 Aug. 17.—V. C.

Wheeler, R. (*Hon. M.-Gen., retd., Ben. S.C.*)—
India, 1858-59.—Action of Sitka Ghat.—Medal.
N.-W. Frontier of India, 1863.—2nd Eusofzai Expedition; operations in the Umbeyla Pass.—Medal with clasp.
Afghanistan, 1879-80.—Zaimusht Expedition; assault of Zawa. Mentioned in Despatches.—Medal.
N.-W. Frontier of India, 1881, Mahsud-Waziri Expedition.

Wheler, C. S. (*Col., retd., I. A.*)—
Egypt, 1882—Action of Tel-el-Kebir.—Medal with clasp. Bronze star.
N.-W. Frontier of India, 1897-98.—Operations on the Samana and in the Kurram Valley during August and September 1897. Medal with 2 clasps.
Tirah, 1897-98.—Clasp.

Whiffin, H. E.—
Soudan, 1885.—Medal with clasp.—Bronze star.
Burma, 1895-96.
South African War, 1902.—Operations in the Transvaal, March 17 to end of May 1902. Despatches *London Gazette,* 18 July 1902. Queen's medal with 2 clasps.

Whitler, A. E.—
Egypt, 1882.—Medal. Bronze star.
Burma, 1885-86.—Taking of Minhla. Operations of the 6th Brigade.—Medal with clasp.

Whitaker, N. R.—
N.-W. Frontier of India, 1908.—Operations in the Mohmand country. Medal with clasp.

Whitaker, S. S.—
The War of 1914—18.—Despatches, *Lond. Gaz.*, 5 Apl. 16.—M. C.

Whitamore, V. N.—
The War of 1914—18.—Despatches, *Lond Gaz.*, 1 Jan. 16.

Whitcombe, E. G. R. (*Lt.-Col., retd., I.M.S.*)—
Mekran, 1898.—Action of Gok Parosh. Despatches. G. G. O. 648 of 98.
China, 1900.—Relief of Pekin. Medal with clasp.

White, G. E.—
N.-W. Frontier of India, Hazara, 1888.—Medal with clasp.
N.-W. Frontier of India, Isazai, 1892—
N.-W. Frontier of India, 1897-98—Buner.—Medal with clasp.
Tirah, 1897-98.—Actions of the Sampagha and Arhanga Passes. Operations against the Khani Khel Chamkanis. Operations in the Bara Valley, 7th to 14th December 1897—Despatches, G. G. O. No. 244 of 1898.—Clasp. Promoted Sub-Conductor.
East Africa, 1903-04.—Operations in Somaliland.—Medal with clasp.
Aden, 1903-04.—Operations in the interior.
N.-W. Frontier of India, 1908.—Operations in the Zakka Khel country.—Medal with clasp.
The War of 1914—18.—Despatches, *Lond. Gaz.*, 13th July and 19th Oct. 16.

White, H. P. M.—
N.-W. Frontier of India, 1908.—Operations in the Zakka Khel country. Operations in the Mohmand country.—Medal with clasp.

White, O. W. (*Capt., retd., I. A.*)—
N.-W. Frontier of India, Waziristan, 1901-02.—Medal with clasp.

White, S. C. (*Sr. Asst. Surgn. and Hony. Capt., retd. I. S. M. D.*)—
N.-E. Frontier of India, Chin-Lushai, 1889-90.—Served in the Gangaw Column.—Specially promoted.—Medal with clasp.

White, W. E.—
The War of 1914—18—Despatches, *Lond. Gaz.*, 19 Oct. 16.

White, W. W.—
N.-W. Frontier of India, Hazara, 1891.—Medal, with clasp.
Chitral, 1895.—Relief of Chitral.—Medal with clasp.
N.-W. Frontier of India, 1897-98.—Operations on the Samana and the Kurram Valley during Aug. and Sep. 97—*Buner.*—2 clasps.
Tirah, 1897-98.—Clasp.
China, 1900.—Relief of Pekin. Despatches, *Lond. Gaz.*, 14 May 01.—Medal with clasp.

The War of 1914.—18—Despatches, *Lond. Gaz.* 22 Jun. 15 and 19 Oct. 16.—Brev. of Col.

Whitehead, J.—
The War of 1914—18.—Despatches, *Lond. Gaz.*, 1 Jan. and 15 Jun. 16, 4 Jan. and 11 Dec. 71. Brev. of Lt.-Col. D.S.O.

Whitford-Hawkey, E. H.—
The War of 1914—18.—Despatches, *Lond. Gaz.* 19 Oct. 16.

Whittall, F. V. (*Bt. Col., retd., I. A.*)—
Burma, 1887-88.—Medal with 2 clasps.
China. 1900.—Special Service Officer. Despatches, *London Gazette*, 13th Sept. 1901.—Medal.
Aden, 1903-04.—Operations in the interior.

Whittell, H. M.—
The War of 1914—18.—Despatches, *Lond. Gaz.*, 15 Jun. 16.

Wickham, A F. (*Commy. and Hony. Major retd. Ordce. Dept.*),
South African War, 1899-1900.—Defence of Ladysmith. Despatches, *Lond. Gaz.*, 10 Sep. 01. Promoted Dy. Commissary with hon. rank of Captain.—Medal with clasp.

Wickham, W. J. (*Col, retd., I. A.*)—
Afghanistan, 1879-80.—As Asst. Road Comdt., S. Afghanistan F. F.—Medal.
Soudan, 1885.—*Suakin.*—Staff Officer to the Director of Indian Transport with Home Forces.—Medal with clasp; bronze star.
N.-W. Frontier of India, Hazara, 1888.—As Transport Officer.—Medal with clasp.
Chitral, 1895.—Relief of Chitral Served as Base Transport Officer and Inspecting Transport Officer, Lines of Communication.—Medal with clasp.

South African War, 1899—1902.—As Assistant Director of Transport Operations in Natal 99, including actions at Talana (20 Oct.) and Lombard's Kop (30 Oct.) Defence of Ladysmith, including action of 6 Jan. 00. As Depy. Assistant Adjutant-General, VIII Division. Operations in the Orange Free State, April and May 00. Operations in Orange River Colony (May to 29 Nov. 00, including actions at Biddulphsberg (29 May), Wittebergen (1 to 29 July). As Assistant Adjutant-General for Transport Head-Quarters. Operations in the Transvaal, Cape Colony and Orange River Colony between 30 Nov. 00 and 31 May 02. Despatches (3 times), *Lond. Gaz.*, 8th Feb. 01 and 29 July 02. Queen's medal with 5 clasps. King's medal with 2 clasps; Brev. of Lt.-Col. C.B.
Tibet, 1903-04.—Medal.

Widdicombe, G. T.—

N.-E. Frontier of India, Lushai, 1889.—Medal with clasp. *Chin-Lushai, 1889-90.*—(Chittagong Column.)—Clasp.
N.-W. Frontier of India, Hunza-Nagar, 1891.—Despatches, G. G. O. 397 of '97.—Clasp.
N.-W. Frontier of India, 1897-98.—Mohmand. Medal with clasp.
Tirah. 1897-98.—Clasp.
The War of 1914—18.—Despatches, *Lond. Gaz.*, 22 Jun. 15. C. B.

Wigley, P. J. R.—

The War of 1914—18.—Despatches *Lond. Gaz.*, 5 Apl. 16 and 15 Aug. 17. M. C.

Wigram, C.—

N.-W. Frontier of India, 1897-98.—Operations on the Samana and in the Kurram Valley during Aug. and Sep. 97. Operations of the Flying Column in the Kurram Valley under Colonel Richardson, 20 Aug. to 1 Oct. 1897.—Medal with 2 clasps.
Tirah, 1897-98.—Clasp.
South African War, 1899-1900.—Relief of Kimberley—Operations in the Orange Free State, Feb. to May 00, including operations at Paardeberg (17 to 26 Feb.), actions at Poplar Grove (7 March) Driefontein (10 March), and Karee Siding (29 March); action at Hountnek (1 May), Vet River (25 and 6 May) and Zand River (10 May).—Operations in the Transvaal in May and June 00, including actions near Johannesberg (29 May), Pretoria (4 June), and Diamond Hill (11 and 12 June).—Operations in the Transvaal west of Pretoria (July to 29 Nov. 00) including actions at Elang River (4 to 16 Aug.)—Operations in the Orange River Colony (May to 29 Nov. 00) including action at Wittebergen (1 to 29 July).—Despatches, *Lond. Gaz.*, 10th Sep. 01. Queen's medal with 6 clasps.

Wigram, K.—

N.-W. Frontier of India, 1897-98, Mohmand.—Medal with clasp.
Tirah, 1897-98.—Operations at and around Dwatoi, and action of 24 Nov. 97. Operations in the Bara Valley, 7 to 14 Dec. 97.—Clasp.
N.-W. Frontier of India, Waziristan, 1901-02.—Clasp.
Tibet, 1903-04.—Action at Niani, 26th June 1904 Operations at and around Gyantse, between 5th May and 6th July 1904. March to Lhassa, 24th July to 3rd August 1904—Medal with clasp.
The War of 1914—18.—Despatches, *Lond. Gaz.*, 1 Jan. and 15 Jun. 16, 4 Jan. 15 May and 11 Dec. 17. Brevs. of Lt.-Col. and Col. and Legion of Honour (5th class). D. S. O.

Wikeley, J. M.—

N.-E. Frontier of India, Chin-Lushai, 1889-90.—Medal with clasp.

Tirah, 1897-98.—Orderly Officer to G. O. C. 1st Division—Actions of the Sampagha and Arhanga Passes.—Reconnaissance of the Kharmana Defile and action of 7 Nov. 97.—Reconnaissance of the Saran Sar and action of 9 Nov. 97.—Operations against the Khani Khel Chamkannis.—Operations in the Bazar Valley, 25th to 30 December 97.—Medal with 2 clasps.

Wilbond, J.H. (*Hony. Capt. Comsy., retd., I.O.D.*)—
Afghanistan, 1878-79-80.—With Kurram Field Force. Advance to Kabul. Action of Charasiah. Operations at and around Kabul. Medal with 2 clasps.

Wilcox, H. B. F.—

Chitral, 1895.—Relief of Chitral.—Medal with clasp.
South African War, 1902.—Operations in the Transvaal, 23 Apl. to 31 May 1902.—Queen's medal with 2 clasps.

Wildes, C. S. D.—

Operations in the Abor country, 1911-12.—Meda with clasp.
The War of 1914-18; *Lon. Gaz.*, 21 Jun. 16.

Wildford, E. E.—

The War of 1914—18.—Despatches, *Lond. Gaz.*, 4 Jan 17. D. S. O.

Wilkieson, C. W. (*Maj., retd. Indian Army.*)—
Burma, 1887-89.—Medal with 2 clasps.
China, 1900.—Relief of Pekin.—Medal with clasp.

Wilkins, T. J. H. (*Col., retd., I. M. S.*)—

Burma, 1885-86.—Tounghoo-Nyingyan Column. Present at storming of Aylah and action at Neybeyagoun.—Medal with clasp.

Wilkinson, C. R.—
West Africa, 1900.—Operations in Ashanti.—Medal.
East Africa, 1903-04.—Operations in Somaliland Action at Jidballi. Medal with 2 clasps.

Wilkinson, E. (*Lt.-Col., retd., I. M.S.*)—
N.-W. Frontier of India, Waziristan, 1894-95.—Medal with clasp.
N.-W. Frontier of India, 1897-98.—Mohmand. Buner. Action of the Tanga Pass.—Medal with clasp.

Wilkinson, G. H.—

S. African War 1900—02.—Operations in Cape Colony South of Orange River, Feb. to Apl, 00. Operations in the Transvaal Feb. to 31 May 02. Queens Medal with 3 clasps
The War of 1914—18.—Despatches *Lond. Gaz.*, 6 Dec. 16.

Wilkinson, J. E. (*Dy. Comy. and Hony. Capt. retd.*)—

N.-W. Frontier of India, Waziristan, 1901-02.—Medal with clasp.

Will, D. A. E.—
British East Africa, 1898.—Operations against the Ogaden Somalis.—Medal with clasp.

Willans, T. J.—
N.-W. Frontier of India, 1897-98.—Malakand. Buner Action of the Tanga Pass.—Medal with clasp
The War of 1914—18.—Despatches, *Lond. Gaz.*, 22 Jun. 15. D. S. O.

Willes, G. F. O. S. O. (*Maj., retd. I. S. C.*)—
Afghanistan, 1878-79-80.—Medal.
Chitral, 1895.—Relief of Chitral.—Medal with clasp.

Williams, A. F. C.—
South African War, 1899-00.—Operation in Orange Free State from March to May 900 including operations near Wepener. Operations in Cape Colony, south of Orange River, December 1899 to March 1900. Operations in Orange River Colony, May to July 1900, including actions near Lindley and Bethlehem. (Dangerously wounded.) Despatches, *Lond. Gaz.*, 16th April 01. Queen's medal with 3 clasps. D.S.O.

The War of 1914—18.—Despatches *Lond. Gaz.*, 4 Jan. 17. Order of St. Stanislas (3rd class with Swords).

Williams, A. H. (*Surg.-Lt.-Col., retd., Ind. Med. Serv.*)—*Afghanistan, 1880.*—Medal.

Williams, A. H.—
Burma, 1886-87.—Operations of the 1st Brigade—Medal with clasp.

N.-E. Frontier of India, Manipur Expedition, 1891.—Clasp.

Williams, C. E.—
N.-W. Frontier of India, 1897-98.—Operations on the Samana.—Medal with 2 clasps.

Williams, C. H. (*Major, retd., I. A.*)—
Burma, 1885—89.—Medal with 2 clasps
Burma, 1895-96.
China, 1900.—Medal.

Williams, C. L. (*Major, retd., I.M.S.*)—
Burma, 1891-92.—Operations on the N.-E. Frontier.

Williams, C. S. (*Maj., retd., I. A.*)—
Burma, 1886-87.—Operations of the 1st Brigade, Ruby Mines Expedition, including advance from Khannyat.—Medal with 2 clasps.

N.-E. Frontier of India, Manipur, 1891.—Clasp.

N.-W. Frontier of India, Hunza-Nagar, 1891.—Despatches, G. G. O. 397 of '92.—Clasp.

Williams, F. I.—
Operations in the Abor country, 1911-12.—Despatches *Gaz. of India*, Notn. 480, dated 23 May 12.—Medal with clasp.—Promoted Asst. Comy. with hony. rank of Lieut.

Williams, G. A. (*Lt.-Col., retd., I. A.*)—
Afghanistan, 1878-79.—Action of Ali Musjid. Despatches. Lt.-Genl. Sir S. J. Browne's Despatch, 8 July 79. (1st) Bazar Valley Expedition. Despatches, G. G. O. 610 of '79.—Medal with clasp.
N.-W. Frontier of India, Hazara, 1888.—Medal with clasp.

Williams, G. R. K.—
N.-W. Frontier of India, Waziristan, 1901-02.—Medal with clasp.

Williams, H. A.—
China, 1900.—Despatches, *Lond. Gaz.*, 13 Sep. 01—Medal. D.S.O.

Williams, L. G.—
South African War, 1900-01.—Served as a trooper in Lumsden's Horse, 26th Feby. to 16th October 1900. Operations in Cape Colony, 20th to 31st March 1900. Operations in Orange River Colony 1st to 30th April 1900, and action at Hout Nek (1st May). Operations in the Transvaal, 6th June 1900 to 11th Feby. 1901, including action at Zilikat's Nek (2nd August 1900) and defence of Eland's River (6th August 1900). Queen's medal with 5 clasps.

The War of 1914—18.—Despatches *Lond. Gaz.*, 5 Apl., 10 May, 13 July, 19 Oct. and 14 Nov. 16 and 15 Aug. 17. D. S. O. Brev. of Lt.-Col.

Williams, W. P. G.—
N.-W. Frontier of India, 1908.—Operations in the Mohmand country.—Medal with clasp.

Williamson, C. V. W. (*Col., retd., I. A.*)—
N.-E. Frontier of India, Lushai, 1890.—Medal with clasp. *Manipur, 1891.*—Clasp.
N.-W. Frontier of India, 1897-98.—Tochi Base Commissariat Officer. Despatches, G. G. O., 1432 of 97.—Medal with clasp.
Tirah, 1897-98.—Chief Transport Officer, Line of Communications.—Clasp.
N.-W. Frontier of India, 1908.—Operations in the Zakka Khel country. Div. Supply Offr.—Medal with clasp.

Williamson, M. (*Lt.-Col., retd., I. A.*)—
Burma, 1887-88.—Medal with clasp.
Burma, 1891-92.
N.-E. Frontier of India, Kachin Hills, 1892-93.—Clasp.
N.-W. Frontier of India, 1897-98.—Malakand-Buner. Action of the Tanga Pass.—Medal with clasp.
East Africa, 1902-04.—Operations in Somaliland, Medal with clasp.

Willis, E. S. C.—
South African War, 1899—1901.—Relief of Ladysmith, including action at Colenso (15th December 1899). Operations of 17th to 24th January 1900, and action at Spion Kop; operations of 5th to 7th February 1900, and action at Vaal Krany. Operations on Tugela Heights (14th to 27th February 1900), and action at Pieters Hill (27th February 1900). Operations in Natal, March to June 1900. Operations in the Transvaal, east of Pretoria, July to 29th November 1900, including action at Lydenburg (5th to 8th September). Operations in the Transvaal, between 30th November 1900 and December 1901.—Despatches. *London-Gaz.*, 17th January 1902.—Queen's medal with 3 clasps. —King's medal with 2 clasps.—D.S.O.
The War of 1914—18,—Despatches *Lond. Gaz.*, 17 Feb. 15.

Willis, G. C. W.—
The War of 1914—18.—Despatches *Lond. Gaz.*, 21 Jun. 16. D.O.S.

Willmore, W. S.—
China, 1900.—Relief of Pekin. Action of Peitsang. Action at Yangtsun. Medal with clasp.

Willmott, E. H. B. (*Deputy Commy. and Hony. Capt., retd., S. and T. Corps*).—
Soudan, 1885.—With the Indian Contingent (Camel Corps).—Medal with clasp; Khedive's bronze star. Mentioned in despatches by Genl. Lord Wolseley.
Burma, 1888-90.—With the Mogoung Column and with the Tonhon Column.—Honourable mention by the C.-in-C. in India.—Medal with clasp.
N.-W. Frontier of India, 1897-98.—Medal with clasp. Promoted Deputy Assistant Commissary.

Willoughby, D. V.—
The War of 1914—18.—Despatches, *Lond. Gaz.*, 19 Oct. 16 and 15 Aug. 17. D.S.O.

Willoughby, J. F. (*M.-Gen., retd., Bo. Ins.*)—
Abyssinia, 1867-68.—Capture of Magdala. Medal.
Afghanistan, 1880.—Defence and battle of Kandahar. Medal with clasp.

Willoughby, M. E.—
N.-E. Frontier of India, Chin-Lushai, 1889-90.—Served as Transport Officer.—Medal with clasp.
China, 1900.—Special Service Officer. Despatches, *Lond. Gaz.*, 14 May 01.—Medal.
The War of 1914—18.—Despatches, *Lond. Gaz.*, 15 Aug. 17.

Willoughby, M. G. P.—
The War of 1914—18.—Despatches, *Lond. Gaz.*, 5 Apl. and 19 Oct. 16, and 12 Jun. 17. M.C.

Willoughby, M. W. (*Lt.-Gen., u. s. l., Ind. S. C.*)—
Persia, 1856-57.—Landing at Hallilah Bay, surrender of Bushire, bombardment and capture of Forts of Mohumera. Despatches, *Lond. Gaz.*, 30 Jan. 57. Medal with clasp.
Abyssinia, 1867-68.—Despatches, *Lond. Gaz.*, 3 June 68.—Medal.

Wilmer, A. H. (*Lt.-Col., retd., I. A.*)—
N.-W. Frontier of India, Hazara, 1891.—Medal with clasp. (*2nd*) *Miranzai, 1891.*—Clasp.

Wilsey, H. W.—
Afghanistan, 1878-79.—Capture of Ali Musjid. Operations in the Khyber and Bazar Valley—Medal with clasp.
N.-W. Frontier of India, 1881.—Mahsud-Waziri Expedition.
Burma.—1885-86-87-88.—Medal with 2 clasps.
Chitral, 1895.—Relief of Chitral.—Medal with clasp.

Wilson, A. M.—
The War of 1914—18.—Despatches, Lond. Gaz., 1 Jan. 16.

Wilson, A. T.—
The War of 1914—18.—Despatches, *Lond. Gaz.*, 5 Apl. 16. 14 Nov. 16 and 15 Aug. 17. C.I.E., D.S.O.

Wilson, D. D.—
South African War, 1899-1901.—Relief of Ladysmith including operations of 17 to 24 Jan. 00, and action at Spion Kop. Operations of 5 to 7 Feb. 00 and action at Vaal Krans. Operations on the Tugela Heights (14 to 27 Feb. 00) and action at Pieter's Hill (27 Feb. 00). Operations in Natal, March to June 00. Operations in the Transvaal, 30 July 00 to July 01. Queen's medal with 7 clasps.
The War of 1914—18.—Despatches, *Lond. Gaz.*, 1 Jan. 16. M. C.

Wilson, D. G.—
N.-W. Frontier of India, Waziristan, 1901-02.—Medal with clasp.

Wilson, G. E. H.—
Tibet, 1903-04.—Medal.
N.-W. Frontier of India, 1908.—Operations in the Mohmand country. Medal with clasp.
Operations in the Abor country, 1911-12.—Clasp.
The War of 1914—18.—Despatches, *Lond. Gaz.*, 17 Feb. 15.

Wilson, H. M.—
The War of 1914—18.—Despatches, *Lond. Gaz.*, 21 July 17.

Wilson, J. A.—
N.-E. Frontier of India, Manipur, 1891.—Medal with clasp.
N.-E. Frontier of India, Abor Expedition, 1894.
Burma, 1895-96.—Served as Staff Officer in Expedition to Kiaruma.
N.-W. Frontier of India, Waziristan, 1901-02.—Served as D. A. A. G. of Punjab Frontier Force and Mahsud Blockade operations. Served as Staff Officer with Lieut.-Colonel V. C. Tonnochy's column. Despatches, G. G. O. 611 of 02.—Medal with clasp.
Operations in the Abor country, 1911-12. D.S.O. Despatches, *Gaz. of India*, Notn. No. 480, dated 23rd May 1912.—Medal with clasp. Brevet of Lieut.-Colonel.

Wilson, J. D.—
The War of 1914—18.—M.C.

Wilson, J. F. (*Lt.-Col., retd., I. A.*)—
Burma, 1885-86.—Taking of Minhla and occupation of Mandalay.—Medal with clasp.

Wilson, N. M.—
The War of 1914—18.—Despatches, *Lond. Gaz.*, 5 Apl., 13 July and 19 Oct. 16. D.S.O.

Wilson, R. C.—
The War of 1914—18.—Despatches, *Lond. Gaz.*, 19 Oct. 16. 14 Nov. 16 and 15 Aug. 17. Brevet of Lt.-Col.—M.C.

Wilson, W. B. (*Col., retd., I. S. C.*)—
Afghanistan, 1878-79-80.—Actions of the Peiwar Kotal and Charasiah.—Operations at and around Kabul.—Medal with 3 clasps.
Burma, 1886-87.—Despatches, G. G. O. 434 of 87. Medal with clasp. Brev. of Maj.

Wilson, W. H. (*Col., u.s.l., Ind. S. C.*)—
India, 1858.—Operations against Dessaies in the North Canara and Bedse Districts. Medal.

Wilson, W. H. D.—
South African War, 1901-02.—Operations in Transvaal, Jan. to Feb. 01. Operations in the Orange River Colony, Feb. 01 to 31st May 02.—Queen's medal with 5 clasps.

Wilson-Johnston, W. E.—
China, 1900.—Medal.
The War of 1914—18.—Despatches, *Lond. Gaz.*, 27 July and 19 Oct. 16. D.S.O.

Wimberley, C. N. C.—
Chitral, 1895.—Relief of Chitral.—Medal with clasp.
N.-W. Frontier of India, 1897-98.—Operations on the Samana and in the Kurram Valley during Aug. and Sep. 97.—Operations of the Flying Column in the Kurram Valley under Colonel Richardson, 20 Aug. to 1 Oct. 97.—2 clasps.
Tirah, 1897-98.—Actions of the Chagru Kotal and Dargai and of the Sampagha and Arhanga Passes. Reconnaissance of the Saran Sar and action of 9 Nov. 97. Operations in the Waran Valley and action of 16 Nov. 97.—Clasp.
Tibet, 1903-04.—Action at Niani, 26th June 1904. Operations at and around Gyantse, between 5th May and 6th July 1904, March to Lhassa, 14th July to 3rd August 1904. Despatches, *Gazette of India*, No. 1065 of 1904.—Medal with clasp.
The War of 1914—18.—Despatches, *Lond. Gaz.*, 15 Aug. 17. C.MG.

Wimberley, F. G. A.—
N.-W. Frontier of India, 1897-98.—Tochi.—Medal with clasp.

Windross, M. (*Sr. Asst. Surgn. and Hony. Capt., retd., I.S.M.D.*)—
Afghanistan, 1878-79-80.—Battle of Ahmed Khel, 19th April 1880. Action of Urzu, 25th April 1880. Medal with clasp.

Windsor, F. N.—
N.-W. Frontier of India, 1897-98.—Malakand.—Operations in Bajaur and the Mamund country.—Medal with clasp.

Wingate, A. W. S. (*Col., retd., I.A.*)—
China, 1900.—Special Service Officer, H. Qrs. Staff, I. B., Relief of Tientsin. Action of Peitsang, action at Yangtsun. Despatches, *London Gazette*, 14th May 1901.—Medal with clasp. Brevet of Lieut.-Colonel.
The War of 1914—18.—Despatches, *Lond. Gaz.*, 15 Jun 16. C.M.G.

Wingate, G.—
N.-E. Frontier of India, Naga Hills, 1879-80.—Despatches, G. G. O. 405 of '80.—Medal with clasp.
N.-W. Frontier of India, 1897-98.—Tochi. Chief Commt. Offr. (28 June to Nov. 97). Despatches *Lond. Gaz.*, 11 Feb. 98. Clasp.

Winter, C. B. (*Capt., Supy. List., I. A.*)—
N.-W. Frontier of India. Isazai. 1892.
The War of 1914—18.—Despatches, *Lond. Gaz.*, 5 Apl, 10 May, 19 Oct. and 14 Nov. 16 and 15 Aug. 17. C. M. G., D. S. O.

Wintle, C. E. H.—
N.-W. Frontier of India, Waziristan, 1894-95.—Medal with clasp.
The War of 1914—18.—Despatches, *Lond. Gaz.*, 27 July 16 and 15 Aug. 17. Brevet of Lt.-Col.

Wise, A. D.—
The War of 1914—18.—Despatches, *Lond. Gaz.*, 31 Oct. 17.

Withers, A. R. (*Maj., retd., I. A.*)—
N.-W. Frontier of India, Waziristan, 1901-02.—Medal with clasp.

Withers, C. McG.—
East Africa, 1902-04.—Operations in Somaliland—Medal with clasp.
Nandi, 1905-06.—Despatches, *London Gazette*, 18th September 1906.—Clasp.

Wodehouse, F. W. (*Lt.-Col., Supy. List., I. A.*)—
N.-E. Frontier, India, Chin-Lushai, 1889-90.—Medal with clasp

Wolley, H. S. L.—
South African War, 1901-02.—Operations in the Transvaal, June 01 to May 02. Despatches, *Lond. Gaz.*, 29 July 02.—Queen's medal with 4 clasps.

Wood, C. A.—
The War of 1914—18.—Despatches, *Lond. Gaz.*, 1 Jan. 16. M. C. Clasp to M. C.

Wood, C. E. (*Major, retd., I. A.*)—
Burma, 1889.—Medal with clasp.

Wood, E. J. F. (*Lieut.-Col., retd., I. A.*)—
N.-W. Frontier of India, Jowaki, 1877-78.—Medal with clasp
Afghanistan, 1878-79-80.—Occupation of Kandahar and Khelat-i-Ghilzai.—Affairs with the Ghilzais a and near Jagdalak. Despatches. Col. F. B. Norman's Despatch, 7 Jan. '80.—Medal.
N.-W. Frontier of India, Zhob Valley, 1884.

Wood, E. J. M.—
The War of 1914—18.—Despatches, *Lon. Gaz.*, 15 Aug. 17. D. S. O.

Wood, F. J. (*Simla Vol. Rifles*)—
N.-W. Frontier of India, 1908.—Operations in the Mohmand country.—Medal with clasp.

Wood, H. C.—
South African War, 1900.—Operations in Cape Colony, north of Orange River, 28th March to 4th April 1900. Operations in the Orange Free State, 5th April to 26th May 1900, including actions at Houtnek (1st May), Vet River (5th and 6th May) and Zand River (10th May). Operations in the Transvaal, 27th May to 26th June 1900, including actions near Johannesburg; (29th May), and Pretoria (4th June). Operations in the Transvaal west of Pretoria, 22nd to 28th August 1900. Operations in the Transvaal east of Pretoria, 29th August to 3rd November 1900.—Queen's medal with 3 clasps.

Wood, H. O. B.—
South African War, 1900-02.—Operations in Orange River Colony (May to 29th Nov. 1900) including action at Wittebergen (1 to 29 July). Operations in the Transvaal in May and June 1901. Operations in Orange River Colony 30 Nov. 1900 to April 1901 and from July 1901 to 31 May 1902. Queen's medal with 3 clasps; King's medal with 2 clasps.
N.-W. Frontier of India, 1908.—Operations in the Zakka Khel country. Operations in the Mohmand country. Engagement of Kargha, 24th May 1908.—Medal with clasp.
The War of 1914—18.—Despatches, *Lond. Gaz.*, 21 Jan. 16 (Sir J. Maxwell. 19 Aug. 15 and 16 Mar. 16) and 15 Aug. 17. Brevet of Maj.

Wood, R. B. M. (*Lt., retd., I. A.*)—
South African War, 1900-02.—Operations in the Transvaal, east of Pretoria, July to 29 Nov. 00. Operations in Orange River Colony, May to June 00. Operations in the Transvaal, 30 Nov. 00 to Jan. 02.—Queen's medal with 3 clasps King's medal with 2 clasps.

Woodcock, E. E.—
Soudan, Nile Expedition, 1898.—Battles of the Atbara and Khartoum.—Medal; Khedive's medal with 2 clasps.

Woodcock, W. C. M. (*Lt.-Col., retd., I.A.*)—
Burma, 1885—87.—Operations of the 6th Brigade.—Medal with clasp.
N.-W. Frontier of India (1st and 2nd) Miranzai, 1891.—Clasp.
N.-W. Frontier of India, Waziristan, 1901-02.—Despatches, G. G. O. No. 611 of '02.—Medal with clasp.

Woodhouse, C. G.—
China, 1900—Relief of Pekin.—Medal with clasp.
East Africa, 1902-04.—Operations in Somaliland. Action at Jidballi.—Medal with clasp.

Woodhouse, H. O. (*Brev. Col., retd., Ind. S. C.*)—
Afghanistan, 1889.—With the Kabul Line Force and 2nd Brigade, Northern Afghanistan Field Force, Logar Valley.—Medal.
N.-E. Frontier of India, 1889. Lushai Expedition.—Medal with clasp.
N.-W. Frontier of India, 1897-98.—Mohmand.—Medal with clasp.
Tirah, 1897-98.—Operations in the Bara Valley, 7 to 14 Dec. 97.—Clasp.

Woodhouse, P. St. J. R.—
South African War, 1902.—Operations in the Transvaal between 8th April and 31st May 1902.—Queen's medal with 4 clasps.
N.-W. Frontier of India, 1908.—Operations in the Zakka Khel country. Operations in the Mohmand country.—Medal with clasp.
The War of 1914—18.—Despatches, *Lond. Gaz.*, 15 Aug. 17. Brevet of Lt.-Col.

Woodward, C. F.—
Tirah, 1897-98.—Medal with 2 clasps.
N.-W. Frontier of India, Waziristan, 1901-02.—Clasp.

Woodward, J. A. H.—
Burma, 1885-87, 1887-89.—Medal with 2 clasps.
N.-W. Frontier of India, Hazara, 1891.—Clasp.
Chitral, 1895.—Relief of Chitral. Action of the Malakand Pass. Action at Khar.—Medal with clasp.
N.-W. Frontier of India, 1897-98.—Present at Fort Jamrud.—Clasp.

Woodwright, W. H. E.—
N.-W. Frontier of India, Hazara, 1888.—Medal with clasp. *1st Miranzai, 1891.*

Woodyatt, N. G.—
The War of 1914—18.—Despatches, *Lond Gaz*, 4 July 16. Promd. Maj.-Gen. C. B.

Wooldridge, H. C.—
The War, 1914—18.—Despatches, *Lond Gaz.*, 4 July 16.

Wooldridge, W. H.—
East Africa (British), 1896.—Operations against the Mazrui Rebels.—Medal.

Woolley, J. M.—
N.-W. Frontier of India, 1897-98.—Malakhand. Relief of Chakdara. Operations in Bajaur and in the Mamund country.—Medal with 2 clasps.
China, 1900.—Medal.

Woon, J. B.—
Afghanistan, 1879-80.—With Kandahar Field Force.—Medal.
N.-E. Frontier of India, Chin-Lushai, 1890.—Relief of Changsil and Aijab.—Medal with clasp.
N.-W. Frontier of India, 1897-98.—Engagement near Shabkadr, 9 Aug. 97.—Commanded the column.—Despatches, G. G. O. 1188 of 97. —Malakand.—Mohmand.—Despatches, G. G. O. 1318 of 97.—Buner. Action of the Tanga Pass —Despatches, G. G. O. 217 of 98.—Medal with clasp.—Brev. of Lt.-Col.
China, 1900.—Despatches, *Lond. Gaz.*, 14 May 01.—Medal. Brev. of Col.

Worgan, R. B.—
South African War, 1900-02.—Operations in Orange River Colony, December 1900 and October 1901 to 31st May 1902. Operations in Cape Colony, December 1900 to October 1901.—Queen's medal with 2 clasps; King's medal with 2 clasps.
The War of 1914—18.—Despatches, *Lond. Gaz.*, 1 Jan. 16, 15 Jun. 16, 4 Jan. 17 and 28 Dec. 17. Brevs. of Maj. and Lt.-Col. D. S. O.

Worlledge, J. F.—
Afghanistan, 1878-79.—With Kandahar Field Force. —Medal.
Egypt, 1882.—Action of Tel-el-Kebir.—Medal with clasp. Bronze star.
Soudan, 1896.—*Dongola Expedition.*—Medal.
N.-W. Frontier of India, 1897-98.—Malakand—Operations in Bajaur and in the Mohmand Country —Utman Khel.—Medal with clasp.

Worsley, C. F. M.—
N.-W. Frontier of India, Waziristan, 1901-02.—Medal with clasp.
N. W. Frontier of India, 1902.—Operations against the Darwesh Khel Waziris.
The War of 1914—18.—Despatches, *Lond. Gaz.*, 15 Aug. 17.

Wortabet, H. G. L. (*Lieut.-Col., retd., I.M.S.*)—
Afghanistan, 1880.—Medal.
Burma, 1886-88.—Operations of the 4th and 6th Brigades.—Medal with 2 clasps.

Wredden, J. L.—
Tirah, 1897-98.—Medal with 2 clasps.

Wright, E. G.—
Burma, 1886-89.—Ruby Mines Expedition. Northern Shan Column. Expedition against the Chins.—Medal with 2 clasps.
N.-E. Frontier of India, Chin-Lushai, 1889-90.—Burma Column.—Clasp.
Tibet, 1903-04.—Medal.

Wright, E. L.—
South Africa, 1879-80.—Zulu Campaign and the operations against Sekukuni, 1879.—Medal with clasp.
Burma, 1886-87.—Medal with clasp.

Wright, F. W. (*Lt.-Col., retd., I.M.S.*)—
Afghanistan, 1878-79-80.—Affairs with the Ghilzais at and near Jugdalak. Action of Chihildakhteran. March from Kabul to the relief of Kandahar, and battle of Kandahar. Operations against the Maris.—Medal with clasp. Bronze star.
Burma, 1886-88.—Operations of the 3rd and 4th Brigades. Despatches, G. G. O. 434 of 87.—Medal with 2 clasps. D. S. O.

Wright, H. C. (*Col., u. s. l., Ind. S. C.*)—
India, 1858.—Operations in the Raichore Dooab district.
Afghanistan, 1879-80.—Action of Charasiah and occupation of Kabul, Chief Commissariat Officer. Despatches, G. G. O. 1139 of '79.—Medal with clasp.
Burma, 1887.—Present at the taking of Minhla and subsequent occupation of Mandalay under Major-Genl. H. N. D. Prendergast. Operations of the 2nd Brigade under Brigr.-Genl. Foord.—Medal with clasp.

Wright, R.—
N.-W. Frontier of India, 1897-98.—Operations in the Mamund country.—Medal with clasp.
Tirah, 1897-98.—Operations in the Bara Valley. Operations in Waran Valley.—Clasp.
China, 1900.—Medal.
Tibet, 1903-04.—Operations at Niani, 26th June 1904. Operations at and around Gyantse between 5th May and 6th July 1904. Medal with clasp.
N.-W. Frontier of India, 1908.—Operations in the Zakka Khel Country. Operations in the Mohmand Country.—Medal with clasp.
Operations in the Abor country, 1911-12.—Despatches, *Gaz. of India*, Notn. 480, dated 23rd May 1912.—Medal with clasp.—Promoted Asst. Comy. with hony rank of Lieut.
The War of 1914-18.—Despatches, *Lond. Gaz.*, 5 Apl. and 19 Oct. 16.

Wright, R. E.—
The War of 1914—18.—Despatches, *Lond. Gaz.*, 5 Apl. 16. Brev. of Maj.

Wyatt, E. R. C.—
East Africa, 1903-04.—Operations in Somaliland. —Medal with clasp.
The War of 1914—18.—Despatches, *Loud. Gaz.*, 4 Jan. and 11 Dec. 17. D. S. O.

Wylie, H. (*Major-General, retd., Bengal Inf.*)—
N.-W. Frontier of India, (2nd) Eusufzai, 1863. —Umbeyla. Storming of Conical Hill.—Medal with clasp.
N.-E. Frontier of India, Bhutan, 1865.—Capture of Nagode stockade.—Clasp.
Abyssinia, 1868.—Medal.
N.-W. Frontier of India, Hazara, 1868.—Clasp.
Afghanistan, 1878-79-80.—Advance from Quetta to Kandahar. Chief Political Officer. Thanked by Her Majesty's Government and Government of India. Despatches. Lieutenant-General *Sir* M. A. S. Biddulph's Despatch, 16 May '79, and G. G. O. 693 of '80.—Medal. C.S.I.

Wyllie, F. (*Lieut.-Col., retd., I.A.*)—
South Africa, 1879, Zulu War.—Medal with clasp.
Burma, 1888-89.—Medal with clasp.

Wyllie, R. J. H. (*Lt.-Col., retd., Ind. S.C.*)—
Afghanistan, 1879-80.—With the Kurram Field Force. Mentioned in Despatches.—Medal.

Wylly, G. G. E.—
South African War, 1899-1900.—(Twice wounded slightly, wounded dangerously.) Queen's medal with 3 clasps. V. C.
The War of 1914—18.—Despatches, *Lond. Gaz.*, 15 June 16, 15 May 17 and 11 Dec. 17. D. S. O.

Wynch, F. J. H.—
Burma, 1887.—Operations of the 1st Brigade.—Medal with 2 clasps.
N.-W. Frontier of India, Hazara, 1888 and 1891. —2 clasps.
Chitral, 1895.—Relief of Chitral.—Action of the Malakand Pass. Action at Khar. (Wounded.)—Medal with clasp.

Wyndowe, S. J. (*Dep. Surg. Gen. retd. Ind. Med. Serv.*)—*India, 1857—59.*—Engagement near Jumbulpure.—Medal.

Wynter, C. P.—
N.-W. Frontier of India, 1897-98.—Malakand.—Operations in Bajaur and in the Mohmund Country.—Medal with 2 clasps.
East Africa, 1903-04.—Operations in Somaliland. Action at Jidballi. Despatches, *Lond. Gaz.*, 2 Sept. '04. Medal with 2 clasps.
The War of 1914-18.—Despatches, *Lond. Gaz.*, 20 Mar 16.

Wynter, J. R.—
The War of 1914-18.—Despatches, Lond. Gaz., 31 May 16. D. S. O.

Y

Yate, A. C. (*Lt.-Col., u. s. l., I.A.*)—
Afghanistan, 1880.—March from Quetta to relief of Kandahar.—Medal.
Burma, 1886--88.—With Northern Shan Column as Intell. Officer, '87-88.—Medal with 2 clasps.

Yate, C. E. (*Col., u. s.l. I. A.*)—
Afghanistan, 1880.—As Poll. Officer, Khelat-i-Ghilzai, and with the force under Lt.-Genl. *Sir* F. Roberts. March to the relief of Kandahar. Battle of Kandahar.—Despatches, G. G. O. 709 of '81.—Medal with clasp; Bronze star.
Mekran, 1901.—Attack on and capture of Nodiz Fort.

Yate, F. H.—
N.-W. Frontier of India, Waziristan, 1901-02.—Medal with clasp.

Yate, W. G. (*Col., retd., I. A.*)—
Afghanistan, 1878-79-80.—Advance on Khelat-i-Ghilzai and Ghazni. Battle of Ahmed Khel. Action of Urzu March from Kabul to the relief of Kandahar, and battle of Kandahar.—Medal with 2 clasps. Bronze star.

Yates, J. A.—
South African War, 1901-02.—Served with 5th Battn. Rifle Bde. Operations in Cape Colony and Orange River Colony, Dec. 1901 to 31st May 1902.—Queen's medal with 3 clasps.
The War of 1914-18.—Despatches, Lond. Gaz., 15 Aug. 17. C. I. E.

Yates, R. J. B.—
The War of 1914-18.—Despatches, Lond. Gaz., 15 May and 11 Dec. 17. D. S. O.

Yearsley, R. A.—
N.-W. Frontier of India, 1908.—Operations in the Zakka Khel country. Operations in the Mohmand country.—Medal with clasp.

Yeats Brown, F. C. C.—
The War of 1914 18.—Despatches, Lond, Gaz., 5 Apl. 16.

Yelding, W. R.—
Afghanistan, 1879-80.—Occupation of Khyber and Jellalabad. Operations in the Logar Valley. March from Kabul to the relief of Kandahar, and battle of Kandahar. Operations against the Maris.—Medal with clasp. Bronze star.
N.-W. Frontier of India, Mahsud Waziri, 1881. Hazara, 1888.—Served as Divisional Transport Officer with the Head-Quarters of the Force. Despatches, G. G. O. 978 of '88.—Medal with clasp D.S.O.
Tirah, 1897-98.—Divisional Commissariat Officer, 1st Division.—Actions of the Sampagha and Arhanga Passes.—Operations in the Bazar Valley, 25 to 30 Dec. 97.—Despatches, G. G. O. 244 of 98.—Medal with 2 clasps.
East Africa, 1903-04.—Operations in Somaliland Director of Supply and Transport. Despatches *Lond. Gaz.*, 2 Sept. 04. Brev. of Col. Medal with clasp.

Younan, A. C. (*Lt.-Col., retd., I. M. S.*)—
Burma, 1886—89.—Medal with 2 clasps.
N.-W. Frontier of India, Hazara, 1891.—Clasp, *Isazai, 1892.*
Chitral, 1895.—Relief of Chitral.—Medal with clasp.
N.-W. Frontier of India, 1897-98.—Tochi.—Clasp.
Mekran, 1898.
N.-W. Frontier of India, 1908.—Operations in the Mohmand country. Medal with clasp.

Young, A. W. C.—
Tibet, 1903-04.—Operations at and around Gyantse between 5th May and 6th July 1904. March to Lhassa, 14th July to 3rd August 1904. Severely wounded. Medal with clasp.
The War of 1914-18. Despatches, Lond. Gaz., 19 Oct. 16. Brev. of Lt.-Col.

Young, C. F. G.—
South Africa, 1879, Zulu Campaign.—Action of Gingindlovo.—Medal with clasp.
China, 1900.—Medal.

Young, C. W. (*Capt., retd., Ind. S. C.*)—
Afghanistan, 1878-79.—Medal.

Young, E. A. (*Col., retd., I. A.*)—
Afghanistan, 1878-79-80.—Occupation of Kandahar and Khelat-i-Ghilzai. Battle of Ahmed Khel. (Severely wounded. Horse shot.)—Medal with clasp.
N.-W. Frontier of India, (2nd) Miranzai, 1891 Medal with clasp.

Young, E. W. (*Surg. Lt.-Col. retd. Ind. Med. Serv.*)—
N.-W. Frontier of India, Zhob Valley Expedition, 1884.—Affair of Dowlatzai.

Young, F. deB.—
N.-W. Frontier of India, Hazara, 1888.—Medal with clasp.
N.-W. Frontier of India, 1897-98.—Operations on the Samana and in the Kurram Valley during Aug. and Sep. '97.—Medal with 2 clasps.
Tirah, 1897-98.—Actions of Chagru Kotal, Dargai, and of the Sampagha and Arhanga Passes.—Despatches, G. G. O. 244 of 98.—Clasp.
South Africa War, 1900-01.—Operations in the Oransge Free State, '00. Operations in the Transvaal, west of Pretoria, '00-01.—Operations in the Transvaal between 30 Nov. '00 and 31 May 02. Operations in the Orange River Colony between 30 Nov. '00 nd 31 May '02. Operations in Cape Colony between 30 Nov. '00 and 31 May '02.—Quenols medal with 4 clasps.
The War of 1914 18.—Despatches, Lond. Gaz., 5 Sept. 16. C. M. G.

Young, G. (*Col., u. s. l, Ind., S. C.*)—
N.-E. Frontier of India, 1865-66. Bhootan Expedition.—Medal with clasp.
N.-W. Frontier of India, 1868. Hazara Campaign.—Clasp.
Perak, 1875-76.—Clasp.
Afghanistan, 1880.—Medal.

Young, G. F. (*Hony. Br. Genl., retd., I. A.*)—
Afghanistan, 1879-80.—As Transport Officer and as Intelligence Officer, Quarter Master General's Department; Deputy Assistant Quarter Master General, Khyber Brigade. Action on the Gara heights near Kam Daka. Despatches—Genl. *Sir* F. Roberts' Despatch, No. 741, 13 Feb. '80.—Medal.
N.-W. Frontier of India, Mahsud Waziri, 1881.—D. A. Q. M. G., 2nd Brigade. Despatches, Br.-Genl. J. J. H. Gordon's Despatch, 26 May '81.—*Hazara, 1888.* Despatches, G. G. O. 978 of '88.—Medal with clasp. Brev. of Lt.-Col.

Young, G. H.—
N.-W. Frontier of India, Waziristan, 1901-02.—Medal with clasp.

Young, H. G.—
Chitral, 1895.—Relief of Chitral—Medal with clasp.
N.-W. Frontier of India, 1897-98.—Buner—Clasp.
South African War, 1902—Served with 9th Battalion, Imperial Yeomanry, from 6 May to 5 Aug. 1902.—Queen's medal with 2 clasps.
The War of 1914-18.—Despatches, Lond. Gaz., 15 Aug. 17. D. S. O.

Young, H. H. (*Lt.-Col., retd., I.A.*)—
Burma, 1887-8.—Medal with clasp.

Young, H. N.—
China, 1900.—Relief of Pekin. Medal with clasp.
The War of 1914-18.—Despatches, Lond. Gaz., 22 Jun. 15 and 1 Jan. 16.

Yung, H. P. (*Maj., retd., Ind. S. C.*)—
Afghanistan, 18,9.—Served with Transport Train, Scinde Reserve Division, Southern Afghanistan Field Force.
South African War, 1899-1902. Asst. Press Censor (graded as a D. A. A. G.) from 5 Apl. 00 to 6 Dec. 00. Afterwards attd. to A. S. Corps from 13 May 01. Queen's medal with 3 clasps. King's medal with 2 clasps.

Young, J. (*Lt.-Col., retd., I. M. S.*)—
Buma, 1885-86.—Occupation of Mandalay.—Medal with clasp.

Young, J. H. (*Lt.-Col., retd., I. A.*)—
Afghanistan, 1878-79-80.—With Khyber and Kurram Field Forces.—Medal.

Young, J. W. L. (*Dy. Comsy. and Hony. Capt., retd.*)—
Afghanistan, 1880.—March from Quetta to Relief of Kandahar.—Medal.
Burma, 1885—87.—Medal with 2 clasps.
N.-W. Frontier of India, Hazara, 1888.—Action of Kotkai.—Clasp.
Zhob Valley, 1890.
N.-E. Frontier of India, Lushai, 1890-91.—Clasp.
N.-W. Frontier of India, Hunza Nagar, 1891-92.—Promoted Sub-Conductor.
N.-W. Frontier of India, 1897-98.—Defence and Relief of Malakand.—Medal with 2 clasps.
The War of 1914-18.—Despatches, Lond. Gaz., 21 Jun. 16.

Young, K. deL.—
The War of 1914-18.—M. C.

Young, T. H. B. (*Col., retd., Ind. S. C.*)—
Operations in the Rumpa and Golgonda Hills.
Burma, 1885-87.—Medal with clasp.

Young, T. S. (*Major, retd., I. A.*)—
Burma, 1887-88.—Port Transport Officer, Htigaing.—Medal with clasp.

Young, W. B.—
Soudan, 1884-85.—Nile.—Medal with clasp.
Burma, 1886-87.—Operations of the 6th Brigade under Brig.-Genl. R. C. Low. Despatches. G. G. O. 864 of 1887.—Medal with clasp.

Young, W. E.—
N.-W. Frontier of India, 1897-98.—Malakand Utman Khel.—Buner.—Medal with clasp.
The War of 1914-18.—Despatches, Lond. Gaz., 15 Aug. 17.

Young, W. Hope (*Lt.-Col., retd., Ind. S. C.*)—
Afghanistan, 1878-80.—Affair of Matoon. Affair at and near Shuturgardan Pass.—Medal.
N.-W. Frontier of India, 1881.—Mashud Waziri Expedition.
Burma, 1888-89.
N.-W. Frontier of India, 1891.—Hazara Expedition. Despatches, G. G. O. 816 of 1891.—Clasp.

Youngerman, E. P. (*Lieut.-Col., retd., (I.M.S.)*—
Burma, 1885-86.—Medal with clasp.

Younghusband, Sir F. E. (*Lt.-Col., retd., I. A.*)—
Tibet, 1903-4.—As British Commr. Operations at and around Gyantse. Medal with clasp. K.C.I.E.

Younghusband, G. J.—
Afghanistan, 1878-79-80.—Action of Ali Musjid (2nd) Bazar Valley Expedition. Action of Futehabad.—Medal with clasp.
Soudan, 1885.—Action of Tamai. Affairs at Thakool and round Suakin.—Medal with clasp. Bronze star.
Burma, 1886-87.—Medal with clasp.
Chitral, 1895.—Relief of Chitral. Served as Brigade-Major of Divisional Troops. Action of the Malakand Pass. Action at Khar. Despatches, G. G. O. 998 of 95.—Medal with clasp.—Brev. of Maj.
South African War, 1899-1902.—Commanded 3rd Imperial Yeomanry, 1st Yeomanry Brigade and Mobile Column. Operations in the Orange Free State, Feb. to May 00. Operations in the Transvaal in May and June 00. Operations in the Transvaal, west of Pretoria, July to 29 Nov. 00, including action at Venterskroon (7 and 9 Aug.) Operations in Orange River Colony (May to 29 Nov. 00), including actions at Lindley (1 June), Rheroster River (11 June). Operations in Cape Colony, south of Orange River, 99-00. Commanded 26th Imperial Yeomanry (Younghusband's Horse) and Mobile Column. Operations in Cape Colony from 3rd December 1901 to 31st May 1902. (Severely wounded.) Despatches, *Lond. Gaz.*, 16 April 01. Brev. of Lt.-Col. C. B. Queen's medal with 3 clasps; King's medal with 2 clasps.
N.-W. Frontier of India, 1908.—Operations in the Mohmand country. Post Comdt., Galanai.—Medal with clasp.
The War of 1914-18.—Despatches, Lond. Gaz., 21 Jun. and 19 Oct. 16. K. C. M. G.

Younghusband, L. N.—
Tirah, 1897-98.—Operations in the Bara Valley, 7 to 14 Dec. 97.—Medal with 2 clasps.
South African War, 1902.—Operations in Cape Colony between 25th April 1902 and 31st May 1902. Commandant, 26th Imperial Yeomanry, from 27th April 1902.—Queen's medal with 2 clasps.
The War of 1914-18.—Despatches, Lond. Gaz., 19th Oct. and 14 Nov. 16 and 15 Aug. 17, Brev. of Col. C. B.

Zorawa, Singh—
The War of 1914-18.—Despatches, Lond. Gaz., 21 Jun. 16. M. C.

WAR SERVICES OF DEPARTMENTAL AND MEDICAL WARRANT OFFICERS.

Acott, H. T.—
N.-W. Frontier of India, 1897-98.—Malakand.—Medal with clasp.—Promoted Sub-Conductor.
British East Africa, 1901.—Operations against the Ogaden Somalis in Jubaland.—Medal with clasp.

Alderton, W. (*Condr., retd., I.O.D.*)—
Tirah, 1897-98.—Medal with 2 clasps.

Archer, J. E.—
Chitral, 1895.—Relief of Chitral.—Medal with clasp.
Nile Expedition, 1898.—Egyptian medal with 2 clasps. Queen's medal.
Tibet, 1903-04.—Medal.

Armstrong, W. A.—
South African War, 1879-80.—Zulu Campaign.
Chitral, 1895.—Relief of Chitral.—Medal with clasp.

Baker, C. H.—
Tirah, 1897-98.—Medal with 2 clasps.
East Africa, 1903-04.—Medal with clasp.

Ballin, F.—
Tibet, 1903-04.—(*i*) Action at Niani, 26th June 1904. (*ii*) Operations at and around Gyantse, between 5th May and 6th July 1904. (*iii*) March to Lhassa, 14th July to 3rd August 1904.—Medal with clasp.
Operations in the Abor country, 1911-12.—Despatches, Gaz. of India, Notn. 480, dated 23rd May 1912.—Medal with clasp.—Promoted Conductor.

Bamford, W.—
N.-W. Frontier of India, 1897-98.—Medal with clasp.
Tirah, 1897-98.—Actions of Chagru Kotal, Dargai and of the Sampagha and Arangha Passes. Operations at and around Dwatoi and action of 24th November 1897. Operations in the Bara Valley 7th to 14th December 1897.—Clasp.

Baptist, A. A. E.—
Tirah, 1897-98.—Medal with 2 clasps.

Bartram, A.—
Chitral, 1895.—Relief of Chitral.—Medal with clasp.
East Africa, 1903-04.—Operations in Somaliland. Action at Jidballi.—Medal with 2 clasps.

Benson, H. E.—
N.-W. Frontier of India, 1897-98.—Tochi.—Medal with clasp. Promoted G. G. O. 1161 of 1898.

Bickley, W. (*Condr., retd., I. O. D.*)—
N.-W. Frontier of India, 1897-98.—*Malakand.*—Medal with clasp.

Bidwell, J. (*Conductor, retd.*)—
Chitral, 1895.—Relief of Chitral.—Medal with clasp.

Bodfish, J. J. (*Conductor, retd.*)—
China, 1900.—Relief of Pekin.—Medal with clasp.

Bosworth, J. A.—
N.-W. Frontier of India, 1908.—Operations in the Zakka Khel country.—Medal with clasp.

Boyd, R. W.—
The War of 1914-17.—Gazette of India, Army Department. Notification No. 1-86, dated 2nd November 1917. Promoted Sub-Conductor.

Braganza, F. B. A.—
N.-W. Frontier of India, 1908.—Operations in the Mohmand country.—Medal with clasp.

Bray, T.—
N.-W. Frontier of India, 1897-98.—Operations on the Samana.—Medal with 2 clasps.
Tirah, 1897-98.—Clasp.

Bristow, R. L.—
Chitral, 1895.—Relief of Chitral. Action of the Malakand Pass.—Medal with clasp.
N.-W. Frontier of India, 1897-98.—Malakand. Operations in Bajaur and in the Mamund country Utman Khel, Buner. Action of the Tanga Pass.—Clasp.
N.-W. Frontier of India, 1901-02.—*Waziristan.*—Clasp.
East Africa, 1903-04.—Operations in Somaliland. Despatches, *London Gazette*, 2nd September 1904.—Promoted Conductor.—Medal with clasp.

Brooks, T. H.—
N.-W. Frontier of India, 1897-98.—Medal with clasp.
Tirah, 1897-98.—Clasp.

Brown, W. H.—
N.-W. Frontier of India, 1897-98.—Malakand. Buner.—Medal with clasp.

Bryce, J.—
South African War, 1899—1901.—Advance on Kimberley, including actions at Belmont, Enslin, Modder River, and Magersfontein. Operations in the Orange Free State, February to May 1900, including operations at Paardeburg, action at Poplar Grove, Karee Siding, Vet River and Zand River. Operations in the Transvaal in May and June 1900, including actions near Johannesburg and Pretoria. Operations in the Transvaal 30th November 1900 to July 1901 and October to November 1901. Operations in Orange River Colony July to October 1901. Queen's medal with 5 clasps.

Byrne, H. R.—
East Africa, 1903-04.—Operations in Somaliland.—Medal with clasp.

Callaghan, G. M.—
China, 1900.—Medal.—Promoted 3rd class Asst. Surgeon.

Calvert, A.—
China, 1900.—Medal.

Cearns, G. W.—
N.-W. Frontier of India, 1897-98.—Medal with clasp.
Tirah, 1897-98.—Clasp.

Charters, W. C. M.—
Chitral, 1895.—Relief of Chitral.—Medal with clasp.
Tirah, 1897-98.—Action of the Sampagha Pass.—Despatches, G. G. O. 244 of 1898.—Clasp. Promoted 1st Class Assistant Surgeon.

Chisholm, A. W.—
N.-W. Frontier of India, Hazara, 1888.—Medal with clasp.

Chisholm, C. (*Sub-Condr., retd.*)—
Burma, 1885-87.—Medal with clasp.
Tirah, 1897-98.—Medal with 2 clasps.

Coombs, M. H.—
South Africa, 1899-1902.—Advance on Kimberley, including actions at Belmont (23rd November 1899), Enslin (25th November 1899), Modder River (28th November 1899) and Magersfontein (11th December 1899). Relief of Kimberley. Operations in the Orange Free State, July to May 1900, including operations at Paardeburg (17th to 26th February), actions at Poplar Grove (7th March 1900) and Karee Siding (29th March 1900).

Operations in the Transvaal in May and June 1900, including actions near Johannesburg (29th May 1900), Pretoria (4th June 1900) and Diamond Hill (11th and 12th June 1900). Operations in the Transvaal, West of Pretoria, July to November 1900. Operations in Orange River Colony, May to 29th November 1900, including actions at Bethlehem (6th and 7th July 1900), Wittebergen and Caledon River. Operations in Orange River Colony, 30th November 1900 to May 1901. Operations in Cape Colony, June 1901 to March 1902. Queens Medal with 8 clasps, King's medal with 2 clasps.

Coates, W.—
Chitral, 1895.—Relief of Chitral.—Medal with clasp.
South African War, 1899-1900.—Defence of Ladysmith.—Medal with clasp.

Coshan, F. W. A.—
Tirah, 1897-98.—Medal with 2 clasps.

Courtney, J.—
Burma, 1889-92.—Served with Maingkhwan and Irrawaddy columns.—Medal with clasp.

Cox, H. W. V.—
N.-W. Frontier of India, 1897-98.—Malakand.—Medal with clasp.

Craggs, H. C.—
South African War, 1899-1902.—Operations in Natal, 1899, including actions at Rietfontein (24th October) and Lombard's Kop (30th October). Defence of Ladysmith, including action of 6th January 1900.—Queen's medal with 2 clasps. King's medal with 2 clasps.

Creagh, C. J.—
Burma, 1889.—Medal with clasp.
Tibet, 1903-04.—Medal.

Cresswell, J.—
N.-W. Frontier of India, 1897-98.—Medal with clasp.

Crowhurst, D. (*Conductor, retd.*)—
N.-W. Frontier of India, Hazara, 1888.—Medal with clasp.
N.-W. Frontier of India, Hazara, 1891.—Clasp.
Chitral, 1895.—Relief of Chitral.—Medal with clasp.
N.-W. Frontier of India, 1897-98.—Malakand. Operations on the Samana.—3 clasps.
Tibet, 1903-04.—Medal with clasp.
N.-W. Frontier of India, 1908.—Operations in the Mohmand country.—Medal with clasp.

Crumley, T. J.—
Tirah, 1897-98.—Medal with 2 clasps.

Culpeper, E. J.—
China, 1900.—Medal.

Cutler, F. G.—
Tirah, 1897-98.—Medal with 2 clasps.

Daly, S. (*Conductor, retd.*)—
Afghanistan, 1878—80.—Served with the force under Major-General *Sir* F. S. Roberts, V.C.—Medal with 2 clasps.
Soudan, 1885.—Served with the Force under Major-General *Sir* Gerald Graham.—Medal with clasp; Bronze star.
Burma, 1885—87.—Medal with clasp.
N.-E. Frontier of India, Lushai, 1891-92-93.
Chitral, 1895.—Relief of Chitral.—Medal with clasp.

Davidson, D. F. C.—
North-East Frontier of India, Chin Hills, 1892-93.—Medal with clasp.
East Africa, 1903-04.—Operations in Somaliland.—Medal with clasp.

Davis, R. A. (*Conductor, retd.*)—
China, 1900.—Medal.

Davison, W. (*Conductor, retd.*)—
Burma, 1885—87.—Medal with clasp.

Deeb'e, G.—
Afghanistan, 1878-80.—With Kurram Valley Field Force.—Medal.
N.-W. Frontier of India, Hazara, 1891.—Medal with clasp.
Chitral, 1895.—Relief of Chitral.—Medal with clasp.
N.-W. Frontier of India, 1897-98.—Operations on the Samana.—Clasp.
Tirah, 1897-98.—Clasp.

DeGruyther, C. G.—
N.-W. Frontier of India, 1908.—Operations on the Mohmand country. Engagement of Kargha, 24th May 1908.—Medal with clasp.

DePenning, H. F.—
N.-W. Frontier of India, 1908.—Operations in the Zakka Khel country. Operations in the Mohmand country.—Medal with clasp.

DeSouza, P. N.—
N.-W. Frontier of India, (1st) Miranzai, 1891—Promoted 3rd Class Assistant Surgeon.

Dewey, H. V.—
Chitral, 1895.—Relief of Chitral.—Medal with clasp.
N.-W. Frontier of India, 1897-98.—Clasp.
Tirah, 1897-98.—Clasp.

Dodd, W.—
China, 1900.—Medal.
Tibet, 1903-04.—Medal.

Douglas, A. E.—
Tirah, 1897-98.—Actions of Dargai and Sampagha and Arhanga Passes. Operations in the Waran and Bara Valley.—Medal with 2 clasps.—Promoted Sub-Conductor.

D'Souza, L.—
N.-W. Frontier of India, 1897-98.—Operations on the Samana.—Medal with 2 clasps.
Tirah, 1897-98.—Actions of Dargai and of the Sampagha and Arhanga Passes. Reconnaissance of the Saran Sar and action of 9th November 1897. Operations at and around Dwata and action of 24th November 1897. Operations in the Bara Valley, 7th to 14th December 1897.—Clasp.

Duckworth, A. R.—
N.-W. Frontier of India, 1897-98.—Mohmand. Medal with clasp.
Tibet, 1903-04.—Medal.

Duckworth, E. V.—
Chitral, 1895.—Relief of Chitral.—Medal with clasp.
Aden, 1903-04.—Operations in the interior.
The War of 1914-16.—Despatches, *London Gazette*, 4th July 1916.

Duckworth, P. T.—
South African War, 1899—1902.—Operations in Natal, 1899, including action of Talana (20th October). Defence of Ladysmith, including sorties of 7th and 10th December 1899 and action of 6th January 1900. Operations in Natal, March to June 1900, including action at Laing's Nek (6th to 9th June). Operations in the Transvaal, east of Pretoria, July to 29th November 1900, including actions at Belfast (26th and 27th August 1900) and Lydenberg (5th to 8th September). Operations in the Transvaal, between 30th November 1900 and 31st May 1902.—Queen's medal with 5 clasps. King's medal with 2 clasps.

Dunn, F. J.—
East Africa, 1903-04.—Operations in Somaliland.—Medal with clasp.

Dyer, W. L. E.—
South African War, 1900.—Queen's medal with clasps.

Edgerley, W.—
N.-W. Frontier of India, Waziristan, 1894-95.—Medal with clasp.
Tirah, 1897-98.—Medal with clasp.
East Africa, 1903-04.—Operations in Somaliland.—Medal with clasp.

Ellis, S. N.—
Tirah, 1897-98.—Medal with 2 clasps.

Elloy, A. L.—
N.-W. Frontier of India, 1908.—Operations in the Mohmand country.

Ewing, J.—
Tirah, 1897-98.—Medal with two clasps.
Tibet Mission, 1903-04.—Medal.

Fannon, J. T.—
South African War, 1899—1902.—Relief of Kimberley. Operations in Orange Free State, February to May 00, including operations at Paardeberg and action at Driefontein; actions at Vet River (5 and 6 May) and Zand River (10 May). Operations in the Transvaal, May and June 00, including actions near Johannesburg (29 May), Pretoria (4 June) and Diamond Hill (11 and 12 June). Operations in the Transvaal, east of Pretoria, including action at Belfast. Operations in Cape Colony, including actions at Colesberg. Operations in the Transvaal (30 Nov. 00 to 31 May 02.)—Queen's medal with 6 clasps. King's medal with 2 clasps.

Ferris, H. (*Conductor, retd.*)—
Burma, 1885—87—89.—Medal with 2 clasps.

FitzGerald, G.—
South African War, 1899-1902.—Relief of Kimberley. Operations in the Orange Free State, February to May 1900, including operations at Paardeberg (17th to 26th February), actions at Poplar Grove (7th March), Driefontein (10th March), Karee Siding (29th March), Vet River (5th and 6th May), and Zand River (10th May). Operations in the Transvaal, May and June 1900, including actions near Johannesburg (29th May), Pretoria (4th June), and Diamond Hill (11th and 12th June). Operations in the Transvaal, east of Pretoria, July to 29th November 1900, including actions at Reit Vlei (16th July), Belfast (26th and 27th August, 1900). Operations in Cape Colony south of Orange River, 1899-1900, including actions at Colesburg (1st January to 12th February 1900.) Operations in the Transvaal, west of Pretoria, October to November 1900. Operations in the Transvaal, 30th November 1900 to March 1902 and May 1902. Operations in the Orange River Colony, March to May 1902.—Queen's medal with 6 clasps. King's medal with 2 clasps.
N.-W. Frontier of India, 1908.—Operations in the Zakka Khel country.—Medal with clasp.

Fitzgerald, L.—
South African War, 1899-1902.—Advance on Kimberley, including actions at Belmont (23rd November 1899), Enslin (25th November 1899), Modder River (28th November 1899), and Magersfontein (11th December 1899). Relief of Kimberley. Operations in the Orange Free State, February to May 1900, including operations at Paardeberg (17th to 26th February); actions at Poplar Grove (7th March), and Karee Siding (29th March); Houtnek (1st May), Zand River (10th May). Operations in the Transvaal in May and June 1900, including actions near Johannesburg (29th May), Pretoria (4th June), and Diamond Hill (11th and 12th June). Operations in the Transvaal, west of Pretoria, July to 29th November 1900. Operations in Orange River Colony (May to 29th November 1900), including actions at Bethlehem (6th and 7th July), Wittebergen (1st to 29th July), Witpoort (15th August), Ladybrand (2nd to 5th September), Caledon River (27th November to 29th November 1900). Operations in Orange River Colony between 30th November 1900 and 31st May 1902. Operations in Cape Colony between 30th November 1900 and 31st May 1902. Queen's medal with 7 clasps. King's medal with 2 clasps.

Fleming, J. F.—
Chitral, 1895.—Relief of Chitral.—Action of the Malakand Pass.—Medal with clasp.

Fox, H. A.—
N.-W. Frontier of India, 1908.—Operations in the Mohmand country.—Medal with clasp.

Foy, F. H.—
N.-W. Frontier of India, 1908.—Operations in the Mohmand country.—Medal with clasp.

Foy, J. R.—
N.-W. Frontier of India, 1897-98.—Operations on the Samana and in the Kurram Valley during August and September 1897.
Tirah, 1897-98.—Action of the Chagru Kotal and of the Sampagha and Arhanga Passes. Reconnaissance of the Saram Sar and action of the 11th November 1897. Operations at and around Dwatoi and action of 24th November 1897. Operations against the Khani Khel Chamkannis. Operations in the Bazar Valley, 25th to 30th December 1897.—Medal with 3 clasps.

Franklin, C.—
Tibet, 1903-04.—Medal.

Gardner, G.—
Tirah, 1897-98.—Medal with 2 clasps.
China, 1900.—Medal.

Gibbon, D. A.—
East Africa, 1902-04.—Operations in Somaliland.—Medal with 4 clasps.

Gillson, E. H.—
The War of 1914-18.—Despatches, *Lond. Gaz.*, 4 July 16.

Gillson, W. J.—
N.-W. Frontier of India, 1897-98.—Malakand. Medal with clasp.

Ginns, L. B.—
South African War, 1901-02.—Operations in the Transvaal, February 1901 to 31st May 1902. Operations in Orange River Colony and Cape Colony, February 1901. Queen's Medal with 5 clasps.

Gleeson, F. H.—
N.-W. Frontier of India, 1897-98.—Malakand. Operations in the Mamund country.—Medal with clasp.
Tirah, 1897-98.—Clasp.

Goodair, A. G.—
N.-W. Frontier of India, Waziristan, 1894-95.—Medal with clasp.
China, 1900.—Relief of Pekin.—Medal with clasp.

Goodburn, F.—
N.-W. Frontier of India, 1897-98.—Malakand. Action at Landakai. Operations in Bajaur. Operations in the Mamund country.—Buner. Action of the Tanga Pass.—Medal with clasp.

Grant, W.—
N.-W. Frontier of India, 1897-98.—Mohmand.—Medal with clasp. Tirah, 1897-98.—Clasp.

Grayson, G. R.—
Tirah, 1897-98.—Medal with clasp.

Grensmith, J.—
South African War, *1900-02.*—Operations in Cape Colony, June 00 to July 01. Operations in Orange Free State, Augt. to Oct. 01. Operations in the Transvaal, Nov. 01 to 31 Mar. 02.—Queen's medal with 3 clasps. King's medal with 2 clasps.

Gregory, T. M.—
N.-W. Frontier of India, Hazara, 1891.—Medal with clasp.

Greson, E. J.—
N.-W. Frontier of India, Hazara, 1891.—Medal with clasp.
N.-W. Frontier of India (2nd Miranzai), 1891.—Clasp.
China, 1900.—Medal.

Griffiths, E. A. C.—
N.-W. Frontier of India, Hazara, 1891.—Medal with clasp.
Chitral, 1895.—Relief of Chitral.—Medal with clasp.

Grimley, H. B.—
Tibet, 1903-04.—Medal.

Grout, T.—
Burmah, 1889-92.—Medal with clasp.
N.-W. Frontier of India and Tirah, 1897-98.—Medal with two clasps.

Hall, E. S.—
Soudan, 1896.—Dongola Expedition.—Khedive's medal.
China, 1900.—Relief of Pekin.—Medal with clasp.

Harding, B. T.—
South African War, 1899-00.—Despatches, *London Gazette*, 8th February 1901. Queen's medal with 5 clasps. Promoted Sub-Conductor.
Aden, 1903-04.—Operations in the interior.

Harding, G. H. (*Conductor, retd.*)—
Burma, 1891-1892.—Bungshe column.—Medal with clasp.

Harrison, E. J.—
N.-E. Frontier of India, Chin Hills, 1892-93.—Medal with clasp.

Hart, R. H. W.—
N.-W. Frontier of India, 1897-98.—Operations on the Samana during August and September 1897. Medal with 2 clasps.
Tirah, 1897-98.—Actions of the Sampagha and Arhanga Passes. Reconnaissance of the Saran Sar and action of 9th November 1897. Operations at and around Dwatoi and action of 24th November 1897. Operations against the Khani Khel Chamkannis. Operations in the Bazar Valley 25th to 30th December 1897.—Clasp.
South African War, 1899-1902.—Operations in Natal, 1899, including actions at Rietfontein (24th October) and Lombard's Kop (30th October). Defence of Ladysmith, including sortie of 10th December 1899 and action of 6th January 1900. Operations in Natal from March to June 1900. Operations in the Transvaal from July 1900 to March 1901. Operations in the Orange River Colony from April 1901 to February 1902. Operations in the Transvaal from March 1902 to 31st May 1902.—Queen's medal with 3 clasps. King's medal with 2 clasps.

Harvey, J. A. F.—
Chitral, 1895.—Relief of Chitral.—Medal with clasp.
N.-W. Frontier of India, 1897-98.—Malakand. Operations in Bajaur and in the Mamund country. Clasp. Promoted 1st class Asst. Surgeon.

Harvey, N. S.—
Tirah, 1897-98.—Medal with 2 clasps.

Hayes, W.—
East Africa, 1903-04.—Operations in Somaliland. Action at Jidbali.—Medal with clasp.

Hefferman, W. St. M.—
Chitral, 1895.—Relief of Chitral.—Medal with clasp.

Hendricks, St. J. E.—
South African War, 1899-1902.—Operations in Natal, 1899, including actions at Elandslaagte (21st October), Rietfontein (24th October), and Lombard's Kop (30th October). Defence of Ladysmith, including sortie of 10th December 1899, and action of 6th January 1900. Operations in Natal, March to June 1900, including action at Laing's Nek (6th to 9th June). Operations in the Transvaal, east of Pretoria, July to 29th November 1900, including actions at Reit Vlei (16th July), Belfast (26th and 27th August 1900), Lydenberg (5th to 8th September), and Rhenoster Kop (29th November). Operations in the Transvaal, between 30th November 1900 and 31st May 1902.—Queen's medal with 4 clasps. King's medal with 2 clasps.

Hill, A. J.—
N.-W. Frontier of India, 1897-98.—Tochi.—Medal with clasp.
N.-W. Frontier of India, 1908.—Operations in the Zakka Khel country.—Medal with clasp. Promoted Conductor.

Holdaway, F.—
Tibet, 1903-04.—Action at Niani, 26th June 1904. Operations at and around Gyantse, between 5th May and 6th July 1904. March to Lhassa, 14th July to 3rd August 1904.—Medal with clasp.
N.-W. Frontier of India, 1908.—Operations in the Mohmand country.—Medal with clasp.

Holmes, J. A. H.—
East Africa, 1902-04.—Operations in Somaliland. Action at Jidbali.—Medal with 2 clasps.

Holmes, R.—
N.-W. Frontier of India, 1908.—Operations in the Zakka Khel country.

Humphreys, E.—
South African War, 1899-1900.—Defence of Ladysmith.—Medal with clasp.

Hurst, W.—
Tirah, 1897-98.—Medal with 2 clasps.

Hussey, W. G. St. J.—
N.-W. Frontier of India, 1897-98.—Malakand.
Tirah, 1897-98.—Despatches, G. G. O. No. 244 of 98.—Medal with clasp.

Ingledew, G. H. (*Condr., I. O. D., retd.*)—
N.-W. Frontier of India, Hazara, 1891.—Medal with clasp.
N.-W. Frontier of India, 1897-98.—Malakand.—Medal with 2 clasps.

Jessiman, R. (*Condr., retd.*)—
China, 1900.—Medal.

Johnson, E. O.—
N.-W. Frontier of India, 1897-98.—Medal with clasp.

Johnstone, J. G.—
N.-W. Frontier of India, 1908.—Operations in the Mohmand country.

Jolley, H. J. L.—
China, 1900.—Relief of Pekin.—Medal with clasp.

Jones, G. T.—
South African War, 1899—1902.—Operations in Natal, 1899, including action at Lombard's Kop (30th October). Defence of Ladysmith, including sorties of 7th and 10th December 1899 and action of 6th January 1900. Operations in Natal, March to June 1900, including action at Laing's Nek (6th to 9th June). Operations in the Transvaal from December 1900 to January 1902.—Queen's medal with 4 clasps.—King's medal with 2 clasps.

Kalkhoven, F. (*Condr., retd., S. and T. Corps.*)—
Burma, 1885-87.—Medal with clasp. Promoted Conductor.
Tirah, 1897-98.—Medal with clasp.

Kaye, F. W.—
Tirah, 1897-98.—Medal with clasp.

Kelly, C. C.—
South African War, 1899-1902.—Operations in Natal, 1899, including actions at Rietfontein (24th October) and Lombard's Kop (30th October). Defence of Ladysmith, including actions of 6th January 1900. Operations in the Transvaal, 10th June 1900 to 4th March 1902.—Queen's medal with 3 clasps. King's medal with 2 clasps.

Kelly, J. (*Conductor, retd.*)—
South Africa, 1879, Zulu Campaign.—Medal with clasp.
Chitral, 1895.—Relief of Chitral.—Medal with clasp.
Tirah, 1897-98.—2 clasps.

Kemp, J.—
Egypt, 1882.—Medal, Bronze Star.

Kendall, W. J.—
The War of 1914–18.—Gazette of India, Army Department Notification 1334, dated 3 Nov. 16, promd. Deputy Commissary.

Kenyon, L. P.—
N.-W. Frontier of India, 1897-98.—Malakand.

Kidby, W. T.—
N.-W. Frontier of India, 1897-98.—Operations in the Mamund country.—Medal with clasp.
Tirah, 1897-98.—Medal with 2 clasps.

Knight, W.—
South African War, 1899-1900.—Defence of Ladysmith.—Medal with clasp.
East Africa, 1903-04.—Operations in Somaliland.—Medal with clasp.

Langhorne, S. G.—
N.-W. Frontier of India, 1897-98.—Operations in the Mamund country. Medal with clasp.
Tirah, 1897-98.—Clasp.

Lawrence, J. L.—
N.-W. Frontier of India, 1897-98.—Medal with clasp.
Tirah, 1897-98.—Actions of Chagru Kotal, Dargai and of the Sampagha and Arhanga Passes. Reconnaissance of the Saran Sar and action of 9th November 1897. Operations in the Warran Valley and action of 16th November '97. Operations at and around Dwatoi and action of 24th November 1897. Operations in the Bara Valley 7th to 14th December 1897.—Clasp.

Lilburn, R. (*Sub.-Condr., retd., I. M. L.*)
N.-W. Frontier of India, 1897-98.—Malakand. Operations in Bajaur and in the Mamund country. Buner. Action of the Tanga Pass.—Medal with clasp.

Lorez, A. P.—
Tirah, 1897-98.—Medal with 2 clasps.
Tibet, 1903-04.—Action at Niani, 26th June 1904. Operations at and around Gyantse, between 5th May and 6th July 1904. March to Lhassa, 14th July to 3rd August 1904.—Medal with clasp.

Lynch, T. B.—
China, 1900—Medal.

McCurtis, A.—
N.-W. Frontier of India, 1908.—Operations in the Mohmand country.

McCurtis, L.—
South African War, 1899–1902.—Advance on Kimberley, including actions at Belmont (23rd November 1899), Enslin (25th November 1899), Modder River (28th November 1899) and Magersfontein (11th December 1899). Relief of Kimberley. Operations in the Orange Free State, February to May 1900, including operations at Paardeberg (17th to 26th February); actions at Poplar Grove (7th March), Karee Siding (29th March) and Zand River (10th May). Operations in the Transvaal in May and June 1900, including actions near Johannesburg (29th May), Pretoria (4th June) and Diamond Hill (11th and 12th June). Operations in the Orange River Colony, May to 29th November 1900, including actions at Bethlehem (6th and 7th July), Wittebergen (1st to 29th) July), Ladybrand (2nd to 5th September) Caledon River (27th to 29th November 1900., Operations in the Orange River Colony between 30th November 1900 and 5th March 1902. Operations in Cape Colony between 30th November 1900 and 5th March 1902.—Queen's medal with 7 clasps. King's medal with 2 clasps.

McDonald, J. J.—
N.-W. Frontier of India, 1897-98.—Medal with clasp.
Tirah, 1897-98.—Clasp.

McGinn, J. J.—
N.-W. Frontier of India, 1897-98.—Operations in Kurram Valley.
Tirah, 1897-98.—Actions at Dargai, Sampagha and Arhanga Passes. Operations at and around Dwatoi, and action of 24th November 1897. Operations in the Bara Valley, 7th to 14th December 1897.—Medal with 2 clasps.
Tibet, 1903-04.—Medal.

McHale, J. E. (*Conductor, retd.*)—
N.-W. Frontier of India (2nd) Miranzai, 1891.—Medal with clasp.
China, 1900.—Medal.

McIntyre, P.—
N.-W. Frontier of India, 1897-98.—Malakand. Operations in Bajaur and in the Mamand country. Buner. Action of the Tanga Pass.—Medal with clasp.

McKay, G. W. (*Condr., retd. Ord. Dept.*)—
N.-W. Frontier of India, Hazara, 1888.—Medal with clasp.
N.-W. Frontier of India, Hazara, 1891.—Clasp.

McNamara, J. D.—
Tirah, 1897-98.—Operations in the Waran Valley and action of 16th November 1897. Operations in the Bara Valley 7th to 14th December 1897. Medal with 2 clasps.

Mackay, D. L.—
N.-W. Frontier of India, 1908.—Operations in the Mohmand country.

Malbene, W. E.—
China, 1900.—Medal.

Maine, C. W.—
Tirah, 1897-98.—Action of the Sampagha Pass. Medal with 2 clasps.

Mann, H. V.—
Chitral, 1895.—Relief of Chitral. Medal with clasp.
N.-W. Frontier of India, 1897-98.—Malakand Operations in Bajaur and in the Mamund country. Utman Khel.—Clasp.
Tirah, 1897-98.—Action of the Sampagha and Arhanga Passes. Reconnaissance of the Saran Sar and actions of the 9th and 11th Novr. 1897. Operations against the Khani Khel, Chamkannis. Operations in the Bazar Valley, 25th to 30th Decr. 1897.—Clasp.

Mansfield, H.—
Chitral, 1895.—Relief of Chitral. Medal with clasp

Marks, F. C.—
N.-W. Frontier of India, 1897-98.—Tochi.—Medal with clasp.
The War of 1914-18.—Despatches, Lond. Gaz., 22 Jun 15. Promd. Conductor.

Marner, T.—
N.-W. Frontier of India, 1897-98.—Medal with clasp.
Tirah, 1897-98.—Clasp.

Martin, W. J. P.—
N.-W. Frontier of India, 1897-98.—Malakand. Medal.
Tirah, 1897-98.—Clasp.

Martingale, J. J. W. (*Condr., retd., Ord. Dept.*)—
South Africa, 1899—1900.—Defence of Ladysmith—Queen's Medal with clasp.

Mason, P. J.—
N.-W. Frontier of India, 1908.—Operations in the Zakka Khel country.—Medal with clasp.
Operations in the Abor country, 1911-12.—Clasp.

Masterson, W. J.—
N.-W. Frontier of India, 1897-98.—Medal with clasp.
Tirah, 1897-98.—Clasp.

Matthews, E. G. A.—
N.-W. Frontier of India, 1897-98.—Malakand. Operations in the Mamund country.—Medal with clasp.
Tirah, 1897-98.—Clasp.

Matthews, J. G.—
Tirah, 1897-98.—Medal with 2 clasps.

Meneaud, C. J. A.—
Chitral, 1895.—Relief of Chitral.—Medal with clasp.

Mendis, F. X.—
N.-W. Frontier of India, 1897-98.—Malakhand.—Medal with clasp.
Tirah, 1897-98.—Clasp.

Mersh, W. G.—
Chitral, 1895.—Relief of Chitral.—Medal with clasp.

Miller, F. E.—
Burma, 1885—87.—Medal with clasp.
N.-W. Frontier of India, 1897-98.—Malakand.—Medal with clasp.
China, 1900.—Medal with clasp.

Mills, G.—
Burma, 1885—87.

Mills, H. G. C.—
Chitral, 1895.—Relief of Chitral.—Medal with clasp.

Mills, P. B.—
N.-W. Frontier of India, 1897-98.—Malakand. Buner. Action of the Tanga Pass.—Medal with clasp.

Milwright, C.—
N.-W. Frontier of India, 1908.—Operations in the Zakka Khel country. Operations in the Mohmand country. Medal with clasp.

Montgomery, J. P.—
Tirah, 1897-98.—Actions of the Sampagha and Arhanga Passes. Operations against the Khani Khel Chamkannis. Operations in the Bazar Valley, 25th to 30th December 1897.—Medal with two clasps.
South African War, 1899-1902.—Operations in Natal 1899, including actions at Rietfontein (24th October) and Lombard's Kop (30th October). Defence of Ladysmith, including action of 6th January 1900. Operations in the Transvaal, 11th December 1900 to 9th July 1901. Operations in the Orange River Colony from 10th July 1901 to 31st May 1902.—Queen's medal with 3 clasps. King's medal with 2 clasps.

Morris, E.—
Tibet, 1903-04.—Action at Niani, 26th June 1904. Operations at and around Gyantse between 5th May and 6th July 1904.
March to Lhassa, 14th July to 3rd August 1904.—Medal with clasp.
N.-W. Frontier of India, Waziristan, 1901-02.—Medal with clasp.
The War of 1914—18.—Despatches, *Lond Gaz.*, 4 July 16.

Morris, J. A.—
Burma, 1889-92.—Medal with clasp.
Chin Hills, 1892-93.—clasp.

Mullins, C.—
N.-W. Frontier of India, 1897-98.—Medal with clasp.
Tirah, 1897-98.—Clasp.
South African War, 1899.—Queen's medal with clasp.

Murphy, W. R.—
South African War, 1899-1900.—Operations in Natal, 99. Defence of Ladysmith. Queen's Medal with clasp.

Nagle, E. H. B.—
Aden, 1903-04.—Operations in the interior.

Nash, A. E.—
China, 1900.—Medal.
The War of 1914—18.—Despatches, *Lond Gaz*, 4 July 16.

Netscher, W. A. C.—
N.-W. Frontier of India, 1897-98.—Malakand.—Medal with clasp.

Newington, F. E.—
Chitral, 1895.—Relief of Chitral.—Medal with clasp.
N.-W. Frontier of India, 1897-98.—Clasp.
Tirah, 1897-98.—Clasp.

Nicholl, W. (*Condr., retd., I. O. D.*)—
Soudan, 1896.—Dongola Expedition.—Medal; Khedive's Medal.

Noble, J. F. (*Conductor, retd.*)—
Tirah, 1897-98.—Medal with 2 clasps.

Noronha, D.—
Aden, 1903-04.—Operations in the interior.

O'Connor, C. E. B.—
The War of 1914—18.—Despatches, *Lond Gaz.*, 4 July 16.

Oliver, F.—
Burma, 1885—87.—Medal with clasp.
N.-W. Frontier of India, Hazara, 1891.—Clasp.

Paes, A.—
Tirah, 1897-98.—Action of Chagru Kotal. Action of Dargai. Actions of the Sampagha and Arhanga Passes. Operations in the Waran Valley, and action of 16th November 1897. Operations in the Bara Valley, 7th to 14th December 1897.—Medal with 2 clasps.

Page, A. F. A.—
East Africa, 1902—04.—Operations in Somaliland.—Medal with clasp.

Payne, C. J.—
Chitral, 1895.—Relief of Chitral.—Medal with clasp.

Pell, E. J.—
N.-W. Frontier of India, 1908.—Operations in the Mohmand country.—Medal with clasp.

Perdrian, A. D. C.—
N.-W. Frontier of India, 1897-98.—Operations in Bajaur and in the Mamund country.—Medal with clasp.
Tirah, 1897-98.—Clasp. Promoted 2nd class Asst. Surgeon.

Pereira, D. F.—
Mekran, 1898.

Pereira, R. B.—
East Africa, 1903-04.—Operations in Somaliland.—Medal with clasp.

Perry, J. (*Conductor, retd.*).—
Chitral, 1895.—Relief of Chitral.—Medal with clasp.

Pleachy, E. A.—
Chitral, 1895.—Relief of Chitral.—Medal with clasp.
China, 1900.—Medal.

Pinto, M. C.—
South African War, 1899—1902.—Operation in Natal, 1899, including actions at Rietfontein, (24th October) and Lombard's Kop (30th October). Defence of Ladysmith, including action of 6th January 1900. Operations in Natal, March to June 1900, including action at Laing's Nek (6th to 9th June). Operations in the Transvaal, July 1900. Operations in the Orange River Colony, December 1901.—Queen's medal with 4 clasps. King's medal with 2 clasps.

Pitchford, S. (*Conductor, retd.*)—
Afghanistan, 1878-79-80.—Defence of Kandahar Sortie from Kandahar and Battle of Kandahar.—Medal with clasp. Medal for distinguished conduct in the field.
Burma, 1885—87.—Medal with clasp.
N.-E. Frontier of India, Chin-Lushai, 1889-90.—Clasp.

Pollock, C. R.—
Tirah, 1897-98.—Action of Dargai. Actions of the Sampagha and Arhanga Passes. Reconnaissance of the Saran Sar and action of 9th Nov. 1897. Operations in the Waran Valley. Operations in the Bara Valley, 7th to 14th Decr. 1897.—Clasp.
N.-W. Frontier of India, 1897-98.—Medal with clasp.

Prior, C. J.—
N.-W. Frontier of India, 1897-98.—Tochi.—Medal with clasp.

Proctor, W. D.—
N.-W. Frontier of India, 1897-98.—Operations in the Samana during August and September 1897.—Medal with 2 clasps.
Tirah, 1897-98.—Clasp.

Quick, A. N.—
The War of 1914—18.—Despatches, *Lond Gaz*, 4 July 16.

Raymond, C.—
Aden, 1903-04.—Operations in the interior.

Riyazuddin—
N.-W. Frontier of India 1897-98.—Malakand, Utman Khel, Buner. Medal with clasp.
Central Africa 1899.—Operations in Uganda,—Medal with clasp.
East Africa 1902-04.—Operations in Somaliland. Clasp.
Operations in the Abor Country 1911-12.—Medal with clasp.

Rebeiro, W. R.—
N.-W. Frontier of India, 1897-98.—Medal with clasp.
Tirah, 1897-98.—Action of the Sampagha Pass. Operations in the Bazar Valley.—Clasp.

Reeks, E. J. (*Condr., retd.*)—
N.-W. Frontier of India, 1892, Isazai Expedition.
Central, 1895.—Relief of Chitral. Action of the Malakand Pass. Action at Khar.—Medal with clasp.
South African War, 1899-1900.—Defence of Ladysmith.—Medal with clasp.

Reilly, J. H. J.—
N.-W. Frontier of India, 1908.—Operations in the Zakka Khel country. Operations in the Mohmand country. Engagement at Matta 24th April 1908.
The War of 1914-16.—Despatches, *London Gazette*, 4th July 1916. Promoted Sub-Conductor.

Richardson, H. A.—

N.-W. Frontier of India, 1897-98.—Medal with clasp.
Tirah, 1897-98.—Action of Dargai. Actions of the Sampagha and Arhanga Passes. Operations at and around Dwatoi and in the Bara Valley.—Clasp.

Roberts, H. W. (*Sub-Condr., retd., B.*)—

Tirah, 1897-98.—Medal with 2 clasps.—Promoted Sub-Condr.

Robertson, A.—

Chitral, 1895.—Relief of Chitral.—Medal with clasp.

Robinson, J. R.—

Nile Expedition, 1898.—Egyptian medal with 2 clasps, Queen's medal.

Rodgers, R. T.—

Chitral, 1895—Relief of Chitral. Action of the Malakand Pass.—Medal with clasp.

Rogers, J. (*Condr., retd., I. O. D.*)—

Tirah, 1897-98.—Medal with 2 clasps.

Rose, J.—

The War of 1914—18.—Despatches, *Lond Gaz*, 4 July 16.

Ruddock, W. B.—

Burma, 1893-94.
N.-W. Frontier of India, 1897-98.—Tochi.—Medal with clasp.
Aden, 1903-04.—Operations in the interior.

Sanderson, E. F.—

South African War, 1899—1901.—Advance on Kimberley, including actions at Belmont (23rd November 1899), Enslin (25th November 1899), Modder River (28th November 1899), and Magersfontein (11th December 1899). Operations in the Orange Free State, February to May 1900, including operations at Paardeberg (17th to 26th February). Actions at Poplar Grove (7th March), Karee Siding (29th March), Vet River (5th and 6th May), and Zand River (10th May). Operations in the Transvaal in May and June 1900, including actions near Johannesburg (29th May), and Pretoria (4th June). Operations in the Transvaal, between 30th November 1900 and 2nd December 1901.— Queen's medal with 5 clasps.

Sayers, C.—

N.-W. Frontier of India, Waziristan, 1894-95.—Medal with clasp.
Expedition to Dongola, 1896.—Medal.
Egyptian Medal.
China, 1900.—Medal.

Searle, A. T.—

N.-W. Frontier of India, 1897-98.—Medal with clasp.

Secombe, A.—

Tirah, 1897-98.—Medal with 2 clasps.
N.-W. Frontier of India, 1901-02, Waziristan.—Clasp.
Tibet, 1903-04—Medal.
Operations in the Abor country, 1911-12.—Medal with clasp.

Shave, E. D.—

China, 1900.—Medal.
N.-W. Frontier of India, 1908.—Operations in the Mohmand country.

Shaw, G. E.—

Chitral, 1895.—Relief of Chitral.—Medal with clasp.
N.-W. Frontier of India, 1897-98.—Action of the Ublan Pass, 27th August 1897.—Clasp.
Tirah, 1897-98.—Medal with 2 clasps.

Sheppard, W. (*Sub.-Condr., retd.*)—

South African War, 1899-1900.—Operations in Cape Colony, south of Orange River, 1899-1900.—Queen's medal with clasp. Specially promoted to Sub-Condr., G. G. O. No. 73 of 1902.

Shortridge, J. J. F.—

Tirah, 1897-98.—Medal with 2 clasps.

Smith, T.—

South African War, 1899-1902.—Relief of Ladysmith, including action at Colenso (15th December 1899). Operations on Tugela Heights (14-27th February 1900). Operations in Natal, March to June 1900, including action at Laings Nek (6-9th June).—Queen's medal with 4 clasps, King's medal with 2 clasps.

Smith, T. W.—

South African War, 1899—1901.—Operations in Natal, Orange Free State and the Transvaal.—Queen's medal with 5 clasps.

Smith, J. N.—

China, 1900.—Relief of Pekin.—Medal with clasp.
N.-W. Frontier, 1897-98.—Medal with clasp.

Southam, E. (*Condr., retd., S. and T.*)—

Burma, 1891-92.
Chitral, 1895.—Medal and clasp.
Tirah, 1897-98.—2 clasps.

Spinks, C. W. F.—

N.-W. Frontier of India, 1908.—Medal with clasp.

Stanfield, T. E.—

Tirah, 1897-98.—Medal with clasp.

Stanley, J. C.—

China, 1900.—Medal.
N.-W. Frontier, 1897.—Medal with clasp.

Stephens, E. B. (*Condr., retd., I. O. D.*)—

Tirah, 1897-98.—Medal with clasp.

Stewart, D. D.—

N.-W. Frontier of India, 1897-98.—Malakand.—Medal with clasp.
British East Africa, 1903.—Sotik Expedition.—Medal with clasp.

Stone, R. J. K.—

N.-W. Frontier of India, 1897-98.—Medal with clasp.
Tirah, 1897-98.—Clasp.

Stone, W. J. K.—

Tibet, 1903-04.—Medal.

Stuart, W. W.—

N.-W. Frontier of India, 1897-98.—Operations on the Samana and in the Kurram Valley.—Medal with 2 clasps.
Tirah, 1897-98.—Clasp.

Swain, J. H.—

Tirah, 1897-98.—Clasp.
Tirah, 1903-4.—Medal with clasp.

Talor, B. C. (*Condr., retd., B.*)—

Afghanistan, 1878—80.—Medal.
Burma, 1885—87.—Actions of Fort Minhla, Myingyan, and occupation of Mandalay.—Medal with clasp.

Tapp, W.—

Tibet Mission, 1903-04.—Medal.

Tate, F. R.—

South African War, 1899-02.—Operations in Natal 1899, including actions at Elandslagate (21st Oct.) Lombard Kop (30th Oct.) Defence of Ladysmith including action on 6th January 1900. Operations in Natal, March to June 1900 including action at Laing's Nek (6th to 9th June). Operations in Transvaal East of Pretoria, July 1900. (Wounded at Rooi Kopjies, 24th July 1900.)

Queen's Medal with 5 clasps. King's Medal with 2 clasps.

Teather, R. T.—

Tibet, 1903-04.—Medal.

N.-W. Frontier of India, 1908.—Operations in the Zakka Khel country.—Medal with clasp.

Tebby, H.—

N.-W. Frontier of India, 1897-98.—Operations on the Samana.—Medal with 2 clasps.

Tirah, 1897-98.—Actions of Chagru Kotal, Dargai, and of the Sampagha and Arhanga Passes. Reconnaissance of the Saran Sar and action of 9th November 1893. Operations in the Bara Valley, 7th to 14th Dec. 1897.—Clasp.

Thompson, C.—

Aden, 1903-04.—Operations in the interior.

Thompson, H. C.—

N.-W. Frontier of India, 1897-98.—Tochi.—Despatches, G. G. O. No. 1432 of 1897.—Medal with clasp. Promoted 2nd class Assist. Surgn.

Torpy, A. T.—

East Africa, 1902-03.—Operations in Somaliland.—Medal with clasp.

Turner, J.—

N.-W. Frontier of India, 1897-98.—Malakand.—Medal with clasp. Promoted Sub-Conductor.

Turner, W. W.—

N.-W. Frontier of India, 1897-98.—Malakand. Utman Khel, Buner. Action of the Tanga Pass—Medal with clasp.

Tibet, 1903-04.—Medal.

Twells, T. W.—

N.-W. Frontier of India, 1897-98.—Malakand.—Medal with 2 clasps.

Tyler, M. W. (*Conductor, retd., S. and T. Corps*)—

N.-W. Frontier of India, Hazara, 1891.—Medal with clasp.

South African War, 1899—1902.—Mentioned in despatches. Queen's medal with 4 clasps and King's medal with 2 clasps. Promoted Conductor.

Vahey, G. F. (*Sub-Condr., S. and T. Corps, retd.*)—

Tirah, 1897-98.—Actions of the Sampagha and Arhanga Passes. Operations against the Khan Khel Chamkanis.—Operations in the Bazar Valley, 25th to 30th December 1897.—Medal with 2 clasps.

East Africa, 1903-04.—Operations in Somaliland. Despatches, *London Gazette,* 2nd September 1904. Promoted Sub-Conductor.—Medal with clasp.

Vicars, W. (*Condr., retd., I. O. D.*)—

East Africa, 1903-04.—Operations in Somaliland.—Medal with clasp.

Vincent, G. W.—

N.-W. Frontier of India, 1897-98.—Medal with clasp.

Tirah, 1897-98.—Clasp.

Wallingford, W.—

Chitral, 1895.—Relief of Chitral.—Medal with clasp

Wallis, C.—

N.-W. Frontier of India, 1897-98.—Mohmand Tochi.—Medal with clasp.

Walsh, P.—

N.-W. Frontier of India, Tochi 1897-98. Medal with clasp.

Warmington, W. T.—

Tirah, 1897-98.—Medal with clasp.

Waugh, R. T.—

South African War, 1899-1902.—Advance on Kimberley, including action at Magersfontein (11th December 1899).

Relief of Kimberley. Operations in Orange Free State, February to May 1900, including operations at Paardeberg (17th to 26th February); actions at Poplar Grove (7th March), Driefontein (10th March), Karee Siding (29th March), Vet River (5th and 6th May), and Zana River (10th May). Operations in the Transvaal in May and June 1900, including actions near Johannesburgh (29th May), Pretoria (4th June), and Diamond Hill (11th and 12th June). Operations in the Transvaal, East of Pretoria, July to 29th November 1900, including action at Reit. Vlei (16th July). Operations in the Transvaal, 30th November 1900 to February 6th 1901, and 5th March 1902 to 31st May 1902. Operations in Cape Colony, 7th February 1901 to 4th March 1901. Operations in the Orange River Colony, 5th March 1901 to 4th March 1902. Queen's medal with five clasps. King's medal with two clasps.

Webb, G.—

N.-W. Frontier of India, 1897-98.—*Malakand.*—Medal with clasp.

Chitral, 1895.—Relief of Chitral. Action of the Malakand Pass. Action at Khar.—Medal with clasp.

Wells, S. H. (*Condr., Ind. Tel. Dept., retd.*)

Afghanistan, 1878—80.—Served with Khyber Force.—Medal.

Burma, 1885.—Medal with 2 clasps.

N.-W. Frontier of India, Hazara, 1891.—Clasp.

Chitral, 1895.—Relief of Chitral.—Medal with clasp.

West, J.—

N.-W. Frontier of India, Waziristan, 1901-02.—Medal with clasp.

Tibet, 1903-04.—Action at Naini, 26th June 1904. Operations at and around Gyantse, between 5th May and 6th July 1904. March to Lhassa, 14th July to 3rd August 1904.—Medal with clasp.

White, A. A. W.—

N.-W. Frontier of India, 1897-98.—Engagement near Shabkadr, 9th Augt. 1897, and operations in the Mamund country.—Medal with clasp.

Tirah, 1897-98.—Action of Chagru Kotal and of Dargai. Actions of the Sampagha and Arhanga Passes. Reconnaissance of the Saran Sar and operations at and around Dwatoi and against the Khani Khel Chamkanis Operations in the Para and Bazar Valleys.—Clasp.

Whittenbury, J. H.—

N.-W. Frontier of India, 1897-98.—Medal with clasp.

Tirah, 1897-98.—Actions of Chagru Kotal, Dargai, Sampagha and Arhanga Passes. Reconnaissance of the Saran Sar and actions of the 9th and 11th November 1897. Operations in the Waran Valley and action of 16th November 1897. Operations at and around Dwatoi. Operations in the Bara Valley 7th to 14th December 1897.—Clasp.

Willes, H. J.—

China, 1900.—Medal.

Williams, A. (*Conductor, retd., Ord. Dept.*)—

Chitral, 1895.—Medal with clasp.

Tirah, 1897-98.—2 clasps.

Williams, C. H.—

Tirah, 1897-98.—Medal with clasp.

Williams, H. (*Conductor, retd.*)—

Afghanistan, 1878-79-80.—Action at Ali Musjid.—Medal with clasp.

N.-W. Frontier of India, 1897-98.—Operations on the Samana and in the Kurram Valley during Aug. and Sep. 97.—Medal with 2 clasps.

Tirah, 1897-98.—Clasp.

Wilson, A.—

Soudan, 1888.—Action at Gamaizah.—Medal with clasp; Khedive's bronze star.

Wood, J. H.—

N.-W. Frontier of India, Isazai, 1892.

Chitral, 1895.—Relief of Chitral.—Medal with clasp.

Wood, S. G.—

Chitral, 1895.—Relief of Chitral.—Action at Panjkora River, 13th April 1895.—Medal with clasp.

Aden, 1903-04.—Operations in the interior.

WAR SERVICES OF INDIAN OFFICERS ON THE ACTIVE LIST.

Governor-General's Body Guard.

RISALDARS—

Jafar Husain—

N.-W. Frontier of India, 1908.—Operations in the Mohmand Country. Engagement of Kargha, 24 May, 1908.—Medal with clasp.

Governor's Body-Guard (Bombay).

Yusuf Ali Khan—

South Africa, 1900.—Medal with 6 clasps.

Governor's Body-Guard (Bengal).

RISALDAR—

Malik Mehr Khan—

N.-W. Frontier of India, 1897-98.— Samana and Kurrum Valley operations 1897.

Tirah, 1897-98.—Medal with 3 clasps.

JEMADAR—

Simrath Singh—

China, 1900.—Medal.

1st Duke of York's Own Lancers (Skinner's Horse).

RISALDAR-MAJOR—

Muhammad Akram Khan, *Bahadur*—

N.-W. Frontier of India, 1908.—Operations in the Mohmand country.

RISALDARS—

Ghulam Kadir Khan—

China, 1900.—Relief of Pekin. Action of Peitsang. Action at Yangtsun.—Medal with clasp.

Ghulam Muhammad Khan—

N.-W. Frontier of India, 2nd Miranzai, 1891.—Medal with clasp.

Faiz Muhammad Khan, I.O.M.—

South African War, 1899-1902.—Operations in Cape Colony, South of Orange River, 1899-1900.

Operations in the Orange Free State (February to May 1900), including operations at Paardeberg (17th to 26th February).—Operations in Orange River Colony. (May to 29th November 1900.)

Operations in Cape Colony, North of Orange River, —Queen's Medal and 3 clasps. King's Medal and 2 clasps.

Muhammad Ali Khan—(I)—

China, 1900.—Relief of Pekin. Action of Peitsang. Action at Yangtsun.—Medal with clasp.

Muhammad Ali Khan—II—

China, 1900.—Relief of Pekin. Action of Peitsang. Action at Yangtsun.—Medal with clasp.

Shaikh Abdul Majid—

China, 1900.—Relief of Pekin. Action of Peitsang. Action at Yangtsun.—Medal with clasp.

Zabte Khan—

China, 1900.—Relief of Pekin. Action of Peitsang. Action at Yangtsan.—Medal with clasp.

RESSAIDARS—

Faujdar Khan—

China, 1900.—Relief of Pekin. Action at Peitsang. Action at Yangtsan.—Medal with clasp.

Nathe Khan, *Bahadur*—

China, 1900.—Relief of Pekin. Action of Peitsang. Action at Yungtsun.—Medal with clasp.

East Africa, 1903-04.—Operations in Somaliland. Action at Jidballi.—Medal with clasp.

Habibur Rahman Khan, *Khan Sahib*—

China, 1900.—Relief of Pekin. Action of Peitsang. Action at Yangtsun.—Medal with clasp.

Barkat Khan—

China, 1900.—Relief of Pekin. Action of Peitsang. Action at Yangtsun.—Medal with clasp.

Muhammad Ali Khan—(III)—

China, 1900.—Relief of Pekin. Action of Peitsang. Action at Yangtsun.—Medal with clasp.

Ruknuddin, I.D.S.M.—

South African War, 1899-1902.—Operations in Cape Colony, South of Orange River, 1899-1900.

Operations in the Orange Free State (February to May 1900) including operations at Paardeberg (17th to 26th February).—Operations in Orange River Colony. (May to 29th November 1900.)

Operations in Cape Colony, North of Orange River —Queen's Medal and 3 clasps. King's Medal and 2 clasps.

JEMADARS—

Murtaza Khan—

China, 1900.—Relief of Pekin. Action of Peitsang. Action at Yangtsun.—Medal with clasp.

Sher Muhammad Khan (I)—

China, 1900.—Relief of Pekin. Action at Yaungtsan.—Medal with clasp.

Abdul Hamid Khan—

China, 1900.—Relief of Pekin. Action at Peittsang. Action at Yangtsan.—Medal with clasp.

Wali Muhammad Khan—

China, 1900.—Relief of Pekin. Action of Peitsang. Action of Yangtsun.—Medal with clasp.

Bahawal Khan—

China, 1900.—Relief of Pekin. Action of Peitsang. Action at Yangtsun.—Medal with clasp.

Abbulla—

China, 1900.—Relief of Pekin. Action at Peitsang. Action at Yangtsun.—Medal with clasp.

2nd Lancers (Gardner's Horse).

RISALDARS—

Bhaju Singh—

East Africa, 1901–04.—Operations in Somaliland. —Medal.

Suraj Singh—

East Africa, 1901-04.—Operations in Somaliland. Officially mentioned for good service in the field, *vide* D. G. of S. and T., No. 6144-T., dated 23rd June 1904.—Medal and clasp.

RESSAIDARS—

Kehri Singh—

N.-W. Frontier of India, 1897-98.—Operations in Samana and Kurram Valley '97. Relief of Gulistan '97 and with Kurram Valley Flying Column '97.—Medal and 3 clasps.

Molar—

N.-W. Frontier of India, 1897-98.—Operations in Tirah—Medal and 2 clasps.

JEMADAR—

Anokh Singh—

East Africa, 1901-04.—Operations in Somaliland.

3rd Skinner's Horse.

RISALDAR-MAJOR—

Balwant Singh—*Bahadur, order of Br. India, 2nd class.*

N.-W. Frontier of India, 1897-98.—Operations on the Samana and in the Kurram Valley during August and September 1897. Relief of Gulistan. —Medal with 2 clasps.

Tirah, 1897-98.—Clasp.

RISALDARS—

Mir Shamshad Ali—

N.-W. Frontier of India, 1897-98.—Operations on the Samana and in the Kurram Valley during August and September 1897. Relief of Gulistan.—Medal with 2 clasps.
Tirah, 1897-98.—Clasp.

Sobha Singh—

N.-W. Frontier of India, 1897-98.—Operations on the Samana and in the Kurram Valley during August and September 1897. Relief of Gulistan. —Medal with 2 clasps.

Tirah, 1897-98.—Clasp.
South Africa, 1900-02.—Medal with 3 clasps.

Nur Muhammad Khan—

N.-W. Frontier of India, 1897-98.—Operations on the Samana and in the Kurram Valley during August and September 1897. Relief of Gulistan. —Medal with 2 clasps.

Tirah, 1897-98.—Clasp.

RESSAIDAR—

Harnarayan Singh—

Afghanistan, 1879-80.—March from Kabul to Kandahar with force under Sir Frederick Roberts. Battle of Kandahar 1st September 1880. Afghanistan Medal with clasp for Kandahar, Bronze Star for Kabul-Kandahar March.

N.-W. Frontier, India, 1897-98.—Operations on the Samana and in the Kurram Valley during August and September 1897.—Relief of Gulistan —Medal with 2 clasps.

Tirah, 1897-98.—Clasp.

JEMADAR—

Imam Ali Khan—

N.-W. Frontier of India, 1897-98.—Operations on the Samana and in the Kurram Valley during August and September 1897.—Relief of Gulistan. —Medal with 2 clasps.

Tirah, 1897-98.—Clasp.

Muhammad Shaffi Khan—

N.-W. Fontier of India, 1897-98.—Operations on the Samana and in the Kurram Valley during August and September 1897. Relief of Gulistan. —Medal with 2 clasps.
Tirah, 1897-98.—Clasp

4th Cavalry.

RISALDAR—

Ram Singh—

Tibet, 1903-04.—Medal.

RESSAIDAR—

Musharraf Khan—

China, 1900.—Medal.

5th Cavalry.

RISALDAR-MAJOR—

Chandar Singh—

N.-W. Frontier of India, Waziristan, 1901-02.—Medal with clasp.

RISALDAR—

Yakub Ali Khan—

N.-W. Frontier of India, Waziristan, 1901-02.—Medal with clasp.

RESSAIDAR—

Lukman Khan—

N.-W. Frontier of India, Waziristan, 1901-02.—Medal with clasp.

JEMADAR—

Sundar Singh—

South Africa, 1899-1902.—2 Medals with 5 clasps.

Muhammad Hussain Khan—

N.-W. Frontier of India.—Waziristan, 1901-02. Medal with clasp.

Chander Singh (I)—

India Medal of 1895.—Clasps. Tirah, 1897-98. - Clasps. Samana, 1897.—Clasps. Samana, 1897.—Clasps. Punjab Frontier, 1897-98.

Chandar Singh (II)—

N.-W. Frontier of India.—Waziristan, 1901-02 —Medal with clasp. India Medal of 1895.—Clasp. Tirah, 1897-98.—Clasp. Samana, 1897. —Clasp. Punjab Frontier, 1897-98.

6th King Edward's Own Cavalry.

RISALDAR-MAJOR—

Fateh Singh—

N.-W. Frontier of India, 1897-98.—Operations on the Samana and in the Kurram Valley during August and September 1897.—Medal with 2 clasps.

Tirah, 1897-98.—Operations against the Khani Khel Chamkanis.—Clasp.

RISALDARS—

Dayal Singh—

N.-W. Frontier of India, 1897-98.—Operations on the Samana and in the Kurram Valley during August and September 1897.—Medal with 2 clasps.

Tirah, 1897-98.—Clasp.

Sher Singh—

N.-W. Frontier of India, 1897-98.—Operations on the Samana and in the Kurram Valley during August and September 1897.—Medal with 2 clasps.

Tirah, 1897-98.—Operations against the Khani Khel Chamkanis.—Clasp.

RESSAIDARS—

Jai Singh—

N.-W. Frontier of India, 1897-98.—Operations on the Samana and in the Kurram Valley during August and September 1897.—Medal with 2 clasps.

Tirah, 1897-98.—Operations against the Khani Khel Chamkanis.—Clasp.

JEMADARS—

Bishan Singh—

N.-W. Frontier of India, 1897-98.

Tirah, 1897-98.—Medal with 2 clasps.

7th Hariana Lancers.

RISALDAR-MAJOR—

Muhammad Akbar Ali Khan, I.O.M.—*Sardar Bahadur. Order of British India, 1st class.*

Burma, 1885—87-88.—Medal with 2 clasps.

RISALDARS—

Gopal Singh—*Bahadur—Order of Br. Ind., 2nd class.*

Burma, 1885—87-88.—Medal with 2 clasps.

Mam Chand—

China, 1900.—Medal.

Suram Chand—

Sikkim, 1888.—Medal with clasp.

9th Hodson's Horse.

RISALDAR-MAJOR—

Malik Khan Mahammad—*Bahadur—Order of Br. Ind., 2nd class.*

Chitral, 1895.—Relief of Chitral.—Medal with clasp.

N.-W. Frontier of India, 1897-98.—Clasp.

Tirah, 1897-98.—Operations in the Bara Valley 7th to 14th December 1897.—Clasp.

RISALDARS—

Mir Dad Khan—*Bahadur—Order of Br. Ind., 2nd class.*

Chitral, 1895. Relief of Chitral.—Medal with clasp.

N.-W. Frontier of India, 1897-98.—Clasp.

Tirah, 1897-98.—Operations in the Bara Valley, 7th to 14th December 1897.—Clasp.

Muhammad Akram Khan—*Bahadur—Order of Br. Ind., 2nd class.*

Chitral, 1895.—Relief of Chitral.—Clasp.

N.-W. Frontier of India, 1897-98.—Clasp.

Tirah, 1897-98.—Operations in the Bara Valley, 7th to 14th December 1897.—Clasp.

Tek Singh, I.D.S.M.—

Chitral, 1895.—Relief of Chitral.—Medal with clasp.

N.-W. Frontier of India, 1897-98.—Clasp.

Tirah, 1897-98.—Operations in the Bara Valley, 7th to 14th December 1897.—Clasp.

Nur Ahmad Khan—*Indian Order of Merit, 1st class.*

Mewa Singh—

Tirah Expeditionary Force, 1897-98.—Medal with clasp.

Punjab Frontier, 1897-98.—Tirah.

JEMADARS—

Ahmad Shah—

Tirah Expeditionary Force, 1897-98.—Medal with clasps.

Punjab Frontier, 1897-98, and Tirah.

Mir Alam Khan—

Tirah Expeditionary Force, 1897-98.—Medal with clasp.

Punjab Frontier, 1897-98.—and Tirah.

Tohid. Gul- Khan, I.D.S.M.—

Tirah Expeditionary Force, 1897-98.—Medal with clasp.

Punjab Frontier, 1897-98 and Tirah.

South Africa, Medl.

Sultan Ahmad Khan—

Tirah Expeditionary Force, 1897-98.—Medal with clasp.

Punjab Frontier, 1897-98.—Clasp.

10th Duke of Cambridge's Own Lancers (Hodson's Horse).

RISALDAR-MAJOR—

RISALDARS—

Sant Singh—

N.-W. Frontier of India, 1897-98.—Medal with clasp.

Mahan Singh—

N.-W. Frontier of India, 1897-98.—Medal with clasp.

Labh Singh—

N.-W. Frontier of India, 1897-98.—Medal with clasp.

Muhammad Ashraf Shah—

N.-W. Frontier of India, 1897-98.—Medal with clasp.

Nur Khan—

N.-W. Frontier of India, 1897-98.—Medal with clasp.

JEMADAR—

Bishn Singh—

N.-W. Frontier of India, 1897-98.—Medal with clasp.

11th King Edward's Own Own Lancers (Probyn's Horse).

RISALDAR-MAJOR—

Muhammad Aslams Khan, *Sardar Bahadur—Order of Br. India, 1st class.*

N.-W. Frontier of India, Hasara, 1891.—Medal with clasp.

Chitral, 1895.—Relief of Chitral.—Medal with clasp.

N.-W. Frontier of India, 1897-98.—Defence of Malakand, Malakand. Action at Landikai. Operations in Bajour and in the Mohmand country.—Mohmand.—2 clasps.

South Africa, 1901-02.—Medal with 3 clasps.

Tibet, 1903-04.—Medal.

RISALDARS—

Bichitar Singh—

N.-W. Frontier of India, Isazai, 1892.

Chitral, 1895.—Relief of Chitral.—Medal with clasp.

N.-W. Frontier of India, 1897-98.—Relief of Malakand. Malakand.—Operations in Bajour and in the Mohmand country. Mohmand—2 clasps.

Rab Nawaz—*Bahadur—*

Order of British India, 1st class, for services in Persia 1916.

Wasawa Singh—

Chitral, 1895.—Medal with clasp.

N.-W. Frontier of India, 1897-98.—Relief of Malakand and Chakdara.—Operations in Bajour.—Mohmand.—2 clasps.

Ram Singh—*Indian Distinguished Service Medal, Waziristan Field Force, 1917.*

RESSAIDARS—

Kahan Singh—

N.-W. Frontier of India, 1897-98.—Defence of Malakand. Relief of Chakdara. Malakand. Action at Landikai. Operations in Bajour and in the Mohmand country. Mohmand.—Medal with 2 clasps.

Rajindar Singh—

Chitral, 1895.—Relief of Chitral.—Medal with clasp.

N.-W. Frontier of India, 1897-98.—Malakand. Action at Landikai. Operations in Bajour and in the Mohmand country. Mohmand.—2 clasps.

Prem Singh—

Chitral, 1895.—Relief of Chitral.—Medal with clasp.

N.-W. Frontier of India, 1897-98.—Relief of Malakand and Chakdara Operations in Bajour.—Mohmand.—2 clasps. Order of Merit, 3rd class.

War, 1914—Star.

JEMADARS—

Partab Singh—

Chitral, 1895.—Relief of Chitral.—Medal with clasp.

N.-W. Frontier of India, 1897-98.—Defence of Malakand. Relief of Chakdara. Operations in Bajour and in the Mohmand Country, Mohmand.—2 clasps.

Sawan Singh—

N.-W. Frontier of India, 1897-98.—Operations in the Mohmand Country.—Medal with clasp.

King's South African medal with clasp.

Mula Singh—

N.-W. Frontier of India, 1897-98.—Defence of Malakand. Relief of Chakdara. Malakand. Action at Landiki. Operations in Bajour and in the Mohmand Country. Medal with 2 clasps.

12th Cavalry.

RISALDAR-MAJOR—

Sukhdayal Singh—

N.-W. Frontier of India, 1897-98.—Medal with clasp.

RISALDARS—

Udham Singh—

N.-W. Frontier of India, 1897-98.—Medal with clasp.

Bakhshish Singh—

N.-W. Frontier of India, 1897-98.—Medal with 2 clasps.

RESSAIDAR—

Kishun Singh—

South Africa, 1901-02.—Queen's Medal with 3 clasps. King's Medal with 2 clasps.

JEMADARS—

Partab Singh—

N.-W. Frontier of India, 1897-98.—Medal with clasp.

Karam Singh—

N.-W. Frontier of India, 1897-98.—Medal with 2 clasps.

Labh Singh—

N.-W. Frontier of India, 1897-98.—Medal with 1 clasp.

13th Duke of Connaught's Lancers. (Watson's Horse.)

RISALDAR-MAJOR—

Fazil Husain—

N.-W. Frontier of India, 1897-98.—Engagement near Shabkadr, 9th August 1897.—Mohmand.—Medal with clasp.
Tirah, 1897-98.—Clasp. Indian Order of Merit, 3rd class.

RISALDARS—

Pahlwan Khan—

N.-W. Frontier of India, 1897-98.—Operations on the Samana and in the Kurram Valley during August and September 1897.—Medal with clasp.
Tirah, 1897-98.—Clasp.

RESSAIDARS—

Bhola Singh—

N.-W. Frontier of India, 1897-98.—Mohmand.—Medal with clasp.

Jiwan Singh—

N.-W. Frontier of India, 1897-98.—Mohmand.—Medal with clasp.

JEMADARS—

Hazara Singh—

Sudan-Dungala Expedition, 1896.—Medal also the Khedive's Medal.
N.-W. Frontier of India, 1897-98.—Mohmand.—Medal with clasp.

14th Murray's Jat Lancers.

RESSAIDAR—

Kure Singh—

South Africa, 1900-02.—Queen's medal with 3 clasps. King's Medal with 2 clasps.

JEMADAR—

Ratti Ram—

South Africa, 1902.—Queen's Medal.

15th Lancers (Cureton's Mooltanis).

RISALDAR—

Ata Muhammad Khan—

War, 1914.—Star.

Muhammad Afzal Khan—

War, 1914.—Star.

RESSAIDAR—

Muhammad Nawaz Khan—

War, 1914.—Star.

Didari—

War 1914.—Star.

Shiudan—

Chiua, 1900.

JEMADAR—

Malik Mehr Khan, I.O.M.—

War, 1914.—Star.

Muhammad Afzal Khan—

War, 1914.—Star.

Hakdad Khan—

War, 1914.—Star.

Feteh Sher Khan—

War, 1914.—Star.

Jan Muhammad Khan—

War, 1914.—Star.

Amir Abdulrahman Khan—

War, 1914.—Star.

Rupa Singh—

War, 1914.—Star.

16th Cavalry.

RISALDAR-MAJORS—

Agia Ram—

China, 1900.—Medal.

Warlam Singh—

China, 1900.—Medal.

RISALDAR—

Raghunath Singh—

China, 1900.—Medal.

RESSAIDARS—

Sher Singh—

China, 1900.—Medal.

Munshi Ram—

China, 1900.—Medal.

Ishar Singh—

China, 1900.—Medal.

Rattan Singh—

China, 1900.—Medal.

JEMADARS—

Ajaib Singh—

China, 1900.—Medal.

Bachittar Singh—

China, 1900.—Medal.

Ran Singh—
China, 1900.—Medal.

Gurdit Singh—
China, 1900.—Medal.

Daya Ram—
China, 1900.

17th Cavalry.

RISALDAR—

Rukan Din—
Mesopotamia Expy. Force, 1916-17.

Rahmat Sher—
N.-W. Frontier of India, 1897-98.—Capture of the Sampagha and Arhanga Passes. Operation in the Bazar Valley, December 25th to 30th, 1897. —Medal with clasps.
Indian Expy. Force, 1915-17.

Bahadur Sher—
East African Expy. Force, 1916-17.—Mentioned in despatches.

Sultan Khan—
Mesopotamia Expy. Force, 1916-17.—Mentioned in despatches.

Sajid Gul—
Chitral, 1895.—Relief of Chitral.—Medal with clasp.
East African Expy. Force, 1915-17.

RESSAIDARS—

Tikka Khan—
N.-W. Frontier of India, 1897-98.—Engagement near Shabkadr, 9th August 1897.—Medal with clasp.

Ata Mohammad Shah—
Mesopotamia Expy. Force, 1918.

Wazir Khan, I.O.M.—
East African Expy. Force, 1915-17.—Awarded I.O.M.

Mommad Raza—
Indian Expy. Force, France, 1915.

Malik Alam Sher Khan, I.D.S.M.—
Indian Expy. Force, France, 1914-17.—Awarded I.D.S.M.

JEMADARS—

Mohammad Sarwar—
Indian Expy. Force, France, 1915-16.

Wazir Mohammad—
East African Expy. Force, 1915-17.

Inayatullah Khan—
East African Expy. Force, 1915-16.

Mohammad Akbar Khan—
East African Expy. Force, 1915-17.

Mohammad Hassan Khan—
Mesopotamia Expy. Force, 1916.

Sharif Khan—
Mesopotamia Expy. Force, 1916.

Sher Mohammad—
Indian Expy. Force, France, 1915-18.

Pir Dad -
Mesopotamia Expy, Force, 1916.

Alam Khan—
Mesopotamia Expy. Force, 1916.

18th King George's Own Lancers.

RISALDAR-MAJOR—

Hazi Gul Mawaz Khan—
N.-W. Frontier of India, 1897-98.—Operations on the Samana and in the Kurram Valley during August and September 1897. Operations of the Flying Column in the Kurram Valley under Col. Richardson, 20th August to 1st October 1897.—Medal with 2 clasps.
Tirah, 1897-98.—Clasp.

Malik Ahmadyar Khan *Bahadur*—
N.-W. Frontier of India, 1897-98.—Operations on the Samana and in the Kurram Valley during August and September 1897. Operations of the Flying Column in the Kurram Valley under Col. Richardson, 20th August to 1st October 1897.—Medal with 2 clasps.
Tirah, 1897-98.—Clasp.

RISALDARS—

Muhammad Khan—
N.-W. Frontier of India, Zhob Valley, 1890.
N.-W. Frontier of India, 1897-98.—Operations on the Samana and in the Kurram Valley during August and September 1897. Operations of the Flying Column in the Kurram Valley under Col. Richardson, 20th August to 1st October 1897.—Medal with 2 clasps.
Tirah, 1897-98.—Clasp.

Lakha Singh *Bahadur*—
N.-W. Frontier of India, 1897-98.—Operations on the Samana and in the Kurram Valley during August and September 1897.—Operations of the Flying Column in the Kurram Valley under Col. Richardson, 20th August to 1st October 1897.—Medal with 2 clasps.
Tirah, 1897-98.—Clasp.

Dost Muhammad Khan—*Bahadur Indian Distinguished Service Medal*—
N.-W. Frontier of India, 1897-98.—Operations on the Samana and in the Kurram Valley during August and September 1897. Operations of the Flying Column in the Kurram Valley under Col. Richardson, 20th August to 1st October 1897.—Medal with 2 clasps.
Tirah, 1897-98.—Clasp.
South Africa, 1899—1902.—Queen's medal with 6 clasps. King's medal with 2 clasps. *Indian Order of Merit, 3rd Cl.*

RESSAIDARS—

Muhamad Inayat Khan *Bahadur*—
N.-W. Frontier of India, 1897-98.—Operations on the Samana and in the Kurram Valley during August and September 1897. Operations of the Flying Column in the Kurram Valley under Col. Richardson, 20th August to 1st October 1897.—Medal with 2 clasps.
Tirah, 1897-98.—Clasp.

Tawakil Khan *Bahadur*—

N.-W. Frontier of India, 1897-98.—Operations on the Samana and in the Kurram Valley during August and September 1897. Operations of the Flying Column in the Kurram Valley under Col. Richardson, 20th August to 1st October 1897.—Medal with 2 clasps.
Tirah, 1897-98.—Clasp.

Dhuman Khan—

N.-W. Frontier of India, 1897-98.—Operations on the Samana and in the Kurram Valley during August and September 1897. Operations of the Flying Column in the Kurram Valley under Col. Richardson, 20th August to 1st October 1897. Medal with 2 clasps.
Tirah 1897-98.—Clasp.

JEMADARS—

Fateh Ali Khan—

N.-W. Frontier of India, 1897-98.—Operations on the Samana and in the Kurram Valley during August and September 1897. Operations of the Flying Column in the Kurram Valley under Col. Richardson, 20th August to 1st October 1897. Medal with 2 clasps.
Tirah 1897-98.—Clasps.

Khuda Baksh Khan—

N.-W. Frontier of India, 1897-98.—Operations on the Samana and in the Kurram Valley during August and September 1897. Operations of the Flying Column in the Kurram Valley under Col. Richardson, 20th August to 1st October 1897.—Medal with 2 clasps.
Tirah, 1897-98.—Clasp.

Allahyar Khan—

N.-W. Frontier of India, 1897-98.—Operations on the Samana and in the Kurram Valley during August and September 1897. Operations of the Flying Colnmn in the Kurram Valley under Colonel Richardson 20th August to 1st October 1897.—Medal with 2 clasps.
Tirah, 1897-90.—Clasp.

19th Lancers (Fane's Horse).

RISALDAR-MAJORS—

Hira Singh' *Bahadur* I.D.S.M.—

N.-W. Frontier of India, 1908.—Operations in the Zakka Khel country.

N.-W. Frontier of India, 1908.—Operations in the Mohmand country.—Medal with clasp.

RISALDARS—

Khushal Khan, *Bahadur*—

N.-W. Frontier of India, 1908.—Operations in the Zakka Khel country.

N.-W. Frontier of India, 1908.—Operations in the Mohmand country.—Medal with clasp.

Bakhshish Singh—

N.-W. Frontier of India, Miranzai, 1891.—Medal with clasp.

N.-W. Frontier of India, 1908.—Operations in the Zakka Khel country.

N.-W. Frontier of India, 1908.—Operations in the Mohmand country. Engagement of Matta, 24th April 1908.—Medal with clasp.

Ali Shan—

N.-W. Frontier of India, 1908.—Operations in the Zakka Khel country.

N.-W. Frontier of India, 1908.—Operations in the Mohmand country.—Medal with clasp.

Moti Singh, I.D.S.M.—

N.-W. Frontier of India, 1908.—Operations in the Zakka Khel country.

N.-W. Frontier of India, 1908.—Operations in the Mohmand country.—Medal with clasp.

Bhagwan Singh—

N.-W. Frontier of India, 1908.—Operations in the Mohmand country. Engagement of Matta, 24th April 1908.—Medal with clasp.

Ghulam Hussain, I.D.S.M.—

N.-W. Frontier of India, 1908.—Operations in the Zakka Khel country.

N.-W. Frontier of India, 1908.—Operations in the Mohmand country.—Medal with clasp.

RESSAIDARS—

Bishan Singh, I.O.M.—

N.-W. Frontier of India, 1908.—Operations in the Mohmand country. Engagement of Matta, 24th April 1908.—Medal with clasp.

Indar Singh—

N.-W. Frontier of India, 1908.—Operations in the Mohmand country. Engagement of Kargha 24th May 1908.—Medal with clasp.

Sawal Singh—

N.-W. Frontier of India, 1908.—Operations in the Zakka Khel country.
N.-W. Frontier of India, 1908.—Operations in the Mohmand country. Engagement of Matta, 24th April 1908.—Medal with clasp.

JEMADARS—

Juma Khan—

N.-W. Frontier of India, 1908.—Operations in the Mohmand country. Engagement of Matta, 24th April 1908.—Medal with clasp.

Hasham Khan I.O.M.—

N.-W. Frontier of India, 190.—Engagement of Matta, 24th April 1908.—Medal with clasp.

Yusaf Ali I.O.M.—

N.-W. Frontier of India, 1908.—Operations in the Zakka Khel country.
N.-W. Frontier of India, 1908.—Operations in the Mohmand country. Engagement of Matta, 24th April 1908.—Medal with clasp. Indian Order of Merit, 3rd class.

Narain Singh—

N.-W. Frontier of India, 1908.—Operations in the Zakka Khel country.
N.-W. Frontier of India, 1908.—Operations in the Mohmand country.—Medal with clasp.

Ram Singh—

N.-W. Frontier of India, 1908.—Operations in the Zakka Khel country.—Medal with clasp.

Hari Singh—

N.-W. Frontier of India, 1908.—Operations in the Zakka Khel country.—Medal with clasp.

Man Singh—

N.-W. Frontier of India, 1908.—Operations in the Zakka Khel country.—Medal with clasp.

20th Deccan Horse.

RISALDARS—

Konsal Singh—

Tirah, 1897-98.—Medal with 3 clasps.

Mir Sadik Ali—

Tirah, 1897-98.—Medal with 2 clasps.

RESSAIDARS—

Richpal Singh—

N.-W. Frontier of India, Waziristan, 1901-02.—Medal and clasp.

N.-W. Frontier of India, 1902.—Operations against the Darwesh Khel Waziris.

N.-W. Frontier of India, 1908.—Operations in the Mohmand country. Engagement of Kargha, 24th May 1908.—Medal with clasp.

21st Prince Albert Victor's Own Cavalry (Frontier Force. Daly's Horse).

RISALDAR-MAJOR—

Sundar Singh—

N.-W. Frontier of India, Waziristan, 1901-02.—Medal with clasp.

RISALDARS—

Kunda Singh—

N.-W. Frontier of India, Waziristan, 1894-95. Action at Wano, 3rd November 1894.—Medal with clasp. Indian Order of Merit, 3rd class.

N.-W. Frontier of India, Waziristan, 1901-02.—Medal with clasp.

N.-W. Frontier of India, 1908.—Operations in the Mohmand country. Engagement of Kargha, 24th May 1908.—Medal with clasp.

Latif Khan—

N.-W. Frontier of India, 1897-98.—Tochi.—Medal with clasp.

N.-W. Frontier of India, 1908.—Operations in the Mohmand country. Engagement of Kargha, 24th May 1908.—Medal with clasp.

Mahbub Khan *Bahadur, Order of Br. India, 2nd class*—

N.-W. Frontier of India, Waziristan, 1894-95—Medal with clasp.

N.-W. Frontier of India, 1897-98.—Tochi.—Medal with clasp.

N.-W. Frontier of India, Waziristan, 1901-02.—Clasp.

N.-W. Frontier of India, 1908.—Operations in the Mohmand country.—Engagement of Matta, 24th April 1908.—Medal with clasp.

Narayan Singh—

N.-W. Frontier of India, 1897-98.—Tochi.—Medal with clasp.

N.-W. Frontier of India, Waziristan, 1901-02.—Clasp.

N.-W. Frontier of India, 1908.—Operations in the Mohmand country.—Medal with clasp.

RESSAIDARS—

Atma Singh—

N.-W. Frontier of India, 1897-98.—Tochi.—Medal and clasp.

N.-W. Frontier of India, Waziristan, 1901-02.—Clasp.

Muhammad Hasham—

N.-W. Frontier of India, 1908.—Operations in the Mohmand country. Engagement of Kargha, 24th May 1908.—Medal with clasp.

Farzand Ali—

N.-W. Frontier of India, Waziristan, 1901-02.—Medal and clasp.

N.-W. Frontier of India, 1908.—Operations in the Mohmand country.—Engagement of Matta, 24th April 1908.—Medal and clasp.

Jahangir Beg—

N.-W. Frontier of India, 1908.—Operations in the Mohmand country. Engagement of Kargha, 24th May 1908.—Indian Order of Merit, 3rd class.

Arbela Singh—

N.-W. Frontier of India, 1897-98.—Defence of Chakdara.—Operations in Bajour and in the Mohmand country.—Medal and 2 clasps.

JEMADARS—

Dewa Singh—

N.-W. Frontier of India, 1897-98.—Tochi.—Medal with clasp.

N.-W. Frontier of India, Waziristan, 1901-02.—Clasp.

N.-W. Frontier of India, 1908.—Operations in the Mohmand country—Medal and Clasp. *Indian Distinguished Service Medal.*

Surayan Singh—

N.-W. Frontier of India, 1902.—Operations against the Darwesh Khel Waziris.

N.-W. Frontier of India, 1908.—Operations in the Mohmand country. Engagement of Kargha, 24th May 1908.—Medal and clasp.

Santa Singh—

N.-W. Frontier of India, Waziristan, 1901-02.—Medal and clasp.

N.-W. Frontier of India, 1908.—Operations in the Mohmand country.—Medal and clasp.

Chain Singh—

N.-W. Frontier of India, 1908.—Operations in the Mohmand country.—Medal and clasp.

Shahbaz Khan—

N.-W. Frontier of India, 1908.—Operations in the Mohmand country.—Medal with clasp.

Harnam Singh—

N.-W. Frontier of India, 1908.—Operations in the Mohmand country.—Medal and clasp.

Masten Singh—

N.-W. Frontier of India, Waziristan, 1901-02.—Medal and Clasp.

N.-W. Frontier of India, 1908.—Operations in the Mohmand country.—Medal and clasp.

Haidar Khan—

N.-W. Frontier of India, 1897-98.—Tochi.—Medal and clasp.

N.-W. Frontier of India, 1902.—Operations against the Darwesh Khel Waziris.

N.-W. Frontier of India, 1908.—Operations in the Mohmand country.—Medal with clasp.

Hazara Singh—

N.-W. Frontier of India, 1897-98.—Tochi, Medal and clasp.

N.-W. Frontier of India, Waziristan, 1901-02.—Clasp.

Wakil Singh—

N.-W. Frontier of India, Waziristan, 1901-02.—Medal and clasp.

N.-W. Frontier of India, 1908.—Operations in the Mohmand country.—Medal with clasp.

Abdul Majid—

N.-W. Frontier of India, 1908.—Operations in the Mohmand country.—Medal and clasp.

22nd Sam-Browne's Cavalry (F. F.).

RISALDARS—

Mir Alam—

N.-W. Frontier of India, Waziristan, 1894-95.—Medal with clasp.

N.-W. Frontier of India, 1897-98.—Operations on the Saman a and in the Kurram Valley during August and September 1897.—Medal with clasp.

N.-W. Frontier of India, Waziristan, 1901-02.—Clasp.

RESSAIDAR—

Harnam Singh—

N.-W. Frontier of India, Waziristan, 1901-02.—Medal and clasp.

23rd Cavalry (Frontier Force).

RISALDAR-MAJOR—

Hisamuddin Khan—

N.-W. Frontier of India, 1897-98. Tirah.—Medal with 2 clasps. Served as Orderly to G. O. C., Khyber Brigade.

N.-W. Frontier of India, 1908.—Operations in the Zakka Khel country. Operations in the Mohmand country.—Medal with clasp. *Indian Distinguished Service Medal.*

RISALDARS—

Shibditt Singh—

N.-W. Frontier of India, Waziristan, 1894-95.—Medal with clasp.

N.-W. Frontier of India, 1897-98.—Operations on the Samana and in the Kurram Valley during August and September 1897.—Operations of the Flying Column in the Kurram Valley under Col. Richardson, 20th August to 1st October 1897.—Medal with 2 clasps.

Tirah, 1897-98.—Clasp.

Muhammad Yakub Khan—

N.-W. Frontier of India, 1897-98.—Action of the Ubian Pass, 27th August 1897. Operations on the Samana and in the Kurram Valley during August and September 1897. Operations of the Flying Column in the Kurram Valley under Col. Richardson, 20th August to 1 1897.—Medal with 2 clasps.

Tirah, 1897-98.—Clasp.

Dost Muhammad—

N.-W. Frontier of India, Zhob Valley, 1890.

N.-W. Frontier of India, 1897-98.—Operations on the Samana and in the Kurram Valley during August and September 1897. Operations of the Flying Column in the Kurram Valley under Col. Richardson, 20th August to 1st October 1897.—Medal with 2 clasps.

N.-W. Frontier of India, 1902.—Operations against the Darwesh Khel Waziris.

Sher Singh—

N.-W. Frontier of India, Waziristan, 1894-95.—Medal with clasp.

N.-W. Frontier of India, 1897-98.—Operations on the Samana and in the Kurram Valley during August and September 1907. Operations of the Flying Column in the Kurram Valley under Col. Richardson, 20th August to 1st October 1897.—Medal with 2 clasps.

Sant Singh—

N.-W. Frontier of India, 1897-98.—Tochi.—Medal with clasp.

RESSAIDARS—

Phula Singh—

N.-W. Frontier of India, 1897-98.—Operations on the Samana and in the Kurram Valley during August and September 1897. Operations of the Flying Column in the Kurram Valley under Col. Richardson, 20th August to 1st October 1897.—Medal with 2 clasps.

Bhagat Singh—

N.-W. Frontier of India, 1897-98.—Tochi.—Medal with clasp.

N.-W. Frontier of India, 1902.—Operations against the Darwesh Khel Waziris.

JEMADARS—

Sahib Dad—

N.-W. Frontier of India, 1897-98.—Operations on the Samana and in the Kurram Valley during August and September 1897. Operations of the Flying Column in the Kurram Valley, under Col. Richardson, 20th August to 1st October 1897.—Medal with 2 clasps.

N.-W. Frontier of India, 1902.—Operations against the Darwesh Khel Waziris.

Rur Singh—

N.-W. Frontier of India, Waziristan, 1894-95.—Medal with clasp.

N.-W. Frontier of India, 1897-98.—Tochi.—Medal with clasp.

N.-W. Frontier of India, 1902.—Operations against the Darwesh Khel Waziris.

Usman Khan—

N.-W. Frontier of India, 1902.—Operations against the Darwesh Khel Waziris.

Natha Singh—

N.-W. Frontier of India, 1897-98.—Operations on the Samana and in the Kurram Valley during August and September 1897. Operations of the Flying Column under Col. Richardson, 20th August to 1st October 1897.—Medal with 2 clasp.

N.-W. Frontier of India, 1902.—Operations, against the Darwesh Khel Waziris.

Jamal Uddin—

N.-W. Frontier of India, 1897-98.—Tochi.—Medal with clasp.

N.-W. Frontier of India, 1902.—Operations against the Darwesh Khel Waziris.

Jamadar Mohkam Chand—

N.-W. Frontier of India, 1897-98.—Operations on the Samana and in the Kurram Valley during August and September 1897.—Operations of the Flying Column under Col. Richardson, 20th August to 1st October 1917.—Medal with 2 clasp.

Dittu Singh—

N.-W. Frontier of India, 1897-98.—Operations on the Samana and in the Kurram Valley during August and September 1897. Operations of the Flying Column under Col. Richardson, 20th August to 1st October 1917.—Medal with 2 clasps.

Chanda Singh—

N.-W. Frontier of India, 1897-98.—Operations on the Samana and in the Kurram Valley during August and September 1897. Operations of the Flying Column under Col. Richardson, 20th August to 1st October 1897.—Medal with 2 clasps.

25th Cavalry (Frontier Force).

RISALDAR-MAJOR—

RISALDARS-MAJOR—

Kale Khan—

N.-W. Frontier of India, 1st and 2nd Miranzai, 1891.—Medal with clasp.

N.-W. Frontier of India, Waziristan, 1901-02.—Medal with clasp.

Indian Distinguished Service Medal.

RISALADARS—

Surayan Singh—

N.-W. Frontier of India, Waziristan, 1901-02.—Medal with clasp.

Siripat Singh—

N.-W. Frontier of India, Waziristan, 1901-02.—Medal with clasp.

Amir Mahammad Khan—

N.-W. Frontier of India, Waziristan, 1901-02.—Medal with clasp.

RESSAIDARS—

Hakim Singh—

N.-W. Frontier of India, Waziristan, 1901-02.—Medal with clasp.

Sher Ali Khan—*Indian Distinguished Service Medal.*

Harditt Singh—

N.-W. Frontier of India, Waziristan, 1901-02.—Medal with clasp.

JEMADARS—

Kirath Singh—

N.-W. Frontier of India, Waziristan, 1901-02.—Medal with clasp.

Mumtaz Mohammad Khan—

N.-W. Frontier of India, Waziristan, 1901-02.—Medal with clasps.

Natha Singh—

N.-W. Frontier of India, Waziristan, 1901-02.—Medal with clasp.

26th King George's Own Light Cavalry.

RISALDAR-MAJORS—

Malik Sher Bahadur Khan, *Sirdar Bahadur—Order of Br. India, 1st class.*

Afghanistan, 1878-79-80.—Medal.

Chitral, 1895.—Relief of Chitral.—Medal with clasp.

N.-W. Frontier of India, 1897-98.—Clasp.

Tirah, 1897-98.—Clasp.

Thakur Mul Singh—

China Expedition, 1900.—Medal.

RISALDAR—

Sher Baz Khan—

N.-W. Frontier of India, 1897-98.—Medal with clasp.—*Indian Order of Merit, 3rd class.*

27th Light Cavalry.

RISALDARS—

Muhammad Khan—

Chitral, 1895.—Relief of Chitral.—Medal with clasp.

N.-W. Frontier of India, 1897-98.—Malakand. Action at Landikai. Operations in the Mohmand country Land Buner.—Clasp.

Mul Singh—

Tirah, 1897-98.—Clasp.

N.-W. Frontier of India, 1897-98.—Medal with clasp.

South Africa, 1900-02.—Queen's, medal with clasps. King's medal with 2 clasps,

Agar Singh—

China, 1900.—Medal.

Ghulam Rasul Khan—

N.-W. Frontier of India, 1897-98.—Operations on the Samana during August and September, 1897. Operations on the Flying Column in the Kurram Valley under Col. Richardson.—Medal with 2 clasps.

Tirah, 1897-98.—Clasp.

Bhagwan Sahai—

N.-W. Frontier of India, Waziristan, 1901-02.—Medal with clasp.

RESSAIDARS—

Muhammad Khan—

N.-W. Frontier of India, 1897-98.—Malakand operation in Bajond and Mamund valley.—Medal.

JEMADARS—

Ghulam Muhammad Khan—

Waziristan, 1901-02.—Medal with clasps.

28th Light Cavalry.

RISALDAR-MAJOR—

Miyan Din Khan—

South Africa, 1901-02.—Medal with 2 clasps.

RESSAIDARS—

Harnam Singh—

N.-W. Frontier of India, 1897-98.—Tochi.—Medal with clasp.

China, 1900.—Medal.

Malik Muhammad Latif—

N.-W. Frontier of India, 1908.—Operations in the Zakka Khel country.

N.-W. Frontier of India, 1908.—Operations in the Mohmand country. Engagement of Matta, 24th April 1908.—Medal with clasp.

JEMADARS—

Abdur Rahman Khan—

South Africa, 1900-02.—Queen's medal with clasps. King's medal with 2 clasps.

Shaikh Haider—

South Africa, 1900-02.—Queen's medal with 3 clasps. King's medal with 2 clasps.

29th Lancers (Deccan Horse).

RISALDAR-MAJOR—

Chanda Singh—

Soudan, 1896.—Medal.

Suakim, 1897.—Medal.

RISALDAR—

Thakur Singh—

China, 1900.—Medal.

30th Lancers (Gordon's Horse).

RISALDARS—

Partab Singh *Bahadur*—

N.-W. Frontier of India, 1897-98.—Medal with clasp.

Mir Mumtaz Hussain—

Burma, 1888—Medal with clasp.

31st Duke of Connaught's Own Lancers.

RISALDAR-MAJORS—

Muhi-ud-din Khan, C.I.E., *Sardar Bahadur—Order of Br. India, 1st class.—Indian Distinguished Service Medal.*

Burma, 1885—89.—Medal with 2 clasps.
Soudan, 1896.—Dongola Expedition.—2 Medals.
South Africa, 1900—02.—Medal with 4 clasps.

Asaram—

Soudan, 1896.—Dongola Expedition.—2 Medals.

RISALDARS—

Sher Singh—

Soudan, 1896.—Dongola Expedition.—2 Medals.

RESSAIDARS—

Ajun Khan—

South Africa, 1900-02—Medal with 2 clasps.

JEMADARS—

Shaikh Muhammad Ishaq—

Burma, 1885-89.—Medal with 2 clasps.
Soudan, 1896.—Dongola Expedition.—2 Medals.

32nd Lancers.

RISALDAR,-MAJOR—

Thakur Jor Singh—

N.-W. Frontier of India, Hazara, 1891.—Medal with clasp.

JEMADARS—

Zafuryab Khan—

N.-W. Frontier of India, 1897.—Medal with clasps.

Tirah, 1897-98 and Samana, 1897.

33rd Queen Victoria's Own Light Cavalry.

RISALDAR-MAJORS—

Rup Chand, *Bahadur—Order of Br. India, 2nd class.*

China, 1900.—Medal.

Santa Singh, *Bahadur, Order of Br. India, 2nd class,* I.O.M.

China, 1900.—Medal.

RISALDARS—

Aladad Khan—

China, 1900.—Medal.

Udham Singh—

China, 1900.—Medal.

Hidayat Ali Khan—*Indian Distinguished Service Medal.*

Tirah, 1897-98.—Medal with 3 clasps.
China, 1900.—Medal.

Muhammad Arsad Khan—

N.-W. Frontier of India, 1897-98.—Medal with 2 clasps.

Tirah, 1897-98.—Clasp.

China, 1900-01.—Relief of Pekin. Action of Peitsang. Action at Yangtsun.—Medal with clasp.

RESSAIDARS—

Ala-uddin Khan—*Indian Distinguished Service Medal.*

China, 1900.—Medal.

Prem Singh—

China, 1900.—Medal.

Sundar Singh—

China, 1900.—Medal.

JEMADARS—

Bhag Mal (I)—

China, 1900.—Medal.

Bhure Khan—

China, 1900.—Medal.

Bhag Mal (II)—

China, 1900.—Medal.

Rur Singh—

China, 1900. Medal.

Muhammad Shafi Khan (II)—

China, 1900.—Medal.

Muhammad Ismail Khan—

China, 1900—Medal.

Siriram—

China, 1900—Medal.

Alam Ali Khan—

China, 1900.—Medal.

34th Prince Albert Victor's Own Poona Horse.

RISALDARS—

Badan Singh—

South Africa, 1902.—Medal with clasp.

RESSAIDAR—

Balwant Singh—

China 1900.—Medal.

35th Scinde Horse.

JEMADARS—

Khuda Baksh—

South Africa, 1901-02.—Medal with clasp.

Sangat Singh—

Tirah Expeditionary Force, 1897-98.—Medal with 2 clasps.

36th Jacob's Horse.

RESSAIDAR—

Muhammad Hasham—

N.-W. Frontier of India, 1908.—Operations in the Mohmand country. Engagement of Kargha, 24th May 1908.—Medal with clasp.

37th Lancers (Baluch Horse).

RISALDARS—

Nur Sher Khan—

East Africa, 1902—04.—Operations in Somaliland.—Medal with clasp.

N.-W. Frontier of India, 1908.—Operations in the Zakka Khel country.—Medal with clasp.

Khan Alam Khan—

N.-W. Frontier of India, 1908.—Operations in the Zakka Khel country.—Medal with clasp.

Yakub Khan—

Somaliland, 1902-04.—Medal with clasps.

JEMADARS—

Kutab Sher Shah—

N.-W. Frontier of India, 1908.—Operations in the Zakka Khel Country.—Medal with clasp.

Mir Alam Khan—

N.-W. Frontier of India, 1908.—Operations in the Zakka Khel Country.—Medal with clasp.

Mir Baz Khan—

N.-W. Frontier of India, 1908.—Operations in the Zakka Khel Country.—Medal with clasp.

Ali Ahmad Khan—

N.-W. Frontier of India, 1908.—Operations in the Zakka Khel Country.—Medal with clasp.

Ghulam Haidar Khan—

N.-W. Frontier of India, 1908.—Operations in the Zakka Khel Country.—Medal with clasp.

Abdulla Khan—

N.-W. Frontier of India, 1908.—Operations in the Zakka Khel Country.—Medal with clasp.

Gul Hassan Khan—

N.-W. Frontier of India, 1908.—Operations in the Zakka Khel Country—Medal with clasp.

Abdul Hafiz—

N.-W. Frontier of India, 1908—Operations in the Zakka Khel Country—Medal with clasp.

Fazal Karim—

N.-W. Frontier of India, 1908.—Operations in the Zakka Khel country.—Medal with clasp.

38th King George's Own Central India Horse.

RISALDAR-MAJOR—

Amar Singh—*Bahadur—Order of British India, 2nd class.*

Tirah, 1897-98.—Operations against the Khani Khel Chamkanis.—Medal with 3 clasps.

RESSAIDAR—

Lihaz Gul Khan—

Tirah, 1897-98.—Reconnaissance of the Kharmana Defile and action of 7th November 1897.—Operations against the Khani Khel Chamkanis.—Medal with 3 clasps.

Mal Singh—

South Africa, 1902.—Queens' Medals with clasp.

39th King George's Own Central India Horse.

RISALDAR-MAJORS—

RISALDARS—

Bachan Singh—

N-W. Frontier of India, 1897-98.—Medal with 2 clasps.

Tirah, 1897-98.—Operations against the Khani Khel Chamkanis.—Clasp.

Surayan Singh—

N.-W. Frontier of India, 1897-98.—Operations on the Samana and in the Kurram Valley during August and September 1897. Medal with 2 clasps.

Tirah, 1897-98.—Operations against the Khani Khel Chamkanis.—Clasp.

South Africa, 1900-02.—King's Medal with clasp.

RESSAIDARS—

Abbas Ali Khan—

N.-W. Frontier of India, 1897-98.—Operations on the Samana and in the Kurram Valley during August and September 1897.—Medal with 2 clasps.

Tirah, 1897-98.—Operations against the Khani Khel Chamkanis.—Clasp.

Fauja Singh—

N.-W. Frontier of India, 1897-98.—Medal with clasps.

Malik Khuda Baksh, Khan—

Tibet, 1903-04.—Medal.

JEMADARS—

Malik Khan Mohamed Khan—

N.-W. Frontier of India, 1897-98.—Medal with 2 clasps.

Tirah, 1897-98.—Operations against the Khani Khel Chamkanis.—Clasp.

Suleman Khan—

South Africa 1900-02.—King's Medal with clasp.

Abdul Ghani Khan—

N.-W. Frontier of India, 1897-98.—Operations on the Samana and in the Kurram Valley during August and September 1897.—Medal with 2 clasps.

Tirah, 1897-98.—Operations against the Khani Khel Chamkanis—Clasp.

Tibet, 1903-04.—Medal.

41st Cavalry.

Muhammad Ibrahim Khan—

N.-W. Frontier of India, 1908.—Operations in the Zakka Khel country.—Medal with clasp.

Aden Troop.

RISALDAR—

Malikdad Khan—*Khan Sahib, Sardar Bahadur, Order of British India, 1st class.* I.D.S.M.

Soudan, 1896.—Dongola Expedition.—Medal, also Egyptian Medal.

Tirah, 1897-98.—Medal with 3 clasps.

Aden Hinterland, 1903-04.—Operations.

Queen's Own Corps of Guides (Cavalry).

RISALDAR-MAJOR—

Bahadur Singh, *Sardar Bahadur—Order of Br. India, 1st class.*

N.-W. Frontier of India, Hazara, 1888.—Medal with clasp.

Chitral, 1895.—Relief of Chitral.—Medal with clasp.

N.-W. Frontier of India, 1897.—Defence of Malakand.—Relief of Malakand.—Relief of Chakdara.—Malakand.—Action of Landakai.—Operations in Bajaur and in the Mamund country.—Buner.—2 clasps. *Indian Order of Merit, 3rd class.*

RISALDARS—

Khawaja Muhammad Khan, *Sardar Bahadur—Order of Br. Ind., 1st class.—Indian Distinguished Service Medal.*

Chitral, 1895.—Relief of Chitral.—Medal with clasp.

N.-W. Frontier of India, 1897-98.—Engagement near Shabkadr, 9th August 1897. Malakand. Mohmand. Buner. Action of the Tanga Pass.—Clasp.

China, 1900.—Medal.

N.-W. Frontier of India, 1908.—Operations in the Mohmand country.—Medal with clasp.

Abnashi Ram—

N.-W. Frontier of India, 1897-98.—Defence of Malakand.—Relief of Malakand.—Relief of Chakdara.—Malakand.—Action of Landakai. Operations in Bajour and in the Mamund country—Medal with 2 clasps.

Nur Khan—

Chitral, 1895.—Relief of Chitral.—Medal with clasp.

N.-W. Frontier of India, 1897-98.—Tochi.—Clasp.

Zardad Khan—

N.-W. Frontier of India, 1908.—Operations in the Mohmand country.—Medal with clasp.

RESSAIDARS—

Jiwand Singh—

Chitral, 1895.—Relief of Chitral.—Medal with clasp.

Abdullah Khan—

N.-W. Frontier of India, 1897-98.—Operations on the Samana and in the Kurram Valley during August and September 1897.—Medal with 2 clasps.

Tirah, 1897-98.—Clasp.

Bhagwan Singh—

N.-W. Frontier of India, 1897-98.—Action at Landakai.—Medal with clasp.

JEMADARS—

Natha Khan—

Chitral, 1895.—Relief of Chitral.—Medal with clasp.

N.-W. Frontier of India, 1897-98.—Buner Clasp.

Ganda Singh—

N.-W. Frontier of India, 1907.—Action of Landakai.—Medal and 1 clasp.

Zarif—

Chitral, 1895.—Relief of Chitral.—Medal with clasp.

N.-W. Frontier of India, 1897-98.—Defence of Malakand. Action of North Camp. Relief of Chakdara. Operations in Upper Swat. Action of Landakai. Operations in Mammund Valley. Action of Agra and Gat. Action of Badelai Operations in Buner.—2 clasps. "Malakand, 1897."

Punjab Frontier, 1897-98.

Aman Khan—

N.-W. Frontier of India, 1897.—Action of Landakai. Operations in Mammund Valley. Action ofBadelai. Operations in Buner.—Indian Medal, 1895.—Clasp.

Punjab Frontier, 1897-98.

Wasawa Singh—

Chitral, 1895.—Relief of Chitral.—Medal with clasp.

N.-W. Frontier of India, 1907-98.—Action of Landakai and Ayraput. Operations in Upper Swat, Mohmand and Buner Valley.—Medal with clasps.

Queen Victoria's Own Corps of Guides (Infantry).

SUBADAR-MAJOR—

Man Bir Thapa, I.O.M.—

N.-W. Frontier of India, 1897-98.—Defence of Malakand.—Relief of Malakand.—Malakand.—Operations in Bajour and in the Mamund country.—Utman Khel.—Buner.—Medal with 2 clasps.

N.-W. Frontier of India, 1908.—Operations in the Mohmand country.—Medal with clasp.

SUBADARS—

Taj Mahammad—

N.-W. Frontier of India, 1908.—Operations in the Mohmand country.—Medal with clasp.

Afzal—

Chitral, 1895.—Relief of Chitral.—Medal with clasp.

N.-W. Frontier of India, 1897-98.—Malakand. Operations in Bajour and in the Mamund country.—Clasp.

Tibet, 1903-04.—Action at Niani, 26th June 1904. Operations at and around Gyantse, between 5th May and 6th July 1904. March to Lhassa, 14th July to 3rd August 1904.—Medal with clasp.

Bahadur Khan Bahadur—

Chitral, 1895.—Relief of Chitral.—Medal with clasp.

N.-W. Frontier of India, 1897-98.—Defence of Malakand. Relief of Malakand.—Relief of Chakdara.—Malakand. Operations in Bajour and in the Mamund country. Utman Khel, Buner.—2 clasps.

N.-W. Frontier of India, 1908.—Operations in the Mohmand country.—Medal with clasp.

Gulab Singh—

N.-W. Frontier of India, 1908.—Operations in the Mohmand country.—Medal with clasp.

Ran Bir Thapa—

Chitral, 1895.—Relief of Chitral.—Medal with clasp.

N.-W. Frontier of India, 1897-98.—Buner.—Clasp.

Nidhan Singh—

Chitral, 1895.—Relief of Chitral.—Medal with clasp.

N.-W. Frontier of India, 1897-98.—Utman Khel, Buner.—Clasp.

N.-W. Frontier of India, 1908.—Operations in the Mohmand country.—Medal with clasp.

Kehr Singh—

N.-W. Frontier of India, 1908.—Operations in the Mohmand country.—Medal with clasp.

Bishan Singh, I.D.S.M.—

N.-W. Frontier of India, 1908.—Operations in the Mohmand country.—Medal and clasp.

Fauja Singh—

N.-W. Frontier of India, 1897-98.—Tochi.—Medal with clasp.

N.-W. Frontier of India.—1901-02.—Waziristan. Clasp.

Hayat Khan—

N.-W. Frontier of India, 1908.—Operations in the Mohmand country.—Medal with clasp.

Chandar Bir—

N.-W. Frontier of India, 1897-98.—Defence of Malakand, Relief of Malakand, Operations in Bajour and in the Mamund country—Utmankhel—Buner.—Medal with 2 clasps.

N.-W. Frontier of India, 1908.—Operations in the Mohmand country.—Medal and clasp.

Gujar—

N.-W. Frontier of India, 1908.—Operations in the Mohmand country.—Medal with clasp.

Chuni Lal—

Chitral, 1895.—Relief of Chitral.—Medal with clasp.

N.-W. Frontier of India, 1897-98.—Utman Khel.—Buner.—Clasp.

N.-W. Frontier of India, 1908.—Operations in the Mohmand country.—Medal with clasp.

Shadi Khan—

Chitral, 1895.—Relief of Chitral.—Medal with clasp.

N.-W. Frontier of India, 1897-98.—Operations in Bajour and in the Mamund country.—Utman Khel.—Buner.—Clasp

N.-W. Frontier of India, 1908.—Operations in the Mohmand country.—Medal with clasp.

Ismail—

Chitral, 1895.—Relief of Chitral.—Medal with clasp.

Sher Singh—

N.-W. Frontier of India, 1908.—Operations in the Mohmand country. - Medal and clasp.

Wazir—

N.-W. Frontier of India, 1908.—Operations in the Mohmand country.—Medal and clasp.

JEMADARS—

Khushal—

Chitral, 1895.—Relief of Chitral.—Medal with clasp.

N.-W. Frontier of India, 1897-98.—Defence of Malakand. Relief of Malakand. Relief of Chakdara.—Malakand. Operations in Bajour and in the Mamund country. Utman Khel, Buner.—2 clasps.

Khalsta Khan—

Tibet, 1903-04.—Action at Niani, 26th June 1904. Operations at and around Gyantse, between 5th May and 6th July 1904. March to Lhassa, 14th July to 3rd August 1904.—Medal with clasp.

N.-W. Frontier of India, 1908.—Operations in the Mohmand country.—Medal with clasp.

Kaman Sing (*re-employed*)—

N.-W. Frontier of India, 1897-98.—Defence of Malakand.—Relief of Malakand, Malakand—Operations in Bajour and in the Mamund country. Utman Khel.—Buner.—Medal with 2 clasps.

N.-W. Frontier of India, 1908.—Operations in the Mohmand country.—Medal with clasp.

Sardar Khan—

N.-W. Frontier of India, 1908.—Operations in the Mohmand country.—Medal with clasp.

Mangtu—

N.-W. Frontier of India, 1908.—Operations in the Mohmand country —Medal with clasp.

N.-W. Frontier of India, 1897-98.—Operations in Bajour and in the Mamund country —Utman-Khel, Buner.—clasp.

N.-W. Frontier of India, 1908.—Operations in the Mohmand country.—Medal with clasp.

Mula—

N.-W. Frontier of India, 1908.—Operations in the Mohmand country.—Medal and clasp.

Samundar—

N.-W. Frontier of India, 1908.—Operations in the Mohmand country.—Medal and clasp.

Narain Singh—

N.-W. Frontier of India, 1908.—Operations in the Mohmand country.—Medal and clasp.

Ghaffur Shah—

N.-W. Frontier of India, 1908—Operations in Zakka Khel country.

N.-W. Frontier of India, 1908.—Operations in the Mohmand country.—Medal with clasp.

Karamdad—

N.-W. Frontier of India, 1897-98.—Tochi.—Medal with clasp.

N.-W. Frontier of India, 1908.—Operations in the Mohmand country. Engagement of Kargha, 24 May 1908.

Jamal—

N.-W. Frontier of India, 1897-98, Tochi.—Medal with clasp.

N.-W. Frontier of India, Waziristan, 1901-02.—clasp

Ahmed Khan—

N.-W. Frontier of India, 1908.—Operations in the Mohmand country.—Medal with clasp.

Siraj Din—

N.-W. Frontier of India, 1908.—Operations in the Mohmand country.—Medal with clasp.

Mohammad Khan—

N.-W. Frontier of India, 1908.—Operations in the Mohmand country.—Medal with clasp.

Taj Mohammad—

N. W. Frontier of India, 1908.—Operations in the Mohmand country.—Medal with clasp.

Lachho—

N.-W. Frontier of India, 1908.—Operations in the Mohamand country.—Medal with clasp.

Mangal Singh—

N.-W. Frontier of India, 1908.—Operations in the Mohamnad country.—Medal with clasp.

3rd Battalion Guides Infantry Frontier Force.

SUBADAR-MAJOR—

Taj Mohammad Khan—

N.-W. Frontier of India, 1907.—Operation in Mohmand country.—Medal with clasp.

SUBADAR -

Chuni Lal—

Chitral Expedition 1895.—Action at Malakand pass 3rd April and Action at Punjkora (SADO) 13th April 1895.—Medal with clasp.

N.-W. Frontier of India, 1908—Operation in the Mohmand country.—Medal the clasp.

N.-W. Frontier of India, 1897.—Utmankhel Buner clasp.

Nadir Khan—

N.-W. Frontier of India, 1908.—Operation in the Mohmand country.—Metal with clasp.

Mohammad Akbar—

Somali Land 1903-04.—Action at Jidballi.—Medal with clasp.

JEMADAR—

Karam Dad—

Somali Land 1903-04.—Action at Jidballi.—Medal with clasp.

Wadhawa Singh—

Punjab N.-W. Frontier, 1897-98.—Medal with clasp.

Somali Land, 1903-04.—Action at Jidballi.—Medal with clasp.

Chattar Sing—

N.-W. Frontier of India, 1908.—Operation in the Mohmand country.—Medal with clasp.

Sajawal Khan—

Somali Land, 1903-04.—Medal with clasp.

Mohamad Khan—

N.-W. Frontier of India, 1908.—Operation in the Mohmand country.—Medal with clasp.

Lal Bahadur Gurung—

N.-W. Frontier of India, 1908.—Operation in the Mohmand country.—Medal with clasp.

21st Kohat Mountain Battery (Frontier Force).

SUBADAR-MAJOR—

SUBADAR—

Chanda, I.O.M.—

N.-W. Frontier of India, Waziristan, 1894-95.—Medal with clasp.

Tirah, 1897-98.—Medal with 2 clasps.

N.-W. Frontier of India, 1902.—Operations against the Darwesh Khel Waziris.

JEMADAR—

Rajha—

Indian Distinguished Service Medal.

22nd Derajat Mountain Battery (Frontier Force).

SUBADARS—

Santa Singh, *Bahadur*—

N.-W. Frontier of India, 1897-98.—Operations on the Samana and in the Kurram Valley during August and September 1897. Relief of Gulistan. Operations of the flying column in the Kurram Valley under Colonel Richardson, 20th August to 1st October 1897.—Medal with 2 clasps.

Tirah, 1897-98.—Actions of the Sampagha and the Arhanga Passes.—Operations in the Bazar Valley, 25th to 30th December 1897.—Clasp.

N.-W. Frontier of India, Waziristan, 1901-02.—Clasp.

Wasan Singh—

Chitral, 1895.—Relief of Chitral.—Medal with clasp.

N.-W. Frontier of India, 1897-98.—Operations on the Samana and in the Kurram Valley, during August and September 1897. Relief of Gulistan. Operations of the flying column in the Kurram Valley under Col. Richardson, 20th August to 1st October 1897.—2 clasps.

Tirah, 1897-98.—Actions of the Sampagha and Arhanga Passes. Operations in the Bazar Valley, 25th to 30th December 1897.—Clasp.

N.-W. Frontier of India, Waziristan, 1901-02.—Clasp.

N.-W. Frontier of India, 1908.—Operations in the Zakka Khel country.

N.-W. Frontier of India, 1908.—Operations in the Mohmand country. Engagement of Matta, 24th April 1908.—Medal with clasp.

Operations against Kachins, 1914-15.

JEMADARS—

Muhammad Amin—

N.-W. Frontier of India, 1897-98.—Operations on the Samana and in the Kurram Valley during August and September 1897. Relief of Gulistan. Operations of the flying column in the Kurram Valley under Col. Richardson, 20th August to 1st October 1897.—Medal with 2 clasps.

Tirah, 1897-98.—Actions of the Sampagha and the Arhanga Passes. Operations in the Bazar Valley, 25th to 30th December 1897.—Clasp.

N.-W. Frontier of India, Waziristan, 1901-02.—Clasp.

N.-W. Frontier of India, 1908.—Operations in the Zakka Khel and Mohmand countries. Engagement of Matta, 24th April 1908.—Medal with clasp.

Nur Hasan—

N.-W. Frontier of India, 1908.—Operations in the Zakka Khel country.—Medal with clasp.

Sherkhan—

N.-W. Frontier of India, Waziristan, 1901-02.—Medal with clasp.

N.-W. Frontier of India, 1908.—Operations in the Zakka Khel country.—Medal with clasp.

Jagat Singh—

N.-W. Frontier of India, Waziristan, 1901-02.—Medal with clasp.

N.-W. Frontier of India, 1908.—Operations in the Zakka Khel country.—Medal with clasp.

Operations against Kachins, .9 4-.5.

23rd Peshawar Mountain Battery (Frontier Force).

SUBADAR-MAJOR—

SUBADAR—

Ghulam Jilani—

N.-W. Frontier of India, 1897-98.—Tochi.—Medal with clasp.

N.-W. Frontier of India, 1908.—Operations in the Mohmand country.—Medal with clasp.

Bagga Singh—

N.-W. Frontier of India, 1908.—Operations in the Mohmand country.—Medal with clasp.

JEMADARS—

Mohammad Din—

N.-W. Frontier of India, 1908.—Operations in the Mohmand country.—Medal with clasp.

Harnam Singh—

N.-W. Frontier of India, 1908.—Operations in the Mohmand country.—Medal with clasp.

24th Hazara Mountain Battery (Frontier Force).

SUBADAR-MAJOR—

Gurditt Singh—

N.-W. Frontier of India, Zhob Valley, 1884.

N.-W. Frontier of India, Hazara, 1888.—Medal with clasp.

N.-W. Frontier of India (1st Miranzai), 1891.

Central Africa, 1895-97.—Medal with clasp inscribed "Central Africa, 1894-98."

East Africa 1917-18.

JEMADARS—

Fazl Khan—

N.-W. Frontier of India, Hazara, 1888.—Medal with clasp.

N.-W. Frontier of India, 1891.

Chitral, 1895.—Relief of Chitral.—Medal with clasp.

Mesopotamia Expdy. Force, 1916-17.

Jalal Singh—

Chitral, 1895.—Relief of Chitral.—Medal with clasp.

Mekran F. F. 1898.—Action of Gok Prosh.

East Africa, 1917-18.

25th Mountain Battery.

SUBADAR—

Kehar Singh—

N.-W. Frontier of India, 1897.—Medal with clasp.

Tirah, 1897-98.—Clasp.

JEMADARS—

Lal Khan—

Suakin Force, 1896.—2 medals (Khedive's and Queen's).

N.-W. Frontier of India, 1897.—Medal with clasp.

Tirah, 1897-98.—Clasp.

Tibet, 1904-05.—Medal.

Haider Khan—

Tirah, 1897-98.—Medal with clasp.

26th Jacob's Mountain Battery.

SUBADAR—

Hem Sing—*Indian Distinguished Service Medal.*

N.-W. Frontier of India, 1897-98.—Medal with 3 clasps.

(*a*) Punjab Frontier, 1897-98.
(*b*) Samana, 1897.
(*c*) Tirah, 1897-98

N.-W. Frontier of India, 1901-02.—Clasp. Waziristan, 1901-02.

JEMADAR—

Mohd. Baksh—*Indian Order of Merit, 2nd class.*

Ghulam Hassan—*Indian Distinguished Service Medal.*

27th Mountain Battery.

SUBADAR—

Sher Baz—

Chitral, 1895.—Relief of Chitral.—Medal with clasp.

N.-W. Frontier of India, Waziristan, 1901-02.—Clasp.

N.-W. Frontier of India, 1902.—Operations against the Darwesh Khel Waziris.

JEMADARS—

Bhan Singh—

N.-W. Frontier of India, Waziristan 1901-02.—Medal with clasp.

N.-W. Frontier of India, 1902.—Operations against the Darwesh Khel Waziris.

Tibet, 1903-04.—Medal.

Shah Nawaz—

N.-W. Frontier of India, Waziristan, 1901-02.—Medal and clasp.

N.-W. Frontier of India, 1902.—Operations against Darwesh Khel Waziris.

28th Mountain Battery.

SUBADAR—

Lal Din—

N.-E. Frontier of India, Manipur, 1891.—Medal with clasp.

N.-E. Frontier of India, Chin Hills, 1891-92.—Clasp.

N.-W. Frontier of India, Waziristan, 1894-95.—Clasp.

N.-W. Frontier of India, 1897-98.—Malakand. Operations in Bajour and in the Mamund country. Utman Khel. Buner.—Attack and capture of the Tanga Pass.—Medal with clasp.

East Africa, 1903-04.—Operations in Somaliland.—Action at Jidballi.—Medal with 2 clasps.

N.-W. Frontier of India, 1908.—Operations in the Mohmand country.—Medal with clasp.

JEMADAR—

Sant Singh—

N.-W. Frontier of India, 1897-98.—Defence of Malakand, Action at Landakai. Operations in Bajour, in the Mamund Country. Utmankkel, Buner, Attack and Capture of Tonga pass.—Medal with 2 Clasps.

N.-W. Frontier of India, 1908.—Medal and clasp.

29th Mountain Battery.

SUBADAR—

Jiwan Khan—

N.-W. Frontier of India, Waziristan, 1901-02.—Medal with clasp.

JEMADARS—

Ghunna Singh—

N.-W. Frontier of India, Waziristan, 1894-95.—Medal with clasps.

Tirah, 1897-98.—Action of Dargai. Capture of Samphaga and Arhanga Passes. Operations against the Khani Khel.

Cham Kannis.—Operations in the Bazar Valley, 26th to 30th December 1897.—Medal with two clasps.

N.-W. Frontier of India, Mahsud Waziri Blockade, 1901-02.—Clasp.

30th Mountain Battery.

SUBADAR—

Daya Singh—*Indian Distinguished Service Medal.*

Aden, 1903-04.—Operations in the Interior.

Tibet, 1903-04.—Medal with clasp.

Order of white Eagle, 5th class (with sword).

JEMADAR—

Kishan Singh—*Indian Distinguished Service Medal.*

Aden, 1903-04.—Operations in the Interior.

Allahyar Khan—

Aden, 1903-04.—Operations in the Interior.

No. 31 Mountain Battery.

SUBADARS—

Khem Singh, *Bahadur—Order of Br. India, 2nd class.*

N.-W. Frontier of India, 1897-98.—Operations on the Samana and in the Kurram Valley during August and September 1897. Relief of Gulistan. Operations of the Flying Column in the Kurram Valley under Col. Richardson, 20th August to 1st October 1897.—Medal with 2 clasps.

Tirah, 1897-98.—Capture of the Sampagha and Arbanga Passes. Operations in the Bazar Valley.—Clasp.

N.-W. Frontier of India, Waziristan, 1901-02.—Clasp.

Mihar Khan—

Chitral, 1895.—Relief of Chitral.—Medal with clasp.

British East Africa, 1895-96.—Operations against the Mazrui Rebels.—Medal.

Lubwa's, Uganda, 1897-98.—Medal with clasp.

N.-W. Frontier of India, Waziristan, 1901-02.—Clasp.

N.-W. Frontier of India, 1908.—Operations in the Zakka Khel country.—Medal with clasp.

JEMADARS—

Prem Singh—

N.-W. Frontier of India, 1897-98. Tochi.—Medal with clasp.

No. 32 Mountain Battery.

SUBADAR—

Hari Singh—

Merkan field force, 1911.

JEMADARS—

Harnam Singh—

N.-W. Frontier of India, Waziristan, 1897-98.—Medal and clasp.

Tirah.—Malakand. Punjab Frontier.—2 clasps.

Gul Muaz—

Waziristan Expedition, 1901-02.—Medal with clasp.

N.-W. Frontier of India, 1902.—Operations against the Darwesh Khel Waziris.

Bara Singh—

N.-W. Frontier of India, 1902.—Operation against the Darwesh Khel Waziris.

No. 33 Mountain Battery.

SUBADAR—

Karm Singh—

Chitral, 1895.—Relief of Chitral.—Medal with clasp.

N.-W. Frontier of India, 1897-98.—Operations on the Samana and in the Kurram Valley during August and September 1897. Relief of Gulistan. Operations of Flying Column in the Kurram Valley under Col. Richardson, 20th August to 1st October 1897.—2 clasps.

Tirah, 1897-98.—Capture of the Sampagha and Arhanga Passes. Operations in the Bazar Valley.—Clasp.

N.-W. Frontier of India, Waziristan, 1901-02.—Clasp.

No. 35 Mountain Battery.

SUBADAR—

Abbas Khan—

N.-W. Frontier of India, Waziristan, 1894-95.—Medal with clasp.

Aden, 1903-04.—Operations in the Hinte lan d.

JEMADARS—

Basant Singh—

Tirah Expedition, 1897-98.—Medal with 2 clasps.

N.-W. Frontier of India, 1902.—Operations against the Darwesh Khel Waziris.

Expeditionary Force Persian Gulf, 1911—

Kehr Singh—

Aden, 1903-04.—Operation in the interior.

Tibet, 1903-04.—Medal with clasp.

Mesopotamia, 1914-16.—Operation in Waziristan, March to August 1917.

The Frontier Garrison Artillery.

SUBADAR—

Amir Singh—

Tirah, 1897-98.—Action of Chagru Kotal.—Medal with two clasps.

Mohmand, 1908.—Engagement of Kargha, 24th May 1908.—Medal with clasp.

JEMADARS—

Rahim Ali—

N.-W. Frontier of India, 1892.—Isaz Expedition.

Chitral, 1895.—Relief of Chitral.—Medal with clasp.

Tirah, 1897-98.—Action of Chagru Ketal and Bara Valley.—Two clasps.

Mohmand, 1908.—Engagement of Kargha, 24th May 1908.—Medal with clasp.

Attar Singh—

Tibet, 1903-04.—Action at Kiam—Opertions at Gyantse—March to Lhassa.—Medal and clasp.

Lall Khan—

Tirah, 1897-98.—Medal and two clasps.

37th Mountain Battery.

JEMADARS—

Baryam Singh—

Aden Column, 1902—

Tibet Expedition, 1903-04—

Mesopotamia Expedition, 1914-16—

South Waziristan Field Force, 1917—

Nur Mohammad—

Chakdara Expedition, 1915—

Sistan (Persian) Field Force, 1916 -

Madad Khan—

Jandola Expedition, 1901—

East Africa Expeditionary Force, 1914-15—

36th (Reserve) Mountain Battery.

SUBADAR—

Karm Singh—

Chitral, 1895.—Relief of Chitral.—Medal with clasp.

N.-W Frontier of India, 1897-98.—Operations on the Samana and in the Kurram Valley during August and September 1897. Relief of Gulistan. Operations of Flying Column in the Kurram Valley under Col. Richardson, 20th August to 1st October 1897—2 clasps.

Tirah, 1897-98.—Capture of the Sampagha and Arhanga Passes Operations in the Bazzar Valley,—Clasp.

N.-W. Frontier of India—Waziristan, 1901-02.—Clasp.

JAMADARS—

Latif Khan—

N.W. Frontier of India, 1908,—Mohmand,—Medal with clasp.

Sher Khan—

N.-W. Frontier of India—Waziristan, 1901-02.—Medal and clasp.

N.-W. Frontier of India, 1908.—Operations in the Zakka Khel Country.—Medal with clasp.

No. 38th (Reserve) Mountain B ttery.

SUBADAR—

Udhe Singh—

N.-W. Frontier of India, 1908.—Operations in the Mohmand Country.—Medal with clasp.

JEMADARS -

Muhammad Khan—

N.-W. Frontier of India, 1908.—Operations in the Mohmand Country.—Medal with clasp.

Ali Akbar—

N.-W. Frontier of India, 1908.—Operations in the Mohmand Country Medal with clasp.

No. 39th Mountain Battery.

SUBADAR-MAJOR—

Muhammad Ismail—

N.-E. Frontier of India, Lushai, 1889-92.—Medal with clasp.

Burma, 1889-92.—Clasp.

N.-W. Frontier of India, 1897-98.—Tochi.—Medal with clasp. *Indian Odrer of Merit, 3rd class.*

Aden, 1903-04.—Operations in the Interior.

42nd Reserve Mountain Battery.

JEMADAR—

Sher Ali—

N.-E. Frontier of India, Manipur, 1891.—Medal with clasp.

N.-E. Frontier of India, Chin Hills, 1891-92.—Clasp.

N.-W. Frontier of India, Waziristan, 1894-95.—Clasp.

N.-W. Frontier of India, 1897-98.—Malakand.—Medal with 2 clasps.

N.-W. Frontier of India, 1908.—Mohmand operations.—Medal with clasp. Operations in Tochi, 1913.

1st King George's Own Sappers and Miners.

SUBADAR-MAJOR—

Akbar Ali, *Bahadur.—Order of Br. India, 2nd class.*

Sword of Honour.

N.-E. Frontier of India, Sikkim, 1888.—Medal with clasp.

N.-W. Frontier of India, (1st and 2nd) Miranzai, 1891.—Clasp.

N.-W. Frontier of India, Waziristan, 1894-95.—Clasp.

N.-W. Frontier of India, 1897-98.—Engagement near Shabkadar, 9th August 1897.—Mohmand.—Medal with clasp.

Tirah, 1897-98.—Operations in the Bara Valley, 7th to 14th December 1897.—Operations in the Bazar Valley, 25th to 30th December 1897.—Clasp.

N.-W. Frontier of India, 1908.—Operations in the Mohmand country.—Medal with clasp.

Mihr Din, *Sardar Bahadur*, I.D.S.M.—*Order of Br. Ind., 1st class.*—

N.-E. Frontier of India, Sikkim, 1888-89.—Medal with clasp.

N.-W. Frontier of India (1st and 2nd) Miranzai, 1891.—Clasp.

N.-W. Frontier of India, Waziristan, 1894-95.—Clasp.

N.-W. Frontier of India, 1897-98.—Engagement near Shabkadar, 9th August 1897.—Mohmand.—Medal with clasp.

Tirah, 1897-98.—Operations in the Bara Valley, 7th to 14th December 1897. Operations in the Bazar Valley, 25th to 30th December, 1897.—Clasp.

China, 1900.—Medal.

Abor, 1911-12.—Medal and clasp.

SUBADARS—

Habibullah Khan, *Sardar Bahadur.—Order of Br. India, 1st class.*

Burma, 1885-86-87.—Medal with clasp.

N.-E. Frontier of India, Sikkim, 1888.—Clasp.

N.-W. Frontier of India, (1st and 2nd) Miranzai, 1891.—Clasp.

N.-W. Frontier of India, Waziristan, 1894-95.—Clasp.

N.-W. Frontier of India, 1897-98.—Engagement near Shabkadar, 9th August 1897.—Mohmand.—Medal with clasp.

Tirah, 1897-98.—Operations in the Bara Valley, 7th to 14th December, 1897. Operations in the Bazar Valley, 25th to 30th December, 1897.—Clasp.

Ram Autar Patak—*Indian Distinguished Service Medal—*

N.-W. Frontier of India, 1908.—Operations in the Zakka Khel country.

N.-W. Frontier of India, 1908.—Operations in the Mohmand country.—Engagement of Kargha, 24th May 1908.—Medal with clasp.

Gauri Shankar Dube, *Bahadur.—Order of Br. India, 2nd class.—Military cross.*

N.-W. Frontier of India, Hazara, 1888.—Medal with clasp.

Tibet, 1903-04.—Operations at and around Gyantse between 5th May and 6th July 1904.—March to Lhassa, 14th July to 3rd August 1904.—Medal with clasp.

Chanda Singh Bahadur,—*Order of Br. India, 2nd class.*

N.-W. Frontier of India, Waziristan, 1894-95.—Medal with clasp.

N.-W. Frontier of India, 1897-98.—Tochi.—Clasp.

Abor, 1911-12.—Medal and clasp.

Fazl Shah,—*Bahadur,—Order of Br. India, 2nd cl.*

Indian distinguished Service Medal and clasp.

N.-W. Frontier of India, Waziristan, 1894-95.—Medal with clasp.

N.-W. Frontier of India, 1897-98.—Engagement near Shabkadar, 9th August 1897.—Mohmand.—Clasp.

Tirah, 1897-98.—Operations in the Bazar Valley, 7th to 14th December 1897. Operations in the Bara Valley, 25th to 30th December 1897.—Clasp.

Abor, 1911-12.—Medal and clasp.

Niamat Ulla Bahadur,—*Order of Br India, 2nd class.*

Indian Distinguished Service Medal.

China, 1900.—Relief of Pekin.—Medal with clasps.

Sultan.—*Indian distinguished Service Medal. Order of Karagcorge 3th Class (with Swords.)*

Chitral, 1895.—Relief of Chitral—Medal with clasp.

Tirah, 1897-98.—Operations in the Bazar Valley 25th to 30th December 1897—Clasp.

N.-W. Frontier of India, 1897-98.—Operations in the Mohmand country.—Clasp.

N.-W. Frontier of India, 1908.—Operations in the Mohmand country—Medal with clasp.

Abor, 1911-12.—Clasp.

Abdul Aziz Khan—*Bahadur.*

Order of British India 2nd class.

Indian distinguiehed Service Medal.

N.-W. Frontier of India, Waziristan, 1901-02.—Medal with clasp.

Tibet, 1903-04.—Medal.

Shiu Shankar Misr—

Chitral, 1895.—Relief of Chitral.—Medal with clasp.

Bhan Singh—

N.-W. Frontier of India, Waziristan, 1894-95.—Medal with clasp.

N.-W. Frontier of India, 1897-98.—Engagement near Shabkadar, 9th August 1897. Mohmand.—Medal with clasp.

Tirah, 1897-98.—Operations in the Bara Valley, 7th to 14th December 1897. Operations in the Bazar Valley, 25th to 30th December 1897.—Clasp.

Nand Singh—

N.-W. Frontier of India, Waziristan, 1894-95.—Medal with clasp.

Tibet, 1903-04.—Operations at and around Gyantse between 5th May and 6th July 1904.—Medal with clasp.

Sada Singh—*Bahadur, order of British India, 2nd class.*

N.-W. Frontier of India, 1897-98.—Medal with clasp.

China, 1900.—Medal.

Nawab Khan Bahadur—*Order of Br. India, 2nd class.*

Indian Distinguished Service Medal.

N.-W. Frontier of India, 1894-95.—Medal with clasp.

N.-W. Frontier of India, 1897-98.—Engagement near Shabkadar, 9th August 1897. Mohmand.—Medal with clasps.

Tirah, 1897-98.—Operations in Bara Valley, 7th to 14th December 1897. Operations in the Bazar Valley, 25th to 30th December 1897.

Abor Expedition, 1911-12.—Medal and clasp.

Ishar Singh—

Tochi F. F., 1897-98.—Medal with clasp.

Yakhudin—*Indian distinguished Service Medal.*

Order of white Eagle, 5th class (with Swords).

Chitral, 1895.—Relief of Chitral.—Medal with clasp.

N.-W. Frontier of India, 1897-98.—Malakand.—Buner.—Clasp.

China, 1900.—Relief of Pekin.—Medal with clasp.

Diwan Ali—

Indian Distinguished Service Medal.

N.-W. Frontier of India, 1897-98.—Medal with clasp.

Tirah, 1897-98—Operations in the Bara Valley, 7th to 14th December 1897. Operations in the Bazar Valley, 25th to 30th December 1897.—Clasp.

N.-W. Frontier of India, 1908.—Operations in the Mohmand country.—Medal with clasp.

Abor Expedition, 1911-12.—Medal and clasp.

Bhola Singh—

Chitral, 1895.—Relief of Chitral.—Medal with clasp.

Tirah, 1897-98.—Operations in the Bazar Valley, 25th to 30th December 1897.—Clasp.

N.-W. Frontier of India, 1897-98.—Operations in the Mohmand country.—Clasp.

N.-W. Frontier of India, 1902.—Operation in the Moumcnd Cou ntry—Medal with clasp.

Abor, 1911-12.—Clasp.

Ram Rup Singh Bahadur—*Order of Br. India 2nd Class Indian Order of Merit.*

Tibet, 1903-04.—Action at Najni, 26th June 1904.—Operations at and around Gyantse, 5th May and 6th July 1904.—March to Lhassa 14th July to 3rd August 1904.—Medal with clasp.

N.-W. Frontier of India, 1908.—Operations in the Zakka Khel country.—Medal with clasp.

Fazal Haq—

Tirah, 1897-98.—Operations in the Bazar Valley, 25th to 30th December 1897.—Medal with clasp.

N.-W. Frontier of India, 1897-98.—Operations in the Mohmand country.—Clasp.

N.-W. Frontier of India, 1908.—Operations in the Mohmand country.—Medal with clasp.

Abor, 1911-12.—Clasp.

Alam Khan—*Indian Order of Merit, 2nd class. Indian distinguished service medal.*

N.-W. Frontier of India, 1908.—Operations in the Mohmand country.—Medal with clasp.

Abor, 1911-12.—Clasp.

JEMADAR—

Chagatta, I. O. M.—

2nd class Indian Order of Merit. Indian Distinguished Service Medl.—Cross of the Order of St. George, 4th Class.

Tibet, 1903-04.—Operations at and around Gyantse between 5th May and 6th July 1904.—March to Lhassa, 14th July to 3rd August 1904.—Medal with clasp.

Gujar Singh—

N.-W. Frontier of India, 1897-98.—Buner.—Medal with clasp.

China, 1900.—Relief of Pekin.—Medal with clasp.

Muhammad Din—

N.-W. Frontier of India, Waziristan, 1894-95.—Medal with clasp.

N.-W. Frontier of India, 1897-98.—Engagement near Shabkadar, 9th August 1897.—Mohmand.—Medal with clasp.

Tirah, 1897-98.—Operations in the Bara Valley, 7th to 14th December, 1897. Operations in the Bazar Valley, 25th to 30th December 1897.—Clasp.

Kishan Singh—

N.-W. Frontier of India, 1897-98.—Engagement near Shabkadar, 9th August 1897.—Mohmand.—Medal with clasp.

Tirah, 1897-98.—Operations in the Bara Valley 7th to 14th December, 1897. Operations in the Bazar Valley, 25th to 30th December 1897.—Clasp.

Abdul Ghafur—

N.-W. Frontier of India, 1908.—Operations in the Mohmand country.—Medal with clasp.

Abor, 1911-12.—Clasp.

Alla Rakha—*Indian Distgd. Ser. Medal*—

N.-W. Frontier of India, 1908.—Operations in the Mohmand country.—Medal with clasp.

Fakir Muhammad—

N.-W. Frontier of India, 1908.—Operations in the Mohmand country.—Medal with clasp.

Manowar Shah—

N.-W. Frontier of India, 1908.—Operations in the Mohmand country.—Medal with clasp.

Sitla Baksh Singh—

Tibet, 1903-04.—Operations at and around Gyantse between 5th May and 6th July 1904.—March to Lhassa 14th July to 3rd August 1904. Medal with clasp.

Rahmat Ullah Khan—

N.-W. Frontier of India, 1908.—Operations in the Zakka Khel country.

N.-W. Frontier of India, 1908.—Operations in the Mohmand country.—Engagement of Kargha, 24th May 1908—Medal with clasp.

Ishar Din Pande—

Tibet, 1903-04.—Operations at and around Gyantse between 5th May and 6th July 1904. March to Lhassa 14th July to 3rd August 1904.—Medal with clasp.

Ramayah Singh.—*Indian Distgd. Ser. Medal*

Tibet, 1903-04.—Medal.

N.-W. Frontier of India, 1908.—Operations in the Zakha Khel Country.—Medal and clasp.

Basant Singh.—*Indian D.stgd. Ser. Medal.*—

N.-W. Frontier of India, 1908.—Operations in the Mohmand Country.—Medal with clasp.

Abor, 1911-12.—Clasp.

Firoz Khan—*Indian Distgd. Ser. Medal*—

N.-W. Frontier of India, 1908.—Operations in the Mohmand Countv.—Medal with clasp.

Abor 1911-12.—Clasp.

Isher Singh.—*Indian Distgd. Ser. Medal.*

Abor, 1911-12.—Medal and clasp.

Thakur Singh—

N.-W. Frontier of India, 1908.—Operations in the Mohmand country.—Medal and clasp.

Sahib Din.—

N.-W. Frontier of India, 1908.—Operations in the Mohmand country.—Medal and clasp.

Fateh Muhammad—

N.-W. Frontier of India, 1908.—Operations in the Mohmand country.—Medal and clasp.

Hashim Shah—

N.-W. Frontier of India, 1908.—Operations in the Mohmand country.—Medal and clasp.

Jhande Khan—

N.-W. Frontier of India, 1908.—Operations in the Zakha Khel Country.

N.-W. Frontier of India, 1908.—Operations in the Mohmand Country. Engagement of Kargha 24th May 1908.—Medal and clasp.

Guman Singh—

N.-W. Frontier of India, 1908.—Operations in the Zakka Khel country.

N.-W. Frontier of India, 1908.—Operations in the Mohmand country. Engagement of Kargha, 24th May 1908.—Medal with clasp.

Chaudhri—

N.-W. Frontier of India, 1897-98.—Malakand Buner.—Medal and clasp.

China, 1900.—Medal.

N.-W. Frontier of India, 1908.—Operations in the Mohmand country.—Medal and clasp.

Hardit Singh, I.O.M.—

Abor, 1911-12.—Medal with clasp.

Feroz Khan—

N.-W. Frontier of India, 1903.—Operations in the Mohmand country.—Medal and clasp.

Mahabir Pathak—

Tibet, 1903-04.—Medal.

Sheoratan Dube—

N.-W. Frontier of India, 1902.—Operations against the Darwesh Khel-Waziris.

Tibet, 1903-04.—Operations at and around Gyantse between 5th May and 6th July 1904. March to Lhassa 14th July to 3rd August 1904.—Medal with clasp.

N.-W. Frontier of India, 1908.—Operations in the Zakka Khel country.—Medal and clasp.

Abdul Rasul—

Abor, 1911-12.—Medal and clasp.

Bhulan Khan I.O.M., I.D.S.M.—

Tibet, 1903-04.—Operations at and around Gyantse between 5th May and 6th July 1904—March to Lhassa 14th July to 3rd August 1904—Medal clasp.

Bagga Singh—

N.-W. Frontier of India, 1908.—Operations in the Mohmand country.—Medal and clasp.

Dasaundha Singh—

N.-W. Frontier of India, 1897-98.—Engagement at Shabkadar 9th August 1897—Medal and clasp.

China, 1900.—Medal and clasp.

Fazalnur—

China, 1900.—Medal and clasp.

Nazar Muhammad—

N.-W. Frontier of India, 1908.—Operations in the Mohmand country.—Medal and clasp.

Shamasdin—

N.-W. Frontier of India, 1908.—Operations in the Mohmand country.—Medal and clasp.

Nica Singh—

N.-W. Frontier of India, 1908.—Operations in the Mohmand country.—Medal and clasp.

Abor, 1911-12.—Clasp.

Fazal Ahmad—

N.-W. Frontier of India, 1908.—Operations in the Mchmand counury.—Medal and clasp.

Abor, 1911-12.—Clasp.

2nd Queen Victoria's Own Sappers and Miners.

SUBADAR MAJOR.—

Michae (II).—

N.-W. Frontier of India—Chin Hills, 1893.—Medal with clasp.

Soudan, 1896.—Dongola Expedition.—Medals with clasp (Queen's and Khedive's).

Egypt, 1914-1918.—Order of British India 2nd class. Serbian Order of White Eagle, 5th Class.

SUBADARS—

Sunderamurthi—

N.-W. Frontier of India, 1897-98.—Samana—Medal with 2 clasps.

Tirah, 1897-98.—Clasp.

Ramasami (I)—

Chitral, 1895.—Relief of Chitral.—Medal with clasp.

Mesopotamia 1915-18—Order of British, India 2nd Class.

Hanumantu—

N.-W. Frontier of India, 1897-98.—Samana—Medal with 2 clasps.

Tirah, 1897-98.—Clasp.

Tibet, 1903-04.—Medal.

Mesopotamia, 1915-18—Mentioned in despatches—I. D. S. M.

Joseph—

Soudan, 1896.—Dongola expedition.—Medals (Queen's and Khedive's).

France, 1914-18—Mesopotamia, 1916—Order of British India—I. D. S. M.

Burland—

Mesopotamia, 1915-18—I. O. M., 2nd Class.

Ramasami (II)—

Soudan, 1896.—Dongola Expedition.—Medals (Queens and Khedives).

N.-W. Frontier of India, 1908.—*Zakka Khel*—Medal and clasp.

Kothandapani—

N.-W. Frontier of India, 1897-98.—*Malakand Buner*—Medal and clasp.

Tirah, 1897-98.—Clasp.

France, 1914-18—Mesopotamia—Egypt—I. D. S. M.

Ponnusami—

N.-W. Frontier of India, 1897-98.—*Malakand—Buner*—Medal clasps.

Tirah, 1897-98.—clasp.

Tibet, 1903-04.—Medal.

Mesopotamia 1915-18—Mentioned in despatches—I. D. S. M.

Varadaraju—

Tibet, 1903-04.—Medal.

Antoni—

China, 1900-01—Medal and clasp.

Mesopotamia, 1915-18—Mentioned in despatches—I. D. S. M.

Rangasami (I)—
Somaliland, 1902-04—Medal and clasp.

Padaivittan—
Burma Chin Hills, 1893-96—Medal and clasp.
China, 1900-01.—Medal.

Krishnasami (I)—
Mesopotamia, 1915-18—I. D. S. M.

Krishnasami (II)—
Tibet 1903-04—Medal and clasp.
Egypt, 1914-18—Cross of Kanageorge, 1st class.

Subramoni (II)—
N.-W. Frontier of India, 1908—Zakka Khel—Medal and clasp.
France, 1914-18—Mesopotamia—I. D. S. M.

JEMADARS—

Kesavalu—
N.-W. Frontier of India, 1897-98.— Medal an clasp.
Tirah, 1897-98.—Clasp.

Narayansami—
China, 1900—Medal and clasp.

Gurusami—
N.-W. Frontier of India, 1897-98.—Malakand-Buner—Medal and 2 clasps.
Tirah, 1897-98.—Clasp.
Tibet, 1903-04.—Medal.

Rajendram—
Tibet 1903-04—Medal.

Subramoni—
Tibet, 1903-04.—Medal.

Viraraghavelu—
China, 1900—Medal and clasp.
Somaliland 1902-04—Medal and clasp.
East Africa 1914-18—Mentioned in despatches M. S. M.

Abdul Rahim—
N.-W. Frontier of India, 1908.—Zakka Khel—Medal and clasp.

Ramasami—
Mesopotamia, 1915-18.—Medal of St. George, 2nd Class (Russian).

Mahomed Hyath—
Tibet, 1903-04—Medal and clasp.

Dharmalingam—
China, 1900—Medal and clasp.
Mesopotamia 1951-18—Mentioned in despatches (twice).

Michael—
N.-W. Frontier of India, 1908—Zakka Khel—Medal and clasp.

Ayyemperumal—
Mesopotamia 1915-18—South Persia—I. D. S. M.

Rasundram—
Egypt 1914-18—Mentioned in despatches M.S.M.

Raman—
Mesopotamia, 1915-18—I. D. S. M.

Sundaram—
China, 1900—Medal and clasp.
Tibet, 903-04—Medal.

Saiyid Yusuf—
Mesopotamia, 1915-18— Mentioned in despatches.

Mariyan—
N.-W. Frontier of India, 1908—Zakka Khel—Medal and clasp.

Raju—
N.-W. Frontier of India, 1908—Zakka Khel—Medal and clasp.

Gopalsami—
Soudan, 1896—Dongola Expedition—Medal (Queen's and Khedive's).
N.-W. Frontier of India, 1908.—Zakka Khel—Medal and clasp.

Loganadan—
Mesopotamia, 1915-18—I. O. M., 2nd class.

Innasimuttu—
Mesopotamia 1915-18—M. S. M.

Ponnusami—
Tibet, 1903-04—Medal.
Sikkim, 1903—Clasp.

Mariasusai—
Chitral, 1895.—Medal and clasp.
Soudan, 1896—Dongala Expedition—Medal (Queen's and Khedive's.)

Munisami—
N.-W. Frontier of India, 1908—Zakka Khel—Medal and clasp.

Rangasami—III.
Egypt, 1914-18—M. S. M.

3rd Sappers and Miners.

SUBADAR-MAJOR—

Arjun Powar—
Tirah, 1897-98.—Medal with 2 clasps.
Mekran, 1901.—Attack and capture of Nodiz Fort.

SUBADARS—

Poshati Mari—*Bahadur, 2nd class Order of Br. India.*
Burma, 1885—87.—Medal with clasp.
East Africa, 1902-04.—Operations in Somaliland. Action at Jidballi.—Medal with 2 clasps.

Bhawani Singh—
East Africa, 1903-04.—Operations in Somaliland —Medal with clasp.

Malla Singh—M. C., I. O. M.

N.-W. Frontier of India, 1897-98.—Operations on the Samana and in the Kurram Valley during August and September 1897. Relief of Gulistan, 14th September 1897.—Medal with 2 clasps.

Tirah, 1897-98.—Capture of the Sampagha and Arhanga Passes. Operations in the Waran Valley.—Clasp.

East Africa, 1903-04.—Operations in Somaliland —Medal with clasp.

Ganpat Mahadeo—*Bahadur.—Order of Br. India, 2nd class.*

N.-W. Frontier of India, 1897-98.—Operations on the Samana and in the Kurram Valley during August and September 1897. Operations of the Flying Column in the Kurram Valley under Col. Richardson.—Medal with 2 clasps.

Tirah, 1897-98.—Capture of the Sampagha and Arhanga Passes. Reconnaissance of the Saran Sar and action of 11th November 1897. Operations against the Khani Khel Chamkanis. Operations in the Bazar Valley, 25th to 30th December 1897.—Clasp.

Ismail Khan—*Bahadur,* I.D.S.M.—*Order of Br. India, 2nd class.*

N-W. Frontier of India, 1897-98.—Malakand.—Operations in Bajour and, in the Mohmand country.—Medal with clasp.

Tirah, 1897-98.—Capture of the Sampagha and Arhanga Passes. Operations at and around Dwatoi and operations against the Khani Khel Chamkanis. Operations in the Bazar Valley.

East Africa, 1903-04.—Operations in Somaliland. —Medal with clasp.

Shah Sawar—

N.-W. Frontier of India, 1897-98.—Malakand Operations in Bajour and in the Mamund country, Mohmand.—Medal with clasp.

Tirah, 1897-98.—Capture of the Sampagha and Arhanga Passes. Operations at and around Dwatoi and action of 24th November 1897.—Operations against Khani Khel Chamkanis.—Operations in the Bazar Valley, 25th to 30th December 1897.—Clasps.

China, 1900.—Relief with Pekin.—Medal with clasp.

East Africa, 1902-04.—Operations in Somaliland. Action at Jidbali.—Medal with 2 clasps.

Muhammad Din—I.O.M.

East Africa, 1902-04.—Operations in Somaliland Action at Jidballi.—Medal with 2 clasps.

Bir Singh—I. O. M.

N.-W. Frontier of India, 1897-98.— Operation on the Samana and in the Kurram Valley during August and September 1897. Relief of Gulistan, 14th Sepember 1897.—Medal with two clasps.

Tirah, 1897-98.—Capture of the Sampagha and Arhanga Passes. Reconnaissance of the Saran Sar and action of 11th November 1897. Operations in Waran Valley. Operation against the Khani Khel Chamkanis. Operations in Bazar Valley, 25th to 30th December 1897.—Clasp.

Mekran, 1901.—Attack and capture of Nodiz Fort.

Dhondibarao Barge—I.O.M.

East Africa, 1902-04.—Operations in Somaliland. Action at Jidballi.—Medal with 2 clasps.

Gujar Singh—

N.-W. Frontier of India, 1897-98.—Operations on the Samana and in the Kurram Valley during August and September 1897. Relief of Gulistan, 14th September 1897.—Medal with 2 clasps.

Tirah, 1897-98.—Capture of the Sampagha and Arhanga Passes. Reconnaissance of the Saran Sar and action of 11th November 1897. Operations in Waran Valley. Operations in Bazar Valley, 25th to 30th December 1897.—Clasp.

East Africa, 1903-04.—Operations in Somaliland.—Medal with clasp.

Khan Muhammad Khan—

China, 1900-01.—Relief of Pekin.—Medal with clasp.

Baryam Singh—

Mekran, 1901.—Attack and capture of Nodiz Fort.

Lakshiman Powar—I.D.S.M.

N.-W. Frontier of India, 1897-98.—Operations on the Samana and in the Kurram Valley during August and September 1897. Operations of the Flying Column in the Kurram Valley under Col. Richardson.—Medal with 2 clasps.

Tirah, 1897-98.—Capture of the Sampagha and Arhanga Passes. Reconnaissance of the Saran Sar and action of 11th November 1897. Operations in the Waran Valley. Operations against the Khani Khel Chamkanis. Operations in the Bazar Valley, 25th to 30th December 1897.—Clasp.

Ghulam Nabi—I.O.M.

East Africa 1903-04—Operations in Somaliland.

Kailas Singh—*Indian Distinguished Service Medal.*

N.-W. Frontier of India, 1897-98.—Malakand. Operations in Bajour and in the Mamund country.—Mohmand.—Medal with clasp.

Tirah, 1897-98.—Capture of the Sampagha and Arhanga Passes, 18th to 31st October 1897.—Operations at and around Dwatoi. Operations against the Khani Khel Chamkanis, 1st to 26th November 1897. Operations in the Bazar Valley, 25th to 30th December 1897.—Clasp.

East Africa, 1902-04.—Operations in Somaliland, action at Jidballi.—Medal with 2 clasps.

JEMADARS—

Ganpatrao Jadhao—

N.-W. Frontier of India, 1897-98.—Operations on the Samana and in the Kurram Valley during August and September 1807. Operations of the Flying Column in the Kurram Valley under Col. Richardson.—Medal with two clasps.

Tirah, 1897-98.—Capture of the Sampagha and Arhanga Passes. Reconnaissance of the Saran Sar and action of 11th November 1897. Operations against the Khani Khel Chamkanis. Operations in the Bazar Valley, 25th to 30th December 1897.—Clasp.

Shirpati Barge—

East Africa, 1903-04.—Operations in Somalil and —Medal with clasp.

Datadin Badhal—*Indian Distinguished Service Medal.*

Aden, 1903-04.—Operations in the Interior.

Tek Singh—

China.—Relief of Pekin.—Medal with clasp.

Gangajirao Khanvilkar—*Indian Distinguished Service Medal.*

N.-W. Frontier of India, 1897-98.—Operations on the Samana and in the Kurram Valley, August and September 1897. Relief of Gulistan, 14th September 1897.—Medal and two clasps.

Tirah, 1897-98.—Capture of the Sampagha and Arhanga Passes. Reconnaissance of the Saran Sar and action of the 11th November 1897. Operations in Waran Valley. Operations against the Khani Khel Chamkanis. Operations in the Bazar Valley, December 1897.—Clasp.

Shankram Pille—

N.-W. Frontier of India, Zhob Valley, 1890.

N.-W. Frontier of India, 1897-98.—Tirah 1898—Medal with 2 clasps.

East Africa, 1902-04.—Operations in Somaliland.—Medal with clasp.

Krishnajirao Bhonsle—*Indian Distinguished Service Medal.*

East Africa, 1902-04.—Operations in Somaliland, Action of Jidballi—Medal with 2 clasps.

Bir Singh—

East Africa, 1902-04.—Operations in Somaliland. Action at Jidballi—Medal with clasp.

Habibullah Khan—I.D.S.M.

East Africa, 1902-04.—Operations in Somaliland, action at Jidballi—Medal with 2 clasps.

Abdullah Shah—

East Africa, 1902-04—Operation in Somaliland. Action at Jidballi.—Medal with 2 clasps.

Sidram Pille—

N.-W. Frontier of India, Zhob Valley 1890.—Waziristan Field Force.—1894-1895—Medal with clasp.

N.-W. Frontier of India, 1897-98.—Samana—Tirah, 1897-98.—Capture of the Sampagha and Arhanga Passes. Operations at and around Dwatoi. Operations against the Khani Khel Chamkanis. Operations in the Bazar Valley.—Medal with 3 clasps.

Aden, 1903-04.—Operations in the interior.

East Africa, 1903-04—Operations in Somaliland—Action at Jidballi—Medal with 2 clasps.

Shaikh Hassan—

N.-W. Frontier of India, Zhob Valley 1890.—

Ali Hussain—

East Africa, 1903-04.—Operations in Somaliland, Action at Jidballi.—Medal with 2 clasps.

Khana—

China, 1900.—Medal.

Krishna Swami—

N.-W. Frontier of India, 1897-98.—Malakand Operations in Bajour, and in the Mamund country. Mohmand.—Medal with clasp.

Tirah, 1897-98.—Capture of Sampagha and Arhanga Passes. Operations against the Khani Khel Chankans. Operations In the Bazar Valley 25th—30th Dec. 1897.—Clasp.

Lalu Rajenna—I.O.M.—

East Africa, 1903-04.—Operation in Somaliland. Medal with clasp.

No. 26 (Railway) Company.

SUBADAR—

Sher Baz—*Bahadur*, I.D.S.M.

N.-W. Frontier of India, Hazara, 1891.—Medal with clasp.

N.-W. Frontier of India, Isazai, 1892.

Chitral, 1895.—Relief of Chitral.—Medal with clasp.

No. 31 (Divisional Signal) Company.

SUBADAR-MAJOR—

Kasim Ali—*Sardar Bahadur.*

Hazara, 1891.—Medal with clasp.

Chitral, 1895.— Ditto.

Mohmand, 1908.— Ditto.

France, 1914-15.—

M. E. Force, 1917-18.—

No. 32 (Divisional Signal) Company.

SUBADARS—

Lehru—*Bahadur*—

N.-W. Frontier of India, 1897-98.—Medal and 2 clasps.

N.-W. Frontier of India, 1901-02.—Clasp.

War 1914—Barah, Marselles, Suez, 1914—18.

Allarakha Khan—I.O.M.

East Africa, 1902-04.—Medal and clasp.

War 1914.—France 1914-15. Basrah 1916-17.

No. 34 (Divisional Signal) Company.

SUBADARS—

Narasingaparumal—

N.-W. Frontier of India, 1897-98.—Operation on the Samana and in the Kurram Valley, during August and September 1897.—Medal and two clasps.

Tirah, 1897-98.—Actions of Chagrie Kotal, Dargai, the Sampagha Pass and the Arhanga Pass. Reconnaissance of the Saran Sar and action of 9th November 1897. Operations in the Waran Valley and action of the 16th November 1897. Operations in the Bara Valley, 7th to 14th December 1897—Clasp.

Tibet Mission Force from 9th November 1903 to 6th October 1904.—Medal.

Signal Service Depot.

SUBADARS—

Dewan Singh—*Bahadur.*—(30th Punjabis)
N.-W. Frontier 1897.—Medical and 3 clasps.
Indian Ex. Force A. (*France*)—*12th Oct. 14 to 10 Dec. 15.*—2nd class Order of British India, with title of Bahadur 4th Aug. 15.
Mesopotamia Ex. Force.—*19th Jan. 16 to 13th Feb. 17.*

SUBEDAR—

Sriniyasulu—(34th Div. Sig. Coy. I.).
Mesopotamia Ex. Force.—16th Nov. 14 to 30th Sept. 15.

SUBEDAR—

Narayan Singh—(38th Dogras. I.)
N.-W. Frontier Waziristan 1901-02.
Somaliland 1902-03.

SUBEDAR—

Swaminathan—(33rd Divl. Sig. Co. I.)
Mesopotamia Ex. Force.—18th Dec. 15 to 22nd Aug. 17.

JEMADAR—

Ghulam Ali Shah—(32nd Divl. Sig. Co.)
Indian Ex. Force A. (*France*)—26th Sept. 14 to 15th Feb. 16.
Mesopotamia Ex. Force.—24 Dec. 17. to 19 May 18.
China Expedy. Force. 1900-01—Medal and clasp.

JEMADAR—

Puran Singh—(31st Divl. Sig Co. I.)
China Ex. Force, 1900-01—Medal.
Indian Ex. Force A. (*France*) *12th Oct. 14 to 10th Dec. 15.*—Awarded Dist. S. Medal.
Mesopotamia Ex. Force 19th Jan. 16 to 13th Jan. 18.

JEMADAR—

Mangal Singh—(33rd Cavalry.)
Indian Ex. Force A. (*France*)—12th Oct. 14 to 10th Dec. 15. Awarded I. Dist. Service Medal.
Mesopotamian Ex. Force, 19th Jan. 16 to 3rd July 16 & 27th June 18 to 2nd Aug. 18.

1st Brahmans.

SUBADARS—

Mahesar Dube—
Tibet, 1903-04.—Medal.

Ramcharan Pande—
Tibet, 1903-04.—Medal.

2-1st Brahmans.

SUBADAR-MAJOR—

SUBADARS—

Ramjiawan Sukul—
Egypt 1914-15.
Mesopotamia 1915-16.—

JEMADARS—

Ramdulare Tewari—
France 1914-15.—
Mesopotamia 1917—

Hukum Chand—
Egypt 1914-15.—
Gallipoli 1915-16.—
France 1915.—
Aden 1916-17.—
Mesopotamia 1917-18.—

Jagdammaparshad Misir—
Mesopotamia 1917.—

Net Ram—
Mesopotamia 1916-18.—

Ramshankar Tewari—
France 1914-15.—
Egypt 1915.—
Mesopotamia 1915-16-17.—

2nd Queen Victoria's Own Rajput Light Infantry.

SUBADAR-MAJORS—

Loknath Singh—
China, 1900.—Medal.

SUBADARS—

Assa Singh—
China, 1900.—Medal.

Pahlad Singh—
China, 1900.—Medal.

Chattar Singh—
China, 1900.—Medal.

Kushiram Singh—
China, 1900.—Medal.

Hari Singh—
China, 1900.—Medal.

JEMADARS—

R. J. Jacob—
China, 1900.—Medal.

Puran Singh—
China, 1900.—Medal.

Naurang Singh—
China, 1900.

Bhura Singh—
China, 1900.—Medal.

Puran Singh—
China, 1900.—Medal.

Sarjit Singh—
China, 1900.—Medal.

2-2nd Qeen Victoria's Own Rajput Light Infantry.

SUBADAR-MAJOR—

Jaswant Singh—
China, 1900.—Medal.
Egypt, 1914-15—
Mesopotamia, 1915-17—

SUBADARS—

Man Singh—
East Africa, 1914-16—
Mesopotamia, 1916-17—

Shiuram Singh—
East Africa, 1914-15—
Mesopotamia, 1916-17—

Loknath Singh—
China, 1900.—Medal.
Egypt, 1914-15—
Mesopotamia, 1915-17—

Balwant Singh—
Mesopotamia, 1915-17—
South Persia, 1917—

Arjan Singh—
Egypt, 1914-15.—
Mesopotamia 1915-17—

Kham Singh—
Egypt, 1914-15—
Mesopotamia, 1915-17—

Sispal Singh—
East Africa, 1914-16—
Mesopotamia 1916-17—

JAMADARS—

Phulandaer Singh—
France, 1914-15—
Egypt 1915—
Mesopotamia, 1915-17—

Surajpal Singh—
Mesopotamia, 1915-17—

Sardar Singh—
Mesopotamia, 1915-17.

Ramsarup Singh, I.D.S.M.—
Egypt, 1914-15—
Mesopotamia, 1915-17—

Chirani Singh—
Mesopotamia, 1915-17—

Arjan Singh—
Egypt, 1914-15—
Mesoptam a, 1915-17—

Chander Singh—
Mesopotamia, 1916-17—

3-2d Queen Victoria's Own Rajput Light Infantry.

SUBADAR MAJOR—

Umrao Singh—
China, 1900.—Medal.

SUBADARS—

Shiu Charan Singh—
China, 1900.—Medal.

Hardwari Singh—
China, 1900.

1-5th Light Infantry.

SUBADARS—

Wahid Ali Khan, I.M.O.—
China, 1900.—Medal.

Suleman Khan—
China, 1900.—Relief of Pekin.—Medal with clasp.

Azizud Din—*Ind. Distgd. Service Medal.*

Rahmat Khan—*Ind. Distgd. Service Medal.*

JEMADARS—

Sikhdar—*Indian Distgd. Service Medal.*

Gulam Nabi Khan—*Indian Distgd. Service Medal.*

Ahmad Din Khan—*Meritorious Service Medal.*

Manzur Ahmad Khan—*Meritorious Service Medal.*

Sazawar Khan—*Meritorious Service Medal.*

6th Jat Light Infantry.

SUBADARS—

Shiulal (I)—
N.-W. Frontier of India, 1897-98.—Tochi.—Medal with clasp.
China, 1900.—Medal.

Lakhi Ram, M.C., I.D.S.M.—
China, 1900.—Medal.

Shiu Lal, I.D.S.M. (II)—
N.-W. Frontier of India, 1897-98.—Tochi.—Medal with clasp.
China, 1900.—Medal.

Mula. I.D.S.M.—
China, 1900.—Medal.

2-6th Jats.

JAMADARS—

Ram Kala—

N.-W. Frontier of India, 1908.—Operations in the Zakka Khel country.—Medal and clasp.

7th Duke of Connaught's Own Rajputs.

SUBADAR-MAJOR—

Shiuambar Singh *Bahadur*, I. O. M.—*Order of Bititsh India, 2nd class.*

China, 1900.—Action of Peitsang. Action at Yang tsun. Relief of Pekin.—Medal with clasp.

SUBADARS—

Ramdayal Singh—

Burma, 1891-92.
China, 1900.—Action of Peitsang. Relief of Pekin.—Medal with clasp.

Bhawani Parshad Singh—

China, 1900.—Action of Peitsang. Action at Yang-tsun. Relief of Pekin.—Medal with clasp.

Brijmohan Singh, I.D.S.M.—

China, 1900.—Action of Peitsang. Action at Yangtsun. Relief of Pekin.—Medal with clasp.

Chandrapal Singh, I.D.S.M.—

China, 1900.—Action of Peitsang. Action at Yangtsun. Relief of Pekin.—Medal with clasp.

Gopal Singh—

China, 1900.—Action of Peitsang. Action at Yangtsun. Relief of Pekin.—Medal with clasp.

Ramball Singh—

China, 1900.—Action of Peitsang. Action at Yangtsun. Relief of Pekin.—Medal with clasp.

JEMADARS—

Baijnath Singh—

China, 1900.—Medal.

Shiudhari Singh—

China, 1900.—Action of Feitoang, Relief Pekin.—Medal with clasp.

Shiu Ratan Singh—

China, 1900.—Action of Peitoang, Relief of Pekin.—Medal with clasp.

2-7th Duke of Connaught's own Rajputs.

SUBADAR MAJOR—

Ramball Singh—

China, 1900.—Relief of Peking—Medal and clasp.
Persian Gulf, 1911.—

SUBADARS—

Debi Bakhsh Singh—

China, 1900.—Relief of Peking—Medal and clasp.

Surajpall Singh—

Persian Gulf 1911.—

Surajball Singh—

Persian Gulf, 1911.—

Sheobodhan Singh—

Persian Gulf, 1911.—

Besdeo Singh—

Persian Gulf, 1911.—

JEMADARS—

Sukhram Singh—

Persian Gulf, 1911.—

Sheodarshan Singh—

Persian Gulf, 1911.—

Ujagar Singh—

Persian Gulf, 1911.—

8th Rajputs.

SUBADARS—

Ram Ikbal Singh—

N.-W. Frontier of India, 1901-02.—Medal with clasp.

M. E. Force 1916.—

Bije Bahadur Singh—

N.-W. Frontier of India, 1897-98.—Medal with clasp.

M. E. Force 1916-17.—

Daulat Singh—

N.-W. Frontier of India, 1897-98.—Medal with clasp.

Girwar Singh—

N.-W. Frontier of India, 1897-98.—Medal with clasp.

M. E. Force 1916.—

Jung Bahadur Singh—

N.-W. Frontier of India, 1897-98.—Medal with clasp.

M. E. Force 1916.—

Ramanand Singh—

N.-W. Frontier of India, 1897-98.—Medal with clasp.

M. E. Force 1916.—

Mothura Singh—

N.-W. Frontier of India, 1897-98.—Medal with clasp.

M. E. Force 1916.—

Inderjit Singh—

China, 1900.—Medal.

M. E. Force 1916—

Chauharja Pershad Misir—

Burma, 1885-87.—Medal with clasp.
Burma, 1887-89.—Clasp.
N.-W. Frontier of India, 1897-98.—Medal with clasp.

2-8th Rajputs.

SUBADRAS—

Ramnarain Singh—
N.-W. Frontier of India, 1897-98.—Medal with clasp.

Jagesar Singh—
N.-W. Frontier, 1897-99.—

Raibhan Singh—
N.-W. Frontier, 1897-99.—

Sukhraj Singh—
China, 1900-02.—

Ajodhya Singh—
China, 1902-02.—

SUBEDAR MAJORS—

Bhure Singh, I. D. S. M.—
France, 19'4-15.—

Govind Singh B. I., *Bahadur.*
France, 1914-15.—
Egypt, 1915.—
Mesopotamia, 1915-16-17-18.—

SUBEDARS—

Sheodarshan Tiwari—
France, 1914-15.—
Egypt, 1915.—
Mesopotamia, 1915-16-17-18.—

Shaikh Abdul Majid—
France, 1914-15.—
Egypt, 1915.—
Mesopotamia, 1915-16-17-18.—

Thakur Sihgh—
France, 1914-15.—
Egypt, 1915.—
Mesopotamia, 1915-16-17-18.—

Abdul Aziz—
France, 1914-15.—
Egypt 1915.—
Mesopotamia, 1915-16-17-18.—

Sher Mohammad Khan, I. D. S. M.—
France, 1914-15.—
Egypt, 1915.—
Mesopotamia, 1915-16-17-18.—

Ganesh Singh—
France, 1914-15.—
Egypt, 1915.—
Mesopotamia, 1915-16-17-18.—

Ramadin Tiwari—
France, 1914-15.—
Mesopotam'a, 1917-18.—

Sheo Nath Singh—
France, 1914-15.—
Egypt, 1915.—
Mesopotamia, 1915-16.—

Hukam Chand—
Mesopotamia, 1916-17.

Amar Singh, I. D. S. M.—
France, 1914-15.—
Mesopotamia, 1916-17-18.

Alam Ali Khan—
Tirah, 1897-98.—

Imam Bakhsh Khan—
Afghan, 1879-80.

JEMADARS—

Sawal Singh—
France, 1914-15.—

Husain Ali Khan—
France, 1914-15.—
Egypt, 1915.—
Mesopotamia, 1915-16-17.—

Jai Lal—
Mesopotamia, 1916-18.—

Rikkhi Singh—
France, 1914-15.—
Egypt, 1915.—
Mesopotamia, 1915-16-17.—

Deo Bakhsh—
France, 1914-15.—
Egypt, 1915.—

Sayyid Muhummad Yakub—
France, 1914-15.—
Mesopotamia, 1917-18.—

Shiv Balak Singh—
Egypt, 1915.—

Basant Singh—

Harjiwan Lal—
Mesopotamia, 1916-17-18.—

Ismail Khan—
France, 1914-15.—
Mesopotamia, 1916-17-18.—

Chandgi Ram—
Mesopotamia, 1916-17-18.—

Sundar Singh—
Mesopotamia, 1916-17-18.—

Sripal Singh—
Fraace 1915.—
Egypt, 1915.
Mesopotamia, 1915-16-17-18.—

Sukhdeo Pande—
Mesopotamia, 1916-17-18.—

Suraj Prasad Misra—
France, 1914-15.—
Egypl, 1914.—
Mesopotamia, 1915-16-17-18.—

Tuhi Ram—
Mesopotamia, 1916-17-18.—

Rajmal Singh—
France, 1914-15.—
Egypt, 1914.—
Mesopotamia, 1915-16-17.—

Neki Ram—
Mesopotamia, 1916-17.—

Hira Lal—
Mesopotamia, 1916-17-18.—

Gumani Singh—
Mesopotamia, 1916-17-18.

Jan Muhammad Khan—
France, 1914-15.—
Egypt, 1915.—

Asghar Khan—
France, 1914-15.—
Egypt, 1915.—

2-9th Delhi Infantry.

SUBADARS—

Behari Lall—
Aden, 1903-04.—Operations in the interior.

Bhagat Ram—
Aden, 1903-04.—Operations in the Interior.

3-9th Bhopal Infantry.

SUBADAR—

Abhaidatt Singh—
N.-W. Frontier of India, Waziristan, 1901-02.—Medal with clasp.

Bhora Singh—
China, 1900.—Medal.

10th Jats.

SUBADAR MAJOR—

Nand Ram *Bahadur*—
Tochi, 1914-15.—
N.-W. F.F., 1917.—

SUBADARS—

Kunwar Chiranji Singh—
Tochi, 1914-15.—

Raje Ram—
Tochi, 1914-15.—
N.-W. F. F., 1917.—

Umrao Singh—
Tochi, 1914-15.—
N.-W. F. F., 1917.—

Molar, I—
N.-W. F. F., 1917.—

Tirkha Ram—
Tochi, 1914-15.—

Sheochand—
Tibet. 1903-04.—Medal.
N.-W. F. F., 1917.—

Molar, II—
N.-W. F. F., 1917.—

Bhola—
Tochi, 1914-15.—

Kanhaiya—
Tochi, 1914-15.—
N.-W. F. F., 1917.—

Molar, III—
Tibet, 1903-04.—Medal.
Tochi, 1914-15.—
N.-W. F. F., 1917.—

Digh Ram—
Tibet, 1903-04.—Medal.
Tochi, 1914-15.—

Jotu—
Tochi, 1914-15.—
N.-W. F. F., 1917.—

JEMADARS—

Gopal—
Tochi, 1914-15.—

Gyani Ram, I—
Tochi, 1914-15.—

Mamraj—
Tochi, 1914-15.—

Lok Ram—
Tibet, 1903-04.—Medal.
Tochi, 1914-15.—
N.-W. F. F., 1917.—

Dani—
N.-W. F. F., 1917.—

Sudhan—
Tochi, 1914-15.—
N.-W. F. F., 1917.—

Sheochand, I—
Tochi, 1914-15.—
N.-W. F. F., 1917.—

Raje Ram—
Tochi, 1914-15.—
N.-W. F. F., 1917.—

Ram Dour Lal—
N.-W. F. F., 1917.—

Chunni Lal—
Tochi, 1914-15.—
N.-W. F. F., 1917.—

Harchand—
N.-W. F.F., 1917.—

Gyani Ram, II—
Tochi, 1914-15.—
N.-W. F. F., 1917.—

Bhola—
Tochi, 1914-15.—
N.-W. F. F., 1917.—

Kalu—
Tochi, 1914-15.—
N.-W. F. F., 1917.—

Bhagwat—
Tochi, 1914-15.—
N.-W. F. F., 1917.—

Digh Ram—
N.-W. F. F., 1917.—

Sheochand, II—
N.-W. F. F., 1917.—

2-10th Jats.

SUBADAR MAJOR—

Shib Lal—*Bahadur. Order of British India, 2nd Class.*
N.-W. Frontier of India, 1897-98.—
Tochi.—Medal with clasp.
China, 1900.—Medal.
France 1917.—
Mesopotamia, 1917.—

11th Rajputs.

SUBADAR-MAJOR—

Jugraj Singh—
N.-W. Frontier of India, Waziristan, 1901-02.—Medal with clasp.

SUBADARS—

Matapalt Singh—*Bahadur.*
Burma, 1887-89.—Medal with clasp.
N.-W. Frontier of India, Waziristan, 1901-02.—Medal with clasp.

Durga Singh—
N.-W. Frontier of India, Hazara, 1891.—Medal with clasp.
N.-W. Frontier of India, Waziristan, 1901-02.—Medal with clasp.

Narayan Singh—
N.-W. Frontier of India, Waziristan, 1901-02.—Medal with clasp.

Lalbahadur Singh—
China, 1900-01.—Medal.
N.-W. Frontier of India, Waziristan, 1901-02.—Medal with clasp.

Bhagwanbakeh Singh—
N.-W. Frontier of India, Waziristan, 1901-02.—Medal with clasp.

Mahatam Singh—
N.-W. Frontier of India, Waziristan, 1901-02.—Medal with clasp.

Jagatpal Singh—
N.-W. Frontier of India, Waziristan.—Medal with clasp.

JEMADARS—

Rikhi Singh—
N.-W. Frontier of India, Waziristan, 1901-02.—Medal with clasp.

Bhagi Singh—
N.-W. Frontier of India, Waziristan, 1901-02.—

Ram Singh—
N.-W. Frontier of India, Waziristan, 1901-02.—

Pertab Singh—
N.-W. Frontier of India, Waziristan, 1901-02.—Medal with clasp.

Jagesar Singh—
N.-W. Frontier of India, Waziristan, 1901-02.—Medal with clasp.

Ramesar Singh—
N.-W. Frontier of India, Waziristan, 1901-02.—Medal with clasp.

Parbhu Singh—
N.-W. Frontier of India, Waziristan, 1901-02.—Medal with clasp.

Jalesar Singh—
N.-W. Frontier of India, Waziristan, 1901-02.—Medal with clasp.

Ram Biswas Singh—
N.-W. Frontier of India, Waziristan, 1901-02.—Medal with clasp.

Beni Singh—
N.-W. Frontier of India, Waziristan, 1901-02.—Medal with clasp.

Dan Bahadur Singh—
N.-W. Frontier of India, Waziristan, 1901-02.—Medal with clasp.

Suraj Bali Singh—
N.-W. Frontier of India, Waziristan, 1901-02.—Medal with clasp.

Bhawani Baksh Singh—
N.-W. Frontier of India, Waziristan, 1901-02.—Medal with clasp.

Deshpati Singh—
N.-W. Frontier of India, Waziristan, 1901-02.—Medal with clasp.

Mahabir Singh—
N.-W. Frontier of India, Waziristan, 1901-02.—Medal with clasp.

Dhenuk Singh—
N.-W. Frontier of India, Wazirstan, 1901-02.—Medal with clasp.

Inderpal Singh—
N.-W. Frontier of India, Wazirstan, 1901-02.—Medal with clasp.

12th Pioneers (The Kelat-i-Ghilzai Regiment).

SUBADARS—

Gyan Singh—
N.-W. Frontier of India, 1897-98.—Defence of Chakdara.—Malakand.—Operations in the Upper Swat and Bajour.—Medal with clasp.
Tirah, 1897-98.—Clasp.
China, 1900-01.—Medal with clasp.

Binda Singh—
Mekran Field Force, 1898.—Action of fort Drosh.

Bhola Singh—
Punjab Frontier, 1897-98.—Medal with clasp.

Pala Singh—
Tibet Mission, 1903-04.—Medal.

JEMADARS—

Mala Singh—
Tibet Mission, 1903-04.—Medal.

Har Narain—
China, 1900-01.—Medal.

Sultan—
China, 1900-01.—Medal.

Naraiyan Singh—
Punjab Frontier, 1897-98.—Medal with clasp.

2nd Battalion 12th Pioneers (The Kelat-i-ghilzai Regiment).

SUBADAR MAJOR—

Man Singh—
War 1914.—Mesopotamia Expeditionary Force, 1916-17.

SUBADARS—

Mohammad Hassan, *Bahadur—Order of Br. India, 2nd class, I.D.S.M.*
China, 1900.—Medal with clasp.
War, 1914.—Mesopotamia Expeditionary Force, 1916-17.

Hassan Khan—
War, 1914.—Mesopotamia Expeditionary Force, 1916-17.

Radha Singh—
War, 1914, France.—Indian Expeditionary Force "A", 1915-16.
War, 1914.—Mesopota mi Expeditionary Force, 1916-17.

Labh Singh—
N.-W. Frontier Mohmand Blockade, 1916.
N. Waziristan F. F., 1917.

Nihala—
China, 1900-01.—Medal.
N.-W. Frontier Mohmand Blockade, 1916.
N. Waziristan F. F., 1917.

Maru—
Mohmand Blockade, 1916.
N. Waziristan Field Force, 1917.

JEMADARS—

Nabi—
Mesopotamia Expeditionary Force, 1916-17.

Abdullah—
Mesopotamia Expeditionary Force, 1916-17.

Jai Ram—
Mohmand Blockade, 1916.

Ram Sarup—
N. Waziristan Field Force, 1917.

13th Rajputs (The Shekhawati Regiment).

SUBADAR-MAJOR—

Bakhtawar Singh—
Chitral, 1895.—Relief of Chitral.—Medal with clasp.
N.-W. Frontier of India, Waziristan, 1901-02.—Clasp.
I. E. F. "B" East Africa, 1914-15.—Indian Order of Merit, 2nd class.
Mesopotamia, 1917-18.

SUBADARS—

Dhan Singh—
I. E. F. "B" East Africa, 1914-15.
Mesopotamia, 1917.

Mangal Singh—
South Africa, 1902.—Medal with clasp.
China, 1900.—Medal.
I. E. F. "B" East Africa, 1914-15.
Mesopotamia, 1917.

Dal Sing—
N.-W. Frontier of India, Waziristan, 1901-02.—Medal and clasp.
I. E. F. "B" East Africa, 1914-15.
Mesopotamia, 1917.

Arjan Singh—
N.-W. Frontier of India, Waziristan, 1901-02.—Medal and clasp.

Bhur Singh—
N.-W. Frontier of India, Waziristan, 1901-02.—Medal and clasp.
I. E. F. "B" East Africa, 1914-15.
Mesopotamia, 1917.

Indar Singh—
N.-W. Frontier of India, Waziristan, 1901-02.—Medal and clasp.
Mesopotamia, 1918.

Rukam Singh—
I. E. F. "B" East Africa, 1914-15.
Mesopotamia, 1917.

Fateh Din Khan—
Mesopotamia, 1918.

JEMADARS—

Rur Singh—
N.-W. Frontier of India, Waziristan, 1901-02.—Medal with clasp.
Mesopotamia, 1917.

Mamand Singh—
N.-W. Frontier of India, Waziristan, 1901-02.—Medal and clasp.
I. E. F. "B" East Africa, 1914-15.
Mesopotamia, 1917.

Hukam Singh—
I. E. F. "B" East Africa, 1914-15.
Mesopotamia, 1917.

Santa Singh—
N.-W. Frontier of India, Waziristan, 1901-02.—Medal and clasp.
Mesopotamia, 1917.
France, 1914-16.

Mohar Singh—
N.-E. Frontier of India, Waziristan, 1901-02.—Medal with clasp.
I. E. F. "B," East Africa, 1914-15.
Mesopotamia, 1917.

Shimbhu Singh—
N.-W. Frontier of India, Waziristan,, 1901-02.—Medal and clasp.

Umrao Singh—
Punjab Frontier, 1897-98.—Medal and clasp.
N.-W. Frontier of India, Waziristan, 1901-02.
Mesopotamia, 1915.

14th King George's Own Ferozepore Sikhs.

SUBADAR-MAJOR—

Sham Singh—
N.-W. Frontier of India, Hazara, 1888.—Medal with clasp.
N.-W. Frontier of India, 1897-98.—Tochi.—Medal with clasp.
China, 1900.—Medal.
Indian Distinguished Service Medal. Medaille Militaire.

SUBADARS—

Narain Singh—*Indian Distinguished Service Medal.*

Jaimel Singh—
N.-W. Frontier of India, Waziristan, 1894-95.—Medal with clasp.
N.-W. Frontier of India, Tochi, 1897-98.—Medal with clasp.
British Central Africa, 1899-1900.—Medal with clasp.
Asanti, 1900.—King's Medal.

Wazir Singh—
N.-W. Frontier of India, Waziristan, 1894-95.—Medal with clasp.
N.-W. Frontier of India, Tochi, 1897-98.—Medal with clasp.
China, 1900.—Medal.
East Africa, Nandi F. F., 1905-06.—Medal and clasp

(African General Service.)

Sant Singh—
N.-W. Frontier of India, Tochi, 1897-98.—(Medal and clasp.)
China, 1900.—Medal.

Phuman Singh—
N.-W. Frontier of India, Tochi, 1897-98.—(Medal and clasp.)
China, 1900.—Medal.

Ghamna Singh—
N.-W. Frontier of India, Tochi, 1897-98.—(Medal and clasp.)
China, 1900.—Medal.

Jaswant Singh—
N.-W. Frontier of India, Tochi, 1897-98.—(Medal and clasp.)
China, 1900.—Medal.
East Africa, Nandi F. F., 1905-06.—Medal.

JEMADARS—

Partap Singh—
China, 1900.—Medal.

Mihan Singh—
N.-W. Frontier of India, Waziristan, 1894-95.—(Medal and clasp.)
N.-W. Frontier of India, Tochi, 1897-98.—(Medal and clasp.)
China, 1900.—Medal.

Mewa Singh—
N.-W. Frontier of India, Tochi, 1897-98.—(Medal and clasp.)
China, 1900.—Medal.

Sarmukh Singh—
China, 1900.—(Medal).

Mastan Singh—
China, 1900-01.—Medal.

Amar Singh—
China, 1900-01.—Medal.

Sawan Singh—
China, 1900-01.—Medal.

Ralla Singh—
China, 1900-01.—Medal.

15th Ludhiana Sikhs.

SUBADAR-MAJOR—

SUBADARS—

Phuman Singh—
Chitral, 1895.—Relief of Chitral.—Medal with clasp.
British East Africa, 1897-98.—Juba Expedition.—Medal with clasp.
Uganda, 1897-98.—2 clasps. *Indian Order of Merit, 3rd class.*

Bhagwan Singh—
Chitral, 1895.—Relief of Chitral.—Medal with clasp.
N.-W. Frontier of India, 1897-98.—2 clasps.
Tirah, 1897-98.—Clasp.

Bishn Singh—
N.-W. Frontier of India, 1897-98.—Medal with 2 clasps.
Tirah, 1897-98.—Clasp.

Gurdit Singh—

N.-W. Frontier of India, 1897-98.—Medal with 2 clasps.

Tirah, 1897-98.—Clasp.

Dhana Singh—

N.-W. Frontier of India, 1897-98.—Medal with 2 clasps.

Tirah, 1897-98.—Clasp.

JEMADARS—

Sundar Singh—

Chitral, 1895.—Relief of Chitral.—Medal with clasp.

N.-W. Frontier of India, 1897-98.—2 clasps.

Tirah, 1897-98.—Clasp.

Mewa Singh—

N.-W. Frontier of India, 1897-98.—Medal with 2 clasps.

Tirah, 1897-98.—Clasp.

Gopal Singh—

Chitral, 1895.—Relief of Chitral.—Medal with clasp.

N.-W. Frontier of India, 1897-98.—2 clasps.

Tirah, 1897-98.—Clasp.

Sardara Singh—

N.-W. Frontier of India, 1897-98.—Medal with 2 clasps.

Tirah, 1897-98.—Clasp.

2-15th Ludhiana Sikhs.

SUBADAR-MAJOR—

Narayan Singh—

N.-W. Frontier of India, 2nd Miranzai, 1891.—Medal with clasp.

Chitral, 1895.—Relief of Chitral.—Medal with clasp.

N.-W. Frontier of India, 1897-98.—2 clasps.

Tirah, 1897-98.—3 clasps.

Ashanti, 1901.—Medal with clasp.

SUBADARS—

Maima Singh—

Chitral, 1895.—Relief of Chitral.—Medal with clasp.

N.-W. Frontier of India, 1897-98.—2 clasps.

Tirah, 1897-98.—Clasp.

Bir Singh, M.C.—

N.-W. Frontier of India, 1897-98.—Medal with 2 clasps.

Tirah, 1897-98.—Clasp.

Albel Singh—

N.-W. Frontier of India, 1897-98.—Medal with 2 clasps.

Tirah, 1897-98.—Clasp.

Basant Singh—

N.-W. Frontier of India, 1897-98.—Medal with 2 clasps.

Tirah, 1897-98.—Clasp.

16th Rajputs (The Lucknow Regiment).

SUBADARS—

Narain Singh—

N.-W. Frontier of India, Waziristan, 1901-02.—Medal with clasp.

Sakalip Singh—

Burma, 1887—89.—Medal with 2 clasps.

N.-W. Frontier of India, 1897-98.—Malakand. Utman Khel.—Buner. Action of the Tanga Pass.—Medal with clasp.

Bulaki Singh—

Burma, 1887-89.—Medal with 2 clasps.

N.-W. Frontier of India, 1897-98.—Malakand.—Utman Khel.—Buner. Action of the Tanga Pass.—Medal with clasp.

Kan Singh—

Chitral, 1895.—Relief of Chitral.—Medal with clasp.

Surju Singh—

N.-W. Frontier of India, 1897-98.—Malakand.—Utman Khel.—Buner. Action of the Tanga Pass.—Medal with clasp.

Mohan Singh—

N.-W. Frontier of India, 1897-98.—Malakand. Utman Khel.—Buner. Action of the Tanga Pass.—Medal with clasp.

Mahadeo Singh (I)—

N.-W. Frontier of India, 1897-98.—Malakand. Utman Khel.—Buner. Action of the Tanga Pass.—Medal with clasp.

Amresar Singh—

N.-W. Frontier of India, 1897-98.—Malakand. Utman Khel.—Buner. Action of the Tanga Pass.—Medal with clasp.

Mahadeo Singh (II)—

N.-W. Frontier of India, 1897-98.—Malakand. Utman Khel.—Buner. Action of the Tanga Pass—Medal with clasp.

JEMADARS—

Shankar Singh—

N.-W. Frontier of India, 1897-98.—Malakand. Utman Khel.—Buner. Action of the Tanga Pass.—Medal with clasp.

Shiunath Singh—

N.-W. Frontier of India, 1897-98.—Malakand. Utman Khel.—Buner. Action of the Tanga Pass.—Medal with clasp.

Sarabjit Singh—

N.-W. Frontier of India, 1897-98.—Malakand. Utman Khel. Buner.—Action of the Tanga Pass.—Medal with clasp.

17th Infantry (The Loyal Regiment).

SUBADAR-MAJOR—

Muhammad Sulaiman Khan, I.O.M.—

N.-W. Frontier of India, Waziristan, 1901-02.—Medal with clasp.

SUBADARS—

Ashraf Ali Khan—
N.-W. Frontier of India, Waziristan, 1901-02.—Medal with clasp.

Amir Khan—
N.-W. Frontier of India, Waziristan, 1901-02.—Medal with clasp.

Hayat Khan—
N.-W. Frontier of India, 1897-98.—Medal with clasp.
Tirah, 1897-98.—Clasp.
South African War, 1899-1902.—Queen's Medal with 3 clasps.—King's Medal with 2 clasps.

JEMADARS—

Muhammad Ismail Khan—
N.-W. Frontier of India, Waziristan, 1901-02.—Medal with clasp.

Fajar Ali Khan—
N.-W. Frontier of India, Waziristan, 1901-02.—Medal with clasp.

Imam Ali Khan—
China, 1901.—Medal.

Faiz Muhamed Khan—
N.-W. Frontier of India, Waziristan, 1901-02.—Medal with clasp.

Qudratullah Khan—
N.-W. Frontier of India, Waziristan, 1901-02.—Medal with clasp.

Tungal Khan—
N.-W. Frontier of India, Waziristan, 1901-02.—Medal with clasp.

Yusuf Khan—
N.-W. Frontier of India, Waziristan, 1901-02.—Medal with clasp.

Anwar Khan—
N.-W. Frontier of India, Waziristan, 1901-02.—Medal wich clasp.

Abdul Ghafur Khan—
N.-W. Frontier of India, Waziristan, 1901-02.—Medal with clasp.

Karam Khan—
N.-W. Frontier of India, Waziristan, 1901-02.—Medal with clasp.

2-17th Infantry (the Loyal Regiment).

SUBADAR-MAJOR—

Yakub Ali Khan—
German East Africa, 1915-18.—Indian Distinguished Service Medal.
Mentioned in despatches.

SUBADARS—

Rahmat Khan—
France, 1914-17.
Mesopotamia, 1917.

Allah Bakhsh—
German East Africa, 1915-17.

Wazir Khan—
N.-W. Frontier of India, Waziristan, 1901-02.—Medal with clasp.
German East Africa, 1915-18.

Muhammed Abdul Karim—
Burma, 1887-89.—Medal with clasp.
Burma, 1889-92.—Clasp.
N.-W. Frontier of India, Waziristan, 1901-02.—Medal with clasp.
German East Africa, 1915-17.

Umrao Ali Khan—
N.-W. Frontier of India, Waziristan, 1901-02.—Medal with clasps.
German East Africa, 1915-17.

JEMADARS—

Arshad Khan—
N.-W. Frontier of India, Waziristan, 1901-02.—Medal with clasps.
France, 1914.

Mansad Khan—
German East Africa, 1915-17.

Mubarik Ali Khan—
N.-W. Frontier of India, Waziristan, 1901-02.—Medal with clasp.
France, 1915.
German East Africa, 1916-18.

Faiz Muhammed Khan—
German Eaet Africa, 1915-18.

Muhammed Yasin Khan—
German East Africa, 1915-18.

Karim Ali Khan—
China, 1900.—Medal.
German East Africa, 1915-18.

Wali Muhammed Khan—
German East Africa, 1915-18.

Nur Ahmad Khan—
German East Africa, 1917-18.

Nabi Bakhsh—
German East Africa, 1915-17.

1-18th Infantry.

SUBADAR-MAJOR—

Karam Sher Khan, *Bahadur, Order of Br. Ind., 2nd cl.*

N.-E. Frontier of India, Chin-Lushai, 1891.—Medal with clasp.

SUBADAR—

Boga Khan—

China, 1900.—Medal and clasp.

2-18th Infantry.

SUBADAR—

Nathu Khan—

France, 1915.

Mohammad Ali—

France, 1915-16.

G. E. Africa, 1916-17.

1-19th Punjabis.

SUBADAR-MAJORS—

Fateh Khan, *Sardar Bahadur—Order of Br. India, 1st class—*

Burma, 1885—87.—Medal with clasp.

N.-W. Frontier of India, Isazai, 1892.

Chitral, 1895.—Relief of Chitral.—Medal with clasp.

N.-W. Frontier of India, 1897-98.—Tochi.—Clasp.

Tibet, 1903-04.—Medal.

N.-W. Frontier of India, 1908.—Operations in the Mohmand country.—Medal with clasp.

Teja Singh—

N.-W. Frontier of India, 1908.—Operations in the Mohmand country.—Medal with clasp.

SUBADARS—

Isar Singh—

N.-W. Frontier of India, Hazara, 1891.—Clasp.

N.-W. Frontier of India, Miranzai, 1891.—Medal with clasp.

Tibet, 1903-04.—Medal.

Ralla Singh, I.D.S.M.—

N.-W. Frontier of India, Hazara, 1891.—Clasp.

N.-W. Frontier of India, Miranzai, 1891.—Medal with clasp.

N.-W. Frontier of India, Waziristan, 1894-95.—Clasp.

Tibet, 1903-04.—Medal.

N.-W. Frontier of India, 1908.—Operations in the Mohmand country. Engagement of Kargha, 24th May 1908.—Medal with clasp.

Jalal Khan—*Indian Distinguished Service Medal.*

N.-W. Frontier of India, Hazara, 1891.—Clasp.

N.-W. Frontier of India, Miranzai, 1891.—Medal with clasp.

N.-W. Frontier of India, Waziristan, 1894-95.—Clasp.

Tibet, 1903-04.—Medal.

N.-W. Frontier of India, 1908.—Operations in the Mohmand country. Engagement of Kargha, 24th May 1908.—Medal with clasp.

Maula Bakhsh—

Tibet, 1903-04.—Medal.

Lal Khan—

Tibet, 1903-04.—Medal.

Hukam Singh—

N.-W. Frontier of India, Waziristan, 1894-95.—Medal and clasp.

Tibet, 1903-04.—Medal.

Fazal Khan—

Ogaden Expedition, 1899.—Medal and clasp.
Tibet, 1903-04.—Medal.

Sohan Singh—

N.-W. Frontier of India, Hazara, 1891.—Clasp.
N.-W. Frontier of India, Miranzai, 1891.—Medal with clasp.

N.-W. Frontier of India, Waziristan, 1894-95.—Clasp.

Tibet, 1903-04.—Medal.

Bal Singh—

Tibet, 1903-04.—Medal.

JEMADARS—

Bak Khan—

Tibet, 1903-04.—Medal.

Suhel Singh—

Tibet, 1903-04.—Medal.

Budha Singh—

Tibet, 1903-04.—Medal.

Nihal Singh—

Tibet, 1903-04.—Medal.

Harnam Singh—

Tibet, 1903-04.—Medal.

Fateh Mohd.—

Tibet, 1903-04.—Medal.

Fazal Khan—

Tibet, 1903-04.—Medal.

Azizullah—

N.-W. Frontier of India.—Operations in the Mohmand country.—Medal and clasp.

Madan Khan—

Tibet, 1903-04.—Medal.

Sahib Shah—

Tibet, 1903-04.—Medal.

Sundar Singh—

Tibet, 1903-04.—Medal.

Saida Khan—
Tibet, 1903-04.—Medal.

Santa Singh—
Tibet, 1903-04.—Medal.

Chaghata Khan—
Tibet, 1903-04.—Medal.

Faiz Talab—
Tibet, 1903-04.—Medal.
N.-W. Frontier of India, 1908.—Operations in the Mohmand country.—Medal with clasp.

Rajwali—
• *Tibet, 1903-04.*—Medal.

Zarulla Khan—
Tibet, 1903-04.—Medal.

2-19th Punjabis.

SUBADAR-MAJOR—

Teja Singh—
N.-W. Frontier of India, 1908.—Operations in the Mohmand country.—Medal with clasp.

SUBADARS—

Jalal Khan, *Bahadur,* I.D.S.M.—2nd Class Order of Br. India.
Hazara Expedition, 1891.—Clasp.
Miranzai Expedition, 1891.—Medal and clasp.
Waziristan Expedition, 1894-95.—Clasp.
Tibet Mission, 1903-04.—Medal.
N.-W. Frontier of India, 1908.—Operations in the Mohmand country.—Medal with clasp.
Indian Distinguished Service Medal, 1908.

Sohan Singh—
Hazara Expedition, 1891.—Clasp.
Miranzai Expedition, 1891.—Medal and clasp.
Waziristan Expedition, 1894-95.—Clasp.
Tibet Mission, 1903-04.—Medal.

Azizullah—
N.-W. Frontier of India, 1908.—Operations in the Mohmand country.—Medal with clasp.

Budha Singh—
Tibet Mission, 1903-04.—Medal.

Khuda Bakhsh—
Tibet Mission, 1903-04.—Medal.

Ghafur Khan—
Tibet Mission, 1903-04.—Medal.
N.-W. Frontier of India, 1908.—Medal and clasp.

Mardan Khan—
Tibet Mission, 1903-04.—Medal.

Sundar Singh—
Tibet Mission, 1903-04.—Medal.

JEMADARS—

Sahib Shah—
Tibet Mission, 1903-04.—Medal.

Sundar Singh—
Tibet Mission, 1903-04.—Medal.

Talok Singh—
Tibet Mission, 1903-04.—Medal.

Nand Singh—
Tibet Mission, 1903-04.—Medal.

Surayan Singh—
Tibet Mission, 1903-04.—Medal.

Ghulam Haidar—
Tibet Mission, 1903-04.—Medal.

Ismail Khan—
Tibet Mission, 1903-04.—Medal.
N.-W. Frontier of India, 1908.—Indian G. S Medal and clasp 1908.

Lal Khan—
Tibet Mission, 1903-04.—Medal.
Indian Distinguished Service Medal War, 1914.

Waryam Singh—
Somaliland Expedition, 1902—04.—Medal and 2 clasps.

Sarwar Khan—
Tibet Mission, 1903-04.—Medal.

Kasim Khan—
N.-W. Frontier of India, 1908.—India G. S. Medal and clasp 1908.

Madat Ali—
Tibet Mission, 1903-04.—Medal.
Somaliland, 1908-10.—Medal.

Rakhim Shah—
Tibet Mission, 1903-04.—Medal.
N.-W. Frontier of India, 1908.—India G. S. Medal and clasp.
N.-W. Frontier of India, 1897-98.—Engagement near Shabkadr, 9th August 1897.—Mohmand.—Malakand.—Buner. Action of the Tanga Pass.—Medal with clasp.
China, 1900.—Medal.

20th Duke of Cambridge's Own Infantry (Brownlow's Punjabis).

SUBADARS—

Moa Sinhg—
N.-W. Frontier of India, Waziristan, 1894-95.—Medal with clasp.
N.-W. Frontier of India, 1897-98.—Eugagement near Subkadar, 9th August 1897. Mohmand.—Malakund.—Boner. Action of the Tanga Pass.—Medal with clasp.

SUBADARS—

Ganda—

N.-W. Frontier of India, Waziristan, 1894-95.—Medal with clasp.

N.-W. Frontier of India, 1897-98.—Engagement near Shabkadr, 9th August 1897.—Malakand. Buner. Action of the Tanga Pass.—Medal with clasp.

Sant Singh—

N.-W. Frontier of India, Waziristan, 1894-95.—Medal with clasp.

Soudan, 1896.—Dongola Expedition.—Medals (Queen's and Khedive's).

N.-W. Frontier of India, 1897-98.—Engagement near Shabkadr, 9th August 1897. Mohmand. Malakand. Buner. Action of the Tanga Pass.—Medal with clasp.

China, 1900.—Medal.

Brahwan—

N.-W. Frontier of India, Waziristan, 1894-95.—Medal with clasp.

N.-W. Frontier of India, 1897-98.—Engagement near Shabkadr, 9th August 1897. Mohmand. Malakand. Buner. Action of the Tanga Pass.—Medal with clasp.

China, 1900.—Medal.

Kashmir Singh—

N.-W. Frontier of India, 1897-98.—Buner. Action of the Tanga Pass.—7th January 1898.—Medal with clasp.

China, 1900-01.—Medal.

Ladha Singh—

N.-W. Frontier of India, 1897-98.—Tochi.—Medal with clasp.

China, 1900-01.—Medal.

Khushala—

N.-W. Frontier of India, 1897-98.—Buner. Action of the Tanga Pass.—Medal with clasp.

China, 1900-01.—Medal.

Habib Khan—

N.-W. Frontier of India, Waziristan, 1894-95.—Medal with clasp.

N.-W. Frontier of India, 1897-98.—Mohmand.—Malakand.—Buner. Action of the Tanga Pass.—Medal with clasp.

China, 1900.—Medal.

JEMADARS—

Udham Singh—

N.-W. Frontier of India, 1897-98.—Buner. Action of the Tanga Pass.—7th January 1898.—Medal with clasp.

China, 1900-01.—Medal.

Hasan Khan—

N.-W. Frontier of India, 1897-98.—Engagement near Shabkadr, 9th August 1897. Mohmand Malakand. Buner. Action of the Tanga Pass—Medal with clasp.

China, 1900-01.—Medal.

Niaz Gul—

China, 1900.—Medal.

Bua Ditta—

N.-W. Frontier of India, Waziristan, 1894-95.—Medal and clasp.

N.-W. Frontier of India, 1897-98.—Engagement near Shabkadr, 9th August 1897. Malakand. Mohmand. Buner. Action of the Tanga Pass.—Medal with clasp.

China, 1900.—Medal.

Alam Khan—

China, 1900.—Medal.

Sant Singh—

N.-W. Frontier of India, Waziristan, 1894-95.—Medal with clasp.

N.-W. Frontier of India, 1897-98.—Engagement near Shabkkadr, 9th August 1897. Malakand, Mohmand. Buner. Action of the Tanga Pass.—Medal with clasp.

China, 1900.—Medal.

Lakha Singh—

China, 1900.—Medal.

1-21st Punjabis.

SUBADAR-MAJOR—

Waryam Singh—

Soudan, 1896.—Dongola Expedition.—2 Medals.

N.-W. Frontier of India, 1897-98.—Malakand Utman Khel. Buner. Action of the Tanga Pass.—Medal with clasp.

N.-W. Frontier of India, 1908.—Operations in the Mohmand country.—Medal with clasp.

SUBADARS—

Allah Dad Khan—

N.-W. Frontier of India, 1908.—Operations in the Mohmand country.—Medal with clasp.

Jowala Singh—

Soudan, 1896.—Dongola Expeditions.—2 Medals.

N.-W. Frontier of India, 1897-98.—Malakand, Utman Khel, Buner. Action of the Tanga Pass.—Medal with clasp.

N.-W. Frontier of India, 1908.—Operations in the Mohmand country.—Medal with clasp.

Nadir Ali—

N.-W. Frontier of India, 1897-98.—Malakand, Utman Khel. Buner. Action of the Tanga Pass.—Medal with clasp.

China, 1900.—Medal.

Lachman Singh—

N.-W. Frontier of India, 1897-98.—Malakand. Utman Khel. Bunner. Action of the Tanga Pass.—Medal with clasp.

Shad Muhammad—

N.-W. Frontier of India, 1908.—Operations in the Mohmand country.—Medal with clasp.

Achhar Singh—

N.-W. Frontier of India, 1908.—Operations in the Mohmand country.—Medal with clasp.

JEMADARS—

Fateh Ali—

N.-W. Frontier of India, 1908.—Operations in the Mohmand country.—Medal with clasp.

Makhana Khan—

N.-W. Frontier of India, 1908.—Operations in the Mohmand country.—Medal with clasp.

Kirpa Singh—

N.-W. Frontier of India, 1908.—Operations in the Mohmand country.—Medal with clasp.

Sohbat Khan—

N.-W. Frontier of India, 1908.—Operations in the Mohmand country.—Medal with clasp.

Sadhu Singh—

N.-W. Frontier of India, 1908.—Operations in the Mohmand country.—Medal with clasp.

Ghagga Singh—

N.-W. Frontier of India, 1908.—Operations in the Mohmand country.—Medal with clasp.

2-21st Punjabis.

SUBADAR-MAJOR—

Wazir Singh—

N.-W. Frontier of India, 1897-98.—Malakand. Utman Khel. Buner. Action of the Tanga Pass.—Medal with clasp.

The War, 1914. Waziristan Field Force, 1917.

SUBADAR—

Hira Singh—

Soudan, 1896.—Dongola Expedition.—2 Medals.

N.-W. Frontier of India, 1897-98.—Malakand.—Medal with clasp.

The War, 1914. Waziristan Field Force, 1917.

Piran Ditta—

N.-W. Frontier of India, 1897-98.—Malakand. Utman Khel. Buner. Action of the Tanga Pass.—Medal with clasp.

N.-W. Frontier of India, 1908.—Operations in the Mohmand country.—Medal with clasp.

The War 1914—France, 1914.—Waziristan Field Force, 1917.

Nur Mast—

N.-W. Frontier of India, 1897-98.—Malakand.—Utman Khel. Buner. Action of the Tanga Pass.—Medal with clasp.

N.-W. Frontier of India, 1908.—Operations in the Mohmand country.—Medal with clasp

The War, 1914. Waziristan Field Force, 1917.

Amir Khan—

N.-W. Frontier of India, 1897-98.—Malakand.—Utman Khel. Buner. Action of the Tanga Pass.—Medal with clasp.

The War, 1914. France, 1915-16.—Waziristan Field Force, 1917.

Isar Singh—

N.-W. Frontier of India, 1908.—Operations in the Mohmand country.—Medal with clasp.

The War, 1914 Waziristan Field Force, 1917.

Mohammad Latif—

N.-W. Frontier of India, 1908.—Operations in the Mohmand country.—Medal with clasp.

The War, 1914. Waziristan Field Force, 1917

Bhag Singh—

The War, 1914. Waziristan Field Force, 1917.

JEMADARS—

Damodar—

N.-W. Frontier of India, 1897-98.—Malakand. Utman Khel. Bunner. Action of the Tanga Pass.—Medal with clasp.

N.-W. Frontier of India, 1908.—Operations in the Mohmand country.—Medal with clasp.

The War, 1914. Mesopotamia, 1915.—Waziristan Field Force, 1917.

Salam Din—

N.-W. Frontier of India, 1908.—Operations in the Mohmand country.—Medal with clasp.

The War, 1914. Waziristan Field Force, 1917.

Daurani—

N.-W. Frontier of India, 1908.—Operations in the Mohamad country.—Medal with clasp.

Isar Singh—I. D. S. M.

N.-W. Frontier of India, 1908.—Operations in the Mohmand country.—Medal with clasp.

The War, 1914. France, 1914-17.—Awarded I. D. S. M. for gallantry in the field.—Waziristan Field Force, 1917.

Sher Baz—

N.-W. Frontier of India, 1908.—Operations in the Mohmand country.—Medal with clasp.

The War, 1914. Mesopotamia, 1916-17.

Hassan Shah—

N.-W. Frontier of India, 1897-98.—Malakand Utman Khel. Buner. Action of the Tanga Pass.—Medal with clasp.

N.-W. Frontier of India, 1908.—Operations in the Mohamand country.—Medal with clasp.

The War, 1914. Waziristan Field Force, 1917.

Dulip Singh—

N.-W. Frontier of India, 1897-98.—Malakand. Utman Khel. Bunner. Action of the Tanga Pass.—Medal with clasp.

N.-W. Frontier of India, 1908.—Operations in the Mohamand country.—Medal with clasp.

The War, 1914. Waziristan Field Force, 1917.

Hakim Singh—

N.-W. Frontier of India, 1897-98.—Malakand. Utam Khel. Bunner. Action of the Tanga Pass.—Medal with clasp.

N.-W. Frontier of India, 1908.—Operations in the Mohmand country.—Medal with clasp.

The War, 1914. Waziristan Field Force, 1917.

22nd Punjabis.

SUBADAR-MAJOR—

Raja Khan—

N.-W. Frontier of India, 1897-98.—Relief of Malakand. Operations in Bajour. Mohmand—Medal with 2 clasps.

Tirah, 1897-98.—Clasp.

N.-W. Frontier of India, 1898-99.—Operations against the Para Chamkanis.

N.-W. Frontier of India, 1902.—Operations against the Darwesh Khel Waziris.

N.-W. Frontier of India, 1908.—Operations in the Mohmand country. Engagement of Kargha, 24th May 1908.—Medal with clasp.

Rala Singh—

N.-W. Frontier of India, 1898-99.—Operations against Para Chamkanis.
N.-W. Frontier of India, 1902.—Operations against the Darwesh Khel Waziris.
Tibet, 1903-04.—Clasp.
N.-W. Frontier of India, 1908.—Operations in the Mohmand country. Engagement of Kargha, 24th May 1908.—Medal with clasp.

Badan Singh—

N.-W. Frontier of India, 1902.—Operations against the Darwesh Khel Waziris.—Tibet Mission, 1903-04—Medal.

N.-W. Frontier of India, 1908.—Operations in Mohmand country. Engagement of Kargha, 24th May 1908.—Medal with clasp.

Mehar Singh—

N.-W. Frontier of India, 1897-98.—Relief of Malakand, Malakand. Operations in Bajour, Mohmand.—Medal and 2 clasps.

Tirah, 1897-98.—Clasp.

N.-W. Frontier of India, 1902.—Operations against the Darwesh Khel Waziris.
Tibet, 1903-04.—Medal.
N.-W. Frontier of India, 1908.—Operations in the Mohmand country. Engagement of Kargha, 24th May 1908.—Medal and clasp.

Gurdit Singh—

N.-W. Frontier of India, 1897-98.—Relief of Malakand. Operations in Bajour, Mohmand.—Medal and 2 clasps.
Tirah, 1897-98.—Clasp.
China, 1900.—Relief of Pekin.—Medal with clasp.
N.-W. Frontier of India, 1902.—Operations against the Darwesh Khel Waziris.
N.-W. Frontier of India, 1908.—Operations in the Mohmand country. Engagement of Kargha 24th May 1908. Medal with clasp.

Roda Singh—

N.-W. Frontier of India, 1897-98.—Malakand· Operation in Bajour, Mohmand.—Medal and clasp.
Tirah, 1897-98.—Clasp.
N.-W. Frontier of India, 1898-99.—Operations against the Para Chamkanies.
N.-W. Frontier of India, 1902.—Operations against the Darwesh Khel Waziris.
N.-W. Frontier of India, 1908.—Operations in Mohmand country. Engagement of Kargha, 24th May 1908.—Medal and clasp.

Ali Beg—

N.-W. Frontier of India, 1897-98.—Relief of Malakand. Operations in Bajour, Mohmand—Medal with 2 clasps.
Tirah, 1897-98.—Clasp.
N.-W Frontier of India, 1902.—Operations against the Darwesh Khel Waziris.
N.-W. Frontier of India, 1908.—Operations in the Mohmand country. Engagement of Kargha, 24th May 1908.—Medal with clasp.

Ahmed Din—

N.-W. Frontier of India, 1902.—Operations against the Darwesh Khel Waziris.

Azim Khan—

N.-W. Frontier of India, 1902.—Operations against the Darwesh Khel Waziris.
N.-W. Frontier of India, 1908.—Operations in the Mohmand country. Engagement of Kargha, 24th May 1908.—Medal and clasp.

Suhel Singh—

N.-W. Frontier of India, 1897-98.—Relief of Malakand, Malakand. Operations in Bajour, Mohmand.—Medal and 2 clasps.
Tirah, 1897-98.—Clasp.
N.-W. Frontier of India, 1898-99.—Operations against the Para Chamkanies.
China, 1900.—Relief of Pekin.—Medal and clasp.
N.-W. Frontier of India, 1902.—Operations against the Darwesh Khel Waziris.
N.-W. Frontier of India, 1908.—Operations in the Mohmand country. Engagement of Kargha, 24th May 1908.—Medal with clasp.

Ali Akbar Khan—

N.-W. Frontier of India. 908.—Operations in the Mohmand country. Engagement of Kargha, 24th May 1908.—Medal and clasp.

Bachitar Singh—

N.-W. Frontier of India, 1908.—Operations in the Mohmand country. Engagement of Kargha, 24th May 1908.—Medal and clasp.

JEMADARS—

Katha Singh—

N.-W. Frontier of India, 897-98.—Medal and clasp.
Tirah, 1897-98.—Clasp.
N.-W. Frontier of India, 902.—Operations against the Darwesh Khel Waziris.
N.-W. Frontier of India, 1908.—Operation in the Mohmand country. Engagement of Kargha, 24th May 1908.—Medal and clasp.

Qasin Khan—

N.-W. Frontier of India, 908.—Operations in the Mohmand country. Engagement of Kargha, 24th May 1908.—Medal and clasp.

Harnam Shing—

N.-W. Frontier of India, 1897-98.—Malakand operations in Bajour, Mohmand.—Medal with clasp.
Tirah, 1897-98.—Clasp.
N.-W. Frontier of India, 1898-99.—Operations against the Para Chamkanis.
China Expedition, 1900-01.—Relief of Pekin.—Medal with clasp.
N.-W. Frontier of India. 1902.—Operations against the Darwesh Khel Waziris.
N.-W. Frontier of India. 1908.—Operations in the Mohmand country. Engagement of Karghas 24th May 1908.—Medal with clasp.

Indar Singh—

N.-W. Frontier of India, 1897-98.—Relief of Malakand, Malakand and operations in Bajour, Mohmand.—Medal and clasp.
Tirah, 1897-98.—Clasp.
China Expedition, 1900-01.—Relief of Pekin.—Medal with clasp.
N.-W. Frontier of India, 1902.—Operations against the Darwesh Khel Waziris.
N.-W. Frontier of India, 1908.—Operations in the Mohmand country. Engagement of Kargha 24th May 1908.—Medal with clasp.

Lal Singh—

N.-W. Frontier of India, 1902.—Operation against the Darwesh Khel Waziris.
N.-W. Frontier of India, 1908.—Operations in the Mohmand country. Engagement of Karghar, 24th May 1908.—Medal and clasp.

Hakim Khan—

N.-W. Frontier of India, 1902.—Operations against the Darwesh Khel Waziris.
N.-W. Frontier of India, 1908.—Operations in the Mohmand country Engagement of Kargha, 24th May 1908.—Medal and clasp.

Fatohi Knan—

N.-W. Frontier of India, 1908.—Operations in the Mohmand ocuntry.—Engagement of Kargha, 24th May 1908.—Medal and clasp.

Muhammad Sher—

N.-W. Frontier of India, 1908.—Operations in the Mohmand country. Engagement of Kargha, 24th May 1908.—Medal and clasp.

Firoz Khan—

N.-W. Frontier of India, 1897-98.—Relief of Malakand, Operations in Bajour, Mohmand.—Medal with 2 clasps.
Tirah, 1897-98.—Clasp.
N.-W. Frontier of India, 1898-99.—Operations against the Para Ghamkanis.
China Expedition, 1900-01.—Relief of Pekin.—Medal with clasp.
N.-W. Frontier of India, 1902.—Operations against the Darwesh Khel Waziris.
N.-W. Frontier of India, 1908.—Operations in the Mohmand Country. Engagement of Kargha, 24th May 1908.—Medal and clasp.

Udam Singh—

N.-W. Frontier of India, 1908.—Operation in the Mohmand country. Engagement of Kargha, 24th May 1908.—Medal and clasp.

Wazir Singh—

China Expedition, 1900-01.—Relief of Pekin.—Medal with clasp.
N.-W. Frontier of India, 1902.—Operations against the Darwesh Khel Waziris.
N.-W. Frontier of India, 1908.—Operations in the Mohmand Country. Engagement of Kargha, 24th May 1908.—Medal and Clasp.

Said Ahmad—

N.-W. Frontier of India, 1897-98.—Relief of Malakand, Malakand Operations in Bajour, Mohmand. Medal with 2 clasps.
Tirah, 1897-98.—Clasp.
N.-W. Frontier of India, 1908.—Operations in Mohmand Country. Engagement of Kargha, 24th May 1908.—Medal and clasp.

Dewi Raj—

1st Miranzai Expedition, 1891.
N.-W. Frontier of India, 1897-98.—Relief of Malakand, Malakand. Operations in Bajour, Mohmand.—Medal with 2 clasps.
Tirah, 1897-98.—Clasp.
N.-W. Frontier of India.—Operations against the Darwesh Khel Waziris.
N.-W. Frontier of India, 1908.—Operations in the Mohmand Country. Engagement of Kargha, 24th May 1908.—Medal and clasp.

Labh Singh—

N.-W. Frontier of India, 1897-98.—Malakand, Operations in Bajour, Mohmand. Medal and clasp.
Tirah, 1897-98.—Clasp.
N.-W. Frontier of India, 1902.—Operations. against the Darwesh Khel Waziris.
N.-W. Frontier of India, 1902.—Operations in the Mohmund Country. Engagement of Kargha, 24th May 1908.—Medal and clasp.

Fateh Singh—

N.-W. Frontier of India, 1897-98.—Medal and Clasp.
Tirah, 1897-98.—Clasp.
N.-W. Frontier of India, 1898-99.—Operations against the Para Chamkanis.
China Expedition, 1900-01.—Relief of Pekin. Medal and clasp.
N.-W. Frontier of India, 1902.—Operations against the Darwesh Khel Waziris.
N.-W. Frontier of India, 1908.—Operations in the Mohmand Country. Engagement of Kargha, 24th May 1908.—Medal and clasp.

Rahim Ali—

Somaliland Field Force, 1904.—Medal and clasp.
N.-W. Frontier of India, 1908.—Operations in the Mohmand country. Engagement of Kargha, 24th May 1908.—Medal and clasp.

Saida Khan—

N.-W. Frontier of India, 1902.—Operations against the Darwesh Khel Waziris.
N.-W. Frontier of India, 1908.—Operations in the Mohmand country. Engagement of Kargha, 24th May 1908.—Medal and clasp.

Mohabat Khan—

N.-W. Frontier of India, 1908.—Operations in the Mohmand country. Engagement of Kargha, 24th May 1908.—Medal and clasp.

Muhammad Amir—

N.-W. Frontier of India, 1902.—Operations against the Darwesh Khel Waziris.
N.-W. Frontier of India, 1908.—Operations in the Mohmand country. Engagement of Kargha, 24th May 1908.—Medal and clasp.

Ghulam Rasul—

N.-W. Frontier of India, 1908.—Operations in the Mohmand country. Engagement of Kargha, 24th May 1908.—Medal and clasp.

Hardit Singh—

N.-W. Frontier of India, 1897-98.—Medal and clasp.
Tirah, 1897-98.—Clasp.
N.-W. Frontier of India, 1898-99.—Operations against the Para Chamkamis.
N.-W. Frontier of India, 1902.—Operations against the Darwesh Khel Waziris.
N.-W. Frontier of India, 1908.—Operations in the Mohmand country. Engagement of Kargha, 24th May 1908.—Medal and clasp.

2-22nd Punjabis.

Dasaundha Singh—*Bahadur.*

Order of Br. India, 2nd class.
N.-W Frontier of India, (1st) Miransai, 1891.
N.-W. Frontier of India, 1897-98.—Malakand, operations in Bajour, Mohmand.—Medal with clasp.
Tirah, 1897-98.—Clasp.
China, 1900.—Relief of Pekin.—Medal with clasp.
N.-W. Frontier of India, 1902.—Operations against the Darwesh Khel Waziris.
Tibet, 1903-04.—Medal.
N.-W. Frontier of India, 1908.—Operations in the Mohmand country. Engagement of Kargha, 24th May 1908.—Medal with clasp.

Ghulam Hussain—

N.-W. Frontier of India, 1902.—Operations against the Darwesh Khel Waziris.
N.-W. Frontier of India, 1908.—Operations in the Mohmand country. Engagement of Kargha, 24th May 1908.—Medal with clasp.

Jiwan Singh—

N.-W. Frontier of India, 1902.—Operations against the Darwesh Khel Waziris.

Nur Khan—

N.-W. Frontier of India, 1897-98.—Malakand operations in Bajaur, Mohmand.—Medal and clasp.

Tirah, 1897-98.—Clasp.

N.-W. Frontier of India, 1902.—Operations against the Darwesh Khel Waziris.

Nijabit Ali—

N.-W. Frontier of India, 1908.—Operations in the Mohmand country. Engagement of Kargha, 24th May 1908.—Medal and clasp.

Haji Khan—

N.-W. Frontier of India, 1908.—Operations in the Mohmand of Kargah, 24th May 1908.—Medal and clasp.

JAMADARS—

Chajja Singh—

N.-W. Frontier of India, 1902.—Operations against the Darwesh Khel Waziris.

N.-W. Frontier of India, 1908.—Operations in the Mohmand country. Engagement of Kargah, 24th May 1908.—Medal and clasp.

Khushal Singh—

N.-W. Frontier of India, 1908.—Operations in the Mohmand country. Engagement of Kargah, 24th May 1908.—Medal and clasp.

Kajir Khan—

N.-W. Frontier of India, 1902.—Operations against Darwesh Khel Waziris.

N.-W. Frontier of India, 1902.—Operation in the Mohmand country. Engagement of Kargha, 24th May 1908.—Medal and clasp.

ur Mohammed—

N.-W. Frontier of India, 1908.—Medal with clasp.

Abdull Khan—

N.-W. Frontier of India, 1908.—Operations in the Mohmand country. Engagement of Kargha, 24th May 1908. Medal and clasp.

Sardar Khan—

N.-W. Frontier of India, 1908.—Operations in the Mohmand country. Engagement of Kargha, 24th May 1908. Medal and clasp.

Khan Muhamad—

N.-W. Frontier of India, 1908.—Operations in the Mohmand country. Engagement of Kargha. 24th May 1908. Medal and clasp.

1/23rd Sikh Pioneers.

SUBADAR-MAJOR—

Man Singh—

N.-W. Frontier of India, (1st) Miranzai, 1891.

Chitral, 1895.—Relief of Chitral.—Medal with clasp.

N.-W. Frontier of India, Waziristan, 1901-02.—Clasp.

Tibet, 1903-04.—Medal.

N.-W. Frontier of India, 1908.—Operations in the Zakka Khel country.—Medal with clasp.

Kala Singh—

China, 1900.—Medal with clasp.

N.-W. Frontier of India, Waziristan, 1901-02.—Medal with clasp.

Tibet, 1903-04.—Medal.

N.-W. Frontier of India, 1908.—Operations in the Zakka Khel country.—Medal with clasp.

Bhola Singh—

Chitral, 1895.—Relief of Chitral.—Medal with clasp.

N.-W. Frontier of India, Waziristan, 1901-02.—Clasp.

Tibet, 1903-04.—Action at Niani, 26th June 1904.—Operations at and around Gyantse, between 5th May and 6th July 1904.—Medal with clasp.

N.-W. Frontier of India, 1908.—Operations in the Zakka Khel country.—Medal with clasp.

Sohan Singh—

N.-W. Frontier of India, Waziristan, 1901-02.—Medal with clasp.

Tibet, 1903-04.—Action at Niani, 26th June 1904. Operations at and around Gyantse, between 5th May and 6th July 1904.—Medal with clasp.

N.-W. Frontier of India, 1908.—Operations in the Zakka Khel country.—Medal with clasp.

Chanda Singh—

N.-W. Frontier of India, Waziristan, 1901-02.—Medal with clasp.

Tibet, 1903-04.—Medal.

N.-W. Frontier of India, 1908.—Operations in the Zakka Khel country.—Medal with clasp.

Bawa Singh—

N.-W. Frontier of India, Waziristan, 1901-02.—Medal with clasp.

Tibet, 1903-04.—Medal.

N.-W. Frontier of India, 1908.—Operations in the Zakka Khel country.—Medal with clasp.

Basawa Singh—

N.-W. Frontier of India, Waziristan, 1901-02.—Medal with clasp.

Tibet, 1903-04.—Action at Niani, 26th June 1904.—Operations at and around Gyantse, between 5th May and 6th July 1904.—Medal with clasp.

N.-W. Frontier of India, 1908.—Operations in the Zakka Khel country. Medal with clasp.

Jawahir Singh—

N.-W. Frontier of India, (1st) Miranzai, 1891.

Chitral, 1895.—Relief of Chitral.—Medal with clasp.

N.-W. Frontier of India, Waziristan, 1901-02.—Clasp.

Harnam Singh (I)—

Chitral, 1895.—Relief of Chitral. Medal with clasp.

N.-W. Frontier of India, Waziristan, 1901-02.—Clasp.

Tibet, 1903-04.—Action at Niani, 26th June 1904.—Operations at and around Gyantse, between 5th May and 6th July 1904.—Medal and clasp.

N.-W. Frontier of India, 1908.—Operations in the Zakka Khel country. Medal with clasp.

Bishan Singh—

N.-W. Frontier of India, (1st) Miranzai, 1891.

Chitral, 1895.—Relief of Chitral.—Medal with clasp.

Teja Singh, I.O.M.—

N.-W. Frontier of India, 1908.—Operations in the Zakka Khel country.—Medal with clasp.

Africa, Shimber Berris, 1914-15.—Medal with clasp.

JEMADARS—

Harnam Singh (III)—

N.-W. Frontier of India, Waziristan, 1901-02.—Medal with clasp.
Tibet. 1903-04.—Actions at Niani, 26th June 1904. —Operations at and around Gyantse, between 5th May and 6th July 1904.—Medal with clasp.
N.-W. Frontier of India, 1908.—Operations in the Zakka Khel country.—Medal with clasp.

Bhagat Singh—

N.-W. Frontier of India, Waziristan, 1901-02.—Medal with clasp.
Tibet, 1903-04.—Action at Niani, 26th June 1904. —Operations at and around Gyantse, between 5th May and 6th July 1904.—Medal with clasp.
N.-W. Frontier of India, 1908.—Operations in the Zakka Khel country.—Medal with clasp.

Sadhu Singh—

N.-W. Frontier of India, 1908.—Operations in the Zakka Khel country.—Medal with clasp.

Inda Parshad—

N.-W. Frontier of India, Waziristan, 1901-02.—Medal with clasp.
N.-W. Frontier of India, 1908.—Operations in the Zakka Khel country.—Medal with clasp.
Tibet, 1903-04.—Action at Niani, 26th June 1904.—Operations at and around Gyantse, between 5th May and 6th July 1904.—Medal with clasp.

Kapur Singh—

N.-W. Frontier of India, Waziristan, 1901-02.—Medal with clasp.
N.-W. Frontier of India, 1908.—Operations in the Zakka Khel country.—Medal with clasp.
Tibet, 1903-04.—Action at Niani, 26th June 1904.—Operations at and around Gyantse, between 5th May and 6th July 1904.—Medal with clasp.

Sundar Singh—

N.-W. Frontier of India, Waziristan, 1901-02.—Medal with clasp.
N.-W. Frontier of India, 1908.—Operations in the Zakka Khel country.—Medal with clasp.
Tibet, 1903-04.—Action at Niani, 26th June 1904. —Operations at and around Gyantse, between 5th May and 6th July 1904.—Medal with clasp.

Sant Singh—

N.-W. Frontier of India, 1908.—Operations in the Zakka Khel country.—Medal with clasp.
Tibet, 1903-04.—Action at Niani, 26th June 1904.—Operations at and around Gyantse, between 5th May and 6th July 1904.—Medal with clasp.

Jhanda Singh—

N.-W. Frontier of India, Waziristan, 1901-02.—Medal with clasp.
N.-W. Frontier of India, 1908.—Operations in the Zakka Khel country.—Medal with clasp.
Tibet, 1903-04.—Action at Niani, 26th June 1904.—Operations at and around Gyantse, between 5th May and 6th July 1904.—Medal with clasp.

Balwant Singh—

N.-W. Frontier of India, 1908.—Operations in the Zakka Khel country.—Medal with clasp.

Atma Singh—

N.-W. Frontier of India, Waziristan, 1901-02.—Medal with clasp
N.-W. Frontier of India, 1908.—Operations in the Zakka Khel country.—Medal with clasp.
Tibet, 1903-04.—Medal.

Jawala Singh—

Tibet, 1903-04.—Medal.
N.-W. Frontier of India, 1908.—Operations in the Zakka Khel country.—Medal with clasp.

Sundar Singh—

Chitral, 1895.—Relief of Chitral.—Medal with clasp.
N.-W. Frontier of India, Waziristan, 1901-02.—Clasp.
Tibet, 1903-04.—Medal.
N.-W. Frontier of India, 1908.—Operations in the Zakka Khel country.—Medal with clasp.

Nihal Singh—

N.-W. Frontier of India, 1908.—Operations in the Zakka Khel country.—Medal with clasp.

Bassawa Singh—

N.-W. Frontier of India, 1908.—Operations in the Zakka Khel country.—Medal with clasp.

Katha Singh—

N.-W. Frontier of India, 1908.—Operations in the Zakka Khel countries.—Medal with clasp.

2/23rd Sikh Pioneers.

SUBADAR-MAJOR—

Mota Singh—

Chitral, 1895.—Relief of Chitral.—Medal with clasp.
N.-W. Frontier of India, Waziristan, 1901-02.—Clasp.
Tibet, 1903-04.—Action at Niani, 26th June 1904.—Operations at and around Gyantse, between 5th May and 6th July 1904.—Medal.
N.-W. Frontier of India, 1908.—Operations in the Zakka Khel country.—Medal with clasp.

SUBEDARS—

Hardit Singh—

Chitral, 1895.—Relief of Chitral.—Medal with clasp.
N.-W. Frontier of India, Waziristan, 1901-02.—Clasp.
Tibet, 1903-04.—Action at Niani, 26th June 1904. Operations at and around Gyantse, between 5th May and 6th July 1904.—Medal with clasp.
N.-W. Frontier of India, 1908.—Operations in the Zakka Khel country.—Medal with clasp.

Arjan Singh—

Tibet, 1903-04.—Medal.
N.-W. Frontier of India, 1908.—Operations in the Zakka Khel country.—Medal with clasp.
Abor, 1911-12.—Medal and clasp.

Sundar Singh—

N.-W. Frontier of India, 1897-98.—Defence of Malakand. Relief of Chakdara. Malakand. Action at Landikai. Operations in Bajour and in the Mamund country.—Medal with 2 clasps.
British Central Africa, 1899-1900.—Medal with clasp.
Ashanti, 1900—Medal with clasp.
N.-W. Frontier of India, 1908.—Operations in the Zakka Khel country.—Medal with clasp.

Chanda Singh—

Chitral—Relief of Chitral, 1895.—Medal with clasp.
N.-W. Frontier of India, Waziristan, 1901-02.—Clasp.

Tibet, 1903-04.—Medal.
N.-W. Frontier of India, 1908.—Operations in the Zakka Khel country.—Medal with clasp.

Hira Singh—
N.-W. Frontier of India, Waziristan, 1901-02.—Medal with clasp.
Tibet, 1903-04.—Action at Niani, 26th June 1904. Operations at an around Gyantse, between 5th May and 6th July 1904.—Medal with clasp.
N.-W. Frontier of India, 1908.—Operations in the Zakka Khel country.—Medal with clasp.

Jhanda Singh—
N.-W. Frontier of India, Waziristan, 1901-02—Medal with clasp.
Tibet, 1903-04. Action at Niani, 26th June 1904. Operations at and around Gyantse, between 5th May and 6th July 1904.—Medal with clasp.
N.-W. Frontier of India, 1908.—Operations in the Zakka Khel country.—Medal with clasp.

Bir Singh—
N.-W. Frontier of India, 1908.—Operations in the Zakka Khel country.—Medal with clasp.

JEMADARS—

Harnam Singh—
Tibet, 1903-04.—Medal.

Sadda Singh—
N.-W. Frontier of India, Waziristan, 1901-02.—Medal with clasp.
Tibet, 1903-04.—Action at Niani, 26th June 1904.—Operations at and around Gyantse, between 5th May and 7th July 1904.—Medal with clasp.
N.-W. Frontier of India, 1908.—Operations in the Zakka Khel country.—Medal with clasp.

Atma Singh—
N.-W. Frontier of India, Waziristan, 1901-02.—Medal with clasp.
Tibet, 1903-04.—Action at Niani, 2 th June 1904.—Operations at and around Gyantse, between 5th May and 7th July 1904.—Medal with clasp.
N.-W. Frontier of India, 1908.—Operations in the Zakka Khel country.—Medal with clasp.

Kehar Singh—
N.-W. Frontier of India, Waziristan, 1901-02.—Medal with clasp.
Tibet, 1903-04.—Action at Niani, 26th June 1904, Operations at and around Gyantse, between 5th May and 7th July 1904.—Medal with clasp.
N.-W. Frontier, 1908.—Operations in the Zakka Khel country.—Medal with clasp.

Tilok Singh, I.O.M.—
Chitral, 1895.—Relief of Chitral.—Medal with clasp.
N.-W. Frontier of India, Waziristan, 1901-02.—Clasp.
Tibet, 1903-04.—Action at Niani, 26th June 1904.—Operations at and around Gyantse, between 5th May and 7th July 1904.—Medal with clasp.
N.-W. Frontier of India, 1908.—Operations in the Zakka Khel country.—Medal with clasp.

Mali Singh—
N.-W. Frontier of India, 1908—Operations in the Zakka Khel country.—Medal with clasp.

Ala Singh—
N.-W. Frontier of India, 1908.—Operations in the Zakka Khel country.—Medal with clasp.

Hira Singh—
N.-W. Frontier of India, 1908.—Operations in the Zakka Khel country.—Medal with clasp.

Bir Singh—
N.-W. Frontier of India, 1908.—Operations in the Zakka Khel country.—Medal with clasp.

Pala Singh—
Chitral, 1895.—Relief of Chitral.—Medal and clasp.
N.-W. Frontier of India, Waziristan, 1901-02.—Clasp.
Tibet, 1903-04.—Action at Niani, 26th June 1904.—Operations at and around the Gyantse, between 5th May and 7th July 1904.—Medal with clasp.
N.-W. Frontier of India, 1908.—Operations in the Zakka Khel country.—Medal with clasp.

24th Punjabis.

SUBADAR-MAJOR—

Sakt Chand, *Sardar Bahadur—Order of British India, 1st class.*
N.-W. Frontier of India, 1897-98.—Defence of Malakand. Relief of Chakdara. Malakand. Action at Landikai, Operations in Bajour and in the Mamund country.—Medal with 2 clasps.
China, 1900.—Action of Peitsan. Action at Yangtsun. Relief of Pekin.—Medal with clasp.

Diwan Singh, *Sardar Bahadur—Order of British India, 1st class.*
N.-W. Frontier of India, 1897-98.—Defence of Malakand. Relief of Chakdara. Malakand. Action at Landikai. Operations in Bajour and in the Mohmand country.—Medal with 2 clasps.
China, 1900.—Action of Peitsang. Action at Yangtsun. Relief of Pekin.—Medal with clasp.

SUBADARS—

Gul Akbar, *Bahadur,* M.C.—*Order of British India, 2nd class. Indian Distinguished Service Medal.*
N.-W. Frontier of India, 1897-98.—Malakand.—Operations in Bajour and in the Mohmand country.—Medal with clasp.
China, 1900.—Action of Peitsang. Action at Yangtsun. Relief of Pekin.—Medal with clasp.
Tibet, 1904.—Action of Gyantse.—Medal with clasp.

Labh Singh, *Bahadur—Order of Br. India, 2nd class—*
N.-W. Frontier of India, 1908.—Operations in the Mohmand country. Engagement of Kargha, 24th May 1908.—Medal with clasp.

Kishan Singh—
N.-W. Frontier of India, 1897-98.—Defence of Malakand. Relief of Chakdara. Action at Landikai and operations in Bajour. Operations in Mamund country.—Medal with two clasps.

Umar Khan—
N.-W. Frontier of India, 1897-98.—Defence of Malakand. Action at Landikai. Bajour and in the Mohmand country.—Medal with 2 clasps.—*Indian Order of Merit, 3rd class.*
China, 1900.—Action of Peitsang. Action at Yangtsun. Relief of Pekin.—Medal with clasp.

Amir Khan—
N.-W. Frontier of India, 1897-98.—Malakand.—Operations in Bajour, and in Mohmand country.—Medal with clasp.
China, 1900.—Action at Pietsang. Action at Yangtsun. Relief of Pekin.—Medal with clasp.

Fateh Singh, *Bahadur—Order of Br. India, 2nd class.*

N.-W. Frontier of India, 1897-98.—Defence of Malakand. Relief of Chakdara. Action at Landikai and operations in Bajour. Operations in Mamund country.—Medal with 2 clasps.—China, 1900. Action at Piet Sung. Action at Yangtsun. Relief of Pekin. Medal with clasp.
Africa, Nandi, 1905-06.—Indian Order of Merit.

Ujagar Singh—

N.-W. Frontier of India, 1897-98.—Defence of Malakand. Relief of Chakdara. Action at Landakai. Operations in Bajour and in the Mamund country.—Medal with clasps.
China, 1900.—Medal.

Lalmir—

N.-W. Frontier of India, 1897-98.—Defence of Malakand. Relief of Chakdara. Action at Landakai. Operations in Bajour and in the Mamund country.—Medal with 2 clasps.
China, 1900.—Relief of Pekin.—Medal with clasp.

Narain Singh (2)—

N.-W. Frontier of India, 1897-98.—Defence of Malakand. Relief of Chakdara. Action at Landakai. Operations in Bajour and in the Mamund country.—Medal with 2 clasps.
China, 1900-01.—Medal with clasp.

Sher Nur—

N.-W. Frontier of India, 1897-98.—Defence of Malakand. Relief of Chakdara. Action at Landakai.—Medal with 2 clasps.
China, 1900-01.—Medal with clasp.

Sarfaraz Khan—

N.-W. Frontier of India, 1908.—Medal with clasp' Action at Gyantse, Tibet.—Medal with clasp.

Kehar Singh—

N.-W. Frontier of India, Hazara, 1888.—Medal with clasp.
N.-W. Frontier of India, 1897-98.—Defence of Malakand. Relief of Chakdara. Action at Landikai. Operations in Bajour and in the Mamund country.—Medal with 2 clasps
China, 1900-01.—Medal with clasps.

Feroz Khan—

N.-W. Frontier of India, 1897-98.—Defence of Malakand. Action at Landakai. Operations in Bajour and in Mamund country.—Medal with clasp.
China, 1900-01.—Medal with clasp.

Hakam Singh—

N.-W. Frontier of India, 1897-98.—Defence of Malakand. Relief of Chakdara. Action at Landakai. Operations in Bajour and in Mamund country.—Medal with 2 clasps.

Ala Singh—

N.-W. Frontier of India, Hazara, 1888.—Medal with clasp.
N.-W. Frontier of India, 1897-98.—Defence of Malakand. Relief of Chakdara. Action at Landakai. Operations in Bajour and in Mamund country.—Medal with 2 clasps.

Suba—

N.-W. Frontier of India, 1897-98.—Defence of Malakand. Relief of Chakdara. Action at Landakai. Operations in Bajour and in Mamund country.—Medal with 2 clasps.
China, 1900-01.—Medal with clasp.

Said Ahmad Khan—

China, 1900-01.—Medal.

Lachman Singh—

N.-W. Frontier of India, 1897-98.—Defence of Malakand. Relief of Chakdara. Action at Landikai. Operations in Bajour and in Mamund country.—Medal with 2 clasps
China, 1900-01.—Medal with clasp.

Wadhawa Singh—

China, 1900-01.—Medal.

25th Punjabis.

SUBADAR-MAJOR—

Sudama—

N.-W. Frontier of India, Isazai, 1895.
Chitral, 1895.—Relief of Chitral.—Medal with clasp.
N.-W. Frontier of India, 1897-98.—Tochi.—Clasp.
N.-W. Frontier of India, 1908.—Operations in the Zakka Khel country.—Medal with clasp.

SUBADARS—

Mawaz Khan—

China, 1900.—Medal.
N.-W. Frontier of India, 1908.—Operations in the Zakka Khel country.—Medal with clasp.

Indar Singh—

N.-W. Frontier of India, Isazai, 1892.
Chitral, 1895.—Relief of Chitral.—Medal with clasp.
N.-W. Frontier of India, 1897-98.—Tochi.—Clasp.
N.-W. Frontier of India, 1908.—Operations in the Zakka Khel country.—Medal with clasp.

Sher Mohamed Khan (II)—

N.-W. Frontier of India, 1897-98.—Tochi.—Medal with clasp.
N.-W. Frontier of India, 1908.—Operations in the Zakka Khel country.—Medal with clasp.

Ghulam Muhammad—

Chitral, 1895.—Relief of Chitral.—Medal with clasp.
N.-W. Frontier of India, 1897-98.—Tochi.—Clasp.
N.-W. Frontier of India, 1908.—Operations in the Zakka Khel country—Medal with clasp.

Harnam Singh—

Relief of Chitral, 1895.—Medal and clasp.
N.-W. Frontier of India, 1897-98.—Clasp.
N.-W. Frontier of India, 1908.—Operations in the Zakka Khel country.—Medal with clasp.

Muhammad Khan—

N.-W. Frontier of India, 1908.—Operations in the Zakka Khel country.—Medal with clasp.

Sadhu Singh—

Chitral, 1895.—Relief of Chitral.—Medal with clasp.
N.-W. Frontier of India, 1897-98.—Tochi clasp.
N.-W. Frontier of India, 1908.—Operations in the Zakka Khel country.—Medal with clasp.

Basant Singh—

Africa, G. S. Nandi, 1905-06.—Medal and clasp.
N.-W. Frontier of India, 1897-98.—Operations in the Zakka Khel country.—Medal with clasp.

JEMADARS—

Lal Khan (I)—

N.-W. Frontier of India, 1908.—Operations in the Zakka Khel country.—Medal and clasp.

Lal Khan (II)—

N.-W. Frontier of India, 1908.—Operations in the Zakka Khel country.—Medal and clasp.

Shah Sawar—

N.-W. Frontier of India, 1908.—Operations in the Zakka Khel country.—Medal and clasp.

Gurdas Singh—

N.-W. Frontier of India, 1908.—Operations in Zakka Khel country —Medal and clasp.

Sahib Gul—

N.-W. Frontier of India, 1908.—Operations in the Zakka Khel country.—Medal with clasp.

Harnam Singh—

N.-W. Frontier of India, 1908.—Operations in the Zakka Khel country.—Medal with clasp.

Adalat Khan—

N.-W. Frontier of India, 1908.—Operations in the Zakka Khel country.—Medal with clasp.

Ghulam Rasul—

N.-W Frontier of India, 1908.—Operations in the Zakka Khel country.—Medal with clasp.

Maula Dad—

N.-W. Frontier of India, 1908.—Operations in the Zakka Khel country.—Medal with clasp.

Johil—

N.-W. Frontier of India, 1908.—Operations in the Zakka Khel country.—Medal with clasp.

Hira—

N.-W. Frontier of India, 1908.—Operations in the Zakka Khel country.—Medal with clasp.

Kabul Shah—

N.-W. Frontier of India, 1908.—Operations the Zakka Khel country.—Medal with clasp.

Harnam Singh (II)—

N.-W. Frontier of India, 1908.—Operations in the Zakka Khel country.—Medal with clasp.

Bhikho—

N.-W. Frontier of India, 1908.—Operations in the Zakka Khel country.—Medal with clasp.

Mir Afzal—

N.-W. Frontier of India, 1908.—Operations in the Zakka Khel country.—Medal with clasp.

Zaman Khan—

N.-W Frontier of India, 1908.—Operations in the Zakka Khel country.—Medal with clasp.

Chet Ram—

N.-W. Frontier of India, 1908.—Operations in the Zakka Khel country.—Medal with clasp.

1-26th Punjabis.

SUBADAR-MAJORS—

Harnam Singh—

Soudan, 1896.—Dongola Expedition.—2 medals.
N.-W. Frontier of India, 1897-98.—Mohmand.—Medal with clasp.

Ishar Singh—

Soudan, 1896.—Dongola Expedition.—2 medals.
N.-W. Frontier of India, 1897-98.—Mohmand.—Medal with clasp.
N.-W. Frontier of India, 1908.—Operations in the Mohmand country. Engagement of Kargha, 24th May 1908.—Medal with clasp.

SUBADARS—

Buta Singh—

N.-W. Frontier of India, 1897-98.—Mohmand.—Medal with clasp.

Sher Akhmad—

N.-W. Frontier of India, 1897-98.—Mohmand.—Medal with clasp.

JEMADARS—

Dalip Singh—

Soudan, 1896.—Dangola Expedition.—2 medals.
N.-W. Frontier of India, 1897-98.—Mohmand.—Medal with clasp.

Gonda Singh—

N.-W. Frontier of India, 1897-98.—Mohmand.—Medal with clasp.

2-26th Punjabis.

SUBADAR-MAJOR—

Parab Dial—

Egypt. 1904-1915.—
Galipoli, 1915.—
France, 1915.—
Egypt, 1915-1916.—
Aden, 1916.—

SUBADARS—

Mehr Chand—

I. E. F. "D," 1917.—Wounded.

Imam Ali—

Somaliland, 1903-04.—Medal.
N.-W. Frontier of India, 1914.—

Sam Undar Khan—

Egypt, 1914-15.—
I. E. F. "D," 1915.—

27th Punjabis.

SUBADAR-MAJORS—

Mir Akbar—

N.-W. Frontier of India. Hazara, 1891.—Medal with clasp.
N.-W. Frontier of India, Miranzai, 1891.—Clasp.
East Africa, 1903-04.—Operations in Somaliland. Action at Jidbali.—Medal with 2 clasps.
N.-W. Frontier of India, 1908.—Operations in the Mohmand country.—Medal with clasp.

Khushal Singh—

N.-W. Frontier of India, Hazara, 1891.—Medal with clasp.
N.-W. Frontier of India, Miranzai, 1891.—Clasp.
Central Africa, 1895-97.—Medal with clasp.
N.-W. Frontier of India, Waziristan, 1901-02.—Medal with clasp.
East Africa, 1903-04.—Operations in Somaliland. Action at Jidbali.—Medal with 2 clasps.
N.-W. Frontier of India, 1908.—Operations in the Zakka Khel country.—Medal with clasp.

SUBADARS—

Arjun Singh—

N.-W. Frontier of India, Waziristan, 1901-02.—Medal with clasp.
East Africa, 1903-04.—Operations in Somaliland.—Medal with clasp.

Bhagat Singh—

N.-W. Frontier of India, Waziristan, 1901-02.—Medal with clasp.
East Africa, 1903-04.—Operations in Somaliland.—Medal with clasp.

Aman Gul—

N.-W. Frontier of India, Waziristan, 1901-02.—Medal with clasp.

Sher Shah—

N.-W. Frontier of India, Waziristan, 1901-02.—Medal and clasp.
East Africa, 1903-04.—Operations in Somaliland.—Medal with clasp.

Rama—

East Africa, 1903-04.—Operations in Somaliland and action at Jidbali.—Medal with 2 clasps.

Naim Gul—

N.-W. Frontier of India, Waziristan, 1901-02.—Medal and clasp.
East Africa, 1903-04.—Operations in Somaliland and action at Jidbali.—Medal with 2 clasps.
N.-W. Frontier of India, 1908.—Operations in the Mohmand country.—Medal with clasp.

Fateh Khan—

East Africa, 1903-04.—Operations in Somaliland.—Medal with clasp.

Sher Ali—

N.-W. Frontier of India, Waziristan, 1901-02.—Medal with clasp.
East Africa, 1903-04.—Operations in Somaliland and action at Jidbali.—Medal with 2 clasps.

Ishar Singh—

N.-W. Frontier of India, Waziristan, 1901-02.—Medal with clasp.
East Africa, 1903-04.—Operations in Somaliland and action at Jidbali.—Medal with 2 clasps.

Labh Singh—

N.-W. Frontier of India, Waziristan, 1901-02.—Medal with clasp.
East Africa, 1903-04.—Operations in Somaliland and action at Jidbali.—Medal with 2 clasps.

Zain Din, I.D.S.M.—

East Africa, 1903-04.—Operations in Somaliland.—Medal with clasp.

JEMADARS—

Tor Khan—

N.-W. Frontier of India, Waziristan, 1901-02.—Medal with clasp.
East Africa, 1903-04.—Operations in Somaliland and action at Jidbali.—Medal with 2 clasps

Chanda Singh—

N.-W. Frontier of India, Waziristan, 1901-02.—Medals with clasps.
East Africa, 1903-04.—Operations in Somaliland.—Medal with clasp.

Hayat Khan—

N.-W. Frontier of India, Waziristan, 1901-02.—Medal with clasp.
East Africa, 1903-04.—Operations in Somaliland and action at Jidbali.—Medal with 2 clasps.

Muhommad Azim—

East Africa, 1903-04.—Operations in Somaliland.—Medal with clasp.

Mustafa—

East Africa, 1903-04.—Operations in Somaliland.—Medal with clasp.

Nand Singh—

East Africa, 1903-04.—Operations in Somaliland and action at Jidbali.—Medal with 2 clasps.

Surjan—

East Africa, 1903-04.—Operations in Somaliland and action at Jidbali.—Medal with 2 clasps.

Badri Das—

Afghanistan, 1878-80.—Medal with Alinasjid clasp.
Burma, 1886-88.—Medal with 2 claps.
N.-W. Frontier of India, Hazara, 1891.—Clasp.
N.-W. Frontier of India, Miranzai, 1891.—Clasp.
N.-W. Frontier of India, Waziristan, 1901-02.—Medal with clasp.
East Africa, 1903-04.—Operations in Somaliland. Medal with clasp.

Hikmat—

East Africa, 1903-04.—Operations in Somaliland and action at Jidbali.—Medal with 2 clasps.

Fateh Khan—

Tibet, 1903-04.—Medal.

2-27th Punjabis.

SUBADARS—

Khan Gul—

N.-W. Frontier of India, Waziristan, 1901-02.—Medal with clasp.
East Africa, 1903-04.—Operations in Somaliland and action at Jidbali.—Medal with 2 clasps.

Tara Singh, M.C., I.D.S.M.—

N.-W. Frontier of India, 1908.—Operations in the Zakka Khel and Mohmand countries.—Medal with clasp.

Sher Bahadur Khan—

N.-W. Frontier of India, 1908.—Operations in the Zakka Khel country.—Medal with clasp.

JEMADARS—

Munshi, I.O.M.—

N.-W. Frontier of India, 1908.—Operations in the Zakka Khel and Mohmand countries.—Medal with clasp.

Harnam Singh—

East Africa, 1903-04.—Opreations in Somaliland and action at Jidbali.—Medal with 2 clasps.

Khadim Shah—

N.-W. Frontier of India, 1908.—Operations in Zakka Khel and Mohmand countries—Medal with clasp.

Khada Bux—

East Africa, 1903-04.—Operations in Somaliland and action at Jidbali.—Medal with 2 clasps.

28th Punjabis.

SUBADAR-MAJOR—

Sundar—

N.-W. Frontier of India, Waziristan, 1901-02.—Medal with clasp.

N.-W. Frontier of India, 1908.—Operations in the Zakka Khel country.
N.-W. Frontier of India, 1908.—Operations in the Mohmand country.—Medal with clasp.

SUBADARS—

Muhammad Nawaz Khan—

N.-W. Frontier of India, Waziristan, 1901-02.—Medal with clasp.
N.-W. Frontier of India, 1908.—Operations in the Zakka Khel country.
N.-W. Frontier of India, 1908.—Operations in the Mohmand country.—Medal with clasp.

Amar Singh—

N.-W. Frontier of India, Waziristan, 1901-02.—Medal with clasp.

Hasan Muhammad—

N.-W. Frontier of India, Hazara, 1891.—Medal with clasp.
Uganda, 1898.—Medal with clasp.
N.-W. Frontier of India, Waziristan, 1901-02.—Medal with clasp.

Mangal Singh—

Nile Expedition, Suakim Force, 1896.—Sudan medal granted by the British Government.—Sudan medal granted by the Khedive of Egypt.
N.-W. Frontier of India, Waziristan, 1901-02.—Medal with clasp.
N.-W. Frontier of India, 1908.—Operations in the Zakka Khel and Mohmand countries.—Medal with clasp.

Kaka Singh—

N.-W. Frontier of India, Waziristan, 1901-02.—Medal with clasp.
N.-W. Frontier of India, 1908.—Operations in the Mohmand country.—Medal with clasp.

Mirza Khan—

N.-W. Frontier of India, Waziristan, 1901-02.—Medal with clasp.
N.-W. Frontier of India.—Operations in the Zakka Khel country.—Medal with clasp.

Thakur Singh—

N.-W. Frontier of India, Waziristan, 1901-02.—Medal with clasp.
N.-W. Frontier of India, 1908.—Operations in the Zakka Khel and Mohmand countries.—Medal with clasp.

Gulam Sarwar—

N.-W. Frontier of India, Waziristan, 1901-02.—Medal with clasp.
N.-W. Frontier of India, 1908.—Operations in the Zakka Khel and Mohmand countries.—Medal with clasp.

Fazal-Mehdi-Khan—

N.-W. Frontier of India, Waziristan, 1901-02.—Medal with clasp.
N.-W. Frontier of India, 1908.—Operations in the Zakka Khel and Mohmand countries.—Medal with clasp.

Gulab Din—

N.-W. Frontier of India, Waziristan, 1901-02.—Medal with clasp.
N.-W. Frontier of India, 1908.—Operation in the Zakka Khel and Mohmand countries.—Medal with clasp.

Sultan Surkhru—

N.-W. Frontier of India, 1908.—Operations in the Zakka Khel and Mohmand countries.—Medal with clasp.

Dalip Singh—

N.-W. Frontier of India, Waziristan, 1901-02.—Medal with clasp.
N.-W. Frontier of India, 1908.—Operations in the Zakka Khel and Mohmand countries.—Medal with clasp.

Imam Ali—

N.-W. Frontier of India, 1908.—Operations in the Mohmand country.—Medal with clasp.

JEMADARS—

Azim Khan—

N.-W. Frontier of India, 1908.—Operations in the Mohmand country.—Medal with clasp.

Mohammed—

N.-W. Frontier of India, Waziristan, 1901-02.—Medal with clasp.

Mehtab Din—

N.-W. Frontier of India, Waziristan, 1901-02.—Medal with clasp.

Wasawa Singh—

N.-W. Frontier of India, Waziristan, 1901-02.—Medal with clasp.

N.-W. Frontier of India, 1908.—Operations in the Zakka Khel and Mohmand countries.—Medal with clasp.

Sunder Singh—

N.-W. Frontier of India, Waziristan, 1901-02.—Medal with clasp.

N.-W Frontier of India, 1908.—Operations in the Zakka Khel and Mohmand countries.—Medal with clasp.

Diwan Ali Khan—

N.-W. Frontier of India, Waziristan, 1901-02.—Medal with clasp.

N.-W. Frontier of India, 1908—Operations in the Zakka Khel country,—in the Mohmand country.—Medal with clasp.

Ranya—

N.-W. Frontier of India, 1901-02.—Medal with clasp.

N.-W. Frontier of India, 1908.—Operations in the Mohmand country.—Medal with clasp.

Niamat Khan—

N.-W. Frontier of India, 1901-02.—Medal with clasp.

N.-W. Frontier of India, 1908.—Operations in the Zakka Khel country—in the Mohmand country.—Medal with clasp.

Azmat Khan—

N.-W. Frontier of India, 1908.—Operations in the Zakka Khel and Mohmand countries.—Medal with clasp.

Allah Lok—

N.-W. Frontier of India, Waziristan, 1901-02.—Medal with clasp.

N.-W. Frontier of India, 1908.—Operations in the Zakka Khel and Mohmand countries.—Medal with clasp.

Teja Singh—

N.-W. Frontier of India, Waziristan, 1901-02.—Medal with clasp.

N.-W. Frontier of India, 1908.—Operations in the Zakka Khel and Mohmand countries.—Medal with clasp.

Nawab Ali—

N.-W. Frontier of India, Waziristan, 1901-02.—Medal with clasp.

N.-W. Frontier of India, 1908—Operations in the Zakka Khel and Mohmand countries.—Medal with clasp.

Operations against the Derweshis at Shimber Berris between 2nd and 9th February —Medal with clasp.

2-28th Punjabis.

SUBADAR-MAJOR—

Allah Baksh—

N.-W. Frontier of India, Waziristan, 1901-02.—Medal with clasp.

N.-W. Frontier of India, 1908.—Operations in the Zakka Khel Country. Operations in the Mohmand Country.—Medal with clasp.

I. E. F. "D" 1915.—Mentioned in Despatches, *wounded 21st July 17.*

SUBADARS—

Wazir Singh—

N.-W Frontier of India, Waziristan, 1901-02.—Medal with clasp.

N.-W. Frontier of India, 1915.—Action at Shabkadr.

E. Africa, 1917.—Slightly wounded.

Sundar—

N.-W. Frontier of India, 1897.—Mohmand Expedition.—Medal and clasp.

East Africa, 1902 and 1904, Somaliland.—Action at Jidballi.—Medal and 2 clasps.

War, 1915-1, with 41st Dogras.

Force "A" 1915.—

" "E" 1915-16.—

" "D" 1915-16.—

JEMADARS—

Hazara Singh—

Chitral Relief Force, 1895.—Medal and clasp.

Tirah, 1897-98.—2 clasps.

War, 1914-16—Egypt, France and Aden.

Wounded in France, 1915.

Alam Sher—

N.-W. Frontier of India.—Operations in Zakka Khel country, 1908, also in Mohmand country, 1908. Medal and clasp. Operations against the Derwishes at Shimber Berris. Feb. 1915.

29th Punjabis.

SUBADAR-MAJOR—

Mehr Khan, *Sardar Bahadur, Hony. Capt. (re-empld.)*

Hazara F. F., 1888.—Medal with clasp.

1st and 2nd Miranzai Expedition, 1891.—Clasp.

Relief of Chitral, 1895.—

N.-W. Frontier of India, Waziristan, 1901-02.—Medal with clasp.

Man Singh—

N.-W. Frontier of India, (2nd) Miranzai, 1891.—

Chitral, 1895.—Relief of Chitral.—Medal with clasp.

N.-W. Frontier of India, Waziristan, 1901-02.—Clasp.

SUBADARS—

Rai Singh—

N.-W. Frontier of India, Waziristan, 1901-02.—Medal with clasp.

Fazl Muhammad, *Sardar Bahadur—Order of British India, 1st class.*

N.-W. Frontier of India, Hazara, 1888.—Medal with clasp.

N.-W. Frontier of India, (1st and 2nd) Miranzai, 1891.—Clasp.

Chitral, 1895.—Relief of Chitral.—Medal with clasp.

N.-W. Frontier of India, Waziristan, 1901-02.—Clasp.

Roshan Khan—

Chitral, 1895.—Relief of Chitral.—Medal with clasp.

N.-W. Frontier of India, Waziristan, 1901-02.—Clasp.

Gurdit Singh—

N.-W. Frontier of India, (1st and 2nd) Miranzai, 1891.—Medal with clasp.

Chitral, 1895.—Relief of Chitral.—Medal with clasp.

N.-W. Frontier of India, Waziristan, 1901-02.—Clasp.

Ali Bahadur—

Chitral Expedition, 1895.—Medal with clasp.—Relief of Chitral, 1895.

N.-W. Frontier of India, Waziristan, 1901-02.

Kesar Singh—

N.-W. Frontier of India, Waziristan, 1901-02.—Medal with clasp.

Indar Singh—

N.-W. Frontier of India, Waziristan, 1901-02.—Medal with clasp.

Wazir Singh—

N.-W. Frontier of India, 1901-02.—Operations in Waziristan.—Medal with clasp.

Ganda Singh—

N.-W. Frontier of India, Waziristan, 1901-02.—Medal with clasp.

JEMADARS—

Lachman Singh—

N.-W. Frontier of India, Waziristan, 1901-02.—Medal with clasp.

Basawa Singh—

N.-W. Frontier of India, Waziristan, 1901-02.—Medal with clasp.

Kaldip Singh—

N.-W. Frontier of India, Waziristan, 1901-02.—Medal with clasp.

Gokal—

N.-W. Frontier of India, Waziristan, 1901-02.—Medal with clasp.

Labh Singh—

N.-W. Frontier of India, Tirah, 1897-98.—Medal with clasp.
N.-W. Frontier of Waziristan, 1901-02.—Clasp.

Muhammad Khan—

N.-W. Frontier of India, Waziristan, 1901-02.—Medal and clasp.

Isar Singh—

Chitral. Relief Force, 1895.
N.-W. Frontier of India, Waziristan, 1901-02.—Medal with clasp.

Kamal Khan—

N.-W. Frontier of India, Waziristan, 1901-02.—Medal with clasp.

Bhagwan Singh—

N.-W. Frontier of India, Waziristan, 1901-02.—Medal with clasp.

Baz Khan, I.D.S.M.—

N.-W. Frontier of India, Waziristan, 1901-02.—Medal with clasp.

Mangtu—

N.-W. Frontier of India, Waziristan, 1901-02.—Medal with clasp.

Balla Singh—

N.-W. Frontier of India, Waziristan, 1901-02.—Medal with clasp.

Lakhu—

N.-W. Frontier of India, Waziristan, 1901-02.—Medal with clasp.

30th Punjabis.

SUBADARS—

Diwan Singh—

N.-W. Frontier of India, 1897-98.—Engagement near Shabkadr, 9th August 1897. Operations on the Samana and in the Kurram Valley during August and September 1897.—Medal with 2 clasps.
Tirah, 1897-98.—Operations in the Bazar Valley, 25th to 30th December 1897.—Clasp.

Chajja Singh—

Chitral, 1895.—Relief of Chitral.—Medal with clasp.
N.-W. Frontier of India, 1897-98.—Engagement near Shabkadr, 9th August 1897. Operations on the Samana and in the Kurram Valley during August and September 1897—2 clasps.
Tirah, 1897-98.—Operations in the Bazar Valley, 25th to 30th December 1897.—Clasp.

Karm Dad—

Chitral, 1895.—Relief of Chitral.—Medal with clasp.
Tirah, 1897-98.—Operations in the Bazar Valley, 25th to 30th December 1897.—Clasp.

Allah Ditta (II).—

N.-W. Frontier of India, Isazai, 1892.
Chitral, 1895.—Relief of Chitral.—Medal with clasp.
N.-W. Frontier of India, 1897-98.—Engagement near Shabkadar, 9th August 1897. Operations on the Samana and in the Kurram Valley during August and September 1897.—2 clasps.
Tirah, 1897-98.—Operations in the Bazar Valley, 25th to 30th December 1897.—Medal with clasp.

Gurdit Singh—

N.-W. Frontier of India, 1897-98.—Engagement near Shabkadr, 9th August 1897. Operations on the Samana and in the Kurram Valley during August and September 1897.—2 clasps with medal.
Tirah, 1897-98.—Operations in the Bazar Valley, 25th to 30th December 1897.—Clasp.

Hamam Singh, II—

Chitral, 1895.—Relief of Chitral.—Medal with clasp.
N.-W. Frontier of India, 1897-98.—Engagement near Shabkadr, 9th August 1897. Operations on the Samana and in the Kurram Valley during August and September 1897.—2 clasps.
Tirah, 1897-98. Operations in the Bazar Valley, 25th to 30th December 1897.—Clasp.

JEMADARS—

Lakha—

N.-W. Frontier of India, 1897-98.—Engagement near Shabkadar, 9th August 1897. Operations on the Samana and in Kurram Valley during August 1897.—Medal with 2 clasps.
Tirah, 1897-98.—Operations in the Bazar Valley, 25th to 30th December 1897.—Clasp.

Ram Saran—

Somaliland, 1913.—Medal.

2-30th Punjabis.

SUBADARS—

Tota Ram—

N.-W. Frontier of India, Isazai, 1892.
Chitral, 1895.—Relief of Chitral.—Medal with clasp.
N.-W. Frontier of India, 1897-98.—Engagement near Shabkadr, 9th August 1897. Operations on the Samana and in the Kurram Valley during August and September 1897.— 2 clasps.
Tirah, 1897-98.—Operations in the Bazar Valley, 25th to 30th December 1897.—Clasp.

Sultan Alam—

Tirah, 1897-98—Operations in the Bazar Valley, 25th to 30th December.—Medal with clasp.

Bagh Ali—

N.-W. Frontier of India, 1897-98.—Engagement near Shabkadr, 9th August 1897.—Operations on the Samana and in the Kurram Valley during August and September 1897.—2 clasps.
Tirah, 1897-98.—Operations in the Bazar Valley 25th to 30th December 1897.—Clasp.

3-30th Punjabis.

Harnam Singh—

N.-W. Frontier of India, 1908.—Operations in the Zakka Khel country.—Operations in the Mohmand country.—Engagement at Kasgha 24th May 1908.—Medal and clasp.

31st Punjabis.

SUBADARS—

Mangal Singh—

N.-W. Frontier of India, 1897-98.—Defence of Malakand. Relief of Malakand. Action at Landikai. Operations in the Mamund country. Buner.—Medal with 2 clasps.

Kor Singh—

N.-W. Frontier of India, 1897-98.—Medal with clasp.
N.-W. Frontier of India, Waziristan, 1901-02. Clasp.
East Africa, 1902-04.—Operations in Somaliland. —Medal with clasp.

Dewa Singh—

N.-W. Frontier of India, 1897-98.—Defence of Malakand. Relief of Malakand. Malakand. Action at Landakai. Operations in the Mamund country. Buner.—Medal with 2 clasps.

Bishan Singh—

N.-W. Frontier of India, 1897-98.—Malakand. Action at Landakai.—Operations in the Mamund country. Buner.—Medal with clasp.

Prem Singh—

N.-W. Frontier of India, 1897-98.—Defence of Malakand. Relief of Malakand. Malakand. Action at Landakai. Operations in the Mamund country. Buner.—Medal with clasps.

Beli Ram—

N.-W. Frontier of India, 1897-98.—Defence of Malakand. Relief of Malakand. Malakand. Action at Landakai. Operations in the Mamund country. Buner.—Medal with 2 clasps.

Muhammad Feroz Khan—

N.-W. Frontier of India, 1908.—Operations in Zakha Khel country.—Medal with clasp.

Kishn Singh—

N.-W. Frontier of India, 1897-98.—Defence of Malakand. Malakand. Action at Landakai. Operations in the Mamund country. Buner.—Medal with 2 clasps.

Karam Khan—

N.-W. Frontier of India, 1897-98,—Defence of Mala Kand, Relief of Malakand. Malakand. Action at Landakai. Operations in the Mamund country. Buner.—Medal with 2 clasps.

JEMADARS—

Anar Khan—

N.-W. Frontier of India, 1897-98.—Defence of Malakand. Relief of Malakand. Malakand. Action at Landakai. Operations in the Mamund country. Buner.—Medal with 2 clasps.

Mihan Singh—

N.-W. Frontier of India, 1897-98.—Defence of Malakand. Relief of Malakand. Malakand. Action at Landakai. Operations in the Mamund country. Buner.—Medal with 2 clasps.

Niamat Khan—

Tibet, Expedition, 1903-04.—Medal and clasp.

Mohammad khan—

N.-W. Frontier of India, 1897-98—Defence of Malakand. Relief of Malakand. Malakand. Action at Landakai. Operations in the Mamund country. Buner.—Medal with 2 clasps.

Shahbaz Khan—

N.-W. Frontier of India, 1897-98.—Tochi. —Medal and clasp.

Sadda Singh—

N.-W. Frontier, 1897-98.—Defence of Malakand. Relief of Malakand. Malakand. Action at Landakai. Operations in the Mamund country. Buner.—Medal with 2 clasps.

32nd Sikh Pioneers.

SUBADAR-MAJOR—

Bishn Singh, *Sardar Bahadur.—Order of British India, 1st class.*
N.-E. Frontier of India, Sikkim, 1888.—Medal with clasp.
N.-W. Frontier of India, Hazara, 1891.—Clasp.
Chitral, 1895.—Relief of Chitral.—Medal with clasp.
N.-W. Frontier of India, Waziristan, 1901-02.—Clasp.
India General Service Medal with clasp, 1908.
Abor, 1911-12.

SUBADARS—

Wasawa Singh, *Bahadur.—Order of Br. India, 2nd class.*
N.-E. Frontier of India, Sikkim, 1888.—Medal with clasp.
Chitral, 1895.—Relief of Chitral.—Medal with clasp.—Indian Order of Merit, 3rd class.
Tibet, 1903-04.—Action at Niani, 26th June 1904. Operations at and around Gyantse, between 5th May and 6th July 1904.—Medal with clasp. *—Indian Order of Merit, 2nd class.*
Abor, 1911-12.- India General Service.—Medal with clasp. *Indian Distgn. Service Medal.*

Lachhman Singh, *Bahadur—Order of Br. India, 1st class.*
N.-E. Frontier of India, Sikkim, 1888.—Medal with clasp.
N.-W. Frontier of India, Hazara, 1891.—Clasp.
Chitral, 1895.—Relief of Chitral.—Medal with clasp.
N.-W. Frontier of India, Waziristan, 1901-02.—Clasp.
Tibet, 1903-04.—Operations at and around Gyantse between 5th May and 6th July 1904. March to Lhassa, 14th July to 3rd August 1904.—Medal with clasp.
Abor, 1911-12.—India General Service Medal with clasp.

Gopal Singh—

Waziristan, 1901-02.—Medal with clasp.
Tibet, 1903-04.—Medal with clasp.
Abor, 1911-12.—Medal with clasp.

Arjan Singh—

Waziristan, 1901-02.—Medal with clasp.
Tibet, 1903-04.—Medal with clasp.
Abor, 1911-12.—Medal with clasp.

Budha Singh—*Indian Distinguished Service Medal.*
Chitral, 1895.—Medal with clasp.
Waziristan, 1901-02.—Medal with clasp.
Tibet, 1903-04.—Medal with clasp.
Abor, 1911-12.—Medal with clasp.

Kesar Singh—

Waziristan, 1901-02.—Medal with clasp.
Tibet, 1903-04.—Medal with clasp.
Abor, 1911-12.—Medal with clasp.

Sant Singh—*Indian Distinguished Service Medal.*
Waziristan, 1901-02.—Medal with clasp.
Tibet, 1903-04.—Medal with clasp.
Abor, 1911-12.—Medal with clasp.

Jhanda Singh—
Waziristan, 1901-02.—Medal with clasp.
Tibet, 1903-04.—Medal with clasp.
Abor, 1911-12.—Medal with clasp.

Sadda Singh—
Waziristan, 1901-02.—Medal with clasp.
Tibet, 1903-04.—Medal with clasp.
Abor, 1911-12.—Medal with clasp.

Assa Singh—
Tibet, 1903-04.—Medal with clasp.
Abor, 1911-12.—Medal with clasp.

Sawan Singh—
Waziristan, 1901-02.—Medal with clasp.
Tibet, 1903-04.—Medal with clasp.
Abor, 1911-12.—Medal with clasp.

JEMADARS—

Sher Singh—
Waziristan, 1901-02.—Medal with clasp.
Tibet, 1903-04.—Medal with clasp.
Abor, 1911-12.—Medal with clasp.

Rup Singh—
Waziristan, 1901-02.—Medal with clasp.
Tibet, 1903-04.—Medal with clasp.
Abor, 1911-12.—Medal with clasp.

Surain Singh—
Waziristan, 1901-02.—Medal with clasp.
Tibet, 1903-04.—Medal with clasp.
Abor, 1911-12.—Medal with clasp.

Natha Singh (I)—
Tibet, 1903-04.—Medal with clasp.

Sundar Singh—
Waziristan, 1901-02.—Medal with clasp.
Tibet, 1903-04.—Medal with clasp.
Abor, 1911-12.—Medal with clasp.

Ishar Singh—
Waziristan, 1901-02.—Medal with clasp.
Tibet, 1903-04.—Medal with clasp.
Abor, 1911-12.—Medal with clasp.

Kirpa Singh—
Waziristan, 1901-02.—Medal with clasp.
Tibet, 1903-04.—Medal with clasp.
Abor, 1911-12.—Medal with clasp.

Arjan Singh—
Waziristan, 1901-02.—Medal with clasp.
Tibet, 1903-04.—Medal with clasp.
Abor, 1911-12.—Medal with clasp.

Mahla Singh—
Abor, 1911-12—Medal with clasp.

Ganesha Singh—
Waziristan, 1901-02.—Medal with clasp.
Tibet, 1903-04.—Medal with clasp.
Abor, 1911-12.—Medal with clasp.

Hira Singh—
Waziristan, 1901-02.—Medal with clasp.
Tibet, 1903-04.—Medal with clasp.
Abor, 1911-12.—Medal with clasp.

Lehra Singh—
Waziristan, 1901-02.—Medal with clasp.
Tibet, 1903-04.—Medal with clasp.
Abor, 1911-12.—Medal with clasp

2-32nd Sikh Pioneers.

SUBADARS—

Ram Singh, *Bahadur—Order of Br. India, 2nd class.*—
N.-E. Frontier of India, Sikkim, 1888.—Medal with clasp.
N.-W. Frontier of India, Hazara, 1891.—Clasp.
Chitral, 1895.—Relief of Chitral.—Medal with clasp.
N.-W. Frontier of India, Waziristan, 1901-02.—Clasp.
Tibet, 1903-04—Action at Niani, 26th June 1904.—Operations at and around Gyantse, between 5th May and 6th July 1904. March to Lhassa, 14th July to 3rd August 1904.—Medal with clasp.
N.-W. Frontier of India, 1908.—Operations in the Zakka Khel country.—Medal with clasp.
Abor, 1911-12.—India General Service Medal with clasp.

Nihal Singh—
N.-E. Frontier of India, Sikkim, 1888.—Medal with clasp.
N.-W. Frontier of India, Hazara, 1891.—Clasp.
N.-W. Frontier of India, Waziristan, 1901-02.—Medal with clasp.
Tibet, 1903-04.—Action at Niani, 26th June 1904. Operations at and around Gyantse, between 5th May and 6th July 1904. March to Lhassa, 14th July to 3rd August 1904.—Medal with clasp.
N.-W. Frontier of India, 1908.—Operations in the Mohmand country.—Medal with clasp.

Mal Singh—
N.-W. Frontier of India, Waziristan, 1901-02.—Medal with clasp.
Tibet, 1903 04.—Action at Niani, 26th June 1904. Operations at and around Gyantse, between 5th May and 6th July 1904. March to Lhassa, 14th July to 3rd August 1904.—Medal with clasp.
Abor, 1911-12.—India General Service Medal with clasp.

Gurbha Singh—
Chitral, 1895.—Medal with clasp.
Waziristan, 1901-02.—Clasp.
Tibet, 1903-04.—Medal with clasp.
Abor, 1911-12.—Medal with clasp.

Jaggat Singh—
Chitral, 1895.—Relief of Chitral.—Medal with clasp.
Tirah, 1897-98.—Operations in the Bara Valley, 7th to 14th December 1897.—2 clasps.
China, 1900.—Medal.
N.-W. Frontier of India, 1908.—Operations in the Mohmand country.—Medal with clasp.

JEMADARS—

Bishan Singh—
Tibet, 1903-04.—Medal with clasp.
Abor, 1911-12.—Medal with clasp.

Mehnga Singh—
Waziristan, 1901-02.—Medal with clasp.
Tibet, 1903-04.—Medal with clasp.
Abor, 1911-12.—Medal with clasp.

Natha Singh—
Waziristan, 1901-02.—Medal with clasp.
Tibet, 1903-04.—Medal with clasp.
Abor, 1911-12.—Medal with clasp.

Santokh Singh—
Abor, 1911-12.—Medal with clasp.

3-32nd Sikh Pioneers.

SUBADARS—

Gurdit Singh—
Chitral, 1895.—Relief of Chitral.—Medal with clasp.
N.-W. Frontier of India, Waziristan, 1901-02.—Clasp.
Tibet, 1903-04.—Action at Naini, 26th June 1904.—Operations at and around Gyantse between 5th May and 6th July 1904.—Medal with clasp.
N.-W. Frontier of India, 1908.—Operations in the Zakka Khel country.—Medal with clasp.

33rd Punjabis.

SUBADARS—

Atar Khan—
N.-W. Frontier of India, Hazara, 1888.—Medal with clasp.
N.-W. Frontier of India, Waziristan, 1894-95.—Clasp.
N.-W. Frontier of India, 1897-98.—Tochi.—Medal with clasp.

Khan Gul—
N.-W. Frontier of India, Waziristan, 1894-95.—Medal with clasp.
N.-W. Frontier of India, Tochi, 1897-98.—Medal with clasp.

Abdullah—
N.-W. Frontier of India, Waziristan, 1894-95.—Medal with clasp.
British East Africa, 1895-96.—Medal with clasp.
Uganda Relief Expedition, 1897-98.—Medal with clasp.

Gholam Hassan—
Somaliland, 1908-10.—Medal.

JEMADARS—

Waris Khan—
N.-W. Frontier of India, Tochi, 1897-98.—Medal with clasp.

Hazara Singh—
N.-W. Frontier of India, Chitral 1895.—Relief of Chitral.—Medal with clasp.

Allah Ditta—
N.-W. Frontier of India, 1897-98.—Tochi.—Medal and clasp

2-33rd Punjabis

SUBADARS—

Muhammad Hussan—
Tibet, 1903-04.—Action at Naini, 26th June 1904. Operations at and around Gyantse, between 5th May and 6th July 1904. March to Lhassa, 14th July to 3rd August 1904.—Medal with clasp.

Khan Gul—
Waziristan, 1894-95.—
N.-W. Frontier of India, Tochi, 1897-98.—
Egypt, 1914-15.—
France, 1915.—
Egypt, 1916.—
East Africa, 1916-17.—

EMADARS—

Jawala Singh—
Hazara, 1888.—
Relief of Chitral, 1895.—
Punjab Frontier, 1897-98.—

Bachan Singh—
Somaliland, 1904-05.—
Egypt, 1914-15.—
France, 1915.—
Aden, 1916.—
East Africa, 1917.—

Lal Khan—
Somaliland, 1904-05.—
Tibet, 1903-04.—
Hong Kong, 1914-15.—

Said Hussain—
Tibet, 1903-04.—
N.-W. Frontier of India, 1908.—Operations in the Mohmand country.
Hong Kong, 1914.—
France, 1915.—
East Africa, E. Force, 1916.—

Gheba Khan—
Egypt, 1914-15.—

Chan Pir Shah—
Egypt, 1914-15.—
France, 1915.—
Egypt, 1915-16.—
Aden, 1916-17.—

Hukan Dad—
Tibet, 1914-15.—
N.-W. Frontier of India, 1908.—
France, 1915.—
East Africa, 1916-17.—

Mohammad Aslam Khan—
N.-W. Frontier, 1915.—Mohmand country.

34th Sikh Pioneers.

SUBADAR-MAJOR—

Jwala Singh, *Sardar Bahadur—Order of Br. India, 1st class and I. O. M., 2nd class.*
N.-W. Frontier of India, Hazara, 1888.—Action of Kotekai.—Medal with clasp.
Chitral, 1895.—Relief of Chitral.—Medal with clasp.
Tirah, 1897-98.—Operations in the Bara Valley, 7th to 14th December 1897.—2 clasps.
China, 1900.—Medal.
N.-W. Frontier of India, 1908.—Operations in the Mohmand country.—Medal with clasp.

SUBADARS—

Natha Singh (II), *Sardar Bahadur.—Order of Br. India, 1st class and I. O. M., 2nd class.*
N.-W. Frontier of India, Hazara, 1888.—Medal with clasp.
N.-W. Frontier of India, 1908.—Operations in the Mohmand country.—Medal with clasp.

Sant Singh, M.C.—
Chitral, 1895.—Relief of Chitral.—Medal with clasp.
Tirah, 1897-98.—Operations in the Bara Valley, 7th to 14th December 1897.—2 clasps.
China, 1900.—Medal.
N.-W. Frontier of India, 1908—Operations in the Mohmand country.—Medal with clasp.

Mundar Singh—
Chitral, 1895.—Relief of Chitral.—Medal with clasp.
Tirah, 1897-98.—Operations in the Bara Valley, 7th to 14th December 1897.—2 clasps.
China, 1900.—Medal.

Thamman Singh—
Chitral, 1895.—Relief of Chitral.—Medal with clasp.
Tirah, 1897-98.—Operations in the Bara Valley, 7th to 14th December 1897.—2 clasps.
Tibet, 1903-04.—Medal.
N.-W. Frontier of India, 1908.—Operations in the Mohmand country.—Medal with clasp.

Malla Singh—
Tirah, 1897-98.—Operations in the Bara Valley, 7th to 14th December 1897.—Medal with 2 clasps.
China, 1900.—Medal.
N.-W. Frontier of India, 1908.—Operations in the Mohmand country.—Medal with clasp.

Bhanga Singh—
Chitral, 1895.—Relief of Chitral.—Medal with clasp.
Tirah, 1897-98.—Operations in the Bara Valley 7th to 14th December 1897.—2 clasps.
China, 1900.—Medal.
N.-W. Frontier of India, 1908.—Operations in the Mohmand country.—Medal with clasp.

Dewa Singh—
Chitral, 1895.—Relief of Chitral.—Medal with clasp.
Tirah, 1897-98.—Operations in the Bara Valley, 7th to 14th December 1897.—2 clasps.
China, 1900.—Medal.
N.-W. Frontier of India, 1908.—Operations in the Mohmand country.—Medal with clasp.

Ganesha Singh—

China, 1900.—Medal.
N.-W. Frontier of India, 1908.—Operations in the Mohmand country.—Medal with clasp.

Mangal Singh—

China, 1900.—Medal.
Tibet, 1903.—Medal.
N.-W. Frontier of India, 1908.—Operation in the Mohmand country.—Medal with clasp.

JEMADARS—

Tara Sing—

Chitral, 1895.—Relief of Chitral.—Medal with clasp.
Tirah, 1897-98.—Operations in the Bara Valley, 7th to 14th December 1897.—2 clasps.
N.-W. Frontier of India.—Operations in the Mohmand country.—Medal with clasp.

Bishn Singh—

N.-W. Frontier of India, 1908.—Operation in the Mohmand country.—Medal with clasp.

Dhanna Singh—

N.-W. Frontier of Indian, 1908.—Operation in the Mamund countsy.—Medal with clasp.

Sundar Singh—

N.-W. Frontier of India, 1908.—Operation in the Mohmand country.—Meadl with clasp.

Gian Singh—

N.-W. Frontier of India, 1908.—Operation in the Mohmand country·medal with clasp.

Sohan Singh—

N.-W. Frontier of India, 1908.—Operation in Mohmand country. Medal with clasp.

Warriam Singh—

Chitral, 1895.—Relief of Chitral, medal with clasp.
Tirah, 1897-98.—Operation in the Bara Valley, 7th to 14th Dcember 1897.—2 clasps.
China, 1900—Medal.
N.-W. Frontier of India, 1908.—Operation in the Mohmand country.—Medal with clasp.

Mihan Singh—

Tirah, 1897-98.—Operation in the Bara Valley, 7th to 14th December 1897.—2 clasps. *China, 1900.*—Medal.
N.-W. Frontier of India, 1908.—Operation in the Mohmand country.—Medal with clasp.

Bisakha Singh—

Chino, 1900—Medal.
N.-W. Frontier of India, 1908.—Operation in the Mohmand country. Medal with Clasp.

Mool Singh—

N.-W. Frontier of India, 1908.—Operation in the Mohmand country, Medal with clasp.

Harnam Singh—

Tirah, 1897-98.—Operation in the Bara Valley, 7th to 14th December 1897.—2 clasps.
China, 1900.—Medal.

Chain Singh—

N.-W. Frontier of India, 1908.—Operation in the Mohmand country.—Medal with clasp.

Gurdit Singh—

Chitral, 1895.—Relief of Chitral.—Medal with clasp.
Tirah, 1897-98.—Operation in the Bara Valley 7th to 14th December 1897.—2 clasps.
China, 1900.—Medal.
N.-W. Frontier of India, 1908.—Operation in the Mohmand country—Medal with clasp.

Sadda Singh—

N.-W. Frontier of India, Hazra, 1888.—Action of Kotikai.—Medal with clasp.
Chitral, 1895.—Relief of Chitral Medal with clasp.
Tira, 1897-98.—Operation in the Bara Valley, 7th to 14th December 1897.—2 clasps.
China, 1900.—Medal.
N.-W. Frontier of India, 1908.—Operation in the Mohmand country.—Medal with clasp.

2/34th Sikh Pioneers.

SUBADAR-MAJOR—

Fateh Singh—

N.-W. Frontier of India, Waziristan, 1901-02.—Medal with clasp.
Tibet 1903-04.—Action at Niani 26th June 1904.—Operations at and around Gyantse between 5th May and 6th July 1914. March to Lhassa 14th July to 3rd August 1904.—Medal with clasp.
Operations in the Abor country, 1911-12.—Medal with clasp.

SUBADARS—

Kishen Singh—

Chitral, 1895.—Relief of Chitral.—Medal with clasp.
N.-W. Frontier of India, Waziristan, 1901-02.—Clasp.
Tibet, 1903-04.—Action at Niani 26th June 1904. Operations at and around Gyantse between 5th May and 6th July 1904.—March to Lhassa 14th July to 3rd August 1904.—Medal with clasp.
N.-W. Frontier of India, 1908.—Operations in the Zakha Khel country.—Medal with clasp.
Operations in the Abor country, 1911-12.—Clasp.

Ala Singh—

N.-W. Frontier of India, Waziristan, 1901-02.—Medal with clasp.
Tibet, 1903 04.—Action at Niani 26th June 1904.—Operations at and around Gyantse between 5th May and 6th July 1904. March to Lhassa 14th July to 3rd August 1904.—Medal with clasp.

Teja Singh—

Chitral, 1895.—Relief of Chitral.—Medal with clasp.
N.-W. Frontier of India, Waziristan, 1901-02.—Clasp.
Tibet, 1903-04—Action at Niani 26th June 1904. Operations at and around Gyantse between 5th May and 6th July 1904.—Medal with clasp.
N.-W. Frontier of India, 1908.—Operations in the Zakka Khel country.—Medal with clasp.

Atma Singh—

N.-W. Frontier of India, Waziristan, 1901-02.—Medal with clasp.
Tibet, 1903-04.—Action at Niani 26th June 1904.—Operations at and around Gyantse between 5th May and 6th July 1904. March to Lhassa 14th July to 3rd August 1904.—Medal with clasp.
Operations in the Abor country, 1911-12.—Medal with clasp.

Jagat Singh—

N.-W. Frontier of India, Waziristan, 1901-02.—Medal with clasp.
N.-W. Frontier of India, 1908.—Operations in the Zakka Khel country. Medal with clasp.

Curditt Singh—

Chitral, 1895.—Relief of Chitral.—Medal with Clasp.

N.-W. Frontier of India, Waziristan, 1901-02.—clasps.

Tibet, 1903-04.—Action at Niani 26th June 1904. Operations at and around Gyantse between 5th May and 6th July 1904.—Medal with clasp.

Operations in the Abour country, 1911-12.—Medal with clasp.

Bir Singh—

N.-W. Frontier of India, 1908.—Operations in the Zakka Khel country.—Medal with clasp.

JEMADARS—

Thakur Singh—

N.-W. Frontier of India, 1908.—Operations in the Mohmand country.—Medal with clasp.

The War 1914.—Star.

Uttam Singh—

N.-W. Frontier of India, Waziristan, 1901-02—Medal with clasp.

Tibet, 1903 04.—Action at Niani 26th June 1904. Operations at and around Gyantse between 5th May and 6th July 1904—Medal with clasp.

N.-W. Frontier of India, 1908.—Operations in the Zakka Khel country—Medal with clasp.

Daya Singh—

Operations in the Abor conntry, 1911-12.—Medal with clasp.

Kishen Singh—

N.-W. Frontier of India, 1908.—Operations in the Zakka Khel country—Medal with clasp.

Partap Singh—

N.-W. Frontier of India, 1908.—Operations in the Mohmand country.—Medal with clasp.

The War 1914.—Star.

Jagat Singh—

N. W. Frontier of India, 1908.—Operations in the Zakka Khel Country—Medal with clasp.

Mundar Singh—

Operations in the Abor country, 1911-12.—Medal with clasp.

Isar Singh—

N.-W. Frontier of Indio, 1908—Operation in the Mohmand country—Medal and clasp.

The War 1918.—Star.

Warriam Singh—

Operation in the Abor country, 1911-12—Medal with clasp.

Sohan Singh—

N.-W. Frontier of India, 1908—Operations in the Zakka Khel country.—Medal with clasp.

Ganga Singh—

Operations in the Abor country, 1911-12.—Medal with clasp.

Teja Singh—

Operations in the Abor country, 1911-12.—Medal with clasp.

3-34th Sikh Pioneers.

SUBADAR-MAJOR—

Maghar Singh, I.D.S.M.—

N.-W. Frontier of India, 1908.—Operations in the Mohmand country.—Medal with clasp.

SUBADARS—

Indar Singh—

Tirah, 1897-98.—Medal with clasp.

China, 1900.—Medal.

N.-W. Frontier of India, 1908.—Operations in the Mohmand country.—Medal with clasp.

Karam Singh—

Tirah, 1897 98.—Operations in the Bara Valley, 7th to 14th December 1897.—Medal with 2 clasps.

China, 1900.—Medal.

N.-W. Frontier of India, 1908.—Operations in the Mohmand country.—Medal with clasp.

JEMADARS—

Jagat Singh—

China, 1900.—Medal.

N.-W. Frontier of India, 1908.—Operations in the Mohmand country.—Medal with clasp.

35th Sikhs.

SUBADAR-MAJOR—

Mangal Singh *Bahadur, Indian Distinguished Service Medal—Order of British India, 2nd class.*

Soudan, 1896.—2 medals (Queen's and Khedive's).

N.-W. Frontier of India, 1897-98.—Malakand. Operations in Bajour and in the Mamund country. Utman Khel—Medal with clasp.

N.-W. Frontier of India, Waziristan, 1901-02.—Clasp.

SUBADARS—

Dula Singh—

Soudan, 1896.—2 medals (Queen's and Khedive's).

N.-W. Frontier of India, 1897-98.—Relief of Malakand. Relief of Chakdara. Malakand. Operations in Bajour and in the Mamund country. Utman Khel.—Medal with 2 clasps.

N.-W. Frontier of India, Waziristan, 1901-02.—Clasp.

Partap Sing—

N.-W. Frontier of India, 1897-98.—Malakand. Operations in Bajour and in the Mohmand country. Utman Khel. Tirah.—Medal with 2 clasps.

N.-W. Frontier of India, Waziristan, 1901-02.—Clasp.

Attar Singh (II)—

N.-W. Frontier of India, Waziristan, 1901-02.—Medal with one clasp.

Sundar Singh—

Soudan, 1896.—2 medals (Queen's and Khedive's).
N.-W. Frontier of India, Waziristan, 1901-02.—Medal with clasp.

Harnam Singh—

N.-W. Frontier of India, 1897-98.—Relief of Malakand. Relief of Chakdara. Malakand. Operations in Baiour and in the Mamund country. Utman Khel. Tirah.—Medal with 2 clasps.
N.-W. Frontier of India, Waziristan, 1901-02.—Clasp.

Hardit Singh—

N.-W. Frontier of India, 1897-98.—Relief of Malakand. Relief of Chakdara. Malakand. Operations in the Mamund country. Operations in the Salarzai Valley. Utman Khel.—Medal and 2 clasps
N.-W. Frontier of India, Waziristan, 1901-02.—Clasp.

Buta Singh—

N.-W. Frontier of India, 1897-98.—Relief of Malakand.—Relief of Chakdara. Operations in Bajour and in the Mohmand country. Utman Khel.—Medal with 2 clasps.
N.-W. Frontier of India, Waziristan, 1901-02.—Clasp.

Arjan Singh—

N.-W. Frontier of India, Waziristan, 1901-02.—Medal with clasp.

JEMADARS—

Labh Singh—

Soudan, 1896.—2 medals (Queen's and Khedive's).
N.-W. Frontier of India, 1897-98.—Operations in Mahmund and Salerzai. Valleys, Utman Khel Expy. Force.—Medal with clasp.
N.-W. Frontier of Inaia, 1901-02.—Waziristan, one clasp.

Gonda Singh—

N.-W. Frontier of India, 1897-98.—Operations in the Mahmund Valley.—Medal and clasp.
N.-W. Frontier of India, Waziristan, 1901-02.—1 clasp.
East Africa, 1902-04.—Operations in Somaliland.—Medal with clasp.

Sahib Singh—

N.-W. Frontier of India, 1897-98.—Relief of Malakand. Relief of Chakdara. Operations in the Salerzai and Mahmund Valleys. Utman Khel Expy. Force.—Medal with 2 clasps.
N.-W. Frontier of India, Waziristan, 1901-02.—1 clasp.

Gainda Singh—

N.-W. Frontier of India, 1897-98.—Utman Khel Expy. Force.—Medal with clasp.
N.-W. Frontier of India, Waziristan, 1901-02.—Clasp.
East Africa, 1902-04.—Medal and 2 clasps.

2-35th Sikhs.

SUBADAR MAJOR—

Attar Singh—

Central Africa, 1893-94.—Storming of Kiwassaurus town Ritu Expedition.—Miranjiras Campaign and action of Ugombo.—Medal with clasp.
Soudan, 1896.—2 medals (Queen's and Khedive's).
N.-W. Frontier of India, 1897-98.—Relief of Malakand, Relief of Chakdara, Malakand. Operations in the Mamund country, wounded.—Medal with 2 clasps.
N.-W. Frontier of India, Waziristan, 1901-02.—Clasp.
East Africa 1902-04.—Operations in Somaliland.—Medal with clasp.

SUBADARS—

Bhola Singh—

N.-W. Frontier of India, 1897-98.—Relief of Malakand, Relief of Chakdara, Malakand. Operations in the Mamund country, Utman Khel.—Medal with 2 clasps.
N.-W. Frontier of India, Waziristan, 1901-02.—Clasp.

Katha Singh—

N.-W. Frontier of India, Waziristan, 1901-02.—Medal with clasp.

Kour Singh—

N.-W. Frontier of India, Waziristan, 1901-02.—Medal with clasp.

Thakur Singh—

N.-W. Frontier of India, 1897-98.—Malakand. Operations in the Munmad country and in Bajour, Utman Khel.—Medal with clasp.
N.-W. Frontier of India, Waziristan, 1901-02.—One clasp.

Ganda Singh—

N.-W. Frontier of India, 1901-02.—Medal with clasp.

Hatti—

Tochi, 1914-15.—
Dirdoni, 1915-16.—
Mesopotamia, 1917-18.—

Ji Ram—

Tochi, 1914-15.—
Mesopotamia, 1917-18.—

Sukh Dev—

Aden, 1903-04.—
Operations in interior Kabul, 1915-16.—
Muscat, 1916.—
Mesopotamia, 1917-18.—

Sheo Chand (I)—

Tochi, 1914-15.—

Sheo Chand (II)—

France, 1914-15.—
Mesopotamia, 1916.—Wounded.
Mesopotamia, 1917-18.—

JEMADARS—

Kirpal Singh—

Frontier Waziristan, 1901-02.—Medal and clasp.

Partab Singh—

Frontier Waziristan, 1901-02.—Medal and clasp.

Badlu—

Tochi, 1914-15.—
Dirdoni, 1915-16.—
Mesopotamia, 1917-18.—

Pritam Singh—

Frontier Waziristan, 1901-02.—Medal and clasp.
Tirah Expedition, 1898.—Medal and clasp.

Sant Singh—

Mesopotamia, 1914-15 —Wounded.

Ami Chand—

Kabul, 1915-16.—
Muscat, 1916.—

Bakhshish Singh—

Uganda Contingent, 1907.—

Harbant Singh—

Frontier Waziristan, 1901.—
Mohmand Expedition, 1908.—Indian Distinguished Service medal.

Nand Singh—

France, 1914.—

Dip Chand—

Force "D," 1916.—

Siri Ram—

Force "D," 1917-18.—

Chhote Lal—

Kabul, 1915-16.—
Muscat, 1916.—
Mesopotamia, 1917-18.—

36th Sikhs.

SUBADAR-MAJOR—

Kishn Singh—

Soudan, 1896.—2 medals (Queen's and Khedive's).
N.-W. Frontier of India, 1897 98.—Operations on the Samana and in the Kurram Valley. Relief of Gulistan.—Medal with 2 clasps.
Tirah, 1897-98.—Actions of the Sampagha and Arhanga Passes.—Reconnaissance of the Saran Sar. Operations in the Warran Valley. Operations at and around Dwatoi. Operations in the Bara Valley, 7th to 14th December 1897. Affairs at Shinkamar.—Clasp.

SUBADARS—

Khem Singh—

N.-W. Frontier of India, 1897-98.—Operations on the Samana and in the Kurram Valley. Defence of Gulistan.—Medal with 2 clasps.
Tirah, 1897-98.—Actions of the Sampagha. Capture of the Arhanga Passes. Reconnaissance of the Saran Sar. Operations in the Warran Valley. Operations at and around Dwatoi and in Bara Valley.—Clasp.

Pertab Singh—

N.-W. Frontier of India, 1897-98.—Operations on the Samana and in the Kurram Valley.—Medal with 2 clasps.
Tirah, 1897-98.—Capture of the Sampagha and Arhanga Passes. Reconnaissance of the Saran Sar. Operations in the Warran Valley. Operations at and around Dwatoi. Operations in the Bara Valley.—Clasp.

Sher Singh—

N.-W. Frontier of India, 1897-98.—Operations on the Samana and in the Kurram Valley. Defence of Gulistan. Relief of Gulistan.—Medal with 2 clasps.
Tirah, 1897-98.—Capture of the Arhanga Pass. Reconnaissance of the Saran Sar. Operations in the Warran Valley. Operations at and around Dwatoi. Operations in the Bara Valley. Affair at Shinkamar.—Clasp.

Phuman Singh—

N.-W. Frontier of India, 1897-98.—Operations on Samana and Kurram. Defence of Gulistan —*3rd Class Order of Merit* for gallantry at Fort Gulistan.—Medal with 2 clasps.
Tirah, 1897-98.—Capture of the Sampagha Pass. capture of the Arhanga Pass. Reconnaissance of the Saran Sar. Operations in Warran Valley. Operations at and around Dwatoi. Operations in the Bara Valley. Affairs at Shinkamar—Clasp.

Fakir Singh—

N.-W. Frontier of India, 1897-98.—Operations on Samana and Kurram. Relief of Gulistan. —Medal and 2 clasps.
Tirah, 1897-98.—Capture of the Sampagha Pass. Capture of the Arhanga Pass. Reconnaissance of the Saran Sar. Operations at and around Dwatoi. Operations in the Bara Valley. Affairs at Shinkamar.—Clasp.

Bara Singh—

N.-W. Frontier of India, 1897-98.—Operations on Samana and Kurram. Relief of Gulistan.—Medal with 2 clasps.
Tirah, 1897-98.—Capture of Sampagha Pass. Capture of Arhanga Pass. Reconnaissance of the Saran Sar. Operations at and around Dwatoi. Operations in Bara Valley. Affairs at Shinkamar. —Clasp.

Naurang Singh—

N.-W. Frontier of India, 1897-98.—Operations on Samana and Kurram. Relief of Gulistan.—Medal and 2 clasps.
Tirah, 1897-98.—Capture of the Sampagha Pass. Capture of the Arhanga Pass. Reconnaissance of the Saran Sar. Operations at and around Dwatoi. Operations in the Bara Valley. Affairs at Shinkamar.—Clasp.

JEMADARS—

Chhaja Singh—

N.-W. Frontier of India, 1897-98.—Operations on Samana and Kurram. Defence of Gulistan. *3rd Class Order of Merit* for gallantry at Fort Gulistan.—Medal with 2 clasps.
Tirah, 1897-98.—Capture of the Sampagha Pass. Capture of the Arhanga Pass. Reconnaissance of the Saran Sar. Operations in the Warran Valley. Operations at and around Dwatoi. Operations in the Bara Valley. Affairs in Shinkamar.—Clasp.

Nand Lall—

N.-W. Frontier of India, 1897-98.—Operations on Samana and Kurram—Medal and 2 clasps.

Ram Lall—

N.-W. Frontier of India, 1897-98.—Operations on the Samana and Kurrum Valley. Defence of Gulistan.—Medal with cla p.
Tirah, 1897-98.—Action of Sanipyha P ss, capture of Arhunga Passes, Reconnaissance of Saran Sar. Operations in the Warran Valley. Operations at and round Dwatoi and in Bara Valley.—Clasp.

2-36th Sikhs.

Surain Singh—
N.-W. Frontier of India, Waziristan, 1901-02.—clasp.

37th Dogras.

SUBADAR MAJOR—

SUBADARS—

Bhag Singh, I.O.M.—*Bahadur order of Br. India 2nd class.*
N.-W. Frontier of India, 1897-98.—Mohmand.—Medal with clasp.

Chagar Singh—*Bahadar—Order of Br. Ind., 2nd class.*
East Africa, 1902-04.—Operations in Somaliland. Action at Jidballi.—Medal with 2 clasps.

Tega—*Bahadur—Order of Br. Ind., 2nd class.*
N.-W. Frontier of India, Hazara, 1891.—Medal with clasp.
Chitral, 1895.—Relief of Chitral.—Medal with clasp.
N.-W. Frontier of India, 1897-98.—Mohmand.—Clasp.

Udho—
Chitral, 1895.—Relief of Chitral.—Medal with clasp.

Janka—
Chitral, 1895.—Relief of Chitral.—Medal with clasp.
N.-W. Frontier of India, 1897-98.—Mohmand.—Clasp.

Ran Singh—
Chitral, 1895.—Relief of Chitral.—Medal with clasp.

Jai Singh—
Chitral, 1895.—Relief of Chitral.—Medal with clasp.
Punjab Frontier—Waziristan—1894-95.—Medal and clasp.
N.-W. Frontier of India, 1897-98.—Mohmand—Clasp.

Ral Singh—
Chitral, 1895.—Relief of Chitral.—Medal with clasp.
Punjab Frontier, 1897-98.—Clasp.

Hakim Singh—
N.-W. Frontier of India, 1897-98.—Mohmand.—Medal with clasp.

Sundar—
N.-W. Frontier of India, 1897-98.—Mohmand.—Medal with clasp.
East Africa, 1902—04.—Operations in Somaliland—Action at Jidballi.—Medal with 2 clasps.

Kahna—
Chitral, 1895.—Relief of Chitral.—Medal with clasp.

Phandi—
N.-W. Frontier of India, 1897-98.—Mohmand.—Medal with clasp.

JEMADARS—

Munshi—I. O. M.
N.-W. Frontier of India, 1897-98.—Mohmand.—Medal with clasp.

Lachman—
East Africa, 1902-04.—Operations in Somaliland, action at Jidballi.—Medal with 2 clasps.

Sohanu—
N.-W. Frontier of India, 1897-98.—Medal and clasp.

38th Dogras.

SUBADAR-MAJOR—

Unkar Chand—
N.-W. Frontier of India, Waziristan, 1894-95.—Medal with clasp.
N.-W. Frontier of India, 1897-98—Malakand Operations in Bajour and in the Mamund country.—Medal with clasp.
N.-W. Frontier of India, Waziristan, 1901-02.—Clasp.
N.-W. Frontier of India, 1908.—Operations in the Mohmand country. Engagement of Kargha, 24th May 1908.—Medal with clasps.

SUBADARS—

Kapur Singh—
N.-W. Frontier of India, Waziristan, 1894-95.—Medal with clasp.
N.-W. Frontier of India, 1897-98.—Relief and defence of Malakand. Relief of Chakdara.—Malakand. Operations in Bajour and Mamund country.—Medal with 2 clasps.

Ushnaki—
N.-W. Frontier of India, 1897-98.—Malakand. Operations in Bajour and in Mamund country.—Medal with clasp.

Nar Singhu—
N.-W. Frontier of India, Waziristan, 1894-95.—Medal with clasp.
N.-W. Frontier of India, 1897-98.—Malakand.—Operations in Bajour and in the Mamund country.—Medal with clasp.
N.-W. Frontier of India, Waziristan, 1901-02.—Clasp.

Hako—
N.-W. Frontier of India, 1897-98.—Relief and defence of Malakand. Relief of Chakdara. Malakand. Operations in Bajour and in the Mamund country.—Medal with 2 clasps.

Jewan Singh—
N.-W. Frontier of India, Waziristan, 1894-95.—Medal with clasp.
N.-W. Frontier of India, 1897-98.—Relief and defence of Malakand. Relief of Chakdara. Malakand. Operations in Bajour and in the Mamund country.—Medal with 2 clasps.
N.-W. Frontier of India, Waziristan, 1901-02.—Clasp.

Kanhaiya—
N.-W. Frontier of India, 1897-98.—Malakand.—Operations in Bajour and in the Mamund country.—Medal with clasp.
N.-W. Frontier of India, Waziristan, 1901-02.—Clasp.

Teg Singh—
N.-W. Frontier of India, Waziristan, 1901-02.—Medal and 1 clasp.

Gujar Singh—*I. O. M.*
N.-W. Frontier of India, 1897-98.—Relief and defence o Malakand. Relief of Chakdara, Malakand. Operations in Bajour and in the Mohmand country. Medal with 2 clasps.
N.-W. Frontier of India, Waziristan, 1901-02.—Medal and 1 clasp.

JEMADARS—

Zalim Singh—
N.-W. Frontier of India, 1897-98.—Relief and defence of Malakand. Relief of Chakdara. Malakand. Operations in Bajour and in the Mamund country.—Medal with 2 clasps.
N.-W. Frontier of India, Waziristan, 1901-02.—Medal and 1 clasp.

Bhagta—
N.-W. Frontier of India, Waziristan, 1901-02.—Medal with clasp.

Nathu—
N.-W. Frontier of India, Waziristan, 1894-95.—Medal with clasp.
N.-W. Frontier of India, 1897-98.—Malakand operations in Bujour and in the Mamund country.—Medical with clasp.

Mohar Singh—
N.-W. Frontier of India, 1897-98.—Relief and defence of Malakand. Relief of Chakdara and Malakand. Operations in Bajour and in the Mamund country.—Medal with 2 clasps.

Lachhman—
N.-W. Frontier of India, Waziristan, 1901-02.—Medal and 1 clasp.

Mohr Singh—
N.-W. Frontier of India, Waziristan, 1901-02.—Medal and 1 clasp.

39th Garhwal Rifles.

SUBADAR-MAJORS—

Kedar Sing Rawat—*Indian Distgn. Service Medal.*
Burma, 1891-92.—Medal with clasp.
Burma, 1892-93.
N.-E. Frontier of India, Chin Hills, 1892-93.—Clasp.
N.-W. Frontier of India, 1897-98.—Malakand. Operations in Bajour. Mohmand.—Medal with clasp.
Tirah, 1897-98.—Clasp.
France, 1914-15.

SUBADARS—

Makhar Sing Kawar—
N.-E. Frontier of India, Lushai, 1892.—Medal with clasp.
Burma, 1892-93.
N.-E. Frontier of India, Chin Hills, 1892-93.—Clasp.
N.-W. Frontier of India, 1897-98.—Malakand. Operations in Bajour. Mohmand.—Medal with clasp.
Tirah, 1897-98.—Clasp.

Ganga Sing Rawat—
Tirah, 1897-98.—Actions of Chagru Kotal and Dargai. Capture of the Sampagha and Arhanga Passes. Operations in the Waran Valley and action of the 16th November 1897. Operations at and around Dwatoi and action of 24th November 1897. Operations in the Bara Valley, 7th to 14th December 1897. Affair at Shinkamar.—Medal with 2 clasps.

Galthi Sing Negi—
N.-E. Frontier of India, Chin Hills, 1892-93.—Medal with clasp.
N.-W. Frontier of India, 1897-98.—Malakand. Operations in Bajour. Mohmand.—Medal with clasp.
Tirah, 1897-98.—Clasp.

Jagat Sing Rawat—Bahadur.
Order of Br. India, 2nd cl. Bahadur.
Burma, 1891-92.—Medal with clasp.
Indian Order of Merit, 3rd Class. Bahadur.
N.-E. Frontier of India, Chin Hills, 1892-93.—Clasp.
N.-W. Frontier of India, 1897-98.—Malakand Operations in Bajour. Mohmand.—Medal with clasp.
Tirah, 1897-98.—Clasp.
France, 1914-15.

JEMADARS—

Shib Sing Negi—
Burma, 1891-92.—Medal with clasp.
N.-E. Frontier of India, Chin Hills, 1892-93.—Clasp.
Tirah, 1897-98.—Medal with 2 clasps.

Sangram Sing Negi—
N.-W. Frontier of India, 1897-98.—Malakand. Operations in Bajour.
Tirah, 1897-98.—Clasp.

Balbhadar Sing Gusain—
N.-W. Frontier of India, 1897-98.—Malakand, Operations in Bajour. Mohmand.—Medal with clasp.
Tirah, 1897-98.—Clasp.

Jura Sing Negi—
N.-W. Frontier of India, 1897-98.—Malakand Operations in Bajour. Mohmand.—Medal with clasp.
Tirah, 1897-98.—Clasp.

2-39th Garhwal Rifles.

SABADAR MAJORS—

Nathu Sing Sunar *(Retd.) Sardar Bahadur and Hon. Capt. Order of British India 1st Cl ss.*
Burma, 1890-91.
N.-W. Frontier of India, 1897-98—Malakand. Operations in Bajour. Mohmand.—Medal with clasp.

Main Sing Chinwarh, M.C., *Bahadur, Order of British India, 2nd Class.*
Burma, 1891-92.—Medal with clasp.
N.-E. Frontier of India, Chin Hills, 1892-93.
N.-W. Frontier of India, 1897-98.—Malakand. Operations in Bajour. Mohmand.—Medal with clasp.
Tirah, 1897-98.—Clasp.
Indian Expy. Force "A" France, 1914-15.

SUBADARS—

Kishan Sing Bisht. (*Re-employed*).

Burma, 1891-92.—Medal with clasp.
N.-E. Frontier of India, Chin Hills, 1892-93.—Clasp.
N.-W. Frontier of India, 1897-98.—Malakand. Operations in Bajour. Mohmand.—Medal with clasp.
Tirah, 1897-98.—Clasp.

Lal Sing Gusain. (*Re-employed*).

Burma, 1891-92.—Medal with clasp.
Burma, 18 2-93.
N.-E. Frontier of India, Chin Hills, 1892-93.—Clasp.
N.-W. Frontier of India, 1897-98.—Malakand. Operations in Bajour. Mohmand.—Medal with clasp.
Tirah, 1897-98.—Clasp.

Gopi Sing Rawat—

N.-W. Frontier of India, 1897-98.—Malakand. Operations in Bajour. Mohmand.—Medal with clasp.
Tirah, 1897-98.—Clasp.
Indian Expy. Force "A", France, 1914-15.
Indian Expy. Force "E", Egypt, 1915-16.
Mesopotamia Expy. Force "D", 1917-18.

Indar Sing Negi—

Indian Expy. Force "A", 1914-15.
Indian Expy. Force "E", 1915-16.
Mesopotamia Expy. Force "D", 1917-18.

Amar Sing Negi—

Indian Expy. Force "A", 1914-15.
Indian Expy. Force "E", 1915-16.
Mesopotamia Expy. Force "D", 1917-18.

Tilok Sing Sauntiyal—

Indian Expy. Force "A", 1914-15.
Indian Expy. Force "E", 1915-16.
Mesopotamia Expy. Force "D", 1917-18.

Meharwan Sing Butola—

Indian Expy. Force "A", 1914-15.
Indian Expy. Force "E", 1915-16.
Mesopotamia Expy. Force "D", 1917-18.

Ratan Sing Butola—

Indian Expy. Force "A", 1914-15.
Mesopotamia Expy. Force "D", 1917-18.

Bir Sing Danu—

Indian Expy. Force "A", 1914-15.
Mesopotamia Expy. Force "D", 1917-18.

Gundar Sing Bhandari—

N.-E. Frontier of India, Lushai, 1892.—Medal with clasp.
Burma, 1892-93.
N.-E. Frontier of India, Chin Hills, 1892-93.—Clasp.
N.-W. Frontier of India, 1897-98.—Malakand. Operations in Bajour. Mohmand.—Medal with clasp.
Tirah, 1897-98.—Clasp.

JEMADARS—

Daulat Sing Sunar—

Indian Expy. Force "A", 1914-15.
Indian Expy. Force "E", 1915-16.
Mesopotamia Expy. Force "D", 1917-18.

Dewan Sing Padhiyar—

Indian Expy. Force "A", 1914-15.
Indian Expy. Force "E", 1915-16.
Mesopotamia Expy. Force "D", 1917-18.

Bahadur Sing Mahar—

Indian Expy. Force "A", 1914-15.
Mesopotamia Expy. Force "D", 1917-18.

Gaur Sing Rawat—

Indian Expy. Force "A", 1914-15.
Indian Expy. Force "E", 1915-16.
Mesopotamia Expy. Force "D", 1917-18.

Mohan Sing Negi—

Indian Expy. Force "A", 1914-15.
Indian Expy. Force "E", 1915-16.
Mesopotamia Expy. Force "D", 1917-18.

Har Sing Gusain—

Indian Expy. Force "A", 1914-15.
Indian Expy. Force "E", 1915-16.
Mesopotamia Expy. Force "D", 1917-18.

Rup Sing Rawat—

Indian Expy. Force "A", 1914-15.
Indian Expy. Force "E", 1915-16.
Mesopotamia Exp. Force "D", 1917-18.

Jaman Sing Chamola—

Indian Expy. Force "A", 1914-15.
Indian Expy. Force "E", 1915-16.
Mesopotamia Expy. Force "D", 1917-18.

40th Pathans.

SUBADAR-MAJOR—

Malapa, *Sardar Bahadur—Order of Br. India, 1st class*

N.-W. Frontier of India, Hazara, 1891.—Medal with clasp.
Chitral, 1895.—Relief of Chitral. Action of the Malakand Pass.—Medal with clasp.
N.-W. Frontier, 1897-98.—Defence and Relief of Malakand. Malakand. Operations in Bajour and the Mamund country.—2 clasps.
Tibet, 1903-04.—Medal.
N.-W. Frontier of India, 1908.—Operations in the Mohmand country.—Medal with clasp.

SUBADARS—

Ghulam Ali—

Tibet, 1903-04.—Action at Naini, 26th June 1904.—Operations at and around Gyantse, between 5th May and 6th July 1904. March to Lhassa, 14th July to 3rd August 1904.—Medal with clasp.
N.-W. Frontier of India, 1908.—Operations in the Mohmand country.—Medal with clasp.

Ata Muhammad—

N.-W. Frontier of India, 1908.—Operations in the Mohmand country.—Medal with clasp.

Abdullah Ji—

Tibet, 1903-04.—Medal.

Sher Afzal—

N.-W. Frontier of India, 1908.—Operations in the Mohmand country.—Medal with clasp.

Ghulam-Muhiyuddin—

Tibet, 1903-04.—Action at Naini, 26th June 1904. Operations at and around Gyautse, between 5th May and 6th July 1904. March to Lhassa, 14th July to 3rd August 1904.—Medal with clasp.
N.-W. Frontier of India, 1908.—Operations in the Mohmand country.—Medal with clasp.

JEMADARS—

Fazal-Dad—
Tibet, 1903-04.—Action at Naini, 26th June 1904. Operations at and around Gyantse, between 5th May and 6th July 1904. March to Lhassa, 14th July to 3rd August 1904.—Medal and clasp.

Jamdad-Khan—
N.-W. Frontier of India, 1908.—Operations in Mohmand country.—Medal with clasp.

Multani-Singh—
Tibet, 1903-04.—Action at Naini, 26th June 1904. Operations at and around Gyantse between 5th May and 6th July 1904. March to Lhassa, 14th July to 3rd August 1904.—Medal with clasp.

Surjan—
Tibet, 1903-04.—Medal.
N.-W. Frontier of India, 1908.—Operations in the Mohmand country.—Medal with clasp.

Sultan-Khan—
Tibet, 1903-04.—Action at Naini, 26th June 1903. Operations at and around Gyantse, between 5th May and 6th July 1904. March to Lhassa, 14th July to 3rd August 1904.—Medal with clasp.

Ali Akbar—
Tibet, 1903-04.—(I) Action at Niani, 26th June 1004. (II) Operations at and around Gyantsi, between 5th May and 6th July 1904. (III) March to Lhassa, 14th July to 3rd August 1904.—Medal and clasp.

Gandaki—
N.-W. Frontier of India, 1908.—Operations in the Mohmand country.—Medal with clasp.

Daurani—
N.-W. Frontier of India, 1908.—Operations in the Mohmand country.—Medal and clasp.

2-40th Pathans.

SUBADARS—

Gul Zaman—
N.-W. Frontier of India, 1908.—Operations in the Mohamand country.—Medal with clasp.

Malook-Singh—
Tibet, 1903-04—March to Lhassa, 14th July to 3rd August 1904.—Medal.
N.-W. Frontier of India, 1908.—Operations in the Mohmand country.—Medal and clasp.

Nikku—
Tibet, 1903-04.—Medal.
N.-W. Frontier of India, 1908.—Operations in the Mohmand country.—Medal with clasp.

41st Dogras.

SUBADAR-MAJOR—

SUBADARS—

Dulo—
N.-W. Frontier of India, 1897-98—Buner.—Medal with clasp and gratuity.
France and Mesopotamia, 1914-16.

2-41st Dogras.

SUBADAR-MAJOR—

Dhani Ram—
N.-W. Frontier of India, 1897-98.—Malakand.—Medal with clasp.
East Africa, 1902-04.—Operations in Somaliland. Action at Jidballi.—Medal with 2 clasps.

SUBADARS—

Kara. N.-W. Frontier of India, Waziristan, 1901-02.—Medal with clasp.

Kansi Ram -
N.-W. Frontier of India, 1897-98.—Malakand. Operations in Bajour and in the Mamund country.—Medal with clasp.
N.-W. Frontier of India, Waziristan, 1901-02.—Clasp.

Charat Singh—
N.-W. Frontier of India, 1897-98.—Relief and defence of Malakand. Relief of Chakdara, Malakand.—Operations in Bajour and in the Mamund country.—Medal with 2 clasps.
N.-W. Frontier of India, Waziristan, 1901-02.—Medal with clasp.

Ran Das—
N.-W. Frontier of India, Waziristan, 1901-02—Medal with clasp.

JEMADARS—

Makhan Singh—
N.-W. Frontier of India, Waziristan, 1901-02.—Medal with clasp.

Lahnu—
N.-W. Frontier of India, 1896-98.—Malakand.—Medal and 2 clasps.
N.-W. Frontier of India, Waziristan, 1901-02.—Clasp.

42nd Deoli Regiment.

SUBADAR-MAJOR—

Chanda—
Aden, 1903-04.—Operations in the Interior.

1-43rd Erinpura Regiment.

SUBADAR-MAJOR—

Goma—
War, 1915-18.—Mentioned in despatches.

SUBADARS—

Wazira, 1915-18.—Mentioned in despatches.

Jala Singh—
War, 1915-18.—Mentioned in despatches.

Rughnath Singh—
War, 1915-17.

Sulaiman Khan—
War, 1915-18.—Mentioned in despatches.

Udhan Singh—
War, 1915-18.

Bagh Singh—
War, 1915-18.

Aladad Khan—
War, 1914-18.

Sadhu Singh—
War, 1915-18.

JEMADARS—

Harnam Singh—
War, 1915-18.—Mentioned in despatches.

Mota—
War, 1915-18.

Umar Khan—
War, 1918.

Nanne Khan—
War, 1915-18.

Bhur Singh—
War, 1915-18.

Gulaba—
War, 1915-18.

Nishan Ali—
War, 1914-16.

Samarah Singh—
War, 1915-18.

Karim Khan—
War, 1914-18.

Gopal Singn—
War, 1915-18.

Salab Singh—
War, 1915-18.

Sanwat Singh—
War, 1915-18.

2-43rd Erinpura Regiment.

SUBADAR-MAJOR—

Abdul Wahab Khan—
Mesopotamia, 1915-17.

SUBADARS—

Sobha Ram—
Aden, 1903-04.—Operations in the Interior.

Harnarayan—
Aden, 1903-04.—Operations in the Interior.

Mendu—
Mesopotamia, 1916-17.

Ram Singh—
Mesopotamia, 1914-17.

JEMADARS—

Teja—
Mesopotamia, 1915-17.

Kana—
Mesopotamia, 1915-17.

Bhura—
France, 1914-16.
Mesopotamia, 1916-17.

Duja—
Mesopotamia, 1915-17.

Chhaga—
Mesopotamia, 1915.

Rupa—
Mesopotamia, 1915-17.

Kehar Singh—
B. E. F. & Persian Gulf, 1915-16.

Shambhu Lall—
Mesopotamia, 1916.

45th Rattray's Sikhs.

SUBADAR-MAJOR—

Sundar Singh (I), I.O.M.—*Bahadur.—Order of Br. India, 2nd Class.*

N.-W. Frontier of India, 1897-98.—Operations in Bajour and in the Mamund country.—Medal with clasp.
Tirah, 1897-98.—Operations in the Bara Valley, 7th to 14th December 1897.—Clasp.
N.-W. Frontier of India, 1908.—Operations in the Zakka Khel country.—Medal with clasp.
Mentioned in despatches Ex. G. of I., No. 1740 of 1917.

Bhagat Singh, *Sardar Bahadur—Order of Br. India, 1st class.*
N.-W. Frontier of India, 1897-98.—Defence of Malakand. Relief of Chakdara.—Malakand. Action of Landikai.—Operations in Bajour and in the Mamund country.—Medal with 2 clasps.
Tirah, 1897-98.—Operations in the Bara Valley, 7th to 14th December 1897.—Clasp.
N.-W. Frontier of India, Waziristan, 1901-02.—Clasp.
N.-W. Frontier of India, 1908.—Operations in the Zakka Khel country.—Medal with clasp.

SUBADARS—

Nidhan Singh—

N.-W. Frontier of India, 1897-98.—Defence of Malakand. Relief of Chakdara. Operations in Bajour and in the Mamund country.—Medal with 2 clasps.

Narain Singh, I.O.M.—

N.-W. Frontier of India, 1897-98.—Malakand. Action of Landikai. Operations in Bajour and in the Mamund country.—Medal with clasp.
Tirah, 1897-98.—Operations in the Bara Valley, 7th to 14th December 1897.—Clasp.
N.-W. Frontier of India, Waziristan, 1901-02.—Clasp.
N.-W. Frontier of India, 1908.—Operations in the Zakka Khel country.—Medal with clasp.
Mentioned in despatches Ex. G. of I., No. 1740 of 1917.

Mall Singh—*Bahadur, Order of Br. India, 2nd class.*
N.-W. Frontier of India, Hazara, 1888.—Medal with clasp.
N.-W. Frontier of India, 1897-98.—Defence of Chakdara. Malakand. Action of Landakai. Operations in Bajour and in the Mamund country.—Medal with 2 clasps.
Tirah, 1897-98.—Operations in the Bara Valley, 7th to 14th December 1897.—Clasp.
N.-W. Frontier of India, Waziristan, 1901-02.—Clasp.
N.-W. Frontier of India, 1908.—Operations in the Zakka Khel country.—Medal with clasp.

Sundar Singh (II)—

N.-W. Frontier of India, 1897-98.—Defence of Malakand. Malakand. Action at Landikai. Operations in Bajour and in the Mamund country.—Medal with 2 clasps.
Tirah, 1897-98.—Operations in the Bara Valley, 7th to 14th December 1897.—Clasp.
N.-W. Frontier of India, Waziristan, 1901-02.—Clasp.
N.-W. Frontier of India, 1908.—Operations in the Zakka Khel country.—Medal with clasp.

Kehar Singh—

N.-W. Frontier of India, 1897-98.—Malakand. Action of Landikai. Operations in the Bajour and in the Malakand country.—Medal with clasp.
Tirah, 1897-98.—Operations in the Bara Valley, 7th to 14th December 1897.—Clasp.
N.-W. Frontier of India, 1908.—Operations in the Zakka Khel country.—Medal with clasp.

Labh Singh—

N.-W. Frontier of India, 1897-98.—Defence of Malakand. Relief of Chakdara Malakand. Action of Landikai. Operations in Bajour and in the Mamund country.—Medal with 2 clasps.
Tirah, 1897-98.—Operations in the Bara Valley, 7th to 14th December 1897.—Clasp.
N.-W. Frontier of India, Warisistan, 1901-02.—Clasp.
N.-W. Frontier of India, 1908.—Operations in the Zakka Khel country.—Medal with clasp.

Mehar Singh, I.D.S.M.—
N.-W. Frontier of India, Waziristan, 1901-02.—Medal and clasp.
Mentioned in Despatches.

Lal Singh II—
N.-W. Frontier of India, 1897-98.—Defence of Malakand.—Relief of Chakdara.—Malakand. Action of Landikai.—Operations in Bajour and the Mamund country. Medal with 2 clasps.
Tirah, 1897-98.—Operations in the Bara Valley, 7th to 14th December 1897.—Clasp.
N.-W. Frontier of India, Waziristan, 1901-02.—Clasp.
N.-W. Frontier of India, 1908.—Operations in the Zakka Khel country.—Medal and clasp.

Abdul Hamid—
N.-W. Frontier of India, 1897-98.—Medal.
Tirah, 1897-98.—Operations in the Bara Valley, 7th to 14th December 1897.—Clasp.
Meritorius Service Medal.

Partab Singh—
N.-W. Frontier of India, 1897-98.—Malakand. Action of Landikai. Operations in Bajaur and in the Mohmand country. Medal with clasps.
Tirah, 1897-98.—Operations in the Bara Valley, 7th to 14th December 1897.—Clasp.
N.-W. Frontier of India, Waziristan, 1901-02.—Clasp.
N.-W. Frontier of India, 1908.—Operations in the Zakka Khel country.—Medal and clasp.

Kesar Singh—
N.-W. Frontier of India, Waziristan, 1901-02.—Medal and clasp.
N.-W. Frontier of India, 1908.—Operations in the Zakka Khel country.—Medal and clasp.

JEMADARS—

Lehna Singh—
N.-W. Frontier of India, 1897-98.—Operations in Bajour and in the Mamund country.—Medal and clasp.
Tirah, 1897-98.—Operations in the Bara Valley, 7th to 14th December 1897.—Clasp.
N.-W. Frontier of India, Waziristan, 1901-02.—Clasp.
N.-W. Frontier of India, 1908.—Operations in Zakka Khel country.—Medal and clasp.

Buta Singh—
N.-W. Frontier of India, 1897-98.—Operations in Bajour and in the Mamund country.—Medal and clasp.
Tirah, 1897-98.—Operations in the Bara Valley, 7th to 14th December 1897.—Clasp.
N.-W. Frontier of India, Waziristan, 1901-02.—Clasp.
N.-W. Frontier of India, 1908.—Operations in the Zakka Khel country.—Medal and clasp.

Attar Singh—
N.-W. Frontier of India, Waziristan, 1901-02.—Medal and clasp.
N.-W. Frontier of India, 1908.—Operations in the Zakka Khel country.—Medal and clasp.

Harnam Singh I—
N.-W. Frontier of India, Waziristan, 1901-02.—Medal and clasp.
N.-W. Frontier of India, 1908.—Operations in the Zakka Khel country.—Medal and clasp.

Harnam Singh II—
N.-W. Frontier of India, 1908.—Operations in the Zakka Khel country.—Medal and clasp.

Sadhu Singh—
N.-W. Frontier of India, Waziristan, 1901-02.—Medal and clasp.
N.-W. Frontier of India, 1908.—Operations in the Zakka Khel country.—Medal and clasp.

Kunda Singh—
N.-W. Frontier of India, Waziristan, 1901-02.—Medal and clasp.
N.-W. Frontier of India, 1908.—Operations in the Zakka Khel and Mohmand country.—Medal and clasp.

Wadhawa Singh—
N.-W. Frontier of India, Waziristan, 1901-02.—Medal and clasp.
N.-W. Frontier of India, 1908.—Operations in the Zakka Khel country.—Medal and clasp.

Abdul Aziz—
N.-W. Frontier of India—Darwesh Khel Operations.
N.-W. Frontier of India, 1908.—Operations in the Zakka Khel country and Mohmand country.—Medal and clasp.
Somaliland, 1911-1916.—Medal and clasp.

Sewa Singh—
East Africa, 1902-03.—Operations in Somaliland.—Medal with one clasp.
N.-W. Frontier of India, 1908.—Operations in the Zakka Khel country.—Medal and clasp.

Kishan Singh—
N.-W. Frontier of India, 1897-98.—Defence of Malakand. Relief of Chakdara, Malakand. Action of Landaikai. Operations in Bajaur and in the Mamund country.—Medal with 2 clasps.
Tirah, 1897-98.—Operations in the Bara Valley, 7th to 14th December 1897.—Clasp.
N.-W. Frontier of India, Waziristan, 1901-02.—Medal and clasp.
N.-W. Frontier of India, 1908.—Operations in the Zakka Khel country.—Medal and clasp.
Meritorious service medal.

Bishan Singh II
N.-W. Frontier of India, Waziristan, 1901-02.—Medal and clasp.
N.-W. Frontier of India, 1908.—Operations in the Zakka Khel country.—Medal and clasp.

Sundar Singh III—
N.-W. Frontier of India, Waziristan, 1901-02.—Medal and clasp.
N.-W. Frontier of India, 1908.—Operations in the Zakka Khel country.—Medal and clasp.

Sant Singh—
N.-W. Frontier of India, Waziristan, 1901-02.—Medal and clasp.
N.-W. Frontier of India, 1908.—Operations in the Zakka Khel country.—Medal and clasp.

46th Punjabis.

SUBADAR-MAJOR—

Habibullah Khan *Bahadur—Order of Br. Ind., 2nd Class.*
Indian Distg. Service Medal.—
N.-W. Frontier of India, Waziristan, 1894-95.—Medal with clasp.
N.-W. Frontier of India, 1897-98.—Tochi.—Medal with clasp.
War, 1914-16.—Despatches.

SUBADARS—

Jota Singh—
N.-W. Frontier of India, Miranzai, 1891.
Chitral, 1895.—Medal with clasp.

Abdul Ghafur Khan—*Indian Distinguished Service Medal.*
War, 1914-16.—Despatches.

Badi-ul-Zaman—
Indian Distinguished Service Medal.
War, 1914-16.—Despatches.
Tibet Mission, 1903-04.—Medal with clasp.

Amir Khan—
N.-W. Frontier of India, 1897-98.—Medal with clasp.

Khushal Singh—
N.-W. Frontier of India, 1897-98.—Malakand.—Medal with 2 clasps.

Ghulam Mohammad I—
N.-W. Frontier of India, 1897-98.—Tochi.—Medal with clasp.

Budha Singh—
N.-W. Frontier of India, 1897-98.—Tochi.—Medal with clasp.

Juma Gul—
Tibet Mission, 1903-04.—Medal and clasp.

Firoz Khan—
N.-W Frontier of India, Waziristan, 1894-95.—Medal with clasp.
N.-W. Frontier of India, 1897-98.—Tochi.—Medal with clasp.

Ghulam Mohammad II—
N.-W. Frontier of India, 1897-98.—Medal with clasp.

Karam Din—
N.-W. Frontier of India, 1897-98.—Tochi—Medal with clasp.

Ahmed Khan—
N.-W. Frontier of India, 1897-98.—Tochi.—Medal with clasp.

Mohamed Bakhash—
N.-W. Frontier of India, 1897-98.—Tochi.—Medal with clasp.
Tibet Mission, 1903-04—Medal with clasp.

Nur Alam—
N.-W. Frontier of India, 1897-98.—Tochi.—Medal with clasp.

Ghulam Mohiuddin—
N.-W. Frontier of India, Waziristan, 1894-95.—Medal with clasp.
N.-W. Frontier of India, 1897-98.—Tochi.—Medal with clasp.

Punjab Singh—
Chitral, 1895.—Relief of Chitral.—Medal with clasp.
N.-W. Frontier of India, 1897-98.
Tochi.—Medal with clasp.

Imam Din—
N.-W. Frontier of India, 1897-98.—Tochi.—Medal with clasp.

JEMADARS—

Habib Khan—
N.-W. Frontier of India, 1897-98.—Tochi.—Medal with clasp.
Somaliland, 1908-10.—Medal with clasp.

Sher Muhammad—
N.-W. Frontier of India, 1897-98.—Tochi.—Medal with clasp.
Tibet Mission, 1903-04.—Medal with clasp.

47th Sikhs.

SUBADAR-MAJOR—

Thakur Singh, *Bahadur,* M. C. *Order of Br. Ind., 2nd Cl.*
Tibet, 1903-04.—Action at Gyantse. March to Lhassa.—Medal with clasp.

SUBADARS—

Ram Singh—
Soudan, 1896.—Dongola Expedition.—Medal Khedive's Medal.
N.-W. Frontier of India, 1897-98.—Relief of Malakand. Relief of Chakdara.—Malakand Operations in Bajour and in the Mamund country. Utman Khel.—Wounded.—Medal with 2 clasps.

Bakhshi Singh, I.O.M.—
N.-W. Frontier of India, 1897-98.—Relief of Malakand. Relief of Chakdara. Operations in Bajour and in the Mamund country. Utman Khel.—Medal with 2 clasps.

Mit Singh, M.C., I.D.S.M.—
N.-W. Frontier of India, 1908.—Operations in the Zakka Khel and Mohmand countries.—Engagement at Kargha, 24th May 1908.—Medal with clasp.

JEMADARS—

48th Pioneers.

SUBADAR-MAJOR—

Dula Singh, *Bahadur, Order of Br. Ind. 2nd Cl.*
N.-W. Frontier of India, Hazara, 1888.—Medal with clasp.
N.-W. Frontier of India, Hazara, 1891.—Clasp.
Chitral, 1895.—Relief of Chitral.—Medal with clasp.

SUBADARS—

Lehna Singh, *Bahadur, Order, of Br. Ind. 2nd Cl.*
N.-W. Frontier of India, Waziristan, 1897.—Medal with clasp.
N.-W. Frontier of India, 1897-98.—Tochi.—Medal with clasp.

Shib Lal—
N.-W. Frontier of India, 1897-98.—Medal with clasp.
China, 1900.—Medal.

Pala Singh—
Chitral, 1895.—Medal with clasp.
N.-W. Frontier of India, 1897-98.—Operations in Bara Valley.—Clasp.
Tirah, 1897-98.—Clasp.
China, 1900.—Medal.
Tibet, 1903-04.—Medal.
N.-W. Frontier of India, 1908.—Operations in Mohmand country.—Medal with clasp.

JEMADARS—

Natha Singh—
N.-W. Frontier of India, 1897-98.—Medal with clasp.
China, 1900.—Medal.

Sohan Singh—
N. W. Frontier of India, 1897-98.—Medal with clasp.
China, 1900.—Medal.

2-48th Pioneers.

SUBADAR—

Hari Ram—
N.-W. Frontier of India, 1897-98—Tochi.—Medal with clasp.
China, 1900-01.—Medal.

2-50th Kumaon Rifles.

SUBADARS—

Chandri Chand, I.O.M.—
Mesopotamia, 1916-17.—Advance on Baghdad. Indian Order of Merit, 2nd class.

Ram Singh Bhandari—
France, 1914-15.

JEMADARS—

Nain Singh Chand—
Mesopotamia, 1917-18.

Joga Chand—
Mesopotamia, 1917.

Tara Da t—
Mesopotamia, 1917.

Pertab Singh Bist—
Mesopotamia, 1917.

Lal Singh Rawat—
France, 1914-15.

Deb Singh Chupal—
France, 1914-15.

Jai Singh Bhandari—
Mesopotamia, 1916-18.

51st Sikhs (Frontier Force).

SUBADAR-MAJOR—

Mirbaz Khan, *Bahadur*—
N.-W. Frontier of India, Zhob Valley, 1890.
N.-W. Frontier of India, Waziristan, 1894-95.—Medal with clasp.
N.-W. Frontier of India, 1897-98.—Tochi.—Medal with clasp.
China, 1900.—Relief of Pekin. Action of Peitsang. Action at Yangtsun.—Medal with clasp.
N.-W. Frontier of India, 1902.—Operations against the Darwesh Khel Waziris.
N.-W. Frontier of India, 1908.—Operations in the Zakka Khel country.—Medal with clasp.

SUBADARS—

Kesar Singh—
N.-W. Frontier of India, Waziristan, 1894-95.—Medal with clasp.
N.-W. Frontier of India, 1897-98.—Action at Maizar, 10th June 1897.—Tochi.—Medal with clasp.
China, 1900.—Relief of Pekin.—Action of Peitsang. Action at Yangtsun.—Medal with clasp.
N.-W. Frontier of India, 1902.—Operations against the Darwesh Khel Waziris.

Sher Gul, *Bahadur*—
N.-W. Frontier of India, Waziristan, 1894-95.—Medal with clasp.
N.-W. Frontier of India, 1897-98.—Action at Maizar, 10th June 1897.—Tochi.—Medal with clasp.

Akbar Khan, I.D.S.M.—
N.-W. Frontier of India, 1897-98.—Action at Maizar, 10th June 1897.—Medal with clasp.
China, 1900.—Relief of Pekin. Action of Peitsang. Action at Yangtsun.—Medal with clasp.
N.-W. Frontier of India, 1902.—Operations against the Darwesh Khel Waziris.

Chanda Singh, I.O.M.—
China, 1900.—Relief of Pekin. Action at Peitsang. Action at Yangtsun.—Medal with clasp.
N.-W. Frontier of India, 1902.—Operation against the Darwesh Khel Waziris.

Labh Singh, *Bahadur*; G.D.S.M.—
N.-W. Frontier of India, Waziristan, 1894-95.—Medal with clasp.
N.-W. Frontier of India, 1897-98.—Tochi.—Medal with clasp.
China, 1900.—Medal.
N.-W. Frontier of India, 1902.—Operation against the Darwesh Khel Waziris.

Arjan Singh, I.O.M., I.D.S.M.—
China, 1900.—Relief of Pekin.—Action of Peitsang.—Action of Yangtsun.—Medal with clasp.
N.-W. Frontier of India, 1902.—Operations against the Darwesh Khel Waziris.

Tika Khan, I.O.M.—
N.-W. Frontier of India, 1897-98.—Action at Maizar, 10th June 1897.—Tochi.—Medal with clasp.
China, 1900.—Relief of Pekin. Action of Peitsang. Action at Yangtsun.—Medal with clasp.

Gokal—
N.-W. Frontier of India, Waziristan, 1894-95.—Medal with clasp.
N.-W. Frontier of India, 1897-98.—Action at Maizar, 10th June 1897.—Tochi.—Medal with clasp.
China, 1900.—Relief of Pekin.—Action of Peitsang. Action at Yangtsun.—Medal with clasp.

Makhmad Din, I.O.M.—
N.-W. Frontier of India, Waziristan, 1894-95.—Medal with clasp.
N.-W. Frontier of India, 1897-98.—Tochi.—Medal with clasp.
China, 1900.—Relief of Pekin. Action of Peitsang. Action at Yangtsun.—Medal with clasp.

Jagat Singh—
N.-W. Frontier of India, Waziristan, 1894-95—Medal with clasp.
N.-W. Frontier of India, 1897-98.—Tochi.—Medal with clasp.
China, 1900.—Relief of Pekin. Action of Peitsang. Action at Yangtsun.—Medal with clasp.

Surayan Singh—
N.-W. Frontier of India, 1897-98.—Tochi.—Medal with clasp.
China, 1900.—Relief of Pekin. Action of Peitsang. Action at Yangtsun.—Medal with clasp.
N.-W. Frontier of India, 1902.—Operation against the Darwesh Khel Waziris.

JEMADARS—

Ghulam Shah—
N.-W. Frontier of India, 1897-98.—Tochi.—Medal with clasp.
China, 1900.—Relief of Pekin. Action of Peitsang. Action at Yangtsun.—Medal with clasp.
N.-W. Frontier of India, 1902.—Operations against the Darwesh Khel Waziris.

Niazbat Khan—
China, 1900.—Relief of Pekin.—Action of Peitsang. Action at Yangtsun.—Medal with clasp.
N.-W. Frontier of India, 1902.—Operations against the Darwesh Khel Waziris.

Gurmukh Singh—
N.-W. Frontier of India, Waziristan, 1894-95.—Medal with clasp.
N.-W. Frontier of India, 1897-98.—Tochi.—Medal with clasp.
China, 1900.—Relief of Pekin. Action of Peitsang.—Action at Yangtsun.—Medal with clasp.

Prem Singh—
N.-W. Frontier of India, 1897-98.—Tochi.—Medal with clasp.
N'qamba Expedition B. S. Africa, August 1899.
Ashanti Expedition, 1900. Actions of 8th August.—Central African Medal and clasp.
N.-W. Frontier of India, 1902.—Operations against the Darwesh Khel Waziris.

Sundar Singh—
N.-W. Frontier of India, Waziristan, 1894-95.—Medal with clasp.
N.-W. Frontier of India, 1897-98.—Tochi.—Medal with clasp.
China, 1900.—Relief of Pekin. Action of Peitsang. Action at Yangtsun.—Medal with clasp.

Mangal Singh, I.O.M.—
China, 1900.—Relief of Pekin. Action of Peitsang. Action at Yangtsun.—Medal with clasp.

Harnam Singh—
N.-W. Frontier of India, 1897-98.—Tochi.—Medal with clasp.
China, 1900.—Relief of Pekin. Action of Peitsang. Action at Yangtsun.—Medal with clasp.

Basawa Singh—
N.-W. Frontier of India.—Operations against the Darwesh Khel Waziris.

Zaman Ali—
N.-W. Frontier of India, 1897-98.—Tochi.—Medal with clasp.
China, 1900.—Relief of Pekin.—Medal and clasp.
N.-W. Frontier of India, 1902.—Operations against the Darwesh Khel Waziris.
Operations against the Darwesh at Shimir Bervis between 19th and 25th November 1914 and between 2nd and 9th February 1915. Medal and Clasp.

1-52nd Sikhs (Frontier Force).

SUBADAR-MAJOR—

Dayal Singh—
East Africa, 1903-04.—Operations in Somaliland. Action at Jidballi.—Medal with 2 clasps.

SUBADARS—

Karam Khan—
N.-W. Frontier of India, 1898.—Medal with clasp.
East Africa, 1903-04.—Operations in Somaliland. Action at Jidballi.—Medal with 2 clasps.

Bhagat Singh—
N.-W. Frontier of India, Hazara, 1888.—Medal with clasp.
East Africa, 1903-04.—Operations in Somaliland. Action at Jidballi.—Medal with 2 clasps.

Parmod Singh—
N.-W. Frontier of India, Hazara, 1888.—Medal with clasp.
East Africa, 1903-04.—Operations in Somaliland. Action at Jidballi.—Medal with 2 clasps.
N.-W. Frontier of India, 1908.—Operations in the Mohmand country. Engagement of Karghar, 24th May 1908.—Medal with clasp.

Ata Khan—*Indian Distinguished Service Medal.*
N.-W. Frontier of India, Waziristan, 1901-01.—Medal with clasp.
N.-W. Frontier of India, 1902.—Operations against the Darwesh Khel Waziris.
N.-W. Frontier of India, 1908.—Operations in the Mohmand country. Engagement of Kargha, 24th May 1908.—Medal with clasp.

Bachitter Singh—
East Africa, 1903-04.—Operations in Somaliland.—Medal with clasp.
N.-W. Frontier of India, 1908.—Operations in the Mohmand country.—Medal with clasp.

Mirza—
N.-W. Frontier of India, Hazara, 1888.—Medal with clasp.
East Africa, 1903-04.—Operations in Somaliland.—Medal with clasp.

Jhanda Singh—
N.-W. Frontier of India, 1898.—Medal with clasp.
East Africa, 1903-04.—Operations in Somaliland.—Medal with clasp.

Karmu—
East Africa, 1903-04.—Operations in Somaliland.—Medal with clasp.

Khalas Khan—
East Africa, 1903-04.—Operations in Somaliland. Action at Jidballi.—Medal with 2 clasps.

Rasam Singh—
East Africa, 1903-04.—Operations in Somaliland.—Medal with clasp.

JEMADARS—

Natha Singh—
East Africa, 1903-04.—Operations in Somaliland, action at Jidballi. Medal with 2 clasps.

Nihal Singh—
East Africa, 1903-04.—Operations in Somaliland. Action at Jidballi. Medal with 2 clasps.

Muhammad Akbar—
East Africa, 1903-04.—Operations in Somaliland.—Medal with clasp.

Lorinda—
East Africa, 1903-04.—Operations in Somaliland.—Medal with clasp.

Bir Singh (1)—
East Africa, 1903-04.—Operations in Somaliland.—Medal with clasp.

Khalas Khan—
East Africa, 1903-04.—Operations in Somaliland.—Action at Jidballi.—Medal with 2 clasps.

Bir Singh (2)—
East Africa, 1903-04.—Operations in Somaliland.—Medal with clasps.

Ahmad Shah—
East Africa, 1903-04.—Operations in Somaliland. Action at Jidballi.—Medal with 2 clasps.

Sahib Gul—
East Africa, 1903-04.—Operations in Somaliland. Action at Jidballi.—Medal with 2 clasps.

2-52nd Sikhs (Frontier Force).

SUBADARS—

Suhel Singh, *Sardar Bahduar - Order of Br Ind a, 1st class.*
N.-W. Frontier of India, Zhob Valley, 1890.
N.-W. Frontier of India (1st and 2nd) Miranzai, 1891.—Medal with clasp.
N.-W. Frontier of India, Waziristan, 1894-95.—Clasp.
N.-W. Frontier of India, 1897-98.—Medal with clasp.
Tirah, 1897-98.—Operations in the Bara Valley. 7th to 14th December 1897.—Clasp.
N.-W. Frontier of India, Waziristan, 1901-02.—Clasp.

Hardit Singh—
N.-W. Frontier of India, 1908.—Operations in the Zakka Khel country.
N.-W. Frontier of India, 1908.—Operations in the Mohmand country. Engagement of Kargha, 24th May 1908.—Medal with clasp.

Sapuran Singh—
N.-W. Frontier of India, Mahsud Waziri Blockade medal and clasp, Waziristan, 1901-02.

Hardial—
N.-W. Frontier of India, 1908.—Bazaar Valley, Mohmand Expedition.—Medal and clasp.

Lachhman—

N.-W. Frontier of India, 1897-98.—Medal and 3 claps.

53rd Sikhs (Frontier Force).

SUBADAR-MAJORS—

Mangal Singh—

N.-W. Frontier of India, Waziristan, 1894-95.—Action at Wana, 3rd November 1894.—Medal with clasp.

N.-W. Frontier of India, 1897-98.—Operations on the Samana and in the Kurram Valley during August and September 1897.—Medal with 2 clasps.

Tirah, 1897-98.—Actions of Dargai and of the Sampagha and Arhanga Passes. Operations against the Khani Khel Chamkanis. Operations in the Bazar Valley, 25th to 30th December 1897.—Clasp.

N.-W. Frontier of India, Waziristan, 1901-02.—Clasp.

N.-W. Frontier of India, 1908.—Operations in the Zakka Khel country.

N.-W. Frontier of India, 1908.—Operations in the Mohmand country. Engagement of Matta, 24th April 1908. Engagement of Kargha, 24th May 1908.—Medal with clasp.

SUBADARS—

Arsla Khan—

N.-W. Frontier of India, 1st and 2nd Miranzai, 1891.—Medal with clasp.

N.-W. Frontier of India, Waziristan, 1894-95.—Clasp.

Action at Wana, 3rd November 1894.

N.-W. Frontier of India, 1897-98.—Operations on the Samana and in the Kurram Valley during August and September 1897.—Medal with 2 clasps.

Tirah, 1897-98.—Action of Dargai. Action of the Sampagha Pass. Operations against the Khani Khel Chamkanis.—Clasp.

N.-W. Frontier of India, Waziristan, 1901-02.—Clasp.

N.-W. Frontier of India, 1902.—Operations against the Darwesh Khel Waziris.

Tibet, 1903-04.—Medal with clasp.

N.-W. Frontier of India, 1908.—Operations in the Zakka Khel country.

N.-W. Frontier of India, 1908.—Operations in the Mohmand country. Engagement of Matta, 24th April 1908. Engagement of Kargha, 24th May 1908.—Medal with clasp. Indian Order of Merit, 3rd class.

Mula Singh—

N.-W. Frontier of India, Waziristan, 1901-02.—Medal and clasp.

N.-W. Frontier of India, 1902.—Operations against the Darwesh Khel Waziris.

N.-W. Frontier of India, 1908.—Operations in the Zakka Khel country.—Medal and clasp.

Chanan Singh—

N.-W. Frontier of India, Waziristan, 1894-95.—Medal and clasp. Action at Wana, 3rd November 1894.

N.-W. Frontier of India, 1897-98.—Tochi.—Operations on the Samana and in the Kurram Valley during August and September 1897.—Medal with 2 clasps.

Tirah, 1897-98.—Action of Sampagha Pass. Action of Arhanga Pass. Reconnaissance of the Saransar and action of the 9th November 1897. Operations at and around Dotoi and action of 24th November 1897. Operations in the Bara Valley, 7th to 14th December 1897. Operations in the Bazar Valley, 25th to 30th December 1897.—Cla

N.-W. Frontier of India, Waziristan, 1901-02.—Clasp.

N.-W. Frontier of India, 1902.—Operations against the Darwesh Khel Waziris.

N.-W. Frontier of India, 1908.—Operations in the Zakka Khel country. Operations in the Mohmand country. Engagement at Matta, 24th April 1908. Engagement at Kargha, 24th May 1908.—I. D. S. Medal.—Medal with clasp.

Gulam Muhammad—

N.-W. Frontier of India, 1897-98.—Medal and clasp.

N.-W. Frontier of India, Waziristan, 1901-02.—Clasp.

N.-W. Frontier.—Operations against Darwesh Khel Waziris, 1902.

N.-W. Frontier of India, 1908.—Operations in the Zakka Khel country. Operations in the Mohmand country. Engagement of Kargha, 24th May 1908.—Medal and clasp.

Hazara Singh—

N.-W. Frontier of India, 1908.—Operations in the Zakka Khel country. Operations in Mohmand country. Engagement at Matta, 24th April 1908. Engagement at Kargha, 24th May 1908.—Medal with clasp.

Gagan Singh—

N.-W. Frontier of India, 1897-98.—Operations on the Samana and in the Kurram Valley during August and September 1897.—Medal and 2 clasps.

Tirah, 1897-98.—Action of Dargai. Action of Sampagha Pass. Action of Arhanga Pass. Operations against the Khani Khel Chamkanis. Operations in the Bazar Valley, 25th to 30th December 1897.—Clasp.

N.-W. Frontier of India, Waziristan, 1901-02.—Clasp.

N.-W. Frontier of India, 1902.—Operations against Darwesh Khel Waziris, 1902.

N.-W. Frontier of India, 1908.—Operations in the Zakka Khel country. Operations in the Mohmand country. Engagement of Kargha, 24th May 1908.—Medal with clasp.

Hira Singh—

N.-W. Frontier of India, Waziristan, 1901-02.—Medal and clasp.

N.-W. Frontier of India, 1902.—Operations against Darwesh Khel Waziris.

N.-W. Frontier of India, 1908.—Operations in Mohmand country. Engagement at Kargha, 24th May 1916. Medal and clasp. I. O. M, 3rd class.

JEMADARS—

Shah Baz—

N.-W. Frontier of India, Waziristan, 1894-95.—Medal and clasp.

Action at Wana, 3rd November 1894.

N.-W. Frontier of India, 1897-98.—Operations on the Samana and in the Kurram Valley during August and September 1897.—Medal with 2 clasps.

Tirah, 1897-98.—Action of Dargai. Action of Sampagha Pass. Action of the Arhanga Pass. Operations against Khani Khel Chamkanis. Operations in the Bazar Valley, 25th to 30th December 1897.—Clasp.

China, 1900-01.—Relief of Peking.—Medal with clasp.

N.-W. Frontier of India. 1901-02.—Clasp.

N.-W. Frontier of India, 1908.—Operations in the Zakka Khel country. Operations in the Mohmand country. Engagement of Matta, 24th April 1908. Engagement of Kargha, 24th May 1908.—Medal with clasp.

Lal Khan—
N.-W. Frontier of India, 1908.—Operations in the Mohmand country. Engagement of Kargha, 24th May 1908.—Medal and clasp.

Sapuran Singh—
N.-W. Frontier of India, 1897-98.—Tochi. Operations on the Samana and in the Kurram Valley during August and September 1897.
Tirah, 1897-98.—Capture of Sampagha Pass. Capture of Arhanga Pass. Reconnaissance of the Saran Sar and action of the 9th November 1897. Operations in the Bazar Valley, 25th to 30th December 1897.—Medal with 3 clasps.
N.-W. Frontier of India, Waziristan, 1901-02.—Clasp.
N.-W. Frontier of India, 1902.—Operations against Darwesh Khel Waziris.
N.-W. Frontier of India, 1908.—Operations in Zakka Khel country. Operations in Mohmand country. Engagement at Matta, 24th April 1908. Engagement at Kargha, 24th May 1908.—Medal and clasp.

Shamir Singh—
N.-W. Frontier of India, Waziristan, 1901-02.—Medal and clasp.

Samundar Khan—
N.-W. Frontier of India, Waziristan, 1901-02.—Medal and clasp.
N.-W. Frontier of India, 1908.—Operations in Zakka Khel country. Operations in Mohmand country. Engagement of Matta, 24th April 1908. Engagement of Kargha, 24th May 1908.—Medal and clasp.

Mir Awas—
N.-W. Frontier of India, 1908.—Operations in the Zakka Khel country. Operations in the Mohmand country. Engagement of Matta, 24th April 1908. Engagement of Kargha, 24th May 1908.—Medal and clasp.

Sajan Singh—
N.-W. Frontier of India, 1908.—Operations in the Zakka Khel country. Operations in the Mohmand country. Engagement of Kargha, 24th May 1908.—Medal and clasp.

Kehr Singh—
N.-W. Frontier of India, 1908.—Operations in Zakka Khel country. Operations in Mohmand country. Engagement at Kargha on 24th May 1906.—Medal and clasp. I. O. M., 2nd class.

Santokh Singh—
N.-W. Frontier of India, 1894.—Action at Wana, 3rd November 1894, Waziristan, 1894-95.—Medal and clasp.
N.-W. Frontier of India, 1901-02.—Medal and clasp.
N.-W. Frontier of India, 1908.—Operations in the Zakka Khel country. Operations in the Mohmand country. Engagement of Kargha, 24th May 1908.—Medal and clasp.

Fateh Muhammad—
N.-W. Frontier of India, Waziristan, 1901-02.—Medal and clasp.
N.-W. Frontier of India, 1902.—Operations against the Darwesh Khel Waziris.
N.-W. Frontier of India, 1908.—Operations in the Zakka Khel country. Operations in the Mohmand country. Engagement at Kargha, 24 May 1908.—Medal with clasp.

Muhd. Khan—
N.-W. Frontier of India, Waziristan, 1901-02.—Medal and clasp.
N.-W. Frontier of India, 1902.—Operations against Darwesh Khel Waziris.
N.-W. Frontier of India, 1908.—Operations in the Mohmand country. Engagement of Kargha, 24th May 1908.—Medal and clasp.

1-54th Sikhs (Frontier Force).

SUBADAR-MAJOR—

Sher Ali—
N.-W. Frontier of India, Isazai, 1892.
Chitral, 1895.—Relief of Chitral.—Medal with clasp.
N.-W. Frontier of India, 1908.—Operations in the Zakka Khel country.
N.-W. Frontier of India, 1908.—Operations in the Mohmand country.—Medal with clasp.

SUBADARS—

Janas Khan—
Tibet, 1903-04.—Action of Naini, 26th June 1904. Operations at and around Gyantse, between 5th May and 6th July 1904. March to Lhassa, 14th July to 3rd August 1904.—Medal with clasp.
N.-W. Frontier of India, 1908.—Operations in the Mohmand country.—Medal with clasp.

Hukam Dad—
N.-W. Frontier of India, Isazai, 1892.
Chitral, 1895.—Relief of Chitral.—Medal with clasp.
N.-W. Frontier of India, Waziristan, 1901-02.—Clasp.
N.-W. Frontier of India, 1902.—Operations against the Darwesh Khel Waziris.
N.-W. Frontier of India, 1908.—Operations in the Zakka Khel country.
N.-W. Frontier of India, 1908.—Operations in the Mohmand country. Engagement of Kargha, 24th May 1908.—Medal with clasp.

Kesar Singh—
N.-W. Frontier of India, Hazara, 1891.
N.-W. Frontier of India, Isazai, 1892.
Chitral, 1895.—Relief of Chitral.—Medal with clasp.
N.-W. Frontier of India, 1908.—Operations in the Zakka Khel country.
N.-W. Frontier of India, 1908.—Operations in the Mohmand country.—Medal with clasp.

Bhikham Singh—
N.-W. Frontier of India, 1902.—Operations against the Darwesh Khel Waziris.
N.-W. Frontier of India, 1908.—Operations in the Mohmand country.—Medal with clasps.

Hira Sing—
N.-W. Frontier of India, 1902.—Operations against the Darwesh Khel Waziris.
N.-W. Frontier of India, 1908.—Operations in the Zakka Khel country.
N.-W. Frontier of India, 1908.—Operations in the Mohmand country.—Medal with clasp.

Teja Singh—
N.-W. Frontier of India, 1908.—Operations in the Zakka Khel country.
N.-W. Frontier of India, 1908.—Operations in the Mohmand country.—Medal with clasp.

Indar Singh—
N.-W. Frontier of India, 1908.—Operations in the Zakka Khel country.
N.-W. Frontier of India, 1908.—Operations in the Mohmand country. Medal with clasp.

Ragha—
N.-W. Frontier of India, Waziristan, 1901-02.—Medal with clasp.
N.-W. Frontier of India, 1902.—Operations against the Darwesh Khel Waziris.
N.-W. Frontier of India, 1908.—Operations in the Zakka Khel country.
N.-W. Frontier of India, 1908.—Operations in the Mohmand country.—Medal with clasp.

Ibrahim Khan—
East Africa, 1903-04.—Operations in Somaliland. Action at Jibdali.—Medal with two clasps.

Rur Singh—
N.-W. Frontier of India, 1908.—Operations in the Zakka Khel country. Operations in the Mohmand country.—Medal with clasp.

JEMADARS—

Dan Singh—
N.-W. Frontier of India, 1902.—Operations against the Darwesh Khel Waziris.
N.-W. Frontier of India, 1908.—Operations in the Zakka Khel country. Operations in the Mohmand country.—Medal with clasp.

Attar Singh—
N.-W. Frontier of India, Isarai, 1892, Chitral 1895.—Relief of Chitral.—Medal with clasp.
N.-W. Frontier of India, 1902. Operations against the Darwesh Khan Waziris.
N.-W. Frontier of India, 1908.—Operations in the Zakha Khel country. Mohmand operations, 1908.—Medal with clasp.

Chanda Singh—
N.-W. Frontier of India.—1908.—Operations in the Zakka Khel country.
N.-W. Frontier of India, 1908.— Operations in the Mohmand country.—Medal with clasp.

Bir Singh—
N.-W. Frontier of India, 1908.—Operations in the Zakka Khel country.—Medal with clasp.

Mukarrab Khan—
N.-W. Frontier of India, 1908.—Operations in the Zakka Khel country.
N.-W. Frontier of India, 1908.—Operations in the Mohmand country.—Medal with clasp.

Gurnam Singh—
N.-W. Frontier of India, 1908.—Operations in the Mohmand country.—Medal with clasp.

Pratab Singh—
N.-W. Frontier of India, 1908.—Operations in the Mohmand country.—Medal with clasp.

Isa Khan—
N.-W. Frontier of India, 1908.—Operations in the Mohmand country.—Medal with clasp.

Surain Singh—
N.-W. Frontier of India, 1902.—Operations against the Darwesh Khel Waziris.
N.-W. Frontier of India, 1908.—Operations in the Zakka Khel country.
N.-W. Frontier of India, 1908.—Operations in the Mohmand country.—Medal with clasp.

Sher Dil—
N.-W. Frontier of India, 1908.—Operations in the Zakka Khel country.
N.-W. Frontier of India, 1908.—Operations in the Mohmand country.—Medal with clasp.

Bir Singh—
N.-W. Frontier of India, 1908.—Operations in the Zakka Khel country.
N.-W. Frontier of India, 1908.—Operations in the Mohmand country.—Medal with clasp.

Harnam Singh—
N.-W. Frontier of India, 1908.—Operations in the Zakka Khel country.
N.-W. Frontier of India, 1908.—Operations in the Mohmand country.—Medal with clasp.

Nur Mohammad—
N.-W. Frontier of India, 1908.—Operations in the Mohmand country.—Medal with clasp.

Lal Khan—
N.-W. Frontier of India, 1908.—Operations in Zakka Khel country.
N.-W. Frontier of India, 1908.—Operations in the Mohmand country.—Medal with clasp.

2-54th Sikhs (Frontier Force).

SUBADAR-MAJOR—

Kehar Singh—
Issazai Field Force, 1892.
Chitral Relief Force, 1895.—Medal and clasp.
Tirah, 1897-98.—Two clasps.
N.-W. Frontier of India, 1902.—Darwesh Khel Operations.
Bazar Valley Field Force, 1908.
Mohmand Field Force, 1908.—Medal and clasp.
Waziristan Field Force, 1917.

SUBADARS—

Allah Dad—
Bazar Valley Expedition, 1908.
Mohmand Field Force, 1908.—Medal and clasp.
Waziristan Field Force, 1917.

Sewa Singh—
Waziristan, 1901-02.—Medal and clasp.
N.-W. Frontier of India, 1902.—Darwesh Khel Operations.
Bazar Valley Field Force, 1908.
Mohmand Field Force, 1908.—Medal and clasp.
France and Force "D", 1915-17.

Mohammad Alam—
I. E. F. "D", 1916.

Ali Bahadur, I.O.M.—
Bazar Valley Expedition, 1908.
Mohmand Field Force, 1908.—Medal and clasp.
France "A" Expeditionary Force and Mesopotamia, 1914-16.
Waziristan Field Force, 1917.—Awarded I.O.M. 2 Class.

JEMADARS—

Mangal Singh—
France, 1915.
Mesopotamia, 1915-17.

Sardar Khan—
France and Mesopotamia, 1915-17.

Ilam Din—
Bazar Valley Expedition, 1908.
Mohmand Field Force, 1908.—Medal and clasp.
France and Mesopotamia, 1915-17.

Katar Singh—
France and Mesopotamia, 1915-17.
Waziristan Field Force, 1917.

Sundar Singh—
Bazar Valley Expedition, 1908.
Mohmand Field Force, 1908.—Medal and clasp.

Muhammad Khan—
Bazar Valley Expedition, 1908.
Mohmand Field Force, 1908.—Medal and clasp. Engagement of Kargha 24th May 1908. Operations against the Darweshes Shimbar Buris between 2nd and 9th February 1915.—Medal and clasp.
Waziristan Field Force, 1917.

1st 55th Coke's Rifles (Frontier Force).

SUBADAR-MAJOR—

Gheba Khan—
N.-W. Frontier of India, 1897-98.—Tochi.—Medal with clasp.
N.-W. Frontier of India, Waziristan, 1901-02.—Clasp.
N.-W. Frontier of India, 1908.—Operations in the Mohmand country.—Medal with clasp.
N.-W. Frontier of India, Waziristan, 1917.
East African Expy. Force, 1917.
N.-W. Frontier of India, Mohmand Blockade, 1917.

SUBADARS—

Dugla—

N.-W. Frontier of India, 1897-98.—Tochi.—Medal with clasp.

N.-W. Frontier of India, Waziristan, 1901-02.—Clasp.

N.-W. Frontier of India, 1908.—Operations in the Mohmand country. Engagement of Kargha, 24th May 1908.—Medal with clasp.

N.-W. Frontier of India, Waziristan, 1917.

N.-W. Frontier of India, Mohmand Blockade, 1917.

N.-W. Frontier of India, Marri Punitive Expedition, 1918.

Rupa Singh—

N.-W. Frontier of India, 1897-98.—Tochi.—Medal with clasp.

N.-W. Frontier of India, 1908.—Operations in the Mohmand country. Engagement at Kargha 24th May 1908.—Medal with clasp.

Indian Expy. Force "B," 1914-15.

Indian Expy. Force "E," 1914-15.

Dheru—

N.-W. Frontier of India, 1897-98.—Tochi.—Medal with clasp.

N.-W. Frontier of India, Waziristan, 1901-02.—Clasp.

N.-W. Frontier of India, 1908.—Operations in the Mohmand country. Engagement of Kargha, 24th May 1908.—Medal with clasp.

Tikhu—

N.-W. Frontier of India, 1897-98.—Tochi.—Medal with clasp.

N.-W. Frontier of India, 1908.—Operations in the Mohmand country. Engagement of Kargha, 24th May 1908.—Medal with clasp.

Anar Gul—

N.-W. Frontier of India, 1897-98.—Tochi—Medal with clasp.

N.-W. Frontier of India, 1908.—Operations in the Mohmand country. Engagement of Kargha, 24th May 1908.—Medal with clasp.

Indian Expy. Force "E", 1914-15.

East African Expy. Force, 1917.

Gul Din—

N.-W. Frontier of India, 1897-98.—Tochi.—Medal with clasp.

N.-W. Frontier of India, 1901-02, Waziristan.—Clasp.

N.-W. Frontier of India, 1908.—Operations in the Mohmand country. Engagement of Kargha, 24th May 1908.—Medal with clasp.

Indian Expy. Force "E", 1914-15.

Hardit Singh—

N.-W. Frontier of India, 1897-98.—Tochi.—Medal with clasp.

British Central Africa, 1899-1900.—Medal with clasp.

N.-W. Frontier of India, Waziristan, 1917.

East African Expy. Force, 1917.—Indian Distinguished Service Medal.

Ibraim Khan—

Indian Expx. Force "A", 1914-15.

Indian Expy. Force "E", 1914-15.

N.-W. Frontier of India, Mohmand Blockade, 1917.

N.-W. Frontrier of India, Waziristan, 1917.—Despatches.

East African Expy. Force, 1917.—Indian Distinguished Service Medal.

Isa Khan—

N.-W. Frontier of India, 1897-98.—Tochi.—Medal with clasp.

N.-W. Frontier of India, Waziristan, 1901-02.—Clasp.

Tibet, 1903-04.—Medal with clasp.

N.-W. Frontier of India, 1908.—Operations in the Mohmand country. Engagement of Kargha, 24th May 1908.—Medal with clasp.

N.-W. Frontier of India, Muhmand Blockade, 1917.

N.-W. Frontier of India, Waziristan, 1917. Despatches.

East African Expy. Force, 1917.

N.-W. Frontier of India Marri Punitive Expedition, 1918.

Shera Ban—

N.-W. Frontier of India, 1897-98.—Tochi.—Medal with clasp.

N.-W. Frontier of India, Waziristan, 1901-02.—Clasp.

Indian Expy. Force "A", 1914-15.

Karam Khan—

N.-W. Frontier of India, 1897-98.—Tochi.—Medal with clasp.

N.-W. Frontier of India, Waziristan, 1901-02.—Clasp.

N.-W. Frontier of India, 1908.—Operations in the Mohmand country. Engagement of Kargha, 24th May 1908.—Medal with clasp.

Hussain—

N.-W. Frontier of India, Waziristan, 1901-02.—Medal with clasp.

N.-W. Frontier of India, 1908.—Operations in the Mohmand country. Engagement of Kargha, 24th May 1908.—Medal with clasp.

East African Expy. Force, 1917.

N.-W. Frontier of India, Marri Punitive Expedition, 1918.

Fazaldad—

N.-W. Frontier of India, 1897-98.—Tochi.—Medal with clasp.

N.-W. Frontier of India, Waziristan, 1901-02.—Clasp.

Mesopotamia Expy. Force, 1918.

Haidar Khan—

N.-W. Frontier of India, 1908.—Operations in Mohmand country. Engagement of Kargha, 24th May 1908.—Medal with clasp.

N.-W. Frontier of India, Mohmand Blockade, 1917.

N.-W. Frontier of India, Waziristan, 1917.

East African Expy. Force, 1917.

N.-W. Frontier of India, Marri Punitive Expedition, 1918.

Abdullah Khan—

N.-W. Frontier of India, 1908.—Operations in the Mohmand Country.—Engagement of Kargha, 24th May 1908.—Medal with clasp.

N.-W. Frontier of India, Mohmand Blockade, 1917.

N.-W. Frontier of India, Waziristan, 1917.

East African Expy. Force, 1917.

N.-W. Frontier of India.—Marri Punitive Expedition, 1918.

Teja Singh—

N.-W. Frontier of India, 1897-98.—Tochi Medal with clasp.

Indian Order of Merit 3rd class.

N.-W. Frontier of India, 1908.—Operations in the Mohmand Country.—Engagement of Kargha, 24th May 1908. Medal with clasp.

East African Expy. Force, 1917.

Mohammad Khan—

N.-W. Frontier of India, 1897-98.—Tochi Medal with clasp.

N.-W. Frontier of India, 1908.—Operation in Mohmand Country.

Engagement at Kargha, 24th May 1908.—Medal with clasp.

Indian Expy. Force "A", 1914-15.

N.-W. Frontier of India, Mohmand Blockade, 1917.

N.-W. Frontier of India, Waziristan, 1917.

N.-W. Frontier of India, Marri Punitive Expedition, 1918.

Eaast African Expy. Force, 1917.

Jahan Khan—

N.-W. Frontier of India, 1908.—Operations in the Mohmand country. Engagement of Kargha, 24th May 1908.—Medal with clasp.

East African Expy. Force, 1917.
N.-W. Frontier of India, Waziristan, 1917.

Gobinda—
N.-W. Frontier of India, Waziristan, 1901-02.—Medal with clasp.
N.-W. Frontier of India, 1908.—Operations in the Mohmand country. Engagement at Kargha, 24th May 1908. Medal with clasp.
Mesopotamian Expy. Force, 1915.
Indian Expy. Force "A", 1914-15.
N.-W. Frontier of India, Waziristan, 1917.
East African Expy. Force, 1917.

JEMADARS—

Shah Baz Khan—
N.-W. Frontier of India, 1897-98.—Tochi.—Medal with clasp.
Tibet, 1903-04.—Medal with clasp. Operations at and round Gyantse, 5th to 6th July 1904.
N.-W. Frontier of India, 1908.—Operations in the Mohmand country. Engagement at Kargha, 24th May 1908.—Medal with clasp.
N.-W. Frontier of India, Mohmand Blockade, 1917.
N.-W. Frontier of India, Waziristan, 1917.
East African Expy. Force, 1917.

Delal Singh—
N.-W. Frontier of India, Waziristan, 1917.
East African Expy. Force, 1917.—Despatches.

Mota Singh—
N.-W. Frontier of India, 1897-98.—Tochi.—Medal with clasp.
British Central Africa, 1899-1900.—Medal with clasp.
N.-W. Frontier of India, 1908.—Operations in the Mohmand country. Engagement at Kargha, 24th May 1908.—Medal with clasp.
East African Expy. Force, 1917.
N.-W. Frontier of India, Marri Punitive Expedition, 1918.
N.-W. Frontier of India, Mohmand Blockade, 1917.
N.-W. Frontier of India, Waziristan, 1917.

Ayub Khan—
N.-W. Frontier of India, 1897-98.—Tochi.—Medal with clasp.
N.-W. Frontier of India, 1908.—Operations in the Mohmand country. Engagement at Kargha, 24th May 1908.—Medal with clasp.
Indian Expy. Force "A", 1914-15.
Indian Expy Force "E", 1914-15.
Tibet, 1903-04.—Gyantse—Medal with clasp.
East African Expy. Force, 1917.—Despatches.

Kishan Singh—
N.-W. Frontier of India, 1901-02, Waziristan.—Medal with clasp.
N.-W. Frontier of India, 1908.—Operations in the Mohmand Country.—Engagement at Kargha, 24th May 1918.—Medal with clasp.
N.-W. Frontier of India, Waziristan, 1917.
East African Expy. Force, 1917.
N.-W. Frontier of India.—Marri Punitive Expedition, 1918.

2-55th Coke's Rifles.

SUBADAR-MAJOR—

Bela Singh—Order of Br. India. 2nd Class, I.O.M.
N.-W. Frontier of India, 1st Miranzai, 1891.—Medal with clasp.
Samana, India, 1895.—Medal with clasp.
Punjab Frontier, 1897-98.
N.-W. Frontier of India, Waziristan, 1901-02.—Clasp. Admitted to 3rd class Order of Merit.

Pala Singh—
N.-W. Frontier of India, 1908.—Operations in the Mohmand Country.—Engagement of Kargha.—Medal with clasp.

N.-W. Frontier of India, Tochi, 1897-98.—Medal with clasp.
N.-W. Frontier of India, Waziristan, 1901-02.—Clasp.

Murafat Shah—
N.-W. Frontier of India, Tochi, 1897-98.—Medal with clasp.
Punjab Frontier, 1897-98.
Waziristan, 1901-02.—Clasp.
N.-W. Frontier of India, 1908.—Operations in the Mohmand Country.—Engagement of Kargha, 24th May 1908.—Medal with clasp.
Field Service overseas, 1916-17.

Hamid Khan—
N.-W. Frontier of India, 1901-02.—Operations against the Darwesh Khel.
Waziris.—Medal and clasp.
France, 1914-15.—Mentioned in despatches, Indian Distinguished Service Medal, Star 1914.

Hukmat Khan—
Field Service overseas, 1914-15.

Hazrat Gul—
N.-W. Frontier of India, Tochi, 1897-98.—Medal with clasp.
N.-W. Frontier of India, Waziristan, 1901-02.—clasp.
N.-W. Frontier of India, 1908.—Operation in the Mohmand country,—Engagement of Kargha 24th May 1918.—Medal with clasp.

Gur Charan—
N.-W. Frontier of India, Tochi, 1897-98.—Medal with clasp.
N.-W. Frontier of India, 1908.—Operations in the Mohmand country. Engagement of Kargha, 24th May 1908.—Medal with clasp.

JEMADARS—

Sundar—
N.-W. Frontier of India, 1901 02.—Waziristan.—Medal with clasp.
N.-W. Frontier of India—Operations in the Mohmand country.—Engagement of Kargha, 24th May 1918.—Medal with clasp.

Muhammad Sharif—
N.-W. Frontier of India, 1897-98.—Tochi. Medal with clasp.

Afzal Khan—
N.-W. Frontier of India, Tochi 1897-98.—Clasp.
N.-W. Frontier of India, Waziristan, 1901-02.—Clasp.
N.-W. Frontier of India, 1908.—Operations in Mohmand country.—Engagement of Kargha, 24th May 1908.—Medal with clasp.

Ghazi Khan—
N.-W. Frontier of India, Waziristan, 1901-02.—Medal with clasp.
N.-W. Frontier of India, 1908.—Operations in the Mohmand country. Engagement of Kargha, 24th May 1908.—Medal with clasp.

Namdar Khan—
Field Service overseas, 1917-18.

56th Punjabi Rifles (Frontier Force).

SUBADAR-MAJOR—

Harnam Singh—*Bahadur, Order of Br. India, 2nd Cl.*, I. O. M.
N.-W. Frontier of India, Zhob Valley, 1890.
N.-W. Frontier of India (1st and 2nd) Miranzai, 1891.—Medal with clasp.

N.-W. Frontier of India, Waziristan, 1894-95.—Clasp.

N.-W. Frontier of India, 1897-98.—Action of the Ublan Pass. Operations on the Samana and in the Kurram Valley during August and September 1897. Relief of Gulistan.—Medal with 2 clasps.

Tirah, 1897-98.—Operations in the Bara Valley, 7th to 14th December 1897.—Clasp.

N.-W. Frontier of India, Waziristan, 1901-02.—Clasp.

N.-W. Frontier of India, 1902.—Operations against the Darwesh Khel Waziris.

SUBADARS—

Mardan Ali—

N.-W. Frontier of India, Zhob Valley, 1890.

N.-W. Frontier of India (1st and 2nd) Miranzai, 1891.—Medal with clasp.

N.-W. Frontier of India, 1897-98.—Medal with clasp.

N.-W. Frontier of India, 1902.—Operations against the Darwesh Khel Waziris.

Nanak Singh—*Bahadur, Order of Br. India, 2nd Class.*

N.-W. Frontier of India, Zhob Valley, 1890.

N.-W. Frontier of India (1st and 2nd) Miranzai, 1891.—Medal with clasp.

N.-W. Frontier of India, Waziristan, 1894-95.—Clasp.

N.-W. Frontier of India, 1897-98.—Action of the Ublan Pass. Operations on the Samana and in the Kurram Valley during August and September 1897. Relief of Gulistan.—Medal with 2 clasps.

Tirah, 1897-98.—Operations in the Bara Valley, 7th to 14th December 1897.—Clasp.

N.-W. Frontier of India, Waziristan, 1901-02.—Clasp.

Jagat Singh—

N.-W. Frontier of India, Waziristan, 1901-02.—Medal with clasp.

Khwaja Muhammad—

N.-W. Frontier of India, Waziristan, 1894-95.—Medal with clasp.

N.-W. Frontier of India, 1897-98.—Action of the Bolan Pass. Operations on the Samana and in the Kurram Valley during August and September 1897. Relief of Gulistan.—Medal with 2 clasps.

Tirah, 1897-98.—Operation in the Bara Valley, 7th to 14th December 1897.—Clasp.

N.-W. Frontier of India, Waziristan, 1901-02.—Clasp.

Sangara—

N.-W. Frontier of India, 1897-98.—Medal and 3 clasps.

N.-W. Frontier of India, 1902.—Operations against the Darwesh Khel Waziris.

Sher Singh—

N.-W. Frontier of India, 1897-98.—Medal and 3 clasps.

Mahsud Waziri Blockade.—*Waziristan, 1901-02.*—Clasp.

Bakhshis Singh—

N.-W. Frontier of India, Mahsud Waziri Blockade medal with clasp, Waziristan, 1901-02.

N.-W. Frontier of India, 1902.—Operations against the Darwesh Khel Waziris.

Lachhman Singh—

N.-W. Frontier of India, 1897-98.—Medal and 3 clasps.

N.-W. Frontier of India, Mahsud Blockade, Waziristan, 1901-02.—Clasp.

N.-W. Frontier of India, 1902.—Operations against the Darwesh Khel Waziris.

Mitt Singh—

N.-W. Frontier of India, Waziristan, 1901-02.—Medal and clasp.

N.-W. Frontier of India, 1902.—Operations against the Darwesh Khel Waziris.

JEMADARS—

Rakhim Gul—

N.-W. Frontier of India, 1897-98.—Medal with clasp.

Tirah, 1897-98.—Operations in the Bara Valley, 7th to 14th December 1897.—Clasp.

N.-W. Frontier of India, Waziristan, 1901-02.—Clasp.

N.-W. Frontier of India, 1902.—Operations against the Darwesh Khel Waziris.

Wazir Singh—

N.-W. Frontier of India, 1897-98.—Medal and 3 clasps.

N.-W. Frontier of India, Mahsud Blockade, Waziristan, 1901-02.

N.-W. Frontier of India, 1902.—Operations against the Darwesh Khel Waziris.

Spin Gul—

N.-W. Frontier of India, Waziristan, 1901-02.—Medal and clasp.

N.-W. Frontier of India, 1902.—Operations against the Darwesh Khel Waziris.

Nadir Khan—

N.-W. Frontier of India, Waziristan, 1894-95.—Medal and clasp.

N.-W. Frontier of India, 1897-98.—Action of the Ublan Pass. Operations on the Samana and in the Kurram Valley. Relief of Gulistan.—Medal with 2 clasps.

Tirah, 1897-98.—Operations in the Bara Valley. Clasp.

N.-W. Frontier of India, Waziristan, 1901-02.—Clasp.

Gul Khan—

N.-W. Frontier of India, Waziristan, 1901-02.—Medal and clasp.

N.-W. Frontier of India, 1902.—Operations against the Darwesh Khel Waziris.

2-56th Rifles.

SUBADARS—

Bishn Singh—

N.-W. Frontier of India, 1908.—Operations in the Zakka Khel country.

N.-W. Frontier of India, 1908.—Operations in the Mohmand country. Engagement of Matta, 24th April 1908. Engagement of Kargha, 24th May 1908.—Medal with clasp.

Jiwand Singh—

N.-W. Frontier, 1897-98.—Medal with 3 clasps.

Mahsud, Waziristan, 1901-02.

Faqir Mohammad—

N.-W. Frontier of India, 1908.—Mohmand Expedition.—Medal and clasp.

57th Wilde's Rifles (Frontier Force).

SUBADAR-MAJOR—

Arsla Khan, M.C., I.O.M., *Sardar Bahadur*—
N.-W. Frontier of India, Waziristan, 1894-95. —Medal with clasp.
China, 1900.—Medal.
N.-W. Frontier of India, 1908.—Operations in the Mohmand country. Engagement of Kargha, 24th May 1908. Mentioned in despatches.—Medal with clasp. Indian Order of Merit, 3rd class.
War, 1914-17.—Mentioned in Despatches (twice wounded), Military Cross and Order of Br. India, 1st class.

SUBADARS—

Madaman, *Sardar Bahadur—Order of Br. India, 1st class.*
N.-W. Frontier of India, Hazara, 1888.—Medal with clasp.
N.-W. Frontier of India, 1st Miranzai, 1891.
N.-W. Frontier of India, Waziristan, 1894-95.—Clasp.
China, 1900.—Medal.
N.-W. Frontier of India, 1902.—Operations against the Darwesh Khel Waziris.
N.-W. Frontier of India, 1908.—Operations in the Mohmand country. Engagement of Matta, 24th April 1908. Engagement of Kargha, 24th May 1908.—Medal with clasp.

Bhagat Sing, I.D.S.M.—
N.-W. Frontier of India, 1908.—Operations in the Mohmand country. Engagement of Matta, 24th April 1908. Severely wounded.—Medal with clasp.
War, 1914-15.—Severly wounded.

Chur Singh—
China, 1900.—Medal.
Tibet, 1903-04.—Medal.
N.-W. Frontier of India, 1908.—Operations in the Mohmand country. Engagement of Kargha, 24th May 1908.—Medal with clasp.

Jagat Singh, I.O.M., I.D.S.M.—
N.-W. Frontier of India, Waziristan, 1894-95.—Medal with clasp.
China, 1900.—Medal.
N.-W. Frontier of India, 1908.—Operations in the Zakka Khel country.
N.-W. Frontier of India, 1908.—Operations in the Mohmand country. Engagement of Matta, 24th April 1908. Engagement of Kargha, 24th May 1908.—Medal with clasp.
Indian Order of Merit (Civil).
War, 1914-16.

Amar Singh—
N.-W. Frontier of India, Waziristan, 1894-95.—Medal with clasp.
China, 1900.—Medal.
War, 1914-16.

Ram Ditta—
N.-W. Frontier of India, 1908.—Operations in the Mohmand country. Engagement of Kargha, 24th May 1908.—Medal with clasp.
War, 1914-17.

Imam Ali, *Bahadur*—
N.-W. Frontier of India, Waziristan, 1894-95.—Medal with clasp.
China, 1900.—Medal.
War, 1914-16—Order of Br. India, 2nd class.

Shah Baz—
China, 1900.—Medal.
N.-W. Frontier of India, 1902.—Operations against the Darwesh Khel Waziris.
N.-W. Frontier of India, 1908.—Operations in the Mohmand country. Engagement of Kargha, 24th May 1908.—Medal and clasp.
War, 1914-17.

Rabel Sing—
N.-W. Frontier of India, 1908.—Operations in the Zakka Khel country.—Medal and clasp.
War, 1914-17.

JEMADARS—

Ahmad Khan—
N.-W. Frontier of India, Waziristan, 1894-95.—Medal and clasp.
War, 1914-17.

Yar Ghulam—
China, 1900.—Medal.
N.-W. Frontier of India, 1908.—Operations in the Mohmand country. Engagement of Matta, 24th April 1908. Engagement of Kargha, 24th May 1908.—Medal and clasp.
War, 1914-16.

Fazal Ahmad—
N.-W. Frontier of India, 1908.—Operations in the Mohmand country. Engagement of Kargha, 24th May 1908.—Medal with clasp.
War, 1914-17.

Mangal Singh, I. O. M.—
China, 1900.—Medal.
N.-W. Frontier of India, 1908.—Operations in the Mohmand country. Engagement of Matta, 24th April 1908. Engagement of Kargha, 24th May 1908.—Medal. 7 clasps.
War, 1914-17.—Mentioned in Despatches, Indian Order of Merit, 2nd class.

Bur Singh, I.D.S.M.—
China, 1900.—Medal.
N.-W. Frontier of India, 1908.—Operations in the Mohmand country. Engagement of Matta, 24th April 1908. Engagement of Kargha, 24th May 1908.—Medal and clasp.
War, 1914-17.—Mentioned in Despatches.—Indian Distinguished Service Medal.

Ganga Ram—
China, 1900.—Medal.
N.-W. Frontier of India, 1908.—Operations in the Mohmand country. Engagement of Kargha, 24th May 1908.—Medal and clasp.
War, 1914-15.

Budhawa Sing—
China, 1900.—Medal.
War, 1914-17.

Pars Ram—
N.-W. Frontier of India, 1908.—Operations in the Mohmand country. Engagement of Kargha, 24th May 1908.—Medal and clasp.
War, 1914-17.

Hira Sing, I.D.S.M.—
N.-W. Frontier of India, 1908.—Operations in the Mohmand country. Engagement of Matta, 24th April 1908. Engagement of Kargha, 24th May 1908.—Medal and clasp.
War, 1914-17.—Mentioned in Despatches, Indian Distinguished Service Medal.

Zargir—
N.-W. Frontier of India, 1908.—Operations in the Mohmand country. Engagement of Kargha, 24th May 1908.—Medal with clasp.
War, 1914-17.

Sultan Khan—
N.-W. Frontier of India, 1908.—Operations in the Mohmand country. Engagement of Kargha, 24th May 1908.—Medal and clasp.
War, 1914-17.

58th Vaughan's Rifles (Frontier Force).

SUBADAR-MAJOR—

Tika Khan—
N.-W. Frontier of India, 1897-98.—Operations of the Flying Column in the Kurram Valley under Colonel Richardson, 20th August to 1st October 1897.
Tirah, 1897-98.—Medal with 2 clasps.
N.-W. Frontier of India, Waziristan, 1901-02.—Clasp.
N.-W. Frontier of India, 1902.—Operations against the Darwesh Khel Waziris.
Tibet, 1903-04.—Medal.

SUBADARS—

Gujar Si gh—
N.-W. Frontier of India, 1897-98.—Operations of the Flying Column in the Kurram Valley under Colonel Richardson, 20th August to 1st October 1897.—Medal with 2 clasps.
N.-W. Frontier of India, Waziristan, 1901-02.—Clasp.
N.-W. Frontier of India, 1902.—Operations against the Darwesh Khel Waziris.

Raj Talab—*Bahadur, Order of Br. India, 2nd cl.*
N.-W. Frontier of India, 1897-98.—Operations of the Flying Column in the Kurram Valley under Colonel Richardson, 20th August to 1st October 1897.—Medal with 2 clasps.
N.-W. Frontier of India, Waziristan, 1901-02.—Clasp.
N.-W. Frontier of India, 1902.—Operations against the Darwesh Khel Waziris.

Karm Singh, I.O.M.—
N.-W. Frontier of India, 1897-98.—Operations of the Flying Column in the Kurram Valley under Colonel Richardson, 20th August to 1st October 1897.
Tirah, 1897-98.—Medal with 2 clasps.
N.-W. Frontier of India, Waziristan, 1901-02.—Clasp.
N.-W. Frontier of India, 1902.—Operations against the Darwesh Khel Waziris.

Indar Singh, M.C.—
N.-W. Frontier of India, Waziristan, 1901-02.—Medal with clasp.
N.-W. Frontier of India, 1902.—Operations against the Darwesh Khel Waziris.

Harchand Singh, I.O.M.—
N.-W. Frontier of India, Waziristan, 1901-02.—Medal with clasp.
N.-W. Frontier of India, 1902.—Operations against the Darwesh Khel Waziris.

Muhmmed Arbi, I.O.M.—
N.-W. Frontier of India, Waziristan, 1901-02.—Medal with clasp.
N.-W. Frontier of India, 1902.—Operations against the Darwesh Khel Waziris.

Hamesh Gul—
N.-W. Frontier of India, Waziristan, 1901-02.—Medal and clasp.
N.-W. Frontier of India, 1902.—Operations against the Darwesh Khel Waziris.

Suba—
N.-W. Frontier of India, 1897-98.—Operations of the Flying Column in the Kurram Valley under Colonel Richardson, 20th August to 1st October 1897.
Tirah, 1897-98.—Medal with 2 clasps.
N.-W. Frontier of India, Waziristan, 1901-02.—Clasp.
N.-W. Frontier of India, 1902.—Operations against the Darwesh Khel Waziris.

JEMADARS—

Rasila—
N.-W. Frontier of India, 1897-98.—Operations of the Flying Column in the Kurram Valley under Colonel Richardson, 20th August to 1st October 1897.
Tirah, 1897-98.—Medal with 2 clasps.
N.-W. Frontier of India, Waziristan, 1901-02.—Clasp.
N.-W. Frontier of India, 1902.—Operations against the Darwesh Khel Waziris.

Santu—
N.-W. Frontier of India, Waziristan, 1901-02.—Medal with clasp.
N.-W. Frontier of India, 1902.—Operations against the Darwesh Khel Waziris.

Madat—
N.-W. Frontier of India, 1897-98.—Operations of the Flying Column in the Kurram Valley under Colonel Richardson, 20th August to 1st October 1897.
Tirah, 1897-98.—Medal with 2 clasps.
N.-W. Frontier of India, Waziristan, 1901-02.—Clasp.
N.-W. Frontier of India, 1902.—Operations against the Darwesh Khel Waziris.

Diwana, I.O.M.—
N.-W. Frontier of India, Waziristan, 1901-02.—Medal with clasp.
N.-W. Frontier of India, 1902.—Operations against the Darwesh Khel Waziris.

Rang Baz Khan—
N.-W. Frontier of India, Waziristan, 1901-02.—Medal and clasp.

Kehr Singh—
N.-W. Frontier of India, 1902.—Operations against the Darwesh Khel Waziris.

Harnam Singh—
N.-W. Frontier of India, Waziristan, 1901-02.—Medal and clasp.
N.-W. Frontier of India, 1902.—Operations against the Darwesh Khel Waziris.

Thakar Singh—
N.-W. Frontier of India, Waziristan, 1901-02.—Medal and clasp.
N.-W. Frontier of India, 1902.—Operations against the Darwesh Khel Waziris.

Budh Singh—
N.-W. Frontier of India, Waziristan, 1901-02.—Medal and clasp.
N.-W. Frontier of India, 1902.—Operations against the Darwesh Khel Waziris.

59th Scinde Rifles (Frontier Force).

SUBADAR-MAJOR—

Nasir Khan, Sardar Bahadur—
Order of British India, 1st class.
Indian Order of Merit, 2nd class.

N.-W. Frontier of India, 2nd Miranzai, 1891.—Medal with clasp.
N.-W. Frontier of India, Waziristan, 1894-95.—Clasp.
China, 1900.—Relief of Pekin.—Medal with clasp.
N.-W. Frontier of India, 1902.—Operations against the Darwesh Khel Waziris.
N.-W. Frontier of India, 1908.—Operations in the Zakka Khel country.
N.-W. Frontier of India, 1908.—Operations in the Mohmand country. Engagement of Kargha, 24th May 1908.—Medal with clasp.

Parbhat Chund, M. C.—
N.-W. Frontier of India, 1908—Operations in the Mohmand country—Medal and clasp.
War, 1915.—Military Cross.

SUBADARS—

Mukhmad Jan, *Sardar Bahadur.—Order of Br. India 1st class. Indian Order of Merit, 2nd class.—Indian Distinguished Service Medal.*
N.-W. Frontier of India, Miranzai, 1891.—Medal with clasp.
N.-W. Frontier of India, Waziristan, 1894-95.—Clasp.
N.-W. Frontier of India, 1902.—Operations against the Darwesh Khel Waziris.
East Africa, 1903-04.—Operations in Somaliland. Action at Jidballi.—Medal with 2 clasps.
N.-W. Frontier of India, 1908.—Operations in the Zakka Khel country.
N.-W. Frontier of India, 1908.—Operations in the Mohmand country. Engagement of Kargha, 24th May 1908. Mentioned in despatches.—Medal with clasp.

Zarin Khan—
N.-W. Frontier of India, 1908.—Operations in the Zakka Khel country.
N.-W. Frontier of India, 1908.—Operations in the Mohmand country. Engagement of Kargha, 24th May 1908.—Medal with clasp.

Gauri Charan, Bahadur—
Order of British India, 2nd class.
Indian Order of Merit, 2nd class.
N.-W. Frontier of India, 1908.—Operations in the Mohmand country. Engagement of Kargha, 24th May 1908.—Medal with clasp.

Muzaffar Khan—
N.-W. Frontier of India, 1908.—Operations in the Zakka Khel country.
N.-W. Frontier of India, 1908.—Operations in the Mohmand country. Engagement of Kargha, 24th May 1908.—Medal with clasp. Order of Merit, 2nd class (as Havildar).

Shahwali—(Retired). Re-employed 29th March 1915.
N.-W. Frontier of India, 2nd Miranzai, 1891.—Medal with clasp.
N.-W. Frontier of India, Waziristan, 1894-95.—Clasp.

Wali Khan—(Retired). Re-employed 10th October 1915.
N.-W. Frontier of India, 2nd Miranzai, 1891.—Medal with clasp.
N.-W. Frontier of India, Waziristan, 1894-95.—Clasp.
N.-W. Frontier of India, 1897-98.—Medal with clasp.

Makhmad Ali—
N.-W. Frontier of India, 2nd Miranzai, 1891.—Medal with clasp.
N.-W. Frontier of India, Waziristan, 1894-95.—Clasp.
N.-W. Frontier of India, 1908.—Operations in the Zakka Khel country.
N.-W. Frontier of India, 1908.—Operations in the Mohmand country. Engagement of Kargha, 24th May 1908.—Medal with clasp.

Sahib Haq—
N.-W. Frontier of India, 1908.—Operations in the Zakka Khel country.
N.-W. Frontier of India, 1908.—Operations in the Mohmand country. Engagement of Kargha, 24th May 1908.—Medal with clasp.

Ali Akbar—
N.-W. Frontier of India, 1908.—Operations in the Zakka Khel country.
N.-W. Frontier of India, 1908.—Operations in the Momand country. Engagement of Kargha, 24th May 1908.—Medal with clasp.

Santa Singh—
N.-W. Frontier of India, 1908.—Bazaar Valley and Mohmand Expedition.—Medal and clasp.

JEMADARS—

Chunnak Singh—
N.-W. Frontier of India, 1908.—Operations in the Zakka Khel country.
N.-W. Frontier of India, 1908.—Operations in the Mohmand country. Engagement of Kargha, 24th May 1908.—Medal with clasp.

Gulab Singh—
N.-W. Frontier of India, 1908.—Operations in the Zakka Khel country. Operations in the Mohmand country. Engagement of Kargha, 24th May 1908.—Medal with clasp.

Jabbar Khan—
N.-W. Frontier of India, 1902.—Operations against the Darwesh Khel Waziris.
N.-W. Frontier of India, 1908.—Operations in the Zakka Khel country. Operations in the Mohmand country. Engagement of Kargha 24th May 1908.—Medal with clasp.

Karam Din—
N.-W. Frontier of India, 1908.—Operations in the Zakka Khel country.—Medal with clasp.

Chattar Singh (Retired).—Re-employed, 10th Sep. 1915.
N.-W. Frontier of India, Miranzai, 1891.—Medal with clasp.
N.-W. Frontier of India, Waziristan, 1894-95.—Clasp.
East Africa, 1903-04.—Operations in Somaliland. Action at Jadbali.—Medal with 2 clasps.

Ram Singh—
N.-W. Frontier of India, 1908.—Operations in the Zakka Khel country. Operations in the Mohmand country. Engagement of Kargha, 24th May 1908.—Medal with clasp.

Rukkan Din—
N.-W. Frontier of India, 1908.—Bazaar Valley. Mohmand Expedition.—Medal and clasp.

61st King George's Own Pioneers.

SUBADAR-MAJOR—

Rangayya—
China, 1900.—Relief of Pekin—Medal with clasp.

SUBADARS—

Muhammad Abdul Hafiz—
China, 1900.—Relief of Pekin.—Medal with clasp.

Veerappan—
China, 1900.—Relief of Pekin.—Medal with clasp.

Saiyid Abdul Ghaffur—
China, 1900.—Relief of Pekin.—Medal with clasp

Kallayaperumal—
China, 1900.—Relief of Pekin.—Medal with clasp.

Shaikh Muhiuddin—
N.-W. Frontier of India, 1897-98.—Medal with clasp.
China, 1900.—Relief of Pekin.—Medal with clasp.

Paparayudu—
China, 1900.—Relief of Pekin.—Medal with clasp.

Duraisami—
China, 1900.—Relief of Pekin.—Medal with clasp.

JEMADARS—

Rathnasabapathi—
China, 1900.—Relief of Pekin.—Medal with clasp.

Katha Nadan—
China, 1900.—Relief of Pekin.—Medal with clasp.

Murtuza Sharif—
China, 1900.—Relief of Pekin.—Medal with clasp.

William Cotter—
China, 1900.—Relief of Pekin.—Medal with clasp.

2-61st King George's Infantry.

SUBADAR-MAJOR—

Gangayya—
China, 1900.—Relief of Pekin.—Medal with clasp.

SUBADARS—

Sivasami—
China, 1900.—Relief of Pekin.—Medal with clasps.

Abdur Rahim—
N.-W. Frontier of India, 1897-98.—Medal with clasp.
China, 1900.—Relief of Pekin.—Medal with clasp.

Rajagopaul—
China, 1900.—Relief of Pekin.—Medal with clasp.

JEMADAR—

Sankerapondian—
China, 1900-01.

62nd Punjabis.

SUBADAR-MAJOR—

Muhammad Zaman—*Bahadur, Order of Br. India, 2nd class.*
N.-W. Frontier of India, Zhob Valley, 1890.
N.-W. Frontier of India, 1st and 2nd Miranzai 1891.—Medal with clasp.
N.-W. Frontier of India, Waziristan, 1894-95.—Clasp.
N.-W. Frontier of India, 1897-98.—Action of the Ublan Pass, 27th August 1897. Operations on the Samana and in Kurram Valley, during August and September 1897. Relief of Gulistan.—Medal with 2 clasps.
Tirah, 1897-98.—Operations in the Bara Valley, 7th to 14th December 1897.—Clasp.
N.-W. Frontier of India, Waziristan, 1901-02.—Clasp.

SUBADARS—

Kalu Khan—*Bahadur, Order of Br. India, 2nd class.*
N.-W. Frontier of India, Hazara, 1891.—Medal with clasp.
N.-W. Frontier of India, Samana, 1891.—Clasp.

Sundar Singh—
N.-W. Frontier of India Hazara, 1888.—Medal with clasp.

Sher Zaman—
China, 1900.—Medal.

Jahandad Khan—
N.-W. Frontier of India, 1897-98.—Tochi.—Medal and clasp.
N.-W. Frontier of India, Waziristan, 1901-02.—Clasp.

Sohna Singh—
Soudan, 1885.—Medal with 2 clasps. Khedive's Bronze Star.
N.-W. Frontier of India, 2nd Miranzai, 1891.—Medal with clasp.
Chitral, 1895.—Relief of Chitral.—Medal with clasp.

Abdulla Khan—
China, 1900.—Medal.

63rd Palamcottah Light Infantry.

SUBADARS—

Kadir Sharif—
China, 1900.—Medal.

Sankarapandian Pillai—
N.-W. Frontier, Tochi, 1897-98.—Medal with clasps.

Muhammad Sharif—
China, 1900.—Medal.

Maduramutla—
China, 1900.—Medal.

Devasagayam—
China, 1900.—Medal.

Shaik Ali—
China, 1900.—Medal.

Hari Singh—
China, 1900.—Medal.

JEMADARS—

Abdul Latif—
China, 1900.—Medal.

Arokiasami—
China, 1900.—Medal.

Saiyid Ali—
China, 1900.—Medal.

Narayanasami—
China, 1900.—Medal.

Abdul Karim—
China, 1900.—Medal.

64th Pioneers.

SUBADARS—

Anantadri Nayudu—
N.-E. Frontier of India, Chin Hills, 1891-92.

Durgayya—
N.-E. Frontier of India, Chin Hills, 1891-92.—
N.-W. Frontier of India, Tirah, 1898.—Medal with clasp.

Krishnamurti—
N.-W. Frontier of India, Tochi, 1897-98.—Medal with clasp.

66th Punjabis.

SUBADARS—

Jamal Khan—
Br. East Africa, 1898.—Operations against the Ogaden Somalis.—Medal with clasp.

Akbar Khan—
Chitral, 1895.—Medal.
Waziristan, 1901-02.—Clasp.

Muhammad Khan—
East Africa (Somaliland), 1908—10.—Medal and clasp.

Sultan Ali—
N.-W. Frontier of India, 1895-96.—Medal with one clasp.
China, 1900.—Relief of Pekin.—Medal with one clasp

Girdhari Singh—
Chitral, 1895.—Medal.
N.-W. Frontier of India, Waziristan, 1901-02.—Clasp.
Aden, 1903-04.—Operations in the Interior.

Hazara Singh—
Egypt (Sanakin), 1885.—Medal and 2 clasps.
Sanakin, 1885.—Bronze Egyptian Star.
Chitral, 1895.—India Medal.
N.-W. Frontier of India, 1897.—Clasp.

Labh Din—
East Africa (Moela), 1895-96.—Medal.
East Africa (Uganda), 1897-98.—Medal with clasp.
East Africa (Labus), 1897-98.—Clasp.
China, 1900.—Relief of Pekin.—Medal with clasp.

JEMADARS—

Bishn Singh—
China, 1900.—Relief of Pekin.—Medal with clasp

Gulzara Singh—
Soudan, 1896.—Medal.
N.-W. Frontier of India, 1897.—Medal with clasp.

Ahmed Khan—
N.-W. Frontier of India, 1897-98.—India medal with clasp.

67th Punjabis.

SUBADAR-MAJOR—

Lachhman Singh—I.D.S.M.
N.-W. Frontier of India, Chitral, 1895.—Relief of Chitral.—Medal with clasp.
N.-W. Frontier of India, 1897-98.—Engagement near Shabkadar on 9th August 1897. Operations on the Samana and in the Kurram Valley during August and September 1897.—2 clasps.
Tirah, 1897-98.—Operations in the Bazar Valley, 25th to 30th December 1897.—Clasp.

SUBADARS—

Shihab Din—
N.-W. Frontier of India, Hazara, 1888.—Medal with clasp.
N.-W. Frontier of India, 1897-98.—Defence of Malakand.—Relief of Chakdara. Malakand. Action at Landikai. Operations in Bajour and in the Mahmund country.—Medal with 2 clasps.
China, 1900.—Relief of Pekin.—Medal with clasp.

Mahomed Din—
East and Central Africa, 1898.—Medal with clasp.

Sher Khan—
N.-W. Frontier of India, Waziristan, 1894-95.—Medal with clasp.
N.-W. Frontier of India, 1897-98.—Tochi.—Medal with clasp.

Budh Singh—
N.-W. Frontier of India, Waziristan, 1901-02.

Gopal Singh—
N.-W. Frontier of India, Waziristan, 1894-95.—Medal with clasp.
N.-W. Frontier of India, 1897-98.—Tochi.—Medal with clasp.
China, 1900-01.—Medal.

Ahmud Khan—I.D.S.M.
N.-W. Frontier of India, Waziristan, 1891-02.—Medal with clasp.

Indar Singh—I.D.S.M.,
N.-W. Frontier of India, 1908.—Medal with clasp.

JEMADARS—

Kaloo—
N.-W. Frontier of India, 1897-98.—Malakand. Defence of Malakand—Medal with 2 clasps.

Rur Singh—
N.-W. Frontier of India, Waziristan, 1901-02.—Medal and clasp.

2-67th Punjabis.

SUBADAR-MAJOR—

SUBADARS—

Bal Mukand—
N.-W. Frontier of India, 1897-98.—Tochi—Medal with clasp.

Ghulam Hassan—
N.-W. Frontier of India, 1897-98.—Defence of Malakand. Relief of Malakand. Malakand. Action at Landikai. Operations in the Mamund country. Buner.—Medal with 2 clasps.

Jahandadkhan—
N.-W. Frontier of India, 1908.—Operations in the Zakka Khel country.—Medal with clasp.

Sham Singh—
N.-W. Frontier of India, 1897-98, Tochi.—Medal with 3 clasps.

Dalip Singh—
N.-W. Frontier of India, Waziristan, 1901-02.—Medal and clasp.

JEMADARS—

Alla Dad Khan—
N.-W. Frontier of India, Tochi.—Medal with clasp.

Bahadur Singh—
Tibet, 1903-04.—Medal.

Tabiyat Singh—
N.-W. Frontier—"Tochi," 1897-98.—Medal and clasp.
N.-W. Frontier of India, 1908.—Operations in the Zakka Khel Country—Medal and clasp.

69th Punjabis.

SUBADAR-MAJOR—

Muhammad Khan—*Indian Order of Merit.*
China, 1900.—Medal.

SUBADARS—

Rahmat Khan—
China, 1900.—Medal.

Fazal Dad—
China, 1900.—Medal.

Rajwali—
China, 1900.—Medal.

Munsha Singh—
N.-W. Frontier of India, 1897-98.—Operations in Samana and Kurram. Relief of Gulistan.—Medal and 1 clasp.
Tirah, 1897-98.—Capture of the Sampagha Pass. Capture of the Arhanga Pass. Reconnaissance of the Saran Sar. Operations in the Waran Valley. Operations at and around Dwatoi. Operations in the Bara Valley. Affair at Shinkamar.—One clasp.

JEMADARS—

Nabi Bakhsh—
Somaliland, 1909-10.—Medal with clasp.

Sardar Khan—*Indian Order of Merit.*

Labh Singh—*Indian Distinguished Service Medal.*

Gheba Khan—*Indian Distinguished Service Medal.*

72nd Punjabis.

SUBADAR-MAJOR—

SUBADARS—

Gyan Singh—*Bahadur—Order of British India, 2nd class.*
Soudan, 1885.—Actions of Hasheen Tofrek and Tamai.—Medal with 2 clasps. Bronze star.
Burma, 1887-89.—Medal with clasp.
N.-E. Frontier of India, Chin-Lushai, 1889-90.—Clasp.
Chinbok Column, 1889-92.—Clasp.
N.-E. Frontier of India, Manipur, 1891.—Clasp.

Chand Khan—
Burma, 1887-89.—Medal with clasp.
N.-E. Frontier of India, Manipur 1891.—Clasp.
N.-E. Frontier of India, Chin Hills, 1892-93.—Klang-Klang Column.

Latif Khan—
Burma, 1887-89.—Medal with clasp.
N.-E. Frontier of India, Chin Hills, 1892.—Pimpi Column.

Fazil Din—
N.-E. Frontier of India, Chin Hills, 1894.—Bangshe Column.

Murad Baksh—
N.-E. Frontier of India, Chin Hills, 1894.—*Bangshe Column.*

JEMADAR—

Jhanda Singh—
N.-E. Frontier of India, Chin Hills, 1894.—Bangshe Column.
Nquamba Expedition, 1899 (B. C. A.).—Medal and clasp.
Ashantee Expedition, 1900.—Medal.
Action of Kumassi.—Clasp.
Action of Danossi.—1900.
Action of Abuassu.—1900.

2-72nd Punjabis.

SUBADAR-MAJOR—

Gyan Singh—*Bahadur—Order of Br. India, 2nd class.*
Soudan, 1885.—Actions of Hasheen Tofrek and Tamai.—Medal with 2 clasps. Bronze star.
Burma, 1887-89.—Medal with clasp.
N.-E. Frontier of India, Chin-Lushai, 1889-90.—Clasp.
Chinbok Column, 1889-92.—Clasp.
N.-E. Frontier of India, Manipur, 1891.—Clasp.

73rd Carnatic Infantry.

SUBADAR-MAJOR—

Innasimuttu—
China, 1900.—Medal and clasp.

74th Punjabis.

SUBADARS—

Mihr Khan—
N.-W. Frontier of India, 1894-95.—Medal with clasp.
China Expeditionary Force, 1900—02.—Medal.

Bhagat Singh—
N.-W. Frontier of India, Waziristan, 1894-95.—Medal with clasp.
N.-W. Frontier of India, 1897-98.—Defence of Malakand. Relief of Chakdara. Action at Landikai. Operations in Bajour and in the Mohmand country.—Medal with 2 clasps.
China, 1900-01.—Medal with clasp.

JEMADARS—

Harnam Singh—
N.-W. Frontier of India, 1897.—Mohmand Expedition.—Medal and clasp.

Ghulam Muhammad—
N.-W. Frontier of India, Waziristan, 1901-02.—Medal and clasp.

Nawab Khan—
N.-W. Frontier of India, Waziristan, 1901-02.—Medal and clasp.

1-75th Carnatic Infantry.

SUBADARS—

Ramana—*Indian Distinguished Service Medal.*
Burma, 1891.—Wuntho. Medal with clasp.
N.-W. Frontier of India, 1897-98.—Tochi.

76th Punjabis.

SUBADAR-MAJOR—

Hussain Shah—

N.-W. Frontier of India, Miranzai, 1891. China, 1900.—Medal.

SUBADARS—

Ahmad Khan—

China, 1900.—Medal.

Indur Singh—

N.-W. Frontier of India, Miranzai, 1891.
N.-W. Frontier of India, 1897-98.—Malakand. Operations in Bajour.—Medal with clasp.
Tirah, 1897-98.—Clasp.

Bahadur Shah—D.S.M.

N.-W. Frontier of India, Waziristan, 1894-95.—Medal with clasp.
N.-W. Frontier of India, 1897-98.—Tochi.—Medal with clasp.

Sundar Singh—D.S.M.

N.-W. Frontier of India, 1897-98.—Malakand. Defence of Malakand. Relief of Malakand. Action at Landakai.—Medal with 2 clasps.
Tirah, 1897-98.—Clasp.

Ghulam Muhammad—

China, 1900.—Medal.

Shan Khan—D.S.M.

Somaliland, 1911-14.—Action at Sheinber Berris.—Medal with clasp.

JEMADARS—

Kapur Singh—

N.-W. Frontier of India, Waziristan, 1894-95—Medal with clasp
N.-W. Frontier of India, 1897-98.—Tochi. Operations on the Samana and in the Kurram Valley during August and September 1897.
Tirah, 1897-98.—Actions Sampagha and Arhanga Passes. Operations against the Khani Khe Chamkanis. Operations in the Bazaar Valley 25th to 30th December 1897.—Medal with 3 clasps.

Hukam Dad—

N.-W. Frontier of India, 1897-98.—Medal with clasp.
China, 1900.—Medal with clasp.

80th Carnatic Infantry.

SUBADARS—

Abdul Majid—

Burma, 1891.—Wunthc. Medal with clasp.
N.-W. Frontier of India, 1897-98.—Tochi.

Gurusami—

N.-W. Frontier of India, 1897-98.—Tochi.

Venkatasami—

N.-W. Frontier of India, 1897-98.—Tochi.

Shaikh Abdur Rahman—

N.-W. Frontier of India, 1897-98.—Tochi.

Narayanasami—

N.-W. Frontier of India, 1897-98.—Tochi.

Abdur Razaq—

N.-W. Frontier of India, 1897-98.—Tochi.

JEMADARS—

Chandika Singh—

N.-W. Frontier of India, 1897-98.

Muhammed Husain—

N.-W. Frontier of India, 1897-98.—Tochi.

1-81st Pioneers.

SUBADAR-MAJOR—

SUBADARS—

Ramasami—

N.-E. Frontier of India, Chin Hills, 1892-93.—Medal with clasp.
N.-W. Frontier of India, 1897-98.—Operations on the Samana and in the Kurram Valley during August and September 1897.—Medal with 2 clasps.
Tirah, 1897-98.—Actions of Chagru Kotal and Dargai.—Operations in the Bazar Valley, 25th to 30th December 1897.—Clasp.

Shaikh Ismail—

N.-E. Frontier of India, Chin Hills, 1892-93.—Medal with clasp.
N.-W. Frontier of India, 1897-98.—Operations on the Samana and in the Kurram Valley during August and September 1897.—Medal with 2 clasps.

Tirah, 1897-98.—Actions of Chagru Kotal and Dargai. Operations in the Bazar Valley, 25th to 30th December 1897.—Clasp.

Palayam—

N.-W. Frontier of India, 1897-98.—Operations on the Samana and in the Kurram Valley during August and September 1897.—Medal with 2 clasps.

Tirah. 1897-98.—Actions of Chagru Kotal and Dargai. Operations in the Bazar Valley, 25th to 30th December 1897.—Clasp.

Kondayya—

N.-W. Frontier of India, 1897-98.—Operations on the Samana, and in the Kurram Valley during August and September 1897.—Medal with 2 clasps.
Tirah, 1897-98.—Actions of Chagru Kotal and Dargai. Operations in the Bazar Valley, 25th to 30th December 1897.—Clasp.

Ramasami—

N.-W. Frontier of India, 1897-98.—Operations on the Samana, and in the Kurram Valley during August and September 1897. Medal with 2 clasps.

Tirah, 1897-98.—Action of Chagru Kotal and Dargai. Operations in the Bazar Valley, 25th to 30th December 1897.—Clasp.

2-81st Pioneers.

Shaikh Daud—

N.-E. Frontier of India, Chin Hills, 1892-93.—Medal with clasp.

N.-W. Frontier of India, 1897-98.—Operations on the Samana and in the Kurram Valley during August and September 1897.—Medal with 2 clasps.

Tirah, 1897-98.—Action of Chagru Kotal and action of Dargai.—Operations in the Bazar Valley, 25th to 30th December 1897.—Clasp.

Shaikh Hamid—

N.-W. Frontier of India, 1897-98.—Operations on the Samana, and in the Kurram Valley during August and September 1897.—Medal with 2 clasps.

Tirah, 1897-98.—Actions of Chagru Kotal and Dargai. Operations in the Bazar Valley, 25th to 30th December 1897 —Clasp.

Rungasami—

N.-W. Frontier of India, 1897-98.—Medal with clasp.

China, 1900.—Relief of Pekin.—Medal with clasp.

JEMADARS—

Ponnusami—

China, 1900-01.—Medal.

Chinnasami—

China, 1900-01.

82nd Punjabis.

SUBADAR-MAJORS—

Afzal Khan, *Bahadur—Order of Br. Ind., 2nd Class.*

China, 1900.—Medal.

Ram Singh, *Bahadur—Order of Br. Ind., 2nd Class,* I.O.M.

Chitral, 1895.—Relief of Chitral —Actions at Malakand, Khar and Swat River.—Medal with clasp.

N.-W. Frontier of India, 1897-98.—Operations on the Samana and in the Kurram Valley.

N.-W. Frontier of India, Tirah, 1897-98.—Actions of Chagru Kotal and Dargai and of the Sampagha and Arhanga Passes. Reconnaissance of the Saran Sar and action of 9th November 1897. Operations in the Waran Valley and action of 16th November 1897.—Severely wounded in 3 places.—Indian Order of Merit, 3rd class, for gallantry at the Chagru Kotal.—Medal with 2 clasps.

SUBADARS—

Mastan Singh—

Soudan, 1896.—Dongola Expedition.—Queen's medal and Khedive's medal.

N.-W. Frontier of India, Utman Khel, 1897-98.—Medal with clasp.

Ghulam Muhiyuddin—

China, 1900.—Medal.

Phuman Singh—

N.-W. Frontier of India, Wasiristan 1894—Medal with clasp.

China, 1900.—Medal.

Harnand—

China, 1902.—Medal.

Fazal Rahman—

N.-W. Frontier of India, (Mohmand) 1908.—Medal with clasp.

Haider Khan—

China, 1900.—Medal.

Amir Khan—

China, 1900.—Medal.

Nawais Ali—

N.-W. Frontier of India, 1897-98.—Operations on the Samana and in the Kurram Valley, during August and September 1897, Medal with 2 clasps.

N.-W. Frontier of India, Tirah, 1897-98.—Actions, of Dargai and Arhanga Passes. Reconnaissance of the Saran Sar of the Sampagha on 24th November 1897. Operations against Khan Khel Chamkanis and action of 9th November in Bara Valley and Bazaar Valley.—Clasp.

N.-W. Frontier of India, Wasiristan, 1901-02—Clasp.

Kirpal Singh—

China, 1900.—Medal.

Sukh Ram—

China 1900-01.—Medal.

JEMADARS—

Khan Muhammad—

China, 1900.—Relief of Pekin—Medal with clasp.

Mausam Shah—

China, 1900.—Medal.

83rd Wallajahbad Light Infantry.

SUBADARS—

Madurai—

China, 1900-01.—Relief of Pekin.—Medal with clasp.

Saiyid Kadir—

Burma, 1889—92.—Medal with clasp.

Abdul Wahab—

Somaliland, 1902—04.—Medal with clasp.

JEMADAR—

Abdul Azim—

Somaliland, 1902—04.—Medal with clasp.

84th Punjabis.

SUBADAR-MAJOR—

Habib Shah—

N.-W. Frontier of India, 2nd Miranzai, 1891.—Medal with clasp.

N.-W. Frontier of India, 1897-98.—Tochi.—Medal with clasp.

N.-W. Frontier of India, 1901-02.—Clasp.

SUBADARS—

Hari Singh—

British East Africa, 1898.—Operations against the Ogaden Somalis —Medal with clasp.

Uganda, 1899.—Operations against Kabarega.—Clasp.

Uganda, 1900.—Medal with clasp.

Firoz Khan—*Distinguished Service Medal, 1915.*

N.-W. Frontier of India, 2nd Miranzai, 1891.—Medal with clasp.

Waziristan, 1894-95.—India Medal of 1854 with clasp inscribed Waziristan, 1894-95.

N.-W Frontier of India, Waziristan, 1901-02.—Medal with clasp.

Harnam Singh—

China, 1900.—Medal.

JEMADARS—

Arjan Singh—

B. E. Africa, 1898.—Operations against the Ogaden Somalis.—Medal with clasp.

Uganda, 1899.—Operations against Kabarega.—Clasp.

N.-W. Frontier of India, Waziristan, 1901-02.—Medal with clasp.

Gopal Singh—

N.-W. Frontier of India, Waziristan, 1901-02.—Medal with clasp.

86th Carnatic Infantry.

SUBADARS—

Tirumalai Nayudu—

Burma, 1885—87.—Medal with clasp.

Abdul Razzak—

Burma, 1885—87, 87—89.—Medal with 2 clasps.

87th Punjabis.

SUBADAR-MAJOR—

Sher Muhammad—

N.-W. Frontier of India, 1897-98.—Medal with clasp.

Tirah, 1897-98.—Clasp.

N.-W. Frontier of India, 1898-99.—Operations against the Para Chamkanis.

N.-W. Frontier of India, 1902.—Operations against the Darwesh Khel Wazirs.

SUBADARS—

Lachhman—

N.-W. Frontier of India, 1897-98.—Tochi Field Force.—Medal with clasp.

China, 1900-01.—Medal.

Fazl Khan—

N.-W. Frontier of India, Waziristan, 1894-95.—Medal with clasp.

British Central Africa, 1895-96.—Medal.

Sarban Singh—*Indian Distinguished Service Medal*—

N.-W. Frontier of India, Waziristan, 1894-95.—Action at Wana, 3rd November 1894.—Medal with clasp.

N.-W. Frontier of India, 1897-98.—Operations on the Samana and in the Kurram Valley during August and September 1897.

Tirah, 1897-98.—Actions of the Sampagha and Arhanga Passes. Reconnaissance of the Saran Sar and action of the 9th November 1897. Actions at and around Dwatoi, and action of 24th November 1897. Operations against the Khan Khel Chamkanis. Operations in the Bazar Valley, 25th to 30th December 1897. Medal with 3 clasps.

N.-W. Frontier of India, Waziristan, 1901-02.—Clasp.

N.-W. Frontier of India, 1902.—Operations against the Darwesh Khel Wazirs.

N.-W. Frontier of India, 1908—Operations in the Mohmand country—Engagement of Matta, 24th April 1908. Engagement of Kargha, 24th May 1908—Medal with clasp.

Muhammad Ali—

China, 1900.—Relief of Pekin.—Medal with clasp.

Ghulam Muhammad—

N.-W. Frontier of India, 1897-98.—Operations on the Samana and in the Kurram Valley during August and September 1897.—Medal with 2 clasps.

Sham Singh—

Tibet, 1903-04—Medal.

Mehar Singh—

Soudan, 1896.—Dongola Expedition, 2 medals.

N.-W. Frontier of India, 1897-98.—Mohmand. Medal with clasp.

Karamdin—

N.-W. Frontier of India, 1897-98.—Tochi Medal with clasp.

JEMADARS—

Dewak Ram—

China, 1900.—Medal.

Sohaj Khan—

China, 1900-01.—Medal.

Narain Singh—

Tirah, 1897-98.—Actions of the Sampagha and Arhanga Passes. Reconnaissance of the Saran Sar and action of the 9th November 1897. Actions at and around Dwatoi, and action of 24th November 1897. Operations against the Khan Khel Chamkanis. Operations in the Bazar Valley, 28th and 30th December 1897.—Medal with 3 clasps.

88th Carnatic Infantry.

SUBADARS—

Abdul Ghafur—

N.-E Frontier of India, Chin Hills, 1892-93.—Medal with clasp.

N.-W. Frontier of India, 1897-98.—Operations on the Samana and in the Kurram Valley during August and September 1897.—Medal with 2 clasps.

Tirah, 1897-98.—Actions of Chagru Kotal and Dargai. Operations in the Bazar Valley, 25th to 30th December 1897.—Clasp.

Muttusami (III)—
China, 1900.—Medal.

Ellayya—
China, 1900.—Medal.

JEMADARS—

Mohiyuddin Khan—
China, 1900.—Medal.

Francis Anthoney—
China, 1900.—Medal.

Kuppusami—
China, 1900.—Medal.

Kalingaraju—
China, 1900.—Medal.

V. Sabagtullah—
China, 1900.—Medal.

Kuppusami—
China, 1900.—Medal.

Chinnasami—
China, 1900.—Medal.

Murugesam—
China, 1900.—Medal.

2-88th Carnatic Infantry.

SUBADAR-MAJOR—

Muhamad Sheriff—
China, 1900—Medal.
East African Expeditionery Force—1914-16.
Battle of Tunga (G. E. A.) 1914.
Capture of Mafia (G. E. A.) 1915.
German East Africa—1915.
Mentioned in Despatches.

1-89th Punjabis.

SUBADARS—

Sobha Singh—
Burma, 1885-89.—Medal with 2 clasps.
N.-E. Frontier of India, Kachin Hills, 1892-93.—Clasp.

2-89th Punjabis.

SUBADAR-MAJOR—

Sher Zaman—
France, 1914-16.

SUBADARS—

Bhan Singh—
China, 1900—Medal.

Mohan Lall—
N.-E. Frontier of India, Chin Hills, 1892-93.—Medal with clasp.

Hazra Singh—
Burma, 1887-89—Medal with clasp.

Jagdeo Dube—
Egypt Dardanelles, France, Mesopotamia, 1914-16.

Nur Khan—
Mesopotamia, 1916-17.

Rhoda Khan—
Mesopotamia, 1916-17.

JEMADARS—

Sewa Singh—
Mesopotamia, 1915-17.

Dyal Singh—
Mesopotamia, 1915-16.

Bahadur Khan—
Egypt, Dardanelles, France, Mesopotamia, 1914-16.

Jai Govind—
Egypt, Dardanelles, France, Mesopotamia, 1914-16.

Fazil Din—
Egypt, Dardanelles, France, Mesopotamia, 1914-16.

Fazal Dad—
Egypt, Dardanelles, France, Mesopotamia, 1914-16.

Accha Singh—
Mesopotamia, 1916-17.

Janki Singh—
Egypt, Dardanelles, France, Mesopotamia, 1914-16.

Sheo Narayan Singh—
Mesopotamia, 1916-17.

Mohamad Khan—
Egypt, Dardanelles, France, Mesopotamia, 1914-16.

90th Punjabis.

SUBADAR-MAJOR—

Ilam Din—*Bahadur.*
N.-E. Frontier of India, Chin Hills, 1903-04.—Mesopotamia, 1916-17—Order of British India 2nd class.

Hazura Singh, Hony. Capt., *Sardar Bahadur*
N.-E. Frontier of India, Burma, 1885-87, 1887, 89.—Medal with 2 clasps.
China, 1900.—Medal.

SUBADARS—

Ratan Singh—
Burma, 1887—89.—Medal with clasp.
N.-E. Frontier of India, Chin Hills, 1892-93.—Clasp.

Harnam Singh—
East Africa, 1902—04.—Operations in Somaliland.—Medal with clasp.
Mesopotamia, 1916.—Indian Distinguished Service Medal.

Narayan Singh—
Mesopotamia, 1916.—Indian Distinguished Service Medal.

Mula Singh—
N.-E. Frontier of India, Chin Hills, 1892-93 and 93-94.—Medal with clasp.
China, 1900.—Medal.
Mesopotamia, 1917—Indian Order of Merit 3rd Class.

Bir Singh—
Mesopotamia, 1916.—Military Cross.

1-91st Punjabis (Light Infantry).

SUBADAR-MAJOR—

Kalandar Khan—
Soudan, 1898.—Operations against the Ogaden Somalis.—Medal with clasp.
N.-E. Frontier of India, Chin Hills, 1894-95-96.
E. Africa, Nandi Punitive Expedition, 1900.—Medal with clasp.
East Africa, 1903.—Mentioned in despatches.
Somaliland Expedition, 1908-1910.—Medal. Wounded in action by arrows.

SUBADARS—

Bijai Singh—
Burma, 1885—87—1887—89.—Medal with clasps.
Burma, 1891.—Wuntho Expedition.—Clasp.
N.-E. Frontier of India, Chin Hills, 1892-93.—Clasp.
China, 1900.—Medal.

Thakur Singh—
N.-E. Frontier of India, Kachin Hills, 1892-93.—Medal with clasp.
N.-E. Frontier of India, Chin Hills, 1893—96.
China, 1900.—Medal.

Warriam Singh—
N.-E. Frontier of India, Chin Hills, 1894-95-96.
China, 1900.—Medal.

Fiazali Khan—
N.-W. Frontier of India, 1897-98.—Medal with clasp.
Tirah, 1897-98.

Khazana—
China, 1900.—Medal.

Saudagar Singh—
China, 1900.—Medal.

Kalu Khan—
China, 1900.—Medal.

Sarju Singh—
China, 1900.—Medal.

Lal Khan—
China, 1900.—Medal.

JEMADARS—

Sundar Singh—
China, 1900.—Medal.

Ghulam Rasul—
China, 1900.—Medal.

Mangal Singh -
China Medal—

Qudir Baksh—
China, 1900.—Medal.

Bishan Singh—
China, 1900.—Medal.

Sher Singh—
China, 1900—Medal.

Ramshakal Rao—
China, 1900—Medal.

2-91st Punjabis.

SUBADAR—

Mirza Khan—
M. E. Force, 1916-18.

Ghulam Mahomad—
M. E. Force, 1916-17.

Ibrahim Khan—
M. E. Force, 1916-18.

Ghulam Rasul—
China Expeditionary Force, 1900-02.
M. E. Force, 1916-18.

Nur Khan—
M. E. Force, 1917-18.

Khan Mohamad—
China Expeditionary Force, 1901-02.
I. E. Force, 1915.
M. E. Force, 1915-16.
M. E. Force, 1916-18.

Sher Yar Khan—
M. E. Force, 1915-16.

JEMADARS—

Mohamad Khan—
M. E. Force, 1915-16.
M. E. Force, 1917-18.

Ahmad Khan—
M. E. Force, 1917-18.

Fakir Mohamad—
M. E. Force, 1916.
M. E. Force, 1917-18.

Mohmood Khan—
M. E. Force, 1916-18.

Abdul Rahman—
M. E. Force, 1916-18.

Mohamad Akbar—
M. E. Force, 1916-18.

92nd Punjabis.

SUBADAR-MAJOR—

Farman Ali—
N.-W. Frontier of India, 1897-98.—*Tochi*—Medal with clasp.

93rd Burma Infantry.

SUBADAR-MAJOR.

Kirpal Singh—
Burma, 1887—89—1889—92.—Medal with 2 clasps.

SUBADARS—

Jawand Singh—*re-employed Indian Order of Merit, 3rd class.*
Burma, 1887—89—1889—92.—Medal with 2 clasps.

94th Russell's Infantry.

SUBADAR-MAJOR—

Duli Chand—
Aden, 1903-04.—Operations in the Interior.

SUBADAR—

Shaikh Muhammad Ali—
Aden, 1903-04.—Operations in the Interior.

JEMADARS—

Sukhdeo—
Aden, 1903-04.—Operations in the Interior.

Shaikh Muhammad Umar—
Aden, 1903-04.—Operations in the Interior.

Ramji Lal—
Aden, 1903-04.—Operations in the Interior.

2-94th Infantry.

SUBADAR-MAJOR—

Shaikh Kalandar—
Aden, 1903-04.—Operations in the Interior.
Swat Valley, 1915.
Persian Gulf, 1916.
Waziristan, Field Force, 1917.

SUBADARS—

Shadi Ram—
China 1900—Medal.

95th Russell's Infantry.

SUBADAR-MAJOR—

Shaikh Ahmadulla—
Tirah, 1897-98.—Medal with clasp.

SUBADARS—

Amjad Khan—
China, 1900.—Relief of Pekin.—Action of Peitsang Action at Yangtsun.—Medal with clasp.

Bhawani Singh—
China, 1900.—Medal.

JEMADARS—

Mohammed Wazir—
Tirah, 1897-98.—Medal with clasp.

Fateh Singh—
China, 1900.—Medal.

2-95th Russells Infantry.

SUBADAR-MAJOR—

Alam Khan—
N. W. Frontier of India, 1914-18.—
Mesopotamia, 1918.—

SUBADARS—

Sukh Deo Singh—
N.-W. Frontier of India, 1914-17.—

Harjiwan Singh—
N.-W. Frontier of India, 1916-17.—
Mesopotamia, 1917-18.—

Hazari Singh—
N.-W. Frontier of India, 1914-16.—
Mesopotamia, 1916-18.—

Jagram Singh—
N.-W. Frontier of India, 1913-16.—
Mesopotamia, 1916-18.—

Sewa Ram—
N.-W. Frontier of India, 1913-16.—
Mesopotamia, 1916-18.—

JEMADARS—

Nihal Singh—
N.-W. Frontier of India, 1916-17.—

Mahomed Shafi Khan—
N.-W. Frontier of India, 1914-16.—
Mesopotamia, 1917-18.—

Amrit Singh—
N.-W. Frontier of India, 1913-16.—
Mesopotamia, 1916-18.—

Shiudas Singh—
N.-W. Frontier of India, 1914-16.—
Mesopotamia, 1916-18.—

Mehar Chand—
N.-W. Frontier of India, 1913-16.—
Mesopotamia, 1916-18.—

Latif Khan—
N.-W. Frontier of India, 1914-16.—
Mesopotamia, 1917-18.—

Deokaran Singh—
N.-W. Frontier of India, 1913-16.—
Mesopotamia, 1916-18.—

Ranjit Singh—
N.-W. Frontier of India, 1913-17.—
Mesopotamia, 1917-18.—

Mahomed Ishaq—
N.-W. Frontier of India, 1914-17.—
Mesopotamia, 1917-18.—

Ghan Shyam—
N.-W. Frontier of India, 1913-17.—
Mesopotamia, 1917-18—.

2-96th Infantry.

SUBADAR—

Nohar Singh—
China, 1900.—Medal.

2-97th Deccan Infantry.

SUB-MAJOR -

Mohamed Ayub Ali Khan—
Mesopotamia, 1915–17.—Twice wounded.
Palestine, 1918.—

SUBEDARS—

Hardhan Singh—
Mesopotamia, 1917.—

Mirza Wilayat Beg.—
Mesopotamia, 1917.—

JEMADARS—

Phekn Singh—
East Africa, 1914.—

Gulzar Singh—
Mesopotamia, 1916-17.—

Ramsarup—
Mesopotamia, 1917.—

Shaeik Abdul Razak—
Mesopotamia, 1916-17.—
Palestine, 1918.—

1-98th Infantry.

SUBADAR-MAJOR—

Jiya Ram—
China, 1900.—Medal.

SUBADARS—

Shaikh Shahamat Ali—
Tirah, 1897-98.—Medal with clasp.
China, 1900.—Medal.

Gopal—
China, 1900—Medal.

Maha Singh—
China, 1900.—Medal.

JEMADARS—

Faizu Khan—
China, 1900.—Medal.

Abdul Rashid—
China, 1900.—Medal.

2-98th Infantry.

SUBADAR-MAJOR—

Bhagwan Singh—
Mesopotamia, 1915-16.—

SUBADARS—

Abdul Rahman Khan—
East Africa, 1914—17.—

Bansgopal Tewari—
China, 1900—02.—

Gumani Ram—
Tirah, 1897-98.—
China, 1900—02.—
Persian Gulf, 1914-15.—
N.-W. Frontier, 1916.—

Nathoo Ram—
Mesopotamia, 1915-16.—

Hanmant Singh—
Mesopotamia, 1915—17.—

Sheikh Ummed—
East Africa, 1914—17.—

Ram Ratan Singh—
Mesopotamia, 1915-16.—

JAMADARS—

Madho Singh—
Mesopotamia, 1915—18.—

Ferzand Khan—
East Africa, 1914—17.—

Gazadhar Singh—
East Africa, 1914—17.—

Rameshar Singh—
East Africa, 1914—17.—

Kamor Singh—
East Africa, 1915—17.—

Sheoji Singh—
East Africa, 1914—17.—

Dalu Singh—
East Africa, 1915—17.—

Govind Singh—
Mesopotamia, 1916-17.—

Bhagwat Singh—
Mesopotamia, 1915-16.—

99th Deccan Infantry.

SUBADAR-MAJOR—

SUBADAR—

Talib Hussain Khan—
Chitral, 1895.—Relief of Chitral.—Medal with clasp.

JEMADAR—

Sada Sukh—
China, 1900.—Medal.

The 101st Grenadiers.

SUBADAR-MAJOR—

Martand Rao Mohite—
Soudan (Suakin), 1896.—Medals (Queen's and Khedive's).

SUBADARS—

Ahmed Din—
East Africa, 1902—04.—Operations in Somaliland.—Medal with clasp.

Bare Khan—
East Africa, 1902-03.—Operations in Somaliland —Medal with clasp.

Daji Sawant—
East Africa, 1903-04.—Operations in Somaliland. —Medal with clasp.

Nur Khan—
East Africa, 1902—04.—Operations in Somaliland. —Medal with clasp.

Wazir Khan—
East Africa, 1902—04.—Operations in Somaliland.—Medal with clasp.

Murar-rao Salunke—
East Africa, 1902—04.—Operations in Somaliland—Medal with clasp.

Rahim Khan—
East Africa, 1902—04.—Operations in Somaliland —Medal with clasp.

JEMADARS—

Ghulam Mahammad—
East Africa, 1903.—Operations in Somaliland—Medal with clasp.
Aden, 1903-04.—Operations in the Interior.

Sakharam Kalbhar—
East Africa, 1902—04.—Operations in Somaliland.—Medal with clasp.

Sakharam Gaikwad—
East Africa, 1902—04.—Operations in Somaliland. —Medal with clasp.

Khaju Khan—
East Africa, 1904.—Operations in Somaliland.—Medal with clasp.

Manowar Khan—
Aden, 1903.—Operations in the Interior.
East Africa, 1903-04.—Operatoins in Somaliland. —Medal with clasp.

Hussain Baksh—
East Africa, 1902—04.—Operations in Somaliland —Medal with clasp.

2-101st Grenadiers.

SUBADARS—

Shaik Abdul Sulaiman—
East Africa, 1902—04.—Operations in Somaliland —Medal with clasp.

Allah Ditta Khan—
N.-W. Frontier of India, Waziristan, 1901-02.—Medal with clasp.
East Africa, 1903-04.—Operations in Somaliland. —Medal with clasp.

Ramalrao Kadam—
East Africa, 1902—04.—Operations in Somaliland.—Medal with clasp.

Kondaji Ghanwat—
East Africa, 1902—04.—Operations in Somaliland. —Medal with clasp.

JEMADAR—

Wahed Ali—
Aden, 1903.—Operations in the Interior.
East Africa, 1903-04.

102nd King Edward's Own Grenadiers.

SUBADAR-MAJOR—

SUBADARS—

Juglal Singh—

Aden, 1903-04.—Operations in the Interior.

Jafar Ali—I.O.M.

British East Africa, 1896.—Medal.
British East Africa, 1898.—Uganda Relief Expedition. Action at Jass Camp and Mruli.—Medal, with 2 clasps.

Bhagwat Singh—

Aden, 1903-04.—Operations in the Interior.

Tota Singh—

Aden, 1903-04.—Operations in the Interior.

Girdhari Singh—

Aden, 1903-04.—Operations in the Interior.

Ganga Ram Singh—

Aden, 1903-04.—Operations in the Interior.

Yar Muhammad—

Aden, 1903-04.—Operations in the Interior.

JEMADARS—

Nadir Khan—

Aden, 1903-04.—Operations in the Interior.

Abraham Daniel—

Aden, 1903-04.—Operations in the Interior.

Roshan Khan—

Somaliland Expedition, 1908-10—Medal with clasp.

Sahiram Singh—

Aden, 1903-04.—Operations in the Interior.

Shiudan Singh—

Aden, 1903-04.—Operations in the Interior.

Chandan Singh—

Aden, 1903-04.—Operations in the Interior.

Tunda Singh—

Aden, 1903-04.—Operations in the Interior.

Rup Singh—

Aden, 1903-04.—Operations in the Interoir.

2-102nd K. E. O. Grenadiers.

SUBADAR-MAJOR—

Ganga Ram—I. O. M., I. D. S. M.

Persian Gulf, 1913—15.—
Mesopotamia, 1915—18.—

Awarded :—I. D. S. M.—19-3-15—I. O. M. 2nd Class 17-3-17.

SUBADARS—

Mangal Singh—

Persian Gulf, 1913—15.—
Mesopotamia, 1915-16.—Wounded 7-1-16.
Mesopotamia, 1917-18.—

Imam Din—

Persian Gulf, 1913—15.—
Mesopotamia, 1915-16.—
Mesopotamia, 1916-17.—
Mesopotamia, 1917-18.—

Kanhiya Singh—

Persian Gulf, 1913—15.—
Mesopotamia, 1915—17.—

Fateh Khan—

Mesopotamia, 1915-16.—

Raghbir Singh—

Persian Gulf, 1913—1915.—
Mesopotamia, 1915-16.—Wounded 1-12-15.—

Ram Diyal Singh—

Persian Gulf, 1913—15.—
Mesopotamia, 1915-18.—Awarded I. D. S. M., 17-3-17.

JEMADARS—

Burhan Ali—

Persian Gulf, 1913-15.—Highly commended for good work, at Muscat, 11-1-15.
Mesopotamia, 1916.—
Mesopotamia, 1917-18.—

Balla Ram—

Persian Gulf, 1913—15.—
Mesopotamia, 1915-16.—

Kumha Ram—

Mesopotamia, 1917-18.—

Jhasi Singh—

Persian Gulf, 1913—15.—
Mesopotamia, 1915—18.—

Kamal Khan—

Persian Gulf, 1913—15.—
Mesopotamia, 1915—17.—Wounded 7-1-16 and 12-2-17.

Chakarpan Singh—

Persian Gulf, 1914-15.—

Karam Ali Khan—

Mesopotamia, 1916.—

Shafait Ali—

Persian Gulf, 1913—15.—
Highly commended Muscat, 11-1-15—
Mesopotamia, 1916.—
Mesopotamia.—*1917-18.*—

103rd Mahratta Light Infantry.

SUBADAR-MAJOR—

Shaikh Imam, *Khan Sahib Sardar Bahadur.*—
Order of British India, 1st class.
N.-W. Frontier of India, 1897-98.—Malakand—Buner. Attack and capture of the Tanga Pass.—Medal with clasp.

Bapurao Gaekwad—*Bahadur*, I.D.S.M.

N.-W. Frontier of India, 1897-98.—Malakand—Buner. Attack and capture of the Tanga Pass.—Medal with clasp.

SUBADARS—

Deoji Rao Bhosle—

N.-W. Frontier of India, 1897-98.—Malakand—Buner. Attack and capture of the Tanga Pass.—Medal with clasp.

Shaikh Abdul Kadir—

N.-W. Frontier of India, 1897-98.—Malakand—Buner. Attack and capture of the Tanga Pass.—Medal with clasp.

Eshwant Bhonsle—

N.-W. Frontier of India, 1897-98.—Malakand—Buner. Attack and capture of the Tanga Pass.—Medal with clasp.

Krishna Ghatge—

N.-W. Frontier of India, 1897-98.—Malakand—Buner. Attack and capture of the Tanga Pass.—Medal with clasp.

Sripati Kadam—

N.-W. Frontier of India, 1897-98.—Malakand—Buner. Attack and capture of the Tanga Pass.—Medal with clasp.

Ilyasdar Khan—

South Africa, 1900-01.—Queen's Medal.

helkh Jaffar—

N.-W. Frontier of India, 1897-98.—Malakand—Buner.—Attack and capture of the Tanga Pass.—Medal with clasp.

Khandu Ghadge—

N.-W. Frontier of India, 1897-98.—Malakand Buner.—Attack and capture of the Tanga Pass.—Medal with clasp.

2-103rd Mahratta Light Infantry.

SUBADAR-MAJOR—

Yeshwantarao Bhosle—

Mesopotamia, 1914-15—Awarded D. S. M.

SUBADARS—

Jai Singrao Sinde—

Mesopotamia, 1914—16.—Shakharam Nikane.

Mesopotamia, 1914—16.—

JEMADARS—

Ram Chandar Rao More—

Mesopotamia, 1915-16.—

Ragho Parab—

Mesopotamia, 1915-16.—

104th Wellesley's Rifles.

SUBADAR-MAJOR (*Tempy.*)—

Rahim Baksh—

British East Africa, 1898.—Operations against the Ogaden Somalis.—Medal with clasp.

SUBADARS—

Sabal Singh, I.O.M.—

British East Africa, 1898.—Operations against the Ogaden Somalis.—Medal with clasp.

Harchand Jat—

British East Africa, 1898.—Operations against the Ogaden Somalis.—Medal with clasp.

Karam Dad—

British East Africa, 1898.—Operations against the Ogaden Somalis.—Medal with clasp.

Muhammad Khan—

British East Africa, 1898.—Operations against the Ogaden Somalis.—Medal with clasp.

JEMADARS—

SHARAF KHAN—

British East Africa, 1898.—Operations against the Ogaden Somalis.—Medal with clasp.

GULAM MUHAMMAD—

British East Africa, 1898.—Operations against the Ogaden Somalis.—Medal with clasp.

Shiuram Sajjan—

British East Africa, 1898—Operations against the Ogaden Somaio—Medal with clasp.

105th Mahratta Light Infantry.

SUBADAR—

Chimaji Garud—

N.-W. Frontier of India, 1897-98.—Medal with 2 clasps.

106th Hazara Pioneers.

SUBADAR-MAJOR—

SUBADARS—

Gird Ali—

Br. East Africa, 1896.—Operations against the Mazrui rebels.—Medal.

China, 1900.—Medal with clasp.

Ali Dost—*Indian Distinguished Service Medal*—

Br. East Africa, 1896.—Operations against the Mazrui rebels.—Medal.

China, 1900.—Medal with clasp.

Sharif—

British East Africa, 1896.—Operations against the Mazrui rebels.—Medal.

China, 1900.—Medal with clasp.

Muhammad Hasan (I)—*Indian Distinguished Medal.*

China, 1900.—Medal with clasp.

Muhammad Hasan (II)—

China, 1900.—Medal with clasp.

Qurban Ali—

China, 1900.—Medal with clasp.

Shafa—

British East Africa, 1896.—Operations against the Mazria rebels.—Medal.

China, 1900.—Medal with clasp.

Ali Juma—*Indian Distinguished Medal.*

China, 1900.—Medal with clasp.

JEMADARS—

Naji Bulla—

British East Africa, 1896.—Operations against Mazria rebels.—Medal.

107th Pioneers.

SUBADAR-MAJOR—

Labh Singh—*Bahadur*, I.D.S.M.

East Africa, 1902—04.—Operations in Somaliland.—Medal with clasp.

Sharam Singh—(*retd.*)

East Africa, 1902—04.—Operations in Somandlila.—Medal with clasp.

SUBADARS—

Babaji Sivaji—
N.-W. Frontier of India, 1897-98.—Operations on the Samana and in the Kurram Valley during August and September 1897. Operations of the Flying column in the Kurram Valley under Col. Richardson —Medal with 2 clasps.
Mekran, 1901.—Attack and capture of Nediz Fort.

Sundar Singh—
British East Africa, 1897-98.—Operations against the Ogaden Somalis.—Medal with clasp.
East Africa, 1902—04.—Operations in Somaliland. —Medal with clasp.

Fateh Muhammad Khan—
East Africa, 1902—04.—Operations in Somaliland. —Medal with clasp.

Hari Singh—
East Africa, 1902—04.— Operations in Somaliland.—Medal with clasp.

JEMADARS—

Lehna Singh—
East Africa, 1902—04.— Operations in Somaliland. —Medal with clasp.

Abdul Rahman—
East Africa, 1902 - 04.—Operations in Somaliland. —Medal and clasp.

Gani Khan—
East Africa, 1902—04.—Operations in Somaliland. —Medal with clasp.

Dial Singh—
East Africa, 1902 - 04.—Operations in Somaliland. —Medal with clasp.

Nanak Singh—
East Africa, 1902—04.—Operations in Somaliland. —Medal with clasp.

Habib Khan—
East Africa, 1902—04.—Operations in Somaliland. —Medal with clasp.

Karim Dad Khan—
East Africa, 1902—04—Operations in Somaliland—Medal with clasp.

Kishan Singh—
East Africa, 1902-04.—Operations in Somaliland.—Medal with clasp.

Samandar Khan—
East Africa, 1902—04.—Operations in Somaliland. Medal with clasp.

2-107th Pioneers.

SUBADARS—

Rajab Khan—
East Africa, 1902-04.—Operations in Somaliland. —Medal with clasp.

Karam Dad—
Tochi Field Force, 1897-98.
Mesopotamia, War, 1914.
Field Force (with 1st Labour Corps).

JEMADAR—

Bishan Singh—
Relief of Chitral, 1895.
Punjab Frontier and Tirah, 1897-98.
China, 1900.
N.-W. Frontier, 1908.
Mesopotamia, War, 1914.
Field Force (with 1st Labour Corps).

Biru Ram—
Somaliland 1902—04.
Mesopotamia and France F. F. War 1914.

108th Infantry.

SUBADARS—

Niaz Ali—
East Africa, 1903-04.—Operations in Somaliland —Medal and clasp.
South Africa, 1901-02.—2 Medals - 3 clasps.

Lakshman Rao Sinde—
Soudan, 1896.—Dongola Expedition.—2 Medals. (Queen's and Khedive's.)

Rao Jadhav—
Tirah, 1897-98.—Medal with 2 clasps.

JEMADARS—

Jogi Naik—
East Africa, 1902—04.—Operations in Somaliland. —Medal with clasp.

Antu Naik—
East Africa, 1902—04.—Operations in Somaliland. —Medal with clasp.

109th Infantry.

SUBADAR-MAJOR—

Lakshiman Kadam—
N.-W. Frontier of India, Waziristan, 1901-02.— Medal with clasp.

SUBADARS—

Amar Ali Khan—
N.-W. Frontier of India, Waziristan, 1901-02.— Medal with clasp.
East Africa, 1902—04.—Operations in Somaliland. —Medal with clasp.

Muhammad Baksh—
N.-W. Frontier of India, Waziristan, 1901-02.— Medal with clasp.

Amrutrao Ghorpade—
N.-W. Frontier of India, Waziristan, 1901-02.— Medal with clasp.

Abbas Khan—
N.-W. Frontier of India, Waziristan, 1901-02.— Medal with clasp.
East Africa, 1902—04.—Operations in Somaliland. —Medal with clasp.

Chhote Khan—
N. W. Frontier of India, Waziristan, 1901-02.— —Medal with clasp.
East Africa, 1902—04.—Operations in Somaliland. —Medal with clasp.

JEMADARS—

Rahimatullah Khan—
N.-W. Frontier of India, Waziristan, 1901-02.— Medal with clasp.
East Africa, 1902-04.—Operations in Somaliland. —Medal with clasp.

Nadir Khan—
N.-W. Frontier of India, Waziristan, 1901-02.— Medal with clasp.
East Africa, 1902—04.—Operations in Somaliland. —Medal with clasp.

Narayan Dhuri—
N.-W. Frontier of India, Waziristan, 1901-02.— Medal with clasp.

2-109th Infantry.

SUBADARS—

Ibrahim Khan—
N.-W. Frontier of India, Waziristan, 1901-02.—Medal with clasp.
East Africa, 1902-04.—Operations in Somaliland.—Medal with clasp.
War, 1915-16—Aden.

Yashwantrao Powar—
War, 1915-16—Aden.

Dattaram Sawant—
War, 1915-16—Aden.

Ahmed Khan—
N.-W. Frontier of India, Waziristan, 1901-02.—Medal with clasp.
East Africa, 1902-04.—Operations in Somaliland.—Medal with clasp.
War, 1915-16—Aden.

Abdul Rahim Khan—
N.-W. Frontier of India, Waziristan, 1901-02.—Medal with clasp.

Umar Baksh—
War, 1915-16—Aden.

Govindrao Sinde—
War 1915-16—Aden.

JEMADARS—

Hassan Muhammad—
N.-W. Frontier of India, Waziristan, 1901-02.—Medal with clasp.

Shivram Naik—
War, 1915-16—Aden.

Bala Parte—
War, 1915-16—Aden.

Ganu Powar—
War, 1915-16—Aden.

Ilam Din—
War, 1915-16—Aden.

Murad Ali—
War, 1915-16—Aden.

111 Mahars.

SUBADAR—

Muhammad Ghouse—
Burma 1890.

Muhammad Yusuf—
Aden,—1915—1917.

Abdul Latif—
China, 1900.

Hiru Ramji—
Tirah Expdy., 1897.
Samana, 1898.
Punjab Frontier, 1899.
Mesopotamia, 1915—1917.

JEMADARS—

Ramji Sajanaji—
Khandhar, 1879-1880.

Namdeo Krishnaji—
Aden, 1903-1904—Operation in the interior.

112th Infantry.

SUBADARS—

Mirchi Singh—
N.-E. Frontier of India, Manipur, 1891.
Abor Expedition, 1894.
Chitral, 1895.—Relief of Chitral.—Medal with clasp.

Rajwali Khan—
Somaliland, G. S. 1909-10.—Medal with clasp.

1-113th Infantry.

SUBADAR-MAJOR—

Kana, Ram—
East Africa, 1908—10.—Medal with clasp.

SUBADARS—

Ganeshi Ram—
East Africa, 1908—10.—Medal with clasp.

Brijlal Singh—
East Africa, 1908—10.—Medal with clasp.

Shama Ram—
East Africa, 1908—10.—Medal with clasp.

JEMADAR—

Pat Ram—
East Africa, 1908—10.—Medal with clasp.

Rasal—
East Africa, 1908—10.—Medal with clasp.

Mat Ram—
East Africa, 1908—10.—Medal with clasp.

2-113th Infantry.

SUBADARS—

Dani Ram—
N.-W. Frontier of India, 1897-98.—Medal with clasp.

Solu Singh—
East Africa, 1908—10.—Medal with clasp.

JEMADARS—

Risal—
East Africa, 1908—10.—Medal with clasp.

Mana Ram—
East Africa, 1908—10.—Medal with clasp.

Naini—
East Africa, 1908—10.—Medal with clasp.

Ganeshi Ram—
East Africa, 1908—10.—Medal with clasp.

Imam Din—
East Africa, 1908—10.—Medal with clasp.

Inoman Ram—
East Africa, 1908—10.—Medal with clasps.

114th Mahrattas.

SUBADAR-MAJOR—

Mahadeorao Khanvilkar—

War, 1914.—Overseas. 19 5—18—N.-W. F. of India, 1915.

SUBADARS—

Dajirao Pawar—

War, 1914.—Overseas, 1915—16.

Shaikh Ahmed—

War, 1914.—Overseas, 1915—18—N.-W. F. of India, 19 5.

Sakharam Pawar—

War, 1914.—Overseas, 1915—18.

Shaikh Abdul—

War, 1914.—Overseas, 1915—18—N.-W. F. of India, 19 5.

Dhanaji Khanvilkar—

War, 1914.—Overseas, 1915—18—N.-W. F. of India, 1915.

Ramehandra Bhosle—

War, 1914.—Overseas, 1915—18.

Dattaram Chavan—

War, 1914.—Overseas, 1914—18.

Balajirao Ghag—

War, 1914.—Overseas, 1917—18.

JEMADARS—

Krishna Holkar—

War, 1914.—Overseas, 1915—18.

Shaikh Aminuddin—

War, 1914.—Overseas, 198.

Gapal Dhanote—

War, 1914.—Overseas ,1914—16.

Shaikh Gul Mohmad—

War, 1914.—Overseas, 1,15—18—N.-W. F. of India, 1915.

Krishna Gaode—

War, 1914.—Overseas, 1914—16—1918.

Shaikh Bandu—

War, 1914.—Overseas, 1914—18.

Krishna Wejre—

War, 1914.—Overseas, 1915—18.

Shaikh Mohmad Khalil—

War, 1914.—Overseas, 1915—18—N. W. F. of India, 1918.

Pandu Naik—

War, 1914.—Overseas, 1915—18—N. W. F. of India, 1915.

Bala Gaade—

War, 1914.—Overseas, 1915—18.

Dhondi Savant—

War, 1914.—Overseas, 1915—18.

Balwant Nholte—

War, 1914.—Overseas, 1915—18.

Ramchandra Chavan—

War, 1914.—Overseas, 1 15—18—N.-W. F. of India, 1915.

Dadu Ingle—

War, 1914.—Overseas, 1915—16.

Wasudeo Rane—

War, 1914.—Overseas, 1915—18.

116th Mahrattas.

SUBADARS—

Mahadorao Sondkar—

Soudan, 1885.—Suakin.—Medal with clasp. Bronze star.
N.-E. Frontier of India, Chin-Lushai, 1889-90.—Medal with clasp.
N.-W. Frontier of India, 1897-98.—Mohmand.—Medal with clasp.
Tirah, 1897-98.—Capture of the Sampagha Pass. Operations at and around Dwatoi and action of 24th November 1897. Operations in the Bara Valley, 7th to 14th December 1897.—Clasp.
Jubaland, 1901.—Medal with clasp.

Kamaruddin—

Jubaland, 1901.—Medal with clasp.

Daji Sinde,—

Jubaland, 1901.—Medal with clasp.

Syed Dastagir—

East Africa, 1902-04.—Operation in Somaliland. Action at Jidballi.—Medal with 2 clasps.

Ramchander Rao Powar—

Jubaland, 1901.—Medal with clasp.

JEMADARS—

Sheikh Mohomed Hussain—

Jubaland, 1901.—Medal with clasp.
East Africa, 1902-04.—Operation in Somaliland. Action at Jidballi.—2 clasps.

117th Mahrattas.

SUBADAR—

Murarirao Sinde—

Somaliland, 1904.—Medal with 2 clasps.

2-117th Mahrattas.

JEMADAR—

Lakhshiman Bhagat—

China, 1900-01.—Medal with clasp.

2-119th Infantry (The Mooltan Regt.)

SUBADAR-MAJOR—

Moti Rawat—

Mesopotamia, 1916-17.

SUBADARS—

Chander Singh—

Mesopotamia, 1917.

Sardara—

Mesopotamia, 1916-17.

Wali Mohammad.

JEMADARS—

Lal Singh—

Mesopotamia, 1916-17.

Khurshed Khan—

Mesopotamia, 1914-17.

Ghulam Ali Khan—

Mesopotamia, 1916-17.

Panna Rawat—

Mesopotamia, 1916.

Sheo Bakhsh—

Mesopotamia, 1914-16.

Ram Jiwan—
Mesopotamia, 1916-17.

Moti Singh—
Mesopotamia, 1916-17.

Umar Ali Khan—
East Africa, 1914-17.

Deda Ram—
Mesopotamia, 1916-17.

Chajju Ram—
Mesopotamia, 1916-17.

Farzand Khan—
East Africa, 1914-17.

Ram Lal—
Mesopotamia, 1916-17.

Jai Mohammad—
Mesopotamia, 1916-17.

120th Rajputana Infantry.

SUBADAR-MAJOR—

Khitab Gul, *Sardar, and I.O.M. 2nd class Bahadur.—Order of Br. India, 1 class.*
China, 1900.—Medal.

121st Pioneers.

SUBADAR-MAJOR—

Sardar Khan—
N.-W. Frontier of India, Isazai, 1892.
Chitral, 1895.—Relief of Chitral—Medal with clasp.
N.-W. Frontier of India, 1897-98. Tochi.—Clasp.

SUBADARS—

Faiel Khan—
N.-W. Frontier of India, 1908—Operations in the Mohmand country.—Medal with clasp.

Gendajat—
East Africa, 1902—04.—Medal with clasp.

Ruda Ram—
East Africa, 1902—04.—Medal with clasp.

Kheta Ram—
East Africa, 1902-04.—Medal with clasp.

Daula Ram—
East Africa, 1902-04.—Medal with clasp.

JEMADARS—

Manulla—
N.-W. Frontier of India, 1897-98—Mohmand. Medal with clasp.
Tirah, 1897-98—Action with Sampagha Pass. Operations at and around Dwatoi and action of 27th November 1897. Operations in the Bara Valley, 7th November to 14th December 1897.—Clasp.

Gharsi Ram—
East Africa, 1902-04.—Medal with clasp.

Tirka Singh—
East Africa, 1902-04.—Medal with clasp.

Narayan Singh—
East Africa, 1902-04.—Medal with clasp.

Magha Ram—
East Africa, 1902-04.—Medal with clasp.

Sewa Ram—
East Africa, 1902-04.—Medal with clasp.

122nd Rajputana Infantry.

SUBADAR-MAJOR—

Shahbaz Khan (*re-employed*)—
N.-W. Fontier of India, 1897-98.
Tirah, 1897.—
China, 1900.—Medal.

SUBADARS—

Behara Rawat—
China, 1900.—Medal.

Pyar—
East Africa, 1902—04.—Operations in Somaliland, Action of Sadbalti—Medal with 2 clasps.

Gugan Ram—
China, 1900.—Medal.

Hira Rawat—
China, 1900.—Medal.

Atbal Singh—
China, 1900.—Medal.

Kutbi khan—
China, 1900.—Medal.

Sheonath—
China, 1900.—Medal.

Teja Rawat—
China, 1900.—Medal.

Net Ram—
China, 1900.—Medal

JEMADARS—

Hari Lakshman—
China, 1900.—Medal.

123rd Outram's Rifles.

SUBADAR-MAJORS—

SUBADARS—

Bhura Ram—
N.-W. Frontier of India, Waziristan, 1901-02.—Medal with clasp.
Aden, 1903-04.—Operations in the Interior.

Nizam-ud-din—
N.-W. Frontier of India, Waziristan, 1901-02.—Medal with clasp.

Malooka Ram—
Aden, 1903-04.—Operations in the Interior.

Bagh Singh—
Aden, 1903-04.—Operations in the Interior.

Ume Singh—
N.-W. Frontier of India, Waziristan, 1901-02.—Medal with clasp.
Aden, 1903-04.—Operations in the interior.

Burhan Ali—
N.-W. Frontier of India, Waziristan, 1901-02.—Medal and clasp.
Aden, 1903-04.—Operations in the Interior.

Hans Ram—
N.-W. Frontier of India, Wasiristan, 1901-02.—Medal and clasp.
Aden, 1903-04.—Operations in the Interior.

Nijabat Ali—
N.-W. Frontier of India, Wasiristan, 1901-02.—Medal and clasp.

Painda Khan—
Aden, 1903-04.— Operations in the Interior

JEMADARS—

Jhuja Ram—
N.-W. Frontier of India, Wasiristan, 1901-02.—Medal with clasp.
Aden, 1903-04.—Operations in the Interior.

Ilam Din—
N.-W. Frontier of India, Wasiristan, 1901-02.—Medal with clasp.
Aden, 1903-04.—Operations in the Interior.

Bhola Ram—
N.-W. Frontier of India, Wasiristan, 1901-02.—Medal with clasp.
Aden, 1903-04.—Operations in the Interior.

Harnath Singh—
Aden, 1903-04.—Operations in the Interior.

Bostan Khan—
N.-W. Frontier of India, Wasiristan, 1901-02.—Medal with clasp.
Aden, 1903-04.—Operations in the Interior.

Arjan Ram—
Aden, 1903-04.—Operations in the Interior.

Ramkaran Ram—
Aden, 1903-04.—Operations in the Interior.

Partab Singh
N.-W. Frontier of India, Wasiristan, 1901-02.—Medal with clasp.
Aden, 1903-04.—Operations in the Interior.

Sukh Singh—
Aden, 1903-04.—Operations in the Interior.

Nop Singh—
N.-W. Frontier of India, Wasiristan, 1901-02.—Medal and clasp.
Aden, 1903-04.—Operations in the Interior.

Dana Ram—
N.-W. Frontier of India, Wasiristan, 1901-02.—Medal with clasp.
Aden, 1903-04.—Operations in the Interior.

Karam Dad—
N.-W. Frontier of India, Wasiristan, 1901-02.—Medal and clasp.
Aden, 1903-04.—Operations in the Interior.

Zaida Khan—
N.-W. Frontier of India, Wasiristan, 1901-02.—Medal and clasp.
Aden, 1903-04.—Operations in the Interior.

Fazaldad Khan—
N.-W. Frontier of India, Wasiristan, 1901-02.—Medal and clasp.
Aden, 1903-04.—Operations in the Interior.

2—123rd Outram's Rifle.

SUBADAR-MAJOR—

Jalta Ram—
N.-W. Frontier of India, Wasiristan, 1901-02.—Medal and clasp.
Aden, 1903-04.—Operations in the Interior.

SUBADARS—

Rura Ram—
N.-W. Frontier of India, Wasiristan, 1901-02.—Medal and clasp.
Aden, 1903-04.—Operations in the Interior.

Feroz Khan I—
N.-W. Frontier of India, Wasiristan, 1901-02.—Medal and clasp.
Aden, 1903-04.—Operations in the Interior.

Gobind Singh—
Aden, 1903-04.—Operations in the Interior.

Chandra Ram—
Aden, 1903-04.—Operations in the interior.

Pahallwan Khan—
Aden, 1903-04.—Operations in the Interior.

JAMADARS—

Nazam Khan—
Aden, 1903-04.—Operations in the Interior.

Gidha Ram—
N.-W. Frontier of India, Wasiristan, 1901-02.—Medal and clasp.
Aden, 1903-04.—Operations in the Interior.

Irtoji Rao Patole—
N.-W. Frontier of India, Wasiristan, 1901-02.—Medal and clasp.
Aden, 1903-04.—Operations in the Interior.

1—124th Duchess of Connaught's Own Baluchistan Infantry.

SUBADAR-MAJOR—

Tota Singh—
China, 1900.—Medal with clasp.

SUBADARS—

Muhammad Ali—
British East Africa, 1896.—Operations against Mazrui rebels.—Medal.

Ali Nazar—
British East Africa, 1896.—Operations against the Mazrui rebels.—Medal.

Hassan Khan—
China, 1900.—Medal with clasp.

JEMADARS—

Ghulam Haidar—
China, 1900.—Medal with clasp.

Bichant Singh—

Central Angonoland Expedition, British Central Africa, December, 1900.—Medal with clasps.

2-124th Duchess of Connaught's Own Baluchistan Infantry.

SUBADARS—

Mohammad Ali—

British East Africa, 1896—Operations against Mazrai rebels—Medal.

Hassan Khan—

China, 1900.—Medal with clasp.

Chanda Singh—

Suakin Force, 1896.—Medal.
Usmankhel Expeditionry Force, 1897.—Medal with clasp.
Tirah Expedition Force, 1898.—

Kishan Singh—

Malakand Expedition Force, 1897.—
Tirah Expedition Force, 1897-1898.—

JEMADARS—

Ghulam Haidar—

China, 1900.—Medal with clasp.

Bachint Singh—

Central Augono and Expedition, British. Central Africa, December 1900.—Medal with clasp.

Charan Singh—

China, 1900-01.—Medal with clasp.
Somaliland, 1903-04.—Medal with clasp.

Wasawa Singh—

Somaliland Field Force, 1902-04.—Medal with clasp.

Ali Zaffar—

China, 1900-01.—Medal with clasp.

3-124th Duchess of Connought's Own Baluchistan Infantry.

SUBADAR-MAJOR—

Kishan Singh—

Burma Field Force, 1885-87.—Medal and clasp.
1st and 2nd Miranzai Expedition, 1891.—clasp.
Waziristan Field Force, 1894-95.—clasp.
N.-W. F. Tirah and Mohmand Field Force, 1897-98.—Medal and 2 clasps.

SUBADARS—

Mota Singh—

Field Service China, 1900.—Medal and clasp.

JEMADARS—

Mohd Din—

Operation against the Mazein rebels in B. E. Africa, 1896.—Medal.
Ogaden Expedition, 1898.
Operations against mutineres and rebels Unyoro including the final defeat and capture of Kabarega and Mwanga.—Medal and clasp.
Mandi Punitive Expedition, 1900—

Bhagwan Singh—

Operations in Somaliland Field Force, 1903-1904.—Medal and 2 claps.
Awarded Indian Order of Merit 2nd class.

Diwan Ali—

Miranzai Expedition 1st and 2nd phase.—Medal and clasp.
Waziristan, 1894-95.—clasp.
N.-W. Frontier, 1897-98.—Medal and 3 clasps.
B. E. Africa, 1903.—Kamiblo Expedition.

1-125th Napier's Rifles.

SUBADAR-MAJOR—

Umar Din Bahadur—

Uganda, 1898.—Medal with clasp.

SUBADAR—

Anna Singh—*Indian Distinguished Service Medal.*

JEMADARS—

Hakim Khan—*Indian Distinguished Service Medal.*

Sher Khan, I.O.M.—

Tirkha Ram—*Indian Distinguished Service Medal.*

2-125th Napier Rifles.

SUBADARS—

Bostan Khan—

N.-W. Frontier of India, Waziristan, 1901-02.—Medal with clasp.
Aden, 1903-04.—Operations in the Interior.

126th Baluchistan Infantry.

SUBADAR-MAJOR—

Mul Singh, *Sardar Bahadur—Order of Br. Ind., 1st class.*

N.-E. Frontier of India, Hazara, 1888.—Medal with clasp.
N. E. Frontier of India, Lushai, 1889-90.
N.-W. Frontier of India, 1897-98.—Defence of Malakand. Relief of Chakdara.—Medal with 2 clasps.
China, 1900.—Medal with clasp.

SUBADARS—

Umar Khan—

China, 1900.—Medal with clasp.

Gulzada, *Sirdar Bahadur—Order of Br. India, 1st class.*

N.-W. Frontier of India, Zhob Valley, 1884.
N.-W. Frontier of India, Hazara, 1888.—Medal with clasp.
N.-W. Frontier of India, 1st and 2nd Miranzai, 1891.
N.-W. Frontier of India, Waziristan, 1894-95.—Clasp.
China, 1900.—Medal with clasp.
Tibet, 1904.—Medal.

Labh Singh—

Indian Order of Merit, 2nd class.

N.-W. Frontier of India, 1897-98.—Medal and clasp.
China, 1900.—Medal and clasp.

Ahmed Khan—
China, 1900.—Medal and clasp.

Ali Muhammad—
China, 1900.—Medal with clasp.

JEMADARS—

Said Muhammad—
China, 1900.—Medal and clasp.

Ghulam Shah—
China, 1900.—Medal and clasp.

Tulsa Singh—
China, 1900.—Medal and clasp.

Shah Niaz Hussain—
China, 1900.—Medal and clasp.

Mehr Singh—
China, 1900.—Medal with clasp.

Gheba Khan—
Somaliland, 1903-04.—Medal with clasp.

Pir Bux—
China, 1900.—Medal with clasp.

Santa Singh—
China, 1900—Medal with clasp.

Gurdit Singh—
China, 1900.—Medal with clasp.

1-127th Queen Mary's Own Baluch Light Infantry.

SUBADAR-MAJOR—

Gulab Shah—
Uganda, 1897-99.—Operations against Soudanese Mutineers and Waganda rebels.
Actions of Jerubz Kiwsri and Kijanguti.—Medal with Clasp.
Mekran, 1901—*Attack andcapture of Nodiz Fort.*
British East Africa (Somaliland), 1908—10.—Meial and Clasp.
Waziristan, 1917.—
German East Africa, 1917-18.—

SUBADARS—

Mahammad Gul—
British East Africa (Somaliland), 1908—10.—Medal and clasp.
Seistan Robat, 1907.—
Mekran Mission, 1916-17.—
Waziristan, 1917.—
German East Africa, 1917-18.—

Mirbad Shah I—
France, 1914.—
German East Africa, 1917-18.—

Sirbuland—
East and Central Africa, 1897-99.—Medal and 2 clasps.
Mekran, 1901.—Attack and capture of Nodiz Fort.
British East Africa (Somaliland), 1908—10.—Medal and clasp.
German East Africa 1916.—

Nazar Muhammad—
British East Africa and Uganda, 1897—99.—Medal and clasp.
British East Africa (Somaliland), 1908—10.—Medal and clasp.
Mekran, 1901.—
Waziristan, 1917.—
German East Africa, 1917-18.—

Mirbad Shah II—
German East Africa, 1917-18.—

Muhammad Yusuf—
Mekran, 1901.—Attack and capture of Nodiz Fort.
British East Africa (Somaliland), 1908—10.—Medal and clasp.
Force "D", 1916.—
German East Africa, 1917-18.—

Shaikh Abdul Kadir—
British East Africa (Somaliland), 1908—10.—Medal and clasp.
Mekran Mission, 1916.—
Waziristan, 1917.—
German East Africa, 1917-18.—

Khan Bahadur—
Somaliland Contingent, 1906—12.—
Mekran Mission, 1916-17.—
Waziristan, 1917.—
German East Africa, 1917-18.—

Amir Ali—
France, 1914.—
Waziristan, 1917—
British East Africa (Somaliland), 1908—10.—Medal and clasp.
German East Africa, 1917-18.—

Yar Muhammad—
Robat, 1907.—
Mekran Mission, 1916-17.—
Waziristan, 1917.—
German East Africa, 1917-18.—

Godar Khan—
British East Africa, 1915-16.—

Sahib Alam—
Robat, 1907.—
Somaliland, 1908-10.—Medal and clasp.

JEMADARS—

Asghar Ali—
Mekran, 1901.—Attack and capture of Nodiz Fort.
British East Africa (Somaliland), 1908—10.—Medal and clasp.

Habib Khan—
Robat, 1907.—
Somaliland, 1908-10.—Medal and clasp.
Mesopotamia, 1916-17.—
German East Africa, 1917-18.

Wasil Khan—
Somaliland, 1908-10.—Medal and clasp.
Waziristan, 1917.—
German East Africa, 1917-18.—

Tora Khan—

Somaliland, 1908—10.—Medal and clasp.
France, 1915.—
Waziristan, 1917.—
German East Africa, 1917.—

Issa Khan—

East Africa, 1915-11.—
German East Africa, 1917-18.

Ahmad Khan—

Robat, 1907.—
Somaliland, 1908—10.—Medal and clasp.
Waziristan, 1917.—

Kalandar Khan—

British East Africa, 1897-99.—Medal and clasp.
Somaliland, 1908-10.—Medal and clasp.
Waziristan, 1917.—
German East Africa, 1917-18.—

Ghulam Muhammad—

Persia Fort, 1903.—
Robat, 1907.—
Somaliland, 1908-10.—Medal and clasp.
France, 1914—16.—
Waziristan, 1917.—
German East Africa, 1917-18.

Jahan Khan—

France, 1915.—
German East Africa, 1916-17.—

Nurdad—

British East Africa (Somaliland), 1908—10.—Medal and clasp.
Mekran Mission, 1916.—
Waziristan, 1917.—
German East Africa, 1917-18.—

Jan Gul—

France, 1915.—
Mohmand Expeditionary Force.—
Waziristan, 1917.—
German East Africa, 1917-18.—

Bahadur Khan—

Seistan, 1906-10.—
Mekran, 1916.—
Waziristan, 1917.—
German East Africa, 1917-18.

Shaikh Muhammad—

Waziristan, 1917.—

Said Ahmad—

Force "D" 1916.—
German East Africa, 1917-18.—

Mirzadah—

Mesopotamia, 1915.—
Waziristan, 1917.—

1-128th Pioneers.

SUBADAR-MAJOR—

Krishna Bhosle—*Indian Distinguished Service Medal.*

N.-W. Frontier of India, 1897-98.—Mohmand.—Medal with clasp.
Tirah, 1897-98.—Action of Sampagha Pass. Operations at and around Dwatoi and action of the 24th November 1897. Operations in the Bara Valley, 7th to 14th December 1897.—Clasp.

SUBADARS—

Maroti Ghorpade—

N.-W. Frontier of India, 1897-98.—Mohmand.—Medal with clasp.

Tirah, 1897-98.—Action of Sampagha Pass. Operations at and around Dwatoi, and action of the 24th November 1897. Operations in the Bara Valley, 7th to 14th December 1897.—Clasp.

Gurditt Singh—

N.-W. Frontier of India, Waziristan, 1894—95.—Medal with clasp.
N.-W. Frontier of India, 1897-98.—Tochi.—Medal with clasp.
China, 1900.—Relief of Pekin. Actions of Peitsang and of Yangtsun.—Medal with clasp.

Sulaiman—

N.-W. Frontier of India, 1897-98.—Mohmand.—Medal with clasp.
Tirah, 1897-98.—Action of Sampagha Pass. Operations at and around Dwatoi, and action of the 24th November 1897. Operations in the Bara Valley, 7th to 14th December 1897 (wounded).—Clasp.

Khushal Khan—

N.-W. Frontier of India, Hazara, 1891.—Medal with clasp.

Chitral, 1895.—Relief of Chitral.—Medal with clasp.
N.-W. Frontier of India, 1897-98.—Tochi—Clasp.

Tukaram Jadhu—

N.-W. Frontier of India, 1897-98.—Mohmand.—Medal with clasp.
Tirah, 1897-98.—Action of Sampagha Pass. Operations at and around Dwatoi and action of the 24th November 1897. Operations in the Bara Valley, 7th to 14th December 1897.—Clasp.

Abdul Latif Khan—I.O.M.

East Africa, 1902-04.—Operations in Somaliland.—Medal with clasp.

JEMADARS—

Mahadu Chowdri—

N.-W. Frontier of India, 1897-98.—Mohmand.—Medal with clasp.

Tirah, 1897-98.—Action of Sampgha Pass. Operations at and around Dwatoi and action of the 24th November 1897. Operations in the Bara Valley, 7th to 14th December 1897.—Clasp.

Bijai Bahadur Singh—

N.-W. Frontier of India, 1897-98.—Mohmand.—Medal with clasp.
Tirah, 1897-98.—Action of Sampagha Pass. Operations at and around Dwatoi and action of 24th November 1897. Operations in the Bara Valley, 7th to 14th December 1897.—Clasp.

Ganda Singh—I. O. M.

Chitral, 1895.—Relief of Chitral. Medal with clasp.
N.-W. Frontier, 1897-98.—Operations in the Bara Valley, 7th to 14th December 1897.—2 clasps. Punjab Frontier, 1897-98, and Tirah, 1897-98.
China, 1900.—Medal.

SUBEDAR MAJOR—

Khushal Khan—

N.-W. Frontier of India, Hazara, 1891.—Medal with clasp.
Chitral, 1895—Relief of Chitral.—Medal with clasp.
N.-W. Frontier of India, 1897-98.—Tochi.—Clasp.
Egyptian E. F. 1914 15.
Mesopotamia E. F., 1915-16.

SUBEDARS—

Nishan Ali Khan—

East Africa, 1902-04.—Operations in Somaliland—Medal with clasp.
France. 1914-15.
Mesopotamia E. F. 1915-16.
Indian E. F. R., 1915.
Waziristan Field Service, 1917.
Mari E. F., 1918.

Malle Khan—

Mesopotamia, 1914—18.

Baburao Mohite—

Gubaland, 1901.—Medal with clasp.
France, 1914-15.—Bronze Star. *1914.*
Mesopotamia E. F., 1915-16.
Mari E. F., 1918.

Sitaram Barge—

Mesopotamia E. F., 1916-17.

Jota Singh—

N.-W. Frontier of India, 1897-98.—Operations in Bara Velley.—Medal with 2 clasps.
Punjab Frontier, 1897-98.
Tirah, 1897-98.
East Africa, 1902-04.—Operations in Somali land. Medal with elasp.
France. 1914-15.
Egypt, 1915.
Mesopotamia, 1915-16.

Mal Singh—

Mesopotamia, 1914—16.
Mesopotamia, 1917—18.

Aziz Khan—

Egyptian E. F., 1914-15.
Mesopotamia E. F., 1915—18.

JEMADARS—
Haidar Khan—

Egyptian E. F., 1914-15.
Mesopotamia E. F., 1915—18.

Mohamed Shofi Khan—

France, 1914-15.
Mesopotamia E. F., 1916.
India E. F. 1915.
Waziristan F. F., 1918.
Mari F. F., 1918.

Konde Jagtap—

Egyptian E. F., 1914-15.
Mesopotamia E. F. 1915-16.
Mentioned in Despatches—Awarded Meritorious Service Medal for Service in the Field.

Krishna Waghode—

France, 1914-16.—Bronze Star *1914.*
Mesopotamia E. F., 1916.

Ghena Singh—

Mesopotamia, 1914-15.

1-129th Duke of Connaught's Own Baluchis.

SUBEDAR MAJOR—

Zaman Khan, *Bahadur*, M.C.—
War 1914.—*(France)*—Military Cross & Order of Br. India, 2nd Class.

SUBADARS—

Mir Badshah—

War 1914—*France and East Africa*—I. O. M. 2nd Class.

Nur Khan—

War, 1914—*France and East Africa*—I. D. S. M. and I. O. M. 2nd Class.
War, 1914—*France and East Africa.*

Sikandar Khan—

War, 1914.—*France and East Africa.*

Rajwali Khan—

East Africa (British), 1895-96—Operations against the mazarine rebels. Medal.
Uganda Relief Expedition, 1897-98.—Action of Changuru 8th February 898. Kabegambe 24th February 189. *Jasscamp 26th* April 189 *and* Amruli to May 1898 Medal with 2 clasps.
East Africa War, 1914.

JEMADARS—

Muhammad-Baksh—

British East Africa, 1898—Expedition against Ogaden Somalis.—Medal with clasp.
Uganda, 1899.—Operations against Kabarega.—Clasp.
Uganda, 1900.—Operations against Nandi Tribes—Medal with clasp.
East Africa War, 1914.

Id Mohamed—

War, 1914—.France.

Muhammad Jan—

War, 1914.—*France.*

Bagh Khan—

War 1914—France and East Africa.

Reshmin—

War 1918—France and East Africa.

Ghufran Khan—

War 1914—France and East Africa.

Gaman Khan—

War 1914—East Africa.

Nawab Ali—

War 1914—France.

Majhi Khan—

War 1914—France and East Africa.

Nawab Khan—

War 1914—East Africa.

Mir Zaman—

War 1918—France.

Nur Muhammad—

War 1914—France.

130th King George's Own Baluchis.

(Jacob's Rifles.)

SUBADAR-MAJOR—

Muhammad Khan—*Bahadar—Order of British India, 2nd class.*
N.-W. Frontier of India, Zhob Valley, 1890.
Mekran, 1898.—Action of Gok Parosh.
China, 1900.—Medal.

SUBADARS—

Kudrat Shah—*Bahadar—Order of British India, 2nd class.*
China, 1900.—Medal.

Muhammad Din—
Mekran, 1898.—Action of Gok Parosh.
China, 1900.—Medal.
East Africa (Somaliland), 1908—10.—Medal with clasp.

Jan—Muhammad—*Bahadur, Order of Br. India, 2nd cl.*
N.-W. Frontier of India, Hazara, 1891.—Medal with clasp.
China, 1900.—Medal.

Saif Ali—
Mekran, 1898.
China, 1900.—Medal.

Saidullah Khan—

Mekran, 1898.—Action of Gok Parosh.
China, 1900.—Medal.

Taj Muhammad—
Mekran, 1898.
China, 1900.—Medal.

Hoshnak Khan—
China, 1900.—Medal.

Nowsher—
China, 1900.—Medal.
Indian Distgd. Service Medal.

Ilam-Din—
China, 1900.—Medal.

Allah-Dad—
China, 1900.—Medal.

JEMADARS—

Sultan-Muhammad.—
China, 1900.—Medal.

Shaikh Sarajuddin—
China, 1900.—Medal.

Shakar Mehdi—
China, 1900.—Medal.

Muhammad Ali—

Indian Distgd. Service Medal.

Bahadur Khan—

China 1900—Medal.

2-130th K. G. O. Baluchis.

JEMADARS—

Gul khan—

China 1900—Medal.

1-150th Infantry.

SUBADAR-MAJOR—

Multan Singh—
Egypt, 1914—16.
Mesopotamia, 1916—18

SUBADARS—

Jiwan Singh—
Mesopotmia, 1915—18.

JEMADARS—

Buti Singh—
Egypt, 1914—17.
Mesopotamia, 1917-18.

Raghubir Singh—
Egypt, 1914—16.
Mesopotamia, 1916—18.

Chuni Singh—
Egypt, 1914—16.
Mesopotamia, 1916—18.

Chhotte Singh—
Mesopotamia, 1915—18.

Phul Singh—
East Africa, 1914—16.
Mesopotamia, 1916—18.

Puttu Sinh—
Mesopotamia, 1915—18.

Chattar Singh—
Mesopotamia, 1915—18.

Harnath Singh—
Mesopotamia, 1915—18.

Shulmbar Singh—
Mesopotamia, 1915-16, 1917-18.

Shulmangal Singh—
Mesopotamia, 1916—18.

Chaltan Singh—
East Africa, 1914-15.
Mesopotamia, 1917-18.

Ramchandar Singh—
Mesopotamia, 1916—18.

Naurang Singh—
East Africa, 1914-15.
Mesopotamia, 1917-18.

Bhagwant Singh—
East Africa, 1914-15.
Mesopotamia, 1917-18.

Sheoji Singh—
East Africa, 1914-15.
Mesopotamia, 1917-18.

JEMADAR—

Danbahadur Singh—
Mesopotamia 1917-18.

2-150th Infantry.

SUBADAR-MAJOR—

Shib Lal—
Tochi, 1915.
Miran Shah, 1917.
Northern Waziristan Field Force, 1917.
Indian Expeditionary Force ' D", 1917.

SUBADARS—

Hardhan Singh—
Field Force, Aden, 1903-04.—Operations, in the interior attack.
M. E. F., 1917-18.

Sri Ram—
Mesopotamia Expeditionary Force, 1917-18.

Kishan Lal—
N.-W. F., Tochi Valley, 1914, 1915, 1916.
M. E. F., 1916-18.

Khubi Ram—
Sikkam, 1903-04.—Tibet—Medal.
Waziristan Field Force, 1917.
Tochi, 1914-15.
Miran Shah, 1917.
M. E. F., 1916—18.

Ganesh Singh—
France M. E. F., 1914.—Wounded.
Mesopotamia, 1916—18.—Wounded.

Gopal Singh—
Mesopotamia "E." Force, 1915, 1917-18.—Wounded.

JEMADARS—

Ramji Lal—
Tochi, 1915.
I. E. Force "D", 1917-18.

Salig Ram—
N.-W. F. India, 1914-16.
M. E. F., 1916-18.—Wounded.

Ujaba Singh—
France, 1914-15.
Egypt, 1915.
M. E. F., 1915-16.—Wounded.
M. E. F., 1917-18.—Battle Shekhsaad.—Awarded Mons Star.—Order of Kara George of Servia, with Sword and Gold Medal, 1st class.

Parsadi Singh—
I. E. F. "D", 1917.

Gulab—
Tochi, 1914-15.
Tochi, 1915.
M. E. F. "D", 1917-18.

Sri Pal Singh—
France, 1915.
Egypt, 1915.
Mesopotamia, 1915.

Lahri—
N. W. F. India, 1914—16.
Mesopotamia Expeditionary Force, 1916—18.

Lal Chand—
Northern Waziristan Field Force, 1917.
Tochi, 1914-15.
I. E. F. "D", 1917-18.

Tota Singh—
Indian E. Force "D", 1917-18.

Zahrla—
Indian E. Force "D", 1917-18.

Mohar Singh—
Mesopotamia E. Force, 1915-16, 1917-18.—Wounded at Awaz.

Ratti Ram—
Mesopotamia E. Force, 1916—18.

3-150th Infantry.

SUBADAR-MAJOR—

Chanda—
Aden 1903-04.—Operations in the Interior.
Baluchistan, 1915-16.
Mesopotamia, 1917-18.

SUBADARS—

Pahara—

Mesopotamia 1915—18.

Sawana—

Mesopotamia, 1915—18.

Khangara—

Mesopotamia, 1915—18.

Dawarika Singh

East Africa, 1914—17.
Mesopotamia, 1917-18.

Subkharan Singh—

Waziristan, 1917.
Mesopotamia, 1917-18.

Jetha—

Mesopotamia, 1915—18.

JEMADARS—

Gulla—

Mesopotamia, 1917-18.

Khuma—

Mesopotamia, 1915—18.

Lachman—

Mesopotamia, 1917-18.

Lakha—

Mesopotamia, 1915—18.

Birda—

Mesopotamia, 1915—18.

Baldeo Singh—

Waziristan, 1917.
Mesopotamia, 1917-18.

Prithipal Singh—

Waziristan, 1917.
Mesopotamia, 1917-18.

Kishna—

Mesopotamia, 1915—18.

1-151st Sikh Infantry.

SUBADAR MAJOR—

Waryam Singh—

N.-W. Frontier of India, 1897-98.—Defence of Malakand-Relief of Chakdara Malakand. Action of Landi-Kai. Operations in Bajaur, and in the Mamund country.—Medal with two clasps.
Tirah, 1897-98.—Operations in the Bara valley 7th to 14th December 1897.—Clasp.
N.-W. Frontier of India, 1908.—Operations in Zakka Khel country.—Medal with Clasp.
Mesopotamia Expeditionary Force, 1916.—Operations before Kut 1916-17.—Wounded 4th February 1917. Invalided to India, 4th June 1917.
Mesopotamia Expeditionary Force, 1917-18.

SUBADARS—

Khalas Khan I—

Zhob Valley, 1890.—African Contingent, 1898-1900.
East Africa 1903-04.—Operations in Somaliland. Action at Jidballi.—Medal with two clasps.—*France, 1914—15.*
Mesopotamia, 1917-18.—Neuve Chapelle and 2nd battle of Ypres.

Prem Singh—

N.-W. Frontier of India, Tochi, 1897-98.—Meda and clasp.
China, 1900.—Medal.
East Africa, Nandi F. F., 1905-06.—Medal with clasp.
Mediterranean Expeditionary Force "E" Gallipali.—Wounded.
Mesopotamia, 1916—18.

Rur Singh—

N.-W. Frontier of India, Tochi, 1897-98.—Medal and clasp.
China, 1900.—Medal.
Mediterranean Expeditionary Force "E" Gallipoli.—Wounded, 4th July 1915.
Mesopotamia, 1916-18.

Phuman Singh—

N.-W. Frontier of India, 1897-98.—Operations on Samana and Kurram. Defence of Gulistan. —3rd Class Order of Merit for gallantry at Fort Gulistan.—Medal with two clasps.
Tirah, 1897-98.—Capture of the Sampagha Pass. Capture of the Arhanga Pass. Reconnaissance of the Sarnsar. Operations at and around Dwatoi.—Operations in the Bara Valley. Affairs at Shinkamar.—Clasp.
Mesopotamian Field Force, 1916—18.

Fakir Singh—

N.-W. Frontier of India, 1897-98.—Operations on Samana and Kurram. Relief of Gulistan.—Medal and two clasps.
Tirah, 1897-98.—Capture of the Sampagha Pass. Capture of the Arhanga Pass. Reconnaissance of the Saran Sar. Operations at and around Dwatoi. Operations in the Bara valley. Affairs at Shinkamar.—Clasp.
Mesopotamia Field Force, 1916—18.

Jiwan Singh—

N.-W. Frontier of India, Tochi, 1897-98.—Medal clasp.
China, 1900.—Medal.
Mediterranean F. F. (Gallipoli Peninsula), 1915-18.—*Mesopotamian F. F., 1916—18.*—I.D.S.M.

Wattan Singh—

Mesopotamia Expeditionary Force, 1916.—Operations before Kut.

Khalas Khan II—

East Africa, 1903-04.—Operations in Somaliland. Action at Jadballi.

JAMADARS—

Gurdit Singh—

Mesopotamian F. F., 1917-18.

Natha Singh—

Mesopotamian F. F., 1917-18.

Ishar Singh—

N.-W. Frontier of India, 1908.—Operations in the Zakha Khel country.—Medal with clasp.

Mesopotamia Expeditionary Force, 1916.—Operations before Kut.

Kunda Singh—

N.-W. Frontier of India, Waziristan, 1901-02.—Medal with clasp.

N.-W. Frontier of India, 1908.—Operations in the Zakha Khel Country.—Medal with clasp and in the Manmud country.

Mesopotamia Expeditionary Force, 1917.

Jiwa Singh

N.-W. Frontier of India, Tochi, 1897-98.—Medal and clasp.

China, 1900.—Medal.

France, First Indian Expeditionary Force, 1915.—Wounded.

Adalat Khan—

Mesopotamia, 1917-18.

Sardar Khan—

East Africa, 1903-04.—Operations in Somaliland. Action at Jadballi.

2-151st Indian Infantry.

SUBADARS—

Latkan Singh—

N.-W. Frontier of India, Waziristan 1894-95.—Medal with clasp.

N.-W. Frontier of India, 1897-98.—Tochi.—Medal with clasp.

China, 1900.—Relief of Pekin. Action of Peitsang. Action at Yangtsun.—Medal with clasp.

Sajawal Khan—

N.-W. Frontier of India, Waziristan, 1901-02.—Medal with clasp.

N.-W. Frontier of India, 1902.—Operations against the Darwesh Khel Waziris.

Hira Singh—

N.-W. Frontier of India, 1902.—Operations against the Darwesh Khel Waziris.

N.-W. Frontier of India, 1908.—Operations in the Zakka Khel country.

N.-W. Frontier of India, 1908.—Operations in the Mohmand country. Medal with clasp.

N.-W. Frontier of India, Waziristan, 1901-02.—Medal and clasp.

Dalel Khan—

China, 1900.—Relief of Pekin. Action of Peitsang. Action at Yangtsun.—Medal with clasp.

Bostan Khan—

N.-W. Frontier of India, Hazara, 1888.—Medal with clasp.

N.-W. Frontier of India, 2nd Miranzai, 1891.—clasp.

N.-W. Frontier of India, Waziristan, 1894-95.—Clasp.

N.-W. Frontier of India, 1897-98.—Operations on the Samana and in the Kurram valley, during August and September, 1897.—Tochi.—Medal with 2 clasps.

Tirah 1897-98.—Action of Dargai. Capture of the Sampagha and Arhanga Passes. Operations against the Khani Khel Chamkanis. Operations in the Bazar Valley, 25th to 30th December 1897.—Clasp.

N.-W. Frontier of India, Waziristan, 1901-02.—Clasp.

Galedu Ram—

Indian Distinguished Service Medal.—

N.-W. Frontier of India, Isazai, 1892.—

Chitral, 1895.—Relief of Chitral.—Medal with Clasp.

N.-W. Frontier of India, 1902.—Operations against the Darwesh Khel Waziris.

N.-W. Frontier of India, 1908.—Operations in the Zakka Khel Country.

N.-W. Frontier of India, 1908.—Operations in the Mohmand Country.—Medal with Clasp.

Shibu—

N.-W. Frontier of India, Waziristan, 1894-95.—Medal with clasp.

N.-W. Frontier of India, 1897-98.—Tochi.—Medal with clasp.

China, 1900.—Relief of Pekin. Action of Peitsang. Action at Yangtsun.—Medal with clasp.

Sadda Singh—

Chitral, 1895.—Relief of Chitral.—Medal with clasp.

Tirah, 1897-98.—Operations in the Bazar Valley.—2 clasps.

N.-W. Frontier of India, Waziristan 1901-02.—Medal with clasp.

N.-W. Frontier of India, 1902.—Operation. against Darwesh Khel Waziris.

N.-W. Frontier of India, 1908.—Operations against Zakka Khel.

N.-W. Frontier of India, 1908.—Mohmand Operations.—Medal with blasp.

Karam Dad—

N.-W. Frontier of India, 1897-98.—Tochi.—Medal with clasp.

N.-W. Frontier of India, 1908.—Operations in the Mohmand country. Engagement of Kargha, 24th May 1908.

Bostan Khan—

N.-W. Frontier of India, 1902.—Operations against the Darwesh Khel Waziris.

N.-W. Fronteir of India, 1908.—Operations in the Zakka Khel country.—Medal with clasp.

JEMADARS—

Dost Mohd.—

N.-W. Frontier of India, Waziristan, 1901-02.—Mahsud Waziris Blockade.—Medal with clasp.

N.-W. Frontier of India, 1902.—Operation against the Darwesh Khel Waziris.

Karam Sher—

N.-W. Frontier of India, Waziristan, 1901-02.—Mahsud Waziris Blockade.—Medal with clasp.

N.-W. Frontier of India, 1902.—Operation against the Darwesh Khel Waziris.

Gul Muwaz—

Operations in the Bazar Valley Expedition, 1908.

N.-W. Frontier of India, 1908.—Operations in the Mohmand country.—Medal with clasp.

Narain Singh—

Operation in the Bazar Valley Expedition, 1908.

N.-W. Frontier of India, 1908.—Operation in the Mohmand country—Medal with clasp.

Harnam Singh—

N.-W. Frontier of India, Waziris, 1897.

N.-W. Frontier of India, 1897-98.—Tochi.—Action at Maizar 10th June 1897.—Medal and clasp.

China, 1900.—Relief of Pekin. Action of Peit sang. Action at Yangtsun.—Medal with clasp.

Faiz Talab—

N.-W. Frontier of India, 1897-98.—Tochi Operations on the Samana and in the Kurram Valley, during August and September 1897.—Medal with 2 clasps.

Tirah, 1897-98.—Action of Dargai. Action of Sampagha Pass. Action of Arhanga Pass. Operation against Khani Khel Chamkanis. Operations in the Bazar Valley, 25th to 30th December 1897.—Clasp.

N.-W. Frontier of India, Waziristan, 1901-02.—clasp.

N.-W. Frontier of India, 1902.—Operations against Darwesh Khel Waziris.

Bhola Singh—

N.-W. Frontier of India, 1908.—Operations in the Mohmand country. Engagement Kargah, 24th May 1908.—Medal with clasp.

Hukam Dad—

N.-W. Frontier of India, 1908.—Operation in the Mohmand country. Engagement Kargha, 24th May 1908.—Medal with clasp.

N.-W. Frontier of India, Waziristan, 1901-02.—Operations in the Zakka Khel country.

Sher Baz Khan—

Operations in the Bazar Valley Expedition, 1908.

N.-W. Frontier of India, 1908.—Operations in the Mohmand country.—Medal with clasp.

Akbar Khan—

N.-W. Frontier of India, Waziristan, 1901-02.—Medal with clasp.

Operations in the Bazar Valley Expedition, 1908.

N.-W. Frontier of India, 1908. Operations in the Mohmand country. Engagement Kargah, 24th May 1908.—Medal with clasp.

Jowahir Shah—

Operations in the Bazar Valley Expedition, 1908.

N.-W. Frontier of India, 1908.- Operations in the Mohmand country.—Medal with clasp.

Gul Mehdi—

N.-W. Frontier of India, Waziristan, 1901-02.—Medal with Clasp.

Gul Mehdi—

N.-W. Frontier of India, 1908.—Operations in the Zakka Khel country. Operation in the Mohmand country. Engagement of Kargha on the 24th May 1908.—Medal with clasp.

Indar Singh—

Chitral, 1895.—Relief Chitral.—Medal with clasp.

Tirah, 1897-98.—Operation in the Bazar Valley.—2 clasps.

Operations in the Bazar Valley Expedition, 1908.

N.-W. Frontier of India, 1908.—Operations in the Mohmand country.—Medal with clasp.

Bishan Singh—

N.-W. Frontier of India, Waziristan, 1894-95.—Medal with clasp.

N.-W. Frontier of India 1897-98.—Operations on the Samana and in the Kurram Valley, during August and September 1897.—Tochi.—Medal with 2 clasps.

Tirah, 1897-98.—Action of Gargai. Capture of the Sampagha and Arhanga Passes. Operations against the Khani Khel Chamkanis. Operations in the Bazar Valley, 25th to 30th December, 1897.—Medal with 3 clasps.

N.-W. Frontier of India, Waziristan, 1901-02.—Clasp.

N.-W. Frontier of India, 1908.—Operations in the Mohmand country. Engagement Kargah, 24th May 1908.—Medal with clasp.

3—151st Punjabi Rifles.

Samundar Khan—

N.-W. Frontier, 1908.
Force "D", 1917.—
Force "E", 1918.—

Santa Bir—

N.-W. Frontier of India, 1908.—

Amir Ali—

Force "A," 1914—16.—
Force "D," 1916—18.—

Ram Lagan Singh—

Force "D," 1918.—

Durga Singh—

Force "D,".1916.—

1-152nd Punjabis.

SUBADAR-MAJOR—

Mushtak Singh—

N.-W. Frontier of India, 1908.—Operations in Zakka Khel country.—Medal and clasp.
North Waziristan Field Force.—

SUBADARS—

Hasham Ali Khan—

N.-W. Frontier of India, 1897-98.—
China, 1900-01.—Medal and clasp.

Ghulam Muhammed—

N.-W. Frontier of India, 1897-98.—Malakand.—Medal and clasp.
China, 1900-01.—Medal with clasp.

Bahadar Singh—

China, 1900-01.—Medal with clasp.

Sadhu Singh—

N.-W. Frontier of India, 1908.—Operations Zakka Khel country.—Medal and clasp.
North Waziristan Field Force.—

Amar Khan—

N.-W. Frontier of India, 1897-98.—Defence of Malakand.
Burma.—Medal with 2 clasps and gratuity.

JEMADARS—

Niamat Khan—

Tibet Expediton.—Medal with clasp.
Somaliland, 1901.—Medal.

Jahan Khan—

Somaliland, 1901.—Medal and clasp.

Amar Singh—

N.-W. Frontier of India, 1897-98.—Tochi.—Medal and clasp.
N.-W. Frontier of India, 1908.—Operations in Zakka Khel country.—Medal and clasp.
North Waziristan Field Force.—

Bhagat Singh—

N.-W. Frontier of India, 1897-98.—Medal and 2 clasps.
China, 1900-01.—Medal and clasp.

2—152nd Punjabis.

SUBADAR-MAJOR—

Shah Nawaz Khan—

Burma Field Force, 1885—87.—Medal and clasp.
Waziristan, 1894-95.—
Tochi, 1897-98.—Medal and clasp.
France, 1917.—

SUBADARS—

Budhi Singh—

Field Service, 1916.—

Motichand—

Field Service, 1915.—

Fazal Ahmad—

B. E. Africa 1898.—Operations against Mazrur Rebels.—Medal.
Field Service, 1915.—

Jai Singh—

Relief of Chitral, 1895.—Medal with clasp.
Punjab Frontier, Waziristan, 1894-95.—Medal and clasp.
Punjab Frontier, 1897-98.—Clasp.
Field Service, 1915.—

JEMADARS—

Lehna Singh—

Field Service, 1915.—

Bakktawar—

Field Service, 1915.—

Rahim Dad—

Field Service, 1915.—

Ahmad Khan—

Field Service, 1915—18.—

Munshi—

Field Service, 1915—17.

Amar Singh—

Field Service, 1916—18.—

3—152nd Punjabis.

SUBADAR-MAJOR—

Sultan Khan—

Somaliland, 1903-04.—Jidholi-Nagul. *Valley—Daras Valley—Gurlo Jid Ali.*—Medal and clasp.

SUBADARS—

Bishen Singh—

Punjab Frontier, 1899.—Medal with clasp.
China, 1902-05.—
Force "E," 1918.—

Mohd. Khan—

China, 1900-01.—Medal.
Aden, 1903-04.—Operations in the interior.

Sunder Singh—

Waziristan, 1901-02.—Medal with clasp.
Somaliland, 1903-04.—

Sulakhan Singh—

N.-W. Frontier of India, Waziristan, 1901-02.—Medal with clasp.
N.-W. Frontier, 1908.—Operations in the Mohmand country.

Imam Ali—

N.-W. Frontier India, 1908.—Operations in the Mohmand country.

Ishan Singh—

North China, 1902—05.—
Mohmand Field Force, 1908.—Medal with clasp.

JEMADARS—

Allah Din—

N. China, 1902-03.—

Force "E", 1918.—

Ishan Singh—

N.-W. Frontier, 1895.—Medal with clasp.
Punjab Frontier, 1897-98.—Medal.
Waziristan, 1901.—
Br. Central Africa, 1907.—
Force "E", 1918.—

Nizam Din—

N.-W. Frontier of India, Waziristan, 1901-02.—Medal with clasp.
N.-W. Frontier of India, 1908.—Operations in the Zakka Khel country—Operations in the Mohmand country.
Force "E", 1915.—

Tahl Singh—

Somaliland, 1903-04.—Goladi.

Sultan Ahmed—

Somaliland, 1903-04.—Nagal Valley.

Latif Shah—

Force "B", 1918.—

Sunder Singh—

N.-W. Frontier of India, Waziristan, 1901-02.—Medal with clasp.
N.-W. Frontier of India, 1908.—Operations in the Zakka Khel country—Operations in the Mohmand country.
Force "D", 1915.—Wounded.
Force "A", 1916.—

Ishan Singh—

Somaliland, 1903-04.—Jidholi.
Wagal Valley.—Daras Valley.
Garlo Jid Ali.—Medal and clasp.

Allah Ditta—

Force "A"—

Ahmed Khan—

N.-W. Frontier of India, 1908.—Operations in the Zakka Khel country.
I. E. F. "D", 1916.—Wounded.

Dalel Khan—

N.-W. Frontier of India, 1908.—Operations in the Zakka Khel country.
N.-W. Frontier of India, 1908.—Operations in the Mohmand country.
I. E. F. "D", 1916.—Wounded.

Binda Singh—

Punjab Frontier, 1897-98.—
N.-W. Frontier of India, 1895.—Medal with clasp.
N.-W. Frontier of India, 1908.—Operations in the Mohmand country.
North China, 1902—05.—

1—153rd Punjabis.

SUBADAR MAJOR—

Baboo Singh—

Mesopotamia, 1915-18.—Wounded.

SUBADARS—

Bakhtawar Singh—

North-West Frontier of India, 1897-98.—Tochi.—Medal with clasps.
Punjab Frontier, 1897-98.
North-West Frontier of India, 1908.—Operations in the Mohmand country. Engagement of Kargha, 24th May 1908.—Medal with clasp.

Ram Baksh—

Force "D", 1916-17.

Dhuman Khan—

Samana, August and September, 1897.—Under Major-General Yeatman Biggs, Flying Column, Kurram Valley, under Colonel Richardsoln September, 1897.
Tirah, 1897-98.—India with 3 clasps.

Ghulam Mohd.—

Kohat Karram Moveable column, 1895.—Medal with clasps.
Punjab Frontier, 1897-98.
Samana, 1897.

Suraj Pande—

Field Service, 1915—18.

Ghulab Khan—

Field Service, 1915—18.—Mentioned in Despatches

Kishna Ram—

Field Service, 1916—18.

Nadhan Singh—

Nquamba Expedition, B. C. A., 1899.
Central Augomiland, B. C. A., 1900.
Field Service, 1917-18.

Nawas Ali—

N.-W. Frontier of India, 1897-98.—Operations on the Samana Valley during August, September 1897.—Medal with 2 clasps.
Kurram, N. W. Frontier of Indin, Tirah, 1897-98.—Actions of Dargai and Arhanga passes. Reconnaissance of the Saran Sar of the Sampagha and 24th November 1897. Operations against Khari Khel Channel and action of 9th November in Bara Valley and Bazar Valley.—Clasp.
N.-W. Frontier of India, Waziristan, 1901-02.—Clasp.
Field Service, 1916—18.

JEMADARS—

Kishen Singh—

Mesopotamia, 1914—18.—Wound.

Lahir Khan—

War Service, 1917-18.

Chowdri Mangal Singh—

Field Service, 1916—18.

Baldeo Singh—

Mahsud Waziri Blockade, 1900-02, Aden, 1903-04.—Operations in the interior.
N.-W. Frontier of India, Waziristan, 1901-02.—Medal and clasp.
Field Service, 1915—18.

Ram Krishna Sukal—

Field Service, 1915—18.

Mir Afzal Khan—

Force "D", 1915-16.—Wounded.
Field Service, 1916—18.

Baldeo Singh—

Field Service, 1915—18.

Nizam Din—
East Africa, 1914-15 and 1916.—Meritorious Service Medal.
Field Service, 1914—18.

Nur Mohd—
Sheik Sayed, 1914.—Wounded.
Egypt, 1914-15.—Dardenelles, 1915, France, 1915.—Wounded.
Mesopotamia, 1916-17.

3.—153rd Rifles.

SUBADARS—

Ghulam Mohd, I.D.S.M.—
N.-W. Frontier of India, Waziristan Blockade, 1900—02.—Medal and clasp.
Aden, 1903-04.—Operations in the interior.—Indian Distinguished Service Medal.
I. E. F. "A", "D". & "E", 1914-16.—Wounded on *1st January 1916.*—Invalided.

JEMADARS—

Sheikh Mohiu-din—
Mesopotamia, 1916—17.—Wounded.

Krishna Bagwe—
Mesopotamia, 1914—17.

Sajawal Khan—
I. E. F. "A", 1914-15.—Wounded.
I. E. F. "E", "D", 1915—18.—Invalided.

Harda Ram—
I. E. F. "D", "A", "E", 1914—16, 1917-18.

Sikander Khan—
I. E. F. "A", 1915.

1—154th Indian Infantry.

SUBADAR-MAJOR—

Chandan Lal—
Persian Gulf, 1913—16.
M. E. F. from 1916—18.

SUBADARS—

Dukhan Singh—
N.-W. F. Operations, Swaat Valley, 1915.
Gulf of Oman, Muscat, 1916.
N.-W. F. Operations, S. W. F. F., 1917,
M. E. F., 1918.

Man Singh—
N.-W. F. Operations, Swaat Valley, 1915.
N.-W. F. Operations, S. W. F. F., 1917.
M. E. F., 1918.

Bhup Singh—
Persian Gulf, F. F., 1913—15.
W. F. F. from 1917.
M. E. F., 1917-18.

Dalsinghar Singh—
Chakdarrah, F. F., 1915.
M. E. F., 1915—18.

Kauleshar Singh—
M. E. F., 1915—18.

JEMADARS—

Beni Singh—
M. E. F., 1918.

Nihal Singh—
N.-W. F. Operations, Swaat Valley, 1915.
Maskat, 1916.
S. W. F. F., 1917.
M. E. F., 1918.

Matadin Singh—
Persian Gulf, F. F., 1914-15.
M. E. F., 1917-18.

Tula Ram—
P. G. F. F., 1913—15.
W. F. F., 1917.
M. E. F., 1917-18.

Mohan Singh—
France, 1914-15.
M. E. F., 1917-18.

Jhunda Singh—
M. E. F., 1915-18.

Ram Rekha Singh—
M. E. F., 1916—18.

Raj Narayan Singh—
M. E. F., 1918.

Ram Pal Singh—
M. E. F., 1915—1918.

Basdeo Singh—
Chakdarrah F. F., 1915.
M. E. F., 1915—18.

2-154th Indian Infantry.

SUBADAR-MAJOR—

Ganga Ram Singh—
Persian Gulf, 1915.
I. E. F. "D", 1915.
Mesopotamia Expeditionary Force from 1917.—Indian Order of Merit.

Shikari—
Mesopotamia Expeditionary Force, 1916-17.—Indian Distinguished Service Medal on 17th August 1917.
Invalided to India, 1917.—Second class Order of Br. India, 1914.

SUBADAR—

Gugan Ram—
China Expeditionary Field Force, 1900.—Medal.
Mesopotamia Expeditionary Force from 1918.

Aungad—
Mesopotamia Expeditionary Force on 1915.

Mubarak Ali—
Aden Field Force on 1915.

JEMADARS—

Bepi Singh—
Egyptian Expeditionary Force on 1918.

Mohamed Ali Khan—
Mesopotamian Expeditionary Force on 1915.
Mesopotamian Expeditionary Force on 1916—18.

Kaloo Khan—
Mesopotamian, 1918.

Munshi—
Mesopotamian Expeditionary Force, 1916—18.

Jokhi Ram—
Mesopotamian Expeditionary Force, 1915.

Ladhu Ram—
Mesopotamian Expeditionary Force, 1916.

Gisa Singh—
Mesopotamian Expeditionary Force, 1917-18.
Egyptian Expeditionary Force, 1918.

Ram Het—
Laskhorai-Somaliland, 1909.—Medal and clasp.

Abdul Majid Khan—
Aden from 1915-16.

Hargiyan—
Mesopotamian Expeditionary Force from 1916—18.

Hardyal—
Mesopotamian Expeditionary Force, 1915—18.

Sirikishan—
Mesopotamian Expeditionary Force, 1915.
Mesopotamian Expeditionary Force, 1917.

Gulab Singh—
Persian Gulf, 1914-15.
I. E. F. "D", 1915-16.—Wounded 21st January 1916.
I. E. F. "D", 1916.

Bhairu Singh—
Aden Operations in the interior, 1903 and 1904.
Persian Gulf, 1914-15.
I. E. F. "D", 1915-16.
Mesopotamian Expeditionary Force, 1916.—Wounded, 11th March 1917.

Safidar Khan—
Muscat on 1916.

1-155th Indian Pioneers.

SUBADARS—

Ewaz Khan—
France and Mesopotamia, 1914—16.
Mesopotamia and Egypt, 1917-18.

Ranjit—
N. W.-Frontier, Tochi, 1897-98.—Medal with clasp.
N. China, 1900-01.
Persian Gulf, 1914-15.
Mesopotamia and Egypt, 1918.

Udmi—
Mesopotamia and Egypt, 1916.

Yasin Khan—
China Expeditionary Force, 1900-01.—Medal.
East Africa (Somaliland), 1904.—Medal and clasp.
Mesopotamia and Egypt, 1914-16.

JEMADARS—

Makhmul Khan—
Mesopotamia, 1915-16.
Egypt and Mesopotamia, 1917.

Chand Khan—
France and Mesopotamia, 1915-16.
Mesopotamia and Egypt, 1918.

Male Khan—
Mesopotamia and Egypt, 1917.

Toka—
Mesopotamia, 1914-15.
Mesopotamia and Egypt, 1917.

Nanta—
Mesopotamia and Egypt, 1917.

Paritam Singh—
Waziristan, 1901-02.—Medal and clasp.
Tibet, 1903-04.—Medal and clasp.
Abhor, 1911-12.—Medal and clasp.
France and Mesopotamia, 1915—17.

Sohan Singh—
Waziristan, 1901-02.—Medal and clasp.
Tibet, 1903-04—Medal and clasp.
Abhor. 1911-12.—Medal and clasp.
France and Mesopotamia, 1915-16, 1917-18.

Naurang Khan—
Mesopotamia and Egypt, 1914.

2-155th Indian Pioneers.

SUBADARS—

Arjan Singh—
N.-W. Frontier of India, Waziristan, 1901-02.—Medal with clasp.
N.-W. Frontier of India, 1908.—Operations in the Zakka Khel country.—Medal with clasp.

Binda Singh—
Tibet, 1903-04.—Action at Niani, 26th June 1900.—Operation at and around Gyantse, between 5th May and 6th July 1904.—Medal with clasp.
N.-W. Frontier of India, 1908.—Operations in the Zakka Khel country.—Medal with clasp.

Bir Singh—
N.-W. Frontier of India, 1908.—Operations in the Zakka Khel country.—Medal with clasp.

Jaggat Singh—
Tibet, 1903-04.—Medal.
N.-W. Frontier of India, 1908.—Operations in the Zakka Khel country.—Medal with clasp.

Narayan Singh—

N.-W. Frontier of India, Waziristan, 1901-02.—Medal with clasp.

Tibet, 1903-04.—Action at Niani, 26th June 1904.—Operations at and around Gyantse, between 5th May and 6th July 1904.—Medal with clasp.

N.-W. Frontier of India, 1908.—Operations in the Zakka Khel country.—Medal with clasp.

JEMADARS—

Sewa Singh—

N.-W. Frontier of India, 1908.—Operations in the Zakka Khel country.—Medal with clasp.

Mitt Singh—

N.-W. Frontier of India, 1908.—Operations in the Zakka Khel country.—Medal with clasp.

Surain Singh—

N.-W. Frontier of India, Waziristan, 1901-02.—clasp.

1-156th Indian Infantry.

SUBADAR-MAJOR—

Tiravengadam—

War, 1916—18.—Force "D."

SUBADAR—

M. Bhagavati—

War, 1916—18.—Barbera, 1916—18.—Force "D", 1918.

Abdus Samad—

War, 1917-18.—Force "D."

Muhamed Waizullah—

War, 1916—18.—Persian Gulf, 1916-17.—Force "D," 1918.

Abdul Hussain—

War, 1916—18.—Persian Gulf, 1916-17.—Force "D," 1918.

Hussain Khan—

War, 1916—18.—Force "D."

Manavazhi Mahadeva Menon—

War, 1918-18.—Force "D."

Munirathnam—

War, 1917-18.—Force "D."

JEMADAR—

Shaikh Hussain—

War, 1917-18.—Force "D."

Abdur Rahimon—

War, 1916—18.—Force "D."

Muhammed Ghaus—

War, 1914—18.—East Africa, 1914—16.—Force, 'D" 1917-18.

Ratna Mudali—

War, 1916—18.—Force "D."

Abdul Jabbar—

War, 1918.—Force "A", 1915-16.—Force "D" 1917-18.

Shaikh Abdul Khader—

War, 1916—18.—Force "D."

Maduraimuthu—

War, 1914—18.—Force "A", 1914-16.—Force "D", 1916—18.

Rungasami—

War, 1917-18.—Force "D."

Nagalingam—

War.—Force "D."

Syed Hussain—

War, 1917-18.—Force "D."

Shaikh Farid—

War, 1916—18.—Force "D."

Abdul Ghani—

War, 1917-18.—Force "D."

Periasami Pillai—

War, 1917-18.—Force "D."

Arumugam—

War, 1916—18.—Force "D."

1st King George's Own Gurkha Rifles.

(The Malaun Regiment.)

SUBADAR-MAJORS—

Bhagatabir Gurung *Bahadur—Order of Br. India 2nd class.*

N.-W. Frontier of India, Waziristan, 1894-95.—Medal with clasp.

Tirah, 1897-98.—Medal with clasp.

SUBADARS—

Kalu Gurung I. O. M.—

N.-W. Frontier of India, Waziristan, 1894-95.—Medal with clasp.

Tirah, 1897-98.—Actions of the Sampagha Pass and Arhanga Pass.—Operations in the Bazar Valley from 25th to 30th December 1897.—Clasp.

Balram Rana—

N.-W. Frontier of India, Waziristan, 1894-95.—Medal with clasp.

Santbir Rana—

Waziristan, 1894-95.—Medal with clasp.
Tirah, 1897-98.—Medal with clasp.

Tulla Thapa—

Waziristan, 1894-95.—Medal with clasp.

Puran Singh Gurung—

Tirah, 1897-98.—Medal with clasp.

Birkh Sing Thapa—

N.-W. Frontier of India, 1897-98.—Mohmand.—Medal with clasp.

Tirah, 1897-98.—Action of the Sampagha Pass and Arhanga Pass.—Operations in the Bazar Valley, from 25th to 30th December 1897.—Clasp.

JEMADARS—

Bahadur Sing Gurung—
N.-W. Frontier of India, 1897-98.—Mohmand.—Medal with clasp
Tirah, 1897-98.—Action of the Sampagha Pass and Arhanga Pass. Operations in the Bazar Valley, from 25th to 30th December 1897.—Clasp.

21—st King George's Own Gurkha Rifles (The Malaun Regiment.)

SUBADAR MAJOR—

Kaluram Thapa, I.D.S.M.—
Operations in Swat, Buner and Mohmand, 1915.
Operations in South Waziristan Field Force, 1917.—Awarded Indian Distinguished Service Medal.

SUBADARS—

Bhagatbir Thapa, I.O.M.—
N.-W. Frontier of India, 1897-98.—Mohmand.—Medal with clasp.
Tirah, 1897-98.—Actions of the Sampagha Pass and Arhanga Pass.—Operations in Bazar Valley, from 25th to 30th December 1897.—Clasp.

Operations in Swat, Buern and Mohmand, 1915.
European War, 1916-18.—Awarded Indian Order of Merit.

Bhimsing Gharti—
N.-W. Frontier of India, 1897-98.—Mohmand.—Medal and clasp.
Tirah, 1897-98.—Actions of the Sampagha Pass and Arhanga Pass.—Operations in Bazar Valley, from 25th to 30th December 1897.—Clasp.
Operations in Swat, Buner and Mohmand, 1915.
European War, 1916.

Kabindra Thapa—
N.-W. Frontier of India, 1897-98.—Mohmand.—Medal with clasp.
Operations in Swat, Buner and Mohmand, 1915.
Operations in South Waziristan Field Force, 1917.

Birbahadur Limbu—
N.-W. Frontier of India, 1897-98.—Mohmand—Medal with clasp.
Tirah, 1897-98.—Actions of the Sampagha Pass and Arhanga Pass.—Operations in Bazar Valley, from 25th to 30th December 1897. Clasp.

Operations in Swat, Buner and Mohmand 1915.
Operations in South Waziristan Field Force, 1917.

Bhagirat Limbu—
N.-W. Frontier of India, 1897-98.—Mohmand.—Medal with clasp.
Tirah, 1897-98.—Actions of the Sampagha Pass and Arhanga Pass.—Operations in Bazar Valley, from 25th to 30th December 1897.—Clasp.
Operations in Swat, Buner and Mohmand, 1915.
European War, 1916-17.

Gangram Khatri—
N.-W. F. of India, 1897-98.—Mohmand.—Medal with clasp.
Tirah, 1897-98.—Actions of the Sampagha Pass and Arhanga Pass. Operations in Bazar Valley, from 25th to 30th December 1897.—Clasp.
Operations in Swat, Buner and Mohmand, 1915.
Operations in South Waziristan Field Force, 1917.

dursing Thapa—
Operations in Swat, Buner and Mohmand, 1915.
Operations in South Waziristan Field Force, 1917.

JEMADARS—

Maubir Rana—
Operations in Swat, Buner and Mohmand, 1915.
Operation in South Waziristan Field Force, 1917.

Jagatsher Gurung I.D.S.M.—
Operations in Swat, Bnner and Mohmand, 1915.—Awarded Indian Distinguished Service Medal.
Operations in South Waziristan Field Force, 1917.

Jitjang Thapa—
Operations in Swat, Buner and Mohmand, 1915.
Operations in South Waziristan Field Force, 1917.

Dilbahadur Ale—
Operations in Swat, Buner and Mohmand, 1915.
Operations in South Waziristan Field Force, 1917.

Mandhoj Rana—
Operations in Swat, Buner and Mohmand, 1915.
Operations in South Waziristan Field Force, 1917.

Fatehsing Thapa—
Operations in Swat, Buner and Mohmand, 1915.
Operations in South Waziristan Field Force, 1917.

Dharamraj Gurung—
Operations in Swat, Buner and Mohmand, 1915.
Operations in South Waziristan Field Force, 1917.

Ramu Thapa—
Operations in Swat, Buner and Mohmand, 1915.
Operations in South Waziristan Field Force, 1917.

Kharakbahadur Pun—
Operations in Swat, Buner and Mohmand, 1915.
European War 1916.

Manbir Gurung—
Operations in Swat, Buner and Mohmand, 1915.
European War 1916-17.

Nandu Gurung—

N.-W. Frontier of India, 1897-98.—Mohmand.—Medal with clasp.
Tirah, 1897-98.—Actions of the Sampagha Pass and Arhanga Pass.—Operations in Bazar Valley, from 25th to 30th December 1897.—Clasp.
European War, 1914-15.

Harkdhoj Limbu—

N.-W. Frontier of India, 1897-98.—Mohmand.—Medal with clasp.
Tirah, 1897-98.—Actions of the Sampagha Pass ad Arhanga Pass.—Operations in Bazar Valley, from 25th to 30th December 1897.—Clasp.
Operations in Swat, Buner and Mohmand, 1915.

Dalbir Rana—

Operations in Swat Buner and Mohmand, 1915.
Operations in South Waziristan, Field Force, 1917.

Narainsing Thapa—

Operations in Swat, Buner and Mohmand, 1915.
Operations in South Waziristan Field Force, 1917.

Kalu Gurung—

Operations in Swat Buner and Mohmand, 1915.
Operations in South Waziristan Field Force, 1917.

Budhiman Ghale—

Operations in Swat Buner and Mohmand 1915.
Operations in South Waziristan, Field Force, 1917.

Puransing Gurung—

Operations in Swat, Buner and Mohmand, 1915.
Operations in South Waziristan, Field Force, 1917.

Amarsing Thapa—

European War—1914-17.

3rd Battalion 1st K. G. O. Gurkha Rifles. (The Malaun Regiment.)

SUBADAR-MAJOR—

Kishan Sing Karki—

N.-W. Frontier of India, 1897-98.—Mohmand—Medal with clasp.
Tirah, 1897-98.—Actions of the Sampagha Pass and Arhanga Pass—Operations in the Bazar. Valley from 25th to 30th December 1897—Claspr
Waziristan, 1915.

SUBADARS—

Dewan Sing Thapa—

China 1900-02.—Medal.
I. E. F. 'A', 1915-16.
I. E. F. 'D', 1916-17.

Lilbir Thapa—

I. E. F. 'A', 1914-15.
I. E. F. 'D', 1916-17.

Bahadur Sing Gurung—

N.-W. Frontier of India, 1897-98.—Mohmand—Medal with clasp.
Tirah, 1897-98.—Actions of the Sampagha Pass and Arhanga Pass—Operations in the Bazar Valley from 25th to 30th December 1897—Clasp.
Waziristan 1915.

Manbir Thapa—

I. E. F. 'A', 1914-1916.
I. E. F. 'D', 1916.

Narla Thapa—

I. E. F. 'D', 1916-17.
Waziristan 1915.

Bhawansing Gurung—

Waziristan 1915.

Tikaram Thapa—

Waziristan 1915.

JEMADARS—

Lalla Pun—

I. E. F. 'A', 1914-16.
I. E. F. 'D', 1917-18.

Dhansing Thapa—

I. E. F. 'D', 1916-17.

Bakhanbahadur Gurung—

Waziristan 1915.

Haste Gurung—

Waziristan 1915.

Harkesor Limbu—

Waziristan 1915.

Bahadu Sing Gurung—

I. E. F. 'D' 1915-16.

Damarsing—

I. E. F. 'A', 1914-15.
I. E. F. 'D' 1917.

Bhadarbir Thapa—

Waziristan 1915.

Thamansing Gurung—

Waziristan 1915.

Karbir Thapa—

Waziristan 1915.

2nd (King Edward's Own) Gurkha Rifles (The Sirmoor Rifles).

SUBADAR-MAJORS—

Tulsiram Gharti—
N.-E. Frontier of India, Manipur, 1891.—Medal with clasp.
N.-W. Frontier of India, 1897-98.—Operations on the Samana and in the Kurram Valley during August and September 1897. Medal with 2 clasps.
Tirah, 1897-98.—Actions of Dargai, Capture of the Sampagha and Arhanga Passes. Operations in the Warran Valley and action of 16th November 1897. Operations at and around Dwatoi and action of 24th November 1897. Operations in the Bara Valley 7th to 14th December 1897.—Clasp
N.-W. Frontier of India, Waziristan, 1901-02.—Clasp.
Abor Expedition, 1911-12.—Medal with clasp.
War Overseas 1914.—Mentioned in Despatches.

SUBADARS—

Manjit Gharti—
N.-E. Frontier of India, Manipur, 1891.—Medal with clasp.
N.-W. Frontier of India, 1897-98.—Operations on the Samana during August and September 1897. Relief of Gulistan.—Medal with 2 clasps.
Tirah, 1897-98.—Actions of Chagru Kotal and Dargai. Actions of Sampagha and Arhanga Passes. Operations in the Waran Valley, and action of 16th November 1897. Operations at and around Dwatoi, and action of 24th November 1897. Operations against the Khani Khel Chamkanis. Operations in the Bara Valley, 7th to 14th December 1897.—Clasp.
N.-W. Frontier of India, Waziristan, 1901-02.—Clasp.
Abor Expedition, 1911-12.—Medal with clasp.

Sarabjit Gurung—M. C.
N.-W. Frontier of India, 1897-98.—Operations on the Samana, during August and September. Relief of Gulistan.—Medal with 2 clasps.
Tirah, 1897-98.—Actions of Chagru Kotal, Dargai, Sampagha, and Arhanga Passes. Operations in the Waran Valley and action of 16th November 1897. Operations at and around Dwatoi and action of 24th November 1897.—Operations in the Bara Valley, 7th to 14th December 1897.—Clasp.
N.-W. Frontier of India, Waziristan, 1901-02.—Clasp.
Abor Expedition, 1911 12.—Medal with clasp.

War 1914-17—Military Cross.

Kharakbir Thapa—
N.-W. Frontier of India, Waziristan, 1901-02.—Medal with clasp.
Tibet, 1903-04.—Medal with clasp.

Abor Expedition, 1911-12.—Medal with clasp.

Dan Sing Gurung, I.O.M.—
N.-W. Frontier of India, 1897-98.—Operations on the Samana during August and September. Relief of Gulistan.—Medal with 2 clasps.
Tirah, 1897-98.—Actions of Chagru Kotal, Dargai, Sampagha and Arhanga Passes. Operations in the Waran Valley and action of 16th November 1897. Operations at and around Dwatoi and action of 24th November 1897.—Operations in the Bara Valley, 7th to 14th December 1897. Clasp.—*Indian Order of Merit, 3rd class.*
N.-W. Frontier of India, Waziristan, 1901-02.—Clasp.
Abor Expedition, 1911-12.—Medal with clasp.

Champasing Gurung—
Abor Expedition, 1911-12.—Medal with clasp.

Lachman Sing Khattri—
N.-W. Frontier of India, 1897-98.—Operations on the Samana during August and September. Relief of Gulistan.—Medal with 2 clasps.
Tirah, 1897-98.—Actions of Chagru Kotal, Dargai, Sampagha, and Arhanga Passes. Operations in the Waran Valley and action of 16th November 1897. Operations at and around Dwatoi and action of 24th November 1897. Operations in the Bara Valley, 7th to 14th December 1897.—Clasp.
N.-W. Frontier of India, Waziristan, 1901-02.—Clasp.
Abor Expedition, 1911-12.—Medal with clasp.

Guman Sing Gurung—
N.-W. Frontier of India, 1897-98, etc., etc.—Medal with 2 clasps.
Tirah, 1897-98, etc., etc.—Clasp.
N.-W. Frontier of India, Waziristan, 1901-02.—Clasp.

Tibet, 1903-04.—Medal.

Abor Expedition, 1911-12.—Medal with clasp.

Narain Sing Thapa—
N.-W. Frontier of India, 1897-98, etc.—Medal with 2 clasps.
Tirah, 1897-98, etc.—Clasp.
N.-W. Frontier of India, Waziristan, 1901-02.—Clasp.
Abor Expedition, 1911-12.—Medal with clasp.

Ahlman Rana—
N.-W. Frontier of India, Waziristan, 1901-02—Medal with clasp.
Abor Expedition, 1911-12.—Medal with clasp.

Man Sing Gurung—
N.-W. Frontier of India, Waziristan, 1901-02.—Medal with clasp.
Abor Expedition, 1911-12.—Medal with clasp.

Man Sing Gurung—
Abor Expedition, 1911-12.—Medal with clasps.

JEMADARS—

Narbahadur Thapa—
N.-W. Frontier of India, 1897-98, etc., etc.—Medal with 2 clasps.
Tirah, 1897-98, etc.—Clasp.
Abor Expedition, 1911-12.—Medal with clasp.

2nd K. E. O. Goorkha Rifles.

SUBADAR-MAJOR—

Arjun Rana, I.O.M.—
Punjab Frontier, 1897-98.
Operations on the Samana in the Kurram Valley, 1897.
France, 1914.—November 2nd Neuve Chapelle. December 20th Festubert.—Mentioned in Despatches I. O. M.

SUBADARS—

Tirthe Gurung—
France, 1914.—November 2nd Neuve Chapelle. Wounded.

Bahadur Ghale—
Tibet, 1903-04.—Medal with clasp.
France, 1914.—November 2nd Neuve Chapelle. Wounded.

Dan Sing Lama, I.O.M.—

Abor Expedition, 1911-12.—Medal with clasp.
France, 1914-15.—November 2nd Neuve Chapelle, December 20th Festubert, March 10th-12th Neuve Chapelle, May 9th Richebourg. September 25th Fanquissait. October 13th Givenchy. Mentioned in despatches, I.O.M.

Gamer Sing Gurung—

N.-W. Frontier of India Waziristan, 1901-02.—Medal with clasp.
Abor Expedition 1911-12.—Medal with clasp.
France, 1914-15.—November 2nd Neuve Chapelle. December 20th Festubert. Wounded. March 10th-12th Neuve Chapelle. May 9th Richebourg. September 25th Fanquissait, October 13th Givenchy. Egypt. Marri Field Force 1918.

Manbir Thapa—

Punjab Frontier 1897-98.—Samana Tirah—Action of Bhagru Kolat. Action of Dargai. Opperations on the Samana and in the Kurram Valley. Relief of Gulistan.—1 medal with 3 clasps.
France, 1914-15.—November 2nd Neuve Chapelle. December 20th Festubert. March 10th-12th Neuve Chapelle. September 25th Fanquissait. October 13th Givenchy. Egypt, 1916. Marri Field Force, 1918.

Dewan Sing Gurung—

Abor, 1911-12.—Medal with clasp.
France, 1914-15.—November 2nd Neuve Chapelle, December 20th Festubert, March 10th-12th Neuve Chapelle. May 9th Richebourg. October 13th Givenchy. Egypt, 1916. Marri Field Force, 1918.

Santbir Gurung—

Abor Expedition 1911-12.—Indian General Service Medal, 1908.
France, 1914-15.—November 2nd Neuve Chapelle. December 20th Festubert. March 10th-12th Neuve Chapelle. May 9th Richebourg. Egypt 1916. Dardenelles, 1915.

Lachman Gurung—

Punjab Frontier 1897-98.—Samana. Tirah. Operations on Samana and in Kuram Valley.—1 Medal and 3 clasp.—
France, 1915.—March 10th-12th Neuve Chapelle. May 9th Richebourg. September 25th Fanquissait. October 13th Givenchy, Egypt. Marri Field Force 1918.

Nar Sing Thapa—

Sikkim Tibet Expedition 1903-04.
Tibet Mission Escort, 1903-04.—(i). Action at Naini 26-6-04.—(ii). Operations at and around Gyantse between 5-5-04—6-7-04. Medal and clasp.
Abor Expedition, 1911-12.—Medal and clasp 1908.
France, 1914-15.—November 2nd Neuve Chapelle. December 20th Festubert. March 10th-12th Neuve Chapelle. Wounded September 25th Fanquissait. October 13th Givenchy. Egypt. Marri Field Force 1918.

Gamer Sing Bura, I.D.S.M.—

France, 1914-15.—November 2nd Neuve Chapelle. December 20th Festubert. March 10th-12th Neuve Chapelle. May 9th Richebourg. September 25th Fanquissait. October 13th Givenchy. Egypt. I. D. S. M. on 2-11-14. Mentioned in Despatch 2-2-15.

JEMADARS—

Suba Sing Gurung—

France, 1914.—November 2nd Neuve Chapelle. December 20th Festubert.

Pirthilal Gurung—

N.-W. Frontier of India, Waziristan, 1901-02.—Medal with clasp.
Abor Expedition 1911-12.—Medal with clasp.
France, 1915.—May 9th Richebourg. September 25th Fanquissait. October 13th Givenchy. Egypt. Marri Field Force, 1918.

Badal Sing Gurung—

N.-W. Frontier of India, Waziristan, 1901-02—Medal with clasp.
Tibet, 1903-04.—Medal with clasp.
Abor, 1911-12.—Medal with clasp.
France, 1914-15.—November 2nd Neuve Chapelle. December 20th Festubert. Wounded. March 10th-12th Neuve Chapelle. May 9th Richebourg. Wounded. September 5th Fanquissait.

Harak Sing Thapa—

France, 1914-15.—November 2nd Neuve Chapelle. December 20th Festubert. March 10th-12th Neuve Chapelle. May 9th Richebourg. October 13th Givenchy.
Egypt.—Marri Field Force, 1918.

Guman Sing Gurung, I.D.S.M.—

Mesopotamia, 1917.—Granted I.D.S.M. for gallantry.

Bhimbahadur Thapa, I.D.S.M.—

France, 1914-15.—November 2nd Neuve Chapelle. December 20th Festubert. March 10th-12th Neuve Chapelle. May 9th Richebourg. September 25th Fanquissait. October 13th Givenchy.
Egypt.—Marri Field Force, 1918. I.D.S.M. 27th February 1915.

Hira Sing Thapa—

Punjab Frontier, 1897-98.—Samana. Tirah.—Medal with 3 clasps.
Operation on the Samana and Kurram Valley, 1897. Tibet Mission. Action at Naini Operation at and arround Gyantse. March to Lhassa.—Medal and clasp.
France, 1914.—Action November 2nd Neuve Chapelle. December 20th Festubert. Wounded.

Kharkia Gharti—

Tibet, 1903-04.—Medal with clasp.
France, 1914-15.—November 2nd Neuve Chapelle. December 20th Festubert. March 10th-12th Neuve Chapelle. May 9th Richebourg. September 25th Fanquissait. October 13th Givenchy.
Egypt, 1916.—Marri Field Force, 1918.

Jangbir Thapa—

N.-W. Frontier of India, Waziristan, 1901-02.—Medal with clasp.
Abor Expedition, 1911-12.—Medal with clasp.
France, 1914-15.—November 2nd Neuve Chapelle December 20th Festubert. March 10th-12th Neuve Chapelle. May 9th Richebourg. September 25th Fanquissait. October 13th Givenchy.
Egypt.—Marri Field Force, 1918.

Keshbir Thapa—

France, 1914-15.—November 2nd Neuve Chapelle. December 20th Festubert. March 10th-12th Neuve Chapelle. May 9th Richebourg. September 25th Fanquissait. October 13th Givenchy.
Egypt.—Marri Field Force, 1918.

Rampershad Thapa, I.D.S.M.—

France, 1914-15.—November 2nd Neuve Chapelle. December 20th Festubert. March 10th-12th Neuve Chapelle. May 9th Richebourg.—Mentioned in despatches, I.D.S.M. Wounded.

Judhia Sarki, I.D.S.M.—

Punjab Frontier, 1897-98.—Samana.—Tirah.—Medal with 3 clasps.
Operation on the Samana and in the Kurram Valley, 1897.
France, 1914-15.—November 2nd Neuve Chapelle. December 20th Festubert. March 10th-12th Neuve Chapelle. Granted I.D.S.M. for gallantry. Mentioned in despatches.

Gobind Sahi—

Chin Lushai Expedition, 1889-90.—Medal with clasp.
Punjab Frontier, 1897-98.—Samana, Tirah.—Medal with 3 clasps.

Balbir Thapa—

France, 1914-15.—November 2nd Neuve Chapelle. December 20th Festubert. March 10th-12th Neuve Chapelle. May 9th Richebourg. Marri Field Force, 1918.

Ramser Gurung—

France, 1914-15.—December 20th Festubert. March 10th-12th Neuve Chapelle. May 9th Richebourg. October 13th Givenchy.
Egypt.—Marri Field Force, 1918.

Padamdhoj Gurung—

France, 1914.—November 2nd Neuve Chapelle. Wounded. I.O.M., November 2nd.

Battu Gurung—

France, 1914.—November 2nd Neuve Chapelle. Wounded.

Bhasker Gurung—

N.-W. Frontier of India, Waziristan, 1901-02.—Medal with clasp.
Abor, 1911-12.—Medal with clasp.
France, 1914-15.—November 2nd Neuve Chapelle. December 20th Festubert. March 10th-12th Neuve Chapelle. September 25th Fanquissait. October 13th Givenchy.

Parsa Sing Thapa—

France, 1914-15.—November 2nd Neuve Chapelle. December 20th Festubert. March 10th-12th Neuve Chapelle. May 9th Richebourg. September 25th Fanquissait.

Tekbahadur Gharti—

France 1914-15.—November 2nd Neuve Chapelle. December 20th Festubert. Marri Field Force, 1918.

Bharat Sing Thapa—

France, 1914-15.—November 2nd Neuve Chapelle. Wounded. March 10th-12th Nueve Chapelle. May 9th Richebourg. September 25th Fanquissait. October 13th Givenchy.

Dalbahadur Gurung—
N.-W. Frontier of India, Waziristan, 1901-02.—Medal with clasp.
Abor Expedition, 1911-12.—Medal with clasp.

Champha Sing Gurung—
N.-W. Frontier of India, Waziristan, 1901-02.—Medal with clasp.
Abor Expedition, 1911-12.—Medal with clasp.

Gagan Sing Gurung—
N.-W. Frontier of India, Waziristan, 1901-02.—Medal with clasp,
Abor Expedition, 1911-12.—Medal with clasp.

Lal Sing Rana—
N.-W. Frontier of India, Waziristan, 1901-02.—Medal with clasp.
Abor Expedition, 1911-12.—Medal with clasp.

War, 1916-17—Indian Distinguished Service Medal Mentioned in despatches.

Harak Sing Gurung—
N.-W. Frontier of India, Waziristan, 1901-02.—Medal with clasp.

Dila Sing Gurung—
N.-W. Frontier of India, 1897-98, etc.—Medal with 2 clasps.
N.-W. Frontier of India, Waziristan, 1901-02.—Clasp.

Nain Sing Thapa—
N.-W. Frontier of India, Waziristan, 1901-02.—Medal with clasp.
Abor Expedition, 1911-12.—Medal with clasp.

Manbahadur Gurung—
N.-W. Frontier of India, Waziristan, 1901-02.—Medal with clasp.
Abor Expedition, 1911-12.—Medal with clasp.

Lalbahadur Sahi—
N.-W. Frontier of India, Waziristan, 1901-02.—Medal with clasp.
Abor Expedition, 1911-12.—Medal with clasp.

Nandbir Thapa—
N.-W. Frontier of India, 1897-98, etc., etc.—Medal with 2 clasps.
Tirah, 1897-98, etc.—Clasp.
N.-W. Frontier of India, Waziristan, 1901-02.—Clasp.
Abor Expedition, 1911-12.—Medal with clasp.

Maniram Khattri—

N.-E. Frontier of India, Manipur, 1891.—Medal with clasp.
N.-W. Frontier of India, 1897-98, etc., etc.—Medal with 2 clasps.
Tirah, 1897-98, etc.—Clasp.

Dalbir Gurung—

N.-W. Frontier of India, Waziristan, 1901-02.—Medal with clasp.

Kharakbir Thapa
N.-W. Frontier of India, Waziristan, 1901-02.—Medal with clasp.
Abor Expedition, 1911-12.—Medal with clasp.

Bala Sing Gurung—
Abor Expedition, 1911-12.—Medal with clasp.

Lachman Sing Thapa—
N.-W. Frontier of India, 1897-98.—Operations in the Samana during August and September. Relief of Gulistan.—Medal with 2 clasps.
Tirah, 1897-98.—Actions of Chagru Kotal, Dargai, Sampagha and Arhanga Passes. Operations in the Waran Valley and action of 16th November 1897. Operations at and around Dwatoi and action of 24th November 1897. Operations in the Bara Valley, 7th to 14th December 1897.—Clasp.
N.-W. Frontier of India, Waziristan 1901-02.—Clasp.

Parman Sing Gurung—
Abor Expedition 1911-12.—Medal with clasp.

3/-2nd King Edward's Own Gurkhas.

SUBADAR-MAJORS—

Lal Singh Thapa—
Relief of Chitral, 1895.—Medal and clasp.
Abor Expedition 1911-12.
France 1914-15.
Egypt, 1916.

SUBADARS—

Gamer Singh Gurung—
Konat Kurum Force and Tirah Expeditionary Force, 1897 98.—Actions of Dargai, Samana and Tirah.—Medal with 2 clasps.
Punjab Frontier, 1897-98.—Clasp.
Mesopotamia, 1916-17.

Jitman Gurung—
France, 1914-15.
Egypt, 1915.

Bag Sing Thapa, I.D.S.M.—
France, 1914-15.—Awarded Indian Distinghuished Service Medal. Wounded.
Egypt, 1915-16.

Tilbir Thapa—
N.-W. Frontier of India, Waziristan, 1901-02.—Medal with clasp.
Abor Expedition 1911-12.
Mesopotamia, 1916-17.—Wounded.

Bahadur Thapa—
N.-W. Frontier of India, Waziristan, 1901-02.—Medal with clasp.
Abor Expedition, 1911-12.
France, 1915.
Egypt, 1916.

Kalu Gurung—
France, 1914.—Wounded.

Lachman Singh Khattri—
France, 1915.
Egypt, 1915-16.

Jemadars.

Bhadra Sing Gurung—
Abor Expedition, 1911-12.—Medal with clasp.
France, 1914-15.
Egypt, 1915-16.

Narbir Gherti—
France, 1914-15.
Egypt, 1915-16.

Gopal Sing Rana—
Abor Expedition, 1911-12.
Mesopotamia, 1916—18.

Manbir Thapa—
France, 1914-15.
Egypt, 1915-16.

Jaman Sing Ale—
Mesopotamia, 1916.

Santok Pun—
Abor Expedition, 1911-12.
Mesopotamia, 1917 18.

Gambhir Sing Thapa—
Abor Expedition, 1911-12.

Gajman Rana—
Kohat and Kurram Force and Tirah Expeditionary force, 1897-98.
Action of Dargai 18 and 20 10-97.
Samanal 1897.—Medal with 3 clasps.
Tirah 1897 98-Punjab, Frontier, 1897-98.
N.-W. Front er of India Waziristan, 1901-02.—Medal and clasp.
France and Mesopotamia, 1914-16.

3rd (Queen Alexandra's Own) Gurkha Rifles.

SUBADAR-MAJORS—

Gunjbir Ale—
Burma, 1886-87.—Medal with clasp.
N.-W. Frontier of India, 1897-98.—Medal with clasp.
N.-W. Frontier of India, Waziristan, 1901-02.—Clasp.

Bhim Sing Thapa—
N.-W. Frontier of India, Waziristan, 1901-02.—Clasp.
War, 1914.—Indian Order of Merit, 2nd class, Army Department, No. 646, dated 9th July 1915.

SUBADARS—

Kirpa Ram Pun—
Burma, 1886-87.—Medal with clasp.
N.-W. Frontier of India, 1897-98.—Medal with 2 clasps.

Tul Sing Thapa—
N.-W. Frontier of India, Waziristan, 1901-02.—Medal with clasp.

Bakht Sing Rana—
N.-E. Frontier of India, 1897-98.—Medal with 2 clasps.
Tirah, 1897-98.—Operations in the Bara Valley, 7th to 14th December 1897.—Clasp.
N.-W. Frontier of India, Waziristan, 1901-02.

Lachhiman Gurung I.O.M.—
N.-E. Frontier of India, Lushai, 1892-93.
N.-W. Frontier of India, 1897-98.—Operations on the Samana and in the Kurram Valley during August and September 1897. Relief of Gulistan.—Medal with 2 clasps.
Tirah, 1897-98.—Actions of Chagru Kotal and Dargai. Capture of the Sampagha and Arhanga Passes. Operations at and around Dwatoi and action of the 24th November 1897. Operations in the Bara Valley, 7th to 14th December 1897.—Clasp.
N.-W. Frontier of India, Waziristan, 1901-02.

Dalkesar Gurung—
N.-E. Frontier of India, Lushai, 1892-93.
N.-W. Frontier of India, Waziristan, 1901-02.—Medal with clasp.
War 14-17.—Awarded the cross of the Order of St. George, 3rd class (*Russia*).

Shibcharan Burathoki—

N.-E. Frontier of India, Lushai, 1892-93.

N.-W. Frontier of India, 1897-98.—Operations on the Samana and in the Kurram Valley during August and September 1897. Relief of Gulistan.—Medal with 2 clasps.

Tirah, 1897-98.—Actions of Chagru Kotal and Dargai. Capture of the Sampagha and Arhanga Passes. Operations at and around Dwatoi and action of the 24th November 1897. Operations in the Bara Valley, 7th to 14th December 1897.—Clasp.

N.-W. Frontier of India, Waziristan, 1901-02.—Clasp.

Shamsher Gurung—

N.-W. Frontier of India, 1897-98.—Operations on the Samana and in the Kurram Valley during August and September 1897. Relief of Gulistan.—Medal with 2 clasps.

Tirah, 1897-98.—Action of Chagru Kotal and Dargai. Capture of the Sampagha and Arhanga Passes. Operations at and around Dwatoi and action of the 24th November 1897. Operations in the Bara Valley, 7th to 14th December 1897.—Clasp.

N.-W. Frontier of India, Waziristan, 1901-02.

N.-W. Frontier of India, 1897-98.—Clasp.

Chandrabir Thapa—

N.-W. Frontier of India, 1897-98.—Operations in the Samana.—Medal with 2 clasps.

N.-W. Frontier of India, Waziristan, 1901-02.—Clasp.

JEMADARS—

Puran Ale—

N.-W. Frontier of India, 1897-98.—Operations on the Samana and in the Kurram Valley during August and September 1897. Relief of Gulistan.—Medal with 2 clasps.

Tirah, 1897-98.—Action of Chagru Kotal and Dargai.

N.-W. Frontier of India, Waziristan, 1901-02.—Clasp.

Dhan Sing Ale—

N.-W. Frontier of India, 1897-98.—Operations on the Samana and in the Kurram Valley during August and September 1897. Relief of Gulistan. Medal with 2 clasps.

Tirah, 1897-98.—Actions of Chagru Kotal and Dargai. Capture of the Sampagha and Arhanga Passes. Operations at and around Dwatoi and action of the 24th November 1897. Operations in the Bara Valley, 7th to 14th December 1897.—Clasp.

Nathu Thapa—

Indian Medal of 1895.—Clasp.

Tirah, 1897-98.—Action of Chagru Kotal. Action of Dargai. Action of Sampagha Pass. Action of the Arhanga Pass. Reconnaissance of Saran Sar, and action of 11th November 1897. Operations in the Bara Valley, 7th to 14th December 1897.

N.-W. Frontier. of India, 1897-98.—Clasp.

Operations on the Samana, 1897.—Clasp.

N.-W. Frontier of India, 1901-02.—*Waziristan.*

Samsher Thapa—

India Medal of 1895.—With clasps.

Punjab Frontier, 1897-98—Samana, 1897—Tirah, 1897-98.

Tulbir Pun—

India Medal of 1895.—With clasps.

Tirah, 1897-98.—Action of Chagru Kotal. Action of Dargai. Action of Sampagha Pass. Action of the Arhanga Pass. Reconnaissance of the Saran Sar, and action of the 11th November 1897. Operations in the Bara Valley, 7th to 14th December 1897.

N.-W. Frontier of India, 1897-98.—Clasp.—Operations in the Samana and in the Kurram Valley during August and September 1897.

Samana, 1897.—Clasp Relief of Gulistan. Operations of the flying Column in the Kurrum Valley, 20th August to 1st October 1897.

Ram Chandra Alé—

India Medal of 1895.—With clasp.

Tirah, 1897-98.—Action of Chagru Kotal. Action of Dargai. Action of Sampagha Pass. Action of the Arhanga Pass. Reconnaissance of the Saran Sar, and action of 11th November 1897. Operation at and around Dwatoi and action of 24th November 1897. Operations in the Bara Valley, 7th to 14th December 1897.

N.-W. Frontier of India, 1897-98.—Clasp—Operations on the Samana and in the Kurram Valley during August and September 1897.

Samana 1897.—Clasp—Operations in the Samana and Kurrum Valley during August and September 1897.

N.-W. Frontier of India, 1901-02.—*Waziristan.* Clasp.

Padam Sing Thapa—

N.-W. Frontier of India, Waziristan, 1901-02.—Medal with clasp.

2-3rd Queen Alexandra's Own Gurkha Rifles.

SUBADARS—

Arjun Sing Thapa—

Waziristan, 1901-02.

France, 1914-15.

Egypt, 1915—18.

Dhanbir Thapà—

Chitral, 1895.—Relief of Chitral.—Medal with clasp.

France, 1914-15.

Balbir Thapa, I.O.M.—

N.-W. Frontier of India, 1897-98.—Operations on the Samana and in the Kurram Valley during August and September 1897. Relief of Gulistan.—Medal with 2 clasps.

Tirah, 1897-98.—Actions of Chagru Kotal and Dargai. Capture of the Sampagha and Arhanga Passes. Operations at and around Dwatoi and action of the 24th November 1897. Operations in the Bara Valley, 7th to 14th December 1897.—Clasp.

France, 1914-15.

Egypt, 1915—18.

Dalbir Gurung—

France, 1914-15.

Egypt, 1915—18.

Hari Chand—

Chitral, 1895.—Relief of Chitral.—Medal with clasp.
Tibet, 1903-04.—Medal.
France, 1914-15.

Gopal Rana—

Burma, 1886-87.
Chitral, 1895.
France, 1915.
Egypt, 1915—17.

An Sing Thapa—

Chitral, 1895.—Relief of Chitral.—Medal with clasp.
N.-W. Frontier of India, 1897-98.—Clasp.
France, 1915.
Egypt, 1915-16.

Bhawan Sing Sahi—

Egypt, 1916.

Tula Gurung—

France, 1914-15.
Egypt, 1915-16.

Gajbir Thapa—

Chitral, 1895.—Relief of Chitral.—Medal with clasp.
N.-W. Frontier of India, 1897-98.—Clasp.

Damar Sing Gurung, M.C.—

France, 1914-15.
Egypt, 1915—18.

Bagbir Gharti—

Egypt, 1916—18.

Dhanbar Gurung—

France, 1914-15.
Egypt, 1915—18.

JEMADARS—

Bhagatbir Thapa—

France, 1914-15.
Egypt, 1915—18.

Dalbahadur Nagarkoti—

Chitral, 1895.

Dhanbir Thapa—

France, 1914-15.
Egypt, 1917-18.

Bahadur Gurung—

France, 1914-15.
Egypt, 1915—18.

Churamani Thapa—

Chitra, 1895.
Egypt, 1918.

Harak Sing Gharti, I.O.M.—

France, 1914-15.
Egypt, 1915—18.

Kharke Rana—

Waziristan, 1901-02.
France, 1914-15.
Egypt, 1917.

Gane Gurung, I.O.M.—

France, 1914-15.
Egypt, 1917-18.

Bhawansing Rana—

Chitral, 1895.

Man Sing Basnet—

France, 1914-15.
Egypt, 1915—18.

Man Sing Mall—

France, 1914-15.
Egypt, 1915—18.

3-3rd Queen Alexandra's Own Gurkha Rifles.

SUBADAR-MAJOR—

Bahadur Thapa—

India Medal of 1895.—Clasp.
Tirah, 1897-98.—Action of Chagru Kotal. Action of Dargai. Action of Sampagha Pass. Action of the Arhanga Pass. Reconnaissance of the Saran Sar, and action of 11th November 1897.—Clasp.
N.-W. Frontier of India, 1897-98.—Operations on the Samana and in the Kurram Valley during August and September 1897.—Clasp.
Samana, 1897.—Relief of Gulistan. Operation of the flying column in the Kurram Valley under Colonel Richardson, 20th August to 1st October 1897.
France, 1914-15.
Egypt, 1915—18.
Waziristan, 1901-02.
France, 1914-15.
Egypt, 1917-18.

SUBADARS.

Karan Sing Thapa—

Chitral, 1895.
Egypt, 1917.

Rupdhan Pun—

France, 1914-15.
Egypt, 1915—18.

Damar Sing Gurung—

Chitral, 1895.
France, 1914-15.
Egypt, 1915—18.

Hari Sing Thapa—

France, 1914-15.
Egypt, 1917—18.

Nain Sing Thapa—

France, 1914-15.
Egypt, 1915-18.

Durgamani Thapa—

Chitral, 1895.—Relief of Chitral.—Medal with clasp.
Tirah, 1897-98.—Clasp.
France, 1915—18.
Egypt, 1915—18.

Birkhdhoz Limbu—

Hpimow Column, December 1910 *to April* 1911.
Egypt, 1916—18.

JEMADARS—

Lalbahadur Gurung, I.O.M.

France, 1914-15.
Egypt, 1915—18.

Kalu Gurung—

France, 1914-15.
Egypt, 1915-18.

Ajab Sing Gurung, I.D.S.M.—

France 1915.
Egypt 1915—18.

Narain Sing Sahi—

France, 1914-15.
Egypt, 1917-18.

Jaman Sing Thapa, I.O.M.—

France, 1914 15.
Egypt, 1916—18.

Birbhan Sing Thapa—

France, 1914-15.
Egypt, 1915—18.

Jabar Sing Thapa—

France, 1914-15.
Egypt, 1915—18.

Haraksing Gurung—

France, 1914-15.
Egypt, 1915—18.

Rup Sing Thapa—

France, 1914-15.
Egypt, 1915—18.

Dalbir Thapa—

Egypt, 1917.

Judbir Ale—

Chitral, 1895.
Egypt, 1917-18.

Damar Sing Gharti—

France, 1914-15.
Egypt, 1915.
Egypt, 1917-18.

Balbahadur Thakur—

France, 1914-15.
Egypt, 1915.
Egypt, 1917-18.

Chakrabahadur Gurung—

Egypt, 1917-18.

Kulbang Sing Sunwar—

Egypt, 1917-18.

4-3rd Queen Alexandra's Own Gurkha Rifles.

SUBADAR-MAJORS—

Kharaksing Rana, M.C.—

N.-W. Frontier of India, 1897-98.—Operations on the Samana and the Kurram Valley during August and September 1897.—Medal with 2 clasps.

Tirah, 1897-98.—Action of Chagru Kotal.—Clasp.

China, 1900.—Medal.

France, 1914.—("1914" Star and Military Cross.

SUBADARS—

Indrajit Thapa—

Gallipoli, 1915.
Egypt, 1914-15-16.

Narbir Thapa—

N.-W. Frontier of India, 1897-98.—Operations in the Samana during August and September, Relief of Gulistan.—Medal with 2 clasps.

Tirah, 1897-98.—Actions of Chargu Kotal, Dargai, Sampagha and Arhanga Passes. Operations in the Waran Valley and action of 16th November 1897. Operations at and around Dwatoi and action of 24th November 1897. Operations in the Bara Valley, 7th to 14th December 1897. —Clasp.

N.-W. Frontier of India, Waziristan, 1901-02.—Clasp.

Abor Expedition, 1911-12.—Medal with clasps.

War of 1914.—General Service, France, 1915.

Gopal Gurung—

N.-W. Frontier of India, 1897-98.—Medal and clasp.
Gallipoli, 1915.
Egypt, 1914-15-16.

Dhanraj Pun—

Gallipoli, 1914-15.

Puran Ale—

N.-W. Frontier of India, 1897-98.—Operations on the Samana and in the Kurram Valley during August and September 1897.

Relief of Gulistan.—Medal with 2 clasps.

Tirah, 1897-98.—Action of Chagru Kotal and Dargai.

N.-W. Frontier of India, Waziristan, 1901-02.—Clasp.

Sarabjit Gurung—

Cross of the Order of St. George, 4th Cl.

N.-W. Frontier of India, Waziristan, 1901-02.—Medal with clasp.

Abor Expedition, 1911-12.—Medal with clasp.

France, Egypt, 1914—16.—"1914—Star."

Partiman Rana—

N.-W. Frontier of India, 1897-98.—Operations on the Samana and in the Kurram Valley during August and September 1897.
Operations of the Flying Column in the Kurram Valley under Col. Richardson, 20th August to 1st October 1897—Medal with 2 clasps.

Tirah, 1897-98.—Reconnaissance of the Kharmana Defile and action of 7th November 1897. Operations against the Khani Khel Chamkanis. —Clasp.

Gallipoli, 1915.

Egypt, 1914-15-16.

JEMADARS—

Ranpati Gurung—

N.-W. Frontier of India, 1897-98.—Operations on the Samana during August and September 1897. Relief of Gulistan.—Medal with 2 clasps.

Tirah, 1897-98.—Action of Chagru Kotal and Dargai. Actions of Sampagha and Arhanga Passes. Operations in the Warran Valley, and action of 16th November 1897. Operations at and around Dwatoi, and action of 24th November 1897. Operations against the Khani Khel Chamkanis. Operations in the Bara Valley, 7th to 14th December 1897.—Clasp.

Granted Indian Distg. Service Medal for gallantry in the Field on 2nd November 1914.

General Service in France, 1914.—"1914—Star" —Mentioned in Despatches.

Bajiram Pun—

N.-W. Frontier of India, 1897-98.—Operations on the Samana and in the Kurram Valley during August and September 1897. Relief of Gulistan.—Medal with 2 clasps.
Tirah, 1897-98.—Actions of Chagru Kotal and Dargai. Capture of the Sampagha and Arhanga Passes. Operations at and around Dwatoi and action of the 24th November 1897. Operations in the Bara Valley, 7th to 14th December 1897.—Clasp.

Dhansing Ale—

N.-W. Frontier of India, 1897-98.—Operations on the Samana and in the Kurram Valley during August and September 1897. Relief of Gulistan.—Medal with 2 clasps.

Tirah, 1897-98.—Actions of Chagru Kotal and Dargai. Capture of the Sampagha and Arhangha Passes. Operations at and around Dwatoi and action of the 24th November 1897. Operations in the Bara Valley, 7th to 14th December 1897.—Clasp.

Kumsing Gurung—

Gallipoli, 1914.

Shamsher Gurung—

Gallipoli, 1915.
Egypt, 1914-15-16.

Singbir Gurung—

Egypt, 1914.
Gallipoli, 1915.

Bhabesur Gurung—

Egypt, 1916.

Harkraj Gurung—

Egypt, 1916.

Damarsing Thapa—

Egypt, 1916.

4th Gurkha Rifles.

SUBADAR-MAJORS—

Sher Sing Rana—

N.-W. Frontier of India, 1891.—Medal and clasp.
N.-W. Frontier of India, 1894-95.—Clasp.
China, 1900-01.—Medal.
France, 1914.

Chamu Thapa—
N.-W. Frontier of India, and 1894-95.—Medal and clasp.
China, 1900-01.—Medal.
France, 1914.

SUBADARS—

Rugbir Gurung—
N.-W. Frontier of India, 1894-95.—Medal and clasp.
China, 1900-01.—Medal.

Pirthi Singh Rana—
N.-W. Frontier of India, 1891.—Medal and clasp.
N.-W. Frontier of India, 1894-95.—Clasp.
China, 1900-01.—Medal.
Tirah, 1897-98.—Medal with 2 clasps.
France, 1914.

Ramkishan Rana, I.D.S.M.—
China, 1900.—Medal.
France, Egypt, Gallipoli, 1914-15.
S. Waziristan. Force, 1917.
Marri, P. F. F., 1918.—Indian Distinguished Service Medal.

Surdal Thapa—
N.-W. Frontier of India, 1897-98.—Medal with clasp.
Tirah, 1897-98.—Clasp.
China, 1900.—Medal.
France, 1914-15.
S. Waziristan Force, 1917.

Harikishen Gurung—
China, 1900-01.—Medal.
France, Egypt, 1914-16.
S. Waziristan Force, 1917.
Marri, P. F. F., 1918.

Ramkisen Dura—
China, 1900-01—Medal.
France, Egypt Gallipoli, 1914-16.
S. Waziristan Force, 1917.
Marri P. F. F. 1918.

Gunjbir Gurang, I. O. M.—
France, Egypt, Gallipoli, 1914-16.
S. Waziristan Force, 1917.
I. O. M., 2nd Class.

JEMADARS—

Pahal Sing Gurung—
China, 1900-01.—Medal.
France, Egypt Gallipoli, 1914-16.
S. Waziristan Force, 1917.
Marri P. F. F. 1918.

Balbahadur Gurung—
N.-W. F., 1897-98.—Medal and clasp.
Tirah, 1897-98—Clasp.
France, Egypt, Gallipoli 1914-16.
S. Waziristan Force, 1917.

Narbir Thapa—
N.-W. Frontier, 1897-98.—Medal and clasp.
Tirah, 1897-98.—Clasp.
China. 1900-01.—Medal.
Gallipoli, 1915.

Budhibal Thapa—
China, 1900-01.—Medal.
France, Egypt, Gallipoli, 1915-16.

Lohasur Gurung—
France, Egypt, Gallipoli, 1914-16.
S. Waziristan Force, 1917.
Marri P. F. F., 1918.

Hira Sing Thapa—
France, Egypt, Gallipolli, 1914-16.
S. Waziristan, F. F. 1917.
Marri, P. F. F, 1918.

Chanderbir Thapa—
France, Egypt, Gallipolli, 1914-16.
S. Waziristan Force, 1917.
Marri P. F. F., 1917.

Jitbahadur Rana—
France, 1914-15.

Ganj Sing Gurung—
France Egypt, Gallipoli, 1914-16.
S. Waziristan Force, 1917.

Ragbir Thapa, I.D.S.M.—
France, Egypt, Gallipoli, 1914-16.
S. Waziristan Force, 1917.
Marri, P. F. F., 1918.—Indian Distinguished Service Medal.

Lokbir Ale, I. D. S. M.—
France, Egypt, Gallipoli, 1914-16.
S. Waziristan Force, 1917.
Marri, P. F. F., 1918—Indian Distinguished Service Medal.

Jaspati Gurung—
France Egypt, 1914-15.
S. Waziristan Force, 1917.

Randhoj Gurung—
France, Egypt, 1914-15.
Marri P. F. F., 1918.

Amer Sing Pun—
France, Egypt, Gallipoli, 1914-16.
S. Waziristan Force, 1917.—Indian Distinguished Service Medal.

Harimard Rana—
France, Egypt, Gallipoli, 1914-16.
Marri, P. F. F., 1918.

Jitman Gurung—
France, Egypt Gallipoli, 1914-16.
Marri, P. F. F., 1918.

Shamsher Jung Thakur—
France, Egypt, Gallipoli, 1914-16.
S. Waziristan Force, 1917.

SUBADAR-MAJOR—

Debi Chand, Sardar Bahadur—
N.-E. Frontier of India, Chin Lushai, 1889-90.—Medal with clasp.
N.-E Frontier of India, Manipur, 1891.—Clasp.
Chitral, 1895.—Relief of Chitral—Medal with clasp.
N.-W. Frontier of India, 1897-98.—Clasp.

SUBADARS—

Bansraj Gurung Bahhdur—
N.-E. Frontier of India, Chin-Lushai, 1889-90.—Medal with clasp.
N.-E. Frontier of India, Manipur, 1891.—Clasp.
Chitral, 1895.—Relief of Chitral.—Medal with clasp.
N.-W. Frontier of India, 1897-98.—Clasp.
Tirah, 1897-98.—Clasp.
China, 1900.—Medal.

Madan Sing Rai—
N.-E. Frontier of India, Chin-Lushai, 1889-90.—Medal with clasp.
N.-E. Frontier of India, Manipur, 1891.—Clasp.
Chitral, 1895.—Relief of Chitral.—Medal with clasp.
N.-W. Frontier of India, 1897-98.—Clasp.
Tirah, 1897-98.—Clasp.

Dhan Lal Gurung—
Chitral, 1895.—Relief of Chitral.—Medal with clasp.
N.-W. Frontier of India, 1897-98.—Clasp.
Tirah, 1897-98.—Clasp.

Harisingh Thapa—
Chitral, 1895.—Relief of Chitral.—Medal with clasp.
N.-W. Frontier of India, 1897-98.—Clasp.
Tirah, 1897-98.—Clasp.

Antbir Gurung, I. O. M.—
N.-W. Frontier, 1897-98.—Medal and clasp.

Pahal Sing Thapa—
China, 1900-02.—Medal.

Mahabir Thakur Bahadur—
N.-W. Frontier of India, 1897-98.—Medal with clasp.
Tirah, 1897-98.—Clasp.

Ramsing Sahi—
Chitral, 1895.—Relief of Chitral.—Medal with clasp.
N.-W. Frontier of India, 1897-98.—Clasp.
Tirah, 1897-98.—Clasp.

Jaman Sing Gurung—
China, 1900-02.—Medal.

JEMADARS—

Kharkbir Ale—
China, 1900-01.—Medal.

Narbir Thapa—
N.-W. Frontier of India, 1897-98.—Medal with clasp.
Tirah, 1897-98.—Clasp.
China, 1900-02.—Medal.

5th Gurkha Rifles (Frontier Force).

SUBADAR MAJOR—

Dhanraj Gurung, I. O. M.—
N.-W. Frontier of India, Isazai, 1892.
Tirah, 1897-98.—Reconnaissance of the Kharmana Defile and action of 7th November 1897. Operations against the Khani Khel Chamkanis.—Medal with 2 clasps.

SUBADARS—

Lal Sing Gurung—
N.-W. Frontier of India, Hazara, 1891.—Medal with clasp.
N.-W. Frontier of India, Waziristan, 1894-95.—Clasp.
N.-W. Frontier of India, 1908.—Operations in the Zakka Khel country.—Medal with clasp.

Mansaram Pun—
N.-W. Frontier of India, Hazara, 1891.—Medal with clasp.
N.-W. Frontier of India, Waziristan, 1894-95.—Clasp.
N.-W. Frontier of India, 1897-98.—Medal with clasp.
N.-W. Frontier of India, 1908.—Operations in the Zakka Khel country.—Medal with clasp.

Hastbir Thapa—
N.-W. Frontier of India, Hazara, 1891.—Medal and clasp.
N.-W. Frontier of India, Isazai, 1892.
N.-W. Frontier of India, Relief of Chitral, 1895.—Medal and clasp.
N.-W. Frontier of India, Punjab Frontier, 1897-98.—Clasp.
N.-W. Frontier of India, Tirah, 1897-98.—Clasp. Actions of the Chagru Kotal and of the Sampagha and Arhanga Passes. Reconnaissance of the Saran Sar and action of 11th November 1897. Operations at and around Dwatoi and action of 24th November 1897. Operations against the Khani Khel Chamkanis. Operations in the Bara Valley, 7th to 14th December 1897.—Operations in the Bazar Valley, 25th to 30th December 1897.
N.-W. Frontier of India, 1908.—Operations in the Zakka Khel country.—Medal with clasp.

Dal Sing Thapa—
N.-W. Fronttier of India, Hazara, 1888.—Medal with clasp.
N.-W. Frontier of India, Hazara, 1891.—Clasp.
N.-W. Frontier of India, Isazai, 1892.
N.-W. Frontier of India, Waziristan, 1894-95.—Clasp.

Budhiman Gurung—
N.-W. Frontier of India, 1897-98.—Medal with 2 clasps.

JEMADARS—

Balbir Rana—
N.-W. Frontier of India, Waziristan, 1901-02.—Clasp.
N.-W. Frontier of India, 1908.—Operations in the Zakka Khel country.—Medal and clasp.

Sukhlal Rana—
N.-W. Frontier of India, 1897-98.—Medal with 2 clasps.

Tikaram Thapa—
N.-W. Frontier of India, 1908.—Operations in the Zakka Khel country.—Medal and clasp.

Kumbh Sing Gurung—
N.-W. Frontier of India, 1897-98.—Medal and clasp.

Tejbir Thapa I.O.M.—
N.-W. Frontier of India, 1908.—Operation in the Zakha Khel country.—Medal and clasp.

Bhawanand Thapa—
N.-W. Frontier of India, 1908.—Operations in the Zakha Khel country.—Medal and clasp.

Narpati Bura—
N.-W. Frontier of India, 1908.—Operation in the Zakha Khel country.—Medal and clasp.

Goria Gurung—
N.-W. Frontier of India, 1908.—Operations in the Zakha Khel country.—Medal and clasp.

Pirthibir Burathoki—
N.-W. Frontier of India, 1908.—Operations in the Zakha Khel country.—Medal and clasps.

2-5th Gurkha Rifles F. F.

SUBADAR-MAJOR—

Amarsing Thapa, M. C.—
Sardar Bahadur.—Order of British India, 1st class.
N.-W. Frontier of India, Hazara, 1888.—Medal with clasp.
N.-W. Frontir of India Hazara, 1891.—Clasp.
N.-W Frontier of India, Isazai, 1892.
N.-W. Frontier of India, Waziristan, 1894-95.—Clasp.
Tirah, 1897-98.—Actions of Chagru Kotal and of the Sampagha and Arhanga Passes.—Reconnaissance of the Saran Sar and action of 11 November 1897.—Operations at and around Dwatoi and action of 24th November 1897.—Operations against the Khani Khel Chamkanis. Operations in the Bara Valley, 7th to 14th December 1897.—Medal with 2 clasps.
N. W. Frontier of India 1908.—Operations in the Zakka Khel country.—Despatches.—Medal with clasp.

SUBADARS—

Ramsing Thapa—
N.-W. Frontier of India, Hazara, 1891.—Medal with clasp.
N.-W. Frontier of India, Isazai, 1892.
N.-W. Frontier of India, 1897-98.—Medal with clasp.
N.-W. Frntier of India 1908.—Operstions in the Zakka Khel country.—Medal with clasp.

Bilbikram Rana—
N.-W. Frontier of India, Hazara 1888.—Medal with clasp.
N.-W. Frontier of India, Hazara, 1891.—Clasp.
N.-W. Frontier of India, 1897-98.—Medal with clasp.
N.-W. Frontier of India, 1908.—Operations in the Zakka Khel country.—Medal with clasp.

Pahal Sing Rana—
N.-W. Frontier of India, 1897-98.—Medal and clasp.
N.-W. Frontier of India, Tirah, 1897-98.—Clasp. Actions of Chagru Kotal and the Sampagha and Arhanga Passes. Reconnissance of the Saran Sar and action of 11th November 1897. Operations at and arround Dwatoi and action of 24th November 1897. Operations against the Khani Khel Chamkanis. Operations in the Bara Valley 7th to 14th December 1897. Operations in the Bazar Valley, 25th to 30th December 1897.
N.-W. Frontier of India, Waziristan, 1901-02—Clasp.
N.-W. Frontier of India, 1908.—Operations in the Zakka Khel country.—Medal and clasp.

Arjun Thapa—
N.-W. Frontier of India, 1908.—Operation in the Zakka Khel country.—Medal and clasp.

Puranbir Rana—
N.-W. Frontier of India, 1897-98.—Medal and clasp.
N.-W. Frontier of India, 1908.—Operations in the Zakka Khel country.—Medal and clasp.

Maidhar Gharti—
N.-W. Frontier of India, 1908.—Operations in the Zakka Khel country.—Medal and clasp.

JEMADARS—

Padamsing Thapa—
N.-W. Frontier of India, Hazara, 1891.—Medal and clasp.
N.-W. Frontier of India, Isazai, 1892.
N.-W. Frontier of India, 1908.—Operations in Zakha Khel country.—Medal with clasp.

Ransur Pun—
N.-W. Frontier of India, 1908.—Operations in the Zakka Khel country.—Medal with clasp.

Nainsing Thapa—
N.-W. Frontier of India, 1908.—Operations in the Zakka Khel country.—Medal with clasp.

Dilmansing Thapa—
N.-W. Frontier of India, 1908.—Operations in the Zakka Khel country.—Medal with clasp.

Puransing Thapa—
N.-W. Frontier of India, 1908.—Operations in the Zakka Khel country.—Medal with clasp.

Chandarsing Gurung—
N.-W. Frontier of India, 908.—Operations in the Zakka Khel country.—Medal with clasp.

3rd Battalion 5th Gurkha Rifles (F. F.).

SUBADARS—

Ty. Sub.-Maj. Kalu Gurung—
N.-W. Frontier, Waziristan, 1894-95.—Medal and clasp.
China, 1900.—Medal.
The War of 1914-18.
France, 1914-15.

Bhagtabir Thapa—
N.-W. Frontier, Waziristan, 1894-95.—Medal with clasp.
China, 1900.—Medal.
The War of 1914-18.
France, 1914-15.
Gallipoli, August 1915.

Hastabir Thapa—
N.-W. Frontier of India, 1897-98.—Mohmad—Medal with clasp.
Tirah, 1897-98.—Action of the Sampagha Pass and Arhang Pass. Operation in the Bazar Valley from 25th to 30th December 1897—Clasp.

Karbir Gurung—
N.-W. Frontier of India, 1897-98.—Mohmad.—Medal with clasp.
Tirah, 1897-98.—Action of the Sampagha Pass and Arhang Pass. Operation in the Bazar Valley from 25th to 30th December 1897.—Clasp.

Embahadur Sahi—
France, 1914-15.

Lalitbahadur Mal—
France and Mesopotamia, 1914-16.

Sarune Gurung—
China, 1900.
The War, 1914-18.
France and Gallipoli, 1915-16.

Dharamdhoj Limbu—
"Sewla Bay" Gallipoli, 1915.

JEMADARS—

Tekbahadur Sahi—
The War, 1914-18.
France, 1914-15.—Wounded.

Dharamraj Gurung—
The War, 1914-18.—France.

Kubir Sing Adhikari—
The War, 1914-18, France.
Mesopotamia, 1914-16.—Wounded.

Hemlal Khattri—
The War. 1914-18.
France and Mesopotamia, 1914-16.

1-6th Gurkha Rifles.

JEMADAR—

Ahmed Nur—
N. W. Frontier of India, Waziristan, 1901-02—Medal with clasp.

2-6th Gurkha Rifles.

SUBADARS—

Parbir Thapa—
N.-E. Frontier of India, Manipur, 1891.—Clasp.

Gamprasad Thapa—
N.-W. Frontier of India, 1908.—Operations in Zakha Khel country.—Medal with calasp.

Subadar Major.

BHADRABIR THAPA.—

Subadars—
Dardanelles, 1915.

Mohantbir Rana—
Tachi Valley 1914-15.
N. W. Frontier Tirah and Samman, 1897-98.
Mesopotamia, 1915.

Ranbir Thapa—
N. W. Frontier Tirah and Samman, 1897-98.
Abor Expendition, 1911-10.
Mesopotamia 1916.

Motilal Limbu—
Dardanelles, 1914.

Birbahadur Lama—
Dardanelles, 1914.

Pirthibir Bura—
N. W. Frontier, 1908—Medal with clasp.

Bompahadur Gurung—
France, 1914.

Jamansing Karki—
Chitral Relief. 1895.
Tirah, 1897-98.
France, 1915.
Mesopotamia, 1916.

Jemadars—

Mahabir Thapa—
France, 1914.

Kesar Ale—
Dardanelles 1915.

Pirthilal Guraug—
France, 1914.

Khimu Rana—
Dardanelles, 1914.

Motilal Thapa—
Dardanelles, 1914.

Partasing Gurung,—
Dardanelles, 1914.

Daljit Pun—
France 1915.
France 1915.
Dardanelles, 1915.

Gaganbahadur Sahi—
France 1915.
Mesopotamia, 1915.

Narbahadur Gurung—
Dardanelles, 1915.

Bhujram Sahi—
France, 1915.
Mesopotamia 1915.

1-7th Gurkha Rifles.

SUBADAR-MAJORS—

Hasturam Gurung—
Tirah, 1897-98.—Medal with clasp.
Mesopotamia. 1917-18.

SUBADARS—

Gopal Singh Rai—
N.-E. Frontier of India, Chin Hills, 1892-94.—Medal with clasp.
Seistan Field Force. 1917.
Mesopotamia, 1917-18.

Rajpal Rai Bahadur—
N.-W. Frontier of India, Waziristan, 1894-95.—Medal with clasp.
N.-W. Frontier of India, 1897-98.—Mohmand.—Medal with clasp.
Tirah, 1897-98.—Clasp.
Tibet, 1903-04.—Operation at and around Gyantse between 5th May and 6th July 1904. March to Lhassa, 14th July to 3rd August 1904.—Medal with clasp.
Dardanelles, 1915.
Mesopotamia, 1915-18.
Palestine, 1918.—Order of British India 2nd Class.

Agam Limbu—
N.-W. Frontier of India, Waziristan, 1894-95.—Medal with clasp.

Padamdhoj Gurung—
N.-W. Frontier of India, Waziristan, 1894-95.—Medal with clasp.
Tirah, 1897-98.—Clasp.
N.-W. Frontier of India, Waziristan, 1901-02.—Clasp.
Me Kron Mission, 1916-18.

Makardhoj Rai—
Seistan Field Force 1917.
Mesopotamia 1917-18.

Mandhoj Rai—
Seistan Field Force, 1917.
Mesopotamia, 1917-18.

Gajarj Rai—
Mesopotamia, 1916-17-18.
Palestine, 1918.

Jitbahadur Rai—
France, 1915.
Mesopotamia, 19 6-17.
Palestine, 1918.

Judhbir Thapa—
Operations in the Punjab Frontier, 1897-98, including the Mohmand and Tirah Expeditions.—Medal with 2 clasps.
Mesopotamia, 1917-18.
Palestine, 1918.

Singbahadur Limbu—
France, 1915.
Mesopotamia, 1916-18.
Palestine, 1918.

KulmanRai—
Mesopotamia, 1216-17.

Lahang Limbu—
Dardanelles, 1915.
Mesopotamia, 1916-18.
Palestine, 1918.

Chehar Singh Thapa—
N.-W. Frontier of India, Waziristan, 1901-02.—Medal with clasp.
Seistan Field Force 1917.
Mesopotamia, 1917-b8.

Lachman Limbu—
Mesopotamia 1916-18.
Palestine 1918.

Balandhar Rai—
Egypt 1915.
Mesopotamia 1916-18.
Palestine 1918.

Harkabahadur Limbu—
Mesopotamia 1917-18.

Bhogbir Limbu—
Seistan Field Force 1917.
Mesopotamia 1917-18.

Kharaksing Limbu—
Mesopotamia 1917-18.

JEMADARS—

Lalbir Thapa—
India Medal, 1895, with clasps.
"Punjab Frontier 1897-98", "Samana 1897," "Tirahi 1893-98".
N.-W. Frontier of India, Waziristan, 1901-02.—Clasp. Delhi Durbar Coronation Medal, 1912.

Purandhoj Limbu—
Served with Chitral relief, Tirah and Frontier India.—Medal with clasps for relief of Chitral 1895.
Tirah, 1897-98.—Action of the Sampagha Pass. G. G. O. 58 of 1898. Action of the Arhanga Pass. G. G. O. 58 of 1898. Reconnaissance of the Saran Sar and action of the 9th November 1897, G. G. O. 244 of 1898. Operations against the Khani Khel Chamkannis. G. G. O. 244 of 1898. Operations in the Bazaar Valley 25th to 30th December 1897. G. G. O. 244 of 1898.
China, 1900-02.—1 Medal.
Mesopotamia, 1917-18.

Gristiram Rai—
France, 1915.
Mesopotamia, 1916-18.
Palestine, 1918.

Narbir Limbu—
France, 1915.
Mesopotamia, 1916.

Surjman Rai—
Seistan Field Force, 1917.
Mesopotamia, 1917-18.

Jaspatti Limbu—
Sikkim Tibet, 1903-04.—Taking of Gyantse Medal.
France, 1915.
Mesopotamia, 1916-18.
Palestine, 1918.

Jasbir Lama—
France, 1915.
Mesopotamia, 1916-18.
Palestine, 1918.

Hastabir Limbu—
Seistan Field Force, 1917.
Mesopotamia, 1917-18.

Padamlal Limbu—
Seistan Feld Force, 1917.
Mesopotamia, 1917-18.

Narsing Rai—
Seistan Field Force, 1917.
Mesopotamia, 1917-18.

Rajhdhoj Rai—
Mekran Mission, 1916-17.
Mesopotamia, 1917-18.

Guruising Gurung—
Seistan Field Force, 1917.
Mesopotamia, 1918.

Janaklal Lama—
France, 1915.
Mesopotamia, 1916-18.
Palestine, 1918.

Rupdal Rai—
Seistan Field Force, 1917.
Mesopotamia, 1917-18.

Jogbir Limbu—
France, 1915.
Seistan Field Force, 1917.
Mesopotamia, 1917-18.

Lalbahadur Rai—
Mesopotamia, 1917-18.

Hastabir Limbu—
Mesopotamia, 1917-18.

Amberdhoj Limbu—
Mekran Mission, 1916-17.
Mesopotamia, 1917-18.

Harkabahadur Limbu—
Seistan Force, 1917.
Mesopotamia, 1917-18.

2-7th Gurkha Rifles.

SUBADARS—

Karkbir Ale—
N.-E. Frontier of India, China Hills, 1892-93.—Medal with clasp.
Mesopotamia, 1914-16.

Balbahadur Limbu.—Order of British India 2nd class.
Relief of Chitral—Medal with clasp.
Tirah, 1897-98.
Saran Sar, 1907.
Mesopotamia, 1914-16.

Birihang Limbu—
Mesopotamia, 1914-16.

Gangabir Rana, M.V.O.
India Medal, 1895.
Punjab Frontier-Tirah-1897-98.
N.W.F. Frontier of Waziristan, 1901-02.
Mesopotamia 1915-17.

Daulbir Rai—
Mesopotamia, 1914.

Kubersing Limbu, I.O.M., I.D.S.M.
N.E. Frontier of India Chin Lushai, 1889-90.
Chitral 1895.
N.W. Frontier, 1897-98.
Tirah, 1897-98.
War, 1914-15.

Bhudbir Thapa—
Sinia, 1892.
Sana Column, 1895-96.
War, 1916-17.

Shyamlal Khettri—
N.W. Frontier, 1897-98.—Medal with clasp.
War, 1914-15.

Bagbir Rai—
N.W. Frontier Manipur, 1891.—Medal with clasp.
N.W. Frontier, 1897-98.
N.W. Frontier of India, 1897-98.
N.W. Frontier of India, Waziristan, 1901-02.
War, 1914-15.

Narjit Rana—
War, 1914-15.

Tekbir Limbu—
War, 1914-15.

JEMADARS—

Birdhoj Limbu—
Mesopotamia, 1914-15.

Sahabir Thapa, I.O.M.
Mesopotamia, 1914-15.

Bahardhan Rai—
Mesopotamia, 1914-16.

Mahadol Gurung—
S. Lushai Hills, 1892-93.

India Medal with clasp 1895—
Punjab Frontier, 1897-98.
War, 1914-15.

3-7th Gurkhas.

SUBADAR MAJOR—

Mansing Thapa—
Operations in Persian Gulf, 1912-13.
Mesopotamia Expedy. Force, November, 1915—July 1917.

SUBADARS—

Kulman Rai—
Mesopotamia Expedy. Force, June 1916—April 1918.

Birbahadur Limbu—
Punjab Frontier, 1897-98—Mohmand and Tirah. Expeditions.—1 Medal and 2 clasps.
Mesopotamia, Expedy. Force Oct. 1914—June 1917. Wounded.
Marri Field Force, Feb. 1918—May 1918.

Tilokbahadur Limbu—
Operations in Persian Gulf, 1912-13.

Birbahadur Limbu—
Mekrán (Seistan), 1916-17.

Rarot Ram Rava—
Naga Expedy. Force March 1906—July 1906.
Aishankuki Promenade Feby. 1910—May 1910.
Makware Expedition, Jany 1911—April 1911.
Aka Expedition, 1913-14.
Mesopotamia Expedy. Force May 1916—June 1917.

JEMADARS—

Narjit Gurung—
Mesopotamia Expedy. Force April 1916—April 1918.

Megraj Limbu—
China, 1900.—Medal.
Mesopotamia Expedy. Force, 1914-16 and 1916-17.

Bhaktabir Limbu—
Marri Field Force Feb. 1918—May 1918.

Bahadur Gurung—
Operation in Persian Gulf, 1912-13.

Dambardhoj Limbu—
Operations in Persian Gulf, 1912-13.

Gunjaman Rai—
Marri Field Force Feb. 1918—May 1918.

Digbahadur Rai—
Mesopotamia Expedy. Force Oct. 1914—May 1917.

Ronlal Rai—
France June 1915—Nov. 1915.
Mesopotamia Expedy. Force Jany. 1916—June 1917.

Jitman Rai—
Mesopotamia Expedy. Force.

Malhang Limbu—
Mesopotamia Expedy. Force Jany. 1916—June 1917.

Bhagesor Gurung—
Mesopotamia Expedy. Force Oct. 1914—April 1916.
Mesopotamia Expedy. Force March 1917—Feb. 1918.

1-8th Gurkha Rifles.

SUBEDAR-MAJORS—

Rukman Sahi—"*Bahadur.*"—
Order of British India, 2nd class.
Burma, 1887-89.—Medal with clasp.
Tibet, 1903-04.—Operations at and around Gyantse, 5th May to 6th July 1904.—March to Lhassa, 14th July to 3rd August 1904.—Medal with clasp.
Abor Expedition, 1911-12.—Medal with Clasp.

SUBADARS—

Indrabir Thapa—
N.-E. Frontier of India, Manipur, 1891.—Medal with clasp.
Tibet, 1903-04.—Operations at and around Gyantse, 5th May to 6th July 1904.—March to Lhassa, 14th July to 3rd August 1904.—Medal with clasp.
Abor Expedition, 1911-12.—Medal with clasp.

Ransur Rana—
Tibet 1903-04.—Operations at and around Gyantse 5th May to 6th July 1904.—March to Lhassa, 14th July to 3rd August 1904.—Medal with clasp.

Bagbir Thapa—
Tibet, 1903-04.—Medal.
Abor Expedition, 1911-12.—Medal with clasp.

Dalsing Thapa—
Military Cross—M.C. Mentioned in despatches, of
Tibet, 1903-04.—Medal.
Abor Expedition, 1911-12.—Medal with clasp.

Ranbir Ghale—
Tibet, 1903-04.—Operations at and around Gyantse, 5th May to 6th July 1904.—March to Lhassa, 14th July to 3rd August 1904.—Medal with clasp.
Abor Expedition, 1911-12—Medal with clasp.

Dhanbir Thapa—
Tibet, 1903-4.—Operations at and around Gyantse, 5th May to 6th July 1904.—March to Lhassa 14th July to 3rd August 1904.—Medal with clasp.
Abor Expedition, 1911-12.—Medal with clasp.

Tulbir Gurung—
Tibet, 1903-04.—Operations at and around Gyantse 5th May to 6th July 1904 and march to Lhassa.—Medal and clasp.
Abor Expedition, 1911-12.—Medal and clasp.

Narbahadur Gurung—
Tibet, 1903-04.—Operations at and around, Gyantse, 5th May to 6th July 1904 and march to Lhassa.—Medal and clasp.
Abor Expedition, 1911-12.—Medal and clasp.
Totok punitive column, 1913.

Narotan Jaisi—
Tibet, 1903-04.—Operations at and around Gyantse, 5th May to 6th July 1904 and march to Lhassa.—Medal and clasp.
Abor Expedition, 1911-12.—Medal and clasp.
Totok punitive column, 1913.

Kharagsing Thapa,—I.D.S M.—
Tibet, 1903-04.—Operations at and round Gyantse, 5th March to 6th July 1904 and march to Lha ssa.—Medal and clasp.
Abor Expedition, 1911-12.—Medal and clasp.

Manbir Thapa—
Tibet, 1903-04.—Operations at and around Gyantse, 5th March to 6th July 1904 and march to Lhassa from 14th July to 3rd August 1904.—Medal and clasp.
Abor Expedition, 1911-12.—Medal and clasp.

Gomu Basnet—
Tibet, 1903-04.—Operations at and around Gyantse, 5th March to 6th July 1904 and march to Lhassa from 14th July to 3rd August 1904.—Medal and clasp.
Abor Expedition, 1911-12.—Medal and clasp.

Damdar Thapa—
Tibet, 1903-04.—Medal.

Kharakbahadur Rana, I.D.S.M.—
Abor Expedition, 1911-12.—Medal and clasp.

Parbir Thapa—
Tibet, 1903-04.—Operations at and around Gyantse, 5th May to 6th July 1904 and march to Lhassa.—Medal and clasp.
Abor Expedition, 1911-12.—Medal and clasp.

Bahadur Thapa—
Tibet, 1904.—Actions at Naini, Sechen Karola and advance on Lhassa.—Medal and clasp.
Abor Expedition, 1911-12.—Medal and clasp.

Dalbahadur Thapa—
Abor Expedition, 1911-12.—Medal and clasp.

Rupsing Thapa—
Tibet, 1903-04.—Operations at and around Gyantse, 5th March to 6th July 1904, march to Lhassa 14th July to 3rd August 1904.—Medal and clasp.

Birbhan Gurung—
Tibet, 1903-04.—Operations at and around Gyantse, 5th March to 6th July 1904.—March to Lhassa, 14th July to 3rd August 1904.—Medal and clasp.

JEMADARS—

Manirej Thapa—
Abor Expedition, 1911-12.—Medal and clasp.

Basurey Thapa—
Tibet, 1903-04.—Operations at and around Gyantse, 5th March to 6th July 1904.—March to Lhassa, 14th July to 3rd August 1904.—Medal and clasp.
Abor Expedition, 1911-12.—Medal and clasp.

Ashu Gurung—
Tibet, 1903-04.—Medal.
Abor Expedition, 1911-12.—Medal and clasp.

Garbhe Pun—
Tibet, 1903-04.—Operation at and round Gyantse, 5th March to 6th July 1904, march to Lhassa, 14th July to 3rd August 1904.—Medal and clasp.

Mandhoj Burathoki—
Tibet, 1903-04.—Operation at and round Gyantse, 5th March to 6th July 1904, march to Lhassa, 14th July to 3rd August 1904.—Medal and Clasp.

Budhibal Thapa—
Tibet, 1903-04.—Operations at and round Gyantse, 5th March to 6th July 1904, march to Lhassa, 14th July to 3rd August 1904.—Medal and clasp.
Abor expedition, 1911-12.—Medal and clasp.

Lalbahadur Thapa—
Tibet, 1903-04.—Operations at and round Gyantse, 5th May to 6th July 1904.—Medal and clasp.

Kaharsing Thapa—
Tibet, 1903-04.—Actions at Naini, Sechen, Gyantse-sejong., Guru, Karola and Karola 1st Action.—Medal and clasp.

Gambhirsing Gurung—
Abor Expedition, 1911-12.—Medal and clasp.

Gajbir Thapa—
Abor Expedition, 1911-12.—Medal and clasp.
Totok Column, 1913.

Tikaram Thapa—
Sikkim Tibet, 1903-04.—Medal.
Abor Expedition, 1911-12.—Medal and clasp.

2-8th Gurkha Rifles.

SUBADAR-MAJOR—

Rimant Thapa—
France, 1914-15.
Egypt, 1915-16.

SUBADAR—

Bagdal Rana—
N.-E. Frontier of India, Manipur, 1891.—Medal with clasp.
N.-E. Frontier of India, Abor, 1911-12.—Medal with clasp.

Basante Sahi—
France, 1914-15.
Egypt, 1915-16.

Abir Thapa—
France, 1914-15.

Dhanbir Thapa—
Egypt, 1915-16.

Jasbir Gurung—
France, 1914-15.
Egypt, 1915-16.

Dhanbahadur Gurung—
France, 1914-15.
Egypt, 1915-16.
Mesopotamia, 1917.

Ramsing Rana—
France, 1914-15.
Egypt, 1915-16.

Krishanlal Chettri—
France, 1914-15.
Egypt, 1915-16.

Lalbir Pun—
France, 1914-15.
Egypt, 1915-16.

JEMADARS—

Nawalbir Gurung—
Abor, 1911-12.
France, 1914-15.
Mesopotamia, 1917-18.

Chabilal Gurung—
France, 1914-15.
Egypt, 1915-16.

Hastalsing Gurung—
France, 1914-15.
Egypt, 1915-16.

Balbhadra Thapa—
France, 1914-15.
Egypt, 1915-16.

Sarbasing Gurung—
France, 1914-15
Egypt, 1915-16.
Mesopotamia, 1917
Egypt, 1918.

Tikaram Mal—
France, 1914-15.
Egypt, 1915-16.

Jitman Pun—
France, 1914-15.
Egypt, 1915-16.

Subhansing Gurung—
France, 1914-15.
Egypt, 1915-16.

Chandrasing Gurung—
France, 1914-15.
Egypt, 1915-16.

Maniraj Gurung—
France, 1914-15.
Egypt, 1915-16.

Bahadur Mal—
Abor, 1911-12.
France, 1914-15.
Egypt, 1915-16.

Kumaraj Thapa—
France, 1914-15.

Dhansur Gurung—
Egypt, 1915-16.
Mesopotamia, 1917.

9th Gurkha Rifles.

SUBADAR-MAJORS—

Bishn Sing Kathait, *Sardar Bahadur—Order of British India, 1st class.*
Afghanistan, 1879-80.—Medal.
N.-W. Frontier of India, 1897-98.—Mohmand.—Medal with clasp.
Tirah, 1897-98.—Operations in the Bara Valley, 7th to 14th December 1897.—Clasp.

Prem Singh Bisht—
N.-W. Frontier of India, 2nd Miranzai, 1891.—Medal with clasp.
N.-W. Frontier of India, Hunza Nagar, 1891.—Clasp.
N.-W. Frontier of India, 1897-98.—Mohmand.—Medal with clasp.
Tirah, 1897-98.—Operations in the Bara Valley, 7th to 14th December 1897.—Clasp.
Tibet, 1903-04.—Medal with clasp.

Mehar Sing Khattri—I.O.M.
N.-W. Frontier of India, 1897-98.—Mohmand.—Medal with clasp.
Tirah, 1897-98.—Operations in the Bara Valley, 7th to 14th December 1897.—Clasp.
France, 1914-Star .4.

SUBADARS—

Haribhagat Sahi—
N.-W. Frontier of India, 2nd Miranzai, 1891.—Medal with clasp.
N.-W. Frontier of India, Waziristan, 1894-95.—Clasp.
China, 1900.—Medal.

Surbir Karki, *Bahadur—Order of British India, 2nd class.*
N.-W. Frontier of India, 1897-98.—Mohmand.—Medal with clasp.
Tirah, 1897-98.—Operations in the Bara Valley, 7th to 14th December 1897.—Clasp.

Jhagarnand—
Burma, 1885-87.—Medal with clasp.
Burma, 1887-89.—Clasp.
N.-W. Frontier of India, 1897-98.—Mohmand.—Medal with clasp.
Tirah, 1897-98.—Operations in the Bara Valley, 7th to 14th December 1897.—Clasp.

Bhimlal Khattri—
N.-W. Frontier of India, 1897-98.—Mohmand.—Medal with clasp.
Tirah, 1897-98.—Operations in the Bara Valley, 7th to 14th December 1897.—Clasp.

Indra Sing Thapa—
N.-W. Frontier of India, 1897-98.—Mohmand.—Medal with clasp.
Tirah, 1897-98—Operations in the Bara Valley, 7th to 14th December 1897.—Clasp.

Jaibhan Karki—
N.-W. Frontier of India, 1897-98.—Mohmand.—Medal with clasp.
Tirah, 1897-98.—Operations in the Bara Valley, 7th to 14th December 1897.—Clasp.

Madho Sing Khattri—
N.-W. Frontier of India, 1897-98.—Mohmand.—Medal with clasp.
Tirah, 1897-98.—Operations in the Bara Valley, 7th to 14th December 1897.—Clasp.

Ratan Sing Basniet—
N.-W. Frontier of India, 1897-98.—Mohmand.—Medal with clasp.
Tirah, 1897-98.—Operations in the Bara Valley, 7th to 14th December 1897.—Clasp.

Sher Sing Adhikari—
N.-E. Frontier of India, Lushai, 1888-89.—Medal with clasp.
N.-E. Frontier of India, Chin-Lushai, 1889-90.—Clasp.
France, 1914.—Star, 1914.

Rupsing Bhandari—
N.-W. Frontier of India, 1897-98.—Mohmand.—Medal with clasp.
Tirah, 1897-98.—Operations in the Bara Valley, 7th to 14th December 1897.—Clasp.

Kishenbahadur Khattri—
N.-W. Frontier of India, 1897-98.—Mohmand.—Medal with clasp.
Tirah, 1897-98.—Operations in the Bara Valley, 7th to 14th December 1897.—Clasp.

Tikaram Majhi—*Bahadur.—Order of British India, 2nd class.*
N.-W. Frontier of India, 1897-98.—Mohmand.—Medal with clasp.
Tirah, 1897-98.—Operations in the Bara Valley, 7th to 14th December 1897.—Clasp.
France 1914.—Star 14.

Gobindram Khattri—
N.-W. Frontier of India, 1897-98.—Mohmand.—Medal with clasp.
Tirah, 1897-98.—Operations in the Bara Valley, 7th to 14th December 1897.—Clasp.
France, 1914.—Star, 1944.

Narbir Khattri—
N.-W. Frontier of India, 1897-98.—Mohmand.—Medal with clasp.
Tirah, 1897-98.—Operations in the Bara Valley, 7th to 14th December 1897-98.—Clasp.
France, 1914.—Star, 1914.

Damodar Khattri, I.D.S.M.—
N.-W. Frontier of India, 1897-98.—Mohmand.—Medal with clasp.
Tirah, 1897-98.—Operations in the Bara Valley, 7th to 14th December 1897.—Clasp.
Tibet, 1903-04.—Medal with clasp.
France 1914.—Star 14.

JEMADARS—

Dilbahadur Khattri—
N.-W. Frontier of India, 1897-98.—Mohmand.—Medal and clasp.
Tirah, 1897-98.—Operations in the Bara Valley, 7th to 14th December 1897.—Clasp.

Kharak Sing Khattri—
N.-W. Frontier of India, 1897-98.—Mohmand.—Medal and clasp.
Tirah, 1897-98.—Operations in the Bara Valley, 7th to 14th December 1897.—Clasp.

Ganesh Bahadur Khattri—
N.-W. Frontier of India, 1897-98.—Mohmand.—Medal and clasp.
Tirah, 1897-98.—Operations in Bara Valley, 7th to 14th December 1897.—Clasp.

Champa Singh Thapa—
N.-W. Frontier of India, 1897-98.—Mohmand.—Medal and clasp.
Tirah, 1897-98—Operations in the Bara Valley, 7th to 14th December 1897.—Clasp.

Jaman Sing Karki—
Chitral, 1895.—Relief of Chitral.—Medal and clasp.
N.-W. Frontier of India, 1897-98.—Clasp.

Gambhir Sing Khattri—
N.-W. Frontier of India, 1897-98.—Mohmand.—Medal and clasp.
Tirah, 1897-98.—Operations in the Bara Valley, 7th to 14th December 1897.—Clasp.

Him Bahadur Sahi—
N.-W. Frontier of India, 1897-98.—Mohmand.—Medal and clasp.
Tirah, 1897-98.—Operations in the Bara Valley, 7th to 14th December 1897.—Clasp.

Durgadutt Sahi—
Chitral, 1895.—Relief of Chitral.—Medal with clasp.
N.-W. Frontier of India, 1897-98.—Clasp.

Bhagatram Khattri—
N.-W. Frontier of India, 1897-98.—Medal with clasp.
Samana, 1897.—Clasp.
Tirah, 1897-98.—Clasp.

Karan Bahadur Khattri—
N.-W. Frontier of India, 1897-98.—Mohmand.—Medal with clasp.
Tirah, 1897-98.—Operations in the Bara Valley, 7th to 14th December, 1897.—Clasp.

Arjun Khattri—
N.-W. Frontier of India, 1897-98.—Mohmand.—Medal and clasp.
Tirah, 1897-98.—Operations in the Bara Valley, 7th to 14th December, 1897.—Clasp.
France, 1914.—Star, 1914.

Chandra Bir Khattri—
N.-W. Frontier of India, 1897-98.—Mohmand.—Medal and clasp.
Tirah, 1897-98.—Operations in the Bara Valley, 7th to 14th December 1897.—Clasp.
France, 1914.—Star, 1914.

Jag Bahadur Khattri—
N.-W. Frontier of India, 1897-98.—Mohmand.—Medal and clasp.

2-9th Gurkha Rifles.

SUBADARS—

Wazir Singh Mahtra—
N.-W. Frontier of India.—Mohmand, 1897-98.—Medal with clasp.

Lalsing Basnet—
N.-W. Frontier of India, 1897-98.—Mohmand.—Medal with clasp.
Tirah 1897-98.—Operations in the Bara Valley, 7th to 14th December 1897.—Clasp.

Padam Sing Thapa, *Bahadur*—
N.-W. Frontier of India, 1897-98.—Mohmand.—Medal with clasp.
Punjab Frontier, 1897-98.—Operations on the Samana and in the Kurram Valley.—Medal and three clasps.
Tirah, 1897-98.—Operations in the Bara Valley, 7th to 14th December 1897.—Clasp.

Keshar Mall—
Tirah, 1897-98.—Medal with clasp.
Wasiristan, 1901-02.—Clasp.

Puran Sing Khandka—
N.-W. Frontier of India, Waziristan, 1894-95.—Action at Wana.—Medal with clasp.

Jogichand—
N.-W. Frontier of India, 1897-98.—Mohmand.—Medal with clasp.
Tirah, 1897-98.—Operations in the Bara Valley, 7th to 14th December 1897.—Clasp.

Gumani Sahi—
N.-W. Frontier of India, Waziristan, 1894-95.—Medal and clasp.

JEMADARS—

Lachman Khandka—
N.-W. Frontier of India, Mohmand, 1897-98.—Medal with clasp.
Tirah, 1897-98.—Operations in the Bara Valley.—Medal with clasp.

Harak Singh Khattri—
Waziristan, 1901-02.—Medal with clasp.

Bahadur Chand—
N.-W. Frontier of India, Mohmand, 1897-98.—Medal with clasp.
Tirah, 1897-98.—Bara Valley.—Clasp.

Moti Parsad Sahi—
N.-W. Frontier of India, Tirah, 1897-98.—Medal with 2 clasps.

3-9th Gurkha Rifles.

SUBADARS—

Durga Datt Sahi—
Mesopotamia, 1916-17.

Tika Ram Khattri—
France, 1914-15.
Mesopotamia, 1915-16.

Shib Parsad Thapa—
Mesopotamia, 1916-17.—Meritorious service medal without annuity.

Wazir Singh Khattri, I.D.S.M.—
Mesopotamia, 1916-17.—Indian Distinguished Service Medal.

Bhagwan Singh Chand—
Mesopatamia, 1916-17.

Combahadur Khattri—
France, 1915-16.
Mesopotamia, 1915-16.

JEMADARS—

Ram Bhagat Adhikari—
France, 1914-15.—Star, 1914.
Mesopotamia, 1915-16.

Kharak Bhadur Khattri—
France, 1915.
Mesopotamia, 1915-16.

Dalip Singh Karki—
France, 1914-15.—Star, 1914.
Mesopotamia, 1915-16.

Tikaram Kanwar—
France, 1914-15.—Star, 1914. Indian Order of Merit 2nd class.
Mesopotamia, 1915-16.—Medal, St. George, 3rd class.

Amar Sing Karki—
France, 1914-15.—Star, 1914.
Mesopotamia, 1915-16.

Balbir Thapa—
Mesopotamia, 1916-17.

Lachman Singh Oli—
France, 1915-16.
Mesopotamia, 1916-17.

Kabiraj Khattri—
France, 1915.
Mesopotamia, 1915-16.

Padam Dhoj Mall—
Mosopotamia, 1916-17.

Deoraj Thapa—
Mesopotamia, 1916-17.

1-10th Gurkha Rifles.

SUBADAR-MAJORS—

Chittahang Limbu—
N.-E. Frontier of India, Chin Lushai, 1889-90.—Medal with clasp.
Chitral, 1895.—Relief of Chitral.—Medal with clasp.
Tirah, 1897-98.—Actions of Sampagha and Arhanga Passes. Reconnaissance of Saran Sar and action of 9th November 1897.—Operations in Bazaar Valley, 25th—30th December 1898.—2 clasps.

Bhola Singh Kandari—
Burma, 1885-91.—Wuntho Expedition.—Medal with 3 clasps.
N.-E. Frontier of India, North Chin Hills, 1892-93-94.—Clasp.

SUBADARS—

Ramjit Remi—
Burma, 1887-89.—Medal with clasp.
N.-E. Frontier of India, Chin Hills, 1892-93.—Clasp.
Burma, 1893-94.

Makhandhoj Rai—
Burma, 1887-89.—Medal with clasp.
N.-E. Frontier of India, Chin Hills, 1892-93.—Clasp.
Burma, 1893-94.

Mandhoj Lama—
N.-E. Frontier of India, Chin Hills, 1892-93.—Medal with clasp.
Burma, 1893-94.

Chakrabahadur Limbu—
N.-E. Frontier of India, Chin-Hills, 1892-93.—Medal with clasp.
Burma, 1893-94.

Jitman Gurung—
N.-E. Frontier of India, Chin Hills, 1892-93.—Medal with clasp.

Chettradhoj Limbu—
N.-E. Frontier of India, North Chin Hills, 1892-93.—Medal with clasp.
Order of British India, 2nd class.

Pharsaldhoj Rai—
N.-E. Frontier of India, North Chin Hills, 1893-94.—Medal with clasp.

Ajitia Rai—
N.-W. Frontier of India, Waziristan, 1894.—Wana.—Medal with clasp.
N.-W. Frontier of India, 1897-98.—Mohmand.—Medal with clasp.

Asbir Rai—
Abor, 1911-12.—Medal with clasp.

Lakhdhoj Rai—
Punjab Frontier, 1897-98.—Mohmand and Tirah expeditions.—Medal and 2 clasps.
Tibet, 1904.—Medal and clasp.

Bajirdhoj Rai—
Abor, 1911-12.—Medal with clasp.

Ram Sing Burathoki—
N.-W. Frontier of India, Waziristan, 1901-02.—Medal with clasp.

JEMADARS—

Shersing Rawat—
Tirah, 1897.—Medal with 3 clasps.
Tibet, 1904.—Medal with clasp.

2-10th Gurkha Rifles.

SUBADAR-MAJOR—

Chittahang Limbu—

Sardar Bahadur, I O.M.
N.-E. Frontier of India, Chin Lushai, 1889-90.—Medal with clasp.

Chitral, 1895.—Relief of Chitral.—Medal with clasp.

Tirab, 1897-98.—Actions of Sampagha and Arhanga Passes Reconnaisance of S ran Sar and action of 9th November 1897.—Operations in Bazar Valley, 25th-30th December 1898.—2 clasps.

SUBADARS—

Chitradhoj Limbu.—*Bahadur I.O.M.*

N.-E. Frontier of India, North Chin Hill, 1892-93.—Medal with clasp.

Lakhdhoj Rai—
Punjab Frontier, 1897-98.—Mohmand and Tirah expeditions.—Medal and 2 clasps.
Tibet, 1904.—Medal and clasp.

Bajirdhoj Rai, I.O.M.—
Abor, 1911-12.—Medal with clasp.

Ramsing Burathoki, I.O.M.—
N.-W. Frontier of India, Waziristan, 1901-02.—Medal with clasp.

JEMADARS—

Shersing Rawat—
Tirah, 1897.—Medal with 3 clasps.
Tibet, 1904.—Medal with clasp.

Udahang Limbu—
Tibet, 1903-04.—Medal.

1-11th Gurkha Rifles.

SUBADAR-MAJOR—

Dhansing Gurung—
M.E.F. 1917-18.
Gallipoli 1915-16.

SUBADARS—

Haricharan Thapa—
M. E. F. 1916-18.

Kharak Bahadur Rana—
I. E. F. "E' & "F" 1914-15.
M. E. F. 1917-19.

Nandbir Rana—
I. E. F. "E' & "F" 1914-15.
M. E. F. 1917-18.

JEMADARS—

Narbir Pun—
M. F. E. 1916-17.

Bhowanand Thapa—
M. E. F. 1916-18.

Narpatti Bura—
M. E. F. 1916-18.

Asbahadur Gurung—
M. E. F. 1916-18.

Dhanraj Thapa—
Gallipoli 1915-16.
M. E. F. 1917-18.

Rabidhoj Rana—
M. E. F. 1917-18.

Mahesor Gurung—
Gallipoli 1915-16.

Gore Pun—
Gallipoli 1915.
M. E. F. 1917-18.

Parsabir Rana—
Gallipoli 1915.
M. E. F. 1917-18.

Karna Bahadur Rana—
I. E. F. "F" 1915-16.
M. E. F. "D" 1917-18.

Shamsher Thapa—
B. E. F. 1915.
I. E. F. "E" 1915.
M. E. F. "D" 1917-18.

Ramprasad Gurung—
M. E. F. "D" 1917-18.

2-11th Gurkha Rifles.

SUBADAR MAJOR—

Nar Patti Rai, I.D.S.M.—
War 1914-18.—Twice wounded. France, Mesopotamia—Indian Distinguished Service Medal.

SUBADARS—

Sirilal Thapa—
Tirah 1897-98.—Action oftDargai. Medal with 2 clasps.
N.-W. Frontier, Waziristan 1901-02.—Clasps.
Abor 1911-12.—Medal with clasp.
France 1914-16.
Mesopotamia 1917-18.

Autar Sing Gurung—
War 1914-18.—France, Dardanellis and Mesopotamia.

Jungraj Limbu—
War 1914-18.—Mesopotamia.

Deobahadur Gurung—
Kohat, Kurram and Tirah, 1897-98.—Actions of Dargai 18th and 20th Oct. 1897. Samana, Tirah, and Punjab Frontier.
1895—Medal with clasp.
N.-W. Frontier, Waziristan, 1901-02.—Clasp.
Abor 1911-12.—Medal with clasp.
Dardanellis—1915.
Egypt—1916.
Mesopotamia—1917-18.

Thaman Sing Thapa—
War 1914-18—
Dardanellis and Mesopotamia.

JEMADARS—

Sukbir Limbu—
War, 1914—18.—Mesopotamia.

Diraj Gurung—
Punjab Frontier, 1897-98.—Medal with clasp.
Samana, 1897, Tirah, 1897-98—
N.-W. Frontier of India, Waziristan, 1901-02.
War, 1914-18.—Mons Star. Wounded Mesopotamia.

Sukbir Limbu—
War, 1914-18.—Mesopotamia.

Santbir Bura—
N.-W. Frontier Waziristan, 1901-02.—Medal with clasp.
Abor, 1911-12.—Medal with clasp.
Mesopotamia, 1916-18.

Sirikishan Thakur—
War 1914—18.—Mesopotamia, France. Wounded.

Parmandhoj Thapa—
War, 1914—18.—Mesopotamia.

Budhiram Roka—
Abor, 1911-12 —Medal with clasp.
Egypt, 1915-16—
Mesopotamia, 1917-18—

Bhanbir Gharti—
Punjab Frontier, 1897-98.—Medal—clasps.
Samana, 1897.—
Tirah, 1897-98.—
N.-W. Frontier of India—Waziristan, 1901-02.—Clasp.
War, 1914—18—

Bhirban Gharti—
N.-W. Frontier of India, Waziristan, 1901-02.—Medal with clasp.
War, 1914—18—

Ude Sing Gurung—
N.-W. Frontier of India, Wazristan, 1901-02.—Medal with clasp.
War, 1914—18.

Darmi Rai—
War, 1914—18.

3-11th Gurkha Rifles.

SUBADAR—

Bhagatram Khattri—
Tirah, 1897-98.—

JEMADAR—

Amarsing Bohara—
Tibet, 1903-04.—

4-11th Gurkha Rifles.

SUBADARS—

Ganga Ram Thapa—
Egypt, 1914.—
France, 1914-15.—
Mesopotamia, 1916.—Wounded 1915.

Budhiman Gurung—
Egypt, 1914.—
France, 1914-15.—
Mesopotamia, 1916-17.—

Ramnath Cachari—
Mishmi Expedition, 1911-12.—
Chinglong Expedition and Totak *column Naga Hills, 1913.*

Kharagbahadur Pun—
Abor Expedition, 1911-12.—Medal and clasp.

Anbir Pun—
Tibet 1903-04—Operations at and around Gyantse, 5th March to 6th July 1904. March to Lhassa, 14th July to 3rd August 1904.—Medal and clasp.
Abor Expedition, 1911-12.—Medal and clasp.

Mandhoj Gurung—
Abor Expedition, 1911-12.—Medal and clasp.

Hastabir Limbu—
Egypt, 1915.—
Mesopotamia, 1915-16.—Wounded.

Paharsing Gurung—
Waziristan, Punjab Frontier, 1894-95.—Medal with clasp.
Tirah, 1897-98.—Medal and 2 clasps.
Egypt, 1914.—
France, 1914-15.—
Mesopotamia, 1916.—Mentioned in despatches.

Bilbahadur Gurung—
Abor Expedition, 1911-12.—Medal and clasp.

1st Mule Corps.

RESSAIDAR—

Hashmat Ali—*Sardar Bahadur, Order of Br. Ind., 1st Class, Indian Order of Merit, 2nd class.*
N.-W. Frontier of India, Wazristan, 1894-95.—Medal with clasp.
N.-W. Frontier of India, 1897-98.—Operations in the Samana and Kurram Valley.—Medal with clasp.
Tirah, 1897-98.—Clasp.
N.-W. Frontier of India, 1908.—Operations in Zaka Khel and Mohmand country.—Engagement of Kargha, 24th May 1908.—Medal with clasp.

2nd Mule Corps.

JEMADAR—

Sardar Khan—
Indian Expdy. Force "A", 1914.
Mesopotamia Expdy. Force, 1916-18.
Egyptian Expdy. Force, 1918.

3rd Mule Corps.

JEMADAR—

Mohd. Ismail—
War, 1915-18.—Mentioned in Despatches.

5th Mule Corps.

RESSAIDAR—

Mohamed Shah—
N.-W. Frontier of India, Waziristan, 1901-02.—Medal with clasp.
N.-W. Frontier of India, 1908.—Operations in the Zakka Khel country.
N.-W. Frontier of India, 1908.—Operations in the Mohmand country.—Medal with clasp.
Abor Survey Party, 1912-13.
M. E. Force, 1916-17.—I.D.S.M.

Karam Khan—
Sikkim, 1888-90.—Medal and clasp 88.
Waziristan, 1894-96.—Clasp 94-95.
Tochi, 1897-99.—Clasp P. F. 97-98.
Tirah, 1879-98.

JEMADAR—

Painda Khan. (*Temp. Ressaidar.*)—
Tirah, 1897-98.—Medal and clasp.
Sikkim-Tibet Mission, 1903-04.—Medal.
War, 1914—I. E. F. "D".—Order of Br. Ind., 2nd class.
War, 1914—France.—Star.

6th Mule Corps.

JEMADARS—

Khuda Dad Khan—
Bazar Valley Field Force, 1908.
Mohamed Field Force, 1908.
Abor Survey Party, 1912-13.
Mesopotamia Expdy. Force since 13-7-16.

Mohd Fazil—
Tibet Mission Force, 1903-04.
N.-W. F., 1908.
M. E. Force since 13-7-16.

Khan Zaman—
Tibet Mission Force, 1903-04.
N.-W. F., 1908.
M. E. Force since 13-7-16.

7th Mule Corps.

RESSAIDAR—

Ditta—
China Lushai, 1888-89.
Samana, 1891.
Waziristan, 1897.
Tirah Expdy. Force, 1897-98.
Mahsud Blockade, 1901-02.
Sikkim Tibet Mission Force, 1903-04.
North-Western Frontier, 1908.

8th Mule Corps.

JEMADAR—

Alam Sher—
N.-W. Frontier of India—Waziristan, 1901-02.

9th Mule Corps.

RESSAIDAR—

Gholam Mohamed—
Chitral Relief, 1895.—Medal with clasp.

10th Mule Corps.

SUBADAR—

Umar Khan, *Bahadur—Order of Br. India.*
N.-W. Frontier of India, Hazara, 1888.—Medal and clasp.

JEMADARS—

Fateh Shah—
N.-W. Frontier of India, 1897-98.—Operations in the Samana and in the Kurram Valley during August and September 1897.—Medal with 3 clasps.
Somaliland, 1904.

12th Mule Corps.

JEMADAR—

Sunder Singh—
Indian General Service, 1908.—Medal with clasp.

15th Pack Mule Corps.

RESSAIDAR—

Nasir Din—
Somaliland, 1902-04.

16th Mule Corps.

RESSAIDAR—

Sham Singh—
N.-W. Frontier of India, 1908.—Medal with clasp.

JEMADAR—

Nathoo Khan—
N.-W. Frontier of India, 1895-98.—Relief of Chitral, Defence of Malakand, and Punjab Frontier Expedition.—Medal with 3 clasp.

17th Mule Corps.

JEMADAR—

Ghulam Nabi—
China, 1905-08.—
Thal & Kohat, 1915-16.—
Bannu, Tank & Miran Shah, 1916-18.—

18th Mule Corps.

JEMADAR—

Harbhagwan—
Somaliland Field Force, 1904.—

19th Mule Corps.

JEMADAR—

Khan Alam—
War, 1914-16.—

20th Mule Corps.

JEMADAR—

Fatteh Khan—
Bazar Valley Field Force, 1908.—

22nd Mule Corps.

JEMADAR—

Kehr Singh—
M. E. Force, 1917-18.—Mentioned in despatches.

24th Mule Corps.

Ramzan—
Soudan, 1885-86.—
N.-W. Frontier of India, 1889-90.—
Chitral, 1895.—
Soudan, 1896.—
N.-W. Frontier of India, Waziristan, 1901-02.—
War, 1917-18.—

26th Mule Corps.

RESSAIDAR—

Gama—
Relief of Chitral, 1895.—Medal and clasp.
Tochi Field Force, 1897.
Tibet, 1903-04.—Operations in and around Gyantse between 5th May and 6th July.—March to Lhassa, 14th July to 3rd August.—Medal and clasp.
N.-W. Frontier of India, 1908.—Operations in the Mohmand country. Engagement at Matta, 24th April 1908. Engagement at Kargha, 24th May 1908.
Abor Expedition, 1911-12.—Clasp.

27th Mule Corps.

RESSAIDARS—

Sham Singh—
Zhob Expedition, 1890—
Samana, 1897.—Medal with 3 clasps.
Flying Column, Kurram Valley.—Medal with clasp.
Tirah, 1897-98.—Medal with 3 clasps.

Toorsum Khan, *Bahadur—Order of Br. India, 2nd class.*
Chitral, 1895.—Relief of Chitral.—Medal with clasp.
N.-W. Frontier of India, 1897.—Defence of Malakand. Relief of Malakand. Relief of Chakdara. Operations in Bajour and in the Mohmand country. Buner.—2 clasps, Indian Order of Merit, 3rd class.

Ali Mardan Khan—
Black Mountain, 1891.—Medal with one Bar.
Chitral Relief, 1895.—Medal with one Bar.
Malakand Force, 1897-98.—One Bar.
South Africa, 1900-02.—Medal.
Mohmand Force, 1908.—Medal with one Bar.

Ram Singh—
Hazara, 1891.—Medal with clasp.
Punjab Frontier, 1897-98.—Medal with clasp.
Tirah, 1897-98.—Clasp.

JEMADAR—

Diwan Chand—
Chitral, 1895.—Relief of Chitral.—Medal with clasp.

29th Mule Corps.

JEMADAR—

Delal Singh—
N.-W. Frontier of India, 1908.—Operations in the Zakka Khel country.
N.-W. Frontier of India, 1908.—Operations in the Mohmand country.—Medal with clasp.

30th Mule Corps.

JEMADAR—

Kam Din—
Hazara Expedition, 1891.—Medal with clasp.

31st Pack Mule Corps.

RESSAIDAR—

Muhammad Ali Nabi—
N.-W. Frontier of India, 1897-98.—Operations on the Samana and in the Kurram Valley during August and September 1897. Relief of Gulistan.—Medal with 2 clasps.
Tirah, 1897-98.—Clasp.
I. E. F. "A., 1914-15.—
I. E. F. "G." & "E.," 1915-17.—

32nd Mule Corps.

JEMADARS—

Fatteh Mohammed—
N.-W. Frontier of India, Waziristan, 1901-02.—
Chitral Relief 1907.—

Samundar Khan—
Chitral Relief, 1895.—
Punjab Frontier Force, 1897-98.—
South Africa, 1902.—
Tibet, 1903-04.—

33rd Mule Corps.

JEMADARS—

Kaim Din—
M. E. Force, 1914.—

Nathoo—
M. E. Force, 1914.—

34th Mule Corps.

JEMADAR—

Karam Ali—
Bazar Valley & Mohmand, 1908.—
War, 1914–18.—

35th Mule Corps.

JEMADARS—

Jaffar Ali Khan—
Chin Lushai, 1889-90.—Medal with clasp.

Jahan Khan, I.D.S.M.—
War 1914–18.—Indian Distinguished Service Medal.

38th Mule Corps.

JEMADAR—

Labh Singh—
France, 1914-15.—

40th Mule Corps.

RESSAIDAR—

Mohd. Shah—
Waziristan, 1901-02.—
Mohmand, 1908.—
War, 1914–18.—

JEMADAR—

Sher Ali Khan—
Mohmand, 1908.—
Abor Survey Party, 1912-13.—
War, 1916-17.

41st Mule Corps.

RESSAIDAR—

Sham Singh—
Zhob Expedition, 1890.—
Samana, 1897.—Medal with 3 clasps.
Flying Column, Kurram Valley.—Medal with 3 clasps.
Tirah, 1897-98.—Medal with 3 clasps.

43rd Pack Mule Corps.

RESSAIDAR—

Karam Khan—
Chitral Expedition, 1895.—Medal.
Tirah, 1897-98.—Medal.
South Africa, 1900-01—Medal.
I. E. F. "D.," 1916-18.—

45th Mule Corps.

JEMADAR—

Nazif Khan—
Thibet, 1887.
Waziristan, 1894-95.
Rief. of Chitral, 1895.
Tirah, 1897-98.
Punjab Frontier, 1897-98.
South Africa, 1899-1902.

46th Mule Corps.

JEMADAR—

Saif Ali—
Tirrah, 1897-98.—Medal.
China, 1900.—Medal.
Abor Survey Party, 1912-13.—
War, 1914-16.—Medal.

48th Mule Corps.

JEMADAR—

Gulab Khan—
Tibet-Sikkim Mission, 1904.—Medal.

52nd Mule Corps.

Mohamed Ali—
Hazara, 1890.—
Somaliland, 1903-04.—
M. E. Force, 1914-15.—

58th Pack Mule Corps.

JEMADAR—

Karim Din—
War, 1914-17.—

65th Mule Corps.

RESSAIDAR—

Mehdi Khan—
Ogaden Expedition.—Operations against mutineers and rebels.
Unyoro, 1899.
Uganda 1898.—Queen's medal with 2 clasps.
Uganda, 1900.—King's medal and two clasps.

50th Silladar Camel Corps.

RISALDAR—

Nawab Khan—
N.-W. Frontier of India, 1908.—Operations in the Mohmand country.—Medal with clasp.

51st Silladar Camel Corps.

RISALDARS—

Rajwali Khan—
Tirah, 1897-98.
Samana, August and September 1897.
Flying Column, Kurram Valley, September 1897.—Medal with 3 clasps.
Indian Remount Depôt, South Africa, 1902.—Medal with clasp.
N.-W. Frontier of India, 1908.—Operations in the Mohmand country.—Medal with clasp.

Shahwali Khan—
N.-W. Frontier of India, 1908.—Operations in the Zakka Khel country. Operations in the Mohmand country.—Medal with clasp.

52nd Silladar Camel Corps.

RISALDARS—

Zaman Khan—
East Africa, 1903-04.—Operations in Somaliland.—Medal with clasp.
N.-W. Frontier of India, 1908.—Mohmand Field Force.—Medal with clasp.

Jamroz Ali Khan—
N.-W. Frontier of India, 1908.—Operations in the Mohmand country.—Medal with clasp.

RESSAIDAR—

Sultan Khan—
N.-W. Frontier of India, 1908.—Operations in the Mohmand country.—Medal with clasp.

53rd Silladar Camel Corps.

RISALDARS—

Mohamed Afzal Khan—
Marri Punitive Operations, 1918.—

Jai Ram—
N.-W. Frontier—Samana, 1897-98.—Medal with 2 clasps.
N.-W. Frontier—Mohmand, 1908.—
Marri Punitive Operations, 1918.—

RESSAIDAR—

Abdulla Khan—
Marri Punitive Operations, 1918.—

54th Silladar Camel Corps.

RISALDAR—

Mehr Ali Khan—
Chitral, 1895.—Relief of Chitral.—Medal with clasp.
N.-W. Frontier of India, 1897-98.—Operations on the Samana and in the Kurram Valley during August and September 1897.—2 clasps.
Tirah, 1897-98.—Clasp.
South Africa, 1900—02.—Transvaal.—Medal with clasp.

Bhai Khan, *Sardar Bahadur.—Order of Br. India, 1st class.*
N.-W. Frontier of India, Hazara, 1888.—Medal with clasp.
East Africa, 1903-04.—Operations in Somaliland. Action at Jidbali.—Medal with 2 clasps.
N.-W. Frontier of India, 1908.—Operation in the Mohmand country.—Medal with clasps.

RESSAIDARS—

Harnam Singh—
East Africa, 1903-04.—Operations in Somaliland.—Medal with clasp.
N.-W. Frontier of India, 1908.—Mohmand Field Force.—Medal with clasp.

55th Silladar Camel Corps.

RISALDARS—

Nasir Khan—
N.-W. Frontier of India, 1908.—Operations in the Mohmand country.—Medal with clasp.

Muhammad Ayub—
N.-W. Frontier of India, 1897-98.—Operations on the Samana and in the Kurram Valley during August and September 1897. Relief of Gullistan.—Operations of the Flying Column in Kurram Valley under Col. Richardson, 20th August to 1st October 1897.—Medal with 2 clasps.

Tirah, 1897-98.—Clasp.
N.-W. Frontier of India, 1908.—Operations in the Mohmand country.—Medal with clasp.

RESSAIDAR—

Sundar Singh—

East Africa, 1902—04.—Operations in Somaliland.—Medal with clasp.

56th Silladar Camel Corps.

RISALDAR—

Zabr Khan—

Buner Field Force, 1898.—Medal and clasp.
Punjab Frontier, 1897-98.
Served with Mohmand Field Force, 1st May 1908 to 5th June 1908.—Medal with clasp.

RESSAIDAR—

Ghulam Mohammed Khan—

Tirah, 1897-98.—Reconnaissance of the Kharmana Depot action of 7th November 1897. Operations Khani Khel Chamkanis.—Medal with 3 clasps.

57th Silladar Camel Corps.

RISALDAR—

Munir Khan, *Khan Sahib, Bahadur.—Order of Br. India, 2nd class.*
East Africa, 1903-04.—Operations in Somaliland.—Medal with clasp.
War, 1914-17.—Mentioned in Despatches.

65th Camel Corps.

RISALDARS—

Ali Asghar Khan—
War, 1916.—

Dilbagh Sing—
War, 1916.—

RESSAIDARS—

Mian Mardan Ali—
War, 1916.—

Rahim Baksh—
War, 1916.—

2nd Government Camel Corps.

RISALDAR—

Nanak Singh—
Tirah, 1897-98.—Medal with 3 clasps.
Persian Gulf, 1914.—
E. P. C. Force, 1918.—

Fateh Khan—

E. P. C. Force, 1917.—

Niaz-Ahmed Khan.—
Muree Field Force, 1918.—
E. P. C. Force, 1918.—

3rd Government Camel Corps.

RISALDAR—

Ashraf Ali Khan—
Egyptian Expdy. Force, 1916.—
Waziristan Field Force, 1917.—

RESSAIDARS—

Fazal Dad Khan—

Tirah, 1897-98.—
M. E. F. 1916.—
Waziristan F. F. 1917.—

5th Government Camel Corps.

RESALDAR—

Noor Din—

Waziristan F. F., 1918.—

Gopi Chand—

Chitral relief, 1895.—
Tochi, 1897-98.—
Tirah, 1898.—
Mohmand Country, 1908.—
E. F. Force, 1916.—

RESSAIDARS—

Dadan Khan—

Mohmand Country, 1908.—
E. F. Force, 1916.—
Waziristan, 1917.—

Fazal Dad Khan—

Tirah, 1897-98.—Usmana.
M. F. Force, 1916.—
Waziristan, 1917-18.—

Grantee Camel Corps.

RISALDARS—

Alam Sher Khan—
Waziristan Blockade, 1901-02.—Medal with clasp.

Mir Alum Khan—
N.-W. Frontier of India, 1897-98.—Buner. Medal with clasp.

JEMADARS—

Mohd. Shafi Khan—

Indian Expdy. Force "D.", 1914-16.—

Abdulla Khan—

Chitral relief, 1895.—
Tochi field Force, 1897-98.—

WAR SERVICES OF SENIOR SUB-ASSISTANT SURGEONS, FIRST AND SECOND CLASS, INDIAN MEDICAL DEPARTMENT.

Abdur Rahim Khan. (*Senior Sub-Asst. Surgn., 1st class.*)—
Burma, 1889-92.—Medal with clasp.
East Africa, 1903-04.—Operations in Somaliland. —Medal with clasp.

Alah Banda, *Bahadur*—(*Senior Sub-Asst. Surgn., 1st class.*)—
Chitral Relief Force, 1895.—Medal with clasp.
N.-W. Frontier of India, Malakand, 1897-98.—Clasp.
China, 1900.—Medal.

Atma Ram Bapuji. (*Senior Sub-Asst. Surgn., 1st class.*)
Soudan, 1886.—Suakin.—Medal with clasp. Bronze Star.
Soudan, 1896.—Dongola Expedition.—Queen's medal and Khedive's medal.
China, 1901-02.

Azhar Hussain. (*Senior Sub-Asst. Surgn., 1st class.*)—
N.-W. Frontier of India, Waziristan, 1894-95. —Medal with clasp.
Chitral, 1895.—Relief of Chitral.
China, 1900.—Relief of Pekin.—Medal with clasp.
Tibet, 1903-04.—Operations at and around Gyantse. Promoted to Senior Hospital Assistant, Second Class.—Medal with clasp.

Barkat Ram *B hadur.* (*Senior Sub-Asst. Surgn., 1st class.*)—
Chitral Relief Force, 1895.—Medal with clasp.
N.-W. Frontier of India, 1897-98.—Malakand. Operations in Bajaur and Mamund country, Buner. Action of the Tanga Pass.—Clasp.

Gurditt Singh. (*Senior Sub-Asst. Surgn., 1st class.*)—
N.-W. Frontier of India, 1897-98.—Mohmand.—Medal with clasp.
Tibet, 1903-04.—Medal.

Ihsan Ali. (*Senior Sub.-Asst. Surgn., 1st class.*)—
N.-W. Frontier of India, 1891.—Medal with clasp.
Miranzai, 1891.—*Clasp.*
Waziristan, 1894-95.—Clasp.
N.-W. Frontier of India, 1897-98.—Defence and Relief of Chakdara.—Medal with 2 clasps.
N.-W. Frontier of India, Waziristan, 1900-01.

Itu Chawan. (*Senior Sub-Asst. Surgn., 1st class.*)—
Chitral, 1895.
East Africa, 1902-04.—Operations in Somaliland. —Medal with clasp.

Karam-chand *Bahadur.* (*Senior Sub-Asst. Surgn., 1st class.*)—
N.-W. Frontier of India, 1897-98, Tochi.—Medal with clasp. Specially promoted to 1st class Hospital Assistant.
China, 1900.—Medal.
N.-W. Frontier of India, Waziristan, 1901-02.—Clasp.
N.-W. Frontier of India, 1908.—Operations in the Zakka Khel country.—Mentioned in despatches. Specially promoted to 2nd class Senior Hospital Assistant.

Mudaliar, P. Sivaprakasa. (*Senior Sub-Asst. Surgeon, 1st class.*)—
Burma, 1887-89.—Medal with clasp.
N.-E. Frontier of India, Chin Hills, 1891.—Clasp.
N.-W. Frontier of India, 1897-98.—Medal with clasp.
Tirah, 1897-98.—Clasp.
China, 1900.—Medal.

Muhammad Khan *Bahadur.* (*Senior Sub-Asst. Surgn., 1st class.*)—
N.-W. Frontier of India, 1897-98.—Defence of Malakand. Relief of Chakdara. Malakand Action at Landikai. Operations in Bajaur and the Mamund country.—Medal with 2 clasps. Promoted to Senior Hospital Assistant.
China, 1900.—Relief of Pekin. Actions of Peitsang and Yangstun.—Medal with clasp. Promoted to Senior Hospital Assistant.

Muhammad Mihr-din. (*Senior Sub-Asst. Surgn., 1st class.*)—
N.-W. Frontier of India, Hazara, 1888.—Medal with clasp.
N.-W. Frontier of India, 1897-98, Tochi.—Medal with clasp.
China, 1900-01.—Relief of Pekin. Action at Peitsang.—Medal with clasp.

Nayudu, T. Ramanujulu. (*Senior Sub-Asst. Surgn., 1st class.*)—
Burma, 1887-89.—Medal with clasp.
Burma, 1889-90.—*Wuntho Field Force.*—Clasp.

Niaz Ali Khan *Bahadur.* (*Senior Sub-Asst. Surgn., 1st class.*)—
Burma, 1887-89.—Medal with clasp.
N.-W. Frontier of India, Miranzai, 1891.
N.-E. Frontier of India, 1894.—Abor. Service brought to the notice of the Government of India.
Tirah, 1897-98.
N.-W. Frontier of India, Waziristan, 1901-02.—Medal with clasp.

Niranjan Dass (*Senior Sub-Assistant Surgeon, 1st class.*)—
N.-W. Frontier of India, Hazara, 1891.—Medal with clasp.
Chitral, 1895.—Relief of Chitral. Mentioned in despatches.—Medal with clasp.
Tibet, 1903-04.—Action at Naini, 26th June 1904. Operations at and around Gyantse between 5th May and 6th July 1904. March to Lhassa 14th July to 3rd August 1904.—Medal with clasp.
Abor, 1911-12.—India General Service.—Medal with clasp.

Nurul Hasan. (*Senior Sub-Asst. Surgn., 1st class.*)—
N.-W. Frontier of India, Chilas, 1893.
N.-W. Frontier of India, Waziristan, 1894-95.—Medal with clasp.
China, 1899-1901.—Operations at Peyang, Psiku. Siege of Tientsin. Relief of Admiral Seymour's Column and Foreign Concessions. Brought to the notice of the Government of India, and promoted to Senior Hospital Assistant, 2nd class.—Medal.
N.-W. Frontier of India, 1908.—Operations in the Mohmand country.

Pillal, J. Sundaram. (*Senior Sub-Asst. Surgn., 1st class.*) –
Burma, 1887-89.—Medal with clasp.
China, 1900.—Medal.

Ram Rao Gangadhar. (*Senior Sub-Assistant Surgeon, 1st class.*)—
Soudan, 1896.—Dongola Expedition.—Queen's medal and Khedive's medal.
Br. East Africa, 1898.—Expedition against the Ogaden Somalis.—Medal with clasp.
Uganda, 1898.—Operations against Kaberega—Clasp.
Uganda, 1900.—Operations against the Nand-tribes.—Medal with clasp.

Riyazuddin, I.O.M. (*Senior Sub-Asst. Surgeon 2nd class.*)—
Malakand, Swat Bunnar, Utmankhel-1897-98.—Metal with clasp.
British Central Afric,—1899-1900.—Medal with clasp.
British East Africa, Somaliland, 1902-04.—Clasp.
Abor, Expedition, 1911-12.—Medal with clasp.

Saiyid Ali Naki *Khan Sahib Bahadur.* (*Senior Sub-Assistant Surgeon, 1st class.*)—
N.-W. Frontier of India, Hazara, 1891.—Medal with clasp.
N.-W. Frontier of India, Samana, 1891.—Clasp.
South African War, 1899-1902.—Defence of Ladysmith and operations in the Transvaal Despatches. Promoted to 2nd class Senior Hospl. Asst.—Queen's medal with 2 clasps. King's medal with 2 clasps.

Shah Mir Khan. (*Senior Sub-Assistant Surgeon 1st class.*)—
N.-W. Frontier of India, Hazara, 1888.—Medal with clasp.
N.-W. Frontier of India, 1897-98.—Malakand.—Medal with clasp. Promoted to Senior Hospital Assistant.
British East Africa, 1898.—Expedition against the Ogaden Somalis.—Medal.
Uganda, 1900.—Operation against the Nandi tribe.—Clasp.
N.-W. Frontier of India, Waziristan, 1901-02.—Medal with clasp.

Shaikh Husain Ali *Bahadur.* (*Senior Sub-Asst. Surgeon, 1st class.*)—
Burma, 1885-89.—Medal with 2 clasps.
British East Africa, 1896.—Medal.

Sher Muhammad *Bahadur.* (*Senior Sub-Assistant Surgeon, 1st class.*)—
Burma, 1887-89.—Medal with clasp.
N.-W. Frontier of India, 1897-98.—Malakand Buner.—Medal with clasp.
China, 1900-01.—Medal.
Abor Expedition, 1911-12.—Medal with clasp.

Solomon, S. (*Senior Sub-Assistant Surgeon, 1st class.*)—
N.-E. Frontier of India, Chin Hills, 1892-93.—Medal.
China, 1900.—Medal.

Thakur Das. (*Senior Sub-Assistant Surgeon, 1st class.*)—
Burma, 1887-89.—Medal with clasp.
Soudan, 1896.—Suakin.—Queen's medal and Khedive's Medal.
China, 1900-01.—Medal.

Tilok Singh *Bahadur.* (*Senior Sub-Assistant Surgeon, 1st class.*)—
N.-W. Frontier of India, Waziristan,—1894-95.—Medal with clasp.
China, 1900.—Relief of Pekin. Actions at Peitsang and Yangtsun.—Medal with clasp. Promoted to Senior Hospital Asst.
N.-W. Frontier of India, 1902.—Operation against the Darwesh Khel Waziris.

INDEX.

NAMES AND RANK. Page.

THE KING-EMPEROR, 1, 577, 660, 691, 743, 875, 885, 895, 909, 923, 947, 949, 985, 1074, 1263, 1411, 1517, 1595 1600
QUEEN ALEXANDRA 712
H. R. H. The PRINCESS ROYAL . 583

A

A'Brook, D., 2nd-Lt. 331
Aaron Ezekiel Ziradker, Sub-Asst. Surgn. 1830
Aba Duduskar, Subdr. . . . 1479
Abaji Rao Mahadik, Jemdr. . . 934
Abajirao Shind, Jemdr. . . . 1705
Abbas Ali, Sub-Asst. Surgn. . . 1809
—— Subdr. 1142
—— Jemdr 1107
—— **Khan,** Ressdr. . . 950
—— **Khan,** Jemdr. . . . 486
—— Jemdr. 1577
—— Subdr. 1439
—— Subdr. 980
Abbay, B. N., Maj. . . . 223, 935
Abbe Ram, Ressdr. . . . 1682, 1686
Abbey, J. B., Lt. . . . 543, 987
—— W. B. T., Lt.-Col. 107, 206, 1894, 1912
Abbi Ram, Jemdr. 1165
Abbon, B. G. E. W., 2nd-Lt. . . 1613
Abbott, A. A., Lt. 304, 885
—— A. S., 2nd-Lt. 363
—— C. P., 2nd-Lt. . . 476, 1439
—— E. G., Capt. 656
—— E. J., Lt. . . . 354, 1441
—— F. B., Capt. . . . 255, 1623
—— G., 2nd-Lt. 1273
—— H. A., Brig.-Gen. 179, 1078, 1917
—— H. J. E., 2nd-Lt. . . . 807
—— J., 2nd-Lt. 701
—— J. F., Capt. . . 273, 1635
—— J. W., Lt. . . . 538, 1311
—— L. H., Lt.-Col. 55, 97, 137, 159, 201, 868, 1046
—— N. J., 2nd-Lt. 788
—— O. A., Maj. 608
—— P. D., 2nd-Lt. . . 552, 1054
—— R. S., Bt.-Maj. . 32, 254, 947
—— S. H. L., Maj. 1748
—— T. W., Capt. . . . 38, 666
—— W., Lt. 691
Abbott-Young, E. C., Lt. . . . 512
Abdoola Khan Subdr. 1268
—— Subdr.-Maj. . . 1515
—— Akbar, Jemdr. . . . 1510
—— **Aleem, A.,** Sub.-Asst. Surgn. . 1828
—— **Azim,** Jemdr. . . . 1340
—— **Khan,** Subdr. . . . 1307
—— **Aziz,** *Hony.* Capt. . . 2, 2102
—— Jemdr. 1001
—— Subdr. 1050
—— Jemdr. 1204
—— **Khan,** Subdr. . . . 988
—— Subdr. . 1035, 1036
—— Subdr. . . . 1307
—— Sr. Sub-Asst. Surgn. 1808
—— Ty. Sub-Asst. Surgn. 1833
—— Jemdr. . . . 1351
—— **Basit,** Jemdr. . 1332
—— Durrani, Sub.-Asst. Surgn. 1817
—— Ghafar, Subdr. 1323
—— Khan, Jemdr. . . 882
—— **Ghafur,** Subdr. 1307
—— Subdr. 1359
—— Subdr. 1327
—— Jemdr. 988
—— Jemdr. 1087
—— Sr. Sub-Asst. Surgn. 1808
—— Buxie, Sub.-Asst. Surgn. 1817
—— **Khan,** Subdr. . . 1207
—— Ressldr. . 893
—— Jemdr. . 1268
—— Jemdr. . 1034

NAMES AND RANK. Page.

—— **Ghaffoor Khan,** Ressaldr. 1683, 1684
—— **Ghani,** Sr. Sub-Asst. Surgn. 1807
—— Ressdr. . . 1684, 1686
—— Subdr.-Maj. . . . 1698
—— **Khan,** Jemdr. . . . 876
—— Jemdr. 1484
—— **Khan,** Jemdr. . . 950
—— Jemdr. . 912, 957
Abdul Ahmad, Jemdr. 1391
—— Alim, Subdr. 1317
—— **Aziz,** Jemdr. 1515
—— Khan 989
—— Ghani, Jemdr. 1591
—— **Ghani Hafiz,** Jemdr. . . 946
—— **Hakim Faruqi,** Sub-Asst. Surgn. . 1813
—— **Khan,** Sub-Asst. Surgn. 1812
—— **Hakim,** Sub-Asst. Surgn. . 1813
—— Khan, Jemdr. . . 1444
—— **Hamid,** Subdr. 1203
—— Jemdr. 1336
—— **Khan,** Sub-Asst. Surgn. 1809
—— Jemdr. . . 875
—— Haq, Jemdr. 1143
—— Khan, Jemdr. . . . 1051
—— **Hussain,** Subdr. . . . 1591
—— **Jabber,** Jemdr. 912
—— Jabar, Jemdr. 1591
—— Kadir, Jemdr. 1332
—— Kalum Shah, Jemdr. . . 884
—— **Karim,** Subdr.-Maj. . . 1315
—— Sub-Asst. Surgn. . 1813
—— Tempy. Sub-Asst. Surgn. 1833
—— Jemdr. 1272
—— Sub-Asst. Surgn. . 1815
—— **Khan,** Risldr.-Maj. 2, 2103
—— Subdr. . . 1052
—— *Hony.* Capt. 2104
—— Jemdr. . . 993
—— Subdr. . . 1405
—— Khal, Jemdr. 1336
—— Khuddus, Subdr.-Maj. . 1359
—— Khan, Subdr. 1343
—— Jemdr. 1305
—— **Latif,** Jemdr. 922
—— Jemdr. 994
—— Subdr. 1447
—— Jemdr. 1391
—— **Khan,** Ressdr. . . 878
—— Subdr. . . 1509
—— **Lais Sahduddin Muhammad, Maulvi, 2nd-Lt.** 481
—— **Majid,** Subdr.-Maj. . . 1327
—— Sub-Asst. Surgn. . 1812
—— Jemdr. . . 912, 916
—— **Khan,** Subdr. . . 1351
—— Jemdr. . . 1583
—— Jemdr. . . 1310
—— *Hony.* Capt. . . 1892
—— Khan Haphysi, Sub.-Asst. Surgn. . . . 1819
—— Shah, Jemdr. . . . 904
—— Shirwani, Sub. Asst. Surgn. . . . 1820
—— Jemdr. . . 1436
—— Manaf, Jemdr. . . . 1385
—— Qadir, Jemdr. . . . 918
—— Jemdr. 994
—— Qadus, Jemdr. 1241
—— Qayum, Jemdr. 1336
—— **Rahim, S.,** Sub-Asst. Surgn. 1827
—— **Jemdr.** 1340
—— Jemdr. 944
—— **Jemdr.** 888
—— Jemdr. . . 930, 1702
—— Khan, Subdr. 1441
—— **Ressdr.** 953
—— Jemdr. 1705
—— **Rahiman,** Subdr. . . . 1435
—— **Rahman,** Jemdr. . . . 1432
—— Jemdr. . . . 968

NAMES AND RANK. Page.

Abdul Rahman, Jemdr. . . . 876
—— Jemdr. . . . 1265
—— Jemdr. . . . 1373
—— Subdr. . . . 1280
—— **Khan,** Ressldr. . 950
—— **Khan,** Jemdr. . 1404
—— Jemdr. . 930
—— Sub. . . 993
—— Jemdr. . . . 928
—— ***Effendi, Sardar,*** *Hony.* Lt. 111, 883
—— **Rasool,** Jemdr. . . . 1682
—— **Rasul,** Jemdr. 990
—— **Rauf Khan,** Ressdr. . . 882
—— Jemdr. 884
—— Jemdr. 876
—— Sub-Asst. Sur. 1817 1828
—— **Razak,** Subdr. 1702
—— Khan, Ressdr. 1684
—— Razaq, Khan, Jemdr. 876, 914
—— Jemdr. 1400, 1704
—— Ressdr. . . . 1684
—— **Raza,** Jemdr. 994
—— Razool, Jemdr. . . . 1702
—— **Said,** Sub-Asst. Surgn. . 1813
—— **Samad,** Jemdr. . . . 922
—— **Samad Shah,** *Hony.* Capt. 56, 111
—— **Satar,** Jemdr. 892
—— **Sattar,** Subdr. . . . 1323
—— Jemdr. . . . 1340
—— **Khan,** Ressdr. . 938
—— **Shakur Khan,** Subdr. 471, 1483
—— Khan, Jemdr. 882
—— Beg, Jemdr. 882
—— Sitar Khan, Ressdr. . .
—— Subhan Khan, Jemdr. . 1351
—— Jemdr. . 1360
—— Jemdr. . 1264
—— **Wahab,** Subdr. . . . 1339
—— Subdr. 1315
—— **Khan,** Subdr.-Maj. 1197
—— Subdr. . . 1263
—— **Wahid,** Sub-Asst. Surgn. 1813
—— Jemdr. . . . 1272
—— **Khan,** Jemdr. . . 878
Abdulla, M., Capt. . . . . .
—— Tempy. Lt. . . . 1769
—— Khan, Subdr. 1707
—— Subdr. 1267
—— Jemdr. 1068
Abdulla, Subdr. 1702
—— Khan Risaldr. . . . 965
—— Jemdr. 1268
—— Risaldr. . . . 934
—— Jemdr. . . . 946
—— Jemdr. . . . 1232
—— Jemdr. . . . 908
Abdullah Khan, Subdr. Maj. . 1143
—— Sr. Sub-Asst. Surgn. . 1808
—— Tr. Vety. Asst. . . 1687
—— Jemdr. 876
—— Ji, Subdr. 1183
—— Khan, Jemdr. . . . 1285
—— Jemdr. . . . 1090
—— Jemdr. . . . 1244
—— *Hony.* Capt. . . 2104
—— Jemdr. . . . 1288
—— Subdr. . . . 1288
—— Subdr. . . . 1435
—— Jemdr. . . 1092, 1107
—— Tr. Vety. Asst. . 1681
—— *Hony.* Capt. . . 2104
—— Ressdr. . . . 1681
—— **Khuttak,** Jemdr. . . 965
—— **Shah,** Jemdr. . . . 999
—— **Sharif,** Risdr. . . . 873
Abdulmia, A. B., Sub. Asst. Surgn. 1832
Abdur Rahman Khan, Ressdr. . 953
Abdur Rahim, Subdr. 1265
—— Sub-Asst. Surgn. 1819
—— Rahiman, Jemdr. 1591
—— **Khan,** Sr. Sub-Asst. Surgn. 1826
—— **Rohaman,** Jemdr. 1308, 1476
—— **Rahman, Khan, Jemdr.** . 900

NAMES AND RANK. Page.

Abdur Rahman Khan, Jemdr. 1373
—— Subdr. 1464
—— Subdr. 993
—— Subdr. 140–
—— Sub.Asst. Surgn. 1818
—— **Rashid**, Subdr. 1011
—— **Rauf Khan**, Sub.-Asst. Surgn. 1828
—— **Razzak**, Sr. Sub-Asst. Surgn. 1808
—— Subdr.-Maj. 1299
—— Razack, Subdr. Maj. 1307
—— Razaq, Jemdr. 1316
—— Subdr. 1327
Abdus Salam, Subdr. 1591
—— **Sattar Khan**, Sub-Asst. Surgn. 1813
—— **Shakur Khan**, Sr. Sub-Asst. Surgn. 1808
Abel, W. C., Maj. 814
Abell, N., Ty. Capt. 619
—— W. H., Lt. 514
Abercrombie, D. E., 2nd-Lt. 331, 893
—— J. R., Lt. 487, 909
Aberigh-Mackay, J. L., Lt.-Col. 179, 917
Abernethy, P. P., Lt. 315, 1255
—— R. C., Lt. 1475
Abhaidatt Singh, Jemdr. 1054
Abhai Ram, Jemdr. 1384
Abhyankar, S. R., Sub-Asst. Surgn. 1832
Abigail, R. P., Lt. 410
Abigail, R. P., 2nd-Lt. 1630
Abiram Gurung, *Hony.* Capt. 2102
Abir Thapa, Subdr. 1523, 1636
—— Jemdr. 1639
Abnashi Ram, Risldr. 965
Abrahall, O. F. T. E., Lt. 516, 1667
Abraham, Capt. A. C. 67, 698, 795
—— C. E. S., Lt. 393
—— R. C., Lt. 196*a*, 1439
—— Daniel, Jemdr. 1412
Abrahams, Paul, Jemdr. 1012
Abram, G. F., Lt. 311, 1191
Abreu, A. J., Tempy. Capt. 1770
Abrocrombie, J. B., Lt. 909
A'Brook, 2nd-Lt. D. 332, 1020, 1557
—— W. R. C., 2nd-Lt. 324, 1239
Abu Khan, Ressdr. 944
—— **Taleh**, Subdr. 1409
Acharmal, Sub-Asst. Surgn. 1813
Achari, K. A. R., Sub-Asst. Surgn. 1827
Acharya, B. V., Ty. Lt. 1777
Acheson, I. St. G., Lt. 310, 505, 1507
—— J. G., Esq. 42
Acheison Grey, C. G. A., 2nd-Lt. 739
Achchra Singh, Jemdr. 1165
Achhar Singh, Jemdr. 1102
—— Jemdr. 1216, 1365
—— Jemdr. 1369
—— Subdr. 1141
Achhaibar Singh, Jemdr. 1051
Achhru Ram, Sub-Asst. Surgn. 1813
Ackland, H. J., Capt. 280—287, 1016
Acland, H. E. P. D., Capt. 895
Acott, H. T., Sub-Condr. 1867
Acton, H. L., Ty. Asst. Surgn. 1804
—— **H. W.**, Capt. 1754
Acworth, D. H., Capt. 54, 258, 1243
—— J. P., Lt. 313, 939
Adair, D. H., Lt. 583
—— R. A. S., 2nd-Lt.
Adalat Khan, Jemdr. 910
—— Jemdr. 920
—— Jemdr. 1121
—— Jemdr. 955
—— Jemdr. 1563
Adalat Zar Khan, Jemdr. 1133
Adam, C. G., Lt. 523
—— F. J. S., Brig.-Gen. 177
—— G. M., Capt. 1724
—— **J. C.**, Capt. 801, 802
—— **Khan**, Jemdr. 936
—— **N. M.**, Capt. 615
Adames, F. H., 2nd-Lt. 328, 1223
Adams, A. E. St. J., Lt. 470, 1611
—— **A. G.**, Lt. 299, 1399
—— **A. K.**, Lt. 700
—— A. W., Lt. 393, 1431
—— **C.**, Lt. 617
—— C. W., Sub-Condr. 1857
—— D. R., Lt. 368, 1164
—— E. F., 2nd-Lt. 534, 1668
—— E. H., Condr. 1861
—— E. J., 2nd-Lt. 360, 1621
—— **F.**, Lt.-Col. 136, 158, 196, 867, 913
—— F. A., 2nd-Lt. 557, 979

NAMES AND RANK. Page.

Adams, F. H., Lt. 758
—— G., Lt. 507, 1007, 1044
—— G. A., Capt. 708
—— G. A., Hon.-Maj. 783
—— G. G. P. D., Lt. 522
—— G. P., Lt. 734, 1007
—— H. A., Lt. 763
—— H. R., Lt. 463, 1305
—— H. T., Capt. 759
—— **H. V.**, Lt. 305, 1064, 1070
—— J. A., Lt. 544, 1365
—— K. L., Lt. 8, 15, 668
—— P., 2nd-Lt. 549
—— P. H., 2nd-Lt. 549
—— P. T., Lt. 508, 1020
—— R. B., *Sir*, Maj.-Genl. 1925, 1955
—— R. F. G., Lt. 300, 1094
—— R. J., 2nd-Lt. 553, 871, 1523
—— R. P., Lt. 9, 717
—— T., Lt. 405
Adamson, M. A. W., Lt. 1585
Addams-Williams, L., Lt.-Col. 133
Adderley, Lt. A. H. 380, 1164
Addinall, C. R., Lt. 541, 1385
—— F. W., Sub-Condr. 1874
Addinsell, Lt. T. A. 1197
Anderson, H.W., 2nd-Lt. 1693
Addington, R. A., Capt. 264, 923
Addis, N. F., 2nd-Lt. 610
—— R. B., Capt 610
—— W. H., Lt. 67, 617
Addison, **A. W. N.**, Capt. 5, 731, 791
—— H. L., Capt. 1717
—— L. J., Sub-Condr. 1863
Adhar Singh, *Hony.* Capt. 2103
Adhya, J. K., Ty. Lt. 1780
Adie, T. D., Lt. 510
Adinarayana, K., Sub-Asst. Surgn. 1826
Adkins, E. C., 2nd-Lt. 354
—— E. C. A., 2nd-Lt. 1060
—— E. W., 2nd-Lt. 1679
Adlam, E. T., Sub-Condr. 1864
Adlard, F. A., Lt. 526, 801, 992
Adler, B. I. H., Maj. 225, 1455
Adlercron, G. R. L., Capt. 589
Admans, H. G., Lt. 793
Advani, G. B., Ty. Lt. 1776
Adye, D. R., Br.-Gen. 12, 23, 105, 121, 126, 153, 190, 870, 1399
Adye-Curran, S. M., Maj. 1711
Aelen, Rt. Revd. Dr. J. 652
Afsar Khan, Jemdr. 1280
Afzal, Subdr. 966
—— Hussain, Jemdr. 878
—— **Khan**, Subdr.-Maj. 1335
—— Jemdr. 1476
—— Jemdr. 1245
—— Subdr. 1371
Aga Oassim Shah, Capt. 278, 879, 951
—— **Murtaz Khan**, Lt. 481
Agam Limbu, Subdr. 1630
Agar, C. S., Lt. 574
—— C. S., 2nd-Lt. 689
—— E. W. T., Lt. 718
—— W., Capt. 1718
—— Singh, Risldr. 925
Agarwala, G. C., Ty. Lt. 1773
Aggarwal, S. N., Ty. Lt. 1772
Agate, A. R., Asst. Surgn. 1802
—— V. N., *Tempy.* Capt. 1765
Agdi Singh, Jemdr. 1452
Agharkar, N. P., Ty. Lt. 1781
Agia Ram, Risldr.-Maj. 906
Agnew, H. C., Maj. 664, 991
Agtey, R. N., Ty. Capt. 1771
Ahamed Baig, Jemdr. 1309
Ahdul Rasul, Tr. Vety. Asst. 1687
Ahibaran Singh, Subdr. Maj. 1043
—— Jemdr. 1043
Ahiman Rana, Subdr. 1602
Ahmad, D. U., Tempy. Lt. 1774
—— **Ali Khan**, Jemdr. 882
—— Jemdr. 908
—— Jemdr. 1227
—— Baig, Jemdr. 1309
—— **Baksh**, Sub-Asst. Surgn. 1814
—— Jemdr. 942
—— **Din**, Subdr. 1407
—— Subdr. 1106
—— Jemdr. 1035
—— Jemdr. 999
—— Husain, Sub-Asst. Surgn. 1817
—— Khan, Sub-Asst. Surgn. 1820
—— Khan, Jemdr. 1351
—— **Jan**, Tr. Vety. Asst. 1687
Ahmad Khan, Resal. Maj. 930

NAMES AND RANK. Page.

Ahmad Khan, Jemdr. 1518
—— Jemdr. 1285
—— **Khan**, Subdr. 1288, 1320
—— **Subdr.** 1207
—— **Ressdr.** 926
—— Subdr. 1441
—— Jemdr. 1304
—— Jemdr. 1121
—— Jemdr. 1373
—— Jemdr. 1240
—— Jemdr. 1682
—— Ressaidr. 1682
—— Ressdr. 1684
—— Ressaidr. 1685
—— Ressaidr. 1686
—— Jemdr. 1502—04
—— **Shah**, Jemdr. 892, 990
—— Subdr. 1249
—— **Gilani**, Jemdr. 1300
—— **Khan**, Subdr. 1501
—— Jemdr. 924, 1133
—— **Kutti**, Jemdr. 1308
—— **Nawaz Khan**, Hony. Capt. 111, 941
—— Yar Khan, Jemdr. 1064
—— Jemdr. 968
—— Jemdr. 968
Ahmed, Z. A., Lt.-Col. 1786
—— **Din**, Subdr. 1303
—— **Khan**, Jemdr. 926, 1510
—— Jemdr. 1283
—— Subdr. 1252
—— Jemdr. 1280, 1284
—— Ressdr. 959
—— **Nabi Khan**, Jemdr. 1012
Ahmud Nur, Jemdr. 1625
Ahsun Ali Khan, Jemdr. 890
Aich, N. B., *Tempy.* Capt. 1766
Aidan Singh, Jemdr. 924
Aikin, R. W. D., 2nd-Lt. 545, 1569
—— V. J. D., 2nd-Lt. 553, 1453
Aikman, E. H., Lt. 59, 504, 832
—— J. H., Capt. 1723
Ainscow, A., 2nd-Lt. 359, 1567
Ainsley, J. E., Ty. Capt. 1725
Ainslie, C., Col. 40, 130, 660
—— C. J., 2nd-Lt. 799
—— **H. P.**, Lt.-Col. 132, 156, 192, 1271
—— **J. A.**, Capt. 718
—— **K.**, Lt. 516*a*, 1385
Ainsworth, H., Lt.-Col. 135, 1744
—— R., Lt. 542, 1189
Aird, M., 2nd-Lt. 1674
Aird-Smith, W., 2nd-Lt. 1487
Airth, O. F. D. Lt. 1718, 1726
Airy, G., Maj. 10, 81, 227
Aitabahadur Limbu, Jemdr. 1649
Aitcheson, H. M., 2nd-Lt. 630
Aitchison, C. T., Genl. 171
—— W., Lt. 78, 492
Aitken, A. B., Capt. 666, 991
—— A. B., Lt. 506
—— A. E., Col. 123, 149, 181
—— C. F., Lt. 389, 949
—— C. S., Lt. 538, 1635
—— R. G., 2nd-Lt. 365, 899
—— W. G., 2nd-Lt. 77, 1884
—— W. N. G., 2nd-Lt. 545
Aitkenhead, J. D., Lt. 112
Aitkins, G. R., Asst. Surgn. 1796
—— **G. V.**, **Asst. Surgn.** 1796
—— L. D. R., Asst. Surgn. 1796
—— S. R., Asst. Surgn. 1803
Aiton, R. L., Lt. 1056
Aiyanna, C. B., Subdr. 1361
Aiyar, N. S., Ty. Lt. 1780
Ajab Khan, *Hon'ble Hony.* Capt. 2104
—— **Sing Gurung**, Jemdr. 1608, 1609
Ajaib Singh, Senr. Sub-Asst. Surgn 1807
—— Ressdr. 888
—— Jemdr. 906
—— Ressdr. 892
—— Ressdr. 886
Ajinkya, R. N., *Tempy.* Capt. 1766
Ajit Singh, Rissdr. 899
—— Rissdr. 1168, 1686
Ajmir Singh, Jemdr. 932
Ajodya, Bakhsh Singh, Jemdr. 1397
—— Singh, Subdr. 1048
Ajudhia Nath Anand, Tempy. Sub-Asst. Surgn. 1833
Ajudhya Parshad Misir, Sr. Sub-Asst. Surgn. 1807
—— **Singh**, Subdr. 1046
Ajunkhan, Ressdr. 934
Ajun Khan, Jemdr. 1256
Akalkot, Capt. *Raja Saheb of* 111
Akbar Ale, Subdr. 1596

NAMES AND RANK. Page.

Akbar Ale, Subdr. 989
—— Jemdr. 1312
—— *Hony*. Capt. . . . 2101
—— **Khan**, Risldr. 931
—— Subdr. 1280
—— Jemdr. 1681
—— Jemdr. . . . 1142, 1232
—— Subdr. 1227
—— Jemdr. 1 43
—— Jemdr. 1090
—— Rissdr. 1683
Akehurst, C. H. L., 2nd-Lt. 332, 1158
Akerman, W. P. J., Capt. . . . 615
Akhmad Khan, Jemdr. 1260
Akhtar Manir Khan, Jemdr. . . 942
Akhurst, W. H., *Hony*. Lt. . . 588
Akram Khan, Jemdr. 1585
Aki Khan, Jemdr. 1280
Aksar Khan, Jemdr. 1117
Ala Singh, Subdr. . . . 1158, 1433
—— Jemdr. 1117
—— Subdr. 1112
Aladad Khan, Risldr. 938
Alagappan, S. C., Ty. Lt. . . . 1776
Alan-banda, Sr. Sub-Asst. Surgn. 1807
Alahdad Khan, *Hony*. Lt. . . . 2104
Alah-ditta, Sub-Asst. Surgn. . 1816
Alah Yar Khan, Subdr. . . . 1363
Alam Dim, Jemdr. 1376
—— **Ali Khan**, Subdr. 1050
—— Jemdr. . . . 938
Alam Gir, ... 946
—— **Khan**, Jemdr. 1679
—— Subdr.-Maj. . . . 1389
—— Subdr. 988
—— *Hony*. Capt. . . . 2104
—— Subdr. 966
—— Jemdr. 908
—— Jemdr. 988
—— Jemdr. 1510
—— Jemdr. 1099
—— Subdr. 1375
—— **Shah**, Subdr. 1507
—— Risldr. 924
—— **Shah**, Ressdr. . . . 1684, 1685
—— **Sher**, Jemdr. 1681
—— Jemdr. 1249
—— Subdr. 1134
—— **Khan**, Risldr. . . . 1684
—— Jemdr. . . . 946
Alau-din Khan, Ressdr. . . . 938
Alavi, H. A., Tempy. Capt. . . . 1772
Alban, R. G. E. W., 2nd-Lt. 1613, 333, 1633
Albel Singh, Ressdr. 874
Alberga, E. M., 2nd-Lt. . 472, 1604
Albert, E. F., 2nd-Lt. . . . 494, 896
Albinson, J. H., Capt. 1718
Albrecht, C. L., Lt. . . . 541, 1673
Albuquerque, C. J., Lt. 1361
—— J. W. F., Ty. Lt. . 1780
Alcock, G. H., Ty. 2nd-Lt. . . . 707
Alder, W. G., 2nd-Lt. . . 559, 1447
Alderman, R. E., Capt. 501, 848, 1419
Alder-Peters, H., 2nd-Lt. . . . 530
Aldersmith, C. A., Lt. 852
Alderson, F. H. 2nd-Lt. . . 556, 1423
—— H. W., Lt. 571
—— M. G., 2nd-Lt. . . 330, 1647
—— R. C., Lt. . . . 528, 1042
—— V. C., Lt. . . . 401, 1515
Alderton, F. W., 2nd-Lt. . . . 1579
—— J. R., Sub-Condr. . . 1872
Aldous, A. M., 2nd-Lt. 721
—— T. R., Lt. 623
—— W. T., Lt. 826
Aldred, T. R., Lt. 821
Aldridge, N. C., Lt. 730
—— W. M., Nursing Sister . 1734
Aldworth, E. E., 2nd-Lt. . . 787, 1134
—— T. P., Capt. 750
Alexander, *Hony*. Capt. . . . 2103
—— A., Lt. 724
—— A. G. S., 2nd-Lt. . . 332, 947
—— A. P., 2nd-Lt. . . . 766
—— A. V., Lt.-Col. 134, 157, 194, 870, 1311
—— C., Maj. 241, 1265, 1331
—— C., Sub-Asst. Surgn. 1819
—— C. G., Capt. . . . 501*a*
—— C. H., Lt.-Col. 102, 135, 158, 195
—— C. St. G. E., 2nd-Lt. 743, 766
—— E. C., Lt.-Col. 97, 137, 159, 210, 1243
—— G., Capt. 503
—— H. B., Lt. 492

NAMES AND RANK. Page.

Alexander, H. G., Capt. . . . 1758
—— H. M., Maj. 101, 228, 1662, 1885
—— H. M., Condr. . . . 1861
—— H. S., Lt.-Col. 133, 156, 193, 869, 1199
—— J. L., 2nd-Lt. . . . 363
—— J. R. J., Lt. . . . 425
—— L. A., 2nd-Lt. . 325, 1617
—— L. F. L., 2nd-Lt. 692, 784
—— L. W., Capt. . 581, 599
—— P. S., Capt. . . . 589
—— R. C. J., Maj. . . . 633
—— R. D., Maj. 2, 222, 1605
—— R. J., Tempy. Asst. Surgn. 1804
—— S., Lt. . . . 724, 738
—— V. S., Capt. . . . 1892
—— S. W., 2nd-Lt. . . . 638
—— W., Lt. . . . 724, 739
—— W., 2nd-Lt. . . 553, 1621
—— W., 2nd-Lt. . . . 818
—— W. A., Ty. Capt. . . 1724
—— W. L., 2nd-Lt. . 558, 1124
—— W. P., Maj. 101
—— Subdr. 1307
—— **Hari Singh**, Jemdr. . 1299
Alfieri, F. J., Lt. . . . 322, 529, 1156
Alford, B. T., Lt. 705
—— E. N., Capt. 39
—— L. H. M. McN., 2nd-Lt. . 595
Algeo, L. W. A., Capt. 768
Aliafsar Khan, Jemdr. 1232
Ali Ahmad, Ty. Sub-Asst. Surgn. . 1833
—— **Ahmad Khan**, Risldr. Maj. 903
—— Jemdr. 946
—— **Akbar**, Subdr. 1106, 1259
—— **Akbar**, Jemdr. 1281
—— Jemdr. 977
—— **Asghar**, Jemdr. 918
—— **Ashgar**, Ressdr. 1685
—— **Akbar**, Jemdr. 1336
—— Jemdr. 1684
—— Ressdr. 1683
—— **Shah**, Ressdr. 1684
—— **Bahadur**, Subdr. 1136
—— Subdr. 1241
—— Jemdr. 1681
—— **Bakhsh**, Jemdr. 1124
—— **Khan**, Subdr. . . . 1086
—— **Bakhsh Khan**, Jemdr. 936, 999
—— Subdr. 1106
—— **Dost**, Subdr. 1427
—— **Gauhar Khan**, *Hony*. Capt. . 2102
—— Tr. Vety. Asst. . . 1687
—— **Haidar Khan**, Risldr.-Maj. . 948
—— *Hony*. Capt. . 2104
—— **Hasan**, Jemdr. 1428
—— **Husain Khan**, Sub-Asst. Surgn. 1812
—— **Husain**, Jemdr. 1428
—— Jemdr. 999
—— **Juma (I)**, Subdr. 1427
—— (II), Jemdr. 1428
—— **Mardan**, Ressdr. 1681
—— **Khan**, Jemdr. . . . 948
—— **Muhammad**, Subdr. . . 1502—04
—— Ressdr. . . . 1633
—— Jemdr. . . . 1533
—— Jemdr. . . . 1518
—— Jemdr. . . . 1117
—— Jemdr. . . . 1682
—— Jemdr. . . . 1681
—— Tr. Vety. Asst. . . 1687
—— Khan, *Hony*. Capt. 2102
—— Jemdr. 1087
—— Jemdr. 1682
—— Tr. Vety. Asst. . . 1687
—— **Nijat**, Jemdr. 1428, 1502
—— **Panah**, Subdr. . . . 1427, 1502
—— Jemdr. 1351
—— Subdr. 1427
—— **Shan**, Risldr. 912
—— Jemdr. 922
—— **Sher**, Jemdr. 1087
—— Khan, Jemdr. . . . 1404
—— Risldr. . . .
—— Risldr.-Maj. . . 955
Allah Din, Jemdr. 1681
Alif Khan, Jemdr. 1001
Aling, G. E., Asst. Surgn. . . . 1802
Alington, N. S., 2nd-Lt. . . . 324
Alison, T. F. A. R., Lt. . . 312, 1407
Alladad Khan, Subdr. 1102
—— Subdr. 1195
Allah Bell Khan, Jemdr. . . . 882
Alladia Khan, Jemdr. 1092
Allah-Din, Jemdr. 1682

NAMES AND RANK. Page.

Alla Ditta, Jemdr. 989, 1681
—— Jemdr. 1457
Allah Baksh, Subdr. 1452
—— Subdr. 1585
—— Subdr.-Maj. . . . 1134
—— Subdr. 1088
—— **Dad**, Jemdr. 1241
—— Jemdr. 1571
—— Subdr. 1375
—— Subdr. 1518
—— **Khan** Jemdr. . . . 1285
—— Jemdr. . . 1498, 1698
—— Subdr. . . . 1207
—— Subdr. . . . 1404
Allah-Din Khan, Jemdr. . . . 1498
Allah Ditta, Jemdr. (II) . . . 1141
—— Jemdr. 1453
—— Jemdr. 1252
—— Jemdr. 1152
—— Subdr. 1363
—— Jemdr. 1285
—— **Khan**, Subdr. 1247
—— Subdr. 1409
Allah Rakha Khan, Subdr. . . . 1011
Allahyar Khan, Jemdr. 910
Allan, A., Lt. 504, 1665
—— A., 2nd-Lt. 715
—— D. H., Capt. 848
—— D. S., 2nd-Lt. 564
—— E. J. A., 2nd-Lt. . . . 327, 1189
—— H. L., Capt. 827
—— H. K. S., 2nd-Lt. 1036
—— J. C., Lt. 504, 986
—— J. M., Lt. 90, 535
—— J. N., 2nd-Lt. 723
—— J. R. N., 2nd-Lt. 765
—— K. D., Capt. . . . 255, 1162
—— L. K., Asst. Surgn. . . . 1799
—— P. A., Capt. 254, 1279, 1575
—— R. A. B., 2nd-Lt. 1068
—— R. G., Capt. 803
—— R. J., Lt. 410, 1635
—— R. M., 2nd-Lt. 796
—— T. L., 2nd-Lt. 554
—— W. A., 2nd-Lt. 365
—— W. S., Capt. 651, 861
Allanson, C. J. L., Bt.-Lt.-Col. 51, 97, 160, 213 1623
—— G. H., Lt. . . . 303, 1006
Alla Rakha, Jemdr. 989
Allard, G. H.. 2nd-Lt. 696
Alam Khan., Jemdr. 1389
Allaud Din, Subdr. 1391
Allaway, G. A., Condr. 1867
Allbrook, W. R., Lt. . . 536, 1491
Allen, A., 2nd-Lt. 7
—— A. H. C., Lt. 1255
—— A. E., Lt. 387
—— A. I. L., Lt. 1708
—— A. J. L., 2nd-Lt. 560
—— A. L., 2nd-Lt. 557
—— A. L. G., Asst. Surgn. . . 1802
—— B., Lt. 81, 307
—— B. C., 2nd-Lt. 531
—— B. M., Lt.-Col. 56
—— C. A., Lt. 311, 1191
—— C. E, 2nd-Lt. . . 26, 356, 1677
—— C. H. P., Capt. 1758
—— C. W., Capt. 608
—— D. J., Lt. 738
—— D. S., 2nd-Lt. 1120
—— E., 2nd-Lt. 735*a*
—— E. C., 2nd-Lt. . . . 559, 1457
—— E. F., 2nd-Lt. 568
—— E. R., Lt. 544
—— E. R. L., 2nd-Lt. 715
—— F., Lt. 520, 1132
—— F. J., Lt. 315
—— F. O. S., Capt. 767
—— G., Sub-Condr. 1857
—— G. C., Lt. 423
—— G. F. S., Capt. 1910
—— G. H., Lt. 534
—— G. M. H., Lt. . . . 540*a*, 1673
—— G. O., 2nd-Lt. . . . 554, 1642
—— H., Maj. 79, 1838
—— H., 2nd-Lt. . . . 609, 996
—— H. I., Capt. . . . 60, 1004
—— H. K., 2nd-Lt. 1114
—— J., Lt 637
—— J. F., Maj 87, 222
—— J. H., Capt. 1843
—— J. H. S., Lt. . . 487, 603, 947
—— J. S., Lt. 303, 1092
—— N. C., Lt. 435
—— P. V., Lt. 1695

NAMES AND RANK. Page.

Allen, R. A., 2nd-Lt. . . 548, 1622
—— R. F. C., Major . . . 1615
—— R. F. W. K., 2nd-Lt. 324, 1038, 1058
—— S. W. J., Sub-Condr. . . 1874
—— W. A., 2nd-Lt. 1251
—— W. B., Lt. . . . 492, 953
—— W. F., 2nd-Lt. 1473
—— W. G., Lt. 695, 697
—— W. J., Esq. 89
—— W. T., Maj. . 102, 244, 919
—— V. W., 2nd-Lt. 723a
Allequen, E. J., 2nd-Lt. . . . 1197
Allison, D., 2nd-Lt. 630
—— D. W., 2nd-Lt. 699
—— G., Tempy. Capt. . . . 1724
—— G., 2nd-Lt. 327, 1598
—— J., Lt. 1720
—— T., 2nd-Lt. 816
Allman, L., Nursing Sister . . 1734
—— R., Lt. 619
Allmand, T. R., 2nd-Lt. . . . 491
Allott, A. D., Lt. 425, 1583
—— H. W. L., Capt. 1714
Allport, D. H., 2nd-Lt. . . . 1130
Allpress, H. V., Capt. . . . 636
Allson, T. F. R. Lt. . . 312, 1407
Allsop, J. J. W., 2nd-Lt. . 554, 1056
Allsebrook, M. B., Lt. . . 310, 1641
Allum, H. G., Lt. 486
Allum, R. W., 2nd-Lt. . . 551, 1044
Allwood, E. S., Rev. 1908
Allwright, S. R., *Hony.* Maj. . . 1843
Almeida, A. E., Senr. Asst. Surgn. 1789
Almond, H. R., 2nd-Lt. . . 568, 1124
—— J., Lt. 95, 524
Alphonso, A. G., Senr. Asst. Surgn. 1787
Alsop, J. Esq. . . . . . .
—— B. P., Lt. 430, 1639
—— W. W. N., The Revd. . . . 1907
Alston, W. L., 2nd-Lt. . 329, 1148
Altaf Husain, Sub-Asst. Surgn. . 1811
—— —— Khan, Subdr. . . . 1403
Altafulla Khan, Subdr. . . . 1403
Alteridge, L. F., Lt. 1080
Altekar, V. K., Sub-Asst. Surgn. . 1832
Altham, E. A., Lt.-Genl. *Sir* 9, 16, 97, 118
Afuhdin, Hony. Capt. 2103
Alves, D. A., 2nd-Lt. 737
Alwar, H. H. Sawai *Sir* Jey Singh, Bahadur, Maharaja of, *Hony.* Lt.-Col. 110
Amal Baz, Jemdr. 966
Aman Gul, Subdr. 1128
—— Khan, Jemdr. 966
Amanat Khan, Jemdr. 1241
Amar Ali Khan, Subdr. . . . 1439
—— —— Ressdr. 953
—— —— Shah, Jemdr. . . . 1513
—— Bahadur Gurung, Jemdr. . 1625
—— Chand, Sub-Asst. Surgn. . 1812
—— Din, Jemdr. 1228
—— —— Sub-Asst. Surgn. . 1815
—— Nath Puri, Sub-Asst. Surgn. 1814
—— Nath Chopra, Sub-Asst. Surgn. 1817
—— Sing Karki, Jemdr. . . . 1645
—— Sing Negi, Subdr. . . . 1179
—— Singh, Subdr. 1011, 1152, 1252
—— —— Jemdr. 1571
—— —— Jemdr. 882
—— —— Jemdr 906
—— —— *Hony.* Capt. 2103, 2104
—— —— Ressdr. 880
—— —— Ressdr. 910
—— —— Jemdr. 894
—— —— Jemdr. 1122
—— —— Jemdr. 944
—— —— Subdr. 1152
—— —— Subdr. 1132
—— —— Ressdr. 955
—— —— Rsldr.-Maj. 948
—— —— Ressdr. 940
—— —— Sub-Asst. Surgn. . . 1815
—— —— Subdr.-Maj. 1894
—— —— Jemdr. 968
—— —— Subdr. 1050
—— —— Subdr. 1343
—— —— Ressdr. . . 1682, 1685
—— —— Jemdr. 1075
—— —— Jemdr. 951
—— —— Jemdr. 989
—— —— Jemdr. 1012
—— —— 2nd-Lt. 482
—— Maruti, Sub-Asst. Surgn. . 1817
—— Pun, Jemdr. 1615

NAMES AND RANK. Page.

Amar Pun Rana, Jemdr. . . . 1625
—— —— Thapa, Subdr.-Maj. . 1618
—— Sing Thapa, Subdr. 1597, 1618
Amarsing Bohra, Jemdr. . . . 1655
Ambalal, Devshanker Bhatt, Sub-Asst. Surgn. 1831
Ambarjit Rai, Jemdr. 1012
Amba Shankar Morarji, Sr. Sub-Asst. Surgn. 1829
Amber Khan, Subdr. 1305
Ambica Kunwar, Subdr. . . . 1363
Ambler, C. T. M., Lt. . . 531, 1225
—— F. B., Capt. 1765
—— H. L., 2nd-Lt. 622
—— M. J., Capt. 593
—— R., Lt. 305, 877
Ambrose, A., Sub-Condr. . . . 1874
—— J., Capt. 9, 779
—— R. D., Lt. . 306, 1429, 1487
Armers, T., 2nd-Lt. 566
Ames, C. G., Maj. 212, 859, 1231, 1249
—— L. H. R., Maj. . 207, 1315, 1889
—— W. R., Maj. 239, 1030
Amesbury, S. J. R. 2nd-Lt. . .
—— W. L. R., Lt.-Col. 67, 132, 156, 193, 1660, 1685
Amey, F. G., 2nd-Lt. 530
Amichan, Jemdr. 924
Amichand, Jemdr. 1165
—— Jemdr. 1456
Amies, B. J., Lt. . . 299, 1056, 1102
—— E. L., 2nd-Lt. 786
Amilal, Jemdr. 1385
Amin, B. D., Sub-Asst. Surgn. . 1832
—— J. L., Sub-Asst. Surgn. . 1832
—— K. V., *Tempy.* Capt. . . 1766
—— Chand, Sub-Asst. Surgn. . 1812
—— Gul, Subdr. 1098
—— Khan, Subdr. 1249
Amir Abdul Rahman Khan, Jemdr. 904
—— Ahmed, Tr. Vety. Asst. . . 1687
—— Ali, Jemdr. 1145, 1681
—— —— Subdr. 1371, 1506
—— —— Jemdr. 1260
—— Ali, Jemdr. 1495
—— —— Khan, Subdr.-Maj. . . 1036
—— —— —— Jemdr. 1240
—— —— Jemdr. 1208
—— Baksh, Jemdr. 1408
—— Baz, Jemdr. 1154
—— Chand, Capt. 1355
—— Haidar, Jemdr. 1453
—— Hamza, Sr. Sub-Asst. Surgn. 1898
—— Khan, Subdr. 1034, 1086, 1207, 1335, 1513
—— —— Subdr. 1090
—— —— Subdr. 1311
—— —— Ressdr. . . 1001, 1682
—— —— Subdr. . . . 1104, 1117
—— —— Subdr. . . . . 1502—04
—— —— Ressdr. 940
—— Lall, Jemdr. 880
—— Muhammad Khan, Ressaidar 921
—— —— —— Jemdr. . . . 884
—— —— —— Jemdr. . . . 979
—— Singh, Jemdr. 886
Amor, L. E., Lt. 533, 1207
—— J. A., Asst. Surgn. . . . 1803
—— W., Maj. 655
Amos, C., 2nd-Lt. . . . 359, 1032
—— T. E., 2nd-Lt. 1018
Amphlett, F. G., 2nd-Lt. . . . 742
Amratlal Jaskaramkothari, Sub-Asst. Surgn. 1829
Amrik Singh, Jemdr. 886
—— —— Jemdr. 918
Amrit Balwant, R., Sub-Asst. Surgn. 183
—— Singh, Ressldr. 924
—— —— Jemdr. 1389
Amritraa, C. M., Sub-Asst. Surgn. 1832
Amritrao Ghorpade, Subdr. . . 1439
Amsden, B. H., Lt. . . 320, 365, 1251
Anand, H. S., Capt. 1760
—— R. R. M., *Tempy.* Capt. . 1770
Ana Ram, Subdr. 1497
Ananda Gopal Bandyopadhyay, Sub-Asst. Surgn. . 1815
—— Thorat, Jemdr. 1460
—— Nambiar, Jemdr. . . . 1308
Anantadri Nayudu, Subdr. . . 1276
Anantan, K. M., Ty. Capt. . . 1772
Anantanarayana Aiyar, C. P., Esq. 89
Ananthaya, U., Ty. Capt. . . . 1772
Anantprasad, S. M., Sub-Asst. Surgn. 1831

NAMES AND RANK. Page.

Anant Ram, Jemdr. 1171
—— —— Ungle, Sub-Asst. Surgn. . 1812
Anantthaya, U., Tempy. Lt. . . 1773
Anar Gul, Subdr. 1243
—— Khan, Jemdr 1145
Anastasiadi, S. E., Maj. . . . 835
Anbir Pun, Jemdr. 1764
—— Subdr. 1657
Ancill, G. C., 2nd-Lt. 690a
Ancrum, H. H., Condr. . . . 1853
Andeen, G. F., Sr. Asst. Surgn. 1791
Anderson, A., 2nd-Lt. 360
—— A. C., Capt. 1755
—— A. G., 2nd-Lt. 788
—— A. J., Maj. 653
—— A. L. B., Capt. 367, 769, 945
—— A. M., Lt.-Col. 130, 155, 190
—— A. R., Lt. . . 543a, 820, 993
—— A. R., Lt. 111
—— A. T. D., Lt. 516
—— A. V., Lieut.-Col. . . 1786
—— A. W., Lt. 725
—— A. W. L., Maj.-Genl. . 1271
—— B., Nursing Sister . . 1735
—— B. E., Maj. 98, 102, 233, 1259
—— C. A., Lt.-Genl. *Sir* 2, 31, 118, 612
—— C. B., Lt. 1241
—— C. E. L., Lt., 55, 503, 835, 1507
—— C. F., Lt.-Col. 102, 135, 662, 1881
—— C. H., Nursg. Sister . 1735
—— C. J., Col. 176
—— C. K., Maj. 1919
—— C. L., Asst. Surgn. . . 1797
—— C. O. D., Lt. 797
—— C. S., Lt.-Col. . . 5, 835
—— C. W., Lt. . 401, 818, 851
—— D. A. T., Ty. 2nd-Lt. . 725
—— D. C., Capt. 710
—— D. C., Lt. . 300—01, 1483, 1583
—— D. M., Lt. . . . 371, 1026
—— Dy., Lt. 834
—— E. D., Lt. . . . 524, 1287
—— E. J. M., Nursg. Sister 1735
—— E. P., Maj . . . 644, 1882
—— E. S. J., Maj. . 26, 87, 214
—— F. E., 2nd-Lt. . . . 61, 818
—— F. G., 2nd-Lt. . . 365, 1351
—— F. J., Capt. 1757
—— F. MacK., Lt. 493
—— F. M. 1698
—— F. S., Capt. 617
—— F. S., Lt 541, 1158
—— G., Lt. 446—449, 879, 1077
—— G., 2nd-Lt. 472, 1161
—— G. A., Lt. 525, 1164
—— G. D., Capt. 269, 1267, 1573
—— G. H., Maj. . 212, 1889, 1913
—— G. K., 2nd-Lt. . 361, 1678a
—— G. W., Capt. . . 252, 1124
—— H. B., 2nd-Lt. 805
—— H. C., Genl. . . . 171, 1031
—— H. C., 2nd-Lt. 669
—— H. J., Lt. 533, 1001
—— J., Maj. 1751
—— J., *Hony.* Capt. . . . 744
—— J. D., Lt. 542
—— J. F. H., Bt. Lt.-Col. 62, 78, 161, 240
—— J. H., 2nd-Lt. 77
—— J. J., Lt. 827
—— J. McK., Capt. 271, 1409, 1439
—— J. W., Lt. 715
—— J. W. 1009
—— K. E., Maj. . 227, 1509, 1883
—— L. A. P., Capt. . . . 1757
—— L. E., Maj.-Genl. 31, 118, 1709
—— M. G., Capt. 102 105, 252, 1629
—— M. L., Lt. 526, 1671, 1699
—— N. K., Revd. 1908
—— N. M., 2nd-Lt. 31
—— O., 2nd-Lt. 325
—— P., B., Lt. 294, 1409, 1463
—— P. J., Capt. 808
—— R. A., Tempy. Capt. . . 1715
—— R. C., Lt. . . 405, 538, 1319
—— R. C., Lt. 470
—— R. E. H., Col. 59
—— R. H., Maj. . . . 214, 895

NAMES AND RANK. Page.

Anderson, B. H., Maj. 234, 1203
——— B. H. B., Lt.-Col. 32, 87, 138, 160, 198
——— B. W., 2nd-Lt. 531, 608
——— B. W. A., Lt. 77, 531, 1884
——— S., Lt.-Col. 134, 1744
——— T. B., Capt. 58, 615
——— W. B., Esq.
——— W. C., Maj.-Genl. 125, 153, 193, 872, 1595
——— W. C., Bt.-Col. 121, 199, 871, 883, 1912
——— W. H., Maj. 78, 220
——— W. H., Lt. 537, 1034
——— W. L., Maj.-Genl. 1271
——— W. M., Maj. 1747
——— W. T., Lt. 830
——— 2nd-Lt. 1676, 1695
Anderson-Morshead, R. Y., Capt. 696, 795
Anding, J. B., 2nd-Lt. 557, 1705
Andras, J. B., 2nd-Lt. 721
Andreae, E. B., Lt. 1720
Andrew, A. W., Col. 50, 104, 108, 121, 126, 150, 189
——— B. H., 2nd-Lt. 722
——— D. C. A., Col. 12, 33, 121, 129, 154, 187
——— F. A., Brig.-Genl. 50, 122, 133, 156, 193, 871, 1509
——— G. P., Maj. 833
——— J. A., 2nd-Lt.
——— H. W., Lt. 423, 1162
——— Jemdr. 1299
Andrews, C. L., Capt. 267, 933
——— H. W., Capt. 268, 1148
——— A. C., Lt. 618
——— A. S., Maj. 808
——— C. G., Capt. 700, 760
——— C. W., 2nd-Lt. 796
——— D. C., 2nd-Lt. 561, 1082
——— D. E., Lt. 516a
——— F. J., Condr. 1853
——— E. W., Capt. 730
——— F. L., *Hony.* Maj. 588, 599
——— G. D., 2nd-Lt. 729a
——— G. O., 2nd-Lt. 595
——— H., Lt. 727
——— H. A., Capt. 66, 272 1170, 1189
——— H. J., Tempy. Capt. 1771
——— H. W., Lt. 7, 776
——— J. A., 2nd-Lt. 325 366, 1367
——— J. P., Sub-Asst. Surgn. 1830
——— J. S. T., Lt. 405, 1070
——— L. J., Col. 104, 108, 123, 149, 182
——— M., Lt. Qr. Mr. 1729
——— W., 2nd-Lt. 362, 567
——— W. E., 2nd-Lt. 807
——— W. M., Lt. 533, 1050
Andrus, T. A., Br.-Genl. 50, 122, 135
Aneill, G. C. R., 2nd-Lt. 689
Angad, Jemdr. 1456
Angad Ram, Subdr. 1483
Angel, B. G., Lt. 544, 1349
Angell, J. W., Lt. 681, 785
Angelo, C. H. C., Lt. 817
——— F. W. P., Maj-Genl. 891
——— M. O., Lt. 44, 513, 1660
——— N. L., Lt. 299, 1391
——— S. S., Lt. 505
Angler, C. T., Capt. 17, 40, 778
Anger, A. E., Lt. 90
Angus, H. C., Lt. 581, 1670
——— G. W., Lt. 798
——— H. F., Lt. 739, 755
——— M. A., Lt. 815
——— T. H., 2nd-Lt. 365, 1203
——— W. C. S., Lt. 686, 725
Anil Chandra Sen, Sub-Asst. Surgn. 1818
Anna, Jemdr. 1200
——— Jemdr. 1202
——— Chawan. Jemdr. 1464
Annamalai, Subdr. 994
Anna Singh, Subdr. 580
Annakin, A. H., 2nd-Lt. 629, 79
Annandale, F. W., Lt. 793
Annear, W. C., 2nd-Lt. 565
Anne Thapa, Subdr. 1596
Annely, F. R. C., Lt. 521, 1678
Annesley, A. S. B., Bt.-Col. 60, 125, 153, 191, 1660
——— F. C., Capt. 604
——— I. D. C., Capt. 827
Annett, A. E., Lt. 591

Annett, J., Capt. 836
——— M. W., Lt. 714
Anno Khan, Jemdr. 940
Anokh Singh, Subdr. 1268
——— Jemdr. 878
———, Jemdr. 950
Anscomb, R. T. J., 2nd-Lt. 357, 1048
Ansdell, R. C., Capt. 583
Ansell, A. M., Capt. 610
——— C. E. O., Lt. 306, 1569
——— J. W., Capt. 280, 287, 1046
An Sing Thapa, Subdr. 1607
Anson, C. Lt. 536, 1191
——— E. E., Lt. 304, 939
——— G. F. W., Maj. 225, 1900, 1913
——— H. A., Lt. 717
——— R. R., Lt. 578, 599
——— T. G., Lt. 583
Anstey, H. C. S., Revd. 1903
——— D., Lt. 420, 1144
——— J. C. K., 2nd-Lt. 561
Anstice, J. H., 2nd-Lt. 585
Anstiss, L. A., 2nd-Lt. 557, 1054
Anston, Lt., G. T. 1674
Anstruther, P. G., Bt.-Lt.-Col. 137, 764
Antbir Gurung, Subdr. 1614
Antelme, P. F., Lt. 578
Anthonisz, C. F., Tempy. Capt. 66, 1723
Anthony, Jemdr. 993, 994
——— Subdr., 1704
——— R. W., Lt.-Col. 139, 1745
——— W. S., Lt.-Col. 135
——— David, Subdr. 1276
Anti Ram, Jemdr. 886
Antony, Subdr.-Maj. 1317
Antonymuthu, S. B., Sub-Asst. Surgn. 1828
Antram, E. W., Capt. 1596, 1892
Antu, Subdr. 1385
——— Naik, Jemdr. 1436
——— Burge, Jemdr. 1480
Anup Singh Jemdr. 966
——— Capt.
Anwar, Subdr. 1035
——— Khan, Jemdr. 1705
Apcar, M. M. O. J., Senr. Asst. Surgn. 1789
——— T. S., Hony. Lt. 112
Apergis, H. D., Lt. 1720
Aplin, P. J. H., Col. 105, 122, 153, 181, 1918
——— S. L., Lt.-Col. 184, 848, 1894, 1991
Appagiri-Thevar, Jemdr. 1328
Appalasami, Jemdr. 994
——— M. N., Sub-Asst.-Surgn. 1826
Appalaswami, M., Sub-Asst. Surgn. 1828
Appa Rao, Jemdr. 993
Appa Salunke, Subdr. 1423
Appalsami, Subdr. 1011
Appasami, Subadar 993
Appasamy, Jemdr. 1205
Appavu, Jemdr. 1332
Appleby, T., Sub-Condr. 1863
Appleton, H. T., 2nd-Lt. 799
Applewhaite, G. R., Lt. 368, 1086
Appleyard, S. W., 2nd-Lt. 788
Applin, R. V. K., Maj. 56, 593
Apte, G. D., *Ty.* Lt. 1780
——— V. K., *tempy.* Capt. 1769
Apthorp, S. E., Maj. 236, 1391
Apthorpe, E. H. F., Maj. 222, 1367, 1369
Aquino, F., Asst. Surgn. 1797
Arangil Chandu, Sub-Asst.-Surgn. 1827
Arbab Muhammad Hussain, Jemdr. 908
Arbela Singh, Risldr. 915
Arbel Singh, *Hony.* Capt. 2103
Arbery, J., *Hony.* Maj. 1850
Arbuthnot, A. D. S., Capt. 664, 996
——— A. H., Maj. 209, 872, 1604, 1629
——— C. W. E., Capt. 501, 835, 997
——— G. H., Lt.-Col. 57
——— P. B., Lt.-Col. 381, 854
——— J. St. C., Lt. 319, 1259, 1567
Abzaa, Hony. Capt., 2102
Archard, Lt. 1676
Archer, A. S., 2nd-Lt. 77, 1884
——— B., 2nd-Lt. 692
——— C., Lt.-Col. 56
——— C. J., 2nd-Lt. 564
——— G., Lt. 699, 796
——— W. A., Lt. 319, 968

Archbold, R., Capt. 619
——— F. H. W., Maj. 58, 663
Archdale, F. A. Lt. 509, 1517
——— T. M., Lt.-Col. 135, 613
Archer, A. J., Senr. Asst. Surgn. 1787
——— A. S., 2nd-Lt. 1884
——— B., 2nd-Lt. 784
——— E. J., Sr. Asst. Surgn. 1789
——— J. E., Condr. 1853
——— W. A., 2nd-Lt. 365, 321
Ardagh, E. V., 2nd-Lt. 561
——— F. D., Capt. 681
Ardoino, J., Maj. 225, 1311
Ardron, H. G., 2nd-Lt. 727
——— J. S., 2nd-Lt. 622
Argent, J., 2nd-Lt. 668
Argles, E. V., Lt. 375, 1020
——— R. L., Lt.-Col. 1711
Argyle, R. E. V., Capt. 846
——— V. L., 2nd-Lt. 303
Ariss, V. H., Asst. Surgn. 1797
Arjandas Gosain, Sub-Asst. Surgn. 1810
Arjani, N. K., Tempy. Capt. 1771
Arjan, Subdr. 1145
——— Bhagat, Subdr. 1707
——— Dass, Jemdr. 1300
Arjan Singh, Subdr. 1148, 1299
——— Subdr. 1070, 1163, 1589
——— Jemdr. 1216
——— Jemdr. 1281
——— Subdr. 1227
——— Jemdr. 1022, 1344
——— Jemdr. 1707
——— Jemdr. 896
——— Jemdr. 1163
——— Jemdr. 1149
——— Sub-Asst. Surgn. 1817
——— Ressdr. 1682, 1686
Arjan Ram, Jemdr. 1488
Arjim Singh, Jemdr. 1698
Arjun Atmaji Redear, Sub-Asst. Surgn. 1830
——— Lukshuman, Jemdr. 1472
——— Khattri, Jemdr. 1643
——— Rama, Jemdr. 1601
——— Maudlekar, Jemdr. 1416
——— Singh, Jemdr. 1280
Arjunlal Gagabhai Jotania, Sub-Asst. Surgn. 1831
——— Powar, Subdr.-Maj. 998
——— Singh, Sub-Asst. Surgn. 1812
——— Sub-Asst. Surgn. 1815
——— Subdr. 1022, 1128
——— Jemdr. 1047
——— Resldr.-Maj. 919
——— Deoi, Sub-Asst. Surgn. 1818
——— Dhillon, Sub-Asst. Surgn. 1819
——— Sekhon, Sub-Asst. Surgn. 1817
Arjunlal, Thapa, Subdr. 1618
——— Subdr. 1607
Arkell, J. G., Capt. 755
——— T. N., Lt. 755
Arklie, R. N., Asst. Surgn. 1803
Arkwright, E. H., Maj. 2, 615
Armitage, A. S., Lt. 315, 1606
——— C. C., Lt. 506, 1665
——— H. M. A., Lt. 637, 972
——— J., *Hony.* Lt. 1849
——— K. L. F., Capt. 501a, 1104
Armitstead, H., Capt. 835
Armour, J. S. D. M., Capt. 18, 40, 719
——— J. S. 526
Armstrong, A., Lt.-Col. 755
——— A. C., Lt. 543
——— A. G., Capt. 267, 923
——— A. H., Lt. 705, 798
——— A. H. H., Lt. 401, 919
——— A. J. E., Lt. 351, 1110
——— C. F., 2nd-Lt. 494, 601, 899
——— D. J. H., Lt. 913
——— E. R., Capt. 10, 1756
——— F. E. J., 2nd-Lt. 565
——— F. H. C., Capt. 274, 1283
——— G. B., Lt. 1056
——— G. G., 2nd-Lt. 752
——— H., Lt. 420, 1443
——— H. F. C., Lt. 322, 1604
——— J. A., 2nd-Lt. 1022
——— J. A. E., Lt. 313
——— J. D. H., Lt. 913
——— J. R. B., Lt. 589
——— J. W., 2nd-Lt. 805
——— K., Lt. 435, 1493-94
——— M. H., 2nd-Lt. 545

NAMES AND RANK. Page.

Armstrong, P. A., Lt. . . . 542, 1007a
—— R. A., Capt. 636
—— R. H., Lt. . . 420, 1591
—— T. T. W., 2nd-Lt. . . 690a
—— V. O., Asst. Surgn. . 1803
—— W. Mc G., Lt.-Col. 46, 710
—— W. L., Lt. . . 534, 1653
Armour, J. S., Lt. 526
Arnall, F. T., Sub-Condr. . . . 1858
Arndt, C. E., 2nd-Lt. . . 566, 1056
Arnell, O. R., Lt. 668
Arnell, L., Capt. 629
Arnold, A., Capt. 740
—— A S., Lt.-Col. 129, 154, 188
—— F., Lt. 1577
—— G., 2nd-Lt. . . . 682, 720
—— H., 2nd-Lt. 1235
—— L. F. C., Lt. . . 470, 1335
—— L. P., 2nd-Lt. 472, 1493-94, 1507
—— N., Lt. 686, 700
—— R. T., 2nd-Lt. 473, 1471, 1473
—— T. F., Capt. 79, 1839
—— V. D., 2nd-Lt. . . . 473, 1040
—— W. W., Lt. . . . 687, 802
Arnott, A. M., Capt. . 263, 1613, 1657
—— H. W., Capt. 850
—— H. W., Lt. 111
—— J., Lt.-Col. 1955
—— S. L., Capt. 687
Arnould, F. G., Capt. 825
—— H. L., Revd. 1907
Arogyasami, Jemdr. 994, 1276, 1699
Aroklam, Jemdr. 1360
Aroklasami, Jemdr. 1273
—— Subdr. 1307
Aron, D. J., Lt. . . . 410, 1642
Arratoon, J., 2nd-Lt. . . 563, 1164
Arridge, H. C., Lt. 535
Arrindell, C.D.S.M., Lt. . . 488, 935
Arshad Khan, Jemdr. . . 1087, 1088
Arsla Khan, Subdr.-Maj. . . . 1351
—— *Hony.* Capt. . . . 2104
—— Subdr. 1235
Arthur, A. S., Maj. 1712
—— C. G., Capt. 603
—— E. J., Capt. 500
—— G., Capt. 1725
—— J. R. H., 2nd-Lt. . 552, 1140
—— L. F., Maj. 66, 98, 105, 211, 913
—— W. G. A., 2nd-Lt. . 552, 1505
Arumugam Mudali, Jemdr. . . 1591
Arunachellam, Jemdr. . . . 1332
Arundell, W. R. H., Lt. . . . 625
Arz Muhammad, Jemdr. . . . 1502
Asa Ram, Subdr. 1011
Asa Singh, Subdr. . . . 1157, 1587
—— *Hony.* Capt. 2104
—— Jemdr. 1232
—— Jemdr. 1114
Asad Ali, Sr. Sub-Asst. Surgn. 1808
Asandhoj Limbu, Subdr. . . .
Asaram, Risldr.-Maj. 934
Asbahadur Gurung, Jemdr. . . 1651
—— Limbu, Jemdr. . . . 1655
Asbir Rai, Gurung, Jemdr. . . 1608
Ascroft, P. T., 2nd-Lt. . . . 567
Aseervathum, J. C., Sub-Asst. Surgn. 1827
Asghar Ali, Jemdr. . . . 1476, 1506
—— Jemdr. 936
—— Jemdr. 1126
——Khan, Jemdr. . . 880, 1051
Ash, F., 2nd-Lt. 756
—— L., 2nd-Lt. 563
—— L., 2nd-Lt. 1678
Ashak Ali Khan, Risldr. . . . 898
Ashanna Poshati, Jemdr. . . . 999
Ashburner, E. H., Lt. 725
—— H. W., Maj. 211, 871, 1427, 1511
Ashby, H. G. N., Lt. 921
—— J. S., Lt.-Col. 56
Ashcroft, J. M., Lt. . . . 488, 911
—— W., Lt. 587
Ashe, A. E., 2nd-Lt. 1695
—— C. H., 2nd-Lt. 556
—— H. D., 2nd-Lt. . . . 55, 929
—— J. W. K., Asst. Surgn. . . 1798
—— W. St. G., 2nd-Lt. . . . 608
Ashfield, R., Condr. 1854
—— F., Maj. 87, 227
Ashford, E., 2nd-Lt. . . . 355, 1511
—— J., Capt. 804
Ashley-Brown, Revd., W. . . . 1907
Ashman, C. H., 2nd-Lt. . . 726, 799
Ashley-Smith, R., Lt. 415, 445, 513, 1396

NAMES AND RANK. Page.

Ashmore, E. J. C., Capt. . 277, 1647
Ashraf Ali Khan, Subdr. . . . 1086
—— Khan, Jemdr. 880
—— Jemdr. 932
—— M. S., Tempy. Lt. . . 1778
Ashton, A., Lt. . . 522, 821, 1032
—— B. C., Capt. 1759
—— C. T., Lt. . . . 352, 1193
—— E. M., Lt. . . . 511, 1665
—— E. M., Lt. (*Hon. Capt.*) . 577
—— F. C., Lt. . . . 354, 1674
—— F. J., Lt. . . . 61, 513
—— G. R., Lt. . . 61, 506, 1666
—— P. W. R., Lt. . . 295, 1255
—— R. W., 2nd-Lt. 91
Ashu Gurung, Jemdr. 1637
Ashworth, C. H., Lt. . . 514, 1508
—— H. G., 2nd-Lt. . . 465, 1471
—— R. Major. 759
—— S. H., Maj. 656
—— S. L., 2nd-Lt. . . 545, 1678a
—— C. H., Lt. . . . 515, 1505
Asirvadam, N., Sub-Asst. Surgn. 1827
Asjit Rai, Subdr. 1648
Askew, J., 2nd-Lt. . . . 356, 1164
—— T. E., 2nd-Lt. . 90, 530, 715
Askey, S. G., Capt. 1723
Askwith, A. V., Lt. . . . 490, 927
—— H. F., Col. 124
Aslam Beg, Jemdr. 916
—— Khan, Ressdr. 1683
Aslett, W. C., Lt.-Col. . . 179, 1918
Aspinall, A. W., Lt. . . . 322, 1178
—— H. H. H., Lt.-Col. 57, 180, 1918
Aspinwall, W. H. E., 2nd-Lt. 522, 1667
Asquith, E. J., Lt. . . . 533, 1044
—— F. H., Asst. Surgn. . . 1800
Assa Singh, Subdr. 1021
—— Subdr. 1151
—— Jemdr. 1681
Assanulla Khan, Jemdr. . . . 1104
Assheton-Smith, G., Lt. . . . 903
—— T., Capt. 517a, 1841
Assig, H. L., Lt. 389a
Astell, I. W., 2nd-Lt. 766
Astler, C., Lt. 917
Aston, H. R., 2nd-Lt. 546
—— R. W., Lt. 529
Asu Gurung, Jemdr. 1637
Aswini Kumar Sarkar, Sub-Asst. Surgn. 1813
Ata Muhammad, Sub-Asst Surgn. 1811
—— Sub-Asst. Surgn. 1815
—— Jemdr. . 1284, 1378
—— Subdr. . 1124, 1183
—— Jemdr. . . . 1090
—— Jemdr. . . . 1249
—— Jemdr. . . . 1428
—— Ressdr. . . . .
—— Khan, Ressaidr. 903
—— Risldr. . . . 955
—— Khan, Sub-Asst. Surgn. 1816
—— Jemdr. . . 945
—— Jemdr. . . 908
—— Jemdr. . . 875
—— Shah, Jemdr. . 908
Ata Singh, Jemdr. 979
Atar Khan, Subdr. 1152
—— Singh, Sub-Asst. Surgn. . 1813
—— *Hony.* Capt. 2101
—— Subdr. 1163
—— Jemdr. 934
—— Jemdr. 1051
—— Chatwal, Sub-Asst. Surgn. 1817
—— Ilmawadi, Sub-Asst. Surgn. 1814
Atbal Singh, Subdr. 1483
Atibal Singh, Jemdr. 1648
Atherton, P., 2nd-Lt. 362
Athol-Liddell, R. E., 2nd-Lt. . . 546
Atkin, E. E., Ty.-Lt. 1729
Atkins, B. S., Capt. . . 260, 1062
—— C. G., Lt. 535
—— C. R. 2nd-Lt. 1193
—— E. A., Lt. . . . 518, 1279
—— E. D., Lt. . . . 489, 885
—— E. R. N., Lt. 1068
—— F. R., Rev. 1908
—— G. F. B., Lt. 536
—— G. W., Maj. 239, 1116, 1120
—— H. R., Maj. 679
—— P. C., Lt. . 307a, 307, 1299
—— R. F., Capt. . . . 260, 1283

NAMES AND RANK. Page.

Atkins, W. V., 2nd-Lt. . . 415, 1164
Atkinson, A. A., 2nd-Lt. . 560, 1082
—— A. J. M., 2nd-Lt. 546, 1044
—— A. W., Lt. . . 529, 1219
—— C. M., Lt. . . 410, 1463
—— C. V., 2nd-Lt. . . . 556
—— D. J., Lt. . . 25, 393, 1614
—— E. D., Lt. 511
—— E. G., Capt. . 263, 907
—— E. H. de V., Col 58, 119, 124, 600, 1889
—— E. R., Sub-Condr. . . 1863
—— F. G., Brig.-Genl. . . 57
—— G. C., Lt.-Col. . . . 56
—— G. P., Capt. 748
—— G. R., Lt. 846
—— H. O., 2nd-Lt. . . . 564
—— J. R. B., Maj.-Genl. . 1955
—— J. T., Sub-Condr. . . 1857
—— L. R., 2nd-Lt. . 355, 1407
—— N. E., Lt. . . . 371, 1419
—— S., Condr. 1873
—— V. H., 2nd-Lt. . . . 788
Atkinson-Willes, G. A., Lt. . 589
Atkints, A. T., Lt. 798
Atma Ram Paul, Sub.-Asst. Surgn. 1818
Atma Singh, Subdr. 1151
—— Subdr. 1303
—— Ressdr. 916
—— Jemdr. 950
—— Jemdr. 1054
—— Subdr. 1151
—— Jemdr. . . 928, 1111
Atmaji, Rao, Ghag, Subdr. Major 1436
Atma-Ram Bapuji, Sub-Asst. Surgn. 1829
—— Paul, Sub-Asst. Surgn. 1818
Atmaram Bhuken-Das, Sub-Asst. Surgn. 1830
Atta Khan, Subdr. Maj. . . . 1249
—— Mohammed, Subdr. . . 1683
—— Jemdr. . . . 1142
—— Jemdr. . . . 1433
Attaullah Khan, Jemdr. . . .
—— Jemdr. . . . 1305
Attar Singh, Subdr.-Maj. . . . 1164
—— Subdr. 1171
—— Jemdr. 1204
—— Jemdr. 1240
Atteridge, L. F., Lt. . . 397, 785
Attfield, C. J., Lt. . . 9, 311, 1113
Attru, *Hony.* Lt. 2104
Attwood, W. W., 2nd-Lt. . . . 734
Auber, J., Lt. 430
Auchinleck, C. J. E., Capt. 243, 1207
Aucutt, W., Lt. 489, 941
Audain, Maj., G. M. 53
Audley, W. D., Lt. . . . 537, 1174
Audsley, A. E. M., Lt. . . . 487
—— C. E., 2nd-Lt. . . . 559
Aue, W. C. R., 2nd-Lt. . . 465, 1291
Auger, A. E., Lt. . . 89, 509, 778
Augier, D. E., Lt. . . 514, 658, 1042
—— E. H., 2nd-Lt. . . . 986
August, F. R. B., Lt. . . . 838
Augustian, Jemdr. 994
Augustine, V. O., Esq. . . .
Aulia Khan, Jemdr. . . 926, 1682
Aulton, F. H., Capt. 735
Aulya Khan, Jemdr. 1288
Auret, A. S., Maj. . 242, 1401, 1505
Austen, E. J., Lt. . . . 540a, 1319
—— W. G., 2nd-Lt. . . . 508
Austin, A. C., Lt. 513, 986
—— A. McN., Lt. 28
—— A. P. M., 2nd-Lt. . . 720
—— E.R., Capt. 779
—— H. H., Brig.-Genl. 50, 97, 103, 121, 126, 661
—— H. N., Capt. 687
—— J. B. G., Lt. 493
—— J. C., Lt. . . . 519, 1066
—— J. M., Lt. 533
—— L. A., Lt. . . . 463, 1028
—— R. F. E., Maj. . . . 1731
—— T., 2nd-Lt. . . . 557, 1359
—— W. A. G., Lt. 797
Auston, G. T., Lt. . . . 522, 1307
Autar Sing Negi, Subdr. . . . 1179
—— Ty. Sub-Asst. Surgn. 1833
—— Gurung, Subdr. . . 1653
Autur Singh Bhangu, Sub-Asst. Surgn. . 1819
—— Butola, Jemdr. . . 1179
Auton, J. H., 2nd-Lt. . . 331, 1120

Names and Rank.	Page.
Auty, S. C., 2nd-Lt. 474,	1567
Avent, J. G., Lt. 304,	1495
Averdeick, G. U., 2nd-Lt.	593
Averill, J. E., Lt. 463,	1636
Avery, A. H., Lt. 68,	797
——— A. H. St. J., Lt.	490
Avill, S. N.,	1781
Avison, J. L., Capt.	686
Avrill, S. M. 2nd-Lt.	838
Awadh Bihari Lall, Sub-Asst. Surgn.	1811
Awal Singh, Jemdr.	1385
Awan, K. B., M.B., Ty. Lt.	1773
Awdry, N. J., Capt.	750
Awtar Singh Bhangu, Sub-Asst. Surgn.	1819
Awatar Singh Grewal, Sub. Asst. Surgn.	1817
Awtar Singh Grewal, Sub-Asst. Surgn.	1817
Axtell, L. H. W., Lt. 405,	1052
Ayckbourne, F. S., 2nd-Lt.	567
Aydon, W. W., Condr.	1853
Aylard, B. F. C., 2nd-Lt.	685
Ayling, C., 2nd-Lt.	754
——— J. C., Capt.	652
Aylmer, F. J., Lt.-Genl. 118, 660,	1955
——— G., Maj. 67, 231,	1662
Aylward, D., Lt. 505,	1275
——— T. B., 2nd-Lt. 473,	949
Ayre, W. H., 2nd-Lt.	788
Ayres, G., 2nd-Lt.	619
Ayscough, G. M., Maj. 240,	1463
Ayub Khan, Jemdr. 1208,	1244
———, Jemdr.	1493
Ayya, Y. V., tempy.-Capt.	1767
Ayyakannu, Jemdr.	1264
Ayyamperumal, Jemdr.	994
Ayyar, R. S. H., Ty. Lt.	1777
Ayyar, G. S. A. S., Ty. Lt.	1779
Ayyulu Naido, Jemdr.	1316
Azam Ali, Risldr.	918
——— Beg, Jemdr.	882
——— Khan, Sub-Asst. Surgn.	1816
Azhagirisami, Jemdr.	1308
Azhar Husain, Sr. Sub-Asst. Surgn.	1807
——— Subdr.	1106
——— Jemdr.	1133
——— Jemdr.	910
Azim Khan, Subdr. 1106,	1249
Azimullah, Jemdr.	1500
——— Khan, Jemdr.	963
Aziz-ud-din Khan, Subdr.	1035
Azizullah, Jemdr.	1695
——— Subdr.	1096
Aziz Khan, Jemdr.	1699
——— Subdr.	1511
——— Jemdr.	1336
Aziz-ul-Hakk, Sub-Asst. Surgn.	1830
Azmatullah Khan, Ressdr.	
Azmat Khan, Jemdr.	1133
B	
Baba Harcharan Singh, Sub-Asst. Surgn.	1819
Baba Kartar Singh Bedi, *Hony.* Lt. 111,	948
Babajirao Dalvi, Subdr.-Maj.	1415
Babaji Sivaji, Subdr.	1432
Baba Rao Ghorpude, Jemdr.	934
Bad Din, Jemdr.	1492
Babington, A. H., Maj. 60, 207,	1661
——— D. M., Col. 71, 124, 612,	1884
——— J. W. H., Maj.	1751
Babonau, A. F., Capt. 42,	1754
——— R. G., Sr. Asst. Surgn.	1789
Babtie, J. W. A., Lt. 425,	1255
Babu, Jemdr.	1276
Babu Lal Sharma, Sub. Asst. Surgn.	1816
——— Parsad Bhargava	1820
Babu Singh, Subdr.	
——— Ty. Sub-Asst. Surgn.	1833
——— Jemdr.	1386
Baburao Ghorpade, Jemdr.	1464
——— Sinde, Jemdr.	1440
——— Pisal Jemdr.	1424
——— Ghag, Jemdr.	1469
——— Mohit, Jemdr.	1514
——— Jadhao, Jemdr.	1469
Bachan Singh, Jemdr.	1154
——— Jemdr. 896,	932
——— Jemdr.	1035
——— Jemdr.	1336
——— Jemdr.	1079
——— Jemdr.	1498
——— Ressdr.	950
Bachan Singh, Uppal, Sub. Asst. Surgn.	1817
Bachan Singh Uppal, Sub-Asst. Surgn.	1817
Bachant Singh, Risldr.	914
Bachint Singh, Jemdr.	1240
Bachanu Dichit, Subdr.	1026
Bachhulal Agnihotri, Jemdr.	1078
Bachint Singh, Sub-Asst. Surgn.	1814
Bachitar Singh, Ressdr.	1682
Bachittar Singh, Subdr. 1231,	1589
——— Jemdr.	950
——— Jemdr.	906
——— Jemdr.	886
Backhouse, B. S., Lt.	1689
——— G. M., Lt.	630
——— J., 2nd-Lt.	724
——— J. R. C., Lt. 532,	1054
——— P. B., Lt. 503,	1665
——— R. S., Capt.	657
Backhurst, L. G., Tempy. Asst. Surgn.	1804
Backworth, 2nd-Lt. G.	90
Bacon, A. L., Capt.	848
——— C. G., Capt. 55, 261,	929
——— R. G., Capt. 267, 1299,	1491
——— R. J., Maj. 27,	732
——— S. F., Capt. 679,	681
——— W. D., 2nd-Lt.	1193
——— W. F., Capt.	24
Badal Sing Gurung, Jemdr.	1602
——— Singh, Hony. Capt.	2104
——— Subdr.	1106
——— Khan, Jemdr.	1515
Badar Bakhsh, Jemdr.	1268
Badar Din, Jemdr.	1456
Badar Uddin Khan, Jemdr.	1087
Badcock, F. F., Col. 14, 22, 126, 150,	188
——— G. H., Maj.	1919
——— W. B., Lt. 525,	1668
Baddeley, C. E., Col. 123,	660
——— G. T. B., Lt.	571
——— H., Lt. 21,	597
——— R. J. H., Maj 218, 903,	929
——— S. E. L., Capt. 263,	911
Bab Din, Jemdr.	1492
Baden-Powell, R.S.S., Lt.-Genl. *Sir*	591
Badekar, V. H.,	1781
Badhams, O. G., 2nd-Lt. 359,	1249
Badi-ul-Zaman, Subdr.	1207
Badlu, Jemdr. 1164,	1385
Badlu Singh, Jemdr.	
———, Subdr.	1403
Badham J. R. C., 2nd-Lt. 494,	947
———, R. A. A., Lt. 321, 365,	1263
Badhamo, O. G., 2nd-Lt.	1249
Badhawa Singh, Jemdr.	934
———, Jemdr.	1252
Badman, F. H., Lt. 210,	1335
Badrinarayan Singh, Subdr.	1403
——— Jemdr.	1405
Badri Das, Jemdr.	1129
——— Parshad, Sub-Asst. Surgn.	1810
———, Sub-Asst. Surgn.	1813
——— Sharma, Sub-Asst. Surgn.	1820
——— Sharma, Jemdr.	1107
Badruddin, Jemdr.	1092
Badsha Khan, Jemdr.	1404
Badshah Husain, Jemdr.	880
——— Ressdr.	1684
Bag Khan, Jemdr.	990
Bagat Singh, Jemdr.	1037
Bagavathi, Jemdr.	1699
Bagbir Gharti, Subdr.	1607
Bagbir Thapa, Subdr.	1636
Bagchee, K., Ty.-Lt.	1778
Bagchi, P. N., Ty. Lt.	1780
Bagehi, K. P., Ty. Lt.	1777
Bagdal Rana, Subdr.	1636
Bagg, H. A., 2nd-Lt.	640
——— P. E., Sub-Condr.	1864
——— R. P. J., Lt.	731
——— W. F., 2nd-Lt. 491,	949
Bagga Singh, Risldr.	944
——— Jemdr.	1364
——— Jemdr. 886,	959
——— Subdr.	974
——— Khan, Jemdr.	1488
——— Jemdr.	1518
——— Jemdr.	1452
Baggar, F. A., Lt.	629
——— J. G., 2nd-Lt.	630
Bagh Ali, Subdr.	1335
——— Subdr. 1011,	1142
——— Khan, Jemdr.	1514
Bagh, Shah, Subdr.	1379
——— Singh, Subdr.	1193
——— Jemdr.	1368
——— Jemdr.	1484
——— Jemdr.	1079
——— Jemdr.	1216
——— Subdr.	1488
——— Negi, Jemdr.	1214
Baghdeo Singh Rawat, Subdr.	
Bagla Parshad Banerjee, Sub. Asst. Surgn.	1818
Bagley, A. A., 2nd-Lt. 7,	546
——— A. H., Lt. 291,	1046
Bagnall, C. E., Maj.	747
——— H. S., Capt. 712,	713
——— R., Maj. 243, 1283,	1311
Bagram, C. E., Capt.	832
Bagshawe, F. W., Lt.-Col. 87, 104, 108, 132, 156,	192
Bag Sing Gurung, Jemdr.	1025
Bag Singh Thapa, Subdr.	1604
——— Jemdr. 1597,	1602
Bahadur Ali, Jemdr.	977
——— Rai Jemdr.	1655
——— Jemdr.	1344
——— Ali Khan, *Hony.* Capt.	2103
——— Jemdr.	908
——— Khan, Jemdr. 966,	1365
——— Ressdlr.	944
——— Jemdr.	1506
——— Jemdr.	1518
——— Jemdr.	910
——— Subdr.	968
——— Ghale, Subdr.	1602
——— Ghasti, Jemdr.	1625
——— Gurung, Jemdr.	1633
——— Malla, Jemdr.	1637
——— Rai, Jemdr.	
——— Shah, Subdr.	1320
——— Subdr.	1699
——— Sher, Risldr.	908
——— Jemdr.	916
——— Sher, Subdr.	1495
——— Khan, Subdr.	1305
——— Jemdr.	946
——— Sing Gurung, Jemdr.	1598
——— Bartwal, Jemdr.	1182
——— Singh, Risldr.-Maj. 5,	905
——— Ressldr.	944
——— Subdr.	1117
——— Jemdr. 932,	1142
——— Jemdr.	1285
——— Thapa, Jemdr.	1637
——— Subdr. 1601, 1607,	1609
——— Jemdr.	1643
——— Jemdr.	1598
——— Jemdr. 1608,	1614
———, Singh, Mahar, Jemdr.	1179
Bahadursing Thapa, Subdr.	1596
Bahawal Khan, Jemdr.	876
——— Sher, Jemdr.	1204
Bahirshal Singh, Jemdr.	940
Bahori Singh, Jemdr.	1397
Baia, Gooda, Jemdr.	1460
Baidullah, Jemdr.	1456
Bailerham, D. J., Lieut.	1715
Bailram Khan, Jemdr.	918
Baichand, Jemdr.	1183
Baij Nath Lal, Sub-Asst. Surgn.	1817
Bailnath Singh, Subdr.	1387
——— Jemdr.	1043
Balaji Rao Kadani, Jemdr.	1469
Baikie, H. McA., Lt. 520,	992
Bailey, A. H., Lt. 543,	1295
——— B., 2nd-Lt.	790
——— C., Col. 125, 150,	185
——— D. C., 2nd-Lt. 537,	1108
——— E. H., Col.	720
——— F. M., Maj. 227,	1913
——— F. R., Lt. 519,	1241
——— G. W. B., Condr.	1873
——— H. E., Lt.	623
——— H. G., 2nd-Lt. 701,	703
——— J. H., Lt. 443,	1351
——— J. W. S., 2nd-Lt.	302
——— L. D., Capt.	682
——— P. H. H., Maj. 241, 1150,	1150
——— R. J., 2nd-Lt. 546, 803,	1463
——— W., Condr.	1807
——— W., Sub-Condr.	1872
——— W. A., Condr.	1853
——— W. A., Lt. 526,	1630
——— W. B., Maj. 221, 1000,	1611
——— W. G., Lt. 354,	779
Baillie, S. M. H., Maj. 239, 1355,	1573

Names and Rank.	Page.
Baillie, W. H., Asst. Surgn.	1801
Ballu Singh, Jemdr.	1071
Bally, B. W. S. S., Lt.	1263
—— J., 2nd-Lt. 359,	1659
—— J. D., Sub-Asst. Surgn.	1830
—— L. H., 2nd-Lt.	465
—— P. E. C., 2nd-Lt. 356,	1263
—— E. M., Maj. 702,	1885
Baillie, E. A., Lt.	780
—— J. C. P., Genl.	169
—— K, temp. 2nd-Lt.	796
—— P., Lt.	905
—— R. Lt.-Col.	180
Bailly, O. H., Lt.	1042
Bain, G. A., Lt. 305, 1600,	1604
—— G. W., Lt. 542,	1227
—— K. D., Hony. Lt.	780
—— L. T., 2nd-Lt.	559
—— R. B., 2nd-Lt.	781
Bainbridge, D., Capt. 267, 970,	1251
—— W. F., Lt.-Col. 51, 97, 131, 156, 195, 869,	1239
Baines, E. F. E., Lt.-Col. 137, 998,	1744
—— F. J. T., Capt.	702
—— G. D., Capt. 278, 937,	959
—— J., Lt.	622
—— R. H., Lt. 295,	1479
Baint Singh, Jemdr.	955
Baird, A. E., Lt. 61, 489,	604
—— C. J., Lt. 425,	1636
—— F., Sergt.-Maj.	1879
—— G. H., 2nd-Lt.	1695
—— H. B. D., Bt. Lt.-Col. 42, 97, 137, 159, 213,	897
—— J., 2nd-Lt.	776a
—— J. W., 2nd-Lt.	763
—— R. E. W., Maj.	715
—— R. P. H., Lt. 430,	1056
—— R. F., Lt.-Col. 138,	1745
—— R. V., Lt. 506,	1665
—— T. W., 2nd-Lt.	560
Bairsto, F. J., Lt. 541,	992
Baisakhi Ram, Jemdr.	1681
Bajijirao Shinde, Jemdr.	1444
Bajji Rao More, Jemdr.	1469
Bajiram Pun, Jemdr. 1608,	1611
Bajirdhoj Rai, Subdr.	1648
Bajpayee, A., Ty. Lt.	1781
Bajranga, Jemdr.	1193
—— Singh, Jemdr.	1047
Bak-Khan, Jemdr.	1095
Bakaji Gadwe, Jemdr.	1464
Bake, A. T. T., Lt. 321,	1501
Baker, A., Lt.	858
—— A. F., Sub-Condr.	1868
—— A. H., Lt. 440,	1433
—— A. V., Capt. 720,	785
—— C. A., *Hony.* Lt.	831
—— C. H. N., Capt.	1758
—— C. H. O'N., Capt.	857
—— C. N., Lt.-Col 87, 132, 156,	193
—— C. S., Lt. 307,	1617
—— D. G., Lt.	595
—— D. G. B. S., Maj.	1748
—— D. M., Lt.	507
—— E. B., 2nd-Lt.	602
—— F., Hony. Lt.	656
—— F. C., Lt. 524,	1082
—— G. F., Capt.	831
—— G. M., Lt. 352,	988
—— G. M., Lt. 296a,	1070
—— H. 2nd-Lt.	1491
—— H. A. J., Ty. 2nd-Lt.	752
—— H. L., Lt. 443,	1351
—— H. N., Capt. 262,	1315
—— H. R., Lt.-Col. 131, 156, 192, 870,	1307
—— J., 2nd-Lt.	700
—— J. N. R. L., Lt. 393,	1471
—— P. N. H., Lt.	826
—— R. A. C., Lt. 463,	1046
—— R. C. N., 2nd-Lt. 700,	723a
—— R. H., 2nd-Lt.	828
—— R. H. 2nd-Lt. 556,	1473
—— S., Lt.	683
—— S., Lt. 521,	1509
—— S., Ty. 2nd-Lt.	742
—— S. J., 2nd-Lt. 666,	996
—— T. MacD., Capt.	721
—— V. A., Lt. 354,	1108
—— V. H. E., 2nd-Lt.	723a
—— W. B., Maj. 234,	1647
—— W. H. G., Capt. 269,	933
—— W. G., Lt.	723
—— W. S., Lt.	516a
Ba-Ket, M., Maj.	1750
Bakewell, W. B., Lt. 752, 783,	798
Bakhan Sing Thakur, Hony. Capt.	2101
Bakhanbahadur, Jemdr.	1598
Bakhatbahadur Limbu, Subdr.	1648
Bakhat Bali Singh, Subdr.-Maj.	1043
Bakhle, C. R., Lt.-Col. 134,	1744
Bakhsh Muhammad, Jemdr.	1107
Bakhshi Autar Singh, Subdr.	1283
Bakhsi Chand, Subdr.-Maj.	1247
Bakhshi Karam Chand, Jamdr.	910
—— Ram, Sub-Asst. Surgn.	1817
—— Ram, Subdr.	1457
—— —— Jemdr.	890
—— Rai, Sub. Asst. Surgn.	1817
—— Singh, Subdr.	1211
—— —— Jemdr.	1165
Bakhshish Singh, Resldr.	912
—— —— Jemdr.	934
—— —— Subdr.	1247
—— —— Ressdr.	898
Baksho, Jemdr.	1427
Bakhtawar, Jemdr.	1060
—— Jemdr.	1571
—— Khan, Subdr.	1509
—— Singh, Subdr.	1021
—— —— Subdr.-Maj.	1070
—— —— Jemdr 930, 936,	951
—— —— Rawat, Jemdr.	1181
Bakht Sing Rana, Subdr.	1607
Bakshi JaswantSingh, Sub-Maj.	2104
Bakta Ram, Subdr.	1488
Bal, J. N., Tempy. Lt.	1774
—— N. K., Capt.	1757
—— N. V., Sub-Asst. Surgn.	1832
Bal Singh, Subdr.	1095
—— Jemdr.	884
Bala Singh, Chanhan, Subdr.	1223
Balwanta, Subdr.	1403
Bala Bhor, Subdr.	1423
—— Dat Joshi, Sub. Asst. Surgn.	1819
—— Gaoda, Jemdr.	1460
—— Ghorpade, Jemdr.	1510
—— Parsad Shaivastav, Sub. Asst. Surgn.	1818
—— Parte, Jemdr.	1441
—— Ram, Jemdr.	1413
—— Parshad Syiaastas Sub-Asst. Surgn.	1818
—— Datt Joshi, Sub-Asst. Surgn.	1819
—— Sahib Daphle, Capt. 278, 1020,	1022
—— Singh Gurung, Jemdr. 1602,	1625
Balaji Mule, Subdr.	1471
—— Rao Ghag, Subdr.	1460-62
—— Singh, Tempy. Sub-Asst. Surgn.	1833
Balakram Misr	1019
Balak Singh, Subdr.-Maj.	2104
Balaraman, Jamdr.	1276
Balandhar Rai, Jemdr.	
Balasundaram, Jemdr.	1276
Balasing Gurung, Jemdr.	1598
Balbahadur Limbu, Subdr.	1630
Balbahadur Thapa, Subdr.	1639
—— Jemdr.	1637
Balbernie, A. G., Capt.	749
Balbir Basent, Jemdr.	1704
Balbir Rana, Jemdr.	1619
—— —— *Hony.* Capt.	2104
—— —— Thapa, Subdr.	1607
—— —— Jemdr.	1615
—— —— Subdr.	1636
—— —— Jemdr. 1602	1645
Balbir Singh, Rao Bahadur, 2nd-Lt.	481
Balbhadar Parshad Misra, Sub-Asst. Surgn.	1816
—— Thapa, Jemdr.	1637
Balch, C. H., Lt. 446—49,	887
—— H. G., Sub-Condr.	1864
Balck, C. A. J. A., Maj.	1711
Bald, A., 2nd-Lt. 77,	494
—— E. A., Capt. 268, 1647,	1655
—— J. A., Maj. 212, 871, 1263,	1417
Baldeo Parsad Gupta, Sub. Asst. Surgn.	1819
Baldeoparshad Dube, Jemdr.	1028
Baldeo Singh, *Hony.* Capt.	2101
—— —— Jemdr.	1396
—— —— Jemdr.	1037
—— Sahi, Jemdr. 1643,	1655
—— —— Jemdr.	1404
Baldewa, I, Jemdr.	1192
—— II, Jemdr.	1192
Baldhoj Limbu, Jemdr.	1649
Balding, C. D., Lt. 297a, 505,	1455
Baldock, W. H., 2nd-Lt. 692,	784
Baldrey, C. H., Lt. 517,	1666
—— F. J., *Hony.* Lt.	1852
—— T., Sr. Asst. Surgn.	1757
—— W. G., 2nd-Lt.	558
Baldu Ram, Jemdr.	886
Baldwin, A. C. J., Lt.	512
—— C., 2nd Lt.	794
—— D. F., 2nd-Lt. 530,	1702
—— E., Tempy. Lt.	1724
—— E. S., 2nd Lt.	556
—— F. E. W., Maj. 53, 283,	1479
—— G. M., Col. 12, 21, 108, 120, 125, 150,	188
—— G. W., Lt. 526,	1668
—— J. E. A., Capt.	589
—— J. W., Lt. 420,	1132
—— L. W., Lt.	1130
—— M. C., Maj. 246, 1595,	1653
—— N., 2nd-Lt.	748
—— P. G. M., Lt. 340,	895
—— R. A. C. 2nd-Lt. 1028,	1702
—— R. H., 2nd-Lt. 547,	1702
—— S., *Hony.* Lt.	730
Bale, W., 2nd-Lt.	711
—— Tewari, Subdr.-Maj.	1016
Bale Thapa, Jemdr.	1615
Bale Rana, Jemdr.	1639
Balfour, A., Lt. 543a,	1411
—— A. S., Capt.	1892
—— D., Lt.	1893
—— J. H., Col. 57, 123, 153, 180,	1917
—— W. M., Lt.	638
Ball, E., Capt.	834
Baliga, B. P., Ty. Lt.	1781
Balindhar Rai, Subdr.	1657
Baliram Rana, Subdr.	1596
Balkaran Singh, Jemdr.	1396
Balkishen, Esq.	91
Balkrishna, Tempy. Capt.	1766
—— Khause, Subdr.	1444
—— Rao, Subdr.-Maj. 1443,	1892
—— Trimbak Lele, Sub-Asst. Surgn.	1831
Ball, A. H., 2nd-Lt. 494,	953
—— C. F., Maj. 15, 25,	829
—— C. F., Lt. 725,	1007
—— E. P., Capt. 16, 40, 261,	1443
—— F. C., Sub-Condr.	1868
—— F. G., Serjt.-Maj.	1879
—— G. H., 2nd-Lt. 21,	712
—— G. W. M. 2nd Lt.	554
—— H. F. 722,	1006
—— H. P. Lieut.	825
—— J. W., 2nd-Lt. 546,	1333
—— L. P., Maj. 239, 1034,	1397
—— T. L., Maj. 218,	1662
—— W. C., 2nd-Lt. 531,	1668
—— W. M., 2nd-Lt. 687,	724
—— W. V., 2nd-Lt.	354
Bala Singh, Jemdr.	1137
Ballamy, G. R. 2nd-Lt.	696
Balla Sawant, Jemdr.	1436
Ballard, G. H., *Hony.* Lt.	693
—— V., 2nd-Lt. 463,	1098
Ballance, S. H., 2nd-Lt. 318,	1653
Ballanden, H. McH., 2nd-Lt.	625
Ballantine, C. L., 2nd-Lt.	597
—— E. W., Lt.	753
—— J., 2nd-Lt.	1086
—— J. R., Lt. 540a,	1696
—— R. W. H., Lt. 435,	1331
Ballantyne, W. J., 2nd-Lt. 475,	1020
Ballentine, G. C., Lt.	294
—— J. S., Lt. 311a,	907
Ballin, J., Lt.	1845
Ballinger, C. A., 2nd-Lt.	365
Balloch, A. C., 2nd-Lt.	544
Ballu Ram Suchadeva, Sub-Asst. Surgn.	1816
Bally, F. K., Maj. 215,	901
—— H. G., Maj. 211, 868, 1054,	1058
—— O. H.	470
Balmes, J. A., Lt.	986
Balmforth, J., *Hony.* Lt.	1843
Balmokand, Jemdr.	1143
Balmukand, Sr. Sub-Asst. Surgn.	1868
—— Sub-Asst. Surgn.	1814
—— Nehraw, Sub-Asst. Surgn.	1817
—— Vaid, Sub-Asst. Surgn.	1818
Bal Mudann, Subdr.	1385
Bal Mukand Singh, Subdr.	1385
—— Singh, Jemdr.	884
Bainas Singh Khattri, Jemdr.	1643

NAMES AND RANK. Page.

Bairaj Nadar, G. J., Sub-Asst. Surgn. 1826
Bal Ram, Sub-Maj. 1040
Balram Das, Sub-Asst. Surgn. . 1816
Balsing Thapa, Subdr. . . . 1624
Bal Singh, Jemdr. 1364
—— Jemdr. 974
Balu Mhadeek 1702
Balu Singh, Jemdr. 936
Balwantrao Dalsukhram Vyas, Sub-Asst. Surgn. . . 1830
Balwant, Jemdr. 1067
—— **Bholle**, Jemdr. 1460
—— **Narayen Masurker**, Sub-Asst. Surgn. . . 1830
—— **Rai Datt**, Tempy. Sub-Asst. Surgn. . . . 1833
—— **Rana**, Jemdr. 1619
—— **Rao**, Ressdr. 896
—— **Rao Gore**, Sr. Sub-Asst. Surgn. 1829
—— **Rao Sawant**, Subdr.-Maj. 1467
Balwant Rao Sinde, Jemdr. . . 934
—— **Singh**, Ressaldr. . . . 888
—— —— Sub-Asst. Surgn. 1817
—— —— Jemdr. . . . 884
—— —— Jemdr. 1356
—— —— Jemdr. . . . 1111
—— —— Subdr. . . . 1024
—— —— Jemdr. . . . 1192
—— —— Jemdr. . 1305, 1321
—— —— Ressdr. . . . 940
—— —— Jemdr. . . . 920
—— —— Jemdr. . . . 948
—— —— Jemdr. . . . 896
—— —— Risldr.-Maj. . . 880
—— —— **Sub-Asst. Surgn.** 1817
—— —— Bindra, Sub-Asst. Surgn. . 1818
Balwant Singh Bindra, Sub-Asst. Surgn. 1818
—— —— Jemdr. . 880, 1054
—— —— Subdr. . . . 1022
Balwanta, Jemdr. 1403
Bamber, C. C. Capt. . . . 367, 1645
Bamboat, N. H., Tempy. Capt. . 1764
Bamfield, H. J. K., Lt.-Col.. 132, 1743
Bamford, A. V. W., Asst. Surgn. 1801
—— C. W., **Asst. Surgn.** . 1802
—— R. G., Lt.-Col. . . 844
—— W., **Sr. Asst. Surgn.** . 1790
Bampfield, G. C., Capt. . 274, 1367
—— R., 2nd-Lt. . . 317, 1162
Bampton, J. A. H., Lt. . . 58, 504
—— R. E. F., Lt. 508
—— R. W., Lt. . . 505, 1830
—— W. J. DeB., **Asst. Surgn.** 1892
Banarji, J. K., Ty. Lt.. . . . 1774
—— K. K., Ty. Capt. . . . 1772
Banarsidas, Verma, Sub.-Asst. Surgn. 1819
Banarsidas Virma, Sub-Asst. Sugn. 1819
Banatvala, H. E., *The Hon.* Col. 1, 124, 1738, 1898
Banbury, W. E., Brig. Gen. 50, 104, 108, 121, 126, 153, 183
Banee, W. J., 2nd-Lt. 1413
Bancroft, C. R. W., Sr. Asst. Surgn. 1789
Bancroft, 2nd-Lt., G. R. C. . . . 1407
Bandbir Gurung, Jemdr. . . . 1602
Bandey, A. E., 2nd-Lt. . . 548, 1327
Bandu Gurung, Jemdr. . . . 1622
Banerjee, A. N., Tempy. Capt. . 1770
—— D. N., Ty. Lt. . . . 1782
—— H. D., Ty. Lt. . . . 1776
Banerjee, H. S., Tempy. Capt. . 1770
—— L. G., Ty. Lt. . . . 1777
—— L. M., Ty. Capt. . . 1771
—— P., Tempy. Capt. . . 1765
—— S. C., Ty. Lt. . . . 1780
—— S. N., Ty. Lt. . . . 1774
—— U. N., Tempy. Capt. . 1765
—— R. N., Esq. 91
Banerji, R., Ty. Lt. 1778
—— S. C. 1052, 1778
Baness, H. E., Lt. . . . 470, 1595
Banerjie, R. C., Ty. Lt. 1782
Banham, G. F., Lt. 1839
—— H., Lt. . . . 18, 23, 534
Bani Singh, Jemdr. 1581
Banister, F., Lt. . . . 518, 1275
—— H., Capt. . . . 485, 877
Banji Singh, Jemdr. 1071
Banker, S. S., Tempy. Capt. . . 1765

NAMES AND RANK. Page.

Bankhead-Browne, A. E., Capt. . 793
Bankim Chandra Mukherjee . 1818
Banks, A. T., 2nd-Lt. . . . 365, 1179
—— H. K., Capt. 656
—— J. R., Capt. 1718
—— R. E., Lt. . . . 380, 1563
—— R. M., Lt. 515
—— T. M., Capt. . . 18, 19, 724
Banna, Jemdr. 1196
Bannatyne, F. C., Maj. 16, 29, 235, 1439
—— N. C., Maj. 8, 15, 98, 219, 1509
—— N. E., Lt. . . 728, 1678*a*
—— T., Capt. 1839
Banner, J. W., 2nd-Lt. . . . 712
Bannerman, A. D'A. G., Lt.-Col. 187, 1889, 1911
—— H. D'A., Lieut. . 309, 1231
—— W. P., Lt.-Col. 133, 157, 193, 933
Bannester, W. C., 2nd-Lt. . . . 1359
Bannon, J. D., Lt. . . . 322, 1283
—— H. M., Lt. . . . 505, 1179
—— H. L., Lt. . . . 519, 1142
Bans Bahadur Singh, Jemdr. . 1051
Bansgopal Sukul, Jemdr.. . . 1016
—— **Tiwari**, Subdr. . . 1401
Bansi Dhar Tawari, Sub-Asst.-Surgn. 1819
—— Subdr. 1455
—— **Ram Sharma**, Sub-Asst. Surgn. 1818
—— **Lal Kockhar**, Sub-Asst. Surgn. 1819
Bansidhar Tripatti, Sub-Asst. Surgn. 1817
Bansraj Gurung, Subdr. . . . 1614
—— **Limbu**, Jemdr. . . . 1657
Bantock, A. E., Sub-Condr. . . 1863
Banwari Lal Sharma, Sub-Asst. Surgn. 1820
Banwell, J. A., 2nd-Lt. 1483
Banyad Ali, Subdr.-Maj. . . . 1054
Banyard, E. F., 2nd-Lt. . 564, 1108
Bapalal Himmatlal Yagnik, Sub-Asst. Surgn. 1830
Bapat, V. B., Tempy. Capt. . . 1770
Ba Pe, Subdr. 1295
Ba phone, Jemdr. 1347
Bapu Bargi, Subdr. 1407
Bapurao Gaekwad, Subdr.-Maj.. 1415
Baptist, A. A. E., Sr. Asst. Surgn. 1790
Bapty, P. A., Lt. 543, 1187
—— P. S., Lt. 486
Baquar, Khan Jemdr. 1704
Bara Singh, Jemdr. 978
Barada Kanta Ray, Asst. Surgn.. 3
Baramdeo Singh, Jemdr. . . 1384
Barbari Khan, Subadar . . . 1122
Barber, A. L., Lt. 700
—— A. H. J., 2nd-Lt. . . 552, 1457
—— A. J., 2nd-Lt. . . 529, 1674
—— C. G., Lt. 525
—— C. H., Maj. 1751
—— E., 2nd-Lt. 531
—— Hon. Sir E. F., Major . 611
—— F. F., 2nd-Lt. . . 565, 1696
—— F. H., 2nd-Lt. 720
—— H. C. L., 2nd-Lt. . . . 859
—— J. F. A., 2nd-Lt. . . 554, 1024
—— L. 2nd-Lt. 819
—— P. T., 2nd-Lt. . . 358, 1493-94
—— V. J., Lt. . . . 40, 712
—— W. N., 2nd-Lt. . . . 640
Barbour, K. D., Capt. . . 258, 903
Barboza, A. H. V., Asst. Surgn. . 1799
Barbton, A. G. A., Lt. 1150
Barchard, L. B., 2nd-Lt. . . 90, 707
Barckley, D. C., Lt. . . . 520, 1038
Barclay, A. G. F., Lt. . . . 525, 992
—— C. N., 2nd-Lt. 763
—— F., Lt. 505, 1098
—— H. L., Lt. . . . 446-49, 1441
—— J. B., 2nd-Lt. 527
—— **P. C. R.**, Lt.-Col.. . 136, 158, 197, 871, 1483
—— T. M., 2nd-Lt. 679
Barclay-Brown, R., 2nd-Lt. . . 1403
Barden, E. A., 2nd-Lt. . . 465, 1279
Bardens, A. T., Condr. . . . 1853
Bardhan, P. K., Tempy. Capt. . 1772
Barfield, F. H., Lt. . . . 435, 1195
Barford, G. E., Lt. 24, 629
—— L. H., Lt. 486
Barge, K., Capt. . . . 51, 243, 907
Barham, C. M., Revd. 1907

NAMES AND RANK. Page.

Barham, W. C., Lt. . . . 375, 1024
Bari Khan, Subdr. 1307
Baring, D. W., Capt.. . . 730, 798
—— E. B., Lt. 734
Baring Durban, E. C., 2nd-Lt. . 1678*a*
Barkat Ali, Sub-Asst. Surgn. . . 1816
—— —— Ralander Bux, Ty. Vet. Asst. 1687
—— Jemdr. 1452
—— Khan, Subdr.-Maj. . . . 1705
—— **Khan**, Jemdr. 876
—— Jemdr. 1088
—— **Ram**, Sr. Sub-Asst. Surgn. 1807
—— —— Sub-Asst. Surgn. . . 1812
—— —— Sub-Asst. Surgn. . . 1819
—— **Shah**, Jemdr. 1285
Barkatullah, Sr. Sub-Asst. Surgn. 1809
—— Khan, Jemdr. . . . 1389
Brake, C. R., Capt. 739
Barkeley, D. C., Lt. 1038
Barker, A. R., Lt. 312
—— A. V., Asst. Surgn. . . 1802
—— C. G., Lt. . . . 504, 577
—— D., Lt. 310, 1303
—— E. H., Lt. 725
—— F., 2nd-Lt. . . . 614, 630
—— F. A., Capt. 1754
—— F. R., 2nd-Lt. . . 545, 1152
—— G. G., Lt. . . . 597, 1006
—— G. M., Lt. . . . 541, 1197
—— H., 2nd-Lt. 363
—— H. A. E., Lt. . . . 332, 1624
—— H. E. R., Lt. . . 507, 1026
—— H. L., Capt. 1756
—— H. W., Lt. 801
—— J. S., Maj. 663
—— J. W., Lt. 628
—— J. W. Serjt.-Maj. . . . 1879
—— K. C., ty. 2nd-Lt. . . . 709
—— P., 2nd-Lt. . . . 553, 1323
—— R., Lt. 536, 1513
—— R. A., Lt. . . . 371, 1611
—— R. E., Capt. . . 741, 1004
—— R. R., Capt. 628
—— R. W., Capt. . . 261, 1894
—— S., 2nd-Lt. 835
—— S. G., Lt. 830
—— T. G., 2nd-Lt. 559
—— T. L., Lt. . . . 292*a*, 1433
—— W. D., Lt. . . . 538, 1028
Barkway, A. P. C., 2nd-Lt. . . 1024
Barlee, A., Esq. . . . . . .
Barley, M. de M. 2nd-Lt. . . . 790
Barlow, A. E., Lt. . . . 293, 1621
—— C. B., Lt. . . . 293, 1231
—— C. P., Maj. 211, 872, 1621, 1635
—— J. H., Lt. . . . 321, 881
—— L., Lt. 527, 1667
—— R. F., 2nd-Lt. 626
—— R. G., Capt. 764
—— W. N. C., Lt. 302, 1132, 1140
Barltrop, A. B., Lt. . . . 318, 1613
Barna Bir, Jemdr. 966
Barnard, A. A., Esq. 73
—— A. E., Lt.-Col. 94, 131, 156, 192
—— A. W. S., 2nd-Lt. . . 794
—— G. S., 2nd-Lt. . . . .
—— J. S. E., Asst. Surgn. . 1802
—— L. G., Lt. . . . 542, 1285
—— T. W. Capt. 1732
—— W., Lt. 527
—— W. O., Maj.-Genl. . . 758
Barnardiston, E., Maj. . . 661, 1883
Barnardo, F. A. F., Bt.-Lt.-Col. 40, 137, 1749
Barnby, J. E., Lt. 720
Barnder, E. W., ... 1838
Barne, G. D., Revd. Maj. . 806, 1904
Barnes, A. E. C., 2nd-Lt. . 358, 1092
—— B. D., 2nd-Lt. 821
—— B. D. J., Lt. 581
—— B. L., 2nd-Lt. 776*a*
—— C. S., Lt. 1130
—— D. H., Lt. . . 55, 505, 1559
—— E., Maj. 1919
—— E. C., 2nd-Lt. . . 330, 1227
—— E. C., Maj. . . 82, 225, 1094
—— H. C., Lt. . . . 57, 532
—— J. F. F., Lt. . . 321, 877
—— J. H., 2nd-Lt. . . 328, 1293
—— J. P., Lt.-Col. 128, 154, 185
—— K. R., Capt. 734
—— L. E., Maj. . 6, 13, 98, 664
—— L. F., 2nd-Lt. . . 669, 997
—— O., Col. 176
—— O. G. R., Capt. . . . 749

Names and Rank.	Page.
Barnes, R. M., Lt. 297,	1102
—— S., Lt.	629
—— T., 2nd-Lt. 555,	1088
—— T. J., Lt. 420,	1275
—— T. L., Lt.	767
—— T. W. C., 2nd-Lt. 365,	1327
—— W. A., Lt.	622
—— W. C., Lt. 511,	1483
Barnet, D. G., Lt.	624
Barnett, A. C., Lt.	756
—— A. H., Lt. 534, 1666,	1704
—— B. C., 2nd-Lt.	1701
—— C. E., 2nd-Lt. 359,	1321
—— D. A., 2nd-Lt. 557,	1054
—— J. W., Capt.	1754
—— S. H., 2nd-Lt.	763
—— W., 2nd-Lt.	713
—— W. E., 2nd-Lt. 355,	1473
—— W. H., Maj. 234,	1371
Baron, F. C., Asst. Surgn.	1800
—— W. W., 2nd-Lt.	787
Barr, A., Lt. 353,	1657
—— A. G., Esq.	91
—— E.N.D., Lt.	623
—— G., Ty. Lt.	579
—— J. H. V., Maj. 104, 217, 1439,	1475
—— M. C., Lt. 320, 365,	1303
Barrack, J. C., Lt. 512,	1110
Barraclough, A. E., Lt. 353,	1676
—— M. W. R., Lt. 312,	887
—— W., Lt.	585
Barrand, H. J., 2nd-Lt. 759,	789
Barratt, B. J., Col. 26,	1731
—— W.C., Maj.-Gl 12, 37, 119, 147,	679
Barrenger, H. E., Lt.	174
Barreto, A. daG., Ty. Capt.	1771
Barrett, A. A., Genl. *Sir* 1, 12, 19, 118, 145, 171,	1917
—— A. L., Lt.-Col. 139, 160, 207, 1154,	1501
—— B. H. H., Capt. 9,	1664
—— C. C. J., Maj. 240, 1893,	1913
—— C. L. R., Lt.	879
—— E. D. C., 2nd-Lt.	1086
—— E. D. E., 2nd-Lt.	476
—— F., 2nd-Lt.	363
—— J. H., Ty. Lt.	1780
—— J. L. M., Lt.	591
—— L. J. S., 2nd-Lt.	640
—— L. W., Capt.	1732
—— M. L., Lt. 300,	964
—— R. C., 2nd-Lt.	626
—— R. O., Lt.	710
Barrie, W. G., 2nd-Lt.	563
Barrington, A., Lt.	540
—— A. C., 2nd-Lt.	591
—— A. H. M., 2nd-Lt. 494, 827,	1293
Barroll, W. S., Lt. 303,	1383
Barron, A. F. N., Maj. 42, 76,	634
—— A. H. D., Capt. 267,	1086
—— A. S. N., Lt. 405, 445,	1614
—— C. A., Lt.-Col.	605
—— F. H., 2nd-Lt.	563
—— F. H., 2nd-L	1678a
—— F. W., Bt Lt.-Col. 53,	633
—— R.M., Brevt Lt.-Col. 139, 935,	1749
—— S., Ty. Lt.	1729
—— S. W. J., Lt.	537
—— W. E., Maj 662,	1883
—— W. G., Lt.	517
—— W. J., 2nd-Lt.	359
Barrons, L. B. A., 2nd-Lt.	752
Barrow, A, 2nd-Lt. 363,	1026
—— E. G., Genl. *Sir* 118, 145, 171, 1042, 1917,	1955
—— G. DeS., Maj.-Genl. 97, 102, 119, 147, 174,	941
—— G. E., Lt. 522,	1195
—— H. G., 2nd-Lt.	729a
—— H. P. W., Lt.-Col.	1710
—— R., Lt. 297a, 1195,	1351
—— R. L., Maj 245,	1174
—— R. H. B., tempy. Capt.	1725
—— R. L., Lt. 487,	953
—— J. C. R., Sub-Condr.	1656
Barru, Jemdr.	1171
Barry, C. C. S., Lt.-Col. 130,	1742
—— C. A., Lt.	622
—— C. B., Lt. 494,	907
—— D. T., 2nd-Lt.	571a
—— E., 2nd-Lt.	769
—— E. H., Lt. 506,	1164
—— E. P., Capt.	66
—— F., 2nd-Lt.	1705
—— F. C., Lt. 307a, 492, 1116,	1231
Barry, G., Capt. 79,	1839
—— J., 2nd-Lt. 361,	1678
—— J. F., Lt.-Col. 57, 123, 154,	185
—— J. J. M., Capt.	1839
—— J. L., Lt. 311a, 968,	1595
—— S. J., Capt. 35,	1715
—— W. E., Lt.	1839
—— W. T., Maj. 208, 1020, 1883,	1913
Barry-Briant, G. S. B., Lt.	90
Barry-Smith, R. C., Lt.-Col. 138, 159, 197, 867,	907
Barson, A. H., *Hony.* Lt.	1849
Barstow, A. E., Capt. 69, 265,	1078
—— H., Brig.-Genl. 8, 14, 122, 136, 159, 211,	1174
—— H. L., Lt. 397,	1614
—— J. A., Capt. 743,	830
—— T. C. R., Capt. 272,	1623
Bartels, G. K. A., Asst. Surgn.	1795
Barter, E. G., Lt. 543a,	987
—— F., Lt. 307a,	1611
Bartholomew, A. G., Maj. 57,	753
—— G., 2nd-Lt. 523,	986
—— H. L., Lt. 505,	992
Bartlect, 2nd-Lt.	1677
Bartleman, E. L., Lt. 425,	1447
—— W. E., Maj.-Genl.	169
Bartler, A. S., 2nd-Lt.	684
Bartlett, A. C., 2nd-Lt. 529,	1399
—— A. H., 2nd-Lt.	1701
—— A. H. S., 2nd-Lt.	551
—— A. O., Lt.	763
—— C. E., Lt.	704
—— D. F., Lt. 9,	527
—— E. L., *Hony.* Ma. 55,	1849
—— G., 2nd-Lt.	797
—— G. D. N., Lt. 516a,	1132
—— H. V., Lt. 309,	1379
—— P. J. A., Lt. 430,	899
—— R. F., Capt.	1689
—— T. E., 2nd-Lt.	491
Bartley, A. St. C., Asst. Surgn.	1801
—— E. S., 2nd-Lt. 475,	1443
—— J. H., Asst. Surgn.	1801
Bartle-Pye, M. F., 2nd-Lt.	1515
Barton, A. E., Lt.-Col. 187, 1896,	1911
—— A. P., Lt.-Col.	608
—— E. R., Lt.	532
—— F. V. G., Lt. 300,	1016
—— G. B. B., tempy. Lt.-Col.	668
—— H. E., Lt.	629
—— H. H., Ty. 2nd-Lt.	687
—— H. J., Lt.-Col. 28, 103, 131,	661
—— J. C., Capt.	818
—— J. E., 2nd-Lt.	819
—— J. E. B., Lt. 440,	1441
—— L. E., Capt. 95, 271,	1140
—— N. F., Lt.	185
—— P., Lt.-Col. 101, 135,	613
—— P. H., Lt.-Col.	628
—— R. F., Lt. 371, 781,	1249
—— R. G., Lt.	1701
—— R. M. S., Lt. 506,	1624
—— R. W.	193
—— V. G. J., Lt. 320,	1323
—— W., 2nd-Lt.	1092
—— W., Lt.	736
Bartram, A., *Hony.* Lt.	1846
Bartrop, C. G., Sub-Condr.	1873
Barty, J. E., 2nd-Lt. 551,	1703
Baruo, G. N., 2nd-Lt.	1054
Barwani, *His Highness*, Sir, Rana of—Capt.	11
Barwis, C. W. A., Capt. 279, 1239,	1293
Barwise, A. H. O., Sub-Condr.	1867
Barya, *Raja of* 111,	1892
Baryam Singh, Jemdr.	999
Baryani Singh, Jemdr.	999
Basakha Singh, Jemdr.	1111
Basan Singh, Subdr.	1577
Basant Singh, Sub-Asst. Surgn.	1811
—— Ressldr.	950
—— Subdr.	1043
—— Jemdr.	1134
—— Jemdr.	1080
—— Subdr.	1120
—— Sub-Asst. Surgn.	1815
—— Sub-Asst. Surgn.	1814
—— Jemdr.	980
—— Jemdr.	989
—— Jemdr.	1320
—— Jemdr. 1051,	1175
—— Jemdr.	1321
Basante Sahi, Subdr.	1636
Basawa Singh, Subdr.	1111
—— Jemdr.	1493
Basawa Singh, Jemdr.	1228
—— Jemdr.	1137
—— Jemdr.	1698
Basawan Singh, Subdr.	1403
Basdeo Singh, Subdr. 1044,	1581
Base, E. H., Lt.	738
Baseley, H. C., Ty. 2nd-Lt.	780
Basevi, W. H. F., Lt.-Col. 132, 156, 192, 870,	1371
BasharatAliKhan, Sub-Asstt. Surgn.	1818
Basheshar Nath, Jemdr.	1698
Bashi Ram, Sub-Asst. Surgn.	1814
—— Sharma, Sub-Asst. Surgn.	1815
Bashir Ahmed Khan, Jemdr.	1515
Basist Singh, Jemdr.	1064
Basith Khan, Jemdr.	1273
Bassett, B. J., 2nd-Lt.	565
—— F. S., Lt. 399,	1285
Baskett, O. W. B., 2nd-Lt.	552
Bassawa Singh, Jemdr.	1111
—— R. P. H., Lt. 67,	722
—— T. P., Maj. 664,	991
—— T. R., Lt.	353
—— W. S., Lt.	46
Basti Khan, Subdr. 1479,	1589
—— Singh, Jemdr.	878
Bastin, H. E., Lt. 405,	903
Bastow, H., Lt.	624
Bastyan, S. J., Hony. Maj. 78,	626
Basu, A., Ty. Lt.	1782
—— B., Ty. Lt.	1774
—— H. N., Ty. Lt.	1782
—— J. N., Temy. Lt.	1775
—— K. M., Ty. Lt.	1774
—— K. S., Ty. Capt.	1767
—— N., Ty. Lt.	1772
—— P. N., Ty. Capt.	1772
—— P. L., Ty. Lt.	1781
—— R. C., Ty. Capt.	1772
—— R. K., Ty. Lt.	1775
—— S. C., Tempy. Capt.	1769
—— S. C., Ty. Lt.	1778
Basurey Thapa, Jemdr.	1637
Batan Singh, Jemdr.	1699
Batchelor, A. E. F., 2nd Lt. 78,	619
—— A. J., Lt. 401,	1134
—— G. F. F., 2nd-Lt. 546,	1275
—— V. A., Maj. 103,	614
Bateman, C. G., Lt.	527
—— C. W., Lt. 90,	514
—— F. W., Lt. 534,	1351
—— G. O., 2nd-Lt.	540
—— G. W., 2nd-Lt.	
—— H. E. R., 2nd-Lt. 329,	328
—— L. C., 2nd-Lt.	357
—— N. G., Lt. 517,	1036
—— T. E., 2nd-Lt.	732
Bateman-Champain, C E., Maj. 208,	1617
—— H. F., Brig.-Gen. 50, 104, 122, 134, 157, 94, 872,	1641
Bates, A. T. T., 2nd-Lt. 327,	1596
—— F. P. E., 2nd-Lt. 331,	1239
—— G. E. L., 2nd-Lt. 330,	947
—— H. S., Lt. 503,	1090
—— H. V. R., Lt. 435,	1128
—— J., 2nd-Lt.	640
—— L. F., Capt.	1732
—— L., Lt.	179
—— R. C. S., Lt. 318,	1363
—— R. S. P., Lt. 320, 365,	1543
—— S. F., Ty. 2nd-Lt.	741
Ba Thaw, Subdr.	1295
Batey, G. H. M., 2nd-Lt.	328
Batley, W., 2nd-Lt.	363
Batra, H. L., Capt.	1750
—— K. R., Capt.	1758
Batt, A. E., Capt.	1841
—— B. J., Sub-Condr.	1857
—— R. H., 2nd-Lt. 699,	795
Batten, A. E., 2nd-Lt.	724
—— E. F., Capt. 52,	502
—— F. G., Col. 125, 150,	185
—— H. H., Maj. 82, 233, 1207,	1894
—— J. F., Capt. 67,	617
—— W. D. G., Capt. 274,	1606
Batterbury, F. H., Lt. 536,	988
—— G. H. Lt.	988
Batterham	1720
Battersby, R. S., 2nd-Lt. 560,	1050
Battershill, W. D., Lt.	721
Batteshill, W. D., 2nd-Lt.	750
Battie, J. W., 2nd-Lt. 463,	1219
—— W. A. A., Lt.	543a
Battine, R. St. C., Lt.-Col. 97, 192, 103, 105, 136, 158, 196,	915

Names and Ranks.	Page.
Battishill, J. H., 2nd-Lt. . 77,	746
Battman, A. R., 2nd-Lt. . 21,	666
Battrick, E. V. B., Lt. .	622
Batty, A. D. G. S., Lt. 529, 651,	957
——— R. E., 2nd-Lt.	567
Battye, A., Asst. Surgn.	1803
——— A. B., Lt.-Col. 138, 160, 199, 885,	959
——— A. H., Lt.-Col. 128, 154,	185
——— B. C., Maj. . . 56, 663,	1883
——— C. F., Capt.	749
——— I. D., Maj. . . 209,	954
——— R. D., Lt. . . . 415,	1393
——— T. H., Capt. . . 267,	1647
——— W. R., Lt.-Col. . . 139,	1745
Ba Gyan, Jemdr.	1295
Ba Min, Jemdr.	1295
Ba Nyum, Jemdr.	1295
Ba Oh, Jemdr.	1295
Ba Shin, Jemdr.	1295
Baum, E. F., Lt.	828
Baumber, J., Lt. . . . 513,	1598
Bavin, L. N., Lt.	490
Bawa Sant, Subdr.	1111
Bawar khan, Jemdr.	1432
Baxter, C. W. W., Capt. . . .	1759
——— D. St. J., Capt. . . 264,	1371
——— E. E., Lt. . . . 33,	450-54
——— G. D., 2nd-Lt. . . 669,	987
——— E. G., Lt.	506
——— G. F., Sub-Condr. . . .	1868
——— R. H. N., Lt. 61, 510,	871
——— W., Sub-Condr.	1873
——— W. C., 2nd-Lt. . . 692,	784
——— W. G. E., Lt. . . 533,	1677
Bay, G. J., 2nd-Lt.	1695
Bayett, H. M., Lt.	1894
Baylay, E. J. L., Maj.	614
Bayldon, R. C., Lt.	487
Bayley, A. S. B., Esq.	91
——— A., Lt.	851
——— E. C., Lt.-Col. . 131, 155, 190,	1896
——— G. A., Maj.	857
——— H. E., Lt.	1885
——— J. P., 2nd-Lt. . . 365,	1150
——— L. C. L., Maj. . 218, 870, 885,	1369
——— L. S., Lt.-Col. . . 134,	6333
——— S. F., Lt.-Col. . . 185, 1889,	1911
——— V., Capt. . . . 545,	806
——— W. H., 2nd-Lt. . . 357,	1032
Bayley-de Castro, A., Asst. Surgn.	1795
Baylis, R. L., 2nd-Lt.	805
Bayliss, B., Lt.	304
——— C. L., Lt.	423
——— H. T., Lt.	817
——— C. E., Condr.	1867
——— C. L., Capt.	1164
Bayly, A. W. L., Maj.-Genl. Sir 1501,	1955
——— H., 2nd-Lt.	562
——— P. F. B., Lt. . . . 507,	1020
Bayne-Jardine, C. W., Capt. . .	615
Bayne, A., Esqr.	93
Baynes, C. E., Lt.-Col. 128, 154,	183
Bazalgette, E., Lt.	846
——— H., Mrs., N. Sister .	1735
Bazaz, M. S., Ty. Lt.	1773
Bazeley, W. N., Maj.	698
Bazett, L. Y., Lt.	518
——— H. M., Lt. . . . 536,	1894
——— S., Maj. . . . 211,	1894
Baz Khan, Subdr.	1311
——— Jemdr.	1565
——— Subdr.	1507
——— Jemdr.	1681
——— Subdr. . . . 1137,	1488
——— Jemdr.	1681
Baz Singh, Jemdr.	1393
Beach, G., Maj.	753
——— W. H., Brig.-Genl. . 51, 121, 134,	661
Beadel, C. F., Maj.	603
Beadle, R. D., Maj. . 238, 1107,	1505
Beadon, H. C., Lt.-Col. 195, 1889, 1896,	1912
——— V., Maj. . 244, 1293, 1641,	1894
——— W. de C., Lt. . . 317a,	1052
——— W. R. C., Capt. . . .	820
Beal, H. E. J., Lt.	515
——— H. G., Lt.	799
Beale, A., Senr. Asst. Surgn. . .	1789
——— A., Bt.-Col.	57
——— C. J., 2nd-Lt.	630
——— E. J., 2nd-Lt. . . 559,	1271
——— G. H., Hony. Capt. . 33, 78,	585

Names and Rank.	Page.
Beale, G. Mc. L., 2nd-Lt. . . .	560
——— R. E. C., 2nd-Lt. . . .	1689
——— W., Lt.-Col. . . 92, 1892,	1912
——— W. F. B., 2nd-Lt. . . 552,	1082
Beales, W. L., 2nd-Lt.	1383
Beall, J. S., Lt. . . . 443,	1152
Beaman, A. A. H., Capt. . 258,	875
——— G. W. D., Lt.	832
——— G. P. R., Maj.	1919
Beamish, R. P., Lt. . . 95, 309,	1102
——— R. R., 2nd-Lt.	828
——— W. S., Maj. . 74, 101,	633
Beamount, R., 2nd-Lt.	1058
Beams, R. H., Lt. 526,	1607
Bean, B. C. L., Capt.	806
——— H., Lt.	764
——— J. A., Maj. . 223, 1052,	1407
——— S. M. L., Capt.	814
——— T. L. L., 2nd-Lt.	554
Beanland, D., Capt. 66, 278, 1106,	1249
Beant Singh, Ressdr.	916
——— Jemdr.	965
Bear, R. M., 2nd-Lt.	112
Beard, A. S., Lt.	769
——— A. W., 2nd-Lt.	666
——— F., Lt. 440,	1239
——— H., 2nd-Lt.	562
——— J. W., 2nd-Lt.	637
——— N. T., 2nd-Lt. . . 546,	1265
——— R., ...	599-600
——— R. H. A., Lt.	1678
——— R. H. F., 2nd-Lt. . . .	548
——— W. H. G., 2nd-Lt. . 475,	1070
Beardall, C. H., Hony.-Lt. . . .	588
Bearne, L. H., Lt. . . . 385,	1567
Beart, P. D., 2nd-Lt. . . 549,	1701
Beasant, C. A., Lt. . 76, 535,	1884
Beath, K. C., 2nd-Lt.	552
Beaton, D. J., Lt. . . . 389,	1046
——— J.S. C., 2nd-Lt. . . 532,	1668
Beatson, B. F., Capt.	1758
——— E. B., Capt.	838
——— J., 2nd-Lt. . . 455-59	1070
Beatson Bell, A., Lt. . . 712,	1006
——— J., Lt.-Col., 19, 131, 156,	191
Beattie, J. G., Lt.	88
——— J. O., Capt. . . 260,	1664
——— M. H., 2nd-Lt.	1696
——— B., Lt.	1839
Beatty, G. A. H., Br.Genl. . 134, 157, 194, 867,	891
——— H., Lt. 493,	903
——— H. P., Lt. . . . 505,	1295
——— L. N., Col. 50, 121, 127, 150,	189
——— P. V., Maj.	1838
Beaty, A. R., Lt. . . . 465,	1232
——— G. H. M., 2nd-Lt. . . 329,	1495
——— P. C. D., Lt. . . . 531,	1672
Beauchamp, K. G., Lt. . . 534,	1335
——— W. G., 2nd-Lt. 539,	1239
Beaufort, P. M., Lt. . . . 524,	1016
Beaumont, G. A. H. Maj. Sir . .	781
——— H., Lt. 527,	1624
——— D. C., Lt.	1718
——— H. S., Lt.	90
——— J. W., Lt.	1841
——— L., 2nd-Lt.	788
——— S. J. G., Maj. 237,	1407
——— T. C., Capt.	657
——— W. B. L., 2nd-Lt. . . .	787
Beausire, R. E., Lt.	629
Beauvais, L., 2nd-Lt.	737
Beaver, G. D. C., Lt. . . 55,	509
——— H. A., Ty. Capt. . . .	1723
Beazeley, G. A., Lt.-Col. 138, 661,	1880
——— G. E. A., 2nd-Lt. . 320,	1655
Beazley, W. E., Capt. . . 258,	1239
Bebbington, H. J., 2nd-Lt. . 328,	1383
Becha Misr, Jemdr.	1027
Becher, G. A., Lt.-Col. 57, 133, 157, 193,	889
Beck, D'A. S., Capt. . . . 62,	1838
——— F. G., 2nd-Lt.	545
——— H. C., Lt.	505
——— J. P., 2nd-Lt. . . . 540,	1699
——— R. C., Capt. . . . 242,	1663
——— T., Maj.	24
——— Lt.	1676
Beek, W. G., 2nd-Lt.	640
Beckerson, R. E., Lt. . . . 397,	1241
Beckett, A., Lt. . 313, 1635, 1657,	1657
——— C. G., Ty. 2nd-Lt. . . .	728
——— C. T., Capt.	615
——— D. Sub.-condr.	1868
——— E. A., Lt.	759

Names and Rank.	Page.
Beckett, E. G., Capt.	851
——— H. J., 2nd-Lt.	618
——— J., 2nd-Lt.	555
——— J. H., Capt.	1732
——— R. S. B., Lt. . . 291,	1635
——— W. H., Col. . . . 161,	1916
Beckham, L. G., Lt.-Col. . 180,	1918
Beckingham, F., Tempy. Capt. F.	629
Beckley, T. H., Maj.	615
Beckwith, L. K., 2nd-Lt. . . .	545
Beckworth, C., 2nd-Lt. . . 739,	791
Beddek, E. E., Maj.	57
Beddy, E., Col.	177
——— P. L., Lt.-Col. . 136, 158, 213,	1217
Bedekar, V. H., Ty. Lt. . . .	1780
Bedell, E. C., Senr. Asst. Surgn. .	1783
——— G. W., Lt. 529,	1022
——— H. C., Asst. Surgn. . .	1799
——— P. S., Asst. Surgn. . .	1798
Bedford, E. W., 2nd-Lt. . . 547,	998
——— H., 2nd-Lt. . . . 360,	1044
Bedford Pim, E. H., Lt.-Col. . .	628
Bedingfeld, R. H. N., Lt. . 540a,	1673
Bedwell, P., Condr.	1853
Bee, J., 2nd-Lt.	360
Beechey, O. P., Lt.	705
Beeching, C. F., Hon. Maj. . .	693
Beecroft, W., 2nd-Lt.	851
Beeden, A. W., Condr. . . .	1853
Beedham, J. W., Lt. . 549, 1143,	1431
Beeley, B. D., Revd.	1906
Beer, E. E., 2nd-Lt. . . . 548,	1265
——— G. E., 2nd-Lt.	361
——— G. H. B., Lt.	699
——— H. R. E., 2nd-Lt. . . .	713
——— J. R. E., Col.	896
——— V. L., Maj.	633
——— W. A., Asst. Surgn. . .	1796
——— W. R., Lt. 291,	939
Beere, H. C., Lt. 731,	791
Beesley, F., Lt.	354
Beeson, R. G., 2nd-Lt.	507
Bega Ram, Jemdr.	1498
——— Jemdr.	1577
——— Jemdr.	1579
Begbie, E. W., Maj.-Genl. . 170,	1916
——— G., Lt.	622
——— T. N., Capt.	821
Begg, P. H., Lt. 316,	1443
Behara Rawat, Subdr. . . .	1483
Behari Lal, Subadar	1051
——— Ty. Sub.-Asst. Surgn.	1833
Behari Singh, Subdr.	1031
Behere, G. H., Sub-Asst. Surgn. .	1831
Behenna, W. H., Sub-Condr. . .	1856
Behrend, E. S., Capt.	852
Behrens, D. H., 2nd-Lt. . . .	624
Beit, F. V. G., Maj.	1716
Bejoy Parsad Singh Roy, Babu, 2nd-Lt.	481
Bela Singh, Subdr. Maj. . . .	1245
——— Jemdr.	1152
——— Jemdr.	1163
——— Jemdr.	1130
——— Jemdr.	888
Belbahadur Gurung, Jemdr. . .	1615
Belben, H. R., Lt. . . . 351,	1487
——— H. A., Lt. . . 318, 1391,	1487
Belchem, A. F., Condr. . . .	1851
Belcher, A. L., 2nd-Lt. . . . 77,	693
——— A. R., 2nd-Lt.	1884
——— D. W., Lt.	446-49
——— E. B., Capt.	485
——— G. W., Lt.	1914
Belder, H. B., Lt.	1263
Belfield, H. E., Lt.-Genl. Sir. .	713
——— G. H., 2nd-Lt.	1199
Belford, H. D., Lt. . . . 420,	1537
Belgrave, H. D., Lt.-Col. 53, 97, 136,	749
Beli Ram, Subdr.	1145
Belk, A., 2nd-Lt.	787
Bell, A. H., Maj. . . 29, 103, 105,	663
——— A. J., 2nd-Lt.	490
——— A. M., Lt.	779
——— A. R., Asst. Surgn. . . .	1800
——— A. R., 2nd-Lt. . . . 531,	1702
——— C. B., Lt.	833
——— C. F. J., 2nd-Lt. . . .	537
——— C. W. A., 2nd-Lt. . . .	765
——— E. W., 2nd-Lt. . . . 780,	1006
——— F. S. W., 2nd-Lt. . . .	728
——— G. F. St. J., 2nd-Lt. . .	691
——— G. d. H., Col. . . 125, 1738,	1894
——— H., Lt. 38, 508,	1566

NAMES AND RANK. Page.

Bell, H. C., Lt. . . . 488, 929, 1331
— H. C., 2nd-Lt. . . . 540a
— H. C. F., Maj 238, 1307
— H. G., Lt.. 1844
— H. G., 2nd-Lt. 1331
— H. G., Lt. 305, 1007
— H. H., Lt. 542, 1001
— H. J., Lt. 519
— J., Lt. . . . 368, 705, 805, 1050
— J. A., Lt. 516a, 986
— J. A., Maj.-Genl. *Sir.* 118, 146, 173
— J. G., Capt. 55, 501a
— J. G. A., Lt. 587
— J. V., 2nd-Lt. . . . 326, 366, 1641
— J. N., 2nd-Lt. . . . 550, 1703
— L., Lt. 514, 1267
— L. C., Lt. 425, 1066
— L. M., Lt.-Col. . . 129, 154, 186
— M. D.. Major 634
— N. D. B., The Hon. Maj.. . . 859
— N. E., 2nd-Lt. 361
— N. G., Tempy. Lt. . . . 45, 767
— N. K., Lt. 7, 405, 1427
— P., Asst. Surgn. 1795
— R., 2nd-Lt. 8 26
— R M., Lt.-Col. .129, 154, 187, 1660
— R. S., Sub-Asst. Surgn. . . 1830
— R. S., 2nd-Lt. 553, 1223
— T. W., 2nd-Lt. . . . 77, 715, 1884
— W., Capt. 815
— W. A., Lt. 681
— W. A. J., 2nd-Lt. . . . 354, 1018
— W. C., 2nd-Lt. 1016
— W. F., Lt. 352, 1197
— W. J., Lt. 626
— W. L., Revd.. 1903
— W. S. B., 2nd-Lt. . . 359, 1104
Bellairs, R. G., Lt. . . . 535, 1696
— S. S., 2nd-Lt. . . . 494, 931
Bellamy, C. S., 2nd-Lt. 577
— E. A., Lt. 353, 1379
— F. H. H., 2nd-Lt. . . 555, 1321
— L. C. M., Lt.. 317a, 365, 1469
Bell-Kingsley, H. E. W., Capt. 256, 1613
— Smyth, J. A., Br.-Genl. 97, 121, 126
Bellars, A. E., Lt. 827
Belle, A., Lt. 711
Bellers, E. V. R., Capt. . . 279, 1595
Bellgard, S. J., Ty. Asst. Surgn. 1804
Bellhouse, A. P., Lt. . 28, 389, 1046
Belli-Bivar, B. F., Capt. . 263, 1505
— P. F., 2nd-Lt. . . . 1489
Bellingham, W. F., 2nd-Lt. . . 1489
Belton, P., 2nd-Lt. . . . 559, 1483
Benares Khan, Jemdr. 1682
Bendall, A. F., 2nd-Lt. . . 358, 1427
— J., Capt. 89, 763
Bender, W. E. G., Lt. . . 520, 605, 986, 1889
Bendle, L. A., Lt. 724, 739
— M. S., Lt. 315
Benegal, V. N. (Ty. Lt.). . . . 1781
Beney, G. E., Capt. . . 502, 747, 1028
Benfield, C. W. C., Capt. . . 485, 909
Beniram, Jemdr. 1063
Beni Singh, Jemdr. 1063
— Sudbr. 1063
Benimadho Singh (I), Subdr. . 1043
— (II), Jemdr.. . 1043
Benjamin, A. L., 2nd-Lt. . . . 597
— David, Sr. Sub-Asst. Surgn. 1819
— G. I., Ty.-Lt. . 1776
— R., Sub-Asst. Surgn. . . 1821
— S. P., Sub-Asst. Surg. . 1830
Benn, F., 2nd-Lt. 713, 1678
— J. P., Maj (Ty. Lt.-Col.) 216, 945
— R. A. E., Lt.-Col. 189, 1901, 1911
Bennedik, S. R. W., Capt. . . . 712
Bennet, G. L., 2nd-Lt. 579
— L. G., 2nd-Lt. 562
— W. R. M., Capt. . . . 501
Bennett, A., 2nd-Lt. 689
— A. D., Capt. . . . 262, 1094
— A. E., Capt. 652
— A H., Lt. 698
— A. J., 2nd-Lt. 362
— A. R., Lt. 821
— A. W., 2nd-Lt. 553
— C. G., Maj. 755
— C. R., 2nd-Lt. 363
— E., Sub-Condr. 1863
— E. E. S., Maj. . 6, 98, 221, 1279
— F. L., Maj. . . . 103, 633
— G., Lt. 752, 784
Bennett, G., 2nd-Lt. . . . . .
— G. E., Lt. 857
— H., Lt.-Col. . . . 130, 1741
— H. B., 2nd-Lt. 749
— H. B., Capt. . . 509, 997, 1707
— H. D., Capt. . . . 668, 996
— J., Condr. 1853
— J. B., Lt. 508
— J. D., 2nd-Lt. 713
— J. F., Maj. . . . 206, 877
— J. H., Maj. . . 105, 222, 879
— J T., Maj. 79
— J. W., 2nd-Lt. . . . 494, 1417
— L. A., Capt 502, 1541, 1497
— M. C., 2nd-Lt. . . 473, 1279
— N., Revd. 1904
— N. B., Lt. 393, 933
— O. D., Maj. . . 95, 251, 1124
— S. B., 2nd-Lt. . . . 367, 1058
— S. G., 2nd-Lt. . . . 669, 997
— T. G., Capt. . . . 728, 798
— V. B., Lt.-Col. . . 131, 1741
— W., 2nd-Lt. 828
— W., Sub-Condr. 1864
— W. A., 2nd-Lt. 1016
— W. G., Lt. . . . 306, 1211
— W. H., Lt. . . . 397, 1016
— W. J., 2nd-Lt. 558
— W R., Asst. Surgn. . . . 1795
Bennetts, G. H., 2nd-Lt. . . . 858
Benni, Risldr. 1684
Bennion, W. O., 2nd-Lt. . . . 356
Bennison, A. J., 2nd-Lt . . . 547
— J. J., 2nd-Lt. 554
Benns, A., Condr. 1854
Benny, H. F., 2nd-Lt. 567
Bensley, C. H., Lt.-Col.. . . 132, 1743
— L. F., 2nd-Lt 786
Bensly, H. F., Lt. . . . 530, 974
Benson, A., Lt. 541
— A. H., 2nd-Lt. 679
— A. J., 2nd-Lt. 728
— C. R., 2nd-Lt. . . 365, 1199
— F. W., 2nd-Lt. 712
— G. F., Hony. 2nd-Lt. . . . 112
— G. R., Capt.. 615
— G. W., 2nd-Lt. 571
— H. E., Asst. Surgn. . . . 1791
— J. W. H., Lt. . 505, 808, 1407
— W. J. P., Maj. . . 707, 787
Benson-Cooke, P. G. 103, 235, 1327, 1439
Bentham, V. H., Lt. . . 440, 1621
— C. P., 2nd-Lt. . . 549, 988
— W. G., Capt. 1839
Bentley, C. W., 2nd-Lt. 362
— J., 2nd-Lt. 683
— J. F., 2nd-Lt. 72
— T., Lt. 521, 998
Benton, H. G., Capt.. . . 255, 881
— G. W., Lt. . . . 542, 1203
— W. B., Maj. . 242, 1417, 1423
Benzie, J. H. S., Lt. . . . 315, 911
Berendt, H., Ty. 2nd-Lt.. . 728, 799
Beresford, G. de la P., Capt. 54, 260, 893
— J. H. B. de la P. . . . 50
— N., Nursing Sister . 1735
Beresford-Barber, H., 2nd-Lt. . 1696
Berey, C. E., Lt. 521, 1642
— W., Lt. 777
Berg, C. L., 2nd-Lt. 557
— G. C., Ty. Capt. . . . 1724
Bergin, F. A., Asst Surgn. . . 1794
Beringer, B., 2nd-Lt. 728
Berkeley, A. M., Lt.-Col. 500, 857, 986
— C. S. J., Lt. . . 298, 1606
— E. L. F., Lt. . . 292, 1120
— G. E., Lt. 818
— G. F., Lt. 1637
— H. F., Lt. 513, 1134
— J. C., Maj.-Genl.. . . . 170
— M. F., 2nd-Lt. . . 535, 1515
— M. H., 2nd-Lt. 328
— R. E., Major 748
— R. E. G., Maj. 53, 237, 1351
— S. S. H., Lt. . . 313, 1635
— V. C. F., Lt.. . . 527, 1355
Berkley-Hill, O. A. R., Maj. . 1753
Berlie, H. C., Asst. Surgn. . . . 1797
Bernard, E. A., Ty. Lt. . . . 1729
— E. A., Ty. Capt. . . . 1727
— J. L., Lt. 578
— P., 2nd-Lt. 91-92, 553
— R. P. St. V., Capt. 103, 264, 1623
Berners, G. H., Lt. . . . 692, 783
Berners, R. A., Brig.-Genl. 12, 22
Berridge, H., Maj. 860
— H. H., Lt. 1689
Beriff, A. W., 2nd-Lt. 562
Berrill, E. F., 2nd 2nd-Lt. . . . 545
— S. C., 2nd-Lt. 553
Berry, A., 2nd-Lt. 1287
— A. C., Sub-Condr. . . . 1857
— A. E., Lt.-Col. . . 130, 1741
— E. F., 2nd-Lt. 545
— E. S., Lt. 523
— F., 2nd-Lt. 564
— H. S., Tempy. Capt. . . . 1723
— O. D. W., Lt. . . . 532, 1299
— P. N., Tempy. Capt. . . . 1764
— W. J., Lt. 514, 755
Berrisford, E. A. 2nd-Lt. . . 474, 1267
Berryman, E. R P., Maj. 54, 255, 1 78
— R. C. P., Capt. . . . 1764
Best, J. H., Lt. 435
Berthon, H. W., Lt.-Col. 196, 1891, 1912
Berthond, L. A., 2nd-Lt. . . . 1679
Bertie, T. P., 2nd-Lt. 560
Bertie-Clay, N. S., Lt.-Col. . 129, 632
Bertin, H., Capt. . . . 17, 45, 707
Bertini, M. A., 2nd-Lt. 720
Bertram, G. F., Sub-Condr. . . 1855
Besley, F., 2nd-Lt. 1369
Best, E., Lt. . . . 389, 1120, 1569
— F. J., Lt.. 986
— G. H., 2nd-Lt. . . . 564, 1387
— J. F., *Hony.* Lt. 593
— *Hon.* J. W., 2nd-Lt. . . . 803
— W. R. P., 2nd-Lt. 758
— W. J., Lt. 420, 1120
Bester, G. C., Lt. 540
Bestford, T., 2nd-Lt. 567
Beswall, A. P., 2nd-Lt. 682
Beswick, L., 2nd-Lt. . . . 491, 945
Betham, G. L., Capt. . 15, 269, 1078
— R. M., Br.-Genl. .12, 26, 120, 123, 149, 182
Bethell, C. A., 2nd-Lt. 715
— C. V., Lt. 352, 1633
— H. K., Brig.-Genl. 50, 97, 119, 122, 137, 587
— L. A., Maj. . . . 220, 1047
Bett, F. J., H., Lt. . . . 430, 1627
Betteridge, G. W., Lt. . . 528, 1707
— R. A., 2nd-Lt. . 359, 1439
— R. C., 2nd-Lt. . 539, 1672
Betterton, F. A., 2nd-Lt. . . 545, 988
Bettles, A. L., Lt. 492, 949
Bettlheim, H., Capt. 56
Bettison, L. W., Lt. 75
Betts, A. J. V., Maj 1749
— D. M., 2nd-Lt. 728
Bevan, A. E., Ty. 2nd-Lt. . . . 709
— A. H. R., Lt.. 299
— G. J. B., 2nd-Lt. 833
— J. D., 2nd-Lt. 1008
Bevan-Petman, B. H., Lt. 298, 915
Beveridge, A., Lt. 638
— E. W., St. C., 2nd-Lt. . 558
— J. W. P., 2nd-Lt. 476, 1174
— R., Lt. 527, 1001
— R., Lt. . . . . . .
— R. L. W., Asst. Surgn. 1800
Beverley, R., Lt. 629
Beville, C. H., Lt.-Col. 129, 154, 187
— E. L. G., Lt. 760
— F. G., Lt.-Col.. 187, 1899, 1911
— H. G. P., Maj. 1919
Bevington, L. F., Capt. . . 258, 1479
Bevis, E. C., Lt. 729
— F. G., Lt. 303, 1395
Beynon, A. C., Lt. . 308, 1399, 1567
— G., 2nd.-Lt.. . . . 548, 779
— F. C., Lt. . . . 806, 1676
— W G L., Maj.-Gen. 12, 24, 119, 147, 174
Beyts, C. A., Capt. 849
— M. S., Lt. 523
— R. F. S., Maj. . 218, 1299, 1883
— W. G., Col. 1709
Bhabesur Gurung, Jemdr. 1603, . 1611
Bhadarir Rana, Subdr. 1595
Bhadra, Sarat Chandra, Esq. .
Bhadrabir Thapa, Jemdr.. . . 1598
—, Subdr.-Maj. . 1627
Bhadra Sing Gurung, Jemdr. . 1604
Bhadria Khattri, Jemdr. . . . 1619
Bhaduri, D. N., Ty. Lt. . . . 1774
Bhag Ali, Sub-Asst. Surgn. . . 1816
— Chandra, Sub-Asst. Surgn. 1820
— Mal, Jemdr. 938
— Singh, Subdr.-Maj. . . . 1171

NAMES AND RANK. Page.

Bhag Singh, Sub-Asst. Surgn. 1815
—— Jemdr. 1212
—— Subdr. 1104
Bhag Singh, Jemdr. 1702
Bhagga Singh, Ressidr. 896
—— Jemdr. 990
Bhagat, M. L., tempy. Capt. 1767
—— Jemdr. 1154
Bhagat Ram, Subadar 1052
—— Sub-Asst. Surgn. 1815
—— Khattri, Subdr. 1642, 1655
—— Singh, Subdr. 1231
—— *Hony.* Capt. 2104
—— Subdr. Maj. 1203
—— Jemdr. 942
—— Ressaldar. 953
—— Jemdr. 1163
—— Jemdr. 1114
—— Jemdr. 948
—— Subdr. 1252
—— Subdr. 1128, 1311
—— Jemdr. 1312
—— Jemdr. 1111
—— Jemdr. 1160
—— Ressdt. 970
—— Jemdr. 1167
Bhagatbir Gurung, Subdr.-Maj. 1596
—— Thapa, Subdr. 1596, 1621
—— Jemdr. 1608, 1609, 1614
Bhagautidin Dube, Jemdr. 1019
Bhagavati, Subdr. 1307, 1591
Bhagel Singh, Jemdr. 1249
Bhageo Singh Rawat, Subdr. 1181
Bhagesur Gurung, Jemdr. 1633
Bhagiman Limbu, Jemdr. 1649
—— Subdr. 1630
Bhagirath Limbu, Subdr. 1596
Bhagirath Singh 1471
Bhagirathi Singh, Jemdr. 1404
Bhagi Singh, Jemdr. 1062
Bhagmal, Jemdr. 938
—— Jemdr. 1039
Bhagta Singh, Jemdr. 946
—— Jemdr. 1175
Bhagu Singh, Jemdr. 1087
Bhaguta, Subdr. 1191
Bhagat Bir Ale, Jemdr. 1608
Bhagwaiparshad Tewari, Subdr. 1018
Bhagwana, Jemdr. 1059
—— Jemdr. 1320
Bhagwan-Bakhsh Singh, Subdr. 1061
Bhagwandas, Sub-Asst. Surgn. 1814
Bhagwan Sahai Ressdr. 925
Bhagwan Sing Chand, Subdr. 1645
—— Singh, Sub-Asst. Surgn. 1812
—— *Hony.* Capt. 2104
—— Senr. Sub-Asst. Surgn. 1807
—— Jemdr. 888
—— Subdr. 1078
—— Risldr. 918
—— Sr. Sub-Asst. Surgn. 1808
—— Jemdr. 912
—— Jemdr. 880
—— Jemdr. 880
—— Jemdr. 894
—— Jemdr. 1495
—— Subdr. 1385
—— Sub-Asst. Surgn. 1815
—— Ressdr. 965
—— Sub.-Maj. 1074, 1401, 1493
—— Jemdr. 1122, 1137
—— Tr. Vety. Asst. 1687
—— Jemdr. 892
—— Jemdr. 1495
—— Jemdr. 1031, 1075, 1122
Bhagwant, Jemdr. 1059
Bhagwant Lal Medan, Sub-Asst. Surgn. 1817
Bhagwant Rao, D., Sub-Asst. Surgn. 1825
—— Singh, Jemdr. 895
Bhagwat Singh, Subdr. 1412
Bhagwant Singh Lal, 2nd-Lt. 481
—— Tewari, Jemdr. 1018
Bhai Kartar Singh Gaga, Sub-Asst. Surgn. 1818
Bahira, Jemdr. 1200
Bhairab Datt Joshi, Sub.-Asst. Surgn. 1819
—— Singh Gurung, Jemdr. 1619
Bhaironparshad Pande 1017

NAMES AND RANK. Page.

Bhairu Singh, Jemdr. 1583
Bhajan Singh, Jemdr. 888
—— Jemdr. 951
Bhajjan Singh, Jemdr. 1369
Bhal, N. S., Tempy. Capt. 1767
Bhailal Dahyabhai Amin, Sub-Asst. Surgn. 1832
Bhailal Motilal Shah, Sub-Asst. Surgn. 1832
Bhairam Singh, Jemdr. 1396
Bhairo Singh, Sub-Maj. 2104
Bhajekar, V. N., Ty. Lt.-Col. 1763
Bhaiya Lal, Sub-Asst. Surgn. 1811
Bhaju Singh, Risldr. 878
Bhaktabir Limbu, Jemdr. 1633
Bhal, P. L., Tempy. Capt. 1768
Bhalchandra Sakharam Shede, Sub-Asst. Surgn. 1830
Bhalla, D. R., Tempy. Capt. 1772
Bhinbir Gharti, Jemdr. 1608, 1653
Bhandare, V. G., tempy. Capt. 1769
Bhandari, M. G., Tempy. Capt. 1765
—— S. L., Ty. Lt. 1774
Bhandarkar, P. R., Ty. Lt. 1778
Bhan Kadam, Jemdr. 1424
Bhansali, N. P., Lt. 1295
Bhan Sase, Jemdr. 1699
Bhan Singh, Jemdr. 1698
—— Jemdr. 1698
—— Subdr. 989
—— Jemdr. 898
—— Jemdr. 1368
—— Jemdr. 912
—— Jemdr. 895
—— Jemdr. 976
—— Jemdr. 1260
Bhanga Singh, Subdr. 1157
—— *Hony.* Capt. 2101
Bhani Singh, Jemdr. 902
Bhansali, N. P., Ty. Lt. 1780
Bhanwar Singh, Jemdr. 934, 1498
Bhao, Rao-savant, Jemdr. 1012
Bharatpur, Lt., *Maharaja of* 112
Bharat Singh, Subdr. 1060
—— Jemdr. 906, 914
—— Jemdr. 1024
—— Ressdr. 926
—— Jemdr. 963
—— Bhandari, Subdr. 1624
—— Risldr. 906
—— Jemdr. 902
—— Jemdr. 890
—— Thapa, Jemdr. 1602
Bhargava, D. P., Ty. Lt. 1771
—— M. P., Ty. Capt. 1780
Bhargava, M. L., Tempy. Capt. 1767
Bharpur Singh, Jemdr. 878
Bharucha, D. H., Ty. Lt. 1781
—— J. C., Capt. 1757
—— J. C., Ty.-Lt. 1443, 1773
—— K. B., Tempy. Capt. 1764
—— P. B., Capt. 1756
—— P. C., Ty. Capt. 1769
Bharucha, R. H. Capt.
Bharu Singh, Risldr 1683
Bhasin, D. R., Ty. Lt. 1776
Bhasker Gurung, Jemdr. 1602
Bhat, A. R. Ty. Lt. 1780
—— R. P., Ty. Lt. 1780
—— S. V., Ty. Capt. 1765
Bhatavadekar, D. M., Tempy. Capt. 1771
Bhathena, S. J., Capt. 1754
Bhatia, K., Tempy. Capt. 1770
—— S. L., Tempy. Lt. 1773
Bhattacharji, S. N., Ty. Capt. 1780
Bhatji, B. R., Ty.-Lt. 1778
Bhattacharya, K. C., Ty.-Lt. 470, 1777
—— S. C., Tempy. Lt. 1772
Bhattacharyya, R., Ty.-Lt. 1774
Bhatt, R. B., Sub-Asst. Surgn. 1832
—— S. T., Sub-Asst. Surgn. 1832
Bhau Lole, Jemdr. 1416
Bhaulra, Sing Gurung, Jemdr. 1602
Bhavnagar, Maharaja of, Lt.-Col. 110
Bhavaneshwar Singh Rajnet, Sub-Asst. Surgn. 1819
Bhawa Singh, Subdr. 1379
Bhawal Din, Jemdr. 1412, 1681
Bhawan Sing Chettri, Jemdr. 1649
—— Singh Kunwar, Jemdr. 1701
—— Nazarkoti, Jemdr. 1224
—— Rana, Jemdr. 1608
Bhawanibaksh Singh, Jemdr. 1063
Bhawan Sing Thapa, Subdr. 1618
—— Rawat, Subdr. 1179

NAME AND RANK. Page.

Bhawani Dat, Jemdr. 990
—— Datt Misra, Sub-Asst. Surgn. 1820
—— Din Pande, Jemdr. 1054
—— Singh, Subdr. 999
—— Subdr. 1387
—— Jemdr. 914
Bhawani Parshad Singh, Subdr. 1043
Bhewa Sarafdar, Subdr. Major 1435
Bhikham Singh, Subdr. 1239
Bhiko, Jemdr. 1121
Bhikhuji Thakur, Jemdr. 1416
Bhima Rawat, Subdr. 1483
Bhimbahadur Bania, Subdr. 1642
Bhim Bahadur Thapa, Jemdr. 1602
Bhimraj Gharte, Jemdr. 1608
Bhimrao Nagojirao, *Hony.* 2nd-Lt. 481
Bhim Sing Gharti, Subdr. 1596
Bhim Sing, Subdr.-Maj. 1607
—— Singh, Jemdr. 1488
—— Subdr. 1483
—— Jemdr. 1044
—— Jemdr. 1035
—— Jemdr. 1384
—— Jemdr. 1258
—— Rawat, Jemdr. 1179
Bhiru Ram, Jemdr. 1433
Bhiwa Mhapdi, Jemdr. 1468
Bhiu Singh, Jemdr. 1420
Bhogbir Limbu, Subdr. 1630
Bholanath Awasthi, Subdr. 1026
—— Pande, Subdr. 1018
Bhola, Subdr. 1059
—— Jemdr. 1059
—— Dat, Jemdr. 1241
—— Nauth, Lt.-Col. 104, 108, 130, 1740
—— Ram, Jemdr. 1488
—— Singh, Subdr. 1066
—— Subdr. 988
—— Subdr. 1111
—— Kandari, Subdr. Maj. 1648
—— Ressdr. 900
—— Jemdr. 1299
—— Subdr. 1165
Bholar, Jemdr. 989
Bhollu, Jemdr. 1193
Bholu, Jemdr. 1192
Bhondu Sawanth, Jemdr. 1424
Bhoop Singh, Jemdr. 1052
Bhopal, Muhammad Nasrulla Khan, *Nawab* of 110, 1050
Bhopat Ram, Subdr. 1419
Bhowan Singh Gurung, Subdr. 1598
—— Sahi, Subdr. 1007
Bhowanand Thapa, Jemdr. 1619
—— Jemdr. 1651
Bhuja Ram Saha, Jemdr. 1627
Bhulan Khan, Jemdr. 990
—— Singh, Jemdr. 930, 959
Bhumbhwal, R. L., Tempy. Lt. 1780
Bhumendra Mohan Roy, Jemdr. 1219
Bhupal Thapa, *Hony.* Capt. 2101
Bhupendra Narayan Choudhary, Srijut, 2nd-Lt. 481
—— Nath Ghosal, Jemdr. 1219
Bhupendranath Lahiri, Sub-Asst. Surgn. 1815
Bhot bahadur, Pande, Jemdr. 1643
Bhup Chand, Sub-Asst. Surgn. 1813
—— Sing Chauhan, Jemdr. 1182
—— Singh, Jemdr. 1312
—— Jemdr. 1232
—— Jemdr. 1165
—— Subdr. 1581
Bhupwarian Dube, Jemdr. 1019
Bhur Singh, Subdr. 1070
—— Subdr. 1193
—— Jemdr. 926
—— Jemdr. 1054, 1190
Bhura, Subdr-Maj. 1199
—— Subdr. 1199
—— Jemdr. 1197
—— Ram, Subdr. 1488
—— Singh, Subdr. 1034
—— Jemdr. 1455
—— Jemdr. 1021
—— Subdr. 1031, 1452
—— Khan, Jemdr. 938
—— Singh, Subdr. Maj. 1055
Bhurke, G. M., Tempy. Lt. 1773
Bhushan, V., Tempy. Capt. 1768
Bhuwandiwala, K. S., Ty. Lt. 1779

NAMES AND RANK. Page.

Bhutal Singh, ... 1597
Bibby, E., 2nd-Lt. . . . 356, 1126
—— J. W. 2nd-Lt. 360
Bichant Singh, Jemdr. 1493
Bichha Ram, Sub-Asst. Surgn. 1811
Bichitar Singh, Rissaldar. . . 896
—— —— Jemdr. 936
—— —— Subdr. 1106
Bickel, T. H., 2nd-Lt. . . 358, 1157
Bickell, E. G., Lt. 681
Bicker, H. J., Lt. 352
—— L. J., 2nd-Lt. 362
Bickers, A. A. D., 2nd-Lt. . 561, 1110
—— S. M. 2nd-Lt. . . 559, 1677
Bickersteth, E. H., Lt. . . 368, 1417
Bickford, M. H., Capt. 94, 262, 1174
—— W. W., Lt.-Col. 15, 21, 138, 160, 199, 871, 1431
—— Smith, W.N.V. Capt. 789
Bickley, G. W., 2nd-Lt. . . . 666
Biddle, W. C., Lt. 526, 1671
Biddlecombe, G. L., Lt. . . . 353
Biddulph, J., Col. . . . 178, 1917
—— S. F., Col. . . 33, 78
Biden, S. W., Lt. . . 506, 857, 1215
Bidhi Chand, *Hony.* Capt. . 5, 2102
—— —— Jemdr. 922
Bidhu Bhusan Bose, Sub-Asst. Surgn. 1815
Bidie, A. G. C., Capt. . . . 279, 893
—— G., Lt.-Col. . . . 130, 1742
Bidwell, R. F., Lt.-Col. 136, 158, 196, 1407
Biggam, A. G., Capt. 1714
Bigge, C. F. E., 2nd-Lt. . . . .
Bigger, T. A. L., Lt. . . . 311*a*, 1263
Biggerstaff, T. J., Sub-Condr. . 1863
Biggin J. V., 2nd-Lt. 707
Bigham, R. A., Lt. 787
Biggs, F. W., Lt. 752
—— J. B., Sub-Condr. . . 1855
—— W. 2nd-Lt. 568, 1678*a*
Biggs-Atkinson, T. B., Capt. 46, 767
Bigg-Wither, F., Maj. . 104, 108, 206, 1894, 1912
—— F. L., Lt.-Col. . . 832
Bignell, B. H., Maj. . . . 210, 1641
—— G. N., Capt. . . . 260, 1136
—— R. L.. Maj 222, 869, 1187, 1189
Bignold, C. A., Capt. . . . 35, 1717
Bigsby, S. H., Lt. . . . 29, 532, 1694
Bigwood, W., Lt. 405, 905
Bihari Sing, Jemdr. 1022
Bijai Singh, Subdr. Maj. . . . 1371
—— Jemdr. 1579
—— Jemdr. 1032
—— Rawat, Jemdr. . . 1181
——, Sidut Singh, Lt. 112
—— Bahadur Singh, Subdr. . 1395
—— —— Jemdr. . . . 1510
Bijan Singh, Jemdr. 1472
Bije Bahadur Singh, Subdr. 1046
Bikaner, H. H., Maharajah *Sir* Ganga Singh Bahadur, *Hony.* Maj. Gen. 1, 110, 877
—— Maharaj Kumar of, Lt. . 112
Bikarma Singh, Jemdr. . . . 1702
Bikram Singh, Jemdr. 1480
—— Jit Singh, Jemdr. . . . 1043
Bilaspur, Rajatof, Capt. . . 111, 1187
Bilderbeck, A. C. L. O'S., Cap. 1758
Bill, E. L., Lt. 525, 1698
—— S. A., Revd. 1904
Billbikram Rana, Subdr. . . 1618
Billimoria, S. D., Tempy. Capt. . 1765
Billing, E. W., Lt. . . . 322, 1098
Billingham, W. F., Lt. . . 1537, 1497
Billington, P. R., 2nd-Lt. . . . 818
—— T. F., 2nd-Lt. . . . 756
Billson, H. G., Capt. 609
—— J. V., Lt. 543, 1279
Bilsborrow, W. G., 2nd-Lt. . . . 788
Binda Singh, Subdr. 1066
—— Subdr. 1157
—— (I) Subdr.. . . . 1589
—— *Hony.* Capt. . . . 2104
Binda Prasad, Jemdr. 1111
Bindeu Pandhari-nath Gobdale, Sub-Asst. Surgn.. 1830
Bindra, S. Ty.-Lt. 1781
Bindu Singh, Subdr. 1419
Binge, F. G. H., 2nd-Lt. . 559, 1144
Bingemann, L. S., 2nd-Lt. . 449, 905
Bingham, C. D'A., Capt. 68, 272, 1439
—— E. G., *Hony.* Lt. . . . 1847
—— W. E., Maj. 103, 105, 106, 107, 211, 1287

NAMES AND RANK. Page.

Bingley, A. H., Maj.-Genl. *Sir* 4, 118, 146, 173
Binney, A. L., 2nd-Lt. 630
—— E. V., Maj. . . 664, 996
Binning, A., Ty. Capt. 1726
——, H. A., Lt. . . . 18, 42. 492
——, J. A., Capt. 1715
Binns, B. O., 2nd-Lt. . . . 540, 1052
Binny, A. C. M., Capt. . . 258, 1090
—— A. J M., Maj. 15, 29, 98, 215, 875
Binstead, V. J., Lt. . . . 534, 1355
Bion, A. E., Lt. 490
—— C. R. F., 2nd-Lt. . . 562, 1601
—— R. H., Lt. 310, 1251
Bir Ali, Jemdr.. 966
Bir Bal Malhantra, Sub-Asst. Surgn. 1819
Bir Sing Danu, Subdr. . . . 1179
—— Singh, Subdr. 999
—— Sub-Asst. Surgn. . 1808
—— Subdr. 1589
—— Subdr. 1367
—— (I) Jemdr. 1232
—— (II) Jemdr. 1232
—— Jemdr. 932
—— Senr. Sub-Asst. Surgn. 1808
—— Subdr. 1267
—— Subdr. 1078
—— Jemdr. 1240
—— Subdr. 1112
—— Jemdr. 1112
—— Jemdr. 1589
Birad Singh, Jemdr. 1497
Biragi Ram, Jemdr. 1312
Biraj Mohan Ghosh, Sub-Asst. Surgn. 1816
Biraj Gurung, Jemdr. 1619
Birbal, Subdr. 1452
—— Nagarkoti, *Hony.* Capt. . 2103
—— Singh, Jemdr. 1031
Birbahadur Limbu, Subdr. 1596, 1623
—— Subdr. . . . 1633
—— Lama, Subdr. . . . 1627
Bir Bal Malkanutra, Sub.-Asst. Surgn. . 1818
Birban Gurung, Jemdr. 1637
Birbhan Gurung, Jemdr. . 1608, 1653
—— Subdr. . . . 1636
Birch, A., 2nd-Lt. 352
—— A. R., 2nd-Lt. 652
—— C., 2nd-Lt. 565
—— C. F. M., Maj. . 17, 34, 246, 1595
—— O. W., Revd. 1904
—— V. K., Maj. 108, 209, 872, 929, 1559
—— W. S., Capt. 1717
Birch-Reynardson, H. T., Capt. . 745
—— M. H., Lt. . 585
Birchall, J. R., Capt. 697
—— S., 2nd-Lt. 788
Bircumshaw, F. R., Lt. . . . 707
Bird, A. . ., Capt. . . . 255, 1663
—— A. G., 2nd-Lt. . . . 545, 1241
—— A. J. G., Maj. . . 664, 985
—— C. A., Maj. . . . 665, 985
—— C. F. A., 2nd-Lt. . . . 686, 986
—— C. J. G., Maj. . . . 87, 217
—— E. A., Lt. 826
—— D. R., 2nd-Lt. 703
—— E. K., Capt. . 278, 1098, 1136
—— G. G., Lt. 302, 1307
Bird, W. H., 2nd-Lt. 788
—— H., 2nd-Lt.. 1653
—— H. W., 2nd-Lt. 361
—— J. G., Ty. Lt. 1759
—— M., Lt. 423, 1479
—— R., Lt . . . 18, 37, 696, 796
—— R. J., Lt. 486, 941
—— T. E. H., 2nd-Lt. . . . 90, 751
—— W. A. H., Capt. . 252, 1110, 1150
—— W. F., Maj. . . . 242, 1663
—— W. H., Lt. 760
—— W. T., Capt. . . 272, 931
Birda, Subdr. 1193
Birdhoj Limbu, Subdr. . . . 1630
Birdwood, C. B., 2nd-Lt. . . 51, 366
—— F. T., Lt. . . 299, 1166
—— G. T., Lt.-Col. . 130, 1742
—— H. B., Col. 12, 42, 121, 126, 150, 196, 867, 925
—— W. R., Genl. *Sir* 1, 101, 118, 145, 171
Birendra Nath Das, Sub-Asst. Surgn. . 1816
Bireswar Mukerji, Jemdr. . . 1219
Birihang Limbu, Subdr. . . . 1630

NAMES AND RANK. Page.

Birkett, A. N., Lt.. 537
Birkett, W. G. N., 2nd-Lt. . . . 741
——, S., Lt. 523, 1479
Birkha Singh, Jemdr. 930
Birkhadoj Limbu, Jemdr. . . 1609
Birkhsing Thapa, Subdr. . . . 1596
—— Sub-Maj. 1657
Birkin, W. N., Lt. . . . 435, 1443
Birkmyre, Hon. Sir A. 603
—— A., 2nd-Lt. 788
Birks, O. M. M., 2nd-Lt. . . . 759
Birley, F., Lt. 524
—— N. A., 2nd-Lt. 595
Birman Rai, Jemdr. 1649
Birmingham, W. R., Lt. . . . 637
Birnarain, Jemdr. 926
Birsing Khauka, Jemdr. . . . 1622
Bir Singh, Subdr. . 1268, 1231, 1367
Bir Sing Danu, Subdr. . . . 1179
Bir Singh, Jemdr. . . . 999, 1250
Birney, C. F., Bt. Lt.-Col. 60, 139, 663 1883
—— H., Lt. 443
Birnie, A., Lt. 544, 1189
Birrell, N., Capt. 859
Birt, A. G., Lt. 511, 1319
Birthistle, R. G. N., 2nd-Lt. . . 1070
Birtles, H., Lt. 617
Birtwistle, J., Lt. 68, 308, 1007, 1060
—— R. K., Lt. 542
Bisakha Singh, Jemdr. . . . 1157
—— Jemdr. . . . 1151
Bisal Sing, Jemdr. 884
Biscoe, H V., Maj. 53, 102. 221, 1898, 1913
—— J. V. M., Maj. 237, 1453, 1883
—— W. T., Lt. 61
—— W. W., Lt.-Genl. . 911, 1955
Bisesar Tewari, *Hony.* Capt. . . 2104
Bishan, G., Tempy. Lt. 1777
—— Das Minocha, Sub. Asst. Surgn. 1819
Bishanbahadur Chand, Jemdr. 1643
Bishanbir Thapa, Subdr. . . . 1624
Bishanda, Sharma, Sub-Asst. Surgn. 1817
Bishan Das, Ty. Sub-Asst. Surgn. 1833
Bishan Das Minocha, Sub-Asst. Surgn. 1819
Bishan Ran, Jemdr 1027
Bishan-Singh, Sub-Asst. Surgn. . 1812
—— Sub-Asst. Surgn. . 1817
—— Jemdr. 1204
—— Jemdr. 1012
—— Jemdr. 906
—— Jemdr. 1704
—— Subdr. 1111
—— Subdr. 1157
—— *Hony.* Capt. 1896, 2104
—— Jemdr. 1151
—— Jemdr. 1114
—— Jemdr. 1150
—— Ressdr. 944
—— Subdr. 1211
—— Subdr. 1145
—— Jemdr. 990
—— Jemdr. 1565
—— Jemdr. 898
—— Jemdr. 1204
—— Jemdr.. 1240
—— Jemdr. 912
—— Subdr. 1078
—— Jemdr. 1372
—— Kathait, *Hony.* Capt. 2103
—— Jemdr. 1078
—— Subdr. 1395
—— Subdr.-Maj. . . . 1642
—— Jemdr. 886
—— Subdr. 1249
—— Subdr. 966, 1166, 1433
—— Bisht, Jemdr. . . . 1608
—— Jemdr. 936
—— Ressdr. 888
Bishen Dass, Jemdr. 1171
Bishen Singh, Risldr. 893
—— *Hony.* Capt. . . . 1896
Bisher, D. R., 2nd-Lt. . . 359, 1088
Bishn Singh, Subdr.-Maj. . 754, 1148
—— Subdr.-Maj. . . . 754
—— Jemdr. 1280
Bishnath Saran Singh, 2nd-Lt. . 482
Bishnu Thapa, *Hony.* Capt. . . 2101
Bishop, E. B., Col. 56, 178
—— C. H., 2nd-Lt. 786
—— E. C. G., Lt. . . 455—59, 883
—— F. J., 2nd-Lt. . . 358, 1569
—— H. C., Lt. . . 533, 682, 1271
—— H. C. W., Lt. . . 508, 1885

Names and Rank. Page.

Bishop, G. J., Hony. Lt. 653
—— G. W. R., Capt. . . 702, 737
—— H. P., 2nd-Lt. . . 359, 1469
—— H. G. S., 2nd-Lt. . . . 1187
—— L. A., 2nd-Lt. 523
—— T. C., 2nd-Lt. 626
—— T H., Surgn.-Capt. . 1763
—— W. E., 2nd-Lt. . . . 359, 1016
—— W. H.. A, Lt. 736
—— Smith, J., 2nd-Lt. . 559, 1678
Bishun Das Sawhny, Sub-Asst. Surgn. . 1814
—— Datt Dube, Subdr. . . 1363
—— Singh, Jemdr. 1256
—— Jemdr. 1236
Bisley, F. de B., Lt. . . . 518, 1144
Bisni, S. F., Tempy. Capt. . . . 1770
Bispham, G., Lt. . . . 520, 1666
Biss, E. E., Capt. 538, 807
—— J. C. de V., Capt. . . . 485
Bissambar Singh, Subdr. . . 1403
Bissesar Singh, Jemdr. . . 1403
Bisset, E., Maj. 1748
—— J. D., 2nd-Lt. 1195
—— W. S. S., Capt. 844
Biswas, S. C., *Rai Sahib* . . . 93
Bisweswar Bhattacharji . . . 91
Bitton, D., 2nd-Lt. 360
Blaber, H. L., 2nd-Lt. 1291
—— H. K., 2nd-Lt. 328
Black, A. B., Maj. 693
—— A. G., Lt. 1537
—— A. S., 2nd-Lt. 1674
—— D. G., Lt. 629
—— G. B., Lt. 595
—— G. C. S., Maj. . . 238, 1016
—— H., Lt. 533
—— J. A., Lt.-Col. . . 133, 1743
—— J. G. McF., Lt. . . . 732
—— J. M., Lt. . . . 440, 1509
—— J. S. C. Capt. 935
—— K. Tempy. Capt. . . . 1725
—— R., Lt. . . . 298a, 1174, 1187
—— R. G., 2nd-Lt. 531
—— W. C., Maj.-Genl. . . . 1407
—— W. C., Col. . 69, 97, 105, 121, 126, 127, 150, 192
Blackburn, B. M., Lt. 300, 1263, 1265
—— D. A., 2nd-Lt. . 562, 1703
—— F., 2nd-Lt. 363
—— H., 2nd-Lt. 361
—— J., Capt. 1839
—— M. E. M., Lt. . 525, 1671
Blackburne, G. I., Capt. . . . 787
—— H. H., 2nd-Lt. . 494, 911
Blackden, J. F., 2nd-Lt. . . 329, 933
Blackhurst, J. J., 2nd-Lt. . . 788
Blacker, H. A. C., Lt. . 540, 805
—— L. V. S., Capt. . . 263, 964
—— W. F., Capt. . . 264, 943
Blackett, C. P., Capt. . . 260, 1623
—— W. S., Lt. . . . 528, 1704
Blackham, J. C., Lt. . . 493, 901
Blackie, F. B., Lt. 528
Blackley, J., 2nd-Lt. 781
Blackman, J. G., Lt. 720
Blackmore, E. W. H., Lt. 446—49, 1064
—— H. S., Capt. . . . 1713
—— J. P., Lt. . . 353, 1455
—— L. A., Capt. 1718
—— L. W. S., Lt. . 731, 798
Blackwell, C. T., Lt.-Col. . . 1718
—— B., Capt. . . 256, 1427
—— J. M., Capt. 706
—— J. N., 2nd-Lt. . . 332, 881
—— P. W., 2nd-Lt. . . . 729
Blackwood, A., *Hony.* Maj. . . 1844
—— C. V. R., Lt. . 322, 1629
—— J. R., Capt. . . 488, 830
—— J. R., 2nd-Lt. . . . 856
—— J. R., 2nd-Lt. . . . 856
—— N. P. O'R., Lt. . 522, 841
Bladwork, H. M., 2nd-Lt. . . 1052
Bladworth, H. M., 2nd-Lt. . . 1024
Blagg, F. O., Lt. . . . 731a, 791
Blaikie, C. J., Capt. 1715
—— H. M., Lt. 525
Blair, A., 2nd-Lt. 763
—— A. E., 2nd-Lt. 744
—— D. A., Maj. . . . 220, 1178
—— F. R., Lt. . . . 526, 1074
—— G. L., Maj. 103, 104, 105, 225, 1166
—— H. A., Capt. 628
—— J. M. 720, 1006, 1507
—— J. R., Lt. . 18, 21, 510, 931
—— R. W. C., Maj. . . 106, 205, 1487

Blair- R. W., Lt. 1693
—— White, J H., Lt. 375
Blake, A. G., Lt. 783
—— A. G., Lt. 293, 1311
—— A. L., Lt. 1004
—— D., Lt. 323, 543a
—— G. A., Capt. 1715
—— G. R., Capt. 756
—— H. W., Lt. . . . 529, 1699
—— J. B., 2nd-Lt. 851
—— L. F. M., Lt. . 528, 638, 1587
—— N. D., Lt. 705
—— R. A. E., Lt. . . . 731, 1004
—— S. B., 2nd-Lt. 361
—— W., Lt. 505, 825
—— W. J., 2nd-Lt. 568
Blakeney, E. C. W., Lt. 323, 959, 1303
—— S. R. St. L., Lt. . 317, 1335
—— W. E. A., Col. . . 123, 149, 183, 708
Blaker, G. H., Asst. Surgn. . . 1707
—— H. B., Asst. Surgn. . . 1796
—— H. B., Asst. Surgn. . . 1800
—— P. S., Tempy. Capt. . . 1764
—— W. F., Maj. 52
Blakesley, W. P., 2nd-Lt. . . . 754
Blakeway, D. B., Lt.-Col. 192, 1900, 1912
—— J. P., Col. . 125, 660, 1881
Blakiston, L. W., Capt. . . 691, 783
—— J. F., Lt. . . . 488, 947
Blakiston Houston, J. E., Capt. 589
Blampied, H. J., 2nd-Lt. . . . 744
—— T., Leg., Lt. 1839
Blanc, L. G., Lt. . . . 311a, 1311
Blanch, N. H., 2nd-Lt. 630
Blanche, F. E., 2nd-Lt. 742
—— R., 2nd-Lt . . . 358, 788
Blanchford, M. W., 2nd-Lt. 356, 1032
Bland, E. G., Capt. . . . 501, 1375
—— E. M., Lt. . . . 410, 1367
—— G. H., Lt. . . . 307, 1423
Blane, A. P., 2nd-Lt. 786
Blandy, R., Capt. . 261, 1641, 1657
—— R. D., Lt. 511, 965
Blanford, A. W., Lt. . . 513, 1517
—— P. T., Capt. . 96, 252, 1343
Blank, A. L., Lt. . . 72, 524, 1884
Blaquiere, N. A., 2nd-Lt. . . . 557
Blascheck, A. D., Lt. 605
Blaxall, M. L., 2nd-Lt. . . 685, 784
Blaxland, A. B., Capt. . . 277, 1082
Blaylock, J., 2nd-Lt. . . 552, 1263
Blayney, A., 2nd-Lt. 1265
Blazey, W. H., Asst. Surgn. . . 1798
Bleathman, W. E. P., 2nd-Lt. . 463, 1293
Bleeck, E., 2nd-Lt. 833
Blenkinsop, B. A., 2nd-Lt. . . 558
—— F. L., Lt.-Col. 133. 1743
Bles, D. G., Lt. 493
Bletsoe, W. B., Lt. . . . 389. 909
Blewitt, A.E., 2nd-Lt. . . 560, 1195
Bli Ram, Sub-Maj. 2103
Blick, L. H., Lt. 90, 542
Blinkworth, K. W., Asst. Surgn. . 1794
Bliss, C. V., Capt. 829
—— J. A., Lt.-Col. 135, 158, 195, 870, 1275
—— J., 2nd-Lt. 741
—— R. L., Capt. 857
Blizard, W. E., Lt. 524
Blockley, J. F. G., 2nd-Lt. . . 1676
Blockley, J. F. G., 2nd-Lt. . 356, 921
Blofeld, G. S., Sub-Condr. . . 1855
—— V. A. G., Asst. Surgn. . 1803
Blomeley, C., Lt. . . . 387, 1152
Blomenstock, L. M., Lt. . 42, 622
Blomfield, C., 2nd-Lt. . . 567, 1517
—— D. J., Lt. . . . 524, 988
—— E. V., Capt. 846
—— H. G., Lt. 493
—— H. M., Capt. . 685, 783
Blood, B., *Sir* Col. 985
—— C. D., 2nd-Lt. . . 554, 1032
—— J. A., Lt. . . . 380, 1241
—— W. H., Capt. . . 20, 159, 964
Bloomfield, FitzM. J. C., Lt. 314, 877
—— J. A., *Hony.* Capt. . 1846
Bloor, J., Sub-Condr. 1870
Blore, C. G., Lt. . . . 731a, 734
—— A. V. 2nd-Lt. 796
Blount, C., 2nd-Lt. 90
—— C. B., Lt. . . . 738, 791
Blower, J. W., 2nd-Lt. 753
Bloxham, H. F., Capt. 59, 501a, 755, 1411
Bloxsome, W. A., 2nd-Lt. . 354, 1339

Blucke, R. S., 2nd-Lt. 737
Blunden, A. C., 2nd-Lt. . . 553, 1521
—— H. G., Capt. 779
Blunt, E. A. H., Capt. 834
Blyde, W. A., Lt. . . . 446—49, 875
Blyth, C., 2nd-Lt. 790
—— E. W., 2nd-Lt. 559
—— F. A., Col. . 57, 122, 153, 180
—— J. C., Lt. . . . 540a, 1207
—— W., Lt. 591
Blythe, J., Hony. Lt. 833
Boal, R., Capt. 666 985
Boardman, J. W., 2nd-Lt. . . . 734
Bobbili, Raja of, 2nd-Lt. . . . 481
Bock, C. C., 2nd-Lt. . . . 326, 931
Bocquet, R. N., Maj. . . 11, 664
Bodas, N. M., Tempy. Capt. . . 1769
Boddam, E. H. C., Lt.-Col. 16, 35, 97, 131, 155, 191, 1617
Boddington, E.S.St.C., 2nd-Lt. 546, 1489-90
Boddy, G. W. 2nd-Lt. 691
—— P. C., 2nd-Lt. 567
—— S. M., Lt. 516, 1385
Boden, J. F. W., Lt. . . 519, 1567
Bodenham, A., Lt. . . . 298, 1187
—— K. L., 2nd-Lt. 327, 1094, 1694
Bodh Raj, Subdr. 1126
—— Jemdr. 932
Bodily, G. R., Lt. 622
Bodkin, L. F., Maj. 105, 226, 145, 1455
Bodie, R. T., 2nd-Lt. . . 324, 1058
Bodley, C. A., Lt. 354
—— L. O., 2nd-Lt. 491
—— R. V. C., Hony. Capt. . . 103
Body, E. U., Maj. 685
—— G. A., 2nd-Lt. . 578, 685, 784
—— J., Capt. 685
Boermel, R. A., Sr. Asst. Surgn. . 1789
Bogarede, L. A. F. V., 2nd-Lt. . 566
Boggs, A. B., Capt. . . . 277, 619
—— G. B., 2nd-Lt. 929
—— W. W., 2nd-Lt. . . 666, 996
Bogie, D., Capt. 858
—— J. S., Maj. 206, 867, 964, 958
—— H. C. L., 2nd-Lt. . . . 993
Bogh Singh, Jemdr. 1075
—— Jemdr. 1347
Bogha Singh, Jemdr. 1368
Bogle-Smith, J A., 2nd-Lt. . . 581
Bogue, W., 2nd-Lt. 1695
Bollard, E. H., Asst. Surgn. . . 1795
Boileau, C. M., Lt. . . . 443, 945
—— G. R. P., Lt.-Col. 66, 105, 133, 157, 195, 872, 1600
—— G. H., Lt.-Col. . . 135, 661
—— W. L. S., Lt. . . 292, 1619
Bois, H. E., 2nd-Lt. . . . 491, 929
Boisragon, G. H., Col. . 124, 150, 185
Bo-kin-Po, Subdr. 993
Bo chan, Subdr. 993
Bo Pan Bu, Jemdr. 994
Bo Sugalay 993
Bo Sugyee 993
—— Than, Subdr. 993
Bolam, A. R., Sub-Condr. . . 1855
—— J E., the Revd. . . . 1908
Bolan, W., 2nd-Lt. 564
Bolckow, H. C. R., Lt. 583
Boldero, A. J., Capt. 737
Boles, D. C., Capt. . . . 55, 595
Boling, L. C., 2nd-Lt. 623
Bolitho, E. H. W., Maj. . . . 614
Bolland, J. F., Condr. 1861
—— L. G. S., Lt. . . 375, 1046
Bollen, J. P., 2nd-Lt. 788
Bolster, R. C., Maj. . . 24, 507, 605
Bolton, A. H., Lt.-Col. 129, 154, 187
—— E. F., Lt. 681, 785
—— F., Lt. 626
—— J. S., Sub-Condr. . . . 1856
—— R. H. D., Lt. . . 296, 1409
Bolus, E. J., Capt. 835
Bombahadur Gurung, Subdr. . 1627
—— Khattri, Jemdr. . 1643
—— Subdr. . . . 1645
—— Limbu, Subdr. . 1633
Bomford, T. L., Capt. 1755
Bonar, A. W. T., 2nd-Lt. . . . 765
—— H. H., 2nd-Lt. 1679
Bond, C. L., 2nd-Lt. 779
—— E. C. J., Capt. 815
—— F. S. 558
—— G. W., Maj. . . 32, 224, 1662
—— G. W. 2nd-Lt. 605
—— H. H., Lt. 1694
—— K. H., Lt. . 399, 1513, . 1515

Names and Rank.	Page.
Bond, L. V., Maj.	664, 985
—— O. H., 2nd-Lt.	551
—— R. F. G., Lt.-Col.	27, 132, 661
—— S. J., Lt.	696
Bone, A. F., Capt.	106, 1910
—— C. B., Capt.	38, 746
—— C. C., 2nd-Lt.	669
—— J. H., Lt.	587
—— J. H., 2nd-Lt.	599
—— W. B., 2nd-Lt.	556
Bonham-Carter, B. H., Capt.	271, 1102, 1183
—— G., Capt.	783
Boniface, C. F., Lt.	360
Bonnar, A. H., Sr. Asst. Surgn.	1787
Bonner, E. H., 2nd-Lt.	686
—— F. J., Lt.	352, 1224
—— T. W., Lt.-Col.	818
Bonney, G. S., 2nd-Lt.	1587
Bonnerjee, K. K. E., Capt.	728
Bonniface, B. H., 2nd-Lt.	564, 1124
Bonnyman, F. J. C., Maj.	759, 795
Bonser, A., 2nd-Lt.	623
Bonsor, R. D., Lt.	781
Bonthrone, A., 2nd-Lt.	764
Boodrie, E. H. B., Asst. Surgn.	1802
Booker, G. A., 2nd-Lt.	550, 1702
Bookey, J. T. B., Lt.	291, 1231, 1624
—— A. C., Lt.	1008
Bookless, A. C. 2nd-Lt.	1008
—— G. T., Capt.	789
Booley, G. H., 2nd-Lt.	565, 1088
Boome, E. H., Brig.-Genl.	59, 106, 121, 126, 153, 191
Boorman, G. H., F., 2nd-Lt.	567
—— J., Lt.	759
—— T., 2nd-Lt.	568
Boor Singh, Jemdr.	1681, 1683, 1704
—— Ressdr.	906
Booth, A., Sub-Condr.	1864
—— A. E., Lt.	486, 937
—— A. H., Lt.	399, 1074
—— A. N., 2nd-Lt.	55
—— A. R., 2nd-Lt.	446—49, 1144
—— E. R. C., Capt.	274, 941
—— F., Bt.-Maj.	1003
—— F. N., 2nd-Lt.	531
—— J. J., Lt.	571, 1696
—— R., Lt.	389
—— T., Lt.	769
—— W. M., Lt.	522, 1667
Boothby, G. A. B. C., Lt.	766
—— R. E., Lt.	597
Booty, F. C., Lt.	389, 1365
Borebank, F. W., 2nd-Lt.	560, 1704
Borissow, F. A., Lt.	61, 489
—— R. W., Capt.	89, 747
Borland, G. D., Lt.	693
Borlase, J., Lt.	509, 1589
—— J. J. D., Revd.	1906
Borrett, T. L., 2nd-Lt.	571
—— J. B., Sub-Condr.	1870
Borrowman, C. G., Capt.	66, 276, 1613
Borthwick, C. J., 2nd-Lt.	354, 1235
Barton, E., 2nd-Lt.	363, 1355
—— W. N., Maj.	58, 664
Borwick, R. H., Lt.	455—59, 1195
Bosanquet, H. P., Lt.	708
—— O. V., the *Hon.* Mr.	1899
—— P. C. M., N. Sister	1735
—— W. C., Ty. Maj.	1722
Bose, A. N., Capt.	1758
—— D. P., Ty. Lt.	1776
—— H. N., Tempy. Lt.	1775
—— K. L., Ty. Lt.	1781
—— L. M. V., 2nd-Lt.	803
—— R, C., Ty. Lt.	1782
—— S., Maj.	1748
—— S. K., Tempy. Capt.	1765
—— Y. M., Ty. Lt.	1775
Bosher, F., Lt.	624
Boshier, L., 2nd-Lt.	1679
Boss, T. H., Lt.	367
Bostan Khan, Subdr.	1280
—— Jemdr.	1480, 1489
—— Subdr.	1565
—— Subdr.	1235
—— Jemdr.	948
—— Jemdr.	955
—— Jemdr.	1344
—— Jemdr.	1440, 1589
—— Jemdr.	918
—— Ressad.	918
—— Jemdr.	1705
Bostock, J. A., Lt.	18, 25, 399, 895
Bostock-Wilson, E. L., Lt.	511, 1672
Boswarva, S. J., 2nd-Lt.	360, 1591
Boswell, A. P., 2nd Lt.	786
—— J. E., Condr.	1853
—— R. B., Lt.	491, 935
—— T. K. C., Asst. Surgn.	1803
—— W. R., Capt.	55, 263, 1132
Bosworth, F. J., Lt.	493, 945
—— J. A., Lt.	1845
—— T. J., Capt.	1839
Bosworth Smith, B. N., Esq.	28
Bothwell, P. J., Lt.	527, 1668
Bott, R. H., Maj.	1753
Botterill, P. W., 2nd-Lt.	77, 767, 1884
—— W. R., 2nd-Lt.	821, 553, 1447
Bottrill, E. R., Lt.	512, 1677
Bottomley, J. E., Lt.	755
Bottu Gurung, Jemdr.	1602
Bouche, B. J., Asst. Surgn.	1707
Boucher, A. S. H., Lt.	544, 1673
—— B. H., Capt.	733, 746
—— C. H., Lt.	312, 1600
—— C. L., Tempy. Capt.	668
——, G. J., Asst. Surgn.	1803
Bouchier, A. C. F., Maj.	57
Boudier, E. W., Brig.-Genl.	14, 43, 120, 125, 150, 185
Boughtflowers, J., Condr.	1853
Boughton, A. E., 2nd-Lt.	360
—— E. J., 2nd-Lt.	473, 1343
Bould, C. H., Sub-Condr.	1863
—— F.	723a
—— H. G., 2nd-Lt.	819
—— R. P., Lt.	818
Boulding, E., 2nd-Lt.	685
Boult, N. C., 2nd-Lt.	551
Boulter, F. C., Lt.	619
—— F. H., Capt.	750, 780
Boulton, C. E., Capt.	264, 1034, 1038
—— H., Lt.-Col.	134, 1745
—— H. B., 2nd-Lt.	721
—— R. A., Lt.	728
Bound, C. G., 2nd-Lt.	1008
Bourdillon, B. H., Lt.	571, 608
—— F. H., Lt.	295, 1108, 1166
—— P. A., 2nd-Lt.	993
Bourke, A. J. H., Lt.	295, 970
—— A. J.	
—— D. R. S., Lt.	523, 849, 998
—— J. J., Lt.-Col.	130, 1742, 1889
—— U. J., Capt.	1759
Bourn, H., Lt.	571
Bourne, F. C., Capt.	17, 23, 751
—— P. R., Lt.	696, 697
—— W., 2nd-Lt.	1708
—— W. FitzG., Lt.-Col.	135, 158, 195, 868, 1038
—— W. H., 2nd-Lt.	323, 1052
—— W. K., Lt.-Col.	97, 135, 158, 210, 869, 913, 1164
Bousfield, B., Maj.	724
—— B. N., Lt.	367, 1287
—— H. T. W., Lt.	506, 1050
Boustead, H., 2nd-Lt.	460
Boutflower, E. C., Capt.	723
Bovet, W., Bt.-Lt.-Col.	996
Bovill, W. J., Capt.	52, 254, 1195
Bowden, A. H. M., 2nd-Lt.	1675
—— C. A., Lt.	624, 976
—— C. C. H., Lt.	320, 947, 1678
—— C. G., 2nd-Lt.	546, 1191
—— E. M., Lt.	683, 685
—— F. H., Capt.	15, 37, 801
—— J. H., Lt.	1007
—— N. H. M., 2nd-Lt.	545
—— P. R., Lt.	700
—— S. A., Lt.	472, 1148
—— W. H., Maj.	76, 207, 1884
—— W. R., 2nd-Lt.	472
—— W. J., Lt.	698
Bowder, N. C., Lt.	538
—— W. E., *Hony.* Maj.	1843
Bowditch, C. E., Lt.	351, 1068
Bowdler, W. A., Capt.	628
Bowe, R. W., Lt.	313, 899
Bowell, G. L.	460
Bowen, C. E., Lt.-Col.	191, 1894, 1912
—— C. H. C., Capt.	501a, 844
—— D. H., 2nd-Lt.	556, 1467
—— E. B., Ty. 2nd-Lt.	669
—— G. J., Tempy. Capt.	1724
—— G. L., Lt.	455, 1028
—— G. M., Lt.	698
—— E. C., Capt.	585
—— H. C., 2nd-Lt.	51
Bowen, H. H., 2nd-Lt.	555, 1160
—— H. W., Lt. Col.	62, 74, 130, 632
—— J., 2nd-Lt.	356, 1008
—— J. E., 2nd-Lt.	1524
—— J. A. R. P., Maj.	664
—— J. P., Capt.	1698
—— J. C. G., Esq.	1907
—— W. O., 2nd-Lt.	325, 366, 1523, 1623
Bower, A. P., 2nd-Lt.	737
—— D. R., 2nd-Lt.	765
—— F., 2nd-Lt.	1030
—— S. W., Lt.	425, 1024
Bowerman, J. F., Lt.	1513
—— J. H.	798
Bowering, J. C., Lt.	638
Bowers, H., 2nd-Lt.	638, 658
—— J. A., 2nd-Lt.	608
—— J. F. F., Lt.	380, 1431
—— P. L., Lt.	513, 997
—— S. J., 2nd-Lt.	756
—— T. I., Lt.	315, 1050, 1062
Bowes, E. C., Lt.	1841
Bowhay, A. B., Capt.	1838
Bowie, H., Lt.	743
—— T. C., Tempy. Capt.	1715
Bowker, W. J., Maj.	103
Bowl, W. T. G., Lt.	1720
Bowle, S. C., Maj.	1712
Bowle-Evans, C. H., Lt.-Col.	130, 1742
Bowler, D., 2nd-Lt.	699
—— H. M., Lt.	76, 522, 1884
—— J., 2nd-Lt.	815
Bowles, C. W., 2nd-Lt.	605
—— W. E., Lt.	492, 941
—— F. C., Sub-Condr.	1855
—— R. F., Lt.	507, 1666
Bowley, G. W., 2nd-Lt.	355, 1895, 1894
Bowman, H. E., Tempy. Capt.	795
—— R. S., Lt.	624
—— W., 2nd Lt.	564, 1082
Bown, H., Capt.	787
—— J. E., 2nd-Lt.	563
Bowraj Limbu, Subdr.	1648, 1655
Bowrey, E. W. C., 2nd-Lt.	651
Bowring, A. H., Lt.-Col.	132, 613
Bowron, R. F., 2nd-Lt.	1140
Bowshil, H. V., 2nd-Lt.	588
Bowtell-Harris, J. F., Lt.	430, 1203
Bowyer, F. D., 2nd-Lt.	751
Box, E. G., 2nd-Lt.	550, 1674
Boxall, M. W. J., Lt.	316, 1132, 1569
Boxer, F. P., Capt.	501a, 734, 1668
Boxwell, A., Maj.	219, 872, 1471, 1583
Boyack, J. M., Lt.	640
Boyagis, G., 2nd-Lt.	654
Boyce, C. E., Capt.	787
—— C. J., Maj.	234, 1662
—— C. O., 2nd-Lt.	492, 1517
—— R. B., Maj.	243, 1066, 1156
—— T. W., Lt.	470, 14457
—— W. A., Lt.	353, 1411
Boyd, A., Lt.	640, 653
—— D. J., Maj.	804
—— H. S., Lt.	487
—— J. E. M., Capt.	1712
—— J. F., Maj.	1752
—— J M., Maj.-Genl.	173
—— L. C., Lt.	577
—— N. R., Lt.	1247
—— O. T., Capt.	56, 269
—— R. W., Sub-Condr.	1874
—— S., Maj.	664, 996
—— T. C., Capt.	1755
—— W. H., 2nd-Lt.	843
—— W. S. J., Lt.	520, 1656
Boyden, P., Lt.	630
Boyd-Parselle, F. G., 2nd-Lt.	553
Boyd-Rochfort, H., Capt.	52, 597
Boyes, W. R., Lt.	509, 997
Boyle, C. A., Capt.	102, 104, 108, 263, 805
—— H. McC. Lt.	420, 933
—— J., 2nd-Lt.	1243
—— R. E., Hon. Col.	176
—— R. G., 2nd-Lt.	602
Boylette, W. R. D., 2nd-Lt.	553, 1705
Boyne, A. R. H., Asst. Surgn	1795
Boynton-Lee, F., Lt.	33, 518, 719, 1673
Boyton-Smith, E. S., Lt.	309, 1058
Boys, H. C., Capt.	1919
Bracey, S. W., Lt.	532, 998
Bracher, H. C., Lt.	506, 1156
Brachlo, J. J. A., Sr. Asst. Surgn	1789
Bracken, G. T. H., Lt.	652

Names and Rank. Page.

Brackenbury, W. J. W., Lt.-Col. 137, 159, 197, 869, 1193, 1195
Bradbury, A. L., Lt. 528
—— C. H., 2nd-Lt. . 545, 992
—— W. E., 2nd-Lt. . . . 666
Braddish, G. S., 2nd-Lt. . . .
Bradish-Elames, J. E. M., Lt. . 589
Bradfield E. W. C., Maj. . . . 1751
—— J. T., 2nd-Lt. . . . 703
—— L. G., 2nd-Lt. . . . 624
Bradford, F. R., Lt. . . 440, 1614
—— F. K. 446-49
—— W. H., Capt. 55
Bradlaugh, J. H., Lt. 538
Bradley, C R. S., Maj . 239, 881
—— F G., Lt.-Col. . . . 61
—— G. G., Lt. . 515, 750, 1671
—— K., Nursg. Sister . . 1735
—— R. J., Maj. 1748
—— R. O., Capt. . . 276, 929
—— S. M., Lt. . . 519, 1135
—— T. A. E., Lt. 129
—— V. M., Lt. . . 55, 617
—— W. C., 2nd-Lt. . . . 693
Bradley-Birt, F. B., 2nd-Lt. . . 549
Bradney, E., Maj. . . . 665, 991
Bradshaw, A. T. R., Lt. . 420, 1251
—— A. R., Lt. 1130
—— C. H., 2nd-Lt. . 703, 704
—— C. R., Lt.-Col. 98, 103, 139, 161, 200, 872, 1641
—— F. J., 2nd-Lt. . . . 354
—— G. T., 2nd-Lt. 473, 535, 1479
—— J. B. L., Lt. . 291, 901
—— J. W. H., Capt. . 486, 885
—— L. J. E., Maj.-Genl. 1086, 1955
—— R. J., 2nd-Lt. . . . 710
Bradshaw-Smith, B., Lt. 460, 465, 1604
Brady, J. P., Lt. . . . 533, 1567
—— R. M. E., Lt. 518
—— W. A. H., Lt. 397
Braganza, F. B. A., Asst. Surgn. . 1796
—— J. B. V., Asst. Surgn. . 1796
Bragg, H. V., Lt. . 307, 1058, 1403
Brahmdeo Singh, Subdr. . . . 1403
Braham, A. M. F. 1857
—— S. H. T., Tempy. Capt. . 795
Brahm Swarup Bhatnagar, Sub-Asst. Surgn. 1816
Brahm Dat, Jemdr. 1125
Brahwan, Subdr. 1098
Braide, R. W., Capt. 741
Braidford, F. G., Lt. 1265
Brailsford, R. W., Lt. 629
Brain, W. G., 2nd-Lt. 1567
—— W. T., Hony. Lt. . . . 606
Brainbridge, R. G., 2nd-Lt. . . . 988
Braithwaite, C. V. C., 2nd-Lt. . 779
Brakspear, C. H., 2nd-Lt. . . . 653
Bramall, B., Lt. 754
—— K. G., 2nd-Lt. . . . 1010
—— K. G. McL., Lt. . . . 507
Bramfitt, A. E., 2nd-Lt. . . . 1024
Bramley, P. B. 59, 522
Brampton, C., 2nd-Lt. . 365, 1148
Bramston, E. O., Hony. Lt. . . 856
Bramwell, F. H., 2nd-Lt. . 710, 786
—— W. H., Lt. . . 533, 1207
Branch, W. G. Lt. . . . 82, 556
Brand, A. R., Lt. . . . 539, 992
—— F. J., 2nd-Lt. . . . 1642
—— J. K., Lt. . . . 699, 795
Brandenbourg, L. F., Capt. . . 1765
Brander, A. A., 2nd-Lt. . . . 561
—— E. H. C., Maj. 229, 020, 1048
—— G. K., 2nd-Lt. . . . 1678a
—— R. W., Capt. . 273, 1670
—— W. B., Capt. 810
Brandon, F. J., Lt. 506
—— G. N., tempy. Capt. . 1725
—— R. V., Capt. . 273, 1604, 1613
—— R. W., 2nd-Lt. . . . 567
Brandt, H. G., Lt. . . 540, 1606
Branfoot, D. C., Capt. . 274, 917
Brannon, A. C., 2nd-Lt. . 474, 1287
Bransden, W. J., 2nd-Lt. . . . 622
Branson, C. H., 2nd-Lt. . 358, 1555
—— L. H., Maj. 56, 108, 219, 1443
Brant-Smith, C., Lt. . . 534, 1227
Brash, C., Ty. Lt. 1729
Brasher, J. H., 2nd-Lt. . . . 782
Brassey, L. P., Maj. 1748
Bratton, V. W., Lt. 415
Braudt, A. C., 2nd-Lt. . . . 679
Braund, H. B. L., Capt. . 18, 42, 789

Names and Rank. Page.

Brawn, J. A., 2nd-Lt. . . 474, 1279
Bray, C. F., 2nd-Lt. 1469
—— E. H., Maj. 859
—— E. M., *Hony.* Maj. . . 1847
—— G. F., 2nd-Lt. . . 357, 1413
—— *Sir* H., *Hony.* Brig.-Genl. . 10
—— F. E., Capt. 681
—— N. D., 2nd-Lt. . . . 1156
—— N. N. E., Capt. . 52, 258, 909
—— S. A., Lt. 709
—— T., Hony. Lt. . . . 1844
Brayham, E. F., 2nd-Lt. . . . 548
Brayne, A. F. L., Lt. . . . 821
—— F. L., Ty. Lt. . . . 909
—— W. F., Maj. . . . 1751
Brazier, C. C. H., Lt. . . . 31, 525
—— F. R., 2nd-Lt. . . . 362
—— H. O., Sub-Condr. . . 1857
Breach, J. B., Condr. 1854
Breadon, G. W. D., 2nd-Lt. 531, 1197, 1885
Bream, J. W., Lt. 516
Brear, B. H., Lt. . . . 387, 1279
Brebner, A., 2nd-Lt. 601
Brede, H. G., Lt. 700
Bredin, A., Lt.-Col. 138, 160, 178, 198
—— A., Col. . . . 177, 1918
—— A. N., Maj. . . . 225, 1623
—— J. A., 2nd-Lt. . . 564, 1134
Bredon, A. S., Capt. 759
Bree, P. G., 2nd-Lt. . . . 471, 1279
Breed, F. G., 2nd-Lt. 749
Breeze, C. L., 2nd-Lt. 564
Bregman, T. M., Lt. . . 538, 1331
Breething, H. R., 2nd-Lt. . . . 742
—— W. J., Lt. . . 443, 1331
Breithaupt, E. A., Maj. . 238, 1044, 1271
—— G. W., Lt. . 84, 516, 1307
Breitmeyer, G. C. A., Capt. . . 587
Bremby, 2nd-Lt. 1036
Bremmage, J. C., Lt. 1124
Bremner, A. G., Bt.-Col. 58, 126, 661, 996
—— C. E. U., Capt. . 274, 1096, 1136
—— H. N. K., Capt. 59, 260, 1459
—— S., Lt. . . . 387, 877
Brennam, R., Lt. 1621
Brennan, C. G. O., Lt. . . 293, 1144
—— R., Lt. 401, 1007
Brereton, H., 2nd-Lt 1054
—— R. A., Maj. 832
Breslauer, L., Lt. . . . 519, 1666
Bretherton, H.W.K., Lt.-Col. . 759
Breton, B. O., Lt. . . . 541, 1255
—— D. C., Lt. . . . 731, 798
—— C. G., 2nd-Lt. . . 325, 1287
Brett, A. L. 2nd-Lt. 1439
—— C. E., Lt. . . . 352, 1675
—— C. T., 2nd-Lt. . . . 563, 1524
—— D. A., Lt. . . . 507, 783, 1052
—— E. T. St. M., 2nd-Lt. . . 324
—— E. W., Capt. . . . 277, 1090
—— F. L. S., Maj. . 230, 1629, 1645
—— H., *Hony.* Lt. 695
—— J. A., Maj. . . 226, 1900, 1913
—— V. W., Capt. . . . 81, 257
—— W. H., 2nd-Lt. . . 331, 923
Brewer, A. A., 2nd-Lt. . . 360, 1044
—— A. G., 2nd-Lt. . . . 1373
—— C. E., 2nd-Lt. . . . 1104
—— G. L., Lt. . . . 68, 683
—— J. C., Capt. . 6, 14, 268, 1367
—— J. P., Lt. . . . . .
—— J. R., 2nd-Lt. 555
—— V., Lt. 531, 1130
Brewer-Little, H., 2nd-Lt. 669, 997, 1001
Brewester, F., 2nd-Lt. 1223
Brewin, A. P., 2nd-Lt. . . 544, 1704
—— A. S., 2nd-Lt. . . . 551
—— G. G., Lt. . . . 516, 1667
—— J. P., Col. . . 35, 123, 660
—— N. A. P., Lt. . . . 543, 1401
—— T. J., Capt. 61
Brewis, R. D., Capt. . 16, 26, 752, 783
—— R. S., Lt. . . . 380, 1303
Brewitt, A. C., Lt. . . 387, 1046
Brewster, J. A., Lt. . . . 540, 1567
—— R. F., Lt.-Col. 101, 132, 632
Brian, W. G., Lt. 542
Briant, C. H., *Hony.* Maj. . . 1848
—— R. J., Sub-Condr. . . . 1867
Brice, A., Hony. Lt. 788
Brick, P. J., 2nd-Lt. . . . 551, 1142
Brickell, H. J., Lt. . . . 430, 1351

Names and Rank. Page.

Brickman, J. M., Capt. . 260, 1026
Brickmann, F. L. O., Capt. . 254, 1471
Bridcut, S. H., Lt. 58
Bridge, H. N., Revd. 1906
—— J. H., *Hony.* Lt. . . . 1850
—— R. E. A., Capt. . 254, 1663
—— W. J., 2nd-Lt. 705
Bridgens, A., Lt. 736
Bridges, A. H., Lt.-Col. . 97, 103, 137, 158, 197, 871, 1463
—— A. B. H., Capt. 5, 1715—16
—— B., Capt. 776
—— E. J., Capt. 593
—— F. L., Revd. 1904
—— J. J., *Hony.* Lt. . . 770
—— K. W., Lt. . . 306, 1042
—— R. F., Capt. . . 20, 1713
Bridgewater, A. G. B., Lt. 526, 1699, 1704
—— E. L., 2nd-Lt. . . . 363
Bridgnell, E., 2nd-Lt. . . 554, 1048
Bridson, W. E., Sub-Condr. . . 1856
Brierley, A. S., 2nd-Lt. . . . 687
—— C. I., Maj. . . 1751, 1900
—— R., Capt. . . . 502, 919
—— R. J., 2nd-Lt. . . 360, 1469
—— R. W., Lt. . . . 59, 797
—— W. E., Capt. 1754
—— W. R. C., 2nd-Lt. . . . 651
Brigg, C. J., Lt. 387, 889
—— G., Lt. 1018
—— W. L., Lt. 490, 917
Briggs, F. C. C., Capt. . . 17, 25, 689
—— G., Lt. 380, 420
—— H. R., Capt. 280, 287, 1144, 1569
—— N., Capt. 1760
—— R. A., Lt. . . . 423, 1655
—— R. C., Lt. . 534, 1690, 1708
—— R. M., 2nd-Lt. 1056
Brigham, G., Lt. . . . 317a, 1387
Bright, F. C., Lt. . . . 509, 1401
—— R. A., Brig.-Genl. 58, 97, 101, 121, 126, 613
—— W. J., Lt. 624
Brightmore-Armour, C. J., 2nd-Lt. 565
Brigstocke, A. M., Maj. . . . 653
Brij Bahadur, Jemdr. 989
Brijbhukan Singh, Jemdr. . 1063, 1705
Brij Lal Fakey, Sub-Asst. Surgn. 1817
—— Dichhit, Jemdr. . . . 1018
Brij Raj Singh, Sub-Asst. Surgn. 1818
Brijlal, Jemdr. 1142
Brijlall, Jemdr. 878
—— Sub-Asst. Surgn. . . . 1815
Brij Lal Fakar, Sub-Asst. Surgn. . 1817
—— Ty. Sub-Asst. Surgn. . . 1833
—— Gnai, Sub-Asst. Surgn. . 1817
—— Singh, Subdr. 1455
Brijlal Singh, Jemdr. 1047
Brij Mohan Lal, Jemdr. . . . 1260
Brijmohan Singh, Subdr. . . . 1043
—— Sukal, Jemdr. . 1054
Brij Pal Singh, Jemdr 1048
Brij Raj Singh, Sub-Asst. Surgn. 1818
Brilliant, L., Lt. . . 299, 1245, 1513
Brimelow, E., Lt. 1160
Brindaban, Sr. Sub-Asst. Surgn. 1808
—— Sub-Asst. Surgn. . 1811
—— Sharma, Sub-Asst. Sugn. 1818
Brindabon Sharma, Sub-Asst. Surgn. 1818
Brindley, W. W., Lt. . 423, 737, 1513
Brindraban Bakshi, Sub-Asst. Surgn. 1817
Brinkler, C. H., 2nd-Lt. . . . 361
Brinkley, E. G., Lt. 621
Brinsmead, H., Capt. 1839
Brinton, O. W., Maj. . . 597, 599
Briscoe, J. R., Ty. Capt. . . . 1724
Brisley, F. de B., 2nd-Lt. . . .
Brislay, H. F., Esqr. 89
Bristow, A. C., 2nd-Lt. . . 552, 1134
—— C. H., Lt. . . . 511, 1614
—— G. M., 2nd-Lt. . . . 780
—— H. J., Condr. . . . 1870
—— R. L., *Hony.* Capt. . . 1845
Bortfan, N.F., 2nd-Lt. 786
Brito, E., 2nd-Lt. . . . 1561, 1702
Britten, W. E., Capt. . . 665, 991
Britton, C. J. J., 2nd-Lt. . . 828
—— J. A. H., Maj. . 233, 1339, 1361
—— T. B., 2nd-Lt. . . . 361
Britts, M. H. G., Lt. . 521, 1668, 1671

Names and Rank.	Page.
Broad, G. C. W., Lt.	715
—— R. N. D., 2nd-Lt.	506, 1618
Broadbent, E. R., Maj.	589
—— H. R., Lt.	513, 992
—— J. T. C., Maj.	220, 872, 1591
Broadfoot, W. A., Capt.	279, 879
Broadhurst, A. F. B., Capt.	763
Broadley, J. F., 2nd-Lt.	1265
Broadmead, H. H., Lt.	17, 24, 705
Broadway, A. C., 2nd-Lt.	332, 366a, 1247
—— P. R., Lt.	319, 365, 1617
Brock, A. W. S., Bt.-Lt.-Col.	138, 711
—— B. deL., Maj.	235, 1501
—— C. H., Maj.	500, 830
—— H. N., Lt.	818
—— W., Lt.	1689
Brockes, E. H., Lt.	515, 1357
Brockbank, G. R., Capt.	748
Brockebank, A. C., 2nd-Lt.	555
—— G. S., 2nd-Lt.	1447
Brocklebank, H. J., 2nd-Lt.	549, 1675
Brockliss, A. A., 2nd-Lt.	476, 953
Brockman, F. H. D., 2nd-Lt.	683
—— P. W. D., Lt.-Col.	54
Brodie, A. H., Asst. Surgn.	1801
—— H. C., *Hony.* Capt.	711
—— J. W., Lt.	543a, 1040
—— K., Lt.	708
—— R., 2nd-Lt.	559, 1020
—— R. P., 2nd-Lt.	1193
—— W. H. H. C., Capt.	501a
—— W. S. C., 2nd-Lt.	522, 1570
Brodie-Boque, W., 2nd-Lt.	
Brodribb, C. H., Maj.	1750, 1900
Brodrick, *Hon.* A. G., Lt.-Col.	681
—— H., 2nd-Lt.	695
—— W. LeC., Capt.	267, 1664
Brohier, C. P., 2nd-Lt.	551, 1195
Broke-Smith, P. W. L., Maj.	663
Bromage, J. A. R., Capt.	754
—— W. A., 2nd-Lt.	793
Bromby, T. A., 2nd-Lt.	1036
Bromham, A. C., Lt.	299, 1030, 1082
—— A. W., 2nd-Lt.	493, 879, 951
Bromhead, B. P., Col. *Sir*	177
Bromilow, D. G., Maj.	243, 901
Bromley, B. G., 2nd-Lt.	1676
—— C. G., 2nd-Lt.	356, 992
—— R. E., Lt.	493, 935
Brommage, L. C., Lt.	
Brook, B. W. C., Capt.	1664
—— H. L., 2nd-Lt.	630
—— L. A., 2nd-Lt.	623
—— T. F., Maj.	237, 1663
—— T. F., 2nd-Lt.	1678a
Brook-Ascough, T. H., Lt.	776
Brooke, A., Capt.	258, 873, 909
—— A. E., 2nd-Lt.	325, 905
—— A. H., Capt.	252, 909
—— A. S., Capt.	268, 1090
—— B. B., Lt.	638
—— B. G., Lt.	595
—— H. A., Lt.	51, 593, 686
—— H. M. M., Lt.-Col.	42, 131, 156, 191, 1650
—— J. A., Capt.	585
—— L. H., 2nd-Lt.	473, 1495
Brooke-Hunt, R. C., Lt.	322, 964, 1567
Brookes, A., Lt.	59, 504, 829, 1287
—— J. A. R., Revd.	1903
—— J. R., 2nd-Lt.	324, 915
—— T. L., Lt.	352, 1263
—— W. L., Sr. Asst. Surgn.	1790
Brooking, H. T., Maj.-Genl. *Sir*	104, 119, 147, 174
Brookman, J., Lt.	455—59, 1331, 1556
Brooks, A. G., Ty. Sub-Asst. Surgn.	1804
—— C. W., 2nd-Lt.	566, 1132
—— E. C., Tempy. Capt.	1770
—— F. L., Lt.	415, 1311
—— G. H., 2nd-Lt.	825
—— L. G., 2nd-Lt.	752
—— J., *Hony.* Lt.	725
—— J. B., Lt.	543a, 1624
—— J. R., 2nd-Lt.	323
—— P., 2nd-Lt.	359
—— R. W., Lt.	353, 1045
—— T. B., Asst. Surgn.	1793
Brooks-King, M., 2nd-Lt.	705
Broom, J. T., Lt.	493, 879
Broome, H. B., Maj.	1750
—— L. R., Lt.	1708
—— W. G. K., Maj.	107, 256, 1363, 1894
Broomfield, R. S., 2nd-Lt.	825
Broon, C. W., Lt.	1016
Brothers, H. J., 2nd-Lt.	753
—— J. A., Lt.	397, 1164
Brotherston, G. T., Lt.	86, 540a, 1702
Brotherton, J. B., Lt.	503, 1665
—— L., Lt.	1321
—— L. P., Capt.	268, 1026, 1028
Brough, A., Brig.-Genl.	122, 136, 662, 1883
—— W. P., Maj.	785
Broughton, A. H., Revd.	1905
—— G., Capt.	258, 1152
—— T. D., Maj.	662
—— C. E., 2nd-Lt.	325, 366, 939
Brown, A., *Hony.* Capt.	737
—— A., Lt.	524, 1323, 1724
—— A., Capt.	1724
—— A., Lt.	582
—— A., Lt.	351, 927
—— A., 2nd-Lt.	363, 1199
—— A. A., 2nd-Lt.	725
—— A. D., 2nd-Lt.	1401
—— A. D., Rev.	1908
—— A. E., 2nd-Lt.	714
—— A. E., Lt.	307, 1433
—— A. F. W., Lt.	371, 1642
—— A. G., Sr. Asst. Surgn.	1790
—— A. H., Capt.	1759
—— A. W. C., 2nd-Lt.	358, 1579
—— B. H., Maj.	71, 101, 633, 1884
—— B. W., Capt.	1717
—— C., 2nd-Lt.	359, 1383
—— C. A. Major	234, 871, 1343, 1543
—— C. E., 2nd-Lt.	733
—— C. E. M., Lt.	1179
—— C. J., 2nd-Lt.	565
—— C. M., Lt.	843
—— D. A., Lt.	720, 1006
—— D. C., Capt.	582
—— D. H., Lt.	738, 740, 1305
—— D. H., Lt.	531
—— D. J., Lt.	523
—— D. S., Lt.	410, 1178
—— E., Lt.	681
—— E. B., Lt.	593
—— E. C., Capt.	55, 257, 1623
—— E. H., Lt.	493, 955
—— E. J., 2nd-Lt.	564
—— E. J., Capt.	679
—— E. R., Lt.	327, 354, 1621
—— F. G., Lt.	629
—— F. H., 2nd-Lt.	691
—— F. H. J., 2nd-Lt.	363, 1124
—— F. J., 2nd-Lt.	357, 1373
—— F. L., 2nd-Lt.	717
—— F. L., 2nd-Lt.	465, 1453
—— G., 2nd-Lt.	666
—— G., 2nd-Lt.	362
—— G. B., 2nd-Lt.	704
—— G. J., 2nd-Lt.	361
—— G. J. F., Sub-Condr.	1867
—— G. L., Lt.	728
—— G. L., Lt.	530, 988
—— G. M., 2nd-Lt.	758
—— G. R., Lt.	317a, 1657
—— H., Capt.	854
—— H., 2nd-Lt.	555, 1140
—— H. A., Capt.	
—— H. A., 2nd-Lt.	828
—— H. C., Maj.	1750
—— H. G., Lt.	797
—— H. H., Capt.	1758
—— H. H., 2nd-Lt.	562
—— H. M., Maj.	1749
—— H. N., Lt.	393
—— H. R., Lt.-Col.	135, 1744
—— H. W., 2nd-Lt.	681
—— H. W., 2nd-Lt.	551, 1678
—— I. C., 2nd-Lt.	840
—— I. T. J., 2nd-Lt.	732
—— J., Maj.	628
—— J., Lt.	296, 435, 1183
—— J. A., Lt.	487
—— J. A., Capt.	859
—— J. A., Capt.	696
—— J. A. H., 2nd-Lt.	558, 1707
—— J. B., 2nd-Lt.	494, 881
—— J. F., 2nd-Lt.	1413
—— J. H., 2nd-Lt.	751
—— J. M. M., Asst.-Surgn.	1799
—— J. R., 2nd-Lt.	566
—— J. R., Lt.	399, 781
—— J. R. R., 2nd-Lt.	905
—— J. W., 2nd-Lt.	599—600
—— K. H., Lt.	578
Brown, K. N., 2nd-Lt.	761
—— L. C., 2nd-Lt.	540, 1034
—— L. C. H., Lt.	321, 1283, 1521
—— L. N., Lt.	511, 1046
—— M. A., Lt.	410, 891
—— N. A. S., Lt.	731, 791
—— N. L., 2nd-Lt.	73
—— R., Senr. Asst. Surgn.	1787
—— R., Lt.	320, 1539
—— R. F., 2nd-Lt.	58, 692
—— R. J. R., Col.	124, 149, 183
—— R. S., 2nd-Lt.	1134
—— R. S. C., Maj.	849
—— R. T., 2nd-Lt.	332, 1271
—— S. B., 2nd-Lt.	879
—— S. M., 2nd-Lt.	1679
—— T. H. C., Lt.	352, 1642
—— V. H., 2nd-Lt.	735
—— W., Bt. Lt.-Col.	60, 139, 160, 225, 1662
—— W., 2nd-Lt.	1453
—— W. A., Capt.	755
—— W. B., Lt.	1885
—— W. B., Lt.	1695
—— W. C., Lt.	637
—— W. F., 2nd-Lt.	331, 355, 1058
—— W. G., 2nd-Lt.	786
—— W. H., Lt.-Col.	108, 129, 155, 188
—— W. H., Capt.	1718
—— W. H., Sr. Asst. Surgn.	1790
—— W. H., Sub-Condr.	1864
—— W. J., Condr.	1867
—— W. P., 2nd-Lt.	787
—— W. S., 2nd-Lt.	557, 1359
—— W. W., Lt.	1720
Browne, A. E., Asst. Surgn.	1795
—— A. G. F., Lt.-Genl. *Sir*	1613, 1955
—— A. J. W., Lt.-Col.	129, 154, 187
—— A. M. F., Asst. Surgn.	1798
—— A. P., Lt.-Col.	130, 155, 189
—— C. A. R., Col.	122, 660
—— C. C., Lt.	535
—— C. E. M., Lt.	508
—— C. H., Maj.	844
—— C. L., Lt.	518, 1563
—— C. M., Lt.-Col.	134, 663, 1886
—— D. R. H., Lt.	523, 998
—— E. W., Maj.	1748
—— F. G., 2nd-Lt.	640
—— G. De La R., Lt.	539
—— H., Lt.	508, 1116
—— H., 2nd-Lt.	859
—— H. F. M., Maj.	213, 1894
—— H. H. G., Maj.	56
—— H. J. H., Lt.	701
—— H. J. P., Br.-Genl.	13, 31, 97, 121, 126, 153, 198, 1617
—— J. C., 2nd-Lt.	692, 784
—— J. F., 2nd-Lt.	555
—— J. G., Maj.	593
—— J., Esq.	89
—— J. W., 2nd-Lt.	593
—— L. S., Lt.-Col.	139, 161, 199, 869, 1170
—— P. W. R., Asst. Surgn.	1795
—— R. B., Capt.	276, 1156
—— R. F., Asst. Surgn.	1794
—— R. G., Maj.	758
—— St. J. A., Capt.	107, 256, 1235
—— T. G. H., 2nd-Lt.	557
—— T. L., 2nd-Lt.	558
—— V. S. T., 2nd-Lt.	1001
—— W. A., Asst. Surgn.	1800
—— W. R. H., Lt.	299, 1505
—— W. S., Esqr.	89
—— W. S. W., Maj.	59, 220, 1199
—— W. W., Maj.	10, 1711
Browning, B. W., Lt.	291, 1050
—— E. R. G., Maj.	105, 228, 1015
—— G., 2nd-Lt.	619
—— H. A., Lt.-Col.	56
—— J. M., 2nd-Lt.	551, 1032
—— L. H., Lt.	618
—— S. H., Tempy. Capt.	1723
—— T. C., Lt.-Col.	22, 195, 1883, 1912
Brownlie, J. K. C., Lt.	393, 1103
—— T. A. M., Lt.	805
Brownlow, C. C., Col.	178, 1917
—— C. C. S., Lt.	1919
—— d'A. C., Brig.-Genl.	59, 122, 134, 157, 194
—— J., Capt.	274, 927
—— W. V., Maj.-Genl.	581
Brownrigg, E. G., the Revd.	1906

Names and Rank. Page.

Browse, G., Lt.-Col. . . . 134, 1745
Bruce, A. McC., Col. . . 169, 1916
—— A. N., 2nd-Lt. . . . 766
—— C E., Maj. . 212, 1898, 1913
—— C. G., Brig.-Genl. *Hon.* 12, 29, 120, 125, 153, 190
—— C. M. A., 2nd-Lt. . . . 820
—— D. W., Capt. . . . 1714
—— E., 2nd-Lt. . . . 544, 653
—— E., 2nd-Lt. 561
—— E. A., Col. 177
—— E. D. B., 2nd-Lt. . . . 545
—— G. D., Bt.-Col. 6, 98, 136, 159, 162, 200, 870, 1265
—— G. E., Maj. . . . 243, 1235
—— H., Sub-Condr. . . . 1863
—— J., Maj. 28 212
—— J. G., Lt. 293, 1627
—— M. E. L., Lt.-Col. . . 87, 140, 161, 200
—— R. M., Capt. . . . 275, 1617
—— R. W., 2nd-Lt. . . 553, 1704
—— R. W. V., Maj . . . 505
—— S. M., Maj. 57
—— W., 2nd-Lt. 1197
—— W. R., Capt. 11
Bruce Kingsmill, J. C. deK., Lt.-Col. 9, 628
Bruce-Kerr, G. H., the Revd. . 1907
Bruce-Smith, R., Lt. . . 321. 023
Bruckman, R. T., Capt. . . . 725
Brunger, B., 2nd-Lt. . . . 359, 1523
Brunker, J. R., Maj. . 43, 215, 1661
Brunlees, L. H., Capt. . 279, 1613
Brunskill, B. A. S., Capt. 271, 1323, 1573
—— G. S., Capt. . 273, 1211
Brunt, P. C., Revd. 1908
Brunton, F. H., 2nd-Lt. . 68, 707
—— J. E. C., Capt. . . 501*a*
—— T. S., Lt. . . . 543, 998
Bryan, F., Ty. Capt. 1723
—— G. M., Tempy. Asst. Surgn. 1804
Bryant, B., 2nd-Lt. 727
—— C. E., Capt. 587
—— E. J. A., 2nd-Lt. . . 365, 1183
—— G. D., 2nd-Lt. . . 329, 964
—— J. F., 2nd-Lt. 802
—— J. M., Maj. . . . 74, 634
—— L., 2nd-Lt. . . . 358, 1275
—— R. J. C., Lt. 1066
—— S. C., 2nd-Lt. 714
Bryce, Condr. J. 1872
—— W., 2nd-Lt. . . 553, 859, 1054
Bryceson, D. J., 2nd-Lt. . . 325, 1239
Bryer, P. J., 2nd-Lt. . . . 548, 1884
Bryson, A. B., Capt . . 22, 259, 1664
—— H. E., 2nd-Lt. 36
—— R., Lt.-Col. . . . 136, 1744
Bua-Ditta, Jemdr. 1099
Bu-Ali Khan, Sub-Asst. Surgn. . 1812
Buchanan A., Lt.-Col. . 127, 1742
—— A., Lt. 637
—— A. T., 2nd-Lt. 561
—— A. W., Capt. 977
—— D. S., Lt. 507
—— F. H. T., Lt.-Col. . . 820
—— H., Lt. 511
—— J. P., 2nd-Lt. . . 558, 1295
—— J. V., Capt. 1723
—— K., 2nd-Lt. 744
—— K. J., Br. Genl. 14, 35, 120, 123, 149, 182
—— R., Lt. 756
—— W., 2nd-Lt. 828
—— W. J., Lt.-Col., *Sir* 128, 1740
Buchan-Hepburn, T. E. A., Capt. 766
Buchanan-Meikle, A., 2nd-Lt. . 544
Buchanan-Riddell, W. R., Lt. . 734
Bucher, F. R. R., Capt. 280—287, 933
Buck, A. G., 2nd-Lt. . . . 76, 1884
—— C H., Lt.-Col. . 197, 1896, 1912
—— F. A., Lt. 318
—— H., 2nd-Lt. . . 18, 21, 696, 796
Buckingham, F., Capt. 721
—— T. N., Lt. 700
Buckland, G. C. B., Maj 14,42,244,1635
—— G. N., Maj. . . 634, 972
—— L. J. E., 2nd-Lt. . . . 1569
—— L. L., Capt. . 268, 921, 925
—— R. J., 2nd-Lt. . . 692, 784
Buckler, F. W., Lt. 490
Buckley, A. W., the Revd. . . 1908
—— E. S., Capt. 617
—— F., Lt. 401, 1203
Buckley, H. C., Maj. 1752
—— L., Capt. 1714
—— L. F., Lt.-Col. . . . 611
—— P. W., 2nd-Lt. 1299
Bucknall, H. L., Lt. . . . 318, 964
Bucknall, R., Lt. 1701
—— R. D. H., Lt. . . 295, 1600
—— W. B., Capt. 735
Buckner, H. M., Lt. 830
—— C. S., 2nd-Lt. 1319
Buckney, F., Capt. 652
Buckton, G. N., 2nd-Lt. . . 325, 1058
Buckwell, F. C. Revd 1904
—— H. L., 2nd-Lt. 588
Budd, F., 2nd-Lt. 362
—— N. A. H., Lt.-Col. 128, 154, 185
Budden, F. H., Maj. . . 665, 1881
—— F. V., Lt. 628
—— J. A., Capt. . . . 81, 254
Budge, G. S., 2nd-Lt. 638
Budgen, H. V., Capt. 252, 1451, 1453
Budh Ram, Subdr. . . . 1283, 1502
—— Singh, Jemdr 1080
—— Jemdr. 1256
—— Jemdr. 1083
Budha Ram, Jemdr. 1484
—— Jemdr. 1457
—— Singh, Subdr. 1207
—— Subdr. 1096
—— Subdr. . . . 1148, 1455
—— Jemdr. 1207
—— Subdr. 1096
Budhi Singh Negi, Jemdr. . . . 1182
Budhbir Thapa, Jemdr. 1631
Budhe Shah, Jemdr. 1420
Budhibal Rana, Jemdr. 1614
—— Thapa, Subdr 1607,1611
—— Thapa, Jemdr. 1637
Budhiman Ghale, Jemdr. . . . 1597
—— Gurang, Jemdr. . . . 1596
—— Subdr. 1618, 1657
—— Singh, Subdr. 1283
—— Gurung, Subdr. 1618, 1657
Budhichand Bura, Subdr. . . . 1618
Budhiram Roka, Jemdr. . 1602, 1658
Budh Ram, Jemdr. 1388
—— Singh, Subdr. . . . 1502-04
Budhi Singh, Jemdr. 1571
Budir Singh, Jemdr. 1132
Budruddin, Risdr. . . . 1683, 1686
Bugg, W. A., 2nd-Lt. 361
Bugler, J., 2nd-Lt. . . . 355, 979
—— E. F., Lt. . . . 351, 1162
Bukhari, A. A., Sub-Asstt. Surgn. 1832
Buist, A. A., Lt. 397
—— A. R., Lt.-Col. 136, 152, 105, 867, 954
—— A. W. T., Lt.-Col. . 129, 1741
—— C N Capt 269 1227
Bujan, Jemdr 1060
Bulaka Singh, Subdr. 1370
Bulaki Singh, Subdr. 1082
—— Subdr. 1321
Buland Khan, Subdr. . . 1367, 1707
Buley, S. H., Lt 352, 1181
Bulfin, E. S., Maj.-Genl. . . . 712
Bulgin, J., 2nd-Lt. . . . 330, 1178
Bull, A. H., 2nd-Lt. . . 544 736, 1677
—— C. A., 2nd-Lt. 685
—— C H., 2nd-Lt. . . . 527, 1667
—— E Revd 1906
—— F. F., Revd. 1906
—— G. G. C., Lt. . . . 209, 1399
—— H., Lt. 544
—— L. C., Lt. 401, 1207
—— W. J., Lt 1007
—— Bryan, Ordce. N. . . . 2110
Bullard, D. A., 2nd-Lt. . . 328, 1293
—— H., 2nd-Lt. . . . 327 1347
Bullied, J. F., Lt. . . . 697, 1255
Bullen, C., 2nd-Lt. 691
—— J. H., Lt.-Col. 862
Buller, F. E., Capt. . . . 665, 996
—— J H G Maj. 1910
Bulley, H. R., Lt. 1331
Bullingham, W. F. 537
Bullmore, F. E., 2nd-Lt. . . 465, 925
Bullock, H., Lt. 312, 1387
—— J. H., Lt. 691
—— J. C. C., Lt. 316
—— J. W. H., 2nd-Lt. . . 554, 1088
—— P. G. W., Lt. . . . 302, 917
—— H. P. S., Lt. . . . 317*a*, 1363
—— R. S., Lt. 104
Bullocke, J. G., Lt. . . . 540*a*, 1333
Bullock-Webster, A. P., Lt. 425, 877
Bulram Singh, Subdr. 1288
Bulteel, C. E., Maj 1750
Bulwant Singh, Ressdr. . . . 940
—— Jemdr. . . . 925
Bunbury, G. B. St. P., Maj 238, 1070
—— G. F., Capt. . 80—287, 1074
—— N. L. St. P., Capt. 272 1427
—— W. C. H., Maj. . . . 766
—— W. E., Maj.-Genl. 97, 118, 146, 173, 1917
Bunce, A., Condr. 1853
—— L. L., Lt. 351, 1451
Bunch, J. B. S., 2nd-Lt. . . 560, 1678
Bundy, F. C., Lt. 541, 1674
Bunn, A. E., 2nd-Lt. . . . 629, 784
—— C. W., 2nd-Lt. . . 366*a*, 1162
—— W., Capt. 1614
Bunnett, F., Condr. 1853
Bunning, W. H., Lt. 1613
Bunt, H., 2nd-Lt. 1078
Bunting, S A. S., Lt. . . . 511, 1001
Bur Singh, Jemdr. 1252
—— Resdr. 891
—— Jemdr. 1364
Burbidge, A. H., 2nd-Lt. . . . 848
—— W. H. M., Lt. . 19, 727
Burbridge, R. P. D., 2nd-Lt. . .
Burbury, F. W., Lt.-Col. . . . 778
Burch, A. P., Lt. . . . 315, 1287
—— F. W., Lt. . . . 295, 1633
—— G. W., Lt. . . . 380, 889
—— P. D., Lt. . . . 313, 1595
Burchell-Crookes, W., Lt. . . 638
Burchett, Y. W. K., 2nd-Lt. 357, 1393
Burcher, F. D. H., Lt. 1689
Burd, E., Maj. . 219, 871, 1347, 1379
Burden, G. N., 2nd-Lt. 325, 366, 1629, 1657
—— H., Lt.-Col. 62, 130, 1742, 1889
—— M. N., 2nd-Lt. 1315
Burdett, E. W., Capt. . . 264 1215
Burdon, R. M., Lt. . . . 405, 1331
—— W. B. C., Lt.-Col. . . 628
Burford-Hancock, H. di S. B., Lt.-Col., 730
Burgan, N., Lt. . . 291, 1613, 1653
Burge, J. T., Lt. 521
—— N. B., 2nd-Lt. . . 330, 1427
Burgess, B., Maj 663
—— B., Lt. 352
—— B. M. M., Lt. . . . 1080
—— C. S., 2nd-Lt. . . . 723*a*
—— F., Lt. 702
—— F. H., 2nd-Lt. . . 476, 1113
—— H. S., 2nd-Lt. 721
—— J. B., tempy. Capt. . . 1723
—— J. F., Lt. 492
—— R A., Lt. 617
—— W. G., 2nd-Lt. 713
—— W. G. A., 2nd-Lt. 535, 1487
—— W. M., Lt. . . 541, 1673
Burghall, C. J. D., Asst.-Surgn. . 1803
Burghope, J. W., 2nd-Lt. . . . 742
Burgoyne, A. H., Maj. . . . 10, 110
Burhan Ali, Jemdr. 900
—— Jemdr. 1514
—— Jemdr. . . . 1231, 1413
—— Subdr. . . . 1327, 1488
Burjan Singh, Jemdr. 902
Burjorjee, B. N., tempy. Capt. . 1767
Burke, A. O. L., Lt. . . . 470, 1630
—— C. H., Asst. Surgn. . . 1800
—— G. T., Capt. . . . 1108, 1755
—— M., Lt. 506
—— N., 2nd-Lt. 578
—— P. H., 2nd-Lt. . . 548, 1032
—— R. H., Lt. 1844
—— R. J. C., Maj. 7, 215, 1891, 1913
—— W. G., 2nd-Lt. . . 552, 1026
Burkett, R. N., 2nd-Lt. 327, 945, 1587
Burkitt, F. H 2nd-Lt. 805
Burland, Jemdr 093
Burleigh, H., 2nd-Lt. 623
Burlingham, D., 2nd-Lt. . . . 530
Burlton, A.B., Lt.-Col. . 24, 129, 155, 188, 1650
—— H. L. G., Lt. . . . . .
—— P. S M., Lt.-Col.1186,1896. 1911
—— W. E. F., Lt.-Col. 40, 76, 129, 154, 186
Burlton-Bennet, J. A., Capt. 260, 1664
Burmester, Z. G., Capt. 254, 933, 1080
Burn, A. F., Lt. 951
—— A. G., Lt.-Col. 57
Burn, A. H., Capt. . . . 253, 1259
—— E. M., Capt. 841

Names and Rank. Page.

Burn, E. M. J., Lt.-Col. . 25, 102, 103, 104, 106, 130, 661
—— G. C., Lt.-Col.. 32, 197, 1883, 1912
—— R. F., Lt. 410
—— R. N., Maj. . . . 663, 1883
—— S. A., Tempy. Capt. . . . 1725
—— W., 2nd-Lt. 568
Burnad, 2nd-Lt. 913
Burnand, H., Capt. 693
—— R. F., 2nd-Lt. . . . 324
Burnard, W., Lt. 1707
Burndred, W. G., Capt. . . . 1839
Burne, F. O. N., Capt. . 264, 1275
—— K. P., Col. 56
—— L. E. L., Maj. 214, 1894, 1913
—— N. A. K., Lt.-Col. . . . 57
—— R. D., 2nd-Lt. 848
—— R. H., Capt. . . 501, 1259
—— S. F., Maj. 614
—— W. G., Lt. . . . 728, 1148
Burnell, R. G., Lt. . . . 460, 1521
Burn-Murdoch, I., Capt . 256, 1148
—— C. T., Capt. 95, 276, 1098
Burnett, A. H., Maj 55, 243, 1303
—— A. H., 2nd-Lt. . . . 562
—— A. S., Capt. . . . 38, 502
—— E. C., 2nd-Lt. 554
—— F. H., Capt. 729
—— F. M. B., Ty. Asst. Surgn. 1804
—— G., 2nd-Lt. . . . 523, 1668
—— R. F. D., Maj 235, 1191, 1193
—— R. R., Lt. . . 420, 1128
—— S. H., Lt.-Col. . . 129, 1741
—— T. H., Capt. 841
Burnett-Hurst, A. R., 2nd-Lt. . 821
Burnham, G. K., Lt. 766
—— G. Le R., Capt. . . 585
Burnley, W. D., Lt. . . . 435, 933
Burns, A. J., Capt. 777
—— A. R., Lt. . . . 534, 1703
—— E. B., Capt. . . . 685, 750
—— F., 2nd-Lt. 1024
—— G. T. A., 2nd-Lt. . . . 560
—— H. J., 2nd-Lt. 750
—— H. P., Asst. Surgn. . . 1799
—— J., 2nd-Lt. 746
—— J. C. H., Capt. 656
—— L. D., Lt. . . . 544, 1704
—— M. M., 2nd-Lt. 791
—— O. H., Sub.-Condr. . . . 1867
—— R. A., Lt. 1695
—— W., Maj. 849
—— W., Lt. 526, 1459
—— W. H., Capt. 843
Burnside, E. A., Col. 124
Burrard, G., Capt. 615
—— S. G., Col. *Sir* 123, 660, 1880
Burrell, E. A., Capt. . 707, 729
—— G. F. D., 2nd-Lt.. . . 759
—— G. P., Maj. 720
—— K., Lt. 420, 1636
—— T. H., Lt. 530
Burrett, F., 2nd-Lt. 362
Burridge, R. A., Lt. . 320, 365, 921
—— R. A., 2nd-Lt. . . 700, 796
—— T. F. S., Capt. 260, 1162, 1305
—— W. T., Capt. . . . 89
Burrough, I. J., 2nd-Lt. 356, 1164
Burroughs, G. H. G., Capt. 501, 807, 1665
—— S. V. G., Capt 262, 1030, 1557
Burrow, S. G., Lt. . . . 538, 1693
Burrowes, P. W., Maj. . 221, 921
Burrows, A. C., Lt. . . 532, 1064
—— G. H. C., Lt. . . 666, 986
—— E., 2nd-Lt. 1271
—— G. R. S. Col. 170
—— G. T., Lt. 1303
—— H. M., Capt. . 253, 1227, 1303
—— L. B., Lt 440
—— S. J., 2nd-Lt. 1676
Burscough, G. E., 2nd-Lt. 360, 1245
Bursey, T. F. Mc.C., Lt. 446—49, 1499—1500
Bur Singh, Jemdr. 1252
—— Jemdr. 896
Burstal, A. C., Lt. 1708
Burt, C. A., Hony. Lt. 779
—— E. H., 2nd-Lt. . . 547, 1705
—— H. C., Capt. . . . 740, 791
—— H. V., Capt. . . 19, 264, 945
—— P. J., 2nd-Lt. . . 366, 911
—— R. F., Lt. 593

Names and Rank. Page.

—— S., 2nd-Lt. 669, 987
Burt, W. G., 2nd-Lt. . . 549, 1178
—— W. H., Capt. . . 704, 746
Burton, C. G., 2nd-Lt. 1187
—— E. B., Col. 57
—— E. R., Lt. . . . 487, 885
—— F., Capt. 658
—— F. C., Col. 56
—— G. H., 2nd-Lt. 474
—— H. A., Lt. 112
—— H. F., Maj. . . . 217, 1279
—— J. A., Lt. 445
—— J. W., 2nd-Lt. 1567
—— R. A., Lt. 1517
—— R. G., Br.-Genl. 12, 38, 105, 120, 124, 149, 183
—— R. O. W., 2nd-Lt. . . . 1451
—— R. W., Lt.-Col. . 194, 1883, 1912
—— T., Lt. 637
—— V. M., 2nd-Lt. . . 357, 1289
—— W. H., 2nd-Lt. 1696
Burvill, H. A., 2nd-Lt. 794
Bury, E. P., Capt. 789
—— J., Lt. 595
—— G. W., Tempy. Capt. . . . 1726
Busar, C. F., Lt. 1583
Busby, T., Condr. 1870
Busch, H. F., Lt.-Col. 821
Bush, A. J., Lt. 536, 1143
—— H. F., Major 821
—— S. L., Lt. 700
—— S. P., Lt. 529, 998
—— W. H., Capt. 1726
—— W. P., 2nd-Lt. 732
—— W. R., Lt. 536
Bushby, W. E., Lt. 523
Bushell, C. W., Maj. . . 664, 991
—— T. O., Lt. 715
Busher, R. C., Capt. 816
Busk, E. W., Ty.-Capt. 619
Buss, B., Capt. 685, 783
—— R. A., Lt. 332, 1600
Bussell, E. W., Asst. Surgn. . . 1803
—— J. C., Lt. . . 312, 440, 1020
Buswell, G., Lt. 508, 997
Buta Khan, Jemdr. 1142
—— Jemdr. 1514
Buta Mal, Sr. Sub.-Asst. Surgn. 1808
—— Sub-Asst. Surgn. . . 1817
—— Resaidar. 948
—— Subdr. 1163
Buta Ram, Jemdr. 966
Butasing, Negi, Subdr
Buta Singh, Jemdr. . 1305, 1376, 1433
—— Jemdr. 950
—— Jemdr. 1149
—— Jemdr. 1204
—— Subdr. 1124
Butchart, A. D., 2nd-Lt. . 560, 1678
Butcher, A. C. S., Capt.. . 717, 1001
—— A. J., 2nd-Lt. 363
—— A. L., Lt. . . . 393, 1203
—— C. R., Lt. . . . 540, 988
—— E. A., 2nd-Lt. . 356, 1215
—— F. C., Lt. 638
—— F. G., Lt. . 66, 295, 1022
—— F. H., Lt. 830
—— G. W. R., 2nd-Lt. . . . 624
—— J., 2nd-Lt. 1008
—— R. A., 2nd-Lt. 354
—— R. E., 2nd-Lt. . . 530, 1042
—— T. R., Asst. Surgn. . . 1795
—— W. H., Lt. 516
Butha Sing Negi, Jemdr. . . . 1182
Butland, R., Lt. 698, 795
Butler, A. E., 2nd-Lt. 619
—— A. G., Lt. . . . 298, 1419
—— A. H., Lt.-Col. 137, 159, 197, 1199
—— A. H., Lt. . . . 410, 905
—— A. H., 2nd-Lt. . . 333, 1505
—— C. C. F., 2nd-Lt. . . . 723a
—— C. J., 2nd-Lt. 749
—— C. T., Lt. . . 450—54 1471
—— C. W., Maj. . . . 87, 220
—— F., Capt. 1729
—— F. H., Lt. 463, 1251
—— F. H. C., Lt. . . . 59, 729
—— G. R., Lt. 315, 1136
—— G. W., Lt. 710
—— H. E., Lt. 492, 939
—— H. M., Maj 213, 870, 1305, 1419
—— H. M., Lt. 387, 887
—— J., Lt.-Col. . . . 179, 1917
—— J. A. N., Lt. 595
—— J. L., 2nd-Lt. 7

Names and Rank. Page

—— M., Lt. 1309
Butler, P. P., Tempy. Capt. . . 1723
—— R. H., 2nd Lt. 1505
—— R. O'N., Lt. 472
—— R. R., Capt. . . 53, 258, 931
—— S. G., 2nd-Lt. . . . 546, 992
—— S. H., Col. *Sir* 1896
—— S. R., Lt. 629
—— W. M., 2nd-Lt. 361
Butt, A. H., tempy. Capt. . . . 1769
—— A. W., 2nd-Lt. 976
—— G. F., 2nd-Lt. . . . 476, 895
—— H., 2nd-Lt. . . . 528, 1675
—— H., 2nd-Lt. 779
—— R. N., 2nd-Lt. 709
Butter, A. H., Lt. . . . .
—— G. R., Lt. 1569
Butt-Gow, H. F. S., Lt. . . . 520
Butterfield, A. R., 2nd-Lt. . . 554
—— E., Maj. 222, 1894, 1913
—— B. F., Lt. . . 40, 435
—— H. R. F., Lt. . . . 693
—— J., Lt. . . . 490, 875
—— L. H., Lt. 516
—— P. H. A., 2nd-Lt. 563, 1524
Butterworth, B., Lt. 758
Buttress, E. S., Lt. . . . 541, 1676
Buxton, V., Lt. 711
Buzzard, A. D., Capt. 501
Byers, C. B., Lt.-Col. . 133, 157, 193, 1660
—— E., 2nd-Lt. 329, 1211
—— G. F., Sr. Asst. Surgn. . . 1789
—— J. A., Lt. 1007
—— R., Condr. 1861
Byesley, R. H., 2nd-Lt. 1473
Bygate, C. M., Lt. . . 446—49 1108
Byles, G. F., 2nd-Lt. . . . 545, 1433
Byne, F., 2nd-Lt. . . . . .
Byramji, P., Ty. Capt. 1770
Byrch, A. E. B., Maj. . . 246, 1663
Byres, E. B. M., Capt. . . 485, 607
Byrne, C., Sub-Condr. 1855
—— E. G. J., Maj. . 227, 871, 1419 1489
—— F., Sub-Condr. 1873
—— F., 2nd-Lt. 505
—— G. L., Lt. 768
—— G. R., Capt. 486
—— H. R., Sr. Asst. Surgn 1792
—— H. S., Lt. 29, 668
—— J., 2nd-Lt. 782
—— T., Lt. 793
Byrnes, W. F., Lt. 544
Byron, F. H., Lt. . . . 525, 1667
—— G., Lt. 516a, 1128
Byura Singh, Sub-Maj. 2104

C

Cabot, C. E., 2nd-Lt. . . 548, 1044
Cabble, E. W., 2nd-Lt. 548
Cable, R. W., Lt. 521, 1283
Cabral, J. R. H., Asst. Surgn. . 1797
—— L. M., Senr. Asst. Surgn 1788
Caddy, A., Maj. 603, 861
—— F., *Hony.* Capt. 690a
Cade, C. M. D., Capt . 95, 501a, 700
—— O. H., Asst.-Surgn.. . . . 1798
Cadell, A., Br.-Gen. . 12, 45, 120, 124, 150, 184
—— A. P. H., Capt. 260, 1383, 1439
—— H. F., Lt.-Col. . . 179, 1918
—— J. G., Maj. . . 98, 217, 1203
—— P. R., Lt.-Col. Hon'ble 821, 1892
—— T., Col. 176
Cadman, A. L., 2nd-Lt. . . . 1676
Cafferata, B. J., 2nd-Lt. . . . 1689
Caffyn, S. C., 2nd-Lt. 556
Cahill, J. J., Asst. Surgn. . . 1800
—— J., 2nd Lt. 318, 545, 1008, 1513
—— M., Hy. Capt. 1843
Cahiran, J., 2nd-Lt. 112
Cahusac, C. F., Capt. . 51, 260, 943
—— S. D. N., Capt. . 272, 893
Caiger, A. B. E., Lt.. . . 311a, 943
Cain, C. W., 2nd-Lt. 725
Caink, A. S., Lt. 629
Caird, D. R., Lt. . . . 533, 1319
Cairn-Duff J. C., Lt. 385
Cairnes, T. A. E., Capt. . . . 583
—— W. E. P., Capt. . . . 589
Cairns, C. A., 2nd-Lt. . . . 324

Names and Rank. Page.

Cairns, J., Maj. 43
—— J., Capt. . . 8, 15, 82
—— J., Tempy. Capt. . . . 1725
Cake, S. F., 2nd-Lt. 791
Calcott, T., Sub-Condr. . . . 1856
Cadell, A. P. H., Capt. . . . 1439
Calder, C C., 2nd-Lt. . . . 568
—— C. K., Capt. 1839
—— J. M., Lt. 375
—— J. T., Lt. 917
—— N. D., Maj. 836
—— O. V., Lt. . . . 541 1223
Calderon, J. J. W., 2nd-Lt. . . . 1695
Caldwell, L. R., 2nd-Lt. . . . 430
—— K. S., 2nd-Lt. . . . 601
Calneck, S. H., Ty. Capt. . . . 1725
Callaghan, G. M., Asst. Surgn. 1792
—— G. F., Maj. . . . 769
Callaway, F. M., 2nd-Lt. . 550, 1701
—— H. W., Lt. 708
—— J., Capt. 756
Callie, G., 2nd-Lt. 761
Callow, J. L., 2nd-Lt. . . 476, 951
Calogreedy, L. H., 2nd-Lt. 548, 1311
Calver, J. E., Lt. 726
Calvert, A. P., 2nd-Lt. 793
—— E., Capt. 1757
—— F. E. R., Lt. . . 513, 1627
—— G. R., 2nd-Lt. . . 492, 949
—— J. T., Lt-Col. . . 128, 1740
—— R. T. C., Lt.-Col. 16, 32, 98, 140, 161, 200, 1483
Calvin, H. F., Lt. 1507
Cam, G., 2nd-Lt. 668
Cambier, V., Capt. . 18, 23, 277, 1148
Cambridge, A. F., Lt. . . 526, 1327
—— E. H., Lt. 579
—— S. J., Lt. . . 533, 998
Cameron, A., Maj. 1752
—— A., 2nd-Lt. 112
—— A. A., Lt. . . . 535, 1413
—— A. J., Lt. . . 76, 1850
—— A. M., 2nd-Lt. . 666, 986
—— A R., Br.-Genl. 98, 120, 125, 743
—— A. W., Lt. . . . 522, 1066
—— B. L., 2nd-Lt. . . . 552
—— C. J. N., Lt.-Col. 78, 485, 827
—— C. S., Maj. . . 224, 919
—— D. A., 2nd-Lt. . . . 558
—— D. H., Maj. . . . 56
—— D. M., 2nd-Lt. . 765, 781
—— E., Lt. . . 295, 1455, 1585
—— E. G., Lt. . . . 493, 915
—— E. H., tempy. Capt. . 1724
—— E. M., 2nd-Lt. . . 558, 1451
—— F. O., Lt. . 446—49, 1433
—— G. S., Lt. . . . 510, 1666
—— G. St. Q., Lt. . . 524, 1347
—— H., Lt. 783
—— H. A., Lt.-Col. 136, 661, 1881
—— H. A., 2nd-Lt. . . . 763
—— H. H., 2nd-Lt. . . 546, 1281
—— H. P., Major. . . . 653
—— J. P., Maj 1749
—— K. K., 2nd-Lt. . . . 786
—— N., Capt. 1717
—— R. H., Lt. 841
—— R. T., Lt. . . . 380, 1617
—— W., Capt. 1732
—— W., 2nd-Lt. 492
—— W. W. L., Lt. 488, 907, 927
Cameron-Smith, G., Capt. . . 619
Cameron-Webb, J. H., Lt. 309, 1074
Caminada, C. B., Lt. . . 354, 1373
Cammell, G A., Capt. . . . 615
Cammiade, L. A., 2nd-Lt. . . 571a
Camm, J., Lt. 836
Campbell, A., Maj . . 103, 663
—— A., Lt. . . 524, 828, 1517
—— A., Ty. Capt. . . . 1771
—— A., Capt. . . . 89, 706
—— A. A., 2nd-Lt, . 545, 1513
—— A. A. E., Br.-Gl. 120, 121, 149, 182
—— A. D., Asst. Surgn 1796
—— A. D. P., Capt. . 279, 1158, 1427
—— A. F., Maj.-Genl . . 172
—— A. F. Capt. . . . 830
—— A. G., 2nd-Lt. . 327, 1347
—— A J., Col. . 126, 150, 185
—— A. J., 2nd-Lt. . . . 728
—— A. R., 2nd-Lt. . . . 727
—— A. Y. G., Maj. . . 611
—— C., Capt. . . 257, 1311
Campbell, C., 2nd-Lt. . . 547, 731a
—— C., Lt. . . . 75, 517
—— C. F., Lt.-Col. . . . 1
—— C. F., Capt. 617
—— C. H., Maj. 95, 241, 964
—— C. M. St. A., Lt. . 313, 1635
—— C. R., 2nd-Lt. . . . 739
—— C. S, Lt. 543a
—— D., 2nd-Lt. . . 742, 764
—— D., 2nd-Lt . . 543, 1621
—— D. D., Lt. . . . 652
—— D. D. H., Maj. . . . 634
—— D. L., Lt. . . . 524, 1042
—— E. C. M., 2nd-Lt. . . 624
—— E. G., Capt. . . . 636
—— F., Lt.-Genl. Sir 12, 20, 118, 172, 145, 1183
—— F. C. G., Capt. 95, 263 1183
—— F. E. D., Capt. . . . 741
—— F. R. M., Capt. . 237, 1371
—— G. C., Maj. 26
—— G E. F., Lt. . . 537, 1668
—— G. P., Bt.-Col. 24, 125, 660
—— G. R., Lt. . . . 489, 933
—— H., Maj. . . . 213, 964
—— H. B. D., Lt.-Col. 141, 661
—— H. D., N. Sister . . . 1735
—— H. G. M., Asst. Surgn. 1801
—— H. N. C., Capt. . . . 611
—— I., Capt. . . . 274, 895
—— I. H., Maj. . . . 57, 1919
—— I. P. F., Capt. . 733, 734
—— J., Maj. . . . 241, 1136
—— J., Lt. . . 425, 828, 1311
—— J., 2nd-Lt. . . 475, 1513
—— J. D., Maj. . . 664, 1880
—— J. E., Col. . . . 177, 1917
—— J. R., Capt. 808
—— J. W., Lt. . 188, 530, 1598
—— K. G., Maj. . . 101, 633
—— L. S., Lt. . . 6, 57, 93
—— L. W. Y., Col. 12, 23, 121, 126, 150, 187
—— M., 2nd-Lt. 1373
—— M. G. G., Capt. . 665, 996
—— M. H. A., Capt. . 277, 1363
—— M. S. C., Col. 71, 74, 124, 612
—— N. D. H . Maj. . . . 583
—— R., 2nd-Lt. 562
—— R. B., Lt. 1695
—— R. C., Lt. 704
—— R. J. F., Asst. Surgn. 1803
—— R. N. R., Maj. . 246, 1110, 1112
—— R. J. T., 2nd-Lt. . . 739
—— R. S. 1009
—— R W., Revd. 1004
—— T., Lt. 691
—— T. E., Lt. . . 516a, 1624
—— T. J., Lt. . . . 26, 528
—— T. P., Capt. . . 76, 518, 807
—— W., Maj. . . 232 1662
—— W., Capt. 1724
—— W., Capt. 1717
—— W. B., Lt. . . 380, 1581
—— W. E. D., Capt. . 262, 939, 1447
—— W. F., Capt. . . 279, 1231
—— W. J., Capt. . . . 830
—— W. J., Lt. . . . 415, 1311
—— W. L., Maj. . 102, 107, 218, 1913
—— W. N., Brig.-Genl. . . 53
Campbell-Harris, C., Lt. 294, 907, 965
Campbell Shaw, H. H., Lt. . . 507
Campbell-Smith, 2nd-Lt. . . . 90
Campion, K. F., Lt. 681
Campfield, O. G., Lt. 630
Candler, L. M., 2nd-Lt. . . . 610
Candy, H. R., Capt. . . . 276, 1307, 1894
—— R. H., Capt. 1757
Cane, C. H., 2nd-Lt. 750
—— C. W., Lt. 725
—— L. B., Ty. Capt. . . . 1723
—— S. P., 2nd-Lt. . 465, 473, 1291
Canesha Singh, Jemdr. . . . 900
Cangley, F. G., Lt. . . 450 54, 1363
Cann, H. D., Capt. 778
—— M., Lt. . . . 534, 1668
Cannell, J. H., Lt. . . . 699, 796
—— J. McA., Hy. Lt. . . 1848
—— M. D., 2nd-Lt. . . 552, 1499
Canney, D. A., Revd. . . . 1903
Canning, J D., Capt. . . . 780
Cant, A., Revd. 1907
—— E. D., Lt. . . 300, 1343, 1587
—— H. D., Condr. 1870
Canteenvalla, J. P., Tempy. Capt. 1764
Cantlie, K., Lt. 512
Cantor, A., Capt. 832
—— C. H., Col. 177
Cantrell, W. C., Capt. 833
Capell, G. C., 2nd-Lt. 689
Capes, C. J., 2nd-Lt. 749
—— G. A., Lt. . 405, 729, 1613
Capon, A J., 2nd-Lt. 361
Capper, A. S., Lt.-Col. . 34, 98, 137, 159, 197, 867, 949
—— G. W., Maj. 707
Capron, G. T. H., 2nd-Lt. . 666, 996
Capstick, F. H., 2nd-Lt. 91—92, 548
Carbery, M., Lt. 619
Carberry, T. B., Lt. . 316, 1409, 1467
Carbould, J. H., 2nd-Lt. . . . 1008
Card, F. W. F., Lt. 581
Carden, F. G. Lt. . . . 430, 1343
—— H. C., Revd. 1004
Carder, C. D., Maj. 628
Cardew, A. M., Maj. 102, 104, 663, 1889
—— D., Capt. 840
—— F. G., Maj. 51
—— H. S., Maj. 245, 1003, 1178
—— J. H., Lt. . . . 504, 833
—— W. F., Lt. 1001
Cardwell, C. R., 2nd-Lt. . . . 740
—— W. A., 2nd-Lt. . 548, 1417
Careless, A. L., Maj. . . . 717
—— J. E., Capt. . . . 611
—— W. E., Revd. . . . 1906
Carew, R. L. O., Capt. 635
—— T. P., Lt. . 536, 1032, 1676, 1678
Carew-Hughes, C. F., Lt. . 415, 1082
Carew-Hunt, A. L., 2nd-Lt. . . 729a
Carew Smith, W., Lt.-Col. . . 841
Carey, A. B., 2nd-Lt. . . . 547, 1275
—— C. O'D., 2nd-Lt. . . 546, 1024
—— C. W., Maj . . . 208, 664
—— F., Condr. 1863
—— H. R. LeM., Lt.-Col. 178, 1917
—— J., Lt. 401, 1108
—— J., Esq. 46
—— J. E., Maj. . . . 234, 1132
—— M. D., Lt.-Col. . . 21, 712
—— O. W., Col. . . 33, 121, 125, 150, 188
—— P. G., Maj. . 238, 1114, 114
—— R. T. O., Lt. 636
—— T. deB., Maj. . . 214, 871, 1471, 1473
—— T. W., 2nd-Lt. . . . 357, 1116
—— W. L., Lt. 859
—— W. L. J., Lt.-Col. . . 10, 74, 101, 103, 138, 632
—— W. R., Capt. 1919
Cargill, A., Lt. 630
—— A. K., Lt. . . . 526, 988
—— J. R., 2nd-Lt. . . 79, 494
—— R. J., Maj. . . . 227, 1124
Cargwell, V. S., 2nd-Lt. . . . 699
Carles, J. R., 2nd-Lt. . . 331, 1080
Carless, H. A., 2nd-Lt. . . 549, 1475
Carleton, A. W., Asst. Surgn. . 1801
—— H. A., Lt.-Col. . . . 1910
—— K. O., Lt. . . 292, 1120, 1541
Carleton-Smith, D. L. G., Lt. 581
Carlile, R. L. H., 2nd-Lt. . 357, 1675
—— T. A., Lt. . . 446—49, 1044
Carlill, C., 2nd-Lt. 640
—— S. R., Lt. 780
Carling, T. E., Lt. . . . 420, 1225
Carlisle, T. R. M., Maj. 103, 614
Carlton, R. S. C. L.C., 2nd-Lt. 553, 1676
Carman, C. R., 2nd-Lt. . . . 723a
Carmichael, C., 2nd-Lt. . . 531, 1699
—— D. G., Maj. . . . 1711
—— T., Lt. . . . 487, 941
—— K. M. R., N. Sister . 1735
—— W. B. G., 2nd-Lt. . 329
—— W. G. H., Lt. . . . 843
Carn, A. T., 2nd-Lt. 559
Carne, E. F., Lt. 617
Carnegie, C. A., Capt. *Lord* . . . 2
Carnegy, C. G., Bt.-Col. 126, 153, 183
—— G. P. O., Lt.-Col. 139, 160, 199, 867, 887
—— P. M., Maj.-Genl. 118, 146, 173
Carney, J., 2nd-Lt. 762
—— M., Sub-Condr. 1856
Carnochan, D. S., Lt. . . . 617

Names and Rank. Page.

Caroe, O. K., Capt. 679
Carpendale, F. M., Bt.-Maj. 53, 262, 1191
—— T. M., Capt 256, 889
—— W. St. J., Capt. 276, 1595
Carpenter, D. B. B., 2nd-Lt. . . 90
—— E. C., 2nd-Lt. . . . 567
—— E. H. O., 2nd-Lt. . . 733
—— H. J., Lt. 387
—— H. R., Capt. . . . 696
—— J. E., Lt. 1720
—— J. H., Capt. 486, 818, 885
—— P. H., Lt. . . 488, 899
Carr, C., Capt. 636
—— C. T., Temp. Maj. . 8, 15, 756
—— C. W., 2nd-Lt. 630
—— D. G., Lt. . . . 423, 1475
—— F. J. B., Lt. 516
—— H. J., Lt. 298a, 1160
—— J. F., Lt. 638
—— P. G., Lt. 583
—— W., 2nd-Lt. 767
—— W. R., Lt. . . . . 455—54
Carral, E. F., Lt. . . . .
Carr-Ellison, H. G., Capt. . . . 598
Carr-Hall, Maj., R. E. 87, 209
Carr-White, G., Capt. . 280,—287 919
—— P., Col. 28, 126, 1738
—— R. A., Capt. . 276, 933
Carreck, F. F. R., Snr. Asst. Surgn. 1769
Carreras, L. A., 2nd-Lt. . . . 560
Carrick, H. E., Lt. 524
—— R. B., Lt. 652
Carrigan. P., *Hony.* Maj. . . 1845
—— R. W., 2nd-Lt. . 622, 973
Carrington, S. S., 2nd-Lt. 547, 1674
Carroll, B. M., Maj. . . 236, 1355
—— A. D., 2nd-Lt. 851
—— C. W. A., Capt. 851
—— E. F., Lt. 505
—— E. W., 2nd-Lt. 554
—— F., Sub-Condr. 1799
—— H. O., Maj. . . . 224, 1662
—— J. E. M., Capt. 707
—— S. R., Lt. 526, 1327, 1662
—— T., Sub-Condr. 1870
Carron, F. P. G., Lt. . . . 638
Carruthers, A. L., 2nd-Lt. . . . 547
—— B. M., Capt. . . . 1724
—— F. W. W., Lt. . 445, 1088
—— C. F. J., Lt. . . . 1720
—— H., Lt. 1653
—— H. St. J., Lt. . 294, 1629
—— J. G. T., Col. . 169, 1916
—— J. H. deW., Maj. . 14, 22, 98, 1178, 1227
—— R. A., Br.-Genl. 53, 120, 123, 149, 182
—— R. J., Lt. . . 318, 1647
—— W. O., Lt. . . 375, 911
Carse, W. M., 2nd-Lt. . . 333, 1028
Carslaw, F. H., Capt. . . . 602
Carson, C. F., Capt. . . 665, 1882
—— C. G., 2nd-Lt. . . 22, 706
—— C. W., Capt. 804
Carstairs, G., 2nd-Lt. 562
—— G. S. de., Lt. . . . 1451
—— N. M., Lt. . . 298, 1397
Carstens, J. E., Lt. . . 463, 1607
Carter, A. B., Lt. . . . 68, 517, 1235
—— A. F., Lt. 732
—— A. C., Lt. 492, 913
—— A. J., Capt. 651
—— A. J., 2nd-Lt. . . 362, 1086
—— A. V., Lt. . 446—49 732, 1453
—— A. W., *Hony.* Lt. . . . 754
—— B. B., Revd. 1904
—— C. C., Lt. 740
—— C. H., Lt. 61
—— E. J., Maj. 591
—— E. J., Lt.-Col. . . 38, 1660
—— E. J., 2nd-Lt. 494
—— E. M., Lt. 380
—— E. N., Capt. . . 267, 1451
—— E. P., Lt.-Col. . 74, 134, 632
—— F., Capt. 696
—— F. G., Capt. . 501, 605, 1664
—— F. N., Lt. . . . 351, 1467
—— F. J., 2nd-Lt. 67
—— F. W., Hony. Capt. . . 859
—— G. E. L., 2nd-Lt. . . . 563
—— G. H. H., Lt. . . 351, 1511
—— G. L., Lt.-Col. 132, 156, 193, 851, 1427
—— G. T., Lt. . . . 538, 1695
Carter, H. B., 2nd-Lt. 739
—— H. L., Lt. 759
—— H. L. M., Lt. . . 143, 927
—— H. P., 2nd-Lt. 732
—— J., 2nd-Lt. 618
—— J., Sub-Condr. 1864
—— J. F. C., Maj. . . 226, 1894
—— J. H., 2nd-Lt. 565
—— J. L., Lt. . . . 311a, 1078
—— J. R. R. G., Lt.-Col. 189, 1892, 1911
—— J. S., Lt. 778
—— L. H., Lt. . . . 353, 976
—— L. J., Maj. 745
—— R. F. F., Lt. . . . 95, 317, 1078
—— R. M., Bt. Lt.-Col. 135, 1748
—— S. N., Lt. . . . 397, 1012
—— T. F., Lt. . . . 532, 988
—— T. L., Lt. . . . . 455—59
—— W. H., Lt. . . . 512, 1898
—— W. T., Lt. . 317a, 365, 1475
—— W. W., 2nd-Lt. 751
Cartland, W. N., Lt. 654
Cartlidge, C., Lt. . . . 313, 911
Cartman, S. A., Snr. Asst. Surgn 1790
Cartmel, A. E., 2nd-Lt. . . 359, 957
Cartner, J. R., Asst Surgn. . . 1802
Cartwright, C. L., Capt. . . . 801
—— C. M., Col. 54, 123, 182, 149
—— E. A., Lt. 748
—— G. N., Brig.-Genl. 24, 122, 132, 613
Cartwright, H., Lt. 785
—— T. H., 2nd-Lt. . . . 1645
—— W. A., 2nd-Lt. . 358, 1405
Caruana, A. J., Col. 8, 125, 150, 185
Carus-Wilson, M. M., Lt. . 738, 740
Caruth, A. C., Lt. . . . 311a, 1629
Carver, F. F., 2nd-Lt. 738
—— L. H. L., Lt. 810
Carvosso, J. P., Capt. 752
—— E. W., Capt. 752
Carwithen, E. T., Lt.-Col. 33 136, 158, 195, 1601, 1883
Carry, W. A., 2nd-Lt. . . 562, 1678a
Case, C. E., 2nd-Lt. . . 555 1614
—— H. A., Bt.-Lt.-Col. . 136, 736
—— J. B., Lt. 488, 901
Case-Morris, C. W., Maj. . 749, 781
Casey, C O., Sub-Condr. . . . 1867
—— J. W., 2nd-Lt. . . 357, 992
—— T. H., 2nd-Lt. 531
Cashmore, A. E., Lt. . 314, 929, 959
—— W., Hony. Lt. . . . 819
Casling, C. T., Lt. 516
—— D. M., Lt. . . 84, 528
—— P. V., Asst. Surgn . . 1800
—— W., Lt. 543a, 1016
Cassell, A. D., Lt. . . . 311, 1343
—— H. E., 2nd-Lt. 776a
Cassels, W. C., Tempy. Capt. . . 1723
—— G. R., Lt.-Col. . 98, 133, 157, 193, 871, 1487
—— K. S., Lt.-Col. 25, 134, 157, 194, 1635
—— R. A., Brig.-Genl. . 50, 97, 121, 125, 153, 211, 935
Casserly, Maj. J. H. 1919
Cassidy, V. C., Capt. . . 269, 1644
Casson, R., Lt. 514, 653
—— L. B., 2nd-Lt. 494
—— W. F. S., Bt. Lt.-Col. 13, 24, 98, 139, 160, 214, 925
Castellino, L., Tempy. Capt. . . 1772
Castells, E., Lt. 313, 1623
Castle, E. D. F., 2nd-Lt. . . . 558
—— J. P., Lt. . . 531, 1164, 1418a
—— W. W., Revd. 1904
Castor, R. H., Lt.-Col. . . 127, 1741
Cater, A. N. L., Lt. 841
—— L. J., Lt. 540
Cates, B. H., Lt. 517
—— K. G. H., Capt. . . 260, 1203
—— W. J., Maj. . . . 206, 1661
Catling, A., Lt. 829
Catron, J. E. T. Lt. . . 371, 1223
Cattanach, A., Lt. . 541, 1106, 1533
Cattell, A. S., Lt. . . . 508, 1656
—— E. E., Lt. 624
—— G. L., Lt.-Col. . 139, 160, 199, 1629, 1883
—— J. G., 2nd-Lt. 571
Catterall, N. C., 2nd-Lt. . . . 787
Cattermole, F. C., 2nd-Lt. . 545, 1181
Catterson-Smith, T. M. O., Capt. 54, 267, 1066
Cattoll, C., 2nd-Lt. . . . .
Catty, T. C., Bt.-Lt.-Col. . 53, 98, 160, 220, 1287
Caudle, H. C. R. Capt. 615
Caulfeild, C. T., Brig.-Genl. 22, 120, 612
Caulfield-Giles, H. R., Lt. . 505, 1673
Caunter, H. E., 2nd-Lt. . . 315, 1183
Cavaye, R. M., 2nd-Lt. 765
Cave, A. M., 2nd-Lt. . . . 751, 1006
—— A. R., 2nd-Lt. 1009
—— C. R., Lt. 514, 1604
—— H., 2nd-Lt. 1305
—— T. S., Hony. Col. *Sir*, . 729
Cave-Allan, W. N., 2nd-Lt. . . 781
Cave-Browne, H., Capt. 16, 25, 919, 258
—— W., Capt. . . 665, 985
Cavell, R. G., Lt. 78, 487
Cavenagh, O., Capt. 619
Cavenaugh, P. H., 2nd-Lt. . . 707
Cavendish-Bentinck, Lord H. C., *Hony.* Col. 724, 725
Cawley, K., Nursing Sister . 1735
Cawthorn, W. J., Lt. . . . 440
Cawthron, E., 2nd-Lt. 567
Cawood, C. W., Lt. . 316, 1327, 1591
Cay, C. S., Lt. 367
Cayford, W., 2nd-Lt. 61
Cayley, P. E., Lt. 314
—— W. de S., Maj.-Genl. *Sir* 119, 124
Cazalet, R. G., Lt. . . . 489, 907
Cazaly, W. H., Maj. 1747
Cearns, G. W., Sr. Asst. Surgn. 1791
Cecil, R. E., Maj. 597
Celestian, C., Sub-Asst. Surgn. 1827
Centeno, J. B., Lt . . . 516a, 1020
Chabilal Gurung, Jemdr. . . 1637
Chabot, V H., Capt. . . . 843
Chadwell, A. E., Lt. . . 696, 697
Chadwick, C. R., Lt. 750
—— P. T., Maj. 714
—— W. H., 2nd-Lt. . . 1327
Chadwyck-Healey, H. P., Lt. . . 720
Chagar Singh, Subdr.-Maj. . . 1171
Chagatta, Jemdr. 989
Chaghta Khan, Jemdr. . . . 1095
Chahat Khan, Jemdr. 1480
Challu, Jemdr. 1404
Chainchal Singh, Jemdr. . . . 1533
Chain Singh, Jemdr. 894
—— Jemdr. 916
—— Jemdr. 1376
—— Jemdr. 896
Chainsukh Ram, Subdr. . . . 1457
Chaltan Singh, Jemdr. 1071
Chajja Singh 1167
—— Jemdr. 1108
Chajjo Ram, Jemdr. 1478
—— Jemdr. 1473
Chajju Ram, Ty. Sub-Asst.-Sugrn. 1833
Chakar Sing Ale, Jemdr. . . . 1597
Chakarpan Singh, ... 1413
Chakrabahadur Limbu, Subdr. 1648
Chakrabatti, D. C., Ty. Lt. . . 1778
Chakrapani, Jemdr. 994
Chakravarti, D. C., Ty. Lt. . .
Chakravarty, N., Ty. Lt. . . . 1775
Chalam, B. S., Capt. 1765
Chaldecott, O. A., Maj. 238, 1491, 1495
—— W. H., Maj. . . 94, 662
Chalk, J., Lt. 536, 1217
—— P., Lt. 748
—— W. G., 2nd-Lt. . . . 555, 1295
Chalke, J. C., Asst. Surgn. . . 1794
Chalker, J. R. 618
Challans, F., Ty. Capt. . . . 1726
Challen, B. S., Lt. . 299, 1303, 1499
—— P., Lt. 293a, 1181
Challice, J., 2nd-Lt. 567
Challis, C. L., Lt. 695
Chalmers, A., Maj. 1747
—— D., Lt. 820
—— H. A., Lt 855
—— T. A., Lt. . . . 76, 528
—— T. C., Lt. . 307a, 432, 923
—— T. H., Lt. 581
Chaman, Jemdr. 1256
—— Singh, Jemdr. 1472
Chamanpal Singh, Subdr. . . 1471
Chamarett, G. R., Sr. Asst. Surgn. 1789
—— R. D., 2nd-Lt. . . . 1704
Chamba Singh, Jemdr. . . . 936

NAMES AND RANK. Page.

Chamberlain, H., 2nd-Lt. . . 859
—— J. A., Lt. . . 10, 76
—— L. L., Lt. . 317, 1128
—— W. G., Lt. . 38, 1273
—— W. H., Lt. . . . 435
—— W. W., Lt. . . 1018
Chamberlayne, A., Maj. . 223, 870, 875, 1321
Chamberlen, E. F., Lt. 294a, 1359, 1469
Chambers, C. C., Capt. . . . 608
—— C. S., Lt. . . . 748
—— G. H., Capt. 31, 273, 1487
—— G. W., 2nd-Lt. . . . 626
—— H. M., Lt. . . 380, 1467
—— J. T., Sub-Condr. . . 1855
—— L. D., Maj. . . 776
—— L. E., Asst. Surgn. . 1794
—— P. R., Maj. 52, 98, 893. 221
—— R., Sub-Condr. . . . 1857
—— R. A., Maj. . . . 1753
—— R. M., Maj.-Genl. . . 170
—— T. B. W., 2nd-Lt. 539, 998
—— W. B., Hony. Lt.-Col. 1663
—— W. J., *Hony.* Lt. . . 1844
—— W. R., 2nd-Lt. . . 816
Chamen, R. T., Lt. 740
Chamier, A., Lt. . . . 530, 1367
—— F. E. A., Maj.-Genl. . 172
—— J. A., Maj. . . 240, 1152
—— R. O., Capt. . . 267, 1443
—— W., Lt.-Col. . . 830
—— W. St. G., Lt.-Col. 19, 138, 160, 198, 1661
Chamingo, R. G., 2nd-Lt. . . . 1150
Chamman Lal, Sub-Asst. Surgn. 1816
Chamman Pal Singh, Jemdr. . 1473
—— Subdr. . 1412
Champa Sing Gurung, Subdr. . 1602
—— Thapa, Jemdr. . 1642
Champha Singh Gurung, Jemdr. 1602
Champion, H. C., Lt. . 311a, 887, 1677
—— F. W., Lt. . . 95, 490
—— O. D., 2nd-Lt. . 558, 858
—— P. H., Capt. . 778, 1899
Champion-de Crespigny, H., Maj. 238, 1245, 1247
Chamu Singh Burathoki, Hony. Capt. 1896, 2101
Chamu Thapa, Subdr. Maj. . . 1614
—— Jemdr. 1615
Chanan Das, Ty. Sub-Asst. Surgn. 1833
Chanan Singh, Subdr. . . . 1235
—— Jemdr. 999
—— Jemdr. 892
—— Jemdr. 1492
—— Deol, Sub-Asst. Surgn. 1819
Chance, F. S., Capt. . . . 724
—— P. V., Lt. 525
—— W. H., Lt. 832
Chand, A., Capt. . . . 1355, 1758
—— Subdr. 1171
—— H., Capt. 1758
—— Khan, Subdr. . . 1303, 1509
—— Pir Shah, Jemdr. . . . 1154
—— Khan, Subdr. . . . 1086
—— Subdr. . . 1479, 1587
—— Subdr. . . . 1479
—— Subdr. 972
—— Singh, Subdr. 989
—— Risldr. 930
—— Jemdr. 1188
—— Subdr. 1473
—— Subdr. . . 1111, 1112
—— Jemdr. 1151
—— Subdr, . . 1227, 1493
—— Jemdr. 1157
—— Subdr. 1160
—— Jemdr. 1129
—— Jemdr. 1240
—— Jemdr. 1473
—— Jemdr. 999
—— Jemdr. 920
Chandan, Subdr. 1471
Chandan Lal, Subdr. 1581
—— Sing, Jemdr. 1412
—— Jemdr. 1056
Chandarbir Thapa, Jemdr. . . 1615
—— Khattri, Jemdr. . . 1643
Chandar Singh, Risaldr-Maj. . . 883
—— Jemdr. . . 1022, 884
—— Jemdr. . . 924, 884
—— Gurung, Jemdr. . . 1637
Chandar Singh Gurung, Jemdr. . 1619
Chanderpal Singh, Subdr. . . 1044
Chander Rao Sinde, Jemdr. . . 1464

Chandgi, Subdr. 1040
—— Jemdr. 1050
—— Ram, Subdr. 1039
—— Jemdr. 1060
—— Jemdr. 1051
Chandi Pershad Mukherji, Lt. 1489
Chandi Singh, *Hony* Capt. . . 2101
Chandika Singh, Jemdr. . . 1043
Chandikaparshad Tiwari, Subdr. 1018
Chandka Singh, Jemdr. . . . 1328
Chandler, F., Condr. 1873
—— H. G. W., Lt.-Col. . 46, 87, 134, 157, 194
—— T. H., Lt. 789
Chandorkar, B. B., tempy. Capt. 1768
Chandos-Pole, G. B., Lt. 587, 713
Chandra, J., tempy. Capt. . . 1769
—— Pali Gupta, Sub-Asst. Surgn 1819
—— Ram, Jemdr. . . . 1537
—— R., Tempy. Capt. . . 1771
—— S. N., Tempy. Lt. . . 1771
Chandra Bir, Subdr. 966
—— Khatri, Jemdr. . . 1643
Chandrabir Burathoki . . . 1622
Chandrabir Thapa, Subdr. . . 1607
Chandra Chur Singh, 2nd-Lt . . 482
Chandrapali Gupta, Sub-Asst. Surgn. 1819
Chandrapal Singh, Subdr. . . 1043
—— Jemdr. . . 1043
Chandra Ram, Subdr. . . . 1489
—— Sing Gurung, Jemdr. . 1619
—— Jemdr. . . . 1604
—— Jemdr. . . 1625, 1637
Chandri Chand, Subdr. . . . 1225
Chandu, A., Sub-Asst. Surgn. . 1827
Chandulal, Jemdr. 1197
Changate Khan, Subdr. . . . 1367
Chanin, P. B., 2nd-Lt. . . . 1189
Chaning Perce, M. S., Lt. 738, 739
Channan Shah, Jemdr., . . . 989
Channan Singh, Jemdr. . . . 894
—— Jemdr. 1212
—— Jemdr. 1256
—— Subdr. 1371
Channell, E. C., Lt. 518
Channer, B. G., Maj. 16, 31, 230, 1239
—— G., Maj. . . . 243, 1074
—— G. K., Maj. 209, 872, 9031, 1609
—— G. O. DeR., Capt. 271, 1629
—— K. F., Capt. . . 274, 943
Chanter, A. B., Capt. 843
—— A. D., Maj. . 28, 237, 1663
Chaplin, E. D., Esq. 91
—— E. L., Lt. . . 318, 1443
—— E. T., 2nd-Lt. . . . 735a
—— R. E., Lt.-Col. 131, 175, 191, 867, 889
Chapman, A., Condr. 1873
—— A., Sub-Condr. . . . 1857
—— A. C., *Hony.* Lt.-Col. 57
—— A. F., Lt. . . 539, 992
—— A. J. B., Lt. . . 526, 1671
—— A. R., 2nd-Lt. . . . 470
—— A. R., Lt. . . 491, 879
—— A. R., Col. 176
—— C. A., Lt. . . . 318, 1299
—— C. W., Lt. . . . 61, 520
—— D. G., 2nd-Lt. . . . 1565
—— E. A. J., Lt. . . 852, 853
—— E. F., Genl. *Sir*. . . 643
—— E. H., Capt. 262, 1150, 1431
—— E. J., Lt. . . . 351, 1641
—— E. S., tempy Capt. . . 1723
—— F. E., Sub-Condr. . . 1868
—— G. P., Lt. . . . 532, 1152
—— H., Maj.-Genl. . 173, 889
—— H. R., Capt. . . 389, 919
—— J. J., 2nd-Lt. 559, 587, 797, 1060
—— L. R. A., Lt.-Col. . . 972
—— P. F., Lt.-Col. . 133, 1743
—— P. H., Revd. . . . 1903
—— R. S., 2nd-Lt. . . . 786
—— R. V., Lt. . . 399, 968
—— V. D., Lt. . . . 787
—— W. L. B., Maj. . 244, 1363
Chappel, B. H., Lt. . . . 385, 722
Chappell, F. S., Lt. . . . 543a
—— H., Lt. . . . 307, 1178
—— W. F., Maj. . . . 583
Chapple, C. H., 2nd-Lt. . . 66, 997
—— G. L., Capt. . . 721, 739
Charan Singh, Subdr. 1395
—— Jemdr. 1493

Charan Singh, Jemdr. 1157
—— Deol, Sub-Asst. Surgn. 1819
Charat Singh, Subdr. 1189
Chard, A. R., Lt. 371, 947
—— R. H., Capt. 628
—— E. H., Lt. 630
—— S. B., 2nd-Lt. . . . 695
Charles, B., Sub-Condr. . . . 1857
—— J. R., E., Bg.-Genl. 51, 97, 119, 126, 662
—— F. D., Lt. . . . 410, 1183
—— P. J., Lt. 387
—— R., Lt. 630
—— Alexander, Sub-Asst. Surgn. 1819
Charlesworth, E. Fitz G., Lt. . 679
—— E. F., Lt. . . . 798
—— F., 2nd-Lt. . . 556
—— F. W., 2nd-Lt. . . 358
—— J. H., Lt. . 389, 1174
—— J M. 723a
Charleton, H. J., Lt. 836
—— J., 2nd-Lt. . . . 765
Charlton, C. E. C. G., Lt.-Col. 133, 613
—— F. J., Lt. . . 375, 1565
Charman, L. C., 2nd-Lt. . . . 780
Charmarette, R. D. C., 2nd-Lt. . 566
Charmichael, W. B. G., 2nd-Lt. . 1487
Charnley, F. H., Lt. 399
Charnalia, G. R., Ty. Capt. 1771, 1789
Charriol, P., Lt. 487
—— R. A., Lt. . . 528, 1007a
Charter, W. F., Lt. . . 397, 1102
Charters, W. C. M., Snr. Asst. Surgn. . . 1790
Chartlon, W. D., 2nd-Lt. . . . 693
Chartres, B. J., 2nd-Lt. . . . 557
—— C. B., Lt. 651
Chartu, Jemdr. 1188
Chase, A. J., Maj. 836
—— H. L., 2nd-Lt. . . . 583
—— J. H., Capt. 836
Chastel De Boinville, G. A., Lt. . 1672
Chatarji, M. N., Ty. Lt. . . . 1776
Chatar Singh, Jemdr. 970
—— Jemdr. 1260
Chatarpal Singh, (I) Subdr. . . 1191
—— Jemdr. . . . 1063
Chater, A., 2nd-Lt. . . . 359, 1405
—— C. W., Lt. . . . 516a, 893
—— A. F., Capt. . . . 665, 985
Chaterjee, S. N., Ty. Lt. . . . 1781
Chatfield, H. S., 2nd-Lt. 35, 539, 1883
Chathotk Kannan, Sub-Asst. Surgn. 1827
Chatrapal Singh, Jemdr. . . 1047
Chatra Ram, Subdr. 1457
Chatru, Jemdr. 1336
Chattar Singh, Risaldar-Maj. . 889
—— Subdr. 1231
—— Jemdr. 1163
—— Jemdr. 896
—— Jemdr. 945
Chatterjee, A. C., Tempy. Lt. . 1773
—— B., Ty. Lt. 1776
—— Bhutnath, Rai Sahib 91
—— B. B., Ty. Lt. . . . 1778
—— J., Tempy. Lt. . . 1775
—— N. N., Tempy. Lt. . 1774
—— N., Tempy. Capt. . 1765
—— R. K., Ty. Lt. . . 1779
—— S. N., Tempy. Lt. . 1782
Chatterji, B., Ty. Lt. 1776
Chatterton, A. L., 2nd-Lt. . . 713
—— B. R., Lt.-Col. . 130, 1742
—— E., Rt. Revd. . . . 1903
—— F., Capt. 611
—— F. W., Col. . . 196, 1916
—— G. D. L., Lt.-Col. 131, 156, 192
—— W. H. H., 2nd-Lt. . 571a
Chatter Singh 1699
Chatur Sing Gurung, Jemdr. . 1615
Chature Rai, Jemdr. 1649
Chaudhari, B. N., Tempy. Lt. . 1775
—— B. R., Ty. Lt. . . . 1776
—— Jemdr. 990
—— Amar Singh, Risaldr. . 1686
—— Khan, *Hony.* Capt. . 2103
—— Nihal Singh, Ressdr. . 1686
Chaudhri, Ram, Sub-Asst. Surgn. . 1816
—— Badan Singh, Jemdr. . 1017
Chaudhri, Jemdr. 1248
Chaudhri Khan, Jemdr. . . . 1492
—— Subdr.-Maj. . . 1375
—— Narain Singh, Jemdr. . . . 1456

Names and Rank. Page.

Chaudhri, Padam Singh, Jemdr. 1384
Chaudhuri, M. N., Maj. 1746
——— B. N., Tempy. Lt. 1775
——— S. K., Ty. Capt. 1768
——— Balwant Gujar Singh, 2nd-Lt. 482
Chaudhury, A. N. R., Ty. Lt. 1775
——— G. C., Ty. Capt. 1771
Chaudri Khan, Jemdr. 1133
Chauharja Pershad Misr, Subdr. 1046
Chauncy, C. H. K., Maj. 210, 871, 1507
Chaundler, O. S., Lt. 371, 1046, 1557
Chavan, N. M., Tempy. Capt. 1767
——— S. G., Ty. Lt. 1537, 1779
Chawand Singh, Ressdr. 951
Cheadle, G., 2nd-Lt. 560, 1678
Cheale, A. R., Maj. 751
Cheape, G. R. H., Lt.-Col. 140, 581
Cheeke, A. M. M., Maj 245, 1419
Cheers, J. McG., Hony. Capt. 7, 1850
Cheeseman, E. J., Lt. 1846
Cheesman, E. J., 2nd-Lt. 740
——— G. W., 2nd-Lt. 598
——— R. E., Lt. 524
Cheetham, H. J., Sub-Condr. 1857
——— R. H., 2nd-Lt. 735
Cheharsing Thapa, Subdr. 1630
Cheke, E. G., Lt.-Col. 102, 628
Chelmsford, *Rt. Hon.* F. J. N. *Lord* 2, 603, 738, 806, 807, 810
Chenery, A. F., 2nd-Lt. 1674
Cheney, J. G., Capt. 1693
Chenevix-Trench, G. H., Lt.-Col. 59
——— H., Lt. 371, 1621
——— R. H., Maj. 211, 1898, 1913
Chengalvarayan, P. A., Sub-Asst. Surgn. 1827
Chenoy, F. B., Capt. 1758
Cheriyan, P. V., Tempy. Capt. 1771
Cherry, J. A., Maj. 40, 60, 519, 821
——— J. W., Capt. 714
Cheshire, H. S., Capt. 501, 986
——— F. M., Lt. 803
——— W. K., 2nd-Lt. 552
Chesney, C. H. R., Maj. 664, 1882
——— D. P., Capt. 55, 103, 261, 110
——— E. C., Capt. 9, 102, 259, 1664
——— H. G., Maj. 245, 1030
Chester, B. F., 2nd-Lt. 725
——— A. J. B., Lt. 507, 1657
——— C. F. L., 2nd-Lt. 750
——— S. C. R., Lt. 75
Chesterton, P. H., 2nd-Lt. 807
Chet Ram, Ressdr. 888
——— Jemdr. 1121
——— Jemdr. 932
——— Singh, Subdr. 1142
——— Subdr. 1367
——— Sing Thapa, Jemdr. 1615
Chetticha, P. K., Jemdr. 1361
Chetwynd-Palmer, E. R., Lt. 313
——— R., Lt. 537, 1705
Chetwynd-Stapylton, G. B., Maj. 52, 720
Chetwyned, C. W. G., Lt. 482, 911
Chevallier, F. T., Lt. 528, 1665
Cheverton, F. P., Lt. 430, 1050
Cheyne, A. Y., Lt.-Col. 136, 158, 197, 867, 903
——— H. B., Maj. 212, 875, 951
——— R. E., Lt.-Col. 104, 134, 157, 194, 889
Chhaba Thapa, Subdr. 1615
Chhagua, Jemdr. 1192
Chajju Ram, Jemdr. 1388
Chhajja Singh, Subdr. 1141
Chhaju Ram, Subdr. 1395
Chhannu Khan, Risaldar 889
Chhatar, Subdr. 1021
Chhatre, K. N., Tempy. Lt. 1773
Chhattar Sing, Sub-Asst. Surgn. 1815
Chhetradhoj Limbu, Subdr. 1648
Chhog Singh, Jemdr. 1571
Chhote Khan, Subdr. 1439
Chholelal Krishnaram Bhatt, Sub-Asst. Surgn. 1831
Chhotelal Daryam Singh Verma, Sub-Asst. Surgn. 1831
Chhotelal Mathurbhai Amin, Sub-Asst. Surgn. 1830
Chhottaji Bedar, Jemdr. 1424
Chhotubhai Bhikubhai Joshi, Sub-Asst. Surgn. 1831
Chhujju Singh, Sub-Asst. Surgn. 1814
Chichester, A. G. deV., Lt.-Col. 131, 155, 191, 1132

Names and Rank. Page.

Chidam Baram, Jemdr. 1699
Chidley, L., 2nd-Lt. 359, 1413
Chiesman, A., Condr. 1854
Chihsolm, D. M. F., Lt. 516
Chilcott, J. S., Lt. 320, 1223
——— W. J., Lt. 354
Child, H. W., 2nd-Lt. 557
——— P., Lt. 691, 783
——— R. H., Maj. 651
——— T. H., Lt. 490, 883
Childers, H. F. E., Capt. 258, 935
Childs, E., 2nd-Lt. 1008
——— F. D., Lt. 533, 1062
——— H., 2nd-Lt. 567
——— H. T., 2nd-Lt. 734
——— T., Capt. 1839
Chill, W. C., Lieut. 816
Chilton, A. H., Lt. 818
Chillu Singh, Rawat, Jemdr. 1181
Chimaji Garud, Subdr. 1423
Chimanrao Anand Rao, Doctor, Sub-Asst. Surgn. 1831
Chimes, W. E. G., 2nd-Lt. 638
Chimna Ram, Subdr. 1419
——— Jemdr. 1412
Chinappa, K. M., Jemdr. 1361
Chinnappan, Jemdr. 1276
Chinnasami, *Hony.* Capt. 2102
——— Jemdr. 1360
——— Jemdr. 1333
——— Naidu, Jemdr. 1012
Chinnayya, Jemdr. 1264
Chinpaw Gam, Subdr. 1347
Chintaman, Gopal, Gogale, Sub-Asst. Surgn. 1831
Chint Ram Burathoki, Subdr. 1618
Chiodetti, V. V., Snr. Asst. Surgn. 1787
Chiragh Singh, Jemdr. 1035
——— Beg, Jemdr. 918
Chiranji Singh, Jemdr. 1052
Chiranji Lal, Sub.-Asst. Surgn. 1815
——— Jemdr. 1039
Chiranji Lal Johar, Sub.-Asst. Surgn. 1819
Chiranji Lal Shah, Sub.-Asst. Surgn. 1818
Chiranji Singh, Jemdr. 1022
——— Subdr. 1021
Chirgwin, S. H., 2nd-Lt. 552, 1271
Chirnside, P. E. F., Lt. 589
Chisholm, A., *Hony.* Maj. 1849
——— D., Esq.
——— D. K. J., Capt. 271, 1459
——— D. M. F., Lt. 516a
——— K. McK., Lt. 541, 1674
Chisholme, A. C., *Hony.* Capt. 1843
Chitale, P. K., Lt.-Col. 139, 1745
Chitmis, D. K., Sub-Asst. Surgn. 1832
Chitrabir Thapa, Subdr. 1607
Chitta, Jemdr. 1391
Chittahang Limbu, Subdr.-Maj 1648
Chitts, M. H. G., 2nd-Lt. 557, 1463
——— T. A. O., 2nd-Lt. 558, 1475
Chitty, A. A. E., Lt. 749
——— A. W., Lt.-Col. 15, 20, 98, 139, 161, 200, 871, 1501
——— C., Lt. 729
——— E. R. L., Lt.-Col. 130, 155, 189
——— W. W., Col. 53, 125, 150, 186
Chivers, E., Lt. 39
——— N. M., Lt. 465, 1327
Choake, F. J., 2nd-Lt. 362
Chob Singh, Jemdr. 1167
Choga, Jemdr. 1197
Chohan, N. K., Tempy. Capt. 1768
Choksy, J. N. H., Tempy. Capt. 1768
Chope, A. J. H., Maj. 244, 1600
Chopra, J. M., Tempy. Lt. 1777
——— P. D., Capt. 1537, 1758
——— R. N., Capt. 109, 1754
Chorley, R. C., 2nd-Lt. 560
Choti Lal, Jemdr. 1165
Choudhury, B. N. S., 2nd-Lt. 1054
Chouler, R., Lt. 512, 712
Chowdhuri, K. B., Tempy. Capt. 1771
Chowdhury, G. S., Ty. Capt. 1772
——— H. C. R., Tempy. Capt. 1766
——— J., tempy Capt. 1768
——— N. N., Ty. Lt. 1776
——— P. C., Tempy. Capt. 1769
Chowns, W. P., Lt. 510, 708, 1668
Chowriappan, Subdr. 1315
Chowrimuthu, Jemdr. 1328
Chrichton, W. R., Revd. 906
Christal, F. A., 2nd-Lt. 476
Christian, A. J., 2nd-Lt. 56

Names and Rank. Page

Christian, G., Brig.-Genl. 12, 22, 103, 120, 124
——— G. C. 2nd-Lt. 987
——— H. D., 2nd-Lt. 564
——— H. N., Capt. 277, 1475
——— J. B., Maj. 1748
——— J. W., Lt. 538
Christie, B., Capt. 268, 1148, 1587
——— D. A., Capt. 280—237, 1074
——— E. L., Maj. 747
——— G., Lt. 405, 1061
——— G. H., 2nd-Lt. 326, 366, 1647
——— H., Lt. 514
——— H. G. F., Maj. 229, 1662
——— J., 2nd-Lt. 715
——— J. I. T. Lt. 986
——— J. McI., 2nd-Lt. 679, 1317
——— J. S., 2nd-Lt. 563, 1217
——— R. C., Capt. 260, 1411, 1413
——— T. M., Lt. 527, 1667
——— W. A. K., Lt. 487
Christison, F. H., Capt. 299, 1110
Christmas, H. R., 2nd-Lt. 730
Christopher, C. deL., Capt. 253, 1663
——— F., Capt. 1839
——— L. W., Maj.-Genl. 1955
Christophers, H. A. S., 2nd-Lt. 564
——— S. R., Maj. 62, 1749
Chrystall, A., Maj. 501, 608, 1696
——— J. I., Lt. 591
Chrystie, G., Col. 177
Chubb, W., Lt. 486, 943
Chucker Butty, S. C., Capt. 1753
Chudleigh, A. C. L., 2nd-Lt. 711
——— C. A. E., Capt. 711
Chumel Singh, Jemdr. 1241
Chunak Singh, Jemdr. 1260
Chunahang Limbu, Jemdr. 1631
Chuhar Khan, Sub-Asst. Surgn. 1814
——— Singh, Sub-Asst. Surgn. 1813
——— Jemdr. 948
Chukerbuti, J. C., Tempy. Capt. 1766
Chunilal, Sub-Asst. Surgn. 1810
——— Jemdr. 1356
——— Subdr. 970
——— Himattram, Bhatt, Sub-Asst. Surgn. 1831
Chunn, G. F., 2nd-Lt. 815
Chunna Ram, Jemdr. 1484—86
Chunni Lal Gandhi, Sub-Asst. Surgn. 1820
Chunni Lal, Jemdr. 1059
Chunni Lal Srivastavr, Sub-Asst. Surgn. 1819
Chur Singh, Subdr. 1252
——— Subdr. 1188
——— Jemdr. 966
Churamani Rana, Jemdr. 1608
Churamarie Thapa, Jemdr. 1608
Church, B. H., Lt. 638
——— C. J. B., Lt. 316, 1074
——— F., 2nd-Lt. 1009
——— D. M. O., Maj. 1752
——— T. R., Maj.-Genl. 172
Churcher, W. D., Lt. 375, 503, 720, 1283
Churchill, A., Lt.-Col. 1885
——— A. P. ff., Capt. 279, 1170
——— E. A. H., Lt. 1004
——— G. R. D., Lt.-Col 138, 159, 198, 868, 1094, 1096
——— J. S., 2nd-Lt. 328
——— W. F., Lt. 776
Cinnamond, H. P., Lt. 410, 1561
Clabby, D. N. J., Lt. 84, 543, 1705
——— W. G., Lt. 1693
Clack, G. H., Lt. 45, 351, 1676
Clague, J., 2nd-Lt. 550, 1297
Clair-Smith, E. I., 2nd-Lt. 561
Clancy, G. D., 2nd-Lt. 690a
Clapham, S., Lt. 516, 1259
Clapp, J. H., Lt. 470, 877
——— C. D., 2nd-Lt. 329, 1203
——— L. J., Lt. 525, 1581
——— W. M., Lt. 524, 1339
Clappen, S. G., Lt. 450—54, 1243
Clapperton, H., 2nd-Lt. 360
Clapton, O., Condr. 1867
Clarabut, F. S., 2nd-Lt. 329, 923
——— R. B., Lt. 401, 1624
Claridge, D. P., Lt. 375
——— E. G. W., 2nd-Lt. 556, 1359
——— J. W. L., Capt. 501, 628
——— P. S. F., Maj. 40, 206
Clark, A., Condr. 1863
——— A., 2nd-Lt. 75
——— A. C., 2nd-Lt. 355

Names and Rank.	Page.
Clark, A. C., 2nd-Lt.	567
—— A. H., 2nd-Lt.	1170
—— A. G. N., Esq.	91
—— A. M., Lt.-Col.	2, 851
—— A. M., Lt.	508, 1630
—— A. N. W., 2nd-Lt.	752
—— A. S., Maj.	228, 1663
—— C., Lt.	7, 690a
—— C. D., Maj.	752
—— C. D. LeG., Lt.	292, 939
—— E. S., Lt.-Col.	133, 1710
—— E. W., Lt.	446—449
—— G., 2nd-Lt.	640
—— G., A., Lt.	528, 1273
—— G. S., Capt.	1724
—— G. S. H., 2nd-Lt.	355, 1417
—— H., 2nd-Lt.	546, 1295
—— H. A., *Hony.* Lt.	111
—— H. W., 2nd-Lt.	354, 1373
—— J., Lt.	21, 304, 1181
—— J. C., Maj.	695, 700
—— J. D., 2nd-Lt.	
—— J. D., 2nd-Lt.	781
—— L. A. F., 2nd-Lt.	551, 1108
—— L. J. L., Lt.-Col.	98
—— R. J. C., Lt.	735
—— R. N., Lt.	490, 834
—— V. P. J., 2nd-Lt.	357
—— V. T. G., 2nd-Lt.	682
—— W. C., Lt.	638
—— W. H., Capt.	259, 883
—— W. J., Lt.	455, 1387
—— W. S., 2nd-Lt.	724
—— W. V., Lt.	306, 1431, 1479
Clarke, A., 2nd-Lt.	761
—— A. A. Mack, Lt.	533
—— A. B., 2nd-Lt.	540, 1704
—— A. C., 2nd-Lt.	355, 1515
—— A. C. K. S., Maj.	245, 903
—— A. R., Asst. Surgn.	1793
—— A. E., Capt.	814
—— A. E., Capt.	276, 1178
—— A. J., Capt.	53, 691
—— A. M. M., 2nd-Lt.	1674
—— A. W., 2nd-Lt.	539, 1511
—— B.G.S., Maj.	9, 16, 103, 233, 1647
—— C. G., 2nd-Lt.	698
—— C. J., Maj	40, 662, 1882
—— C. J. L., Lt.-Col.	97
—— C. P., Lt.	291, 1140, 1142
—— C. V., N. Sister	1735
—— D. G., 2nd-Lt.	921
—— E., Maj.	685
—— E., 2nd-Lt.	769
—— E. H., Capt.	665, 991
—— E. R. 2nd. Lt.	769
—— E. W., 2nd-Lt	557
—— F. A. H., Maj.	32, 1711
—— F. A. S., Lt.	368, 1251
—— F. E., 2nd-Lt.	786
—— F. E. M., Lt.	294, 1096
—— F. G., 2nd-Lt.	807
—— F. H., 2nd-Lt.	551
—— F. I. L., Esq.	88
—— F. R., Sergt.-Maj.	1879
—— F. S., Capt.	7, 276, 1207
—— G., Lt.	537, 992
—— G. A., Maj.	238, 1666
—— G. C. C., Maj.	226, 1662
—— G. R., Capt.	280—287, 1303
—— G. R., Capt.	806
—— G. S., Maj.	209, 869, 1150, 1156
—— H., Maj.	40, 74, 633
—— H., Capt.	257, 1030
—— H., Lt.	1708
—— H., 2nd-Lt.	492, 1435
—— H. A., Sub-Condr.	1864
—— H. B. J., 2nd-Lt.	701
—— J., Sub-Condr.	1864
—— J. A., Lt.	320, 365
—— J. A., Lt.	460—462, 937, 1279
—— J. E., *Hony.* Capt.	55, 1850
—— J. H., Ty. Lt.	1778
—— J. H., 2nd-Lt.	791
—— J. L. J., Br.-Genl.	12, 24, 133
—— J. R. R., Lt.	430, 1086
—— J. W., 2nd-Lt.	567
—— J. W., Lt.	692, 784, 1636
—— J. W. A., Lt.	470, 1102
—— L. C., 2nd-Lt.	929
—— M. F. C., Lt.	540, 1672
—— M. J., Lt.	410, 1291
—— N. St. J. S., 2nd-Lt.	558, 1009
—— N., 2nd-Lt.	1695
—— P. G., 2nd Lt.	692
Clarke, P. S., Lt.	504
—— P. T., Lt.	318, 1259
—— P. T. 2nd-Lt.	549, 1333
—— R. C., Maj.	242, 1663
—— R. C. B., Revd.	1904
—— R. L. H., 2nd Lt.	819
—— R. V., Capt.	1718
—— S. H., Lt.	538
—— T. G., Col	176
—— T. V., 2nd-Lt.	315, 891
—— V. S., Capt.	749
—— V. T. G., 2nd-Lt.	786
—— W. D. M., Lt.	508, 1885
—— W. H., 2nd-Lt.	540, 1028
—— W. H., Lt.	554
—— W. L., Revd.	1903
Clarke-Bell, D. M., 2nd-Lt.	323, 1651
Clark-Kennedy, A. H., Col.	57
—— H. D. G. A., 2nd-Lt.	550
Clarkson, A. S., 2nd-Lt.	1511
—— B. St. J., Lt.-Col.	132, 736
—— R., Asst. Surgn.	1796
—— R. M., Lt.	1614
Claudius, A. H. T., Asst. Surgn.	1800
—— C. J., Asst. Surgn.	1796
Clauson, G. L. M., Lt.	56
Claxton, B., 2nd-Lt.	703
—— E., Lt.	637, 975
Clay, B. G., Maj.	583
—— C. H., Col.	97, 105, 124, 149, 183
—— F. J., *Hony.* Maj	1849
—— N. C., 2nd-Lt.	475, 786, 1418a
—— R. D., Lt.	1273
Clay,-Inman, H., 2nd-Lt.	552
Claydon, S., Capt.	833
Claye, A. M., Lt.	738
Clayton, J., 2nd-Lt.	562
—— J. A., Capt.	826
—— J. H., 2nd-Lt.	492
—— R. V., Capt.	826, 861
—— W. A., Maj.	614
—— W. D., Lt.	712
Clayton-Cowell, H. C. F., Lt.	507
Clayton-Jones, R., Lt.	511, 1090
Clear, T., Lt.	493
Cleary, C. J., Lt.	527, 1343
—— G. W., 2nd-Lt.	561, 1403
—— R. P., Lt.	846
Cleator, A. E., 2nd-Lt.	355, 1143
Clease, E. K. J., 2nd-Lt.	325, 1235
Cleave, J. C. T., Lt.	317, 1042
Cleaver, L., Lt.	517, 996
—— N. R., Lt.	352, 1443
Cleere, A. T. C., 2nd-Lt.	779
Cleeson, J., 2nd-Lt.	564
Cleghorn, J., Surgn.-Genl.	1
Clemence, S., Lt.	292, 1629
Cleland, W. S. C., 2nd-Lt.	715
Clement, J. Lt.	1187
Clementi, J., Maj.	214, 867, 964, 970
Clements, C., 2nd-Lt.	619
—— G. W., Lt.	519, 1407
—— H. S., Condr.	1853
—— I., *Hony.* Lt.	1845
—— J. E., Maj.	1748
Clements-Finnerty, H., Lt.	595
Clemesha, W. W., Lt.-Col.	62, 133, 1743
Clamo, J., 2nd-Lt.	358
Clemson, W., Capt.	736, 737
Clendining, C. H., Lt.	757
Clen-Murphy, E. W., 2nd-Lt.	366, 1809
Clerici, M., Lt.	521
Clery, C. B. L., Brig.-Genl.	50, 120, 125, 191, 153
—— J. A., 2nd-Lt.	552, 1633
—— P. A. A., 2nd-Lt.	327, 1235
Cleveland, E. A. M., Lt.	420, 1607
—— H. F., Col.	24, 104, 109, 123, 1738
Cleverly, G. C., Lt.	681
—— O. S., Capt.	681
Clews, L., 2nd-Lt.	935
—— L. C., Lt.	450—454
Clibborn, J., Lt.-Col.	179, 1917
Cliff, W. P., Lt.	518, 814, 917
Clifford, C. H. W., Lt.	512, 1136
—— E. K., 2nd-Lt.	546, 998
—— G. A., 2nd-Lt.	557, 1701
—— G. G. J., Lt.	589
—— H. H., Lt.	489, 883
—— J. H., Asst. Surgn.	1802
—— R., Maj.	215, 868, 1106, 1108
—— R. C., Capt	1756
—— R. C. R., Lt.-Genl.	172, 917
—— R. E. St. J., Capt.	820
Clifford, W., Lt.	758
—— W. H. R. Capt.	82, 502
Clift, A., 2nd-Lt.	356
—— E., Sub-Condr.	1874
—— F. C. L., 2nd-Lt.	571a
Clifton, A. J., Capt.	761, 791
—— F. R., 2nd-Lt.	358, 1453
—— H. A., Lt.	51
Climo, P. H., Maj.	633, 972
—— S. H., Maj.-Genl.	12, 21, 119, 147, 175
Clinch, H. G., 2nd-Lt.	550
Cline, A. H., Lt.	357, 1101
Clinton, A. G., Lt.	20, 536, 1022
Clive, H. W. F., Maj	244, 1613
—— J. McC. Lt.	511
Cliyde, J. C., 2nd-Lt.	1653
Cloete, H. D., Maj.	107, 233, 1375, 1898
—— I. B. Maj.	98, 213, 1170
Clogg, C. T., 2nd-Lt.	563
Clogstoun, R. C, Lt.	1695
Cloran, D. J. H., Lt.	311, 1331, 1511
Close, J. K., Col.	29, 127, 1738
—— L. H., Lt.-Col.	134, 661
—— W. A., 2nd-Lt.	363
Clough, E R. Revd.	1903
—— T., Lt.	490, 915
Clover, P. T., Lt.	699
—— W. B., Lt.	827
Clow, P., Lt.	712
Clowes, W. P., Lt.	589
Cloy, C. Condr.	1872
—— F. A., 2nd-Lt.	560
Club, A., Lt.	540, 988
Clubbe, C. H., Lt.	540, 1251
Clunie, T., 2nd-Lt.	713
Clutterbuck, C. H., Maj.	56, 105, 230, 1497
—— E. L., Lt.	704
Clyde, Lt., W. T.	401, 1385
—— D., Capt.	1718
Coad, T. H. E., Lt.	537
Coady, J. W., Sub-Condr.	1872
Coalrington, K.D.B., 2nd-Lt.	937
Coaker, V. A., Maj	79, 215, 879
Coape-Smith, H., Maj.-Genl.	172
Coates, H. L., Lt.	443, 1107
—— J., Maj.	817
—— J. C., Lt.	302, 955, 1617
—— N. E., Lt.	725
—— S. B., Maj.	242, 1407, 1411
—— V., Maj.	227, 1355
—— W., Condr.	1854
—— W. J., 2nd-Lt.	88, 638
Coath, P. S., 2nd-Lt.	363
Coats, D. H., Tempy. Capt.	1715
—— J. R., Capt.	801
—— N. G. R., Capt.	267, 1120, 1569
Cobb, C., Capt.	16, 29, 253, 1663
—— H. J. G., 2nd-Lt.	324
—— H. S., Capt.	766
—— H. V., the Hon'ble Mr.	808, 1900
—— R. H., 2nd-Lt.	739
Cobban, C. L., Maj.	33, 104, 105, 226, 1195
—— H. L. N., Capt.	830
—— S. W., Lt.-Col.	833
Cobbe, A. S., Lieut.-Genl. *Sir*	97, 118, 146, 174
—— H. H., Bt.-Col.	53, 132, 156, 192, 899
Cobbett, J. E., Lt.	321, 1319
Cobble, V. G., 2nd-Lt.	562
Cobbold, M. F. D., Capt.	267, 1595
Cobham, H. W., Brig.-Genl.	58
Coburn, F. R., Lt.	17, 38
—— M. R., Esq.	88
Cockburn, R. C. R., Lt.	465
Cockeran, A. E., 2nd-Lt.	360, 1054
Cockerton, G., 2nd-Lt.	640
Cochran, A., Capt.	651
—— A. C. P., Maj.	221, 1183
—— G. W., Bt.-Lt.-Col.	54, 161, 233, 1331
Cochrane, A., Capt.	1724
—— A. W. R., Lt.-Col.	133, 1743
—— C. P., Lt.	927
—— D. W., Lt.	949
—— J. A., 2nd-Lt.	564, 1108
—— R. P. I., Capt.	277, 907
—— T. G. F., Capt.	744
Cockaday, W. G. C., Lt.	626
Cockayne, M. H. D., Lt.	380, 1016
Cockburn, A. B., 2nd-Lt.	805

NAMES AND RANK. Page.

Cockburn, H. A., Genl. . . . 160
—— J., 2nd-Lt. 563
—— J. G. H., Capt. . 89, 727
—— J. L. K., Capt. . . 761
—— T. S., 2nd-Lt . . . 719
Cockcroft, E. F., Capt. . . 32, 717
Cockell, E. L., Capt. . . . 1708
Cocker, R. W., 2nd-Lt. . . . 491
Cockerton, G., 2nd-Lt. . . . 640
Cockie, B. H., 2nd-Lt. . . . 750
Cockman, G. H., Esq. . . . 88
—— G. R., Lt. 544
Cockerill, J. W., Lt. . 351, 883, 1575
Cockram, L. H., Lt. . . . 405
—— L. H., 2nd-Lt. . . 1630
Cockrell, G. H., 2nd-Lt. . . 494, 945
—— L. H., Lt. 405
—— R. A., Lt. 1695
Cockrill, S. R., Lt. . . . 593, 1607
Cocks, C. V. S., Lt. . . . 744
Codd, H. C., 2nd-Lt. . . 558, 1708
Codrington, G. H. F., Col. . . 577
—— H. deB., Lt.-Col 9, 132, 156, 193, 1660
—— H. W., Col. 14, 26, 121, 127, 150. 189
—— K. deB., 2nd-Lt. . . 327
Cody, W. E., Asst. Surgn. . . . 1796
Coen, W. W., Maj. 846
—— W. W., Lt.-Col. 500
Coffey, E. P., Lt. . . 27, 519, 1379
Coffin, A. S., Capt. . . 261, 1132
—— C., Col. . . 10, 97, 126, 160
—— D. M., Ty. Capt. . . . 619
—— S. W., Ty. Capt. . . . 1723
Coghill, N. S., Lt.-Col. 197, 1892, 1912
—— C. C., 2nd-Lt. . . . 684
Coghlan, E. M. E., 2nd-Lt. . . 1007
Cogswell, A., Lt. . . . 753, 785
Cohen, A. J., Lt. 833
—— A. M., Maj. 751
—— N. C., Lt. . . . 372, 1319
—— N. H., Lt. . . 9, 516a, 1677
—— S. L. Capt. 714
Cohil, J. L., 2nd-Lt. 987
Cohn, M. D., Lt. . . . 582, 599
Cohu, C. J., Revd. 1904
Coke, A. H., Lt.-Col 132, 156, 193, 1453
—— A. J. P., Maj. . . 211, 917
Colan, H. N., Maj . . . 236, 1283
—— W. R. H., Lt.-Col. 135, 158, 195, 870, 1283
Colbeck, B. B., Lt.-Col. . . . 633
—— C. E., Capt. 664
—— H. K., Lt. 642
Colborne, T. L. D., 2nd-Lt. . . 455
Colbourne, W. A., Lt. 313, 1116, 1120
Colchester, M. T., Lt. . . 506, 1271
Coldstream, G. A. P., 2nd-Lt. 328, 1143, 1251
—— J. C., Maj. 207, 1896, 1889, 1913
—— W. M., Lt.-Col. 174, 661, 1880
Coldwell, J. A., 2nd-Lt. . . . 549
—— L. R., Lt. 1493
Coldwell-Horsfall, J. H., Maj. 614
Coldham, P. C., 2nd-Lt . 356, 1150
Cole, A. B. F., Revd. . . . 1903
—— A. B. H., Lt. 1732
—— A. C., 2nd-Lt. 566
—— A. F., Maj. 634
—— A. H., 2nd-Lt . . 474, 1112, 1114
—— B. L., Maj. . . . 82, 219, 1070
—— C. W., Lt. 652
—— E. H., Col. . . 124, 149, 183
—— E. J. 2nd-Lt. . . 700, 796
—— F. E., Capt. 817
—— G. R., Capt. 732
—— H. A., Capt. . 260, 1056, 1199
—— H. M., Capt. 697
—— H. W. G., Lt.-Col. 87, 1691, 1696, 1898, 1911
—— J., 2nd-Lt. 799
—— J. E. L., 2nd-Lt. . . . 565
—— J. T., Lt. . . 59, 506, 737
—— L. B., 2nd-Lt. . . 83, 626
—— O. W., 2nd-Lt. . . . 554
—— R. D., Lt. . . . 528, 1201
—— R. F., Condr. 1862
—— R. St. G., Lt. . . 387, 1191
—— W. J., Lt. . . . 520, 1666
—— W. J. L., Lt. 522
Coleburt, J. G., *Hony.* Maj. . . 1844
Colegrave, G. R., 2nd-Lt. . . . 684
Coleman, A., Lt.-Col. . . 127, 1740

NAMES AND RANK. Page.

Coleman, A. J., 2nd-Lt. . 358, 1245
—— C. M., 2nd-Lt. . . . 330
—— G. V. L., 2nd-Lt. . . 332, 336a, 1245
—— H. C., 2nd-Lt. . 361, 887
—— J. W., Sub-Condr. . . 1857
—— P. W., 2nd-Lt. . . . 559
—— R. E., Capt. . . 754, 1004
Colenso, J. E., Maj . . 239, 1629
Coleridge, J. F. S. D. Bt. Lt.-Col. 50, 98, 122, 137, 159, 216, 1635
—— P. L., Maj. . . 222, 1327
Coles, C. E. H., Lt. . . 509, 1665
—— F. H., Lt. 622
—— G. R., Lt. . . . 293a, 911, 732
—— N. R., Lt. . . . 375, 951
—— T. H., Lt. . . . 535 1151
Colfox, W. P., Capt. 615
Colhoun, W. A., Lt. . . . 371, 1136
Colkers, A. B., Esq. 89
Collard, B. H. A., 2nd-Lt. . 365, 1181
—— B. S., Capt. 683
—— E., 2nd-Lt. 618
Colledge, J. E., Hony. Lt. . . 652
Callender, K. G. F., Capt. . . 1732
Collet, A. L., Lt. . . . 518, 1501
—— A. P., Lt. 489
Collett, A. L., Maj. . . . 852, 1501
—— C. M., Capt. 486
—— H. S., Lt. . 533, 1307, 1674
—— N. T., 2nd-Lt. . . 473, 1164
—— J. W., Lt. . . . 532, 1447
—— R. A., Lt. . . . 353, 1559
—— W. A., 2nd-Lt. 354
Collett-Mason, W. W. C., Lt. . 589
Colley, R., Capt. 750
Collier, A. E., Capt. . . 658, 991
—— C. B. G., Revd. . . . 1907
—— E. 624
—— J., 2nd-Lt. 713
—— S., 2nd-Lt. 1008
—— W. J., 2nd-Lt. . . 331, 1483
Collin, E. P. C., Lt. . . 295a, 909
—— F. S., Maj. . . 665, 985
Collinge, J., 2nd-Lt. . 77, 1185, 1885
—— W. C., Lt. . . . 522, 1411
Collingridge, H. F., Maj. . 223, 1541
—— H. V., Lt. . . 293, 1623
Collings, E. G., Capt. 264, 1162, 1166
—— F. W., 2nd-Lt. . . 550, 1676
—— J. S., 2nd-Lt. . . 473, 909
—— T. H. O., Capt. . 685, 746
Collingwood, A. L., Lt. . 311, 1467
—— C. F., Sub-Condr. . . 1856
—— F. W. N., Lt. . . . 55
Collins, A. H. T., 2nd-Lt. . . . 568
—— A. E., Sub-Condr. . . 1857
—— C., *Hony.* Maj. . . . 1848
—— C. F., 2nd-Lt. . . . 591
—— C. W. M., Lt. . . . 825
—— D. J. Col. 126
—— E. G. W., Lt. . . . 789
—— F. C., Lt. . . 460—62, 1371
—— F. J. 2nd-Lt. . . . 1653
—— F. P., 2nd-Lt. . . . 545
—— F. T., 2nd.-Lt. . . 564, 1363
—— G., Sergt.-Maj. . . . 1874
—— G. F. S., 2nd-Lt. . . . 523
—— G. H., 2nd-Lt. . . . 751
—— G. P., Lt. 1606
—— H. M., Capt. . . 1455, 1759
—— H. S., 2nd-Lt. . . . 1054
—— H. W., Lt. . . . 532, 1339
—— H. W. S., 2nd-Lt. . . . 553
—— J., 2nd-Lt., ... 363
—— J., Lt. 689
—— J. C., Lt. 1718
—— J. H., Revd. 1903
—— J. L., 2nd-Lt. . . 560, 703
—— J. R., Capt. 913
—— L., Lt. 504
—— L. P., Maj. . 229, 1521, 1613
—— M. R., 2nd-Lt. . . 327, 911
—— R. E., 2nd-Lt. . . 360, 1088
—— R. T., Tempy. Lt. . . 1178
—— S. St. B., Capt. . 636, 976
—— T. C., 2nd-Lt. . . 562, 1143
—— T. M., Lt. 541
—— T. V., 2nd-Lt. . . . 638
Collinson, C. D. H., Lt. . 491, 903
—— C. P., 2nd-Lt. . . . 560
—— J. C., *Hony.* Maj. . . 1844
—— W. J., Maj. 1750
Collis, F. S., 2nd-Lt. . . 323. 455
—— J. G., Capt. 8
—— M. S., Lt. . . . 403, 1291

NAMES AND RANK. Page.

Collis, W. M., 2nd-Lt. . . 550, 1701
Collishaw, T., 2nd-Lt. . . . 363
Collison, A. E., Lt. . . . 617
—— S. S., 2nd-Lt. . 539, 1407
Collister, H. J., 2nd-Lt. . 553, 829
Collum, J. A., Capt. . 105, 255, 923
Collyer, J., 2nd-Lt. . . 559, 1094
Collyers, H. N., Capt. . . . 808
Collyns, G. G., Lt. . . . 295, 545
Colman, B. A., Lt. . . . 51, 514
—— E., 2nd-Lt. 622
—— L. M., Lt. 628
Colmore, H., Capt. 587
—— J. E., Lt. 1693
Colnett, R. D., Lt. . . 368, 1557
Colomb, G. H. C., Br.-Genl. 120, 124, 149, 183
Colquhoun, A. F. R., Maj. . . 711
—— A. R., Major. 76, 715, 1884
—— J. W. C., Lt. . . 405, 1569
—— S. P., 2nd-Lt. . . . 1044
—— T. J., Sub-Condr. . . 1855
Colsey, J. W. T., 2nd-Lt. . 473, 717
Coltman, Lt. A. C. . . . 515, 1092
Coltham, W. F., 2nd-Lt. . . 668
Colter, P. 1850
Colthup, B. W., 2nd-Lt. . . . 566
Columbine, R. H., Lt. . . 369, 1655
Colven, A. F., 2nd-Lt. . . . 684
Colvin, C. P., Capt. 833
—— C. J. D., Capt . 9, 59, 257, 947, 1889 1899
—— G. L., Maj. 60
—— J. M. C., Col. 101, 137, 661, 996
Colwell, H. H., Ty. Lt. . . . 1778
Colwill, F., Lt. 529
—— R. F., Lt. . . . 537, 1038
Combe, B. A., Maj.-Genl. . . . 593
—— H. W., 2nd-Lt. 90
—— L. M., Lt. . . 292, 1148, 1150
—— S. B., Bt.-Lt.-Col. 104, 138, 159, 215, 1211
Comeron, E. M., N. Sister . . 1735
Comerford, T. J., Lt. . . 294a, 909
Comins, R. I., Lt. 440
Comley, H. M., 2nd-Lt. . . . .
Commeline, F.H.R., Br.-Genl. 12, 37, 120, 123, 149, 183
—— F. T., Lt. . 307, 877
Commen, C., Temp. Lt. . . . 1779
Common, F. B., Lt. . . 488, 883
Comper, G. A., Lt. . . . . .
—— G. Q., Lt. . . . 514, 1068
Compton, C. W. McG., Maj. 14, 37, 225, 1287
—— G. G., 2nd-Lt. 653
Compston, P., Lt. . . 380, 1645
Comyn, G. V., Maj. . . 245, 1663
—— K., Capt. 1713
Conceicao, F., Tempy. Lt. . . 1775
Conder, C. S., Lt. 1839
—— H., Capt. . 260, 1189, 1451
—— J. M., 2nd-Lt. 547
Condon, deV., Lt.-Col. . 138, 1745
—— E. C. H., Capt. . . . 826
—— E. H., Col. . 23, 126, 1709
—— J. K., Lt.-Col. 109, 192, 1892, 1912
—— W. E. H., Capt. 109, 263, 1685, 1321
Cones, H. M. A., Lt. 490
Coney, G. B., 2nd-Lt. 704
—— H. G., 2nd-Lt. 565
Congdon, J. C., 2nd-Lt. . . 563, 1092
Congreve-Pridgeon, R., 2nd-Lt. . 799
Conibear, W. J., Lt. 352
Coningham, F. E., Lt.-Col. 136, 158, 196, 872, 1647
—— M. E., Maj. 240, 1379, 1391
—— R. E., Maj. . 13, 98, 238, 1647
Conley, W. Mac. C., Tempy. Capt. 1723
Conlon, H., Lt. 618
Connaught, *H.R.H.* A. W. P. A., *Duke of*, Field Marshal, 1, 585, 731, 762, 899, 933, 1042, 1513, 1660
Connell, C. B., 2nd-Lt. 854
—— C. M., Lt. 491, 949
—— E. S., Capt. . . . 37, 43
—— E. W. H., 2nd-Lt. . 357, 1441
—— H. W., Lt. . . . 534, 1655
—— J. H., 2nd-Lt. . . 465, 1251
—— T., Hony. Maj. 624
—— V. J. A., Capt. . . 261, 899
—— W. W., 2nd-Lt. 790
Connellan, P. S., Capt. . . . 1756

Names and Rank. Page.

Connell-McDowell, A. E. J., Maj. 235, 1463, 1894
Connelly, J. P., 2nd-Lt. 539
Conner, T. S., Lt. 293, 1359, 1591
Connole, J. S., Capt. 722
Connolly, E. P., Lt.-Col. 1711
—— D. R. V., 2nd-Lt. 326, 366, 1174, 1521, 1641
—— E. T., Sub-Condr. 1857
—— E. T. P., 2nd-Lt. 531
Connop, H. E., Maj. 222, 915, 941
Connor, A. D., Maj. 234, 1191
—— F. P., Maj. 1749
—— G. H., Lt. 33, 300, 1152, 1367
—— G. P., Lt. 543a
—— J. M., Tempy. Asst. Surgn. 1804
—— T. S., Lt. 1307
Connors, J. P., Capt. 691, 784
Conochie, N. C., L. 781
Conolly, J. J. P., Lt. 297a, 1162, 1211
Conquest, C. H. C., Asst. Surgn. 1799
Conran, E. L., Lt. 597
—— W. D. B., Maj. 663, 991
Conry, J. deL., Maj. 230, 1391
—— H. T., 2nd-Lt. 1695
Considine, W. J., Lt. 516(a), 1666
Constable, E. A., Lt.-Col 2, 651
—— R. C., 2nd-Lt. 591
—— W. N. B., 2nd-Lt. 463, 1383
Constant, C. F., Ty. Capt. 1723
Contractor, A. K., Ty. Maj. 1763
—— C. M., Tempy. Lt. 1775
—— K. H., Tempy. Capt. 1769
—— S. C., Tempy. Capt. 1766
Conville, L. H. G., Lt. 637
Conway, C. McC., Capt. 843
—— F. J., Lt. 629
—— H., 2nd-Lt. 1674
Conway-Gordon, E. C. W., Maj. 13, 35, 98, 209, 879
Conyers-Brown, H. P., Lt. 311a, 1271
Cooch Behar, Hony. Lt. 1011
Coode, A. P., Maj. 776a
Cook, A., Sub-Condr. 1856
—— A., 2nd-Lt. 726
—— A. F., Capt. 502, 1030
—— A. W., 2nd-Lt. 1695
—— C. C., Col. 15, 32, 126, 150, 188
—— C. W., 2nd-Lt. 564
—— E. H., Lt. 541, 1108
—— E. J. A., Lt. 543a, 987
—— E. N., 2nd-Lt. 790
—— E. R. M., Lt. 541
—— H., Maj.-Genl. 736
—— H. C., Sub Condr. 1874
—— H. G., Lt. 521, 1323
—— H. O., 2nd-Lt. 488, 937
—— H. P., Maj. 1752
—— H. R., Bg.-Genl. 39, 42, 102, 120, 125, 611
—— H. W., 2nd-Lt. 546, 988
—— J., Maj.-Genl. 1136, 1955
—— J., Capt. 44, 705
—— L., Maj. 1749
—— P. R., Lt. 623
—— R., 2nd-Lt. 1048
—— R. H., 2nd-Lt. 1457
—— R. J., 2nd-Lt. 987
—— S., Lt. 511, 1174
—— T. P., Lt. 511, 1020
—— W., Lt.-Col. 180, 1917
—— W. G., 2nd-Lt. 474, 1614
—— W. K., Maj. 241, 1663
—— W. M., 2nd-Lt. 561, 1355
Cooke, A., Maj. 591
—— A. B., Lt. 455-59, 1473
—— C., 2nd-Lt. 735
—— C. A., Maj. 826
—— C. A., 2nd-Lt. 357, 1054
—— C. B., Lt.-Col. 179, 1918
—— C. W., Lt. 416-49, 1066, 1677
—— E. B., Capt. 819
—— G. H., Lt. 516a
—— G. M. B., Lt. 295, 925
—— G. P., Lt. 506, 1607
—— H. F., Col. 8, 15, 119, 126, 150, 195, 859, 1148
—— H. L., 2nd-Lt. 561, 1323
—— J. C. C., 2nd-Lt. 516a, 1673
—— J. H., *Hony.* Lt. 752, 784
—— K. F. R., Maj. 819
—— R. H., 2nd-Lt. 682, 786
—— R. D., 2nd-Lt. 1114
—— R. L., 2nd-Lt. 472
—— S. F., Lt. 503, 1395

Names and Rank. Page.

Cooke, T. C., Esq. 88
Cookson, G. A., Maj.-Genl. 118, 145, 173, 929
—— S. M., Capt. 258, 1664
Coole, A. E., 2nd-Lt. 332, 1148
Cooling, C. E., Lt. 533, 1007a
Coombe, J. R., *Hony.* Lt. 697
Coomber, D. K., 2nd-Lt. 754
—— F. E., 2nd-Lt. 798
—— H. P., Capt. 636
—— M. G., Asst. Surgn. 1797
—— W. K., Asst. Surgn. 1797
Coombs, A. P., Capt. 702
—— C. C., 2nd-Lt. 630
—— E. E., Lt. 655, 849
—— E. W. G., 2nd-Lt. 360
—— M. E., Lt. 487
—— M. H., Condr. 1873
—— S. H. J., Ty. Asst. Surgn. 1804
—— W. H., 2nd-Lt. 554
Cooper, A., 2nd-Lt. 55, 1467
—— A. C., 2nd-Lt. 1181
—— A. J., Lt.-Col. 857
—— A. P., 2nd-Lt. 568
—— A. T., Sub-Condr. 1864
—— A. V., Lt. 90, 509
—— C. A., Capt. 600
—— C. H., Lt. 510, 1006
—— C. H., 2nd-Lt. 365, 1497
—— C. R. P., Lt. 848, 535, 1295
—— D. G., Capt. 1756
—— E., 2nd-Lt. 758
—— F. A., Lt. 536, 988
—— F. A., 2nd-Lt. 547
—— F. L., 2nd-Lt. 362, 1038
—— F. W., *Hon.* Maj. 766
—— G. B. A., Maj. 246, 1663
—— G. C., Maj. 103, 105, 234, 1433, 1479
—— G. M., 2nd-Lt. 571
—— G. S., Maj. 614
—— H. G., Lt. 516a, 1517
—— H. L., Capt. 501, 1259
—— H. L., 2nd-Lt. 554, 1553
—— H. R., Lt. 528, 1339
—— J. A. W., Capt. 835
—— J. E., *Hony.* Lt. 1846
—— J. H., Maj. 22, 706
—— K. E., Maj 244, 1443
—— L. B., Asst. Surgn. 1801
—— L. G., Lt. 460-62, 1423
—— L. K., Lt. 526, 998
—— N. E., 2nd-Lt. 1152
—— O., Maj. 695
—— P. A., Capt. 519, 806, 998
—— P. T., Lt.-Col. 74, 101, 129, 611
—— R. E., Lt. 530
—— R. E., Lt. 76, 528, 1884
—— R. F., Lt. 526, 1894
—— R. G., Lt. 1425
—— T., Capt. 485, 826
—— T., 2nd-Lt. 723
—— T. W., 2nd-Lt. 689
—— W., 2nd-Lt. 637, 979
—— W. G., Lt.-Col. 132, 156, 192, 867, 939
—— W. H., 2nd-Lt. 535
—— W. P., 2nd-Lt. 76, 757
—— W. R. F., Capt. 561
Cooper-Chadwick, J. L., Lt. 322, 1245
Coote, M. C., Capt. 260, 1431, 1898
Cooter, H., Lt. 533
Copcutt, E. E., Asst. Surgn. 1796
Cope, A. A. R., Hony. Capt. 111
—— R., Ty. Capt. 1729
Copeland, A. G. J., Capt. 265, 883, 1004
—— D. P., Lt. 507
—— J. H., Lt. 506, 1696, 1898
—— T., 2nd-Lt. 493
Copeman, C. N. C., Lt. 711
—— D. H., Lt. 68
Copland, R. W., Lt. 1292a, 1598
Copleston, Lt. R. E. 1124
Coppinger, W. V., Maj. 1747
Copplestone, F. E., Lt. 430, 1893
Corban-Lucas, A. J. L., 2nd-Lt. 1297
Corbett, B. A., Maj. 27, 207, 1070, 1893
—— E. C., 2nd-Lt. 561
—— H. R., 2nd-Lt. 545, 1289
—— J. E., Lt. 535, 1696
—— J. H., 2nd-Lt. 717, 1009
—— R. C., Ty. Lt. 1726
—— T. W., Capt. 109, 263, 891
—— V. W., Lt. 311, 1607

Names and Rank. Page.

Corbitt, J. H. 2nd-Lt. 1008
Corbould, E. C., Lt. 352
—— P. S., Capt. 36, 500, 997
Corbould-Warren, E., Bt.-Lt.-Col. 136, 614
—— W., Lt. 540a, 1279
Corbyn, M. McL., Maj. 221, 1662
Corcoran, B. J., 2nd-Lt. 556, 1104
—— J., 2nd-Lt. 767
Corderio, E., Asst. Surgn. 1799
Cordeiro, P. E., Asst. Surgn. 1801
Cordington, H. deB., Lt.-Col. 1660
Cordon, C. I., Lt. 393, 1108
—— C. G. W., 2nd-Lt. 846
Cordwell, E. G., Sub-Condr. 1857
—— E. T., 2nd-Lt. 361, 1048
Corfield, H. R., Lt. 516a, 1668
—— R. FitzH., Lt. 623
—— V. C. W., Lt. 305, 1647, 1657
Corin, E. J. R., Lt. 521, 1359
—— T., Lt. 1273
Corke, F. S., Lt. 455-59, 1140
—— L. H., 2nd-Lt. 361
—— R. M., Capt. 272, 1383, 1581
Corkery, A. T. M., Lt. 29, 540a
—— J. S. W., 2nd-Lt. 368
Corley, J. W., Lt. 492, 875
Corlett, D. S., Lt. 526, 1666
—— J. S., Lt. Col. 105, 140, 162, 200, 903
—— M. J. J., Capt. 501a, 658, 1665
Cormack, F. E., Lt. 489
—— H. S., Capt. 1756
—— R. K. W., Lt. 43?, 1618
—— W., Condr. 1854
Cormody, E. K., Lt. 66, 726
Cornah, J. R., Lt. 842
Cornedius Sundararaj Gopal, Sub-Asst. Surgn. 1827
Cornelius, A. B., Asst. Surgn. 1792
Cornell, W. E. L., Lt. 371, 375, 937
Corner, A. H. S., Sub-Condr. 1870
—— C. G. S., Asst. Surgn. 1801
—— E. S., 2nd-Lt. 564
—— R. J., Lt. 313, 933
—— W., 2nd-Lt. 782
Cornfield, C. J., 2nd-Lt. 562
Cornforth, E., 2nd-Lt. 605
Cornish, A. S., Lt. 538
—— A. W. D., Maj. 19, 242, 1623
—— H., Lt. 527, 1191, 1601
—— H. D., Lt. 397
Cornish-Trestrail, A. G., 2nd-Lt. 550, 1009
Cornock, W. B., Lt. 1144
Cornwall, J. W., Lt.-Col. 135, 1744
Cornwallis, F. W. M., Lt. 593
Cornwell, R. B., Lt. 528, 1275
—— V. J. S., 2nd-Lt. 1189
—— W. S., 2nd-Lt. 780
Corridon, A., 2nd-Lt. 1701
—— A. F., 2nd-Lt. 545
—— R., Sr. Asst. Surgn. 1787
Corrie, A. B., 2nd-Lt. 545, 1297
—— G. G. W., Lt.-Col. 137, 632
—— W. F. T., Maj. 101
Corrin, T. O., Lt. 308
—— T. S., Lt. 1144
Corry, M., Maj. 1747, 1889
—— P. N., Lt. 526, 815
Corse-Scott, E. J., Capt. 262, 1600, 1604
Corser, H. D., 2nd-Lt. 1591
Corson, A., 2nd-Lt. 798
Cortwright, T. H., Lt. 527
Cory, E. M., Capt. 511, 819, 998
Cory-Smith, L. B., 2nd-Lt. 359, 1050
Cosby, N. R. C., Capt. 501, 1618
Cosens, F. R., Maj. 246, 1030
—— F. R. S., Maj. 1722, 697
Cosgrove, E. T., Lt. 758
Coshan, F. W. A., Sr. Asst. Surgn. 1787
Cosnell, A. G., 2nd-Lt. 599
Cosserat, E. W., 2nd-Lt. 1008
Cossey, E. E., 2nd-Lt. 788
Cosslett, E. T., Lt. 577
Costello, C. P., Condr. 1851
—— E. W., Lt-Col. 50, 98, 122, 135, 15?, 140, 162, 200, 868, 1106
—— J. F., Capt. 818
—— J. W., 2nd-Lt. 330, 919
Costelloe, W. H. G., Lt. 666, 992
Coster, G. W., 2nd-Lt. 360, 1064
Costobadie, L. P., Ty. Capt. 1723
Cotching, A., 2nd-Lt. 684
Cotesworth, C. G., Capt. 826
—— C. H., Capt. 597

Names and Rank. Page.

Oothay, H. W., Capt. .77, 1885
Oottam, C. H., 2nd-Lt. 1367
—— J. S., 2nd-Lt. 1679
Ootter, E., Capt. 1759
—— H. J., Lt.-Col. 137, 632
—— P., Capt. 9, 18, 571
—— R. K., Capt. 615
—— V. R., Maj. 94, 228, 1880
Ootterell, R. J., 2nd-Lt. 565
Ootterill, L., Maj. 1711
Oottle, J. L., Capt. 852
—— S. G., 2nd-Lt. 791
Ootton, A., Condr. 1861
—— A. C., 2nd-Lt. 563, 1443
—— B., Revd. 1904
—— C. W., 2nd-Lt. 552, 1259
—— D. C., Lt. 58, 622
—— E. A., Asst. Surgn. 1795
—— F. W., Lt.-Col 1711
—— H. J., Lt. 352
—— J. C., 2nd-Lt. 324, 1116
—— J. T., Lt. 445, 1393
—— P. V., Capt. 8, 15, 753
—— S. C. N., 2nd-Lt. 561, 1509
—— T. C., Lt. 351
—— W. B., Lt. 95, 487, 608
—— W. P., 2nd-Lt. 652
—— W. L., Lt.-Col. 133, 157, 194, 868, 1034
Ootton-Swanston, W. L., Lt. 425, 423, 1223
Oottrell, H., 2nd-Lt. 1319
Oouche, T. J., 2nd-Lt. 1140
Oouchman, C. C., 2nd-Lt. 685
Oougle, M., 2nd-Lt. 566
—— H. J., Maj. 663, 1880
Oouldrey, O., Lt. 430, 1114
—— T. R., Capt. 1729
Ooulile, A. G., Capt. 1754
Ooulon, F., Lt. 19
Oouison, R. A., Lt. 511, 727, 1507
Oouiston, H. C., Lt. 776a
—— W. B., 2nd-Lt. 539, 1672
Oouiter, A. W., Lt. 1405
—— R., *Hony.* Maj. 1852
Oouiton, G. F., 2nd-Lt. 553, 1521, 1523
Oouitor, G. F., 2nd-Lt. 1523
Ooulthard, J. R., 2nd-Lt. 651
—— G. R., Lt. 514, 1385
Ooupe, A., Capt. 1732
Oouper, J. M., 2nd-Lt. 355
—— T., Lt. 820
—— T., 2nd-Lt. 550
Oourley, J. M. C., 2nd-Lt. 1122
Oourt, A. E., Esqr. 89
—— E. M., Sub-Condr. 1874
—— E. T., 2nd-Lt. 565, 1207
—— R. H., 2nd-Lt. 687
Oourtenay, C. E., 2nd-Lt. 733
—— H. G., 2nd-Lt. 733
—— J. M., 2nd-Lt. 522
—— R. B., *Hony.* Lt. 1849
—— R. H., Capt. 830
—— S. W., 2nd-Lt. 720, 786
Oourtney, F. L., Asst. Surgn. 1797
—— J., *Hony.* Capt. 1846
—— M., Sr. Asst. Surgn. 1787
Oourthope, E. A., Capt. 501a, 609
Oourthope-Munroe, C. H., Capt. 485, 937
Oousens, B. P., Lt. 313, 947
Oousins, C. W., Lt. 517, 1618
—— E. C., Lt. 518
—— M. A. S., Lt. 505, 1575
Ooutts, D., Capt. 1754
—— J., 2nd-Lt. 790
—— J. D., Capt. 1839
—— J. S. P., Asst. Surgn. 1802
—— W. F., 2nd-Lt. 763
Oove, V. D., Lt. 515, 1666
Oovell, B. C., Lt. 504, 826
—— G., Capt. 1758
Ooventon, J. A., 2nd-Lt. 460-62, 889
Ooventry, G. St. J., 2nd-Lt. 790
—— H. F., Lt. 381, 1559
—— W. A., 2nd-Lt. 563, 1703
Ooverdale, G. L. B., Lt. 630
Ooverton, J. W. G., Lt. 318, 1465
—— S. H., Lt. 540a
Oowan, A. H., Lt. 623
—— D. T., Lt. 307a, 1611
—— E. S., Capt. 707
—— G. J., Lt. 489
—— J., Lt.-Col. 133, 1710
—— J. de B., Maj. 614

Oowan, J. L., 2nd-Lt. 745, 1008
—— J. M., Lt. 526
—— N. M., 2nd-Lt. 731a
Ooward, A. J., Ty. Capt. 1850
Oowdell, A. F., Capt. 1670
Oowell, F. J., 2nd-Lt. 25, 539, 1596
—— H. C., 2nd-Lt. 896
—— R. O., 2nd-Lt. 797
Oowen, S. W., 2nd-Lt. 1455
Oowie, A., Lt. 542, 1122
—— C. H., Col. 60, 122, 660
—— D. G., Maj. 71, 633, 1884
—— D. W. G., Lt. 1693
—— F., Lt. 530, 1323
—— H. E. C., Maj. 662, 1882
—— H. K., 2nd-Lt. 623
—— H. McC., Maj. 662, 1880
—— J. M., Lt. 446, -49, 1152
—— P. P., 2nd-Lt. 640
—— T. McK., 2nd-Lt. 744
—— W. A. L., Lt.-Col. 26, 186, 1883, 1911
—— W. P., 2nd-Lt. 493
Oowieson, S., 2nd-Lt. 778
Oowin, D. H. F., Maj. 1746
Oowley-Brown, H. T., 2nd-Lt. 1696, 1702
Oowling, C. S., 2nd-Lt. 1630
—— C. S. C., Lt. 472, 1630
—— E. J. S., Lt. 623
Oownie, J. C., 2nd-Lt. 786
Oowper, A. G., Capt. 1758
—— G. G. St. J., 2nd-Lt. 560, 1447
—— G. J., Asst. Surgn. 1800
—— M., Maj-Gen. 118, 146, 174
—— R. H., 2nd Lt. 750
—— W. E., 2nd-Lt. 1679
Oowper-Smith, S. C., 2nd-Lt. 1225
Oox, A. D., Lt.-Col. 136, 158, 196, 868, 1124
—— A. J., 2nd-Lt. 90, 522, 732
—— A. P., 2nd-Lt. 560, 1677
—— C. E. S., Bt. Lt.-Col. 134, 158, 211, 1283
—— C. G., Asst. Surgn. 1795
—— C. H., Ty. Capt. 1725
—— C. E. C., Lt. 111
—— C. J. B., 2nd-Lt. 566
—— E. O., Lt. 523, 1699
—— F. B., Lt. 533, 1676
—— F. W. H., Col. 16, 37, 123, 149, 182
—— G. W., Lt. 307, 1144
—— G. W., Lt. 539, 988
—— G. W. S., 2nd-Lt. 554, 1122
—— H., 2nd-Lt. 517, 706, 1674
—— H. G., 2nd-Lt. 1020
—— H. R., Capt. 516a, 806, 1247
—— H. T., 2nd-Lt. 588
—— H. V., Lt.-Genl. *Sir* 118, 145, 172, 1287
—— H. W. V., Sr. Asst. Surgn. 1789
—— L. E., Ven'ble, 1906
—— M., Lt. 313, 947
—— O., Lt. 507
—— P. Z., Hony. Maj.-Genl. *Sir*. 184, 1889, 1911
—— R. A., Lt. 710
—— R. C., Maj. 808
—— R. F., Maj. 591
—— R. N., 2nd-Lt. 747, 935
—— Stephen, Capt. 611
—— S. J. S., Lt. 371
—— T., Maj. 808
—— T. S., Maj. 208, 1170
—— T. W., 2nd-Lt. 552, 1379
—— V. M., 2nd-Lt. 1367
—— W. C., Lt.-Col. 702
—— W. J., Lt. 543
Ooxall, G., 2nd-Lt. 637
Ooxe, C. B., Sr. Asst. Surgn. 1792
—— K. H., Lt. 504, 1601
Ooxon, J. O. 1009
Ooy, F. C., Lt. 470, 1271
Oozens, A., 2nd-Lt. 777
—— S. J., 2nd-Lt. 692
Orabbe, C. B., 2nd-Lt. 325, 366, 1515
Orabtree, C., 2nd-Lt. 690a
—— H., 2nd-Lt. 361, 1678
—— H. G., Revd. 1904
Orace, J. H., Lt. 506, 524, 1630
Oracknell, H. W., 2nd-Lt. 566, 1703
Oraddock, A. B., Capt. 367
—— C. W., 2nd-Lt. 750
—— H. E. F., 2nd-Lt. 491
—— *Sir* R. H., Hon. Comdt. 827, 1894
—— S., Lt. 781
Oraft, H. T., Lt. 317a, 365, 1475

Orage, F., Lt. 796
Oragg, F. W., Maj. 35, 1753
Oraggs, H. C., Sr. Asst. Surgn. 1790
Oraig, A. D., Lt. 322, 1647, 1653
—— A. F., Lt. 751
—— C. C., Lt. 302, 1022
—— G., 2nd-Lt. 316, 957
—— H. A., Capt. 833
—— H. T., Capt. 274, 1195, 1425
—— J. A., 2nd-Lt. 1245
—— J. H., Lt. 781, 1225
—— J. W., 2nd-Lt. 1331
—— M. T., 2nd-Lt. 669, 986
—— S. F., 2nd-Lt. 555, 1126
—— S. M. M., Lt. 315, 1487
—— V. F., Lt. 996
—— W., 2nd-Lt. 541
—— W. J., Lt. 1885
Oraig Adams, R. L., Lt. 509, 1668
Oraigen, A., *Hony.* Lt. 75
Oraigie, P. N., Maj. 214, 885
—— J. H., 2nd-Lt. 1193
Oraik, D., Lt. 51, 503
—— J., Maj. 60, 1919
—— W. R., 2nd-Lt. 603
Oraker, P. W., Lt. 509
Oramer-Roberts, M. T., Maj. 1910
—— W. E., Bt.-Lt.-Col. 136
Orampin, R. W., Lt. 463, 1557
Orampton, E. A., 2nd-Lt. 571a
—— G. P., Lt. 293a, 1641
Oran, H. R., Ty.-Lt. 1729
Orandell, W., 2nd-Lt. 360, 1158
Orane, A., 2nd-Lt. 1114
—— B., Nursg. Sister 1734
—— E. R., Lt. 986
—— L. F., Lt. 681
—— M. E., Capt. 723
—— W. J., Sub.-Condr. 1874
Oranswick, W., 2nd-Lt. 465
Oranton, D., 2nd-Lt. 734
Oranswick, H. S., Capt. 735
—— W., 2nd-Lt. 473, 1048
Oranwill, G. B., 2nd-Lt. 472, 1028
Orapp, W. V., Lt. 544, 1187
Orapper, N., 2nd-Lt. 556
Oraster, E. E., Maj. 45, 668
—— G., Maj. 13, 29, 52, 98, 216, 885
—— J. C. B., Lt.-Col. 56
—— R. F., Lt. 415, 731
—— S. L., Col. 123, 660
—— W. S., Maj. 216, 889
Oraufurd-Lindsay, H. W., 2nd-Lt. 492
Oraven, A., *Hony.* Maj. 1844
—— A. G. W. S., Lt. 443, 1092
—— C. H. J., Lt. 535
—— L., 2nd-Lt. 355
Oraven-Thomson, C. A., 2nd-Lt. 556
Oraw, H. H., Capt. 526, 848
Orawford, A., Ty. Capt. 1724
—— A. E., Capt. 81, 634, 1664
—— A. G., Lt.-Col. 136, 158, 197, 870, 1143, 1373
—— A. G., 2nd-Lt. 553, 1523
—— C. G., Lt. 510, 1143
—— C. O., Lt. 435, 1283
—— E. L., Lt. 516a
—— F. R. R., 2nd-Lt. 552, 1495
—— G. A., Lt. 445, 1413
—— H., Capt. 502, 1665
—— J., Capt. 656
—— J. C., Maj. 88, 658
—— J. D., Bt. Maj. 107, 253, 1363
—— J. H., Lt.-Col. 136, 158, 196
—— J. M., Lt.-Col. 128, 1741
—— K. B. S., Capt. 666, 996
—— M. N., Lt. 651
—— N. S., Lt. 515, 1701
—— R. McG. C., Lt. 781
—— S. C. R., Capt. 57, 720
—— T. A., Lt. 623
—— T. W., Sub-Asst. Surgn. 1819
Orawfurd, H. P. B. F., Col. 176
Orawley, C. P., Lt. 736
—— N. L., Lt. 450-54, 1142
Orawshaw, H. E. J., Lt. 529, 1668
Orawshaw, H. H., Lt. 385, 1335, 1575
—— P. W., 2nd-Lt. 522
Orayden, C. C., 2nd-Lt. 535, 1672
Oraze, F., 2nd-Lt. 699
Oreagh, A. H. D., Col. 126, 150, 187
—— C. J., Condr. 1861
—— D. V., Lt. 587

Names and Rank.	Page.

Creagh, E. C., Lt.-Col. 56, 1919
——— O'M., Genl. Sir 1513, 1925, 1995
——— P. H., Maj. 711
Creais, E. J., Asst. Surgn. 1796
Creak, E. H., Revd. 1906
——— E. V., Capt. 8, 571, 1850
Crealock, W. E. W., Lt. 504, 1665
Crean, G. J., Capt. 768
——— L. E., Asst. Surgn. 1803
Crease, E. H., 2nd-Lt. 599
Creasey, A. C., 2nd-Lt. 363
Creasy, E. C., Capt. 623
——— N. E., Lt. 354
——— R. L., 2nd-Lt. 549, 1215
Credland, H. W., 2nd-Lt. 728
Cree, G., Maj.-Genl. 43, 1709
Cremer, G. O. R. 455—59
Creed, H., 2nd-Lt. 749
Creedy, H. D., Lt. 510
Creeke, W. R., 2nd-Lt. 846
Creery, W. F., Lt. 769
Creffield, G., Lt. 475
Cregan, T. C., Capt. 1732
Cremer, G. O. R., Lt. 450
——— H. L. H., Lt. 683, 684
Cressall, H. A., Lt. 17, 45, 754
Cresswell, John 2111
Creswick, M. D. J., Lt. 686
Crew, F. A. E., Capt. 699, 700
——— R. D., Capt. 272, 1231
Crewdson, W. D., Capt. 724
——— W. T. O., Capt. 615
Crewe, E. N., 2nd-Lt. 548
Crichton, A. B., Lt. 526, 1705
——— B. D., Ty. Capt. 1725
——— H. G. L., Sir 730
——— McV., Maj. 212, 1415
——— R. T., Lt.-Col. 185, 1880, 1911
——— T. C., Lt. 317, 883
——— T. E., 2nd-Lt. 444—45
——— W. M., Lt. 1676
Crick, C. C., Capt. 54, 265, 1439
——— W., 2nd-Lt. 798, 841
——— W. H. R., 2nd-Lt. 740
——— W. L., Lt. 538
Crickett, G. W., Lt. 21, 760
Crighton, C., Col. 843
——— D. W., Lt. 814
——— T. C., 2nd-Lt. 883
Crilly, A. J., Lt. 536, 1283
Crimmin, C. C., Lt. 507
——— J., Col. 1, 29, 124, 1738
Criper, S. F., Lt. 512
Cripps, A. W., Col. 124, 149, 183, 1660
——— E. J., Capt. 263, 1664
Crisp, J. H., 2nd-Lt. 786
——— M. F., 2nd-Lt. 1267
——— R. E., 2nd-Lt. 565
——— R. H., Lt. 510, 1195
Crispin, E. H., Maj. 1689
Cristall, B. P., Lt. 306, 1237
Critchell, D. E. R., 2nd-Lt. 550, 1705
Critchley, D. R., Lt. 291, 1148, 1156
——— R., 2nd-Lt. 761
Crittall, P. C., Lt. 380, 905
Croad, A., Lt. 516a
Crossdale, E. W., Lt. 724
Croban-Lucas, A. J. L. 2nd-Lt. 456
Crocker, A., Lt. 523, 1082, 1675
——— E. G. M., Lt. 703, 796
——— G., Lt. 793
——— G. D., Br.-Genl. 12, 21, 120, 125
——— H. E. T., Lt. 515
——— S. F., Maj.-Gl. 124, 149, 183
——— W. J., Lt. 297, 389, 505, 1074
Crockett, N. S., Lt. 541, 1066
Crockwell, C. G., 2nd-Lt. 778
——— L. H. W., Lt. 514, 986
Crocombe, F. R., Lt. 420, 1319
Croft, T. deC., Lt. 518, 1666
——— F. R., Lt. 332, 1046
Crofton, G. H. S., Maj. 710
——— M. R. H., Maj. 614
——— R. M., Lt. 493, 943
Croghan, S. C., 2nd-Lt., 564, 1124
Croil, T. M., Lt. 629
Croken, I. E., Capt. 1839
Croly, R. G. G., Maj. 1752
——— W. C., Maj. 62
Crombie, A., 2nd-Lt. 744
——— D. C., Maj. 15, 35, 98, 215, 919
——— W. M., Lt. 1759
Crompton, J., Capt. 269, 1058
Cronan, C. P., 2nd-Lt. 91, 566
Crouan, L. B., Lt. 532, 1339
Croning, H. L., Asst.-Surgn. 1803
Crook, C. J., 2nd-Lt. 750
Crooke, F. M. W. 2nd-Lt. 475
——— W., 2nd-Lt. 1598
Crooks, B. D., 2nd-Lt. 712
——— F., Ty. Lt. 1726
Crookshank, C. K., Maj. 105, 1910
——— S. D'A., Br.-Genl. 60, 121, 127, 605, 661, 1882
Cropper, H., 2nd-Lt. 669, 1001
——— P. G., Lt. 591
Crosbie, A., Capt. 836
——— C. S., *Hony.* Capt. 1850
——— F. S., 2nd-Lt. 365, 1604
——— H., Lt. 756
Crosby, C., Lt. 748
Crosby-Browne, D., Lt. 528, 1630
Crosleigh, E. L., Maj 225, 1148, 1150
Cross, A. B., Capt. 61, 505, 843, 1407, 1732
——— A. B. R., 2nd-Lt. 7
——— C. H., Maj. 1753
——— C. H., M., 2nd-Lt. 566
——— D. L. C. A., 2nd-Lt. 681
——— E. C., Lt. 690a
——— F. G., Tempy. Capt. 1724
——— F. N., Lt. 380, 937
——— H. E., Lt. 575a, 753
——— H. L., Esq. 91
——— F. L., 2nd-Lt. 548, 1701
——— P. L., 2nd-Lt. 1702
——— P. W., Maj. 768
——— S. D., Lt. 351
——— S. G. N., 2nd-Lt. 324, 1110
——— W., Lt. 354
Crossfield, E. A. K., Capt. 107, 262, 1647
——— C. R. K., Lt. 505
Crossle, H., Maj. 1749, 1900
——— H. C., Lt. 523, 998, 1698
Crossley, A. A., Capt. 582
Crosthwait, H. L., Maj. 661, 1880
——— L. G., Maj. 227, 1880
Crosthwaite, B. M., Maj. 832
——— C. G., Maj. 214 1900, 1913
——— H. B., Lt. 83, 543, 803
——— J. S., Maj. 230, 723, 1900
Crothers, A. V., Lt. 1142
Croucher, F. E., Lt. 317a, 365, 891, 1483
——— E. W., Lt. 493
Croudace, C. G., 2nd-Lt. 544
Crow, C. G., Asst. Surgn. 1794
Crowden, H. C., *Hony.* Lt. 768
Crowdy, C. R., Maj. 633, 974
——— J. D., Maj. 54, 98, 220, 1617
Crowe, H. W., 2nd-Lt. 1018
——— J. A., Lt. 571
——— W. H. F., Capt. 617
Crowhurst, J. D. J., 2nd-Lt. 551, 1128, 1573
——— A. D., 2nd-Lt. 554, 1143
Crowley, B. E., 2nd-Lt. 669, 997
——— C., Sub-Condr. 1864
——— H. S., Lt. 526, 988
——— J. T., 2nd-Lt. 362
Crowsley, F., Lt. 712
Crowther, A. H., Capt 276, 1283
——— C. T. B., Lt. 656, 992
——— H. J., 2nd-Lt. 539, 1387
——— H. O., Lt. 516a, 1471
Croxford, S. H., 2nd-Lt. 1695
Croxson, J. A. F., Lt. 534
Crozier, D., Lt. 351, 1673
——— P. H., Revd. 1904
——— W. H., Lt. 423, 1623
Cruchley, W. H., 2nd-Lt. 693
Crudass, R. H., Lt. 687
Cruddas, H. M., Maj. 62, 1746
Cruickshank, A. J., Capt. 52, 665, 991
——— A. S., Lt. 309, 1074
——— F. C. M., Lt. 513, 905
——— H. St. J., Maj. 101, 633
——— J. A., Maj. 1753
——— J. E., Maj. 33, 231, 871, 1545, 1600
——— J. G., Lt. 435, 1092
——— J. R. H., Lt. 297a, 909
Cruise, W., 2nd-Lt. 1030
Crum, W. E., Maj. 603
Crumblehulme, W., 2nd-Lt. 567
Crump, D. J., Lt. 18, 23, 778
——— E. W. E., Maj. 83, 501, 814
——— R. H., 2nd-Lt. 540
——— S. T., Maj. 1752
Crunden, E. G., Asst. Surgn. 1796
Crundwell, E. F., 2nd-Lt. 776a
Cruttwell, R. G., Lt. 817
Cruttenden, L. C., 2nd-Lt. 360
Cubbon, F. R., Capt. 274
——— R., *Hony.* Capt. 1847
Cubitt, A., Sub.-Condr. 1863
——— F. L., 2nd-Lt. 692
Cuddeford, R. B. G., Capt. 273, 1505, 1515
Cude, C. J., Lt. 353
——— J., 2nd-Lt. 1674
——— T., Lt. 354, 1577
Cuerden, C., Condr 1867
——— H., Sub-Condr. 1867
——— T., *Hony.* Lt. 1848
Cuerdon, F. J., Lt. 525
——— G. H. E., Hony. Capt. 1843
Cuffe, *Sir* O. F. L. W., Hy. Lt.-Col. 848
Cull, A. B. C., 2nd-Lt. 776
——— B. N., 2nd-Lt. 360
Cullen, A. B., Lt. 302, 992
——— D. E., Hony. Lt. 1863
——— E. H. S., Lt.-Col. 133, 157, 193, 869, 1156
——— J. Sub.-Condr. 1868
——— J. A. W., Capt. 22, 1718
——— J. G., Lt. 521, 1066
——— L., 2nd-Lt. 560, 1678
——— P., Lt. 637
——— R. R. B., Sub-Condr. 1863
——— S. E., 2nd-Lt. 548, 1094
——— W. R., 2nd-Lt. 765
——— W. B. A. K., Maj. 1752
Cullenum, J., Capt. 1718
Cullerne, C. P., Maj. 614
Culley, Lt. A. 453, 1319
——— R. H., 542
Cullingford, F. H., Lt. 351, 1676
Cullington, E., 2nd-Lt. 691, 784
Cullinan, J., Sub-Condr. 1857
Cullis, L., Lt. 536, 1110
Cullom, A. E., 2nd-Lt. 599
Culpeck, T. C., 2nd-Lt. 551, 1265
Culpeper, A. G., Asst. Surgn. 1791
Culverwell, J. S., Capt. 501a, 835, 1235
Cumber, H. S., Esq. 9, 88
Cumberlege, A. F., Lt.-Col. 11, 134, 661
——— C. F., Lt. 582
——— C. J., Lt.-Col. 194, 1883, 1912
——— J. H., Lt. 485, 1665
——— R. M., Capt. 629
Cuming, B. S., 2nd-Lt. 768, 781
——— R. J., Lt.-Col. 137, 200, 159, 869, 1215
Cumming, A., 2nd-Lt. 763
——— A. E., Lt. 303, 1255
——— D. B., 2nd-Lt. 534
——— J. F. 455—59
——— J. F., Lt. 1092
——— R. C. R., Lt. 529
——— R. H. R., 2nd-Lt. 315, 891
——— W. J., Lt. 95, 303
——— W. T., Sub-Condr. 1856
Cummings, C. G., 2nd-Lt. 565
——— W. H., 2nd-Lt. 38, 758
Cummins, H. A. V., Bt.-Col. 97, 126, 153, 194, 1116
——— J. E., Lt. 322, 1379
——— W., Lt. 425
Cunliffe, J. B., Maj. 652
——— R. G., Lt. 544, 1601
Cunnah, E. E., Capt. 1732
Cuningham, A. P., Capt. 55, 276, 1505
——— C. A., Maj.-Genl. 173, 1505
——— G. E., 2nd-Lt. 547, 820
Cuninghame, T. B., Lt. 515, 1124
——— A. W., Lt. 59, 703
——— D. S., Col. 179, 1917
Cunningham, A. H., Lt.-Col. 137, 661, 985
——— A. P., Lt. 440, 1116
——— C. C., Maj. 99, 227, 871, 1066, 1433
——— C. J., Condr. 1867
——— E. C. T., 2nd-Lt. 82, 1189
——— E. M., Nursg. Sister 1734
——— G. Capt. 881
——— G. E., Lt. 820
——— J., Maj. 1752
——— J. R., Capt. 1839
——— N., 2nd-Lt. 624
——— P. H., Lt.-Col. 130, 155, 190
——— R. A., Lt.-Col. 133
——— T. C., Lt. 354, 1154

NAMES AND RANK.	Page.
Cunningham, T. S., Lt.	763, 781
——— V., 2nd-Lt.	356, 1419
——— W., Lt.	622
——— W., Lt.	618
——— W. B., Capt.	256, 1174
——— W. E., 2nd-Lt.	325, 366, 1231
——— W. J., Lt.	354, 1104
Cunnington, J., Lt.	701, 733
Curd, A. T., Lt.	537, 1573
Curell, A., Maj.	589
Curl, G. H., 2nd-Lt.	354
Curle, V., Lt.	296, 1275, 1331
Curley, T., *Hy.* Capt.	1848
——— W. L., Lt.	351, 1189
Curll, C. E., Bt.-Col.	57
Curnow, E., Lt.	295, 1283
Curran, J. F., Senr. Asst. Surgn.	1787
——— P. J. E., 2nd-Lt.	1705, 556
Currie, A., Col.	170
——— A. H., Lt.	785
——— A. J., Esq.	91
——— C. J. R. W., Sub-Condr.	1864
——— C. O., Lt.	619
——— F., Maj.-Genl.	169
——— J., Lt.	446-49, 1495
Curry, A. V., Lt.	387, 1022
——— F. R., Capt.	16, 38, 708
——— G. R., Lt.	529, 1671
——— L. H., 2nd-Lt.	741
——— M. J., Capt.	636, 980
——— R. C., Lt.	700
Cursetjee, H. J. M., Capt.	1757
——— H. R., Tempy. Lt.	1773
Curthoys, F. H., 2nd-Lt.	361
Curtin, A. F. S., Lt.	316, 1199
Curtis, A., 2nd-Lt.	554
——— A. C., Capt.	279, 1203
——— A. G. C., Lt.	443, 1132
——— C. M., 2nd-Lt.	554
——— F., Maj.	238, 911
——— G. R., Capt.	731
——— H., Lt.	668, 997
——— H. C., Condr.	1861
——— K. S., 2nd-Lt.	557
——— J. E., Condr.	1863
——— M., 2nd Lt.	567
——— O. O., Capt.	267, 1247
——— P. H., 2nd-Lt.	560, 1411
——— P. V., 2nd-Lt.	725
——— R. J., Lt.	72
——— R. L., Lt.	96, 516
Curtler, A. C. H., Lt.	9
——— W. L., Lt.	367, 1613
Curtoys, O. F., 2nd-Lt.	328, 1158
Cusack, J., Lt.	778
Cuscaden, R. L., Lt.	505, 1409
Cusins, A. F., Maj.	663
Cusse, C. A., Lt.	696
Cuthbert, L. A., Lt.	533, 1096
——— R., *Hony.* Lt.	1848
Cutler, H. C., Maj.	213, 862, 1032, 1046
——— F. G., Asst. Surgn.	1792
——— G. J., Lt.	541, 1673
——— M. H. C., Lt.	300, 1543
——— N. E., Lt.-Col.	679
——— P. H., Lt.	679
Cutting, C. E., Lt.	536
Cutts, E. J., 2nd-Lt.	739
Cuvelier, E. A., Lt.	526, 610
Cuzen, E. McL., Asst. Surgn	1798

D

NAMES AND RANK.	Page.
D'Abreo, F., Esq.	91
D'Abreu, A. R., Asst. Surgn	1800
D'Albani, C. L., Lt.	371, 1487
D'Alessio, E. S. J., Revd.	1907
D'Ambrunenil, W., 2nd-Lt.	18, 25, 722
D'Anglbau, W. E., Lt.	733
——— W. E. D., Lt.	798
D'Arcy, A. F. J., Asst. Surgn.	1800
——— M. A. A., Lt.	353, 1104
——— M. R., Lt.	488
D'Auvergne, V., Sub-Condr.	1867
D'Avray, A. D., Capt.	38, 1717
D'Cruz, G. L., 2nd-Lt.	571a
——— J. B., Sub-Asst. Surgn.	1827
——— H. J., Lt.	543, 1704
D'Esterre, J. C. E., Major.	614, 634
DaCosta, J. A., Tempy. Lt.	1773, 1798
D'Souza, W. M., Ty. Capt.	1771
Dabbs, S. E., 2nd-Lt.	1160
Dabholkar, A. Y., Tempy. Capt.	1766
Daboo, E. B., Tempy. Capt.	1764
Dacres, L. S. L., Lt.	487, 609
Dacre-Tyler, J. L., Lt.	508
Dad, K., Ty. Capt.	1772
Dada More, Jemdr.	1468
Dadabhai, E. ., Sub-Asst. Surgn.	1832
Dadachanji, R. E., Tempy. Capt.	1766
——— R. K., Tempy. Capt.	1765
Dadan Khan, Jemdr.	1344
——— Ressdr.	1684
Dadhibal Pun, Jemdr.	1619
Dadi Gurung, Jemdr.	1590
Dadu Ingle, Subdr.	1460
Dadle, R. T., Lt.	1195
Dady, M., Tempy. Capt.	1767
Dagdu Chawan, Jemdr.	1416
Dagg, A. V., Lt.	401, 1279
Dagge, R. C., Lt.	313, 1020
Daglish, G., Sub.-Condr.	1872
Daick, W. J., 2nd-Lt.	1387
Daimpre, M. R. W., Lt.	700
Daintith, Lt., E. M., St. G.	518
Daintree, G. W., Lt.	1761
Dainty, S., 2nd-Lt.	794
Daji Sawant, Subdr.	1407
Dajiras Rane	1444
Dajirao Powar, Subdr.	1460
Daji Sinde, Subdr.	1463
Dakin, A. E., Lt.	711
Dal Singh, Subdr.	1070
——— Jemdr.	1167
Dalal, H. P., Ty. Lt.	1775
——— K. R., Ty. Lt.	1778
Dalbahadur, Bhandari, Jemdr.	1643
——— Jemdr.	1602
——— Thapa, Subdr.	1674
——— Jemdr.	1622
——— Jemdr.	1636
——— Jemdr.	1597
——— Gurung, Jemdr.	1657
——— Nagar Kote, Jemdr.	1608
——— Rana, Jemdr.	1598
Dalbir Gurung, Jemdr.	1602, 1608
——— Burathoki, Jemdr.	1657
——— Rana, Jemdr.	1597
——— Thapa	1597
——— Jemdr.	1608
——— Jemdr.	1609
Daldy, A. J., Lt.	316, 1263, 1275
——— A. W., Maj.	22, 87, 212
Dale, C. H., Lt.	543, 1098
——— D. H., Lt.	1359
——— E. C. B., Capt.	615
——— G. A., Br.-Gen.	50, 103, 121, 130, 155, 189
——— P. T., Lt.	789
——— R., 2nd-Lt.	546
Dalel Khan,	1375
——— Jemdr.	1573
——— Singh, Subdr.	
——— Jemdr.	1103, 1681
——— Subdr.	1577
——— Jemdr.	1577
——— Subdr.	1565
——— Jemdr.	1244
——— Jemdr.	880
Dale Ram, Jemdr.	1216
Daley, F. M., Lt.	508
——— J. F., Lt.	519
Dalgarno, D. J., 2nd-Lt.	801
Dalgety, A. E., 2nd Lt.	326, 366, 1078
——— H. B., Maj.	587
Dalgleish, A. J., Maj.	42, 1838
——— J., Regtl. Sergt. Maj.	993
Dalgliesh Singh, Jemdr.	1573
Dalkesar Gurung, Subdr.	1607
Dali Ahmad, Subdr. Maj.	1369
Dalip Chand, Jemdr.	1141, 1228
——— Singh, Jemdr.	1533
Dallao Conte, N., 2nd-Lt.	1419
Dalia Thapa, Jemdr.	1637
Dalip Singh, Resdr	914
——— Subdr.	1285
——— Subdr.	896
——— Jemdr.	1099
——— Jemdr.	882
——— Jemdr.	1111
——— Jemdr.	1174
——— Subdr.	1189
——— Subdr.	1132
——— Subdr.	1335
——— Jemdr.	1179
——— Subdr	1321
——— Jemdr.	942
——— Ty. Sub-Asst. Surgn.	1833
——— Karki, Jemdr.	1645
Dalison, J. B., Lt.	305, 1283, 1541
Daljit Singh, Subdr.	1043
Daljit Singh Pun, Jemdr.	1627
Daliyon Singh, Jemdr.	924
Dallas, A. E., Lt.-Col.	32, 134, 157, 194, 1660
——— D. A. G., Capt.	278, 943
——— H. W., Lt.	375, 692, 1335
——— J., Brig.-Genl.	123, 660
——— J. S., Lt.-Col.	131, 156, 192
——— P., 2nd-Lt.	741
Dallas-Conte., N., 2nd-Lt.	330
Dalley, R. P., Lt.	489, 1007
Dallimore, Capt. A. G.	...
Dalling, W. E., 2nd-Lt.	562
Dallison, J. B.	306
Dallison, H. W., Capt.	695, 850
Dallimore, A. G., Capt.	75
Dallmeyer, H. S. T., Lt.	543a
Dalishankar Mulji Trivedi, Sub-Asstt. Surgn.	1830
Dallu Ale, Jemdr.	1625
——— Thapa, Jemdr.	1563
Dalmahov, P. C., Maj.-Genl.	169
Dalman Ale, Hony. Capt.	2103
Dalmir Khan, Jemdr.	1405
Dalpat Singh, Subdr. Maj.	1038
——— Jemdr.	926
——— Subdr.	1391
Dalrymple, B. G. R., Hony. 2nd-Lt.	947
——— W. C., 2nd-Lt.	552, 1379
——— W. L., Maj.-Gen.	769
Dalrymple-Hay, S. F. B., Maj.	207, 195
——— B. G., Lt.	315
——— B. W., 2nd-Lt.	1689
Dalrymple-Waller, G. H. D., Lt.	537
Dal Singh, Jemdr.	1166
——— Jemdr.	1384
Dalsingar, Subdr.	1581
Dalsing Basnet, Subdr.	1630
——— Thapa, Subdr.	1618
Dalsii Rai, Jemdr.	1633
Dalsur, Rana, Jemdr.	1625
Dalton, A. G., Lt.	1006
——— G. M., Asst. Surgn.	1796
——— R. C., Lt.	1383
Dalu Ram, Jemdr.	1577
Dalvi, N. A. Temy. Capt.	1771
Daly, C. D., Capt.	260, 1664
——— C. E., Capt.	257, 1058, 1279,
——— C. K., Capt.	102, 103, 104, 265, 1664
——— D. A., Lt.	536
——— D. J., Lt.	522, 1670, 1705
——— D. W., Lt.	589
——— F. O. W. A., 2nd-Lt.	326, 366, 1126
——— J. T., Col.	43, 125, 1738
——— J. T., Ty. Capt.	1727
Dalyell, J. T., Lt.-Gen.	715
Dalzell, J., 2nd-Lt.	362
Dalziel, K. McL., Maj.	1747
Damarbahadur Sen, Jemdr.	1643
Damar Singh Gurung, Subdr.	1607
——— Jemdr	1608
——— Jemdr.	1609, 1615
——— Thapa, Jemdr.	1608
——— Jemdr.	1611, 1619
Dambardhoj Limbu, Jemdr.	1633
Damdar Thapa, Jemdr.	1636
Damerell, R. C., 2nd-Lt.	786
——— W. J., Lt.	531, 1668
Dames, T. D. L., Lt.	585
Damiano, N. A., 2nd-Lt.	807
Dammers, E. H. F., Lt.	740
Damodar, Jemdr.	1104
——— Janardan Phrabhavalkar, Sub-Asst. Surgn.	1832
——— Singh Chaudhri, 2nd-Lt.	481
Damodhar Khattri, Subdr.	1642
Damodoram, T. V., Sub-Asst. Surgn.	1828
Danagher, A. V., Lt.	353, 1092
Dana Ram, Jemdr.	1488
Danbahadur Singh, Jemdr.	1063
Danby, A. L., Capt.	485, 909
——— E. C., Lt.	488, 935
——— G. P., Lt.	489
——— J. A. MacD., 2nd-Lt.	473, 1319
Dance, A. J., Lt.	351, 1673
——— E., 2nd-Lt.	476, 1363
Dan Cuylenberg, V. A., Asst. Surg.	1802
Dand, J., Lt.	533, 1122
Dandabir Thapa, Jemdr.	1619
Dandi, H. S., Sub.-Asst. Surgn.	1832
Dane, H., Capt.	277, 964, 1004

Names and Rank. — Page.

Dane, R., Capt. 261, 915
Dangerfield, E R., Maj 819
Dangnath Imbu, Jemdr. 1549
Dani, R. G., Ty. Lt. 1774
—— Jemdr. 1059
Daniel, A. C. E., Lt. 602
—— A. R., Lt. 540, 975
—— E. A., Esq. 91
—— F. G., 2nd-Lt. 689
—— F. L., Lt. 540, 1619
—— H. C., Lt. 595
—— N S., Sub-Asst. Surgn. 1837
Daniell, C. I., Lt. 793
—— J. A. S., Maj. 234, 1074
—— M. I. V. G. Q., Ty. Asst. Surgn. 1804
—— P. S. Capt. 858
Daniels, A., 2nd-Lt. 91, 358
—— A. M., Br.-Maj. 54. 256, 879
—— C. J. F. W., Lt. 503, 721, 1609
Daniels, L. S., Tempy. Lt.-Col. 61
Dani Ram Subdr. 1457
—— Ressdr. 886
Danishmand Khan, Ressdr. 1684
Danks, E., Maj. 803
—— R. B., 2nd-Lt. 630
Dann, H C. R., Lt.-Col. 128, 154, 183
—— C. F., Lt. 300, 1132
—— C. W., Lt. 405, 1453
Dan Singh, Jemdr. 1210
Dan Singh, Gurung, Subdr. 1601
Danson, E. T., 2nd-Lt. 325, 366, 1235, 1565
—— T.
—— Rawat, Jemdr. 1223
Danskin, J. W., Ty. Capt. 7, 571, 1850
Danter, A. W., 2nd-Lt. 560, 1092
—— J. E., 2nd-Lt. 363, 1098
Dao Rawat, Jemdr 1472
Daram Sing Thapa, Jemdr. 1598
Daraz Khan, Jemdr. 1320
Darbara Singh, Jemdr. 1533
Darby, B. G., Lt. 656
—— G., 2nd-Lt. 854, 1678
—— G. S., Capt. 503 832
—— W. A., *Hony.* Lt. 1847
Dardier, C. E., Capt 680, 754
Dare, N. F., Lt. 622
Dare Khan, Sr. Sub-Asst. Surgn. 1808
Darell, E. M. D., Lt. 310, 1355
—— R D E., Capt. 265, 1219, 1319
Dargan, P. A., Tempy. Capt. 1142, 1725
Dark, R. H. Lt. 630
Darley, C. G., Capt. 52, 593
—— J. R., Lt.-Col. 133, 157, 193, 871, 1471
Darling, H. J., Lt. 41, 510
—— J. S., Lt. 540a, 1517
—— O. R., Lt. 440, 1092
—— T. N., Ty. Capt. 1725
Darme Rai, Jemdr 1653
Darne, A., Revd. Father 848
Darnell, W. A., 2nd-Lt. 355, 1475
—— W. R., 2nd-Lt. 329, 1074
Darnley-Smith, H. W., 2nd-Lt. 799
Darshan Singh, Jemdr. 890
—— Jemdr. 1385
Dart, J L. C., Revd. 1907
—— K. G., 2nd-Lt. 362
—— L. V., Lt. 368, 968
Darts, W., 2nd-Lt. 618
Darukhanawala, K. A., Capt. 1767
Daruwalla, B. P. Ty. Lt. 1780
—— J. D., Tempy. Capt. 1770
—— P. P., Tempy. Capt. 1770
Darvell, G. G., 2nd-Lt. 554
Darwan Sing Negi, Subdr. 1179
Darweza, Khan 1493
Darwell, G. P., Lt. 303
Darwin, G. M., 2nd-Lt. 819
——, J. H., 2nd-Lt. 559, 1191
Darwood, A. J. S., Capt. 820
Darya Khan, Subdr.
Das, A., Tempy. Capt. 1770
—— B., Tempy. Capt. 1768
—— J., Tempy. Lt. 1777
—— K. K., Ty. Lt. 1780
—— L., Tempy. Capt. 1772
—— M., Capt. 1758
—— M., Tempy. Capt. 1767
—— P., Tempy. Capt. 1769
—— S. P., Tempy. Capt. 1772
—— U. N., Tempy. Lt. 1775
Dasanda Singh, Subdr. 1124
Dasaundha Singh, Subdr. 1106, 1268
—— Jemdr. 990
Dasaundha Singh, Subdr. 1011, 1050
—— Subdr. 1267
—— Subdr. Maj. 1108
—— Sub-Asst. Surgn. 1817
—— Subdr. 1343
Dasaundi Ram, Subdr. 1285
Dasent, L. H., Lt. 493, 937
Dasgupta, K., Tempy. Lt. 1776
Dash, A. J., Lt. 515
Daspati Singh, Subdr. 1702
—— Jemdr. 1053
Das Ram, Sub-Asst. Surgn. 1816
Dasrath Khattri, Jemdr. 1643
Das Gupta, K. C., Lt.
Dastoor, P. A., Tempy. Lt. 1777
Dastur, H. S., Tempy. Capt. 1765
Datadin Badhal, Jemdr. 999
Datar Singh, Jemdr. 1453
Date, M., Lt. 704
—— W. A., Lt 700
Datnarayan Dube, Jemdr. 1027
Datt, B. R., Tempy. Sub-Asst. Surgn. 1833
Dattajirao Khanvilkar, Jemdr. 1414
Datta, R., Tempy. Capt. 1770
—— Jadhow, Jemdr. 1417
—— Ram, Jemdr. 1320
Dattaram, Chauhan, Subdr. 1460
—— Sawant, Jemdr. 1464
—— Karle, Jemdr. 1444
Dattatraya Balkrishna Malliker, Sub-Asst. Surgn. 1829
—— Ganga-dhar Kale, Sub-Asst. Surgn. 1830
—— Mohaniraja Makeshir, Sub-Asst. Surgn. 1831
—— Shripad Mahabal, Ty. Sub-Asst. Surgn. 1833
—— Waradkar, Jemdr. 1469
Datu Bhosle, Subdr. 1423
Datye, V. V., Sub-Asstt. Surgn. 1832
Daubney, R., Capt. 1841
Daud Ali Khan, Ressdr. 882
—— Beg, Subdr. 1327
—— Khan, Jemdr. 1272
Daudbir Thapa, 1608
Daud Khan, Hony. Capt. 2103
—— Jemdr. 938
—— Jemdr. 1305
Daukes, C. T., Maj. 220, 1898, 1913
Daula Ram, Subdr. 1479
—— Singh, Jemdr. 1112
Daulat Bhosle, Subdr. 1423
—— Ram. Sub-Asst. Surgn. 1812
——, Sub-Asstt. Surgn. 1817
—— Rao Bhosle, Subdr. 1423
—— Singh, Sub-Asst. Surgn. 1808
—— Singh, Subdr. 1288
—— Sub-Asst. Surgn. 1818
—— Jemdr. 1150
—— Subdr. 1046
Daunt, A., Lt. 722
—— C. O'B., Capt. 261, 947
—— E. W. F. T., Lt. 77, 513, 1667
—— W. D., Lt.-Col. 79, 129, 155, 188
Davi Singh, Hony. Capt. 2103
Dave, P. D., Sub-Asst. Surgn. 1832
Davenport, E. C., Lt. 487
—— C. V., 2nd-Lt. 355, 1245
—— G. C., Capt. 732
—— R., 2nd-Lt. 355, 1245, 1405
—— T., Lt. 397, 1183
Davern, F. B., Lt. 653
Davey C. C., Lt. 534, 1463
—— C. F., Ty. Capt. 1725
—— F., Lt. 769
—— H. B., Hony. Lt. 654
—— H. R., Lt. 1677
—— H. W., 2nd-Lt. 759
David, J. S., Tempy. Capt. 1767
—— J., 2nd-Lt. 473
—— T, Capt. 501, 1667
Davids, W. J., Esq. 91
Davidson, A., Maj. 724
—— A., Lt. 470, 1323
—— A. J., 2nd-Lt. 558
—— A. N., Maj. 1919
—— A W. E., Condr. 1852
—— D. J., Lt. 76, 524, 1884
—— D MacD., Lt.-Col. 117, 1740
—— F. D., Lt.-Col. 135, 158, 195, 871, 1403
—— F. M. Lt.-Col. 101, 131, 632
—— G. F., Lt. 1265
—— G. S., Lt. 527, 658, 1671
—— H. B., Capt. 264, 1647
Davidson, H. C., Lt. 1080
—— H. H., Lt. 540, 1701
—— H. M. Maj. 222, 943
—— J., Lt.-Col. 23, 105, 130, 1742
—— J., Lt. 510, 1666
—— J. R., Lt. 515 997
—— J. S., Col. 33, 34, 124, 1709
—— J. W., Capt. 280—287, 897
—— M. D., 2nd-Lt. 547
—— R., Capt. 1713
—— S. I., Capt. 1732
—— S. R., Br.-Genl. 50, 121, 125, 153, 195, 869, 1211
—— W. L. L., Sub-Condr. 1864
Davie, J H M., Lt.-Col. 53, 135, 158, 198, 939
Davies, A., Lt. 504, 1668
—— A., 2nd-Lt. 682, 785, 799
—— A. A. M., Capt. 1714
—— A. F., Lt. 26, 1679
—— A. T., Lt. 516a, 1614, 1653
—— C. A., Sub-Condr. 1856
—— C. C. Lt. 445, 1596
—— C. H., Br.-Genl. 50, 121, 125, 153, 189
—— C. H. L., Asst. Surgn. 1799
—— C. R. V., Lt. 1465
—— C. W., 2nd-Lt. 75
—— D. A., Lt. 755
—— D. A. 2nd-Lt. 630
—— D. M., Lt. 626
—— D. S., 2nd-Lt. 571
—— E. A., Asst. Surgn. 1792
—— E. E., Lt. 738
—— E. I., Tempy. Capt. 1724
—— E. J. Sub-Condr. 1864
—— E. L., Lt. 205, 410, 1607
—— E. R. L., 2nd-Lt. 330, 1046
—— F. A., Lt. 304, 964
—— F. B., Lt. 638
—— F. C., Lt. 513, 1024
—— F. G. H., Br.-Genl. 12, 21, 121, 126, 150, 187
—— F. L., Lt. 638, 976, 1698
—— G. A., 2nd-Lt. 559
—— G. B., Capt. 260, 1215, 1898
—— G. C., Lt. 1082, 1579
—— G. P., Lt. 532, 1668
—— G. S., 2nd-Lt. 544, 1393
—— H., Lt. 527, 998
—— H. L., Lt. 304, 1178
—— H. M., Capt. 500
—— H. N., Lt. 843
—— H. T. W., 2nd-Lt. 767
—— H. W., Maj. 60, 208, 1661
—— I. A., Lt. 787
—— J., Lt.-Col. 56
—— J. A., 2nd-Lt. 619
—— J. B., 2nd-Lt. 361
—— J. F., Lt. 34, 504, 1409
—— J. F., 2nd-Lt. 779
—— J. H., Lt. 512
—— J. P., Lt. 36, 507, 1883
—— J. T., Lt. 911, 305
—— J. W., Lt. 686, 1007
—— K. C. J., Lt. 367, 1647
—— L. M., Maj. 534, 975
—— L. S. R. G., Lt. 658
—— M. A., Lt. 516, 1499
—— N. D., Lt. 638
—— O. G., 2nd-Lt. 638
—— P., Capt. 803
—— P. M. A., Capt. 520, 858
—— R. D. Maj. 243, 1505
—— R. J., 2nd-Lt. 361
—— R. M., 2nd-Lt. 622
—— R. M., Lt. 319, 1305, 1419
—— R. T., Lt. 28, 746, 1579
—— S. C., Lt. 540, 1672
—— T. B., Capt. 1717
—— T. W. E., Lt. 626
—— V. K. N., Lt. 304, 1645
—— W., Esq. 89
—— W., Lt. 525
—— W. A., 2nd-Lt. 555, 1166
—— W. B., 2nd-Lt. 39, 560, 706
—— W. J., Lt. 353, 1040
—— W. P. L., Bt.-Col. 58, 97, 122, 126, 613
—— W. V., 2nd-Lt. 473, 751, 1158
Davies-Bunton, A. E. K., Lt. 740
Davin, A. L. V., Capt. 1718
Davis, A. P., 2nd-Lt. 550, 1293
Davis, A. I., 2nd Lt. 548, 993
—— A. S., 2nd-Lt. 767
—— A. T. Lt. 401, 1092

NAMES AND RANK. Page.

Davis, A. W., 2nd-Lt. 610
—— C. J. McN., Lt. . . 531, 1696
—— C. K., Lt. 387, 929
—— C. T., Maj. 14, 46, 230, 1431
—— D., 2nd-Lt. 732
—— A. O. L., Lt. 723a
—— D. A., 2nd-Lt. . . . 463, 1291
—— E., Condr. 1853
—— E., Sub-Condr. 1857
—— F. G. T., 2nd-Lt. 1104
—— F. L., 2nd-Lt. . . . 354, 1413
—— G., 2nd-Lt. 571a
—— G. E. P., Maj. . 95, 221, 1427
—— G. G., Lt. . . . 313, 1183
—— G. H., Maj. . . . 209, 1661
—— G. J., Lt.-Col. 95, 135, 158, 195, 1106
—— G. R., 2nd-Lt. 360
—— H., Lt. . . . 297, 704, 1611
—— H. C. M., 2nd-Lt. . 331, 1606
—— H. J. A., *Hony.* Lt. . . . 1850
—— J., Condr. 1854
—— J. Lt. 1143a
—— J. E. P., Maj. 95
—— J. T., 2nd-Lt. . . . 547, 1471
—— M. A., Lt. 516a
—— M. L., Lt. 630
—— N. E., 2nd-Lt. 1126
—— P. J., Lt. 836
—— R., Lt. 450-54 1641
—— R. W. J., 2nd-Lt. . . 669, 986
—— S., Lt. 524, 1435
—— T. A., Maj. 230, 1020, 1046
—— T. A. W., 2nd-Lt. . 326, 366, 1600
—— T. J., Capt. 20, 1838
—— T. W. G., Lt. 511
—— V. C., 2nd-Lt. 779
—— W., Lt. 508, 1666
—— W., *Hony.* Maj. 1849
—— W. J., Lt. 610
Davison, A., 2nd-Lt. 803
—— D. S., Capt. . . . 265, 877
—— H. T., Lt. . . . 425, 1373
—— K. H. M., Capt. . 272, 1427
—— K. S., Maj.-Genl. .118, 146, 173
—— W. M., 2nd-Lt. 724
Davson, H. J. H., Maj. 221, 872, 1335, 1585
Davy, B. G., Lt. 513, 871
—— C. L., Lt. 793
—— H. N., 2nd-Lt. . . . 328, 897
—— R. M. M., Maj. 57
—— T. A. L., Lt. 622
Davys, G. I., Maj. 1749
—— J. V., Lt. . . 310, 1070, 1545
Daw, H. H., 2nd-Lt. 668
Dawatsering Lama 1649
Dawe, H. R., Lt. . . . 514, 1395
—— T. R., Capt. . 272, 1309, 1323
Dawbarn, J. R., Lt. 759
Dawes, A. M., Capt. . . . 267, 1263
—— C. D., Lt.-Col. . . 134, 1744
—— F. L., Lt. 544, 1698
—— H. A., 2nd-Lt. 1417
—— H. D., Capt. . . . 278, 1150
—— K. C. B., Lt. 300, 1162, 1499
—— L. F. S., Lt. . . . 666, 986
—— R. M., Lt. . 302, 1006, 1038
—— W., 2nd-Lt. 619
—— W. M., Lt.-Col. 57
Dawkes, B. B., Lt. . . . 317, 1363
Dawkins, H., 2nd-Lt. . . . 668, 985
Dawn, A., 2nd-Lt. . . . 539, 1291
Dawson, A. A. F. C. H., Capt. 267, 1098, 1901
—— A. J. E., Lt. . 317, 1020, 1467
—— A. T., Lt. . . . 492, 827
—— A. V., 2nd-Lt. . . . 332, 903
—— B. N., Capt. . . . 10, 76
—— C., 2nd-Lt. 1036
—— C., Lt. 537, 1263
—— C. H., Bt.-Col. 1901
—— C. W., Lt. 720
—— E., Lt.-Col. . . . . 5??
—— E. J., 2nd-Lt. 1399
—— F. D., *Hony.* Capt. . . 1850
—— F. T. G., 2nd-Lt. . . . 668
—— F. W. W., Maj. . . . 1711
—— G. A., 2nd-Lt. 748
—— G. F., Capt. 1712
—— G. T. C., 2nd-Lt. . . . 796
—— H. H., Maj. 15, 19, 226, 1315
—— J., 2nd-Lt. 765
—— J. C., 2nd-Lt. 863
—— J. V., Lt. 591
—— L. A. F. Capt. 1841
—— R. C., Lt. . . . 527, 1708
—— R. M., 2nd-Lt. . . 568, 1227
Dawson, T., Lt. 455-59 1156
—— T. E., Lt. . . 90, 519, 1144
—— W., tempy. Lt. 1729
—— W., tempy. Capt. . . . 172
—— W. C. Lt. 75
—— W. H. H., Capt. . . . 501
Day, A. D., Maj. . . 217, 1371, 1567
—— A. H., Asst. Surgn. . . . 1801
—— C. L., Capt. . . . 636, 1004
—— C. M., 2nd-Lt. 588
—— E. A. P., Lt. 710
—— E. R., 2nd-Lt. 324
—— E. V. G., Lt. . . . 465, 1647
—— F., Lt. 726, 1006
—— F. B., Maj. 691
—— H. M., 2nd-Lt. . . . 330, 883
—— I. H., 2nd-Lt. 1678
—— J., Maj. . . . 7, 14, 663, 1003
—— J. H., 2nd-Lt. 566
—— J. H., Lt. 750
—— L. F. L., Lt. 513
—— L. T. L., Lt. 1678D
—— P. E., Sub-Condr. 1873
—— R., 2nd-Lt. 482
—— R. E., Capt. 695
—— R. F. O' H. F., Lt. . . . 544
—— S. N., Lt. 860
Daya Kishan, Jemdr. 902
Daya, Ram, Sub-Asst. Surgn. . . 1814
—— Sub-Asst. Surgn. . . 1820
—— Jemdr. 906
—— Jemdr. 1039
—— Shankar, Jemdr. . . . 1386
—— **Singh**, Sub-Asst. Surgn. . . 1812
—— Jemdr. 1158
—— Subdr. 977
—— Jemdr. 930
Dayal Singh, *Hony.* Capt. . . . 2103
—— Jemdr. 1365
—— Jemdr. 1240
—— Sr. Sub-Asst. Surgn. 1808
—— Ressdr. 948
—— Subdr. Maj. 1231
—— Ressdr 959
—— Risldr. 886
——, Tempy. Sub-Asstt. Surgn. . 1833
Dayanu Jadhav, Jemdr. . . . 1444
Dayrell, W. S., Maj. . 54, 234, 1303
D'Costa, J. A., Asst. Surgn. . . 1798
D'Cruz, J. B., Sub-Asst. Surgn. . 1827
—— H. J., Lt. 542
D'Este, H. A. M., Lt. . . . 425, 1579
De, J. C., Capt. 1757
—— P. Ty. Lt. 1781
—— S. Temp Lt. 1776
Deacon, A. E., 2nd-Lt. 1008
—— A. F., Maj. 32, 1838
—— C. G., 2nd-Lt. . 487, 939, 1675
—— H., 2nd-Lt. 539, 1696
—— H. R. G., Lt.-Col. 135, 763, 769
Deakin, C. C., Lt. 543, 1285
—— J. S., Lt. 320, 1343
—— R. H., Lt. 292, 1058
Dealy, J. A., Col. . 26, 120, 124, 658
Dean, A., Lt. 319, 321, 1327
—— A. E., Lt. 875
—— C. S. C., 2nd-Lt. 737
—— F. W., Capt. 636
—— G. E., 2nd-Lt. 548
—— G. H., 2nd-Lt. 472
—— T. H., 2nd-Lt. 363
—— V. Lee, Lt. 836
—— W. E., Lt. 305, 1235
Deane, C. H. S., Capt. . . 256, 1191
—— D., Lt.-Col. .98, 103, 136, 159, 207, 897
—— E. S., Lt. 351, 603
—— F. J., Lt. 352, 1088
—— G. A., Sr. Asst. Surgn. . 1791
—— H. H. B., Capt. . 66, 271, 1275
—— J. A. L., Capt. . . 636, 978
—— L. S., Capt. 91, 804
—— N. G., 2nd. Lt. . . 25, 550, 1677
—— S. O. M., 2nd-Lt. 831
—— V. M., Lt. 526
—— W. S., 2nd-Lt. 835
Deans, T. J., Lt. 315, 1375
—— W., Revd. 1905
Dear, G. W. N., 2nd-Lt. . . . 786
Dear, W. T., 2nd-Lt. . . . 357, 1014
Deare, B. H., Lt.-Col. . . 129, 1741
Dearnley, H. S., 2nd-Lt. . . . 361
Deas, L. J. M., Bt. Lt.-Col. 137, 1747, 1889
Dease, C. E., Maj. . . 67, 221, 1371
Deatker, F. G. F., Asst. Surgn. . 1801
—— F. T., Asst. Surgn. . . 1795
Deb, M. L., Tempy. Capt. . . . 1771
—— Singh, Chupal, Jemdr. . . . 1225
Debenham, W. C., Esq. 493
DeBeer, E. S., 2nd-Lt. . . 357, 1164
deBerry, P. P. E., Lt.-Col. 101, 134, 610
Debibakhsh Singh, Subdr. . . . 1040
Debi Chand, Subdr.-Maj. . . . 1614
—— **Singh**, Subr. 1397
—— Jemdr. 1024
—— Jemdr. 926
Debi-dayal, Sr. Sub-Asst. Surgn. 1808
deBoinville, G. A. C., Lt. . . . 515
deBourbel, G., Lt. . . . 521, 1707
deBourcier, H. A., 2nd-Lt. . . . 630
De-Brooy, H. T., 2nd-Lt. . . . 1708
DeBrunner, W. E., Lt. . . 536, 998
DeBrath, E., Lt.-Genl. *Sir* 1148, 1955
—— E. E., Lt. .510, 1671, 1702
—— N. S., Capt. . . 276, 903
DeBretton, D. W., 2nd-Lt. . . . 551
Deb Singh Negi, Jemdr. . . . 1224
deBurgh, E., Lt.-Col. 14, 20, 51, 99, 240, 891
DeButts, F. C., Capt. 264, 1004, 1243
deClermont, A. C., Lt. 61, 508, 1614
deCourcy, G. L., Lt. . . 310, 965
DeCourcy, *Hon'ble* M. W. R., Maj. . . . 17, 22, 242, 1148
De Coverly, A. T., 2nd-Lt. 356, 1048
deCruz, F. G., Sr. Asst. Surgn. . 1788
DeCunha, A., Sr. Asst. Surgn. 1789
—— C. M., Asst. Surgn. . 1801
—— J. H., Asst. Surgn. . . 1800
Deda Ram, Jemdr. 1473
Dedman, J. J., Lt. . . . 302, 1473
—— J. B., Lt. 322, 1423
Dee, J., 2nd-Lt. 75
—— L. J., 2nd-Lt. 556
——, P., Lt.-Col. 135, 1744
Deed, L. C. B., Capt. 664
—— L. H., Lt. 521, 1275
Deedes, R. B., Capt. . 5, 272, 1144
Deedham, J. V., 2nd-Lt. . . . 763
Decks, F. G. H., Sr. Asst. Surgn. . 1790
—— T. B., 2nd-Lt. . 29, 555, 1883
—— W. C. L., Sr. Asst. Surgn. . 1789
Dees, J. S., Asst. Surgn. . . . 1800
DeFonblanque, E. B., Capt. 55, 636
Defreville, F. J., Lt. . . . . .
De Frieville, G. V., Lt. 897
Defries, A. G., Capt. . . . 78, 485
deGale, H. O., Lt. . . . 543, 1104
—— W. A. S., Maj. . . 241, 883
deGama, C. E., Asst. Surgn. . . 1801
deGrey, G., Maj. 691
—— R., 2nd-Lt. 710
DeGruchy, F. A. L., Maj. 234, 1387
deGruyther, A. N., Asst. Surgn. . 1795
—— C. G., Asst. Surgn. . 1794
—— T. D. D., Capt. . . 658
Dehn, C. G., 2nd-Lt. 45, 757
deHoxar, A. R., Lt. 608
Deighton, A. A., Lt. 538
—— F. J., Capt. 267, 1664, 1685
—— H., Lt. 26
Deish, D., Ty. Lt. 1781
deKantzow, C. A., Col. . . . 175
De la Bere, H. P., Lt.-Col. . . . 72
delaBere, R., Lt. 721
deLabilliere, E. G. D., Lt.-Col. 87, 140, 162, 201, 201
De la Coze, F. M., 2nd-Lt. . 25, 779
De la Fargue, D. G. H., Lt. 445, 885
Delafons, R. A., Lt. . 68, 298, 1633
Delaforce, E. F., Col. . . . 58, 121, 126, 613
De la Harpe, S. St. J., 2nd-Lt. . . 557
Delahunty, R. A., Lt. . . . 520, 1106
Delamain, J. H., Lt. . . . 460-62, 1001
—— *Sir* W. S., Maj.-Genl. 50, 118, 146, 173
Dela Motte B. B., Maj. . . 232, 1652
Delaney, C. D., Asst. Surgn. . . 1795
—— M., 2nd-Lt. 637
DeLanty, H. W., Sr. Asst. Surgn. . 1788
Delap, A., Lt. 393, 1598
Delip Singh, Jemdr. 1052
Delmege, C. H., 2nd-Lt. . . 522, 1704
——, C. H., Maj. 597
Delmé-Radcliffe, A., Maj. 103, 217, 1423
—— S. A., Maj. . . . 1919
Delves, N. D., 2nd-Lt. 357
—— T. G., 2nd-Lt. 1441
Dema Parab, Jemdr. 1468
Demetriadi, C. C., Capt. . . . 835
deMonte, A. N., Asst. Surgn. . . 1801

NAMES AND RANK. Page.

DeMorgan, H. A., Capt. . . . 1726
Demorgan, A. C., 2nd-Lt. . . . 556
deMowbray, R. M., Tempy. Capt. 1725
Dempster, D. E., Lt. . . 389, 1241
——— E. R., 2nd-Lt. . . 562
——— F. E., Maj. . . 57
——— H. F., Lt. . . 512, 1152
——— H. S., Lt. . . 508, 1064
——— J., Lt. . . 522, 1673
——— T., Sub-Condr. . . . 1853
den-Bogaerde, L. A. F. V., 2nd-Lt. 566
Denehy, F. P. G., Lt. . . 310, 935
deNetto, F., Ty. Lt. 1776
Dening, D. M., Lt. 622
——— H., Maj. . . 94, 107, 216, 921
——— J. P., Capt. . . . 279, 915
——— L. E., Maj. . . . 212, 937
——— R., Capt. . . . 107, 909
——— S. J. C., Lt. . 460—62, 1112
Denis, A. L., 2nd-Lt. . . . 796
——— W., Lt. 691, 783
Denison, O. G., 2nd-Lt. . . 77, 1884
Denman, F. C., 2nd-Lt. . 618, 619
Denne, L. G. L., Capt. . . . 628
——— R. T., Maj. 628
——— R. W. A., Lt.-Col. 130, 155, 189, 1323
Dennehy, H. G., Lt. . . 514, 1647
Dennes, H. M., 2nd-Lt. . 357, 1373
Denness, R. B., Lt. . . 537, 1050
———S. W. G., 2nd-Lt. . 355, 1064
Dennett, F. C. H., 2nd-Lt. . . . 361
——— G., *Hony.* Lt. 707
Denney, L. E., 2nd-Lt. . . . 693
——— W., 2nd-Lt. 820
Denning, J. DuP., Lt. . 487, 911
Dennison, G. D., Capt. . 55, 734
Denniston, J. G., 2nd-Lt. . . . 764
——— J. G., 2nd-Lt. . . . 765
——— R. D., 2nd-Lt. . . . 652
Dennistoun, J. G., Lt.-Col. . . 613
Denny, J. A., Lt. 2
——— M. P. R., Lt. . . 303, 1195
——— T. H., Lt. . . 385, 700
Dennys, A. H., Lt.-Col. .129, 154, 185
——— C. H. M., Capt. 367, 769, 1613
——— G. T., Maj. . . 245, 1144
——— H. T., Lt.-Col. 128, 154, 187, 1896
——— L. E., Maj. . . 272, 1239
——— W. A. B., Col. . 122, 153, 181
DeNoronha, M. X., Asst. Surgn. . 1803
Dent, R. B., Capt. . . 267, 1605
——— R. C., 2nd-Lt. . 332, 566*a*, 1251
——— W., Bt.-Lt.-Col. 51, 99, 138, 160, 241, 1415
——— W. E., Lt. 589
Denton, T., Condr. 1854
Denyer, 2nd-Lt. P. H. 1114
Deobahadur Gurung, Subdr. 1602, 1653
DeoBaksh, Jemdr. 1031
Deochand, Subdr. 1585
Deodat Thapa, Subdr. . . . 1624
Deoji Gorole, Subdr. 1416
Deoji Rao Bhonsle, Subdr. . . 1416
——— Kadam, Subdr. . . . 1463
——— Parte, Jemdr. . . . 1409
Deokaran Singh, Jemdr. . . . 1389
Deoraj, Sub-Asst. Surgn. . . . 1813
——— Thapa, Jemdr. 1545
Deosi Ram, Jemdr. 1575
dePenning, H. C., Asst. Surgn. . 1796
——— H. F., Asst. Surgn. . 1794
DePledge, O. F., Lt. 629
DePree, H. D., Br.-Genl. 51, 98, 102, 121, 126, 613
dePret, J., Lt. 102, 597
DeRavara, R. W., Asst. Surgn. . 1803
Dereham, H. G., 2nd-Lt. . . . 595
Derham, W., Lt. 712
deRhe-Philipe, G. W. V., Capt. . 838
DeRosario, C. M., Capt. . . . 833
——— V. (Ty.-Lt.) . . . 1781
deRozario, N., Esq. 89
Derrick, L., Lt. 702
Derwent, F. R., 2nd-Lt. . . .
Desai, N. K., Tempy. Capt. . . 1770
——— S. D., Sub-Asst. Surgn. . 1832
——— U. L., Tempy. Capt. . . 1768
De St. Croix, C. Capt. 747
deSalis, R. A., Lt. . . . 310, 881
Desborough, B., Lt. 1267
——— G., Lt. 430
——— J. N. C., 2nd-Lt. 365, 1179
deSelincourt, L., Lt. 733
Deshmukh, G. V., Ty. Maj. . . 1763
Desibra, G. P., Ty. Lt. 1780

NAMES AND RANK. Page.

DeSilva, E. H., 2nd-Lt. . . 559, 1309
——— G. F., Sr. Asst. Surgn. . 1788
Desilva, L. X., Sr. Asst. Surgn. 1789
——— W. R., 2nd-Lt. . . 365, 1116
Desmond, A. E., 2nd-Lt. . . . 618
Desraj Urs, Lt.-Col. 110
——— Singh, Jemdr. . . . 1056
DeSouza, A. B., Sr. Asst. Surgn. 1791
———, DP., Sr. Sub-Asst. Surgn. 1829
——— E., Esq. 88
——— F. B., Tempy. Capt. . . 1769
——— L. J., Sr. Asstt. Surgn. . 1791
——— M. S. J., Sub-Asst. Surgn. 1832
——— P., Ty. Lt. 1778
——— P. J., Tempy. Capt. . . 1766
——— P. N., Sr. Asst. Surgn. . 1789
——— S. A., Asst. Surgn. . . 1802
deSouza, J. F., Tempy. Capt. . . 1771
——— V., Ty. Lt. 1779
Dessa, A. C. W., Ty. Asst. Surgn. 1804
Determes, A. C., 2nd-Lt. . 546, 988
——— M. J., Lt. . . 518, 1271
Deuber, A. G. C., Capt. . . . 1245
Develin, A. A., Sub-Condr. . . 1864
deVesian, N. E., Lt. . . 753, 785
Des-Vœux, H. Lt.-Col. 191, 1895, 1911
D'Esterre, J. C. E., Maj. . 614, 634
Detmold, K. J., Lt. . . . 747, 1007
Deuchars, H., 2nd-Lt. 725
Deung, M. T., 2nd-Lt. . . 357, 1383
Deuskar, V. N., Sub-Asst. Surgn. 1832
Devanasan, Jemdr. 994
Devaraju, Jemdr. 993
——— Subdr. 1327
Devasagayam, A., Ty. Lt. . . 1774
——— Subdr. 1271
Devasahayam, *Hony.* Capt. . 5, 2103
Deverell, J. J., 2nd-Lt. 668
Devereux, W., Lt. 617
Devenish-Meares, B. J., Lt. 302, 1187, 1409
Devi-Bakhsh, Jemdr. 1043
Devi Chand, Subdr. 1343
——— Jemdr. 878
——— Dayal, Ty. Sub-Asst. Surgn. . 1833
——— Das, Jemdr. 1171
——— Raj, Jemdr. 1107
——— Singh, Subdr. 1393
Devindar Singh, Sub-Asst. Surgn. 1818
Devlin, D. G., 2nd-Lt. 557
——— O. G., 2nd-Lt. 1447
Devine, A., Senr. Asst. Surgn. . 1787
——— C., Asst. Surgn. . . . 1801
——— T., Capt. 636
Devonshire, C. H., Lt. . . 533, 998
——— F. V., Lt. 587
——— F. V., Maj. . . . 577
——— N. G., 2nd-Lt. 326, 366, 970
Devos, J. P., 2nd-Lt. 551
Devoy, J. C., Lt. . . . 415, 1469
Dew, A. B., Lt.-Col. . 193, 1898, 1912
——— T. H., Lt. 353
Dewa, Subdr. 1199
——— Singh, Subdr. 1145
——— Jemdr. . . . 1111, 1252
——— Risdr. . . . 1682, 1686
——— Subdr. 1157
——— Jemdr. 1192
——— Jemdr. 916
——— Subdr. 1211
——— Ram, Jemdr. 1452
Dewak Ram, Subdr. 1066
——— Jemdr. 1356
Dewan, R. K., Tempy. Capt. . . 1770
——— Ali, Subdr. 1132
——— Chand, Jemdr. 1683
——— Shah, Jemdr. 1682
——— Singh, Jemdr. 1217
——— Sing Thapa, Subdr. . . 1598
——— Gurung, Subdr. . . 1602
——— Padhiyar, Jemdr. . . . 1179
——— Negi, Jemdr. 1179
——— Singh, Sub.-Asst. Surg. 1813
——— Takuli, Jemdr. . . . 1179
Dewar, F., Lt. 512, 803
——— H. S. L., Lt. . . . 521, 1094
Dewardt, T., 2nd-Lt. 630
Dewat Ram, Risldr. 1686
Dewberry, A., Sub-Condr. . . . 1868
DeWend-Fenton, J. D. L., Capt. . 593
Dewes, F. J., Lt.-Col. . . . 128, 1739
Dewey, C. A. R., Lt. . . . 532, 988
——— H. A. R., 2nd-Lt. . . . 554
——— H. V., Sr. Asst. Surgn. . 1789
——— J. O., Asst. Surgn. . . 1797
Deya Ram, Jemdr. 892

NAMES AND RANK. Page.

DeWilton, J. G. B., Capt. . 277, 889
Dewing, R. H., Capt. . . 52, 666, 991
Dhairyawan, K. R., Tempy. Capt. 1770
Dhal Singh, Subdr. 1024
Dhalbir Burathoki Jemdr. . . . 1657
Dhali Khan, Subdr. 1495
Dhalip Singh, Jemdr. 896
Dhana Singh, Jemdr. 1299
Dhanbahadur Rana, Jemdr. . . 1511
——— Gurung, Subdr. . . 1636
——— Jemdr. 1625
——— Limbu, Jemdr. . . 1649
——— Sen, Jemdr. . . . 1643
Dhanbar Gurung, Jemdr. . . . 1608
Dhanbir Gharti, Jemdr. . . . 1608
——— Gurung, Jemdr. . . . 1347
——— Subdr. 1607
——— Subdr. 1624
——— Pun, Jemdr. 1597
——— Thapa, Subdr. . . . 1607
——— (I) Subdr. . . 1638
——— (II) Subdr. . . 1636
——— Subdr. . . . 1636
——— Jemdr. . . . 1608
Dhane Khan, Subdr. 1409
——— Jemdr. 1480
Dhaniram, Sub-Asst. Surgn. . . 1811
Dhanman Rai, Jemdr. 1649
Dhanna Singh, Jemdr. 1111
Dhanniram Thapa, Jemdr. . . 1625
Dhani Ram, Subdr.-Maj. . . . 1189
——— Subdr. 1122
——— Jemdr. 890
Dhan Lal Gurung, Subdr. . 1614, 1618
Dhanna Singh, Subdr. 1078
——— Jemdr. 1157
——— Jemdr. 1200
——— Ram, Jemdr. . . 1585
Dhanpat Rai, Singh, Jemdr. . . 890
Dhanraj, Sub-Asst. Surgn. . . 1814
——— Gurung, Subdr. . . . 1618
——— Jemdr. 1657
——— Jemdr. 1608
——— Pun, Subdr. 1611
——— Thapa, Jemdr. . . . 1625
Dhan Singh, Subdr. 1070
——— Jemdr. 1391
——— Negi, Jemdr. 1179
——— Thapa, Jemdr. . . 1598
Dhansing Ale, Jemdr. . . 1608, 1611
——— Gurung, Subdr. . . 1618
Dhan Sing Lama, Subdr. . . . 1602
Dhansur Gurung, Jemdr. . . . 1637
——— Rai, Subdr. 1630
Dhanu, Jemdr. 1608
Dhanwat Meo., Jemdr. 1408
Dhar, J., Tempy. Capt. 1769
Dhara Singh, Jemdr. 936
Dharam Singh, Jemdr. . . 970, 1083
——— Jemdr. 1369
Dharam Singh Gurung, Jemdr. . 1611
——— Jemdr. . . 918, 1012
——— Jemdr. 1189
——— Jemdr. 926
——— Subdr. 1011
Dhari RamBhatia, Sub-Asst. Surg. 1818
Dhari Ram Bhatia, Sub-Asst. Surgn. 1818
Dharmanand Pande, Sub-Asst. Surgn. 1819
Dharmarajulu, Subdr. 1327
Dharmraj Gurung, Jemdr. . . 1597
Dharm Singh, Sub-Asst. Surgn. . 1808
——— Subdr. 1280
——— Datt, Jemdr. 1225
Dharmadhoj Limbu, Jemdr. . . 1621
Dharmalingam, Jemdr. 1276
——— Jemdr. . . . 994
Dharmanand Pande, Sub-Asst. Surgn. 1819
Dharmatma Singh, Sub-Asst. Surgn. 1818
Dharmraj Gurung, Jemdr. 1597, 1622
Dharrao Madhik, Jemdr. . . . 1460
Dhaul Bir Rai, Subdr. . 1630, 1657
Dhaunkal Singh, Jemdr. . . . 1193
Dhauraj Gurung, Jemdr. . . . 1619
Dhavle, D. S., Ty. Lt. 1776
Dhawan, M. L., Tempy. Lt. . . 1766
Dhawarial, Jemdr. 1484
Dhenuk Singh, Jemdr. 1063
Dhera Mal, Jemdr. 894
Dheru Khan, Subdr. 1243
Dhig Ram, Jemdr. 1059
Dhingra, C. L., Tempy. Lt. . . 1777
Dhira, Subdr. 1199

Names and Rank. Page.

Dhig Ram, Jemdr. 1561
Dhirendra Nath Das Gupta, Sub-Asst. Surgn. 1815
Dhiraj Gurung, Jemdr. 1608
Dhirajlal Virjibhai Bhatt, Sr. Sub-Asst. Surgn. 1831
Dhirendra Kumar Sen, Subdr. 1219
Dhondibarao Barge, Subdr. 999
—— Kadam, Jemdr. 1447
Dhondi Ghichar, Subdr. 1444
—— Nagooji Vizrae, Sr. Sub-Asst. Surgn. 1829
—— Sawant, Jemdr. 1460
Dhondo Mahadeo, Sr. Sub-Asst. Surgn. 1829
Dhondo Narayan Kadhe, Sub-Asst. Surgn. 1831
Dhondy, B. S., Tempy. Capt. 1493, 1766
Dhonkal Ram, Jemdr. 1488
—— Singh, Jemdr. 1281
Dhulip Singh, Risldr.-Maj. 925
—— Jemdr. 914
Dhuman Khan, Ressdr. 910
Dhum Singh Chauhan, Subdr. 1181
Dhunjibhoy, F. Tempy. Lt. 1773
—— J. E., Tempy. Capt. 1770
Dhura Ram, Jemdr. 1498
Dhyan Singh, Sub-Asst. Surgn. 1810
Diack, W. J., 2nd-Lt. 330
—— H. J. W., 2nd-Lt. 3317, 1617
Diacono, H. V., 2nd-Lt. 546, 1675
Dial, A. E., Sub-Asst. Surgn. 1819
—— Singh, Jemdr. 132
Dias, A. P., Ty. Lt. 1780
—— B. G. V., Tempy. Capt. 1765
Dibben, E. L., Lt. 738
Dibble, H. R., 2nd-Lt. 587
Dibbs, H. J., Capt. 821
Dibdin, A., Capt. 679
—— L. G., Maj. 679, 681
Dick, A. MacD., Bt. Maj. 1755
—— A. R., Col. 56, 102, 123, 149, 184
—— J. G., 2nd-Lt. 713
—— R., 2nd-Lt. 727
—— H. N., 2nd-Lt. 1507
—— F. G., 2nd-Lt. 713
—— R. N., Brig. Gen. 51, 98, 122, 136
Dicker, B. P., Lt. 320, 365, 1339
—— 2nd-Lt. D. E. 357
Dickerson, A. H., Lt. 540a
Dickens, A. C., 2nd-Lt. 323, 1497
—— A. H., Lt. 637
—— C. E., 2nd-Lt. 1379
—— E., Capt. 1704
—— G. W. W., 2nd-Lt. 724
—— H. E., 2nd-Lt. 474, 1401
—— W. S., Lt. 527, 1704
Dickenson, 2nd-Lt. 1677
Dickie, D. H., 2nd-Lt. 90, 724
—— J. C., Capt. 781
—— R. A., Lt. 538
—— R. S., Lt. 653
—— W., 2nd-Lt. 1705
Dickins, C. E., 2nd-Lt. 1379
—— E. A., 2nd-Lt 512
—— E. K., 2nd-Lt. 540, 1699
—— F., Maj. 74, 634
Dickinson, C. E., Lt. 733, 791
—— F. R., Lt. 321, 1587
—— G. H., 2nd-Lt. 705
—— H. B. H., Capt. 485, 951
—— J., Sub-Condr. 1856
—— J. J., *Hony.* Lt. 1843
—— L. D., 2nd-Lt. 561
—— R., Lt. 703
—— R. C., Maj. 94, 228, 927
—— R. S., 2nd-Lt. 557
—— T. M., Capt. 276, 905
—— W., Lt.-Col. 1955
—— W. G., Lt. 78, 491, 895
—— W. H., Maj. 1746
—— W. M. K., Lt. 55, 1910
Dick-Lauder, J. N. D., Maj 14, 43, 243, 905
Dickson, A., Lt. 623
—— A. N., Capt. 1754
—— A. N., Lt. 638
—— E., Maj. 219, 1283
—— E. C. S., 2nd-Lt. 763
—— E. J., 2nd-Lt. 562, 1678a
—— H. R. P., Maj 243, 929
—— J. H., Lt.-Col. 37, 131, 156, 191, 1660
—— J., Capt. 1725
—— J. W., Lt. 767
—— J. W. E., Lt. 765

Names and Rank. Page.

Dickson, R., 2nd-Lt. 651
—— R. C., Capt. 1717
—— T., 2nd-Lt. 75
—— T. H., Capt. 1713
—— W. A., 2nd-Lt. 545, 1148
—— W. C., Lt. 543a, 1347
—— W. E. R., Br.-Genl. 59, 106, 121, 126, 661
—— W. M., Capt. 1717
Didan Singh, Jemdr. 912
—— Khan, Jemdr. 1493
Didari, Ressldr. 903
Didcock, R. S. H., Lt. 542, 1567
Dier, H. V., Lt. 1839
DigbahadurRai, Jemdr. 1633
Digby, C., Hon. Capt. 1843
—— E., 2nd-Lt. 494
—— K. G., 2nd-Lt. 523, 1447
—— P. R., Lt. 527, 1114
—— T. A., Capt. 262, 1140, 1577
Digges La Touche, C. H., Maj. 107, 240, 905
—— H. N., Lt. 1271
Diggory, P. Lt. 387, 1054
—— S. L., Lt. 651
Digh Ram, Subdr. 1054
—— Subdr. 1059
—— Jemdr. 938
Dighton, P. C., 2nd-Lt. 595
Dignam, J. R., 2nd-Lt. 475, 921
Dikshit, H. K., Tempy. Capt. 1771
Dila Sing Gurung, Jemdr. 1615
—— Jemdr. 1602
—— Subdr. 1636
Dilbagh Rai, Ty. Sub-Asst. Surgn. 1833
—— Singh, Risldr. 1685, 1686
Dilawar Khan, Risldr. 961
—— Subdr. 1130
Dilbahadur Ale, Jemdr. 1597, 1615
Dilbahadur Khattri, Jemdr. 1643
—— Rana, Jemdr. 1604
Dilks, A. C., Lt. 420, 1140
Dilke, T. T., 2nd-Lt. 36
Dillingham, C. S. D., 2nd-Lt. 564, 1375
Dillon, D., 2nd-Lt. 856
—— H. V. S., 2nd-Lt. 558
Dillow, D'A. H., 2nd-Lt. 681, 786, 1001
Dillu Ram, Risldr. 1686
Dilman Singh Thapa 1619
Dilsukh Singh, Jemdr. 1452
Dimond, C. F. W., Lt.-Col. 754, 785
Dinadayalan, P. Sub-Asst. Surgn. 1828
Dina Nath, Jemdr. 1175
—— Jemdr. 922
—— Ressdr. 1683
Dinanath, Babu Rao Sirsat, Sub-Asst. Surgn. 1831
Dina Nath Vesudeva, Sub-Asst. Surgn. 1818
Dina Nath Vasodeva, Sub.-Asst. Surg. 1818
Dinarnath Wadhawa Mal, Jemdr. 1299
Dinesh Chandra Sarkar, Sub-Asst. Surgn. 1816
—— Sen, Sub-Asst. Surgn. 1815
Dingaran, G. S., Asst. Surgn. 1799
Dinkarao Yadao, Subdr. 1423
—— Subdr. 1409
Dinker Rao Dareker, Subdr. 1435
Din Muhammad, Jemdr. 989
Dinnage, 2nd-Lt. H. A. 578
Dinwiddie, H. W., 2nd-Lt. 365, 1078
Dinwiddy, M. J., Capt. 749
Dinwoodie, J., Condr. 1861
—— J. F., 2nd-Lt. 844
Dip Chand, Jemdr. 938, 1388
—— Jemdr. 1165
—— Jemdr. 1217
Diplock, A., 2nd-Lt. 796
Dipnarayan Singh, Subdr 1021
Dipram, Jemdr. 1393
Dique, J. S., Asst. Surgn. 1802
Dirks, A. J., Lt. 543a, 1026
Ditmas, A. R., Lt.-Col 39, 129, 154, 185, 1660
Ditta Khan, Subdr. 1207
—— Ram, Ressdr. 1683
—— Sawant, Subdr. 1441
Dittu Singh, Jemdr. 920
Diwan Ali, Jemdr. 1133, 1704
—— Jemdr. 1511
—— Jemdr. 1495—96
—— Jemdr. 1571
—— Subdr. 988
—— Jemdr. 1148

Names and Rank. Page.

Diwan Chand, Sub-Asst. Surgn. 1812
Diwan Chand, Sub-Asst. Surgn. 1814
—— —— Sub-Asst. Surgn. 1814
—— —— Ressdr. 878
—— Singh, Sub-Asst. Surgn. 1816
—— Jemdr. 1384
—— Bharara, Sub-Asst. Surgn. 1818
—— Bahadur, Subdr. 1141
—— Subdr.-Maj. 1116
—— Subdr. 1011, 1054
—— Baharara 1818
—— Padhiyar, Jemdr. 1182
—— R. K., Ty. Lt.
Diwana, Jemdr. 1256
Dix, R. C. St.J., Lt. 779
—— W., 2nd-Lt. 749, 1670
Dixie, W. J., Sub-Condr. 1864
Dixon, A. F. W., Capt. 724, 739
—— C. R. B., 2nd-Lt. 1060
—— E. G., 2nd-Lt. 807
—— G., Lt. 488
—— G., Lt. 532, 988
—— G., 2nd-Lt. 855
—— J. Maj. 852
—— M. J., Lt. 51, 491, 1892
—— R., 2nd-Lt. 1122
—— R. C., Lt. 519, 1666
—— T., Lt. 430, 1134
—— T. H., Revd. 1904
—— V. H., 2nd-Lt. 330, 907
—— W., Lt. 420, 925
Dixson, A. H. 2nd-Lt. 611
—— H. F., Lt. 587, 1007
D'Netto, C. I., Asst. Surgn. 1803
Dobbie, A. G., Lt. 700, 719
—— C. F., Lt.-Col. 128, 154, 185
—— H. H., Col. 122, 153, 181
—— T. B., 2nd-Lt. 539, 1487
—— W. H., Col. 123, 149, 181
Dobbin, F. Le F., 2nd-Lt. 332, 1623
—— J. M., Lt. 767
Dobbs, A. H., Lt.-Col. 57
—— A. F., Lt. 504
—— C. F., Lt.-Col. 98, 134, 157, 197, 870, 1383
—— W. E. J., Lt. 8, 82, 524
Dobell, C. M., Maj.-Gen. *Sir* 12, 22, 118
Dobney, T. A., 2nd-Lt. 475, 1116
Dobson, E. T., Lt. 313, 1311, 1493
—— F. E., Lt. 525
—— G. H., 2nd-Lt. 1678a
—— J. H., 2nd-Lt. 843
—— T., Lt. 849
—— W. E., Condr. 1870—71
—— W. G., Lt. 629
Docking, A. C., Sub-Condr. 1855
Dockrell, M., Capt. 59, 258, 1463
—— M., Ty. Lt. 1726
Doctor, D. B., Ty. Lt. 1781
Dodd, A., Lt. 637
—— A. C. B., Lt. 380, 1154
—— A. H. R., Maj. 54, 241, 907
—— A. V., Lt. 463, 1427
—— B. F., Lt. 425, 1433
—— C. K., Capt. 696
—— F., 2nd-Lt. 624
—— G. E., Revd. 1905
—— H. B., Lt. 844
—— J. J., Lt. 79, 490
—— J. R., Lt. 533, 1614
—— P. C. B., Capt. 95, 253, 933
—— W., Condr. 1853
Dodds, A., Lt. 760
—— B. W. P., Lt. 537, 980
—— H. C., Lt. 313, 1191, 1195
—— H. S., 2nd.-Lt. 618
Dodge, M. W., 2nd-Lt. 548
Dodgson, E. C., 2nd-Lt. 742
Dodkins, L. C., 2nd-Lt. 577, 725
Dods, A. W., Maj. 651
—— E. R. S., Capt. 280—287, 1411
Dodwell, H. H., 2nd-Lt. 802
Dogra, A. S., tempy. Capt. 1769
—— P. N., tempy. Capt. 1255, 1768
—— E. W., Lt. 1203
—— K. A. C., Maj. 1712
Doherty, J. A., 2nd-Lt. 1211
—— J. M.-Lt. 8
—— J. M., Lt. 82
—— P. J., 2nd-Lt. 450—54, 1094
Doig, E. W., Lt. 420, 1263
Dolan, M. J., Lt. 666, 996
Dolben, A. C., Lt. 683, 799
Dolman, H. G., 2nd-Lt. 475, 733, 1387
Dolphin, H. M. B., 2nd-Lt. 331, 1249
—— M. J. X., Asst. Surgn. 1799

NAMES AND RANK. Page.

Dolphin, W. H., Lt. 518
Dominy, A. G., Lt. 538
—— J. C., Maj. 730
Don, P. J., 2nd-Lt. 965
Donald, A., Lt. 516a, 656, 1105
—— A. A., 2nd-Lt. 552
—— C. D., Capt. 56, 485, 653
—— C. H., Capt. 605
—— G., Lt. 410
—— H., Lt. 470, 1327
—— J., 2nd-Lt. 565, 1144
—— J., Maj. 807
—— J. W., 2nd-Lt. 1467
—— McP., 2nd-Lt. 362
—— R. T., Lt. 798
Donaldson, A., Capt. 269, 1156
—— A. H., 2nd-Lt. 360, 1401
—— A. L., Lt. 368, 1601
—— H. V., Lt. 533
—— J., Lt. 399, 1403
—— R., Lt. 826
—— R. C., Lt. 859
—— S., Lt. 515, 1289
Donegan, M., Lt. 637
Donkin, H. J., 2nd-Lt. 58
Donnan, W., Lt.-Col. 191
Donne, A. J., 2nd-Lt. 563
—— B. G., Lt. 629
—— F., Capt. 1839
—— W. J. B., 2nd-Lt. 564, 1355
Donnell, N. A., Lt. 353, 1415
Donnelland, J. H. A., Sr. Asst. Surgn. 1792
Donnelly, H. R., 2nd-Lt. 331, 1651
Donnison, E. G., Lt. 798
Donnithorne-Lutter, C.E., Lt. 537
Donoghue, P. J., Asst. Surgn. 1801
Donovan, B. D., 2nd-Lt. 1375, 539
—— C., Lt.-Col. 129, 1739
—— C. C., Lt.-Col. 19, 74, 137, 632
—— C. H., Capt. 267, 1385, 1485
—— E., Lt. 519, 1020
—— J., Lt. 637
Donowan, A. B., Asst. Surgn. 1801
Doog, J. M. G., 2nd-Lt. 833
Dooley, J. A., Lt. 519
Doorly, A. R. C., tempy. Capt. 1725
Dootson, H., 2nd-Lt. 619
Dopping-Hepenstal, M. E., Lt.-Col. 104, 139, 160, 199, 871, 872, 1595
Dorai, R. B., Raja *Hony.* 2nd-Lt. 48, 84
Doraisami, Jemdr. 1324
—— Nayadu, Jemdr. 1318
Doraisamy, S., Capt. 1757
Doran, P. J., Asst. Surgn. 1802
—— H. F., 2nd-Lt. 567, 826
Dore, A. M., Lt. 443
—— G. B., Lt. 430, 1489
Dorling, L. H. G., Capt. 615
Dorman, C. H., Capt. 636
Dormer, E. J., Lt. 313, 943
—— J. G., Capt. 256, 1419
Dorrell, J. O., 2nd-Lt. 565
Dorrington, G. H., 2nd-Lt. 566, 1678a
DoRosairo, V., Ty. Lt. 1781
Dorward, *Sir* A. R. F. M., Genl. 991
—— W., Lt. 571
Doshi, V. J., Ty. Lt. 1774
Dossett, A. E., Lt. 440, 1311
Dossettsr, A. H. 798
Dost Muhammad, Jemdr. 1428
—— Risldr. 891
—— Subdr. 1567
—— Risldr. 1682
—— Jemdr. 900
—— Risldr. 919
—— Jemdr. 1248
—— Jemdr. 1565
—— Subdr. 1259
—— Subdr. 1235
—— Ressdr. 1683
—— Khan, Ressaidr. 945
—— —— Jemdr. 888
—— —— Ressdr. 908
—— —— Jemdr. 932
—— —— Ressdr. 884
—— —— Risldr. 5, 910
—— —— Subdr. 1092, 1311
—— —— Subdr. 1321
—— —— Ressdr. 1683, 1684
—— —— Subdr.-Maj. 1311
Dotivala, N. M. P., Tempy. Lt. 1773
-oubell, P. S., Lt. 571, 602

NAMES AND RANK. Page

Double Day, 2nd-Lt. 797
Donbney, R., Capt. 1841
Doubt, E. A., Lt. 514
—— H. E., Lt. 534, 1187
Doudney, G. W., Lt. 293a, 947
—— R. V., Capt. 1689
Douetil, P. V., Capt. 485, 874, 1893
Douglas, A. E., *Hony.* Lt. 1847
—— A. E. R., Lt. 310, 1423
—— A. P., Lt.-Col. 10, 74, 107, 132, 632
—— D., Capt. 493, 855, 943
—— F. W., Capt. 1689
—— F. W., 2nd-Lt. 571
—— G. D. E., 2nd-Lt. 358, 1604
—— G. S., Maj. 225, 871, 1090, 1457
—— H. A., Maj. 9, 200, 1661
—— J. A., Br.-Genl. 50, 105, 120, 124, 149, 184
—— J. S., 2nd-Lt. 624
—— J. K., Capt. 665
—— L. N., Lt. 430, 1597
—— M. W., Lt.-Col. 184, 1889, 1895, 1911
—— R., Lt. 524, 1219
—— R. O., Lt. 751
—— T., Lt. 656
—— V. E., 2nd-Lt. 558, 1197
—— W. B., Lt.-Col. 132, 156, 193, 868, 869, 1016
—— W. G., 2nd-Lt. 640
Douglas-Hamilton, J. A. G., Lt. 9, 601
Douglas-Withers, H. H., Capt. 741
Douglass, P. C., Lt.-Col. 1710
Doule, F. McC., Maj. 665, 985
Douse, F. W., Lt. 515, 1473
Dove, E. P., Lt. 492, 848, 883
—— J. H., Capt. 1729
—— M., Lt. 410, 1082
Dover, R. W., Esq. 9, 88
—— W. E., Lt. 352
Doveton, A. C., Capt. 109, 255, 1102, 1140
—— H. J., Maj. 218, 872, 1102, 1573
Dow, J. N., 2nd-Lt. 555
—— D. J., Lt. 304, 1645
—— G., 2nd-Lt. 475, 1383
—— T. M., Lt. 524, 1166
Dowd, H., 2nd-Lt. 778
Dowdall, J. J., tempy. Capt. 1724
Dowdell, R. A., 2nd-Lt. 736
Dowell, G. C., Col. 123, 632
Dower, J. H., 2nd-Lt. 658, 986
Dowie, A. M., 2nd-Lt. 782
Dowland, G. R., Capt. 279, 1255
—— W., 2nd-Lt. 1273
Dowley, F. M., Capt. 501a, 854
Dowling, D. M., Lt. 310, 1435
—— G., Lt. 43, 292a, 1026
—— J. C., 2nd-Lt. 768
—— J. R. H., 2nd-Lt. 739
—— L. E., 2nd-Lt. 818
—— T., 2nd-Lt. 796
—— W. H., 2nd-Lt. 578
Down, E. A., 2nd-Lt. 362, 1195
—— E. C., 2nd-Lt. 833
—— F. D., Lt. 587, 1211, 1495
—— G. O., Lt. 512, 747, 1668
—— H. R., Lt. 351, 1007a
Downes, E. F. F., Sub-Condr. 1846
—— G. E., 2nd-Lt. 799
Downie, A., Lt. 440, 1088
Downing, L. D., 2nd-Lt. 548, 1265
—— P. H., Lt. 735
Downs, G. E., 2nd-Lt. 734
—— R. A., Capt. 691, 783
—— R. W., Lt. 779
—— W. S., Lt. 640
Downton, A. M., Lt. 628
Dowse, B. T. J., 2nd-Lt. 328, 1142
Dowson, V. H. W., 2nd-Lt. 547
Doy, E. P., 2nd-Lt. 619
Doyle, C. R., Lt. 793
—— E. C., Capt. 1838
—— E. E., Capt. 20, 458, 1756
—— F. J. B., Lt. 535, 1695
—— G. W., Asst. Surgn. 1797
—— H., Asst. Surgn. 1800
—— H. H., Asst. Surgn. 1802
—— J., Sr., Asst. Surgn. 1792
—— J. P., Capt. 501a, 653
—— M. J., *Hony.* Maj. 1844
—— P. J., Lt. 306, 1102
D'Oyly, J., Maj. 211, 1471

NAMES AND RANK. Page.

D'Oyly-Hughes, R., Lt. 310, 1595, 1617, 1621
Dracott, F. H., Lt. 543a
Draffen, J. E., 2nd-Lt. 326, 1174, 1225
Drake, A. J. A., 2nd-Lt. 519, 815
—— D. H., Lt. 303, 1040
—— F. H., 2nd-Lt. 90, 722
—— H. R., Maj. 1757, 1889
—— E. T. T., Lt. 598
—— J. A., Lt. 311, 1343
—— J. C. B., 2nd-Lt. 494, 905
—— P. C., Lt. 528, 1487
Drake-Brockman, D. H., Col. 14, 26, 121, 127, 150, 189
—— F. T., Capt. 272, 887
—— G. T., Capt. 252, 1090
—— H. E., Lt.-Col. 128, 1740
—— R. A., Lt. 666, 996
—— V., Maj. 72, 101, 103, 634
Draper, F. H., Sub-Condr. 1873
—— M., 2nd-Lt. 638
Drawbridge, W. H., Revd. 1903
Draycott, C. D., 2nd-Lt. 713
Drayson, Lt.-Col., A. P. 26, 721
—— Fitz A., Capt. 1004
Drayton, P., 2nd-Lt. 90
Dreffen, J. E., 2nd-Lt. 1174
Drever, G. D., 2nd-Lt. 359, 1164
Drew, A., Lt. 72
—— A. B. H., Brig.-Genl. 16, 32, 120, 125, 150, 186
—— B. C. H., Lt.-Col. 51, 98, 137, 159, 217, 1110
—— F. G., Capt. 666, 985
—— F. J., Lt. 531
—— J., 2nd-Lt. 73
—— J. D., Lt. 514, 1106
—— O. W., 2nd-Lt. 363, 1030
Dreyer, G. V., Maj. 634
Dreyfus, B. E., Lt. 490, 935
Drinan, G. C., Maj. 229, 1215
Drinkall, R., 2nd-Lt. 752
Drinkwater, C. R., Lt. 492, 927
Driver, C. E., Capt. 698
—— G. L. S., 2nd-Lt. 548, 1219
—— H. C., Lt. 1841
—— J. C., tempy. Capt. 1769
—— R. A., Lt. 389, 1096
D'Roche, E., Asst. Surgn. 1793
Dron, C. W., Lt. 455, 1060
D'Rose, F. J., Asst. Surgn. 1796
Drought, J. V., Capt. 264, 1491, 1519
—— A. C., 2nd-Lt. 332, 1457
D'Rozario, S., Sub-Asst. Surgn. 1830
Drudge, H. A., 2nd-Lt. 355, 1491
Druett, H. C., Lt. 540a, 1305
Druitt, G. T., Capt. 732
—— T. R., Lt. 491, 925
Drummond, A. B., Lt.-Col. 196, 1912
—— F. H. R., Maj.-Genl. *Sir* 118, 146, 173
—— J. 2nd-Lt. 778
—— J. G., Lt. 490, 893, 955
—— J. G. L., 2nd-Lt. 1689
—— J. G. P., Capt. 269, 1094
—— J. L., Capt. 1004
—— R. H., Lt. 730, 731
Drury, A. N., Tempy. Capt. 1724
—— J., Capt. 526, 808, 992
Dry, R., 2nd-Lt. 653
Drysdale, A. E., Capt. 271, 1174, 1211
—— D., *Hony.* Capt. 1850
—— J., Lt. 1703
—— J. K., Lt. 493
—— J. R., Capt. 693
—— L., Maj. 610
D'Silva, A., Ty. Lt. 1776
—— E., Asst. Surgn. 1798
—— H. A. H., Asst. Surgn. 1800
—— M. J., Asst. Surgn. 1801
—— N. A. T., Ty Asst. Surgn. 1804
D'Silveira, J. L. L., Asst. Surgn. 1801
D'Souza, A. J., Tempy. Capt. 1764
—— J. A. A., Tempy. Capt. 1771
—— J. B., Asst. Surgn. 1796
—— J. F. A., Ty. Asst. Surgn 1804
—— L., Sr., Asst. Surgn. 1789
—— L. J. M., Ty. Asst. Surgn. 1804
—— W. M., Tempy. Capt. 1772
D'Sylva, G. H., Tempy. Asst. Surgn. 180

Names and Rank. Page.

D'Sylva, W. R., Sr. Asst. Surgn. . 1788
Duarte, A. M., Tempy. Lt. . . . 1774
Dubash, M. N., Ty. Lt. . . . 1779
Dube, K. C., Hony. Tempy. Capt. . 1772
——— P. L., Ty. Lt. 1799
Duberly, M. R. W., Capt. . 362, 1110
Dubois, H. A., Lt.. . . . 506, 1533
DuBoulay, J. H., *Sir* Lt.-Col. . 806
Dubs, C. E. D., Lt. 595
Duchesne, R. L., Lt. . . 399, 877
Duckenfield, W., 2nd-Lt. . 359, 1565
Duckworth, A. R., Asst. Surgn. . 1793
——— E., Asst. Surgn. . . 1800
——— E. V., Sr. Asst. Surgn. 1790
——— G. E., Asst. Surgn. . 1794
——— G. F., Asst. Surgn. . 1796
——— G. P., Esq. 89
——— H., Asst. Surgn. . . 1793
——— H. C., Asst. Surgn. . 1802
——— M. B., Lt. . . 541, 1457
——— P. T., Asst. Surgn. . 1794
Dudani, G. L., Ty. Lt. 1780
Dudding, W., Sub-Condr. . . . 1864
Dudgeon, C. W., Nur. Sis. . . 1735
——— M. F., Capt. . . . 585
Dudley, A. E., Asst. Surgn. . . 1880
——— F., Lt. 543, 1187
——— G. W., Asst. Surgn. . . 1795
——— R. S., Lt. . . . 520, 1259
——— S., Lt. 925
——— S., Lt. 371
Dudman, C. H., Sub-Condr. . . 1855
Duer, C., Lt.-Col. 1786
Duesbury, A. A., 2nd-Lt. . 330, 1600
Dufall, J. B., Lt. . . . 538, 1672
Duff, A. C., Lt. 543, 1698
——— B. B. H., Lt. 538
——— G. M., Col. . . 22, 125, 660
——— J. M., 2nd-Lt. . . 776, 1895
——— K. C., Capt. . . . 78, 485
——— S. H., Lt. 533
Duff-Sutherland-Dunbar, G., Maj. *Sir* . . 217, 1144
Duff Torrance, W. B., Lt. . 294a, 1545
Duffus, A. H. H., Lt. . . 520, 1154
Duffy, J. V., Capt. 1724
——— L., Capt. . . . 17, 38, 741
——— S. C. E., 2nd-Lt. . 475, 1389
——— T. A., *Hony.* Maj. 53, 806, 1849
Dugdale, J. A., Lt. 691
Duggal, S. S., Tempy. Capt. . . 1770
Duggan, C. R., Maj. 850
——— E. M., Capt. 849
——— J., *Hony.* Maj. . . . 1844
——— J. N., Ty. Maj. . . . 1763
——— R. D., Lt. . 56, 310, 1120
Dugla, Subdr. 1243
Duguid, T. C., Lt. . . . 302, 1311
Duhan, W. W. T., Brig.-Genl. 17, 101, 120, 123, 612
Duirs, F. D., 2nd-Lt. 753
Duja, Jemdr. 1197
Duke, A. A. G., Maj. . 29, 244, 1663
——— A. B. C., Capt. 628
——— A. L. R., Lt. . . . 321, 1336, 1575
——— H. E., Maj. . . . 738, 791
——— O. A., Capt. 103, 105, 255, 877, 917
——— V. W. H., Maj. 1004
——— W. F., Lt. 534, 1607
——— W. S., Lt. . . 504, 830, 1665
——— W. V., Lt. . . 540a, 854, 1359
Dukes, M. B., 2nd-Lt. 562
Dukhan Singh, Subdr. . . . 1581
Dukhi Singh, Jemdr. 1063
Dula Ram, Sub-Asst. Surgn.. . 1808
——— Jemdr. 1498
——— Singh, Subdr.-Maj.. . . 1215
——— Jemdr. 1107
——— Subdr. 1163
Dulat Singh Sundar, Subdr. . . 1179
Dulcken, H. W., Lt. . . 317a, 911
Dulip Sing, Jemdr. 1104
Duli Chand, Jemdr. 1405
Duli Chand, Subdr. Maj. . . . 1383
Dulla Singh, Jemdr.. 1698
Dulo, Subdr. 1188
Dulsing Thapa, Subdr. . . . 1636
Dumbreck, W. Y., 2nd-Lt. . . 1022
Dumper, E., Lt. 351, 1673
Dumpoor Penchaloo Nayudu, Sub-Asst. Surgn. . . . 1827
Dunbar, J. C., Maj. . . . 53, 614
——— W. C., 2nd-Lt. 696
——— W. R., Lt. 751

Names and Rank. Page

Dunbar-Sutherland, C. F., Lt. 440
Duncan, A. W., Capt. 1756
——— A. W., Lt. . . 410, 537, 1636
——— C. A., 2nd-Lt. 558
——— D. D., 2nd-Lt. 624
——— D. J., Lt. . . . 531, 1702
——— D. L., Lt. . 298, 1600, 1653
——— G., Lt. 605
——— G. H., Capt. 850
——— G. L., Capt. 1756
——— H. C., Maj. . 8, 15, 53, 99, 213, 1641
——— H. C., Lt. . . . 393, 1401
——— J. G., Lt. 55, 509
——— J. G., Lt. . . . 506, 1223
——— L. F., Capt. . . 501a, 1148
——— R. Ty. Lt. 1726
——— R. A., 2nd-Lt. . . 556, 1451
——— R. C., Bt.-Maj. . . 54, 258, 1617
——— S. F., 2nd-Lt. 552
——— T., 2nd-Lt. 719
——— T. G., Lt. 624
——— T. M., Lt. 387
Dunckley, P. J., Lt. . . . 518, 923
Dundas, C. H., Lt. 923
——— P. H., Lt.-Col. . 61, 98, 137, 159, 200, 1038
——— R. H., Lt. 7, 744
——— W. J., Capt. 1732
——— R. M., Maj. 834
——— W. L., Maj. . 209, 872, 1606, 1657
Dundee, W. J. D., Col. 123
Dunford, C. M. P., Lt. . 297, 1579
Dunga Rawat, Subdr. . . . 1476
Dunham, G. A., Sub-Condr. . . 1854
——— H. H., Lt. . . . 542, 1673
Dunk, E. F., Capt. 653
Dunkerley, W., Lt. 654
Dunkerton, E. L. H., Capt. . 1106
Dunley-Owen, A. H. U., 2nd-Lt. . 365, 1227
Dunkin, J. C., Lt.. 738
Dunlop, A., Sub-Condr. . . . 1853
——— C. A. M., Maj. . 220, 1170
——— C. W. D., Asst. Surgn. 1797
——— D. L. C., Revd. . . . 1904
——— H. H., Lt.-Col. . . 129, 154, 187
——— H. J., Maj. 841
——— H. L., 2nd-Lt. . . 550, 1701
——— J., Capt. 861
——— R. W. B., 2nd-Lt. . 561, 1321
——— W. B., Lt.-Col. . 139, 160, 215, 1661
Dunn, C. L., Maj. 1750
——— C. L., Lt. . . . 515, 1207
——— D., 2nd-Lt. . . . 472, 1453
——— D. A. C., Lt. . . . 300, 1046
——— D. T., Lt. 512, 1636
——— E. G., Maj. 54, 99
——— F. E., Capt. 39, 501, 806, 1664
——— F. J., Sr. Asst. Surgn.. . 1788
——— G. L., Capt. 780
——— H. S. A., Capt. 1839
——— J. J. F., Asst. Surgn. . . 1797
——— J. Mc. L., 2nd-Lt. . . . 752
——— N. E. J., Lt. . . . 67, 764
——— O. H. B., Lt. . . . 537, 1255
——— R. L., Asst. Surgn. . . 1796
Dunne, E. L., 2nd-Lt. . . 551, 1034
——— J., Condr. 1863
——— T. F. K., Maj. 633
Dunning, A. A. J., Lt. 529
——— A. G. A., Capt. . . 502, 1195
——— C. W. A., Lt. 506
——— G. K., Lt. . . . 733, 734
——— H. G., Capt.. . 56, 502, 1656
Dunolly, K. J. G., Lt.-Col. 104, 129, 155, 188
Duns, G. S. S., Lt. 623
Dunscombe, N. B., Maj. . . . 1919
Dunsford, E. H., Maj. . . 212, 1289
——— F. P., S., Lt.-Col.. 34, 121, 131, 155, 191
Dunsmure, C. T. W., Lt. . 519, 1559
Dunstan, F. M., Maj. 683
——— G. T., Lt. 579
——— W., Lt. 697
Dunsterville, E. L., Lt.-Col. . 127, 128, 660
——— L. C., Maj.-Genl. 12, 26, 103, 105, 119, 128, 147, 175
Dunt, R. B., Lt.. . . . . 516, 1136
Dunwell, C. N., 2nd-Lt. . . . 1415

Names and Rank. Page.

Dunwoodie, L. B., Sr. Nursg. Sister 1734
Dupe, T. G., 2nd-Lt. 613
Duplessis, G. L. J., Capt. . . . 733
DuPre, W. D., 2nd-Lt. . . . 1030
Dupree, H., 2nd-Lt. 561
Duraisami, Subdr.-Maj. . . . 1307
——— Subdr. 1317
——— Jemdr. 1264
——— II., Subdr. . . . 1264
——— Mudali, Jemdr. . . 1360
Duraisami, Subdr. 1339
Durack, J. J., Lt. 488, 939
Durand, E. P. M., Capt. 55, 264, 1046, 1901
Durani, Jemdr. 1184
Durant, E. G., Lt. 491
——— E. J., 2nd-Lt. 82
——— R. M., Lt. 618
Durban, E. C., 2nd-Lt.. . . . 357
Durga Das Assal, Sub-Asst. Surgn. 1818
Durga Das Asal, Sub-Asst. Surgn. . . . 1818
Durgadut Sahi, Jemdr. . 1643, 1645
Durgaman, Jemdr. 1696
Durgamani Thapa, Subdr. 1607, 1609
——— Jemdr. 1602
Durga Narain Singh, Raja, 2nd-Lt. 481
DurgaiParsad Khatri, Sub-Asst. Surgn. . 1820
Durga Parsad Khatri, Sub-Asst. Surgn. . . 1810
Durga Parshad Misra, Sub-Asst. Surgn.. . . . 1815
——— Jemdr. 906
——— Singh, Subdr. 1062
——— Jemdr. 1281
Durgapoda Benerjee, Subdr. . 1219
Durgayya, Subdr. 1276
Durham, W. G. W., Maj. 104, 109, 242, 925
——— W. R., Sr. Asst. Surgn.. 1788
Durjan Singh, Jemdr.. . . . 1052
——— Subdr. 1387
——— Subdr. 1032
Dur Muhammad Khan, *Hony.* Capt. 2103
Durie, G. A., 2nd-Lt.. 803
Durnan, J. A., Lt. . . . 516, 1006
Durning, J. G., Asst. Surgn. . . 1800
Durrani, Subdr. 1507
Durrant, E. M., Lt.. 769
——— G., 2nd-Lt. . . . 475, 1158
——— H. B., Rev. 1903
——— O. T., Lt. . . . 393, 1567
Durston, E. R., Lt. . . . 541, 1673
Durzan Singh, Jemdr. 961
Dutchman, R. H., Lt. 714
Duthie, A. M., Maj. 614
——— K. J. N., 2nd-Lt. . 327, 1418a
——— T., *Hony.* Maj. . . . 1845
Dutt, A. C., Tempy. Capt. . . . 1766
——— N. B., Ty. Lt. 1782
——— S., Lt. . . . .
——— S., Tempy. Capt. 1768
Dutta, J., Ty. Lt. 1779
——— N., Tempy. Capt. . . . 1767
Dutton, C. C., 2nd-Lt. 748
——— H. R., Maj. . . . 37, 1749
——— R. M. L., Maj. . . 634, 980
——— W. H. R., 2nd-Lt. . . 1243
duVallon, G. C. deJ., 2nd-Lt. . . 624
Duxbury, C. D., Maj. . 101, 209, 1661
——— E. L., Lt. 505, 1116
Dwane, H. M., Lt.-Col. . . . 844
——— L. M. R., 2nd-Lt. . 328, 1038, 1559
Dwarka Singh, Subdr. 1681
——— Jemdr. 1561
Dwelley, W. E., 2nd-Lt. . . . 472
Dwyer, H. M. O'C., Capt. . . . 769
——— J. C. O., 2nd-Lt. . . . 907
——— P., Maj. 10, 1712
Dya Ram, Jemdr. 1702
Dyas, G. E., Capt. 1715
Dyball, E. E., 2nd-Lt. . . 589, 602
Dybell, C. W., 2nd-Lt. . . . 1581
Dyce, A. G., Capt. . . . 261, 1664
——— H. L., Maj. . . . 215, 891
Dye, A., Hony. Lt. 848
Dyer, A. J., Lt. 1519
——— A. J. A., Lt. 834
——— A. W., Asst Surgn.. . . . 1793
——— B. S., Revd. 1904

NAMES AND RANK. Page.

Dyer, C. J. F., 2nd-Lt. 90, 695
—— G. M., Lt. 310, 365, 901
—— G. S., Lt. 628
—— H. R., Maj. 221, 941
—— I. R., Lt. 387, 1235
—— J. C., Asst. Surgn. 1798
—— J. E. F., Maj. 583
—— J. F., Capt. 803
—— R. E. H., Br.-Genl. 12, 25, 97, 120, 125, 150, 185
—— W. L. E., Sub-Condr. 1855
Dyke, O. M., Lt.-Col. 215, 871, 915, 1447
—— P. H., Lt.-Col. 137, 159, 199, 1517
—— H. A. G., Capt. 1718
—— H. B., Lt. 1720
Dykes, S. S., Capt. 1713
Dymott, B. K., Capt. 42, 279, 1144
Dyne, C. F., 2nd-Lt. 1895
Dyson, F., 2nd-Lt. 18, 475, 925
—— H. P., Lt. 351, 1090
—— T. E. Col. 24, 124, 1738
Dyton, A. H., 2nd-Lt. 563
Dzitt, J., Ty. Lt.

E

Eade, J., Sub-Condr. 1857
—— R. O., 2nd-Lt. 546, 1389
Eaden, C. L., 2nd-Lt. 1016
Eadie, J. L., Maj 52, 102, 104, 107, 108, 238, 1395
—— R. B., Tempy. Capt. 1724
Eadwine Franklin Das, Sub-Asst. Surgn. 1818
Eagar, F. M., Lt. 380, 1411, 1583
Eagle, H., Sub.-Condr. 1856
Eagles, J., Condr. 1853
Eaglesome, A., Lt. 851
Ealand, V. F., Lt. 617, 628
Eales, A. C. H. A., Capt. 263, 1054, 1471
—— C. H. H., Capt. 280, 287, 914
—— R. A. G., 2nd-Lt. 618
—— T. F. J., Lt. 380, 1395
Eardley-Wilmot, R., Maj.-Genl. 1955
Earl, A., Capt. 741
—— E. A., Lt. 696, 796
—— E. J., Esq. 88
—— E. J., Lt. 741
Earle, C. O., 2nd-Lt. 560
—— E. S., Lt.-Col. 136, 158, 196, 868, 1074
—— R. G., Bt. Lt.-Col. 662, 991
Earp, A. C., Capt. 502, 1605
Earwaker, G. P., Lt. 544
Easdon, H. V., Asst. Surgn. 1802
—— L. V. O'B., Asst. Surgn. 1794
Easey, R., Tempy. Asst. Surgn. 1804
Eassie, F., Col. 126, 1838
East, C. F., Lt. 320, 1355, 1457
Eastabrooke, I., Lt. 443, 1618
Eastley, C. M., Lt. 579
Eastland, W. H., 2nd-Lt. 555, 1373
Eastmead, C., Lt. 318, 1078, 1563
—— C. S., Lt.-Col. 105, 131, 156, 191, 872, 1606
Eastmond, A., Lt. 512, 986
—— R. C., Lt. 514
Easton, C. A., Lt. 779
—— F. A., Lt.-Col. 137, 632
—— M. H., Maj. 779
—— R. M., Capt. 1758
Eastwick, A. J., 2nd-Lt. 750, 1413
Eastwood, F., Lt. 629
—— G. A., 2nd-Lt. 618
—— V., 2nd-Lt. 739, 791
Eate, S. N., Asst. Surgn. 1802
Eates, A. V., Sr. Asst. Surgn. 1791
—— E. A., Asst. Surgn. 1796
Eaton, H., Lt. 491
—— R. S., 2nd-Lt. 1397
—— R. W., 2nd-Lt. 555
Eatwell, G. H. F., Lt. 807
Ebden, F. T. P., Maj. 227, 1307
—— J. W., Capt. 56, 501a, 608
Eberhardie, R. C., Capt. 261, 1463
—— R. E., Lt. 535, 1443
Ebert, W. E. R. B., 2nd-Lt. 554, 1217
Eccles, G. J., Lt. 304, 879
—— J. G. T., Lt. 470, 1016
—— M., Capt. 271, 1471
—— P., Maj. 652
Eccles-Holmes, J. F., Lt. 310, 1151
Echlin, G. F. W., Lt. 75

NAMES AND RANK. Page.

Eckel, P., Lt. 1068
Eckstein, J. McD., Capt. 1755
—— H. F., Capt. 595
Ecott, H., Lt. 61, 593
Ede, H. S., Lt. 503, 1158
Eddie, R. W., Capt. 1726
Edelston, E., Capt. 502
Eden, A. G., Maj. 614
—— A. E., 2nd-Lt. 556
—— A. W. B., 2nd-Lt. 358, 1096
—— B., Lt. 90, 517
—— E. A., Lt. 511, 974
—— G. M., Lt. 721, 1689
—— H. C. H., Capt. 615
—— H. G. A., 2nd-Lt. 358, 1128
—— J. L., Capt. 256, 1195
—— P. R., temp. Lt. 795
Edenborough, L. A. B., Lt. 679
Edeson, B. J. S., 2nd-Lt. 786
Edgar, D. K., Maj. 58, 664
—— L. A., Capt. 628
Edgcome, G. E. R., Lt. 317, 877
Edge, A. B., Lt. 638
—— E. H., 2nd-Lt. 546, 1389
—— F. L., Lt. 527, 955
—— G. P., 2nd-Lt. 545, 1674
—— J., Lt. 322, 1128
—— V. A., Asst. Surgn. 1802
Edgecombe, F., *Hony.* Capt. 700
Edgerley, H. F., Condr. 1853
—— W., *Hony.* Lt. 1846
Edgson, T. J., Lt. 446, 49, 1034
Edkins, R. H., Capt. 486, 1675
Edlin, C. A. E., 2nd-Lt. 564, 1375
Edmeades, F. D., 2nd-Lt. 475, 1289
Edmiston, A. W., 2nd-Lt. 1267
Edmonds, H. H., Revd. 1906
—— P. N., 2nd-Lt. 622
Edmondson, B. E. W., Lt. 315
—— H. C., 2nd-Lt. 494, 921
—— W. H., 2nd-Lt. 565
Edmunds, H. B., Lt. 629
Ednie, L., 2nd-Lt. 473, 1467
Edridge, Sir F. T., Hony. Col. 679
Edsell, E. V., Capt. 721
Edward, W. D., Lt. 306, 1251
Edward-Collins, C. E., Maj. 223, 1662
—— G., Capt. 17, 26, 27, 178, 937
Edwardes, A. C., Lt.-Col. 135, 158, 198, 871, 1411
—— H. W., Lt. 798
—— J. G., Lt.-Col. 107, 136, 158, 196, 872, 1606
—— S. M., Brig.-Genl. 50, 120, 124, 149, 184
Edwards, A. J., 2nd-Lt. 724
—— A. H. M., 2nd-Lt. 361
—— A. L., Lt. 68, 744
—— A. W., 2nd-Lt. 1048
—— C. G. N., Lt. 294, 964
—— D., 2nd-Lt. 638
—— D., Capt. 1732
—— D. B., Capt. 55, 357, 949
—— D. J., 2nd-Lt. 77, 357, 1884
—— E., Lt. 543, 1080
—— E., Maj. 240, 1136
—— F., 2nd-Lt. 560, 1287
—— F. A., 2nd-Lt. 717
—— F. B., Lt. 90
—— F. D., Lt. 542
—— F. H., Lt. 506, 1383
—— Fitz. J. M., Col. 1, 97, 123, 149, 182
—— F. O., Lt. 425, 1142
—— G. A., Lt. 445, 49
—— H., 2nd-Lt. 1387
—— H. B., 2nd-Lt. 548, 1618
—— H. J., 2nd-Lt. 567
—— H. R., Capt. 1713
—— H. S., Capt. 681, 785
—— J. B., Col. 57
—— J. B., Lt. 717
—— J. C., 2nd-Lt. 630
—— J. W., 2nd-Lt. 565
—— L. E., 2nd-Lt. 834
—— M. F., Lt. 435, 1080
—— N. H., Capt. 276, 1629
—— O. Lt. 1664
—— O. G., Capt. 521, 818
—— P. A., Maj. 818
—— R. C., Lt. 696, 697
—— S. J., Lt. 538, 1699
—— S. W., Lt. 1156
—— T., Lt. 529, 1427
—— W., 2nd-Lt. 545

NAMES AND RANK. Page.

Edwards, W. B., Lt.-Col. 19, 138, 1838
—— W. C., 2nd-Lt. 548
—— W. H., Maj. 9, 610
—— W. H., Ty. Capt. 571, 1850
—— W. M., Capt. 634, 1663
—— W. R., Maj.-Genl. 1, 4, 119, 1737, 1889, 1955
Edye, E. H. H., 2nd-Lt. 571
Egambaram, Jemdr. 893
Egan, E. M., 2nd-Lt. 475, 1001
—— V. G., 2nd-Lt. 552
Egan-Wyer, O. J., Lt. 543a, 1317
—— T. M., 2nd-Lt. 566
Egerton, C. C., Fld. Ml. *Sir* 118, 145, 171, 919, 1925, 1955
—— F., Lt. 595
—— R. G., Lt.-Gen. *Sir* 118, 146, 172, 1925
—— S. W., Maj. 78, 107, 224, 887
—— W., Capt. 500, 1664
Eggar, J. N., Capt. 720
—— W. H., Ty. Lt. 1729
Egginton, J. F., 2nd-Lt. 359, 1064
Elloart, B. H., Capt. 502, 1160
Ekambaram Mudali, C. V., Sub-Asst. Surgn. 1828
Ekin, R. G., Capt. 288, 290, 1243, 1355
—— W. J., 2nd-Lt. 365, 893
Elachola Mohi-ud-din, Jemdr. 1308
Elchidana, K. H., Tempy. Lt. 1774
Elder, E. M., 2nd-Lt. 760
Elderton, M. T., Lt.-Col. 132, 156, 193, 871, 1435
Eldon, A., Lt. 21, 530, 965
Eldred, H. S., 2nd-Lt. 355, 1114
—— W., 2nd-Lt. 622
Eldridge, F. L., Lt. 535, 1672
—— G. T., 2nd-Lt. 558, 1702
—— J. H., Condr. 1854
Elford, D. E., Lt. 796
Elgood, C. L. 105, 108, 371, 754, 901
Elias Wynn, A. H. W., Maj. 99, 222, 1595
Elijah, N., Ty. Lt. 1776
Elijah Abraham, Sr. Sub.-Asst. Surgn. 1829
Eliot, E. E. M., Capt. 850
—— H. S., Maj. 220, 1662
—— M., Lt.-Col. 76, 218, 1885
—— O. H., Lt. 517
Elkington, E. H. O., Capt. 273, 1166
Elkins, F. C. A., Asst. Surgn. 1800
—— T. I. H., Asst. Surgn. 1803
—— W. H., Lt. 526, 1671
Ellaby, G. A., Revd. 1908
Ellappah, Subdr. 1359
Ellayya, Subdr. 1359
Ellcome, J. E., Capt. 1713
Ellen, W., 2nd-Lt. 473, 1243
Ellery, A. C., Lt. 540a, 1239
Elles, F. W. C., Lt. 351
—— P. G. M., Capt. 615
Elliot, A. C., 2nd-Lt. 294a, 494
—— C., Lt. 509
—— C. R., Lt.-Col. 57, 129, 154, 187
—— C. D., Lt. 423, 1207
—— E. L., Lt.-Genl. *Sir* 933, 1925, 1955
—— G., 2nd-Lt. 545
—— G. O., Capt. 703, 704
—— H. R., Maj.-Genl. 170
—— H. W. A., Capt. 755
—— J. J., 2nd-Lt. 1703
—— J. R. McD., Capt. 755
—— J. S., Lt. 608
—— L., 2nd-Lt. 796
—— L. F., Lt. 778
—— R. A. D., Lt. 518
—— R. H., 2nd-Lt. 585
—— S. L. H., 2nd-Lt. 559
—— S. S., 2nd-Lt. 844
—— T., Lt. 490
—— T. R., 2nd-Lt. 788
—— W., 2nd-Lt. 1521
—— W. J., 2nd-Lt. 1453
—— W., 2nd-Lt. 550
Elliott, A. B., 2nd-Lt. 561
—— A. C., Lt.-Col. 195, 1896, 1912
—— A. C., Lt. 380, 1387
—— A. O., 2nd-Lt. 79
—— B. H., Capt. 101
—— C. A., Lt.-Col. 122, 138, 661
—— C. H., 2nd-Lt. 742
—— C. E., Capt. 732
—— E. C., Lt. 731
—— E. D., Maj.-Genl. 643

Names and Rank. Page.

Elliott, E. T., 2nd-Lt. . . 539, 1178
—— F. A. B., Lt. 638
—— F. P., Lt. 538
—— G. F., Lt. 509
—— H. H., Maj. . 32, 81, 207
—— H. J., Condr. 1863
—— J., Hy. Lt. 1850
—— J. F., 2nd-Lt. 555
—— J. G., Lt . . . 313, 1231
—— J. J., 2nd-Lt. 561
—— J. M., Capt. 1713
—— L., 2nd-Lt. 690
—— P., Lt. 765
—— P. W., Maj. . . 238, 1098
—— W., 2nd-Lt. 568
Ellis, A. B., Lt. . . . 308, 1148
—— A. C. S. B., Maj. . 6, 13, 99, 101, 212, 1479
—— A. E., Sub-Condr. . . . 1874
—— A. H. J., Lt. . . . 410 1251
—— A. R., 2nd-Lt. 700
—— A. W., 2nd-Lt. 1197
—— B. G., 2nd-Lt. 686
—— E. I. A., 2nd-Lt. . . 729, 791
—— E. R., Lt. 75, 638
—— E. T. H., Lt. 578
—— F. R., Lt. 737
—— G. A., Lt. 7, 638
—— H. C., Lt. 389, 1183
—— J., Maj. 722
—— J. C., 2nd-Lt. . . . .
—— J. W. L., Capt. 708
—— R., Lt. 490, 921
—— S., Lt. 352, 1070
—— S., Temp., 2nd-Lt. . . . 55
—— S. G. V., Maj. . . . 87, 219
—— S. N., Asst. Surgn. . . . 1792
—— T. J., Hony. Maj 1847
—— T. S., Lt. 90, 491
—— W. J., 2nd-Lt. 697
—— W. J., Lt. 523, 1032
—— W. M., Col. . . . 123, 660
Ellison, D. H., Lt. . 302, 1387, 1439
—— E. C., Lt. . . . 316, 1267
—— H. F., Lt. 1001
Elloy, A. L., Asst. Surgn. . . . 1796
—— K. P., Asst. Surgn. . . . 1798
—— L. M., Lt. . . . 529, 1399
Ellson, P., 2nd-Lt. 67, 776
Ellwood, B. P., Lt.-Col. . 138, 159, 198, 867, 933
Elmer, P. S., Lt. 473, 1557
Elmes, J. P. J., Lt. . . . 488, 897
Elmitt, G. L., Capt. 1732
Elmo, C. E., 2nd-Lt. 1521
Elms, C. E., 2nd-Lt. . . . 553, 1523
Elmslie, C. S., Lt. . . . 525, 758, 1475
—— G. F., Capt. 72
Elphick, H. C., Lt. 389
—— L. S., 2nd-Lt. 752
Elphinston, W. G., Capt. . 259, 939
Elrington, R., 2nd-Lt. . . . .
Elson, E. J., 2nd-Lt. . . . 474, 1447
—— W., 2nd-Lt. 1583
Elsmie, A. M. S., Col. 50, 97, 103, 121, 125, 127, 150, 190, 869
Elsworth, C. W. T., Lt. . 489, 895
Eltham, J. F., Sub-Condr. . . . 1854
—— F. J., Condr. 989
Elton, A. C., Lt. . . . 529, 988
—— G., Lt. 729
—— H. S., Maj.-Genl. . . 173, 1319
—— S., 2nd-Lt. 1674
Elton-William-Jervis, A., 2nd-Lt. 536
Elverson, C. G. L., 2nd-Lt. . . . 750
Elvidge, R. W., Hony. Capt. . . 717
Elwes, F. F., Maj. . . 1746, 1892
Ely, E. S., 2nd-Lt. 723a
Embahadur Sahi, Subdr. . . . 1621
Embury, T. F., 2nd-Lt. . . .
Emeney, A. M., Lt. 493
—— L. H., 2nd-Lt. . . 556, 1583
Emerson, D. B., Lt. . . 487, 543
Emery, T. S., Lt. . 18, 34, 367, 1319
—— W. C., 2nd-Lt. 701
Emile, C. A. A., Asst. Surgn. . 1758
—— V. W. A., Asst. Surgn. . 1758
Eminson, B. F., Capt. 1757
Emmerson, M., Hony. Lt. . . 1846
Emmett, C. H., 2nd-Lt. 561
—— P. W., Asst. Surgn. . . 1800
—— W. H., Hony. Capt. . . 1847
Emmott, B. St. B., 2nd-Lt. . 540, 1627
Empson, A. B. A. Capt. . 264, 889
—— C., 2nd-Lt. 626
Emslie-Smith, H., Maj. . . . 1749

Endley, W. G., 2nd-Lt. . . . 679
Engeldow, G. R., Lt. . . 492, 953
Engineer, S. K., Tempy. Maj. . 1763
England, H. G., Revd. 1904
—— P. H., Condr. 1868
—— R. D., 2nd-Lt. 821
—— R. S., Lt. . . . 503, 1509
Engledow, F. L., Capt. 752
Engledue, R. S., Maj. 7, 107, 242, 1363
Englefied, R. H., 2nd-Lt. . . . 624
Engley, E. R., 2nd-Lt. . . 562, 1707
English, A., Lt. 523, 1698
—— B. A., Lt. 658
—— B. M., Lt. . . . 511, 1674
—— E. E., 2nd-Lt. . . 547, 1705
—— F. W. H., Lt. 314
—— H. H. E., 2nd-Lt. . 475, 889
—— J. W., Capt. 636
Ennes, C., 2nd-Lt. 1699
Ennis, C., 2nd-Lt. 559
—— J. I., Lt. . . 315, 1614, 1647
Enriquez C. M. D., Maj. 246, 1102, 1347
Enticknapp, A. W., 2nd-Lt. . . 568
Entrican, J., Lt.-Col. 129, 1730, 1895
Entwisle, G. R., Lt. . . 530, 1435
Entwistle, J., 2nd-Lt. 363
—— R., 2nd-Lt. 640
Erck, J. C. W., Lt.-Col. . 128, 154, 182
Eric, H., 2nd-Lt. 849
Ernest, A. R., 2nd-Lt. 571
Errington, J. P., Capt. . . 67, 724
—— J. R. F., Capt. . 712, 798
Erskine, C. E. T., Capt . 261, 964
—— D. D. A., 2nd-Lt. . . . 731a
—— K. C. S., Capt. 17, 23, 268, 1617
—— W. R., Lt. 527, 1517
Ersu Nukayya, Jemdr. 1308
Erswell, H. W., Condr. 1872
Eshwant Bhosle, Subdr. . . . 1417
—— Rao Bhosle, Subdr. . . 1416
Eshwanta Shinde, Subdr. . 1423, 1579
Esilman, A., 2nd-Lt. 695
Estill, J., Lt. 1598
Eswaramurthi, Jemdr. . . . 1324
Esse, F. A., Lt. 425, 1601
Esu Vase, Subdr. 1699
Etchells, J. E., 2nd-Lt. 363
Etheridge, F., Maj. . . 99, 233, 1042
—— F. H. 2nd-Lt. 751
Etherington, J., Hony. Lt. . . 778
—— R. A., 2nd-Lt. 330, 1034, 1613
Etherton, P. T., Maj. 103, 244, 1178
Ethirajalu Nayadu, P. C. Sub-Asst. Surgn. 1828
Eustace, A. H., Maj-Gen. 12, 29, 119, 147, 174
—— F. R. H., Maj. . . 662, 1882
—— R., Capt. 280, 1178
Evans, A., 2nd-Lt. . 354, 1008, 1239
—— A. E., 2nd-Lt. 356
—— A. G., Lt. 515, 1667
—— A. H., Condr. 1854
—— A. L., Lt. 368, 1050
—— A. S., Maj. 21, 663
—— B. L., Capt. . . . 17, 45, 679
—— C., 2nd-Lt. 73, 790
—— C. F., Lt. 537
—— C. H., Hony. Lt. 112
—— D. E., Capt. 1720
—— D. G., Capt. 1717
—— D. J., 2nd-Lt. 9
—— D. M., 2nd-Lt. 624, 790
—— D. MacN., Capt. 261, 1243, 1247
—— D. R., 2nd-Lt. 599
—— D. W., Lt. . . . 455—59, 1170
—— E. A., Lt. . . . 371, 746, 1183
—— E. G. H., Lt. . . . 536, 1569
—— E. J., Capt. 707
—— E. M., Lt. 541, 1698
—— E. T., Lt. 755
—— E. W., Lt. 540, 997
—— F. L., 2nd-Lt. 1048
—— G., Lt. 32, 493, 803
—— G., Condr. 1873
—— G. H., Col. . . . 653, 1842
—— H. K., Lt. 756
—— H. M., Genl. Sir 171, 1635, 1917, 1925
—— H. R., 2nd-Lt. 1575
—— H. S., Asst. Surgn. . . . 1803
—— H. W., Capt. 717
—— I. A. Lt. 528, 1671
—— I. B., Lt. 293
—— J., 2nd-Lt. 790
—— J., 2nd-Lt. 1122

Evans, J. B., Lt. 1034
—— J. E., Lt. 31a, 909
—— J. D., 2nd-Lt. . . . 550, 1677
—— J. H., 2nd-Lt. 704
—— J. J., Hony. Lt. 1848
—— J. L., Lt. 511, 1647
—— J. S. P., Capt. 789
—— J. W. D., Capt. . . 52, 597
—— J. W. P., 2nd-Lt. . 561, 1678a
—— L. Maj. 663
—— L. E., 2nd-Lt 1689
—— M. O., 2nd-Lt. 737
—— N. W., Lt. . . . 316, 1211
—— P. B., Lt. 621
—— P. S., Astt. Surgn. . . . 1803
—— R., Capt. 587
—— R. A. G., 2nd-Lt. 362
—— R. E., Lt. 722
—— R. E., 2nd-Lt. 748
—— R. J., Capt 501, 1664
—— R. M., 2nd-Lt. . . . 473, 1263
—— S., Lt. 503, 1517
—— S., Lt.-Col. . . . 132, 1743
—— S. F., 2nd-Lt. 325, 366, 970, 1567
—— T. F., 2nd-Lt. 790
—— T. G., Tempy. Capt. . . . 1726
—— T. J. C., Bt. Maj. . . . 1755
—— T. L., 2nd-Lt. . . 326, 366, 1104
—— T. R., Lt. 298, 1136
—— U. W., Brig.-Genl. 50, 104, 120, 124, 651
—— V. C., 2nd-Lt. . . . 359, 1559
—— W. A. A., Capt. . 510, 830, 1197
—— W. H., Maj. 663
—— W. J., Maj. 216, 1595
—— W. K., 2nd-Lt. 362
—— W. S., Ty. Capt. 1723
Evans-Gordon, K. A. G., Capt. 260, 949, 1889
Evans-Lombe, T. B. A., Lt. . . . 589
Evanson, E. A., Lt. . . . 389, 1187
—— G. A., Lt. . . . 538, 1701
Evatt, C. A., Lt. 530, 1668
—— J. T., Brig.-Genl. . . . 1178
Eve, H. E., Lt. 1585
Eveleigh, P. B. C., Lt. 522
Eveling, W. R., 2nd-Lt. 558
Evenden, J. M., 2nd-Lt. . 357, 1044
Evennett, H. C., Lt. . . . 307a, 897
Everard, B. G., Lt. 544a
—— G. H., Lt. 798
—— K. V., 2nd-Lt. 626
—— R., Condr. 1867
—— T. A., Lt. 538, 1696
Everett., C. F., 2nd-Lt. . . 363, 1094
—— G. E., Capt. 819
—— M., Bt. Maj. . 665, 996, 1882
—— W. H., 2nd-Lt. . . 328, 947
——, W. M. 2nd-Lt. 1573
Evers, T. G., 2nd-Lt. 947
Evershed, W. A., Lt. . . 523, 998
Everton, B., 2nd-Lt. 736
—— R. F., 2nd-Lt. . . . 40, 637
Eves, G. W., Capt. 849
Evett, J. H., 2nd-Lt. 713
Evetts, D. W., Lt. 303, 895
—— E. F., Capt. 681
Evill, C. E., 2nd-Lt. 825
Ewart, H. P., Maj.-Genl. Sir . . 583
—— J. A., Lt. . . . 51, 487, 891
—— R. F., Maj. . . . 226, 1062
—— R. H., Maj.-Genl. Sir 119, 147, 174
Ewaz Khan, Ressdr. 79
—— Subdr. 1587
—— Subdr 1479
Ewbank, R. B., 2nd-Lt. 819
—— W., Col. . . 58, 97, 124, 665
Ewen, A. C., Lt. 393
Ewens, G. P., Lt. 521, 1513
Ewers, L. F., Lt. 724, 919
Ewin, H. L. H., 2nd-Lt. 1046
Ewing J., Col. 176
—— J., Condr. 1851
—— J. A., 2nd-Lt. 558
—— R. B., Capt. . 274, 1263, 1895
—— S. R., Esq. 91
Ewings, J. L. S., Capt. 911
Exham, H., Maj. . . . 54, 245, 1629
Exshaw, P., Nursg. Sister . . . 1735
Extance, T. J., 2nd-Lt. . . 545, 1413
Eyles, C. M. E., Asst. Surgn. . . 1803
Eyre, E. H., Col. 177
—— J. S. G., 2nd-Lt. . . 332, 1130
—— R. G., Maj. 769

NAMES AND RANK. Page.

Eyre, M. H., 2nd-Lt. . 550, 571a, 1523
Eyre-Walker, C., 2nd-Lt. . . . 805
—— W., Capt. 804
Eyton, J. S., Lt. 488, 885
Ezekiel, J., Tempy. Capt. . . . 1771
Ezra, E., Lt. 56, 516a

F

Facon, W. S., 2nd-Lt. 90
Faddy, F. L., Lt. 540, 1365
Fagan, A. W., Lt. . . . 510, 1082
—— B. J., Maj. . . 203, 1046
—— D. P., Capt. . . 787, 788
—— E A., Brig.-Genl. 50, 1 2, 135 158, 195, 867, 943
—— H. A., Lt. . . . 435, 1642
—— H. W. F., Revd. . . . 1904
—— J. L., Maj.-Genl. . . . 173
—— L. E., Ty. Capt. . . 619, 979
Fagnani, H. H., Lt.. 443, 505, 965, 1567
Fahey, W., Lt. 618
Fair, F. W., 2nd-Lt. 566
Fairclough, F. S., 2nd-Lt. 355, 993
—— J. E. B., 2nd-Lt. . . 1275
—— R., 2nd-Lt. 742
Fairhurst, A., Esq. 88
—— H. E. 2nd-Lt. . . . 788
—— J. A., 2nd-Lt. . . . 361
Fairley, J. Capt. 608
—— J. W., Lt. 806
—— R. S., Lt. 1621
Fairlie, J. A., the Revd., . . . 1905
—— J. W., 2nd-Lt. . . . 544
—— W. S., 2nd-Lt. . . 557, 1054
Fairweather, A., 2nd-Lt. . 549, 1297
—— D. O., Ty. Lt. . . 1729
—— H., Capt. . . . 485
—— J. W., Capt. . . 858
Faith, N. W., Maj 821
Faithfull, F. E. N., Lt. . . 510, 1475
—— F. H., Lt. . 61, 506, 1589
—— G. F. H., Lt. . . 269, 1501
Faiyazali Khan, Subdr. . . . 1771
Faizu Khan, Jemdr. 1400
Faizullah Khan, *Hony.* Lt. . . 2104
Faiz Ahmad Khan, Jemdr. . . 1117
—— Jemdr. . . 1154
Faiz Ali Khan, Subdr. . . . 1683
—— Jemdr. . . . 1507
Faiz Buksh, Ressdr. . . . 1683
Faiz Khan, *Hy.* Capt. . . . 2103
—— Muhammad, Jemdr. . . 1056
—— Subdr. . . 1501
—— Subdr. . . 1501
—— Buksh, Ressdr. . 1683
—— Khan, Risaldar 876, 1893
—— Subdr. . 1086
—— Jemdr. . 1256
—— Jemdr. . 1078
—— Subdr. . 1090
—— Chaudhry, Sub.-Asst. Surgn. 1818
Faize Muhammad, Subdr. . . 1427
Faizullah, Snr. Sub-Asst. Surgn. 1807
—— Khan, Risldr. . . . 889
Fajir Ali Khan, Subdr. . . . 1086
Faiz Talab Khan, *Hy.* Capt. . . 2101
—— Jemdr. 1095
—— Talb, Jemdr. 1336
—— Ul-Hasan Khan, Jemdr. . . 1388
Fakharuddin, Jemdr. 1507
Fakir Muhammad, *Hy.* Lt. . . 2104
—— Subdr. . . 1249
—— Subdr. . . 1259
Fakir, Singh, Subdr. 1567
—— Singh, Jemdr. 1216
Falcke, J., Lt. 624
Falcon, R. R. B., Capt. 82, 368, 371, 501a, 679
—— R. W., Lt.-Col. 82
Falconer, J. F., Capt. . . 486, 929
—— G. A. L., 2nd-Lt. . . 465
Falel Khan, Subdr. . . . 1479, 1589
Falgar, R. J., Lt. . . . 529, 1704
Falk, H., Maj 1752
Falkenstein, A. J., Lt. . 352, 1596
Falkiner, J. Mcl., tempy. Capt. . 1763
Falkland, H. L. F., Capt. 26), 1070, 1808
Falkner, J. M., Lt. . . . 26, 485
—— P. H. E., Lt.-Col. . 133, 1710
Fallon, J., Col. . . . 27, 124, 1709
Falloon, P. H., 2nd-Lt. . . 561, 1281

NAMES AND RANK. Page.

Falloon, T. St. A., Asst. Surgn. . 1801
Fanaken, P. F., Asst. Surgn. . . 1800
Fancourt, St. J. M., Col. . . . 179
Fane, A. G. C., Lt. 508
—— V. B., Maj.-Genl. *Sir* 119, 147, 174, 1102
Fanning, R. J., Lt. . . . 443, 1181
Fannon, J. T., Sub-Condr. . . 1855
Fanshawe, L. A., Maj. . 74, 101, 633
—— W., 2nd-Lt. . 494, 825, 939
Faqir Chand, Jemdr. 1232
—— Khan, Subdr. 1320
—— Muhammad, Jemdr. . . 1373
—— Jemdr. . . 1280
—— Jemdr. . . 989
Faqira Singh, Subdr. 1031
Fardell, D. O., Capt. 1636
—— J. G., Capt. 729
Farbrother, A. A., Lt. . . 297, 1351
—— E. S., Lt. . . 571a, 1841
Farebrother, F. H., Capt. 102, 258, 1509
Farewell, M. W., Capt. 4
Farelly, M., Capt. 1841
Farfan, A. J. T., Maj. . 109, 634, 975
Farid Khan, Risldr. 1686
Faridkot, *Raja of*, Maj. . . . 110
Faris, J. G., Maj. 221, 1629
Farley, A. T., 2nd-Lt. 557
—— C. W., Lt. 721
—— E. L., Capt. . . . 665, 985
—— F. A., 2nd-Lt. . . . 1154
—— G. P., Lt. . . . 508, 1255
—— H. J., Lt. . . . 552, 1391
—— L. G., 2nd-Lt. . . . 626
Farlinger, A. C., Ty. Lt. . . . 1729
Farman Ali, Subdr. . . . 1011, 1493
—— Subdr.-Maj. . . . 1375
—— Khan, Subdr. . . 1497
—— Shah, Jemdr. 1364
—— W. S., *Hony.* Lt. . . . 838
Farmar, W. C. R., Lt.-Col. . 135, 632
Farmborough, J. C., Ty. 2nd-Lt.
Farmer, C. A., Sr. Asst. Surgn. . 1787
—— C. R. H., Lt. 798
—— C. R. S., 2nd-Lt. . . . 820
—— E. J., Lt. 848
—— F. S., Lt. 692
—— G. C., Lt. 667
—— J., Col. 126, 1842
—— J. C., Lt. 537
—— N. R., Lt. 797
—— F. M., 2nd-Lt. . . . 568
Farnbourough, J., Capt. . . . 1708
Farn, E. A., Lt. 1703
Farnell, F. B., Lt. 798
Farnon, L. R. St. J., Lt. . 540, 1703
Farnsworth, C., 2nd-Lt. . . . 1629
Farquhar, 2nd-Lt. 798
—— D., Lt. 542
—— F. R., Maj. . . 245, 943
—— W., 2nd-Lt. . . . 696
Farquharson, C. W., Capt. 16, 40, 263, 1058
—— D. A., Tempy. Capt. 1725
—— F. A., Lt. . 509, 986
—— J. C. 2nd-Lt. . . 790
—— J. R, 2nd-Lt. . . 652
—— S., Lt. 730
Farquharson-Roberts, K. F., Capt. 42, 75
Farr, C. A. C., 2nd-Lt. 799
—— C. J., Lt. 528, 1060
—— R. P., Lt. 848
Farran, A. C., Ty. 2nd-Lt. . . . 572
—— G L., Maj. . . 56, 222, 88.
Farrance, S. C. E., Lt. . . 380, 880
Farrar, C. B., Lt. . . 405, 881 1007
—— F. H. 2nd-Lt. . . . 788
Farrar, W. B., Lt. . 315, 1034, 1387
Farrell, G. V., Capt. . . 351, 1443
—— H. H., Lt. 1705
—— J., 2nd-Lt. . . 551, 805, 1227
—— J. K. O'M., Lt. . . 375, 877
—— J. O'M., 2nd-Lt. . . .
—— N. N., Lt. . . . 523, 1699
—— R., Condr. 1872
—— R. A., 2nd-Lt. 790
Farren, D., 2nd-Lt. 563
Farrington, F. L., Lt. . . . 38, 528
—— G., *Hony.* Maj. . . 1848
—— H. R., Lt. 538
—— R. L., 2nd-Lt. . . . 758
Farrow, A. E., Lt. 626
—— A. R., Lt. 629
—— G. U., 2nd-Lt. 561
—— J. K., Lt. . . . 399, 1007
Faruki, S. M. A. Tempy. Capt. . 1771

NAMES AND RANK. Page.

Farwell, A. E., Lt. . 95, 311, 1078, 1515
Farwell, E. W., Capt. 704
—— G. A. L., Capt. . 280-287, 1124
—— W. L., Capt. . . . 267, 1613
Farzand-Ali, Ressdr. 915
—— Khan, Jemdr. . . 876
—— Subdr. . . 1473
Farzand Khan, Jemdr. 999
—— Jemdr. 1401
Fasken, C. G. M., Maj.-Genl. . . 123
—— K. L., Capt. . . . 276, 1387
—— W. H., Col. . . 123, 149, 183
—— Bahadur Singh, Jemdr. . 1181
Fateh Ali, Jemdr. 1396
—— Jemdr. 1103
—— Jemdr. 981
—— Jemdr. 1208
—— Jemdr. 1456
—— Bahadur Sing, Jemdr. 1281
Fateh Chand Prahara, Sub-Asst. Surgn. 1817
—— Din, Jemdr. 980
—— Din Khan, Jemdr. . . . 1010
—— *Hony.* Capt. 2104
—— Ali Khan, Jemdr. . . . 910
—— Jang, Jemdr. 1321
—— Subdr. 1507
—— Khan, Subdr.-Maj. . . . 1094
—— Jemdr. 975
—— Jemdr. 1513
—— Jemdr. 1312
—— Ressdr. 1683
—— Jemdr. 1682
—— Subdr. 1379
—— Jemdr. 1244
—— Subdr. 1413
—— Subdr. 1001
—— Jemdr. 896
—— Jemdr. 1129
—— Jemdr. 968
—— Subdr. 1452
—— Subdr. 1471
—— Subdr. 1128
—— Jemdr. 908
—— Jemdr. 999
—— Mohamed, Ressdr. . . 1685
—— Mohammed, Jemdr. . . 1022
—— Jemdr. . . 1143
—— Jemdr. . . 990
—— Muhammad, Jemdr. . . 1095
—— Jemdr. . . 1370
—— Subdr. . . 1311
—— Subdr. . . 1343
—— Jemdr. . . 1304
—— Jemdr. . . 1518
—— Subdr. . . 137-
—— Jemdr. . . 1236
—— Jemdr. . . 990
—— Jemdr. . . 918
—— Jemdr. . . 1488
—— Beg, Ressdr. 892
—— Subdr. . . 1432
—— Jemdr. . . 876
—— Ty. Sub-Asst. Surgn. 1833
—— Khan, . . 951
—— Subdr. 1085
—— Jemdr. . . 1133
—— Roz, Jemdr. . . . .
—— Khan, Subdr. . . 1256
—— Shah, Ressdr. . . 1682
—— Shah, Jemdr. . . . 1054
—— Jemdr. . . . 1333
—— Jemdr. . . . 1133
—— Jemdr. . . . 1052
—— Khan, Jemdr. . . 904
—— Singh Newar, Sub.-Maj. 1601
—— Subdr. 1114
—— Subdr. 1387
—— Risldr.-Maj. . . 885
—— Jemdr. 1107
—— Jemdr. 1163
—— Jemdr. 1245
—— Subdr. 1046
—— Subdr. 1117
—— Jemdr. 1167
—— Ressdr. 891
—— Jemdr. 932
—— Bisht, Jemdr. . . 1224
—— Singh Gusain, Subdr. . . 1181
—— Sub-Maj. . . . 1158
—— Jemdr. 928
—— Jemdr. 1157
Fatehsingh Thapa, Jemdr. . . 1597
Fateh-ud-din Khan, Jemdr. . . 1090
Fatehi Khan, Jemdr. 1107
Fatteh Din, Jemdr. 1682

Names and Rank.	Page.
Fatteb Khan, Jemdr.	1681
——, Jemdr.	1681
Fauja Singh, Ressdr.	950
—— Jemdr.	894
—— Subdr.	1211
—— Subdr.	968
—— Jemdr.	896
Faujdar Khan, Ressdr.	876
—— Jemdr.	1092
—— Jemdr.	1409
Faujdar Singh, Ressdr.	942
Faulconer, E. A., Lt.	532
Foulder, G. B., Lt.	389, 1383
Faulkner, A. K., Capt.	827
—— E. A., Lt.	22
—— R., Lt.	389, 1624
—— St. J. T., Lt.	723
—— W. G., 2nd-Lt.	539, 1219
Faunce, A. deL., Maj.	206, 1050
—— B. deL., Lt.-Col.	130, 155, 189, 1892
Faunthorpe, J. C., Lt.-Col.	608
Fausset-Baker, A. A., Capt.	280—287, 1004, 1102
Faux, C. E., 2nd-Lt.	637
Fawcett, A. H., Lt.	506, 1665
—— A. M., Lt.	622
—— C. E. W. S., Maj.	32, 1712
—— F. O.D., Maj.	1754
—— H. E. M., Lt.	35, 529, 1671
—— J. F., Maj.	844
—— J. H., Lt.	514, 986
—— R. B., Lt.	405, 1642
—— W. L., Lt.	293a, 1645
Fawcitt, C. S., Capt.	80
Fawcus, L. B., Lt.	490
Fawkes, L. E. Lt.	785
Fawkner, H. S., Lt.	389, 927
Fay, A. T., 2nd-Lt.	560, 1701
Fayrer, F. D. S., Lt.-Col.	139, 1745
Fazal Ahmad, Jemdr.	1252
——, Sub-Asst. Surgn.	1830
—— Jemdr.	1208
—— Khan, Jemdr.	1036
—— Ali Khan, Jemdr.	890
—— Jemdr.	886
—— Dad, Jemdr.	1498
—— Jemdr.	1184
—— Jemdr.	1488
—— Subdr.	1288
—— Subdr.	1369
—— Jemdr.	1244
—— Jemdr.	1365
—— Subdr.	1513
—— Ressdr.	1683
—— Hassan, Jemdr.	1284
—— Husain, Jemdr.	1141, 1143
—— Khan, Subdr. Maj.	1281
—— Ressdr.,	1683
—— Din, Subdr.	1001
—— Jemdr.	1124
—— Jemdr.	1365
—— Jemdr.	1704
—— Jemdr.	999
—— Hasan, Sub-Asst. Surgn.	1817
—— Ilahi, Jemdr.	1122
—— Khaliq, Jemdr.	1567
—— Khan, Subdr.	1335
—— Subdr.	1577
—— Jemdr.	1472
—— Jemdr.	926
—— Subdr.	1288
—— Jemdr.	892
—— Karim, Jemdr.	946
—— Sub-Maj.	1453
—— Shah, Jemdr.	989, 990
—— Jemdr.	1428
Fazar Ali Khan, Jemdr.	936
—— Subdr.	1233
Fazil Din, Jemdr.	1683
Fazili Husain, Risdr.-Maj.	899
Fazl Ahmad, Subdr.	1571
—— Khan, Jemdr.	1143
—— Haq, Ressldr.	942
—— Haqq, Jemdr.	988
—— Husain Chohan, Sub-Asst. Surgn.	1820
—— Ilahi, Sub-Asst. Surgn.	1815
—— Jemdr.	1518
—— ul-Rahman, Sub-Asst. Surgn.	1818
—— Karim, Sub-Asst. Surgn.	1813
—— Khan, Subdr.	1355
—— *Hony.* Capt.	2102
—— Jemdr.	1095
—— Muhammad, Subdr.	1136
—— Nur, Jemdr.	990
Fazl Rahman, Subdr.	1335
—— Shah, Sub-Asst. Surgn.	1814
—— Subdr.	988
Fazl-Din, Sub-Asst. Surgn.	1810
Fazl-i-Ilahi, Sub-Asst. Surgn.	1811
Fazl-i-Khaliq, Jemdr.	970
Fazl Khan, Jemdr.	974
—— Subdr.	1095
Fazlur Rahman, Sr. Sub-Asst. Surgn.	1809
Fazul-ud-din, M., Tempy. Capt.	1771
Fazzal Din, Jemdr.	1456
Fear, F. R., Capt.	18, 47, 386, 730
Fearnside, Speed, K. A., Lt.	385, 889
Fearon, J. G., Capt.	679
Featherstone, B. K., Lt.	319, 1239
—— O., Capt.	679
—— W. A. E., 2nd-Lt.	579
Feegrade, A. A., Asst. Surgn.	1802
—— E. S., Asst. Surgn.	1796
Field, R., Lt.	780
Feiling, K. G., Lt.	8
Fellmann, P. M. J., Lt.	518, 1668
Felce, E., Maj.	709
Felix-Williams, A., Lt.	455-59 1243
Fell, A. L., Lt.	317a, 1604
—— E A. H., Maj.	224, 897
—— R. J. H., Lt.	518, 1026
Fellingham, F. W., Lt.	353, 1702
Fellowes, C. A., 2nd-Lt.	692
—— H. LeM., Maj.	1919
Fellows, B. C., Bt.-Maj.	57, 1919
—— C. S., Lt.-Col.	135, 158, 195, 1439
Feltham, W. I. P., Capt	288—290, 1231
Felton, G. H., 2nd-Lt.	46, 356, 1351
Fenn, C. D., 2nd-Lt.	566
—— E. H. McB., Lt.-Col.	136, 158, 197
Fennell, A E. S., Maj.	232, 1331
—— C. D., Lt.	509
—— W. G., Lt.	450—54 1385
Fenning, W. M., Maj.	207, 1663, 1883, 1913
Fenstone, J., 2nd-Lt.	545
Fenton, A., Lt.-Col.	137, 1745
—— G. A., Lt.	617
—— G. C. V., Maj.	11, 664
—— H. F., Ty. Capt.	1726
—— J. N., Capt.	33, 260, 1086, 1199
—— L. S., Major	45, 232, 1455
Fenwick, F., Lt.	487
—— F. R., 2nd-Lt.	720
—— J. S., 2nd-Lt.	1419, 371
Fenwicke-Clennell, E. C., Lt.	380, 1433
—— G. E., Lt.	300, 1427
Ferguson, D. C. H., Lt.	330, 1397
—— D. H., Capt.	485, 832
—— E. G., Tempy. Asst. Surgn.	1804
—— F. E., Lt.	371, 970
—— G., Lt.	617
—— G. C. S., Capt.	264, 885
—— H. F., Ty. Capt.	1725
—— H. J., Lt.	536, 1030
—— J., Capt.	1004
—— J., Ty. Lt.	1725
—— J., Lt.	828
—— J. A., Capt.	272, 1070
—— J. A., 2nd-Lt.	605
—— J. F., Capt.	761
—— J. H., Lt.	534, 1583
—— J. M., 2nd-Lt.	473, 895
—— R. H., 2nd-Lt.	558, 532, 1703
—— W. A., Maj.	656
—— W. A. M., Lt.	304, 1611
—— W. M., Capt.	1732
Fergusson, A. C., Lt.-Col.	101, 134, 632, 972
Fermer, E. J., 2nd-Lt.	553
Fern, E. A., Lt.	571
Fernandes, C., Ty. Hony. Maj.	111, 176
Fernandez, C. J., Tempy. Capt.	1766
—— M. T., Asst. Surgn.	1801
Fernbach, H. H. 2nd-Lt.	565
Feroz Ali, Sub-Maj.	998
Feroz Khan, Ressdr.	1684
—— Subdr.	1321
Feroze Khan, Ressldr.	924
—— Jemdr.	1569
—— Jemdr.	1569
—— Jemdr.	1683
Ferrar, M. L., Maj.	104, 108, 211, 1896, 1913
Ferrey, E. M., Lt.	785
Ferriday, T. B., 2nd-Lt.	1007
Ferrier, C. D., Lt.	9, 18, 511
—— I., Capt.	269, 1215, 1511
—— W., Capt.	804
Ferris, G. J., Sub-Asst. Surgn.	1830
—— W. A. T., Maj.	231, 937
—— W. B., Lt.-Col.	56, 179, 1918
—— W. H., 2nd-Lt.	740
Ferry, E. S. H., Maj.	834
Ferst, W. J., 2nd-Lt.	619
Festing, J. E. G., Maj.	35, 664
Fetherstonhaugh A. E. H., Capt.	593
—— M. R., Lt.	399, 937
—— W. A., Bt.-Lt.-Col.	15, 42, 98, 137, 159, 211, 889
—— W. E., Capt. (Temp.)	1723
Fettes, D., Capt.	1724
Few, G., Lt.	597
Fewery, J., 2nd-Lt.	362
Fewtrell, A. H. H., Lt.	367, 925
Ffolkes, E. J. P. B., 2nd-Lt.	326, 366, 895
fforde, A. B., Lt.	609
—— G. F., Lt.	528, 1704
—— J. S. B., Lt.	316, 1471, 1483
Flanagan, G., Lt.	838
Ffrench, R. P. T., Capt.	268, 1487
—— T., Lt.-Col.	57
ffrench-Mullen, J. L.W., Lt.-Col.	131, 155, 190, 1894, 1895
Fiaz Muhammad Khan, Subdr.	1086
Fickling, W. A., 2nd-Lt.	754, 799
Fida Mohammed, Ressdr.	1684
Fiddes, J., Lt.	320, 323, 1359
—— J. M., Capt.	1919
Fiddiman, J. E., 2nd-Lt.	1008
Fido, C. B., Lt.	525
—— G. B., Asst. Surgn.	1798
Field, A. V., 2nd-Lt.	723, 1008
—— C. D., Lt-Col.	135, 158, 195, 870, 1315
—— C. D., Capt.	29
—— C. W., Lt.-Col.	184, 1883, 1911
—— D. M., Maj.	227, 1889, 1913
—— F., Sergt.-Maj.	1879
—— F. C., Lt.	68, 292, 935
—— F. M. S., Lt.	509, 1211
—— G. G., 2nd-Lt.	553, 1521
—— G. L., Lt.	505, 1639
—— H., 2nd-Lt.	554
—— J., 2nd-Lt.	1679
—— J. H., Capt.	829
—— L., Capt.	615
——, L., 2nd-Lt.	739
—— R. T., Lt.	450—54, 1309
—— W. H., Lt.	506, 1151
—— W. J., 2nd-Lt.	546, 1431
—— W. P., Capt.	857
Fielder, G. H., Lt.	435, 1633
—— H. G., 2nd-Lt.	601
Fielden, G., Lt.	587
Fielding, A., Lt.	1701
—— C. H., Capt.	1754
—— H., Lt.	527, 1671, 1704
—— J., 2nd-Lt.	363, 1052
Fieldsend, C. E., Lt.	511, 1235
Fife, D. Mc. D., 2nd-Lt.	331, 1040
Figg, W. G., Lt.	955
Figgins, A. P., Lt.	446—49 1144
Figgate, T. R., Maj.	485, 601
Fill, H. H. G., 2nd-Lt.	7, 354
Filleul, R. R., Lt.	738, 740, 791
Filmer, W. G. H., Lt.	686
Filose, A. A. E., Capt.	280, 947, 949
Filtcroft, E. A. S.	455—59
Finch, C., Lt.-Col.	35, 81, 128, 154, 184
—— C., 2nd-Lt.	564, 1144
—— C. H., 2nd-Lt.	493, 875
—— F. G., *Hony.* Lt.	1843
—— H. C., 2nd-Lt.	45, 356, 1018
—— K. H., 2nd-Lt.	365, 897, 1451
—— P. W., Lt.	410, 1431
—— R. E., 2nd-Lt.	630
—— R. H., Lt.	516
Findlay, A. J., Lt.	517, 1589
—— J., Capt.	1757
—— L. F., Lt.	753, 798
—— M. A., Lt.	536, 1333
Finglas, N. H., Lt.	503
—— U. C. N., 2nd-Lt.	1630
Finlay, J., Lt.	526, 1667
—— J. A., Lt.	303, 1203
—— J. R. P., 2nd-Lt.	327, 1078
—— R. F., Maj.	6, 13, 214, 1255
Finlayson, C. S. M., Lt.	90, 516
—— R. E. M., Lt.	460—62, 1174
—— T., Lt.	455—459

NAMES AND RANK. Page.
Finlayson, W. T., Maj. 1751
Finelli, C. W. F., 2nd-Lt. 551, 1249
Finlinson, H. Capt. 785
Finlow, R. S., 2nd-Lt. 856
Finnie, A. McL., Maj. 244, 1303, 1895
Finnigan, L. D., 2nd-Lt. 547, 1702
—— T. F., Lt. 543
—— W. B., 2nd-Lt. 547
Finnis, B. H., Maj. 869, 216, 1098, 1245
—— F. A., Maj. 62, 74, 101, 633
—— H., Capt. 66, 271, 1235
—— H. C., Capt. 256, 1074, 1898
—— S. W., Capt. 256, 1120
Finny, C. M., Capt. 1713
Firbank, T., Condr. 1861
Firdoskhan, Jemdr. 966
Firman Ali, Subdr. 1011
Firminger, W. K., Revd. 1905
Firoz Ali, Jemdr. 1258
Firoz Din, Jemdr. 999
Firoz Khan, Subdr. 1207
—— Subdr. 1343
—— Jemdr. 1001, 1107
—— Jemdr. 939
—— Jemdr. 1336
—— Jemdr. 990
—— Subdr. 1122
—— Jemdr. 1344
—— Jemdr. 1117
—— Jemdr. 990
—— Jemdr. 1117
—— (I) Subdr. 1489
—— Ressaidr. 939
—— (II) Subdr. 1489
—— Jemdr. 900
—— Jemdr. 1130
Firth, C. W., 2nd-Lt. 735
—— D. G., Capt. 730
—— F. J. W., Capt. 276, 885
—— H., 2nd-Lt. 818
—— H. B., Asst. Surgn. 1802
—— H. V., Lt. 532, 1676
—— J. R., Lt. 798, 806
—— R. A., Lt.-Col. 132, 156, 193
Fischer, L. G., Lt.-Col. 127, 1742
—— N. W., Lt. 516, 1164
—— T. G. T., Lt. 67, 309, 1641
Fish, E. W., Capt. 1718
—— F. M., 2nd-Lt. 624
—— W. H., 2nd-Lt 548, 1471
Fishburn, J., Temp. Lt. 351, 939
Fisher, A. H., 2nd-Lt. 757
—— A. R. G., 2nd-Lt. 474, 1211
—— A. W. 1009
—— B. D., Lt.-Col. 98, 136, 595
—— B. O. F. deC., Lt. 692, 783
—— C. J., Capt. 754
—— C. S., 2nd-Lt. 637
—— C. W., Capt. 753
—— D. G., Capt. 1732
—— F., Lt.-Col. 136, 158, 196
—— F. A. B., Lt. 313, 1167
—— F. B., Lt. 516, 1618
—— G., Condr. 1870
—— G., 2nd-Lt. 640
—— G. T., Capt. 267, 1606
—— H. A., Lt. 537
—— J., Col. 57, 105, 123, 149, 182
—— J., Lt.-Col. 131, 1743, 1901
—— J., 2nd-Lt. 764
—— J. W., Lt. 707
—— L. E. R., 2nd-Lt. 729
—— M., Tempy. Asst. Surgn. 1804
—— M. G., Lt. 17, 33, 735
—— P., Lt. 512, 1457
—— S., Lt. 527, 1701
—— S. M., Lt. 780
—— T. A., Capt. 696
Fisk, E. A., Capt. 501a, 1666
—— C. L., 2nd-Lt. 90
—— L. E., Lt. 537, 1299
Fisk-Aldis, A., Lt. 504, 1267
Fison, C. H., Lt. 420, 681, 1116
Fitch, A. A., Capt. 1718
—— A. P., Condr. 1863
—— F., 2nd-Lt. 1695
—— K. S., 2nd-Lt. 328, 1239
Fitch, W. O., Lt. 540a, 724, 1401
Fitchew, H. P., Lt. 1006
Fitkin, S. A., 2nd-Lt. 797
Fitter, J. H., Condr. 1853
Fitton, C. V., Lt. 1154
Fitz, E. W., Lt. 733
Fitzsimons, G., 2nd-Lt. 720
Fitz Gerald, A., Capt. 101, 257, 1664
—— C. R. L., Maj. 56, 239, 1501
—— D. C. V., Maj. 79, 1752

NAMES AND RANK. Page.
Fitz Gerald, E. A., Maj. 585
—— G., Sub-Condr. 1873
—— G. M., Capt. 272, 911
—— J., Sergt.-Maj. 1879
—— J., Lt. 525
—— J. D., 2nd-Lt. 1475
—— J. J., Lt. 543
—— J. P., 2nd-Lt. 365
—— K. P., 2nd-Lt. 1287
—— L., Sub-Condr. 1855
—— L. A. W., Asst. Surgn. 1797
—— L. C., Lt. 294, 1457
—— P., Condr. 1861
—— S. G. V., 2nd-Lt. 552
—— W. R., Lt. 668, 996
Fitz-Gibbon, A. Fitz G., Lt. 513, 1156
—— L. B., Lt. 513
Fitzhardinge-Berkeley, G., Lt. 531
FitzHerbert, G. A., Lt. 528, 1691
—— R. A., 2nd-Lt. 563
—— R. F., Lt. 505, 1219
Fitz-Hugh, A., 2nd-Lt. 793
Fitz-Maurice, W. H., 2nd-Lt. 328, 1104
Fitzpatrick, C., Condr. 1854
—— C., Lt. 527, 915
—— C. J., Lt. 528
—— E. J., Lt. 518, 1668
—— G. V., Lt. 56, 505, 1630
—— H. J., Lt. 542
—— J. J., 2nd-Lt. 546, 1475
—— J. O'C., Lt. 488, 957
—— J. T., Lt. 40, 540, 1403
—— L., 2nd-Lt. 1701
—— L. F. J., Lt. 549
—— L. J., Lt. 514, 1211
—— L. V., Lt. 293, 1006, 1096
—— M. J. J., Lt. 532, 1215
—— N. T., Capt. 665, 1882
—— P. J., 2nd-Lt. 566
—— P. W., Lt. 523, 1178
—— R. J., Asst. Surgn. 1794
—— R. H. J., 2nd-Lt. 550, 1126
Fitz-William, M. R., Lt. 75
Fitzsimons, G., 2nd-Lt. 785
Flanders, C. M., Capt. 1839
Flanagan, H. J. C., Asst. Surgn. 1795
Flashman, H. C., Lt. 506, 1665
—— T. C., Ty. Capt. 559, 1883
Flatt, L. N., Lt. 519
Flawn, A. H. T., 2nd-Lt. 356
Flaxman, H. J. M., Lt. 503, 1050
Fleet, A. L. E., Capt. 615
Fleming, A. M., Lt.-Col. 140, 1746
—— A. N., Lt.-Col. 135, 1744
—— D. J., 2nd-Lt. 562, 1495-96
—— E. G., Lt. 544
—— F., *Hony.* Maj. 77, 625
—— F., 2nd-Lt. 465, 933
—— G. G., 2nd-Lt. 560
—— G. R., Lt. 504
—— G. T., Lt. 700, 723
—— H. L. O., Asst. Surgn. 1793
—— H. M., Capt. 581
—— J. F., Sr. Asst. Surgn. 1789
—— J. G. G., Lt. 443, 1104
—— J. H., 2nd-Lt. 556, 1195
—— J. K. S., Maj. 10, 1618, 1748
—— M. J., 2nd-Lt. 561, 1473
—— N. B. B., Capt. 1718
—— R. P., Lt. 816
—— R. S. T., Lt. 315, 1355, 1379
—— T., Lt. 515, 1098
—— T., Lt. 1720
—— V. J. P., Lt. 516
—— W. E., Maj 56, 237, 1187
Fleming-Madge, D., 2nd-Lt. 551
Flemington, R. R., Capt. 28, 501a, 1675
Flemming, R. L. V., Capt. 288-290, 1407, 1463
Flemying, B. W. B. T., 2nd-Lt. 854
Fletcher, A. B., Capt. 279, 1038, 1058
—— A. F., Maj. 595
—— F. D., 2nd-Lt. 1693
—— H., Capt. 501a
—— J. A., Lt. 509, 723, 1667
—— R. C., Capt. 279, 1110
—— T. R., Lt.
—— W., Lt. 986
—— W. T., Capt. 738
—— W. T., Capt. 267, 1395
Flett, F., 2nd-Lt. 765
—— R., 2nd-Lt. 324, 1509
Fleuret, F. S., Lt. 685
Fleury-Teulon, C. A., Capt. 585
Flewett, W. E., Lt. 513, 997
Flick, C. L., Lt.-Col. 698

NAMES AND RANK. Page.
Flight, E., Condr. 1870
—— W. E., 2nd-Lt. 618
Flint, A. L. A., Maj. 230, 1662
—— R. P., 2nd-Lt. 796
Flitcroft, E. A. S., Lt. 450, 1150
Flood, F. W. S. P., Lt. 425
—— J. B., 2nd-Lt. 909
—— J. D. M., Maj. 226, 1255
—— J. W. S. P., 2nd-Lt. 761
Florance, D. C. H., Lt. 622
Florence, G. D., 2nd-Lt. 331, 1193
Flower, A. C., Capt. 513, 843
—— E. H., 2nd-Lt. 818
—— H. G. S., Lt. 704, 1034
—— L. T., 2nd-Lt. 540, 1285
Flowerdew, J. L., Condr. 1873
—— R. E., Capt. 1755, 1889
—— S. P., Capt. 836
Flowers, E., Maj. 628
Fliose, A. A. E. 279
Floyd, A. B., Capt. 691
—— H., 2nd-Lt. 90
Floyer, E. A., Lt. 486
Flye, T. H., Lt. 354, 1289
Flynn, A. A. L., Lt. 835
—— A. E., 2nd-Lt. 769, 1245
—— G. E. C., Lt. 506
—— H. H., Revd. 1906
—— J. W., 2nd-Lt. 354, 998
—— T. F. A., Lt. 316
Flyter, M. J., Lt. 84, 542
Focken, F. J. C., 2nd-Lt. 554
Foddy, A. H., Lt. 527
Foden, W. H., Hony. Lt. 855
Fogarty, P. C., Lt. 529, 827, 991
Fogerty, W. P., 2nd-Lt. 327, 1251, 1255
Fogwell, L. B., 2nd-Lt. 367
Folconer, G. A., Lt. 927
Folds, J., 2nd-Lt. 549
Foley, B., Lt.-Col. 855
—— J. E., Capt. 1717
—— M., Sub-Condr. 1867
—— P. F., 2nd-Lt. 666, 996
Folkard, F. T., Lt. 510, 1032
—— W. F. D., Lt. 401
Folkestone, W. P. B., Capt. 755
Fonseca, E. I. A., Asst. Surgn. 1803
—— I., Tempy. Capt. 1767
Fontain, P., 2nd-Lt. 640
Fookes, T. G., Lt. 530, 1219
Fooks, G. E., Lt.-Col. 1786
—— H., Col. 46, 115, 1738
—— H. E., Capt. 636
—— O. J. F., Capt. 593
Foort, F., Sub-Condr. 1863
Foott, F. D., Capt. 769
Foote, B. A. M., Lt. 709
—— H., Revd. 1907
—— H. B., Lt.-Col. 71, 74, 101, 129, 632
Footner, F. L., Maj. 729
Foot, A. B., 2nd-Lt. 556
—— D. W., *Hon.* Lt. 706
—— H. O., *Hon.* Lt. 443, 1604
—— P. B. D. H., Capt. 68, 1004
—— W. J. H., 2nd-Lt. 548, 1439
Foottit, J. A. W., Capt 265, 907, 1126
Foran, E., 2nd-Lt. 770
Forbes, A., Lt. 727, 798, 1054-5
—— A., Maj. 821
—— A. D. C., Lt 488, 935
—— A. E. B., Capt. 528, 818
—— A. F. G., Capt. 277, 909
—— A. G., Capt. 729
—— A. W., Br.-Genl., 61, 120, 122, 153, 181, 1918
—— B. C., Lt. 82, 524
—— C. A., Capt. 485, 887
—— E. E., Maj. 60, 209, 1661
—— E. R. C., Lt. 537, 1026
—— G. H. N., Lt. 521
—— J., Lt. 764
—— K. D. M., 2nd-Lt. 558
—— L., Maj. 236, 1251
—— L. W. Mck., Lt. 304, 1606
—— R. B., Asst. Surgn. 1802
—— R. M. N., Maj. 58, 634
—— T. C., 2nd-Lt. 547, 1673
—— W. G. H., Lt. 785
Forbes-Leith, H. L., 2nd-Lt. 731
Force-Jones, R., 2nd-Lt. 560
Ford, A., 2nd-Lt. 359, 1563
—— A. R., 2nd-Lt. 726
—— B. H., 2nd-Lt. 567, 1678a
—— C. B., Lt. 393, 1415
—— C. W. W., Lt. 291, 1162
—— C. W., 2nd-Lt. 357, 1303

NAMES AND RANK. Page.

Ford, D. O'C., 2nd-Lt. 565
—— E. E., 2nd-Lt. 790
—— E G., Capt. 257, 1435
—— F. E., 2nd-Lt. 1475
—— F. O'N., Lt. 528, 1347
—— F. P., 2nd-Lt. 355
—— G., Lt. 986
—— G. H, 2nd-Lt. 551, 1427
—— G. N, Maj. 105, 240, 1423
—— H., Condr. 1854
—— H. B., Lt.-Col. 131, 155, 192, 868, 1136
—— H. J., Lt. 806
—— J., Sub Condr. 1857
—— J. M. R., Capt. 252, 1261, 1307
—— L. D., Lt. 508
—— M. A., 2nd-Lt. 362
—— R., Sergt.-Maj. 1879
—— R. A., Lt. 353, 980
—— R. S., 2nd-Lt. 734
Forde, A., Capt. 8, 571, 1550
—— F. L., Lt. 737
Forder, H. N., 2nd-Lt. 355, 1511
Fordham, F. P. G., Asst. Surgn. 1799
—— H., Lt. 526
—— H. J. J., Asst. Surgn. 1795
—— W. C. W., Asst. Surgn. 1795
—— W. M., Lt.-Col. 98, 139, 162, 199, 868, 1098
Fordyce, H L. D., Bt.-Col. 125, 153, 190
—— J. F. D., 2nd-Lt. 323, 1151, 1156
Fordyce-Harvey, C. A., 2nd-Lt. 546
Foreman, C. L., Capt. 502, 1665
—— F. C., Lt. 491, 941
Forester, F. J., Lt. 531
Forge, F. W., 2nd-Lt. 710
Forjett, F. H., Lt.-Col. 178, 1918
Formby, R. F. R., Lt.-Col. 57
Forrest, A. E. G., Lt. 320, 365, 875
—— C., 2nd-Lt. 708, 1596
—— F. M. F., Asst. Surgn. 1803
—— G., 2nd-Lt. 476
—— J., Maj. 1749
—— J. D. D., Lt. 455–59, 369
—— L. H., Lt. 302, 1211
Forrester, G. D., 2nd-Lt. 494, 848, 913
—— H. E., Lt. 541, 1189
—— H. W., Capt. 72, 777
—— R. D., Lt. 739
—— W. R., 2nd-Lt. 623
Forshaw, H. W., 2nd-Lt. 556
—— W. T., Lt. 368, 1319
Forster, D., Bt.-Lt.-Col. 51, 98, 107, 663
—— F. J., Lt. 1677
—— G. H., 2nd-Lt. 1008
—— H. R., 2nd-Lt. 1034
—— P. B. B., Lt.-Col. 130, 155, 189
—— R. W. H., 2nd-Lt. 593
—— T., 2nd-Lt. 669, 987
—— W. C. H., Maj. 1746
Forsyth, C. E. P., Lt. 606, 801
—— C. S., Lt. 927
—— D. G., Sub-Condr. 1867
—— I. M., Lt. 322, 1639
—— J. A., 2nd-Lt. 640
—— W. L., Capt. 1755
Forsyth-Forrest, R. S., Lt. 536
Forteath, A. M., Maj. 233, 943, 961
—— F. W. H., Lt.-Col. 43, 133, 157, 193, 1660
Fortescue, A., Lt. 1677
—— L. S., Lt. 505, 1667
Forth, C. O., Lt. 450–54, 1160
Fortt, N H., Lt. 704
—— W. G., Lt. 704
Fortune, F. R., 2nd-Lt. 765
Forward, F. C. M., 2nd-Lt. 326, 366, 1489, 1611
—— W. G., Maj. 695
Foscholo, H. W., Senr. Asst. Surgn. 1787
Foskett, R. C., Lt. 10, 550
—— R. M., 2nd-Lt. 330, 1653
—— W., Lt. 446-49, 1309
Fossett, F. C., 2nd-Lt. 599
Fossey, R. H., Lt. 531, 704
Foster, A. E., 2nd-Lt. 360, 1561
—— C. S., Maj. 16, 37, 241, 945
—— E. C. H., 2nd-Lt. 524, 1343
—— E. L. P., Lt. 318, 1181
—— F. L. 723a
—— G. R., Lt. 582, 599
—— H. B., Maj. 1748
—— H. B., Lt. 492, 965
—— H. E., 2nd-Lt. 1323
—— H. M., Maj. 14, 42, 729
Foster, J. W., Lt. 307, 921, 1007
—— L. A., Lt. 316, 1595
—— R., Maj. 234, 1142, 1880
—— R., 2nd-Lt. 539, 1309
—— R. B. B., Maj. 1747
—— R. C., Lt. 708
—— R. F., Lt. 445, 1291
—— R. G., Lt. 808
—— T., Condr. 1854
—— T. D., 2nd-Lt. 354, 943
—— T. F. V., Bt. Maj. 53
—— W. M. A., Maj. 17, 22, 54, 242, 893
Fothergill, W., 2nd-Lt. 549
Fotheringham, J. A. R., Lt. 720
Foulkes, G. F. F., Maj. 854
—— T. H., Lt.-Col. 130, 1741
Fourdrinner, C. D., 2nd-Lt. 786
Fournier, A. E., Lt. 353, 1676
Foweraker, H. A., 2nd-Lt. 551, 1164
Fowle, J. LeC., Capt. 277, 903
Fowle T. C. W., Capt. 102, 105, 250, 1183, 1889
—— W. T., Lt. 846
Fowler, A. C. O., Lt. 798
—— A. G., 2nd-Lt. 746
—— C. A., Maj.-Genl. 12, 23, 119, 147, 175
—— C. W., 2nd-Lt. 559
—— E. K., Bt.-Maj. 29, 262, 1251
—— E. O., Bt.-Maj. 532, 1695
—— F. J., Maj.-Genl. 12, 39, 119, 147, 174
—— G., Maj. 1748
—— G. H. A., Lt. 445, 968
—— G V., Lt. 530, 1158
—— H. O. W., 2nd-Lt. 324
—— J. C. H., Capt. 854
—— J. T. F., Lt. 705
—— R. St. L., Lt. 595
Fowles, S., *Hony.* Lt. 1847
Fox, B. F., Lt. 351, 1447
—— B. H., Maj. 29, 664
—— C. B., Lt. 542, 1673
—— C. J., Sr. Asst. Surgn. 1788
—— C. J. J., Maj. 655
—— E., Condr. 1854
—— E. B., Capt. 665, 996
—— E. B., Lt. 516, 1166
—— E. C. R., Asst. Surgn. 1793
—— E. L. W., Lt. 322, 1178
—— E. V., Lt.-Col. 760
—— F. E., Capt 702
—— F. N., Capt. 267, 1178
—— F. W. J., Lt. 796
—— G. C., Condr. 1870
—— G. L., Lt. 753
—— H. B., Maj. 501, 602, 1695
—— H. V., Lt. 526, 1505
—— H. W., Ty. Capt. 1723
—— J., Condr. 1852
—— J. McD., 2nd-Lt. 25, 553
—— L. W., Lt.-Col. 35, 137, 159, 197, 651
—— M., 2nd-Lt. 821
—— R. G, 2nd-Lt. 1489
—— R. G., Lt. 581
—— R. deV. R., Lt. 318, 1203
—— R. J., 2nd-Lt. 1489
—— R. P., Maj. 689
—— S. J. V., Asst. Surgn. 1793
—— T. F., 2nd-Lt. 545, 988
—— W. B. O., Lt. 307, 1617
Foxlee, F. J., Lt. 489, 1007
Fox-Male, W. C. E. T., Maj. 107, 239, 1036, 1379
Fox-Pitt, C. Capt. 1731
Fox-Strangways, V., Lt. 318, 968
Foy, F. H., Asst. Surgn. 1795
—— J. R., Asst. Surgn. 1793
Foyster, E. R., Lt. 732
—— F., 2nd-Lt. 564
—— G., Lt. 696
Fradd, L. C. W., Capt. 787
Framji, A., Esq. 89
Framurz Khan, Jemdr. 1368
France, E., Lt. 55, 298a, 1178, 1678a
Francies, J. V., Lt. 516
Francis A. K., 2nd-Lt. 778
—— A. S., 2nd-Lt. 362
—— Antony, Jemdr. 1360
—— C. A., 2nd-Lt 779
—— E. E., Tempy. Capt. 1763
—— G. N., 2nd-Lt. 553
—— H. C., Lt. 721
—— H. F., Capt. 635
Francis H. V., Tempy. Asst. Surgn. 1804
—— J. E., Lt. 697
—— J. L., Sub-Condr. 1857
—— L. A., Asst. Surgn. 1795
—— L. V. DaC., Asst. Surgn. 1797
—— M. H., 2nd-Lt. 324
—— P. E. C., Lt. 450, 721
—— P. V., Ty. Lt. 1776
—— R. F., Capt. 260, 1211
—— R. G., Lt. 722
—— T V., 2nd-Lt. 355, 1247
—— W., 2nd-Lt. 328, 879
—— W. D., Lt. 541, 1251
—— W. J., Lt. 636
—— W. L., 2nd-Lt. 362
Frankau, G. N., Capt. 501a, 753, 1668
Frankish, W. M., Lt. 405, 1054
Frankland, A., Lt. 306, 1407
Franklin, A. D., 2nd-Lt. 719
—— C., Condr. 1861
—— E. A. McA., 2nd-Lt. 303, 1397
—— F. J., Lt. 538
—— G. D., Maj. 1749, 1889
—— H. J., Lt. 1311a, 1070, 1086
—— H. S. E., Br.-Genl. 60, 122, 136, 158, 214, 1003, 1074
—— R. J., Maj. 1711
Franks, H. S., 2nd-Lt. 545, 905, 1371
—— K. F., Capt. 259, 1467
—— S. H., Lt. 493
—— S. W., 2nd-Lt. 568, 1130
—— W. A., Maj.-Genl. 169
Frank, C. M., Lt. 799
—— E. B. P., 2nd-Lt. 572
Fraser, A., Lt. 531
—— A. 2nd-Lt. 653
—— A., Lt. 787
—— A. B., 2nd-Lt. 551, 1415
—— A. C., Maj. 217, 1661
—— A. D., Capt. 485, 903
—— A. D., Capt. 1717
—— A. G. L., Asst. Surgn. 1799
—— A. H. H., Lt. 95, 505, 1677
—— C. D., 2nd-Lt. 552, 1219
—— C. G., 2nd-Lt. 711
—— C. J. S., Lt. 291, 1007, 1239
—— D., Lt. 533, 1331
—— D. DeM. S., Capt. 269, 949
—— D. F., Lt. 617
—— D. F. S. 2nd-Lt. 362
—— D. H., Lt. 536, 1699
—— D. S., Lt. 493
—— E., Lt. 518, 986
—— E. R. D. 2nd-Lt. 1583
—— F. C., Maj. 1753
—— F. J., Lt.-Col. 132, 155, 192
—— G. H., Ty. Capt. 1725
—— H., Capt. 844
—— H. A D., Col. 20, 126, 660, 1900
—— H. A. H., Lt. 598
—— H. F., Lt.-Col. 101, 134, 597
—— H. F. Capt. 1493-94
—— H. J., Lt. 513, 1666
—— H. W. M., Lt. 518, 587
—— I. S., 2nd-Lt. 652
—— J., Capt. 1732
—— J., 2nd-Lt. 618
—— J., 2nd-Lt. 90
—— J. A., Lt. 583
—— J. D., Capt. 280-287, 935
—— J. W., 2nd-Lt. 363, 1026
—— K. M., Lt. 533, 1606
—— R., Lt. 352, 1092
—— R A. J., 2nd-Lt. 548, 724, 1557
—— R. J. L., Lt. 1720
—— R. T., Capt. 602
—— S. G. G., Capt. 265, 1331, 1509, 1587
—— S. M., the Hon. Sir 841, 1900
—— T., Lt.-Col. 97, 102, 121, 126, 660
—— T., Capt. 818
—— T., Ty.-Lt. 1726
—— T. C, Maj. 15, 71, 633, 1884
—— T. F. G., 2nd-Lt. 326, 366, 879
—— W A. K., Bt.-Maj. 104, 258, 949
—— W. C., Lt 537, 1132, 1676
—— W. G., 2nd-Lt. 556
—— W. J, Maj. 1752
—— W. L., Lt. 540a
—— W. S., Capt. 826
Fraser Tytler, W. K., Capt. 269, 921
Fratel, E., Asst. Surgn. 1797
—— W., Asst. Surgn. 1799

NAMES AND RANKS. Page.

Frater, W. M., 2nd-Lt. 562
Frattini, G. N., Lt. . . . 530, 998
Frazer, D. S., Capt. . . . 279, 909
—— F. A., Lt.-Col. . . 752, 783
—— G. S., Col. . 40, 125, 150, 185
—— W. G., 2nd-Lt. 1708
Frederick, C. W. E., Asst. Surgn 1797
Free, C. W., Lt. . . . 368, 970
Freeborn, F. V., 2nd-Lt. . . . 720
Freed, E., Condr. 1854
Freegard, S. E. L., 2nd-Lt. . . 362
Freel, J., *Hony.* Maj. . . . 761
Freeland, H. F. E., Maj.-Gen. 119, 133, 662, 1881
—— J. C., Bt. Lt.-Col. 51, 99, 160, 213, 1162
—— R. A. B., Lt. 380
Freeman, A. B., 2nd-Lt. . . . 555
—— A. F. A., Lt. 729
—— E. C., Lt. . . . 351, 1361
—— J. St. Q., Tempy. Asst.-Surgn. 1804
—— K. C., 2nd-Lt. . . . 624
—— M. L., 2nd-Lt. . . . 802
—— O. R. C., Asst. Surgn. . 1798
—— T. H., *Hony* Lt. . . 1850
—— W. E., Sub-Condr. . . 1874
Fremantle, H. E., Lt. . . . 533
Freeman-Thomas, *The Hon.* I. B., Lt. 309, 879
Freer, R. C., Capt. 615
Freer-Smith, D. F., Lt. . . . 585
Freeston, W. L., Sub-Condr. . . 1856
Freeth, C.J.D., Lt.-Col. 72, 101, 137, 632
Fregaskis, G., 2nd-Lt. . . . 1331
Freitas, A. M., Tempy. Lt. . . 1774
Freke, C. G., Lt. 655
Fremantle, A. F., Lt. . . . 488
Fremlin, H. J. L., Lt. . . 487, 911
—— R. W., Capt. . . 486, 889
French, A. G., Nursing Sister . 1735
—— C. R., 2nd-Lt. . . 473, 1385
—— F. G., Lt. . . 514, 1895
—— F. H., Lt. . . . 308, 1611
—— H. J., Capt. . . . 135, 789
—— L., 2nd-Lt. 361
—— V. C., Maj. 852
Frenchman, J. E., Lt. . . . 1780
Frend, G. V. R., Lt. . 533, 1120, 1569
Frere, A. G., Maj. . 235, 1351, 1559
—— R. T., Capt. . 36, 691, 783
Fressanges, G. F., Hony. Lt. . 1850
Freston, T. A., 2nd-Lt. . . 327, 366
Fretz, W. L. E., Maj. . 37, 1713
Frew, J. P., 2nd-Lt. 559
Frewin, A. R., 2nd-Lt. . . . 760
Fricker, E. S. G., Lt. . 521, 1585
—— L., Hony. Capt. . . . 790
Friedlander, O. A., Lt. . . 669, 997
—— W., 2nd-Lt. . . . 626
Friel, B. H., 2nd-Lt. 786
—— R., 2nd-Lt. 554
Friend, A. L. I., Capt. . . . 583
—— E. A., 2nd-Lt. . . 551, 1333
—— O. F., Lt. 623
—— R. H., 2nd-Lt. . . . 1007a
Frier
Frisby, F., 2nd-Lt. . . . 780, 799
—— W. T., 2nd-Lt. . . 565, 1678a
Fritchley, G. B., Lt. . . 534, 998
Frith, F. C. A., Capt. . . . 1718
—— H. E., 2nd-Lt. . . . 709
—— J. G., Lt. 302, 1501
—— J. W., Lt. . . 314, 1635, 1657
—— N. B. C., 2nd-Lt. . . 90, 779
—— W. H., Lt. . . . 759, 1885
Frizelle, J., Lt.-Col. 140, 162, 200, 1896, 1912
Frizzel, W. G., 2nd-Lt. . . . 1090
Froom, C., Lt. 1679
Frossard, C. B. S., Lt. . . 696, 795
Frost, A. T., Maj. 1712
—— C. C. J., Lt. . . 310, 1007, 1647
—— D. M., Asst. Surgn. . . 1797
—— E. G., Capt. 679
—— F. D., Bt. Maj. . . 259, 1664
—— F. H., Lt. . . . 543, 1674
—— F. J. T., Lt. . . . 78, 491
—— G. H., 2nd-Lt. 558
—— G. H., Lt.-Col. . . 129, 1742
—— H. B., Lt. . . . 543, 1152
—— R. L., Lt. . . 317, 1255, 1287
—— W. A., Capt. 1713
Froud, H. G., 2nd-Lt. . . . 637
Froy, W. A., Lt. . . . . . .
Fruvall, H. A., Asst. Surgn. . 1801

NAMES AND RANKS. Page.

Fry, A. B., Maj. 1746
—— A. S., Capt. 1759
—— C. I., Maj.-Genl. 118, 146, 173, 1497
—— D. C., Capt. 798
—— E. N., 2nd-Lt. . . . 557, 1078
—— F. A., Sub-Condr. . . . 1874
—— G. C., Lt. . . 313, 943, 961
—— H. R., Lt. . . . 533, 1102
—— O. M., Lt. 749
—— W. J., Lt. 525, 1658
Fryer, F. W., *Hony.* Lt. . . . 710
—— G. W. S., Lt.-Col. . 181, 1918
—— J., Maj. 587
—— P. J., 2nd-Lt. 76
Fuge, R. D., 2nd-Lt. 546
—— T. L. M., Capt. . . 502, 1517
Fulcher, E. W. B., Lt. . . . 1007
—— G. W., Lt. . . . 286, 909
Fulford, G. L., Lt. 700
Fuller, A. G., Lt. 537
—— B. H., 2nd-Lt. 1283
—— C. A., Lt.-Col. . . . 608, 861
—— E. T., 2nd-Lt. . . . 760
—— G., 2nd-Lt. 1627
—— H. F., Maj. . . . 244, 1663
—— J. B., 2nd-Lt. . . . .
—— R. W., 2nd-Lt. . . . 547, 1705
—— T. A., Lt. 780
Fullerton, G. A., 2nd-Lt. . . 326, 1614
—— J. P., Capt. . . 279, 1187
—— L. W., Lt. . . . 353, 980
—— R. A. D., Lt. 587
Fullford, H., Sub-Condr. . . . 1864
Fulton, C. F. V., Capt. . . 367, 689
—— C. M., Capt. . 103, 273, 937
—— E. J., Capt. . . . 279, 875
—— E. A. C., Lt. . . 7, 397, 1613
—— F. C., Col. 35
—— F. C. R., Bt.-Col. 31, 125, 153, 192, 1660
—— G. K., Lt. 1720
—— J. D., Capt. . . . 267, 1124
—— J. O., Lt. . . . 294, 1595
—— R., Lt. . . 399, 683, 1235
—— T., Lt. 529, 1427
Furey, H. G., Asst. Surgn. . . 1800
—— T. F., 2nd-Lt. . . 562, 1703
Furguson, W. E., 2nd-Lt. . 550, 1701
Furlong, G. O., Capt. 623
—— M., Col. 177
Furminger, H. J., Lt. . 455—59, 1126
Furneaux, T. E., Lt. . . 515, 1042
Furner, A. S., 2nd-Lt. . . . 1008
Furnell, Ty. H. A., Condr. . . 1868
Furness, H. G., Lt. . . . 523, 1459
—— W. S., Lt. . . . 294, 1026
Furney, J. L., Bt.-Lt.-Col. 135, 158, 207, 1106, 1116
—— S. K., Lt. 389, 1241
Furnivall, J. S., Lt. 827
Fursdon, A. A., Lt. . . . 698, 700
Furze, G., 2nd-Lt. 835
—— T. E., Capt. 501a
Fussell, H. S., Lt. 393
Futers, S. H., 2nd-Lt. . . . .
Fyfe, W. S., 2nd-Lt. 550
Fyffe, E. L., Capt. 1713
—— F. C., 2nd-Lt. 978
—— R. S., Lt. Revd. . . . 1908
Fyldes, G. B., Lt. . . . 380, 1203
Fyson, P. F., 2nd-Lt. 801
Fyzee, A. A. H., Tempy. Capt. 1765

G

Gabb, J. D., Lt. . . 292, 681, 1311
Gabbett, A. C., Hony. Lt.-Col. 57, 1919
—— K. J., Capt. 76, 273, 1110, 1886
Gabdhoo, Jemdr. 1320
Gabriel, C. H., Maj. . 237, 1889, 1913
—— E. V., Maj. 52, 826
Gaby, C. O., Capt. 843
—— V. F., 2nd-Lt. 740
Gade, O. H., Asst. Surgn. . . 1798
Gadgil, B. B., Ty. Capt. . . . 1771
—— S. B., Ty. Capt. . . . 1766
Gadney, H. G., Lt. 778
Gadsden, C. H., Lt. 541
Gadsdon, H. A., Lt. . . 531, 1668
Gafur Khan, Jemdr. 882
Gagan Sing Gurung, Jemdr. . 1602
Gagan Bahadursahi, Jemdr. . . 1627
Gagan Singh, Jemdr. 1235
Gaggan Ram, Subdr. 1583
Gage, A. T., Lt.-Col. . . 138, 1745
Gagliardi, F. B., Lt. . . . 82, 532

NAMES AND RANKS. Page.

Gahan, C. G. D., Capt. 1664
—— E. C. J., Lt. 850
—— J., Lt. 380
—— P. J., Lt. 850
Gal, N. J., *Tempy.* Capt. . . . 1765
Gaillard, P., 2nd-Lt. . . . 754, 799
Gain, C. T., 2nd-Lt. . . . 561, 1678a
—— J. C., Lt. 397, 1187
Gains, A. T., 2nd-Lt. 750
Gainda Singh, Risldr. 925
—— Jemdr. 1163
—— Jemdr. 1288
Gainda, Jemdr. 1476
Gainsford, D., 2nd-Lt. . . 562, 1511
Gaisford, P., Capt. . . . 274, 1144
Gait, E. A., Lt.-Col. *Hon. Sir* 601, 855, 1894
Gajbahadur Thapa, Jemdr. . . 1897
—— Subdr. . . . 1596
Gajbir Rana, Jemdr. 1625
—— Jemdr. 1649
Gajbir Thapa, Subdr. 1607
Gajendar Singh, Jemdr. . . . 1252
Gajman Rana, Jemdr. 1604
Gajraj Rai, Subdr. 1630
Gajraj Singh, Subdr. 1031
—— Gorkha, Sub-Asst. Surgn. 1820
Gakhar, Jemdr. 1252
Galbraith, E. D., Maj. 235, 1243, 1567
—— D. H. A. 1493
—— I. W., 2nd-Lt. . . 315, 911
Gale, A. C., Maj. 644
—— B., Capt. 1755
—— B. M. B., Capt. . . . . .
Gale, E. S., Lt.-Col. 42, 139, 161, 200, 872, 1647
—— H. C., 2nd-Lt. 1371
—— H. J. G., Maj. 634
—— J. F., Sub-Condr. . . . 1863
—— R. C., Asst. Surgn. . . 1802
Gale, S. T., 2nd-Lt. 684
Galodu, Subdr.-Maj. 1565
Gall, L., Maj. 243, 921
Galland, J. F., 2nd-Lt. . . . 1088
Gallichan, H. N., 2nd-Lt. . . 710
Galliene, J. W. H., Lt. . . 540, 1677
Galloway, F. W., Capt. . . . 816
—— H. J., Sr. Asst. Surgn. . 1789
—— J., Capt. . . . 659, 781
—— W. C., Lt. . . . 296, 1403
Galvin, A., Lt. 506, 1170
—— C. M., Lt. . . . 526, 1701
—— H. Fitz M., Lt. . . . 504
—— J. S., Ty. Capt. . . . 1770
—— L. E., Lt. . . . 504, 1375
—— M., Sr. Asst. Surgn. . . 1788
—— O. H., Lt. . . 76, 528, 1884
—— P. H., Lt. 516a
—— W. H., 2nd-Lt. 552
Gama, Jemdr. 1681
Guman Khan Subdr. 1413
Gambir Sing Pun, Subdr.-Maj. 1624
—— Sing Thapa, Jemdr. . . 1604
Gambhir Thapa, Jemdr. . . . 1625
Gamble, A., Lt. 692
—— H. C., 2nd-Lt. 565
—— L., Lt. 423, 1614
—— M. F., Lt. 541, 1676
—— W. M. F., Lt. . . 67, 400—62
—— V. F., Lt. . . 506, 622, 1678
Gamblin, W. H., Ty. 2nd-Lt. . . 623
Game, S. W., Lt. 535, 1331
Gamer Sing Bura, Jemdr. . . . 1602
—— Gurung, Subdr. . . 1602
—— Thapa, Jemdr. . . 1608
Gammon, F. G., 2nd-Lt. . . 557, 1707
—— J. C., Capt. . . . 501, 997
Gamon, R. B., Lt. 534, 1399
Gamparsad Thapa, Jemdr. . . 1625
Ganapathy, C. M., Ty. Capt. . . 1765
Ganda, Subdr. 1098
—— R. C., Ty. Lt. 1781
Gandaki, Jemdr. 1184
Ganda Ram Chauhan, Sub-Asst. Surgn. 1817
Ganda Singh, Resaldr. 898
—— Resaldr. 918
—— Subdr. 1136
—— Ressdr. 886
—— Jamdr. 1117
—— Jemdr. 966
—— Subdr. 1702
—— Ressdr. 1686
—— Jessds. 900
—— Subdr. 1215
—— Jemdr. 1510
—— Risaldr. 893

Names and Rank.	Page.
Ganda Ram, Sub-Asst. Surgn.	1817
Ganda Ram Chuhan, Sub.-Asst. Surgn.	1817
Gandarbh Sing, Jemdr.	893
Gane Gurung, Jemdr.	1608
Ganes, Jemdr.	1476
Ganesaiyer, R., Jemdr.	1012
Ganesh Bahadur Khattri, Subdr.	1642
Ganesh Das, Sub-Asst. Surgn.	1813
—— Das Bhandhari, Sub-Asst. Surgn.	1819
—— Hari Patange, Sub-Asst. Surgn.	1832
—— Raghunath Shirodker, Sub-Asst. Surgn.	1829
—— Singh, Subdr.	...
—— Jemdr	1044
—— Subdr.	1559
—— Jemdr.	1272
—— Subdr.	1157
—— Vasudeo Sakrikar, Sub. Asst. Surgn.	1832
Ganesha, Jemdr.	1583
Ganesha Singh, Jemdr.	1149
—— Subdr	1111
—— Jemdr.	900
Ganesh Ram, Subdr.	1472
—— Jemdr.	1583
Ganeshi Ram, Subdr.	1455
—— Jemdr.	1457
—— Singh, Jemdr.	1032
Ganga Bishan, Resdr.	914
—— Baksh Singh, Jemdr.	1043
—— Dat, Risldr.-Maj.	878
—— Parshad Sukul, Subdr. Maj.	1026
—— Parsad, Sub-Asst. Surgn.	1815
—— Pershad, Risldr.	924
—— Tewari, Jemdr	1018
—— Ram, Subdr.	1188
—— Subdr.-Maj.	1413
—— Jemdr.	1192
—— Subdr.	1193
—— Jemdr.	1476
—— Jemdr.	1365
—— Jemdr.	1039
—— Jemdr.	1480
—— Bhaguji, Jemdr.	1472
—— Kadam, Jemdr.	1436
—— Ram, Ressdr.	1705
—— Ram Hariba, Sr. Sub-Asst. Surgn.	1829
—— Ram Subdr	1252
—— Ram Singh, Subdr.-Maj.	1412, 1586
—— Singh, Jemdr.	158
—— Jemdr.	1385
—— Jemdr.	1577
—— Sub-Asst. Surgn.	1815
—— Sub-Asst. Surgn.	1818
—— Sub-Asst. Surgn.	1816
—— Jemdr.	1344
—— Punn, Sub.-Asst. Surgn.	1818
—— Sewak Pande, *Hony.* Capt.	2102
—— Sing Rawat, Subdr. Maj.	1225
Gangabir Rana, Subdr.	1630
Ganga Dhar, Jemdr.	1476
Gangadhar Singh, Subdr.	457
Gangaji Rao Khanvilkar, Jemdr.	999
Gangaram Khattri, Subdr.	1596
—— Bhaguji, Jemdr.	1472
—— Ramkrishna Narba, Sub-Asst. Surgn.	1831
—— Thapa, Subdr.	1657
Gangayya, Subdr.	1265
Ganguli, P. P., Lt.	1774
Gani Ishmael Nayarene, Sub.-Asst. Surgn.	1819
Ganj Sing Gurung, Jemdr.	1615
Ganja Singh Gurung, Jemdr.	1657
Ganniclift, A.L., Lt.	415, 1193
Gannon, J. C. R., Maj.	77, 242, 919
Ganpatdayal, Tewari, Jemdr.	1017
Ganpat Kanoj Rao Rana, Sub-Asst. Surgn.	1831
—— Krishnaji, Subdr.	1471
—— Mahadeo, Subdr.	999
—— Ramaya Guru, Sub.-Asst. Surgn.	1832
—— Rao Dalwe, Jemdr.	1416
—— Rao More, Jemdr.	999, 1444
—— Rae, Sub-Asst. Surgn.	1814
—— Rao Raghunath Rajwade, Col.	2, 111
Ganpat Singh, Ressdr.	974
Ganpat Singh, Ressdr.	940
—— Jemdr.	1499
—— Jemdr.	1031
—— Jemdr.	1024
—— Ressdr.	1682
—— Risaldar	1686
—— Jemdr.	1489
—— Jemdr.	1420
Ganpatrao Jadhao, Jemdr.	999
Gant, F. C., Maj.	778
—— T., 2nd-Lt.	743
Gantzer, J. F., Ty. 2nd-Lt.	571a
Ganuji Bagwe, Jemdr.	1444
Ganu Panaskar, Jemdr.	1424
Gannu, Jemdr.	1567
Gapper, J. E. B., Lt.	536, 1407
Garbett, C. H. V., Col.	179, 1917
—— D. A., Lt.	302, 1391
—— H. C., Maj.	608
—— R. B. L., 2nd-Lt.	355
Garbha, Jemdr.	1117
—— Singh, Subdr.	1150
Garbhe Limbu, Subdr.	1648
Garbhey Pun, Jemdr.	1637
Gard, H. M., 2nd-Lt.	562
Garcia-Henry, A., 2nd-Lt.	666, 1009
Garden, R. L., Lt.	435, 1517
Gardener, G. O. T., 2nd-Lt.	560, 1678
—— H. W., Capt.	1725
Gardiner, A. B., Senr. Asst. Surgn.	1787
—— A. L., 2nd-Lt.	919
—— D., 2nd-Lt.	788
—— D. A., Capt.	636
—— D. I., 2nd-Lt.	540, 1499
—— E. J. R., 2nd-Lt.	852
—— H. W., Maj.	74, 101, 633
—— H. J., 2nd-Lt.	735a
—— R., Lt.-Col.	99, 103, 139, 160, 199, 869, 1235
—— R. C., Lt.	508, 1665
—— W. A., Capt.	260, 1613
—— W. F., Ty. Lt.	
Gardner, A., Capt.	636
—— C. F., Capt.	263, 1162
—— C. P., Capt	67, 256, 1307
—— G., *Hony.* Lt.	1843
—— J. Capt.	27
—— J., Ty. 2nd-Lt.	669, 997
—— J. H. D., 2nd-Lt.	332, 1636
—— J. M. S., Capt.	279, 1491, 1495
—— J. W. B., Sub-Condr.	1874
—— K. A. R., 2nd-Lt.	629
—— L., 2nd-Lt.	754
—— O. J., 2nd-Lt.	550, 1223
—— R. G., Capt.	263, 1120
—— R. J., Capt.	1732
—— T. A. A., 2nd-Lt.	731a
Garg, G. C. Ty. Lt.	1782
Garewal, A. S., Ty. Lt.	1777
Garforth, W., Maj.	52, 664, 1882
Garkha, *Hony.* Capt.	2103
Garland, A., Revd.	1908
—— A. G. L., Lt	1575
—— A. S T., 2nd-Lt.	740
—— E. A., 2nd-Lt.	709
Garle, H. A., Lt.	628
—— J. A. B., Capt.	485, 907
Garner, J. D., Lt.	738
——, E. M., 2nd-Lt.	475, 1263
Garnett, C. L. S., Capt.	658
—— R. T., Ty. Lt.	638
Garrad, A. T., Sub-Condr.	1864
—— W. E., Capt.	776a
—— W. F., the Revd.	1907
Garratt, B. F., Maj.	238, 1082
—— G. T., Lt.	488
—— H. S., Lt.-Col.	134, 137, 194, 1415
—— R., Maj.	209, 1900, 1913
—— R. B.	1009
Garratty, M. C., 2nd-Lt.	563, 1589
Garrett, A. ff., Maj.	662, 1881
—— C. S., 2nd-Lt.	788
—— E. C. McN., Capt.	508, 1698
—— H. L. O., Lt.	82, 540a, 805
—— J. E., Capt.	720
—— J. N., 2nd-Lt.	357, 1271
—— J. T., 2nd-Lt.	328, 903
—— K. A., Capt.	279, 1227
—— R. C., Maj.	634, 972
—— R. S., Capt.	826
—— R. V., Lt.-Col.	57
Garrity, W. H., Sub-Condr.	1857
Garside, H., 2nd-Lt.	754
Garrod, H. J. J., Asst. Surgn.	1793
—— W. F., Lt.	440, 1122
Garrow, R. K., 2nd-Lt.	332, 566a, 1054
—— V. J., Lt.	387, 1044
Garson, I. G. M., Lt.	76, 715, 1884
Garstin, E. L. L., Lt.	528, 1705
—— H. A., Capt.	267, 915
—— H. E., Lt.-Col.	74, 134, 613
—— W. A MacD., Maj.	227, 1901
Garthwaite, E. J. K., Lt.	515, 1567
—— T. E., Asst. Surgn.	1802
Gartside-Tipping, R. F., Lt.-Col.	179, 1917
Garvie, J., Col.	45, 126, 1638
—— K. C., 2nd-Lt.	324, 1617
Garwood, F. S., Maj.	662, 996
Gaskell, H. S., Maj.	663
—— R. W., Maj.	224, 1050
—— T. K., Maj.	7, 206
—— W., Capt.	806
Gaskill, T., Lt.	642
Gaskin, A. E., 2nd-Lt.	708
Gasper, B. C., 2nd-Lt.	556 1219
—— H. LeG., Lt.	543a
Gass, F. T., Lt.	320, 1457
Gasson, H. J., Lt.	543a, 993
—— H T., *Hony.* Maj.	993, 1848
—— L. B., 2nd-Lt.	931
Gastrell, E. T., Col.	57, 122, 149, 181
—— E. H., Lt.	313, 957
Gatacre, J., Maj.-Genl. *Sir.*	173, 1487
Gate, H. J., Ty. Lt.	622
Gatehouse, R. O. A., Lt.	310, 927
Gateley, T. P., 2nd-Lt.	563
Gates, T., Condr.	1863
Gatherer, A. H. N., Capt.	267, 1004, 1207
—— R. G. T., Maj.	224, 872, 1647, 1653
Gauhar Ali, Ressdr.	908
—— Rahman Khan, Ressdr.	934
Gaul, A. C., the Revd.	1907
—— A. D., Lt.	537, 1596
Gault, C., Lt.	317a, 1116
—— J., Lt.	399, 1164
Gaunt, A. F., 2nd-Lt.	565
—— C. H., Capt.	827
—— C. R., Lt.-Col.	766
Gaur Sing Negi, Subdr.	1181
—— Jemdr.	1181
—— Rawat, Jemdr.	1182
Gauri Charan, Subdr.	1259
—— Prasad Varma, Ty. Sub-Asst. Surgn.	1833
—— Saran, Jamader	1189
Gauri Shankar Dube, Subdr.	989
Gaus Muhammed, Subdr.	1515
Gaussen, J. R., Lt.-Col.	139, 162, 199, 867, 879, 951
Gavaghan, E. L., Lt.	489, 929
Gavin, W. L., 2nd-Lt.	784
Gavin Jones, T. F., Lt.-Col.	608
Gaya Parshad Saksena, Sub-Asst. Surgn.	1820
Gawld, M. A.	1009
Gawler, A. E. J., 2nd-Lt.	705
Gawthorne, C. H., 2nd-Lt.	546, 1297
Gaya Singh	1384
—— Bux Singh, Jemdr.	1047
Gayadin Sharma, Sub-Asst. Surgn.	1810
Gayand Singh, Jemdr.	1396
Gayer, E. C., Capt.	850
—— J. H. C., Lt.	295, 1665
—— W. M., Lt.	637, 972
Gaynor, S. S., Asst. Surgn.	1800
Gayton, C., 2nd-Lt.	356, 1471
Gazder, J. D., Ty. Lt.	1782
Gazzard, T., Lt.	300, 1148, 1156
Geach, E. S., Lt.	687
—— P., Lt.	686
Geard, R. C., Capt.	273, 903
Gear, L. E. H., Lt.	1463
Geary, A. B., 2nd-Lt.	56
—— F., 2nd-Lt.	629
—— H. N. G., Capt.	665
—— H. V., Lt.	505
Gebbie, F. St J., 2nd-Lt.	849
Geddes, T., Lt.	776, 785
—— T. J. R., 2nd Lt.	324
Geddie, D. W., 2nd-Lt-	473
Gee, F. W., Lt.-Col.	61, 128, 1740
—— T., Ty. Lt.	638
—— W. C., Lt.	536
—— W. H. S., 2nd-Lt.	324, 1215
Geere, A. J. K., 2nd-Lt.	331
Geeson, I., 2nd-Lt.	754
Gega, Jemdr.	1192
Geidt, E. W., Capt.	274, 1431
Gell, H. V., Capt.	273, 1287
—— P. F., Maj.	224, 901
—— W. T., *Hony.* Lt.	21, 1848
Gelston, J. E. H., Lt.	380, 875

Names and Rank. Page.

Gem, G., Lt.-Col. . . 138, 726
Gemmell, A., 2nd-Lt. 474
—— G., Lt. . . . 517, 1509
—— A., 2nd-Lt. . . . 552, 1497
—— G. W., Lt. . . 300, 1030
—— J. C., 2nd-Lt. 1140
Gemmill, W., Capt. 1726
Gemnill, A., Lt. 699
Gena Jadhow, Jemdr. 1440
Gena Rawat, Subdr. 1483
—— Subdr. 1471
Genda Jat, Subdr. 1479
—— Singh, Jemdr. 1031
Geddie, D. W., 2nd-Lt. 1413
Gehna Singh, Jemdr. 1887
Gennak Lakhnak, Jemdr. . . . 1447
Gentle, F. W., 2nd-Lt. 701
—— C. E., The Revd. . 1908
Genu Pawar, Jemdr. 1443
—— Jadhav, Jemdr. 1436
Geoghegan, F.E., Lt.-Col. 40, 133, 157, 193, 1660
—— G., 2nd-Lt. . 69, 997
—— N. M., Lt.-Col. 107, 136, 159, 211, 1363
—— O. A., Maj. 101, 220, 1662
—— S., Lt.-Col. . . 98, 105, 129, 154, 187
George, A., Lt. 445, 1279
—— A. B., Lt. 515, 1670
—— A. B., Esq. 91
—— A. C., 2nd-Lt. 356
—— A. C., 2nd-Lt. 710
—— A. G., 2nd-Lt. . . 560, 1895
—— A. W., Ty. Asst.-Surgn. . 1804
—— D. S., Maj. . . 33, 212, 1191
—— **G. Aseeravatham,** Jemdr. 1308
—— H. J., Lt. 789
—— H. S., Lt. . . . 539, 988
—— J. G., 2nd-Lt. . . 567, 1678a
—— N. M., Lt. . . 513, 1665
—— R., Lt. . . . 314, 947
—— R. E. G., Lt. . . 491, 949
—— S. P., 2nd-Lt. . . 328, 1164
—— W. J., 2nd-Lt. . 549, 1417
Georges, H. W.E., Col. 181
Geraghty, P., Condr. 1853
Gerard, E. L., Lt. 543
—— M. F. M. J., Capt. 89, 747
Gerard-Smith, V., 2nd-Lt. . 551 1604
Gernon, A. W., Asst. Surgn. . . 1794
—— L. P., Asst. Surgn. . . 1795
Gerrard, C. B., Ty. Lt. . . . 1726
—— D. M., 2nd-Lt. 1371
—— E. F., Lt. . 521, 1673, 1702
—— E. F. H., 2nd-Lt. 540, 1703
—— F. W., Lt. 511
—— H. J. F., Lt. 450-54, 624, 916
—— R. J., Lt. . 316, 1152, 1289
Gerrey, G. W., 2nd-Lt. . . 360, 1114
Gervers, F. R. S., Maj 662
Gething, L. R., Lt. . . . 537, 1363
Ghafaran Khan, Jemdr. 1514
Ghafoor Khan, Jemdr. 1035
Ghafur Khan, Subdr. . . 1096, 1432
—— Jemdr. 1133
—— **Shah,** Jemdr. 7, 966
Ghagga Singh, Jemdr. 1103
Ghamandi, Jemdr. 1462
Ghamna Singh, Subdr. 1074
Ghanaiya Singh, Jemdr. . . . 1114
—— Jemdr. . . . 1432
Ghani Khan, Subdr. 1432
Ghansar Sing, Risdr. 1898
Ghansham, Jemdr. 1321
—— Parab, Jemdr. . . . 1460
Ghansyam, Jemdr. 1388
Gharam Singh, Ressdr. 890
Gharpurey, K. G., Maj. . . . 1752
Gharsi Ram, Jemdr. 1480
Ghasi, Jemdr. 1193
Ghaus Muhammad, Sr. Sub-Asst. Surgn. 1808
—— Khan, Ris. Maj. . . 31
Ghazanfar Ali, Tr. Vety. Asst. . 1687
Ghazi Khan, Jemdr. 1208
—— Jemdr. 1245
Ghebe Khan, Jemdr. 1288
—— Jemdr. 1336
—— Jemdr. 1502
—— Subdr. 1154
—— Jemdr. 1409
Gheba Khan, Sub-Maj. 1243
—— Jemdr 1132

Names and Rank. Page.

Gheba Khan, Jemdr. 1514
Ghena Singh, Jemdr. 1511
Ghirla Thapa, Jemdr. 1619
Ghisa Singh, Jemdr. 1452
—— Jemdr. 1583
Gholam Ali Khan, Mirzada . . 111
Gholam Kadar Khan, Ressdr. . 1686
Ghose, K. M., Ty. Lt. 1779
Ghosh, A. M., Ty. Capt. . . . 1767
—— B. C., Ty. Lt. 1777
—— J. M., Ty. Lt. 1775
—— L. M. 1781
—— N. N. Ty., Lt. 1782
—— M. C., Ty. Capt. . . . 1768
—— N., Ty. Lt. 1775
—— N. N., Ty. Lt. 1774
—— N. N., Ty. Lt. 1781
—— N. N., Ty. Lt. 1782
—— P., Ty. Lt. 1775
—— P. M., Ty. Lt. 1781
—— P. N., Ty. Capt. . . . 1765
—— S., Ty. Capt. 1772
—— S. M., Ty. Capt. . . . 1764
—— S. N., Ty. Lt. 1774
—— S. N., Ty. Lt. 1775
Ghoshal, L., Ty. Lt. 1782
Ghulab Khan, Subdr. 1577
Ghulam Ahmad, Sub-Asst. Surgn. 1811
—— Jemdr. . . . 1510
—— **Akbar Shah,** Jemdr. . . 942
—— **Ali,** Subdr. 1183
—— **Ali,** Subdr. 1427
—— Shah, Jemdr. 10 2
—— **Asghar Khan Khalidie,** Sub. Asst. Surgn. 1818
Ghulam Ali Khan, Jemdr. . . . 1473
—— Subdr.-Maj. . . 1142
—— Jemdr. . . . 958
—— **Dastgir Khan,** Sr. Sub-Asst Surgn. 1808
—— Risldr.-Maj. . 930
—— **Habib,** Subdr. 1347
—— **Haidar,** Jemdr. 1096
—— Jemdr. I. 1096
—— Jemdr. II 1095
—— Jemdr 1344
—— Jemdr. 918
—— **Haidar Khan,** Subdr. . 1035
—— Jemdr. . . . 999
—— Ressdr. 1684, 1686
—— Subdr. . . 1256
—— Sub-Asst. Surgn. . 1811
—— **Jemdr.** 1493
—— **Jemdr.** 946
—— **Hasan Khan,** Jemdr. . . 940
—— Jemdr. . . . 975
—— **Hassan,** Subdr. 1285
—— **Shah,** Jemdr. . . 942
—— **Shah, Subdr.** . . 1379
—— **Husain,** Sub-Asst. Surgn. 1811
—— Subdr. 1680
—— **Khan,** Ressdr. 1683
—— Subdr. 1108
—— Khan, Jemdr. . . 1289
Ghulam Husain, Jemdr. 1124
—— Subdr. 1333
—— Jemdr. 1493
—— *Hony.* **Capt.** . . 2102
—— Ressldr. . . . 912
—— Subdr. 1320
—— Subdr. 1507
—— Jemdr. 1122
—— **Jafir Shah,** Jemdr. 894
—— **Jafer Shah,** Jemdr. 1104
—— **Jilani,** Jemdr. 974
—— Subdr. 1707
—— Ressdr. 1687
—— **Khan,** Jemdr. . . 1395
Ghulam Kadir Khan, Ressdr. 1616, 1682, 1684
—— Jemdr. . . . 1104
—— **Kadir,** Ressdr. . . . 918
—— **Khan,** Rissaldr. . . . 876
—— **Khan,** Subdr. . . . 1283
—— **Mahomed,** Subdr. . . 1575
—— Ressdr. 1682
—— **Mohammad Khan,** Ressdr. 1684
—— **Mohluddin,** Jemdr. . . 1404
—— **Mohi-ud-din,** Subdr. . . 1183
—— Jemdr. . . 951
—— Jemdr. . . 1208
—— **Mohiuddin,** Tr. Vety. Asst. 1687
—— Jemdr. . . . 1332
—— **Muhammad,** Ressdr. . 1683
—— Sub-Asst. Surgn. 1810

Names and Rank. Page.

Ghulam Muhammad, Jemdr. . . 1281
—— Jemdr. . . 974
—— Subdr. . . 1518
—— *Hony.* Capt. 1303
—— Subdr. (i) . 1307
—— Jemdr. . . 1288
—— Jemdr. . . 1408
—— Subdr. . . 1371
—— Jemdr. . . 1492
—— Subdr. . . 1569
—— Subdr. . . 1379
—— Subdr. . . 1120
—— Subdr. . . 1320
—— Subdr. . . 1117
—— Subdr. (ii) . 1207
—— Subdr. . . 1095
—— Subdr. . . 1373
—— Subdr. . . 1493
—— *Hony.* Capt. 2102
—— *Hony.* Capt. 2102
—— Subdr. . . 1235
—— Jemdr. . . 999
—— Jemdr. . . 1364
—— Subdr. . . 1311
—— Jemdr. . . 1420
—— **Khan,** Ressdr. 944
—— *Hony.* Capt. 1896
—— (I) Ressdr. . 940
—— (II) Jemdr. . 926
—— Risldr. . . 876
—— Ressdr. . . 1684
—— Jemdr. . . 1258
—— Jemdr. . . 1443
—— Jemdr. . . 999
—— **Khan,** Jemdr. 950
—— Jemdr. 940, 946
—— Jemdr. 1258
—— Ty. Sub-Asst. Surgn. 1833
—— **Imamdin,** Tr. Vety. Asst. 1687
—— Subdr. . 1335
—— Subdr. . 1207
—— **Mehdi,** Jemdr. 1236
—— **Nabi,** Jemdr. 999
—— Jemdr. 1320
—— Jemdr. 1245
—— Jemdr. 1232
—— Jemdr. 1012
—— **Khan,** Jemdr. 1035
—— **Rasul,** Jemdr. 1121
—— Subdr. . . . 1373
—— Jemdr. . . . 1107
—— Jemdr. . . . 1356
—— Jemdr. . . . 1376
—— Jemdr. . . . 1098
—— Jemdr. . . . 1681
—— **Khan,** Risldr. . 925
—— Jemdr. . . . 912
—— Sadiq, Subdr. . . . 1289
—— Subdr. 1132
—— **Sarwar Khan,** Ressdr. 1684, 1686
—— **Shah,** Jemdr. . . . 1502
—— Jemdr. 940
—— Jemdr. 1228
—— **Yunis,** Subdr. . . . 1495
Ghunna Singh, Jemdr. 977
Ghure, Subdr. 1455, 1585
Ghure Singh, Jemdr. 1404
Gian Chand, Sub-Asst. Surgn. . 1815
—— Singh, Ty. Sub-Asst. Surgn. 1833
Giband, A. B., Lt. 711
—— H. A., Maj. 1919
Gibb, F. G., Lt. 530, 1335
—— H. M., Lt. 629
—— J. B., Lt. . . . 45, 511, 1665
—— L. J., 2nd-Lt. 791
—— W. H., Capt. 744
Gibby, D., Lt. 375
Gibbings, O. D., 2nd-Lt. 725
Gibbon, C. B. B., Asst. Surgn. . . 1796
—— D. A., Asst. Surgn. . . 1794
—— H. A., Esq. 33
—— S. H., 2nd-Lt. 838
Gibbon, S. R., Lt. 623
Gibbons, A. J., 2nd-Lt. 544, 1886—88
—— J. H., 2nd-Lt. . . . 571
—— S., 2nd-Lt. 769
—— W. R., 2nd-Lt. . . . 588
Gibbs, A. A., Lt.-Col. 79, 101, 132, 1743
—— A. E. T., 2nd-Lt. . . 548, 1351
—— A. R., Lt. 534
—— E. W. B., Lt. 315, 1219, 1443
—— F. J., Capt. . . 502 746, 1668
—— H. H., 2nd-Lt. 815
—— H. R. K., Lt. 322, 1351

Names and Rank. Page.

Gibbs, J. W., Asst. Surgn. . . 1798
—— R. V., Asstt. Surgn. . . 1803
—— W., Capt. 587
Gibby, D., Lt. 1607
Giblett, J. V., 2nd-Lt. . . . 1096
Giblin, H. T., Asst. Surgn. . . 1801
Gibson, A., Sub-Condr. 1857
—— A. B., Lt. 375
—— A. D., 2nd-Lt. 686
—— A. J., 2nd-Lt. . . 548, 1587
—— D. A., Lt. . . . 524, 1668
—— D. E., Lt. . . . 440, 1339
—— D. V., Lt. . . . 511, 1136
—— F. A. S., Lt. 731
—— F. C., Lt. . . . 423, 1557
—— F. J. R., 2nd-Lt. . . . 331
—— G., *Hony.* Lt. 859
—— G. M. L., Lt. 1405
—— H. A., 2nd-Lt. 565
—— H. E. B., Capt. . . . 1755
—— J. F., Sr. Asst. Surgn. . 1788
—— J. G., Capt. . . . 254, 901
—— J. R., Lt. . . . 321, 1443
—— L. B., 2nd-Lt. 624
—— L. G., Maj. . . . 103, 1712
—— P. G. H., Lt. . . 351, 1241
—— P. R., 2nd-Lt. 1309
—— R., Sub-Condr. 1857
—— R. J., 2nd-Lt. 567
—— R. J. T., Lt. . . 696, 697
—— T. B., Lt. 807
—— T. J., Asst. Surgn. . . . 1795
—— W. C., Ty. 2nd-Lt. . .
—— W. G. 2nd-Lt. 987
—— W. R., Maj. 844
Gibsone, H. C., Capt. 679
Gidden, J. A., Lt. . . . 535, 1307
Giddens, H. L., 2nd-Lt. . . 553, 1305
Gidha Ram, Jemdr. 1489
Gidley, C. de B., Maj. 614
—— D. J., 2nd-Lt. . . 560, 1030
—— J., Capt. 746
Gidley-Kitchin, E. G., Capt. 665, 985
Gidney, L. A., 2nd-Lt. 563
Gidwani, B. B., Ty. Capt. . . . 1768
Giffard, D. H. R., Maj. . 29, 224, 872, 1635, 1639
—— G. G., Maj.-Genl. . . 119, 1737
—— J. F., Maj. 102
Giffen, L. E., Lt. 1223
—— S., Lt. 781
Giffin, R. A., Capt. 500
Gifford, F. R. Lt. . . . 292a, 1647
—— R. E., Capt. 703
Gigg, F. J., Lt. 353
Gig Sing, Jemdr. 916
Gilbanks, C. B., 2nd-Lt. 556
Gilbee, S. J., Lt. 796
Gilbert, A. S. A. S., 2nd-Lt. . . 557
—— B. J., Capt. 75, 746
—— C., 2nd-Lt. . . . 327, 1247
—— C., 2nd-Lt. 358
—— C. H. F., 2nd-Lt. . . . 1418a
—— C. T., 2nd-Lt. 703
—— E. C., Lt. . . . 440, 1038
—— E. O., Lt. 543
—— F., 2nd-Lt. 722
—— H., 2nd-Lt. 640
—— H. H., Capt. . . . 268, 1664
—— J. R., 2nd-Lt. 1363
—— L. E., Maj. 1746
—— R. S., Lt. . . . 520, 1667
—— S. W., Lt. 696
—— T., Lt. 489, 578
—— V. J., 2nd-Lt. 626
—— W. G. L., Lt. 509
—— W. H., Lt. . . . 538, 997
—— W. L., 2nd-Lt. 786
Gilbertson, G., *Hony.* Maj. . . 1845
—— J. G., 2nd-Lt. 549
Gilchrist, T. E., 2nd-Lt. . . . 567
—— W. F. C., Maj. 53, 99, 217, 1231
Gildea, E. P., 2nd-Lt. 1573
—— J. P., 2nd-Lt. 1283
—— R. B., Lt. 512
Gilder, M. D. D., Ty. Maj. . . . 1763
—— T. N., 2nd-Lt. 365
Giles, A., Lt.-Col. . . 127, 154, 182
—— C. D., Revd. . . . 652, 1906
—— E. D., Bt. Lt.-Col. 98, 139, 160, 220, 941
—— E. P., Lt. 700
—— H. E., Lt. 504
—— L., Lt. 531, 1668
—— L. R., 2nd-Lt. 701
—— M. G., 2nd-Lt. 1679

Names and Rank. Page.

Gibs, V., Bt.-Lt.-Col. 102, 103, 139, 663
—— W., Lt.-Col. 56, 1852
Gilfillan, N., Lt. 770
Gilja, Jemdr. 1188
Gillham, B. C., Lt. 720
—— J. E., 2nd-Lt. 752
Gilholme, E., Lt. 597
Gilkes, C. H., 2nd-Lt. 1441
—— G. H., Lt. 637
Gill, A. G., Capt. 47, 665
—— A. W. R., 2nd-Lt. 558
—— B. S., Ty. Capt. 1772
—— C. A., Lt.-Col. . . . 140, 1750
—— C. A. G., 2nd-Lt. . 556, 828, 1295
—— E. H. N., Lt. 512
—— G. T., Lt. 510, 1624
—— H., Lt. 642
—— H. W., Esq. 91
—— J. A., 2nd-Lt. 801
—— J. G., Lt.-Col. . . . 133, 1710
—— L., Lt. 1720
—— R. B., Lt. 543, 1418a
—— S. S., Sub-Asst. Surgn. . . 1819
—— W. H. I., 2nd-Lt. 362
Gillan, D. H., Rev. 1906
—— D. H., 2nd-Lt. 802
—— G. V. B., Capt. . . 276, 1641
—— J. B., 2nd-Lt. 47
Gillanders, C. F. Ty. Lt., . . 1599, 600
Gille, R. B., 2nd-Lt. . . . .
Gillespie, A. D., Lt. . . . 491, 925
—— D. S., Lt. 299, 1255
—— H. R. S., Asst. Surgn. . 1797
—— J., 2nd-Lt. . . . 358, 1677
—— J. K., Lt. 371
—— R. St. J., Maj. . . 25, 652
Gillespy, F. R., Revd. 1905
Gillett, B. G., 2nd-Lt. . . 535, 1693
—— C. E. S., 2nd-Lt. . . . 731
—— E. S., Maj. . . . 9, 79, 1838
—— N. C., Capt. . . 18, 29, 779
Gillham, J. C. E., Lt. 75
Gilliat, L., 2nd-Lt. 1028
Gillie, K. M., 2nd-Lt. . . . 548, 988
Gillies, F. G., Maj. . . 99, 222, 937
—— H., Maj. . . 15, 32, 242, 875
—— J. E., 2nd-Lt. 555
Gillies-Reyburn, D., Lt. 68, 298a, 1379
Gilligan, C. H., Lt. . . . 425, 1142
—— W. B., Lt. . . 311, 425, 1469
Gilling, H. T., Sub-Condr. . . . 1874
Gillitt, W., Maj. 801, 1751
Gillman, B. B., 2nd-Lt. . . . .
Gillmon, J. H., 2nd-Lt. . . 332, 1311
Gillon, G. K., 2nd-Lt. . . 558, 1459
—— J. B., Lt. 541, 1271
—— W. D., Lt. 111
Gillott, A., Lt. 681
Gillson, E. H., Asst. Surgn. . . 1795
—— W. J., Sr. Asst. Surgn. . 1790
Gilman, H. C. R., Lt. . . . 521, 1627
Gilmore, A. L., Lt. . . . 533, 1351
—— J. E., Lt. 1696
—— A. M., Nursing Sister . 1734
—— H. F., Lt. 380, 1062
—— M. G., Sr. Nursg. Sister, 1734
Gilmore-Ellis, P. D., Lt. . 397, 1241
Gilmour, A. C., Capt. 766
—— C. D., Capt. 743
—— J. A., 2nd-Lt. 763
Gilpin, G., *Hony.* Maj. 727
—— E. L. P. P., Lt. . . . 385, 933
—— F. E., Lt. 578
Gilroy, A. W., 2nd-Lt. 566
—— P. K., Capt. 1756
Gimi, H. D., Ty. Lt. 1774
Gimlette, G. H. D., Lt.-Col. . . 1786
Gimson, C., 2nd-Lt. 547
—— A., 2nd-Lt. 798
—— W. M., Lt. 508
Ginger, L. S., Lt. 778
—— A. H., 2nd-Lt. 709
Gingell, G. H., Sub-Condr. . . 1870
Ginni, H. D., Lt. . . . . . .
Ginns, L. B., Capt. . . . 8, 571, 1850
—— R., 2nd-Lt. 1106
Gipps, W. F., Lt. 527, 1094
Gird Ali, Subdr. 1427
Girdhari, *Hony.* Capt. Subdar Bahadur, 1191, 2104
—— Jemdr. 1212
—— Singh, Jemdr. 1388
—— Subdr. 1412
—— Singh, Negi, Jemadr. . 1181
Girdhar Lal, Sub-Asst. Surgn. . 1813
Girdhari Pande, Subdr. . . . 1022

Names and Rank. Page.

Girdhari Pande, Singh, Subdr.-Maj. 1279
Girley, A., 2nd-Lt. 1144
Girling, J. B., Lt. 405, 1068
—— G. F. H., Lt. . . . 543a, 1154
—— H. M., Capt. 602
—— W. H., 2nd-Lt. . . . 1678a1
Girsham, W. S., 2nd-Lt. . . 563, 1387
Girwar Singh, Subdr. 1046
—— Jemdr. 1032
Gisborne, H. P., 2nd-Lt. . . . 748
Gissing, A. C., Lt. 638
Gitaram Paras Ram Shyngla, Sub-Asst. Surgn. . 1818
Gittings, T. A., Lt. 617
—— F. S., 2nd-Lt. 707
Giyan Chand, Jemdr. 1244
Gladstone, A. C., Lt. . . . 506, 1618
—— H. L., Lt. 724
—— S. D., Lt. 572, 1641
—— T. H., Lt. . . 18, 25, 581
Glaister, J. D., 2nd-Lt. 763
Glancey, J. W., Esq. 89
Glanville, E. A., Lt. 512
—— F., Br.-Gen. 16, 43, 120, 123, 660
Glascott, J. R. D., Lt. 832
Glasfurd, A. I. R., Lt.-Col. 98, 135, 158, 195, 869, 1183
Glasse, M. St. G., Maj. 1838
Glasson, J., Lt. 725
Glasspool, L. A., Lt. . . . 486, 913
Glaysher, W. E., Capt. 636
Glazebrook, E. J., *Hony.* Lt. . . 786
Gladstanes, A. U., Maj. . . 214, 931
Gleeson, F. H., Asst. Surgn. . . 1793
—— J., 2nd-Lt. 1134
Glegg, J. A., Capt. 279, 1251
Gleig, D. D. W., 2nd-Lt. 618
Glen, J. C. G., 2nd-Lt. 562
—— J. V. J., Lt. . . . 306, 1038
—— R., Lt.-Col. . . . 856, 1893
—— R. J. G., Lt. . . . 521, 1347
Glendening, C. J., Lt. 1271
Glendinning, H. J., Capt. . . . 615
Glennie, E. A., Capt. 665
—— J. W., 2nd-Lt. . . 474, 1297
Gloria, J. S., Asst. Surgn. . . . 1799
Gloster, T. H., Maj. 1748
Glover, A. C., Sub-Condr. . . . 1854
—— B. T. J., Capt. 1722
—— C. A. W., 2nd-Lt. . . 551, 1179
—— E. H., 2nd-Lt. 564
—— M., Lt. . . . 423, 787, 1108
—— R. C., Capt. 827
—— W. M., Lt. 651
Glyani, Jemdr. 1059
Glynne-Jones, R. R., Capt. . 55, 717
Glynton, G. M., Maj. 99, 218, 1606, 1609
Gnanabishegam, R., Jemdr. . . 1328
Gnanamuthu, G. D., Ty. Lt. . . 1780
Gnanaprakasam, Jemdr. . . . 1276
Gnanasiromani, D., Sub-Asst. Surgn. 1827
Goad, C. E. G. B., Capt. . . 258, 1509
Gobar Singh, Butola Subdr. . . 1179
Goberdhan Gurun, Jemdr. . . . 1649
Gobind Ram, Ty. Sub-Asst. Surgn. 1833
Gobindram Khattri, Subdr. . . 1642
Gobind, Jemdr. 1244
—— Jemdr. 1171
—— Sahi, Jemdr. 1602
Gobind Singh, Subdr. 1489
—— Jemdr. 928
Gobiraj Gurung, Jemdr. . . . 1625
Govinda, Subdr. 1188
Goby, J. H., Lt. 538, 1707
Goda, Subdr. 1195
Godar Khan, Subdr. 1506
—— Subdr. 1569
Godber, A. R., Capt. 848
—— J., The Revd. . . 1903, 1904
Godbold, O. M., Condr. 1853
Godby, J. G., Lt. 781
Goddall, 2nd-Lt., F. E. 1707
Goddard, A. G. I. A., Lt. . 380, 1389
—— E. N., Lt. . 300-01, 1509, 1431
—— F. C., Lt. 310, 1459
—— L. A., 2nd-Lt. 567
Godden, A. L., 2nd-Lt. 856
Godfrey, A. S., Lt. 913
—— D. W., Capt. 597
—— E. K. M., Lt. 516a
—— F., Lt. 493
—— G. C., Lt.-Col. 850
—— R., *Hony.* Lt. 1850
—— R. E., Lt. 423, 1579

Names and Rank. Page.

Godfrey, B. L., 2nd-Lt. 789
—— S. H., Lt.-Col. 57
—— W. E., 2nd-Lt. 522, 1299
Godinho, P. X., Ty. Capt. 1765
Godkin, S. R., Maj. 1747
Godly, B. W. T., 2nd-Lt. 356, 1457
Godley, B. R., Lt. 321, 1038
—— E. L., Lt. 543a, 1273
—— J. L., 2nd-Lt. 778
—— T., 2nd-Lt. 787
Godsland, C. H., Capt. 367, 1094
Godlonton, B. G., Lt. 681, 785
Godson, C. A., Maj. 1752
Godwin, A. H. V., 2nd-Lt. 791
—— A. R., Lt. 628
—— C. A. C., Lt.-Col. 50, 97, 121, 127, 153, 158, 209, 919
—— T., Sub-Condr. 1873
—— W. E., Lt. 525, 1561
—— W. J. R., Asst. Surgn. 1803
Goepel, J. A., Lt. 513, 992
Goff, C. A., 2nd-Lt. 568
—— H. H., 2nd-Lt. 552, 1036
Goffin, R. C., 2nd-Lt. 556, 1335
Goll, D. P., Maj. 1750
—— J. N., Ty. Lt. 1780
Gokal, Subdr. 1227
—— Jemdr. 1137
—— Chand, Sub-Asst. Surgn. 1809
—— Singh, Jemdr. 884, 1400
—— —— Jemdr. 1453
—— Ty. Sub-Asst. Surgn. 1833
Gokhale, V. B., *Ty.* Capt. 1765
Gokul Chand, Sub-Asst. Surgn. 1815
—— Prasad Dube, Jemdr. 1027
—— Rai, Jemdr. 1631
—— Singh, Sub-Asst. Surgn. 1815
Golam Mohamed, Tr. Vety. Asst. 1687
Golak Nath Chatterji, Subdr. 1307
Goldberg, G. H. A., Lt. 519, 1371
Golden, C. H., 2nd-Lt. 693
—— H., Lt. 726, 1007
—— H. A., Lt. 435
Goldfrap, H. W., Capt. 59, 105, 256, 1415
Goldie, E. A. M. J., Ty. Capt. 1765
—— J. O., Col. 177
—— K. O., Maj 53, 234, 893
Golding, G. C., Capt. 751
—— G. V., Capt.
Goldingham, D. D., Capt. 16, 39, 693
Goldman-Monk, M. E. deB., Capt. 102, 107, 262, 1128, 1371
Goldney, A. P., Capt. 818
Goldring, H., 2nd-Lt. 683
Goldsack, A. G. R., Lt. 354, 1305
Goldschmidt, G. F., 2nd-Lt. 761
Goldsmith, E. O., Capt.
—— H. A., Bt.-Maj. 260, 1387
—— M. L., 2nd-Lt. 551, 1281
—— W. S., Lt. 504, 1515
Goldthorp, F. H., Maj. 1919
—— H. D., Lt. 423, 917
Goldsworthy, L. J., Lt. 493
Gole, G. B., Ty. Lt. 1782
Gollan, R., Maj. 610
Goma, Subdr. 1195
Gomer, L. W., 2nd-Lt. 365, 957
Gomes, D. V., Asst. Surgn. 1798
—— A. L., Asst. Surgn. 1798
Gomez, A. E., Asst. Surgn. 1795
—— A. F., Asst. Surgn. 1800
—— A. J., Asst. Surgn. 1797
—— S. C., Asst. Surgn. 1797
Gomme, D. E., 2nd-Lt. 549, 988
Gompertz, A. V., Capt. 6, 14, 103, 665
—— A. V. H., 2nd-Lt. 577
—— A. W. B., Capt. 269, 1670
—— G. H. M., Lt. 528, 992
—— M. C. T., Lt. 14, 29, 306, 1046, 1143a
—— M. L. A., Capt. 28, 103, 108, 255, 1435
Gomu Basnet, Jemdr. 1636
Gonda Singh, Risldr. 898
—— Jemdr. 999
—— Jemdr. 1150
—— Jemdr. 1124
Gonsalves, A., Ty. Lt. 1775
—— C., Esq. 88
—— D. R., Sr. Asst. Surgn. 1788
—— E. F., Esq. 88

Names and Rank. Page.

Gooch, F. C., 2nd-Lt. 568
—— H. W. 619
—— K. T., Lt. 617
Good, C. A., 2nd-Lt. 737
—— F. N., 2nd-Lt. 638
—— J., Maj. 1747
—— O. N. H., Lt. 535, 980
—— S. B., Lt. 313, 947, 1162
—— W. J., Lt. 529
Goodair, A. G., Condr. 1854
Goodall, A., *Hony.* Lt. 1844
—— F. E., 2nd-Lt. 551
—— G. M., Lt. 504, 1614
—— J. H. W., 2nd-Lt. 754
—— L. M., Lt. 387, 1143
Goodban, L., 2nd-Lt. 746
Goodbody, C. M., Lt.-Col. 135, 1746, 1889
Goode, F. V., Lt. 736
—— H. H., 2nd-Lt. 1678a
—— W. H. E. N., 2nd-Lt. 686
—— C. H., Lt. 759
—— G. G., 2nd-Lt. 760
—— H. H., 2nd-Lt. 563
—— R. A., Sub-Condr. 1868
—— R. L., 2nd-Lt. 324, 1247
—— T. C., Lt. 701, 781, 782
—— W., 2nd-Lt. 784
—— W. W., Capt. 45, 705
—— L. C. H., Lt. 537, 1235
Goodenough, H. L., Lt.-Col. 56, 128, 154, 185
Gooderidge, R. A., Capt. 1838
Goodfellow, N. G. B., Lt.-Col. 138, 159, 217, 1509
—— R. C., Maj. 209, 947
Goodger, A., 2nd-Lt. 564
Goodnall, E. J., *Hony.* Maj. 38, 1846
Goodhind, M. G., temp. 2nd-Lt. 796
Gooding, W. P., 2nd-Lt. 723
Goodland, C. H., Capt. 15, 46, 705
Goodman, E. 599—600
Goodman, J. G., Asst. Surgn. 1802
—— R., Lt. 795
—— R. H. B., Lt. 618
—— W. J., 2nd-Lt. 637
Goodson, D. F. J., Lt. 589
Goodwillie, D., 2nd-Lt. 653
Goodwin, A. E., Lt. 525, 1667
—— F., Col. 1
—— F. A., 2nd-Lt. 361
—— J., 2nd-Lt. 357, 1473
—— W. E., 2nd-Lt. 1675
Goodwyn, A., Nursing Sister 1734
Goody, G. R., 2nd-Lt. 740
—— R., 2nd-Lt. 690a
Goolden, A. W., Lt. 304, 1495
Gopal, Subdr. 1399
—— Jemdr. 1059
—— Chandra Ghose, Jemdr. 1219
—— Chandra, Sub-Asst. Surgn. 1818
—— Dass, Jemdr. 1288
—— Dhanaote, Jemdr. 1460
—— Gurung, Subdr. 1611
—— Luxuman, Jemdr. 1447
—— Narain Saran, *Maharaj Kumar* of Tikari 111
—— Parab, Jemdr. 1469
—— Ram Lama, *Hony.* Capt. 2103
—— Jadhao, Subdr. 1418a
—— Rao Palande, Jemdr. 1416
—— Sing Rai, Subdr. 1630
—— Rana, Subdr. 1609
—— Rana, Subdr. 1607
—— Sajan, Jemdr. 1447
—— Singh, Negi 1347
—— *Hony.* Capt. 2104
—— Subdr. 1148
—— Subdr. 1559
—— Subdr. 1283
—— Subdr. 1483
—— Subdr. 1217
—— Risaldr. 888
—— Subdr. 1043
—— Jemdr. 1707
—— Jemdr. 1344
—— Subdr. 1056
—— Jemdr. 1079
—— Jemdr. 1248
—— Jemdr. 1196
—— Jemdr. 1376
—— Hari Behare, Sub-Asst. Surgn. 1832
—— Dhaliwal, Sub-Asst. Surgn. 1819
—— Ittu, Jemdr. 1510

Names and Rank. Page.

Gopal Dhaliwal Waje, Subdr. 1423
—— Jemdr. 1285
—— Rana, Jemdr. 1504
Gopala, *Hony.* Capt. 2103
—— Menon, K. P., Jemdr. 1360
Gopalan Nayar 91
—— K., Jemdr. 1309
Gopaldas, Sub-Asst. Surgn. 1810
Gopal-datt Joshi, Sub-Asst. Surgn. 1814
Gopalan, C. K., Sub-Asst. Surgn. 1825
—— N., Sub-Asst. Surgn. 1828
Gopalaswami, K. S., Ty. Lt. 1779
Gopalsami, Jemdr. 1308
—— Subdr. 1276
—— Jemdr. 1327
—— Jemdr. 994
Gopi, Jemdr. 1193
Gopi Chand, Ressaldar 1683
—— Nath Agarwal, Sub-Asst. Surgn. 1815
—— Sing Rawat, Subdr. 1179
Gorakh Bahadur Mul, Jemdr. 1643
Gordeiro, E., Asst. Surgn. 1799
Gordhan, Jemdr. 1321
Gordhandas Jethabhai Desai, Sub-Asst.-Surgn. 1819
Gordon, A. D., Capt. 276, 905, 955
—— A. H., Capt. 485
—— A. R., 2nd-Lt. 1323
—— B. M. M., 2nd-Lt. 549
—— C. A., Lt. 504
—— C. C., Capt. 102, 615
—— C. F., Lt. 22, 778
—— D. B., 2nd-Lt. 360
—— D. H., Capt. 67, 279, 1156
—— D. St. V., Capt. 279, 897, 949
—— E. R. Mac. I., 2nd-Lt. 546, 1359
—— F. L., Lt. 317, 1007a
—— G., Lt. 543a
—— G., Ty. Lt. 622
—— G., 2nd-Lt. 1695
—— G. A. S., Bt. Lt.-Col. 139, 162, 205, 899, 957
—— G. S., Maj 219, 941
—— G. S., Ty. Lt. 1729
—— G. W., 2nd-Lt. 776a
—— H. F., Maj 10, 78, 218
—— H. S., Capt. 279, 1635
—— H. V., Lt. 543a, 1293
—— H. W., Col. 176
—— J. D., Revd. 825, 1907
—— J. de la H., Maj. 262, 1889
—— J. G. B., Maj. 8, 15, 221, 1231
—— J. H., Col. 177
—— J. K., Maj. 614
—— J. L. R., Br.-Genl. 12, 42, 132, 156, 192, 868, 1078
—— J. M., Lt. 519, 965
—— J. P., Lt. 450—54
—— J. W. S., Lt. 508, 1195
—— L. A., Col. 180
—— L. C., Condr. 1853
—— L. G. F., Br.-Gen. 26, 43, 101, 120, 124, 612
—— M., Lt. 387, 1676
—— N. H., Lt. 21, 540a, 1611, 1883
—— P. J., Lt.-Col. 52
—— R., Col. 179, 1917
—— R., 2nd-Lt. 329, 1080
—— R. D. L., 2nd-Lt. 332, 366a, 1315
—— R. E., Capt. 665, 1001
—— R. E., Capt. 617
—— R. F. C., Lt.-Col. 118, 154, 185
—— R. N. S., Capt. 1919
—— S., Capt. 1757
—— S. H., Lt. 314, 1219, 1403
—— S. V. L. deM., Lt. 540a, 1158
—— T. J., Lt. 375, 1217
—— W., Genl. 171
—— W., 2nd-Lt. 959
—— W. D., *Hony.* Lt. 34, 1846
—— W. A. L., Lt. 806
Gordon-Brown, D. S., Lt. 743
Gordon-Ives, L. E., Lt. 2
Gordon-Powell, D. H., Lt. 310, 1427
Gordon-Smith, W., Lt. 375, 513
Gore, A., 2nd-Lt. 751
—— C. H. C., Capt. 66, 279, 1178
—— C. H., Lt. 410, 955
—— F. L., Capt. 254, 1455
—— F. W., Lt. 608
—— F. W. B., 2nd-Lt. 531
—— R., Lt. 591

NAMES AND RANK. Page.

Gore, R. M., Lt. 450-54, 1359
—— S. N. D., Ty. Lt. 1776
Goria, Gurung, Subdr. 1618
—— Pun, Jemdr. 1619, 1651
—— Thapa, Jemdr. 1625
Goring-Jones, M. D., Br.-Gen. 122, 132
Gorman, J., Lt. 1689
—— J. T., *Hony.* Capt. 770
—— T. S., Lt. 624
Gormanly, P. J., Asst. Surgn. 1794
Gorrie, G., 2nd-Lt. 547
—— H. T., Major. 821
Gorringe, W. P. H., 2nd-Lt. 324
Gorst, R., Lt. 524
Gorton, W. H., 2nd-Lt. 817
Goru Ram, Jemdr. 1488
—— Subdr. 1193
Gosden, A. G., 2nd-Lt. 788
—— W. I., Lt. 470, 1411
Gosling, E. A. C., 2nd-Lt. 903
—— G. T. D., Lt.-Col. 683
Gosnell, K. A., Lt. 304, 1070
Gosney, C. G., 2nd-Lt. 786
—— H. J., Maj. 679
—— L. L., Lt. 679, 681
Goss, B. G., Lt. 533
—— E. S., Capt. 1757
—— E. C., 2nd-Lt. 761
—— J., Capt. 126, 1725
—— J. E. S., Lt. 440, 1042
—— J. L. L., 2nd-Lt. 653
—— J. S., Capt. 817
—— W. A., Lt. 541
Gosse, P. H. G., Capt. 1722
Gossage, A. F. W., 2nd-Lt. 595
Gosselin, H. de V., Lt. 488, 1675
Gosset, H. H. E., Capt. 665, 1882
Gossling, F. F., Lt. 734
Gostling, D. E., 2nd-Lt. 332, 943
Gotch, C. W., 2nd-Lt. 557, 1703
Gothaskar, S. B., Ty. Capt. 1767
Gott, G. A., Bt.-Col. 57, 122, 153, 180
Gottwaltz, B. L., Capt. 617
Goudge, B. N., 2nd-Lt. 577
Gouge, W. G., 2nd-Lt. 565
Gough, E. H., Ty. Lt. 638
—— F. B., Lt. 535, 1026
—— H. A. K., Lt.-Col. 195, 1889, 1912
—— J. A., 2nd-Lt. 796
—— W. G., H., Capt. 288-290, 1600
Gould, B. J., Lt. 806
—— E. F., 2nd-Lt. 554
—— G., Maj. 234, 877
—— H. R., Lt. 18, 38, 523
—— J., Lt.-Col. 130, 1742
—— J. G., Capt 1009
—— P. R., 2nd-Lt. 358, 1120
—— S. C., Capt. 253, 1347, 1895
—— W. G., Lt. 7, 543, 710, 793
Goulden, C. F., Lt. 1676
Goulding, C. C., 2nd-Lt. 493, 1008
—— H. P. R., 2nd-Lt. 836
Gouldsbury, C. A., Capt. 55, 268, 1617
—— C. P., Lt. 518, 1651
Goulstone, W. H., Lt. 536, 968
Gourdhan, Jemdr. 906
Gourlay, C. A., Maj. 1749
Gourlie, J., Maj. 219, 947
—— N. E. J., 2nd-Lt. 474, 1158
Gout, P. J., Maj. 232, 1383
Gutiers, E. M. de M., 2nd-Lt. 1289
—— G. F. C. de M., 2nd-Lt. 561, 1289
Govan, H. M., Ty. Lt. 1695
Gove, B. V., Lt. 489, 578, 883
Gover, A. C., Capt. 254, 1479
Govill, S. N., Ty. Lt. 1781
Govind Adhikari, Jemdr. 1643
—— Ghorpade, Jemdr. 1468
—— Rao Bargwe, Jemdr. 1464
—— Rao Sinde, Subdr. 1441
—— Lal, Sub-Asst. Surgn. 1819
—— Palao, Jemdr. 1416
—— Parab, Subdr. 1467
—— Ram, Jemdr. 1356
—— Ramchandra Jejuraker, Sub-Asst. Surgn. 1829
—— Ramchandra Upaleker, Sub-Asst. Surgn. 1831
—— Sakpal, Jemdr. 1464
—— Sawant, Jemdr. 1464
—— Subdr. 1444
Govind, Singh, Subdr.-Maj. 1050
—— Singh, Jemdr. 1476
Govindrao Power, Jemdr. 1470
—— Shindhi, Jemdr. 1409
Govinda, Jemdr. 1171
—— Dhamne, Subdr. 1423
—— Jadhaw, Jemdr. 1436
Govindarajalu, Subdr. 1339
—— Jemdr. 1316
Govindasami, I, Jemdr. 993
—— II, Jemdr. 994
—— Pillai, T. R., Sub-Asst. Surgn. 1827
—— Subdr. 1351
Govinderajulu, Jemdr. 1351
Gow, A. L., Lt. 729a
—— H., Lt. 544
—— P. F., Capt. 175
Gowland, C. J., Lt. 410, 1007
Gowan, C. H., Lt. 591
—— H. C., Maj. 803
Gower, J. M. N., Lt. 523, 1433
Gowing, D. S., Lt. 410
Goyal, K. N., Ty. Capt. 1772
Goyder, G. B., Capt. 59, 485
Grabham, H. M., 2nd-Lt. 1094
Graburn, N. E. B., 2nd-Lt. 566
—— N. E. B., 2nd-Lt. 1678a
Grace, F., Lt. 617
—— H. G., Capt. 254, 915, 959
—— O. G., Lt. 714
Gracey, D. D., Capt. 280-287, 1595
—— H. M. K., Capt. 277, 1407, 1409
Gracias, C. M., Ty. Lt. 1779
—— H. D., Esq. 88
Gradidge, J. H., Lt. 295, 901
Gradwell, J., Sub.-Condr. 1856
Græme, J. A., Maj. 664, 1882
—— N. F., Capt. 257, 1595
Graham, A. C., 2nd-Lt. 567, 1457
—— B. C., Maj. 214, 1497
—— C. G., 2nd-Lt. 547, 818
—— C. T., Lt.-Col. 501, 827, 124, 1696
—— D., Lt. 560, 798, 1056
—— D. L., Maj. 1359, 1753
—— D. S., Maj. 222, 1509
—— E. C., Capt. 658
—— F., 2nd-Lt. 465, 1152
—— F. M., Asst. Surgn. 1802
—— F. R. V., Lt. 589
—— G., Condr. 1854
—— G. D., 2nd-Lt. 534, 1696
—— G. F., Capt. 1754
—— H., Lt. 512, 1641
—— J. D., Bt. Lt.-Col. 62, 137, 1747
—— J. E. C., Maj. 725
—— J. M., Capt. 793
—— J. M., Lt. 585
—— J. P., Lt. 95, 317, 1231
—— J. S., Maj. 87, 240
—— J. S., 2nd-Lt. 561, 1321
—— K., N. Sister 1735
—— L. C. T., Capt. 272, 891
—— M. D., Lt. 1636
—— M. J., Capt. 1717
—— M. T., Lt. 517, 798
—— N. I., Lt. 303, 1371
—— P. H., Lt. 521, 1911
—— R., Hony. Lt. 1850
—— R. A., Capt. 274, 1144
—— R. B., Bt. Lt.-Col. 53, 139, 160, 208, 1207
—— R. J. D., Lt. 543a, 803
—— R. L., Ty. 2nd-Lt. 623
—— R. M., 2nd-Lt. 365, 1235
—— S. B., Col. 84, 124, 150, 184
—— T., 2nd-Lt. 363
—— T. C. V., Lt. 534, 1668
—— T. R., 2nd-Lt. 717
—— W. H., Ty. Lt.
Grahame, D., Lt. 720
Graham-Brown, E. D., 2nd-Lt. 618
Grahamshaw, J. G., 2nd-Lt. 563, 1112
Grainge, B., Sub-Condr. 1864
—— D. G., 2nd-Lt. 1479
Grainger, R. L., 2nd-Lt. 753, 786
—— S. N., 2nd-Lt. 330, 1565
—— T., Surgn.-Genl. 1, 1955
Gramble, W. M. F., Lt. 1371
Gramson, D. T. R., 2nd-Lt. 362
Grandhi, M. R., Ty. Capt. 1770
Grandison, R. McK., 2nd-Lt. 566
Grange, D. G., 2nd-Lt. 563
Granger, A. P., Capt. 1722
—— H. H., Lt. 514
—— T. A., Lt.-Col. 3, 10, 131, 1743
Grant, A., Lt.-Col. 40, 128, 154, 184
—— A., Esq. 91
—— A. G., Capt. 501a
—— A. E., 2nd-Lt. 546, 1022, 1044
—— A. G. C., Lt. 515
—— A. J., 2nd-Lt. 473, 893, 1088
—— A. O., Lt. 1655
—— A. R. 2nd-Lt. 361
—— A. W. H., 2nd-Lt. 708
—— C. A., Lt. 511, 992
—— C. D., 2nd-Lt. 637, 638, 976
—— C. F., Lt. 488, 827, 1347
—— C. J. P. MacA., Capt. 252, 1403
—— C. T., 2nd-Lt. 558
—— C. W., 2nd-Lt. 609
—— D. St. J. D., Lt.-Col. 57
—— D. M., 2nd-Lt. 551
—— D. S., 2nd-Lt. 566
—— E. P., Maj. 15, 20, 222, 921
—— E. R., 2nd-Lt. 550, 1701
—— E. R., Asst. Surgn. 1798
—— F. C., Maj. 778
—— F. R. E., Capt. 269, 1090, 1483
—— F. W., Sub-Condr. 1854
—— G. N., Lt. 540, 1701
—— G. P., Maj. 99, 212, 1427
—— G. R. E., 2nd-Lt. 805
—— G. S., 2nd-Lt. 1321
—— I., Lt. 95, 520
—— I. D., Capt. 1758
—— J., Col. 179, 1918
—— J., 2nd-Lt. 564, 1140
—— J. C., Lt. 420, 1050
—— J. D., Maj. 99, 214, 872, 1035, 1655
—— J. D., 2nd-Lt. 619
—— J. F., Capt. 500, 720
—— J. L., Lt. 523
—— J. W., Lt.-Col. 132, 1743
—— L. B., Capt. 688
—— L. I. 2nd-Lt. 1008
—— N. M., 2nd-Lt. 564, 1203
—— P., Capt. 276, 964
—— P. F. A., Lt. 1720
—— P. W., 2nd-Lt. 557, 1199
—— R., Lt. 516, 953, 1307
—— R., Lt. 489, 715, 1487
—— R. G., Capt. 850
—— R. A. P., Capt. 267, 1451
—— T., Sub-Condr. 1868
—— T. E., Lt. 806
—— T. G. L. E., Lt. 295a
—— W. C., Ty. Capt. 519
—— W. O., Maj. 208, 871, 1401, 1505
Grantham, G. J., 2nd-Lt. 626
—— V. A., 2nd-Lt. 654
Grant, Peterkin, M. H., Lt. 542, 1673
Grant-Smith, H. G. H., Maj. 226, 1419
Grassby, C. T., 2nd-Lt. 562
Grassick, W. J., 2nd-Lt. 742
Grassie, J. Lt. 743
Grattan, H. W., Col. 126
Gratton, C. K., 2nd-Lt. 561, 1321
—— J. S., 2nd-Lt. 494, 923
Graves, D. H. McD., Lt.-Col. 129, 1742
—— E., Lt. 158, 891
—— J., 2nd-Lt. 362
—— L. C., Lt. 776, 1894
Graveson, H., *Hony.* Maj. 1847
Gravestock, G. T., Lt. 776a
Graveston, G. J. B., 2nd-Lt. 561, 1367
—— H. B., Lt. 509
Gravett, G. W., Lt. 540, 1672
Gray, A. H., 2nd-Lt. 640
—— A. H., Ty. Lt. 1726
—— A. L., Lt. 1369
—— A. R., 2nd-Lt. 327
—— C. E., Lt. 307, 1606
—— D. B., Capt. 252, 1215
—— D. P. G., Capt. 698, 795
—— E. H., Capt. 276, 1311
—— E. St. C., Maj. 17, 42, 220, 939
—— F. W., Lt. 84, 553, 1653
—— F. W. B., Br.-Genl. 12, 27, 121, 126, 153, 192
—— G. C., Lt. 763, 781
—— G. C. G., Capt. 14, 31, 54, 267, 879

Names and Rank. Page.

Gray, G. E., 2nd-Lt. 1060
—— H. A., Ty. Capt. 1723
—— J., Capt. 695
—— J. B., 2nd-Lt. 860
—— J. E., 2nd-Lt. 743
—— J. K., *Hony.* Capt. 24, 728
—— J. M., 2nd-Lt. 566
—— J. Mc C. T., Lt. 735
—— J. W., Lt. 509, 1347
—— I. G., 2nd-Lt. 323
—— M. E., Sr. Nursing Sister 1734
—— N. R., 2nd-Lt. 656
—— O. E., Capt. 716
—— R., Capt. 720
—— S., Capt. 485
—— S. M., 2nd.-Lt. 656
—— S. T., Capt. 502, 1255
—— T., 2nd-Lt. 77, 785, 1884
—— W. A., 2nd-Lt. 828
—— W. C., Maj. 1751
—— W. D., 2nd-Lt. 558
—— W. du G., Maj.-Genl. 1243, 1955
Grayson, G. R., Condr. 1853
—— N., 2nd-Lt. 844
Greany, M. P. M., Lt. 511, 1355
Greatbatch, G., Ty. 2nd-Lt. 622
Greatoret, J., 2nd-Lt. 619
Greatwood, H. E., 2nd-Lt. 330, 1489-90
Greaves, E., Maj. 852
—— G. R., Genl. *Sir*
—— J., Lt. 368, 1060
—— R. B., 2nd-Lt.
—— W. H. *Hony.* Lt. 1849
Green, A. J., 2nd-Lt. 258, 1473
—— A. L., Lt. 720, 731a
—— A. M., Capt. 654
—— A. W., 2nd-Lt. 473, 1389
—— C. H. V., Capt. 816
—— C. R. M., Col. 126, 1738, 1899
—— C. W. T., Lt. 510, 997
—— D. C., Lt. 533, 1255
—— E., Nursing Sister 1735
—— E. B., 2nd-Lt. 720
—— E. C., Capt. 17, 26, 683
—— E. F., Sub.-Condr. 1857
—— E. G. H., 2nd-Lt. 328, 1617
—— E. J., Lt. 90, 505
—— E. W. K., Lt. 542, 1698
—— F., 2nd-Lt. 554, 1678a
—— G. H. B., 2nd-Lt. 91–92
—— H., Sub-Condr. 1854
—— H., 2nd-Lt. 354, 1601
—— H. D., Capt. 836
—— H. L., Ty. 2nd-Lt. 623
—— H. S., Lt. 532
—— H. T., 2nd-Lt. 355, 1511
—— H. W., 2nd-Lt. 668
—— J., 2nd-Lt. 552, 1110
—— J. D., Capt. 838
—— J. E. S., Capt. 1895
—— J. G. J., Capt. 1759
—— J. H., Lt. 440, 1347
—— P. A., Lt. 463, 970, 1143
—— P. J., Lt. 68, 617, 797
—— R., Lt. 777
—— R. K., 2nd-Lt. 652
—— S. A. C., 2nd-Lt. 555
—— T. G., Maj. 803
—— T. L., Lt. 780
—— T. W., 2nd-Lt. 602, 1126
—— V. N., 2nd-Lt. 359, 1140
—— W., Condr. 1863
—— W., 2nd-Lt. 566
—— W., Lt. 425, 1054
—— W. E., *Hony.* Maj. 1844
—— W. G., Lt. 308, 1082
—— W. G. K., Maj. 210, 943
—— W. J., Sub-Condr. 1873
—— W. W., Maj. 615
—— W. W., *Hony.* Capt. 21, 1847
Green-Armytage, P. D., Maj. 7, 103, 127, 1167
—— V. B., Capt. 1754
Greenacre, E., Lt. 778
—— G. A., 2nd-Lt. 539
Greenaway, H., Maj. 221, 1287
—— W. J., Esq. 89
Greene, A. C. B., Maj. 99, 633
—— E. A. J., 2nd-Lt. 567
—— H., Capt. 636
—— H. A., Sr. Asst. Surgn. 1792
—— J., Capt. 583
—— J., 2nd-Lt. 768
Greener, N. B., 2nd-Lt. 679

Names and Rank. Page.

Greener, W. J., Lt. 529
Greenfield, A. C., Lt. 507
Greenford, C., Condr. 1854
Greengrass, R. P. W., Lt. 440
Greenham, L., Maj. 850
Greenhaugh, F. H., Temp. Maj. 61
Greenhouse, F. S., Capt. 108, 256, 1383
—— R. E., Capt. 280-287, 1339
Greenhow, E. L., Lt. 375, 935
Greening, A. H., Lt. 530, 1707
Greenshields, D. J., Maj. 53, 581, 615
Geenslade, E. W., Maj. 777
Greenop, M. E., Lt. 399, 1090
Greensmith, J., Sub-Condr. 1873
Greenstreet, C. B. L., Lt.-Col. 37, 135, 661
—— W., 2nd-Lt. 589
Greenway, A. L., Asst. Surgn. 1800
—— C. D. K., Maj. 76, 1884
—— C. K., Capt. 265, 874, 923, 1892
—— C. M., 2nd-Lt. 833
—— L. A. N., Asst. Surgn. 1803
Greenwood, A., Sr. Asst. Surgn. 1788
—— A. B., Lt. 541
—— C. J. E., Capt. 666
—— E. M., Ty. 2nd-Lt. 623
—— E. T., Lt. 697
—— H., 2nd-Lt. 363, 1052
—— H. C., Lt. 531
—— J., Lt. 687
—— J. A., Lt. 518, 1016
—— N., 2nd-Lt. 356
—— R. S., Lt. 1299
—— W. Lt. 1007
—— W. L., 2nd-Lt. 545, 1365
Greer, A., Lt. 729a
—— E. R., Lt. 293a, 1427
—— J., *Hony.* Maj. 679
—— R. D., Lt. 380, 1613
—— R. E., Maj. 45, 662
—— U. McG., 2nd-Lt. 328, 1611
—— W. N., Capt. 45, 1718
Greer-Spencer, B., 2nd-Lt. 549
Greeves, S., Lt. 470 1571
Greffield, G., 2nd-Lt. 1369
Greg, L. H., Maj. 854
Gregory, A., Condr. 1863
—— A., Ty. Lt. 1726
—— C. J. E., 2nd-Lt. 1577
—— C. L., Br.-Gen. 50, 97, 121, 126, 153, 194, 867, 911
—— G. Lt. 923
—— G. M., Lt. 84, 532
—— H. G., Lt. 539, 1007a
—— J. R., Lt. 511, 1495
—— M. S., Lt. 509
—— R. V., 2nd-Lt. 841
—— S., Lt. 450-54
—— T., Lt. 460
—— T. M., Sub-Condr. 1870
—— V. W., 2nd-Lt. 562
Gregson, C. D., Maj. 239, 913
—— D. A., 2nd-Lt. 921
—— E. G., Capt. 485
—— H., Lt. 463, 1475
Greig, A. R., 2nd-Lt. 555
—— A. W., Maj. 1749
—— E. D. W., Maj. 1746
—— J. G., Maj. 206, 1479
—— R. M., Capt. 254, 1040, 1191
—— S., Lt. 516
—— W. B., Maj. 82, 1919
Grell, L. G. N., Lt. 397
Grellier, B. M. C., Capt. 1717
—— E. F. W., Capt. 1715
Greson, E. J., Sr. Asst. Surgn. 1788
Greswell, E., Lt. 525, 1886—88
Gretton, G. F., Maj. 216, 887
—— J. C., Maj. 54, 218, 1501
Greville, B. S. A. F., Maj 237, 1082
—— S. E., 2nd-Lt. 776a
Grey, A., Lt.-Col. 56
—— A. J. H., Maj. 107, 217, 1901, 1911
—— A. R., 2nd-Lt. 1621
—— E., Bt.-Col. 57
—— F., Maj. 750
—— F., 2nd-Lt. 1243
—— J. W. A., 2nd-Lt. 700
—— L. J. H., Col. 177
—— P., Capt. 273, 1667

Names and Rank. Page.

Grey, R. C., Lt. 299
—— R. W. H., Lt. 302, 1615
—— W. A. S., Capt. 57, 1895, 1910
—— W. G., Lt.-Col. 102, 104, 108, 189, 1889, 1911
Gribble, D. S., Lt. 368, 503, 1643
—— H. J. C., Lt. 698
Gribbon, H. H., Lt.-Col. 733
Grice, W. T., Lt.-Col. 5, 807
Gridley, R. F., 2nd-Lt. 793
Grier, H. D., Br.-Genl. 58, 120, 124, 632
—— J. A. B., Capt. 849
Grierson, J., Col. 177
—— Jackson, H. C., Lt. 725
Grieve, A. McL., Capt. 744
—— C. C. M., 2nd-Lt. 561
—— W., Lt. 526, 1247
Grieves, A. J. M., Lt. 415, 1435
—— B. B., 2nd-Lt. 324, 1148, 1589
Grielson, G. V., 2nd-Lt. 539
Griffen, F. W. B., Lt. 534, 1219
Griffe-Williams, J. H., Lt. 90, 519
Griffin, A. J. B., Capt. 1718
—— A. M., Lt. 529
—— G., Lt. 541, 1507
—— G. W., Lt. 315, 885
—— J. B., Capt. 15, 43, 754
—— S. F. H., Lt. 506, 1293
—— T., Lt. 55, 97, 505
—— T., Capt. 76
Griffith, A. J. W., 2nd-Lt. 736
—— E. S., Lt. 531, 1158
—— F., Capt. 1717
—— F. G., 2nd-Lt. 836
—— F. J., 2nd-Lt. 1455
—— G. M., Lt.-Col. 633
—— H. G., 2nd-Lt. 776a
—— H. G. B., Lt. 717
—— L., Maj 17, 23, 244, 1431
—— R., Col. 176
—— R. E. H., Maj. 233, 1901
—— W. R. C., Maj. 243, 1020
Griffith-Griffin, F. M., Maj. 265, 1267, 1513
Griffith-Jones, M. L., 2nd-Lt. 858
Griffiths, C., Lt.-Col. 62
—— C. E., Lt. 472, 1473
—— C. P., Lt. 443
—— E. A. C., Sr. Asst. Surgn 1789
—— E. G., Lt. 756
—— F. J., 2nd-Lt. 331
—— H., Lt. 520, 1471
—— H. M., Lt. 532, 1693, 1696
—— J. C., Lt. 970
—— J. Y., 2nd-Lt. 727
—— R. W., 2nd-Lt. 357, 1573
—— R. J. F. 2112
—— T. A., 2nd-Lt. 539, 1463
—— W. A., 2nd.-Lt. 753
—— W. S., Capt. 440, 907, 1886—88
—— W. S., Lt. 445, 538
Grigg, A. C., Capt. 683, 686
—— C. A., 2nd-Lt. 754, 799
—— E. C., Lt. 493, 947
Griggs, A., Lt. 618
Grigson, P. St. J. B., Lt. 492, 1291
Grille, F. L., Lt. 26, 542, 1606
Grimble, H. N. H., Lt. 711
Grime, G. M. 2nd-Lt. 715
—— H. B., Lt. 303, 1339
Grimes, C. J., 2nd-Lt. 547
—— C. J., Revd. 852-53 1904
Grimley, H. B., *Hony.* Lt. 1846
Grimgross, R. P. W., Lt. 1509
Grimmond, J. A., 2nd-Lt. 568
Grimshaw, R. W. W., Maj. 94, 219 939
—— J. E., Lt. 63, 1425
Grimson, G. R., Esq. 89
Grimston, L. A., Hony.-Col. 2, 606
—— S. B., Br.-Gen. 12, 27, 120, 124, 150, 184
Grindlay, A. P. R., 2nd-Lt. 840
Grindley, W. H., *Hony.* Maj. 1845
Grinnol, H. R., Esq. 88
Gripper, H. B., Lt. 435, 955
—— H. E., 2nd-Lt. 752, 784
Grisewood, A. E., Maj. 1753
Grisdale, W., 2nd-Lt. 717
Gristiram Rai, Jemdr. 1631, 1657
Grogan, T., Sub-Condr. 1857
Groocock, G. H., Lt. 463, 1676

Names and Rank. Page.

Groom, C. O., 2nd-Lt. 1193
—— C. R. N., 2nd-Lt. 793
—— J. P., Lt. 529
Groombridge, K. C., Lt. 526, 679, 1675
Gronow, I. A., 2nd-Lt. 1181
Grose, E., Capt. 253, 1082, 1307
—— J. S., 2nd-Lt. 557, 1104
—— N. A., Lt. 700
Grose-Hodge, H., Lt. 524
Grosnell, K. A., Lt. 1070
Gross, A. A., Lt. 668, 996
Grosvenor, C. J. P., Lt. 423, 909
—— V. L., 2nd-Lt. 686
Groube, G. P. T., Maj. 1746
Grout, T., Condr. 1853
Grove, A., 2nd-Lt. 778
—— C., Maj.-Gen. *Sir* 707
—— E. W., Lt.-Col. 132, 613
—— G. R., Lt. 315, 1623
—— P. L., Bt. Lt.-Col. 33, 68, 101, 141
—— W. E. U., 2nd-Lt. 1695
Grover, A. C., Capt. 1068
—— G. H., 2nd-Lt. 568
—— M. H. S., Genl. *Sir* 105, 145, 171, 1955
—— W., 2nd-Lt. 359, 1066
Groves, G. L., Capt. 679
Growse, H. E., Capt. 1910
Gruar, R. F., Lt. 492, 875, 1339
Gruer, H. G., Lt. 543a
Grummett, W. J., 2nd-Lt. 691
Grummitt, W. C., Lt. 492
Grundy, S. E., Lt. 293
Grylls, T. B., Lt. 711
—— W. M., Capt. 258, 1130, 1231
Gude, N. C., Lt. 490, 915
Gudal Sing Gusain, Subdr. 1182
Gudalore Ranganayakalu Nayadu Sub-Asst. Surgn. 1827
Gudgeon, P. E. 2nd-Lt. 1453
Gueizelar, J. A., Asst. Surgn. 1795
Guest, F. F., 2nd-Lt. 360
—— S. V., 2nd-Lt. 547
—— R., 2nd-Lt. 362
—— R. E., 2nd-Lt. 753, 786
—— W. W. K., Lt. 817
Gueterbock, E. A. L., Lt. 666, 996
Gugan, Hony. Capt. 2104
—— Jemdr. 1068
Gugan Sing, Ressdr. 914
Guha, P. C., Ty. Capt. 1453, 1772
Guider, M. E., Lt. 540a, 1673
Guidon, G. J. R., 2nd.-Lt. 735a
Gulab, Jemdr. 1559
Guilford, E., 2nd-Lt. 561
Guilfoyle, J. M., Ty. Capt. 1770
Gulichand Singh, Subdr. 1585
Guinee. J. A. 2nd-Lt. 472, 1297
Guiness, J. F. G., Ty.-Lt. 622
Guinness, V. E. G., Capt. 666, 667
—— W., Capt. 66
Guise, C. P., 2nd-Lt. 327, 1078
—— F. E. B., Capt. 276, 1191
—— H. B. C., Lt. 52, 487
—— J. W., Capt. 280—287, 1203, 1245
—— V. R., Capt. 636
Guiton, L. W., Lt. 306, 1283, 1513
Gujar, Jemdr. 966
Gujar Singh, Subdr. 1256
—— Ressdr. 1686
—— Jemdr. 1336
—— Jemdr. 989
—— Jemdr. 1702
—— Subdr. 999
—— Subdr 1174
—— Ressldr. 898
Gujjar Mal Varma, Sub-Asst. Surgn. 1815
Gul, Jemdr. 1507
—— Akbar, Subdr. 1117
—— Badshah, Jemdr. 1103
—— Din, Subdr. 1243
—— Hasan Khan, Jemdr. 946
—— Husain, Subdr. 1435
—— Khan, Jemdr. 1248
—— Jahan Khan, Ressdr. 944
—— Mehdi, Jemdr. 1565
—— Muaz, Jemdr. 978
—— Mawaz Khan, Subdr. 1231

Names and Rank. Page.

Gul Mawaz Khan, Jemdr. 1565
—— Muhammad, Jemdr. 950
—— Zaman, Subdr. 1183
Gulaba, Jemdr. 1196
Gulabdin, Sub-Asst. Surgn. 1810
Gulab Din, Subdr. 1132
—— Jemdr. 1305
—— Jemdr. 1248
—— Khan, Subdr. 1375
—— Subdr. 1701
—— Jemdr. 1681
—— Jemdr. 999
—— Jemdr. 1698
—— Jemdr. 1404
—— Jemdr. 1440
—— Jemdr. 1380
—— Subdr. 1705
—— Rai, Ty. Sub-Asst. Surgn. 1833
—— Singh Jemdr. 1192
—— Subdr. 1583
—— Jemdr. 1260
—— Shah, Subdr.-Maj. 1506
—— Subdr. 966
—— Subdr. 1585
—— Jemdr. 930, 1022
—— Jemdr. 1215
—— Jemdr. 1452
—— Ressdr. 1684
—— Negi, Subdr. 1182
Gulam Hassan, Jemdr. 1507
—— Haider, Jemdr. 1702
—— Kadar. 1436
—— Khan, Subdr. 1299
—— Jalani, Subdr. 1515
—— Muhammad, *Hony.* Capt. 2103
—— Jemdr. 1507
—— Khan, *Hony.* Capt. 2104
—— Nabi, Jemdr. 1681
—— Rasul, Jemdr. 1107
—— Khan, Jemdr. 876, 912
Gulbadan Singh, Jemdr. 1368
Guldar Shah, Subdr. 1502—04
Gulla Ram, Jemdr. 1472
Gulland, J. P., Capt. 269, 1287
Gulichand Singh Jemdr. 1452
Gulistan Khan, Jemdr. 926
—— Jemdr. 1122
Gulliland, Lt. C. C. 489
Gulliver, J., 2nd-Lt. 637
Gulpham Khan, Subdr. 1391
Gulsher Khan, Jemdr. 1070
Gulzada, Subdr. 1502—04
Gulzar Singh, Jemdr 1397
Gulzara Singh, Jemdr. 1320
—— Jemdr. 1280
Guman Sing, Jemdr. 990
—— Dangwal, Jemdr. 1181
—— Singh Gurung, Subdr. 1602
—— Jemdr. 1602
—— Jemdr. 1608, 1609
—— Jemdr. 924
Gumani Singh, Jemdr. 1051
Gumani, Jemdr. 1060
—— Ram, Subdr. 1401
Gumani Sahi, Subdr. 1642
Gumming, R. H. R., Lt. 315
Guna Nand, Jemdr. 990
Gundar Sing Bhandari, Subdr. 1182
—— Singh, Subdr. 1583
Gundry, A. R., Lt. 817
Ganga Man, Ressdr. 938
Gunindranath Barua, Srijut, 2nd-Lt. 481
Gunjaman Rai, Jemdr. 1633
Gunjbir Ale, Subdr.-Maj. 1607
Gunjirao Ghag, Jemdr. 1464
Gunn, A., 2nd-Lt. 859, 1122
Gunn, A. D., Capt. 254. 1443
—— A. R., Ty. Capt. 1723
—— C. F. 2nd-Lt. 1653
—— J. P., Lt. 575, 986
Gunner, H., 2nd-Lt. 682
Gunning, C. S., 2nd-Lt. 566
—— G. H., Maj. 213, 915
—— H. R., Ty. Capt. 578
Gunjbir Gurang, Subdr. 1614
Gunpat Rao Umre, Risldr.-Maj. 79
Gunson, G. H., 2nd-Lt. 919
Gunter, E. E., Capt. 835
—— C. P., Bt. Lt.-Col. 137, 662, 1880

Names and Rank. Page.

Guppy, S. J., Lt. 351, 1585
Gupta, A. C., Esq. 91
—— A. N., Ty. Lt. 1781
—— B. L., Ty. Capt. 1771
—— B. N., Ty. Lt. 1782
—— B. S., Ty. Lt. 1774
—— D. N., Ty. Capt. 1768
—— I. D., Ty. Capt. 1767
—— I. N. S., Ty. Lt. 1774
—— J. C., Ty. Lt. 1774
—— J. N. S., Ty. Lt. 1782
—— K. C. D., Ty. Lt. 1774
—— K. C. S., Ty. Lt. 1776
—— M. R., Ty. Capt. 1768
—— P., Ty. Lt. 1782
—— P. K., Ty. Capt. 1765
—— P. R., Ty. Lt. 1772
—— P. R. D., Ty. Lt. 1781
—— R. S., Ty. Capt. 1774
—— S. C. S., Ty. Capt. 1766
—— S. D. 91
—— S. N. S., 2nd-Lt. 482
—— U. M., Ty. Capt. 1771
—— U. S., Ty. Capt. 1772
—— V. P., Ty. Lt. 1777
Gupte, P. S., Ty. Lt. 1780
—— V. B., Ty. Capt. 1769
Guran Ditta, Sub-Asst. Surgn. 1814
Gurbachan Singh, Risaldar 1684
—— Grewal, Sub-Asst. Surgn. 1818
—— Rail, Sub-Asst. Surgn. 1818
Gurbaksh Singh, Jemdr. 918
—— Jemdr. 928
Gurcharan, Subdr. 1245
Gurdas, Jemdr. 1175
Gurdas Singh, Jemdr. 1121
Gurdatt Singh, Subdr.-Maj. 1459
—— *Hony.* Capt. 2101
—— *Hony.* Capt. 2102
—— *Hony.* Capt. 2103
Gur-dayal Kashop, Sub-Asst. Surgn. 1814
Gur-dayal, Singh, Sub-Asst. Surgn 1815
—— Jemdr. 1236
Gurdial, Subdr. 1320
—— Singh, Jemdr. 1112
Gurdit Singh, Subdr. 1141
—— Subdr. 1151
—— Jemdr. 936
—— Jemdr. 1502
—— Jemdr. 1117
—— Jemdr. 950
—— Subdr. 1215
—— Subdr. 1078
—— Jemdr. 1288
—— Jemdr. 1141
—— Ressdr. 1686
—— Jemdr. 1367
—— Jemdr. 906
—— Jemdr. 1150
—— Ressdr. 1683, 1684
Gurditt Singh, Sr. Sub-Asst. Surgn. 1809
—— Sr. Sub-Asst. Surgn. 1807
—— Subdr. 1106
—— Subdr. 1136
—— Jemdr. 1157
—— Jemdr. 1160
—— Jemdr. 1493
—— Jemdr. 1563
—— Subdr. 1163
—— Jemdr. 1158
—— Subdr. 1509
—— Subdr. Maj. 974
—— Saini, Sub-Asst. Surgn. 1810
Gurdon, E. P., Maj.-Genl. 173
—— P. R. T., Lt.-Col. 183, 1898, 1911
—— W. N., Lt. 1006
Gurmukh Shah, Ty. Sub-Asst. Surgn. 1833
—— Singh, Subdr. 1054
—— Jemdr. 1228
—— Jemdr. 1587
—— Jemdr. 999
—— Jemdr. 1212
Gurnam Singh, Jemdr. 1290
—— Maj. 110
Gurnett, W. V., 2nd-Lt. 556, 1459
Gurney, A., 2nd-Lt. 548, 1122
—— A. H., 2nd-Lt. 556, 1471
—— C. R., Capt. 668, 991
—— S., Sub-Condr. 1868

Names and Rank. Page.

Gursaran Singh Grewal, Sub-Asst. Surgn. 1819
Gurton Hobbs, A. G., 2nd-Lt. 356, 1675
Gurudayal Singh, B. H., Sub-Asst. Surgn. 1827
Gurul Sing Gurung, Jemdr. . . 1631
Gurusami, Subdr. 1327
——— Jemdr. (I) 994
——— Jemdr. (II) 994
Gaustab Khan, Jemdr. . . . 908
Guthrie, A., Capt. 351, 652, 1886–88
——— D. T. G., 2nd-Lt. 561, 1020
——— F. C., Lt. . . . 486, 913
——— I. D., Capt. . 267, 907, 1122
——— J. A., Lt. 797
——— W., Lt. . . . 68, 726
Gutteridge, B. F., Maj. . . . 730
Guy, G. R., Capt. 781
—— K., Lt. 387, 1489
—— N. G., Lt. 305, 1251
Guyton, G., Lt. 476, 1395
Gwalior, Maharaja of, *Hony.* Lt.-Genl. 1, 110
Gwatkin, F., Maj. . . 55, 245, 909
——— F. S., Col. 56
——— W. L. S., Lt. 111
Gwinnett, B. C. H., 2nd-Lt. 357, 1435
Gwyer, V. E., Lt-Col. . 6, 13, 98, 138, 159, 221, 1058
Gwyn, A. H., Maj. 241, 1030, 1880
——— P. E. C. J., 2nd-Lt. . . 324, 1116
Gwyn-Thomas, G., Br.-Genl. 50, 98, 122, 137, 159, 197, 867, 877
Gwynn, C. H., Lt. . . 10, 523, 1660
——— F. E., Sub-Condr. . . 1868
Gwyn-Williams, Lt. R. H. . . 55
Gwynne, C. W., 2nd-Lt. . . 4, 547
——— J. R., Ty. Capt. . 1507, 1766
Gwyther, E. N., 2nd-Lt. . . . 708
Gyann Jadhao, Jemdr. . . . 1424
Gyan Singh, Subdr.-Maj. . . 1305
——— Subdr. 1303
——— Subdr. . . 1066, 1215
——— Singh, Jemdr. 1157
Gyani, Jemdr. 1059
Gyani Sing, Subdr. 1359
Gyll, G. W. F., Lt. 727
Gypson, J. C., Sub.-Comdr. . . 1856
Gyton, G. H., Lt. 1007

H

Habgood, H. W., 2nd-Lt. . . . 735
Habib Gul, Ressdr. 892
——— Khan, Jemdr . . . 918, 930
——— Subdr. 1098
——— Jemdr. 1812
——— Subdr. 1102
——— Jemdr. 1506
——— Jemdr. 1051
——— Subdr. 1207
——— Jemdr. 1432
——— **Shah**, Subdr. Maj. . . . 1343
Habibul Rahman Khan, Jemdr. 920
Habib-ur-Rahman Khan, Jemdr. 1400
Habibullah, Ty. Sub-Asst. Surgn. 1833
Habibullah Khan, Subdr. . . . 989
——— **Khan**, Subdr.-Maj. . 1207
——— Jemdr. . . . 999
Habibur Rahman Khan, Ressdr. 876
——— Jemdr. . 1035
Hack, W. D., Lt. 1693
——— T. W., 2nd-Lt. 796
Hackett, H. F., 2nd-Lt. . . 494, 1009
——— H. M. M., Capt. 55, 254, 623
——— T. D. H., Maj. . 229, 1895
Hacking, H., Revd. 1906
Hadaway, J. C. S., Lt. . . 317, 1098
Hadayatullah Khan, . . . 1495-96
Hadi Hussain Shah, Jemdr. . . 1143
Hadden, A. J., Lt. 507
——— D. H., Capt. 1715
Haddick, V. A., Capt. . . . 1670
Haden, W. R., 2nd-Lt. . . . 551
Hadfield, W., 2nd-Lt. . . . 790
——— H. W., 2nd-Lt. . . . 911
Hadow, D. S., Maj. 501*a*, 605, 1143*a*

Names and Rank. Page.

Hadow, F. A., Capt. 836
——— H. R., Maj. 1910
——— R. H., Capt. 765
Haffenden, E. D., 2nd-Lt. . 546, 992
Haggar, J. M. C., 2nd-Lt. . . 692
Hagger, E. R., Maj. . . 78, 219
Haggett, H. G., 2nd-Lt. . . 9, 554
Haidar Ali, Subdr. 1141
——— Jemdr. 1428
——— **Beg**, Jemdr. . . . 1316
——— **Khan**, Subdr. . . 1335
——— Jemdr. 1035
——— Jemdr. 1511
——— Jemdr. 1369
——— Jemdr. 1141
——— Jemdr. 916
——— Jemdr. 1244
——— Jemdr. 976
——— Jemdr. 1492
——— Jemdr. 980
—— Rahman, Jemdr. . . . 1515
——— Shah, Jemdr. . . 1364
Haldipur, S. V., Esq. 8
Haig, A. B., Capt. . . . 256, 1116
——— D., Genl. *Sir* 595
——— D. deH., Maj. . . . 33, 656
——— D. P., Lt.-Col. . . 26, 719
——— H. G., Capt. . . 16, 46, 485, 826
——— P. B., Lt.-Col. . 129, 1742, 1889, 1901
——— T. W., Lt.-Col. 104, 108, 184, 1889, 1911
——— W. deH., Maj. 53, 664, 1882
Haige, H. C., Asst.-Surgn. . . . 1803
Haigh, F. F., Lt. 510, 1433
——— F. H., Lt. 726
——— P. B., Maj. . 68, 509, 849, 1459
Hailes, M. H., Lt. 832
——— W. L., Capt. . . . 262, 1038
Hailey, W. M., *Hon.* Lt.-Col. 804
Hailstone, G. R., Lt. . 318, 1635
——— R., Lt. . . 305, 1647, 1657
Haimard Rana, Jemdr. . . . 1615
Haines, E., Lt. 777
——— G., Nursg. Sister . . . 1735
——— J. B., Capt. . 105, 261, 1058
——— M. A., Lt. . . . 371, 1178
——— R. L., Ty. Lt. 1729
——— W. E., Lt. . . . 530, 988
Hainsworth, H. M., Lt. . 529, 1034
Haji Hafiz Obeidulla Khan, Maj. 2, 110
Haji, H. S. G., Capt. 1757
—— Khan, Subdr. 1108
Haji Ahmed Khan, Jemdr. . . 944
——— Gul Mawaz Khan, Rsedr. Maj. . 910
——— KhanJemdr. . . 1108
Hajra, B. B., Ty. Lt. . . . 1775
Hakamdad Khan, Jemdr. . . 1320
——— Jemdr. . . . 1117
Hakam Khan, *Hony.* Lt. . . 1303
——— **Singh**, Subdr. . . . 1117
Hake, H. D., Lt. 733
Hakikat Singh, Jemdr. . . . 934
Hakim Khan, Risldr. 934
——— Jemdr. 1356
——— Subdr.-Maj. . . . 1141
——— Jemdr. 1498
——— Hony. Lt. . . . 2104
——— Jemdr. 1107
——— Jemdr. 1510
——— Risldr. 1684
Hakim Singh, Sr. Sub-Asst Surgn. 1809
——— Jemdr. . 1103, 1104
——— Subdr. 1585
——— Subdr. 1141
——— Jemdr. 1228
——— Jemdr. 938
——— Ressdr. 922
——— Subdr. 1171
Hak-Newaz, Jemdr. 922
Hako, Subdr. 1174
Hakum Singh, Jemdr. . . . 980
Haldane, C. L., Lt.-Col. . 106, 132, 156, 193, 872, 1629
Halden, K., Lt. 629, 975
Haldipur, S. V., Esq. . . . 8, 91
Hale, R. G. S., Capt. 1892
——— E. C., Lt. 405, 1062
——— J. W. 2113
——— T. H., 2nd-Lt. 684
——— W. G., 2nd-Lt. . . 330, 1459
Hales, G., Lt. 95, 509
——— G., Lt. 521, 1411

Names and Rank. Page.

Hales, T. C., Capt. 838
Halford, E. H. J., Lt. 697
——— E. S., Capt. 101
Halford Watkins, J. F., Lt. . . 504
Halket F. G., 2nd-Lt. 687
Hall, A. C. S., Capt. 636
—— B. A. M., Maj. 633
—— B. K. B., Lt. 755
—— C. B. 2113
—— C. B. Lt. 695
—— C. E., 2nd-Lt. 619
—— C. F., Lt. 542
—— C. S., 2nd-Lt. 737
—— D. K., Lt 656
—— E. A. M., sub-Condr. . . . 1857
—— E. A. W., Lt.-Col. . 127, 1739
—— E. G., Maj. 9, 15, 238, 1467
—— E. J., Asst. Surgn. . . . 1803
—— E. R. M., Lt. 292, 1174
—— E. S., Lt.-Col. . 129, 155, 188
—— E. S., *Hony.* Capt. . 1702, 1847
—— F. A., Lt. 443, 1393
—— F. G., Lt. . . . 39, 59, 737
—— F. R., 2nd-Lt. . . 359, 708, 1052
—— F. W., 2nd-Lt. 559
—— G. C., Sub.-Condr. 1855
—— G. C. O., Lt. 420, 1405
—— G. H., Lt. 517, 1178
—— H., 2nd-Lt. 475, 1431
—— H., 2nd-Lt. 669, 986
—— H. C., Maj. . . . 26, 229, 1662
—— I. F. G., Capt. 278, 921
—— J. C., Revd. 1907
—— J. F., 2nd-Lt. 831
—— J. W., 2nd-Lt. 1056
—— P. A., Ty. Capt. 1726
—— P. M., 2nd-Lt. 629
—— R. F., 2nd-Lt. 1611
—— R. F. 2nd-Lt. 609
—— R. G., 2nd-Lt. . . . 494, 887
—— R. M., 2nd-Lt. . . . 472, 1521
—— R. S., Lt. 737
—— R. S., 2nd-Lt. 552
—— R. V., 2nd-Lt. 788
—— R. W. M., Lt. 311*a*
—— S. O., Lt.-Col. . . . 133, 1710
—— T. C., Lt. 22, 1848
—— T. H., 2nd-Lt. . . . 359, 1373
—— W. A., Lt. 516*a*, 1407
—— W. D., Capt. 459, 1034
—— W H., Lt. 389, 1399
Hallahan, R. J., Lt. 543*a*
——— R J., 2nd-Lt. 1695
Hallam, D. E., Capt. 746
——— E. H., Lt. . . . 352, 1557
——— H., 2nd-Lt. 727
——— S. Capt. 691
Halland, G. H. R., Lt. . 6, 14, 513
Halles, W. L., Capt. 262
Hallett, P., Maj. . . . 14, 24, 721
——— S. C., 2nd-Lt. 618
Halley, M. W., Lt. 504
——— D. H., 2nd-Lt. 786
Halliday, G. T., Lt.-Genl. . 169, 905
——— H. I., Lt. 504, 1898
——— H. M., Lt.-Col. 188, 1702, 1898, 1911
——— J. G., Genl. 170
——— W. J. F., Maj. 614
Hallifax, H. F., Maj . . . 803, 913*a*
——— H. F., 2nd-Lt. 929
——— J. S., 2nd-Lt. 315
Hallilay, H., Maj. 1750
Hallilley, W. S., Lt. . . . 297*a*, 1042
Halloran, J., Sub-Condr. . . . 1863
——— G. B., 2nd-Lt. . . 472, 1385
——— H., Lt. 24
Hallowes, F. W., Col. 10, 81, 126, 150, 188, 1650
———J. H., Maj. . 212, 903, 953
Hallum, S. S., Lt. 735
Hallums, W. B., Asst.-Surgn. . . 1800
Halsey, G. E.C, 2nd-Lt. . . 563, 1701
Halstead, J., 2nd-Lt. . . . 355, 1299
Halton, F. W., Maj. 725
——— H., Maj. 1720
——— S. C., 2nd-Lt. 1467
Ham, D. M., Lt. . . 291, 1315, 1591
——— P. S., Lt. 368, 1676
Hamber, L. G. W., Capt. . . 279, 1595, 1614
Hamblen, G., 2nd-Lt. 555
Hambling, V. F., 2nd-Lt. . 540, 1401
Hambly, R. L., Lt. . . 510, 779, 1068

NAMES AND RANK. Page.

Hamby W. E., 2nd-Lt. 355, 1008
—— W. P., 2nd-Lt. 696
Hamblyn, H. H., 2nd-Lt. 1689
Hamel Singh, Jemdr. 1683
—— Jemdr. 1702
Hamer, F. R., Lt. 472, 1359
—— G. M., Lt. 593, 599
—— M. A., Maj. 99, 215, 871, 1499, 1513
—— R. A., Lt. 630
—— R. F., 2nd-Lt. 626
—— R. G., Lt. 50, 532
Hamerton, F. W. M., Retd. 1907
Hamesh Gul, Subdr. 1256
Hamidullah, Jemdr. 945
—— Khan, Jemdr. 1495-96
Hamid Husain, Subdr. 1476
—— Sub-Asst. Surgn. 1820
Hamid Khan, Subdr. 1245
—— Jemdr. 1416
—— G. E., 2nd-Lt. 1271
Hamill, G. E., Lt. 415
—— H., Lt. 521, 1413
—— P., Capt. 1723
—— W. J. W., 2nd-Lt. 565
Hamilton, A. F., Maj. 1752
—— A. F., Capt. 254, 1263, 1273
—— A. F., Lt. 729
—— A. H. C., Capt. 45, 560, 1883
—— A. J., Lt. 532, 1581
—— A. S., Lt. 493, 905
—— A. W. B., 2nd-Lt. 557
—— C., Maj. 57
—— C. A. B., Maj. 221, 1606
—— C. B., 2nd-Lt. 564, 1363
—— D. S., Lt. 460-62, 1557
—— E. G., Bt. Lt.-Col. 52, 138, 769
—— E. W. S., 2nd-Lt. 1096
—— F. A., Maj. 53, 56, 230, 889
—— F. M., Capt. 642
—— G. H., Lt. 993
—— H. M., Capt. 279, 1259, 1567
—— Surg. Genl., *Sir* H. 1925
—— H. W. R., Capt. 666
—— I. S. M., Genl., *Sir* 766
—— J. A., Lt.-Col. 132, 1743
—— J. A. deC., 2nd-Lt. 729
—— L. S., Lt. 310, 1275, 1331
—— O. P., Capt. 268, 1311
—— P. S., Lt. 14, 20, 725
—— R., Capt. 665, 991
—— R., Lt. 511, 1335
—— R. E. A., Lt.-Col. 196, 1901, 1912
—— S. W. S., Maj. 664, 1880
—— W., *Hony.* Lt. 1848
—— W. A., Lt.-Col. 13
—— W. G., Brig.-Genl. 50, 97, 120, 122
—— W. G., Maj. 139, 1747
—— W. H., Bt.-Lt.-Col. 62, 1752
—— W. H., Maj. 722, 729
—— W. W., Asst. Surgn. 1803
Hamilton-Kerr, D., 2nd-Lt. 781
Hamilton-Smith, R., Lt. 1676
Hamir Singh, Risldr. 939
Hamira, Subdr. 1199
Hamlet, R. O., Lt. 512, 1673
Hamley, A. E. J., 2nd-Lt. 729, 791
—— D. W., 2nd-Lt. 729
—— H. R., 2nd-Lt. 654
—— W. S. C., Capt. 485, 913
Hamlin, A. E., Snr. Asst. Surgn. 1790
Hamlyn, C., Esq. 1908
—— E. A., Lt. 311a, 1407
—— H. W. A., Lt. 293a, 1114, 1359
—— W. T., Capt. 651
Hammersley-Smith, R. H., Maj. 16, 32, 231, 901
Hammick, St. V. F., Capt. 94, 262, 1070
Hammill, A. E., Capt. 858
Hammond, A. G., Col. *Sir* 178, 1917, 1925
—— A. V., Capt. 274, 964
—— C. F. B., Lt. 1693
—— F. A. L., Lt.-Col. 135, 1744
—— G. T., 2nd-Lt. 751
—— J. F., Capt. 1724
—— W. H., Lt. 658

NAMES AND RANK. Page.

Hammond, W. P., Maj. 234, 1090
Hampson, A., 2nd-Lt. 564, 1443
—— J., Lt. 591, 599
—— J. H. J., Lt. 304a, 1597
Hampton, A. V. F., Lt. 293a, 1066, 1156
—— C. H., Lt. 353, 1018
—— H. V., 2nd-Lt. 667
—— L. C., Lt. 529, 1427
—— R. *Hony.* Capt. 589
—— S. H., 2nd-Lt. 848
Hamzah Khan, Jemdr. 1424
Hamzullah Khan, Hony. Capt. 2103
Hanafin, J. B., Capt. 1754
Hanbury Williams, *Sir* J., Maj.-Genl. 745
Hance, H. M., Capt. 501a, 985
—— J. B., Capt. 1757
Hancock, A., 2nd-Lt. 1038
—— C. P., Lt. 291, 1459
—— D. L., Lt. 695
—— E. S., Maj. 237, 1367
—— F. B., Lt. 25, 780
—— G. A., 2nd-Lt. 1140
—— G. E., Col. 177
—— G. R., Lt. 700
—— H. F., 2nd-Lt. 566
—— J., 2nd-Lt. 714
Handcock, A. G., Lt.-Genl. 172, 1038
—— W. J., 2nd-Lt. 27, 1677
Hancox, G. R. 686
Handley, F., *Hony.* Lt. 1843
—— F. 2nd-Lt. 790
—— H., 2nd-Lt. 720, 786
—— H. G., 2nd-Lt. 568
—— L. M., Capt. 273, 923, 963
—— P., 2nd-Lt. 551, 1009
Handpal Sing, Subdr. 1567
Hands, A. S., 2nd-Lt. 737
—— P. L. J., 2nd-Lt. 552
Handscomb, L. G., Lt. 543, 1575
Handscombe, H. T., 2nd-Lt. 681
Handu Singh, Jemdr. 1453
Handy, F. R., Lt. 353, 1361
Hankin, D. L., Maj. 242, 1090, 1092
—— E. L., Lt.-Genl. 172
—— E. M., *Hony.* Lt. 826
—— H. M., Capt. 264, 964
Hankins, R. F. H., Sr. Asst. Surgn. 1787
Hanlon, G. J. P., *Hony.* Lt. 1844
Hanley, M. P., 2nd-Lt. 551, 1323
Hanly, J., *Hony.* Maj. 1850
Hanmant Dattatrya Bedeskar, Sub-Asst. Surgeon 1832
—— Shrinivas Dandi, Asst. Surgn. 1832
Hanmer, C. G., Lt. 318, 915, 955
Hanna, A. L., Lt. 322, 895
—— G. B., Capt. 1759
—— H., Lt. 375, 1415
—— J., 2nd-Lt. 566
—— J. R. M., Lt. 1515
Hannagan, L., 2nd-Lt. 546, 1473
Hannah, A. J., Lt. 514, 1283
—— G. K., Lt. 487, 951
—— J. S., Lt. 524, 1575
—— W. F., 2nd-Lt. 782
Hannam, H. L., Lt. 455-59
Hannay, R. J., 2nd-Lt. 543a
Hannen, F. B., Lt. 608
—— R., Lt. 834
Hanney, R. J., Lt. 1267
Hannigan, A., 2nd-Lt. 1363
—— H., Capt. 1724
Hannyngton, J., Capt. 808
Hanooman Singh, Rissaldar Maj. 914
—— Jemdr. 1043
Hanrahan, M. W., Lt. 525, 1699
Hansard, A. J. St. L., Lt. 302, 1365
—— J. H., Capt. 730, 1004
Hansen, G. E., Lt. 299, 1231, 1259
—— H. E., 2nd-Lt. 555
Hansford, E. V., Lt. 538, 1601
Hanson, A. D., Asst. Surgn. 1800
—— F. A., Lt. 445, 1265
—— F. J. C., Lt. 844
—— G. J. G., Maj. 234, 1287, 1901
—— H., 2nd-Lt. 1088
Hans Raj, Subdr. 1035
—— Madhok, Ressdr., 1684, 1686
Hans Ram, Jemdr. 1059
—— Raj Madhok, Subdr. 1488
Hanse Singh, Jemdr. 1412
Hanumant Singh, Subdr. 1401

NAMES AND RANK. Page.

Hanumantu, Subdr. 993
Hanut Singh, *Rao-Raja* 112
Hanwant Singh, *Hony.* Capt. 2103
Hanvey, G. R. J., Sr. Asst. Surgn. 1791
—— M. P., Lt. 1060
Hanwell, J. O., Capt. 288, 290, 895
Happell, A. J., Lt. 536
—— W. H., Capt. 271, 1403
Haqdad Khan, Jemdr. 904
Har-bakhsh Singh, Ressdr. 890
Harakh Bahadur Singh, Jemdr. 1062
Harakbahadur Limboo, Jemdr. 1631
Harakbir Limboo, Subdr. 1597
—— Thapa, Jemdr.
Harak Sing Gurung, Jemdr. 1608
—— Jemdr. 1619
—— Gharti, Jemdr. 1608
—— Khattri, Jemdr. 1643
—— Thapa, Jemdr. 1602
Hara Singh, Hony. Capt. 2103
Harbakhsh Singh, Subdr. 1064
Harband Singh, Ressdr. 892
Harbans Lal, Ty. Sub-Asst. Surgn. 1833
—— Singh, Jemdr. 1452
—— Subdr. 1166
Harbant Singh, Jemdr. 1165
Harbhaj, Jemdr. 900
—— Singh, Jemdr. 880
Harbhan, Jemdr. 1476
Har-bhawan, Jemdr. 1687
Harbord, C. B., Lt.-Col. 122, 140, 162, 200, 857, 931
—— G., Lt. 624
—— G. M., Bt.-Maj. 615
—— K. B., Capt. ...
—— L. B., Capt. 6, 14, 257, 1199
—— R. E., Capt. 16, 35, 683
Harborow, J. E. W. P. A., 2nd-Lt. 559
Harbottle, R. C., Maj. 205, 887
Harchand, Subdr. 1455
—— Jemdr. 1404
—— Jemdr. 1059
—— Jat, Subdr. 1419
—— Sing, Subdr. 1355
—— Singh, Subdr. 1256
—— Sub-Asst Surgn. 1817
Harcourt, C. B., Maj. 223, 1132
—— E. S., Maj. 66, 103, 108, 244, 1629
—— H., Capt. 804
—— J. S. M., Lt.-Col. 140, 161, 200, 1600
Hardaker, A. J., Asst. Surgn. 1791
—— C. B., *Hony.* Maj. 1843
—— F. G., Asst. Surgn. 1799
—— H. L. W., 2nd-Lt. 551, 1701, 1704
Hardam Singh, Risaldar 1684
Hardas Singh, Ressdr. 1682
Hardat, Jemdr. 1397
Hardayal, Jemdr. 1171
—— Jemdr. 1583
Har-Dayal, Subdr.-Maj. 1403
—— Jemdr. 1260
Hardcastle, A. E. L., Lt. 685
—— J., Maj. 239, 1207
—— R. N., Lt-Col. 136, 758
Harden, F. G., Capt. 689
—— H. A., 2nd-Lt. 363, 1189
Harder, J. R., 2nd-Lt. 522, 1651
Hardhan, Jemdr. 1216
—— Singh, Subdr. 1397
—— Subdr. 1559
Hardie, F. M., Lt. 745
—— G. E., Maj. 212, 868, 1040, 1487
—— J. D., Lt. 399, 879
—— J. F., Lt. 785
—— J. J., 2nd-Lt. 957
—— W. A., 2nd-Lt. 563
—— W. B., Lt. 629
—— W. J. 2nd-Lt. 729a
Hardiman, J. P., Capt.
Harding, A. G. L., 2nd-Lt. 356, 1505
—— A. W., Lt. 541
—— E. A., 2nd-Lt. 828
—— C. N., Capt. 274, 1433, 1539
—— F., 2nd-Lt. 566
—— G., *Hy.* Maj. 40, 59, 1843
—— G. T. H., 2nd-Lt. 523
—— H., Lt. 695, 697
—— J. A., Lt. 780
—— K. O'B., Capt. 279, 899
—— L. O'B., 2nd-Lt. 556, 1292
—— M. F., Lt.-Col. 34, 103, 130, 155, 190

Names and Rank. Page.

Harding, R., 2nd-Lt. 794
—— R. A. C., Lt. 48
—— W. F., Lt. 517, 1667
—— W. J., Lt. 318, 1375
Hardinge, Lt. *Hon.* A. H. L. 507
—— G. W. P., 2nd-Lt. 549, 1703
—— J. G., Maj. 882
—— H. R. von Donop, Maj. 87, 215
Hardingham, H. 89
Hardit Singh, Jemdr. 990
—— Jemdr. 1699
—— Jemdr. 1107
—— Subdr. 1259
—— Subdr. 1243
Harditt Singh, *Hony.* Capt. 2013
—— Sub-Asst. Surgn. 1816
—— Subdr. 1163
—— Ressdr. 922
—— Sub-Maj. 1151
Hardman, A., 2nd-Lt. 1557
—— E. F. C., 2nd-Lt. 327, 1635
—— P. L., Lt. 514, 1624
—— R. C., Capt. 720
—— R. G., 2nd-Lt. 561
Hardwari Singh, Subdr. 1024
—— Jemdr. 1021
—— Jemdr. 1028
—— Lal, Jemdr. 1056
Hardwick, N. DeC., Capt. 59, 505, 832
Hardwicke, C. B., Lt. 510, 1665
—— S. M. B., 2nd-Lt. 551
Hardy, A. E., Capt. 749
—— E. P., Lt. 389, 725, 1303
—— G. W., Asst. Surg. 1800
—— H. DeL., 2nd-Lt. 531
—— H. J., Lt. 638
—— R. H., Lt. 832
—— T. H., Maj.-Genl. 12, 42, 119, 147, 174
—— V., 2nd-Lt. 55
—— W. E., Col. 130, 1709
—— W. K., Br.-Genl. 58, 121, 130, 632
Hardayal, Sub-Maj. 1403
—— Jemdr. 1484
Hardyal Singh, Sr. Sub-Asst. Surgn. 1809
Hare, A. N., The Revd. 1905
—— E. H. 2nd-Lt. 713
—— G. A., Maj. 71, 101, 614, 1884
—— I. L. O'H., Capt. 265, 1433
—— P. V., 2nd-Lt. 786
—— S. St. G., Lt. 68, 405, 1321
—— T., 2nd-Lt. 571a
—— W. F., temy. Capt. 1724
Hare-Duke, A. R. A., 2nd-Lt. 558
Harenc, R. E., Maj. 243, 881
Harendra Kumar Guha, Sub-Asst. Surgn. 1816
Harford, E. B., Capt. 704
—— G. A., 2nd-Lt. 597
—— H. H., Capt. 615
Hargian, Jemdr. 1484
—— Singh, Jemdr. 924
Hargrave, J. D., 2nd-Lt. 595
—— J. F. H., 2nd-Lt. 328, 1523, 1606
—— L. O., 2nd-Lt. 557
Hargrave-Carroll, F., Capt. 502
Hargreaves, C. K., 2nd-Lt. 554
—— F., Lt. 521
—— F. B., Lt. 353
—— F. W., Sub-Condr. 1861
—— H., Capt. 527, 804, 1702
—— H., Lt. 420, 1507
—— J., Lt. 797
—— P. S., Lt. 315, 901
—— R., Capt 89, 725
Hargyan Singh, Risldr. 901
Harhet, Subdr. 1455
Hari Lakhsman, Jemdr. 1484
Haribakan Rana, Subdr. 1609
Haribans Lal, Sr. Sub-Asst. Surgn. 1807
Haribhagat Sahi, Subdr. 1642
Haricharan Thapa, Subdr. 1651
Hari Baran Rana, Subdr. 1607
—— Chand, Sub-Asst. Surgn. 1816
—— Subdr 1607
—— Ressdr. 1682
—— Das Gupta 89
—— Jadhao, Jemdr. 1468
—— Kishan Gupta, Sub-Asst. Surgn. 1818
—— Ram, Subdr. 1066
—— Subdr.-Maj. 1217
—— Subdr. 1215

Names and Rank. Page.

Hari Ram, Subdr-Maj. 1066
—— Mehta, Asst. Surgn. 1820
—— Mehta, Sub-Asst. Surgn. 1820
—— Parasar, Sub-Asst. Surgn. 1818
—— Singh, Ressdr. 936
—— *Raja Sir.* 111, 909
—— Sub-Asst. Surgn. 1817
—— Jemdr. 884, 959
—— Jemdr. 912
—— Subdr. 1021
—— Jemdr. 894
—— Subdr. 1343
—— Subdr. 1082
—— Subdr. 978
—— Subdr. 1021
—— Subdr. 1215
—— Subdr. 1114
—— Jemdr. 1121
—— Jemdr. 1212
—— Jemdr. 1051
—— Jemdr. 1347, 1388
—— Jemdr. 946, 953
—— Jemdr. 999
—— Jemdr. 1683
—— Subdr. 1031
—— Subdr. 1432
—— Subdr. 1271
—— Jemdr. 1040
—— Subdr. 1367
—— Burathoki, Subdr.-Maj. 1607, 1609
—— Bhandari, Jemdr. 1643
—— Gurung, Jemdr. 1625
—— Parasa, Asst. Surgn. 1888
—— Sing Thapa, Subdr. 1614
—— Thapa, Jemdr. 1408
—— Warute, Jemdr. 1469
Harihar Pershad, Jemdr. 1369
Hari Krisan Gupta, Asst. Surgn. 1818
Harikishan Gurung, Subdr. 1615
Hari Lal Narsinh Pancholi, Asst. Surgn. 1832
Harington, A. W. D., Lt.-Col. 24, 139, 160, 199, 1661
—— H. R., Capt. 265, 1635
Haris Chandra Govindras Shirke, Sub-Asst. Surgn. 1830
Harjiwan Singh, Subdr. 1389
Harjiwan Lal, Jemdr. 1051
Harkabahadur Limbu, Jemdr. 1631
—— Subdr. 1630
Harkabir Limbu, Subdr. 1648
Harkadhoj Limbu, Jemdr. 1597
—— Thapa, Jemdr. 1631
Harka Sing Gurung, Jemdr. 1602
—— Thapa, Jemdr. 1649
Harkbir Ale, Subdr. 1630
—— Gurung, Jemdr. 1619
Harker, A. W. A., Capt. 636
—— G. A., 2nd-Lt. 742
—— J. T. H., Lt. 111
Harkesor Limbu, Jemdr. 1598
Harkess, A. E., 2nd-Lt. 566, 1678a
Harkishan, Jemdr. 1472
—— Jemdr. 1583
—— Jemdr. 1453
Harkishan-Das, Sub-Asst. Surgn. 1813
Harkness, H. L., Maj. 1910
—— E. B., 2nd-Lt. 547
Harkraj Gurung, Jemdr. 1611
Harku Thapa, Subdr. 1624
Harlal Singh, Jemdr. 1537
Harland, D. N., 2nd-Lt. 714
—— F. W., Lt. 317a, 881
—— G. B., Capt. 1757
—— H., 2nd-Lt. 758
Harley, A. B., Maj. 57, 230, 1479
—— G. T., 2nd-Lt. 761, 1008
—— I., Nursing Sister 1735
—— J. E., 2nd-Lt. 790
—— J. R. H., Lt. 763
—— N., Lt. 1219
—— T. W., Maj. 1749
—— W., 2nd-Lt. 1361
Harley-Manson, R. J., Ty. Capt. 1726
Harlow, C. M., Lt 525, 1148
Harma Singh, Jemdr. 1698
Harman, B. B., 2nd-Lt. 465, 1443
—— C. W. T., Lt. 517
—— F. A., 2nd-Lt. 356, 1074
—— O. A., Lt. 513, 858, 1666
—— S. H., Lt. 506, 1054
—— T. E., *Hony.* Lt. 747

Names and Rank. Page.

Harmer, F. L., 2nd-Lt. 546, 1339
—— S. H., Lt. 352
Harna Singh, Subdr. 1571
—— Subdr. 1452
Harnam Singh, Sub-Asst. Surgn. 1813
—— Sub-Asst. Surgn. 1812
—— Jemdr. 1196
—— Jemdr. 1107
—— Subdr. 1163
—— Jemdr. 1240
—— Jemdr. 1121
—— Subdr.-Major 1124
—— *Hony.* Capt. 2104
—— Subdr. 1143
—— Jemdr. 882
—— Jemdr. 1380
—— Subdr.
—— Ressdr. 918
—— Jemdr. 1256
—— Risldr. 925
—— Jemdr. 1167
—— Subdr. (I) 1111
—— Subdr. (II) 1111, 1112
—— Subdr. 1343
—— Subdr.-Major 1247
—— Jemdr. 1212
—— Jemdr. 1130
—— Jemdr. 1112
—— Subdr. 1157
—— Ressdr. 1683
—— Ressdr. 1684
—— Jemdr. 918
—— Jemdr. 1095
—— Jemdr. 888
—— Subdr. 1379
—— Jemdr. 936
—— Ressdr. 1686
—— Jemdr. 1565
—— Jemdr. 953
—— Subdr. 1311
—— Resldr.-Maj. 935
—— Subdr. 1375
—— Sub-Asst. Surgn. 1815
—— Sub-Asst. Surgn. 1816
—— Sub-Asst. Surgn. 1817
—— Jemdr. 978
—— Jemdr. 928
—— Jemdr. 1157
—— Jemdr. 1079
—— Jemdr. 916
—— Jemdr. 1228
—— Jemdr. 989
—— Jemdr. 974
—— Subdr. 1320
—— Subdr. 1114
—— Jemdr. 1112
—— Subdr. 988
—— Jemdr. (I) 1204
—— Jemdr. (II) 1204
—— Jemdr. 1701
—— Subdr. 1120
—— Subdr. (II) 1121
—— Gill, Sub-Asst. Surgn. 1819
Harnand, Subdr. 1335
—— Subdr. 1060
Harnarain, Jemdr. 1067
—— Subdr. 1056
—— Jemdr. 1391
—— Jemdr. 938
—— Singh 1071
Harnarayan Singh, Ressdr. 880
—— Jemdr. 1583
—— Jemdr. 1388
—— Jemdr. 1035
Harnath Singh, Jemdr. 1483
—— Subdr. 1488
Harnett, W. F., 2nd-Lt. 544, 817
—— W. L., Maj. 1753
Harold, C. F., Lt.-Col. 194, 1893, 1912
—— C. H. H., Capt. 1713
Har-parshad, Sub-Asst. Surgn. 1807
Harper, A. F., Maj. 6, 14, 227, 1343
—— A. H., Lt. 533, 1571
—— C. O., 2nd-Lt. 814
—— C. S., Lt. 527, 1419
—— E., Esq. 88
—— G., Lt. 510, 1665
—— G. L., 2nd-Lt. 689
—— G. M., 2nd-Lt. 560, 1050
—— H. P., Ty. Capt. 1723
—— J. K., 2nd-Lt. 492
—— J. P., 2nd-Lt. 561, 1433
—— K. B., Lt. 807
—— K. G., 2nd-Lt. 561
—— K. J., 2nd-Lt. 897

NAMES AND RANK. Page.

Harper, R., Lt. 510
——— R. K., Lt. . . . 463, 887
Harper-Collins, R. S., Lt. 543a, 1469
Harphool, Subdr. 1393
Harphul Singh, Ressdr. . . 930, 959
——— Subdr. 1320
——— Jemdr. 1397
——— Jemdr. 1028
——— Jemdr. 1212
Harpul, Jemdr. 1356
——— Jemdr. 1404
——— Subdr. 1403
Harraway, C. G., 2nd-Lt. . . . 622
Harricks, N., Lt. 515, 1672
Harridence, R. T., 2nd-Lt. . . . 545
Harries, D. G., Ty. 2nd-Lt. . . . 623
——— D. J., tempy. Capt. . . 1725
——— J. F., Lt. 721
Harrington, F., Capt. . . . 818, 986
Harriott, W. G., Capt. 615
Harris, A. A. F., 2nd-Lt. . 312, 1120
——— A. E. C., Capt. 69, 268, 887, 951
——— A. F., 2nd-Lt. . . . 579, 734
——— A. M., 2nd-Lt. 565
——— A. T. R., Lt. . . 540, 1674
——— A. M., Nursing Sister . 1734
——— A. W., Lt. 514, 1565
——— A. W., Lt. 943
——— B. A., Esq. 91
——— B. B., 2nd-Lt. 559
——— C., Lt. 626
——— C. E., Lt. . . . 521, 1259
——— C. E. L., Lt. . . . 314, 877
——— C. G., Maj. 608
——— C. J. W., Lt. 835
——— C. P., Lt. . . . 515, 756, 1667
——— C. R., Lt. 753
——— C. J. W., Lt. 835
——— D. B. B., Lt. . . 540a, 1255
——— D. G., Lt. 806
——— E., Capt. 1004
——— E. J., 2nd-Lt. . . 1197, 1676
——— E. J. B., 2nd-Lt. . . . 693
——— E. N., Lt. . . . 292, 1613
——— E. R., Lt. 756
——— E. T., Maj. 1751
——— F. J., Capt. . . . 703, 705
——— F. O., Lt. . . . 515, 1375
——— F. W., Lt. 608
——— F. W., 2nd-Lt. 571a
——— G. G., Capt. . . . 693, 1427
——— G. L., Lt. . . . . .
——— G. R. C., Col. *Lord* . . 683
——— G. W., Lt. 510, 1667
——— G. W., Sub-Condr. . . . 1874
——— H., Lt. 915
——— H. F. E., Lt. . . . 538, 1705
——— H. G., Sub-Condr. . . . 1855
——— H. J., Lt. . . 548, 730, 1373
——— J. E., Lt. 711
——— J. H., Capt. 729
——— J. N. A., 2nd-Lt. . . . 324
——— L., Lt. 34, 532, 1668
——— O. M., Maj. 614
——— P. V., Lt. 595
——— S., 2nd-Lt. 760
——— T. B., Capt. . . . 664, 991
——— T. G. M., Maj. 230, 1898, 1913
——— T. T., Rev. 1908
——— U. A., 2nd-Lt. 556
——— W., Capt. 606, 861
——— W. H., 2nd-Lt. 1009
Harrison, A. F., Lt. . . . 725, 1007
——— A., 2nd-Lt. 1104
——— A. A., 2nd-Lt. . . . 560, 1120
——— A. F. L. H., Lt. . . 29, 525
——— A. G., Maj. 804
——— A. H. P., Lt.-Col. 8, 14, 82, 134, 157, 201, 869, 1152
——— A. J., 2nd-Lt. 567
——— A. M. L., Lt. . . . 299, 1613
——— A. P., Maj. 226, 1463
——— A. S., 2nd-Lt. . . . 355, 1417
——— A. S., 2nd-Lt. 552
——— A. W., Capt. 485
——— C., 2nd-Lt. 56
——— C. C., Capt. 1718
——— C. E., Tempy. Capt. . . . 1724
——— C. H., Capt. . . . 255, 1663
——— C. W., Lt. 726
——— D. H., Lt. 518
——— D. J., Lt. 519
——— E. G., 2nd-Lt. 683
——— E. G. W. W., Capt. . . 636
——— E. J., *Hony.* Lt. . . . 1843

NAMES AND RANK. Page

Harrison, E. L., Lt. . . . 293a, 1283
——— E. P., Lt. 514
——— F. M. K., 2nd-Lt. . . . 760
——— F. S., Lt. 516a, 1624
——— G. B., Lt. 681
——— G. E., 2nd-Lt. 1497
——— G. K., Capt. . . . 686, 720
——— H., Lt.-Col. . 199, 1893, 1912
——— H., 2nd-Lt. 626
——— H., Sub-Condr. 1855
——— H., 2nd-Lt. 358, 1104
——— H. B., 2nd-Lt. . . . 665, 996
——— H. H., Lt. 618
——— H. L., Lt. 440, 1056
——— H. L., 2nd-Lt. . . . 561, 1363
——— H. S., Lt. . . 493, 941, 1676
——— H. T., Capt. 615
——— I. L., Nursing Sister . 1735
——— J., Capt. 62
——— J. A., Lt. 307, 1016
——— J. E., ... 1843
——— J. M. R., Maj. 614
——— J. S., 2nd-Lt. 1678
——— J. W. H., 2nd-Lt. . . . 327
——— K. H., Lt. 529, 1611
——— L. G. B., Maj. . . . 220, 1497
——— M. K., 2nd-Lt. 1517
——— P. F., Maj. 614
——— P., 2nd-Lt. 1040
——— R., Lt. 749
——— R. D., Lt. 816
——— R. S., 2nd-Lt. 724
——— R. St. C., Lt. . . . 533, 1060
——— T., 2nd-Lt. 756
——— T. A., Col. . . 87, 123, 149, 181
——— T. D., Lt. 734
——— T. R., 2nd-Lt. 326, 366, 1621
——— V. M., 2nd-Lt. . . 692, 784
——— W., Condr. 1873
——— W. A., Capt 683, 686
——— W. E. P., 2nd-Lt. . . 543, 970
——— W. H., 2nd-Lt. 1596
——— W. N., Lt. 58, 505
Hari Singh, *Raja Sir*, Capt. . . . 111
Hari Ram, Subdr. 1217
Harriss, S. A., Lt.-Col. . . 131, 1743
Harrold, J. H., Lt. 487, 897
——— P. H., 2nd-Lt. 1122
Harrop, E., 2nd-Lt. 808
——— H. R., Lt. 513, 1223
——— T. G., Lt. 617
Harrup, W. D., Lt. 738, 791
Harr , F. L., Lt. 371, 1098
——— H. T., 2nd-Lt. . . . 356, 1088
——— W. T., Lt. 537, 1007a
Harsing Gusain, Jemdr. 1179
Har Sing Rawat, Subdr. 1181
Harston, J., Maj. . . . 39, 752, 783
Hars Ram, Jemdr. 1682
Hart, A. C., Lt. 55, 749
——— A. E., Capt. 45, 754
——— A. M., Nursing Sister . . 1735
——— E. G., Maj. . . . 60, 105, 214, 1661
——— E. J. A., Capt. . . . 502, 1704
——— F., Revd. 1908
——— F. A. A., Lt. 623
——— F. W., *Hony.* Lt. 1848
——— H. A., 2nd-Lt. 566
——— J. A., Asst. Surgn. . . . 1794
——— J. H. B. B., Lt. . . . 507, 1154
——— J. H. W., Lt. 758
——— L. W., Lt. . . . 415, 1066, 1533
——— M. R. W., Asst. Surgn. . . 1794
——— N. B., Lt. 508
——— N. F., Lt. 525
——— N. V., Lt. 317, 1094
——— S., Capt. 615
——— V., Capt. 831
——— W., 2nd-Lt. 358, 1389
——— W. C., 2nd-Lt. . . . 557, 1701
——— W. E. F., Asst. Surgn. . . 1797
——— W. P. B., Senr. Asst. Surgn. 1787
Hart-Synnot, A. H. S., Maj. . . 104
Hartigan, G. E. R. S., Lt. . . . 314
——— K. L. S., 2nd-Lt. . . . 365
Hartland, B. S., Lt. . . 318, 941, 1641
——— E. W., 2nd-Lt. 708
Hartle, J. C., Lt. 798
——— J. W. L., 2nd-Lt. 561
Hartles, W. E., Lt. 23, 1677
Hartless, L. C., 2nd-Lt. . . 553, 1303
Hartley, A. F., Maj. . . . 230, 895
——— C. R. B., Lt. . . . 463, 1187
——— E. R., 2nd-Lt. . . . 78, 494,

NAMES AND RANK. Page.

Hartley, F., 2nd-Lt. 712
——— G. C., 2nd-Lt. 1675
——— G. McG., Asst. Surgn. . 1802
——— G. R., *Hony.* Maj. . . 1844
——— J. M., Esq. 91
——— R. B., Lt. 787
Hartman, V. J., 2nd-Lt. . . 474, 1645
Hartmann, W. H., Lt. . . . 415, 881
Hartnett, G. W., Sub-Condr. . . 1873
——— W. F., 2nd-Lt. 535
Hartnoll, E. S., Lt. . . . 542, 1291
——— F. B., Capt. . . . 520, 1633
——— F. H., Lt. 375, 1651
——— P. E., Sub-Condr. . . . 1875
Harton, J. G., Condr. 1867
Hartree, K., Capt. 751
Hartridge, P., Lt. 761
Hartshorn, E. P., Lt. . . 425, 1636
Hartwell, B. C. D. A., Maj. *Sir*.
——— J. R., Capt. 260, 1596, 1613
Harty, E. P., 2nd-Lt. 742
Harun Khan, Jemdr. 1424
Harvest, H. deV., Lt.-Col. 108, 190, 1883, 1911
Harvey, A., Sub-Condr. 1867
——— A. B., Lt.-Col. . 138, 154, 185
——— A. E. D., Asst. Surgn. . 1794
——— A. F., Sub-Condr. . . . 1870
——— A. F. B., Maj. . . . 101, 614
——— A. G., 2nd-Lt. 832
——— A. J., 2nd-Lt. 1677
——— A. W. M., Bt. Lt.-Col. 62, 138, 1751
——— B. L., Lt. 517
——— C., Lt. 313, 929
——— C. A. F., 2nd-Lt. . . . 1032
——— C. O., Capt. . . 54, 267, 957
——— C. W., 2nd-Lt. . . . 559, 1678
——— C. W. L., Lt. . 387, 734, 1596
——— D. V., Lt. . . . 24, 527, 1677
——— E. F., Lt. 425
——— E. E., 2nd-Lt. 93, 793
——— E. L., 2nd-Lt. 780
——— F., Lt.-Col. 62, 133
——— F. J., Maj. 697
——— G., Capt. 603
——— G. B., Capt. . . . 278, 883
——— G. F., Revd. 1907
——— G. J., Asst. Surgn. . . . 1797
——— G. T. B., Maj. . 489, 852, 965
——— H., Lt. 352
——— J., Tempy. Capt. . . . 1724
——— J. A. F., Sr. Asst. Surgn. 1789
——— J. F. B., Lt. . . . 508, 997
——— J. S., Lt. 375, 1505
——— M. E., Nursing Sister . 1734
——— M. P., 2nd-Lt. 1060
——— N. S., Sr. Asst. Surgn. . 1788
——— R. H. M., 2nd-Lt. 527, 610, 1671
——— R. L. . . . . . 599—600
——— R. K., Lt.-Col. 720
——— R. O., 2nd-Lt. 56
——— S. E., Capt. 581
——— W., *Hony.* Maj. 1849
——— W. H., 2nd-Lt. . . . 77, 1884
——— W. F., Maj. 1746
——— W. N., Lt. 493
——— W. S., Capt. 852
Harvey-Hurst, A. J., Lt. . 487, 998
Harvey-Kelly, C. H. G. H., Bt.-Maj. 52, 256, 1505
——— G. I., Lt. . . 8, 82, 540a
——— H. St. G. H., Maj. 221, 1275
Harward, A. J. N., Col. 125, 153, 191
——— R. B., Capt. . . 272, 1307, 1457
Harwood, A., Lt. 761
——— G., Lt. 747
——— G. E., Lt. 1122
——— W. A., Lt. . . 519, 829, 986
——— W. F., 2nd-Lt. 565
Hasham Khan Hayat Khan Pathan, Sub-Asst. Surgn. . . 1530
Hasham Ali Khan, Subdr. . . . 1569
Hasan Ali Khan, Ressdr. 936
——— Jemdr. 1012
——— Din Dar, Sub-Asst. Surgn. 1816
——— Shah, Ressdr. 892
——— Jemdr. 1104
——— Muhammad, Subdr.-Maj. . 1419
Haseler, A. R., Lt. 488
Hasell-Wright, E. L. C., Lt. 312, 1375
Hashan Khan, Jemdr. 912
Hashim Ali, Sub-Asst. Surgn. . 1817
——— Din, Jemdr. 1435

Names and Rank.	Page.
Hashim Shah, Jemdr.	990
Hashmat Ali Khan, Ressdr.	1681
Haskard, D. D., Capt.	634
Haskew, L. C., Lt. 524, 1666,	1702
Haslam, B. E., Lt. 503,	885
—— B. H., 2nd-Lt.	539
Haslehust, G. W., Maj. 99, 206,	1136
—— H. M., Lt. 84—86, 507,	1489
Hasler, H. J., Lt. 295,	1479
Haslett, R. S., 2nd-Lt.	324
Hasluck, R. F., Lt. 68, 517,	1005
Hasnip, G. H., Lt. 353,	1491
Hassamkhan Hayatkhan Pathan, Sub-Asst. Surgn.	1830
Hassall, T. C. D., Lt.	711
Hassan, Subdr.	1068
Hassan Muhammad, Subdr.	1132
—— Jemdr.	1441
—— Jemdr.	1368
—— Shah, Jemdr.	1208
Hassells-Yates, G. A., Maj. 8, 15,	633
Hasso Khan, Ressdr.	1683
—— Jemdr.	1681
Hastal Sing Gurung, Jemdr.	1637
Hastabir Limbu, Jemdr.	1657
—— Jemdr.	1653
—— Jemdr.	1631
——Thapa, Jemdr.	1627
Hastbir Thapa, Subdr.	1621
Hastibir Limbu, Jemdr.	1631
—— Rana, Jemdr.	1619
Hastdal Thapa, Jemdr.	1619
Hasted, A. W. O., Lt. 389,	1465
—— J. O. C., Capt.	761
Haste Gurung, Jemdr.	1598
Hastie, H. B., Lt.	536
Hastings, A. W., Esq.	88
—— C., 2nd-Lt.	733
—— E. S., Maj.-Genl. 118, 146, 173, 1375,	1918
—— E. W., 2nd-Lt.	788
—— F. J. B., 2nd-Lt.	668
—— G. F., Maj. 229,	917
—— G. L., Lt.	583
—— J. E., 2nd-Lt. 565,	1255
—— J. M., Lt. 375,	933
—— P. G., 2nd-Lt. 365,	947
—— W. H., Bt.-Maj. 107, 243,	1375
Hasturam Gurung, Subdr.	1630
Haswell, C. H., Bt.-Lt.-Col. 29, 139,	663
—— F. W., Lt. 465,	1431
—— J. F., Lt.-Col. 725,	1722
—— P. L., Lt. 529,	1042
Hatch, Col. A. V.	57
—— K., Lt. 297,	939
—— W. E. J., Capt.	779
Hatchard-Smith, W. H., Lt.	720
Hatchell, C. F. W., Revd.	1907
—— D. T., Col.	177
Hatchman, G. H. S., 2nd-Lt.	561
Hatfeild, H. S., Capt.	581
Hathaway, C. J., Lt.	1106
—— H., Lt.	597
—— J. R., Lt. 492,	933
Hathiramani, H. T., Tempy. Capt.	1772
Hathornthwaite, J. C., Maj. 226,	1479
Hatti, Subdr.	1165
Hattiangadi, S. K., Tempy. Capt.	1769
Hatton, F., Capt.	760
—— S. O., 2nd-Lt.	552
—— W. A., Lt. 352,	1018
Haughton, C. S., Lt.	777
—— E. J. H., Maj. 16, 21, 238,	1423
—— H., 2nd-Lt.	
—— H. L., Maj. 107, 206,	1379
—— H. L., Maj. 69, 241,	1166
—— S. G. S., Maj.	1752
—— W. B., Lt. 523,	986
—— W. H., 2nd-Lt.	1467
Haulton, W. F., Capt.	
Hauser, S. B., Lt. 514,	1395
Haveli Ram Ghei, Sub-Asst. Surgn.	1818
Havelock, L. A., 2nd-Lt.	556
—— R. A. S., 2nd-Lt. 557,	1347
—— W. A. C., Lt. 522,	1403
Haviland, L. P., Capt. 253,	899
Haward, E., Lt.	805
Hawes, A. J., Ty. Capt.	1725
—— C. H., Lt.-Col. 54, 138, 162, 198, 867,	919
—— C. M., Maj. 52, 243,	1098
—— E. H., 2nd-Lt.	571a
—— E. L., Lt.	720
Hawes, E. J., Lt.	975
—— H. F., Lt.	752
—— W., 2nd-Lt. 637,	978
Hawke, A. W., Lt. 68,	761
—— W., Lt.	630
Hawken, B. R. P., Lt. 440,	1457
—— G. H., Lt. 440,	1078
Hawker, G. C., Lt. 22,	623
—— G. V., 2nd-Lt. 550,	1675
Hawkes C. C., Capt.	726
—— F. R., Lt. 516a,	1451
—— G., Col. 179,	1918
—— G. E., Lt. 208, 300, 1239,	1513
—— H. Capt.	1217
—— H. M. P., *Sir*, Maj.-Genl. 1925,	1955
—— J. S., 2nd-Lt. 354,	1515
—— L. V. C., Lt. 357,	1317
—— M. H., Lt. 291, 1066,	1215
—— R., Lt.-Col. 137, 159, 197, 868,	1026
—— T. Y., 2nd-Lt. 316,	927
—— W. B., Lt.	540a
—— W. C. W., Lt.-Col. 139, 162, 201, 868, 1066,	1068
Hawkeswood, H. W., Lt. 492, 925,	1517
Hawkins, A., *Hony.* Lt.	685
—— A. B., Capt.	1717
—— A. E., Capt. 736,	1004
—— A. S., Lt.	638
—— E. J., 2nd-Lt.	564
—— E. N., Lt.	776a
—— F., Col. 180,	1917
—— F. H., Capt. 608,	1181
—— F. L., 2nd-Lt. 363,	1166
—— F. T., Lt. 304,	1467
—— G. A., Lt.-Col. 129, 155, 188,	1660
—— H. R., 2nd-Lt.	591
—— H. W., Lt. 540,	1673
—— J., 2nd-Lt.	742
—— M., 2nd-Lt. 475,	929
—— R. I., Lt.	785
—— R. P., Capt. 486,	1675
—— R. S., Col. 857,	1893
Hawks, G. A., Maj. 43, 81,	208
Hawksworth, P., Lt. 314,	1199
Hawley, N. F., Lt.	534
—— N. W., Lt.	630
—— W. G. B. I., Maj. 214,	927
Haworth, A. N., Tempy. Capt.	1725
—— G. M., 2nd-Lt. 476,	1411
—— J. K., Capt.	1717
—— L. B. H., Maj. 53, 104, 207, 1889,	1913
—— R., 2nd-Lt. 566,	1070
Haworth-Booth, M., 2nd-Lt.	624
Haws, A. H.	582
Hawthorn, D. C., Lt.	470
Hawthorne, C. F. D., Lt.	710
Hawtrey, H. C., Lt.-Col. 137, 664,	1001
Hay, A., Capt. 105,	636
—— A. S., Maj. 56, 217,	933
—— C. E., Lt. 312, 899,	931
—— C. E. E., Lt.	595
—— C. M., Lt.	445
—— C. J. B., Lt.-Col. 51, 98, 102, 104, 137, 159, 214,	954
—— D. J., Lt.	715
—— E. B., Lt. 522,	1231
—— E. F. A., Lt. 493,	1291
—— E. J. A., 2nd-Lt.	326
—— F. S., 2nd-Lt.	530
—— F. W., Capt.	1756
—— J. A. C., 2nd-Lt. 555,	1413
—— J. B. W., Capt. 254, 1165,	1211
—— J. H., Capt.	752
—— L. T., Lt.-Col. 134, 157, 194, 867,	883
—— R., Capt.	1760
—— R. W. M., 2nd-Lt.	1335
—— T. W. D., Lt. 415,	1305
—— W., Capt.	1838
—— W. N., Lt.-Col. 98, 103, 104, 105, 138, 159, 197, 871,	1513
—— W. R., Lt. 380, 503, 738,	1116
Hay Jahans, P. F., 2nd-Lt.	563
Hayat Ali, Jemdr.	976
Hayat Ali Beg, Ressdr.	930
—— Jemdr.	1515
—— Gul Khan, Jemdr.	948
——Khan, Jemdr.	1129
—— Jemdr.	1412
—— Subdr.	1086
—— Subdr.	1283
Hayat Ali Khan, Jemdr.	966
——Khan, Jemdr.	894
——Muhammad, Jemdr.	1436
—— Jemdr.	1124
——Khan, Jemdr.	1452
—— Mir Khan, Ressdr.	914
Haycraft, W. C. S., Capt. 273,	1379
Haycroft, J. S., Lt. 526,	1671
Haydon, J. R., 2nd-Lt.	
—— L. R., Capt.	1732
—— W. P., Lt.-Col. 60, 136, 158, 196,	1661
Haye, N., 2nd-Lt. 551,	1162
Hayes, A. H., Maj.	1711
—— A. W., 2nd-Lt. 357,	1164
—— B. H., 2nd-Lt.	710
—— C. R., 2nd-Lt. 549,	1699
—— E. C., Lt.	692
—— E. J. T., Lt. 397,	1223
—— E. R. C., Lt. 529,	1698
—— F., Lt.	617
—— F. B., Capt. 22,	1838
—— G. F. L., Hony. Lt.	746
—— H. C., 2nd-Lt.	566
—— J. W., Capt.	1839
—— L., Lt. 508,	1098
—— M. E., 2nd-Lt.	757
—— P., Capt.	1713
—— T., Lt. 712,	1658
—— W., Condr.	1861
Hayes-Flanagan, J., 2nd-Lt.	562
Hayes-Newington, E. A., Lt. 321, 1136,	1411
Hayes-Sadler, J., Maj. 633, 972,	975
Hayes-Watkins, A. S., Lt.	492
Hayfield, B., Lt.	503
Haygarth, R. N., 2nd-Lt.	564
Hayhurst, J. H. J., *Hony.* Maj.	1844
Hayland, A. K., Maj.	1889
Hayling, A. A., 2nd-Lt. 563,	1696
Haylock, S. W., Lt. 455,—459, 1335,	1575
Hayman, F. K., Capt.	1718
—— J. R., Capt.	1717
—— W. C., *Hony.* Lt.	1846
Haymes, G. N., 2nd-Lt.	354
—— T. F., Lt. 354,	1577
—— W. R., 2nd-Lt.	96
Hayne, A. R. S., Lt. 512,	1283
—— C. W., Lt. 506,	1178
—— M. L., Lt. 300,—301,	1170
Haynes, A. E., Lt. 430,	1062
—— A. H., 2nd-Lt.	362
—— F. J., 2nd-Lt.	776a
—— G. W., Lt.	776a
—— W. F., Capt. 259,	1102
—— S. J., Lt. 29, 353,	1639
—— W., 2nd-Lt.	618
Hays, H., 2nd-Lt. 545,	987
—— W., 2nd-Lt.	359
Haysey, R., 2nd-Lt.	1465
Haysom, R. A., Lt. 515,	1144
Hayter, L. C. D., 2nd-Lt. 554,	1060
—— L. N., 2nd-Lt.	362
—— R. T. M., Asst. Surgn.	1801
Hayward, C. D., 2nd-Lt.	1469
—— C. J., 2nd-Lt.	808
—— E. 2nd-Lt.	786
—— E. J., 2nd-Lt.	735a
—— F. G., Lt.-Col.	628
—— H. B., 2nd-Lt.	1455
—— H. H., 2nd-Lt. 566,	1523
—— L. G., 2nd-Lt.	779
—— M. H. W., Lt.-Col.	654
—— R. D., Lt. 541,	988
—— R. H. P., Lt.	617
—— S. D., 2nd-Lt.	681
—— T. E., Lt. 300, 1148,	1332
—— W. D., Lt.-Col. 79, 132,	1743
Haywood J. D., Lt.	1839
—— J. S., Lt. 537,	1668
—— W. G., Sub-Condr.	1868
—— W. T., Sub-Condr.	1864
Hazar Singh Rai, Jemdr.	1631
Hazara Singh, Jemdr.	938
—— Subdr.	1533
—— Jemdr.	1134
—— Jemdr.	1154
—— Subdr.	1533
—— Jemdr.	1702
—— Risldr.	1684
—— *Hony.* Capt.	2101
—— Jemdr.	1577
—— Subdr.	1303
—— Jemdr.	916
—— Jemdr.	944
—— Jemdr.	1103

NAMES AND RANK. Page.

Hazara Singh, Jemdr. 932
—— Sub-Asst. Surgn. 1817
—— Jemdr. 1698
—— Subdr. 1235, 1280
Hazarat Gul, Khan, Jemdr. 1260
Hazari Mall, Sub-Asst. Surgn. 1816
—— Singh, Subdr. 1389
—— Jemdr. 1032
—— Jemdr. 1035
Hazdas Singh, Ressdr. 1686
Hazell, A. J., 2nd-Lt. 723
Hazle, A. W., Asst. Surgn. 1796
Hazrat Gul, Subdr. 1245
Hazledine, H. St. J. C., Capt. 768
Hazur Ahmed, Ressdr. 884, 959
Hazura Singh, Risldr. 1686
—— Subdr.-Maj. 1367
—— Jemdr. 950
—— Grewal, Asst. Surgn. 1818
Hazzard, F., 2nd-Lt. 686
—— P. H., 2nd-Lt. 564, 1255
Head, E. B., 2nd-Lt. 732
—— F. C., Lt. 505
—— H. K., Capt. 505, 830, 1667
—— W. R., Lt. 543
—— H., 2nd-Lt. 1695
—— H. R., 2nd-Lt. 473
Headrick, W. S., 2nd-Lt. 790
Heald, B. H., Capt. 848
Heale, A. S., Capt. 1714
—— B. J. W., Maj. 108, 226, 1901, 1913
Healing, N. C., Maj. 634
Healis, C. F., Lt. 732
Healy, H. H., Capt. 747
—— J. J., Lt. 522
Heanly, P. W., Lt. 776
Heape, B. R., Lt. 617
—— C. H., 2nd-Lt. 491
Heaps, W., 2nd-Lt. 733
Heapy, G., Sub-Condr. 1855
Heard, R., Lt.-Col. 3, 10, 125, 1742
Hearn, C. R. T., 2nd-Lt. 769
—— E. S., Lt. 505, 815, 1048
—— E. R., Capt. 502, 683, 1667
—— G. R., Lt.-Col. 139, 662, 1881
—— G. S., Lt. 517, 1453
—— J., Lt. 508, 1110
—— L. A., 2nd-Lt. 696, 1008
—— W. S. H., Lt. 295, 939
Hearnden, H. J., 2nd-Lt. 618
Hearring, T. C., Lt. 435
Hearsey, A. J., Capt. 816
—— G. A. C., Lt. 489
—— G. C. M., 2nd-Lt. 877
—— J. T. M., 2nd-Lt. 362
—— L. D. W., Lt. 489, 608, 877
—— N. N., Lt. 507, 1613
Hearson, H. F. P. 604
Hearth, L. A., 2nd-Lt. 796
Heath, A., 2nd-Lt. 568
—— C. J., Lt. 533, 1170
—— D. F., 2nd-Lt. 758
—— E., 2nd-Lt. 751
—— F., 2nd-Lt. 693
—— F. M., Lt. 692
—— F. S., Lt. 489, 925
—— H. C. S., Lt. 410, 1231
—— H. L., Capt. 804
—— H. N., Lt. 399
—— L. J., 2nd-Lt. 561
—— L. M., Capt. 66, 256, 1094
Heathcote, C. N., Lt. 514, 1565
—— F. C., Lt. 519
—— G., Tempy. Capt. 1724
—— J. R. C., Maj. 577
Heaton, A. D., Lt. 420, 1040
—— R. B., 2nd-Lt. 324
—— T. B., Capt. 1717
Heaton-Armstrong, J. D., Capt. 486
Heaton-Ellis, J. S., Capt. 636, 980
Heaver, L. G. H., Lt. 304, 1187
Hebberd, E. St. J., Lt. 528, 1699
Hebbert, G. K. P., Capt. 636
—— P. B., Capt. 273, 1199, 1898
Hebron, C. W., 2nd-Lt. 475, 1513
Hebson, T. E., 2nd-Lt. 541, 1319
Heckle, E. W., 2nd-Lt. 713
Hector, S. A., Lt. 1630
Hedger, H., 2nd-Lt. 363, 1074
Hedley, I. M., Lt. 595
—— R. W., 2nd-Lt. 358, 1447
Heelis, G. H., Maj. 725
—— G. H., Lt. 724, 739
Heenan, J., 2nd-Lt. 20, 666
Heffernan, W. St. M., Sr. Asst. Surgn. 1789

Heffernan, H. W., Lt.-Col. 131, 156, 191
—— P., Maj. 1752
Hegarty, E., 2nd-Lt. 554, 1417
—— T. C. V., 2nd-Lt. 668
Heggart, J. E., Lt. 318, 1339
Heggs, T. B., Capt. 683
Hehir, P., Maj.-Genl. 19, 120, 1736
Heilgers, F. W., 2nd-Lt. 855
Hele, J. W., Capt. 724
—— G. C., Lieut. 516, 1363
Helliwell, S., Sub-Condr. 1864
Hellier, J., Maj. 658
—— J., Esq. 89
Helm, F., Lt. 505
Helme, C. G. G., 2nd-Lt. 559
—— F. S., Lt. 518, 1154
Helmore, G. D., 2nd-Lt. 1676
Helps, A. O., 2nd-Lt. 702
—— M. D., Capt. 66, 679
Helson, G., 2nd-Lt. 703, 706, 1489
Hely, T. F., Lt. 518, 1343
Hely Hutchinson, J. W., Capt. 856
Hema Ram, Jemdr. 1456
Hemeon, C. R., 2nd-Lt. 313, 1199
Hem-raj, Jemdr. 1687
—— Jemdr. 1039
Hem Singh, Subdr. 975
—— Jemdr. 1031
Hemming, C. H., Revd. 1904
Hempsall, E., Sub-Condr. 1864
Hemsley, C., Maj. 54, 245, 1275
Hemsted, F., 2nd-Lt. 78, 112, 486
Hemsworth, D. D., Lt. 1886
—— N. E. C., Lt. 298a, 1136
—— T. G. F., Lt. 669, 997
Henderson, A. G. R., 2nd-Lt. 571
—— A. G., 2nd-Lt. 1000
—— A. N., Lt. 531, 1668
—— C., 2nd-Lt. 362
—— C. A., Capt. 802
—— C. G. H., Maj. 107, 220, 871, 929, 1465
—— C. H., Lt. 491, 913
—— C. R. B., Maj. 743
—— D. R., 2nd-Lt. 560, 1311
—— D. S., Lt. 536, 1673
—— D. T., Lt. 491
—— E. G. S., 2nd-Lt. 330, 1499, 1500
—— G., Maj. 240, 947
—— G. B., Capt. 272, 1078
—— G. H., Lt. 766
—— G. P., 2nd-Lt. 568
—— G. S., Capt. 758
—— H. W. W., Lt. 534, 998
—— H. W. W., Lt. 1679
—— J., Condr. 1853
—— J. H., Capt. 1919
—— J. H., Maj. 746
—— J. L., Capt. 830
—— J. S. W., 2nd-Lt. 976
—— J. T., 2nd-Lt. 18, 26, 719
—— J. W., 2nd-Lt. 555, 1441
—— K., Maj. 99, 209, 1178
—— K. D. M., Maj. 221, 868, 1036
—— L. D., Lt. 75
—— M. H., Bt.-Col. 137, 159, 197, 867, 923
—— R. E., Capt. 827
—— R. S., Lt. 521, 1082
—— R. W., Maj. 208, 907, 913
—— S. H., Lt.-Col. 127, 1739
—— T. A., 2nd-Lt. 744
—— T. H., Lt.-Col. 128, 154, 183
—— W., Lt. 818
—— W. L., Capt. 485, 827, 903
—— W. S. R., Lt. 399, 1415
Hendley, C. E., Maj.-Genl. 12, 28, 119, 147, 175
—— H., Maj.-Genl. 1, 42, 119, 1738
Hendrick, C. A., Lt. 318, 1187
—— U. St. J., Asst. Surgn. 1802
Hendricks, St. J. E., Sr. Asst. Surgn. 1792
—— W. St. A., Asst. Surgn. 1800
Hendry, C. V., 2nd-Lt. 744
—— F. C., Lt. 504, 827, 1030
—— J. M., 2nd-Lt. 763
Henegan, J., Lt.-Col. 128, 154, 183
Heneghan, W. J., Lt. 770
Henke, F. E., Lt. 537, 1648
Henley, C. F., Maj. 745
Henn, J. F., Lt. 750
Hennell, A. M., Maj. 209, 935
—— F. J., Lt. 515

Hennessey, D. M. L., Lt. 317, 1052
Hennessy, E. P., Sub-Condr. 1856
—— J., Lt.-Col. 130, 1710
—— J. A., 2nd-Lt. 766
—— J. M. R., Capt. 1758
—— L. J., 2nd.-Lt. 703
—— J. P. C., Col. 127, 150, 189, 1660
—— T., 2nd-Lt. 764
—— T. F., 2nd-Lt. 637, 972
Henriques, C. Q., 2nd-Lt. 544, 987
Henrotin, H. V., Lt. 389, 933
Henry, A. G., 2nd-Lt.
—— G. C., 2nd-Lt. 668, 1008
—— J. B. A., Capt. 288, 290, 1058
—— M., Lt. 312, 1459, 1467
—— R., *Hony.* Capt. 1846
—— T. R., Lt. 371, 1056
—— W., 2nd-Lt. 554
—— W. D., Col. *Sir* 806
—— W. E. C., Revd. 1904
—— W. R. P., Maj. 843, 897
Henshaw, A. H., Sub-Condr. 1864
—— A. G., 2nd-Lt. 329, 1305
—— R. L., 2nd-Lt. 761
Hensley, F. K., Maj. 107, 230, 964, 968
Henslowe, F. J. D., Lt.-Col. 79, 132, 156, 192
Hensman, H. S., Tempy.-Lt. 1772
Henson, F. J., Maj. 751
—— P. H., Ty. Capt. 1726
—— R. K., Lt. 302, 1120, 1203
Henstock, P., Capt. 827
—— R. P., Capt. 267, 1629
Henwood, F. H., Lt. 455
—— F. W., Lt. 1283
Hepburn, C. A., Lt. 622
—— R., 2nd-Lt. 626
—— W. R., 2nd-Lt. 653
—— W. S., 2nd-Lt. 559, 1387
Heppenstall, C. H., Capt. 1759
Hepper, A. W., Maj. 663
—— E. C., Maj. 1748
Heppolette, F. H., 2nd-Lt. 562
—— J. E. R., Asst. Surgn. 1800
Hepworth, M. A., Lt. 307, 1102
—— N., Lt. 90, 514
—— S. M., Capt. 1756
Herald, A. T. A., Lt. 492
Herapath, C. B., Capt. 263, 1082, 1664
—— D. K. MacK., Capt. 54, 583
Heraper, J. C., Lt. 300, 1315, 1455
Herbert, C., Lt.-Col. 27, 185, 1883, 1911
—— D., Lt.-Col. 187, 1898, 1911
—— D. M. A., Capt. 288, 290, 1335
—— G. M., Bt. Lt.-Col. 736
—— G. S., *Hon.* Capt. 755
—— H., Lt.-Col. 1786
—— H. A., Lt. 318, 1020, 1048
—— H. R., Capt. 367, 1676
—— Lynch, E., Capt. 1910
—— L., Maj.-Genl. 57
—— L. N., Br.-Genl. 14, 32, 103, 120, 123, 149, 181
—— R. L., Lt. 519, 1263
—— V. A., Capt. 486, 929
—— W. C., 2nd-Lt. 776
—— W. C., 2nd-Lt. 759
Herbison, L. J., 2nd-Lt. 881
Herd, W. L., Capt. 787
Herda Ram, Jemdr. 1579
Herdon, A. H., Lt. 522, 1701
—— B. L., Lt.-Col. 533, 1505
—— H. E., Lt.-Col. 98, 139, 162, 199, 869, 1243
—— S. P. N., Maj. 82, 509, 804
Hereford, E. H., Lt. 538
—— R. J., 2nd-Lt. 559, 1104
Hering, M. O. L., Capt. 707
Heriz-Smith, G. V., Capt. 257, 1128
Herman, C. R., 2nd-Lt. 546, 1505
—— G. L., Capt. 750
Herne, E. Eu. P., 2nd-Lt. 759
—— R. V., Lt. 1483, 521
Heron, D., Maj. 107, 1750, 1889
—— E. F., 2nd-Lt. 558
—— G. C. H., Lt. 595
—— J., 2nd-Lt. 1519
—— J. M., Capt. 1722
—— R., Lt. 617
Heron-Jones, C. V., Lt. 506, 1443
Herpreaves, C. K., 2nd-Lt. 555
Herrick, J. R. W., Lt. 389, 1604
—— R. deS. B., Capt. 1756
—— R. L. W., Capt. 485, 929

Names and Rank. Page.

Herridge, W. J., Lt. . . . 544, 1385
Herring, C. S., Capt. . . . 747
—— R. L., Lt. . . 292a, 1596
—— T. C., Lt. . . 425, 1487
Herriot, T. R., Tempy. Capt. . . 1723
Hervaiwil-Cox, V. M., Lt. . . . 320
Hervey, E. G., Lt. . . . 506, 1108
—— W. F. G., Lt. . . 309, 1647
Herschorn, D. F., Lt. . . . 622
Hervyy, N. L., 2nd-Lt. . . 84, 566
Heselton, R. T., 2nd-Lt. . . 548, 1305
Hesketh, R. W., Col. . . . 177
—— M., 2nd-Lt. 571a
—— W., Maj. . . 209, 897
Heslop, G. D., Lt. . . . 490, 951
—— W. T. B., Lt. . . . 619
Hesterlow, A. M. V., Tempy. Lt. 1773
Hetson, G. Lt. . . . . 1489—90
Het Ram, Risldr. 1682
—— Subdr. 1215
—— Jemdr. 1216
—— Jemdr. 1388
—— Ressdr. 878
—— Jemdr. 1400
—— Risaldar 1686
Hetram Singh, Subdr.-Maj. . . 1030
Hetherington, C., 2nd-Lt. . . . 553
Heughan, J. K., 2nd-Lt. . . 556, 1455
Hewat, J. P. D., Capt. . . 781, 798
Hewer, J. A., Lt. . . . 519, 1667
Hewes, E. E., *Hony.* Capt. . . 1844
Hewetson, E. F., 2nd-Lt. . . . 599
—— F. E. A., 2nd-Lt. 90, 516a
—— J. J., Lt. . . 305, 1635
—— R. P., Lt. . . . 622
Hewett, D. P., Lt. . . 493, 1295
—— G., Maj. . 232, 1215, 1217
—— H., 2nd-Lt. . . . 548, 1307
—— H. P., Lt. . . . 487, 919
—— H. S., Lt. . . . 512, 1203
—— J. P. O'C., 2nd-Lt. 317, 1170
—— M. S., Maj. . . 227, 1662
—— W. G. O'C., Lt. . 310, 1215
—— W. P., Maj. . . . 753
—— W. S., Lt.-Col. . . 179, 1917
Hewick, J. S., Lt. . . . 736
Hewitt, B. E., 2nd-Lt. . . . 750
—— D. J., 2nd-Lt. . . 545, 1217
—— C. L. R., Esq., Asst. Insp. 72
—— D. R., Maj. . . 78, 225
—— G., Revd. 1907
—— H. T., Capt. . . 769, 770
—— J., 2nd-Lt. 567
—— J., *Hony.* Maj. . . . 1845
—— S. H., Lt. . . . 352, 1674
—— W. R., Sub-Condr. . . 1874
Hewlett, A., Lt.-Col. 139 162, 199, 867, 947
—— E. A., Bt.-Lt.-Col. 138, 160, 218, 901
—— J. H., Capt. 753
—— K., Vety. Lt. . 40, 604, 861
—— L. M., Capt. . . 753, 785
Hewson, J. W., Sub. Condr. . . 1864
—— F. M., Sub-Condr. . . 1867
—— G. C., Tempy. Asst. Surg. 1804
Hext, C. W., Maj. . . . 225, 1662
—— G. T. B., Maj. . . 225, 1662
—— J. E., Maj. . . . 225, 1662
Hey, B. S., Lt. . . . 420, 1195
—— N. H., 2nd-Lt. 563
Heycock, C. H., Lt.-Col. 129, 661, 867, 991
—— F. W., Revd. 1906
Heyland, A. K., Maj. . . 207, 1661
—— G. D., Maj. . 239, 1479 1589
—— W. F., Lt. 1919
Heyne, G. A. F., 2nd-Lt. . 549, 1675
Heywood, W. H. V., Capt. . . 843
—— H. C. L., Lt. . . 293a, 1311
Hezlet, R. K., Maj. . . . 614
Hiatt, A., *Hony.* Capt. . . . 583
Hibberd, A. S., Lt. . . 298, 1122
—— E., Lt. 528
Hibbert, C. Capt. 1722
—— O. Y., Capt. . 54, 100, 749
—— W. J. C., Capt. . . . 89
Hickey, H. G., Lt. . . . 373, 1509
—— P. F. B., Lt. . . 513, 1433
—— S. G. M., Lt. . . 571a, 1841
Hickie, A. D., Capt. . 313, 843, 1251
—— C. C., Maj. . 239, 1483, 1895
—— F. C., Capt. . . 208, 1664
—— H. W., Maj. . . 243, 1343
—— J. C., Capt. . . 502, 1665
—— J. M., 2nd-Lt. . . 558, 1339
Hickley, A. C., Lt.-Col. . 130, 155, 189
—— V. N., Lt.-Col. . . . 2
Hickman, H. T. D., Capt. 52, 262, 1156
—— W. T., 2nd-Lt. . . . 568
Hickmott, R. R. J., 2nd-Lt. 553, 1705
—— W. H., 2nd-Lt. . 358, 1571
Hickox, S E., Revd. . . . 1903
Hicks, A. N., 2nd-Lt. . . . 626
—— E. S., 2nd-Lt. . 548, 1126, 1675
—— G. A., Capt. 832
—— H. E., Lt. . . . 541, 1048
—— L. J., Lt. . . . 512, 1647
—— M. B., Capt. 653
—— M. G., Lt. 731
—— W., Lt. 1843
Hickson, B. E., Capt. . 78, 267, 925
—— E. B., Capt. 1718
—— J. C. A., 2nd-Lt. . . . 1633
Hidayat Ali Khan, Rsldr. . . 938
Hidayatullah, Sub-Asst. Surgn. . 1813
—— Jemdr. . . . 1103
—— Khan, Jemdr. . . 1495
Hide, T. G. S., Lt. . . 68, 389, 1136
Higby, H. V., Lt. . . . 519, 1573
—— R. W. E., 2nd-Lt. . . . 1001
Higgens, C. R., Lt. . . . 526, 1471
Higgin, H. S., Lt. . . . 505, 1315
Higginbotham, P. P., 2nd-Lt. 328, 1291
Higgins, D., Ty. Lt. . . . 1726
—— F. M. P., 2nd-Lt. . . . 619
—— G. L., Lt. . . 753, 785
—— H., Lt. 776a
—— R. F. R., 2nd-Lt. . . . 1471
Higginson, G. St. G., Lt. . . 505
—— J., 2nd-Lt. 361
—— N., Lt. . . . 524, 1112
Higgs, H. J., Lt. 1006
—— J. V., Lt. . . . 552, 1038
—— P. R., Lt. 95, 111
—— R. H., 2nd-Lt. 1032
Higham, B., Maj. 1752
—— L. A., 2nd-Lt. . . . 91, 563
Highet, H., Ty. Capt. . . . 1724
—— H., Capt. 724
Highett, C. T., Capt. . . . 736
Highley, H. F., 2nd-Lt. . . . 784
—— W. M. B., Lt. . . 535, 1579
Higman, J. W., Lt. 785
Hignett, H. A. Du C., Capt. . . 652
Higson, F., Capt. 691
—— W., 2nd-Lt. . . . 558, 1708
Hikmat, Jemdr. 1129
Hilabrand-Todd, H. V., 2nd-Lt. . 554
Hilbourne, G. M., Lt. . . 638, 1979
Hildebrand, H. A., Maj. . 239, 939
Hilder, E. C., Lt. 749
—— T. M. M., Lt. 585
Hildick-Smith, J. N., Lt. . 380, 1046
Hildreth, G. A., Tempy. Capt. . 1769
Hildyard, R. J. T., Brig.-Genl. 50, 122, 134
Hildyard-Marris, R. W., Lt. . . 543
Hiles, M., Capt. 755
Hill, A., Sub-Condr. 1873
—— A. E., 2nd-Lt. . . . 551, 1521
—— A. H., Capt. . . 522, 848, 1295
—— A. H. C., Capt. 1757
—— A. J., Capt. 6
—— A. J., *Hony.* Lt. . . . 1850
—— A. L., Lt. 59, 504
—— A. H., Lt. 1295
—— A. R., Maj. 850
—— B., Lt. 637
—— B. B., 2nd-Lt. . . . 475, 1607
—— C., Lt. 470, 1365
—— C. E., 2nd-Lt. 1695
—— C. F., Sr. Nursing Sister . . 1734
—— C. F., 2nd-Lt. . . . 561, 1046
—— C. I. H., 2nd-Lt. 708
—— C. P., Capt. . . . 7, 827, 828
—— C. R., Lt. 536
—— D. G., Lt. 520, 1666
—— D. N., Lt. 435, 1166
—— D. R. O., Lt. . . . 311, 1629
—— E. D., Lt. 512, 1060
—— E. E., Revd. 1907
—— E. F. J., Bt.-Lt.-Col. 58, 138, 663, 985
—— E. G., Capt. 1732
—— E. R., Asst. Surgn. . . . 1800
—— E. R., Lt. 697
—— E. W., 2nd-Lt. . . . 542, 992
—— F., Lt. 597
—— F. G. A., 2nd-Lt. . . 543a, 1673
—— F. O. R., 2nd-Lt. . . 329, 1170
—— F. R., Lt. . . . . 455—59, 887
Hill, G. G., Lt. 1567
—— G. D., Lt. 587
—— G. N., Capt. 633
—— H. A., Maj. 831, 1183
—— H. B. C., Lt. . . . 20, 492, 931
—— H. C., Lt.-Col. 136, 158, 196, 1441
—— H. G., 2nd-Lt. 625
—— H. H., 2nd-Lt. . . . 556, 1058
—— H. L., Lt. 506, 1465
—— H. M., Condr. 1855
—— H. P., Maj. 730
—— H. V. J., Lt. 310, 1136
—— H. W. D., Maj. . . . 219, 905
—— J., 2nd-Lt. 618, 713
—— J., Maj.-Genl. . . 119, 147, 175
—— J., Serjt.-Maj. 1879
—— J. R., Lt.-Col. . . 128, 154, 185
—— J. H., Lt. 668, 996
—— K. M. R., 2nd-Lt. . . . 550, 1677
—— L. E., Lt. 506, 1479
—— N. F., Lt. . . . . . .
—— P. E., Lt. 629
—— R. A. C., Lt. 798
—— R. C., Lt. 591
—— R. C. B., Maj. 663, 985
—— R. D. O., Maj. 29, 243, 1600, 1639
—— R. F., Lt. 699, 796
—— R. L., Lt. 799
—— T. W., Condr. 1872
—— W., Condr. 1873
—— W. A., 2nd-Lt. . . . 557, 1046
—— W. A., Maj. 206
—— W. D., Lt. 622
—— W. E., Lt. 556, 1293
—— W. L., 2nd-Lt. 1463
Hill Dickson, J. E., Lt. . . . 768
Hiller, L. O., Lt. 1008
Hill-Forster, R. W., 2nd-Lt. . . 599
Hill-Murray, H. J., 2nd-Lt. . 552, 1375
—— N. J. G., 2nd-Lt. 762, 782
—— S. W. H., 2nd-Lt. . . . 755
Hills-Johnes, Sir J., Lt.-Genl. 1925, 1955
Hill-Whitson, T. E. L., Lt.-Col. 51, 99, 132, 593
Hill-Workman, J., Capt. . . . 579
Hilliard, F. H., 2nd-Lt. . . . 790
—— H. N., Lt.-Col. . 128, 154, 185
—— J. J., Capt. 1838
—— L. M., Lt. 887, 296
—— P. J., Lt. 322, 911
—— R. B., 2nd-Lt. 553
—— W. A., 2nd-Lt. 539
Hillier, F. N., Lt. 623
—— L. O., Lt. 352, 1007a
—— V. F., Lt. 513, 1098
Hillman, A. J. H. S., Lt. . . . 5, 628
—— A. K., Lt. 380, 1315
—— G. E., Lt. 779
Hills, A. C., Capt. 485, 885
—— A. E., Maj. . . . 6, 752, 783
—— G. G., Lt. 511
Hilsdon, H., 2nd-Lt. 579
Hilson, R. J., Maj. 208
Hilton, F. L., *Hony.* Maj. . . . 1849
—— J. G. M., 2nd-Lt. . . 547, 1673
—— W. H., 2nd-Lt. 362
Hiluludin, Ressdr. 1686
Himayat Ali Khan, Jemdr. . . . 1705
Himbahadur Sahi, Jemdr. . . . 1642
Hime, E. H., 2nd-Lt. . . . 365, 1642
Himlal Karki, Jemdr. 1643
Himlal Thapa, Jemdr. 1654
—— Khattri, Jemdr. . . 1622
Himmat Singh, Sub-Asst. Surgn. 1814
Himta, Jemdr. 1202
Hinch, T. W., 2-Lt. 562
Hinchliffe, A. H. S. *Hony.* 2nd-Lt. 188
Hind, F. M. J., 2nd-Lt. 1020
—— H. W., Lt. 385, 1106
—— J. H., Capt. 591
Hind-Fletcher, T. R., Lt. . 413, 1585
Hinde, N. G., Capt. . . . 274, 1600
—— H. T. L., Maj. . . . 246, 1403
—— R. G., Capt. . 263, 1491, 1513
Hinder, B. A. P., Asst. Surgn. . 1802
Hindle, M. D., 2nd-Lt. 329
Hindley, C. E. H., Lt. . . 530, 1671
—— D. C. R., Capt. . . 486, 911
Hindmarsh, F., 2nd-Lt. 695
—— J. H. L., Capt. 276, 1046, 1586
Hindpal Singh, Subdr. 1399
Hinds, W. A. G., Capt. . 19, 277, 1415

NAMES AND RANK. Page

Hingston, A. T., Lt. 380
—— C. A. F., Maj. . . . 1750
—— C. V., Lt. . . . 728, 731
—— F. P. B., Lt. . . 541, 1624
—— H., Capt. 1757
—— J. C. L., Capt. 26, 1713
—— R. W. G., Capt. . . 1756
Hinkley, A. M., Condr. . . . 1854
Hinston, A. T., Lt. 375
Hinton., A. A., 2nd-Lt. 788
—— F. E., 2-Lt. . . . 559, 1060
—— L., Lt. 1007
—— N. R. 619
—— T. A., Sub-Condr. . . . 1855
Hipsey, A. W. L., 2nd-Lt. . . . 1048
Hira, Subdr. 1066
—— Jemdr. 1121
—— Jemdr. 1171
—— Jemdr. 1202
—— Jemdr. 1400
—— Lal, Sr. Sub-Asst. Surgn. 1808
—— Ranch Hodlal James, Assst.-Surgn. 1832
—— Jemdr. 1051
—— Subr. 1327
—— Sub-Asst. Surgn. . . 1810
—— P., Sr. Sub-Asst. Surgn. 1826
—— Jemdr. 1420
—— Singh, Sub-Asst. Surgn. 1813
—— Capt. 1894
—— Sub-Asst. Surgn. 1808
—— Sub-Asst. Surgn. 1810
—— Jemdr. . . . 1252
—— Jemdr. . . . 882
—— Jemdr. . . . 892
—— Sub-Asst. Surgn. 1811
—— Risldr. Maj. . 912
—— Subdr. . . . 1682
—— Jemdr. . . . 1079
—— M., Sr. Sub-Asst. Surgn. 1826
—— Ressdr. 906
—— Jemdr. 1111
—— Jemdr. 1304
—— Subdr. Maj. . . 1101
—— Risldr.-Maj. . . 931
—— Singh, Subdr. . 1565
—— Singh, Jemdr. . . 1149
Hirajit Thapa, Jemdr. 1615
Hira Singh, *Hony*. Lt. 2103
—— Subdr. 1335
—— Subdr. 1239
—— Subdr. 1112
—— Jemdr. 1151
—— Jemdr. 1022
—— Jemdr. 1252
—— Jemdr. 1112
—— Jemdr. 1067
—— Jemdr. 1376
—— Jemdr. 1256
—— Jemdr. 1228
—— Jemdr. 1433
—— Subdr. 1235
—— Jemdr. 1071
—— Subdr. 1189
—— Bisht, Jemdr. . . . 1223
—— Sing Thapa, Jemdr. 1637
—— Subdr. 1211
—— Sing, *Hony*. Capt. 2101, 2103
—— Subdr. . . . 1104, 1699
Hiralal Parshotamdash Sheth, Asst. Surgn. 1832
Hira Rawat Subdr. 1471
Hira Rawat, Jemdr. 1472
—— Subdr. 1483
Hirarup Sahi, Jemdr. 1625
Hirasing Thapa, *Hony*. Capt. 2104
Hird, S., 2nd-Lt. 750, 784
Hirendra Nath Sarkar, Subdr. . 1219
Hirjeebhai Narbhoram Dave, Sub-Asst. Surgn. 1830
Hirons, J. T. C., 2nd-Lt. . . . 691
Hirsch, J. H., Lt. 591
—— L., Maj. 1749
—— V. R., Capt. 1725
Hirst, A. H., Lt. 440, 1399
—— F. C., Maj. 45, 209, 1016, 1893, 1913
—— F. H., Lt. 576a, 1367
—— G. G., Maj. 26 1751
—— H., 2nd-Lt. 630
—— J., 2nd-Lt. . . . 77, 735, 1884
—— J. B., 2nd-Lt. . . . 947, 961

NAMES AND RANK. Page.

Hirst J. E., Lt. 299, 1066
—— W. E. L., 2nd-Lt. 798
Hirtes, A. E., Sub.-Condr. . . . 1873
Hiru Ramji, Jemdr. 1447
—— Bhatose, Jemdr. . . . 1469
Hisco, R., 2nd-Lt. 549, 1479
Hislop, A. F., Maj. . . . 1663, 1919
—— J. F. G., Capt. 741
—— J. H., Capt. 1757
—— R. G. G., Lt. . . . 536, 1367
Hitchcock, F. C., 2nd-Lt. . . 535, 1672
—— L. 1009
—— L., 2nd-Lt. 733
—— L. W., Lt. . 446—49, 1106
—— R. H., Lt. . . . 542, 1309
—— R. V., Capt. . . 520, 818
—— W. H. H., 2nd-Lt. . . 563
Hitchens, A. P., 2nd-Lt. . . . 638
Hitchin, F. E., 2nd-Lt. . . 564, 1335
Hitchins, C. S., Lt. 514
—— H. E., Lt.-Col. 131, 156, 192
Hixon, A. J., Asst. Surgn. . . . 1802
Hla Bu, Jemdr. 1295
Hoad, H. R., 2nd-Lt. 1142
Hoade, J. R., Lt. . . . 509, 1143
Hoare, A. S., Capt. 756
—— A. W., Lt. . . . 353, 1305
—— C. G., Maj. . 122, 234, 949
—— C. E., Maj. 730
—— E. R., Lt. 597
—— F. C., Capt. 628
—— G. G., Lt. 756
—— H. P. B., Lt. . . . 423, 1130
—— J. G., Lt. . . . 430, 1144
—— L. J., 2nd-Lt. 712
—— W., Capt. 1704
—— W. I., *Hony*. Capt. . . 1846
Hobart, P. C. S., Capt. . . 665, 985
—— R. C., 2nd-Lt. 658
Hobbs., A. E. S., Ty. Capt. . . . 494
—— A. J., Lt. . . . 322, 1110
—— A. W., 2nd-Lt. 623
—— C. C. A., Capt. . 509, 852, 997
—— C. E., Lt. 303, 905
—— C. F. G., Lt. 798
—— E. V., Capt. 1839
—— G. M., Lt. 1618
—— H. R., 2nd-Lt. 618
—— J. M., Lt. 423
—— W. C., *Hony*. Maj. . . . 1843
Hobday, E. M., Maj. . . . 245, 1187
—— J. R., Col. 56
—— W. E., Lt. . 17, 28, 514, 1379
Hobgen, D. J., Lt. 628
Hobhouse, R. O., Lt. 705
Hobkirk, R., Capt. 89, 758
Hobley, A. J., Lt. . . . 415, 1182
Hobson, C. M., Lt. 741
—— E., Lt. 293, 931
—— J. R. Capt. 986
—— J. W. S., Capt. . . . 54, 269
—— L. C., Capt. 844-45
—— P. W., Lt. 795
—— R. J., 2nd-Lt. . . . 549, 988
Hockaday, P. H., Capt. 292a, 545, 1351
Hocken, C. A. F., Lt.-Col. 54, 98, 102, 135, 158, 195, 941
—— N. A., Lt. . . . 514, 1335
Hockin, F. W., Lt. . . . 534, 1323
Hocking, E. A. E., 2nd-Lt. . . 630
Hodder, E. M., Lt. . . . 351, 1633
Hodding, C., Lt. . . . 518, 1427
—— C. F., Capt. 755
—— F., 2nd-Lt. . . . 556, 611
Hodge, E. H. V., Capt. . . 10, 1755
—— H. G., 2nd-Lt. 965
—— J. F., Lt. . . 317, 1419, 1497
—— W. E., 2nd-Lt. 722
Hodgen, G. W., Capt. 82, 280-287 1487
—— H. 2nd-Lt. 1487
Hodges, H. H., Lt. . . . 699, 700
—— H. S., Capt. 501a, 1665
—— R. W., Maj 111
—— J. E. C., 2nd-Lt. 547
—— T. O., 2nd-Lt. 801
—— W. H., 2nd-Lt. 557
—— W. L., Maj. . . . 681, 739
Hodgetts, W., Lt. 589
Hodgins, J. N. M., Lt. 299, 638, 968
Hodgkin, C. J., Lt. . . . 538, 1693
—— N. F., Lt. 521
Hodgkinson, C., Lt.-Col. . 139, 161, 199, 885
—— H. G., Maj. 78
—— J., Lt.-Col. 139, 161, 199, 883

NAMES AND RANK. Page

Hodgman, S. F. J., 2nd-Lt. . . 1585
Hodgson, A. I., Capt. 841
—— C. H., 2nd-Lt. . . . 765
—— E. C., Maj. 1751
—— G. B., Col. 122, 153, 181, 1917
—— G. C., Maj. 213, 869, 1110 1151
—— G. H., Capt. 830
—— H. C., 2nd-Lt. . . . 315, 909
—— H. E. A., Lt. 693
—— H. L., Lt. 42, 536
—— J. F., 2nd-Lt. 1679
—— J. L., 2nd-Lt. . . 554, 1469
—— P. E., Bt.-Lt.-Col. . 139, 662
—— W. G., Lt.-Col. . 28, 188, 1883, 1911
—— W. H., Maj. . . . 236, 1894
—— W. M., Lt. . . . 724, 739
Hodgston, F. W., Lt. . . . 450, 1004
Hodkin, W., Lt. 401, 1439
Hodson, C., Lt. 525, 1668
—— D., Lt. 307, 1463
—— E., 2nd-Lt. 554
—— J. S. F., Lt. . . . 508, 1215
—— S. S., Capt. . . . 501a, 1379
—— V. C. P., Maj. . . . 240, 893
Hoerder, P. E. R., 2nd-Lt. . 534, 1672
Hoernle, E. S., Lt. . . . 535, 1369, 1695
Hoff, S. R., Lt. 529, 1699
Hoffman, J. A., Lt. 723
Hofford, R. McV., 2nd-Lt. . . . 519
Hoffton, P. N. 2nd-Lt. 1009
Hogan, E. J. P., Lt. 517
—— E.M. A. J., Maj. . . 236, 1662
—— E. O., Capt. . . . 257, 1663
—— H. F., Asst. Surgn. . . . 1800
—— H. F., Lt. 533
—— P. C., Capt. . . . 528, 1702
—— P. T., Capt. . . . 265, 1391
—— R. J., Lt. 532, 1705
—— W. E., 2nd-Lt. . . . 1702, 562
—— W. J. A., Senr. Asst. Surgn. 1787
Hog, R. D., 2nd-Lt. . . 24, 323, 366, 895
Hogbin, C. J., 2nd-Lt. 1030
—— F. R., 2nd-Lt. 784
Hogg, A. G. M., Maj. . 208, 1191, 1195
—— C. C. H., Bt. Lt.-Col. . 501, 1882
—— C. M. T., Maj. 14, 32, 99, 217, 1613
—— D. McA., Capt. . . 103, 665, 991
—— F., Capt. 1838
—— F. A., Sub-Condr. 1873
—— G. C., Maj.-Genl. . . . 170, 939
—— G. E. M., Maj. . . . 217, 1661
—— G. P. 2nd-Lt. . . . 554, 1219
—— H. C. G., 2nd-Lt. 361
—— J. McK. T., Lt.-Col. . 135, 158, 195, 869, 1178
—— P. G. H., Maj. . . . 663, 1882
—— R., 2nd-Lt. 363
—— R. E. T., Brig.-Genl. 50, 122, 138, 162, 212, 947
—— W. L., Maj. 214, 1026
—— W. P., Capt. 1759
—— T. W., Col. 177
Hoggan, G. W., Lt. 511
Hoggarth, L. S., Capt. 724
Hoggart, D., 2nd-Lt. 765
Hogge, A. H. F., Capt. 276, 1110, 1112
—— L. R., Maj. . 104, 107, 217, 1211
Hoggett, F. A. J., 2nd-Lt. . 356, 1044
—— J. E., 2nd-Lt. 362
Hogh., A. L. V., 2nd-Lt. 564
Hodgkino, N. F., Lt. 1573
Hogshaw., B. W., 2nd-Lt. . . 558, 1058
—— F. H., Lt. 524, 988
Hoile, G. W. C. H., 2nd-Lt. . . . 618
Hojel, J. G., Lt.-Col. . . . 128, 1739
Holbrook, H. F., Lt. . . . 351, 1501
—— S. D., Lt. 410, 1577
Holbrooke, B.F. B., Maj. 207, 871, 1491, 1495
Holburn, W. C., Capt. 1732
Holdaway, F., Condr. 1863
Holdcroft, G. E., Lt. . . . 541, 1443
Holden, E. A. T. A., Lt. 738
—— G. G., Lt. 368, 911
—— G. P. C., 2nd-Lt. . . 358, 1587
—— H., *Hony*. Maj. 1844
—— H. D., Lt. 758
—— S., Sub-Condr. 1868
—— W. S., Lt. 293, 1050
Holder, E. D., 2nd-Lt. . . . 365, 917
Holderness, H., Maj. 224, 1223, 1595
—— W., Capt. 726, 727
Holdich, H. A., Brig.-Genl. 9, 15, 98, 122, 133, 157, 208, 1617

Names and Rank. Page.

Holdsworth, B. G., Lt. . 543a, 1587
—— E. R., Capt. . . . 828
—— M. J., Lt. . . 425, 1447
—— P. B. C., Lt. . . . 1630
Hole, G. B., Temp. Capt. . . . 795
—— H. L., Lt.-Col. 135, 158, 195, 870, 1327
—— H. E. M., Lt. 629
Holehouse, A., Asst. Surgn. . . 1703
—— W., Condr. . . 1870
Holford, W. F., 2nd-Lt. . 693, 1106
Holford Walker, L., Lt. . . . 622
Holgate, M. J., Capt. 1755
Hollamby, W. C., 2nd-Lt. . 356, 1895
Holland, A., Lt. 737
—— A. P., 2nd-Lt. . . 1499, 1500
—— D. M., 2nd-Lt. . . . 626
—— E F., Maj. 1919
—— E. T., Capt. 1717
—— E. W., 2nd-Lt. 754
—— G., 2nd-Lt. 619
—— G. L., Lt.-Col. 57, 128, 154, 182
—— G. M., 2nd-Lt. . . 556, 1263
—— G. R., Lt. . . 530, 1668
—— G. W., Capt. . 367, 689, 1104
—— H. T., Capt. 835
—— J. A., 2nd-Lt. 713
—— J. R., Capt. 692
—— R. E., Maj. 806
—— T. G., 2nd-Lt. 548
—— W. T. F., Lt. 597
Hollands, E. W., Lt. 508, 979, 1705
Holland-Pryor, P., Brig.-Genl. 50, 121, 126, 153, 191
Hollebone, H. C., 2nd-Lt. . . . 626
Hollick, G., 2nd-Lt. 735
Holliday, J. C. H., Maj. 53, 244, 1483
—— J. M., Ty. 2nd-Lt. 622, 980
Hollidge, S. G., Lt. 798
Hollingbery, G. R. V., 2nd-Lt. 551, 1020
—— N. H., 2nd-Lt. . 562, 1130
Hollinrake, R., 2nd-Lt. 760
Hollis, F. S., Lt. 533, 1645
Hollom, G. H. S., Lt. 712
Holloway, B., Brig.-Genl. . 97, 105, 120, 123, 149, 182
—— B. T., Lt. 753
—— C. A., 2nd-Lt. 360
Hollwey, J. B., Capt. 617
Holwell, R. F., 2nd-Lt. 1561
Holman, A. T., 2nd-Lt. 623
—— H. C., Maj.-Genl. 53, 97, 102, 103, 105, 119, 125, 153, 194, 905
Holman-Hunt, H. L., Lt. . 534, 1695
Holme, A. S., Maj. 103, 104, 662, 1881
—— H. B., Lt. . . 530, 848, 1293
—— H. L., Maj. 633
Holmes, A. C., 2nd-Lt. 564, 658, 1678a
—— A. H. R., Lt. . 520, 1670, 1702
—— A. S., Lt. 503
—— B., Lt. 371, 947
—— B. O. S. H., Asst. Surgn. 1802
—— B. W., Surgn. Maj. . . . 111
—— C., 2nd-Lt. 1126
—— C. W. A., Maj. 99, 215, 871, 1463, 1469
—— E., Lt. 455, 1295
—— E. B., Capt. 758
—— E. J., 2nd-Lt. 699
—— F. W., Asst. Surgn. . . 1795
—— G., Lt. 834
—— G. F., Sr. Asst. Surgn. 1789
—— G. H., Condr. 1867
—— G. L., Lt. . . . 311a, 1561
—— G. V., Lt.-Col. 47, 130, 155, 189
—— H. B., Col. 840
—— H. G., Lt. . . . 11, 666
—— H. L., Maj. . 235, 1473, 1465
—— H. R., Lt. . . . 534, 1690
—— H. R. J., Capt. . . . 728
—— H. W. J., 2nd-Lt. . . . 562
—— J., temp. Lt. 795
—— J. A., Lt. 472, 1197
—— J. A. F., Sr. Asst. Surgn. 1790
—— J. F., Capt. 1757
—— J. F., Lt. . . . 544, 1613
—— J. M., Maj. . 62, 1751, 1889
—— L. S., Asst. Surgn. . . 1795
—— L. V., 2nd-Lt. . . 365, 1187
—— R., Asst. Surgn. . . . 1799
—— R. E. J., Lt. . . 316, 1178
—— R. J., *Hony.* Maj. . . . 1845
—— S. B., Lt. . . . 443, 1369
—— S. J., 2nd-Lt. 717
—— W., Sub-Condr. . . . 1861
—— W. G., 2nd-Lt. . . 564, 1283

Holmes, W. K., Lt. 624
Holness, H. J., Maj. 1838
Holroyd, F. H., Capt. 1733
—— G., Capt. 1753
—— J. H. B., Capt. 1729
Holt, B. G., Lt. 82, 581, 1008, 1068
—— D. G., Asst. Surgn. . . . 1802
—— D. R., 2nd-Lt. 1143
—— D. R. W., 2nd-Lt. . . . 548
—— E., Lt. . 76, 532, 1886—88
—— E. B., Asst. Surgn. . . . 1800
—— E. M., 2nd-Lt. 945
—— G. H., 2nd-Lt. . . 546, 1295
—— V., Lt. 753, 1006
Holton, W. E., 2nd-Lt. 565
Holworthy, A. W. W., Lt. . 318, 1606
Holyoak, A. H. P., 2nd-Lt. 356, 1675
Holyoake, R., Lt.-Col. . . . .
Homan, A. R., 2nd-Lt. . . . 840
—— C. F. W. B., Capt. . . 755
—— E. M., N. Sister .
—— R., 2nd-Lt. . . 356, 1387
Home, A. C., 2nd-Lt. 1463
—— D. C., Maj. . . 37, 81, 230
—— J. E., Maj. . 45, 216, 1661
—— J. M., Bt.-Col. 54, 97, 103, 105, 126, 153, 188
—— P. M., Maj. . . . 206, 1178
—— W., Capt. 765
—— W. E., Capt. 1919
Homer, N. H. T., Lt. . . 412, 487
Homfray, N. E., 2nd-Lt. . . . 790
Homi, Kawasji Kanga, Sub-Asst. Surgn. 1832
Hone, A. C., Lt. 522
—— H. L., 2nd-Lt. 788
—— E. J., Capt. 720
Honeboon, H., 2nd-Lt. 75
Honey, B. St. C., Asst. Surgn. . 1800
Honeyball, F. R., Lt. . 298a, 1495
Honeyman, T. J., Capt. . . . 1718
Honeywell, B., 2nd-Lt. . . 553, 1303
Honnor, A. J., Lt. . . 534, 1181
Honour, W. 619
Hony, H. C., Lt. 56
Honywood, W. W., 2nd-Lt. . . . 555
Hood, A., Capt. 22, 1714
—— D. A., 2nd-Lt. 1217
—— F. W., Lt. 1299
—— H. M., Lt. . . . 490, 927
—— T. C., Lt. 818
—— T. H., *Hony.* Capt. . . . 705
Hoogwerf, V. H., 2nd-Lt. . . . 547
Hook, 2nd-Lt. T. Y. B. . . . 704
—— W. J., 2nd-Lt. 726
—— C. W. T., Lt. 685
Hooker, J. S., Maj. . . . 222, 1066
Hooper, A. F., Capt. . . . 750, 752
—— C. A., Lt. 56
—— E. G., Lt. 748
—— E. J., Lt. . . . 534, 1898
—— F. E., Capt. 801
—— H. G., Lt. 720
—— W. J., 2nd-Lt. 360
—— W. H., *Hony. Col. Sr* . . 1
Hoore, J., 2nd-Lt. 1393
Hooton, A., Lt.-Col. . . 133, 1743
—— A. C., Capt. 59, 712
Hopcroft, R. T., Lt. . . . .
Hope, A. A., 2nd-Lt. . . 325, 366, 1164
—— A. C., Lt. 515
—— A. V. W., Maj. . . 207, 1148
—— J. C., Lt. 827
—— J. W., Lt. . . . 520, 1670
—— P. M., Capt. 724
Hopewell, G. M., 2nd-Lt. . 550, 1701
Hopkins, A. C., Capt. . . . 43, 75
—— C. J. W., Lt. 530
—— E., Lt. 520
—— F., Lt. 430
—— G. B. E., 2nd-Lt. . . . 781
—— F. L., 2nd-Lt. 925
—— H. E. E. J., Lt. . . . 517
—— J. A. S., Lt. . . . 527, 1701
—— L. E., Maj. . . . 662, 1882
—— W. J., 2nd-Lt. . . 315, 913
Hopkinson, G. G., 2nd-Lt. . 359, 1106
—— G. H., Capt. . . 263, 1463
—— W. D., Lt. 624
Hopper, F. C., 2nd-Lt. . . 324, 1102
—— G., Condr. 1871
—— R. McN., Ty. Asst. Surgn. 1804
—— S., *Hony.* Lt. 577
Hopps, O. D., Lt. 21, 536
Hopton, W. J., 2nd-Lt. . . . 21, 687

Hopwood, H. B., Col. . . 51, 97, 126, 153, 198, 937
Horan, K., Capt. 859
Hordern, R. D. E. L., 2nd-Lt. 552, 1678
Hore, A. T., Lt. 352, 1397
Horman, J. C. L., Lt. 974
Hormasji Manekji Limbu-walla, Hospl. Asst. 1830
Hornby, M. L., Maj. 1919
—— R. F., Lt. 589
—— J. A., 2nd-Lt. 1060
Horne, A. E. M., 2nd-Lt. . . . 709
—— A. W. M., Lt. 578
—— C. W., Capt. 485
—— G. G., 2nd-Lt. 492
—— G. H., Lt. 541
—— J. H., Capt. 1755
Horner, B. F., Lt. 692
—— B. S., Capt. 721
—— E. T., Capt. . . 46, 691, 1142
—— R. 1009
—— W. J., Lt. . . . 46, 509, 1281
Hornett, W. G., 2nd-Lt. 850
Hornor, F. H. F., Capt. . . 265, 1431
Hornsby, F. H., Maj. 614
—— H. F. P., Capt. . 265, 883
—— M. E. T., Nursing Sister 1735
—— N., Capt. . . 501, 830, 1665
—— R. W., Capt. 52, 54, 259, 1094
—— W., Tempy. Capt. . . 1724
Horris, J. P. A., Lt. 1839
Horsfall, C. M., Lt. 507
—— T. M. M., Ty. Capt. . . 1724
Horseman, E. V., 2nd-Lt. . . . 760
Horsfield, G., Lt. 707
—— H. E., Lt. . . . 17, 45, 490
Horsley, C. D., Capt. . . . 53, 597
Horsman, H., Lt. 491
Horton, K., Lt. 514, 986
—— R. A., 2nd-Lt. . . 553, 1521
—— J. H., Bt.-Lt.-Col. . . . 62
Horton-Bennett, A. G. R., Capt. 76, 501a, 1886—88
Horwill, L. C., Lt. 489
Horwood, J. C., Lt. . . 504, 1665
—— K. C., Revd. 1904
Hosain, M. N. A., 2nd-Lt. . . . 1054
—— M. S. M., 2nd-Lt. . . . 1064
Hosbons, P. R., Revd. 1905
Hoseason, W. S., Lt. 821
Hosgood, J. W., 2nd-Lt. 546, 1598, 1607
Hoshank Khan, Subdr. 1518
Hosan Khan, Jemdr. 1248
Hoshan Khan, Jemdr. 1515
Hoshiar Singh, Ressdr. 1086
—— **Singh Mahar Bisht**, Jemdr. 1225
Hosie, A. L., Capt. 531
Hosken, A. F., Maj. 841
—— C. C. F., Maj. 754
Hoskin, F., 2nd-Lt. 786
Hoskin, J. G., 2nd-Lt. . . 357, 1030
Hosking, B. C., Lt. 516
—— F. M., Lt. 515
Hoskings, R. V. T., 2nd-Lt. . . 549
Hoskins, C. J., Lt. . . 415, 1207
—— G. D., Lt. . . . 532, 1351
—— J., *Hony.* Capt. . . . 740
Hoskyn, J. C. M., Maj. 54, 104, 108, 209, 1199
—— R., Maj. 633
Hossack, I. F., 2nd-Lt. 324
Hotblack, H. C., Lt. 624
Hotham, G. I. M., 2nd-Lt. . . 332, 881
—— J. C., Maj. . . 76, 244, 881
Hotine, M., 2nd-Lt. . . . 666, 992
Hotson, J. E. B., Capt. . . 501, 815
Hottinger, E. F., Asst. Surgn. . 1793
Hotton, H. O., 2nd-Lt. . . 331, 1551
Hotz, R., 2nd-Lt. 555
Houchan, J. B., Lt. 946
Houghton, A. T., 2nd-Lt. . 7, 703, 796
Houdey, J. E., Capt. . . 501a, 1297
Hounsell, R. H. S., Capt. . . 11, 665
—— T. H., 2nd-Lt. . . 589, 599
Hourihane, G. O'B., 2nd-Lt. 476, 1299
—— T. C., 2nd-Lt. . 327, 1187, 1521
House, B. H., Lt. 589
—— E. H., Lt. . . 311, 491, 885
—— E. N., Lt. 491, 951
—— M. H. N., Maj. 628
—— P. J., Maj. 618
Houseman, E. A., Maj. . . . 603, 861
Houston, A. L., Capt. 801
—— A. M., Lt.-Col. 138, 162, 199, 870, 1287
—— B., 2nd-Lt. 743

Names and Rank.	Page.
Houston, C., Lt.	533, 1351
—— C., 2nd-Lt.	1674
—— E., Col.	123, 660
—— J. F., Lt.	731
—— J. J., Asst. Surgn.	1803
—— W. M., Maj.	1747
How, A. P., Lt.-Col.	57
Howard, A., 2nd-Lt.	630
—— A., 2nd-Lt.	805
—— C., *Hony.* Maj.	72, 1844
—— G., Sub-Condr.	1863
—— G., 2nd-Lt.	540, 1385
—— G. A. L., Capt.	96, 252, 935
—— G. W., 2nd-Lt.	365, 929
—— D. H. The Revd.	1905
—— E., Lt.	638
—— E. S. G., Capt.	617
—— G. B., Lt.	542, 1696
—— H. E., 2nd-Lt.	362
—— H. G., Lt.	624
—— H. H. G., 2nd-Lt.	545
—— J. E., Asst. Surgn	1798
—— J. G., Lt.	528, 1667
—— R., 2nd-Lt.	66, 741
—— S. W., Capt.	769
—— W. H. K., Lt.	534, 1696
Howard-Vyse, G. A., Maj.	685
Howarth, H. S., Lt.	528, 998, 1699
—— R.	723a
—— R. A. S., 2nd-Lt.	697
Howatson, C. N., Lt.	492a, 911
—— G. A. S., Asst. Surgn.	1793
Howden, G. H., 2nd-Lt.	556
Howdle, T. G., Lt.	303, 1475
Howe, A., Capt.	749
—— A. J., Sub-Condr.	1863
—— E., Lt.	528
—— E. G., Lt.	734
—— G., 2nd-Lt.	536, 1090
—— G. E., 2nd-Lt.	965
—— R., 2nd-Lt.	852
—— T. M., 2nd-Lt.	554
—— W., Lt.	618
—— W. G. T., 2nd-Lt.	551
—— W. T., Capt.	701, 1004
Howell, C. B., Maj.	228, 909
—— C. J., Capt.	40, 683, 783
—— G. B., Capt.	54, 256, 911
—— G. C. L., 2nd-Lt.	605
—— H. F., Lt.	420, 1642
—— H. W. H., 2nd-Lt.	548, 1705
—— I. B., Lt.	679, 681
—— J., *Hony.* Maj.	1845
—— R. B., Maj.	11, 221, 903
—— V., Lt.	389, 1126
—— W. E., Lt.	799
Howells, P. I., Lt.	700
—— T. R. R., 2nd-Lt.	356
—— W. M., temp. Capt.	1723
Howes, H. C., 2nd-Lt.	540, 1577
—— S., Capt.	54, 597
—— W. T., *Hony.* Capt.	713
Howey, W., Maj.-Genl.	169
—— J. R., Sub-Condr.	1874
Howfield, J. A. J., 2nd-Lt.	1183
Howgrave-Graham, H. M., Capt.	734
Howie, G., Capt.	79, 1839
—— H. C., Lt.	538, 849
—— J. R. S., 2nd-Lt.	884
—— J., Capt.	776, 793
Howison, G., 2nd-Lt.	559, 1497
Howitt, C. N., Lt.	297a, 1435
—— G., Lt.	352, 1677
—— G. M., 2nd-Lt.	1475
Howland, C., Lt.	515, 1666
—— C. C., Lt.	313, 939
Howlett, A. H. J., Lt.	460, 62, 455, 1215
—— C. G., Capt.	1756
—— L. W., Ty. Capt.	1723
—— E. G., Capt.	254, 1303
Howling, F., *Hony.* Capt.	724
Howman, R. C., 2nd-Lt.	327, 1347
Howson, G., Maj	68, 243, 797, 881
—— W. G., 2nd-Lt.	564, 1367
Hoyle, A. L., Capt.	489, 1705
—— R., 2nd-Lt.	767
Hozier, F. R., 2nd-Lt.	555, 1417
Hpau, Lu La, Subdr.	1347
Hubab Khan, Jemdr.	1433
Huban, J. P., Capt.	1717
Hubball, J. L., 2nd-Lt.	549, 1028
Hubble, L. U., Lt.	34, 747
—— S. L., Lt.	683
Hubbard, A. E., Lt.	538
—— D. S., 2nd-Lt.	559
—— E. A., 2nd-Lt.	759
—— G. R., Lt.	1720
Hubbard, H., 2nd-Lt.	359, 1130
Hubert, J., Sub-Condr.	1874
Huddleston, G. B. G., Capt.	500, 1106
—— J. K., 2nd-Lt.	475, 1636
—— E. W., Capt.	1892
Hudgesdon, H. R., Lt.	1501
Hudson, A. L., Asst. Surgn.	1795
—— A. R., Lt.	446, 49 889
—— B. R., Capt.	78, 259, 889
—— C., Lt. Col.	140, 62, 1745
—— C. T., Lt.-Col.	128, 1739
—— E., 2nd-Lt.	1642
—— E. H., Capt.	841
—— E. J. M., 2nd-Lt.	548, 1435
—— H., Lt.-Genl. *Sir*	8, 14, 118, 146, 172, 1925
—— J., 2nd-Lt.	59, 770
—— J. E., Lt.	528, 1417
—— J. H., Lt.-Col	87, 103, 104, 108, 137, 159, 197
—— J. R., *Hony.* Maj.	998, 1849
—— L., Lt.	531
—— P. B., Lt.	78, 493
—— R. J. H., 2nd-Lt.	33, 561
—— T., Lt.	317a, 933
—— W. G., Lt.	698
Hudson-Heaven, L. G., Capt.	278, 1267
Huelin, W., Lt.	306, 1070
Hufton, J. H. S., Asst Surgn.	1795
—— P. N., 2nd-Lt.	550
Huggan, G. H., Lt.	750
Huggard, E. H., Lt.	767
—— J. C., Ty 2nd-Lt.	678
Huggins, F. M., 2nd-Lt.	492
—— G. F., Lt.	652
—— H. D., 2nd-Lt.	494, 943
—— H. W., Capt.	615
Hugh-Jones, N., Lt.	291a, 1235, 1565
Hughes, A. H., 2nd-Lt.	365, 1148
—— A. P., Capt.	56, 1919
Hughes, A. S., Lt.	624
—— C. E., Lt.	506, 1315
—— C. F., Maj.-Genl.	173
—— C. F. W., Maj	215, 1078
—— C. R., Lt.	420, 968
—— E. D. G., 2nd-Lt.	758
—— E. E., tempy. Capt.	1725
—— E. W., Lt.	521, 1575
—— E. W., 2nd-Lt.	695
—— F., Lt.	73, 698, 796
—— F., Capt.	629
—— F. E. C., Lt.	311a, 1182, 1613
—— F. S., 2nd-Lt.	851
—— F. T. C., Lt.-Col.	186, 1889, 1911
—— F. W. T., 2nd-Lt.	679
—— G. R., Maj.	941, 233, 1195
—— G. V., Capt.	267, 1654
—— H., Lt.	326, 537, 1565
—— H., 2nd-Lt.	366
—— H. B. W., Capt.	665
—— H. C., Lt.	628
—— H. D., Lt.	32, 717
—— H. T., Lt.	410, 1126
—— I., 2nd-Lt.	748
—— J., Lt.	506
—— J. E., Maj.	62, 1919
—— J. E., Lt.	517
—— L. W., Capt.	274, 1497
—— L. W., Capt.	1718
—— O., 2nd-Lt.	577
—— P., Capt.	848
—— R. H., Asst. Surgn.	1795
—— R. M., Maj.	850
—— R. R., Lt.	531, 1699
—— R. W., *Hony.* Lt	1843
—— S., Lt.	352, 1389, 1415
—— S. A., Capt.	709
—— T., 2nd-Lt.	474, 1121
—— T., Condr.	1872
—— T. A., Capt.	1756
—— T. E. 2nd-Lt.	1676
—— W., 2nd-Lt.	356
—— W. B., Condr.	1861
Hughes Rowlands, Lt.	945
Hughes-D'Aeth, C. C., Maj.	738
Hughes-Hallett, R. L. Bt. Lt.-Col.	139, 162, 212, 1661
Hugo, E. V., Lt.-Col.	3, 229, 1742
—— J. H., Lt.-Col.	136, 1744, 1889
Huish, L., Lt.	697
Hukam Chand, Subdr	1050
—— Jemdr.	1019
Hukam Chand, Dad, Jemdr.	1320
—— Dad, Subdr.	1565
—— Dad, Jemdr.	1498
—— Dad, Jemdr.	1068
—— Jemdr.	1565
—— Jemdr.	1154
—— Subdr.	1239
—— Khan, Jemdr.	1236
—— Singh, Sub-Asst. Surgn.	1814
—— Subdr.	1095
—— Sub-Asst. Surgn.	1812
—— Jemdr.	502
—— Jemdr.	1124
—— Ressdr.	945
—— Subdr.	1150
—— Subdr.	1445
—— Jemdr.	989
—— Esq.	91
—— Jemdr.	1412
—— Jemdr	1071
Hukmat Khan, Subdr.	1345
—— Jemdr.	1095
Hukmi, Subdr	1215
Hukm Tej Partab Singh, Raja 2nd-Lt.	481
Hukum Dad, Jemdr.	1260
Hulas Singh, Risldr.	889
Hulatt, C., Lt.	491, 923
Hulbert, H. B., Lt.	9, 793
—— J. G., Lt.-Col.	1786
—— L., Lt.	543
—— M. L., Lt.	903
—— T. E., Maj.	36, 68, 223, 879
Hull, A., Sub-Condr.	1856
—— B., Lt.	401, 1517
—— J. R., Lt.	518, 1108
Hullock, E. R., 2nd-Lt.	552, 1614
Hulme, H., 2nd-Lt.	623
Hulseberg, H. Lt.-Col.	138, 162, 198, 871, 1505
Humby, L. J., Lt.	297, 1148
—— H. R., Capt.	1732
Hume, A. H. B., Lt.-Col.	42, 104, 133, 661
—— E. G., Maj.	241, 887, 955
—— E. J., 2nd-Lt.	363
—— R. N., Capt.	1752
—— R., Lt.	352, 974
—— S. G. W., Lt.-Col.	139, 161, 200, 1042
—— W. P., Lt.-Col.	1708
Hume-Spry, F., 2nd-Lt.	622
—— W. E., Maj.	31, 28, 1662
Hume-Wright, F. J. O., Bt. Maj.	262, 1042, 1082
—— G., Capt.	57
Humfress, H. O., Lt.	111
Humfrey, B., Lt.	401, 759, 1007
—— F. G. C., Lt.-Col.	132, 156, 192, 867, 897
—— F. T., Col.	177, 1918
—— H. B., Lt.	313, 913
—— J. C., Capt.	585
—— L. E. H., Maj.	735
Humm, P. S., Capt.	1732
Hummell, J. J., Lt.	1722
Humpherus, E. H., Lt.	1489 90
Humphrey, D. A. R., Lt.	504, 1666
—— H. P., Lt.	617
—— J., Capt.	16, 39, 501a, 657, 1363
—— G. R. D., 2nd-Lt.	703
Humphreys, A., *Hony. Lt.*	10, 1847
—— C. E., Lt.	318, 1287, 1311
—— D. W. G., Lt.	534, 1074
—— E. A. F., 2nd-Lt.	1708
—— R. M., Capt.	1720
—— E., Condr.	1854
Humphries, A., Lt.	68, 530, 797
—— F., 2nd-Lt.	1008
Humphris, T. H., Lt.	617
Humphrise, E. E., Lt.	491
Humphrys, B. H., Lt.	517
—— D. H., Lt.	697
—— E. F. A., 2nd-Lt.	558
—— F. H., Maj.	225, 1901, 1913
Hungerford, S. A. H., Capt.	272, 1463
Hunnebun, B., 2nd-Lt.	563, 1295
Hunnybun, C., Lt.	506, 1475
Hunsworth, E., 2nd-Lt.	559
Hunt, A. E., Lt.	628
—— C. E., Capt.	55, 252, 1156

Names and Rank. Page.

Hunt, C. O. C., Maj 614
—— C. V., 2nd-Lt. . . . 494, 959
—— E., Capt. 838
—— E. C., Condr. 1861
—— F. D., Lt.-Col. . . 79, 134, 1838
—— F. E., Maj. . . . 241, 919
—— G. E. D., Capt. . . 9, 735, 1677
—— H. G., Lt. . . . 353, 1473
—— H. R. A., Maj. 99, 103, 218, 1120
—— J. L., Lt. 435, 1170
—— J. M., Lt. 291, 1355
—— J. V., Capt. 78, 485
—— J. W. A., 2nd-Lt. . . 327, 1427
—— L. F., Lt. . . . 446—449
—— N. G., Capt. 821
—— R. A. C., Maj.-Genl. . . . 170
—— R. E., Lt. . . . 387, 1144
—— R. S., Maj. 581
—— R. V., Maj. . . . 224, 1487
—— S., Lt.-Col. . 138, 1745, 1889, 1899
—— V. de V., Lt.-Col. 66, 108, 138, 159, 197, 1883
—— W. E., Lt. 705
—— W. M., Maj. 633
Hunter, A. G. M., 2nd-Lt. . . . 840
—— A., 2nd-Lt. 360
—— C. F., 2nd-Lt. 555
—— C. G. W., Lt.-Col. . 134, 662
—— D. A., 2nd-Lt. 1048
—— D. C., 2nd-Lt. 555
—— E., Lt. 1768
—— F. F., Maj. . . 56, 217, 1880
—— G. G. R., Lt. 543
—— H. C., Lt. . 311a, 1459, 1463
—— H. C. A., Lt. 527, 1178, 1655
—— I. G. C., 2nd-Lt. . . . 851
—— J., *Hony.* Capt. 56
—— J., Lt. 430, 1098
—— J. B. D., Maj. . . 1751, 1889
—— J. C., Lt. 836
—— J. G., Maj.-Genl. 57, 119, 147, 174
—— J. H. D., Lt. . . 311, 1383
—— J. M., Lt.-Col. . . 178, 1918
—— J. P., Lt. 682
—— R. G. P., Capt. 664
—— R. J., 2nd-Lt. 854
—— R. J. L., 2nd-Lt. . . . 642
—— R. T., Lt. . . . 696, 697
—— S. L., 2nd-Lt. 744
—— T., Lt.-Col. . . . 139, 1745
—— T. C. L., Lt. 1142
—— T. J., Sub-Condr. . . . 1867
—— T. S., Capt. 1839
—— W. C., 2nd-Lt. 744
—— W. J. H., Lt.-Col. 139, 160, 199, 871, 1467
—— W. M., 2nd-Lt. . . 633, 638
Hunter-Gray, G. T., Capt. . . 1919
Hunter-Jones, L., Capt. . . . 593
Huntingfield, W. C. A., Capt. . 591
Huntley, W. B., Lt. . . . 308, 1487
Hurburgh, L. H., Lt. . . . 90, 779
Hurford-Jones, M., Lt. . . 293, 1371
Hurlbatt, D. G., 2nd-Lt. . . . 112
Hurley, F. S., 2nd-Lt. . . 539, 1630
—— H. A. J., Sub-Condr. . . 1864
—— J. M., 2nd-Lt. 540
—— W. C., 2nd-Lt. 362
Hurly, M. F. F., Lt. . . 306, 1156
—— M. R., Lt.-Col. . 129, 154, 188
Hurn, A. S., Lt. 709
Hurndall, B. K. F., Lt. . . . 797
Hurrell, A. E., 2nd-Lt 692
Hurst, C., 2nd-Lt. 757
—— F. W., Capt. . . 60, 502, 1665
—— F. G., Asst. Surgn. . . . 1793
—— G. E., Lt. 1720
—— G. S., 2nd-Lt. 324
—— G. T., 2nd-Lt. 362
—— I., Lt. 504, 1665
—— N., Lt. 423, 1642
—— R. H., Lt. 347, 1657
—— T. R., 2nd-Lt. . . . 365, 1611
—— V. de la M., Lt. . . 517, 1347
—— W., Condr 1872
Husain, M. A., Ty. Lt. 1781
Husain, Jemdr. 1244
—— Ali Khan, Jemdr. 1051
—— Bakhsh, Jemdr. 1452
—— Baksh, Jemdr. 1409
—— Khan, Jemdr. 1285

Names and Rank. Page.

Husain, Bakhsh Khan, Risldr.-Maj. 939
—— —— Khan, Jemdr. 1087
—— Bakhsh Khan, Jemdr. . . 1090
—— Muhammad Khan, Jemdr. 1308
Husain Shah, Subdr.-Maj. . . 1319
Husamuddin Khan, Risldr.-Maj. 919
Husband, J., Maj. 1748
—— R., Lt. 90, 487
—— R. H., 2nd-Lt. 818
—— W. F., Lt. . . . 351, 883
Hushyar Singh, Risldr. 906
Huskisson, A. E., Condr. . . . 1861
Huss, S. A., Capt. . . . 501a, 722
Hussain, Baksh, Jemdr. . . . 1409
—— Khan, Subdr. 1591
—— —— Subdr. 1142
—— —— *Hony.* Capt. . . 2102
—— Shah, Jemdr. . . . 1305
Hussey, D. C., Lt. 1203
—— E. W., *Hony.* Col. . . . 685
—— G. O. G., Lt. . 397, 508, 881
—— H., *Hony.* Capt. 8
—— N. G. F., Lt. . . . 529, 1219
—— W. G. St. J., Sr. Asst. Surgn. . 1791
Hussey-Freke, P. E., 2nd-Lt. . . 819
Hutchens, F. K., Lt. . . 528, 1234
—— H. J., Lt. . 1229a, 733, 791
Hutcheson, G., Lt.-Col. . 139, 1745
—— J. H., Lt. . . . 521, 985
—— S., Maj. 220, 578, 1026, 1187
—— W. C., Maj. . . 237, 1026
—— W. R., 2nd-Lt. . . . 628
Hutchesson, J. F., Lt. 56
Hutchings, C. M., 2nd-Lt. . . . 324
—— D. A., 2nd-Lt. . . . 425
—— F. S., 2nd-Lt. . . 560, 1143
Hutchins, A. G., Col. . . 170, 1916
—— E. E., *Hony.* Capt. . . 1847
Hutchinson, A. G. C., Maj. 99, 217, 859, 1110, 1160
—— A. McI., 2nd-Lt. . . . 568
—— C. A. R., Col. 66, 57, 125, 150 193, 1157
—— C. R. M., Lt.-Col. 131, 156, 198
—— E., Lt. 530
—— F. H., 2nd-Lt. 562
—— F. H. G., Lt.-Col. 157, 1744
—— F. P., Col. 105, 122, 149, 181, 1917
—— G. W., 2nd-Lt. 1110
—— G. S. M., Maj. 1919
—— G. W., 2nd-Lt. 332
—— H. D., Lt.-Genl. 1606, 1955
—— J. B., Lt.-Col. 178
—— L., Ty. Lt. 1726
—— L. T. R., Lt.-Col. 140, 1745
—— W. G., Maj. 212, 1899, 1913
Hutchison, B. O., Capt. 59, 587, 1006
—— E., 2nd-Lt. 1305
—— F. H., Capt. 1703
—— G., Maj. . . 59, 213, 1483
—— H. S., Maj. 1752
—— J. B., Maj. . . 9, 235, 947
—— J. R. H., Lt. 292a, 268, 907
Huth, F. L., 2nd-Lt. . 327, 366, 901
Huthwaite, H. Y., Lt. . . 367, 1247
Hutson, H. W. P., Lt. 56
Hutton A. R., 2nd-Lt. . 325, 366, 1601
—— C. T., Capt. 832
—— D. J., 2nd-Lt. 577
—— E. A., 2nd-Lt. . . . 359, 1295
—— E. D., Lt. 463, 937
—— G. C., Lt. 435, 1577
—— G. M., Capt. . . . 269, 917
—— H. O., 2nd-Lt. 1657
—— H. W., 2nd-Lt. 776a
—— T. A., Capt. 793
Hutton-Squire, R. H. E. . . . 633
Huxa, H. J., Asst. Surgn. . . . 1802
Huxford, H. J., Capt. 261, 1497, 1499
Huxtable, A. E. V., *Hony.* Capt. . 583
Huzura Sing, Subdr. 1705
Hyatt, P. T., 2nd-Lt. 702
Hyde, C. B., Condr. 1853
—— G. L., Lt. 423, 1577
—— G. W. T., Ty. 2nd-Lt. . . 571a
—— H. S., 2nd-Lt. . . . 548, 1096
—— J. A., Lt. 527, 1239
—— J. A., Asst. Surgn. . . . 1802
—— J. K., 2nd-Lt. 742

Names and Rank. Page.

Hyde, R. W., Capt. . . . 695, 1893
—— W. C., Lt. 579
Hyde-Smith, V. G., Lt. . . 524, 1699
Hyder, H. S., 2nd-Lt. . . 359, 1678a
Hyderabad, *His Highness* Sir Usman Ali of, Lt.-Genl. . . 110
Hyland, H. F., 2nd-Lt. 1689
—— J. P., Lt. . . 446—449, 1058
Hynes, B. M., Lt.-Col. 779
—— C. E., Asst. Surgn. . . . 1803
—— C. F. J., Condr. 1854
—— H. G. C., Capt. . . 106, 1919
—— W., 2nd-Lt. 562
Hypher, P. P., Esq. 91
Hyslop, A. J., Lt. . . . 681, 785
—— F., Maj. 689
—— G. G., Lt. 1060

I

Ibad-Ullah Khan, Jemdr. . . . 1391
Ibbotson, A. W., Lt. . 18, 42, 487, 608
—— G., Lt. . . . . .
—— H., 2nd-Lt. 793
Ibrar Hasan Khan, Jemdr. . . . 876
Ibrahim, Jemdr. 1052
—— Ali Khan, Nawab, 2nd-Lt. 481
Ibrahim, Beg, Subdr. 1315
—— Khan, Subdr.-Maj. . . . 1441
—— —— Subdr. 1239
—— —— Subdr. 1373
—— —— Subdr. 1243
—— —— Jemdr. . . 924, 930
—— —— Jemdr. 944
—— —— Jemdr. 1036
—— —— Jemdr. 890
Iceton, E. A., Lt. 445, 1641
Id Buskh, Jemdr. 1698
Idar, Maharaja of, Lt.-Col. . . . 110
Id Mohamed, Subdr. . . . 1502—04
—— Jemdr. 1514
Idris Khan, Ressdr. 918
Igglesdon, G. H., Lt. 683
Iggulden, H. A., Brig.-Genl. 120, 123
Ignatius Wellington, Asst. Surgn. 1818
Insan Ali, Sr. Sub-Asst. Surgn. . 1807
Ijay Ali Khan, Jemdr. 936
Ilahi Bakhsh, Daniel Joseph, Asst. Surgn. 1827
—— Jemdr. 1208
Ilam Din, Jemdr. 1241
—— Jemdr. 1441
—— Subdr. 1518
—— Jemdr. 1681
—— —— Subdr. 1117
—— Subdr. 1117
Iles, A. G., Lt. 486, 939
—— F. W., Maj. . . . 500, 1092
—— J. H., Ty. Lt. 1726
Iliff, W. A. B., Lt. . . . 317, 1251
Illnan, S. G., 2nd-Lt. 556
Illington, E. M., Lt.-Col. . 132, 1743
Illingworth, A. E., Lt. . . . 714
—— G. R., Lt. . . . 320, 365
—— J. R., Lt. . . . 460, 1499
—— R. E., *Tempy.* Capt. . . 1724
Illius, H. W., Maj. 1748
—— J. W., Maj. 1749
Iltaf Hussain Shah, Ressldr . . 944
Ilyasdar, Khan, Subdr. 1416
Imam Ali, Subdr. 1573
—— Jemdr. 1079
—— Subdr. 1252
—— Subdr. 1126
—— Din, Subdr. 1413
—— Khan, Jemdr. 880
—— —— Jemdr. 1436
—— —— Jemdr. 1086
—— —— Jemdr. 1479
—— Shah, Jemdr. 1304
Imamuddin, Jemdr. 1704
—— Jemdr. 1036
Imam Bakhsh, Din, Subdr. . . 1380

Names and Rank.	Page.

Imam Baksh, Din, Jemdr. . . . 1704
Iman Baksh Khan, Subdr. . . . 1050
—— Jemdr. 1036
Iman Din, Subdr. Maj. . . . 1367
—— Jemdr. 1380
—— Jemdr. 1459
Imdad Khan, Jemdr. . . . 1705
—— Jemdr. . . . 878
—— Ali Khan, Jemdr. . . 884
Impey, L., Lt.-Col. . . . .
Imrat Sing, Jemdr. . . . 934
Imrie, B. C., Capt. . . . 1732
—— W., Lt. . . . 387, 1243
Imtiyaz Muhammad Khan, Jemdr. . . . 1036
Inayat Khan, Jemdr. . . . 1565
Inayat Ali, Subdr. . . . 1001
Inayat Masih, Jemdr. . . . 1299
Inayatullah Khan, Jemdr. . . 908
Inaytullah, Sub-Asst. Surgn. . 1813
Ince, H. T., Asst. Surgn. . . 1802
Inda Rawat, Jemdr. . . . 1476
Indar Sing Bisht, Subdr. . . 1182
Indar Singh, Sub-Asst. Surgn. . 1812
—— Sub-Asst. Surgn. . 1814
—— Sub-Asst. Surgn. 1814, 1817
—— Jemdr. 1336
—— Jemdr. 892
—— Ressdr. 873
—— Jemdr. 1510
—— Subdr. 1150
—— Subdr. 1120
—— Subdr. 1367
—— Subdr. 1160
—— Subdr. 1375
—— Subdr. 1066
—— Ressdr. 912
—— Jemdr. 1336
—— Jemdr. 1067
—— Jemdr. 1122
—— Jemdr. 916
—— Jemdr. 1129
—— Jemdr. 890
—— Subdr. 1239
—— Jemdr. 1452
—— Ressdr. . . 945, 953
—— Subdr.-Major . . 1020
—— Subdr. 1070
—— Jemdr. 1260
—— Jemdr. 1565
—— Jemdr. 1585
—— Subdr. 1705
—— Jemdr. . . 902, 934
—— Jemdr. 1031
—— Jemdr. 1215
—— Subdr. 1288
—— Subdr. 1286
—— Subdr.-Maj. . . . 1255
—— Jemdr. 1476
—— Subdr. 1577
—— Subdr. 1136
—— Jemdr. 977
—— Jemdr. I . . . 1024
—— Jemdr. II . . . 1024
—— Jemdr. 1142
—— Jemdr. 1231
—— Jemdr. 1152
—— Jemdr. 1107
—— Asstt. Surgn. . . 1817
—— Jemdr. 892
—— Jemdr. 1012
—— Risaldar . . . 1686
—— Kathait, Jemdr. . 1179
—— Negi, Subdr. . . 1179
Indradhoj Limbuy, Jemdr. . . 1649
Indarjit Sing, Subdr. . . . 1046
Indrapal Singh, Jemdr. . . . 1063
Indra Singh, Subdr. . . . 1235
Inder, B. W., Lt. . . . 525, 1463
Inder Singh, Ressaldr. . 1684, 1685
Indesri Parshad Awasthi, Subdr. 1018
Indrabir Gurung, Jemdr. . . . 1625
—— Khan, Jemdr. . . . 1619
—— Thapa, Subdr. . . . 1636
Indra Datt Dixit, Sub-Asst. Surgn. 1817
Indraj Singh, Jemdr. . . . 882
—— Ressdr. 926
Indrajit Thapa, Subdr. . . . 1611
Indur Singh, Subdr. . . . 1320
Ingen, C. T. V., 2nd-Lt. . . 494
Ingam, H., 2nd-Lt. . . . 1499—1500
Ingham, B., 2nd-Lt. . . . 788
—— H., 2nd-Lt. . . . 758
—— H., 2nd-Lt. . . 355, 1499
Ingle, L. S., Lt. 317, 1094
Inglefield-Dall, B. M., Lt. . 530, 1704
Ingles, H. N. G., Lt. . . 304, 1144
Inglet, B. A., Lt. . . . 542, 1415
Inglis, A., 2nd-Lt. 1653
—— A. F., Maj. . . . 236, 1595
—— F. A. G., 2nd-Lt. . . . 786
—— H. A., Lt. . . . 542, 1122
—— H. G., Lt. . . . 507, 1565
—— J. P., 2nd-Lt. . . 560, 1020
—— T. A., Capt. . . . 501, 1666
—— W. M., Lt. . . 508, 1068, 1676
—— Lt. 1675
Ingoldby, E., Lt. 638
Ingman, G. E., Lt. 1191
Ingram, A. C., Maj. . . . 1751
—— A. G., *Hony* Lt. . . . 606
—— A. I., Lt. 704
—— A. L., 2nd-Lt. . . . 1068
—— C. D., Lt. 720
—— C. R., Maj. . . . 53, 749
—— F. E., Lt. . . . 352, 1674
—— G., Lt. . . 39, 669, 1884
—— G. E., Lt. . . . 504, 1665
—— H., 2nd-Lt. 758
—— J., Sub-Condr. . . . 1863
—— J. M., Maj. 614
—— W. J. S., Capt. . . . 1760
Igram-Johnson, R. E. S., 2nd-Lt. 329, 970, 1567
Inkson, N. L., Maj. . . . 501, 849
Inman, H. D., Lt. 538
—— H. M., Maj. 1753
—— W. F., Lt. . . . 543a, 901
Innasimuttu, Subdr.-Maj. . . 1307
—— Jemdr. 994
Innes, E. A. R., Lt. . . . 491, 881
—— G. A., Lt. . . . 353, 998
—— H., Lt.-Col. . . . 140, 1745
—— H. D.' A., Capt. . . . 1695
—— H. W., Condr. . . . 1867
—— J. H., Lt. . . . 380, 1635
—— R. J., Lt. . . . 486, 1537
Innes-Lillingston, F.F., Maj . . 213, 1661
Inniss, B. J., Col. 124
Innocent, A., 2nd-Lt. . . . 925
Inrig, L., Lt. 535, 1144
Inskip, R. D., Capt. . . 55, 258, 1259
Inward, A. A., Condr. . . . 1867
Irani, M. S., Maj. 1751
Iredell, F. S., Lt.-Genl. . . 170, 1459
Ireland, B. J., 2nd-Lt. . . 552, 1142
—— C. E., 2nd-Lt. . . . 561
—— G., Capt. . . 262, 1050, 1070
—— R. C., Asst. Surgn. . . 1797
Ironside, O. D., Capt. . . . 629
Irshad Ali, Ressdr. 1683
Irvin, A. M. N. Suster . . . 1735
—— D., Lt. 380, 1413
Irvine, A., 2nd-Lt. . . . 556, 1311
—— A. A., Lt.-Col. . 197, 1896, 1912
—— C. P. J., Lt. 767
—— H. E. G., 2nd-Lt. . . . 1679
—— M. L. C., Capt. . . . 1756
—— T. W., Lt.-Col. . 129, 1740, 1901
—— W. D., 2nd-Lt. 626
—— C. E., 2nd-Lt. . . . 358, 1573
—— D. F., Lt. 705
—— E. P., Lt. 1720
—— G., Capt. 1718
—— J. H. R., Lt. . . . 488, 1883
—— M., 2nd-Lt. . . . 538, 1636
—— M., Maj. . . 14, 31, 501, 804
—— M. C., 2nd-Lt. . . . 358, 1154
—— R., 2nd-Lt. 362
—— S. E., 2nd-Lt. . . . 555, 1373
—— B. C. B., Revd. . . . 1904
—— C., 2nd-Lt. 68, 760
—— C. D., Maj. 758
—— C. G., Lt. 1720
—— C. J., Capt. . . . 491, 803
—— G. W., Lt. 520
—— H. N., Lt. . . . 298, 1609
—— H. R. B. H., Capt. . . 272, 1423
—— J., Condr. 1854
—— N. L. C., Lt. . . . 1399, 1636
—— R. J. B., Revd. . . . 1904
—— S. F., Lt. . . 397, 397, 1319
—— W. J. A., 2nd-Lt. . . . 742
—— Carter, E., 2nd-Lt. . 360, 1289
Isaac, A. H., 2nd-Lt. . . . 781
—— A. J., 2nd-Lt. . . . 705, 1463
Isaacs, R. L., Lt. 307
Isac, Jemdr. 994
Isacke, H., Brig.-Genl. . 6, 13, 97, 103, 121, 126
Isa Charan, Mr. 3
—— Khan, Jemdr. 1240
—— Subdr. . . . 1090, 1092
—— Subdr. 1243
Isar Singh, Subdr. . 989, 1095, 1104
—— Jemdr. 1137
—— Jemdr. 1158
—— Jemdr. 1104
Isbell, H., 2nd-Lt. 700
Isbister, R., Capt. 1718
Isemonger, F. C., 2nd-Lt. 552, 869, 1143
Ishar Din Pande, Jemdr. . . . 989
—— Singh, Sub-Asst. Surgn. . 1812
—— Sub-Asst. Surgn. . . 1816
—— Subdr.-Maj. . . . 1124
—— Subdr. 988
—— Subdr. 1128
—— Subdr. 1573
—— Jemdr. 1151
—— I, Jemdr. 1573
—— II, Jemdr. 1573
—— Jemdr. 990
—— Jemdr. 1248
—— Jemdr. 1111
—— Ressdr. 906
—— Tr. Vety. Asst. . . . 1687
—— Jemdr. 1149
—— Jemdr. 894
Isher Singh, Jemdr. 1563
Isherwood, R., Capt. . . 179, 1839
Ishwaridatt, Sub-Asst. Surgn. . 1812
Ishwari-Parshad Sharma, Sub-Asst. Surgn. 1813
Ismail, Subdr. 968
—— Khan 1090
—— Jemdr. 1381
—— Jemdr. 938
—— Jemdr. 1096
—— Jan, Jemdr. 920
Ismail Khan, Subdr. 999
—— Jemdr. 1704
—— Resldr. 927
—— Sharif, Jemdr. . . . 928
Ismay, H. L., Bt.-Maj. . 96, 258, 915
Issa Khan, Jemdr. 1506
Issaac, Y., Jemdr. 1340
Isvar Rao, Subdr. 1332
Ittoji Rao Patole, Jemdr. . . . 1489
Ittycheria, C. C., Ty. Lt. . . . 1775
Itu Chawan, Sub-Asst. Surgn. . 1829
Ivatts, E., Asst. Surgn. . . . 1802
Ivens, H. T. C., Maj. . . 223, 1124
Ives, C. St. J., Lt. 487
—— E. C., Lt. . . . . .
—— G. M. P., Capt. 501
—— R. G., Sr. Asst. Surgn. . . 1791
—— V. G. Lav. N., 2nd-Lt. . . 707
—— W. P., Condr. 1853
Iwey, T., *Hony.* Capt. . . . 745
Iyengar, B. R., Ty. Lt. . . . 1780
—— C. N. Ramaswami, Esqr., M. A. . . . 91
Iyyasami, Jemdr. 1317
—— Pillai, Jemdr. . . . 1316
Iyer, A. R. 91
—— C. K. A., Ty. Lt. . . . 1780
—— P. V. V., Sub-Asst. Surgn. . 1828
—— R. S., Ty. Lt. 1776
—— T. S. S. 91
Izat, J., Lt. 514, 985
—— J. R., Lt. 517, 985
—— W. B., Maj. 663, 1882

J

Jabarbahadur Gurung, Jemdr. . 1609
Jabar Khan, Jemdr. 1260
Jabdal, Jemdr. 1386
Jack, A. J., Lt. 1020
—— H. A., 2nd-Lt. . . . 474, 1140
—— L. A., Lt. 530
—— M. M., Lt. 1082
—— W., Lt. 538, 1761

NAMES AND RANK. Page.

Jack, W. A. M., Capt. 1735
—— W. B., 2nd-Lt. 599
Jackman, H. R., Lt. . . 317, 1207
—— J., 2nd-Lt. 736
Jacks, W. R., 2nd-Lt. 1193
Jackson, A. B., Maj. 611
—— A. H. C., Lt. . 17, 40, 524
—— A. I., Lt. 846
—— A. L., 2nd-Lt. . . . 561, 1367
—— A. W. J., 2nd-Lt. . 546, 998
—— B. B., Capt. 1732
—— C., Maj. 767
—— C., Lt. 530, 1170
—— C. G., Capt. . . 9, 18, 571, 1850
—— C. H., Capt. . 95, 264, 1102
—— C. W., Capt. 705
—— D. B. F., Maj. 44
—— D. R. H., Maj. . . 231, 1623
—— D. S., Asst. Surgn. . . . 1800
—— E. F., Lt. 765
—— F. A., Maj. . . . 211, 915
—— G., Sub-Condr. 1868
—— G. O'N. B., 2nd-Lt. . . . 324
—— G. T. T., Maj. . . . 37, 1838
—— H., Lt. 508, 1666
—— H., Lt. 531, 1672
—— H. C., 2nd-Lt. . . . .
—— H. J. R., Lt. . . 307, 1255
—— H. L., 2nd-Lt. 1571
—— H. S., Lt. 520, 1895
—— H. W., Lt.-Col. 16, 42, 137, 159, 199, 870, 1363
—— H. W., Lt. 797
—— J., Lt.-Col. . . . 128, 1740
—— J., Lt. 540, 1677
—— J., Lt. 725
—— J. D., Lt. 524, 1699
—— J. D., Lt. 319, 1375
—— J. E., Sub-Condr. . . . 1855
—— J. W. C., Lt. 540a
—— K. H., Lt.-Col. 20, 87, 138, 159, 197
—— L., Lt. 617
—— L. E. W., Lt. 305
—— L. F., Capt. 840
—— L. H., Capt. 271, 1058, 1070, 1880
—— L. T., Lt. 488
—— M. H., Capt. . . . 257, 929
—— N. C., Lt. 397, 1152
—— O., Lt. 695
—— P. S. E., Lt. 652
—— P. W., 2nd-Lt. 643
—— R., 2nd-Lt. 557
—— R. J., 2nd-Lt. 803
—— R. C., 2nd-Lt. . . 548, 1596
—— R. J., 2nd-Lt. . . 566, 1132
—— S. B., Sub-Condr. . . . 1874
—— S. G., Asst. Surgn. . . . 1798
—— T., 2nd-Lt. 1094
—— T., Lt.-Col. . . . 129, 1741
—— T. H., Tempy. Capt. . . 1725
—— T. L., 2nd-Lt. 566
—— V. L. F. P., Maj. . . 237, 925
—— W. E., Lt. . . 410, 533, 1120
—— W. F. B., 2nd-Lt. . . . 568
—— W. G., 2nd-Lt. 568
—— W. J., Lt. 543a
Jacob, A. L., Lt.-Col. . . 197, 1899, 1912
—— A. LeG., Col. 1, 54, 126, 150, 187
—— C. W., Lt.-Genl. *Sir* . 107, 118, 145, 1427, 1925
—— E., Lt. 505, 801
—— G., Lt. 544, 1495
—— G. A., Col. 177
—— G. W. B., Lt. 420
—— H. B., Col. 177
—— H. F., Lt.-Col. . . . 108, 190, 1893, 1911
—— H. H., 2nd-Lt. . . . 476, 1573
—— J. R., Lt. 493, 1447
—— L. A., Asst. Surgn. . . . 1799
—— L. W. R., 2nd-Lt. . . 355, 1217
—— R., Sub-Asst. Surgn. . . . 1828
—— R. J., Jemdr. 1021
—— R. M., Capt. . . 53, 260, 1517
—— V., Jemdr. 1340
—— W., Maj.-Genl. 172
Jacobs, A., 2nd-Lt. 595
—— E. J., 2nd-Lt. 77, 786, 1844
Jacobson, E. J. H., Lt. . . . 504
Jacomb, F. B. W., Capt. . 260, 1664

NAMES AND RANK. Page.

Jacques, A., Lt. 581
—— F., 2nd-Lt. . . . 787, 1844
—— G. P. R., Capt. 911
—— S., Capt. 759
Jadunandan Singh, Subdr. Maj. 1363
Jaensch, L. V., Asst. Surgn. . . 1795
Jafar Ali, Subdr. 1412
—— **Husain**, Risldr. 873
—— **Khan**, Subdr. 1323
—— **Shar**, Subdr. 1074
Jafarey, M. A., Ty. Lt. 1775
Jaffar Ali, Jemdr. 1681
Jaffray, *Sir* W. E., Lt. 581
Jafri, Z. H., Ty. Capt. 1772
Jagan Nath Pande, *Hony.* Capt. . 2102
Jag Ram, Jemdr. 906
—— **Singh**, Subdr. 1193
—— —— Jemdr. 930
Jaganaya Kulu, Jemdr. 994
Jagannath Singh, Jemdr. . . . 1063
—— —— Jemdr. 1192
Jagannatham, Jemdr. 1264
Jagan Singh, Jemdr. 930
Jagatbahadur Rai, Jemdr. . . . 1649
Jagat Ram, Ty. Sr. Sub-Asst. Surgn. 1832
—— **Singh**, Sub-Asst. Surgn. . . 1809
—— —— Subdr. 1247
—— —— Subdr. 1252
—— —— Ressdr. . . . 898, 910
—— —— Ressdr. 920
—— —— Subdr. 1237
—— —— Jemdr. 910
—— —— Jemdr. 1112
—— —— Jemdr. 1212
—— —— Jemdr. 1134
—— —— Jemdr. 1151
—— —— Jemdr. 973
—— —— Jemdr. 1433
—— —— Subdr. 896
—— —— Jemdr. 1160
—— —— Subdr. II. 1158
—— —— *Hony.* Capt. . . . 2102
—— —— Jemdr. 1380
—— —— Jemdr. 1698
—— —— Subdr. I 1158
—— —— Ressdr. 1682
—— —— Jemdr. 1320
Jagatpal Singh, Subdr. 1062
—— —— Jemdr. 1391
Jagatsher Gurung, Subdr. . . . 1596
Jagbahadur Thapa, Jemdr. . . . 1643
Jagdama Sukul, Jemdr. 1018
—— —— Parshad Misr, Jemdr. 1019
Jagdeo Singh, Subdr. 1391
—— —— Jemdr. 934
—— —— Jamdr. 1388
—— **Dube**, Jemdr. 1365
—— **Sukul**, Subdr. 1026
Jagdish Chandra Madan, Ty. Sub-Asst. Surgn. 1818, 1833
—— —— Mohan, Sub-Asst. Surgn. 1818
Jagesar Singh, Jemdr. 1063
—— —— Jemdr. 1048
Jaggan Singh, Jemdr. 930
Jaggat Singh, Subdr. . . 1150, 1589
—— —— Jemdr. 1589
—— —— Jemdr. 902
Jaggar, J. J., Lt. 538
Jagir Singh Grewal, Sub-Asst. Surgn. 1819
Jagmohan Singh, Jemdr. 1054
Jagu Singh, Jemdr. 1192
Jagger, R., Lt. 588
Jagmal Singh, Jemdr. 1031
—— —— Jemdr. 1383
Jagnaths, Subdr. 1191
—— Jemdr. 1193
Jagpal Singh, Jemdr. 900
—— —— Jemdr. 890
Jagram Singh, Subdr. 1389
Jagratan Singh, Jemdr. 1048
Jag Singh, *Hony.* Lieut. . . . 2104
Jahandad, Jemdr. 1280
Jahan Dad, Jemdr. 1681
Jahan Khan, Jemdr. 1569
—— —— Jemdr. 1506
—— —— Jemdr. 1356
—— —— Jemdr. 900
—— **Shah**, Subdr. 1497
Jahandad Khan, Subdr. Maj. . . 1457
—— —— Subdr. 1285
—— —— Jemdr. 1280
—— —— Subdr. 1268

NAMES AND RANK. Page.

Jahandad Khan, Jemdr. 894
—— —— Jemdr. 1681
Jahan Khan, Jemdr. 1244
Jahangir Beg, Resldr. 915
Jahangir Khan, *Hony.* Capt. . . 2101
—— —— Subdr. 1467
—— —— Ressdr. 900
Jahans, F. R., 2nd-Lt. 557
—— L. E. H., Lt. . 304, 513, 1439
Jahar Singh Gurung, Jemdr. . . 1625
Jai Bahadur Mal, Jemdr. 1643
Jadhan Rai, Jemdr. 1649
Jaipatra Singh, Subdr. 1062
Jai Gopal, Sub-Asst. Surgn. . . 1812
—— —— **Govind Upadhya**, Jemdr. 1365
—— **Ram**, Jemdr. 1457
—— —— **Lal**, Jemdr. 1405
—— —— Risdr. 1684
—— —— **Dalvi**, Jemdr. 1440
—— —— Jemdr. 1040
—— **Singh**, Jemdr. 1188
—— —— Jemdr. 1175
—— —— Jemdr. 959
—— —— **Bhandari**, Jemdr. . . . 1225
—— —— *Hony.* Capt. 2102
—— —— Ressdr. 886
—— —— Ressdr. 888
—— —— Subdr. 1571
—— —— Risldr. 886, 959
—— —— **Rao Scinde**, Jemdr. . . 1416
—— —— Ty. Sub-Asst. Surgn. . 1833
Jaikaran Singh, Jemdr. 1023
Jailal, Jemdr. 1051
Jaimal, Jemdr. 968
—— **Din**, Subdr. 1011
—— **Singh**, Subdr. . . 1074, 1079
—— —— Subdr. 1074
—— —— Jemdr. 968
—— —— Subdr. 1165
—— —— Bachher, Sub-Asst. Surgn. . 1818
Jair, B. R., Ty. Lt. 1780
Jaipur, Maharaja of—*Hony.* Maj.-Genl. 110, 1070
Jairam Singh, Subdr. 1031
—— —— **Chauhan**, Jemdr. . . . 1407
Jairam Rao Rane, Jemdr. . . . 1416
—— —— **Thakur**, Jemdr. . . . 1424
Jaisi Ram, Jemdr. 930
Jaising Rao Mahadik, Ressdr. . . 934
—— —— **Sinde**, Subdr. 1417
Jaita Ram, Subdr. 1488
—— —— Subdr. Maj. 1489
Jai Singh, Subdr. 1052
—— —— Jemdr. 1052
Jalal Din, Subdr. Maj. 1122
—— —— Jemdr. 999
—— **Khan**, Subdr. 1096
—— —— Jemdr. 1312
—— —— Jemdr. 1249
—— —— Jemdr. 1376
—— **Singh**, Subdr. 1195
—— —— Jemdr. 974
Jalalud-din, Jemdr. 1436
Jalesar Singh, Jemdr. . . 1063, 1064
Jalim, Subdr. 1471, 1483
Jaloo Ram, Jemdr. 1537
Jalu Ram, Subdr. 1488
Jalpaparshad Misir, Subdr. . . 1016
Jamal Khan, Subdr. 1280
—— **Ressdr.** 918
—— **Din**, Jemdr. 1440
—— —— Jemdr. 1433
—— —— Subdr. 1299, 1573
—— —— Jemdr. 1682
Jamal-ud-din, Capt. . . . 108, 1758
Jamaiyat Singh, Ressdr. 890
Jaman Singh, Subdr. 1268
—— —— Subdr. 1571
Jaman Sing Ale, Jemdr. 1604
—— —— **Bisht**, Jemdr. 1179
—— —— Subdr. 1615
—— —— **Karki**, Jemdr. . . . 1627
James, A., Sr. Asst. Surgn. . . . 1791
—— A. A., Lt.-Col. 139, 160, 199, 808, 1140, 1457
—— A. E., 2nd-Lt. . . . 356, 1160
—— A. V., Lt. 706, 1160
—— B., Condr. 1854
—— B. S., Capt. . . 66, 367, 921
—— B. S., 2nd-Lt. 787
—— C. A., Maj. 213, 872, 1501, 1577
—— C. H., Lt.-Col. . . . 123, 1741, 1889
—— C. H., Esq. 91

NAMES AND RANK. Page.

James, C. M., Lt. 445, 1467
——— D., 2nd-Lt. 709
——— D. H., Maj. 653
——— E., Sub-Asst. Surgn. . . 1830
——— E., Lt. 696
——— E H S., Maj. 209, 1500, 1913
——— E. J. T., Lt. 90
——— E. R., Asst. Surgn. . . 1803
——— F., Capt. 255, 927
——— F., 2nd-Lt. . . . 549, 1674
——— F., Revd. 1908
——— F. A., *Hony.* Lt. . . . 610
——— F H, Maj . . 240, 1419
——— G. G., Bt. Maj. 1755
——— H., Lt. 1373
——— H., Maj. 628
——— H. C., Lt. . . . 387, 1467
——— H. K., Ty. Asst. Surgn. 1804
——— H. M., Lt. 602
——— H. O. B., Lt. . . . 310, 1343
——— H. P., 2nd-Lt. . . 359, 1351
——— H. R., Sub-Asst. Surgn. . 1832
——— H. R., 2nd-Lt. 560
——— H. V., Lt. . . . 410, 1114
——— W. H., 2nd-Lt. 1678a
——— I. S., Tempy. Capt. . . . 1724
——— J. F., Capt. 1754
——— J. H., 2nd-Lt. 720
——— J. R. Lt., . . . 430, 1036
——— J. L., 2nd-Lt. . . 763, 781
——— J. W. L., Esq. 1903
——— L., 2nd-Lt. . . . 365, 1211
——— P. W., Revd. 1908
——— R., The Revd. 1906
——— R. E., 2nd-Lt. . . 593, 599
——— R. F., Lt. 544
——— R. V., 2nd.-Lt. 564
——— T. A., Lt. 528
——— W., Sub.-Condr. . . . 1857
——— W. A. L., Lt. . . 302, 1487
——— W. B., Maj.-Genl. 16, 31, 119, 147, 175
——— W. E., Lt. 515
——— W. H. Capt. 62
——— W. H., 2nd-Lt. 561
——— W. R., Capt. . 6, 14, 107, 258, 1363
——— W. R., Lt. 352, 1611
——— W. B., Sub-Condr. . . . 1856
——— W. T., Lt. . . . 522, 1483
Jameson, A. McC., Maj. . 224, 899
——— A. N., Sub-Condr. . . 1867
——— C. E. E., Lt. . . 371, 1088
——— G. G., Lt . . . 435, 1142
——— J. H., Lt. . . . 294, 970
Jamieson, A. J., Lt.-Col. 130, 155, 189
——— D. J. T., Lt. . . . 826
——— G., Lt. . . . 515, 1667
——— H. W., Lt. 818
——— J. W. E., Lt. . 298, 1255
——— R. G., Revd. 1905
——— W. H., Lt. 1056
Jamin Sing Gurung, Subdr. . . 1615
Jamison, A. W., 2nd-Lt. . . . 735a
Jamit Singh, Subdr. 1141
——— Subdr. 1171
Jamkhandi, Capt., Chief of . . 111
Jammu and Kashmir, Lt.-Genl. *Sir* Partab Singh, *Maharaja* of., 110, 1170
Jamna Sahai, Jemdr. 1583
Jamroze Ali Khan, Ressdr. . . 1684
Jamsetji Pestonji Patel, Asst. Surgn. 1832
Janakial Lama, Jemdr. 1631
Janak Singh, Jemdr. 890
——— Jemdr. 1171
Jan Gul, Subdr. 1124
——— Jemdr. 1506
Jan Muhammad, Jemdr. . . . 922
——— Jemdr. . . . 1289
——— Subdr. . . . 1518
——— Khan, Jemdr. . . 1051
——— Jemdr. 942
——— Jemdr. 890
——— Jemdr. 951
——— Jemdr. 903
——— Jemdr. 1492
Janardana Reddi, P., Sub-Asst. Surgn. 1828
Jangbahadur Newar, Jemdr. . . 1598
Jang Bahadur Singh, Subdr. . . 1046
——— Subdr. . . 1397
——— Jemdr. . . 1400
Jang Bahadur Singh, Jemdr. . . 878
Jangbir Thapa, Subdr. . . . 1596
——— Jemdr. . . . 1602
Jangia Rana, Jemdr. 1596
Jangrai Limbu, Subdr. . . . 1653
Jani Mirza Ghazanfar Beg, Sub-Asst. Surgn. 1819
Janka, Subdr. 1171
Janki Parsad Bhatnagur, Asst. Surgn. 1820
Janki Singh, Jemdr. 1365
Janmeja Singh, Jemdr. . . 918, 955
Janmeji Singh, Ressdr. . . 1682, 1686
Janojirao Bhave, Subdr. . . . 1409
Jansi Singh, Jemdr. 1413
Janson, S. J., Lt. 515
Jansz, E. O., 2nd-Lt. 568
——— W. H. S., 2nd-Lt. . . . 563
Jant Singh, Ressdr. 894
Janus Khan, Subdr. 1239
Jaora, *Nawab* of, *Hony.* Maj. 110, 950
Jaques, A. B., Lt. . . . 490, 965
Jaquet, J. M., 2nd-Lt. . . 522, 1355
Jardine, A., Lt. 410
——— C. H., Bt.-Lt.-Col. 108, 138, 159, 219, 870, 1391, 1393
——— D. P., 2nd-Lt. 492
——— E., 2nd-Lt. 794
——— J. D., Lt. 430, 1493
——— J. E. B., Capt. 681
——— L. W., Lt. 681
Jarman, A. J., 2nd-Lt. . . . 742
Jarrad, E. D. D., Capt. 82, 505, 859, 1259
——— F. L. D., Maj. . . 230, 1501
——— V. E. D., Capt. . . . 851
Jarrett, A. F. V., Maj. . . 634, 975
——— H. C. D., Maj. . . 212, 1094
——— H. S., Col. 177
Jarrett-Kerr, H. J., Maj. 33, 331, 243, 1663
Jarvis, C., Maj. 238, 913
——— E. D., Lt. 623
——— H. L., 2nd-Lt. . . 553, 1676
Jasbahadur Limbu, Jemdr. . . 1631
——— Jemdr. . . 1649
Jasbir Gurung, Subdr. . . . 1636
——— Lama, Jemdr. . . . 1631
——— Thapa, Jemdr. . . . 1625
Jaskeram Pande, Jemdr. . . . 1054
Jasmal Singh, Subdr. 1368
Jasmer Singh Gill, Sub-Asst. Surgn. 1817
Jaspati Gurung, Jemdr. . . . 1615
Jasmir Sing Gill, Asst. Surgn. . 1817
Jasaptti Limbu, Jemdr. . . . 1631
Jasjit Singh Sirdar, 2nd-Lt. . . 905
Jas Ram, Jemdr. 1456
——— Subdr. 1457
——— Resdr. 1686
Jasudasam, F., Sub-Asst. Surgn. 1826
Jasudasan, Sr. Sub-Asst. Surgn. . 1825
Jaswant Singh, Sub-Asst. Surgn. 1812
——— Sub-Asst. Surgn. 1813
——— Jemdr. 1114
——— Jemdr. 940
——— Jemdr. . . 906, 959
——— Risldr. 896
——— Subdr. Maj. . . 1022
——— Subdr. 1032
——— Subdr. 1074
——— Sethi, Asst. Surgn. . 1818
Jatar Shankar Jeshta Ram Bhatt, Sub-Asst. Surgn. . . . 1830
Jatar, N. S., Capt. 1758
Jathar, R. G., Ty. Sub-Asst. Surgn. 1833
Jatindra Nath Neogi, Sub-Asst. Surgn. 1816
Jati Ram, Subdr. 1307
Jatni Sing Negi, Subdr. . . . 1182
Jattan Singh, Resldr. 940
Jawahir Singh, Subdr. 1230
——— Subdr. 1383
——— Subdr. 1111
Jawala Parshad Pande, Subdr. 1016
——— Singh, Subdr. . 1102
——— Singh, Jemdr. I 1151
——— Singh, Jemdr. II 1151
Jawand Singh, Subdr. 1379
——— Subdr. 1379
——— Ressdr. 948
Jawande Singh, Sub-Asst. Surgn. 1816
Jay, Lt. C. E. 399, 1016
Jayajirao Palande, Jemdr. . . 1424
Jaynes, A. G. H. 537
Jeans, F., 2nd-Lt. 554, 1056
Jecks. E. D., 2nd-Lt. 363, 1192
Jeevaratnam, Jemdr. 1699
Jefcoate, H. M., 2nd-Lt. . 326, 1251
Jeff, J., 2nd-Lt. 744
Jefferis, A. N., Lt. 879
——— A. R., Lt. . . . 368, 679
——— C. E., Lt. 509
Jefferiss, F. B., Maj. . . 752, 1720
Jeffery, C. F., Lt. . . . 534, 1692
——— F. B., 2nd-Lt. . . 363, 1203
——— H. E., Lt. . . 540, 700, 988
——— R., 2nd-Lt. 756
——— W. H., Maj. 6, 13, 107, 216, 1307
Jefford, A., 2nd-Lt. 553
——— P. E., 2nd-Lt. . . 556, 1211
Jeffrey, E. J. B., Lt. . . 700, 785
——— J. A., Lt. 591
——— J. E., Condr. 1854
Jeffreys, H. F., Lt. . . . 473, 1389
——— J. F. D., Lt. . . 308, 1391
——— K., 2nd-Lt. . . . 355, 1393
——— R. C., Lt. . . . 6, 14, 543a
——— W. J., 2nd-Lt. . . 355, 1040
Jeffries, H., Sub-Condr. . . . 1855
——— H. R. M., Lt. . . . 533, 970
Jeffs, W., Sergt.-Maj. 1879
——— H. S., Lt. 513, 1447
Jelbart, W. A., Maj. 1838
Jells Joseph, Jemdr. 1331
Jellett, H. H., Lt.-Col. . . . 850
Jelley, C. A., 2nd-Lt. . . 540, 1086
Jellicoe, J. A., Lt. . . . 508, 1158
Jemadar Khan, Jemdr. . . . 1489
Jenkin, A. J., 2nd-Lt. 728
——— F. C., Bt.-Lt.-Col. 74, 137, 633
——— H. J., Revd. 1908
——— R. T., Lt. 511, 1666
Jenkins, A. G., Lt. . . . 401, 1223
——— A. J., 2nd-Lt. . . 642, 974
——— C. H., Lt. 352, 1515
——— C. M., Lt. 706
——— D. L., Lt. 505
——— E. M., Lt. 517, 1371
——— F. A. M., 2nd-Lt. . . . 737
——— F. A. M. B., Lt. . . 415, 1303
——— G. D., Lt. 732
——— J. W. P. J., 2nd-Lt. . . 1387
——— L. C., 2nd-Lt. 700
——— N. L., Lt. . . 305, 1152, 1533
——— R. E., Capt. 732
——— R. G., Lt. 638
——— S. E., Lt. 531, 1668
——— T. G., Lt. 1007
——— W., Capt. 834
——— W. A., 2nd-Lt. . . 555, 1627
——— W. M., Capt. 636
——— W. M., Lt. 73, 537
Jenks, A. I., 2nd-Lt. 737
——— R. L., 2nd-Lt. 852
Jenner, C. H., Lt. . . . 385, 1112
Jennings, A. R., 2nd-Lt. . . . 362
——— E., Bt.-Col. . . . 125, 1740
——— C. K., 2nd-Lt. . . 450, 1144
——— G. S., Asst. Surgn. . . 1793
——— H. A. K., Br.-Genl. 10, 120, 124, 612
——— H. W., 2nd-Lt. . . 355, 1056
——— J. M., Lt. . . . 537, 1674
——— J. W., Sub-Condr. . . 1864
——— L. W. B., 2nd-Lt. . 330, 1275
——— P. W., 2nd-Lt. 597
——— R., Sub-Condr. . . . 1864
——— R. D., Maj. . . . 215, 899
——— R. M., Genl. *Sir* 169, 885, 1916, 1925
——— S. P., Lt. 543, 1295
——— W. E., Maj.-Genl. . 35, 126, 1738
——— W. G., 2nd-Lt. 723a
Jennison, S. A., Lt. . . . 294, 881
——— F. G., Lt. 517, 1215
Jensen, C. L. R., Capt. . . 729, 791
Jephcott, A. H., Lt. 517
Jepson, L. W. 2nd-Lt. 558
——— S., Lt. 300, 1183
Jermy, W. F., 2nd-Lt. 691
Jermyn, K., Lt. 463
——— T., Bt.-Col. 57
Jerram, M. R. K., Lt. 509, 1601, 1653
Jerrom, J. H., Capt. 1839
Jervis, A. E. W., Lt. . . 841, 1507
——— C. W. J., Lt. . . 518, 1158
——— E. O., Revd. 1906

Names and Rank. Page.

Jervis, T. S., Lt. . . . 302, 1391
—— W. W., Lt. . . 368, 795, 1601
Jesse, W., Lt. . . . 530, 1670
Jeshanker Manishanker Rekai, Asst. Surgn. . . . 1818
Jessel, G., Lt. . . . 685
—— R. H., Lt. . . . 733
Jessop, C. R., Lt. . . 304, 1575
—— F. T., Lt. . . 522, 1184
—— J. C., Capt. . . .
Jeste, D. G., Esq. . . . 89
Jesudasan, F., Asst. Surgn. . . 1828
Jetha, Subdr. . . . 1561
—— Singh, Jemdr. . . . 918
—— —— Jemdr. . . . 891
Jeudwine, W. W., Maj. . . 1749
Jewan Das, Jemdr. . . . 1681
—— Singh, Subdr. . . . 1174
—— —— Jemdr. . . . 1698
—— —— Jemdr. . . . 1698
Jewell, C. J. S., Lt. . . . 628
—— E. C., Lt. . . 460, 787, 1207
—— H. A., Capt. . . . 700
—— J. W., Lt. . . 313, 921
Jewett, A. F., Maj 212, 872, 1627, 1635
Jeynes, A. G. H., Lt. . . 437, 1271
Jeypore Kumara, Raja of, 2nd-Lt. 481
Jhabbar, Jemdr. . . . 1035
—— Ram, Jemdr. . . . 1372
Jhabu Singh, Jemdr. . . . 1268
Jhagarnand, Subdr. . . . 1642
—— Subdr.-Maj. . . . 1645
Jhagar Sing Bisht, Subdr. . . 1182
Jhanda, Jemdr. . . . 1481
Jhanda Singh, Ressdr. . . . 1682
—— —— Ressdr. . . . 1683
—— —— Subdr. . . . 1303
—— —— Subdr. . . . 1112
—— —— Jemdr. . . . 1698
—— —— Subdr. . . . 1117
—— —— Subdr. . . . 1148
—— —— Jemdr. . . . 1188
—— —— Jemdr. . . . 1304
—— —— Jemdr. . . . 900
—— —— Jemdr. . . . 914
—— —— Jemdr. . . . 1111
—— —— Jemdr. . . . 1043
—— —— Subdr. . . . 1587
Jhande Khan 990
—— Singh, Jemdr. . . . 989
Jhandu Singh Man, Ressdr. . . 1686
—— Lal Jiwari, Jemdr. . . 1018
Jhonda, Jemdr. . . . 1397
Jhonson, G., Condr. . . . 1862
—— W. E. C., 2nd-Lt. . . .
Jharmal Singh, Risaldar-Maj. 914
Jhutha Singh, Jemdr. . . 1031
Jhujha Ram, Subdr. . . 1488, 1497
Jhup Sing Pun, Jemdr. . . 1615
Jia Lal, Sub-Asst. Surgn. . . 1819
Jimby, C. W., 2nd-Lt. . . 67, 701
Jind, *Maharaja of*, Lt.-Col. . . 170
—— Singh, Subdr. . . . 1124
Ji Ram, Subdr. . . . 1164
Jisukh Singh, Subdr. . . . 1395
Jita Gurung, Jemdr. . . . 1008
Jitan Singh, Subdr. . . . 1383
Jitbahadur Gurung, Jemdr. . . 1602
Jitbahadur Rai, Subdr. . . . 1530
—— Rana, Jemdr. . . . 1515
—— Jemdr. . . . 1625
—— Thapa, Jemdr. . . . 1619
—— Jemdr. . . . 1543
Jitendra Narain Bhup, *Maharaja*, *Sir Hony.* Lt. 610
Jitu Limbu, Jemdr. . . . 1597
Jitjung Thapa, Jemdr. . . . 1597
Jitman Gurung, Subdr. . . . 1648
—— —— Jemdr. . . 1603, 1604
—— —— Subdr. . . . 1624
—— —— Jemdr. . . . 1615
—— Khattri, Subdr. . . . 1602
—— Pun, Jemdr. . . . 1637
—— Rai, Subdr. . . . 1648
—— —— Jemdr. . . . 1633
Jit Singh, Sub-Asst. . . . 1813
—— —— Resldr. . . . 899
—— —— Subdr. . . . 1395
—— —— Pundir, Subdr. . . 1179
Jivabhai Llallubhai Amin, Asst. Surgn. . . . 1832
Jivaji Rao, Jemdr. . . . 1447
Jiwa Singh, Jemdr. . . . 1533
—— —— Jemdr. . . . 1569
—— —— Jemdr. (1) . . . 989

Names and Rank. Page.

Jiwa Singh, Jemdr. . . . 882
Jiwan Khan, Subdr. . . . 977
Jiwan Khan, Jemdr. . . . 1440
—— —— Subdr. . . . 1409
—— Lal, Sub-Asst. Surgn. . . 1819
—— Mal, Sub-Asst. Surgn. . . 1810
—— Ram, Jemdr. . . . 1420
—— Singh, Subdr. . . . 1509
—— —— Subdr. . . . 1108
—— —— Jemdr. . . . 878
—— —— Subdr. . . . 1363
—— —— Subdr. . . . 1563
—— —— Subdr. . . . 1299
—— —— Subdr. . . . 1122
—— —— Jemdr. . . . 1171
—— —— Jemdr. . . . 920
—— —— Jemdr. . . . 1285
—— —— Ressdr. . . . 900
—— —— Sub-Asst. Surgn. . . 1817
—— —— Sub-Asst. Surgn. . . 1817
Jiwand Sing, Ressaidar . . . 965
—— —— *Hony.* Capt. . . 2101
Jiya Ram, Subdr. . . . 1399
Joachim, B. I. H., Tempy. Capt. . 1770
Joaquim, B. J. P., 2nd-Lt. 546, 1114
Joblin, G. W., Lt. . . 535, 1299
Jobbing, G. A., Maj. . . . 628
Jobson, T. S., Capt. . 78, 163, 921
Jodhbir Singh, Subdr. . . . 1231
Jodh Singh, Sub-Asst. Surgn . 1811
—— —— Sub-Asst. Surgn. . 1817
—— —— Jemdr. . . . 1280
—— —— Jemdr. . . . 1167
—— —— Jemdr. . . . 1071
—— Khan, Subdr. . . . 1247
—— —— Jemdr. . . . 1565
Jodha, Jemdr. . . . 1196
—— Jemdr. . . . 1188
—— Khan, Jemdr . . . 968
—— Singh, Jemdr. . . . 1047
—— —— Jemdr . . . 1452
Jodrell, F. J., 2nd-Lt. . 555, 1249
Joel, B. T., Sub-Asst. Surgn. . 1827
Jog, S. C., Tempy. Capt. . . 1765
Joga Chand, Jemdr. . . . 1347
—— —— Jemdr. . . . 1225
Jogbir Limbu, Jemdr. . . . 1631
Jogendra-nath Khan, Sub-Asst. Surgn . 1815
Jogendra Singh, Sirdar of Alra . 111
Jogesar Singh, Jemdr. . . . 1388
Jogesh Chandra Chackravati, Asst. Surgn. . . . 1820
Joglekar, R. G., Sub-Asst. Surgn. 1831
Jogi Chand, Subdr. . . . 1615
Jogichand, Subdr. . . . 1642
Jogi Naik, Subdr. . . . 1435
Johanson, J. L., Capt. . 477 1629
Johil, Jemdr. . . . 1121
John Lloyd George, Asst. Surn. . 1819
John, A. W. U., Maj. . 37, 500, 829
—— A. W., Lt. . . . 611
—— D., Tempy. Capt . . 1771
—— D. C., 2nd-Lt. . . 329, 1267
—— J. C., Capt. . . . 1757
—— W. D., Lt. . . . 1839
—— Francis, Jemdr. . . . 1273
—— K. E., 2nd-Lt. . . . 324
—— K. I., Ty. Lt. . . . 1775
—— W. H., 2nd-Lt. . . 564, 1166
—— W. R., Lt. . 17, 22, 536, 1385
Johns, A. B., 2nd-Lt. . . . 583
—— B. C. W., Lt. . . 321, 1375
—— T. S., Capt. . . . 777
—— W. H., Lt. . . . 885
Johnson, A., Lt. . . . 543a
—— A. E., Maj. 207, 872 1595, 1651
—— A. H. G., 2nd-Lt. 328, 1174, 1567
—— A. G., Lt. . . . 1056
—— C. A. K., Maj. . . . 57
—— C. E., Col. . 123 149, 181
—— C. F., Lt. . . . 728
—— C. G., Nursing Sister . . 1735
—— C. P., Ty. Asst. Surgn. . 1804
—— C. R., Lt.-Col. 134, 157, 94, 871, 1617
—— D. G. K., 2nd-Lt. . . 679
—— E., 2nd-Lt. . . . 551
—— E., Lt. . . 507, 1183
—— E., *Hony.* Lt. . . . 1849
—— E. C., Lt. . . . 410
—— E. C. 2nd-Lt. . . . 324
—— E. G., Lt. . . . 443
—— E. J., Lt. . . . 431
—— E. O., Asst. Surgn. . . 1792
—— E. P., Brig.-Genl 58, 120, 124, 660

Names and Rank. Page.

Johnson, E. P., Lt. . . 506, 1657
—— E. V., Condt. . . . 1854
—— F. B., Capt. . . . 728
—— F. E., Maj.-Genl. . 101, 104, 119, 123, 612
—— F. H., Asst. Surgn. . . 1801
—— F. H., 2nd-Lt. . . . 548
—— F. L., 2nd-Lt. . . . 1586
—— F. W. F., Lt.-Col. . . 728
—— G., Condr. . . . 1862
—— G. M. W., Lt. . . 404, 1339
—— G. S., Capt. . . 272, 1263
—— G. T., Capt. . . 269, 1118
—— H., 2nd-Lt. . . . 687
—— H. A. B., Capt. 67, 259, 923
—— H. B., Lt. . . 371, 1182
—— H. E., Capt. . . 666, 1066
—— H. E., Lt. . . 540a, 1072
—— H. G., Condr. . . . 1853
—— H. H., Lt. . . 430, 1231
—— H. M., 2nd-Lt. . . . 787
—— J., Capt. . . . 832
—— J., *Hony.* Lt. . . . 735a
—— J., 2nd-Lt. . . . 618
—— J. C., Capt. 280, 287, 1395
—— J. E. B., Lt.-Col. 139, 161, 260, 867, 915
—— J. F., 2nd-Lt. . . 549, 1705
—— J. N. G., 2nd-Lt. . . 556
—— J. R., Lt. . . 541, 1219
—— K. H., Lt. . . . 1487
—— M. E. S., Maj. 55, 99, 217, 1215
—— P. E., 2nd-Lt. . . 549, 1360
—— P. V., 2nd-Lt. . . . 750
—— R. A., Lt. . . 393, 1143
—— R. A., Lt.-Col. . . . 734
—— R. B., Asst. Surgn. . . 1803
—— R. H. K., 2nd-Lt. 699, 700, 796
—— R. S., Lt. . . 320, 1557
—— S. H., 2nd-Lt. . . 329, 877
—— T. F., Lt. . . 428, 891
—— T. G., Lt. . . . 931
—— V. A., Lt. . . 90, 776
—— W., Asst. Surgn. . . 1801
—— W. H. C., (Revd. . 828, 1908
—— W. J., Lt. . . . 543a
Johnson-Coombs, K. S., 2nd-Lt. 578, 1696
Johnston, A. H., Lt. . . 528, 1667
—— A. F., Lt. . . . 838
—— A. J. S., 2nd-Lt. . 475, 1082
—— A. L., Capt . . 27, 1501
—— C. *Hony.* Lt. . . . 1844
—— C. A., Lt.-Col. 62, 128, 1739
—— C. E. L., Maj. 10, 74, 102, 107, 634
—— D. D. H., Lt. 308, 1662, 1541
—— D. L., Capt. . . . 609
—— D. S., Capt . . 665, 1882
—— E. A., 2nd-Lt. . . . 366
—— E. F., Capt. . . 656, 906
—— E. N. W., Capt. 66, 485, 929
—— F., 2nd-Lt. . . 32, 337
—— H. M., Col. . 123, 149 183
—— H. W., Lt.-Col. 57, 129, 154, 187
—— J. A., 2nd-Lt. . . 557, 1285
—— J. A., Lt. . . . 532
—— J. H., Lt. . . 518, 992
—— J. S., Lt. . . . 536
—— M., 2nd-Lt. . . 562, 1068
—— M. A. B., Capt. . . 636
—— M. St. G. C., 2nd-Lt. . 531
—— P. D. C., Maj. 237, 1003, 1016
—— R., Maj. 8, 14, 82, 216, 881
—— R. B., 2nd-Lt. . . . 561
—— R. H. G., Lt. . 96, 109, 1002
—— T. E., 2nd-Lt. . . 669, 986
—— W., 2nd-Lt. . . . 786
—— W., Maj . . . 228, 1629
—— W. C., Lt. . . 536, 1303
—— W. H., Esq. . . . 69
—— W. H., 2nd-Lt. . . . 556
—— W. H. H., Lt. . . . 801
—— W. S., 2nd-Lt. . . . 91
Johnstone, A. A. J., Brig.-Genl. 120, 123, 149, 181
—— A. M., 2nd-Lt. . 545, 1160
—— B. A., Maj. . . 135, 779
—— C. H. C., Capt. . . . 628
—— C. W. H., 2nd-Lt. 544, 1375
—— D. P., Maj. . . . 1712
—— E., Lt. . . 450-54, 1355
—— F. A. B., Maj 83, 232, 1407
—— J., Sgt. Maj. . . . 1879

Names and Rank. Page.

Johnstone, J. G., Asst. Surgn. . 1796
—— J. M., Ty. Capt. . . 1725
—— N. W. W., Lt. 512, 1618
—— R. B., 2nd-Lt. . 561, 1130
—— T. S., Asst. Surgn. . 1797
—— W. W., Capt. . 501, 609
Joicey, H. E., Bt.-Lt.-Col. *Hon.* . 136, 593
Joiner, B. H. T., Lt. . . . 695, 755
Jokhi Ram, Jemdr. 1583
Joll, A. E., 2nd-Lt. 731a
Jolley, H. J., *Hony.* Lt. . . . 1846
Jollie, F., Lieut.-Col. 57, 129, 154, 188
Jolliffe, G. W. V., 2nd-Lt. . . . 713
Jolly, G. A., Maj. . . 104, 1750
—— A. W., Lt. 1732
—— E. H. P., Lt. 522
—— G. G., Capt. 1754
—— W., 2nd-Lt. 834
Jollye, H. C. B., Lt. 490
Joly de Lotbiniere, A. C. deL., Br.-Genl. 125, 660
—— H. A., Capt. 066, 985
Jonas, L. G., Lt. 516, 1001
—— R. M., Lt. 296a, 1215
—— A. F., Lt.-Col. . . 179, 1917
—— A., Revd. 1908
—— A., 2nd-Lt. 758
—— A. E., 2nd-Lt. . . 564, 653
—— A. G., 2nd-Lt. 535
—— A. G., Lt. 777
—— A. H., Lt. 543, 1391
—— A. N. G., Maj. 103, 233, 943, 1305
—— A. N. P., 2nd-Lt. 551
—— B. C., Lt. 1418a
—— B. H., Capt. 4
—— B. I., Capt. . 271, 1174, 1571
—— B. L., Capt. . . 265, 1664
—— B. L., 2nd-Lt. 746
—— B. M., Lt. 371, 1182
—— C., 2nd-Lt. 1501
—— C. A., 2nd-Lt. . . . 742, 1676
—— C. A., 2nd-Lt. . . . . .
—— C. D., Lt. 423, 1189
—— C. E., Lt. 542
—— C. R., Lt. 311, 1343
—— C. G. P., Capt. 636
—— C. Y., Lt. 435, 1140
—— C. W., 2nd-Lt. 797
—— D. J., Lt. 536, 1143
—— D. L., 2nd-Lt. 789
—— D. V., 2nd-Lt. 731a
—— E., *Hony.* Maj. 1844
—— E. B., Capt. 1717
—— E. G., Lt. 511
—— E. H., Lt. 516
—— E. H., 2nd-Lt. . . . 587, 1008
—— E. H., 2nd-Lt. . . . 475, 1469
—— E. J., 2nd-Lt. 1319
—— E. L., Lt. 752
—— E. O., Lt. . . . 90, 524, 1467
—— E. W., Capt. 748
—— F. B., Lt. 504
—— F. H., Capt. 502, 827
—— F. L., 2nd-Lt. . . . 531, 1668
—— F. L., 2nd-Lt. 563
—— F. L., Lt.-Col. 180
—— F. M. L., 2nd-Lt. 624
—— F. T., Lt. 815
—— G. E., 2nd-Lt. 1271
—— G. E., 2nd-Lt. 566
—— G. F., Lt. . . 312, 335, 443, 887
—— G. H., 2nd-Lt. 530
—— G. P., 2nd-Lt. 1018
—— G. R., Lt. 858
—— G. T., Lt.-Genl. 160
—— G. T., Asst. Surgn. 1793
—— G. W., Lt. 397, 1263
—— H., Revd. 1906
—— H., Sub-Condr. 1863
—— H., Lt. . . . 423, 515, 1219
—— H. C., Lt. 488
—— H. E., Ty. 2nd-Lt. 571a
—— H. H., Lt.-Col. 19, 131, 155, 190, 1660
—— H. H., 2nd-Lt. . . 360, 1160
—— H. I. K., 2nd-Lt. . . 326, 1102
—— H. J., Lt.-Col. . . 128, 154, 183
—— H. L., Bt.-Maj. 501
—— H. P., Capt. 825
—— H. R., Lt. 638
—— H. R., 2nd-Lt. 566, 637, 1143
—— I. D., Maj. 1751
—— I. E., Lt. 789, 790
—— J., Lt. 701
—— J. A., Capt. 825

Jones, J. B., Capt. 1713
—— J. D., 2nd-Lt. 530, 1671
—— J. E., Lt. 21, 1670
—— J. H., 2nd-Lt. 1142
—— J. H., Lt. 519
—— J. K., Lt. 297, 1627
—— J. L., Lt. 320, 1108
—— J. L., Lt. 1247
—— J. S., Revd. 1904
—— J. S., 2nd-Lt. . . . 355, 1245
—— J. W., Capt. 1751
—— K. C. J., Lt. 307a, 1645
—— L., Lt. 311a, 1487
—— L. A. M., Maj. . 47, 228, 1062
—— L. B., Lt. 1419
—— L. C., Maj.-Genl. . . 97, 103, 119, 126, 153, 195, 887
—— L. W., Lt. . . 460, 697, 1152
—— L. W. N., Lt. 1621
—— M. P., 2nd-Lt. 669, 997
—— M. R., 2nd-Lt. . . . 328, 1094
—— M. S. H., Capt. . . . 277, 949
—— O. C., 2nd-Lt. 472, 1611
—— O. E., Lt 516, 1601
—— P. H., Lt. 631, 975
—— P. J. D., Lt. . 597, 1048, 1062
—— P. P., Lt. 389, 1347
—— R. A., Lt. . . . . 446—49, 1655
—— R. B., 2nd-Lt. 1699
—— R. C., 2nd-Lt. 797
—— R. E., Lt. 1672
—— R. G., Lt.-Col. . . . 180, 1918
—— R. I., Lt. 410, 1359
—— R. I., Lt. 353
—— R. J. P., 2nd-Lt. . . . 331, 917
—— R. M., 2nd-Lt. 1455
—— R. N., Sub-Condr. 1855
—— R. R., Lt. 443, 1355
—— R. S., Lt. 385, 1641
—— R. T., Capt. . . 501, 986, 1889
—— S. C., 2nd-Lt. 73, 768
—— S. V., 2nd-Lt. . . . 359, 1094
—— S. W., Maj. 62, 1751
—— S. W. E., 2nd-Lt. 703
—— T., tempy. Capt. 1726
—— T. C., Major 829
—— T. J., 2nd-Lt. 717
—— T. J. W., Lt. . 519, 779, 1671
—— T. L., tempy. Capt. 1725
—— T. M., Capt. 832
—— T. N. B., Lt. 387
—— T. P., Lt. 357, 1195
—— T. P., Condr. 1854
—— T. W. H., Maj. 59, 1919
—— V., Capt. 830
—— V., 2nd-Lt. 788
—— V. H., Lt. 593
—— V. S., Capt. . . 17, 21, 725
—— V. W., 2nd-Lt. . . . 329, 1217
—— W. A., Lt. 598
—— W. E., Lt. 618, 654
—— W. E. R., 2nd-Lt. . . . 356, 1361
—— W. H., 2nd-Lt. 361
—— W. H. D., Lt.-Col. . . 178, 1918
—— W. H. C., 2nd-Lt. 331
—— W. J., Sub-Condr. 1856
Jopp, C. H. K., Capt. 95, 257, 901, 1385
—— J., Brig.-Genl. 177
Joques, B. W., Lt. 889
Jora, Jemdr. 1202
Jorawar Singh, Jemdr. . . . 940
Jordan, C., 2nd-Lt. . . . 475, 1042
—— E. G., Lt. 430, 1088
Joscelyne A. H., Capt. 836
Joseph, Jemdr. 1264
—— Ezekiel, Jemdr. 1498
—— A., Tempy Capt. 1766
—— A. F., Capt. . . . 17, 27, 502
—— B. A., Jemdr. 1299
—— D. P., Sub-Asst. Surgn. . 1827
—— E. V., Lt. 622
—— Jemdr. 1331
—— Subdr. 993
—— I. D., Sub-Asst. Surgn. . 1827
—— **Appaji Karvekar**, Sub-Asst. Surgn. 1830
—— **Chinnapen Ascerrathun**, Sub-Asst. Surgn. 1827
—— J. P., Ty. Lt. 1785
—— S. P., Sub. Asst. Surgn. . 1819
Joshada Kinkar Ghosh, Jemdr. . 1219
Joshi, K. G., Ty. Lt. 1778
—— M. S., Tempy. Capt. . . . 1765
—— N. N., Tempy. Capt. . . . 1771
Joshipura, R. K., Sub-Asst. Surgn. 1832

Joshua, W. J., 2nd-Lt. 1678
Josland, G. C., Lt. . . . 351, 992
Jot Ram, Jemdr. 884, 959
—— Jemdr. 1533
Jot Singh Gusain, Jemdr. . . . 1179
—— Rawat, Jemdr. 1179
Jota Singh, Subdr. 1207
—— Subdr. 1511
Jotu, Subdr. 1059
Joti More, Jemdr. 1416
Jourdain, C. E. A., Lt.-Col. 131, 748
—— P. F. C., Capt. . . . 830
Jowahir Shah, Jemdr. 1565
Jowara Ram, Jemdr. 1489
Jowhar Singh, Jemdr. 1476
—— Subdr. 1259
—— G. A., Lt. 785
Joy, H. R. L., Ty. Maj. 1722
Joyce, A. H. B., Lt. . 315, 1042, 1267
—— B. J. M. R., 2nd-Lt. . . . 1032
—— C. E., 2nd-Lt. . . . 559, 1379
—— G., Sub-Asst. Surgn. . . . 1827
—— H. W., Lt. 850
—— P. E., 2nd-Lt. 669, 992
—— T. H., Lt. 525, 1763
—— W. D., Lt. 307a, 1106
Joyes, A., Lt. . . . 516, 1601, 1678
Joyner, H., *Hony.* Lt. 1850
Joynes, A., 2nd-Lt. 623
Joynson, K B., Capt. 1910
Judah, D., Ty. Maj. 1763
Judbir Thapa, *Hony.* Capt. . 5, 2101
—— Ali, Jemdr. . . . 1608
Judd, E. L., Lt. . . 321, 365, 891
Judge, C. E. M., Lt. . 9, 317, 1666
—— C. H., 2nd-Lt. 554
Judah, D. A., 2nd-Lt. . 545, 1673, 1763
Judhabir Thapa, Subdr. . . . 1630
Judhia Sarki, Jemdr. 1602
Judkins, B. E. H., Capt. . . 502, 1672
Jug Lal, Subdr. 1395
—— Subdr. 1039
—— Jemdr. 1396
—— Jemdr. 902
Jugal Kishor, Sub-Asst. Surgn. . 1816
Juglal Singh, Subdr. 1412
Jugraj Singh, Subdr.-Maj. . . . 1062
—— Subdr. 1403
Jukes, A. H., Bt.-Maj. 257, 1143, 1641
Julian, A. L., Lt. 353, 1001
Julius, S. de\` A., Maj. . . . 52, 99
—— Revd. Leo. 854
Juma Gul, Subdr. 1207
—— Khan, Jemdr. 948
—— Jemdr. 1245
—— Jemdr. 912
—— Jemdr. 1256
Jumma Khan, Jemdr. 1698
Jummah Khan, Jemdr. 1124
Jumman Khan, Jemdr. 1433
Jung Bahadur Singh, Jemdr. . . 1268
Jungbluth, V. A., Lt. 617
Juniper, A., 2nd-Lt. 638
Jupp, E. P., 2nd-Lt. 735
Jusbahadur Singh Rai, Jemdr. . 1631
Justice, C. LeG., Col 122, 153, 181, 1917
—— W. A., Maj 1748
Juxon-Jones, J. H., Lt.-Col. 179, 1918
Jwala Parshad Aginhorti . . . 1017
Jwala Singh, Subdr.-Maj. . . . 1156
—— *Hony.* Capt. . . . 2101
—— Jemdr. 1031
—— Jemdr. 1154
—— Jemdr. 910
—— Jemdr. 1150
—— *Hony.* Capt. . . . 2104
—— *Hony.* Lt. 2104
—— Sub-Asst. Surgn. . 1810, 1813
—— Jemdr. . . . . .
—— **Sing Mahar**, Subdr. . . 1181
Jwana Singh, Jemdr. 1083

K

Kabindra Thapa, Subdr. . . . 1596
Kabiraj Khattri, Jemdr. . . . 1645
Kabiram Gurung, Jemdr. . . . 1619

NAMES AND RANK. Page.

Kabir Thapa, Jemdr. 1622
Kaburam Rai, Jemdr. 1657
Kabul Shah, Jemdr. 1121
Kabul Singh, Risldr. 930
Kabzair Khan, Ressdr. 959
Kacker, R. K., Capt. 1769
Kader Bakhsh, Ressdr. . . 1682, 1686
Kadir Bakhsh 1456
—— Khan, N., Sub-Asst. Surgn. 1827
—— Nawaz Khan, Subdr. . . . 1092
—— M., Sr. Sub-Asst. Surgn. 1826
Kahan Singh, Ressdr. 895
—— Jemdr. 1142
—— Jemdr. 930
Kahar Sing Ali, Jemdr. . . . 1596
Kahna, Subdr. 1170
—— Ram, Jemdr. 1587
Kali, Jemdr. 994
Kalkini, V. M., Capt. 1766
Kalkobad, H. D., Sub-Asst. Surgn. 1831
Kailas Singh, Jemdr. 999
Kaim Din, Jemdr. 1681
Kajiman Limbu, Jemdr. . . . 1649
Kajimn Lama, Jemdr.. 1649
Kajir Khan, Subdr. 965
—— Jemdr. 1108
Kajju Khan, Jemdr.. 1196
Kaka, D. L., Temp. Capt. . . . 1770
Kaka, Ram, Sub-Asst. Surgn. . 1809
—— Sub-Asst. Surgn. . . 1811
—— Sub-Asst. Surgn. . . 1813
—— Singh, Subdr.. . . . 1132
—— Singh, Subdr. 1132
Kale Khan, Jemdr. 890
Kakoobhai Jayantilal Trivadi, Sub-Asst. Surgn. 1831
Kaku Singh, Hony. Capt. . . . 2101
Kalab Hussain, Tr. Vety. Asst. . 1687
Kala Chaku, V., Sub-Asst. Surgn. 1831
Kala Khan, Ressdr. 1684
—— Singh, Subdr. 1111
—— Subdr. 1707
—— Hony. Capt. . . . 2103
—— Ressaidar 906
—— Jemdr. 1585
—— Jemdr. 1704
—— Jemdr. 1305
—— Jemdr. 1285
—— Jemdr. 1456
—— Jemdr. 1012
Kalandar Khan, Risldr.-Maj.. . 887
—— Jemdr. 1506
Kalberer, J. F., Lt. 524
Kalbi Hussain, Jemdr. 1068
Kalbir Gharti, Jemdr. 1622
Kaldip Singh, Jemdr. 1137
Kalia Rawat, Jemdr. 1484
Kalian Singh, Jemdr. 1472
Kalapesi, H. D., Ty. Lt. . . . 1782
Kalingaraju, Jemdr. 1360
Kaliyana, Subdr. 1191
Kalathil Gopalan, Sub-Asst. Surgn. 1831
Kale Khan, Risldr.-Maj. . . . 921
—— Jemdr. 890
—— Jemdr. 940
—— Jemdr. 1533
Kalibahadur Rai, Jemdr. . . . 1649
Kaliaperumal, Subdr. 1263
Kalicharan Shiuratan Misr, Sub-Asst. Surgn. 1829
—— Dutt. 88
Kali-das Chhote-Lal, Sub-Asst. Surgn. 1829
Kali Mohan Chanda, Sub-Asst. Surgn. 1815
Kalimuttu Thevar, Jemdr. . . . 1332
Kalkhoven, F. G., 2nd-Lt. . 558, 1501
Kalka Singh, Subdr.-Maj. . . . 1043
Kallu Singh, Jemdr. 1011
Kaloo Khan, Jemdr.. 1583
Kalsavala, N. B., Capt. 1767
Kalu, Jemdr. 1284
—— Jemdr. 1059
—— Thapa, Subdr.-Maj. . . . 2101
—— Subdr. 1191
—— Sing Gurung, Subdr. . . 1633
—— Gurung, Jemdr. . . 1608, 1619
—— Subdr. 1596
—— Subdr. 1604
—— Jemdr. 1597
—— Khan, Subdr. 1168
—— Subdr. 1371
—— Khattri, Subdr. 1642

NAMES AND RANK. Page.

Kalu, Ram, Jemdr. 1420
—— Thapa, Subdr.. . . 1596
—— Singh, Jemdr. 1031
Kalyanasundram, V. R., Rao Saheb 88
Kalyan Singh, Subdr. 1471
—— Jemdr. 936
Kama, Jemdr. 1702
Kamakaka, B. H., Ty. Lt.. . . 1773
Kamal Khan, Jemdr. 1280
—— Jemdr. 1413
—— Shah, Ressdr. 1684
Kamaluddin, Hony. Capt. . . . 2104
—— Khan, Risldr. . . . 948
Kaman Singh, Jemdr. 966
Kamardin, Hony. Capt. . . . 2104
Kamar-ud-din, Subdr. 1403
Kamal Khan, Jemdr. 1413
—— Jemdr. 1137
Kamat, D. D., Capt.. 1753
Kamata Kurmi, Jemdr. . . . 1468
Kameshvar Narayan Singh, Bahu, 2nd-Lt. 481
Kamil Sing Rawat, Subdr. . . 1179
Kamta Singh, Jemdr. 1064
Kanagasabapaty, R., Lt. . . . 1776
Kana, Jemdr. 1197
Kan Singh, Subdr. 1082
Kanthimathinathan, S. P., Lt. . 1777
Kana Ram, Subdr.-Maj. . . . 1455
Kanal, V. M., Capt. 1766
Kanaiya, Jemdr. 1040
Karan Singh, Risldr. 896
Kan Beg, Subdr. 1247
Kanchan Singh, Hony. Capt. . 2104
Kandasami, Subdr.. 993
—— Jemdr. 1316
Kandhai Singh, Hony. Capt. . . 2102
Kandhyalal, Jemdr. 1028
Kane, G. B., Lt. 399, 1187
—— J. L. K., Capt. 271, 1439, 1475
Kanga, J. S., Ty. Capt. 1770
Kanhai Parshad Dube, Hony. Capt. 2101
Kanhaiya, Subdr.-Major . . . 1395
—— Subdr. 1174
—— Subdr. 1059
—— Jemdr. 1067
—— Jemdr. 1067
—— Jemdr. 1068
—— Subdr. 1217
—— Lal, Sub-Asst. Surgn. . 1812
—— Sub-Asst. Surgn. . 1819
—— Lal Sukul, Sub-Asst. Surgn. 1814
—— Singh, Jemdr. . . . 1032
Kanhiya Singh, Subdr. 1413
—— Jemdr. 878
Kann, J. H. C., Lt. 808
Kannan, C., Sub-Asst. Surgn. . 1827
Kannayram, Jemdr. 1264
Kannayya Thevar, Subdr. . . . 1327
Kansbir Rana, Jemdr. 1619
Kanshi Ram, Subdr. 1189
—— Jemdr. 896
—— Sub-Asst. Surgn. . . 1833
Kantiprasad B. Pathak, Jemdr. . 1197
Kanwal Singh, Jemdr. 906
Kanwar Amar Singh, Capt. 40, 278, 877, 903
—— Dig Bijay Singh, Ressaldr. 1686
—— Lachmi Singh, Jemdr. . 1188
—— Pirthi Singh, Capt. 108, 278, 883
—— Udham Singh, Subdr. . 1136
Kapadia, F. N., Ty. Lt. 1781
—— M. S., Lt. 1775
—— P. R., Capt. 1769
—— R. N., Capt. 1766
Kapila, B. B., Capt. 1765
Kapoor J., Lt. 1781
—— Singh, Subdr. . . . 1698
Kapur, A. N., Lt. 1775
—— D. D., Ty. Capt.. . . 1770
—— G. D., Capt. 1768
—— K. L., Capt. 1765
—— N. C., Capt. 1757
—— M. K., Lt. 1778
—— S. N., Ty. Capt.. . . 1772
—— Chand, Sub-Asst. Surgn. 1807
—— Khan, Jemdr. 888
—— Singh, Grewal, Sub-Asst. Surgn. 1819
—— Jemdr. 1121
—— Jemdr. 932
—— Subdr. 1143

NAMES AND RANK. Page.

Kapur Singh, Jemdr. 1111
—— Subdr. 1174
—— Jemdr. 1320
Kapura Ram, Jemdr. 1288
Kapurthala, Maharaja of, Lt.-Col. 110
Kar, A. K., Lt. 1780
Kara, Subdr. 1189
Karaka, D. P., Capt. 1766
Karaksing Thapa, Jemdr. . . 1597
Karam, Ali, Jemdr. 1681
—— Khan, Rissdr. . . . 1685
—— Jemdr. 1090
—— Bakash, Jemdr. 1129, 1356
—— Jemdr. 1129
—— Subdr. 1507
—— Subdr. 1355
—— Chand, Snr. Sub-Asst. Surgn. 1807
—— Tempy. Sub-Asst. Surgn. 1833
—— Kapur, Sub-Asst. Surgn. 1814
—— Dad, Subdr. 1419
—— Subdr. 1141
—— Subdr. 1379
—— Jemdr. 1380
—— Jemdr. 1488
—— Subdr. 1565
—— Jemdr. 970
—— Subdr. 1452
—— Jemdr. 1248
—— Jemdr., ... 1681
—— Khan, Subdr. . . . 1052
—— Subdr. . . . 1152
—— Ressldr. . . 1681
—— Jemdr. . . . 1320
—— Subdr. . . . 1283
—— Jemdr. . . . 948
—— Din, Subdr. 1355
—— Jemdr. 1408
—— Jemdr. 1228
—— Subdr. 1207
—— Elahi, Jemdr. 1265
—— Jemdr. . . . 1260
—— Husain, Sr. Sub-Asst. Surgn 1809
—— Ilahi, Jemdr. 1117
—— Khan, Risaldar. . . 945
—— Khan, Subdr. . . . 1243
—— Subdr. 1439
—— Jemdr. 1440
—— Dad Khan, Subdr. . . 1145
—— Ressdr. 1681
—— Jemdr. 1579
—— Jemdr. 1087
—— Jemdr. 1704
—— Khan, Jemdr. 1682
—— Sher, Jemdr. 1248
—— Jemdr. 1565
—— Khan, Subdr.-Maj. . 1090
—— Singh, Jemdr.. . . . 898
—— Subdr. 1157
—— Jemdr. 1260
—— Jemdr. 882
—— Thapa, Subdr.. . 1609
Karamdad Khan, Subdr. . . . 1283
Karamdhoj Limbu, Jemdr. . . 1649
Karamyar Khan, Jemdr. . . . 1405
Karan Bahadur Thapa, Subdr. . 1614
—— Sing Thapa, Jemdr. . . 1607
Karbir Limbu, Jemdr. . 1631, 1657
—— Thapa, Jemdr. 1598
—— Rana, Jemdr. 1597
Karkhabadur Rana, Jemdr. . . 1636
Karnabahadur Rana, Jemdr. . . 1619
—— Jemdr. . . 1651
Karandhoj Ghalle, Jemdr. . . 1631
Karan Sing Gurung, Hony. Capt. 2103
Karbir Gurung, Subdr. 1621
—— Rai, Subdr. 1648
Karera Singh, Jemdr. 1498
Karim Ali Khan, Jemdr. . . . 1088
—— Baksh, Sub-Asst. Surgn. 1812
—— Subdr. 1507
—— Sub-Asst. Surgn.. 1815
—— Bux, Jemdr. . . 1704
—— Dad Khan, Jemdr. 1432, 1480
—— Subdr. 1433
—— Din, Jemdr. 1682
—— Khan, Subdr. 999
—— Jemdr. 1495
—— Jemdr. 1196
—— Nawaz Khan, Resldr. . . 903
Khagendra Chandra Nag, 2nd-Lt. 482
Kharki, K. B. S., Capt. 1767
Karmadhoj Limbu, Jemdr. . . 1639

Names and Rank.	Page.
Karmeswar Das, Sub-Asst. Surgn.	1830
Karm Singh, Risldr.-Maj.	2
—— Subdr.	1509
—— Subdr.	1160
—— Sub-Asst. Surgn.	1810
—— Subdr.	1256
Karna Singh Thapa, Subdr.	1607
Karnabir Gurung, Jemdr.	1347
Karndhoj Limbu, Jemdr.	1649
Karode, R. N., Capt.	1766
Karshasaf Khan, Jemdr.	898
Kasi Ram Gharti, Subdr.	1602
Karta Ram, Sub-Asst. Surgn.	1813
Kartar Bakhsh, Subdr.	1355
—— Chand, Jemdr.	928
—— Singh Bewal, Sub-Asst. Surgn.	1820
—— Grewal, Asst. Surgn.	1818
—— Jauhal, Asst. Surgn.	1818
—— Bedi, Asst. Surgn.	1818
—— Singh, Sub-Asst. Surgn.	1812
—— Risldr.	942
—— Jemdr.	1111
—— Jemdr.	944, 1167
—— Jemdr.	1433
—— Jemdr.	1241
—— Jemdr.	882
—— Risldr.	942
—— Jemdr.	898
—— Jemdr.	948
—— Ressldr.	948
—— Sub.-Asst. Surgn.	1833
Kartara Singh, Jemdr.	1228
Karumbayram, T. B., Sub-Asst. Surgn.	1828
Karve, S. D., Lt.	1779
Kasiru Din, Jemdr.	1095
Kashi Nath, Pandit, 2nd-Lt.	481
—— Ty. Lt.	
Kashinath Dalvi, Subdr.	1463
—— Mane, Subdr.	1407
—— Ram Chandra Kalamkar, Sub-Asst. Surgn.	1830
—— Zine, Jemdr.	1444
Kashi Ram, Sub-Asst. Surgn.	1811
—— Jemdr	1054
Kashmir Singh, Subdr.	1098
—— Jemdr.	1164
Kashy, P. K., Lt.	1777
Kasim Ali Bahadur, Subdr.-Maj.	1011
—— Jemdr.	930
—— Khan, Jemdr.	1488
Kasim Raza, Ressdr.	918
—— Shah, Subdr.	1457
Kasiram Thapa, Jemdr	1619
Kasturi, Jemdr.	1340
Kashturirungam Pillai, Jemdr.	1316
Katar Singh, Ressdr.	910
—— Jemdr.	882
Katha Nadan, Jemdr.	1264
—— Singh, Subdr.	1164
—— Jemdr.	1107 1249
—— Jemdr.	1111
Kaulesar Singh, Subdr.	1581
Kauntze, B. C., Lt.-Col. 141, 161, 200, 870,	1323
—— C. R., Lt.-Col. 131, 156,	192
Kaur Singh, Subdr.	1164
Kaushash, R., Capt.	1766
Kavanagh, C., Lt. 519	1707
Kay, A. H., Lt.	1007a
—— A. N., Lt.	789
—— A. T., 2nd-Lt	563
—— H. C., Maj. 215,	889
—— R. H., 2nd-Lt.	1705
—— T. A., Lt. 527,	1886—88
Kay-Mount, C. G., 2nd-Lt.	556
Kaye, C., Lt.-Col. 7, 56, 107, 133, 156,	193
—— F. W., *Hony.* Capt.	1844
—— G. F., Sergt.-Maj.	1879
Kazi Jalolud-Din Ahmed, Asst. Surgn.	1818
—— Muhammad Nazir Hasan, Sub-Asst. Surgn.	1820
Kazi Saiyd Abdul Kayum, Sub-Asst. Surgn.	1830
Kealy, P. H., Maj. 43,	663
Keane, M., Maj.	1712
Kear, J. E., Lt.	683
Kearney, P. W., 2nd-Lt.	722
Kearns, R. W., 2nd-Lt. 1, 326, 366,	1343
Kearns, W. I., Lt. 541,	1707
Kearsey, A. R., 2nd-Lt.	849
—— E. G., 2nd-Lt.	
—— E. S., Lt.	538
—— F. W., Lt. 529,	1393
—— W., Sub-Condr.	1868
—— S. H., Lt. 39,	515
Kearton, A. N. M., Lt	638
Keary, H. D'U., Lt.-Genl. *Sir* 12, 46, 118, 146, 172, 1371,	1918
—— H. P., Lt. 380,	1142
Keates, H. C., Maj.	1750
—— J. A., Lt.	525
Keating, J. W., Lt.	617
Keatinge, C. A., Lt. 503,	1078
—— F. W. G., Lt. 508,	1668
—— G., Lt. 487,	1667
Keats, A. W., 2nd-Lt 566,	1098
Keays, R. L. C., Capt.	748
—— R. W. C., Lt.-Col., 130, 155, 190, 870	1339
Kedarnath Atrish, Sub-Asst. Surgn.	1810
—— Dube, Subdr.	1028
Kedar Nath, Sub.-Asst. Surgn.	1833
—— Sing Rawat, Subdr.-Maj.	1109
Keddie, C. M., Lt. 512,	1166
Kee, W J, *Hony.* Lt.	1846
Keeble, I., Nursing Sister.	1735
Keegan, D., Lt. 306, 1152,	1563
—— B. A., 2nd-Lt. 554	1705
—— R. J., 2nd-Lt. 561,	1245
Keelan, A. P., Lt. 450,	1158
—— D. H., Capt.	840
—— H. P., Lt.-Col. 135, 158, 196, 871,	1479
—— R. S., Asst. Surgn	1795
Keeler, F. H., *Hony.* Capt.	684
Keeley, W. A., 2nd-Lt.	475
Keeling, E. J., 2nd-Lt.	760
—— E. P., Capt. 502,	1666
—— E. W. P., Retd.	1903
—— E. H., Capt. 501a, 653,	1471
—— F. G., 2nd-Lt. 559,	1287
—— R. F., Mr.	10
Keely, I. E., Lt. 508,	1601
Keen, E. W., Lt.	1624
—— F. S., Maj 51, 99, 207,	1203
—— H. C., Lt	1453
—— P H., Maj 215 869, 1142,	1247
—— W. J., Lt. Col., 201, 1601,	1912
Keenan, F. W., Lt. 450,	1303
—— T. J., Lt.	525
—— T. P. C. S., 2nd-Lt, 333,	1517
Keene, A. H., 2nd-Lt.	858
—— B. G., Capt. 277, 1375,	1581
—— C. W., Maj. 206,	1132
—— G. N. S. Maj 212,	1140
—— H., 2nd-Lt.	785
—— H., Capt.	732
—— M. R., Capt. 288,	1395
Keer, W. H., Lt. 549,	854
Keesham, M. J., 2nd Lt.	6902
Kehar Singh, Subdr. Maj.	1241
—— Subdr.	975
—— Jemdr.	1197
—— Jemdr.	1299
—— Subdr.	1203
—— Ressdr.	934
—— Risldr.	1085
—— Jemdr.	934
—— Jemdr.	912
—— Jemdr.	1347
—— Jemdr.	976
—— Subdr	1707
—— Subdr.	1074
—— Jemdr.	950
—— Jemdr.	1111
—— Jemdr.	888
—— Sur. Sub-Asst Surgn.	1867
—— Sub-Asst. Surgn.	1811
—— Sub-Asst. Surgn.	1814
—— Chandail, Sub-Asst. Surgn.	1813
—— Gill, Sub-Asst- Surgn.	1818
Kehar, D., Ty. Lt.	1781
—— Subdr.	966
—— Jemdr.	986
—— Jemdr.	1236
—— Jemdr.	1256
—— Subdr.	1114
—— Jemdr.	1117
—— Subdr.	1256
—— Subdr.	1577
—— Jemdr.	1081
—— Jemdr.	1336
Kehr Singh, Subdr.	1311
—— Subdr.	1074
—— Jemdr.	1117
Kehri Singh, Ressdr.	878
—— Subdr.-Maj.	1052
Kehir Sing, Subdr.	1217
Kehru, Jemdr.	1312
Keiger, F. G., Sub-Condr.	1874
Keighley, F. W., Lt. 515,	923
—— G. C., Capt.	692
—— H. D. S., Capt.	947
—— K., 2nd-Lt.	719
—— V. A. S., Lt.-Col. 138, 160, 198, 867, 909,	139
Keightley, M. F., Lt.	315
Keir, C. Mc. H., 2nd-Lt.	554
—— G., Lt. 530,	1483
Kelly, F. P. C., Lt.-Col. 158, 196, 870,	1497
—— H. G., Maj.	818
Keith, C. N., Maj.	858
—— N. Y., Lt. 541,	1604
—— S., Capt.	1723
Kekewich, S., Lt.	597
Kelappan, P. K., Sub-Asst. Surgn.	1826
Kelk, B. M., Lt. 443,	1333
Kellaway, G. D., 2nd-Lt.	1189
Keller, R. C., Lt.	798
Kellett, M. L., Nursing Sister	1735
—— W. R., 2nd-Lt.	555
Kelley, H. H., 2nd-Lt.	1886
Kellie, C. C. J., Lt. 291, 1016,	1034
—— G. J. D., Capt. 17, 23, 271,	1140
Kelly, A., 2nd-Lt.	1134
—— A. W. J., 2nd-Lt.	998
—— C. C., Sr. Asst.-Surgn.	1791
—— C. J., Lt. 528, 808,	1397
—— E. H., Maj. 57, 664	985
—— F. P. C., Lt.-Col. 136,	871
—— F. C., Lt.	699
—— F. H., Maj.-Genl. 107, 118,	660
—— G. A., Capt. 29,	1833
—— H. H., 2nd-Lt.	789
—— H. R., Asst. Surgn.	1795
—— J. A., Sub-Condr.	1857
—— J. G., Col. 179,	1917
—— M. G., Lt. 460,	1399
—— P. I., Lt. 397,	1351
—— P. A., Lt. 518,	1705
—— P. D. L., 2nd-Lt. 539,	1325
—— R. G., Lt. 358,	1152
—— R. J., 2nd-Lt. 554,	1702
—— R. W., Asst. Surgn.	1800
—— R. W., Lt.	1203
—— R. W. L., Lt.	543a
—— T., Condr.	1862
—— T. R., Lt.-Col. 137,	1744
—— V. S., Lt. 318,	1517
—— W. A. D., Lt. 503	1038
—— W. G., 2nd-Lt. 535,	1695
—— W. J., 2nd-Lt. 546,	803
—— W. S., Lt. 470,	1385
Kelsall, H. W., Lt.-Col. 23, 107, 135,	651
—— J., 2nd-Lt. 700,	1796
—— R., Maj.	1750
Kelsey, A. R., Capt.	751
—— T., 2nd-Lt.	769
Kelso, E., Nursing Sister	1735
—— W. D., 2nd-Lt. 455, 473,	933
Kelu, M., Sub-Asst. Surgn.	1826
Kemball, A. G., Brig.-Genl. 50, 122, 135, 158, 195, 869,	1144
—— G. V., Maj.-Genl. *Sir* 12, 32, 97, 101, 118,	612
—— J. S., Lt.-Col. 95, 131, 156,	192
Kemble, F. A., Lt.-Col 74, 136,	613
—— P. B., Maj. 60, 107, 22,	1362
—— T. A., Capt. 262, 1331,	1509
Kemm, A. H., Lt. 524,	968
—— St. J. G., Lt.	618
Kemmis, A. W. M., Maj. 107, 221,	893
—— W. D. O., Lt.	585
Kemp, A. H., *Hony.* Lt.	75
—— C. A., Lt.	629
—— D. C., Lt.-Col. 141,	1746
—— G. C., Capt.	615
—— J., *Hony.* Capt.	1849
—— K. McI., Capt. 501, 821, 1116,	1501
—— N. W., Capt.	821
—— T. N. C., Capt. 277,	1641
—— W. G. J., Lt. 526,	1667
—— W. J., *Hony.* Capt. 28,	1845

Names and Rank.	Page.
Kemp, W. N., 2nd-Lt. . . . 494,	923
Kempe, E. M., Lt.	520
—— H. F. C., Capt. . . .	615
Kemplay, W., 2nd-Lt.	545
Kempson, C. J., Lt. . . . 513,	1491
—— E. C. R., Maj. 230, 1016,	1018
Kempster, A. C., Lt. . . 475,	1333
—— J. C., Capt. . . .	658
Kemp-Welch, H. A., Lt. . . 732,	798
Kenaelly, W., Condr.	1872
Kendall, H. J., Lt. . . . 510,	1671
—— S., 2nd-Lt.	731a
—— S. B. G., Maj. . . 219,	1662
—— W. J. *Hony.* Maj. . . .	1844
Kenderdine, W. N., 2nd-Lt. 561,	1303
Kendle, F. W., Maj.	1720
—— W. E. C., Lt. . . 507,	1667
Kendrick, T. L., Capt. . . .	780
—— V. J. M., Lt. . . 430,	1090
Kenlston, L. R. H., Sub-Condr. .	1857
Kennard, D. C., 2nd-Lt. . . .	786
—— W. A., Capt. . . .	591
—— W. G. L., 2nd-Lt. .	619
Kennaway, J., Capt.	696
—— J. H., *Rt. Hon. Sir* 696,	697
Kennedy, A., Capt.	1757
—— A. C., Lt.-Col. 10, 74, 76, 137,	632
—— A. M., Capt. . . 271,	1635
—— A. P., 2nd-Lt. . . 474,	1497
—— A. S. C., Capt. . . .	615
—— B., Lt.	59
—— D. D., 2nd-Lt. . . 735,	786
—— D. Mc. K., Lt. . . .	385
—— D. W., Capt.	819
—— E. E., Capt. . . 277,	941
—— E. G., Capt.	1756
—— E. R., Capt. . . 288,	1090
—— H., Col. . . 123, 183,	149
—— H. K., 2nd-Lt. . . 555,	1373
—— J., Capt.	743
—— J., Lt.	779
—— J. C., Lt.-Col. . . .	1711
—— J. P., Lt. . . 486,	1665
—— M. G., Nursing Sister	1735
—— K. H. H., Capt. 501a, 857,	1367
—— K. M., Lt. . . 490,	897
—— M. C., Lt.	591
—— R. L., Lt.	589
—— R. S., Capt.	1752
—— S. V., Lt. . . 51,	541
—— V., Nursing Sister . .	1735
—— W. D., 2nd-Lt. . . 10,	494
—— W. M., Lt.-Col. 195, 1898,	1912
Kennedy-Craufurd-Stuart, C., Maj. 52, 107, 232,	1505
Kennelly, D. McK., Lt. . 295a,	1467
—— W. J. McK., Revd. . .	1907
Kenneth, G., 2nd-Lt.	
Kenney, B., Lt.	764
—— M. J., Capt.	768
Kennion, R. L., Lt.-Col. 189, 1889	1911
Kenny, E. C., Capt. . 259, 1191,	1399
—— E. J., Lt. . 309, 1423,	1463
—— H. T., Col.	57
—— H. T., Lt.	692
—— J. L., Condr.	1872
—— R., *Hony.* Lt.	1852
—— W. D., 2nd-Lt. . . 330,	1182
Kenrick, W. H., Lt.-Col. . 136,	1744
Kensington, E. C., Bt. Lt.-Col. 99, 138, 159, 217, 1371,	1517
Kent, A. R., Lt. 353,	1319
—— C. R., Lt. . . . 18, 35,	689
—— D. G., temp Lt. . 18, 35,	796
—— H. B., Lt. 505,	1403
—— H. H., Lt.-Col. . . .	687
—— H. S., Lt. 518,	1667
—— H. T., 2nd-Lt. . . . 325,	1217
—— J. B., 2nd-Lt. . . . 355,	1255
—— L., 2nd-Lt.	1040
—— L. M., Maj . . . 654,	1882
—— N. G., 2nd-Lt.	735
—— W., Condr.	1854
—— W. H., Lt. . . . 354,	1267
Kentish, E. G., 2nd-Lt. . . .	619
Kenward, N. H., Lt. . . 768,	785
—— P. E., 2nd-Lt. . . .	1489
Kenworthy, L. A., Lt. . 298a,	1263
—— W., Capt. . . 16, 28, 253,	937
Kenyon, H. E., Maj. . . 58,	634
Kenyon, L. P., Sr. Asst.-Surgn.	1792
—— L. R., Brig.-Genl. 71, 101, 120, 125,	632
Keogh, E. H., 2nd-Lt.	357
—— F. deC., Capt.	1724
—— J. B., Lt.-Col. . . 138, 159, 198, 867,	935
—— J. H., Lt.-Col. . . 104,	632
Keppel, T. R. H., Maj. . . 230,	1140
Keppel-Palmer, C. M., Lt. . .	597
Ker, E. C., Lt.	111
—— J. S., Lt. 492,	965
—— T. G., Lt.	504
—— T. M., 2nd-Lt. . . 329,	1158
—— W. E., 2nd-Lt.	623
Kerans, G. C. L., Maj. . . 43,	1748
Keri, Jemdr.	1036
Kerley, S. W., 2nd-Lt. . . 357,	1046
Kernahan, J. A. A., Capt. . .	1756
Kernick, J. E., 2nd-Lt. . . .	561
Kerr, A. N., Maj. . . 94, 238,	1174
—— C. W. E., Senr. Asst. Surgn.	1788
—— C. M. C. H., 2nd-Lt. . .	1483
—— D. H., 2nd-Lt.	38
—— F. A., Ty. Capt.	1724
—— H. N., Lt.	470
—— J. C., 2nd-Lt.	1611
—— J. N., 2nd-Lt. . . . 332,	1170
—— N., Lt. 401,	1335
—— P. J., Capt. . . 28, 571,	1841
—— T., Lt. 533,	1571
—— W. B., 2nd-Lt.	713
—— W. H., 2nd Lt.	83
Kerridge, H. J., The Revd. . .	1907
—— S. C., Lt. . . . 446,	1126
—— T. H., 2nd-Lt. . . 43,	705
Kersey, A. O., Lt. . . . 440,	1505
—— J., 2nd-Lt.	803
Kershaw, C. H., Lt.	626
—— H. O., 2nd-Lt. . . .	802
—— H. W. P., Lt. . . 322,	1431
Kerslake, S. S., 2nd-Lt. . . .	714
Kesar Ale, Jemdr.	1627
Kesar Chand, Sub-Asst. Surgn.	1812
Kesar Lal, Subdr.	1193
—— Mal, Subdr.	1642
—— Ram Minocha, Asst. Surgn.	1819
—— Singh, Sub-Asst. Surgn. .	1810
—— Subdr.	1189
—— Subdr.	1239
—— Subdr.	1227
—— Shah, Jemdr.	1364
—— Singh, Subdr.	1136
—— Jemdr.	918
—— Jemdr.	1141
—— Jemdr.	1212
—— Subdr.	1203
—— Subdr.	1148
—— Dann, Jamdr	1179
—— Thapa, Jemdr. . . .	1608
Kesavelu, Jemdr.	994
Kesharya, Jemdr.	1456
—— Jemdr.	1585
Keshav. Waman Khuperkar, Sub-Asst. Surgn.	1831
Keshan, Singh, Subdr.	1699
Keshavlal, H. B., Sub-Asst. Surgn.	1831
—— D. U., Sub-Asst. Surgn. .	1832
Keshbir Gurung, Subdr. . . .	1607
—— Thapa, Jemdr.	1502
Kesheo Abajee Pradhan, Ty. Sub-Asst. Surgn.	1833
—— Talekar, Jemdr	1460
Kesra, Jemdr.	1192
Kesri Singh, Ressdr.	906
Kestin, C. A., Lt.	638
Keston, T. H., Lt. . . . 491,	925
Ketley, E. W., Lt. . . . 7, 99,	715
Kettleband, H. W. B., 2nd-Lt. 567	1287
Kettlewell, A.M., Maj 53, 99 207,	949
—— H. H., Revd.	1904
—— J., Condr.	1862
—— L., Capt.	789
Kevill-Davies, E. F. W.H., Lt. 420,	1627
Kewal Singh, Jemdr.	1453
Kewley, J., 2nd-Lt. . . . 56,	1708
Key, B. W., Capt. . . 280-287,	1103
—— A. R., 2nd-Lt.	752
—— C. T., Capt.	829
—— J. H., 2nd-Lt. . . 814,	1008
Key-Mowat, C. G., Lt. . . .	1895
Keyes, R. Capt.	752
—— T. H., Bt.-Lt.-Col. 105, 213,	1913
Keyms, J., Capt.	1725
Keys, W. D. A., Maj.	1747
—— F. H., 2nd-Lt.	790
Keyte, P. E., Lt. 529,	1889
Keywood, H. A., 2nd-Lt. . 90,	742
Keyworth, W. D., Capt. . . .	1755
—— W., Lt.-Col.	1702
Khadam Khan, Jamdr.	1117
Khader Patcha, Ressldr. . . .	1683
Khadir Sharief, Subdr. . . .	1273
Khageswar Bhuyan, Sub-Asst. Surgn.	1816
Khairati Singh, Jemdr. . 876,	926
Khairuddin Khan, Jemdr. . . .	951
Khalya Mohamed Khan, Jemdr.	1122
Khaju Khan, Jemdr.	1408
Khaista Khan, Jemdr. . . 966,	970
Khalas Khan, (I) Subdr. . . .	1563
—— (II) Subdr. . . .	1563
Khalik Dad, Subdr.	1507
Khalil-ul-Rahman Khan, Jemdr.	892
Khaliqdad, Khan, Ressdr. . .	945
Khan, A. S., Capt. . . . 108,	1754
—— G. A., Capt.	1769
—— J. N., Sub-Asst. Surgn. .	1813
—— L. H., Capt.	1756
—— M. A., Capt.	1774
—— M. K. Y., Sub-Asst. Surgn.	1827
—— M. M., Capt.	1758
—— P. M. T., Ty. Capt. . . .	1770
—— S. A., Lt.	1780
—— W., Lt.	1778
—— Abdus Sattar, Capt. . .	1754
—— Alam, Jemdr.	1681
—— Alam Khan, Risaldr. . .	945
—— Bahadur Khan, Jemdr. . .	918
—— Bahadur, Subdr.	1256
—— Jemdr.	1577
—— Jemdr. . . 1311,	1505
—— Jemdr.	1682
—— Subdr.	1080
—— Jemdr.	1241
—— Jemdr.	966
—— Jemdr.	942
—— Jemdr.	972
—— Dilawat Khan, Risaldr. .	898
—— Gul, Subdr.-Maj. . . .	1154
—— Subdr.	1259
—— Subdr.	1130
—— Muhammad, Jemdr. . .	1012
—— Subdr.	1373
—— Khan, Jemdr.	904
—— Khan, Jemdr.	938
—— Jemdr. . . .	1336
—— Jemdr. . . .	1092
—— Jemdr. . . .	1107
—— Jemdr. . . .	888
—— Jemdr . . .	890
—— Jemdr. . . .	950
—— Akbar Khan, Capt. 51, 278,	1016
—— Subdr. .	999
—— Tr. Vety. Asst. . .	1687
—— Singh, Jemdr.	1704
—— Singh, Jemdr.	1372
—— Jemdr.	924
—— Jemdr.	1704
—— Shirin Khan, Jemdr. 898,	944
—— Sikandar, Hayat Khan Khattar.	481
—— Zaman, Jemdr.	966
—— Jemdr.	1380
—— Subdr.	1011
—— Khana, Jemdr.	999
Khandu, Ghadge, Subdr. . . .	1416
Khandwalla, M. T., Capt. . . .	1767
Khangara, Subdr.	1561
Khanna, A. R., Lt.	1775
—— C., Ty. Capt.	1771
—— D. D., Ty. Capt. . . .	1772
—— G. N., Lt.	1778
—— L. C., Capt.	1768
—— T. R., Capt.	1768
Khansing Gurung, Subdr.-Maj. .	1651
Khan Zaman, Subdr.	1375
Khani Zaman, Jemdr.	1344
Kharakbahadur Pun, Jemdr. 1597,	1637
Khanolkar, S. R., Sub-Asst. Surgn.	1829
—— R. G., Sub-Asst. Surgn.	1830
Khanraj Thapa, Jemdr. . . .	1651
Khanzada, Subdr.	1343
Kharak, Bahadur Khattri, Jemdr.	1645
—— Mal, Jemdr. . .	1643
—— Rana, Subdr. .	1624
—— Subdr. . .	1651
—— Singh, Subdr.	1157

Names and Rank.	Page.
Kharak Singh Gurung, Subdr	1607, 1609
—— Subdr.	1624
—— Limbu, Jemdr.	1631
—— **Sing Rana**, Subdr.	1601
—— Subdr. Maj.	1611
—— **Thapa**, Subdr.	1636
—— —— Jemdr.	1639
Kharakbir, Thapa, Subdr.	1601
—— —— Jemdr.	1602
Karkal, Charti, Jemdr.	1602
Khadim Shah, Jemdr.	1130
Kharegat, R. M., Capt.	1760
Khareghat, M. P. M., Lt.-Col.	1786
Kharjit Gurung, Jemdr.	1619
Kharka Sing Thapa, Jemdr.	1625
—— Limbu, Subdr.	1630
Kharkbahadur pun, Jemdr.	1657
Kharkabir, Pun, Jemdr.	1625
Kharkbir Ale, Jemdr.	1615
Kharke, Thapa	1608
—— Rane, Jemdr.	1608
Khairpur, Mir of, Lt.-Col.	110
Kharrak, Sing, Khattri, Subdr.	1642
Khashaba Barge, Jemdr.	1432
Khaw, I. J. O. K., Capt.	1764
Khawaj Bakhsh Khan, Jemdr.	1087
Khawja Muhammad Khan, Risldr.	51, 965
Khazan Singh, Risldr.	901
—— Jemdr.	894
—— Jemdr.	944
—— Jemdr.	924
—— Jemdr.	924
—— Jemdr.	890
—— Subdr.	1379
—— Sub-Asst. Surgn.	1819
—— Subdr.	1299
Khazana, Subdr.	1371
Khem Singh, Subdr.	1166
—— Subdr.	978
—— Jemdr.	990
—— Subdr.	1022
—— Jemdr.	1157
—— Jemdr.	1299
—— Resaldr.	882
—— **Ram**, Subdr.	1387
Khera, R. L., Lt.	1773
Kheru, Jemdr.	1312
—— **Ram**, Subdr.	1479
—— Subdr.	1433
—— Jemdr.	1336
Khilali Ram, Subdr.	1289
Khilanda Ram Diwan, Sub-Asst. Surgn. I.	1811
Khimu Rana, Jemdr.	1627
Khin, H . A., Capt.	1765
Khitab Gul, Subdr.-Maj.	1476
Khiwan Singh, Jemdr.	1368
Khiyal Gul, Subdr.	966
Khizar Mohammad Hayat Malik, 2nd-Lt.	481
Khosla, R. N., Capt.	1766
Khub Chand, Subdr.	1453
—— **Ram**, Jemdr.	1400
Khubi, Jemdr.	1320
—— **Ram**, Subdr.	1559
—— Jemdr.	930
—— **Dad**, Jemdr.	1428
Khuda Bakhsh, Asst. Surgn.	1812
—— Asst. Surgn.	1808
—— **Baksh**, Jemdr.	1232
—— Jemdr.	1130
—— **Khan**, Jemdr.	910
—— Subdr.	1707
—— **Bakhsh**, Jemdr.	910
—— Jemdr.	1126
—— **Baksh**, Subdr.	1096
—— Risldr.	1683
—— **Bux**, Jemdr.	1687
—— **Khan**, Jemdr.	942
—— **Dad**, Subdr.	1698
—— **Yar**, Jemdr.	1375
Khudadad Khan, Subdr.	1455
—— Jemdr.	1681, 1515
—— Jemdr.	1501
Khuma, Jemdr.	1561
—— Jemdr.	1195
—— **Ram**, Jemdr.	1413
Khunda Singh, Jemdr.	1563
Khursaid Alikhan, Sub-Asst. Surgn.	1811
Khurshed Ali Khan, Risaldr. Maj	887
Khurshed Khan, Jemdr.	1473
Khusal, Jemdr.	956
Khusal Singh, Jemdr.	1108
Khushal Khan, Risaldr.	912
—— Subdr. Maj.	1511
—— **Singh**, Subdr. Maj.	1028
—— **Singh**, Jemdr.	1114, 1128, 1207
—— Jemdr. I	1114
—— Jemdr. II	1114
Khushala, Subdr.	1098
Khushia, Jemdr.	1284
Khushi Mohammad, Subdr.	1305
—— Jemdr.	1092
Khushial *Hony.* Capt.	2103
Khusi Ram, Jemdr.	1189
Khusial Singh, Subdr.-Maj.	2104
Khushiram Singh, Subdr.	1021
—— Subdr.	1497
Khwaja Muhammad, Subdr.	1247
—— Risaldr.	912
—— Subdr.	1239
—— **Khan**, Risaldr.	19, 965
Khyali Ram, Subdr.	1335
Khyat Singh Negi, Jemdr.	1182
Kidar Singh, Jemdr.	1047
Kidby, W. T., Sr. Asst. Surgn.	1700
Kidd, B. G. B., Maj.	209, 872, 1497, 1579
—— G. R., Capt.	278, 1098
—— H. B., 2nd-Lt.	494, 1469
—— R. J., 2nd-Lt.	465, 473, 1651
Kiddle, J. St. A., Asst. Surgn.	1799
—— J. A. C., Lt.	513, 1493
Kidarnath Tewari, Jemdr.	1027
Kidner, W. E., Maj.	7, 664, 991
Kidson, R. A. B., Lt.	534, 1112
Kidsten, A. H., 2nd-Lt.	1068
Kiernander, C. D., 2nd-Lt.	557
—— F. D., 2nd-Lt.	1695
—— O. G., Lt.	509, 1319
Kifayat Allah, Jemdr.	978
Kikani, C. M., Ty. Capt.	1770
Kilbee, F. W., 2nd-Lt.	686
Kilbourn, W. G., 2nd-Lt.	794
Kilburn, J. M., Lt.	543, 901
Kilby, H. J., Sub-Condr.	1868
Kilcoin, C. L., 2nd-Lt.	666
Killeen, O. J. H. P., Lt., Lord	595
Killery, St. J. B., Lt.-Col.	133, 1710
Kilgour, F., Col.	177
—— R., Lt.	541, 1673
Killick, A. E. H., 2nd-Lt.	560, 1295
—— A. S., Lt.	683, 1006
—— C. F. R., 2nd-Lt.	549, 1439
Kilkelly, P. P., Lt.-Col.	1786, 1901
Kilner, F. W., Lt.	541, 1674
—— G. A., 2nd-Lt.	1273
Kilpin, J. G., Lt.	300, 1120
Kilroy, A. J. M., Lt.	317, 1070, 1483
Kincaid, J., 2nd-Lt.	875
—— Kinch, E. G. N., Capt.	820
Kind, F., Lt.	352, 1495
King, A., 2nd-Lt.	555
—— A. F. W., Lt.-Col.	132, 1743
—— A. H., 2nd-Lt.	977
—— A. J., Lt.	630
—— A. J. J., 2nd-Lt.	622
—— B. C., Lt.	589
—— C., Lt.	491
—— C. A., 2nd-Lt.	27, 1679
—— C. J. S., Capt.	665, 991
—— C. T. W., Lt.	636
—— E. H., Lt.	491, 577, 1007
—— E. O., Capt.	84, 515, 611
—— E. R., Lt.	491, 937
—— E. St. J., Lt.	317
—— E. T., 2nd-Lt.	549, 1333
—— E. W., Lt.	544, 735, 1513
—— F. J. H., 2nd-Lt.	355, 1056
—— F. H., Lt.	389, 560, 929
—— F. J. M., Capt.	664, 1880
—— G., Sub-Condr.	1863
—— G. E., Lt.	487
—— G. H., Asst. Surgn.	1794
—— G. P., Asst. Surgn.	1801
—— H. A., Asst. Surgn.	1795
—— H. H., Capt.	1755
—— H. L., 2nd-Lt.	728
—— H. T, Col.	1918, 181
—— J., Lt.	529, 1150
—— J. A., *Hony.* Maj.	1846
—— J. A., Lt.	539, 1361
—— J. B., Revd.	1904
—— J. G., Lt.	629
—— J. H., 2nd-Lt.	361, 1125
—— J. H. L., Lt.	420, 732, 1178
—— J. L., Lt.	1007
—— J. R., Lt.	535, 1699
King, J. B., *Esq.*	88
—— J. St. A., Maj.	56, 230, 1267
—— L. C., 2nd-Lt.	463, 1128
—— R. B., 2nd-Lt.	475
—— R. G., Maj.	662
—— R. M., 2nd-Lt.	565
—— R. M. E., Lt.	293a, 1645
—— R. S., Lt.	315, 929
—— S. J., Lt.	443, 1108
—— W. B., 2nd-Lt.	559, 1090
—— W. G., *Hony.* Lt.	1846
—— W. H Condr.	993, 1867
Kingdon, J., Capt.	260, 1351
—— S. C. K., 2nd-Lt.	565
Kingett, A. E., Lt.	542, 1636
Kingham, P., 2nd-Lt.	550, 1677
Kinghorn, E. C., 2nd-Lt.	725
—— P., Sub-Condr.	1864
King-Harman, M. J., Col.	178, 1917
King-Mason, A. R. L., Maj.	219, 1435
King-Salter, N. H., Capt.	264, 1623
Kingsbury, W. G., 2nd-Lt.	565
Kingsford, G. M., Capt.	683
Kingsley, C. H., 2nd-Lt.	560
—— H. H., 2nd-Lt.	690a
—— W. E., 2nd-Lt.	571, 1695
Kingsmill, W. J., 2nd-Lt.	548, 1455
Kingston, C. G., Lt.	747, 785
—— F., Lt.	544, 1443
—— G. E., 2nd-Lt.	552
—— W. V., Capt.	1732
Kingwell, G. L., Lt.	696, 697, 795
Kinloch, G., Maj.	656
—— J., 2nd-Lt.	796
—— R. I. H., Capt.	72, 101, 103, 635, 1884
—— W., Lt.	488
Kinnear, N. B., Capt.	40, 821
—— T., 2nd-Lt.	743
Kinneen, J. F, 2nd-Lt.	568
Kinnes, G. W., Lt.	629
Kinns, H. E., Lt.	816
—— P. C., Capt.	502, 1144
Kirath Singh, Jemdr.	922
Kirby, A. D., Br.-Genl.	58, 121, 126, 613
—— A. S., Lt.	793
—— E. T. R., Lt.	721
—— H., Lt.	368, 903
—— H. A. Maj.	634
—— J. L., 2nd-Lt.	577
—— J. N., 2nd-Lt.	638, 977
—— J. T., Bt. Lt.-Col.	104, 160, 233, 1662
—— N., Col.	32, 125, 660
—— N. C., Lt.	375, 1164
—— P. R. E., Capt.	1718
—— S. J., 2nd-Lt.	669, 996
Kirk, C. G. P., Lt.	595
—— F. C. deL., Lt.	691
—— G. B., 2nd-Lt.	748
—— J. C. M., Lt.	319, 1263
—— R. T. F., 2nd-Lt.	553, 1676
Kirkbride, G., Lt.	375
Kirkby, R. A., 2nd-Lt.	546, 992
Kirke, E. St. G., Maj.	664, 1001
—— K. St. G., Lt.-Col.	101, 135, 613
Kirkham, H., 2nd-Lt.	363
—— H., 2nd-Lt.	362
—— J. L., Lt.	445, 1060
Kirkhope, K. M., Capt.	8, 15, 82, 523, 817
—— J. L., 2nd-Lt.	465, 473, 1018
Kirkman, H. G., Lt.	470, 1479
—— W. A., Lt.	535, 1672
Kiron Kumar Mitra, Jemdr.	1219
Kirpashanker Girdharilal, Dave Sub-Asst. Surgn.	1820
Kirkpatrick, A. R., 2nd-Lt.	425, 1607
—— C., Maj.	52, 99, 139, 160, 217, 964
—— C. D., Capt.	634
—— E., Lt.-Col.	135, 158, 195, 1259
—— G. M., Maj.-Genl. *Sir*	6, 13, 97, 118, 660
—— H., Lt.-Col.	1745
—— H., Capt.	37, 830
—— K. E., Maj.	43, 223, 1016
—— W., Lt.	56, 523
—— W. E., Sr. Asst. Surgn	1791
Kirkwood, A. S., Maj.	81, 219
—— A. T., Lt.-Col.	47, 198, 1883, 1895, 1912
—— C. H. K., Maj.	104, 108, 234, 919

Names and Rank. Page.

Kirkwood, H. W., 2nd-Lt. . . . 708
—— J., Maj. 1750
—— J. N. S., Col. . 179, 1918
—— T. M., Lt.-Col. 130, 155, 189
—— T. W., Maj. . 243, 907
—— W. C., Maj. 236, 1395
Kirpa Ram, Sub-Asst. Surgn. . . 1810
—— Jemdr. 1312
—— Jemdr. 1188
—— Jemdr. . . . 886, 1145
—— Pun, Subdr. . . . 1607
—— Thapa, *Hony.* Capt. 2102
—— Singh, Risaldar. . . . 945
—— Sing, Rsldr. 1699
—— Jemdr. . . 1103, 1149
—— Jemdr. 990
—— Sub-Asst. Surgn. . 1813
—— Chand, Jemdr. . . 1608
Kirpashanker Girdharilal Dove, Sub-Asst. Surgn. . . . 1820
Kirpal Singh, Subdr.-Maj. . . 1379
—— Subdr. 1335
—— Subdr. 1681
—— Subdr. 1363
—— Subdr. 1152
—— Jemdr. 1682
—— Jemdr. 990
—— Jemdr. 1164
Kirtiman Khattri, Subdr. . . 1645
Kirtiman Rai, Subdr. . . . 1567
Kirwan, E. W. O'G., Capt. . . 1756
—— J. W. K., 2nd-Lt. . 476, 1443
—— L. E., Maj. . . 519, 801
—— N. G. B., Lt. . . 516, 1455
Kisaram Gharti, Subdr. . . . 1602
Kisch, F. H., Capt. 665
—— H. A., Lt. 1723
Kishan Chand, Sr. Sub-Asst. Surgn. . 1807
—— Lal, Sub-Asst. Surgn. . . 1812
—— Jemdr. 1052
—— Subdr. 1559
—— Sing Bohra, Jemdr. 1643
—— Bisht, Subdr. . . 1182
—— Singh, Jemdr. . . . 1111
—— Sumra, Sub-Asst. Surgn. 1818
—— Jemdr. 1502
—— Jemdr. 1484
—— Jemdr. 1344
—— Jemdr. 946
—— Subdr. 1493
—— Sub-Asst. Surgn. 1814
—— Sub-Asst. Surgn 1815
—— Sub-Asst. Surgn 1811
—— Sr. Sub-Asst. Surgn. . . . 1808
—— Jemdr. . . . 1561
—— Jemdr. 989
—— Sub-Asst. Surgn. . 1812
—— Subdr.-Maj. . . 1495
—— Risldr. . . . 918
—— Subdr. 1371
—— Jemdr. 1432
—— Subdr. 1179
—— Jemdr. 1145
—— Jemdr. 1160
—— Jemdr. 1244
—— Jemdr. 1204
—— Jemdr. 1079
Kishandhoj Rai, Subdr.-Maj. . 1630
Kishangarh, *Maharaja* of, *Hony.* Lt.-Col. 110
Kishena Ram, Subdr. 1335
Kishenbahadur Khattri, Jemdr. . 1643
Kishensai, Jemdr. 1472
—— Subdr 1059
Kishen Singh, Subdr. 1117
—— Jemdr. 1158
—— Jemdr. 1698, 1704
—— Subdr. 975, 1158
—— Sing Kharki, Subdr.-Maj 1598
Kishna Ram, Jemdr. 1420
Kishn Singh, Subdr.-Maj. . . 1106
—— Subdr. 1145
Kishnaswami Naidu, Risldr. . . 1683
Kishor, N., Ty. Capt. . . . 1771
Kishori Lal, Jemdr. 894
Kishor Singh, Sub-Asst. Surgn. . 1811
Kishun Singh, Risldr. 898
—— Subdr. Maj. . . 1303
Kissane, R., Lt. 521, 1657
Kissick, T., 2nd-Lt. 715
Kitchen, A. C., 2nd-Lt. 327, 1130, 1251
—— J. E., Capt. 1718

Names and Rank. Page.

Kitchen, W. H., 2nd-Lt. . 357, 1401
Kitchingman, H., Capt. . . . 730
Kitching, B., 2nd-Lt. . . 555, 1369
—— W. L. W., Revd. . . . 1903
Kite, P., Lt. 90, 733
—— E. F., Sub-Condr. . . . 1870
Kithawardin Pande, Jemdar. . . 1027
Kitson, C. E., Bt. Lt.-Col. 16, 35, 134, 749, 783
—— G. H. W., 2nd-Lt. . . . 324
—— H. L., Maj. 738
—— W. K., Capt. . . 699, 795
Kittoe, M. C. S. H., Capt 502, 779, 1667
Kiy, D. W., 2nd-Lt. 1361
Knaggs. C. F., Maj. . . 78, 219
Knapman, A. A., 2nd-Lt. . . . 787
—— E. H., 2nd-Lt. . 548, 1062
Knapp, A. F. P., Capt. . . . 748
—— H. H. G., Maj. . . 1748, 1895
—— M. D., Nursing Sister . 1734
Knapton, H. A. F., Lt.-Col. 133, 1743
—— J. H., 2nd-Lt. . . . 1034
Knatchbull, G. W. C., Br.-Genl. 121, 128, 154, 183
—— R. E., 2nd-Lt. 1457
Kneen, A. G., 2nd-Lt. . . . .
Knight, A., 2nd-Lt. 640
—— A. B., Lt. . . . 292, 881, 891
—— A. F. J., Lt. 571
—— C. C. A., 2nd-Lt. . . . 1573
—— C. R. B., Lt. . . . 686, 783
—— E. E., Capt. 619
—— E. H., 2nd-Lt. 668
—— F. E., Sr. Asst. Surgn. . 1791
—— F. P., Sub-Condr. . . . 1870
—— H., 2nd-Lt. . . . 549, 1704
—— H., Lt. 66
—— H. F., Lt. 523, 579
—— H. W., Lt. 1775
—— J. H., Bt. Maj. . 60, 634, 1003
—— K. A., Lt. 798
—— L. C., 2nd-Lt. 626
—— L. R., Lt. . . 380, 970, 455
—— M. M., Lt. . . . 530, 1701
—— R., 2nd-Lt. . . . 330, 964
—— R. A. W., Lt. 589
—— R. H., Lt. 755
—— S. J., 2nd-Lt. 782
—— W., 2nd-Lt. 1674
—— W. C., Maj.-Genl. 12, 40, 97, 102, 119, 147, 174
—— W. F., Lt. 538
—— W. N., 2nd-Lt. . . 358, 1678a
—— W. R., Sub-Condr. . . . 1863
Knighteaton, H.D., H. U. M., Capt. 785
Knights, R., Lt. . . . 460—62, 1636
Kniveton, F., 2nd-Lt. . . 564, 1191
Knobel, W. B., Capt. 1725
Knock, J., Lt. 76, 617
Knocker, J. P., 2nd-Lt. . 365, 1497
Knollys, D. E., Maj. . . 245, 1094
—— R. W. E., Maj. 207, 1896, 1913
Knott, A. E., Rev. 1908
—— H., Lt. 585, 608
—— V., 2nd-Lt. . . . 356 1489
—— W. J., Lt. 90, 512
Knowland, G. H., Capt. . . 501, 1471
Knowles, C. G., Lt. 90
—— C. J., Esq. 8, 93
—— G., Maj. 207, 877
—— G. B., Lt. 523, 1667
—— G. J., Lt. 507, 1611
—— H. P., Lieut. . . 536, 1587
—— J. A., Maj. 25
—— J. B., Lt. 747
—— J. F., Lt. 521, 1391
—— J. G., Col. 602
—— J. K., Maj. . 215, 1883, 1913
—— L. T., 2nd-Lt. . . 559, 1678
—— R., Capt. 1754
—— W. S., 2nd-Lt. . . 315, 1597
Knowlton, H. T., Capt. . . . 804
Knox, A. E., Lt. 524, 1699
—— A. W. F., Brig.-Genl. . 97, 103, 104, 105, 121, 126, 153, 196, 870, 1255
—— E. B., Lt.-Col. 1710
—— E. F., Lt.-Col. . 136, 158, 197, 869, 1106
—— J. H. G., Lt. 488, 941
—— R. W., Lt.-Col. . . . 137, 1745
—— S. G., Lt.-Col. 193, 1889, 1912
Knox-Niven, H. W., Maj. . . . 1919
Knox-Ord, D., Capt. 807

Names and Rank. Page.

Knyvett, W. A., Lt. 511
—— F. B., Maj. 619
Knyvett-Hough, C. A., 2nd-Lt. 547, 1673
Koch, E. M., Lt. 533, 1383
Kochhar, J. R., Capt. 1769
Koebel, F. E., Maj. . 105, 224, 1227
Koester, P. E., Capt. 785
Kolapore, F. J., Capt. 1756
Kolaporewalla, P. J., Capt. . . 1764
Kolhapur, H. H., *Sir* Shahu Chhatrapati Maharaj, Maharaja of, *Hony.* Col. 110, 1415
Kolu Gam, Jemdr. 1347
Konayya, Subdr. 1323
Kondaji Ghanwat, Subdr. . . . 1409
Kondaji Bedar, Jemdr. 1424
Kondayya, Subdr. 1331
Kondiba Shelke, Jemdr. . . . 1409
—— Power, Jemdr. 1469
Kondi Jagtap, Jemdr. 1511
Konsal Singh, Risldr. 914
Korb, G. L., 2nd-Lt. . . . 564, 1696
—— P. A., 2nd-Lt. . . . 556, 1702
Korke, V. T., Lt. 1773
Kor Singh, Subdr. 1145
Koshi, E. J., Lt. 1639, 1776
Kotah, *Hony.* Lt.-Col. *Sir* Umed Singh, Maharaja of . . . 110, 1191
Kotalayya, Jemdr. 1350
Kothandapani, Subdr. 993
Kothari, A. J., Sub-Asst. Surgn. . 1829
Kotwal, N. S., Lt. 1777
Kozundaivelu, Mudali, Sub.-Asst. Surgn. 1826
Kreyer, J. A. C., Maj 243, 927
Kripalani, J. K., Capt. 1769
Kripal Singh, Jemdr. 882
Krishna Chauhan, Jemdr. . . . 1424
—— Bagwe, Jemdr . . . 1424
—— Jemdr . . . 1579
—— Bhosle, Subdr.-Maj. . 1509
—— Chandra, Varma, Asst. Surgn. 1818
—— Chandra Singh, Ressdr. 878
—— Changu, Jemdr . . . 1447
—— Kalubha, 2nd-Lt. . . 112
—— Chavan, Jemdr. . . . 1416
—— Gaoda, Jemdr. . . . 1460
—— Ghorpade, Jemdr. . . 1432
—— Rao Khanvilkar, Jemdr. 934
—— Rao Gadke, Subdr. . . 1707
—— Ladoo Matker, Sub-Asst. Surgn. 1830
—— Naik Kadam, Jemdr. . 1408
—— Sinde, Jemdr. 1469
—— Krishna Swami, Jemdr. 999
—— Wajande, Jemdr. . . 1511
—— Wejre, Jemdr. . . . 1460
—— Lal Chattri, Subdr. . 1636
Krishnaji Rao Bhonsle, Jemdr. . 999
Krishnajirao Ghorpade, Jemdr. . 1444
—— Bhosle, Jemdr . . . 1424
Krishnamurti, A. S., Capt. . . 1768
—— Subdr. 1276
Krishnan, B. T., Lt. 1778
—— H., Lt. 1777
—— N. N., Capt. 1768
Krishnasami, (I) Jemdr. . . . 994
—— (III) Jemdr. . . . 993
—— (II) Jemdr. . . . 994
—— Subdr.-Maj. . . 1275
—— Subdr. 1323
—— (IV) Jemdr. . . . 994
—— (V) Jemdr. . . . 994
—— Subdr. 1323
—— Pillai, T. G., Sub-Asstt. Surgn. 1828
Krishnamoorthy, Y. V., Ty. Lt. . 1780
Krishnaswami, P., Capt. . . . 1768
—— C. V., Lt. 1781
Krishna Sami, Jemdr. 1317
Kubir Sing Adlikar, Jemdr. . . 1622
—— Limbu, Subdr . . . 1630
Kudrat Ali, Sub-Asst. Surgn. . . 1814
—— Shah, Subdr. . . . 1518
Kukday, K. V., Lt.-Col. . . 134, 1744
Kulbahadur Gurung, Subdr. . . 1618
—— Thapa, Jemdr. . . . 1645
Kulbir Rana, Jemdr. 1619
Kulkani, N. H., Lt. 1780
Kulla Thapa, Subdr. 1624
Kulman, Bhajoo, Subdr. Maj. . . 1347
—— Khattri, Jemdr. . . . 1643
—— Rai, Subdr. 1633
Kulwant Singh, Sub-Asst. Surgn. 1817

NAMES AND RANK. Page.

Kumar Adhikram Majumdar Subdr. 1219
Kumar, D. R., Capt. 1769
—— M., Lt. 1779
—— Shri Chandra Kalcondoha, *Hony.* 2nd-Lt. 112
——Kalcondorha Shivinghji Karan Singhji, *Hony.* 2nd-Lt. 112
Kumarasami, Subdr. 1307
Kumbh Singh Gurung, Subdr. 1618
Kumja Naw, Jemdr. 1347
Kumraj Thapa, Jemdr. 1637
Kumsing Gurung, Jemdr. 1611, 1625
Kunda Sing, Risldr. 915
Kundal Sukul, Jemdr. 1659
Kundan, Jemdr. 1018
Kundan Lal, Sub-Asst. Surgn. 1814
—— Lal, Jaini, Asst. Surgn. 1818
—— Singh, Jemdr. 1381
—— —— Jemdr. 1202
—— Lall, Sub-Asst. Surgn. 1811
—— Singh, Ressdr. 1683
—— —— Subdr. 1199
—— —— Nagi, Jemdr. 1179
Kunhardt, F. G., Maj. 234, 1311
—— J. C. G., Maj. 1749
Kunhi, Jemdr. 1309
Kunhi Ali, Jemdr. 1360
—— Raman, K. V., Sub-Asst. Surgn. 1828
—— Menon, Jemdr. 1309
—— —— O., Sub-Asst. Surgn. 1827
—— —— T., Sub-Astt. Surgn. 1827
Kunning, A. B., Esqr. 93
Kunniyil Raman, Sub-Asst. Surgn. 1827
Kunwar Digbajai Singh, Rissldr. 1682
—— Chiranji Singh, Subdr. 1059
—— Daulat Sing, 2nd-Lt. *Hony.* 482
—— Nand Lall, Ty. *Hony.* 2nd-Lt. 482
—— Saleh Singh, Subdr. 1059
—— Savai Sinhji, Lt. 311a, 1030
—— Shiudatt Singh, Subdr. 1387
Kuppa, P. P., Sub-Asst Surgn. 1828
Kuppuswami, P. P. Sub-Asst. Surgn. 1832
Kuppusami, Subdr. 993
—— Jemdr. 994
—— Jemdr. 994
—— Jemdr. 994
—— Jemdr. 1308
—— Jemdr. 1272
—— Jemdr. 1360
Kuppu Swami Pillai, M., Sub-Asst. Surgn. 1827
Kurban Ali Khan, *Hony.* Capt. 2104
Kurban Hussain Shah, Jemdr. 1493
Kurda, Jemdr. 1040
—— Subdr. 1040
Kure Singh, Risaldr. 901
—— Jemdr. 902
Kureishi, M. D. A., Capt. 108, 1755
Kuriyan, A. T., Capt. 1768
Kurra, Subdr. 1476
Kurshed Ali Khan, Sub-Asst. Surgn. 184
Kurshed Muhammad Khan, Risldr. 961
Kushal Singh Burathoki, *Hony.* Capt. 2103
—— Jemdr. 1472
Kushiram Singh, Subdr. 1497
Kutab Sher Shah, Jemdr. 946
Kutbi Khan, Subdr. 1483
Kutubuddin, T. R., Vet. Asst. 1687
Kutumbiah, P., Lt 1780
Kybett, C. H., Condr. 1853
Kydd, W. A., 2nd-Lt. 1028
Kynaston, N. E. V., Capt. 367, 1243

L

Labanti, V. W., 2nd-Lt. 567, 1102
Labh Din Khan, Subdr. 1280
Labh Singh, Subdr. 1379
—— Subdr. 1106
—— Jemdr. 1107
—— Jemdr. 1068
Labh Singh, Ressdr. 893
—— Jemdr. 1288
—— Subdr.-Maj. 1431
—— Jemdr. 1217
—— Subdr. 1502—04
—— Subdr. 1188
—— Subdr. 1128
—— Jemdr. 1704
—— Subdr.-Maj. 1116
—— Subdr. 1227
—— Jemdr. 950
—— Jemdr. 948
—— Subdr. 1203
—— Jemdr. 1143
—— Subdr. 1285, 1106
—— Jemdr. 894
—— Jemdr. 1137
—— Jemdr. 898
—— Jemdr. 1510
—— Jemdr. 1299
—— Jemdr. 1299
—— Jemdr. 1681
Labhu Ram, Sub-Asst. Surgn. 1814
—— Jemdr. 1142
La Bouchardiere, B. R., Asst. Surgn. 1799
—— M. P., Lt. 537, 1351
Laborda, F. E. L., Ty. Capt. 1723
Lace, E. W., Capt. 501a, 854, 1699
Lacey, C. W., Capt. 547, 611, 1699
—— F. A., Lt. 486, 1675
—— G., Lt. 542, 988
—— J. M., Lt.-Col. 802
—— J. S., 2nd-Lt. 357
—— O. B., Lt. 840
—— R., 2nd-Lt. 760
—— R. H. D., Esq. 91
Lachham Sing Rawat, Subdr.-Maj. 1223
Lachhi Ram, Subdr. 1188
Lachhiman Thapa, Jemdr. 1607
Lachhman, Subdr. 1355
—— Jemdr. 1175
—— Jemdr. 1143
—— Jemdr. 1189
—— Subdr. 1247
—— Jemdr. 1171
—— Das, Sub-Asst. Surgn. 1809
—— —— Jemdr. 1369
—— Gurung, Subdr. 1007
—— Sunwar, Subdr. 1648
—— Singh, Subdr.-Maj. 1405
—— —— Sub-Asst. Surgn. 1810
—— —— Subdr. 1379
—— —— Subdr.-Maj. 1082
—— —— Jemdr. 1248
—— —— Jemdr. 1476
—— —— Subdr.-Maj. 1078
—— —— Sub-Asst. Surgn. 1813
—— —— Jemdr. 884
—— —— Jemdr. 1158
Lachho, Jemdr. 936
Lachman Gurung, Subdr. 1615
—— Limbu, Jemdr. 1631
—— —— Jemdr. 1649
—— Khandkai, Jemdr. 1643
—— Singh, Oli, Jemdr. 1645
—— —— Khattri, Subdr. 1602
—— —— Subdr. 1604
—— Thapa, Jemdr. 1602
—— Singh, Subdr. 1148
—— —— Jemdr. 1137
—— —— Jemdr. 1579
—— —— Jemdr. 1488
—— —— Jemdr. 898
—— —— Sub.-Maj. 1283
—— —— Jemdr. 1372
—— —— Jemdr. 1117
—— —— Jemdr. 1068
—— —— Subdr. 1102
Lachmana, Jemdr. 1191
—— Jemdr. 1561
Lachmi Singh, Jemdr. 1385
Lack, L. A. H., Maj. 1752
Laddha Ram, Sub-Asst. Surgn. 1815
Ladha Singh, Subdr. 1098
—— Subdr. 1114
Ladhu Ram, Jemdr. 1583
—— Jemdr. 1420
Ladu, Jemdr. 1192
—— Sawant, Subdr. 1467
La Frenais, A. C. L., Tempy. Capt. 1767
—— H. M., Asst. Surgn. 1799
—— L. N., Asst. Surgn. 1803
Lahang Limbu, Subdr. 1630
Lahiri, P. N., Tempy. Lt. 1775
Lahora, Singh, Jemdr. 945
Lahore, Subdr. 1471
Lhhoru, Jemdr. 968
Lahrasib Khan, Jemdr. 896
Lahri Ali Khan, Jemdr. 914
Laidlaw, D., Lt.-Col. 709
—— H. A. L., 2nd-Lt. 828
Laidler, T., Lt. 463, 1501
Laik Ram, Jemdr. 902
Laihra Singh, Jemdr. 1149
Laing, A. D. M. G., Lt. 299, 1323, 1459
—— F. C., Lt.-Col. 128, 154, 185
—— H., 2nd-Lt. 626
—— H. J. G., 2nd-Lt. 474, 1319
—— S. van B., Maj. 243, 1319
Laird, A. B., 2nd-Lt. 1187
—— E. H., 2nd-Lt. 1393
—— H. G. C., Capt. 268, 1407
—— R. B., Lt. 656
—— W. B., Capt. 1713
—— W. R., Lt. 298a, 1197
Laithwaite, A., Lt. 1160
Lajje Ram, Jemdr. 888
—— Jemdr. 1040
Lajmi, V. S., Ty. Lt. 1781
Lajpat Rae, Sub-Asst. Surgn. 1814
Lake, A. E., 2nd-Lt. 357, 1056
—— C. Lt. 808
—— C. H., Lt. 1441
—— C. H. F. L., Lt. 519
—— C. H. W., 2nd-Lt. 561, 1447
—— E. A. W., Capt. 259, 1505
—— F. T. M., Lt. 380, 1024
—— M. C., Capt. 253, 1439
—— W. G., Lt. 524
Lakeland, W. J., Lt. 522, 848
Lakeman, W. H., 2nd-Lt. 731a
Lakha, Jemdr. 1561
—— Jemdr. 1141
—— Jemdr. 1699
—— Rawat, Jemdr. 1476
—— —— Jemdr. 1484
—— Singh, Risdr. 910
—— —— Jemdr. 1099
—— —— Sub-Asst. Surgn. 1814
—— —— Tr. Vety. Asst. 1687
Lakhan Singh, Jemdr 1404
Lakhbahadur Limbu, Subdr. 1648
Lakhdhoj Rai, Subdr. 1648
Lakhi Ram, Subdr. 1039
—— Singh, Ressdr 1682, 1680
Lakhmi Das, Jemdr. 894
Lakhpat Singh, Rissdr. 924
—— Jemdr. 902
—— Sub.-Asst. Surgn. 1819
Lakh Ram, Jemdr. 1035
Lakhu, Jemdr. 1137
Lakin, C. A., Maj. 500, 820
—— J. H. F., Maj. 53, 99, 107, 216, 1629
Lakpat Singh, Jemdr. 926
Lakshiman, Kadam, Subdr.-Maj. 1439
—— Powar, Subdr. 999
—— Rao Sinde, Jemdr. 1440
—— —— Subdr. 1435
Lakshman, Prasad Jurawan Singh, Sr. Sub-Asst. Surgn. 1829
Laksmanparshad, Sub-Asst. Surgn. 1811
Lakshmi Narayan Chaudri, Babu 3
Lal, S., Tempy. Capt. 1772
—— S. B., Tempy. Capt. 1768
—— Akbar Khan, Subdr. 1320
Lala, Subdr. 1199
—— Jemdr. 1200
—— Ram, Jemdr. 1498
—— Rawat, Subdr. 1483
—— Singh, Subdr. 1021
Lal Bahadur Singh, Subdr. 1062
—— Jemdr. 1397
—— Singh Jemdr. 1212
—— —— Jemdr. 930
Lalbahadur Bura, Jemdr. 1608
—— Gurung, Jemdr. 1603
—— —— Jemdr. 970
—— Rai, Jemdr. 1631
—— Sahi, Jemdr. 1602
—— Singh
—— Thapa, Jemdr. 970
—— —— Jemdr. 1637
Lal Beg, Jemdr. 979
Lalbir Thapa, Subdr. 1631
—— —— Jemdr. 1625
—— Limba, Jemdr. 1622

Names and Rank. — Page.

LalbiriLimba, Pun, Subdr. . . . 1636
—— Rai, Jemdr. . . . 1622
Lal Chand, Jemdr. 906
—— Jemdr. 1559
—— Jemdr. 1040
——Chandra Meara, Sub.-Asst. Surgn. 1818
Lalkaka, K. A. J., Ty. Lt. . . 1777
Lal Khan, *Hony.* Capt. . . . 2101
—— Jemdr. 920
—— Jemdr. 1407
—— Jemdr. 968
—— Jemdr. (I) 1121
—— Jemdr. 1305
—— Jemdr. (II) 1121
—— Jemdr. 1134
—— Jemdr. 1231
—— Jemdr. 1240
—— Jemdr. 978
—— Jemdr. 1682
—— Subdr. 1154
—— Jemdr. 948
—— Subdr. 1235
—— Subdr. 979
—— Subdr. 1095
—— Jemdr. 1288
—— Subdr. 1333
—— Jemdr. 1036
—— Jemdr. 9183
—— Jemdr. 1573
Lal Chandra Meara, Sub-Asst. Surgn. 1818
—— Din, Subdr. 976
—— Mir, Subdr. 1117
—— Muhammad, Sub-Asst. Surgn. 1827
—— Shah Gul, Jemdr. 1502
—— Sher, Jemdr. 1507
—— Sing Gusain Subdr. . . . 1179
—— Rana, Jemdr. 1692
—— Thapa, Subdr. 1602
—— Sub.-Maj. . . 1504
—— Jemdr. . . . 1615
—— Singh, Subdr. 1141
—— Subdr. 1301
—— Jemdr. 1107
—— Jemdr. 1163
—— Jemdr. 1281
—— Jemdr. 1472
—— Jemdr. 1308
—— Jemdr. 922
—— Subdr. 1203
—— Jemdr. 1473
—— *Hony.* Capt. 2102
—— Jemdr. 1698
—— Sub-Asst. Surgn. . . 1816
—— Jemdr. 1268
—— Jemdr. 1117
—— Jemdr. 1054
—— Jemdr. 1145
—— Subdr. 1301
—— Basnet, Subdr. . . . 1642
—— Bedi, Sub-Asst. Surgn. 1814
—— Gill, Sub-Asst. Surgn. 1819
—— Rawat, Jemdr. . . 1225
—— Sandhu, Sub-Asst. Surgn. . . . 1813
—— Subdr. 1419
Lale Pun, Jemdr. 1598
Lalitbahadur Mal, Subdr. . . 604
—— Subdr. . . 1021
Lalji Ram, Jemdr. 1386
Lall Khan, Jemdr. 979
—— Jemdr. 1143
—— Ressdr. 1683
—— Mast, *Hony.* Lt. 2104
Lalor, D. C., Lt. . . . 425, 949
——, J. F. V., Lt. . 300, 1379, 1383
Lalta Singh, Jemdr. 1401
—— Pershad Upadhya, Jemdr. 1019
Lalu Khattri, Jemdr. 1643
Lamacraft, H., 2nd-Lt. . . . 91, 547
Lamb, A. B., Esq. . . . 1702, 91
——, B., Maj. 633
—— D. O. W., Capt. 8, 15, 252, 893
—— G., 2nd-Lt. . . . 550, 618
—— P., Lt. 531
—— R. J., 2nd-Lt. . . 701, 786
—— T., Lt. 656
—— V. I., Nursing Sister . . 734
—— W., Maj. 801
Lambagraon, Jaichand, *Raja* of— *Hony.* Lt.-Col. . 110, 1170
Lambden, C. B., Lt. . . 351, 1007a
Lambe, G., 2nd Lt. . . 550, 1702
Lambert, A., 2nd-Lt. 780
—— A. F., Maj. 615

Lambert, A. G., Lt. . . 522, 1367
—— D. H. C., Lt. . 486, 893
—— E. W., Lt. 510
—— H., Lt. 848
—— H., 2nd-Lt. . . 357, 1158
—— J. H., Capt. . . . 843
—— V. G., 2nd-Lt. . . . 637
—— W. G. A., Lt. . . 371, 1150
—— W. J., Lt.-Col. . 208, 929
—— W. R., Capt. 279, 1052, 1395
Lamberton, N. C., 2nd-Lt. . . 779
Lambie, J., 2nd-Lt. . . . 472, 1032
—— S. H., Lt. 354
Lambourne, A. E., Condr. . . 1863
Laming, S. A., 2nd-Lt. 927
Lamonby, L., Maj. . . 724, 739
Lamond, J. D. M., Asstt. Surgn. 1802
Lamont, G. A., Ty. Capt. . . . 1724
—— J., 2nd-Lt. 553
—— J. C., Lt.-Col. 1785
—— J. N., Capt. 753
—— J. W. F., Br.-Genl. 121, 126, 613
Lampard, L. A., 2nd-Lt. . 559, 611
Lancashire, J. K., 2nd-Lt. . . 571a
Lancaster, A. S., Lt. . . 420, 1130
—— C. E. A., Lt. . 21, 532, 1668
—— E. H., Maj. . . 221, 1662
—— G. C., 2nd-Lt. . . . 566
—— M. P., Lt. 883
Lance, F. Fitz. H., Maj. . . 209, 911
Landale, C. H., Capt. 274, 905, 917
—— H. R., Capt. . . 486, 943
Landeman, H. G., 2nd-Lt. . 540, 1695
—— Capt. 1702
Lander, A., 2nd-Lt. 56
—— F. J., 2nd-Lt. 750
—— E. T. W., 2n-Lt. . . . 728
—— J. H., Lt. . 425, 722, 1007
—— K. J. M., Lt. . . . 530
Landon, C., 2nd-Lt. 826
—— C. R.H.P., Maj. 107, 218, 870, 941, 1297
—— R. P., Capt. . 634, 914, 974
—— S. L., Lt. . . . 309, 1427
Landragin, L. C., Lt. 754
Lane, A. D., Lt. . . . 446-49, 1018
—— C. A., Col. 126
—— C. A., Lt.-Col. . . 137, 1744
—— C. M., 2nd-Lt. . . 329, 1124
—— C. R. C., Capt. 54, 104, 272, 911
—— C. T., Col. 177
—— D. T., Lt.-Col. . . . 127
—— F. B., Maj. . . 53, 219, 889
—— F. C., Lt.-Col. . 72, 132, 613, 1884
—— F. M., Maj. 1919
—— F. C. T., 2nd-Lt. 779
—— G. H., Capt. . 278, 1415, 1467
—— H. A., Brig.-Genl. 14, 28, 29, 131, 155, 190
—— H. E., Maj. 101
—— H. P., Col. . . 125, 150, 185
—— H. R. C., Capt. 256, 1617, 1621
—— H. S., Lt. 846
—— H. V., Maj. 33, 225, 868, 1016, 1028
—— J. T. H., Maj. 221, 869, 1178, 1182
—— J. W., Capt. 1713
—— N. M., Lt. 629
—— P. G. C., Lt. 316
—— S. T. H., 2nd-Lt. . . . 608
—— V. C. T., 2nd-Lt. . . 465, 937
—— W. B., Lt.-Col. . . 128, 1740
—— W. E., 2nd-Lt. . . 546, 1405
—— W. G., 2nd-Lt. 752
—— W. H., Maj. . . 208, 1383
Lane-Brown, H., 2nd-Lt. . . . 658
Lane-Ryan, S. E., 2nd-Lt. . . . 332
Lanfear, E. W., 2nd-Lt. . . . 753
Lang, A., *Hony.* Capt. . . . 1845
—— A. G. B., Lt.-Col. . . . 57
—— C. F. G., Maj. . . 209, 1661
—— D. W., Lt. 577
—— E. B., Col. . . 59, 124, 183
—— E. M., Maj. . . 869, 1223
—— H. F., Capt. . . . 485, 911
—— H. G., Lt. . . . 330, 1389
—— H. S., Lt. 754
—— H. W., 2nd-Lt. 1497
—— J. Lt. 532
—— J. H., Lt. 310
—— J. T., Lt. . . 450-54, 1397
—— L. E., Capt. . . 256, 1893
—— W. F., Capt. . . 9, 18, 500
—— W. H., Bt.-Maj. . 52, 232, 905
—— W. I., Lt. 1517
—— W. W. Capt. . . . 24, 1839
Langdle-Smith, R. M., Revd. . 1906

Langdon, A. G., 2nd-Lt. . . . 361
Langdon, F. G., Lt. 818
Lange, B. D., Lt. 1287
Langer, W. ., Lt. . . . 521, 1886
Langford, G. W., 2nd-Lt. . . . 546
—— R., Lt. 617
—— W., Lt. . . 18, 27, 581
Langford-James, J. W., Esq. . 1903
—— R. H. L., 2nd-Lt. 651
Langhorne, C. C., Capt. . 272, 1116
—— J. A. D., Maj. 53, 101, 633
—— S. G., Sr. Asst. Surgn. . 1791
Langlands, A. R., Lt. . . 443, 1585
—— E. W., Lt. . 373, 1607
Langley, E. E., 2nd-Lt. 561
—— L. C., 2nd-Lt. . . 475, 1098
—— W. V., 2nd-Lt. . . . 360
Langley-Taylor, W. L., Capt. . . 628
Langrish, F., Sub-Condr. . . . 1855
Langrishe, R. B., Maj. . . 245, 1603
Langton, J., Sub-Condr. . . . 1856
Langshaw, C. M., 2nd-Lt. . . . 563
Langstaff, J. W., Lt.-Col. 10, 133, 1710
Langston, K. E. W., 2nd-Lt. . . 724
—— T. A. O., Lt.-Col. . . 1786
Langtry, H. V. M., Lt.-Col. 197, 1895, 1912
Langworthy, V. U., *Hony.* Col. 702, 704
Lanktree, H. R., Lt. . . . 541, 1297
Lansdale Ruthven, H. P., 2nd-Lt. 1696
Lansley, G. E., Sub.-Condr. . . 1867
Lapsley, W., Maj. 1747
—— J. B., Capt. 1754
Larcomb, J. N. 2nd-Lt. . . . 788
Large, D. T. M., Capt. 1714
—— O., 2nd-Lt. 564
—— G. S., Capt. 1058
Larking, H. S., Lt. . . 321, 365, 1395
Larkins, E. P., Lt. . . . 358, 913
Larlham, P. E., Lt. 302, 323, 1399, 1401
Larminie, J. A., 2nd-Lt. . . 328, 1483
Larmour, L. C., Maj. . . . 101, 634
Larnder, E. W., Col. . . . 31, 1838
Larymore, P. A. D., Lt. . 323, 1647
Larymore, M. A., Lt. 489
Lascelles, H., 2nd-Lt. 784
—— H. B., 2nd-Lt. 750
—— R. H., Maj. 614
Lash, A. O., Maj. 1919
Laskari, P. F., Tempy. Capt. . . 1691
Laskie, A. L., Lt. 1245
Last, F. B., Capt. . . . 732, 785
LaTerriere, F. J. B. des, Lt. . . 597
Latham, A., Maj. . . 226, 1223, 1594
—— C. R. C., 2nd-Lt. . 554, 1339
—— E., Capt. 615
—— E. B., Lt. . . . 516, 1090
—— F., Bt.-Maj. 71
—— H. B., Capt. . . . 58, 617
—— H. C., Lt. . 311, 410, 1104
—— H. W., Ty. Capt. . . . 1726
—— P. D., Lt. 489, 915
Lathbury, H. O., Lt.-Col. . 37, [illegible], 001
Latif Khan, Subdr. 1301
—— Subdr. 1235
—— Jemdr. 1389
—— Rissdr. 915
—— Sher, Jemdr. 1373
Latimer, H., Lt. 493
—— R. A., 2nd-Lt. . . 475, 1413
Latkan Singh, Subdr. 1565
Latte Singh, Jemdr. . . . 1452, 1585
Latter, F. A., Lt. 313, 855
Lattey, H. P. T., Lt. . . . 511, 1630
—— J. T., Lt. 518
—— W. T., 2nd-Lt. 622
Leuddle, A. K., Maj. 1750
—— F. P., Lt.-Col. 1711
—— T. C., Lt.-Col. . . 133, 1710
Laughland, A. P., Lt. 763
Laughlin, G. E., 2nd-Lt. . . .
—— G. E., 2nd-Lt. . . . 786
Laughton, F. S., 2nd-Lt. . 446-49, 1032
—— G. C., 2nd-Lt. . . . 840
—— G. G., Lt. 1695
—— J. V., *Hony.* Maj. . . 598
—— J. W., Lt. . . . 299, 1647
Laul Khan, Subdr. 1203
Laurasib Khan, Jemdr. 894
Laurie, G., 2nd-Lt. . . . 541, 1653
—— J., Lt. 401, 1191
Lavender, G. A., 2nd-Lt. . . . 652
—— S. S., Lt. . . . 401, 1207
Laver, W. E., Lt. . . 351, 1509, 1587
Laverton, G. A., Lt. . . . 756, 1892
—— W. R. C., Capt. . . 685, 68

Names and Rank. Page.

Laville, F. A. J., Lt. . . 291, 1082
—— T. T., Capt. . 272, 1062, 1070
Lavin, J. F., 2nd-Lt. . . . 555, 1521
Lavis, L. E., Lt. 797
Law, C. H., Capt. 279, 931
—— C. E., Lt. 820
—— E. L., Lt. 445, 1561
—— G. H. M., 2nd-Lt. . 473, 1451
—— H. R. McK., 2nd-Lt. . . . 766
—— H. S., Lt. 630
—— R. O., Capt. 603
—— T. S., Ty. Capt. 1725
Lawes, G. W., 2nd-Lt. 1009
Lawford, E. E., Lt. . . . 486, 913
—— E. E. M., Col. . . 179, 1918
—— N. B., Lt. . . . 517, 1505
Lawler, G. F., 2nd-Lt. 316
—— J. R., Lt. 834
Lawlor, P. M., Lt. 815
Lawn, A. J., 2nd-Lt. 637
Lawrance, E. R., 2nd-Lt. . . . 717
—— G. H., Asst. Surgn. . 1797
Lawrence, A. E., Lt. . . 387, 1613
—— A. G., Asst. Surgn. . . 1795
—— A. G., 2nd-Lt. . . . 561
—— C. A., Ty. Lt. . . . 1716
—— E., Lt. 526, 998
—— E. D., 2nd-Lt. . 331, 1317
—— G. C., Lt. 1839
—— G. E., Maj. 707
—— H. A., 2nd-Lt. . . .
—— H. R., Maj 224, 1343, 1889, 1913
—— H. S., Hon. Col. . . 657
—— H. S., Lt. 1006
—— J. A., Lt. . . . 415, 1343
—— J. L., Sr. Asst. Surgn. 1791
—— N. S., Lt. . 514, 1641, 1701
—— R. R. S., Lt. 722
—— R. D., Capt. 1718
—— R. T., Capt. . . 273, 893
—— T. B., Lt. . . 399, 1359
—— W. A., Maj.-Genl. .173, 907
—— W. H. A., 2nd-Lt. . . 778
Lawrence-Archer, J. H., 11, 72, 136, 632
Lawrenson, T. G. P., Lt.-Col. 45, 194, 1883, 1912
—— W. R., Maj. 40, 219, 870, 1309, 1351
Lawrie, G., Lt. 531, 1698
—— L. J., Asst. Surgn. . . 1801
—— T. M., Lt. 1036
—— W. W., 2nd-Lt. 356
Laws, M. W. P., 2nd-Lt. . . . 561
—— W., 2nd-Lt. . . . 561, 1166
Lawson, A. G., 2nd-Lt. 567
—— C., 2nd-Lt. 618
—— D., Lt. 656
—— E., 2nd-Lt. 746
—— E. F., Lt. . 522, 1455, 1701
—— H. A. C., Lt. . . 523, 1355
—— H., Lt.-Col. 180
—— J., Lt. 465, 781
—— J. E., 2nd-Lt. 578
—— J. F., Lt. 443
—— R. A., Lt. . . 513, 1613
—— W. G., 2nd-Lt. . 739, 791
—— W. G., 2nd-Lt. 560
—— W. W., Sub-Condr. . . . 1864
Lawton, A. C., Lt. . . . 519, 1046
—— C. A., 2nd-Lt. 361
—— J., 2nd-Lt. 750
—— S. P., 2nd-Lt. 553
—— T. D. V., Lt. 318
Laxman Prasad Pandy, Jemdr. . 1705
Laximan Jaukar, Subdr. 1432, 1479
Laxuman Narayan Gogate, Sub-Asst. Surgn. 1832
—— More Jemdr. 1469
—— Bhagat, Jemdr. . . . 1469
Layard, C. W. C., 2nd-Lt. . . . 848
Layfield, F., 2nd-Lt. 1018
Laying, T. M., Capt. . 67, 277, 1058
Laythorpe, F. W. H., Lt. 33, 531, 1668
Layton, E. C., 2nd-Lt. 750
Lazaro, T., Asst. Surgn. . . . 1803
Lea, C. M., Lieut. 798
—— E. H., Lt. 375, 1413
—— E. T. H., Capt. 1717
—— W. H., Sub-Condr. . . . 1864
—— W., 2nd-Lt. . . . 473, 1018
Leach, A., Tempy. Capt. . . . 1723
—— A. G., Lt. . . . 84, 542, 1642
—— A. H. L., 2nd-Lt. . . 558, 1455
—— E., Maj.-Genl. Sir, . . . 749
—— F. B., 2nd-Lt. . . . 494, 1297

Lees, H. J., Lt. 492
—— R. B., Lt. 368
Leach, R. B., Capt. 798
Leachman, G. E., Lt.-Col. . 136, 726
Leader, H. P., Maj.-Gen. . 7, 14, 119
—— S., 2nd-Lt. 955
Leader-Williams, B. C., 2nd-Lt. . 776
Leaf, C. S., Lt. 685
Leah, T. C., Maj. 633
Leake, H., 2nd-Lt. 1064
—— H. M., Capt. 608
—— J. I., Lt. 317a, 365, 1148, 1587
Leamon, H W., Lt. . . . 534, 968
Lean, R., Lt. . . . . . 844—45
—— R. L., 2nd-Lt. 1693
—— W. W., Lt.-Col. . 179, 725, 1911
Leaning, A., Maj. 1838
Leapman, H M , Maj. 16, 21, 242, 1070
Learmond, A. W., Lt. 511, 814, 1227
Learner, J. H., 2nd-Lt. . . . 756
Leat, H. G., Lt. 351, 1062
Leatham, H. H., Esq. 89
Leaver, J. B., 2nd-Lt. 1679
Leawood, R. R., Lt. . . . 415, 1653
Le-Breton, E. P., Maj. . . 664, 985
Le Brooy, H. T., 2nd-Lt. . 549, 1601
Leckie, A., Lt. 622
—— F. W. V., Col. . . 178, 1918
Leckie-Ewing, W. C., Maj. . . 702
Lecky, J. G., Maj. 68, 243, 797, 1471
—— R. L., Capt. . 507, 854, 1475
—— R. St. C., Maj.-Genl. 7, 14, 58, 119, 612
Lecky-Thompson, W., 2nd-Lt. . 561
LeCornu, C. J. S., Capt. 262, 1004, 1247
Ledgard, R. G., Revd. 1904
—— R. S., Capt. 712
Ledger, A. V., Ty. Lt. 1729
Ledgerwood, F. H., Lt. . 294, 1363
—— T., 2nd-Lt. 567
Ledingham, J. J., 2nd-Lt. . 356, 1219
Ledward, R. B., Lt. 367
Ledwich, G., 2nd-Lt. 551
Lee, A. E., Lt. 503, 1164
—— A. J., Capt. 1755
—— A. J., 2nd-Lt. . . . 551, 1886
—— A. K., Lt. . . . 59, 763, 704
—— A. W., Lt. 513, 1397
—— A. W. H., Lt.-Col. 132, 156, 193, 1319, 1895
—— C. G., 2nd-Lt. 1678a
—— D. H., Lt. 1066
—— D. J. N., Lt. . . . 638, 978
—— E. O., Lt. . . . 525, 1160, 1219
—— F. R., Maj. 848
—— G., Maj. 68
—— G. B., Capt. 17, 23
—— G. J., Lt. 503, 778
—— G. St. C., Lt. 518
—— H , Lt. 538, 1696
—— H., Lt. 714
—— J., Sr. Asst. Surgn. . . . 1787
—— J. F., Lt. 508
—— J. S. S., 2nd-Lt. 559
—— M. G., Maj. 231, 1183
—— R., Capt. 1479, 1758
—— R. E., the Revd. 1907
—— R. H., Maj. 1752
—— S. S., Capt. . . . 102, 104, 615
—— T. E., 2nd-Lt. 1060
—— T. R., Lt. . . . 425, 705, 1102
—— W. C. N., Maj. . . 244, 915, 955
—— W. C. N., 2nd-Lt. 1677
—— W. M., Lt. 463, 1058
Lee-Mitchell, H. A., 2nd-Lt. . . 545
Lee-Warner, J., Lt. . . . 512, 687
Leech, A. B. R., Lt. . . . 298, 1399
—— A. J., 2nd-Lt. 611
—— J. C., Capt. 589
Leedam, W. D., Lt. . . . 669, 1007
Leedham, W., Lt. 509, 1665
Ledward, R. B., Lt. . . . 368, 1539
Leeds, H. J., Lt. 622, 1174
—— T. L., Lt.-Col. 133, 157, 197, 870, 1259
Leeke-Roe, D. C., 2nd-Lt. . . . 732
Leeman, J. A., 2nd-Lt. 567
Leembruggen, B. C. F., 2nd-Lt. . 568
Leeming, A., Lt. 425, 1509
Lees, A. S., Capt. 707
—— C. H. B., Lt.-Col. 96, 98, 139, 160, 199, 869, 1227
—— E. A., 2nd-Lt. 1108
—— G. W. M., 2nd-Lt. 595
—— L., Lt. 463, 1343
—— R. M., Lt. 749

Leese, F. W., Lt. . 307, 1130, 1136
Leeson, G. W., 2nd-Lt. 855
Leete, B. T., 2nd-Lt. . . 539, 1581
LeFeuvre, C. E., 2nd-Lt. . . . 554
—— C. O., 2nd-Lt. . 549, 1701
—— E. E., Asst. Surgn. . 1802
LeFleming, J. N., Lt 780
—— R. E., Capt. 280, 287, 1411
LeFranc, G. A., 2nd-Lt. . . . 558
Lefroy, A. L. M., Lt. 697
—— G. A., Most Revd. . . . 1903
Leftwich, W. M., Lt. . . 518, 1411
Legard, D'A., Br.-Genl. . 50, 98, 104, 122, 132, 595
LeGeyt, P. H., Maj.-Genl. . . . 172
—— P. S., Lt. 523, 1551
Legg, E. D. B., Lt. 353
—— P. C., Lt. 352, 1106
Leggat, J. W., 2nd-Lt. 558
—— W., Lt. 463, 1559
—— W. K., Lt. 785
Leggatt, A. F., Lt. 704
—— G., Lt. 622
—— H. A., Lt. 302, 895
Legge, A. G., 2nd-Lt. 566
—— G. E., 2nd-Lt. . . . 759, 1678
—— W. E., Lt. . . 536, 1618, 1676
Lego, J. T., 2nd-Lt. 1009
Lehane, M. J. A., 2nd-Lt. . . . 690a
Lehana Singh, *Hony.* Capt. . . 2101
—— Jemdr. 946
—— *Hony.* Capt. . . . 2102
—— Subdr. 1215
—— *Hony.* Capt. . . . 2102
—— Jemdr. 1432
—— Subdr. 1150
Lehna Singh, Jemdr. 1571
Lehnu, Jemdr. 1189
Lehri, Subdr. 1215
—— **Singh**, Jemdr. 818
—— Jemdr. 1559
Lehru, Subdr. 1011
Leicester, B., Maj. . . . 1663, 1684
—— C. B. W., Lt. 581
—— J. C. H., Lt.-Col. 138, 1745
—— O. H. Sub-Condr. . . . 1864
Leigh, A. P., Lt. 405, 921
—— H., 2nd-Lt. 1008
—— J., *Hony.* Lt. 714
—— J. T., 2nd-Lt. 750
—— P. C. J., 2nd-Lt. 539
Leighton, A. V., 2nd-Lt. . . . 715
—— J. H. L., Lt. 517, 73
—— W., 2nd-Lt. 363
—— W. W. M. F., 2nd-Lt. 553, 1359
Leigh-Bennett, P. R., Lt. . . . 512
Leigh-Clare, H. J. L., Lt. 487, 885, 959
Leiper, J. A., Capt. 1724
Leir, J. R. C. V., Lt. . . 307, 1651
Leishman, A. R., Capt. 857
Leitch, G., *Hony.* Capt. 1845
—— R. A., Lt. 541, 1642
Leith-Ross, W., Capt. . 52, 253, 1243
Lekh Ram, Subdr. 1385
—— Jemdr. 1391
—— Subdr. 1056
Lekhi Ram, Jemdr. 1052
Lekhu Ram, Jemdr. 1175
Lekraj, Jemdr. 1052
Lele, S. C., Tempy. Lt. 1778
Lely, W. G., Lt. 828
Leman, C. E., Lt. 860
—— F B., Maj. . . . 243, 1351
—— J. D , Lt. 1919
Lemarchand, F. E., Capt. 278, 1182, 1017
—— M. M., Capt. 844
LeMauug, Jemdr. 1295
Lemay, A. G., Lt. 1263
LeMessurier, G. D. S., Maj. 38, 247, 1359
Le Mesurier, H. G., Maj. . 11, 662
—— T. H. F., Lt. 320, 365, 970
Lemon, G. T., Capt. . 11, 18, 501, 1419
—— R. E., Lt. 542, 987
—— R. L., Capt. . . . 273, 1647
—— W. S., Lt. . . 76, 522, 1885
Lenanton, G. F. R., Lt. 618
Leney, R. J. B., Tempy. Capt. . . 1726
Lenfestey, L. d'E., Maj. 72, 633, 1884
Leng, C., 2nd-Lt. 668, 1008
Lenox, G. H., 2nd-Lt. . . 35, 933
Lennox, A., 2nd-Lt. 1417
—— G., Capt. 277, 1170
—— G. M., Maj. . . . 208, 1467
—— J., Lt. 859

Names and Rank. Page.

Lenox-Conyngham, G. P., Col. 124, 660, 1880
Lentaigne, E. C., Capt. 59, 254, 1613
——— J. I. N., Lt. 518, 1630
——— W. D. A., 2nd-Lt. 332, 1613
Lenton, H. T. F., 2nd-Lt. 565
Leonard, A. W., Col. 123, 153, 181, 1660
——— D. J., Lt. 293, 1251
——— D. W., Capt. 852
——— D. W., 2nd-Lt. 1373
——— G. R., 2nd-Lt. 854
——— G. T. I., Lt. 529, 1096
——— H. B., Lt. 748
——— H. W., Lt. 470, 931
——— J. A., 2nd-Lt. 578
——— J. J., 2nd-Lt. 562
——— N., 2nd-Lt. 794
——— P. D., Sub-Condr. 1868
——— P. J., Lt. 332, 802, 1339
——— R. E., 2nd-Lt. 1032
——— R. G., 2nd-Lt. 328, 1621
——— W. H., Bt. Lt.-Col. 139, 1747
Leopold, R. T., Asst. Surgn. 1800
LePatourel, E. C., Capt. 502, 1471
——— H. J. B., Lt. 571
Lepper, J. G., Capt. 52, 752
Leplar, W., Sub-Condr. 1857
Lescher, R., 2nd-Lt. 564
Leslie, A. S., Maj. 1753
——— D., Capt. 850
——— D. W., Lt. 7
——— D. W. R., Lt. 491
——— F. D., Capt. 265, 1215, 1587
——— F. M., Lt.-Col. 852
——— F. S., Capt. 763, 781
——— G. A. J., Bt. Lt.-Col. 107, 119, 125, 661
——— J. C. R., Esq. 93
——— N. R. A. D., *Bart.*, Bt.-Maj. *Sir* 268, 947
——— P. N., Lt.-Col. 197, 1883, 1912
——— T. C. C., Maj. 43, 1712
——— T. L., Capt. 269, 941
——— W., Capt. 105
——— W. C. C., Lt.-Col. 26, 183, 1883, 1911
——— W. D., Lt. 7
——— W. H. H., 2nd-Lt. 563
——— W. S., Bt. Col. 51, 98, 136, 158, 211, 1144
———-Jones, L. H., 2nd-Lt. 19, 143 571
Leslie-Moore, A., Lt. 507, 1247
Leslie-Smith, B. R., Lt. 397, 1513
——— C., Capt. 262, 1106, 1533
Lesmond, C. A., Lt. 507, 826, 1665
LeSouza, S. A., Asst.-Surgn. 1802
Lesse, F. W., Lt. 1136
Lesser, S., 2nd-Lt. 545
Lesslie, W. B., Bg.-Genl. 50, 121, 126, 661
Lester, A. R., Tempy. Capt. 1770
——— C., 2nd-Lt. 359
——— H. M., Lt. 81 1321
——— H. N., Lt. 415, 747
——— L., Lt. 423, 480, 1181
——— R., Lt. 506, 1665
Lester-Garland, A. G., Lt. 367, 729
——— G. H., Lt. 740
LEstrange, E. B., Maj. 55, 229, 1311, 1323
——— T. W., Lt. 541, 1297
Le Sueur, J. H. R., Lt. 666, 992
Lethbridge, A., Maj. 245, 1239, 1895
——— D. J. L., 2nd-Lt. 782
——— E. A. E., Lt.-Col. 131, 745
——— J. S., Lt. 666, 986
——— M. H. B., Lt. 29, 490
——— S., Lt.-Col. 74, 134, 632
——— W., Lt.-Col. 138, 1745
——— W. B., Lt. 521, 1263
Lethern, W. C., 2nd-Lt. 1469
Lett, J. W. H., Lt. 536
——— T., Lt. 762
Lettington, A. W., Sub-Condr. 1868
Letts, F. C., Sub-Condr. 1864
Levack, J. S., Capt. 39, 1713
Leventon, A., Lt.-Col. 132, 1743
Levett, P., 2nd-Lt. 550, 1415
Levick, A. C., Capt. 689
——— H., 2nd-Lt. 568
Levings, F. G., Sub-Condr. 1863
Levingston, T. C., Lt. 315, 1471, 1483
Levinson, H., 2nd-Lt. 302
Levy, M. P., Lt. 77, 747, 791, 1885
——— R., 2nd-Lt. 50
——— R. M., 2nd-Lt. 550

Names and Rank. Page.

Levey, J., 2nd-Lt. 708
Lewellin, D. J., Lt. 443, 733, 1423
Lewin, W. H., Lt.-Col. 56
Lewington, R., 2nd-Lt. 794
Lewis, A. E., Asst. Surgn. 1800
——— A. E., Lt. 8, 15, 529
——— A. G. L., 2nd-Lt. 557
——— A. H., 2nd-Lt. 781
——— A. M., Lt. 297*a*, 1231
——— A. N., Lt. 718
——— C., *Hony.* Capt. 1846
——— C. B., Lt. 47, 322, 1596
——— C. C., 2nd-Lt. 1695
——— C. C. C., Lt. 298, 759, 1667
——— C. G., Capt. 665, 1880
——— C. L., Lt. 540, 1672
——— D. G., Lt. 423, 1007
——— E., Lt. 311, 1355
——— E. H., Lt. 703
——— E. S., 2nd-Lt. 1674
——— F., Lt.-Col. 136, 711
——— F. D., Lt. 532
——— F. H., Lt. 296*a*, 1543, 1611
——— F. W. H., Lt. 530, 1668
——— G., Maj. 225, 947, 1365
——— G. W., Lt. 521, 1668
——— H., Capt. 32
——— H. A. C., Lt. 435, 1457
——— H. C., 2nd-Lt. 350, 1639
——— H. D., Capt. 43, 1838
——— H. J., 2nd-Lt. 361
——— H. M., Revd. 806
——— H. P., Lt. 929
——— H. V., Capt. 267, 1507, 1513
——— J. A., 2nd-Lt. 465, 1443
——— J. S., 2nd-Lt. 350, 1361
——— J. W., *Esq.* 88
——— L., 2nd-Lt. 549
——— L. R., 2nd-Lt. 539
——— N. C., Lt. 724
——— P. H., Lt. 1314
——— P. S., Lt. 632
——— R. E., 2nd-Lt. 568, 1056
——— R. F., Capt. 264, 9 5
——— R. M., 2nd-Lt. 555
——— R. N., 2nd-Lt. 551, 1351
——— R. P., Asst. Surgn. 1800
——— R. R., Maj. 1712
——— S., Sub-Condr. 1857
——— T., Lt. 789
——— T. C., 2nd-Lt. 790
——— T. G., Capt. 96, 279, 1058
——— T. M., Lt. 443, 1323
——— T. W. L., Lt. 654
——— W., Capt. 702, 703
——— W., 2nd-Lt. 618, 1523
——— W. H., 2nd-Lt. 567
——— W. H., Capt. 634
——— W. J., 2nd-Lt. 619
——— W. R., Capt. 717
——— W. T., Sub-Condr. 1873
——— W. T., 2nd-Lt. 1677
Lewis—Lewis, S., Lt. 470
Lewthwaite, R., Lt. 724
Ley, A. E. H., Maj. 226, 913
Leys, J. F., Lt. 393, 1020
——— W. G., *Esq.* 88
Leyshon, W., 2nd-Lt. 356, 1399
Leyland, F. B., Maj. 587
L'Fleur, J. F., Asst. Surgn. 1797
——— M. J. P., Asst. Surgn. 1802
Libby, W. H., 2nd-Lt. 355, 1090
——— H. J., 2nd-Lt. 1095
Lichfield, S. J., 2nd-Lt. 552, 1319
Lichmore, B. C., *Khan Sahib* 88
Lickman, E. W., Lt. 307*a*, 857
——— B. S., Lt. 637
Liddell, C. O., Lt. 540
——— J. A., Capt. 485, 1675
——— R. E. A., 2nd-Lt. 1231
——— W. S. 986
——— W. T., Lt. 559, 848
Liddle, A., Lt. 685
——— F. A., Lt. 685
Liddy, J., Condr. 1870
Lidierth, J. E., 2nd-Lt. 842
Light, H., Capt. 280—287, 1343
——— F., Lt. 789
Lightbody, J. G., 2nd-Lt. 465, 473, 943
——— W. A., 2nd-Lt. 974
Lightfoot, G. S., Lt. 371, 1407
——— K., Capt. 58
Lighthall, W. S., 2nd-Lt. 737
Lihaz Gul Khan, Risaldr. 948
Likhi Ram, Jemdr. 1476
Limakh Ram, Jemdr. 1433

Names and Rank. Page.

Lilbir Thapa, Jemdr. 1598
Lilico, W., Ty. Capt. 1726
Lilley, U., Lt. 816
——— W. G., Lt. 455, 1030
Lillie, C. J. W., Lt. 315, 1613
Lillies, A. R. G., Lt. 817
——— G. L., Lt. 311, 1112, 1339
Lillingston, W. E. G., Lt.-Col. 132, 156, 193
Lilly, A. N. I., Capt. 54, 272, 1042
——— A. T. I., Lt.-Col. 1731
Limaye, G. G., Ty. Lt. 1779
——— V. P., Tempy. Capt. 1771
Limb, H. E., 2nd-Lt. 1271
Limby, S. O., Lt. 513
Limpenny, F. R., Lt. 622
Linaket, A. F. R., 2nd-Lt. 748
Linberry-Kerr, G. W., 2nd-Lt. 552, 1343
Lincke, S. H. P., Lt. 818
Lincoln, A. E. 23
——— C. H., Asst. Surgn. 1796
——— C. J., Asst. Surgn. 1798
——— E. H., 2nd-Lt. 558, 805, 1883
——— N., Lt. 505, 825
——— S. W., 2nd-Lt. 1321
——— S. W., Lt.-Col. 37, 1660
Lind, A. G., Maj. 214, 1255
——— E. N. A., Lt. 515, 1094
Lindesay, A. L., Col. 123, 149, 181
——— F. S., Maj. 15, 46, 211, 1883
——— G. W. G., Maj. 1919
——— V. E. H., Lt.-Col. 133, 1744
Lindley, G. W., Lt. 776*a*
Lindop, A. H. E., Lt. 506
Lindquist, W. H. H., Lt. 309, 1647
Lindsay, A., 2nd-Lt. 362
——— D. M., Lt. 315, 1259
——— H. A. P., Bt.-Col. 22, 104, 108, 125, 153, 189, 1660
——— H. C. C., Lt. 517, 1678
——— H. N., Lt. 455—59 1259
——— J. G., Lt. 541, 992
——— M. E., Capt. 583
——— P. L., 2nd-Lt. 328, 1517
——— W. D., Maj. 634, 981
Lindsay-Young, E., Capt. 278, 1094, 1096
Lindsley, R. M., 2nd-Lt. 322, 1070
Line, R., 2nd-Lt. 858
Lineker, L. L., Lt. 352
Linell, C. J. B., Lt. 352, 1417
Lines, R. E., Lt. 534, 1007*a*
Linfield, W. A., Lt. 686, 783
Ling, F. G., Lt. 76, 535, 1885
Lingard, H., Capt. 844
Lingeman, A. A., Lt. 622
Lingwood, C. P., Lt. 754
Linker, J. E., Lt. 353
Linney, W. W., Lt. 312, 1475
Linskill, H., Lt. 544
Linton, E. C., Capt. 1717
——— J., Lt. 544, 1189
——— R., 2nd-Lt. 760
Lintott, P. J., Lt. 532, 1062
Lipscomb, F. M., Capt. 1715
——— L., Lt. 492, 877
Lirks, A. J., 2nd-Lt. 543
Lisby, V. C., Lt. 368, 732
Lisle, P. C., Capt. 754, 798
Lisles, R. A., Condr. 1868
Lismore, T., Lt. 798
Lisney, J., 2nd-Lt. 710
List, F. J., 2nd-Lt. 355
——— J. N., Capt. 521, 820, 992
Lister, A. E. J., Maj. 1748
——— C. C., Capt. 597
——— C. E. L., 2nd-Lt. 565, 1443
——— J. G., The Revd. 1908
——— T., *Esq.* 1895
Lister-Jackson, A., Lt. 516, 998
Liston, J. J., Capt. 1758
——— R. C. C., Capt. 94, 268, 1411
——— W. D. F., Lt. 489, 578
——— W. G., Lt.-Col. 139 1745
Liswell, J. A., Lt. 440, 709, 1189
Litt, J., Lt. 1839
Little, A. G., Lt. 617
——— C. W. R., Capt. 753
——— D. D., 2nd-Lt. 475, 1150
——— E. G., Lt. 430, 1018
——— E. M., Maj. 635, 977
——— G. L. C., Capt. 1755
——— J. A., Lt.-Col. 178, 1917
——— J. O., Capt. 485, 610
——— R. Capt. 485, 691

NAMES AND RANK. Page.

Little, W. C., Lt.-Col 139, 161, 199, 872, 1623
—— W. J., 2nd-Lt. . . . 357, 1064
Littlebury, A. T. B., 2nd-Lt. 326, 366
Littledale, A. G., Lt. . . 519, 1627
Littlehailes, R., 2nd-Lt. . . . 802
Littlejohn, A., Lt. . . 539, 992
Littlewood, F., 2nd-Lt. 1693
—— R. V., Lt. . . 542, 1698
—— S., Ty. Capt. 1724
Litton, G. E., Lt. 296, 1151
Livermore, W. C., Lt. 779
Livesay, E., 2nd-Lt. . . . 358, 1225
—— J., Maj. . . 234, 1058, 1411
—— T. R., Lt. 487
Livesey-Wardle, R. M., Capt. . . 502
Livingstone, D. A., 2nd-Lt. . . 358
—— H. L., Esq. 88
—— L., Lt. 532
—— P., Lt. 1655
Liyakat Ali, Subdr. 999
Llewellin, W. W., Lt. . 738, 740, 791
Llewellyn, B., Lt. 487
—— R. G., 2nd-Lt. . 359, 1024
Llewhellin, G. E. Capt. . . . 485
—— N. G. R., *Hony.* Capt. 111
Lloyd, A. C., Lt. . . . 504, 1667
—— A.H., Capt. 807
—— C. F., Capt. 776a
—— C. G., Capt. . . . 255, 1663
—— C. H., 2nd-Lt. . . . 708, 1678a
—— C. P. W., Lt. 488
—— C. R., Bt.-Lt.-Col. 140, 160, 233, 1662
—— E., *Hony.* Lt. 67
—— E. T., 2nd-Lt. . . . 356, 1044
—— *Sir* G. A. 1892
—— G. T., Lt. . . 39, 526, 1193
—— H. C., Lt. 712
—— H. S., 2nd-Lt. . . . 43, 668
—— H. S. J., Lt. . . . 303, 1598
—— I. G., Maj. 827
—— I. M.A., Lady Supdt. . . 1734
—— J. E., Lt. 410, 1164
—— J. H., Lt.-Col. 16, 43, 134, 157, 194, 623
—— J. H., 2nd-Lt. 1674
—— J. S., Lt. . . . 291, 1600
—— J. T., 2nd-Lt. . . . 494, 941
—— K. H., Maj. 206
—— L. N., 2nd-Lt. 826
—— L. W., Lt. 597
—— N. R., 2nd-Lt. . . 742 788
—— O. S., Maj. 614
—— R. A., Maj. 1749
—— R. B., Capt. 1754
—— R. E., Maj. 1748
—— R. H., Lt.-Col. . . 133, 1710
—— R. H., 2nd-Lt. 560
—— R. S., Lt. 692
—— S., Lt. 490, 909
—— T. C., 2nd-Lt. 790
—— T. F., Maj.-Genl. 760
—— T. W., *Hony.* Capt. . 8, 93, 1845
—— T. W., Capt. 1841
—— W., 2nd-Lt. 1134
—— W. E. E., Lt.-Col. . 131, 156, 192, 1052, 1391
—— W. H., 2nd-Lt. . . . 754, 1009
—— W. L., Lt. 375, 1225
Lloyd-Evans, H. J., Lt. . . 53, 297
Lloyd-Jones, F. L., Lt.-Col. 131, 156, 191, 871, 1455
—— R. W. A., 2nd-Lt. 76, 684
—— W. R., Lt. . . 306, 1509
Lloyd-Williams, J. E., Lt. 487, 935
Loasby, H. C., Revd. . . . 1906
Lobo, C., Asst. Surgn. 1796
—— E. D., Asst. Surgn. . . . 1796
—— M. V. J., Asst. Surgn. . . 1801
Loch, E. C., Maj. . . . 78, 503
—— G. C. B., Maj. . . 58, 99, 663
—— G. H., Lt.-Col. 57
—— H. F., Col. . . 97, 123, 149, 181
—— H. N., Capt. . . 280—287, 1017
—— J. C., Maj. . 54, 214, 1600, 1519
—— P. G., Capt. . . 103, 259, 1898
—— S. G., Br.-Genl. 13, 19, 97, 121, 120, 662
Lochan Singh, Subdr. . . . 1120
Lochner, C. B., Lt. . . . 385, 933
Lochrie, C., 2nd-Lt. 715
Lock, C. F., 2nd-Lt. 599
—— F. R. E., Br.-Genl. 50, 121, 126, 150, 187
—— H. G., Lt. 503, 1609
—— H. O., Maj. 740
—— J., Sub-Condr. 1867

NAMES AND RANK. Page.

Lock, P. G., Tempy. Capt. . . . 1723
Locke, D. W., Lt. . . . 697, 1255
—— L., Nursg. Sister 1735
Locke-Elliott, Lt.-Genl. Sir E. . . 1691
Locket, L. G., Capt. 751
Lockett, G. B., 2nd-Lt. 623
—— V. N., Capt. 595
Lockhart, J. G., Capt. 755
—— R. McM., Capt. 277, 1116, 1227
—— W., 2nd-Lt. . . . 55, 1164
Lockie, H. R., Maj 830
Locke, G. M., Lt. 622
Lockstone, J., Lt. 354, 1012
Lockwood, R. U., 2nd-Lt. . . . 566
Loder, L. N., 2nd-Lt. . . 358, 1160
Lodge, A. B., Lt. 405, 968
—— E. R., Lt. . . 460—62, 1327
—— F. C., Maj. 691
Loewe, H. M. J., 2nd-Lt. 76, 695, 1885
Lofthouse, A., 2nd-Lt. 567
—— J. D. 2nd-Lt. 786
Loftus, G. M., 2nd-Lt. 1102
Loftus-Tottenham, F. J., Lt. 307, 1600
Logan, A. F., Capt. 94, 280—287, 915
—— A. T., 2nd-Lt. . . . 24, 1679
—— E. W., Lt. 531, 1143
—— J., 2nd-Lt. . . . 357, 1675
—— J. S., 2nd-Lt. . . . 357, 1287
—— R. H., Maj. 748
—— T., Lt. 544, 1611
—— W., Capt. 696
Logan-Home, W. M., Capt. 255. 1451
Loganadan, A. D., Lt. 1693
—— Jemdr. 994
Lohasur Gurung, Jemdr. . . . 1615
Lokbahadur Thapa, Jemdr. . . 1619
Lokendra Sah, Raja, 2nd-Lt. . . 481
Lokhbir Ale, Jemdr. 161
Lok Ram, Jemdr. 1059
Loknath Singh, Subdr.-Maj. . . 1030
—— Subdr. 1012
Lole, E. F., 2nd-Lt. 746
Lollingdale, E. T., 2nd-Lt. . . . 599
Lomas, H. A., Capt. 609
—— L., Lt. 520, 1595
Lomax, C. E., Lt. 807
—— G. W., 2nd-Lt. 1018
London, H. R., Lt. . . . 529, 1158
Long, C. A., 2nd-Lt. . 548, 732, 1048
—— D. T., Lt. 313
—— E. W., Capt. . . . 779, 797
—— F. A., 2nd-Lt. . . . 729, 791
—— H. A., 2nd-Lt. . . 540, 970
—— H. W., 2nd-Lt. . . 326, 366
—— P., Lt. . . . 423, 733, 1006
—— W. E. L., Capt. . 74, 101, 634
—— W. H., Lt. . . 317, 1219, 1501
Longbotham, C. M., Capt. 272, 1301
Longbottom, A. W., Lt. . . 380
—— J. L., Lt. . . 299 1283
Longden, A. B., Lt.-Col. 131, 156, 191, 869, 1174
—— D. J., Lt. . . . 375, 1110
Longdin, W. G., Maj. . . 518, 804
Longfellow, G. D., Lt 577
Longford, G. W., 2nd-Lt. . . . 1056
Longhurst, A. L., Lt.-Col. 98, 103, 135, 158, 207, 1629
Longland, R. H. B., 2nd-Lt. 666, 992
Longley, A. H., Lt. . . . 435, 1038
Longstaff, T. G., Capt. . . . .
Longstaffe, J. W., Lt. . . 532, 1668
Longthorn, J., 2nd-Lt. . . 363, 1180
Longstreet, R. W., 2nd-Lt. . . . 599
Lonsdale, A. T., 2nd-Lt. . . . 359
Looney, D., Lt. 723
Lopez, A. P., Asst. Surgn. . . . 1793
—— A. V., Ty. Lt. 1780
—— J. W. C., Asst. Surgn. . . 1798
—— V. J., Asst. Surgn. . . . 1794
Lord, A. J., 2nd-Lt. . . . 363, 1074
—— H. P. M., Maj. . 105, 215, 1895, 1913
—— J. A., Lt. 599 600
—— J. W., Lt. . . 297a, 1491, 1495
—— R. C., Capt. 36, 665
Lordan, L. P., 2nd-Lt. 953
—— W. L., Lt. 734
Lorimer, A. P. G., Maj. 1753
—— D. L. R., Maj. 211, 1889, 1913
—— E., Maj. . . . 234, 945, 953
—— J. M., Capt. . . . 260, 879
Lorinda, Jemdr. 1249
Loring, E. J., Maj. 8, 15, 664, 669, 996
Lorrayne-Wadley, G. C., Lt. 507, 1363
Lory, F. B. P., Lt. . . 489, 849, 929

NAMES AND RANK. Page.

Losasso, M. R., 2nd-Lt. 562
Lothian, A. G. 34
—— J., Lt. . . . 2920, 927
Lottus, G. M., 2nd-Lt. 551
Loudon, J. A., Lt.-Col. 40, 128, 153, 184, 869, 1197
Loudoun, R. F., Lt. . . . 401, 1052
Lough, A. T., Lt.-Col. . . 136, 613
Lougheed, Lt. A. 76, 789
Loughlin, J. A., 2nd-Lt. 565
Loughman, G., Lt. 517
Loughten, J. A., 2nd-Lt. . . . 1703
Loup, L. A., Lt. 304, 1647
Louw, D. L., 2nd-Lt. 770
Lovat-Fraser, W. A., Lt. 299, 1215, 1375
Love, G. W., 2nd-Lt. 530
—— R. M., 2nd-Lt. . . . 365, 935
—— S. H., 2nd-Lt. 714
Loveday, F. W., Maj. 033
—— S. G., Capt. 754
Loveland, F., Lt. 354
Loveless, G. J., 2nd-Lt. . . 356, 1417
—— W. G., 2nd-Lt. 360
Lovell, I. R., 2nd-Lt. 784
—— Maj., W. C. L. 608
—— W. D., Lt. 488, 899
Lovering, A. J., 2nd-Lt. . . 355, 1032
Loverock, R. C., Capt. . . 61, 745
Lovett, O. deT., Lt. . . . 318, 1641
Lovick, R. D., Sub-Condr. . . . 1873
Low, F., Lt. 731
—— J., Lt. 413, 1465
—— J. H., Lt. 578
—— M. M. E. Nursing Sister . 1735
—— R. B., Lt.-Col. . . 129, 154, 186
—— S. P., Capt. 734
—— W. H., 2nd-Lt. 1389
Lowe, C. G., Lt. . . . 492, 893, 955
—— E. A. H., Lt. 397, 913
—— F. G., Lt. 56, 518
—— H. A., Lt. 1007
—— J. S., Lt. 521
—— P. W. C. S., 2nd-Lt. . . . 728
—— T. W., 2nd-Lt. . . . 356, 1674
Lowcock, D. R., 2nd-Lt. . 567, 1678a
Lowes, G. W., 2nd-Lt. 551
Lowis, F. C., Lt. 517, 848
—— H. D., Capt. 150
—— H. E., Br.-Genl. 14, 43, 120, 125, 1851
—— H. R., 2nd-Lt. . . . 327, 1034
—— P. S., Lt.-Col. . . . 131, 632
Lowndes, D. G., 2nd-Lt. . 333, 1178
—— R. C., Capt. 636
—— R. C., Lt. 491, 897
—— W. G., 2nd-Lt. 598
Lownes, J. R., Lt. 527, 1704
Lowrie, A. E., Lt. 62, 505
Lowry, A., Lt. 543a
—— Maj. 84
—— H. W., Lt.-Col. 50
—— J., 2nd-Lt. 557, 1295
—— W. E., 2nd-Lt. 565
Lowry-Corry, W., Capt. 255, 919, 1898
Lowsley, G. L., 2nd-Lt. . . 332, 1613
—— W. A., Revd. 1906
Lowson, C. S., Maj. 1746
—— H J. 472
—— J. M., 2nd-Lt. 1024
—— J. McE., 2nd-Lt. . . . 472, 925
—— J. McK. A., Lt. 638
Lowther, A. W., 2nd-Lt. . . 360, 1319
Lowther, H., Lt.-Col. . 125, 154, 185
—— R. A. N., Capt. 71
—— T. E., *Hony.* Capt. . . . 760
Loyd, E. B. K., Lt. 308, 895
—— P. H., the Revd. 1907
Luard, B. E., Capt. 610
—— C. C., Br.-Genl. 50, 120, 125
—— C. E., Lt-Col. 199, 1889, 1899, 1912
—— G. E., Lt. 610
Lubbock, G., Bt. Col. 60, 121, 127, 601
Lucas, C. A., Capt. . . . 262, 1664
—— C. A., Lt. 730
—— C. St. J., Lt. . . . 526, 1671
—— E. W., 2nd-Lt. 571
—— F., Lt. 581
—— F. G., Col. 50, 121, 126, 150, 187
—— H. W. C., Maj. . . . 216, 939
—— H. D. N., Capt. 16, 19, 259, 931
—— H. T. 1009
—— J., Lt. 793
—— J. deB. T., Maj. 615
—— M. H., Maj. . . 238, 945, 1088
—— P. N. A., 2nd-Lt. . . 562, 843
Luce, E. P. W., 2nd-Lt. 630
—— G. H., 2nd-Lt. 555

Names and Rank. Page.

Lucey, F. E., 2nd-Lt. 563
Luchesne, A. E., 2nd-Lt. . . . 1693
Luck, C. A., Col. . 57, 125, 150, 185
—— C. P., 2nd-Lt. . . 553, 1521
—— J., Lt. 90, 517, 766
—— T., Maj. . . 222, 1283, 1285
Ludgate, W., Maj. . . 26, 1838
Ludlow, C. W. P., Lt. . . . 701
—— F., Lt. 523, 1395
—— H., 2nd-Lt. . . . 356, 1483
—— L., Capt. 709
Lugard, H. I., Col. 176
Luigi, P., Lt. 425
Luke, N. R., Lt. 651
—— O. E., Lt. 492, 1575
—— T. M., Br.-Genl. 8, 15, 122, 234, 632
Lukin, H. C. W., Lt.-Col. 16, 22, 122, 135, 158, 195, 867, 941
Lukis, S., 2nd-Lt 851
Lukman Khan, Ressdr. . . . 883
Lukshiman Raoji, Jemdr. . . . 1436
Lumb, F. G. E., Maj. . . 214, 1178
—— T. F., Capt. 748
Lumba, Rawat, Jemdr. 1473
Lumby, A. F.R., Capt. 7, 14, 274, 1287
—— L. J., Lt. 797
Lumley, J. N., Capt. . . 52, 591
Lumley-Smith, T. G. L., Capt. 59, 597
Lumsdaine, A. A. M., Lt. . 695, 733
Lumsden, E., Lt. 617
—— E. S., Lt. 532
—— H. R. W., Lt.-Col. .180, 1917
—— P. J., Lt.-Col. 35, 128, 1740
—— P. S., Genl. Sir . . . 171
—— W., Commodore . . . 2
Lund, C., 2nd-Lt. 325, 366
—— C. E., 2nd-Lt. 1636
—— J. H., 2nd-Lt. 758
Lundy, C. R. ., Lt. . 460—62, 1068
Lunham, J. L., Maj. 1750
—— T. M., Lt. . . 304, 939
Lunn, A. J., Capt. . . . 272, 1535
—— B. H., 2nd-Lt. 744
—— E. D., 2nd-Lt. 742
Lunnon, Lt. J. D. K. 1036
Lunt, R. M., Lt. . . . 450—54, 1355
Lupprian, E. V., Asst. Surgn. . 1801
Lupton, G., 2nd-Lt. 585
—— W. M., Capt. 1758
—— F. E., Lt. . . . 514, 1667
Lurcott, R. G., Lt. 1571
Luscombe, A. M. P., Lt. 385, 705, 1088
—— F. W., 2nd-Lt. . . . 548
Lushington, A. N., Lt. . 46, 507, 1501
—— H. E. C., Lt. 493
—— W. N., Lt.-Col. 31, 134, 157, 197, 1661
Lusty, G. H., 2nd-Lt. 1695
Luthra, B. N., Ty. Lt. 1775
—— J. N., Tempy.-Capt. . . 1766
Lutman, J. B., Lt. 854
Lutter, W. F., Lt. . . 508, 1665
Luttman-Johnson, H. W., Capt. 276, 891
Luxmoore, C. de J., Maj. 103, 1208, 1403
—— C. T. P., Col. 177
Luxton, F.C., 2nd-Lt. 776
—— F. W., Capt. 818
Lyall, C. H. G., Ty. Lt. 1726
—— C. R., Maj. . 222, 1166, 1563
—— J. H., Lt. 526, 1569
—— R. A., Maj. . . 212, 1889, 1913
—— W. H., Lt. . . . . . .
—— W. J., Lt. 666, 945
Lyell, A. G., Maj. 230, 1178
—— G. D., Lt. 491, 887
—— T. R. G., Lt. 535
Lyle, A., 2nd-Lt. 712
—— D. R., 2nd-Lt. 829
—— G. H., Lt-Col. 846
—— P., Lt. 721
—— P., 2nd-Lt. 362
—— T. McB., 2nd-Lt. . . . 555
Lynch, C. M., Sub-Condr. . . . 1873
—— C. St. J., Maj . 60, 664, 1882
—— H. E. H., Lt. 1585
—— J., Capt. 78, 626
—— J. A., Lt. 68, 533
—— J. S., 2nd-Lt. . . . . . .
—— J. V., 2nd-Lt. 557
—— P. W., 2nd-Lt. . . 359, 1036
—— R. J., Lt. 95, 318, 1501
Lynch-Blosse, C. E., Lt. . . 493, 939
—— H. R., Capt. . . . 841
Lynch-Staunton, H. V. C., Capt. 259, 1491, 1493
Lynden-Bell, L. A., Capt. . . . 764
Lynes, F. C., 2nd-Lt. 692
—— R. H., Lt. 514, 1624
Lyngdoh, H., Ty. Lt. 1774
Lynn, G. R., Capt. 1755
—— G., Lt. 543, 653
—— J. A. G., Capt. . . . 367, 921
—— J. C., Lt 629
—— T. P., 2nd-Lt. 713
—— T. V., 2nd-Lt. 547
Lynsdale, A. H., 2nd-Lt. . 557, 1361
—— A. W. J., Senr. Asst. Surgn. . 1787
—— B. St. C.F., Asst. Surgn. 1802
—— C. B. F., 2nd-Lt. . . . 563
—— H. H., Asst. Surgn. . 1794
Lyon, C. B., 2nd-Lt. 598, 1893
—— C. W., Maj 221, 1191
—— H. B., Lt. 536, 1255
—— H. N., 2nd-Lt. 532
—— J. W. H., Lt.-Col. 193, 1883, 1912
—— W. A., Lt. 305, 1563
—— W. A., 2nd-Lt. . . . 669, 986
Lyons, A. J., Lt. 521, 1667
—— C. P., Lt. 527, 1007a
—— H. H., Lt. 371, 879
—— I. A., Capt. 787
—— J., 2nd-Lt. 711
—— J. F., Capt. 1717
—— P., Condr. 1853
—— R. W. S., Maj.-Genl. 1, 119, 1737, 1955
Lys, A. J., Sub-Condr. 1868
Lysaght, D. R., 2nd-Lt. 779
Lysaght-Griffin, E. H. L., Lt. . 1036
Lysons, K. C., 2nd-Lt. . . 549, 1701
Lyster, A. St. G., Lt. . 509, 986, 1698
—— C. J. H., Maj. . 67, 219, 1662
—— H. H., Lt.-Genl. . . . 172
Lyte, E. F. N., 2nd-Lt. 1243
Lyttle, W. J., *Hony.* Maj. . . . 1845
Lyus, A. O., Lt. 692

M

McAdam, J., Lt. 538, 1701
McAlister, N. H., Capt. . . . 1839
—— S. G., Ty. 2nd-Lt. . . 742
McAllister, H. M., Capt. . . 719, 300
—— J., 2nd-Lt. 828
McAlpin, D. F. C., Lt. . 525, 1670
McAlpine, E. G., Lt. 808
McAndrew, A., 2nd-Lt. 1491
—— G. D., 2nd-Lt. . 669, 997
McAndrews, T., *Hony*-Lt. . . . 835
McArdle., R. H., 2nd-Lt. . . . 566
Mc Arthur, A. G., 2nd-Lt. . . . 554
—— D. F., 2nd-Lt. . . . 1696
—— J., Lt. 785
McArthy, S. E. F. D'Arcy, Lt. . 834
McAuliffe, J. P. T., 2nd-Lt. 545, 998
—— E. C., 2nd-Lt. . . . 1693
Mc Avoy, W., Lt. 709
McBain, J. A. D., Lt. 821
McBride, J., Lt. . . 514, 719, 1463
McBrien, C. H., Lt. 488, 1705
McCall, B., Lt. 90, 527
—— R. G., Lt. 470, 1044
—— W. C., 2nd-Lt. 551
McCallum, D. V., Lt. . . 505, 1195
—— R. E., Capt. . . . 1663
McCalmont, D. H. B., Capt. 59, 587
—— H., Maj.-Genl. *Sir* . 587
—— W. V., Capt. 277, 1058, 1060
McCammon, J. A., Lt. 638
McCanlis, W. J., Capt. 825
McCann, J. S., 2nd-Lt. 707
—— W. M., 2nd-Lt. . . . 562
McCarrison, R., Lt.-Col. . . 139, 1747, 1890
McCarthy, A. J., 2nd-Lt 1008
—— A. W., Lt. 706
—— C. E. A., Lt. 617
—— C. L. J., Lt. . 423, 1359
—— E. A. R., Sub-Condr. . 1855
—— E. D., Lt. . . . 472, 1068
—— J. McD., Lt.-Col. . . . 133
—— J. W., Lt. . . . 463, 1190
—— M. P., 2nd-Lt. . 563, 1046
McCarthy, P., Sr. Asst. Surgn. . 1788
—— R. W., Lt. . . . 536, 1581
—— T. C., Lt. . . . 310, 1038
—— W. O., Lt. . . . 415, 1086
—— W. T., Sub-Condr. . . 1874
McCaskie, H. B., Ty. Capt. . . 1726
McCaskill, J. C. H., Lt.-Col. . . 138, 159, 197, 870, 1359
McCaughey, F. G., Lt. 315, 1283, 1311
McCaul, M. W., Revd. 1905
McCausland, E. T. W., Capt. 253, 1606
McCay, D., Maj. 1746
—— R. C., Lt. 23
McClay, A. D. G., 2nd-Lt. . . . 565
McCleery, H. W. F., 2nd-Lt. 465, 473
—— R. J., Lt. 319
McClellan, H. T., Maj. 589
McClelland, A. A., 2nd-Lt. . 669, 992
—— W., Lt. 465, 1405
McClenaghan, H. C., Lt. . 304, 1197
—— H. E., Lt. 1515
—— P. J. W., Lt. 306, 1397, 1513
McCleverty, A. H., Maj. 216, 868, 1048 1070
—— G. M., Capt. . 259, 1600
—— P. H., Maj 225, 1054, 1095
McClintock, R. L., Bt.-Lt.-Col. 137, 662, 991
McCloughin, B. G., Lt. 843
McClumpha, C. D., Lt. . 509, 1451
McClymont, A. J., 2nd-Lt. . . . 705
McColl, J. S., Lt. 112
McCollough, R. L., Lt. 582
McClurkin, T., Capt. 1717
McCloskey, F. J., Lt. . . 470, 1092
McCluskey, J. J. P., 2nd-Lt. 473, 1321
Mc Combe, A., Lt. . . . 540a, 1415
—— F. C., Lt. . . . 606, 861
—— J. S., Capt. . . 10, 1713
—— R., Lt. 513, 1409
McConaghey, A., Lt.-Col. . 185, 1911
—— F., Lt.-Col. 197, 1899, 1912
McConaghy, C. B., Maj. . . 1890, 1901
—— J. G., Maj. 15, 22, 217, 921
McConkey, A. I. G., Capt. . 278, 1255
McConnell, C. H. Q., Lt. . . . 111
—— J. K., Lt. . . . 311, 1311
McConnochie, J. S., Lt. 25
McCord, J., 2nd-Lt. 550, 1702
McCormack, R. B., Capt. 856
McCormick, A., Capt. 777
—— A. I. C., Lt.-Col. 135, 661, 1890
—— G. D., Maj. . . 215, 1303, 1453
—— J., 2nd-Lt. 734
McComy, T. F. E., 2nd-Lt. . . 1674
McCorquodale, A. R., Lt. . . . 1670
McCowen, W. T., Maj. 1751
McCoy, J. W., Maj. 1748
McCoy-Hill, L. McG., Lt. . . 397, 1245
McCracken, A., Lt. 95, 523
—— G., 2nd-Lt. 555
McCrea, A. C., Col. 14, 32, 126, 150, 189
—— T., Sub-Condr. 1836
McCready, D., 2nd-Lt. 553
McCreath, A. F., 2nd-Lt. . . . 653
McCririck, D. H. G., Lt. . . . 701
McCrossan, A. D., 2nd-Lt. . 23, 714
McCudden, J. H., Maj. . 230, 915
McCue, W. T., Ty. Asst. Surgn. . 1804
McCuish, A., Lt. 351, 1007a
McCulloch, A. J., Maj. . . 98, 283
—— E. A., 2nd-Lt. . . . 1008
—— H. S., Lt. . . . . 455—59
McCully, A. J., Sub-Condr. . . . 856
McCurtis, A., Asst. Surgn. . . 1795
—— L., Asst. Surgn. . . 1703
McCutcheon, G. H., Lt. . 314, 1249
—— O. E., Maj. . 610, 861
McDermott, B. J., Sub-Condr. . 1874
—— C. H., Lt. . 504, 1664
—— F., Lt. . . 440, 1191
—— J. W., Lt. . . . 509
—— M. D., 2nd-Lt. . . 553
—— N. W. E., 2nd-Lt., 545, 1309
—— R. K., Capt. 765
—— R. M., Ty. Asst. Surgn. 1804
—— W., 2nd-Lt. . . . 618

NAME AND RANK. Page.

McDonald, A., Capt. 825
—— A. J. P., Asst. Surgn. 1803
—— C. K., 2nd-Lt. . 559
—— E. J. C., Maj. . . 1752
—— D. C., 2nd-Lt., 548, 1419
—— D. W., Lt. . . 535, 1672
—— H. G., Lt. 726
—— H. P., Lt. . . 526, 1667
—— J. B., 2nd-Lt. . 1110, 1160
—— J. H., Lt.-Col. 132, 1743
—— J. J., Hony. Lt. . . 1190
—— J. S., Maj. . . . 859
—— M., 2nd-Lt. . . . 1050
—— R. H., Capt. . . . 1004
—— S. H., Lt. . 77, 522, 1885
—— W., Lt. 745
—— W., 2nd-Lt. . . . 763
McDonnell, F. A., Lt. 316, 1034, 1389
—— T. F. R., 2nd-Lt. . . 828
McDonogh, C. L., Ty. Capt. . . 1726
McDonough, A. D., Lt. . . 45, 1848
McDougall, A., Lt. . . . 463, 931
—— A. H., Capt. . . . 1839
—— D. M., 2nd-Lt. . 321, 1407
—— F. A., 2nd-Lt. . 494, 925
—— J., 2nd-Lt. 540
—— J. L. N., Condr. . . 1854
—— P., Lt. . . . 534, 1062
—— R. M., 2nd-Lt. . 329, 1078
—— W. F., Capt. . . . 1839
McDowall, J. P. K., Lt. . . . 687
McDowell, G. A., Lt. 769
—— R. E., Maj. 871
McElligot, Capt. 1839
McElwaine, E. J. D., Capt. 273, 1319
—— E. J. Lt. 796
McEnery, R. T., Maj. . 60, 222, 662
McEntee, J., *Hony.* Capt. . 1849
McEntire, H. S. M., Lt. 67, 389, 487, 1487
McErvel, D. S., 2nd-Lt. . . . 559
McEwn, H. C., 2nd-Lt. 558
McFadyen, A., Lt. . . . 504, 1443
McFarland, W. C. D., Lt. . . 762
McFarlane, A. H., Lt. . . 515, 970
—— A. McP., Capt. . . 1839
—— F. N., 2nd-Lt. . . . 624
—— W., 2nd-Lt. . 763, 782
—— W. E., Revd. . . . 1907
McGann, H. B., Maj. 239, 1287, 1895
McGarry, J., 2nd-Lt. . . 530, 1007a
McGarth, P. J., Asst. Surgn. . 1794
McGarvey, B. V., Lt. . . . 487, 1383
McGavin, A. E. C., Lt. . . 514, 1701
—— D. G. K., Lt. . . 532, 1219
McGee, G. L., 2nd-Lt. 624
McGeorge, D. J., Lt. . . 515, 1548
McGill, D. S., Lt. 542, 1705
—— E. A., Lt. 621
—— H. R., Maj. . . 206, 1307
McGillivray, W. S., Maj. . . . 1751
McGinn, J. J., Lt. 1846
—— T. G., Lt. 816
McGlashan, E. G., Lt. . 491, 941
McGorry, T. E. E., 2nd-Lt. . . 1674
McGovern, F., Lt. . . . 712, 1006
—— J. B., Lt. 730
McGovren, F. G., *Hony. Lt.* . . 611
McGowan, C. S., Sr. Nursg. Sister 1734
—— J. A. B., Maj. . . . 1838
—— T., Maj. 614
—— T. C. S., 2nd-Lt. . . 1373
—— T. E., 2nd-Lt. . . 474, 1166
—— W. H., Maj. . . . 634
McGrath, J., Lt. 821
—— P. J., Asst. Surgn. . 1194
—— W. H., 2nd-Lt. . 545, 1704
McGregor, A., 2nd-Lt. . . 546, 1295
—— F. H., 2nd-Lt. . . . 653
—— G. N., 2nd-Lt. . . . 705
—— K. S., 2nd-Lt. . 568, 1696
—— R. S., Capt. . . . 692
—— W. D., 2nd-Lt. 329, 1653
McGuire, C. W., Asst. Surgn. . . 1802
—— G., Asst. Surgn. . . 1799
—— J., Asst. Surgn. . . 1803
—— J., 2nd-Lt. 362
—— J. P., Asst. Surgn. . . 801
—— P. J., 2nd-Lt. . . . 1698
—— T. B., Lt. 516
McGurk, N. A., Lt. . 505, 801, 1666, 1702
—— O., Lt. 1839
McHaffie, J. P. 2nd-Lt. . . 556, 1315
McHale, F. F., 2nd-Lt. . . . 757
McHardly, L., 2nd-Lt. . . . 724
McHardy, W. S., Lt. 1319

McHarg, A. A., Lt. Col. . 103, 139, 662, 1880
McIlroy, H. D., Capt. 1718
McIlwaine, A., 2nd-Lt. 1323
—— W. A. St. J., Lt. . . 533
McInerney, F. E., 2nd-Lt. . . . 711
McInnes, P., Condr. 1853
McInroy, C., Col. 176
McIntosh, A., Capt. . 521, 843, 1667
—— E. P., Lt. 490
—— J. A., Capt. . 14, 37, 719
—— K. H., 2nd-Lt. . . . 816
McIntyre, D. P., Lt. 491
—— H. F., Lt. . . . 318, 1170
—— H. M. J., Capt. . . . 637
—— J., Lt. 731a
—— J., 2nd-Lt. . . . 630, 977
—— J. McD., Lt. . . . 445
—— J. P., 2nd-Lt. . . . 1678
McIver, A. C., 2nd-Lt. . . 549, 1674
—— C., Capt. . . . 1193, 1757
McIyer, N. A. H., Lt. 832
McKay, C. H. 2nd-Lt. 828
—— C. I., Capt. 615
—— G. S. M., 2nd-Lt. . 763, 782
—— J., 2nd-Lt. 724
—— T., Lt.-Col. 57
—— T. A., 2nd-Lt. 465, 473, 1489
McKeag, H. T. A., Lt. . 509, 768
McKean, H., Lt. 652
—— M. C., Capt. 776
McKeand, A., Capt. 827
McKechnie, A., 2nd-Lt. . . . 1009
—— V. A., Lt. 1036
—— W. E., Maj. . . . 1740
McKee, R., 2nd-Lt. F. F. . 786, 1885
McKelvie, R. F. S., Revd. . . . 1904
McKelvey, J. W., Capt. . . . 759
McKendrick, A. G., Maj. . . . 1747
McKenna, C., 2nd-Lt. . . 547, 1674
—— J. C., Maj. 54, 99, 217, 1020
—— J. W., Capt. 196
—— W., 2nd-Lt. 790
—— A. T., 2nd-Lt. . 543, 1110
—— H. N., 2nd-Lt. . 539, 1225
—— R. D., Lt. . . . 527, 1618
McKenzie, A. T., Lt. 542
—— R. E., 2nd-Lt. . . . 537
McKenzie-Muirson, K. B., Lt. 446-449, 875
McKernan, F. P., 2nd-Lt. 318, 1311
—— R. L., 2nd-Lt. . . . 547
McKerron, P. A. B., Lt. . . . 1098
McKerrow, H. B., 2nd-Lt. . 767, 1886
—— J. A., Lt. . . . 492, 883
McKew, P. J., Sub-Condr. . . . 1870
McKie, R., Sr. Asst. Surgn. . . . 1788
McKillop, A., Lt. 793
McKinney, J., 2nd-Lt. . . 355, 1385
McKinnon, J. A., 2nd-Lt. . . . 1078
McKinstry, H., Lt. . . . 425, 1515
McKnight, M. Capt. 1723
McLagan, C. A. G., 2nd-Lt. . . 597
McLaren, D. A. J., Tempy. Capt. 1723
—— J. A., Lt. 387, 1563
—— W. D., Lt. 815
Mcharty, A. M., 2nd-Lt. . . . 1393
McLaughlin, H. D., Maj. 216, 1623, 1899
—— W., Maj. 768
—— W. F. M. D., Maj. 99, 221, 1431
McLaurin, D. J., Lt. 1325
McLay, D. B., 2nd-Lt. . . 650, 992
McLean, A., 2nd-Lt. 760
—— A. D., 2nd-Lt. . . 681, 786
—— A. E. J., Lt. 1223
—— A. M., Lt. . . . 515, 1006
—— D. A., Lt. . . 446-449, 1158
—— F. G. S., Capt. 271, 1096, 1283
—— G. C., Lt. 859
—— K., Lt. 525, 1313
—— L., Revd. 1905
—— R. R. B., Capt. . . 501, 1415
—— T. D., 2nd-Lt. . . . 557
—— W., Lt. 519, 1403
—— W. A. G., 2nd-Lt. . . . 561
McLeod, D. J. S., Genl. *Sir* 927 1925, 1955
—— D. K., Maj. . . . 244, 904
—— D. Mc K., 2nd-Lt. . . 788
—— H. T., Capt. 844
—— K., *Hony.* Col. 1
—— K., Maj. . . . 230, 1141
—— N. C., Lt. 840
—— R. D., 2nd-Lt. . . . 786
—— T. J., Lt.-Col. 136, 158, 196, 1629

McLeod, W., Tempy. Capt. . . 1724
—— W. K., Brig.-Genl. . 19, 120, 123, 612
McLeavy, H. E. B., Asst. Surgn. 1800
McLellan, A. J., Lt. 508
—— G. H. H., Lt. . . . 767
—— J. S., Lt. 89, 778
McLennane, G. N., 2nd-Lt. . . 568
McMahon, F. A., 2nd-Lt. . . . 556
—— H. A., Lt. . 321, 365, 1397
McMahon O'Byrne, D., Lt. . . 399
McManus, A. A., 2nd-Lt. . 356, 1297
—— E. A., Lt. 489
—— J., *Hony.* Lt. . . . 1844
McMaster, W., Maj. 615
McMath, R. F., Lt. 701
McMillan, P. K. T., Ty. 2nd-Lt. . 762, 782
—— W., Capt. 852
—— T. B., Lt. . . . 533, 1627
—— W., Lt. 537
McMorran, J., Lt. 518, 1189
McMullu, W. K., Lt. . . . 445, 953
McMurray, F. C., 2nd-Lt. . 549, 1009
McMurrough Kavanagh, A. T., Capt. 52, 587
McNab, A., Lt. 352, 1048
McNair, A. L., Col. 178
—— A. W., Maj. 815
—— D. C., Ty. Lt. . . . 1777
—— E. J., Lt.-Genl. . . . 159
—— G. D., Lt. . . . 523, 1670
McNally, E. W., Asst. Surgn. . . 1801
—— I., Nursg. Sister . . 1735
McNamara, J. D., Condr. . . . 1883
—— M., Sub-Condr. . . . 1864
—— M. L., Lt. 805
McNaught, D., Lt. . . 90, 543, 1648
—— J. G., Lt.-Col. . . . 1710
McNaughton, J., Lt. . . 38, 764
McNeight, A. A., Maj. 10, 104, 1752
McNeil, J., 2nd-Lt. 653
—— W. A., Lt. . . . 401, 1120
McNeile, D. H., Lt.-Col. 105, 137, 159, 197, 911
McNeill, A. A. C., Capt. . . . 1754
—— D., Col. 176
McNeill, D. S. E., 2nd-Lt. . . 315, 961
—— H. L., Capt. 822
—— J. H. R., Revd. . . . 1905
—— J. S., Col. 819
—— K. J., Capt. 818
—— M. D., Lt. 1
McNeill Love, R. T., Tempy. Capt. 1724
McNeish, A. W., 2nd-Lt. . . . 357
McNevin, F. L., Sub-Condr. . . 1868
McNicol, W. M., 2nd-Lt. . . . 550
McOstrich, A. H. M., 2nd-Lt. . 330, 1178
McPherson, A. B., Capt. 9, 15, 261, 1238
—— D. L., Capt. 817
—— E. M., N. Sister . . 1735
—— G., Lt.-Col. . . 138, 1745
—— H., Lt. 513
—— J., Lt. 535, 1703
—— J., Maj. 1759
—— W. H., 2nd-Lt. . 360, 1034
McQuade, J. L., 2nd-Lt. . . 550, 1677
McQueen, J. A., Lt. . . . 540a, 1475
McRae, A. St. G. M., Maj. 99, 218, 868, 1080, 1203
McRoberts, B. A. K., Capt. . . . 763
—— P. F., Lt. . . . 520, 1666
McSwiney, H. F. C., Capt. 53, 262, 1606
—— S. A., Tempy. Capt. . 1724
McSwiny, P. M., Lt. 507
McTavish, F. H. C., Capt. . 78, 766
McVean, D. A. D., Lt.-Col. . 138, 159, 160, 198, 869, 1203
—— E. D., 2nd-Lt. . . . 887
McWatter, J. W., 2nd-Lt. . 763, 782
McWalters, W., *Hony.* Maj. . . 1844
McWatters, G. A., Maj. . . 209, 1331
—— H. C., Bt. Lt.-Col. 8, 15, 99, 138, 159, 216, 1116
McWhinne, R. S., 2nd-Lt. . 565, 1678a
—— W., Lt. . . . 517, 1067
McWhirrer, R. J., Lt. 767
Mabbott, P. W., Maj. . 55, 500, 814
McAdam, A. Mc. F., 2n-Lt. . . 719
MacAlister, A. S., Lt. 610
Macalister, S. C., Lt. . . 317, 1423
Macallum, A. H., Capt. . . 501, 1359
Macandrew, H. J. M. Maj.-Genl. 101, 119, 147, 174
Macann, A. E. H., Lt. . . 320, 318
MacArthur, A. G., 2nd-Lt. . . 1707

Names and Rank. Page.

MacArthur, D. C., Lt. 638
—— D. F., 2nd-Lt. . . 550
—— D. T. S., 2nd-Lt. . . 563
—— R. G., Lt. . . 541, 1707
Macara, J., 2nd-Lt. 1695
Macartney, F. A., Maj 5, 246, 1235
—— H. J., Lt. . . 510, 1618
Macaulay, A. D. C., Lt., 420, 513, 630
—— D. I. M., Maj. 102, 105, 208, 875
—— F. G. A., Lt. . . 652
—— J., Maj. 33, 715
Macausland, R. A., Capt. 26, 264, 1203
MacBain, J. R. G., 2nd-Lt. . 719
Macbay, W. G. W., Lt.-Col. 179, 1918
Macbean, J. A., Lt. 610
MacBean, H. A., 2nd-Lt. . . . 806
—— J. S., Lt. 830
MacBeth, A., Lt. 818
—— C. J. S., Lt. . . 541, 1078
MacBey, S. G., 2nd-Lt. . . 560, 1187
—— W. D., 2nd-Lt. . . . 801
MacBrayne, R. J., Maj. 94, 233, 1078
MacCay, R. C., Lt. 907
Maccoy, N., Lt. 687
MacCulloch, A. F., Capt. . . . 514
—— H. S., Lt. 450
MacDermott, M. J., Capt. . . 502
Macdonald, A., Lt. . . . 528, 998
—— A. J., 2nd-Lt. . . . 1052
—— A. St. J., Lt. . . . 528
—— A. S., Lt. 701
—— C. K., 2nd-Lt. . . . 1164
—— D., 2nd-Lt. 1601
—— D., Maj. 1838
—— D. T., Capt. . 267, 1497
—— E. G., 2nd Lt. . . . 666
—— F. W. P., Lt.-Col. 105, 182, 1911
—— G. G., Ty. Lt. . . . 1729
—— H., Maj. . 54, 261, 895
—— H. I., Sub-Condr. . 1874
—— I. M., Capt. . . . 762
—— I. P., Lt. 111
—— J., Lt. . . . 727, 731a
—— J., Hony.-Lt. . . . 843
—— J. B., Lt. 322
—— J. K., 2nd-Lt. . . . 766
—— J. R. L., Maj.-Genl. 1955
—— J. V., Capt. . . . 1756
—— L. J., Lt. . . 511, 1006
—— L. T., Lt. . . 505, 1654
—— M., Lt. 542
—— N., Lt. . . . 317, 1187
—— O. S., Capt. . . . 833
—— P. S., Lt. 1630
—— R. H., Capt. . 741, 1004
—— R. R. M., Capt. . . . 687
—— R. W., Maj. 226, 1293, 1895, 1913
—— S., Capt. 658
—— S., Lt. . . . 435, 1239
—— S. R., Lt. . . 513, 714
—— W., Maj. . . 238, 1144
—— W. C., 2nd-Lt. . . . 578
—— W. R., Capt. . . . 861
—— W. R., 2nd-Lt. . . . 640
Macdonald-Clark, A. F., Lt. 668, 986
Macdonell, O. S., Capt. . . . 832
MacDonnell, H. N. F., Lt.-Col. 35, 87, 137, 159, 217
—— R. G., Lt. 683
MacDougal, J., 2nd-Lt. . . . 1154
MacDougall, D., Capt . . 501, 1604
—— D. K., Ty. Capt. . 752
—— R. McI., Lt. . . . 68
Mace, A. C., Lt. 737
—— F. C., 2nd-Lt. 618
—— J. M., Lt.-Col. 783
—— P. H., 2nd-Lt. 357
Macey, C. T., 2nd-Lt. 705
—— T. Y., *Hony.* Capt. . . . 57
MacEwan, A. K., Capt. . . 619, 1892
MacEwen, D. W., Capt. . . . 744
MacFadden, R. B., Lt. . . 516a, 986
Macfadyen, M., Capt. 859
—— W. A., Lt. 685
MacFarlan, J. B., Maj. . . . 633
Macfarlane, A., Revd. 1907
—— E. A., 2nd-Lt. . . . 363
—— H., 2nd-Lt. . . 473, 1447
—— J. W., 2nd-Lt. . . . 801
Macfarlane-Black, J. G., Lt. . . 785
Macfarlane-Grieve, R. W., Lt. . 744
MacFarquhar, J., 2nd-Lt. . . . 1695

Macfie, J., Lt. . . . 31, 513, 777
MacGeorge, H. K., Col. 40, 126, 150, 188
—— J. B., Maj. . 38, 652, 1882
MacGilchrist, A. C., Maj . . 1746
MacGillivary, W., 2nd-Lt. . . 787
MacGillivray, D. G., Capt. 502, 1598
MacGlasson, L., Lt. 725
Macgregor, A., Capt. . . 636, 981
—— A., Maj. 651
—— A., Lt. . . . 532, 973
—— A. D., Capt. . 571a, 1841
—— A. H., Sr. Asst. Surgn. 1791
—— D. R., Lt. . . 520, 1667
—— J., Lt.-Col. . . . 1886
—— J. S. I., 2nd-Lt. . . 1134
—— L. E., Lt. . . 507, 1227
—— R. F. D., Capt. . . 1756
—— R. G., Capt 268, 938, 961
—— W. L., 2nd-Lt. . 547, 1705
Macgrigor, G. N., Lt. . 517, 1670
MacGuckin, C. E., 2nd-Lt . 326, 1154
Machado, L. S., Tempy. Capt. . 1766
Machin, F. W., Lt. 543
—— H. S., Tempy. Lt. 695, 1006
Macintosh, I. D., 2nd-Lt. . . 597
—— K. J., Lt. . . 309, 1617
MacIntyre, A., Lt. 490
—— D., Capt. 1910
—— D. C. F., Maj.-Genl. 57, 118, 146, 173
—— D. I., Lt. 585
—— D. P., Lt. . . . .
—— F. P., Lt. 593
—— J., Lt. 719
—— J. T., 2nd-Lt. . . . 465
—— I. H., Lt. . . 299, 1187
—— W., 2nd-Lt. . . . 1154
—— W. F., Lt. 629
Mack, C. H., Lt. . . . 518, 1487
—— F. C., 2nd-Lt. 566
Macllwaine, A. H., Capt. . . . 615
MacIver, N., Lt. 640
—— P. 2nd-Lt. 602
—— R. G., Maj. 651
MacIvor, I., Capt. 1910
—— R. J., Lt. . . 440, 1648
—— R. S. P., Capt. . 278, 1513
Mackaness, S. G., 2nd-Lt. . 547, 1673
Mackarness, C. G. M., Lt. . . . 517
—— R. S. P., Lt. . . . 380
Mackay, A. G. D., 2nd-Lt. . . . 630
—— A. J., Lt. . . 410, 1239
—— A. S., Capt. . . 274, 1629
—— C. H., 2nd-Lt. . . . 555
—— D. A., 2nd-Lt. . . . 560
—— D. H., 2nd-Lt. . . 494, 941
—— D. L., Asst. Surgn. . . 1795
—— D. M., Asst. Surgn. . . 2803
—— D. S., Lt.-Col. . . . 854
—— D. S., 2nd-Lt. . . . 764
—— D. S., Lt. . . . 393, 903
—— E. A., Maj. 755
—— E. H., 2nd-Lt. . . . 765
—— F. F., Lt. . . 516, 611, 1701
—— G. S., Capt. . . 501a, 1365
—— G. S., Ty. Lt. . . . 622
—— J., 2nd Lt. . . 539, 1279
—— J., 2nd-Lt. 1695
—— J. M., Lt. 714
—— J. W., 2nd-Lt. . . 547, 571
—— R., Sub-Condr. . . . 2872
—— R., 2nd-Lt. 547
—— R. D., 2nd-Lt. . 560, 1405
—— R. J., Lt. 393
—— T. I. S., Lt. 519
MacKelvie, Lt.-Col. M. 139
MacKenzie, A., Sub-Condr. . . 1855
—— A. E., Capt. 852
—— A. H., Capt. 814
—— A. R. D., Col. *Sir* 163, 1925
—— A. S. C., Lt. . . . .
—— C., Lt.-Col. . 134, 157, 194, 867, 899
—— C. A., 2nd-Lt. . 558, 1094
—— C. A. C., Capt. 252, [illegible]
—— C. F., Maj. 10[illegible], [illegible], 1890, 1913
—— C. F., Capt. . 269, 1102
—— D., 2nd-Lt. 77, 786, 1685
—— D., Capt. 729
—— D. A. L., 2nd-Lt. 365, 1227
—— D. B., Capt. . 277, 1255
—— D. F., Capt. . . . 1712
—— D. G., I. C. S. . . . 2
—— E. A. H., Capt. 265, 1026

Mackenzie, E. L., Lt.-Col. 134, 726
—— F., Lt. . . . 470, 1140
—— F. McD., Capt. . . 1839
—— G. M., 2nd-Lt. . . 744
—— G. P., Capt. . . . 859
—— H. C., Lt. . . 387, 1044
—— H. H., Capt. . 636, 972
—— H. J., Maj. 15, 22, 235, 1207
—— H. M., Maj. . . . 1746
—— I. D., Lt. 875
—— J., 2nd-Lt. . . 553, 1467
—— J., Maj. . . 2, 212, 1162
—— J., Maj. . . . 24, 1711
—— J. D., Capt. 16, 35, 269, 1187
—— J. D. H., 2nd-Lt. . . 785
—— J. F., Lt. 1674
—— J. F. O. P., Lt. . 415, 1363
—— J. McA., Capt. 486, 935
—— J. M. H., Capt. . . 1910
—— K. L., Capt. . . . 610
—— K. S., 2nd-Lt. 557, 1703
—— L. H. L., Maj . . . 1755
—— R. F., Asst. Surgn. . 1800
—— R. H. H., Capt. . .
—— R. H. T., Capt. . 17, 754
—— R. T. H., Capt. . . 821
—— S. N., 2nd-Lt. . . . 563
—— T., Condr. . . . 1867
—— T. C., Capt. . 265, 885
—— T. C., 2nd-Lt. . . . 791
—— W., Lt. 601
—— W. R., 2nd-Lt. 558, 1395
—— W. S., Lt. 296, 719, 1543
Mackenzie-Kennedy, E. C. W., Maj.-Genl. 118, 143, 173
Mackenzie-Rogan, H., Capt. 501a, 1517
Mackernan, F. P., Lt. . . 318, 1355
Mackey, C. E., Sub-Condr. . . 1864
—— D. A., 2nd-Lt. . . . 560
—— D. D., 2nd-Lt. . 359, 1285
—— F., Sub-Condr. . . . 1854
—— G., Asst. Surgn. . . 1799
—— H. J. A., Lt.-Col. 58, 122, 137, 613
—— R., 2nd-Lt. 1674
Mackichan, K. W., Lt. . . . 618
Mackie, A. H., Lt. . . 294a, 909
—— C., Capt. . . . 39, 1839
—— F. P., Maj. 1749
—— J., 2nd-Lt. 1092
—— J. B., Lt. 704
—— J. R., Capt. 704
—— R., Lt.-Col. 38, 214, 870, 1275, 1333
—— R., Condr. 1873
—— R. H., Lt. 832
Mackinnon, A., 2nd-Lt. . . 560, 1407
—— A. C. B., Maj. 213, 872, 1641, 1645
—— R. A., 2nd-Lt. 465, 473, 2457
—— N. D., Rev. . . . 1905
—— W. H., Genl. . . . 689
Mackintosh, A., Capt. . . . 1892
—— G. H. T., Maj. . 229, 883
—— G. N., Lt. . . 299, 1166
—— J. B., Lt.-Col. 101, 135, 632
—— J. D., Lt. 761
—— W., 2nd-Lt. . . . 859
Mac'Intosh, J. N., Capt. . . 1717
MacKnight, J., Lt. 602
Mackrell, A., Maj. . 236, 1439, 1505
—— G., Capt. 578
Mackrow, G. H., Lt. 528
Mackworth, N. W., Maj. . . . 1749
MacLachlan, A., Lt. 306, 1227, 1565
—— A. B., Lt. 401, 1046, 1557
—— D., 2nd-Lt. . 95, 523
—— T. R., Bt.-Col. 50, 121, 126, 153, 193, 870, 1375
Maclagan, R. S., Col. 42, 123, 660, 1899
Maclaren, C. S., Lt. 318, 1251, 1255
—— W. de B., Lt. . 405, 1126
Maclarron, C. M., Lt. . . 530, 1674
Maclean, A. D. D., Capt. . . . 715
—— C. H. M., 2nd-Lt. . . 998
—— C. H. Mac K., 2nd-Lt. . 568
—— C. S., Maj.-Genl. 172, 915
—— D., 2nd-Lt. 1032
—— G. G. C., Maj. 5, 235, 1419
—— H. A., 2nd-Lt. . . . 828
—— H. G., Lt. . . 524, 1703
—— J. R., 2nd-Lt. . 763, 782
—— L. L., Capt. . . 273, 1635

NAMES AND RANK. Page.

Maclean, M. E., Nursg. Sister . 1735
——— W., Nursg. Sister . . 1735
Maclear, G. D'O., Lt. . . . 513
Macleman, W. D., 2nd-Lt. .715, 764
MacLennan, N. L., Lt. . . . 679
MacLeod, A., Lt. 730
——— C., Lt. 602
——— D. G., 2nd-Lt. . . . 562
——— D. N., Lt. 1715
——— E. C., Lt.-Col. . 130, 1742
MacLeod, H. T. 844-45
——— I. N., 2nd-Lt. . 549, 1624
——— L. C., 2nd-Lt. . . . 565
——— M. N., Maj. . . 663, 1880
——— N., Maj. 17, 47, 99, 218, 1078
——— N., Lt. 718
——— N., 2nd-Lt. 764
——— N. M., Lt. . . 318, 1595
——— R., Capt. 615
——— R. C., Capt. 714
——— R. D. C., Maj. . 233, 911
——— W. M., Maj. 13, 19, 99, 215, 933
MacMahon, G. D. R., Maj 233, 871, 1491, 1533
——— H. F. E., Bt. Lt.-Col. 34, 139, 160, 222, 1662
——— J. J., Lt. 537
——— O'Byrne, D., Tempy. 2nd-Lt. 623, 921
Macmillan, D. MacL., Lt. . 317, 1156, 1158
——— E. J. A., Lt. . 489, 877
——— H. A., 2nd-Lt. 723a, 732, 1885
——— H. R. M., Lt. . . . 825
——— J., Lt. 352
——— J. McC. A., Maj . 1750
——— K. W., 2nd-Lt. 522, 1565
Macmullen, C. N., Br.-Genl. 51, 97, 121, 126, 153, 214, 1078
——— W. H. F., Lt.-Col. 179, 1917
MacNab, A. J., Col. .125, 1741, 1890
Macnab, Br.-Genl., C. L. . . .
Macnabb, D. J. C., Lt.-Col. 185, 1895, 1911
——— H., Lt. 690a
——— R. J., Maj... . 234, 1890
Macnaghten, A. W., Col. . . 170
Macnair, R. H., Lt. 803
Macnamara, J. P., Capt. . . . 1718
Macnaughen, H. P. W., 2nd-Lt. 821
Macnee, E. A., Lt. . . 503, 893
Macnicol, A. C., Lt. 751
Macniven, G. A., Lt. . . 530, 988
Macomish, R., the Revd. . . . 1907
Maconachie, G. W., Maj. . . 1751
——— R. R., Lt. . 542, 1235
Maconchy, G. A., Lt. . . 291, 1617
Maconochie, A. E. F., Lt. . 509, 897
MacPhail, D., 2nd-Lt. . . . 361
Macpherson, A., 2nd-Lt. . . . 786
——— A. D., Lt.-Col. 198, 1890, 1901, 1912
——— A. I., 2nd-Lt. 556, 1271
——— A. K., Maj. 267, 1066, 1160
——— C. F., Lt. 56
——— D., 2nd-Lt. . 550, 1701
——— D. I., Maj. 242, 1058, 1893
——— G. C., Revd. . . . 1905
——— H. A., Lt. . 371, 1293
——— H. R., Lt.. . . . 623
——— J., Maj. 211, 1491, 1493
——— J. D., Col. . . . 177
——— J. D. G., Lt.-Col. . 1710
——— J. E., Lt.. . 375, 1367
——— R. G., Lt.-Col. . 1910
——— R. V., Lt. . . . 885
——— W., Capt. . . . 585
Macqueen, E. T. L., Capt. . 367, 1064
——— J. E. B., Sr. Asst. Surgn. 1787
——— K. G. S., Senr. Asst. Surgn. 1789
——— P., 2nd-Lt. . . . 571a
——— R. B., 2nd-Lt. . 365, 879
——— S. W., 2nd-Lt. . . . 1126
Macquoid, C. E. E. F. K., Lt.-Col. 131, 156, 191, 867, 881
Macrae, A. C., Capt.. . . . 1757
——— D. E., Lt. 1839
——— I. M., Maj. 1751
——— J. C., Maj. 99, 227, 1094
——— J. L., Lt.-Col. 129, 1742, 1890
——— J. N., Maj. . . . 764, 765

NAMES AND RANK. Page.

Macrae, M., 2nd-Lt. . . . 763, 782
——— W., Maj. . . 663, 1881
Macrostie, E. J., 2nd-Lt. . 29, 779
Macroy, W. M., Lt. 765
MacSwiney, E. V. F., Capt. . . 789
Mactaggart, C., Col. . 124, 1738, 1898
Mactavish, A., Lt.Col. 45, 213, 868, 1018, 1026
MacVay, S., 2nd-Lt. . . . 465, 1651
Macvean, E. D., Lt. . 320, 365, 1397
Macwatt, R. C., Col. . 127, 738, 1896
MacWatters, M. R. C., Maj. . . 1752
Macwhirter, D., Capt. 485, 601, 885
MacWilliam, T. A., Lt. . . 487, 937
Madad Khan, Subdr. . . . 1521
Madan, F. F., 2nd-Lt. . . 563, 1524
——— K. R., Tempy. Capt. . . 1769
——— Sing Rai, Subdr. 1614
Madaman, Subdr. 1252
Mada Rawat, Subdr. . . . 1483
Madan Gopal, Jemdr. . . . 1126
——— Singh Panwar, Subdr. . 1223
——— Gusain, Jemdr. . . . 1179
Mardar Khan, Jemdr. . . . 1248
——— Jemdr. 1324
Madat Ali, Jemdr. . . . 1096
——— Subdr. 1255
——— Jemdr. 1380
——— Khan, Ressdr. . . . 944
Madden, J. W., Lt. 1707
——— S. F., 2nd-Lt. . . 561, 826
——— W. J., Lt. 543
Maddison, J. G., 2nd-Lt. . . 351
Maddock, E. C. G., Maj. . . 1746
Maddox, R. H., Lt.-Col.. . 129, 1742
Madeley, J. W., Capt. . . 802
——— W. Y., 2nd-Lt. . 557, 1283
Madgavkar, V. D., Tempy. Lt. . 1771
——— V. B., Lt. . . . .
Madge, H. E., 2nd-Lt. . . 324, 1614
——— K. A., Lt. . . 536, 1639
Madgett, C., Capt. . . . 749
Madhava Menon, Subdr. . 1307, 1309
Madhavan, O. C., Ty. Lt. . . 1777
Madhaveshvarendra Sahi Babu, 2nd-Lt. 481
Madhok, A. N., Ty. Lt. . . . 1778
——— M. C., Tempy. Lt. . . 1776
Madhoparshad, Sr. Sub-Asst. Surgn. 1809
Madho Sing Rana, *Hony.* Capt. 2, 2102
——— Khattri, Subdr. . 1642
——— Rawat, Subdr. . . 1179
——— Jemdr. 1114
——— Jemdr. 1401
Madhu Jagdale, Jemdr. . . . 1444
——— Sudan Pal, Sub-Asst. Surgn. 1820
——— Wogmode, Jemdr. . . 1444
Madland, C., 2nd-Lt 626
Madley, J., 2nd-Lt. 1674
——— J. W., Capt. . . . 801
Madurai, Jemdr. 1351
——— Subd 1339
Maduraimuthu, Jemdr. . . . 1308
——— Jemdr. . . . 1591
Maduraimuttu, Subdr. . . . 1331
Maduraimuttu, Padaiyachi, Subdr. . 1331
Maduramuttu, Subdr. . . . 1271
——— Jemdr. . . . 994
Maduranayakam, Subdr. . . 1359
——— Jemdr. . . . 994
Madurai, Jemdr. 993
——— Jemdr. 994
Maer, C. A., Lt. 1074
Maffat, J., 2nd-Lt. 1030
Maffett, D. B., Lt. . 314, 1136, 1144
——— R. E., Lt.-Col. . . 134, 723
Maffey, J. L., 2nd-Lt. . . . 2, 493
Maflin, A. J. G., Lt. . . . 534
Mag Singh, Jemdr. 1476
Magan Lal Dhanji Udani, Sub-Asst. Surgn. 1830
——— Kanji, Sub-Asst. Surgn. 1830
——— Singh, Jemdr. 936
Magar Ram, Jemdr. 1452
——— Sing, Hony. Capt. . . 2101
——— Singh, Jemdr. 1075
Magenty, E. R., Lt. . . 470, 1164
Magee, J. H., 2nd-Lt. . . 359, 1265
——— R., Hony. Lt. . . . 828
Maggs, C. R., 2nd-Lt. . . . 733
——— C. W., Capt. . . . 764, 456

NAMES AND RANK. Page.

Magh Singh, Jemdr. 990
——— Jemdr 1067
Maghan Singh, Subdr. 1268
Magha Ram, Jemdr. 1483
Maghar Singh, Subdr. . . . 1160
——— Jemdr. 1376
——— Subdr. 1121
——— Jemdr. 1160
Magill, R. H., 2nd-Lt. . . . 785
Magnay, A. D., Capt. . 37, 277, 875
Magner, J., Lt. 368, 1677
Magniac, C. J. L., Lt.-Col. 60, 139, 662, 1881
——— F. A., Maj. 210, 858, 1106, 1125
Magrath, B. H. W., Maj.-Genl. . 172
——— H. A. F., Lt.-Col. 129, 154, 188
Maguire, J. J., Lt. . . . 401, 1507
——— P. J., Tempy. Capt. . . 1724
——— P. J., Lt. 527
Mahabir Parshad, Sub-Asst. Surgn. . 1812
——— Misir, Jemdr. 1019
——— Pathak, Jemdr. . . 990
——— Singh, Jemdr. . . . 1003
——— Singh, Subdr. . . . 1061
——— Singh, Subdr. . . . 1020
——— Tewari, Jemdr. . . . 1017
——— Thakur, Bahadur, Subdr. 1614
——— Thapa, Jemdr. . . . 1627
——— Jemdr. 999
Maha Singh, Jemdr. 1400
Mahar, F. M., 2nd-Lt. . . . 1509
Mahadal Limbu, Subdr. . . . 1630
Mahadbir Thapa, Jemdr. . . 1637
Mahadeo Chowdry, Jemdr. . . 1510
——— Parshad, Sr. Sub-Asst. Surgn. . 1807
——— Tewari, Jemdr. . . 1027
——— Ramchandra Yaswantrao, Jemdr. . 1468
——— Singh (I), Subdr. . 1082
——— (II), Subdr. . 1082
——— Jemdr. . . . 1044
——— More, Jemdr. . . . 1424
——— Nalaude, Subdr. 1107
——— Sondkar, Subdr. . 1403
Mahadevan, V., Capt. . . . 1758
Mahadewa, Jemdr. (I) . . . 1193
——— Jemdr. 1561
——— Jemdr. 1193
——— Jemdr. 1191
Mahadset Kodale, Jemdr. . . 1436
Mahadu Jadhao, Jemdr. . . 1416
Mahamad Yasin, Jemdr. . . . 918
Mahammad Abdul Rashid, Jemdr. 1400
——— Alam, Jemdr. . . 1284
Mahan Singh, Risldr. Maj. . . 893
——— Jemdr. 930
——— Jemdr. 934
Mahant Bir Rana, Subdr. . . 1627
Mahaparsad, Rai, Jemdr. . . . 1633
Mahbub Alam, Jemdr. . . . 1375
Maharaj Kishan, Sub-Asst. Surgn. 1810
——— Singh, Subdr. . . . 1452
——— Subdr. . . . 1585
——— Sub-Asst. Surgn. . 1810
Maharam Singh, Subdr.. . . . 1031
Mahasher Limbu, Jemdr. . . . 1639
Mahatab Singh, *Hony.* Capt. 2101
Mahatam Singh, Subdr. . . . 1062
Mahoob Khan, Jemdr. 908
Mahbub Ali Khan, Sub-Asst. Surgn. . 1811
——— Alikhan, Jemdr. 930, 953
——— Khan, Ressdr. . . . 914
——— Risalr. . . . 915
——— Jemdr. . . . 1088
——— Ressdr. . . . 932
——— Subdr. . . . 1052
Mahbubali Badulla Shaikh, Sub-Asst.-Surgn. 1831
Mahbullah, Jemdr. 1682
Mandi Hassan, Sub-Asst. Surgn. 1814
Mahendar Singh, Ressdr. . . 912
Mahendra Man Singh, Raja, 2nd-Lt. . 481, 1537
Mahendu Lal Garg, Sr. Sub-Asst. Surgn. . 1808
Maher, C. B., Lt.. . . . 527, 1699
——— F., *Hony.* Maj. . . . 1843
——— F. M., 2nd-Lt. . . . 560
——— W. H., Asst. Surgn. . . 1794
Mahesar Dube, Subdr. 1010

NAMES AND RANK. Page.

Mahesh Narain Dube, Subdr. . 1026
Mahesur Gurung, Jemdr. . 1625, 1651
Mahfuz Ali Khan, Jemdr. . . 932, 961
Mahindar Singh, Ressdr. . . . 1686
Mahipal Singh, Jemdr. . . . 1047
Mahipat Singh, Subdr. . . . 1405
Mahiuddin, A. K. M., Ty. Lt. . 1778
Mal Sukh, Ressdr. 1686
Mahman Singh, Sub-Asst. Surgn. 1816
Mahmood Mian Sahib, S. Khan Sahib 89
Mahmud Khan, Ressdr. . . . 921
—— Jemdr. 938
—— Jemdr. . . 1088, 1537
—— Jemdr. 1373
—— Jemdr. 1460
—— Jemdr. 1321
—— Jemdr. 1681
—— Shah, Jemdr. 948
Mahomed Hayat, Jemdr. . . . 994
Mahon, A. E., Maj . . 223, 1243
—— B. MacM., Capt. . 367, 1122
—— B. T., Lt.-Genl. *Sir* . . 589
—— H. J., Maj. . . 16, 39, 835
Mahoney, A. J., Lt. 68, 514, 1355
—— C. C., Asst. Surgn. . . 1803
—— E. R., Asst. Surgn. . . 1794
—— J., Lt. 761
—— J. S., Lt. 836
—— R. S., 2nd-Lt. . . . 563
—— T., Sub.-Condr. . . . 1857
—— T., 2nd-Lt. 564
Mahony, B. C., 2nd-Lt. . . 557, 988
—— C., Lt. 531
—— G. H., Capt. 1758
—— J. S., Condr. . . . 1868
—— N. E. B. C., Lt. . . . 701
Mahr Ali Khan, Jemdr. . . . 1705
Mahrab-Ali Khan, *Hony.* Capt. . 2102
Mahsow, Jemdr. 1681
Mahta Ram, Jemdr. 1457
Mahtab, Jemdr. . . . . .
—— Khan, Subdr. 1479
—— Subdr. 1457
—— Singh, Jemdr. 1031
Maiden, C. J. D., 2nd-Lt. . . 784
Maidhar Gharti, Subdr. . . 1618
Maidlow, J. F., Lt. . . . 73, 695
Maidment, W. T., 2nd-Lt. . . 619
Maiku Lal, Sub-Asst. Surgn. . 1809
Mailya, G. B., Tempy. Lt. . . 1764
Maima Singh, Subdr. . . . 1080
Main, B. W. C., 2nd-Lt. . . . 757
Mainga Singh, Jemdr. . . . 1145
Maingha Sing, Subdr. . . . 1056
Mainprise, G. D., Lt. . . . 843
Mainwaring, A. J., Lt. . . . 490
—— C. V., Col. 123, 149, 182
—— E. P., Col. . 169, 1916
—— G. B., Capt. . 259, 1178
Mainwaring-White, E P., Maj. 230, 1475
Mair, A., 2nd-Lt. 834
—— A., 2nd-Lt. . . 544, 722, 1678*a*
—— B. V., Capt. 55
—— R., Capt. . . 505, 986, 1884
Maisey, F. C., Col. 179
Mal Sukh, Ressdr. 1683
Maltby, G., Lt. 1007
Maite Bura, Jemdr. 1627
Maitland, A., Lt. . . . 73, 508
—— G. R., Maj. 40, 99, 228, 901
—— P. J. Maj. Gen. . . . 1955
Maitra, A. P., Ty. Lt. . . . 1776
—— G. C., Tempy. Capt. . . 1772
—— M. M., Tempy. Capt. . . 1772
—— N., tempy. Capt. . . . 1770
—— S. S., Tempy. Lt. . . . 1776
Majid Gul, Jemdr. 934
Majhi Khan, Jemdr. 1514
—— Sing, Subdr. 1577
Majumdar, D. C., Ty. Lt. . . 1781
—— P. T., Tempy. Capt. 1767
Major, C. R., 2nd-Lt. . . . 559
—— D. D. J., Asst. Surgn. . . 1802
—— E., 2nd-Lt. 567
—— F. F., Lt.-Col. 131, 156, 191 870, 1387
—— F. W., Col. 177
Majumdar, N. N., Ty. Lt. . . 1779
Makar Singh Chauhan, Jemdr. . 1182
Maker, L., Lt. 533, 1573
Makhan, Jemdr. 1189
—— Lal, Jemdr. 932
—— Banerjee, Jemdr. . 1012
—— Khan, Subdr. . . . 1357
—— Singh, Jemdr. . . . 1167

Makhan Singh, Resdr. 1682
—— Jemdr. . . . 892
—— Jemdr. . . . 1163
—— Jemdr. . . . 1240
—— Jemdr. . . . 950
——Gurung, Sub.-Maj. . 1639
Makhana Khan, Jemdr. . . . 1103
Makhane Khan, Jemdr. . . . 1122
Makhar Sing Kawar, Subdr. . 1181
—— Jemdr. 950
Makhmad, Subdr. 1098
—— Ali, Subdr. 1257
—— Din, Subdr. 1227
——Janab, Jemdr. . . . 1096
Makhmul Khan, Jemdr. . . . 1480
——Jemdr. . . . 1587
Makin, R., 2nd-Lt. 742
Makhandhoj Rai, Subdr. . . 1630
Mal Singh, Subdr. 1511
—— Ressdlr. 948
—— (I) Risldr. 894
—— Subdr. 1150
—— (II) Jemdr. . . . 894
——Jemdr. (III) . 894, 955
—— Jemdr. 1304
Mala Khan, Sub.-Maj. . . . 1513
—— Jamdr. 1565
—— Rawat, Subdr. 1471
—— Singh, Jemdr. 1493
—— Subdr. 1587
Malak Din, Jemdr. 1493
—— **Muhammad Hayat Khan**, Ressdr. 65*b*, 891
—— **Sher Khan**, Jemdr. . . . 889
Malan, L. N., Lt.-Col. . 137, 663, 1001
Malang Khan, Subdr. . . . 1259
Malapa, Subdr.-Maj. 1183
Malcolm, A. A., 2nd-Lt. . . . 595
—— A. R. H., 2nd-Lt. . . 619
—— B. E., Lt.-Col. . . . 61
—— F. A., 2nd-Lt. . . . 54
—— J., Condr. 1854
—— J. D., Lt. . . . 525, 1050
—— J. F., Lt. . . . 81, 405
—— W. L., Lt.-Col. . 130, 155, 189
Malcomson, G. E., Capt. . . . 753
Malden, C. C., Capt. 726
—— C. M., Lt. . . . 305, 1259
—— W. J., 2nd-Lt. . . 330, 968
Male, E., 2nd-Lt. 1383
—— Khan, Subdr. 1511
—— Jemdr. 1587
Malerbi, J. W. B., Lt. . . . 730
Malerkotla, Nawab of . . 110, 998
Malet, A. W., Capt. 258, 1174, 1291, 1895
—— R. J., Maj. . 215, 870, 1275, 1361
—— W. deC., Maj. 14, 32, 246, 901
Maley, J., Condr. 1853
Malhang Limbu, Jemdr. . . . 1633
—— R. C., Tempy. Capt. . . 1765
Malhoutra, G. D., Tempy. Capt. . 1767
Mali, Jemdr. 1175
Malik, Ahmadyar Khan, Risldr.-Maj. 910, 2101
—— **Ahmad Yar Khan**, Jemdr. . 908
—— **Alam Sher Khan**, Ressdr. . 908
—— **Aman**, Subdr. 1507
—— **Ghulam Hasain Khan**, Jemdr. 908
—— **Khan**, Jemdr. 968
—— **Khan Muhammad**, Risldr.-Maj. 891, 957
—— **Khan**, Risldr. 914
——Jemdr. . . 950
—— **Khizar Mohammad** Hayt, 2nd-Lt. 481, 484
—— **Khuda Bakhsh Khan**, Jemdr. 950
—— **Mehrkhan**, Resldr. . . . 874
—— **Mihr Khan**, Jemdr. . . . 904
—— **Mohammad Akbar Khan**, Khan Sahib, 2nd-Lt. . . 481
—— **Mohammad, Sher Khan**, Tiwana, 2nd-Lt. . 481—84
—— **Muhammad Bahadur Khan**, Ressdr. 950
——, Jemdr. 950
—— **Latif**, Ressdr. 928
—— **Mubaraz Khan**, *The Hon., Hony.* Capt. 111, 891
—— **Sher Khan**, Jemdr. . 892
—— **Mumtaz Muhammad Khan**, Capt. . 82, 278, 881
—— **Muzzaffar Khan**, Ressdr. . 942

Malik Sher Bahadur Khan, Risldr.-Maj. 873, 924
—— **Bisaldar**. 950
——Sardar Khan Nun, 2nd-Lt. . 481—84
—— 82
—— **Sikandar Khan**, Jemdr. . 928
—— **Sher Muhammad Khan**, Jemdr. . 950
—— **Sultan Khan**, Risldr. . 910
—— **Ahmad Khan**, Ressdr. 910
—— Mahmmud Khan, Jemdr. . . . 910
—— **Umar Hayat Khan, Tiwana**, *Hony.* Maj *Sir* . 110, 909
Malikdad Khan, Jemdr. . . . 963
Malim, A. W., Capt. 732
Malir, J., Sr. Asst. Surgn. . . . 1790
Malkani, B. R., Ty. Lt. . . . 1777
Malkhan Singh, Jemdr. . . . 1457
Mallam, G. L., Lt. 297*a*, 877, 959
—— R. K., Capt. 1714
Mall Singh, Subdr. 1203
—— Jemdr. 1112
Malla Singh, Subdr. 999
—— Jemdr. 1067
—— Subdr. 1157
Mallaby, A. W. S., 2nd-Lt. 332, 565*a*, 1128
Mallalieu, W., Capt. 617
Mallam, Lt. G. L. . . . 297, 877
Malleson, W., Maj.-Gen. . 119, 147, 174
Mallet, W. H., Lt. . . 455–59, 1501
Mallett, G. W. L., 2nd-Lt. . . . 745
Mallik, B. B., Tempy. Capt. . . 1772
—— P. N., Tempy. Lt. . . . 1774
Mallinson, E. H. P., 2nd-Lt. 327, 1142
Mallis, J. E., 2nd-Lt. 560
Mallock, A. R. O., Capt . 23, 253 1192
—— H. A., Col. . . . 176, 1917
Mallya, G. B., Tempy. Capt. . . 1764
Malone, A., Lt. 624
—— E. M., Capt. . . 273, 1034
—— H. A. P., Lt. 583
—— M., Sub-Condr. . . . 1857
—— R. H., Capt. 1760
Maloney, H. J., 2nd-Lt. . . . 1693
—— J. C., 2nd-Lt. . . . 530
Malook, Jemdr. 1152
Malooka Ram, Subdr. . . . 1488
Maloy, T. H., Lt. . . . 523, 1668
Mal Singh, Jemdr. 1114
Maltby, C. M., Capt. . . . 273, 1387
—— F. N., Capt. . 279, 1497, 1579
—— G., Lt. . . . 506, 1007, 1665
—— G B., 2nd-Lt. 622
Maluk Khan, Ressdr. 1682
—— **Singh**, Subdr.-Maj. . . .
——, Subdr. 1183
Malvern, J. F., Temp. Lt. . . . 742
Malyon, F. H., Capt. 55, 102, 265, 1102
Mam Chand, Risldr. . . . 888, 959
—— Raj, Jemdr. 1067
Maman Singh, Jemdr. . . . 1071
Mamraj, Jemdr. 880
—— Jemdr. 1433
—— Jemdr. 1412
—— Jemdr. 1059
Mama Thakur, Jemdr. . . . 1416
Man Singh, ... 1268
—— Subdr. 1581
Mana Khan, Hony. Capt. . . 2101
Manajirao Palande, Subdr. . . 1423
Mana Ram, Jemdr. . . . 1420, 1457
Mana Rawat, Jemdr. . . 1472, 1484
Manavazhi Madavhavan Menon, Jemdr. 1308, 1591
Manawar Khan, Jemdr. . . . 1408
Manbahadur Gurung, Jemdr. . 1602
—— Jemdr . 1611
—— **Limbu**, Jemdr. . . 1657
—— **Mal**, Jemdr. . . 1627
—— **Rai**, Subdr. . . 1648
—— **Thapa**, Subdr.-Maj. 1624
Manbhari Singh, Subdr. . . . 1062
Manbher, Jemdr. 1039
Manbir Gurung, Jemdr. . 1597, 1525
—— Subdr. 1657
—— **Limbu**, Subdr. . . . 1630
—— **Thapa**, Subdr.-Maj. . . 966
—— **Subdr.** 1602
—— **Jemdr.** 1619
—— **Jemdr.** 1604
—— **Subdr.** 1636
—— **Rana**, Jemdr. . . . 1597
—— **Jemdr.** 1598

Names and Rank.	Page.
Manbir Rana, Jamdr.	1596
Mandar Singh Thapa, Subdr.	1624
Mandasa, Zamindar of 2nd-Lt.	481
Mander, F. D., Maj.-Genl.	1415
——— T. J., 2nd-Lt.	565
Manders, H. C., 2nd-Lt.	805
Manderson, R. W., Capt.	253, 879
Mandeville, W., Lt.	318, 1387
Mandhle, Ty. Lt. M. B.	949, 1779
Mandhoj Burathoki, Jemdr	1637
——— **Gurung**, Jemdr.	1637, 1657
——— **Lama**, Subdr.	1648
——— Jemdr.	1631
——— **Rai**, Subdr.	1530
——— **Rana**, Jemdr.	1597
——— **Sen**, Jemdr.	1643
Mandlik, G. S., Tempy. Lt.	1773
Manduell, J. S., Lt.	77, 530, 1885
Maneckshaw, H. F., Ty. Capt.	1771
Mangal Dass, Jemdr.	1369
——— **Sen**, Tr. Vety. Asst.	1687
——— **Singh**, Ressaidar	1683
——— Subdr.	1145
——— Subdr.	1070
——— Subdr.	1132
——— Jemdr.	1212
——— Jemdr.	1241
——— Subdr.	1070
——— Subdr.	1157
——— Jemdr.	1372
——— Jemdr.	1112
——— Jemdr.	1252
——— Jemdr.	1228
——— Jemdr.	1114
——— Jemdr.	1032
——— Jemdr.	1284
——— Subdr.-Maj.	1162
——— Subdr.-Maj.	1235
——— Jemdr.	1001
——— Jemdr.	880
——— Jemdr.	966
——— **Limbu**, Subdr.	1648
——— Hony. Capt.	2101
——— Jemdr.	1368
——— Subdr.	1413
——— Jemdr.	892
——— Jemdr.	1533
——— Jemdr	1012
——— Ressdr.	914
——— Ty. Sub-Asst. Surgn.	1833
——— **Bisht**, Subdr.	1181
Mangapati Raju, Subdr.	1331
Mangat Singh, Jemdr.	1404
Manghat, Subdr.-Maj.	1391
Manghu Lal, Ty. Sub-Asst. Surgn	1833
Mangin, E. B., Capt.	55, 255, 1431
——— **F. M.**, Lt.-Col.	133, 1710
——— **P. C.**, Lt.	57, 487
——— **P. H.**, Lt.	509
Mangtu, Jemdr.	966
Manibhai Rambhai Desai, Sub-Asst. Surgn.	1832
Manico, E., 2nd-Lt.	838
Manifold, C. C., Maj.-Genl.	37, 119, 1737
Manik Ram, Subdr.	1453
Manikkam, Jemdr.	1332
——— C., Sub-Asst. Surgn.	1827
——— S., Sub-Asst. Surgn.	1826
Manikial Motilal Talati, Sub-Asst. Surgn.	1831
Manikoth Narayanan Nambair, Jemdr.	1328
Manir Shah, Jemdr.	1268
Maniraj Guraung, Jemdr.	1637
——— Rai, Jemdr.	1649
——— Thapa, Jemdr.	1637
Mani Ram, Jemdr.	979
——— Jemdr.	1122
Maniram Khattri, Jemdr.	1602
Manishanker Karunshanker Shukla, Sub-Asst. Surgn.	1832
Manisty, G. J. E., Capt.	264, 1283
——— G. S., Lt.	111
Manjit Gharti, Subdr.	1601
Mankad, N. P., Ty. Lt.	1781
Mankodi, K. K., Ty. Lt.	1778
Manley, E. N., **Maj.**	662, 1882
——— H., Lt.	506, 1671
——— R. T., Capt.	628
——— W. B., Lt.	492, 933
——— W. J., Lt.	401, 1104
Manlove, J. E. D., Lt.	375, 1255
Mann, A., 2nd-Lt.	539, 965
——— A. C. S., Asst. Surgn.	1800
Mann, A. E., Capt.	827
——— A. R., Lt.	624
——— D. B. U., Lt.	704
——— E. A. Capt.	58
——— E. W., Capt.	1717
——— G. C., Lt.	844
——— G. E., 2nd-Lt.	758
——— G K., Lt.	834
——— H. V., Sr. Asst. Surgn.	1780
——— J. R., Lt.	638
——— R. L., Lt.	583
——— S. H., Lt.	323, 460, 520
——— S. G., Lt.	720
——— W. H., Lt.	778
Mannering, G., 2nd-Lt.	731a
Manners, C. M., Maj.	1708
——— C. M. S., Capt.	256, 1419
——— R., Lt.	1695
——— R. H. H., Maj.	17, 33, 237, 1427
——— W. McK., Lt.	789
Mannes-Abbott, W. S. E., Lt.	543, 1199
Manning, C. L., 2nd-Lt.	752
——— E., Capt.	1571
——— J., Lt.	303, 1351
——— T. W., Lt.	1677
Manning-Press, C. A., 2nd-Lt.	1104
Mannock, R. C., Lt.	309, 1635
——— W. A., Lt.	309, 1647
Mannooch, G. H., Lt.	537
Manohar, Jemdr.	1040
——— Jemdr.	1192
——— **Lal Ghei**, Sub-Asst. Surgn.	1817
——— M., Lt.	1781
——— **Singh**, Subdr.	1320
Manook, J. S., Lt.	540a
Manowar Shah, Jemdr.	989
Manphul, Jemdr.	1028
Mansa Ram, Ressdr.	902
——— Risldr.	901
——— Jemdr.	888
——— **Singh**, Jemdr.	888
——— Jemdr.	1215
Mansab Ali Khan, Risldr.	1707
Mansar Ali, Subdr.	1412
Mansel, J., Sub-Condr.	1855
——— J. L., Capt.	583
Mansfield, C. B. H., Maj.	229, 889, 953
——— H., Maj.-Genl. *Sir*	1955
——— F. A., 2nd-Lt.	750
——— H., Sr. Asst. Surgn.	1789
——— L. G., 2nd-Lt.	1679
——— M. G., Lt	528, 1705
——— P. T., 2nd-Lt.	550, 1524
——— W. S., Lt.	589
Man Singh, Subdr.-Maj.	1110
——— Hony. Capt.	2101
——— Subdr.-Maj.	1136
——— Subdr.	1022
——— Bosnet, Jemdr.	1608
——— Subdr.	1607
——— **Gurung**, Subdr.	1602
——— **Rawat**, Subdr.	1223
——— Jemdr.	885
——— Mal., Jemdr.	1608
——— Thapa, Subdr.-Maj	1633
——— Jemdr.	990
——— Jemdr.	884
——— Jemdr.	944
——— Jemdr.	950
——— Jemdr.	936
——— Jemdr.	1571
——— Jemdr.	912
——— Jemdr.	1268
——— Jemdr.	1059
——— Subdr.-Maj.	1068
——— **Bisht**, Jemdr.	1182
Mansab Khan, Jemdr.	1088
Manser, A. R., 2nd-Lt.	783, 791
Mansha Singh, Subdr.	1288
——— Jemdr.	1083
——— Jemdr.	1121
——— Jemdr.	1163
——— Subdr.	1287
——— Jemdr.	1493
Manson, C. E. F., Lt.	520, 1671
——— D., Ty. Capt.	1724
——— F. P., 2nd-Lt.	818
——— J. A. F., Maj.	28, 60, 1210, 1661
——— R. G., 2nd-Lt.	857
——— T. W., 2nd-Lt.	1761
——— W. A., 2nd-Lt.	578
——— W. M., 2nd-Lt.	359, 1333
Manstu Singh, Jemdr.	1011
Mansur Khan, Hony. Capt.	2104
Manthel, W. J., Lt.	351, 1030
Mantle, W. E., 2nd-Lt.	1034
Manton, L., Capt.	60, 103, 665, 1881
Manuk, T. M., Lt.	295a, 917
Manulla, Jemdr.	1480
——— Subdr.	1432
Manzur Ahmad Khan.	1035
Maola Baksh Khan, Subdr.	1409
Mapp, C. L., Lt.	507
Mappin, J., Lt.	624
Maqsud Ali Khan, Jemdr.	936
March, A. M., Sub-Condr.	1856
——— G. J., 2nd-Lt.	710
——— J. C., Lt.	514, 1618
Marchant, C. H., Asst. Surgn.	1797
——— D. R., Lt.	514, 1483, 1583
——— F. O., Capt.	685
——— G., Maj.	90, 215, 919, 957
——— H. L. MacL., 2nd-Lt.	563
——— J. D., Sub. Cond.	1857
——— H. L. M., 2nd-Lt.	1704
——— L. B., 2nd-Lt.	551, 1678
——— T. H. S., Maj.	591
——— T. J., Asst. Surgn.	1799
Mardall, F. H., Capt.	272, 1187
Mardan Ali, Subdr.	1247
——— Ali, Jemdr.	1376
——— Jemdr.	1488
——— Jemdr.	1304
——— Khan, Subdr.	1096
——— Shah, Jemdr.	1681
Marden, J., 2nd-Lt.	558, 1321
Mare, E. J., Lt.	90, 513, 746
Margerison, L. A., 2nd-Lt.	992
Maria Cruz, Subdr.-Maj.	993
Mariadas, M. S., Sub-Asst. Surgn.	1828
Mariasusai, Jemdr.	994
Marigold, A. J., 2nd-Lt.	473, 1074
Mari Muthu, Jemdr.	1316
Marindin, J. F., Lt.	297a, 1617
Mariyan, Jemdr.	994
——— Jemdr.	994
Marjad Singh, Jemdr.	1386
Marjoribanks, J. L., Lt.-Col.	137, 1745
——— R. D., Maj.	54, 99, 217, 1431
——— S. D., 2nd-Lt.	293a, 947
Marjoribanks-Marcar, E., tempy. Capt.	1723
Markasagayam, Jemdr.	994
Markoy, F. E., Revd.	1903
Marketis, W. P., Hony. Lt.	1850
Markham, J. H., Capt.	1732
——— R. W., Capt.	761, 1732
Markiew, E. G., 2nd-Lt.	1217
Marks, E. B. M. F., Asst. Surgn.	1803
——— F. C., Ty. Capt.	7, 571, 1850
——— G. A., Lt.	544, 1385
——— R. M., Lt.	516a, 1128
Marksman, L. B. C., Capt.	1717
Markwick, H. W., Capt.	9, 261, 1664
——— W. L., Lt.	695, 1886
Marley, R. W., Lt.	797
——— V. D. K., Capt.	758
——— W. H., Lt.	514
Marklew, E. G., 2nd-Lt.	554
Marlow, B. W., Col.	187, 123, 149, 181
——— T., Lt.	536, 1647
——— W., 2nd-Lt.	1581
Marmion, W. J., Capt.	1732
Marner, T., Sub-Condr.	1854
Maroti Fartare, Jemdr.	1444
——— **Ferande**, Subdr.	1439
——— **Ghorpade**, Subdr.	1509
——— **Jadhao**, Jemdr.	999
——— **Jadhow**, Jemdr.	1480
——— **Rao Chawan**, Jemdr.	1440
Marple, R. G., Lt.	1289
Mar. C. F., Maj.	79, 1748
——— H. R., Lt.	1273
Marrable, E. D., Lt.	638
Marrian, A G., Lt.	695
Marriette, F. W., Lt.	597, 598
Marriott, A. E., Lt.	520, 1667
——— A. S., Maj.	24, 81, 213
——— B. M., Lt.	541
——— C. F., Capt.	276, 901
——— E. F., Lt.	487, 897
——— E. M. L., Col.	169, 1916
——— F. C., Lt.	637
——— J. H. G., Maj.	115, 223, 1066, 1151
——— J. R., Lt.	320, 1403
——— K. M. H., 2nd-Lt.	723a
——— L. J., Lt.	540, 857
——— N. E., Capt.	280-287, 919, 921
——— R. E., Lt.	511, 840, 1001

NAMES AND RANK. Page.

Marriott, S. A., 2nd-Lt. 758
—— W. G., Lt. 58
Marris, S. P. Revd. 826
Marrisami. Jemdr. 1284
Marrs, F. W., 2nd-Lt. . . . 554, 849
Marryat, J. R., Capt. . . 664, 1882
—— N., Capt. 825
Marsden, A. P., Capt. 831
—— G., 2nd-Lt. 567
—— R. T., Lt.-Col. . 132, 643
—— W. G., tempy. Capt. . 1725
—— P., 2nd-Lt. . . 494, 891
Marsh, B. C., Lt. 1287
—— C. C. A., Lt. . 303, 516, 1211
—— C. H., Maj. 219, 868, 909 1060
—— C. H., Lt. 527
—— C. H., Lt. 965
—— C. K., Lt. 295
—— B. C., Lt. . . . 375, 1287
—— E. W., tempy. Capt. . . 1768
—— F. C., 2nd-Lt. . . 328, 1217
—— F. G., Maj. 96, 98, 105, 136, 158, 211, 1641
—— F. H., Ty. 2nd-Lt. . . . 748
—— F. H. R., Maj.-Genl. 169, 1916
—— G. F., Lt. 706
—— G. P., Lt. 723
—— G. S., 2nd- Lt. . . 545, 998
—— H. R., 2nd-Lt. 828
—— H. W., Sub-Condr. . . . 1857
—— J. J., 2nd-Lt. . . 474, 1048
—— N. O. C., Lt. . . 487, 919
—— R. B., 2nd-Lt. 765
—— R. H., Maj. . . 221, 909
—— T., Lt. . . . 72, 629
—— T., Sub-Condr. 1870
—— W. R., *Hony.* Capt. . . 75
Marshal, R. D., 2nd-Lt. . . . 835
Marshall, A., Maj. 13, 42, 239, 927
—— A., 2nd-Lt. 619
—— A., *Esq.* 72
—— A. D., Lt. . . 519, 1708
—— A. E., 2nd-Lt. 563
—— A. T., 2nd-Lt. 732
—— A. W., 2nd-Lt. 835
—— C., Lt. . . . 318, 945
—— C. H. T., Col. . 177, 1917
—— D. F., Lt. . . 520, 1696
—— E. C., 2nd-Lt. . . 539, 1672
—— E. N. C., 2nd-Lt. . 669 991
—— E. O., Nurg. Sister . . 1735
—— F., Lt. . . . 523, 992
—— F. H., Maj. . . 101, 614
—— F. S., 2nd-Lt. . . 720, 786
—— G. W., Maj. . . . 4, 806
—— H., 2nd-Lt. 1008
—— H. D., Maj 602
—— H. F., Ty. Lt. . . . 1729
—— H. H., Capt. . 274, 1215
—— H. G., Capt. . 17, 21, 725
—— H. J. M., Brig.-Genl. 58, 121, 131, 661
—— J., Lt. 56, 489
—— J. D., Capt. 807
—— J. D. C., Lt. . . 311a, 1178
—— J. G., Lt. . . . 516a, 998
—— J. S. B., -Maj 14, 43, 254, 1162
—— J. T., Lt. 714
—— K. R., Lt. 740
—— N. E., 2nd-Lt. 803
—— N. E., Revd. 1904
—— P. H., Ty. Capt. 6, 14, 571, 1850
—— R., Maj. . . . 733, 791
—— R. H., Lt. . . . 440, 881
—— R. W., 2nd-Lt. 765
—— T. D., Capt. . . 750, 752
—— T. R., 2nd-Lt. . 316, 947 961
—— W. D., Lt. . . . 420, 1415
—— W. D. L., Lt. 111
—— W. H., Capt. 685
—— W. H., Capt. 1724
—— W. J., Sr. Asst. Surgn. 1791
—— W. M. C., Capt. . 501, 1497
—— W. R., Maj.-Genl. *Sir* . 118
—— W. S., Esq. 91
Marsham, R. H. P., Lt. . . . 692
Marsland K. D., Lt. 1056
Marston, E. N., Lt. . . . 523, 1471
Martand Rao Mohite, Subr.-Maj. 1407
Martelli, C. I. H., 2nd-Lt. . . . 826
—— N. C., Lt.-Col. . 179, 1917
Marten, S. A., 2nd-Lt. . . . 1673
Martin, A., Asst. Surgn. . . . 1758
—— A., Lt. 532, 1311
—— A. D. DeB., Capt. . . . 665
Martin, A. E., Asst. Surgn. . . 1801
—— A. J., Lt. 746
—— A. R., Lt.-Genl. *Sir* . . 1955
—— A. S. W., Lt. . 84, 538, 1178
—— A. W., 2nd-Lt. . . 658, 798
—— C. A., Asst. Surgn. . . 1802
—— C. de C., Capt. . . . 1758
—— C. J. S., 2nd-Lt. . . . 362
—— D. D., 2nd-Lt. . . 474, 1604
—— E. B., Lt. . . . 511, 1665
—— E. H. Hony. Capt. . . . 1850
—— E. H., deC., Lt. 292, 1455, 1475
—— E. L., 2nd-Lt. . . . 90, 683
—— E. T., Capt. 615
—— F., Maj. 205
—— F. B., Lt. 836
—— F. E., 2nd-Lt. 753
—— F. G., Asst. Surgn. . . 1805
—— F. R., Lt. 656
—— F. W., Revd. 1903
—— F. W., Revd. 1905
—— G., Lt. 695
—— G. F., 2nd-Lt. 1114
—— G. D., Maj. . . . 242, 1050
—— G. H., 2nd-Lt. 584
—— G. N., 2nd-Lt. 705
—— G. N. C., Capt. 615
—— G. St. J., Lt. . . . 435, 1505
—— H., Lt. 541, 1439
—— H. C., Revd. 1907
—— H. E. L. 1009
—— H. G., Maj. 615
—— H. J., Lt. . . . 312, 1411
—— H. J., 2nd-Lt. 780
—— H. L., 2nd-Lt. 1281
—— H. S., Capt. 722
—— H. T., Capt. 857
—— J., 2nd-Lt. 363
—— J., Lt. . . 90, 509, 776, 1487
—— J., Lt. 375, 1489
—— J., Maj. . . . 224, 870, 1383
—— J. A., Lt. 1720
—— J. E. L., 2nd-Lt. 555
—— J. G., Lt. 656
—— J. H. A., Asst. Surgn. . 1798
—— J. H. A., Lt. . . . 353, 977
—— J. M., Capt. . . . 668, 986
—— J. S. S., Capt. 1757
—— L. J., Lt. 826
—— M., Sub-Condr. 1863
—— N. M., Capt. . . . 269, 947
—— O. M., Lt. . . . 490, 901
—— P. B., Maj. . . . 242, 1339
—— R., Maj. 791
—— R., Lt. 626
—— R. F., Lt. 505
—— R. H., Col. . . . 32, 693
—— R. H., Lt. 843
—— R. Mc. G., Lt. . . . 47, 710
—— R. V., Capt. 1758
—— S. A., 2nd-Lt. 539
—— S. F., Lt. 317, 1435, 1441, 1565
—— T., Lt. 517 1517
—— T. S. S., Capt. . . 486, 833
—— V. J., Lt. . . 389, 777, 1114
—— W. E., Lt. 748
—— W. D., Lt. 493
—— W. H., Lt. 68, 797
—— W. J., 2nd-Lt. 626
—— W. J. P., Sr. Asst. Surgn. 1790
—— W. J. R., Lt. . . . 375, 1225
—— W. M., Lt. . . . 311, 947
—— W. S., Asst. Surgn. . . 1798
Martindale, C. S. deB., Maj. . . 57
—— H., Revd. 1907
—— J. B., Lt. 748
Martyn, J. K., Lt. . . 508, 997, 1701
—— R. O. F., Lt. 1160
—— W. W., Lt. . . . 507, 1463
Martyr, H., Lt. 463, 1641
Marwood G., 2nd-Lt. 630
—— S. L., Capt. . . . 502, 1004
Maru, Jemdr. 1039
—— Jemdr. 1040
—— Subdr. 1068
—— **Ram,** Subdr. 1412
Maruti Rao Ghorse, Jemdr. . . 934
Marutirao Kadam, Jemdr. . . 1468
—— Dhmual, Jemdr. . . 1469
—— Ghag, Jemdr. . . . 1468
Marutiraoji Vipat, Sub-Asst. Surgn. 1832
Mascall, F. E., Major 614
—— M. E., Maj. 634
Mascarenhas, C. J., Asst. Surgn. 1800
—— R., Asst. Surgn. . . 1803
—— V. E., Ty. Lt. . . . 1775
Mashiter, E., Lt. 724
Masilamoni, Jemdr. 1328
Masina, P. M., Ty. Lt. 1773
Mason, A., Capt. 666, 985
—— A., 2nd-Lt. 712
—— A. E., Lt. 745
—— A. S., Capt. . 16, 24, 698, 795
—— C. A., Capt. 1717
—— C. D. T., Revd. . . . 1907
—— C. H., Asst. Surgn. . . 1795
—— C. M., Lt. 699
—— D. R. W., Lt. . . . 306, 1020
—— E. W. R., 2nd-Lt. . . . 1657
—— G. H., Asst. Surgn. . . 1800
—— G. K. M., Capt. 593
—— H., 2nd-Lt. 679
—— H. C., Lt. . . . 58, 530
—— H. C. E., Lt. 1465
—— H. M., Maj.-Genl. . . . 931
—— H. St. J., Asst. Surgn. . 1802
—— J. C., 2nd-Lt. 361
—— J. L., 2nd-Lt. 331
—— J. S., 2nd-Lt. . . . 360, 1405
—— K., Bt.-Maj. . . 52, 665, 1880
—— N., Esq. 88
—— P. J., Condr. 1861
—— R. D., Asst. Surgn. . . 1801
—— S., 2nd-Lt. 637
—— T. K., 2nd-Lt. 669
—— T. W., 2nd-Lt. 549
—— W., Maj. 602
—— W. F., Capt. 1718
Mason-MacFarlane, F. N., Capt. 7, 615
Mass, L., Lt. 1143a
Massey, A. P., Lt. 354, 1052
—— J. W., Capt. . . . 25, 706
Massey-Dawson, K. H., 2nd-Lt. .
Massie, R. H., Col. 126, 632
Masson, J., Maj. 1747
——, K., Lt. 1720
Massy, D. F., Capt. . 276, 578, 937
—— F. S., Maj. . 230, 1613, 1627
—— I. P., Lt. . . 298a, 1263, 1331
—— S. D., Maj. . . . 230, 1136
—— W. I. J., Maj. 239, 1066, 1297, 1895
Mastan Singh, Subdr. (I) . . 1335
—— Subdr. (II) . . 1336
—— Jemdr. 1167
—— Jemdr. 1079
—— Jemdr. 916
—— Jemdr. 1075
Master, A., Capt. . . . 560, 825
—— F. S., Tempy. Capt. . . 1765
—— **K. S.,** Tempy. Lt. . . 1765
Masterman, C. H., Lt. 492
—— G. V., Lt. 630
Masters, A. M., Tempy. Capt. . 1724
—— E., Capt. . . 268, 1435, 1583
—— H. F., Lt. . . . 516, 1667
—— J., Maj. 241, 1082
—— N. F., Ty. Lt. 1772
—— O., Capt. . . . 260, 1403
Masterson, W. J., Sr. Asst. Surgn. 1790
Mastu Sing Gusain, Jemd. . . 1182
—— Rawat, Jemdr. . . 1182
Masum Shah, Jemdr. 1336
Masurekar, V. R., Ty. Lt. . . 1776
Mataram, Jemdr. 1456
Matabakhsh Singh, Jemdr. . . 1043
Matadin Tiwari, Subdr. . . . 1026
—— Singh, Jemdr. . . 1581
—— Jemdr. . . 1143
—— Quananjaya, Sub-Asst. Surgn. . 1820
Matan Singh, Jemdr 1043
Matapalt Sing, Subdr. . . . 1062
Matayi Govindan, Sub-Asst. Surgn. 1828
Matbar Singh Thapa, *Hony.* Capt. 2102
Matc ham, 2nd-Lt. 1439
Mathai, P. M., Ty. Lt. . . . .
Mathajone, M., 2nd-Lt. . . 550, 1674
Mather, J. J., Lt. 522, 1070
—— O., Capt. . . 268, 1435, 1439
—— O. H., Lt. . . . 505, 1674
—— R. W., 2nd-Lt. . . 330, 1459
—— W. M., 2nd-Lt. 494
Matheros, F. C., 2nd-Lt. . . . 1009
Matheson, B. H., Capt. . . 272, 1239
—— M., 2nd-Lt. . . . 686, 784
—— S. F., 2nd-Lt. 987
Mathew, C. M., Lt.-Col. 107, 132, 1743
—— H. B., 2nd-Lt. 544
—— J. D., Condr. 1867
—— K. M., Tempy. Lt. . . . 1778

NAMES AND RANK. Page.

Mathew, W. C., Lt. . . . 516, 1066
—— Jemdr. 1300
Mathewes, J. R., Col. 54, 123, 149, 182
Mathews, A. E., Asst. Surgn. . 1798
—— A. 2nd-Lt. 568
—— A. H. Lt. 463
—— C. W., Lt. . . 668, 992
—— E., 2nd-Lt. 568
—— F., Sub-Condr. . . . 1864
—— F. W., Asst. Surgn. 1792
—— G. L., Lt. 1460-62
—— J. C., Revd. 1907
—— J. S., 2nd-Lt. . . . 553
—— L. D., Lt 698
—— S. L., Lt. 935
—— T. C., Lt. . . . 1591, 1780
—— V., Lt. 536
—— V. de R., Lt. 1455
—— V. G., Asst. Surgn. . 1795
—— W. A., 2nd-Lt. . . . 681
—— W. F., Maj. . 663, 1882
—— W. H., 2nd-Lt. . . . 941
Mathias, L. J., Lt.-Col. . . 40, 1660
—— L. W. H., Capt. 272, 509, 1667
Mathic, W. A. S., 2nd-Lt. . 1. . 556
Mathulla, M. M., Ty. Lt. . . . 1777
Mathur, H. G. D., Ty. Lt. . . . 1773
Mathura Das, Sub-Asst. Surgn. 1812
—— Jemdr. 1035
—— Parshad Sotl, Sub-Asst. Surgn. . 1819
—— Singh, Jemdr. . . . 1043
—— *Hony.* Capt. . 2102
—— Subdr. . . . 1046
—— Jemdr. . . . 1404
Matiullah Khan, Jemdr. . . . 894
MatlabKhan, Ty. Sub-Asst. Surgn. 1833
Matley, C. A., Esq. 988
Matlocks, F. S., Lt. 39
Mat Ram, Jemdr. 888
Matson, H. S., Maj. 1751
—— J., Maj. . . . 81, 225
Mattey, A. G., 2nd-Lt. . . . 640
Matthai, P. M., Tempy. Capt. 1365, 1767
Matthewman, A. S., Lt. . 295a, 1383
—— G. B., Lt. 295a, 1455, 1585
Matthews, A., 2nd-Lt. . . . 8, 567
—— A. B., Capt. . . 666, 996
—— A. G. T., Ty. Asst. Surgn. 1804
—— B. F., Capt. . . 19, 501a
—— C., 2nd-Lt. 711
—— C. J., Lt.-Col. . . . 729
—— C. R., Col. 176
—— E. A. C., Maj. . . 3, 1746
—— E. G. A., Sr. Asst. Surgn. 1790
—— F., Lt. 737
—— F. C., 2nd-Lt. . . . 552
—— F. J., Lt. . . 321, 1604
—— F. M., Capt. 40, 61, 261, 1166
—— G., 2nd-Lt. 851
—— G. L., 2nd-Lt. . . . 1243
—— H., Capt. 1704
—— H., Lt. . . . . . .
—— J., Sub-Condr. . . . 1856
—— J. G. T., Sr. Asst. Surgn. 1790
—— J. M., Lt. . . 506, 1667
—— L. G., Lt. . . 401, 1052
—— R., Lt. . . . 397, 949
—— R. C., Maj. 1838
—— R. W. Y., Lt. . . . 629
—— S. A., 2nd-Lt. . . . 640
—— S. F. W., Lt. . . 353, 1162
—— S. L., Lt. 488
—— V., Lt. 542
—— V. M., Asst. Surgn. . . 1802
—— W. H., 2nd-Lt. . . . 494
Mattey, A. G., Capt. 642
Mattock, F. S. Lt. 39
—— L. A., Lt. 746
Mattocks, F., 2nd-Lt. 666
Maturin, H. G., Capt. . . 105, 108, 255, 1215, 1263
—— F. H., Lt.-Col. . . . 1722
Matwali, Subdr. 1379
Maud, W. M., Capt. 679
Maude, E. A., Maj. . . . 226, 1124
—— E. W., Lt. . . 293a, 1249
—— G. W. E., Capt. . . . 712
—— H., Sub-Condr. . . . 1874
Maudult, R. F. S., Lt. . . . 583
Maughan, N., Lt. 654
—— W. C., 2nd-Lt. . . . 1373
Mauji, Subdr. 1395
Maukad, N. P. L. 1782
Maula Bakhsh, Subdr. 1095
—— Jemdr. 1142
—— Sub-Asst. Surgn. . 181

NAMES AND RANK. Page.

Maula, Dad, Subdr. 1303
—— Khan, Jemdr. 898
—— Jemdr. 912
—— Subdr. 1321
—— Jemdr. 1569
—— Jemdr. 1121
—— Jemdr. 1457
—— Khan, Jemdr. . . . 1681
Maule, H. N. St. J., Lt.-Col. 58, 101, 133, 613
—— J. H., Condr. 1861
Mauleverer, W. M. M. G., Lt. 740, 1006
Maulvi Muazzim Husein Mohomed Furrokh, 2nd-Lt. 482
Maumesar Gharti, Jemdr. . . . 970
Maun, J. W., 2nd-Lt. 668
Maund, J. A. H., 2nd-Lt. . . . 328
Maunder, G. H., Lt. . . . 397, 1433
—— G. H. O., 2nd-Lt. . . 1058
—— R. A., 2nd-Lt. . 34, 78, 491
Maung Aung Bwin, Subdr. . . 1894
—— Ba Gyan, Jemdr. . . . 1291
—— Jemdr. . . . 994
—— Ba Pe, Subdr. 1347
—— Ba Sein, Jemdr. . . . 1291
—— Ba Than, Jemdr. . . . 1191
—— Ba Tin, 2nd-Lt. . . 84, 481
—— Ba Zan, Jemdr. . . . 1291
—— Gyee 1297
—— Gyine, Jemdr. 993
—— Khin, Jemdr. 1297
—— Kin Maung, Jemdr. . . 1291
—— Lu Oh Jemdr. 1698
—— Lugloy, Jemdr. 994
—— Nee, Jemdr. 1297
—— Nyein, Jemdr. 1295
—— Po Shwe, Jemdr. . . . 1297
—— Po Thaw, Jemdr. . . . 994
—— Pu, Jemdr. 1297
—— San Mya Jemdr. . . . 1291
—— Sao Ohu, Jemdr. . . . 1297
—— Shwe, Jemdr. 1297
—— U., Subdr. 1291
—— Tin, Jemdr. 1291
—— Tit, Jemdr. 1347
—— Tun Aung 1297
—— Tun Gyaw, Jemdr. . . . 129
—— Zan Kin, Jemdr. . . . 994
Maungpe, 2nd-Lt. . . . 481, 1695
Maunsell, A. P. G., Revd. . . 1904
—— E. B., Maj. . 214, 941, 943
—— E. C. C., Maj. . . 38, 1749
—— I. O'G., Capt. 252, 1365, 1371
Maurice, J. O. F., 2nd-Lt. . 522, 1341
Mauulla, Jemdr. 1480
Maver, A. M., 2nd-Lt. 793
Mawaz Din, Jemdr. 1612
Mawaz Khan, Subdr. Maj. . . 1120
—— Subdr. 1407
—— Subdr. 1124
—— Jemdr. 1502
Mawdesley, J. L., Capt. . . . 75
Mawe, M. D., Rev. 1908
Mawer, E. S., 2nd-Lt. 623
Mawhood, J. C., 2nd-Lt. . . 332, 964
Mawhinny, B. J. W., Lt.-Col. 22, 130, 1710
Mawson, G. H., Lt. . . . 668, 1509
—— L. R., 2nd-Lt. 567
Maxted, 2nd-Lt. F. S. 885
Maxwell, A., Capt. 615
—— A. G., Maj. 59
—— B. C. D'A., Tempy. Maj. 9, 668
—— C. W., 2nd-Lt. . . 491, 1577
—— D. W., Maj. 25, 99, 209, 869 1225, 1613
—— F. S., 2nd-Lt. . . 365, 909
—— F. W., Maj. 8
—— G. W., Col. . 122, 153, 180
—— H., 2nd-Lt. 552
—— H. B. Lt. . . . 490, 915
—— H. G., Lt.-Col. 130, 155, 190
—— H. St. P., Lt.-Col. . 179, 1917
—— J. F., Lt. . . . 487, 913
—— J. G., Genl. Sir . . . 743
—— J. McC., Col. . . 128, 612
—— L. L., Bt.-Col. 66, 121, 127, 153, 193
—— M. T., Lt. 624
—— P. A., Maj. . 61, 238, 1026
—— R., Lt. 509, 1309
—— R. M., Lt. . . . 487, 911
—— W. E., 2nd-Lt. . 327, 1505
—— W. F., Maj. . . 663, 1003
—— W. R., 2nd-Lt. . 18, 23, 790
May, A. H., 2nd-Lt. 630

NAMES AND RANK. Page.

May, A. J., 2nd-Lt. 825
—— B. H., Lt. 505
—— C. D., Capt. . 264, 1034, 1664
—— C. R. A., 2nd-Lt. . . 329, 1249
—— E., Lt. 53, 749
—— F. W., Capt. . . . 486, 911
—— H. R., Lt. 731a
—— H. V., Lt. 1689
—— L. H., 2nd-Lt. 1211
Maya Ram, Jemdr. 883
—— Jemdr. 904
Maybury, M. A. K., 2nd-Lt. 565, 1702
Maycock, H. W., 2nd-Lt. . . . 1263
Mayes, A. T., Hony. Lt. . . . 807
—— F. A., 2nd-Lt. . . 540, 1441
May-Somerville, J. A. C., Capt. 277, 895, 957
Maydwell, H. S. L., Capt. . . . 252
Mayer, C. A. L., Tempy. Lt. . . 1707
—— G. H., Lt. . . . 533, 1707
Mayers, E. W., Lt. 544
—— H. S., 2nd-Lt. 567
—— N., 2nd-Lt. 731, 791
—— N. G., Lt. 1471
—— W. E., 2nd-Lt. 1293
Mayhew, A. G., 2nd-Lt. 567
Mayles, R. W. J., Lt. . . 528, 1671
Maynard, B. M., the Revd. . . 1904
—— C. I. F., Lt. . . 293, 1102
—— F. H., Maj. 105, 237, 1142, 1497
—— F. P., Lt.-Col. . 127, 1740
—— N. C., 2nd-Lt. . . 550, 1677
—— P. A., Capt 17, 33, 685, 783
—— K., 2nd-Lt. 1614
—— R. deK., Lt. 846
Mayne, A. G. O. M., Capt. . 268, 899
—— C. E. M., Maj. . . . 78, 218
—— P., Lt. 410
Mayo, A. R., Sub-Condr. . . . 1856
—— G. L. L., Maj . . . 9, 226, 1622
—— H. R., Ty. Capt. 1725
Mazagonwala, E. B., Ty. Lt. . . 1775
Mazal Singh, Jemdr. 1024
Mazumdar, D. C., Lt. 1782
—— I. B., Lt. 1781
Mazhar Ali Khan, Jemdr. . . . 938
Meaburn, M., 2nd-Lt. 619
Mead, A. W., 2nd-Lt. 778
—— B., Lt. 733, 1007
—— H. R., Col. . . 36, 123, 149, 182
—— H. R. C., Maj. 27
—— P., Lt. 637
—— R., Lt. 55, 729
Meade, H. R. C., Capt. 272, 1223, 1635
—— G. A., Maj. 852
—— J. W., Capt. 745
—— J. W. B., Lt.-Col. . . 57, 183
—— M. E. M., Capt. . 69, 268, 1497
—— P. A., Capt. . . . 277, 1451
—— W. L., Maj. . . . 241, 1387, 1894
Meadows, C. A. G. P., Maj 13, 218, 868, 1042, 1044
—— D., Capt. . 62, 571a 1841
—— G. S., Maj. 722
Meadmore, C. R. A., 2nd-Lt. 780, 1596
—— T. B. R., Lt. 460-62, 1143
Meager, C. J., Lt. . . . 528, 1501
Meakins, H. W., Lt. 817
Meares, G. C. K., Lt. . . 527, 1671
—— G. W., Capt. 836
—— J. W., Lt. 806
—— W. K., Lt. 526, 1667
Mearns, H. G., 2nd-Lt 850
—— J. A., Ty. Lt. 1726
—— J. P., Lt. . . . 321, 1020
—— W. A., Maj. 1752
Mears, A., Lt.-Col. 105, 193, 1880, 1912
—— C. D., Lt.-Col. 135, 158, 198, 867, 905
—— H., Lt. 506, 1106
Meates, W. W., 2nd-Lt. 361
Meatyard, W. E., Lt. . . . 78, 492
Mecredy, C. C., Capt. 1759
—— G. W., 2nd-Lt. 557
—— H., Maj. . . . 74, 101, 614
—— W. H., 2nd-Lt. . . 556, 1703
Medill, P. M., Maj. 634
Medland, J. W., Esq. 93
Medley, A. G., Lt.-Col. 57, 105, 128, 154, 184
—— A. F., 2nd-Lt. 360
—— F. J., Lt. 534, 592
—— N. R., Maj. 825
Medlicott, H. E., Maj. . 54, 235, 879
—— S. J. E., 2nd-Lt. . . . 1417

Names and Rank.	Page.
Medwell, P. R., Lt. . 493, 889,	941
Mee, E. A., Lt.	470
Meehan, G. F., Lt. . 423,	1193
—— M., Lt.	531
Meek, A. S., Maj. . 243, 1893,	1913
Meeks, F. S., Lt.	1233
Meers, Lt., P. J. . 420,	1289
Meese, W. A. B., Lt. . 77,	1885
Megaw, J. W. D., Maj.	1746
Meggett, H. A., 2nd-Lt.	1289
Megraj Limbu, Jemdr.	1633
Mehar Ali Khan, Jemdr.	1682
—— Baksh Subdr.	1111
—— Singh, Subdr.-Maj.	1187
—— —— Ressdr.	932
—— —— Subdr.	1355
—— —— Jemdr.	1175
—— —— Jemdr. . 912,	1142
—— —— Subdr.	
—— —— Subdr.	1211
—— —— Jemdr.	1157
—— —— Subdr.	1106
—— —— Khattri, Subdr.-Maj.	1642
—— Singh Chawal, Ty. Sub-Asst. Surgn.	1833
—— Khan, Jemdr.	950
Meharman Thapa, Jemdr.	1625
Mehdi Hasan, Subdr.	1387
—— Khan, Subdr.	1495
—— —— Ressdr. . 1682,	1684
Meher Singh, Subdr.	1698
Mehew, R. S., 2nd-Lt.	1225
Mehnga Singh, Jemdr.	1150
Mehra, A. C., Tempy. Lt.	1769
Mehrab Khan, Jemdr. . 1488,	1579
Mehr Ali Khan, Resldr.	1684
—— Chand, Jemdr.	1124
—— —— Subdr.	1369
—— —— Jemdr.	1389
—— —— Subdr.-Maj.	1136
—— —— Singh, Jemdr.	1158
—— Ilahi Jemdr.	1507
—— Khan, Jemdr. . 977,	1495
—— —— Jemdr.	1569
—— Muhammad, Jemdr.	1232
—— Singh, Jemdr.	1502
—— —— Jemdr.	1158
—— —— Subdr.	1367
—— —— Jemdr.	900
—— Mohammad Khan Bahadur, of Maler Kotla, Lt.	111
Mehrban Sing Butola, Subdr.	1179
Mehta, G. D., Tempy. Capt.	1768
—— K. K., Tempy. Capt.	1765
—— N. B., Tempy. Capt.	1765
—— N. D., Ty. Lt.	1776
—— N. M., Capt.	1757
—— S. B., Capt. . 108,	1753
—— Abinasi Ram, Sub.-Asst. Surgn.	1818
—— Gyan Singh.	892
Mehtab Din, Jemdr.	1133
Mela Singh, Jemdr.	1217
Melck, G. H., Lt.	1839
Melkle, H. G. W., Capt.	806
—— D. M., 2nd-Lt.	557
—— J. A. S., 2nd-Lt.	742
—— R., Capt.	733
—— W. A., 2nd-Lt. . 472,	1243
—— W. Q. R., Lt.	986
Meiklejohn, J. F., Capt. 9, 16, 271,	937
—— R., 2nd-Lt.	566
—— W., 2nd-Lt. . 546,	1645
—— W. L. S., Maj 99, 105, 216, 870, 1385,	1427
Mein, D. B., Capt. . 271,	1243
—— J. E., Bt.-Col.	56
—— J. J., Lt.	622
Melhuish, H. M. H., Bt.-Lt.-Col. 137,	1750
—— H., Lt. . 450-54	1018
Mell, F. O. N., Lt.-Col. . 135,	1744
Meller, F. R., Lt. . 489,	875
Mellersh-Jackson, L., Capt. 55,	753
—— W., Capt. . 753,	785
Mellis, A. R., Capt.	500
—— J. H. S., 2nd-Lt. . 356,	1674
Mellish, A., Lt. . 506,	1666
Melliss, C. J., Maj.-Gen., *Sir* 50, 105, 118, 146, 173,	1925
—— H, Col. *Sir* . 178, 1918,	1955
Mellor, A., Maj.	614
—— F. R., Lt. . 490,	875
—— G. F., Maj. . 26, 81,	231
—— J. E. P., Lt.	733
—— J. R., Lt. . 446-49	1533
Mellor, J. S., Capt.	583
—— J. S. P., Lt.	705
—— L. A. St. J., Lt. 314, 1108,	1399
—— R. W., Capt. . 652,	1886
Meloney, J. C., 2nd-Lt.	1411
Melvill, T. P., Maj	595
Melville, A. M., Lt.	1369
—— C. W. F., Maj	1747
—— D. W. L., 2nd-Lt.	597
—— E. C., 2nd-Lt.	787
—— E. P. A., Capt.	1910
—— H. B., Lt.-Col. . 128, 1740,	1890
—— H. J., Lt. . 296,	917
—— J. S., Bt.-Col.	76
Melvin, J. W., Lt. . 540,	1677
—— W. G., Lt. . 524,	938
Mendes, M., 2nd-Lt. . 546,	1415
Mendi Khan, Jemdr.	1130
Menois, F. X., Asst. Surgn.	1793
—— J. C., Asst. Surgn.	1801
Mendoze, A. L., 2nd-Lt.	363
Mendu, Subdr.	1197
Meneaud, C. J. A., Sr. Asst. Surgn.	1790
Menezes, J. F., Sr. Asst. Surgn.	1791
—— J. S., Asst. Surgn.	1800
—— L. D. C., Asst. Surgn.	1801
Menga Sing, Jemdr.	1217
Menhinick, J. S., Lt.	679
Mennie, J. O., Col. 6, 13, 124, 150,	184
Menon, A. K., Ty. Lt.	1777
—— C. K., Ty. Lt.	1778
—— C. G., Lt. . 1295,	1777
—— C. V. A., Capt.	1767
—— E. K., Ty. Lt.	1774
—— E. K., Ty. Lt.	1778
—— K. G., Ty. Lt.	1777
—— K. P., Ty. Lt.	1778
—— M. P. K., Ty. Capt.	1765
—— M. V., Ty. Capt.	1770
—— P. B., Ty. Lt.	1777
—— P. K. K., Sub-Asst. Surgn.	1828
—— T. K., Ty. Capt.	1765
—— V. C. G., Ty. Lt.	1776
—— V. K. K., Ty. Lt.	1776
Meny-Gibert, M. A., 2nd-Lt.	793
Menzies, F. A., Capt. 9, 18, 502, 835,	986
—— H., Revd. . 658,	1903
—— T., Lt.	1717
—— W. F., Lt. . 294*a*,	887
Meo Singh, Jemdr.	1396
Meral Singh, Jemdr.	1083
Mercer, A. A., Lt.	929
—— A. A., Capt.	1515
—— A. V. A., Lt. . 302,	1629
—— A. R., Lt. . 544,	1611
—— A. W., 2nd-Lt.	552
—— A. W., 2nd-Lt. 28, 510, 869,	1143*a*
—— E. C., 2nd-Lt. . 739,	791
—— J. A., 2nd-Lt. . 28,	748
—— J. L. C., Capt. 681, 783,	785
—— L. F., Capt. . 280-289,	1623
—— S. V., Lt.	761
—— W. D. F., Lt.	1806
Meredith, A. E., Capt. 264, 1004,	1459
—— D. F., 2nd-Lt. . 546,	1639
—— F. B. S., 2nd-Lt. . 562,	1265
—— H. T., 2nd-Lt.	1601
—— K., Lt.	834
Merewether, J. W. B., Lt.-Col. 189, 1893,	1911
Merrett, A. B., Lt.	637
Merrill, W. E., Lt. . 353,	945
Merriman, A. B., Maj. 215, 871 1505,	1515
—— S. T., Lt. . 463,	1389
Merritt, B. T. B., Capt. . 260,	1647
Merry, D. G., 2nd-Lt. . 683,	686
—— N. Mc. I.	1036
—— R., Lt. . 540,	1327
Merrylees, S. B., Lt. . 37?,	1321
Mersh, W. G., Sr. Asst. Surgn.	1790
Merton, J., 2nd-Lt.	746
Messenger, H. T. K, Maj. 225,	1662
—— J. W., Lt.	1689
—— W., 2nd-Lt.	619
Messervy, F. W., Capt. . 278,	891
Messum, H. R., Lt. . 493,	881
—— L. R., Capt. 264, 1102, 1120,	1569
Meston, A., 2nd-Lt. . 549,	970
Metcalf, F., Lt. . 529,	1668
Metcalfe, A. N., Lt.	1839
—— C., 2nd-Lt.	638
—— C. J., Lt. . 435,	1048
—— E., 2nd-Lt.	1695
—— E. D., Capt. . 264, 879,	1004
Metcalfe, E. D. T., Capt. 268,	1664
—— G. C., Ty. Capt.	1723
—— H. A. F., Lt.	605
—— H. J., 2nd-Lt.	626
—— J. H., Lt. . 534,	1707
—— J. S., Ty. 2nd-Lt.	571*a*
—— M. R., Lt. . 318,	1193
—— T., 2nd-Lt. . 571,	1693
Methven, D. M., Capt.	591-592
Meudell, E., 2nd-Lt.	547
Mewa Khan, Jemdr.	1507
—— Ram Vashisht, Sub-Asst. Surgn.	1815
—— —— Subdr.	1419
—— Singh, Jemdr. . 892,	1075
—— Subdr. Maj.	1699
—— Thapa, Subdr.	1639
Meyer, B., 2nd-Lt. . 68, 315,	1178
—— C. H. L., Lt.-Col.	1786
—— H. A., Lt.	788
—— S. S., 2nd-Lt.	640
—— W. R. M., 2nd-Lt. . 331,	915
Meyers, J. W., Asst. Surgn.	1803
Meyler, E. M., Revd.	1904
Meynell, H., Maj. . 79, 213,	929
—— R., Condr.	1861
Meyrick, R., Lt. . 492,	931
Meyrick-Jones P., 2nd-Lt. . 324,	1600
Meysey-Thompson, A. de C. C., Lt.	619
Mg Aung Din, Subdr.	1293
—— Aye, Jemdr.	1291
—— Ba Chaen, Jemdr.	1293
—— Han, Jemdr.	1291
—— Ket, Jemdr.	1291
—— On Subdr.	1297
—— Shwe, Jemdr.	1293
—— Saw, Jemdr.	1293
—— Gyi, Subdr.	1293
—— Htang Law, Jemdr.	1291
—— Ba than, Jemdr.	1297
—— —— Jemdr.	1293
—— Kyaw Zan, Subdr.	1293
—— Kyaw Nyun, Jemdr.	1291
—— Mg Gale, Jemdr.	1293
—— Maung, Jemdr.	1291
—— —— Jemdr.	1293
—— Myit, Jemdr.	1293
—— Nyan Gyi, Jemdr.	1291
—— Nyo, Jemdr.	1291
—— Nyun U, Subdr. Maj.	1293
—— Pe Gyi, Jemdr.	1297
—— On Kyaw, Subdr. Maj.	1291
—— Pa Pon, Subdr.	1291
—— Pe Thein, Jemdr.	1293
—— Po Byu, Jemdr.	1293
—— San, Subdr.	1291
—— Po Chu, Subdr.-Maj.	1297
—— Po Myit, Subdr.	1297
—— Po Pyu, Jemdr.	
—— San Tun, Subdr.	1297
—— Saw, Jemdr	1293
—— Shwe Gyaw, Jemdr.	1293
—— So Hlaing, Jemdr.	1297
—— So Maung, Jemdr.	1297
—— Thein Maung, Jemdr.	1291
—— Aung Gyaw, Jemdr.	1293
—— Maung, Jemdr	1293
—— Win, Jemdr.	1297
Mhadu Wagmode, Subdr.	1444
Mhatre, S. W., tempy. Capt.	1769
Mian Khan, Jemdr.	1241
—— Mardan Ali, Ressdr.	1685
—— Nur, Jemdr.	968
—— Shamsher Singh, Jemdr.	1188
Michael, L. H. F., Asst. Surgn.	1801
—— (I) Subdr.	993
—— (II) Subdr.	993
—— (III) Jemdr.	994
—— D. F., Asst. Surgn.	1799
—— N. A., Asst. Surgn.	1803
—— W., 2nd-Lt.	358
—— Jemdr.	1265
—— T. J., Lt. . 353,	1289
Michaelson-Yeetes R. de M. Capt.	1704
Michell, J. W. A., Col.	177
—— J. H., Lt. . 415,	949
—— R. J. V., Maj.	68
Michie, G. F., Lt.	753
—— J. B., 2nd-Lt.	1241
Micholls, M. G., Capt.	595
—— R. M. B., Lt.	709
Michelmore, R. G., Ty. Capt.	1725
Middlemass, G. deS. H., Capt. 254, 1108,	1483
—— R. W. H., Maj. 132,	1661

Names and Rank. **Page.**

Middleton, C., 2nd-Lt. 843
—— C. W., 2nd-Lt. . . 683
—— H. J. J., Col. Ty. . . 180
—— J., Lt. 785
—— J. E., 2nd-Lt. . . . 848
—— L. W., Lt. . 534, 1505
—— M., Capt. . . 274, 1355
—— P. A. H., 2nd-Lt. . . 819
—— R. H., 2nd-Lt. . . 825
—— T. B., Lt. . 318, 1611
Middleton-Stewart, C. E., 2nd-Lt. 566
—— J. C., 2nd-Lt. 552, 1463
Middleton-West, S. H., Capt. . 1754
Middleweek, L. W. S., Lt. . 351, 1674
Midelton, E. W., Lt. 514
Midgley, E. A., Lt. . 503, 830, 1315
Midwood, D. N., Lt. 629
Miers, W. H., Sergt.-Maj. . . 998
Mieville, A. F. H., Lt. . . . 753
Mihan Singh, Jemdr. 1145
—— Jemdr. 1075
—— Jemdr. 1157
—— *Hony.* Capt. . . . 2103
Mihar Khan, Jemdr. 978
Mihr Ali, Ty. Sub-Asst. Surgn. . 1833
—— **Din**, Subdr. Maj. 989
—— **Khan**, Subdr. . . . 1311
—— **Khan**, *Hony.* Capt. . . . 2103
—— **Singh**, Subdr. 1311
Mihrab Khan, Sub-Asst. Surgn. . 1812
Mikram Singh, Ressd. 1683
Milburn, J. C., Asst. Surgn. . . 1793
Milburne, W., Lt. 516
Mildmay, A. B. C., Lt. . . 505, 1666
Mildon, R. K., 2nd-Lt. . . 329, 1495
Mileham, C. A., Lt. . 508, 728, 1666
Milchem, H. G., Lt. . . . 526, 998
Miles, C. W., Capt. *Sir* . . . 5, 795
—— E., 2nd-Lt. . . . 356, 998
—— H., 2nd-Lt. 619
—— H. H., 2nd-Lt. 825
—— J. W., 2nd-Lt. . . 363, 1267
—— P. J., Brig.-Genl. 12, 29, 120, 124, 149, 184
—— W. H., Capt. 736
—— W. H., Capt. 704
—— W. H., Lt. 504, 1501
Milford, E., Maj. 215, 1319
Mikha Singh Dhindsa, Sub-Asst. Surgn. 1819
Mikhi Ram, Sub-Asst. Surgn. . 1814
—— Sub-Asst. Surgn. . . 1819
—— Singh, Ressdr. . . . 951
Mill, G. S., Lt. 534, 1621
Millar, A. B., Lt. 306, 1247
—— A. S., Capt. . . . 501a, 1399
—— A. J. F., Lt. . . . 513, 843
—— E. G., Lt. 683
—— E. L., Maj. 577
—— G. McG., Maj. . . 1239, 1753
—— J. G., Lt. . . . 526, 1094
—— J. H. B., Capt. 615
—— P. P., 2nd-Lt. 637
—— W. B., Lt. 1653
—— W. H., Col. . . 123, 149, 182
—— W. W. C., Lt. 1369
Milbourne, C. K., Capt. . . . 709
Milburne, W., Lt. 504
Millen, J. A., 2nd-Lt. 752
Miller, A., Lt.-Col. . . . 135, 1744
—— A. B., Lt. 420, 1211
—— A. C., 2nd-Lt. 559
—— A. D., 2nd-Lt. . . . 476, 1351
—— A. D., Lt. . 521, 1670, 1704
—— A. G., 2nd-Lt. 1708
—— A. R., 2nd-Lt. 637
—— A. R. C., Lt. . . . 534, 1672
—— B. L. St. C., 2nd-Lt. . . . 558
—— C. B., Lt. 541, 1427
—— C. R., 2nd-Lt. 1009
—— C. H., 2nd-Lt. 550
—— E. C., 2nd-Lt. 637
—— E. V., 2nd-Lt. . . . 562 1418a
—— F., Maj. 239. 1435
—— F. B., Lt. 352, 1493
—— F. E., *Hony.* Maj. . . . 1843
—— F. N., Lt. 597
—— G. N., Capt. . . 55, 257, 367
—— G. R. B., Lt. . 310, 1020, 1082
—— H., Capt. 1919
—— H., 2nd-Lt. 624
—— H., 2nd-Lt. 546, 1373
—— J. A. T., Capt. 593
—— J. F., Lt. 764
—— L. H., Lt. 405, 1114

Names and Rank. **Page.**

Miller, M. R., Capt. 731
—— O. G., 2nd-Lt. . . . 562, 1114
—— R. A., 2nd-Lt. 793
—— R. G., 2nd-Lt. . . . 356, 1116
—— R. S. C., 2nd-Lt. 1676
—— R. W. H., Capt. 1718
—— R. W., Ty. Capt. 175
—— S., Lt. 736
—— T., Lt. 531, 1668
—— T. S., 2nd-Lt. . . . 77, 1885
—— W., Lt.-Col. 834
—— W. C. W., Maj. 500
—— W. G., Ty. Capt. 1770
—— W. J. B., Lt. . . . 517, 986
—— W. V., 2nd-Lt. 790
Miller-Hallet, J. L., Capt. 95, 280-287 1600
Miller-Stirling, A. E. S., Capt. 280-287
Millet, P., Maj. 628
Millett, T. W., 2nd-Lt. 567
Milligan, H. C., Lt. . . . 38, 1507
—— J. R., Lt. 505
—— J. W., Lt. 1114
Milliken, E. K., Lt. . . . 322, 1245
Millin, D. J., 2nd-Lt. . 359, 1130
Milling, R. C., Lt. . . 320, 365, 1595
Millinship, E. H., Lt. 637
Mills, A. E., 2nd-Lt. 361
—— A. M., Maj. . . . 232, 909
—— A. W. F., Lt. . . . 507, 1630
—— B. A. J., Lt. 700
—— C. F., Lt. 353, 1225
—— E. B., Capt. 808
—— E. G., Lt. 353, 1042
—— F. C. R. Lt. . . . 540, 1275
—— G., 2nd-Lt. 618
—— G. A., 2nd-Lt. 618
—— G. R., 2nd-Lt. 550
—— H., 2nd-Lt. 754
—— H. A. Lt. 24
—— H. A. Capt. 75
—— H. C., 2nd-Lt. . . . 545, 1704
—— H. G. C., Sr. Asst. Surgn. 1789
—— J., Lt. 749
—— J. E., Lt. . 446-49, 707, 1561
—— L. P., Lt. 708
—— P. B., Sr. Asst. Surgn. . . 1789
—— P. S., Capt. 1752
—— R. G., Lt. 709
—— R. T., 2nd-Lt. . . . 532, 1611
—— S. D., Capt. 367, 1197
—— S. G., Nursing Sister . . 1734
—— S. H. H., 2nd-Lt. . . 560, 1219
—— S. W., 2nd-Lt. 1104
—— W. S., Lt. 510, 1333
Millward, L. J. M., 2nd-Lt. . . 728
Milward, A. G., 2nd-Lt. 543a
—— R. G., Lt. 368
Milwood, A. M., Lt. 526
Milne, A. F., Capt. 1732
—— A. S., Lt. 535
—— A. Y., Lt. 1720
—— C., Lt.-Col. . . . 130, 1742
—— C. A., 2nd-Lt. . . 555, 1489-90
—— C. I., Ty. Capt. 1726
—— E. M., Lt. 506, 1050
—— E. O., Capt. 51, 702
—— G. E., Capt. 815
—— J. L., Lt. 801
—— J. S. W., Lt.-Col. . . . 654
—— J. W., Lt.-Col. 137, 159, 197, 870, 1335
—— O. J. T., Lt. . . . 317a, 909
—— P., Ty. Capt. 1725
—— R. B., 2nd-Lt. 781
—— T., Maj. 234, 1243
—— W. G., Lt. 515
Milner, C. E., Lt. . . . 488, 947
—— E. J., Sub-Condr. . . . 1864
—— F. D., 2nd-Lt. . . . 552 1132
Milnes, A. V., Capt. 1732
Milram Singh 1380
Milsom, E. G. D., tempy. Capt. . 1724
Milton, A. M., 2nd-Lt. 494
Milward, A. G., Lt. . . . 544, 1265
—— A. S., Lt. 787
—— C. A., Bt. Lt.-Col. 51, 99, 160, 214, 1235
—— H. H., Lt. . . . 367, 1223
—— R. C., Lt. 511
—— R. G., Lt. . . . 368, 1074
Milwright, C., Condr. 1851
Mims, C. C., Lt. 622
Minahan, D. J., 2nd-Lt. 561
Minbahadur Mal, Jemdr. . . . 1702
Minch, J. B., Capt. 1717

Names and Rank. **Page.**

Minch, W., Lt. 769
Minchin, A. B., Lt.-Col. 197, 1899, 1912
—— A. A. F., 2nd-Lt. . . 611
—— C. F., Lt.-Col. 183, 1901, 1911
—— H. C. S., Capt. . 277, 1231
—— R. H. L., Capt. 280,-827 1203, 1287
—— V. R., 2nd-Lt. . 729, 1009
Minchinton, H. D., Capt. . 263, 1595
Minhinnick, J. B., 2nd-Lt. . 546, 1008
Minner, C., 2nd-Lt. 669
Minniken T. B., Capt. . 96, 252, 1319
Minniss, A. O., 2nd-Lt. 1001
Minnitt, G. C., Lt. . . 516a, 998
Minns, A., Lt. 703
—— E. W., Sub-Condr. . . . 1856
Minshall, A. O., 2nd-Lt. . . 355, 1092
—— C. H., Lt. 796
Minshull, W. T., 2nd-Lt. 561
Minter, G., Lt. 534
—— J. P. S., Capt. . 273, 1517, 1571
Mintey, A. H., 2nd-Lt. . . . 358, 1511
Minto, A. H. 2nd-Lt. . . . 554, 1273
Mintoft, T. C., Capt 102
Mir Akbar, Subdr.-Maj. 1128
—— Shah, Jemdr. 1432
—— **Afzal**, Jemdr. 1336
—— Jemdr. 1121
—— *Hony.* Capt. 2102
—— **Khan**, Rissldr. . . . 918
—— **Alam Khan**, Jemdr. (I) . . 892
—— Jemdr. (II) . . 892
—— Jemdr. 946
—— **Ali**, Jemdr. 1408
—— Jemdr. 1256
—— **Khan**, Risdr. 1685
—— Jemdr. 1495
—— **Asghar Ali**, Subdr.-Maj. . 1271
—— **Aslam Khan**, Jemdr. . . . 934
—— Jemdr. . . . 948
—— **Aulad Ali**, Jemdr. 1092
—— **Awas**, Jemdr. 1236
—— **Badshah**, Subdr. (I) . . . 1506
—— Subdr. (II) . . . 1506
—— Subdr. 1513
—— **Baz**, Jemdr. 1001
—— **Khan**, Subdr. 1227
—— Jemdr. 946
—— **Dad**, Jemdr. 1336
—— **Dad Khan**, Risaldr. 891
—— Jemdr. 1090
—— **Dast**, Jemdr. 1124
—— **Haidar**, Jemdr. 1571
—— **Haider Shah Khan**, *Hy.* Capt. 2103
—— **Hussain**, Jemdr. 1320
—— **Ishan Ali**, Jemdr 1328
—— **Jafar Khan**, *Hony.* Capt. . 2104
—— **Kambir Khan**, Subdr. Maj. 1507
—— **Mohamad**, Jemdr. 1567
—— **Mumtaz Hussain**, Risldr. . 931
—— **Padshah**, Ressdr. 889
—— **Sadik Ali**, Risldr. 914
—— **Shamshad Ali**, Risldr. . . 880
—— **Zaman**, Jemdr. 1518
—— **Khan**, Jemdr. . . 895
—— Jemdr. 1514
—— Jemdr. . . . 948
Miraj-ud-din, Sub-Asst. Surgn. . 1811
Mirchandani, D. H., Ty. Lt. . . 1779
Mirajkar, V. R., Tempy. Capt. . 1764
Miran Bakhsh, Subdr. 1070
Miran Baksh, Subdr. 1122
—— Jemdr. 1393
—— Jemdr. . . . .
—— Shah, Jemdr. 1124
Mirasadulla, Subdr. 1307
Mirchi Singh, Subdr. 1452
Miri, G. A., Tempy. Capt. . 1285, 1770
Mirtassa Duk Hussain, Jemdr. . 1436
—— Dad, Jemdr. 1506
Mirzadad, Jemdr. 1506
Mirza Khan, Subdr. . . . .
—— Jemdr. 1495
—— Jemdr. 957
—— Subdr.-Maj . . . 1373
—— **Khan**, Subdr. 1132
—— **Muhammad Beg**, Subdr. 1407
—— Sr. Sub-Asst. Surgn. 1829
—— **Azim Barlas**, Sub-Asst. Surgn. . . . 1819
—— **Muhammad Yusuf Beg**, Sub-Asst. Surgn. . . . 1809
—— **Naqiulla Beg**, Jemdr. . 1333
—— **Riyaul, Hussain**, Risaldar 889
—— **Shamshad Beg**, Sub-Asst. Surgn. 1819

NAMES AND RANK. Page.

Mirza, Umra o Beg, *Hony.* Capt. 2103
—— Wali Jan Beg, Jemdr. 1088
—— Dahab Beg, Jemdr. 1404
—— Wilayat Beg, Subdr. 1397
Mirzaman, Jemdr. 1681
—— Khan, Jemdr. 948
—— Ressdr. 896
Misa, J. M., 2nd-Lt. 595
Miskin, W. L., Capt. 260, 1475
Misri Khan, *Hony.* Capt. 2104
—— Jemdr. 1488
—— Jemdr. 1256
Missing, C. M., 2nd-Lt. 739
—— C. M. S., 2nd-Lt. 791
Misson, W., Condr. 1853
Mistri, A. B., Tempy. Capt. 1769
Mitchell, A. C., Lt. 666, 992
—— A. H., 2nd-Lt. 682, 785
—— A. J., Lt. 751
—— A. J., Condr. 1867
—— B. E., Lt. 543a
—— B. M., Condr. 1854
—— C. A., Capt. 697, 1004
—— C. H., Lt. 534, 968
—— C. R. S., 2nd-Lt. 558
—— D. G., Lt. 803
—— E. A., Lt. 618
—— E. J. R., Lt. 531, 1505, 1708
—— F. M., Lt. 507, 1501
—— G. A., 2nd-Lt. 692
—— G. G., Lt. 530
—— H., 2nd-Lt. 1401
—— H. C. B., Capt. 779, 822
—— H. F., 2nd-Lt. 358
—— H. H. G., Col. 2, 652
—— H. J., 2nd-Lt. 562
—— H. R. W., Lt. 540a, 1014
—— H. S., Maj. 225, 1148, 1613
—— J., Lt. 617
—— J., 2nd-Lt. 566, 1032, 1696
—— J., Sub-Condr. 1803
—— J., Capt. 605
—— J. D., Revd. 1906
—— J. H., 2nd-Lt. 323, 1166
—— J. R. V., 2nd-Lt. 599
—— K. G., 2nd-Lt. 553
—— K. W. S., 2nd-Lt. 686
—— L., Lt. 440, 446-49 1459
—— N., 2nd-Lt. 1068
—— P. T., 2nd-Lt. 554
—— P. W., Lt. 540a, 1701
—— R. A., Lt. 542
—— R. H., 2nd-Lt. 557, 1359
—— R. P. J., Capt. 205, 945
—— R. R., Lt. 351
—— R. St. J., Lt. 523, 998
—— S. A., 2nd-Lt. 333, 1183
—— T. G., Capt. 715
—— W., Lt. 492
—— W. G., Lt. 736
—— W. G., Lt. 638, 817
—— W. H., 2nd-Lt. 712
—— W. J., Lt.-Col. 138, 160, 198, 871, 1491, 1493
—— W. P. S., Asst. Surgn. 1799
Mitchell-Carruthers, N. L., Capt. 208, 1140
Mitchelson, G., 2nd-Lt. 742
Mitchley, E. C. T., Lt. 405, 1597
—— G. A., Lt. 303, 1311
Mitford, B. V., Maj. 746
—— J. P., Maj. 229, 1399
—— W. S., 2nd-Lt. 595
Mithanlal Vaisha, Sub-Asst. Surgn. 1818
Mithavala, M. A., tempy-Capt. 1769
Mitra, B. M., Capt. 1758
—— B. N., Esq. 91
—— J. M., Ty. Capt. 1769
—— K. C., Ty. Lt. 1775
—— N. C., Ty. Lt. 1776
—— P. N., Ty. Capt. 1770
—— S. C., *Hony.* Ty. Capt. 1771
—— S. L., Ty. Capt. 1765
Mitralal Thapa, Subdr. 1648
Mit Singh, *Hony.* Capt. 2, 2104
—— Jemdr. 1280
—— Sub-Asst. Surgn. 1814
—— Subdr. 1211
—— Jemdr. 1589
—— Jemdr. 1149, 1587
—— Bhasauria, Sub-Asst. Surgn. 1819
Mitt Singh, Subdr. 1247
—— Jemdr. 895
—— Subdr. 1347
—— *Hony.* Capt. 2101

NAMES AND RANK. Page.

Mitter, H. C., Ty. Capt. 1772
—— B. K., Lt.-Col. 105, 129, 1740
—— S. C., Ty. Lt. 1775
Mitthu Lal, Sr. Sub-Asst. Surgn. 1808
Mitton, F., Sub-Condr. 1864
Miyan Din Khan, Risldr. Maj. 927
—— Jan, Sub-Asst. Surgn. 1809
Mizen, H. A., Lt. 299, 1046
Moase, W., 2nd-Lt. 358, 1405
Moat, C. A., 2nd-Lt. 706, 785
Moazzam Ali Talukdar, Sub-Asst. Surgn. 1820
Moberly, B. R., Bt.-Lt.-Col. 6, 13, 99, 139, 160, 214, 1247
—— C. N., Maj. 1440, 822
—— F. J., Brig.-Genl. 6, 13, 98, 121, 130, 155, 189, 808, 1120
—— H. S., Lt.-Col. 13, 43, 98, 135, 158, 199, 870, 1279
—— R. S., Lt. 525, 819, 1148
Mobta, Jemdr. 1583
—— Jemdr. 1476
Mobin Khan, *Hony.* Capt. 2104
Mocatta, C. H., Lt. 393, 1606
—— V. E., Maj. 54, 103, 593
Mockett, V., 2nd-Lt. 55, 686
Mockler, E. C., Capt. 268, 1595, 1629
—— G. H. G., Lt.-Col. 105, 107, 130, 155, 189
—— G. S., 2nd-Lt. 492, 953
Mockridge, A. P., Lt. 353, 1621
Modak, N. W., Ty. Capt. 1771
—— Y. V., Ty. Lt. 1776
Modget, H. R., Lt. 520, 1614
—— L. D., Lt. 508
Modi, L. S., Ty. Capt. 1764
Mody, M. S. H., Lt. 1780
Moens, A. W. H. M., Bt.-Lt.-Col. 51, 98, 137, 159, 217, 1231
Moffat, J., Lt. 503, 560, 691, 1624
—— L. P., Lt. 443, 959
—— S. H., Lt. 543a
Moger, W., Capt. 702, 795
Mogul, D. D., Tempy. Capt. 1667
Moghul Khan, Subdr. 1142
—— Jemdr. 1433
Mohadset Kokate, Jemdr. 1436
Mohammad Abdul Latif, Ty. Sub-Asst. Surgn. 1833
—— Karim, Subdr. 1088
—— Hassan, Subdr. 1068
Mohd. Azam., Subdr. 1704
—— Azizuddin, Subdr. 1704
—— Hussain Khan, Ressdr. 1683
—— Ishaq., Jemdr. 1704
—— Taqi Shah, Ressdr. 1682, 1686
Mohr Singh, Jemdr. 1032
Mohamed, L. G., Ty. Lt. 1780
Mohamed, Ali, Jemdr. 1082
—— Khan, Ressdr. 955
—— Abas Khan, Subdr. 1707
—— Amir Khan, Risldr. 1686
—— Beg, Sr. Sub-Asst. Surgn. 1829
—— Haji, Ressdr. 1683
—— Ibrahim Khan, Jemdr. 1360
Mohamad Yusuf Ali Shah, Ressdr. 1683
—— Shafi Khan, 1389
—— Beg, Sub.-Asst. Surgn. 1809
—— Khan, Jemdr. 1705
Mohamed Akram Khan, Jemdr. 1704
—— Anwan Khan, Jemdr. 1573
Mohbat Singh, Subdr. 1517
Mohomed Khan, Subdr. 1479
—— Risldr. 1707
Mohamed Khan, Jemdr. 1702
—— Mustaq Khan, Jemdr. 890
—— Yasin Khan, Jemdr. 1126
Mohammed, Subdr. 1507
—— Abdul Hakim, Subdr. 1447
—— Aslam Khan, Jemdr. 1957
—— Afzal, Jemdr. 1012
—— Ghouse, Subdr. 1447
—— Faizullah, Subdr. 1707
—— Fazil, Jemdr. 1092
—— Hanif, Jemdr. 1583
—— Ismail, Jemdr. 1681
—— Yousuf, Subdr. 1447
Mohommed Yakuob, Subdr. 1702
—— Khan, Subdr. 1573
Mohammad Khan, Ressdr. 1685
—— Ressdr. 1683
—— Mohsin, Jemdr. 1683
—— NasirKhan, Risldr. 1685
—— Quadir, Jemdr. 1705

NAMES AND RANK. Page.

Mohammad, Said, Jemdr. 1240
—— Sher Khan Malik, 2nd-Lt. 481
Mohan, G. V. R., Tempy. Capt. 1765
—— Lall, Sub-Asst. Surgn. 1809
—— Sr. Sub-Asst. Surgn. 1808
—— Jemdr. 1585
—— Subdr. 1365
—— Lal Nagesar Shukla, Sub-Asst. Surgn. 1829
—— Fulji Thakur, Sub.-Asst. Surgn. 1820
—— Jemdr. 1452
—— Singh, Subdr. 1082
—— Jemdr. 1940
—— Jemdr. 872
—— Negi, Jemdr. 1182
—— Jemdr. 880
—— Jemdr. 951
—— Subdr. 1082
—— Jemdr. 948
—— Jemdr. 1581
—— Jemdr. 1031
—— Jemdr. 880
—— Jemdr. 880
—— Rawat, Jemdr. 1182
—— Jemdr. 1224
Mohar Singh, Jemdr. 902
—— Jemdr. 1175
—— Jemdr. 1559
—— Jemdr. 1393
—— Jemdr. 1071
—— Jemdr. 1054
Mohil, E. K. G. Ty. Lt. 1780
Mohiuddin Shariff, Jemdr. 1698
Mohkam Chand, Jemdr. 920
Moinet, L. V., 2nd-Lt. 603
Moinuddin Bilgrami, Sub-Asst. Surgn. 1819
Moji Ram, Jemdr. 1453
Moir, E. T., Capt. 827
—— G., Lt. 518, 1383
—— J. H., 2nd-Lt. 568, 1070
—— J. S. S., Maj. 234, 923
Molar, Ressdr. 878
—— Subdr. 1059
—— I, Subdr. 1059
—— II, Subdr. 1059
Mold, G. L., Lt. 292, 1128, 1183
Mole, J. A., Lt. 517
—— P. C., Esq. 91
Molesworth, A. L. M., Maj. 230, 1635
—— A. O., 2nd-Lt. 567
—— F. C., Maj. 663
—— G. E. N., Lt. 66, 75, 700, 795, 1287
—— G. N., Capt. 701
—— H. E., Lt.-Col. 137, 632
—— R. C., Lt. 311, 1611, 1633
—— R. E., Lt.-Col. 1731
—— R. P., Lt.-Col. 125, 632
—— W., Col. 3, 10, 126, 1738
Moll, W. B., Capt. 279, 1050
Mollison, E. J., Maj. 96, 225, 1497
Molloy, C. P., Capt. 844, 846
—— G. M., Maj. 6, 13, 99, 207, 939
—— H. T., Maj. 238, 1617, 1624
Moloney, J. T., Lt. 740
—— W. J., Ty.-Capt. 1724
Molony A. D., Maj. 216, 1629, 1707
—— A. W., Maj. 797
—— C. W., Capt. 267, 1251
—— H. J. C., 2nd-Lt. 552, 1521
—— J. C., Lt. 801
—— J. B. deW., Capt. 1758
—— J. T., Lt. 738, 746
—— O. H. C., Maj. 40, 228, 1883, 1383
—— P. J., Rev 1903
Molyneux, E. Lt. 748
—— H. J., 2nd-Lt. 1474, 887
—— P. L., Lt. 298, 1178
—— J. J., Capt. 739
Moman Ram, Subdr. 1457
Monaghan, T., Lieut. 691
Monbahadur Singh, Subdr. 1219
Monck-Mason, G. N. G., Maj. 242, 1343
—— H. E., Lt.-Col. 180
Monckton, C. A. W., Capt. 707
—— F. E., Lt. 1657
Moncrieff, A. B., Capt. 585
—— I. J. S., Capt. 1839
—— R. L., Lt. 530, 978
Moncur, J. O., 2nd-Lt. 562, 1130
Money, E. A., Col. 178, 1917
—— E. D., Brig.-Genl. 14, 42, 122 131, 156, 192

NAMES AND RANK — Page

Money, E. F. D., Maj. 33, 210, 871, 1559, 1613
—— G. A., Col. . . . 179, 1917
—— G. W. P., Capt. . 265, 1606
—— H. D. K., Capt. 280-287, 964
—— W. G.M., Lt. . . 311a, 1407
—— W.S. E., Capt. . 256, 917, 1892
Money-Simons, J.J., Lt.-Col. 179, 1917
Monga Singh, Jemdr. 1151
Moniadurai, Jemdr. 1360
Monier-Williams, L., Lt. . 294a, 1606
Monies, S. O., Lt. . . . 526, 1219
—— V. A., Lt. . . . 313, 909
Monins, J. E., Lt. 683
Monisse, W. T. J., Asst. Surgn. . 1803
Monk, F. F., Lt. 534, 1239
—— F. H., 2nd-Lt. . . . 551, 1361
Monks, E. W., 2nd-Lt. . . . 554
—— H. G., Lt. . 486, 953
Monro, C. C., Genl. Sir 1, 4, 5, 97, 118
Monro, D., Lt. . . . . . .
—— D. C., Capt. 1715
Monsell, H. H., 2nd-Lt. . . . 767
Mont-Thomas, E. B., Lt. . . . 521
Montagu J. W. E., Lord Lt.-Col. . 10
Montagu-Smith, E., Capt. . 252, 917, 941
Montague, J. E. F., Capt. . 636
Monteath, G., 2nd-Lt. 849
Montefiore, C. E., Capt. 255, 1443, 1898
—— L. N., Capt. . . . 734
Monteith, M. E. C. Von S., Maj. 241, 1423
Montford, D. R., Capt. . . 272, 1399
Montgomerie, H. S., Maj. 751, 785
—— L. J., Lt. 312
—— N. E., N. Sister 1735
Montgomery, C. A. S., Bt.-Col. . 776
—— C. W. T., Asst Surgn 1798
—— E. M., 2nd-Lt. . . .
—— H., 2nd-Lt. . . 568
—— H., 2nd-Lt. . . 563
—— H. R. G., Capt. 697, 783
—— J., Capt. . . 104, 583
—— J. P., Sr. Asst. Surgn 1790
—— L. J., Lt. . . 311a, 1058
—— S. M., Lt. . . . 589
Montray, F. C., Lt. . . . . .
Montrose, E., Sub-Condr. . . . 1870
Mony penny, R.d'A., Lt. 367, 1263
Moodie, A. E. S., Lt. 1245
—— A. L. M., Lt. 854
—— E. H., 2nd-Lt. . 326, 366, 881
—— J., 2nd-Lt. 791
Moody, B., Lt. 60, 510
—— C. G., Lt. 67, 722
—— C. M., Asst.-Surgn. . . 1795
—— F. H., Maj. . 9, 16, 240, 899
—— F. W., 2nd-Lt. . . 360, 1140
—— J., 2nd-Lt. 980
—— W. E., Asst. Surgn. . . 1794
—— W. P., Lt. 770a
Moojen, L. G., Lt. . . . 618, 651
Moolky, D. M., Ty. Lt. . . . 1776
Mool Singh, Jemdr. 1157
Moomtaz Ali Khan, Ressdr. . . 1084
Moon, E. J. 1009
Moone, A. J., Lt. 513
Mooney, H. A., Hony Lt. 1849, 1849
—— H. F., 2nd-Lt. . . 473, 897
—— W. J., Lt. 520
Mooni Rathnam, Subdr. . . . 1099
Moor, G. W., 2nd-Lt. . . . 624
—— J. A., Lt. . . 515, 796, 1189
Moorcroft, E., Lt. . . 460-62, 1251
Moore, A., Lt.-Col. 133, 157, 193, 1279
—— A. C., Lt. . . 305, 1375, 1581
—— A. E. duPre, Capt. . 832
—— A. G. W., Lt.-Col. 35, 78, 129, 154, 186
—— A. H. W., 2nd-Lt. . . 640
—— A. J., Lt. 520, 1614
—— A. J., Lt. 351, 1299
—— A. N. FitzG., Lt. . 489, 893
—— A. W., 2nd-Lt. . . . 1695
—— A. W. U., Capt. 758
—— C. A., Maj.-Genl. . . . 170
—— C. F. F., Capt. . 55, 262, 1487
—— C. H., Lt. 1679
—— C. H. G., Lt.-Col. 31, 131, 561, .94
—— C. S., Lt. 322, 1635
—— E. A., Lt. 685
—— E. C. F., Lt. . . . 56, 488
—— E. Du P. H., Maj. . . . 701
—— E. R., 2nd-Lt. . . 549, 1441
—— F. F., 2nd-Lt. 1008

Moore, F. G., Maj. . . . 4, 213, 1038
—— F. L., Lt.-Col. 61, 129, 154, 186
—— F. M., Lt. 303, 1231
—— G., 2nd-Lt. . . 465, 473, 1305
—— G., Lt. 808
—— G. D., Lt. 825
—— G. D. M., Maj. 1919
—— G. W., Capt. 843
—— H. A., Lt.-Col. 129, 155, 843, 188
—— H. A., 2nd-Lt. 1166
—— H. D., Maj . . . 599, 1517
—— H. E., Lt. 90
—— H. H., Lt. 514, 1399
—— H. L., Lt. . . . 312, 1052
—— H. W., Lt. 636
—— J., 2nd-Lt. 548
—— J. N., Capt. 789
—— P., Sub-Condr. 1855
—— P. J., 2nd-Lt. 350, 1369
—— R. J., Capt. 56
—— R. N., Lt. . . . 544, 1239
—— W. E. C., Lt. 629
—— W. G., Lt.-Col. 780
—— W. G. T., Lt. . . . 536, 1699
—— W. J. A., 2nd-Lt. . . . 825
—— W. S., Lt. 300, 1170
—— W. W. T., Lt. . . . 515, 1347
Moore-Lane, V. W. C., Capt. 267, 1427
Moorhead, E. J., Capt. . . . 615
—— H. D., Lt. . . 310, 1144
Moorhouse, E. G., Lt. . . 465, 931
—— F. H., Lt. . . . 476, 1044
Mooring, L. L., Lt. . . . 354, 1038
Moran, Capt. E. W. 1054
—— J. P. 2nd-Lt. 608
—— J., Maj . 38, 225, 1664, 1883
—— L. A., Tempy. Capt. . . 1723
—— T., Lt. 517, 1046
Morbey, E. H., Lt. . . 509, 1199
Morby, J. C. H., 2nd-Lt. . . . 568
Mordaunt, L. J. P., Ty. Lt. . . 1777
Morde, R. W., Lt. 493
Mordecai, J., Esq. 89
More, A. H., 2nd-Lt. 630
—— J. C., Maj. 52, 102, 238, 1227
—— J. S., Lt. 571
—— L. P., Bt.-Col. 38, 126, 161, 170, 206, 1709
—— P. St. C., Lt.-Col. . 130, 1742
Morel, H. J., 2nd-Lt. . . 492, 893
Morell, D. W., Lt . . . 443, 1678
—— J. A. C., Sub-Condr. . . 1864
Moreland, J. A., 2nd-Lt. . 473, 889
Moresby, C., Maj. 801
Moreshawar Balwant Bhagwat, Sub-Asst. Surgn. 1831
Moreton, B. M., 2nd-Lt. . 330, 1170
—— C. A., Lt. 718
—— E. A., Lt. . . 455-59, 1169
—— J., Hony. Lt. 841
—— J. P., Lt. 581
—— T., 2nd-Lt., . . . 557, 1701
—— Marshall, F. W., Maj. . 1249
Morfey, R. J., 2nd-Lt. . . . 623
Morford, E. St. C., Lt. . . . 529
Morgan, D. E., Maj. 219, 870, 1295, 1379
—— C. C., Maj. 651
—— C. G., 2nd-Lt. 1096
—— E., Ty. Lt. 1726
—— E., Sub-Condr. 1803
—— E., 2nd-Lt. 552
—— E. H., 2nd-Lt. 702
—— E. J., Lt. 527, 1022
—— E. J. L., Lt. . . . 529, 1673
—— E. W., 2nd-Lt. 1009
—— F., Lt. 536
—— F. A. S., Maj. 653
—— F. L., Capt. 746
—— F. R., Lt. 840
—— F. T., Lt. 704
—— G. A., 2nd-Lt. 1678a
—— G. H., Maj. . . . 87, 227
—— G. J., 2nd-Lt. 505
—— G. N. R., Lt. 63
—— H., 2nd-Lt. 640
—— H. A., Maj. . . . 2.8, 1090
—— H. A., Capt. 717
—— H. G., Lt. . . . 703, 796
—— H. H., Lt. 61
—— H. S., Lt. 492, 891
—— H. W., Lt. 500
—— I. M., 2nd-Lt. . . . 564, 1311
—— J. B. W., Capt. 1919
—— K. S., Lt. . . . 292, 1114
—— L., 2nd-Lt. . . . 358, 1285

Morgan, L. E., Lt. . . . 308, 1600
—— M. H. L., Maj. 105, 227, 1267
—— P. B. J., Lt. . . 375, 988
—— R. W., 2nd-Lt. . . 355, 1008
—— T.H., Capt. 681
—— W. D., 2nd-Lt. . . 561, 1433
—— W. G. C., Lt. . . 397, 1233
—— W. L., Lt. . . . 510, 1158
—— W. M., Lt. . . . 405, 119
Morgan-Owen, L. I G., Lt.-Col. 51
Morgans, C. G., 2nd-Lt. . 563, 815
Moriarty, D., Lt. . . . 403, 1135
—— E., Nursg.-Sister . . . 1735
—— J., Lt. 506
—— J. J., Tempy. Capt. . . 1725
—— P. P., 2nd-Lt. 1121
Morice, R. C. L., Maj. . 230, 1170
—— J. D., 2nd-Lt. 1695
Morier, W. E., Capt. 714
Morin, A. H., Maj. . . 500, 611
Morison, J., Maj. 1753
—— H. Mc. D. Lt. . . 509, 1533
Moritz, F. E. B., Lt. . . 393, 1299
Morland, W. E. T., Capt. . 58, 745
Morley, A. D., Lt. 732
—— A. R., 2nd-Lt. 693
—— B. E., Lt. 46, 770
—— E. A., 2nd-Lt. . . 559, 1092
—— H., Sub-Condr. 1857
—— H. G., Lt. 59
—— L. V., 2nd-Lt. 302
—— M. H. A., Lt. . 32, 527, 1071
—— W. B. S., Lt. 518
Moroney, W. M., Sub-Condr. . . 1850
Morphett, G. C., Bt. Lt.-Col. . . 53
Morphy, E. C. O., Lt. . . 425, 1271
Morrell, H. G., Capt. 258, 1003, 1471
—— S. J., Lt. 579
—— W. T., 2nd-Lt. 599
Morris, A. E. C., 2nd-Lt. . 558, 1707
—— A. L., Maj. . . . 244, 1080
—— A. R., 2nd-Lt. 558
—— C. E., Mag. . 237, 964, 1347
—— C. J., Lt. 311
—— C. M., Lt. . . . 507, 1148
—— C. O., Maj. . . 23, 216, 1061
—— C. T., Maj. . . . 213, 871, 1335, 1413
—— D., Capt. 600, 990
—— D. N., Lt. 785
—— D. O., Lt.-Col. 192, 1899, 1912
—— E., Condr. 1801
—— F., Lt. 531, 1395
—— G., Sub-Condr. 1803
—— G. H. J. G., Capt. . . . 604
—— G. L., Lt. 785
—— G. M., Bt.-Col., 50, 105, 121, 135, 158, 196, 870, 1207
—— G. F., Maj. 17, 42, 235, 931
—— H. C., Lt. . . . 489, 909, 1075
—— H. E., 2nd-Lt. 787
—— H. E. A., Lt. . . . 399, 1583
—— H. G., Lt. . . 310, 1215, 1072
—— H. L., Lt. 537, 1008
—— H. M., Lt. 371, 1333
—— H. McG., Lt. 1884
—— H. S., Lt. . . . 314, 870, 1375
—— J., Hony. Maj. 1047
—— J. A., Hony. Lt. 1840
—— J. C. T., 2nd-Lt. . . 511, 1901
—— J. H., Capt. . . . 608, 1001
—— J. J., 2nd-Lt. 1657
—— J. S., 2nd-Lt. . . . 545, 1704
—— L. J., Lt. 489
—— M. G., Lt . . . 541, 998
—— N. B., Tempy. Capt. . . 1708
—— N. M., Capt. . . . 485, 891
—— P. G., Ty. Asst. Surgn. . 1804
—— R. G., Lt. 793
—— T., Sub-Condr. 1803
—— W., Sub-Condr. 1874
—— W. G., Lt. . . 518, 814, 1701
—— W. H., Lt. 767
—— W. M., Capt. 1718
——Jones, J., Lt. . . . 445, 953
Morrish, J. H., 2nd-Lt. . . . 750
Morrison, A., Lt. . . . 519, 1487
—— A., 2nd-Lt. 719
—— A. O'H., Lt. . . 529, 998
—— B. H., Lt. . . . 312, 1106
—— C. C., Capt. . 277, 915, 941
—— D., Lt. 787
—— D. C. A., Maj. . . . 755
—— G. A., Lt. 420
—— J., Maj. . . . 71, 633, 1886

NAMES AND RANK. Page.

Morrison J., 2nd-Lt. 790
——— J. W., 2nd-Lt. . . . 1225
——— M. D., 2nd-Lt. . 563, 1703
——— R., 2nd-Lt. 788
——— R. G., Lt. . . 489, 1677
——— R. V., Capt. 1757
——— W. H., Lt. 111
Morrogh, V. M., 2nd-Lt. . . . 769
——— **Bernard**, J. A. D., Capt. . . 872, 917, 955
Morrow, A., Hony. Lt. . . . 844
——— C. T., Capt. . . 5, 571, 1850
——— J. S., Lt. 307a
——— W. J. A., 2nd-Lt. . 545, 1701
Morse, A. H., Capt. 665
——— B. S., Lt. 703
——— C. J., Lt.-Col. . . . 9, 129
——— L. H., Capt. 258, 1003, 1267
——— R. N. S., Capt. . . . 7 0
——— T. B. M., 2nd-Lt. . . .
——— W. S., Lt. 732, 1008
Morshead, H. T., Maj. . 664, 1880
——— V. H., 2nd-Lt. . . . 51
Mort, G. M., Lt.-Col. 589
Mortimer, A. C., 2nd-Lt. . . . 555
——— A. E., Lt. 1703
——— C.O., 2nd-Lt. . . . 1095
——— D. C., 2nd-Lt. 539
——— H., 2nd-Lt. . 549, 1359
——— P., Maj. . 230, 1602
——— W. G., Lt. . . 492, 927
——— R., Lt. 601
Mortleman, C. C., Lt. 732
Mortlock, J., Condr. . . . 1862
Morton, A. F., Lt. . 371, 1046, 1090
——— B. M., Revd. 1906
——— H. W. D., Lt.-Genl. . . 172
——— C. E., Lt. . 321, 1363, 1575
Morton, G. C., Capt. 681
——— E. H., 2nd-Lt. . 559, 1311
——— E. R., Lt. 779
——— E. T., 2nd-Lt. . . . 1655
——— **F.**, Asst. Surgn. . . . 1800
——— H. K., Lt. . . . 318, 1503
——— J. I., 2nd-Lt. . . 553, 1271
——— N., Capt. 1910
——— R. C., 2nd-Lt. . . . 728
——— T. F., 2nd-Lt. . . . .
——— T. M., Lt. . . . 292, 1611
——— W., 2nd-Lt. 91-92
——— W. D., Lt. 537
——— W. R., Col. 102, 106, 123
Morton-Marshall, F.W., Maj. 225, 1517
Moseley, B., Lt. . . 293, 730, 1443
——— F. S. H., Lt. . . 540, 826
——— L. A., 2nd-Lt. . . . 28, 779
Moses-Daniel, Jemdr. 1412
Moses, C. R., 2nd-Lt. 96
——— J. G. O., Capt. . . . 1758
——— O. St. J., Maj. . . . 1747
——— Solomon Vincent, Sub-Asst. Surgn. 1827
Mosley, A. D., 2nd-Lt. . . .
Moss, A. G., 2nd-Lt. . . 360, 1022
——— B. H., 2nd-Lt. . . . 558
——— C. H. de V., Lt. 315, 1235, 1505
——— C. E. R., 2nd-Lt. . . . 618
——— D., 2nd-Lt. 974
——— E. de V., Lt. . . 292, 1451
——— E. M., Lt. . . 540a, 1673
——— L. Lt., ... 367
——— R. C., Lt. 817
——— R. L., Capt. . . 679, 798
——— S. G., 2nd-Lt. . . . 1679
——— T., Capt. . . . 256, 1140
——— W. H., 2nd-Lt. . . . 568
Moss-Blundell, R. S., Maj. . 68, 244, 925
Mosse, A. H. E., Maj. 214, 1893, 1913
——— C. B., Capt. . . . 277, 1463
——— C. O. R., Capt. . . 273, 1475
Mossop, J. C. C., Maj. . 501a, 653
Mostyn-Owen, Lt. H. L. . 304, 923
Motabhai Pirozshah Bharucha, Sub-Asst. Surgn. 1837
Mota, Jemdr. 1196
——— **Singh**, Sub-Asst. Surgn. . 1810
——— Jemdr. 1244
——— Subdr.-Major. . . 1112
——— Subdr. Major . . 1098
——— Subdr. 1495
——— Jemdr. 1502
Motha Khan, Jemdr. 1488
Moti Chand, Subdr. . . . 1171
——— Subdr 1567
———, Subdr. 1199

NAMES AND RANK. Page.

Moti Jemdr. 1060
——— **Negi**, *Hony.* Capt. . . . 2101
——— **Persad Sahi**, Jemdr. . . 1643
——— **Ram**, Sub-Asst. Surgn. . 1813
——— Jemdr. 882
——— **Rawat**, Subdr. 1473
——— **Singh**, Ressdr. . . . 912
——— Jemdr. 1683
——— Jemdr. 1473
——— Jemdr. 1573
——— Subdr. 1280
——— Jemdr. 950
Motilal Limbu, Subdr. . . . 1627
——— **Thapa**, Jemdr. . . . 1027
Mott, G. E. T., Lt. 582
Mottram, P. J. Mac. Q., Capt. 500, 1130
Mouat, G. E. D., Maj 38, 219, 870, 1271, 2173
Modcll, E., 2nd-Lt. 1705
Moul, S. W. A., Asst.-Surgn. . 1803
Mouladad Khan, Jemdr. . . . 953
Mould, A. E., Lt. . 518, 818, 1699
——— B. S., Lt. . . . 538, 1001
——— G. B., Lt. 519
——— H., Revd. 1907
Moulding, E. J., 2nd-Lt. . 625, 976
Moule, G. G., Capt. 593
Moulson, G., Lt. 1715
Moung Ba Kin, Jemdr. 994
——— Gyi, Jemdr. 994
——— Hman, Jemdr. 994
Mount, A. H. ... Maj. 60, 664, 1882
——— T., 2nd-Lt. 1251
Mountbatton, L. A. L., Capt. . 1892
Mountain, R. G., Lt. 300, 1505, 1579
Mountford, E. W., Capt. . 681, 785
Mouritz, C. E. A., 2nd-Lt. . . . 629
Mousley, E. O., Ty., 2nd-Lt. . 622
Moutray A. G., Capt. 769
Mowat, J. S., Maj. . . . 231, 923
Mowels, J. H. 32, 75
Mowforth, H., Lt. . . . 522, 1367
Moxon, R. J.-Lt. 1701
——— R., Lt. 524
Moyes, E. M., Lt. . . 397, 1319
Moylan, W. D., Lt. . . . 510, 1001
——— T J, 2nd-Lt. . . 547, 1673
Moyles, S., 2nd-Lt. . . . 578, 723a
Moynan, R. N. O., Capt. . . . 1717
Moysey, R. A. L., Lt. . . 316, 1106
Mubarak Ali, Subdr. . . . 1435
——— Subdr. 1583
——— Subdr. 1088
——— Subdr. 1035
——— Sub-Asst. Surgn. . . 1817
——— **Shah, Khan**, Sub-Asst. Surgn. 1816
Mubarik Ali, Subdr. . . 1488, 1579
——— Khan, Jemdr . . . 1507
Mubarz Khan., Jemdr. 912
Mucklow, J. 608
Mudali, C. V.E., Sub-Asst. Surgn. 1828
——— N. V., Sub-Asst. Surgn. . 1815
——— P. S. K., Sub-Asst. Surgn. 1826
Mudaliar, P. S., Sr. Sub-Asst. Surgn. 1826
Muddi, E. E. 1009
Mudge, J. B., Capt. 1718
Mudliar, S. B., Sub-Asstt. Surgn. 1832
Mudhol, Chief of 112
Mudie, D., Maj. 656
——— P. U., Lt. . . . 537, 1672
——— R. F., Capt. . . . 68, 501a,
Mugford, J., 2nd-Lt. 630
——— P. W., Sub-Condr. . . 1864
——— T. G., Lt. . . . 352, 1162
Muggeridge, H. C., Lt. . . 521, 986
Muggridge, H. G., Lt. . . 521, 1693
Mughal Khan., Jemdr. 1807
Muh Smmad Saib, Hony. Capt. 2103
Muhabat Khan, Jemdr. . 1107, 1249
Muhabbat Singh Jemdr. . . . 1083
Muhammad, Jemdr. 1133
——— **Abbas**, Tr. Vety. Asst. 1687
——— Jemdr. 1380
——— **Abdul**, Subdr. . . . 1379
——— **Abdul Azim**, Jemdr. . 1264
——— **Subdr.** 1359
——— **Abdul Azim Khan**, Sub-Asst. Surgn. . . 1827
——— **Abdulazim Khan**, Subdr. 1359
——— Abdul Rashid, Jemdr. 1400
——— **Abdul Ghafur**, Sub-Asst. Surgn. . . 1826
——— Abdul Haq, Jemdr. . 1265
——— Karim, Jemdr. . 1276

NAMES AND RANK. Page.

Muhammad Wahab, Jemdr . 1328
——— **Amin**, Jemdr. . . . 973
——— **Amir**, Jemdr. . . . 1107
——— Ressdr. 916
——— **Qadir**, Jemdr. . . 925
——— Jemdr. . . . 1396
——— Ashraf Khan, Jemdr. . 910
——— Ayub Khan, Jemdr. 894
——— Ayyub, Sub-Asst. Surgn. 1817
——— Hakim, Jemdr. . . . 1328
——— **Azim**, Jemdr. . . . 1129
——— **Beg**, *Hony.* Capt. . 2101
——— **Abdul Hafiz**, Subdr. . 1263
——— **Karim**, Sr. Sub-Asst. Surgn. 1836
——— **Munaf**, Subdr. . . 1331
——— Jemdr. 1233
——— **Rahman**, Sub-Asst. Surgn. . . . 1816
——— **Abdul Rashid**, Jemdr. 1324, 1400
——— **Rahim**, Sub-Asst. Surgn. 1828
——— **Abdullah**, Sub-Asst. Surgn. . 1814
——— Jemdr. . 1578
——— **Abdur Rauf**, Sub-Asst. Surgn. . . 1827
——— **Abdus Subhan**, Subdr. . . . 1307
——— Jemdr. . . 1506
——— Jemdr. . . . 920
——— **Afzal Khan**, Risldr. 1684
——— Ressdr. 903
——— Jemdr. 1284
——— Jemdr. 903
——— **Ahmad Shamshi**, Sub-Asst. Surgn. . 1820
——— Ty. Sub-Asst. Surgn. 1833
——— **Ajmeri Khan**, Jemdr. 950
——— Subdr. 970
——— **Hussain**, Subdr. 1395
——— **Khan**, Ressdr. 928
——— Ressdr. 876
——— Jemdr. . 898
——— Jemdr. 2, 898
Muhammad Khan, Jemdr. . . . 908
——— Jemdr. . . . 948
——— Jemdr. . . . 1305
——— **Akbar**, Jemdr. . . . 1373
——— **Akbar Ali Khan**, Risldr.-Maj. . 2, 887
——— Khan, Jemdr. . . . 1154
——— Risldr. 79
——— **Ahmad Said Khan**, Ty. Hony. 2nd-Lt. . . . 482
——— **Akram Khan**, Risaldr. 876
——— Jemdr. 1344
——— Risldr. 891
——— **Alam**, Jemdr. . . . 1284
——— Subdr. . . . 1241
——— Jemdr. . . . 1579
——— Jemdr. . . . 1488
——— **Ali**, Jemdr. . . . 1518
——— Subdr. . . . 1355
——— Ressdr. . . . 894
——— Risldr. . . . 893
——— Subdr. . . . 1407
——— **Khan**, Jemdr. . . . 1409
——— Jemdr. . . . 1472
——— Jemdr. . . . 1583
——— **Ally**, Tr. Vety. Asst. . 1587
——— Jemdr. . . . 989
——— Jemdr. . . . 994
——— **Beg, Nawab** *Hony.* Lt.-Col. *Sir* 51, 110, 913
——— **Khan**, Risldr. (I) 876, 882,
——— Jemdr. . . . 975
——— Risldr. (II) . 876
——— Subdr. . . . 1373
——— Jemadar . . . 1090
——— Jemdr. . . . 1365
——— Jemdr. . . . 1579
——— Jemdr. . . . 1092
——— Jemdr. . . . 970
——— Jemdr. . . . 966
——— Khurshaid Jan, Sub-Asst. Surgn. . 1820
——— **Alum**, Jemdr. . . 1408
——— **Beg**, *Hony.* Capt. 2101
——— Ressdr. . 876
——— **Afzal**, Jemdr. . . 920
——— Subdr. . . . 1493
——— **Afzal Khan**, Risldr. 903

Names and Rank.	Page.

Muhammad Nabi, Ressdr. 1682
—— Ressdr. 894
—— Amin Khan, Jemdr. 884
—— *Hony*. Capt. 2101
—— Amir, Sub-Asst. Surgn. 1809
—— Khan, Jemdr. 876
—— Jemdr. 910
—— Arbi, Subdr. 1256
—— Arshad Khan, Ressdr. 938
—— Asadullah Saiyid, Subdr. 1359
—— Ashraf Shah, Ressdr. 893
—— Aslam Khan, Risaldar Maj. 895
—— Jemdr. 1154
—— Subdr. 1502–04
—— Ataullah Khan, Sub-Asst. Surgn. 1827
—— Ayub, Risaldr. 1684
—— Khan, Resldr. 942
—— Ali Khan, Subdr. 1397
—— Subdr.-Major 1397
—— Subdr. 1367
—— Jemdr. 1248
—— Azim, Subdr. 1375
—— Khan, Subdr. 1252
—— Risldr.-Major 942
—— Bahaud-din, Sub-Asst. Surgn. 1827
—— Bakhsh, Subdr. 1513
—— Subdr. 1207
—— Subdr. 1086
—— Jemdr. 1288
—— Jemdr. 999
—— Jemdr. 975
—— Jemdr. 1356
—— Subdr. 1439
—— *Hony*. Capt. 2102
—— Barkat-Ullah, Sub-Asst. Surgn. 1814
—— Basbir Khan 1397
—— Bashir Khan, Jemdr. 950
—— Burhan, Jemdr. 1276
—— Chirag-ud-din Khan, Subdr. 1416
—— Din, Subdr. 1283
—— Subdr. 1518
—— Subdr. 999
—— Jemdr. 1573
—— Jemdr. 974
—— Jemdr. 1452
—— Jemdr. 1492
—— Sub-Asst. Surgn. 1815
—— Sub-Asst. Surgn. 1814
—— Sub-Asst. Surgn. 1814
—— Jemdr. 1380
—— Jemdr. 1001
—— Jemdr. 1495
—— Faiyaz Khan, Sr. Sub-Asst. Surgn. 1809
—— Fakhr-ud-din Sahib, Sr. Sub-Asst. Surgn. 1826
—— Fakruddin Khan, Lt. 481, 1693
—— Gul Khan, 2nd-Lt. 1693
—— Feroz Khan, Subdr. Maj. 7, 1145
—— Ghalib, Subdr. Maj. 1323
—— Ghani Khan, Subdr. 1399, 1591
—— Gul, Subdr. 1506
—— Hanif Khan, Subdr. 1393
—— Ressdr. 912
—— Hafiz Khan, Sub-Asst. Surgn. 1811
—— Hanif 1436
—— Hasna Subdr. II 1427
—— Hashim Khan, Subdr. 1268
—— Hashum, Ressdr. 916
—— Jemdr. 908
—— Subdr. 1052
—— Subdr. 1488
—— Subdr. 1056
—— Hayat Khan, Ressdr. 936

Muhammad Hayat Khan, Jemdr. 910
—— Husain, Sr. Sub-Asst. Surgeon 1808
—— Ty. Sub-Asst. Surgeon 1833
—— Sahib, Jemdr. 1316
—— Shah, Jemdr. 1232
—— Khan, Jemdr. 1312
—— Subdr.-Maj. 1287
—— Jemdr. 892
—— Jemdr. 892
—— Jemdr. 1324
—— Jemdr. 900
—— Jemdr. 1372
—— Jemdr. 1145
—— Jemdr. 1137
—— Jemdr. 79
—— Jemdr. 1141
—— Jemdr. 1036
—— Jemdr. 926, 1152
—— (I) Jemdr. 1285
—— Resldr. 925
—— Jemdr. 999
—— (II) Jemdr. 1285
—— Jemdr. 944
—— Jemdr. 1373
—— *Hony*. Capt. 2101
—— Husain, Jemdr. 1579
—— Hussain, Sub-Asst. Surgn. 1811
—— Jemdr. 1428
—— Jemdr. 1400
—— Jemdr. 1328
—— Jemdr. 1264
—— Sub-Asst. Surgn. 1814
—— Subdr. 132
—— Subdr. 1333
—— Jemdr. 1079
—— Sub-Asst. Surgn. 1816
—— Sub-Asst. Surgn. 1816
—— Khan, Ressdr. 1686
—— Khan, Jemdr. 884
—— Jemdr. 1305
—— Hasan Jemdr. 1035
—— Jemdr. 951
—— Ibrahim, Jemdr. 1360
—— Subdr. 1467
—— Jemdr. 1308
—— *Hony*. Capt. 2102
—— Ibrahim Khan, Ressdr. 945, 953
—— Jemdr. 1035
—— Jemdr. 884
—— Illiaskhan, Risldr. 876
—— Inayat Khan, Ressdr. 910
—— Inayatullah Khan, Subdr. 1357
—— Isa Khan, Ressdr. 963
—— Ishak, Sub-Asst. Surgn. 1810
—— Ishaq, Jemdr. 934
—— Ishaq Siddiqi, Sub-Asst. Surgn. 1817
—— Khan, Jemdr. 1391
—— Jemdr. 1389
—— Islam Khan, Jemdr. 912 934
—— Ismail, Sub-Asst. Surgn. 1817
—— Ismail Khan Subdr. 1399
—— Jemdr. 1087
—— Subdr. 1327
—— Ismail, Ty. Sub-Asst. Surgn. 1833
—— Jemdr. 938
—— Ismail Khan, Jemdr. 890
—— Jemdr. 1701
—— Jalaluddin, Subdr. 1315
—— Jafar Husain, Jemdr. 1351
—— Sharif, Jemdr. 1332
—— Jamiluddin, 2nd-Lt. 481
—— Juma, Subdr. 1427
—— Jemdr. 999
—— Jan, Jemdr. 1514
—— Khan, Jemdr. 1397
—— Kazim, Jemdr. 970
—— Jemdr. 1280
—— Kazim Khan, Jemdr. 946

Muhammad Khan, Subdr. 1280
—— Jemdr. 1440
—— Jemdr. 1571
—— Senr. Sub-Asst. Surgn. 1807
—— Jemdr. 976
—— Jemdr. 966
—— Resldr. 919
—— Jemdr. 1236
—— Jemdr. 1244
—— Jemdr. 1433
—— Jemdr. 1506
—— Ressdr. 914
—— Subdr.-Maj. 1517
—— Risldr. 910
—— Jemdr. 910
—— Jemdr. 926
—— Jemdr. 1248
—— Jemdr. 1241
—— Jemdr. 1320
—— Subdr. 1399
—— Jemdr. 912
—— Subdr. 1419
—— Jemdr. 9 2
—— Subdr. 1120
—— Ressdr. 1684
—— Jemdr. 1228
—— Khwaja Hussain, Jemdr. 1308
—— Latif, Subdr. 1104
—— Maksud Ali Khan, Ressdr. 1686
—— Mihirdin, Sr. Sub-Asst. Surgn. 1807
—— Mir Afzal Khan, Subdr. 1268
—— Mohsin, Sub-Asst. Surgn. 1816
—— Munir, Ty. Sub-Asst. Surgn. 1833
—— Mustafa Khan, Jemdr. 1393
—— Murtaza Khan, Jemdr. 1391
—— Jemdr. 936
—— Muzaffar Hussain, Jemdr. 1396
—— Sher, Jemdr. 1356
—— Nasar Nazir, Jemdr. 1305
—— Nawaz, Jemdr. 942, 953
—— Khan, Subdr. 1132
—— Jemdr. 876
—— Jemdr. 903
—— Nazir Khan, Jemdr. 932
—— Nur-ul-Hak, Ressdr. 1083, 1086
—— Nurkhan, Ressdr. 944, 961
—— Panah, Jemdr. 1428
—— Rafi, Sub-Asst. Surgn. 1816
—— Riazuddeen, Sub-Asst. Surgn. 1826
—— Rasul, Jemdr. 1320
—— Jemdr. 1126
—— Raza, Ressdr. 908
—— Raza, Ressdr. 910
—— Raza Khan, Risldr. 878
—— Subdr. 1387
—— Sr. Sub-Asst. Surgn. 1809
—— Roshan, Subdr. 1492
—— Subdr. 1499
—— Sadikali Quraishi, Sub-Asst. Surgn. 1820
—— Sadiq, Subdr. 1379
—— Jemdr. 1361
—— Khan, Jemdr. 1356
—— Jemdr. 1404
—— Sahib, Tr. Vety. Asst. 1687
—— Said, Sub-Asst. Surgn. 816
—— Jemdr. 1316
—— Saiyid, Subdr. 1335
—— Saiyid Khan, *Hony*. Capt. 2103
—— Jemdr. 1405
—— Sakhawatullah Khan, Sub-Asst. Surgn. 18 1
—— Salim, Sub-Asst. Surgn. 1816
—— Sanah Khan, Jemdr. 876
—— Saraj, Jemdr. 1249
—— Sarwar, Jemdr. 908
—— Sattar Sharif, Ty. Sub-Asst. Surgn. 1833
—— Shaffi, Tr. Vety. Asst. 1687
—— Jemdr. 1473
—— Sayyid, Jemdr. 1304
—— Jemdr. 1236
—— Shafi, Jemdr. 1092

Names and Rank.	Page.

Muhammad, Shafi Khan, Ressdr. 876
—— Jemdr. 880
—— Jemdr. 938
—— Shafiq, Sub-Asst-Surgn. 1817
—— Sarwar Khan, Risldr. 965
—— Jemdr. 1208
—— Shafe Khan, Jemdr. 938
—— Jemdr. 938
—— Shah, Jemdr. 909
—— Shamsuddin, Jemdr. 1424
—— Shiabuddin, *Hony.* Capt. 2104
—— Sharif, Jemdr. 1435
—— Subdr.-Maj. 1361
—— Sub-Asst. Surgn. 1811
—— Khan, Jemdr. 946
—— Sharif, Jemdr. 1245
—— Jemdr. 1332
—— Sher, Jemdr. 1236
—— Jemdr. 1256
—— Sher, Jemdr. 1107
—— Jemdr. 924
—— Khan, Jemdr. 946
—— Jemdr. 966
—— Sulaiman Khan, Subdr. 1086
—— Jemdr. 1447
—— Suleman, Tr. Vety. Asst. 1687
—— Taqi Khan, Jemdr. 876
—— Tosawar Khan Afridi, Sub-Asst. Surgn. 1820
—— Umar, Sub-Asst. Surgn. 1811
—— Jemdr. 1384
—— Faruk Khan, 876
—— Khan, Jemdr. 888
—— Jemdr. 1035
—— Jemdr. 918
—— Usman, Sub-Asstt. Surgn. 1828
—— Khan, Jemdr. 1388
—— Jemdr. 882
—— Wali, Jemdr. 1288
—— Wasil Khan, Sub-Asst. Surgn. 1820
—— Wazir, Jemdr. 1388
—— Resldr. 944
—— Yakub, Sub-Asst. Surgn. 1817
—— Yakub Khan, Risldr. 919
—— Jemdr. 928
—— Jemdr. 1344
—— Jemdr. 1099
—— Sub-Asst. Surgn. 1820
—— Yar Khan, Subdr. 1036
—— Yasin, Jemdr. 940
—— Subdr. Maj. 1273
—— Jemdr. 1358
—— Yasin Khan, Jemdr. 932
—— Yunus, Jemdr. 1351
—— Yusuf, K., Sub-Asst. Surgn. 1827
—— Subdr. 1195
—— Khan, Jemdr. 938
—— Subdr. 1506
—— Khan, Ressdr. 888
—— Risldr. 882
—— Jemdr. 878
—— Bhatti, Sub-Asst. Surgn. 1813
—— Zainul Abedin, Sub-Asst. Surgn. 1825
—— Zahir-ud-din, Jemdr. 928
—— Zaki, Sub-Asst. Surgn. 1813
—— Zaman, Subdr. 1343
—— Subdr. 1259
—— Subdr.-Maj. 1207
—— Subdr. 1567
—— Subdr. 1393
—— Jemdr. 1488
—— Subdr. 1128
—— Sr. Sub-Asst. Surgn. 1807
—— Subdr. 896
—— Zanib, Jemdr. 1616
—— Zarif, Jemdr. 1380
Muhiyuddin, Jemdr. 1360
Muhi-ud-din, Subdr. 1307
Muhi-ud-din Khan, Rissldr.-Maj. 934
Muir, A., 2nd-Lt. 743
—— A. C. 2nd-Lt. 1501
—— A. H. H., Maj. 14, 66, 250, 1078
Muir, A. P., Lt. 859
—— J. H., Capt. 595
—— J. L., Lt. 509, 818, 986
—— W., 2nd-Lt. 763
—— W. H., Lt. 581
—— W. W., Maj. 5, 217, 1078
Muirhead, A., Hon. Comdt. 843
—— A. K., 2nd-Lt. 817
—— J., Capt. 764
—— J., Lt. 927
—— J. A., Maj. 51, 99, 226, 875
—— J. I., Lt. 368
—— J. M., 2nd-Lt. 1499, 1500
—— J. W., Lt. 491, 935
—— T. 788
Muizud-din Khan, *Hony.* Capt. 2101
Muizzullah Khan, Subdr. 1078
Mukand, S. N., Ty. Lt. 1775
—— Lal Varma, Sub-Asst. Surgn. 1819
—— Ram, Sub-Asst. Surgn. 1816
—— Singh, Risldr. 878
—— Ressdr. 880
—— Jemdr. 880
Mukarrab Khan, Jemdr. 1480
—— Jemdr. 1240
Mukerjee, C. P., Tempy. Lt. 1774
—— H. P., Tempy. Capt. 1768
—— H. P. (*Ty. Lt.*) 1782
—— M., Tempy. Lt. 1525, 1776
—— P., Ty. Capt. 1772
—— P. C., Tempy. Capt. 1768
—— S. B., Tempy. Capt. 1766
—— S. C., Tempy. Capt. 1768
—— S. K., Ty. Lt. 1774
—— S. N., Capt. 1758
Mukerji, A. K., Ty. Lt. 1774
—— J. N., Tempy. Lt. 1778
—— A. P. Ty. Lt. 1781
Mukharji, J. G., Tempy. Capt. 1767
—— J. M., Ty. Lt. 1774
—— K. D., Ty. Lt. 1780
—— N. B., Ty. Capt. 1771
Mukhdad Singh, Jemdr. 1364
Mukhdam Bakhsh Khan, 1388
Mukherji, M. 1776
Mukh Ram, Risldr. 878
—— Subdr 1419
Mukkatilla Kaiappa Uttaya, Subdr. 1561
Mukhmad Jan, Subdr 1359
Mukhtiar Singh, Ressaidr. 1168
Mukhtiara, Jemdr. 1184
Mukhtiyar Singh, Jemdr. 1031
Muktibahadur Rana, Jemdr. 1619
Mulchandra, Sub-Asst.-Surgn. 1820
—— Sub-Asst. Surgn. 1820
Mul Raj, Ressdr. 1685
—— Singh, Subdr. 1203
—— Ressdr. 932
—— Jemdr. 1433
—— Subdr. Maj. 1501
—— Sub-Asst. Surgn. 1814
—— Jemdr. 1488
—— Risldr. 925
Mula, Subdr. 1039
—— Jemdr. 960
—— Singh, Subdr. 1367
—— Jemdr. 1142
—— Sub-Asst. Surgn. 1814
—— Subdr. 1235
—— Jemdr. 942
—— Jemdr. 895
Mulcahy, W. J. P., Lt. 531, 1702
Mulchand Sharma, Sub-Asst. Surgn. 1811
—— Jemdr. 1028
Mulgrue, J. A., Lt. 509, 1667
Mulholland, H. H., Capt. 1715
Mulik Ghulam Muhammad, *Hony.* Capt. 2104
Mulk Raj Singh, Sub-Asst. Surgn. 1814
—— Puri, Sub-Asst. Surgn. 1810
Mulk Shah, Tr. Vety. Asst. 1687
Mull, A., Lt. 304, 1030
Mullan, H. D., Ty. Lt. 1775
Mullaly, B. R., Capt. 276, 1647
—— H., Maj.-Genl. 600
—— H., Maj. 246, 1641
—— H. A., Lt. 321, 305, 1403
—— J. C. D., Capt. 273, 1050
Mullarkey, P. S., Lt. 352, 1677
Muller, C. G., Asst. Surgn. 1799
—— C. J., 2nd-Lt. 790
Mulliken, E. M., Capt. 259, 1664
Mullin, R. H., Lt. 541
Mulliner, A. R., Lt. 589
Mullinger, R. O., 2nd-Lt. 577
Mullings, J. R., Maj. 615
Mullins, C., Sr. Asst. Surgn. 1791
—— C. W., Lt. 75
—— H., Asst. Surgn. 1798
—— H. L., Lt. 820
Mulloy, N. F. C., Capt. 1256, 935
Mulvany, J., Lt.-Col. 130, 1742
Mumford, E. C., 2nd-Lt. 565
—— E. M., Maj. 687, 710
—— J. J., Lt. 66, 760
—— W. J. T., Capt. 719
—— S. E. A., Lt. 518, 1199
Multani Singh, Jemdr. 1184
Mumtaz Khan, Subdr. 1086
—— Jemdr. 1476
—— Jemdr. 1511
—— Bhatti, Sub-Asst. Surgn. 1807
—— Jemdr. 1475
—— Jemdr. 884
—— Muhammad Khan, Jemdr. 922
Munckton, F. E., 2nd-Lt. 460, 462
—— V. C. A., Lt. 492, 879
Mundar Singh, Subdr. 1157
—— Jemdr. 1158
Munden, S. E., Lt. 55, 489
Mundock-Smith, M., 2nd-Lt. 1698
Munay, S. C. B., Capt. 745
—— H. A. G., Lt. 17, 36, 728
Munford, C. J. H., Capt. 754
Mungavin, H. St. J., Lt. 511
Mungli Singh, Jemdr. 1453
Munim, M. D., Tempy. Capt. 1767
Munir Khan, Risldr. 1084
Munirathnam, Subdr. 1328, 1591
—— Jemdr. 994
Munisami, Jemdr. 1276
—— Jemdr. (I) 994
—— Subdr. 1315
—— Jemdr. (II) 994
—— Jemdr. (III) 994
—— Jemdr. (IV) 994
—— Govander, Subdr. 131[illegible]
Munisami Naidu, Sr. Sub-Asst. Surgn. 1829
—— Raju, Jemdr. 1139
Munn, A., 2nd-Lt. 1509
—— A. E., 2nd-Lt. 561, 1132
—— F. L. R., Capt. 21, 269, 1207
—— J., 2nd-Lt. 783
—— L., Capt. 841
—— R. G., Lt.-Col. 133, 156, 193 1601
Munna Singh, Subdr. 1031
Munns, F. A. C., Lt. 492
—— P. G., 2nd-Lt. 494, 905
Munnsmy, Jemdr. 1704
Munro, A., 2nd-Lt. 472, 1389
—— A. C., Capt. 1754
—— A. R. Lt. 798
—— A. S., 2nd-Lt. 562, 1418*a*
—— D., Maj. 1748
—— D., Lt. 525
—— G. B., Maj. 1753
—— G. C. G., Lt. 715
—— G. D., 2nd-Lt. 640
—— H., 2nd-Lt. 355, 1088
—— J. C., Lt. 719, 759
—— R. C., Lt. 727
—— R. P., Capt. 843
—— T. M., Lieut. 651
—— T. O. A., 2nd-Lt. 1678*a*
Munroe, J., 2nd-Lt. 819
Munrowd, H. G. H., Asst. Surgn. 1794
Munsabdar Khan, Risdr. 899
Munsey, S. T. H., Lt. 515, 980
Munsna Singh, Jemdr. 1493
Musharraf Khan, Jemdr. 1304
Munshi, B. C., Ty. Lt. 1775
Munshi, Jemdr. 1130
—— Khan, Jemdr. 1142
—— Ram, Jemdr. 1103
—— Jemdr. 1571
—— Jemdr. 1571
—— Jemdr. 1145
—— Jemdr. 1505
—— Ressdr. 906
—— Ressdr. 1082, 1086
—— Singh, Jemdr. 1404
—— Shaikh Ahmad Shaikh Lal, Sub-Asst. Surgn. 1830
Munton, V. R., Capt. 253, 1415
Munnsomi, Jemdr. 1324
Muqarrab Khan, Jemdr. 953
Murad Ali Khan, Ressdr. 79

NAMES AND RANK. Page.

Murad Ali Khan, Jemdr. 1441
—— **Bakhsh**, Subdr. 1303
Murafat Shah, Subdr. . . . 1245
Murari Sinde, Subdr. 1467
Murar Rao Khore, Sub-Asst. Surgn. 1815
Murar Rao Salunke, Jemdr. . . . 1407
Murch, R. W., Lt. 778
Murcott, S. H. D., Lt. . . 425, 1431
Murcutt, F. A., 2nd-Lt. 568
Murdoch-Smith, M., Lt. . . . 530
Murgatroyd, J. O., Lt. . . . 687
Murie, J. C. W., 2nd-Lt. . 630, 976
Murison, C. O., Maj. 1747
—— C. A. P., Lt. 617
—— G. de H., Capt. . . 293, 1595
—— R. E. Capt. 1839
—— T. D., Capt. 1755
Murland, H. F., Maj 15, 37, 240, 1275
Murley, P. A., Lt. . . . 535, 1672
Murli Singh, Jemdr. 1384
—— Jemdr. 1488
Murphy, A., Maj. . . . 1415, 1748
—— A., Lt. . . . 8, 15, 82, 516a
—— C. C. R., Lt.-Col. 102, 133, 157, 207, 1140
—— D., Lt. 745
—— D. F., Capt. 1756
—— D. W., Lt. . 77, 617, 1885
—— E. J., Sr. Asst. Surgn. . 1789
—— E. J., Lt. . . . 597, 1009
—— E. L., Lt. . . . 487, 959
—— F., 2nd-Lt. 1636
—— G. P., Capt. 265, 1401, 1893
—— H. A., Asst. Surgn. . . 1803
—— H. N., Asst. Surgn. . . 1800
—— H. W., 2nd-Lt. . . 553, 1703
—— J., Lt. 543, 1305
—— J. E., Lt. 798
—— J. J., 2nd-Lt. . . 544, 1674
—— J. J. C., 2nd-Lt. . 455-59, 1401
—— J. L., Maj. . . 231, 1323
—— J. P., Capt. 777
—— J. R. K., Lt. . . 313, 891
—— L. A., 2nd-Lt. . . 545, 1677
—— R. S., Lt. . . 23, 579, 1128
—— R. T., Sr. Asst. Surgn. . 1787
—— R. W., Ty. Capt. . . 66, 1723
—— S. B. W., Asst. Surgn. . 1799
—— S. K., 2nd-Lt. . . . . .
—— W. O'S., Maj. 1747
—— W. R., *Hony*, Lt. . . . 1844
Murr, C., 2nd-Lt. 640
Murray, A. B., Lt.-Col. . 107, 131, 156, 192
—— A. E., Lt. . . . 516a, 1407
—— A. G., Maj. . . . 87, 220
—— A. J. Lt. 785
—— A. R., Capt. 764
—— B. J., 2nd-Lt. 363
—— C., Jemdr. 1308
—— C. A., Capt. 1838
—— C. A. G. O., Capt. . . . 718
—— C. F. T., Lt.-Col. 128, 154, 184
—— C. H., Capt. 720
—— C. H., 2nd-Lt. 745
—— C. J., Capt. . . 8, 571, 1850
—— C. R. B., Lt. . . 487, 1525
—— C. W., Asst. Surgn. . . 1796
—— E. M., Capt. . . . 262, 904
—— E. R. B., Brig.-Genl. 50, 121, 131, 156, 191, 870, 1367
—— F., Col. 149
—— F. M., Maj. . . . 74, 614
—— F. B. J., Maj. . . 213, 1661
—— G., Capt. 614
—— G., Lt. 524, 988
—— G. A., Sub-Condr. . . . 1835
—— G. B., Maj. . . . 229, 1002
—— G. E., 2nd-Lt. 530
—— G. W., Capt. 654
—— H., Maj. 87, 225
—— H., Lt. 709
—— H. A., Maj. . 227, 1140, 1164
—— H. B., Maj. 57
—— H. E., Capt. 1758
—— H. F., Maj. 885
—— H. K., Lt. . . . 668, 991
—— H. L., 2nd-Lt. 561
—— J., Capt. 715
—— J. A., 2nd-Lt. 656
—— J. G. P., Lt.-Col. . 137, 1744
—— J. H., Maj. . . 1749, 1890
—— J. H. S., Lt.-Col. . 22, 129, 601
—— J. R., 2nd-Lt. . . 365, 1166
—— J. W., Lt.-Genl. *Sir* . . 844

NAMES AND RANK. Page.

Murray, J. W., 2nd-Lt. . . . 1174
—— K. D. B., Maj 13, 26, 99, 219, 1259
—— K. M., Capt. 741
—— K. R., Lt. 708
—— M. J., Lt. . . 519, 1451
—— M. J., Lt. 489
—— M. J., Capt. 1723
—— N., 2nd-Lt. 832
—— N. T. G., Lt. . . 420, 1136
—— P., Lt. . . . 465, 1120
—— P. F., 2nd-Lt. . . 357, 1088
—— R., Capt. 1369
—— R. E., Asst. Surgn. . . 1794
—— R. G., Lt. . . . 443, 1575
—— R. G. H., Capt. . . 269, 1641
—— R. H., *Hon.* Maj. Genl. . 764
—— R. T. G., Capt. *Hon'ble* . 744
—— S. G. C., Capt. . . 255, 1060
—— S. J., Bt.-Col. . . . 769
—— S. K., 2nd-Lt. . . 329, 1647
—— T., Capt. 1722
—— T., Lt. 1068
—— T. H., Lt. 585
—— T. W., Lt. . . 314, 1463, 1407
—— V., Col. . . 60, 126, 660, 1881
—— V. R., Lt. . . 450-54, 1601
—— V. W., Lt. . . . 304, 961
—— W. G. P., Lt.-Col. 98, 139, 160, 199, 868, 1102
—— W. P., Lt. . . . 828, 861
—— W. H., Lt. 805
—— W. W. M., Capt . . . 485
Murray-Johnson, F. K., Ty. Lt. 581
Murray-Shirreff, B. G., Lt. 310, 1152, 1563
Murrell, A. F., Lt. 352
—— F. E., 2nd-Lt. 807
—— P. A., 2nd-Lt. 754
Murrian, Jemdr. 1332
Murtaza Khan, Jemdr. . . . 876
—— Jemdr. . . . 1130
—— Subdr. . . . 1493
—— Sharif, Jemdr. . . . 1264
Murty, P. V. R., Tempy. Lt. . 1773
Murugayya, Subdr. 1271
Murrugasan, Jemdr. . . . 1272
Murrugesam, Jemdr. . . . 1360
Murugesan, Jemdr. . . . 1012
—— (I) Jemdr. . . . 993
—— (II) Jemdr. . . . 994
—— (III) Jemdr. . . . 994
—— Subdr. 1011
Musa Khan, Subdr. 1509
Muschamp, W. A. H., 2nd-Lt. . 1122
Muscroft, W., Bt. C., Lt.-Col 20, 131, 156, 206, 1661
Musgrave, A. D., Br.-Genl. 121, 126, 613
—— T. H., Lt. 630
Mushak Ali Khan, Subadar Maj. 1435
Musharraf Khan, Ressdr. . . 882
—— Jemdr. . . . 1305
—— **Ali Khan**, Jemdr. . 920
Muspratt, C. K., Capt. . . . 733
—— F. C., Col. 120, 123, 149, 182
—— G., Lt. . . . 528, 1391
—— R. A. L., 2nd-Lt. . 356, 1391
—— S. F., Lt.-Col. 98, 138, 159, 215, 897
—— V. E., Maj. . 9, 16, 99, 103, 210, 931
—— W., Revd. 1903
Muspratt-Williams, C. A., Lt.-Col. 128, 632, 1890
Mussadi, Subdr. . . . 510, 1171
Mussallam Khan, Subdr. . . . 1086
Mussenden, F. W., Lt.-Col. 131, 589
Mustafa, Jemdr. 968
—— Subdr. . . . 1128, 1129
—— Khan, Ressdr. . . . 926
—— **Sharif**, Jemdr. . . 932
Mustak Singh, Subdr.-Maj. . . 1589
Musto, A. A., 2nd-Lt. . . 61, 523
—— C. R., Lt. . . 353, 1197
—— W. R., Lt. 521
Mutch, C. H., Lt. 1689
Muter, R. S., Major. . . 662, 1881
Muthu Karappan, Jemdr. . . 1699
Muthukrishnan, A., Ty.-Lt. . . 1778
Muthusami, Subdr. 1359
—— Jemdr. 1340
—— Nayudu, Jemdr. . . 1304
Muthuviru, Subdr. 1327
—— Jemdr. 1351

NAMES AND RANK. Page.

Muttusami, T. A., Sr. Sub-Asst. Surgn. 1826
—— Subdr. 1323
Muttuvell, Subdr. 1276
Muzaffar Khan, *Hony.* Capt. 1894, 2101
—— Jemdr. 1143
—— Jemdr. 1228
—— Jemdr. 950
—— Jemdr. 942
Muzhar Hussain, Sub-Asst. Surgn. 1828
Muzzammil Hussain, Sr. Sub-Asstt. Surgn. 1807
—— Jemdr. . . 1308
Myburgh, P. B., Lt. 797
Myer, B. D., Ty. Lt. . . . 1781
—— H. A., 2nd-Lt. . . . 788
Myers, F. G., Lt. 1583
—— N. C., Lt. . . . 519, 998
—— N. G., Lt. 397
Myles, A. V. (Tempy. Capt.) . . 1691
—— G. S., 2nd-Lt. 625
—— R. B., Capt. 1715
—— R. H. S., Lt. 1674
—— A. V., Lt. . . . 1693, 1919
Mylles, C. C., Capt. 763
Mylne, R. H., Capt. . . 265, 1547
Mylon, D., 2nd-Lt. 361
Mysore, *Maharaja* of, *Hony*. Col. 110, 923

N

Nabi, Jemdr. 1068
—— **Ahmad, Sadiki**, Sr. Sub-Asst. Surgn. 1807
—— **Bakhsh**, Jemdr. 1092
—— Jemdr. 1288
—— **Buksh**, Jemdr. 1681
—— **Bux**, Jemdr. 1702
—— Jemdr. 1088
—— **Jan Hussain Ali Shaikh**, Sub-Asst. Surgn. 1831
Nabi-Yar Khan, Jemdr. . . . 914
Nadar, G. J. B., Sub-Asst. Surgn. 1828
—— **Khan**, Jemdr. 948
—— Jemdr. 1507
—— II, Jemdr. . . . 1691
Nadesan Pillai, Subdr. . . . 1331
—— Subdr. . . . 1334
Nadhan Sing, Jemdr. . . . 1260
Nadin, G., Lt. . . . 300-01, 897
Nadir Ali, Subdr. 1102
—— Jemdr. 1054
—— **Khan**, Jemdr. . . . 1312
—— Subdr. 1122
—— Jemdr. 1412
—— Jemdr. 1248
—— II, Jemdr. . . . 1208
—— Subdr. 972
—— Jemdr. 1681
—— **Shah**, Jemdr. . . . 1248
—— **Ali Khan**, Jemdr. . . . 898
—— Ressdr. 896
Nag, S., Tempy. Capt. . . . 1707
—— Jemdr. 1171
Nagabhushanam, Subdr. . . . 1705
Nagakar, V. R., Tempy. Capt. . . 1770
Nagendra Nath Choudhary, 2nd-Lt. 481
—— **Das Gupta**, Sub-Asst. Surgn. . 1815
—— **Vehari Mukerjee**, Jemdr. . 1219
Nagesar Sing, Jemdr. 1047
—— Tewari, Subdr. . . . 1018
Nagina Singh, Sub-Asst. Surgn. 1808
—— Jemdr. . . . 1121
—— Ressdr 896
Nagle, E. H. B., *Hony.* Lt. . . 1847
Nagindar Singh, Sub-Asst. Surgn. 1813
Naido, R. G., Sr. Sub.-Asst. Surgn. 1829
Naidu, B. P. B., Tempy. Capt. . 1766
—— C. R., Ty. Lt. 1828
Naik, N. R., Tempy. Capt. . . 1766
—— **Din Muhammad**, Jemdr. . 1412
—— **Mirya**, Jemdr. . . . 1079
Naim Gul, Subdr. 1128
Naimjirao Kadam, Jemdr. . . 1469

Names and Rank. — Page.

Nain Singh, Jemdr. 1225
—— Bhandari, Jemdr. . 1182
—— Subdr. . 1223
—— Bisht, Jemdr. . 1179
—— Ohinwarh, Subdr.-Maj. 1179
—— Gharti, Subdr. . . . 1615
—— Gurung, Jemdr. . . . 1615
—— Thapa, Jemdr. . . . 160
—— Jemdr. . . . 1602
—— Jemdr. . . . 1618
—— Thakur, Subdr.-Maj. 1223
Nain Lama, Subdr. 1630
Naini, Jemdr. 1457
—— Lama, Jemdr. 1630
Nairn, A. M., Lt. 622
Nairne, E. G., Capt. . . . 261, 1895
Naismith, W. N., Lt. . . . 537, 1098
Naiz Ali, Jemdr. 1453
—— Ahmed Khan, Ressdr. . 1683
Najab Khan, Jemdr. . . . 1681
Najabat Ali, Subdr. 1108
Najibullah Khan, Subdr. . . . 1207
Najib Khan, Ressdr. 926
—— Jemdr. 1380
Najibullah, Jemdr. 1428
Najif Khan, Jemdr. . . . 1681, 1682
Nakched, *Hony.* Capt. 2102
Naider, F. W., Lt. 702
Nalini Mohan Roy, Chaudhuri Babu, 2nd-Lt. . . 481
Nalwa, I. S., Ty. Lt. . . . 1780
Nambiar, P. K. R., Ty. Lt. . . 1779
Nambyar, K. K. Ty. Lt. . . . 1779
Namdar Khan, Jemdr. . . . 1245
Namdia Krishnaji, Jemdr. . . 1433
Nana Nikarr, Subdr. 1699
Nanajirao Palande, Jemdr. . . 1468
Nanak Chand, Subdr. 1171
—— Jemdr. 1683
—— Ram, Jemdr. 1356
—— Singh, Bahadur, Jemdr. 1247
—— Jemdr. 1432
—— Subdr. 1217
—— Jemdr. 1208
—— Risaldar. 1683
—— Jemdr. 1216
—— (I) Jemdr. . . . 1067
Nance, F., Sub-Condr. 1857
—— W. J., Capt. . . 265, 331
Nanda, Jemdr. 1060
—— Subdr. 1193
Nandabir Chetri, Jemdr. . . . 1347
—— Rana, Jemdr. . 1624, 1651
—— Thapa, Jemdr. . . 1602
Nandi, A. K., Ty. Lt. 1775
—— C. L., Ty. Lt. 1774
Nandlal Misr, Jemdr. 1027
Nand Kumar Tawai, Sub-Asst. Surgn. . 1819
—— Lal, Sub-Asst. Surgn. . 1814
—— Jemdr. 1400
—— Jemdr. 1167
—— Jemdr. 1054
—— Jemdr. 1188
—— Gurung, Subdr. . . . 1624
—— Gupta, Sub-Asst. Surgn. 1819
—— Ram, Subdr.-Maj. . . 1058
—— Sahi, Jemdr. . . . 1619
—— Sing Gurung, Jemdr. . . 1625
—— Jemdr. 1096
—— Singh, *Hony.* Capt. . . 2102
—— II, Jemdr. 1129
—— Jemdr. 1068
—— Jemdr. 1165
—— Subdr. 989
—— Jemdr. 1096
—— Jemdr. 1075
—— Jemdr. 1241
—— Jemdr. 1160
Nandu, Subdr. 1320
—— Jemdr. 1232
—— Gurung, Jemdr. . . . 1597
Naneh Singh, Jemdr. 901
Nanhe Singh, Resdr. 902
Nanga, Jemdr. 1192
—— Rawat, Jemdr. 1484
Nangle, K. E. Lt.-Col. 136, 158, 196, 871
—— M. C., Maj. 99, 104, 870, 1375, 1405
Nangpaul, D. C., Tempy. Capt. . 1768
Nanha Singh, Jemdr. 888
Nankarvis, T. J., 2nd-Lt. . . . 360
Nanla, S. D., Ty. Lt. 1774
Nanig Ram, Jemdr. 1385
—— Jemdr. 1488
Nanjappa, Subdr. 1276
Nanjappa, Naidu, Jemdr. . . . 1324
Nanne Khan, Jemdr. 1196
Nanti, Jemdr. 1587
Nanton, H. C., Brig.-Genl. 31, 120, 122, 663
Nanwa, Jemdr. 1456
—— Jemdr. 1585
—— Singh, Subdr. 1455
Napier, A. H., Maj. 1753
—— C. S., Lt. 996
—— G., Lt. 636
—— J. F., Sub-Condr. . . . 1864
—— M., Lt. 66, 730
Napier-Ford, G. S., Lt. . . 493, 953
Nar Singh Thapa, Jemdr. . . . 1637
—— Subdr. . . 1602
—— Rai, Jemdr. 1631
Narain Kadam, Jemdr. . . . 1416
—— Das, Sub-Asst. Surgn . . 1809
Narainsami, Subdr. 1011
—— Jemdr. . . 1264, 1351
Narain Dutta, Jemdr. 1225
—— Mal, Jemdr. 1604
—— Sing Thapa, Subdr. . . 1602
—— Jemdr. . . 1597
—— Singh, Subdr. 1203
—— Jemdr. 1565
—— Pawar, Jemdr. . 1181
—— Subdr. 1117
—— Ressdr. 1684
—— Jemdr. 1321
—— Jemdr. 1165
—— Risldr 918
—— Jemdr. 1375
—— Subdr. 1699
—— (I), Jemdr. . . . 1117
—— Sub-Asst. Surgn. . 1813
—— Jemdr. 850
—— Jemdr. 990
—— Jemdr. 900
—— Rana, Jemdr. 1608, 1609
Navainsawmy, Jemdr. 1272
Narang Singh, Subdr. 1166
Naranjan-das, Sub-Asst. Surgn. . 1812
Nara Rawat, Jemdr. 1476
Narasimulu, Subdr. 1339
—— Subdr. 1323
Narasimhan, T. C., Sub-Asst. Surgn. 1828
Narasinhula Bahadur, Jemdr. . 1327
Narasingaperumal, Subdr. . . 1011
Narasingasami, Jemdr. . . . 994
Narayanaras, D., Ty. Lt. . . . 1778
Narayan, R., Lt. 1691
Narayan Das, Sub-Asst. Surgn. . 1808
—— Sub-asst. Surgn. 1808
—— Dat, Jemdr. 1175
—— Dhuri, Jemdr. 1440
—— Giri, Subdr. 1699
—— Kadam, Jemdr. . 1416, 1464
—— Khonne, Jemdr. . . . 1479
—— Moreshvar, Sr. Sub-Asst. Surgn. . . . 1829
—— Parshao Sukul, Sub-Asst. Surgn. 1808
—— Rao Goghle, Jemdr. . 1705
—— Sinde, Jemdr. . . 1430
—— Sindhe, Subdr. . . 1699
—— Sadashiv, Jemdr. . . 1472
—— Singh, Subdr. . . . 1062
—— *Hony.* Capt. . . 2102
—— Subdr.-Maj. 1078, 1080
—— *Hony.* Capt. . . 2102
—— Jemdr. 1111
—— Subdr. 1379
—— Subdr. 1080
—— Jemdr. 966
—— Jemdr. 1480
—— Jemdr. 1124
—— Subdr. 1082
—— Jemdr. 1356
—— Subdr. 1367
—— Subdr. 1074
—— Ressdr. 915
—— Subdr. . . 1111, 1589
—— Subdr. 1145
—— Subdr. 1011
—— Jemdr. 1587
—— Jemdr. 922
—— R, Subdr. 1315
—— Subdr. 1476
—— Jemdr. 1124
—— Jemdr. 1240
—— Jemdr. 1166
—— Jemdr. 989
—— Subdr. 1216
Narayan Singh, Subdr. 1375
—— Jemdr. 1114
—— Jemdr. 1112
—— Jemdr. 873
—— Kadam, Jemdr. . . . 1432
—— Singh Dhariwal, Sub-Asst. Surgn. . 1818
—— Shetge, Jemdr. . . . 1468
Narayansami, Jemdr. 1333
Narayanan Nair, M., Sub-Asst. Surgn. . . . 1828
—— Nambier C., Jemdr. . . 1309
—— Nainbujar, C., Jemdr. 1308
Narayanasami, Subdr. (II) . . 1276
—— Subdr.-Maj. . . 1331
—— Subdr. (I). . . 1276
—— Subdr. 1327
—— Naidu, Subdr. . 1323
—— Jemdr. . . . 1276
—— Jemdr. 994
—— (I) Jemdr. . . 994
—— (II) Jemdr. . . 994
—— (III) Jemdr. . . 994
—— Jemdr. 1275
—— Jemdr. 1275
—— P. Jemdr. . . . 1275
Narayandat Dube, Jemdr. . . 1026
Narbahadur Gurung, Subdr. . 1653
—— Subdr. . . 1618
—— Jemdr. . . 1627
—— Jemdr. . . 1649
—— Subdr. . . 1636
—— Thapa, Jemdr. . 1602
—— Limbu, Jemdr. . 1649
—— Ram, Subdr. . . 1653
Narbir Gurung, Jemdr. . . . 1622
—— Khatri, Subdr. 1642
—— Limbu, Jemdr. . . . 1631
—— Pun, Subdr. 1615
—— Jemdr. 1637
—— Thapa, Subdr. 1602
—— Subdr. 1611
—— Jemdr. 1615
—— Gharti, Jemdr. 1604
Narendra Nath Ghatak, 2nd-Lt . 482
Narbhu Thapa, Jemdr. 1625
Narhar Balwant Salunke, Sub-Asst. Surgn . 1831
NariaThapa, Jemdr. 1598
Narian Sing Thapa, Jemdr. . . 1002
Narikuti Ramunni, Sr. Sub-Asst. Surgn 1826
Nariman, J. K., Capt. 1704
Narindar Singh, Sub-Asst. Surgn. 1814
—— Jemdr. 1163
Narinjan Singh, Jemdr. . . . 1103
Narjit Gurung, Jemdr. 1631
—— Jemdr. 1633
—— Jemdr. 1639
Narpati Rai, Subdr.-Maj. . 1030, 1653
—— Lama, Jemdr. 1649
Narpatti Bura, Jemdr. 1651
Narpat Singh, Jemdr. 1043
Narsing Thapa, Jemdr. 1002
—— Jemdr. 1639
—— Dat, Jemdr. 1016
Narsu Jadhav, Jemdr. 1444
Narsinghu, Subdr. 1174
Narsingh Das Khurana, Sub. Asst. Surgn. 1817
Naru Panre, Jemdr. 1642
Nash, A. S., *Hony.* Lt. 1845
—— A. V., Lt. 305
—— G. F., Lt. 543, 1401
—— H., 2nd Lt. 560, 1524
—— H. J., 2nd-Lt. 820
—— J. A., Capt. . . . 485, 608
—— S. G. C., Lt. . . . 543a, 1323
—— W., Capt. 751
Nasir Khan, Subdr.-Maj. . . . 1259
—— Risldr. . . . 1682, 1051
Nasiruddin, Ressdr. 1683
Nasir-ud-din, V. K., Sub-Asst. Surgn. . . . 1827
—— Jemdr. 993
Nasir-ulla Khan, Subdr. . . . 1492
Nasmith, S. U., Lt. 1069
Nasurullah Khan, Ressdr. . . 1084
—— Jemdr. 1432
Natesan, V. R., Ty. Lt. 1777
Nath, B., Ty. Lt. 1777
—— V., Ty.-Lt. 1773
Natha, Jemdr. 1561
—— Khan, Jemdr. 966
—— Ram, Jemdr. 1056

Names and Rank.	Page.
Natha Singh, Subdr.	1299
——— Jemdr.	1702
——— Subdr.	1157
——— Subdr.	1150
——— Jemdr.	1151
——— Risldr.	950
——— Jemdr.	922, 930
——— Subdr. (II)	1156
——— Jemdr.	920
——— Hony. Capt.	2104
——— Jemdr.	1112
——— Jemdr.	1589
——— Ressdr.	948
——— Jemdr.	1228
——— Jemdr.	1563
——— Jemdr.	1698
——— Jemdr.	1449
——— Jemdr.	1216
——— Sub-Asst. Surgeon.	1811
——— Jemdr.	920
——— Jemdr.	1260
Nathan, Jemdr.	1484
——— Singh, Jemdr.	1472
——— M., 2nd-Lt.	719
——— S. K. Ty. Lt.	1777
——— S. L., Capt.	502, 1247
Nathe Khan, Ressdr.	876
——— Subdr.	1108
——— Subdr.	1001
Nathu, Subdr.	1395
——— Jemdr.	1117
——— Khan, Subdr.-Major	1367
——— Subdr.	1092
——— Jemdr.	1052
——— Jemdr.	1090
——— Ram, Subdr.	1401
——— Singh, Jemdr.	940
——— Jemdr.	1585
——— Jemdr.	884
——— Sunar, Hony. Capt.	1179, 2104
——— Thapa, Jemdr.	1608
Nathwa Ram, Jemdr.	1143
Naubat Singh, Ressdr.	953
——— Jemdr.	1054
Naughton, F. A., 2nd-Lt.	559
——— T. H., Condr.	1870
Naulakh Singh, Jemdr.	1391-71
Naurang Singh, Sub-Asst. Surgn.	1814
——— Khan, Jemdr.	1480
——— Singh, Subdr.	1070
——— Jemdr.	1021
——— Subdr.	1074
——— Jemdr.	1011
Naurata Ram, Sub-Asst. Surgn.	1815
Nauroz Khan, Jemdr.	892, 1587
Nausher Khan, Jemdr.	990
Navanagar, *Maharaja of*, Lt.-Col.	110
Nawa Singh, Jemdr.	1456
Nawab Ali, Jemdr.	1514
——— Jemdr.	990
——— Ali Khan, Jemdr.	892
——— Jemdr.	1087
——— Khan, Jemdr.	1241
——— Jemdr.	1311
——— II Jemdr.	1336
——— Subdr.	989
——— Risldr.	1684
——— Jemdr.	1514
——— Jemdr.	894, 955
——— Subdr.	1095
——— Jemdr.	946
——— Ressdr.	884
——— Singh, Jemdr.	1510
Nawab-ud-Din, Tempy. Lt.	1773
Nawab Khajeh Habibullah, Jemdr.	1219
Nawab Singh, Resdr.	884
Nawal Singh, Jemdr.	1083
——— Jemdr.	926, 1299
——— Rana, Subdr.	1607
Nawalbir Gurung, Jemdr.	1637
Nawaz, M., Tempy. Lt.	1773
Nawazil Ali, Jemdr.	1335
Nawazish Ali, Jemdr.	902
——— Subdr.	1122
Nayagar, C. P., Sub-Asst. Surgn.	1827
Nayakar, P. T. V., Sub-Asst. Surgn.	1828
——— P. P. V. R., Sub-Asst. Surgn.	1827
Nayanar Pillai, Jemdr.	1317
Nayaran, B., Tempy. Capt.	1768
Nayar, T. K., Tempy. Capt.	1770
——— T. M., Tempy. Lt.	1773
Naye, B. A., Lt.	440, 1389
Naylor, G., Lt.	745
——— H. W., Maj.	66, 620
Naylor, T., 2nd-Lt.	90, 767
Nayudu, C. K. K., Sub-Asst Surgn.	1827
——— D. P., Sub-Asst. Surgn.	1827
——— E. M. A., Sub-Asst. Surgn.	1827
——— E. V., Sub-Asst. Surgn.	1827
——— G. R., Sub-Asst. Surgn.	1827
——— J. A., Sub-Asst. Surgn.	1827
——— M. A. N., Sub-Asst. Surgn.	1828
——— M. A., Sub-Asst. Surgn.	1825
——— M. V., Sub-Asst. Surgn.	1828
——— T. R., Sub-Asst. Surgn.	1825
——— V. V., Sr. Sub-Asst. Surgn.	1826
Nazam Khan, Jemdr.	1489
Nazar Muhammad, Subdr.	1006
——— Jemdr.	1122
——— Khan, Jemdr.	1484
Nazir Khan, Sub-Asst. Surgn.	1812
——— Subdr. Maj.	1056
——— Jemdr.	1372
——— Ali, Jemdr.	1701
——— Khan, Jemdr.	1088
——— Muhammad, Jemdr.	990
Nazru Khan, Jemdr.	1480
Neal, G. E., 2nd-Lt.	
——— E. A. C. 2nd-Lt.	455, 1181
——— G. T. St.-J., Lt.	521
——— H. A., Lt.	302
——— J. H., 2nd-Lt.	556
——— W. D., Sr. Asst. Surgn.	1787
Neale, F. C. L., 2nd-Lt.	472, 895
——— K. G., 2nd-Lt.	828
——— M. R., Maj.	22, 75
——— R. K., Lt.	306, 1120
——— R. T., 2nd-Lt.	1088
——— T., Sub-Condr.	1863
——— W. G., Maj.	239, 1890, 1913
Nealor, W. S., Maj.	1732
Neame, E. M., Lt.	752
——— L. V., Maj.	752, 783
Neate, W. H., Lt.	535, 1092
——— T. H., Lt.	525
Neare, L. H., Cap	1699
Needham, J., Lt.-Col.	532,
——— J. O., Lt.	321, 1120, 1519
——— B. A., Ma	4, 107, 1750, 1890
Neel, K. H., Lt.	297, 1433
Neely, A. T., Capt.	851
——— F. R., Lt.	850
Nehal Singh, Subdr.	1698
——— Subdr.-Maj	1698
——— Sing, Jemdr.	1581
——— Subdr.-Maj.	1691
Neighbour, P. M., Lt.	1720
Neil, J. A., Lt.	528, 1895
Neild, R., Lt.	501
Neill, F. H., 2nd-Lt.	996
——— J. F. B., Lt.	50, 531
Neilson, E. G., Lt.	1699
——— W. H., Capt.	835
Nek Alam, Jemdr.	1347
——— Subdr.	1698
Nek Muhammad, Jemdr.	1518
——— Jemdr.	1440
Neki Ram, Subdr.	1320
——— Jemdr	1433
——— Jemdr.	1397
——— Jemdr.	1320
Nelson, A. M., Revd.	1905
——— C., Capt.	17, 21
——— H., 2nd Lt.	548
——— J. I., 2nd-Lt.	556, 828, 1293
——— J. J. H., Capt.	1755
——— J. W., Maj.	749
——— S., 2nd-Lt.	556, 1703
——— W. C., Sub-Condr.	1873
Nepal, *Sir* Chandra Shamsher Jang, Prime Minstr. of, *Hony.* Lt.-Genl.	170, 1613
Nepal, Singh, Jemdr	924
Nepean, H. D. B. Y., Lt.	279, 1617
——— H. E. C. B., Col.	56, 121, 126, 150, 187
Ne'Roy, W. C., Asst. Surgn.	1798
Nery, P. E., Asst. Surgn.	1798
Nesbitt, E. A., Lt.	501a, 1665
——— H. K., Lt.	537
——— H. U. S., 2nd-Lt.	1114
——— J. D. W., 2nd-Lt.	728
Nesfield, V. B., Maj.	1749
Nesh, H., 2nd-Lt.	506
Netar Singh Nagi, Jemdr.	1179
Nether Sing Bisht, Jemdr.	1179
Nethersole, F. R., Lt.-Col.	193, 1895, 1912
——— J., Capt	263, 921
Nether wood, E., 2nd-Lt.	565
——— R. J., Sub-Condr.	1870
Net Ram, Subdr.	1453
——— Subdr.	1483
——— Jemdr.	1019
——— Subdr.	1039
Netscher, C. S., Asst. Surgn.	1803
——— E. G., 2nd-Lt.	566, 1387
——— F. H. D., Sr. Asst. Surgn.	1791
——— W. A. C., Sr. Asst. Surgn.	1791
Neumann, C. W., Bt. Lt.-Col.	2.6, 1215
Neve, E., Lt.	778
Nevile, R. L., Nursg. Sister	1735
Nevill, E. P., Lt.	608
——— H. R., Lt.-Col.	8, 15, 82, 485, 608
——— J. B., Lt.	524, 1143a
Neville, E. A., Lt.-Col.	1896
——— R. B., Lt.	724
——— W. E., Lt.	577
——— Rolfe, A., Capt.	1617, 264
Nawat Khan, Jemdr.	1134
Newall, C. L. N., Capt.	122, 259, 1600
——— G., 2nd-Lt.	788
Newal Singh, Jemdr.	1583
Newbery, C. A., Revd.	651
——— E. A., 2nd-Lt.	328, 1567
——— J., Sub-Condr.	1857
——— J. E., 2nd-Lt.	608
Newbold, A. W., Col.	123, 149, 182
——— A. W., 2nd-Lt.	1585
——— J. W., Asst. Surgn.	1798
——— W., 2nd-Lt.	758
Newcomb, C., Capt.	1756
——— H. R., Lt.	430
Newcombe, G. H., Maj.	105, 218, 1895, 1913
Newell, E. L., 2nd-Lt.	640
——— H. A., Lt.-Col.	104, 136, 158, 195, 1661
——— S. L., Lt.-Col.	1009
Newey, R. H., Lt.	761
Newfield, M., Lt.	747, 791
——— M. M. Lt.	1006
Newham, H. H., The Revd.	1908
——— F. G., 2nd-Lt.	351
Newhouse, A. W., Lt.	726
——— C. G., Lt.	817
Newill, J. H., Lt.-Col.	178, 1918
——— W. M., Lt.	299, 937
Newing, F. J. E. L'E., Snr. Asst. Surgn.	1751
Newington, F. E., Condr.	1872
——— G. H., 2nd-Lt.	328
——— M. T. L., Lt.	487, 885
Newitt, D. M., Capt.	501a, 1235
Newland, B. E. M., Maj.	1759
——— C. D., Lt.	371, 1597
——— C. J. Lt.	1263
——— J. H. F., 2nd-Lt.	316, 366, 1108, 1151
Newling, A. B., Capt.	589
Newlyn, E. W. G., Lt.	354, 1439
Newman, B. L., 2nd-Lt.	669, 1884
——— C. E. P., 2nd-Lt.	1483
——— C. J., Lt.	759
——— C. J., Lt.	789
——— C. A. R., Lt.-Col.	130, 1742
——— E. B., 2nd-Lt.	713
——— E. J., Lt.	491, 925
——— E. W., Sub-Condr.	1864
——— F. C., 2nd-Lt.	693
——— G. J., 2nd-Lt.	357, 1088
——— H. 2nd-Lt.	717
——— H., Maj.	852
——— H., Capt.	501a, 1559
——— H., 2nd-Lt.	67
——— H. J., Ty. 2nd-Lt.	707
——— J., Lt.	353
——— J. H., 2nd-Lt.	549
——— R. A. 2nd-Lt.	40, 793
——— T., 2nd-Lt.	736
——— T. D., 2nd-Lt.	568
——— W. C., 2nd-Lt.	361
Newman-Saunders, C., 2nd-Lt.	802
Newnham, C. C., Lt.-Col.	98, 134, 157, 195, 867, 885
——— H. J., Lt.	352, 1126
Newsam, E. S., 2nd-Lt.	1393
——— W. O., Lt.	493, 903
Newsom, A. C., Col.	130, 1838
Newson, W. 2nd-Lt.	561, 1123
Newstead, F., 2nd-Lt.	619
Newton, A., 2nd-Lt.	1009
——— A. B. G., Lt.-Col.	826
——— E., Capt.	1718
——— G. H., Capt.	843
——— J. F., 2nd-Lt.	742
——— L., Condr.	1872
——— P. L., Capt.	636
——— R. P., the Revd.	1904

NAMES AND RANK. Page.

Newton, Davis, C., Capt. . . . 1756
Newton-King, R. M., Lt. . . . 300
Neylan, P. J., 2nd-Lt. . . . 1679
Newquist, H. H., 2nd-Lt. . . . 535
Nhanu Sawant, Jemdr. . . . 1468
Niamat Khan, Jemdr. . . . 1133
—— Jemdr. . . . 1117
—— Subdr. . . . 1391
—— Jemdr. . . . 1569
—— Subdr. . . . 1571
—— Jemdr. . . . 1575
—— Rsssdr. 890
—— Ullah, Subdr. . . . 989
—— Khan, Tr. Vety. Asst. . 1687
Niamatullah Khan, Subdr. . . 1092
—— Jemdr. . . 1192
Niaz Ahmed Khan, Ressaidr. . . 1684
—— Ali Khan, Sr. Sub-Asst. Surgn. 1807
—— Jemdr. . . . 1232
—— Jemdr. . . . 1435
—— Mohammed, Subdr. . . 1088
—— Khan, Jemdr. . . 1056
Niazbat Khan, Jemdr. . . . 1128
—— Jemdr. . . 1567
Niaz Gul, Jemdr. 1099
—— Mir, Jemdr. 1502
—— Muhammad Khan, Jemdr. 957, 1192
Nibaran Chandra Roy, Sub-Asst. Surgn. 1820
Niblett, L. A., 2nd-Lt. 361
—— R. H., Lt. 562, 814
Niblock, W. J., Lt.-Col. . 134, 1743
Nicachi, J. C., Lt. . . . 534, 1699
Nice, C. G., 2nd-Lt. 681
—— G. L., Lt. 785
Nichol, D. S. F., 2nd-Lt. . . . 557
—— W., 2nd-Lt. 712
Nicholas, A. A., Lt. . . . 302, 1128
—— B. G., Capt. 597
—— C. A. N., Lt. . 302, 1493, 1501
—— C. F., 2nd-Lt. . 360, 1245
—— C. H., Lt. . . . 393, 1090
—— D. H. J., Asst. Surgn. 1797
—— J., Maj. 1838
—— J. A., 2nd Lt. . . . 1293
—— J. K. B., Lt. . . . 619
—— J. R., 2nd-Lt. . . . 1054
—— J. W., Lt 507
—— M. J., Asst. Surgn. 1801
—— O. R., 2nd-Lt. . . 358, 1241
—— R. H., Capt. . . 501, 1078
—— S. H. E., Lt.-Col. . 8, 136, 158, 196
—— T., Condr. 1853
—— T., 2nd-Lt. . . 564, 1211
—— W. A., Lt. . 507, 1359, 1675
Nicholl, C. K., Capt. . . 415, 939
—— E. McK., Surgn.-Maj. 62, 1838
—— E. McK., Lt. . . 317, 901
—— H., Condr. 1853
—— M. C., Lt. . . 322, 1239
—— N. F., Capt . . . 7, 1239
Nicholl-Carne, E. D. S., Capt. . 818
Nicholls, A., 2nd-Lt. 797
—— A. W., 2nd-Lt. 332, 1273, 1361
—— C. W., Lt. . 717, 797, 1007
—— E. H., 2nd-Lt. . . . 1583
—— E. M., Lt. 537
—— H. B. C., Lt. . . . 721
—— J. W. D., 2nd-Lt. . . 1447
—— R. L., Lt. . . . 522, 1062
Nichols, C. E., 2nd-Lt. 659
—— D., 2nd-Lt. . . . 358, 1180
—— F. W., Lt. . . . 307, 1007
—— H., Maj. 852
—— H. J., Lt. . . . 515, 997
—— H. L., Lt. . . 493, 1293
—— J. A., 2nd-Lt. . . 560, 848
—— J. E., Lt. . . 7, 534, 1885
Nicholson, A. R., Lt. . 33, 511, 1411
—— A. T., 2nd-Lt. . . . 717
—— C. F. S., Lt. . . 385, 1606
—— G. A., 2nd-Lt. . 475, 1401
—— E. M. F., Lt. 512, 821, 1636
—— E. A., Lt. 1513
—— F. D., Capt. . . . 1729
—— F. L., Maj. . . 246, 1170
—— J., 2nd-Lt. . . 564, 1303
—— J. P., 2nd-Lt. . . . 554
—— J. R., 2nd-Lt. . . . 758
—— J. W., Lt. . . 490, 897
—— M. A., Capt. . . . 1755
—— R. B., Maj. . . . 1753
—— R. L., Lt. . . . 353, 1183

NAMES AND RANK. Page.

Nicholson, R. W., Lt.-Col. 35, 128, 154, 1660
—— S., 2nd-Lt. 981
—— W. C., Lt.-Col. 134, 157, 194, 868, 1086
—— W. H., Lt. 515, 707, 1058
Nickalls, J. W., Lt. 536
—— J. W. D., 2nd-Lt. . . . 356
Nickels, T., Lt. . . 76, 77, 528, 1885
Nicklin, G. N., Lt. . . . 504, 1397
—— S. G., 2nd-Lt. . . 691, 784
Nickolds, A., 2nd-Lt. 567
Nicol, W. B., Maj 1838
Nicolas, C. P., Maj. 213
—— F. C., Maj. . 207, 1895, 1913
—— R., Maj 213, 872, 1110, 1589
Nicolay, B. U., Lt.-Col. . 200, 139, 161, 872, 1611, 1613
Nicoll, N. F., Capt. . 215, 1239, 1317
—— W. A., 2nd-Lt. . . 553, 814
—— W. F., Lt . . . 540a, 1673
Nicolle, A. V., Capt. . . 502, 997
Nicolls, B. E., Capt. . . . 17, 27
Nicolson, D. O., Lt. . . . 306, 1435
—— H. G., Lt. . . . 536, 1647
Nidhan Singh, Subdr . . . 1203
—— Subdr.-Maj. . . .
—— Subdr. . . 965, 970
Nield, H. M., 2nd-Lt. 767
Nielson, E. G., Lt 527
Nierses, W. L. M., Lt. . . 520, 1666
Nigah Ali Shah, Jemdr. 900
Nigam, K. S., Lt. 1781
Nightingale, A. E., The Revd. . 1908
—— A. W., Lieut. 511, 854
—— G. O., Lt. . . 296, 1052
—— C. G., 2nd-Lt. . . . 748
—— M. R. W., Brig.-Genl. 50, 105, 122, 135, 158, 105, 872, 1617
—— V. R. J., 2nd-Lt. 356, 1217
Nihala, Subdr. 1068
—— Jemdr. 1393
Nihal, Jemdr. 1052
—— Chand, Tr. Vety. Asst. . 1687
—— Singh, Ressdr. 948
—— Jemdr. 1389
—— Jemdr. 1433
—— Jemdr. 1705
—— Jemdr. 1095
—— Jemdr. 1022
—— Jemdr. 1111
—— Jemdr. 936
—— Jemdr. 1067
—— Jemdr. 1216
—— Jemdr. 889
—— Jemdr. 898
—— Jemdr. 930
—— Subdy 1231
Nijat, Jemdr. 1428
Nijabut Husain, Jemdr. . . . 942
Nijabut, Ali, Subdr. 1488
Nijhawan, G., Capt. 1769
Nilamegham, Jemdr. 1323
Nilkhant Ramchandra Mauskar, Sub-Asst. Surgn. . 1831
Nikka Singh, Jemdr. 989
Nikku, Jemdr. 1184
Nimmo, A., Lt. . . 430, 1241, 1369
Nimonkar, S. A., Ty. Lt. . . . 1776
Ninan, V. T., Ty. Lt. 1779
Ninis, G. W., 2nd-Lt. . . 293, a 891
—— R. D., Revd. 1903
Niranjan-Das, Sr. Sub-Asst. Surgn. . 1807
Niranjan Singh, Jemdr. 950
Nirmal Singh, Ty. Sub-Asst. Surgn. 1833
Nisar Husain, Subdr.-Maj. . . 1323
Nisbet, J. D., Lt. . . . 763, 781
—— T., Bt. Lt.-Col. 54, 140, 160, 232, 927
Nisbett, W. G., Lt.-Col. 131, 155, 191
Nishan Ali, Jemdr. 1196
—— Khan, Subdr. . . 1511
—— Subdr. . . . 1288
Nissen, F. H. C., 2nd-Lt. . . . 491
—— G. A., Lt. 529
—— T. A., Lt. . . . 508, 1606
Niven, D. S., Maj. . . 15, 240, 921
Nivet, E. L. A., 2nd-Lt. 722
Nixon, A. I., Capt. 75
—— B. G. M. F., Capt. . 261, 1187
—— C. H. F., Capt. 252, 1371, 1585
—— C. N., Capt. . . . 485, 885
—— E. J., Capt. 630
—— E. M., Maj. . . . 318, 943

NAMES AND RANK. Page.

Nixon, F. B., Maj. 585
—— J. E., Genl. Sir 105, 145, 171, 1925, 1945
—— J. T., 2nd-Lt. 727
Niyaz Ahmad Khan, Subdr. . . 1092
—— Genl. Jemdr. 1260
—— Khan, Hony. Lt. . . . 2104
—— Muhammad Khan, Sub-Asst. Surgn. 1812
Nizam Din, Jemdr. . . . 1103, 1305
—— Jemdr. 1573
—— Ressdr. 1684
Nizammud-din, Sr. Sub-Asst. Surgn. 1808
—— Subdr. . . . 1488
—— Jemdr. . . . 1304
Nizam-ud-din, Jemdr. 908
—— Subdr. . . . 1497
—— Khan, Subdr. . 1052
Njanapragasam, Sub-Asst. Surgn. D. I. 1828
Noad, C. H. C., Lt. . . . 753, 806
—— C. K., Lt. . . . 66, 297, 1641
Noake, A. R., the Revd. . . . 1904
—— A. S., Maj. . . . 217, 1661
Noakes, E. J. W., Lt. . . 399, 923
—— S. W., 2nd-Lt. 360
—— T. C., Capt. 1405
Noble, A., Ty. Lt. 1778
—— A. G., Lt. 517
—— G., Lt. . . . 460, 62, 937
—— H. B., 2nd-Lt. 361
—— H. E., 2nd-Lt. 714
—— J. W., 2nd-Lt. 793
—— R., Capt. 583
—— T. L., Lt. . . 77, 520, 1886
—— W. J. W., Maj. . . 32, 663
Noel, E. F. H., 2nd-Lt. . . 327, 897
—— K. H., Lt. 297
—— O. C., 2nd-Lt. . 325, 366, 1598
—— W. H. M., 2nd-Lt. . . . 714
Nohar Singh, Subdr. 1387
Nokes, G. B. L., Lt. . . . 385, 941
Nolan, A. A. M., Ty. Asst. Surgn. 1804
—— M. C., 2nd-Lt. 325, 366, 1122
—— N. G., Ty. Lt. 1726
—— T. R., Maj. 857
Nolan Ferrall, H. J., Capt. . . 769
Noorullah Khan, Jemdr. . . . 1289
Nop Singh, Jemdr. 1488
Norbury, A., Condr. 1853
—— P. FitzG., Maj. 102, 220, 939
Nordon, B. F., 2nd-Lt. 599
Norendro Singh, Ressdr. . . . 1683
—— Ressdr. . . . 1686
Norfor, F. F., 2nd-Lt. 801
Norie, C. E. deM., Bg.-Genl. 12, 35, 97, 103, 120, 124, 150, 185
Norkett, H. W., Condr. . . . 1854
Norman, A. C., Maj. . 223, 883, 1725
—— A. L., Lt. 733
—— C., Lt. 776
—— C. A., Lt. 1565
—— C. E. R., Capt. . . . 1765
—— C. G., Lt. 770
—— C. L., Brig.-Genl. 1, 6, 13, 97, 121, 127, 153, 210, 1964
—— C. L., Capt. 751
—— D., Lt 1299
—— D. R., 2nd-Lt. 523
—— F., 2nd-Lt. 112
—— F. E. B., 2nd-Lt. 475, 1024
—— F. W., 2nd-Lt. 1693
—— G., Lt. 617
—— G. E., 2nd-Lt. 768
—— G. R., Lt. 597
—— H. J., Capt. . 94, 263, 1411
—— J. M., Lt. 1720
—— S. H., Lt. 353, 1054
—— W., 2nd-Lt. 1060
—— W. H., Bt.-Col. 51, 98, 136, 153, 196, 867, 895
Normand, C. W. B., Lt. . 519, 806
Nornabell, H. M., Maj. . . . 614
Noronha, D., Sr. Asst. Surgn. . 1791
Norrie, C. M., Capt. . . . 58, 668
—— F. H. B., Capt. 1717
—— J., Maj. 696, 795
Norris, A. K., Maj. . . . 226, 1487
—— C. E., Lt. 543, 1007
—— C. E., Capt. 1702
—— C. P., 2nd-Lt. 730
—— C. W., Lt. . . . 538, 1693
—— H., Maj. 226, 1327
—— H. F., 2nd-Lt. 326, 366, 1319
—— J. W., Lt. . . . 455, 591, 90

NAMES AND RANK. Page.

Norris, R. J. N., Capt. 277, 1042
—— R. V., Lt. 516p
—— W. A., Capt. 5, 571, 1850
—— W. C. G., Lt. 304, 1425
—— W. J., Lt. 463, 1189
Norrish, E. H. B., Lt. 514, 1132
—— F., 2nd-Lt. 455, 1399
North, A. E., 2nd-Lt. 1423
—— F. E., 2nd-Lt. 1088
—— F. L., 2nd-Lt. 1405
—— H. N., Bt. Lt.-Col. 664
—— K. A., Capt. 273, 1148, 1156
—— L. V., Lt. 401, 1401
—— The Hon. W. F. J., Lt.-Col. 706
Northam, R. M., Lt. 538, 1672
Northcote, C. C., 2nd-Lt. 562
Northey, H. S., Lt. 525
—— W. B., Capt. 104, 256, 1225, 1595
Northfield, E. W., Lt. 470, 1471
Norton, C. E. G., Brig. Gen. 50, 121, 133, 587
—— C. E., Lt. 375
—— D. J. E., Lt. 591
—— E. C., 2nd-Lt. 754
—— G. A., Lt. 526, 1667
—— W. J. M., Asst. Surgn. 1804
—— W. R. T., Lt. 1919
—— W. O., 2nd-Lt. 546, 1295
—— W. D., 2nd-Lt. 1371
—— W. D., Lt. 1677
Norwood, W. S., Lt. 538, 1668
Nosworthy, F. P., Capt. 665, 1880
—— J. L., Lt 629
Notley, R. A., 2nd-Lt. 630
—— T., Sub-Condr. 1857
—— W., Lt. 322, 1431
Nott, A. H., Lt.-Col. 127, 1740
Nottage, J. B., Lt. 541
Nott-Bower, Lt. 1006
—— J. R. H., 2nd-Lt. 553, 1523
Nottidge, G., Capt. 664
Nougerede, V. A. Dela, Lt. 532
Nour Ahmed Hossain Maulvi, 2nd-Lt. 481
Novis, T. S., Lt.-Col. 139, 1745
Nowsher, Subdr. 1518
Noyes, C. D., Capt. 256, 1020, 1022
—— H., 2nd-Lt. 739
Noyle, H. W., Lt. 533, 1108
Nugent, C. H.H., Lt.-Col. 11, 134, 661
—— J. F. H., Capt. 16, 42, 269, 1132
—— L. St. D. H. V., Ty. Asst. Surgn. 1804
—— O. I. W., Maj.-General 12, 97, 119
—— O. S. B., Maj.-Gen. 118
—— T. D., Lt. 504, 830, 1581
Nunan, W., Ty. Capt. 1772
Nulkor, A. K., Capt. 1769
Nunes, J. M., Sr. Asst. Surgn. 1789
Nunn, C. A., Sub-Condr. 1870
—— F. M., 2nd-Lt. 1465
—— H., 2nd-Lt. 463, 1144
—— J. N., Lt. 540
—— R. E., Lt. 844
—— R. L. H., Lt. 629
—— R. N., Lt. 296, 1279
Nurana Singh, Subdr.-Maj. 1704
Nur Ahmad Khan, Resldr. 891
—— Jemdr. 1088
—— Sher Khan, Resdr. 945
—— Akhmad, Subdr. 7, 1256
—— Jemdr. (I) 999
—— Jemdr. (II) 999
—— Alam, Hony. Capt. 2104
—— Jemdr. 1208
—— Jemdr. 1373
—— Jemdr. 1368
—— Subdr. 1207
—— Sub-Asst. Surgn. 1820
Nur Ali, Jemdr. 1420
—— Dad, Jemdr. 1506
—— Din, Risldr. 1683, 1684
—— Jemdr. 1075
—— Hasan, Sub-Asst. Surgn. 1807, 1814
—— Jemdr. 973
—— Husain, Jemdr. 1280
—— Khan, Jemdr. 1087
—— *Hony.* Capt. 2103
—— Jemdr 1407
—— Jemdr. 1288
—— Risldr. 965
—— Risldr. 893
—— Subdr. 1373
—— Subdr. 1407

NAMES AND RANK. Page.

Nur Khan, Jemdr. 1365
—— (I) Subdr. 1513
—— Subdr. 1515
—— Subdr. 1108
—— Mast, Subdr. 1104
—— Mohammad, Ty. Sub. Asst. Surgn. 1833
—— Mohomed, Jemdr. 850
—— Muhammad, Subdr. 1495—96
—— Muhammad Khan, Jemdr. 890, 938
—— Jemdr. 975
—— Jemdr. 1240
—— Subdr. 966
—— Subdr. 1492
—— Jemdr. 1514
—— Risldr. 883
—— Jemdr. 1108
—— Tr. Vety. Asst. 1687
—— Jemdr. 989
—— Jemdr. 942
—— Sub-Asst. Surgn. 1809
—— Jemdr. 1268
—— Jemdr. 892
—— Risldr. 880
—— Tr. Vety. Asst. 1687
—— Sher Khan, Risldr. 945
Nurbal Gurung, Jemdr. 968
Nurdad, Subdr. 1506
Nurul Hasan, Sr. Sub-Asst. Surgn. 1807
Nussey, H. S., 2nd-Lt. 626
Nute, J. F. A., 2nd-Lt. 532
Nuthall, C. E., Brig.-Gen. 9, 16, 120, 123, 1838
Nutt, H. R., Maj. 1747
Nuttall, B. A., Lt. 90, 781
—— F., Revd. 1906
—— J. R., Maj. 1919
—— N., Lt. 536, 1491
—— R. W., Sub-Condr. 1855
Nutting, H. S., Capt. 595
Nyan, K., Ty. Lt. 1778
Nye, A. R., Lt. 307a, 420, 514, 1614
Nyss, A. N., Asst. Surgn. 180

O

O'Beirne, J. J., Lt. 617
O'Brian, H. E. W., *Esq.* 88
—— J. A., Esq. 91
O'Brien, A. J., Lt.-Col. 105, 109, 198, 1896, 1912
—— B. P. T., Lt. 84, 309, 1635
—— C., Nursing Sister. 1735
—— C. P., Asst. Surgn. 1796
—— C. R., Capt. 1752
—— C. W., Lt.-Col. 56
—— D. P., Lt. 519, 961
—— E., Lt.-Col. 199, 1893, 1912
—— E. C., Lt. 375, 1639
—— G. P., Sr. Asst. Surgn. 1789
—— H., 2nd-Lt. 685
—— J., Capt. 500
—— J., Capt. 90
—— J. G., 2nd-Lt. 90
—— J. J., Lt. 368, 1453
—— J. M. D., 2nd-Lt. 565
—— M. C., Lt. 504, 1607
—— M. J., 2nd-Lt. 565, 1493—94
—— P., Maj. 758
—— R. W., Lt. 8, 527, 1671
—— T., 2nd-Lt. 563, 1245
—— W., Lt. 769
—— W. G., 2nd-Lt. 560, 1703
—— W. J., Lt. 320, 1543
O'Bryan, J. R., Lt. 885
O'Bryen, C. W., Lt.-Col. 55, 104, 127, 154, 182
—— R. E., 2nd-Lt. 1695
O'Callaghan, C., Asst.-Surgn. 1803
—— C. H. D., Lt. 307, 399, 1335
—— F. J. O., 2nd Lt. 1676
—— L. M., 2nd-Lt. 712
—— T. P. M., Lt. 510, 1038, 1557
O'Connell, W. R. F., Lt. 353
O'Collin, P. F. 2nd-Lt. 1120
O'Collins, P. F., 2nd-Lt. 476
O'Connor, A. B., 2nd-Lt. 557
—— A. J., Lt. 509
—— A. T., 2nd-Lt. 559, 1703
—— C. E. B., *Hony.* Lieut 1848
—— G. B., Sub-Condr 1857
—— F. P., Capt. 486
—— G. B., Capt. 502, 1665

NAMES AND RANK. Page.

O'Connor, J. C. J., Lt. 487, 929
—— K. K., Lt. 309, 1074
—— M. J., Lt. 316, 1038
—— M. J., 2nd-Lt. 358, 1557
—— M. S. S., Maj. 60, 485, 1663
—— N. F. M., N. Sister 1735
—— P., *Hony.* Lt. 1849
—— R. A., Lt. 488
—— R. L., Lt. 297a, 1038
—— T. G., 2nd-Lt. 560, 1289
—— T. J., Lt. 420, 933
—— W., Asst. Surgn. 1792
—— W. W. Asst. Surgn. 1797
—— W. F. T., Lt.-Col. 195, 1890, 1912
—— George, J. M. 2nd-Lt. 599
O'Conor, W. O., Revd. 1904
O'Day, T., 2nd-Lt. 637, 979
O'Dea, M. L. B. P., Lt. 503
—— M. L., Lieut. 1665
O'Doherty, T. E., 2nd-Lt. 539
O'Donald, E., *Hony.* Maj. 1844
O'Donnell, B. H., Capt. 272, 923
—— F. G., Lt. 450—54, 1122
—— I. A., 2nd-Lt. 319
—— O., 2nd-Lt. 1696
—— S. P., Lt.-Col. 816
—— T. F., Capt. 829
—— T. J., Lt.-Genl. Sir 10, 118, 1709
O'Donoghue, E. J., 2nd-Lt. 1060
—— P. E., Sr. Asst. Surgn. 1790
O'Donovan, D. F., 2nd-Lt. 530
O'Dowd, G. R., *Esq.* 88
—— J. H. B., Lt. 423
O'Dwyer, M. F., *Hon.*, Sir 605, 804, 836, 838, 1896
O'Ferrall, J. L. F., 2nd-Lt. 332, 1211, 1596
O'Flynn, A. J., 2nd-Lt. 563, 1315
O'Gilry, J. D., Capt. 1635
O'Gogarty, M. G., Capt. 1839
O'Gorman, A. R., 2nd-Lt. 554
—— C. J., Lt.-Col. 1710
—— C. J., 2nd-Lt. 82, 564
—— P. W., Lt.-Col. 128, 1742
—— M. K., Lt. 310, 1463
O'Grady, H. DeC., Brig.-Genl. 11, 42, 97, 127, 153, 207, 1231
—— J., 2nd-Lt. 83
—— W. J., Lt. 516a
O'Hagan, R., Lt. 503, 1565
O'Halloran, H., Lt. 450—54, 1491
O'Hara, A. E., *Esq.* 88
—— E. H., Lt. 534
—— C. H., Lt. 525, 1705
O'Keeffe, A. R. W. M., 2nd-Lt. 425, 925
—— D. S. A., Maj. 1749
O'Kinealy, F., Lt.-Col. 128, 1741
O'Leary, B. J., Capt. 636
—— D. F. K., 2nd-Lt. 551, 1052
—— F. H., Asst. Surgn. 1793
—— H. H., Lt. 1321
—— M. G., 2nd-Lt. 532, 1355
—— M. P., Lt. 314
—— R. F., 2nd-Lt. 553, 1701
O'Malley, D. V., Ty. Capt. 1725
O'Mealy Piggot, F. M., Lt. 1442
O'Meara, A. E., Maj. 758
—— A. J., 2nd-Lt. 1455
—— C. A. E., Lt.-Col. 76, 132, 156, 193
—— E. J., Lt.-Col. 138, 1745
—— M. J., 2nd-Lt. 358, 1373, 1678
O'Morchoe, N. C., Lt. 296a, 1136
O'Neill, A. Capt. 1839
—— E. M. M., Sub-Condr. 1857
—— F. O., Lt. 26, 543, 608
—— G. P., Sub-Cond. 1874
—— H. J. D., Capt. 252, 1505
—— J. S., Maj. 1752
—— P. L., Maj. 1748
—— W. H. L., Capt. 271, 1227
O'Neil-Ready, C. W., 2nd-Lt. 524, 1633
O'Neil-Shaw, J. C. L., Lt. 541
O'Regan, W. F., Tempy. Capt. 1724
O'Reilly, C. C. G., Asst. Surgn. 1798
—— C. M., Lt-Col. 46, 135, 138, 195, 870, 1351
—— D. O., Lt. 504
—— H. W. H., Lt.-Col. 133, 1710
—— H. R. H., Capt. 37, 229—290, 1060
—— J. M., 2nd-Lt. 359
—— K. W. R., Lt. 296a, 1471

Names and Rank. Page.

O'Reilly, R., 2nd-Lt. 1483
O'Riordan, K. M., 2nd-Lt. . 553, 1521
—— P. M., Lt. 516
O'Rorke, C. H., Capt. . . 28, 1713
—— J. M. W., Capt. 52, 256, 921
—— M. D., Lt. 769
O'Shaughnessy, H. P., Capt. . . 843
O'Shea, P. J., 2nd-Lt. 547
O'Sullivan, A. G. J., 2nd-Lt, 568, 1708
—— E. T., Lt. . 317, 1908, 1279
—— H. P., Lt. 317, 1016, 1026
—— J., Capt. . . 272, 1030
—— K. L., Capt. . . 43, 1717
—— O. D., 2nd-Lt. . . . 567
O'Toole, A. H., 2nd-Lt. . . . 741
—— N. D. G., Lt 517
Oakden, R., Maj. 776a
Oakes, G. B., Lt.-Col. 84, 130, 155, 190
—— H. C., Lt.-Col. . 56, 105, 138
—— J. O., Capt. 160, 199, 599, 867, 927
—— T. T., Maj. . 9, 16, 227, 1411
Oakeshott, D. B., 2nd-Lt. . 565, 1267
Oakley, C. B., Lt. 503
—— C. H., Maj. 808
—— G. C., Lt. 64
—— G. U., Asst. Surgn. . . 1797
—— J., 2nd-Lt. 733
Oakman, A. S., 2nd-Lt. . . . 357
Oaten, E F., Lt. . . . 490, 895
Oates, A. T., Lt. . . . 307a, 893
—— A., Lt. 405, 1285
—— S. J., 2nd-Lt. 571
—— W. J. S., Lt. . 304, 1110, 1112
Oatey, G. H., 2nd-Lt. 618
Oatts, H. A., Lt. . .310, 1187, 1561
Oatway, S. H., Lt. . . . 515, 1379
Obbard, E. N., Lt.-Col, 20, 22, 81, 131, 155, 190
—— H. N., 2nd-Lt. . 666, 986
—— O. J., Lt.-Col. 187, 1895, 1911
Obee, J., 2nd-Lt. 618
—— L. E., 2nd-Lt. 1674
Oberal, G. R., Tempy. Capt. . . 1769
Obourn, N. A., 2nd-Lt. . . 356, 1199
Occ..n, C. K. 2nd-Lt. 552
Ocock, W. P., Capt. . . . 754, 795
Ockelford, A., Sub-Condr. . . . 1857
Odan Govindan, Sub-Asst. Surgn. 1827
Oddie, B. E., Lt. 313
—— E., 2nd-Lt. . . . 762, 782
—— W. P., Lt. . . . 515, 1132
Oddin Taylor, H. W., Lt. . . . 655
Odlum, W. H., Maj. 1750
Odling, B. E., Lt. 541
—— T. M., Lt. . . . 492, 1677
Oecken, C. A. J., Lt. 889
Oehme, C. N., Capt. 827
Officer, H. R., 2nd-Lt. . . 329, 1241
Ogborn, E. F., 2nd-Lt. 669
Ogbourne, T., Sub-Condr. . . . 1856
Ogden, A. W., Sub-Condr. . . . 1864
—— H. G., Capt. 801
—— H. L., Lt. . . . 463, 887
—— J. L., Asst. Surgn. . . . 1103
Ogg, A. C., Maj. . . 99, 216, 1042
—— G. S., Col. . 11, 71, 126, 612
Oggier, S. L., 2nd-Lt. 545
Ogier, L. L., Lt. 389
Ogilvie, D., Maj. 651
—— E. C., Bt.-Col. . 42, 125, 661
—— G. D., Maj. . 4, 227, 1890, 1913
—— G. H., Lt. 507
—— J. F. W., Capt. . 256, 964
—— P. G., Lt. . . . 73, 508
Ogilvie, W. H., Lt.-Col. . . 130, 1742
—— W. N., Lt. . . . 537, 973
Ogilvy, A. G., Capt. . 271, 1164, 1274
—— D., Maj. 663
—— G. M. H., Capt. . . . 718
—— J. D., Capt. . . 263, 1595
Ogle, J. E., Lt. 738
—— N., Bt. Lt.-Col. 13, 66, 98, 161, 234, 1283
—— J. W., Lt. 836
Oglethorpe, J. A., Lt. 579
—— R. S., Lt. . . . 425, 1223
Ogley, E. F., Lt. 486
Oka, M. G., Ty. Lt. 1776
Oke, H., 2nd-Lt. . . . 357, 1197
Okell, G. F., 2nd-Lt. 1007
Old, R. E., Lt. 435, 1333
Oldacre, V. G., 2nd-Lt. . . . 1040
Oldall, C. H., Lt. . . . 543a, 1673
Oldbury, Lt. A. E. 35, 75
Olden, S. M., 2nd-Lt. 1559
Oldendorff, A. J. A., Maj. . . . 780
Oldfield, G. E. B., Maj. 20, 213, 870, 1311, 1365
—— M. H., Lt. 850
—— T. A. F. R., Lt.-Col. 193, 1893, 1912
Oldham, A. W., Lt. 435
—— G. M., Lt.-Col. . 130, 663
—— J. B., 2nd-Lt. 722
—— J. S., Maj. . . 241, 885
Oldridge dela Hey, C. J., Lt. 294, 925
Oldroyd, O. E. 2nd-Lt. . . . 1736
Oliffe, F., Lt. 300, 1321
Oliphant, D. A., Capt. 520, 829, 1601
—— I. M. B., Lt. . 425, 759, 1046
—— R. W., Lt. . . . 542, 1303
Oliver, A. J., Maj. 852
—— C., Capt. 723a
—— D. R. G., Maj. 105, 221, 1870, 1913
—— E. S., Lt. 542, 1636
—— E. W., Maj. 608
—— F., Sub-Condr. 1870
—— F. J. 2nd-Lt. 557
—— F. P., Lt. 489, 885
—— J. S., Lt. 961
—— L. C., 2nd-Lt. 808
—— N. E., 2nd-Lt. 1557
—— R. E., Maj. 843
—— St. J. L., Lt. . . 18, 23, 491
—— T. W., Lt. . . . 526, 1611
—— W. B., Asst. Surgn. . . 1803
Olivier, R. T., Lt. 1130
Olley, A. H., Condr. 1863
Olney, J. W. C., 2nd-Lt. 555
Olorenshaw, A. J., Lt. . . . 515
Ommanney, A. H., Maj. 226, 1359, 1702
—— C. V., Maj. 139, 160, 199, 871, 1439
—— F. R., Lt. . . 318, 1048
—— W. M., Capt. 262, 1148, 1589
Onraet, P. T., Lt. . . . 489, 919
Onslow, R. C., Col. . . 31, 180, 1917
Onyon, R. R., Lt. 721
Oonawala, J. H., Ty. Capt. . . 1764
Openshaw, E. R., Lt. 702
—— S. N., Lt. . . . 401, 1279
Ople, A. L., 2nd-Lt. . . . 360, 1321
Oppenheimer, K. M., Lt. . . . 511
—— P. H. M., 2nd-Lt. . 736
Orchard, A. J., Maj. . 223, 871, 1046, 1557
—— D. G., 2nd-Lt. 547
—— D. S., Capt. 239, 1182, 1635
—— G. F., 2nd-Lt. 746
—— H. C. W., 2nd-Lt. . . . 548
—— J. W., Capt. . 17, 38, 696, 795
—— M. F., Lt. 699
Ord, B., 2nd-Lt. 56
—— B. K., Lt. . . . . 91—92, 528
Orford, C. J., Lt. 776a
Organ, P. C., Lt. 638
Orgill, A. W., 2nd-Lt. . 326, 366, 1143
—— T. C., Lt. 514
Orman, C. H., Senr. Asst. Surgn. 1787
—— F. L., Lt.-Col. 107, 135, 158, 196, 1647, 1805
Orme, B. F. P., Lt. 727
Ormerod, H. E., Capt. 821
Ormesher, C. I., Lt. 699
Ormiston, T. L., Lt.-Col. 196, 1895, 1912
Ormond, E. C., Lt. 623
Ormrod, L. A., Lt. . . . 599—600
Ormsby, M. H., Lt . . 491, 611, 1708
Orpen, G. E., Lt. . . . 317a, 1181
Orr, A. E., 2nd-Lt. 1695
—— C. H. H., Lt. 1016
—— E. P., 2nd-Lt. 720
—— G. B., 2nd-Lt. 638
—— G. M., Lt.-Col. 13, 32, 97, 127, 153, 211, 895
—— H. D., Lt. 440, 943
—— J., Lt. 534, 988
—— J. H. R., Lt. . . . 317a, 1178
—— I. M., Lt. 294a, 1471
—— R., Hony. Lt.-Col. . . . 1845
—— W., Lt. 86, 540a
Orr-Ewing, M. B., Capt. . . . 615
Orridge, F. F., Sub-Condr. . . 1870
Orris, W. E., 2nd-Lt. 362
Orton, E. F., Lt.-Col. 99, 140, 162, 201 228, 945
—— S. B., Bt. Lt.-Col. . 99, 140, 228, 1178
Orwin, W., Capt. 9, 707
Ory, Revd. S. L. (Hony. Capt.) . . . . .
Osborn, C. R., Lt. 696
—— H. H., Capt. 858
Osborne, C. A., Lt. . . . 445, 1080
—— C. C. G., Lt. 533
—— C. E. F., Lt. 1361
—— D. G. H., Capt. . . . 729
—— F. W., Capt. 502, 733, 1004
—— F. W., 2nd-Lt. . . . 581
—— G. F. F., Maj. . 651, 1882
—— G. H., Maj. . . . 721
—— H., Lt. 317, 1411
—— H. M., 2nd-Lt. 545, 581, 1008
—— M. S., Lt. 589
—— S., 2nd-Lt. . . . 555, 1281
Osbourne, A. B., Asst. Surgn. . 1801
—— C. E. F., Lt. 387
Osmond, F. E., 2nd-Lt. . 682, 786
—— W. F. R., Lt. . . . 8, 618
Ostler, C., Lt. 489
—— R. S., 2nd-Lt. 362
Ostrehan, F. G. R., Lt.-Col. . 56, 181, 1917
—— M., Capt. . 256, 1641, 1898
Oswald, F., Capt. . . 7, 262, 913
—— J. A. R., Lt. 765
—— R. A., 2nd-Lt. 324
Oswell, G. C., Lt. . . . 463, 1225
Otter, R. B., 2nd-Lt. . . 355, 1080
Ottley, A. R. C., 2nd-Lt. . 550, 1265
—— W. J., Lt.-Col. 139, 160, 199, 868, 1110, 1112
—— V. W., Lt. 538
Ottmann, G. W., Asst. Surgn. . 1798
Otto C. R., 2nd-Lt. . . . 669, 997
—— E. C., 2nd-Lt. 561
—— L. A., 2nd-Lt. 1140
—— D. S., Ty. Lt. 1777
—— H. F., Asst. Surgn. . . . 1794
Ouseley, J. W. G., Maj. 94, 485, 609
—— R. G., Br.-Genl. 58, 101, 126, 613
Outerson, W., 2nd-Lt. 73
Outlaw, W. H., Lt. 738
Outram, J. D., Maj. 719
Ovenden, E. A., 2nd-Lt. . . 355, 1034
Ovens, J., 2nd-Lt. 1679
—— T. L., Maj. . . . 238, 1062
Over, J., 2nd-Lt. 555, 1707
Overbeck-Wright, A. W., Maj. 1707
Overbury, J. G., Capt. . . . 754
Overend, H. F., Ty. Lt. . . . 1726
Overton, A. G., 2nd-Lt. . . 325, 366
—— H., Lt. 445, 1026
Overy, A. E. R., Sub-Condr. . . 1855
Owden, J. S., Lt. . . . 488, 578
Owen, A. C., Lt.-Col. 825
——, A. J., Lt. . . 723a, 988, 1533
—— A. T. J., Lt. . . . 302, 1247
—— A. W., Lt. 423, 1585
—— D. P., Lt. 491, 88
—— L. G., Lt. 525
—— L. W. G., Lt. 789
—— O., Lt. 1919
—— R. H. W., Capt. 723
—— R. J., Sr. Asst. Surgn. . . 1787
—— S. F., 2nd-Lt. 1678
—— T., Lt. 789
—— W., 2nd-Lt. 548, 1453
—— V. E. O'N., Lt. . . . 511, 1596
—— Smith, R., Lt. 1703
Owens, D. J., Lt. 701
—— D. W. B., Lt. . . . 533, 1050
—— F. C., Lt.-Col. 199, 1695, 1895, 1912
—— T. D. C., Lt. . . . 322, 1613
—— T. F., Maj. 1751
Owers, F. F., Asst. Surgn. . . 1799
Oxenham, S. H., Lt. . . 28, 530, 1671
Oxley, H. H., 2nd-Lt. 758
—— J. C. S., Maj. 1747
Oxley-Brennan, C. G., Lt. . .
Oza, S. T., Sub-Asstt. Surgn. . . 1832
Ozanne, E. H. B., Capt. . 271, 1479
—— G. D., Capt. . . 271, 1403
Ozzard, F. B., Lt.-Col. 104, 128, 1740

NAMES AND RANK. Page.

P

Pacheco, J. N. J., Asst. Surgn. . 1800
Pachiappan, Jemdr. 1308
Packer, H. D., Lt.-Col. . . 133, 1710
—— E. V., Lt. 731a
—— N. E., Capt. 1724
Packington, W. H. I., Lt. 637, 981
Packwood, I., *Hony.* Lt. . 1009, 1849
Padaivittan, Jemdr. 993
Padam Singh Thapa, Jemdr. 1608, 1619
—— Karki, Jemdr. . . 1223
Padamdhoj Gurung, Subdr. . 1630
—— Jemdr. . 1602
Padamhoj Mall, Jemdr . . . 1645
Padam Sing Rawat, Subdr. . 1182
—— Thapa, Subdr. . 1642
Padamlal Limbu, Jemdr. . . 1631
Padbidri, G. R., Ty. Lt. . . . 1778
Padbury, G. F., 2nd-Lt. . . 690a
Paddison, S., 2nd-Lt. . . 681, 786
Paddock, F. A., 2nd-Lt. . . . 788
—— W. G., Lt. 808
Paddon, S. S. W., Capt. . . . 1919
Paddon-Row, D. A. T., Lt. 538, 1293
Padfield, C. R., 2nd-Lt. . . . 1142
—— G. A. S., Revd. . . 1904
—— R. C. H., Lt. . 61, 486
Padgham, T. H., 2nd-Lt. . . 358 1491
Padley, A. C., 2nd-Lt. 768
Padki, S. C., Sub-Asst. Surgn. . 1832
Padma, Subdr. 1099
Paes, A., Sr. Asst. Surgn. . . 1791
Page, A. F. A., *Hony.* Lt. . . 1847
—— A. J., Lt. . 446-49, 719, 915
—— A. J., 2nd-Lt. 848
—— C., Lt. 352, 1116
—— C. E., Lt. 974
—— E. R., Lt. 518, 1667
—— F. J., Capt 825
—— H. F., 2nd-Lt. 728
—— H. J., Lt. . . . 542, 1391
—— H. V. S., Lt. 295a, 1487, 1579
—— J. A., 2nd-Lt. 829
—— J. F., ... 1009
—— L. A., Lt. 90
—— L. H., Lt. 703
—— T. S., 2nd-Lt. . . . 561 1678a
—— W. H., Lt. . . . 535, 1672
—— W. K., Lt. 947
Page-Turner, E. H. M., Capt. . 777
Paget, B. J., 2nd-Lt. 737
—— C. N., Capt. 779
—— G. H., Lt. . . . 368, 1607
—— G. H., Capt. 1038
—— H. E. G., Capt. . . 264, 1207
—— P., Maj. 684
Paget, Tomlinson, W., Capt. . 587
Pahalman Gurung, Jemdr. . . 1625
Pahal Sing Rana, Subdr. . . 1618
—— Thapa, Subdr. . . 1615
Pahalsing Gurung, Subdr. .
—— Jemdr. . 1615
Paher Singh, Subdr. 1031
Pahara, Subdr. 1561
Paharsingh Gurung, Jemdr. . .
Pahlad Sing, Subdr. . . . 1021
—— Singh, Jemdr. . 902, 1063
—— Jemdr. 1054
Pahlwan Khan, Risldr. . . . 899
—— Subdr. . . . 1489
—— Subdr. 1143
—— Jemdr. . 886, 999
Paice, A. H., Lt. 519
Paige, C. P., Maj. . 6, 14, 239, 1439
—— R. H., Lt. . . . 77, 515, 1885
Paillad Ram, Jemdr. 1488
Pain, H. B., Sub-Condr. . . . 1853
—— H. W., Lt. . . . 514, 1630
—— S. A., 2nd-Lt. 358
—— S. C. A. J., Lt. . . 532 1707
Painda Khan, Ressdr. 1683
—— Jemdr. 1364
—— Subdr. 1488
Paip Singh, Subdr. 1191
—— Jemdr. 924
Paine, R. L., 2nd-Lt. 712
—— W. H., 2nd-Lt. 1631
Painter, R. F., 2nd-Lt. . . 356, 1331
Parsley, G., 2nd-Lt. 1591
Pakenham-Walsh, W. P., Maj. 664, 996
Pakhar Singh, Jemdr. 1699
Pakes, W. C., Lt. 624

Pal, A. C., Ty. Lt. 1774
—— B. N., Ty. Lt. 1782
—— M. N., Ty. Capt. 1772
—— N. B., Ty. Capt. 1772
—— S. C., Capt. 1751
Pala, Jemdr. 1129
—— Ram, Jemdr. 1288
Pal Singh, Subdr. 1124
—— Jemdr. 873
—— Jemdr. 950
—— Jemdr. 975
Palairet, C. R., Lt. 841
—— R. C. N., Capt. 10, 74, 700, 795, 1886
Pala Singh, Subdr. 1311
—— Subdr. 1056
—— Subdr. 1215
—— Jemdr. 1111
—— Jemdr. . . 1001, 1699
—— Subdr. 1112
—— Subdr. 1245
Palavasa, Thevan, Jemdr. . . 1328
Palayam, Subdr 1311
Palethorpe, C. N. deW., Lt. . . 730
Palin, C. E. E., Lt. . . . 291, 1195
—— C. W., Capt. . . . 279, 1513
—— D. V., 2nd-Lt. 693
—— G. W., Col 26, 123, 149, 181, 1560
—— P. C., Maj.-Genl. 56, 119, 147. 175
—— B. H., Maj. . . 207, 1883, 1913
Palit, A. N., Capt. . . . 1705, 1756
—— S. K., Ty. Lt. 1774
Palk, C. H. L., Maj. 1786
—— L. C., Lt. 401, 1074
Pallant, H. C. 2nd Lt. 1674
Pallin, R. H., Maj. 1693
Pallock, A. H. Lt. 1125
Palmer, A. C., Capt. . 260, 1387, 1395
—— A. R. B., Lt. . . . 493, 941
—— A. V., Qr.-Mr. Sergt. . . 1879
—— C. C., Lt.-Col. 71, 101, 134, 632, 1886
—— C. E., Maj. 1752
—— C. H., Maj. 654
—— C. W. G., Maj. 733
—— E. J., Rt. Revd. . . . 1907
—— E. de S. Lt. 692
—— F., Lt. 527, 1327
—— F. C. Capt. 759
—— F. J., Lt.-Col. . . . 133, 1710
—— F. O., Lt. 697
—— G. F., Bt Lt.-Col. 57, 105, 108, 139, 160, 209, 1395
—— G. W., 2nd-Lt. . . . 330, 1187
—— H. A., Capt. . . . 1501a, 1371
—— H. I. R., Lt.-Col. 128, 154, 184
—— H. R., 2nd-Lt. 732
—— J. H. G., Maj. 54, 103, 245, 1102
—— P. R., 2nd-Lt. 790
—— S. V., Capt. . . . 279, 1343
—— W. A. L., 2nd-Lt. . . . 558
—— W. E., Lt. 739
—— W. J. Lt. . . . 450-54, 1630
—— W. J. L., *Hon.* Lt. . . 731
Palmes, W. P., Lt. 490
—— W. T., Lt. 1293
Pancham Singh, Jemdr. . . .
—— Negi, Jemdr. . . . 1182
Panchanadan, Jemdr. 1012
Panchbir Mal, Jemdr. 1643
—— Singh, Gorkha, Sub. Asst. Surgn. . . . 1819
Pancholi, H. N., Sub-Asst. Surgn. 1832
Panckridge, H. R., Lt. . . 505, 1665
Pandalai, K. G., Capt. 1756
Pande Khan, Subdr.-Maj. . . . 1001
Pandit Balkishan Kaul. . . . 3
—— Dewat Ram, Risdr. . . 1682
—— Natthu Ram, Sr. Sub-Asst Surgn. 1807
—— Shankardas, Sr. Sub-Asst. Surgn. 1807
Pandnak Lakknak, Jemdr. . . 1447
Pandu Naik, Jemdr. 1460
Pandu Vithal, Jemdr. 1473
Pandurang Parab, Subdr. . . . 1439
—— Ramchandra Dhavle, Sub-Asst. Surgn. 1831
Pandya, P. R. Y., Sub-Asst. Surgn. 1832
Pane Sing, Jemdr. 961
Panet, A. E., Br.-Gen. . 56, 27, 661
Paney Singh, Ressdr. . . . .
Panikkar, K. P., Tempy. Capt. . 1766
Panji Sihg, Jemdr. 1071

Pank, P. E. D., Lt. 1715
Panna, Subdr. . . . . . .
Panna, Lal, Sub-Asst. Surgn. . 1813
—— Sur., Sub. Asst. Surgn. . 1820
—— Sur., Sub-Asst. Surgn. . 1820
—— Rawat, Jemdr. 1473
—— Singh, Jemdr. 1196
Parnell, W. G. A., 2nd-Lt. . . 753
—— N. D., 2nd-Lt. . . 1499-1500
Pannet, C. A., Capt. 1722
Pannett, W. J., Lt. 543a
Pan Sing Ghariya, Subdr. . . 1179
Pant, P., Tempy. Capt. . . . 1770
Pantlin, H. C., 2nd-Lt. . . . 554
Panton, A. W., Tempy. Capt. . 1766
Paonaskar, S. G., Tempy. Capt. 1768
Paparayadu, Subdr. . . . 1263
Papillon, A. H. B. Capt. . 666, 991
Papps, F. J., 2nd-Lt. 710
Papworth, F. A. G., Lt. . 533, 1225
Papworth, W. F., Capt. . . . 807
Par Singh Rawat, Jemdr. . . 1223
Parab, V. K. Ty. Capt. . . . 1770
—— Dayal, Subdr. 1126
Paramoor, C. H., 2nd-Lt. . . . 556
Paramer, C. H., 2nd-Lt. . . . 1415
Paramsukh Misr, Jemdr. . . . 1027
Paranjpe, V. M., Sub-Asst. Surgn. 1832
Parasabir Rana, Jemdr. . . . 1651
Paras Ram, Jemdr. 1252
Parasuram, A. R., Ty. Lt. . . 1776
—— G. R. P., Ty. Lt. . . 1778
Parasuraman, Jemdr. . . 994, 1699
Parbal Thapa, Jemdr. 1604
Parbat Chand, Subdr.-Maj. . . 1259
Parbati Jagtap, Jemdr. . . . 1416
Parbhu Dyal, Jemdr. 1197
—— Singh, Jemdr. 1063
—— Subdr. 1054
Parbhudayal Sukul, Jemdr. . . 1017
Parbir Burathoki, Subdr. . . . 1596
—— Pun, Subdr. 1639
—— Rana, Subdr. 1639
—— Thapa, Jemdr. . . 1611, 1637
Parbury, E., Maj. . . 72, 614, 1884
—— H. F., Capt. 595
Pardington, G. E. L., Capt. 751, 797
Purduman Singh, Risldr. . . . 1684
Parem Singh, Subdr. 1563
Pargick, C. E., 2nd-Lt. . . . 1703
Pargiter, J. G., 2nd-Lt. 332, 556a, 957
—— B. B., Capt. 636
Paris, A. I., Maj. . . . 663, 996
—— A. H., *Hony.* Capt. . . . 22
—— G. N. A., Lt. 410
—— W. E., Lt. 714
Parish, G. N. O., Lt. 1367
Parisotti, L. Lt. 425, 1607
Paritam Singh, Jemdr. 1158
Parja, Subdr. 1143
—— Gurung, Subdr. 1607
Park, D., Lt. 1092
—— E. R., Lt. 540, 988
—— L. Lt. 1179
—— H. H., 2nd-Lt. . . 566, 1678a
—— M. E., Capt. 743
—— R. C. W., 2nd-Lt. . . . 736
—— T. W. U., Lt. 783
—— W. B., 2nd-Lt. . . . 551, 1678
—— W. J., Lt. 540
—— W. R., Revd. 1908
Parke, C., Lt. 628
—— L. S., Capt. . . 18, 27, 732
—— R., Condr. 1854
Parker, A. H., Lieut. 804
—— C. G., 2nd-Lt. . . . .
—— C. J., 2nd-Lt. . . . 328, 1373
—— D. R., Lt. . . . 1, 521, 1323
—— E., Lt. . . . 430, 970, 1718
—— F. J. H., 2nd-Lt. . . . 729
—— G., Lt. 59
—— G. T. M., Lt. . . . 489, 963
—— H., 2nd-Lt. 1561
—— H. E., Lt. 605
—— H. N., Lt. 817
—— H. R. C., Lt. . . . 530, 1387
—— J. I., Br.-Genl. 120, 124, 612
—— L., 2nd-Lt. 567
—— L., Lt. 841
—— L., 2nd-Lt. 1678a
—— R. N., Lt. . . 10, 542, 1673
—— T. C., 2nd-Lt. 752
—— V. L., Lt. 311a, 885
—— W., Cap. 748
—— W. A. H., Revd. 1903

NAMES AND RANK. Page.

Parker, W. L., Capt. *Sir* 734
—— W. S. H., Lt. 581
Parkes, E. E., Maj. 1711
—— R. G., Lt. 733
—— R. S., 2nd-Lt. 618
Parkin, J. F., Major. . 233, 1287, 1455
—— J. H. D., Lt. . . 319, 1066
—— R. H., Lt. 623
Parkinson, A. G., 2nd-Lt. . 535, 1143
—— C. E., 2nd-Lt. 555
—— E., 2nd-Lt. 557
—— H. L., Lt. . . . 525, 1110
—— J. T., Sr. Asst. Surgn. 1787
—— J. W., 2nd-Lt. . . . 552
—— W. H., Ty. Capt. . . 1729
—— R. H. F., Asst. Surgn. 1799
Parks, A. H., Lt. 463
—— H. C., Capt. . . . 6, 14, 1850
—— H. J., Lt. . . . 470, 1479
Parlakimedi, Sri Krishna, 2nd-Lt. 481
Parlby, J., Esq. 88
Parmanand, Ty. Capt. 1772
—— Jemdr. 1400
—— Jemdr. 1567
—— Sub-Asst., Surgn. . 1813
—— Misra, Sub-Asst. Surgn. 1812
—— Sharma, Sub-Asst. Surgn. 1813
Parmandhoj Thapa, Jemdr. . . 1653
Parman Thapa, Jemdr. . . . 1608
—— Subdr. 1657
Parman Singh Gurung, Jemdr. . 1602
Parmenter, A. S. P., Lt. . . . 542
—— F., Lt. . 304, 1126, 1433
Parmeshwardin Misr, Jemdr. . 1027
Parmer, S. W., Lt. . 78, 487, 885
Parminter, R. H. R., Capt. . . . 55
Parmod Singh, Risldr. . . . 898
Parnell, F. R., 2nd-Lt. . . . 571a
Parr, A. E., Lt. 488
—— A. 2nd-Lt. 798
—— C. E., Capt. 729
—— H. O., Bt.-Col. 53, 125, 153, 191, 868, 1042
—— T. W., Condr. 1870
Parrish, A., 2nd-Lt. 779
Parrott, A., Lt. 700, 712
—— W. Fitz A., Asst. Surgn. 1792
Parrington, J. W., Maj. . 39, 101, 614
Parry, A., 2nd-Lt. 351
—— B. K., 2nd-Lt. . . . 560, 1419
—— D. F., 2nd-Lt. 1571
—— D. G. 519
—— E. A., 2nd-Lt. . . . 67, 757
—— E. R., Lt.-Col. . . . 130, 1742
—— H. T., 2nd-Lt. 776
—— J. G., Lt. . . 534, 1641, 1855
—— J. H., *Ty.* Capt. 1765
—— J. K., 2nd-Lt. . 326, 366, 1363
—— M. J. 2nd-Lt. 787
—— N. E., Lt. . . 524, 1630, 1898
—— R. E., Lt. 543, 1080
—— T. O., 2nd-Lt. 798
—— W. L., 2nd-Lt. 825
—— Jones, W. H., 2nd-Lt. 474, 1641
Pars Ram, Jemdr. 1175
Parsa Singh, Thapa, Jemdr. . . 1602
Parsadi Singh, Jemdr. 1559
Parselle, J., Lt. . . . 11, 18, 666
Parshadi Singh, Subdr. . . . 1455
Parshad Gurang, Subdr. . . . 968
—— Singh, Jemdr. . . . 1052
—— Thapa, Subdr. . . . 1618
Parsick, C. E., 2nd-Lt. . . 552, 1696
Parsley, W., Capt. . . . 61, 691
Parsons, A. E. B., Capt. 95, 108, 261, 1231
—— A. H., Lt. 578
—— C. H., Sub-Condr. . . . 1855
—— D. S., 2nd-Lt. 703
—— F. F., Lt. . . . 523, 1671
—— F. G., Lt.-Col. . . . 756
—— F. S., Lt. 540a, 1162
—— J., 2nd-Lt. . . . 568, 1327
—— J. H., Lt. . . . 423, 933
—— J. P. D., 2nd-Lt. . . 549, 1675
—— J. W. A., Lt. . . . 367, 1589
—— L. C., Lt. . . . 703, 1677
—— L. J., Lt. . . . 354, 978
—— P. J. 2nd-Lt. . . . 551, 1463
—— R., Esq. 2
—— R. G., 2nd-Lt. . . . 638
—— W. G. L., Lt. . . . 542, 1271
Parsons-Smerdon, C., 2nd-Lt. . 561
Parsons-Smith, S. H., Maj. . . 45
Parta Singh Gurung, Jemdr. . 1627

NAMES AND RANK. Page.

Partabner, Raja of, 2nd-Lt. . . 1048
Partab Singh, Risldr. 931
—— Jemdr. 1164
—— Jemdr. 934
—— Jemdr. 1158
—— Subdr. 1259
—— Subdr. 1241
—— Sub-Asst. Surgn. 1814
—— Subdr. . . . 1335
—— Jemdr. . . . 896
—— Subdr. . . . 1021
—— Subdr. . . 1075, 1343
—— Jemdr. 1225
—— Jemdr. 898
—— Jemdr. 1063
—— Subdr. 1145
—— Subdr. 1163
—— Jemdr. 1241
—— Subdr. 1203
—— Jemdr. 1272
—— Subdr. 1488
Partal Singh, Jemdr. 905
Partap Singh, Sub.-Asst. Surgn. 1819
—— Sub-Asst. Surgn. 1818
—— Jemdr. . . 898, 1312
Parte Gurung, Jemdr. . . . 1639
Parthasaradhi Nayakar, C., Sub-Asst. Surgn. 1827
Parthasarathy, Jemdr. . . . 994
Parliman Rana, Subdr. . 1611, 1523
Partridge, C., 2nd-Lt. . . 354, 993
—— C. H., 2nd-Lt. . . . 604
—— E., Lt. 542
—— E. C., Lt. . . 84, 517, 1614
—— H. V., 2nd-Lt. 326, 366, 1521
—— J. W., Lt. . . 303, 1387
—— T. G. Lt. . . . 534, 987
Parsee, Jemdr. 1260
Pascoe, E. H., Lt. 512
—— F. J., Lt. . . . 722, 1007
—— J. J., 2nd-Lt. 722
Paske, C. A. M., Maj. . . 246, 919
Passanha, W. H., Snr. Asst. Surgn. 1791
Passmore M., 2nd-Lt. 736
Passy, F. H. B. Maj. . . 78, 233
—— W. DeL., Capt. . 276, 1052, 1455
Patandin Sukul, Jemdr. . . . 1017
Patch, B. G., Maj. 1712
—— F. R., Br.-Gen. 58, 122, 132, 632
Patch, R., Col. 1955
Patchappan, Subdr. 1339
Patchett, E., 2nd-Lt. . . . 567
—— R. T., Ty. Lt. . . . 1729
Patel, J. P., Sub-Asst. Surgn. . 1832
—— K. K., tempy. Capt. . . 1766
—— K. S., *Hony.* Capt. . . 1770
Pateman, E. A. T., Asst. Surgn. . 1803
Paterson, A. D., Lt. . . 536, 1299
—— A. G., Capt. . . . 718
—— A. H., 2nd-Lt. . . 77 1885
—— A. M., Lt 531
—— A. R., Senr. Asst. Surgn. 1787
—— A. R., Capt. 739
—— C. R. S., Lt. . 368, 720, 1150
—— D. C., 2nd-Lt. . . . 556
—— E., Lt.-Col. 585
—— E. A., *Hony.* Lt. . . 588
—— F. W. J., Maj. . 242, 1279
—— G. F., Lt. . . . 470, 1618
—— G. F. J., Capt. . 253, 1156
—— G. K., Capt. . . . 31, 790
—— G. M., 2nd-Lt. . . 763, 781
—— G. S. V., Lt. 1179, 18, 26, 512
—— H., 2nd-Lt. 623
—— H. E., Lt. 1056
—— H. F. W., Maj. 241, 1883, 1158
—— J., 2nd-Lt. 785
—— J. C., 2nd-Lt. . . . .
—— J. D., Lt. 656
—— J. J. C., 2nd-Lt. . . . 825
—— K., Lt. . . . 425, 1641
—— L. M., Lt. . 291, 1148, 1156
—— P. J., Lt. Col. . . 104, 613
—— R. H., O'D., Maj. 235, 939, 1048
—— S., 2nd-Lt. 61
—— T. G. F., Maj. . 62, 1748
—— T. S., Capt. . . 256, 911
Pat Ram, Subdr. 1403
—— Subdr. 1393
—— Jemdr. 1289
—— Jemdr. 1455
—— Singh, Jemdr. . . . 981

NAMES AND RANK. Page.

Pathak, M. V., Ty. Capt. . . . 1771
—— S. M., Ty. Lt. 1775
Patiala, Maj.-Genl. *Maharaja of,* . 110, 1078
Patiram Singh 1384
Patmore, F. J. M. E., Lt. . . . 732
—— S. H., 2nd-Lt. . 566 1702
Patney, S. L., Ty. Lt. . . . 1772
Paton, E. W., 2nd-Lt. 802
—— H. G., Lt. . . . 313, 1521
—— J., Lt. 658, 986
—— J. D., Lt. 589
—— J. E. F., Lt. . . 380, 933, 1006
—— N. F., Capt. . . . 811, 1892
—— W., Capt. 834
—— W. C., Capt. 1757
Patrick, H., 2nd-Lt. . . . .
—— P. J., Lt. 519, 1657
—— R. M. F., Maj. . 105, 246, 1191, 1895
Patrickson, D. M., Maj. . . 227, 1351
Partridge, E., 2nd-Lt. 1187
—— T. G. Lt., ... 988
Pattamada Chetticha, Jemdr. . 1324
Pattanath, Sankaran, Ty. Sub-Asst. Surgn. 1833
Patterson, A. A. C., 2nd-Lt. 539, 1052
—— A. H., 2nd-Lt. . . . 794
—— C., Lt. 55
—— H. E., 2nd-Lt. . . 669, 997
—— S. B. A., Lt.-Col. 140, 162, 200, 1901, 1912
Pattinson, E., Lt. 541
—— H., Lt. 294, 1020
—— H. A., Lt. . . . 317, 1040
—— R. I., 2nd-Lt. . . 540, 1563
Pattison, H. A., Lt. . . . . .
—— F. R., Lt. 380
Pattisson, C. A., Lt. 750
Patton, A. G., 2nd-Lt. . . 546, 1639
—— S. L. S., 2nd-Lt. . 560, 1199
—— W. F., Lt. 445
—— W. S., Maj. 1749
Pattullo, H. A., 2nd-Lt. . 474, 1040
Paul, A. E., Capt 843
—— A. R., Sub-Asst. Surgn. . 1818
—— G. E., Ty. Lt. 1773
—— R. S., Lt.-Col. . 95, 132, 156, 193, 1501
—— J. B., Sub-Condr. . . . 1870
Paulie, E., Esq. 89
Paull, J. R., Lt.-Col. . . . 43, 705
Paul Singh, Jemdr. 1300
Paulson, P., 2nd-Lt. 750
—— P. Z., Maj. 1003
Paupiah, Jemdr. 1388
Pavton, A. W. Capt. . . . . .
Pavey, G. P., 2nd-Lt. . . 701, 1007
Pawson, F. A., Lt 518
Paxton, W. M., Capt. . 501a, 1678a
Pay, E., Sub-Condr. 1874
Paymaster, B. B., Maj. . . . 1749
—— S. A. B., Capt. 1001, 1764
Payne, C. C., 2nd-Lt. 692
—— C. Condr., ... 1854
—— C., *Hony.*-Lt. 1846
—— C. A. W., Maj. 652
—— C. H., Capt. 775
—— D. W., Maj. 634
—— E. C., 2nd-Lt. 362
—— E. H., Lt.-Col. 87, 134, 157, 194
—— G. D., Capt. . . . 278, 1397
—— H. L. H., Lt. 617
—— J. H., Lt. 59, 736
—— P., Lt. 617
—— R. L., Maj.-Genl. . . 701, 705
—— W., 2nd-Lt. 551
—— Gallwey, L. P., Lt. . . . 587
Paynter, W. P., Maj. 614
Payton, W. H., Lt. 95, 523
Peace, A. C. 528
—— J. N., 2nd-Lt. . . 359, 1255
—— L. R., 2nd-Lt. 629
Peach, E. N., Lt. 445
—— W. A., 2nd-Lt. 362
Peacock, A. D., 2nd-Lt. . . . 600
—— E. B., Maj. . . . 87, 207
—— G. E., Asst. Surgn. . . 1800
—— G. I. C., 2nd-Lt. . . 332, 1435
—— H. B., Col. . . 189, 1901, 1911
—— L. V., Lt. . . 311, 1092, 1207
—— R. P. M., 2nd-Lt. . . . 356
Peacocke, A., 2nd-Lt. 1096
—— H. L., Lt. 691
—— L. G. L., Lt. 581
—— T. G., Col. 78, 125

NAMES AND RANK. Page

Peak, C. A., Lt. 295, 1058
Peake, A. J., Lt. . . 515, 1607
—— H. G., Tempy. Capt. . . 1723
—— S. C., 2nd-Lt. . . 691
Peake-Cottam, H., Lt. . . 302, 559
Peal, B. A., 2nd-Lt 554
—— E. F., Lt. 789
Pearce, A., Capt. 741
—— A. G., 2nd-Lt. 845
—— A. H. B., Ty. Capt. . . 1725
—— C. F. B., Lt. 389, 705, 1088
—— C. M., Col. 1893
—— C. B., Lt.-Col. . 131, 1741
—— E. A., Capt. 1839
—— E., Sub-Condr. 1856
—— G. A., 2nd-Lt. . . 361, 1305
—— J., 2nd-Lt. . . . .
—— J. C., Lt. 519
—— R. G. F. Lt. 1006
—— R. E. S., Lt. 733
—— T., Lt. . . 510, 997, 1215
—— T., Sub-Condr. . . . 1856
—— W. K., Capt. . . 730, 868
—— W. R., Capt. 817
Pearcey, R. A., 2nd-Lt. . 473 1267
Pearcy, F. G., Lt. . . . 541, 1676
Peareth, J. T., Lt. 766
—— W. F. E., Capt. . . . 751
Pearman, H. A., Lt. . . 530, 1671
—— W. T., 2nd-Lt. 784
Pears, T C., Lt.-Col. . 56, 179, 1917
Pearse, C. J., Col. 176
—— F. N., 2nd-Lt. . . 550, 1702
—— G., Maj. . . 60, 227, 1662
—— G. E., Lt. . . 669, 997
—— G. G., Maj. . . 391, 852, 1219
—— J. C., Maj. . . 224, 881
—— T. F., Lt. . . . 302, 1475
—— W. P., Maj 22
Pearson, A. A., Genl. *Sir* 145, 171, 1491, 1955
—— A. F., Lt. . . . 28, 666
—— C. McM., Lt. . . 513, 1667
—— C. V B., 2nd-Lt. . 474, 875
—— E. A., 2nd-Lt. . . . 657
—— E. G., 2nd-Lt. . . . 604
—— G. T., Lt. . . 322, 1255
—— H., 2nd-Lt. 1008
—— H. C., Lt. . 40, 397, 1601
—— H. F. A., Lt.-Col. 130 155, 190
—— H. F., 2nd-Lt. . 494, 949
—— H. G., Esq. . . . 1903
—— H. G., Lt. . . 534, 1331
—— H. G. A., Capt. . 267, 1235
—— H. L., Maj. . . 58, 827
—— H. S. L. Capt. 253, 1443, 1459
—— H. S. 2nd-Lt. . . 328, 939
—— J. B., Lt. 524
—— J. R., Lt. 608
—— L. C., Lt. 515
—— L. M., 2nd-Lt. . 558, 1457
—— R. F., 2nd-Lt. . 324, 931
—— T. F., 2nd-Lt. . 540, 1140
—— W. MacM., Lt.-Col. 140, 1746
—— W. G., 2nd-Lt. . 356, 1673
Peart, C. L., Maj. 53, 99, 104, 107, 108, 210, 1427, 1890
—— D. G., Maj. 27, 1883
—— G. R., Col. 177
Pease, C. N., Lt. 585
—— G. B. R., Capt. 623
—— H. T., Col. . . 1842, 1895
Peat, H. R., Lt. 1561
—— W. E., Lt. . . . 470, 925
Peattie, G. D., 2nd-Lt. . . . 1624
Peck, A. W., Bt.-Col. 119, 126, 153, 195, 867, 921
—— H. E., 2nd-Lt. . . 555, 1413
—— J. H., Lt.-Col. 36, 193, 1883, 1912
—— L. J., Lt. . . . 488, 877
Pecker, C. B., Lt. . . . 313, 921
Peckover, R. G., 2nd-Lt. . 328, 1305
Pedder, G. R., Lt. 591
—— J. W. R., 2nd-Lt. . . . 623
Peddis, G., Maj. 841
—— R. M., Lt. 1425
Peden, W., 2nd-Lt . 38, 762, 781
Pedgrift, A. E., 2nd-Lt. . . . 751
Pedly, J. E., Lt. . . . 506, 1251
Peebles, A., Lt. 844
—— A. S. M., Maj. . . . 1751
—— E. C., Bg.-Genl. 12, 22, 120, 125
—— J. H., 2nd-Lt. . . . 844—45
—— R. E., Maj. 614

NAMES AND RANK. Page

Peek, B. M., Lt. 1006, 732
—— W., Capt. *Sir* 59
Peel, A. W., Condr. 1872
—— B. G., Maj. . . 52, 222, 1331
—— W., Lt. 517, 1160
—— W. R. B., Lt. . . . 405, 937
Peele, C. R. DeC., Lt. 503, 738, 1575
Peerman, T. C., Lt. . . 524, 1667
Peers, A., 2nd-Lt. 1321
Peek, L. M., Capt. . 264, 1038, 1893
Peffers, A. Capt. 763
—— J., Lt. . . . 470, 1036
Pegg, A. C., Capt. . . 278, 1379
—— T., Sub-Condr. 1856
Pegge, F. H., Lt. . . . 489, 917
Pelham-Clay, E.C. G., Lt. . 527, 1561
Pelle, H. D., Maj 1745
Peirse, C. E., Col. . . . 180, 1918
Pelhwan Khan, Jemdr. . . . 1515
Pell, E. J., Asst. Surgn. . . . 1799
—— L. R., 2nd-Lt. 739
—— T. O., Lt. 530, 1668
Pelly, A., Sub-Condr. 1855
—— C. B. N., 2nd-Lt 652
—— C. T., 2nd-Lt. . . . 551, 1645
—— G. S., 2nd-Lt. 810
—— H. A., Lt. 587
—— H. N., Capt. 1722
—— H. R., Capt. 107, 262, 004, 1805
Pelmear, A., 2nd-Lt. 361
Pema Ram, Subdr. 1419
—— Jemdr. 1456
Pemberton, A. L., Capt. . 975, 636
—— E. H., Maj. . . 245, 1049
—— H., Condr. 1853
—— J. D., Lt. . . . 455, 59
—— J. S., Lt. 1132
—— S., Maj. . . . 664. 991
Pembroke, S. K., Lt. 758
Pembrook, J. G., Lt. 780
Penberthy, W. F., 2nd-Lt. . 722, 799
Pender, G. T., Lt. . . . 505, 1152
—— W. S., Lt. . . . 299, 865
Pendery, G. B., Condr. . . . 1854
Pendlebury, J. W., Capt . . . 745
—— R., 2nd-Lt. . . 669, 997
Pendrigh, T. G., Lt. 535
Penfold, H. de L., Capt. . 278, 1641
Pengelley, G. H., Capt. . 254, 1663
Peniston-Bird, E. L., 2nd-Lt. . 475
Penlington, A. T., Sub-Condr. . 1856
Penley, H. O., Revd. 1904
Penn, C. E., Lt. 533, 952
—— C. L., Lt. 1511, 465
—— G. G., 2nd-Lt. 794
—— W. E., *Hony.*-Capt. . . . 1845
Penn, Simpkins G. Lt. 307*a*
Pennack, C. Lt. 710
Pennant, W. A. T., Temp, Lt.-Col. 67
Pennefather, E. M., Maj. . . . 1711
—— E. M., Lt. . . 698, 1006
Pennefather-Evans, G., Lt.-Col. 135, 158, 96, 868, 1094, 1106
Pennell, C. S., 2nd-Lt. 833
—— G. E., 2nd-Lt. 1008
—— K. E. L., Capt., . . 668, 996
Penney, J. D., 2nd-Lt. . . . 7, 552
—— N. A., Lt. . . . 322, 1120
Penniger, R. C., Lt. 1124
Pennington, A. W., Lt.-Col. 131, 156, 191
—— C. E. R., Lt. . . . 73, 410
—— R. W. B., Maj. . . . 654
Penny, C. E., 2nd-Lt. 1116
—— E., Lt. 583
—— E. A., Capt. . . . 25, 1756
—— J., Lt.-Col. . . . 129, 1742
—— J. D. 2nd-Lt. . . . 6, 14
Penrose-Welsted, R. H., Capt. 273, 1030
Pentham, J., 2nd-Lt. . . . 557, 1703
Pentland of Lyth, *Baron* 611, 801, 652, 1892
—— A., 2nd-Lt. . . . 362, 1199
Penton, B. C., Maj. . . 99, 218, 872, 1.20, 1569
—— H. E., Maj.-Genl. . . . 1431
Pentoney, L. R. 2nd-Lt. . 558, 1497
Pentreath, G. A., Lt. 538
Peppain, S. H., 2nd-Lt. . . . 1487
Peppe, A. T., Maj. 1894
—— G. T. V., Lt. 487
—— J. H. H., Lt. 595
Pepper, G. H. N., Lt. 518
—— H., *Hony.* Maj. . . . 1850

NAMES AND RANK. Page.

Pepper, H., Maj. 806
—— J. W., Sub-Condr. . . . 1864
Pepperall, R. A., Lt. . 450—54, 1285
Pepperett, W., Ty. 2nd-Lt. . . . 725
Peppin, S. H., Lt. 536
Pepys, G. L., Maj. . 6, 13, 99, 102, 104, 217, 1251
Peradon, F. L., 2nd-Lt. 1066
Perbhu Sing, Subdr. 1064
Perch, H. L., Lt. 1008
Percy, A. C., Lt. 520, 1886
—— J., Condr. 1868
Percy-Jones, C.M., Capt. 280—287, 1102, 1122
Percy-Smith, E. S. Bt.-Lt.-Col. 59, 107, 220, 1895, 1913
—— H. K., Lt. 317*a*, 1143, 1183
Perdriau, A D. C., Sr. Asst Surgn 1790
Perdue, C. G., Lt. . . . 578, 1403
Pereira, A. X., Capt. . . . 1769
—— D. F., Asst. Surgn. . . 1791
—— E J. C., Sr. Asst. Surgn. 1788
—— J. A. 1781
—— J. C., Asst. Surgn. . . 1801
—— J. E., Lt. . . 493, 579, 883
—— J. G., Asst. Surgn. . . 1799
—— J. M., Asst. Surgn. . . 1797
—— J. P., Asst. Surgn. . . 1797
—— R. B, Sr. Asst. Surgn . 1788
Pereyra, A., Esq. 88
Perfect, S. C., 2nd-Lt. 32
—— S. R., 2nd-Lt. 558
Perigo, J., Lt. 637
Periyasami, Jemdr. 1333
—— Pillai, Jemdr. 1591
Perkin, C. R. 2nd-Lt. . . 563, 1483
Perkins, B. W., 2nd-Lt. . . . 820
—— G. H., Asst. Surgn. . . 1793
—— H., 2nd-Lt. 637
—— J. W., Asst. Surgn. . . 1800
—— N. K., Lt. 734
—— W. H., Lt. . . . 348, 1359
Perks, G. A. C., 2nd-Lt. . . . 1375
—— H. B, Lt. . . . 540, 1696
Perley, J., Sub-Condr 1873
Permeshwari Parshad Tewari, Jemdr. 1019
Perram, H. C., Capt. . 271, 279, 1343
Perraton, C. W., Lt. . . . 535, 1249
Perreau, A. M., Col. . . . 126, 613
Perreira, W. A. C., Asst. Surgn. . 1796
Perrett, W. A., Lt. . . . 399, 1321
Perrin, F. E., 2nd-Lt. . . 361, 1052
—— F. H., Lt. . . . 519, 1667
—— L. C., 2nd-Lt. . . 557, 1614
—— N. J., 2nd-Lt. 626
Perrins, H. G., 2nd-Lt. . . . 1413
Perriton, W. L, Sub-Condr. . . 1855
Perrott, H. R. T. S., Lt. . 532, 1691
—— L. C. P., Lt. . . 528, 1243
Perry, A. C., Lt. 529
—— A. E. P., 2nd-Lt. . . . 815
—— A. S. H., Lt. . . . 524, 1124
—— C. A. S., Lt. . . 547, 608, 1673
—— D. G., Ty. 2nd-Lt. . . . 725
—— E. L., Lt.-Col. . . . 134, 1744
—— E. L., 2nd-Lt. 356
—— E. G., Lt. 310, 1227
—— E. W., Lt. 543*a*
—— F., Sub-Condr. 1863
—— F. F. Lt.-Col. 1931
—— F. H., 2nd-Lt. 790
—— F. W. R., Lt. . . . 536, 1060
—— H. A. P., 2nd-Lt. . . . 559
—— H. S. J., 2nd-Lt. . . . 352
—— H. W., Brig. Gen. *Sir* 10, 74, 76, 120, 123
—— J. P., 2nd-Lt. 696
—— P. E., 2nd-Lt. . . . 545, 1673
—— S., Maj. 633, 978
—— T. T., the Revd. 1907
—— W. E., Capt. . . . 665, 880
—— W. H. G., Lt. 587
Persa, Jemdr. 1700
Persse, S. H., Capt. . . . 485, 907
Pert, C. E., Lt. . . . 318, 317, 907
—— J. E., 2nd-Lt. 686
Pertab Singh, Subdr. . . 1166, 1497
—— Jemdr. 1193
—— Risldr. . . . . .
Pertiman Gurung, Jemdr. . . . 1602
Pertwee, J. W. 2nd-Lt. . . . 788
—— L., Lt. 465, 524
Perumal, V. G., Subdr. . . . 1351
—— Jemdr. 1994
—— Reddi, Jemdr. 1254

Names and Rank.	Page.
Perumal, Subdr.	1359
Perumal, Pillai, Jemdr.	1270
——— Thevar, Jemdr.	1332
Pescod, R. S., 2nd-Lt.	560, 1678
Peshaurilal Mehta, Sub-Asst. Surgn.	1819
Pesk, C. A.	295a
Pestonjee, R., Capt.	1768
Pestonji, A. B., Capt.	1766
Petch, P. B., 2nd-Lt.	561, 1678
Peterkin, G., 2nd-Lt.	
——— R. C., 2nd-Lt.	606
Peters, A. S. C., 2nd-Lt.	550, 1674
——— C. W. E., Asst. Surgn.	1798
——— F. A., 2nd-Lt.	1393
——— F., 2nd-Lt.	637, 1469
——— G. L., Capt.	844-45
——— H., Ty. 2nd-Lt.	724
——— H. A., Lt.	
——— J., Maj.	219, 871, 893, 1541
——— J. H. C., Asst. Surgn.	1795
——— J. H. G., Lt.	293, 685, 1227
——— N. M., Sub-Condr.	1872
——— W., 2nd-Lt.	1575
Petersen, R. G., 2nd-Lt.	555, 1205
Peterson, A. H., 2nd-Lt	1885
——— F. H., Brig.-Genl.	12, 23, 106, 120, 123, 149, 184
Pethick, R. S., Capt.	1839
Petit, G., Capt.	1712
Peto, V. F. C., Lt.	111
Petre, A. H. C., Lt.	593
Petri, A. T., *Hony.* Capt.	1843
Petrie, A. G., Lt.	315
——— C. A. B., Lt	707
——— C. L. R., Maj.	735
——— J. F., Lt.	95, 310, 968, 1617
——— J. C. E., 2nd-Lt.	539, 1708
——— W. S., Lt.	1839
——— W. T., MacG., Lt.	815
Petter, L. St. G., Revd.	1905
——— P. E., 2nd Lt.	787
Pettersen, S., 2nd-Lt.	530, 1483
Petters, H. D., Lt.	77, 512
——— H.D.,	1885
Pettifer, F. G., Lt.	571
——— J., Lt.	1525
Pettigrew, F. C., Lt.	445, 1636
——— J. D., 2nd-Lt.	750
——— R. W., Sr. Asst. Surgn.	1791
——— W., 2nd-Lt.	548, 1696, 1701
Pettitt, A. H., Lt.	623, 1471
——— N. 2nd-Lt.	562
Pettman, F. L., Lt.	410, 410, 512, 1403
Petts, C. E., Lt.	1036
——— H. J., Lt.	531, 1672
Petty, H. G., 2nd-Lt.	1677
——— J., 2nd-Lt.	730
Peyton, A. G., Col.	59
——— A. H. Ty. Lt. Col.	229, 1662
——— W. H., 2nd-Lt.	826
Phadke, S. K., Ty. Capt.	1766
Phaga Singh, Jemdr.	1112
Phaltan, Chief of, Hony. 2nd-Lt.	481
Phandi, Subdr.	1171
Phani, N., Ty. Capt.	1772
Phani Bushan Datta, Subdr.	1219
Paharsing Gurung, Jemdr.	1597
Phatak, S. A., Ty. Capt.	1768
——— V. M., Ty. Capt.	1764
Phaudasing Limbu, Subdr.	1648, 1655
——— Jemdr.	1655
Phaure, A. E. A., Asst. Surgn.	1802
Phayre, A., Lt.-Genl. Sir	118, 145, 172, 1918
——— R. B., Bt. Maj	105, 258, 1613
Phelan, F., Capt.	1757
——— J. O'C., *Hony.* Capt.	1848
——— P. G., 2nd-Lt.	327, 1415
——— P. J., 2nd-Lt.	327, 1413
Phelips, H. V. M., Lt.	
Phelps, A., Lt.-Genl.	170
——— A. A., Capt.	501, 1311
——— A. E., Lt.	387, 1249
——— H. H. Y., Lt.	1160
——— J. N., 2nd.-Lt.	330, 1174
——— J. W., Capt.	755, 756
——— J. W., Capt.	746
Pheku Sing, Jemdr.	1397
Phera Singh, Jemdr.	1051
Phibbs, H., 2nd-Lt.	547
Philbrick, H. A., Lt.	538, 1158
Philip, C. L., 2nd-Lt.	557, 889
——— J. G., Revd.	1905
——— J. L. R., Capt.	1760

Name and Rank.	Page.
Philip, Poonnoos, Subdr.	
Philips, A. A., Lt.	836
Philips, J., Col.	177
——— J. C., Maj.	223, 1131, 1249
——— J. L., Maj.	614
——— J. W., Capt.	257, 1663
——— N. R. G., Lt.	1886
——— R. L., 2nd-Lt.	321, 1423
Phillipson, H., 2nd-Lt.	1571
Phillipsz, B. S., 2nd Lt.	557, 1702
Phillimore, R. H., Maj.	663, 1880
Phillips, A. A. H., Lt.	629
——— A. J., Lt.	721
——— A. L. Col.	57
——— C. A., Lt.	405, 490, 1563
——— C. A., 2nd-Lt.	562
——— C. E., Capt.	529, 830
——— C. E. V., Lt.	300, 1427
——— C. H. K., Lt.	503, 1116
——— C. H. W., Lt.	502, 1665
——— C. S., Capt.	17, 27, 780
——— E., Lt.	532, 1668
——— E. C., 2nd-Lt.	356, 1161
——— E. J., 2nd-Lt.	1576
——— F., Hony. Lt.	846
——— F. E., Lt.	655
——— G. A., Maj.	9, 16, 500, 803
——— G. C., Capt.	485, 961
——— G. H., *Hony.* Capt.	1849
——— G. H., 2nd-Lt.	549, 1044
——— G. S., Condr.	1854
——— G. W., 2nd-Lt.	788
——— H. A., Maj.	9, 755
——— H. C., Asst. Surgn.	1796
——— H. J. L., Lt.	509, 1611
——— H. L., Lt.	507, 1655
——— H. S., Capt.	43, 262, 925
——— H. T., 2nd-Lt.	564, 1303
——— H. W., Lt.	541, 1673
——— I. P., Asst. Surgn.	1802
——— J. A. S., Maj.	1479, 1753
——— J. H., Lt.	527, 1423
——— J. R. C., 2nd-Lt.	763
——— L., Maj	223, 1271
——— P. F., 2nd-Lt.	77, 553, 1885
——— P., Hony. Capt.	1847
——— P. B., 2nd-Lt.	547, 1705
——— P. J., Lt.	322, 1373
——— R. G. T., 2nd-Lt.	1886
——— R. S., Lt.	291, 1178
——— T. E., 2nd-Lt.	734
——— W. E., Capt.	702
——— W. G., 2nd-Lt.	727
——— W. H., Lt.	533, 998, 1007a
——— W. J., Capt.	1722
——— W. J. C., Lt.	405, 768, 1457
——— W. W., Maj.	611
——— W. W., Lt.	507, 1116
Philp, R. E., Lt.	668, 996
——— W. H. S., Lt.	353, 1008
Philpot, H. C. V., 2nd-Lt.	553
Philpott, F. L., Lt.	292, 925, 1587
——— G. R., Lt.	491, 353, 1110
——— D. A., Lt.	423, 1397
Phipson, E. S., Capt	1755
Phipps, A. H., 2nd-Lt.	554
——— C. C., Capt.	19, 655, 1681
——— H. C., 2nd-Lt.	618
——— J. P., 2nd-Lt.	328, 881
Phonda Sing Limbu, Subdr.	1648
Phulander Singh, Jemdr.	1022
Phulchand, Subdr.	1040
Phul Singh, Subdr.	1497
——— Jemdr.	1039
——— Jemdr.	1032
——— Resaldr.	901
——— Jemdr.	1047
——— Khan, Jemdr.	1480
Phula Ram, Jemdr.	1476
——— Singh, Ressdr.	910
——— *Hony.* Capt.	1066, 2103
Phuman Singh, Subdr.	1078
——— Subdr.	133
——— Subdr.	1563
——— Subdr.	1145
——— Subdr.	1074
——— Jemdr.	1304
Phumman Singh, Jemdr.	1157
Phusa Ram, Jemdr.	1336
Phusan Singh, Jemdr.	1585
Phythian-Adams, E. G. P., Capt.	107, 253, 1327, 1501
Piaru, Jemdr.	1284
Pibworth, A. H., 2nd-Lt.	351, 1299

Names and Rank.	Page.
Picachy, F. A., Sr. Asst. Surgn.	1796
Pickard, B. de H., Lt.	397, 941
——— L. A., Lt.	544, 1241
Pickersgaill, R., Lt.	420, 1154
Pickering, F. C., 2nd-Lt.	565
——— M. B., 2nd.-Lt.	361, 1141
Pickford, A. D., Maj.	603
——— H. V. F., 2nd-Lt.	362
Pickmore, C. E., Lt.	510, 1577
Pickthall, C. M., Lt.	511, 1642
Picton, A. G. C., 2nd-Lt.	717
——— R. E., Col.	37, 125, 660
Pierce, F.G., Lt.-Col	9, 28, 84, 131, 156, 192, 1411
——— H. J. J., 2nd-Lt.	1483
——— I. G. F., Lt.	315
——— P. A., Ty. Asst. Surgn.	1804
Peircy, C. H. Z., Lt.	690a, 798
Pibworth, F., 2nd-Lt.	353
Pierpoint, H. W., Maj.	1712, 1899
Piers, S. O., Revd.	1906
——— Lt.-Col. W., B.	181
Pigeon, J. W., Capt.	1758
Piggford, H. G., Lt.	506
Piggott, A. H., Capt.	749
——— J. T., Maj.	686
Pignon, H. L., 2nd-Lt.	551, 1098
Pigot, G., Lt.	291, 1094
——— H. G., Lt.	314, 947
——— W. T., Lt.	506
Pigott, B., *Sir* Lt.	597
——— F. M., Lt.	399
——— G. B., Asst. Surgn.	825, 1803
——— G. T., Maj.	783
——— R. E. P., Col.	825
——— T. F., Capt.	521, 830, 1026
Pigou, Maj. F. H.	52
Pike, A., Lt.	489, 911
——— F. F., 2nd-Lt.	815
——— G. S., 2nd-Lt.	252
——— J., Lt.	629
——— J. P., 2nd-Lt.	669
——— W. H., Lt.	380, 511, 1581
Pilbeam, F. W., 2nd-Lt.	359, 1309
Pilcher, A. H., Lt.	529
——— A. J., Lt.-Col.	26, 130, 611
——— A. S., Lt.-Col.	597
——— H. R., Capt.	485, 939
——— T. D., Maj.-Genl.	708, 709
——— V. D., Lt.	460, 1596
Pilditch, I., Maj.	780
——— D., 2nd-Lt.	553
Pildith, D., 2nd-Lt.	1523
Pilgrim, H. G. E., Lt.	512
Pilkington, J., Sub-Condr.	1855
——— T. L., Lt.	405, 1511
Pili, G. C. R., Lt.	455, 1501
Pillai, B. V. S., Sub-Asst. Surgn.	1827
——— C. V. R., Sub-Asst. Surgn.	1817
——— J. S., Sr. Sub-Asst. Surgn.	1816
——— K. P., Sub-Asst. Surgn.	1832
——— M. A., Sr. Sub-Asst. Surgn	1825
——— M. A. T., Sr. Asst. Surgn.	1816
——— M. K., Capt.	1769
——— M. K. S., Sub-Asst. Surgn.	1827
——— M. T., Sub-Asst. Surgn.	1817
——— N. S., Sr. Sub-Asst. Surgn.	1816
——— P. A., Sub-Asst. Surgn.	1827
——— P. B., Sub-Asst. Surgn.	1816
——— S. C. D., Sub-Asst. Surgn.	1827
——— S. K., Ty. Lt.	1778
——— S. R., Sub.-Asst. Surgn.	1828
——— T. G. K., Sub-Asst. Surgn.	1828
——— T. M. S., Sub-Asst. Surgn.	1828
——— T. R. G., Sub-Asst. Surgn.	1827
Pillans, J. E., 2nd-Lt.	631
Pillar, C. H., 2nd-Lt.	557
——— E. J., 2nd-Lt.	357, 1038
——— F. V., 2nd-Lt.	1263
Pillay, A. P., Ty. Capt.	1770
——— C. K. Ty., Lt.	1780
Pilling, R. W., 2nd-Lt.	475
Pim, G. A., Capt.	264, 1517
——— G. E., 2nd-Lt.	561, 1311
——— G. R., Lt.	666, 986
——— H. M., Capt.	274, 1116
Pinch, R. P., 2nd-Lt.	778
Pinches, A. G., Lt.	455, 1621
——— H. I., Ty. Capt.	1726
Pindi Das Ball, Sub-Asst. Surgn.	1819
Pine, F. C., Lt.	543, 1321
Pine-Coffin, E. C., Lt.	302, 1106, 1130
Pinhey, H. T., Lt.	686, 700
——— R. A. Capt.	367, 1006

Names and Rank.	Page.
Pinker, F. R., 2nd-Lt.	565
Pinkerton, A. T. H., Lt.	768
—— G. E., Lt	353
—— S. R., 2nd-Lt. 739,	791
Pinney, J. C. D., Maj.	59
Pinnock, V. St. L., Ty. Capt.	1726
Pinniger, R. C., Lt.	111
Pinto, J. A., Asst. Surgn.	1798
—— M. C., Asst. Surgn.	1794
Pipe, H. W., Lt.	532
—— T. S., Lt.	846
Piper, A. B. C., Lt. 470,	1475
—— N. L., 2nd-Lt. 720,	786
—— W. J. F., Lt. 317,	1187
Pippet, G. R., Lt.	543a
Piran-Ditta, Subdr.	1104
—— Jemdr.	1682
Pir Baksh, Jemdr.	1502
—— Jemdr.	1391
—— *Khan Sahib*, Sub-Asst. Surgn.	1807
—— Dad Khan, Jemdr.	908
—— Dost, Jemdr.	1433
—— Subdr.	1493
—— Mahi, Jemdr.	1436
—— Khan, Jemdr. 1683,	1702
—— Muhammad, Jemdr.	1701
—— Khan, Jemdr.	1268
—— Khan, Sub.-Maj.	1120
Pirbhu Singh, Subdr.	1032
Pirie, C P W, Maj.-Gen. 118 146,	173
—— D. G. L., Lt. 528,	998
—— G. B., Capt.	752
—— G. W., Lt. 538,	1672
—— J. P., Lt.	636
Pirouet, H. N., 2nd-Lt. 330,	1160
Pirrie, F. W., Bt.-Col. 190, 1880, 1911,	1931
—— J. V., 2nd-Lt.	611
Pirta Sing Thapa, Subdr.	1639
Pirthibir Bura, Subdr.	1627
Pirthi Chand, Subdr.	1117
—— Lal Gurung, Jemdr.	1627
Pirthipal Singh, Jemdr.	1561
Prithi, Singh, Jemdr.	1192
—— Rana, Subdr.	1614
Pitambar Singh, Subdr.	1021
Pitcairn, D. N., Lt. 294,	1611
—— G. D., 2nd-Lt.	766
—— J. E., Lt. 763,	810
—— J. E. Lt.	781
Pitcher, D. G., Col. 177,	1917
—— D. LeG. Br.-Genl. 50, 122, 137, 159, 215, 949,	1931
Pitchford, H., 2nd-Lt.	354
Pitha Rawat, Subdr.	1436
Pithawala, H. E., Lt. 1497,	1779
—— H. E., Ty. Lt.	1779
Pitkeathly, J. S., Maj. 58,	605
Pitman, C. R. S., Capt. 271,	1128
—— F., 2nd-Lt.	637
Pite, R. W., 2nd-Lt.	1008
Pitt, B. I., Lt. 443, 1495,	1507
—— C. A., Capt.	
—— G., 2nd-Lt. 465,	1305
—— F. Lt.	1677
—— M. F., Lt.	26
—— H. F. G., 2nd-Lt. 539,	1459
—— R. H., Lt.	351
—— W. A., Lt.	785
—— W. H., 2nd-Lt. 84, 492,	887
—— W. O., Lt.	10
Pittar, A. H., 2nd-Lt. 558,	1703
Pittock, J. A., Lt.	685
Pitts, G., Lt. 533,	1120
Place, N. D., Capt. 253,	1046
—— W., 2nd-Lt.	571
Plaford, R. B., Lt. 425,	1427
Planche-Hearn, C. R., Lt.	760
Plant, I. T., 2nd-Lt. 360,	1369
Plant, J. A., Capt. 76,	778
—— J. C. G., Lt. 652,	783
Plater, W. E., 2nd-Lt.	302
Platt, E. H. R. C. R., Col. 127,	701
—— E. P., Lt. 68, 317,	1187
—— H. C., 2nd-Lt. 472, 1034,	1466
—— J., 2nd-Lt.	723a
—— P., 2nd-Lt.	723a
—— W. P., Lt. 90,	687
Platts, A., Lt. 528,	1668
—— C. S., Lt. 517,	1007
—— J. C., Lt. 294a,	907
—— M. A., Lt. 322,	1617
—— S. A., 2nd-Lt. 331,	1641
—— T. C., 2nd-Lt. 699,	796
Playfair, A., Lt.-Col. 57, 192, 1691, 1898,	1912
Playfair, H. R., Lt. 310,	1355
Playne, H. F., 2nd-Lt. 88,	848
Pleasance, A. G., 2nd-Lt. 357,	1114
Plewman, T. L., 2nd-Lt. 365,	1303
Plineke, J. F., Capt. 721,	739
Plinston, G. H., Capt. 14, 38, 254,	1062
Plomer, G., Lt. 300,	1479
—— E. M. 2nd Lt.	815
Plommer, G. F., 2nd-Lt.	568
Plowden, A. H. C., Lt.	697
—— B. E. C., Capt. 277,	1335
—— C. H., Col.	176
—— C T. C., Maj	1910
—— H. R. H., Capt	595
—— J. M. B. C., Lt. 321,	1178
—— P. P. M. C., Lt.	819
—— T. C., Lt.-Col. 193, 1899,	1912
—— W. F. C., Lt.-Col.	56
Pluck, H., Sub-Condr.	1873
Plumb, W. C., 2nd-Lt. 554,	1181
Plumer, C. G. M., Maj. 237,	1263
—— G. W. H., 2nd-Lt.	1679
—— T. H., Lt.-Col. 127, 154, 181	1918
Plumby, A. G. C., Capt. R.A.M.C.	913
Plummer, E. W., Lt.-Col. 101, 133,	613
Plumpton, R. R., Lt. 543a,	1331
Plumptre, C. M., Capt.	1756
Plunkett, A. L. B., 2nd-Lt.	844
—— H. R. S., 2nd-Lt.	3 0
—— R. F. D., Lt.	317
Plymen, F. J., Capt.	803
Pocklington, J. F. J., 2nd-Lt. 475,	1601
Pocock, H. S., 2nd-Lt.	1008
—— J. A, 2nd-Lt.	666
—— J. G., Lt.	293a
—— L. J., Lt. 543, 1289,	1289
—— N. E. I., Lt. 316,	1030
—— P. F., Lt.-Col. 137, 159, 197,	1475
—— R. N., Lt.	722
Pogose, R. Lt.	1695
Pogson, C. A., Capt. 17, 27, 261,	1467
—— C. A., Lt. 531,	1707
—— F. V., Capt. 105, 253,	1663
—— L. V. J., Capt. 367,	1431
Poh, L. H., Ty. Capt.	1768
Bhkar, Jemdr.	1060
Pohlo, Jemdr.	1188
Pohlo Ram, Sr. Sub-Asst. Surgn.	1808
Pohp Sing, Jemdr.	932
Pokhar, Jemdr.	1391
Poingdestre, C. Le G., Lt. 516a, 1607,	1653
Pola Khan, Jemdr.	1408
Pole, G. E., Lt. 354,	1676
Pole Carew, G. A., Capt.	722
Pollaco, R. H., Lt. 508,	1666
Pollacco, R. G., 2nd-Lt. 559,	1094
Pollard, A. R., Lt. 511,	986
—— C. C., 2nd-Lt. 356,	1499
—— E. A., Capt.	43
—— E A., Lt.	695
—— G. L. G., Capt. 273, 1004,	1497
—— I. E., Lt.	534
—— J. H., Lt.-Col. 128, 154,	184
—— L., Maj	696
—— R. C., 2nd-Lt.	957
Pollard-Lowsley, H. deL., Maj. 663, 1882,	1890
Pollen, H. S., Lt. 523,	1347
Pollett, J. O., S. K. C., 2nd-Lt.	808
Polley, S. T., Maj. 78, 242,	879
Pollicot, H. E., 2nd-Lt. 474,	1495
Pollitt, H. T., Lt. 315,	913
—— J., 2nd-Lt.	360
Pollock, C. A., 2nd-Lt. 753,	781
—— C. R., Sr. Asst. Surgn.	1791
—— H. C., Maj. 664,	1882
—— J. A H., Maj.-Genl. 1527,	1955
—— K. C., 2nd-Lt.	359
—— R. C. G., Maj. 237, 968,	1227
—— V. D., 2nd-Lt. 332,	1162
Pollock-Gore, W. A. J., Lt. 309,	1509
Pollok, A. B., Capt. 56,	587
Polu Ram, Subdr.	1412
Polwhele, J. A., Lt.	610
Po Mya, Jemdr.	1295
Pomphrey, C., Lt.	859
Ponnusami, Subdr.	993
—— Naidu, Jemdr.	1332
—— Subdr.	1351
Ponnusami, Jemdr. (I)	994
—— Jemdr. (II) 994,	1333
Ponnuswmny Dinadayalan, Sub-Asst. Surgn.	1828
Ponsonby, D. G., Capt. 250,	1156
—— H. P., Lt. 629,	976
Pont, L. V., Lt.	489
Ponting, T. L., Capt. 263,	1375
Pook, E. R., Lt. 342,	1507
—— J. de C., Lt. 504, 610,	1665
Pool, W. A., Capt. 57a,	1841
—— W G, 2nd-Lt.	818
Poole, A. R., 2nd-Lt.	361
—— E. J. R., Capt. 259,	1207
—— E. R. L, Lt. 314,	935
—— F. A., Condr.	1861
—— F. G., Maj.	102
—— G. E., 2nd-Lt.	
—— G. W., Lt. 352,	1453
—— I. M. C., Ty. Lt.-Col. 108, 215,	1661
—— R. L. G., Lt.	375
Pooley, J. P. F., Lt. 521,	1667
—— J. S., Capt.	1717
Poor, W. A. J., 2nd-Lt.	638
Poore, L. L. 2nd-Lt.	728
—— R. M., Brig.-Genl. 12, 34, 120,	124
Pope, A. V., Capt.	503
—— D., Lt.	589
—— H. L., Capt.	500
—— J. A., 2nd-Lt. 480,	1495
—— H. W., Lt. 583,	1207
—— H. W., 2nd.-Lt.	357
—— P. F., Maj. 744, 1016,	1018
—— S. B., Maj. 52, 99, 233,	1255
—— S. F., Capt. 53,	700
—— W. V., Asst. Surgn.	1800
Pope-Hennessy, L. H. R., Brig-Gen. 51,	122
Popham, E. L., Bt. Lt.-Col. 160, 213,	923
—— R. L., Maj.	1711
Porch, E. A., Bt. Lt.-Col. 60, 139, 160, 219,	1662
Porchester, H. G. A. M. V F., *Lord*, Lt.	587
Porral, F. J. W., Lt. 397	917
Porrest, J. D. D., Lt.	1369
Porrett, C. R., 2nd-Lt. 8, 82, 545,	801
Porri, L. P., Rev.	1908
Porritt, A. H., Lt. 537,	1668
Portal, A E., Lt. 368,	883
Portas, J., Lt. 304,	1501
Porteous, C. A., Col.	177
—— C. McL., Lt.-Col. 131, 155,	191
—— D. G., Capt. 15, 31, 261,	1367
—— E. R., Capt.	830
—— H. A., Lt. 446—49,	1044
—— R., Maj.	1838
Porter, A. B., Capt. 259,	885
—— A. G., Lt. 304,	1026
—— D. A., Nursing Sister	1735
—— F G., Maj 29,	1086
—— G. H. J., Lt. 533, 1415,	1678
—— H., Lt. 691,	783
—— L. L., Maj. 78, 485,	830
—— M. T., Maj. 664,	1882
—— N., 2nd-Lt.	565
—— R. R. M., Capt.	1757
—— W., 2nd-Lt. 666,	996
—— W. D., Lt. 415,	1060
Portsm outh, F. V., Lt.	691
Poshati Mari, Subdr.	999
Postance, J. F. A., Lt.	507
Posth, C., Condr.	1855
Postlethwaite, J. L., 2nd-Lt.	669
Poston, G., Capt. 720,	785
Postwalla, R. M., Capt. 1347,	1764
Potbury, H. W., Lt. 697,	1255
PoThan, Sub-Maj.	1295
Pothan, K. E., Ty. Lt.	1777
Pott, D., Capt. 262, 899,	1004
—— E. H., Maj. 14, 35, 230,	875
—— H. P., Maj. 220, 1883, 103,	1299
Pottekat, K. K. M., Sub-Asst. Surgn.	1828
Potter, A. G., Capt. 679,	681
—— G. E., Lt.	587
—— J. C., 2nd-Lt.	768
—— J. W., Ty. Capt.	1724
—— L. R., Capt. 501a, 807,	1106
—— R. J. K., Maj. 239, 1409,	1467
—— R. W., Maj.	679
—— S., Lt. 673,	974

Names and Rank. Page.

Potter, W. N. 2nd-Lt. 1689
Pottinger, E. G., Lt. 317
——— J. A., Maj. 246, 1140, 1166
——— R. S., Lt.-Col. 194, 1893, 1912
Pottle, A. E., Lt. 25, 514
Potts, F., Br.-Genl. 101, 121, 126, 611
——— T., Sub-Condr. 1856
——— W., 2nd-Lt. . . . 476, 1423
Poulter, H. W., Lt. 93
——— J. Condr. 1861
Poulton, H. M., Lt. . . 95, 318, 1162
Pouncey, G. R., Lt. . . 318, 1102
——— T. C., 2nd-Lt. . . 564, 1255
Pound, F., 2nd-Lt. 637
Poulter, H. W., Lt. 93
Povah, F., Lt. 103
——— J. W., Maj. 614
Pow, G. S., 2nd-Lt. . . . 539, 1443
——— J. S. 27, 1679
Powe, H., 2nd-Lt. 626
Powell, A., 2nd-Lt. 565
——— A. E., Lt. 505
——— A. L. H., Lt. 536
——— C. A., 2nd-Lt. 746
——— C. B., Maj.-Genl. *Sir* 105, 118, 145, 173, 1595, 1955
——— C. T., 2nd-Lt. 1305
——— D., Capt. 288, 890, 1140
——— D. H., Bt. Lt.-Col. 15, 37, 140, 160, 231, 1423
——— E. E., Lt.-Col. 135
——— E. H., 2nd-Lt. . . 472, 1343
——— F., Capt. 279, 1178
——— F. E., T., 2nd-Lt. . . . 565
——— J., *Hony*. Lt. . . 25, 1845
——— F. N. H., Lt. . . 423, 1120
——— F. T., 2nd-Lt. . . 326, 366
——— G., 2nd-Lt. 840
——— G. A., 2nd-Lt. . . 558, 1703
——— G. M., 2nd-Lt. 752
——— G. W., 2nd-Lt. 739
——— H. D., Sub-Condr. . . . 1873
——— H. L., Lt.-Col. 628
——— H. N., Lt. . . . 393, 949
——— J. A., Ty. Lt. 1720
——— J. J., Capt. 745
——— L. F., Lt. . . 44, 302, 1086
——— L. K. M., Lt. . . . 304, 911
——— R. E., 2nd-Lt. . . 327, 1170
——— S. A., Surgn. Capt. . . . 111
——— S. H., Br.-Genl. 58, 103, 120, 124, 660
——— S. H., 2nd-Lt. 611
——— T. H. G., 2nd-Lt. . . . 707
——— T. M., 2nd-Lt. 776a
——— V. N. ff., Lt. . . . 487, 826
——— W., Lt. 90, 537
——— W. B., Lt.-Col. 131, 155, 190, 872, 1629, 1931
——— W. G., 2nd-Lt. 669
——— W. I. Fitz. . ., Lt. . . . 1720
——— W. J., Maj 1752
——— W. N., Lt. . . . 371, 923
——— W. N., Lt. 623
——— W. R., Lt. 740
——— W. S., Lt. . . . 493, 885
Powell-Price, J. C., Lt. 503
Power, A. F., Lt. 668
——— C. C., 2nd-Lt. 1624
——— F. W., 2nd-Lt. . . 331, 1187
——— H. R., Lt. . . . 306, 1170
——— J. A., Capt. . . . 79, 1839
——— J. E., Maj . . . 254, 911
——— M., Hony. Lt. 1848
——— O. E. S., Lt. 526
——— P., Maj. 38, 1712
——— R. T., Lt. 832
——— T. V., Condr. 1861
——— W., *Hony*. Maj. 1844
——— W. A., Lt. . . . 446, 49 1573
——— W. N., 2nd-Lt. 330, 1231, 1521
Powers, G. C., 2nd-Lt. 506
Powis, H. B., Lt. 491, 1668
——— L. J., Lt. . . . 520, 817, 986
Pownall, J. F., Sub-Condr. . . . 1873
Pownes, H. C., Asst. Surgn. . . 1800
Powter, H. F., 2nd-Lt. 685
Poynder, F. S., Capt. . . . 276, 1641
Poynor, M. P., Lt. . . . 445, 1385
Poynter, H. G., Lt. 622
Prabhaker, C. J., Ty. Lt. . . . 1774
Prabh Dayal, Jemdr. 1141
——— Jemdr. 1252
Pradhan, K. A., Ty. Sub-Asst.-Surgn. 1833

Names and Rank. Page.

Pradyumn Singh Bedi, Sub-Asst. Surgn. . . 1818
Preed, G. A. M., Maj. . . . 235, 1034
Pragnell, J., Lt. 731a
Pragnarain, arvasthi 1017
Prag Singh, Jemdr. 1083
——— Jemdr. 1048
——— Jemdr. 878
Prahlad, Jemdr. 1400
Prain, D., Lt.-Col. 1931
——— W., 2nd-Lt. 1674
Prakas Rae, *Hony*. Capt. . . . 2102
Prall, S. E., Lt.-Col. . . . 128, 1739
——— S. R., Capt. 1724
Pramatha Krishna Santra, Jemdar 1219
Pran Shankar Purusottam Raval, Sub-Asst. Surgn. . . . 1830
Prance, B. C., Lt. . . . 534, 1695
Prasad, B., Ty. Capt. 1764
——— K., Lt.-Col. 1786
——— M., Ty. Capt. 1771
——— R., Ty. Capt. 1772
——— R., Ty. Lt. 1782
Prashara, F. C. Sub. Asst.-Surgn. 1817
Praslow, J. E., 2nd-Lt. 1028
Pratap Singh, H.H. *Sir*. 1,110,939, 1170
Pratchitt, W. M., Capt. 724
Pratt, A. C., 2nd-Lt. 1695
——— A. J., Condr. 1861
——— A. R., Lt. . . 315, 1505, 1517
——— C., Lt. 585
——— C. T. E., 2nd-Lt. . . 494, 887
——— E. B., Lt. 491, 953
——— E. G. W., Lt.-Col. 106, 868, 1092, 1387
——— G. A., Ty. Lt. 1726
——— G. H., 2nd-Lt. 701
——— H. E. H., Capt. . . . 655, 657
——— H. M., Col. 177
——— H. R. E., Maj. 209, 868, 1122, 1163
——— J., Lt. 300, 883
——— L. H. Lt. 1007a
——— M. P., Capt. 278, 1479
——— S. S. T., 2nd-Lt. 355
——— W. P. R., Capt. 832
——— W. W., Capt. 1714
Pratt-Barlow, R. A. L. 638
Pratul Chandra Chakravarti, Sub-Asst. Surgn. . 1815
Prayag Das Dvivedi, Sub-Asst Surgn. . . 1820
Prebble, B., 2nd-Lt. 355
——— C. C., 2nd-Lt. . . . 545, 1393
——— C. E. 2nd-Lt. . . . 474, 1068
——— J., 2nd-Lt. . . . 567, 1521
Preece, C. B., Lt. 519, 998
——— D. E. G., Lt. 1104
Preedy, C., Capt. 605
Preddie, A. R., Lt. . . 502, 440, 1160
Preiss, R. B., 2nd-Lt. 475, 733, 791, 1090
Prem Narain Misra, Sub-Asst. Surgn. . 1817
——— **Chandra Singh**, Sub-Asst. Surgn. . . 1819
——— **Singh**, Subdr. . . . 1043
——— **Nath**, Sub-Asst. Surgn. . . 1815
——— **Singh**, Risldr.-Maj. . . . 914
——— Sub-Asst. Surgn. . . 1817
——— Risldr. . . . . . .
——— Ressdr. 896
——— Jemdr. 1182
——— Jemdr. 1682
——— Jemdr. 1207
——— Jemdr. 1196
——— Jemdr. 1228
——— Subdr. 1122
——— Subdr. 1150
Premendu Mukerji, Sub-Asst. Surgn. . 1820
Prem Narayan Singh, Sub-Asst. Surgn. 1820
——— **Singh**, Jemdr. 978
——— Subdr. 1145
——— Subdr. 1207
——— Subdr. . . . . .
——— Ressdar. 938
——— Ressdr. 1682
——— Jemdr. 945
——— *Hony*. Capt. 5, 2103
——— Sisht, Subdr.-Maj . . . 1642
——— Jemdr. 1228
——— Ty. Sub.-Asst. Surgn. 1833
Prendergast, C. G., Br.-Genl. 12, 33, 120, 124, 149, 184
——— C. . ., Maj.-Genl. . 1132
——— G. A., Maj.-Genl 172, 903

Names and Rank. Page.

Prendergast, L. C., Lt. . 311a, 1178
——— N. H., Capt. 95, 260, 964
——— R. V., Lt. . . . 72, 623
Prentice, J., Lt. 656
——— J., 2nd-Lt. 782
——— J. F., Lt. 1060
Prentis, W. S., Lt.-Col. . 134, 157, 194, 870, 1303
Prescot, C. P., Lt. . . . 668, 1007
Prescott, C. W., Maj. . 231, 1901, 1913
——— H. C., Maj. . . . 236, 1895
——— H. W. DeB., Senr. Asst. Surgn. 1787
——— J. A., Lt. 51
——— J. E., 2nd-Lt. 562
——— J. J. W., Lt.-Col. . . . 1710
Presgrave, E. R. J., Col. 56, 180, 1918
Presswell, T. G., Lt. . . . 528, 1707
Preston, A. B., Capt. 1715
——— A. K., 2nd-Lt. . . . 494, 927
——— B. H., Col. 176
——— E. A., 2nd-Lt. 73
——— E. B. C., Capt. . . . 82, 502
——— E. E., Lt.-Col. 39, 102, 138, 159, 198, 1660
——— Lt.-Col., F. G. 179
——— G. A., Lt.-Col. 138, 160, 198, 872, 1623
——— G. A., Lt. 317, 1187
——— J. E., Lt.-Col. . . . 180, 1918
——— K., Lt. 525
——— K. H., Lt. . 302, 1203, 1563
——— L. C. E. W., 2nd-Lt. 566, 1702
——— N., Lt. 1172
——— P. H. H., Maj. . . . 101, 633
——— T., Lt. 492
——— W. J. P., Lt.-Col. 140, 162, 200, 870, 1395
Prestwick, F., Lt. 1239
Prevost, G. H., Lt.-Col. 135, 158, 195, 870, 1355
Price, A., Lt. 726
——— A., The Revd. 1906
——— A. R. P., Lt. . . . 508, 997
——— A. W., 2nd-Lt. . . . 358, 1678a
——— C. A., 2nd-Lt. 559
——— C. E., Ven'ble . . . 803, 1903
——— C. E., 2nd-Lt. . . . 357, 1469
——— C. F., *Hony*. Lt. 708
——— C. H., 2nd-Lt. . . . 333, 1158
——— C. H. C. Col. . 57, 123, 149, 181
——— C. J., 2nd-Lt. . . . 327, 1152
——— C. N., Lt. 705, 1321
——— C. S., 2nd-Lt. . . 330, 970, 1567
——— C. T., Lt. 692
——— C. U., Bt.-Col. . . 126, 153, 192, 871, 1517, 1931
——— E. J., 2nd-Lt. 564, 1287
——— E. W. A., Lt. 111
——— F., 2nd-Lt. 618
——— F. W. J., Sub-Condr. . . . 1863
——— G., 2nd-Lt. 356
——— G. W., 2nd-Lt. . . . 553, 1143
——— H., Lt. 515, 1241
——— H. E., Revd. 1908
——— H. W., Maj. 223, 1562
——— J., Capt. 634
——— J. C., 2nd-Lt. . . . 330, 1479
——— J. D., Lt. 512, 1601
——— J., Revd. 1905
——— R., 2nd-Lt. 360
——— R. H., Lt.-Col. . . . 136, 1744
——— S. H., 2nd-Lt. 753, 786
——— T. J. P., 2nd-Lt. . . . 656, 986
——— T. W., 2nd-Lt. . . . 355, 1601
——— W. D., Capt. 734
——— W. H., Lt. 352, 1507
Prichard, H. M., 2nd-Lt. . . 562, 1283
——— L. E., Capt. 1839
——— W. J., Capt. 717
Prichat Singh, Jemdr. 1452
Priday, A. K., Lt. . . . 399, 1122
Priddey, A. H. P., 2nd-Lt. . 548, 1459
Prideaux, C. H., Lt. 700
——— C. W. A., Lt. 511
——— F. B., Lt.-Col. 195, 1890, 1912
——— G. V. A., Lt. . . 515, 1291
——— V. H., Lt. 700
Pridgeon, A. F., *Hony*. Maj. . . 593
——— R. C., 2nd-Lt. 733
Pridmore, W. G., Col. 37, 129, 1738, 1031
Priestley, C. K. N., Lt.-Col. 130, 155, 189, 859, 1191
——— E. C., Lt. 293
——— E. C., Lt. . . 516a, 1579, 1671
——— G. W., Lt. Col. 57

Names and Rank. **Page.**

Priestley, G. W., 2nd-Lt. . 563, 1056
—— H. E. B., Lt. . . . 519
—— H. S., 2nd-Lt. . . . 725
—— W., Capt. 619
—— W. H., 2nd-Lt. . . . 564
Priestley-Evans, Capt. M. . 5012, 1018
Priestman, H. E., Capt. . . 5, 787
Prikity Kumar Ghose, Jemdr. . 1329
Primalani, C. H., Ty. Capt. . . 1766
Prince, A., 2nd-Lt. 733
—— H., Lt. 425, 1181
—— P. E., Maj. . . . 664, 991
—— R., Maj. . . . 87, 219
—— T. G., Lt. 806
—— W. C. R., Lt.-Col. 45, 105, 105, 108, 191, 883, 1912
—— W. J., 2nd-Lt. . . . .
Pringle, J., 2nd-Lt. . . . 669, 987
—— J. A., 2nd-Lt. 858
—— J. C., Lt . . 664, 991
—— J. E., 2nd-Lt. . . 669, 992
—— J. H. S., Lt. . . 196, 1501
—— I. T., St. C. Lt. 817
—— R., 2nd-Lt. 602
—— W., 2nd-Lt. . . . . .
—— W. B., 2nd-Lt. 723a
Prins, E. G. A., Sr. Asst. Surgn. 1789
—— G. E. A., Ty. Sub.-Asst. Surgn. 1804
Prinsep, D. W. McL., Capt. 280-287, 179, 1675
—— E. S. MacL., Capt. 274, 895
Printer, M. D., Ty. Lt. 1779
Prioleau, C. P. J., Lt. . . 318, 904
Prior C. G., Lt. . . . 295, 1505
—— C. J., Capt . 10, 18, 571, 1850
—— E. G., Lt. 73, 746
—— E., 2nd-Lt. 618
—— H. C., Lt. 516a, 1898
—— H. H., 2nd-Lt. 760
Prissick, C., Lt.-Col. . 137, 159, 201, 869, 1247
Pritam Singh, Ty. Sub.-Asst. Surgn. 1833
—— Subdr. 1340
—— Sub. Asst. Surgn. . . 1818
—— Jemdr. 1163
—— Jemdr. 1587
Pritchard, A. G., Bt.-Col. 12, 45, 121, 126, 153, 94, 677, 1931
—— A. J. L., Lt. . . . 385, 570
—— C. D., Lt . . . 296, 1011
—— C. H., Lt.-Col. . . . 106
—— D. J., Capt. 1919
—— E. J., Sub-Condr. . . . 1857
—— H. E., Lt.-Col. 128, 154, 184
—— H. R. N., Maj. . 28, 1899, 1913
—— J. M., Capt. 752
—— P. A. R., Maj 233, 1601, 1893
—— T. A., Lt. . . . 465, 1143
Pritchett, H. N., Lt. 1720
Prithi Chand., Subdr. 1117
Prithpal Singh, Jemdr. . . . 1321
—— Jemdr. 1320
Prithilal Gurung, Jemdr. . . . 1627
Prithipal Singh, Jemdr. . . . 1404
—— Jemdr. 1054
Proby, R. S. B., the Revd. . . 1954
Probyn, D. M., Genl. Sir 1, 169, 895, 1925
Proctor, A. H., Maj. 1751
—— A. L., Lt. 789
—— D. P., Lt. . . . 520, 1128
—— F. O., Revd. 1906
—— G. N., Capt. 261, 1323, 1339
—— J., Maj. 703
—— N. C., Lt. 530
—— P., 2nd-Lt. . . . 330, 1319
—— R. H., 2nd-Lt. 752
—— W. D., Condr. . . . 1853
—— W. J. E., Lt. 415
Prodger, S. E., Lt. 793
Proes, E. M., Maj. . . . 500, 986
Prophit, G. C., Lt. . . 319, 1178
Prosser, P. H., 2nd-Lt. . 5612, 1056
Prothero, J. E. S., Capt. . . . 1895
Proudlock, R., Lt. 836
—— R. V., Lt. 309, 1287, 1375
Proust, R., 2nd-Lt. 551
Provan, A., 2nd-Lt. 362
Provis, F., Lt. 528, 1052
Pruce, C. A., Senr. Asstt. Surgn. 1787
Prussia, H. B., 2nd-Lt. 544
Pryce, D. D., Lt.-Col. . 98, 179, 1918

Pryce, H. E. ap. R., Lt.-Col. 50, 98, 122, 135, 158, 140, 162, 200, 1868, 1090, 1931
—— P. L., 2nd-Lt. 706
—— W. W. R., 2nd-Lt. . 330, 1475
Pryde, A. W., Lt. . . . 537, 1455
—— J., 2nd-Lt. 557
—— R. W., 2nd-Lt. . . 548, 1459
Prynn, C. C., Lt. 742
Prynne, H. F., 2nd-Lt. 598
Pryor, E. D., Lt. 516
—— J. W., 2nd-Lt. . . 360, 1511
—— O. J., Lt. . . . 296, 1691
—— T., Capt. . . 58, 501, 858
Puckle, F. H., Lt. . . . 95, 511
—— G. H., Lt. 487
Pudan Singh, Jemdr. 1281
Pudi Khan, Jemdr. 1471
Puech, A. G., Lt.-Col. . 81, 351, 815
Pugh, A. J., *Hony* Col. . . 2, 603
—— C. L., 2nd-Lt. 790
—— F. M., Lt. 551
—— H. O., Lt 111
—— H. V., 2nd-Lt. . . . 558, 1197
—— J. C., 2nd-Lt. . . . 360, 1676
—— J. H., Lt. 696, 727
—— L. B., Lt. 304, 1641
—— R. A. C., 2nd-Lt. . . 78, 494
—— T. E., 2nd-Lt. . . . 539, 1419
Pughe, O. L., Capt. . 261, 1303, 1347
Pugsley, W. H., 2nd-Lt. . . . 739
Pullar, E. J., Maj. . 31, 501, 653
Pulleine, H., Lt. . . . 420, 889
Pullen, A. F., Col. . . 101, 102, 125, 612
—— A. P., 2nd-Lt. . . . 73, 759
—— F., Lt. 602
Pulley, H. C., Maj. 103, 107, 108, 228, 1066
—— O. C., Capt. . . . 255, 1082
Pulling, F. E. S., Lt. 731
—— G. H., Lt. . . . 430, 1627
—— H. D., Lt. . . . 387, 1627
Pullman, G. C., Capt. . . 55, 711
Pulman, T. M., Lt. . . . . .
Pulvermann, A. W., Capt. . 268, 1004, 1509
Pumfrey, A. L., Lt. 1305
Pumffrey, R. J., 2nd-Lt. . . . 844
Puna Singh Lama, Subdr. . . 1347
Puncharo, F. N., Lt. 519
Punjab Singh, Jemdr. . . 1163, 1208
—— Subdr. . . 1145, 1207
Punn, G. S., Sub-Asst. Surgn. . . 1818
Punnai Singh, Ressdr. 928
Punnusami, Jemdr. 994
Puniakoti Mudali, Jemdr. . . 1276
Punyakoti, Subdr. 1276
Pura Rawat, Jemdr. 1484-86
Puran Ale, Jemdr. 1608
—— **Bahadur Pun**, Subdr. . . 1615
—— **Sunwar**, Jemdr. . 1633
—— **Singh Gharti**, Jemdr. . 1625
—— Subdr. 1082
—— Jemdr. 890
—— Subdr. 1052
—— Subdr. 1021
—— Jemdr. 972
—— Jemdr. 1143
—— Jemdr. 1079
—— Jemdr. 1111
—— Sub-Asst. Surgn. . . 1819
—— Jemdr. 882
—— Jemdr. 1021
—— Jemdr 1012
—— **Gurung**, Subdr. . . 1596
—— Jemdr. . . 1597, 1102
—— **Khancka**, Subdr. . 1642
—— **Rana**, Subdr. . .
Puranbir Rana, Subdr. . . . 1618
Puran Singh Thapa, Jemdr. . . 1619
Purcell, C. H. C., 2nd-Lt. . . . 790
Purcell-Gildea, E. J., Lt. 56, 528, 1678a
—— E. T. H., Lt. 541
—— J., 2nd-Lt. 557
Purandhoj Limbu, Jemdr. . . . 1531
Purchase, R. G., 2nd-Lt. 548, 723a, 1094
Purcon, D. W., Lt.-Col. . 181, 1918, 131
Purdumman Singh, Ressdr. . . 1686
Puri, M. L., Capt. 1754
—— N. D., Capt. 1758
—— S. R., Tempy. Lt. . . 1291, 1776
Purkis, A. E., Sub-Condr. . . . 1854
Purry P. S., Lt. 753
Purser C. R., 2nd-Lt. 552
Pursey, C. E., 2nd-Lt. 739

Purslow, F. G., Sub-Condr. . . 1864
Purusottam Saran, Vaish, Sub-Asst. Surgn. . . . 1819
Purushottam, D. M., Sub-Asst. Surgn. 1831
Purves, J. J., Lt. . . . 300-01, 1102
—— P. F., Lt. 503
Purvis, M., Capt. 1756
—— R. G., Lt. . . . 420, 1058
Pusey, A. G., 2nd-Lt. . . 550, 1291
Puthenpurackal Mathoonuy Mathew, Jemdr. . 1360
Puthussery Kunhi, Raman, Sub-Ast. Surgn. . . 1827
Putman, T. C., 2nd-Lt. . . . 1207
Putnam, G. W., 2nd-Lt. . . . 1359
—— W. A., 2nd-Lt. 327, 1162, 1543
Puttick, A. W., Lieut. 321
Pyar, Subdr. 1483
Pyara Singh, Jemdr. 890
Pyare Jan, Siddigi, Sub-Asst. Surgn. 1818
—— **Lal, Gahlor**, Sub-Asst. Surgn. . . . 1820
Pyare Singh, Sub-Asst. Surgn. . 1820
Pybus, G. D., Lt. . . 306, 1074, 1203
—— H., Capt. 628
Pye, H. B., Lt. 470, 1515
—— J., Sub-Condr. 1863
—— K. W., Maj. 604, 1880
—— L. K., 2nd-Lt. 624
—— W. E., Lt.-Col. . 139, 160, 199, 1399
Pyett, C. F., 2nd-Lt. 547
Pye-Smith, T. E. B., Capt. . . . 725
Pyke, D. P., Lt. 706
Pyle, S. T., 2nd-Lt. 757
Pym, F. W., 2nd-Lt. 554
Pymm. J. M., 2nd-Lt. 557
Pyne, F. S., Maj. 606
—— G. D., Esq. . . . . . .
—— H. P., Lt. 490
Pyper, H. Mc. D., Lt. . . 415, 1604
—— J. C., Capt. 1717
Pyra Singh, Subdr., ... 1699
Pyster, N. M., 2nd-Lt. 1008

Q

Qadir Bakhsh, Khan, *Hony*. Lt. 111
Qalandar Khan, Jemdr. . 1260, 1320
—— Jemdr. 1433
Qananjaya, M. D. S., Sub-Asst. Surgn. 1820
Qasm, Jemdr. 1428
Qasim Ali, Subdr. 1239
—— Jemdr. 1145
—— **Khan**, Jemdr. . . . 1092
—— Jemdr. 1252
—— Jemdr. 1096
—— Jemdr. 1096
—— Jemdr. 1376
—— Jemdr. 918
—— Jemdr. 950
—— Jemdr. 1388
Qazi Khan, Jemdr. . . . 946, 953
Quadros, L. G., Sr. Asst. Surgn 1767
Qualtrough, E. F., Lt. . . 405, 1627
Quanbrough, F. R. B., Asst. Surgn. . 1801
—— F. G., Capt. 609
Quarry, P. S., Lt. 573
Quayle, P. R., Capt. 103, 271, 1505, 1507
Quazi Mahomed Akbar, Ressdr. 1686
Qudrat Khan, *Hony*. Risaldr.-Major 1896
Qudratullah Khan, Subdr. . . 1686
Queripel, L. H., Bt.-Lt.-Col. 136, 614
Quest, T. P., Lt. 393, 1473
Quick, A. N., Asst. Surgn. . . 1800
—— C. F. E., Asst. Surgn. . . 1799
—— N., Lt. 527
Quig, A., 2nd-Lt. 782
Quigley, J., Condr. 1867
Quilter, C. F. G., Lt. . . . 516a, 1066
—— G. J. H., Lt. 523
—— T. H., Lt. 531, 1702
Quillinan, L. P., Lt. . . . 538, 1491
Quin, L. M., Lt. 292a, 1359
—— R. B., Lt. 435, 1223
—— R. F. W., Lt-Col. 103, 141, 162, 200, 1187
—— **T. J.**, Maj.-Genl. . . 109, 1910

NAMES AND RANK. Page.

Quinan, E. P., Bt. Maj. 253, 1128, 1469
Quine, B. H., 2nd-Lt. 744
Quinlan, P. B., Lt. . . 490, 935
—— W. T., Capt. 1717
Quinton, A. W., 2nd-Lt. . 358, 1143
—— F. W. D., Bg.-Genl. 32, 35, 37, 121, 126, 612
—— L. M., 2nd-Lt. . . . 568
Quirke, M. J., Maj. 1751
—— J. P. F., 2nd-Lt. 751
Quraishi, M. S. A., Sub-Asst. Surgn. 1820
—— N. A., Tempy. Capt. . 1770
Qurban Ali, Subdr. 1427
Qusim, Khan, Jemdr. . . . 1107
Qutab Khan, Jemdr. . . . 1502
Qutbi Khan, Jemdr. . . . 1090
Qutbuddin Khan, Subdr. . . 1092

R

Rabbidge, M. D., Nursing Sister . 1735
Rabbit, C., 2nd-Lt. . . . 685, 784
Rabbitt, J. B., 2nd-Lt. 565
Rabel Singh, Subdr. 1252
Rab Nawaz, Risldr. 896
Rab Nawaz, Khan, Jemdr. . . . 942
Rabidhoj Rana, Jemdr. 1651
Race, T. T. G., Lt. . 314, 1403, 1407
—— W. C., 2nd-Lt. 736
Rachpal Singh, Ressdr. . . . 914
—— Jemdr. . . . 1596
Rachha Ram, Jemdr. 878
Rackett, G. L., Lt. 354
Rackstraw, G., Lt. . . . 511, 1215
Radcliff, H. G., 2nd-Lt. . 326, 1249
—— S. G., Lt.-Col. 57
Radcliffe, C. 2nd-Lt. . . . 475, 923
—— C. E. St. J., Lt. . . . 111
—— H., Lt. 1074
—— W., 2nd-Lt. 637
—— W. C. A., Lt.-Col. 74 128, 613
Radcliffe-Barrett, C. L., 2nd-Lt. 491
Radcliffe-Smith, N. M. B., Maj. 234, 1199
Radford, E. A., 2nd-Lt. 556
—— E. J., Lt. . . . 90, 777
—— O. B., Maj. . . 243, 1351
Radha Singh, Jemdr. 1068
Radice, W. A., Capt. 807
Radley, H. P., Capt. 274, 1004, 1303
—— S. B., Tempy. Capt. . . 1723
Radwell, J., Maj. 731
Rae, A. D., 2nd-Lt. 747
—— A. E. P., 2nd-Lt. 1695
—— C., Lt. 519, 1699
—— F. L., Lt. 529, 1702
—— G. B., Capt. . 260, 1042, 1664
—— H. P. M., 2nd-Lt. 802
—— J. G., Maj. 243, 1487
—— K. W. R., Lt. . . . 320, 1441
—— L. F. D., 2nd-Lt. 562
—— M. L., Maj. 221, 1913
—— R., Lt. 596, 727
Raeburn, J. P., Lt. 354
Raeside, D. A., 2nd-Lt. . . . 782
Rae Singh, *Hony.* Capt. . . . 2102
Rafiud-din Khan, Sr. Sub-Asst. Surgn. 1807
Ragha, Subdr. 1239
Raghavulu Chettiar, *Rao Sahib* . 89
Raghbir Limbu, Jemdr. . . . 1631
—— Singh, Bedi, Sub-Asst. Surgn. 1820
—— **Singh**, Subdr. 1031
—— Jemdr. . . . 890
—— Jemdr. . . . 918
—— Sub-Asst. Surgn. 1817
—— Thapa, Jemdr. . 1615
Raghubardayal Tewari, Jemdr. . 1017
Ragho Parab, Jemdr. . . . 1416
Raghoji Bagwe, Jemdr. . . . 1464
Raghojirao, Jemdr. 999
Raghubir Singh, Jemdr. . . . 926
—— Jemdr. . . . 924
—— Subdr. . . . 1391
—— Jemdr. . . . 1031
—— Subdr. . . . 1413
Raghuel Isudas Harry, Sub-Asst. Surgn. 1831
Raghu-nandan Parsad, Sub-Asst. Surgeon 1820
Ragunandanparahad, Awasthe 1017
Raghubardayal, Tewari, . . . 1017
Raghunandan Singh, ... 1043
Raghu-nandan, Parsad Geol, Sub-Asstn. Surgn. 1820
Raghunath, Sr. Sub-Asst. Surgn. 1807
—— —— **Balkrishna Alate**, Sub-Asst. Surgn. . 1829
—— **Balwant Limaye**, Sub-Asst. Surgn. . 1831
—— Das Vadara, Sub-Asstn. Surgn. . . 1819
—— **Rao Chawan**, Jemdr. 1435
—— **Sahai**, Sub-Asst. Surgn. . 1815
—— **Singh**, Sub-Asst. Surgn. . 1814
—— Risaldar . . . 906
—— Subdr. . . . 1195
—— Jemdr. . . . 1024
—— Jemdr. 1064
—— **Sitaram Patki**, Sub-Asst. Surgn. 1830
Raghunathrao Kadam . . . 1444
Raghuraj Singh, Jemdr. . . . 1063
Raghuram Sajjan, Jemdr. . . . 1420
Raharadan Rai, Subdr. . . . 1630
Rahim Ali, Jemdr. 979
—— Subdr. 1515
—— Jemdr. 1314
—— **Khan**, Jemdr. . . . 1107
—— **Baksh**, Subdr. 1419
—— Sub-Asst Surgn. . 1814
Rahim Baksh, Sub-Asst. Surgn. . 1814
—— Ressdr. 918
—— Jemdr. . . 1420
—— Jemdr. . . 916
—— **S. A.**, Sub-Asst. **Surgn.** 1827
—— **Khan**, Subdr. 1401
—— *Hony.* Capt. . . 2102
—— **Shah**, Jemdr. . . 1096
—— Sub-Asst. Surgn. . 1814
—— Resdr. 1685
Rahim Dad Khan, Subdr. . . . 1571
Rahim Sher Khan, Jemdr. . . 914
—— Jemdr. . . 1036
Rahimullah Khan, Risaldar. . 1684
—— Jemdr. 953
—— Khan, Jemdr. . . 1440
Rahles-Rahbula, E. H., 2nd-Lt. . 475, 1068
Rahman, M. A., Capt. 1755
Rahamat Ali Beg, Tr. Vety. Asst. 1087
—— Khan, Jemdr. . . 1470
—— Jemdr. . . . 1284
Rahmat Ali Khan, Risaldar. . . 1684
—— Jemdr. . . 1312
—— Subdr. . . . 1035
—— Jemdr. . . . 1453
—— Subdr. . 1035, 1035
Rahmat Khan, Subdr. 1208
—— Subdr. . . . 1088
—— Jemdr. . . . 1252
—— Subdr. . . . 1086
—— Jemdr. . . . 1372
—— **Sher**, Resldr. . . . 908
Rahmatullah, Jemdr. 1428
—— **Khan**, Jemdr. . . 989
Rai, B. R., Sub-Asst. Surgn. . . 1817
—— Baij Nath Vyas, Bahadur . . 3
—— D. G., Maj. . . 105, 108, 1748
—— D. H., Capt. 108, 1754
—— H. S., Tempy. Lt. 1773
—— K., Ty. Lt. 1777
—— R. J., 2nd-Lt. 1509
—— R. N., Ty. Lt. . . 1495, 1777
—— Baij Nath Vyas 3
—— Singh, Subdr. 1136
—— Singh, Subdr. 1171
—— Sihgh, Jemdr. 1189
—— Singh, Jemdr. 1232
—— **Singh**, Subdr. 1399
—— Jemdr. 938
Rakhan Singh, Jemdr. 1021
Raibhan Singh, Jemdr. 1048
Rail, G. S., Sub-Asst. Surgn. . . 1820
Raikes, L. T., Maj. 614
—— A. F., 2nd-Lt. 1044
—— M. H., Capt. . 486, 724, 1672
Raines, H., 2nd-Lt. 355
Rainford, H., 2nd-Lt. 362
Rainier, N. B. J., Lt.-Col. . 134, 1744
—— W. H. C., Lt. . . 415, 1251
Rainsbath, J. G., Lt. . . 440, 1120
Rainsford, J. A. G., Lt.-Col. 57, 128, 154, 164
Rainsford-Hannay, D., Capt. 250, 1235
Rainy-Robinson, R. M., Col. 57, 1931
Raiside, D. A., 2nd-Lt. . . . 703
Rai Singh, Jemdr. 894
Rait, H. A. M., Sr. Nursing Sister 10, 1734
—— J. W. F., Lt.-Col. . . 138, 1745
Rait Kerr, R. S., Capt. . . 665, 996
Raitt, H. A., Maj.-Genl. . . . 118
—— **Mal** 1249
—— **Singh**, Ressdr. 880
—— —— Sub-Asst. Surgn. . 1816
Raj Talab, Subdr. 1256
—— **Wali Khan**, Jemdr. 1513
Raja, Jemdr. 972
—— R. N., Tempy. Capt. . . . 1768
Raja Gopal, Subdr. 1265
—— **Khan**, Subdr.-Maj. . . . 1106
—— Jemdr. 1249
—— Jemdr. 126
—— Khan, Jemdr. 1567
Raja Ram, Jemdr. 966
—— **Singh**, Risaldar. 1686
Raja Hari Singh, Genl. 2
—— Khan, Subdr. 1289
Rajab Khan, Subdr. Maj. . . . 1433
Rajagopal, Jemdr. 994
Rajam, R. V., Ty. Lt. 1781
Rajan, B. S., *Tempy.* Capt. . . 1766
Rajaram Jadhow, Jemdr. . . . 1432
—— Pande, Jemdr. . . . 1027
—— Singh, Jemdr. . . . 1047
Rajaran, G. K., Sub-Asst. Surgn. 1830
Rajaratnam, T. V. P., Tempy. Capt. 1768
—— Jemdr. 994
Rajbali Singh, Subdr. 1387
Rajdhoj Rai, Jemdr. 1631
Rajdeo Singh, Jemdr. 1391
Raje Khan, Jemdr. 1385
Raje Ram, Subdr. 1059
—— Jemdr. 1059
—— **Singh**, Subdr. 1452
Rajendra Kishore Sen Gupta, Sub-Asst. Surgn. . . . 1815
—— Lal Mukherjee, Jemdr. . 1219
Rajindram, Jemdr. 994
Rajindar Singh, Jemdr. . . . 920
—— 2nd-Lt. 481
Rajinder Singh, Ressdr. . . . 896
Rajkishor Singh, Jemdr. . . . 1383
Rajmal Singh, Jemdr. 1051
Rajman Thapa, Jemdr. . . . 1657
Rajnande Limbu, Subdr. . . . 1655
Rajnarayan Singh, Jemdr. . . 1581
Rajogopal, Jemdr. 1772
Rajpal Rai, Subdr. 1630
Raju, Jemdr. (I) 994
—— Jemdr. (II) 994
Raju Singh, Jemdr. 1472
Rajwali, Subdr. 1288
—— Jemdr. 1095
—— **Khan**, Risldr. 1684
—— Jemdr. 922
—— Subdr. 1513
—— Subdr. 1452
Rajwant Singh, Sub-Asst. Surgn. 1817
—— Jemdr. 1044
Rakhim Gul, Jemdr. 1248
—— Shah, Jemdr. 1095
Rakhman Shah, Jemdr. . . . 1262
Rala Ram, Sub-Asst. Surgn. . . 1813
Rala Singh, Sub-Asst **Surgn.** . . 1815
—— Subdr. 1106
—— Jemdr. 886
Ralla Singh, Subdr. 1095
—— Jemdr. 1150
Ralph, A. C., Lt.-Col. 132, 156, 192, 868, 1030
—— Revd. Fr. 826
—— M. M., Lt. . . . 18, 33, 719
—— S. A., Capt. . . . 501*a*, 1665
—— S. L., Maj. 233, 1040
Ralphs, F. G., 2nd-Lt. 472
Ralston, A. H. B., Capt. . 255, 1050
—— W. H., Maj. . 238, 107, 1211
—— W. J. B., Maj. . . 243, 1030
Ram, A., Ty. Lt. 1779
—— G., Tempy. Capt. 1768
—— G. S., Ty. Lt. 1777
—— H., Ty. Capt. 1772
—— M. A., Sub-Asst. Surgn. . . 1818
Rama, Subdr. 1128
—— **Chor**, Subdr. 1416
—— **Limbalkar**, Jemdr. 999
—— **Methe**, Jemdr. 1436
—— **Rao**, *Rao Sahib* 3
Ramadhar Misr. 1017
Ramalingam, Jemdr. 1360
—— **Pillai**, Subdr. . . . 1702

NAMES AND RANK. Page.

Ramalingam Pillai, ... 1683
Ramanan Singh, Jemdr. . . 1393
Ramanathan, P. A. A., Ty. Lt. . 1777
Ramanand Singh, Subdr. . . . 1046
Ramannah, Jemdr. 1340
Ramasami, Jemdr. 1308
—— Subdr. 1331
—— Jemdr. 1332
——(I) Jemdr. 993
—— Jemdr. 1270
—— Jemdr. . . 1332, 1699
——(III) Jemdr. 994
——(II) Jemdr. 993
——(IV) Jemdr. . . 994
——(V) Jemdr . . 994
Raman, Jemdr. 994
Raman, Kavakappara Muppil, Nayar, 2nd-Lt. 481
Ramautar Singh, Subdr. . . 1043
Ramana, Subdr. Maj. 1315
Ramayan Singh 989
Rambahadur Rana, Jemdr. . . 1597
Ram Adhin Tiwari, Subdr. . . 1050
—— Autar Patak, Subdr. . . . 989
—— Bakhsh, Subdr. 1335
—— Bux Singh, Jemdr. . . 1567
—— Chabila Singh, Sub-Asst. Surgn. 1814
—— Chandra, B. S., Sub-Asst. Surgn. 1831
—— Parab, Jemdr. . 1409
—— Gopal Thitte, Sub-Asst. Surgn. . . 1831
—— Das, Jemdr. 1189
—— Datta Kumar, Sub-Asst. Surgn. . . 1815
—— Ditta, Jemdr. 1189
—— Din Dube, Jemdr. . . . 1051
—— Dullag, Jemdr. 968
—— Ikbal Singh, Subdr. Maj. . 1040
—— Jiwan, Jemdr. 1473
—— Kishan Dura, Subdr. . . 1615
—— Jemdr. 1103
—— Sukul, Jemdr. . . 1368
—— Krishna Anant Parab, Sub-Asst. Surgn. . . . 1830
—— Ganadat Shinde, Sub-Asst. Surgn. . . 1830
—— Kunwar Singh, Ressdr. . . 918
—— Lagan Sing, Jemdr. . . 1390
—— Jemdr. . . . 1567
—— Lal, Sub-Asst. Surgn. . . 1511
—— Jemdr. 1453
—— Abrole, Sub-Asst. Surgn. 1813
—— Jemdr. . . 1004
—— Jemdr. . . 1004
—— Narain, Sub-Asst. Surgn. . 1809
—— Nath, Ressdr. 889
—— Oachar, Subardar . . 1657
—— Tewari, Subdr. . . . 1383
—— Pershad Singh, Subdr. . 1387
—— Sing Bhandari, Subdr. 1181, 1223
—— Singh, Ressdr. 902
—— Jemdr. . . . 898
—— Dharamwal, Sub. Asst. Surgeon . 1818
—— Rana, Jemdr. . . 1636
—— Chandra, Tempy. Lt. . . 1199
—— Chandra Ale, Jemdr. . . 1608
—— Bhiku, Jemdr. 1460
—— Chandra Chawan, Jemdr. 1460
—— Chandar Chaike, Jemdr. . 1414
—— Chandrarao Bhonde, Jemdr. 1468
—— Chandra Rao More, Jemdr. 1417
—— Chandra Goli, Jemdr. . . 1510
Rambrich Singh, Jemdr. . . . 1004
Rambal, Jemdr. 1050
Rambali Singh, Subdr. Maj. . . 1044
Rambiswas Singh, Jemdr. . . 1003
—— Ressdr. . . 1083
Ramchandar Sawant, Jemdr. . . 1424
—— Rao Ekunder, Jemdr. . 1423
Ramchandra, Gangadhar Jather, B. Sub-Asst. Surgn. . 1833
Ramcharan Pande, Jemdr. . . 1010
—— Singh, Rsldr. . . 934
Ramcharakh Misr, Jemdr. . . 1010
Ramdass, Jemdr. 1099
Ramdas Singh, Jemdr. . . . 1395
Ramdayal Ram, Jemdr. . . . 1452
Ramdayal Singh, Subdr. . . . 1191
—— Misr 1010
Ramdihal Pande, Jemdr. . . . 1018
Rhani Singh, Subdr. . . . 1383
Ram Dilla, Jemdr. 1082

NAMES AND RANK. Page.

Ramhet Singh, Subdr. 1452
Ramjahan Singh, Jemdr. . . 1364
Ramjee Dass, Esq. 89
Ramjilal, Jemdr. 1216
—— Jemdr. 1559
—— Jemdr. 1008
—— Subdr. 1040
Ramjirao Kadam, Subdr. . . 1409
—— Jiwan, Jemdr. . . . 1420
Ram Lal, Subdr. 1211
—— Tewory Jemdr . 1575
Ramlal Singh, Subdr. . . . 1497
—— Subdr. Maj. . . 1032
—— Jemdr. . . . 1064
—— Subdr. 1082
—— Jemdr. . . . 961
Rampal Singh, Jemdr. . . 1064
Ram Biswas Singh, Ressdr. . . 1684
—— Parshad Nayaran Sahi, Babu, 2nd-Lt. 481
—— Phal, Jemdr. 890
—— Rao Anand Rao, Doctor, Sub-Asst. Surgn. . . . 1830
—— Rao Gangadhar, Sr. Sub-Asst. Surgn. 1829
—— Rawat, Jemdr. 1484
—— Rup Singh, Subdr. . . . 988
Ram Saran, Jemdr. 1141
Ram Saran Mehra, Sub-Asst. Surgn. 1814
Ram Sarup, Jemdr. 926
—— Jemdr. 1401
—— Jemdr. 1401
—— Jemdr. 1039
—— Sing, *Hony.* Capt. . 2104
—— Singh, Subdr. 1211
—— Jemdr. . . . 1456
—— Subdr. . . . 1502—04
—— Subdr. 2102
—— Rsldr. 882
—— Jemdr. 922
—— Sub-Asst. Surgn. . . 1810
—— Subdr. 1335
—— Jemdr 1380
—— Sub-Asst. Surgn. . 1812
—— Subdr. . . . 1082
—— Sub-Asst. Surgn. . 1810
—— Subdr. 1142
—— Jemdr. 1250
—— Subdr. 1150
—— Ressdr. 895
—— Jemdr. 1003
—— Ressdr. 1681
—— Subdr. 1343
—— Subdr. 1132
—— Jemdr. 1312
—— Subdr. 1399
—— Jemdr. 1141
—— Subdr. 1197
—— Jemdr. 1698
—— Rsldr. 896
—— Jemdr. 912
—— Bhandari, Jemdr. . 1181
—— Subdr. 1225
—— Burathoki, Subdr. . 1648
—— Dharmwal, Sub-Asst. Surgn. . . . 1818
—— Thapa, Subdr. . . 1018
—— Sohan Tewari, Jemdr. 1017
—— Sing Sahi, Subdr. . . . 1614
—— Burathoki, Subdr. . 1648
—— Sudh Tewari, Jemdr. . 1017
—— Sukal Rai, Jemdr. . . . 1372
Rama Nand, Sub-Asst. Surgn. . 1810
Ramnand, Jemdr. 1476
Raman, A., Sub-Asst. Surgn. . 1827
—— O. K., Sub-Asst. Surgn. . 1827
—— P. K., Sub-Asst. Surgn. . 1827
—— T. K., Sub-Asst. Surgn. . 1827
Ramanjulu, Subdr. 1317
Ramanna, Subdr. 1315
Ramappa, S. P. Jemdr. . . . 1361
Ramasere Singh, Jemdr. . . . 1561
Ramaswami, N. S., Ty. Lt. . . 1779
—— Iyengar, R. Esq. . . 89
Rambakhsh Singh, Subdr. . . 1399
Rambaran Singh, Subdr. . . . 1043
Rambarat Singh, Jemdr. . . . 1385
Rambhagat Adhikari, Jemdr. . 1645
Rambharosa Tiwari, Subdr. . . 1023
Rambharose Misr, Jemdr. . . 989
Ramchandra Balkrishna Kusurkar, Sub-Asst. Surgn. . . . 1831
Ramchandra Bhogijal Bhatt, Sub-Asst. Surgn. 1832
Ramchandra Bhosle, Subdr. 1460—62
—— Ghorepade, Subdr. 1463

NAMES AND RANK. Page.

Ramchandra Roa Erundkar, Subdr. . . . 1579
—— Rao Mahadik, *Hony.* Capt. . . 2101
—— Rao Pawar, Jemdr. 1423
—— Sondhle, Jemdr. . 1460
—— Sing, Jemdr. . . 1047
Ramchander, Gaikwad, Jemdr. . 1424
Ramchandra Gangadhar Jathar, Sub.-Asst. Surgn. . . . 1833
—— Singh, Jemdr. . . 1581
Ramchandra Ganu, Naidoo., Sr., Sub-Asst. Surgn. 1829
Ramchandrarao, *Hony.* 2nd-Lt. . 482
Ramdas Singh, Jemdr. . . . 1400
—— Pathak, Jemdr. . . . 1027
Ramdat Singh 1022
Ram Ditta, Jemdr. 1563
Ramdayal Singh, Subdr. . . . 1191
Ramdeo, Subdr. 1455
Ramdhani Singh, Jemdr. . . . 1384
Ramdhari Singh, Jemdr. . . 1395
Ramdihah Pande, Subdr. . . . 1018
Ram Ditta, Subdr. 1252
Ramditta Mal, Sr. Sub-Asst Surgn 1809
Ramdiyal Singh, Jemdr. . . . 1413
Ramdour Lal, Jemdr. 1059
Ramdulare Tewari, Jemdr. . . 1019
Ramesar Dyal, Sub-Asst. Surgn . 1816
—— Pathak, Jemdr. . . . 1017
—— Singh, Jemdr. . . . 1047
—— Jemdr. . . . 1048
Ramesardat Pande, Subdr. . . 1018
Ramesardth Misr 1017
Rameswar Singh, Subdr. . . . 1054
—— Jemdr. 1004
Ram Gopal Singh, 2nd-Lt. . . 482
Ramisetty Subbayya, Sub-Asst. Surgn. 1827
Ramji Bhonsle, Jemdr . . . 999
—— Garud, Jemdr. 1424
—— Lal, Jemdr. . . . 1480
Ramjidas, Jemdr. 1312
—— Lal, Jemdr. . . . 1216
—— 1067
—— Jemdr. 1384
—— Jemdr. 1060
—— Jemdr. 914
—— Singh, Jemdr. . . 1385
—— Misr, Jemdr. 1304
—— Sajanaji, Jemdr. . . 1447
Ramji Lal Singh, Jemdr. . . 1032
Ramjiawan Sukul, Subdr. . . 1018
—— Tiwari, Jemdr. . . 1027
Ramji Lal Jemdr. 901
—— Jemdr. 1336
Ram Kala, Subdr. (I) 1040
—— Jemdr. (II) 1040
Ramkaram Ram, Jemdr. . . . 1488
Ramkishan Thapa, Jemdr. . . 1597
Ramkishen Rana, Subdr. . . 1014
Ramkrishna Mahadeo Haldankar, Sub-Asst. Surgn. . . . 1831
Ramkumar Singh, Subdr. . . . 1395
Ram Lagan Singh, Jemdr. . . 1054
Ram Lal, Tr. Vety. Asst. . . . 1087
—— Jemdr. 1167
—— Jemdr. 1473
—— Jemdr. 1100
—— Tewari, Jemdr. . . 1575
—— Jemdr. 1032
—— Singh, Jemdr. . . . 1387
Ramnad, Raja of, 2nd-Lt. . . . 481
Ramnath Oachari, Jemdr. 1657, 1030
—— Jemdr. 1443
Ram Nath, Rsldr. . . . 889, 901
—— Jemdr. 1583
—— Singh, Jemdr. . . 1032
—— Sharma, Jemdr. . . 1051
—— Tiwar, Jemdr. . . . 1018
Ramnarayan Singh, Subdr. . . 1387
Ramniklal Karasanlal Joshipura, Sub-Asst. Surgn. . 1832
—— Jemdr. . . 1048
Ramnawaz Singh, Jemdr. . . 1047
Rampal Singh, Jemdr. . . . 1043
—— Jemdr. . . . 1047
—— Jemdr. . . . 1521
Rampargat Pande, Jemdr. . . 1026
Ramparshad Singh, Subdr . . 1031
Rampartab Singh, Jemdr. . . 1003
—— Gurung, ... 1051
Rampersad Tiwari, Subdr. . . 1319
Rampershad Thapa, Jemdr. . . 1604
—— Singh 1498
Rampezotti, E. A., 2nd-Lt . . 553

NAMES AND RANK. Page.

Ramphal Tiwari, Subdr. 1026
Ram Phal, Ressdr. 890
——— **Sukul,** Jemdr. . . . 1024
Ram Prasad Naryan, Sahi., 2nd-Lt. 1013
Ramprith Singh, Subdr. . . . 1405
Rampur, *Sir* Muhammad Hamid Ali Khan, *Bahadur, Hony.* Col., *Nawab* of . . . 1, 110, 891
Ram-rachhpal, Sr. Sub-Asst. Surgn. 1807
Ramradhe Singh, Jemdr. . . . 1048
Ramraj Singh, Jemdr. . . 1043, 1385
Ram Ratan Singh, Jemdr. . . 1401
Ram Rao Powar, Jemdr. . . . 934
——— Sub-Asst. Surgn. 1830
Ram Rakhu Singh, Jemdr. . . . 1344
Ramrekha Singh, Jemdr. . . . 1581
Ramsamujh Upadhya, Jemdr. . 1027
Ram Saran Gurung, Subdr. . 1615
——— **Sahar,** Subdr. 1056
——— **Sharma,** Sub-Asst. Surgn. 1820
——— **Saran,** Jemdr. 1117
———**Singh,** Jemdr . .
Ram Sarup, Jemdr. 1052
——— Jemdr. . . . 1443
——— **Singh,** Subdr. . . 1395
——— Jemdr. . .
——— Jemdr. . . 1164
——— Jemdr. 904
——— Subdr. 1401
——— Subdr. . . . 1371
——— Jemdr. . . . 1068
——— Subdr. 1052
Ramsay, A. D. G., Lt.-Col. 200, 1899, 1912
——— A. H. W., 2nd-Lt. . . . 568
——— A. Fitz G., Maj. . 223, 917
——— A. Macl., Tempy. Capt. . 1766
——— B. F., Capt. 740
——— J. G., Maj.-Genl. *Sir* . 1116
——— W. F. C., 2nd-Lt. . 550, 1677
——— W. J., *Hony.* Capt. . 1843
Ramsbotham, R. B., Lt. 515, 856, 1203
Ramsden, A. R., Lt. . . . 931, 375
——— G. C. F. Lt. 726
——— H. F. S., Col. . . . 87
——— J. E., 2nd-Lt. . . . 640
Ramsey, G. A. S., Capt. . . . 1760
——— H. L., 2nd-Lt. . . . 1679
——— L. A., Lt. 685, 784
Ramshankar Tiwari, Jemdr. . . 1018
Ramsidh Singh, Jemdr. . . . 1047
Ram Singh Rawat, Jemdr. . . 1181
——— Jemdr. 894
——— Jemdr. 966
——— Jemdr. 901
——— Jemdr. 1130
——— Jemdr. 1064
——— Jemdr. 1248
Ramsor Gurung, Jemdr. . . 1602
Ram Suchet Singh, Jemdr. . . 1384
Ramsudh Tewari, ... 1016
Ramsuman, Tewari 1017
Ramsumer Sukul 1017
Ramswarath Singh, Jemdr. . .
——— Jemdr. . . 1048
Ramulu, Jemdr. 994
Ramuni Narikuti, Sub-Asst. Surgn. 1825
——— **Nayar, P. P. V.,** Sub-Asst. Surgn. 1817
Ramu Singh, Subdr. 1499
——— **Thapa,** Jemdr. 1597
Ramzar, Jemdr 1681
Ran Singh, Jemdr. 905
Ranaday, S. G., Tempy. Capt. . 1763
Rana Jodha Jang Bahadur, Lt. 2, 311a, 1026
Rana Sher Muhammed, Sub-Asst. Surgn. 1820
Rana Sher Mohamed Khan, Risldr. . 1684, 1686
Ranade, Y. B., Ty. Lt. 1781
Ranbahadur Gurung, Jemdr. . 1025
——— **Limbu,** Jemdr. . . 1522
Ranbaz Khan, Jemdr. 928
Ran Bir, Subdr 968
——— **Ghaley,** Subdr. . . . 1636
——— **Thapa,** Subr. . . . 1602
——— Subdr. . . . 1627
Rance, C. J., Lt. 629
Ranchhdal Popatlal Parikh, Sub-Asst. Surgn. 1832
Ranchor Singh, Jemdr. . . . 926
Rand, T. A., Lt. 629
Randall, B. Fitz. H., Capt. 274, 879, 951
——— B., T., 2nd-Lt. 550

NAMES AND RANK. Page.

Randall, E., 2nd-Lt. 860
——— H. V., 2nd-Lt. . . 551, 1009
——— L. B., 2nd-Lt. 794
——— L. E., Lt. . . . 504, 1197
Randell, J. W., 2nd-Lt. . 358, 1434
Randhir Singh, Ressdr. . . . 884
——— Jemdr. . . . 890
——— Jemdr. . . . 904
Randhoj Rana, Subdr. . 1607, 1609
——— **Gurung,** Jemdr. . . . 1615
Randles, G. W., 2nd-Lt . . 731, 1009
Rang Baz Khan 1682
Rangabhashyam, C., Tempy. Lt. 1768
Ranganayakulu, Subdr. . . . 1351
Rangasami, Subdr. 993
——— Jemdr. 994
——— **Mudali,** Jemdr. . . . 1591
——— Subdr. 1333
——— Jemdr. . . . . .
Rangayya, Subdr.-Maj. . . . 1263
Rangbaz Khan, Subdr. . . . 1256
Rang Baz Khan, Jemdr. . . . 1256
Ranger, C. A., Lt. . . . 510, 1148
——— W. J., Lt. . . . 31, 67, 701
Ranishwar Singh, Residr. . . 1683
Ranjan Rana, Jemdr. 1625
Ranjit, Subdr. 1587
——— **Ram** Rissdr. 686
——— Jemdr. 1193
——— **Singh,** Subdr. 1399
——— Subdr. 1032
——— Jemdr. 974
——— Subdr. 1371
——— Jemdr. 1389
——— Subdr. 1397
——— Jemdr. 961
——— Sub-Asst. Surgn. . 1818
——— **Sing, Rawat,** Subdr. . . 1179
Ranjoer Singh, Subdr. 1171
Ranjor Singh, Jemdr. 1047
Ranken, L. D., Maj. 765
Rankin, C. H., Br.-Genl. 121, 127, 587
——— H. W., Lt. 617
——— J. T., Maj. 856
——— R. M., 2nd-Lt. 577
——— W. B., Maj. 715
Ranking, R. P. L., Lt. . . 299, 883
——— W. L., Capt. . . . 485, 61.
Ranlal Rai, Jemdr 1633
Rannu Thapa, Subdr.-Maj. . . 1883
Ranoda Prosad Saha 1319
Ranoji Beluse, Jemdr. 1416
Ranpati Gurung, Jemdr. . . . 1611
Ran Singh, Subdr. 1171
——— Jemdr. 888
——— Jemdr. 906
——— Subdr. 1280
Rans, C. A., Sub-Condr. . . . 1879
Ranser Limbu, Jemdr. 1631
Ran Sing Thapa, Jemdr. . . . 1619
Ransford, E. M., 2nd-Lt. . . . 793
Ransford, F. M., Maj. 211, 1126, 1623
——— W. B., Lt. . . . 510, 1651
Ransom, A. J., Sub-Asst. Surgn. 1819
——— J. M., Bt.-Col. . . . 57
——— R. A., Capt. . . . 741
Ransome, F. D., 2nd-Lt. . . . 797
Ranson, A. J., Sub-Asst. Surgn. . 1819
——— H., Lt. 789
——— H. E., Lt. . . . . .
Ransur Rana, Subdr. 1636
——— **Pun,** Jemdr. 1619
Rao, B. D., Ty. Lt. 1778
——— C. S. R. V., Ty. Lt. . . . 1778
——— H. S., Ty. Lt. 1782
——— K. B., Capt. 1758
——— M. N., Ty. Lt. 1779
——— N. M., Tempy. Capt. . . 1768
——— S. R., Tempy. Capt. . . 1765
——— T. S. H., Sub-Asst. Surgn. . 1828
——— Jemdr. 1171, 1188
——— **Jadhow,** Subdr. 1435
Raoji Daji Gholap, Sub-Asst. Surgn 1829
——— **Mahamunkar,** Jemdr. . . 1444
Raper, C. F., 2nd-Lt. . . 330, 1098
Raper, J. H. F., Lt. 818
Raphael, A. J., Asst. Surgn. . 1797
——— S. C., Asst. Surgn. . . 1795
Rapot Ram Rava, Subdr. . . . 1633
Rappa, A. A., Sub-Asst. Surgn. . 1804
Rappoport, F. G., Lt 544
Rapson, R. A., 2nd-Lt. . . . 788
Raquam Din, Jemdr. 1260
Rasal, Subdr. 1455
Rasam Singh, Subdr. 1231

NAMES AND RANK. Page.

Raschen, G. H., Lt. . . . 507, 1098
Rashe, E. Capt. 1321
Rashid, M. A., Tempy. Capt. . 1769
Rasila, Jemdr. 1256
——— Jemdr. 1232
Rassia Ram, Jemdr. 896
Rasul, S. G., Tempy. Capt. . . 1769
——— **Khan,** Jemdr. 926
——— **Shah,** Sub-Asst. Surgn. . 1816
Rasundram, Jemdr . . . 994, 1699
Ratan Chand, Ressdr. 965
——— **Chand Marwah,** Sub-Asst. Surgn. 1812
——— **Rao Bhofse,** Jemdr. . . . 1464
——— **Sing Butola,** Jemdr. . . 1179
——— **Tulsara,** Jemdr. . . . 1181
——— **Singh,** Ressdr. 931
——— Jemdr. 880
——— Subdr. . . . 1357
——— Ressdr. . . 936, 950
——— *Hony.* Capt. . . . 2102
——— Risldr. 880, 1682, 1686
——— **Basnet,** Subdr. . . . 1042
Ratansinh, Shambhusinh., 2nd-Lt. 48
Ratcliffe, J., 2nd-Lt. 788
——— W., Lt. 637
Rgtipal Singn, Jemdr. 1064
Rati Ram, Risldr. 901
——— Jemdr. 1557
Rati-lal Keshav-lal Modi, Sub-Asst. Surgn. 1815
Ratlam, *Rajaol*, *Hony.* Col. 110, 950
Ratna, Subdr. 1039
Ratnam, Jemdr. 1332
Ratnamalal, Subdr. 1317
Ratna Mudali, Jemdr. 1591
Ratnasabapathy, Jemdr. . . . 1264
Rattan Singh, Jemdr. 906
——— Risldr. . . . 1684
——— **Thapa,** Jemdr. . . . 1625
Rattray, C., Br.-Genl. 53, 98, 121, 125, 191
Rattu Singh, Jemdr. 1031
Rau, B. S., Ty. Lt. 1778
——— G. R. D., Ty. Capt. . . . 1772
——— M. S., Ty. Lt. 1776
——— R. V., Ty. Lt. 1778
Rauf Khan, Jemdr. 918
Raunaq Singh, Jemdr. 1284
Raval, J. M., Sub-Asst. Surgn. . 1818
Raven, T. F., 2nd-Lt. . . . 559, 1118
Ravenhill, C., Lt.-Col. . . 134, 613
——— M. H., Lt. . . . 515, 1893
Raven-Hill, A. L., Lt. . . 393, 1126
——— L. E., Lt. 399, 780, 1333
Ravenscroft, S., Lt. 835
——— G. E., 2nd Lt. . . . 687
Ravenshaw, H. S. L., Maj.-Genl. 119, 125, 759
Ravi Partab Narain Singh, Sahib Zada, 2nd-Lt. 481
Rawail Singh, Sr. Sub-Asst. Surgn. . 1809
Rawat Singh, (I) Risldr. . . . 539
——— Jemdr. 940
——— Jemdr. 928
Rawbone, D. B. M., Lt. . . 385, 1034
Rawcliffe, J., 2nd-Lt. 687
Rawlands, W. S., 2nd-Lt. . . . 552
Rawlence, M., Capt. . . . 665, 996
Rawlings, D. A. B., 2nd-Lt. 556, 828, 1347
——— G. N., Capt. . . 501a, 1667
Rawlingson, H. G., 2nd-Lt. . . 502
Rawlins, A. K., Lt.-Col. . 134, 157, 194, 1116
——— E. F., Lt. 490
Rawlinson, C. A. W., 2nd-Lt. . . 354
——— H., Maj. 846
Rawson, C. D., Maj. . . . 58, 634
Ray, C. B., Sub-Condr. . . . 1858
——— G. C., Ty. Capt. 1781
——— G. G., Lt. 535
——— H. M., Maj. 500, 858
——— J. N., Ty. Lt. 1772
——— K. M., Ty. Lt. 1774
——— L. K., Ty. Lt. 1774
——— R. S. A., Lt. 538
——— S. K., Ty. Lt. 1797
Rayment, W. M., 2nd-Lt. . . . 714
Raymond, C., Sr. Asst. Surgn. 1789
——— E. D., Maj. 16, 43, 224, 931
——— M. C., Capt 54, 254 939
Raynbird, H. 2nd-Lt. . . 465, 1459
Rayne, P. C., Lt. . . . 317a, 365, 1211

Names and Rank. Page.

Rayneau, J. R., Lt. 111
—— L. A., Capt. 486, 1048
Rayner, A., Sub-Condr. 1856
—— A. G., 2nd-Lt. 360
—— A. J., 2nd-Lt. 546
—— C. S. W., Lt. 387, 1387, 1215
—— G. E., Lt. 539, 1007
—— G. H., Lt. 1136
—— J., Lt. 430, 1152, 1489
—— R. H., Lt. 380, 1006
—— T. A., Lt. 353, 1054
Raynor, C. A., Capt. 271, 1215
—— E. V., Lt. 310, 1243
—— V. H., Capt. 280, 287, 1355
—— W. D., Lt. 526
Raywood, W., 2nd-Lt. 733
Raza Ali Shah 1514
Rea, A., Lt. 393, 1046
—— A. E. P., 2nd-Lt. 1595
—— A. N., Lt. 420, 1022
—— F. A. R. R., Lt. 508
—— F. W., Lt.-Col. 128, 154, 185
Read, C. E. L., 2nd-Lt. 703
—— E. H. 2nd-Lt. 739
—— E. W., 2nd-Lt 552
—— H., Brig.-Genl. 179, 1917
—— H. W. K., Lt.-Col. 133, 1710
—— J., Lt. 313, 544, 905, 1104
—— J. E., Lt. 380, 731, 1613
—— J. F. W., 2nd-Lt. 691
—— M. A., 2nd-Lt. 669
—— N. J., Lt. 722
—— P. E., 2nd-Lt. 730
—— R. F., Lt. 852
—— R., 2nd-Lt. 353, 1537, 1419
—— S. A., Ty. 2nd-Lt 577
—— W. D. B., 2nd-Lt. 679
—— W. R., Lt. 581
—— W. T., Lt. 518
Reade, P. V. J., 2nd-Lt. 474, 1475
Reader, E. J. W., *Hony.* Lt. 682
—— F. E. M., Capt. 17, 37, 734
—— N., Tempy. Capt. 1723
Reardon, W. A., Ty. Capt. 1767
Reavell, H. S., Capt. 72, 698, 795
Rebbeck, T. V., Maj. 732
Rebeiro, C. W., Asst. Surgn. 1794
—— E. F., Asst. Surgn. 1793
—— F. V. deV., Asst. Surgn. 1799
—— W. R., Sr. Asst. Surgn. 1791
Rebsch, W. K., Capt. 29, 107, 255, 1063
Recknell, D. L., Lt. 443, 1068
Recton, T., Lt. 25, 527, 1668
Reddick, G. P., Lt. 537, 883, 1034
Redding, J., Sub-Condr. 1874
—— J. E., Lt. 371
Reddi, M. S., Sub-Asst. Surgn. 1828
Reddiar, M. S., Sub-Asst. Surgn. 1827
Redfarn, F. W., 2nd-Lt. 362
Redfern, T., 2nd-Lt. 1110
—— A. R. 798
Redington, D. C., 2nd-Lt. 599, 600
Redkar, A. A., Sub-Asst. Surgn. 1830
Redl, E. A. F., Lt.-Col. 102, 135, 158, 1455, 195
Redman, H. E., Maj. 229, 1403
—— H. G., Capt. 755
Redmond, C. J., 2nd-Lt. 532, 1699
—— L. G., Lt. 818
Redshaw, T. A., Lt. 776a
Redwood, J., Lt. 711
Relief, P. J., 460, 462
Reece, J. D., Maj. 206
Reed, C., Bt. Lt.-Col. 62, 74, 633
—— E. W., 2nd-Lt. 713
—— F. W., 2nd-Lt. 568
—— G. S., Lt. 56, 524
—— H. J. F., Lt. 430
—— H. R. B., Maj. 221, 1178, 1181
—— H. S., Lt.-Col. *Sir* 2, 604
—— H. S., Capt. 16, 697
—— J. H., Lt. 415
—— L., Capt. 1705
—— L. N., Lt. 514
—— M. W., Lt. 488
—— R. B., Capt. 1722
—— R. R., Lt. 305, 1215
—— S. P. K. J. H., Lt. 401, 1495
—— T., Condr. 1855
—— W. A., 2nd-Lt. 786
—— W. L. L., Lt. 491
Reekie, J., 2nd-Lt. 965
Reeks, F., Lt. 538, 1707
—— N., Lt. 729
Reel, A. G. W., 2nd-Lt. 1695
Rees, C. G. A., 2nd-Lt. 550, 1702
—— C. H., 2nd-Lt. 558, 1678
—— F. A., 2nd-Lt. 554
—— G. H., Maj. 1712
Rees, H. J. V., Lt.
—— H. V., 2nd-Lt. 551, 1110
—— I. D., Lt. 512
—— J. C., 2nd-Lt. 658
—— O. M., 2nd-Lt. 653
—— R., Condr. 1867
—— R., Capt. 520, 602, 1696
—— T. O., 2nd-Lt. 553, 828, 1407
—— T. W., Lt. 300, 301, 1307
Reeve, A., Lt. 721
—— A. J., Lt. 445, 1140
—— E. G., 2nd-Lt. 692
—— F., 2nd-Lt. 769
—— G. A., 2nd-Lt. 754
—— G. B., Ty. Lt. 796
—— H. L. 2nd-Lt. 1032
—— M. B. P., Lt. 318, 1096, 1573
Reeves, A. J. M., Lt. 298a, 1267
—— F. H., 2nd-Lt. 363
—— F. K., Lt. 630
—— G. L., 2nd-Lt. 750
—— H. Mac.D., 2nd-Lt. 567
—— S. G., 2nd-Lt. 21, 626
—— S. R., Maj. 72, 633
—— W. A., Lt. 786
—— W. H., 2nd-Lt. 549, 1064
Regan, C. H., Lt. 1443
—— G. H., 2nd-Lt. 397, 1442
—— W. A. V., 2nd-Lt. 723a
Regnart, N. S., Capt. 589
Rego, A. S., Asst. Surgn. 1802
Rehling, G. C., Asst. Surgn. 1794
Reiche, A. D. E., Capt. 280, 287, 1189
Reichwald, M. B, Ty. Lt. 1726
Reid, A. D., Lt. 1919
—— A. G., 2nd-Lt. 715
—— A. M., Ty. Asst. Surgn. 1804
—— C. L., Capt. 111, 485
—— D., Sub.-Condr. 1864
—— D. C. G., Lt. 766
—— D. E., Lt. 508
—— D. W., Lt. 401, 1417
—— E. J., Sr. Nursg. Sister 1734
—— G., Lt. 763, 781
—— H. A., Lt. 529
—— J., 2nd-Lt. 566
—— J. B., Lt. 510, 1006
—— K. B. 2nd-Lt. 1008
—— L. H., Lt. 624
—— N., *Hony.* Capt. 765
—— R., Capt. 628
—— R. N., Lt. 490
—— S. D. Capt. 1715
—— W. C., Maj. 241, 935
—— W. D., Maj. 27, 211, 883
—— W. F., Lt. 415, 1144
—— W. P., 2nd-Lt. 571
Reiffenstein G. P., Capt. 1693
Reilly, A. F., Lt. 95, 490
—— B. R. Maj. 238, 1893, 1913
—— C., Esq. 91
—— C. O. C., Lt. 488, 987
—— E. F. J., 2nd-Lt. 494, 938
—— H. D. B., Capt. 679
—— H. L., Bt.-Maj. 258, 1335
—— J. H. J., Sub-Condr. 1874
—— N. E., Maj. 95, 241, 1901
—— S. G., 2nd-Lt. 552
—— T. J., Lt. 465, 1287
Reiner, G. V., 2nd-Lt. 652
—— N. A., Lt. 534, 1627
Reinhold, C. H., Maj. 1255, 1752
—— H. E., Maj. 104, 105, 222
Renshaw, C. M., Lt.
Reiss, H. B., Lt. 489
Reith, C. E. W., Lt. 56, 507, 1026, 1570
Rekha Ram, *Hony.* Capt. 2103
—— Jemdr. 1412
Remfry, D. H., 2nd-Lt. 851
—— M., Lt. 852
Remington, G. H. L., Lt. 826
—— J. B., Capt. 818
Remrekha Singh, Jemdr. 1384
Rendall, C., Lt. 351, 1156
—— F. C. E., Lt. 420, 1361
—— J. C. S., Maj. 40, 230, 1662
Rendel, A. J., Capt. 615
Rendell, A. C., Capt. 844
—— C. B., Lt. 516, 1517
—— E. A., 2nd-Lt. 735
—— F., 2nd-Lt. 757
Rendle, J. M., 2nd-Lt. 531, 1671
Rennick, A. de C., Lieut.-Col. 179, 1917
—— A. de C., Maj. 79, 217, 895
—— C. M., 2nd-Lt. 826
—— D., *Hony* Maj. 1849
Rennick, G. B., Lt. 314, 1363
—— R. W. de C., Maj. 219, 868, 1056, 1387
Rennie, C. A., Lt. 587, 797
—— F. P., 2nd-Lt. 805
—— J. O., Lt. 520
—— P. M., Maj. 39, 1753
—— S. J., Col. 1731
—— T., Maj. 653
—— W. L., Lt. 465, 1363
Rennison, A. J., Maj. 225, 1662
Renny, A. McW., Bt.-Col. 57
—— G. B., Bt.-Col. 56
—— G. S., Maj. 218, 872, 1174, 1567
—— H. B., Maj. 239, 1110
—— S. C., Capt. 718
—— S. E., Capt. 578
—— S. M., Col. 11, 71, 119, 123, 612
Renny-Tailyour, J. W., Maj. 614, 1880, 660
—— T. F. B., Col. 122
Renshaw, C. M., Lt. 401
Renton, H. D., 2nd-Lt. 565
Renwick, H., Lt. 308, 1181
—— W. B., 2nd-Lt. 610
Reoti, Parshad, Jemdr. 1028
Rerrie, E. S. E., Lt. 318, 1415
Reshmin, Jemdr. 1514
Retallack, C. W., Lt. 304, 1407
—— P. H., Maj. 654
Retallick, J. M. A., Lt.-Col 57
Retief, P. J., Lt. 455, 1319
Rettie, A. H., 2nd-Lt. 619
—— W. J. K., Lt.-Col. 122, 132, 613
Reuben, W., *Hony.* Maj. 1849
Revell, A. J., Lt. 310, 1263
—— F., 2nd-Lt. 705
—— L., 2nd-Lt. 1675
—— P. J., Lt. 721
Revnell, A. J., Rev. 1908
Rewa, H. H. *Sir*, Venkat Raman-Singh, *Bahadur*, Maharaja of, *Hony.* Lt.-Col. 110
Rewantia Thapa, Jemdr. 940, 1625
Rewat Singh, Ressdr. 936
Reyazuddin Ahmad Khan, Ty. 2nd-Lt. 482
Reyne, G. Van R., Capt. 258, 1319
Reyner, 2nd-Lt., A. G. 1197
—— B., Maj. 67, 1059
Reynolds, A. J., Maj. 221, 945
—— A. D. M. 1679
—— A. W., Lt. 309, 1473
—— B. O., Maj 38, 652
—— B. T., Capt. 617
—— C., Lt. 617
—— C. E. J., Lt. 1008
—— C. C. R., Lt. 312 911
—— C. G., 2nd-Lt. 546
—— C. W., 2nd-Lt. 732
—— D., Capt. 24, 1713
—— E. W., Capt. 256, 1391
—— F. J., Sub-Condr. 1857
—— G. N., Capt. 597
—— H. J., Sr. Nursing Sister 1734
—— H. L., *Hony.* Capt. 1050
—— H. V., Maj. 244, 1335, 1423
—— J. E., Lt. 40, 60, 519, 815
—— J. R., 2nd-Lt. 324
—— J. W., 2nd-Lt. 1565
—— L., Maj. 1750
—— N., 2nd-Lt. 77, 723a, 1885
—— R. J., Ty. Lt. 1729
Rheam, P. M., 2nd-Lt. 725
Rhenius, E. E., 2nd-Lt. 1693
Rhodes, B. B., Lt. 492, 927
—— C. K., Lt. 512, 1487
—— G. D., Capt. 665, 1882
—— G. E., Maj. 668, 1884
—— G. E., 2nd-Lt. 549, 1675
—— H. H., Lt. 351
—— H. S., Capt. 820
Riach, A. H. D., Lt.-Col. 37, 134, 651
Riadore, A. F., 2nd-Lt. 788
Riazuddeen, M., Sub-Asst. Surgn. 1826
Ribbins, W. V., 2nd-Lt. 530
Ricardo, W. H., Lt. 27
Riccard, J. S., Maj. 62, 74, 633
Riccomini, J. A., 2nd-Lt. 20, 669, 992
Rice, A. K., Lt. 303, 1621
—— B. A. McH., Maj. 212, 1623
—— E. M., Lt. 520, 819
—— G. D., Maj. 237, 1663
—— H. D., 2nd-Lt. 808
—— H. A. H., Maj. 1919

Names and Rank.	Page.
Rice, H. G., Lt.	353, 1140
—— J. P., Lt.	1839
—— P. H., Lt.	511, 1227
—— R. A., Capt.	485, 929
—— R. V., Lt.	685, 783
—— R. W. 2nd-Lt.	363
—— S. K. B., Lt.-Col.	139, 160, 199, 869
—— S. M., Maj.	15, 19, 99, 207, 1275
—— T., Sub-Condr.	1856
—— T. W., Esq.	88
Rich, C. N., Lt.	520, 821
—— E. P., Lt.	299, 964
—— E. T., Maj.	662, 1880
—— G., Lt.	401
—— H. H., Col.	11, 123, 612
—— H. H., Capt.	274, 1475
—— H. W. W., Condr.	1857
Richard, C. M., 2nd-Lt.	493
—— T. J. R., Lt.	998
Richards, A., Capt.	502, 858, 1665
—— A. P., Capt.	89, 787
—— A. R. T., Capt.	734, 798
—— B. W., Capt.	465
—— C. E., Capt.	738
—— C. D., Lt.	319, 889
—— C. G., 2nd-Lt.	77, 356, 1885
—— C. H., Lt.-Col.	5, 858
—— E. V., Lt.	849
—— F. J., Capt.	830
—— H., 2nd-Lt.	361
—— H. G., Col. Sir	814
—— H. L., 2nd-Lt.	704
—— J., 2nd-Lt.	790
—— J. C., Lt.	1653
—— R. W., 2nd-Lt.	358, 1160
—— S. H., Capt.	852
—— T. J. R., Lt.	519
—— W., 2nd-Lt.	361
—— W. E., Esq.	89
—— W. J., Sub-Condr.	1874
—— W. N., Lt.	304, 1475
—— W. V., Lt.	709
—— W. V., Maj.	87, 213
Richardson, A., Lt.	819
—— A. B. H. N., 2nd-Lt.	581
—— A. E., *Hony.* Lt.	818
—— A. J., Lt.	491
—— A. J. C., Lt.	368, 1517
—— A. M., 2nd-Lt.	692
—— A. V., 2nd-Lt.	831
—— C., 2nd-Lt.	566
—— C. W. G., Maj.-Genl. 9, 16, 97, 102, 105, 119, 147,	175
—— D. T., Capt.	1714
—— E., Asst. Surgn.	1800
—— F., Lt.	735a
—— F. B. W., Col.	122, 149, 181
—— F. C., Lt.	816
—— F. J., 2nd-Lt.	357, 1114
—— F. L., Capt.	11, 18, 570, 1850
—— G., Lt.	793
—— G., 2nd-Lt.	355, 1561
—— G., Sub-Condr.	1857
—— G. A., Asst. Surgn.	1798
—— G. G., Maj	46, 244, 1347, 1423, 1895
—— G. L. R., Lieut.-Genl. *Sir*,	909, 1955
—— H., 2nd-Lt.	528, 1383
—— H., Capt.	761
—— H. A., Asst. Surgn.	1792
—— H. F., Lt.	629
—— H. L., Br.-Genl.	12, 23, 120, 125, 185, 150
—— H. L., Capt.	752
—— J. F., Ty.-Capt.	1772
—— J. H. S., 2nd-Lt.	355, 1062
—— J. J., Lt.-Col.	591
—— J. M., Asst. Surgn.	1798
—— J. S., Lt.	529
—— L., 2nd-Lt.	725
—— R., Lt.	291, 1243
—— R., Sub-Condr.	1856
—— R. A., Lt.	512
—— R. C., Lt.	542
—— R. J. D., Lt.	22, 512
—— R. K., Lt.	493, 923
—— S. G. P. L., Lt.	1839
—— S. H., Lt.	624
—— T. W., Lt.-Col.	859
—— W., 2nd-Lt.	546
—— W. H., Lt.	29
—— W. N., Esq.	39
Richardson Gibson, F. J., 2nd-Lt.	330, 1249
Richdale, C. R., Lt.	308, 1385
Richer, R. A., Lt.	544, 1092
Riches, C. A., Lt.	537
—— C. F. A., 2nd-Lt.	1359
—— G. H., Lt.	637, 975
Richmond, M. G., Lt.	352, 1642
—— J. P., Lt.	397, 1152
—— R. H., Condr.	1873
—— R. D., Lt.	611
—— R. Y., Lt.	1732
Richpal Singh	1052
Rickard, J. H., 2nd-Lt.	1415
Rickcard, C. R., Capt.	628
Rickell, G. G., Capt.	61, 721
Ricketts, C. H., Esq.	91
—— E. L., Capt.	274, 1391
—— E. M., Lt.	520, 1666
—— E. W. C., Capt.	253, 1110, 1156
—— F. R. R., *Hony.* Maj.	1845
—— F. W., Capt.	47
—— G. A. McG., Lt.	401, 1034
—— H. C., Lt.-Col.	78, 130, 155, 189
—— P., 2nd-Lt.	362
—— P. E., Lt.-Col.	135, 158, 195
—— R. L., Lt.-Col.	98, 139, 162, 195, 199, 867, 893
—— T. C. D., Lt.	302, 1335
—— T. F. W., Capt.	768
Rickie, J. H., Lt.	832
Ricks, G., *Hony.* Lt.	1850
Riddell, A., Maj.	238, 1311
—— B., Lt.	540
—— R. B., Lt.-Col.	137, 632
—— R. C. H., 2nd-Lt.	705, 1321
—— W. H., Capt.	1754
Riddle, T. E., Lt.	491, 931
Riddock, J. D., 2nd-Lt.	1678
Rideout, J. E. W., Lt.	491
Ridge, J. H., Lt.	776a
—— K. J. T., Lt.	293, 1407
Ridgeway, D. G., Maj.	99, 217, 1606
Ridgway, R. T. I., Lt.-Col.	127, 153, 190
Ridsdale, A. C., Revd.	1904
—— C. G., Capt.	851
Ridler, W., 2nd-Lt.	356, 1405
Ridley, A. G., 2nd-Lt.	640
—— E. R., Lt.	296, 1327
—— G. A., Lt.	380, 931
Ridlington, H. T., Lt.	541, 1162, 1678
Rieley, S. D., Asst. Surgn.	1797
Ries, J. L., Lt.	322, 1170
Rieu, E. V., 2nd-Lt.	559, 1423
Rigby, G. H., Lt.	64
—— J. P., 2nd-Lt.	554
—— L., 2nd-Lt.	758
—— W., 2nd-Lt.	581, 605
Rigg, A. C. P., Revd.	1905
—— H., Capt.	608
—— H. M., Maj.	218, 870, 1293, 1375
—— J. H., Lt.	522, 1379
Riggall, H. B., Capt.	695
—— P. R., Capt.	1718
Riggs, A. J., Lt.	352, 1673
Rigold, B. F., Lt.	493
Rikhesor Khattri, Jemdr.	1615
Rikhi Ram Khattri, Jemdr.	1637
Rikhya, M. L., Tempy. Capt.	1768
Rikkhi Singh, Jemdr.	1051
—— Subdr.	1064
Riley, J. E., 2nd-Lt.	986
—— C. B., Maj.	1919
—— C. L., Lt.	415
—— D., Lt.	77, 507, 1885
—— E. C., 2nd-Lt.	545, 1493-94
—— E. H., 2nd-Lt.	821
—— E. S., 2nd-Lt.	552, 821, 1459
—— H. J., 2nd-Lt.	788
—— M. H., 2nd-Lt.	679
—— R. C., Lt.	510
Riley-Irving, C. T., 2nd-Lt.	568
—— J. H., 2nd-Lt.	33, 1883
Riley-Lloyd, C., Lt.	1189
Riley-Smith, W., 2nd-Lt.	599
Rimell, F. J., Lt.	304, 1042, 1070
Rimane Thapa, Subdr. Maj.	1636
Rimington, J. C., Maj.-Genl.	119, 585, 660
—— M. B., Lt.	1919
Rimington, R. G. W., Lt.	585
Rimmer, H. A., Capt.	89, 777
Rind, A. T. S. A., Bt. Col.	1931
—— G. B. A., Bt. Lt.-Col.	8, 15, 98, 141, 162, 217, 1491
Ring, H. J., Lt.	540, 1696
—— J. S. H., Capt.	253, 1207
Ringrow, N. V., 2nd-Lt.	547
Rintoul, C. R., Revd.	1903
—— J. P., 2nd-Lt.	653
Ripley, T., Capt.	610
Rippen, A. C., 2nd-Lt.	1679
Rippon, P. I. V., Lt.	323, 455, 525, 1459
Risal, Jemdr.	1457
—— Singh, Resldr.	902
—— Jemdr.	906
Rishi Lal Vaid, Ty. Sub-Asst. Surgn.	1833
Rishworth, H. R., Asst. Surgn.	1802
Risley, C. G., Maj.	236, 909
Ritchie, A. A., 2nd-Lt.	550, 907
—— A. B., Maj.-Genl.	119, 125
—— A. B., 2nd-Lt.	828
—— C. Macl., Lt.-Col.	103, 134, 613
—— D. D., Ty. Lt.	1726
—— G. S., 2nd-Lt.	817
—— J. H., Lt.	525
—— J. L., Capt.	1714
—— N. M., Lt.	743, 744
—— T. K., Lt.	719
—— W. D., Maj.	1748
Ritson, F., Lt.	536, 1110
—— R. G., Capt.	585
Ritter, J. C., Lt.	514, 1375
Rivaz, V., Col.	177, 1917
Rivers, W. E., Lt.	366, 1647
—— W. E., 2nd-Lt.	326
Rivers-Smith, P. E., Lt.	622, 977
Rivett, E. G., Capt.	28, 516a, 605, 1668
—— H. H., Lt.	551, 852, 1533
—— W. C., Lt.	537, 1068
Rivett-Carnac, H. G., Capt.	59, 274, 1323
—— J. C. T., Lt.	490
—— J. S., Maj.	13, 37, 99, 209, 901
—— P. K., Capt.	273
—— S. G., Lt.-Col.	132, 601, 1881
Rix, C. H. B., Lt.	76, 573
—— H. S., Capt.	605
—— H. V., Lt.	508, 1081
—— W. H. B., *Hony.* Lt.	1843
Rixon, F., 2nd-Lt.	548, 1699
Riyazuddin, Sub-Asst. Surgn.	1808
Rizvi, S. F. H., Sub-Asst. Surgn.	1819
Roach, I. N., Lt.	1513
—— J. O., 2nd-Lt.	326, 1255
—— J. P., 2nd-Lt.	530, 1096
—— L. J., Lt.	543a, 1331
—— L. N., Lt.	443
Roan, W., Lt.	524, 1331
Roark, C. T. I., Lt.	380, 939
Robarts, C. B. H., Maj.	233, 1315
Robathan, K. M., Lt.	516, 1601
Robb, A. A., Lt.	84, 535, 830, 1361
—— A. H. D., 2nd-Lt.	546, 1361
—— A. L., tempy. Capt.	1715
—— C. B., Maj.	15, 26, 751
—— G., 2nd-Lt.	796
—— J. J., Maj.	1747
—— J. M., 2nd-Lt.	782
—— T. R., Lt.	304, 1359, 1591
Roberson, A. R. G., Capt.	728
Robert, V. A., 2nd-Lt.	554
—— V. G., Lt.	295, 1515
Robertson, H. E. 2nd-Lt.	1315
Roberts, A., 2nd-Lt.	362
—— A. B., Revd.	1904
—— A. C., Lt.	352, 1677
—— A. C., 2nd-Lt.	355, 1409
—— A. S. B., Maj.	216, 827, 1895, 1913
—— A. W., 2nd-Lt.	587
—— C., Lt.	75
—— C., Sub-Condr.	1874
—— C. E. L., 2nd-Lt.	703
—— C. F., 2nd-Lt.	623
—— C. J., Lt.	351, 1675
—— C. L. D., Tempy Capt.	1770
—— C. R. J., Lt.	527, 1671
—— C. R., Ty.-Lt.	1331
—— C. S., Lt.	793
—— C. W., Sub-Condr.	1856
—— D. E., 2nd-Lt.	558

Names and Rank. Page.

Roberts, E. A., Maj. 1751
—— E. C., 2nd-Lt. . . 557, 1009
—— E. F., *Hony.* Lt. . . . 75
—— F. C., Lt. . . . 514, 1050
—— F. L., Capt. 259, 1042, 1383
—— G. B., 2nd-Lt. . . 359, 1533
—— G. E., 2nd-Lt. . 325, 366, 968
—— G. H., Lt. . . . 463, 1147
—— H. A., Lt. 617
—— H. B., Lt. . 296, 1315, 1317
—— H. C. W., Lt. 405
—— H. J., Sub-Condr. . . 1655
—— H. L., Brig.-Genl. 12, 27, 105, 120, 124, 149, 184
—— J. O., 2nd-Lt. 1008
—— J. R., Lt.-Col. *Sir* 128, 1740, 1890
—— J. R., Capt. 665, 1001, 1882
—— J. R., 2nd-Lt. . . 473, 1154
—— J. W. B. T., 2nd-Lt. . . 848
—— M. B., Capt. . 44, 279, 1455, 1457
—— N. d'E. Maj. . . 79, 1838
—— O. L., 2nd-Lt. . . 666, 985
—— R. E., Ty. Capt. . . . 1723
—— R. F., Lt. 518
—— R. G., Lt. 583
—— R. J., 2nd-Lt. 786
—— R. R. P., Lt. 1695
—— S. A., Lt. . . . 525, 988
—— S. F., Sub-Condr. . . . 1857
—— S. N., Lt. . . . 351, 1676
—— V. G., Lt. 1513
—— V. H., Lt.-Col. . . 140, 1746
—— V. L., 2nd-Lt. 683
—— V. P., 2nd-Lt. . . 331, 925
—— W. B., Maj. . 104, 218, 1407
—— W., 2nd-Lt. 1225
—— W. E., 2nd-Lt. 807
—— W. H., Maj. . 664, 1882, 1890
—— W. H., Lt. . . . 531, 1704
Robertson, A., *Hony.* Maj. . . 1843
—— A., Lt. 841
—— A., 2nd-Lt. 637
—— A., 2nd-Lt. 1695
—— A. MacL., Lt. . 512, 1086
—— A. C., Capt. 817
—— A. L., 2nd-Lt. . 561, 1678a
—— B., *The Hon. Sir* 803, 1899
—— C., 2nd-Lt. 653
—— C. A., Maj. . . 236, 1333
—— C. L., Lt.-Col. 131, 661, 1880
—— D., Lt.-Col. *Sir* 178, 1918
—— D., 2nd-Lt. 557
—— D., 2nd-Lt. . . 539, 1455
—— D. E., Maj. 53, 221, 895
—— D. G., Lt. . . 317, 1058, 1585
—— D. K., 2nd-Lt. . . . 534
—— D. N., Capt. 785
—— E., Lt. . 368, 1197, 1701
—— E. A., Lt. 538
—— E. D. S., Maj. 239, 915, 1104
—— F., 2nd-Lt. 814
—— F. A. deV., Maj. . . . 111
—— F. E., Capt. . 60, 486, 939
—— F. W. R., 2nd-Lt. . . 854
—— G. A., Lt.-Col. . . . 103
—— G. G., Capt. 852
—— H. B., Lt. . . 537, 1598
—— H. E., 2nd-Lt. 560
—— H. E., Capt. 67, 279, 1431
—— H. L. C., Lt. . 315, 1124
—— H. S., Capt. 652
—— J., Capt. 1839
—— J., Lt. 1397
—— J., 2nd-Lt. . . 493, 937
—— J., Lt. . . . 293, 1054
—— J., Senr. Asst. Surgn. 1767
—— J., Lt. 1054
—— J., 2nd-Lt. 937
—— J. C., Lt.-Col. 134, 1744, 1890, 1931
—— J. F., Maj. 651
—— J. H., Capt. 579
—— J. K., 2nd-Lt. . . . 782
—— J. McI., Lt. . 543a, 1614
—— J. P. M., 2nd-Lt. . . 1441
—— K., Maj. . . . 236, 877
—— K. E. R., Asst. Surgn. 1802
—— M., Capt. 760
—— M. K., Ty. Capt. . . 1725
—— M. R., Capt. 783

Robertson, R., 2nd-Lt. 1056
—— R. A., Lt. . . 508, 1607
—— R. A. H., Maj. 10, 76, 227, 1140
—— R. B., 2nd-Lt. . 530, 1677
—— R. H., Capt. . 89, 766
—— R. H., 2nd-Lt. . 551, 1098
—— S., 2nd-Lt. . . 355, 1457
—— S. R., Lt. 719
—— W., Condr. 1873
—— W. V., Lt. . . 532, 998
Robertson-Glasgow, N., Maj. 74, 107, 108, 633
Robey, A. E., 2nd-Lt. . . . 769
—— C. A., 2nd-Lt. 559
Robin, N. E., Lt.-Col. . 129, 154, 186
—— M. L., 2nd-Lt. 1040
Robins, E., Lt. 544, 970
—— F. J., Capt. 807
—— J. H., *Hony.* Lt. . . . 696
—— R. T., Lt. . . . 470, 1040
Robinson, A. E., 2nd-Lt. 357, 821, 1078
—— A. E. W., Capt. . 501a, 1132
—— A. G., 2nd-Lt. 1693
—— A. J., Sub-Condr. . . . 1855
—— A. L. C., 2nd-Lt. . . . 843
—— A. M., Capt. 1335
—— A. W., Lt. 543a
—— C., 2nd-Lt. 535
—— C., Lt. 1708
—— C. P. W., 2nd-Lt. 325, 366, 1241
—— C. T., Lt. . . . 538, 1603
—— C. W., Lt. . . . 393, 883
—— D. B., Lt. . . . 534, 1207
—— D. C., Capt. . . 272, 1166
—— D. G., Maj. . 99, 215, 1207
—— D. G., Lt. . . . 515, 986
—— E. A., 2nd-Lt. 1695
—— E. C., Lt. . . 299, 1178
—— E. C., Lt. 490
—— E. L., Maj. 101, 238, 1315, 1447
—— F. C., Lt. . . 415, 512, 1489
—— F. C., 2nd-Lt. . . 358, 1009
—— F. G., L. 1489-90
—— F. J., Capt. 29, 75
—— F. R. H., 2nd-Lt. . 317, 913
—— F. W. O., Lt. 36, 320, 929
—— G., Lt. 618
—— G. A., Sub-Condr. . . . 1856
—— G. D., 2nd-Lt. 55
—— G. J., Lt. 1699
—— G. St. J., Capt. . 278, 1359
—— G. W., Lt.-Col. 46, 130, 155, 190, 763
—— H., Lt. . . . 450-54, 1623
—— H., 2nd-Lt. 1054
—— H. I., Capt. . . . 699, 751
—— H. N., Lt. 617
—— H. R., Lt. 299, 1369, 1371
—— J. A., Lt. . 309, 1339, 1591
—— J. C., Lt. . . . 528, 1339
—— J. G., Lt. . . . 530, 988
—— J. G., 2nd-Lt. . . 331, 1128
—— J. G. K., 2nd-Lt. 326, 365, 1082
—— J. H., Sub-Condr. . . . 1855
—— J. McM., Lt. . . 410, 1455
—— J. P., Lt. 589
—— J. R., Condr. 1853
—— J. S., Lt. 986
—— L. R. C., Lt. 487
—— N. P., Lt. . . . 313, 1263
—— S. W., Br.-Genl. 53, 121, 127, 632
—— W. A., 2nd-Lt. 567
—— W. A., Lt. 734
—— W. H., 2nd-Lt. 1696
—— W. H. B., Maj.-Genl. 62, 120, 1737
—— W. P., Maj. . . . 9, 1663
—— W. W., Lt. 697
Roblin, C. H., 2nd-Lt. . . 549, 1170
—— M. L., 2nd-Lt. . . 546, 1040
Robotham, W. D. N., 2nd-Lt. . . 90
Robson, A., Lt. 682, 720
—— A., Maj. 734
—— C. G., Lt.-Col. 105, 131, 155, 191, 868, 1058
—— E., Lt. . . . 415, 787, 1020
—— H. A. B., Capt. 1756
—— H. W. C., Capt. . . 261, 1629
—— M., Lt. 512
—— R. N., 2nd-Lt. 1621
Roche, G. P., Lt. 420
—— G. R. M., Revd. 1905
—— H. J., Col. . . 124, 150, 184

Roche, I. A., 2nd-Lt. . . . 331, 1641
—— J. M., Lt. 623
—— R. T. H., Capt. . . . 668, 1001
—— V. W., Capt. 701
Rochfort, A. P. G., Lt. . . . 857
—— H. G., Lt. 29, 542
Rockford-Boyd, H. C., Maj. . . 614
Rockley, F. M., Lt. . . 77, 528, 1885
Roda Ram, Ressdr. . . 1681, 1685
—— Singh, Subdr. 1106
—— Jemdr. 1268
—— Khan, Jemdr. 1365
Rodd, N. J., 2nd-Lt. . . . 356, 1164
Rodda, W. J., 2nd-Lt. 566
Roddy, F. A., Capt. 1717
—— H. H., Lt.-Col. 10, 130, 155, 189, 1560
Roderick, C. B., Sub-Condr. . . 1854
Rodger, A., Lt. 848
—— A. J., Capt. 501a
—— F. McG., 2nd-Lt. . . 61, 679
—— T. T., 2nd-Lt. 765
Rodgers, E. W. F., Lt. . . 543, 1699
—— F. D., 2nd-Lt. . . 503, 1315
—— F. McG. 61
—— J., Lt. 380, 1128
—— J. J., Lt. . . . 508, 1665
—— J. F., 2nd-Lt. . . 76, 550
—— M. C. R., Asst. Surgn. . 1798
—— R. T., Sr. Asst. Surgn. 1789
—— W. E., Asst. Surgn. . 1801
Rodgerson, A. P., Capt. . 274, 1178
Rodrigues, A. C., Ty. Lt. . . . 1780
—— F., Tempy. Capt. . . . 1767
—— G. D., Asst. Surgn. . . 1802
—— H. A., Esq. 89
—— J. B., Tempy. Capt. . . 1770
—— J. M., Asst. Surgn. . . 1795
—— P. H., Sr. Asst. Surgn. 1789
—— R. H. S., Asst. Surgn. 1799
Rodriguez, L. A., Sr. Asst. Surgn. 1788
Rodwell, E. H., Col. 97, 105, 122, 149, 181
—— E. H., Lt. 307, 1235
—— E. H., Lt. 112
—— L. H., Lt. 314, 1050
Roe, A. E., Lt. 738
—— B. O., Lt.-Col. . 194, 1896, 1912
—— C. D., Maj. 239, 1613
—— C. H., Br.-Genl. 58, 108, 120, 113, 660
—— F. F., 2nd-Lt. 640
—— G. A. M., Maj. 720
—— J. H., 2nd-Lt. 699, 796
—— K. T., Lt. 425, 1647
Roff, C. W., 2nd-Lt. 589
—— J. M., 2nd-Lt. 763
Roffey, H. R., *Hony.* Lt. . . . 1850
—— E. A., Lt. 606
—— L. S., 2nd-Lt. 606
Roger, G. B., Capt. . . . 501, 1065
—— K. B., Lt. 487
—— W. W., 2nd-Lt. 763
—— A. C. C., Lt. 1919
—— A. G., Lt. 689, 1567
—— A. L., 2nd-Lt. . . . 540, 1624
—— C. D., Capt. 1722
—— E., Sub-Condr. 1861
—— E. W., 2nd-Lt. . . 15, 22, 723
—— F. C., Maj. 1750
—— G. D., 2nd-Lt. . . . 552, 1183
—— G. G., Capt. 277, 1595
—— G. H., Maj. 6, 13, 99, 217, 1062
—— G. N., Lt. 506, 1403
—— G. S., Capt. . . 257, 1178, 1225
—— H., Lt. 536, 1606
—— H. B., Lt. 533
—— H. O., Sub-Condr. . . . 1864
—— J., Sub-Condr. 1870
—— J. A., Asst. Surgn. . . . 1800
—— J. R. D., Lt. 76, 510
—— J. S., Lt. 543a, 1675
—— J. W., 2nd-Lt. 1191
—— L., Lt.-Col. *Sir* . . 130, 1742
—— P. H., Lt.-Col. . . . 26, 722
—— R., 2nd-Lt. 937
—— R. H., Lt. 490
—— R. L., 2nd-Lt. 1160
—— T., *Hony.* Lt. 687
—— T. G., *Hony.* Capt. . . . 753
—— T. J., Lt. 311, 1363
—— T. P. W., Lt. . . . 393, 1496
—— W. G., Lt. 312, 1070

NAMES AND RANK. Page.

Rogers, W. P., Lt. . . . 506, 1667
——— W. S., Sub-Condr. . . . 1864
Rogerson, A. M., 2nd-Lt. . . . 828
——— J. A., Lt. . . . 507, 1411
——— J. F., Lt. . . . 537, 1672
Rohde, J. H., Capt. . . 665, 996
Rohlu, Jemdr. 1189
Rohtan Singh, Jemdr. 1021
Roland-Price, A., Lt. . . 39, 509
Rolfe, A. T., Sub-Condr. . . . 1857
——— F. W., 2nd-Lt. . . . 619
——— R. H., Esqr. . . . 8, 88
Rolinson, W. J., Lt. . . 450, 454
——— W. S., Lt. . . . 1309
Rolland, A., Lt.-Col. . . . 661
——— S. E., Brig.-Genl. 178, 1351, 1918
Rollins, C., Sub.-Condr. . . . 1857
——— W., Condr. 1872
Rollinson, R. G., 2nd-Lt. . . . 1267
Rollo, A. D., Lt. . . . 503, 825
——— J. C., 2nd-Lt. . . . 808
——— L. D., Maj. . . . 242, 1391
Rolt, P. J., Lt. . . . 507, 1128
Rombulow-Pearse, A. B., Maj 233, 1621, 1623
Romilly, H. A., Lt. . . 701, 1006
Ronaldshay, *Earl of*, *Hony.* Col. 610, 651, 656, 856, 859, 1893
Ronaldson, E. C. S'J., Lt. 503, 1042
——— G. W., Ty. Capt. . . 1725
Ronayne, C. R. L., Maj. . . 1711
Roney-Dougal, C. N., Capt. . 615
Ronnfeldt, W. E., Lt. . 307, 1267
Rood, L. F., Lt. . . . 353, 1397
Rooke, D. M., Lt. . . . 579
——— H., Nursing Sister. . . 1735
——— I., 2nd-Lt. . . 530, 1026
Rooker, G., Capt. . 501, 751, 1006
Rooks, R., Sub-Condr. . . . 1867
Roome, H. E., Capt. . . 665, 985
——— H. N., Lt.-Col. . 128, 154, 185
——— R. E., Lt.-Col. 79, 130, 155, 189
Roop Singh, Jemdr. . . . 892
Roos-Keppel, G. O., *Hon.* Lt.-Col. *Sir* 95, 102, 188, 1900, 1911
Roos-naie-Cocq, C. A., Lt.-Col. 44, 133, 157, 193, 624, 872, 1635
Rootes, A. H., 2nd-Lt. . 326, 365, 1038, 1523
Rooth, A. V., Capt. 501a
Rope, C., 2nd-Lt. 361
Roper, C. R., Sub-Condr. . . 1856
——— C. R., Lt. 731
——— E. W., 2nd-Lt. . . . 740
——— G. E., 2nd-Lt. . . 550, 1677
——— J., Capt. . . 36, 738, 791
——— R. B., Lt. . . 310, 1128, 1573
——— T. P., Lt. . . . 375, 1002
Ropner, G. V., 2nd-Lt. . . 325, 1043
Rorison, W. M., Lt. . . . 489
Rosair, H. B., Asst. Surgn. . . 1795
Rose, A. G., 2nd-Lt. 361
——— A. M., *Hony.* Lt. . . . 1846
——— A. S., 2nd-Lt. . . 77, 554, 1885
——— B. A., Lt. 843
——— C., Lt. 540
——— C. G., Lt. . 533, 1062, 1074
——— C. S., Col. . . 123, 650
——— D'A. L., Maj. . . . 734
——— E. D., Lt. 628
——— E. R., Lt. 652
——— F. U., 2nd-Lt. . . 543a, 1463
——— F. W., 2nd-Lt. . . 1463, 1678
——— G. A., Capt. 1839
——— G. C. G. F., 2nd-Lt. . 559, 1403
——— G. H., 2nd-Lt. . . 549, 1675
——— G. W. N., Lt. . 510, 840, 986
——— G. X. A., Asst. Surgn. . . 1803
——— H., Col. 57
——— H. A., Lt.-Col. . . . 110
——— J., Condr. 1853
——— J. L., Brig.-Genl. 121, 131, 155, 191
——— L. W., Ty. Asst. Surgn. . 1804
——— P. T., Lt. . . . 520, 656
——— R., 2nd-Lt. 1701
——— R. E., Lt. . . . 540a, 1701
——— R. P., 2nd-Lt. . . . 361
——— S. G. S., Lt. . . 351, 968
——— T. S., Sub-Asst. Surgn. . . 1820
——— W., 2nd-Lt. 1265
——— W. A., Lt. . . . 489, 875
Rose-Gordon, G., 2nd-Lt. . . . 727
Rose-Meyer, J. F. C., Asst. Surgn. 1801
Rose-Smith, W., Capt. . . . 485
Roseveare, E., Capt. . . 61, 698

NAMES AND RANK. Page.

——— R. V. H., Lt. . . . 623
Rosher, E. M., Capt. . . . 61
Roshan Khan, Subdr. . . . 1136
——— Jemdr. 1412
——— Lal, Sub-Asst. Surgn. . 1816
——— Ali, Subdr. 1371
——— Lal, Jemdr. 1051
Roskilly, S. T. H., Lt. . . 73, 695
Roskin, H. H., Lt. 714
Ross, A. C., Maj . . 51, 219, 913
——— A. D., Capt. 636
——— A. D., Asst. Surgeon . . 1803
——— A. J. H., Capt. . 288, 290, 1619
——— A. L., 2nd-Lt. . . . 1679
——— A. N., Asst. Surgn. . . 1796
——— A. T., 2nd-Lt. . . . 723a
——— C., Lt. 1136
——— D. A., Lt. . . . 542, 1655
——— D. B., Maj. . . . 238, 1094
——— D. J., 2nd-Lt. 557
——— C. E. W., 2nd-Lt. . . 539, 1512
——— C. G., Lt. 787
——— E. F., Maj. 240, 1197, 1359
——— E. J., Capt. . . . 253, 1635
——— G. F., 2nd-Lt. 802
——— G. W., Maj. . . . 87, 215
——— H., Maj. . . . 4, 1749, 1890
——— H., Lt.-Col. 135, 158, 195, 871, 1415
——— H. S., 2nd-Lt. . 556, 921, 1455
——— J. A., Lt. 805
——— J. A., 2nd-Lt. 763
——— J. B., 2nd-Lt. 1677
——— J. C., 2nd-Lt. 1106
——— J. D., 2nd-Lt. . . 553, 1371
——— J. H. C., 2nd-Lt. . . . 494
——— J. M., 2nd-Lt. 763
——— J. R., 2nd-Lt. . . 332, 1114
——— J. W., Lt. 9, 1689
——— K. M., Lt. 517
——— N. H., Lt.-Col. . . 133, 1710
——— N. McL., Maj. . . . 651
——— R., *Hony.* Maj. . . . 1844
——— R. C., Bt.-Lt.-Col. 9, 16, 108, 141, 162, 227, 1038
Ross, R. D., Ty. Asst. Surgn. . . 1804
——— R. D., 2nd-Lt. 731a
——— R. E., Lt. 974
——— R. P., 2nd-Lt. 1028
——— R. W., Capt. 765
——— S. B., Asst. Surgn. . . . 1804
——— T. R., Lt. . . . 518, 1670
——— T. B., Lt.-Col . . 140, 1746
——— W. B., 2nd-Lt. . . 763, 782
——— W. C., Maj. 1749
——— W. E. T., 2nd-Lt. . . . 693
——— W. H., Col. 176
——— W. J., 2nd-Lt. . . 548, 1702
——— W. R., Lt. 1919
Ross-Counsell, L., 2nd-Lt. . . 548
Ross Lewin, F. H. W., Lt. 530, 784
Ross-Hurst, B. P., 2nd-Lt. 327, 1243, 1245
——— W. J. O. B., Lt. 309, 1255, 1543
Rossel, H. G. T., Lt. . . 296, 1275
Rosser, J. C., Condr. . . . 1872
Rossiter, B. E., Condr. . . 1861
——— F. N. C., Capt. . 8, 636
Rost, E. B., Lt.-Col. . . 134, 1744
Rostron, H., Lt. . . 532, 1675, 1283
Roth, E. J. H., Lt. . . 410, 931
Rothe, C. W., Lt. 624
Rothera, G. H., Lt. . . 489, 909
——— P., Capt. 843
Rotheram, W., Ty. Lt. . . . 1726
Rothwell, A. D., Maj. . . . 634
Rotton, J. G., Br.-Genl. 121, 126, 613
Rouch, E. A., Lt. 507
Rouche, G. P. Lt. 423
Roughead, F. A., 2nd-Lt. . . 595
——— A. O. B., Lt.-Col. . 1710
Roughton, F. A. G., Maj. . 244, 1454
——— G. E., Lt. . . 354, 1673
Round, G. E. G., Lt. . . 401, 1007
Rountree, A. Fitz G., Lt. . 90, 5381
Rourke, D., *Hony.* Lt. . . . 761
Routhledge, L., G. F., 2nd-Lt. . 669
Routliff, J. J., Lt. 1667
Rouquette, H. R. H., Capt. . . 636
——— L. P., Lt. 981
Rouse, A., Capt 1840
——— A. H. T., Maj. . . . 1919
——— H., Col. . . . 124, 612
——— R. P., Lt. . . . 322, 1629
——— T. A., 2nd-Lt. . . . 1673
Roussel, P. B., Capt. . . . 769
Routh, C. F. R., 2nd-Lt. . . . 56

NAMES AND RANK. Page.

——— C. I., 2nd-Lt. 539
——— G. M., Maj. . 74, 102, 105, 108, 634
——— G. S. F., Maj. 212, 869, 1241, 1247
——— L. M., Capt. 1717
Routleff, J. L., Lt. . . . 47, 524
Row, C. K., Tempy. Lt. . . . 1780
——— G., Lt.-Col. 44
——— G., R., Lt.-Col. . . . 1898
——— R., Ty. Lt.-Col. . . . 1763
Rowan, T., Capt. 849
Rowan-Robinson, J. R., Bt.-Lt. Col. 160, 225, 1662
Rowbotham, A. S., 2nd-Lt. . 67, 786
——— C. J., Lt. 524
Rowcliffe, N. H. B., Lt. 40, 523, 1309
Rowcroft, C. H., Lt.-Col., 139, 162, 199, 891
——— G. C., Maj.-Genl. . . 172
——— G. F., Tempy. Maj. . . 1763
Rowe, G. C., 2nd-Lt. 986
——— H. C., 2nd-Lt. 1263
——— J. B., Asst. Surgn. . . . 1803
——— M., 2nd-Lt. . . . 696, 697
——— S. A., 2nd-Lt. . . . 551, 1375
——— S. G., Lt. 514, 1104
——— V. G., Lt. 1050
——— W. C., Lt. . . . 397, 1303
——— W. H., Lt. . . . 353, 1030
——— W. J., Asst. Surgn. . . . 1802
——— W. T. M., Lt. . . 543, 1493
Rowechapple, W. C., 2nd-Lt. . . 759
Rowell, J. W., Lt. 516
Rowland, C., Lt. 723
——— C. J., Lt. . . . 401, 1413
——— J. E. M., Lt. 832
——— J. W., Lt. . . . 507, 1120
——— T. A., 2nd Lt. 1693
Rowland-Hill, T., Capt. . . . 658
Rowlands, I. B. H., Lt. . . . 397
——— W. S., 2nd-Lt. . . . 803
Rowlandson, F., Esq. . . . 1906
——— H. W., Maj. 105, 219, 1335, 1893
——— M. A., Col. . . . 171
——— M. S. D., Lt.-Col. 51, 98, 105, 136, 158, 215, 1174
Rowley, B. G., Lt. . . . 295, 1002
——— E. D. T., Lt. . . 7, 718, 1894
——— F. H. J., Asst. Surgn. . . 1800
——— G. S., Lt. 589
——— J. T., 2nd-Lt. . . . 530, 1701
——— R., *Hony.* Lt. . . . 748
——— W. J. D., 2nd-Lt. . . 539, 1267
Rowlinson, O. A., Lt. . . 360, 1327
——— P. J., 2nd-Lt. 1207
Rowney, F., Lt. 637
Rowntree, H. K., Capt. . . . 1757
Rowse, H. C., 2nd-Lt. . . . 722
——— T. A., Lt. 540a
Rowsell, V. N., 2nd-Lt. . . . 825
Rowson, C. M., 2nd-Lt. . . . 561
——— J. C., Lt. . . . 311a, 1120
——— W. S., Lt. 522
Rowton, J., 2nd-Lt. 563
Roxburgh, G. A., Lt. . . 298, 947
——— T. J. Y., Lt. 525
——— W., Lt. 31a, 1120
——— W. A., 2nd-Lt. . . . 790
Roy, A. C., Tempy. Lt. . . . 1780
——— C. B., 2nd-Lt. 560
——— J. N., Esq. 91
——— K. S., Ty. Lt. 1774
——— M., 2nd-Lt. 1405
——— M. H. 2nd-Lt. 561
——— N. C., Sub-Asst. Surgn. . . 1820
——— P. C., Capt. 1758
——— R. M., Lt. . . 291, 1307, 1339
Royal, A., Esq. 89
Royal Dawson, F. G., Lt.-Col. . 836
——— H., Capt. 654
Royan, A., 2nd-Lt. 1703
Royds, G. F., Capt. 738
Royston, C. G., Capt. . 276, 909, 957
Rozario, C. J., Asst. Surgn. . . 1801
——— G. S., Asst. Surgn. . . 1801
Rozdon, S., *Hony.* Tempy. Lt. . 1773
Rozier, C., Esq. 88
Rubie, C. B., Capt. 39
Ruck, C. F. L., Lt. . . . 292a, 1000
——— O. L., Maj. . . . 139, 1239
——— R. C., Capt. 108, 267, 1387, 1401
Ruda Ram, Subdr. . . . 1456, 1479
Rudd, C., 2nd-Lt. 714
——— C. H. C., Lt. . . . 397, 1187
——— N. B., Lt. 509, 1215

NAMES AND RANK. Page.

Ruddle, W. H., Capt. 841
Ruddock, W. B., *Hony.* Maj. 1847
Rudduck, W., 2nd-Lt. 362
Rudkin, B. A., Capt. 264, 1327
Rudman, F. B. B., Lt. 803
Rudolph, W., Lt. 465, 1126
Rudrnarain Nagarkoti, Subdr.-Maj. 1596
Ruegg, A. J., *Hony.* Maj. 6, 14, 571, 1850
Ruel, W. G., Maj. 221
Ruffell, B., Lt. 309, 1379
Ruffer, R., 2nd-Lt. 683
Ruga Ram, Jemdr. 1585
Ruggles, J., Maj.-Genl. 1094
Rugha Singh, Subdr. 1039
Ruglys, H., 2nd-Lt. 787
Rukam Singh, Jemdr. 1070
Rukan Din, Risldr. 908
Rukman Sahi, Sub.-Maj. 1636
Ruknuddin, Jemdr. 876
Rule, D., 2nd-Lt. 1702
—— D. D. P. 2nd-Lt. 567
—— D. G., Capt. 274, 1128, 1573
—— G. C. S., 2nd-Lt. 722
—— W. N. P., Lt. 543, 1699
Rulya, Jemdr. 1092
Rumal Singh, Jemdr. 900
Rumbold, H., 2nd-Lt. 579
—— W., *Hony.* Capt. 1645
Runcorn, J. E., Sub-Condr. 1870
Rundall, F. M., Col. 179, 1917
—— J. W., Lt. 321, 365, 1595
—— L., Capt. 1910
Rundle, A. L. de V., 2nd-Lt. 330, 1595
—— C. A. G., Capt. 270, 1509
—— F. P., Lt.-Col. 138, 661
—— H., Lt. 543
Rundlett, L. St. C., 2nd-Lt. 506
—— R. C., 2nd-Lt. 556, 1054
Rudr Singh, Subdr. 1497
Rungasami, Subdr. 1323
Rung Shah, Subdr. 1355
Runnalls, H. G., 2nd-Lt. 360, 1361
Rup Chand, Subdr. 1039
—— Jemdr. 902
—— Bisht, Jemdr. 1179
—— Singh Bisht, Jemdr. 1179
Rupa, Jemdr 1195
—— Singh, Subdr. 1243
—— Jemdr. 904
Rupdal Rai, Jemdr. 1631
Rupdhan Pun, Jemdr. 1008
Rup Sing Khandka, Subdr. 1642
—— Jemdr. 1412
—— Jemdr. 1149
—— Singh, Jemdr 898, 1151
—— Negi, Jemdr. 1224
—— Thapa, Jemdr. 1008
—— Subdr. 1636
—— Sing Rawat, Jemdr. 1179
—— Sing, Jemdr. 1192
—— Jemdr. 1192
Ruqam Din, Subdr. 1259
Rui Singh, Ressdr. 932
—— Subdr. 1503
—— *Hony.* Capt. 2103
—— Subdr. 1239
—— Jemdr. 1075
—— Jemdr. 1284
—— Jemdr. 920, 930
—— Jemdr. 938
—— Jemdr. 1095
—— Jemdr. 1347
—— Jemdr. 989
Rura Ram, Subdr. 1489
—— Jemdr. 1212
Rure Khan, Subdr. 1363
Rurh Singh, Risldr. 1084
Rusby, L. S., Lt. 397, 1611
Rush, J. S., *Hony.* Maj. 1492, 1844
—— W., Condr. 1853
Rushbrooke, A. V., 2nd-Lt. 428, 1009
—— R. B. W., Capt. 708
Rushforth, F. V., Lt. 90, 512
Rushton, E. H., 2nd-Lt. 563, 1678a
—— H. W., Brig.-Genl. 58, 122, 132, 661, 1881
—— J. P., 2nd-Lt. 545, 1673
—— S. K., *Hony.* Lt. 1850
Russell, A. A., Lt. 315, 1195
—— A. G., Lt. 393, 1365
—— A. J. H., Capt. 1754
—— A. M. H., Lt. 42, 488, 879
—— B. H., Lt. 518, 683, 1668

NAMES AND RANK. Page.

Russell, C., 2nd-Lt. 1609
—— C. A., Capt. 636
—— C. J., *Hony.* Lt. 777
—— C. R., 2nd-Lt. 1048
—— C. S. C., Lt. 1708
—— D., Lt. 380, 1397
—— E. C., Maj. 79, 1838
—— E. P., Lt. 529, 998
—— E. T., 2nd-Lt. 550, 1701
—— E. W., Lt. 818
—— F., 2nd-Lt. 587, 1008
—— F. D., Lt.-Col. 53, 98, 102, 135, 158, 196, 867, 875
—— G. A., 2nd-Lt. 360, 1389
—— G. H., Maj. 95, 240, 1501
—— H. B., Lt. 330, 1433
—— H. G., 2nd-Lt. 545, 1673
—— H. L., 2nd-Lt. 331
—— J. B., Capt. 501
—— J. G., Lt. 311a
—— J. M., Lt. 519, 1046
—— J. S., 2nd-Lt. 355
—— M., Lt. 789
—— O. B. P., Lt. 300, 893
—— P., Lt. 697
—— R. B., 2nd-Lt. 1008
—— R. E., Lt. 84, 516a
—— R. T., Lt. 521, 1641
—— R. W., Capt. 271, 1641
—— T. G., 2nd-Lt. 818
—— V. H., Lt. 752
—— W. H., 2nd-Lt. 568
—— W. K., Maj. 60, 662, 1882
Rust, F. H., Capt. 486
—— L. D., Lt. 487, 911
—— N. A., Lt. 525
—— T. H. 1675
Rustam Ali, Sr. Sub-Asst. Surgn. 1808
—— Khan, *Hony.* Capt. 2103
—— Singh, Jemdr. 1021
Rustom Khan, Jemdr. 950
Rutherfoord, C., Tempy. Capt. 1724
Rutherford, A., Tempy. Capt. 1724
—— E. V., 2nd-Lt. 598
—— H. R., 2nd-Lt. 564, 1395
—— J. S., Lt. 430, 1078
—— R., Maj. 1711
—— T. G., Lt. 493, 947
—— T. W., Maj.-Genl. 172
Ruthven, H. P. L., Lt. 111
—— J. L., 2nd-Lt. 547
—— W. J., *Hony.* Capt. 35
Rutledge, C. J., Lt. 401, 1501
Ruttledge, L. H. N., Lt. 401, 1636
—— R. F., 2nd-Lt. 330, 939
Ruzzak, S. A., Maj. 1307, 1747
Ryall, C. R., Revd. 1907
—— H. H., Lt. 1166
—— T., Sub-Condr. 1855
—— W. D., Capt. 767
Ryan, B. A., Lt. 314, 1613, 1621
—— C. C. L., 2nd-Lt. 315, 913, 961
—— D. G. J., Capt. 257, 1623
—— E. J., *Hony.* Maj. 804, 1845
—— E. J., Lt. 1335
—— H. W., Maj. 709
—— P., Condr. 1867
—— P., 2nd-Lt. 668, 1008
—— P. C., Condr. 1851
—— P. C. H., Ty. Capt. 1724
—— R., 2nd-Lt. 1383
—— T., 2nd-Lt. 355
Rybot, N. V. L., Maj. 212, 1319
Rycroft, H. Z., Lt. 860
Ryde, A. A., 2nd-Lt. 356, 1241
Ryder, C. H. D., Col. 103, 130, 660, 1880
—— F., 2nd-Lt. 754
—— E. N., Lt. 17, 21, 597
—— C. F., 2nd-Lt. 599
—— R. C., Lt. 514, 1614
—— W. L., Lt.-Col. 42, 118, 154, 185, 872, 1598
Rylands, J., 2nd-Lt. 553
Ryle, C. D., Capt. 843
Ryley, L. F. M., 2nd-Lt. 365, 937
Ryles, C., Capt. 22, 1713
Rynd, F. C., Lt.-Col. 57
—— F. F., Maj. 102, 633
Ryper, G. H., 2nd-Lt. 328, 879
—— H. J., 2nd-Lt. 551, 1054
Ryrie, G., 2nd-Lt. 744
Ryves, B. H., Maj. 209, 1661
—— C. H., 2nd-Lt. 556, 1265

NAMES AND RANK. Page.

S

Saadat Ali, S., Sub-Asst. Surgn. 1814
—— Husain, Jemdr. 966
Sabaji Rao Rani, Jemdr. 1469
Sabai Singh, Subdr. 1419
Sabbatella, F., 2nd-Lt. 786
Sabhapathi, Subdr.-Maj. 1699
Sabawala, B. P., Ty.-Capt. 1768
Sabhesan, D. K.I.K., Tempy. Lt. 1778
Sabin, W. H., 2nd-Lt. 794
Sabine, H. E., 2nd-Lt. 355, 1433
Sabit Khan, Ressdr. 945
Sachin, Nawab of, *Hony.* Lt. 35, 111
Sachindra Kumar Roy, Jemdr. 1219
Sackett, F. C., 2nd-Lt. 566
Sacre, L. G. B., 2nd-Lt. 691
Sada Ram, Sub-Asst. Surgn. 1819
—— Singh, Subdr. 989
—— Sukh, Subdr. 1403
Sadanundan Singh, Ressdr. 1686
Sadar Din, Jemdr. 953, 989
—— Jemdr. 1260
—— Khan, Jemdr. 1460
—— Ud Din Shah, Ressdr. 953
Sadashio Parab, Jemdr. 1469
Sadashiv Rao Surve, Jemdr. 1444
—— Waman Gupte, Ty. Sub-Asst. Surgn. 1833
Sadashiva Rau, K. R., Esq. 91
Sadasiva Naidu 1340
Sadda Singh, Subdr. 1112
—— Subdr. 1587
—— Subdr.
—— Jemdr. 1587
—— Jemdr.
—— Jemdr. 1145
Saddha Singh, Risldr. Maj. 882
Sadhoo Singh, Jemdr. 906
Sadhu, Jemdr. 1171
—— Singh, Ressldr. 922
—— Jemdr.
—— Jemdr. 1103
—— Subdr. 1120
—— Jemdr. 1151
—— Jemdr. 932
—— Jemdr. 1252
—— Jemdr. 1195
—— Jemdr.
Sadik Aaron, Sr. Sub-Asst. Surgn. 1829
—— Muhammad Khan, Risldr. 944
Sadajirao Kadam, Jemdr. 1510
Sadleir, A. W. W., Sr. Asst. Surgn. 1788
Sadler, F., Sub-Condr. 1863
—— J. W. V., 2nd-Lt. 1629
Sdlier, J. W. V., 2nd-Lt. 365
—— R. W. C. T., Sub-Condr. 1861
Sadullah Khan, Tr. Vety. Asst. 1687
Safidar Khan, Jemdr. 1583
Safi Muhammad, F., Sub-Asst. Surgn. 1815
Sagar Singh, Jemdr. 1032
Sagat Singh, Rao-Raja 112
Sage, H., 2nd-Lt. 358, 1036
Sagon Naik, Jemdr. 1444
Saha, N. N., Tempy. Lt. 1870
Sahabir Gurung, Subdr. 1618
—— Thapa, Subdr. 1630
—— Limbu, Jemdr. 1649
Sahab-ud-din, Jemdr. 1141
Sahadeo Ramchandra Khanolkar, Sub-Asst. Surgn. 1829
Sahdeo Sawant, Subdr. 1417
—— Singh, Jemdr. 1047
Sahaman Thapa, Jemdr. 1649
Saharmukhi Limbu, Subdr. 1648, 1655
Sahbir Gurung, Subdr. 1618, 1619
Sahedeo, R. K., Sub-Asst. Surgn. 1829
Sahib Alam, Jemdr. 1506
—— Bakhsh Singh, Subdr. 1046
—— Dad, Jemdr. 920
—— Din, Jemdr. 990
—— Gul, Jemdr. 1231
—— Jemdr. 1121
—— Haq, Subdr. 1259
—— Khan, Subdr.-Maj. 1145
—— Shah, Subdr. 1096
—— Singh, Jemdr. 1163
—— Jemdr. 916, 944
—— Jemdr. 1380

NAMES AND RANK — Page.

Sahib Singh, Jemdr. 955
—— Ressdr. 890
—— Sub-Asst. Surgn. . 1815
Sahi Ram, Jemdr. 1391
—— Singh, Jemdr. . . . 1412
Sahu, R., Sr. Sub-Asst. Surgn. . 1826
Saich, H. C., 2nd-Lt. 793
Said Ahmad, Jemdr. 1506
—— Ghulam Khan, Jemdr. . 1344
—— Jemdr. . . 128
——, Subdr. . . 1305
—— Gul, Subdr. 1098
—— Hussain, Jemdr. . . . 1154
—— Habib, Jemdr. 1518
—— Muhammad, Subdr. . . . 1280
—— Jemdr. . . . 1502
—— Jemdr. . . . 1506
—— Jemdr. . . . 1071
—— Raja, Jemdr. 908
Said-ud-din Ahmad, Sub-Asst. Surgn. 1820
Saidulla Khan, Jemdr. . . . 895
Said Alam, Jemdr. 999
Saka Ram Niram, Jemdr. . . . 1416
—— Patole, Jemdr. . . . 1469
Saida Khan, Subdr. 1363
—— Jemdr. 1095
—— Jemdr. 1107
—— Jemdr. 1137
—— Subdr. 989
—— Jemdr. 894
Saidal Khan, Jemdr. . . . 1571
Saiidullah Khan, Subdr. . . . 1518
—— Risldr. . . . 934
Saif Ali, Subdr. 1518
—— Ullah, Jemdr. 1495
—— Jemdr. 1681, 1682
—— Khan, Jemdr. . . . .
Saifud-din, Sr. Sub-Asst. Surgn. . 1807
—— Saifur, Jemdr. 1244
Saifullah, Jemdr. . . . . 1495—96
Saifur Khan, Jemdr. 1012
Saigol, R. D., Maj. 1750
Saileswar Singh Roy, Babu, 2nd-Lt. 481
St. Aubyn, J. H. A. G., Lt. . . 415, 1591
St. Clair, L. F. C., Lt. 597
St. John, C. R., 2nd-Lt. 1104
—— F. O., Capt. 1004
—— H. B., Maj 207, 1890, 1899, 1913
—— H. B., Capt. 741
—— O., Maj. . . 235, 1315, 1383
—— P. G., 2nd-Lt. 556
—— R. S., Br.-Genl. 41, 98, 122, 136, 158, 210, 1098
St. Maur, F. P., Lt. 598
—— R. W. M., Lt. . . 367, 901
St. Maur-Wynch, H. St. M, Col. 176
St. Pourcain, E. A., 2nd-Lt. 560, 1289
St. Romaine, E. A., Senr. Asst. Surgn. . . 1787
—— E. C., 2nd-Lt. . 547, 1701
Sainsbury, H., Maj. 708
Saint, J. B., 2nd-Lt. 725
Sainter, J. D., Lt. . 307, 307a, 1279
Saiyad, Ghulam Ali, 2nd-Lt. Hony. 482
Saiyid Abbas Haidar, Sub-Asst. Surgn. 1819
—— Abdul Basit, Sr. Sub-Asst. Surgn. 1808
—— Akhtar Husain Naqvi, Sub-Asst. Surgn. 1820
—— Abbas Haider, Sub-Asst. Surgn. 1819
—— Abdul Jafar, Jemdr. . . 994
—— Jalil, Jemdr. . . . 924
—— Kadir, Moulvi, Sr. Sub-Asst. Surgn. . 1829
—— Latif, Sub-Asst. Surgn. . . 1827
—— Abdul Sattar, Jemdr. .
—— Wahab, Sub-Asst. Surgn. . 1816
—— Ghafur, Subdr. 1263
—— Abdullah, Risldr. . . .
—— Subdr. . . . 1423
—— Jemdr. . . . 1273
—— Ahmad, Sub-Asst. Surgn. . 1812
—— Jemdr. . . . 1480
—— Subdr. . . . 1273
Saiyed Khan, Jemdr. 1117
—— Akhmad Shah, Jemdr. 916, 955
—— Akhtar, H. N., Sub-Asst. Surgn. 1820
—— Ali, Hony. Capt. 2104
—— Naki, Sub-Asst. Surgn. 1807
—— Asghar Ali, Sub-Asst. Surgn. 1816

Saiyed Faiyaz, H. R., Sub-Asst. Surgn. 1819
—— Husain, Hony. Capt. . . 2101
—— Hashim, Jemdr. 1384
—— Irshad Husain Alavi, Sub-Asst. Surgn. 1817
—— Ishtiyaq Ali, Sub-Asst. Surgn. 1820
—— Ismail, Tr. Vety. Asst. . . 1687
—— Jafar Husain, Sub-Asst. Surgn. . . 1816
—— Kadir, Subdr. 1339
—— Kamal, Tr. Vety. Asst. . . 1687
—— Mohi-ud-din, Subdr. . . . 1339
—— Muhammad, Jemdr. . . 1502
—— Ali, Sub-Asst. Surgn. . . 1816
—— Ejaz, Sub-Asst. Surgn. 1812
—— Ahmad Naqvi, Sub-Asst. Surgn. 1819
—— E. Z., Sub-Asst. Surgn. . . 1820
Saiyid Munsin Shah, Risldr. . 945
—— Nazim Ali, Sr. Sub-Asst. Surgn. 1808
—— Padshah, Subdr.-Maj. . . 1447
—— Subdr. . . . 1383
—— Qamer Ali, Sub-Asst. Surgn. 1818
—— Sa-adat Ali, Sub-Asst. Surgn. 1811
—— Sikandar, Jemdr. . . . 135
—— Suliman, Subdr. . . . 1423
—— Turab Ali, Sr. Sub-Asst. Surgn. 1807
—— Yasin, Subdr. 1323
Sajan Singh, Jemdr. 1236
Sajawal Khan, Jemdr. 1344
—— Jemdr. 1579
—— Subdr. 1001
—— Jemdr. 974
Sajid Gul, Risldr. 908
Sakaramrao Surwe, Jemdr. . . 1464
Sakaram Nimak 1417
Sakaldip Singh, Subdr. . . . 1082
Sakhawat Khan, Jemdr. . . . 1090
Sakt Chand, Sub-Maj. 1116
Sakat Singh, Subdr. 1371
Sakawat Khan, Jemdr. 1092
Sakharam Gaikwad, Jemdr. . . 1408
—— Kalbhar, Jemdr. . . 1408
Sakka Ram Patole, Subdr. . . 1496
—— Powar, Subdr. . 1460—62
—— Rao Jagtap, Subdr.-Maj. 1467, 1469
—— Sanu Parab, Sub-Asst. Surgn. 1831
Sakhawat Khan, Jemdr. . . . 890
Sakhi Muhammad, Jemdr. . . . 1364
—— Zaman, Jemdr. . . . .
Salab Singh, Jemdr. 1196
Salagram, Sub-Asst. Surgn. . . 1810
Salah Muhammad, Subdr. . . . 1495
Salam Din, Jemdr. . . . . .
Salberg, F. J., Lieut. 506
Saldanha, A., Ty. Lt. 1773
—— J. L., Tempy. Lt. . . 1776
——, M. J., Ty. Lt. . . 1773
Sale, A. L., Maj. 793
—— C. A., 2nd-Lt. . . . 474, 1639
—— J. L., Lt. 510, 986
—— W. G., Lt. . . . 320, 365, 1497
Sales, H. D., 2nd-Lt. 1227
—— J. J., 2nd-Lt. 505
Salew, G. A., Hony. Capt. . 1453, 1885
Salig, Jemdr. . . . . . .
—— Ram, Sub-Asst. 1816
—— B. M. V., Sub-Asst. Surgn 1818
—— Singh, Jemdr. 1397
Salim Khan, Jemdr. 916
—— Jemdr. 1480
—— Jemdr. 1423
Salisbury, F. H., Maj. 1753
Salkeld, H. Y., Maj. . 227, 877, 1289
Salkield, T., Maj. 805
Sallai, C. G. 2nd-Lt. . . . 465, 887
—— F., 2nd-Lt. 1491
Salmon, F. H., 2nd-Lt. . . . 365
—— G. H., Lt. 698
—— T. F., Hony. Capt. . . 1850
—— W. H., Col. . . . 179, 1917
—— W. H. B., Maj. . . 220, 1662
Salmond, H. A. B., Lt.-Col. . . 1009
Salt, A. C. J., 2nd-Lt. . . 470, 888
—— G., Lt. 303, 1307
—— J., 2nd-Lt. 360

Salt, J. J., Lt. 302, 1032
—— J. S., Asst. Surgn. . . . 1795
—— R. F., Lt. 397, 1178
Saltanat Bahadur Singh, Subdr. 1082
Saltanat Singh, Jemdr. . . . 1063
Salter, E., Lt. 318, 1606
—— M. G., Lt. 491
Salusbury, R. T. G., Maj. . 41, 223, 1124
Salusbury-Trelawny, Lt., J. M. 510, 986
Salvesen, F. M. R., Lt. 302, 1335, 1539
—— H. K., Lt. . . 293, 1191
Salvin, H. C. J., Maj. 634
Salway, H., Lt. 1044
Saman Singh, Jemdr. 1150
Samand Khan, Jemdr. 1492
Samandar Khan, Jemdr. . . . 1268
—— Jemdr. . . . 1681
—— Jemdr. . . . 1236
—— Jemdr. . . . 1432
Samarbahadur Jemdr. 1643
Samardhoj Rai, Jemdr. . . . 1649
Samborne-Palmer, F. C. S., Lt.-Col. 20, 134, 157, 194, 868, 1052
Samdu La, Jemdr. 1347
Samih-Ullah, Jemdr. . . 1510, 1589
—— Jemdr. 1587
Saminathan Pillai, Jemdr. . . 1316
Samler, W. H. G., Lt. . . . 702
Sammunder Khan, Ressdr. . . 1684
—— Shah, Jemdr. . . . 1682
Sampey, A. W., Maj. 1711
Sampson, C. A., 2nd-Lt. . . 363, 1116
—— E. D., Maj. 846
—— H. S., Lt. 723
—— J. C., Tempy. Capt. . . 1767
—— J. E., 2nd-Lt. . . . 563
—— R. W. K., Lt. . . 531, 1672
Sampuran Singh, Sub-Asst. Surgn. 1814
Samrath Singh, Jemdr. . . . 874
Samrat Singh, Subdr. 1488
Sams, H. A., Lt. 803
Samuel, C. M., Lt. 622
—— David, Subdr. . . . .
—— Jemdr. 1308
—— Subdr. Maj. . . . 1333
—— G. R., Sub-Asst. Surgn . 1827
—— Manikkam, Sr. Sub-Asst. Surgn 1826
—— P., Jemdr. 1351
—— Peter Joseph, Sub-Asst. Surgn. 1819
—— Jemdr. 1308
—— Aide, Jemdr. 1456
—— Subdr. 1315
Samuels, C. J. H., 2nd-Lt. . . 1413
—— L., Lt. 698, 796
Samund Sing, Jemdr. 918
Samundar, Subdr. 1432
—— Jemdr. 956
—— Khan, Jemdr. . . . 990
—— Jemdr. 966
—— Subdr. 1126
—— Jemdr. . . . .
Samways, J., Sub-Condr. . . . 1863
Sanak Sing Limbu, Jemdr. . . 1649
Sandanasami, L. D., Jemdr. . . 1012
Sandeman, A. P., Maj. 216, 1895, 1913
—— C. S., Capt. . . . 1717
—— D. G., Maj. . . 243, 964
—— J. E., Col. . . 178, 1917
—— R. F., 2nd-Lt. . . . 315
Sanders, A., Sub-Condr. . . . 1856
—— B. G. T., Esq. . . . 88
—— C. F., 2nd-Lt. . . . 790
—— C. W., Capt. 272, 1016, 1026
—— D. B., Capt. 1009
—— E. T., 2nd-Lt. . . . 90
—— G. A. F., Br.-Genl. 50, 98, 122, 133, 661
—— G. L. H., Col. 125, 150, 184, 1660
—— G. P., Maj. . . 221, 1617
—— H., Lt. 515
—— H. H., Lt. . 546, 852, 1537
—— H. R., Lt. 779
—— H. W., Lt. . . . 541, 1696
—— J. H., Lt. 489
—— R. G., Lt. . 291, 1174, 1585
—— W. G., Lt. . . . 514, 1677
Sanderson, E. F., Condr. . . . 1853
—— F. G. O., Maj. . 224, 1501
—— F. H., Lt. 1839
—— G., 2nd-Lt. 552
—— G. M., 2nd-Lt. . . . 693

Names and Rank. Page.

Sanderson, G. W., Condr. . . 1867
—— J. F. B., Lt. . . . 1898
—— P. C., Lt. . . 317, 1613
—— R., Maj. 806
—— R. J., 2nd-Lt. . . 626
—— R. J., Condr. . . . 1854
—— T., Sub-Condr. . . 1856
Sandes, E. W. C., Capt. 664, 815, 1898
—— F., 2nd-Lt. 1009
—— J. D., Maj. 1753
Sandford, H. R., Capt. 665
—— W. S. B., Capt. . 82, 485
Sandham, W. F. N., Lt. 318, 1279, 1305
Sandhya Das, *Hony*. Capt. . . 2103
Sandiford, V. V. V., Maj. 227, 869, 1219, 1267
Sandison, J., 2nd-Lt. . . 90, 695
—— W., 2nd-Lt. 784
Sandover, W. L., Maj. . . . 721
Sands, E. 2nd-Lt. 1008
—— E. S., 2nd-Lt. 637
Sandways, P. N., Asst. Surgn. . 1794
—— W. G., Asst. Surgn. . 1798
Sandy, A. E. L., 2nd-Lt. . . . 1419
—— W. J. W., Lt. 1579
Sandys, M. K. T., Lt. 722
—— U. C., Capt. 836
Sanford, G. B., Lt.-Col. . 7, 196, 1896, 1912
—— S. A., Capt. 583
Sangara, Subdr. 1247
Sangat Singh, Jemdr. 942
Sangili, Jemdr. 1264
Sangster, D. B., 2nd-Lt. . 325, 366, 1227
—— P. B., Lt.-Col. 136, 158, 196, 867, 929
——Green, 2nd-Lt. G.C., L. 1679
Sanimathan, Jemdr. 1317
Sanjeeva Menon, Jemdr. . . . 1309
Sanjam Singh, Jemdr. 878
Sankaran, T., Sub-Asst. Surgn. 1826
Sankarapandian Pillai, Subdr.
—— —— Kadam Jemdr. 1440
—— —— Khan, Jemdr. 1087
Sankar Singh Kathait, Jemdr. . 1182
Sankara Pondian, Jemdr. . . . 1265
Sankey, G. B., Capt. . 8, 15, 780
Sanman Singh, Jemdr. . . . 1054
Sanowar Khan, Jemdr. . . . 936
San Pe, Jemdr. 1295
Sansar Chand, Subdr. . . . 1231
—— Sub-Asst. Surgn. 1815
Sansaru, Jemdr. 1099
Sansom, A., Condr. 1854
—— A. E., Lt. . . 318, 949
—— H. J., Lt. . . 311a, 1487
Sant Ram Varma, Sub-Asst. Surgn. 1819
Sant Singh, Ressdr. 1684
—— Risldr. 893
—— Jemdr. 888
Santa Bir, Jemdr. 968
Santbir Gurung, *Hony*. Capt. 5, 2103
—— Subdr. . . . 1602
—— Rai, Subdr. . . . 1648
—— Rana, Subdr. . . . 1624
—— Jemdr. 912
Sant Singh, Sub-Asst. Surgn. . 1814
—— Sub-Asst. Surgn. . 1818
—— Jemdr. 1149
—— Risaldar. 893
—— Jemdr. 1336
—— Subdr. 1148
—— Subdr. . . . .
—— Subdr. 1157
—— Subdr. 1074
—— Jemdr. 1165
—— Jemdr. 1012
—— Subdr. 1166
—— Jemdr. 1204
—— Jemdr. 888
—— Jemdr. 976
—— Jemdr. 1099
—— Risdr. 919
Santa Singh, Jemdr. 1111
—— Jemdr. 1493
—— Jemdr. 916
—— Jemdr. 1071
—— Subdr. 973
—— Jemdr. 1502
—— Sub. Asst. Surgn. . 1817
—— Subdr. 1365
—— Jemdr. 1356

Names and Rank. Page.

Santa Singh, Risldr. Maj. . . 938
—— Subdr. 1259
—— Jemdr. 1095
—— Ressdr. 898
—— Jemdr. 1368
Santan Singh Bisht, Sub-Asst. Surgn. . 1820
—— Sub-Asst. Surgn. . 1810
Santbir Bura, Jemdr. 1602
—— Gurung, Subdr. . . . 1614
—— Rana, Subdr. . . . 1596
—— Thapa, Jemdr. . . . 1615
—— Subdr. 1624
—— Santiago, S., Jemdr. . . 1324
Santok Pun, Jemdr. 1604
Santokh Singh, Sub-Asst. Surgn 1811
—— Jemdr. . . . 1304
—— Jemdr. . . . 2031
—— Jemdr. . . . 1236
—— Subdr. . . . 1150
Santu, Jemdr. 1256
—— Chawan Jemdr. . . . 1436
San Tun, Subdr. 1297
—— U. Kaing, Jemdr. . . . 1293
—— Tha, Jemdr. 1293
Sanwal, Jemdr. 1060
—— Khan, Jemdr. 950
—— Singh, Jemdr. 1024
Sanwalia Singh, Subdr. . . . 1453
Sanwal Singh, Jemdr. . . . 1021
—— Jemdr. 1196
Sanwar Khan, Jemdr. . . . 1096
Sanyal, D., Ty. Lt. 1773
—— S., Ty. Lt. 1777
—— S. K., Lt. 1691
Sapuran Singh, Jemdr. . . . 1236
—— Subdr. 1114
—— Jemdr. 1075
—— Jemdr. 1143
—— Jemdr. 1376
—— Jemdr. 1577
—— Jemdr. 1248
—— Jemdr. 1158
—— Sub-Asst. Surgn. 1817
Sarabjit Gurung, Subdr. . . . 1601
Sarat Kumar Roy, Jemdr. . . . 1219
Sarbaz Sing Gurung, Jemdr . . 1637
Sarbadhoj Rai, Jemdr. . . . 1633
Sarbaland Khan, Jemdr. . . . 912
Sarban Singh, Subdr. . . . 1355
Sarab Singh, Jemdr. 916
Sarabjit Gurung, Jemdr. . . . 1602
—— Subdr. . . . 1611
—— Jemdr. . . . 1608
—— Singh, Jemdr. . . . 1083
—— Thapa, *Hony*. Capt. . . 2103
Sarbuland, Subdr. 1506
Sardar, Abdul Rahman Khan, *Hony*. Lt. 883
—— Ali, Jemdr. 912
—— Hira Singh, Risldr. . . 1883
—— Janmeja Singh, *Hony*. Capt. 2103
—— Khan, Jemdr. 1440
—— Kumar Prem Singh, Ressdr. 1686
Sardara (II), Jemdr. 1030
—— Muhammad, Jemdr. . . 1012
—— Raghbir Singh Sindhanwala . 481
—— Khan Subdr.-Maj. . . 1479
—— Subdr. 1405
—— Jemdr. 1681
—— Subdr.-Maj. . . . 1883
—— Jemdr. 1493
—— Jemdr. 1107
—— Subdr. 999
—— Jemdr. 999
—— Subdr. 1124
—— Subdr. Maj. . . . 1092
—— Sub-Asst. Surgn. . 1815
—— Ressdr. 892
—— Jemdr. 1241
—— Jemdr. 1321
—— Jemdr. 1682
—— Jemdr. 904
—— Subdr. 1288
—— II, Jemdr. 1288
—— Jemdr. 942
—— Jemdr. 966
—— Jemdr. 1563
—— Sahib Chowdhery
—— Tachitar Singh, Ressdr. 1686
—— Shah, Subdr. 1507
—— Jemdr. 1134

Names and Rank. Page.

Sardara, Singh Gill, Sub-Asst. Surgn. 1819
—— Singh, Ressldr. . . . 901
—— Jemdr. . . . 1022
—— Subdr. Maj. . . 1064
—— Sujan Singh, Ressdr. . 1683
—— Gill, Sub-Asst. Surgn. . 1819
—— Jemdr. 1039
—— Singh Kokari, Sub-Asst. Surgn. 1819
—— Jemdr. 1472
—— Jemdr. 1079
—— Subdr. 1039
—— Subdr. 1473
—— Singh Kokori, Sub-Asst. Surgn. 1819
Sarel, G.B. M., Lt.-Col. 136, 158, 197, 895
Sare-Soar, M. W. R., Lt. 538, 577
Sarfaraz Khan, Subdr. . . . 1247
—— *Hony*. Capt. . . 2102
—— Jemdr. 1117
—— Jemdr. 1436
—— Subdr. . . . .
Sargeant, J. W., 2nd-Lt. . . . 548
—— W., 2nd-Lt. 563
Sargent, A. E. E., Maj. . 224, 1662
—— E. L. K., Tempy. Capt. 1725
—— H. H., Lt. . . 540a, 977
—— H. L., Asst. Surgn. . 1800
—— W. N., Asst. Surgn. . 1800
—— W. P. M., Maj. 214, 868, 1064, 1419
Sarginson, G., 2nd-Lt. 793
Sargisson, W. F., Lt. . . . 488
Sargon, A. I., Capt. . . 501, 1459
Sarjerao Supekar, Jemdr. . . 1460
Sarjit Singh, Jemdr 1021
—— Jemdr. 878
Sarju, Subdr. 1189
—— Singh, Subdr. 1371
—— Subdr. 1047
—— Jemdr. 1405
—— Jemdr. 1281
Sarkar, D. L., Ty. Lt. 1775
—— S., Capt. 1755
—— S. C., Ty. Lt. 1778
Sarkies, O. St. C., Lt. . . . 846
—— C. S. S. Ty. Lt. . . . 1780
—— T. R. V., Esq. 91
Sar Mukh Singh, Jemdr. . . . 1075
Sarsfield, T. H., Capt. . . . 1715
Sartorius, E. H., Maj.-Genl. . . 741
—— G., Capt. 103, 260, 797, 885
Sanuria Gurung, Subdr. . . . 1621
Sarratt, N. D. 2nd-Lt. 706
Sarup Singh, Ressdr. 936
—— Ressdr. 894
—— Jemdr. . . . .
—— Jemdr. . . . .
Sarwan Singh 914
Sarwana, Subdr. 1191
Sarwar Khan, Subdr. 1497
Sasidhar Thapa, Subdr. . . . 1618
Sasiram Thapa, Subdr. . . . 1630
Sassi Mal, Jemdr. . . . 1643, 1655
Sassoon, A. M., Capt. . . . 55, 591
—— S. J., Capt. 585
Sastry, N. S. N., Ty. Lt. . . . 1779
—— T. S., Capt. 1758
Satal Singh Thapa, Subdr. . . 1624
Satar Muhammad, Subdr. . . 1489
Satchell, W., 2nd-Lt. . . . 78, 619
Sathe, V. L., Tempy. Capt. . . 1766
Satis Chandra, Mr. 3
Satnaram Singh, Jemdr. . . . 1047
Satpute, N. R., Tempy. Capt. . 1770
Sattar Shah, Ressdr. 939
Satterthwaite, R. E., Capt. 17, 42, 725, 783
Sattidin Sukul, Subdr. . . . 1026
Satyaranjan Bhowmick, Sub-Asst. Surgn. 1815
Satyendra Nath Aditya, *Rai Sahib* 89
Saubolle, C. E., 2nd-Lt. . . 559, 1191
Saudagar Singh, Subdr. . . . 1371
Saul, E. R., Capt. 1732
—— G. W. O., 2nd-Lt. . . . 794
—— J. B., Capt. 798
Sau Mun, Jemdr. 1347
Sau tu, Jemdr. 1347
Saunders, A., Capt. . . . 55, 501
—— A. R., Lt.-Col. 130, 155, 190
—— C., Capt. 710
—— C. D., Capt. 502
—— C. H. S., Capt. . 846, 847

Names and Rank. Page.

Saunders, C. J. G., Revd. . . . 1904
—— C. W., Lt. . . 310, 1435
—— E. E. L., Lt. . . 527, 1403
—— E. G., Ty. Capt. . . 1725
—— E. H., Maj. . . . 214, 1661
—— E. J., Lt. . . . 90, 532
—— E. T., 2nd-Lt. . 753, 754
—— F. W., Lt.-Col. 41, 134, 661
—— G. N., Lt. . 399, 683, 1140
—— H. A. C., Lt. . 545, 1674
—— H. F., 2nd-Lt. . . . 553
—— H. F., Capt. . . . 819
—— H. W. B., 2nd-Lt. 766, 731a
—— J., 2nd-Lt. . . 545, 1673
—— J. St. C., 2nd-Lt. . . 555
—— M., Capt. 52, 105, 252, 1166
—— M. C., 2nd-Lt. . . . 754
—— P. C., Maj. . . 221, 1662
—— R. A. R., Lt. . . . 517
—— T. H., 2nd-Lt. . 354, 1249
—— W., Lt. . . . 528, 988
—— W. E. P., 2nd-Lt. 327, 1189
—— W. H., Lt. . . 352, 1038
Saunderson, R. G., Lt. . . . 624
Saunders-Jones, D. W., 2nd-Lt. 328, 1321
Saurinayagam, Jemdr. . . . 1276
Sauriyappan, Jemdr. . . . 1276
Savage, C. T. 2nd-Lt. . 77, 742, 1885
—— E. G., Capt. . 280, 1413, 1451
—— E. G. Lt. . . 505, 1094, 1151
—— F. G. 2nd-Lt. . . . .
—— P., Capt. 1758
—— P. G., 2nd-Lt. 1066
—— R., Lt. 537, 1672
—— R. J., 2nd-Lt. 733
—— T. A., Capt. 636
—— W. C., Lt. . . . 543a, 1886
—— W. H., Col. 123, 149, 182, 280 —287, 1931
Savantvadi, Chief of, 2nd-Lt. . . 112
Savanur, Nawab Abdul Majid Khan of, Lieut. . 111, 874, 930
Saveall, E. W., 2nd-Lt. . . . 1679
Savary, J. R., Lt. 445, 1604
Savege, C. R., Lt. 532
Savereux, E., Lt. 682
Savi, R. J. T., Lt.-Col. . 81, 155, 190
Savile, A. R. C., Lt.-Col. 141, 162, 201, 870, 933, 1389
Savill, E. M., Lt. 731
Savory, A. B., 2nd-Lt. . . . 760
—— R. A., Capt. . . 278, 1074
Saw, G. A., 2nd-Lt. . . 547, 1765
Sawa Singh, Sub-Asst. Surgn. . 1812
Sawal Singh, Jemdr. . . . 912
—— Subdr. 1231
Sawali Singh, Jemdr. . . . 1051
Sawan, Singh, *Hony.* Capt. . . 2104
—— Sub-Asst. Surgn. . 1814
—— Subdr. 1148
—— Jemdr. . . 1699, 1704
—— Jemdr. 999
—— Jemdr. 1075
—— Jemdr. 896
—— Jemdr. 1117
—— Jemdr. 990
—— Jemdr. 1702
—— Tang, Jemdr. . . . 1347
Sawant Ram, Jemdr. . . . 1457
Sawar Khan, Jemdr. . . . 1142
—— Jemdr. 1433
—— Subdr. 1497
Saward, M. H., Maj.-Genl. . . 643
Sawdy, A. E. L., 2nd-Lt. . . . 365
Sawariya Khan, Jemdr. . . . 999
Sawers, C. M. O., 2nd-Lt. . 315, 929
Sawford, A. N., 2nd-Lt. . . . 357
—— Sawle, J. F., 2nd-Lt. . . 1104
Sawhney, M. R. Ty. Lt. . . . 1782
Sawle, J. F., 2nd Lt. . . . 476
Sawminathan, Subdr. . . . 1011
Sawtell, W. A., Revd. . . . 1907
—— W. G., 2nd-Lt. . . 560, 1427
Sawyer, A. J., 2nd-Lt. . . . 361
—— C. P. B., Lt. . . 389, 893
—— H. H., Maj. 657
—— H. T., Lt.-Col. . . 39, 1838
—— N. H., 2nd-Lt. . . . 29
—— P. S., Lt. . . . 543a, 980
—— W. H., 2nd-Lt. . . . 554
Saxby, H. B. Lt. 825
—— W., 2nd-Lt. . . . 637, 978
Saxena, R. B., Ty. Lt. . . . 1774
—— V. D., Ty. Capt. . . . 1772
Saxton, A. C., Lt. . . . 425, 1054
—— P. D., Maj. . . . 237, 1098

Names and Rank. Page.

Saxton, W. T., Lt. . . . 525, 1102
Sayad Shams-ud-din, Jemdr. . . 1464
—— Sam Suddin, Jemdr. . . 1464
—— Umar, Subdr. 1444
Sayed Raunak Ali Khan, Risldr. 951
Sayee, F., 2nd-Lt. 790
Sayers, C., Hony. Maj. . . 35, 1847
—— W. W. Lt. . . 55, 700, 727
Sayres, A. C., Lt. . . 491, 929
—— E. H., Lt. 491
Saywell, J. R., Capt. 1126
—— R., 2nd-Lt. 557
Saywer, Lt. C. P. B. 1048
Sayyed Aftab Shah, Jemdr. . . 894
Sayyid, Abdur Rahim, Subdr. . 1323
—— Akbar, Jemdr. 990
—— Ali, Jemdr. 1272
—— Ahmad, Subdr. 1351
—— Jemdr. 914
—— Jemdr. 1107
—— Jemdr. . 994, 1380
—— Jemdr. . . . 876
—— Amir, *Hony.* Capt. . . 2102
—— Azimuddin 1316
—— Hussain, Jemdr. . . . 1316
—— Jemdr. . . . 1360
—— Ibrahim, Jemdr. . . . 1012
—— Jalal, *Hony.* Capt. . . 2102
—— Khan, Ressdr. 895
—— Mastan, Jemdr. . . . 1324
—— Muhammad Yakub, Jemdr. 1051
—— Muhammed, Jemdr. . . 1444
—— Musa, Jemdr. 994
—— Qadir, Jemdr. 1316
—— Quadir, Jemdr. . . . 1333
—— Qudratullah, Jemdr. . . 928
—— Sikander, Jemdr. . . . 1360
—— Usman, Jemdr. . . . 1328
—— Yakub, *Hony.* Capt. . . 2102
—— Yusaf, Jemdr. . . . 994
Sazawar Khan, Jemdr. . . . 1035
Sazewar Khan, Jemdr. . . . 1035
Scale, J. D. Maj . . 105, 234, 1355
Scales, A. M., 2nd-Lt. . . . 696
Scallon, R. I., Genl. *Sir* . 145, 171, 1955
Scanlon, P. A., Asst. Surgn. . 1794
Scar, J., 2nd-Lt. 567
Scarff, A., 2nd-Lt. 563
Scarr, K. R., Lt. 685
Schaffter, A. J., Sub-Asst. Surgn. . 1826
Scheidweiler, F. C., 2nd-Lt. . . 1008
Schmidt, C. T., Lt. . . . 440, 893
Schneider, A., Lt. . . . 298a, 1174
—— L. W., Capt. . . . 628
Schocktee, S. H., 2nd-Lt. . . . 752
Schofield, A. J. A., 2nd-Lt. . 315, 913
—— F. W., *Hony.* Lt. . . 782
—— H. W. M., Lt. . . . 311a
—— W. H. S., Capt. 272, 1090
Schokman, C. C., Lt. . . 321, 921
Scholefield, M. T., Lt. . . 509, 1665
Scholes, W., Lt. 470, 1427
Scholfield, C. G., Lt. 748
Scholtz, C. J., Ty. Capt. . . . 1723
Schomberg, R. C. F., Maj. . . . 764
Schrader, G. J., 2nd-Lt. . . . 564
Schrottky, C. E. G., Maj. 72, 101, 614
Schulz, B. B., Lt. 683
Scobie, A., Lt. 90
Scobles, J., Lt. 707
—— I. H., McK., Capt. . . . 764
Scoggins, H. R., 2nd-Lt. 796
Sconce, J., Col. 176, 1917
Scoones, E. H. C., 2nd-Lt. . . . 640
—— G. 2nd-Lt. 1693
—— G. A. P., Capt. 17, 42, 276, 1600
Scorey, J. B., 2nd-Lt. 558
Scorgie, N. A., 2nd-Lt. . . 359, 1513
Scorrsby-Jackson, T., Capt. . . 861
Scotcher, W. J., 2nd-Lt. . . . 566
Scotland, A. G., Lt. . . . 310, 1166
Scotman, R., 2nd-Lt. 568
Scotson, R., 2nd-Lt. . . . 507, 1096
Scott, A. A. B., Ty. Capt. . . . 1724
—— A. B., Maj.-Genl. *Sir* 12, 43, 119, 612
—— A. K. S., Maj. . . 53, 221, 945
—— A. G., Lt. 307, 1385
—— A. H, 2nd-Lt., . . . 549, 1411
—— A. L., Lt.-Col. 133
—— A. N. De V., Maj. . . 221, 877
—— A. W., Lt. 700
—— B., Col. . . . 124, 150, 183
—— B. B., Capt. 848
—— B. C., Maj. 843
—— C. H., Maj.-Gen. *Sir* . . 1955

Names and Rank. Page.

Scott, C. W. F., Lt. . 298a 393, 507, 1641 1655
—— E., 2nd-Lt. 753
—— E. A. 567, 836
—— E. H., Lt. 530, 1275
—— E. J. W., Lt. . . . . 599—600
—— E. L., Capt. 836
—— E. S., 2nd-Lt. 1704
—— F. B., Capt. 256, 1024, 1106, 1880
—— F. H., Revd. 1907
—— F. J., Capt. 583
—— F. L., Lt. 430
—— F. W. A., Lt.-Col. . . . 740
—— G. B., Maj. . . . 95, 211, 1128
—— G. C., Capt. 29, 1664
—— G. D., Lt. 509, 1596
—— G. E., Capt. 842
—— G. E. B., 2nd-Lt. 808
—— G. M., Capt. 1717
—— G. S., Lt. 735
—— H. B., Maj. 1753
—— H. G., 2nd-Lt. 727
—— H. L., Maj. . . 99, 238, 1595
—— H. L., Capt. 829
—— H. St. G. S., Maj. . . 231, 1613
—— I. J., 2nd-Lt. 67
—— J., Capt. 818
—— J., Capt 1756
—— J., 2nd-Lt, ... 750, 784
—— J. A., Lt. 523, 1379
—— J. B., Capt. . 276, 1134, 1152
—— J. B., Lt. 587
—— J. E., Lt. 535, 1691
—— J. G., 2nd-Lt. 332, 1613, 1653
—— J. W., Lt. 723
—— K. D., 2nd-Lt. 1207
—— K. G., Lt. 781
—— K. R., Capt. . . 52, 276, 1144
—— L., Sr. Asst. Surgn. . . . 1791
—— L. B., Maj. 21, 1749
—— M., Lt. . . . . 455—59, 905
—— M. A. H., Capt. . . . 21, 665
—— M. D., Nursing Sister . . 1735
—— M. L., Nursing Sister . . 1735
—— M. G. S., Capt. 798
—— N. E. H., Maj. . . 1748, 1890
—— O. R., Lt. 527, 998
—— R., Maj. . . . 216, 1883, 1913
—— R. A., Capt. 939, 367
—— R. E., Lt. 533, 1096
—— R. J. L., Lt. 295
—— R. L., 2nd-Lt. 1263
—— R. M., 2nd-Lt. 1008
—— R. N. G., Capt. . . . 269, 1132
—— R. S., Maj. 78, 238
—— S., Lt. 701
—— S. C., Lt. 385, 1439
—— S. K., Lt. 656
—— S. V. J., 2nd-Lt. . . . .
—— T. A., Col. 177
—— T. E., Maj.-Genl. 5, 119, 147, 175
—— T. J., Lt. 686, 783
—— T. T., Lt. 524, 1231
—— W., Lt. 399, 1231
—— W., Lt. 1367
—— W., Col. 177
—— W., Lt.-Col. 138, 160, 198, 871, 1419
—— W., Lt. 399, 1104
—— W. D., 2nd-Lt. 563
—— W. H., 2nd-Lt. 803
—— W. H. S., Maj. 818
—— W. R. A., Capt. . . . 273, 1670
—— W. S., 2nd-Lt. 881, 494
Scott-Brown, E. G., Lt. . . 527, 1431
Scott-Coward, C. W., 2nd-Lt. 539, 1395
Scott-Elliot, C. R., Lt.-Col. . 57, 135, 158, 197, 870, 1263
Scotter, H. H. G. D., Lt. . . . 537
Scott-O'Connor, P. C., Lt. . . 1693
Scott-Oldfield, J. L., Lt. . . . 538
Scott-Ruffle, C. F., Lt. . . 509, 992
Scourfield, W. H., 2nd-Lt. . . 717
Scovell, A. M., Lt. 102, 106
—— G. V. F., Maj. . . . 501, 808
Scowcroft, H. D., Lt. . . . 11, 577
Scrafton, J. E. B., Maj. . . 240, 1899
Scragg, S., Sub-Condr. . . . 1856
Scratton, G. W. B., Capt. . 271, 911
Scrivener, A. H., Lt. 511
—— R. F. S., Lt. 1001
Scriven, E. G., 2nd-Lt. . . . 362
—— R. G., 2nd-Lt. 363
Scrivens, A. T., 2nd-Lt. . . . 363
Scroggie, J. A., Lt. . . . 523, 1164
—— W. R. J., Maj. 1748
Scroope, C. F., Capt. . 268, 1116, 1279

Names and Rank. Page
Scruby, J., Lt. 509, 817
—— W. S. J., Maj. 107, 238, 1363
Scudamore, P. C., Capt. 20, 274, 1062
—— W. V., Col. 124, 660, 1881
Scully, H., 2nd-Lt. 562
Scutt, C. C., 2nd-Lt. 549, 1435
Seabroke, C., Maj. 748
Seager, J. E. B., Lt. 463, 1066
—— R. L., Lt. 295, 909
Seagrim, C. B., Lt. 303, 1561, 1403
Seal, C. E. B., Maj. 610, 861
—— J. R., Lt. 90 511,
—— L. S., Lt. 537, 1327
—— R. C., Lt. 463, 1491
Seale, C. I., Lt. 529
—— J. H., 2nd-Lt. 683
Sealy, A. E., Lt.-Col. 42, 135, 158, 195, 872, 1633
—— D. C., Ty. Asst.-Surgn. 1804
—— G. O. F., Lt.-Col. 137, 1744
—— H. G., Maj. 228, 1162, 1517
—— H. N., Capt. 1715
—— S. C., Sub-Condr. 1857
Seaman, C. K., 2nd-Lt. 560, 603, 1373
Seamark, F., Lt. 701
Searight, A. K., Capt. 105, 749
Searl, A. E. G. 2nd-Lt. 300, 1295
Searle, A. E. S., Bt.-Col. 57
—— A. T., Maj.-Genl. 172
—— C. S., Capt. 279, 1395
—— G. A. H., Lt. 543a
—— H., Lt. 686
—— H. F., 2nd-Lt. 546, 1297
—— H. G., 2nd-Lt. 551, 1677
—— R. P., Sub-Condr. 1864
—— W. A. A., Lt. 313, 957
Searls, A. G., 2nd-Lt. 637
Seater, G. H., Lt. 318, 1046
Seaton, F. D. R., Maj. 227, 1405, 1451
—— O. C., Lt. 445, 1193
—— W. A., 2nd-Lt. 555
Seaver, C. D. K., Capt. 1713
Sebag-Montefiore, G. E., Lt. 59, 597
Sebastian, F. Jemdr. 1447
Seccombe, J. B. P., 2nd-Lt. 327, 1259
Secker, V. H., Capt. 593, 599
Secluna, G., Tempy. Capt. 1767
Secombe, A., Condr. 1861
—— H. S., Lt. 399, 1195, 1537
Secondé, E. C., Capt. 107, 108, 261, 1082
Seddon, A., Capt. 1757
—— T. Y., Col. 126, 150, 186
Sedgeley, C. H., Lt. 516, 133
Sedgley, C. N. Lt. 1670
Sedu, Jemdr. 1193
Seebar, H. C. R., Lt. 455, 59, 1443
Seed, C. J., 2nd-Lt. 707
—— J. A. T., Lt. 387, 1443
—— R. D., Lt. 1307
—— R. B., 2nd-Lt. 465
Seeds, A. A., Lt.-Col. 133, 1710
Seeley, A. C., Lt. 353, 1674
—— B. J., Lt. 618
—— C. H., Revd. 848
—— G. H., Revd. 1908
Seers, F. W., 2nd-Lt. 1696
Segar, E., Capt. 259, 1154, 1183
—— W. J., 2nd-Lt. 1678
Seikh Amer Ahmed, Jemdr. 1705
Sein Po, Jemdr. 1295
Selby, H. R., Condr. 1854
—— W. R., Lt. 296, 1399
Selfe, G. V. H., 2nd-Lt. 790, 796
Sell, C. E., Revd. 1906
—— W. H., 2nd-Lt. 802
Seller, C., 2nd-Lt. 326, 366, 1170
—— J. R., 2nd-Lt. 787
Sellers, S. R., 2nd-Lt. 567
—— W. L., Capt. 76, 77, 1886
Selley, W. A., 2nd-Lt. 565
Sellick, H. D., Capt. 501a
—— W. H., 2nd-Lt. 567
Sellier, P. J., Lt. 516
Selves, H. S., 2nd-Lt. 579
Selway, H., Lt. 463
Selwyn, A. L. H., Revd. 1904
—— C. H., Col. 97, 103, 105, 124, 149, 184
—— H. G., Lt. 691
Semini, E. J., Capt. 265, 1323, 1883
Semins, A. M., Lt. 1001
Semon, H. C., *Tempy.* Capt. 1763
Semple, A. G., 2nd-Lt. 790
—— R. E., 2nd-Lt. 719, 759
Sen, A. K., Ty. Lt. 1773
—— A. K., Ty. Lt. 1782

Name and Rank Page.
Sen, G. C., Ty. Lt. 1776
—— G. D., Ty. Lt. 1782
—— H. C., Ty. Lt. 1774
—— H. K., Ty. Lt. 1773
—— J. L., Capt. 1060, 1756
—— L. M., Ty. Lt. 1782
—— N. K. Ty.-Lt. 1776
—— P. N., Ty.-Lt. 1774
—— R., *Tempy.* Lt. 1773
—— S. C., Temp. Capt.
—— S. K., Ty.-Lt. 1766, 1778
Sendall, W. H., Lt. 445, 1393
Sen-Gupta, A. K., Ty. Capt. 1772
—— B. C., *Tempy.* Lt. 1775
Senior, A. T., Sub-Condr. 1857
—— E. B. G., Lt. 354, 1575
—— E. W., Sub-Condr. 1864
—— H. F., Capt. 619
—— H. W. R., Col. 53, 97, 103, 126, 150, 188
Senneck, F. H., 2nd-Lt. 571a
Seole, Subdr. 1476
Seoraj Pun, Subdr. 1624
Sequeira, J. C., Ty.-Lt. 1779
Sergeant, A. H., Lt. 353, 1020
Serjeant, J. M., 2nd-Lt. 1675
Serres, E. D., 2nd-Lt. 556
Seth, E. R., 2nd-Lt. 1088
Sethi, B. N. S., Ty. Lt. 1781
Seton, H. W., Capt. 255, 1641
Seton-Browne, C. L. S., Lt.-Col. 137, 158, 197
Sevenoaks, C. L., Lt. 306, 1116 1569
——, P. L., 2nd-Lt. 1124
Sevorn, F., 2nd-Lt. 1189
—— P., 2nd-Lt. 1124
Seward, J. B., 2nd-Lt. 1140
Sewa Ram, Subdr. 1389
—— Subdr. 1403, 1480
—— Ty. Sub Asst. Surgn. 1833
Sewa Singh, Sub-Asst. Surgn. 1817
—— Bajwa, Sub-Asst. Surgn. 1817
—— Jemdr. 1150 1189
—— Subdr. 1241
—— Subdr. 1166
—— Jemdr. 1336
—— Jemdr. 1204
—— Jemdr. 1365
—— Bajwa, Sub-Asst. Surgn. 1817
Sewal, R. N., Tempy. Capt. 1767
Sewell, D. L., Ty. Capt. 1724
—— H. K., Lt. 522
—— L. H., Capt. 841
—— R. A. D., Maj. 841
—— R. B. S., Capt. 1754
—— S. E., Lt. 72, 652
Sexton, C. B., Lt. 513, 1283
—— D. C., 2nd-Lt. 623
—— E. G., Maj. 213, 875
—— F. W. O., Maj. 1712
—— J. A., Ty. Asst. Surgn. 1804
—— J. J. O'B. Maj. 57
Seymour, A. H., Lt. 722, 1614
—— C. G., Maj. 1601, 1750
—— C. L., Lt. 534, 1217
—— C. R. F., Maj. 99, 105, 223, 1070
—— E. V. F., Capt. 267, 873, 189, 957
—— H. C., Lt. 518
—— M. H., Maj. 243, 1647
——, W. W., Lt. 1471
Seymour-Higgins, A. C. P., Lt. 512, 708, 1671
Seys, R. C., Maj. 102, 634
Shabir Hussain, Jemdr. 1432
Shabaz Khan, Subdr. Maj. 1483
Shackell, A., 2nd-Lt. 361, 1116
Shackleford, B. G. V., 2nd-Lt. 713
Shadam Khan, Subdr. 1117
Shad Man Khan, Jemdr. 926
Shad Muhammad, Subdr. 1102
—— Jemdr. 1133
—— Jemdr. 1681
—— Khan, Resldr.
—— Resldr. 915
Shaddock, J., Lt. 7, 791
—— J. C., 2nd-Lt. 729a
Shadi, Subdr. 1066
—— Khan, Subdr. 968
—— Jemdr. 970
—— Ram, Jemdr. 1385
Shafa, Jemdr. 1428
Shafi Muhammad Khan, Ressldr. 890

Names and Ranks Page.
Shah, B. B., Ty. Capt. 1772
—— B., Ty. Lt. 1764
—— J. M., Lt. 1759, 1766
—— S. H. U., Ty. Lt. 1768
Shahamad Ali, Jemdr. 900
Shahane, N. R., Ty. Lt. 1773
Shahbaz Khan, Jemdr. 1480
Shah Baz, Subdr. 1252
—— Jemdr. 1152
—— Jemdr. 1236
Shah Gul, Jemdr. 908
—— Khan, Jemdr. 1244
—— Jemdr. 1145
—— Subdr. 1299
—— Jahan, Jemdr. 1103
—— Mir Khan, Sub-Asstt. Surgn. 1807
—— Mirza Beg, *Hony.* Maj. 110
—— Muhammad, *Hony.* Capt. 2103
—— Jemdr. 1456
—— Khan, Jemdr. 1090
Shah Nawaz, Subdr. 1409
—— Niwaz, Jemdr. 1433
—— Nawaz Khan, Jemdr. 896, 1439
—— Subdr. 1571
—— Jemdr. 976
—— Niaz Husain, Jemdr. 1502
—— Pasand, Jemdr.
—— Sawar, Jemdr.
—— Jemdr. 1122
—— Subdr. 999
—— Sowar, Jemdr. 1121
—— Wali, Subdr. 1259
—— Wali Khan, *Hony.* Lt. 2104
—— Jemdr. 910
—— Jemdr.
—— Jemdr. 948
—— Zaman, Subdr. 1335
—— Subdr. 1247
Shahbaz Khan, Jemdr. 916
—— Subdr. 1299
Shahzad Khan, Jemdr. 940, 1388
Shahbaz Khan, Jemdr.
—— Jemdr. 1145
Shahibullah, Subdr.-Maj.
Shah Qadam, Jemdr.
Shahryar Khan, Jemdr. 1404
Shahwali Khan, Risldr. 1684
Shahzad Khan, Jemdr. 1387
Shahzada Khan, Subdr. 1492
Shahzullah Khan, Jemdr. 1001
Shaikh, A. H., Lt.
—— Haidar, Jemdr. 928
—— Abdul, Subdr. 1460-62
—— A. H., Tempy. Capt. 1764
—— Hakim, Jemdr. 1404
—— Kadir, Subdr. 1351
—— Khadir, Jemdr. 1591
—— Abdul Qadir, Jemdr.,
—— Subdr. 1506
—— Sulaiman, Subdr. 1409
—— Samad, Jemdr. 1424
—— Majid, Subdr. 1050
—— Abdulla, Jemdr. 994
—— Abdullah, Subdr. 1379
—— Subdr. 1383
—— Aminuddin, Jemdr. 1460
—— Dada, Jemdr. 1416
—— Abdul Karim, Subdr. 1385
—— Majid, Risldr. 876
—— Abdulla Shaikh Rahim, Sub-Asst. Surgn. 1830
—— Abdur Rahman, Subdr. 1327
—— Abdur Razaq, Jemdr. 1397
—— Adam, Tr. Vety. Asst. 1687
—— Ahmad, Subdr. 1460-62
——, *Hony.* Capt. 2102
—— S. H. S., Sub-Asst. Surgn. 1820
—— Sub-Maj. 1471
—— Subdr. 1331
—— Ahmadullah, Subdr.-Maj. 1387
—— Ahmed, Subdr. 1444
—— Ahmed Ali, Ressdr.
—— Ahmed, Sahib 89
—— Ali, Subdr. 1271
—— Jemdr. 1328
—— Tr. Vety. Asst. 1687
—— Ali Jan, Sub-Asst. Surgn. 1815
—— Muhammad, Subdr.-Maj. 1050
—— Jemdr. 1506
—— Amarbhai Daudbhai, Sub-Asst. Surgn. 1830
—— Amir, Jemdr. 1324
—— *Hony.* Capt. 2102

Names and Rank. Page.

Shaikh Anser, Jemdr. 1324
—— **Azmud-din Shaikh Ismail**, Sub-Asst. Surgn. 1830
—— Bab, Subdr. Maj. 1459
—— Babu, Hony. Capt. 2104
—— **Bade Sahib**, Jemdr. 1444
—— Bandu, Jemdr. 1460
—— Chand, Jemdr. 1460
—— Daud, Subdr. 1333
—— Jemdr. 1328
—— Jemdr. 1328
—— Jemdr. 1579
—— Jemdr. 1424
—— Subdr. 1339
—— **Faiz-ud-din**, Ressdr. 914
—— **Fakir Muhammad**, Subdr. 1479
—— **Farid**, Sub-Asst. Surgn. 1815
—— Jemdr. 1591
—— Subdr.-Maj. 1395
—— *Hony.* Capt. 2101
—— G. M. L. S., Sub-Asst. Surgn. 1831
—— **Gul Muhammad**, Subdr. 1460
—— **Haider**, Subdr. 1423
—— Jemdr. 928
—— **Hamid**, Subdr. 1333
—— **Hasain**, Risldr. 935
—— Jemdr. 1318
—— Ali, Sr. Sub-Asst. Surgn. 1807
—— Hasan, Jemdr. 999
—— **Hussain**, Jemdr. 1591
—— Jemdr. 1591
—— **Ibrahim**, *Hony.* Lt. 2104
—— **Hussain**, *Hony.* Capt. 2101
—— **Imam**, Subdr. Maj. 1415
—— **Inayatullah**, Jemdr. 1308
—— Isa, Jemdr. 1400
—— Ishaq Muhammad, Jemdr. 1051
—— Ismail, Subdr. 1333
—— *Hony.* Capt. 2101
—— *Hony.* Capt. 2102
—— Subdr. 1359
—— Jemdr. 1436
—— Jemdr. 1480
—— Jaffar, Subdr. 1418a
—— **Jafar Ali**, Jemdr. 932
—— **Jamal**, Jemdr. 1444
—— Kalandar, Subr.-Maj. 1385
—— Uddin, Jemdr. 1276
—— **Lal Abdul Karim**, Sub-Asst. Surgn. 1830
—— **Mahatabdin**, *Hony.* Capt. 2102
—— **Mahbub**, *Hony.* Capt. 2101
—— Jemdr. 1324
—— Jemdr. 1317
—— Subdr. 1052
—— **Mahibob Muhammad Sahib**, Sub-Asst. Surgn. 1830
—— **Mahomed Usman**, Subdr. 1423
—— Mansur, Jemdr. 1391
—— **Maktum**, Jemdr. 1416
—— Masud Ali, Jemdr. 1388
—— **Mohi-uddin**, Jemdr. 1424
—— Jemdr. 1579
—— **Miran**, Shaikh Rahim, Sr. Sub-Asst. Surgn. 1829
—— Ressdr. 914
—— Risldr. 935
—— **Muhammad Abdul Mattalib**, Sub-Asst. Surgn. 1816
—— **Muhammad**, *Hony.* Capt. 2102
—— Jemdr. 1506
—— Jemdr. 1416
—— Ali, Subdr. 1383
—— Ishak, Jemdr. 1447
—— Khalil, Jemdr. 1460
—— Jemdr. 1444
—— **Dadasahib**, Sub-Asst. Surgn. 1831
—— **Shafaat Hussain Siddegi**, Sub.-Asst. Surgn. 1820
—— **Muhammad Umar**, Subdr. 1476
—— **Umar**, Subdr. 1391
—— **Muhiuddin Subdr.** 1263
—— **Naruddin**, Jemdr. 1333
—— **Nur-Allah** 1819
—— **Qadir Bakhsh**, Jemdr. 1472
—— **Rafiuddin**, Jemdr. 1472
—— **Ramrag**, Jemdr. 1472
—— **Sadiq Ali**, Jemdr. 1404
—— **Shahamat Ali**, Jemdr. 1399
—— **Sirajuddin**, Jemdr. 1518

Shaikh Sultan, Jemdr. 1476
—— **Usman**, Jemdr. 1468
—— Jemdr. 1468
—— **Usmanmia Hussainmia**, Sub-Asst. Surgn. 1830
—— Wada, Ressdr. 1683
—— **Wali Ahmad**, *Hony.* Capt. 2102
—— **Yasin**, Subdr.-Maj. 1443
—— **Yakub**, Subdr. 1423
—— **Yusuf**, Jemdr. 1464
—— **Ali**, Subdr. 1092
Shairp, A., Maj. 4, 103, 209, 1661, 1931
—— H. F., Maj. 42, 87, 211
Shakar, Mehdi, Jemdr. 1518
—— Khan, Subdr. 1283
Shakeil, A., 2nd-Lt. 363
Shakesnaft, A. J., Capt. 103, 104, 691
Shakespear, G. F. C., Bt.Maj. 52, 271, 1359
—— G. R. J., Maj.-Genl. 173
—— H. T., Maj. 230, 897
—— L. W., Col. 123, 149, 181, 1898
—— W. B., Capt. 25, 268, 1600
Shakrulla Khan, Jemdr. 1683
Shallcroff, L. E., 2nd-Lt. 1122
Shallcross, E. P., Capt. 1839
Shalless, E. G., Ty. Lt. 795
Shallom Soloman, Jemdr. 1444
Shallow, C. R., Lt. 68, 410, 1311
—— D. S., 2nd-Lt. 640
—— G., Lt. 517, 1247
Sham Lal Basva, Sub.-Asst. Surgn. 1819
Shama Rao, Subdr. 1455
—— Singh, Jemdr. 1533
Shamaldas, D. D., Sub.-Asst. Surgn. 1832
Shamas Din, Jemdr. 1681
—— Jemdr. 1284
Shambhudayal Tiwari, Jemdr. 1027
Shambhulal Jethallal Shah, Sub-Asst. Surgn. 1831
Shambhu, C. P., Sub-Asst. Sugn. 1832
Shambhu Nath, Sub-Asst. Surgn. 1814
—— Sub-Asst., Surgn. 1817
Shamir Singh, Jemdr. 1236
Shamrath Singh, Jemdr. 1196
Shams Din, Jemdr. 990
Sham Nang, Jemdr. 1248
Sham Lal Bawa, Sub-Asst. Surgn. 1819
Shamsul Haq, Ressaidr. 1686
Shamsher Jang Thaker, Jemdr. 1615
Sham Singh, Subdr.-Maj. 1074
—— Subdr. 1285
—— Ressdr. 1681
—— Jemdr. 1156
—— Subdr.
—— Jemdr. 890
—— Jemdr. 1124
—— Jemdr. 930, 1141
—— Jemdr. 1681
—— Jemdr. I 957
—— II Jemdr. 957
—— Ressdr. 1681
—— Subdr. 1148, 1587
—— Sub-Asst. Surgn. 1810
—— Tr. Vety. Asst. 1687
—— Sub-Asst. Surgn. 1813
Sham Lal Singh, Subdr. 1268
—— Jemdr. 1484
Shams Khan, Jemdr.
Shamsher Gurung, Subdr. 1618
—— Jemdr. 1608
—— Jemdr. 1611
—— **Ali Khan**, Jemdr. 898
—— Ressdr. 940
—— **Jang**, Sub-Asst. Surgn. 1814
—— **Singh**, Jemdr. 1248
—— Jemdr. 1117
—— Jemdr. 1143
—— Thapa, Jemdr. 1608
—— Jemdr. 1651
Shamsul Khan, Jemdr.
Shamsuddin, Sub-Asst. Surgn. 1816
—— Sub-Asst. Surgn. 1817
—— Subdr. 1276
—— Jemdr.
—— Jemdr. 1304
—— Jemdr. 1067
—— Jemdr.
Sham Sukh, Subdr. 1355
Shamsul Haq, Ressdr. 1683
—— Ressdr. 1686

Shamzada Khan, Jemdr. 1571
Shan Khan, Subdr. 1320
Shand, G., Capt. 485
—— J. G. B., Capt. 1755
—— N. P., Capt. 691, 692
—— W. K. W., Lt. 691, 783
Shankar, P. U., Ty. Lt. 1774
Shanks, G., Capt. 1760
Shankar Ashenna, Jemdr. 999
Shankar Das, Sr. Sub-Asst. Surgn. 1807
Shankar Kerkar, Subdr. 1469
—— Savant, Jemdr. 1469
Shankardial Misir, Jemdr. 1018
Shankar **Gopal Patwardhan**, Sub-Asst. Surgn. 1831
—— **Krishna Gole**, Sub-Asst. Surgn. 1831
—— **Lal**, Sub-Asst. Surgn. 1813
—— Lawande, Jemdr. 1416
Shankarrao B., Ingle Hony. 2nd-Lt. 482
Shankar, R. A., Sub-Asst. Surgn. 1832
—— Baber, Jemdr. 1480
—— Singh, Jemdr. 1083
—— Sing Rawat, Jemdr. 1182
—— Rawat, Jemdr.
—— Vishnu Gole, Sr. Sub-Asst. Surgn. 1829
Shanker, Laxman Matker, Sub-Asst. Surgn. 1830
Shanker Nath Dutt, Jemdr. 1012
Shankram Pille, Jemdr. 999
Shanley, T., Lt. 591
Shanmugham, Jemdr. 1276
Shan Muhammad, Jemdr. 1376
Shannon, A. G. L., 2nd-Lt. 551, 1116
—— E. C. B., Lt. 465, 903
—— G. C., 2nd-Lt. 544, 1407
—— J. A., Capt. 1839
—— W. T., Lt. 789
Shanur Khan, Subdr. 1509
Shan Sowar, Subdr. 1409
Shapter, L. J., Lt. 696, 697
Sharaf Ali Khan, Jemdr. 1687
Sharaf Ali, Risldr. 1683
Sharafat Ali Khan, Sub-Asst. Surgn. 1815
Sharaf Din, Subdr. 1518
—— Khan, Jemdr. 1420
Sharaf-ud-din, Subdr. 1035
—— Jemdr. 1702
Sharam Singh, Subdr.-Maj. 1431
Sharbat Alikhan, Jemdr. 906
Sharekar Karkar, Subdr. 1459
Sarfaraz, Jemdr. 1433
Sharfu Khan, Jemdr. 1432
Sharhab Khan, Jemdr. 1235
Sharim Khan, Subdr. 1698
Sharif, Subdr. 1427
—— Jemdr. 1099
—— Khan, Jemdr. 1480
—— Subdr. 1283
—— Jemdr. 908
Sharland, W., 2nd-Lt. 705
Sharma, A. N., Ty. Capt. 1251, 1771
—— J. K., Ty. Lt. 1778
—— R. N., Ty. Lt. 1777
—— R. S., Ty. Capt. 1771
Sharman, E. P., 2nd-Lt. 77, 719, 1885
—— C. T., 2nd-Lt. 722
Sharp, C. J., Lt. 294, 1070
—— J. E., tempy Capt. 1726
—— J. C., Capt. 1397
—— J. C., 2nd-Lt. 794
—— J. M. C., 2nd-Lt. 568
—— P. C., Maj. 719
—— T. G., 2nd-Lt. 362
—— S. H., 2nd-Lt. 818
Sharpe, A. G., Lt. 389, 881
—— A. J., Lt. 679
—— A. R., Lt. 844
—— G. F., Lt. 543, 1154
—— H., Lt. 721
—— J., Capt. 1696
—— J. C., 2nd-Lt.
—— J. C., Sub-Condr. 1856
—— J. N., 2nd-Lt. 550
—— P., 2nd-Lt. 67, 723
—— R., *Hony.* Maj. 1843
—— W., Condr. 1854
—— W. R. S., Capt. 821
—— W. W. S., Capt. 1717
Sharples, J. L., Capt. 759
—— R., Sr. Asst. Surgn. 1787
Sharpus, A. F., L. 531

Names and Rank.	Page.
Shattock, E. K., 2nd-Lt.	556, 887
Shave, E. D., Asst. Surgn.	1793
Shaw, A., Condr.	1853
—— A. A. St. J., 2nd-Lt.	475
—— A. N. St. F., 2nd-Lt.	1469
—— B. V. D. H., Lt.	313, 1431, 1479
—— C. C. C., Maj.	1343, 1751
—— C. H., Nursg. Sister	1735
—— C. H., Lt.	88
—— C. H., Lt.	519
—— C. J., 2nd-Lt.	562
—— C. W., Lt.	799
—— D. G. L., Maj.-Gen	118, 144, 173
—— D. H., Lt.	981
—— E., Col.	177
—— E., Lt.	638, 977
—— E. B., Tempy. Capt.	524
—— F. C., Lt.	415
—— F. J. F., 2nd-Lt.	601
—— E. L. H., Asst. Surgn.	1798
—— F. McL., Lt.	489
—— G. F., Lt.	521, 1453
—— G. R., 2nd-Lt.	556, 998
—— G. W., Lt.	571, 1703
—— H. H., 2nd-Lt.	554
—— H. M. D., Maj.	216, 1595, 1609
—— H. P. J., Maj.	26, 244, 1663
—— J., 2nd-Lt.	554
—— J. Lt.	798
—— J. A., 2nd-Lt.	539, 1671
—— J. F., Lt.	405, 511, 1483
—— L. B., Lt.	460—62, 1040
—— L. H., Lt.	787
—— M. H., 2nd-Lt.	355, 1215
—— N., Lt.	741, 742
—— P. B., 2nd-Lt.	1124
—— R., Lt.	543
—— R., Lt.	534, 1231
—— R. B., Lt.	78, 624
—— R. C., Lt.	738, 740, 1007
—— R. L., Lt.	302, 1215, 1333
—— S.	1005
—— S. C., Ty. Lt.	1774
—— S. G. N., 2nd-Lt	825
—— T. A., Lt.	1839
—— T. W., Lt.	1720
—— V. E., Lt.	415, 1092
—— W., Lt.	529
—— W., 2nd-Lt.	1207
—— W. H., 2nd-Lt.	356, 1006
—— W. S. J., Maj.	1750
—— W. W., Maj.	244, 1653
Shawali Khan, Jemdr.	1702
Shawe, E. B., 2nd-Lt.	553, 1054
—— R. A., 2nd-Lt.	549, 1405
Shea, A. G., Maj.	8, 15, 218, 1227
—— J. S. M., Maj.-Genl.	97, 101, 119, 147, 174
—— T. L., Capt.	62, 1838
—— V. G., 2nd-Lt.	722
Shead, C. G., 2nd-Lt.	552, 1293
Sheard, C., Capt.	485, 943
Shearer, E. J., Capt.	274, 933
—— F. C., 2nd-Lt.	1181
—— J. E., Capt.	272, 1124
—— W., 2nd-Lt.	1295
Shearman, W. W., Lt.	450—54, 522, 1659
Shears, R. H., Capt.	577
Sheath, F. G., Condr.	1867
Shedu Ram, Jemdr.	1420
Shee, J. R. L., Lt.	488, 911
Sheehan, A. F. C., Lt.	389, 1607
—— T. P., Lt.	530, 1056
Sheehy, J. F., Lt.	542, 1295
Sheen, A. W., Lt.-Col.	1720
Sheepshanks, E., Lt.	493
—— R. H., Capt.	263, 897
Sheffield, D. E., 2nd-Lt.	780
—— J. C., 2nd-Lt.	1056
Shebryar Khan, Subdr.	1373
Sheikh Ahmad, Jemdr.	1698
—— Subdr.	1379
—— Dada, Jemdr.	1416
—— ImtiazRasulKhan, 2nd-Lt.	482
—— Mamuddin, Jemdr.	14 6
—— Masnad Ali, Jemdr.	1388
—— Mohomed Usman, Jemdr.	1464
—— Muhomed Husain, Jemdr.	1464
—— Subdr.	1707
—— Ibrahim Siri Bah, Jemdr.	1464
—— Jaffer, Subdr.	1416
—— Jamal, Jemdr.	1416
—— Shahid Husain, 2nd-Lt.	482
—— Taj Mohammad	91
—— Ummed, Subdr.	1401
—— Wilayat-Ali, Jemdr.	1091
Sheikh Amir Ahmed, Jemdr.	1705
Sheddon, L. H., 2nd-Lt.	779
Shell, P. J. H., Lt.	935, 405
Sheldon, H. G., Lt.	295, 893, 323
—— N. L., 2nd-Lt.	831
Sheldrick, A. H. W., Capt.	485
Shelford, E. J. H., 2nd-Lt.	708
Shelley, P., Lt.	455—59, 1036
Shellick, H. D., Lt.	1006
—— W. A., 2nd-Lt.	567
Shelly, C. E., Lt.	1467
Shelswell, O. H. C., Capt	501a, 1183
Shelton, R. C. M., Lt.	585, 974
Shelton-Palmer, A. L., 2nd-Lt.	564
Shelverton, E. M., Lt.	487, 1077
—— St. J. A., Lt.	111
—— H. M., Lt.	536, 633, 1633
Shemain, N. P., Sr. Asst. Surgn	1791
Shenton, E. W., Lt.	543a, 1373
Sheobakhsh Singh, Subdr.	1363
—— Jemdr.	1498
Sheo Chand, Risldr. Maj., Ty.	951
—— Subdr.	1165
—— Subdr.	1161
—— Jemdr.	1059
—— Jemdr.	1388
—— Dat, Jemdr.	1039
—— Karan, Ressdr.	888
—— Jemdr.	902
Sheodan Singh, Jemdr	890
—— Jemdr.	1192
—— Jemdr.	1385
Sheodarshan Tiwari, Subdr.	1050
Sheodath Singh, Jemdr.	1368
Sheodhan Singh, Jemdr.	1388
Sheoji, Jemdr.	1472
Sheoji Singh, Jemdr.	
Sheolal, Subdr.	1059
—— Jemdr.	1385
Sheonath, Subdr.	1471
—— Jemdr.	1059
—— Singh, Subdr.	1050
Sheoparash Rai, Jemdr.	1369
Sheoratan Dube, Jemdr.	990
Sheo Ram, Ressdr.	1683, 1686
Sheorey, N. L., Ty. Capt.	1764
Sheo Singh, Subdr.-Maj.	1193
—— Sankar Pathak, Ressdr.	1683
Shepard, A. J., Lt.	352, 1305
—— R. W., 2nd-Lt.	618
Sheperd, A. V. W., Lt.	304, 1627
Shepherd, C. W., Asst. Surgn.	1799
—— F. E., Lt.	
—— G. C., Lt.	1655
Shepherd, C. H., Col.	695
—— C. E., Col.	178
—— C. I., Maj.	241, 1235
—— F. A., Lt.	535, 1235
—— F. A., 2nd-Lt.	91
—— F. E. Lt.	1499
—— F. M., Lt.	623
—— G. F., Ty. Lt.	1726
—— H. C. B., 2nd-Lt.	1441
—— H. V., Lt.	512, 1178
—— J., Lt.	793
—— J. D., Lt.	668, 986
—— R., 2nd-Lt.	562
—— R. J., Sub-Condr.	1855
—— S., Sub-Condr.	1864
—— W. I., Lt.	530, 1707
—— W. S., Lt.	488
—— W. S. W., Esq.	88
Sheppard, A. L., Capt.	1755
—— B. K., Lt.	312, 1596
—— C. B., Lt.	724
—— E. A., 2nd-Lt.	355, 1086
—— F., Sub-Condr.	1857
—— F. E., Lt.	399, 1489
—— G. S., Bt.-Col.	87, 125, 153, 193
—— S. H., Lt.-Col	97, 120, 125, 661
—— S. T., Lt.	41, 821
—— S. V., Lt.	90—707
—— W. H., 2nd-Lt.	18, 36
Sheppard-Capurro, R. M. F. F. de S., Lt.	380, 917, 1495
Shepperd, C. B., 2nd-Lt.	724
Shepton, H. McL., Lt.	543
—— W. S., 2nd-Lt.	629
Sherard, W. G., Asst. Surgn.	1798
—— A. F. L., 2nd-Lt.	1070
Sher Afzal, Subdr.	1509
—— Subdr.	1183
—— Khan, Subdr.	1268
—— Ahmad Khan, Jemdr.	1248
Sher Akhmad, Jemdr.	1124
—— Alam, Jemdr.	970
—— Ali, Subdr. Maj.	1239
—— Jemdr.	976
—— Ressdr.	1686
—— Jemdr.	1498
—— Ressaidr.	1682
—— Subdr.	1128
—— Jemdr.	1498
—— Khan, Jemdr.	1681
—— Jemdr.	910
—— Ressdr.	922
—— Bahadur, Jemdr.	908
—— Jemdr.	924
—— Subdr.	1130
—— Khatti, Subdr.	1630
—— Khan, Jemdr.	957
—— Jemdr.	940
—— Baz, Jemdr.	
—— Baz Khan, Jemdr.	1145
—— Dil, Jemdr.	1507
—— Jemdr.	1240
—— Dil Khan, Subdr.	1492
—— Baz, Subdr.	976, 1001
—— *Hony.* Capt.	2101
—— Jemdr.	1104
—— Jemdr.	1240
—— Jemdr.	1117
—— Khan, Ressdr.	924
—— Jan Khan, Jemdr.	948
—— Gul, Jemdr.	1268
—— Subdr.	1227
—— Jang, Jemdr.	
—— Jemdr.	966
—— Khan, Subdr.	1283
—— Jemdr.	1256
—— Jemdr.	1511
—— Jemdr.	892
—— *Hony.* Capt.	2103
—— Jemdr.	1432
—— Jemdr.	1515
—— Jemdr.	1122
—— Jemdr.	973
—— Subdr.	1305
—— Mohomed, Jemdr.	
—— Mohamed Khan, Ressdr.	1684
—— Muhammad, Senr. Sub-Asst. Surgn.	1807
—— Jemdr.	924, 966
—— Subdr.-Maj.	1355
—— Jemdr.	1208
—— Jemdr.	1143
—— Jemdr.	1075
—— Jemdr.	953
—— Jemdr.	1489
—— Auli, Sub-Asst. Surgn.	1819
—— Khan, Resdr.	1684
—— Khan, Subdr.	1050
—— Jemdr.	876
—— Jemdr.	908, 910
—— Khan Jemdr.	876
—— Muhammad, Khans, Jemdr.	1090
—— Subdr.	1120
—— Jemdr.	1492
—— Shah, Subdr.	1128
—— Khan, Risldr.	927
—— Nur, Jemdr.	1117
—— Singh, *Hony*, Capt.	5, 2101
—— Jemdr.	936, 955
—— Jemdr.	1028
—— Jemdr.	920
—— *Hony.* Capt.	2101
—— Ressdr.	906
—— Ressadr.	906
—— Jemdr.	1149
—— Subdr.	1148
—— Subdr.	1305
—— Jemdr. (I)	928
—— Jemdr. (II)	928
—— Risldr.	919
—— Risldr.	885
—— Subdr.	1166
—— Jemdr.	966
—— Jemdr.	1151
—— Jemdr.	914
—— Singh, Jemdr.	1372
—— Jemdr.	1385
—— Jemdr.	1681
—— Jemdr.	1111
—— Subdr.	
—— Jemdr.	1682
—— Subdr.	1247
—— Liddar, Sub-Asst. Surgn.	1819
—— Jemdr.	932
—— Jemdr.	1320

Names and Rank.	Page.
Sher Singh, Bhullar, Sub-Asst. Surgn.	1818
—— Subdr.	1321
—— Jemdr.	1433
—— Bhullar, Sub-Asst. Surgn.	1818
—— Liddar, Sub-Asst. Surgn.	1819
Sher Sing Adhikari, Subdr.	1642
—— Jemdr.	1433
—— Jemdr.	1054
—— Khattri, Jemdr.	1639
—— Rana, Subdr.-Maj.	1614
—— Rawat, Subdr.	1648
—— Zaman, Subdr.	1268
—— Subdr. Maj.	1365
—— Subdr.	1267
—— Subdr.	1577
—— Jemdr.	1493
—— Jemdr.	
—— Subdr.	1280
Sheraban, Jemdr.	1244
Sherard, A. F. L., 2nd-Lt.	540
Sherbahadur Thapa, Jemdr.	1625
—— Gurung, Jemdr.	1611
Sherbulak Singh, Ressdr.	1683
Sherelen, J., 2nd-Lt.	1696
Sherer, J. C., Lt.-Col. 16, 41, 131, 156, 191,	867
—— J. D., Lt.-Col. 136,	632
Sheridan, 2nd-Lt. F. L.	1044
—— F. S., Lt.	511
—— R. J., *Hony.* Lt.	1848
—— R. S., Condr.	1858
Sheriff, G., Lt. 700,	719
—— T. H., the Revd.	1908
Sherjan Khan, Jemdr.	948
Sheringham, A. T., Maj. 105, 108, 229, 1479,	1587
Sheriston, J. A., 2nd-Lt.	1140
Sherlock, G. W. S., Maj. 66, 99, 213, 868, 1104,	1623
—— W. W., Lt. 510,	1665
Sherman, C. B., Capt.	765
—— D. M. H., 2nd-Lt.	324
—— F. J., 2nd-Lt.	565
Sherrard, G. C., Capt.	1886
—— J. W., Bt.-Lt.-Col. 60, 137, 159, 213,	1651
Sherson, D. B. N., 2nd-Lt. 566,	1060
Sherston, J. R. V., Capt. 17, 21, 264	895
Sherwell, S., Capt.	766
Sherwin, C. E., Lt.	578
—— C. E., Lt.	579
Sherwood, H., Lt.	753
—— J., 2nd-Lt.	683
—— W. O., Lt. 489,	1667
Shettle, F. B., Maj. 62,	1751
Shevlin, W., Sub-Condr.	1868
Shewell, E. F., Lt.-Col.	614
Shewen, D. G. P. M., Capt. 273, 1128,	1473
—— M. T., *Hony.* Brig.-Genl.	56
Shewn, D. G. P. M., Capt.	1154
Shiambahadur Limbu, Subdr.	1548
Shiam Sing Rana, Subdr.	1614
Shiam Singh, Sub-Asst. Surgeon	1817
Shiama Singh, Jemdr.	1021
Shihbu, Subdr.	1228
Shibcharan Burathoki, Subdr.	1607
Shib Charan Das, Sub Asst. Surgn.	1817
Shibdhoj Mal, Subdr.	1642
Shibsita Singh, Risaldr.	919
Shib Dyal, Jemdr.	1171
Shib Lal, Subdr.	1039
—— Jemdr.	1405
—— Subdr.	
—— Subdr.	1215
—— Subdr. Maj.	1060
—— Singh, Ressdr	1685
—— Danu, Jemdr.	1214
Shibu Gharti, Jemdr.	1608
Shibprasad Thapa, Subdr.	1645
Shield, G. E. R., Capt.	697
Shields, C. J., 2nd-Lt.	363
—— F. T., Lt.	415
Shier, M. B., Lt. 389,	1655
Shidrawapa Sangapa Motgi, Jemdr.	1417
Shinab Din, Subdr.	1283
Shikari Gujar, Subdr.-Maj	1471
Shikandar Khan, Jemdr.	1579
Bahadur, Subdr. Maj.	1583
Shillingford, C. A. N., 2nd-Lt.	552
Shillong, J. H., Asst. Surgn.	1794
Shillitoe, G. R., 2nd-Lt. 539,	1672
Shilstone, W. C. N., Lt.	511
Shilu Singh, Jemdr.	1021
Shim Bhoo, Jemdr.	886
Shimbu Lal, Jemdr.	1197
—— Singh, Jemdr.	1071
Shine, E. P. F. Maj. 226, 1662,	1685
Shipley, H., 2nd-Lt.	1511
Shipman, H. J., *Hony.* Maj.	73
—— H. L., Lt. 460—462,	1495
Shipp, G. R. G., 2nd-Lt.	568
—— W. E., Maj.	826
Shippard, H. C., Lt.	756
—— S. S. A., Lt. 318,	964
Shipster, G. C., Capt.	653
—— W. N., Capt.	758
Shipton, G. L., Lt. 322,	1425
Shir Singh, Ressldr.	1683
Shiraz Gul, Jemdr.	1228
Shirgaokar, S. R., Ty. Maj.	1763
Shirgaonkar, J. V., Tempy. Capt. 14, 39,	1766
Shirilal Thapa, Subdr.	1602
Shirley, E. A., 2nd-Lt.	637
—— G. S., 2nd-Lt.	555
—— S. R., Capt 258,	1239
Shirreff, A. G., 2nd-Lt. 554,	949
Shishpal Singh, Subdr.	1021
Shiuambar Singh, Subdr.	1046
Shiubalak Singh, Subdr.	1383
Shiubakhsh Singh, Jemdr.	1021
Shiubaran Misr, Jemdr.	999
Shiubardan Singh, Jemdr.	1064
Shiubhajan Dube, Subdr.	1016
Shiubodan Singh, Subdr.	1044
Shiuchand, Subdr.	1059
—— Jemdr.	1059
Shiu Balak Singh, Jemdr.	1051
—— Dat, Subdar	1039
—— Datt Singh, Jemdr.	999
—— Dube.	1017
—— Singh, Subdr.	1215
—— Lal, Jemdr.	1052
—— Nath Singh, Subdr.	1521
—— Narayan Ram, Jemdr.	1488
—— Singh, Jemdr.	1365
—— Pal Singh, Jemdr.	1048
—— Pershad Singh, Jemdr.	1047
—— Ratan Singh, Jemdr.	1581
—— Rattan Dube, Jemdr.	1017
—— Shankar Misr, Subdr.	989
—— Ram Singh, Subdr.	1452
—— Jemdr.	1396
—— Singh, Jemdr.	1385
Shiucharan Singh, Subdr.	1452
—— Jemdr.	1024
—— Jemdr.	1043
Shiudan, Ressdr.	903
—— Jemdr.	1336
Shivdan Singh, Jemdr.	1412
Shivdas Singh, Jemdr.	1389
Shiudarshan Singh, Jemdr.	1044
Shinhad Singh, Jemdr.	1400
Shiumbar Singh, Subdr.-Maj.	1043
Shiugopal Pande	1019
Shiuji Ram	1419
Shiukaran Singh, Jemdr.	1047
Shiulal, Subdr. (I)	1039
Shiumangal Tiwary, Jemdr.	1027
Shiunandan Misr, Subdr.	1026
Shiunarain Singh, Jemdr.	1043
Shiunath Misr	1027
Shiunath Singh, Jemdr.	1047
—— Jemdr.	1083
—— Jemdr.	
—— Jemdr.	1064
—— Subdr.	1046
Shiupal Singh, Jemdr.	1047
Shiuparshad, Sub-Asst. Surgn.	1810
—— Krishnalal Vyas, Sub-Asst. Surgn.	1830
—— Tewari, Subdr.	1025
—— Jemdr.	1017
Shiuram Singh, Subdr.	1022
Shiurutan Singh, Jemdr.	1043
Shiv Lal, Jemdr.	1417
Shivram Sajjan Jemdr.	1420
Shiushanker Singh, Subdr. 988,	1397
Shivalal T. O., Sub-Asst. Surgn.	1832
Shivchand, Jemdr.	1388
Shiv Charan Das, Sub.-Asst. Surgn.	1817
Shivram Naik, Jemdr.	1441
Shiv Nath Tewari, Jemdr.	1575
Shiv Ram, Sub-Asst. Surgn.	1814
—— Nath, Subdr.	1583
Shivram Vithal Patekar, Sub-Asst. Surgn.	1831
—— Babaji, Patel, Sub-Asst. Surgn.	1829
Shivramhinjad, Subdr.	1460-62
Shiv Ramdas, Sub-Asst. Surgn.	1813
—— Singhji Karan Singhji, 2nd-Lt.	111
Sholdham, W. F. Q., Lt.	1419
Shore, N. B. P., Lt.	542
—— O. B. S. F., Col. 97, 105, 123, 149,	181
—— R., Lt.-Col. 1786,	1901
Shores, C. D. T., 2nd-Lt.	854
Shorrock, A. G., Lt.	825
—— J. C., Lt.	787
Short, H. E., Capt.	1756
—— F. P., Lt.	697
—— J. McL., Lt. 310,	1211
—— R. L. B., 2nd-Lt. 546,	1008
—— S. E., Lt.	1676
—— V. C. G. O'B., Lt. 317, 1020,	1070
Shorten, J. A., Capt.	1754
Shorter, L. V. H., 2nd-Lt.	683
—— R. M., Lt.	718
Shortland, C. V. N., Maj	851
—— R. B., 2nd-Lt. 548,	1307
Shortridge, G. C., Lt.	486
—— J. J. F., Condr.	1853
Shortt, G. A., Capt. 89,	791
—— N. R., Lt.	508
—— W. H. O., 2nd-Lt. 548,	988
Shoubridge, C. A. G., Maj. 208,	1451
Shoulder, H. C., Lt.	68
—— H., Lt.	1006
Shout, W. G., Capt. 485,	854
Shove, H. L., Lt. 540a,	1672
—— N. A., Capt. 502,	1479
Showers, F., 2nd-Lt.	549
Shrader, G. J., 2nd-Lt.	1489—1490
Shrager, A. L., Lt. 504,	1511
Shrewsbury, O. L., 2nd-Lt.	730
Shrimpton, S. V., Capt.	1732
Shriniwas, R.B., Sub-Asst. Surgn.	1832
Shripad Hari Sathe, Sub-Asst. Surgn.	1831
Shrirang Ramchandra, Sub-Asst. Surgn.	1829
Shroff, A. D., Tempy. Capt.	1768
—— J. D., Ty. Lt.	1781
Shrubsall, P., 2nd-Lt.	1008
Shubart, S. D., 2nd-Lt. 564,	1678a
Shuell, C. A. J., 2nd-Lt. 355,	1044
Shujat Ali, Jemdr.	1583
Shujait Ali, Transport Vety. Asst.	1687
Shuker, H., Lt. 420,	1627
Shukhmandar, Subdr.	1492
Shukul Chand, Jemdr.	1208
Shuldham, W. F. Q., Lt. 371,	1419
Shunker, C. P. V., Asst. Surgn	1796
—— W. H. D., Asst. Surgn.	1800
Shunmagan, Jemdr.	1012
Shurey, H. R., 2nd-Lt.	626
Shute, A. F., Lt.	657
—— C. A., Capt. 53, 259,	1431
—— G. E. F., Capt. 238—290,	1094
—— P. B., Lt. 387,	1511
Shuttlewood, A. E., 2nd-Lt.	640
Shuttleworth, A. R. B., Maj. 99, 211,	1651
—— B. W., Maj. 99, 220,	1203
—— D. I., Bt.-Lt.-Col. 51, 98, 137, 159, 212,	1606
—— E. H. S., Lt. 318,	1235
—— F., 2nd-Lt.	353
—— J. T., Lt. 835,	491
Shyamlal Khattri, Jemdr.	1631
Sialu Singh, Jemdr.	918
Sibbald, . W., Lt.	761
Sibban Singh, Jemdr.	1483
Sio Lal, Subdr.	1355
Sibson, A. J. B. M., 2nd-Lt.	591
Siddall, J. B., *Hony.* Maj.	585
—— T., Capt.	793
Siddiq Ahmad, Sub-Asst. Surgn.	1813
—— Y. M.,	1781
Side, A. P. B., 2nd-Lt.	630
Sidebottom, S. R., Lt.	628
Sidey, C. L., *Hony.* Capt.	111
Sidgwick, C. K. ., Maj.	825
Sidley, C. L., 2nd-Lt.	1693
Sidmouth, G. W., *Visct.* Capt. 698,	795
Sidney, C. S., Sub-Condr.	1873
Sid Ram Nikam, Jemdr.	1440
Sidram Pille, Jemdr.	999

NAMES AND RANK. Page.

Siedle, C. E., Lt. 519, 1026
—— F. J., Lt. 505, 1665
Sienesi, A. R., Lt. . . . 445, 1614
Sievwright, A. G. H., 2nd-Lt. . 547
—— F. J. H., Capt. . 825
Sigrist, A. E., Lt. 736
Sikandar Khan, Jemdr. . . . 1320
—— —— Subdr. . . . 1513
—— —— *Hony.* Capt. . . 2102
—— —— Shah, Ressdr. . 1683
Sikes, F. G., Capt. 9, 1839
Sikdar, J. M., Ty. Capt. . . . 1772
Sikhdar, Jemdr. 1035
Silas San Wa 1291
Silberbauer, H. L., Lt. . 297, 1068, 1617
Sil Singh, Subdr. 1174
Silver, A., Revd. 1905
—— G. J., 2nd-Lt. 317
—— B. J., Capt. 768
—— H. E., Capt. . . . 274, 1090
—— J. R. R., Maj. 608
—— P. W., 2nd-Lt. 359, 1365, 1545
Silvester, C. J., Esq. 73
—— F. H., 2nd-Lt. . . . 781
Silvester Mayars, H., 2nd-Lt. . 1678a
Sim, A. J. A., Lt. 793
—— D., Lt. 1604
Simcox, A. H. A., Maj. 654
Simbhu Singh, Subdr. . . . 1031
Sime, E. A., 2nd-Lt. . . . 539, 1203
Simeon, G. N., Lt. 516a
Simkin, F. E., 2nd-Lt. 1679
Simkins, G. P., Lt. . . . 927, 1223
Simmon, W. W. J., Lt. 785
Simmonds, H., Lt. . . . 542, 1624
—— R., 2nd-Lt. . . . 721
—— V. B., Lt. . . 521, 1517
—— W. G., 2nd-Lt. . . . 618
—— W. W., 2nd-Lt. . . 754
Simmons, B. H., Asst. Surgn. . . 1795
—— C. D., Lt. 695
—— C. E., Lt. . . . 516a, 1667
—— D. A., Capt. 736
—— F., Lt. . . . 449, 903
—— F. D., Lt. 1571
—— F. S., 2nd-Lt. . . 359, 1273
—— J. M. R., 2nd-Lt. . 365, 1463
—— R. D. H., Lt. . . . 628
Simms, G. H., Lt. . . . 311a, 1259
—— J., Lt. . . . 443, 1403
—— F. H., Lt. 1036
Simon, A. I., Ty. Lt. 1774
—— A. O., Lt. . . . 524, 1483
—— C. E., Lt. 506
—— D. C. M., *Hony.* Capt. . 742
—— P. P., Condr. 1862
—— W. M., 2nd-Lt. 741
—— Jemdr. 2012
——**Simon-Peter,** G., Jemdr. 1012
Simonds, J. de L., Capt. . . . 634
—— J. N., Maj. . . 244, 941
—— M. H., Maj. . . 230, 887
Simons, A. L., 2nd-Lt. 563
—— G. E. L., Capt. . . . 1718
—— F. H., Lt. . . . 445, 925
—— J. G., Lt. 393, 1595
Simpson, A., Lt. . . . 735, 1886
—— A. A., 2nd-Lt. . . . 560
—— A. F., Capt. . . 258, 1387
—— A. R., Lt. . . . 305, 1007
—— C. A., 2nd-Lt. 1679
—— C. A. L., Sub-Condr. . . 1864
—— C. E., Lt. . . 536, 1618
—— C. N., Lt. . . . 82, 508
—— C. R., Maj.-Genl. . . . 695
—— D. T., 2nd-Lt. 630
—— E. S., 2nd-Lt. . 320, 1309
—— F. C., Lt. . . . 306, 1170
—— F. F., Capt. 1726
—— G. A. D., 2nd-Lt. 326, 1439
—— G. G., 2nd-Lt. . 759, 798
—— G. H., 2nd-Lt. . . . 558
—— G. P., Capt. 615
—— H. B., 2nd-Lt. . 359, 1287
—— H. C., Capt. 1722
—— H. J. A., Lt. . . 401, 1606
—— H. Mc. K., 2nd-Lt. . . 568
—— J., 2nd-Lt. 750
—— J., Maj. 107, 222, 1598, 1647
—— J. A., Lt. . . . 530, 1495
—— J. C., Bt.-Lt. Col. 137, 159, 211, 1020, 1082
—— J. H., Capt. 852
—— J. N., Lt. 440
—— J. S. C., 2nd-Lt. . 324, 1497
—— J. N., Lt. 913

NAMES AND RANK. Page.

Simpson, K. S., Lt. 1839
—— L. C., Lt. . . . 491, 1886
—— N. S., Maj. 1753
—— P. W., Lt. 785
—— R. C., Capt. . . 263, 1517
—— R. G., Capt 1718
—— R. M., 2nd-Lt. 653
—— R. W., Capt. 1839
—— S. R., Lt. 955
—— T. W., 2nd-Lt. . . . 67
—— V. E. W., Capt. . 357, 1193
—— W. A., Lt. 450—54, 1611
—— W., *Hony.* Capt. . . . 718
—— W. J., Capt. 1754
Simrat Singh, Jemdr. 1280
Sins, A. W., 2nd-Lt. 794
—— E. A., 2nd-Lt. 817
—— J. C., 2nd-Lt. 1008
—— J. H., 2nd-Lt. 760
—— R. J., 2nd-Lt. 637
Simson, A., Maj. 656
—— G. N., Lt. 1609
—— G. O., Capt. . . 274, 939
—— R., Capt. 1919
Sinclair, A. C., Asst. Surgn. . 1798
—— A. V., 2nd-Lt. . . 562, 1303
—— G., 2nd-Lt. 565
—— J. R., 2nd-Lt. 693
—— H. D. B., Maj. 846
—— M. C., 2nd-Lt. . . 332, 1513
—— N., Lt. 523, 1517
—— R. J., Capt. . . . . .
—— W., 2nd-Lt. 561
—— W. C. B., Capt. 18, 47, 776
—— W. V. S., Lt. . . . 33, 747
—— **Sinduraju,** Jemdr. . . . 994
Sinclair-Day, A. R., 2nd-Lt. . . 561
Singaram, Jemdr. 994
Singaravelu, Jemdr. 1333
Singer, C. A., Lieut. . . 668, 1006
—— E. O. A., Lt. . . . . .
Singh Bahahadur Limbu, Subdr. 1630
Singbir Gharti, Subdr. . . . 1615
Singbaj Limbu, Jemdr. . . . 1055
Singbir Rai, Subdr. 1648
—— —— Gurang, Jemdr. . . 1611
—— —— Ghale, Hony. Capt. . 2103
Singha, Singh, Jemdr. 1452
Singh, A. B., Ty. Sub-Asst. Surgn. 1833
—— B. Ty. Lt. 1775
—— B. *Hony.* Ty. Lt. . . . 1782
—— B. H., Capt. 1758
—— B. H. G., Sub-Asst. Surgn.
—— B. J., Lt.-Col. . . . 128, 1641
—— D. R. R., Ty. Capt. . . . 1769
—— G., Ty. Capt. 1771
—— Hukum 91
—— K. S., Maj. 1753
—— M. A., Tempy. Capt. . . . 1766
—— N., Tempy. Capt. . . . 1768
—— R., Tempy. Lt. 1777
—— S., Tempy. Capt. . . . 1772
—— S., Tempy. Lt. 1763
Singha, N. L. 91
Singleton, *C. Actg. Capt.* . . . 1006
—— J. 2nd-Lt. 1689
Singbhaj Limbu, Jemdr. . . . 1649
Sinha, J. C., Tempy.-Lt. . . . 1778
—— J. D. Ty.-Lt. 1778
—— N. P., Maj. 1786
Sinker, R. C., 2nd-Lt. . . 326, 1624
Shinkandar Khan, Jemdr. . . . 1493
Sinton, C. McD. Maj. 747
—— J. A., Capt. . 1639, 1756, 1925
—— J. S., Lt. 371, 1397
Sirajuddin, Sr. Sub-Asst. Surgn. 1809
—— Jemdr. 966
Sircar, G., 2nd-Lt. Hony. . . . 482
Sirbaland Khan, Jemdr. . . . 946
Sirdar Autar Singh, Risaldar . 918
—— **Hamid Hasan,** Sub-Asst. Surgn. 1818
—— **Jasjit Singh,** 2nd-Lt. . 481
—— **Khan,** Jemdr. 926
—— **Norendra Singh,** Ressdr. 1686
—— **Piar Singh,** Subdr. . . 1135
—— **Sant Singh,** Risaldr.-Maj 918
—— **Singh,** Jemdr. 914
Sirichand, Jemdr. . . . . . .
Sirigange Rai, Jemdr. 1631
Sirikishn Thakur, Jemdr. . . . 1653
Siri Lal, Ressd. 884
—— Lala, Jemdr. 1193
—— Ram, Jemdr. 938
—— Jemdr. 1165

NAMES AND RANK. Page.

Sirilal Thapa, Subdr. 1653
Siria Gharti, Subdr. 1618
Sirkar, K. K., Ty. Lt. . . . 1779
Sirpat Rao Ghag, Jemdr. . . . 1447
Sirput Pudwal, *Hony.* Capt. . . 2031
Sirpati Kadam, Subdr. . . . 1416
—— **Powar,** Jemdr. . . . 1416
Sirmur, Bir Bikram Singh of—— *Hony.* Lt.-Col. 110, 986
—— H. H. Raja *Sir* Amar Parkash of . 110
Sirrell, E., 2nd-Lt. 1008
Sispal Singh, Subdr. 1031
Sisley, F. W., 2nd-Lt. . . . 476, 1351
—— J. L., Sub-Condr. . . . 1874
Sis Ram, Subdr. . . . . . .
Sis Ram, Singh, Jemdr. . . . 930
—— Jemdr. 1039
—— Jemdr. . . . 1039, 1385
Sital Pershad Singh, Jemdr. . . 1063
—— **RamVarma,** Sub-Asst. Surgn. 1807
—— **Singh Lama,** Hon. Lt. 2104
—— Subdr. 1036
Sita Ram Barge, Subdr. . . . 1511
—— Gaode, Jemdr. 1424
—— Jemdr. 1579
—— Ghogle, *Hony.* Lt. . . 2104
—— Misr, Supdt. 1016
—— Parab, Subdr. . . . 1444
—— Sabarwal, Sub-Asst. Surgn. 1817
—— Singh, Jemdr. . 930, 1384
—— Subdr. Maj. 1891
—— Jadho, Subdr. 1467
Sitaram Sellar, Jemdr. . . . 1468
Sitaramrao Vichare, Jemdr. . . 999
Sitaram Vishnu Sathe, Sub-Asst. Surgn. 1821
Sitlabaksh Singh, Jemdr. . . . 1063
—— —— Jemdr. . . . 989
—— —— Jemdr. . . . 1044
—— —— Subdr. . . . .
Sitladin, Misr. 1017
Siunath Nath Singh, Jemdr. . . 1476
Sitwell, N. S. H., Maj. . 171, 633, 1884
Sitzler, E. A., Lt. . . 513, 1323, 1591
Sivaramulu, Jemdr. 1328
Sivalingam, Jemdr. 1272
Sivansankaran, A. J. Ressdr. . . 1683
Sivary, H. S., Lt. . . . 440, 1150
Sivasami, Subdr. . . . 1264, 1265
Siwa Moduck, *Hony.* Capt. 1407, 2101
Skead, B., Lt. 430 1143a
Skeen, A., Maj.-Genl. 6, 13, 97, 119, 147, 175
Skehan, W. J., Condr. 1851
Skellon, P. E., 2nd-Lt. 567
Skelton, J. W. A., 2nd-Lt. . . .
—— W., Lt. 797
Skemp, F. W., Mr. 22
Skempton, A. W., 2nd-Lt. . . . 556
Skene, J. G., Maj. . . . 230, 1635
—— S. G. S., Lt. . . 306, 1635
—— W. F., 2nd-Lt. . . . .
Sketchley, B. T., Lt. . . 758, 761
Skey, L. G., Lt. 1299
Skillter, W. I., Lt. 307
—— W. M., Lt. 1497
Skinner, A. B., Maj. . 67, 218, 883
—— A. L., Capt. . . 268, 1483
—— E. M., Capt. 262, 1489, 1497
—— F. E., 2nd-Lt. 552
—— F. H., 2nd-Lt. . 460—62, 1068
—— F. J., *Hony.* Maj. . . . 1847
—— F. M., 2nd-Lt. 1371
—— F. S. D., Lt. 725
—— G. E. M., Capt. . . 66, 689
—— G. J., Col. 177
—— G. S., Maj. . . . 233, 1247
—— H. T., Maj. . . . 218, 1136
—— H. W., Lt. 629
—— J. MacG., Capt. 1752
—— J. R. R., 2nd-Lt. . 494, 1678
—— L. S. C., Lt. . . . 435, 949
—— R. H., Maj. . . . 243, 1371
—— R. McK., Maj. 1711
—— R. O., Capt. 636
—— S. E., *Hony.* Capt. . 111, 875
—— T. B., Maj. 236, 1275, 1587
—— W., Sub.-Condr. . . . 1857
—— W. E., Lt. . . . 517, 1677
Skipwith, C. G. Y., Capt. . 271, 907
—— F., Maj. . . 99, 214, 1617
—— F. E., Capt. 826
—— P. A., Maj. 78
Skone, H. A., 2nd-Lt. . . 329, 1606

Names and Rank. Page.

Skrimshire, F. B. B., Capt. . . 1714
Skrine, P. B. H., Lt. . . . 309, 1070
Slack, H. B., 2nd-Lt. . . . 715
Slacke, G. C., Maj. . . 244, 905, 1893
Slade, F., *Hony.* Lt. 1849
—— G. E. R., 2nd-Lt. 825
—— J. W. C., Lt. . . 318, 1409
—— M., Lt. . . . 300, 1275
Sladen, D. R., Brig.-Genl. 12, 37, 38
—— W. D., Lt. 716
Slancy, W. G. 33
Slark, J., 2nd-Lt. 563
Slater, A., Esq. 88
—— A., Lt. . . . 506, 1427
—— A. E., 2nd-Lt. . . 82, 556
—— A. C. N., Lt. . . 535, 1705
—— A. F., Capt. 815
—— A. F. M., Maj. . . 1233, 1895
—— A. W., Maj. 807
—— D. McL., Maj. 239, 1032, 1052
—— H., Capt. 638
—— O., Capt. . . . 55, 665
—— R. A., Lt. . . . 380, 1355
—— S. H., 2nd-Lt. 547
Slatter, C. B., 2nd-Lt. 564
—— J. C., 2nd-Lt. 1489
Slaughter, C. M., Lt. . . 525, 1507
—— G. A., Capt. . . 68, 797
—— H. L., Lt. 522
Sleigh, A. L., Capt. 806
—— G. P., Lt. . . . 367, 907
Slenesi, A. R. . . . . . 446—49
Slesser, W. D. V., Lt. . . . 95, 490
Slessor, J. G., Lt. 796
Sleven, E. J., 2nd-Lt. 637
Slim, W. J., Lt. . . . . .
Slingsby, T. W., Maj. 211, 872, 917, 1581
Sloan, R., 2nd-Lt. 567
Sloane, W. H. St. J., Lt. . . 529, 1058
Slocombe, W., 2nd-Lt. . . 355, 1034
Sloper, H. C., 2nd-Lt. . . 474, 1889
Sloot, J. N. C., 2nd-Lt. . . . 740
Sly, F. C., Lt. 1846
Slynn, H. J., 2nd-Lt. 689
Smales, C. B., Lt. 848
—— R., 2nd-Lt. 791
Small, E., Lt. 710
—— C. G., 2nd-Lt. . . 548, 1009
—— G. A., Lt. . . 489, 907, 1007
—— J. H., 2nd-Lt. . . 357, 1008
Smalley, J., Maj. 1753
—— S., Lt. . . . 514, 1287
Smallman, W. F., 2nd-Lt. . . . 1195
Smallpeice, F. W., Maj. 19, 681, 785
Smallwood, E. S., Ty. Lt. . . 638
—— F. G., Lt.-Col. 31, 74, 132, 632
—— H. St. C., Lt. . . . 486
—— L. S., Lt. 302, 1383, 1581
—— R., 2nd-Lt. . . . 324
Smart, A. W., Lt. . . . 608, 1585
—— C. N. B., Lt. . . 517, 998
—— E. L., 2nd-Lt. 357
—— G. B., 2nd-Lt. 744
—— G. M., Lt. 624
—— G. R., Lt. 681, 785
—— J. H., Lt. . . . 536, 988
—— J. N., Lt. . . 46, 353, 1676
—— R., *Hony.* Lt. 715
—— R. L., Lt. 543, 1601
—— R. W. L. De B., Maj. 336, 915
—— 2nd-Lt. 1676
Smedley, E. A., 2nd-Lt. . . . 721
Smeeth, W. C., 2nd-Lt. . . 996, 1170
Smellie, C. S., Lt. 492
Smelt, M. V., Lt. . . 297, 365, 964
Smerdon, H. M., Lt. 697
—— C. P., 2nd-Lt. 1265
—— H. E., 2nd-Lt. 1243
Smethurst, S. J., Lt. 527
Smith, A., Lt.-Col. . . . 1842, 1893
—— A., 2nd-Lt. . . . 360, 1473
—— A. (*temp. Capt.*) . . . .
—— A. A., Maj. 101, 214, 872, 1255, 1565
—— A. A., Lt. 399, 933
—— A. B., Lt. . . 291, 1191, 1493
—— A. C., Lt. 630
—— A. D., Capt. . . . 257, 1600
—— A. F., 2nd-Lt. 1391
—— A. F., Lt. 741
—— A. F., 2nd-Lt. . . . 780, 799
—— A. G. C., 2nd-Lt . . 741, 1678a
—— A. H., Maj. 653
—— A. H. D., Capt. 1724
—— A. H., 2nd-Lt. 358
—— A. K., Lt. . . 489, 604, 897
Smith, A. L., Lt. 311a
—— A. L. F., Lt. . . . 733, 734
—— A. Le F., Lt.-Col. 137, 158, 197, 868, 1020, 1042
—— A. R., 2nd-Lt. . . 547, 640, 1156
—— A. S. D., 2nd-Lt. 565
—— A. T. H., 2nd-Lt. . . . 754
—— B E., *Hony.* Maj. . 10, 74, 1844
—— B. G., Lt. 817
—— B. K. Lt. 986
—— B. L., Sr. Asst. Surgn. . . 1791
—— B. L. S., Lt. . . 305, 1493, 1501
—— B. St. J., Capt. 502, 807, 1435, 1583
—— C., 2nd-Lt. . . . 73, 767, 790
—— C. A., Lt.-Col. . . . 198, 1912
—— C. A. B., Lt. 543, 1094
—— C. A. J., Maj. 224, 439, 871, 1441
—— C. C. H., Lt. . . 311a, 1030, 1521
—— C. G., Maj 849
—— C. G., 2nd-Lt. 734
—— C. H., Capt. 1755
—— C. J., Asst. Surgn. . . . 1802
—— C. M., Capt. . . . 500, 1670
—— C. P., Asst. Surgn. . . . 1799
—— C. S., Lt.-Col. 1710
—— C. T., Lt. 525
—— C. T. M., Bt. Maj. . . 261, 1355
—— C. W. J., Maj. . 227, 1136, 1241
—— D., 2nd-Lt. 764
—— D. B'Arcy, 2nd-Lt. . . . 836
—— D. E., Lt. . . . 488, 873, 1892
—— D. F., Lt. 479, 1225
—— D. G. S., Lt. 1130
—— D. R., Lt. . . . 415, 1156, 1589
—— E., Lieut. 561, 804
—— E., Lt. . . . 524, 856, 1645
—— E., 2nd-Lt. . . . 565, 1678a
—— E. C., Maj. 119, 869, 1106, 1134
—— E. C., Lt. 697
—— E. D., Tempy. Capt. . . . 1724
—— E. D. D., Capt. . . . 253, 1128
—— E. H., 2nd-Lt. 1034
—— E. K., Lt. 1839
—— E. L., Condr. 1861
—— E. S., 2nd-Lt. 549
—— E. T., Lt. 316, 1641
—— E. W., Sub-Condr. . . . 1874
—— E. W., Lt. 638
—— F., 2nd-Lt. 552, 1249
—— F. A., Lt.-Col 69, 135, 1744, 1899
—— F. A., Br.-Genl. 12, 37, 120, 125, 150, 185
—— F. A. L., Nursing Sister . 1735
—— F. C., Maj. 1786
—— F. E., Capt. 1732
—— F. F. S., Capt. 1755
—— F. G., Lt. . . . 446—49, 1491
—— F. G., 2nd-Lt. 561
—— F. H., Revd. 1904
—— F. H., 2nd-Lt. . . . 700, 761
—— F. J., Capt. 738, 740
—— F. M., Lt. 463, 1455
—— F. N., Lt. 73, 695
—— F. R. Sr. Asst. Surgn. . . 1791
—— G., Col. 779
—— G. A., Lt. 489, 903
—— G. B., Brig.-Genl. 58, 120, 123, 612
—— G. C. A., Revd. 1906
—— G. E., Lt. 759
—— G. J., *Sir Hony.* Col. . . 722
—— G. L. S., Capt. 636
—— G. M., Lt. 520, 987
—— G. R., Lt. . . 508, 1379, 1577
—— G. McL. C., Lt.-Col . 120, 1618, 1742
—— G. W., Lt. 1413
—— H., 2nd-Lt. 557
—— H., Lt.-Col. 128
—— H., Capt. 1717
—— H. A., Lt.-Col. 2, 129, 1742, 1890
—— H. A., Lt. 527, 1667
—— H. B. Lt.-Col . . . 135, 613
—— H. C. E., 2nd-Lt. . . . 568
—— H. F., Sub.-Condr. . . . 1857
—— H. D., 2nd-Lt. 700
—— H. C., Lt. . . 543a, 794, 1291
—— H. H., Bt. Lt.-Col. 160, 215, 1050
—— H. M., Esq. 88
—— H. M., Lt. . . . 425, 709, 1601
—— H. McM., Lt. 540, 1701
—— H. P., Lt. 543
—— H. N., 2nd-Lt. 1417
—— H. R. W. M., Lt.-Col. . . 67
—— H. S., Lt. 511, 1006
—— H. T. P., Revd. 1903
Smith, H. V., 2nd-Lt. . . 568, 1245
—— H. W., Sub-Condr. . . . 1873
—— J., 2nd-Lt. 18, 42, 328, 547, 1182
—— J. A. Capt. 84
—— J. A., 2nd-Lt. . . . 552, 1086
—— J. B., Col. . . . 8, 23, 127, 1738
—— J. B., Maj.-Genl. 173
—— J. C., 2nd-Lt. . . . 473, 1275
—— J. C., 2nd-Lt. 790
—— J. F., Revd 1903
—— J. F. S., 2nd-Lt. 363
—— J. H., Capt. 1758
—— J. H., Lt. 731
—— J. H., Lt. 523, 1601
—— J. I., Capt. 854
—— J. L., Maj. 209, 1661
—— J. L. L., 2nd-Lt. 714
—— J. L. V., 2nd-Lt. . . 328, 1066
—— J. M., Lt.-Col. 183, 1890, 1911, 1925
—— J. M., *Hony.* Maj. 1847
—— J. MacD., Col. 176
—— J. N., Condr. 1867
—— J. P., 2nd-Lt. 757
—— J. W., Sub-Condr. 1864
—— K. C., Lt. . . . 522, 1670, 1702
—— L., Capt. 268, 885
—— L. A., Lt.-Col. . 58, 122, 132, 613
—— L. A., 2nd Lt. 726
—— L. A. H., Lt. 543a, 975
—— L. B. S., Lt. 375
—— L. E., Lt. 321, 1451
—— L. F., Lt. 1044
—— L. R., Lt. 1140
—— L. W., Lt. 636
—— L. W., Lt. 491, 923
—— M. C., Maj. . . . 68, 244, 1275
—— M. H., Lt. 489
—— M. H. H., Capt. 288-290, 1223, 1623
—— M. I. L., Lt. 510
—— M. L. C., Lt. 507, 1042
—— M. R., Lt. 696
—— N. V., Lt. 380, 1160
—— O. F., Maj. 7, 230, 891
—— O. L., Lt.-Genl. 172
—— P., Condr. 1853
—— R., Ty. Lt. 622
—— R., 2nd-Lt. 848
—— R., *Hony.* Lt. 1843
—— R. A., Lt 624, 976
—— R. C. K., 2nd-Lt. 1223
—— R. C., 2nd-Lt. . . 552, 1499—1500
—— R. E., Lt. 516a, 1667
—— R. G., Lt. 312, 1495
—— R. G., Asst. Surgn 1799
—— R. H., Asst. Surgn. . . . 1802
—— R. H., Lt. 526
—— R. H., 2nd-Lt. 474 913
—— R. H. V., 2nd-Lt. 595
—— R. I., Lt. 725
—— R. J., 2nd-Lt. 1106
—— R. N., 2nd-Lt. 705
—— R. N. D., Lt. 513, 1665
—— R. O., Lt. 540, 1696
—— R. O. A., Asst. Surgn. . . 1802
—— R. S., Lt. 425, 740
—— R. S., Lt. 425, 1018
—— R. T. W., Capt. 780
—— R. W., Lt. 55, 798
—— R. W. H., Lt. 516, 1678
—— S., 2nd-Lt. 354, 1674
—— S., 2nd-Lt. 563, 1367
—— S. A., Lt. 541, 1022
—— S. B., Lt. 491, 949
—— S. B., Lt.-Col. . . . 39, 129, 1741
—— S. C., 2nd-Lt. 361
—— S. H. P., Maj. . . . 60, 225, 662
—— S. H. M., Lt. 516a
—— S. J. 2nd-Lt. 553
—— S. W., Lt. 793
—— T., Sub-Condr. 1855
—— T. A., 2nd-Lt. . . . . . .
—— T. A., 2nd-Lt. 1624
—— T. N., 2nd-Lt. 330, 1623
—— T. G., Esq. 88
—— T. R., 2nd-Lt. 629
—— T. P., Lt. 538
—— T. S., 2nd-Lt. 568
—— T. St. C., Ty. Capt. . . . 1726
—— T. S., Lt. 571, 1704
—— V. C. E., Lt. . . . 14, 31, 791
—— V. H. S., Capt. . . 260, 877, 935
—— V. P., Maj. 663
—— V. S., Lt. 509, 723, 1668
—— V. T., Lt. 435, 1158
—— V. W., Lt. 487
—— W., Esq. 81

Names and Rank. Page.

Smith, W., Condr. 1873
—— W., 2nd-Lt. . . . 703, 779
—— W. A., Lt. 512
—— W. D., Maj. 801
—— W. D., Lt.-Col. 131
—— W. F., Lt.-Col. 101, 133, 157, 194, 1660
—— W. F., 2nd-Lt. 630
—— W. F. Capt. 801
—— W. G., 2nd-Lt. . . . 472, 1403
—— W. H., 2nd-Lt. 1195
—— W. H., 2nd-Lt. . 354, 560, 1387
—— W. H. R., Tempy. Capt. . 1724
—— W J. 619
—— W. M., Condr. 1854
—— W. M., N. Sister 1735
—— W. P., 2nd-Lt. 548
—— W S., Lt. . . . 292a, 1096
—— W. S., 2nd-Lt. . . 548, 1243
—— W. T., Lt. . . . 538, 602, 1701
—— Y. E., Maj. . . 238, 1279, 1281
Smith-masters, E. L. Lt. . . . 1007
Smitherman, H. S., 2nd-Lt. . . . 1008
Smithers, H. E., Lt. . 310, 1407, 1435, 1583
—— L. S. H., Maj. 216, 868, 1086, 1088
—— K. O., Lt. . 33, 7.2, 783
—— M. H. W., Capt. . 728, 798
Smithett, W. H., Lt. . . . 311, 1303
Smithson, E. A., 2nd-Lt. . . . 1475
Smithwick, D. J., Maj. 693
Smyth, C. A., 2nd-Lt. 1678
—— C. A. W., Maj. 14, 22, 236, 1016
—— E. G., 2nd-Lt. . . . 547, 1705
—— F. E., 2nd-Lt. 570
—— F., 2nd-Lt. 790
—— F. G. A., Capt. 1715
—— G. C. G., Lt. . . . 506, 1665
—— G. H., Lt. Capt. 82, 524, 814
—— G. J. W., Maj. . . 662, 1882
—— H. A., 2nd-Lt. 626
—— H. M., Lt. . . . 322, 1641
—— J., Lt. 1243
—— J. G., Capt 17, 24, 277, 1078, 1925
—— R., Sub-Condr. 1874
—— S. Lt. 512, 1243
—— S. G., Asst Surgn. . . . 1794
—— W. C., Lt.-Col. . 132, 661, 1881
—— W. J., Lt. 848
Smythe, A. C. G., Capt. 259, 883, 1293
—— A. H., Capt. . . . 843
—— L. G. F., 2nd-Lt. . 328, 1293
Smyth-Osbourne, D. G., Lt. 84, 505, 1641
Smyth-Richards, F. G., Capt. 698, 795
Snatt, S. G., 2nd-Lt. 356
Snell, W., Capt. 696, 737
—— W. J. B., Lt. . . . 371, 1255
Snelling, A. H. J., Lt. . 300, 1030, 1557
—— C., Lt. . . . . . 460—61
—— C. G., Lt. . . . 385, 1419
—— O. W., Lt. 1333
—— W. T., Sub.-Condr. . . . 1863
Sneyd, R. T. S., Lt. . . . 510, 997
Sniders, A., 2nd-Lt. 1048
Snodgrass, J., Lt. . . 508, 1624 1657
Snook, C. G., 2nd-Lt. . . . 473, 1197
—— F., Condr. 1853
Snow, A. E., Lt. 701
—— C. A., Lt. . . . 511, 1601
—— D. G., 2nd-Lt 779
—— F. W., Lt. 814
—— G. H., 2nd-Lt. 725
Snowden, W. C., Lt. 521
Snowdon, H. S. K., Maj. . . 100, 633
—— J. H., Sub-Condr. . . . 1855
Snoxall, C. E., Lt. 754
Soady, G. J. Fitz M., Br.-Gl. 120, 124, 149. 183
Soanes, L. J. W., Lt. 793
Soar, J., 2nd-Lt. 566
—— S. F., Lt. . . 527, 1667, 1671
—— T., 2nd-Lt. 1678a
Sobail Singh, Jemdr 922
Sobha Ram, Subdr. 1197
—— **Singh**, Risaldar 880
—— Subdr. . . . 1363
Sobh Narain Sukul, Subdr. . . 1659
Sodalamuthu, Jemdr 1265
Soden, E. E., Lt. 841
—— H. L. B., Lt. 1674
—— J. N., Lt. . . 300, 1493, 1513
Soganadan, A. D., Ty. Lt. . . 1775
Sohaj Khan, Jemdr. 1356
—— Jemdr. 1514
Sohan, Subdr.-Maj. 1321
Sohan Lal, Jemdr. 1052
—— **Ohohra**, Sub-Asst. Surgn. 1818
—— **Ram**, Jemdr. 1433
—— Subdr. 1289
—— **Singh**, Subdr. 1096
—— Jemdr. 1698
—— Jemdr. 1433
—— Jemdr. 1587
—— Jemdr. 942
—— Jemdr. 1157
—— Jemdr. 894
—— Subdr. 1111
—— Jemdr. 1083
—— Subdr. 1303
—— Jemdr. 168
—— Jemdr. 1158
—— Jemdr. . . . 1111, 1149
—— Jemdr. 1163
—— Jemdr. 1232
—— Ressdr. 878
—— Subdr. 1395
—— Subdr. 1303
—— **Sidhu**, Sub-Asst. Surgn. . . . 1818
—— Risdr. 899
—— Subdr. 1375
—— **Pannun**, Sub-Asst. Surgn. 1820
—— Jemdr. 950
—— Jemdr. 1216
—— Jemdr. 902
—— Ty. Sub.-Asst. Surgn. . 1883
—— **Pannun**, Sub-Asst. Surgn. 1820
—— **Sidhu**, Sub-Asst. Surgn. 1818
Sohann, Jemdr. 1171
Sohna Singh, Subdr. 1268
Sohawa Singh, Subdr. . . . 1157
—— Subdr. . . . 1160
—— *Hony.* Capt. . 2104
Sohbat, Jemdr. 1040
—— Jemdr. 1103
—— **Khan**, Jemdr. 1507
—— Jemdr. 920
Sohel Sing, Jemdr. 1095
Sohnu, Jemdr. 1260
Sokhey, S. S., Capt. 1757
Solano, B. A., Capt . 262, 1092, 1207
Solly, A. R., Lt. 302, 1463
—— P. R., Lt. 514, 1409
Solomon D'Rozario, Sub-Asst. Surgn. 1830
—— E. V., 2nd-Lt. 40
Solomon, M., Ty. Lt. 1775
—— E. V., Ty. 2nd-Lt. . . . 579
—— S., Senr. Sub-AsstSurgn 1826
—— W E. G., Capt. . . . 789
Solomon Benjaman, Jemdr. . . 1452
Solu, Subdr. 1157
Somaling, Jemdr. 1447
Somasundaram, T. P., Senr. Sub-Asst. Surgn. 1826
Somerfield, G. J., Sub-Condr. . . 1874
Somayaji, S. L., Ty. Lt. . . . 1777
Somerlad, H., 2nd-Lt. 693
Somerset, A., *Hony.* Col. *Duke of* 755, 756
—— C. W., Br.-Gen. . 14, 37, 120, 124, 150, 184
—— I., 2nd-Lt. . . . 473, 1423
Somervaille, G., Capt. 827
Somervell, D. B., Capt. . . . 753
Somerville, A. P., 2nd-Lt. . . 559, 1383
—— E. A., Capt. . . 263, 917
—— G. A , *Hony.* Lt. . . 1850
—— J A H. R., Maj . 634, 977
—— T., 2nd-Lt. . . . 566, 1339
Sommers Cox, S. J., Lt. . . . 1575
Sommerville, R., Lt. . . 440, 1091
Somnath Mahadeo Rawal, Sub-Asst. Surgn. . . . 1832
Sona Singh, Jemdr. 1493
Sondhi, R. C., Ty. Lt. 1773
—— S. D., Capt. 1760
Soni, B. D., Lt. . . . . .
—— **Singh**, Jemdr. 1457
Sonpal, Subdr. 1191
Sonpar, K. N. Tempy. Capt. . . 1770
Sonu Sawant, Jemdr. 1424
Sopari Singh, Subdr 1303
Sopana Thopth, Jemdr. 999
Sopana, The Watkar, Jamdr. . . 999
Soper, V., Lt. 738, 740
—— O. N., 2nd-Lt. . . . 548, 1407
Soper, W. T., 2nd-Lt. . . . 356, 1395
Sopper, E., Capt. 595
Soppit, H. J., 2nd-Lt. 748
Sopwith, G. E., Maj. . . . 664, 1882
—— R. J. 2nd-Lt. 626
Sorabjee Pallanjee 89
Sorabji Fardunji Gandhi, Sr. Asst. Surgn. 3
Sorby, W. J., Lt. 509, 1095
Soroshi Ranjan Sen, Sub-Asst. Surgn. . . 1815
Sorrell, F. G., 2nd-Lt. 687
Sorsbie, R. F., Brig.-Gen. 19, 108 120, 124, 660
Sotham, R. C., Lt. 752
Sothers, D. B., Lt. 511
Sotton, H. G., Lt.-Col. 776a
Souchon, G. H., Lt. 595
Soundy, P. R., Lt. 821
Sourinayagum, Subdr. 1139
Soutar, C. J. S., Lt. 629
Souter, A. B., Lt.-Col. 105, 137, 158, 197, 915
—— H. M. W., Lt.-Col. 136, 158, 196, 867, 901
—— J. H., Lt. . . . 435, 1441
—— L. F., Lt. . . . 510, 1215
—— W. D., Lt. . . 455—59, 1199
Southan, J. H. P., Lt. 750
Southcombe, J. C., Capt. . 73, 795
Southern, W., Lt. . . . 443, 1453
Southey, A., Lt. 858
—— B. I., Temp. Lt.-Col. . . 61
—— W. M., Col. . 12, 24, 25, 120, 125, 150, 188
Southgate, C., Lt. . . . 375, 1333
—— E. H. D., 2nd-Lt. . . 359
—— E. J., *Hony.* Lt. . . 1844
—— T. D., Lt. . . 375, 911
Southon, C. E., Maj. 1748
—— H. M. Lt. 799
Southron, A. H., 2nd-Lt. . . . 717
Southwell, W. B., 2nd-Lt. . . . 684
Souza, A., Lt. 658, 861
Sowar Khan, Jemdr. 1256
Sowerby, M. H., Maj. 654
Sowler, J., Lt. 312, 1457
Sowton, B. S., Lt. 304, 1509
Spackman, W. C., Capt. . . . 1757
Spafford, D. N., 2nd-Lt. . . . 630
Spain, G. A. R., Capt. . . . 267, 1415
Spalding, W. B. Maj. 72, 633
Spanier, L., Sub-Condr. . . . 1857
Spankie, C. G., Capt. . . . 259, 1191
—— H. V., Lt. . . . 302, 1074
Spargo, F. W., Lt. 543a, 1291
Sparham, A. C., 2nd-Lt. . . 359, 1581
Spark, A. J., 2nd-Lt. 614
—— F. E., 2nd-Lt. 799
—— J. A., Lt. . . . 312, 1095
Sparke, H. C., *Capt.* . . 60, 501, 838
—— W., Lt. 832
Sparkes, H. B, Lt. . . 353, 1355, 1575
—— H. W., 2nd-Lt. 1567
—— J. P., Lt.-Col. 56
Sparrow, R., Lt.-Col. . . . 132, 583
Spaull, C. M., Lt. 300, 1351
Spear, C. R., Lt. 305, 1255
Speed, A. V., 2nd-Lt. 704
—— B, 2nd-Lt. 704
—— W. A., Lt. 529
Speer, C. H., Lt. 1475
Speke, W. H., Maj. 705
Speldewinde, C. G. O., 2nd-Lt. 564, 1183
Spence, A. D. Maj. 828, 868
—— W. M. 2nd-Lt. 324
—— A. H. O., Lt.-Col. 4, 103, 104, 135, 158, 195, 949
—— A. L., 2nd-Lt. 565
—— J., Lt. 455, 1561
—— J. M. . . . . . . 460—67
—— P. T. A., Lt.-Col. 1871, 1899, 1911
—— R. A., Lt 604
—— R. B., Maj. . . 54, 244, 1391
—— R. M., Lt. 655
—— W. J. F., 2nd-Lt. 561
—— W. M. G., 2nd-Lt. . . . 324
Spencer, B. G., 2nd-Lt. 1701
—— E. C., 2nd-Lt. 315
—— F. R. B., Esq. 24
—— G. W., Lt. . . . 425, 1140
—— H., Lt. 681
—— H. A., 2nd-Lt. 618
—— H. J., Lt. 304, 1026
—— H. Y., 2nd-Lt. 563

Names and Rank. Page.

Spencer, J. C., Lt. . . . 435, 1267
—— J. S., Lt. 708
—— L., Capt. . . . 265, 1046
—— L. A., 2nd-Lt. . . 332, 939
—— R. B., Tempy. Capt. . . 1765
—— R. D., Lt. 630
—— R. R., Lt. 789
—— W., 2nd-Lt. . . 494, 959
—— W. E. C., Lt. . . 401, 1365
—— W. G., Lt. . . 352, 1676
—— W. N., Lt. . . 669, 997
—— J., 2nd-Lt. 748
—— W. J. F. 2nd-Lt. . . . 91
Spens, W. P., Capt. 681
Spicer, N., 2nd-Lt. 709
Spiers, A. J. O., Ty. Lt. . . . 578
Spires, J., Sub-Comdr. . . . 1874
Spiller G. C., 2nd-Lt. . . . 753
—— T. R., Lt. . . 508, 1665
—— W., Sub-Condr . . 1861
—— W. M. H., Lt.-Col. . . 1710
Spin Gul, Subdr. 1247
Spink, F., 2nd-Lt. . . . .
—— H. H. M., Capt. . 501a, 986
Spinks, C. W. F.. *Hony*. Lt. . . 1846
—— J. H. C., 2nd-Lt. . . . 564
—— W., Lt. 617
Spitteler, A., Maj. 1747
Spittal, J., Lt. 588
Spivey, J. A., 2nd-Lt. 362
Splane, E. G., Lt. . . . 520, 1001
Spong, W. A., Capt. . . . 39, 1713
Spooner, A., Condr.. 1853
—— A. F., 2nd-Lt.. . 554, 1273
—— F. V., Capt. . . 277, 931
—— H., Revd. 1904
—— H., 2nd-Lt. 324
—— J. D., 2nd-Lt. . 545, 1008
Spottiswoode, A. U. M., Capt. 287, 1617
Spowart, W. C., Lt. . 506, 1671
Spragge-White, T. E., Lt. 77, 521, 1885
Sprawson, C. A., Maj. 1747
Spreadbrow, F. J., 2nd-Lt. . 359, 1078
Spring, G. C., Lt. . . . 513, 1708
Sprot, E. W. H., Capt.. . 581, 582
Sprott, F. H., Lt. 537
Sproule, C. L., Ty. Capt. . . . 1724
—— H. H., Maj. . 21, 207, 881
—— R. F., Lt. . . 318, 925
Spruce, P. A., Lt.. 798
Spurgeon, C. E., Lt. 838
—— F. W., 2nd-Lt. . . . 1327
Spurgen, J. C., 2nd-Lt. . . . 739
Spurgin, A. R., 2nd-Lt. . 330, 1066
—— E. W. Capt. . . 269, 929
—— H. S., 2nd-Lt. . . . 754
Spurrier, J. R., 2nd-Lt. . . . 581
—— R. S., Capt. 581
Spurway, F. E., Lt. 705
—— W. R. P., Lt. . 315, 964
Squire, J. H., Capt. 628
—— L. G., Lt. 1673
Squires, A. W. T., 2nd-Lt. . 356, 988
—— E. K., Maj. . . 58, 664
—— J. H., Sub-Condr. . . 1855
Squirrell, E. A., Capt. . . . 760
Sri Bahadur Singh, Subdr. . . 1476
Sri Chand, Jemdr. 888
—— Subdr. 1453
—— Jemdr. 1320
Srikishn Singh, Jemdr. . . . 1412
Srikishna, Subdr. 1191
—— Singh, Jemdr. . . . 1583
Srikrishna Vasishtha, Sub-Asst. Surgn. 1818
Srinivasamurti, G., Ty. Lt. . . 1777
Srinivasalu, Subdr. 1011
Sripal Singh, Jemdr. 1051
Sripat, Ressdr. 921
—— Barge, Jemdr. 999
—— Tiwari, Jemdr. . . . 1368
Sri Ram, Sr. Sub-Asst. Surgn.. . 1808
—— Jemdr. 1391
—— Sub-Asst. Surgn. . . . 1816
—— Misr, Jemdr.. 1017
Srish Chandra Chakravarti, Sub-Asst. Surgn. 1818
Stabb, C. T., Lt.. . . 685, 700, 795
Stubbings, R. A., Lt., ... 1140
Stable, H. H., Lt. . . . 385, 740
—— R. H., Capt. . 277, 949, 1483
Stables, A. D., 2nd-Lt.. . . . 1695
—— F. H. A., 2nd-Lt. . 333, 1617
—— J. H., Lt. 514
Stace, R. E., Capt. 664
Stacey, A. F., 2nd-Lt. . . 530, 1030
Stacey, L. B., 2nd-Lt. . . 332, 1134
Stack, C. E., Col. 170
—— C. S., Lt.-Col. 131, 156, 191, 867, 937
—— G. H., Bt.-Lt.-Col. 58, 138, 663
—— H. T., Maj. . . . 46, 1718
Staddon, C. E., Lt. 752
—— H. E., Lt.-Col. . 133, 1710
—— T. W., 2nd-Lt. 696
Stafford, H. C. C., Lt. 519, 808, 1667
—— W. S., Maj. 779
—— W., Lt.. 987
Stagg, M., Capt. 11, 665
Staig, B. M., 2nd-Lt. 550
Stahnard, G. E., 2nd-Lt. . . . 1678a
Stainforth, L. C. H., Bt. Col.. . 57
Stainton, T. H., Maj. . . 843, 1663
Staines, F. J., 2nd-Lt. 1678a
Staley, G., Lt. 543a, 1673
Stall, C. J., 2nd-Lt. . . . 564, 1050
Stallard, G. B., 2nd-Lt.. . 568, 1120
—— H. E., 2nd-Lt. 362
—— J. P., Capt. 1717
—— R. H., Capt.. 60, 664, 1882
—— R. T., Lt. 731
—— S. F., Lt.-Col. . 134, 613
Stamp, L. D., Ty. Lt 1729
—— R. G., Lt. 533
—— W. A. E., Capt. 696
Stampe, W. L., Lt. . . . 539, 988
Stanbury, S. J., Lt. . . . 517, 1367
Standage, R. F., Lt.-Col. . 132, 1743
Standbridge, C. H., Capt. . 252, [illegible] 1459, 1467
Standen, J. D., Capt. . . 272, 1058
—— W. H., Condr. . . . 1854
Standing, H., Lt. . . . 443, 931
Standish, A., 2nd-Lt. 565
—— N., Lt. 322, 1170
Standley, R., Sub-Condr.. . . . 1856
Stanes, F. J., 2nd-Lt. 567
Stanfield, E. A. F., Lt. . 307a, 901
—— T. E., Sub-Condr. . . 1870
Stanford, W. H. C., 2nd-Lt. 553, 1708
Stanforth, L. H., Lt. .692, 747, 783
Staniforth, J. R., Lt. 751
Stanger, F. R., Lt. . . . 31, 1493
Stanger-Leathes, H. E., Maj. . 1751
Staniland, D. L., 2nd-Lt.. . . . 758
—— E. A., Lt. 589
Stanley, A., 2nd-Lt. 618
—— A. E., Revd. 1903
—— C. H., 2nd-Lt. 1561
—— D. M., Lt. . . . 593, 599
—— D. R., Lt. 317a
—— E. H. B., Maj. 1748
—— G., Capt. . . . 755, 1004
—— G. C., 2nd-Lt. 1319
—— J. C., *Hony*. Lt . . . 1843
—— P. T., Lt. . . . 425, 1385
—— R. C. H. S., Capt. . . . 733
—— S. P., 2nd-Lt. 1581
Stannard, G. E., 2nd-Lt. . . . 358
—— P. V., 2nd-Lt.. . . . 362
Stansbury, C. A., Lt.. . . 387, 1090
Stansfeld, V., 2nd-Lt. 651
—— C. G., Lt.-Col. 99, 139, 162 200, 869, 1181, 1635
—— W., 2nd-Lt. . . 355, 1331
Stanton, C. A., Lt. . . 489, 1890
—— F. W. S., Bg.-Genl. 121, 129, 632
—— H. M. A., 2nd-Lt. . . 624
—— P. E., 2nd-Lt. 722
Stanyforth, R. T., Lt. 595
Staples, E. C., Capt. . 81, 267, 1052
—— E. H., Capt. 259, 1393, 1483
Stapleton, F., Lt. 858
—— F. H., Lt.-Col. 51, 98, 136, 745
—— H. E., Capt. 501a 1116, 5010
—— J., Lt. 622
Stapylton, A., 2nd-Lt. 857
—— J. M., Lt. . 299, 875, 919
Stark, C. W., Lt. . . . 353, 1243
—— J., 2nd-Lt. 1702
—— J. D., Lt. 538
Starkey, H. A., Maj. 654
Starky, W. B., Lt. 1695
Starley, J. R., Lt. 793
Startin, H. A., Lt. 525, 1674
Statham, W., 2nd-Lt. 1689
Staughton, A. J., 2nd-Lt. . . 178, 494
Staunton, J. W., 2nd-Lt. . . 525, 689
Staveley, J. P., 2nd-Lt. 530
—— R., Capt. 615
Staveley, W., *Hony*. Maj. . . 1844
Stavert, G. H., 2nd-Lt. . . . 730
Stayner, J. W., 2nd-Lt. 112
Steabben, J. O., 2nd-Lt. . . . 352
Stead, E. A., Capt. 288, 290, 1114, 1162
—— J., 2nd-Lt. 691
Steane, S. W., Lt. . . . 94, 509
Stear, B. T., *Hony*. Lt. 722
Stearn, C., 2nd-Lt. 1417
Stearns, H. D., Lt. . . . 504, 1467
Stebbing, H. T., Capt. . 68, 636, 1004
Stedman, F., Lt. 506
—— C. R., Lt. 112
—— H., Lt. 504, 1895
—— J. F., Lt. . . . 534, 1383
—— R. G., 2nd-Lt. 814
Steed, F. C. W., 2nd-Lt. . . . 327
Steeds, F. G., 2nd-Lt. 776a
Steedman, C. D., Lt. . . 39, 622
—— J. A., Lt. . . . 525, 988
—— L., 2nd-Lt. 921
—— L., Lt. 443
—— M. T. W., Ty. Capt . . . 1726
—— W. McQ., 2nd-Lt. 476, 1211
Steel, C. E., Maj. 1838
—— C. N., Maj. . . 245, 1120, 1537
—— E., 2nd-Lt. 361, 1080
—— E. G., Lt. 445, 1513
—— F. R., Lt. 530, 1243
—— G., 2nd-Lt. 357
—— G. W., 2nd-Lt. 1321
—— R. A., ... 201
Steele, R. P., ... 307a
—— J., 2nd-Lt. 1034
—— J. G., Lt. 514
—— R. A., Lt.-Col. 51, 96, 97, 103, 104, 127, 153, 201, 907
—— R. F., Maj. 1751
Steele, A. H., 2nd-Lt. 802
—— C., Capt. . . 15, 35, 591, 599
—— C. E. B., Maj. . . 42, 213, 1661
—— F. Lt., ... 636
—— G. B. E., Lt. 1028
—— H., Lt. 354, 1082
—— J. L., 2nd-Lt. 705
—— J. S., Lt. 380, 1487
—— J. T., Lt. 511, 1672
—— L., Lt. 463, 945
—— M. L. A., Asst. Surgn.. . 1802
—— R. L., 2nd-Lt. 808
—— R. P., Lt. 307, 1397
—— St. G. L., Col. 105, 122, 149 181
Steele, Perkins, C. S., Lt. . . 603
Steen, H. B., Maj. . . . 871, 1748
—— H. C., Maj. 207, 870, 1425, 1467
—— R., Maj. 1746
Steer, J. M. B., Lt. . . 298a, 1339
Stein, J. A., Lt. 524, 988
Steinhoff, B. G., 2nd-Lt. . . . 571
—— F. G., Asst. Surgn. . 1795
Steirn, F., Lt. 636
Stenson, L. H. E., Lt. . 450—54, 961
—— W. S., Sub-Condr. . . . 186
Stent, D. H., Lt. 828
—— P. J. H., Lt. 5256
Stephen, A. A., Lt. . . . 351, 889
—— A. T., Lt. 317a, 365, 1247, 1523
—— J. R. M., 2nd-Lt. . . 324
—— L. P., Maj. 1746
—— P. W. R., Lt. . . 352, 1279
—— R. L., Lt.. 571
Stephens, E. J., Esq. 191
—— E. J., 2nd-Lt.. . . . 722
—— F. L., Capt. 843
—— G. H., Condr. 1872
—— G. R. T., Sub-Condr. . . 1863
—— H. D., Capt. 1732
—— H. F., *Hony*. Lt. . . 1844
—— J., Lt. 858
—— J. W., Capt. 695
—— K. T., 2nd-Lt. . . . 1285
—— R., Capt. . . . 59, 745
Stephenson, B., Capt. 827
—— C. B., Capt. 760
—— E. L., Lt. 736
—— F. H. A., Capt. 501, 858, 1666
—— G. E. B., 2nd-Lt. 328, 1164
—— G. T., Lt. 638
—— H., Capt. 1838
—— H. S., Revd. 605, 1903, 1904
—— J., Lt.-Col. 105, 133, 1743
—— J., Lt. . . 322, 1433
—— J. E., Capt. . . 703, 704
—— N. A., 2nd-Lt. . . . 624

NAMES AND RANK. Page.

Stephenson, P., 2nd-Lt. . . . 733
——— P. W., 2nd-Lt. . 565
——— R. W., Lt. . . . 720
——— T., Lt.. . . . . 540a
Stepto, E., Lt. 399, 885
Sterl, T. H., 2nd-Lt. . . 357, 1052
Sterling, E., Esq.. 88
——— J. McN., Asst. Surgn. 1802
Stern, C. E., Lt. . . . 751
Sterndale-Bennett, J., Maj. 68, 226, 1431
Sterwin, R. J., 2nd-Lt. . . . 1140
Steuart, A., 2nd-Lt. 652
——— F. G., Capt.. . . 501a, 1199
Stevenage, E. A., Asst. Surgn. . 1794
——— J. J., Capt. 808
Steven, A. R., Capt. 1718
——— G. D., Lt. 781
——— W. S. R., Capt.. . . . 1713
Steveni, O. K., 2nd-Lt. . . 329, 1431
Stevens, A., Lt. 527
——— A. E., Lt. 587, 588, 599—600
——— A. E., Lt. . . . 666, 991
——— A. F., Lt.-Col. . . 132, 1743
——— C., Maj. 56
——— C., Lt. 778, 1007
——— C. F. L., Capt. 276, 891, 893
——— C. G. B., Esq. 43
——— C. R., Lt.-Col. . . 130, 1742
——— C. W., 2nd-Lt. . . 562, 1335
——— C. W., Lt. . . 399, 1134
——— D. T., Lt.. . . . 470, 933
——— E. P., 2nd-Lt. 470
——— E. T. D., Lt. 776a
——— F. W.. 2nd-Lt. 357
——— G. B., Lt.-Col . . 181, 1918
——— G. C. N., 2nd-Lt. . . . 554
——— H. C., 2nd-Lt. 545
——— H. L., Lt. 9, 1689
——— I. M., Capt.. 707
——— J. A., Lt.-Col. 827
——— J. R., Lt. . . . 508, 1665
——— K. G., 2nd-Lt. 722
——— L. C., Capt. . . . 628, 1890
——— M. W., Lt.-Col. . 170, 1916
——— N. M. C., Lt.-Col. 105, 131, 156, 191, 1702
——— N. S., 2nd-Lt. 1359
——— N. W., Capt.. 1714
——— R. H., Lt. . . . 506, 1030
——— R. P., Lt. . . . 537, 1072
——— S., Capt. 10
——— S., Esq. 44
——— S. G., Capt. 817
——— S., Hon. Capt. 1715
——— T. H. G., Lt. . . 58, 523
——— T. M., Lt. . . . 490, 921
——— W. D., 2nd-Lt. . . . 640
Stevenson, A., 2nd-Lt. . . . 768
——— A. F. M., 2nd-Lt. . . 595
——— A. M., Lt. . . 393, 1321
——— C. Y., Capt. . . . 668
——— F., Capt. 1754
——— H. A., Lt. . . 319, 1595
——— H. D. M., Maj. 246, 1663
——— H. L., Lt. 848
——— J., Lt. . . 399, 516, 1581
——— J., 2nd-Lt. . 763, 782, 1676
——— J. F. B., Lt. . 28, 415, 1199
——— J. G., Lt. . 669, 827, 997
——— J. H. M., Capt. 630, 973
——— L. J., Lt. . . 533, 988
——— M. M., Capt. . 272, 875
——— P. N., 2nd-Lt. 366a, 1439
——— P. R., 2nd-Lt. . . . 1679
——— R. C., Lt. . . 520, 998
——— T. L. Lt. . . 430, 1636
——— W., Capt. . 46, 111, 571a, 827, 1841
——— W., Condr. 1851
——— W. B., Capt. . . 28, 1714
——— W. D. H., Maj. . . 4, 42, 1752, 1890
——— W. H., Lt. . . 423, 1152
Stevenson-Hamilton, O. C. J., Maj. . 233, 1662
Steward, A. W., 2nd-Lt. . . . 562
——— C. W. d'A., Capt. . 274, 1279, 1451
——— D. H., 2nd-Lt. 1130
——— E. M.. Maj. . . 19, 223, 1662
——— J. R. O'B., Lt. . . . 397
——— M. E. C., 2nd-Lt. . . . 618
——— R. O. B., 2nd-Lt. . . . 885
Stewart, A. C., Lt.-Col. 139, 160, 199, 964
——— A. D., Maj. . . . 41, 1753

Stewart, A. E., Maj. 57, 226, 1491, 1495
——— A. F., Lt.-Col 140, 162, 201, 1459
——— A. F., Maj. . . . 218, 1662
——— A. M., Lt. 435
——— A. R., Lt. . . . 532, 1585
——— A. W., Lt. . . 493, 941, 1499
——— A. W., 2nd-Lt. . . 1499—1500
——— C., Revd. 1903
——— C., Lt. 544, 1676
——— C. D. A., 2nd-Lt.. . 365, 895
——— C. E., Lt. . . . 313, 1050
——— C. H., Lt. . 307, 1427, 1509
——— C. K., 2nd-Lt. 55
——— C. N., Asst. Surgn. . . 1795
——— C. W., Lt. . . . 522, 1667
——— D., 2nd-Lt. 705
——— D. D., Sr. Asst. Surgn. . 1791
——— D. M., Lt. 515
——— F. G., 2nd-Lt. . . 327, 1243
——— F. H., Sir, Maj. . 859, 1751
——— F. T.. Lt.-Col. . 130, 155, 189
——— F. W., Lt. 518, 1094
——— G., Maj.-Genl.. . . . 964
——— G., Lt. 608
——— G. E., Lt.-Col.. . 139, 1745
——— G. H., Maj 1746
——— G. V., 2nd-Lt.. . . . 744
——— H., Maj. . . 207, 1890, 1913
——— H., 2nd-Lt.. . . . 637
——— H., Tempy. Capt. . . . 1724
——— H. C., Capt. . . . 35, 1838
——— H. J. L., 2nd-Lt. . 768, 781
——— H. L. G., Lt. 628
——— H. N. Sr. Asst. Surgn. . 1791
——— H. S., Maj. . . . 226, 907
——— H. W., Lt. . . . 36, 535
——— I., Lt. . . . 367, 915, 1174
——— J., Lt. 492
——— J. Ty. Lt. 1729
——— J., Maj. 744
——— J. A., Lt.-Col.. 132, 156, 192, 868, 1070
——— J. A., Lt. . . . 512, 992
——— J. C., Maj.-Genl. 173, 887
——— J. H. K., Br.-Genl. 97, 121, 126 153, 196, 869, 1178
——— J. J., 2nd-Lt.. . . 568, 1678a
——— J. M., Lt.-Genl. . . . 169
——— J. M., Maj.-Genl. 50, 118, 146, 173
——— J. M. V., Maj. 84, 215, 1647
——— J. R., 2nd-Lt. . 563, 790, 1517
——— J. St. C. D., Maj. 225, 945
——— M. Mc. D., Lt. . . 543, 1385
——— M. J., Lt. . . . 387, 1193
——— N. R., Maj.-Genl. *Sir* . 1955
——— N. T., 2nd-Lt. 987
——— R. H., Lt. . . . 506, 1110
——— R. J. T., Lt.-Col. 42, 138, 159, 197, 1663, 1883
——— R. R., Lt. . . . 541, 1106
——— T., 2nd-Lt. 719
——— T. A., 2nd-Lt. . . . 555, 1509
——— T. W., Lt.-Col. . . 128, 1742
——— T. W., 2nd-Lt. 744
——— W., 2nd-Lt. . . . 559, 1271
——— W., Capt. 1714
——— W. F., 2nd-Lt. 723a
——— W. G., 2nd-Lt. 1695
——— W. H., 2nd-Lt. 61
——— W. R., Capt. 1758
Stewart-Martin, T. S., 2nd-Lt. . 1675
Stewart-Smith, D. C., Lt. . . 744
Stick, L. T., Lt. 1265
Stickland, H., Sub-Condr. . . . 1864
Stidston, D. D., Ty. Asst. Surgn. 1804
Stiff, R. 619
Stiffe, A. F. E., Lt.-Col. 26, 132, 632
Stigings, J. C., Condr. . . . 1855
Stileman, C. E. C., Lt. . . 385, 1245
——— R. F., Lt. 702
Stiles, J. H., 2nd-Lt. 363
——— S., 2nd-Lt. 730
Still, C. H., 2nd-Lt. 599
Stilliter, W. M., 2nd-Lt. . . . 1497
Stilling, N. A., Lt. . . . 723, 1007
Stillingfleet, R., 2nd-Lt. . . 666, 992
Stillman, A. C. H., Lt. . . 352, 1305
——— R. C. B., Capt. . . . 615
Stillwell, E. H., Maj. 777
Stilwell, J. G., Lt. 729
——— W. B., Lt.-Col. . . . 791
Stirling, C. Mc K., 2nd-Lt. . . 1679
——— D. A., Lt. 591
——— F., Lt. 523, 1667
——— J. H., 2nd-Lt. 826

Stirling, J. M., Lt. . . . 430, 1259
——— R., Temp. 2nd-Lt. . . . 598
——— S. W., Lt. . . . 537, 1672
——— W. G. H., Lt.-Col. 127, 154, 182
Stiven, R., Lt. 317a, 1279
Stobbart, J., 2nd-Lt. . . 540, 1672
——— R. Mc. I., Lt. . . 389, 1020
Stoble, D., Sub-Condr. 1870
Stock, C. A., 2nd-Lt. . . 359, 1040
——— S. M., Lt. 316
Stockdale, S. R., Lt. 729a
——— T. C., Lt. 752
Stocker, A., 2nd-Lt. 723a
——— C. J., Capt. 1716
——— C. L., Lt. . . . 691, 783
——— E. H., Capt. . . 54, 591
Stockley, C. H., Maj. . 235, 1279
——— C. I., Ty. Capt. . . . 1726
——— H. G., 2nd-Lt. . . .
——— H. R., Lt.-Col. 7, 14, 132, 661
Stoddard, A. A., Lt. 111
——— R. H., Ty. Capt. . . 1725
Stodart, D. E., Capt. 578
——— T., Col. . . 29, 127, 1738
Stoddart, G., Capt. . . 54, 256 419
——— H., Capt. . 252, 1062, 1319
——— J. S. P., Lt. . . 518, 1267
——— R. A., Lt. . . 529, 1705
——— W. T., Lt.. 536
Stoehr, C. F., Capt. . . 664, 996
Stokes, A., Brig.-Genl. 101, 120, 123, 612
——— C. B., Bt. Lt.-Col. . 51, 99, 103, 104, 108, 160, 210, 879
——— C. L. L., Capt. 501, 752, 1004
——— E. V., Lt.. 752
——— H., Capt. . . . . .
——— H. A., Lt. . . . 423, 1018
——— H. F. S., Lt. 729a
——— H. G., Lt. 520
——— L. E., Lt. 843
——— L. W., Capt. 818
——— R. G., 2nd-Lt. . . 358, 1524
——— S. E., Lt. . . . 82, 542
——— T. A., Lt. 752
——— T. G. N., Maj. 1746
Stokoe, Revd. C. G. 1903
Stone, A. G., Maj . . 52, 244, 1606
——— A. P., Capt. 502
——— A. P. G., Lt.. . . 352, 1451
——— A. S., *Hony* Lt. . . . 822
——— B. S., Capt. . . . 261, 1359
——— C. A., Lt.-Col., . 129, 1710
——— C. A., Lt. . . . 610, 861
——— E. R., Tempy. Capt. . . 1723
——— E. W. R., 2nd-Lt. . . 731
——— F. L., Capt. . . . 686, 751
——— H. C., Lt. 679
——— H. C. B., Revd. . . . 1906
——— H. O'D., 2nd-Lt. . 1547, 1034
——— J. E., 2nd-Lt. 708
——— J. F., 2nd-Lt. . . 354, 1674
——— J. P., Condr. 1857
——— N. H., Maj. . . . 701, 703
——— R. J. K., Sr. Asst. Surgn. . 1792
——— T. W., Lt. 470, 1431
——— R. C., 2nd-Lt. . . . 551, 1704
——— W., Sub.-Condr. 1856
——— W. J. K., Sr. Asst. Surgn. 1789
——— W. R., Lt.-Col. . 191, 1896, 1912
Stoneham, T. E., Condr. . . . 1861
Stonehewer, E. H., Lt. . 506, 1275
Stonehouse, E. J., Nursing Sister 1734
Stones, W., 2nd-Lt. . . . 797, 1321
Stoney, E. C., Capt. 1713
——— P. S., Maj. 52, 99, 218, 870, 1124
Steinhoff, B. G., 2nd-Lt. . . . 571
——— F. G., Asst. Surgn. . . 1795
Stonex, W., 2nd-Lt. 1679
Stonor, E. C., Capt. 695
——— W. T., 2nd-Lt. . . 565, 808
Stooks, G. S., Maj.. 806
Stoop, A. D., Lt. 681
——— B. L., 2nd-Lt. 626
——— F. G. F., 2nd-Lt. . . 536, 1433
——— T. M. M., Lieut. . 393, 598
Strey, F. G. F., Lt. . . . 82, 536
Storey-Cooper, E. S., Lt. . 470, 1036
Stork, C. H., 2nd-Lt. 558
——— H. C., Lt. . . 53, 524, 1026
Storm, W., 2nd-Lt. 546 1417
Storr, C. L., Maj. . 51, 99, 207, 1239
Storrar, A. Y., 2nd-Lt. 826
——— J., 2nd-Lt. 987
Storrie, W., Lt. 1360
Storrow, S. L., Lt. . . 91—92, 445
Storrs, B. St. J., 2nd-Lt. . . . 623

NAMES AND RANK. Page.

Storrs, G. N., Lt. 509, 1609
——— W. G., 2nd-Lt. . . . 1591
Story, F. E., 2nd-Lt. . . . 330, 913
——— G. A., Lt. 724
——— H. F., Capt. 16, 29, 253, 1162
——— H. W., Lieut. . 494, 808
——— J. A., Capt. . 258, 1217, 1263
——— W. G. T., Tempy. Capt. . 1724
Stotherd, E. A. W., Lt.-Col. . . 57
Stott, F. L., Lt. 308
——— H., Capt. 1754
——— H. W. M., 2nd-Lt. . 350, 1614
——— R., Esqr. 88
Stourton, Hon. W. M., Lt. . . 589
Stout, J. C., 2nd-Lt. 705
Stovold, H. W., Maj. 651
Stow, V. A. S., 2nd-Lt. . . 557, 803
Stowell, A. T., Capt. . . . 831
——— J. C., Lt. . . . 56, 510
——— P. C., Lt. . . 45, 509, 1016
——— R. S., 2nd-Lt. 816
Stowers, R. Lt. 1715
Stoyle, E. G., Lt. 617
Stracey, P. St. C., Snr. Asst. Surgn. 179
——— S. R., Asst. Surgn. . . 1799
Strachan, A. B., 2nd-Lt. . . . 763
——— A. W., Lt. 513
——— E. A., Capt. . . . 1714
——— R. S. Ty. Capt. . . . 1726
Strachen, G. R., Lt. . . 455, 1515
Strachey, E., *the Hon.* Maj. 17, 702, 795
Strahan, C., Lt.-Genl. . . . 671
——— G. A., Lt.-Col. 132, 156, 193
——— G. C., Capt. 269, 1623, 1629
Straker, F. E., 2nd-Lt. . . 332, 1375
——— W. R., Lt. . . 455-59, 927
Stranack, F. W., Capt. 805
——— H. R., Capt. . . 260, 1136
Strang, J. L., Lt. 725
——— D. A., Lt. 1407
——— W., 2nd-Lt. 748
Strange, A. J., Lt.-Col. 136, 158, 196, 1623
———F., Lt. . . . 353, 1614
——— F., 2nd-Lt. 555
——— R. G., Br.-Gl. 12, 120, 123, 612
Strangeman, C. V., 2nd-Lt. 360, 1499-1500, 3419
Stranger, F. R., Lt. . . . 393, 1152
Stratton, E. B., Lt., . 1451, 1543, 321
——— F. C. G., Capt. . . . 753
——— J. G., Sub-Condr. . . 1872
——— L. E. 2nd-Lt. . . . 1451
——— W. C. R., Lt.-Col. 182, 1911
Strawbridge, H. G., 2nd-Lt. . . 358
Streatfeild, S. F. Revd. . . . 1907
Streatfield, H. G. C., Lt. . . . 732
——— H., Col. 780
——— N., Lt. . . . 487, 885, 959
Street, A., Lt.-Col. . . 128, 1740
——— A. R., Lt. . . 66, 701
——— W., Lt. . 543, 848, 1695
——— W., 2nd-Lt. 1695
——— W. F., Lt. . . . 544, 1673
Strellett, D. L., 2nd-Lt. . . . 362
Strettell, C. B. D., Maj. . 222, 919
Strickland, E. P., Maj.-Genl. . 119, 691
——— H. B., 2nd-Lt. 328, 909
——— H. F., Capt. (Temp.) . 1724
——— P. C. B., Col. 124, 1738, 1896
Stringer, B. de H., Lt. . . 543, 1645
——— F. W., Capt. 732
Strip, R. A., 2nd-Lt. . . . 552, 1010
Strologo, R. C., Lt. . . 687, 731*a*
Stronach, J. L., 2nd-Lt. . . 553, 1583
Strong, A. D., Bt.-Lt.-Col. 160, 391, 209, 872, 893, 1563
——— C. A., Lt. . . . 312, 1598
——— E. S., Maj. 836
——— G., 2nd-Lt. 568
——— H. P., Lt.-Col. 139, 160, 871, 1407, 1409
——— H. S., Maj. . . 207, 1893, 1913
——— H. W., 2nd-Lt. 796
——— J. D., Maj. 236
——— J. R., 2nd-Lt. 722
——— N. B. G., Lt.-Col. . . 57, 1919
——— R. G., Maj. . . 224, 868, 1022, 1062
Stroud, G. M., Lt. . . . 313, 943
——— T. J., 2nd-Lt. 760
Stroughill, M. A., Sr. Nursg. Sister . . . 1734
Strover, E. J., Capt. . . . 263, 1026

Strover, W. G., Capt. . . . 262, 1034
Struben, L. F., 2nd-Lt. . . . 583
Strudwick, G., Sub-Condr. . . 1873
———W. A. J., Lt. . . 354, 1007*a*
Strudwicke, M. A., Lt. . . 443, 1423
Strump, R., 2nd-Lt. . . . 328, 1647
Stuart, A., 2nd-Lt. . . . 356, 1130
——— A. R., 2nd-Lt. 844
——— A. F., 2nd-Lt. 602
——— C. H., 2nd-Lt. 806
——— D., Lt. 430
——— D. F., Lt.-Col. 129, 154, 186
——— D. N., Maj. 229, 870, 1291, 1896
——— D. S., Lt. . . . 18, 29, 1153
——— F. J., Capt. 1712
——— H. H., *Hony.* Lt . . . 1848
——— H. R., Asst. Surgn. . . . 1802
——— J., 2nd-Lt. . . . 559, 1090
——— J. D., 2nd Lt. 848
——— J. M. B., Lt. . 511, 848, 992
——— J. P., Lt. 1359
——— J. R., Lt.-Col. 1731
——— L., Lt.-Col. 658
——— L. A., Capt. . . . 278, 1487
——— L. J., Capt. 577
——— M., Capt. 859
——— P. P., Lt. 95, 487
——— R., 2nd-Lt. . . . 357, 1677
——— R. C. O., Maj-Genl. 118, 612, 1886
——— R. P., Lt.-Col. 707
——— R. W. H., Revd. . . . 1903
——— T. C., Lt. 522, 1639
——— T. J., Asst. Surgn. . . . 1796
——— W., Maj. . . . 220, 917
——— W. W., Sr. Asst. Surgn. . . 1791
——— Menteth, A. H. Maj. 103, 234, 1415, 1451
Stuart-Prince, C. E., Capt. . 278, 1259, 1565
——— D., Lt. . 77, 510, 1885
Stuart-Shepherd, U. G. S., 2nd-Lt. 358, 1837
Stuart-William, W. L., Lt. . 380, 899
Stubbings, R. A., Lt. 368
Stubbs, H. A., 2nd-Lt. . . . 1695
——— H. G. P., Lt. . . 318, 1116
——— H. H., Lt. . . . 525, 1702
——— J. B., Capt. . . . 274, 899
——— R. H., Lt. . . . 375, 1022
——— S. G., 2nd-Lt. 554
——— S. T., Lt. 524, 1699
Studd, E., Capt. . . . 43, 351, 603
Studdy, H., Lt. 492
Sturdy, A. E., Lt. 711
——— E. C., 2nd-Lt. . . . 927, 494
——— J. R., Lt. . . 9, 18, 393, 1611
Sturge, C., Lt. 778
——— E., Capt. 754, 1004
Sturges, F. G., 2nd-Lt. . . . 797
Sturrock, G. C., Lt.-Col. 71, 101, 137, 632, 1884
——— L. A., Lt. 510
Sturt, G. L., Capt. 778
Stutter, J. E., Lt. . . . 543, 1636
Styles, A. F., Sub-Condr. . . . 1856
——— H. W., Capt. 751
Suba, Jemdr. 1117
——— Subdr. 1256
——— Khan, Jemdr. 1119
——— Ram, Jemdr. . . . 1288
——— Sing Gurung, Jemdr. . . . 1602
——— Subdr. . . . 1188
Subbayya Naidu, Jemdr. . . . 1316
——— Reddiar, M., Sub-Asst. Surgn. 1827
Subhanji Ellappa Bhorgay, Sr. Sub-Asst. Surgn. . . . 1829
Subh Ram, Ressdr. 930
——— Subdr. 1320
——— Subdr. 1060
——— Subdr. 1403
Subha Chand, Jemdr. 1321
Subhan, Jemdr. 1035
Subhan Singh Gurung, Jemdr. . 1637
Subramanian, N. S., Ty. Lt. . . 1775
Subramanyam, K. S., Ty. Lt. . 1778
Subramoni, (I) Jemdr. 993
——— (II) Jemdr. 994
——— Jemdr. 1012
Subramanta Pillai, Resdr. . . 1683
Subramanyam, M., Ty. Capt. . . 1766
——— R.B., Ty. Capt. . . 1769
Subrayan, Subdr. 1339
Such, J., Lt. 509
Sucha Singh, Subdr. 1281
——— Jemdr. 990

Sucha Singh, Jemdr. 1212
——— Jemdr. 1079
——— Jemdr. 944
——— Sub-Asst. Surgn. . 1817
——— Jemdr. 888
Suchet Singh, Subdr. 1011
——— Jemdr. 1175
——— Risaldar 1683
Sud, R. N., Ty. Lt. 1773
Sudam Balu, Jemdr. 1447
Sudama, Subdr.-Maj. 1120
——— Subdr. 1117
——— Ram, Sub-Asst. Surgn. . 1814
Sudan Singh, Jemdr. 1699
Sudha Singh, Jemdr. 1157
Sudhan, Subdr. 480
——— Jemdr. 1059
——— Rao Nimbalkar, Subdr. 1444
Sudhanya Kumar Ghosh, Sub-Asst. Surgn. 1820
Sudhiriswar Sen Gupta, Sub-Asst. Surgn. 1816
Sudh Singh, *Hony.* Lt. 2104
Sudlow, F. S., 2nd-Lt. 362
Sufi, I. A., Ty. Capt. 1767
Sugden, A. S., Capt. 26
——— H., 2nd-Lt. 560
——— J. W., 2nd-Lt. . 355, 1413
——— K. G. J., 2nd-Lt. . 763, 782
Sughar Singh, Jemdr. . . . .
——— Jemdr. 1032
——— Jemdr. 1064
——— Jemdr. 1043
Suhel Singh, Subdr. 1247
——— Subdr. 1105
Sulnaik Singh, Jemdr. 1397
Sujrda, Jemdr. 1175
Suja Khan, Jemdr. 916
Sujan Singh, Subdr. Maj. . . 1483
——— Jemdr. 989
Sujjan Singh, Risldr. 1686
Sukbir Limbu, Jemdr. 1653
Sukdhe, Subdr. 1040
Sukha, Subdr. 1191
——— Jemdr. 1193
Sukha Ram, Subdr. 1412
Sukhbasi Lal Saksena, Sub-Asst. Surgn. 1820
Sukhbir Gurung, Jemdr. . . . 1619
——— Subdr. . . . 970
Sukhdayal Singh, Risldr. Maj. . 898
——— Subdr. . . . 1164
Sukhdeo Singh, Subdr. . . . 1389
——— Jemdr. . . . 1056
——— Pande, Jemdr. . . . 1051
Sukhdev, Subdr. 1165
——— Jemdr. 1028
Sukhi Ram, Risldr. Maj. . . .
Sukh Lal, Jemdr. 1040
——— Saksena, Sub-Asst. Surgn. 1820
Sukhlal Rana, Jemdr. 1619
Sukhmandar, Jemdr. 1492
Sukhnandan Singh, Jemdr. . . 1054
Sukhpal, Dube, Jemdr. 1027
Sukhpal Singh, Resldr. 918
Sukhraj Singh, Subdr. Maj. . . 1048
Sukh Ram, Subdr. 1391
——— Subdr. 1335
——— Singh, Jemdr. 1488
——— Jemdr. 1044
Sukhramdas, Sub-Asst. Surgn. 1812
——— Sub-Asst. Surgn. . . 1819
Sukman Rai, Subdr. 1648
Sukumbar Thapa, Subdr. . . . 1624
Sulaiman Khan, Jemdr. . . . 1412
——— G., Sub-Asst. Surgn. . 1829
——— Khan, Jemdr. 950
——— Sub-Asst. Surgn. . 1820
Sulakhan Singh, Jemdr. . . .
Suleman Khan, Subdr. 1035
——— Resldr. 882
——— Subdr. 1195
——— Jemdr. 1035
——— Jemdr. 1035
——— Jemdr. 950
——— Jemdr. 1707
Sulley, J. P., Lt. . . . 20, 511, 1665
Sullivan, A. S., Lt. 532
——— C. S., 2nd-Lt. . . 546, 1297
——— E. L., Br.-Genl. 120, 125,
——— G. M. R., 2nd-Lt. 354, 1673
——— J. H. B., Lt. 489
——— M. B., Lt. 785
——— R. E., Col. . 124, 149, 183
——— R. T., ... 544

NAMES AND RANK. Page.

Sullivan, T. E., Condr. 1867
——— W. C., Condr. 1853
Sully, T. D., 2nd-Lt. 829
Sultan, Subdr. 989
——— Subdr. 1040
——— Jemdr. 1040
——— Ahmad, Jemdr.
——— ——— Jemdr. 1376
——— Jemdr. 1216
——— Ahmed Khan, Jemdr.
——— ——— Jemdr. 892
——— Alam, Jemdr. 1142
——— Ali, Sr. Sub-Asst. Surgn. 1808
——— ——— Jemdr. 1433
——— ——— Subdr. 1281
——— ——— Jemdr. 1515
——— Asghar Ali Khan, Subdr. 1285
——— Jan, Risldr. Major 79
——— Khan, Subdr. 1519
——— ——— Ressldr. 908
——— ——— Jemdr. 1435
——— ——— Subdr. 1128
——— ——— Subdr. 1409
——— ——— Jemdr. 1096
——— ——— Jemdr 1252
——— ——— Jemdr 1154
——— ——— Jemdr. 1435
——— ——— Subdr. 1311
——— ——— Jemdr. 942
——— ——— Jemdr. 944
——— ——— Jemdr. 910
——— ——— Jemdr. 1134
——— ——— Jemdr. 1440
——— ——— Resadr. 1684
——— Mahmud, Jemdr. 1124
——— ——— Jemdr. 1208
——— Muhammad, Jemdr. 920
——— ——— Jemdr. 894
——— ——— ——— Jemdr. 1518
——— ——— Khan, Subdr. 1052
——— ——— ——— Jemdr. 892
——— Sarkhru, Jemdr. 1133
——— Sikandar Khan, Jemdr. 908
——— Singh, Ressdr.
——— ——— Risaldr. 939, 951
——— ——— Jemdr. 1364
——— ——— Subdr. 1032
——— Ahluwalia, Sub-Asst. Surgn. 1818
Sultano, P. S., 2nd-Lt. 546
Sulton, H. G., Maj. 1585
Summanwar, N. R. C., Capt. 1768
Summers, E. W., 2nd-Lt. 561
——— G. H., Capt. 96, 213, 923
Sumner, F. W., Maj. 1747
——— Jemdr. (I), (II) 1189
——— G. A., Lt. Qr.Mr. 1729
——— W., 2nd-Lt. 831
Sumption, F., Lt. 805
Sumanbir Limbu, Jemdr. 1649
Sumandar Shah, Jemdr. 1682
Sundar, Subdr.-Maj. 1132
——— Jemdr. 1171
——— Subdr. 1171
——— Subdr. 1134
——— Jemdr. 1245
Sundarachari, R., Esq. 91
Sundaramurti, Subdr. 993
Sundaram, Jemdr. 994
Sundardas, Sub-Asst. Surgn. 1810
Sundaresan, V. R., Ty. Lt. 1777
Sundar Khan, Jemdr. 1376
——— ——— Jemdr. 1577
——— Raju, Jemdr. 1351
——— Singh, Sub-Asst. Surgn. 1816
——— ——— Jemdr. 1112
——— ——— Jemdr. 1368
——— ——— Subdr. 1151
——— ——— Jemdr. 1150
——— ——— Jemdr. 1096
——— ——— Risldr. 935
——— ——— Subdr. 1699
——— ——— Jemdr. 1228
——— ——— Risaldr.-Maj. 915
——— ——— Sr. Sub-Asst. Surgn. 1808
——— ——— Sr.Sub-Asst.Surgn. 1808
——— ——— Sub-Asst. Surgn. 1817
——— ——— Subdr. 1031
——— ——— Risaldr. 944
——— ——— Subdr. 1188
——— ——— Subdr. 1163
——— ——— Subdr. 1375
——— ——— Jemdr. 1433
——— ——— Jemdr. 1111
——— ——— Jemdr. 1204

NAMES AND RANK. Page.

Sundar Singh, Jemdr. 978
——— ——— Subdr.-Maj. 1203
——— ——— Subdr. 1203
——— ——— Jemdr. 1681
——— ——— Subdr. 1148
——— ——— Subdr. 1320
——— ——— Ressldr. 938
——— ——— Jemdr. 1149
——— ——— Subdr. 1112
——— ——— Jemdr. 1111
——— ——— Jemdr. 1157
——— ——— Jemdr. 1165
——— ——— Subdr. 1432
——— ——— Jemdr. 1163
——— ——— Jemdr. 1372
——— ——— Jemdr. 1031
——— ——— Jemdr.
——— ——— Jemdr. 1079
——— ——— Subdr. 1208
——— ——— *Hony.* Capt. 2101
——— ——— Subdr. 1096
——— Sing, Sub-Asst. Surgn. 1833
Sunder, C. E., Lt.-Col. 127, 1739
——— Singh, Jemdr. 1701
——— ——— Jemdr. 1211
——— ——— Ressdr. 1684
——— Vishnu Vedak, Sub-Asst. Surgn. 1832
Sundarlal Tribhuvan Bhatt, Sub-Asst. Surgn. 1832
Sunderland, W., Lt. 583
Sunit Chand, Jemdr.
——— ——— Subdr. 1231, 1247
Sunka, Jemdr. 1188
Sunkha Singh, Subdr. 1288
Sunkunni, V. Moon Jemdr. 1309
Sunwat Singh, Jemdr. 1192
——— ——— Jemdr. 1111
Surajbaksh Singh, Jemdr. 1281
Surajbali Sing, Subdr. 1046
——— ——— Jemdr. 1044
——— ——— Subdr. 1063
Suraj Bux, Risldr. 888
——— Din, Subdr. 1124
——— Mal, Risaldr.
——— ——— Jemdr 1035
——— Pal, Sub-Asst. Surgn. 1810
——— ——— Dube, Jemdr.
——— ——— ——— Jemdr. 1523
——— ——— Misra, Jemdr. 1054
——— ——— Singh, Subdr. 1022
——— ——— ——— Jemdr. 1182
——— ——— Pande, Subdr.
——— ——— ——— Resldr. 878
Surajpal Tewari, Subdr. 1026
——— Singh, Subdr. 1044
Suraj Prasad Misra, Jemdr. 1031
——— ——— Mal *Hony.* Capt. 2104
Suram Chand, Ressdr. 888
Suran Singh, Jemdr. 890
——— ——— Singh, Jemdr 1192
——— ——— Subdr. 1032
Surat Ram, Jemdr. 902, 926
——— ——— Jemdr. 1401
——— Singh, Ressdr. 882
——— ——— Jemdr. 1396
Surain Singh, Jemdr. 1114
——— ——— Jemdr. 1589
——— ——— Jemdr. 1587
Surayan Singh, Risaldr. 921
——— ——— Sub-Asst. Surgn. 1816
——— ——— Ressdr. 950
——— ——— Subdr. 1095
——— ——— Jemdr. 1252
——— ——— Jemdr. 1142
——— ——— Jemdr. 1376
——— ——— Subdr. 1218, 1227
——— ——— Jemdr. 916, 1096
Surbir Karki, Subdr. 1642
——— Gurang, Subdr. 1639
——— Limbu, Subdr. 1648
——— Thapa, Jemdr. 1608
——— ——— Jemdr. 1609, 1657
Surbuland Khan, Risdr. 1682
Surdal Thapa, Subdr. 1614
Surja, Subdr.-Maj. 1215
——— *Hony.* Capt. 1896, 2104
Surja Narayan Ghosh, Sub-Asst. Surgn. 1816
Surjan, Jemdar. 1129
Surjan, Subdr. 1256
——— Jemdr. 1184
——— Jemdr.
——— Singh, Subdr. 1493
Surjiman Rai, Subdr. 1630
Surju Singh, Subdr. 1082

NAMES AND RANK. Page.

——— Subdr. 1371
Surrey, H. P., Lt. 538, 1672
——— H. R., 2nd-Lt. 556
Sussex, H. G. V., 2nd-Lt. 1369
Sutcliffe, H. Lt. 1007
——— J. E., Lt. 463, 1391
——— L. B., Lt. 789, 1886
——— O. D., Lt. 306, 1355, 1575
——— W., Serg. Maj. 1879
Sutherlnd, A. O., Maj. 222, 1106
——— A. W., Lt. 435, 1443
——— D., 2nd-Lt. 794
——— D. W., Lt.-Col. 131, 1743
——— D. W., Lt. 711
——— J. W., Lt. 353, 980
——— R. O., Maj. 237, 901
——— W., Lt.-Col. 806
——— W. D., Lt.-Col. 128 1739, 1890
Sutherland-Hawes, J. V., 2nd-Lt. 877
Sutherland-Stevens, R. G. 2nd-Lt. 557
Suttill, K., Capt. 738
Sutton, C. B., 2nd-Lt. 692, 784
——— F. G H., Lt.-Col. 42, 131, 155, 191
——— F. K., Lt. 618
——— E. S., 2nd-Lt. 630
——— G. F., Lt. 401, 1583
——— S. H., Lt. 527, 1491
——— H. G., Maj. 45, 244, 1191, 1585
——— J. C., Lt. 463, 1355
——— R. W. E., Lt. 1355
——— V. C., Lt. 534
——— W. R., 2nd-Lt. 565
Suys, P. W., 2nd-Lt. 768, 1096
Swaffer, H. P., Lt. 534, 1672
Swain, G. A., 2nd-Lt. 766
——— J. H., Sub-Condr. 1863
——— G. H., Lt. 721
——— G. V., 2nd-Lt. 361
Sussex, H. G. V., 2nd-Lt. 1583
Swaine, A. L. B. 780
Swainson, C. A., Lt. 529, 804
Swales, F. W., 2nd-Lt. 561, 1110
——— H. G., 2nd-Lt. 750, 784, 1008
——— T., Lt. 537
Swallow, J. C., 2nd-Lt. 737
Swaminathan, V., Sub-Asst. Surgn. 1828
——— ——— Subdr. 1009
Swan, C. T., Lt.-Col. 129, 154, 187
——— E., 2nd-Lt. 686
——— H. S., Capt. 1732
——— J. C. K., Lt. 516, 1651
——— J. G. G., Maj. 1747
——— J. H., Lt. 380, 1191
Swan, J. S., Lt.-Col. 46, 134, 157, 194, 1660
——— J. S., 2nd-Lt. 652
——— L. M., Lt. 738, 740
——— S. R., Lt. 540*a*, 1132
Swanborough, F. C. V., Lt. 352
Swann, A. E., 2nd-Lt. 329, 857, 947
——— C. J. H., Maj. 101
——— L. G., 2nd-Lt. 692
Swanston, C. B. R., Lt. 506, 1471
——— E. R. 1009
——— J. F. A., Lt. 751
Swanzy, R. E., Lt. 488
——— E. J. E. Bt.-Col., *Sir* 57
——— F. G., Maj. 225, 1122, 1162
——— R. P. N., Lt. 487
Sweeney, E. R. S., Asst. Surgn. 1803
——— J. E., Asst. Surgn. 1802
——— M., 2nd-Lt. 624
——— S., Lt. 534, 977
——— W. J., 2nd-Lt. 684
Sweeny, R. L. C., Maj. 219, 1662
Sweet, E. H., Lt.-Col. 137, 159, 197, 872, 1600
——— R., Capt. 1757
Sweet-Escot, W. W., 2nd-Lt. 740
Swift, G. A., 2nd-Lt. 95, 708
Swifte, E. L., Lt.-Col. 59, 137, 159, 197, 1251
Swinburn, H. R., 2nd-Lt. 359, 1078
Swindells, H. R., 2nd-Lt. 787
——— T. A., 2nd-Lt. 734
Swinehatt, P. H., 2nd-Lt 658
Swiney, A. J. H., Br.-Genl. 58, 120, 124, 660
Swinhoe, C., Col. 176
Swinnerton, H., Maj. 779
——— J. C., *Hony.* Capt. 1850
Swinson, C. A., Capt. 90
Swithenbank, R. S., Lt. 537, 998
Swinton, F. E., Lt.-Col. 78, 129, 1740
——— T. S., 2nd-Lt. 567
Swinton-Skinner, E., Col. 178, 1918
Swords, W. J., 2nd-Lt. 567
Swynnerton, F. R., 2nd-Lt. 549, 1249

NAMES AND RANK. Page.

Sybray, J., Capt. 793
Syddall, J., 2nd-Lt. . . . 545, 1673
Sydney-Smith, D. C., Lt. . 311a, 1275
Syed Abbas, Subdr... 1704
—— Ali P. Subdr-Maj. . . . 1309
Syed Ahmad Khan, Jemdr. . . 1681
—— **Budruddin**, Sub-Asst. Surgn. 1828
—— Dastagir, Subadar . . . 1464
—— Subdr. 1307
—— Issrail Shah, Ressdr. . . . 1683
—— Mohammad Hossain Maulvi, 2nd-Lt. . . . 481
—— **Muzzaffar Ali Shah**, Ressdr. 1682
—— —— Ressdr. 1686
Syed Timur Ali Shah., Ressdr. . 1683
—— Umar, Ressdr. 1683
S. Mohamed Jamen, Resldr . 1707
—— —— Ressdr. . 1686
Sydenham, A. F., Lt. . . 399. 907
—— J. P., *Hony.* Capt. 1843
—— P. S. B., 2nd-Lt. 666, 996
Syer, H. H., Maj. . . 214, 1661
—— L. B., Lt. 505
Sykes, A. A., 2nd-Lt. 624
—— A. C., Capt. . . . 666, 1004
—— A. W., Lt. . . . 777, 794
—— A. W., 2nd-Lt. 566
—— C., 2nd-Lt. 640
—— C. A., Lt.-Col. . 122, 132, 613
—— E., **Sr. Nursg. Sister** . . 1734
—— E. F., 2nd-Lt. 564
—— F. C., Lt. 393, 1136
—— J., **Hony. Lt.** . . 45, 1848
—— H., 2nd-Lt. 565
—— H. J., 2nd-Lt. 545
—— J. H., Lt. 885
—— M. G., Lt. 515
—— P. M., Lt.-Col. 181, 192, 890, 1912
—— S. P., Capt. 1,14
—— W. D., 2nd-Lt. 315
Sym, J. M., Maj.-Genl. *Sir* 173, 1617
Syme, J. G. S., Revd. . . . 1903
—— K., Lt. 1355
Symes, A. J., Capt. . . . 1755
—— F. S., 2nd-Lt. 554
—— J. D., Lt. . . . 738, 791
—— L., 2nd-Lt. 565
—— R. W., 2nd-Lt. 731
Symington, K. B., Lt. . . . 1689
Symns, J. M., 2nd-Lt. . . 6, 14, 559
Symon, G., 2nd-Lt. 545
—— C., Asst. Surgn. . . . 1803
—— E., 2nd-Lt. 354
—— E. T., Lt. 709
—— F. H., Lt. 728
—— H. Tempy. Asst. Surgn. 1804
—— P. R., Lt. . . 668, 992
Symonds, A. H. B., Lt. . . 78, 488
—— D. G., Capt. 738
—— E., Lt. 1395
—— J. D. B., 2nd-Lt. . . . 1463
—— W. F., Lt. . 89, 738, 740
—— W. F., Lt. . 89, 738, 740
Symons, A., Col. . . 97, 103, 126
—— E. W., 2nd-Lt. . . . 758
—— H. G., 2nd-Lt. . . . 359
—— H. J. H., Ty. Capt. . . 1726
—— T. H., Lt.-Col. . 134, 1744
Synge, M., Maj. . . 101, 209, 1661
—— R. F. M. F. M., Lt.-Col. 687
Synnott, N. W., Lt. . . . 840
Syrett, F. G., Lt. 626
Szczepanski, H. C., Maj. 37, 87, 108, 214
—— H. C. A., Maj.-Genl. 173

T

Taaffe, W. A., Lt. 769
Tabuteau, T. R., Col. . . . 177
Tablat Singh, Jemdr. . . . 1681
Tabib, M. E. A. K., Sub-Asst. Surgn. 1832
Tabiyat Singh, Jemdr. . . . 1285
Tacchella, C. F., Capt. . . . 808
Taffs, L. H., Lt. 493
Tafts, L. H., 2nd-Lt. 73
Tagetmeier, P. A., 2nd-Lt. . . 352
Taggart, M. J., 2nd-Lt. . . . 354
—— P. D., 2nd-Lt. . 355, 1022
Tahourdin, R. J. L., Lt. . 310, 1211
Taihal Sing, Subdr. 1305
Tainsh, J. R., Maj. 501
—— J. R., Maj. 834
Tait, A., Lt. 559, 828
—— J., 2nd-Lt. 494, 821

NAMES AND RANK. Page.

Tait, R., 2nd-Lt. 563
—— T. R., Lt. 443
—— T. R., 2nd-Lt. 1413
—— J., 2nd-Lt. 907
Taitt, N. R., Lt. 393
—— W. R., Lt. 1036
Taj Muhammad, Sub.-Maj. . . 970
—— **Muhammad**, Subdr. . . . 966
—— —— Subdr. . . . 1518
—— —— Jemdr. . . . 1472
—— —— Jemdr. . . . 970
—— —— Subdr. . . . 1435
—— —— Khan, Jemdr. . 940
—— —— Risldr. . . . 938
—— —— Jemdr. . . 890, 966
—— —— Jemdr. . . . 1643
Tekbahadur Mal, Jemdr. . . 1483
Taj Din, Jemdr. 999
Takhat Singh, Subdr. . . . 1031
Talbot, A. C., Lieut.-Col., *Sir* 179, 1917
Talbot, A. D., the Revd. . . . 1904
—— C. C., Lt. 733
—— C. C., 2nd-Lt. 362
—— D. H., Capt. 595
—— E. F., Nursg. Sister . . 1735
—— F. E. G., Bt.-Lt.-Col. 98, 137, 159, 216, 868, 1074, 1114
—— G. R. FitzR., Lt.-Col. 28, 628
—— J., Lt.-Col. . 57, 131, 155, 190
—— J. E. L., 2nd-Lt. . . . 358
—— W. E. H., Lt. 295a, 1339, 1435
—— W. L., Lt. 32, 717
Tale Muhammad Khan, *Hony.* Capt. . 111
Talib Husain Khan, Subdr. . . 1403
Tait, T. R., Lt. 435
Talok Singh, Jemdr. . . . 968, 1096
Talwar, K. C., Tempy. Capt. . . 1769
Tams, G. E., 2nd-Lt. 1501
Tambling, F. C. H., The Revd. 1904
Tambusami, Jemdr. 1264
Tamiz Gul, Jemdr. 1502
Tampi, G. S., Ty. Lt. 1779
Tancock, A. C., Maj. 21, 95, 221, 1144
—— O. K., Lt.-Col. . 132, 632
Tancred, G. M. W., Lt. . . 543, 1673
—— J. Y., Maj. 218, 869, 1094, 1130
—— N. G. W., 2nd-Lt. . . . 1193
Tendavarayan, Jemdr. 1351
Tandan, P. L., Tempy. Capt., . 1769
Tandy, **E. A., Maj.** 662, 1880
—— H. S., Col. . . 178, 1918
—— M. O'C., Maj. . . . 662, 1880
—— O. C. S., Capt. 1717
Tangul Khan, Subdr. . . . 1085
Tankard, B. L., Lt. . 393, 750, 1463
Tann, E. J., Capt. 778
Tanner, C. O. O., Col., . 12, 23, 121, 126, 150, 189,
—— G., Lt. . . . 18, 27, 708
—— G., Hy. Lt. 68, 797
—— G. W., Lt. . . . 375, 1369
—— H., Lt. . . 504, 733, 1666
—— H. R., Capt. 702
—— J. T., Lt. . . . 492, 798
—— P., Sub-Condr. 1864
Tanser, H. A., 2nd-Lt. . . . 1469
Tansley, L. R. H., Ty. Lt. . . 795
Tapp, A. V., 2nd-Lt. 76
—— D. H., Lt. 320, 1459
—— W., Sub-Condr. . . . 1861
Tappin, W. H., Lt. 793
Tapsell, C. B., Lt. . . . 522, 1391
—— E. B., Lt. 491
—— E. G., Lt. 1674
—— N. B., Lt. . . . 308, 1247
Tara, Jemdr. 1189
Tarachand, Sub-Asst. Surgn. . 1810
Tara Chand, Subdr. 1188
—— Dutt, Jemdr. 1347
—— —— Jemdr. 1225
—— **Singh**, Jemdr. 888
—— Jemdr. 918
—— Jemdr. 1157
—— Jemdr. 1112
—— Ressdr. 1684
—— Subdr. 1403
—— Jemdr. 1356
—— Subdr. 1130
—— Jemdr. 924
—— Jemdr. 1344
—— Jemdr. . . . .
Tarapore, B. S., Ty. Capt. . . 1772
—— P. K., Capt. . . . 1753
Tarif Singh, Subdr. . . . 1024
Tarling, T. D., 2nd-Lt. . . . 734

NAMES AND RANK. Page.

Tarr, **W., Maj.** 1751
—— W. G., Lt. . . . 354, 1497
Tarran, F. P., Lt. . . . 529, 658
—— W., *Hony.* Maj. . . . 720
Tarver, A. L., Bt.-Col., 1, 13, 20, 97, 127, 153, 196, 871, 1491
Tasadduq, Hussain, Sub-Asst. Surgn. . 1828
Tassie, J. A., Capt. 859
Tatam, A. C., Lt. 491
Tate, **A. E.**, Col. . . . 32, 1731
—— **A. R. W.**, Lt. 689
—— **F. R.**, Sub-Condr. . . . 1855
—— **G.**, Capt. 79, 1755
—— **G**, Lt.-Col . . 138, 1745, 1890
—— J., 2nd-Lt. 566
—— **J.**, Lt. . . . 511, 854, 992
—— **J. C.**, Capt. . 256, 1495, 1893
—— J. W., Lt. . . . 543, 891
—— T. B., Lt. 490, 965
—— W. M., Capt. 40, 258, 1243, 1497
Tatem, E. G., 2nd-Lt. . . 1489—90
Tatham, A. M., 2nd-Lt. . . . 905
—— C. W., 2nd-Lt. . . . 1399
—— C. W., Lt. 1281
—— W. H., 2nd-Lt. . . . 987
—— W. M., 2nd-Lt. . 324, 901
—— P. R., 2nd-Lt. . 327, 919
Tatoba Amre, Subdr. . . . 1469
—— Ambre, Subdr. . . . 1469
Tatojirao Sinde, Jemdr. . . . 1436
Tattersall, O., Condr. . . . 1863
Tattie, J. B., 2nd-Lt. 1379
Tatton-Winter, C. W., Ty. 2nd-Lt. 730
Tatya Jadhav, Jemdr. 1416
Taunton, H. L., Lt. 683
—— I. H., Lt. 543a
Tayner, E. J. 1696
Tavernor, G. V., Capt. . . . 844
—— G. V., Capt. . . . 846
Taverner, D. T., Lt. . . . 353
Tawakil Khan, Ressdr. . . . 910
Tawney, A. E., Capt. 636
Tay, B., Capt. 79, 1839
Tayler, C. H., Lt. . . . 485, 827
—— E. R. N. A. W., Lt. 490, 883
—— F. L., Maj. . 242, 1191, 1193
—— H. F., Lt. 638
—— **H. H.**, Capt. . . . 500, 842
—— H. J. D., *Hony.* Lt. . 21, 1848
—— R. G. A'W., Lt. 528
—— S. E., Lt. . . . 809, 1178
Tayleur, G., Lt. . . 380, 697, 1154
Taylor, A. A., *Hony*, Capt. . . 755
—— A. C., Lt. . . . 440, 1024
—— A. D., Sub.-Condr. . . . 1864
—— A. G., Lt. 629
—— **A. G. V.**, Lt. 310
—— A. H., Maj.-Gen. . . . 297
—— A. H. S., 2nd-Lt. . . . 1611
—— A. J., Lt. 681, 785
—— **A. O.**, Asst. Surgn. . . 1799
—— A. T., Lt. . . . 298, 1467
—— A. T. W., Capt. 263, 1004, 1195
—— A. W. D. C., Capt. . . 1701
—— A. W. N., Lt.-Col. 119, 155, 188
—— B., 2nd-Lt. 1467
—— B. C., Lt. 511
—— C. A., 2nd-Lt. 588
—— C. G., Asst. Surgn. . . 1795
—— C. H., Lt. . . 486, 695, 737
—— C. H., 2nd-Lt. 551
—— C. W., Lt. 352, 1447
—— D., Lt. 593
—— D. E. D., Lt. 577
—— D. Mc. N., Capt. . . . 1756
—— E., Lt. 637
—— E. B. N., Lt. 836
—— E. C., Maj. 1750
—— E. J., 2nd-Lt. . . . 90, 727
—— E. T. N., Ty. Lt. . . . 1779
—— E. M., Lt.-Col. 99, 207, 917, 961
—— E. R. L., 2nd-Lt. . . . 564
—— F., Lt. 761
—— F., Lt. 803
—— F. H., 2nd-Lt. 785
—— G. C. B., Capt. 15, 24, 692, 783
—— G. D., Lt. . . . 294, 1629
—— G. F., 2nd-Lt. . . . 465, 933
—— G. L., Lt. 385, 1633
—— G. L., 2nd-Lt. 858
—— G. T., 2nd-Lt. . . . .
—— H., Lt. 485, 923
—— H. A., Lt. 733
—— H. A., Lt. 797
—— H. B., Lt. 1369

Names and Rank. Page.

Taylor, B. D., Lt. 1042
—— H. D., Lt. 380
—— H. P., Lt. 752
—— H. S., Sub.-Condr. 1868
—— H. S., 2nd-Lt. 566
—— H. W. D. C., Lt. 529, 1696
—— J., Maj. 1753
—— J., Capt. 1717
—— J., Maj. 621
—— J., Capt. 636, 979
—— J. A., Lt. 542
—— J. B., Lt. 684
—— J. F. S., 2nd-Lt. 550, 1275
—— J. H., Capt. 279, 895
—— J. H. C., Ty. 2nd-Lt. 669, 997
—— J. McL. G., Maj. 245, 1471
—— J. N., Lt. 371, 1335
—— J. N. C., Lt. 389
—— J. S., Lt. 420, 889
—— L., Lt. 751
—— M. D., 2nd-Lt. 1695
—— N., Maj. 654
—— N. C., Maj. 229, 1052, 1375
—— N. S., Capt. 269, 1148
—— P. H. M., Maj. 212, 935, 955
—— R., 2nd-Lt. 587
—— R. H., 2nd Lt. 1591
—— R. E. S., Lt.-Col. 57
—— R. M., Asst.-Surgn. 1802
—— R. N., 2nd-Lt. 560
—— R. O'B., Bg.-Genl. 16, 37, 98 102, 122, 138, 160, 200, 911
—— R. W., 2nd-Lt. 817
—— R. V., Capt. 748
—— S., 2nd-Lt. 1299
—— S. A., Lt. 1501
—— S. G., Lt. 525, 1219
—— S. H., Maj. 608
—— S. T. R., 2nd-Lt. 354, 1315
—— T., *Hony.* Maj. 6, 1849
—— V. C. L., Capt. 273, 1303, 1379
—— W., Lt. 351
—— W., Lt. 354, 1088
—— W. D., 2nd-Lt. 1499—1500
—— W. F. 2nd-Lt. 565
—— W. J., 2nd-Lt.
—— W. R., Lt. 9, 18, 516a, 720, 1607
—— W., Sub-Condr. 1863
—— W. H., 2nd-Lt. 788
—— W. W., Lt. 687
Teagle, C. H., Lt. 755
Teague, J., Lt. 293a, 1495
—— L. E. S., Lt. 543a
—— R. J. D., Lt. 302, 1066
—— S. W. M., 2nd-Lt. 72, 622
Teague-Jones, R., Lt. 532
Teal, F. H. D., 2nd-Lt. 549, 1056
Teale, D. B. S., 2nd-Lt. 550, 1337
Teape, C. C. T., Lt. 397
Tearce, A. T., Lt. 304, 1495
Teather, R. T., Sub-Condr. 1861
Tebbutt, H. J., 2nd-Lt. 735a
Tebby, H., Sub-Condr. 1861
Tedman, C., Lt. 351, 818, 997
Tee, S. E., 2nd-Lt. 551, 1064,
Teed, A. B. H., Lt.-Col. 178, 136, 158, 197
Teehan, P., Lt. 522, 1455
Teeling, A. R. B., 2nd-Lt. 560
Teemey, S. P., 2nd-Lt. 1569
Teg Sing, Subdr. 1174
Teg Singh Kaphola, Subdr. 1182
Tega, Subdr. 1171
Tegetmeier, P. A., Lt. 352, 974
Tehri, Raja of 111, 1179
Teire, D., 2nd-Lt. 564, 1367
Tej Singh, Jemdr. 928, 1476
Teja, Jemdr. 1197
—— Bajwa, Sub-Asst. Surgn. 1318
—— Rawat, Jemdr. 1483
—— Singh, Jemdr. 878, 1216
—— —— Jemdr. 1244
—— —— Subdr. 1158
—— —— Jemdr. 1364
—— —— Jemdr. 910, 957
—— —— Jemdr. 912
—— —— Jemdr. 900
—— —— Jemdr. 1133
—— —— Subdr. 1509
—— —— Jemdr. 1158
—— —— Jemdr. 1111
—— —— Subdr. 1239
—— —— Subdr.-Maj. 1096
Tejbir Thapa, *Hony.* Capt. 2101
Tej Muhummad Khan, Jemdr. 1473
Tekbahadur Gharti, Jemdr. 1602
—— Pun, Jemdr. 1637
—— Sahi, Jemdr.
Tek Chand, Sub-Asst. Surgn. 1871
—— Sub Asst. Surgn. 1820
—— Ram, Jemdr. 1167
—— Singh, Ressdr. 878
—— —— Risldr. 891
—— —— Jemdr. 621
—— —— Jemdr. 999
—— —— Jemdr. 1143
—— —— Jemdr. 896
—— —— Jemdr. 1376
—— —— Subdr. 1171
Teku, Jemdr. 1171
Telang, V. D., Ty. Lt. 1782
Telfer, J. W., Lt. 387
Telford, B. D., Lt. 315, 1287, 1379
Tellery, J. A., 2nd-Lt. 557
Telprole, Zamidar of, 2nd-Lt. 481
Temperley, H. R., 2nd-Lt. 77
Temple, B., Lt. 531, 1671
—— E., Capt. 617
—— F. C., Lt. 601
—— J. A., Col. 178, 1917
—— P. DeL., Maj. 237, 944
—— T., Maj. 74, 101, 614
Templer, C. F., Brig.-Genl. 9, 16, 78, 122, 132, 156, 193
—— H., Col. 123, 153, 181, 1917
—— H. E., Lt. 368, 754, 763, 1651
Templeton, C. A. S., Lt. 520, 1455
—— R. B., Lt. 26, 798
—— W. G. H., Lt. 509, 1561
Tennant, A. W., Lt. 1493
—— C. A. M., Capt. 265, 1064
—— C. S., Ty. Capt. 1725
—— E., Lt.-Col. 78, 105, 132, 156, 193
—— F. O., 2nd-Lt. 1405
—— H. L., Lt.-Col. 98, 101, 129, 613
—— J. B. H., Lt. 820
—— J. E., Maj. 11
—— J. H., 2nd-Lt. 332, 1225
—— W. R., Lt. 638, 972
Tennent, A. K., Lt. 728
Terrell, F. J. A., 2nd-Lt. 539
Terrot, C. R., Maj. 585
Terry, A. D. R., Capt. 710
—— C. E., Sub-Condr. 1854
—— E., 2nd-Lt. 568, 1696
—— H. F., 2nd-Lt. 630
—— H. J., 2nd-Lt. 618
—— J. H. S., 2nd-Lt. 357, 1235
—— L. R., Lt. 516a, 1565
—— P. E., 2nd-Lt. 571
—— S. E., 2nd-Lt. 1703
—— T., Sub-Condr. 1874
—— W., Condr. 1861
Tervitt, J. F. T., 2nd-Lt. 325, 1614
Tester, A. F., Lt. 7, 751
—— C. K., Lt. 307, 1038
Tetley, J. A., 2nd-Lt. 1473
Teversham, M. S., Capt. 280—287, 1148
—— R. K., Bt.-Col. 56, 180
Teviot-Kerr, A. E. K., Lt. 514
Tew, G. C., 2nd-Lt. 653
Tewson, R. G., Lt. 399, 1657
Teyen, C. St. L., Lt. 816
Tezu Khanka, Capt. 2102
Tha, B. R. H. O., Lieut. 1773
Thacker, H. J., Lt.-Col. 129, 155, 188
—— N., Lt. 399, 1078
—— W. L. H., Capt. 485
Thackwell, A. C. A., Maj. 240, 1331
—— C. G. R., Col. 122, 149, 181
Thaker, V. P., Sub-Asst. Surgn. 1832
Thain, E. W., Maj. 776
Thakar Singh, Jemdr. 1493
—— Chand, Risldr. 921
—— Das, Sub-Asst. Surgn. 1807
—— Dip Singh, Rao Bahadur, *Hony.* Capt. 111
—— Jor Sing, Risldr. Maj. 935
—— K. S., Capt. 1755
—— Karan Singh, Ty. 2nd-Lt. 482
—— Kundan Singh, Ressdr. 1686
—— Mul Singh, Risldr. Maj.
—— Parshad Singh, Jemdr. 1391
—— Sawant Singh, Risldr. 927
—— Shiva Dhyan Sinha, Ressdr. 1686
—— Singh, Subdr.-Maj. 1211
—— —— Subdr.-Maj. 1114
—— —— Subdr.-Maj. 1141
—— —— Subdr. 1164
Thakar Singh, Risldr. 882
—— —— Risldr. 951
—— —— Risldr.
—— —— *Hony.* Lt. 2104
—— —— Jemdr. 1256
—— —— Jemdr. 1158
—— —— I. Subdr. 1217
—— —— II. Subdr. 1217
—— —— Jemdr. 898
—— —— Subdr.-Maj. 1371
—— —— Jemdr. 1150
—— —— Jemdr. 990
—— —— Risdr. 882
—— —— Jemdr. 948
—— —— Subdr. 1050
—— —— Subdr. 1132
Thakur Bhagwant Singh, *Hony.* Lt. 2104
—— Surendra Bahadur Singh, Jemdr. 1085
Thakurdin Misr, Jemdr. 1027
Thal Singh, Jemdr.
Thaman Singh, Subdr. 1305
—— Rai, Jemdr. 1631
—— Gurung, Jemdr. 1598
Thambiah, S., Ty. Lt. 1780
Thamu Limbu, Jemdr. 1631
Thamburan, Jemdr. 1308
Than Singh, Jemdr. 930
Thane, C. S., Lt.-Col. 820
Thapar, D. R., Ty. Lt. 1781
—— K. S., Tempy. Lt. 1768
—— K. R., Ty. Lt. 1781
Tharle-Hughes, R., Capt. 531, 806
Thatcher, A. F. B., Capt. 380, 704
—— C., 2nd-Lt. 365
Thavamony Joel, B., Sub-Asst. Surgn. 1827
Theak, L. F., Lt. 440, 1309
Theaker, H., Lt. 67
Theal, C., 2nd-Lt. 640
Thear, H. T., 2nd-Lt. 693
Thein Kin, Subdr. 1295
Theobald, A. C. L., Maj. 614
—— J. M., 2nd-Lt. 58
Theodore, J. B., Asst. Surgn. 1800
Theokaram Singh, Jemdr. 1705
Thepar Singh Negi, Jemdr. 1223
Thew, I. A., 2nd-Lt. 324, 968
Thewlis, N. E., 2nd-Lt. 539, 1624
Thiele, C. E., Lt. 440, 1391
Thierry, E. H., *tempy.* Lt. 622
Thipthorp, E. E. Sr. Asst. Surgn. 1791
—— W. B., Asst. Surgn. 1793
Thirka Ram, Jemdr. 1400
—— Jemdr. 1498
Thirsk, F. A., 2nd-Lt. 359, 1515
Thiselton Dyer, G. H., 2nd-Lt. 849
Thirupod, K., Ty. Lt. 1779
Thirkell-White, E. B., Capt. 501, 842, 1557
Thoday, F. A., Capt. 696
Tholasy, C. M., Sub-Asst. Surgn. 1828
Thom, A., 2nd-Lt. 713
—— J. B., Maj. 634
Thomas, A. E., Lt. 89, 698
—— A. F. F., Lt. 445, 1207
—— A. H., 2nd-Lt. 559, 1164
—— A. N., Capt. 1754
—— A. R., Lt. 36, 516, 1670
—— A. W., Snr. Asst Surgn. 1789
—— B. L., Lt. 780
—— B. S., Lt. 703
—— C. A., 2nd-Lt.
—— C. E., Capt. 501a, 685, 1140
—— C. J. H., Maj. 628
—— C. S. W., Maj. 72, 633
—— C. V., 2nd-Lt. 797
—— D., 2nd-Lt. 587
—— D. E., Capt. 78, 486
—— D. J., Tempy. Capt. 1723
—— D. R., Capt. 1759
—— D. V., 2nd-Lt. 559, 1158
—— D. W. E., Lt.-Col. 717
—— E., Lt. 533, 1007a
—— E. A., 2nd-Lt. 1405
—— E. F., Lt. 8, 82, 538
—— E. K., Lt. 1777
—— E. S., Revd. 1903
—— E. W., Capt. 817
—— F., *Hony.* Lt. 1848
—— F. A. S., Lt. 526, 1459
—— F. G. S., Lt. 463
—— F. G., 2nd-Lt. 555, 1024
—— F. H. S., Col. 123, 149, 183
—— G. A., Lt. 731
—— G. A., Lt. 704

NAMES AND RANK. Page.

Thomas, G. D. E., Capt. . . . 717
—— G. E. V., 2nd-Lt. . . 240
—— G. H., Lt. 302, 1606, 1609
—— G. I., Capt. . . . 617
—— G. S., 2nd-Lt. . 355, 1068
—— H., 2nd-Lt. . . . 551
—— H. G. S., Capt. . . . 832
—— H. P., Capt. . . 277, 1501
—— H. R. D., Col. . 179, 1918
—— H. St. G., Lt.-Col. 128, 154, 183
—— J. D., Sr. Asst. Surgn. . 1791
—— J. E., Lt. . . . 446, 1124
—— J. H., Condr. . . . 1868
—— J. P., Capt. 777
—— K. F. W., Lt. . 312, 1597
—— K. M., Ty. Lt. . . . 1779
—— L. A. S., Lt. . . 435, 1034
—— M. L., Lt. . . 541, 1673
—— M. P., Lt. . . 513, 1199
—— N. S., Lt. 1689
—— P. E., Lt. . . 641, 1001
—— R. C. P., Asst. Surgn. . 1802
—— R. D., Lt. 519
—— R. E., 2nd-Lt. . . . 1695
—— R. G., Capt. 791
—— R. H., Maj. . . 663, 1880
—— R. H., 2nd-Lt. . 546, 1507
—— R. L., Lt. 579
—— R. M. B., Col. . . . 177
—— R. S., Maj. . . 209, 1661
—— R. S., Capt. . . . 501a
—— R. T., Ty. Lt. . . . 1779
—— S., 2nd-Lt. 567
—— S. S. L. St. G., 2nd-Lt. . 719
—— T. I. G., Lt. . . . 1010
—— T. L., 2nd-Lt. . 23, 706
—— W., Condr. 1861
—— W., Capt. 858
—— W. C., Ty. Lt. . . . 1776
—— W. H. G., Asst. Surgn. . 1798
—— W. R., Hony. Capt. . . 1723
—— W. W., Rev. . . . 1508
Thomas-Pillai, Sr. Sub-Asst. Surgn. 1826
Thomas-Peter, G. F., Lt. . . . 722
Thomason, A. D. F., Lt. 307, 897, 899
—— A. F., Lt.-Col. 26, 131, 191, 155, 1660
—— R. M., Lt.-Col. . . . 825
Thompson, A., Maj. 245 1451, 1457
—— A., Lt. . . . 583, 1525
—— A., Lt. 1701
—— A., Lt. 571
—— A. B., 2nd-Lt. . . . 1020
—— A. G., Ty. 2nd-Lt. 720, 786, 799
—— A. K., 2nd-Lt. . 549, 1293
—— B. E., Lt. 760
—— B. N., Lt. . . 519, 1667
—— C., Sr. Asst. Surgn. . 1792
—— C. B., 2nd-Lt. . . . 566
—— C. C., 2nd-Lt. . . . 698
—— C. G., Sr. Asst. Surgn. 1789
—— C. G., Lt. . . 516, 1038
—— C. L., Asstt. Surgn. . 1803
—— C. M., Maj. 243, 1283, 1880
—— C. N., Lt. . . 521, 577
—— C. P., Lt.-Col. 187, 711, 1895
—— D., Lt. . . . 443, 953
—— D. C., Capt. . . . 696
—— E., Capt. . 538, 852, 1695
—— E. C. A., Lt. . . . 535
—— F. G. A., Lt. . . . 1303
—— F. T., Maj. 1748
—— F. S. C., Lt.-Col. 139, 1745
—— G. L., 2nd-Lt. . 354, 1243
—— G. N., Maj. . . 233, 1195
—— G. V. L., 2nd-Lt. 559, 1678
—— H., Condr. 1854
—— H. C., Sr. Asst. Surgn. 1791
—— H. E., Hon. Lt. . . 794
—— H. V. J., Asst. Surgn. 1794
—— H. W., Lt. . . . 75, 753
—— J., 2nd-Lt. 565
—— J. P., Maj. . . 234, 941
—— J. P., Lt. 605
—— J. S., Lt. . . . 393, 1042
—— J. S., Lt. 515
—— J. S., Lt. . . 487, 1667
—— J. W., 2nd-Lt. . . . 689
—— L. M., Lt. . . 543, 1467
—— N. L., 2nd-Lt. . . . 1154
—— N. R., Capt. . . 66, 767
—— P., 2nd-Lt. . . 555, 1319

NAMES AND RANK. Page.

Thompson, P. T. L., Capt. 258, 1523, 1327
—— R., Lt. . . . 25, 777
—— R. C., Ty. Capt. . . 56
—— R. D., Lt. . . 507, 1267
—— R. D., Capt. . . . 844
—— R. H., Lt. . . 401, 1060
—— R. M., Lt. . . 389, 1559
—— R. M., 2nd-Lt. . . . 709
—— R. H. G., Asst. Surgn. 1795
—— S. H., 2nd-Lt. . . . 566
—— S. L., Capt. . . 273, 1455
—— S. M., Lt. 1689
—— T. R. L., Capt. *Sir* 100, 587
—— W., 2nd-Lt. 1062
—— W. G. M., 2nd-Lt. 329, 909
—— W. H., 2nd-Lt. . . . 571a
—— W. J., Lt. 516a
—— W. S., Lt. . . 533, 1601
Thoms, A. K., Capt. 656
—— R. S., Maj. 656
Thomson, A. A. Mc. D., 2nd-Lt. . 626
—— A., 2nd-Lt. . . . . .
—— A. G., Bt., Lt.-Col. 51, 52, 99, 139, 160, 207, 1255
—— A. J. D., Maj. 241, 1090, 1896
—— A. R., Capt. . . 202, 1042
—— A. P. Q., Lt. . . 309, 1617
—— C., Maj. 1786
—— C. G., Lt. 308
—— D., 2nd-Lt. . . . . .
—— D. A., Maj. . . 663, 1003
—— G. C., 2nd-Lt. . . . 708
—— G. Y., Capt. 1757
—— H., Capt. . . . 725, 1732
—— H. H., Lt. . . 524, 1667
—— H. J., Lt. . . . 512, 857
—— J., 2nd-Lt. 361
—— J., Capt. 715
—— J. L., Lt. 79, 491
—— J. R., 2nd-Lt. . . . 687
—— J. S., Capt. 1702
—— J. S., 2nd-Lt. . . . 725
—— P. R., 2nd-Lt. . . . 748
—— R. F., Lt. . . . 401, 927
—— R. J., Lt. . . . 703, 706
—— R. S., Lt. . 505, 825, 1199
—— R. St. C., Capt. . . 27, 723
—— S. H., 2-nd-Lt. . . . 567
—— T. A., the Revd. . . . 1904
—— T. R., 2nd-Lt. . . . 1441
—— W., Revd. 1905
—— W., 2nd-Lt. 355
—— W. B., Capt. 736
—— W. D., Col. 57, 122, 149, 181, 1917
—— W. M., Br.-Genl. . 119, 126, 764
—— W. B., Lt.-Col. . . . 732
—— W. R., Lt. 530
—— W. R., 2nd-Lt. . . 331, 1439
—— Glover, J. W., Capt. 262, 1162, 1898
Thorburn, H. H., Maj. . . 1753, 1890
—— J. M., Capt. . . . 848
—— M. M., Capt. . . 68, 744
—— R., Lt. . . . 508, 1092
—— R., 2nd-Lt. 1263
Thorn, A. P., Lt. . . . 393, 1104
—— G., 2nd-Lt. 359
—— H. T. C., Lt. . . 506, 992
Thornback, A. H., Lt. . . . 642
Thorne, A. C. D., Maj. 223, 1415, 1883
—— A. M., 2nd-Lt. . . 35, 1265
—— E. P., 2nd-Lt. . 359, 1559
—— G., Condr. 1853
—— H., Lt. . . . 514, 1042
—— M., Lt. . . . 367, 1231
—— S. G., Lt. . . . 527, 965
—— T., *Hony.* Maj. . . . 1843
Thornely, B., 2nd-Lt. 1189
—— M. H., Maj. . . . 1746
Thornewill, A. S., Capt. 501a, 760, 1666
Thornhill, C. H., 2nd-Lt. . . 624
—— C. J. M., Bt. Lt.-Col. 57, 105, 139, 161, 238, 1038
—— C. M., Capt. . . . 262, 1116
—— H. B., Lt.-Col. *Sir* . . 53
—— M. B., Lt. 1124
Thornilley, Lt., J. H. . . 402, 915
Thornton, A. M., Lt. . . 89, 578
—— C. A., 2nd-Lt. . . 323, 1178
—— C. A. F., 2nd-Lt. . . 472
—— C. E., Lt.-Col. 131, 155, 190
—— E. A., Lt. . . . 380, 1020
—— G. E., Lt. . . 352, 1673

NAMES AND RANK. Page.

Thornton, H. E., Lt. . . 470, 1393
—— R. T., Lt. . . . 1507, 1183
—— T. A., Capt. 587
—— W. W., Lt. . . . 430, 1020
Thorogood, P. W., Lt. 397, 509, 1058
Thorp, M., Lt. 512, 1395
—— F. W., 2nd-Lt. 817
Thorpe, A. J., Lt. . . . 530, 1678
—— C. W. G., Lt. . . 308, 1231
—— H. S., Lt. 638
—— J. A., Lt. . . . 518, 1363
—— P., Asst. Surgn. . . . 1797
—— T. W., Sub-Condr. . . . 1856
Thoy, A. R., Asst. Surgn. . . . 1799
—— G., Asst. Surgn. 1791
Threle, C. E. Lt. 445
Threlfall, C. H., 2nd-Lt. 695
—— V. R. S., Lt. 543a
Threlkeld, S., Lt. . . . 489, 1674
Thring, R. H. D., Col. . 57, 149, 182
Thrupp, G. A., Lt. . . . 513, 1573
—— H. F., 2nd-Lt. . . 567, 1411
—— W. G., Lt. 490
Thubron, J. B. S., 2nd-Lt. . . 860
Thuillier, H. F., Maj.-Genl. 119, 126, 661
—— H. S., Capt. . . 637, 975
—— L. C., Maj. 213, 1880, 1913
Thunder, H. F. A., 2nd-Lt. . 324, 1583
Thurgar, H. M., Lt. 755
Thurgood, W. J., Lt. . . 465, 1309
Thurley, M., Lt. 521
Thurlow, B. W., 2nd-Lt. . . . 548
Thursfield, G. A. R., Revd. . . 1908
Thurston, C. W., Lt. . . 354, 1674
—— E. O., Maj. 1745
—— E. T., Lt. 658a
—— E. T., 2nd-Lt. . . 542, 1028
—— J. W., Capt. 807
Thurton, B. W., 2nd-Lt. . . . 1699
Thwaites, A. J., 2nd-Lt. . . . 362
—— W. St. C., Ty. Capt. . . 1775
—— W. St. C., Tempy. Lt. . 1765
Thwaytes, L. L., Capt. . . 279, 917
—— R. C. G., Capt. . . 62, 1838
Thyer, W. E., Ty. Asst. Surgn. . 1804
Thyne, W., Capt. 253, 1367
—— W. H., Asst. Surgn. . . 1801
Tiarks, J. G. E., Lt. 581
Tibbits, W., Maj. 1711
Tibbitt, W. T., 2nd-Lt. . . . 735a
Tibbs, D. M., 2nd-Lt. 564
—— J., *Hy.* Capt. 1845
—— P. G., Revd. 1907
Tice, C. A., Lt. 508, 1665
Ticehurst, C. B., Ty. Lt. . . . 1726
—— W., Lt. 67
Tickell, G. W., Lt. . . . 397, 1166
Tidd, G. M., 2nd-Lt. 739
Tidmarsh, C. F., 2nd-Lt. 358, 1151, 1509
—— J. M., Lt. 723
Tidswell, E. C., Br.-Genl. 50, 120, 123
Tierney, H., Lt. 516, 1395
Tiffen, W. L., 2nd-Lt. . . 524, 1211
Tighe, M. J., Maj.-Genl. *Sir* 12, 35, 118, 146, 173
Tika Khan, Subdr. Maj. . . . 1255
Tikam Singh, Jemdr. 1021
—— Jemdr. 1212
Tika Ram, Sub-Asst. Surgn. . . 1809
—— Joshi, Sub-Asst. Surgn. 1819
—— Thapa, Subdr. . . . 1618
—— Khattri, Subdr. . . . 1645
—— Kanwar, Jemdr. . . . 1645
—— Mal Jemdr. 1637
—— Malhi, Subdr. . . . 1642
—— Thapa, Subdr. . . . 1598
Tikam, Sing, Jemdr. 1456
Tikaram, Jemdr. 1396
Tikhu, Subdr. 1243
—— Ram, Subdr. 1419
Tikka Khan, Ressdr. 908
—— Subdr. 1227
—— Brij Mohan Pal, Ressdr. . 900
Tilak Singh, Subdr. 1698
Tilbikram Rana, Jemdr. 1597
Tilbir Thapa, Jemdr. 1602
—— Subdr. . . . 1604, 1637
—— Jemdr. 1607
Tilden, E. D., 2nd-Lt. . . 365, 945
Tilfer, A. F., Lt. 1098
Tillard, A. B., Lt.-Col. 137, 158, 197, 872, 1613
—— E. D., Maj. 663, 996
—— J. A., Maj.-Genl. . . . 1955
Tilly, R., Capt. 255, 1487

NAMES AND RANK. Page.

Tilley, F. L., Condr. 1868
—— H., Lt. 577, 761
—— H. R., Lt. 353
—— J. S., Lt. 518, 1479
—— R. H., 2nd-Lt. 1447
—— W. E. C., Lt. . . . 112, 503
Tilokbahadur Limbu, Subdr. . 1633
Tilok Sing, Jemdr. 1181
—— **Limbu**, Jemdr. 1657
—— **Sauntiyal**, Subdr. . . . 1179
—— **Singh**, Snr. Sub-Asst. Surgn. 1807
—— —— Subdr. 1114
—— —— Jemdr. . . . .
—— —— Jemdr. 1653
Timbrell, T., Maj. . . 101, 216, 1661
Timmins, L. M., 2nd-Lt. . . . 705
—— T. B., Maj. . . 700, 702, 795
Timms, G. R., Capt. 706
Tindall, A., Lt. 750
—— J. W. B., Capt. . . 278, 1042
Tindle, T. A., Lt. 756
Tingey, J. M., Sub-Condr. . . . 1864
Tingley, A., 2nd-Lt. 565
Tinley, F. B. N., Capt. . . 274, 913
Tinker, H., 2nd-Lt. 814
Tinley, R. G. W. N., 2nd-Lt. 737, 1134
Tinney, G. E., Lt. . . . 300, 1132
—— L. H., Capt. . . 367, 1158
Tinsley, W. S., Lt. . . . 523, 988
Tipper, G. H., Lt. 511
Tippetts-Aylmer, G. A., Lt. 536, 1699, 1701
Tippetts, M. E., Nursg. Sister . 1735
Tipple, A., 2nd-Lt. 548
Tipping, F. B., 2nd-Lt. . . 692, 784
—— R. P. M., Lt. . 302, 1331, 1587
Tirat Singh, Jemdr. 1028
Tirath Ram, Hony. Capt. . . . 2102
Tirebuck, C. T., Lt. . . 294, 1613
Tirkha, Subdr. 1059
Tireman, H., 2nd-Lt. 808
—— **Ram**, Jemdr. 1052
Tirka Singh, Jemdr. 1480
Tirloki Singh, Jemdr. 1064
Tirney, H., Lt. 516a
Tirthe Gurung, Subdr. . . . 1602
—— **Pun**, Jemdr. 1639
Tirumalay, Jemdr. 1276
Tiruvangadam, Jemdr. 1272
—— Jemdr. 1351
Tiruvengadam, Subdr.-Maj. . . 1591
Tirwa, Raja of, 2nd-Lt. . . . 1048
Tisdall, A. J. Lt. 1919
Titchmarsh, E. A., Lt. 300, 1042, 1044
—— F. L., Lt. 796
Titherington, G. W., Bt.-Maj. . . 745
Titterington, J. H., 2nd-Lt. 549, 1359
Tittle, D. E., 2nd-Lt. . . . 328, 1122
—— D. R., Lt. 769
Titterton, C. M., Lt. 1720
Tivey, W. H., *Hony.* Lt. . . . 1848
Tiwana Khan, Subdr. 1492
Tiwari, G. P., Lt. . . . .
Tobin, A. E., Condr. 1868
—— E. J., Lt. . . . 536, 1367
—— H. W., Maj. . 14, 39, 99, 218, 1509
—— J., 2nd-Lt. . . . 566, 1703
—— P. W., Asst. Surgn. . . . 1800
—— W. J., Condr. 362
Tobit, P. K., Capt. 500
Tod, A. F., 2nd-Lt. 618
—— A. K., Lt. 708
—— J. K., Col. 13, 46, 97, 105, 124, 150, 184
Toda Singh, Subdr. 1451
Todar Singh, Ressdr. 932
—— Jemdr. 934
Todd, A. C., 2nd-Lt. 715
—— A. E., 2nd-Lt. . . . 549, 1699
—— A. S., Lt. 73, 652
—— E. J., Lt 1843
—— E. T. T., Capt. . . 55, 261, 1479
—— F. R. P., Lt. . . . 514, 1395
—— F. W., Lt. 389
—— H. C., 2nd-Lt. 1896
—— H. J., Lt. 492, 895
—— J., Lt. 744
—— J. F., Capt. . . . 252, 949
—— N. C., Asst.-Surgn. . . . 1803
—— R., Capt. . . 162, 1128, 1441
—— S. E. R., 2nd-Lt. 652
—— T. B., *Hony.* Maj. . . . 1843
—— W., Lt. 524
Toder Singh, Jemdr. 924
Todman, A. McN., Lt. . . 529, 1671

Tofts, C. F., Capt. 500, 818
Tohid Gul Khan, Jemdr. . . . 892
Toker, A. C., Maj.-Genl. *Sir* 173, 1090
Tokha, Subdr. 1215
—— Jemdr. 1587
Tolat, K. V., Ty Capt. . . . 1771
Tollemache, D. P., Capt. *Hon'ble* 100, 587
—— H. V., 2nd-Lt. . . . 292
Toller, G. G. T., Lt. 29, 504, 832
Tollit, W. E., Lt. . . . 410, 1217
Tolmie, J. A., Capt. 1718
Tolputt, W. L., Lt. 769, 770
Tomkins, C. J. D., Lt. . . . 463, 1225
—— E. L., Lt.-Col. 71, 101, 134, 632, 1882
—— G. L., Lt. . . . 506, 1509
—— H. L., Lt.-Col. 133, 157, 196, 1132
—— W. P., Genl. 671
Tomkyns, R. A. G., 2nd-Lt. . . . 557
Tomlin, W., Condr. 1854
Tomlinson, C., 2nd-Lt. 723a
—— A. E., Lt. 735
—— E. W., Lt. 111
—— F., Lt. . . . 294, 1423
—— H. W., Capt. 664
—— J., Condr 1854
—— R. A. H., 2nd-Lt. 330, 1225
—— W. G., Hon. Lt. . . . 802
Tompkins, R. H., Lt. 721
Toms, R. G., Lt. 474, 1181
—— W. T. 2nd-Lt. 1009
Tomsett, E. W., Lt. . 308, 1403, 1606
Tonkin, O. P., 2nd-Lt. 588
—— R. S., Lt. . . . 389, 1437
Tonkinson, H., Lt. 848
Toogood, C. G., Lt. . . . 389, 1602
—— L. G. C., 2nd-Lt. . 885, 1008
Toohey, J., *Hony.* Capt. . 501, 656
Toolmin, A. F. H., 2nd-Lt. . .
Toomey, A. T., 2nd-Lt. 559
—— S. P., Lt. 544,
—— W. G., Lt. . . . 430, 1130
Toon, H. E., 2nd-Lt. . . . 554, 1676
Tooth, H. C. A., Lt. . . . 399, 1491
Toovey, C. W., Capt. . 292a, 753, 1335
—— E. H., Lt. . . . 515, 1668
Topham, G. H., Lt. 725
—— L. H. C., 2nd-Lt. 546, 1389
—— S., Lt. 460
Topley, A., Lt. 739
Toplis, G A., 2nd-Lt. 693
Toppin, A. J., Lt. 767
Topping, W. D., Lt. . . . 397, 1026
Tor, Jemdr. 1184
Tora Baz Khan, Jemdr. . . . 1152
—— Jemdr. 1134
Tora Khan, Jemdr. 1500
—— Jemdr. 1457
Tor Khan, Subdr. 1118
Torkington, G. S., Maj. . 217, 1287
Torpy, A. T., Asst. Surgn. . . . 1794
Torrance, I. S., Lt. . . . 397, 1193
—— W. R. D., Lt. . . . 294
Torrens, R. G., Lt. . . . 518, 1493
Torrie, C. J., Maj. . 219, 1003, 1140
—— L. J., Capt. . . . 265, 1124
Tosh, C. W., Lt.-Col. 656
Tostevin, F. P., Lt. . . 560, 803, 1363
Tota Ram, Subdr. 1142
—— Jemdr. . . . . . .
—— Jemdr. 1452
—— Subdr. 1335
—— Subdr. 1412
Tottenham, J. L., Lt. 318, 1096, 1166
Totterdell, J. L., Lt. 585
Tottle, J. B., 2nd-Lt. 565
Tough, F. W. K., Capt. 1722
Toulmin, V. F., Ty. 2nd-Lt. . . . 577
Toussaint, H. W., Asst. Surgn. . 1799
—— J. B., 2nd-Lt. . 564, 1183
Tovell, A. F., Sub-Condr. . . . 1867
—— G. A. C., 2nd-Lt. 555
Tovey, R. C. D., 2nd-Lt. . . . 790
—— W., Sub-Condr. 1864
Towell, A. F. H., 2nd-Lt. . . 546, 1271
Tower, H. G. E., 2nd-Lt. . . 323, 1251
—— J. A., 2nd-Lt. 819
Towers, F., Lt. 757
Towerton, H. J., Lt. . 503, 707, 1323
Towler, R. A. G., Lt. 651
Towndrow, F. E., 2nd-Lt. . . . 1673
Townend, H. D., Lt. 821
Towner, C., 2nd-Lt. 361, 1106

Towning, W. E., Ty. Asst. Surgn. 1804
—— C. F., Lt. . . . 543, 1355
Townley, J. S., Capt. . . 717, 1722
—— N. P., Lt. . 305, 1170, 1137
Townsend, F. C., Lt. 638
—— F. J., 2nd-Lt. . . . 757
—— F. O., Capt. 501
—— G., 2nd-Lt. 554
—— H., Maj. . . . 696, 795
—— R. R., Lt. 75
—— R. S., Capt. 1754
—— T., 2nd-Lt. . . . 357, 1151
—— W. A., 2nd-Lt. . . . 559
—— Smith, W., Capt. . . 1702
—— 1886
Townshend, C. A., Lt. 1886
—— C. V. F., Maj.-Genl. *Sir*, 118
—— C. M. W. L., 2nd-Lt. 328 891
—— F. H. E., Capt. 665, 1882
—— R. P. L., Capt. . . 850
Toy, L., Lt. 75
Toyle, R. C., Ty. Lt. 1776
Toyne, C. J., 2nd-Lt. . . 330, 1606
Tozer, D. C. E., Lt. . . . 492, 935
—— H. S., 2nd-Lt. . . . .
—— P. H. S., Lt. 519
Trafford, E. A., Maj. . 25, 235, 1883
—— W. E., Tempy. 2nd-Lt. . 622
—— W. L., Maj. 1749
Trail, R. M., Capt. . . . 268, 1174
—— W. S., Capt. . . 5, 258, 1251
Traill, H. B., Capt. . 288, 1355, 1373
Tranchell, H. G., Capt. 1910
Tranter, F. G., Lt. . . . 75, 490
Tranthen, J. R., 2nd-Lt. . . . 618
Trapman, A. H., Capt. . 735, 795
Travers, F. G., Lt. 835
—— O. A., Capt. . 501, 803, 1666
—— R. E., Lt.-Col. 133, 156, 193
—— W. L., Capt. 610
Travers-Drapes, E. H. A., 2nd-Lt. 559
—— V. L., 2nd-Lt. . . 563
Travis, G. R., Lt. 577
Traylen, H. E., 2nd-Lt. . 549, 1507
Traynor, T. W., Asst. Surgn. . 1796
Treadgold, G. W. R., 2nd-Lt. . . 790
Treanor, R. F. S., Esq. 73
Treasure, F. L., 2nd-Lt. 362
Trechmann, A. D., Lt. . . 410, 1285
Tredaway, G. H., Capt. . 59, 504, 852, 1026
Tredinnick, N. W., 2nd-Lt. . . . 623
Tree, G. C., 2nd-Lt. 723a
—— H. B. leD., Capt. . 54, 254, 1451
—— P. B. LeD., Lt. 668
Treeby, H. J., 2nd-Lt. . . . 562, 1678a
—— J. W. C., 2nd-Lt. 700
Tregale, J. W. W., Lt. . . 522, 1618
—— P. M. 2nd-Lt 361
Tregaskis, G., Lt. . . . 539, 1331
—— O., Lt. 528, 1487
Tregear, F. C., Lt.-Col. 137, 159, 197, 1042
—— V. F. W., Lt.-Col. 57, 137, 159, 197, 1411
—— V. W., Maj.-Genl. *Sir* 169, 1916
Tregenza, E. L., 2nd-Lt. . . 720, 786
Tregoning, E. A., Capt. 17, 22, 687
Trehane, C. H., Capt. . . 277, 921
Treharne, T. W., Ty. 2nd-Lt. . . 689
Treherne, F. H., Maj.-Genl. *Sir* 3, 118
Trelawny, L. C., Maj . . 24, 244, 1663
Treleaven, C. V., Lt. 491
—— W. J., 2nd-Lt. . . 720, 786
Treloar, C. C. M., 2nd-Lt. . . . 565
—— R. M., Capt. . 502, 858, 1110
Tremaine, Lt. J. 90
Tremenheere, C. W., 2nd-Lt. . . 571a
—— H. S., 2nd-Lt., 549, 1419
Tremlett, C. P., Maj. 696
Trench, A C., Maj. 58
—— A. H. C., Capt. 664
—— C. F., Capt. . 14, 29, 256, 887
—— J. S., Maj. 485
—— R. H., Capt. 653
Trenchard, G. B. B., Capt. . . 634
—— O. H. B., Maj . 664, 1880
Trent, A. R., 2nd-Lt. 1507
Tresawna, H., Lt. 55, 722
Tresham, B. F., Lt. . . . 513, 1178
Tresidder, A. G., Capt. . 1355, 1754
Tresler, K. T. B., *Hony.* Lt. . . 112
Treston, M. L., Capt. 1758
Trethowan, R. F., Lt. 720
Trevaskis, H. K., Lt. 538

NAMES AND RANK. Page.

Trevelyan, W. R. F., Capt. . 278, 1379
Trevenen, W. B., Maj. 89, 698, 795
Trevett, H. V., 2nd-Lt. 360
Trevor, A. C. H., Capt. . 252, 1116
—— A P., Lt.-Col. 103, 197, 1890, 1912
—— C. P., Capt. 367 1601
—— D. B., Lt. 836
—— G H., Col. 177
Trevor-Roper, A. M., Capt. 696, 697
Trewin, J. W., 2nd-Lt. 787
Trewnella, M., Sub-Condr. . . . 1873
Tribe, A. G., 2nd-Lt. . . . 332, 1431
——, C. J., Capt. 61, 706
Tribble., N. V., 2nd-Lt. . . 560, 1407
Triebler, F. C. 2nd-Lt. 799
Trigg, T. R., 2nd-Lt. 355
Trimbakha Hargovind, Trivedi, Sub-Asst. Surgn. 1817
Trimmings, A. E., 2nd-Lt. . . . 561
Trinkle, P., Lt. 533, 1114
Triphook, O. L., 2nd-Lt. 89
Tripp, H. C., 2nd-Lt. 841
—— W. H., Sub-Condr. . . . 1867
Tristram, C. E., Maj. 1919
Tritton, G. C., 2nd-Lt. . . . 549, 1497
Trivedi, C. P., Ty. Lt. 1778
—— D. M., Sub-Asst. Surgn. . 1830
—— J. G., Ty. Lt. 1780
—— S. V., Ty. Lt. 1776
Troath, W. A., 2nd-Lt. 739
Trobridge, F. M., 2nd-Lt. 626
Trotman, E. W., Lt. . . . 470, 1419
Trott, W. A., Lt. 435, 1054
Trotter, E. G. S., Maj. . 103, 104, 106, 208, 1052, 1883, 1913
—— H., Lt. 507, 1601
—— J. P., 2nd-Lt. 719
Trought, T., Capt. . . 14, 19, 751
Troughton-Dean, G. P., Lt. 302, 1116
Trounsen, W., 2nd-Lt. 722
Troup, F. C. A., Capt. . . 636, 978
—— B. S., Maj. 806
—— J. R. Capt. 9867
Trousdell, C. F., Lt. 624
—— J. P., Lt. 486, 897
Troutbeck, J. M., 2nd-Lt. . . . 5
Trow, C. E. A., Ty. Capt. . . . 1725
—— V. H., Capt. 387, 1475
Trowbridge, B. B., 2nd-Lt. . . . 623
Trowsdale, H. J., 2nd-Lt. . . . 1225
Truell, E. G. S., Maj. 769
Truelove, C. W., 2nd-Lt. 753
Trueman, H. J. H., Maj. 219, 872, 1195, 1561
Truesdale, H. V., 2nd-Lt. . . . 788
Truman, M. R. Lady Supt. . . 1734
Truscott, N. P. L. 737
Trusler, C. H., Lt. 527, 988
—— W. J., Lt. 525
Trusted, H. H., Lt. . . 17, 29, 722
Truter, A. W., Snr. Asst. Surgn. 1790
Trutch, H. F., 2nd-Lt. . 474, 1391
Trutwein, P. St. C., Snr. Asst. Surgn. 1791
Trydell, B., Lt.-Col. . . 129, 154, 187
Trythall, W. J., Lt. 858
Tryhorne, M. W., Lt. . . . 529, 1487
Tuck, C. H. A., Maj. . . . 228, 1605
—— S. A., Capt. 760
Tucker, A. C., Capt. . . . 738, 740
—— A. C., Lt.-Col. 714
—— B. B., Lt. Hon. . . . 403, 1601
—— B. H. G., Lt. . . . 304, 1645
—— C., Condr. 1872
—— E. F. G., Lt.-Col. . 139, 1745
—— F. G., 2nd-Lt. . . . 722, 1708
—— G. E., Lt. 755
—— H. B., Maj. 244, 1395, 1399, 1883
—— J. D., Lt. 460-62 1295
—— K. D., 2nd-Lt. 1565
—— N. P., Lt. 487
—— V., Lt. 1134
—— W. A., Esq. 34
—— W. H., Maj. 1746
—— W. L., Lt. 756
Tuckey, G., 2nd-Lt. 1689
—— R. E. S., Lt. . . . 516a, 1676
Tudball, W., Lt.-Col. the Hon'ble 814
Tudhope, M. A., 2nd-Lt. . . . 78[illegible]
Tudor, W. J. E., Lt. . . 618, [illegible]
Tudor-Craig, P. C., Lt. . [illegible], 16[illegible]
Tudor-Owen, W. C., 2nd-Lt . . . [illegible]553
Tuffin, R., 2nd-Lt. . . . . [illegible], 1152
Tugwell, W. B. P., M[illegible]. . . 229, 1479
Tuhi Ram, Jemd[illegible] 1051
Tuite, H. G. S., 2nd-[illegible] 1696

NAMES AND RANK. Page.

Tuite-Dalton, F. G. T. Capt. 254, 1606
Tukaram Bhogle, Subdr. . . . 1467
—— Dharo, Jemdr. 1480
—— Jadhu, Subdr. 1509
—— Jagtap, Jemdr. 1416
—— Kokate, Jemdr. 1436
—— More, Jemdr. 1436
Tuke, A. W., Maj. 1746
Tuker, F. I. S., Capt. 278, 1600, 1898
Tul Ram, Jemdr. 938
—— Gurung, Jemdr. 1608
—— Gurung, Subdr. 1607
—— Ram Rana, Subdr. . . . 1624
Tulbar Pun, Jemdr. 1608
Tularam, Jemdr. 1581
—— Joshi, Sub-Asst. Surgn. 1810
Tulbir Gurung, Subdr. . . . 1636
Tule Thapa, Jemdr. . . . . .
Tuljashanker Hargovind Trivedi, Sub-Asst. Surgn. 1817
Tulla Thapa, Subdr. 1595
Tulloch, D., Lt. 523, 1473
—— D. A., 2nd-Lt. . . . 557, 1696
—— J. C., Lt. 1007
—— H. M., Lt. . . . 304, 937
—— S. S. G., Maj. 500
Tully, E. J., 2nd-Lt. 1104
—— E. J. S., 2nd-Lt. . . 476, 959
Tulsa Singh, Jemdr. 1502
Tulsi Ram, Sub-Asst. Surgn. . 1816
—— Jemdr. 1145
—— Jemdr. 1289
—— Gharti, Subdr.-Maj. . 1601
Tul Sing Thapa, Subdr. . . . 1607
—— Jemdr. 1608
Tumber., W. A., 2nd-Lt. . 363, 1203
Tun, A., Ty. Lt. 1774
Tun Zam, Jemdr. 1698
Tunaley, A. E., Lt. 73
Tunda, Subdr. 1471
—— Singh, Jemdr. 1412
—— Subdr. 1453
Tungate, W. C., 2nd-Lt. . . . 1641
Tunkniss, C. S., 2nd-Lt. . . . 1417
Tunks, G. P. D'A. G., Capt. . . 711
Tunnicliffe, Lt., F. D . . . 503, 993
Tun Sein, Jemdr. 1295
Tunstall, J. E., 2nd-Lt. . . 546, 1297
Tupper, G. W. H., Capt. . . . 753
Turketine, A. W. H., Lt. . . . 720
—— T L., 2nd-Lt. 720
Turkestan, Subdr. 1513
Turl, G. H., Lt. 1441
Turle, H. B., Lt. 492
Turnbull, A. W., 2nd-Lt. . . . 788
—— B., Maj. 227, 869, 1110, 1158
—— D., Lt. 353
—— G. O., Maj. . . 214, 1124
—— J., *Hony.* Capt. 589
—— J. N., Ty. Lt. 1729
—— P., Maj. 614
—— P. S., Surgn.-Maj.-Genl. 1
—— T. E., Lt.-Col. 777
—— W. G., 2nd-Lt . . 669, 997
—— W. J., Lt. . . 669, 997, 1886
—— W. J. U., Capt. 611
—— W. R., 2nd-Lt. 879
Turner, A., 2nd-Lt. 723a, 1009
—— A. C, Lt. 844-45
—— A. F. St. A., Lt. 686
—— A. G. B., Lt.-Col. 57, 130, 155, 189
—— A. H., Col. 178, 1917
—— A. M., Maj. 581
—— A. P., Capt. 714
—— A. W., Asst. Surgn. . . 1803
—— A. W., Capt. . . . . . 1[illegible]
—— B. G., 2nd-Lt. . . . . . [illegible]30
—— C., Lt. . . . . [illegible]
—— C. A. G., Lt. . . 340a, 1701
—— C. D. L., Lt. . 318 [illegible], 1467
—— C. J. B., Ca[illegible]. 27[illegible], 1004, 1078
—— C. L., 2nd-Lt. . . . 554, 1676
—— C. W. A. F[illegible]., ... 35
—— D., Con[illegible]. 1861
—— D. [illegible], Lt. 1,83
—— [illegible] M., Lt. 503, 997
—— D. V., 2nd-Lt. 566
—— E., Lt. 410, 965
—— E. L., Lt. 292a, 679
—— E. L., Lt. 875, 901
—— F. R., Lt. 515, 1004
—— F. C., Lt.-Col. 54, 98, 102, 103
—— F. W. C., Lt.-Col. 78, 129, 155, 188
—— G. F., Capt. . 96, 263, 1335

NAMES AND RANK. Page.

Turner, G. F., Lt. . . . 492, 923
—— G. H., Lt. 623
—— G. H., Brig.-Genl. . . . 57
—— G. K., Lt. 542, 1303
—— G. R., 2nd-Lt. 568
—— H., Capt. 768
—— H., Lt. 617
—— H. A., 2nd-Lt. . . 554, 1676
—— H. F. C., Lt. 927
—— H. F. G., Lt. 410
—— H. G., Maj. . . . 244, 1427
—— H. G., 2nd-Lt. 787
—— H. H., Lt.-Col. 138, 661, 1880
—— H. J. C., Lt. . . . 520, 1701
—— H. L., 2nd-Lt. . . 692, 784
—— H. R., Lt. 624
—— J., Capt. . . 255, 1459, 1896
—— J. C., Lt. 692
—— J. C. H., Capt. . . . 1696
—— J. F., Maj. 58, 663
—— J. G., Maj.-Genl. 57, 105, 108, 118, 146, 173, 881
—— J. H. P., 2nd-Lt. . . 545, 988
—— J. M., Lt. . . . 91, 92, 529
—— J. N., Sr. Asst. Surgn. . 1789
—— J. R., Capt. 1723
—— J. R., Capt. 827
—— J. T., Lt. . . . 95, 487, 927
—— J. W., Lt. 368, 925
—— M., Capt. 608
—— M. R., 2nd-Lt. . . . 331, 921
—— N. T. B., Lt. . . . 520, 1667
—— O., Sub-Condr. 1855
—— P. H., 2nd-Lt. . . . 551, 1088
—— R. G., Lt.-Col. . . . 130, 1742
—— R. H. F., Lt. . . . 389, 877
—— R. L., Lt. 513, 1007
—— R. V., Lt. . . 389, 885, 1702
—— V. E., Lt. 535
—— W. 1009
—— W., Lt. 719
—— W. C., 2nd-Lt. 552
—— W. F., Lt. 739
—— W. T., Lt. . . . 460-62 1389
—— W. W., Asst. Surgn. . . . 1792
Turnor, H. B., Capt. 595
Turpin, C. F., 2nd-Lt. 712
—— R. J., Lt. 686
Turra Baz Khan, Subdr.-Maj. . 1227
Turrell, J. W., Lt. 89. 759
Turton, A. E., Lt. 589, 609
—— H. M., Maj. 1919
—— O., 2nd-Lt. 657
—— W. H., 2nd-Lt. . . . 535, 1271
Tustin, C. V., Lt. . . 435, 784, 951
Tute, C. S., Maj. . 71, 101, 634, 1882
Tutt, A. R., Lt. 776, 1699
Tutte, E., 2nd-Lt. 564, 1395
Tutten, C., Sub-Condr. 1856
Tutton, H. K., 2nd-Lt. . . 360, 1241
—— P. C., Lt. 423, 1367
Tweddell, H., Lt.-Col. . . 130, 155, 189, 1335
Tweed, J. R. H., Lt. . . . 389, 1383
Tweedie, F. F., Capt. . . . . . 8[illegible]
—— M., Col. 177
—— N. C. L. B., Capt. . . . 1[illegible]
Tweedy, H. R. F., 2nd-Lt [illegible] 32[illegible] 1363
—— R. J., Capt. [illegible], 42 [illegible]09, 911
Twelftree, A. E., 2nd-L[illegible] . . . 1183
Twells, T. W., Asst. Surgn. . . . 1792
Twentyman, W. P., the Revd. . 1907
Twiddy, [illegible] A., [illegible]. . 18, 33, 513, 1052
—— G. [illegible], 2nd-Lt. 693
—— [illegible] W., *Hony.* Lt. . . . 1852
Twig[illegible]n, Capt. 814
Tw[illegible]g, C. H., Lt. 533
—— E F., Lt.-Col. 130, 155, 189, 1491
—— H. D., Capt., ... 1696
—— R. H., Br.-Genl. 14, 26, 110, 123, 149, 181, 1917
Twinberrow, J. D., Lt. . 302, 1259
Twining, P G., Maj.-Genl. . 119, 660
Twisleton-Wykeham-Fiennes, N. I. E., Maj. 614
Twiss, E. K., Maj. 16, 39, 236, 1058
—— H. W. F., Maj. 9, 102, 219, 1662
—— W. L. O., Bt. Lt.-Col. 6, 13, 98, 102, 103, 138, 159, 215, 1641
Twist, E. F., Maj. 591
Twynam, H. J., Lt. 512, 1489
Tye, H. J., 2nd-Lt. . . 68, 474, 911
—— J., Condr. 1854
Tyer, A. M., 2nd-Lt. 734
Tyers, W. F., Lt. . . . . . . [illegible]
Tyfe, W. S., 2nd-Lt. . . . . .

Names and Rank. Page.

Tylden-Pattenson, E. C., Lt.-Col. 129, 662, 985
Tyler, A., Condr. . . . 1851
—— A. T., Capt. . . . 727
—— B. M. C., Lt. . . 501a, 1665
—— E. W., 2nd-Lt. . . 79, 494
—— F. R., Lt. . . . 727, 1007
—— H. H. F. M., Lieut. . . 516
—— J. A., Br.-Genl. 58, 121, 126, 613
—— L. J., 2nd-Lt. . . . 763
—— N. E., 2nd-Lt. . . . 710
—— R. C., Lt. . . . 540
—— V., 2nd-Lt. . . . 724
—— W., 2nd-Lt. . . 563, 1524
Tynan, P., Sub-Condr. . . 1853
Tyndale, A. C., Lt. . . 88, 821
—— H. E., 2nd-Lt. . . 359, 1511
Tyndale-Biscoe, Maj. E. C. . . 500
Tyndall, C. T. A., Lt. . . 539, 1653
—— H. S., Maj. . . 209, 1183
—— J. W. H. D., Maj. . 235, 1034
—— W. W., Maj. . . . 656
Tyne, C. J., 2nd-Lt. . . . 329
Tyres, F. W., Lt. . . . 1228
Tyrer, R., Maj. . . 53, 212, 1343
Tyrie, D. A., Col. . . . 656, 1893
Tyrrell, A., 2nd-Lt. . . 552, 1385
—— A. C. L., Maj. 15, 31, 99, 216, 921
—— C. H., Maj. . . 227, 1120
—— F. H., Lieut.-Genl. 170, 1311
—— H. E., Lt. . . 445, 1142
—— J. F., Lt.-Col. 10, 74, 138, 632
—— J. R. J., Maj. . 1748, 1890
—— R. C., Sub-Condr. . . 1868
Tysoe-Smith, Lt. C. . . . 1007a
Tyson, E. W., Lt. . 302, 1427, 1509
Tyte, A. N., Lt. . . . 401, 1211
—— W. L., 2nd-Lt. . . . 547
Tytler, H. C., Col. . 126, 150, 186
—— S., Sub-Condr. . . . 1868

U

Uberoi, B. D., Tempy. Capt. . . 1768
—— T. S., Tempy. Capt. . . 1769
Ubhaya, N. R. R., Tempy. Capt. 1766
Ubsdell, A. R., Capt. . . . 502
Ubdul Hakim Khan, Jemdr. . . 936
Uda, Jemdr. 1200
Udahang Limbu, Jemdr. . . . 1649
Udai Singh, Ressdr. . . . 880
—— Subdr. . . . 1068
—— Jemdr. . . 1021
Udani, M. L. D., Sub-Asst. Surgn. 1830
—— K. D., Sub-Asst. Surgn. . 1832
Udehbir Gurung, Jemdr. . . . 1618
Ude Ram, Jemdr. . . . 1197
—— Sing, Jemdr. . . . 1060
—— Jemdr. . . . 878
—— Gurung, Jemdr. 1608, 1653
—— Ghale, Jemdr. . . . 1608
Udey Singh, Jemdr. . . . 1420
Udham Singh, Risldr. . . . 938
—— Risldr. . . . 898
—— Jemdr. . . . 1189
—— Subdr. . . . 1195
—— Jemdr. . . . 1099
—— Subdr. . . . 1160
—— Jemdr. . . . 1137
—— Jemdr. . . . 957
—— Jemdr. . . . 912
—— Jemdr. . . . 1107
—— Sub-Asst. Surgn. 1818
—— Ty. Sub-Asst. Surgn. 1833
Udhav Sakha-Ram Shindge, Sub-Asst. Surgn. . 1830
Udho, Subdr. . . . 1171
Udit Misr, Subdr. . . . 1026
—— Narain Singh, Jemdr. . 1054
Udmi, Risldr. . . . 882
—— Subdr. . . . 1587
—— Subdr. . . . 1217
—— Ram, Subdr.-Maj. . . .
—— Jemdr. . . . 1054
—— Singh, Jemdr. . . . 1031
Uff, G. T. R., Lt. . . 399, 1239
Ugan Singh, Jemdr. . 940, 1393
Ujagar Singh, Subdr. . . . 1117
—— Jemdr. . . . 990
—— Jemdr. . . . 1044
—— Jemdr. . . . 1533
—— Sub-Asst. Surgn. . 1817
—— Jemdr. . . . 1376
Ujala, Jemdr. 1039
Ujirhoj Rai, Jemdr. . . . 1649
Ujir Sing Limbu, Jemdr. . . 1649
Ulloh, M. S., Tempy. Lt. . . 1773
Uloth, G., Capt. . . . 278, 927
Umapathi Pande, Jemdr . . 1224
Umar Ali, Subdr . . . 150, 204
Umar Rawat, Subdr. . . . 1471
—— **Daraz Khan**, Ressdr. . . 1683
—— Ressdr. . . . 1684
—— Ressdr. . . . 1686
—— **Din**, Subdr.-Maj. . . 1497
—— **Khan**, Subdr. . . . 1117
—— Sing, Jemdr. . . . 1216
Umdah Singh, *Hony.* Capt. . . 2101
Umed Singh, Jemdr. . . . 940
Ume Singh, Subdr. . . . 1488
Umar Ali, Jemdr. . . . 1321
—— Khan, Jemdr. . . . 1196
—— Jemdr. . . . 1480
Umda, Jemdr. 1193
Umer Khan, Ressdr. . . . 1682
Umfreville, A. T., 2nd-Lt. . . 623
Umrao Ali Khan, Subdr. . . 1088
—— Jemdr. . . . 1473
—— **Singh**, Subdr.-Maj. . . 1024
—— Rawat, *Hony.* Capt. 2103
—— Jemdr . . . 1071
—— Tr. Sub-Asst. Surgn. 1833
—— Jemdr. . . . 1193
—— Singh, Jemdr. . . . 1452
—— Subdr. . . . 1059
—— Chaudhari, Jemdr. . . 1179
Umar Baksh, Subdr. . . . 1441
Unad Singh, Subdr. . . . 1471
Unao Singh, Jemdr. . . . 940
Underdown, A. H., . . 534, 1707
Underhill, G. F. C., Maj. 99, 227, 1267
—— N., Capt. . . . 636
—— W. E., 2nd-Lt. . . 295
—— **Faithorne**, C. F., Capt. 369
Underwood, G. F., Maj. . 228, 903
—— H. J., Lt. . . 375, 1136
—— J. E. A., Capt. . . 1724
—— W. R., Lt. . . 443, 1293
Unger, O. R., Capt. . . . 1758
—— V. C. W., Capt. . 526, 852, 1667
Unkar Chand, Subdr.-Maj. . . 1174
—— **Singh**, Jemdr. . . . 1583
Unkara, Jemdr. . . . 1193
Unsworth, N., Lt. . . 516, 1096
Unwin, G. B., Bt.-Col. . 123, 153, 181
Upcher, R., Maj.-Genl. . . 761
Upchurch, L. C., Lt. . . 353, 1007a
Upendra Kumar Ganguli, Sub-Asst. Surgn. . . 1815
Uppal, K. C., Ty. Lt. . . . 1782
Upperton, J., Maj.-Genl. . . 173
Upshon, C. O., Asst. Sugn. . . 1803
—— L. H., 2nd-Lt. . . 557, 843
Upson, P. C., 2nd-Lt. . . . 361
Upton, G. H. M., 2nd-Lt. . . 1698
—— J. A. E., Lt. . . 322, 1606
—— R. B. E., Capt 280, 1178, 1595
Ure, H. O., 2nd-Lt. . . 712, 1008
Uridge, A. J. B., Capt. . . . 778
—— H. J., Lt. . . 73, 488
Urmson, D. G. S., Capt. . 271, 1275
Urmston, H. N., Capt. . . . 1919
Urquhart, A. M., Maj. . 71, 633, 1884
—— D. D. M., Lt. . . 298, 1271
—— D. I., 2nd-Lt. . . . 766
—— D. S., Lt. . . . 765
—— F. M., Capt. . . 763, 781
—— W., Lt. . . . 1839
Urwin, J. J., Maj. . . . 1746
—— S. E., 2nd-Lt. . . . 359
Usher, J. T., Lt. . . . 585
Ushnaki, Subdr. . . . 1174
Ushnak Singh, Subdr. . . . 1699
Usman Khan, Jemdr. . . . 1324
—— Jemdr. . . . 920
—— Jemdr. . . . 1468
—— Jemdr. . . . 1464
Uttam Chand Malhotra, Ty. Sub-Asst. Surgn. . . . 1833
—— **Singh**, Sub-Asst. Surgn. 1812
—— Subdr. . . . 1211
—— Subdr. . . . 1565
—— Jemdr. . . . 936
—— Subdr. . . . 1211
—— Jemdr. . . . 1158
—— Jemdr. . . . 1142
—— Jemdr. . . . 1305
—— Jemdr. . . . 1167
Uttram Singh, Jemdr. . . . 999
Utterson, H. K., Maj. . . . ...
Utting, F. J., *Hony.* Maj. . . 1843
—— F. A. K., 2nd-Lt. . 566, 696
—— L. E., 2nd-Lt. . . . 361
Uzielli, H. R., Capt. . . . 854

V

Vadamalai, Jemdr. . . . 1276
Vagnelini, S. V., 2nd-Lt. . . 707
Vaid, H. B., Ty. Lt. . . . 1781
Vaidalingam, Jemdr. . . . 1265
Vaidanadasami Naidu, Jemdr. . 1317
Vaidya, J. B., Capt. . . . 1758
Vakil, N. H., Tempy. Capt. . . 1769
—— P. R., Tempy. Capt. . . 766
Valett, 2nd-Lt. . . . 553, 1054
Valentine, A., Maj. . . . 719
—— C. R., Capt. . . . 808
—— S. W., 2nd-Lt. . . 742, 788
—— W. C., 2nd-Lt. . . . 1040
—— W. S., Lt. . . 389, 1182
Vallance, G. A., Lt. . . 17, 23, 705
—— J. P., 2nd-Lt. . . . 781
Vallis, E. W. H., Lt. . . . 703
—— R. W. H., Lt. . . . 704
Valladares, W., Ty. Capt. . . 1771
Vaman Balwant Unde, Sub-Asst. Surgn. 1831
Vance, H. R., Lt. . . . 793
—— R. L., Capt. . . . 1759
Van Cuylenberg, V. A., Asst. Surgn. 1802
—— Der Beek, A. G., 2nd-Lt. . . 833
Vander Byl, J., Maj. . . . 589
Vander Gucht, G. T., Maj. 33, 216, 879, 1537
Vanderhide, R. F., Asst. Surgn. 1798
Vanderwart, S. C., Asst. Surgn. . 1803
Vanderwert, J. A. H., Ty. Capt. . 1725
Vanes, J. A., Revd. . . . 1908
Vania, S. D., Lt. . . . 1774
Van Langenberg, J. F., 2nd-Lt. 554, 1513
Van Ingan, C. T., 2nd-Lt. . . 875
Vankaner, Capt. *Raj Saheb of* . 111
Van Houten, F., 2nd-Lt. . 352, 1678
Vankatasami, Subdr. . . . 1315
—— Jemdr. . . . 1316
Van Mulken, Rev. Peter . . 817
Van-Reenen, J. W., Capt. . . 1758
Vanrenen, D. H., Maj. . 78, 227
—— G. R., Lt.-Col. 131, 156, 191, 868, 1082
—— J. R. D., 2nd-Lt. 365, 1355
—— L. F. E., Lt. . 491, 881
Vansittart, E., Bt.-Col. . . 180
—— G. N., Capt. . 277, 949
Vanstone, H. W., 2nd-Lt. 354, 1674
Vanston Rumney, W. F., 2nd-Lt. 363, 1203
Van Someren, A. G. V., Capt. 734, 1722
—— L. W., Capt. . . 817
—— W. W., Maj. 56, 103, 105, 212, 1203
Varadachari, B. V., Ty. Lt. . . 1777
Varadaraju, Subdr. . . . 993
—— **Nayathar**, P. T., Sub-Asst. Surgn. . 1828
Varde, D. N., Tempy. Capt. . . 1770
Vardya, B. K. Ty. Lt. . . . 1782
Vardon, A. C., Asst. Surgn. . . 1798
Varlava, D. D., Tempy. Capt. . . 1767
—— H. E., Ty. Lt. . . . 1779
Varma, A. S., Ty. Capt. . . 1772
Varvill, J. K., Capt. . . . 787
—— A. M., Capt. . . . 1001
Varyam Singh, Sub-Asst. Surgn. 1810
Varyan Singh, Jemdr. . . . 1699
Vasant, G. A., Sub-Asst. Surgn. . 1831
—— **Sakharam Gaikwad**, Sub-Asst. Surgn. 1831
Vasey, G. H., Capt. . . . 787
Vasudev Ramchandra Mudhekar, Sub-Asst. Surgn. . . 1831
Vaswani, T. P., Ty. Lt. . . . 1778
Vaughan, A. A., 2nd-Lt. . . 739
—— C. C., 2nd-Lt. . . . 552
—— E., 2nd-Lt. . . . 554
—— E. A., 2nd-Lt. . 558, 1678
—— E. G., Lt.-Col. 46, 131, 156, 191, 1660
—— E. N. L., Lt. . . . 704

NAMES AND RANK. Page.

Vaughan, E. W. D., Capt. . 278, 877
—— F. B., 2nd-Lt. . . . 1183
—— H. B., Capt. 297, 830, 1187, 1896
—— J., 2nd-Lt. 560
—— J. C. S., Lt.-Col. 128, 1741, 1894
—— L., Lt. 728
—— L. R., Maj.-Genl., 97, 119, 126, 153, 209, 1629
—— R. E., Col. 9, 16, 120, 125, 150, 188, 1660
—— R. R., Lt.-Col. 23, 189, 1883, 1911
—— R. W., Lt. . . 535, 1699
—— S. G., Lt. . . 528, 1036
Vaughan-Jones, H., Lt. . 291, 1455
Vaughan-Sheehan, G. M.R., Lt. . 503
Vaughan-Stevens, D. L., Lt.. . 490
Vaux, H. G., Capt. 1893
—— R., 2nd-Lt. 701, 1008
Vawdrey, R. H., 2nd-Lt. . . . 729
Vazialingam, Subdr. 1317
Vazifdar, S. S., Capt. . . 108, 1755
Veale, J. F., 2nd-Lt. 1008
—— P. J., Capt. 1757
—— R. D. S., 2nd-Lt. 1160
Vears, L., Lt. 622
Vedak, S. V., Sub-Asst. Surgn.. . 1832
Veech, A. G., Nursg. Sister . . 1735
Veerappa, Subdr. 1307
Veerappen, Subdr. 1263
Velamurgam, Subdr. 1339
Velanganni, Jemdr. 993
Velankar, S. V., Ty. Lt. . . . 1782
Vellinayakam, Subdr. 1351
Vellore Gangadara Ranganthem, Sub-Asst. Surgn. 1828
Velu Kutty Nair, Jemdr. . . . 1309
Velunayar, P., Jemdr. 1360
Venables, A. V., Capt. . . 501, 997
—— E. P., Lt. . . 504, 1263
Veneer, F. A., 2nd-Lt. . . 552, 1009
Vengalil Sankunni Menon, Jemdr. 1309
Vengu Aiyar, *Rao Bahadur* P. R. 88
Venkatachalam, K., Ty. Lt. . . 1776
—— —— Subdr. . . . 1307
—— —— Jemdr. . . . 1351
Venkatagir, Raja of, 2nd-Lt. . . 481
Venkataramanan, C. S., Ty. Lt. 1776
Venkatasami, Subdr.-Maj. . . . 1339
—— (i) Jemdr. . . . 994
—— (ii) Jemdr. . . . 994
—— Subdr. 1327
—— Subdr. 1315
Venkatachellam, Jemdr. . . . 1308
—— Jemdr. . . . 1360
Venketachalam, Jemdr. . . . 1361
Venkatanarasayya, Subdr. Maj. . 1327
Venkatesape-Rumal, Jemdr. . . . 994
Venktesh Anant Chandragiri, Sub-Asst. Surgn. 1831
Venn, E. N. L., Lt. 595
—— T. W., Lt. 503
—— W. G. W., 2nd-Lt.. . . . 728
Venning, F. E. W., Maj. . . 234, 1144
Venters, W. B., Capt. 857
Ventham, W. G., Condr. . . . 1872
Venton, W. F., 2nd-Lt. . . 549, 1144
Ventris, E. F. V., Maj. . 10, 76, 634
Venugopal, S. B., Capt. . . . 1758
Venagopaul, K., Ty. Lt. . . . 1776
Vercoe, F. G., 2nd-Lt. . . 750, 784
Verdon, P., Lt. 1759
Verede Matthews, R. Lt. . . . 45
Verghese, G., Ty. Lt. 1775
Vermont, C. L., 2nd-Lt . . . 568
Vernal, E. E. L., Asst. Surgn.. . 1802
—— M. S., Lt.. 528
Verner, E. A., 2nd-Lt. 567
Verney, J. H. P., *Hon.* Lt. . . . 595
—— R., Lt.-Col. 2
Vernieux, C. O., Lt. . . . 525, 1701
Vernon, G. F. D., 2nd-Lt. . 666, 992
—— H. B., Maj. . . . 242, 1128
—— O., Nursg. Sister . . . 1735
—— R. C. 1009
—— T. F., Lt. 893
Vernon Brown, E. C., 2nd-Lt. . . 1355
Verschole, F. H. H. S., Capt. . . 767
Vertue, H. St. H., Capt. . . . 1718
Vessey, S. J., Lt. . . 77, 516a, 1885
Vetch, W. H., Lt. 380
Veysey, G., 2nd-Lt. . . . 549, 1705
Vibart, E. D. H., Col. 176

NAMES AND RANK. Page.

Vicary, E. S., Bt. Maj. 53, 257, 905
Vicary, G. E., Capt. 755
Viccars, L. J., Lt. 754
Vick, R. I., Lt. 791
Vickers, A., Maj. 219, 1215, 1898
—— C. K., 2nd-Lt. . 326, 366, 889
—— H. J., Lt. 490
—— S., 2nd-Lt. 690a
—— S. G., 2nd-Lt. 705
—— T. B., 2nd-Lt. . . 668, 992
—— W. G. H., Capt. . 272, 933
Vickery, L. M., Lt., . . . 532, 988
—— R. M., Lt. . . . 529, 1671
Victor, A. C., Asst. Surgn. . . . 1799
Vidler, C. H., Lt.. 630
Vieyra, C. F., Asst. Surgn. . . . 1802
—— P., Capt. 1758
Vigar, W. A., 2nd-Lt. 361
Vigne, F., Lt. 797
Vigors, M. D., Capt. . . 260, 891
Vijaya Narain Gupta, Sub-Asst. Surgn. 1819
Vilvaraghava Mudali, Sr. Sub-Asst. Surgn. 1826
Vikramshah, Raja Hony. 2nd-Lt. 482
Vijaya Naryan Gupta, Sub-Asstt. Surgn. 1819
Villar, A. W., 2nd-Lt.. 739
Villiers, A., *Hony.* Maj. 11, 71, 91, 1843
—— A., Condr. 1854
—— E. H., Lt. 763
—— S. A., Lt. 617
—— T. L., Lt.. 776
Villiers-Stuart, J. P., Bt. Lt.-Col. 6, 13, 98, 139, 160, 218, 1243
—— P. G., Capt. 367, 1395
—— —— W. D., Brig. Genl. 7, 74, 141, 163, 208, 1617
Vincent, B., Brig.-Genl. 50, 98, 127, 585
—— B., Asst. Surgn. . . . 1801
—— F. L., Lt.-Col. 34, 78, 129, 155, 188
Vincent, G., Lt. 759
—— G. W., Sr. Asst. Surgn. . 1791
—— H., Asst. Surgn . . . 1795
—— H. O., Brig.-Genl. . 7, 14, 612
—— J. A., 2nd-Lt. 358, 1365, 1545
—— J. F. des., Asst. Surgn. . 1801
—— J. W. M., Capt. . . . 705
—— L. F., 2nd-Lt. 740
—— M. S., Sub-Asst. Surgn. . 1827
Vine, H., Capt. 651
Vine-Stevens, L. T., Lt. . . . 750
Vines, C. E., Maj. . 72, 101, 634, 1882
—— D. F., Capt. 1893
—— R. G., Lt. 445, 927
Vinen, J. E., Lt. 508, 1665
Vining, C., Ty. 2nd-Lt. 760
—— L. E., Capt. 840
Vintom, J., 2nd-Lt.. 717
Vipan, A., 2nd-Lt. 808
—— R. S., 2nd-Lt. 857
Vir Singh, Jemdr. 1212
Viranna, Jemdr. 1315
Viraraghavelu, Jemdr. 994
Viraraghavulu, Subdr. 1275
Virasmai Nayudu, V., Sr. Sub-Asstt. Surgn. 1826
Virgin, A. H., Lt. . . . 293, 1066
—— C. D., 2nd-Lt. 565
Virgo, F. R., Lt. 543a, 1636
Virgoe, C. G. B., Lt. 705
Virrayya, Jemdr. 1332
Vishnu Aire, Subdr. 1467
—— Ghone, Subdr. 1444
—— Jirao Ghag, Jemdr. . . 1464
—— Malhar Bandevadikar, Tr. Vety. Asst. . . . 1687
—— Narvekar, Subdr. . . . 1707
—— Palao, Subdr. 1407
—— Sakha Ram, Jemdr. . . 1444
—— Sakha Ram Panjre, Sub-Asst. Surgn. . . 1816
Vishram Rao Chowan, Subdr. 1463
—— Gangaram, Jemdr. . . 1472
Vishvanath Narayan Divshikar, Sub-Asst. Surgn. . 1831
Visvanathan, S., Ty. Lt.. . . 1779
Vishwanath Bhagawantrao Mohite, Sub-Asst. Surgn. . . 1830
Vishwa Nath Shukla, Sub-Asst. Surgn. 1817
Visvalingam, Jemdr. 994
Vittal Ghande, Jemdr. 1467
Vithal Mahadeo Satpute, Sub-Asst. Surgn. 1830

NAMES AND RANK. Page.

Vithoba Jadhow, Subdr. . . . 1479
Vithoba, Khule, Jemdr. . . . 1424
Vittle, D. H., 2nd-Lt. . . . 700, 796
Vivian, F. G., Bt.-Col. 56
—— G. L., Lt. 493
—— J. F. G., 2nd-Lt. . . . 565
—— O. H., 2nd-Lt. . . . 554
—— S. C. E., 2nd-Lt. . . . 1693
—— V. P. T., 2nd-Lt.. . . . 551
—— Sheriff, Jemdr. 1299
Vyvyah, R. T., Capt. 776a
Vizard, W. G., 2nd-Lt. . 55, 738, 740
Vizialingam, Subdr. 1316
Vizrae, D. N., Sub-Asst. Surgn. 1829
Vlieland, C. P., 2nd-Lt. 557
V. Jacob, Jemdr. 1273
Voelcker, G. G., Capt. 94, 280—287, 1606
Vogwill, J. W., Lt. . . . 538, 1672
Voit, R. A., 2nd-Lt. . 325, 366, 1162
Vokes, F. L., 2nd-Lt. 556
Von Bibra, F. H., Maj. . . . 817
Von-Braun, C. R. B., Tempy. Capt. 1723
—— W. A. W. G., Lt. . . 733
Von Hogt, A. L., 2nd-Lt. . . . 1207
Vos, J. P. de., 2nd-Lt. 1395
—— S., Lt. 699, 796
Vosper, E. E., Lt. . . . 313, 943
Vost, W., Lt.-Col. . . . 127, 1739
Vousley, F., Ty. Capt 18, 24, 571, 1850
Vowles, A. C., 2nd-Lt. . . 739, 791
—— A. S., Lt. 1455
—— R. D., 2nd-Lt. 788
Voyle, S. G., Lt. 445, 1623
Voysey, G. C. A., Lt. . 96, 306, 1162
Vroj Lal Umed Ram Pandit, Sub-Asst. Surgn. 1830
Vrooman, P. H., Lt. . . . 311, 937
Vyall, F. S., Lt. 523, 992
Vyner, R. J., Maj.. 595
Vyasam Venkataramayya, Sub-Asst. Surgn. 1827

W

Wace, D. C., Lt. 1919
—— E. W. C., Lt. 111
—— F. B., Lt. 525, 1305
—— M., Maj. . 66, 245, 1074, 1563
—— M. H., Lt. . . . 311a, 1507
—— S. L., Maj. . . . 228, 1662
Wacher, H. S., Ty. Lt. 1773
Waddell, C. M., Lt. 767
—— G., 2nd-Lt. . . . 553, 1523
—— J. H., 2nd-Lt. 356
—— J. W., 2nd-Lt. . . . 494, 961
—— W., 2nd-Lt. 719
Waddingham, C. A., Lt. . . 519, 1636
Waddington, C. W., Maj. . 501, 825
Waddy, J. M. E., Lt.-Col. . . . 705
Wade, E. G., 2nd-Lt. . 539, 1042, 1469
—— H., Lt. 293a, 1249
—— H. R., Capt. 628
—— J., Sub-Condr. 1855
Wade-Evans, D.G., Lt., . . . 525, 1403
—— H., Capt. 89
Wadeson, F. F., Lt. 90, 506
—— F. W. G., Maj.-Genl. 107, 118, 146, 174
—— H. H., Capt. . . . 501a, 731
Wadhawa Singh, Sub-Asst. Surgn. . 1814
—— Subdr. 970
—— Jemdr. 1204
—— Jemdr. 898
—— Jemdr. 932
—— Jemdr. 1117
—— Jemdr. 1228
—— Jemdr. 1305
Wadia, J. R., tempy. Capt.. . . 1767
—— M. D., Capt. . . 105, 1754
Wadilal Mohan-Lal, Sub-Asst. Surgn. . 1830
Wadsworth, S., Lt. 801
Wagget, J. L., Lt. 581
Waggott, R. L., 2nd-Lt. 794
Waghorn, W. D., Br.-Genl. 58, 121, 126, 661, 1881
Waghray, D. C., Ty. Lt. 1780
Wagle, S. G., Ty. Lt. 1780
Wagstaff, C. M., Br.-Genl. 98, 122, 137, 663
—— H., Lt. 1223
—— H. W., Capt. . . 666, 667
—— L. C., Maj. 52, 55, 105, 238, 1020

NAMES AND RANK. Page.

Wagstaffe, H., Lt. 450
Wahid, Ali Khan, Sub-Maj. . . 1034
——— Jemdr. 1409
——— Baksh 39
Wahidyar Khan, Sub-Asst. Surgn. 1810
Waight, G., Lt. 352, 1285
Wailling, W., Lt. . . 510, 808, 1702
Wainwright, J. G., Lt. . . 310, 964
——— J. R., Lt. . . . 1567
——— V. P., Lt. . . 513, 1666
Waite, E. W., Lt. . . 297, 788, 1279
——— F. E., Lt. 835
——— H. S., 2nd-Lt. . . . 566
——— J. J., Lt. . . . 781, 1259
——— W. H., 2nd-Lt. . . 356, 1183
Waiton, C., Lt. 761
Waitt, W. E., Lt. . . . 352, 1303
Wajid Ali, *Hony.* Capt. . . 2102
Wajid Ali, Khan, Resldr. . . . 935
——— Jemdr. 882
Wake, E. St. A., Lt.-Col. . . . 56
——— T. A., 2nd-Lt. . . 692, 784
Wakefield, A. G., 2nd-Lt. . 557, 1703
——— G. E. C., Capt. . . . 841
——— L. V., Lt. 776
——— W. E., 2nd-Lt. . . . 1695
Wakeham, A. R., Lt. . . . 522
——— C. E., 2nd-Lt. . . . 315
Wakeling, A., 2nd-Lt. . . . 618
——— H. E., 2nd-Lt. . . . 619
Wakil Singh, Jemdr. . . . 916
Wakila Dhian Singh, Jemdr. . . 1145
Walait Shah, Jemdr. . . . 1305
Walati Ram Nehraw, Sub.-Asst. Surgn. 1820
Walayat Khan, Jemdr. . . . 1312
——— Shah, Jemdr. . . . 1364
——— Jemdr. 1364
Walcott, E. S., Col. . . 698, 700
Walcroft, T. C., 2nd-Lt. . . 552
Walden, A. E., Lt. . . . 638, 974
——— A. E., 2nd-Lt. . . . 562
——— G. P., 2nd-Lt. . . . 557
——— H. L., 2nd-Lt. . . 561, 1351
——— W. C., 2nd-Lt. . . 568, 1678a
Waldram, H. G., Capt. . . 698, 737
Waldron, C. B., 2nd-Lt. . . . 555
Wale, C. C. A., Sr. Asst. Surgn. . 1789
——— C. W., Asst. Surgn. . . . 1798
——— N. W., Lt. . . . 516, 579
Waley, R. P. S., 2nd-Lt. . . . 752
Walford, H. G., 2nd-Lt. . . 547, 1702
——— J. N., 2nd-Lt. . . . 1689
Wali Dad Khan, Risldr. . . . 951
——— Jemdr. 1256
——— Khan, Subdr. . . . .
Wali Mohammed, Subdr. . . . 1092
——— Khan, Jemdr. . 1472
Wali Muhammad, *Hony.* Capt. 2, 2101
——— Subdr. . . . 1473
——— Jemdr. . . . 1092
——— Jemdr. . . . 1436
——— Tr. Vety. Asst. 1687
——— Ressdr. . . . 910
——— Jemdr. . . . 876
——— Khan, Jemdr. . 886
——— Jemdr. . 1088
——— Jemdr. . 880
Wali Mohomed Khan, Jemdr. . 1249
Walke, W. F., Sub-Condr. . . . 1870
Walker, A., 2nd-Lt. 1289
——— A., 2nd-Lt. 555
——— A. D., Maj. . 102, 663, 1882
——— A. C., Lt. 708
——— A. M., 2nd-Lt. . . . 545
——— A. M. E. H. T., Maj. 108, 220, 1883, 1913
——— A. R., Lt. . . . 305, 1080
——— A. W., Lt. . 504, 835, 1665
——— B. N., Condr. . . . 1854
——— B. W. G., Capt. . . 279, 919
——— C. A. P., Lt. . . 455, 1199
——— C. C., Maj. . . . 239, 927
——— C. W. G., Maj. 6, 14, 53, 99, 240, 1170
——— E. A., Maj. 1749
——— E. B., Asst. Surgn. . . 1798
——— E. C., Maj. . 31, 103, 664
——— E. C., 2nd-Lt. . . 571, 1695
——— E. J. P.T., Capt. 7, 14, 261, 935
——— E. M., Lt. . . . 506, 1675
——— F. H., Lt. . . . 522, 998
——— F. K., Lt. . . . 521, 830
——— F. L., 2nd-Lt. . . . 494
——— F. McF., Lt. . . 637, 980
——— F. R., Lt. . . . 300, 1144
Walker, F. S., Lt.-Col. 34, 195, 1883, 1912
——— F. W., Lt. 748
——— G. B., Capt. . . 253, 1899
——— G. C., the Rev. . . . 1908
——— G. D., Lt. . . . 526, 1898
——— G. K., Lt.-Col. 138, 849, 1841, 1893
——— G. K., 2nd-Lt 652
——— G. W., Genl. 671
——— G. W. H., Lt. 642
——— H., Sub-Condr. . . . 1873
——— H. L., Lt. . . . 528, 1335
——— H. R. O., Capt. 252, 1098, 1120
——— H. T., Maj. . . . 219, 931
——— H. W., Maj. 614
——— J., Maj. 1720
——— J., 2nd-Lt. . . . 356, 1227
——— J., 2nd-Lt. . . . 550, 1676
——— J. B., 2nd-Lt. 791
——— J. C., Capt. . . . 268, 883
——— J. C., Lt. 704
——— J. C., 2nd-Lt. 654
——— J. D., Capt. 708
——— J. E., Lt. . . . 314, 1617
——— J. G. D., Maj.-Genl. . . 170
——— J. H. C., Lt. 1715
——— J. N. Col. . . . 179, 1908
——— J. N., Lt.-Col. . . 140, 1746
——— J. R. H., Tempy. Capt. . 1763
——— L., 2nd-Lt. . . . 545, 1058
——— L. H. C., 2nd-Lt. . 91—92, 550
——— L. W., Lt. . . . 329, 1026
——— M., Lt.-Col. . . . 128, 632
——— M. A., Lt. 844
——— N., Lt. 490
——— M., Lt. 758
——— M., 2nd-Lt. 818
——— N, 2nd-Lt. 582
——— N. W., 2nd-Lt. . . . 725
——— P. J., 2nd-Lt. 717
——— P. L. C., Lt. 706
——— P. L. E., Capt. . . . 587
——— R., Lt. . . . 685, 786, 1007
——— R. E., Lt. . . . . .
——— R. G, Lt. 622
——— R. Y. K., Lt. 617
——— S. E., Lt. 302, 1211
——— S. G., 2nd-Lt. 857
——— T., Capt. . . 18, 43, 44, 754
——— T. H., Lt. 626
——— W., Nursg. Sister . . . 1735
——— W., 2nd-Lt. . . . 365, 1058
——— W. A., 2nd-Lt. 802
——— W. B., Maj. . . . 74, 633
——— W. G., Maj.-Genl. . . 146, 174
——— W. O., Capt. 1756
——— W. R., Lt.-Col. . 131, 156, 192
——— W. R., Lt. 8, 725
Walkinshaw, C. C., Lt. . . 729, 732
——— I. C., Lt. . . 401, 1289
Wall, A. C., Lt.-Col. . 139, 160, 799, 872, 1635
——— C. E., Capt. . . . 32, 693
——— E. W., Col. . . 126, 150, 187
——— F., Bt.Lt.-Col. . 62, 131, 1743
——— H. J. G., 2nd-Lt. . . . 887
——— R. H., Lt. . . 405, 1006, 1513
——— W. H., 2nd-Lt. . . 860, 1245
Wallace, A., Maj.-Genl. 118, 146, 173, 1128
——— A. A., Lt. . 298a, 968, 1677
——— A H., Lt. 1677
——— A. J. L., Capt. 273, 1231, 1335
——— A. D, Hony. Capt. . . . 703
——— A. M., Capt. 485
——— A. R., Lt. 490
——— C. L., 2nd-Lt. . . . 7, 547
——— C. R. D., 2nd-Lt. . . . 1730a
——— C. W., Capt. . . . 253, 1106
——— D., Lt. 540, 1359
——— D. J. C., 2nd-Lt. . . 329, 1243
——— E., Capt. 1841
——— E. C. L., Maj. . 104, 209, 1140
——— E. G., Lt. 302, 1391
——— E. H., Lt. 352
——— G. S., Maj. 37, 1712
——— H. C., 2nd-Lt. 567
——— J., 2nd-Lt. 769
——— J. R. K., Lt. . . . 527, 1387
——— M. D., Capt. 1405
——— R. W. A., 2nd-Lt. . . 327, 1223
——— S., Lt. 729a
——— V., Tempy. Capt. . . . 1724
Waller, A. A. P., Lt.-Col. . 139, 160, 199, 871, 1423, 1443
——— A. G., Capt. . . . 262, 1199
Waller, B. C., Lt. . . . 311a, 943
——— C., 2nd-Lt. 618
——— D. G., Lt. 830
——— E., Lt.-Col. . 186, 1883, 1911
——— E. I. A., Lt.-Col. . . . 1707
——— E de W, Capt. 262, 1235, 1303
——— F. C., 2nd-Lt. 553
——— G. D., Capt. 43
——— H. W. L., Maj. 615
——— J. F., Capt. . . . 256, 1203
——— L. G., 2nd-Lt. . . 357, 1044
——— R. de W., Maj. 70, 235, 1255, 1435
——— R. S. deW., Maj. . 11, 72, 634
——— S. S., 2nd-Lt. 1693
——— W. A. C., Lt. . . . .
Wallers, R. W., 2nd-Lt. . . 476, 1158
Walley, A. F., 2nd-Lt. 833
Walling, A. H. B., Lt. . . 506, 1689
Wallingford, W., *Hony.* Lt. . . 1813
Wallington, G., 2nd-Lt. 780
——— J. M., Lt. . . 297a, 937
Wallis, A. R., Capt. . . . .
——— R. H., Capt. . . . 268, 1431
——— C., Lt. 295a, 964
——— H., Lt. 492
——— H. J. K., Maj. . 219, 913, 1890
——— S., Lt. 729
——— S. A., Lt. . . 668, 996, 1886
——— S. I., Lt. 306, 1110
Wallullah, Sr. Asst Surgn. . . 1808
Walpole, P. W., Lt. . . . 505, 1674
Walpole-Simmons, C. R., Lt. . 735
Walmesley, A. M., 2nd-Lt. . . . 1677
——— H. J., Lt. . . 512, 1170
Walmsley, A. M., 2nd-Lt. . . . 551
——— D. C., Lt. . . 528, 1459
——— H. E., Lt. . . 544, 1648
——— E. S., Capt. . 18, 28, 780
——— J. B., Lt. . . 297a, 1034
——— R., *Hony.* Maj. . . 1844
Walmisley, A. L., Lt. . . 544, 1839
Walsh, A. P., Lt. . . . 321, 365, 1467
——— C. L., Lt. 911
——— D. T., 2nd-Lt. . . . 77, 1885
——— E., 2nd-Lt. . . . 562, 1418a
——— E. T., Sub-Condr. . . . 1864
——— G. G., Capt. . . 42, 500, 1666
——— G. J. T., Lt. . . 59, 487, 899
——— H., 2nd-Lt. . . . 530, 815
——— H., Sub-Condr. 1874
——— H. P., Capt. 748
——— J., Lt. 513
——— J., Hony. Lt. . . . . .
——— J. E., Sub-Condr. . . . 1874
——— J. E., Lt. 544, 1702
——— L. E., Lt. 521, 1614
——— L. W., Lt. . . 368, 534, 1178
——— L. W., Lt. 1102
——— M. C., Lt. 311, 1287
——— P., Condr. 1853
——— P. G., Lt. 430
——— T. E., Sub-Condr. 1874
——— V. P. T., 2nd-Lt. . . . 1034
Walsha, J. W., Lt. . . . 692, 783
Walshe, J. J., Tempy. Capt. . . 1725
Walter, A. E., Lt.-Col. 140, 1745, 1890
——— C. A., Lt. 353
——— E., Maj. . 10, 104, 220, 1662
——— E. F., Lt. 313, 1199
——— G. B., Capt. . . . 486, 947
——— J. A., 2nd-Lt. . . . 669, 997
——— J. Mac N., Maj.-Genl. 16, 19, 97, 119
Walters, E. C., Lt. . . 405, 511, 1132
——— B., *Hony.* Capt. 1844
——— R., Lt. 522, 1007
——— R. W., 2nd-Lt. . . . .
——— T. J. G. S., Lt. . . 507, 830
Waltlake, J. H., 2nd-Lt. . . . 1679
Walton, C., Lt.-Col. . 139, 653, 1882
——— E. W., Col. . . 123, 660, 1207
——— G., Lt.-Col. 135, 158, 198, 869, 1207
——— G., Capt. 665, 1881
——— G., Lt. 1136
——— G. W., Lt. . . . 351, 1636
——— H., Capt. 515, 848
——— H. B. G., Lt.-Col. . 133, 1710
——— H. J., Lt.-Col. . . 134, 1744
——— J. H., Lt. 512, 1630
——— L. C. J. B., Capt. 288, 290, 1613, 1645
——— W. C., Brig.-Gen. 7, 120, 124, 149, 183
Walze, C. E., 2nd-Lt. 553
Waman, Hari Supekar, Sub-Asst. Surgn. 1829

NAMES AND RANK. Page.

Waman, Ramkrishna Rajpathak, Sub-Asst. Surgn. 1830
Wambeek, C. L., 2nd-Lt. . . . 561
Wands, R. D., 2nd-Lt. . . . 859
Wania, H. J., Tempy. Capt. . . 1766
Wapshare, R., Lt.-Genl. 12, 42, 118, 146, 174
Warburton, H. G., Lt.-Col. . 2, 500, 658
—— J. D. MacL I., 2nd-Lt. 563
—— J. F., Lt. . . 410, 1215
—— P. D., Capt. . . . 1715
—— R. P., Lt. . . 82, 535
—— W., Capt. . . . 844
Ward, A., Lt.-Col. . 53, 131, 156, 191
—— B. M., Lt. 67, 581
—— B. S., Maj 16, 40, 104, 240, 1403
—— B. S., Maj. 614
—— C., Lt. 760
—— C. H., Lt.-Col. . 134, 157, 194, 870, 1475
—— C. H., 2nd-Lt. 362
—— E., 2nd-Lt. 1401
—— E., 2nd-Lt. . . . 357, 1323
—— E. L., Lt.-Col. 1746
—— F. D., 2nd-Lt 835
—— F. H. T., 2nd-Lt. . 545, 1265
—— F. K., Lt. 506, 1463
—— F. R., 2nd-Lt. 611
—— F. W., Maj.-Genl. 643
—— G. A., Lt.-Col. 57, 128, 154, 185
—— G. L. S., Maj. . . . 55, 1919
—— G. R., Lt. 322, 1624
—— G. S. La. C., 2nd-Lt. . . . 554
—— G. S. W., 2nd-Lt. . . . 1369
—— H., Condr. 1853
—— H. A., Lt. 785
—— H. C. S., Bt. Lt.-Col. 13, 22, 98, 101, 138, 159, 217, 877
—— H. F., Lt. 610
—— H. G. L., Lt. . . . 367, 1413
—— H. J., Capt. . . 111, 696, 697
—— J., Sub-Condr. 1863
—— J. A., Capt. 62, 1838
—— J. E. 1009
—— J. M., 2nd-Lt. 1373
—— K. E., Lt. 540, 1702
—— L. H., Lt. 522, 1896
—— P. G., Lt. 1499
—— R., *Hony.* Capt. 598
—— R. I., Lt. 581
—— R. L., Lt. . . . . . .
—— T., Lt. 522, 1479
—— T. E., *Hony.* Lt. 1844
—— T. E., 2nd-Lt. . . . 558, 1473
—— T. M., Lt.-Col. . 129, 154, 187
—— V. D., Lt.-Col. 38
—— W., Lt.-Col. . . . 500, 830
—— W. D., Capt. 741
—— W. G., 2nd-Lt. . . 360, 1361
—— W. J., Lt. . . . 510, 827
—— W. N., 2nd-Lt. 757
—— W. T., Capt. 708
—— W. W., *Hony.* Lt. . . . 759
—— Jackson, C., Capt. 79
Wards, G. T., 2nd-Lt. . . 473, 1343
Wardell, A. F. G., Revd. . . . 1904
—— C., Sub-Condr. 1874
—— C. G. V. M., Maj. . 230, 1102
—— F. W., 2nd-Lt. 1040
—— G. F. E., Maj. 40, 60, 209, 1661
—— M., Nursg. Sister . . . 1735
Warden, A. W., Lt.-Col. 57
—— S. B., Tempy. Capt. . . . 1764
—— W. B., Capt. 256, 1003, 1315
Wardle, F., 2nd-Lt. 360
—— F. M., Lt. . . . 305, 1042
—— F. MacL., Lt. . . . 525, 998
—— H. M., 2nd-Lt. 1265
—— J. L., 2nd-Lt. . . 330, 913
—— R. M. L., Lt. 1373
—— R. P., 2nd-Lt. 1441
—— M. K., Capt. 711
—— W. F., 2nd-Lt. . . 559, 1707
Wardley, A. N., Lt. . . . 559, 859
Wardrop, D., Capt. 1719
Ware, F. C. W., Lt.-Col. . . 191, 1912
—— J. R., Lt. 755
—— W. J. K., 2nd-Lt. 752
Warhurst, A. E., Lt. 306, 1601, 1623
Wariam Singh, Rissldar.-Maj. . . 906
—— —— Subdr. . . 1203, 1563
—— —— Subdr. 1311
—— —— Jemdr. 1103
Waring, E. H., Lt.-Col. 140, 162, 200, 870, 1271, 1367
Waring, H. S., Lt. 624
—— J. L., Lt. 307, 1419
Waris Khan, Subdr. 1152
Wariyam Singh, Jemdr. . . . 1364
Wark, R. G., Capt. 66, 626
Warliker, D. P., Lt.-Col. . . . 1786
Warlow, A. M., Lt. 630
Warma, J. D., Ty. Capt. . . . 1771
Warmington, J., 2nd-Lt. . . . 557
—— K. P., Lt. 801
—— W. E., Lt. . . . 423, 1335
—— W. T., Condr. . . . 1853
Warn, J. E., 2nd-Lt. . . . 563, 1524
Warne, F. E., Lt. 806
Warner, A. C., 2nd-Lt. 760
—— A. H., Lt. 628
—— B. M., Capt. 732
—— C. M. E., Asst. Surgn. . . 1795
—— C. T., Capt. . . . 269, 1106
—— E. P., 2nd-Lt. 619
—— F. C., Sub-Condr. . . . 1855
—— G. R., 2nd-Lt. 756
—— H. R., 2nd-Lt. 27, 556, 871 1521
—— J., 2nd-Lt. . . . 618, 979
—— W. G. H., Asst. Surgn. . . 1796
Warnford-Davies, R. D., 2nd.-Lt. . 738, 740, 791
Warr, C. H., 2nd-Lt. . . 360, 1134
Warrack, C. L., Sr. Nursing Sister 1734
Warran, W. S. E., Lt. . . 318, 1647
Warre, F. R., 2nd-Lt. 714
Warren, A. A. G., 2nd-Lt. . . . 658
—— A. G., Lt. 322, 1106
—— D. F. W., Lt. . . . 415, 1110
—— D. S., Lt. 515, 1124
—— E. H., 2nd-Lt. . . . 560, 1678
—— E. T., Lt. 493, 1104
—— G. M., Lt. 756
—— H. F., 2nd-Lt. 730
—— M. C., Capt. 720
—— P., 2nd-Lt. 331, 905
—— S., 2nd-Lt. 562
—— S. H., Capt. 1722
—— V. A., Lt. 306, 1170
—— W. J., Lt. 490, 919
—— W. L., Lt.-Col. . . . 130, 632
—— W. R., Maj. 614
Warren-Boulton, C. G., Capt. . . 852
Warren-Codrington, H. O., Lt.-Col. 198, 1415, 1883, 1912
Warrender, G., 2nd-Lt. 762
Warrener, J. V., Lt. 443
Warriam Singh, Subdr. 1371
—— —— Jemdr. 1372
—— —— Jemdr. 1158
—— —— Subdr. 1142
Warrior, D., 2nd-Lt. 1056
Warton, C. P. F., Maj. . . . 234, 1411
—— R. B., Maj. 614
Warwick, D. V., Ty. Capt. . . . 619
—— E. A., 2nd-Lt. 640
—— G. N., 2nd-Lt. 365
—— T., Condr. 1853
—— T. G., Lt. 351
Warry, C. F. G., 2nd-Lt. . . . 630
—— H. E., Lt. 375, 1439
Waryam Singh, Subdr.-Maj. . . 1102
—— —— Jemdr. . . 892, 894, 957
—— —— Jemdr. 1080
—— —— Jemdr. 1288
—— —— Jemdr. 967
Warryan Singh, Resildr. . . . 935
—— —— Jemdr. 896
—— —— Jemdr. 1232
—— —— Jemdr. 1157
Waryan Singh, Sub-Asst. Surgn 1813
—— —— Sub-Asst. Surgn. . 1816
—— —— Sub-Asst. Surgn. 1814
—— —— *Hony.* Capt. . . 2101
—— —— Jemdr. 1096
Wasal Khan, Jemdr. 1506
Wasan Singh, Subdr. 973
Wasborough, L. S., Lt. . . . 703
Wasudeo Kadem, Jemdr. . . . 1468
—— —— Rana, Jemdr. . 1460
Wasudeo Ghag, Jemdr. . . . 1409
Wasawa Singh, Subdr. . . . 1148
—— —— Jemdr. 1133
—— —— Jemdr. 966
Wassawah Singh, Risldr. . . . 896
Watan Singh, Subdr. 1563
Waterfall, 2nd-C. F. 803
—— J., 2nd-Lt. . . 476, 1032
Waterfield, A. C. M., Lt.-Col 51, 98, 103, 129, 155, 189
—— B. C., Maj. 59
Waterfield, C. H., 2nd-Lt. . . . 729a
—— S., Esq. 33
—— R., Esq. 91
Waterhouse, A. M., Sr. Lady Supdt. 10, 1734
—— J., Maj.-Genl. . . . 173
Wateridge, G. H., 2nd-Lt. . . . 637
Waterlow, J. F., Lt.-Col. . . .
Waters, A., Condr. 1861
—— E. E., Lt.-Col. . . 132, 1743
—— R. S., Maj. . . . 225, 1183
—— S., Capt. 776
—— W. H., Lt. . . . 435, 1405
Waterworth, J. W., ty. 2nd-Lt. 748, 788
—— T. L., Capt. 844
Wathawa Mall, Subdr. 1299
Wathen, F. B., Capt. 844
—— L. W. D., Lt. 589
Watkin, H. G., Lt. . . . 368, 1134
—— W., 2nd-Lt. . . . 742, 788
Watkins, A. N. M., Lt. . . 302, 1211
—— A. W., Lt. 517
—— B. S., 2nd-Lt. . . 332, 1241
—— E. W., 2nd-Lt. 622
—— F. J., Lt. . . . 77, 522, 1885
—— F. W. S., Lt. . . . 313, 943
—— G. D., 2nd-Lt. 769
—— H., Lt. 737
—— H., Lt. 510, 997
—— L. T., Capt. . . 535, 797, 804
—— P. J., Tempy. Lt. 1001
—— V. C., Lt. 401, 919
Watkis, G. L., 2nd-Lt. 111
—— H. B. B., Genl. *Sir* 145, 171, 1144, 1955
—— H. L., Capt. . . . 273, 945
Watling, F. H., Lt.-Col. . . 132, 1743
—— G., Lt.-Col. 107, 131, 155, 190, 1062
—— L. D., Maj. . 37, 38, 215, 1661
—— M. C. G., Lt. 510
Watmore, A., Sub-Condr. . . . 1857
Watney, C. N., Lt.-Col. . . . 751
—— C. W., Maj. . . . 228, 1662
—— R. D., Capt. . . . 686, 751
—— R. W., Lt. 1677
Watson, A., Maj. 608
—— A., Capt. 729a
—— A., 2nd-Lt. 713
—— A. C., Lt.-Col. . 129, 155, 188
—— A. C., Capt. 587
—— A. C., Lt. . . . 405, 1383
—— A. E., Lt. . . . 351, 1565
—— A. G., Capt. 754
—— A. G., Lt. 532
—— A. G., 2nd-Lt. 1383
—— A. O., 2nd-Lt. . . . 549, 1699
—— A. W., Lt. 538
—— B. S., Lt. 758
—— C., 2nd-Lt. 1399
—— C. H., Lt.-Col. . . 137, 1744
—— C. S. M. C., Maj. . 663, 1881
—— D., Tempy. Capt. . . . 1723
—— D. G., Lt. 542
—— E. C., Capt. 583
—— E. C., Lt. 543a, 987
—— E. E., Maj. 274, 1267
—— E. L., Lt. 797
—— E. L., Capt. 700
—— E. L., 2nd-Lt. . . . 494, 887
—— E. McI., 2nd-Lt. 790
—— E. V., Maj. 615
—— E. T., 2nd-Lt. 726
—— F., 2nd-Lt. 770
—— F. R. J. MacD., 2nd-Lt. . . 768
—— G. A., 2nd-Lt. 562
—— G. D., Capt. . . . 666, 985
—— G. H., Lt.-Col. 56
—— G. L., Lt. 522, 1203
—— G. N., Capt. . . . 269, 899
—— G. T. C., Capt. . . 367, 909
—— G. W., Capt. 1717
—— H., Capt. 501a
—— H. A., 2nd-Lt. 550
—— H. B., Capt. . . 68, 485, 879
—— H. D., Brig.-Genl. 1, 50, 119, 124, 150, 184
—— H. D., Maj. . . . 87, 105, 238
—— H. E., 2nd-Lt. 808
—— H. R., Maj. 1216, 870, 1317, 1331
—— H. R. W., Capt. . . 486, 909
—— H. S., 2nd-Lt. 1491
—— H. T., 2nd-Lt. 767
—— J., Genl. *Sir* 899
—— J., Capt. 9, 485, 1064
—— J., Lt. 848

NAMES AND RANK. Page.

Watson, J., Lt. 535, 1231
—— J., 2nd-Lt. 786
—— J. A. M., Capt. . . . 856
—— J. F., Capt. 811
—— J. G., Lt. . . . 537, 1333
—— J. H., Maj. . . 212, 899
—— J. H., Sub.-Condr.. . . 1874
—— J. H., Lt. 787
—— J. H. L., Lt. 423
—— J. J., 2nd-Lt. . . 560, 1678
—— J. L., Capt. 814
—— J. V., 2nd-Lt. . . . 562
—— J. W., Lt.-Col. 139, 1745, 1890
—— L. A., Lt.-Col.. 98, 103, 105, 131, 155, 191
—— L. D.. Lt. 583
—— N., 2nd-Lt. . . . 330, 891
—— O., Lt. 81, 298
—— P., Lt. 19, 67
—— P. C., Lt. 463
—— P. H., Hony. Lt. . . . 1845
—— R., 2nd-Lt. 834
—— R. A. D., Capt. . . . 666, 985
—— R. M., Lt. 29, 543
—— S. B., Bt.-Lt.-Col. 135, 158, 208, 872, 1275, 1587
—— S. G., 2nd-Lt. 742
—— S. J., Sub-Condr. . . . 1870
—— T. G., Lt. 589
—— T. N., Bt.-Maj. . 262, 897
—— T. T. B., Ty. Capt. . . 1725
—— W. A., Maj.-Genl. 97, 146, 174, 949
—— W. B., 2nd-Lt. 1433
—— W. G., Sub.-Condr. . . . 1870
—— W. L., Capt. . . 1411, 1754
—— W. M., Br.-Genl. 105, 120, 125
Watson-Armstrong, the Hon'ble, W. J. M. Capt. 1056
Watson-Smyth, R. M., 2nd-Lt. . 807
Watt, E. G., 2nd-Lt. 630
—— D. M., Br.-Genl. . 98, 122, 135, 158, 198, 1600
—— G. T., 2nd-Lt. 669
—— J. M., 2nd-Lt. 1393
—— J. O., 2nd-Lt. 752
—— W. B. F., Lt. 785
—— W. D. B., Capt. . . 656, 975
—— W. R., Lt. . . . 311, 1621
—— W. P., Lt. . . . . . .
—— W. W., Lt. 828
Watters, G. H., Lt. 450
Watterson, J. W., Lt. 704
—— J. S., Lt. . . 703, 704, 796
Wattom, I. C., 2nd-Lt. 567
Watts, A. D., Asst. Surgn. . . . 1803
—— A. E., 2nd-Lt. 363
—— A. H., Maj. . . . 216, 1661
—— A. T., Lt. 637
—— B., Lt.-Col. 1710
—— B., 2nd-Lt. . . . 355, 1090
—— E. P., Capt. . . . 267, 1235
—— F. H., 2nd-Lt. . . . 356, 988
—— F. C., 2nd-Lt. 1441
—— G. A. R., Lt.-Col. . . 137, 159, 200, 921
—— G. S., 2nd-Lt. 363
—— H., Maj. 1751
—— H. C., Lt. . . 380, 1435, 1583
—— H. E., 2nd-Lt. . . . 354, 1195
—— H. J., 2nd-Lt.. 360
—— H. P., Maj. . 37, 214, 1661
—— J. W., Lt. 732, 785
—— L. G., 2nd-Lt. . . . 332, 1595
—— M. E., Esq. 91
—— N. H. L., Capt. 67, 107, 108, 253, 1375
—— P. J., 2nd-Lt.. 567
—— R. B., Lt. 401
—— R. P., Capt. . . 367, 1403
—— W. H., 2nd-Lt. 56
—— W. W., 2nd-Lt. . . 552, 1018
Wauchope, A. G., Br.-Genl. 50, 121, 126, 743
—— R. S., Capt. 29, 262, 1880
Waugh, A. D., 2nd-Lt. . . 669, 997
—— C. R., Sub-Condr. . . . 1863
—— J. N., 2nd-Lt. 558
—— R. T., Condr. 1872
Waumsley, T. H., Lt. . . 541, 1439
Way, J. D., Lt.. 525
—— T. A. H., Maj. 816
Waycott, A., Lt. 629
—— E. W. J., Maj. 614
Waylen, G. H. H., Capt. . . . 755
Wayman, W. L., Capt. 252, 1383, 1415

Waymouth, O. S., Lt. . . . 521, 577
Wayte, V. W., Lt. . 46, 463, 758, 1351
Wazid Khan, Jemdr. 1404
Wazir, Jemdr. 966
—— Beg, Risaldar 1684
—— Khan, Ressdr. 908
—— Subdr. 1088
—— Subdr. 1407
—— Subdr. 1268
—— Jemdr. 1436
—— Muhammad, Jemdr. . . . 908
—— Singh, Subdr. 1104
—— Singh, Subdr. 1074
—— Subdr. 1136
—— Ressdr. . . . 944, 961
—— Subdr. 1698
—— Jemdr. 1248
—— Jemdr. 1051
—— Subdr. 1134
—— Jemdr. 1245
—— Jemdr. 1107
—— *Hony.* Capt. . . . 2102
—— Khattri, Subdr. . . . 1645
—— Jemdr. . . . 1643
—— Mahatra, Subdr. . . . 1642
Wazira, Subdr. 1195
—— Subdr. . . . .
—— Jemdr. 966
—— Jemdr. 1200
Weallens, W. R. W., Lt. . . 291, 1613
Wearing, W., 2nd-Lt. . . 354, 1199
Wearn, A. W., 2nd-Lt. 494
Weatherall, N. E., Capt. . . 55, 587
Weatherdon, H. E., Lt. . 322, 1653
Weatherhead, E. S., Lt. . . . 721
—— L. D., 2nd-Lt. . . 1339
—— W. A. Lt. . . . 1720
Weatherly, E. F. C. F., Lt. 300, 1170, 1571
—— J. E., 2nd-Lt. . . . 734
Weathrall, C. H. G., Capt. . 571*a*, 1841
—— I. A., Mc. D., Lt. 509, 1050
Weaver, B. H., 2nd-Lt. . . 735*a*, 1678
—— J. Y., Lt.. . . . 385, 899
—— M. G., Lt. . . . 445, 915
Weber, H. G., Lt. 654
Webb, A., Sub-Condr. 1856
—— A. B., Lt. 507, 1665
—— A. C., 2nd-Lt. 561
—— A. E., Lt.-Col. 20, 131, 156, 191, 1660
—— A. O.'B., 2nd-Lt. . . . 610
—— A. W. T., Capt. . . 269, 1893
—— B. H., 2nd-Lt. 732
—— C. A., Lt. 527
—— C. C., Lt. 789
—— C. E., Sub-Condr.. . . . 1870
—— C. E., Lt. 1886
—— C. I., N. Sister 1735
—— E. A., 2nd-Lt. 361
—— F. B., Lt. . . . 68, 303, 1639
—— G., Hony. Capt. 1847
—— G. A., Maj. 610
—— G. R., Maj. 818
—— G. S. R., Lt. 41, 68, 420, 945
—— H. G., 2nd-Lt. . . 77, 707, 1885
—— H. G. S., Maj. 1749
—— H. J., Capt. 731
—— H. P., 2nd-Lt. 361
—— J. E. N., Lt. . . 504, 516, 1166
—— J. E. T., 2nd-Lt. 690*a*
—— J. R. D., Capt.. 1757
—— N. D., 2nd-Lt. 794
—— N. G., 2nd-Lt. 610
—— R., 2nd-Lt. 707
—— R. A., Lt. 375, 1399
—— R. M., 2nd-Lt. 1677
—— S. D., Lt. 446, 1495
—— S. R., Lt. 1189
—— T., 2nd-Lt. 357, 1331
—— T. H., Sub-Condr. . . . 1863
—— W. 2nd-Lt. 567
—— W. F., Lt. 393, 1419
—— W. F. R., Maj. . . 233, 1106
—— W. H., Lt. 537, 630
—— Johnson, S., Capt. . . 67, 721
—— C., Capt. . 754, 1720
Webbe, D. B., 2nd-Lt. 550
Webber, A., Capt. 767
—— C. J., 2nd-Lt. . . . 357, 1144
—— C. D., Lt. . . . 415, 1098
—— M. R. H., Maj. 214, 879, 1893
—— W. G., 2nd-Lt.. 987
Webley, E. A., 2nd-Lt. 599
Webster, A., 2nd-Lt. 21, 549
—— C. D., Maj. 107, 222, 1140, 1573
—— C. G., Capt. 1219

Webster, C. G. G., 2nd-Lt. 566, 1463
—— C. H., 2nd-Lt. . . 553, 815
—— F. G., Lt.. . . 487, 909, 957
—— G. W., Lt. . . . 516*a*, 1068
—— H., Lt. 692
—— H. P., Lt. . . . 460, 1150
—— J., 2nd-Lt. 1271
—— J. F., 2nd-Lt. 692
—— P. S., Lt. 531
—— R., 2nd-Lt. 724
—— R. O., 2nd-Lt. 757
—— T. G. C., 2nd-Lt. . 546, 1295
—— W. H. A., . Lt. 541, 1293
—— S., 2nd-Lt. 547
Weedon, C. W., 2nd-Lt. 751
—— F. F., Col. . 43, 105, 126, 160
—— H. H., 2nd-Lt. 1581
Weekes, H. E., Maj. . . 28, 228, 1647
Weeks, A. H., Sub-Condr. . . . 1873
—— C. A., 2nd-Lt. . 325, 367, 1613
—— H. A., 2nd-Lt. 474
—— R. F. J., Maj. 858
—— V. H., Lt. . . . 542, 1511
Weidenbach, L. P., Lt. . . . 1839
Weigall, E. T., Lt. 617
—— G. J. V., Capt. 709
—— G. S. C., 2nd-Lt. . . . 324
—— L. A. F., Maj. 735
Weil, R. G., 2nd-Lt. . . . . .
Weinman, A. N., 2nd-Lt. . 567, 1056
—— L. O., Ty. Capt. . . . 1729
Weir, A. D., Lt. 542, 1624
Weir, D. L., Capt. 55
—— G., Lt. 455, 1347
—— G. T., Lt. 1639
—— H. M., 2nd-Lt. 640
—— I. E. B., Capt. . . 278, 1501
—— J. L. R., Maj. 103, 229, 723, 883, 1890
—— R. S., Lt. 814
Weitzal, R. H., Lt. . . . 493, 913
Welby, B. C., Lt. 535, 1703
Welch, A. J., Maj. 718
—— F. E., Capt. . 257, 1423, 1443
—— G. A. J., Sub-Condr.. . . 1835
—— J. F., 2nd-Lt. 473
Weinman, T., Ty. Lt. 1729
Welch, H. C., Maj. 1838
—— J. F., 2nd-Lt. . . . 476, 1191
Welchman, E. W., Bt.-Col. . . . 56
—— G. de V., Lt. . . . 55
—— G. H., Lt.. . . 540*a*, 1702
Weld, C. J., Capt. . . . 277, 1247
Weldhen, R. B., Lt. . . . 420, 1495
Weldon, E. G., Capt. 589
—— F., Col. 176
—— S. W., Lt. 685
—— W. E., 2nd-Lt. 815
Wellborne, C. deM., Maj. . 243, 1896
Weller, H. O., Lt. 836
—— the Revd. J. R. . . . 1905
Wellesley, H. A., Capt. . . 274, 1617
Wellington, I., Sub-Asst. Surgn. . 1818
Wellings, D., 2nd-Lt. . . . 356, 1166
Wellman, F. J., 2nd-Lt. . . . 1381
Wells, A., 2nd-Lt. . . . 563, 750, 1517
—— A. M., Lt. 313, 1505
—— C. A.. 2nd-Lt. . . . 358, 1575
—— C. A., Asst. Surgn. . . . 1796
—— C. A., 2nd-Lt. . . . 358, 1453
—— C. M., 2nd-Lt. 735*a*
—— C. Mc, 2nd-Lt. 90
—— E. H., 2nd-Lt. 720
—— E. T. W., 2nd-Lt. . . 721, 742
—— F. W. A., Maj. . . . 234, 1038
—— G. E. H., Lt. 779
—— H. A. T., Asst. Surgn. . . 1801
—— H. J., Sub-Condr. . . . 1873
—— H. J. G., Capt. 1714
—— H. V., 2nd-Lt. . . . 328, 1676
—— J. R. P., Capt. . . . 827, 1162
—— N. S., Maj. 1747
—— R. H., Lt. 622, 972
—— R. L. P., Lt. 489, 923
—— R. T., Maj. 1751
—— R. W., Lt. . . 523, 1670, 1702
—— T. R., Capt. . . . 261, 1152
—— W., 2nd-Lt. . . . . . .
—— W. G., Esq. 88
—— W. G. H., Lt. . . . 425, 1624
—— W. T. J., *Hony.* Capt. . . 582
Wells-Smith, F. T., 2nd-Lt. . 562, 1339
Welman, W. H. L. J., Capt., 263, 1606, 1633
Welsford, E. P., 2nd-Lt. 328, 808, 875
Welsh, A. C. R., Lt. . . . 295*a*, 1473

NAMES AND RANK. Page.

...sh, G. J., 2nd-Lt. 752
...R. H. W., Capt. 277, 893, 905
—— T. V., 2nd-Lt. 327, 897
—— W. M. M. O'D., Maj. 615
...sman, C. H., 2nd-Lt. 1090
...stead, G. L. M., Lt. 591
...on, F. E., Capt. 1732
—— J. H., 2nd-Lt. 1018
—— J. J. G., Lt. 423, 1591
—— W., Sub-Condr. 1885
...myss, A. B., Lt. 77, 502, 740, 791, 1885
—— A. F., Lt. 298a, 1267
...mouth, H. H., Lt. 547, 1385
...nyon, Revd. F. A. 1908
...tworth-Hyde, R. R., Lt. 529, 998
...rden, L. H., Ty. Lt. 1726
...rnicke, A. E., Lt. 307, 1136
—— F. P., Capt. 1752
...skin, H. K., Asst.-Surgn. 1803
...sson, R. A., Lt. 410, 1401
...st, A. H. D., Lt.-Col. 139, 614
—— A. P. J., Lt. 511, 1666
—— C., Capt. 617
—— C., 2nd-Lt. 357
—— C. H., 2nd-Lt. 793
—— D., Nurs. Sister. 1735
—— E., 2nd-Lt. 465, 1211, 1657
—— E. M., Capt. 288, 1606
—— G. F., Lt. 543, 1673
—— J., Sub-Condr. 1862
—— M. P., 2nd-Lt. 556
—— L. C., Lt. 846
—— P. S. 2nd-Lt. 332, 1152
—— W. G., 2nd-Lt. 355, 1040
...stcott, G. H., Revd. 1903
...stern, C. E. M., Capt. 257, 945
—— C. M. T., Maj. 43, 242, 1663
—— F. H., Lt. 686, 784
—— J. S. E., Col. 51
...stgarth, S. H., 2nd-Lt. 1102
...stlake, A. P., Bt.-Col. 122, 153, 181
—— A. R. C., Lt. 722
—— T., 2nd-Lt. 358
...stland, D., 2nd-Lt. 539, 1439
...stmacott, P. G., Capt. 501a, 1144
—— R., Lt.-Col. 603
—— R., Capt. 485
—— R., Maj.-Genl. *Sir* 173, 1509
—— T. H., Maj. 59, 603
—— T. R., Capt. 947, 1467
...stoby, C. B., Lt. 460, 1166
...ston, A., 2nd-Lt. 355, 1499
—— A. F., Lt.-Col. 1711
—— E. B., Lt. 446, 1469
—— F. H., Lt. 525, 1150
—— G. E., Col. 170, 1916
—— G. O., Sr. Asst. Surgn. 1788
—— J. R., Lt. 321, 1471
—— M., Lt. 540
—— N. A., Lt. 513, 1676
—— J. S. N., 2nd-Lt. 549, 1701
—— W., Sub-Condr. 1873
—— 2nd-Lt. 1676
...ston Foster, A. W., 2nd-Lt. 357, 1647
...strap, A. E., Asst. Surgn. 1797
...strope, F. E., Lt. 617
...stropp, F. M., Lt.-Col. 136, 661, 1881
...strupp, P., 2nd-Lt. 474, 1633
...twood, J. D., Capt. 834
...herall, C. R., 2nd-Lt. 696
—— F. G. B., Capt. 265, 1495, 1505
—— G. F. J., Lt. 420
—— G. A. C., Maj. 6, 244, 875
—— K. L., 2nd-Lt. 729
—— W. A., Col. 179, 1918
...on, J. L., Sub-Condr. 1874
...mouth, J. C., 2nd-Lt. 553, 1523
...dcoat, C. C., Lt. 593
...ll, G. H. B., Lt. 1018
...ley, C. A., 2nd-Lt. 564, 1263
...lley-Kelly, J. H., Capt. 798
...rton, A. F., Capt. 730
—— F., Lt. 522, 1701
—— J., Lt. 832
—— R. F. H., 2nd-Lt. 564
...atley, A. H. B., Capt. 52, 260, 885
—— A. R., Capt. 720
—— C. K., 2nd-Lt. [illegible]
—— G. R. P., Maj. 209, 925?
—— H. S., Col. [illegible], 1917
—— T. P., Capt. 32, 714

Wheatley, W. A., Lt. 637
—— W. P. R., Maj. 237, 1663
Wheeler, A. G., Lt. 533
—— C., Lt. 723, 1006, 1585
—— C. D., Lt. 526, 1559
—— E. O., Capt. 665, 985
—— F. S., Maj. 22, 81, 241
—— F. O., Lt. 629
—— G. A. S. d'E., Lt. 316
—— G. C., Maj. 222, 1641
—— G. E., Lt. 401, 1624
—— H. T., Revd. 1903
—— P. J., Lt. 72, 622, 1884
—— T. W., 2nd-Lt. 780
—— W., 2nd-Lt. 358, 1413
Wheelwright, T. H., Capt. 9, 15
Whelan, A. A., Esq. 4
—— H. C., Lt. 316
—— J., 2nd-Lt. 562
—— J., Capt. 637
—— J. P., 2nd-Lt. 769
Wheler, G. B. B., Capt. 597
—— *Sir* T. W., Capt. 728
Whelton, J. H., 2nd-Lt. 413
Whetham, L. C., 2nd-Lt. 555, 1676
Wheway, T., Ty.-Lt. 1729
Whicher, J. D' A., Capt. 750
Whidborne, C. S. L., Lt. 593
Whiffen, G. G., Maj. 8, 720, 785
—— G. S. L., 2nd-Lt. 1383
—— N. H., Lt. 720
Whiffin, H. E., Lt.-Col. 128, 154, 183
—— L. D., 2nd-Lt. 545, 1627
Whig, P. L. Ty.-Lt. 1782
Whigham, G. C., Maj. 817
—— R. D., Maj. 718
Whigham-Teasdale, D. G., Lt. 14, 24, 509, 1183
Whillier, E. C., Lt. 795
Whillier, E. C., 2nd-Lt. 703
Whinyates, S., Lt. 492, 933
Whistler, A. E., Bt.-Col. 122, 153, 181, 1917
—— A. B., Capt. 279, 893
Whistondale, F., 2nd-Lt. 564
Whitaker, B., Capt. 486, 1675
—— C. F., Lt. 507, 1020
—— C. L. D. H., Maj. 227, 1090
—— F. G., Lt. 734
—— F. N., 2nd-Lt. 558
—— H. C., Bt.-Maj 2775?
—— S. S., Capt. 255, 1729
—— S. W., Lt. 435, 905
—— W. A., 2nd-Lt. 360, 1160
Whitamore, A. G., Lt. 56, 516a, 1666
—— E. O., Lt. 380
—— V. N., Maj. 1752
Whitbread, F. G., Asst. Surgn. 1798
Whitburn, W. W., Lt. 538
Whitby, E. V., Capt. 1714
—— H. F., Maj. 221, 911
—— R. B., Lt. 527, 1618
Whitchurch, R. H. S., Maj. 87, 223
Whitcombe, C. E., Lt. 541, 992
—— L. C., Lt. 304, 1311
White, A. A. W., Asst. Surgn. 1793
—— A. D., Maj. 1752
—— A. E., Condr. 1854
—— A. G., Capt. 254, 1501, 1517
—— A. L. P., Sub-Condr. 1868
—— B. A., Capt. 807
—— B. G., Capt. 278, 1629, 1633
—— B. K., Lt. 380
—— C. E. P., 2nd-Lt. 791
—— C. F., Lt. 838
—— C. F. O., Tempy. Capt. 1723
—— C. T. F., Lt. 789
—— C. R., Lt. 512, 843, 1001
—— C. M., Lt. 787
—— C. T., 2nd-Lt. [illegible]
—— C. W. G., 2nd-Lt. 324
—— D. E. E., Asst. Surgn. 1795
—— D. G., Lt. 1886
—— D. O., Lt. [illegible], 1026, 1034
—— E. D., Asst. Surgn. 1793
—— E. W., 2nd-Lt. 330, 917
—— F., Lt. 293a, 1120
—— F. G., 2nd-Lt. 1701
—— F. A. I. M., 2nd-Lt. 549
—— F. J., 2nd-Lt. 472, 1648
—— F. N., Maj. 1750, 1890
—— F. R., 2nd-Lt. 742
—— G., 2nd-Lt. [illegible]
—— G., 2nd-Lt. 618
—— G., 2nd-Lt. 354
—— G. E., *Hony.* Maj. 1845

White, G. F., Lt. 450
—— G. F., Brig.-Genl. 57, 120, 124, 612
—— G. F., 2nd-Lt. 608, 899
—— G. H., 2nd-Lt. 554
—— G. L., 2nd-Lt. 721
—— G. W. W., Lt. 292a, 883
—— H. E. N., 2nd-Lt. 739
—— H. G. N., Lt.-Col. 500, 836
—— H. M., 2nd-Lt. 794
—— H. P. G., 2nd-Lt. 1695
—— H. P. M., Maj. 213, 1661
—— H. R., 2nd-Lt. 559, 1283
—— H. S., Lt. 460, 988
—— J., 2nd-Lt. 760, 1008
—— J. H. B., Lt. 1018
—— J. H., Sub-Condr. 1856
—— J. H. G., Capt. 1759
—— J. H. M., Capt. 1841
—— J. M., Capt. 729
—— K. W. E., Asst. Surgn. 1798
—— L., Lt. 629
—— L. A., Nursing Sister 1734
—— L. G., 2nd-Lt. 728
—— L. N. W., Lt. 455, 1455
—— L. S., 2nd-Lt. 555, 1645
—— M., Lt. 530
—— M. F., Maj. 1751
—— M. J., Lt. 540, 1674
—— N., 2nd-Lt. 361
—— O. J., Lt. 296, 1621, 1657
—— O. W., Capt. 1919
—— P. R., Lt. 352, 1150
—— R. E., Lt. 617
—— R. S., Lt. 90, 704
—— R. S. M., Lt. 375, 1054
—— R. W., Capt. 102
—— R. Y., 2nd-Lt. 626
—— S. E. C., Lt. 423, 1189
—— S. H., Maj. 17, 28, 754
—— T. H., Lt. 992
—— T. N., 2nd-Lt. 473, 1162
—— W., Lt. 75
—— W. B., Maj. 59, 234, 949
—— W. E., Lt.-Col. 129, 154, 186
—— W. F., 2nd-Lt. 750
—— Rev. W. E. 602
—— W. H., Lt. 1720
—— W. H., 2nd-Lt. 358, 1561
—— W. J., Lt. 536, 1365
—— W. T., Snr. Asst. Surgn. 1789
—— W. T. H., Lt. 529
—— W. W., Bt.-Col. 62, 124, 882, 1740
Whitehead, E. C., Lt. 318, 1399, 1409
—— F. W., 2nd-Lt. 1611
—— G., Capt. 7
—— H., Rt. Revd. 1906
—— H. E., 2nd-Lt. 750
—— J., Bt.-Lt.-Col. 53, 98, 137, 159, 220, 1016
—— J., Capt. 724
—— J., Esq. 89
—— R. B., Lt. 1007
—— T. A., Lt. 493, 951
—— W., Sub-Condr. 1855
—— W. A., Lt. 302, 1319, 158[illegible]
—— W. A. B., 2nd-Lt. [illegible]
Whitehill, R. D., Lt. 368, 1479
Whitehorn, A. D., Lt. 852
Whitehouse, A. L. W., Maj. 721
—— R. H., Capt. 805
Whitelaw, D., Lt. 763
Whiteley, F. N. [illegible] 531, 577
Whitelock, T. [illegible] 2nd-Lt. 1350
Whiteman, W. M., 2nd-Lt. 638
—— W. H., 2nd-Lt. 707
Whiteside, A., 2nd-Lt. 1032
—— H. H., 2nd-Lt. 1130
—— J., 2nd-Lt. 638
—— M. B. D., Lt. [illegible]
Whitestone, B. F., Lt. 1004
Whitfield, H. E., Lt. 683
Whitford-Hawkey, E. H., Capt. 272, 905, 953
Whiting, W. C., Lt. 465, 1080
—— H. N., 2nd-Lt. 705
Whitley, A. H., Asst. Surgn. 1792
—— A. P., 2nd-Lt. 330, 1158
—— C., Sub-Condr. 1863
—— H. C., Lt. 368, 1565
—— L. G. M., Lt. 1674
Whitlock, D. K., Lt. 730
—— P. O., Lt. 540
Whitman, H., Sub-Condr. 1870
Whitmore, A., Maj. 1750
Whitney, W. W., Lt. [illegible]

NAMES AND RANK. Page.

Whittaker, A. P., 2nd-Lt. . 568, 1271
—— E. H., 2nd-Lt. 325, 883
—— J., Condr. 1868
—— G., 2nd-Lt. . . 776a
—— T., *Hony.* Maj. . 1845
Whittall, G. E., Bt.-Maj. 61, 745
—— J. M., Capt. . 52, 272, 1623
Whittell, H. M., Capt. . . 255, 1663
Whitten, A. M., Lt. . . . 571, 1695
Whittenbury, J. H., Sr. Asst. Surgn. 1791
Whittick, E. G., Sub-Condr. . . 1874
—— H. D., Lt. 316
Whitting, E. LeG., Maj. . . . 634
Whittingdale, W. L., 2nd-Lt. . . 740
Whittington, E. F., 2nd-Lt. . . 730
—— H., Lt. . . 354, 1142
—— G. W., Lt. . . 405, 1395
—— L., Capt. 18, 23, 681, 785
—— P. R., Capt. . . . 681
Whittington-Ince, C. H., Lt. . . 732
Whittle, H. E., 2nd-Lt. . . . 474
Whittome, A. G., 2nd-Lt. . . . 363
—— J. E., Lt. . . 508, 1158
Whitty, N. I., Capt. 749
Whitwain, R., 2nd-Lt. 1166
Whitwell, H. E., 2nd-Lt. 692, 784, 1521
—— H. W., Maj. . . 239, 1070
Whitworth, D. E., Capt. . 272, 877
—— H. J., 2nd-Lt. . 564, 1235
—— T. S., Maj. 17, 21, 238, 1303
Whyte, D., Capt. 1724
—— D. E., Lt. 656
—— H. A., 2nd-Lt. 805
—— L. J. deB., 2nd-Lt. . . . 1679
—— W., 2nd-Lt. 828
Wibby, E. J., 2nd-Lt. . . 358, 1309
Wicher, J. D'A., Capt. 1900
Wick, F. G., 2nd-Lt. 351
Wickham, E. T. R., Capt., 105, 272, 949, 1890
—— H. F., Lt.-Col. . 137, 581
—— J., Lt. . . 292, 1102, 1537
—— J. C., Capt. . . 664, 991
—— J. J. O., Lt. . 450—54, 1513
Wickham, W. C., 2nd-Lt. 91—92, 539
Wickham-Jones, E., Lt. . . . 622
Wickins, W. J., Revd. 1
Wicks, E. L., 2nd-Lt. . . 357, 1211
—— F. G., Lt. 351
—— W. M., Capt. . . . 89, 754
—— G., 2nd-Lt. 723a
Wide, L. A. C., Lt. . . . 533, 1355
Widdicombe, E. P., Lt. . . 505, 986
—— F. S., Lt.-Col. 129, 154, 187
—— G. T., Lt.-Col. . . 57, 129, 154, 188
—— J. L. T., Lt. . . 312, 1641
Widdowfield, C. S., 2nd-Lt. . . 980
Widdows, F. M., Lt. . . . 393, 1108
Wienholt, E. A., Capt. . . . 581, 599
Wiffin, G. S. W., 2nd-Lt. . . . 356
Wigan, C. R., Lt. 681
—— S. R. B., Capt. 1583
—— T. K., Capt. 698
Wigden-Janks, J. C., 2nd-Lt. . . 1417
Wigfall, C. S., Lt. . . 519, 779, 1671
—— R. H., Capt. . 500, 843, 1665
Wigg, E. C., 2nd-Lt. 653
Wiggins, D. G., Lt. . . . 540, 1701
—— R., Capt. 1718
Wightwick, H. M., Capt. . 259, 1893
Wigley, A. B., Maj. . . 16, 717, 726
—— A. J., Lt. 1001
—— F. G. M., Capt. 7, 14, 263, 1359
—— P. J. R., Capt. 167, 108, 265, 1367
Wigner, G. D., 2nd-Lt. . . 361, 1677
Wigram, C., Bt. Lt.-Col. 1, 96, 98, 133, 157, 207, 909
—— R., Brig.-Gen. 51, 97, 121, 126, 211, 1600
Wikeley, J. M., Lt.-Col. 78, 130, 155, 189
Wilayat Ali, Tr. Vety. Asst. . . 1687
—— Jemdr. 1284
—— Shah, Jemdr. . . . .
—— Resdr. 1684
—— Sub-Asst. Surgn. . . . 1815
Wilberforce-Bell, H., Capt. 5, 108, 260, 1596, 1893
—— P. F., Capt. 749
Wilbraham, A. E., *Hony.* Lt. . . 588
Wilby, V. G., 2nd-Lt. 555
Wilcock, F. W., Lt. . . . 536, 1032
—— A. G. E., Lt. 1720
—— C. A., 2nd-Lt. 1679
—— H. A., 2nd-Lt. 1317
Wilcox, F. J., Lt. . . . 503, 1245
—— H. B. F., Maj. . . 239, 1371

Wild, C. E., Lt. 608
Wildeblood, J. P., Capt. . 27, 501
Wilder, B. L., 2nd-Lt. . . 326, 1231
—— J. M., Lt. 692
—— P. E., 2nd-Lt. . . 326, 366, 1600
—— C. S. D., Maj. . . 20, 211, 1661
Wildy, H. A., Capt. . . . 779, 797
Wiles, B. H., Lt. 488, 893
—— F., Sub-Condr. 1857
—— G., Lt. 849
—— G., Lt. 821
—— H. H., Lt. . . . 71, 509, 1884
—— J. J., Lt. 756, 757
Wiley, W. J. G., Lt. 1589
Wilford, E. E., Maj. . . . 212, 931
Wilken, W., Lt. 534, 1299
Wilkes, B. St. C., 2nd-Lt. . . .
—— G. S., Maj. 74, 101, 633, 1884
Wilkie, H. D., 2nd-Lt. 652
Wilkins, C. E. W., Sub-Condr. . . 1808
—— E. G. B., Lt.-Col. 105, 128, 154, 184
—— F. B., Lt. 571
—— G. A. W., 2nd-Lt. . . 560, 1245
—— H. W. J., 2nd-Lt. . . 357, 1259
—— J. F., Lt. 531
—— J. P. K., 2nd-Lt. . . 540, 1068
—— L. J., 2nd-Lt. 687
—— R., 2nd-Lt. 329, 1243
—— T., Lt. 617
—— W. H., Col. 177
—— W. S., 2nd-Lt. . . . 560, 1678
Wilkinson, A. C. L., 2nd-Lt. . . 571a
—— A. E., 2nd-Lt. . . . 699, 766
—— A. S., 2nd-Lt. . . . 331, 895
—— C. K., Maj. 213, 872, 1231, 1571
—— C. R., Capt. 848
—— C. W., Lt.-Col. 135, 661, 1881
—— D. A., Lt. 507
—— D. S., Capt. . . . 685, 763
—— E. J., Lt. . . . 47, 531, 1668
—— E. M., Lt. 1383
—— E. R., Revd. 1904
—— F. B., Lt. 532
—— F. R., Lt. 415, 1110
—— F. S., 2nd-Lt. . . . 668, 992
—— G. H., Maj. 221, 1662
—— G. N., Lt. 293a, 1178
—— H., Lt. 748
—— H. A., Lt. 302, 1247
—— H. A., Lt. 504, 1285
—— H. E., Sub-Condr. . . . 1868
—— J. B., Lt. 516, 1497
—— J. F., Lt. 526
—— J. H., Lt. 317
—— J. V. S., Lt. 540a
—— K. W., Lt. 526, 1435
—— N., Lt. 654
—— P. S., Maj.-Genl. . . . 687
—— R., Lt. . . . 508, 1195, 1565
—— R. C., Lt. 492
—— R. J., Lt. 305, 1007
—— R. R., 2nd-Lt. . . . 327, 970
—— S. A., Lt. 820, 861
—— S. P., 2nd-Lt. . . . 546, 1433
—— T. H. A., Lt. . . . 638, 978
—— T. O., Maj 107, 242, 1371, 1579
—— T. W., 2nd-Lt. 555
—— W. D., Lt. 393, 1650
—— W. E., 2nd-Lt. . . . 552, 1219
—— W. J., 2nd-Lt. 558
Wilks, B. St. C., Lt. . . . 543a, 905
—— C. S., Sub-Condr. 1864
—— D. J., Lt. 533, 1219
—— E., Lt. 489, 893
Will, D. A. E., Maj . . . 43, 87, 208
—— G. D., 2nd-Lt. 1281
—— H. St. C. O., 2nd-Lt. 330, 1203
—— R. R., Capt. 656
—— W. M., Ty. Lt. 1759
Willard, F. W., Lt. . . . 536, 1335
Willan, H. C., 2nd-Lt. 640
Willans, J. F., Lt. 544
—— N. J., Lt. 375
—— T. J., Lt.-Col. 139, 160, 199, 870, 1251
Willcock, A., 2nd-Lt. 626
—— F. W., 2nd-Lt. 536
—— H. A., 2nd-Lt. 1193
Willcocks, J., Genl. *Sir* . . . 748
—— R. L., Maj. 1748
—— W. G., 2nd-Lt. . . . 82, 571
Willcox, G. F., Capt. 637
Willett, H. G., Lt. . . . 523, 1323
—— H. L., Maj. . . . 691, 785
Willey, V. S., 2nd-Lt. 1703
William Cotter, Jemdr. . . . 1264

NAMES AND RANK. Page.

Williams, A., Lt. 520, 804
—— A., Lt. 591
—— A., Lt. 5404, 1160
—— A., Tempy. Capt. 1786
—— A. B., Lt. . . . . 488, [illegible]
—— A. C., Lt. . . 33, 528, 1667
—— A. deC., Lt. 327
—— A. E., Capt. 819
—— A. E., Sub.-Condr. . . . 1864
—— A. F. C., Maj. 99, 211, 933
—— A. H., Capt. 271, 847
—— A. H., Lt. 544
—— A. H., Lt.-Col. 128, 154, 183
—— A. H. K., Lt. . . . 390, 953
—— A. J., Maj. 62, 1838
—— A. J., 2nd-Lt. . . . 742, 980
—— A. J., 2nd-Lt. 359
—— A. T., Lt. 535, 1703
—— A. V., Capt. 816
—— B. F., Lt. . . 314, 947, 959
—— B. T., Lt. 1629
—— B. W., Lt. 245, 683
—— C., Lt. 729
—— C. A., Capt. 721
—— C. A. H., Lt. . 77, 516, 1808
—— C. B., 2nd-Lt. 547
—— C. G., Lt. 487, 907
—— C. E., Lt.-Col. . . 130, 1742
—— C. G., Hon. Lt. 709
—— C. H., Condr. 1852
—— C. J., Sr. Asst. Surgn. . 1791
—— C. S., Col. 751
—— C. W., Lt. 536, 1058
—— D., Lt. 543, 1673
—— D. H. J., 2nd-Lt. . . 329, 1162
—— D. J., Lt. 727
—— D. M., Lt. 443, 1591
—— E. C., Lt. 538, 1677
—— E. J., Capt. 638
—— E. O., Ty. Asst. Surgn. . 1804
—— E. P., 2nd-Lt. 564
—— E. P., Lt. 526, 1668
—— E. W., Lt. 399, 1134
—— F., Condr. 1854
—— F. C., 2nd-Lt. 717
—— F. P. J., Capt. 636
—— F. I., *Hony.* Lt. . . . 1843
—— F. S., ... 986
—— F. T., 2nd-Lt. 630
—— F. W., R. S. M. 988
—— G., 2nd-Lt. . 539, 794, 1419
—— G., Maj.-Gen. *Sir* 11, 119, 660
—— G. A., 2nd-Lt. . . . 750, 784
—— G. B., Maj. 807
—— G. B., Lt. 295, 1046
—— G. E. M., Lt. 455—59, 933
—— G. G., Lt. 738, 740
—— G. G. R., Lt. 748
—— G. H., 2nd-Lt. 1289
—— G. M., Lt.-Col. 62
—— G. R. K., Maj. 210, 1896, 1915
—— G. S., Lt. 656
—— H., 2nd-Lt. 986
—— H. A., Maj. 1747
—— H. C., Condr. 1863
—— H. E., 2nd-Lt. 817
—— H. F., Revd. 1904
—— H. G. H., Lt. 541
—— H. J., Maj. 581
—— H. M., Lt. 303, 1311
—— H. R., Capt. 107, 254, 1647
—— H. S., 2nd-Lt. 572
—— H. T., 2nd-Lt. 986
—— J., Capt. 18, 52
—— J., 2nd-Lt. 1569
—— J., *Hony.* Lt. 753
—— J., Lt. 715
—— J. B., Lt. 1839
—— J. C., Lt. 521, 1667
—— J. E. H., Revd. 1904
—— J. F. H., Lt. 489, 925
—— J. G., Asst. Surgn. . . . 1794
—— J. G., 2nd-Lt. 62
—— J. H., Capt. . . . 276, 1641
—— J. H., Lt. 488, 92
—— J. H., Lt. . . . 68, 699, 79
—— J. K., Lt. 52
—— J. M., 2nd-Lt. 56
—— J. P., 2nd-Lt. 180
—— J. W., Lt. . . 292, 1287, 135
—— K. L., Lt. 50
—— L. G., Bt. Lt.-Col. 51, 138, 159, 228, 88
. . . . 126
—— L. J., Lt. 85
—— M. J., 2nd-Lt. 5
—— N. M., 2nd-Lt. . . . 699, 79
—— N. T., Lt. . . . . .

NAMES AND RANK. Page.
ams, P. D., Capt. 777
—— R., Capt. 776a
—— P. M., 2nd-Lt. . 568, 1098
—— R., Col. *Sir* . . 739, 740
—— R. C. B., Maj. 17, 25 238, 1162
—— R. deV., Maj. . . . 628
—— R. F., Lt. 1279
—— R. G., Lt. 389, 505, 1453, 1585
—— S., Lt. 710
—— S., Capt. . . 253, 1890
—— S. C., Capt. 16, 42, 271, 1664
—— S. J., 2nd-Lt. . 551, 1657
—— S. M., 2nd-Lt. 326, 1225, 1613
—— S. N., 2nd-Lt . .548, 1559
—— S. P., Capt. . . 259, 227
—— T. L., Lt. . . 26, 722
—— T. P., Capt. 1718
—— T. S. B., Bt. Lt.-Col. 137, 1748, 1901
—— T. T., 2nd-Lt. . . . 556
—— V. P. B., Maj. 56, 57, 230, 881
—— V. S., Lt. . . 491, 923
—— V. S. M., Capt. . 502, 1078
—— W., *Hony.* Maj. . . . 1845
—— W. A., Lt.. 704
—— W. D. C., Lt. . 493, 951
—— W. E. B., Capt. . 62, 1754
—— W. H., Lt. . . 542, 1007a
—— W. J., Lt. . . . 759, 1006
—— W. M., 2nd-Lt. . . 790
—— W. P. G., Capt. 1639, 1753
—— W. R. B., Capt. 280—287, 1629
—— W. T., Lt. . . 415, 1453
ams-Taylor, T., Lt. . . 591
iamson, A., 2nd-Lt. . . 356, 1106
—— A., 2nd-Lt. . . . 1020
—— A., 2nd-Lt. . . 546, 1295
—— A. A., Capt. . 269, 1315
—— A. B., 2nd-Lt. . . . 566
—— A. F., 2nd-Lt. . . . 782
—— A. G., Maj. 6, 14, 108, 242, 887
—— A. U., 2nd-Lt. . 358, 1577
—— C. P. F., Capt. . 277, 1387
—— F. Capt. . 501, 753, 1657
—— G. A., Lt. . . 313, 1195
—— H., Lt. 571
—— H. E., Lt. 785
—— J., Capt. 608
—— J. B., Capt. . . . 1717
—— J. R., 2nd-Lt. . . . 618
—— P. R. C., Lt. 513, 843, 1339
—— U. G., 2nd-Lt. . . . 559
—— W. J. 2nd-Lt. . . . 788
iamson Oswald, O. C., Col. 124, 632
ie, E. E., 2nd-Lt. . . .358, 1144
ingdon of Ratton, *Baron* 604, 654, 821, 849
is, A. C., Lt. 538, 1695
—— C. R., 2nd-Lt. 622
—— E. H., Br.-Genl. 120, 125, 613
—— E. S. C., Maj. . . 219, 1255
—— G. C. W., Maj. 34, 78, 233, 933
—— G. H., Maj. . . 662, 1890
—— G. P., Lt. 681
—— H. A., 2nd-Lt. . 557, 551, 1024
—— H. G., Lt. 491
—— H. H., Maj. . . . 755, 795
—— J. H. P. 2nd-Lt. 325
—— J. L. N., Col. . . 177, 1917
—— N. O., Lt. 701
—— R. D., Lt. 488, 901
—— R. G., Lt. 746
—— R. H., Lt. 524
—— S. G. R., Lt.-Col. . . 103, 614
—— T. A., 2nd-Lt. 754
—— V. J., Ty. 2nd-Lt. . 669, 997
—— W., Condr. 1870
—— W. E., 2nd-Lt. 709
ison-McQuillen, D., 2nd-Lt. . 805
ier, G. I., Capt. 1707
nore, P. A., 2nd-Lt 359
—— W. S., Lt.-Col. . 140, 1745
nott, M. J, 2nd-Lt. . 554, 1223
—— N. R., Lt. . . 528, 1415
ughby, B. D., Lt. . 297, 1102
—— D. V., Maj. 54, 228, 1016
—— F. A., 2nd-Lt. . . . 563
—— J. F., Maj.-Genl. 937, 1955
—— J. J., Lt.. . 440, 1130
—— M. E., Br.-Genl. 53, 97, 102, 107, 120, 124, 150, 184
—— M. G. P., Capt. 265, 893
—— M. W., Lt.-Genl. 172, 1419
—— -Harris, A. J. 2nd-Lt. 557, 1705
ghby-Osborne, A. de V., Col. 135, 741

NAMES AND RANK. Page.
Wills, E. V., Capt. . . 82, 254, 1050
—— J. H. P., 2nd-Lt. 326
—— L. K., Lt. 740
—— P. G., 2nd-Lt. 555
—— R. D., Lt. 901
—— W. A., Capt. 778
—— W. R., Lt. 510
Willison, W. L. P., Capt. 90, 501a, 731
Willson, C. B., Lt. . . 505, 1666
Willmer, E. J., 2nd-Lt. . . 565, 1070
Wilmer, E. B. G., Bt. Lt.-Col. 58, 137, 614
—— D. H., 2nd-Lt. . . 494, 921
—— J. R., Col. . . . 177, 1917
—— T. G., Lt. 666
Willmore, P. A., 2nd-Lt. . . . 1056
Wilmot, R. C., Maj. 1711
Wilmshurt, A. P. Lt. 1007
Wilsey, F. H. W., Lt. . . 311, 1641
—— H. W., Hony. Maj. . . 1846
Wilshin, H. L., Lt. 756
Wilson, A., Hony. Capt. . . . 1844
—— A. D., Lt. . . 299, 875, 933
—— A. F., Sub-Condr. . . . 1863
—— A. F., 2nd-Lt. 844
—— A. H., Capt. . . 89, 768
—— A. K., Lt.. . . . 543a, 974
—— A. P., Capt. . . . 17, 29, 724
—— A. T., Bt. Lt.-Col. 103, 244, 1890, 1885
—— A. T., 2nd-Lt. . 77, 715, 1885
—— C. C., Lt. . . . 491, 897
—— C. E., Lt. 581
—— C. H., Lt. 531
—— C. H., Lt. . . . 512, 1672
—— C. H. E., Capt. . . . 619
—— C. T., Sub-Asst. Surgn. . 1819
—— D. C., Maj. 615
—— D. D., Maj. . . 99, 220, 907
—— D. G., Maj. .220, 1901, 1913
—— D. G., Lt. 758
—— D. McI., 2nd-Lt. . . . 739
—— D. S., 2nd-Lt. 799
—— D. W., Capt. . . 57, 485, 604
—— E. C., Lt. 397
—— E. H. E., 2nd-Lt. . 546, 1297
—— E. R. C., Maj. . . 634, 980
—— F., Ty. 2nd-Lt. . . 717, 1703
—— F., 2nd-Lt. 561
—— F. A., Lt.-Col. . 58, 129, 661
—— F. E., Maj. . . . 1749, 1901
—— F. H., Capt. 801
—— F. G., 2nd-Lt. 597
—— F. H., 2nd-Lt. . . 550, 1701
—— F. W., Capt. 521
—— F. W. B., Capt. 615
—— G., Capt. 1713
—— G., Lt. 488, 891
—— G., Sub-Condr. . . . 1873
—— G. B., 2nd-Lt. 553
—— G. E. R., Maj. . .217, 1156
—— G. E., 2nd-Lt. 544
—— G. P., 2nd-Lt. 653
—— G. S., Capt. 1718
—— H. M., Capt. 14, 26, 260, 875
—— H. M., Tempy. Capt. . . 1724
—— H. R., Lt. 793
—— H. W. D., Lt. 512
—— I. C., Capt. 636
—— L. L. D., 2nd-Lt. . . . 1696
—— J., 2nd-Lt, ... 563
—— J. C. M., 2nd-Lt. . . . 564
—— J., 2nd-Lt. . . 91, 690a, 1275
—— J. A., Lt.-Col. 130, 155, 191
—— J. A. M., Maj. 1894
—— J. C. J., 2nd-Lt. . . . 624
—— J. D., Capt. . . 852, 947
—— J. F. S., Lt. . . . 313, 913
—— J. H. A., 2nd-Lt. . . 365, 907
—— J. K. R., 2nd-Lt. . . . 689
—— J. L., Capt. . . 668, 996
—— J. L. D., 2nd-Lt. . . . 558
—— J. M., Lt. . . . 520, 1698
—— J. M., Lt. . . . 488, 891
—— J. McL., Lt. . . 518, 1247
—— J. P., Lt. . . . 492, 927
—— J. S., Ty. Lt. 1726
—— J. S. R., Nursing Sister . 1734
—— J. T., 2nd-Lt. 357
—— J. W., 2nd-Lt. 763
—— J. W., Lt. . . . 410, 1132
—— J. W., Sub-Condr. . . . 1868
—— K. B., Lt. 504
—— K. H. 1009
—— L. E., Lt. . . . 380, 1128
—— L. M., 2nd-Lt. . . 328, 1066
—— N. B., Capt. . . . 486, 877

NAMES AND RANK. Page.
—— N. F. J., Capt. 4
Wilson, N. J., 2nd-Lt. 765
—— N. M., Bt. Maj. 55, 254, 1629
—— N. M., Maj. . . . 1596, 1752
—— O., Capt. 1758
—— P. F. H., 2nd-Lt. 763
—— R., Sub.-Condr.. 1867
—— R. B., Lt.-Col. . . 507, 1886
—— R. B., Lt.. 629
—— R. C., Bt. Lt.-Col. 53, 98, 138, 159, 234, 1459
—— R. E., Lt. 465, 1259
—— R. E., Maj. . . 31, 634, 979
—— R. H., Capt. . 95, 262, 1335
—— R. H. B., Capt. 280—287, 1641
—— R. P., Lt.-Col. . . 133, 1743
—— R. P., Capt. . . . 751, 1004
—— R. R., Capt. 636
—— R. V., 2nd-Lt. 726
—— S., Capt. 702
—— T., Condr. 1854
—— T. A., 2nd-Lt. 723
—— T. A. A., Maj. 8, 15, 233, 881
—— T. C., Lt. 509
—— T. G., Ty. Capt. . 9, 571 1850
—— T. K., Capt. . 255, 913, 961
—— W., 2nd-Lt, . . . 362. 1657
—— W., Maj. 605
—— W., Lt. 711, 925
—— W. C. F. A., Bt. Maj. 271, 1419
—— W. H., Col. 177
—— W. H., Lt. 699
—— W. H. D., Maj. 9, 16, 236, 1016
—— W. K. P., Maj. . 242, 1140
—— W. L., Lt. . . 460—62, 1183
—— W. M. C., Lt. . . 425, 1267
—— W. R., Lt. . 699, 700, 795
Wilson-Green, M.A., Nurs. Sister 1735
Wilson-Haffenden, E. L., Lt. 313, 885
Wilson-Johnston, J., Lt. . . . 605
—— W. E., Bt. Lt. Col. 51, 99, 160, 217, 1166
Wilton, A. J. M., Lt. . . . 320, 1469
—— E. F., Lt. 629
—— C. E. P., Lt. . . 420, 1183
—— J. S., Lt.. 72
—— T., Lt. 1189
—— V. R., Lt. . . 460—62, 1195
Wimberley, C. I., Lt.-Col. . . . 57
—— C. N. C., Col. 128, 1741
—— F. G. A., Lt.-Col. 21, 198, 1883, 1912
—— R. J., Lt.. 1006
Wimbush, A., 2nd-Lt. 609
—— N. C., Lt. 504
Wimpey, S. G., 2nd-Lt. . . 551, 1024
Wimshurst, C. R., 2nd-Lt. . . 547
Winckworth, A. R., 2nd-Lt. . . 786
—— H. C., Maj. 1712
—— W. N., Lt. 111
Winckler, C. P., Lt. . . 518, 1375
Winckley, S. F., Lt. . . 322, 897
Windeler, W. T., Capt. . 269, 1271
Windham, A. W. G., Lt. . . . 585
—— C. J., Lt.-Col. 189, 1901, 1911
—— J. B. R., Lt. 311, 909, 1675
Windsor, E. P., Lt. 797
—— F. N., Lt.-Col. . .132, 1743
—— H. A., Esq. 88
—— H. C. W., Asst. Surgn. 1801
WindsorAubrey, W. A. 2nd-Lt.. 331, 1150
Winfield, L., 2nd-Lt. 692
Wing, R., Lt. 435. 905
Wingate, A. W. S., Col. . . . 54
—— G., Col. . . . 179, 1917
—— G. F. R., Capt. . . . 58
Wingfield, C. A. F., Lt. . . . 591
—— C. M. H., Lt. . 306, 1600
—— R. W. D., Lt. . . 304, 1090
—— W. J. R., Maj. 53
Wingfield-Digby, K. E., Capt., . 615
Wingfield-Stratford, E. C., Capt. 751
Winkfield, W. B., Lt.-Col. . 133, 1710
Winn, F. B., Lt. . . 540a, 1618, 1621
—— F. G., 2nd-Lt. 1679
—— H., Lt. 61, 700
—— H. E., Lt. 504, 610
—— W. J., Lt. 624
Winnall, E. I., Lt. . . . 425, 1154
Winnington-da-Costa, A. F., Tempy. Capt. 1767
Win Pe, Jemdr. 1295
Winsland, F. E., Lt. 1695
Winsloe, A. R., Lt.-Col. . .132, 661
—— H. E., Bt.-Lt.-Col. . 135, 662

NAMES AND RANK. Page.

Winslow, L. C., 2nd-Lt. 619
Winstanley, B. J., 2nd-Lt. 549
Winter, C. B., Bt.-Col. 53, 105, 108, 135, 158, 195, 871, 1451
—— G. W., 2nd-Lt. . . 563, 1203
—— H. B., Lt. 76, 530
—— H. E., Lt.-Col. 1731
—— J. H., 2nd-Lt. . . . 354, 1678
—— J. W., Lt. . . . . .
—— W. E., 2nd-Lt. 831
Winters, A. E., Lt. . . . 537, 1672
—— L. P., Lt. . . . 435, 1179
Winthrop, H. E., Lt. . . . 306, 1074
Wintle, C. E. H., Lt.-Col. . 137, 159, 199, 871, 1459
Winton, C., 2nd-Lt. 540
—— M., 2nd-Lt. 629
—— W. C., 2nd-Lt. 1515
Wintour, E., Lt.-Col. . 130, 155, 190
Wir Singh, Jemdr. 1575
Wise, A. A., Lt. 530, 1375
—— A. D., Maj. 244, 915
—— A. F., Lt. 530
—— H., 2nd-Lt. 68, 726
—— H. H., 2nd-Lt. . . . 567, 1050
—— J. H., Lt. 511, 1375
—— S. A., Capt. 628
Wiseham, W. B. K., Asst. Surgn. 1793
Wiseman, W., Lt. . . . 1443, 1064
Wishart, D. G., Lt. 1839
—— G., Lt. 505
—— W. G., 2nd-Lt. 638
—— W. H., Capt. 1724
—— W. J., Capt. 1004
Wisher, J. D., Lt. . . . 543a, 1367
Witchell, H., Lt. 492, 905
Witherow, T. H., Lt. 1401
Withers, A. R., Capt. 1919
—— B. H., Capt. 748
—— C. McG., Maj. 99, 213, 372, 1355, 1575
—— F. W., Lt. 1457
—— J. R., Capt. 696, 760
—— P. L., 2nd-Lt. 640
Withinshaw, L., 2nd-Lt. . . 494, 907
Withnell, J., 2nd-Lt. . . . 358, 1511
Witney, W. W., Lt. 857
Witthaus, G. T., 2nd-Lt. 592
Witts, C. A., Lt. . . . 450—54, 1281
—— C. S. F., Lt. . . . 90, 368, 683
—— F. V. B., Capt. . . . 665, 985
Woakes, R. B., Lt. . . . 510, 992
Wodehouse, C. G., Capt. 20, 512, 1279
—— F. W., Lt.-Col. 188, 1893, 1911
Wogan, H. J., Lt. 666
Wolcroft, T. C., 2nd-Lt. 988
Woledge, A. R., Sub-Condr. . . . 1864
—— C. G. C., 2nd-Lt. . . 546, 1293
—— F. L., Lt. . . . 297, 1050
—— H. W., Hony. Capt. . . . 1846
Wolfe, G. C., Esq. 91
Wolfe-Murray, D. K., Lt. . . . 637
Wolfenden, J. C., Lt. 593
Wolverson, T. B., Capt. 1732
Wolverton, F., Lord, Col. . . . 754
Wollaston, J. W., Lt. 764
Wolley, H. S. L., Maj. 82, 235, 1247
Wolley-Dod, O. C., Brig.-Genl. 12, 38, 97, 120, 123
Wollstein, L. J., Ty. 2nd-Lt. 753, 786
Wolters, N. C., Lt. 524
Womersley, F. E., 2nd-Lt. . . . 61
Wonnacott, A. J., Lt. 353
Wontersz, E. P. T., 2nd-Lt. . . . 552
Wood, A. D., Lt. 766
—— A. E. F., Lt. 488, 927
—— A. H., Bt.-Col. 45, 767
—— A. H., 2nd-Lt. 591
—— A. H., Condr. 1853
—— A. R. B., 2nd-Lt. 679
—— A. W., Capt. 500, 658
—— C., Lt. 353, 1007a
—— C. A., Capt. 1756
—— C. F., Lt. 95, 490
—— D. G., Lt. 527
—— E., Lt. 389, 1020
—— E. C., Lt. 490
—— E. J. M., Col. 12, 35, 121, 126, 150, 188, 870
—— E. R., 2nd-Lt. . . . . .
—— E. W. M., Lt. 1674
—— F., 2nd-Lt. . . . 788, 1423
—— F. E., Maj. 610
—— F. J., Capt. 1407
—— F. M. A., Lt. 623

Wood, G. H., Ty. Capt. 1724
—— H., Lt.-Col. . . 139, 662, 1880
—— H. C., Lt. 808
—— H. D., 2nd-Lt. 859
—— H. O. B., Maj. 224, 868, 1024, 1046
—— H. S., Lt.-Col. 129, 1742
—— J., 2nd-Lt. 787
—— J. C., Lt. 789
—— J. E., 2nd-Lt. 1557
—— J. H., *Hony.* Lt. 1844
—— J. H., 2nd-Lt. 763
—— M., 2nd-Lt. 1443
—— M. H., Lt. 351
—— N. R. T., 2nd-Lt. 1001
—— P., Lt. 818
—— R., Lt. 637
—— R., 2nd-Lt. 723
—— R. B. M., Lt. 59
—— R. G. P., Lt. 583
—— R. R., Capt. 754
—— R. T., Capt. 767
—— T., Lt. 583
—— T., Sub-Condr. 1855
—— T. B., Lt. 848
—— W. C. S., Tempy. Capt. . . 1723
—— W. M. P., Lt.-Col. 198, 1893, 1916
—— W. L., 2nd-Lt. 550
Woodall, H. W., Maj. 740
Woodcock, E. E., Maj. 214, 1030, 1064
—— F. B., 2nd-Lt. 779
—— F. D., 2nd-Lt. 743
—— R. C., Lt. 292
—— T., 2nd-Lt. 1393
—— W. D. G., 2nd-Lt. 539, 1507
Woodcroft, J., Sub-Condr. . . . 1863
Woodd, P. C., Lt. 844
Woodfall, H. A. J., Capt. . . 55, 721
Woodfield, A. W., 2nd-Lt. . . . 723a
Woodford, A.S., Lt.445, 450—54, 763, 1405
Woodgate, H. P., Lt. 748
—— H. W., 2nd-Lt. . . 355, 1673
Woodger, J. H., Lt. 691
Woodham, J. F., Maj. . . . 215, 1319
Woodhams, D., 2nd-Lt. . . . 700, 728
Woodhead, A. W., Capt. 501, 610, 1598
Woodhouse, A. C., 2nd-Lt. . . . 830
—— C. G., Maj. . . 211, 1501
—— C. H., 2nd-Lt. . . . 618
—— F. S., Capt. . . 258, 1664
—— F.V. R., Lt. 300—01, 1243
—— G., 2nd-Lt. 599
—— H., Col. 169
—— H. L., Capt. 665, 1001, 1882
—— H. S., Lt. 738
—— J. D. F., Capt. 593
—— L. J., Lt. 736
—— P. St. J. B., Bt. Lt.-Col. 138, 159, 238, 1663
—— R. A. G., Lt. . . 51, 1593
—— S., 2nd-Lt. 618
—— Adolphus, W. H., 2nd-Lt. 559
Wooding, W.W., Tempy. Capt. . . 745
Woodland, P. F., Capt. 1841
Woodliffe, H. D., Maj. . 18, 25, 717
—— L. F., Lt. 425, 1050
Woodman, F. C., 2nd-Lt. 724
—— G. S, Tempy. Capt. . . 1723
—— J., 2nd-Lt. 1679
—— J. B., Lt. 319, 937
—— R. S., Lt. 320, 1463
—— W., 2nd-Lt. 701
—— W. F., Lt. 533, 1641
—— W. H. L., 2nd-Lt. . 305, 1150
Woodrow, F. H., Lt. 668, 1006
—— T. W., 2nd-Lt. 619
Woodruff, K. M., Ty.-Lt. . . . 1779
Woods, A., 2nd-Lt. 558
—— A. F., Lt. 492
—— A. J., 2nd-Lt. 790
—— C. F., 2nd-Lt. 547, 1674
—— D. F. St. J., 2nd-Lt. . . . 567
—— G. G., Lt.-Col. 74, 107, 135, 632
—— G. H., 2nd-Lt. 548, 1126
—— H., Lt. 352, 1140
—— H. A., Lt. 352, 1197
—— H. K., Lt.-Col. . 134, 741, 1197
—— H. S., Lt. 537, 1672
—— H. W., 2nd-Lt. 735, 1885
—— J., 2nd-Lt. 67, 701
—— J. McB., Lt. 445, 1074
—— L. H. W., 2nd-Lt. 77, 1885
—— N., Lt. 535, 1696
—— W. L. J., 2nd-Lt. 1008
—— P. W. H., Lt. 526, 1701

Woodsell, J. W., Asst. Surgn. . . [illegible]
Woodside, R., Capt. . . . . . [illegible]
Woods-Jennings, H. W., 2nd-Lt. . [illegible]
Woods-Scawen, C. E., 2nd-Lt. . [illegible]
—— P. N., 2nd-Lt. . . . . 551, [illegible]
Woodward, C. F., Maj. . . . . [illegible]
—— C. P., Maj. . . . 9, 211, [illegible]
—— D. L. J., 2nd-Lt. . . . 227, [illegible]
—— E. M., Maj.-Genl., *Sir*, [illegible]
—— J. A. H., Col. 123, 149, [illegible]
—— M., Sub-Condr. . . . . [illegible]
—— R. F., Maj. . . 103, 146, [illegible]
—— R. G., Lt. . . . . 309, [illegible]
—— W. J., Lt. . . . . . 303, [illegible]
Woodwright, W. H. E., Lt.-Col. 127, [illegible]
Woodyatt, N. G., Maj.-Genl. 12, 29, 119, 147, 174, [illegible]
Woodyer, J. M., Lt. . . . . . 375, [illegible]
Wooff, S. H., 2nd-Lt. . . . . 257, [illegible]
Wooldridge, H. C., Lt.-Col. 137, 158, [illegible]
—— J. H. C., Lt. . . . . 304, [illegible]
—— L. E., 2nd-Lt. . . . . . [illegible]
—— W. H., Lt.-Col. 130, 155, 190, [illegible]
Woolf, A. E. M., Tempy. Capt. . . [illegible]
—— S. H., Lt. . . . 291, 1094, [illegible]
Woolgar, C., Lt. . . . . . . 68, [illegible]
Woollam, J. L., Lt. . . . . . 303, [illegible]
Woollat, P. C., Ty. Maj. . . . . [illegible]
Woollcombe, C. L., Lt.-Genl. *Sir* [illegible]
—— G. H. D., Lt. . 313, [illegible]
—— L. D., Maj. . . . . . [illegible]
Woolley, J., Lt. . . . . . . 498, [illegible]
—— J. H. C., Lt. 314, 852, 957, [illegible]
—— J. M., Lt.-Col. 130, 1744
Woolmer, F. J., 2nd-Lt. . . . . . [illegible]
—— H. E., Lt. . . . . [illegible]
Woolner, A. C., Capt. . . . . . [illegible]
—— P. C., 2nd-Lt. 1668
—— R. J., 2nd-Lt. 1265
Woolsey, C. H., 2nd-Lt. . . . . 359, [illegible]
Woolrich, W. G., Capt. 685
Woon, J. B., Genl. *Sir* . 145, 171, [illegible]
Woosnam, A. G., Lt. . . . 507, 830, [illegible]
—— R. W., Capt. . . . . . . [illegible]
Wootten, E. H. W., Maj. . . . 501a, [illegible]
—— R. M., Capt. 585, 1512
Wootten, W. J. G., 2nd-Lt. 550
Wooton, H. E., 2nd-Lt. 717
Worden, C. N., Lt. 544, 1451
—— G. M., 2nd-Lt. 465, 1203
—— L., 2nd-Lt. 722, 1070
Wordsworth, R. H., Lt. 373, 905
Worgan, R. B., Br.-Genl. 50, 122, 139, 160, 223, 913
Worke, F. E. L., 2nd-Lt. 802
Workman, D. J. G., Lt. 73, 708
Worlledge, F. H., Capt. . 96, 277, 923
—— J. F., Col. . . 195, 179, 1917
—— L. H., Lt. 310, 1479
Wormald, R. L., Revd. 1907
Worrall, E. W., Maj. 102, 701
Worsdell, G. B., Capt. 1004
Worsley, C. F. M., Maj. 51, 215, 915
—— R. E., Lt. 740
Warssam, C., Lt. 1156
Worswick, E. T., Capt. 1732
Worster, A. H., Capt. 511, 852
Worth, A. M., 2nd-Lt. 363
—— J. G., Ty. 2nd-Lt. 732
Wortham, O. O., Maj. 11, 1013
Worthington, R. F., 2nd-Lt 330, 1375
Wotherspoon, W. S., Lt. . . 528, 998
—— C., 2nd-Lt. 685
Woulfe Flanagan, R. J., Lt.-Col. 104, 155, 749
Woutersz, E. W. C., Asst. Surgn. 180
Woyka, A. G., 2nd-Lt. 744
Wrafter, G. T., Asst. Surgn. . . 1800
Wragg, W. F., 2nd-Lt. 455
Wraight, G. J. R., 2nd-Lt. . . 682, 740
Wrangham, D. G., Maj. 393
Wrangham-Hardy, G., Lt. 614
Wratislaw, J. M. B., Maj. 769
Wray, D. W., 2nd-Lt. 968
—— G. H., Lt. 1703
Wredden, J. L., Asst. Surgn. . . 1793
Wreford, A. P., 2nd-Lt. . . . 555, 1151
—— C. F., 2nd-Lt. 543
—— R. G., Lt. 507, 1067
—— V. S. J., Lt. 689
Wren, P. C., Capt. 111
Wrench, G. T., Ty. Capt. 1749

Names and Rank.	Page.
Z	
Zabat Khan, Risldr.	1684
Zabte Khan, Jemdr.	876
Zacharias, P. T., Ty.-Lt.	1779
Zafar Husain, Sub-Asst. Surgn.	1813
—— Muhammad Khan, Jemdr.	884
Zafar Hussian, Jemdr.	1117
Zargir, Jemdr.	1252
Zafuryab Khan, Jemdr.	936
Zahirud-din, Jemdr.	884
Zahirud-din Khan, Sub-Asst. Surgn.	1811
Zahur Ali Khan, Jemdr.	1090
Zahur Khan, Subdr.	1086
Zaid, Gul, Subdr.	1120
Zaidullah Khan, Jemdr.	1589
—— —— Jemdr.	1480
Zaiman, B. A., Lt.	542, 1295
Zain-Din, Subdr.	1128
Zain Khan, Jemdr.	1336
Zainulabdin, Jemdr.	1332
Zainuladin, Jemdr.	1134
Zalim Singh, Jemdr.	
—— —— Jemdr.	
—— —— Jemdr.	1567
—— —— Ressdr.	
Zaman Ali, Subdr.	1207
—— —— Jemdr.	1228
—— —— Jemdr.	1565
—— —— Jemdr.	1369
Zaman Khan, Subdr.-Maj.	1513
—— —— Risldr.	
—— —— Risldr.	1684
—— —— Subdr.	1518
—— —— *Hony.* Capt.	2101
—— —— Jemdr.	1492
—— —— Jemdr.	1121
—— —— Shah, Subdr.	1347
—— Mehdi, Khan, Jemdr.	922
Zarif Khan, Jemdr.	966
Zardad Khan, Ressdr.	975
—— —— Subdr.	1108
Zarin Khan, Subdr.	1259
Zarullah Khan, Jemdr.	1095
Zauki Ram, Subdr.	1476
Zeigler, Lt. J. H.	610
Ziada Khan, Jemdr.	1488
Zekhraj, Jemdr.	1609
Ziarud Din Khan, Subdr.	1483
Ziegler, J. H., Lt.	622
Zikriya Khan, Risldr. Maj.	908
Ziladar Khan, Sub-Asst. Surgn.	1813
Ziman, S. N., Lt.	489, 941
Zobel, M., Lt.	514, 1667
Zorawar Singh, Capt.	278, 875
—— —— Sr. Sub-Asst. Surgn.	1809
—— —— Jemdr.	1193
—— —— Subdr.	1191
Zulfi, Jemdr.	1312
Zul-fikar Hyder, Sub-Asst. Surgn.	1815

www.ingramcontent.com/pod-product-compliance
Ingram Content Group UK Ltd.
Pitfield, Milton Keynes, MK11 3LW, UK
UKHW010157180926
475500UK00003B/63

9 781843 422969

By Authority.

The Quarterly

INDIAN ARMY LIST

JANUARY 1919.

Volume 3

A DISTRIBUTION LIST OF OFFICERS ON THE ACTIVE LIST OF THE ARMY IN INDIA, INDIAN DEFENCE FORCE, ETC.

TO WHICH IS ADDED, HALF-YEARLY, ORDERS OF KNIGHTHOOD, ETC., INDIAN ORDERS, FOREIGN ORDERS, NON-EFFECTIVE LISTS, AND WAR SERVICES OF OFFICERS OF THE INDIAN ARMY.

ARMY HEADQUARTERS, INDIA.

CALCUTTA
SUPERINTENDENT GOVERNMENT PRINTING, INDIA
1919

NOTE.—The Indian Army List being an Official publication, and the Press having no power to make alterations in it, all communications regarding deaths, errors, or omissions should be addressed to "Military Secretary to the Commander-in-Chief" with the words "Army List" written on the upper left-hand corner of the cover.

All correspondence regarding the War Services of Officers to be addressed to the Adjutant-General in India ***through Brigade or Divisional Commanders.*** (***Vide*** India Army Order 31 of 1911.)

Regt. Centre—Secunderabad.
Linked with—75th & 86th Carnatic Infy.

Uniform—Scarlet.
Facings—Yellow.

79th Carnatic Infantry.

Muirhead-ki-paltan.—Formed at Trichinopoly in 1777, by Capt. D. Muirhead, of drafts from the 1st, 3rd, 8th and 16th Carnatic Battalions, and was originally designated the 20th Carnatic Battn. Became the 20th Madras Bn., 1784; the 2nd Bn., 7th Regt. of Madras Native Infantry, 1796; the 19th Regt. of Madras N. I., 1824; the 19th Regt. of Madras Infy., 1885; the 19th Madras Infy., 1901; present designation, 1903.

"Sholinghur" "Carnatic" "Mysore" "Seringapatam" "Pegu" "Central India."

Composition—2 Companies of Madrasi Musalmans, 1 of Tamils, 1 of Paraiyans and Christians.

1st Commn. or date of entering service.	Names and Rank.	Army rank.	Present appointment in Regiment.	Remarks.
	Commandant.			
7 Mar. 94	Kauntze, Lt.-Col. B. C.	31 July 18	31 July 18	
	Company Commanders. (4)			
5 Jan. 01	L'Estrange, Maj. E. B.	31 Dec. 15	19 Aug. 18	2nd-in-Comd. With 74th Punj.
17 Jan 00	Murphy, Maj. J. L.	7 Feb. 16	2 Apl. 12	
5 Aug. 05	Thompson, Capt. P. T. L.	5 Aug. 14	8 June 18	With 80th Infy.
11 Dec. 07	Semini, Capt. E. J.	1 Sep. 15	19 Aug. 18	With 111th Mahars.
	Company Officers. (5)			
8 Sep. 09	Brunskill, Capt. B. A. S.	1 Sep. 15	23 Nov. 10	With 3-152nd Infy.
29 Jan. 10	Dawe, Capt. T. R.	1 Sep. 15	11 Mar. 11	With 2-73rd Infy.
6 Sep. 11	Rivett Carnac, Capt. H. G.	6 Sep. 15	21 Nov. 12	
15 Nov. 15	Laing, Lt. A. D. M. G.	15 Nov. 16	26 Nov. 15	Attd. With 114th Mahrattas.
	Attached.			
23 Aug. 06	Proctor, Capt. G. N. (*Actg. Maj.*)	29 Aug. 15	28 Dec. 10	83rd L. Infy.
4 Sep. 16	Lewis, Lt. T. M.	4 Sep. 17	9 Nov. 18	M. G. Corps, I. A. (*on probn.*)
1 Dec. 17	Davidson, Lt. A.	1 Dec. 18	29 Apl. 18	I. A. (*on probn.*)
5 Nov. 14	Towerton, Lt. H. J.	5 Nov. 15	1 Oct. 16	I. A. R. O.
10 May 16	Cook, Lt. H. G.	10 May 17	6 Oct. 16	I. A. R. O.
10 June 16	Parker, Lt. D. R.	10 June 17	21 Aug. 16	I. A. R. O.
4 Aug. 16	Willett, Lt. H. G.	4 Aug. 17	11 Aug. 16	I. A. R. O.
10 Aug. 16	Brown, Lt. A. (*Actg. Capt.*)	10 Aug. 17	12 Sep. 17	I. A. R. O. Adjt.
1 Mar. 17	Cowie, Lt. F.	1 Mar. 18	26 Sep. 17	I. A. R. O.
5 May 17	McIlwaine, Lt. A.	5 May 18	13 Oct. 17	I. A. R. O.
5 May 17	Hockin, Lt. F. W.	5 May 18	15 Oct. 17	I. A. R. O.
8 Aug. 17	Kelly, Lt. P. D. L.	8 Aug. 18	....	I. A. R. O.
12 Sep. 17	Barton, Lt. V. G. J.	12 Sep. 18	....	75th Infy.
5 Nov. 17	Nash, Lt. S. G.	5 Nov. 18	....	I. A. R. O.
29 May 18	Ottley, 2nd-Lt. A. R. C.	29 May 18	15 June 18	I. A. R. O.
28 May 18	Foster, 2nd-Lt. H. E.	28 May 18	29 Aug 18	I. A. R. O. (*on probn.*)
28 May 18	Gordon, 2nd-Lt. A. R.	28 May 18	29 Aug. 18	I. A. R. O. (*on probn.*)
5 Nov. 17	Blake, 2nd-Lt. D.	5 Nov. 17	14 Aug. 18	I. A. R. O.
29 Mar. 18	Ward, 2nd-Lt. E.	29 Mar. 18	14 Aug 18	I. A (*T. C.*)
7 July 18	Hanley, 2nd-Lt. M. P.	7 July 18	8 Oct. 18	I. A. R. O.

Subadar-Major.

		Jemadar.	Subadar.	Subadar-Major.
17 Dec. 88	Muhammad Ghalib (56, 66*a*)	1 Mar. 01	Aug. 06	20 Oct. 15.
	Nisar Husain (*Retd.*)	....	....	Re-employed 17 Feb. 17.

Subadars.

10 June 89	Jatar Khan (56, 66*a*)	1 June 03	27 Feb. 08
1 June 91	Rungasami (56)	8 May 06	7 Aug. 12
3 Sep. 98	Narayanasami Naidu (58)	11 May 11	1 June 14
6 Mar. 95	Abdul Sattar (56, 72*a*)	24 Dec. 07	22 July 15
11 May 97	Saiyid Yasin (56, 66*a*)	5 June 13	20 Oct. 15
20 May 99	Krishnasami (I) (56, 66*a*)	7 Aug. 12	21 Sep. 16
9 Nov. 97	Nilamegham (56, 66*a*)	3 Aug. 11	20 Apl. 17
1 Dec. 99	Abdul Ghafar (66*a*)	20 Oct. 15	20 Apl. 17
24 June 99	Krishnasami (II) (56)	31 Oct. 12	9 Nov. 17
18 Oct. 00	Sayyid Abdur Rahim	21 Sep. 16	4 Mar. 18
19 May 98	Muttusami	21 Sep. 16	16 Mar. 18
19 Mar. 99	Konayya	21 Sep. 16	1 June 18
23 May 01	Narasimulu	21 Sep. 16	1 June 18

79th Carnatic Infantry—*contd.*

1st Commn. or date of entering service.	Names and Rank.	Army rank.	Remarks.
		Jemadars.	
27 Dec. 94	Shaikh Amir (66*a*)	19 June 13	
16 Nov. 98	Sayyid Mastan	19 July 16	
10 Dec. 98	Marrisami	19 July 16	
10 Apl. 04	Muhammad Khan	20 Apl. 17	
19 Apl. 04	Shaikh Mahbub	20 Apl. 17	
14 July 00	Eswaramurthi	11 Nov. 17	
16 Sep. 04	Muhammad Abdul Rashid	3 Dec. 17	
6 Sep. 07	Nanjappa Naidu	1 June 18	
26 Mar. 08	S. Santiago	1 June 18	
	Munusami	1 June 18	
1 July 01	Shaikh Anser	1 June 18	
14 May 01	Madar Khan	2 Sep. 18	
16 July 07	Doraisami	2 Sep. 18	
	Usman Khan	1 Sep. 18	

Regt. Centre—St. Thomas' Mount.
Linked with—73rd & 88th Carnatic Infy.

Uniform—Scarlet.
Facings—Emerald Green.

80th Carnatic Infantry.

Bagot-ki-paltan.—Formed at Tanjore in 1777, by Capt. Thomas Bagot, of drafts from the 2nd, 6th, 12th and 15th Carnatic Battalions and originally styled the 21st Carnatic Bn. Became the 21st Madras Bn., 1784; the 2nd Bn., 2nd Regt. of Madras Native Infantry, 1796; the 20th Regt. of Madras N. I., 1824; the 20th Regt. of Madras Infy., 1885; the 20th Madras Infy., 1901; present designation, 1903.

Carries an honorary colour, with the inscription "Hyder Ally, Sholinghur, Hejira 1195."
"Sholinghur" "Carnatic" "Mysore" "Seringapatam."

Composition—2 Companies of Madrasi Musalmans, 1 of Tamils, 1 of Paraiyans and Christians.

1st Commn. or date of entering service.	Names and Rank.	Army rank.	Present appointment in Regiment.	Remarks.
	Commandant.			
3 May 90	Hole, Lt.-Col. H. L.	3 May 16	30 Dec. 16	
	Company Commanders. (4)			
20 Jan 00	Coleridge, Maj. P. L. (*Actg. Lt.-Col.*)	1 Sep. 15	30 Dec. 16	2nd-in-Comd.
4 Dec. 00	Norris, Maj. H.	1 Sep. 15	4 Nov. 12	2nd-in-Comd. (*temp.*)
11 May 01	Benson-Cooke, Maj. P. G. [l]	11 May 16	25 Aug. 14	With 109th Infy.
27 Jan. 04	Phythian Adams, Capt. E. G. P.	27 Jan. 13	5 June 17	With 1-156th Infy.
	Company Officers. (6)			
16 Feb. 07	*Rudkin, Capt. B.A.*	*1 Sep. 15*	*3 Dec. 12*	
29 Jan. 10	Wright-Warren, Capt. C. T. [l]	1 Sep. 15	21 Mar. 11	
11 Aug. 15	Ridley, Lt. E. R. (*Actg. Capt.*)	11 Aug. 16	30 Oct. 15	Attd.
15 Nov. 15	Rowlinson, Lt. O. A. (*Actg. Capt.*)	15 Nov. 16	23 Nov. 15	Attd. Qr.-Mr.
30 Jan. 17	Cawood, Lt. C. W.	30 Jan. 18	6 Feb. 17	Attd. With 1-156th Infy.
12 Sep. 17	Dean, Lt. A.	12 Sep. 18	5 Dec. 17	Attd.
	Attached.			
5 Aug. 05	Thompson, Capt. P. T. L. (*Actg. Maj.*)	5 Aug. 14	1 Feb. 18	79th Infy.
5 Feb. 15	Lovat-Fraser, Lt. W. A.	5 Feb. 16	1 Dec. 16	I. A. R. O.
15 Sep. 15	Merry, Lt. R. (*Actg. Capt.*)	5 May 16	28 Jan. 18	I. A. R. O. (*on probn.*)
6 Oct. 16	Cambridge, Lt. A. F.	6 Oct. 17	27 Nov. 16	I. A. R. O.
6 Oct. 16	Carroll, Lt. S. R.	6 Oct. 17	30 Nov. 16	I. A. R. O.
21 Nov. 16	Palmer, Lt. F.	21 Nov. 17	4 Jan. 17	Adjt. (*Offg.*)
20 June 17	Seal, Lt. L. S.	20 June 18	20 Oct. 17	
16 June 17	Spurgeon, 2nd-Lt. F. W.	16 June 17	23 Jan. 18	I. A. R. O. (*on probn.*)
20 Sep. 17	Chadwick, 2nd-Lt. W. H.	20 Sep. 17	28 Jan. 18	I. A. R. O. (*on probn.*)
29 Mar. 18	Bandey, 2nd-Lt. A. E.	29 Mar. 18	20 July 18	I. A. R. O.
29 Mar. 18	Wooff, 2nd-Lt. S. H.	29 Mar. 18	20 July 18	I. A. (*T. C.*)
1 Dec. 17	Donald, Lt. H.	1 Dec. 18	4 July 18	North'd. Fus., I. A. (*on probn.*)
23 Aug. 17	Chivers, Lt. N. M.	23 Aug. 18	....	Glouc. R., I. A. (*on probn.*)
1 June 18	Lodge, Lt. E. R.	1 June 18	....	R. F. A., I. A. (*on probn.*)
24 Apl. 18	Barnes, 2nd-Lt. T. W. C.	24 Apl. 18	....	I. A. (U. L.)
7 June 18	Turner, 2nd-Lt. D.	7 June 18	....	I. A. (*T. C.*)

Subadar-Major.

		Jemadar.	Subadar.	Subadar-Major.
3 May 89	Abdul Majid (52, 56, 66*a*)	1 July 05	1 Jan. 11	29 Dec. 15
....	Venketanarasayya Bahadur (Ret.)	....	....	Re-employed 30th Oct. 16.

Subadars.

28 Jan. 90	Abdul Ghafur (56, 66*a*)	15 Oct. 09	16 Oct. 13	
19 Nov. 92	Gurusami (66*a*)	17 Nov. 11	19 May 14	
14 Jan. 95	*Venkatsami*	*23 Nov. 11*	*29 Sep. 15*	
8 May 99	Hira Lal (66*a*)	15 Apl. 12	29 Dec. 15	
1 Apl. 92	Narayanasami (56)	19 May 14	17 Mar. 16	
4 Sep. 93	Daud Beg (52, 56)	1 Jan. 11	23 Apl. 16	
22 Sep. 94	Shaikh Abdur Rahman (52, 56, 66*a*).	1 July 11	9 Apl. 17	
12 Jan. 95	Abdur Razaq (52, 56, 60*a*, 66*a*).	1 July 14	1 June 18	
12 Jan. 95	Muhammad Ismail (56, 66*a*)	3 Dec. 15	1 June 18	
1 Apl. 01	Devaraju (60*a*)	9 Apl. 17	21 Sep. 18	
15 Mar. 09	Kannayya Thevar (72*a*)	9 Apl. 17	21 Sep. 18	
	Narasimhulu Bahadur (*Ret.*)	....	....	Re-employed.
	Dharmarajulu (*Ret.*)	....	....	Do.
	Gopalsami (*Ret.*)	....	....	Do.

80th Carnatic Infantry—*contd.*

1st Commn. or date of entering service.	Names and Rank.	Army rank.	Remarks.
	Jemadars.		
25 Mar. 01	Munirathnam (57)	29 Dec. 15	
17 Mar. 16	R. Gnanabishegam	17 Mar. 16	(*On probn.*)
11 Aug. 96	Shaik Hussain	23 Apl. 16	
6 Feb. 99	*Muhammad Abdul Hakim*	*18 July 16*	
8 June 00	Chandke Singh (56)	18 July 16	
24 Sep. 99	Sivaramulu	1 Aug. 16	
1 Oct. 16	Manikoth Narayanan Nambiar	1 Oct. 16	*Supy.*
17 Apl. 04	Masilamoni (56)	9 Apl. 17	
8 June 97	Muhammed Hussain (56)	16 Apl. 17	
26 May 04	Appagiri Thevar	1 June 18	
1 July 03	Sayyid Usman	24 June 18	Indian Adjt.
	Shaikh Daud	1 Aug. 18	
	Muhammad Abdul Wahab	7 Aug. 18	
	Chowrimuthu	1 June 18	
	Palavasa Thevan	21 Sep. 18	
	Shaikh Daud	21 Sep. 18	
	Shaikh Ali	21 Sep. 18	
	Doraisami Nayadu	21 Sep. 18	
	Mir Ishan Ali	21 Sep. 18	

The regiment is allowed an extra Jemadar for the honorary colour.

Regt. Centre—Secunderabad.
Linked with—82nd & 87th Punjabis.

Uniform—Scarlet.
Facings—Emerald Green.

1st Battalion, 76th Punjabis.

Lane-ki-paltan.—Formed at Trichinopoly in 1776, by Capt. Thomas Lane, of drafts from the 1st, 3rd and 8th Carnatic Battalions and was originally designated the 16th Carnatic Bn. Became the 16th Madras Bn., 1784; the 2nd Bn., 5th Regt. of Madras Native Infantry, 1796; the 16th Regt. of Madras N. I., 1824; the 16th Regt. of Madras Infy., 1885; the 16th Madras Infy., 1901; present designation, 1903.

"Sholinghur" "Carnatic" "Mysore" "Seringapatam" "Ava" "Burma, 1885-87."

Composition—2 Companies of Punjabi Musalmans, 1 of Sikhs, 1 of Jats.

COLONEL—Major-General Henry S. Elton 13 May 1904.

1st Comm. or date of entering service.	Names and Rank.	Army rank.	Present appointment in Regiment.	Remarks.
	Commandant.			
	Company Commanders. (4)			
5 Aug. 96	*Rybot, Maj. N. V. L.*, D.S.O.	*5 Aug. 14*	*25 Feb. 16*	2nd-in-Comd.
22 Jan. 98	Woodham, Maj. J. F. (*Actg. Lt.-Col.*)	1 Sep. 15	2 Dec. 12	Comdt. (*temp.*).
4 May 98	*Milford, Maj. E.*	*1 Sep. 15*	*24 Nov. 15*	
21 Jan. 03	*Laing, Maj. S. van B.*, M.C.	*21 Jan. 18*	*25 Feb. 16*	
	Company Officers. (9)			
18 Nov. 03	*Minniken, Capt. T. B.*	*18 Nov. 12*	*1 Dec. 05*	*Malay States Guides.*
5 Aug. 05	*Reyne, Capt. G. van R.*, D.S.O.	*5 Aug. 14*	*25 Oct. 06*	
25 Jan. 08	Darell, Capt. R. D. E.	1 Sep. 15	5 Apl. 09	With 49th Bengalis.
17 Jan. 11	McElwaine, Capt. E. J. D. (*Actg. Maj.*)	1 Sep. 15	16 Apl. 13	Adjt., 1 Jan. 17. 2nd-in-Comd. (*temp.*)
15 Nov. 15	Whitehead, Lt. W. A.	15 Nov. 16	18 Nov. 15	With 2-154 Infy.
27 Oct. 16	Cohen, Lt. N. C.	26 Oct. 17	2 Jan. 17	Attd.
12 Sep. 17	Cobbett, Lt. J. E.	12 Sep. 18	....	Attd.
31 Aug. 18	Proctor, 2nd-Lt. P.	31 Aug. 18	....	Attd.
	Attached.			
18 Jan. 03	Stoddart, Capt. H.	12 July 10	7 Oct. 17	11th Rajputs. *Lv. ex I., m. c.*
12 Mar. 15	*Kiernander, Lt. O. G.*	*12 Mar. 16*	*30 Aug. 15*	I. A. R. O.
30 Apl. 15	Birt, Lt. A. G.	30 Apl. 16	28 Jan. 17	I. A. R. O.
5 May 15	Emery, Lt. T. S., M.C.	5 May 16	5 Aug. 17	*Staff Capt., Jhansi Bde.*
13 Mar. 14	𝔒. 𝔈. Forshaw, Lt. W. T.	13 Mar. 15	4 Dec. 17	
27 Oct. 15	Moyes, Lt. E. M.	27 Oct. 16	2 Apl. 18	I. A. (*on probn.*)
9 Nov. 15	Irwin, Lt. S. F.	9 Nov. 16	14 Feb. 18	I. A. (*on probn.*)
19 Sep. 17	Anderson, Lt. R. C.	19 Sep. 18	9 Feb. 18	I. A. (*on probn.*)
26 Mar. 16	Crocombe, Lt. F. R.	26 Mar. 17	11 June 17	
6 May 17	Retief, Lt. P. J.	6 May 18	18 Jan. 18	I. A. (*on probn.*)
5 May 17	Caird, Lt. D. R.	5 May 18	28 Aug. 17	I. A. R. O.
29 Aug. 17	Austen, Lt. E. J.	29 Aug. 18	16 Dec. 17	I. A. R. O.
26 Sep. 17	Hebson, 2nd-Lt. T. E.	26 Sep. 17	22 Jan. 18	I. A. R. O.
31 Oct. 17	Culley, Lt. A.	31 Oct. 18	15 Feb. 18	I. A. R. O.
5 Nov. 17	Kent, 2nd-Lt. A. R.	5 Nov. 18	16 Mar. 18	I. A. (*T. C.*)
26 Jan. 18	Danby, 2nd-Lt. J. A. M.	26 Jan. 18	18 Apl. 18	I.A. (*on probn.*)
27 June 17	Laing, 2nd-Lt. H. G. J.	27 June 17	15 Aug. 18	I. A. R. O.
26 Mar. 18	Buckner, 2nd-Lt. C. S.	26 Mar. 18	16 Aug. 18	I. A. R. O.
21 Dec. 17	Norris, 2nd-Lt. H. F.	21 Dec. 17	16 Aug. 18	I. A. (*U. L.*)
29 Aug. 17	Jones, 2nd-Lt. E. J.	29 Aug. 17	11 Sep. 18	
1 May 18	Cottrell, 2nd-Lt. H.	1 May 18	16 Oct. 18	S. W. Bords, I. A. R. O. (*on probn.*)
5 Jan. 18	Stanley, 2nd-Lt. G. C.	5 Jan. 18	16 Oct. 18	Lond. Yeomanry, I. A. R. O. (*on probn.*)
26 June 17	McHardy, Lt. W. S.	26 June 18	....	K. O. Sco. Bord., I. A. (*on probn.*)
7 June 18	Lichfield, 2nd-Lt. S. J.	7 June 18	12 Dec. 18	I. A. R. O.
22 July 18	Lowther, 2nd-Lt. A. W.	22 July 18	3 Dec. 18	I. A. (*T. C.*)
22 July 18	Thompson, 2nd-Lt. P.	22 July 18	3 Dec. 18	I. A. R. O.

Subadar-Major.

		Jemadar.	Subadar.	Subadar-Major.
1 Apl. 87	Husain Shah, I.O.M., I.D.S.M.	24 Feb. 02	16 May 03	24 July 15

2nd Battalion, 75th Carnatic Infantry.

Formed at Bangalore on 19th October 1918.

Composition :—2 Companies of Tamils. 1 Company of Madrasi Musalmans. 1 Company of Paraiyans and Christians.

1st commn. or date of entering service.	Names and Rank.	Army rank.	Present appointment in Regiment.	Remarks.

Commandant.

20 July 98	Watson, Maj. H. R., *q.* (*Actg. Lt.-Col.*)	1 Sep. 15	19 Oct. 18	1-81st Prs. (*temp.*)

Company Commanders (4).

13 Aug. 04	Nicoll, Capt. N. F.	13 Aug. 13	4 Nov. 18	1-54th Sikhs 2nd-in-Comd. (*temp.*)
12 June 15	Hawkes, Lt. L. V. C.	12 June 16	31 Oct. 18	York R., I. A. (*on probn.*)
11 Aug. 15	Roberts, Lt. H. B.	11 Aug. 16	19 Oct. 18	1-75th Infy.
15 Sep. 16	Christie, Lt. J. McI.	15 Sep. 17	26 Nov. 18	R. Scots., I. A. (*on probn.*)

Company Officers.

31 Oct. 17	Wilcoek, 2nd-Lt. H. A.	31 Oct. 17	19 Oct. 18	Genl. List., I. A. R. O. (*on probn.*)
5 Nov. 17	Egan-Wyer, Lt. O. J.	5 Nov. 18	19 Oct. 18	I. A. R. O. Adjt. (*temp.*)
31 Aug. 18	Lawrence, 2nd-Lt. E. D	31 Aug. 18	19 Oct. 18	1-75th Infy.

Subdar Major.

		Jemadar.	Subadar.	Subadar. Major.	
30 Nov. 97	Antony (56, 66*a*)	1 Oct. 10	1 Jan. 14	19 Oct. 18	(*temp.*)

Subdars.

13 Oct. 98	Ratnamalai (51, 66*a*)	1 Jan. 14	23 Dec 15
11 Oct. 99	Ramanjulu (56, 106)	3 Feb. 12	2 Nov. 16
1 Jan. 01	Duraisami (66*a*)	23 Dec. 15	14 Nov. 16
21 Aug. 96	Abdul Alim (56, 66*a*)	23 Mar. 17	21 Oct. 18
22 May. 04	Vazialingam (61, 66*a*)	14 Nov. 16	24 Apl. 18

Jemadars.

12 July 99	Nayanar Pillai (66*a*)	23 Mar. 17	
14 Sep. 96	Krishnasami (66*a*)	1 Apl. 17	
7 Apl. 97	Shaikh Mahbub (56, 66*a*)	1 July 17	Indian Adjt.
24 Apl. 04	Vaidanadasami Naidu	18 Mar. 18	
4 Mar. 01	Sanimathan	18 Mar. 18	
12 Oct. 10	Iyyasami	22 Apl. 18	

1st Battalion, 75th Carnatic Infantry—*contd.*

1st Commn. or date of entering service.	Names and Rank.	Army rank.	Remarks.
	Jemadars.		
21 June 96	Muhammad Said (56, 66*a*)	2 June 14	
29 May 00	Kandasami	28 July 16	
15 Sep. 99	Sayyid Azimuddin	1 July 17	
20 May 04	S. Ayyulu Naidu	1 Nov 17	Indian Adjt.
30 Jan. 00	Abdur Razaq	1 Jan. 18	
18 Nov. 02	Muhammad Sahib	1 Feb. 18	Indian Qr. Mr.
	Haidar Beg	8 Aug. 18	
	Abdul Majid Khan	8 Aug. 18	
	Iyyasami Pillai	12 Sep. 18	
	Saminathan Pillai	12 Sep. 18	
	Govindarajalu	12 Sep. 18	
	Mari Muthu	12 Sep. 18	
	Sayyid Hussain	1 Aug. 18	
	Sayyid Qadir	18 Sep. 18	
	Subbayya Naidu	18 Sep. 18	
	Kasturirungam Pillai	18 Sep. 18	

Regt. Centre—Secunderabad.
Linked with—79th & 86th Carnatic Infy.

Uniform—Scarlet.
Facings—Yellow.

1st Battalion, 75th Carnatic Infantry.

Davis-ki-paltan.—Formed at Tanjore in 1776, by Capt. John Davis, of drafts from the 2nd, 6th and 12th Carnatic Battalions and was originally designated the 15th Carnatic Battn. Became the 15th Madras Bn., 1784; the 2nd Bn., 4th Regt. of Madras Native Infantry, 1796; the 15th Regt. of Madras N. I., 1824; the 15th Regt. of Madras Infy., 1885; the 15th Madras Infy., 1901; present designation, 1903.

"Sholinghur" "Carnatic" "Mysore"

"Afghanistan, 1879-80" "Burma, 1885—87."

Composition—2 Companies of Madrasi Musalmans, 1 of Tamils, 1 of Paraiyans and Christians.

1st Commn. or date of entering service.	Names and Rank.	Army rank.	Present appointment in Regiment.	Remarks.
	Commandant.			
1 Mar. 90	Field, Lt.-Col. C. D.	1 Mar. 16	10 May 17	
	Company Commanders. (4)			
30 Aug. 93	Ames, Maj. L. H. R.	30 Aug. 11	10 May 17	2nd-in-Comd.
23 May 00	*Dawson, Maj. H. H.*	*1 Sep. 15*	*27 Dec. 15*	*D. A. A. G., N. Comd.*
8 May 01	Robarts, Maj. C. S. H.	8 May 16	4 Nov. 13	With 80th Infy.
18 Jan. 02	Robinson, Maj. E. L.	18 Jan. 17	10 May 17	With 111th Mahars.
	Company Officers. (5)			
18 Jan. 05	*Warden, Capt. W. R.*	*18 Jan. 14*	*28 June 13*	*With Sig. Service.*
29 Aug. 06	Baker, Capt. H. N. (*temp. Maj.*)	29 Aug. 15	12 Nov. 08	Adjt., 15 June 15.
29 Jan. 09	Williamson, Capt. A. A.	1 Sep. 15	5 Apl. 12	
22 Jan. 15	Ham, Lt. D. M.	22 Jan. 16	21 Mar. 15	With 1-156th Infy.
11 Aug. 15	Roberts, Lt. H. B.	11 Aug. 16	27 Oct. 15	Attd. With 2-75th Infy.
15 Nov. 15	Heraper, Lt. J. C.	15 Nov. 16	24 Sep. 16	Attd.
12 Sep. 17	Barton, Lt. V. G. J.	12 Sep. 18	....	Attd. With 79th Infy.
31 Aug. 18	Lawrence, 2nd-Lt. E. D.	31 Aug. 18	....	Attd. with 2-75th Infy.
	Attached.			
8 Feb. 02	St. John, Maj. O.	29 June 16	7 Jan. 17	94th Infy.
24 Dec. 14	Hughes, Lt. C. E.	24 Dec. 15	4 Aug. 16	I. A. R. O.
12 Feb. 15	Midgley, Lt. E. A.	10 Nov. 15	7 Jan. 17	I. A. R. O.
11 Dec. 14	Higgin, Lt. H. S.	11 Dec. 15	25 Mar. 16	I. A. R. O.
27 June 17	Burder, 2nd-Lt. M. N.	27 June 17	....	I. A. R. O. (*on probn.*).
22 June 15	Hillman, Lt. A. K. (*Actg. Capt.*)	22 June 16	....	I. A. (*on probn.*).
5 Nov. 17	Taylor, 2nd-Lt. S. T. R.	5 Nov. 17	....	I. A. (*T. C.*)
4 Oct. 18	Rodgers, 2nd-Lt. F. D.	4 Oct. 18	....	I. A. R. O.
29 July 18	Mchaffie, 2nd-Lt. J. P.	29 July 18	....	I. A. R. O.
30 Aug. 18	Roberton, 2nd-Lt. H. E.	30 Aug. 18	....	I. A. R. O.
27 Sep. 18	O'Flynn, 2nd-Lt. A. J.	29 Sep. 18	....	I. A. R. O.

Subadar-Major.

		Jemadar.	Subadar.	Subadar-Major.	
19 July 90	Ramana (51, 66*a*)	1 July 10	15 Apl. 12	20 June 18	
	Abdul Kadir (*Ret.*)	....	....	....	Re-employed, 24 Dec. 15.

Subadars.

2 Mar. 94	Abdul Wahab (56, 58, 66*a*, 105)	1 Oct. 10	16 Oct. 15
July 96	Ibrahim Beg (106)	29 Jan. 15	23 Mar. 17
25 Mar. 04	Viranna	21 Oct. 16	16 Dec. 17
19 Feb. 00	Vanketasami	2 Nov. 16	18 Mar. 18
11 Apl. 97	Munisami Govander	3 Nov. 16	18 Mar. 18
25 Nov. 15	R. Narayan Singh	6 Nov. 15	8 Aug. 18
21 Jan. 08	Y. Samuel	3 Dec. 17	12 Sep. 18
21 Mar. 14	Muhammad Jalaluddin	8 Aug. 17	18 Sep. 18
	Chowriappan	8 Aug. 18	3 Oct. 18
	Munisami	8 Aug. 18	4 Nov. 18

Regt. Centre—Bangalore.
Linked with—67th and 69th Punjabis.

Uniform—Scarlet.
Facings—Emerald Green.

74th Punjabis—*contd.*

1st Commn. or date of entering service.	Names and Rank.	Army rank.	Present appointment in Regiment	Remarks.
	Jemadars.			
5 Feb. 01	Biragi Ram	30 May 14		
5 Jan. 94	Kehru	20 May 15		
16 Nov. 02	Zulfi	25 Sep. 15		
10 July 00	Partap Singh	25 July 16		
5 Sep. 03	Muhammad Khan	3 Apl. 17		
22 Dec. 02	Bhagat Singh	1 July 17		
27 Nov. 02	Walayat Khan	9 Oct. 17		
3 Jan. 03	Rahim Ali	15 Dec. 17		
9 May 06	Rahmat Khan	15 Dec. 17		
30 Sep. 15	Habib Khan	15 Dec. 17		
26 Nov. 02	Fateh Khan	18 Dec. 17		
16 Nov. 02	Ram Singh	18 Dec. 17		
16 Nov. 02	Bhup Singh	18 Dec. 17		
1 Feb. 01	Kirpa Ram	18 Dec. 17		
1 Jan. 13	Ramjidas	25 Jan. 18		
17 Dec. 02	Akbar Ali	8 Feb. 18		
11 Aug. 04	Jalal Khan	16 June 18		
2 Dec. 02	Nadir Khan	13 July 18		
7 Sep. 97	Nadir Khan (*Retd.*)	....		Re-employed 1 Nov. 18.

Regt. Centre—Bangalore.
Linked with—57th & 69th Punjabis.

Uniform—Scarlet.
Facings—Emerald Green.

74th Punjabis.

Wahab-ki-paltan.—Formed at Vellore in 1776, by Capt. Donald Campbell, of drafts from the 5th, 9th and 10th Carnatic Battalions, and was originally styled the 14th Carnatic Battn. Became the 14th Madras Bn., 1784; the 2nd Bn., 6th Regt. of Madras Native Infantry, 1796; the 14th Regt. of Madras N. I., 1824; the 14th Regt. of Madras Infy., 1885; the 14th Madras Infy., 1901; present designation, 1903.

THE DRAGON with the Motto تیار و وفادار "*Ready and True.*"

"Sholinghur" "Carnatic" "Mysore" "Maheidpoor" "China" "Burma, 1885—87."

Composition—2 Companies of Punjabi Musalmans, 1 of Sikhs (other than Jats and Mazbis), 1 of Punjabi Hindus.

COLONEL—Lieutenant-General Francis H. Tyrrell . 1 January 1906.

1st Commn. or date of entering service.	Names and Rank.	Army rank.	Present appointment in Regiment.	Remarks
	Commandant.			
21 Dec. 89	Alexander, Lt.-Col. A. V. . .	21 Dec. 15	13 Nov. 17	
	Company Commanders. (4)			
3 Mar. 97	Oldfield, Maj. G. E. B. (*Actg. Lt.-Col.*)	3 Mar. 15	13 Nov. 17	2nd-in-Comd. with 2-89th Punj.
21 Apl. 00	Ardoino, Maj. J.	1 Sep. 15	9 May 13	
8 May 01	Kunhardt, Maj. F. G.	8 May 16	26 Sep. 15	
18 Jan. 02	Riddell, Maj. A., D.S.O. . . .	18 Jan. 17	26 Sep. 15	With 2-153rd Infy.
	Company Officers. (9)			
21 Jan. 03	Bagnall, Maj. R.	21 Jan. 18	19 Apl. 04	With 1-67th Punj.
20 May 05	Campbell, Capt. C.	20 May 14	9 Aug. 06	
14 Oct. 08	Hamilton, Capt. O. P. . . .	1 Sep. 15	4 Mar. 14	
20 Jan. 12	Gray, Capt. E. H.	20 Jan. 16	15 Mar. 13	
19 Mar. 15	Duguid, Lt. T. C. (*Actg Capt.*)	19 Dec. 16	19 Mar. 15	
19 July 16	McConnell, Lt. J. K.	19 July 17	8 Dec. 16	Attd.
14 Nov. 16	Dobson, Lt. E. T.	14 Nov. 17	. . . .	Attd. with 2-124th Infy.
30 Jan. 17	McCaughey, Lt. F. G. . . .	3 Jan. 18	30 Jan. 17	Attd. with 67th Punj.
18 June 17	Humphreys, Lt. C. E. . . .	18 June 18	. . . .	Attd. with 69th Punj.
18 June 17	Mckernan, Lt. F. P.	18 June 18	. . . .	Attd. with 87th Punj.
5 May 15	Gabb, Lt. J. D.	5 May 16	6 July 17	Attd.
22 Aug. 16	Blanc, Lt. L. G.	22 Aug. 17	18 June 17	Attd.
3 Jan. 16	Mitchley, Lt. G. A.	3 Jan. 17	22 June 17	Attd.
26 May 15	Heywood, Lt. H.C.L. (*Actg.Capt.*)	26 May 16	1 Aug. 17	Attd. Adjt.
19 Jan. 16	Williams, Lt. H. M. (*Actg. Capt.*)	19 Jan. 17	19 Jan. 17	Attd
8 Mar. 16	Whitcombe, Lt. L.C. (*Actg.Capt.*)	8 Mar. 17	1 Aug. 17	Attd.
31 Aug. 18	Gillmon, 2nd-Lt. J. H. . . .	31 Aug. 18	14 Sep. 17	Attd.
	Attached.			
5 Jan. 01	L'Estrange, Maj. E. B. . . .	31 Dec. 15	31 Dec. 17	79th Infy. 2nd-in-Comd. (*temp.*)
20 Feb. 16	Campbell, Lt. W. J. (*Actg Capt.*)	20 Feb. 17	22 July 17	I. A. (*on probn.*)
12 Aug. 16	Dossett, Lt. A. E., M.C. . .	12 Aug. 17	29 Nov. 17	I. A. (*on probn.*)
5 May 17	Abbott, Lt. J. W.	5 May 18	18 Aug. 17	I. A. R. O.
5 May 17	Martin, Lt. A.	5 May 18	18 Aug. 17	I. A. R. O.
21 May 16	Campbell, Lt. J. (*Actg. Capt.*) .	21 May 17	29 Nov. 17	K. O. Sco. Bord., I. A. (*on probn.*)
28 Mar. 18	Calogreedy, 2nd-Lt. L. H. . .	28 Mar. 18	15 July 18	I. A. R. O.
16 Feb. 16	Shallow, Lt. C. R. (*Temp Capt.*)	16 Feb. 17	22 Jan. 18	Essex R., I.A. (*on probn.*)
10 June 15	Brooks, Lt. F. L.	10 Mar. 17	2 Sep. 18	M. G. Corps, I.A. (*on probn.*)
18 Sep. 18	Christal, 2nd-Lt. F. A. . . .	18 Sep. 18	8 July 18	Linc R., I.A. (*on probn.*)
16 Apl. 15	Blake, Lt. A. G.	16 Apl. 16	13 Nov. 18	I. A. R. O.
26 July 18	Irvine, 2nd-Lt. A.	26 July. 18	29 Aug. 18	I. A. R. O.
27 Aug. 18	Morton, 2nd-Lt. E. H. . . .	27 Aug. 18	9 Sep. 18	I. A. R. O.
4 Sep. 18	Pim, 2nd-Lt. G. E.	4 Sep. 18	25 Sep. 18	I. A. R. O.
30 Aug. 18	Henderson, 2nd-Lt. D. R. . .	30 Aug. 18	3 Oct. 18	I. A. R. O.

		Jemadar.	Subadar.	Subadar-Major.
16 Nov. 02	Dost Muhammad Khan, I.O.M. (50, 72*a*)	16 Nov. 02	25 Jan. 10	25 Jan. 18.
	Subadars.			
18 Sep. 88	Mihr Khan (56, 66*a*)	22 Oct. 04	1 Nov. 09	
27 Nov. 02	Baz Khan (56, 66*a*) . . .	27 Nov. 02	16 Dec. 11	
5 Apl. 90	Bhagat Singh (56)	1 Nov. 09	20 May 15	
13 Feb. 04	Wariam Singh (50, 53, 66*a*, 162)	17 June 11	1 Oct. 16	With 67th Punj.
10 May 93	Pala Singh (56, 66*a*)	16 July 11	3 Apl. 17	
1 Feb. 05	Mihr Singh (56)	1 June 12	3 Apl. 17	
29 Jan. 95	Dost Muhammad	29 Apl. 15	1 May 17	
16 Dec. 98	Fateh Muhammad	20 May 15	1 July 17	
1 Jan. 96	Harnam Singh	20 May 15	18 Dec. 17	
1 July 99	Nawab Khan (56)	18 Oct. 16	25 Jan. 18	
28 Sep. 07	Khan Bahadur	3 Apl. 17	25 Jan. 18	
22 May 98	Ghulam Muhammad	1 Oct. 16	16 June 18	
26 Nov. 02	Amir Khan (51)	3 Apl. 17	16 June 18	
3 Feb. 04	Sultan Khan	1 May 17	13 July 18	

Regt. Centre—Trichinopoly. *Uniform*—Scarlet.
Linked with—61st & 64th Pioneers. *Facings*—White.

1st Battalion, 81st Pioneers.

Roberts-ki-paltan.—Formed at Chicacole in 1786, by Capt. G. Roberts, from the Ganjam Sebundy Corps, and drafts from the 11th and 18th Madras Battalions, and was originally designated the 28th Madras Bn. Became the 1st Bn., 11th Regt. of Madras Native Infantry, 1796; the 21st Regt. of Madras N.I., 1824; the 21st Regt. of Madras Infy., 1885; the 21st Regt. of Madras Infy. (Pioneers), 1891; the 21st Madras Pioneers, 1901; present designation, 1903.

"Mysore" "Seringapatam" "Nagpore" "Afghanistan, 1878-80."
"Burma, 1885-87" "Tirah" "Punjab Frontier."

Composition—2 Companies of Tamils, 1 of Madrasi Musalmans, 1 of Paraiyans and Christians.
COLONEL—Major-General *Sir* H. T. Brooking, K.C.B., K.C.M.G.

1st Commn. or date of entering service.	Names and Rank.	Army rank.	Present appointment in Regiment.	Remarks.
	Commandant.			
20 May 93	Bruce, Lt.-Col. G. D., D.S.O., *p.s.c.*	1 Jan. 17	13 Aug. 18	With 2-61st Prs.
	Company Commanders. (4)			
14 Aug. 95	McWatters, Maj. G. A., *q.* (*Actg. Lt.-Col.*)	14 Aug. 13	5 May 17	2nd-in-Comd.
20 July 98	Watson, Maj. H. R., *q.*	1 Sep. 15	14 Nov. 14	With 2-75th Infy.
18 Apl. 00	*Peel, Maj. B. G.*, D.S.O., *q.*	*1 Sep. 15*	*3 Jan. 16*	
8 Jan. 01	Fennell, Maj. A. E. S.	31 Mar. 16	5 May 17	
	Company Officers. (9)			
28 May 01	*Cochran, Bt. Lt.-Col. G. W.*, D.S.O., *q.*	*3 June 18*	*30 Nov. 02*	
28 Jan. 03	Thackwell, Maj. A. C. A., M.C., *q.*	15 June 17	13 Aug. 05	
27 Aug. 02	Alexander, Maj. C., M.C., *q.*	27 Aug. 17	18 Jan. 04	With 2-61st Prs.
29 Aug. 06	*Kemble, Capt. T. A., q.*	*29 Aug. 15*	*26 Sept. 08*	*With 1-128th Prs.*
17 Aug. 07	*Fraser, Capt. S. G. G., q.*	*1 Sep. 15*	*24 Nov. 08*	*With 1-155th Prs.*
2 Jan. 08	Nance, Capt. W. J. *q.*	1 Sep. 15	8 Apl. 09	Adjt., 7 Nov. 17.
11 Aug. 15	Curle, Lt. V. (*Actg. Capt.*)	11 Aug. 16	28 Oct. 15	Attd. With 64th Prs.
20 Oct. 15	Massy, Lt. I. P.	20 Oct. 16	11 Dec. 15	Attd.
29 June 16	Tipping, Lt. R. P. M.	26 Dec. 16	. . . .	Attd. With 1-155th Prs.
29 June 16	Hamilton, Lt. L. S.	29 June 17	. . . .	Attd. With 64th Prs.
19 July 16	Cloran, Lt. D. J. H.	19 July 17	9 Nov. 16	Attd. With 2-128th Prs.
	Attached.			
9 Sep. 15	Breething, Lt. W. J. (*Actg. Capt.*)	9 Sep. 16	. . . .	I. A. (*On probn.*) Qr.-Mr. (*temp.*)
5 May 17	Fraser, Lt. D.	5 May 18	. . . .	I. A. R. O.
20 July 16	Ballantine, Lt. R. W. H. (*Actg. Capt.*)	20 July 17	. . . .	I. A. (*On probn.*)
19 May 15	Roberts, (*temp. Capt.*) C. R.	25 July 17	. . . .	I. A. R. O.
22 July 17	Bregman, 2nd-Lt. T. M.	22 July 17	. . . .	I. A. R. O.
5 May 17	Pearson, Lt. H. G.	5 May 18	. . . .	I. A. R. O.
29 Aug. 17	Bell, Lt. H. C. (*Actg. Capt.*)	29 Aug. 18	6 Dec. 17	I. A. R. O.
15 Jan. 18	Painter, 2nd-Lt. R. F.	15 Jan. 18	. . . .	I. A. (*T. C.*)
15 Jan. 18	Stansfeld, 2nd-Lt. W.	15 Jan. 18	26 Jan. 18	I. A. (*T. C.*)
26 Sep. 17	Craig, 2nd-Lt. J. W.	26 Sep. 17	21 June 18	I. A. R. O. (*On probn.*)
29 Mar. 18	Webb, 2nd-Lt. T.	29 Mar. 18	19 July 18	I. A. (*T. C.*)
30 May 17	Game, Lt. S. W.	30 May 18	5 June 18	I. A. R. O.
1 Aug. 17	Tregaskis, Lt. G.	1 Aug. 18	31 Jan. 18	I. A. R. O.
5 Nov. 17	Roach, Lt. L. J.	5 Nov. 18	25 May 18	I. A. R. O.
5 Nov. 17	Plumpton, Lt. R. R.	5 Nov. 18	2 June 18	I. A. R. O.
18 Jan. 16	Burdon, Lt. R. M., M.C.	18 Jan. 17	10 Nov. 18	R. W. Surr. R., I. A. (*on probn.*)
	Bulley, Lt. H. R.	. . .	11 Nov. 18	R. F. A., I. A. R. O. (*on probn.*)

Subadar-Major.

		Jemadar.	Subadar.	Subadar-Major.
13 Mar. 95	Narayanasami (52, 56, 66*a*)	26 Aug. 09	17 Apl. 15	26 Aug. 16

Subadars

26 Apl. 92	Ramasami (52, 56, 66*a*)	5 May 04	1 Dec. 09
17 Apl. 93	Palayam (52, 56, 66*a*)	1 Dec. 09	1 Feb. 16
7 Sep. 98	Maduraimuttu (56, 66*a*)	24 July 13	18 Feb. 17
15 Apl. 98	Muhammad Abdul Munaf (50, 52)	9 Dec. 14	18 Dec. 17
19 Nov. 02	Mangapati Raju, (51, 52)	9 Dec. 14	18 Dec. 17
5 Feb. 03	Nadesan Pillai (56, 66*a*)	17 Apl. 15	18 Dec. 17
24 May 94	Maduraimuttu Padaiyachi	7 Jan. 15	20 Apl. 18
17 July 00	Shaikh Ahmad (46, 66*a*)	1 Feb. 16	20 Apl. 18
28 Oct. 96	Kondayya	21 Feb. 16	17 Sep. 18

1st Battalion, 81st Pioneers—*contd.*

1st Commn. or date of entering service.	Names and Rank.	Army rank.	Remarks.
	Jemadars.		
17 June 91	*Ratnam (52, 56, 66a)*	*25 June 09*	*With 61st Prs.*
2 Oct. 98	Manikkam (48, 51)	30 June 14	
18 Mar. 99	Joseph (66a)	17 Apl. 15	
18 Mar. 96	Ramasami (51, 66a)	1 July 16	
23 Jan. 03	Kalimuttu Thevar (51, 66)	1 July 16	
24 Dec. 02	Abdul Basid	21 July 16	Indian Adjt., 18 Dec. 17.
2 Jan. 03	Perumal Thevar (56)	1 Feb. 17	
1 Nov. 02	Virrayya	1 Feb. 17	
29 July 99	Ghulam Mohuyid-din (51)	18 Feb. 17	
1 Mar. 03	Zainulabdin (51, 60a)	6 Apl. 17	
23 May 17	Jelis Joseph	23 May 17	(*On probn.*)
10 July 99	Arunachellam	15 Oct. 17	
6 Dec. 93	Appavu (57)	18 Dec. 17	
7 Mar. 03	Abdul Kadir (50, 57)	30 Mar. 18	
19 Apl. 99	Ramasami	20 Apl. 18	
5 Aug. 04	Muhammad Sharif	1 Sep. 18	
20 Feb. 11	Muhammad Jafar Sharif	1 Sep. 18	
29 June 08	Ponnusami Naidu (48a, 50)	1 Sep. 18	
5 Aug. 04	Murrian	1 Sep. 18	

2nd Battalion, 81st Pioneers.

Formed at St. Thomas Mount.

Composition—2 Companies of Musallis, 2 companies Madrassis.

1st Commn. or date of entering service.	Names and Rank.	Army rank.	Present appointment in Regiment.	Remarks.
	Commandant.			
4 Aug. 97	Mackie, Maj. R. (*Actg. Lt. Col.*)	4 Aug. 15	16 Feb. 18	64th Prs. (*temp.*) *Comdg. St. Thomas Mont.*
	Company Commanders. (4)			
28 Aug. 01	Robertson, Maj. C. A.	28 Aug. 16	16 Feb. 18	61st Prs. 2nd-in-Comd. (*temp.*)
15 Mar. 15	Mills, Lt. W. S. (Actg. Capt.)	15 Mar. 16	7 Mar. 18	I. A. R. O.
15 Nov. 15	Hayward, Lt. T. E. (Actg. Capt.)	15 Nov. 16	18 Feb. 18	32nd Prs.
19 Sep. 14	Southgate, Lt. C., M.C. (*Actg. Capt.*)	19 June 16	5 Sep. 18	North'd. Fus., I.A. (*on probn.*)
	Company Officers.			
15 Nov. 15	Shaw, Lt. R. L.	15 Nov. 16	28 Feb. 18	48th Prs.
23 Nov. 15	Morgan, Lt. W. G. C.	23 Nov. 16	21 May 18	S. Wales Bord., I. A. (*on probn.*)
17 Dec. 15	Raven-Hill, Lt. L. E.	17 Dec. 16	20 July 18	Kent Cyclist Bn., I. A. (*on probn.*) Qr.-Mr. (*Offg.*)
19 July 16	Old, Lt. R. E.	19 July 17	21 May 18	Devon. R., I. A. (*on probn.*)
5 May 17	Snelling, Lt. C. W.	5 May 18	10 Sep. 18	Aust'n. I.F., I.A. (*on probn.*)
4 Sep. 16	Kelk, Lt. B. M.	4 Sep. 17	21 May 18	Notts. & Derby. R., I. A. (*on probn.*)
28 Dec. 17	Ralphs, 2nd-Lt. F. G.	28 Dec. 17	7 Sep. 18	York & Lanc. R., I.A. (*on probn.*)
20 June 17	Findlay, Lt. M. A.	20 June 18	7 Mar. 18	I. A. R. O.
27 June 17	Watson, Lt. J. G., M.M. (*Actg. Capt.*)	27 June 18	20 July 18	I. A. R. O. Adjt. (*Offg.*)
29 Aug. 17	Bullocke, Lt. J. G.	29 Aug. 18	26 Apl. 18	I. A. R. O.
15 Jan. 18	Ball, 2nd-Lt. J. W.	15 Jan. 18	3 June 18	I. A. R. O.
29 Mar. 18	Clarke, 2nd-Lt. P. T.	29 Mar. 18	27 July 18	I. A. R. O.
8 Apl. 18	King, 2nd-Lt. E. T.	8 Apl. 18	9 Nov. 18	I. A. R. O.
29 May 18	Kempster, 2nd-Lt. A. C.	29 May 18	3 Oct. 18	North'n. R., I.A. (*on probn.*)
22 May 18	Teale, 2nd-Lt. D. B. S.	22 May 18	6 Oct. 18	I. A. R. O.
2 June 18	Friend, 2nd-Lt. E. A.	2 June 18	6 Oct. 18	I. A. R. O.
10 July 18	Manson, 2nd Lt. W. M.	10 July 18	4 Dec. 18	I. A. (T. C.)

Subadar-Major.

		Jemadar.	Subadar.	Subadar-Major.
22 Mar. 94	Samuel David (52, 56)	1 June 06	13 July 11	26 Feb. 18

Subadars.

30 Jan. 99	Ghulam Hussain	11 Sep. 14	1 Aug. 16
19 June 91	Shaikh Daud (52, 56, 58)	21 Oct. 13	5 May 17
29 Nov. 95	Shaikh Hamid (51, 58)	17 Apl. 15	26 Feb. 18
15 June 99	Isvar Rao	15 May 14	6 Mar. 18
18 Apl. 98	Rangasami	. . . .	6 Mar. 18
18 Dec. 99	Lal Khan	1 Oct. 17	1 Sep. 18
3 Jan. 00	Muhammad Hussain	1 Feb. 18	1 Sep. 18

Jemadars.

10 July 99	Periyasami	26 Feb. 18
3 Jan. 06	Narayansami Nayudu	6 Mar. 18
29 Oct. 88	Ponnusami	11 Mar. 18
1 May 18	Singaravelu	11 May 18
5 Oct. 00	Shaikh Naruddin	24 May 18
16 Sep. 97	Chinnasami	1 June 18
14 July 99	Mirza Naqiulla Beg.	20 June 18
2 Aug. 05	Sayyid Qadir	1 Sep. 18
5 Jan. 08	Fateh Sher	1 Sep. 18

Regt. Centre—Secunderabad. *Uniform*—Scarlet.
Linked with—76th & 87th Punjabis. *Facings*—Emerald Green.

82nd Punjabis.

Dalrymple-ki-paltan.—Raised at Ellore in 1788, by Capt. James Dalrymple, as the 29th Madras Battalion. Became the 2nd Bn. 11th Regt. of Madras Native Infantry in 1796; the 22nd Regt. of Madras N.I. in 1824, the 22nd Regt. of Madras Infy. in 1885; the 22nd Madras Infy. in 1901; present designation in 1903.

"Mysore" "Seringapatam" "Ava."

Composition—2 Companies of Punjabi Musalmans, 1 of Sikhs, 1 of Jats.

1st Commn. or date of entering service.	Names and Rank.	Army rank.	Present appointment in Regiment.	Remarks.
	Commandant.			
4 Mar. 91	Milne, Lt.-Col. J. W.	4 Mar. 17	4 Sep. 18	
	Company Commanders (4)			
15 May 97	Morris, Maj. C. T.	15 May 15	4 Sep. 18	2nd-in-Comd. With 2-102nd Grs.
1 Sep. 98	Davson, Maj. H. J. H.	1 Sep. 15	11 Sep. 14	3-154th Infy.
8 Jan. 01	Rowlandson, Maj. H. W. [L]	8 Jan. 16	24 Feb. 17	
22 Apl. 03	Reynolds, Maj. H. V.	22 Apl. 18	4 Sep. 18	With 105th Infy.
	Company Officers. (9)			
5 Aug. 05	*Reilly, Bt.-Maj. H. L.*, D.S.O.	*29 Oct. 15*	*11 Sep. 07*	
29 Aug. 06	*Wilson, Capt. R. H.*	*3 June 15*	*22 Sep. 08*	*Kurram Militia.*
19 Jan. 07	*Turner, Capt. G. F.* M.C.	*1 Sep. 15*	28 Mar. 08	*Malay States Guides*
3 Sep. 10	Luna, Capt. A. J., M.C.	1 Sept. 15	13 Dec. 11	
24 Aug. 12	*Plowden, Cap. B. E. C.*	*24 Aug. 16*	*19 Oct. 13*	
11 Nov. 14	Herbert, Capt. D. M. A.	11 Nov. 18	15 Mar. 15	
3 Apl. 17	Toovey, Lt. C. W.	3 Apl. 18	15 Aug. 16	Adjt.
22 Dec. 15	Ricketts, Lt. T. C. D.	22 Dec. 16	18 Feb. 17	Attd.
29 June 16	Badman, Lt. F. H. (*Actg. Capt.*)	29 June 17	18 Feb. 17	Attd.
15 Nov. 15	Salvesen, Lt. F. M. R.	15 Nov. 16	19 Nov. 15	Attd. With 1-142nd Infy.
30 Jan. 17	Blakeney, Lt. S. R. St. L.	30 Jan. 18	5 Sep. 17	Attd.
20 Feb. 17	Gibb, Lt. F. G.	20 Feb. 18	22 Feb. 18	Attd.
11 Apl. 17	Haylock, Lt. S. W.	11 Apl. 18	29 Nov. 17	Attd. With 1-153rd Infy.
12 Sep. 17	Duke, Lt. A. L. R.	12 Sep. 18	15 Dec. 17	Attd. With 1-153rd Infy.
4 Dec. 15	O'Collaghan, Lt. C. H. D.	4 Dec. 16	1 July 18	Attd.
	Attached.			
17 Jan. 11	Wallace, Capt. A. J. L. (*Actg. Maj.*)	1 Sep. 15	22 June 18	52nd Sikhs.
12 Aug. 14	Taylor, Lt. J. N.	12 May 16	28 July 18	Suff. R., I. A. (*on probn.*)
12 Sep. 14	Dallas, Capt. H. W.	12 June 16	23 Dec. 17	I. A. (*on probn.*)
30 Apl. 15	Hamilton, Lt. R. (*Actg. Capt.*)	30 Apl. 16	24 June 15	I. A. R. O.
20 July 15	Hocken, Lt. N. A.	20 July 16	15 Aug. 15	I. A. R. O. Qr. Mr.
11 Oct. 15	McLaurin, Lt. D. J.	11 Oct. 16	23 Aug. 18	I. A. R. O. (*on probn.*)
27 Dec. 15	Kerr, Lt. N.	27 Dec. 16	23 July 18	R. Sc. Fus., I. A. (*on probn.*)
	Ryan, Lt. E. J.	. . . .	31 July 18	I. A. (*on probn.*)
12 May 16	Warmington, Lt. W. E.	12 May 17	22 July 18	I. A. (*on probn.*)
	Robinson, Capt. A. M.	. . .	9 Nov. 18	E. Lan. R., I. A. R. O. (*on probn.*)
5 Jan. 17	Walker, Lt. H. L.	5 Jan. 18	1 July 18	I. A. R. O.
5 May 17	Beauchamp, Lt. K. G.	5 May 18	15 July 18	I. A. R. O.
20 June 17	Willard, Lt. F. W.	20 June 18	1 July 18	I. A. R. O.
29 Aug. 17	Hay, Lt. R. W. M.	29 Aug. 18	29 Nov. 18	I. A. R. O.
19 Sep. 17	Arnold, Lt. L. F. C., M.C.	19 Sep. 18	15 Aug. 18	Lond. R., I. A. (*on probn*)
31 July 18	Goffin, 2nd-Lt. R. C.	31 July 18	2 Sep. 18	I. A. R. O.
20 Sep. 18	Stevens, 2nd-Lt. C. W.	20 Sep. 18	27 Oct. 18	I. A. R. O.
7 Oct. 18	Taylor, 2nd-Lt. E. R. L.	7 Oct. 18	20 Oct. 18	I. A. R. O.
7 Oct. 18	Hitchin, 2nd-Lt. F. E.	7 Oct. 18	25 Oct. 18	I. A. R. O.

	Subadar-Major.	Jemadar.	Subadar.	Subadar-Major.	
15 Mar. 89	Afzal Khan, *Bahadur* (56)	11 June 02	16 May 03	1 May 12	*Order of Br. Ind.*, [2nd Cl.
24 May 87	Ram Singh, *Bahadur*, I.O.M. (56, 72a)	16 May 03	16 July 07	21 July 16	*Order of Br. Ind.*, *2nd Cl.*
	Subadars.				
18 June 87	Mastan Singh	16 July 07	1 May 12		
29 May 97	Ram-bakhsh	22 Oct. 08	1 Nov. 13		With 1-153rd Infy.
23 July 96	Ghulam Muhiyuddin, I.D.S.M. (56)	7 Jan. 10	5 Jan. 15		
13 Apl. 98	Harnand (56)	1 Nov. 13	1 Feb. 15		
2 Aug. 91	Phuman Singh	1 May 12	6 Apl. 15		
17 July 98	Khyali Ram	1 July 14	15 Apl. 15		
28 Aug. 03	Fazl Rahman	5 Jan. 15	29 Apl. 15		
22 Feb. 98	*Haidar Khan*	*5 Jan. 15*	*17 Dec. 15*		*Instr. Sch. of N.C.O., Belgaum.*
24 Dec. 98	Hira Singh, I.D.S.M.	5 Jan. 15	17 Dec. 15		
5 Aug. 03	Kishena Ram	15 Apl. 15	10 May 16		With 1-153rd Infy.
29 May 97	Amir Khan	6 Apl. 15	12 Dec. 16		
20 Jan. 03	Muhammad Saiyid *Bahadur*	9 Mar. 16	3 Feb. 17		*Order of Br. Ind., 2nd Cl.*
1 June 96	Nawaz Ali	1 June 16	9 Feb. 17		With 1-153rd Infy.
12 Jan. 00	Kirpal Singh (56)	6 Apl. 15	21 May 17		
12 Apl. 98	Mastan Singh (II)	11 July 16	13 Feb. 17		
9 Nov. 03	Fazal Khan	29 Apl. 15	25 Feb. 17		
11 June 04	Bagh Ali	15 July 16	12 Oct. 17		
13 June 00	Partab Singh	6 Apl. 15	8 Feb. 18		
20 Jan. 04	Tota Ram	1 May 15	18 May 18		
	Sukh Ram	. . . .	18 Sep. 18		
20 June 03	Shah Zaman	25 Jan. 17	18 Sep. 18		

82nd Punjabis—*contd.*

1st Commn. or date of entering service.	Names and Rank.	Army Rank.	Remarks.
	Jemadars.		
3 Dec. 95	Khan Muhammad	29 Apl. 15	
4 July 03	Zain Khan	1 June 15	
18 June 00	Kehr Singh	15 July 15	
13 Aug. 02	*Bachan Singh*	*18 July 15*	*Instr. Sch. of N.C.O., Belgaum.*
1 July 03	Sant Singh	18 July 15	
4 Apl. 97	Masum Shah	17 Dec. 15	
30 May 04	Kheta Ram	28 Mar. 16	Indian Adjt., 1 July 17.
1 July 03	Gujar Singh	26 Aug. 16	
6 Nov. 03	Sewa Singh	2 Sep. 16	
20 Oct. 03	Ali Asghar	1 Dec. 16	
12 Dec. 03	Mir Afzal (58)	1 Jan. 17	With 1-155th Infy.
20 June 03	Gheba Khan	27 Jan. 17	
19 Sep. 03	Firoz Khan	9 Feb. 17	
1 Feb. 18	Abdul Hamid	1 Feb. 18	(*On probn.*)
31 Aug. 03	Chatru	1 Mar. 18	
7 May 04	Abdul Qayum	9 Mar. 18	
1 Jan. 06	Nawab Khan	15 Mar. 18	
20 Apl. 05	Mir Dad	18 May 18	
4 Dec. 05	Faiz Talib	20 Aug. 18	
8 Dec. 03	Phusa Ram	20 Aug. 18	
26 Sep. 03	Aziz Khan	2 Sep. 18	
25 Aug. 04	Indar Singh	18 Sep. 18	
31 May 07	Abdul Khai	18 Sep. 18	
10 May 10	Ramji Lal	18 Sep. 18	

Regt. Centre—Trichinopoly. *Uniform*—Scarlet.
Linked with—63rd L. Infantry. *Facings*—Emerald Green.

83rd Wallajahbad Light Infantry.

Tolfrey-ki-paltan.—Raised at Madras in 1794, by Capt. E. Tolfrey, as the 33rd Bn. of Madras Native Infantry. Became the 1st Bn., 12th Regt. of Madras N. I., 1797; the 1st Battn., 12th Regt. of Madras N. I., or Wallajahbad Lt. Infy., 1812; the 23rd Regt. of Madras N. I., or Wallajahbad Lt. Infy., 1824; the 23rd (Wallajahbad) Regt. of Madras (Light) Infy., 1885; the 23rd (Wallajahbad) Madras Lt. Infy., 1901; present designation, 1903

"Now or Never."

"Seringapatam" "Nagpore" "Burma, 1885-87."

Composition—2 companies of Madrasi Musalmans, 1 of Tamils, 1 of Paraiyans and Christians.

1st Commn. or date of entering service.	Names and Rank.	Army rank.	Present appointment in Regiment.	Remarks.
	Commandant.			
14 Sep. 87	Keays, Lt.-Col. R. W. C.	14 Sep. 13	2 Dec. 14	
	Company Commanders. (4)			
28 July 00	LeMessurier, Maj. G. D. S., *q*	1 Sept. 15	25 Feb. 13	
4 May 01	Britton, Maj. J. A. H.	4 May 16	17 Apl. 14	With 2-88th Infy.
22 Oct. 02	Martin, Maj. P. H.	22 Oct. 17	9 Aug. 17	
	Company Officers. (5)			
29 Aug. 06	Proctor, Capt. G. N.	29 Aug. 15	28 Dec. 10	With 79th Infy.
15 Aug. 14	Greenhouse, Capt. R. E.	15 Aug. 18	14 Oct. 14	
4 Aug. 15	Talbot, Lt. W. E. H.	4 Aug. 16	27 Oct. 15	
20 Oct. 15	Steer, Lt. J. M. B.	20 Oct. 16	10 Dec. 15	Adjt.
29 June 16	Robinson, Lt. J.A.	29 June 17	6 July 16	Attd.
7 Jan. 16	Grime, Lt. H. R.	7 Jan. 17	22 May 17	Attd.
19 July 16	Lillies, Lt. G. L.	19 July 17	15 Nov. 16	Attd. with 2-23rd Sikh Prs.
18 June 17	Heggart, Lt. J. E.	18 June 18	24 June 17	Attd.
12 Sep. 17	Dicker, Lt. B. P.	12 Sep. 18	13 Dec. 17	Attd.
15 May 17	Leonard, 2nd-Lt. P. J.	5 Feb. 18	9 Mar. 18	Attd.
	Attached.			
22 Jan. 15	Roy, Lt. R. M.	22 Jan. 16	22 Mar. 15	1-73rd Infy.
18 June 15	Williamson, Lt. P. R. C.	18 June 16	4 Oct. 16	I. A. R. O.
5 Jan. 17	Robinson, Lt. J. C. (*Actg. Capt.*)	5 Jan. 18	27 Mar. 17	I. A. R. O.
5 Jan. 17	Cooper, Lt. H. R.	5 Jan. 18	27 Mar. 17	I. A. R. O.
5 May 17	Cronan, Lt. L. B.	5 May 18	9 Nov. 17	I. A. R. O.
21 Aug. 16	Clapp, Lt. W. M.	21 Aug. 17	10 Oct. 16	I. A. R. O.
8 Aug. 17	Weatherhead, Lt. L. D.	8 Aug. 18	18 Feb. 18	I. A. R. O.
12 Nov. 17	Gruer, 2nd-Lt. H. G.	12 Nov. 17	8 June 18	I. A. R. O. S. S. O., Cannanor.
11 Aug. 16	Gibson, Lt. D. R.	11 Aug. 17	11 July 18	R. Berks. R., I. A. (*on probn.*)
12 July 18	Latham, 2nd-Lt. C. R. C.	12 July 18	14 Aug. 18	I. A. R. O.
15 Aug. 18	Hickie, 2nd-Lt. J. M.	15 Aug. 18	18 Sep. 18	I. A. R. O.
16 Sep. 18	Wells-Smith, 2nd-Lt. F. T.	16 Sep. 18	9 Oct. 18	I. A. R. O.
21 Oct. 18	Somerville, 2nd-Lt. T.	21 Oct. 18	17 Nov. 18	I. A. R. O.

Subadar-Major.

		Jemadar.	Subadar.	Subadar-Major.
20 June 88	Madurai (56)	19 Dec. 09	1 Feb. 12	15 Sep. 18.

Subadars.

9 Oct. 88	Saiyid Kadir (56)	16 May 08	9 Feb. 13
13 Dec. 90	Abdul Wahab (56)	28 Sep. 09	18 Aug. 13
1 Aug. 95	Narasimhulu (56, 66*a*)	12 Nov. 10	10 Dec. 14
20 Oct. 96	Velamurgam (56, 58)	1 Feb. 12	15 Sep. 15
11 Mar. 94	Ellappah (56)	16 Oct. 14	1 Nov. 16
28 Oct. 97	Sourinayagam (56, 66*a*)	22 May 10	12 Mar. 17
19 Mar. 95	Saiyed Mohi-ud-din (56, 66*a*)	16 Oct. 14	12 Mar. 17
10 Apl. 99	Shaikh Daud (56)	18 Nov. 14	11 Apl. 17
28 May 08	Patchappan (52, 56)	7 Dec. 15	24 July 17
26 Oct. 14	Duraisami	7 Dec. 15	31 July 17
29 Sep. 14	Govinderajalu (46)	22 Sep. 16	16 Dec. 17
23 June 09	Subrayan	12 Mar. 17	26 May 18
25 Apl. 11	Munisami Raju (57, 56)	22 Sep. 16	26 May 18

83rd Wallajahbad Light Infantry—*contd.*

1st Commn. or date of entering service.	Names and Rank.	Army rank.	Remarks.
	Jemadars.		
16 Jan. 95	Abdul Azim (56)	1 Aug. 13	
21 Jan. 95	Abdul Rahim (56)	16 Oct. 14	
2 Nov. 04	Y. Issac	2 Sep. 16	Indian Adjt.
23 June 97	Abdul Sattar	12 Mar. 17	
17 May 01	Kasturi (56)	11 Apl. 17	
3 May 10	Sadasiva Naidu	2 Aug. 17	
19 Sep. 06	Muthusami (56)	16 Dec. 17	
18 Nov. 14	Ramannah	26 May 18	

Regt. Centre.—Secunderabad.
Linked with—62nd & 66th Punjabis.

Uniform—Scarlet.
Facings—Emerald Green.

84th Punjabis.

Macdonald-ki-paltan.—Raised at Vellore in 1794, by Capt. D. Macdonald, as the 34th Bn. of Madras Native Infantry. Became the 2nd Bn., 12th Regt. of Madras N. I. in 1797; the 24th Regt. of Madras N. I. in 1824, the 24th Regt. of Madras Infy. in 1885; the 24th Madras Infy. in 1901; present designation in 1903.

THE ELEPHANT.

"Seringapatam" "Assaye" "Bourbon."

Composition—2 companies of Punjabi Musalmans, 1 of Sikhs, 1 of Rajputs from Western Rajputana and the Eastern Punjab.

1st Commn. or date of entering service.	Names and Rank.	Army rank.	Present appointment in Regiment.	Remarks.
	Commandant.			
29 Nov. 90	Crawford, Lt.-Col. A. G.	29 Nov. 16	23 Apl. 18	With 2-91st Punj.
	Company Commanders. (4)			
5 Sep. 96	Tyrer, Maj. R.	5 Sep. 14	23 Apl. 18	2nd-in-Comd.
28 July 00	Harper, Maj. A. F.	1 Sep. 15	5 May 16	Attd. G. S. Br., A. Hdqrs.
8 May 01	*Brown, Maj. C. A.*	*8 May 16*	*11 Mar. 17*	*I. S. Troops.*
30 Apl. 02	Monck-Mason, Maj. G. N. G.	24 Dec. 17	23 Apl. 18	
	Company Officers. (9)			
21 Jan. 03	Hickie, Maj. H. W.	21 Jan. 18	8 Apl. 04	
23 Dec. 03	*Blanford, Capt. P. T.*, M.C.	*23 Dec. 12*	*30 Mar. 09*	*Malay States Guides.*
8 Sep. 09	Perram, Capt. H. C.	1 Sep. 15	26 Jan. 11	With 66th Punj.
8 Aug. 14	Palmer, Capt. S. V.	8 Aug. 18	13 Oct. 14	*Lv., ex I., m.c.*
1 Oct. 14	Light, Capt. H.	1 Oct. 18	16 Dec. 14	
15 Nov. 15	Cant, Lt. E. D.	15 Nov. 16	20 Nov. 15	Attd. With 2-152nd Punj.
29 June 16	James, Lt. H. O. B.	29 June 17	. . . .	Attd.
19 July 16	Cassell, Lt. A. D.	19 July 17	16 Nov. 16	Attd. Qr.-Mr. (*Offg.*)
19 July 16	Drake, Lt. J. A. (*Actg. Capt.*)	19 July 17	16 Nov. 16	Attd.
19 July 16	Jones, Lt. C. R. (*Actg. Capt.*)	19 July 17	16 Nov. 16	Attd. Adjt.
20 Dec. 17	Kearns, 2nd-Lt. R. W.	20 Dec. 17	20 Mar. 18	Attd.
12 Sep. 17	Deakin, Lt. J. S.	12 Sep. 18	9 Dec. 17	Attd.
	Attached.			
28 Mar. 00	Lawrence, Maj. H. R.	1 Sep. 15	6 Mar. 18	Supy. List.
21 May 15	Lawrence, Lt. J. A.	21 Feb. 17	29 Nov. 17	I. A. (*on probn.*)
27 Jan. 16	Hely, Lt. F. F.	27 Jan. 17	1 Apl. 16	I. A. R. O.
1 July 16	Maurice, Lt. J. O. F. (*Actg. Capt*).	1 July 17	9 Sep. 16	I. A. R. O.
10 Nov. 16	Cleary, Lt. C. J.	10 Nov. 17	17 Feb. 17	I. A. R. O.
28 June 16	Carden, Lt. F. G.	28 June 17	19 Aug. 18	R. Suss. R., I. A. (*on probn*).
28 Apl. 17	Boughton, 2nd-Lt. E. J.	28 Jan. 18	31 July 18	Manitoba R., I. A. (*on probn.*)
26 Jan. 18	Wards, 2nd-Lt. G. T.	26 Jan. 18	29 Mar. 18	I. A. (*on probn.*)
22 Aug. 17	Lees, Lt. L.	22 Aug. 18	29 Nov. 17	I. A. (*on probn.*)
28 Dec. 17	Powell, 2nd-Lt. E. H.	28 Dec. 17	4 July 18	R. Ir. Fus., I. A. (*on probn.*)
15 June 18	Linberry-Seer, 2nd-Lt. G. W.	15 June 18	8 July 18	I. A. R. O.
7 Oct. 18	Foster, 2nd-Lt. E. C. H.	7 Oct. 18	26 Oct. 18	I. A. R. O.

Subadar-Major.

		Jemadar.	Subadar.	Subadar-Major.
4 Apl. 89	Habib Shah (56)	16 May 02	1 Oct. 02	6 July 16

Subadars.

30 May 92	Hari Singh (56, 63)	16 May 02	6 Jan. 06	
20 July 02	Muhammad Zaman (56)	20 July 02	27 June 09	
15 Oct. 02	Devi Singh (56)	4 Feb. 12	17 Feb. 15	
9 Jan. 86	Firoz Khan, I.D.S.M. (56)	1 July 05	10 May 15	
22 Jan. 03	Ram Singh	17 Feb. 15	29 Aug. 15	
17 Dec 97	Harnam Singh	17 May 15	26 Oct. 15	
14 July 00	Dasaundha Singh	27 Mar. 15	9 Mar. 16	With 62nd Punj.
20 Dec. 04	Amar Singh	21 Apl. 17	7 Nov. 17	
2 Oct. 03	Abdul Khan	27 Apl. 15	18 May 18	
1 Mar 97	Fateh Muhammad	21 Feb. 18	23 May 18	
2 Apl. 05	Khanzada (48a, 56)	17 Jan. 15	16 July 18	
27 May 02	Partab Singh	7 Nov. 17	26 Sep. 18	

84th Punjabis—*contd.*

1st Commn. or date of entering service.	Names and Rank.	Army rank	Remarks.
	Jemadars.		
30 Dec. 03	Dadan Khan (56)	6 Apl. 15	Indian Adjt., 23 Mar. 17.
14 July 02	Said Ghulam Khan (56) . . .	12 Apl. 15	
1 June 93	Arjan Singh	21 Apl. 15	
28 June 02	*Muhammad Yakub Khan* . .	*10 May 15*	*With Somaliland Contingent.*
13 July 02	Khani Zaman	17 May 15	
12 Oct. 99	Gopal Singh	29 Aug. 15	
2 June 02	Sajawal Khan	29 Aug. 15	
3 Jan. 03	Kundan Singh	26 Oct. 15	
2 Jan. 93	Ganga Singh	27 Nov. 15	
4 Jan. 91	*Tara Singh*	*15 Oct. 16*	
July 02	*Firoz Khan*	*15 Oct. 16*	
1 Nov. 02	Ram Rakku Singh	18 May 18	
4 July 02	Bahadur Ali	18 May 18	
19 Oct. 12	Muhammad Akram	14 Aug. 17	
13 July 02	Bostan Khan	19 July 18	
6 July 02	Kishan Singh	12 Sep. 18	
27 Apl. 06	Yakub Khan	12 Sep. 18	
21 Aug. 02	Ghulam Haidar	25 Sep. 18	

85th Burman Rifles.

Formed at Mandalay on July 1917.

Composition—1 Coy. of Kachins, ½ Coy. of Burmans, ½ Coy. of Karens, ½ Coy. of Sikhs, ½ Coy. of Punjab Mahomedans, 1 Coy. of Gurkhas.

1st Commn. or date of entering service.	Names and Rank	Army Rank.	Present appointment in Regiment.	Remarks.
	Commandant.			
12 Aug. 99	Burd, Maj. E. (*Actg. Lt.-Col.*)	1 Sep. 15	28 July 17	93rd Infy.
	Company Commanders. (4)			
4 Dec. 01	Morris, Maj. C. E. (*Actg. Lt.-Col.*)	4 Dec. 16	1 Sep. 17	Guides, 2nd-in-Comd. (*temp.*)
19 Aug. 03	Richardson, Maj. G. G.	19 Aug. 18	20 June 18	105th Infy.
10 Oct. 03	Enriquez, Capt. C. M. D. (*Actg. Maj.*)	10 Oct. 12	1 Sep. 17	1-21st Punj.
4 Jan. 04	Gould, Capt. S. C.	9 Jan. 13	26 July 17	Civil, Burma.
	Company Officers.			
24 Jan. 06	Pughe, Capt. O. L.	24 Jan. 15	20 Aug. 17	1-72nd Punj.
26 Feb. 15	Gray, Lt. J. W., M.C.	26 Feb. 16	7 Oct. 17	I. A. R. O.
9 July 15	Wyatt, Lt. R. M.	9 July 16	22 July 17	I. A. R. O.
6 Aug. 15	Moore, Lt. W. W. T.	6 Aug. 16	24 July 17	I. A. R. O.
8 Dec. 14	Jones, Lt. P. P.	9 Aug. 16	10 Feb. 18	
10 June 16	Glen, Lt. R. J. G.	10 June 17	....	I. A. R. O.
2 Aug. 16	Pollen, Lt. H. S.	2 Aug. 17	17 July 17	I. A. R. O.
7 Aug. 17	Green, 2nd-Lt. J. H.	7 Aug. 17	....	I. A. (*on probn.*)
10 Aug. 16	Cameron, Lt. G. St. O.	10 Aug. 17	1 Aug. 17	I. A. R. O. Qr.-Mr. (*offg.*)
5 Dec. 15	Hurst, Lt. V. de la M.	5 Dec. 16	11 Dec 17	I. A. R. O.
5 Jan. 17	Ford, Lt. F. O. N. (*Actg. Capt.*)	5 Jan. 18	....	I. A. R O. Adjt.
5 Nov. 17	Dickson, Lt. W. C.	5 Nov. 18	14 Aug. 18	I. A. R. O.
5 May 17	Weir, Lt. G.	5 May 18	10 Feb. 18	I. A. (*on probn.*)
31 Jan. 18	Howman, 2nd-Lt. R. C.	31 Jan. 18	12 Feb. 18	Attd.
31 Jan. 18	Bullard, 2nd-Lt. H.	31 Jan. 18	12 Feb. 18	Attd.
30 July 18	Rawlings, 2nd-Lt. D. A. R.	30 July 18	19 Aug. 18	I. A. R. O.
9 Aug. 18	Havelock, 2nd-Lt. R. A. S.	9 Aug. 18	14 Aug. 18	I. A R. O.
14 Aug. 18	Wright, 2nd-Lt. F. A.	14 Aug. 18	10 Sep. 18	I. A. R. O.

Subadar-Major.

		Jemadar.	Subadar.	Subadar-Major.
9 July 91	Kulman, Bhajoo	25 Oct. 10	1 July 17	20 July 17

Subadars.

7 June 88	Ghulam Habin	25 Sep. 05	1 July 12
20 Apl. 88	Pritam Singh	11 Oct. 11	4 Aug. 18
12 Nov. 00	Chinpaw Gam	26 June 13	1 Apl. 15
20 Dec. 93	Mitt Singh, Sirdar Bahadur, I.O.M.	20 Jan. 15	1 July 17
8 Oct. 93	Gopal Singh Negi	26 Mar. 12	1 July 17
15 Apl. 90	Zaman Shah	23 Feb. 16	1 Aug. 17
2 Dec. 05	Hpan Lu La	21 Mar. 17	1 Aug. 17
30 Oct. 14	Puna Singh Lama	21 Nov. 14	1 Aug. 17

Jemadars.

1 Sep. 92	Bogh Singh	21 Mar. 18
31 July 00	Samdu La	1 Dec. 14
1 Jan. 03	Sau Mun	22 May 18
29 May 07	Kumja Naw	1 Aug. 15
26 July 95	Karnabir Gurung	1 Nov. 15
1 Apl. 03	Hari Singh	16 June 16
10 Sep. 95	Makum Lall Limbu	26 May 18
9 Sep. 91	Amir Singh	31 July 18
9 Oct. 99	Mangal Singh	3 May 18
3 Feb. 96	Nek Alam	15 Sep. 16
23 July 94	Rur Singh	1 Oct. 16
1 Aug. 03	Sawan Tang	1 Jan. 17
31 Oct. 08	Sau Tu	21 Mar. 17
17 Feb. 04	Kolu Gam	1 Apl. 17
2 May 96	Manrath Limbu	21 Apl. 17
3 Apl. 99	Bajbir Limbu	30 Nov. 17
12 Feb. 06	Maung Tit	16 July 17
1 June 10	Nandabir Chetri	1 Aug. 17
23 Mar. 18	Ibrahim Khan	10 Aug. 18
11 May 08	Mitung Tang	29 July 17
30 Aug. 05	Kehar Singh	3 Dec. 17
17 Feb. 00	Dhanbir Gurung	21 Feb. 18

Regt. Centre—Secunderabad. *Uniform*—Scarlet.
Linked with—75th & 79th Carnatic Infy. *Facings*—Emerald Green.

86th Carnatic Infantry.

Innis-ki-paltan.—Raised at Tanjore in 1794, by Capt. James Innes, as the 36th Battalion of Madras Native Infantry. Became the 2nd Bn., 13th Regt. of Madras N. I. in 1798; the 26th Regt. of Madras N. I. in 1824; the 26th Regt. of Madras Infy. in 1885; the 26th Madras Infy. in 1901; present designation in 1903.

"Nagpore" "Kemendine" "Ava" "Pegu" "Burma, 1885-87."

Composition—2 Companies of Madrasi Musalmans, 1 of Tamils, 1 of Paraiyans and Christians.

COLONEL—Brigadier-General Stuart E. Rolland . . . . 15 August 1913.

1st Commn. or date of entering service.	Names and Rank.	Army rank.	Present appointment in Regiment.	Remarks.
	Commandant.			
3 May 90	O'Reilly, Lt.-Col. C. M. . . .	3 May 16	24 May 17	
	Company Commanders. (4)			
20 July 18	Lawrenson, Maj. W. R. (*Actg. Lt.-Col.*)	1 Sep. 15	31 July 18	2nd-in Comd With 2-73rd Infy.
28 July 00	Patrickson, Maj. D. M. . . .	1 Sep. 15	19 May 15	With 119th Infy.
26 June 01	Frere, Maj. A. G.	26 June 16	25 May 17	With 2-150th Infy.
4 Dec. 01	*Berkeley, Maj. R. E. G. q.* . .	*4 Dec. 16*	*31 July 18*	*Lv., ex I., m. c.*
	Company Officers. (5)			
21 Jan. 03	Leman, Maj. F. B. . . .	21 Jan. 18	25 Apl. 04	
29 Nov. 05	*Kingdon, Capt. J.*	*29 Nov. 14*	*4 Mar. 09*	*Bde.-Maj., 59th Bde.*
15 Sep. 15	Barrow, Lt. R.	15 Sep. 16	27 Oct. 15	Attd.
15 Nov. 15	Spaull, Lt. C. M.	15 Nov. 16	30 Nov. 15	Do.
19 Jan. 16	Manning, Lt. J. (*temp. Capt.*) .	19 Jan. 17	. . . .	Attd.
7 Sep. 15	Farbrother, Lt. A. A. . . .	7 Sep. 16	29 Aug. 17	
9 Sep. 16	Baker, Lt. H. L. (*Actg. Capt.*) .	9 Sep. 17	22 Dec. 17	Attd. Adjt. (*offg.*)
	Attached.			
15 May 16	Hockaday, Lt. P. H.	5 May 16	. . . .	Mon. R., I. A. (*on probn.*)
23 Nov. 15	Kelly, Lt. P. I. (*Actg. Capt.*) . .	23 Nov. 16	9 Mar. 18	Leins. R., I. A. (*on probn.*)
21 Aug. 16	Bailey, Lt. J. H.	21 Aug. 17	. . . .	York & Lanc. R., I. A. (*on probn.*) Qr.-Mr. (*offg.*)
10 Mar. 16	Bratton, Lt. V. W.	10 Mar. 17	20 Aug. 17	R. War. R., I. A. (*on probn.*)
13 June 16	Brickell, Lt. H. J.	13 June 17	22 Dec. 17	I. A. R. O. [*Bhamo.*
26 June 17	Wayte, Lt. V. W.	26 June 18	22 June 18	Manch. R., I.A. (*on probn.*) *S.S.O.*,
5 May 17	Hoskins, Lt. G. D.	5 May 18	. . . .	I. A. R. O.
5 May 17	Gilmore, Lt. A. L.	5 May 18	. . . .	I. A. R. O.
5 May 17	Bateman, Lt. F. W.	5 May 18	31 Aug. 17	I. A. R. O.
20 June 17	La-Bouchardiere, Lt. M.P. . .	20 June 18	12 Oct. 17	I. A. R. O.
15 Jan. 18	Felton, 2nd-Lt. G. H. . . .	15 Jan. 18	21 May 18	I. A. (*T. C.*) *S.S.O., Shwebo.*
19 Mar. 18	Gibbs, 2nd-Lt. A. E. T. . . .	19 Mar. 18	25 July 18	I. A. R. O.
28 Aug. 18	Miller, 2nd-Lt. A. D. . . .	28 Aug. 18	19 Nov. 18	Bord. R., I. A. (*on probn.*)
28 Aug. 18	Sisley, 2nd-Lt. F. W. . . .	28 Aug. 18	. . . .	Essex R., I. A. (*on probn.*)
24 May 18	Anderson, 2nd-Lt. F. G. . . .	24 May 18	. . . .	I. A. (U. L.)
7 June 18	James, 2nd-Lt. H. P. . . .	7 June 18	. . . .	I. A. (*T. C.*)
3 June 18	Lewis, 2nd-Lt. R. N.	3 June 18	. . . .	I. A. R. O.

		Jemadar.	Subadar.	Subadar-Major.
	Subadar-Major.			
2 July 01	Shaikh Abdul Kadir (56, 66*a*) .	22 July 01	10 Apl. 06	1 July 17
	Subadars.			
29 June 04	Muhammad Inayat Ullah Khan (52, 56)	29 June 04	21 June 11	
25 Mar. 95	V. G. Perumal (56, 66*a*) . . .	19 June 10	12 Jan. 16	
5 Nov. 95	Abdul Majid Khan (52, 56, 66*a*)	1 June 10	8 June 16	
1 June 93	Govindasami (52, 56, 66*a*) . .	7 Apl. 11	1 Sep. 16	
11 May 98	Ponnusami (56, 66*a*)	20 Feb. 13	26 Sep. 16	
20 May 97	Vellinayakam (56, 60*a*, 66*a*) .	1 Mar. 15	6 Dec. 16	
1 Dec. 98	Ranganayakulu (46, 56, 60*a*) . .	18 June 16	27 July 17	
28 Mar. 99	Abdul Aziz Khan (52, 53, 56, 66*a*)	1 May 16	1 Sep. 17	
10 Feb. 97	Sayyid Ahmad (56, 58, 66*a*) . .	1 Sep. 16	18 Sep. 17	
	Jemadars.			
5 Aug. 97	Samuel, P. (56, 66*a*)	12 Jan. 16		
23 Feb. 97	Sundar Raju	26 Sep. 16		
5 Apl. 99	Ali Raza (56, 66*a*)	29 Sep. 16		
16 Jan. 96	Narainsami	29 Sep. 16		
2 Mar. 00	Ahmad Hussain Khan (56) . .	1 Nov. 16		
6 July 99	Madurai (56)	6 Dec. 16		
1 Aug. 02	Tiruvangadam (56)	24 July 17		
5 May 04	Abdul Subhan Khan (48*a*, 52, 56, 66*a*)	1 Sep. 17	Indian Adjt.	
25 July 03	Venkatachalam	18 Sep. 17		
5 July 04	Govindarajulu (46, 56) . . .	18 Sep. 17		
5 July 04	Saiyid Sikander (58)	1 Mar. 18		
15 Apl. 09	Muhammad Yunus	1 Apl. 18	Indian Qr. Mr.	
	Tandavarayan	1 Sep. 18		
	Muhammad Jafar Husain . .	1 Sep. 18		
	Yusuf	1 Sep. 18		
	Muthuviru	1 Sep. 18		

Regt. Centre—Secunderabad.
Linked with—76th & 82nd Punjabis.

Uniform—Scarlet.
Facings—Emerald Green.

87th Punjabis.

Lindsay-ki-paltan.—Raised at Trichinopoly in 1798, by Major A. Lindsay, as the 1st extra Battalion of Madras Native Infantry. Became the 1st Bn., 14th Regt. of Madras Native Infantry the same year; the 27th Regt. of Madras Native Infantry in 1824; the 27th Regt. of Madras Infy. in 1885; the 27th Madras Infy. in 1901; present designation in 1903.

"Maheidpoor" "Lucknow" "Burma, 1885-87."

Composition—2 Companies of Punjabi Musalmans, 1 of Sikhs, 1 of Jats.

1st Commn. or date of entering service.	Names and Rank.	Army rank.	Present appointment in Regiment.	Remarks.
	Commandant.			
28 June 90	Prevost, Lt.-Col. G. H.	22 Apl. 16	22 Apl. 16	
	Company Commanders. (4)			
20 Jan. 97	Withers, Maj. C. McG., *p.s.c.*	20 Jan. 15	22 Apl. 16	2nd-in-Comd. With 1-153rd Infy.
28 July 00	Coates, Maj. V.	1 Sept. 15	22 Mar. 13	
8 May 01	Scale, Maj. J. D., D.S.O. [L]	8 May 16	3 Oct. 15	
8 Feb. 02	Bailie, Maj. S. M. H.	8 Feb. 17	22 Apl. 16	With 1-153rd Infy.
	Company Officers. (9)			
28 Jan. 03	Carroll, Maj. B. M.	14 Oct. 16	23 Apl. 04	
23 May 06	Smith, Bt. Maj. C. T. M.	3 June 18	16 July 09	
18 Jan. 11	Middleton, Capt. M.	1 Sept. 15	4 Apl. 12	Adjt.
1 Oct. 14	Raynor, Capt. V. H.	1 Oct. 18	22 Dec. 14	*Lv., ex I., m. c.*
1 Oct. 14	Traill, Capt. H. R.	1 Oct. 18	22 Nov. 16	With 2-91st Punj.
16 Dec. 14	Hunt, Lieut. J. M. (*Actg. Capt.*)	16 Dec. 15	12 Feb. 15	
15 Nov. 15	Williams, Lt. J. W. (*Actg. Capt.*)	11 Apl. 16	19 Nov. 15	Attd. With 69th Punj.
18 Apl. 16	Sutcliffe, Lt. O. D.	18 Apl. 17	26 Apl. 16	Attd. With 1-153rd Infy.
29 June 16	Playfair, Lt. H. R. (*Actg. Capt.*)	29 June 16	2 July 16	Attd.
29 June 16	Darell, Lt. E. M. (*Actg. Capt.*)	29 June 16	2 July 16	Attd.
19 July 16	Lewis, Lt. E.	19 July 17	9 Nov. 16	Attd.
12 Sep. 17	East, Lt. C. F.	12 Sep. 18	. . . .	Attd.
	Attached.			
23 Apl. 15	Greany, Lt. M. P. M.	23 Apl. 16	13 July 15	I. A. R. O.
20 July 15	Mahoney, Lt. A. J.	20 July 16	16 July 17	I. A. R. O. *Asst. Instr., Musky. Sch., Pachmari.*
5 July 16	Jaquet, Lt. J. M. (*Actg. Capt.*)	5 July 17	31 Aug. 16	I A. R. O. Qr.-Mr. (*Offg.*)
5 July 16	O'Leary, Lt. M. G.	5 July 17	30 Aug. 16	I. A. R. O.
10 Nov. 16	Berkely, Lt. V. C. F.	10 Nov. 17	15 Feb. 17	I. A. R. O.
5 Aug. 16	Lawson, Lt. H. A. C. (*Actg. Capt.*)	5 Aug. 17	29 Nov. 17	I. A. R. O.
22 June 15	Slater, Lt. R. A., M.C.	22 June 16	15 May 18	I. A. (*on probn.*)
25 July 17	Sutton, Lt. J. C.	25 July 18	3 June 18	Manch. R., I. A. (*on probn.*)
5 Nov. 17	Sparkes, Lt. H. B.	5 Nov. 18	14 Mar. 18	I. A. (*T. C.*)
6 July 15	Crawshaw, Lt. H. H.	6 July 16	. . . .	M. G. Corps. I.A. (*on probn.*)
15 May 15	Sutton, Lt. R. W. E.	. . . .	11 Sep. 18	A.S.C. I. A. R. O. (*on probn.*)
9 Sep. 15	Jones, Lt. R R.	9 Sep. 16	. . .	L'pool R., I. A. (*on probn.*)
2 Mar. 17	Lunt, Lt. R. M.	2 Mar. 18	. . .	L'pool R., I. A. (*on probn.*)
26 June 16	Syme, Lt. K.	. . . .	11 Sep. 18	A.S.C. I. A. R. O. (*on probn.*)
1 May 18	Vernon-Browne, 2nd-Lt. E. C.	1 May 18	3 Aug. 18	Gen. List I. A. R. O. (*on probn.*)
5 May 17	Wide, Lt. L. A. C.	5 May 18	7 Mar. 18	I. A. R. O.
5 May 17	Binstead, Lt. V. J.	5 May 18	2 Mar. 18	I. A. R. O.
18 June 17	Mackernan, Lt. F. P.	18 June 18	18 June 18	74th Punj.
30 Jan. 17	Fleming, Lt. R. S. T.	30 Jan. 18	7 July 18	93rd Infy.
31 Oct. 17	Townley, Lt. C. F.	31 Oct. 18	1 Sep. 18	I. A. R. O.
24 Apl. 18	Van Renen, 2nd-Lt. J. R. D.	24 Apl. 18	14 Oct. 18	I. A. (U. L.)
11 Sept. 18	Walden, 2nd-Lt. H. L.	11 Sep. 18	7 Oct. 18	I. A. R. O.
11 Sep. 18	Cook, 2nd-Lt. W. M.	11 Sep. 18	3 Oct. 18	I. A. R. O.
7 Oct. 18	Borton, 2nd-Lt. E.	7 Oct. 18	22 Oct. 18	I. A. (*T. C.*)
7 Oct. 18	Donne, 2nd-Lt. W. J. B.	7 Oct. 18	. . . .	I. A. R. O.

Subadar-Major.

		Jemadar.	Subadar.	Subadar-Major.
1 June 97	Sher Muhammad (56, 66a)	1 July 00	16 May 03	1 June 11.

Subadars.

7 Mar. 96	Lachhman (56, 66a, 72a)	27 Apl. 06	1 Sep. 08	
6 July 93	Fazl Khan (56, 66a, 72a)	11 May 04	15 Mar. 10	
1 Dec. 91	Sarban Singh, I.D.S.M. (56, 66a)	1 July 06	5 Oct. 10	
5 Apl. 92	Muhammad Ali (56)	21 Jan. 06	17 Apl. 15	
22 Dec. 95	Harchand Singh (56)	27 Apl. 07	15 Apl. 15	
15 Aug. 87	Mehar Singh	1 June 11	29 Apl. 15	
14 Sep. 95	*Sham Sukh* (56, 66a)	*1 Sep. 08*	*12 Feb. 16*	
10 Apl. 96	Karam Din (56)	26 Oct. 14	12 Aug. 16	
15 Aug. 03	Kartar Baksh (56, 66a)	30 Nov. 14	12 Oct. 16	
10 July 96	Sib Lal (56, 66a)	30 Nov. 14	1 Jan. 17	
	Rung Shah (*Retd.*)	. . . .	. . . .	Re-employed, 1 Feb. 18.
2 Sep. 03	Karam Bakash (56)	17 Apl. 15	18 May 18	

87th Punjabis—*contd.*

1st Commn. or date of entering service.	Names and Rank.	Army rank.	Remarks.
	Jemadars.		
16 Apl. 97	Dewak Ram (56, 66*a*)	15 Apl. 15	Indian Adjt., 6 Sep. 15.
8 Dec. 97	Chuni Lal	15 Apl. 15	
5 May 00	Harpul (56, 66*a*)	17 Apl. 15	
30 Sep. 93	Hakim Khan	17 Apl. 15	
9 July 94	Jahan Khan (66*a*)	29 Apl. 15	
5 July 01	Santa Singh	29 Apl. 15	
10 Apl. 99	Ghulam Rasul (66*a*)	1 Jan. 16	
16 Apl. 99	Muhammad Sadiq Khan (57)	1 Jan. 16	
20 Aug. 01	Muhammad Sher	1 Jan. 16	
7 May 97	Nanak Ram (66*a*)	12 Feb. 16	
7 Dec. 95	Muhammad Baksh	12 Feb. 16	
22 Feb. 97	Sohaj Khan	12 Feb. 16	
3 Feb. 93	*Narayan Singh*	*1 July 16*	
13 Mar. 91	Tara Singh (66*a*)	1 Jan. 17	
1 Aug. 98	Govind Ram	1 Apl. 18	
22 Aug. 03	Balwant Singh	29 Sep. 18	

Regt. Centre—St. Thomas' Mount.
Linked with—73rd & 80th Carnatic Infy.

Uniform—Scarlet.
Facings—Yellow.

1st Battalion, 88th Carnatic Infantry.

1st Martin's.—Raised at Vellore in 1798, by Capt. George Martin, as the 2nd extra Battalion of Madras Native Infantry of drafts from the 4th, 5th, 6th, and 12th regiments. Became the 2nd Bn., 14th Regt. of Madras N. I., the same year; the 28th Regt. of Madras N. I. in 1824; the 28th Regt. of Madras Infy. in 1885; the 28th Madras Infy. in 1901; present designation in 1903.

"Maheidpoor" "Nagpore" "Ava" "China, 1900."

Composition—2 Companies of Madrasi Musalmans, 1 of Tamils, 1 of Paraiyans and Christians.

1st Commn. or date of entering service.	Names and Rank.	Army rank.	Present appointment in Regiment.	Remarks.
	Commandant.			
27 Jan. 92	McCaskill, Lt.-Col. J. C. H.	11 June 17	11 June 17	
	Company Commanders. (4)			
6 Jan. 00	Broadbent Maj. J. T. C. (*Actg. Lt.-Col.*)	1 Sep. 15	5 June 17	2nd-in-Comd. With 1-156th Infy.
12 Aug. 02	Ross, Maj. E. F.	12 Aug. 17	15 Feb. 14	With 2-43rd Infy.
23 May 06	Stone, Capt. B. S.	23 May 15	30 Dec. 16	2nd-in-Comd. (*temp.*) [*A. Hdqrs.*
16 Feb. 07	*Wigley, Capt. F. G. M.*	*1 Sep. 15*	*3 May 17*	*Translation Officer, G. S. Br.,*
	Company Officers. (5)			
8 Sep. 09	*Shakespear, Bt.-Maj. G. F. C.,* D.S.O., M.C.	*1 Jan. 17*	*12 Apl. 13*	
26 Sep. 11	Harding, Capt. C. N.	6 Sep. 15	10 Nov. 12	With 105th Infy.
22 Jan. 13	Robinson, Capt. G. St. J.	22 Jan. 17	15 Mar. 14	Adjt.
29 June 16	Conner, Lt. T. S.	10 Apl. 16	5 July 16	With 1-156th Infy.
26 Jan. 16	Robb, Lt. T. R. (*Actg. Capt.*)	26 Jan. 17	17 Mar. 16	With 1-156th Infy.
18 June 17	Perkins, Lt. W. H.	18 June 18	2? June 17	Attd.
5 Dec. 15	Peake-Cottam, Lt. H.	5 Dec. 16	10 May 17	Attd.
1 June 15	Hamlyn, Lt. H. W. A.	1 June 16	20 Aug. 17	Attd. with 3-23rd Infy.
24 June 15	Chamberlen, Lt. E. F.	24 June 16	2 Aug. 17	Attd. with 2-117th Infy.
14 July 16	Gilligan, Lt. W. B.	14 July 17	8 Dec. 17	Attd. with 2-117th Mah.
12 Sep. 17	Fiddes, Lt. J.	12 Sep. 18	8 Dec. 17	Attd.
	Attached.			
11 Nov. 14	Dale, Lt. D. H.	11 Nov. 15	11 Sep. 18	A. S. C., I. A. R. O. (*on probn.*)
5 May 15	Quin, Lt. L. M.	5 May 16	14 May 17	I. A. (*on probn.*)
6 Sep. 15	Stuart, Lt. J. P.	6 Sep. 16	5 Sep. 18	North'd Fus. I. A. R. O. (*on probn.*)
12 May 16	McCarthy, Lt. C. L.	12 May 17	30 Nov. 17	I. A. (*on probn.*)
13 Dec. 15	Lawrence, Lt. T. B.	13 Dec. 16	2 June 18	R. Berks. R., I. A. (*on probn.*)
6 Feb. 16	Jones, Lt. R. I.	6 Feb. 17	25 Aug. 18	S. Wales Bord., I. A. (*on probn.*) Qr.-Mr. (*offg.*).
29 Aug. 17	Wallace, Lt. D.	29 Aug. 18	8 Dec. 17	I. A. R. O. Adjt. (*offg.*)
6 Dec. 17	Hamer, Lt. F. R.	6 Dec. 18	10 Sep. 18	W. Ind. R., I. A. R. O. (*on probn.*)
5 Jan. 17	Gore, Lt. R. M.	5 Jan. 18	2 June 18	I. A. (*on probn.*)
27 Apl. 18	Titterington, 2nd-Lt. J. H.	27 Apl. 18	30 May 18	I. A. R. O.
20 Apl. 18	Mortimer, 2nd-Lt. H.	20 Apl. 18	. . .	I. A. R. O.
29 May 18	Whitelock, 2nd-Lt. T.	29 May 18	7 Sep. 18	I. A. R. O. (*on probn.*)
5 June 18	Woods-Scawen, 2nd-Lt. C. E.	5 June 18	19 June 18	I. A. R. O.
19 June 18	Putnam, 2nd-Lt. G. W.	19 June 18	19 June 18	Lon. R., I. A. R. O. (*on probn.*)
22 June 18	Leighton, 2nd-Lt. W. W. M. F.	22 June 18	28 June 18	I. A. R. O.
31 July 18	Claridge, 2nd-Lt. E. G. W.	31 July 18	22 Aug. 18	I. A. R. O.
1 Aug. 18	Mitchell, 2nd-Lt. R. H.	1 Aug. 18	4 Sep. 18	I. A. R. O.
1 Aug. 18	Brown, 2nd-Lt. W. S.	1 Aug. 18	6 Sep. 18	I. A. R. O.
1 Aug. 18	Austin, 2nd-Lt. T.	1 Aug. 18	11 Sep. 18	I. A. R. O.
9 Sep. 18	Duke, 2nd-Lt. W. V.	9 Sep. 18	19 Sep. 18	I. A. R. O.
25 June 18	Bannister, 2nd-Lt. W. C.	25 June 18	2 Oct. 18	Genl. List, I. A. R. O. (*on probn.*)
25 June 18	Riches, 2nd-Lt. C. F. H.	25 June 18	9 Oct. 18	Genl. List, I. A. R. O. (*on probn.*)

Subadar-Major.

		Jemadar.	Subadar.	Subadar-Major.
24 Feb. 94	Abdul Khuddus (53, 56)	14 May 08	1 Aug. 11	9 Feb. 17

Subadars.

4 Feb. 91	Abdul Ghafur (52, 56, 66*a*)	17 Mar. 10	1 July 13
5 Apl. 94	Muhammad Asadullah Saiyid (56)	14 July 13	18 Feb. 14
15 Jan. 92	Muthusami (43, 56)	6 Jan. 12	3 Nov. 15
22 Apl. 93	Ellayya (56)	1 Sep. 13	8 Apl. 16
11 Aug. 94	Maduranayakam	10 May 13	1 Dec. 16
21 Nov. 95	Shaikh Ismail (56, 66*a*)	18 Feb. 14	1 Feb. 17
11 Sep. 96	Pe[illegible]mal (52, 56)	13 May 14	9 Feb. 17
7 Dec. 97	Muhammad Abdulazim Khan (56, 66*a*)	11 Aug. 14	8 Aug. 17
17 July 94	Muhammad Abdul Azim (56, 66*a*)	18 Feb. 14	8 Aug. 17

1st Battalion, 88th Carnatic Infantry.

1st Commn. or date of entering service.	Name and Rank.	Army Rank.	Remarks.
	Jemadars.		
25 Jan. 15	Kunhi Ali	25 Jan. 15	
18 Oct. 92	*Muhiyuddin*	*30 Sep. 16*	
5 Sep. 95	*Francis Antony*	*30 Sep. 16*	
9 June 17	Kuppusami (56)	1 Dec. 16	
11 Oct. 12	Ramalingam (56)	1 Feb. 17	
9 Feb. 98	Kalingaraju (56)	9 Feb. 17	
12 Feb. 98	V. Sabagtullah (60*a*)	19 Feb. 17	
14 June 94	Mohammed Ibrahim Khan (56, 60*a*)	8 Aug. 17	
27 Feb. 15	P. Ainuddin	1 Dec. 17	
14 June 94	Muhammad Ibrahim	3 Dec. 17	
7 July 97	Kuppusami	1 Mar. 18	
3 Nov. 96	Moniadurai	1 May 18	
1 June 18	Abdul Subhan	1 June 18	
5 Aug. 98	Chinnasami	1 June 18	
3 Apl. 18	Puthenpurackal Mathoonny Matthew	3 Apl. 18	
14 May 04	Arokiam	1 July 18	
13 Jan. 98	Murugesam	1 July 18	
13 Dec. 04	Kotalayya	1 Aug. 18	
25 Nov. 08	Sayyid Husain	1 Aug. 18	
19 Oct. 11	Duraisami Mudali	1 Aug. 18	
19 July 15	K. P. Gopala Menon	1 Aug. 18	

2nd Battalion, 88th Carnatic Infantry.

Formed at Bangalore on 3rd July 1918.

Composition :—2 companies coorgs, 1 company Maravans, 1 company Madrassis.

1st Commn. or date of entering service.	Names and Rank.	Army rank.	Present appointment in Regiment.	Remarks.
	Commandant.			
22 Jan. 98	Malet, Maj. R. J. (*Actg. Lt.-Col.*)	1 Sep. 16	3 July 18	64th Prs. (*temp.*)
	Company Commanders. (4)			
4 May 01	Pritton, Maj. J. A. H. . . .	4 May 16	3 July 18	83rd Infy. 2nd-in-Comd.
13 Apl. 16	Rendall, Lt. F. C. E. (*Act'g. Capt.*)	13 Apl. 17	16 Aug. 18	I. A. (*on probn.*)
1 June 17	Robb, Lt. A. A. (*Act'g. Capt.*) .	1 June 18	1 Sep. 18	I. A. R. O.
27 Sep. 17	Harley, 2nd-Lt. W.	27 Sep. 17	3 July 18	I. A. R. O.
	Company Officers.			
3 Nov. 14	Osborne, Lt. C. E. F. (*Act'g. Capt.*)	3 Aug. 16	3 July 18	Att'd.
29 July 17	King, Lt. J. A.	29 July 18	3 July 18	I. A. R. O. Adjt.
1 Oct. 17	Freeman, Lt. E. C.	1 Oct. 18	26 Aug. 18	I. A. (*T. C.*)
5 Nov. 17	Handy, Lt. F. R.	5 Nov. 18	15 Aug. 18	I. A. (*T. C.*)
15 Jan. 18	Robb, 2nd-Lt. A. H. D. . . .	15 Jan. 18	15 Jan. 18	I. A. R. O.
18 Jan. 18	Jones, 2nd-Lt. W. E. R. . . .	18 Jan. 18	16 Aug. 18	I. A. R. O.
29 Mar. 18	Lewis, 2nd-Lt. J. S.	29 Mar. 18	16 Aug. 18	I. A. (*T. C.*)
1 May 18	Kiy, 2nd-Lt. D. W.	1 May 18	15 Nov. 18	I. A. R. O.
7 June 18	Monk, 2nd-Lt. F. H. . . .	7 June 18	6 Oct. 18	I. A. R. O.
1 Aug. 18	Lynsdale, 2nd-Lt. A. H. . . .	1 Aug. 18	26 Aug. 18	I. A. R. O.
22 July 18	Ward, 2nd-Lt. W. G.	22 July 18	4 Dec. 18	I. A. (*T. C.*)
22 July 18	Runnalls, 2nd-Lt. H. G. . . .	22 July 18	4 Dec. 18	I. A. (*T. C.*)
11 Oct. 18	Nicholls, 2nd-Lt. A. W. . . .	1 Oct. 18	. . . .	Attd.

Subadar Major.

		Jemadar.	Subadar.	Subadar Major.
22 Feb. 92	Muhammad Sharif (52, 56, 66*a*) .	1 June 08	5 Dec. 14	30 July 18

Subadars.

28 Oct. 17	C. B. Aiyanna	. . . .	28 Oct. 17
4 Mar. 18	Mukkatila Kalappa Uttaya . .	4 Mar. 18	3 July 18
4 Mar. 18	K. M. Chinappa	4 Mar. 18	1 Nov. 18

Jemadars.

7 Jan. 01	Venkatachalam	1 July 17	Indian Adjt.
4 Mar. 18	P. K. Chetticha	4 Mar. 18	
22 Nov. 17	S. P. Ramappa	3 July 18	

Regt. Centre—Meiktila.
Linked with—90th & 91st Punjabis.

Uniform—Drab Serge.
Facings—Blue.

1st Battalion, 89th Punjabis.

Macleod-ki-paltan.—Raised at Masulipatam in 1798 by Capt. A. Macleod, as the 3rd Extra Battalion of Madras Native Infantry of drafts from the 2nd, 7th, 8th, 10th and 11th regiments. Became the 1st Bn., 15th Regt. of Madras N. I., 1800; the 29th Regt. of Madras N. I., 1824; the 29th Regt. of Madras Infy., 1885; the 29th Regt. (7th Burma Batt.) of Mad. Infy., 1891; the 29th Burma Infy., 1901; present designation, 1903. Composition—1½ Companies of Sikhs, ½ of Brahmans ½ of Rajputs, 1½ of Punjabi Musalmans.

1st Commn. or date of entering service.	Names and Rank.	Army rank.	Present appointment in Regiment.	Remarks.
	Commandant.			
Dec. 91	Jackson, Lt-Col. H. W., D.S.O., (*temp. Br.-Genl.*)	1 Jan. 17	10 Mar. 18	
	Company Commanders. (4)			
5 Aug. 96	Geoghegan, Bt. Lt.-Col. N. M. (*Actg. Lt.-Col.*)	3 June 16	9 Apl. 18	2nd-in-Comd.
18 Jan. 02	Scruby, Maj. W. S. J.	18 Jan. 17	6 Apl. 16	
22 Oct. 02	Engledue, Maj. R. S.	8 Nov. 15	18 June 16	
15 July 03	Chapman, Maj. W. L. B.	15 July 18	9 Apl. 18	
	Company Officers. (8)			
9 Jan. 04	Crawford, *Bt.-Maj.* J. D., D.S.O., M.C. (*temp. Col.*)	1 Jan. 18	21 July 05	
18 Jan. 05	*Broome, Capt. W. G. K.*	*18 Jan. 14*	*29 May 09*	*Burma Mily. Police.*
5 Aug. 05	*James, Capt. W. R.*, M.C.	*5 Aug. 14*	*1 July 07*	*G. S. O., 3rd gr., A. Hdqrs.*
24 Aug. 12	Campbell, Capt. M. H. A.	24 Aug. 16	25 Dec. 13	Adjt.
19 July 16	Rogers, Lt. I. J.	19 July 17	11 Nov. 16	Attd.
14 Nov. 16	Rennick, Lt. G. B.	14 Nov. 17	20 Nov. 16	Attd.
7 Apl. 17	Dawkes, Lt. R. B. (*Actg. Capt.*)	7 Apl. 18	16 Apl. 17	Attd.
7 Apl. 17	Bullock, Lt. H. P. S.	7 Apl. 18	16 Apl. 17	Attd.
18 June 17	Morton, Lt. C. E.	18 June 18	24 June 17	Attd. with 1-153rd Infy.
18 June 17	Bates, Lt. R. C. S.	18 June 18	24 June 17	Attd. Qr. Mr.
27 Oct. 17	Phillips, Lt. P. J.	27 Oct. 18	4 Nov. 17	Attd.
21 Dec. 17	Parry, 2nd-Lt. J. K.	21 Dec. 17	20 Mar. 18	Attd.
16 June 15	Ledgerwood, Lt. F. H. (*Actg. Capt.*)	16 June 16	21 Aug. 18	Attd.
	Attached.			
20 Aug. 14	Humphrey, Lt. J.	20 Aug. 16	12 Oct. 15	I. A. R. O.
15 May 16	Morris, Lt. H M.	15 May 16	3 Aug. 17	I A. (*on probn.*) *Lv., ex I., m.c.,*
10 Sep. 15	Hele, Lt. G. C.	10 Sep. 16	7 Oct. 17	I. A. R. O.
22 Jan. 15	Lorrayne-Wadley, Lt. G. C. (*Actg. Capt.*)	22 Jan. 16	24 Oct. 15	I. A. R. O.
1 Jan. 16	Thorpe, Lt. J. A.	1 Jan. 17	. . . .	I. A. R. O. *Lv. in I. m. c.*
23 June 17	Gething, Lt. L. R. B.	23 June 18	7 Mar. 18	I. A. R. O.
27 Oct. 17	Tweedy 2nd-Lt. H. R. F.	27 Oct. 17	23 May 18	I. A. R. O.
24 July 16	Smalley, Lt. S.	24 July 16	. . . .	I. A. R. O.
29 May 18	Gilbert, 2nd-Lt. J. R.	29 May 18	29 Aug. 18	I. A. R. O. (*on probn.*)
29 May 18	Hannigan, 2nd-Lt. A.	29 May 18	29 Aug. 18	I. A. R. O. (*on probn.*)
20 Feb. 16	Mackenzie, Lt. J. F. O. P.	20 Feb. 17	. . . .	A. S. C., I. A. (*on probn.*)
23 Feb. 17	Cangley, Lt. F. G., D.S.O., M.C.	23 Feb. 18	16 Aug. 18	L'pool. R., I. A. (*on probn.*)
19 Sep. 17	Rennie, Lt. W. L.	19 Sep. 18	. . . .	
31 July 18	Dance, 2nd-Lt. E.	31 July 18	6 Nov. 18	North'd. Fus., I. A. (*on probn.*)
28 Aug. 18	Tostevin, 2nd-Lt. F. P.	28 Aug. 18	20 Sep. 18	I. A. R. O.
13 Sep. 18	Harrison, 2nd-Lt. H. L.	13 Sep 18	4 Oct. 18	I. A. R. O.
7 Oct. 18	Collins, 2nd-Lt. F. T.	7 Oct. 18	27 Oct. 18	I. A. R. O.
7 Oct. 18	Hamilton, 2nd-Lt. C. B.	7 Oct. 18	27 Oct. 18	I. A. R. O.

Subadar-Major.

		Jemadar.	Subadar.	Subadar-Major.
	Subadars.			
4 Jan. 01	Ambika Kunwar (52, 56, 66*a*)	4 Jan. 01	3 Aug. 03	
10 Nov. 04	Alan-yar Khan (52, 56)	10 Nov. 04	2 Sep. 07	
26 Feb. 87	Sobha Singh (56, 66*a*)	21 July 03	1 Sep. 11	
23 Dec 04	Saida Khan (52, 56, 66*a*)	Jan. 12	21 Sep. 14	
20 Oct. 98	Jasmal Singh	1 July 13	12 Oct. 14	
16 Nov. 94	Bishan Datt Dube	12 Oct. 14	16 June 16	
16 July 94	Allah Ditta	. . . .	15 Dec. 16	
17 Aug. 97	Rure Khan (66*a*)	21 Sep. 14	24 Dec. 17	
	Hazara Singh (*Retd.*)	. . . .	. . . .	Re-employed 2 Apl. 15.
3 Aug. 97	Jiwan Singh	10 May 15	14 Mar. 18	
14 July 98	Kirpal Singh (52, 56)	13 May 15	18 May 18	
9 Feb. 03	Sheobaksh Singh	7 Feb. 16	18 May 18	

1st Battalion, 89th Punjabis—*contd.*

1st Commn. or date of entering service.	Names and Rank.	Army rank.	Remarks
		Jemadars.	
2 Feb. 97	Kesar Shah	1 Oct. 14	
17 June 99	Bagga Singh, I.D.S.M.	13 May 15	
1 Sept. 00	Ghulam Muhammad, I.D.S.M. (52, 56) .	12 Mar. 16	
11 Dec. 03	Sultan Singh (66*a*)	1 May 16	
13 Aug. 02	Bur Singh, I.D.S.M. (58) . . .	1 June 16	Indian Adjt., 18 May 18.
18 Feb. 00	Wariyam Singh (56,58) . . .	10 June 16	
	Painda Khan	20 Oct. 16	
13 Sep. 04	Haidar Shah.	15 Dec. 16	
11 Mar. 04	Sakhi Muhammad (56, 58) . .	15 Dec. 16	
25 May 08	Ramji Misr, I.D.S.M.	1 Sep. 17	
1 Aug. 12	Teja Singh	3 Dec. 17	Indian Qr. Mr.
25 Nov. 01	Walayat Shah	15 Mar. 18	
8 Apl. 05	Mukhdad Singh (51)	18 May 18	
1 Mar. 07	Farman Shah	18 May 18	
17 Apl. 00	Bal Singh	1 Sep. 18	
18 Aug. 07	Ramjanan Singh	2 Sep. 18	
28 May 12	Muhammad Sadiq	2 Sep. 18	

2nd Battalion, 89th Punjabis.

Formed at Ferozepore on 5th June, 1917.

Composition—1½ Companies Punjabi Musalmans, 1½ Sikhs, ½ Brahmans, and ½ Rajputs.

1st Commn. or date of entering service.	Names and Rank.	Army rank.	Present appointment in Regiment.	Remarks.
	Commandant.			
3 Mar. 97	Oldfield, Maj. G. E. B. (*Actg. Lt.-Col.*)	3 Mar. 15	24 Nov. 17	74th Punjabis.
	Company Commanders. (4)			
19 May 00	Lewis, Maj. G.	1 Sep. 15	15 May 18	38th Horse. 2nd-in-Comd. (*temp.*).
18 Jan. 02	Maunsell, Capt. I. O. G.	18 Jan. 11	16 Oct. 17	91st Punjabis.
15 Apl. 15	Mackay, Capt. G. S.	15 Apl. 18	10 Nov. 17	I. A. R. O.
30 Aug. 15	Booty, Lt. F. C. (*Actg. Capt.*)	30 Aug. 16	1 Apl. 18	I. A. (*on probn.*)
	Company Officers (4).			
16 Oct. 15	Russell, Lt. A. G.	16 Oct. 16	23 Sep. 17	I. A. (*on probn.*)
17 Nov. 16	*Hansard, Lt. A. J. St. L.*	*17 Nov. 16*	*15 June 17*	*I. A. (on probn.) Asst. Musky. Instr., Satara.*
31 Dec. 15	Spencer, Lt. W. E. C.	31 Dec. 16	4 Oct. 18	R. Suss. R., I. A. (*on probn.*)
1 Jan. 17	Johnstone, Lt. E.	1 Jan. 18	6 Mar. 18	I. A. (*on probn.*)
20 June 17	White, Lt. W. J.	20 June 18	5 Oct. 17	I. A. R. O. *Lv., ex I., m. c.*
29 Aug. 17	Faddy, Lt. F. L. (*Actg. Capt.*)	29 Aug. 18	9 Dec. 17	I. A. R. O. Adjt., 29 May 18.
21 Dec. 17	Greenwood, 2nd-Lt. W. L.	21 Dec. 17	15 Feb. 18	I.A.R.O. Qr. Mr.
28 Nov. 17	Adams, Lt. J. A.	28 Nov. 18	2 Feb. 18	I.A.R.O.
1 Oct. 17	Hill, Lt. C., M.C.	1 Oct. 18	16 Apl. 18	I. A. (*on probn.*)
7 June 18	Silver, 2nd-L. P. W.	7 June 18	. . . .	I. A. (*T. C.*)
7 June 18	Vincent, 2nd-Lt. J. A.	7 June 18	. . . .	I. A. (*T. C.*)
22 July 18	Yates, 2nd-Lt. G. C. H.	22 July 18	3 Dec. 18	I. A. (*T. C.*)

Subadar-Major.

		Jemadar.	Subadar.	Subadar-Major.
30 Dec. 02	Sher Zaman (56)	30 May 14	7 Apl. 17	4 Aug. 18

Subadars.

29 Sep. 91	Mohan Lall (56)	13 Aug. 14	25 June 17
22 Nov. 00	Nur Khan (66*a*)	12 Nov. 14	25 June 17
17 Feb. 06	Jagdeo Dube, I.D.S.M.	25 June 17	1 June 18
4 Jan. 08	Roda Khan	25 June 17	1 June 18
11 Aug. 02	Santa Singh	25 June 17	1 July 18

Jemadars.

26 Aug. 03	Achhar Singh	25 June 17	
26 Apl. 98	Sewa Singh	25 June 17	
20 Aug. 17	Dayal Singh	20 Aug. 17	
17 Nov. 18	Bahadur Khan	3 Dec. 17	Indian Qr. Mr.
3 Apl. 10	Fazal Dad	1 June 18	Indian Adjt.
22 Sep. 03	Muhammad Khan	1 June 18	
10 Apl. 08	Jaigovind Upadhya	1 June 18	
10 June 15	Ganga Ram	1 June 18	
1 Sep. 97	Janki Singh	1 July 18	
6 Mar. 09	Fazal Din	1 July 18	
1 Apl. 05	Shivnarayan Singh	1 July 18	

Regt. Centre—Meiktila.
Linked with—89th & 91st Punjabis.

Uniform—Drab Serge
Facings—Black.

1st Battalion, 90th Punjabis.

Crewe-ki-paltan.—Raised at Masulipatam in 1799 by Capt. T. K. Crewe, principally of men who had belonged to Raymond's Brigade at Hyderabad, and was originally styled the Masulipatam Battalion. Became the 2nd Bn., 15th Regt. of Madras Native Infantry, 1800; the 30th Regt. of Madras N. I., 1824; the 30th Regt. of Madras Infy., 1885; the 30th Regt. (5th Burma Battn.) of Mad. Infy., 1892; the 30th Burma Infy., 1901; present designation, 1903.

"Ava" "Afghanistan, 1878-80" "Burma, 1885-87."

Composition—2 Companies of Sikhs, ½ of Brahmans, ½ of Rajputs, 1 of Punjabi Musalmans.

1st Commn. or date of entering service.	Names and Rank.	Army rank.	Present appointment in Regiment.	Remarks.
	Commandant.			
1 Feb. 88	Murray, Lt.-Col. E. R. B. (*tempy. Br.-Genl.*)	1 Feb. 14	14 Sept. 16	
	Company Commanders. (4)			
9 Nov. 92	McGill, Maj. H. R. (Q)	9 Nov. 10	30 Jan. 08	
20 Jan. 00	Apthorpe, Maj. E. H. F.	1 Sep. 15	10 Mar. 18	With 2-90th Punj.
	Company Officers. (8)			
4 Dec. 01	Hancock, Maj. E. S.	4 Dec. 16	30 Jan. 05	
9 Jan. 04	Thyne, Capt. W.	9 Jan. 13	30 Apl. 07	
20 May 05	*Miller, Capt. G. N.*	*20 May 14*	*4 Nov. 08*	
29 Aug. 06	*Porteous, Capt. D. G. (temp. Maj.)*	*29 Aug. 15*	*2 Nov 11*	*D. A. A. G., S. Commd.*
9 Oct. 07	Wigley, Capt. P. J. R., M.C.	1 Sep. 15	7 Oct. 11	*Lv., ex I., m.c.*
9 Sep. 08	*Brewer, Capt. J. C.*, M.C.	*1 Sept. 15*	*3 Jan. 10*	*G. S. O., 3rd Gr., Army Hdqrs.*
18 Jan. 11	Bampfield, Capt. G C.	1 Sept. 15	10 Mar. 12	Adjt. (*Offg*).
21 Dec. 17	Andrews, 2nd-Lt. J. A.	21 Dec. 17	....	Attd.
	Attached.			
14 Aug. 95	Waring, Lt.-Col. E. H.	19 Aug. 18	....	63rd Infy.
15 Nov. 15	Connor, Lt. G. H. (*Actg. Capt.*)	15 Nov. 16	....	33rd Punj.
15 July 15	Kennedy, Capt. K. H. H., M.C.	15 July 18	....	I. A. R. O.
17 June 16	Lambert, Lt. A. G. (*Actg. Capt.*)	17 June 17	....	I. A. R. O.
3 Sept. 15	Brockes, Lt. E. H. (*Actg. Capt.*)	3 Sept. 16	....	I. A. R. O.
29 Oct. 15	Hirst, Lt. F. H.	29 Oct. 16	....	I. A. R. O., *Lv., ex I., m.c.*
8 Oct. 15	Stanbury, Lt. S. J. (*Actg. Capt.*)	8 Oct. 16	....	I. A. R. O.
20 June 17	Tobin, Lt. E. J.	20 June 18	....	I. A. R. O. Qr. Mr. 8 Nov. 18.
7 July 17	Mowforth, Lt. H.	7 July 17	....	I. A. R. O.
22 Apl. 16	Tutton, Lt. P. C.	22 Apl. 17	....	R. Suss. R., I. A. (*on probn.*)
5 Nov. 17	Wisher, Lt. J. D.	5 Nov. 18	....	I. A. R. O.
26 Sept. 17	Cottam, 2nd-Lt. C. H.	26 Sept. 17	....	
16 May 15	Parish, Lt. G. N. O.	16 May 16	....	I. A. (*on probn.*)
8 Mar. 17	Chamier, Lt. A.	8 Mar. 18	....	I. A. R. O.
	Cox, 2nd Lt. V. M.		....	
20 June 17	Hislop, Lt. R. G. G.	20 June 18	....	I. A. R. O.
7 Feb. 16	Bland, Lt. E. M., M.C.	7 Feb. 17	....	I. A. (*on probn.*)
1 June 15	Macpherson, Lt. J. E.	1 June 16	....	R. Scots, I. A. (*on probn.*)
7 Sep. 18	Graveston, 2nd-Lt. G. J. B.	7 Sep. 18	....	I. A. R. O.
13 Sep. 18	Jackson, 2nd-Lt. A. L.	13 Sep. 18	....	I. A. R. O.
7 Oct. 18	Teire, 2nd-Lt. D.	7 Oct. 18	....	I. A. R. O.
7 Oct. 18	Howson, 2nd-Lt. W. G.	7 Oct. 18	....	I. A. R. O.
26 Sep. 18	Smith, 2nd-Lt. S.	26 Sep. 18	....	I. A. R. O.

Subadar-Major.

		Jemadar.	Subadar.	Subadar-Major.	
22 Nov. 90	Imam Din *Bahadur* (52, 56)	20 Nov. 12	5 Jan. 15	18 Apl. 18	*Order of Br. Ind., 2nd Cl.*
	Hazura Singh, *Hony. Capt., Sardar Bahadur* (56)	13 Aug. 03	16 Oct. 04	15 Oct. 07	*Order of Br. Ind., 1st Cl.* Re-employed, 23 May 15.

Subadars.

10 Feb. 89	Rattan Singh (56)	5 Aug. 08	1 Jan. 14	
19 Sept. 95	Narayan Singh, I.D.S.M.	1 Oct. 12	1 July 14	
19 Oct. 01	Mula Singh (56)	1 Oct. 12	9 July 14	
23 June 96	Bir Singh (52, 56)	1 Jan. 14	20 Jan. 15	
3 Apl. 96	Muhammad Azam (56, 58)	1 Aug. 13	9 Aug. 17	
20 Oct. 98	Gurdict Singh (56)	20 Jan. 15	31 Aug. 17	
22 Oct. 98	Man Khan	5 Jan. 15	18 May 18	
	Mehr. Singh (*Retd.*)	....	....	Re-employed 26 Sep. 18.
	Changate Khan (*Ret.*)	....	....	Re-employed 8 Feb. 18.
	Indar Singh (*Ret.*)	....	....	Re-employed 24 June 17.
	Hari Singh (*Ret.*)	....	....	Re-employed 1 May 17.
	Bulande Khan (*Ret.*)	....	....	Re-employed 24 Mar. 17.
20 Oct. 98	Chet Singh	30 Sep. 17	1 Aug. 18	

1st Battalion, 90th Punjabis—*contd.*

1st Commn. or date of entering service.	Name and Rank.	Army rank.	Present appointment in Regiment.	Remarks.
	Jemadars.			
	Mangal Singh	15 Sep. 15		
	Khiwan Singh	6 Apl. 16		
	Santa Singh	15 June 16		
	Hagh Singh	12 Sep. 16		
	Sunder Singh	12 Sep. 16		
5 Jan. 99	Muhammad Yasin	25 Sep. 17		
13 June 97	Lal Khan Singh	20 Jan. 15		
10 Aug. 03	Bogha Singh	20 May 18		
27 Oct. 03	Bhan Singh	20 May 18		
20 Aug. 95	Nur Alam	18 May 18		
6 June 04	Framurz Khan	18 May 18		
	Gulb dan Singh	18 May 18		
	Sri Pat Tiwari	18 May 18		

2nd Battalion, 90th Punjabis.

Formed at Dhond on 20th July 1918.

Composition—1 Company Sikhs, 1 Company Punjabi Musalmans, 1 Company Brahmans, 1 Company Ahirs.

1st Commn. or date of entering service.	Names and Rank.	Army Rank.	Present appointment in Regiment.	Remarks.
	Commandant.			
25 Jan. 99	Bayley, Maj. L. C. L. (*Actg. Lt.-Col.*)	1 Sep. 15	25 July 18	6th Cav.
	Company Commanders. (4)			
20 Jan. 00	Apthorpe, Maj. E. H. F.	1 Sep. 15	31 July 18	1-90th Punj. 2nd-in-Comd. (*temp.*)
12 June 17	Hoernle, Lt. E. S. (*Actg. Capt.*) .	12 June 18	1 Aug. 18	I. A. R. O.
9 Sep. 16	Holmes, Lt. S. B. (*Actg. Capt.*) .	9 Sep. 17	30 July 18	R. Muns. Fus., I. A. (*on probn.*)
3 May 17	Nimmo, Lt. A. (*Actg. Capt.*) .	3 May 18	2 Aug. 18	R. Sc. Fus., I. A. (*on probn.*)
	Company Officers.			
22 June 15	Tanner, Lt. G. W., M.C. (*Actg. Capt.*)	22 June 16	1 Aug. 18	Leic. R., I. A. (*on probn.*) Adjt. 2 Aug. 18. (*Offg.*)
15 Nov. 15	Robinson, Lt. H. R.	15 Nov. 16	24 Aug. 18	1-91st Punj. Qr. Mr. 14 Aug. 18. (*Offg.*)
27 Mar. 17	Rolinson, Lt. W. J.	27 Mar. 18	31 Aug. 18	R. Berks. R., I. A. (*on probn.*)
18 Apl. 17	Forrest, Lt. J. D. D.	18 Apl. 18	31 Aug. 18	Sea. Highrs., I. A. (*on probn.*)
5 May 17	Gray, Lt. A. L.	5 May 18	4 Sep. 18	I. A. R. O. (*on probn.*)
5 Nov. 17	Besley, 2nd-Lt. F.	5 Nov. 17	30 July 18	I. A. (*T. C.*)
15 Apl. 18	Greffield, 2nd-Lt. G.	15 Apl. 18	28 Aug. 18	I. A. (*on probn.*)
29 Mar. 18	Johnson, 2nd-Lt. P. E. . . .	29 Mar. 18	30 Aug. 18	I. A. (*T. C.*)
29 Mar. 18	Sawford, 2nd-Lt. A. N. . . .	29 Mar. 18	31 Aug. 18	I. A. (*T. C.*)
1 May 18	Sussex, 2nd-Lt. H. G. V. . . .	1 May 18	14 Nov. 18	Lond. R., I. A. R. O. (*on probn.*)
10 July 18	Ward, 2nd-Lt. G. S. W. . . .	10 July 18	2 Dec. 18	I. A. R. O.
22 July 18	Pla t, 2nd-Lt. I. T.	22 July 18	2 Dec. 18	I. A. (*T. C.*)
22 July 18	Moore, 2nd-Lt. P. J.	22 July 18	2 Dec. 18	I. A. (*T. C.*)
22 July 18	Kitching, 2nd-Lt. B.	22 July 18	2 Dec. 18	I. A. R. O.
6 Mar. 15	Murray, Capt. R.	6 Mar. 17	17 Nov. 18	High. L. I.
31 May 15	Taylor, Lt. H. B.	1 July 17	17 Nov. 18	K. O. Sco. Bord.
1 Nov. 15	Melville, Lt A. M.	1 July 17	17 Nov. 18	High. L. I.
5 Aug. 16	Millar, Lt. W W. C.	5 Feb. 18	17 Nov. 18	High. L. I.
24 Feb. 17	Storrie, Lt. W., D.S.O. . . .	24 July 18	17 Nov. 18	High. L. I.

Subadar-Major.

		Jemadar.	Subadar.	Subadar-Major.
22 Aug. 03	Dali Ahmad	8 Mar. 17	1 Aug. 18	9 Sep. 18

Subadars.

1 Sep. 14	Mehr Chand	1 Oct. 15	5 June 18
1 Jan. 06	Ram Persan Tewari	16 June 16	26 Oct. 18
3 Oct. 01	Fazal Dad	1 Aug. 18	1 Nov. 18

Jemadars.

16 June 17	Lachman Dass	6 June 18
16 June 04	Harihar Pershad	1 Aug. 18
27 Sep. 05	Zaman Ali	1 Aug. 18
23 June 05	Dharam Singh	1 Aug. 18
18 Sep. 11	Haidar Khan	1 Aug. 18
5 Sep. 16	Bhajjan Singh	1 Aug. 18
22 May 09	Shreoharash Rai	1 Nov. 18
1 May 05	Achhar Singh	1 Nov. 18
4 May 18	Mangal Das	1 Nov. 18

Regt. Centre—Meiktila.
Linked with—89th & 90th Punjabis.

Uniform—Drab.
Facings—Cherry.

1st Battalion, 91st Punjabis (Light Infantry).

Jennerett-ki-paltan.—Raised at Trichinopoly in 1800 by Lt.-Col. S. Jennerett, as the 1st Battalion, 16th Regiment of Madras Native Infantry. Became the 1st Bn., 16th Regt. of Madras N. I., or Trichinopoly Light Infy., 1812; the 31st Regt. of Madras N. I., or Trichinopoly Light Infy., 1824; the 31st (Trichinopoly) Regt. of Madras Light Infantry, 1885; the 31st Regt. (6th Burma Battn.) of Madras (Light) Infy., 1892; the 31st Burma Light Infy., 1901; present designation, 1904.

"*Now or Never.*"

"Maheidpoor" "China, 1900."

Composition—1½ Companies of Punjabi Musalmans, ½ of Hindustani Musalmans, 1 of Sikhs, 1 of Dogras.

COLONEL—Lieut.-Genl. *Sir* Henry D'U. Keary, K.C.B., K.C.I.E., D.S.O. . 29 October 1912.

1st Commn. or date of entering service.	Names and Rank.	Army rank.	Present appointment in Regiment.	Remarks.
	Commandant.			
22 Aug. 88	Basevi, Lt.-Col. W. H. F.	22 Aug. 14	28 Apl. 18	
	Company Commanders. (4)			
3 Aug. 98	Day, Maj. A. D.	1 Sep. 15	14 Aug. 10	With 3-151st Infy.
20 Jan. 09	Dease, Maj. C. E. (*temp. Lt.-Col.*)	1 Sep. 15	22 Aug. 14	*Comdt., Sch. of Instr. for Officers, Nasik.*
8 May 01	Barnett, Maj. W. H.	8 May 16	28 Apl. 18	
	Company Officers. (13)			
18 Jan. 02	Maunsell, Capt. I. O. G.	18 Jan. 11	15 June 04	With 2-89th Punj.
12 Feb. 02	Wilcox, Maj. H. B. F.	6 Apl. 17	19 Nov. 03	
22 Oct. 02	Wilkinson, Maj. C. O.	22 Oct. 17	10 Feb. 06	With 3-153rd Infy.
29 July 03	Skinner, Maj. R. H.	11 Jan. 18	18 Nov. 05	
4 Nov. 03	Nixon, Capt. C. H. F., D.S.O.	4 Nov. 12	9 June 10	With 3-154th Infy.
21 June 05	Campbell, Capt. F. R. M.	22 June 14	9 Feb. 11	
20 Dec. 06	Goldman-Monk, Capt. M. E. deB. [L].	1 Sep. 15	10 Sep. 11	With 1-27th Punj.
17 Aug. 07	Baxter, Capt. D. St. J.	1 Sep. 15	20 Sep. 10	
15 Nov. 15	Robinson, Lt. H. R.	15 Nov. 16	28 Nov. 15	Attd. with 2-90th Punj.
29 June 16	Bartlett, Lt. H. V.	29 June 17	20 Oct. 16	
12 Sep. 17	Jones, Lt. J. L.	12 Sep. 18	. . . .	Attd. With 2-22nd Punj.
12 Sep. 17	Illingworth, Lt. G. R.	12 Sep. 18	. . . .	Attd.
17 Jan. 16	Graham, Lt. N. I.	17 Jan. 17	. . . .	Attd.
	Attached.			
17 Feb. 88	Murray, Lt.-Col. A. B., *I. A.*	17 Feb. 14	28 Apl. 18	Comdt. (*temp.*)
27 July 98	Kensington, Bt.-Lt.-Col. E. C., M.C., *p.s.c.*	1 June 17	. . . .	130th Baluchis. Comdt. (*temp.*)
14 Sep. 14	Palmer, Capt. H. A.	14 Sep. 18	. . . .	I. A. R. O.
20 Feb. 16	Goldberg, Lt. G. H. A.	20 Feb. 17	11 Apl. 16	I. A. R. O.
29 May 18	Gale, 2nd-Lt. H. C.	29 May 18	. . . .	I. A. R. O. (*on probn.*)
5 June 17	Collins, Lt. F. C.	5 June 18	24 Dec. 17	R. Ir. Fus., I. A. (*on probn.*)
7 Nov. 15	Jenkins, Lt. E. M.	7 Nov. 16	9 Dec. 17	I. A. R. O.
9 Aug. 16	Gamble, Lt. W. M. F., M.C.	9 Aug. 17	23 Aug. 18	I. A. (*on probn.*)
15 Jan. 18	Franks, 2nd-Lt. H. S.	15 Jan. 18	23 Aug. 18	I. A. R. O.
15 Jan. 18	Norton, 2nd-Lt. W. D.	15 Jan. 18	23 Aug. 18	I. A. R. O.
1 May 18	White, 2nd-Lt. C. J.	1 May 18	12 Aug. 18	I. A. R. O. (*on probn.*)
24 June 18	Ross, 2nd-Lt. J. D.	24 June 18	7 Aug. 18	I. A. R. O.
13 June 18	Skinner, 2nd-Lt. F. M.	13 June 18	15 July 18	I. A. R. O.
1 Aug. 17	Gerrard, 2nd-Lt. D. M.	1 Aug. 17	8 Sep. 18	I. A. R. O. (*on probn.*)

Subadar-Major.

		Jemadar.	Subadar.	Subadar-Major.
20 June 00	Thakur Singh (56)	1 Nov. 03	1 Jan. 11	23 Sep. 17
23 Feb. 87	Bijai Singh, *Sardar Bahadur* (56, 66*a*).	1 Oct. 98	1 June 09	*Order of Br. Ind., 1st Cl.*

Subadars.

4 Feb. 01	Sakat Singh (56, 66*a*, 72*a*)	4 Feb. 01	16 July 09	
30 May 90	Warriam Singh (56, 66*a*)	1 Jan. 06	15 July 12	
12 Oct. 95	Faiyazali Khan (56, 66*a*)	1 Feb. 09	12 Nov. 14	
29 Nov. 95	Khazana (66*a*)	1 Jan. 14	5 Jan. 15	
24 Jan. 98	Kalu Khan (56, 66*a*)	16 July 11	9 Jan. 16	
31 May 97	Saudagar Singh (56, 66*a*)	1 Jan. 11	6 July 16	
8 Sep. 03	Ranjit Singh	5 Jan. 15	5 Jan. 17	
7 July 93	Surju Singh (56)	5 Apl. 15	28 June 17	
2 May 05	Amir Ali	6 July 16	21 Feb. 18	
27 Jan. 08	Ghulam Muhammad	1 Mar. 18	27 May 18	
23 Nov. 00	Kishan Singh	12 July 15	18 June 18	
13 Nov. 03	Ram Singh	1 Jan. 18	1 Aug. 18	
	Roshan Ali (*Ret.*)	. . . .	. . .	Re-employed. 16 Dec. 16.
	Channan Singh (*Ret.*)	. . . .	. . .	Do. 16 Dec. 16.
	Fateh Muhammad (*Ret.*)	. . . .	. . .	Do. 20 Mar. 17.

1st Battalion, 91st Punjabis (Light Infantry)—*contd.*

1st Commn. or date of entering service.	Names and Rank.	Army rank.	Remarks.
	Jemadars.		
18 May 00	Khan Singh	5 Jan. 15	
16 Mar. 00	Sundar Singh (66)	25 Nov. 15	
28 June 99	Bishan Singh	6 July 16	
15 July 97	Mangal Singh	6 July 16	
26 Aug. 03	Warriam Singh	6 July 16	
14 Aug. 99	Sher Singh	3 Dec. 17	
8 May 00	Lachman Singh	1 Jan. 18	
6 May 12	Nazir Khan	21 Feb. 18	
1 Dec. 97	Ram Shukal Rai	1 Mar. 18	
	Rahmat Khan	27 May 18	
	Nur Alam	27 May 18	
	Muhammad Khan	27 May 18	

2nd Battalion, 91st Punjabis (Light Infantry).

Formed at Ferozepore.
Composition.—8 platoons Hindustani Musalmans, 4 platoons Punjabi Musalmans, 4 platoons Ahars.

1st Commn. or date of entering service.	Names and Rank.	Army rank.	Present appointment in Regiment.	Remarks.
	Commandant.			
20 Nov. 90	Crawford, Lt.-Col. A. G.	29 Nov. 16	19 Feb. 18	84th Punj.
	Company Commanders. (4)			
26 Sep. 14	Livesay-Wardle, Capt. R. M. (*Act. Maj.*)	26 Sep. 18	19 Mar. 18	I. A. R. O. (*on probn.*) 2nd-in-Comd. (*temp.*)
1 Oct. 14	Traill, Capt. H. R.	1 Oct. 18	19 Mar. 18	87th Punj. Adjt. 1 Sep. 18.
31 May 15	James, Lt. H. (*Actg. Capt.*)	19 May 18	19 May 18	I. A. R. O. (*on probn.*)
16 May 16	Davison, Lt., H. T.	16 May 17	24 May 18	I. A. (*on probn.*)
	Company Officers.			
5 Nov. 17	Shenton, Lt. E. W.	5 Nov. 18	2 Apl. 18	I. A. R. O. (*on probn.*)
30 May 17	Brewer, 2nd-Lt. A. G.	30 May 17	17 Nov. 18	Durh. L. I.
30 May 17	Ward, 2nd-Lt. J. M.	30 May 17	17 Nov. 18	Durh. L. I.
27 June 17	Maughan, 2nd-Lt. W. C.	27 June 17	17 Nov. 18	Durh. L. I.
29 Aug 17	McGowan, 2nd-Lt. T. C. S.	29 Aug. 17	17 Nov. 18	Durh. L. I.
5 Nov. 17	Caminada, 2nd-Lt. C. B.	5 Nov. 17	23 July 18	I. A (*T. C.*)
5 Jan. 18	Clark, 2nd-Lt. H. W.	5 Jan. 18	5 June 18	I. A. (*T. C.*)
15 Jan. 18	Miller, 2nd-Lt. H.	15 Jan. 18	5 June 18	I. A. R. O. Qr. Mr.
31 Jan. 18	Parker, 2nd-Lt. C. J.	31 Jan. 18	2 Apl. 18	Attd.
19 Mar. 18	Harris, 2nd-Lt. H. J.	19 Mar. 18	16 July 18	I. A. R. O.
29 Mar. 18	Brown, 2nd-Lt. F. J.	29 Mar. 18	13 Aug. 18	I. A. (*T. C.*)
29 Mar. 18	Dennes, 2nd-Lt. H. M.	29 Mar. 18	16 July 18	I. A. (*T. C.*)
2 June 18	O'Meara, 2nd-Lt. M. J.	2 June 18	5 Oct. 18	I. A. (*T. C.*)
7 June 18	Leonard, 2nd-Lt. D. W.	7 June 18	5 Oct. 18	I. A. R. O.
26 June 18	Campbell, 2nd-Lt. M.	26 June 18	17 Nov. 18	Durh. L. I.
10 July 18	Hall, 2nd-Lt. T. H.	10 July 18	2 Dec. 18	I. A. (*T. C.*)
19 July 18	Kennedy, 2nd-Lt. H. K.	19 July 18	18 Oct. 18	I. A. R. O.
23 July 18	Eastland, 2nd-Lt. W. H.	23 July 18	2 Dec. 18	I. A. R. O.
23 July 18	Irving, 2nd-Lt. S. E.	23 July 18	2 Dec 18	I. A. R. O.
28 Aug. 18	Seaman, 2nd-Lt. C. K.	28 Aug. 18	21 Sept. 18	I. A. R. O.

Subadar-Major.

		Jemadar.	Subadar.	Subadar-Major.	
26 Nov. 00	Mirza Khan	18 May 15	10 Jan. 16	23 Mar. 18	(*temp.*)

Subadars.

3 Jan. 08	Ghulam Mohammed	14 Aug. 15	6 July 16
4 Aug. 05	Ibrahim Khan	10 Jan. 16	12 Feb. 17
16 Feb. 95	GhulamRasul	10 May 16	23 Mar. 18
26 Nov. 00	Nur Khan	6 July 16	23 Mar. 18
15 Dec. 09	Khan Mohammed	6 July 16	23 Mar. 18
24 Dec. 05	Shehryar Khan	1 Jan. 18	16 May 18
15 Dec. 00	Muhammad Khan	23 Mar. 18	1 Aug. 18

Jemadars.

7 Oct. 05	Mahmud Khan	23 Mar. 18
30 July 07	Muhammad Akbar	23 Mar. 18
7 July 05	Faqir Muhammad	23 Mar. 18
26 Dec. 00	Ahmad Khan	23 Mar. 18
4 Sep. 05	Abdul Rahman	23 Mar. 18

Regt. Centre—Mandalay. *Uniform*—Drab Serge.
Linked with—72nd Punjabis & 93rd Burma Infy. *Facings*—White.

92nd Punjabis.

Dyce-ki-paltan.—Raised at Madura in 1800 by Lt.-Col. A. Dyce, as the 2nd Battalion, 16th Regiment of Madras Native Infantry. Became the 32nd Regt. of Madras N. I., 1824; the 22nd Regt. of Madras Infantry 1885; the 4th Regt. of Burma Infy., 1890; 32nd Regt. (4th Burma Battalion) of Madras Infy., 1891; the 32nd Burma Infy., 1901; present designation, 1903.

"Ava."

Composition—2 Companies of Sikhs, 2 of Punjabi Mussalmans.

COLONEL—Major-General Edward S. Hastings, C.B., D.S.O. . . 10 June 1910.

1st Commn. or date of entering service.	Names and Rank.	Army rank.	Present appointment in Regiment.	Remarks.
	Commandant.			
23 Mar. 89	Maclachlan, Bt.-Col. T. R., C.M.G. (*temp. Br.-Genl.*) . . .	10 Mar. 17	3 July 15	Bde. Comdr.
	Company Commanders. (4)			
10 Oct. 94	Nangle, Maj. M.C., *p.s.c.* (*l*).(Q) .	10 Oct. 12	24 Sep. 17	2nd-in-Comd. With 2-99th Infy.
4 Jan. 99	Rigg, Maj H. M., (*Actg. Lt.-Col.*)	1 Sep. 15	23 Nov. 12	With 2-70th Rif.
19 Dec. 00	Taylor, Maj. N. C., D.S.O. . .	19 Dec. 15	6 Feb. 14	With .-9th Infy.
8 May 01	*Cloete, Maj. H. D.*, M.C. . . .	*8 May 16*	*24 Se* .	*Assam Rifles.*
	Company Officers. (8)			
21 Jan. 03	Hastings, Maj. W. H., D.S.O. (*Actg. Lt.-Col.*)	1 Jan. 16	6 Apl. 04	
9 Jan. 04	*Watts. Capt. N. H. L. (temp. Maj.)*	*9 Jan. 13*	*14 Mar. 08*	*Asst. Comdt. Sch. of Instn., Bangalore.*
19 Jan. 07	Ponting, Capt. T. J., M.C. . .	1 Sep. 15	29 Mar. 08	
24 Aug. 12	Keene, Capt. B. G.	24 Aug. 16		With 2-153rd Infy.
18 Apl. 16	Moore, Lt. A. C.	18 Apl. 17	. . . .	Attd. With 2-153rd Infy.
29 June 16	Proudlock, Lt. R. V.	29 June 17	. . . .	Attd. With 69th Punj.
1 Oct. 16	Hasell-Wright, Lt. E. L. C. .	1 Oct. 17	. . . .	
30 Jan. 17	Deans, Lt. T. J.	30 Jan. 18	. . . .	Attd.
18 June 17	Harding, Lt. W. J.	18 June 18	. . . .	
12 Sep. 17	Johns, Lt. R. C. W.	12 Sep. 18	6 Dec. 17	Attd.
12 Sep. 17	Jackson, Lt. J. D.	12 Sep. 18	6 Dec. 17	Attd.
20 Nov. 16	Morris, Lt. H. S., M.C. . . .	20 Nov. 17	28 Oct. 16	With 1-117th Mah.
1 Oct. 18	Straker, 2nd-Lt. F. E. . . .	1 Oct. 18	. . . .	Attd.
5 Nov. 15	Lovat-Fraser, Lt. W. A. . . .	5 Nov. 16	. . . .	Attd. with 1-48th Prs.
	Attached.			
26 July 15	*Bland, Capt. E. G.*	*19 Sep. 16*	. . . . .	*I. A. R. O. Adjt, Prisoners of War Camp, Bellary.*
20 Nov. 14	Galvin, Lt. L. E.	20 Nov. 15	11 Aug. 17	I. A. R. O.
9 Apl. 15	Wise, Lt. J. H.	9 Apl. 16	24 Apl. 17	I. A. R. O.
30 July 15	Ritter, Lt. J. C. (*Actg. Capt.*) .	30 July 16	7 Sep. 16	I. A. R. O.
2 Aug. 15	Harris, Lt. F. O., M.C. . . .	2 Aug. 16	7 Sep. 16	I. A. R. O.
1 Jan. 16	Winckler, 2nd-Lt. C. P. . . .	1 Jan. 16	19 Sep. 16	I. A. R. O.
10 Mar. 17	Wise 2nd-Lt., A. A. . . .	10 Mar. 17	. . . .	I. A. R. O.
25 Oct. 16	Perks, 2nd-Lt. G. A. C. . . .	25 Oct. 16	6 Feb. 18	I. A. R. O.
8 Aug. 17	Donovan, 2nd-Lt. B. D. . . .	8 Aug. 17	2 Nov. 17	I. A. R. O.
11 June 18	Hill-Murray, 2nd-Lt. H. J. . .	11 June 18	. . . .	I. A. R. O.
7 June 18	Rowe, 2nd-Lt. S. A. . . .	7 June 18	. . . .	I. A. R. O.
7 Oct. 18	Dillinghan, 2nd-Lt. C. D. S. . .	7 Oct. 18	. . . .	I. A. R. O.
7 Oct. 18	Edlin, 2nd-Lt. C. A. E. . . .	7 Oct. 18	. . . .	I. A. R. O.

Subadar-Major.

		Jemadar.	Subadar.	Subadar-Major.	
Mar. 91	Farman Ali *Bahadur*, M.C., I.O.M., I.D.S.M. (56, 62, 66).	16 Nov. 03	20 July 11	7 July 14	*Order of Br. Ind., 2nd Cl.*
20 Feb. 76	Chaudhri Khan, (*Hony. Capt.*), *Sardar Bahadur* (*Retd.*) . .	. . . . .	Re-employed, 24 Jan. 16.		

Subadars.

28 June 95	Sundar Singh (56)	17 Sep. 12	7 Jan. 16	
17 Aug. 18	Harnam Singh	26 Aug. 15	13 Jan. 16	
19 Jan. 98	Khan Zaman	23 Jan. 15	15 Oct. 16	
28 Apl. 01	Gulab Khan (56)	24 Oct. 14	2 Feb. 17	With 2-153rd Punj.
17 Jan. 96	Muhammad Azim, I.O.M. . .	20 Nov. 15	22 Feb. 17	
3 June 99	Sohan Singh	21 Jan. 16	23 Feb. 17	
19 Aug. 02	Allahdad	14 Jan. 16	23 Feb. 17	With 2-152nd Punj.
24 Aug. 04	Alam Khan	18 Dec. 16	19 Dec. 17	
	Narain Singh	. . . . .	. . . .	Re-employed, 15 Apl. 16.
1 Oct. 03	Khuda Yar	23 Feb. 17	2 Jan. 18	
20 Nov. 02	Dalel Khan	14 Jan. 16	27 May 18	With 2-153rd Punj.
27 Aug. 99	Inder Singh	23 Feb. 17	27 May 18	With 2-153rd Punj.

92nd Punjabis—*contd.*

1st Commn. or date of entering service.	Names and Rank.	Army rank.	Remarks.
	Jemadars.		
9 Apl. 00	Buta Singh	8 Feb. 16	
1 Nov. 05	Qasim Ali	13 Feb. 16	With 2-152nd Punj.
. . . .	Fateh Muhammad	17 Dec. 16	
19 July 98	Mahbub Alam	23 Feb. 17	
12 Sep. 01	Gopal Singh	23 Feb. 17	
30 Jan. 06	Alam Din	23 Feb. 17	
1 Oct. 04	Shan Muhammad	1 Apl. 17	
15 Dec. 99	Narain Singh	14 Apl. 17	
24 Nov. 04	Jalal Khan	6 Nov. 17	
29 July 11	Ata Muhammad	19 Dec. 17	
20 Mar. 08	Ghulam Rasul	19 Dec. 17	
3 Aug. 06	Mardan Ali	1 Jan. 18	With 2-153rd Punj.
2 Oct. 01	Maghar Singh	2 Jan. 18	
17 July 02	Hira Singh	27 May 18	
2 June 05	Ujagar Singh	1 Jan. 18	
7 Mar. 06	Sapuran Singh	1 Jan. 18	
1 Dec. 09	Sundar Khan	1 Jan. 18	
21 Aug. 03	Surayan Singh	27 May 18	
20 Oct. 06	Tek Singh	2 July 18	
	Sultan Ahmad	10 July 18	
18 Oct. 07	Chain Singh	20 Sep. 18	

Regt. Centre—Mandalay.
Linked with—72nd & 92nd Punjabis.

Uniform—Drab Serge
Facings—Yellow.

93rd Burma Infantry.

Wahab-ki-paltan.—Raised at Guntoor in 1800 as the 1st Battalion, 17th Regiment of Madras Native Infantry. Became the 33rd Regt. of Madras N. I. in 1824; the 33rd Regt. of Madras Infy. in 1885; the 3rd Regt. of Burma Infantry in 1890; the 33rd Regt. (3rd Burma Battn.) of Mad. Infy. in 1891; the 33rd Burma Infy. in 1901; present designation in 1903.

Composition—2 Companies of Sikhs, 2 of Punjabi Musalmans.

1st Commn. or date of entering service.	Names and Rank.	Army rank.	Present appointment in Regiment.	Remarks.
	Commandant.			
	Company Commanders. (4)			
17 June 91	Haughton, Maj. H. L.	17 June 09	31 May 16	2nd in-Comd.
20 May [illegible]	Morgan, Maj. [illegible]. F., D.S.O.	1 Sep. 15	10 Apl. 15	With 3-7th Infy.
12 Aug. 99	Burd, Maj. E. (*Actg. Lt.-Col.*).	1 Sep. 15	11 Aug. 15	With 85th Rif.
18 Jan. 02	Fox-Male, Maj. W. C. E. T., M.C.	18 Jan. 17	18 Apl. 16	With 2-5th Infy.
	Company Officers. (8)			
18 Jan. 11	Haycraft, Capt. W. C. S., M.C.	1 Sep. 15	12 Mar. 12	
18 Jan. 11	Taylor, Capt. V. C. L.	1 Sep. 15	1 Sep. 15	With 72nd Punj.
22 Jan. 13	*Pegg, Capt. A. C.*	*22 Jan. 17*	*19 Mar. 14*	Adjt. (*Offg.*).
14 Jan 14	Trevelyan, Capt. W. R. F.	14 Jan. 18	11 Oct. 14	With 2-153rd Infy.
29 June 16	Ruffell, Lt. B. (*temp. Capt.*)	29 June 17		
14 Jan. 15	*Gillies Reyburn, Lt. D. (temp Capt.)*	*14 Oct. 16*	. . . .	*Attd. Asst. Instr. Bombing, Mhow.*
30 Jan. 17	Fleming, Lt. R. S. T.	30 Jan. 18	. . . .	Attd. with 82nd Punj.
30 Jan. 17	Telford, Lt. B. D.	30 Jan. 18	. . . .	Attd. with 69th Punj.
27 Oct. 17	Cummins, Lt. J. E.	27 Oct. 18	4 Nov. 17	Attd.
31 Aug. 18	Ferguson, 2nd-Lt. D. C. H.	31 Aug. 18	5 Sep. 18	Attd.
	Attached.			
27 Aug. 02	Coningham, Maj. M. E.	27 Aug. 17	. . . .	96th Infy. 2nd-in-Comd. (*Offg.*)
31 Aug. 15	Oatway, Lt., S. H.	31 Aug. 16	13 Mar. 16	I. A. R. O.
26 Mar. 14	Hodson, Lt. S. S.	1 Sep. 15	18 Sep. 14	I. A. R. O.—Qr.-Mr., 29 Oct. 15.
26 July 15	*Hobday, Lt. W. E. (temp. Capt.)*	*26 July 16*	. . .	*I. A. R. O.—Staff Capt., Ambala Bde.*
20 Feb. 16	Coffey, Lt. E. P.	20 Feb. 17	31 Aug. 17	I. A. R. O.
7 Nov 17	Dickens, 2nd-Lt. C. E.	7 Nov. 17	5 June 18	I. A. R. O.
5 Nov. 17	Bellamy, Lt. E. A.	5 Nov. 18	14 Mar. 18	I. A. (*T. C.*)
20 June 17	Highley, Lt. W. M. B.	20 June 18	7 Oct. 17	I. A. R. O.
15 Nov. 15	Lalor, Lt. J. F. V.	15 Nov. 16	1 Jan. 17	1-94th Infy.
18 Apl. 16	Mc Clenaghan, Lt. P. J. W.	18 Apl. 17	. . . .	1-129th Baluchis.
13 June 16	Rigg, Lt. J. H.	13 June 17	. . . .	I. A. R. O.
22 July 16	Scott, Lt. J. A.	22 July 17	. . . .	I. A. R. O.
7 Nov. 17	Dickens, 2nd-Lt. C. E.	7 Nov. 17	5 June 18	I. A. R. O.
10 June 15	Dalrymple, 2nd-Lt. W. C.	10 June 18	5 July 18	I. A. R. O.
20 June 18	Cox, 2nd-Lt. T. W.	20 June 18	3 July 18	I. A. R. O.
27 Aug. 18	Joyce, 2nd-Lt. C. E.	27 Aug. 18	. . . .	I. A. R. O.
16 Oct. 18	Tattle, 2nd-Lt. J. B.	16 Oct. 18	25 Oct. 18	I. A. R. O.
	Subadar-Major.			
		Jemadar.	Subadar.	Subadar-Major.
14 July 88	Kirpal Singh (56)	1 June 03	19 Oct. 08	26 Mar. 17.
	Subadars.			
1 Mar. 91	Narayan Singh (56, 105)	1 Nov. 99	1 June 02	
22 Nov. 90	Bagh Shah (56)	1 Apl. 11	10 Oct. 14	
4 Mar. 91	Labh Singh (56)	19 Oct. 08	17 Feb. 15	
16 Oct. 92	Jawand Singh	15 Oct. 14	8 Jan. 16	
21 Nov. 00	Muhammad Sadiq	29 Jan. 16	6 Mar. 16	
19 July 00	Muhammad Abdul	15 Oct. 14	6 Mar. 16	
20 July 02	Karam Dad, I.O.M.	10 Mar. 16	14 Mar. 16	
13 Jan. 96	Lachhman Singh, I.O.M.	8 Mar. 16	17 Apl. 16	
13 Aug. 01	Bulaka Singh	4 Apl. 16	12 Aug. 16	
12 Dec. 02	Ghulam Hussain Shah	8 Mar. 16	20 Feb. 17	
28 Apl. 99	Khazan Singh	14 July 16	26 Mar. 17	
24 Apl. 00	Sheikh Ahmad	23 Mar. 17	30 Jan. 18	
20 Nov. 01	Harnam Singh	15 July 16	12 Apl. 18	
30 June 96	Fateh Khan	3 June 13	19 Apl. 18	
5 Nov. 01	Bhawa Singh	1 Dec. 16	24 May 18	
16 Nov. 04	Shaikh Abdullah, I.D.S.M.	7 Mar. 16	24 May 18	
15 Dec. 06	Ghulam Muhammad	3 Dec. 17	21 May 18	
	Jawand Singh, I.O.M. (*Retd.*)	. . . .	. . . .	Re-employed, 19 Sep. 15.
	Matwali (*Retd.*)	. . . .	. . . .	Re-employed, 27 Sep. 15.

93rd Burma Infantry—*contd.*

1st Commn. or date of entering service.	Names and Rank.	Army rank.	Remarks.
	Jemadars.		
31 Oct. 92	Karam Dad	10 Oct. 14	
30 July 03	Muhammad Din	8 Mar. 16	
15 July 02	Sayyid Ahmad	15 Mar. 16	
1 Aug. 06	Gulab Khan	13 Apl. 16	Indian Qr.-Mr.
26 May 02	Imam Din	2 June 16	
11 Aug. 96	Milran Singh	26 Mar. 17	Indian Adjt.
	Ram Singh (*Retd.*)		Re-employed, 4 Apl. 15. Depôt.
19 Jan. 02	Muhammad Zarif	30 Jan. 18	
1 Mar. 07	Muhammad Abbas	12 Apl. 18	
6 Mar. 02	Khan Singh, I.D.S.M.	19 Apl. 18	
6 Apl. 02	Jagat Singh	3 May 18	
5 June 02	Sahib Singh	24 May 18	
26 Nov. 11	Najib Khan	24 May 18	
15 Aug. 08	Khan Zaman	21 May 18	
16 May 06	Madat Khan	26 Aug. 18	
7 June 99	Harnam Singh	3 Sep. 18	

Linked with—95th & 96th Infy.

Uniform—Scarlet.
Facings—Dark Green.

1st Battalion, 94th Russell's Infantry.

Raised at Hyderabad in 1813, under the superintendence of Mr. Henry Russell, the Resident, and was originally designated the 1st Battalion of the Russell Brigade. Became the 1st Regiment of Infantry, Nizam's Army, in 1826; the 1st Infy., Hyderabad Contgt., in 1854; present designation, 1903.

"Maheidpoor" "Nowah."

Composition—1½ Companies of Rajputs, 1 of Jats, 1½ of Dekhani Musalmans.

1st Commn. or date of entering service.	Names and Rank.	Army rank.	Present appointment in Regiment.	Remarks.
	Commandant.			
18 June 92	Dobbs, Lt.-Col. C. F., D.S.O., *p.s.c.*	1 Jan. 16	4 Feb. 17	
	Company Commanders. (4)			
10 Oct. 94	Lane, Maj. W. H. (*Actg. Lt.-Col.*)	10 Oct. 12	25 Jan. 18	2nd-in-Comd. Comdt. (*temp.*)
17 Mar. 00	Martin, Maj. J., M.C.	1 Sep. 15	17 Nov. 14	With 2-97th Infy.
18 Apl. 00	*Molony, Maj. O. H. C.*	*1 Sept. 15*	*4 Feb. 17*	*Offg. Cant. Magte.*
14 Sep. 01	Gout, Maj. P. J., M.C.	6 Apl. 16	17 Sep. 18	2nd-in-Comd. (*temp.*)
	Company Officers. (8)			
8 Feb. 02	St. John, Maj. O.	29 June 16	7 Feb. 14	With 1-75th Infy.
4 May 01	Wayman, Capt. W. L.	10 Mar. 11	2 Feb. 16	With 1-103rd Mah.
18 Jan. 05	*Greenhouse, Capt. F. S.*	*18 Jan. 14*	*26 Mar. 06*	*Poll. Employ.*
24 Jan. 06	Palmer, Capt. A. C.	24 Jan. 15	15 Oct. 07	With 1-97th Infy.
27 May 08	Donovan, Capt. C. H.	1 Sept. 15	6 Sep. 11	With 2-94th Infy.
29 Jan. 10	Corke, Capt R. M.	1 Sept. 15	10 Mar. 11	With 1-154th Infy.
4 Aug. 15	*Matthewman, Lt. A. S.*	*4 Aug. 16*	*28 Oct. 15*	No. 4 Rem't Sqdr.
15 Nov. 15	Lalor, Lt. J. F. V.	15 Nov. 16	19 Nov. 15	Attd With 93rd Infy.
2 Jan. 16	Barroll, Lt. W. S.	2 Jan. 17	14 May 17	Attd.
18 June 17	Yates, Lt. D. H. P.	18 June 18	24 June 17	With 1-154th Infy.
18 June 17	Bebbington, 2nd-Lt. H. J.	18 June 17	24 June 17	Attd. *Lv., ex I., m. c.*
10 Dec. 15	Smallwood, Lt. L. S.	10 Dec. 16	28 Aug. 17	Attd. With 1-154th Infy.
10 July 16	Hunter, Lt. J. H. D.	10 July 17	28 Aug. 17	Att d.
	Attached.			
16 Aug. 05	Roberts, Capt. F. L.	16 Aug. 14	25 Dec. 17	1-7th Rajputs.
24 Jan. 06	Cadell, Capt. A. P. H. (*Actg. Maj.*)	24 Jan. 15	27 May 18	109th Infy.
26 June 17	Constable, 2nd-Lt. W. N. B.	26 June 17	24 July 17	Essex R., I. A. (*on probn.*)
5 May 17	Koch, Lt. E. M.	5 May 18	18 Aug. 17	I. A. R. O.
5 May 17	Stedman, Lt. J. F.	5 May 18	17 Aug. 17	I. A. R. O.
9 Dec. 14	Tweed, Lt. J. R. H. (*Actg. Capt.*)	9 Sep. 16	2 Aug. 17	Devon. R. I. A. (*on probn.*) Adjt.
19 July 15	Beales, Lt. W. L. (*Actg. Capt.*)	19 July 15	30 Aug. 17	Norfolk R. I. A. R. O. (*on probn.*)
3 Jan. 15	Edwards, Lt. F. H. (*Actg. Capt.*)	3 Jan. 16	31 Oct. 17	I. A. R. O.
24 Aug. 15	Wilkinson, Lt. E. M.	23 Feb. 17	29 Apl. 18	R. Innis. Fus., I. A. R. O., (*on probn., Lv., ex I., m. c.*)
27 Jan. 16	Watson, Lt. A. C.	22 Jan. 17	20 May 18	R. Ir. Fus., I. A. (*on probn.*)
23 Dec. 15	Moir, Lt. G.	23 Dec. 16	27 Mar. 18	I. A. R. O.
19 Dec. 16	Richardson, Lt. H.	19 Dec. 17	20 May 18	I. A. R. O.
26 Jan. 18	Dow, 2nd-Lt. G.	26 Jan. 18	4 Apl. 18	R. Scots., I. A. (*on probn.*)
15 Jan. 18	Wiffen, 2nd-Lt. G. S. W.	15 Jan. 18	5 June 18	I. A. (*T. C.*)
29 Mar. 18	Deung, 2nd-Lt. M. T.	29 Mar. 18	16 July 18	I. A. (*T. C.*)
11 Sep. 15	Faulder, Lt. G. B.	11 Sep. 16	4 July 18	W. Rid. R., I. A. (*on probn.*)
26 Feb. 15	Dalton, Lt. R. C., M. C.	20 Nov. 15	21 Sep. 18	Lond. R., I. A. R. O. (*on probn.*)
20 Aug. 18	Somerville, 2nd-Lt. A. P.	20 Aug. 18	21 Sep. 18	I. A. R. O.
7 June 18	Brown, 2nd-Lt. C.	7 June 18	13 Oct. 18	I. A. (*T. C.*)
26 Apl. 17	Watson, 2nd-Lt. A. G.	26 Apl. 17	18 Oct. 18	Gord Highrs., I. A. R. O. (*on probn.*)
21 Dec. 17	Male, 2nd-Lt. E.	21 Dec. 17	23 Aug. 18	Genl List I. A. R. O. (*on probn.*)
	Ryan, 2nd-Lt. R.	. . . . .	4 Sep. 18	I. A. (*on probn.*)

Subadar-Major.

		Jemadar.	Subadar.	Subadar-Major.
29 June 99	Duli Chand (52, 56, 66*a*)	1 Nov. 05	26 Oct. 08	1 Mar. 17.

Subadars.

18 June 95	Jitan Singh (53, 56, 66*a*)	1 Aug. 09	1 Jan. 13	
16 Feb. 03	Shiupalak Singh (56, 62)	5 Apl. 15	28 July 15	
3 Aug. 08	Jawahir Singh (56, 62)	28 Oct. 15	1 Jan. 17	
4 May 03	Rajkishor Singh (58)	12 Nov. 15	2 May 17	
10 Jan. 07	Shaikh Abdullah (56, 60*a*)	1 July 14	8 Oct. 17	
1 Apl. 00	Shaikh Muhammad Ali	1 Jan. 16	27 Oct. 17	
1 Feb. 03	Ramdhari Singh	2 May 17	18 May 18	
	Rampershad Tewari (*Retd.*)	. . . . .	. . . . .	Re-employed 13 Feb. 18.
	Saiyid Padshah (*Retd.*)	. . . . .	. . . . .	Re-employed 2 June 18.

1st Battalion, 94th Russell's Infantry—*contd.*

Date of entering service.	Names and Rank.	Army rank.	Remarks.
	Jemadars.		
8 Aug. 04	Remrekha Singh (56)	17 Apl. 15	
3 Mar. 07	Saiyid Hashim (66*a*)	1 Jan. 17	
22 Feb. 07	Murli Singh	1 Jan. 17	
8 Jan. 00	Ramji Lal	15 May 17	
1 July 97	Muhammad Umar	26 May 17	
7 Jan. 04	Patiram Singh	3 June 17	
6 Sep. 04	Ramsuchet Singh (58)	9 Sep. 17	
13 July 10	Abhai Ram (60*a*)	9 Sep. 17	
17 July 06	Gaya Singh (56, 60*a*)	27 Oct. 17	Indian Adjt.
19 Dec. 08	Sita Ram Singh (56)	27 Oct. 17	
25 Sep. 14	Dal Singh	3 Dec. 17	
8 Oct. 98	Bhim Singh (62)	18 May 18	
16 Feb. 03	Shiujadik Singh	18 May 18	
19 Dec. 09	Baramdeo Singh	18 May 18	
15 Sep. 11	Ismail Khan (50)	1 Sep. 18	
20 Nov. 11	Kundan Singh	19 Sep. 18	
20 Aug. 06	Diwan Singh	19 Sep. 18	
....	Chaudhri Padam Singh	....	(*temp Hony.*)

2nd Battalion, 94th Russell's Infantry.

Formed at Saugor on 11 June 1917.

Composition ½ Company Dekhani Musalmans, 1½ Companies Rajputs, 1 Company Jats, 1 Company Ahirs.

1st Commn. or date of entering service.	Names and Rank.	Army rank.	Present appointment in Regiment.	Remarks.
	Commandant.			
20 July 98	Meiklejohn, Maj. W. L. S., *p. s. c., q* (*Actg. Lt.-Col.*)	1 Sep. 15	1 Aug. 18	106th Prs. (*temp.*).
	Company Commander. (4)			
5 Aug. 05	Jopp, Capt. C. H. K. (*Actg. Maj.*)	5 Aug. 14	29 Aug. 18	4th Lrs. 2nd-in-Comd. (*temp.*)
30 Sep. 15	Boddy, Lt. S. M.	30 Sep. 16	19 June 18	I. A. R. O.
10 Dec. 15	Clyde, Lt. W. D.	10 Dec. 16	...	I. A. (*on probn.*)
18 July 15	Scott, Lt. A. G.	18 Apl. 17	2 Aug. 17	Attd.
	Company Officers.			
27 May 08	Donovan, Capt. C. H.	1 Sep. 15	15 July 18	1-94th Infy.
13 July 15	Coulthard, Lt. G. R. (*Actg. Capt.*)	13 July 16	27 Sep. 17	I. A. R. O. Adjt.
20 June 17	John, Lt. W. R. (*temp. Capt.*)	20 June 18	3 Oct. 17	I. A. R. O.
20 Mar. 15	Young, Lt. A. D.	20 Mar. 16	29 Jan. 18	I. A. R. O. *S. S. O., Saugor.*
19 May 16	Stanley, Lt. P. T.	19 May 17	16 Jan. 18	I. A. (*on probn.*)
26 Aug. 15	Richdale, Lt. C. R.	26 May 17	14 July 17	Attd.
22 Dec. 15	Poynor, Lt. M. P.	22 Sep. 17	26 Nov. 17	Attd.
16 Feb. 17	Fennell, Lt. W. C.	16 Feb. 18	20 Jan. 18	I. A. (*on probn.*)
26 July 17	Marks, Lt. G. A.	26 July 18	5 Sep. 17	I. A. R. O.
15 Oct. 15	Ainslie, Lt. K.	15 Oct. 16	3 Aug. 17	I. A. R. O. *Lv., ex I., m c.*
29 Aug. 17	Howard, Lt. C.	29 Aug. 18	8 Dec. 17	I. A. R. O. (*on probn.*)
26 Sep. 17	Addinall, Lt. C. R.	26 Sep. 18	11 Feb. 18	I. A. R. O.
31 Oct. 17	Wenmoth, Lt. H. H.	31 Oct. 18	20 Jan. 18	I. A. R. O. (*on probn.*)
17 Oct. 17	Kelley, Lt. W. S.	17 Oct. 18	28 Mar. 18	I. A. (*on probn.*)
31 Oct. 17	Stewart, Lt. M. Mc D.	31 Oct. 18	12 Feb. 18	I. A. R. O.
25 Nov. 17	Herredge, Lt. W. J.	25 Nov. 18	15 Jan. 18	I. A. R. O.
28 Dec. 17	Halloran, 2nd-Lt. G. B.	28 Dec. 17	28 Jan. 18	I. A. (*on probn.*) Qr. Mr.
15 Jan. 18	McKinney, 2nd-Lt. J.	15 Jan. 18	5 July 18	I. A. (T. C.)
28 Feb. 18	French, 2nd-Lt. C. R.	28 Feb. 18	15 Jan. 18	I. A. (*on probn.*)
7 June 18	Tyrell, 2nd-Lt. A.	7 June 18	9 Oct. 18	I. A. R. O.

Subadar-Major.

		Jemadar.	Subadar.	Subadar-Major.
9 July 92	Shaikh Kalandar	17 Aug. 04	3 Oct. 17	11 June 17 (*temp.*)

Subadars.

9 Aug. 02	Antu	13 May 11	11 June 17
2 Dec. 02	Bal Mukand Singh	21 Aug. 12	11 June 17
13 Apl. 95	Shaikh Abdul Karim	8 Aug. 14	11 June 17
4 June 98	Shadi Ram	13 June 16	15 Jan. 18
18 May 06	Bhagwan Singh	11 June 17	1 Oct. 17
25 Nov. 05	Ram Lal Singh	11 June 17	1 Oct. 17
15 Jan. 18	Lekh Ram	15 Jan. 18	14 Feb. 18

Jemadars.

23 Dec. 07	Ganga Singh	11 June 17	
25 Dec. 03	Lachmi Singh	11 June 17	
29 May 03	Shiu Singh	11 June 17	
23 Feb. 06	Amilal	11 June 17	
1 May 10	Badlu	1 Oct. 17	
20 Jan. 14	Raje Khan	23 Jan. 18	
8 July 03	Ramraj Singh	14 Feb. 18	
16 Dec. 01	Lalji Ram	14 Feb. 18	
26 Dec. 00	Nanig Ram	14 Feb. 18	
14 Sep. 17	Awal Singh	21 Mar. 18	(*On probn.*)
3 Dec. 05	Sis Ram	1 May 18	
28 Mar. 10	Jabdal	1 May 18	
9 Mar. 03	Marjad Singh	1 May 18	
16 July 08	Daya Shankar	1 May 18	
23 July 06	Rambarat Singh	15 July 18	
5 Apl. 07	Darshan Singh	15 July 18	
27 Aug. 08	Lekh Ram	15 July 18	
24 Mar. 09	Abdul Manaf	15 July 18	
14 Oct. 14	Sher Singh	15 July 18	

Linked with—94th & 96th Infy.

Uniform—Scarlet.
Facings—Dark Green.

1st Battalion, 95th Russell's Infantry.

Raised at Hyderabad in 1813, under the Superintendence of Mr. Henry Russell, the Resident, and originally designated the 2nd Battalion of the Russell Brigade. Became the 2nd Regiment of Infantry, Nizam's Army, in 1826; the 2nd Infy., Hyderabad Contgt., in 1854; present designation in 1903.

"Maheidpoor" "Nowah" "Burma, 1885-87."

Composition—1½ companies of Rajputs, 1½ of Hindustani Musalmans, 1 of Ahirs of the Eastern Punjab.

1st Commn. or date of entering service.	Names and Rank.	Army rank.	Present appointment in Regiment.	Remarks.
	Commandant			
11 Feb. 88	Major, Lt.-Col. F. F.	11 Feb. 14	11 Oct. 14	
	Company Commanders. (4)			
3 Sep. 92	Pratt, Maj. E. G. W. (Q)	3 Sep. 10	1 May 16	2nd-in-Comd. With 2-18th Infy. *Lv., ex I., m. c.*
20 Jan. 97	Rennick, Maj. R. W. deC.	20 Jan. 15	1 May 13	With 4-9th Infy.
8 May 01	DeGruchv, Maj. F. A. L.	8 May 16	12 May 16	
27 Aug. 02	*Meade, Maj. W. L.*	*27 Aug. 17*	*25 Jan. 18*	*Pte. Secy. to the L. G., Burma.*
	Company Officers. (8)			
5 Aug. 05	Simpson, Capt. A. F. (*Actg. Maj.*)	5 Aug 14	16 May 07	
20 Dec. 05	*Goldsmith, Bt. Maj. H. A.*, M. C.	*3 June 18*	*24 July 08*	*I. S. Troops. Lv., ex I., m. c.*
9 Sep. 08	Ruck, Capt. R. C.	1 Sept. 15	24 Nov. 09	With 2-98th Infy.
3 Sep. 10	Maltby, Capt. C. M., M. C. (*Actg. Maj.*)	1 Sept. 15	17 Dec. 11	
20 Jan. 12	Fasken, Capt. K. L.	20 Jan. 16	10 Mar. 13	
24 Aug. 12	Williamson, Capt. C. P. F.	24 Aug. 16	5 Nov. 13	
15 Nov. 15	Ellison, Lt. D. H.	15 Nov. 16	20 Nov. 15	With 109th Infy.
27 Oct. 16	Bullock, Lt. H. (*Actg. Capt.*)	27 Oct. 17	2 Jan. 17	Attd.
18 June 17	Mandeville, Lt. W. (*Actg. Capt.*)	18 June 18	24 June 17	Attd.
18 June 17	Partridge, Lt. J. W.	27 Dec. 16	24 June 17	Attd.
22 June 15	Elliott, Lt. A. C. (*Actg. Capt.*)	22 June 16	....	Attd.
27 Aug. 15	Rayner, Lt. C. S. W. (Actg. Capt.)	27 Aug. 16	....	Attd. With 1-48th Prs.
	Attached.			
10 Nov. 16	Wallace, Lt. J. R. K. (*Actg. Capt.*)	10 Nov. 17	30 Jan. 17	I. A. R. O. Adjt.
30 Jan. 17	Farrar, Lt. W. B.	30 Jan. 18	3 Feb. 17	5th Infy.
1 Mar. 17	Parker, Lt. H. R. C.	1 Mar. 18	12 Sep. 17	I. A. R. O.
8 Aug. 17	Crowther, 2nd-Lt. H. J.	8 Aug. 17	3 Nov. 17	I. A. R. O.
26 Sep. 17	Jenkins, 2nd-Lt. J. W. P. J.	26 Sep. 17	12 Feb. 18	I. A. R. O.
29 Aug. 17	Edwards, 2nd-Lt. H.	29 Aug. 17	12 Feb. 18	I. A. R. O.
5 May 17	Clark, Lt. W. J.	5 May 18	29 Apl. 18	I. A. (*on probn.*)
5 Jan. 18	Smith, 2nd-Lt. W. H.	5 Jan. 18	4 June 18	I. A. (*T. C.*)
19 Mar. 18	Homan, 2nd-Lt. R.	19 Mar. 18	16 July 18	I. A. (*T. C.*)
	Stuart-Shepherd, 2nd-Lt. U. G. S.	....	....	I. A. (*T. C.*)
19 Apl. 17	Brigham, Lt. G.	19 Apl. 18	18 Aug. 17	
	Daick, 2nd-Lt. W. J.	....	7 Sep. 18	
28 May 18	Dolman, 2nd-Lt. H. G.	28 May 18	29 Aug. 18	Hamps. R., I. A. (*on probn.*)
20 Aug. 18	Hepburn, 2nd-Lt. W. S.	20 Aug. 18	23 Oct. 18	
26 Sep. 18	Girsham, 2nd-Lt. W. S.	26 Sep. 18	27 Oct. 18	
11 Oct. 18	Best, 2nd-Lt. G. H.	11 Oct. 18	10 Nov. 18	
23 Oct. 18	Netscher, 2nd-Lt. E. G.	23 Oct. 18	10 Nov. 18	

Subadar-Major.

		Jemadar.	Subadar.	Subadar-Major.
8 May 95	Shaikh Ahmadullah (52, 56, 66*a*)	11 Dec. 07	11 Nov. 08	1 Apl. 17

Subadars.

28 Oct. 93	Nohar Singh (53, 56, 66*a*)	16 July 10	21 Aug. 12	
23 Jan. 00	Bhawani Singh (53, 56)	16 Apl. 10	1 July 13	
1 Apl. 10	Kunwar Shiudatt Singh (56, 66*a*)	1 Apl. 10	1 Jan. 14	
4 Mar. 03	Mehdi Hasan (46, 51, 66*a*)	29 May 11	11 Nov. 14	
13 Feb. 01	Rajbali Singh (52, 56, 66*a*)	1 Jan. 12	5 May 15	
22 Dec. 01	Ramnarayan Singh (56)	1 July 13	23 Aug. 15	
	Baijnath Sing (*Retd.*)	....	....	Re-employed, 30 Mar. 15. Depôt.
3 Jan. 01	Ram Pershad Singh, I.D.S.M. (66*a*)	1 Jan. 14	1 Oct. 16	
4 Dec. 96	Muhammad Raza Khan	11 Nov. 14	28 Jan. 18	
4 June 98	Fateh Singh (60*a*)	13 June 16	17 May 18	
16 Dec. 01	Khem Ram (46, 66)	1 Aug. 16	17 May 18	
16 Dec. 01	Durjan Singh (60*a*)	1 Aug. 16	23 Sep. 18	

Linked with—94th & 96th Infy.

Uniform—Scarlet.
Facings—Dark Green.

1st Battalion, 95th Russell's Infantry—*contd.*

1st Commn. or date of entering service.	Names and Rank.	Army rank.	Present appointment in Regiment.	Remarks.
	Jemadars.			
17 Nov. 03	Muhammad Usman Khan (56, 66*a*)	28 May 15		
19 Mar. 91	Muhammad Wazir (58) . . .	1 July 16		
26 May 97	*Shahzad Khan*	*11 July 16*		
21 Nov. 01	Harnarayan Singh	1 Apl. 17		
24 Feb. 06	Sheodhan Singh (56)	1 June 17		
11 Feb. 06	Budh Ram (56)	1 Aug. 17		
14 July 08	Hari Singh (51).	1 Aug. 17		
25 Jan. 07	Shaikh Masnad Ali (48*a*, 52, 56, 61) I.D.S.M.	1 Aug. 17		
22 July 04	Sheo Chand	28 Jan. 18		
24 Feb. 97	Qasim Khan	28 Jan. 18		
13 May 12	Girdhari Singh	29 Jan. 18		
30 June 08	Shivchand (56)	1 June 18		
22 Dec. 06	Ghansyam	1 June 18		
21 Mar. 06	Dipchand (56)	1 June 18		
23 July 02	Paupiah	3 Dec. 17		
. . .	Ramlal Singh	1 Sep. 17		
. . .	Babu Singh	17 May 18		
. . .	Faiz-ul-Hasan Khan	17 May 18		
16 Feb. 08	Jagesar Singh	6 Sep. 18		
13 May 12	Chhaju Ram	28 Sep. 18		
15 Aug. 10	Het Ram	28 Sep. 18		
	Mukhdum Bakhsh Khan . .	28 Sep. 18		
	Rambadan Singh	10 Oct. 18		

2nd Battalion, 95th Russell's Infantry.

Formed at Secunderabad on 20th January 1918.

Composition—1 company Rajputs, 1 company Hindustani Musalmans, and 2 companies Ahirs of Western U. P.

1st Commn. or date of entering service.	Names and Rank.	Army rank.	Present appointment in Regiment.	Remarks.
	Commandant.			
10 Nov. 92	Savile, Lt.-Col. A. R. C.	20 Jan. 18	20 Jan. 18	31st Lancers.
	Company Commanders. (4)			
21 July 15	Goddard, Lt. A. G. I. A. (*Actg. Capt.*)	21 July 16	22 Oct. 18	Mon. R., I. A. (*on probn.*) 2nd-in-Comd. (*temp.*)
13 Aug. 15	Telfer, Lt. J. W. (*Actg. Capt.*)	13 Aug. 16	15 Nov. 18	R. Highrs., I. A (*on probn.*)
25 July 17	Merriman, Lt. S. T. (*Actg. Capt.*)	25 July 18	23 Oct. 18	North'n R., I. A. (*on probn.*)
20 June 17	Lang, 2nd-Lt. H. G. (*Actg. Capt.*)	20 Mar. 18	21 Feb. 18	Attd.
	Company Officers.			
28 July 15	Grey, Lt. R. C.	28 July 16	29 Aug. 18	I. A. (*on probn.*) *Lv., ex I., m. c.*
11 Aug. 16	Naye, Lt. R. A.	11 Aug. 17	10 Nov. 18	R. Innis. Fus., I. A. (*on probn.*)
10 May 16	Jeffreys, Lt. H. F. (*Actg. Capt.*)	10 May 17	. . .	Lond. R., I. A. (*on probn.*)
30 Jan. 17	McDonnell, Lt. F. A.	30 Jan. 18	14 Feb. 18	1-5th Infy. *Lv., ex I., m. c.*
5 June 17	Turner, Lt. W. T.	5 June 18	. . .	R. W. R., I. A. (*on probn.*)
5 Oct. 17	Hughes, Lt. S.	5 Oct. 18	11 May 18	I. A. (*T. C.*)
31 Oct. 17	Low, 2nd-Lt. W. H.	31 Oct. 17	2 May 18	I. A. R. O.
28 Dec. 17	Munro, 2nd-Lt. A., D.S.O.	28 Dec. 17	21 May 18	I. A. (*on probn.*)
15 Jan. 18	Eade, 2nd-Lt. R. O.	15 Jan. 18	3 June 18	I. A. R. O. Qr.-Mr. (*temp.*)
15 Jan. 18	Edge, 2nd-Lt. E. H.	15 Jan. 18	3 June 18	I. A. R. O.
15 Jan. 18	Topham, 2nd-Lt. L. H. C.	15 Jan. 18	3 June 18	I. A. R. O.
26 Jan. 18	Green, 2nd-Lt. A. W.	28 Jan. 18	14 May 18	I. A. (*on probn.*)
1 May 18	Duffy, 2nd-Lt. S. C. R.	1 May 18	25 May 18	I. A. (*on probn.*)
10 July 18	Russell, 2nd-Lt. G. A.	10 July 18	3 Dec. 18	I. A. (*T. C.*)
	Hart, 2nd-Lt. W.	. . .	7 Oct. 18	I. A. (*T. C.*)

Subadar-Major.

		Jemadar.	Subadar.	Subadar-Major.
18 May 97	Alam Khan	12 July 14	20 Jan. 18	18 Oct. 18

Subadars.

30 July 15	Harjiwan Singh	. . . .	20 Jan. 18
2 Dec. 99	Jagram Singh	10 Mar. 16	20 Jan. 18
4 Dec. 99	Sewa Ram	10 Mar. 16	20 Jan. 18
21 Aug. 06	Sukh Deo Singh	12 Nov. 14	27 Jan. 18
12 Oct. 08	Hazari Singh	10 Mar. 16	20 Jan. 18

Jemadars.

25 Oct. 15	Nihal Singh	1 Oct. 17
27 Nov. 07	Barkatulla Khan	27 Jan. 18
5 May 04	Mohd. Shafi Khan	20 Jan. 18
22 May 04	Amrit Singh	20 Jan. 18
22 May 04	Shiudas Singh (48)	26 Jan. 18
27 Aug. 08	Mehar Chand	20 Jan. 18
24 Dec. 02	Latif Khan	1 Aug. 18
26 Feb. 06	Deokaran Singh	1 Aug. 18
24 Feb. 06	Ranjit Singh	1 Aug. 18
14 Oct. 12	Muhammad Isaq Khan	1 Aug. 18

Linked with—94th & 95th I: fy.

Uniform—Scarlet.
Facings—Dark Green.

1st Battalion, 96th Berar Infantry.

Raised at Hyderabad in 1797. Was the 2nd Battalion, Aurangabad Division, and afterwards the 2nd Battalion, Berar Infantry. Became the 3rd Regiment of Infantry, Nizam's Army, in 1826; the 3rd Infantry, Hyderabad Contingent in 1854; present designation in 1903.

"Nowah" "Central India" "Burma, 1885-87."

Composition—1½ companies of Rajputs, 1 of Jats, 1½ of Hindustani Musalmans.

1st Commn. or date of entering service.	Names and Rank.	Army rank.	Present appointment in Regiment.	Remarks.
	Commandant.			
8 Oct. 90	Nangle, Lt.-Col. K. E. . . .	8 Oct. 16	3 May 17	
	Company Commanders. (4)			
11 Mar. 99	Jardine, Bt.-Lt.-Col. C. H. . .	3 June 17	21 May 18	2nd-in-Comd. With 2-96th Infy.
8 Jan. 01	Conry, Maj. J. deL.	8 Jan. 16	3 Aug. 17	2nd-in-Comd. (*temp.*)
28 Jan. 03	*Apthorp, Maj. S. E.*, D.S.O. .	*22 Sep. 16*	*29 Mar. 18*	*C. C. Sch. of Instn. for Officers, Sabathu.*
27 Aug. 02	Coningham, Maj. M. E. . . .	27 Aug. 17	21 May 18	With 93rd Infy.
	Company Officers. (8)			
22 Oct. 02	Rollo, Maj. L. D. . . .	22 Oct. 17	7 Feb. 04	
29 July 03	*Spence, Maj. R. B.*	*29 July 18*	*1 May 06*	*D. A. A. & Q. M. G.*
18 Jan. 05	Reynolds, Capt. E. W. (*Actg. Maj.*)	18 Jan. 14	21 Mar. 06	
4 Sep. 07	Hogan, Capt. P. T.	1 Sep. 15	17 Sep. 09	Qr.-Mr., 1 May 16. *Lv., ex I., m. c.*
18 Jan. 11	*Ricketts, Capt. E. L.*	*1 Sep. 15*	*6 Mar. 12*	*Bde. Major.*
12 Feb. 15	Angelo, Lt. N. L. (*Actg. Capt.*) .	12 Nov. 16	18 Apl. 15	Attd.
18 Feb. 15	Jervis, Lt. T. S.	18 Nov. 16	22 Apl. 15	Attd.
15 Nov. 15	Garbett, Lt. D. A. (*Actg. Capt.*)	15 Nov. 16	21 Nov. 15	
5 Dec. 15	Wallace, Lt. E. G. (*Actg. Capt.*)	5 Dec. 16	20 Apl. 15	Attd. Adjt.
8 Sep. 15	Jeffreys, Lt. J. F. D.	8 June 17	3 May 16	Attd.
	Attached.			
10 May 16	Knowles, Lt. J. F.	10 May 17	12 Feb. 17	I. A. R. O.
7 Oct. 17	Page, Lt. H. J.	7 Oct. 18	29 Nov. 17	I. A. R. O.
5 Oct. 17	Farley, Lt. H. J.	5 Oct. 18	18 Jan. 18	I. A. (*T. C.*)
5 Nov. 17	Jones, Lt. A. H.	5 Nov. 18	18 Mar. 18	I. A. R. O.
19 Mar. 18	Muspratt, 2nd-Lt. R. A. L. . .	19 Mar. 18	16 July 18	I. A. (*T. C.*)
1 July 16	Tapsell, Lt. C. B.	1 July 17	26 July 18	I. A. R. O.
8 Oct. 16	Thiele, Lt. C. E.	8 Oct. 17	5 July 18	Welsh R., I. A. (*on probn.*)
27 Mar. 18	Trutch, 2nd-Lt. H. F.	27 Mar. 18	26 July 18	I. A. (*on probn.*)
25 Oct. 16	Sutcliffe, Lt. J. E.	25 July 18	2 Sep. 18	M. G. Corps, I. A. (*on probn.*)
5 Jan. 17	Muspratt, Lt. G. (*Actg. Capt.*) .	5 Jan. 18	3 Mar. 18	I. A. R. O. Qr.-Mr. (*offg.*)
18 June 17	Belben, Lt. T. A.	18 June 18	12 June 18	1-123rd Rif.

Subadar-Major.

		Jemadar.	Subadar.	Subadar Major.
26 May 00	Manghat (53, 56)	1 Feb. 07	13 May 11	1 June 17

Subadars.

23 Mar. 03	Shaikh Muhammad Umar . .	1 May 10	15 June 14
2 Nov. 04	Niamat Khan (51, 52, 56) . . .	1 Sep. 11	5 Mar. 15
6 Jan. 04	Gulpham Khan (52, 56) . . .	1 May 10	26 Apl. 15
1 Apl. 04	Chatarpal Singh (1), (52) . . .	3 Feb. 10	15 Aug. 15
25 June 01	Sukh Ram (52, 56)	16 June 10	15 Aug. 15
12 May 03	Jagdeo Singh (52, 56) . . .	25 June 10	29 June 16
9 Sep. 04	Dalpat Singh	15 Aug. 15	9 July 16
8 July 08	Allaud Din	5 Mar. 15	3 Dec. 17
18 July 06	Raghubir Singh	8 June 15	11 Mar. 18

Jemadars.

10 Apl. 00	Sahi Ram	15 Aug. 15	
7 Mar. 03	Dhan Singh	9 July 16	
4 July 06	Chitta	28 July 16	Indian Adjt.
19 Sep. 05	Sri Ram	22 Aug. 16	
1 Oct. 07	Jagatpal Singh (46)	29 Sep. 16	
9 Mar. 03	Muhammad Murtaza Khan . .	11 June 17	
	Rajdeo Singh	13 Dec. 17	
19 June 03	Pokhar	11 Mar. 18	
18 Apl. 03	Abdul Latif	11 Mar. 18	
20 Oct. 06	Thakurparshad Singh	4 Mar. 18	
6 Jan. 04	Muhammad Ishaq Khan . . .	1 June 18	
4 Oct. 04	Pir Bakhsh	18 May 18	
5 Apl. 03	Naulakh Singh	18 May 18	
28 Feb. 08	Harnarain	18 May 18	
	Abdul Ahmad	18 May 18	
	Ibad-ullah Khan	1 July 18	
29 Apl. 07	Shaikh Mansur.	1 Sep. 18	Indian Qr.-Mr.
17 Apl. 01	Lekhram.	2 Sep. 18	

2nd Battalion, 96th Infantry.

Formed at Ahmednagar on the 4th March 1918.

Composition—2 Companies Hindustani Musalmans, 1 Company Tagas, ½ Company Rajputs, ½ Company United Provinces Jats.

1st Commn. or date of entering service.	Names and Rank.	Army Rank.	Present appointment in Regiment.	Remarks.
	Commandant.			
11 Mar. 99	Jardine, Bt. Lt.-Col. C. H.	3 June 17	7 Mar. 18	1-96th Infy.
	Company Commanders (4)			
16 Aug. 95	Staples, Capt. E. H. (*Actg. Maj.*)	16 Aug. 14	9 Mar. 18	122nd Infy., 2nd in-Comd. (*temp.*)
1 Jan. 15	Cotton, Lt. J. T. (*Actg. Capt.*)	1 July 17	20 Mar. 18	Norf. R., I. A. (*on probn.*)
14 Dec. 15	Hall, Lt. F. A.	1 July 17	17 May 18	R. Ir. R., I. A. (*on probn.*)
30 Sep. 16	Sendall, Lt. W. H.	30 Sep. 17	7 Sep. 18	Lond. R., I. A. (*on probn.*)
	Company Officers.			
	Attached.			
28 Sep. 17	Thornton, Lt. H. E.	28 Sep. 18	8 Sep. 18	York. R., I. A. (*on probn.*) Adjt. (*Actg.*)
1 Feb. 17	Kearsey, Lt. F. W.	1 Feb. 18	2 Apl. 18	I. A. R. O.
12 Mar. 17	Laird, 2nd-Lt. E. H.	12 Mar. 17	5 June 18	Lond. R., I. A. R. O. (*on probn.*)
15 June 17	Peters, 2nd-Lt. F. A.	15 June 17	23 Aug. 18	Suff. R., I. A. R. O. (*on probn.*)
1 Aug. 17	Mclarty, 2nd-Lt. A. M.	1 Aug. 17	19 May 18	Lanc. Fus. I. A. R. O. (*on probn.*) Qr.-Mr.
1 Aug. 17	Watts, 2nd-Lt. J. M.	1 Aug. 17	3 June 18	I. A. R. O.
15 Jan. 18	Prebble, 2nd-Lt. C. C.	15 Jan. 18	3 June 18	I. A. R. O.
15 Jan. 18	Jeffreys, 2nd-Lt. K.	15 Jan. 18	3 June 18	I. A. (*I. C.*)
29 Mar. 18	Burchett, 2nd-Lt. Y. W. K.	29 Mar. 18	17 July 18	I. A. R. O. (*on probn.*)
29 Mar. 18	Hoore, 2nd-Lt. J.	29 Mar. 18	22 July 18	I. A. R. O. (*on probn.*)
3 Mar. 16	Battye, Lt. R. D.	3 Mar. 17	. . . .	A. S. C., I. A. (*on probn.*)
	Newsam, 2nd-Lt. E. S.	. . . .	. . . .	Genl. List, I. A. R. O. (*on probn.*)
	Woodcock, 2nd-Lt. T.	. . . .	. . . .	Genl. List, I. A. R. O. (*on probn.*)
16 Mar. 17	Davies, 2nd-Lt. G. S.	16 Mar. 17	. . . .	I. A. R. O.
	Medical Officer.			

		Jemadar.	Subadar.	Subadar-Major.
	Subadar-Major.			
12 Apl. 03	Jadu Nandan Singh, I.D.S.M.	22 Jan. 10	20 Nov. 12	4 Mar. 18.
	Subadars.			
9 May 08	Muhammad Zaman Khan	15 June 14	4 Mar 18	
17 May 00	Devi Singh	26 Apl. 15	4 Mar. 18	
4 July 08	Muhammad Hanif Khan (48*a*)	26 Apl. 15	4 Mar. 18	
27 July 05	Ramdas Singh	26 Apl. 15	4 Mar. 18	
23 Aug. 06	Harphool	26 Apl. 15	4 Mar. 18	
4 June 98	Pat Ram	1 July 04	20 July 16	
13 June 08	Shamshul Khan	26 Sep. 16	4 Mar. 18	
	Jemadars.			
13 Sep. 04	Miran Bakhsh	29 June 15		
26 Feb. 04	Mohar Singh	1 Apl. 17		
11 Mar. 04	Baz Singh	4 Mar. 18		
10 Oct. 08	Muhammad Mustafa Khan	4 Mar. 18		
19 Jan. 06	Dipram	4 Mar. 18		
22 Dec. 01	Nihala	4 Mar. 18		
1 Nov. 05	Ramanan Singh	4 Mar. 18		
4 July 06	Ugan Singh (46)	4 Mar. 18		

Linked with—98th & 99th Infy.

1st Battalion, 97th Deccan Infantry.

Uniform—Scarlet.
Facings—Dark Green.

Raised at Hyderabad in 1794 by Monsieur Raymond. Was the 3rd Battalion, Aurangabad Division, and afterwards the 3rd Battalion, Berar Infantry. Became the 4th Regt. of Infantry, Nizam's Army, in 1826; the 4th Infantry, Hyderabad Contingent, in 1854; present designation in 1903.

"Nagpore."

Composition—1½ companies of Rajputs, 1 of Jats, 1½ of Dekhani Musalmans.

1st Commn. or date of entering service.	Names and Rank.	Army rank.	Present appointment in Regiment.	Remarks.
	Commandant.			
31 Aug. 92	Preston, Lt.-Col. W. J.P., D.S.O.	31 Aug. 18	5 Nov. 18	
	Company Commanders. (4)			
11 Feb. 88	Lloyd, Lt.-Col. W. E. E. (Q)	11 Feb. 14	3 Oct. 13	2nd-in-Comd. With 2-9th Infy. *Lv. ex. I. mc.*
16 Jan. 95	Palmer, Bt. Lt.-Col. G. S. [L]	1 Jan. 18	26 May 14	
24 July 01	Kirkwood, Maj. W. C.	24 July 16	22 Jan. 16	
18 Jan. 02	*Eadie, Maj. J. I.*, D.S.O. [L]	*18 Jan. 17*	5 Nov. 18	
	Company Officers. (8)			
27 May 08	*Fletcher, Capt. W. T.*	*1 Sep. 15*	*3 Sep. 11*	
8 Aug. 14	Searle, Capt. C. S., M.C.	8 Aug. 18	13 Nov. 15	Attd.
8 Aug. 14	Lambert, Capt. W. R.	8 Aug. 18	12 Oct. 14	With 2-9th Infy.
15 Aug. 14	Johnson, Capt. J. C., M.C.	14 Aug. 18	13 Nov. 15	Attd.
1 Oct. 14	Keene, Capt. M. R.	1 Oct. 18	16 Dec. 14	With 1-144th Infy.
12 Sept. 17	Larkins, Lt. H. S.	12 Sep. 18	8 Dec. 17	Attd.
9 Apl. 15	Bevis, Lt. F. G.	9 Jan. 17	. . . .	Attd.
	Attached.			
15 July 03	Tucker, Maj. H. B.	15 July 18	12 Mar. 17	98th Infy.
20 Sep. 11	Villiers-Stuart, Capt. P. G.	20 Sep. 15	21 May 18	Ches. R., I. A. (*on probn.*) Adjt.
24 Jan. 06	Palmer, Capt. A. C.	24 Jan 15	. . . .	1-94th Infy.
11 Nov. 14	Cooke, Lt. S. F.	11 Nov. 15	13 Dec. 17	I. A. R. O.
7 May 15	Thorp, Lt. M. (*Actg. Capt.*)	7 May 16	30 May 16	I. A. R. O.
13 July 15	Eales, Lt. T. F. J.	13 July 16	16 Nov. 17	Oxf. & Bucks. L. I., I. A. (*on probn.*)
20 July 15	Dawe, Lt. H. R.	20 July 16	20 Aug. 16	I. A. R. O.
20 July 15	Hauser, Lt. S. B.	20 July 16	15 Aug. 15	I. A. R. O.
30 July 15	Todd, Lt. F. R. P.	30 July 16	31 Aug. 18	I. A. R. O.
10 Mar. 16	Ashley-Smith, Lt. R. (*Actg. Capt.*)	10 Mar. 17	20 Sep. 16	R. Lanc. R., I. A. (*on probn.*)
29 Oct. 15	Tierney, Lt. H.	29 Oct. 16	7 Nov. 16	I. A. R. O.
13 Jan. 16	Whittington, Lt. G. W.	13 Jan. 17	. . . .	I. A. (*on probn.*)
22 July 16	Ludlow, Lt. F.	22 July 17	31 Jan. 17	I. A. R. O.
5 Apl. 17	Morris, Lt. F.	5 Apl. 18	7 Apl. 17	I. A. R. O.
8 Aug. 17	Scott-Coward, Lt. C. W.	8 Aug. 18	3 Nov. 17	I. A. R. O.
5 Nov. 17	Symonds, Lt. E.	5 Nov. 18	10 Mar. 18	I. A. (*T. C.*) Qr. Mr.
30 Oct. 18	Guyton, 2nd-Lt. G.	30 Oct. 18	2 July 18	Linc.R., I. A. (*on probn.*)
19 Mar. 18	Soper, 2nd-Lt. W. T.	19 Mar. 18	17 July 18	I. A. (*T. C.*)
5 June 18	de Vos, 2nd-Lt. J. P.	5 June 18	27 June 18	I. A. R. O.
15 Aug. 18	Mackenzie, 2nd-Lt. W. R.	15 Aug. 18	17 Sep. 18	I. A. R. O.
7 Oct. 18	Tutte, 2nd-Lt. E.	7 Oct. 18	20 Oct. 18	I. A. R. O.
7 Oct. 18	Rutherford, 2nd-Lt. H. R.	7 Oct. 18	20 Oct. 18	I. A. R. O.

Subadar-Major.

		Jemadar.	Subadar.	Subadar-Major.	
7 July 97	Kanhaiya, I.D.S.M. (56)	Oct. 03	16 May 07	28 Nov. 15	
	Shaikh Farid (*Retd.*)	. . . .	. . . .	. . . .	Re-employed 23 Nov. 16.

Subadars.

15 Aug. 02	Râm-Sarúp Singh (50, 53)	1 Apl. 10	28 Mar. 15
22 Feb. 97	Charan Singh (56)	10 June 10	15 Apl. 15
7 Dec. 00	Gyani Singh	29 Nov. 10	24 May 15
23 Dec. 00	Sohan Singh (56)	20 Oct. 14	30 Aug. 15
9 Dec. 01	Ramkumar Singh, I.O.M.	24 May 15	8 Jan. 16
28 Mar. 15	Bijai-Bahadur Sing (66)	28 Mar. 15	27 May 16
30 Nov. 01	Chhaju Ram	16 June 15	27 May 16
24 May 07	Mauji (50)	30 Aug. 15	15 July 16
20 Feb. 93	Ramdhari Singh (56)	27 Apl. 12	1 Jan. 17
14 Jan. 02	Jit Singh	15 Apl. 15	7 Jan. 18
24 Feb. 06	Muhammad Akbar Hussain	28 June 15	7 Jan. 18
19 Jan. 02	Jug Lal (I)	30 Aug. 15	7 Jan. 18
14 Mar. 03	Nathu	24 May 16 / 4 July 16	7 Jan. 18
7 May 00	Jisukh Singh	27 May 16	3 Mar. 18
3 Nov. 07	Bishan, Singh.	27 Mar. 16	16 Oct. 18

1st Battalion, 97th Deccan Infantry—*contd.*

1st Commn. or date of entering service.	Names and Rank.	Army rank.	Present appointment in Regiment.	Remarks.
	Jemadars			
21 May 07	Fateh Ali	27 May 16		
19 Sept. 05	Jodha	27 May 16		
16 Nov. 97	Rachpal Singh, I.O.M.	22 Aug. 16		
6 Dec. 01	Jug Lal	13 Oct. 16		
11 Nov. 06	Ghulam Jilani Khan	23 Oct. 16		
4 July 06	Bhairam Singh	2 Nov. 17		
8 Oct. 05	Surat Singh	2 Nov. 17		
21 May 07	Muhammad Muzaffar Hussain	3 Dec 17		
1 July 06	Meo Singh	3 Mar. 18		
18 May 06	Balkaran Singh	11 Mar. 18		
20 Sep. 10	Gayand Singh	11 Mar. 18		
2 Apl. 10	Tikaram	15 Mar. 18		
26 Apl. 18	Muhammad Abdul Karim	26 Apl. 18	(*on probn.*)	
20 Dec. 00	Shiuram Singh	1 July 18		
12 Aug. 00	Mukhram Singh	1 July 18		
...	Baldeo Singh	6 Sep. 18		

2nd Battalion, 97th Deccan Infantry.

Formed at Sitapur on 27th June 1917.

Composition—1½ companies of Rajputs, 1 company of Jats, 1½ companies of Dekhani Mussalmans.

1st Commn. or date of entering service.	Names and Rank.	Army rank.	Present appointment in Regiment.	Remarks.
	Commandant.			
17 Mar. 00	Martin, Maj. J., M.C. (*Actg. Lt. Col.*)	1 Sep. 15	. . . .	1-94th Infy.
	Company Commanders. (4)			
23 Apl. 02	Ball, Maj. L. P., D.S.O. . . .	23 Apl. 17	. . . .	5th L. Infy. 2nd-in-comd. (*temp.*)
14 Dec. 14	Sharp, Capt. J. C.	20 Jan. 16	15 Oct. 17	
20 Nov. 14	Nicklin, Lt. G. N. (*Actg. Capt.*)	2 Apl. 18	. . . .	I. A. R. O.
22 June 15	Russell, Lt. D. (*Actg. Capt.*) .	22 June 16	30 Nov. 17	Attd.
	Company Officers.			
15 May 15	Sinton, Lt. J. S. (*Actg. Capt.*) .	15 May 16	15 Oct. 17	Dorset. R., I.A. (*on probn.*)
23 Sep. 15	Carstairs, Lt. N. M. (*Actg. Capt.*)	23 Sep. 16	30 Oct. 17	Adjt.
29 June 15	Lee, Lt. A. W. (*Actg. Capt.*) .	29 June 16	. . . .	I. A. R. O.
19 Jan. 16	Franklin, 2nd-Lt. F. A. McA., M.C. (*Actg. Capt.*)	19 Jan. 17	26 Sep. 17	Attd. Qr.-Mr.
28 Apl. 16	*Steele, Lt. R. P.*	*28 Apl. 17*	*2. Aug 17*	*Attd. With R. A. Force.*
4 May 17	Philpott, Lt. D. A.	4 May 17	30 Nov. 17	Attd.
22 Nov. 16	Robertson, 2nd-Lt. J.	22 Nov. 16	21 Sep. 17	I. A. R. O.
12 Dec. 16	Lang, Lt. J. T. M.C.	12 Dec. 17	. . . .	Aus. Imp. Force, I. A. (*on probn.*)
12 Sep. 17	Macvean, 2nd Lt. E. D. . . .	12 Sep. 17	. . . .	I. A (*U. L.*)
26 Sep. 17	Eaton, 2nd-Lt. R. S.	26 Sep. 17	. . . .	I. A. R. O.
5 Oct. 17	Hore, Lt. A. T.	5 Oct. 18	. . . .	I. A. (*T. C.*)
5 Nov. 17	Rood, 2nd-Lt. L. F.	5 Nov. 17	. . . .	I. A. (*T. C.*)
5 Jan. 17	Kelly, Lt. C. J.	5 Jan. 18	27 June 17	I. A. R. O.
12 Sep. 17	McMahon, Lt. H. A.	12 Sep. 18	10 Dec. 17	Attd.

Subadar-Major.

		Jemadar.	Subadar.	Subadar-Major.
19 Apl. 06	Muhammad Ayub Ali Khan . .	19 Apl. 07	18 July 14	22 July 17

Subadars.

19 Apl. 06	Muhammad Ayub Ali Khan . .	19 Apl. 07	18 July 14
10 July 06	Shiushanker Singh	26 May 16	5 July 17
29 June 00	Neki Ram	24 May 15	22 July 17
3 Sep. 00	Hardhan Sing	19 Oct. 14	6 Aug. 17
6 June 99	Mirza Wilayat Beg	6 Aug. 17	1 Nov. 17
26 May 04	Jang Bahadur Singh	5 July 17	15 Jan. 18
9 Sep. 05	Debi Singh	5 July 17	15 Jan. 18
14 Sep. 11	Ranjit Singh	28 Nov. 17	14 July 18

Jemadars.

18 Mar. 08	Muhammad Jan Khan . . .	14 Jan. 16
12 Apl. 10	Shams Khan	9 Sep. 16
12 Aug. 00	Salig Singh	22 July 17
2 July 06	Hardat	22 July 17
1 June 09	Ajodhya Bakhsh Singh . . .	6 Aug. 17
15 Mar. 08	Shaikh Abdur Razaq	28 Nov. 17
7 Aug. 11	Muhammad Basbir Khan . .	28 Nov. 17
10 Mar. 02	Bahori Singh	27 Feb. 18
31 Dec. 10	Bhutal Singh	27 Feb. 18
27 Nov. 05	Lal Bahadur Singh	27 Feb. 18
20 Aug. 04	Pheku Singh	4 Mar. 18
6 May 06	Suinaik Singh	15 Jan. 18
16 July 10	Gulzar Singh	15 Jan. 18
	Harphul	14 July 18
6 Aug. 10	Jhonda	13 Aug. 18

Linked with—97th & 99th Infy.

Uniform—Scarlet.
Facings—Dark Green.

1st Battalion, 98th Infantry.

Raised at Ellichpur, 1788, by Nawab Salabat Khan, as Salabat Khan's Regiment. Was afterwards the 1st Battalion, Ellichpur Brigade. Became the 7th Regt. of Infantry, Nizam's Army, in 1826; the 5th Infantry, Hyderabad Contingent, in 1854; present designation in 1903.

"Central India" "China, 1900."

Composition—1½ companies of Rajputs, 1½ of Hindustani Musalmans, 1 of Ahirs of the Eastern Punjab.

1st Commn. or date of entering service.	Names and Rank.	Army rank.	Present appointment in Regiment.	Remarks.
	Commandant.			
14 Sep. 87	**Adye, Bt.-Col. D.R. (L)** (*tem. Br. Gen.*)	1 Jan. 17	3 Oct. 15	*Comdr., 59th Bde.*
	Company Commanders. (4)			
23 May 91	Tregear, Lt.-Col. V. F. W. (Q)	23 May 17	5 Aug. 15	2nd-in-Comd. With 1-102nd Grs.
27 Jan. 92	Pye, Lt.-Col. W. E., C.I.E. (Q)	27 Jan. 18	24 Sep. 03	
5 Jan. 01	Mitford, Maj. J. P., D.S.O.	5 Jan. 16	5 Aug. 15	2nd-in-Comd. (*temp.*)
8 Apl. 00	*Wright, Maj. G. D.*	*4 May 16*	*13 Jan. 18*	*Burma Military Police.*
	Company Officers. (8)			
15 July 03	Tucker, Maj. H. B.	15 July 18	11 June 08	With 97th Infy.
16 Aug. 05	Kenny, Capt. E. C.	16 Aug. 14	30 Jan. 13	With 1-70 Rif.
2 Sep. 10	Montford, Capt. D. R.	1 Sep. 15	15 Oct. 12	
22 Jan. 13	Payne, Capt. G. D.	22 Jan. 17	13 Mar. 14	*Lv., ex I., m. c.*
18 Aug. 15	Selby, Lt. W. R. (*Actg. Capt.*)	18 Aug. 16	. . . .	Attd.
12 Sep. 15	Bull, Lt. G. G. C.	12 Sep. 16	6 July 17	Attd.
15 Sep. 15	Leech, Lt. A. B. R	15 Sep. 16	26 June 17	
29 June 16	Tomsett, Lt. E. W.	29 May 17		
27 Oct. 16	Beynon, Lt. A. C.	18 June 17	2 Jan. 17	Attd. With 3-151st Infy.
14 Nov. 16	Mellor, Lt. L. A. St. J.	14 Nov. 17	. . . .	Attd.
30 Jan. 17	Young, Lt. W. A. L.	30 Jan. 18	30 Jan. 17	Attd.
14 Nov. 15	Adams, Lt. A. G.	14 Nov. 16	8 July 17	Attd.
11 Dec. 16	Larlham, 2nd-Lt. P. E.	11 Dec. 17	. . . .	Attd.
18 June 17	Whitehead, Lt. E. C.	18 June 18	. . . .	Attd.
19 July 16	Moore, Lt. H. L.	19 July 16	4 Aug. 17	Attd.
	Attached.			
17 June 15	Webb, Lt. R. A. (*Actg. Capt.*)	17 June 16	. . . .	I. A. (*on probn.*)
5 May 17	Norrish, Lt. F.	5 May 18	. . . .	I. A. (*on probn.*)
25 Jan. 17	Bartlett, 2nd-Lt. A. C.	25 Jan. 17	1 Oct. 17	I. A. R. O.
1 Feb. 17	Elloy, Lt. L. M.	1 Feb. 18	24 Apl. 17	I. A. R. O. (*on probn.*) Qr.-Mr. (*temp.*)
5 May 17	Gamon, Lt. R. B.	5 May 18	. . . .	I. A. R. O.
26 June 17	Watson, 2nd-Lt. C.	26 June 17	. . . .	I. A. (*on probn.*)
14 Aug. 19	Hirst, Lt. A. H. (*Actg. Capt.*)	14 Aug. 16	30 Mar. 18	I. A. (*on probn.*) Adjt.
	Dawson, 2nd-Lt. E. J.	. . . .	30 Mar. 18	
8 June 17	Kelly, 2nd-Lt. M. G.	8 June 17	. . .	I. A. (*on probn.*)
15 Jan. 18	Leyshon, 2nd-Lt. W.	15 Jan. 18	11 June 18	I. A. (*T. C.*)
	Tatham, 2nd-Lt. C. W.	. . . .	4 Apl. 18	
10 Sep. 15	Millar, Capt. A. S.	10 Sep. 18	. . . .	I. A. R. O.
15 Jan. 18	Richardson, 2nd-Lt. G.	15 Jan. 18	11 June 18	I. A. (*T. C.*)
12 Aug. 15	Hall, Lt. W. H.	12 Aug. 16	. . . .	A. S. C., I. A. (*on probn.*)
28 Nov. 17	Angel, Lt. B. G.	28 Nov. 18	. . . .	Genl. List, I. A. (*on probn.*)

Subadar-Major.

		Jemadar	Subadar.	Subadar-Major.
6 Mar. 98	Jiya Ram (66*a*)	1 May 10	4 Mar. 14	1 Jan. 18

Subadars.

1 Jan. 08	Hindpal Singh (56)	1 Jan. 08	21 Oct. 13
10 Nov. 00	Rai Singh	11 Aug. 14	13 Sep. 14
30 July 95	Shaikh Shahamat Ali	18 Aug. 08	4 Nov. 14
22 May 95	Ram Bakhsh Singh (66*a*)	19 Sep. 13	5 Nov. 15
1 Apl. 05	Muhammad Ghani Khan (50, 52)	1 Oct. 11	29 Feb. 16
26 May 04	Ranjit Singh	4 Nov. 14	17 July 16
17 Dec. 00	Muhammad Khan (46, 56)	5 Aug. 13	15 Jan. 18
12 May 98	Gopal	1 Nov. 15	15 Jan. 18
30 Sep. 12	Muhammad Ismail Khan (56)	13 Sep. 14	8 Feb. 18
13 Apl. 03	Ram Singh	1 Mar. 17	23 Mar. 18

1st Battalion, 98th Infantry—*contd.*

1st Commn. or date of entering service.	Names and Rank.	Army rank.	Present appointment in Regiment.	Remarks.
	Jemadars.			
21 Oct. 13	Jang Bahadur Singh	21 Oct. 13		
25 Dec. 00	Nand Lal	13 Sep. 14		
14 June 97	Faqir Bakhsh Singh	16 Sep. 15		
9 Oct. 99	Maha Singh	10 Nov. 15		
12 Aug. 95	Faizu Khan	20 Mar. 16		
23 Aug. 04	Ramdas Singh	17 July 16		
24 Oct. 00	Hira	14 Jan. 17		
8 Nov. 00	Prahlad	1 Feb. 17		
12 Apl. 17	Muhammad Hussain	12 Apl. 17		
5 Dec. 00	Shiuhad Singh	19 June 17		
7 July 05	Het Ram	22 June 17		
1 Mar. 06	Khub Ram	2 Nov. 17		
9 Dec. 09	Gokal Singh	3 Dec. 17		
	Mahammad Abdul Rashid	3 Dec. 17		
20 Jan. 03	Abdul Razaq Khan	14 Dec. 17		
	Thirka Ram	14 Dec. 17		
	Habib-ur-Rahman Khan	15 Jan. 18		
	Shaikh Isa	15 Jan. 18		
	Parmanand	27 Jan. 18		

2nd Battalion, 98th Infantry.

Formed at Jubbulpore on 20th January 1918.

Composition—One company of Rajputs, one company of Hindustani Musalmans, and two companies of Ahirs.

1st Commn. or date of entering service.	Names and Rank.	Army rank.	Present appointment in Regiment.	Remarks.
	Commandant.			
7 Mar. 94	Grant, Maj. W. O. (*Actg. Lt.-Col.*)	7 Mar. 12	24 Jan. 18	1-127th Bal. L. I.
	Company Commanders. (4)			
22 Oct. 02	Auret, Maj. A. S.	22 Oct. 17	12 Feb. 18 (*temp.*)	1-127th Bal. L. I. 2nd-in-Comd.
6 Sep. 08	Ruck, Capt. R. C. (*Actg. Maj.*)	1 Sep. 15	12 Sep. 18	1-95th Infy.
25 Jan. 08	Murphy, Capt. G. P.	1 Sep. 15	31 Jan. 18	*Poll. Dept.*
	Company Officers.			
4 Aug. 15	Duncan, Lt. T. M. (*Actg. Capt.*)	4 Aug. 16	8 June 18	I. A. (*on probn.*) Adjt.
11 Mar. 15	Larlham, Lt. P. E.	11 Dec. 16	11 Feb. 18	I. A. (*on probn.*)
8 Aug. 17	Hambling, Lt. V. F.	8 Aug. 18	9 Feb. 18	I. A. R. O.—Qr.-Mr.
5 Mar. 15	Bright, Lt. F. C. (*Actg. Capt.*)	5 Mar. 16	13 Oct. 18	I. A. R. O.
23 June 15	Pattison, Lt. F. R.	23 June 16	3 Aug. 18	I. A. (*on probn.*)
23 Sep. 15	Duncan, Lt. H. C. (*Actg. Capt.*)	23 Sep. 16	2 Sep. 18	Sea Highrs. I. A. (*on probn.*)
7 Jan. 16	North, Lt. L. V.	7 Jan. 17	31 July 18	N. Staff. R., I. A. (*on probn.*)
22 Mar. 16	Wesson, Lt. R. A.	22 Mar. 17	29 Aug. 18	R. Lanc. R., I. A. (*on probn.*)
1 May 17	Murphy, Lt. J. J. C.	1 May 18	22 May 18	I. A. (*on probn.*)
	Witherow, Lt. T. H.	....	1 Oct. 18	R. Ir. Rif., I. A. (*on probn.*)
	Fitch, Lt. W. O.	....	14 Oct. 18	Bord. R., I. A. R. O. (*on probn.*)
31 Oct. 17	Nash, 2nd-Lt. G. F.	31 Oct. 17	....	I. A. R. O. (*on probn.*)
27 Feb. 18	Ward, 2nd-Lt. E.	27 Feb. 18	8 Sep. 18	Norf. R., I. A. R. O. (*on probn.*)
5 Nov. 17	Brewin, Lt. N. A. P.	5 Nov. 18	16 July 18	I. A. R. O.
29 Mar. 18	Kitchen, 2nd-Lt. W. H.	29 Mar. 18	16 July 18	I. A. (*T. C.*)
29 Mar. 18	Williams, 2nd-Lt. S. N.	29 Mar. 18	22 July 18	I. A. R. O.
26 June 18	Nicholson, 2nd-Lt. G. A.	26 June 18	22 May 18	I. A. (*on probn.*) *Lv., ex I., m. c.*
26 June 18	Brown, 2nd-Lt. A. D.	26 June 18	26 Sep. 18	Genl. List, I. A. R. O. (*on probn.*)
27 Mar. 18	Dickens, 2nd-Lt. H. E.	27 Mar. 18	1 Oct. 18	Glouc. R. I. A. (*on probn.*)
7 June 18	Mitchell, 2nd-Lt. H.	7 June 18	9 Oct. 18	I. A. (*T. C.*)
22 July 18	Donaldson, 2nd-Lt. A. H.	22 July 18	2 Dec. 18	I. A. (*T. C.*)

Subadar-Major.

		Jemadar.	Subadar.	Subadar-Major.
17 Dec. 94	Bhagwan Singh (56)	16 Jan. 09	27 Apl. 12	14 Mar. 18 (*temp.*)

Subadars.

1 Aug. 95	Abdur Rahman Khan	18 Aug. 08	5 Aug. 13
4 July 91	Bansgopal Tewari	1 Nov. 15	1 Mar. 18
6 June 94	Gumani Ram	30 Oct. 17	18 Mar. 18
17 Apl. 03	Nathu Ram, I.D.S.M.	22 Aug. 16	1 June 18
16 May 06	Hanumant Singh (46)	8 Jan. 16	1 July 18
11 Dec. 17	Ram Sarup	11 Dec. 17	1 Aug. 18

Jemadars.

25 Aug. 02	Madho Singh	27 May 16
27 Jan. 01	Sheikh Ummed	22 Mar. 18
25 Oct. 10	Surat Ram	1 June 18
5 May 06	Ram Ratan Singh	1 July 18
19 Apl. 07	Farzand Khan	1 Aug. 18

Linked with—97th & 98th Infy.

Uniform—Scarlet.
Facings—Dark Green.

1st Battalion, 99th Deccan Infantry.

Raised in 1788 by Nawab Salabat Khan. Was afterwards the 2nd Battalion, Ellichpur Brigade. Became the 8th Regiment of Infantry, Nizam's Army, in 1826; the 6th Infantry, Hyderabad Contingent, in 1864; present designation in 1903.

Composition—1½ companies of Rajputs, 1 of Jats, 1½ of Hindustani Musalmans.

1st Commn. or date of entering service.	Names and Rank.	Army rank.	Present appointment in Regiment.	Remarks.
	Commandant.			
13 July 92	Davidson, Lt.-Col. F. D.	1 April 16	1 April 16	
	Company Commanders. (4)			
10 Oct. 94	Luxmoore, Maj. C. de J. [*l*] (*Actg. Lt.-Col.*)	10 Oct. 12	20 Dec. 17	2nd-in-Comd.
5 Jan. 01	Redman, Maj. H. E.	20 Oct. 15	24 May 16	
	Company Officers. (8)			
4 Apl. 00	Grant, Capt. C. J. P. MacA.	18 Jan. 09	26 Aug. 08	*Lv., ex I., m.c.*
27 Aug. 02	*Ward, Maj. B. S. [l] (temp. Lt.-Col.)*	*1 Jan. 17*	*15 Jan. 04*	*D. A. Q. M. G., Embarkation, Bombay.*
18 Nov. 03	Hinde, Maj. H. T. L.	18 Nov. 18	22 Mar. 06	
13 Aug. 04	Farrell, Capt. G. V.	13 Aug. 13	3 Dec. 09	
29 Nov. 05	Masters, Capt. O. (*Actg. Maj.*)	29 Nov. 14	9 Nov. 08	
8 Sep. 09	*Ozanne, Capt. G. D.*, M.C.	*1 Sept. 15*	*3 Feb. 11*	*With Sig. Ser.*
18 Sep. 09	Happell, Capt. W. H.	1 Sept. 15	1 Apl. 14	Adjt.
12 May 15	Ridge, Lt. K. J. T., M.C.	12 May 16	8 Nov. 15	With 1-101st Grs.
7 Apl. 16	Seagrim, Lt. C. B.	7 Apl. 17	9 Nov. 16	Attd. With 3-150th Infy.
16 Aug. 16	Holmes, Lt. G. L.	16 Aug. 17	11 Nov. 16	Attd. With 3-150th Infy.
14 Nov. 16	Gordon, Lt. S. H.	14 Nov. 17	23 Nov. 16	Attd. With 49th Bengalis.
12 Sep. 17	Marriott, Lt. J. R.	12 Sep. 18	7 Dec. 17	Attd.
12 Sep. 17	Mullaly, Lt. H. A.	12 Sep. 18	7 Dec. 17	Attd.
	Attached.			
23 Jan. 14	Watts, Capt. R. P.	23 July 18	15 Aug. 18	Hamps. R., I. A. (*on probn.*)
15 Dec. 14	Kent, Lt. H. B. (*Actg. Capt.*)	15 Dec. 15	7 Dec. 17	I. A. R. O.
12 Feb. 16	Pettman, Lt. F. L. (*Actg. Capt.*)	12 Feb. 17	. . . .	I. A. (*on probn.*)
10 Dec. 15	Donaldson, Lt. J.	10 Dec. 16	13 Jan. 18	North'd Fus., I. A. (*on probn.*) *S. S. O., Ahmedabad.*
27 Jan. 16	Perdue, Lt. C. G.	27 Jan. 17	15 Mar. 16	I. A. R. O.
20 Feb. 16	McLean, Lt. W. (*Actg. Capt.*)	20 Feb. 17	3 Apl. 16	I. A. R. O.
18 Apl. 16	Bragg, Lt. H. V. (*Actg. Capt.*)	18 Apl. 17	21 Nov. 17	1-10th Jats.
29 June 16	Tomsett, Lt. E. W.	28 May 17	29 June 16	1-3rd G. R.
1 July 16	Havelock, Lt. W. A. C.	1 July 17	3 Aug. 16	I. A. R. O. Qr.-Mr. (*Offg.*)
5 Sep. 16	Wade-Evans, Lt. D. G. (*Actg. Capt.*)	5 Sep. 17	4 Mar. 18	I. A. R. O. Adjt. (*Offg.*)
9 Sep. 16	Simms, Lt. J.	9 Sep. 17	29 Nov. 17	I. A. (*on probn.*)
14 Nov. 16	Race, Lt. T. T. G.	14 Nov. 17	20 Oct. 17	1-101st Grs.
10 Nov. 16	Saunders, Lt. E. E. L.	10 Nov. 17	27 Feb. 17	I. A. R. O.
1 Sep. 17	Fitzpatrick, Lt. J. T.	1 Sep. 18	8 Sep. 17	I. A. R. O.
28 Dec. 17	Smith, 2nd-Lt. W. G.	28 Dec. 17	15 Aug. 18	N. Lan. R., I. A. (*on probn.*)
	Barclay-Brown, 2nd-Lt. R.	. . . .	16 Oct. 18	R. Naval Vol. R., I. A. R. O. (*on probn.*)
27 Aug. 18	Rose, 2nd-Lt. G. C. G. F.	27 Aug. 18	21 Sep. 18	I. A. R. O.
4 Sep. 18	Cleary, 2nd-Lt. G. W.	4 Sep. 18	27 Sep. 18	I. A. R. O.
13 Aug. 15	Galloway, Lt. W. C.	13 Aug. 16	. . . .	I. A. R. O.
8 Jan. 15	Rogers, Lt. G. N.	8 Jan. 16	. . . .	I. A. R. O.
	Subadar-Major.			
		Jemadar.	Subadar.	Subadar-Major.
22 Jan. 94	Brahmdeo Singh, *Bahadur* (56, 58)	16 Oct. 01	1 Jan. 07	1 Aug. 17 *Order of Br. Ind., 2nd cl.*
15 May 93	Hardayal (56, 66*a*) (*Retd.*)			Re-employed, 30 Aug. 17.
	Subadars			
17 Nov. 90	Talib Husain Khan (56)	7 Apl. 03	7 Aug. 11	
4 June 04	Altafulla Khan	16 May 07	7 Aug. 11	
20 Mar. 02	Bissambar Singh (56)	17 Jan. 15	13 Aug. 15	
25 July 94	Sada Sukh	7 Aug. 11	16 Sept. 15	
17 Oct. 93	Badrinarayan Singh (56)	16 Jan. 10	19 Oct. 15	
6 July 00	Subh Ram (56)	15 Aug. 15	23 Nov. 15	With 76th Punj.
31 Oct. 93	Basawan Singh	1 June 10	16 Jan. 16	
13 Apl. 94	Altaf Husain Khan (56, 60*a*)	7 Aug. 11	12 Feb. 16	
28 May 95	Jugraj Singh (58)	1 Oct. 13	13 May 16	
8 Mar. 99	Tara Singh	15 Aug. 15	13 Aug. 17	
8 June 07	Harphul (I) (52, 56)	10 Apl. 17	12 Dec. 17	
9 Nov. 97	Bissesar Singh	19 Oct. 15	18 Feb. 18	
9 Jan. 02	Balwanta (56)	28 May 17	18 Feb. 18	
13 June 00	Badlu Singh	15 Aug. 15	24 May 18	
20 Oct. 02	Sewa Ram	12 Dec. 17	28 Aug. 18	

1st Battalion, 99th Deccan Infantry—*contd.*

1st Commn. or date of entering service.	Names and Rank.	Army rank.	Present appointment in Regiment.	Remarks.
	Jemadars.			
[1]4 Nov. 98	*Shaikh Sadiq Ali*	*1 Dec. 11*		
[1]9 May 95	Allahbaksh Khan	15 July 16		
[1]8 July 07	Muhammad Sadiq Khan (56)	29 Nov. 16		
1 Aug. 07	Badsha Khan	23 July 17		
2 Apl. 97	Mathura Singh	1 Aug. 17		
15 Dec. 06	Shaikh Abdul Hakim (56)	1 Aug. 17		
1 Jan. 01	Mangat Singh	1 Aug. 17		
22 July 00	Harphul (56)	13 Aug. 17		
18 June 01	Abdul Rahman Khan	3 Dec. 17		
21 Nov. 08	Baldeo Singh	18 Feb. 18		
25 Sep. 09	Lalta Singh	20 Feb. 18		
8 June 07	Harchand	20 Feb. 18		
18 July 07	Muhammad Sadiq Khan	4 Mar. 18		
17 Jan. 11	Prithipal Singh	21 Mar. 18		
25 Aug. 11	Ghure Singh	21 Mar. 18		
28 Sep. 02	Ghulam Mohiuddin	21 Mar. 18		
13 Sep. 13	Shahryar Khan	23 Apl. 18		
11 Feb. 09	Munshi Singh	23 Apl. 18		
8 Dec. 08	Chailu (56)	23 Apl. 18		
12 Dec. 06	Mirza Wahab Beg	24 May 18		
	Lakhan Singh	24 May 18		
11 Feb. 07	Alisher Khan	24 May 18		
25 Aug. 11	Ram Sarup	28 Aug. 18		
23 Sep. 12	Wazid Khan	28 Aug. 18		
1 Aug. 11	Gulab Khan	2 Sep. 18		

2nd Battalion, 99th Deccan Infantry.

Formed at Arangaon on 4th March, 1918.

Composition—2 Companies Hindustani Musalmans. 1 Company Rajputs. 1 Company United Provinces Jats..

1st Commn. or date of entering service.	Names and Rank.	Army rank.	Present appointment in regiment.	Remarks.
	Commandant.			
10 Oct. 94	Nangle, Maj. M. C., *p.s.c.* (*L*) (*Actg. Lt.-Col.*)	10 Oct. 12	2 Oct. 18	92nd Punj.
	Company Commanders. (4)			
11 Aug. 00	Seaton, Maj., F. D. R.	1 Sep. 15	20 Mar. 18	12th Infy. 2nd.-in comd. (*temp.*)
17 Sep. 14	Noakes, Capt. T. C.	12 Mar. 17	17 Nov. 18	York. & Lanc. R., I.A. (*on probn.*)
22 Nov. 15	Wallace, Capt. M. D.	23 July 17	17 Nov. 18	I. A. (*on probn.*)
29 Sep. 15	Coulter, Lt. A. W.	29 Sep. 15	16 Oct. 18	C. M. G. Corps, I. A. R. O. (*on probn.*)
	Company Officers.			
8 Apl. 16	Hall, Lt. G. C. O.	8 Apl. 17	19 May 18	I. A. (*on probn.*)
2 June 16	Hutchings, Lt. D. A.	2 June 17	19 May 18	Berk. R., I. A. (*on probn.*)
7 July 16	Waters, Lt. W. H.	7 July 17	. . .	R. Ir. Fus., I. A. (*on probn.*)
15 Aug. 16	Gibson, Lt. G. M. L.	15 Feb. 18	17 Nov. 18	I. A. (*on probn.*)
19 Sep. 17	McCelland, Lt. W.	19 Sep. 18	15 May 18	I. A. (*on probn.*)
1 Aug. 17	Thomas, 2nd-Lt. E. A.	1 Aug. 17	6 Nov. 18	R. W. Kent. R., I. A. R. O. (*on probn.*)
29 Aug. 17	North, 2nd-Lt. F. L.	29 Aug. 17	17 Nov. 18	I. A. (*on probn.*)
26 Sep. 17	Tennant, 2nd-Lt. F. O.	26 Sep. 17	30 Mar. 18	I. A. R. O. (*on probn.*)
17 Dec. 17	Wykes, 2nd-Lt. H. J.	17 Dec. 17	16 Oct. 18	Labour Corps, I. A. R. O. (*on probn.*)
28 Dec. 16	Thornton, 2nd-Lt. C. A. F.	28 Dec. 17	2 June 18	W. York. R., I. A. (*on probn.*)
15 Jan. 18	Davenport, 2nd-Lt. R.	15 Jan. 18	2 June 18	I. A. (*T. C.*)
15 Jan. 18	Lane, 2nd-Lt. W. E.	15 Jan. 18	2 June 18	I. A. R. O.
19 Mar. 18	Ridler, 2nd-Lt. W.	19 Mar. 18	17 July 18	I. A. (*T. C.*)
29 Mar. 18	Shawe, 2nd-Lt. R. A.	29 Mar. 18	17 July 18	I. A. R. O. (*on probn.*)
2 June 18	Cartwright, 2nd-Lt. W. A.	2 June 18	5 Oct. 18	I. A. (*T. C.*) Adjt.
7 June 18	Moase, 2nd-Lt. W.	7 June 18	5 Oct. 18	I. A. (*T. C.*)
7 June 18	Chater, 2nd-Lt. A.	7 June 18	5 Oct. 18	I. A. (*T. C.*)
30 Aug. 18	Mackay, 2nd-Lt. R. D.	30 Aug. 18	22 Sep. 18	I. A. R. O.
10 Sep. 18	Roy, 2nd-Lt. M.	10 Sep. 18	10 Oct. 18	I. A. R. O.
22 July 18	Mason, 2nd-Lt. J. S.	22 July 18	2 Dec. 18	I. A. (*T. C.*)

Subadar-Major.

		Jemadar.	Subadar.	Subadar-Major.	
27 Apl. 97	Lachhman Singh, (56)	16 June 06	16 Jan. 10	23 Apl. 18	(*temp.*)

Subadars.

9 July 04	Abdul Karim Khan (50, 66)	16 May 14	23 Apl. 18
25 July 07	Bhagirathi Singh (56)	1 Jan. 17	23 Apl. 18
1 Nov. 14	Sardar Khan	1 Feb. 17	23 Apl. 18
13 Aug. 00	Maman Singh	1 Aug. 17	23 Apl. 18
14 Sep. 11	Mahipat Singh	20 Feb. 18	23 Apl. 18
21 July 03	Rampirith Singh	4 Mar. 18	23 Apl. 18

Jemadars.

8 Mar. 04	Dalmir Khan	13 Feb. 18
8 Dec. 02	Sarju Singh	23 Apl. 18
1 Oct. 07	Mohammed Sayid Khan	23 Apl. 18
7 Aug. 11	Duli Chand	23 Apl. 18
20 Sep. 10	Badrinarayan Singh	23 Apl. 18
1 Oct. 04	Shib Lal	23 Apl. 18
18 June 05	Jai Lal	23 Apl. 18
. . . .	Karamiyar Khan	17 June 18

Regt. Centre—Ahmednagar. *Uniform*—Scarlet.
Linked with—108th & 109th Infy. *Facings*—White.

1st Battalion, The 101st Grenadiers.

Formed at Bombay in 1779 of drafts of Grenadier companies from the 1st, 2nd, 3rd, 4th, 5th and 6th Battalions of Bombay Sepoys, and of two complete companies from the Marine Bn., and was originally designated the 8th Bombay Bn. Became the Bombay Grenadiers in 1783; the Bombay Grenadier Battn., 1784; the 1st Bombay Bn., or Bo. Grenadiers, 1788; the 1st of Grenadier Bn., 1st Regiment of Bombay Native Infantry, 1796 the 1st or Grenadier Regt. of Bo. N. I., 1824; the 1st Regt. of Bo. Infantry (Grenadiers), 1885; the 1st Bombay Grenadiers, 1901; present designation, 1903.

THE WHITE HORSE.

"Mangalore" "Mysore" "Hyderabad."

"Kandahar, 1880" "Afghanistan, 1878-80" "Burma, 1885-87."

Composition—1 company of Dekhani Mahrattas, 1 of Konkani Mahrattas, 1 of Rajputana Musalmans and 1 of Punjabi Musalmans.

COLONEL—Major-General William C. Black . . . 28 January 1908.

1st Comm. or date of entering service.	Names and Rank.	Army rank.	Present appointment in Regiment.	Remarks.
	Commandant.			
12 Dec. 94	Strong, Lt.-Col. H. P. . . .	30 Mar. 18	30 Mar. 18	With 2-101st Grs.
	Company Commanders. (4)			
25 Jan. 99	Roberts, Maj. W. B. [*l*], D.S.O. (*Actg. Lt.-Col.*)	1 Sep. 15	6 Mar. 18	2nd-in-Comd.
15 Aug. 00	*Johnstone, Maj. F. A. B.* . .	*23 Mar. 16*	*14 Mar. 14*	*Attd. A. G.'s Bch. A. Hdqrs.*
5 Jan. 01	Bean, Maj. J. A.	1 Sep. 15	14 Jan. 16	With 2-9th Infy.
4 Dec. 01	Beaumont, Maj. S. J. G. . . .	4 Dec. 16	6 Mar. 18	
	Company Officers. (9)			
9 Sep. 08	*Laird, Capt. H. G. C.*	*1 Sep. 15*	*20 Dec. 09*	*Adjt., 5 Nov. 14.*
24 Aug. 12	Gracey, Capt. H. M. K. . . .	24 Aug. 16	1 Oct. 14	With 2-101st Grs.
10 Dec. 15	Money, Lt. W. G. M. . . .	10 Sep. 17	27 Dec. 15	Attd.
29 June 16	Retallack, Lt. C. W. . . .	28 Feb. 17	9 Mar. 17	
18 Apl. 16	Frankland, Lt. A.	18 Apl. 17	24 Apl. 16	Attd.
29 June 16	Smithers, Lt. H. E.	29 June 17	. . . .	With 108th Infy.
16 Aug. 16	Hamlyn, Lt. E. A.	16 Aug. 17	3 Feb. 17	Attd.
14 Nov. 16	Race, Lt. T. T. G.	14 Nov. 17	. . . .	Attd. With 1-99th Infy.
12 Sep. 17	McDougall, Lt. D. M. . . .	12 Sep. 18	10 Dec. 17	Attd. *Lv. m. c. ex I., 6 months.*
1 Sep. 18	Mason, 2nd-Lt. J. L.	1 Sep. 18	13 Sep. 18	Attd.
	Attached.			
22 Oct. 02	Coates, Maj. S. B., M.C. . .	22 Oct. 17	3 June 18	102nd Grs.
1 Oct. 14	Flemming, Capt. R. L. V. . . .	1 Oct. 18	3 Feb. 17	116th Mah.
11 Dec. 14	Cross, Lt. A. B.	11 Dec. 15	24 Dec. 14	I. A. R. O.
11 Dec. 14	Benson, Lt. J. W. H.	11 Dec. 15	15 Dec. 14	I. A. R. O.
12 May 15	Lightfoot, Lt. G. S. (*actg. Capt.*)	12 May 16	23 Nov. 17	I. A. (*on probn.*)
12 May 15	Ridge, Lt. K. J. T., M.C. . . .	12 Nov. 15	27 Dec. 17	99th Infy.
29 Oct. 15	Murray, Lt. A. E.	29 Oct. 16	3 Feb. 17	I. A. R. O.
29 Oct. 15	Hall, Lt. W. A.	29 Oct. 16	. . . .	I. A. R. O.
5 Dec. 15	Strang, Lt. D. A.	5 Dec. 15	29 Nov. 17	I. A. R. O.
1 Jan. 16	Wood, Capt. F. J.	1 Jan. 16	1 Mar. 16	Genl. List.
27 Oct. 16	Allson, Lt. T. F. A. R. . . .	27 Oct. 17	. . . .	U. L.
8 Mar 16	Clements, Lt. G. W.	8 Mar. 17	4 Jan. 18	I. A. R. O.
10 June 16	Neal, Lt. G. T. St. J. . . .	10 June 17	27 Dec. 17	I. A. R. O.
20 June 17	Gapper, Lt. J. E. B.	20 June 18	4 Oct. 17	I. A. R. O.
8 Aug. 17	Collison, Lt. S. S.	8 Aug. 18	12 Nov. 17	I. A. R. O.
26 Aug. 17	Bancroft, 2nd-Lt. G. R. S. . .	26 Aug. 17	2 May 18	I. A. R. O. (*on probn.*)
5 Oct. 17	Wykes, Lt. E.	5 Oct. 18	19 Jan. 18	I. A. (*T. C.*)
24 Oct. 17	Lynch-Blosse, Lt. C. E. . . .	24 Oct. 18	5 Oct. 18	
15 Jan. 18	Atkinson, 2nd-Lt. L. R. . . .	15 Jan. 18	23 Sep. 18	I. A. (*T. C.*)
29 Mar. 18	Soper, 2nd-Lt. O. N.	29 Mar. 18	17 July 18	I. A. R. O.
28 June 18	Rees, 2nd-Lt. T. O.	28 June 18	2 Oct. 18	I. A. R. O.
31 Aug. 18	Tribble, 2nd-Lt. N. V.	31 Aug. 18	10 Sep. 18	I. A. R. O.
19 Aug. 18	LeFrank, 2nd-Lt. G. A. . . .	19 Aug 18	10 Sep. 18	
2 Sep. 18	MacKinnon, 2nd-Lt. A. . . .	2 Sep. 18	23 Sep. 18	I. A. R. O.
		Jemadar.	**Subadar.**	**Subadar Major.**
	Subadar-Major.			
15 Oct. 90	Martand Rao Mohite (56) . .	21 Mar. 05	11 Apl. 05	19 Jan. 15.
	Siwa Moduck (*Hon. Capt.*), *Sardar Bahadur, Retd.*	. . . .	. . . .	Re-empld., 3 Jan. 16.
	Subadars.			
5 Mar. 97	Ahmad-Din (52, 56)	18 Jan. 05	1 Aug. 13	
2 Apl. 94	Vishnu Palao (52, 56)	1 Sep. 06	5 Nov. 14	
30 May 98	Bari Khan	1 Sep. 10	5 Dec. 14	
11 Mar. 02	Daji Sawant (56)	11 Sep. 14	19 Jan. 15	
1 Oct. 94	Nur Khan	11 Sep. 14	3 Feb. 17	
25 May 98	Wazir Khan	11 Sep. 14	3 Feb. 17	
5 Jan. 01	Rahim Khan	20 Jan. 15	3 Feb. 17	
22 Jan. 06	Bapu Bargi (66 *a*)	5 Nov. 14	3 Feb. 17	
21 Oct. 01	Murar Rao Salunke (56, 66*a*) .	5 Nov. 14	3 July 17	
16 Jan. 10	Kasinath Mane	3 Feb. 17	19 June 18	
11 Jan. 10	Muhammad Alam	23 Feb. 17	14 Aug. 18	
	Mawaz Khan (*Retd.*)	. . . .	. . . .	Re-empld., 23 Apl. 15, Dep[illegible]

1st Battalion, The 101st Grenadiers—*contd.*

1st Commn. or date of entering service.	Names and Rank.	Army rank.	Remarks.
	Jemadars.		
26 Nov. 94	Ghulam Muhammad	1 Aug. 13	
26 May 96	Sakharam Kalbhar	11 Sep. 14	
18 Feb. 02	Sakharam Gaikwad (*72a*)	5 Dec. 14	
9 Mar. 14	Yeshwant Rao Mane	3 Feb. 17	
15 June 03	Khaju Khan	23 Feb. 17	
9 Jan. 00	Manawar Khan	23 Feb. 17	
30 Mar. 06	Krishna Naik Kadam	3 July 17	
....	Dhanwat Meo	....	Re-empld.
....	Mir Ali	3 Dec. 17	
22 Mar. 02	Pola Khan	24 Jan. 18	
14 Jan. 08	Amir Bakhsh	23 Mar. 18	
9 June 07	Jairam Chauhan	19 June 18	
....	Lal Khan	14 Aug. 18	
7 Dec. 04	Karam Din I	1 Oct. 18	

2nd Battalion, The 101st Grenadiers.

Formed at Suez on 3rd February 1917.

Composition—1 company of Dekhani Mahrattas, 1 of Konkani Mahrattas, 1 of Rajputana Mussalmans, & 1 of Punjabi Mussalmans.

1st Commn. or date of entering service.	Names and Rank.	Army rank.	Present appointment in regiment.	Remarks.
	Commandant.			
12 Dec. 94	Strong, Lt.-Col. H. P.	30 Mar. 18	3 Feb. 17	101st Grs.
	Company Commanders. (4)			
26 Mar. 02	Potter, Maj. R. J. K.	26 Mar. 17	. . . .	117th Mah.
8 Sep. 09	Anderson, Capt. J. McK., M.C.	30 Nov. 11	. . . .	109th Infy.
24 Aug. 12	Gracey, Capt. H. M. K.	24 Aug. 16	3 Feb. 17	1-101st Grs.
	Company Officers.			
13 Nov. 14	Bolton, Lt. R. H. D.	13 Aug. 16	3 Feb. 17	I. A. R. O.
20 Nov. 14	Davies, Lt. J. F.	20 Nov. 15	. . . .	I. A. R. O. Adjt.
18 Dec. 14	Cuscaden, Lt. R. L.	18 Dec. 15	. . . .	I. A. R. O.
18 June 15	McCombe, Lt. R.	18 June 16	3 Feb. 17	I. A. R. O.
22 July 15	Solly, Lieut. P. R.	22 July 16	3 Feb. 17	I. A. R. O.
15 Nov. 15	Anderson, Lt. P. B.	15 Nov. 16	3 Feb. 17	116th Mah.
15 Nov. 15	Deveinish-Meares, Lt. B. J. D.	15 Nov. 16	. . . .	1-41st Dogras.
18 Aug. 16	Slade, Lt. J. W. C.	28 May 18	. . . .	
	Attached.			
30 Jan. 17	Carberry, Lt. T. B.	30 Jan. 18	. . . .	117th Mah.
18 June 17	Whitehead, Lt. E. C.	18 June 18	. . . .	I. A. (*on probn.*)
	Subadar-Major.			
		Jemadar.	Subadar.	
	Subadars.			
25 Mar. 96	Shaikh Abdul Sulaiman (52, 56, 58, 72a)	9 Aug. 02	17 Dec. 04	
3 June 98	Maola Baksh Khan	1 July 11	3 Feb. 17	
30 Dec. 95	Allahditta Khan	25 July 11	3 Feb. 17	
14 Nov. 01	Ramajirao Kadam	11 Sep. 14	3 Feb. 17	
20 Jan. 00	Kondaji Ghanwat	5 Nov. 14	3 Feb. 17	
1 Jan. 97	Sultan Khan	21 June 15	3 Feb. 17	
26 May 02	Dhane Khan	5 Dec. 14	3 Feb. 17	
30 Oct. 02	Shah Sowar	3 Feb. 17	3 Feb. 17	
8 June 07	Dinkarrao Yadhao	4 May 13	3 Feb. 17	
9 Feb. 06	Abu Talah (58)	5 Nov. 14	3 Feb. 17	
8 Oct. 05	Janoji Rao Bhawe	10 Mar. 16	14 Nov. 17	
8 Oct. 05	Shah Nawaz	3 Feb. 17	7 Dec. 17	
24 Apl. 05	Jiwan Khan	3 Feb. 17	23 Apl. 18	
20 Jan. 02	Keshao Dalvi	14 June 15	19 Aug. 18	
	Jemadars.			
28 June 05	Kondiba Selke	19 Mar. 16		
10 Apl. 00	Wahid Ali	3 Feb. 17		
24 Sep. 99	Wasudeo Ghag	3 Feb. 17		
. . . .	Nadir Khan	3 Feb. 17		
9 Mar. 99	Hussain Baksh	3 Feb. 17		
28 Oct. 04	Gheba Khan	14 Nov. 17		
7 Jan. 08	Deoji Parte	7 Dec. 17		
	Muhammad Ali Khan	10 Apl. 18		
. . . .	Govind Rao Sindhi	23 Apl. 18		
. . . .	Ramchandar Parab	19 Aug. 18		

Regt. Centre—Deesa.
Linked with—112th & 113th Infy.

Uniform—Scarlet
Facings—White.

1st Battalion, 102nd King Edward's Own Grenadiers.

Raised at Calicut in 1796 by Capt. D. Cameron, and was originally designated the 13th Battalion of Bombay Native Infantry. Became the 1st Bn., 5th Regiment of Bombay N. I., 1797; the 2nd Bn., 1st Regt. of Bo. N. I., 1798; the 2nd Batt., 1st or Grenadier Regt. of Bo. N. I., 1818; the 2nd or Grenadier Regt. of Bo. N. I., 1824; the 2nd (the Prince of Wales's Own) Bombay N. I. (Grenadiers), 1876; the 2nd (the Prince of Wales's Own) Regiment of Bo. Infantry (Grenadiers), 1885; the 2nd (Prince of Wales's Own) Bombay Grenadiers, 1901; the 102nd Prince of Wales's Own Grenadiers, 1903; present designation, 1st January 1906.

THE PLUME OF THE PRINCE OF WALES—THE SPHINX.
THE ROYAL AND IMPERIAL CYPHER OF KING EDWARD VII.
"Egypt" "Kirkee" "Corygaum" "Abyssinia."

Composition—1 Company of Western Rajputana Jats, 1 of Bagri Jats and Jats from Eastern Rajputana, 1 Rajputana Gujars, 1 of Punjabi Musalmans.

COLONEL-IN-CHIEF . . . . THE KING.
COLONEL—

1st Commn. or date of entering service.	Names and Rank.	Army rank.	Present appointment in Regiment.	Remarks.
	Commandant.			
27 Jan. 92	Edwardes, Lt.-Col. A. C.	3 June 16	12 Oct. 17	
	Company Commanders. (4)			
11 Feb. 88	Pierce, Lt.-Col. F. G. (Q)	11 Feb. 14	23 July 11	2nd-in-Comd. S. S. O., Simla.
11 Aug. 00	*Oakes, Maj. T. T.*	*1 Sep. 15*	*9 Jan. 16*	*D. A. Q. M. G., A. Hdqrs.*
8 May 01	Warton, Maj. C. P. F.	8 May 16	13 Feb. 17	
	Company Officers. (9)			
8 May 01	Livesay, Maj. J.	8 May 16	15 Oct. 06	With 1-10th Jats.
22 Oct. 02	Coates, Maj. S. B., M.C.	22 Oct. 17	11 Aug. 05	With 1-101st Grs.
24 Jan. 06	Christie, Capt. R. C., M.C. . .	24 Jan. 15	24 Mar. 07	With 2-102nd Grs.
19 Jan. 07	*Norman, Capt. H. J.*	*1 Sep. 15*	*8 Mar. 08*	*I. S. Troops, Ambala.*
9 Sep. 08	*Liston, Capt. R. C. C.*	*1 Sep. 15*	*15 Dec. 09*	*I. S. Troops, Gilgit.*
15 Aug. 14	Dods, Capt. E. R. S., M. C. . .	15 Aug. 18	8 Nov. 15	
15 Aug. 14	Le Fleming, Capt. R. E. . . .	15 Aug. 18	7 Nov. 15	
10 May 16	Osborne, Lt. H.	10 Feb. 18	21 Nov. 17	Attd.
	Attached.			
23 May 91	Tregear, Lt.-Col. V. F. W. . .	23 May 17	8 Nov. 18	1-98th Infy. O. C. Depot.
5 Aug. 14	Bloxham, Capt. H. F.	5 Aug. 18	13 Feb. 18	I. A. R. O.
22 Jan. 15	Rogerson, Lt. J. A.	22 Jan. 16	8 Apl. 17	I. A. R. O.
28 May 15	*Nicholson, Lt. A. R.*	*28 May 16*	*26 July 15*	*I. A. R. O. A. D. C. to G. O. C., Poona.*
27 Jan. 16	*Leftwich, Lt. W. M.* . . .	*27 Jan. 17*	. . . .	*I. A. R. O. Asst. Cable Censor.*
10 May 16	Hales, Lt. G.	10 May 17	6 Dec. 17	I. A. R. O.
7 July 16	Collinge, Lt. W. C.	7 July 17	25 Nov. 17	I. A. R. O. (*on probn.*) Qr.-Mr.
27 Oct. 16	Martin, Lt. H. J.	27 Oct. 17	28 Dec. 17	
1 Mar. 17	Meloney, Lt. J. C.	1 Mar. 18	23 Dec. 17	I. A. R. O.
18 June 17	Hayes-Newington, Lt. E. A. . .	18 June 18	24 Dec. 17	29th Punj.
25 Oct. 17	Gosden, Lt. W. I.	25 Oct. 18	6 July 18	I. A. (*on probn.*)
5 Nov. 17	Boyce, Lt. W. A.	5 Nov. 18	15 Mar. 18	I. A. (*T. C.*)
5 Nov. 17	Balfour, Lt. A.	5 Nov. 18	15 Mar. 18	I. A. R. O.
29 Mar. 18	Scott, 2nd-Lt. A. H.	29 Mar. 18	16 July 18	I. A. R. O.
30 Aug. 18	Curtis, 2nd-Lt. P. H.	30 Aug. 18	25 Sep. 18	I. A. R. O.
31 Oct. 18	Thrupp, 2nd-Lt. H. F.	31 Oct. 18	21 Nov. 18	I. A. R. O.
. . . .	Higgins, 2nd-Lt. R. F. R. . .	. . . . .	. . . . .	M. G. C., I. A. (*on probn.*)

Subadar-Major.

	Jemadar.	Subadar.	Subadar-Major.

1st Battalion, 102nd King Edward's Own Grenadiers—*contd.*

1st Commn. or date of entering service.	Names and Rank.	Army rank.	Present appointment in Regiment.	Remarks.
	Subadars.			
7 July 94	Juglal Singh (52, 56, 66*a*)	1 Dec. 08	1 Sep. 12	
3 May 92	Jafar Ali, I. O. M. (56, 66*a*)	6 Jan. 06	1 Apl. 13	
18 June 94	Bhagwat Singh (50, 56)	9 June 10	18 Sep. 15	
2 Jan. 99	Tota Singh (58)	1 Feb. 12	26 Nov. 15	
29 Sep. 02	Girdhari Singh (56)	1 Feb. 15	7 Jan. 16	
30 Oct. 99	Ganga Ram Singh (52, 56, 66*a*)	1 Sep. 12	1 July 16	
27 Apl. 02	Maru Ram	27 Oct. 15	12 Sep. 10	
18 May 01	Yar Muhammad	15 Oct. 16	13 Feb. 17	
20 July 02	Mansar Ali, I.D.S.M.	15 Jan. 17	14 Apl. 17	
18 Jan. 09	Sukha Ram	15 Oct 16	19 Feb. 18	
22 Jan. 05	Polu Ram	13 Feb. 17	24 Sep. 18	
	Jemadars.			
27 Jan. 94	Nadir Khan (56)	27 Oct. 15		
29 Nov. 92	Abraham Daniel	27 Oct. 15		
	Bhawal Din	13 Feb. 17		
11 Aug. 03	Roshan Khan	13 Feb. 17		
19 Apl. 93	Moses Daniel	27 Aug. 17		
30 June 98	Sahiram Singh	27 Aug. 17		
29 Aug. 02	Shiudan Singh	27 Oct. 17		
9 Apl. 03	Chandan Singh	30 Oct. 17		
23 Feb. 07	Rekha Ram	31 Oct. 17		
19 Mar. 07	Hukam Singh	3 Jan. 18		
4 Jan. 06	Tunda Singh	3 Jan. 18		
23 Dec. 99	Sulaiman Khan	3 Jan. 18		
2 Jan. 09	Chimna Ram	19 Feb. 18		
17 Feb. 09	Mamraj Ram	19 Feb. 18		
	Hayat Khan	19 Feb. 18		
25 Oct. 96	Rup Singh	1 June 18		
20 Feb. 06	Naik Din Muhammad	1 June 18		
....	Hanse Singh	24 Sep. 18		

2nd Battalion, 102nd King Edward's Own Grenadiers.

Formed at Arangaon, near Ahmednagar.

Composition—2 Companies Rajputana Gujars, 1 Coy. Rajputana Jats, & 1 Coy. Punjabi Musalmans.

1st Commn. or date of entering service.	Names and Rank.	Army rank.	Present appointment in Regiment.	Remarks.
	Commandant.			
15 May 97	Morris, Maj. C. T. (*Actg. Lt.-Col.*)	15 May 15	17 Feb. 18	82nd Punj.
	Company Commanders. (4)			
24 Jan. 06	Christie, Capt. R. C., M. C. (*actg. Maj.*)	24 Jan. 15	22 Aug. 18	1-102nd Grs., 2nd-in-Comd.
1 Oct. 14	Savage, Capt. E. G.	1 Oct. 18	16 Feb. 18	1-112th Infy. *offg.* Adjt.
19 June 15	Lea, Lt. E. H. (*actg. Capt.*)	19 June 16	19 May 18	N. Staff. R., I. A. (*on probn.*)
29 July 17	Ward, Lt. H. G. L. (*actg. Capt.*)	29 July 18	1 Aug. 18	Worc. R., I. A. (*on probn.*)
	Company Officers.			
7 July 15	Irvin, Lt. D.	7 July 16	19 May 18	M. G. Corps, I. A. (*on probn.*)
7 Jan. 16	Rowland, Lt. C. J.	7 Jan. 17	10 Sep. 18	M. G. Corps, I. A. (*on probn.*)
1 Sep. 16	Tait, Lt. T. R.	1 Sep. 17	15 Aug. 18	A. S. C., I. A. (*on probn.*)
16 Sep. 15	Crawford, Lt. G. A.	16 Sep. 17	19 May 18	Bord. R., I. A. (*on probn.*)
5 Jan. 18	Davis, 2nd-Lt F. L.	5 Jan. 18	3 June 18	I. A. (*T. C.*)
26 Jan. 18	Geddie, 2nd-Lt. D. W.	26 Jan. 18	28 July 18	Sea. Highrs., I. A. (*on probn.*)
31 Jan. 18	Phelan, 2nd-Lt. P. J.	31 Jan. 18	16 Feb. 18	
29 Mar. 18	Cunningham, 2nd-Lt. V.	29 Mar. 18	23 Aug. 18	I. A. (*T. C.*)
29 Mar. 18	Bray, 2nd-Lt. G. F.	29 Mar. 18	23 Aug. 18	I. A. (*T. C.*)
29 May 18	Eastwick, 2nd-Lt. A. J.	29 May 18	3 Dec. 18	R. W. Kent, I. A. (*on probn.*)
26 June 18	Latimer, 2nd-Lt. R. A.	26 June 18	10 Sep. 18	M. G. Corps, I. A. (*on probn.*)
7 June 18	Wheeler, 2nd-Lt. W.	7 June 18	8 Oct. 18	I. A. (*T. C.*)
7 June 18	Woolsey, 2nd-Lt. C. H.	7 June 18	8 Oct. 18	I. A. (*T. C.*)
10 July 18	Chidley, 2nd-Lt. L.	10 July 18	2 Dec. 18	I. A. (*T. C.*)
30 Oct. 18	Burgess, 2nd-Lt. F. H., D.C.M.	30 Oct. 18	7 Sep. 18	4th Linc. R., I. A. (*on probn.*)
	Attached.			
13 Nov. 14	Smith, Lt. G. W.	13 Nov. 15	6 Sep. 18	7th York and Lanc. R., I. A. R. O. (*on probn.*)
15 Dec. 14	Samuels, Lt. C. J. H.	15 Dec. 15	2 Oct. 18	Ir. R., I. A. R. O. (*on probn.*)
6 June 16	Hamill, Lt. H.	6 June 17	16 Mar. 18	I. A. R. O. *offg.* Qr.-Mr.
30 May 17	Cameron, Lt. A. A.	30 May 18	16 Mar. 18	I. A. R. O.
5 Jan. 18	Extance, 2nd-Lt. T. J.	5 Jan. 18	3 June 18	I. A. R. O.
26 June 18	Bance, 2nd-Lt. W. J.	26 June 18	....	Genl. List, I. A. R. O. (*on probn.*)
26 June 18	Perrins, 2nd-Lt. H. G.	26 June 18	....	Genl. List, I. A. R. O. (*on probn.*)
2 June 18	Payne, 2nd-Lt. W.	2 June 18	8 Oct. 18	I. A. R. O.
22 July 18	Brown, 2nd-Lt. J. F.	22 July 18	2 Dec. 18	I. A. R. O.
22 July 18	Hay, 2nd.-Lt. J. A. C.	22 July 18	2 Dec. 18	I. A. R. O.

		Jemadar.	Subadar.	Subadar Major.
	Subadar-Major.			
7 Dec. 05	Ganga Ram, I.O.M., I.D.S.M.	1 Aug. 16	25 Feb. 16	15 Mar. 18
	Subadars.			
1 July 01	Mangal Singh	27 Oct. 15	16 Feb. 18	
29 Oct. 99	Imam Din	15 Oct. 16	16 Feb. 18	
1 June 01	Kanhiya Singh	25 Nov. 16	16 Feb. 18	
14 Nov. 99	Fateh Khan	8 Jan. 17	16 Feb. 18	
18 June 04	Raghubir Singh	8 Jan. 17	16 Feb. 18	
30 June 01	Ramdiyal Singh, I.D.S.M.	13 Feb. 17	16 Feb. 18	
	Jemadars.			
27 Sep. 02	Burhan Ali	16 Feb. 18		
20 Oct. 06	Bala Ram	16 Feb. 18		
15 Jan. 05	Khuma Ram (56)	16 Feb. 18		
4 Dec. 06	Kamal Khan (56)	16 Feb. 18		
8 Sep. 06	Jansi Singh	16 Feb. 18		
10 Nov. 08	Chakarpan Singh (56)	16 Feb. 18		
9 Feb. 03	Karam Ali Khan	1 Apl. 18		
16 May 07	Shafait Ali	1 Apl. 18		

Regt. Centre—Satara. *Uniform*—Scarlet.
Linked with—105th & 110th Mahratta Lt. Infy. *Facings*—Black.

1st Battalion, 103rd Mahratta Light Infantry.

Raised in 1768, and was originally designated the 2nd Battalion of Bombay Sepoys. Became the 1st Bn., 2nd Regiment of Bombay Native Infantry, 1796; the 3rd Regiment of Bombay N. I., 1824; the 3rd Regt. of Bom. Native (Light) Infy., 1871; the 3rd Regt. of Bom. (Light) Infantry, 1885; the 3rd Bombay Lt. Infy., 1901; present designation, 1903.

"Mysore" "Seedaseer" "Seringapatam" "Beni Boo Alli."
"Mooltan" "Goojerat" "Punjaub" "Abyssinia."

Composition—1 Company of Dekhani Mahrattas, 2 of Konkani Mahrattas, 1 of Dekhani Musalmans.

COLONEL—Major-General Frederick D. Mander . . . . 13 May 1904.
HONORARY COLONEL—Hony. Colonel *His Highness Maharaja Sir* Shahu Chhatrapati Maharaj, G.C.S.I., G.C.I.E., G.C.V.O., LL.D., of Kolhapur . . . . 1 January 1915.

1st Commn. or date of entering service.	Names and Rank.	Army rank.	Present appointment in Regiment.	Remarks.
	Commandant.			
1 May 90	Ross, Lt.-Col. H.	31 May 16	10 May 17	
	Company Commanders. (4)			
9 Nov. 89	*Garratt, Lt.-Col. H. S. (Q)* . .	*9 Nov. 15*	*6 July 02*	*Comdt., Prisoners of War Camp, Sumerpur.*
5 Aug. 96	Crichton, Maj. McV.	5 Aug. 14	5 Aug. 05	
21 Feb. 00	Thorne, Maj. A. C. D. . . .	1 Sep. 15	20 May 14	
	Company Officers. (9)			
3 Sep. 02	*Yates, Maj. J. A.*, C.I.E., D.S.O.	*21 Dec. 16*	*9 Oct. 07*	
27 Aug. 02	*Dent, Bt.-Lt.-Col. W.*, D.S.O., *p.s.c.*	*27 Aug. 17*	*22 Jan. 04*	
18 Jan. 05	Goldfrap, Capt. H. W., M.C. [L]	18 Jan. 14	12 Mar. 06	
10 Jan. 07	Munton, Capt. V. R.	1 Sep. 15	27 Mar. 08	
26 June 08	Spain, Capt. G. A. R. . . .	1 Sep. 15	7 Sep. 13	
24 Aug. 12	*Hinds, Capt. W. A. G.*, M.C. .	*24 Aug. 16*	*4 Nov. 13*	*A. D. C. to the G. O. C., N. Comd.*
14 Jan. 14	Lane, Capt. G. H.	14 Jan. 18	10 Oct. 14	With 1-117th Mah.
4 Apl. 16	Marshall, Lt. W. D.	4 Apl. 17	22 Aug. 17	Attd.
18 June 17	Rerrie, Lt. E. S. E.	18 June 18	5 Nov. 17	Qr.-Mr.
18 June 17	Covernton, Lt. J. W. G. . . .	18 June 18	22 June 17	With 2-116th Mah.
31 Jan. 18	Phelan, 2nd-Lt. P. G.	31 Jan. 18	8 Feb. 18	
	Attached.			
9 Jan. 92	Warren-Codrington, Lt-Col. H. O.	9 Jan. 18	25 Apl. 17	Cant. Magte.'s Dept.
8 May 01	Stuart-Menteth, Maj. A. H. [L]	8 May 16	13 Sep. 18	1-112th Infy.
4 May 01	Wayman, Capt. W. L. . . .	10 Mar. 11	. . . .	1-94th Infy.
10 Nov. 14	McLean, Capt. R. R. B. . . .	21 Nov. 17	4 Apl. 18	*Lv., m.c., 6 mths.*
17 June 15	Hanna, Lt. H. (*actg. Capt.*) . .	17 June 16	15 Apl. 18	I. A. (*on probn.*)
13 Dec. 15	Henderson, Lt. W. S. R. . . .	13 Dec. 16	30 Mar. 18	I. A. (*on probn.*)
14 Nov. 16	Norris, Lt. W. C. G.	2 Jan. 17	29 Jan. 18	
5 Jan. 17	Willmott, Lt. N. R.	5 Jan. 18	24 Mar. 17	I. A. R. O.
6 Oct. 17	Inglet, Lt. B. A.	6 Oct. 18	14 June 18	I. A. R. O.
31 Oct. 17	Hughes, Lt. S.	31 Oct. 18	16 Feb. 18	I. A. (*T. C.*)
1 Nov. 17	Donnell, Lt. N. A. (*Actg. Capt.*)	1 Nov. 18	16 Mar. 18	I. A. (*T. C.*) Adjt.
29 Aug. 17	McCombe, 2nd-Lt. A.	29 Aug. 17	. . .	I. A. R. O.
31 Oct. 17	Rickard, 2nd-Lt. J. H. . . .	31 Oct. 17	16 Feb. 18	I. A. R. O.
31 Oct. 17	Dunwell, 2nd-Lt. C. N. (46) .	31 Oct. 17	16 Feb. 18	I. A. R. O.
15 Jan. 18	Mendes, 2nd-Lt. M.	15 Jan. 18	7 June 18	I. A. R. O.
7 June 18	Fraser, 2nd-Lt. A. B.	7 June 18	17 Oct. 18	I. A. R. O.
23 July 18	Levett, 2nd-Lt. P.	23 July 18	2 Sep. 18	I. A. R. O.
23 July 18	Paramor, 2nd-Lt. C. H. . . .	23 July 18	2 Sep. 18	I. A. R. O.

		Jemadar.	Subadar.	Subadar-Major.	
	Subadar-Majors.				
10 Nov. 85	Shaikh Imam, *Khan Sahib Sardar Bahadur* (53, 56) . .	12 Feb. 92	4 July 97	5 June 08	*Order of Br. Ind., 1st Cl.*
4 June 90	Bapurao Gaekwad *Bahadur*, I.D.S.M. (66*a*).	29 July 97	5 June 08	30 Sep. 15	*Order of Br. Ind., 2nd Cl.*
19 Feb. 00	Babajirao Dalvi (56, 57) . . .	24 Aug. 11	24 Aug. 14	16 May 16	

1st Battalion, 103rd Mahratta Light Infantry—*contd.*

1st Commn. or date of entering service.	Names and Rank.	Army rank.		Remarks.
	Subadars.			
2 Mar. 92	Deoji Rao Bhonsle (52, 56, 66) .	3 Apl. 08	5 Nov. 08	
3 May 96	Eshwant Bhosle (53, 56, 66) .	22 Nov. 02	23 Jan. 10	
9 Apl. 03	Muhammad Chiragh-ud-din Khan (53, 56)	9 Apl. 03	9 Sep. 14	
28 May 96	Sirpati Kadam (52, 56, 66) . .	23 Jan. 10	29 Sep. 15	
15 June 99	Ilyasdar Khan (56, 66), I.D.S.M. .	1 Sep. 14	29 Sep. 15	
19 Jan. 03	Shankar Lowande	21 Jan. 16	14 Oct. 17	
	Deoji Gorole	. . . .	. . . .	Re-employed, 4 Mar. 16.
22 Sep. 09	Narayan Kadam (46,52,56,66) .	5 June 16	1 Mar. 18	
15 June 96	Khandu Ghadge (56)	21 Oct. 16	1 July 18	
	Jemadars.			
23 July 97	Sirpati Powar (56, 66, *72a*) . .	1 Nov. 12		
27 Aug. 99	Tukaram Jagtap (56, 57, *72a*) .	8 Nov. 13		
11 Aug. 96	Sheikh Jamal	1 Sep. 14		
13 Apl. 97	Krishna Chavan	9 Sep. 14		
30 Dec. 03	Gopal Rao Palande, I.O.M., I.D.S.M.	29 Sep. 15		
21 Oct. 98	Shaikh Maktum (56) I.D.S.M. .	29 Sep. 15		
29 Jan. 05	Shaikh Muhammad	29 Sep. 15		
20 Dec. 99	Dagdu Chawan	29 Sep. 15		
26 Mar. 02	Mahadu Jadhao 46, 52, 56 . .	29 Sep. 15		
2 Oct. 06	Sheikh Imamuddin (56) . . .	5 June 16		
3 June 01	Sheikh Dada	6 June 16		
22 May 04	Govind Palao (58)	1 Mar. 17		
12 Oct. 07	Jairam Rao Rane	1 Dec. 17		
12 Nov. 01	Joti More	3 Dec. 17		
10 May 02	Mama Thakur	1 Mar. 18		
18 Apl. 04	Hamid Khan	1 July 18		
10 May 02	Bhikaji Thakur	1 July 18		
14 Dec. 04	Parbati Jagtap	1 July 18		
1 Jan. 08	Arjun Mandlekar	1 July 18		
25 June 03	Bhau Lole	1 July 18		
23 Dec. 08	Ranoji Beluse	1 July 18		
11 Sep. 11	Ganpat Rao Dalwi.	1 July 18		
11 Dec. 04	Tatya Jadhow	1 July 18		

2nd Battalion, 103rd Mahratta Light Infantry.

Formed at Kamptee.

Composition—1 Company Lingayats, 1 Coy. Bedars, 1 Coy. Dekhani Mahrattas, 1 Coy. Konkani Mahrattas.

1st Commn. or date of entering service.	Names and Rank.	Army Rank.	Present appointment in Regiment.	Remarks.
	Commandant.			
5 Aug. 96	Bald, Maj. J. A., C.M.G.	5 Aug. 14	23 June 18	61st Prs. (*temp.*)
	Company Commanders. (4)			
24 Dec. 02	Benton, Maj. W. B., M.C.	24 Dec. 17	19 June 18	105th Mah. L. I. 2nd-in-Comd. (*temp.*)
5 May 15	Bickersteth, Lt. E. H.	5 May 16	. . . .	E. York. R., I. A. (*on probn.*)
7 Jan. 16	Reid, Lt. D. W., D.S.O., M.C.	7 Jan. 17	1 Sep. 18	Sea. Highrs., I. A. (*on probn.*)
18 June 17	Turner, 2nd-Lt. C. D. L.	18 June 18	1 Sep. 18	1-117th Mah.
	Company Officers.			
5 Jan. 17	Hudson, Lieut. J. E.	5 Jan. 18	10 Oct. 18	I. A. R. O.
28 July 17	Lennox, 2nd-Lt. A.	28 July 17	20 June 18	I. A. R. O. (*on probn.*) Qr.-Mr.
26 Sep. 17	Smith, 2nd-Lt. H. N.	26 Sep. 17	14 July 18	I. A. R. O. (*on probn.*)
5 Oct. 17	Linell, Lt. C. J. B.	5 Oct. 18	23 June 18	I. A. (*T. C.*) Adjt.
25 Oct. 17	Daws, Lt. H. A.	25 Oct. 17	28 Aug. 18	N. Staffs. R., I. A. (*on probn*
30 Oct. 17	Tuckniss, 2nd-Lt. C. S.	30 Oct. 17	28 Aug. 18	
15 Jan. 18	Harrison, 2nd-Lt. A. S.	15 Jan. 18	1 Sep. 18	I. A. (*T. C.*)
15 Jan. 18	Clark, 2nd-Lt. G. S. H.	15 Jan. 18	20 June 18	I. A. (*T. C.*)
19 Mar. 18	Cardwell, 2nd-Lt. W. A.	19 Mar. 18	17 July 18	I. A. R. O.
19 Mar. 18	Loveless, 2nd-Lt. G. J.	19 Mar. 18	17 July 18	I. A. (*T. C.*)
28 Mar. 18	George, 2nd-Lt. W. J.	28 Mar. 18	26 July 18	I. A. R. O.
1 May 18	Stearn, 2nd-Lt. C.	1 May 18	14 Aug. 18	I. A. R. O. (*on probn.*)
15 Jan. 18	Storm, 2nd-Lt. W.	15 Jan. 18	4 Dec. 18	I. A. R. O.
26 June 18	Medlicott, 2nd-Lt. S. J. E.	. . . . .	. . . .	Genl. List, I. A. R. O. (*on probn.*)
26 June 18	Wigden-Jenks, 2nd-Lt. J. C.	. . . . .	. . . .	Genl. List, I. A. R. O. (*on probn.*)
15 July 18	Bennett, 2nd-Lt. J. W.	15 July 18	. . . .	I. A. R. O.
10 July 18	Hegarty, 2nd-Lt. E.	10 July 18	4 Dec. 18	I. A. R. O.
21 July 18	Hozier, 2nd-Lt. F. R.	21 July 18	4 Dec. 18	I. A. R. O.

		Jemadar.	Subadar.	Subadar-Major.
	Subadar-Major.			
16 Oct. 03	Eshwant Rao Bhonsle	29 Sep. 15	11 Mar. 16	27 July 18
	Subadars.			
23 July 67	Saka Ram Niram	5 June 16	1 July 18	
5 Feb. 02	Jai Singh Rao Sinde	1 Jan. 17	1 July 18	
27 Feb. 06	Sahdeo Sawant	1 Sep. 17	1 July 18	
	Jemadars.			
o July 11	Datta Jadhow	14 Oct. 17		
13 June 04	Ram Chandra Rao More	28 Oct. 17		
3 July 07	Ragho Parab	1 Jan. 18		
	Rama Ghor	. . . .		
	Shidramapa Sangapa Motgi	1 July 18	(*on probn.*)	

3rd Battalion, 103rd (Kolhapur) Mahratta Light Infantry.

Formed at Kolhapur on 5th August 1918.

Composition :—

1st Commn. or date of entering service.	Names and Rank.	Army Rank.	Present appointment in Regiment.	Remarks.
	Commandant.			
11 Oct. 93	Steen, Maj. H. C. (*Actg. Lt.-Col.*)	11 Oct. 11	5 Sept. 18	117th Mahs. (*temp.*)
	Company Commanders. (4)			
18 Jan. 11	Craig, Capt. H. T. (*Actg. Maj.*)	1 Sept. 15	6 Oct. 18	43rd Regt. 2nd-in-Comd.
14 Nov. 16	Norris, Lt. W. C. G.	2 Jan. 17	5 Aug. 18	1-103rd Infy.
2 Jan. 16	Peddie, Lt. R. M.	2 Jan. 17	4 Sept. 18	R. Innis. Fus., I. A. R. O. (*on probn.*)
6 Jan. 16	Cooper, Lt. R. G.	6 Jan. 17	22 Aug. 18	I. A. R. O.
	Company Officers.			
23 June 17	ꝟꝯ Grimshaw, Lt. J. E. . . .	23 June 18	6 Sept. 18	Lan. Fus., I. A. (*on probn.*)
27 Oct. 17	Shipton, Lt. G. L.	27 Oct. 18	22 Aug. 18	1-105th Infy.
31 Oct. 17	Gill, Lt. R. B.	31 Oct. 18	12 Sep. 18	I. A. R., O., Qr.-Mr.
30 Jan. 18	Duthie, 2nd-Lt. K. J. N. . . .	30 Jan. 18	12 Sep. 18	Attd.
7 June 18	Castle, 2nd-Lt. E. D. F. . . .	7 June 18	7 Oct. 18	I. A. (*T. C.*)
7 June 18	Gilbert, 2nd-Lt. C. H. F. . .	7 June 18	7 Oct. 18	I. A. (*T. C.*)
26 June 18	Clay, 2nd-Lt. N. C.	26 June 18	16 Sep. 18	I. A. (*on probn.*)
19 Sep. 18	Munro, 2nd-Lt. A. S.	19 Sep. 18	21 Oct. 18	I. A. R. O.
21 Sep. 18	Miller, 2nd-Lt. E. V.	21 Sep. 18	19 Oct. 18	I. A. R. O.
19 Oct. 18	Walsh, 2nd-Lt. E.	19 Oct. 18	12 Oct. 18	I. A. R. O.
. . . .	Jones, 2nd-Lt. A. N. P. . . .	. . . .	16 Oct. 18	I. A. R. O.

		Jemadar.	Subadar.	Subadar-Major.
	Subadar-Major.			
	Subadar.			
24 Sep. 87	Shaikh Jaffer	1 Apl. 03	5 Aug. 18	
. . . .	Gopalrao Jadhao.	. . . .	12 Mar. 18	
	Jemadar.			

Regt. Centre—Nasirabad.
Linked with—123rd & 125th Rifles.

Uniform—Dark Green.
Facings—Scarlet.

104th Wellesley's Rifles.

Raised in 1775 as the 5th Battalion of Bombay Sepoys. Became the 9th Bn. of Bombay Sepoys, 1778; the 2nd Bn., 2nd Regiment of Bo. Native Infantry, 1796; the 4th Regiment of Bombay Native Infantry, 1824; the 4th Regt. of Bombay N. I., or Rifle Corps, 1841; the 4th Regt. of Bo. Infantry, or Rifle Corps, 1885; unauthorisedly styled the 1st Battalion, Bombay Rifle Regt., 1889, and title corrected to 4th Regt., 1st Battn., Rifle Regt. of Bo. Infantry, 1889; the 4th Bombay Rifles, 1901; present designation, 1903.

"Mysore" "Seringapatam" "Bourbon" "Beni Boo Alli" "Mooltan" "Punjaub"
"Reshire" "Bushire" "Koosh-ab" "Persia" "Central India"
"Kandahar, 1880" "Afghanistan, 1879-80" "British East Africa, 1898."

Composition—2 Companies of Rajputana Jats, 1 of Rajputana Rajputs, 1 of Punjabi Musalmans.

COLONEL—Lieut.-General Michael W. Willoughby, C.S.I. . . 16 June 1914.

1st Commn. or date of entering service.	Names and Rank.	Army rank.	Present appointment in Regiment.	Remarks.
	Commandant.			
7 Nov. 91	Scott, Lt.-Col. W.	7 Nov. 17	6 Feb. 18	
	Company Commanders. (4)			
20 Jan. 97	Butler, Maj. H. M. (*actg. Lt.-Col.*)	20 Jan. 15	6 Feb. 18	2nd-in-Comd. With 2-72nd Punj.
4 Aug. 97	Sargent, Maj. W. P. M. (*actg. Lt.-Col.*)	4 Aug. 15	8 Dec. 13	With 2-11th Rajputs.
19 Jan. 01	Grant-Smith, Maj. H. G. H.	1 Sep. 15	10 July 17	*Offg.* 2nd-in-Comd.
11 Aug. 00	Byrne, Maj. E. G. J. (*actg. Lt.-Col.*)	1 Sep. 15	6 Feb. 18	With 2-123rd Rif.
	Company Officers. (9)			
23 Apl. 02	*Maclean, Maj. G. G. C. (temp. Lt.-Col.)*	*3 June 16*	*20 Dec. 04*	*Dy. Mily. Secy., A. Hdqrs.*
27 Aug. 02	James, Maj. F. H., M.C.	27 Aug. 17	23 Jan. 04	
19 Aug. 03	Cheeke, Maj. A. M. M.	19 Aug. 18	10 Nov. 04	With 125th Rif.
18 Jan. 05	Dormer, Capt. J. G. (*actg. Maj.*)	18 Jan. 14	26 Mar. 06	
18 Jan. 05	Manners, Capt. C. M. S., D.S.O.	18 Jan. 14	8 Mar. 06	
18 Jan. 05	Stoddart, Capt. G.	18 Jan. 14	9 Mar. 06	
8 Sep. 09	Wilson, Bt.-Maj. W. C. F. A.	3 June 18	6 Nov. 10	Poll. Employ.
1 Oct. 15	Butler, Lt. A. G.	1 Oct. 16		
18 Apl. 16	Ambrose, Lt. R. D., M.C.	18 Apl. 17	23 Apl. 16	With 1/123rd Rif.
24 Apl. 16	Waring, Lt. J. L.	24 Apl. 17	10 July 17	Attd.
18 June 17	Davies, Lt. R. M.	18 June 18	. . . .	With 2-72nd Pun'.
	Attached.			
13 Nov. 14	*Lemon, Capt. G. T.*	*15 Sep. 16*	. . . .	*I. A. R. O. Attd. M. W. Bch. A. Hdqrs.*
27 Nov. 14	Alderman, Capt. R. E., C.I.E.	19 Sep. 16	. . . .	I. A. R. O.
30 Jan. 17	Hodge, Lt. J. F.	30 Jan. 18	. . . .	125th Rif.
15 May 15	Atkinson, Lt. N. E.	15 May 16	. . . .	I. A. (*on probn.*) *Lv., ex I., m. c.*
29 Sep. 15	Webb, Lt. W. F.	29 Sep. 16	. . . .	R. N. V. R. *Offg.* Qr.-Mr.
23 July 15	Snelling, Lt. C. G. (*Temp. Capt.*)	23 July 16	. . . .	I. A. (*on probn.*) Platoon Comdr.
29 May 16	Shuldham, Lt. W. F. Q.	29 May 16	. . . .	I. A. (*on probn.*)
16 Aug. 16	Jones, Lt. L. B.	. . . .	. . . .	
10 Nov. 16	Harper, Lt. C. S. (*temp. Capt.*)	10 Nov. 17	. . . .	I. A. R. O. Adjt.
22 May 15	Fenwick, Lt. J. S.	22 May 16	12 Apl. 18	I. A. (*on probn.*)
8 Aug. 17	Williams, Lt. G.	8 Aug. 18	. . . .	I. A. R. O.
8 Aug. 17	Pugh, Lt. T. E.	8 Aug. 18	. . . .	I. A. R. O.
26 Nov. 17	Trotman, Lt. E. W.	26 Nov. 18	. . . .	Glouc. R., I. A. (*on probn.*)
28 Mar. 18	McDonald, 2nd-Lt. D. C.	28 Mar. 18	. . . .	I. A. R. O.
29 Mar. 18	Tremenheere, 2nd-Lt. H. S.	29 Mar. 18	. . . .	I. A. R. O.
15 Sep. 18	Dallas-Conte, 2nd-Lt. N.	15 Sep. 18	. . . .	Attd.
7 June 18	Wilkinson, 2nd-Lt. W. E.	7 June 18	. . . .	I. A. R. O.
. . . .	Sawdy, 2nd-Lt. A. E. L.	. . . .	. . . .	*U. L.*
. . . .	Parry, 2nd-Lt. B. K.	. . . .	. . . .	I. A. R. O.

		Jemadar.	Subadar.	Subadar Major.	
	Subadar-Major.				
10 Aug. 86	Hasan Muhammad, *Bahadur* (52, 56).	1 Jan. 99	1 Dec. 02	2 Nov. 08	*Order of Br. Ind. 2nd Cl.*
21 Sep. 95	Rahim Bakhsh (53, 56) I.D.S.M.	1 Dec. 04	4 Dec. 14	4 July 17	(*Temp.*)
	Subadars.				
1 Jan. 94	Sabal Singh, I.O.M. (52, 56, 66*a*)	14 July 00	20 Feb. 04		
24 Aug. 98	Mukh Ram, I.O.M. (52, 56, 66*a*)	2 Nov. 08	10 June 10		
3 Sep. 95	Harchand Jat (52, 56, 66*a*)	1 July 05	17 Nov. 11		
3 Oct. 03	Pema Ram (52, 56, 58, 66*a*)	25 Aug. 10	1 Feb. 14		
11 Dec. 95	Karam Dad (56)	20 Mar. 13	27 Oct. 15		
5 Oct. 96	Muhammad Khan (56)	15 Aug. 14	1 Jan. 17		
1 Mar. 00	Bindu Singh (56)	15 Aug. 14	1 Jan. 17		
5 Feb. 05	Tikhu Ram (56)	15 Aug. 14	1 Jan. 17		
1 Dec. 10	Chimna Ram	31 Dec. 10	1 Jan. 17		
1 Apl. 07	Mewa Ram, I.D.S.M.	. . . .	1 Jan. 17		
3 June 00	Lal Sing (56)	15 Aug. 14	4 July 17		
5 Mar. 08	Bhopat Ram (52)	1 Jan. 17	4 July 17		

104th Wellesley's Rifles—*contd.*

1st Commn. or date of entering service.	Names and Rank.	Army rank.	Remarks.
	Jemadars.		
16 Dec. 95	Kalu Ram (52, 56)	16 Mar. 12	
11 Aug. 95	Kishna Ram (66*a*)	1 Feb. 14	
17 Mar. 03	Rahimdad Khan	27 Oct. 15	
7 Sep. 96	Sharaf Khan	12 Dec. 15	
9 Nov. 05	Ganpat Singh	16 Jan. 16	
20 Jan. 97	Nur Ali (51)	11 Feb. 16	Indian Adjt., 1 Jan. 17.
7 Aug. 96	Ghulam Muhammad (51)	1 Dec. 16	
10 Nov. 91	Raghoram Sajjan	1 Dec. 16	
22 Oct. 04	Bahawal Khan (51)	1 Jan. 17	
5 July 05	Udey Singh (51)	1 Jan. 17	
25 Feb. 08	Ladhu Ram (51)	1 Jan. 17	
16 Feb. 11	Shedu Ram (51)	1 Jan. 17	
1 Jan. 01	Budhe Shah	4 July 17	
14 Jan. 05	Mana Ram	4 July 17	
19 Aug. 05	Hira Ram	4 July 17	
6 Mar. 02	Shiuram Sajjan	1 Mar. 18	
21 Apl. 07	Jiwan Ram (51)	1 May 18	
6 Jan. 08	Bhiv Singh (51)	1 Aug. 18	Indian Qr.-Mr.

Regt. Centre—Satara. *Uniform*—Scarlet
Linked with— 103rd & 110th Mahratta Light Infy. *Facings*—Black.

105th Mahratta Light Infantry.

Kali-Panchwen Paltan.—Raised in 1788 as the 3rd Battalion of Bombay Sepoys. Became the 1st Bn., 3rd Regiment of Bombay Native Infantry, 1796; the 5th Regt. of Bombay N. I., 1824; the 5th Regt. of Bombay Nat. (Light) Infantry, 1841; the 5th Regt. of Bombay (Light) Infantry, 1885; the 5th Bombay Light Infantry, 1901; present designation in 1903.

"Mysore" "Seedaseer" "Seringapatam" "Beni Boo Alli" "Kahun"
"China, 1860-62" "Afghanistan, 1879-80" "Burma, 1885-87."

Composition—2 Companies of Dekhani Mahrattas, 1 of Konkani Mahrattas, 1 of Dekhani Musalmans.

COLONEL—Major-General Robert A. C. Hunt . . . 23 March 1909.

1st Commn. or date of entering service.	Names and Rank.	Army rank.	Present appointment in Regiment.	Remarks.
	Commandant.			
7 Jan. 92	Waller, Lt.-Col. A. A. P. . . .	27 Jan. 18	28 Feb. 18	With 110th Mah. L. I.
	Company Commanders (4)			
3 Aug. 98	Delmé-Radcliffe, Maj. A. (*actg. Lt.-Col*). [L]	1 Sep. 15	28 Feb. 18	2nd-in-Comd. *Offg.* Comdt.
19 Oct. 01	*Powell, Bt.-Bt. Col. D. H.*, M. C. *(temp. Lt.-Col)*.	*3 Mar. 16*	*3 June 18*	*D. A. G.*
18 Jan. 02	*Haugton, Maj. E. J. H.*, D.S.O..	*18 Jan. 17*	*26 Mar. 17*	*Bde.-Maj., Peshawar Infy. Bde.*
27 Aug. 02	*Ford, Maj. G. N. [L]*	*27 Aug. 17*	*28 Feb. 18*	*G. S. O., 2nd gr. A. Hdqrs.*
	Company Officers, (8)			
27 Aug. 02	Monteith, Maj. M. E. C. V. S. .	31 Aug. 17	17 Feb. 04	*Offg.* 2nd-in-Comd.
24 Dec. 02	Benton, Maj. W. B., M.C. . .	24 Dec. 17	9 Nov. 06	With 2-103rd Mah. L. I.
19 Aug. 03	*Richardson, Maj. G. G.* . . .	*19 Aug. 18*	*12 Jan. 05*	*Burma Mily. Police.*
29 Jan. 10	Irwin, Capt. H. R. B. H. . . .	1 Sept. 15	20 Mar. 11	Adjt.
16 June 15	Tomlinson, Lt. F.	16 June 16	28 Oct. 15	With 2-9th Infy.
18 Apl. 16	Bland, Lt. G. H., M. C. . . .	18 Apl. 17	26 Apl. 16	Attd.
29 June 16	Douglas, Lt. A. E. R. . .	29 June 17	2 June 16	Attd.
30 Jan. 17	Macalister, Lt. S. C.	30 Jan. 18	26 Sep. 17	Attd.
12 Sep. 17	Phillips, Lt. R. L.	12 Sep. 18	8 Dec. 17	Attd.
27 Oct. 17	Dedman, Lt. J. B.	27 Oct. 18	6 Nov. 17	Attd.
27 Oct. 17	Shipton, Lt. G. L.	27 Oct. 18	6 Nov. 17	
	Attached			
22 Apl. 03	Reynolds, Maj. H. V.	22 Apl. 12	1 Jan. 18	82nd Punj.
5 Aug. 05	Welch, Capt. F. E.	5 Aug. 14	. . . .	110th Mah. L. I.
6 Sep. 11	Harding, Capt. C. N.	6 Sep. 15	. . . .	88th Infy.
26 Sep. 15	Mason, Lt. H. C. E.	26 Sep. 16	29 Apl. 18	I. A. R. O. (*on probn.*)
1 Oct. 15	Wright, Lt. W. R. B.	1 Oct. 16	. . . .	I. A. R. O.
20 Feb. 16	Young, Lt. J. M.	20 Feb. 17	. . . .	I. A. R. O.
29 June 16	Kenny, Lt. E. J.	29 June 17	3 July 16	116th Mah.
21 Nov. 16	Phillips, Lt. J. H.	21 Nov. 17	. . . .	I. A. R. O.
5 June 17	Cooper, Lt. L. G.	5 June 18	9 Sept. 18	Sea. Highrs, I. A. (*on probn.*)
6 Sep. 16	Lewellin, Lt. D. J.	6 Sep. 17	. . . .	High. L. I., I. A. (*on probn.*)
28 Mar. 16	Wood, Lt. F.	28 Mar. 17	15 Jan. 18	I. A. R. O. (*on probn.*)
18 Aug. 17	Strudwicke, Lt. M. A. . . .	18 Aug. 18	. . . .	Dorset. R., I. A. (*on probn.*)
30 Sep. 18	Potts, 2nd-Lt. W. . . .	30 Sep. 18	23 Aug. 18	A. S. C., I. A. (*on probn.*)
25 Oct. 16	North, 2nd-Lt. A. E. . . .	25 Oct. 16	10 Sept. 18	4th London R., I.A.R. O. (*on probn.*)
26 Jan. 18	Somerset, 2nd-Lt. I. . . .	26 Jan. 18	24 May 18	I. A. (*on probn.*)
29 May 18	Yealand, 2nd-Lt. W. C. . . .	29 May 18	29 Aug. 18	I. A. R. O. (*on probn.*)
30 July 18	Alderson, 2nd-Lt. F. H. . . .	30 July 18	21 Aug. 18	I. A. R. O.
5 Aug. 18	Lewis, 2nd-Lt. A. G. L. . . .	5 Aug. 18	16 Sep. 18	I. A. R. O., (*on probn.*)
21 Aug. 18	Rieu, 2nd-Lt. E. V.	21 Aug. 18	11 Sep. 18	I. A. R. O. (*on probn.*)
		Jomadar.	**Subadar.**	**Subadar-Major.**
	Subadar-Major,			
	Subadars.			
23 Dec. 95	Dinkarrao Yadao (52, 56) . . .	28 July 04	1 Aug. 11	
21 May 89	Saiyid Suliman, I. D. S. M. (52, 56)	15 Mar. 08	2 Feb. 13	
26 Feb. 97	Chimaji Garud, I. O.M., I.D. S. M. (52, 56)	17 June 07	16 July 13	
27 Nov. 93	Shaikh Mahomed Usman (52, 56)	9 Apl. 10	23 Feb. 15	
13 Feb. 97	Bala Bhor (56)	16 July 13	28 Nov. [illegible]	
21 Feb. 03	Saiyid Abdullah	23 Feb. 15	12 [illegible] 15	
11 Jan. 10	Daulat Rao Bhosle (56)	11 Jan. 10	[illegible] Dec. 15	
5 Oct. 99	Gopal Waje (52, 56)	24 Aug. 14	17 Aug. 16	
18 Jan. 08	Manajirao, Palande	15 June 15	10 Jan. 17	
15 June 97	Sheikh Yakub (56)	1 Sep. 15	12 Jan. 17	
19 Apl. 03	Appa Salunke	[illegible] July 15	14 Jan. 17	
3 Feb. 94	Balkrishna Khanse (56, 66*a*) . .	3 Dec. 08	26 Mar. 17	110th Mah. L. I.
24 Apl. 03	Ramchandar Rao Powar . . .	6 June 15	26 Mar. 17	
20 Sep. 98	Eshwanta Shinde (52, 56) . . .	16 Jan. 15	15 Aug. 17	With 3-153rd Infy.
18 June 98	Shaikh Haider	27 July 16	31 July 17	
4 Sep. 07	Ramchandar Rao Erundkar (56) .	20 Jan. 16	13 Oct. 17	With 3-153rd Infy.
14 Feb. 99	Govinda Dh[illegible]ne (56)	7 Oct. 15	5 Feb. 18	
0 Aug. 93	Datu Bh[illegible] (Retd.)	16 July 03	11 Aug. 09	Re-employed, 1 Jan. 18.
17 June 06	Shaikh Jamal	26 Mar. 17	14 Mar. 18	
27 Dec. 01	Daulat Bhosle, I.D.S.M. . . .	18 Oct. 15	31 July 18	

105th Mahratta Light Infantry—*contd.*

1st Commn. or date of entering service.	Names and Rank.	Army rank.	Remarks.
	Jemadars.		
24 Mar. 02	*Jairam Thakur*	*26 Apl. 15*	
1 Apl. 07	*Jayajirao Palande*	*1 Sep. 15*	
28 Feb. 04	Shaikh Mohi-uddin (56)	18 Oct. 15	With 3-153rd Infy.
13 June 05	Sonu Sawant	28 Nov. 15	
27 Feb. 99	Vithoba Khule	28 Nov. 15	
21 July 02	Muhammad Shamsuddin	28 Nov. 15	
13 June 05	Krishna Bagwe	13 Dec. 15	With 3-153rd Infy.
13 Dec. 01	*Ramchandar Sawant*	*24 Jan. 16*	
18 Feb. 06	Ram Chander Gaikwad	27 July 16	
27 Aug. 02	*Hamzah Khan*	*10 Jan. 17*	
4 Oct. 05	Sitaram Gaode	10 Jan. 17	With 3-153rd Infy.
7 June	Ramji Garud	15 Aug. 17	
1 July 05	Bhau Kadam	13 Oct. 17	
5 Dec. 02	Ganu Panaskar	3 Dec. 17	
17 Mar. 07	Shaikh Abdul Samad	25 Dec. 17	
11 Oct. 07	Dhondu Sawanth	1 Jan. 18	
10 Oct. 93	Kondaji Bedar	1 Jan. 18	
10 Apl. 06	Krisnajirao Bhosle	1 Feb. 18	
14 Oct. 07	Mahade Rao More	14 Mar. 18	
12 Apl. 12	Shaikh Daud	14 Mar. 18	With 3-153rd Infy.
4 May 03	Baburao Pisal	24 May 18	
16 July 07	Krishna Chauhan	24 May 18	
1 May 92	Gyanu Jadhao	16 Aug. 18	
27 Feb. 03	Rajoi Powar	16 Aug. 18	
21 Nov. 08	Harun Khan	16 Aug. 18	
27 Sep. 02	Chhottaji Bedar	1 Oct. 18	

Regt. Centre—Quetta. *Uniform*—Drab. *Facings*—Scarlet.

106th Hazara Pioneers.

Raised at Quetta in 1904, by Major C. W. Jacob; a nucleus being formed by drafts from the 124th Dunchess of Connaught's Own, and 126th Baluchistan Infantry Regiments.
Composition—4 Companies Hazaras.

COLONEL—Lieut.-Genl. *Sir* C. William Jacob, K.C.B.

1st Commn. or date of entering service.	Names and Rank.	Army rank.	Present appointment in Regiment.	Remarks.
	Commandant.			
8 Dec. 88	Carter, Lt.-Col. G. L., C.I.E.	8 Dec. 14	1 Oct. 16	
	Company Commanders. (4)			
22 Jan. 96	*Peart, Maj. C. L., p. s. c. (L)*	*22 Jan. 14*	*20 Oct. 18*	*2nd-in-Comd.*
5 Aug. 96	Ashburner, Maj. H. W., D.S.O., *q.*	5 Aug. 14	15 Jan. 12	With 2-128th Prs.
5 Sep. 96	*Grant, Maj. G. P.*, D.S.O., *p.s.c. (temp. Lt.-Col.)*	*5 Sep. 14*	*2 Aug. 16*	*G. S. O., 1st gr.*
20 July 98	*Meiklejohn, Maj. W. L. S., p. s. c., q. (l) (Actg. Lt.-Col.)*	*1 Sep. 15*	*20 Oct. 18*	With 2-94th Infy.
	Company Officers. (9)			
25 Jan. 99	*Wyatt, Maj. E. R. C.*, D.S.O., *p.s.c. (temp. Lt.-Col.)*	*1 Sep. 15*	*1 Oct. 04*	*G. S. O., 1st gr., A. Hd-Qrs.*
17 Jan. 00	*Davis, Maj. G. E. P., q.*	*1 Sep. 15*	*30 Apl. 07*	*Kurram Militia.*
4 Dec. 01	*Manners, Maj. R. H. H., q.*	*4 Dec. 16*	*20 Oct. 04*	*Bde.-Maj., Jubbulpore Bde.*
22 Apl. 03	Turner, Maj. H. G., *q.*	22 Apl. 18	18 June 05	
18 Jan. 05	Blackwell, Capt. H., *q.*	18 Jan. 14	22 Oct. 06	2nd-in-Comd. (*temp.*)
9 Sep. 08	Moore-Lane, Capt. V. W. C., *q.*	1 Sep. 15	30 Nov. 09	Adjt., 26 June 16.
29 Jan. 10	Davison, Capt. K. H. M.	1 Sep. 15	10 Mar. 11	
3 Sep. 10	Bunbury, Capt. N. L. St. P.	1 Sep. 15	13 Nov. 11	
18 Jan. 11	*Pollard, Capt. G. L. G.*	*1 Sep. 15*	*28 Mar. 12*	*With Sig. Ser.*
8 Aug. 14	Campbell, Capt. A. D. P.	8 Aug. 18	29 Feb. 16	Attd. With 2-34th Prs.
15 Nov. 15	Fenwicke-Clennell, Lt. G. E.	15 Nov. 16	16 Nov. 15	Attd.
15 Nov. 15	Phillips, Lt. C. E. V.	15 Nov. 16	16 Nov. 15	Attd.
22 Dec. 15	Tyson, Lt. E. W.	22 Dec. 16	1 Mar. 15	Attd. With 1-128th Prs.
18 Apl. 16	Stewart, Lt. C. H.	18 Apl. 17	30 Apl. 16	Attd. With 1-128th Prs.
29 June 16	Gordon Powell, Lt. D. H.	29 June 17	30 Apl. 16	Attd.
31 Jan. 18	Hunt, 2nd-Lt. J. W. A.	31 Jan. 18	23 Feb. 18	
....	Burge, 2nd-Lt. N. B.	....	....	Attd.
	Attached.			
24 Dec. 14	Slater, Lt. A.	24 Dec. 15	13 Jan. 17	I. A. R. O. Qr.-Mr.
26 May 15	Greer, Lt. E. R.	26 May 16	28 June 17	I. A. (*on probn.*)
12 June 15	Willans, Lt. N. J.	12 June 16	....	A. S. C., I. A. (*on probn.*)
3 Jan. 16	Woollam, Lt. J. L.	3 Jan. 17	....	I. A. (*on probn.*)
1 Jan. 16	Hodding, Lt. C.	1 Jan. 17	26 Feb. 16	I. A. R. O.
19 Jan. 16	Bell, Lt. N. K.	19 Jan. 17	17 Dec. 17	I. A. (*on probn.*)
17 May 16	Plaford, Lt. R. B.	17 May 17	29 Nov. 17	I. A. (*on probn.*)
25 June 16	Landon, Lt. S. L.	25 June 17	....	R. Fus., I. A. (*on probn.*)
5 Feb. 17	Fulton, Lt. T.	5 Feb. 18	23 Apl. 17	I. A. R. O.
1 Feb 17	Edwards, Lt. T.	1 Feb. 18	23 Apl. 17	I. A. R. O.
1 Feb. 17	Hampton, Lt. L. C.	1 Feb. 18	23 Apl. 17	I. A. R. O.
25 July 17	Dodd, Lt. A. V.	25 July 18	13 Apl. 18	I. A. (*on probn.*)
5 Oct. 17	Miller, Lt. C. B.	5 Oct. 18	22 Jan. 18	I. A. R. O.
25 Oct. 17	Scholes, Lt. W.	25 Oct. 18	25 Sep. 18	N. Lan. R., I. A. (*on probn.*)
2 June 18	Bendall, 2nd-Lt. A. F.	2 June 18	19 Oct. 18	I. A. (*T. C.*)
7 June 18	Ford, 2nd-Lt. G. H.	7 June 18	19 Oct. 18	I. A. R. O.
4 Sep. 18	Sawtell, 2nd-Lt. W. G.	4 Sep. 18	9 Sep. 18	I. A. R. O.
....	Harris, Lt. G. G.	....	20 Nov. 18	W. Kent R., I. A. (*on probn.*)

		Jemadar.	Subadar.	Subadar-Major.
	Subadar-Major.			
2 Mar. 95	Ali Dost (56) I.D.S.M.	22 June 00	1 Jan. 05	
	Subadars.			
6 Apl. 93	Gird Ali (56)	1 Oct. 95	6 Nov. 01	
25 May 93	Sharif (56, 66)	1 Dec. 04	1 May 06	
21 May 01	Yazdan Khan (56, 66)	1 May 06	7 Apl. 15	
27 Aug. 02	Muhammad Juma (56)	1 Apl. 11	7 Apl. 15	
23 June 97	Muhammad Hasan (II) (56)	12 Apl. 13	19 Apl. 15	
4 Mar. 95	Ali Panah	24 Mar. 11	24 June 15	
13 Aug. 99	Qurban Ali (56)	1 Oct. 13	1 Aug. 15	
4 Dec. 96	Faize-Muhammed	12 Oct. 14	1 Oct. 15	
5 Apl. 97	Ali Shafa (66)	28 Jan. 14	29 Dec. 15	
27 Aug. 05	Ali Juma (56), I.D.S.M.	7 Apl. 15	26 May 16	
5 Oct. 99	Ghulam Ali	22 Feb. 18		

106th Hazara Pioneers—*contd.*

1st Commn. or date of entering service.	Names and Rank.	Army rank.	Remarks.
	Jemadars.		
21 Dec. 05	Khuda Dad (56, 66)	7 Apl. 15	
6 Sep. 95	Shafa (56)	19 Apl. 15	
27 Aug. 05	Ali Juma (56) I.D.S.M.	7 Apl. 15	
20 Sep. 05	Rahmatullah (56)	22 June 15	Indian Adjt.
4 Apl. 99	Qasim	1 Aug. 15	
30 Jan. 05	Ata Muhammad	1 Aug. 15	
16 Aug. 92	Najibullah (56)	25 Nov. 15	
1 Oct. 04	Ali Juma (II) (56)	1 Dec. 15	
24 Oct. 04	Salim	29 Dec. 15	
4 Nov. 04	Nijat	3 Jan. 16	
17 Oct. 04	Ali Hasan (46, 56)	12 Jan. 16	
16 May 08	Ali Hussain	1 July 16	
	Dost-Muhammed (*Ret.*) *Hon. Capt.*	....	Re-empld., 17 Dec. 17.
18 Dec. 96	Muhammed-Panah	5 Nov. 17	
25 Aug. 05	Haider-Ali	5 Nov. 17	
4 Oct. 05	Fazal Shah	23 Dec. 17	
30 Apl. 04	Muhammad Ali	1 Apl. 18	
2 Oct. 01	Ali Najat	24 June 15	
....	Muhammed Hassan	....	
....	Bakhsho	....	

Regt. Centre—Kirkee.
Linked with—121st & 128th Pioneers.

1st Battalion, 107th Pioneers.

Uniform—Scarlet.
Facings.—White.

Raised in 1788 as the 4th Battalion of Bombay Sepoys. Became the 1st Batt., 4th Regiment of Bombay Native Infantry in 1796; the 7th Regt. of Bombay N. I. in 1824; the 7th Regt. of Bombay Infantry in 1885 the 7th Bombay Infantry (Pioneers) in 1900; the 7th Bombay Pioneers in 1901; present designation in 1903
"Mysore" "Seedaseer" "Seringapatam" "Beni Boo Alli" "Burma, 1885-87."
Composition—1 Companies of Pathans, 1 of Rajputana Musalmans, 1 of Sikhs, 1 of Dekhani Mahrattas.
COLONEL—Major-General Herbert E. Penton. . 3 March 1914.

1st Commn. or date of entering service.	Names and Rank.	Army rank.	Present appointment in Regiment.	Remarks.
	Commandant.			
5 Dec. 91	Bickford, Lt.-Col. W. W., C.I.E.	17 Sep. 17	27 Feb. 18	A. A. & Q. M. G., 16th Dn.
	Company Commanders. (4)			
3 Aug. 98	*Marjoribanks, Maj. R. D., p.s.c. (temp. Lt.-Col.)*	*1 Sep. 15*	*27 Feb. 18*	*2nd-in-Comd. A. A. & Q. M. G.*
1 Sep. 98	McLaughlin, Maj. W. P. M. D. *q., p.s.c. (Actg. Lt.-Col.)*	1 Sep. 15	Mar. 14	Comdt. (*temp.*)
23 May 00	*Sterndale-Bennett, Maj. J., q.*	*1 Sep. 15*	*19 Feb. 13*	*Comdt., Musky. Schl., Changli Gali.*
8 Jan. 01	*Davis, Maj. C. T.*	*8 Jan. 16*	*27 Feb. 18*	*G.S.O., 2nd gr., Burma Dn.*
	Company Officers. (9)			
22 Apl. 03	*Griffith, Maj. L., q.*	*22 Apl. 12*	*25 Oct. 06*	*Bde.-Maj., Abbottabad.*
13 Aug. 04	*Mangin, Capt. E. B., M.C., q.*	*13 Aug. 13*	*1 Mar. 06*	*Staff Captain.*
10 Aug. 05	*Shute, Capt. C. A., q. (temp. Lt.-Col.)*	*16 Aug. 14*	*11 Mar. 08*	*A. A. G.*
29 Nov. 05	*Coote, Capt. M. C., q.*	*29 Nov. 14*	*11 Apl. 11*	*Assam Mily. Police.*
29 Aug. 06	Chapman, Capt. E. H.	29 Aug. 15	11 May 12	With 2-32nd Prs.
25 Jan. 08	Hornor, Capt. F. H. F., M.C., *q. (Actg. Maj.)*	1 Sep. 15	18 May 11	2nd-in-Comd. (*temp.*)
19 Sep. 08	Wallis, Capt. B. H., M.C.	1 Sep. 15	7 Oct. 11	Adjt., 29 Apl. 18.
6 Sep. 11	*Geidt, Capt. R. W.*	*6 Sep. 15*	*7 Nov. 12*	*With Hazara Levy.*
14 Jan. 14	*Robertson, Capt. H. E.*	*14 Jan. 18*	*10 Oct. 14*	*Platoon Comdr., Young Officers' Schl. of Instrn., Bangalore.*
15 Nov. 15	Goddard, Lt. E. N.	15 Nov. 16	18 Nov. 15	Attd. With 1-128th Prs.
15 Nov. 15	Plomer, Lt. G.	15 Nov. 16	18 Nov. 15	Attd. With 121st Prs.
18 Apl. 16	Clark, Lt. W. V.	18 Apl. 17	24 Apl. 16	Attd. With 121st Prs.
14 Nov. 16	Shaw, Lt. B. V. D. H.	14 Nov. 17	27 Nov. 16	Attd. With 121st Prs.
27 Oct. 17	Kershaw, Lt. H. W. P.	27 Oct. 18	1 Nov. 17	
31 Aug. 18	Steveni, 2nd-Lt. O. K.	31 Aug. 18	1 Sep. 18	Attd.
1 Oct. 18	Tribe, 2nd-Lt. A. G.	1 Oct. 18	10 Oct. 18	Attd.
	Attached.			
6 Sep. 11	Marshall, Capt. H. H., M.C., (*Actg. Maj.*)	6 Sep. 15	4 Oct. 18	
1 Sep. 18	Pogson, Capt. L. V. J., M.C.	8 Aug. 18	10 July 18	Ches. R., I. A. (*on probn.*)
7 July 15	Chappel, Lt. B. H. (*b*)	7 July 16	16 Aug. 17	I. A. (*on probn.*) Offr. Staff Sch., Saugor.
23 July 15	Bowers, Lt. J. F. F.	23 July 16	9 June 18	High. L. I., I. A. (*on probn.*)
15 Oct. 15	Adams, Lt. A. W.	15 Oct. 16	. . . .	R. E., I. A. (*on probn.*)
5 Feb. 16	Finch, Lt. P. W. (95)	5 Feb. 17	21 Feb. 18	York. R., I. A. (*on probn.*)
28 May 16	Murcott, Lt. S. H. D.	28 May 17	18 Apl. 18	Hants. R. E., I. A. (*on probn.*)
10 Nov. 16	Scott-Brown, Lt. E. G.	10 Nov. 17	. . . .	I. A. R. O.
6 Mar. 17	Murray, Lt. G. E.	6 Mar. 18	25 Apl. 17	I. A. R. O. Emb. Staff, Karachi.
19 Sep. 17	Haswell, Lt. F. W.	19 Sep. 18	25 May 18	Bord. R., I. A. (*on probn.*)
25 Oct. 17	Stone, Lt. T. W.	27 Oct. 18	9 June 18	High. L. I., I. A. (*on probn.*)
26 Nov. 17	Notley, Lt. W., M.C. (207*a*)	27 Nov. 18	15 Oct. 17	R. Innis. Fus., I. A. (*on probn.*)
28 Nov. 17	Worden, Lt. C. N.	28 Nov. 18	1 Feb. 18	I. A. R. O.
15 Jan. 18	Field, 2nd-Lt. W. J.	15 Jan. 18	10 June 18	I. A. R. O.
29 Mar. 18	Beedham, 2nd-Lt. J. W.	29 Mar. 18	. . . .	I. A. R. O.
2 June 18	Hall, 2nd-Lt. H.	2 Jan. 18	10 June 18	R. Gn. Art., I. A. (*on probn.*) Offrs. Sch., Nasik.

		Jemadar.	Subadar.	Subadar-Major.	
	Subadar-Major.				
7 Dec. 89	Labh Singh *Bahadur*, I.D.S.M. (56, 66*a*)	1 Sep. 08	1 July 12	5 Dec. 14	*Order of Br. Ind., 2nd Cl.*
30 Sep. 86	Sharam Singh (*Retd.*) (50)	10 July 02	1 Sep. 08	7 Feb. 14	Re-empld. 27 Apl. 15. Depôt.

1st Battalion, 107th Pioneers—*contd.*

1st Commn. or date of entering service.	Names and Rank.	Army rank.		Remarks.
	Subadars.			
19 Aug. 93	Babaji Sivaji	1 Jan. 10	7 Aug. 13	
24 Sep. 94	Sundar Singh (56)	1 May 09	9 Aug. 14	
22 Aug. 00	Hari Singh, *Bahadur* (66a)	1 July 12	5 Dec. 14	*Order of Br. Ind., 2nd Cl.*
16 July 95	Fateh Muhammad Khan, I.D.S.M. (56, 71).	1 July 09	19 Dec. 14	
30 Sep. 05	Samundar	9 Aug. 14	29 Dec. 15	
1 Nov. 00	Ghafur Khan	4 Nov. 10	18 Jan. 16	
26 Jan. 01	Laximan Jankar	28 Dec. 14	7 Mar. 16	With 121st Prs.
1 Apl. 02	Rajaram Jadhow	9 Aug. 14	14 Aug. 18	
12 Oct. 96	Manulla	1 Feb. 17	15 Sep. 18	
19 Feb. 00	Ghani Khan (66a)	1 Nov. 14	15 Sep. 18	
	Jemadars.			
25 July 00	Lehna Singh	9 Aug. 14		
25 Mar. 01	Abdul Rahman, I.D.S.M.	9 Aug. 14		Indian Adjt.
22 Mar. 01	Dial Singh (46)	5 Feb. 15		
17 Aug. 00	Nanak Singh (56)	1 Oct. 15		
14 Nov. 00	Habib Khan (56), I.D.S.M.	16 Feb. 16		
20 Oct. 99	Krishna Ghorpade, I.D.S.M.	7 Mar. 16		
26 Dec. 00	*Bawar Khan*	*29 May 16*		
	Sher Khan (*Retd.*)	. . .		Re-empld. 1 Sep. 15.
1 July 01	*Karim Dad Khan*	*21 Aug. 16*		*With 121st Prs.*
1 Feb. 87	Budri Singh	25 June 18		
21 Aug. 04	Narayan Kadam	27 June 18		
3 Feb. 02	Nasrullah Khan	1 July 18		
17 Nov. 06	Mir Akbar Shah	15 Sep. 18		
16 July 07	Shabir Hussain	15 Sep. 18		
22 Mar. 01	Kishan Singh	15 Sep. 18		
5 Aug. 01	Samundar Khan	15 Sep. 18		
4 Apl. 05	Sharfu Khan	15 Sep. 18		
27 May 05	Yeshwant Sawant	15 Sep. 18		

2nd Battalion, 107th Pioneers.

Formed at Bareilly on 26 July 17.

Composition—4 mixed Companies of Punjabis.

1st Commn. or date of entering service.	Names and Rank.	Army rank.	Present appointment in Regt.	Remarks.
	Commandant.			
28 July 00	Cunningham, Maj. C. C., *p.s.c.* (Q.) (*Actg Lt.-Col*)	1 Sep. 15	13 Aug. 17	12th Prs.
	Company Commanders (4).			
8 May 01	Cooper, Maj. G.C., M.C. . . .	8 May 16	25 Nov. 17	121st Prs.
25 Jan. 08	Hare, Capt. I. L. O'H. . .	1 Sep. 15	16 Aug. 17	64th Prs.
5 May 15	Barker, Lt. T. L.	5 May 16	16 Aug. 17	
18 June 15	Hickey, Lt. P. F. B., D.S.O. . .	8 June 16	26 July 17	I. A. R. O.
	Company Officers.			
2 Sep. 15	Noel, Lt. K. H.	2 Sep. 16	21 Sep. 17	I. A. (*on probn.*)
25 Nov. 15	Maunder, Lt. G. H., M.C. (*Actg. Capt.*)	25 Nov. 16	26 Mar. 18	I. A. (*on probn.*) Adjt.
12 Feb. 16	Parmenter, Lt. F. (*Actg. Capt.*) .	12 Feb. 17	9 Dec. 17	I. A. (*on probn.*)
24 Apl. 16	Brown, Lt. A. E.	24 Apl. 17	30 Oct. 17	I. A. (*on probn.*)
8 July 16	Fenwicke-Clennell, Lt. E. C., M.C.	8 July 16	4 Dec. 17	Attd.
16 Aug. 16	Baker, Lt. A. H.	16 Aug. 17	10 Oct. 18	Midd'x. R., I. A. (*on probn.*)
14 May 16	Dodd, Lt. B. F.	14 May 17	12 Oct. 18	E. Surr. R., I. A. (*on probn.*)
6 Dec. 16	Cameron, Lt. F. O.	6 Dec. 17	4 June 18	Aust. Impl. Forces, I. A. (*on probn.*)
12 Sep. 17	Russell, Lt. H. B.	12 Sep. 18	5 Dec. 17	
27 Oct. 17	Stephenson, Lt. J.	27 Oct. 18	9 Dec. 17	
	Attached.			
2 Apl. 15	Haigh, Lt. F. F.	2 Apl. 16	31 Aug. 17	I. A. R. O.
10 May 16	Wyles, Lt. J. W.	10 May 17	26 July 17	I. A. R. O. Qr.-Mr.
5 Aug. 16	Gower, Lt. J. H. N..	5 Aug. 17	1 Dec. 17	I. A. R. O. (*on probn.*)
5 Jan. 18	Sabin, 2nd-Lt. H. E.	5 Jan. 18	6 June 18	I. A. (*T. C.*)
15 Jan. 18	Wilkinson, 2nd-Lt. S. P. . . .	15 Jan. 18	12 June 18	I. A. R. O.
26 Nov. 17	Watson, 2nd-Lt. W. B. . . .	26 Nov. 17	22 June 18	I. A. R. O. (*on probn.*)
2 June 18	Randell, 2nd-Lt. J. W. . . .	2 June 18	. . . .	I. A. (*T. C.*)
5 Sep. 18	Harper, 2nd-Lt. J. P.	5 Sep. 18	5 Sep. 18	I. A. R. O.
6 Sep. 18	Morgan, 2nd-Lt. W. D. . . .	6 Sep. 18	6 Sep. 18	I. A. R. O.
		Jemadar.	**Subadar.**	**Subadar-Major.**
	Subadar-Major.			
16 Apl. 01	Rajab Khan	7 Feb. 14	21 July 17	23 Nov. 17
	Subadars.			
23 Mar. 93	Ala Singh	13 Oct. 15	1 May 17	
31 Aug. 05	Karim Dad Khan	24 Feb. 17	26 July 17	
10 Sep. 99	Bhiru Ram	26 July 17	1 Nov. 17	
1 May 99	Bishan Sing	1 May 16	23 May 18	
2 Sep. 96	Kartar Singh	1 Sep. 17	5 June 18	
7 Nov. 02	Shah Niwaz	1 Oct. 17	5 June 18	
3 Jan. 00	Neki Ram	15 Mar. 17	6 June 18	
	Jemadars.			
9 Aug. 17	Atta Muhammad	1 Sep. 17		
	Datar Singh	1 Sep. 17		
5 Nov. 04	Pir Dost	22 Nov. 17	. . . .	
16 Sep. 00	Buta Singh	1 Nov. 17	. . . .	
15 Oct. 10	Sohan Ram	1 Nov. 17	. . . .	
30 Jan. 15	Sawan Khan	1 Dec. 17	. . . .	
8 Sep. 17	Ghulam Mahammed	15 Dec. 17	. . . .	
15 Aug. 16	Mul Singh	1 Jan. 18	. . . .	
16 May 03	Sher Singh	1 Jan. 18	. . . .	
20 Nov. 17	Kheta Ram	1 Jan. 18		
22 Dec. 17	Jumman Khan	1 Mar. 18		
9 May 03	Sundar Singh	1 Mar. 18		
4 June 12	Likhma Ram	1 Apl. 18	. . . .	
12 Sep. 05	Hubab Khan.	5 June 18		
15 Apl. 02	Jagat Singh.	5 June 18		
24 Apl. 18	Muhammad Khan	5 June 18		
1 Dec. 08	Mam Raj.	5 June 18		
1 June 92	Hira Singh	6 June 18		
22 Aug. 10	Ram Sarup	22 July 18		
1 Oct. 00	Nihal Singh	15 Sep. 18		
22 Sep. 00	Sohan Singh	15 Sep. 18		
2 Nov. 07	Qalandar Khan	15 Sep. 18		
11 Dec. 17	Jamal Din	15 Sep. 18		
20 Nov. 14	Sultan Ali (38)	15 Sep. 18		
14 June 02	Maghal Khan	19 Sep. 18		
3 Nov. 15	Sha..aray	20 Sep. 18		
13 Jan. 01	[R]am Nath	15 Oct. 18		

Regt. Centre—Ahmednagar.
Linked with—101st Grs. & 109th Infy.

Uniform—Scarlet.
Facings—White.

108th Infantry.

Raised at Bombay in 1768 as the 1st Battalion of Bombay Sepoys. Became the 7th Batt. of Bombay Sepoy in 1788; the 2nd Batt., 4th Regiment of Bombay Native Infantry in 1796, the 8th Regt. of Bombay N. I. in 1824 the 8th Regt. of Bombay Infantry in 1885; the 8th Bombay Infantry in 1901; present designation in 1903.

"Mysore" "Hyderabad" "Afghanistan, 1879-80."

Composition—1 Company of Dekhani Mahrattas, 1 of Konkani Mahrattas, 1 of Punjabi Musalmans, 1 of Rajputana Musalmans.

1st Commn. or date of entering service.	Names and Rank.	Army rank.	Present appointment in Regiment.	Remarks.
	Commandant.			
10 Nov. 88	Elderton, Lt.-Col. M. T.	10 Nov. 14	20 Oct. 15	*Lv., ex I., 6 mos., m. c.*
	Company Commanders. (4)			
20 May 99	King-Mason, Maj. A. R. L. (*Actg. Lt.-Col.*)	1 Sep. 15	30 Mar. 18	2nd-in-Comd. Comdt. (*temp.*).
23 July 01	*Waller, Maj. R. deW.*	*23 July 16*	*21 Oct. 13*	
29 Jan. 02	Miller, Maj. F.	29 Jan. 17	20 Oct. 15	With 2-95th Infy.
3 Dec. 04	*Gompertz, Capt. M. L. A. [l] (temp. Lt.-Col.)*	*3 Dec. 13*	*30 Mar. 18*	*Comdt., Offrs.' Sch. of Instruction, Sabathu.*
	Company Officers. (9)			
25 Aug. 05	Ford, Capt. E. G.	5 Aug. 14	5 Nov. 06	1-109th Infy.
9 Sep. 08	Mather, Capt. O. (*Actg. Maj.*)	1 Sep. 15	....	With 2-154th Infy.
9 Sep. 08	Masters, Capt. E. (*Actg. Maj.*)	1 Sep. 15	2 Feb. 10	Attd.
15 Sep. 15	Howitt, Lt. C. N.	15 Sep. 16	13 Nov. 15	Attd.
18 Apl. 16	Nicolson, Lt. D. O.	18 Apl. 17	22 Apl. 16	Attd.
29 June 16	Dowling, Lt. D. M.	29 June 17	3 July 16	Attd.
29 June 16	Saunders, Lt. C. W.	29 June 17	3 July 16	With 109th Infy.
30 Jan. 17	Martin, Lt. S. F.	30 Jan. 18	....	Attd.
10 Oct. 18	Peacocke, 2nd-Lt. G. I. C.	1 Oct. 18	....	
	Attached.			
14 July 15	Watts, Lt. H. C.	14 July 16	29 Nov. 17	I. A. (*on probn.*)
10 Aug. 16	Davis, Lt. S.	10 Aug. 17		I. A. R. O.
8 Mar. 17	Entwistle, Lt. G. R.	8 Mar. 18		Qr.-Mr.
6 Oct. 16	Wilkinson, Lt. K. W.	6 Oct. 17	....	I. A. R. O.
8 Mar. 16	Grieves, Lt. A. J. M.	8 Mar. 17	29 June 18	Glouc. R., I. A. (*on probn.*)
29 Mar. 18	Hudson, 2nd-Lt. E. J. M.	29 Mar. 18	17 July 18	I. A. R. O.
29 Mar. 18	Scutt, 2nd-Lt. C. C.	29 Mar. 18	17 July 18	I. A. R. O.
29 Mar. 18	Gwinnett, 2nd-Lt. R. C. H.	29 Mar. 18	17 July 18	I. A. (*T. C.*)
	Clarke, 2nd-Lt. H.	....	....	Genl. list, I. A. R. O. (*on probn.*)

		Jemadar.	Subadar.	Subadar-Major.
	Subadar-Major.			
1 Nov. 99	Mushak Ali Khan (56, 66a)	9 Nov. 02	9 Sep. 10	21 Jan. 17.
	Bhewa Sarafdar (*Ret.*)	....	....	Re-empld., 2 Mar. 17.
	Atmaji Rao Ghag (Ret.)	....	....	*Re-empld., 21 Jan. 17.*
	Subadars.			
1 July 91	Lakshiman Rao Sinde (56)	17 Feb. 04	1 Mar. 09	
7 Nov. 98	Dinker Rao Dareker (56, 66a)	1 Nov. 06	30 Oct. 10	
16 Sep. 95	Taj Muhammad Khan	9 Sep. 10	1 Apl. 12	
30 Mar. 97	Gul Husain (56)	16 Sep. 10	21 June 15	
10 Apl. 95	Rao Jadhow (56, 66a)	1 Aug. 11	21 June 15	
11 Apl. 93	Jogi Naik (56, 60a)	30 Oct. 10	2 Aug. 16	
17 Feb. 03	Raghunath Rao Chawan	21 June 15	21 June 17	
2 Aug. 99	Abdulla Khan (56, 66a)	8 July 11	20 Oct. 17	
23 July 99	Mubarak Ali (66a)	1 Apl. 12	22 Oct. 17	With 2-154th Infy.
	Abdul Rahiman (*Ret.*)	....	....	Re-empld., 29 July 17.
28 May 03	Niaz Ali Khan	28 May 03	29 Sep. 07	
26 Sep. 95	Hashan Din	23 Dec. 16	6 Aug. 17	
[illegible] Jan. 97	Sultan Khan	21 June 15		

108th Infantry—*contd.*

1st Commn. or date of entering service.	Names and rank.	Army rank.	Remarks.
	Jemadars.		
11 Apl. 01	Antu Naik (56)	28 May 13	Indian Adjt.
1 Sep. 02	Govinda Jadhaw	10 Dec. 14	
3 Mar. 97	Hayat Muhammad	10 Aug. 15	
29 June 01	Jalal-ud-din	19 Nov. 15	
9 June 85	Tukaram Kokate	19 July 16	
18 Jan. 00	Shaikh Muhammad Ismail	19 July 16	
26 Dec. 98	Abdul Majid Khan	1 Aug. 16	
22 Nov. 02	Tatojirao Sinde	1 Aug. 16	
19 Sep. 99	Mohadset Kokate	2 Aug. 16	
13 Sep. 95	Pir Mahi	2 Aug. 16	
13 Dec. 02	Mir Tassadik Hussain	30 Apl. 18	
21 Jan. 97	Balla Sawant	31 May 17	
13 July 99	Wali Muhammad	21 June 17	
9 June 02	Santu Chawan	1 July 17	
21 Sep. 09	Muhammad Hanif	20 Oct. 17	With 2-154th Infy.
6 Apl. 04	Narayan Rao Sinde	1 Feb. 18	With 2-154th Infy.
	Genu Jadhow	12 Dec. 17	
	Muhammad Sharif	28 Dec. 17	
12 July 07	Gulam Kadar	6 Apl. 17	
3 Dec. 98	Sarfaraz Khan	12 Nov. 17	
24 Feb. 86	Lukshiman Raoji	18 June 18	
	Tukaram More	25 May 18	
	Gangaram Kadam	25 May 18	
	Rama Methe	2 Sep. 18	
	Wazir Khan	2 Sep. 18	
	Imam Ali Khan	20 Oct. 18	

Regt. Centre—Ahmednagar.
Linked with—101st Grs. & 108th Infy.

Uniform—Scarlet.
Facings—Black.

1st Battalion, 109th Infantry.

Raised in 1788 as the 5th Battalion of Bombay Sepoys. Became the 2nd Batt., 1st Regiment of Bombay Native Infantry in 1796; the 1st Batt., 5th Regt. of Bombay N. I. in 1798; the 9th Regt. of Bombay N. I. in 1824; the 9th Regt. of Bombay Infantry in 1885; the 9th Bombay Infantry in 1901; present designation in 1903.

"Mysore" "Seringapatam" "Mooltan" "Punjaub" "Afghanistan, 1879-80."

Composition—1 company of Dekhani Mahrattas, 1 of Konkani Mahrattas, 1 of Rajputana Musalmans, 1 of Punjabi Musalmans.

1st Comman. or date of entering service.	Names and Rank.	Army rank.	Present appointment in Regiment.	Remarks.
	Commandant.			
7 Dec. 95	Ommanney, Lt.-Col. C. V.	29 May 18	29 May 18	
	Company Commanders. (4)			
3 May 90	Fellows, Lt.-Col. C. S. (Q)	3 May 16	20 Feb. 15	2nd-in-Comd. with 2-76th Infy. *Lv., m.c., ex I., 6 mths.*
27 July 98	Barr, Maj. J. H. V. (*I*)	1 Sep. 15	10 Oct. 09	With 120th Infy.
17 Mar. 00	Smith, Maj. C. A. J. (*Actg. Lt.-Col.*)	1 Sep. 15	9 Feb. 10	With 2-109th Infy.
26 June 01	*Bannatyne, Maj. F. C.*	*3 June 15*	*29 May 18*	*D. A. A. & Q. M. G., Kohat Bde.*
	Company Officers. (9)			
24 July 01	Mackrell, Maj. A.	9 Aug. 16	16 Mar. 05	
29 Jan. 02	*Paige, Maj. C. P.*, C.I.E.	*29 Jan. 17*	*30 Sep. 05*	*G. S. O., 2nd gr., A. Hdqrs.*
9 Jan. 04	*Lake, Capt. M. C.*	*9 Jan. 13*	*28 July 06*	*Arab Levy Corps.*
24 Jan. 06	Cadell, Capt. A. P. H.	24 Jan. 15	30 Mar. 07	With 1-94th Infy.
25 Jan. 08	*Crick, Capt. C. C.*	*1 Sep. 15*	*4 Aug. 09*	*Bde.-Maj., Kerman.*
9 Sep. 08	Mather, Capt. O.	1 Sep. 15	13 Dec. 09	
8 Sep. 09	Anderson, Capt. J. McK., M.C.	1 Sep. 15	31 Dec. 10	With 2-101st Grs.
29 Jan. 10	Kane, Capt. J. L. K.	1 Sep. 15	4 Apl. 12	With 120th Infy.
20 Jan. 10	*Bingham, Capt. C. D'A.*	*1 Sep. 15*	*7 Mar. 11*	*Instr., Sch. of Musky., Satara.*
16 Aug. 16	*Abraham, Lt. R. C.*	*20 Aug. 16*	*10 Nov. 16*	*Attd. Asst. Censor, Rangoon.*
18 Mar. 16	Jahans, Lt. L. E. H.	18 Mar. 17	23 June 16	Qr.-Mr.
1 Sep. 18	Thompson, 2nd-Lt. W. R.	1 Sep. 18	5 Sep. 18	Attd.
	Attached.			
11 May 01	Benson-Cooke, Maj. P. G. (*l*)	11 May 16	2 Mar. 18	80th Infy. 2nd-in-Comd. (*temp.*)
22 June 15	Warry, Lt. H. E.	22 June 16	15 Jan. 18	Devon. R., I. A. (*on probn.*) C. C. (*temp.*)
17 July 15	Scott, Lt. S. C., M.C.	17 July 16	2 Mar. 18	Essex. R., I. A. (*on probn.*) C. C. (*temp.*)
15 Nov. 15	Ellison, Lt. D. H. (*Actg. Capt.*)	15 Nov. 16	24 Sep. 16	95th Infy. Adjt.
9 Jan. 16	Hodkin, Lt. W.	9 Jan. 17	19 Apl. 18	Lan. Fus. I. A. (*on probn.*)
8 Aug. 17	Westland, Lt. D.	8 Aug. 18	3 Nov. 17	I. A. R. O. *Offg.* Qr.-Mr.
15 Oct. 17	Martin, Lt. H.	5 Oct. 18	22 Jan. 18	I. A. R. O. Asst. R. T. O., Karachi.
5 Nov. 17	Newlyn, Lt. E. W. G.	5 Nov. 18	15 Mar. 18	I. A. (*T. C.*)
19 Dec. 17	Walmisley, Lt. A. L.	19 Dec. 18	30 Mar. 18	I. A. R. O.
19 Mar. 18	Foot, 2nd-Lt. W. J. H.	19 Mar. 18	16 July 18	I. A. R. O.
29 Mar. 18	Killick, 2nd-Lt. C. F. R.	29 Mar. 18	16 July 18	I. A. R. O.
7 June 18	Betteridge, 2nd-Lt. R. A.	7 June 18	27 Oct. 18	I. A. (*T. C.*)
30 July 18	Brett, 2nd-Lt. A. L.	30 July 18	18 Oct. 18	Genl. List, I. A. R. O. (*on probn.*)
31 July 18	Abbott, 2nd-Lt. C. P.	31 July 18	5 Oct. 18	R. I. Rifles, I. A. (*on probn.*)
1 Oct. 18	Stevenson, 2nd-Lt. P. N.	1 Oct. 18	7 Oct. 18	Attd.
	Matcham, Sub-Lt. L. C.	....	18 Oct. 18	R. N. Divn., I. A. R. O. (*on probn.*)

		Jemadar.	Subadar.	Subadar-Major.	
	Subadar-Major.				
9 Mar. 97	Lakshiman Kadam (56, 58, 66*a*)	5 June 00	1 Aug. 02	23 Mar. 17	
	Subadars.				
27 Apl. 98	Amar Ali Khan (56, 66*a*)	19 Nov. 10	1 Nov. 13		
8 Nov. 94	Muhammad Baksh (56, 66*a*)	20 Nov. 04	14 June 15		With 1-101st Grs.
11 July 98	Amritrao Ghorpade (56, 66*a*)	7 Nov. 08	14 June 15		
16 Feb. 03	Pandurang Parab (56)	1 July 12	27 Aug. 15		
16 Dec. 98	Abbaskhan (56)	20 Jan. 14	5 Nov. 15		
20 Apl. 98	Chhote Khan (56)	1 Nov. 13	10 Jan. 16		
7 Aug. 16	Shah Nawaz Khan, I. O. M. (56)	5 Nov. 15	1 May 18		Asst. Instr., School of Musketry, Sabathu.
5 Nov. 04	Maroti, Rao Chawan (56, 66*a*)	7 Jan. 16	16 May 18		
28 Nov. 04	Karam Khan	1 Aug. 18	30 Aug. 18		

1st Battalion, 109th Infantry—*contd.*

1st Commn. or date of entering service.	Names and Rank.	Army rank.	Remarks.
	Jemadars.		
16 Dec. 98	Jamal Din (56, 66*a*)	13 Mar. 15	With 2-132nd Infy.
26 May 02	Lakshiman Rao Sinde	27 Aug. 15	
29 Aug. 00	Rahimatullah Khan (56)	10 Jan. 16	With 2-101st Grs.
17 Oct. 97	Genu Jadhow	17 Feb. 17	
9 July 00	Narayan Dhuri	2 Apl. 17	
7 July 03	Krishnaji Rao Powar (46)	22 June 17	With 2-101st Grs.
4 June 05	Sultan Khan	8 Aug. 17	
28 June 98	Nadir Khan	3 Feb. 17	
5 Aug. 03	Gulab Khan	1 May 18	
20 June 03	Bostan Khan	1 May 18	
9 July 03	Sid Ram Nikam	1 May 18	
3 Feb. 08	Jai Ram Dalvi	1 May 18	
8 July 02	Sankar Kadam	16 May 18	
. . . . .	Muhammad Khan	12 July 18	
23 June 03	Nek Muhammad Khan	25 July 18	
. . . . .	Karam Khan	1 Aug. 18	
25 Mar. 01	Jiwan Khan	30 Aug. 18	
3 June 05	Baburao Sinde	11 Sep. 18	
7 May 05	Sardar Khan	11 Sep. 18	

2nd Battalion, 109th Infantry.

Formed at Kohat, 1 May 18.

Composition—1 company Dekhani Mahrattas, 1 company Konkhani Mahrattas, 1 company Rajputana Musalmans, 1 company Punjabi Musalmans.

1st Commn. or date of entering service.	Names and Rank.	Army rank.	Present appointment in Regiment.	Remarks.
	Commandant.			
17 Mar. 00	Smith, Maj. C. A. J. (*Actg. Lt.-Col.*)	1 Sep. 15	27 May 18	(*temp.*)
	Company Commanders. (4)			
29 Aug. 06	Todd, Capt. R. (*Actg. Maj.*)	29 Aug. 15	4 Nov. 18	2nd-in-Comd. (*temp.*)
20 Feb. 16	Lake, Lt. C. H.	20 Feb. 17	16 May 18	I. A. R. O.
16 July 16	Souter, Lt. J. H., M.C.	16 July 17	9 May 18	I. A. (*on probn.*)
30 Jan. 17	Martin, Lt. S. F.	30 Jan. 18	15 May 18	Attd.
	Company Officers.			
20 Oct. 16	Barclay, Lt. H. L. (*Actg. Capt.*)	20 Oct. 17	9 May 18	I. A. (*on probn.*) Adjt.
8 Aug. 17	Mayes, Lt. F. A.	8 Aug. 18	16 May 18	I. A. R. O.
12 Sep. 17	Rae, Lt. K. W. R.	12 Sep. 18	20 May 18	I. A. (*on probn.*)
5 Nov. 17	Abbott, Lt. E. J.	5 Nov. 18	9 May 18	I. A. (*T.C.*) Qr. Mr. 9 May 18.
5 Nov. 17	Turl, Lt. G. H.	5 Nov. 18	16 May 18	I. A. (*T.C.*)
26 Sep. 17	Gilkes, 2nd-Lt. C. H.	26 Sep. 17	9 May 18	I. A. R. O. (*on probn.*)
21 Dec. 17	Wardle, 2nd-Lt. R. P.	21 Dec. 17	9 May 18	I. A. R. O. (*on probn.*)
	Attached.			
29 Mar. 18	Moore, 2nd-Lt. E. R.	29 Mar. 18	17 July 18	I. A. R. O.
29 Mar. 18	Pryde, 2nd-Lt. R. W.	29 Mar. 18	17 July 18	I. A. R. O.
29 Mar. 18	Connell, 2nd-Lt. E. W. H.	29 Mar. 18	24 July 18	I. A. (*T. C.*)
29 Mar. 18	Delves, 2nd-Lt. N. D.	29 Mar. 18	16 July 18	I. A. (*T. C.*)
10 Aug. 16	Barton, Lt. J. E. B.	10 Aug. 17	. . . .	Manch. R., I. A. (*on probn.*)
28 Nov. 17	Robertson, 2nd-Lt. J. P. W.	28 Nov. 17	10 Sep. 18	Lanc. R., I. A. R. O. (*on probn.*)
28 Mar. 18	Thompson, 2nd-Lt. T. R.	28 Mar. 18	9 Sep. 18	Norf. R., I. A. (*on probn.*)
26 June 18	Shephard, 2nd-Lt. H. C. B.	26 June 18	1 Oct. 18	Genl. List, I. A. R. O. (*on probn.*)
26 June 18	Watts, 2nd-Lt. F. C.	26 June 18	1 Oct. 18	Genl. List, I. A. R. O. (*on probn.*)
22 July 18	Pickering, 2nd-Lt. M. B.	22 July 18	2 Dec. 18	I. A. (*T. C.*)
22 July 18	Henderson, 2nd-Lt. J. W.	22 July 18	2 Dec. 18	I. A. R. O.
18 Sep. 18	Haworth, 2nd-Lt. G. M.	28 Sep. 18	1 Oct. 18	Shrop. L. I. I. A. (*on probn.*)
		Jemadar.	Subadar.	Subadar-Major.
	Subadar-Major.			
20 May 98	Ibrahim Khan	1 July 08	19 Nov. 10	1 May 18
	Subadars.			
8 Feb. 03	Yeshwant Rao Power	1 Nov. 12	12 June 17	
12 Dec. 02	Dattaram Sawant	14 June 15	1 May 18	
20 Mar. 01	Ahmed Khan	14 June 15	1 May 18	
12 July 00	Abdul Rahim Khan	7 Jan. 16	1 May 18	
12 June 03	Umar Bakhsh	13 Aug. 16	1 May 18	
21 Dec. 02	Govindrao Sinde	12 June 17	1 May 18	
	Jemadars.			
16 Feb. 99	Hassan Mahommed	15 Dec. 16		
9 Jan. 00	Shiveram Naik	1 May 18		
1 Mar. 03	Bala Parte	1 May 18		
10 May 98	Genu Pawar	1 May 18		
3 Nov. 98	Ilam Din	1 May 18		
11 Jan. 03	Murad Ali Khan	1 May 18		

Regt. Centre—Satara.
Linked with—103rd & 105th Mahratta Lt. Infy.

Uniform—Scarlet.
Facings—Black.

110th Mahratta Light Infantry.

Raised at Calicut in 1797, as 2nd Battalion, 5th (or Travancore) Regt. of Bombay Native Infantry. Became the 10th Regt. of Bombay N. I. in 1824; the 10th Regt. of Bombay Native (Light) Infantry in 1871; the 10th Regt. of Bombay (Light) Infantry in 1885; the 10th Bombay Light Infantry in 1901; present designation in 1903.

"Central India" "Abyssinia" "Afghanistan, 1879-80."

Composition—2 companies of Dekhani Mahrattas, 1 of Konkani Mahrattas, 1 of Dekhani Musalmans.

1st Commn. or date of entering service.	Names and Rank.	Army rank.	Present appointment in Regiment.	Remarks.
	Commandant.			
	Company Commanders. (4)			
29 Oct. 90	Hill, Lt.-Col. H. C., D.S.O. (Q)	29 Oct. 10	7 Jan. 00	2nd-in-Comd.
11 Feb. 99	Branson, Maj. L. H.	1 Sep. 15	1 Oct. 10	*Lv., ex I., 6 mos., m. c.*
22 Apl. 03	Cooper, Maj. K. E., M.C.	22 Apl. 18	15 Apl. 15	
2 Mar. 04	Gunn, Capt. A. D.	2 Aug. 13	15 July 17	
	Company Officers. (9)			
13 Aug. 04	*Montefiore, Capt. C. E.*	*13 Aug. 13*	*1 Mar. 06*	*Assam Mily. Police.*
5 Aug. 05	Welch, Capt. F. E.	5 Aug. 14	30 June 07	With 105th Infy.
29 Aug. 05	*Ball, Capt. E. P. (temp. Maj.)*	*29 Aug. 15*	*9 Nov. 07*	*D.A.Q.M.G., Bombay.*
9 Sep. 08	Chamier, Capt. R. O.	1 Sep. 15	13 Nov. 09	
16 Aug. 14	Moseley, Lt. B.	16 May 16	29 Apl. 17	Attd. Actg. Adjt.
30 Jan. 17	Gibbs, Lt. E. W. B.	30 Jan. 18	. . . .	Attd.
30 Jan. 17	Begg, Lt. P. H.	30 Jan. 18	. . . .	Attd.
18 June 17	Chaplin, Lt. E. L.	18 June 18	. . . .	Attd.
	Attached.			
17 Jan. 92	Waller, Lt.-Col. A. A. P.	27 Jan. 18	20 Mar. 18	105th Infy. (*Comdt. temp.*)
19 Jan. 07	Pearson, Capt. H. S. I.	1 Sep. 15	27 Aug. 17	114th Mah.
20 Nov. 14	McFadyen, Lt. A.	20 Nov. 15	. . . .	I. A. R. O.
8 Jan. 15	Heron-Jones, Lt. C. V., M.C.	8 Jan. 16	. . . .	I. A. R. O.
5 Nov. 15	Regan, Lt. C. H. (207a)	5 Nov. 16	16 Jan. 18	R. Fus., I. A. [illegible] *probn.*)
26 Mar. 16	Armstrong, Lt. H.	26 Mar. 17	27 July 17	R. Dub. F[illegible] I. A. (*on* [illegible]*bn.*) Actg. Qr.-M.
7 July 16	Birkin, Lt. W. N., M.C.	7 July 17	26 Jan. 18	Nott[illegible] Derby. R., I. A. (*on probn.*)
22 July 16	Sutherland, Lt. A. W.	22 July 17	5 Mar. 18	[illegible]n Highrs. I. A. (*on probn.*)
18 Aug. 16	Seed, Lt. J. A. T.	18 Aug. 16	6 Aug. 17	[illegible] Fus., I. A. (*on probn.*)
2 June 17	Eberhardle, Lt. R. E.	14 Aug. 18	1 Oct. [illegible]	I. A. R. O. (*on probn.*)
22 Aug. 17	Lewis, Lt. J. A., M.C. (207a)	22 Aug. 18	1 O[illegible] 17	R. Ir. Regt., I. A. (*on prob*[illegible])
27 Aug. 17	Harman, Lt. B. R.	22 Aug. 18	13 [illegible]t. 17	E. Kent R., I. A. (*on* [illegible]*bn.*)
[illegible]5 Oct. 17	Cleaver, Lt. N. R.	5 Oct. 18	[illegible] Jan. 18	W. Kent R., I. A. (T. C.)
21 Nov. 17	Kingston, Lt. F.	21 Nov. 1[illegible]	. . . .	I. A. R. O.
25 Oct. 17	Gibson, Lt. J. R.	25 Oct. [illegible]	30 Oct. 17	R. Scott. Fus., I. A. (*on probn.*)
[illegible]5 Jan. 18	Sugden, 2nd-Lt. J. W.	5 [illegible] 18	21 June 18	I. A. (*T. C.*)
27 July 18	Bartley, 2nd-Lt. E. S.	2[illegible] July 18	. . . .	I. A. (*on probn.*)
1 Oct. 17	Holdcroft, 2nd-Lt. G. E. (*Actg Capt.*)	[illegible] Oct. 17	21 Jan. 18	I. A. [illegible]. O.
1 Oct. 17	Wood, 2nd-Lt. M.	1 O[illegible]. 17	21 Jan. 18	[illegible]5th Queen's, I. A. (*on probn.*)
30 June 18	Kirwan, 2nd-Lt. J. W. [illegible]	30 June 18	6 Nov. 1[illegible]	R. Dub. Fus., I. A. (*on probn.*)
2 Oct. 18	Cotton, 2nd-Lt. A. C.	2 Oct. 18	3 Nov. 18	I. A. R. O.
10 Oct. 18	Hampson, 2nd-Lt. [illegible]	10 Oct. 18	26 Oct. 18	I. A. R. O.
18 Oct. 18	Lister, 2nd-Lt. [illegible] L.	18 Oct. 18	2[illegible] Oct. 18	I. A. R. O.

Subadar-Majors.

		Jemadar.	Subadar.	Subadar-Major.	Remarks.
9 July [illegible]	*Balkri[illegible]na Rao Bahadur.* (56, 72a)	*30 Aug. 97*	*1 May 03*	*6 July 15*	*Order of Br. Ind., 2nd Cl., A. D. C. to H. E. the Govr. of Bomb*[illegible]
Mar. 94	Shaikh Yasin, *Bahadur*, I.D.S.M. (56, 66a).	1 Jan. 09	16 Oct. 12	22 Nov. 15	*Order of Br. Ind*[illegible] *Cl.*

110th Mahratta Light Infantry—*contd.*

1st Commn. or date of entering service.	Names and Rank.	Army rank.	Present appointment in Regiment.	Remarks.
	Subadars.			
26 May 06	Sudhan Rao Nimbalker (56, 72*a*).	26 May 06	1 May 10	
1 Jan. 96	Maroti Ferande (56, 66*a*)	2 Apl. c6	9 Sep. 14	
7 Mar. 94	Vishnu Ghone (56, 66*a*), I.D.S.M.	10 Jan. 15	. . . .	
16 Oct. 12	Dattajirao Khanvilkar, M.C.	16 Oct. 12	22 Nov. 15	
21 Mar. 94	Govind Sawant (56)	11 Jan. 18	22 Nov. 15	
24 Oct. 00	Shaikh Bade Sahib	22 Feb. 15	9 Feb. 16	
10 Apl. 07	Dajirao Rane (46)	6 Oct. 15	28 Jan. 17	
3 Feb. 94	Balkrishna Khanse	3 Dec. 08	26 Mar. 17	
16 Dec. 98	Mhadu Wagmode	9 Sep. 14	11 June 17	
1 Oct. 92	Sayad Umar	9 Sep. 14	11 June 17	
	Sitaram Parab	. . .	11 June 17	
5 Oct. 05	Shaikh Ahmed	14 Oct. 15	10 Feb. 18	
12 Oct. 01	Dhondi Gaichar	18 Jan. 17	10 Feb. 18	
	Jemadars.			
1 Nov. 03	Narsu Jadhav	22 Nov. 15		
16 Nov. 01	Raoji Mahamunkar	22 Nov. 15		
29 Jan. 97	Maroti Fartare	22 Nov. 15		
17 Jan. 07	Dattaram Karle	22 Nov. 15		
11 Aug. 96	*Shaikh Jamal*	*1 Sep. 14*		*With 2-103rd L. Infy.*
25 Mar. 97	Krishnajirao Ghorpade	9 Sep. 14		
26 June 00	Madhu Jagdale	1 July 15		
11 Aug. 95	Shaikh Muhammad	11 June 17		
21 Jan. 02	Ganpat Rao More	11 June 17		
11 Oct. 01	Bajijirao Shiude	11 June 17		
28 Aug. 99	Kashinath Zine	11 June 17		
3 Mar. 94	Vishnu Sakharam	3 Dec. 17		
2 Dec. 06	Yeshwant Rao Nalge	10 Feb. 18		
10 Apl. 01	Dayanu Jadhav	10 Feb. 18		
29 Oct. 07	Ganuji Bagwe	10 Feb. 18		
26 Oct. 05	Sagon Naik	1c Feb. 18		
9 July 04	Sayyid Muhammad	10 Feb. 18		
28 Dec. 17	Sadashiv Rao Surve	10 Feb. 18		
24 June 13	Abdul Hakim Khan	10 Feb. 18		
4 Dec. 12	Raghunathrao Kadam	10 Feb. 18		
24 Sep. 87	Shallom Solomon	1 July 18		

111th Mahars.

Formed at Mhow on 19 June 1917.

Composition—4 Companies of Mahars.

1st Commn. or date of entering service.	Names and Rank.	Army rank.	Present appointment in Regiment.	Remarks.
	Commandant.			
4 May 98	Dyke, Lt.-Col. O. M.	30 June 18	27 June 18	21st Cav.
	Company Commanders. (4)			
18 Jan. 02	Robinson, Maj. E. L.	23 Feb. 17	. . . .	75th Infy.
29 Aug. 06	Campbell, Capt. W. E. D. . .	29 Aug. 15	19 June 18	34th Horse.
19 Dec. 14	Tomkin, Lt. R. S.	18 Sep. 16	1 June 18	
	Company Officers.			
2 July 15	Jeffs, Lt. H. S.	2 July 16	. . . .	I. A. R. O.
14 May 16	Bartleman, Lt. E. L.	14 May 17	27 Mar. 18	I. A. (*on probn.*)
21 July 16	Digby, Lt. K. G.	22 July 17	. . . .	I. A. R. O.
5 May 17	Collett, Lt. J. W.	5 May 18	. . . .	I. A. R. O. Qr.-Mr.
27 May 16	Holdsworth, Lt. M. J., M.C. . .	27 May 17	1 Oct. 18	R. Suss. R., I. A. (*on probn.*)
25 July 17	Roberts, Lt. G. H.	25 July 18	. . . .	M. G. Corps, I. A. (*on probn.*)
1 Oct. 17	Fox, Lt. B. F.	1 Oct. 18	16 Jan. 18	I. A. (*T. C.*)
18 Oct. 17	Jacob, Lt. J. R.	18 Oct. 18	17 Jan. 18	I. A. R. O.
1 Nov. 17	Taylor, Lt. C. W.	1 Nov. 18	16 Mar. 18	I. A. (*T. C.*)
5 Nov. 17	Tilley, 2nd-Lt. R. H.	5 Nov. 17	16 Mar. 18	I. A. (*T. C.*)
20 Nov. 17	Boyce, 2nd-Lt. T. W., M.C. . .	20 Nov. 17	13 Oct. 18	R. Innis. Fus., I. A. (*on probn.*)
19 Dec. 17	Brockel bank, 2nd-Lt. G. S. (*Actg. Capt.*)	19 Dec. 17	27 Mar. 18	I. A. R. O. Adjt.
28 Jan. 18	Macfarlane, 2nd-Lt. H.	28 Jan. 18	12 Oct. 17	I. A. (*on probn.*)
26 Mar. 18	Elson, 2nd-Lt. E. J.	26 Mar. 18	3 Oct. 18	M. G. Corps, I. A. (*on probn.*)
7 June 18	Hedley, 2nd-Lt. R. W. . . .	7 June 18	11 Oct. 18	I. A. (*T. C.*)
17 July 18	Nicholls, 2nd-Lt. J. W. D. . .	17 July 18	17 July 18	I. A. (*T. C.*)
22 July 18	Botterill, 2nd-Lt. W. R. . . .	22 July 18	6 July 18	I. A. R. O.
8 Aug. 18	Devlin, 2nd-Lt. D. G.	8 Aug. 18	12 Sep. 18	I. A. R. O.
20 Aug. 18	Alder, 2nd-Lt. W. G.	20 Aug. 18	9 Sep. 18	I. A. R. O.
28 Aug. 18	Cowper, 2nd-Lt. G. G. St. J. .	28 Aug. 18	18 Sep. 18	I. A. R. O.
10 Sep. 18	Lake, 2nd-Lt. C. H. W. . . .	10 Sep. 18	9 Oct. 18	I. A. R. O.

		Jemadar.	Subadar.	Subadar-Major.
	Subadar-Major.			
27 Nov. 82	Jivaji Rao	1 Feb. 12	28 May 18	
	Subadars.			
5 Oct. 87	Abdul Latif	. . . .	1 Sep. 17	
23 Feb. 97	Muhammad Yousuf	. . . .	7 Sep. 17	
6 Feb. 99	Mohammed Abdul Hakim . .	. . . .	14 Sep. 17	
1 May 98	Namdeo Krishnaji	9 July 17	1 Aug. 18	
21 Jan. 88	Mohammed Ghouse (*Retd.*) . .	. . . .	2 Oct. 17	Re-empld.
7 Apl. 85	Sebastian, F. (*Retd.*) . . .	. . . .	6 Oct. 17	Re-empld.
	Jemadars.			
4 Jan. 80	Ramji Sajanaji	28 July 17		
1 Aug. 89	Krishna Changu	1 Sep. 17		
16 Mar. 86	Shaikh Muhammad Ishak (*Retd.*)	10 Apl. 05	8 Oct. 17	Re-empld.
6 Aug. 94	Hiru Ramji	29 Sep. 17		
. . .	Sirpatrao Ghag	1 Aug. 18		
. . .	Sudam Balu	1 Aug. 18		
. . .	Dhondbarao Kadam	1 Aug. 18		
. . .	Gopal Sajan	1 Aug. 18		
. . .	Gopal Laxuman	3 Oct. 18		
20 Oct. 17	Somalingiah, B.	20 Oct. 17	. . . .	(*t emp.*)
1 Jan. 88	Pandnak Laknak	2 Sep. 17		

Regt. Centre—Deesa.
Linked with—102nd Grs. & 113th Infy.

Uniform—Scarlet.
Facings—Yellow.

1st Battalion, 112th Infantry.

Raised at Bombay in 1798 by Lieut. James McClintock, as the 2nd Battalion, 6th Regiment of Bombay Native Infantry. Became the 12th Regt. of Bom. N. I. in 1824; the 12th Regt. of Bombay Infantry in 1885; the 12th Bombay Infantry in 1901; present designation in 1903.

"Kirkee" "Meanee" "Hyderabad" "Central India."

Composition—1 Company of Western Rajputana Jats, 1 of Bagri Jats and Jats from Eastern Rajputana, 1 of Rajputana Gujars, 1 of Punjab Musalmans.

1st Commn. or date of entering service.	Names and Rank.	Army rank.	Present appointment in Regiment.	Remarks.
	Commandant.			
1 Mar. 90	Winter, Bt.-Col. C. B., C.M.G., D.S.O. [*L*].	3 June 16	23 July 16	A. Q. M. G.
	Company Commanders. (4)			
10 Oct. 94	Shoubridge, Maj. C. A. G. (*Actg. Lt.-Col.*)	10 Oct. 12	10 Apl. 18	2nd-in-Comd.
11 Aug. 00	Seaton, Maj. F. D. R.	1 Sep. 15	11 Jan. 14	With 2-99th Infy.
8 May 01	Stuart-Menteth, Maj. A. H. [*l*]	8 May 16	17 Jan. 14	With 1-103rd Mah.
19 Aug. 03	Thompson, Maj. A.	19 Aug. 18	10 Apl. 18	With 2-113th Infy.
	Company Officers. (9)			
9 Jan. 04	Budgen, Capt. H. V.	9 Jan. 13	5 July 05	With 2-112th Infy.
10 Feb. 04	*Tree, Capt. H. B. LeD.*	*10 Feb. 13*	*14 Oct. 05*	*Bde.-Maj.*
3 Dec. 04	Logan-Home, Capt. W. M.	3 Dec. 13	1 July 08	
24 Jan. 06	Conder, Capt. H.	24 Jan. 15	23 Mar. 07	With 2-41st Dogras.
22 Feb. 08	Carter, Capt. E. N.	1 Sep. 15	28 Nov. 11	*Lv., m.c., ex I., 6 mths.*
9 Sep. 08	Grant, Capt. R. A. P., M.C.	1 Sep. 15	14 Nov. 09	Qr.-Mr., 24 Jan. 15. With 67th [Punj.
24 Aug. 12	Meade, Capt. P. A.	24 Aug. 16	8 Nov. 13	Adjt.
1 Oct. 14	Savage, Capt. E. G.	1 Oct. 18	23 Dec. 14	With 2-102nd Grs.
2 May 15	*Moss, Lt. E. de V. (48, 206)*	*2 May 16*	*2 Dec. 17*	*Staff School, Saugor.*
12 Sep. 17	Smith, Lt. L. E.	12 Sep. 18	....	Attd.
12 Sep. 17	Stratton, Lt. E. B.	12 Sep. 18	....	Attd. with 144th Infy.
	Attached.			
23 May 00	Bodkin, Maj. L. F.	1 Sep. 15	....	113th Infy.
18 Jan. 11	Steward, Capt. C. W. D'A. (*Actg. Maj.*)	1 Sep. 15	1 Sep. 18	66th Punj.
5 Sep. 14	Carstairs, Lt. G. S. (*Actg. Capt.*)	5 Sep. 15	....	Candian Infy. I. A. (*on probn.*)
18 Feb. 15	McClumpha, Lt. C. D.	18 Feb. 16	18 Apl. 15	I. A. R. O.
1 Oct. 15	Hawkes, Lt. F. R.	1 Oct. 16	22 Nov. 15	I. A. R. O.
20 Feb. 16	Murray, Lt. M. J.	20 Feb. 17	3 Aug. 16	I. A. R. O.
30 Sep. 16	Finch, Lt. K. H.	30 Sep. 16	....	I. A. R. O.
1 Dec. 17	Butler, Lt. R. O'N.	1 Dec. 18	....	Leins. R., I. A. (*on probn.*)
1 Oct. 17	Bunce, Lt. L. L.	1 Oct. 18	....	I. A. (*T. C.*)
5 Oct. 17	Stone, Lt. A. P. G., M.M.	5 Oct. 18	....	I. A. (*T. C.*)
28 May 18	Haughton, 2nd-Lt. H.	28 May 18	....	I. A. R. O. (*on probn.*)
26 Jan. 18	Law, 2nd-Lt. G. H. M.	26 Jan. 18	....	S. Staff. R., I. A. (*on probn.*)
30 July 18	Duncan, 2nd-Lt. R. A.	30 July 18	....	I. A. R. O. (*on probn.*)
14 Aug. 18	Cameron, 2nd-Lt. E. M.	14 Aug. 18	....	I. A. R. O. (*on probn.*)

		Jemadar.	Subadar.	Subadar-Major.
	Subadar-Major.			
15 Apl. 98	Toda Singh (53, 56, 58, 66*a*, 72*a*)	31 Jan. 05	14 May 12	2 Jan. 18.

1st Battalion, 112th Infantry—*contd.*

1st Commn. or date of entering service.	Names and Rank.	Army rank.		Remarks.
	Subadars.			
26 May 87	*Mirchi Singh (52, 56, 66a)*	*28 Nov. 08*	*4 Feb. 15*	*With 67th Punj.*
4 Jan. 01	Shiucharan Singh (52, 56, 66a)	9 Oct. 13	14 June 15	
26 July 99	Allah Bakhsh (52, 56, 66a)	1 Oct. 07	14 June 15	With 3-154th Infy.
17 May 93	Fateh Khan (66a)	1 Apl. 12	6 July 15	
19 Oct. 99	Raje Singh (66a)	1 June 10	27 Oct. 15	
16 Dec. 00	Harna Singh	1 Jan. 15	16 May 16	With 2-152nd Infy.
13 Jan. 00	Birbal (56, 66a)	28 Oct. 11	17 Dec. 16	
13 May 97	Ram Het	14 June 15	21 July 16	
21 Aug. 95	Rajwali Khan	14 June 15	10 June 17	
14 May 95	Shiu Ram Singh	14 May 12	16 June 17	
20 Jan. 96	Bhura Singh	27 Oct. 15	10 July 17	
22 Jan. 98	Maharaj Singh	14 June 15	4 Aug. 17	With 3-154th Infy.
1 Aug. 93	Karam Dad	6 July 15	29 Aug. 18	
	Jemadars.			
3 June 99	Husain Bakhsh	6 Apl. 15		
23 Aug. 96	Indar Singh	27 Oct. 15		With 3-154th Infy.
15 May 99	Hayat Muhammad Khan	27 Oct. 15		
10 Mar. 01	Magar Ram	19 Nov. 15		
26 Jan. 94	Latte Singh (66)	3 Mar. 16		
23 Apl. 00	Umrao Singh	8 June 16		With 143rd Infy.
3 Sep. 00	Gulichand Singh	26 July 16		With 3-154th Infy.
5 Apl. 01	Barkat Ali	8 Jan. 17		
16 Nov. 02	Ramdayal Ram	29 Jan. 17		
12 July 03	Prichat Singh	16 Feb. 17		
16 May 04	Agdi Singh	27 Feb. 17		
3 Jan. 97	Mohan Lal	27 Feb. 17		With 3-154th Infy.
8 June 97	*Dilsukh Singh*	*12 Feb. 17*		*With 18th Labour Corps.*
25 Mar. 03	Muhammad Din	10 June 17		
1 Dec. 03	Tota Ram	11 Mar. 18		
12 May 04	Singha Singh	10 Apl. 18		
28 June 99	*Dewa Ram*	*1 Jan. 17*		*With 18th Labour Corps.*
22 June 03	Gulab Singh	5 Aug. 17		
19 Sep. 04	Ghisa Singh	3 Oct. 17		With 108th Infy.
10 Nov. 86	Soloman Benjaman	15 Feb. 18		
30 Aug. 01	Bagga Khan	8 June 18		With 3-154th Infy.
9 Nov. 03	Harbans Singh	13 May 18		
19 July 07	Jodha Singh	13 May 18		
22 July 00	Jhabbar Ram	8 Aug. 18		

2nd Battalion, 112th Infantry.

Formed at Nasirabad on 8 July 1917.

Composition.—2 companies Jats, 1 company Punjabi Musalmans, 1 company Gujars.

1st Commn. or date of entering service.	Names and Rank.	Army rank.	Present appointment in Regiment.	Remarks.
	Commandant.			
23 Feb. 98	McCormick, Maj. G. D.	1 Sep. 16	28 Oct. 18	1-72nd Punj. (*temp.*)
	Company Commanders. (4)			
30 Apl. 02	Biscoe, Maj. J. V. M.	24 Dec. 15	1 July 17	Cant. Magte.'s Dept. 2nd-in-Comd.
9 Jan. 04	Budgen, Capt. H. V..	9 Jan. 15	16 May 18	1st-112th Infy.
5 May 15	O'Brien, Lt. J. J.	5 May 16	16 June 18	Leins. R., I. A. (*on probn.*)
13 Apl. 15	Stuart, Lt. D. S.	13 Apl. 16	1 May 18	Candn. Regt., I. A. (*on probn.*)
	Company Officers.			
19 Apl. 15	Dann, Lt. C. W., M.C. . . .	19 Apl. 16	24 Oct. 17	Suff. R., I. A. (*on probn.*)
8 Dec. 15	Hearn, Lt. G. S.	8 Dec. 16	14 Dec. 17	I. A. R. O., Qr. Mr.
11 Mar. 16	Williams, Lt. W. T., M.C. . .	11 Mar. 17	19 May 18	E. Kent. R., I. A. (*on probn.*)
2 June 16	Shaw, Lt. G. F.	2 June 17	17 Jan. 18	I. A. R. O.
23 June 16	Scott, Lt. F. L.	23 June 17	2 Aug. 17	I. A. (*on probn.*)
6 Sep. 16	Southern, Lt. W.	6 Sep. 17	16 Jan. 18	I. A. (*on probn.*)
20 Nov. 16	Carter, Lt. A. V., M.C., M.M. .	20 Nov. 17	1 Apl. 18	Aust. Infy., I. A. (*on probn.*)
23 Aug. 17	Brown, Lt. F. L.	23 Aug. 18	12 Oct. 17	Attd.
29 Aug. 17	Gudgeon, 2nd-Lt. P. E.	29 Aug. 17	15 Jan. 18	I. A. R. O.
31 Oct. 17	Wroe, Lt. H.	31 Oct. 18	12 Feb. 18	I. A. R. O.
1 Nov. 17	Poole, Lt. G. W.	1 Nov. 18	17 Mar. 18	I. A. (*T. C.*)
28 Dec. 17	Dunn, 2nd-Lt. D.	28 Dec. 17	8 Nov. 18	Arg. & Suth'd. Highrs., I. A. (*on probn.*)
26 June 18	Brown, 2nd-Lt. W.	26 June 18	4 Oct. 18	Genl. List, I. A. R. O. (*on probn.*)
15 June 18	Elliot, 2nd-Lt. W. J.	15 June 18	16 Sep. 18	Genl. List, I. A. R. O. (*on probn.*)
29 Mar. 18	Owen, 2nd-Lt. W.	29 Mar. 18	3 Mar. 18	I. A. R. O.
2 June 18	Clifton, 2nd.Lt. F. R.	2 June 18	16 Oct. 18	I. A. (*T. C.*)
7 June 18	Wells, 2nd-Lt. C. A.	7 June 18	12 Oct. 18	I. A. (*T. C.*)
3 July 18	Aikin, 2nd-Lt. V. J. D. . . .	3 July 18	3 Dec. 18	I. A. R. O.
		Jemadar.	Subadar.	Subadar-Major.
	Subadar-Major.			
	Fazal Karim	. . . . .	1 July 17	
	Subadars.			
13 Apl. 03	Manik Ram	. . . . .	11 July 17	
8 Sep. 02	Sri Chand	. . . . .	26 July 17	
14 May 04	Rahmat Khan	6 Oct. 17	1 Jan. 18	
3 Aug. 03	Net Ram	. . . . .	14 Mar. 18	
8 July 99	Tunda Singh	. . . . .	1 Apl. 18	
13 Sep. 03	Khub Chand	. . . . .	6 June 18	
23 June 99	Sanwalia Singh	. . . . .	6 June 18	
	Jemadars.			
1 Aug. 97	Ramlal	1 July 17		
19 May 07	Sunchri	19 July 17		
5 June 95	Niaz Ali	24 July 17		
15	Handu Singh	6 Oct. 17		
5 Feb. 10	Mungli Singh	6 Oct. 17		
16 May 12	Kewal Singh	6 Oct. 17		
1 Mar. 18	Amir Haidar	1 Mar. 18	. . . . .	*On probn.*
26 Oct. 05	Harkishen	20 Mar. 18		
20 June 11	Moji Ram	20 Mar. 18	. . . . .	Indian Adjt.
14 Dec. 09	Gokal Singh	20 Mar. 18		
30 Sep. 04	Sibban Singh	20 Mar. 18		
	Allah Ditta	8 June 18		

Regt. Centre.—Deesa. *Uniform*—Scarlet.
Linked with—102nd Grs. & 112th Infy. *Facings*—Yellow.

1st Battalion, 113th Infantry.

Raised in 1800, as the 1st Battalion, 7th Regiment of Bombay Native Infantry. Became the 13th Regt. of Bombay Native Infantry in 1824; the 13th Regt. of Bombay Infantry in 1885; the 13th Bombay Infantry in 1901; present designation in 1903.

THE SPHINX.

"Egypt" "Kirkee" "Beni Boo Alli" "Central India" "Afghanistan, 1879-80."

Composition—1 Companies of Western Rajputana Jats, 1 of Bagri Jats and Jats from Eastern Rajputana, 1 of Rajputana Gujars, 1 of Punjabi Musalmans.

1st Commn. or date of entering service.	Names and Rank.	Army rank.	Present appointment in Regiment.	Remarks.
	Commandant.			
11 Feb. 88	Lloyd-Jones, Lt.-Col. F. L.	11 Feb. 14	3 July 16	
	Company Commanders. (4)			
1 Mar. 90	*Redl, Lt.-Col. E. A. F.*, C.I.E. *[L].*	*1 Mar. 16*	*3 July 16*	*2nd-in-Comd. G. S. O., 1st gr.*
5 May 00	Adler, Maj. B. I. H.	1 Sep. 15	9 June 10	2nd-in-Comd. (*temp.*)
23 May 00	Bodkin, Maj. L. F. [L.]	1 Sep. 15	17 Jan. 14	With 1-112th Infy.
8 Jan. 01	*Fenton, Maj. L. S.*	*10 Mar. 16*	*3 July 16*	
	Company Officers. (9)			
1 May 01	*Parkin, Maj. J. F.*, D.S.O.	*1 May 16*	*9 Mar. 04*	
15 July 03	Roughton, Maj. F. A. G.	15 July 18	3 Mar. 08	
2 Mar. 04	Gore, Capt. F. L.	2 Mar. 13	9 Apl. 09	
17 Jan. 11	Thompson, Capt. S. L.	1 Sep. 15	20 Mar. 13	
20 Jan. 12	Passy, Capt. W. DeL.	20 Jan. 16	18 Mar. 12	With 1-11th Rajputs.
8 Aug. 14	Roberts, Capt. M. R.	8 Aug. 18	Dec. 15	Attd. with 2-113th Infy.
29 June 16	Vaughan-Jones, Lt. H.	4 Jan. 16	. . .	Attd. Adjt.
14 July 15	Cameron, Lt. E.	14 July 16	28 Oct. 15	Attd. with 3-154th Infy.
4 Aug. 15	Mathewman, Lt. G. B.	4 Aug. 16	6 Dec. 15	Attd. with 3-154th Infy.
15 Nov. 15	Heraper, Lieut. J. C.	15 Nov. 16	18 Nov. 15	Attd.
31 Aug. 18	Griffiths, 2nd-Lt. F. J.	31 Aug. 18	5 Sep. 18	Attd.
	Attached.			
3 May 15	Martin, Lt. E. H. deC.	3 May 17	. . . .	120th Infy.
14 May 15	Gill, Lt. E. H. N.	14 May 16	. . . .	I. A. R. O.—A. R. O., Allahabad.
9 Sep. 15	Balding, Lt. C. D. (*temp. Capt.*)	9 Sep. 16	26 June 16	I. A. (*on probn.*)
6 Fb. 16	Robinson, Lt. J. McM.	6 Feb. 17	30 June 18	I. A. (*on probn.*)
. . . .	O'Meara, Lt. A. J.	. . . .	21 Aug. 18	I. A. (*on probn.*)
25 Mar. 16	Templeton, Lt. C. A. S.	25 Mar. 17	27 Feb. 17	I. A. R. O.
8 Sep. 15	Kirwan, Lt. N. G. B.	8 Sep. 16	27 Feb. 17	I. A. R. O.
22 May 16	Fricker, Lt. E. S. G.	22 May 17	22 June 16	I. A. R. O.
23 June 16	Teehan, Lt. P. (*temp. Capt.*)	23 June 17	28 July 16	I. A. R. O., T. C. O., Bombay.
23 June 16	Lawson, Lt. E. F.	23 June 17	28 July 16	I. A. R. O.
	Cowan, Lt. S. N.	. . . .	9 Sep. 18	Notts & Derby. R., I.A.R.O. (*on probn.*)
5 May 17	White, Lt. L. N. W.	5 May 18	22 Nov. 18	I. A. (*on probn.*)
. . . .	Mathews, Lt. V. de R.	. . . .	5 Mar. 18	I. A. R. O.
18 June 17	Lawton, Lt. T. D. V.	18 June 18		I. A. R. O. (*on probn.*)
	Vowles, Lt. A. S., D.S.O. (207a)	. . . .	27 Mar. 18	I. A. (*on probn.*)
20 June 17	Pryde, Lt. A. W.	20 June 18	4 Oct. 17	I. A. R. O.
8 Aug. 17	Robertson, Lt. D.	8 Aug. 18	2 Nov. 17	I. A. R. O.
5 Nov. 17	Blackmore, Lt. J. P.	5 Nov. 18	17 Mar. 18	I. A. (*T. C.*)
29 Mar. 18	Kingsmill, 2nd-Lt. W. J.	29 Mar. 18	23 July 18	I. A. R. O.
29 May 18	Hayward, 2nd-Lt. H. B.	. . . .	29 Aug. 18	I. A. R. O. (*on probn.*)
7 June 18	Branson, 2nd-Lt. C. H.	7 June 18	10 Oct. 18	I. A. (*T. C.*)
30 July 18	Henghan, 2nd-Lt. J. K.	30 July 18	7 Sep. 18	I. A. R. O.
31 July 18	Ross, 2nd-Lt. H. S.	31 July 18	16 Sep. 18	I. A. R. O.
16 Aug. 18	Leach, 2nd-Lt. A. H. L.	16 Aug. 18	25 Sep. 18	I. A. R. O.
		Jemadar.	**Subadar.**	**Subadar-Major**
	Subadar-Majors			
8 July 95	Kana Ram	23 Jan. 03	2 Nov. 06	16 July 12.
	Budha Singh (Retd.)	. . . .	. . . .	Re-empld., 27 Mar. 17.
	Subadars.			
15 Feb. 00	Ghure (56)	20 Nov. 06	8 June 07	With 3-154th Infy.
7 Sep. 97	Khudadad Khan (56)	25 Apl. 03	15 May 11	Actg. Subdr.-Maj.
28 July 98	Ganeshi Ram (56)	21 Aug. 04	20 Nov. 12	
8 Jan. 99	Bansi (56)	[illegible] July 12	28 Jan. 16	
26 July 03	*Ramdeo (56)*	*1 July 15*	*10 June 16*	*With 102nd Grs.*
14 June 99	Brijlal Singh (56)	15 Aug. 15	23 Mar. 17	
14 Oct. 03	Harchand	30 Aug. 15	16 July 17	
	Parshadi Singh (*Retd.*)	. . . .	1 Sep. 16	Re-empld.
	Nanwa Singh (*Retd.*)	. . . .	30 June 17	Re-empld.
27 Sep. 04	Shama Ram	1 May 16	13 May 18	
8 Mar. 05	Rasal	7 Oct. 17	13 May 18	
13 May 06	Harch[illegible]	1 Dec. 16	17 Sep. 18	
. . . .	R[illegible]ura Singh (*retd*)	. . . .	. . . .	Re-empld.

1st Battalion, 113th Infantry—*contd.*

1st Commn. or date of entering service.	Names and Rank.	Army rank.	Remarks.
	Jemadars.		
20 Mar. 00	Moman Ram (56)	1 Dec. 15	With 2-113th Infy.
1 Apl. 05	Kadir Bakhsh	1 May 16	Indian Adjt.
1 July 01	Shah Muhammad	1 May 16	
14 Aug. 03	Fateh Ali	1 May 16	
1 May 01	Kesharya	10 June 16	With 3-154th Infy.
28 Feb. 97	*Fazzal Din*	*21 July 16*	
1 June 01	Jas Ram	1 Sep. 16	With 2-113th Infy.
25 Apl. 05	Nanwa	7 Nov. 16	With 3.154th Infy.
29 Sep. 04	Pat Ram	1 Dec. 16	
2 Oct. 98	Kala Singh	1 Dec. 16	
21 Feb. 04	Hema Ram	26 May 17	
28 Dec. 09	Ruga Ram	26 May 17	With 3-154th Infy.
	Mataram (48*a*)	3 Nov. 17	
6 Oct. 95	Samuel Aide	14 Jan. 18	
15 Sep. 98	Badar Din	15 Oct. 16	
	Pema Ram	. . . .	Re-empld. 13 Oct. 17.
	Angad	3 Dec. 17	
	Ram Singh	13 May 18	
	Harnarayan	13 May 18	
	Amichand	25 July 18	
	Tikam Singh	17 Sep. 18	
16 Nov. 98	Sahib Din	2 Sep. 18	
	Chaudhri Narain Singh	. . . .	Temp. Hony.

2nd Battalion, 113th Infantry.

Formed at Bombay on 27th November 1916.

Composition—2 companies of Rajputana Jats, 1 of Rajputana Gujars, 1 of Punjabi Musalmans.

1st Commn. or date of entering service.	Names and Rank.	Army rank.	Present appointment in Regiment.	Remarks.
	Commandant.			
19 May 00	Douglas, Maj. G. S. (*Actg. Lt.-Col.*)	1 Sep. 15	. . . . .	18th Infy.
	Company Commanders. (4)			
19 Aug. 03	Thompson, Maj. A.	19 Aug. 18	18 Dec. 17	112th Infy.
8 Aug. 14	Roberts, Capt. M. R. (*Actg. Maj.*)	8 Aug. 18	18 Dec. 17	1—113th Infy. 2nd-in-Comd.
19 June 15	Fitz Gerald, Lt. L. C. . . .	19 June 16	18 Dec. 17	Attd.
	Company Officers.			
29 Jan. 10	Harward, Capt. R. B. . . .	1 Sep. 15	28 Feb. 18	73rd Infy.
12 May 15	Fisher, Lt. P. (*Actg. Capt.*) . .	12 May 16	12 Feb. 18	Nort'd. Fus., I. A. R. O. (*on probn.*)
17 Aug. 15	Withers, Lt. F. W.	17 Aug. 16	14 Feb. 18	Midd'x. R., I. A. R. O. (*on probn.*)
30 Jan. 16	Philips, Lt. W. J. C.	30 Jan. 17	23 Sep. 18	1st Gn. Bn., R. Ir. Fus., I. A. (*on probn.*)
7 July 16	Lewis, Lt. H. A. C. (*Actg. Capt.*)	7 July 17	31 July 17	I. A. (*on probn.*) Adjt.
1 Aug. 16	Hawken, Lt. B. R. P. . . .	1 Aug. 17	14 Feb. 17	North'd. Fus., I. A. (*on probn.*)
29 Oct. 16	Sowler, Lt. J.	29 Oct. 17	6 Oct. 17	R. Scots., Qr.-Mr., I. A. (*on probn.*)
25 July 17	Gass, Lt. F. T.	25 July 18	26 Sep. 17	I. A. (*on probn.*)
26 Sep. 17	Duckworth, Lt. M. B.	26 Sep. 18	31 Jan. 18	I. A. R. O.
26 Sep. 17	Young, 2nd-Lt. J. C. . . .	26 Sep. 17	31 Jan. 18	I. A. R. O. (*on probn.*)
31 Oct. 17	Cook, 2nd-Lt. R. H.	31 Oct. 17	14 Feb. 18	I. A. R. O. (*on probn.*)
12 Sept. 17	East, Lt. C. F.	12 Sep. 18	23 Sep. 18	87th Punj.
28 Nov. 17	Knatchbull, 2nd-Lt. R. E. . .	28 Nov. 17	1 Feb. 18	I. A. R. O.
15 Jan. 18	Godley, 2nd-Lt. B. W. T. . .	15 Jan. 18	6 June 18	I. A. (*T. C.*)
15 Jan. 18	Robertson, 2nd-Lt. S. . . .	15 Jan. 18	6 June 18	I. A. (*T. C.*)
26 Jan. 18	Mackinnon, 2nd-Lt. R. A. . .	26 Jan. 18	31 Jan. 18	I. A. (*on probn.*)
7 June 18	Barber, 2nd-Lt. A. H. J. . . .	7 June 18	10 Oct. 18	I. A. R. O.
31 July 18	Forshaw, 2nd-Lt. H. W. . . .	31 July 18	2 Sep. 18	I. A. R. O. (*on probn.*)
16 Aug. 18	Parson, 2nd-Lt. L. M.	16 Aug. 18	9 Sep. 18	I. A. R. O. (*on probn.*)
21 Aug. 18	Allen, 2nd-Lt. E. C.	21 Aug. 18	12 Sep. 18	I. A. R. O.
31 Aug. 18	Drought, 2nd-Lt. A. C. . . .	31 Aug. 18	5 Sep. 18	Attd.
31 Oct. 18	Graham, 2nd-Lt. A. C. . . .	31 Oct. 18	Nov. 18	I. A. R. O.
		Jemadar.	**Subadar.**	**Subadar-Major.**
	Subadar-Major.			
25 Sep. 96	Jahandad Khan (56)	24 Apl. 03	1 May 07	1 Dec. 16
	Subadars.			
1 Jan. 94	Dani Ram (56)	1 Mar. 07	5 Jan. 15	
9 Apl. 87	Kasim Ali Shah (56)	1 May 07	5 Jan. 15	
27 Feb. 09	Chatra Ram	20 Nov. 12	1 May 16	
11 Oct. 03	Solu	1 July 15	1 May 16	
21 May 00	Chainsukh Ram	5 Jan. 15	1 Sep. 16	
13 June 96	Gangadhar Singh	5 Jan. 15	1 Dec. 16	
9 Aug. 95	Mahtab Khan	1 Aug. 15	28 May 18	Indian Adjt.
1 June 01	Jas Ram	1 Sep. 16	28 May 18	
20 Mar. 00	Moman Ram	1 Dec. 15	28 May 18	
6 July 03	Bakshi Ram	1 Dec. 16	19 Aug. 18	
	Jemadars.			
16 Nov. 98	Malkhan Singh	1 June 16		
16 June 95	Mauladad (46)	1 Dec. 16		
13 May 07	Risal	1 Dec. 16		
3 Jan. 97	Mana Ram	12 Mar. 17		
24 June 05	Naini	10 Apl. 17		
12 Oct. 03	Tora Khan	24 July 17		
3 Dec. 00	Soni Singh	26 July 17		
1 Mar. 02	Jai Ram	30 Sep. 17		
6 Oct. 17	Sawant Ram	28 May 18		
26 Aug. 06	Ganeshi Ram	28 May 18		
14 Jan. 03	Alah Ditta	8 June 18		
. . . .	Budha Ram	19 Aug. 18		
. . . .	Mahta Ram	19 Aug. 18		
4 July 02	Imam Din	17 July 18		

Regt. Centre—Poona.
Linked with—116th & 117th Mahrattas.

Uniform—Scarlet.
Facings—Yellow.

114th Mahrattas.

Raised at Bombay in 1800, by Lieut. Wood, as the 2nd Battalion, 7th Regiment of Bombay Native Infantry. Became the 14th Regt. of Bombay Native Infantry in 1824; the 14th Regt. of Bombay Infantry in 1885; the 14th Bombay Infantry in 1901; present designation in 1903.

Composition—2 companies of Konkani Mahrattas, 1 of Dekhani Mahrattas, 1 of Dekhani Musalmans.

COLONEL—Lieutenant-General Francis S. Iredell, M. . 13 May 1904.

1st Commn. or date of entering service.	Names and Rank.	Army rank.	Present appointment in Regiment.	Remarks.
	Commandant.			
16 Dec. 96	Wintle, Lt.-Col. C. E. H.	29 Mar. 18	29 Mar. 18	
	Company Commanders. (4)			
28 Sep. 92	Stewart, Lt.-Col. A. F. (Q)	28 Sep. 18	19 Nov. 15	2nd-in-Comd.
8 May 01	Wilson, Bt.-Lt.-Col. R. C., D.S.O., M.C., *p. s. c.*	3 June 17		
13 Aug. 04	*Turner, Capt. J.*	*13 Aug. 13*	*18 Mar. 18*	*Burma Mily. Police.*
24 Jan. 06	Bremner, Capt. H. N. K.	24 Jan. 15	16 Apl. 18	
	Company Officers. (7)			
19 Jan. 07	Pearson, Capt. H. S. I.	1 Sep. 15	11 Mar. 08	With 110th Infy.
17 Aug. 07	Meredith, Capt. A. E., M.C.	1 Sep. 15	4 Nov. 09	
6 Nov. 09	Chisholm, Capt. D. K. J.	1 Sep. 15	27 Feb. 14	
16 Dec. 15	Hancock, Lt. C. P.	16 Dec. 15	18 Feb. 15	
29 June 16	Goddard, Lt. F. C.	29 June 17	10 Oct. 17	
	Attached.			
18 Jan. 02	Standbridge, Capt. C. H.	16 Mar. 12	18 Jan. 18	117th Mah.
15 Nov. 15	Laing, Lt. A. D. M. G.	15 Nov. 16	15 Nov. 16	79th Infy.
8 Jan. 15	Sargon, Capt. A. I.	19 Sep. 16	18 Mar. 15	I. A. R. O.
23 Apl. 15	Sothers, Lt. D. B.	23 Apl. 16	3 July 15	I. A. R. O.
10 Mar. 16	*Haigh, Lt. P. B.*	*10 Mar. 16*	*5 Aug. 16*	*I. A. R. O. Asst. Musky. Instr., Satara.*
16 Aug. 16	Hunter, Lt. H. C.	16 Aug. 17	19 Dec. 17	116th Mah.
21 Oct. 16	Mitchell, Lt. L.	21 Oct. 17	22 Apl. 18	110th Mah. L. I.
27 Oct. 16	Henry, Lt. M.	27 Oct. 17	3 Jan. 18	117th Mah.
19 Mar. 15	Yeld, Lt. R. K.	19 Mar. 16	5 Aug. 16	I. A. R. O.
22 July 16	Furness, Lt. H. G.	22 July 17	3 Jan. 17	I. A. R. O.
6 Oct. 16	Thomas, Lt. F. A. S.	6 Oct. 17	4 Dec. 17	I. A. R. O.
5 Jan. 17	Walmsley, Lt. D. C.	5 Jan. 18	27 Jan. 18	I. A. R. O. Off. Qr.-Mr.
19 Oct. 17	Burns, Lt. W.	19 Oct. 18	1 Nov. 17	I. A. R. O.
10 June 17	Rippon, Lt. P. I. V.	10 June 18	24 Nov. 16	I. A. R. O.
8 Aug. 17	Pitt, Lt. H. F. G.	8 Aug. 18	5 Nov. 17	I. A. R. O.
12 Sep. 17	Tapp, Lt. D. H.	12 Sep. 18	7 Dec. 17	
23 Aug. 17	Raynbird, 2nd-Lt. H.	23 Aug. 18	21 Apl. 18	I. A. (*on probn.*).
29 Mar. 18	Priddey, 2nd-Lt. A. H. P.	29 Mar. 18	17 July 18	I. A. R. O.
23 July 18	Gurnett, 2nd-Lt. W. V.	23 July 18	2 Aug. 18	I. A. R. O.
13 Aug. 18	Gillon, 2nd-Lt. G. K.	13 Aug. 18	7 Sep. 18	I. A. R. O.
31 Aug. 18	Hale, 2nd-Lt. W. G.	31 Aug. 18	14 Sep. 18	Attd.
31 Aug. 18	Mather, 2nd-Lt. R. W.	31 Aug. 18	14 Sep. 18	Attd.
7 June 18	Riley, 2nd-Lt. E. S.	7 June 18	....	I. A. R. O.

Subadar-Major.

Name	Jemadar.	Subadar.	Subadar-Major.	Remarks
Gurdatt Singh Bahadur (*Retd.*)	....	....	....	Re-empld., 6 Nov. 16.
Shaikh Babu (Retd.) *Hon. Capt. Sardar Bahadur*	....	4 May 18		Re-empld.

114th Mahrattas—*contd.*

1st Commn. or date of entering service.	Names and Rank.	Army rank.	Present appointment in Regiment.	Remarks.
	Subadars.			
24 May 95	Dajirao Powar (56)	21 May 03	20 Sep. 08	
12 Apl. 00	Shaikh Ahmad (56)	10 Oct. 07	11 Jan. 09	
18 July 99	Ramchandar Bhosle (56)	17 Apl. 11	29 June 17	
1 Feb. 02	Dattaram Chauhan	1 June 17	1 June 17	
4 Nov. 92	Balajirao Ghag (56)	28 Oct. 14	11 Oct. 17	
	Shaikh Abdul Kadir (*Ret.*)	. . . .	. . . .	Re-empld.
	Shivram Zinjad	. . . .	. . . .	Re-empld. 28 Aug. 18.
	Jemadars.			
20 Aug. 98	Shaikh Aminuddin (56)	23 Aug. 15		With 110 Infy.
1 Nov. 99	Gopal Dhanaote (56)	1 Nov. 15		Jemdr. Adjt.
8 Apl. 01	Krishna Gaoda	8 Feb. 16		With 110 Infy.
24 Dec. 00	Shaikh Bandu	10 Apl. 17		
25 Apl. 01	Krishna Wejre	5 May 17		
2 Mar. 09	Shaikh Mohammad Khalil	14 May 17		
2 Apl. 01	Pandu Naik	14 May 17		
14 Mar. 02	Bala Gaoda	14 May 17		
19 Dec. 04	Dhondi Sawant	1 June 17		
6 May 01	Balwant Bhoite	29 June 17		
21 Nov. 01	Ghansham Parab	25 July 17		
28 Nov. 01	Ram Chandra Chawan	5 Oct. 17		
1 Apl. 02	Shaikh Chand	1 Dec. 17		
11 Sep. 01	Ramchandra Bhiku	10 Apl. 18		Jemadar, Qr.-Mr.
7 May 01	Dadu Ingle (56)	15 Apl. 18		
	Wasudeo Rane	1 June 18		
25 Oct. 05	Mahmud Khan	12 Aug. 18		
27 Mar. 04	Ramchandra Jondhle	12 Aug. 18		
	Ananda Thorat	1 Sep. 18		
19 Dec. 03	Dharrao Mahdik	2 Sep. 18		
19 June 06	Sarjerao Supekar	14 Sep. 18		
20 June 06	Sadar Khan	14 Sep. 18		
16 Sep. 07	Keshao Talekar	14 Sep. 18		

Regt. Centre—Poona.
inked with—114th & 117th Mahrattas.

Uniform—Scarlet.
Facings—Yellow.

116th Mahrattas.

Raised at Mangalore in 1800, by Major R. T. Heath, as the 2nd Battalion, 8th Regiment of Bombay Native Infantry or Gore's Regiment. Became the 16th Regiment of Bombay Native Infantry in 1824; the 16th Regiment of Bombay Infantry in 1885; the 16th Bombay Infantry in 1901; present designation in 1903.

"Afghanistan, 1879-80" "British East Africa, 1901."

Composition—2 companies of Konkani Mahrattas, 1 of Dekhani Mahrattas, 1 of Dekhani Musalmans.

1st Commn. or date of entering service.	Names and Rank.	Army rank.	Present appointment in Regiment.	Remarks.
	Commandant.			
25 Mar. 91	Bridges, Lt.-Col. A. H., D.S.O., P.S.C. [l].	1 Jan. 17	9 Apl. 18	
	Company Commanders. (4)			
20 July 98	Holmes, Maj. C. W. A., p.s.c. (*Actg. Lt.-Col.*)	1 Sep. 15	9 Apl. 18	2nd-in-Comd. with 2-117th Mahs.
23 May 00	Harrison, Maj. A. P.	1 Sep. 15	11 Oct. 13	
28 Jan. 03	Ayscough, Maj. G. M.	18 July 17	19 Nov. 15	
21 Dec. 04	Young, Capt. H. W. [L]	21 Dec. 13	9 Apl. 18	
	Company Officers. (8)			
5 Aug. 05	*Dockrell, Capt. M.*	*5 Aug. 14*	*2 Mar. 08*	*G. S. O., 2nd gr.*
19 Aug. 06	Eberhardie, Capt. R. C.	29 Aug. 15	16 June 08	
19 Jan. 07	Hopkinson, Capt. G. H.	1 Sep. 15	20 Mar. 08	
29 Jan. 10	*Hungerford, Capt. S. A. H.*	*1 Sep. 15*	*5 Apl. 11*	
24 Aug. 12	Mosse, Capt. C. B.	24 Aug. 16	3 Nov. 13	Adjt.
1 Oct. 14	Flemming, Capt. R. L. V.	1 Oct. 18	18 Dec. 14	With 1-101st Grs.
11 Nov. 15	Anderson, Lt. P. B.	16 June 16	2 Jan. 16	Attd. with 2-101st Grs.
22 Dec. 15	Solly, Lt. A. R.	22 Dec. 16	1 Mar. 16	Attd.
18 Apl. 16	Hodson, Lt. D.	18 Apl. 17	25 Apl. 16	Attd.
9 June 16	Kenny, Lt. E. J.	29 June 17	3 July 16	Attd. with 105th.
16 Aug. 16	Hunter, Lt. H. C.	16 Aug. 17	11 Nov. 16	Attd.
9 Oct. 16	Tankard, Lt. B. L.	9 Oct. 17	7 Dec. 17	Attd.
1 Oct. 15	O'Gorman, Lt. M. K. (*temp. Capt.*)	1 July 17	5 Mar. 17	Attd.
20 July 15	Isaacs, Lt. R. L.	20 Apl. 17	. . . .	Attd.
	Attached.			
8 May 01	Connell McDowell, Maj. A. E. J.	8 May 16	17 Dec. 14	Civil.
5 Feb. 16	Atkinson, Lt. C. M.	5 Feb. 17	14 May 18	I. A. (*on probn.*)
14 Nov. 16	Murray, Lt. T. W.	14 Nov. 17	23 Nov. 16	117th Mahs.
7 July 16	Hone, 2nd-Lt. A. C., M.C.	7 July 16	15 Jan. 18	I. A. R. O. (*on probn.*)
22 July 16	Collins, 2nd-Lt. G. F. S.	22 July 16	17 Sep. 16	I. A. R. O.
28 Aug. 16	Inder, Lt. R. W.	28 Aug. 17	4 Nov. 16	I. A. R. O.
31 Dec. 14	Ward, Lt. F. K.	31 Dec. 15	4 July 17	I. A. R. O.
14 Jan. 15	Martyn, Lt. W. W.	14 Jan. 15	21 Mar. 17	I. A. R. O.
1 July 16	Mather, 2nd-Lt. J. J.	1 July 16	. . . .	I. A. R. O.
1 Jan. 17	McBride, 2nd-Lt. J.	1 Jan. 17	24 July 17	I. A. R. O.
8 Aug. 17	Griffiths, Lt. T. A.	8 Aug. 18	4 Nov. 17	I. A. R. O.
16 May 17	Davey, Lt. C. C.	16 May 18	25 Nov. 17	I. A. R. O.
12 Sept. 17	Woodman, Lt. R. S.	12 Sep. 18	7 Dec. 17	Attd.
15 Jan. 18	Bailey, 2nd-Lt. R. J.	15 Jan. 18	2 June 18	I. A. R. O.
2 Aug. 18	Chitty, 2nd-Lt. M. H. C.	2 Aug. 18	9 Aug. 18	I. A. R. O.
4 June 18	Parsons, 2nd-Lt. P. J.	4 June 18	17 June 18	I. A. R. O.
18 June 18	Middleton, Stewart, 2nd-Lt. J. C.	18 June 18	18 June 18	I. A. R. O.
21 Oct. 18	Webster, 2nd-Lt. C. G. G.	21 Oct. 18	25 Nov. 18	I. A. R. O.
23 Apl. 18	Simmons, 2nd-Lt. J. M. R.	23 Apl. 18	25 Oct. 18	I. A. (*on probn.*)
	Hill, 2nd-Lt. W. L.	. . . .	. . . .	R. War. R., I. A. R. O. (*on probn.*)

		Jemadar.	Subadar.	Subadar-Major.
	Subadar-Major.			
1 Jan. 87	Ramchandar Ghorpade	6 Apl. 00	28 Sep. 01	8 Jan. 17
	Subadars.			
1 Apl. 84	Mahadeorao Sondkar *Bahadur* (56).	10 July 02	12 Nov. 04	*Order of Br. Ind., 2nd Cl.*
3 Aug. 01	Vishram Rao Chowan (56)	4 Sep. 06	4 Mar. 09	
14 May 97	Deoji Rao Kadam (56, 66a)	2 Feb. 09	1 Nov. 10	
19 July 96	Kamar-ud-din (56)	22 Oct. 07	2 Sep. 12	
1 Apl. 09	Kashinath Dalvi (56, 66a)	16 Mar. 10	25 Dec. 15	
4 Mar. 00	Daji Sinde (56)	13 Oct. 14	1 May 16	

1st Battalion, 116th Mahrattas—*contd.*

1st Commn. or date of entering service.	Names and Rank.	Army rank.	Subadars.	Remarks.
	Subadars—*contd.*			
10 June 02	Abdurrahaman Khan	5 Aug. 15	2 Jan. 17	
11 July 02	Baburao Ghorpade (56, 66*a*, 72*a*).	5 Apl. 15	9 Feb. 17	
22 Apl. 02	Syed Dastagir	20 June 16	9 Feb. 17	
2 Feb. 04	Usman Khan	8 Jan. 17	12 June 18	
13 Oct. 03	Chander Rao Sinda	8 Dec. 15	30 July 18	
	Sheikh Ibrahim Siri Bah (*Retd.*)	. . . . .	Re-empld.	
	Jemadars.			
11 Jan. 03	Govind Sawant	8 Dec. 15		
28 June 02	Bakaji Gadwe	25 Dec. 15		
4 Apl. 05	Shaikh Yusuf	19 July 16		110th Mah. L. Infy.
1 Jan. 01	Dattaram Sawant (56)	1 Aug. 16		Indian Adjt.
6 May 05	Vishnujirao Ghag (56)	10 Oct. 16		110th Mah. L. Infy.
25 Dec. 00	Anna Chawan	1 Nov. 16		
1 Aug. 08	Sheikh Mohomed Usman . .	17 Jan. 17		
5 June 04	Yadeorao Barge	1 Feb. 17		
3 June 07	Narayan Kadam	8 Feb. 17		
26 June 03	Ramchandar Chalke	9 Feb. 17		
1 Apl. 01	Sakaramrao Surwe	3 Feb. 18		
19 Mar. 94	Govindrao Barge	2 May 18		
1 Feb. 08	Raghoji Bagwe	31 July 18		
1 June 08	Gunajirao Ghag	22 Sep. 18		
20 May 03	Govind Sakpal	23 Sep. 18		
1 Aug. 99	Sheikh Muhomed Hussain . .	12 Jan. 17		
	Sheikh Mohomed (*retd.*) . . .	. . . . .		Re-empld.
	Ratanrao Bosle	1 Sep. 18		

2nd Battalion, 116th Mahrattas.

Formed at

Composition :—

1st Comm n. or date of entering service.	Name and Rank.	Army Rank.	Present Appointment in Regiment.	Remarks.
	Commandant.			
6 Jan. 00	Henderson, Maj. C. G. H. . .	1 Sep. 15	. . .	29th Lrs.
	Company Commanders (4).			
23 July 01	Holmes, Maj. H. L.	23 July 16	. . .	122nd Infy.
. . .	Lough, Lt. J.	. . .	. . .	2nd-in Commd.
5 Sep. 15	Hasted, Lt. A. W. O.	5 Sep. 16	. . .	Wilts. R., I. A. (*on probn.*)
21 Apl. 16	Low, Lt. J.	21 Apl. 17	. . .	Sea. Highrs., I. A. (*on probn.*)
	Company Officers.			
7 Apl. 17	Woodford, Lt. A. S.	7 Apl. 18	. . .	I. A. (*on probn.*)
18 June 17	Covernton, Lt. J. W. G. . . .	18 June 18	. . .	1-103rd Mah. Adjt.
. . .	Davies, Lt. C. R. V., M. M. . .	. . .	. . .	
. . .	Flatt, 2nd-Lt. H. C.	. . .	. . .	I. A. (*T. C.*)
. . .	Haysey, 2nd-Lt. R.	. . .	. . .	
. . .	Nunn, 2nd-Lt. F. M., D. C. M. .	. . .	. . .	
18 Dec. 14	Hill, 2nd-Lt. H. I.	14 Dec. 15	. . .	I. A. R. O., Qr. Mr.
. . .	Jones, 2nd-Lt. R. M.	. . .	. . .	
. . .	Grainger, 2nd-Lt. S. N. . . .	. . .	. . .	
. . .	Mason, 2nd-Lt. H. C. E. . . .	. . .	. . .	I. A. R. O. (*on probn.*)

1 Jemadar. 1 Subadar. 1 Subadar-Major.

Subadar-Major.

Subadars.

28 Oct. 96	Sayad Shamsuddin	2 Sep. 12	. . .	
28 Mar. 06	Kesheorao Sinde	1 Dec. 16	. . .	
. . .	Krishna Holkar	. . .	. . .	
. . .	Shaikh Gul Muhammad . . .	. . .	. . .	
. . .	Eshwanta Jadhao	. . .	. . .	
. . .	Narayan Parab	. . .	. . .	

Jemadars.

Regt. Centre—Poona.
Linked with—114th & 116th Mahrattas:

Uniform—Scarlet.
Facings—Yellow.

1st Battalion, 117th Mahrattas.

Raised at Bombay in 1800 as the Bombay Fencible Regiment. Became the 1st Battalion, 9th Regiment of Bombay Native Infantry in 1803; the 17th Regiment of Bombay Native Infantry in 1824; the 17th Regiment of Bombay Infantry, 1885; the 17th Bombay Infantry in 1901; present designation in 1903.

Composition—2 Companies of Konkani Mahrattas, 1 of Dekhani Mahrattas, 1 of Dekhani Mussalmans.

1st Commn. or date of entering service.	Names and Rank.	Army rank.	Present appointment in Regiment.	Remarks.
	Commandant.			
13 Aug. 92	Hunter, Lt.-Col. W. J. H.	18 Mar. 18	18 Mar. 18	
	Company Commanders. (4)			
11 Oct. 93	Steen, Maj. H. C.	11 Oct. 11	18 Mar. 18	2nd-in-Command. With 3-103rd Mah. L. I.
17 Nov. 94	Lennox, Maj. G. M.	17 Nov. 12	23 Mar. 12	.
1 Sep. 99	*Green-Armytage, Maj. P. D.* [L.].	*1 Sept. 15*	*26 Sept. 15*	*D. A. Q. M. G., A. Hdqrs.*
18 Jan. 02	*Hall, Maj. E. G.*	*18 Jan. 17*	*8 Apl. 16*	*D. A. Q. M. G., A. Hdqrs.*
	Company Officers. (8)			
26 Mar. 02	Potter, Maj. R. J. K.	26 Mar. 17	26 Aug. 06	With 2-101st Grs.
18 Jan. 02	Standbridge, Capt. C. H. . . .	16 Mar. 12	7 Oct. 04	With 114th Mah.
16 Aug. 05	Franks, Capt. K. F. (*Actg. Maj.*)	16 Aug. 14	11 Mar. 08	2nd-in-Comd. (*temp.*)
24 Jan. 06	*Pogson, Capt. C. A.*, M.C. . .	*24 Jan. 15*	*23 Apl. 12*	*Bde.-Maj., Dehra Dun. Bde.*
8 Jan. 15	*Taylor, Lt. A. T.*	*8 Oct. 16*	*28 Mar. 15*	*Attd.*
25 June 15	Chambers, Lt. H. M.	25 June 16	. . . .	Attd.
7 Aug. 15	Kennelly, Lt. D. McKeon, (*Actg. Capt.*)	7 Aug. 16	. . . .	Attd., Adjt.
27 Oct. 16	Henry, Lt. M.	27 Oct. 17	. . . .	With 114th Mah.
14 Nov. 16	Murray, Lt. T. W.	14 Nov. 17	. . . .	Attd.—With 116th Mahs.
30 Jan. 17	Carberry, Lt. T. B.	30 Jan. 18	. . . .	With 2-101st Grs.
30 Jan. 17	Dawson, Lt. A. J. E.	30 Jan. 18	. . . .	With 1-2nd Rajputs.
18 June 17	Turner, Lt. C. D. L.	18 June 18	. . . .	With 2-103rd Mah. L. I.
12 Sep. 17	Walsh, Lt. A. P.	12 Sep. 18	. . . .	Attd.
21 July 17	Collingwood, Lt. A. L. . . .	21 July 17	28 Aug. 17	Attd.
10 Mar. 16	Hawkins, Lt. F. T.	10 Mar. 17	. . . .	Attd.
	Attached.			
14 Jan. 14	Lane, Capt. G. H.	14 Jan. 18	. . . .	1-103rd Mah. L. I.
6 Mar. 14	Westmacott, Capt. T. R. . . .	26 Mar. 17	. . . .	38th Horse.
20 Nov. 14	*Stearns, Lt. H. D.*	*20 Nov. 15*	*26 Apl. 15*	*I. A. R. O.*
14 Aug. 15	James, Lt. H. C.	14 Aug. 16	. . . .	Genl. List, I.A. (*on probn.*)
12 Nov. 15	Shelly, Lt. C. E.	12 Nov. 16	. . . .	I. A. R. O. (*on probn.*)
11 Aug. 16	Jones, Lt. E. O.	11 Aug. 17	. . . .	I. A. R. O.
24 Sep. 16	James, Lt. C. M.	24 Sep. 17	. . . .	I. A. (*on probn.*)
20 Nov. 16	Morris, Lt. H. S., M.C. . . .	20 Nov. 17	. . . .	92nd Punj.
5 Nov. 17	Thomson, Lt. L. M.	5 Nov. 18	. . . .	I. A. R. O. Qr. Mr.
5 Oct. 17	Carter, Lt. F. N.	5 Oct. 18	. . . .	I. A. (T. C.)
28 Aug. 17	Donald, 2nd-Lt. J. W. (46) . .	28 Aug. 17	. . . .	I. A. R. O. (*on probn.*)
15 Nov. 17	Shannon, 2nd-Lt. G. C. . . .	5 Nov. 17	. . . .	I. A. R. O.
28 Nov. 17	Taylor, 2nd-Lt. B.	28 Nov. 17	. . . .	I. A. R. O. (*on probn.*)
27 Jan. 18	Ednie, 2nd-Lt. L.	27 Jan. 18	23 Apl. 18	I. A. (*on probn.*)
	Houghton, 2nd-Lt. W. H. . .	. . . .	. . . .	I.A.R.O. (*on probn.*)
15 June 18	Halton, 2nd-Lt S. C.	15 June 18	. . . .	I. A. R. O.
26 June 18	Mackenzie, 2nd-Lt. J. . . .	26 June 18	. . . .	I. A. R. O.
	Symonds, Capt. J. D. B. . . .	. . . .	. . . .	I. A. R. O. (*on probn.*)
27 July 18	Bowen, 2nd-Lt. D. H.	27 July 18	. . . .	I. A. R. O.

		Jemadar.	Subadar.	Subadar-Major.	
	Subadar-Major.				
16 May 91	*Balwant Rao Sawant, Bahadur*, I.O.M. (72*a*).	*17 Jan. 00*	*1 Oct. 04*	*1 June 13*	*Order of Br. Ind., 2nd Cl.*
	Subadars.				
10 Nov. 91	Sakharam Rao Jagtap, *Sardar Bahadur* (52, 56).	19 Feb. 02	20 Dec. 11	*Order of Br. Ind., 1st Cl.* With 2-117th Mahs.	
14 Jan. 95	Vittal Chande (56)	27 July 02	20 Dec. 11		
22 Sep. 95	Murari Sinde, I.O.M. (56) . . .	15 Feb. 06	1 Aug. 12		
3 Nov. 91	Govind Parab (52, 50, 57, 66*a*) .	3 Nov. 05	16 Nov. 12		
26 Aug. 94	*Ladu Sawant* (52, 56)	*12 Nov. 07*	*1 June 13*		
14 Apl. 97	*Mahadeorao Nalaude*, I.O.M. (52, 56, 72*a*).	*29 June 06*	*14 Apl. 15*		
4 Dec. 99	*Mirza Muhammad Beg* (52, 56, 72*a*).	*7 July 12*	*14 Apl. 15*		
15 June 98	Muhammad Ibrahim	19 Aug. 14	22 Nov. 15		
14 May 97	Tukaram Bhogle (56)	19 Apl. 08	8 Dec. 15		
3 May 03	Vishnu Aire, I.D.S.M.	14 Apl. 15	1 Nov. 17		
2 Aug. 08	Sitaram Jadhao	8 Dec. 15	18 Apl. 18		
9 Jan. 99	Jahangir Khan	1 Oct. 17	1 July 18		

1st Battalion, 117th Mahrattas—*contd.*

1st Commn. or date of entering service.	Names and Rank.	Army rank.	Remarks.
	Jemadars.		
23 Nov. 00	***Sitaram Sellar*, M.C. *(56, 66a)***	*1 June 13*	
7 Nov. 00	*Marutirao Kadam (56, 52)*	*19 Aug. 14*	
16 Dec. 03	Hari Jadhao (52, 56, 57)	19 Aug. 14	
2 June 01	Nhanu Sawant	14 Apl. 15	
25 June 00	Shaikh Usman (I)	14 Apl. 15	
1 Mar. 03	Wasudeo Kadam	1 Oct. 17	
23 Jan. 06	Bhiwa Mhapdi	1 Oct. 17	
3 July 03	Shaikh Usman (II)	1 Oct. 17	
1 Apl. 05	Dema Parab	1 Oct. 17	
15 Sep. 07	Nanajirao Palande	18 Apl. 18	
2 Nov. 08	Dada More	18 Apl. 18	
16 Aug. 98	Ramchandarao Bhonsle	16 May 18	
16 Nov. 03	Marutirao Ghag	16 May 18	
12 June 05	Yenkatrao Salunke	16 May 18	Jemdr. Adjt.
10 May 01	Narayan Shetge	16 May 18	
28 Oct. 07	Mahadeo Ramchandra Yeshwantrao.	16 May 18	
16 Feb. 12	Usman Khan	1 July 18	
7 June 02	Maruti Dhumal	15 July 18	
8 June 03	Govind Ghorpade	16 Sep. 18	
10 Mar. 04	Kamta Kurmi	16 Sep. 18	Jemdr. Qr.-Mr.

2nd Battalion, 117th Mahrattas.

Formed at Arangaon.

Composition—1 Company Bhandaris, 2 Companies Dekhani Mahrattas, 1 Company Konkani Mahrattas.

1st Commn. or date of entering service.	Names and Rank.	Army rank.	Present appointment in Regiment.	Remarks.
	Commandant.			
20 July 98	Holmes, Maj. C. W. A. (*Actg. Lt.-Col.*), *p.s.c.*	1 Sep. 15	4 Apl. 18	116th Mahs. (*temp.*)
	Company Commanders. 4			
9 Jan. 04	Quinan, Bt.-Maj. E. P.	1 Jan. 18	19 Oct. 18	27th Punj. 2nd-in-Comd. (*temp.*)
5 Aug. 14	Wadeson, Capt. H. H.	5 Aug. 18	11 Sep. 18	I. A. R. O.
24 June 15	Chamberlen, Lt. E. F.	24 June 16	26 Aug. 18	1-88th Infy.
5 Mar. 16	Devoy, Lt. J. C.	5 Mar. 17	19 May 18	I. A. (*on probn.*)
	Company Officers.			
14 July 16	Gilligan, Lt. W. B.	14 July 17	4 July 18	1-88th Infy.
26 Oct. 16	Weston, Lt. E. B.	26 Oct. 17	6 June 18	Lond. R., I. A. (*on probn.*)
1 May 17	Bellamy, Lt. L. C. M.	1 May 18	3 June 18	I. A. (*on probn.*)
12 Sep. 17	Wilton, Lt. A. J. M.	12 Sep. 18	3 June 18	*U. L.*
5 Nov. 17	Harper-Collins, Lt. R. S.	5 Nov. 18	8 Apl. 18	I. A. R. O.
5 Jan. 18	Roberts, 2nd-Lt. A. C. (*Actg. Capt.*)	5 Jan. 18	3 June 18	I. A. (*T. C.*) Adjt.
15 Jan. 18	Brown, 2nd-Lt. W. F.	15 Jan. 18	3 June 18	I. A. (*T. C.*)
26 Feb. 18	Tanser, 2nd-Lt. H. A.	26 Feb. 18	6 Sep. 18	I. A. R. O. (*on probn.*)
28 Feb. 18	Wright, 2nd-Lt. P.	28 Feb. 18	30 Sep. 18	R. W. Kent. R., I. A. (*on probn.*)
29 Mar. 18	Price, 2nd-Lt. C. E.	29 Mar. 18	17 July 18	I. A. (*T. C.*) *Actg.* Qr.-Mr.
29 May 18	Shaw, 2nd-Lt. A. A. St. J.	29 May 18	4 Sep. 18	Norf. R., I. A. (*on probn.*)
10 July 18	Bishop, 2nd-Lt. H. P.	10 July 18	2 Dec. 18	I. A. (*T. C.*)
10 July 18	Hodgson, 2nd-Lt. J. L.	10 July 18	2 Dec. 18	I. A.R. O.
15 July 18	Kidd, 2nd-Lt. H. B.	15 July 18	21 Oct. 18	A. R. O.
22 July 18	Brierley, 2nd-Lt. R. T.	22 July 18	2 Dec. 18	I. A. (*T. C.*)
22 July 18	Cooper, 2nd-Lt. A.	22 July 18	2 Dec. 18	I. A R. O.
1 Sep. 18	Price, 2nd-Lt. J. C.	1 Sep. 18	13 Sep. 18	Attd.
....	Hayward, 2nd-Lt. C. D.	....	1 Oct. 18	Notts and Derby. R., I. A. R. O. (*on probn.*)
....	Howells, 2nd-Lt. D. R.	....	2 Oct. 18	I. A. R. O. (*on probn.*)
....	Lethern, 2nd-Lt. W. E.	....	2 Oct. 18	I. A. R. O. (*on probn.*)
		Jemadar.	Subadar.	Subadar-Major.
	Subadar-Major.			
10 Nov. 91	Sakharam Rao Jagtap, *Sardar Bahadur* (51, 52, 56)	19 Feb. 02	26 Dec. 11	*Order of Br. Ind., 1st cl.* (*temp.*)
	Subadars.			
7 June 01	Shankar Kerkar	1 Sep. 17	1 Nov. 17	
31 May 01	Govindrao Powar (51, 52, 56, 66A)	19 Aug. 14	18 Apl. 18	
18 May 01	Tatoba Ambre (66A)	19 Aug. 14	16 May 18	
17 July 05	Sadashiv Parab (51, 52)	28 Jan. 17	16 May 18	
22 May 07	Baburao Jadhao (51)	28 Jan. 17	16 May 18	
12 May 01	Sakharam Patole	1 Oct. 17	16 May 18	
15 Aug. 98	Mhadeo Bhonsle	1 Oct. 17	16 May 18	
	Jemadar.			
15 Nov. 17	Laxuman Bhagat	15 Nov. 17		
7 May 01	Sabajirao Rani	18 Apl. 18		
6 Nov. 07	Krishna Sinde	18 Apl. 18	Ind. Adjt.	
4 Jan. 03	Gopal Parab	16 May 18		
25 May 08	Balaji Rao Kadam	16 May 18		
5 Nov. 07	Shankar Shawant	16 May 18		
25 June 06	Bajirao More	16 June 18		
7 Oct. 13	Laxuman More	27 Aug. 18		
21 Dec. 08	Baburao Ghag	27 Aug. 18		
27 Dec. 13	Nanujirao Kadam	27 Aug. 18		
30 Oct. 12	Hari Warute	27 Aug. 18		
20 Apl. 18	Hiru Bhatose	27 Aug. 18		
27 June 07	Kondiba Power	26 Sep. 18		
1 Aug. 17	Dattatray Waradkar	26 Sep. 18		

Regt. Centre—Mhow. *Uniform*—Scarlet.
Linked with—120th & 122nd Infy. *Facings*—Yellow.

1st Battalion, 119th Infantry (The Mooltan Regiment).

Raised at Bombay in 1817 by Lieut.-Col. Michael Kennedy, as the 1st Battalion, 10th Regiment of Bombay Native Infantry. Became the 19th Regiment of Bombay N. I. in 1824; the 19th Regiment of Bombay Infantry in 1885; the 19th Bombay Infantry in 1901, present designation in 1903.

"Ghuznee, 1839" "Affghanistan, 1839" "Mooltan" "Goojerat" "Punjaub" "Kandahar, 1880" "Afghanistan 1878-80."

Composition—1 Company of Rajputana Gujars, 1 of Mers, 1 of Rajputana Rajputs, 1 of Hindustani Musalmans.

1st Commn. or date of entering service.	Names and Rank.	Army rank.	Present appointment in Regiment.	Remarks.
	Commandant.			
6 July 89	Darley, Lt.-Col J. R., C.I.E., D.S.O.	6 July 15	25 Nov. 16	
	Company Commanders. (4)			
29 Feb. 96	D'Oyly, Maj. J.	1 Mar. 14	22 June 17	2nd-in-Comd.
17 May 99	Boxwell, Maj. A., C.I.E. (*Actg. Lt.-Col.*)	1 Sep. 15	3 Mar. 14	With 2-154th Infy.
18 Apl. 00	Carey, Maj. T. deB. (*Actg. Lt.-Col.*)	1 Sep. 15	25 Nov. 6	With 2-119th Infy.
21 Jan. 03	*Lecky, Maj. J. G.*	*21 Jan. 18*	*7 June 18*	*Asst. Instr., M. G. Trng. Centre., Mhow.*
	Company Officers. (8)			
10 Oct. 03	Taylor, Maj. J. McL. G.	10 Oct. 18	3 Aug. 04	
27 Jan. 04	*Brickmann, Capt. F. I. O.*, M.C.	*27 Jan. 13*	*19 Mar. 08*	
5 Aug. 05	*Morrell, Capt. H. G.*	*5 Aug. 14*	*3 Apl. 14*	*With Sig. Ser.*
19 Jan. 07	Eales, Capt. A. C. H. A.	1 Sep. 15	16 Mar. 08	With 3-9th Infy.
18 Sep. 09	Eccles, Capt. M., M.C.	1 Sep. 15	13 Aug. 14	
20 Aug. 16	O'Reilly, Lt. K. W. R.	20 Aug. 16	1 Dec. 16	With 28th Punj.
30 Jan. 17	Levingston, Lt. T. C.	30 Jan. 18		
30 Jan. 17	Forde, Lt. J. S. B.	30 Jan. 18		
20 Nov. 14	Orr, Lt. I. M.	20 Aug. 16	1 Dec. 16	Attd.
1 Sep. 17	Weston, Lt. J. R.	12 Sep. 18	10 Dec. 17	Attd.
	Attached.			
13 Nov. 14	le Patourel, Capt. E. C., M.C.	13 Nov. 18	29 May 15	I. A. R. O.
2 Dec. 14	Keeling, Capt. E. H.	5 Aug. 18	7 May 15	I. A. R. O.
29 Dec. 14	Swanston, Lt. C. B. R.	29 Dec. 15	. . . .	I. A. R. O.
28 May 15	Knowland, Capt. G. H.	23 Nov. 17	8 Dec. 15	I. A. R. O. Adjt.
22 June 15	Seymour, Lt. W. W.	15 Sep. 17	27 Nov. 17	I. A. R. O.
22 Sep. 15	Baker, Lt. J. N. L.	22 Sep. 16	15 May 18	I. A. (*on probn.*)
23 Oct. 15	Crowther, Lt. H. O.	23 Oct. 16	6 June 16	I. A. R. O.
24 Nov. 15	Mayers, Lt. N. G.	24 Nov. 16	31 July 18	Lanc. Fus., I. A. (*on probn.*)
8 Apl. 16	Griffiths, Lt. H.	8 Apl. 17	3 Sep. 16	I. A. R. O.
5 Aug. 16	Marston, Lt. E. N.	5 Aug. 17	28 Nov. 17	I. A. R. O.
25 Sep. 16	Higgens, Lt. C. R.	25 Sep. 17	19 Feb. 17	I. A. R. O.
2 Feb. 17	Arnold, Lt. R. T.	2 Feb. 18	. . . .	I. A. (*on probn.*)
2 Apl. 17	Butler, Lt. C. T.	2 Apl. 18	20 May 18	I. A. (*on probn.*)
7 Aug. 17	Ashworth, Lt. H. G.	7 Aug. 18	28 June 16	I. A. (*on probn.*)
15 Jan. 18	Gayton, 2nd-Lt. C. (59)	15 Jan. 18	2 June 18	I. A. (*T. C.*)
8 Mar. 18	Davis, 2nd-Lt. J. T.	8 Mar. 18	22 Mar. 18	
29 Mar. 18	Fish, 2nd-Lt. W. H.	29 Mar. 18	17 July 18	I. A. R. O.
27 July 18	Gurney, 2nd-Lt. A. H.	27 July 18	15 Aug. 18	I. A. R. O.
14 Sep. 18	Pettitt, 2nd-Lt. N.	14 Sep. 18	6 Oct. 18	I. A. R. O.

		Jemadar.	Subadar.	Subadar-Major.	
	Subadar-Major.				
	Shaikh Ahmad (Retd.)	. . . .	. . . .	. . . .	Re-empld., 21 Apl. 15.
4 July 93	Shikari *Bahadur*, D.S.M.	11 Mar. 02	31 May 05	1 Jan. 17	*Order of Br. Ind., 2nd. Cl. With 2-154th Infy.*
	Subadars.				
15 Oct. 93	Uma Rawat, *Bahadur* (56)	2 Aug. 03	1 Mar. 09		*Order of Br. Ind., 2nd Cl.*
8 Sep. 96	Unad Singh, I.O.M. (56, 66*a*)	10 Oct. 08	9 Sep. 14		
1 June 98	Fateh Khan	9 Sep. 14	28 Sep. 15		
2 Mar. 98	Tunda (56)	21 Jan. 10	22 Nov. 15		
1 Oct. 95	Hira Rawat	15 June 11	15 Dec. 15		122nd Infy.
9 Feb. 65	Bhagirath Singh	28 Sep. 15	27 Jan. 16		
	Balaji Mule, *Bahadur* (*Retd.*)				Re-empld., 15 Jan. 16.
26 Aug. 04	Abdul Shakur Khan, *Bahadur* (50, 56, 60)	1 Jan. 11	1 July 16		Order of Br. Ind., 2nd Cl.
16 Jan. 93	Jalim	1 Mar. 15	30 Sep. 16		122nd Infy.
18 May 03	Chandan	1 Oct. 15	25 Nov. 16		With 2-154th Infy.
10 Aug. 01	Lahore	1 Jan. 16	25 Nov. 16		
21 Nov. 79	Ganpat Krishnaji	1 Sep. 15	26 Nov. 16		
1 May 99	Sheonath	27 July 15	20 Jan. 17		With 2-154th Infy.
1 June 16	Chamanpal Singh	. . . .	20 Mar. 17		
82 May 11	Lal Singh	4 Apl. 16	29 June 17		
2 Jan. 06	Mala Rawat	1 Aug. 16	17 Dec. 17		
31 May 99	Gena Rawat	4 Nov. 16	26 Jan. 18		122nd Infy.
. . . .	Pudi Khan	. . . .	11 Aug. 18		Re-empld.
3 Sep. 99	Kalyan Singh	11 Jan. 16	23 May 18		

1st Battalion, 119th Infantry (The Mooltan Regiment)—*contd.*

1st Commn. or date of entering service.	Names and Rank.	Army rank.	Remarks.
	Jemadars.		
16 Jan. 97	Shaikh Qadir Bakhsh	28 Sep. 15	
25 June 07	Dao Rawat	11 Nov. 15	
12 Nov. 00	Yadram Singh	22 Nov. 15	
19 Jan. 01	Fazal Khan	11 Jan. 16	
28 Mar. 04	Sheoji	11 Jan. 16	
2 Sep. 08	Chaman Singh	11 Jan. 16	
19 May 02	Raju Singh	19 Apl. 16	
23 Aug. 05	Sardara	1 July 16	
14 Sep. 03	Ganesh Ram	25 July 16	With 2-154th Infy.
16 Sep. 07	Shaikh Ramzan	1 Aug. 16	
17 Jan. 99	Kalyan Singh	9 Sep. 16	
7 May 99	Ghamandi	3 Nov. 16	
18 Feb. 04	Kishensai	26 Nov. 16	
21 Dec. 97	Lal Singh	26 Nov. 16	
13 Aug. 87	Narayan Sadashiv	26 Nov. 16	
13 Oct. 08	Harkishan	20 Jan. 17	With 2-154th Infy.
25 July 09	Gulla Ram, D.S.M.	22 Mar. 17	
7 July 14	Mana Rawat	29 June 17	
13 Feb. 05	Hira Rawat	29 June 17	
12 May 08	Nathan Singh	15 Dec. 17	
12 May 07	Kushal Singh	17 Dec. 17	
11 Nov. 11	Shaikh Rafiuddin	26 Jan. 18	
9 Mar. 96	Arjun Lukshuman	1 Aug. 18	
1 July 07	Bijai Singh	11 Sep. 18	
26 May 14	Muhammad Ali Khan	11 Sep. 18	
18 Aug. 15	Kalu Khan	11 Sep. 18	
1 May 07	Vishram Gangaram	19 May 18	
25 Sep. 07	Wali Muhammad Khan	20 Sep. 17	

2nd Battalion, 119th Infantry. (The Mooltan Regiment.)

Formed at Jubbulpore on 15 June 17.

Composition—1 Company of Rajputana Rajputs, 1 Company of Rajputana Gujars, 1 Company of Mers, 1 Company of Hindustani Musalmans.

1st Commn. or date of entering service.	Names and Rank.	Army rank.	Present appointment in Regiment.	Remarks.
	Commandant.			
18 Apl. 00	Carey, Maj. T. de B. (*Act'g. Lt.-Col.*)	1 Sep. 15	22 May 18	119th Infy.
	Company Commanders (4).			
30 July 15	Douse, Lt. F. W. (*Actg. Capt.*) .	30 July 16	25 Jan. 18	Midd'x. R., I. A. R. O.
8 Aug. 15	Welsh, Lt. A. C. R., M.C. (*Actg. Capt.*).	9 Aug. 16	5 Nov. 17	Attd.
19 Dec. 15	Dedman, Lt. J. J. (*Actg. Capt.*) .	19 Dec. 16	26 Nov. 17	Attd.
16 Feb. 15	Speer, Lt. C. H. (*Actg. Capt.*) .	16 Feb.	28 Mar. 18	Bord. R., I. A. (*on probn.*)
	Company Officer.			
13 Oct. 15	Quest, Lt. T. P.	13 Oct. 16	13 Apl. 18	Yorks & Lancs. R., I. A. (*on probn.*)
22 June 16	Reynolds, Lt. A. W. (*Act'g. Capt.*)	22 June 17	1 Aug. 17	Attd. Adjt.
5 Aug. 16	Tulloch, Lt. D.	5 Aug. 17	1 Oct. 17	I. A. R. O., Qr.-Mr.
11 Apl. 17	Cooke, Lt. A. B.	11 Apl. 18	15 Feb. 18	Lond. R. I. A. (*on probn.*)
5 Nov. 17	Hunt, Lt. H. G.	5 Nov. 18	15 Mar. 18	I. A. (*T. C.*)
6 Nov. 17	Gasper, Lt. H. LeG.	6 Nov. 18	15 Dec. 17	I. A. R. O.
1 Dec. 17	Griffiths, Lt. C. E.	1 Dec. 18	11 Nov. 18	R. Welsh. Fus., I. A. (*on probn.*)
	Tetley, Lt. J. A.	. . . .	22 Oct. 18	A. Cyclist Corps, I. A. R. O. (*on probn.*)
4 Jan. 18	Allen, 2nd-Lt. W. E.	4 Jan. 18	9 Nov. 18	S. Staff R., I. A. (*on probn.*)
15 Jan. 18	Hannagan, 2nd-Lt., L. . . .	15 Jan. 18	10 June 18	I. A. R. O.
15 Jan. 18	Barnett, 2nd-Lt. W. E. . . .	15 Jan. 18	10 June 18	I. A. (*T. C.*)
2 Feb. 18	Arnold, 2nd-Lt. R. T.	2 Feb. 18	10 Dec. 17	Suff. R., I. A. (*on probn.*)
7 June 18	Green, 2nd-Lt. A. J.	7 June 18	25 Oct. 18	I. A. (*T. C.*)
22 July 18	Baker, 2nd-Lt. R. H.	22 July 10	3 Dec. 18	I. A. R. O.
22 July 18	Smith, 2nd-Lt. A.	22 July 18	3 Dec. 18	I. A. (*T. C.*)
	Byerley, 2nd-Lt. R. H. . . .	. . . .	14 Oct. 18	York and Lanc. R., I. A. (*on probn.*)
17 Aug. 18	Ward, 2nd-Lt. T. E.	17 Aug. 18	30 Aug. 18	I. A. R. O.

		Jemadar.	Subadar.	Subadar-Major.	
	Subadar-Major.				
18 Nov. 92	Moti Rawat	22 Oct. 03	8 May 09	15 June 17	(*temp.*)
	Subadars.				
6 Jan. 04	Chander Singh	15 Apl. 15	5 June 17		
3 Apl. 01	Sardara	1 July 16	5 June 17		
16 Sept. 00	Wali Mohammad	24 Nov. 16	5 June 17		
26 July 07	Muhammad Shafi	5 June 17	16 June 17		
	Jemadars.				
1 Dec. 03	Lal Singh	5 June 17			
9 Sep. '04	Khurshed Khan.	5 June 17			
1 Aug. 03	Ghulam Ali Khan	5 June 17			
12 Mar. 07	Lumba Rawat	5 June 17			
7 July 06	Panna Rawat	5 June 17			
23 Jan. 95	Pandurang Vithal	15 June 17			
23 Dec. 03	Sheo Bakhsh	15 June 17			
25 Oct. 09	Ram Jiwan	15 June 17			
23 Apl. 03	Moti Singh	16 June 17			
15 Mar. 07	Umar Ali Khan	16 June 17			
26 May 09	Deda Ram	4 Dec. 17			
24 Nov. 03	Chhaiju Ram	1 Jan. 18			
13 Mar. 05	Chandar Singh	1 Jan. 18			
4 Sep. 07	Faryand Khan	1 Jan. 18			
4 May 10	Ram Lal.	1 Jan. 18			
1 June 12	Taj Mohammud	1 Jan. 18			

Regt. Centre—Mhow.
Linked with—119th Infy. & 122nd Infy.

Uniform—Scarlet.
Facings—Yellow.

120th Rajputana Infantry.

Raised at Baroda in 1817, by Lieut.-Col. D. S. Fallon, as the 2nd Battalion, 10th Regiment of Bombay Native Infantry. Became the 20th Regiment of Bombay Native Infantry in 1824; the 20th Regiment of Bombay Infantry in 1885; the 20th Bombay Infantry in 1901; present designation in 1903. Reformed on January 1st, 1917.

"Reshire" "Bushire" "Koosh-ab" "Persia."

Composition—1 company of Rajputana Gujars, 1 of Mers, 1 of Rajputana Rajputs, 1 of Hindustani Musalmans.

COLONEL—

1st Commn. or date of entering service.	Names and Rank.	Army rank.	Present appointment in Regiment.	Remarks.
	Commandant.			
21 Dec. 89	Ward, Lt.-Col. C. H.	21 Dec. 15	11 Jan. 17	*Lv. m. c. ex. I.*
	Company Commanders. (4)			
25 Mar. 91	*Pocock, Lt.-Col. P. F.*, D.S.O. *(Q)*	*25 Mar. 17*	*3 Mar. 14*	*2nd-in-Comd.*
21 Apl. 00	Mainwaring-White, Maj. E. P. (*temp. Lt.-Col.*)	11 Jan. 16	30 May 12	
	Company Officers. (8)			
24 Jan. 06	*Miskin, Capt. W. L.*	*24 Jan. 15*	*19 Feb. 08*	
3 Sep. 10	Mosse, Capt. C. O. R., M.C.	1 Sep. 15	1 Apl. 12	
18 Jan. 11	*Rich, Capt. H. H.*	*1 Sep. 15*	*1 Apl. 12*	
14 Jan. 14	Christian, Capt. H. N.	14 Jan. 18	21 Oct. 14	
15 Nov. 15	Pearse, Lt. T. F.	15 Nov. 16		
3 May 15	Martin, Lt. E. H. deC.	3 May 16	. . . .	Attd. With 113th Infy.
12 June 15	Richards, Lt. W. N.	12 Mar. 17	27 June 17	I. A. (*on probn.*)
27 Oct. 16	Linney, Lt. W. W.	27 Oct. 17	3 Jan. 17	
1 May 17	Carter, Lt. W. T.	1 May 18	2 Dec. 17	
1 May 17	Craft, Lt. H. T.	1 May 18	2 Dec. 17	
31 Aug. 18	Pryce, 2nd-Lt. W. W. R.	31 Aug. 18	13 Sep. 18	Attd.
	Attached.			
27 July 93	Barr, Maj. J. H. V.	1 Sep. 15	4 Mar. 18	109th Infy. Comdt. (*temp.*)
29 Jan. 10	Kane, Capt. J. L. K.	1 Sep. 15	13 June 18	109th Infy.
18 Jan. 11	Shewen, Capt. D G. P. M. (*actg. Maj.*)	1 Sep. 15	. . . .	27th Punj.
15 Sep. 15	Howitt, Lt. G. N.	15 Sep. 16	6 Aug. 18	108th Infy.
8 Jan. 15	Hunnybun, Lt. C.	8 Jan. 16	27 Sep. 17	I. A. R. O.
29 Jan. 15	*Lecky, Lt. R. L.*	*29 Jan. 16*	. . . .	I. A. R. O.
2 Apl. 15	Faithfull, Lt. F. E. N. (*actg. Capt.*)	2 Apl. 16	20 May 15	I. A. R. O.
15 Sep. 16	Elmslie, Lt. C. S.	15 Sep. 17	7 Mar. 17	I. A. R. O.
11 Aug. 15	Trow, Lt. V. H. (*temp. Capt.*)	11 Aug. 16	16 Feb. 18	Wilts, R., I. A. (*on probn.*) Platoon Comdr.
23 Apl. 15	Howdle, Lt. T. G.	23 Jan. 17	8 Apl. 17	Qr.-Mr. 13 Aug. 18.
4 Aug. 15	Carr, Lt. D. G.	11 May 17	1 Oct. 17	I. A. (*on probn.*)
4 Mar. 16	Abernethy, Lt. R. C.	9 Oct. 17	11 Dec. 17	I. A. R. O. (*on probn.*)
29 Aug. 17	McQueen, Lt. J. A.	29 Aug. 18	17 Jan. 18	I. A. R. O.
31 Oct. 17	FitzGerald, 2nd-Lt. J. J.	31 Oct. 17	15 Feb. 18	I. A. R. O. Adjt.
19 Nov. 17	Smithson, 2nd-Lt. E. A.	19 Nov. 17	28 Dec. 17	I. A. R. O. (*on probn.*)
28 Nov. 17	Jones, 2nd-Lt. B. C.	28 Nov. 17	2 Feb. 18	I. A. R. O. (*on probn.*)
25 July 17	Gregson, Lt. H. (46)	25 July 18	2 June 18	M. G. Corps., I. A. (*on probn.*)
25 Oct. 17	Piper, Lt. A. B. C.	25 Oct. 18	2 June 18	D. of Corn. L. I., I. A. (*on probn.*)
15 Jan. 18	Fitzpatrick, 2nd-Lt. J. J.	15 Jan. 18	5 June 18	I. A. R. O.
15 Jan. 18	Darnell, 2nd-Lt. W. A.	15 Jan. 18	5 June 18	I. A. (*T. C.*)
15 Jan. 18	Ford, 2nd-Lt. F. E.	15 Jan. 18	5 June 18	I. A. (*T. C.*)
27 Mar. 18	Reade, 2nd-Lt. P. V. J.	27 Mar. 18	2 June 18	R. Innis. Fus., I. A. (*on probn.*)
29 Mar. 18	Carless, 2nd-Lt. H. A.	29 Mar. 18	17 July 18	I. A. R. O.
2 June 18	Goodwin, 2nd-Lt. J.	2 June 18	13 Oct. 18	I. A. (*T. C.*)
10 Aug. 18	Chitts, 2nd-Lt. T. A. O.	10 Aug. 18	6 Sep. 18	I. A. R. O.
6 Sep. 18	Fleming, 2nd-Lt. M. J.	6 Sep. 18	27 Sep. 18	I. A. R. O.

120th Rajputana Infantry—*contd.*

1st Commn. or date of entering service.	Names and Rank.	Army rank.	Present appointment in Regiment.	Remarks.
		Jemadar.	Subadar.	Subadar-Major.
	Subadar-Major.			
13 July 87	Khitab Gul, *Sardar Bahadur*, I.O.M. (56).	19 Dec. 91	1 May 95	19 July 01 *Order of Br. Ind., 1st Cl.*
	Subadars			
15 Nov. 95	Dunga Rawat, I.O.M. (56)	16 July 01	4 Mar. 08	
12 Sep. 06	Sri Bahadur Singh *Bahadur* (56)	13 Sep. 06	1 Oct. 11	*Order of Br. Ind., 2nd Cl.*
12 May 99	Pitha Rawat (56, 72*a*), I.D.S.M.	1 Apl. 04	8 Sep. 14	
21 Apl. 01	Shaikh Muhammad Umar (56)	1 Mar. 08	15 June 15	
20 Mar. 03	Seole	14 Apl. 15	1 Jan. 17	
	Narayan Singh (*Retd.*)	.	. " .	Re-empld, 5 Feb. 16.
	Kurra (*Retd.*)	. . .	. . .	Re-empld.
17 Mar. 08	Hamid Husain	1 Jan. 17	1 Dec. 17	
7 Aug. 05	Zauki Ram	1 Jan. 17	23 Mar. 18	
	Jemadars.			
4 Nov. 99	Tej Singh (56)	1 Oct. 11		
27 May 04	Phula Ram (56)	8 Sep. 14		
10 May 06	Nara Rawat	8 Sep. 14		
14 Aug. 94	Indar Singh	14 Apl. 15		
14 Jan. 08	Mag Singh	25 July 15		
14 Sept. 03	Ganes	25 July 16		
22 July 04	Abdur Rahaman (46)	1 Jan. 17		
9 Sep. 01	Rahamat Khan	1 Jan. 17		
10 May 05	Siunath Singh	1 Jan. 17		
28 Jan. 99	Lakha Rawat	1 Jan. 17		
1 Jan. 17	Likhi Ram	1 Jan. 17		
9 May 00	Ashgar Ali, I.D.S.M.	1 Dec 17		
21 June 09	Govind Singh	1 Dec. 17		
20 Nov. 07	Jowhar Singh	1 Dec. 17		
23 Aug. 04	Balu Ram	1 Dec. 17		
1 June 03	Inda Rawat	1 Dec. 17		
22 Aug. 06	Lachhman Singh	1 Jan. 18		
	Mumtaz Khan	17 Aug. 15		*Temp.*
4 Nov. 03	Gainda	23 Mar. 18		
15 June 08	Gangadhar, I.D.S.M.	23 Mar. 18		
27 May 12	Ganga Ram	23 Mar. 18		
3 Apl. 01	Ramnand	27 Mar. 18		
6 Jan. 12	Harbhan	5 June 18		
28 June 09	Mobta	27 Mar. 18		
15 Aug. 06	Balu Singh	27 Mar. 18		
14 Mar. 12	Afzal Khan	27 Mar. 18		
20 May 04	Shaikh Sultan	24 May 18		

Regt. Centre—Kirkee. *Uniform*—Scarlet.
Linked with—107th & 128th Pioneers. *Facings*—White.

121st Pioneers.

The Marine Battalion.—Raised at Bombay, in 1777, as the Marine Battalion. Became the 1st (or Marine) Battalion, 11th Regiment of Bombay Native Infantry, 1818; the 21st or Marine Bn., 1824; "The Marine Battalion," 1824, the 21st Regt. of Bombay Native Infantry (The Marine Battalion), 1851; the 21st Regiment of Bo. Infantry (the Marine Battalion), 1885; the 21st Bombay Infantry (the Marine Battalion), 1901; the 21st Bombay Pioneers, 1903; present designation, 1903.

AN ANCHOR AND LAUREL WREATH, with Motto in Hindustani corresponding with "*Per Mare per Terram.*"
"Persian Gulf" Beni Boo Alli" "Burma" "Aden"
"Hyderabad" "Punjab" "Abyssinia."

Composition—1 company of Dekhani Mahrattas, 1 of Rajputana Mussalmans, 1 of Western Rajputana Jats, 1 of Pathans.

1st Commn. or date of entering service.	Names and Rank.	Army rank.	Present appointment in Regiment.	Remarks.
	Commandant.			
18 June 92	Keelan, Lt.-Col. H. P.	3 June 16	14 Aug. 16	
	Company Commanders. (4)			
19 Nov. 92	Greig, Maj. J. G., C.I.E. [*Lt.-Col.*]	19 Nov. 10	24 Dec. 17	2nd-in-Comd.
5 Aug. 96	*Ellis, Maj. A. C. S. B., p.s.c.) temp.*	*5 Aug. 14*	*1 Jan. 10*	*G. S. O., 1st gr., A. Hdqrs.*
17 Feb. 00	*Baldwin, Maj. F. E. W., q.*	*1 Sep. 15*	*1 Jan. 10*	*D. A. A. G.*
8 July 00	Hathornthwaite, Maj. T. C., *q.*	1 Sep. 15	14 Aug. 16	
	Company Officers. (9)			
5 Jan. 01	Tugwell, Maj. W. B. P., *q*	5 Jan. 16	24 Dec. 03	
8 Jan. 01	*Harley, Maj. A. B., D.S.O., q.* [*L*]	*8 Jan. 16*	*1 Sep. 06*	*D. A. A. G.*
8 Jan. 03	Sheringham, Maj. A. T., D.S.O., *q.* [L]	19 Dec. 15	11 Aug. 05	With 2-155th Prs.
8 May 01	Cooper, Maj. G. C., M.C., *q.* [L]	8 May 16	16 Nov. 02	With 2-107th Prs.
18 Jan. 02	Heyland, Maj. G. D., *q.*	18 Jan. 17	22 July 04	With 2-155th Infy.
3 Aug. 04	Gover, Capt. A. C., M.C., *q.*	13 Aug. 13	7 Nov. 05	With 2-12th Prs.
5 Aug. 05	Bevington, Capt. L. F., q.	5 Aug. 14	7 Oct. 08	
25 July 06	*Todd, Capt. E. T. T., q.*	*25 July 15*	*19 June 10*	*D. A. Q. M. G.*
8 Sep. 09	Ozanne, Capt. E. H. B., *q.*	1 Sep. 15	31 Dec. 10	
13 Jan. 14	Pratt, Capt. M. P.	13 Jan. 18	13 Nov. 15	*Lv. m.c., up to 13 Oct. 19.*
14 July 15	Baines, Lt. R. H., M.C.	14 July 16	27 Oct. 15	
18 Apl. 16	Hasler, Lt. H. J.	16 July 16	27 Apl. 16	
	Attached.			
27 Nov. 14	Shove, Capt. N. A.	27 Nov. 18	28 Sep. 17	I. A. R. O.
31 Dec. 14	Hill, Lt. L. E.	31 Dec. 15	16 Dec. 16	I. A. R. O.
5 May 15	Whitehill, Lt. R. D. (*actg. Capt.*)	5 May 16	....	R. War. R., I. A. (*on probn.*) Adjt.
9 July 15	Youngman, Lt. J. S.	9 Apl. 17	20 Sep. 15	
15 Nov. 15	Plomer, Lt. G.	15 Nov. 16	28 Sep. 17	107th Prs.
1 Jan. 16	Tilley, Lt. J. S.	1 Jan. 17	1 Mar. 16	I. A. R. O.
8 Feb. 16	Kirkman, Lt. H. G.	8 Feb. 17		R. Innis. Fus., I. A. (*on probn.*)
12 May 16	Bird, Lt. M.	12 May 17	....	I. A. (*on probn.*)
29 June 16	Worlledge, Lt. L. H.	29 June 17	3 July 16	
14 June 16	Ward, Lt. J.	14 June 17	20 July 16	I. A. R. O.
18 Aug. 16	Clark, Lt. W. V.	1 Apl. 17	28 Apl. 17	107th Prs.
14 Nov. 16	Shaw, Lt. B. V. D. H.	14 Nov. 17	24 Nov. 17	107th Prs.
5 Aug. 18	Birkett, Lt. S.	5 Aug. 17	1 Dec. 17	I. A. R. O.
1 Feb. 18	Bradshaw, 2nd-Lt. G. T.	1 Feb. 18	7 Dec. 17	Cav. Reserve Regt., I.A. (*on probn.*)
28 Oct. 17	Parks, Lt. H. J., M.M.	28 Oct. 18	12 Jan. 18	York. R., I. A. (*on probn.*)
29 Mar. 18	Hisco, 2nd-Lt. R.	29 Mar. 18	15 July 18	I. A. R. O.
30 Aug. 18	Chorley, 2nd-Lt. R. C.	30 Aug. 18	13 Sep. 18	I. A. R. O.
1 Oct. 18	Grange, 2nd-Lt. D. G.	1 Oct. 18	....	I. A. R. O.
		Jemadar.	Subadar.	Subadar-Major.
	Subadar-Major.			
6 Nov. 90	Sardar Khan (56)	24 Sep. 01	1 Apl. 07	1 July 14
	Subadars.			
....	Chand Khan	....	20 Mar. 14	Re-empld.
....	Juma Bux Khan (*retd.*)	....	15 Oct. 18	Re-empld.
16 Nov. 87	Mohomed Khan (retd.)	1 July 02	....	Re-empld.
12 Sep. 02	Falel Khan (56)	9 July 04	1 Jan. 10	With 2-155th Infy.
13 June 95	Ruda Ram (56, 66)	10 June 10	7 July 12	
5 Mar. 97	Ewaz Khan (56)	10 Oct. 08	17 Oct. 14	With 1-155th Prs.
15 Sep. 99	Kheta Ram (50, 52, 66*a*) (50, 52)	7 July 12	4 Jan. 15	
8 June 97	Aba Duduskar (56)	1 July 14	25 Apl. 15	
16 July 98	Mahtab Khan	18 Oct. 14	1 Aug. 15	
20 July 00	Vithoba Jadhow	25 Apl. 15	20 Nov. 15	
1 Apl. 04	Basti Khan	23 Apl. 15	15 Dec. 15	With 2-155th Infy.
4 Feb. 98	Daula Ram	....	24 Oct. 16	
12 July 98	Chand Khan	15 Apl. 13	23 Aug. 17	With 128th Prs.
26 June 01	Laximan Jankar	28 Dec. 14	7 Nov. 16	07th Prs. attd.
24 Aug. 05	Imam Khan	20 Nov. 15	13 June 18	

121st Pioneers—*contd.*

1st Commn. or date of entering service.	Names and Rank.	Army rank.	Remarks.
	Jemadars.		
12 Oct. 96	*Manulla (56)*	*1 Feb. 13*	*With 107th Prs.*
7 Mar. 98	*Yar Khan (56)*	*15 Apl. 13*	*With 107th Prs.*
3 Mar. 02	Mukarrab Khan (56)	30 Oct. 14	Offg. Indian Adjt.
21 Aug. 99	Gharsi Ram	Jan. 15	
1 Aug. 98	Tirka Singh	4 Jan. 15	
27 Jan. 00	Narayan Singh	4 Jan. 15	
4 Oct. 96	Chahat Khan	23 Feb. 15	
31 Apl. 96	Shaikh Ismail	19 May 15	
27 Feb. 99	Saiyed Ahmed	19 May 15	
26 June 99	Maroti Jadhow	22 Oct. 15	
25 Jan. 98	Magha Ram	20 Nov. 15	
4 June 01	Dhane Khan	20 Nov. 15	
4 Apl. 05	Salim Khan	20 Nov. 15	
10 Sep. 00	Umar Khan	20 Nov. 15	
21 Mar. 01	*Sharif Khan*	*20 Nov. 15*	*With 107th Prs*
16 May 08	Bikram Singh	20 Jan. 16	
10 Feb. 98	Sewa Ram	12 July 16	
19 Aug. 96	Bostan Khan	1 Dec. 16	With 2-155th Infy.
15 Sep. 07	Ganga Ram	1 May 18	
15 Jan. 05	Tukaram Dharo	1 May 18	
25 Dec. 06	Nazru Khan	1 May 18	
4 Mar. 03	Makhmul Khan	11 Mar. 18	With 128th Prs.
24 Mar. 04	Naurang Khan	11 Mar. 18	With 128th Prs.
1 July 01	Karim Dad Khan	21 Aug.	107th Prs. attd.
5 Apl. 02	Zaidulla Khan	12 June 18	With 2-155th Infy.
10 Nov. 06	Shahbaz Khan	13 June 18	
25 Aug. 02	Ramji Lal, I.D.S.M.	13 June 18	
11 Apl. 06	Shankar Baber	13 June 18	
1 Apl. 05	Antu Burge	20 June 18	
11 Feb. 10	Phul Khan	20 Aug. 18	

Regt. Centre—Mhow.
Linked with—119th and 120th Infy.

Uniform—Scarlet.
Facings—Emerald Green.

122nd Rajputana Infantry.

Raised at Bombay in 1818 by Lieut.-Col. John Cunningham, as the 2nd Battalion, 11th Regiment of Bombay Native Infantry. Became the 22nd Regt. of Bombay Native Infantry in 1824; the 22nd Regt. of Bombay Infantry in 1885; the 22nd Bom. Infy. in 1901; present designation in 1903.

"China, 1900."

Composition—1 company of Rajputana Gujars, 1 of Mers, 1 of Rajputana Rajputs, 1 of Hindustani Musalmans.

1st Commn. or date of entering service.	Names and Rank.	Army rank.	Present appointment in Regiment.	Remarks.
	Commandant.			
29 Nov. 90	Barclay, Lt.-Col. P. C. R.	29 Nov. 16	2 Apl. 17	
	Company Commanders. (4)			
13 July 92	*Calvert, Lt.-Col. R. T. C., p.s.c. (Q.)*	*13 July 18*	*11 Nov. 18*	*2nd.in-Comd. A.Q.M.G., 5th Divn.*
20 Jan. 97	Hutchison, Maj. G., *p.s.c.*	20 Jan. 15	1 Apl. 10	With 84th Punj.
23 July 01	Holmes, Maj. H. L.	23 July 16	17 July 16	
15 July 03	*Holliday, Maj. J. C. H.*	*15 July 18*	*9 Aug. 16*	*A. Q. M. G.*
	Company Officers. (8)			
13 Aug. 04	Middlemass, Capt. G. de S. H.	13 Aug. 13	1 Mar. 06	With 2-22nd Punj.
16 Aug. 05	Staples, Capt. E. H.	16 Aug. 14	20 June 08	With 2-96th Infy.
9 Sep. 08	Skinner, Capt. A. L.	1 Sep. 15	10 Dec. 09	
20 Jan. 09	Grant, Capt. F. R. E.	1 Sep. 15	29 Mar. 10	With 1-18th Infy.
22 Jan. 13	Stable, Capt. R. H., D.S.O.	22 Jan. 17	17 Mar. 14	
15 Nov. 15	Anderson, Lt. D. C.	15 Nov. 16	18 Nov. 15	Attd. With 2-154th Infy.
25 Mar. 16	O'Leary, Lt. M. P.	25 Dec. 17	3 Nov. 16	Attd.
30 Jan. 17	Kilroy, Lt. A. J. M.	30 Jan. 18	. . .	With 13th Rajputs.
1 May 17	Croucher, Lt. F. E.	1 May 18	30 Nov. 17	Attd. With 9th Horse.
31 Jan. 18	Larminie, 2nd-Lt. J. A.	31 Jan. 18	16 Oct. 18	Attd.
31 Aug. 18	Collier, 2nd-Lt. W. J.	31 Aug. 18	13 Sep. 18	Attd.
	Attached.			
7 May 02	Hickie, Maj. C. C.	7 May 17	27 Dec. 14	Civil.
30 Apl. 15	Barnes, Lt. W. C.	30 Apl. 16	18 Sep. 18	I. A. R. O.
19 Jan. 16	Shaw, Lt. J. F.	19 Jan. 17	4 Mar. 15	I. A. R. O. Adjt.
11 May 16	Gibson, Lt. F. C.	11 May 17	11 July 17	11th Welsh R., I. A. (*on probn.*)
10 June 16	Herne, Lt. R. V.	10 June 17	10 June 17	I. A. R. O.
28 June 16	James, Lt. W. T.	28 June 17	29 June 17	I. A. R. O.
25 Aug. 16	Simon, Lt. A. O.	25 Aug. 17	6 Sep. 18	I. A. R. O.
30 Jan. 17	Levingston, Lt. T. C.	30 Jan. 18	24 May 17	119th Infy.
30 Jan. 17	Forde, Lt. J. S. B.	30 Jan. 18	24 May 17	119th Infy.
8 Aug. 17	Petterson, Lt. S.	8 Aug. 18	11 Nov. 17	I. A. R. O.
9 Oct. 17	Pierce, 2nd-Lt. H. J. J.	9 Oct. 17	8 Dec. 17	I. A. R. O.
9 Oct. 17	Banwell, 2nd-Lt. J. A.	9 Oct. 17	22 Dec. 17	I. A. R. O.
19 Mar. 18	Ludlow, 2nd-Lt. H.	19 Mar. 18	17 July 18	I. A. (*T. C.*)
1 May 18	O'Reilly, 2nd-Lt. R.	1 May 18	3 Aug. 18	I. A. R. O.
15 July 18	Keir, 2nd.Lt. C. McH.	15 July 18	11 Aug. 18	I. A. R. O.
11 Aug. 18	Newman, 2nd-Lt. C. E. F.	11 Aug. 18	15 Oct. 18	I. A. R. O.
20 Aug. 18	Belton, 2nd-Lt. P.	20 Aug. 18	1 Oct. 18	I. A. R. O.
24 Sep. 18	Perkin, 2nd-Lt. C. R.	24 Sep. 18	16 Oct. 18	I. A. R. O.

		Jemadar.	Subadar.	Subadar-Major.
	Subadar-Major.			
31 Dec. 02	Sujan Singh (56, 72*a*)	31 Dec. 02	16 July 05	1 June 14
. . . .	Shahbaz Khan (*Ret.*)	. . . .	. . . .	Re-empld.
	Subadars.			
23 July 97	Behara Rawat	1 Oct. 03	16 Nov. 17	
20 Apl. 98	Pyar (72*a*)	1 Apl. 13	1 June 14	With 120th Infy.
18 June 04	Zahirud-din Khan	18 June 04	27 July 15	
22 Sep. 01	Mada Rawat	1 Aug. 14	27 July 15	
1 Oct. 95	Hira Rawat	15 June 11	15 Dec. 15	With 119th Infy.
17 July 97	Atbal Singh	1 Aug. 12	14 Jan. 16	
1 Dec. 97	Bhima Rawat (56*a*)	19 Nov. 14	10 July 16	
16 Jan. 03	Jalim	1 Mar. 15	30 Sep. 16	With 119th Infy.
23 Nov. 01	Bhim Singh (72*a*)	7 Aug. 05	1 Aug. 16	
10 Nov. 99	Kutbi Khan	1 Dec. 15	7 Nov. 16	
21 June 03	Angad Ram	8 Aug. 16	7 Apl. 17	Transfd. to 2-154th Infy
4 Jan. 05	Lala Rawat	11 May 15	4 June 17	With 120th Infy.
31 May 99	Gena Rawat	4 Nov. 16	26 Jan. 18	With 119th Infy.
19 Jan. 10	Taj Muhammad Khan	12 Feb. 16	2 Mar. 18	
21 June 97	Teja Rawat	27 July 15	18 May 18	
27 Jan. 06	Harnath Singh	1 Dec. 15	18 May 18	
24 July 02	Gopal Singh (56)	6 Nov. 14	11 Sep. 18	
18 Mar. 99	Net Ram	1 Dec. 15	11 Sep. 18	

122nd Rajputana Infantry—*contd.*

1st Commn. or date of entering service.	Names and Rank.	Army rank.	Remarks.
	Jemadars.		
25 May 03	Budha Ram	27 July 15	
4 Apl. 05	Lakha Rawat	10 July 16	
20 June 04	Abdul Ghani Khan	20 Sep. 16	
19 May 05	Nazar Muhammad Khan	20 Sep. 16	
7 May 94	Hari Lakshman	9 Oct. 16	
17 Dec. 05	Abdul Ghani Khan	1 May 17	
6 May 07	Kalla Rawat	29 June 17	
7 Aug. 14	*Mana Rawat*	*29 June 17*	*With 119th Infy.*
16 Nov. 00	Sham Lal Singh	3 Dec. 17	
12 May 08	*Nathan*	*15 Dec. 17*	*With 119th Infy.*
22 Jan. 05	Bhawarlal Singh	25 Mar. 18	
3 Mar. 01	Nanga Rawat	18 May 18	
27 Feb. 07	Rama Rawat	18 May 18	
6 Aug. 05	Bagh Singh	18 May 18	
29 Aug. 05	Kishan Singh	18 May 18	
2 Mar. 01	Pura Rawat	11 Sep. 18	
6 Nov. 14	Chunna Ram	11 Sep. 18	

Regt. Centre—Nasirabad.
Linked with—104th & 125th Rifles.

Uniform—Dark Green.
Facings—Scarlet.

1st Battalion, 123rd Outram's Rifles.

Raised, in 1820, to a considerable extent from men who had served at the battle of Kirkee in the disbanded Dapuri Battalion (Peishwa's service) under Major Ford, Madras Army, and was originally designated the 1st Bn., 12th Regt. of Bombay Native Infantry. Became the 23rd Regt. of Bo. N. I., 1824; the 23rd Regt. of Bo. Native (Light) Infy., 1841; the 23rd Regt. of Bom. (Light) Infy., 1885; unauthorisedly designated the 2nd Battalion, Bo. Rifle Regt., 1889; title corrected to 23rd Regt. (2nd Battn., Rifle Regt.) of Bombay Infantry, 1889; became the 23rd Bombay Rifles, 1901; present designation, 1903.

"Kirkee" "Persia" "Afghanistan, 1879-80" "Burma, 1885-87."

Composition—2 companies of Rajputana Jats, 1 of Rajputana Rajputs, 1 of Punjabi Musalmans.

COLONEL—Major-General *Sir* John Gatacre, K.C.B. . 13 May 1904.

1st Commn. or date of entering service.	Names and Rank.	Army rank.	Present appointment in Regiment.	Remarks.
	Commandant.			
24 Apl. 89	Cassels, Lt.-Col. G. R., *p.s.c.*	24 Apl. 15	13 Jan. 17	
	Company Commanders. (4)			
2 May 91	Blair, Maj. R. W. C. (Q) [*l*]	2 May 09	26 Sep. 15	2nd-in-Comd.
23 Sep. 96	Hardie, Maj. G. E. (*actg. Lt.-Col.*)	23 Sep. 14	13 Jan. 12	With 2-6th Jat L. I.
11 May 00	Hunt, Maj. R. V.	1 Sep. 15	24 Sep. 16	
8 July 00	Norris, Maj. A. K.	1 Sep. 15	2 Dec. 17	
	Company Officers. (9)			
11 Jan. 03	*Rae, Maj. J. G. (actg. Lt.-Col.)*	*21 Jan. 18*	*20 Apl. 04*	*I. S. Troops.*
3 Aug. 04	Tilly, Capt. R., M.C.	13 Aug. 13	27 Dec. 05	
21 Aug. 06	Moor, Capt. C. F. F., M.C.	29 Aug. 15	13 Mar. 09	Adjt.
19 Sep. 08	Ffrench, Capt. R. P. T., M.C.	1 Sep. 15	18 Mar. 14	
5 Oct. 10	Chambers, Capt. G. H., M.C.	1 Sep. 15	18 Mar. 1	Inspr. Phys[l]. & Bayonet. Tr.
22 Jan. 13	Stuart, Capt. L. A.	22 Jan. 17	29 Apl. 14	Qr.-Mr.
16 Sep. 14	*Hodgen, Capt. G. W.*	*16 Sep. 18*	*11 Nov. 15*	
4 Aug. 15	Page, Lt. H. V. S.	4 Aug. 16	4 Jan. 16	Attd. With 3-153rd Rif.
16 Aug. 16	Sansom, Lt. H. J.	16 Aug. 17	11 Nov. 16	Attd.
16 Aug. 16	Jones, Lt. L.	16 Aug. 17	10 Nov. 16	Attd.
23 June 16	Huntley, Lt. W. B.	23 June 17	. . . .	Attd.
14 Nov. 16	James, Lt. W. A. L.	4 Jan. 17	14 Nov. 16	
1 Jan. 17	Craig, Lt. S. M. M.	1 Jan. 18	. . . .	Attd. With 3-153rd Rif.
18 June 17	Belben, Lt. T. A.	18 June 17	. . . .	With 1-96th Infy.
20 Nov. 16	Wrigley, 2nd-Lt. H., M.C.	20 Nov. 16	. . . .	Attd.
	Attached.			
15 May 15	D'Albani, Lt. C. L.	15 May 16	15 Aug. 15	Suff. R., I. A. (*on probn.*)
26 June 15	Steele, Lt. J. S., M.C. (*actg. Capt.*)	26 June 16	. . . .	R. Ir. Rif., I. A. (*on probn.*) Adjt.
9 Dec. 14	Mc'Entire, Lt. H. S. M. (*temp. Capt.*)	9 Dec. 15	9 Dec. 15	I. A. R. O. Platoon Comdr.
21 May 15	Rhodes, Lt. C. K.	21 May 16	25 July 15	I. A. R. O.
29 May 15	Martin, Lt. J.	29 May 16	. . . .	High. L. I., I. A. (*on probn.*)
. . . .	Charmichael, 2nd-Lt. W. B. G.	. . . .	. . . .	Attd.
28 Aug. 15	Aird-Smith, 2nd-Lt. W.	28 Aug. 15	28 Aug. 15	I. A. R. O.
8 Jan. 17	Mack, Lt. C. H.	8 Jan. 17	6 Jan. 17	I. A. R. O.
15 Feb. 16	Morrison, Lt. A.	15 Feb. 17	. . . .	I. A. R. O.
18 Apl. 16	Ambrose, Lt. R. D., M.C.	18 Apl. 17	. . . .	104th Rifles.
4 July 16	Herring, Lt. T. C., M.C.	4 July 17	27 Aug. 18	E. York. R., I. A. (*on probn.*)
8 Sep. 16	Grant, Lt. R.	8 Sep. 17	. . . .	I. A. R. O.
10 Jan. 17	Tryhorne, Lt. M. W.	10 Jan. 18	12 Jan. 17	I. A. R. O.
19 Dec. 16	Tregaskis, Lt. O.	19 Dec. 17	19 Dec. 17	I. A. R. O.
19 Dec. 16	Drake, Lt. P. C.	19 Dec. 17	19 Dec. 17	I. A. R. O.
26 Apl. 17	Burgess, Lt. W. G. A.	26 Apl. 18	. . . .	I. A. R. O.
20 June 17	Peppain, 2nd-Lt. S. H.	20 Jan. 17	. . . .	I. A. R. O.
8 Aug. 17	Dobbie, 2nd-Lt. T. B.	8 Aug. 17	7 Aug. 18	A. S. C., I. A. R. O.
. . . .	Johnson, Lt. K. H.	. . . .		
. . . .	Hodgen, 2nd-Lt. H.	. . . .	. . . .	

1st Battalion, 123rd Outram's Rifles—*contd.*

1st Commn. or date of entering service.	Names and rank.	Army rank.	Present appointment in Regiment.	Remarks.
		Jemadar.	Subadar.	Subadar-Major.
	Subadar-Major.			
	Subadars.			
2 June 91	Jaita Ram (56)	16 Sep. 02	8 Dec. 03	Subr. Maj. (Temp.) 29 Dec. 15. With 2-123rd Rif.
1 Feb. 96	Bhura Ram, I.O.M. (56)	22 Mar. 08	14 May 08	
3 Oct. 95	Nizam-ud-din (56)	1 Dec. 09	4 Apl. 15	
19 Oct. 02	Malooka Ram	2 Nov. 14	29 Dec. 15	
25 Oct. 12	Bagh Singh (56)	16 Jan. 15	29 Dec. 15	
4 June 05	Samrat Singh, I.D.S.M.	21 Apl. 15	29 Dec. 15	
7 Mar. 04	Jalu Ram (56)	25 Aug. 15	18 Jan. 17	
16 Oct. 00	Ume Singh	22 Sep. 14	18 July 17	
22 Feb. 99	*Partab Singh*	*8 Jan. 16*	*6 May 17*	*With 125th Rif.*
1 Sep. 00	*Jhuja Ram (56)*	*7 Sept. 14*	*24 Feb. 17*	*Do.*
16 Nov. 03	Harnath Singh, M. C., I.O.M.	29 Dec. 15	8 Sep 17	
24 Nov. 00	Burhan Ali	3 Sep. 17	13 Nov. 17	
17 Dec. 00	Bhola Ram	29 Dec. 15	18 Nov. 17	
5 July 05	Bagga Khan	29 Dec. 15	18 Nov. 17	
21 June 05	Shiunarayan Ram	29 Dec. 15	23 Nov. 17	
11 Sep. 00	Hans Ram, I.O.M.	6 Oct. 17	11 Apl. 18	
1 May 99	Nijabut Ali	3 Dec. 17	12 Apl. 18	
1 Dec. 02	Painda Khan	8 Mar. 15	12 Apl. 18	
5 July 09	Baz Khan, I. D. S. M.	19 Mar. 18	22 May 18	
	Jemadars.			
18 Aug. 04	Murli Singh	5 Apl. 15		
1 Nov. 06	Motha Khan	15 Oct. 15		
24 July 03	Pailad Ram	29 Dec. 15		
24 Dec. 03	Ramkaran Ram (66)	29 Dec. 15		
11 Oct. 02	Sukh Singh	23 Sep. 16		
	Goru Ram	1 Dec. 16		
12 Aug. 16	Kasim Khan (56)	26 Jan. 17		
20 Feb. 01	Nop Singh (56)	18 Jan. 17		
16 July 04	Dhonkal Ram	18 Jan. 17		
17 July 05	Bhim Singh, I.O.M.	18 July 17		
18 Aug. 04	Mul Singh	8 Sep. 17		
7 Dec. 00	Dana Ram	18 Nov. 17		
2 July 04	Mardan Ali	19 Mar. 18		
23 Oct. 12	Fateh Muhammad	19 Mar. 18		
25 Oct. 00	Arjan Ram, I.O.M.	11 Apl. 18		
19 July 04	Misri Khan	11 Apl. 18		
1 May 99	Karam Dad	22 May 18		
6 Apl. 98	Ziara Khan	22 May 18		
	Fazal Dad	24 May 18		

2nd Battalion, 123rd Outram's Rifles.

Formed at Baroda on 29th December 1915.

Composition—2 companies of Punjabi Mussalmans, 1 of Rajputana Jats and 1 of Rajputana Rajputs.

1st Commn. or date of entering service.	Names and Rank.	Army rank.	Present appointment in Regiment.	Remarks.
	Commandant. (4).			
11 Aug. 00	Byrne, Maj. E. G. J. (*Actg. Lt.-Col.*)	3 June 15	20 May 18	104th Rif. (*temp.*)
	Company Commanders.			
29 Aug. 06	Skinner, Capt. E. M. (*Actg. Maj.*)	29 Aug. 15	11 Jan. 17	125th Rif. 2nd-in-Commd. (*temp.*)
14 Apl. 15	Haslehust, Lt. H. M. (*temp. Maj.*)	15 Jan. 16	12 Jan. 16	I. A. R. O.—D. R. O., Mah.
7 May 15	Twynam, Lt. H. J. (*temp. Capt.*)	7 May 16	4 Feb. 16	I. A. R. O.
29 May 15	Martin, Lt. J., M.C. (*temp. Capt.*)	29 May 16	12 Oct. 17	Attd.
	Company Officers.			
7 Aug. 15	Guy, Lt. K.	7 Aug. 16	29 Mar. 18	Lanc. Fus., I. A. (*on probn.*) Qr.-Mr.
20 Dec. 15	Shephard, Lt. F. E.	20 Dec. 16	4 Oct. 17	Attd.
20 May 16	Robinson, Lt. F. G.	20 May 16	22 Oct. 17	I. A. R. O.
27 June 16	Rayner, Lt. J., M.C.	27 June 17	27 Aug. 18	Tank Corps, I. A. (*on probn.*)
20 Nov. 16	Wrigley, Lt. H., M.C. (*Actg. Capt.*)	20 Nov. 17	12 Oct. 17	Aus. Impl. Force, I. A. (*on probn.*) Adjt., 3 Mar. 18.
30 Oct. 17	Kenward, 2nd-Lt. P. E.	30 Oct. 17	13 Feb. 18	I. A. R. O. (*on probn.*)
31 Oct. 17	Fox, 2nd-Lt. R. J.	31 Oct. 17	13 Feb. 18	I. A. R. O. (*on probn.*).
28 Nov. 17	Slatter, 2nd-Lt. J. C. (*temp. Capt.*)	28 Nov. 17	3 Feb. 18	I. A. R. O. (*on probn.*) A. R. T. O., Deolali.
21 Dec. 17	Forward, 2nd-Lt. F. C. M.	21 Dec. 17	21 Aug. 18	4-3rd G. R.
15 Jan. 18	Boddington, 2nd-Lt. E. S. St. C.	15 Jan. 18	22 Aug. 18	I. A. R. O.
26 Jan. 18	McKay, 2nd-Lt. T. A.	26 Jan. 18	26 Jan. 18	R. Highrs., I. A. (*on probn.*)
19 May 18	Knott, 2nd-Lt. V.	19 May 18	15 July 18	I. A. (*T. C.*)
22 July 18	Milne, 2nd.Lt. C. A.	22 July 18	3 Dec. 18	I. A. R. O.
31 July 18	Tatem, 2nd-Lt. E. G.	31 July 18	3 Dec. 18	Genl. Ljst., I. A. R. O. (*on probn.*)
1 Sep. 18	Greatwood, 2nd-Lt. H. E.	1 Sep. 18	4 Sep. 18	Attd.
11 Oct. 18	Shrader, 2nd-Lt. G. J.	11 Oct. 18	2 Nov. 18	I. A. R. O.
....	Humpherus, Lt. E. H.	....	14 Oct. 18	E. Kent. R., I. A. R. O. (*on probn.*)
....	Helson, Lt. G.	....	15 Oct. 18	Som. L. I.

		Jemadar.	Subadar.	Subadar-Major.	
	Subadar-Major.				
2 June 91	Jaita Ram (56)	16 Sep. 02	8 Dec. 03	29 Dec. 15	Temp.
	Subadars.				
14 Feb. 92	Rura Ram, I.D.S.M. (63)	8 Dec. 03	22 Mar. 08		
28 Nov. 95	Firoz Khan (I) (56)	1 Sep. 09	1 Dec. 09		
23 Aug. 02	Gobind Singh	5 Jan. 15	5 July 15		
2 July 02	Chandra Ram (56)	2 Nov. 14	11 Dec. 15		
15 Oct. 07	Firoz Khan (II)	15 Oct. 15	29 Dec. 15		
18 Jun. 04	Satar Muhammad	15 Oct. 15	29 Dec. 15		
	Jemadars.				
5 June 02	Nazam Khan (52)	1 Aug. 15			
11 Aug. 00	Gidha Ram (66a)	11 Dec. 15			
6 July 05	Jowara Ram	11 Dec. 15			
1 July 02	Ittoji Rao Patole	29 Dec. 15			
20 Aug. 08	Sher Muhammed (46)	29 Dec. 15			
18 Dec. 14	Jemadar Khan	3 Dec. 17			

Regt. Centre—Quetta.
Linked with—126th Baluch Infy.

Uniform—Drab (Red serge trousers).
Facings—Scarlet.

1st Battalion, 124th Duchess of Connaught's Own Baluchistan Infantry.

Raised, in 1820, by Capt. H. R. Deschamps, as the 2nd (Marine) Battalion, 12th Regiment of Bombay Native Infantry. Became the 2nd Batt., 12th Regt. of Bom. N. I., 1823; the 24th Regt. of Bom. N. I., 1824; the 24th Regt. of Bom. Infy., 1885; the 24th (Baluchistan) Regt. of Bom. Infy., 1891; the 24th (the Duchess of Connaught's Own Baluchistan) Regt. of Bom. Infy., 1895; the 24th (the Duchess of Connaught's Own Baluchistan) Infy., 1901; present designation 1903.

"Aden" "Central India" "Afganistan, 1879-80" "British East Africa, 1896."

Composition—1 Company of Hazaras, 1 of Punjabi Musalmans, ½ of Khattaks, ½ of Mahsud Wazirs, 1 of Sikhs other than Jat Sikhs.

COLONEL—General Sir Alfred A. Pearson, K. C. B. . . . 6 January 1905.

1st Commn. or date of entering service.	Name and Rank.	Army rank.	Present appointment in Regiment.	Remarks.
	Commandant.			
29 Oct. 90	Tarver, Bt.-Col. A. L., D.S.O., A. D. C. *p.s.c.*	3 June 18	22 June 18	G. S. O., 1st gr., Peshawar Dn.
	Company Commanders. (4)			
28 Oct. 91	Mitchell, Lt.-Col. W. J. (Q) . .	28 Oct. 17	21 Mar. 09	2nd-in-Comd., With 2-124th Infy.
28 June 93	Holbrooke, Maj. B. F. R. (Q) (*Actg Lt.-Col.*)	28 June 11	21 Mar. 09	With 3-124th Infy.
7 Dec. 95	Chauncy, Maj. C. H. K. (Q) .	7 Dec. 13	12 May 13	With 2-127th Bal.
	Company Officers. (9)			
22 Jan. 96	Macpherson, Maj. J., D.S.O. . .	22 Jan. 14	16 July 98	With 2-124th Infy.
26 June 01	*Stewart, Maj. A.E.*, M.C. . .	*1 Sep. 15*	*16 Aug. 03*	*With 3-124th Infy.*
28 July 00	*Rind, Bt. Lt.-Col. G. B. A., p.s.c. (temp. Lt.-Col.)*	*20 Dec. 18*	*15 May 03*	*A. A. G., A. Hdqrs*
4 May 01	MacMahon, Maj. G. D. R. . .	4 May 16	30 Sep. 04	With 1-140th Infy.
18 Jan. 02	Chaldecott, Maj. O. A. . . .	18 Jan. 17	15 Aug 03	With 3-124th Infy.
16 Aug. 05	Lynch-Staunton, Capt. H.V.C. .	16 Aug. 14	8 Mar. 09	With 2-124th Infy.
19 Jan. 07	Hinde, Capt. R. G.	1 Sep. 15	24 Mar. 08	With 1-129th Infy.
17 Aug. 07	Drought, Capt. J.V. (*temp. Maj.*)	1 Sep. 15	15 Oct. 09	With 2-130th Infy.
25 Jan. 08	Bacon, Capt. R. G.	1 Sep. 15	13 Mar. 09	With 71st Punj.
9 Oct. 15	Gardner, Capt. J. M. S. . . .	5 Aug. 18	1 Feb. 16	Attd. With 8-124th Infy.
15 Sep. 15	Lord, Lt. J. W.	15 Sep. 16	30 Oct. 15	Attd. With 3-124th Infy.
	Attached.			
5 Feb. 87	Twigg, Lt.-Col. E. F.	5 Feb. 13	22 June 18	Comdt. (*temp.*)
9 July 15	Kempson, Lt. C. J. (*Actg. Capt.*)	9 July 16	5 Sep. 16	I. A. R. O.
21 Oct. 16	Sutton, Lt. S. H. (*Actg. Capt.*) .	21 Oct. 17	24 June 17	I. A. R. O.
20 June 17	Allbrook, Lt. W. R.	20 June 18	10 Oct. 17	I. A. R. O. Qr. Mr.
20 June 17	Nuttall, Lt. N.	20 June 18	10 Oct. 17	I. A. R. O.
8 July 17	Quillinan, 2nd-Lt. L. P. (*Actg. Capt.*)	8 July 17	10 Oct. 17	I. A. R. O. Adjt.
9 Dec. 15	Tooth, Lt. H. C. A.	9 Dec. 16	24 Apl. 18	I. A. (*on probn.*)
29 Oct. 16	Smith, Lt. F. G.	29 Oct. 17	16 May 18	I. A. (*on probn.*)
2 Feb. 17	O'Halloran, Lt. H.	2 Feb. 18	16 May 18	I. A. (*on probn.*)
25 July 17	Seal, Lt, R. C.	25 July 18	1 Oct. 17	I. A. (*on probn.*)
15 Jan. 18	Drudge, 2nd-Lt. H. A. . . .	15 Jan. 18	. . . .	I. A. (*T. C.*)
7 June 18	Padgham, 2nd-Lt. T. H. . . .	7 June 18	. . . .	I. A. (*T. C.*)
15 June 18	Salmon, 2nd-Lt. F.	15 June 18	14 Sep. 18	Genl. List, I. A.
15 June 18	McAndrew, 2nd-Lt. A. . . .	15 June 18	14 Sep. 18	Genl. List, I. A. (*on probn.*)
1 Nov. 18	Bicker, 2nd-Lt. H.	1 Nov. 18	22 Mar. 18	I. A. (*T. C.*)
5 Nov. 17	Hasnip, Lt. G. H.	5 Nov. 18	22 Mar. 18	I. A. (*T. C.*)
. . . .	Watson, 2nd-Lt. H. S. . . .	. . . .	. . . .	Lanc. R., I. A. R. O. (*on probn.*)

		Jemadar.	Subadar.	Subadar. Major.
	Subadar-Major.			
1 Dec. 95	Mohammad Roshan	21 Jan. 15	12 July 15	1 June 18

1st Battalion, 124th Duchess of Connaught's Own Baluchistan Infantry—*contd.*

1st Commn. or date of entering service.	Names and Rank.	Army rank.		Remarks.
		Jemadar.	Subadar.	
	Subadars.			
29 Aug. 96	Sher Dil Khan	13 Feb. 15	13 July 15	
12 Aug. 03	Shahzada Khan	27 Apl. 15	19 May 16	
1 Sep. 07	Tiwana Khan	31 Dec. 15	23 Aug. 17	
14 May 05	Sher Muhammad	23 Aug. 17	1 Jan. 18	
1 May 17	Nasrullah Khan	......	......	Re-employed.
5 Sep. 14	Sukhmandar Singh	6 Nov. 17	1 Oct. 18	
	Jemadars.			
21 Mar. 02	Bab Din, I.D.S.M.	22 Jan. 16		
4 Feb. 02	Samand Khan	11 Aug. 16		
18 Oct. 06	Chaudri Khan	1 Jan. 18		
3 Apl. 11	Jan Muhammad	1 Feb. 18		
1 Sep. 07	Haidar Khan	15 Feb. 18		
28 Nov. 08	Chanan Singh	15 Feb. 18		
20 Apl. 05	Ghulam Muhammad	1 May 18		
20 Apl. 05	Zaman Khan	10 May 18		
	Muhammad Din	1 Sep. 18		

2nd Battalion, 124th Duchess of Connaught's Own Baluchistan Infantry.

Formed at Bushire, 11th August 1916.

Composition—2 Companies Sikhs, 2 Companies Punjabi Musalmans and Khattacks.

1st Commn. or date of entering service.	Names and Rank.	Army rank.	Present appointment in Regiment.	Remarks.
	Commandant.			
28 Oct. 91	Mitchell, Lt.-Col. W. J. (Q)	28 Oct. 17		1-124th Infy.
	Company Commanders. (4)			
22 Jan. 95	Macpherson, Maj. J., D.S.O.	22 Jan. 14	18 Aug. 17	1-124th Infy. 2nd-in-Comd. (*temp.*)
16 Aug. 05	Lynch-Staunton, Capt. H. V. C. (*Actg. Maj.*)	16 Aug. 14	6 Jan. 17	1-124th Infy.
13 Jan. 15	Smith, Lt. A. B. (*Actg. Capt.*)	13 Jan. 16	5 Nov. 17	42nd R.
12 Sep. 15	Ogier, Lt. L. L. (*Actg. Capt.*)	12 Sep. 16	2 July 18	Attd.
	Company Officers.			
4 Dec. 12	Fraser, Capt. H. F.	4 Dec. 16	6 Nov. 18	A. S. C., I. A. (*on probn.*)
9 July 15	Kiddle, Lt. J. A. C., M.C.	9 July 16		I. A. R. O.
23 Sep. 15	Stanger, Lt. F. R. (*Actg. Capt.*)	23 Sep. 16	4 July 18	I. A. (*on probn.*)
15 Nov. 15	Soden, Lt. J. N. (*Actg. Capt.*)	15 Nov. 16	5 Nov. 17	Adjt. (*temp.*) 129th Infy.
29 June 16	Nicholas, Lt. C. A. N.	13 Dec. 16	25 Jan. 17	126th Infy.
18 Apl. 16	Smith, Lt. B. L. S. (*Actg. Capt.*)	18 Apl. 17	16 Dec. 17	126th Infy.
19 June 16	Jardine, Lt. J. D.	19 June 17	30 Sep. 18	L'pool. R., I. A. (*on probn.*)
27 July 16	Coldwell, Lt. L. R.	27 July 17	11 July 18	
14 Nov. 16	Dobson, Lt. E. T.	14 Nov. 17	7 Jan. 18	74th Punj.
20 June 17	Wright, Lt. E. McW.	20 June 18		I. A. R. O.
1 Nov. 17	Miller, Lt. F. B.	1 Nov. 18	17 Mar. 18	I. A. (*T. C.*)
8 Oct. 15	Armstrong, Lt. K.			5th Lrs., I. A. (*on probn.*)
6 Oct. 15	Tennant, 2nd-Lt. A. W.		2 July 18	I. A. R. O. (*on probn.*)
5 Nov. 17	Rowe, Lt. W. T. M.	5 Nov. 18	19 July 18	I. A. R. O.
28 Dec. 17	Arnold, 2nd-Lt. L. P.	28 Dec. 17		Lond. R., I. A. (*on probn.*)
15 Jan. 18	Riley, 2nd-Lt. E. C.	15 Jan. 18		I. A. R. O.
7 June 18	Barber, 2nd-Lt. P. T.	7 June 18		I. A. (*T. C.*)
15 Oct. 18	O'Brien, 2nd-Lt. M. J.	15 Oct. 18	5 Nov. 18	I. A. R. O.
		Jemadar.	Subadar.	Subadar-Major.
	Subadar-Major.			
8 May 15	Bhagwan Singh, M.C., I.D.S.M. (56, 60).	8 May 05	1 Dec. 13	2 June 18
	Subadars.			
5 Sep. 02	Farman Ali (56)	9 June 15	1 Dec. 15	
1 Jan. 97	Muhammad Afzal, I.O.M.	3 Apl. 15		Supy.
1 Sep. 99	Darweza Khan, I.O.M.	21 Jan. 15	1 Jan. 16	
4 May 99	Surjan Singh	1 May 15	27 Sep. 16	
21 May 93	Chanda Singh	1 Jan. 16	12 Jan. 17	
3 Feb. 06	Ghulam Muhammad	22 Jan. 16	26 Mar. 17	
29 Nov. 01	Murtaza Khan (56)	8 Aug. 14	30 Aug. 17	
17 June 95	Kishen Singh	12 Jan. 17	13 Sep. 17	
27 May 02	Pir Bakhsh	1 Mar. 17	6 Nov. 17	
	Jemadars.			
11 Sep. 95	Didan Khan	29 July 10		
11 Sep. 99	Mala Singh	1 May 15		
23 Mar. 94	Bichant Singh (56)	1 May 15		
10 Sep. 06	Ayub Khan, I.O.M.	21 June 15	Supy.	
12 Dec. 00	Mansha Singh	27 Sep. 16		
25 Apl. 00	Basawa Singh	12 Jan. 17		
10 Nov. 99	Sona Singh	26 Feb. 17		
24 Oct. 03	Gurdit Singh	26 Mar. 17	Indian Adjt.	
6 Aug. 10	Sher Zaman	21 July 17		
16 Sep. 08	Ghulam Hussain	21 July 17		
4 May 98	Sardar Khan	13 Sep. 17		
15 Mar. 99	Thakar Singh	25 Sep. 17		
30 May 99	Charan Singh	6 Nov. 17		
14 Dec. 10	Kurban Hussain Shah	6 Nov. 17		
19 July 06	Malak Din	27 Nov. 17	Supy.	
	Santa Singh	3 Dec. 17		
20 Oct. 93	Muhammed Din	1 Sep. 18		

3rd Battalion, 124th Duchess of Connaught's Own Baluchistan Infantry.

Formed at Karachi on 14th August 1917.

Composition—2 Companies Punjabi Mussalmans, 1 Company Khuttaks, 1 Company Baluchis.

1st Commn. or date of entering service.	Names and Rank.	Army rank.	Present appointment in Regiment.	Remarks.
	Commandant.			
28 June 93	Holbrooke, Maj. B. F. R. (*Actg. Lt.-Col.*)	26 June 11	9 Sep. 17	1-124th Infy.
	Company Commanders. (4)			
26 June 01	Stewart, Maj. A. E., M.C.	1 Sep. 15	4 Aug. 17	1-124th Infy. 2nd-in-Comd.
18 Jan. 02	Chaldecott, Maj. O. A.	18 Jan. 17	14 Aug. 17	1-124th Infy.
	Company Officers.			
18 Jan. 05	Tate, Capt. J. C. (*Actg. Maj.*)	18 Jan. 14	20 Mar. 18	Poll. Dept.
25 Jan. 08	Wetherall, Capt. F. G. B.	1 Sep. 15	20 Mar. 18	1-127th Bal. L. I.
1 Sep. 14	Gardner, Capt. J. M. S.	5 Aug. 18	24 Oct. 17	1-124th Infy.
11 Jan. 15	Honeyball, Lt. F. R.	29 Sep. 15	27 Nov. 17	Attd.
30 Apl. 15	Gregory, Lt. J. R.	30 Apl. 16	14 Aug. 17	I. A. R. O.
29 May 15	Teague, Lt. J., M.C. (*Actg. Capt.*)	29 May 16	13 Oct. 17	R. War. R., I. A. (*on probn.*) Adjt.
15 Sep. 15	Lord, Lt. J. W.	15 Sep. 16	14 Aug. 17	
25 Sep. 15	Rogers, Lt. T. P. W., M. C.	25 Sep. 16	28 Aug. 18	W. York. R., I. A. (*on probn.*)
31 Dec. 15	Reed, Lt. S. P. K. J. H.	10 Dec. 16	5 Mar. 18	Conn. Rang., I. A. (*on probn.*)
25 Jan. 16	Tearoe, Lt. A. T.	25 Jan. 17	23 Aug. 17	Lond. R., I. A. (*on probn.*).
9 Feb. 16	Goolden, Lt. A. W.	9 Feb. 17	1 Oct. 17	Lond. R., I. A. (*on probn.*)
30 Jan. 16	Currie, Lt. J.	30 Nov. 16	12 Dec. 17	Attd.
8 Mar. 16	Webb, Lt. S. D., M.C.	10 Dec. 17	27 Nov. 17	Attd.
17 Mar. 16	Avent, Lt. J. G.	17 Mar. 17	17 Apl. 18	
12 Apl. 16	Weldhen, Lt. R. B.	12 Apl. 17	12 Feb. 18	Worc. R., I. A. (*on probn.*)
1 Oct. 16	Smith, Lt. R. G.	1 Oct. 17	5 Oct. 17	Attd.
1 Mar. 17	Simpson, Lt. J. A.	1 Mar. 18	1 Oct. 17	I. A. R. O.
23 June 17	Shipman, Lt. H. L.	23 June 18	11 Oct. 17	M. T., A. S. Corps, I. A. (*on probn.*)
5 Oct. 17	Kind, Lt. F.	5 Oct. 18	21 Jan. 18	I. A. (*T. C.*)
28 Nov. 17	Jacob, Lt. G.	28 Nov. 18	29 Jan. 18	I. A. R. O.
31 Jan. 18	Beaty, 2nd-Lt. G. H. M.	21 Jan. 18	19 Feb. 18	
26 Mar. 18	Pollicot, 2nd-Lt. H. E.	26 Mar. 18	28 Aug. 18	I. A. (*on probn.*)
7 June 18	Crawford, 2nd-Lt. F. R. R.	7 June 18	4 July 18	I. A. R. O.
31 Aug. 18	Mildon, 2nd-Lt. R. K.	31 Aug. 18	2 Sep. 18	Attd.
20 Sep. 11	Fleming, 2nd-Lt. D. J.	20 Sep. 18	7 Oct. 18	I. A. R. O.

		Jemadar.	Subadar.	Subadar-Major.	
	Subadar-Major.				
21 Mar. 84	Kishan Singh	16 Oct. 02	23 Mar. 06	14 Aug. 17	Re-employed on 31 Oct. 15.
	Subadars.				
8 Jan. 98	Mehdi Khan	1 May 15	13 July 15		
9 Mar. 07	Ghulam Yunas	1 Nov. 15	20 May 16		
1 Aug. 00	Saleh Mohd.	22 Jan. 16	20 Jan. 18		
15 May 17	Hamidullah Khan	19 Nov. 17	1 Feb. 18		
1 Feb. 18	Nur Muhammad	....	1 Feb. 18		
11 Jan. 99	Mota Singh	1 July 16	20 Mar. 18		
16 Aug. 04	Mehr Khan	1 Jan. 16	1 Apl. 18		
6 Aug. 10	Dhali Khan	20 Jan. 18	4 Aug. 18		
	Jemadars				
27 Feb. 94	Mohd. Din	11 Aug. 16			
	Karim Khan	*21 Jan. 17			
	Mir Ali Khan	21 Jan. 17			
15 June 99	Bhagwan Singh	24 Mar. 18			
	Bahadur Sher	1 Apl. 18			
13 Oct. 02	Fateh Khan	1 May 18			
11 Mar. 07	Sher Zaman	1 May 18			
	Mirza Khan	1 May 18			
	Hamidullah Khan	4 Aug. 18			
22 Oct. 17	Saifullah	4 Aug. 18			
1 Oct. 17	Diwan Ali	17 Aug. 18			
	Hadayatullah Khan	6 Sep. 18			

Regt. Centre—Nasirabad.
Linked with—104th & 123rd Rif.

Uniform—Dark green
Facings—Scarlet.

1st Battalion, 125th Napier's Rifles.

Raised in 1820 as the 1st Extra Battalion, formed to a considerable extent from men who had served in the disbanded Dapuri Brigade (Peishwa's service), commanded by Major Ford. Became the 25th Regiment of Bombay Native Infantry, 1826; the 25th Regt. of Bombay Nat. (Light) Infantry, 1858; the 25th Regt. of Bom. (Light) Infantry, 1885; unauthorisedly designated the 3rd Battn., Bombay Rifle Regt., 1889; title corrected to 25th Regt. (3rd Battn., Rifle Regt.) of Bom. Infy., 1889; became the 25th Bombay Rifles, 1901; present designation, 1903.

"Meeanee" "Hyderabad" "Central India" "Abyssinia" "Burma, 1885-87."

Composition—2 companies of Rajputana Jats, 1 of Rajputana Rajputs, 1 of Punjabi Musalmans.

COLONEL—Major-General Charles Irwin Fry, C.B. . . 8 Sep. 1910.

1st Commn. or date of entering service.	Names and Rank.	Army rank.	Present appointment in Regiment.	Remarks.
	Commandant.			
25 July 91	Keily, Lt.-Col. F. P. C., C.M.G., D.S.O.	24 Sep. 16	24 Sep. 16	
	Company Commanders. (4)			
6 Jan. 95	Kidd, Maj. B. G. B.	16 Jan. 13	24 Sep. 16	2nd-in-Comd. With 3-153rd Infy.
4 Aug. 97	*Graham, Maj. B. C.*	*4 Aug. 15*	*26 Mar. 12*	*Lv., ex I., m. c.*
18 Oct. 99	Harrison, Maj. L. G. B.	1 Sep. 15	6 Feb. 14	
5 May 00	Mollison, Maj. E. J., C.I.E.	1 Sep. 15	19 Apl. 14	
	Company Officers. (9).			
8 Jan. 01	*Clutterbuck, Maj. C. H.* [L] *(temp. Lt.-Col.)*	*8 Jan. 16*	*15 May 03*	
30 Oct. 01	Maynard, Maj. F. H., M.C. [l]	30 Oct. 16	15 Feb. 03	With 2-30th Punj.
25 Apl. 06	Huxford, Capt. H. J. (*Actg. Maj.*)	25 Apl. 15	31 Mar. 09	With 2-125th Rif.
29 Aug. 06	Skinner, Capt. E. M. (*actg. Maj.*)	29 Aug. 15	1 Aug. 08	With 2-123rd Rif.
25 Jan. 08	Macdonald, Capt. D. T., M. C.	1 Sep. 15	2 Apl. 09	With 1-43rd Rif.
9 Sep. 08	Meade, Capt. M. E. M.	1 Sep 15	11 Dec. 09	Coy. Offr., Tg. Sch., Indore.
18 Jan. 11	Hughes, Capt. L. W.	1 Sep. 15	10 Mar. 12	Qr.-Mr.
14 Jan. 14	Maltby, Capt. F. N. (*Actg. Maj.*)	14 Jan. 18	9 Oct. 14	With 3-153rd Infy.
20 Apl. 16	Skilliter, Lt. W. M.	20 Apl. 17	. . . .	Attd.
30 Jan. 17	Hodge, Lt. J. F.	30 Jan. 18	. . . .	With 104th Rif.
12 Sep. 17	Sale, Lt. W. G.	12 Sep. 18	8 Aug. 18	Attd.
30 Jan. 17	Dickens, 2nd-Lt. A. C.	30 Jan. 17	. . . .	
27 Oct. 17	Simpson, 2nd-Lt. J. S. C.	27 Oct. 17	5 Nov. 17	With 96th Infy.
21 Dec. 17	Long, 2nd-Lt. H. W.	21 Dec. 17	20 Mar. 18	
	Attached.			
9 Sep. 08	Tate, Capt. W. M. (*Actg. Maj.*)	1 Sep. 15	5 Oct. 18	55th Rif.
13 Nov. 14	Marshall, Capt. W. M. C.	13 Nov. 18	15 Feb. 15	I. A. R. O.
13 Nov. 14	Bennett, Capt. L. A.	13 Nov. 18	15 Feb. 15	I. A. R. O., A. R. O., Ajmere.
8 Sep. 15	Wilkinson, Lt. J. B.	8 Sep. 16	. . . .	I. A R. O.
27 June 17	Billingham, Lt. W F.	27 June 17	6 Aug. 18	I. A. R. O.
15 Oct. 17	Harrison, 2nd-Lt. G. E., D.C.M.	15 Oct. 17	25 Jan. 18	I. A. R. O.
5 Nov. 17	Tarr, Lt. W. G.	5 Nov. 18	6 Aug. 18	I. A. (*T. C.*)
27 Mar. 18	Tritton, 2nd-Lt. G. C.	27 Mar. 18	18 July 18	I. A. R. O.
27 Mar. 18	Kennedy, 2nd-Lt. A. P.	27 Mar. 18	25 June 18	R. Irish Fus., I. A. (*on probn.*)
14 Aug. 18	Pentony, 2nd-Lt. L. R.	14 Aug. 18	7 Sep. 17	I. A. R. O. (*on probn.*)
28 Feb. 18	Gemmel, 2nd-Lt. A.	28 Feb. 18	8 Nov. 18	L'pool. R., I. A. (*on probn.*)
24 Apl. 18	Knocker, 2nd-Lt. J. P.	24 Apl. 18	16 Oct. 18	*U. L.*
24 Apl. 18	Cooper, 2nd-Lt. C. H.	24 Apl. 18	16 Oct. 18	*U. L.*
	Howison, 2nd-Lt. G.	. . . .	27 Sep. 18	I. A. R. O. (*on probn.*)

		Jemadar.	Subadar.	Subadar-Major.	
	Subadar-Major.				
3 June 92	Umar Din *Bahadur* (56, 63)	30 Sep. 04	2 Dec. 04	1 Apl. 12	*Order of Br. Ind., 2nd Cl.* Depôt.

	Subadars.			
23 Jan. 95	Ana Rám (56)	9 Jan. 04	22 May 10	
10 May 93	Rud Singh (56, 72*a*)	1 July 07	1 Mar. 14	
6 Dec. 99	Sarwar Khan (56)	18 Feb. 12	17 Apl. 15	
7 June 02	Phul Singh (56)	5 Aug. 09	8 Jan. 16	
3 July 98	Ramlal Singh (52, 56, 66*a*)	8 Aug. 14	8 Jan. 16	
8 Oct. 08	Kushiram Singh	8 Jan. 16	23 Apl. 16	
22 Nov. 98	Nizam-ud-din	1 Jan. 15	23 Feb. 17	
1 Sep. 00	Jhujha Ram (56)	7 Sep. 14	24 Feb. 17	
16 Jan. 01	Birad Singh (56, 66*a*)	28 Mar. 17	3 Mar. 17	
	Pertab Singh	8 Jan. 16	3 May 17	
21 July 02	Anna Singh, I.D.S.M.	3 Apl. 17	9 Apl. 17	
	Farman Al Khan (*Retd.*)	. . . .	1 Jan. 18	Re-employed.

1st Battalion, 125th Napier's Rifles—*contd.*

1st Commn. or date of entering service.	Names and Rank.	Army rank.	Remarks.
	Jemadars.		
13 Nov. 99	*Dhura Ram* (50)	*22 May 10*	
7 July 05	*Fazaldad*	*8 Aug. 14*	
24 June 01	Sher Ali	3 Nov. 14	Indian Adjt., 3 June 15.
10 July 00	Sheobaksh Singh . . . . a	8 Jan. 16	
12 Aug. 05	Hakim Khan, I.D.S.M.	8 Apl. 16	
19 July 05	Dula Ram	8 Apl. 16	
27 Oct. 06	Bhanwar Singh	3 Apl. 17	
28 Oct. 02	Allah Dad Khan	5 Apl. 17	
8 July 06	Lala Ram	6 Nov. 17	
5 June 03	Allahdin Khan	3 Dec. 17	Indian Qr. Mr. 3 Dec. 17.
7 July 97	Joseph Ezekiel	1 Apl. 18	
1 May 06	Tirkha Ram, I.D.S.M.	24 May 18	
. . . .	Hukam Dad	1 June 18	
2 Mar. 04	Ramparsad Singh	28 May 18	
25 Dec. 00	Kaiera Singh	28 May 18	
. . . .	Bachan Singh	9 Aug. 18	

2nd Battalion, 125th Napier's Rifles.

Formed at Anandi.

Composition.—

1st Commn. or date of entering service.	Names and Rank.	Army rank.	Present appointment in Regiment.	Remarks.

Commandant.

22 Apl. 98	Hamer, Maj. M. A., D.S.O., M.C., *p.s.c.* (*Actg. Lt.-Col.*)	1 Sep. 15	1 Aug. 18	I-129th Baluchis.

Company Commanders. (4)

25 Apl. 06	Huxford, Capt. H. J. (*Actg. Maj.*)	25 Apl. 15	2 Aug. 18	1-125th Rif., 2nd-in-Comd.
10 June 16	Walsh, Lt. P. G. (*actg. Capt.*)	10 June 17	22 Aug. 18	I. A. (*on probn.*)
19 June 17	Illingworth, Lt. J. R. (*actg. Capt.*)	19 June 18	8 Sep. 18	I. A. (*on probn.*)
20 Nov. 16	Bursey, Lt. T. F. Mc C., M. C. (*actg. Capt.*)	20 Nov. 17	8 Sep. 18	Austr, Impl. Force, I. A. (*on probn.*)

Company Officers.

27 Oct. 15	Davies, Lt. M. A.	27 Oct. 16	15 Aug. 18	I. A R. O.
15 Nov. 15	Dawes, Lt. K. C. B.	15 Nov. 16	9 Aug. 18	1-22nd Punj. I. A.
15 Nov. 15	Challen, Lt. B. S. (*actg. Capt.*)	15 Nov. 16	3 Aug. 18	1-72nd Punj. Adjt.
29. Aug. 17	Gardiner, Lt. D. I.	29 Aug. 18	31 Aug. 18	I. A. R. O.
15 Jan. 18	Weston, 2nd-Lt. A.	15 Jan. 18	28 July 18	I. A. (*T. C.*) Qr. Mr.
15 Jan. 18	Ingham, 2nd-Lt. H.	15 Jan. 18	8 Oct. 18	I. A. (*T. C.*)
19 Mar. 18	Pollard, 2nd-Lt. C. C.	19 Mar. 18	22 Aug. 18	I. A. (*T. C.*)
7 June 18	Cannell, 2nd-Lt. M. D.	17 June 18	8 Oct. 18	I. A. R. O.
27 May 18	Muirhead, 2nd-Lt. J. M.	27 May 18	9 Nov. 18	R. Lanc. R., I. A. R. O. (*on probn.*)
11 June 18	Smith, 2nd-Lt. R. C.	11 June 18	2 Dec. 18	I. A. R. O.
25 June 18	Taylor, 2nd-Lt. W. D.	25 June 18	26 Sep. 18	I. A. R. O (*on probn.*)
26 June 18	Holland, 2nd-Lt. A. P.	26 June 18	2 Oct. 18	I. A. R. O. (*on probn.*)
22 July 18	Strangeman, 2nd-Lt. C. V.	22 July 18	2 Dec. 18	I. A. (*T. C.*)
31 Aug. 18	Henderson, 2nd-Lt. E. G. S.	31 Aug. 18	13 Sep. 18	Attd.
20 Sep. 18	Stewart, 2nd-Lt. A. W.	20 Sep. 18	15 Oct. 18	I. A. R. O.

		Jemadar.	Subadar.	Subadar-Major.

Subadar-Major.

Subadars.

16 Apl. 15	Ramu Singh	19 Aug. 18	19 Aug. 18	
29 Dec. 15	Lachman Singh	19 Aug. 18	19 Aug. 18	
29 Dec. 15	Bostan Khan	19 Aug. 18	19 Aug. 18	
29 Dec. 15	Ganpat Singh	19 Aug. 18	19 Aug. 18	

Jemadars.

19 Aug. 18	Piru Ram	19 Aug. 18		
19 Aug. 18	Maula Dad	19 Aug. 18		
19 Aug. 18	Mangla Ram	19 Aug. 18		
22 Aug. 18	Ram Nath Singh	22 Aug. 18		
22 Aug. 18	Khuda Bux	22 Aug. 18		
22 Aug. 18	Baksha Ram	22 Aug. 18		

Regt. Centre.—Quetta.
Linked with—124th Baluch, Infy.

Uniform.—Drab (Red serge trousers).
Facings.—Scarlet.

126th Baluchistan Infantry.

Raised, in 1825, by Capt. E. M. Wood, as the 2nd Extra Battalion of Bombay Native Infantry. Became the 26th Regiment of Bombay Native Infantry in 1826; the 26th Regt. of Bom. Infy. in 1885; the 26th (Baluchistan) Regt. of Bombay Infy. in 1892; the 26th Baluchistan Infy. in 1901; present designation in 1903.

"Koosh-ab" "Persia" "China, 1900."

Composition—1 company of Hazaras, ½ of Khattaks, ½ of Wazirs, 1 of Baluchis and Brahuis, 1 of Sikhs other than Jat Sikhs.

COLONEL—Major-General *Sir* Alfred W. L. Bayly, K.C.B., K.C.M.G., C.S.I., D.S.O. . 7 May 1907.

1st Commn. or date of entering service.	Names and Rank.	Army rank.	Present appointment in Regiment.	Remarks.
	Commandant.			
13 July 92	*Chitty, Lt.-Col. A. W., p.s.c.* .	*1 July 18*	*20 July 18*	*A. A. & Q. M. G., 2nd Dn.*
	Company Commanders. (4)			
30 Jan. 89	*Paul, Lt.-Col. R. S. (Q)* . .	*30 Jan. 15*	*20 July 13*	*2nd-in-Comd. Zhob Militia.*
30 Aug. 93	Barrett, Bt.-Lt.-Col. A. L. (Q) (*Actg. Lt.-Col.*)	30 Aug. 11	14 Dec. 05	With 2-33rd Punj.
22 Jan. 96	Woodhouse, Maj. C. G. .	22 Jan. 14	20 July 13	
20 Feb. 97	James, Maj. C.A. (*temp. Lt.-Col.*)	20 Feb. 15	20 July 18	2-153rd Ind. Infy.
	Company Officers. (9)			
20 May 99	*Gretton, Maj. J. C.*	*1 Sep. 15*	*7 Dec. 04*	*Bde.-Maj.*
18 Apl. 00	Sanderson, Maj. F. G. O. .	1 Sep. 15	10 Oct. 03	
8 Jan. 01	Jarrad, Maj. F. L D. . . .	8 Jan. 16	21 Oct. 04	
30 Apl. 02	*Brock, Maj. B. deL.*, M.C. .	*22 June 16*	*9 Oct. 04*	*A. M. S. to G.O.C., S. Comd.*
28 Jan. 03	*FitzGerald, Maj. C. R. L.*, D.S.O.	*29 Jan. 17*	*14 Dec. 05*	*Bde.-Maj.*
27 Aug. 02	*Russell, Maj. G. H.*	*27 Aug. 17*	*10 Jan. 04*	*Kurram Militia.*
13 Aug. 04	White, Capt. A. G. . . .	13 Aug. 13	1 May 06	With 130th Baluchis.
20 Jan. 09	*Faithfull, Capt. G. F. H. (temp. Maj.)*	*1 Sep. 15*	*25 Mar. 10*	*R. A. F.*
18 Sep. 09	Johnston, Capt. A. L. . . .	1 Sep. 15	20 Aug. 14	Adjt., 13 July 15.
24 Aug. 12	Thomas, Capt. H. P.	24 Aug. 16	14 Oct. 13	
22 Jan. 13	Weir, Capt. I. E. R. . . .	22 Jan. 17	8 Mar. 14	
13 Aug. 15	Pringle, Lt. J. H.	13 Aug. 16	. . . .	Attd.
10 Dec. 15	Frith, Lt. J. G.	10 Dec. 16	. . . .	Attd.
29 June 16	Nicholas, Lt. C. A. N. . .	13 Dec. 16	28 Oct. 15	Attd. 2-124th Infy.
26 Jan. 16	Wooldridge, Lt. J. H. C. (*Actg. Capt.*)	26 Jan. 17	21 Mar. 16	Attd. *Offg. Qr.-Mr.*
1 Apl. 16	Portas, 2nd-Lt. J.	1 Apl. 17	. . . .	Attd.
18 Apl. 16	Smith, Lt. B. L. S. . . .	18 Apl. 17	. . . .	Attd. with 2-124th Infy.
14 Nov. 16	Paton, Lt. H. G. (*temp. Capt.*) .	14 Nov. 17		
18 June 17	Bake, Lt. A. T. T. . . .	18 June 18	27 June 17	
18 June 17	*Lynch, 2nd-Lt. R. J.* . . .	*18 June 18*	*27 June 17*	*Attd. with Frontier Militia.*
25 July 17	Laidler, 2nd-Lt. T.	25 July 17	. . . .	M. G. Corps, I. A. (*on probn.*)
	Attached.			
16 Nov. 87	Wooldridge, Lt.-Col. W. H. . .	20 July 13	20 July 18	Comdt. (*temp.*)
24 Nov. 14	Miles, Lt. W. H.	20 Nov. 15	. . . .	I. A. R. O.
14 Jan. 15	Lushington, Lt. A. N. . . .	14 Jan. 16	15 Jan. 15	I. A. R. O.
14 Jan. 15	Mitchell, Lt. F. M. . . .	14 Jan. 16	14 Jan. 15	I. A. R. O.
14 May 15	Kemp, Capt. K. McI. . . .	16 Dec. 17	1 Aug. 15	I. A. R. O.
31 Dec. 15	Rutledge, Lt. C. J.	31 Dec. 16	. . . .	N[illegible]h. Fus., I. A. (*on probn.*)
. . . .	Hudgesdon, Lt. H. R., M.C. .	. . . .	19 Aug. 18	[illegible]. A. R. O.
. . . .	Taylor, Lt. S. A.	. . . .	13 Aug. [illegible]	I. A. R. O.
10 May 16	Long, Lt. W. H.	10 May 17	10 July 16	I. A. R. O.
5 Jan. 17	Collet, Lt. A. L.	5 Jan. 17	16 Mar. 17	I. A. R. O.
5 Jan. 17	Meager, Lt. C. J.	5 Jan. 18	16 Mar. 16	I. A. R. O.
5 May 17	Pill, Lt. G. C. R.	5 May [illegible]	. . . .	M. G. Corps, I. A. (*on probn.*)
5 Oct. 17	Holbrook, Lt. A. F.	5 Oct. 18	22 Jan. 18	I. A. (*T. C.*)
. . . .	Tams, 2nd-Lt. G. F.	. . . .	29 Jan. 18	I. A. R. O.
. . . .	Kalkhoven, 2nd-Lt. F. G. . .	. . . .	13 Aug. 18	I. A. R. O.
. . . .	Jones, 2nd-Lt. C.	. . . .	. . . .	I. A. R. O. (*on probn.*)

Subadar-Major.

		Jemadar.	Subadar.	Subadar-Major.	
12 Mar. 83	Mul Singh, *Sardar Bahadur* (56)	21 Jan. 99	15 May 00	1 Oct. 15	*Order of Br. [illegible] Cl.*

126th Baluchistan Infantry—*contd.*

1st Commn. or date of entering service.	Names and Rank.	Army rank.	Present appointment in Regiment.	Remarks.
	Subadars.			
24 Oct. 98	Umar Khan	17 Oct. 07	17 Mar 11	*Order of Br. Ind., 1st Cl.*
24 Aug. 02	Gulzada, *Sardar Bahadur*	22 June 00	1 Dec. 13	
22 Nov. 02	Muhammad Aslam Khan	12 Nov. 02	12 Oct. 14	
4 Apl. 87	Labh Singh, I.O.M. (56)	16 July 09	11 Mar. 15	
30 June 99	Ahmad Khan (56)	1 Dec. 13	10 May 16	
22 Jan. 02	Id Mohamed (56)	24 Mar. 11	1 July 16	
17 Apl. 00	Guldar Shah, I.D.S.M.	24 June 15	1 Aug. 16	
. . . .	Amir Khan (*Retd.*)	Re-employed	15 Mar. 15	Depôt.
21 July 98	Ali Muhummad	1 Oct. 15	1 Aug. 16	
10 Mar. 12	Ram Singh	10 Mar. 12	11 Aug. 16	
29 Aug. 02	Budh Singh, I.D.S.M.	19 Mar. 15	9 Mar. 18	
	Jemadars.			
1 Apl. 98	Said Muhammad	18 Mar. 14		
9 June 98	Ghulam Shah (56)	12 Oct. 14		
8 Jan. 99	Tulsa Singh	12 Oct. 14		
10 Jan. 93	Shah Niaz Hussin	7 May 15		
25 Jan. 00	Lal Shah Gul	24 June 15		
19 Mar. 05	Niaz Mir	24 June 15		
19 Apl. 95	Arz Muhammad	1 Oct. 15		
4 Feb. 99	Mehr Singh	10 May 16		
1 Aug. 01	Gheba Khan	1 Aug. 16		
1 Jan. 06	Mawax Khan (56)	1 Aug. 16	Ind. Adjt.	
5 Oct. 01	Pir Baksh	1 Jan. 17		
25 Jan. 99	Santa Singh	1 Jan. 17		
20 Mar. 05	Tamiz Gnl	3 Mar. 17		
22 Dec. 98	Gurdit Singh	3 Dec. 17		
18 Aug. 03	Kishan Singh	9 Mar. 18		
7 Nov. 01	Qutab Khan	12 May 18		
5 Aug. 18	Khuda Dad Khan	5 Aug. 18	(*On probn.*)	

Regt. Centre—Karachi.
Linked with—129th & 130th Baluchis.

Uniform—Green (Red serge trousers)
Facings—Scarlet.

1st Battalion, 127th Queen Mary's Own Baluch Light Infantry.

Raised at Karachi, in 1844, by Maj. J. Jackson, by order of Sir C. Napier, and originally designated the Baluch Battalion or the Sind Baluchi Corps. Became the 1st Baluch Bn., 1846; the 1st Baluch Extra Bn. of Bombay Native Infantry, 1858; the 1st Baluch Regiment, 1859; the 1st Bombay Baluoh Regt., 1859; the 27th Regt. of Bombay N. I. or 1st Baluch Regt. 1861; the 27th Regt. of Bombay Nat. (Light) Infantry or 1st Baluch Regt., 1871; the 27th Regt. of Bombay (Light) Infantry (1st Baluch Regt.), 1885; the 27th Regt. of Bombay (Light) Infantry (1st Baluch Bn.), 1888; the 27th (1st Baluch Bn.) Bombay Light Infy., 1888; the 27th Baluch Light Infy., 1901; 127th Baluch Light Infy. in 1903; 127th Princess of Wales's Own Baluch Light Infantry, 1909, present designation, 1910.

IN EACH OF THE FOUR CORNERS THE CYPHER OF QUEEN MARY.
"Delhi, 1857" "Abyssinia," "Afghanistan, 1879-80"
"Burma, 1885-87" "British East Africa, 1897-99."

Composition—1 Company of Punjabi Musalmans, 1½ of Mahsuds, 1½ of other Pathans.

COLONEL—Major-General Charles A. Cuningham . 13 May 1904.

1st Commn. or date of entering service.	Names and Rank.	Army rank.	Present appointment in Regiment.	Remarks.
	Commandant.			
9 Sep. 91	Hulseberg, Lt.-Col. H., D.S.O.	9 Sep. 17	25 Aug. 18	
	Company Commanders. (4)			
27 Mar. 94	Grant, Maj. W. O. (*actg. Lt.-Col*)	7 Mar. 12	25 Aug. 18	2nd-in-Comd., with 2-98th Infy.
22 Jan. 98	Merriman, Maj. A. B. (*actg. Lt.-Col.*)	1 Sep. 15	31 Oct. 14	With 2-129th Baluchis.
4 Apl. 00	*Kennedy-Craufurd-Stuart, Maj. C.*, D.S.O.	*1 Sep. 15*	*13 May 15*	
22 Oct. 02	Auret, Maj. A. S.	22 Oct. 17	25 Aug. 18	With 2-98th Infy.
	Company Officers. (9)			
12 Apl. 03	*Davies, Maj. R. D.*	*22 Apl. 18*	*31 Mar. 09*	
18 Jan. 05	*Harvey-Kelly, Bt.-Maj. C. H. G. H.*	*3 June 17*	*28 Mar. 06*	*G. S. O., 3rd gr.*
15 Aug. 05	Lake, Capt. E. A. W. (*actg. Maj.*)	5 Aug. 14	30 Oct. 06	2nd-in-Comd. (*temp.*)
10 Aug. 06	O'Neill, Capt. H. J. D., M.C.	29 Aug. 15	29 Oct. 07	10th Arm. Motor Bde.
9 Jan. 07	Belli-Bivar, Capt. H. F.	1 Sep. 15	26 Mar. 08	1-69th Punj.
25 Jan. 08	Wetherall, Capt. F. G. B.	1 Sep. 15	30 Mar. 09	With 3-124th Infy.
18 Sep. 09	Quayle, Capt. P. R. [L] (*actg. Maj.*)	1 Sep. 15	16 Dec. 10	With 2-127th Bal. L. I.
28 Jan. 11	Cuddeford, Capt. R. B. G. (*actg. Capt.*)	1 Sep. 15	27 Mar. 12	With 2-129th Bal.
16 Sep. 11	*Cuningham, Capt. A. P.*	*6 Sep. 15*	*4 Nov. 12*	
14 July 15	Prior, Lt. C. G.	14 July 16	28 Oct. 15	Attd. Qr.-Mr.
15 Nov. 15	Browne, Lt. W. R. H.	15 Nov. 16	17 Nov. 15	Attd.
15 Nov. 15	Mountain, Lt. R. G.	15 Nov 16		Attd. with 3-153rd Rif.
19 June 16	Acheson, Lt. I. St. G.	29 June 17	. . . .	Attd. With 2-127th Bal. L. I.
14 Nov. 16	Wells, Lt. A. M.	14 Nov. 17	17 Nov. 16	Attd.
30 Jan. 18	Maxwell, 2nd-Lt. W. E.	30 Jan. 18	18 Feb. 18	Attd.
. . . .	Butler, 2nd-Lt. R. H.	. . . .	1 Oct. 18	Attd.
	Attached.			
24 July 01	Mackrell, Maj. A.	9 Aug. 16	9 Aug. 17	109th Infy.
1 June 15	Harvey, Lt. J. S.	1 June 16	10 June 18	M. G. Corps. I. A. (*on probn.*)
3 Aug. 15	Ashworth, Lt. C. H.	3 Aug. 16	29 Aug. 17	
17 July 16	Martin, Lt. G. St. J.	17 July 17	9 June 18	Lond. R., I. A. (*on probn.*)
1 Aug. 16	Kersey, Lt. A. O., M.C.	1 Aug. 17	. . . .	Shrop. L. I. I. A. (*on probn.*)
6 Oct. 16	Fox, Lt. H. V.	6 Oct. 17	. . . .	I. A. R. O.
30 Jan. 17	Moss, Lt. C. de V.	30 Jan. 18	29 Oct. 17	With 53rd Sikhs.
30 Jan. 17	Pratt, Lt. A. R.	30 Jan. 18	4 Feb. 17	130th Baluchis.
16 Mar. 17	Mitchell, Lt. E. J. R.	16 Mar. 18	1 Apl. 17	I. A. R. O.
5 May 17	Herdon, Lt. B. L.	5 May 18	18 Aug. 17	I. A. R. O.
19 May 17	Middleton, Lt. L. W.	19 May 18	22 Sep. 17	I. A. R. O.
15 Jan. 18	Herman, 2nd-Lt. C. R.	15 Jan. 18	10 June 18	I. A. R. O.
18 Jan. 18	Harding, 2nd-Lt. A. G. L.	18 Jan. 18	10 June 18	I. A. (*T. C.*)
. . . .	Lawford, Lt. N. B.	. . . .	17 Nov. 15	I. A. R. O.
7 June 08	Arthur, 2nd-Lt. W. G. A.	7 Jun '18	20 Oct. 18	I. A. R. O.

1st Battalion, 127th Queen Mary's Own Baluch Light Infantry —*contd.*

1st Commn. or date of entering service.	Names and Rank.	Army Rank.	Present appointment in Regiment.	Remarks.
		Jemadar.	Subadar.	Subadar-Major.
	Subadar-Major.			
10 Sep. 06	Gulab Shah	14 Feb. 00	29 Jan. 01	16 Sep. 13
	Subadars.			
23 Nov. 12	Mir Badshah (I) (56)	23 Nov. 12	9 Dec. 14	
1 Jan. 05	*Muhammad Gul (56)*	*1 Sep. 13*	*7 Mar. 15*	
26 Dec. 93	*Sarbuland (56)*	*16 Oct. 13*	*1 July 15*	*With 129th Bal.*
20 July 97	Nazar Muhammad	5 Feb. 15	16 Aug. 15	
7 Dec. 11	Mir Badshah (II) (66*a*)	1 Jan. 14	14 Nov. 15	
29 Aug. 98	Muhammad Yusuf (56)	2 Nov. 14	14 Nov. 15	
25 July 99	*Shaikh Abdul Qadir*	*1 July 15*	*15 Dec. 15*	
2 May 04	Khan Bahadur	1 Aug. 15	21 Sep. 17	
1 June 04	Amir Ali (46, 56)	7 Feb. 16	1 Oct. 17	
28 Aug. 15	Yar Muhammed	16 Nov. 15	6 Dec. 17	
8 Apl. 01	Sahib Alam	1 Oct. 15	1 Sep. 18	
17 Apl. 14	Godar Khan	16 Aug. 15	1 Sep. 18	
	Jemadars.			
26 Jan. 01	*Asghar Ali*	*7 Mar. 15*		
4 Jan. 05	*Muhammad Khan*	*14 Nov. 15*		
8 Mar. 05	*Habib Khan*	*1 May 16*		
20 May 02	Wasal Khan	28 Sep. 16		
17 June 95	Tora Khan (56)	1 Oct. 16	Ind. Adjt.	
31 May 04	Issa Khan	1 Nov. 16		
29 Mar. 02	Ahmad Khan	13 Jan. 17		
8 Sep. 95	Kalandar Khan	21 Sep. 17		
6 Feb. 99	Gulam Muhammad	1 Oct. 17		
31 Aug. 10	Muhammad Afzal	19 Oct. 17		
25 July 10	Jahan Khan	1 Sep. 18		
30 July 99	Nurdad	1 Sep. 18		
15 Nov. 09	Jan Gul	1 Sep. 18		
24 Jan. 16	Bahadur Khan	1 Sep. 18		
1 Jan. 05	Shaik Muhammed	1 Sep. 18		
24 Aug. 15	Said Ahmad	1 Sep. 18		
18 July 14	Mirzadad	1 Sep. 18		

2nd Battalion, 127th Queen Mary's Own Baluch Light Infantry.

Formed at Multan.

Composition.—6 platoons Baluchis and Derewals, 4 platoons Pathans, 6 platoons Punjabi Musalmans.

1st Commn. or date of entering service.	Names and Rank.	Army rank.	Present appointment in Regiment.	Remarks.
	Commandant.			
7 Dec. 95	Chauncy, Maj. C. H. K. (*actg. Lt.-Col.*)	7 Dec. 13	16 Feb. 18	1-124th Infy.
	Company Commanders. (4)			
9 Sep. 08	Lewis, Capt. H. V., D.S.O., M.C. (*actg. Maj.*)	1 Sep. 15	5 May 18	1-129th Baluchis, 2nd-in-Comd.
18 Sep. 09	Quayle, Capt. P. R. (L) (*actg. Maj.*)	1 Sep. 15	14 May 18	1-127th Bal. L. I.
23 Oct. 14	Anderson, Lt. C. E. L.	15 Oct. 15	6 July 18	I. A. R. O.
20 Nov. 14	Galvin, Lt. H. F. (*actg. Capt.*)	29 Nov. 15	1 Apl. 18	I. A. R. O.
	Company Officers.			
22 Apl. 15	Coulson, Lt. R. A.	22 Apl. 15	29 July 18	I. A. R. O.
29 July 15	Milligan, Lt. H. C. (*actg. Capt.*)	29 July 16	27 Sep. 18	M. G. Corps., I. A. (*on probn.*) Adjt.
26 Sep. 15	Durrant, Lt. O. T.	26 Sep. 16	5 July 18	I. A. (*on probn.*)
7 Apl. 16	Hargreaves, Lt. H.	7 Apl. 17	28 Apl. 18	A. S. C., I. A. (*on probn.*)
19 June 16	Acheson, Lt. I. St. G.	29 June 17	8 Mar. 18	1-127th Bal. L. I.
1 July 16	Pitt, Lt. B. I.	1 July 17	14 Feb. 18	I. A. (*on probn.*)
26 Aug. 16	Wace, Lt. M. A.	26 Aug. 17	6 July 18	
5 Jan. 17	Slaughter, Lt. C. M.	5 Jan. 18	27 June 18	I. A. R. O.
30 Jan. 17	Kelly, Lt. V. S.	30 Jan. 18	27 July 18	I. A. R. O.
15 May 17	Dick, Lt. H. N.	15 May 18	17 Feb. 18	
20 June 17	Jervis, Lt. A. E. W.	20 June 18	8 July 18	I. A. R. O.
5 Oct. 17	Pook, Lt. E. R.	5 Oct. 18	21 June 18	I. A. (*T. C.*)
5 Oct. 17	Griffin, Lt. G.	5 Oct. 18	5 May 18	I. A. R. O.
1 Nov. 17	Price, Lt. W. H.	1 Nov. 18	28 Apl. 18	I. A. (*T. C.*)
27 Nov. 17	Trent, 2nd-Lt. A. R.	27 Nov. 17	21 June 18	I. A. R. O.
30 Oct. 17	Blair, 2nd-Lt. J. M.	31 Oct. 17	2 July 18	I. A. R. O.
15 Jan. 18	Thomas, 2nd-Lt. R. H.	15 Jan. 18	21 June 18	I. A. R. O.
29 Mar. 18	Traylen, 2nd-Lt. H. E.	29 Mar. 18	14 July 18	I. A. R. O.
26 Apl. 18	Brooke, 2nd-Lt. L. H.	26 Apl. 18	4 May 18	I. A. (*on probn.*)
		Jemadar.	Subadar.	Subadar-Major.
	Subadar-Major.			
1 Sep. 02	Mir Kambir Khan	1 Sep. 02	10 Aug. 07	
	Subadars.			
20 July 97	Mohammed (56)	21 June 13	2 Nov. 14	
9 Jan. 94	Baz Khan	28 Feb. 16	7 Feb. 16	
18 Aug. 99	Ghulam Muhamad	2 June 15	8 July 16	
22 Oct. 10	Malik Aman, I.O.M.	18 May 16	23 Nov. 16	
21 Aug. 01	Sardar Shah	1 July 15	2 Jan. 17	
8 Oct. 14	Alam Shah	26 Apl. 15	2 Oct. 16	
6 July 01	Karam Bakhsh, I.D.S.M.	22 Oct. 15	6 Apl. 17	
3 May 01	Durrani	14 Nov. 15	3 Apl. 17	
3 Mar. 01	Sher Dil	22 Oct. 15	19 Oct. 17	
26 Dec. 00	Gul	15 May 18	15 May 18	
12 Apl. 09	Khalik Dad	1 May 16	15 May 18	
	Jemadars.			
29 July 90	Gulam Hassan	12 Mar. 16		
29 May 05	Fakharuddin	5 Apl. 17		
9 Aug. 04	Sohbat Khan	10 Aug. 17		
8 May 18	Fazal Khalig	8 May 18	(*On probn.*)	
1 June 08	Ghulam Mohd.	15 May 18		
13 Mar. 97	Nur Mohammed	15 May 18		
8 Dec. 01	Naukar Khan	15 May 18		
24 Nov. 00	Alla Rakha	15 May 18		
7 Mar. 01	Ibrahim	15 May 18		
1 Oct. 06	Nawab Khan I	15 May 18		
29 July 99	Jalal Din	15 May 18		
4 May 15	Bahram	15 May 18		
3 Sep. 07	Ahmad Shah	15 May 18		
28 Apl. 03	Nawab Khan II	15 May 18		
1 June 08	Abdul Karim	15 May 18		
15 June 18	Mughal Khan	16 June 8		
22 Sep. 97	Faiz Ali Khan	9 Sep. 18		
1 June 04	Mehr Ilahi	17 Sep. 18		
6 Mar. 01	Ghulam Mohammed	20 Sep. 18		
24 July 07	Mubarak Khan	20 Sep. 18		

Regt. Centre—Kirkee. *Uniform*—Scarlet.
Linked with—107th and 121st Pioneers. *Facings*—White.

1st Battalion, 128th Pioneers.

Raised at Ahmednagar in 1846, by Lieut.-Col. R. Taylor, as the 28th Regiment of Bombay Native Infantry. Became the 28th Regiment of Bombay Infantry in 1885; the 28th (Pioneer) Regiment of Bombay Infantry in 1888; the 28th Bombay Pioneers in 1901; present designation in 1903.

"*Progredior.*"

"Kandahar, 1880" "Afghanistan, 1879-80" "Tofrek" "Suakin, 1885" "Tirah" "Punjab Frontier."

Composition—1 company of Yusufzais, 1 of Rajputana Musalmans, 1 of Sikhs, 1 of Dekhani Mahrattas.

COLONEL—Major-General *Sir* Richard Westmacott, 𝕂𝕥., K.C.B., D.S.O. 13 May 1904.

1st Commn. or date of entering service.	Names and Rank.	Army rank.	Present appointment in Regiment.	Remarks.
	Commandant.			
23 Mar. 89	Andrew, Lt.-Col. F. A., D.S.O., *q.* (*temp. Br.-Genl.*)	23 Mar. 15	14 Aug. 16	
	Company Commanders. (4)			
27 July 98	Goodfellow, Bt.-Lt.-Col. N. G. B., *q.* (*Actg. Lt.-Col.*)	3 June 17	8 May 16	2nd-in-Comd.
23 Dec. 98	*Tobin, Maj. H. W.*, D.S.O., *p.s.c.*	*1 Sep. 15*	*9 July 11*	*G. S. O., 1st gr.*
19 July 99	*Bannatyne, Maj. N. C., p.s.c.* (*temp. Lt.-Col.*)	*1 Sep. 15*	*19 Apl. 16*	*G. S. O., 1st gr.*
10 Jan. 00	Graham, Maj. D. S.	1 Sep. 15	8 May 16	
	Company Officers. (9)			
5 Aug. 05	Goad, Capt. C. E. G. B., M.C.	5 Aug. 14	19 Nov. 06	
5 Aug. 05	*Farebrother, Capt. F. H.* [L]	*5 Aug. 14*	*8 Dec. 06*	*Staff Capt.*
9 Sep. 08	*Pulverman, Capt. A. W., q.*	*1 Sep. 15*	*20 Dec. 09*	*With Sig. Ser.*
3 Sep. 10	*Mathias, Capt. L. W. H.*, D.S.O., *q.*	*1 Sep. 15*	*8 Nov. 11*	*S. and T. Corps.*
20 Jan. 12	Rundle, Capt. C. A. G., M.C.	20 Jan. 16	4 Apl. 13	
11 Feb. 16	Sowton, Lt. B. S.	11 Feb. 17	24 Jan. 18	Attd.
18 Apl. 16	Tomkins, Lt. G. L., M.C.	18 Apl. 17	23 Apl. 16	Attd. *Offg.* Adjt.
18 Apl. 16	Lloyd-Jones, Lt. W. R.	18 Apl. 17	23 Apl. 16	Attd.
29 June 16	Pollock-Gore, Lt. W. A. J.	29 June 17	13 Dec. 16	Attd.
27 Oct. 17	Flett, 2nd-Lt. R.	27 Oct. 17	4 Nov. 17	Attd.
	Attached.			
11 Aug. 00	Anderson, Maj. K. E.	1 Sep. 15	28 Mar. 17	Cant. Magte.'s Dept.
29 Aug. 06	Kemble, Capt. T. A., *q.*	29 Aug. 15	9 Dec. 17	81st Prs.
17 Aug. 07	Fraser, Capt. S. G. G., *q.*	17 Aug. 16	31 Jan. 18	81st Prs.
4 May 15	Black, Lt. J. M.	4 Aug. 17	8 Dec. 17	I. A. (*on probn.*)
17 June 15	Hickey, Lt. H. G.	17 June 16	15 May 18	E. York. R., I. A. (*on probn.*)
4 Aug. 15	Leeming, Lt. A.	11 Aug. 17	7 Dec. 17	I. A. (*on probn.*)
25 Oct. 15	England, Lt. R. S.	25 Oct. 15	8 Dec. 17	I. A. R. O.
15 Nov. 15	Goddard, Lt. E. N.	15 Nov. 16	23 Sep. 16	107th Prs.
23 Nov. 15	Gemmell, Lt. G.	23 Nov. 16	30 Nov. 17	I. A. R. O.
22 Dec. 15	Tyson, Lt. E. W.	22 Dec. 16	4 Oct. 17	106th Prs.
18 Apl. 16	Stewart, Lt. C. H.	18 Apl. 17	28 Oct. 17	106th Prs.
10 May 16	Baker, Lt. S.	10 May 17	22 Sep. 16	I. A. R. O.
16 Aug. 16	Greengrass, Lt. R. P. W.	16 Aug. 17	14 Nov. 18	Manch. R., I. A. (*on probn.*)
25 Oct. 17	Lewis-Lewis, Lt. S.	25 Oct. 18	9 Sep. 18	Welsh R., I. A. (*on probn.*)
1 May 18	Rae, 2nd-Lt. R. J.	1 May 18	4 Aug. 18	
29 May 18	Mawson, 2nd-Lt. J. E.	29 May 18	30 Aug. 18	I. A. R. O. (*on probn.*)
7 June 18	Tidmarsh, 2nd-Lt. C. F.	7 June 18	....	I. A. (*T. C.*)
17 July 18	Stewart, 2nd-Lt. T. A.	17 July 18	4 Dec. 18	I. A. R. O.
30 Aug. 18	Mahar, 2nd-Lt. F. M.	30 Aug. 18	23 Sep. 18	I. A. R. O.
5 Sep. 18	Cotton, 2nd-Lt. S. C. N.	5 Sep. 18	27 Sep. 18	I. A. R. O.
23 Oct. 18	Clean-Murphy, 2nd-Lt. E. W.	23 Oct. 18	15 Nov. 18	I. A. R. O.

		Jemadar.	Subadar.	Subadar Major.
	Subadar-Major.			
Dec. 96	Krishna Bhosle *Bahadur* (56, 58) *Order of Br. Ind., 2nd Cl.*, I.D.S.M.	1 Mar. 04	25 Dec. 07	22 Jan. 16
	Subadars			
10 Sep. 90	Maroti Ghorpade (56)	16 Feb. 04	1 Sep. 06	
12 Apl. 99	Karm Singh *Bahadur* (56, *Order of Br. Ind. 2nd Cl.*, I.D.S.M.	4 Dec. 07	25 Aug. 12	
22 May 99	Jiwan Singh (56)	6 May 08	20 Oct. 14	
1 Mar. 04	Shanur Khan (56)	1 Sep. 10	26 Dec. 15	
17 Feb. 12	Sher Afzal, M.C., I.D.S.M.	11 July 12	8 Jan. 16 8 Feb. 16	
19 Aug. 02	Musa Khan, I.O.M. (56)	12 Apl. 1	8 Jan. 16	
20 Oct. 08	Abdul Latif Khan, I.O.M.	20 Oct. 14	8 Jan. 16	
6 Jan. 93	Tukaram Jadhu (56), I.D.S.M.	10 Dec. 95	22 Jan. 16	
11 Apl. 99	Teja Singh (56), I.D.S.M.	25 Aug. 12	7 Feb. 16	
8 Dec. 01	Bakhtawar Khan	26 Dec. 15	12 Dec. 16	
12 July 98	Chand Khan (56)	15 Apl. 13	23 Aug. 17	
	Yassin Khan (*retd.*)	....	1 Feb. 18	Re-empld.

1st Battalion, 128th Pioneers—*contd.*

1st Commn. or date of entering service.	Names and Rank.	Army rank.	Remarks.
	Jemadars.		
11 June 95	Mahadeo Chowdry	20 Oct. 14	
6 Jan. 02	Alam Khan	20 Oct. 14	
3 Aug. 81	Bijai Bahadur Singh	20 Oct. 14	
12 Apl. 99	Indar Singh, I.O.M. (56)	20 Oct. 14	
3 June 00	Bala Ghorpade	22 Jan. 16	
6 Nov. 90	Ganda Singh, I.O.M.	18 Apl. 16	
18 Dec. 03	Hakim Khan	18 Apl. 16	
1 Aug. 94	Gopal Ittu	1 Mar. 18	
6 May 13	Ahmed Khan	11 Mar. 18	
6 Nov. 96	Ramchandar Goli	8 Apl. 18	
12 June 06	Abdul Akbar	5 Jan. 17	
28 May 99	Nawab Singh	21 Apl. 18	
21 Feb. 99	Labh Singh, I.D.S.M.	19 May 18	
14 Aug. 02	Ghulam Ali	19 May 18	
15 Aug. 02	Samiullah	19 May 18	
8 Jan. 03	Sadojirao Kadam	19 May 18	

2nd Battalion, 128th Pioneers.

Formed at Arangaon.

Composition.—4 platoons of Pathans, 4 platoons of Lobana and Saini Sikhs, 2 platoons of Rajput Musalmans, 2 platoons of Dekhani Mahrattas, 4 platoons of Herio.

1st Commn. or date of entering service.	Names and Rank.	Army Rank.	Present appointment in Regiment.	Remarks.
	Commandant.			
5 Aug. 96	Ashburner, Maj. H. W., D.S.O., *q.* (*Actg. Lt.-Col.*)	5 Aug. 14	22 May 18	106th Prs.
	Company Commanders (4).			
20 Jan. 09	Ferrier, Capt. I. *q.*	1 Sep. 15	23 May 18	48th Prs.
20 Nov. 14	Shrager, Lt. A. L. (*actg. Capt.*) .	20 Nov. 15	20 Mar. 18	I. A. R. O.
23 Dec. 14	Woolf, Lt. S. H.(*actg. Capt.*) .	23 Dec. 15	. . .	19th Punj.
	Company Officers.			
13 Aug. 15	Shute, Lt. P. B.	13 Aug. 16	31 May 18	R. E., I. A. (*on probn.*)
26 Jan. 16	Pilkington, Lt. T. L. (*actg. Capt.*)	26 Jan. 17	15 May 18	R. Ir. Regt. I. A. (*on probn.*) Adjt.
19 July 16	Cloran, Lt. D. J. H.	19 July 17	6 June 18	1-81st Prs.
8 Aug. 17	Clarke, Lt. A. W.	8 Aug. 18	30 May 18	I. A. R. O.
11 Sep. 17	Penn, Lt. C. L.	11 Sep. 18	15 May 18	E. York. R., I. A. (*on probn.*)
5 Oct. 17	Weeks, Lt. V. H.	5 Oct. 18	31 May 18	I. A. R. O.
5 Oct. 17	Carter, Lt. G. H. H.	5 Oct. 18	6 June 18	I. A. (*T. C.*)
2 Nov. 17	Shipley, 2nd-Lt. H. (46) . . .	2 Nov. 17	3 June 18	I. A. R. O.
25 Nov. 17	Wootten, 2nd-Lt. R. M. . . .	25 Nov. 17	16 Mar 18	Lond. R., I. A. R. O.
5 Jan. 18	Ashford, 2nd-Lt. E.	5 Jan. 18	3 June 18	I. A. (*T. C.*)
5 Jan. 18	Green, 2nd-Lt. H. T.	5 Jan. 18	3 June 18	I. A. (*T. C.*)
15 Jan. 18	Forder, 2nd-Lt. H. N. . . .	15 Jan. 18	3 June 18	I. A. (*T. C.*)
30 Jan. 18	Clarkson, 2nd-Lt. A. S.	30 Jan. 18	8 Nov. 18	H. L. Infy., I. A. R. O. (*on probn.*)
2 June 18	Withnell, 2nd-Lt. J.	2 June 18	19 Oct. 18	I. A. (*T. C.*)
7 June 18	Mintey, 2nd-Lt. A. H.	7 June 18	8 Oct. 18	I. A. (*T. C.*)
7 June 18	Tyndale, 2nd-Lt. H. E. . . .	7 June 18	10 Oct. 18	I. A. (*T. C.*)
20 Sep. 18	Gainsford, 2nd-Lt. D.	20 Sep. 18	24 Oct. 18	I. A. R. O.

		Jemadar.	Subadar.	Subadar-Major.	
	Subadar-Major.				
5 July 90	Khushal Khan (56) *t*	24 Aug. 07	11 July 12	26 Apl. 18	(*Temp.*)
	Subadars.				
1 July 97	Baboo Rao Mohite	7 Aug. 13	12 Dec. 15		
29 Dec. 99	Nishan Ali Khan (56, 65) . . .	18 Oct. 08	26 Jan. 16		
19 Oct. 96	Jota Singh	9 Aug. 14	29 Feb. 16		
15 Dec. 98	Sita Ram Barge	27 Aug. 10	1 Apl. 18		
6 Nov. 99	Mal Singh	20 Oct. 14	1 Apl. 18		
19 May 03	Aziz Khan	8 Feb. 16	4 Apl. 18		
10 Dec. 01	Male Khan	25 Feb. 17	4 Apl. 18		
	Jemadars.				
8 Dec. 98	Chena Singh	1 Apl. 18			
5 Apl. 99	Kondi Jagtap	1 Apl. 18			
1 Aug. 03	Krishna Wagaude	1 Apl. 18			
20 Aug. 06	Haidar Khan	10 Apl. 18			
31 Mar. 06	Muhammad Shafi Khan . . .	2 May 18			
	Mumtaz Khan	8 July 18	(*On probn.*)		
	Sher Khan	. . . .	(*On probn.*)		
	Diwan Ali	17 Aug. 18			

Regt. Centre—Karachi.
Linked with—127th & 130th Baluchis.

Uniform—Green (Red serge trousers).
Facings—Scarlet.

1st Battalion, 129th Duke of Connaught's Own Baluchis.

Raised at Karachi in 1846, by Capt. Wright as the 2nd Baluch Battalion. Became the 2nd Baluch Extra Bn. of Bombay Native. Infantry 1858; the 2nd Baluch Regt., 1859; the 2nd Bombay Baluch Regt., 1859; the 29th Regt. of Bom. N. I., or 2nd Baluch Regt., 1861; the 29th (the Duke of Connaught's Own) Regt. of Bom. N. I. or 2nd Baluch Regt., 1883; the 29th (the Duke of Connaught's Own) Regt. of Bom. Infy., or 2nd Baluch Regt., 1885; the 29th (the Duke of Connaught's Own) Regt. of Bom. Infy. (the 2nd Baluch Bn.), 1888; the 29th (the Duke of Connaught's Own) Baluch Infy., 1903; present designation, 1903.

"Reshire" "Bushire" "Koosh-ab" "Persia."
"Kandahar, 1880" "Afghanistan, 1878-80" "Tel-el-Kebir" "Egypt, 1882."

Composition—1 Company of Punjabi Musalmans, 1½ of Mahsuds, 1½ of other Pathans. 1 special Coy. of Araino.

COLONEL-IN-CHIEF—Field Marshal *His Royal Highness* Arthur William Patrick Albert, *Duke of* CONNAUGHT and STRATHEARN, K.G., K.T., K.P., G.C.B., G.C.S.I., G.C.M.G., G.C.I.E., G.C.V.O., A. D. C., G.B.E. . . . . 13 May 1904.
COLONEL—General *Sir* O'Moore Creagh, V.C., G.C.B., G.C.S.I. . . 16 February 1912.

1st Commn. or date of entering service.	Names and Rank.	Army rank.	Present appointment in Regiment.	Remarks.
	Commandant.			
24 July 91	Hay, Lt.-Col. W.N., C.I.E., D.S.O., *p.s.c.* [L.]	24 July	22 Aug. 18	
	Company Commanders. (4)			
22 Jan. 98	Hamir, Maj. M. A., D.S.O., M.C., *p.s.c.*	1 Sep. 15	22 Aug. 18	2nd-in-Comd., with 2-125th Rif.
4 July 03	Clive, Maj. H. W. F. . . .	4 July 18	1 Mar. 17	*Offg.* 2nd-in-Comd.
	Company Officers. (9)			
17 Aug. 07	Griffith-Griffin, Bt.-Maj. F. M., M.C.	1 Jan. 17	14 Nov.	With 62nd Punj.
9 Sep. 08	Lewis. Capt. H.V., D.S.O., M.C. .	1 Sep. 15	24 Dec. 09	With 2-127th Bal. L. I.
26 July 13	*MacIvor, Capt. R. S. P.* . . .	*1 Sep. 15*	*8 Oct. 14*	
8 Aug. 14	Palin, Capt. C. W., M.C. . . .	8 Aug. 18	15 Oct. 14	
22 Oct. 14	Roberts, Lt. V. G.	22 July 18	. . . .	With 2-129th Baluchis.
15 Nov. 15	Brilliant, Lt. L..	15 Nov. 16	17 Nov. 15	With 2-55th Rif.
15 Nov. 15	Soden, Lt. J. N.	15 Nov. 16	17 Nov. 15	With 2-124th Infy.
18 Apl. 16	McClenaghan, Lt. P. J. W. . .	18 Apl. 17	27 Apl. 16	With 93rd Infy.
23 May 16	McKinstry, Lt. H.	23 May 17	30 Nov. 17	Attd.
31 Jan. 18	Barnes, 2nd-Lt. J. H.	31 Jan. 18	31 Jan. 18	With 2-70th Rif.
. . . .	Sinclair, 2nd-Lt. M. C.	. . . . .	. . . .	Attd.
	Attached.			
19 Jan. 07	Hinde, Capt. R. G.	1 Sep. 17	. . . .	1-124th Infy.
1 June 15	Gordon-Smith, Lt. W. . . .	1 June 16	29 Jan. 18	*Lv., m.c., ex I.*
1 Aug. 15	Bowerman, Lt. J. F.	1 Aug. 16	. . . .	I. A. (*on probn.*)
15 Nov. 15	Hawkes, Lt. G. E., M.C. . . .	15 Nov. 16	. . . .	54th Sikhs.
23 Nov. 15	Leslie-Smith, Lt. B. R. . . .	23 Nov. 16	. . . .	I. A. (*on probn.*)
18 Apl. 16	Guiton, Lt. L. W. (*Actg. Capt.*)	18 Apl. 17	. . . .	67th Punj. *Offg.* Adjt.
10 June 16	Ewens, Lt. G. P.	10 June 17	. . . .	I. A. R. O.
14 Jan. 16	Wall, Lt. R. H.	14 Jan. 17	. . . .	I. A. (*on probn.*)
24 Apl. 16	Brindley, Lt. W. W.	24 Apl 17	. . . .	I. A. (*on probn.*)
7 Oct. 16	Steel, Lt. E. G.	7 Oct. 17	12 Feb. 18	I. A. (*on probn.*)
8 Dec. 16	Roach, Lt., I. N.	8 Dec. 17	8 June 18	I. A. (*on probn.*)
2 Jan. 17	Wickham, Lt. J. J. O. (207*a*) .	2 Jan. 18	15 Feb. 18	I. A. (*on probn.*)
29 May 17	Hebron, Lt. C. W.	29 May 18	. . . .	I. A. (*on probn.*)
20 June 17	Barker, Lt. R.	20 June 17	. . . .	I. A. R. O. Qr. Mr. 1 Feb. 18.
21 Dec. 17	King, Lt. E. W.	21 Dec. 18	. . . .	I. A. R. O.
15 Jan. 18	Campbell, 2nd-Lt., A. A. . .	15 Jan. 18	8 June 18	I. A. R. O.
. . . .	Nicholson, Lt. E. A. . . .	. . . .	. . . .	D. L. Infy. I. A. R. O. (*on probn.*)
7 June 18	Scorgie, 2nd-Lt. N. A. . . .	7 June 18	. . . .	I. A. (*T. C.*)
15 July 18	Van Langenberg, 2nd-Lt. J. F. .	15 July 18	. . . .	I. A. R. O.

		Jemadar.	Subadar.	Subadar-Major.	
	Subadar-Major.				
24 Mar. 90	Zaman Khan, *Bahadur* M.C. (56)	10 Mar. 05	1 July 11	6 Sep. 18	*Order of Br. Ind., 2nd Cl.*
	Subadars.				
21 July 86	Amir Khan	5 Aug. 03	1 Oct. 10		
11 Dec. 11	Turkestan	14 Jan. 12	18 July 13		
12 Dec. 11	Mir Badshah (56, 66*a*), I.O.M. .	1 Nov. 13	31 Oct. 14		
16 Dec. 95	Muhammad Bakhsh (56) . . .	8 Aug. 14	16 Jan. 16		
19 Nov. 00	Sikandar Khan	20 Dec. 14	17 Dec. 16		
16 Aug. 08	Nur Khan, I.O.M., I.D.S.M. . .	14 Feb. 16	4 Apl. 17		
28 Apl. 94	Rajwali Khan	14 Dec. 16	8 Apl. 17		
7 Mar. 91	Amir Ali Shah	7 Jan. 17	26 June 18		
4 Aug. 11	Fazl Dad	8 Apl. 17	4 Apl. 18		With 2-130th Infy.
15 July 03	Gaman Khan	27 Nov. 17	26 June 18		

1st Battalion, 129th Duke of Connaught's Own Baluchis—*contd.*

1st Commn. or date of entering service.	Names and Rank.	Army rank.	Remarks.
		Jemadars.	
1 Oct. 00	Id-Muhammad	31 Oct. 14	
5 Jan. 11	Muhammad Jan	17 Dec. 16	
30 Oct. 11	Bagh Khan	30 Dec. 16	
4 Aug. 11	Reshmin (46)	3 Apl. 17	
25 Apl. 13	Ghafaran Khan	6 Aug. 17	
3 Dec. 17	Raza Ali Shah	3 Dec. 17	
9 June 93	Buta Khan	4 Apl. 18	
31 Jan. 01	Nawab Ali	26 June 18	
5 Sep. 02	Gheba Khan	26 June 18	
17 Sep. 09	Majhi Khan	26 June 18	
5 Nov. 15	Nawab Khan	26 June 18	
....	Sohaj Khan	26 June 18	
....	Mir Zaman	26 June 18	
....	Nur Muhammad	26 June 18	
....	Burhan Ali	26 June 18	
....	Manowar Din	22 Oct. 18	
....	Ghulam Haidar	22 Oct. 18	

2nd Battalion, 129th Duke of Connaught's Own Baluchis.

Formed at Quetta.

Composition—1 company of Baluchis, 3 companies of Punjabi Musalmans.

1st Commn. or date of entering service.	Names and Rank.	Army rank.	Present appointment in Regiment.	Remarks.
	Commandant.			
22 Jan. 98	Merriman, Maj. A. B. (*Actg. Lt.-Col.*)	1 Sep. 15	9 June 18	1-127th Bal. L. I.
	Company Commanders. (4)			
28 Jan. 11	Cuddeford, Capt. R. B. G. (*Actg. Maj.*).	1 Sept. 15	6 Sept. 18	1-127th Bal. L. I., 2nd-in-Comd.
6 Sep. 16	Mercer, Capt. A. A.	1 May 18	26 May 18	29th Lrs.
21 Nov. 14	Goldsmith, Lt. W. S.	21 Nov. 15	20 Jan. 18	I. A. R. O.
15 May 15	Sommers Cox, Lt. S. J., M.C.	16 May 16	27 Jan. 18	Worc. R., I. A. (*on probn.*)
	Company Officers.			
20 Oct. 14	Robert, Lt. V. G., M.C., (*Actg. Capt.*).	22 June 16	1 Apl. 18	1-129th Baluchis. Adjt.
24 Dec. 15	Alderson, Lt. V. C., M.C.	24 Dec. 16	3 Mar. 18	Aus. Imp. Forces, I. A. (*on probn.*)
16 Mar. 15	Bond, Lt. K. H., M.C.	16 Mar. 16	1 Apl. 18	I. A. *on probn.*)
14 July 15	Hanna, Lt. J. R. M.	14 July 16	29 Jan. 18	I. A. R. O. (*on probn.*)
5 Aug. 16	Strachan, Lt. G. R.	5 Aug. 17	5 Apl. 18	I. A. (*on probn.*)
8 Oct. 16	McClenaghan, Lt. H. E.	8 Oct. 17	26 Nov. 18	R. I. Fus., I. A. (*on probn.*)
1 June 17	Barkeley, Lt. M. F.	1 June 18	4 Jan. 18	I. A. R. O. *Lv., m. c., ex I., 6 mths.*
1 Nov. 17	Jenkins, Lt. C. H.	1 Nov. 18	18 Mar. 18	I. A. (*T. C.*)
1 Dec. 17	Bartle Pye, Lt. H. (207*a*)	1 Dec. 18	20 Feb. 18	Dors. R., I. A. (*on probn.*)
21 Dec. 17	Crabbe, 2nd-Lt. C. B. (207*a*)	21 Dec. 17	29 Mar. 18	I. A. (*T. C.*)
31 Jan. 18	Larminie, 2nd-Lt. J. A.	31 Jan. 18	20 Feb. 18	
26 Apl. 17	Winton, 2nd-Lt. W. C.	26 Apl. 17	23 Apl. 18	I. A. R. O. (*on probn.*)
15 Jan. 18	Clarke, 2nd-Lt. A. C.	15 Jan. 18	6 June 18	I. A. (*T. C.*)
15 Jan. 18	Hawkes, 2nd-Lt. J. G.	15 Jan. 18	6 June 18	I. A. (*T. C.*)
31 Jan. 18	Cahill, 2nd-Lt. J.	31 Jan. 18	20 Feb. 18	Army Signals.
10 Sep. 18	Thirsk, 2nd-Lt. F. A.	10 Sept. 18	2 Dec. 18	I. A. (*T. C.*)

		Jemadar.	Subadar.	Subadar-Major.	
	Subadar-Major.				
19 Sep. 93	Abdullah Khan, I. O. M.	1 Mar. 14	3 June 15	1 Apl. 18	*Actg.*
	Subadars.				
5 June 07	Gulam Jilani, I.D.S.M.	22 Oct. 14	2 Jan. 16		
13 Feb. 01	Rahim Alli	16 Dec. 14	23 Dec. 16		
21 July 94	Nur Khan, I.D.S.M.	2 Jan. 16	1 Feb. 18		
10 July 12	Gaus Muhammad	13 Feb. 17	1 Feb. 18		
	Jemadars.				
26 Oct. 06	Khuda Dad Khan	15 Mar. 17			
4 Apl. 02	Pelhwan Khan	6 Aug. 17			
4 Apl. 02	Hayat Ali	16 June 17			
27 Feb. 90	Sultan Ali	1 Feb. 18			Qr.-Mr. Jemdr.
30 Oct. 11	Sher Khan	1 Feb. 18			Adjt. Jemdr.
25 Mar. 01	Atta Muhammad	1 Feb. 18			
22 Aug. 04	Sardar Khan	1 Feb. 18			
19 Jan. 18	Khan Muhammad	1 Feb. 18			
16 July 18	Muhammad Zanib	16 July 18			
24 July 18	Haidar Rahman	24 July 18			
17 Sep. 18	Abdul Aziz	17 Sep. 18			
1 July 18	Bashir Ahmed Khan	1 Oct. 18			
25 Feb. 08	Badal Khan	1 Oct. 18			

Regt. Centre—Karachi.
Linked with—127th & 129th Baluchis.

Uniform—Green (Red serge trousers).
Facings—Scarlet.

130th King George's Own Baluchis (Jacob's Rifles).

Raised, in 1858, by Major John Jacob, as the 1st Regiment of Jacob's Rifles; became "Jacob's Rifles," 1861, and the 30th Regt. of Bombay Native Infantry or Jacob's Rifles, 1861; the 30th Regt. of Bombay Native Infantry or 3rd Baluch Battalion, 1881; the 30th Regt. of Bombay Infy. (3rd Baluch Battn.), 1885; the 30th Baluch Infantry, 1901; the 130th Baluchis in 1903; the 130th Prince of Wales's Own Baluchis, 1st January 1906; 130th Prince of Wales's Own Baluchis (Jacob's Rifles) and present designation, 1910.

THE PLUME OF THE PRINCE OF WALES. THE ROYAL AND IMPERIAL CYPHER.

"Afghanistan, 1878-80" "China, 1900."

Composition—3 companies of Punjabi Musalmans, 1 of other Pathans.

COLONEL-IN-CHIEF—THE KING.

1st Commn. or date of entering service.	Names and Rank.	Army rank.	Present appointment in Regiment.	Remarks.
	Commandant.			
11 Feb. 88	Price, Bt.-Col. C. U., C.M.G.	1 Jan. 17	17 Apl. 15	*Lv., ex I., m.c., 6 mths.*
	Company Commanders. (4)			
18 May 92	Dyke, Lt.-Col. P. H., D.S.O. (Q) (*temp. Col.*)	1 Jan. 17	17 Apl. 15	2nd-in-Comd. *Offg.* Comdt.
27 July 98	Kensington, Bt.-Lt.-Col. E. C., M.C., *p.s.c.*	. . . . .	21 Nov. 14	With 91st Punj.
19 May 00	Morton-Marshall, Maj. F. W.	1 Sep. 15	17 Apl. 15	With 2-56th Infy.
	Company Officers. (9)			
17 Oct. 00	Sealy, Maj. H. C.	17 Oct. 15	24 June 03	With 1-35th Sikhs. S. S. O., Hyderabad (Sind).
5 Jan. 01	Moore, Maj. H. D., D.S.O.	5 Jan. 16	17 Jan. 04	*Offg.* 2nd-in-Comd.
10 Oct. 03	Woodward, Maj R. F. [*l*]	10 Oct. 18	7 Aug. 07	
24 Jan. 06	*Jacob, Capt. R. M.*, M.C. (*temp. Maj.*)	*24 Jan. 15*	*1 Apl. 08*	D. A. A. G.
16 Jan. 07	*Simpson, Capt. R. C.*	*1 Sep. 15*	*6 Oct. 11*	
22 Feb. 07	Pim, Capt. G. A. (Actg. Maj.)	1 Sep. 15	7 Apl. 10	Comdg. Depôt.
3 Sep. 10	Minter, Capt. J. P. S.	1 Sep. 15	9 Nov 11	With 2-152nd Infy.
30 Jan. 17	Kelly, Lt. V. S.	30 Jan. 18	13 Sep. 17	
30 Jan. 17	Pratt, Lt. A. R.	30 Jan. 18	. . . .	With 1-127th Bal. L. I.
	Attached.			
13 Aug. 04	White, Capt. A. G.	13 Aug. 13	. . . .	126th Bal. L. I.
27 Nov. 14	Fuge, Capt. T. L. M.	17 Oct. 18	6 Dec. 14	I. A. R. O.
12 Sep. 15	Mackenzie-Rogan, Capt. H.	12 Sep. 18	10 Dec. 17	I. A. R. O.
14 Sep. 14	Lang, Lt. W. I. (Actg. Capt.)	14 Sep. 14	30 Jan. 18	I. A. R. O. Adjt.
14 Nov. 14	Burton, Lt. R. A.	14 Nov. 14	26 Jan. 18	I. A. R. O. (*on probn.*)
5 Mar. 15	Archdale, Lt. F. A. (*temp. Capt.*)	5 Mar. 16	5 Nov. 15	I. A. R. O.
8 Apl. 15	Hull, Lt. B.	8 Apl. 16	10 Dec. 17	I. A. (*on probn.*).
5 May 15	Richardson, Lt. A. J. C.	5 May 16	4 Sep. 18	I. A. (*on probn.*)
9 July 15	Blanford, Lt. A. W., M.C. (*temp. Capt.*)	9 July 16	31 Dec. 15	I. A. R. O.
24 Sep. 15	Rendell, Lt. C. B.	24 Sep. 18	17 Jan. 18	I. A. R. O.
28 Oct. 15	Cooper, Lt. H. G., M.C. (*temp. Capt.*)	28 Oct. 16	1 May 16	I. A. R. O.
16 Nov. 15	Evans, Lt. S., M.C.	16 Nov. 15	6 Dec. 17	I. A. R. O.
2 June 16	Simmonds, Lt. V. B.	2 June 17	20 July 16	I. A. R. O.
26 July 16	Garden, Lt. R. L.	26 July 17	12 Dec. 17	I. A. (*on probn.*)
5 Aug. 16	Sinclair, Lt. N.	5 Aug. 17	17 Jan. 18	I. A. R. O.
21 Aug. 16	Campbell, Lt. A., M. C.	21 Aug. 17	23 Jan. 18	I. A. R. O.
23 Nov. 16	Martin, Lt. T.	23 Nov. 16	9 Dec. 17	I. A. R. O.
1 Nov. 16	Erskine, Lt. W. R.	1 Nov. 17	8 Dec. 16	I. A. R. O.
27 Apl. 17	Hawkeswood, Lt. H. W.	26 Apl. 18	16 Sep. 18	I. A. R. O.
1 June 17	Boyce, Lt. C. O.	1 June 18	2 Sep. 18	I. A. R. O.
29 Aug. 17	Darling, Lt. J. S.	29 Aug. 18	17 Jan. 18	I. A. R. O.
28 Aug. 17	Harrison, Lt. M. K.	28 Aug. 18	15 Feb. 18	I. A. R. O. (*on probn.*)
31 Jan. 18	Lindsay, 2nd-Lt. P. L.	31 Jan. 18	6 Feb. 18	I. A. (*on probn.*)
26 June 18	Campbell, 2nd-Lt. J.	26 June 18	2 Oct. 18	R. Highrs., I. A. (*on probn.*)
1 Oct. 18	Keenan, 2nd-Lt. T. P. C. S.	1 Oct. 18	8 Oct. 18	*U. L.*
. . . .	Blomfield, 2nd-Lt. C.	. . . . .	19 Nov. 18	I. A. R. O. (*on probn.*)
. . . .	Wello, 2nd-Lt. A.	. . . . .	2 Nov. 18	I. A. R. O. (*on probn.*)
. . . .	Stewart, 2nd-Lt. J. R.	. . . . .	3 Nov. 18	I. A. R. O. (*on probn.*)

		Jemadar.	Subadar.	Subadar-Major.	
	Subadar-Major.				
Dec. 88	Muhammad Khan *Bahadur* (72a)	1 July 02	2 Aug. 07	1 Feb. 15	*Order of Br. Ind., 2nd Cl.*

130th King George's Own Baluchis (Jacob's Rifles)—*contd.*

1st Commn. or date of entering service.	Names and Rank.	Army rank.	Present appointment in Regiment.	Remarks.
	Subadars.			
12 July 95	Kudrat Shah *Bahadur*	1 June 04	1 Nov. 04	*Order of Br. Ind., 2nd Cl.*
20 Sep. 95	Muhammad Din (72a)	11 Jan. 09	10 Sep. 14	
13 Jan. 96	Saif Ali (52, 56, 66a)	15 Sep. 07	1 May 16	
13 June 96	Saidulla Khan (52, 56, 66a)	10 Sep. 14	7 Aug. 16	
30 Sep. 99	Hoshnak Khan (52, 56, 72a)	10 Sep. 14	7 Aug. 16	
	Jan Muhammad, *Bahadur* (Retd.)	....	....	Re-empld., 27 Apl. 15.
15 Dec. 03	Zaman Khan, (66a, 56)	13 Aug. 15	4 Oct. 16	
10 Dec. 97	Nowsher, I.D.S.M.	1 Jan. 16	14 Oct. 16	
24 Jan. 05	Sharaf Din (56, 72a)	17 May 16	14 Oct. 16	
23 June 96	Ilam Din	26 Jan. 15	26 Jan. 17	
27 Aug. 97	Taj Muhammad	10 Sep. 14	15 Feb. 17	
3 Sep. 98	Allah Dad (56, 66a, 72a)	26 Jan. 15	6 Aug. 17	
9 Mar. 12	Mir Zaman (56)	4 Oct. 16	11 Oct. 18	
	Jemadars.			
5 Dec. 98	Sultan Muhammad	1 May 16		
27 July 97	Shaikh Sirajuddin	1 June 16		
9 Sep. 99	Shakar Mehdi (46, 56, 58)	7 Aug. 16		
22 Sep. 05	Fazl Ilahi	7 Aug. 16		
21 May 03	Muhammad Ali, I.D.S.M. (60a)	14 Oct. 16		
18 Mar. 02	Nek Muhammad (66a)	13 Dec. 16		
5 July 99	Bahadur Khan (72a)	2 Jan. 17		
23 Apl. 03	Ahmad Khan (58)	10 Jan. 17		
30 June 03	Said Habib	15 Feb. 17	....	Indian Adjt.
	Muhammad Abdullah (Retd.)	....	....	Re-empld., 19 Jan. 16.
1 Aug. 96	Bagga Khan	20 Sep. 18		
....	Fateh Muhammad	12 Oct. 18		

2nd Battalion, 130th King George's Own Baluchis.

Formed at on

Composition :—

1st Commn. or date of entering service.	Names and Rank.	Army Rank.	Present appointment in Regiment.	Remarks.
	Commandant.			
4 Aug. 97	Loch, Maj. J. C. (*Actg. Lt.-Col.*)	4 Aug. 15	12 Nov. 18	3rd G. R.
	Company Commanders.			
17 Aug. 07	Drought, Capt. J. V. (temp. Major).	1 Sep. 15	29 Oct. 18	2nd-in-Comd. 124th Infy.
27 July 15	Dyer, Lt. A. J.	27 July 16	1 Nov. 18	I. A. R. O.
	Company Officer.			
28 Nov. 17	Heron, 2nd-Lt. J.	23 Nov. 17	. . . .	I. A. R. O.
	Medical Officer.			
	Subadar-Major.			
		Jemadar.	Subadar.	Subadar-Major.
	Subadars.			
4 Aug. 11	Fazal Dad	8 Apl. 17	4 Apl. 18	
8 June 95	Nurdad, I. O. M.	2 Nov. 14	1 July 15	
	Jemadars.			
. . . .	Zaman Shah	26 June 18		
4 Nov. 10	Naim Khan	6 Aug. 17		
8 Apl. 01	Sahib Alam	1 Oct. 15		
1 Sep. 05	Azimullah	1 Jan. 17		

1st Battalion, 131st United Provinces Regiment.

Formed at on

Composition—3 Companies Rajputs, 1 Company Hindustani Musalmans.

1st Commn. or date of entering service.	Names and Rank.	Army rank.	Present appointment in regiment.	Remarks.
	Commandant.			
24 July 18	Warner, 2nd-Lt. H. R. (*Actg. Lt.-Col.*)	24 July 18	26 July 18	I. A. R. O.
	Company Commanders. (4)			
5 Jan. 01	Collins, Maj. L. P., D.S.O. . .	5 Jan. 16	30 Oct. 18	1-4th G. R., 2nd-in-Comd.
26 Aug. 16	Smith, Lt. C. C. H.	26 Aug. 17	11 Oct. 18	1-4th Raj.
21 June 18	O'Riordon, 2nd-Lt. K. M. (*Actg. Capt.*).	21 June 18	. . . .	I. A. R. O.
21 Jan. 18	Horton, 2nd-Lt. R. A. (*Actg. Capt.*)	21 June 18	. . . .	I. A. R. O.
	Company Officers.			
5 Jan. 17	Burnell, Lt. R. G.	5 Jan. 18	5 Nov. 18	I. A. (*on probn.*)
12 Sep 17	Brown, Lt. L. C. H.	12 Sep. 18	20 Sep. 18	1-67th Punj., Qr. Mr.
27 Oct. 17	Rouse, 2nd-Lt. R. P. . . .	27 Oct. 17	27 Oct. 18	1-7th G. R.
21 Dec. 17	Powell, 2nd-Lt. F. T.	21 Dec. 17	21 Dec. 17	Attd.
21 Dec. 17	Connolly, 2nd-Lt. D. R. V. . .	21 Dec. 17	27 Oct. 18	38th Dogras.
21 Dec. 17	Partridge, 2nd-Lt. H. V. . . .	21 Dec. 17	28 Oct. 18	2-1st G. R.
28 Dec. 17	Hall, 2nd-Lt. R. M.	28 Dec. 17	11 Oct. 18	North'd Fus., I. A. (*on probn.*)
22 May 18	Elliot, 2nd-Lt. W.	22 May 18	3 July 18	I. A. R. O.
7 June 18	Lester, 2nd-Lt. C.	7 June 18	. . . .	I. A. (*T. C.*)
7 June 18	Hill, 2nd-Lt. A. E.	7 June 18	. . . .	I. A. R. O.
21 June 18	Molony, 2nd-Lt. H. J. C. . .	21 June 18	9 June 18	I. A. R. O.
21 June 18	Blunden, 2nd-Lt. A. C. (*temp. Lt.*)	21 June 18	. . . .	I. A. R. O.
21 June 18	Field, 2nd-Lt. G. G. (*temp. Lt.*)	21 June 18	. . . .	I. A. R. O.
21 June 18	Coulton, 2nd-Lt. G. F. . . .	21 June 18	3 July 18	I. A. R. O.
21 June 18	Luck, 2nd-Lt. C. P.	21 June 18	24 July 18	I. A. R. O.
31 Jan. 18	Hourihane, 2nd-Lt. T. C. . . .	31 Jan. 18	31 Jan. 18	1-41 Dogras.
28 Feb. 18	Whittle, 2nd-Lt. H. E. . . .	28 Feb. 18	5 Oct. 18	I. A. (*on probn.*) Adjt., 31 Oct. 18.
22 July 18	Lavin, 2nd-Lt. J. F.	22 July 18	2 Dec. 18	I. A. R. O.
22 July 18	Pryor, 2nd-Lt. J. W.	22 July 18	6 Dec. 18	I. A. (*T. C.*)
28 Oct. 18	Prebble, 2nd-Lt. J.	28 Oct. 18	20 Nov. 18	I. A. R. O.
31 Aug. 18	Power, 2nd-Lt. W. N.	31 Aug. 18	31 Aug. 18	52nd Sikhs.
		Jemadar.	Subadar.	Subadar-Major.
	Subadar-Major.			
	Subadars.			
25 Aug. 88	Bulaki Singh (56, 66*a*) . . .	1 Jan. 07	14 May 15	16th Raj.
19 Dec. 00	Shiunath Singh (56)	5 Aug. 12	1 Jan. 18	1-8th Raj.
26 June 96	Madad Khan (52, 56, 66*a*) . .	1 Aug. 14	25 June 17	2-89th Punj.
	Jemadars.			
1 Aug. 17	Rampal Singh	1 Aug. 17		

2nd Battalion, 131st United Provinces Regiment.

Formed at Bareilly on 2 July 18.

Composition—1 Company of Brahmans, 1 Coy. of Hillmen, 1 Coy. of Hindustani Musalmans, and 1 Coy. of Ahirs.

1st Commn. or date of entering service.	Names and Rank.	Army rank.	Present appointment in regiment.	Remarks.
	Commandant.			
21 June 18	Adams, 2nd-Lt. R. J. (*Actg. Lt.-Col.*).	21 June 18	2 July 18	I. A. R. O.
	Company Commanders (4).			
25 May 18	Eyre, 2nd-Lt. M. H. (*Actg. Maj.*)	25 May 18	25 July 18	I. A. R. O. (*T. C.*) 2nd-in-Comd.
21 July 18	Bell, 2nd-Lt. R. S. (*Actg. Lt.*)	21 July 18	18 July 18	I. A. R. O.
21 June 18	Stronach, 2nd-Lt. J. L. (*Actg. Lt.*)	21 June 18	22 July 18	I. A. R. O.
21 July 18	Weymouth, 2nd-Lt. J. C. (*Actg. Lt.*).	21 July 18	22 July 18	I. A. R. O.
	Company Officers.			
21 June 18	Crawford, 2nd-Lt. A. G. (*Actg. Capt*).	21 June 18	12 July 18	I. A. R. O.
21 June 18	Elms, 2nd-Lt. C. E. (*Actg. Lt.*).	21 June 18	11 July 18	I. A. R. O.
21 June 18	Nott-Bower, 2nd-Lt. J. R. H. (*Actg. Lt.*).	21 June 18	1 July 18	I. A. R. O.
21 June 18	Pilditch, 2nd-Lt. D. (*Actg. Lt.*) .	21 June 18	8 July 18	I. A. R. O.
21 June 18	Waddell, 2nd-Lt. G. (*Actg. Lt.*).	21 June 18	1 July 18	I. A. R. O. Qr.-Mr.
7 June 18	Brunger, 2nd-Lt. B.	7 June 18	10 Oct. 18	I. A. (*T. C.*)
22 July 18	Lewis, 2nd-Lt. W.	22 July 18	2 Dec. 18	I. A. (*T. C.*)
	Attached.			
3 Jan. 16	Woodward, Lt. W. J. . . .	3 Jan. 17	3 Oct. 18	The Buffs, I. A. (*on probn.*)
1 May 17	Stephen, Lt. A. T.	1 May 18	24 Sep. 18	56th Rif.
20 Dec. 17	Bowen, 2nd-Lt. W. O. . . .	20 Dec. 17	23 Oct. 18	1-6th G. R.
20 Dec. 17	Rootes, 2nd-Lt. A. H. . . .	20 Dec. 17	21 Sep. 18	1-6th Jats, Adjt.
31 Jan. 18	Hargrave, 2nd-Lt. J. F. H. . .	31 Jan. 18	2 Oct. 18	1-3rd G. R.
24 Oct. 18	Hayward, 2nd-Lt. H. H. . . .	24 Oct. 18	30 Nov. 18	I. A. R. O.
		Jemadar.	Subadar.	Subadar-Major.
	Subadar-Major.			
	Subadars.			
12 Nov. 94	Partiman Rana	10 Aug. 09	6 Oct. 14	4-3rd G. R.
20 Nov. 00	Abir Thapa	1 Nov. 14	1 Nov. 18	2-8th G. R.
13 July 18	Uma Dat	13 July 18	13 July 18	
13 July 18	Muhammad Akbar Khan . . .	13 July 18	13 July 18	
	Jemadars.			
1 Mar. 18	Suraj Pal Dube	1 Mar. 18		
13 July 18	Lalman Singh	13 July 18		
13 July 18	Tej Singh	13 July 18		
13 July 18	Lachhman Pershad.	13 July 18		
13 July 18	Ram Phal	13 July 18		
13 July 18	Bal Kishen	13 July 18		
13 July 18	Bhairon Singh	13 July 18		
13 July 18	Ram Charan	13 July 18		

1st Battalion, 142nd Jodhpur Infantry.

Formed at

Composition.

1st Commn. or date of entering service.	Names and Rank.	Army Rank.	Present Appointment in Regiment.	Remarks.
	Commandant.			
20 July 98	Money, Maj. E. F. D., D.S.O.	1 Sep. 15	. . . .	2-4th G. R.
	Company Commanders.			
	Company Officers.			
5 May 15	Ledward, Lt. R. B. . . .	5 May 16	. . . . .	I. A. (*on probn.*) Q..Mr.
15 Nov. 15	Salvesen, Lt. F. M. R. . . .	15 Nov. 16	. . . . .	87th Punj. Adjt.
12 Sep. 17	Brown, Lt. R.	12 Sep. 18	. . . . .	I. A. (*on probn.*)

Jemadar. Subadar. Subadar-Major.

	Subadar-Major.			
	Subadars.			
	Jemadars.			

1st Battalion, 141st Bikaner Infantry.

Formed at Bikaner on 9 July 18.

Composition—3½ Companies Rajputana Hindus,

½ Coy. Rajputana Musalmans.

1st Commn. or date of entering service.	Names and Rank.	Army Rank.	Present appointment in Regiment.	Remarks.
	Commandant.			
20 July 98	Vander Gucht, Maj. G. T., (*Actg. Lt.-Col.*)	1 Sep. 15	20 July 18	3rd Horse (*temp.*)
	Company Commanders. (4)			
10 Oct. 03	Steel, Bt.-Maj. C. N. (*Actg. Maj.*)	3 June 17	8 Oct. 18	25th Punj., 2nd-in-Comd. (*temp.*)
4 Dec. 14	Innes, Lt. R. J. (*Actg. Capt.*)	4 Dec. 15	2 Aug. 18	I. A. R. O.
17 Apl. 15	Wickham, Lt. J. (*Actg. Capt.*)	17 Apl. 16	11 July 18	21st Punj.
15 May 15	Black, Lt. A. G. (*Actg. Capt.*)	15 May 16	9 July 18	I. A. (*on probn.*)
	Company Officers.			
13 Dec 15	Secombe, Lt. H. S. (*Actg. Capt.*)	13 Dec. 16	16 Oct. 18	I. A. (*on probn.*) *Offg.* Adjt.
29 Mar. 16	Belford, Lt. H D.	29 Mar. 17	17 July 18	I. A (*on probn.*)
5 Nov. 17	Read, 2nd-Lt. R. S.	5 Nov. 17	10 July 18	I. A. (*T. C.*)
15 Jan. 18	Sanders, 2nd-Lt H. H.	15 Jan. 18	25 July 18	I. A. R. O. *Offg.* Qr.-Mr.
	Attached.			
	Mahendra Man Singh, Raja, 2nd-Lt. (*Hony.*)	10 May 18	....	(*Temp.*)

		Jemadar	Subadar	Subadar-Major.
	Subadar-Major.			
	Subadar.			
13 May 05	Shiuji Ram (52, 56)	26 Sep. 12	1 Jan. 17	104th Rif.
12 Nov. 02	Asgar Khan (56)	1 July 15	27 July 18	109th Infy.
9 Feb. 03	Harlal Singh	22 Feb. 16	9 Aug. 18	62nd Punj.
5 Aug. 95	Shiunath Singh	11 Sep. 14	11 Aug. 18	13th Raj
31 Aug. 03	Chatru	2 Mar. 18	13 Aug. 18	82nd Punj.
25 Oct. 18	Mohbat Singh	....	25 Oct. 18	
	Jemadars.			
	Mahmud Khan	27 July 18	....	
	Chandra Ram	12 Aug. 18	....	
	Jaloo Ram	13 Aug. 18	....	

1st Battalion, 140th Patiala Infantry.

Formed at Patiala, 2nd July 18.

Composition :—2 Companies of Sikhs, 1 Coy. of Musalmans from Patiala State, 1 Coy. of Hindu Jats.

1st Commn. or date of entering service.	Names and Rank.	Army Rank.	Present appointment in Regiment.	Remarks.
	Commandant.			
4 May 01	MacMahon, Maj. G. D. R. (*Actg. Lt.-Col.*)	4 May 16	28 June 18	124th Bal. Infy.
	Company Commanders. (4)			
29 Aug. 06	Leslie-Smith, Capt. C. (*Actg. Maj.*)	29 Aug. 15	26 June 18	2nd-in-Com d., 22nd Punj.
22 Feb. 15	Morison, Lt. H. Mc. D. (*Actg. Capt.*)	30 June 16	5 Aug. 18	I. A. R. O. (*on probn.*)
19 Mar. 15	Owen, Lt. A. T. (*Actg. Capt.*) .	19 Mar. 16	30 June 18	56th Rif.
8 Mar. 16	Hart, Lt. L. W. (*Actg. Capt.*) .	8 Mar. 17	8 July 18	I. A. (*on probn.*)
	Company Officers.			
18 Dec. 14	Dubois, Lt. H. A.	18 Dec. 15	14 Aug. 18	I. A. R. O.
12 Dec. 16	Mellor, Lt. J. R. (*Actg. Capt.*) .	12 Dec. 17	7 Sep. 18	I. A. (*on probn.*) Adjt.
18 Apl. 16	Jenkins, Lt. N. L.	18 Apl. 17	17 Aug. 18	1-33rd Punj.
26 Sep. 17	Cattanach, Lt. A.	26 Sep. 18	2 July 18	I. A. R. O.
7 June 18	Roberts, 2nd-Lt. G. B. . . .	7 June 18	10 Oct. 18	I. A. (*T. C.*)
7 June 18	Rivett, 2nd-Lt. H. H. . . .	7 June 18	10 Oct. 18	I. A. R. O.
		Jemadar.	Subabar.	Subadar-Major.
	Subadar-Major.			
	Subadars.			
16 Dec. 90	Hazara Singh (51)	16 Apl. 11	. . . .	
6 June 92	Badan Singh (56, 66*a*) . . .	5 Aug. 14	. . . .	
13 Apl. 93	Jiwa Singh (66*a*)	25 Nov. 14	2 July 18	
15 Sep. 93	Mangal Singh (66*a*)	10 Mar. 15	2 July 18	
29 Apl. 96	Kale Khan	12 Oct. 15	2 July 18	
6 Apl. 07	Dalip Singh (56)	26 May 17	2 July 18	
	Jemadars.			
7 Jan. 15	Ujjagar Singh	1 Jan. 18		
	Darbara Singh	2 July 18		
	Chainchal Singh	12 Oct. 18		
	Sham Singh	12 Oct. 18		
	Ali Muhammad	12 Oct. 18		
	Fateh Khan	12 Oct. 18		
	Jot Ram	12 Oct. 18		

1st Battalion, 133rd Regiment.

Formed at

Composition :—

1st Commn. or date of entering service.	Names and Rank.	Army Rank.	Present appointment in Regiment.	Remarks.
	Commandant.			
8 May 01	Connell-Mcdowell, Maj. A. E. J.	8 May 16	11 Sep. 18	Civil Employ (*temp.*)
	Company Commanders (4).			
17 Apl. 15	Murrey, Lt. C. R. B.	17 Apl. 15	....	I. A. R. O.
13 Aug. 17	Pertwee, Lt. L.	13 Aug. 18	....	I. A. (*on probn.*)
21 May 17	Pettcifer, Lt. J.	21 May 17	....	I. A. R. O.
23 May 17	Thomson, Lt. A.	23 May 17	....	I A. R. O.
	Company Officers.			
1 May 18	Mansfield, 2nd-Lt. P. T.	1 May 18	....	I. A. R. O.
2 June 18	Stokes, 2nd-Lt. R. G.	2 June 18	....	I. A. (*T. C.*)
....	Bown, 2nd-Lt. J. E.	....	18 Oct. 18	
....	Madan, 2nd-Lt. F. F.	....	....	I. A. R. O.
....	Brett, 2nd-Lt. C. T.	....	....	I. A. R. O. Adjt.
....	Butterfield, 2nd-Lt. P. H. A.	....	11 Oct. 18	
....	Warn, 2nd-Lt. J. E.	....	....	I. A. R. O. Qr.-Mr.
....	Tylor, 2nd-Lt. W.	....	....	I. A. R. O.
....	Nash, 2nd-Lt. H.	....	24 Nov. 18	

	Jemadar.	Subadar.	Subadar-Major.

Subadar-Major.

Subadar

Jemadar

1st Battalion, 143rd Narsingh (Dholpure) Infantry.

Formed at Dholpure on 31st August 1918.

Composition—1 Company Jats, 1 Company Rajputs, 1 Company Musalmans including Meos.

1st Commn. or date of entering service.	Names and Rank.	Army Rank.	Present appointment in Regiment.	Remarks.
	Commandant.			
2 Aug. 99	Peters, Maj. J. (*Actg. Lt.-Col.*)	1 Sep. 15	31 Aug. 18	10 Lrs. (*temp.*)
	Company Commanders.			
13 Nov. 14	Bennett, Lt. I. A. (*Actg. Maj.*)	13 Nov. 15	9 Nov. 18	I. A. R. O. 2nd-in-Comd.
22 Apl. 15	Carleton, Lt. K. C.	22 Apl. 16	8 Sep. 18	1-25th Punj.
7 Apl. 16	Johnston, Lt. D. D. H.	7 Apl. 17	14 Sep. 18	1-11th Raj.
18 Apl. 16	Dalison, Lt. J. B.	18 Apl. 17	31 Aug. 18	67th Punj.
	Company Officers.			
		Jemadar.	Subadar.	Subadar-Major.
	Subadar-Major.			
3 Jan. 01	Ram Pershad Singh, I. D. S. N. (66*a*.)	1 Jan. 14	1 Oct. 16	23 Oct. 18	95th Infy.
	Subadars.			
	Jemadars.			
23 Apl. 00	Umarao Singh	8 June 16		
	Dalip Singh	1 Oct. 18		
	Sabal Singh	1 Oct. 18		
	Nanu Singh	1 Oct. 18		
	Megh Singh	1 Oct. 18		
	Muhammad Shafi Khan	1 Oct. 18		

1st Battalion, 144th Bharatpur Infantry.

Composition—

1st Commn. or date of entering service.	Names and Rank.	Army Rank.	Present appointment in Regiment.	Remarks.
	Commandant.			
8 May 01	Brown, Maj. C. A.	8 May 16	. . . . .	84th Punj.
	Company Commanders.			
1 Oct. 14	Keene, Capt. M. R.	1 Oct. 18	. . . . .	2-97th Infy. 2nd-in-Comd.
11 Aug. 15	Mackenzie, Lt. W. S.	11 Aug. 16	. . . . .	4th G. R.
24 Aug. 15	Lewis, Lt. F. H.	24 Aug. 16	. . . . .	4-3rd G. R.
15 Nov. 15	Cutler, Lt. M. H. C.	15 Nov. 16	. . . . .	Attd.
	Company Officers.			
29 June 16	Ross-Hurst, Lt. W. J. O. B. . .	29 June 17	. . . . .	58th Rif.
5 July 17	Bates, Lt. R. S. P.	5 July 18	. . . . .	Attd.
12 Sep. 17	O'Brien, Lt. W. J.	12 Sep. 18	. . . . .	Attd.
12 Sep. 17	Stratton, Lt. E. B.	12 Sep. 18	. . . . .	Attd.
31 Jan. 18	Putnam, 2nd-Lt. W. A.	31 Jan. 18	. . . . .	Attd.

Jemadars.	Subadars.	Subadar. Major.

Subadar-Major.

Subadars.

Jemadars.

1st Battalion, 145th Alwar (Jai Paltan) Infantry.

Formed at

Composition.

1st Commn. or date of entering service.	Name and Rank.	Army Rank.	Present appointment in Regiment.	Remarks.
	Commandant.			
14 Sep. 01	Cruickshank, Maj. J. E.	15 Feb. 16	. . . .	2-2nd G. R.
	Company Commanders (4).			
	Company Officers.			
. . . .	Duff-Torrance, Lt. W. R.	. . . .	. . . .	Adjt.
29 June 16	Davys, Lt. J. V.	29 June 17	. . . .	13th Raj. Qr.-Mr.
7 June 18	Vincent, 2nd-Lt. J. A.	7 June 18	. . . .	I. A. (*T. C.*)
7 June 18	Silver, 2nd-Lt. P. W.	7 June 18	. . . .	I. A. (*T.C.*)

Jemadar. Subadar. Subadar-Major.

Subadar-Major.

Subadars.

Jemadars.

1st Battalion, 150th Indian Infantry.

Formed at Busrah.

Composition—4 Companies Rajputs.

1st Commn. or date of entering service.	Names and Rank.	Army Rank.	Present appointment in Regiment.	Remarks.
	Commandant.			
5 Jan. 01	Orchard, Maj. A. J. (*Actg. Lt.-Col.*)	1 Sep. 15	5 July 18	8th Rajputs.
	Company Commanders (4).			
6 Oct. 06	Burroughs, Capt. S. V. G. (*Actg. Maj.*)	1 Sep. 15	10 May 18	2nd-in-Comd. 1-4th Raj.
19 Dec. 16	Thirkell-White, Capt. E. H. . .	19 Dec. 16	10 May 18	I. A. R. O.
6 May 15	Chaundler, Lt. O. S.	6 May 16	10 May 18	I. A. (*on probn.*).
15 Nov. 15	Snelling, Lt. A. H. J. . . .	15 Nov. 16	19 May 18	I. A. R. O.
	Company Officers.			
24 Dec. 15	Maclachlan, Lt. A. B., M.C. . .	24 Dec. 16	5 Oct. 18	I. A. R. O.
11 Apl. 16	Johnson, Lt. R. S., M.C . . .	11 Apl. 17	12 Aug. 18	I. A. (*on probn.*).
5 May 17	Hamilton, Lt. D. S.	5 May 18	1 Oct. 18	N. Lan. R., I. A. (*on probn.*).
25 June 17	Crampin, 2nd-Lt. R. W. . . .	25 June 18	20 Oct. 18	R. Berks. R, I. A. (*on probn.*)
	Attached.			
5 Oct. 17	Hallam, Lt. E. H.	5 Oct. 18	28 Sep. 18	I. A. (*T. C.*) Qr. Mr.
26 Sep. 17	Oliver, 2nd-Lt. N. E.	26 Sep. 17	10 June 18	I. A. R. O. (*on probn.*). Adjt.
26 Jan. 18	Elmer, 2nd Lt. P. S.	26 Jan 18	16 Oct. 18	Lond. R., I. A. (*on probn.*).
. . . .	Wood, 2nd-Lt. J. E.	. . . .	20 Sep. 18	I. A. R. O. (*on probn.*).
. . . .	Hardman, 2nd-Lt. A.	. . . .	20 Sep. 18	I. A. R. O. (*on probn.*).
2 June 18	Bickel, 2nd-Lt. T. H.	2 June 18	7 Oct. 18	I. A. (*T. C.*)
7 June 18	O'Connor, 2nd-Lt. M. J. . . .	7 June 18	7 Oct. 18	I. A. (*T. C.*)
16 Mar. 18	Fraser, 2nd-Lt. R. A. J. . . .	16 Mar. 18	. . . .	I. A. R. O.
. . . .	Mason, 2nd-Lt. E. W. R. . . .	. . . .	22 Sep. 18	I. A. R. O. (*on probn.*).
11 May 16	Gibson, Lt. F. C.	11 May 17	20 Sep. 18	I. A. (*on probn.*).
31 Aug. 18	A'Brook, 2nd-Lt. D.	31 Aug. 18	20 Sep. 18	1-2nd Raj. L. I.
		Jemadar.	Subadar.	Subadar-Major.
	Subadar-Major.			
16 Jan. 98	Multan Singh	16 Nov. 10	15 Apl. 15	10th May 18.
	Subadars.			
2 Aug. 99	Jiwan Singh	15 Oct. 14	9 May 16	
3 Feb. 01	Chhotte Singh	8 Nov. 17	. . . .	
26 June 03	Phul Singh	19 Feb. 18	. . . .	
15 Dec. 99	Shuimbar Singh	13 Jan. 08	17 Apl. 18	
27 July 97	Shumangal Singh	1 Apl. 18	. . . .	
26 Mar. 98	Naurang Singh	1 May 14	. . . .	
3 Nov. 99	Bhagwant Singh	3 Nov. 14	. . . .	
	Jemadars.			
2 May 04	Butti Singh	1 Sep. 16		
28 Aug. 00	Puttu Singh	8 Nov. 17		
30 Apl. 03	Chattar Singh	6 Dec. 17	Indian Adjt.	
3 Dec. 07	Ramchandar Singh	1 Apl. 18		
21 Jan. 96	Danbahadur Singh	9 May 17	1-11th Raj.	
17 Jan. 03	Sheoji Singh	18 May 16	Indian Qr.-Mr.	
4 May 07	Chailtan Singh	3 July 17		

2nd Battalion, 150th Indian Infantry.

Formed at Basra, 20 May 18.

Composition—2 companies Jats, 1 coy. Rajputs, 1 coy. Gujars.

1st Commn. or date of entering service.	Names and Rank.	Army rank.	Present appointment in Regiment.	Remarks.
	Commandant.			
16 Jan. 95	Birch, Maj. V. K., (*Act'g. Lt.-Col.*)	16 Jan. 13	20 May 18	29th Lrs.
	Company Commanders. (4)			
26 June 01	Frere, Maj., A. G.	26 June 16	20 May 18	86th Infy.
12 Mar. 15	Newman, Lt. H.	12 Mar. 15	27 June 18	I. A. R. O.
18 Dec. 14	Barnes, Lt. D. H.	18 Dec. 15	24 May 18	I. A. R. O.
2 Apl. 15	O'Callaghan, Lt. P. P. M.	2 Apl. 16	20 May 18	I. A. R. O.
	Company Officers.			
3 Sep. 15	Thompson, Lt. R. M., (*Act'g. Capt.*)	3 Sep. 16	23 May 18	I. A., (*on probn.*) Adjt. 23rd May 18.
20 Feb. 16	Dunsmure, Lt. C. T. W.	20 Feb. 17	. . . .	I. A. R. O.
6 Oct. 16	Wheeler, Lt. C. D.	6 Oct. 17	26 May 18	I. A. R. O., *Offg.* Qr.-Mr.
18 June 17	Coventry, 2nd-Lt. H. F.	18 June 17	26 May 18	
5 Nov. 17	Collett, Lt. R. A.	5 Nov. 18	27 June 18	I. A. (*T. C.*)
	Attached.			
26 June 17	Leggatt, Lt. W.	26 June 18	. . . .	N. Lan. R., I. A. (*on probn.*).
				R. W. Surr. R., I. A. (*on probn.*)
21 Dec. 17	Williams, 2nd-Lt. S. M.	21 Dec. 17	. . . .	1-4th G. R.
26 Apl. 18	Olden, 2nd-Lt. S. M.	26 Apl. 18	3 Oct. 18	I. A.
31 Jan. 18	Dwane, 2nd-Lt. L. M. R.	31 Jan. 18	18 Nov. 18	I. A. R. O.
7 June 18	Evans, 2nd-Lt. V. C.	7 June 18	15 Oct. 18	I. A. (*T. C.*)
10 July 18	Thorne, 2nd-Lt. E. P.	10 July 18	2 Dec. 18	I. A. (*T. C.*)

		Jemadar.	Subadar.	Subadar-Major.	
	Subadar-Major.				
10 June 00	Shib Lal (52, 56)	10 June 12	7 Dec. 14	6 June 18	1-10th Jats.
	Subadar.				
9 June 89	Hardhan Singh	16 July 04	24 June 09		2-9th Infy.
17 Nov. 01	Siri Ram	25 May 14	8 Jan. 16		1-6th Jats.
10 May 98	Kishan Lal (56, 52, 66*a*)	30 Dec. 13	14 May 16		1-6th Jats.
25 July 97	Khubiram (56)	7 Dec. 14	7 Mar. 17		1-10th Jats.
30 Apl. 02	Ganesh Singh (52, 56)	1 Sep. 12	1 Dec. 17		1-9th Infy.
13 June 94	Gopal Singh	1 Nov. 16	1 June 17		1-9th Infy.
17 Jan. 06	Salig Ram	9 Oct. 16	22 May 18		1-6th Jats.
	Jemadar.				
3 Apl. 04	Ramji Lal	1 Apl. 17	. . . .		1-10th Jats.
2 Dec. 06	Ujala Singh	27 June 17	. . . .		1-6th Infy.
2 Jan. 17	Parsadi Singh	14 Nov. 17	. . . .		2-9th Infy.
3 July 08	Gulab, (56)	1 Nov. 17	. . . .		1-10th Jats, Jemdr. Adjt.
9 June 05	Siripal Singh	25 Nov. 17	. . . .		1-9th Infy.
17 Feb. 01	Lehri	20 May 18	. . . .		1-6th Jat., Qr.-Mr., Jemdr.
26 Aug. 05	Lalchand	20 May 18	. . . .		1-10th Jats.
1 Feb. 17	Tota Singh	21 May 18	. . . .		2-9th Infy.
1 Feb. 17	Zeheriya	21 May 18	. . . .		2-9th Infy.
19 June 10	Mohar Singh	19 June 18	. . . .		
. . . .	Rati Ram	1 July 18			

3rd Battalion, 150th Indian Infantry.

Formed at

Composition—2 companies Minas, 1 company Mehro, 1 company Rajputs.

1st Commn. or date of entering service.	Names and Rank.	Army rank.	Present appointment in Regiment.	REMARKS.
	Commandant.			
11 Feb. 99	Trueman, Maj. H. J. H. (*actg. Lt.-Col.*)	1 Sep. 15	22 May 18	43rd R.
	Company Commanders. (4)			
26 Feb. 15	Templeton, Lt. W. G. H. (*actg. Maj.*)	26 Feb. 16	22 May 18	I. A. R. O., 2nd-in-Comd.
7 Apl. 16	Seagrim, Lt. C. B.	7 Apl. 17	. . . .	99th Infy.
29 June 16	Oatts, Lt. H. A.	29 June 17	. . . .	41 Dogras.
10 Aug. 16	Legeyt, Lt. P. S.	10 Aug. 17	. . . .	I. A. R. O.
	Company Officers.			
7 May 15	Cinnamond, Lt. H. P., M.C. . .	7 Feb. 17	. . . .	W. York. R., I. A. (*on probn.*)
16 Aug. 16	Holmes, Lt. G. L.	16 Aug. 17	. . . .	99th Infy.
10 Sep. 16	Godwin, Lt. W. E. (*actg. Capt.*)	10 Sep. 17	. . . .	I. A R. O. Adjt.
6 Jan. 16	Law, Lt. E. L.	6 Jan. 17	. . . .	I. A. (*on probn.*)
10 Nov. 16	Pelham,-Clay, Lt. E. C. G. . .	10 Nov. 17	. . . .	I. A. R. O. Qr. Mr.
25 Jan. 16	Stanley, 2nd-Lt. C. H. . . .	. . . .	. . . .	Essex R., I. A. R. O. (*on probn.*)
23 Oct. 16	Peat, Lt. H R.	. . . .	. . . .	High. L. I.
5 Sep. 16	Spence, Lt. J., M. C.	5 July 18	. . . .	I. A. (*on probn.*.)
30 Oct. 17	Machin, Lt [illegible]. W.	30 Oct. 17	. . . .	I. A. R. O.
31 Oct. 17	Mills, Lt. J. E.	31 Oct. 18	. . . .	I. A. (*on probn.*)
18 Dec. 17	Muir, 2nd-Lt. A. C.	. . .	. . . .	High. L. I.
15 Jan. 18	Richardson, 2nd-Lt. G. . . .	. . . .	. . . .	I. A. (*T. C.*)
. . . .	Parker, 2nd-Lt. H.	. . . .	. . . .	High. L. I.
22 July 18	Foster, 2nd-Lt. A. E.	22 July 18	. . . .	I. A. R. O.
7 June 18	White, 2nd-Lt. W. H. . . .	7 June 18	. . . .	I. A. (*T. C.*)
. . . .	Hollwill, 2nd-Lt. R. F. . . .	. . . .	. . . .	Genl. List. I. A. R. O. (*on probn.*)

		Jemadar.	Subadar.	Subadar-Major.
	Subadar-Major.			
6 Oct. 92	Chanda, I. O. M. (52, 56) . . .	1 Apl. 05	15 July 10	1 Apl. 17.
	Subadar.			
9 Mar. 98	Pahara	1 Jan. 10	23 Jan. 15	
7 Mar. 02	Sarwana	1 Nov. 13	20 Nov. 15	
17 June 96	Khangara (52, 56)	7 Jan. 10	6 Jan. 16	
14 Jan. 05	Dwarka Singh	5 Sep. 16	15 Mar. 18	
16 Sep. 07	Subh Karan Singh	15 Mar. 18	18 Mar. 18	
1 Jan. 04	Jetha	23 July 15	22 May 18	
19 Dec. 07	Lakha	7 July 17	22 May 18	
	Jemadar.			
11 Sep. 04	Gulla	19 June 15		
18 Oct. 01	Khuma	23 June 16		
14 Jan. 02	Lachmana	18 Aug. 16		
24 Oct. 04	Birdha	1 July 17		
21 Nov. 08	Baldeo Singh	18 June 18		
17 Jan. 11	Prithipal Singh	21 Mar. 18		
1 Mar. 99	Kishna	22 May 18		
. . .	Dhira	22 May 18		
. . . .	Natha	22 May 18		
. . . .	Mahadewa	22 May 18		
. . . .	Ramasere Singh	1 Oct. 18		

1st Battalion, 151st Sikh Infantry.

Formed at

Composition—3 companies Sikhs, 1 company Punjabi Musalmans.

1st Commn. or date of entering service.	Names and Rank.	Army rank.	Present appointment in Regiment.	Remarks.
	Commandant.			
15 Jan. 95	Strong, (*Bt. Lt.-Col.*) A. D., D.S.O.	3 June 18	16 May 18	10th Lrs.
	Company Commanders. (4)			
20 Jan. 00	Lyall, Maj. C. R.	1 Sep. 15	. . . .	2nd-in-Comd., 36th Sikhs.
10 Oct. 03	Wace, Maj. M.	10 Oct. 18	. . . .	14th Sikhs.
20 Mar. 15	Preston, Lt. K. H.	20 Dec. 16	. . . .	45th Sikhs.
	Company Officers.			
13 Aug. 15	McLaren, Lt. J. A.	13 Aug. 16	. . . .	Sco. Rif., I. A. (*on probn.*)
18 Apl. 16	Keegan, Lt. D.	18 Apl. 17	. . . .	1-33rd Punj.
20 Jan. 16	Phillips, Lt. C. A.	20 Jan. 17	28 Aug. 18	W. York. R., I. A. (*on probn.*)
29 June 16	Murray-Shirreff, Lt. B. G. (*actg. Capt.*)	29 June 17	24 July 18	1-33rd Punj. (*temp.*) Adjt.
8 Jan. 16	Browne, Lt. C. L.	8 Jan. 17	26 Aug. 18	Essex. R., I. A. R. O.
18 June 17	Eastmead, Lt. C.	18 June 18	. . . .	I. A. 15th Sikhs.
8 Aug. 17	Patinson, Lt. R. I.	8 Aug. 18	. . . .	I. A. R. O. Qr.-Mr. (*Offg.*) 14th Sikhs.
18 June 17	Morton, Lt. H. K.	18 June 18	22 Aug. 18	2-41st Dogras.
7 June 18	Ford, 2nd-Lt. A.	7 June 18	. . . .	I. A. (*T. C.*)
		Jemadar.	Subadar.	Subadar-Major.
	Subadar-Major.			
31 Jan. 93	Wariam Singh	1 Feb. 05	1 Nov. 12	15 May 15 (*Offg.*) 45th Sikhs.
	Subadars.			
18 Oct. 89	Khalas Khan I	1 Nov. 09	27 Feb. 15	52nd Sikhs.
28 June 95	Parem Singh	4 July 14	10 Mar. 16	14th Sikhs.
24 Mar. 96	Rur Singh	16 Oct. 14	15 Mar. 16	14th Sikhs.
24 Aug. 94	Phuman Singh	2 May 14	9 Apl. 16	36th Sikhs.
27 June 96	Jiwan Singh	16 Oct. 14	10 Oct. 16	14th Sikhs.
10 Dec. 98	Watan Singh	25 Aug. 16	2 Feb. 17	45th Sikhs.
3 Jan. 96	Khalas Khan II	26 Apl. 15	1 Oct. 17	52nd Sikhs.
	Jemadars.			
25 June 92	Gurditt Singh	10 May 16	36th Sikhs.	
9 Mar. 01	Natha Singh	24 June 16	Jemdr. Adjt. (*Offg.*) 36th Sikhs.	
6 Nov. 98	Isher Singh	5 Oct. 16	45th Sikhs.	
21 Apl. 99	Khunda Singh	2 Feb. 17	45th Sikhs.	
29 Dec. 97	Jiwa Singh	4 Feb. 17	14th Sikhs.	
7 May 02	Adalat Khan	16 Sep. 17	52nd Sikhs.	
26 Sep. 98	Sardar Khan	19 Oct. 17	52nd Sikhs.	
	Dallu Thapa	24 May 18		
	Ram Ditta	24 May 18		

2nd Battalion, 151st Indian Infantry.

Formed at
Composition—2½ companies, Punjabi Musalmans, ¾ Coy. Dogras, ¼ Jat Sikhs, ½ Malwai Sikhs.

1st Commn. or date of entering service.	Names and Rank.	Army rank.	Present appt. in Regiment.	Remarks.
	Commandant.			
1 Dec. 97	Smith, Maj. A. A. (*act'g. Lt.-Col.*)	1 Sep. 15	8 June 18	58th Rif.
	Company Commanders. (4)			
18 Jan. 02	Beadle, Maj. R. D.	18 Jan. 17	2 July 18	46th Prs.
	Company Officers.			
13 Mar. 15	Hugh-Jones, Lt. N.	13 Mar. 16	10 June 18	53rd Sikhs.
18 Apl. 16	Maclachlan, Lt. A. (*act'g. Capt.*)	18 Apl. 17	25 Apl. 16	51st Sikhs, Adjt.
30 Jan. 17	Martin, Lt. S. F.	30 Jan. 18	10 June 18	108th Infy.
20 Dec. 17	Hughes, 2nd-Lt. H.	20 Dec. 17	9 Sep. 18	56th Rif.
20 Dec. 17	Danson, 2nd-Lt. T. E. . . .	20 Dec. 17	13 Sep. 18	53rd Sikhs.
	Attached.			
22 Jan. 13	Stuart-Prince, Capt. C. E. (*act'g. Maj.*)	22 Jan. 17	24 May 18	59th Rif.
18 Dec. 14	Dunning, Lt. A. G. A., M.C. . .	18 Dec. 15	31 May 18	I. A. R. O.
1 Oct. 13	Whitley, Lt. H. C.	1 Oct. 14	3 Sep. 18	York. R., I. A. (*on probn.*)
14 Jan. 15	Inglis, Lt. H. G., M.C. . . .	14 Jan. 15	31 May 18	I. A. R. O.
18 Feb. 15	Wilkinson, Lt. R.	18 Feb. 16	31 May 18	I. A. R. O.
12 June 15	Charlton, Lt. F. J.	12 June 16	8 Aug. 18	I. A. (*on probn.*)
20 July 15	Heathcote, Lt. C. N.	20 July 16	31 May 18	I. A. R. O.
20 July 15	Harris, Lt. A. W.	20 July 16	31 May 18	I. A. R. O.
1 July 16	Macmillan, Lt. R. W. . . .	1 July 17	8 July 18	I. A. R. O.
17 Oct. 16	Terry, Lt. L. R.	17 Oct. 16	5 July 18	I. A. R. O.
5 Oct. 17	Watson, Lt. A. E.	5 Oct. 18	31 May 18	I. A. (*T.C.*)
5 Sep. 16	Norman, 2nd-Lt. C. A. . . .	. . . .	30 Aug. 18	Glouc. R., I. A. (*on probn.*)
. . . .	Reynolds, 2nd-Lt. J. W. . . .	. . . .	7 July 18	I. A. R. O. (*on probn.*)
7 June 18	Duckenfield, 2nd-Lt. W. . .	7 June 18	21 Oct. 18	I. A. (*T.C.*)
. . . .	Chapman, 2nd-Lt. D. G. . . .	. . . .	17 Oct. 18	R. N. V. R., I. A. R. O. (*on probn.*)
1 May 18	Tucker, 2nd-Lt. K. D. . . .	1 May 18	. . . .	I. A. R. O.

		Jemadar.	Subadar.	Subadar-Major.
	Subadar-Major.			
5 Dec. 93	Sadda Singh	2 May 15	16 Mar. 16	(*temp.*)
	Subadars.			
4 Dec. 86	Latkan Singh	1 Feb. 08	26 Feb. 09	
12 July 91	Galodu Ram	21 Dec. 08	. . . .	
15 Apl. 87	Bostan Khan, *Bahadur* . . .	24 June 05	16 June 11	*Order of Br. Ind. 2nd, Cl.*
12 Nov. 05	Jodh Khan, I.O.M.	24 Oct. 14	27 June 16	1-56th Rif.
28 Dec. 91	Shibu	22 Jan. 16	21 May 18	
19 Sep. 08	Dalel Khan, I.D.S.M.	18 Apl. 16	21 May 18	
30 Dec. 99	Karam Dad	1 Dec. 15	30 May 18	1-56th Rif.
19 Dec. 95	Sajawal Khan	9 Apl. 16	30 May 18	
27 July 98	Hira Sing	14 June 16	31 May 18	
4 May 88	Hukam Dad	22 Sep. 16	19 Oct. 18	
. . . .	Uttam Singh (*retd.*)	. . . .	. . . .	Re-empld.
	Jemadars.			
4 Sep. 94	Faiz Talab Khan, I.O.M. . .	22 June 15	Jemdr. Qr.-Mr.	
21 June 00	Dost Mohd.	8 Mar. 16	1-56th Rif.	
28 Apl. 95	Harnam Singh	7 Apl. 16		
11 Dec. 01	Bhola Singh	10 June 16		
14 May 93	Indar Singh	4 July 16		
22 June 00	Karam Sher	24 Oct. 16	1-56th Rif.	
4 Apl. 06	Hoshan Khan	10 Mar. 17		
12 Apl. 95	Akbar Khan.	22 June 17		
1 Nov. 97	Jowaher Shah	1 May 18		
25 Jan. 01	Inayat Khan	30 May 18		
15 Apl. 03	Sher Baz	30 May 18		
13 Apl. 05	Gul Muwaz	31 May 18		
11 Feb. 02	Munshi Ram	31 May 18	Jemdr. Adjt.	
. . . .	Gul Mehdi	31 May 18		
. . . .	Baz Khan	1 June 18		
16 Feb. 08	Malak Khan	1 June 18		
24 June 94	Bishan Singh	1 July 18		
27 Oct. 06	Narain Singh	1 July 18		
9 Mar. 03	Zaman Ali	5 July 18		

3rd Battalion, 151st Punjabi Rifles.

Formed at

Composition—1½ companies Punjabi Musalmans, ½ Coy. Gurkhas, 1 Coy. Dogras, ½ Coy. Rajputs, ½ Coy. Ahirs.

1st Commn. or date of entering service.	Names and Rank.	Army rank.	Present appt. in Regiment.	Remarks.
	Commandant			
4 Jan. 99	Renny, Maj. G. S. (*Actg. Lt.-Col.*)	1 Sept. 15	8 June 18	38th Dogras.
	Company Commanders (4).			
3 Aug. 08	Day, Maj. A. D.	1 Sep. 15	5 July 18	91st Punj.
14 Jan. 14	Hamilton, Capt. H. M. . . .	14 Jan. 18	5 July 18	59th Rif.
23 Apl. 14	Hill, Lt. G. G., M.C.	23 Apl. 16	24 May 18	I. A. R. O.
13 Sep. 16	Fagnani, Lt. H. H.	13 Sep. 17	24 May 18	I. A. R. O.
	Company Officers.			
5 May 15	Colnett, Lt. R. D.	5 May 16	26 July 18	I. A. (*on probn.*)
. . . .	Rogers, Lt. A. G.	. . . .	30 July 18	I. A. R. O. (*on probn.*)
23 Sep. 15	Carr, Lt. F. B.	23 Sep. 16	8 July 18	I. A. R. O.
29 June 16	Wainwright, Lt. J. R.	29 June 17	24 May 18	1st Guides Infy.
27 Oct. 16	Beynon, Lt. A. C.	18 June 17	24 May 18	98th Infy.
5 May 17	Brady, Lt. J. P.	5 May 18	11 Aug. 18	I. A. R. O.
18 June 17	Handrick, Lt. C. A.	18 June 18	10 Aug. 18	41st Dogras.
18 Aug. 17	Woodcook, Lt. W. D. G.. . .	18 Aug. 18	6 Aug. 18	I. A. R. O.
5 Oct. 17	Didcock, 2nd-Lt. R. S. H. . .	5 Oct. 17	22 July 18	I. A. R. O.
8 Oct. 17	Brain, 2nd-Lt. W. G.	8 Oct. 17	13 July 18	I. A. R. O.
28 Nov. 17	Sparkes, 2nd-Lt. H. W. . . .	. . . .	12 Aug. 18	I. A. R. O. (*on probn.*)
31 Jan. 18	Newbury, 2nd-Lt. E. A. . . .	31 Jan. 18	13 July 18	I. A. R. O.
26 Mar. 18	Auty, 2nd-Lt. S. C.	26 Mar. 18	12 Aug. 18	I. A. (*on probn.*)
	Brewster, 2nd-Lt. J. A. . . .	. . . .	27 July 18	I. A. R. O.
. . . .	Burton, 2nd-Lt. J. W.. . . .	. . . .	2 Sep. 18	I. A. R. O.
	Attached.			
9 June 01	Galbraith, Maj. E. D., D.S.O. .	19 June 16	. . . .	55th Rif. O. C. Depôt.
14 July 15	Bearne, Lt. L. H.	14 July 16	22 Oct. 18	Wilts. R., I. A. (*on probn.*)
6 Aug. 15	Garthwaite, Lt. E. J. K. . . .	6 Aug. 16	22 June 15	I. A. R. O.
12 Sept. 17	Arbuthnot, Lt. J. St. C. . . .	12 Sep. 18	. . . .	59th Rif.
27 Oct. 17	Brooke-Hunt, Lt. R. C. . . .	27 Oct. 18	1 Aug. 18	1st Guides Cav.
21 Dec. 17	Evans, 2nd-Lt. S. F. . . .	21 Dec. 17	30 Aug. 18	3rd Guides Infy.
31 Jan. 18	Johnson, 2nd-Lt. A. H. G. . .	31 Jan. 18	1 Aug. 18	38th Dogras.
7 June 18	Ainscow, 2nd-Lt. A.	7 June 18	7 Oct. 18	I. A. (*T. C.*)
31 Aug. 18	Ingram-Johnson, 2nd-Lt. R. E. S.	31 Aug. 18	4 Oct. 18	3rd Guides Infy.
31 Aug. 18	Price, 2nd-Lt. . S.	31 Aug. 18	4 Oct. 18	3rd Guides Infy.
		Jemadar.	Subadar.	Subadar-Major.
	Subadar-Major.			
11 May 91	Jai Singh	1 Jan. 13	13 Feb. 15	
	Subadars.			
1 Jan. 08	Handpal Singh	1 Jan. 08	21 Oct. 13	
2 June 99	Fatteh Singh	5 Jan. 15	25 Sept. 17	
9 Apl. 96	Fakir Singh	10 July 15	. . . .	36th Sikhs
18 Dec. 00	Mohd. Zaman	1 Dec. 16	10 Jan. 17	59th Rif.
14 Dec. 97	Nur Mohd.	16 Jan. 17	4 Aug. 18	
18 Jan. 97	Kirtiman Rai	5 June 17	12 Oct. 18	
17 Dec. 97	Dost Mohd.	6 Nov. 17	22 June 18	59th Rif.
	Jemadars.			
18 Jan. 97	Zalam Singh	25 June 15	38 Dogras.	
1 Nov. 98	Niazbat Khan	8 Jan. 16	51st Sikhs.	
1 May 00	Samundar	1 Aug. 16		
16 Oct. 16	Totaram	10 Oct. 16		
9 Jan. 96	Gannu	1 Mar. 17		
24 Feb. 09	Bhan Sing	10 Jan. 17	56th Rif.	
5 Dec. 00	Sheolad Singh	19 June 17		
2 Aug. 07	Nadar Khan	6 Nov. 17	59th Rif.	
2 Aug. 07	Parmanand	27 Jan. 18		
	Barna Bir	27 Jan. 18		
13 Oct. 08	Raja Khan	27 Jan. 18	59th Rif.	
28 Nov. 04	Santa Bir	1 Feb. 18		
18 Oct. 01	Ram Lagan Singh.	11 Mar. 18		
23 Aug. 98	Amir Ali	17 June 18	95th Rif.	

1st Battalion, 152nd Punjabis.

Formed at

Composition—2¾ companies Punjabi Musalmans, ¾ company Sikhs, ¼ company Pathans, ¼ company Dogras.

1st Commn. or date of entering service.	Names and Rank.	Army rank.	Present appt. in Regiment.	Remarks.
	Commandant.			
25 Jan. 99	Penton, Maj. B. C. (*Actg. Lt.-Col.*)	1 Sep. 15	. . . .	25th Punj.
	Company Commanders. (4)			
17 Aug. 07	Messum, Capt. L. R. (*Act'g. Maj.*)	1 Sep. 15	24 May 18	21st Punj., 2nd-in-Commd.
15 Aug. 14	Briggs, Capt. H. R.	15 Aug. 18	. . . .	31st Punj.
18 Apl. 16	Ansell, Lt. C. E. O.	18 Apl. 17		
18 Apl. 16	Sevenoaks, Lt. C. L.	18 Apl. 17	. . . .	24th Punj.
	Company Officers.			
6 Oct. 16	Lyall, Lt. J. H. (*Actg. Capt.*) .	6 Oct. 17	24 May 18	I. A. R. O., Adjt.
30 Jan. 17	Boxall, Lt. M. W. J.	30 Jan. 18	24 May 18	28th Punj.
30 Jan. 17	Butter, Lt. G. R.	30 Jan. 18	24 May 18	
5 May 17	Frend, 2nd-Lt. G. V. R. . . .	5 May 17	24 May 18	I. A. R. O.
18 Jan. 17	Needham, 2nd-Lt. J. O. . . .	18 June 17	24 May 18	1-25th Punj.
20 June 17	Evans, 2nd-Lt. E. G. H. . . .	20 June 17	24 May 18	I. A. R. O.
9 Sep. 08	Coats, Capt. N. G. R. (*Actg. Maj.*)	1 Sep. 15	. . . .	2-25 Punj. O. C. Depôt.
. . . .	Best, Lt. E.	. . . .	1 Aug. 18	I. A. (*on probn.*)
23 Jan. 16	Colquhoun, Lt. J. W. C. . .	23 Jan. 17	19 July 18	R. Jr. Fus., I. A. (*on probn.*)
15 Jan. 18	Aikin, 2nd-Lt. R. W. D. . .	15 Jan. 18	10 June 18	I. A. R. O.
28 May 18	Williams, 2nd-Lt. J.	. . . .	8 Nov. 18	Lanc. R., I. A. R. O. (*on probn.*)
7 June 18	Bishop, 2nd-Lt. F. J.	7 June 18	14 Oct. 18	I. A. (*T.C.*)
. . . .	Buckland, 2nd-Lt. L. J. E. . .	. . . .	19 Oct. 18	R. N. Divn., I. A. R. O. (*on probn.*)
. . . .	Martin, 2nd-Lt. G. N.	. . . .	22 Nov. 18	I. A. (*on probn.*)
. . . .	Munn, 2nd-Lt. H.	. . . .	10 Oct. 18	I. A. R. O. (*on probn.*)
. . . .	Teemey, 2nd-Lt. S. P. . . .	. . . .	17 Sep. 18	I. A. R. O.

		Jemadar.	Subadar.	Subadar-Major.
	Subadar-Major.			
4 May 01	Mustak Sing (46, 72*a*) . . .	1 Apl. 12	1 Aug. 15	24 May 18
	Subadars.			
30 Dec. 95	Hasham Ali Khan (51, 56) . .	4 Aug. 12	6 July 15	
1 Nov. 97	Ghulam Mohd.	16 Oct. 14	14 Feb. 17	
1 Nov. 99	Bahadur Singh	12 Mar. 16	14 Feb. 17	
1 Apl. 98	Sadhu Singh (50, 72*a*) . . .	1 May 13	1 May 17	
1 Mar. 92	Anar Khan	28 Mar. 15	14 Oct. 17	
1 Jan. 01	Alum Sher Khan	12 Dec. 16	1 Dec. 17	
10 Oct. 03	Godar Khan	2 Nov. 16	. . . .	
2 Nov. 98	Niamat Khan	4 Sep. 15	24 May 18	
	Jemadars.			
8 June 98	Jahan Khan	4 Sep. 15		
17 Jan. 01	Mehr Khan	17 Dec. 15		
12 June 96	Amar Singh	1 Aug. 16		
14 Aug. 02	Feroze Khan	6 Jan. 16		
24 Apl. 94	Bhagat Singh	25 Sep. 16		
23 July 00	Maula Dad	29 Sep. 17		
1 Mar. 01	Amir Ali	12 Jan. 18		
15 July 12	Feroze Khan	25 Apl. 18		
2 Sep. 02	Burhan Ali	24 May 18		
2 Jan. 06	Niazbat Khan	24 May 18		
29 Sep. 03	Ditta Khan	24 May 18		

2nd Battalion, 152nd Punjabis.

Formed at

Composition—1 company Dogras, 1 company Rajputs, 2 companies Punjabi Musalmans.

1st Commn. or date of entering service.	Names and Rank.	Army rank.	Present appt. in Regiment.	Remarks.
	Commandant.			
20 Jan. 97	Wilkinson, Maj. C. R. (*Actg. Lt.-Col.*)	20 Jan. 15	. . . .	52nd Sikhs.
	Company Commanders. (4)			
25 Dec. 09	Jones, Capt. B. I. (*Actg. Maj.*)	1 Sep. 15	. . . .	38th Dogras, 2nd-in-Comd.
6 Sep. 11	Watson, Capt. E. E.	1 Sep. 15	. . . .	62nd Punj.
15 Nov. 15	Longbottom, Lt. J. L.	15 Nov. 16	. . . .	1-67th Punj.
15 Nov. 15	Cant, Lt. E. D.	15 Nov. 16	. . . .	84th Punj.
	Company Officers.			
15 Nov. 15	Weatherly, Lt. E. F. C. F.	15 Nov. 16	. . . .	37 Dogras, Adjt.
18 Apl. 16	Ronnfeldt, Lt. W. E.	18 Apl. 17	. . . .	62nd Punj.
26 Sep. 17	Northfield, Lt. E. W.	26 Sep. 18	. . . .	I. A. (*on probn.*)
26 Sep. 16	Simmons, Lt. F. D.	. . . .	. . . .	I. A. R. O. (*on probn.*)
5 May 17	Harper, Lt. A. H.	5 May 18	. . . .	I. A. R. O. Qr.-Mr.
5 May 17	Kerr, Lt. T.	5 May 18	. . . .	I. A. R. O.
20 July 15	Philipson, 2nd-Lt. H.	. . . .	. . . .	I. A. R. O. (*on probn.*)
21 Jan. 15	Lurcott, Lt. R. G.	. . . .	. . . .	I. A. R. O. (*on probn.*)
	Attached.			
3 Sep. 10	Minter, Capt. J. P. S. (*Actg. Maj.*)	1 Sep. 15	. . . .	130th Bal.
	Manning, Capt. E.	. . . .	. . . .	R. Berk. R.
24 July 15	Longbottom, Lt. A. W. (*Actg. Capt.*)	24 July 16	. . . .	I. A. (*on probn.*)
...	Jephcott, Lt. A. H.	. . . .	. . . .	
27 Oct. 17	Standish, 2nd-Lt. N.	27 Oct. 17	. . . .	37th Dogras.
25 Oct. 17	Greeves, Lt. S.	25 Oct. 18	. . . .	Lanc. R., I. A. (*on probn.*)
7 June 18	Hickmott, 2nd-Lt. W. H.	7 June 18	. . . .	I. A. (*T. C.*)
	Jackson, 2nd-Lt. H. L.	. . . .	. . . .	Lond. R., I. A. R. O. (*on probn.*)

		Jemadar.	Subadar.	Subadar-Major.	
	Subadar-Major.				
15 July 91	Shahnawaz Khan	. . . .	16 May 02	18 May 18	*Offg.*

1st Commn. or date of entering service.	Names and Rank.	Army rank.	Present appt. in Regiment.	Remarks.
	Subadars.			
15 July 91	Fazl Ahmad (56, *66a*, *66b*, *72a*)	6 Jan. 06	16 May 10	
17 Mar. 90	Jai Singh	2 June 12	26 Mar. 15	
1 Jan. 12	Moti Chand	1 Nov. 14	6 Jan. 16	
14 Oct. 98	Jaman Singh	1 Nov. 13	21 Jan. 16	
19 Aug. 02	Allahdad	14 Jan. 16	2 Feb. 17	92nd Punj.
16 June 98	Budhi Singh	1 Oct. 15	31 July 17	
Nov. 05	Quasim Ali	13 Feb. 16	19 Dec. 17	92nd Punj.
16 Jan. 98	Man Singh	19 Jan. 17	13 May 18	
21 Nov. 99	Rahim Dad Khan	17 Jan. 15	13 May 18	
	Niamat Khan	. . . .	24 May 18	
	Jemadars.			
1 Jan. 96	Munshi	15 Apl. 15		
8 May 99	Bakhtawar	6 Jan. 16		
1 Dec. 99	Amar Singh	12 Feb. 16		
1 Mar. 00	Lehna Singh	27 Feb. 16		
17 Dec. 02	Ahmed Khan	16 May 16	. . . .	
	Dewan Ali	19 Dec. 17	. . . .	
	Shahwali Khan	19 Dec. 17	. . . .	
	Saidal Khan	. . . .		
	Mohammad Khan	. . . .		
	Shamzada Khan	. . . .	13 May 18	
	Mir Haidar	. . . .	13 May 18	
	Chhog Singh	. . . .	13 May 18	

3rd Battalion, 152nd Punjabis.

Formed at

Composition—1½ companies Sikhs, 1½ coy. Punjabi Musalmans, 1 coy. Kashmiri Musalmans, ¼ coy. Pathans.

1st Commn. or date of entering service.	Names and Rank.	Army rank.	Present appointment in Regiment.	Remarks.
	Commandant.			
4 Jan. 99	Doveton, Maj. H. J. (*actg. Lt.-Col.*)	1 Sep. 15	10 June 18	21st Punj.
	Company Commanders. (4)			
14 Feb. 00	Webster, Maj. C. D. . . .	1 Sep. 15	24 May 18	30th Punj. 2nd-in-Comd.
8 Sep. 09	Brunskill, Capt. B. A. S. . .	1 Sep. 15	25 May 18	79th Infy.
6 Sep. 11	Rule, Capt. D. G.	6 Sep. 15	23 May 18	27th Punj.
20 Feb. 16	Higby, Lt. H. V. (*actg. Capt.*)	20 Feb. 17	24 May 18	I. A. R. O.
	Company Officers.			
10 June 16	Hodgkins, Lt. N. F. (*actg. Capt.*)	10 June 17	25 May 18	I. A. R. O. Adjt.
29 June 16	Roper, Lt. R. B.	29 June 17	23 May 18	27th Punj.
5 May 17	Maker, Lt. L.	5 May 18	12 June 18	I. A. R. O.
18 June 17	Reeve, Lt. M. B. P. . . .	18 June 18	3 July 18	19th Sikhs.
. . . .	Parry, Lt. D. F.	. . . .	31 July 18	I. A. (*on probn.*)
9 July 15	Thrupp, Lt. G. A.	9 July 16	23 Sep. 18	I. A. R. O.
. . . .	Knight, 2nd-Lt. C. C. A. . .	. . . .	8 June 18	I. A. (*on probn.*)
	Attached.			
16 June 09	Anderson, Capt. G. D. (*actg. Maj.*)	1 Sep. 15	. . . .	62nd Punj., O. C. Depôt.
28 Oct. 16	Power, Lt. W. A.	28 Oct. 17	24 July 18	
20 June 17	Curd, Lt. A. T.	20 June 18	10 Aug. 18	I. A. R. O.
5 Oct. 17	Gildea, Lt. E. P.	5 Oct. 18	5 Sep. 18	I. A. R. O.
17 Dec. 17	Everett, 2nd-Lt. W. M. . .	17 Dec. 17	1 Sep. 18	I. A. R. O.
29 Mar. 18	Griffiths, 2nd-Lt. R. W. . .	29 Mar. 18	24 July 18	I. A. (*T. C.*)
3 June 18	Crowhurst, 2nd-Lt. J. D. J. .	3 June 18	2 Dec. 18	I. A. R. O.
7 June 18	Irving, 2nd-Lt. C. E. . . .	7 June 18	19 Oct. 18	I. A. (*T. C.*)
24 Aug. 18	Jacob, 2nd-Lt. H. H. . . .	24 Aug. 18	19 Oct. 18	R. Cav., I. A. (*on probn.*)

		Jemadar.	Subadar.	Subadar-Major.
	Subadar-Major.			
31 July 01	Sultan Khan	1 Dec. 10	1 Aug. 14	
	Subadars.			
10 Apl. 92	Bishen Singh (53, 56, 66*a*) .	22 Dec. 09	13 Jan. 16	
9 Feb. 07	Sundar Singh	11 Nov. 15	17 July 17	
16 Aug. 82	Ishar Singh	1 Apl. 99	24 Mar. 06	Re-employed.
19 Apl. 03	Sulakhan Singh	14 Jan. 16	10 Apl. 18	Do.
2 Aug. 05	Mohammed Yacub Khan . .	1 Oct. 15	14 Sep. 16	Do.
3 Nov. 98	Dalip Singh	13 May 17	24 May 18	Do.
26 Aug. 04	Imam Ali	22 Apl. 17	. . . .	Do.
15 Sep. 96	Jamal Din	1 Feb. 16	13 Apl. 16	
	Jemadars.			
27 Oct. 98	Sundar Singh (58)	4 July 13	. . . .	
25 Jan. 97	Allah Din	6 Feb. 15	. . . .	
26 Mar. 93	Mohd. Din	14 Sep. 16	. . . .	
19 Jan. 06	Allah Ditta	1 Nov. 17	. . . .	
11 Mar. 01	Ishar Singh I	1 Nov. 17	. . . .	
6 Jan. 97	Ishar Singh II	12 Dec. 17	. . . .	
25 Jan. 18	Mohd. Anwan Khan . . .	17 Feb. 18	. . . .	
1 Feb. 04	Ahmed Khan	2 Mar. 18	. . . .	
29 May 05	Dalel Khan	2 Mar. 18	. . . .	
26 Jan. 98	Nizam Din	10 Apl. 18	. . . .	
26 Nov. 95	Bindar Singh	25 June 18	. . . .	
15 Nov. 14	Moti Singh	24 May 18	. . . .	Jemdr. Adjt.
14 Apl. 01	Tahal Singh	24 May 18	. . . .	
5 Aug. 01	Sultan Ahmed	24 May 18	. . . .	
20 Aug. 04	Lal Khan	8 Feb. 17	. . . .	
1 July 09	Latif Sher	1 June 18	. . . .	

1st Battalion, 153rd Punjabis.

Formed at

Composition—1½ companies Punjabi Musalmans, ½ coy. Jats, ¾ coy. Sikhs, ¾ coy. Rajputs, ½ coy. Brahmans.

1st Commn. or date of entering service.	Names and Rank.	Army rank.	Present appointment in Regiment.	Remarks.
	Commandant.			
20 Jan. 97	Withers, Maj. C. McG. (*Actg. Lt.-Col.*)	20 Jan. 15	25 Apl. 18	87th Punj.
	Company Commanders. (4)			
8 Feb. 02	Bailie, Maj. S. M. H. . .	8 Feb. 17	. . . .	87th Punj. 2nd-in-comd.
1 Dec. 11	Cousins, Lt. M. A. S. (*actg. Capt.*)	11 Dec. 15	18 May 18	I. A. R. O.
18 Apl. 16	Sutcliffe, Lt. O. D. (*actg. Capt.*)	18 Apl. 17	18 May 18	87th Punj.
14 Nov. 16	Jessop, Lt. C. R. (*actg. Capt.*)	15 Mar. 17	18 May 18	
	Company Officers.			
3 Oct. 14	Peele, Lt. C. R. deC. . .	3 Oct. 15	18 May 18	I. A. R. O.
26 May 16	Hughes, Lt. E. W. . . .	26 May 17	. . . .	I. A. R. O.
14 May 16	Hannah, Lt. J. S., M.C. . .	14 Aug. 17	18 May 18	I. A. R. O.
18 June 17	Morton, Lt. C. E. (*actg. Capt.*)	18 June 18	18 May 18	1-89th Punj. Adjt.
12 Sep. 17	Duke, Lt. A. L. R. . . .	12 Sep. 18	. . . .	82nd Punj.
. . . .	Garland, Lt. A. G. L. (*actg. Capt.*)	. . . .	. . . .	
20 Aug. 16	Murray, Lt. R. G. . . .	20 Aug. 17	14 Oct 18	I. A. (*on probn.*)
26 July 17	Luke, Lt. O. E. . . .	26 July 18	15 Sep. 18	I. A. R. O.
13 Sep. 17	Cockerill, Lt. J. W. . . .	13 Sep. 18	28 Sep. 18	I. A. (*T. C.*)
	Attached.			
2 Mar. 04	Allan, Capt. P. A.	2 Mar. 13	25 Apl. 18	66th Punj. O. C. Depôt.
6 Oct. 14	Crawshaw, Lt. H. H. (*actg. Capt.*)	. . . .	9 June 18	I. A. (*on probn.*) *Actg. Adjt.*
11 Apl. 17	Haylock, Lt. S. W. . . .	11 Apl. 18	1 Aug. 18	I. A. (*on probn.*)
5 Nov. 17	Senior, Lt. E. B. G. . . .	5 Nov. 18	10 July 18	I. A. (*T. C.*)
5 Nov. 17	Handscomb, Lt. L. G. . .	5 Nov. 18	10 July 18	I. A. R. O.
5 Nov. 17	Sparkes, Lt. H. B. . .	5 Nov. 18	. . . .	I. A. (*T. C.*)
7 June 18	Wells, 2nd-Lt. C. A., D. C. M.	7 June 18	11 Oct. 18	I. A. (*T. C.*)
...	Peters, 2nd-Lt. W., M. M. .	. . . .	9 Nov. 18	Bedf. R., I. A. R. O. (*on probn.*)
...	Evans, 2nd-Lt. H. R. . .	. . . .	13 Oct. 18	Genl. List, I. A. R. O. (*on probn.*)

		Jemadar.	Subadar.	Subadar-Major.	
	Subadar-Major.				
26 Aug. 04	Baboo Singh (56)	3 Mar. 07	23 Oct. 10	18 May 18	90th Punj,
	Subadars.				
6 July 91	Bakhtawar Singh (56, 66*a*) .	22 Jan. 11	1 July 11		
29 May 97	Ram Bakhsh	22 Oct. 08	1 Nov. 13		
20 Sep. 08	Thakar Singh	20 Sep. 08	15 July 14		
1 Apl. 91	Dhuman Khan (56, 66*a*) . .	5 Aug. 06	29 Apl. 15		
9 Aug. 94	Ghulam Muhd.	16 Sep. 08	27 July 15		
13 Apl. 97	Suraj Pande	17 Oct. 14	5 Apl. 16		
1 July 00	Ghulab Khan	30 Dec. 14	1 May 16		
5 Aug. 03	Kishena Ram	15 Apl. 15	10 May 16		
1 June 96	Nawaz Ali	1 June 16	9 Feb. 17		
16 Jan. 97	Kishen Singh	6 Apl. 15	20 Sep. 18		
	Jemadars.				
26 Aug. 03	Lahir Khan (52, 56, 66*a*) . .	29 Apl. 15	Ind. Adjt.		
1 Aug. 15	Chawdri Mangal Singh . .	1 Aug. 15			
14 May 98	Baldeo Singh I	11 Feb. 16			
1 June 97	Ram Kishan Sukal . . .	5 Apl. 16			
12 Dec. 03	Mir Afzal	1 Jan. 17			
17 Aug. 97	Baldeo Singh II	30 Aug. 17			
18 Jan. 04	Nizam Din	15 Oct. 17			
1 Jan. 09	Nur Muhd.	15 Oct. 17			
8 Dec. 03	Deosi Ram	18 May 18			
12 Sep. 03	Niamat Khan	3 July 18			
30 Oct. 01	Shivnath Tewari	6 July 18			
20 July 03	Ghulam Muhd.	17 July 18			
14 July 03	Ram Lal Tewari	8 Aug. 18			
...	Wir Singh	20 Sep. 18			

2nd Battalion, 153rd Punjabis.

Formed at

Composition—1½ companies Punjabi Musalmans, 2 coy. Sikhs, ½ coy. Hindustani Musalmans.

1st Commn. or date of entering service.	Names and rank.	Army rank.	Present appointment in Regiment.	Remarks.
	Commandant.			
20 Feb. 97	James, Maj. C. A. (*actg. Lt.-Col.*)	20 June 18	20 June 18	126th Infy.
	Company Commanders. (4)			
18 Jan. 02	Riddell, Maj. A., D.S.O.	18 Jan. 17	23 May 18	74th Punj., 2nd-in-Comd.
24 Aug. 12	Keene, Capt. B. G.	24 Aug. 16	27 May 18	92nd Punj.
14 Jan. 14	Trevelyan, Capt. W. R. F.	24 Jan. 18	1 June 18	93rd Infy.
18 Dec. 14	Pigot, Lt. W. T.	18 Dec. 15	15 June 18	I. A. R. O.
	Company Officers			
12 Feb. 15	Smith, Lt. G. R.	12 Feb. 16	1 June 18	
29 Aug. 15	Hutton, Lt. G. C.	29 Aug. 16	22 June 18	I. A. (*on probn.*)
18 Apl. 16	Moore, Lt. A. C.	18 Apl. 17	27 May 18	I. A. (*on probn.*) Qr.-Mr. 27 May 18.
24 Apl. 16	Hyde, Lt. G. L.	24 Apl. 17	23 May 18	I. A. (*on probn.*) Adjt., 27 May 18.
5 May 17	Hamilton, Lt. A. J.	5 May 18	19 June 18	I. A. R. O.
20 June 17	McCarthy, Lt. R. W.	20 June 18	2 July 18	I. A. R. O.
5 Nov. 17	Wonnacott, Lt. A. J.	5 Nov. 18	22 June 18	I. A. (*T. C.*)
5 Nov. 17	Haymes, Lt. T. F.	5 Nov. 18	22 June 18	I. A. (*T. C.*)
5 Nov. 17	Turnbull, Lt. D.	5 Nov. 18	15 June 18	I. A. (*T. C.*)
	Attached.			
29 Aug. 06	Digby, Capt. T. A.	29 Aug. 15	27 May 18	1-30th Punj. O. C. Depôt.
8 Apl. 15	Pickmere, Lt. C. E.	8 Apl. 16	30 Oct. 18	I. A. R. O.
6 Feb. 16	Holbrook, Lt. S. D.	6 Feb. 17	6 Oct. 18	E. York. R., I. A. (*on probn.*)
29 Aug. 17	Howes, 2nd-Lt. H. C.	29 Aug. 17	22 July 18	I. A. R. O.
5 Nov. 17	Cude, Lt. T.	5 Nov. 18	17 July 18	I. A. (*T. C.*)
28 Nov. 17	Arnold, 2nd-Lt. F.	28 Nov. 17	13 July 18	I. A. R. O.
24 Mar. 17	Maxwell, Lt. C. W.	24 Mar. 18	17 July 18	I. A. R. O.
7 June 18	Williamson, 2nd-Lt. A. U.	7 June 18	15 Oct. 18	I. A. ((*T. C.*)
	Gregory, 2nd-Lt. C. J. E.	....	30 Oct. 18	Genl. List, I. A. R. O. (*on probn.*)

		Jemadar.	Subadar.	Subadar-Major.
	Subadar-Major.			
17 Dec. 92	Jhanda Singh	8 Sep. 15	17 Apl. 16	1 June 18

	Subabars.			
1 Mar. 97	Fazal Khan (56,66a)	1 Nov. 09	1 Oct. 16	
27 Mar. 94	Dalel Singh	3 July 13	1 Sep. 15	
8 July 93	Partab Singh	15 Oct. 14	8 Mar. 16	
16 Oct. 99	Basan Singh	29 Aug. 15	15 July 16	
28 Apl. 01	Ghulab Khan	24 Oct. 14	2 Feb. 17	92nd Punj.
3 June 97	Lal Khan	14 Mar. 16	18 Mar. 17	
30 Dec. 02	Sher Zaman (56)	30 May 14	3 Apl. 17	
3 June 01	Afzal Khan	6 July 16	21 Feb. 18	92nd Punj.
20 Nov. 02	Dalel Khan	14 Jan. 16	27 May 18	92nd Punj.
27 Aug. 99	Indar Singh	23 Feb. 17	27 May 18	
10 July 00	Prem Singh	26 Mar. 17	26 Mar. 17	
10 Jan. 01	Kehr Singh	3 Apl. 17	25 Jan. 18	
12 Nov. 04	Majhi Khan	24 Mar. 17		
18 Dec. 01	Sawan Singh	26 Mar. 17	22 June 18	
23 Nov. 02	Beant Singh	25 Jan. 18	10 Sep. 18	
	Bega Ram	....	19 Sep. 18	
	Jemadar.			
24 July 00	Sher Zaman	1 May 17		
27 Nov. 02	Abdul Ghafur	15 Dec. 17		
19 Apl. 00	Badal Khan	1 Jan. 18		
3 Aug. 06	Mardan Ali	1 Jan. 18	92nd Punj.	
1 Dec. 09	Sundar Khan	1 Jan. 15		
7 Mar. 06	Sapooran Singh	1 Jan. 15		
14 Dec. 02	Amir Ali	25 Jan. 18		
	Maluk Khan	24 Mar. 18	Ind. Adjt.	
	Mirza Khan	27 May 18		
10 May 04	Ganga Singh	27 May 18		
	Abbas Khan	27 May 18		
	Harnam Singh	27 May 18	Ind. Qr.-Mr.	
20 Jan. 04	Narayan Singh	22 June 18		
	Khan Bahadur	12 July 18		
	Hazara Singh	10 Sep. 18		
	Dalu Ram	19 Sep. 18		

3rd Battalion, 153rd Rifles.

Formed at—Palestine.

Composition—½ company Konkani Mahrattas, ¼ coy. Dekhani Mahrattas, ¼ coy. Dekhani Musalmans, 2½ coy. Punjabi Musalmans, ½ coy. Rajputana Jats.

1st Commn. or date of entering service.	Names and Rank.	Army rank.	Present appointment in Regiment.	Remarks.
	Commandant.			
6 Jan. 95	Kidd, Maj. B. G. B. (*temp. Lt.-Col.*)	16 Jan. 13	. . .	125th Rif.
	Company Commanders. (4)			
22 Oct. 02	Wilkinson, Maj. T. O. . .	22 Oct. 17	. . . .	91st Punj., 2nd-in-Comd.
5 Aug. 15	Page, Lt. H. V. S. . . .	4 Aug. 16	. . . .	123rd Rif.
14 Jan. 15	Reith, Lt. C. E. W. . . .	14 Jan. 16	. . . .	I. A. R. O.
1 Oct. 15	Priestley, Lt. E. C. . . .	1 Oct. 16	. . . .	I. A. R. O.
	Company Officers.			
1 Jan. 17	Craige, Lt. S. M. M. . . .	1 Jan. 17	. . .	123rd Rif.
27 Aug. 15	Dunford, Lt. C. M. P. . .	27 Aug. 16	. . .	I. A. (*on probn.*) Adjt.
24 Jan. 16	Darwell, Lt. G. P. . . .	24 Jan. 17	. . .	120th Infy.
15 Nov. 15	Mountain, Lt. R. G. . .	15 Nov. 16	. . . .	1-127th Bal. L. I.
16 May 16	Godfrey, Lt. R. E.	16 May 17	. . . .	I. A. (*on probn.*)
23 May 16	D'Este, Lt. H. A. M. . . .	23 May 17	. . . .	I. A. (*on probn.*)
	Attached.			
14 Jan. 14	Maltby, Capt. F. N. (*Actg. Maj.*)	14 Jan. 18	. . .	125th Rif. O. C. Depôt.
3 Dec. 15	Saunders, Lt. R. A. R. . .	3 Dec. 16	. . .	I. A. R. O.
1 July 16	Dore, Lt. G. B.	1 July 17	. . .	I. A. (*on probn.*)
5 Nov. 17	Bodley, 2nd-Lt. C. A. . . .	5 Nov. 17	. . .	I. A. (*T. C.*)
5 Nov. 17	Smith, 2nd-Lt. H. P. . . .	5 Nov. 17	. . .	I. A. R. O.
. . .	Neville, 2nd-Lt. V. R. W. .	. . .	. . .	
2 June 18	Brown, 2nd-Lt. A. W. C. .	2 June 18	. . . .	I. A. (*T.C.*)
. . . .	Davies, Lt. G. C., M.C. . .	. . . .	. . . .	I. A. R. O. (*on probn.*)
. . . .	Roberts, Lt. R. G. L. . . .	. . . .	. . .	1-6th R. Welsh. Fus.
. . . .	Davies, Lt. R. T. . . .	. . . .	. . . .	I. A. R. O. (*on probn.*)
. . . .	Sandy, Lt. W. J. W. . . .	. . . .	. . . .	R. N. Divn., I. A. R. O. (*on probn.*)
. . . .	Alderton, 2nd-Lt. F. W. . .	. . . .	. . .	Genl. List, I. A. R. O. (*on probn.*)

		Jemadar.	Subadar.	Subadar-Major.
	Subadar-Major.			
	Subadars.			
7 July 97	Ghulam Mahomed, I.D.S.M. .	1 Oct. 09	21 Dec. 14	
25 June 97	Mubarik Ali, I.D.S.M. . . .	22 Sep. 14	7 Aug. 15	
31 July 03	Mahammad Hussain . . .	18 Nov. 17	19 Mar. 18	
20 Sep. 98	Eshwanta Shinde	16 Jan. 15	15 Aug. 17	
4 Sep. 07	Ram Chander Rao Erundkar	20 Jan. 16	13 Oct. 17	
	Jemadars.			
2 July 03	Bega Ram	18 Aug. 14		
4 Oct. 05	Sitaram Gaode	10 Jan. 17		
12 Apl. 12	Shaik Daod	14 Mar. 18		
27 Nov. 00	Mehrab Khan	5 July 15		
15 July 05	Lachman Singh	29 Dec. 15		
15 Mar. 05	Karam Dad Khan	29 Dec. 15		
24 July 04	Muhammad Alam	12 Apl. 18	22 May 18	
5 Oct. 04	Muhammad Zaman . . .	12 Apl. 18		
28 Feb. 04	Shaikh Mohiyuddin . . .	18 Oct. 15		
13 June 05	Krishna Bagwe	13 Dec. 15		
9 July 09	Sajawal Khan	26 Jan. 18		
23 June 10	Herda Ram	26 Jan. 18		
5 July 05	Sikandar Khan	1 July 18		

1st Battalion, 154th Indian Infantry.

Formed at—

Composition—2¼ companies Rajputs, ¾ coy. Jats, 1 coy. Ahirs.

1st Commn. or date of entering service.	Names and Rank.	Army rank.	Present appointment in Regiment.	Remarks.
	Commandant.			
29 Feb. 96	Slingsby, Maj. T. W. (*Actg. Lt.-Col.*)	1 Mar. 14	17 May 18	22nd Cav.
	Company Commanders. (4)			
29 Jan. 10	Corke, Capt. R. M. (*Actg. Maj.*)	1 Sept. 15	26 Aug. 18	1-94th Infy. 2nd-in-Comd. (*temp.*)
24 Dec. 14	Nugent, Lt. T. D. (*Actg. Capt.*)	22 Nov. 14	. . .	I. A. R. O.
	Company Officers.			
10 Dec. 15	Smallwood, Lt. L. S. *Actg. Capt.*)	10 Dec. 16	. . .	1-94th Infy. Adjt.
7 Sep. 16	Clapp, Lt. L. J.	7 Sep. 17	. . .	I. A. R. O.
18 June 17	Yates, Lt. D. H. P.	18 June 18	. . .	1-94th Infy.
8 Aug. 17	Leete, 2nd-Lt. B. T.	8 Aug. 17	. . .	I. A. R. O.
7 June 18	Sparham, 2nd-Lt. A. C.	7 June 18	. . .	I. A. (*T. C.*)
. . .	Weedon, 2nd-Lt. H. H.	. . .	11 Sep. 18	Norf. R., I. A. R. O. (*on probn.*)
. . .	Welman, 2nd-Lt. E. J.	. . .	11 Sep. 18	Genl. List, I. A. R. O. (*on probn.*)
. . .	Stanley, Lt. S. P.	. . .	6 Nov. 18	W. Yorks. R., I. A. R. O. (*on probn.*)
. . .	Marlaw, Lt. W.	. . .	. . .	R. N. V. R., I. A. R. O. (*on probn.*)
. . .	Dybell, 2nd-Lt. C. W.	. . .		Genl. List, I. A. R. O. (*on probn.*)
		Jemadar.	Subadar.	Subadar-Major.
	Subadar-Major.			
8 Oct. 02	Chandan Lal (56)	12 Nov. 14	10 May 16	17 May 18
	Subadars.			
18 May 00	Kaulesar Singh	23 Sep. 14	27 May 16	
19 Nov. 01	Bhup Singh (66*a*)	28 June 15	20 July 16	
12 July 05	Man Singh	1 Jan. 17	3 June 17	
1 Dec. 99	Dalsingar (66*a*)	1 Jan. 13	1 Dec. 17	
4 Sep. 04	Dukhan Singh	29 Mar. 17	8 Oct. 17	
1 Sep. 02	Ramrekha Singh	30 Aug. 15	15 Nov. 18	
. . .	Basdeo Singh	29 June 16	15 Nov. 18	
	Jemadars.			
19 Nov. 07	Rajnarāyan Singh	30 Aug. 15		
3 Apl. 06	Rampal Singh	27 May 16		
19 Aug. 16	Bani Singh	19 Aug. 16		
9 July 12	Nehal Singh (44)	9 Sept. 17		
. . .	Matadin Singh	28 Jan. 18		
. . .	Jula Rām	17 Feb. 18		
. . .	Mohan Singh	11 Mar. 18		
. . .	Jhanda	11 Mar. 18		
. . .	Ramchandar Singh	12 July 18		
. . .	Shiur Ratan Singh	12 July 18		

2nd Battalion, 154th Indian Infantry.

Formed at

Composition—3½ companies Gujars, ½ company Rajput Musalmans.

1st Commn. or date of entering service.	Names and Rank.	Army Rank.	Present appointment in Regiment.	Remarks.
	Commandant.			
17 May 99	Boxwell, Maj. A., C.I.E. (*Actg. Lt.-Col.*)	1 Sep. 15	. . . .	119th Infy.
	Company Commanders. (4)			
20 Jan. 12	Hindmarsh, Capt. J. H. L.	20 Jan. 16	. . . .	8th Raj.
13 Nov. 14	Turner, Lt. D. H.	13 Nov. 15	24 May 18	I. A. R. O.
14 July 15	Watts, Lt. H. C.	14 July 16	. . . .	I. A. (*on probn.*)
22 July 15	Marchant, Lt. D. R.	22 July 16	24 May 18	I. A. R. O.
	Company Officers.			
20 July 15	Pike, Lt. W. H.	20 July 16	9 Sep. 18	I. A. (*on probn.*)
15 Nov. 15	Anderson, Lt. D. C.	15 Nov. 16	31 May 18	122nd Infy.
15 Nov. 15	Whitehead, Lt. W. A.	15 Nov. 16	. . . .	1-76th Punj.
. . . .	Busar, Lt. C. F.	. . . .	. . . .	
1 Dec. 15	Marris, Lt. H. E. A.	1 Dec. 16	6 Aug. 18	I. A. (*on probn.*)
26 May 16	Allott, Lt. A. D.	26 May 17	. . . .	I. A. (*on probn.*)
29 June 16	Smithers, Lt. H. E.	29 June 17	. . . .	I. A. R. O.
5 May 17	Ferguson, Lt. J. H. (*Actg. Capt.*)	5 May 18	24 May 18	I. A. R. O. Adjt. 24 May 18.
29 June 17	Nicholls, 2nd-Lt. E. H.	29 June 17	2 June 18	I. A. R. O. Qr.-Mr.
27 Oct. 17	Thunder, 2nd-Lt. H. F. A.	27 Oct. 17	. . . .	I. A. (*on probn.*).
24 July 18	Emeny, 2nd-Lt. L. H.	24 July 18	9 Sep. 18	I. A. R. O.
	Attached.			
9 Sep. 08	Masters, Capt. E.	1 Sep. 15	29 May 18	108th Infy.
24 Sep. 14	Smith, Capt. B. St. J.	24 Sep. 18	16 Aug. 18	I. A. R. O.
23 June 15	Eagar, Lt. F. M.	23 June 16	20 July 18	R. Muns. Fus., I. A. (*on probn.*)
20 Nov. 15	Myers, Lt. F. G.	. . . .	26 Aug. 18	M. G. C., I. A. (*on probn.*)
3 Jan. 16	Sutton, Lt. G. F.	3 Jan. 17	26 Aug. 18	Lond. R. I. A. (*on probn.*)
25 Nov. 17	Elson, 2nd-Lt. W.	25 Nov. 17	9 June 18	
. . . .	Sussex, 2nd-Lt. H. G. V.	. . . .	. . . .	Lond. R., I. A. R. O. (*on probn.*)
. . . .	Fraser, 2nd-Lt. E. R. D.	. . . .	20 Oct. 18	Essex R., I. A. R. O. (*on probn.*)
		Jemadar.	Subadar.	Subadar-Major.
	Subadar-Major.			
30 Oct. 99	Ganga Ram Singh, I.O.M.	1 Sep. 12	. . . .	
4 July 93	Shikari *Bahadur*, D.S.M.	11 Mar. 02	20 Aug. 13	*Order of Br. Ind., 2nd Cl.*
	Subadars.			
17 July 97	Gaggan Ram	1 June 14	19 June 15	
1 May 99	Sheonath	27 July 15	20 July 17	1-119th Infy.
23 July 99	Mubarak Ali	1 Apl. 12	. . . .	
18 May 03	Chandan	1 Oct. 15	25 Nov. 16	
21 June 03	Angad Ram	8 Aug. 16	7 Apl. 17	
1 May 05	Gulab Singh	15 Dec. 14	31 Oct. 17	
	Jemadars.			
26 Dec. 98	Abdul Majid Khan	20 Aug. 16		
26 May 03	Hurgyan Ram	7 Nov. 16		
26 July 02	Ram Heth	25 Nov. 16		
31 July 03	Hardial	11 Apl. 17		
20 Sep. 05	Ladhu Ram	12 Feb. 16		
13 Oct. 08	Harkishan	20 Jan. 17		
31 Aug. 02	Jokhi Ram	7 Apl. 17		
. . . .	Ghisa Singh	3 Oct. 17		
14 Sep. 03	Ganesh Ram	25 July 16		
4 July 06	Bhairu Singh	2 Nov. 17		
. . . .	Mohammed Hanif	. . . .		
20 Sep. 16	Genesha	20 Sep. 16		
31 Mar. 06	Shrikrishna Singh	24 Feb. 17		
. . . .	Unkar Singh	19 Feb. 18		
26 May 14	Mohammed Ali Khan	10 Sep. 18	119th Infy.	
18 Aug. 15	Kaloo Khan	10 Sep. 18	119th Infy.	
1 July 07	Bijai Singh (56) (61)	10 Sep. 18		
1 June 03	Munsi	24 May 18		
17 Feb. 02	Safidar Khan	24 May 18		
. . . .	Mohbta	24 May 18		

3rd Battalion, 154th Indian Infantry.

Formed at

Composition—1¾ companies Gujars, 2 coys. Jats, ¼ coy. Punjabi Musalmans.

1st Commn. or date of entering service.	Names and Rank.	Army rank.	Present appointment in Regiment.	Remarks.
	Commandant.			
11 Sep. 98	Davson, Maj. H. J. H. (*Actg. Lt.-Col.*)	1 Sep. 15	. . . .	82nd Punj.
	Company Commanders. (4)			
	Company Officers.			
	Attached.			
22 Apl. 03	Sutton, Maj. H. G.	22 Apl. 12	29 May 18	42nd R. O. C. Depôt.
4 Nov. 03	Nixon, Capt. C. H. F., D.S.O. (*Actg. Maj.*)	4 Nov. 12	. . . .	1-91st Punj
7 Oct. 14	Eve, Lt. H. E., M.C.	. . . .	17 Oct. 18	Lond. R., I. A. R. O. (*on probn.*)
11 Sep. 15	Williams, Lt. R. G. . . .	11 Sep. 16	16 June 18	R. Fus., I. A. (*on probn.*)
13 Jan. 15	Sanders, Lt. R. G. . . .	13 Jan. 16	. . . .	38th Dogras.
21 May 15	Lynch, Lt. H. E. H. . . .	21 May 16	24 May 18	I. A. R. O. (*on probn.*)
14 July 15	Cameron, Lt. E.	14 July 16	15 May 18	113th Infy.
14 Aug. 15	Matthewman, Lt. G. B. . .	4 Aug. 16	15 May 18	113th Infy.
30 Apl. 16	Hind-Fletcher, Lt. T. R. . .	30 Apl. 17	19 Sep. 18	M. G. C., I. A. (*on probn.*)
22 May 16	Fricker, Lt. E. S. G. . . .	22 May 17	15 May 18	I. A. R. O.
10 May 16	Owen, Lt. A. W.	10 May 17	30 Sep. 18	I. A. (*on probn.*)
2 Sep. 16	Langlands, Lt. A. R. . .	2 Sep. 17	17 July 18	R. Scots, I. A. (*on probn.*)
1 Jan. 17	Smallwood, Lt. R. . . .	1 Jan. 18	17 July 18	
5 May 17	Stewart, Lt. A. R. . . .	5 May 18	31 Aug. 18	I. A. R. O.
18 June 17	Robertson, Lt. D. G. . . .	18 June 18	30 May 18	1-10th Jats.
. . . .	Newbold, 2nd-Lt. A. W. . .	. . . .	6 Nov. 18	S. Staffs. R., I. A. R. O. (*on probn.*)
1 Oct. 17	Guppy, Lt. S. J.	1 Oct. 18	31 Aug. 18	
1 May 18	Hodgman, 2nd-Lt. S. F. J. .	1 May 18	25 Oct. 18	I. A. R. O. (*on probn.*)
. . . .	Johnson, 2nd-Lt. F. L. . .	. . . .	8 Aug. 18	
. . . .	Wigan, Capt. S. R. D. . .	. . . .	10 Aug. 18	
. . . .	Adamson, Lt. M. A. W. . .	. . . .	. . . .	
		Jemadar.	Subadar.	Subadar-Major.
	Subadar-Major.			
	Subadars.			
16 Feb. 00	Ghure (56)	20 Nov. 06	8 June 07	1-113th Infy.
26 July 99	Allabakhsh (52, 56, 66*a*) . .	1 Oct. 07	14 June 15	1-112th Infy.
22 Jan. 98	Maharaj Singh	14 June 15	4 Aug. 17	1-112th Infy.
17 Aug. 99	Hukim Singh (56)	5 Jan. 15	1 Sep. 16	
. . . .	Binja Ram	. . . .	. . . .	
2 Oct. 98	Kala Singh	1 Dec. 16	1 Sep. 18	
. . . .	Indar Singh	. . . .	1 Sep. 18	
. . . .	Keshariya	. . . .	1 Sep. 18	
. . . .	Latte Singh	. . . .	10 Sep. 18	
	Jemadars.			
3 Jan. 97	Mohan Lal	27 Feb. 17		
3 Sep. 00	Gulichand Singh	26 July 16		
30 Aug. 01	Bagga Khan	8 June 18		
26 Dec. 01	Deochand	21 July 16		
25 Apl. 05	Nanwa	7 Nov. 16	1-113th Infy.	
28 Dec. 09	Ruga Ram	26 May 17	1-113th Infy.	
. . . .	Nathu Singh	10 Sep. 18		
. . . .	Dhanna Ram	. . . .		
. . . .	Jamna Sahai	17 Sep. 18		
. . . .	Sheo Lal	17 Sep. 18		
. . . .	Phusan Singh	22 Sep. 18		
. . . .	Akram Khan	22 Sep. 18		

1st Battalion, 155th Indian Pioneers.

Formed at

Composition—2 companies Mazbhi Sikhs, 1 coy. Jats, 1 coy. Rajput Musalmans.

1st Commn. or date of entering service.	Names and Rank.	Army rank	Present appt. in Regiment.	Remarks.
	Commandant.			
10 Oct. 94	Watson, Bt. Lt.-Col. S. B., *q.*	3 June 16	27 May 18	64th Prs.
	Company Commanders. (4)			
4 July 03	Skinner, Maj. T. B., *q.*	26 Sep. 16	27 May 18	64th Prs. *Actg.* 2nd-in-Commd.
20 Sep. 08	Christie, Capt. B.	1 Sep. 15	19 May 18	C. C. *Temp.* 32nd Prs.
25 Jan. 08	Leslie, Capt. F. D.	1 Sep. 15	19 May 18	48th Prs.
17 Aug. 07	Fraser, Capt. S. G. G., *q.*	1 Sep. 15	20 May 18	81st Prs.
	Company Officers.			
29 June 16	Tipping, Lt. R. P. M. (*actg. Capt.*)	26 Dec. 16	20 May 18	*Offg.* Adjt. 20 May 18, 81st Prs.
	Attached.			
8 Jan. 03	Sheringham, Maj. A. T., D. S. O., *q.*, (L.)	19 Dec. 15	19 May 18	121st. O. C. Depôt 1-155th Prs.
27 Jan. 16	Fraser, Lt. H. W. M.	27 Jan. 17	19 May 18	I. A. R. O.
29 June 16	Phillpott, Lt. F. L.	23 Feb. 16	. . .	27th Cav.
5 Jan. 17	Blake, Lt. L. F. M.	5 Jan. 18	10 June 18	I. A. R. O.
12 Sep. 17	Dickinson, Lt. F. R.	12 Sep. 18	24 Sep. 18	I. A.
20 June 17	Knowles, Lt. H. P.	20 June 18	20 May 18	I. A. R. O. *Offg.* Qr. Mr.
5 Nov. 17	Holdsworth, Lt. B. G.	5 Nov. 18	8 June 18	"
5 Oct. 17	Laver, Lt. E. W.	5 Oct. 18	23 July 18	I. A. (*T. C.*)
1 May 17	Leake, Lt. J. I.	1 May 18	23 July 18	3-32nd Prs.
19 Mar. 18	Gibson, 2nd-Lt. A. J.	19 Mar. 18	16 Aug. 18	I. A. R. O.
31 Jan. 18	Burkett, 2nd-Lt. R. N.	31 Jan. 18	. . .	I. A.
9 June 18	Holden, 2nd-Lt. G. P. C.	7 June 18	9 Sep. 18	I. A. (*T. C.*)
31 July 18	Bonney, 2nd-Lt. G. S.	31 July 18	15 Sep. 18	Genl. list, I. A. R. O. (*on probn.*)
		Jemadar.	Subadar.	Subadar-Major.
	Subadar-Major.			
	Subadars.			
5 Mar. 97	Ewaz Khan	10 Oct. 08	17 Oct. 14	121st Prs.
26 Apl. 96	Jhanda Singh, I. O. M.	1 June 11	30 Nov. 14	32nd Prs.
4 July 96	Asa Singh	1 July 12	16 June 15	32nd Prs.
1 May 95	Ranjit	30 Apl. 16	16 July 16	48th Prs.
27 Feb. 93	Sham Singh	16 July 11	9 Sep. 16	32nd Prs.
23 Apl. 04	Mala Singh		1 Jan. 17	32nd Prs.
25 June 98	Sadha Singh	16 June 15	12 Apl. 17	
12 July 98	Chand Khan	15 Apl. 13	23 Aug. 17	121st Prs.
5 July 87	Yasin Khan	4 Feb. 03	16 Feb. 04	Re-empld.
12 May 01	Udmi	12 July 16	19 May 18	48th Prs.
	Jemadars.			
2 Jan. 97	Surain Singh	26 Nov. 15	32nd Prs.	
12 June 00	Mit Singh	4 Sep. 16	32nd Prs.	
3 Oct. 08	Sadda Singh	12 Apl. 17	32nd Prs.	
20 June 01	Nante	6 Aug. 17	48th Prs.	
25 May 01	Tokha	3 Mar. 18	48th Prs.	
26 Mar. 04	Nauroz Khan	11 Mar. 18	121st Prs.	
4 Mar. 03	Makhmul Khan	11 Mar. 18	121st Prs.	
27 Apl. 96	Pritam Sing	5 Oct. 16		
	Male Khan	20 May 18		
	Narayan Singh	20 May 18		
19 June 99	Sohan Singh	18 June 18		
16 Apl. 09	Samiullah	26 June 18		
3 Nov. 97	Kahna Ram	18 July 18		
. . . .	Gurmukh Singh	1 Sep. 18		
. . . .	Gehna Singh	1 Sep. 18		

2nd Battalion, 155th Indian Infantry.

Formed at

Composition—1 company Yusafzais, 3 coys. Mazbhi Sikhs.

1st Commn. or date of entering service.	Names and Rank.	Army rank.	Present appointment in Regiment.	Remarks.
	Commandant.			
4 Aug. 97	Nicolas, Maj. R. (*actg. Lt.-Col.*)	4 Aug. 15	. . . .	23rd Prs.
	Company Commanders. (4)			
18 Jan. 02	Heyland, Maj. G. D., *q.*	10 Jan. 18	5 July 18	121st Prs.
29 Aug. 06	Ommanney, Capt. W. M., *q.*	29 Aug. 13	12 June 18	1-32nd Prs.
18 Dec. 14	Faithfull, Lt. F. H.	18 Dec. 15	12 June 18	1-23rd Prs.
	Company Officers.			
13 Jan. 15	Parsons, Lt. J. W. A.	13 Jan. 16	12 June 18	I. A. (*on probn.*)
. . . .	Smith, Lt. D. R.	. . . .	12 June 18	
18 Feb. 15	Borlase, Lt. J.	18 Feb. 16	12 June 18	I. A. R. O.
1 Dec. 15	Findlay, Lt. A. J.	1 Dec. 16	12 June 18	I. A. R. O.
30 Oct. 15	Wiley, Lt. W. J. G.	30 Oct. 16	12 June 18	I. A. (*on probn.*)
27 Oct. 17	Grieves, 2nd-Lt. R. B.	27 Oct. 17	3 July 18	1-32nd Prs.
28 Sep. 18	Garratty, 2nd-Lt. M. C.	28 Sep. 18	12 Oct. 18	I. A. R. O.
		Jemadar.	Subadar.	Subadar-Major.
	Subadar-Major.			
	Subadars.			
12 Sep. 02	Falel Khan	9 July 04	1 Jan. 10	
1 Apl. 04	Basti Khan	8 Apl. 15	15 Dec. 15	
27 June 97	Arjan Singh	30 June 07	8 May 15	
6 Oct. 99	Binda Singh	15 Aug. 14	3 Feb. 17	
24 July 03	Bir Singh	6 Nov. 16	2 May 17	
1 July 92	Jaggat Singh	8 Aug. 11	1 June 17	
14 Nov. 98	Bachhittar Singh	11 Aug. 15	12 Feb. 18	
27 Apl. 00	Narain Singh	6 Nov. 16	. . . .	
	Jemadars.			
19 Aug. 96	Bastan Khan	1 Dec. 16		
11 Apl. 04	Sewa Singh	1 June 17		
10 Feb. 05	Mitt Singh	7 Dec. 17		
4 Jan. 90	Surain Singh	7 Dec. 17		
7 Aug. 01	Natha Singh	17 Jan. 18		
2 Feb. 11	Bir Singh	1 Feb. 18		
10 June 03	Jaggat Singh	28 Mar. 18		
5 Apl. 02	Zaidullah Khan	12 June 18		
16 Apl. 09	Samih-Ullah	26 June 18		

Facings—Green.

1st Battalion, 156th Indian Infantry.

Formed at—

Composition—¾ company Tamils, 2 coy. Madrassi Musalmans, ¾ coys. Mayaros, ¼ coy. Moplahs, ¼ Coy. Madrassi Christians.

1st Commn. or date of entering service.	Names and Rank.	Army rank.	Present appointment in Regiment.	Remarks.
	Commandant.			
6 Jan. 00	Broadbent, Maj. J. T. C. (Actg. Lt.-Col.)	1 Sep. 15	24 Oct. 18	88th Infy. (*temp.*)
	Company Commanders. (4)			
27 Jan. 04	Phythian-Adams, Capt. E. G. (*actg Maj.*)	27 Jan. 13	26 May 18	80th Infy. 2nd-in-Comd.
22 Jan. 15	Ham, Lt. D. M. (*actg. Capt.*)	22 June 16	16 Sep. 18	1-75 Infy.
10 Apl. 15	Conner, Lt. T. S. (*actg. Maj.*)	29 June 17	26 May 18	88th Infy. *offg.* 2nd-in-comd.
9 July 15	Sitzler, Lt. E. A. (*actg. Capt.*)	9 July 16	1 June 18	78th Infy.
	Company Officers.			
26 Jan. 16	Robb, Lt. T. R.	26 Jan. 17	9 June 18	88th Infy.
29 June 16	Robinson, Lt. J. A. (*actg. Capt.*)	29 June 17	1 June 18	83rd Infy. Adjt.
9 Sep. 16	Williams, Lt. D. M.	9 Sep. 17	1 June 18	I. A. (*on probn.*) Qr.-Mr.
31 Jan. 17	Cawood, Lt. C. W.	31 Jan. 18	26 May 18	80th Infy. Adjt.
	Attached.			
4 Sep. 15	Campbell, Lt. W. B.	4 Sep. 16	29 Sep. 18	I. A. (*on probn.*)
7 Apl. 16	Armstrong, Lt. R. H.	7 Apl. 18	16 Apl. 18	I. A. (*on probn.*)
26 Apl. 16	Welton, Lt. J. J. G.	26 Apl. 17	23 Sep. 18	I. A. (*on probn.*)
10 Mar. 16	St. Aubyn, Lt. J. H. A. G.	10 Mar. 16	3 Oct. 18	I. A. (*on probn.*)
. . .	Storr, 2nd.Lt. W. G.	. . .	3 Oct. 18	I. A. R. O. (*on probn.*)
. . .	Paisley, 2nd-Lt. G.	. . .	30 Oct. 18	Wilts, R., I. A. (*on probn.*)
. . .	Taylor, 2nd-Lt. R. H.	. . .	10 Nov. 18	I. A. (*on probn.*)
10 July 18	Wright, 2nd-Lt. H.	10 July 18	2 Dec. 18	I. A. (*T. C.*)
22 July 18	Boswarva, 2nd-Lt. S. J.	22 July 18	. . .	I. A. (*T. C.*)
31 Oct. 17	Corser, 2nd-Lt. H. D.	31 Oct. 17	16 Sep. 18	I. A. R. O.
		Jemadar.	Subadar.	Subadar-Major.
	Subadar-Major.			
15 Dec. 95	Tiruvengadam (56, 66*a*)	8 May 10	31 Oct. 12	1 June 18
	Subadars.			
9 May 96	M. Bhagavati (56)	12 Jan. 12	1 Apl. 13	
13 Apl. 92	Abdus Samad (66*a*)	1 Jan. 10	1 July 14	
3 June 92	Muhammad Waizullah (56, 58, 66*a*).	9 Feb 13	16 Oct. 14	
15 Feb. 95	Abdul Hussain (52, 56, 66*a*)	25 Feb. 14	22 Sep. 16	
6 Mar. 95	Hussain Khan (66*a*)	6 Dec. 13	20 Apl. 17	
20 Dec. 15	Manavazhi Mahadeva Menon	20 Dec. 15	21 May 18	(*On probn.*)
25 Mar. 01	Muniratham (57)	29 Dec. 15	30 June 18	
	Jemadars.			
11 Aug. 96	Shaikh Hussain	23 Apl. 16	Ind. Adjt.	
17 May 98	Abdur Rahiman	1 Oct. 16		
24 Jan. 95	Muhammad Ghaus	12 Mar. 17		
1 Aug. 01	Ratna Mudali	20 Apl. 17		
20 Mar. 01	Abdul Jabbar	2 Aug. 17		
17 Aug. 00	Shaikh Abdul Khadir	4 Mar. 18		
5 June 01	Maduraimuthu	5 Apl. 18		
16 Dec. 03	Nagalingam	29 June 18		
10 Oct. 09	Rangasami Mudali	30 June 18		
5 July 00	Saiyed Hussain	15 Aug. 18		
30 June 99	Shaikh Farid	15 Sep. 18		
13 Oct. 10	Abdul Ghani	15 Sep. 18		
9 July 10	Periyasami Pillai	15 Sep. 18		
28 May 08	Arumugam Mudali	15 Sep. 18		

Regt. Centre—Dharmsala.

Uniform—Dark Green.
Facings—Scarlet.

1st King George's Own Gurkha Rifles (The Malaun Regiment.)

1st Battalion
2nd Battalion

Raised at Subathu, in 1815, by Lt. R. Ross, chiefly from Gurkha soldiers of Amar Singh Thapa's army, who took service with the British after the fall of Malaun, and originally designated the 1st Nasiri Battalion, became the 6th or 1st Nasiri Local Bn., 1823; 4th or Nasiri Local Bn., 1826; the 66th or Gurkha Regt., 1850; the 66th or Gurkha Light Infy. Regt., 1858; the 11th Regt. of Bengal Native Infantry, 1861; the 1st Gurkha Regt. (Lt. Infy.), 1861; the 1st Gurkha Regt. (Lt. Infy.), 1886; the 1st Gurkha (Rifle) Regt., 1891; the 1st Gurkha Rifles, 1901; the 1st Gurkha Rifles (the Malaun Regt.) 1903; 1st Prince of Wales's Own Gurkha Rifles (the Malaun Regiment) 1906; present designation, 1910.

The 2nd Battalion was raised at Dharmsala, in 1886, by Major G. W. Rogers.

THE PLUME OF THE PRINCE OF WALES, THE ROYAL AND IMPERIAL CYPHER.
"Bhurtpore" "Aliwal" "Sobraon" "Afghanistan, 1878-80"
"Tirah" "Punjab Frontier."

COLONEL-IN-CHIEF—THE KING.

COLONEL—Maj.-Genl. *Sir* Charles H. Powell, K.C.B., Indian Army.

1st Commn. or date of entering service.	Names and Rank.	Army rank.	Present appointment in Battalion.	Remarks.
	Commandants.			
8 Dec. 88	1 Anderson, Bt.-Col. W. C., C.M.G. (*temp. Br.-Genl.*)	3 June 16	21 May 15	
18 June 92	2 Dopping-Hepenstal, Lt.-Col. M. E., D.S.O. [*l*] (Q)	11 Mar. 18	11 Mar. 18	
	Company Commanders. (8)			
20 July 98	*2 Shaw, Maj. H. M. D.*	*1 Sep. 15*	*11 Mar. 18*	*With 3-3rd G. R.*
20 July 98	1 Evans, Maj. W. J.	1 Sep. 15	3 Sep. 18	2nd-in-Comdt.
27 July 98	2 Johnson, Maj. A. E	1 Sep. 15	9 Aug. 10	With 1-11th G. R.
20 Jan. 00	2 Elias Wynn, Maj. A. H. W., *p.s.c.*	1 Sep. 15	23 Dec. 14	*Lv., ex I., m. c.*
11 Apl. 00	1 Holderness, Maj. H.	1 Sep. 15	9 Mar. 16	With 1-50th Kumaon Rifles.
28 July 00	1 Latham, Maj. A	1 Sep. 15	8 Dec. 17	With 1-50th Kumaon Rifles.
14 Sep. 01	*2 Inglis, Maj. A. F. (actg. Lt.-Col.)*	*14 Sep. 16*	*11 Mar. 18*	*I. S. Troops.*
18 Jan. 03	*1 Scott, Maj. H.L.*, D.S.O., M.C., *p. s. c.*	*3 Sep. 18*	*3 Sep. 18*	
	Company Officers. (18)			
10 Oct. 03	2 Baldwin, Maj. M. C., M.C.	10 Oct. 18	1 Mar. 06	With 2-11 G. R.
18 Nov. 04	*2 Birch, Maj. C. F. M.*	*18 Nov. 18*	*1 Mar. 06*	*Bde. Maj. Jhansi Bde.*
18 Jan. 05	1 Northey, Capt. W. B., M.C. [*l*]	18 Jan. 14	29 July 06	With 2-50th Rif.
5 Aug. 05	2 Græme, Capt. N. F.	5 Aug. 14	15 Sep. 18	2nd-in-Comd. (*temp.*)
19 Jan. 07	1 Minchinton, Capt. H. D.	1 Sep. 15	22 Mar. 08	
19 Jan. 07	2 Ogilvy, Capt. J. D.	1 Sep. 15	26 May 08	With 1-1st G. R.
25 Jan. 08	2 Cobbold, Capt. M. F. D.	1 Sep. 15	19 Mar. 09	
9 Sep. 08	2 Mockler, Capt. E. C.	1 Sep. 15	14 Nov. 09	With 2-7th G. R.
20 Jan. 12	1 Carpendale, Capt. W. St. J.	20 Jan. 16	12 Mar. 13	With Nepalese Cont.
20 Jan. 12	2 Rogers, Capt. G. G.	20 Jan. 16	9 Mar. 13	With Nepalese Cont.
14 Jan. 14	2 Bellers, Capt. E. V. R.	14 Jan. 18	16 Oct. 14	With 1-1st G. R.
8 Aug. 14	1 Hamber, Capt. L. G. W.	8 Aug. 18	10 Nov. 15	With 1-4th G. R.
15 Aug. 14	1 Gracey, Capt. D. D., M.C.	15 Aug. 18	16 Nov. 15	
15 Aug. 14	1 Upton, Capt. R. B. E.	15 Aug. 18	9 Nov. 15	With 2-39th Garh Rif.
26 May 15	2 Murison, Lt. G De H.	26 May 16	1 Apl. 18	Attd. Adj. (*temp.*)
16 June 15	2 Fulton, Lt. J. O.	16 June 16	28 Oct. 15	With 1-1st G. R.
22 Oct. 16	2 Thomas, Lt. K. F. W.	22 Oct. 17	....	Attd.
28 Feb. 16	2 Hampson, Lt. J. H. J.	28 Feb. 17	7 Oct. 17	Attd. With 1-1st G. R.
29 June 16	2 D'Oyly Hughes, Lt. R.	29 June 17	....	With 2-5th G. R.
16 Aug. 16	1 Barry, Lt. J. L.	16 Aug. 17	11 Nov. 16	With 2nd Bn., Guides.
14 Nov. 16	2 Burch, Lt. P. D.	14 Nov. 17	24 Nov. 16	Attd. with 3-5th G. R.
30 Jan. 17	1 Foster, 2nd-Lt. L. A.	30 Jan. 17	....	With 1-5th G. R.
23 Jan. 17	2 Knowles, Lt. W. S., M. C.	23 Jan. 18	29 Nov. 17	Attd.
28 June 17	1 MacLeod, Lt. N. M.	28 June 18	....	
28 June 17	1 Stevenson, Lt. H. A.	28 June 18	....	
	2 Rundle, 2nd-Lt. A. L. de V.	....	....	Attd.
	2 Watts, 2nd-Lt. L. G.	....	....	Attd.
5 July 17	2 Milling, 2nd-Lt. R. C.	5 July 17	1 Dec. 17	Attd. Qr. Mr. (*temp.*)
12 Sep. 17	1 Rundall, Lt. J. W.	12 Sep. 18	10 Dec. 17	Attd.
21 Dec. 17	2 Partridge, 2nd-Lt. H. V.	21 Dec. 17	....	Attd. With 1-131st U. P. R.
15 May 15	2 Newland, Lt. C. D.	15 May 16	10 Dec. 17	Attd.
26 Jan. 16	1 Mitchley, Lt. E. C. T.	26 Jan. 17	4 Dec. 17	Attd.
25 June 16	2 Douglas, Lt. L. N.	25 June 17	1 Dec. 17	Attd.

1st King George's Own Gurkha Rifles—*contd.*

Attached.

1st Commn. or date of entering service.	Names and Rank.	Army rank.	Present appointment in Battalion.	Remarks.
29 Nov. 05	*1 Wilberforce-Bell, Capt. H. (temp. Maj.)*	*29 Nov. 14*	. . . .	*Poll. Dept. A. M. S., A. Hdqrs.*
24 Jan. 06	2 Hartwell, Capt. J. R., D.S.O.	24 Jan. 15	4 July 18	
30 Sep. 15	2 Simons, Lt. J. G., M.C.	30 Sep. 16	3 July 18	North'd Fus. I. A. (*on probn.*)
18 Feb. 15	2 Scott, Lt. G. D.	18 Feb. 16	24 Aug. 15	I. A. R. O. With Nepalese Cont.
	1 Antram, Capt. E. W.	*27 Nov. 14*	*9 Apl. 15*	*Asstt. Comdt. Nayar Bde.*
15 Oct. 16	1 Sheppard, Lt. B. K.	15 Oct. 17	. . . .	Som. L. I. I. A. (*on probn.*)
5 May 15	1 Herring, Lt. R. L.	5 May 16	28 Nov. 17	Lond. R. I. A.
20 June 17	1 Gaul, 2nd-Lt. A. D.	20 June 17	9 Oct. 17	I. A. R. O.
8 Aug. 17	1 Cowell, 2nd-Lt. F. J.	8 Aug. 17	9 Nov. 17	I. A. R. O. S. S. O., Dharmsala.
23 Apl. 15	2 Owen, Lt. V. E. O'N. (*Actg. Maj.*)	23 Apl. 16	28 Nov. 17	I. A. R. O. (*on probn.*)
1 Oct. 16	2 Davies, Lt. C. C.	1 Oct. 17	3 July 18	W. York. R. I. A. (*on probn.*)
27 Oct. 17	2 Lewis, Lt. C. B.	27 Oct. 18	4 Nov. 17	
3 Feb. 15	2 Wilson, Lt. E. C. (*Actg. Capt.*)	3 Nov. 16		Bedf. R. I. A. (*on probn.*)
5 Oct. 17	2 Falkenstein, Lt. A. J.	5 Oct. 18	21 Jan. 18	I. A. (T. C.)
12 Aug. 15	2 Harvey, Lt. C. W. L.	12 Aug. 16	16 Feb. 18	Hamps R. I. A. (*on probn.*)
31 Jan. 18	2 Bates, 2nd-Lt. A. T. T.	31 Jan. 18	6 Feb. 18	With 1-1st G. R.
31 Jan. 18	2 Harrison, 2nd-Lt. W. H.	31 Jan. 18	6 Feb. 18	
1 May 17	1 Brookman, Lt. J.	1 May 18	10 Dec. 17	Hamps R. I. A. (*on probn.*)
29 Mar. 18	Jackson, 2nd-Lt. R. C.	29 Mar. 18	. . . .	I. A. R. O.
30 Apl. 16	Lomas, Lt. L.	30 Apl. 17	. . . .	I. A. R. O.
5 May 17	Pilcher, Lt. V. D.	5 May 17	17 Aug. 18	M. G. Corps. I. A. (*on probn.*)
26 Sep. 17	1 Baness, Lt. H. E.	26 Sep. 18	2 Nov. 18	I. A. (*on probn.*)
30 May 17	2 Meadmore, 2nd-Lt. C. R. A.	1 Mar. 18	23 Nov. 18	E. Kent R. I. A. (*on probn.*)
18 Dec. 17	2 Forrest, 2nd-Lt. C.	18 Sep. 18	17 Aug. 18	High L. I.-I. A. (*on probn.*)
1 Oct. 18	2 O'Ferral, 2nd-Lt. J. L. F.	1 Oct. 18	15 Oct. 18	47th Sikhs.

Subadar-Majors.

		Jemadar.	Subadar.	Subadar-Major.	
20 Jan. 87	1 Bhagatbir Gurung, *Bahadur* (56)	1 June 97	1 Mar. 01	20 Dec. 14	*Order of Br. I., 2nd* [*Cl.*
3 Sep. 02	2 Kalu Ram Thapa, I.D.S.M. (*66a, 72a*)	3 Sep. 02	1 July 05	1 Aug. 18.	

Subadars.

1 Nov. 90	1 Kalu Gurung, I.O.M. (56)	1 June 07	1 Mar. 11	*Order of Br. I., 2nd class.*
27 Oct. 89	2 Bhagatbir Thapa, I.O.M. (56)	1 July 07	16 June 13	I. O. M., 2nd Class.
12 Nov. 94	1 Puransing Gurung (56)	1 Jan. 14	20 Dec. 14	
8 Dec. 90	1 Santbir Rana (56)	1 Sep. 12	19 Jan. 15	
6 Feb. 93	1 Baliram Rana (56)	1 Jan. 11	9 June 15	
20 Nov. 87	2 Bhim Sing Gharti (56)	1 Jan. 08	1 July 15	
26 Oct. 93	1 Tulia Thapa (56, *66a*)	1 Apl. 12	24 Dec. 15	
17 Nov. 91	2 Kabindra Thapa (*66a, 72a*)	11 Feb. 10	12 Feb. 16	
26 Jan. 97	1 Parbir Burathoki	29 Sep. 14	22 Feb. 16	
18 Feb. 99	1 Bhadarbir Rana	19 Jan. 15	21 Mar. 16	
21 Dec. 93	2 Birbahadur Limbu (56, *66a*)	5 Dec. 10	1 Apl. 16	
16 Dec. 96	1 Birkhsing Thapa	15 Apl. 15	18 Apl. 16	With 1st Bn.
21 Jan. 94	2 Bhagirath Limbu (56)	15 Apl. 15	19 Apl. 16	
	1 Jangbir Thapa (Retd.)	. . . .	. . . .	Re-employed, 15 July 15.
4 Aug. 97	2 Gangaram Khattri (50, 53, *72a*)	16 Dec. 13	27 Sep. 16	
16 June 13	2 Bahadursing Thapa (50)	16 June 13	1 Nov. 16	
14 Nov. 96	1 Dadi Gurung	24 Dec. 15	11 Aug. 17	
18 Dec. 96	1 Jangia Rana, I.D.S.M. (46)	8 Aug. 14	24 May 18	
8 Jan. 97	1 Kabar Sing Ale	18 Apl. 16	24 May 18	
26 Nov. 98	2 Manbir Rana	1 July 15	18 Sep. 18	With 4-11th G. R.
21 Nov. 99	2 Jagatsher Gurung	7 Dec. 15	18 Sep. 18	
15 Nov. 98	1 Gajbahadur Thapa	2 Jan. 17	12 Aug. 18	
17 Nov. 01	1 Akbar Ali, I.D.S.M.	26 Mar. 17	18 Aug. 16	
12 Nov. 00	1 Anne Thapa	26 Mar. 17	19 Sep. 18	

1st King George's Own Gurkha Rifles—*concld.*

1st Commn. or date of entering service.	Names and Rank.	Army rank.	Remarks.
		Jemadars.	
5 May 99	2 Jitjung Thapa (59, 66*a*, 72*a*) .	12 Feb. 16	Gurkha Adjt., 29 Sep 16.
10 Nov. 99	2 Dilbahadur Ale (56) . . .	12 Feb. 16	
10 Jan. 99	2 Mandhoj Rana	12 Feb. 16	
12 Nov. 98	2 Fatehsing Thapa	12 Feb. 16	
8 Feb. 99	1 Dalbir Thapa	21 Mar. 16	
2 Nov. 02	2 Dharmraj Gurung (56) . .	29 Mar. 16	
28 Dec. 98	2 Ramu Thapa	1 Apl. 16	
18 Feb. 01	2 Kharakbahadur Pun . . .	18 Apl. 16	
3 Feb. 99	2 Manbir Gurung	18 Apl. 16	
15 Nov. 93	2 *Nandu Gurung*, I.D.S.M. . .	7 *July* 16	
15 July 93	2 *Harkadhoj Limbu*	7 *July* 16	
13 Nov. 04	2 Dalbir Rana	23 July 16	
27 July 01	2 Narain Sing Thapa	27 Sep. 16	
27 Oct. 01	2 Kalu Gurung	27 Sep. 16	
20 Dec. 01	2 Budhiman Ghale	2 Nov. 16	
10 Nov. 01	1 Bag Sing Thapa	8 Apl. 17	
5 Jan. 99	2 Puransing Gurung	18 Apl. 17	
25 Nov. 01	2 Amar Sing Thapa . .	1 July 17	
20 Nov. 02	1 Chakar Sing Ale	29 Aug. 17	
2 Nov. 02	1 Tilbikram Rana, I.D.S.M. . .	24 May 18	
13 June 00	1 Ramkishan Thapa, I.O.M. .	24 May 18	
5 Nov. 03	1 Manbir Gurung	24 May 18	
5 Dec. 01	1 Kharaksing Thapa, I.D.S.M. .	24 May 18	
2 Jan. 06	1 Dhanbir Pun	7 July 18	
11 June 15	2 Bandbir Gurung	11 June 15	
. . . .	Karanbir Rana	12 Aug. 18	
. . .	Rambahadur Rana	18 Aug. 18	
. . . .	Jite Limbu	19 Sep. 18	
		Temporary.	
		Subadar.	
	1 Harakbir Limboo, I.D.S.M. .	28 Nov. 14	

3rd Battalion, 1st King George's Own Gurkha Rifles. (The Malaun Regiment.)

Formed at Dharmsala on 5 June 1917.

Composition—4 companies of Magars and Gurungs.

1st Commn. or date of entering service.	Names and Rank.	Army rank.	Present appointment in Regiment.	Remarks.
	Commandant.			
23 Aug. 84	Ryder, Lt.-Col. W. I.	9 Aug. 10	30 Sept. 17	
	Company Commanders.			
14 Feb. 00	Simpson, Major J.	1 Sep. 15	13 June 17	1-10th G. R. 2nd-in-Comd.
20 Nov. 14	Woodhead, Lt. A. W. (*Actg. Capt.*)	20 Nov. 14	17 June 17	I. A. R. O.
20 Nov. 14	MacGillivray, Capt. D. G.	20 Nov. 18	31 Jan. 18	I. A. R. O.
4 May 15	Copland, Lt. R. W. (*actg. Capt.*)	4 May 16	24 Sep. 17	Attd.
	Company Officers.			
8 July 15	Baumber, Lt. J.	8 July 16	25 Sep. 17	I. A. R. O. (*S. S. O., Fort Sandeman.*)
28 Dec. 14	Delap, Lt. A.	28 Sep. 16	28 Oct. 17	I. A. (*on probn.*)
4 Apl. 15	Lloyd, Lt. H. S. J.	4 Jan. 17	24 Sep. 17	Attd. Qr. Mr.
29 Oct. 16	Strong, Lt. C. A., M.C. (*Actg. Capt.*) (46)	29 Oct. 17	8 Oct. 17	Attd. Adjt.
1 Mar. 17	Campbell, Lt. J. W.	1 Mar. 18	21 Apl. 17	I. A. R. O.
5 July 17	Robertson, 2nd-Lt. H. B. (61)	5 July 17	12 Oct. 17	I. A. R. O.
21 Dec. 17	Noel, 2nd-Lt. O. C.	21 Dec. 17	1 Apl. 18	Attd.
1 Feb. 18	Allison, 2nd-Lt. G.	1 Feb. 18	6 Feb. 18	I. A. (*on probn.*)
15 Jan. 18	Hosgood, 2nd-Lt. J. W.	15 Jan. 18	15 June 18	I. A. R. O.
23 Apl. 18	Crooke, 2nd-Lt. F. M. W. (M.M.)	23 Apl. 18	27 Apl. 18	I. A. (*on probn.*)
25 May 17	Estill, Lt. J.	25 May 18	. . .	S. A. Infy., I. A. (*on probn.*)
		Jemadar.	Subadar.	Subadar-Major.
	Subadar-Major.			
22 Mar. 87	Kishensing Karki	1 May 06	1 July 10	1 July 17
	Subadars.			
37 Jan. 98	Dewan Sing Thapa	11 May 16	2 Oct. 17	
7 Dec. 97	Lilbir Thapa	20 Dec. 14	26 Mar. 17	
27 Nov. 95	Bahadur Sing Gurung	1 July 17	1 July 17	
20 Jan. 01	Manbir Thapa	11 May 16	21 Oct. 17	12 Dec. 17
24 Oct. 02	Naria Thapa	1 July 17	1 July 17	12 Dec. 17
21 Nov. 05	Bhowan Sing Gurung	1 July 17	1 July 17	12 Dec. 17
15 Dec. 98	Tikaram Thapa	1 July 17	1 Apl. 18	
	Jemadars.			
15 Dec. 00	Lale Pun	26 Mar. 17		
13 Feb. 99	Dhan Sing Thapa	1 July 17	1 July 17	
11 Nov. 99	Bakhanbahadur Gurung	1 July 17	1 July 17	
21 Sep. 17	Jangbahadur Newar	21 Sep. 17	21 Sep. 17	Gurkha Adjt
21 Feb. 09	Haste Gurung	12 Dec. 17	. . . .	
7 Sep. 00	Bahadur Sing Gurung	12 Dec. 17	. . .	
14 Sep. 08	Harkesor Limbu	12 Dec. 17	. . .	
10 Dec. 98	Daram Sing Thapa	1 Apl. 18	. . . .	Gurkha Qr. Mr.
18 Nov. 06	Dalbahadur Rana	1 June 18		
18 Dec. 07	Balasing Gurung	1 June 18		
26 Apl. 04	Bhadrabir Thapa	1 Apl. 18		
28 Dec. 07	Thaman Sing Gurung	1 July 18		
13 Nov. 06	Karbir Thapa	1 July 18		

Regt. Centre—Dehra Dun.

Uniform—Dark Green.
Facings—Scarlet.

2nd King Edward's Own Gurkha Rifles. (The Sirmoor Rifles.)

1st Battalion . .
2nd Battalion . .

Raised at Nahan (Sirmoor), in 1815, by Lt. F. Young, from Gúrkha soldiers, who took service with the British on the termination of the first phase of the Nepal War, and originally designated the Sirmoor Battalion. Became the 8th (or Sirmoor) Local Bn., 1823; the 6th (or Sirmoor) Local Bn., 1826; the Sirmoor Bn., 1850; the Sirmoor Rifle Regt., 1858; the 17th Regt. of Bengal Native Infantry, 1861; the 2nd Gurkha Regt., 1861, the 2nd Gúrkha (the Sirmoor Rifle) Regt., 1864; the 2nd (Prince of Wales's Own) Gúrkha Regt. (the Sirmoor Rifles) 1876; 2nd (the Prince of Wales's Own) Gúrkha Regt. (the Sirmoor Rifles), 1886; 2nd (the Prince of Wales's Own) Gurkha (Rifles) Regt. (the Sirmoor Rifles), 1891; 2nd (the Prince of Wales's Own) Gúrkha Rifles (the Sirmoor Rifles), 1901; present designation, 1st January 1906.

Granted a Truncheon for distinguished service at Delhi in 1857.
The 2nd Battalion was raised at Dehra Dun, in 1886, by Col. S. E. Becher.
THE PLUME OF THE PRINCE OF WALES.
THE ROYAL AND IMPERIAL CYPHER OF KING EDWARD VII.
"Bhurtpore" "Aliwal" "Sobraon" "Delhi, 1857"
"Kabul, 1879" "Kandahar, 1880" "Afghanistan, 1878-80"
"Tirah" "Punjab Frontier."

COLONEL-IN-CHIEF—THE KING.

1st Commn. or date of entering service.	Names and Rank.	Army Rank.	Present appointment in Battalion.	Remarks.
	Commandants.			
28 June 90	2 Boileau, Lt.-Col. E. R. P. [L]	3 June 15	2 Apl. 16	Comdt., Cadet College, Quetta.
9 Apl. 92	1 Sweet, Lt.-Col. E. H., D.S.O.	1 Jan. 17	1 Jan. 17	
	Company Commanders. (8)			
7 Nov. 91	2 *Watt, Lt.-Col. D. M.*, D.S.O., *p.s.c. (Q).*	*7 Nov. 17*	*24 Dec. 14*	*2nd-in-Comd.*
13 July 92	1 Harcourt, Lt.-Col. J. S. M. (Q)	13 July 10	30 Mar. 18	2nd-in-Comd.—Comdt. 2nd-Bn. (*temp.*)
22 Jan. 96	2 *Wigram, Bt.-Col. K.*, D.S.O., *p.s.c.* (*temp. Br.-Genl.*)	*1 Jan. 18*	*1 Nov. 14*	*G. S. O., 1st gr.*
17 Jan. 00	1 Bailey, Maj. W. B.	1 Sep. 15	9 Mar. 16	With 4-3rd G. R.
14 Sep. 01	2 Cruickshank, Maj. J. E.	15 Feb. 16	24 Feb. 17	With 1-145th Infy.
21 Jan. 03	1 Hill, Maj. R. D. O.	21 Jan. 18	2 Feb. 17	With 3-8th G. R.
22 Apl. 03	1 Chope, Maj. A. J. H., D.S.O.	22 Apl. 18	1 Jan. 17	Depôt.
5 Aug. 05	2 Smith, Capt. A. D., M.C. (*Actg. Maj.*)	5 Aug. 14	30 Mar. 18	With 1st. Bn.
	Company Officers. (18)			
16 Aug. 05	2 *Newall, Bt.-Maj. C. L. N.* (*temp. Br.-Genl.*)	1 Jan. 18	3 May 10	*R. A. F.*
16 Aug. 05	1 McCleverty Capt. G. M., M.C. D.S.O.	16 Aug. 14	29 Aug. 15	Adjt.
29 Aug. 06	2 Corse-Scott, Capt. E. J., M.C. (*Actg. Maj.*)	29 Aug. 15	17 July 18	2nd-in-Comd. (*temp.*) With 3-2nd G. R.
9 Sep. 08	1 Shakespear, Capt. W. B. (*Actg. Maj.*)	1 Sep. 15	1 Sep. 11	With 1-1st G. R.
6 Sep. 11	2 Hind, Capt. N. G.	6 Sep. 15	5 Dec. 12	Adjt.
20 Jan. 12	2 *Scoones, Capt. G. A. P.*, D.S.O., M.C.	*20 Jan. 16*	*11 Mar. 13*	*Bde.-Maj., Quetta.*
14 Jan. 14	1 *Tuker, Capt. F. I. S.*	*14 Jan. 18*	*10 Oct. 14*	*With Assam Mily. Police.*
15 Aug. 14	1 *Miller-Hallet, Capt. J. L.*	*15 Aug. 18*	*29 Jan. 16*	*With Gilgit Scouts.*
11 Nov. 14	1 Lloyd, Lt. J. S.	11 Nov. 15	14 May 17	Attd.
11 Nov. 14	2 Gough, Capt. W. G. H., M.C. (*Actg. Capt.*)	11 Nov. 18	5 Mar. 18	
14 July 15	1 Bucknall, Lt. R. D. H.	14 July 16	27 Oct. 15	Attd.
1 Oct. 15	1 Duncan, Lt. D. L.	1 Oct. 16	5 June 17	With 2-11th G. R.
7 Apl. 16	2 Bain, Lt. G. A.	7 Apl. 17	6 Nov. 16	Attd.
18 Apl. 16	2 Wingfield, Lt. C. M. H.	18 Apl. 17	25 Apl. 16	
18 Apl. 16	1 Loftus-Tottenham, Lt. F. J.	18 Apl. 17	25 Apl. 16	Attd.
22 Sept. 15	1 Morgan, Lt. L. E.	22 June 17	17 Oct. 15	Attd.
14 Nov. 16	2 Boucher, Lt. C. H.	14 Nov. 17	1 Dec. 16	Attd.
14 Nov. 16	2 Woolcombe, Lt. G. H. D.	14 Nov. 17	21 Nov. 16	Attd.
27 Oct. 17	2 Buss, Lt. R. A.	27 Oct. 18	5 Nov. 17	Attd.
5 Nov. 17	1 Meyrick-Jones, 2nd-Lt. P.	5 Nov. 17	24 Feb. 18	Attd.
21 Dec. 17	2 Davis, 2nd-Lt. T. A. W.	21 Dec. 17	22 Mar. 18	Attd.
21 Dec. 17	1 Wilder, 2nd-Lt. P. E.	21 Dec. 17	21 Mar. 18	Attd.
5 Sep. 18	1 Toyne, 2nd-Lt. C. J.	5 Sep. 18	. . . .	Attd.
31 Aug. 18	2 Duesbury, 2nd-Lt. A. A.	. . . .	3 Sep. 18	Attd.
21 Dec. 17	1 Hutton, 2nd-Lt. A. R.	21 Dec. 17	21 Mar. 18	Attd.

2nd King Edward's Own Gurkha Rifles—*contd.*

1st Commn. or date of entering service.	Names and Rank.	Army rank.	Present appointment in Battalion.	Remarks.

Attached Officers.

4 May 01	1 Pritchard, Maj. P. A.	4 May 16	12 Dec. 14	Civil.—With 2nd Bn. *Lv. ex. I. m. c.*
5 Oct. 10	2 Trevor, Capt. C. P. (48*a*, 51)	25 Sep. 15	1 Apl. 18	I. A. (*on probn.*)
4 May 15	2 Donaldson, Lt. (A. L. (*Actg. Capt.*)	4 May 15	22 Jan. 18	I. A. (*on probn.*)
18 Sep. 15	1 Toogood, Lt. C. G., D.S.O.	18 Sep. 16	23 Oct. 16	Glouc. R. I. A. (*on probn.*)
5 Feb. 15	2 Keely, Lt. I. E.	5 Feb. 16	9 Feb. 15	I. A. R. O.
23 Apl. 15	1 Snow, Lt. C. A.	23 Apl. 16	8 Nov. 15	I. A. R. O. Qr.-Mr. (*offg.*)
15 Jan. 15	1 Trotter, Lt. H.	15 Jan. 16	22 July 15	I. A. R. O., *Lv. ex., Ind. m. c.*
5 Mar. 15	1 Jerram, Lt. M. R. K., M.C.	5 Mar. 16	. . . .	I. A. R. O., With 2-11th G. R.
8 Sep. 15	1 Robathan, Lt. K. M. (*Actg. Capt.*) (52)	8 Sep. 16	3 Jan. 16	I. A. R. O., Adjt.
26 Jan. 15	2 Cornish, Lt. H.	26 Oct. 16	17 May 18	I. A. (*on probn.*)
24 Mar. 17	2 Oliphant, Lt. D. A. (*Actg. Capt.*)	24 Mar. 17	5 Apl. 17	I. A. R. O.
5 May 17	2 Thompson, 2nd-Lt. W. S.	5 May 17	5 May 17	I. A. R. O.
27 Nov. 14	1 Coxe, Lt. K. H., M.C., (51, 52)	27 Nov. 15	5 Mar. 17	I. A. R. O.
18 Apl. 16	1 Warhurst, Lt. A. E.	18 Apl. 17	11 Apl. 17	6th G. R.
2 Aug. 16	1 Smith, Lt. J. H.	2 Aug. 17	. . . .	I. A. R. O.
23 Nov. 16	1 Pearson, Lt. H. C., M.C.	23 Nov. 16	29 Sep. 17	K. R. Rif. C. I. A. (*on probn.*) Staff Capt., Ahmedabad.
8 July 17	1 Hansford, Lt. E. V. (207*a*)	8 July 18	4 Oct. 17	I. A. R. O.
29 Aug. 17	1 Macdonald, 2nd-Lt. D.	29 Aug. 17	28 Jan. 18	I. A. R. O. (*on probn.*)
18 Aug. 15	2 Esse, Lt. F. A. (48*a*, 51)	1 July 17	1 Mar. 18	Manch. R. I. A. (*on probn.*)
27 Sep. 15	2 Joyes, Lt. A. (46)	27 Sep. 16	11 Dec. 17	I. A. R. O. (*on probn.*)
3 June 15	2 Price, Lt. J. D. (46)	3 June 16	8 Dec. 17	I. A. R. O. (*on probn*)
5 May 17	2 Mallet, Lt. W. H.	5 May 17	16 Jan. 18	Suff. R. I. A. (*on probn.*)
27 July 17	1 Mould, 2nd-Lt. B. S.	27 July 17	25 Feb. 18	I. A. R. O.
31 Oct. 17	2 Smart, Lt. R. L.	31 Oct. 18	16 Feb. 18	Attd.
19 Dec. 16	2 Cunliffe, Lt. R. G.	19 Dec. 18	19 Mar. 18	I. A. R. O. (*on probn.*)
5 Jan. 18	2 Green, 2nd-Lt. H.	5 Jan. 18	8 June 18	I. A. (T. C.)
29 Aug. 17	1 Pocklington, 2nd-Lt. J. F. J.	29 May 18	30 Apl. 18	Ches R., I. A. (*on probn.*)
15 Jan. 18	2 Price, 2nd-Lt. T. W.	15 Jan. 18	12 June 18	I. A. (T. C.)
23 Apl. 16	2 Meredith, 2nd-Lt. H. T.	23 Apl. 17	9 June 18	Suff. R. I. A. (*on probn.*)
7 Apl. 17	1 Murray, Lt. V. R.	7 Apl. 18	17 May 18	Ches. R. Cam'n Highrs.—I. A. (*on probn.*)
25 Apl. 18	2 Tucker, Lt. B. B. Hon.	25 Apl. 17	19 Dec. 17	E. Surr. R. I. A. (*on probn.*)
29 Mar. 18	1 Le Brooy, 2nd-Lt. H. T.	29 Mar. 18	14 July 18	I. A. R. O.
5 May 15	2 Jervis, Lt. W. W.	5 May 16	. . . .	Devon. R.—I. A. (*on probn.*)
2 Sep. 15	1 Smith, Lt. H. M. (51, 95)	2 June 17	26 June 18	Bedf. R. I. A. (*on probn.*)
24 Apl. 18	1 Salmon, 2nd-Lt. F. H.	24 Apl. 18	17 Oct. 18	I. A. (U. L.)
16 Sep. 18	1 Bion, 2nd-Lt. C. R. F.	16 Sep. 18	23 Oct. 18	I. A. R. O. (*on probn.*)
8 Sep. 15	2 Jones, Lt. O. E. (*Actg. Capt.*)	8 Sep. 16	. . . .	I. A. R. O.

Subadar-Majors.

		Jemadar.	Subadar.	Subadar-Major.	
20 Jan. 90	1 Tulsiram Gharti *Bahadur* (56, 72*a*)	1 Apl. 07	1 Oct. 11	18 Apl. 16	*Order of Br. I., 2nd Cl.*
13 June 96	2 Arjun Rama, I.O.M. (56)	1 Apl. 12	2 Nov. 14	1 Sep. 18	

Subadars.

15 Jan. 85	1 Manjit Gharti (56, 60*a*)	1 Mar. 07	16 Jan. 10	*Order of Br. I., 2nd Cl.*
1 Apl. 93	1 Sarabjit Gurung, M.C., *Bahadur* (56, 60*a*).	19 Oct. 07	1 Feb. 13	*Order of Br. I., 2nd Cl.*
25 Nov. 91	1 Kharakbir Thapa (56)	1 May 09	14 Feb. 13	
1 Nov. 90	2 Kharak Sing Rana, M.C. (58, 66*a*)	15 June 10	1 Mar. 14	With 4-3rd G R.
11 Apl. 94	1 Dan Sing Gurung, I.O.M. (66*a*)	19 Feb. 11	2 Nov. 14	

2nd King Edward's Own Gurkha Rifles—*concld.*

1st Commn. or date of entering service.	Names and Rank.	Army rank.	Present appointment in Regiment.	Remarks.
	Subadars—*contd.*			
2 Feb. 92	1 Champha Sing Gurung (66a, 27a)	1 Oct. 11	2 Nov. 14	
15 Nov. 92	2 Tirthe Gurung (52)	21 May 13	2 Nov. 14	
25 May 98	2 Bahadur Ghale	9 Aug. 14	10 Dec. 14	
13 Dec. 95	1 Shirilal Thapa (56)	16 Dec. 13	20 Dec. 14	2-11th G. R.
3 June 98	2 Dhan Sing Lama, I.O.M.	9 Aug. 14	7 Jan. 15	
3 Nov. 97	2 Gamer Sing Gurung (52, 56)	20 Dec. 14	9 Mar. 15	
12 Nov. 93	2 Manbir Thapa	1 Mar. 14	17 Mar. 15	
14 Nov. 94	2 Lal Sing Thapa	23 Dec. 14	9 Mar. 15	
1 Jan. 96	1 Lachman Sing Khattri (56, 66a)	1 May 12	22 Sep. 15	
2 Apl. 97	*1 Narbir Thapa (56, 60a)*	*1 Jan. 14*	*31 Oct. 15*	*With 4-3rd G. R.*
12 Feb. 98	*2 Dewan Sing Gurung*	*24 Jan. 15*	*8 Nov. 15*	
1 Dec. 98	2 Santbir Gurung	7 May 15	3 Nov. 15	
11 Nov. 93	*1 Ranbir Thapa*	*14 Feb. 13*	*1 Jan. 16*	*With 3-6th G. R.*
3 Dec. 93	1 Guman Sing Gurung	1 June 14	6 Mar. 16	
2 Apl. 97	1 Narain Sing Thapa (60a)	2 Nov. 14	14 Apl. 16	
1 Jan. 98	1 Ahiman Rana, I.D.S.M.	23 Apl. 15	3 Feb. 17	
9 Nov. 96	1 Deobahadur Gurung	23 Apl. 15	3 Feb. 17	2-11th G. R.
4 Oct. 99	1 Man Sing Gurung	9 Mar. 16	5 Feb. 17	
8 Apl. 99	1 Man Sing Gurung	22 Jan. 16	5 Feb. 17	
	1 Jitman Khattri (*Retd.*)	. . .	. . .	Re-employed, 15 Feb. 16. Depôt.
20 Nov. 99	2 Nar Sing Thapa (56)	8 Nov. 15	15 July 18	
20 Nov. 01	2 Gamer Sing Bura, I.D.S.M. (56)	11 Dec. 15	1 Sep. 18	
	Jemadars.			
21 Nov. 97	2 Suba Sing Gurung	9 Aug. 14		
13 Jan. 94	1 Narbahadur Thapa	20 Sep. 14		
4 Mar. 94	*2 Ranpati Gurung*, I.D.S.M.	*2 Nov. 14*		With 4-3rd G. R.
6 Feb. 92	*2 Hera Sing Thapa*	*2 Nov. 14*		
5 Nov. 00	*2 Jitman Gurung*	*11 May 15*		*With 3-2nd G. R.*
22 Feb. 99	*1 Sarabjit Gurung*	*29 Apl. 15*		*With 4-3rd G. R.*
11 June 15	1 Bandbir Gurung	11 June 15		With 1-2nd G. R.
10 Dec. 00	2 Bag Sing Thapa, I.D.S.M.	11 Oct. 15		*With 3-2nd G. R.*
15 Feb. 99	1 Dalbahadur Gurung (48a, 51)	14 Dec. 15		
17 Feb. 01	2 Pertlal Gurung	26 Dec. 15		
16 Feb. 98	1 Champha Sing Gurung	1 Jan. 16		
16 June 99	1 Gagan Sing Gurung	1 Jan. 16		
16 July 97	1 Lal Sing Rana, I.D.S.M.	9 Mar. 16		
16 Feb. 98	1 Santbir Bura	9 Mar 16		2-11th G. R.
27 Nov. 00	1 Harka Sing Gurung	9 Mar 16		
7 Nov. 97	1 Tilbir Thapa (48a)	14 Apl. 16		3-2nd G. R.
7 June 00	2 Kharki Charti (52, 56)	24 July 16		
17 Nov. 99	2 Jangbir Thapa (56)	24 July 16		
9 Feb. 99	2 Keshbir Thapa	24 July 16		
19 Feb. 99	2 Rampershad Thapa, I.D.S.M.	24 Aug. 16		
1 Mar. 01	2 Badal Sing Gurung (46)	24 Aug. 16		
10 Oct. 99	2 Harak Sing Thapa	24 Aug. 16		
13 Dec. 02	2 Guman Sing Gurung, I.D.S.M.	30 Aug. 16		
14 Mar. 00	1 Nain Sing Thapa	4 Feb. 17		Qr. Mr. Jemdr.
5 Dec. 00	1 Manbahadur Gurung, I.D.S.M.	4 Feb. 17		Gurkha Adjt.
20 Jan. 17	1 Dila Sing Gurung	4 Feb. 17		
8 Nov. 97	1 Lalbahadur Sahi	5 Feb. 17		
3 Dec. 95	1 Nandbir Thapa	26 Feb 17		
7 Feb. 81	1 Maniram Khattri	28 Feb. 17		
19 Nov. 99	2 Bhim Bahadur Thapa, I.D.S.M. (56).	1 Mar. 17		
15 Jan. 99	1 Dalbir Gurung	4 Mar. 17		
16 Nov. 98	1 Budhiram Roka	14 Apl. 17		2-11th G. R.
	1 Puran Sing Gurung	18 Apl. 17		
20 Oct. 95	2 Judhia Sarki, I.D.S.M.	1 May 17		Qr. Mr. Jemdr.
21 Nov. 97	1 Kharakbir Thapa	12 June 17		
19 Nov. 01	1 Bala Sing Gurung	12 June 17		
17 Nov. 82	2 Gobind Sahi	1 Mar. 18		
26 Dec. 98	2 Bhaulra Sing Gurung	1 Mar. 18		With 3-2nd G. R.
1 Jan. 99	2 Balbir Thapa	1 Mar. 18		
22 Nov. 99	2 Ramsor Gurung	1 Mar. 18		
4 Oct. 95	1 Lachman Sing Thapa	24 Dec. 17		Qr. Mr. Jemdr.
	1 Jitbahadur Gurung	24 May 18		
4 Nov. 97	1 Parman Sing Gurung	24 May 18		
	1 Durgamani Thapa (*Retd.*)	. . . .		Re-employed, 26 May 16.
30 Nov. 00	2 Padamdhoj Gurung, I. O. M.	1 May 18		
16 Dec. 03	2 Bottu Gurung	1 May 18		
27 Dec. 99	2 Bhasker Gurung	1 May 18		
11 Nov. 02	2 Parsa Sing Thapa	1 May 18		
16 Feb. 02	2 Tekbahadur Gharti	1 May 18		
18 Oct. 06	2 Bharat Sing Thapa	2 Sep. 18		Gurkha Adjt.

The first Battn. has an addl. Jemdr. for carrying the Truncheon.

3rd Battalion, 2nd King Edward's Own Gurkha Rifles (The Sirmoor Rifles)

Formed at Dehra Dun on June 12th, 1917.

Composition—4 companies of Gurkhas.

1st Commn. or date of entering service.	Names and Rank.	Army rank.	Present appointment in Regiment.	Remarks.
	Commandant.			
6 Mar. 95	Arbuthnot, Maj. A. H. (*temp. Lt.-Col.*)	6 Mar. 13	5 Apl. 18	2-7th G. R.
	Company Commander (4)			
29 Aug. 96	Corse-Scott, Capt. E. J., M. C.	29 Aug. 15	. . . .	2nd G. R., 2nd-in-Comd. (*temp.*)
17 Jan. 11	Brandon, Capt. R. V.	15 Sep. 15	30 Nov. 17	1-4th G. R.
	Company Officers.			
7 Apl. 16	Bain, Lt. G. A. (*Actg. Capt.*)	7 Apl. 17	6 Nov. 16	2nd G. R. Adjt. 15 June 17.
11 Apl. 17	Fell, Lt. A. L.	11 Apl. 18	1 Oct. 17	Attd.
27 Oct. 17	Armstrong, Lt. H. F. C.	27 Oct. 18	5 Nov. 17	Attd.
23 Apl. 18	Crosbie, 2nd-Lt. F. S.	23 Apl. 18	18 Oct. 18	Attd.
22 Dec. 14	Sim, Lt. D.	22 Dec. 15	7 Oct. 17	I. A. R. O.
30 July 15	Cave, Lt. C. R. (*Actg. Capt.*)	30 July 16	6 Feb. 18	I. A. R. O.
5 Sep. 15	Herrick, Lt. J. R. W.	5 Sep. 16	12 Jan. 18	Notts. & Derby. R., I.A. (*on probn.*)
20 Feb. 16	Pyper, Lt. H. McD.	20 Feb. 17	2 Aug. 17	I. A. (*on probn.*)
27 Aug. 16	Foot, Lt. H. O.	27 Aug. 17	29 May 18	L'pool R. I. A. (*on probn.*)
23 Sep. 16	Savary, Lt. J. R.	23 Sep. 17	23 Jan. 18	I. A. (*on probn.*)
30 Sep. 16	Hodgshon, Lt. F. W.	30 Sep. 17	15 Feb. 18	Lond. R. I. A. (*on probn.*)
19 Sep. 17	Bradshaw,-Smith, Lieut. B.	19 Sep. 18	15 Apl. 18	W. York R.—I. A. (*on probn.*)
19 Sep. 17	Matthews, Lt. F. J. (207*a*)	19 Sep. 18	13 Oct. 17	Attd.
26 Sep. 17	Keith, Lt. N. Y.	26 Sep. 18	26 Jan. 18	I. A. R. O.
28 Dec. 17	Alberga, 2nd-Lt. E. M.	28 Dec. 17	31 May 18	Lond. R.—I. A. (*on probn.*)
28 Feb. 18	Martin, 2nd-Lt. D. D.	28 Feb. 18	25 May 18	High. L. I.—I. A. (*on probn.*)
7 June 18	Duglas, 2nd-Lt. G. D. E.	7 June 18	12 Oct. 18	I. A. (*T. C.*)
7 June 18	Gerard-Smith, 2nd-Lt. V.	7 June 18	12 Oct. 18	I. A. R. O.

1st Commn. or date of entering service.	Names and Rank.	Jemadar.	Subadar.	Subadar-Major.
	Subadar-Major.			
14 Nov. 94	Lal Sing Thapa	23 Dec. 14	9 Mar. 15	1 May 18
	Subadars.			
5 Nov. 00	Jitman Gurung	11 May 15	1 May 18	
10 Dec. 01	Bag Sing Thapa	11 Oct. 15	1 May 18	
7 Nov. 97	Til Bir Thapa (51, 52, 48)	14 Apl. 16	1 May 18	
3 Mar. 99	Bahadur Thapa (50, 72*a*, 48*a*)	12 June 17	1 May 18	
18 Oct. 99	Kalu Gurung	12 June 17	1 May 18	
17 Dec. 03	Lachman Sing Khattri (50)	12 June 17	14 May 18	
	Jemadars.			
10 Mar. 03	Narain Mal, I.D.S.M.	12 June 17		
26 Dec. 98	Bhadra Sing Gurung	1 Mar. 18		
7 Jan. 00	Narbir Gherti (72*a*)	1 May 18		
1 Sep. 99	Gopal Sing Rana	1 May 18	Gurkha Qr. M.	
15 Jan. 00	Man Bir Thapa	1 May 18		
29 Nov. 01	Jaman Sing Ale (58)	1 May 18		
6 Nov. 02	Santok Pun (60*a*)	1 May 18		
20 Dec. 03	Gambir Sing Thapa (51, 60*a*, 48*a*)	1 May 18	Gurkha Adjt.	
22 Mar. 94	Gajman Rana	1 May 18		
26 Feb. 03	Dilbahadur Rana (50, 48*a*)	15 Sep. 18		
7 Dec. 02	Parbal Thapa (50, 48*a*)	15 Sep. 18		
20 Jan. 03	Chandra Sing Gurung (46)	15 Sep. 18		

Regt. Centre—Almora.

Uniform—Dark Green.
Facings—Black.

3rd Queen Alexandra's Own Gurkha Rifles.

1st Battalion . . .
2nd Battalion . . .

Raised at Almora, in 1815, by Lt. Sir R. Colquhoun, Bart., from Gúrkha soldiers who took service with the British after the fall of Malaun and conquest of Kamaon, supplemented by transfers from the Gorakhpur Hill Regiment, was originally designated the Kamaon Battalion. Became the Kamaon Provincial Bn., 1816; the 9th (or Kamaon) Local Bn., 1823; the 7th (or Kamaon) Local Bn., 1826; the Kamaon Bn., 1860; the 18th Regt. of Bengal Native Infantry, 1861; the 3rd Gúrkha Regt., 1861; the 3rd (the Kumaon) Gúrkha Regt., 1864; 3rd Gúrkha Regt., 1887; 3rd Gúrkha (Rifle) Regt., 1891; 3rd Gúrkha Rifles, 1901; 3rd The Queen's Own Gúrkha Rifles, 1907; present designation in 1908.

The 2nd Battalion was raised at Lansdowne, in 1891, by Major H. D. Hutchinson.

THE CYPHER OF QUEEN ALEXANDRA.

"Delhi, 1857" "Ahmad Khel" "Afghanistan, 1878-80" "Burma, 1885-87"
"Chitral" "Tirah" "Punjab Frontier."

COLONEL—Lieut.-General Henry D. Hutchinson, C.S.I. 13 May 1904.

1st Commn. or date of entering service.	Names and Rank.	Army rank.	Present appointment in Battalion.	Remarks.
	Commandant.			
11 Feb. 88	2 Eastmead, Lt.-Col. C. S. (Q) [L]	11 Feb. 14	28 Feb. 16	
29 Nov. 90	1 Edwardes, Lt.-Col. J. G. . .	29 Nov. 16	3 July 17	
	Company Commanders. (8)			
16 Jan. 95	2 Channer, *Maj. (Actg. Lt.-Col.)* G. K., D.S.O.	16 Jan. 13	30 July 16	2nd-in-Comd. With 3-3rd G. R.
14 Aug. 95	1 Dundas, Maj W. L. (Q)	14 Aug. 13	3 July 17	With 4-11th G. R. 2nd-in-Comd.
5 Aug. 96	*2 Shuttleworth, Bt.-Lt.-Col. D.I., p.s.c. (Q) (temp. Lt.-Col.)*	3 June 18	. . . .	*G. S. O., 1st gr.*
4 Aug. 97	*1 Loch, Maj. J. C.*	*4 Aug. 15*	. . . .	*D. A. A. & Q. M. G.*
3 Aug. 98	*2 Ridgeway, Maj. D. G., p.s.c. (temp. Lt.-Col.)*	*1 Sep. 15*	*4 Nov. 14*	*G. S. O., 1st gr.*
25 Jan. 99	*2 Glynton, Maj. G. M., p.s.c.*	*1 Sep. 15*	*14 Mar. 15*	
20 Jan. 00	1 Hamilton, Maj. C. A. B. .	1 Sep. 15	30 July 16	
20 Jan. 00	*1 Alexander, Maj. R. D.* . .	*1 Sep. 15*	*3 July 17*	*A. D. C. to H. E. the Viceroy.*
	Company Officers. (18)			
19 Oct. 01	1 Tuck, Maj. C. H. A. . . .	17 Oct. 15	20 Mar. 05	
27 Jan. 04	*1 Stone, Maj. A G., m. c.* . .	*19 July 18*	*4 Sep. 05*	*Bde. Maj.*
27 Jan. 04	1 McCausland, Capt. E. T. W. *(temp. Maj.)*	27 Jan. 13	9 Mar. 09	
13 Aug. 04	2 Tuite-Dalton, Capt. E.G.T., M.C.	13 Aug. 13	20 Apl. 06	
29 Aug. 06	*2 McSwiney, Capt. H. F. C., m.c. (temp. Maj.)*	*29 Aug. 15*	*11 Sep. 08*	*D. A. A. G.*
19 Jan. 07	2 Welman, Capt. W. H. L. J. .	1 Sep. 15	13 Oct. 15	With 3-7th G. R.
17 Aug. 07	1 Money, Capt. G. W. P. . .	1 Sep. 15	13 Nov. 09	
25 Jan. 08	*1 Dent, Capt. R. B.*	*1 Sep. 15*	*9 Apl. 09*	*With Nepalese Contgt.*
9 Sep. 08	*1 Fisher, Capt. G. T.*	*1 Sep. 15*	*1 Nov. 10*	*Poll. Dept.*
5 Sep. 11	*2 Batten, Capt. W. D. G.* . .	*5 Sep. 15*	*6 Apl. 14*	*With Nepalese Contgt.*
1 Oct. 14	*1 Voelcker, Capt. G. G.* . . .	*1 Oct. 18*	*19 Dec. 14*	*Asst. Ispctg. Offr., Kashmir.*
11 Nov. 14	1 West, Capt. E. M.	11 Nov. 18	27 Feb. 15	
1 Aug. 15	2 Nicholson, Lt. C. F. S. *(Actg. Capt.)*	1 Aug. 16	. . . .	
29 June 16	Tomsett, Lt. E. W.	28 May 17	. . . .	With 1-99th Infy.
15 Nov. 15	1 Thomas, Lt. G. H.	15 Nov. 16	20 Nov. 15	With 3-3rd G. R.
7 Jan. 16	2 Simpson, Lt. H. J. A. . .	7 Jan. 17	11 Dec. 17	
26 Jan. 16	1 Forbes, Lt. L. W. McK. *(Actg. Capt.)*	26 Jan. 17	19 Mar. 16	Attd.
18 Apl. 16	1 Gray, Lt. C. E.	18 Apl. 17	25 Apl. 16	Attd.
18 Apl. 16	2 Berkeley, Lt. C. S. J. . . .	6 Oct. 16	18 Apl. 16	Attd.
29 June 16	2 French, Lt. F. H.	28 May 17	. . . .	
27 Oct. 16	1 Monier-Williams, Lt. L. . .	28 June 16	16 Jan. 17	Attd.
30 Jan. 17	2 Armitage, Lt. A. S.	30 Jan. 18	10 Feb. 17	
2 Jan. 16	2 Allanson, Lt. G. H.	2 Jan. 17	10 Oct. 15	Attd.
18 June 17	1 Holworthy, Lt. A. W. W. . .	18 June 18	29 June 17	
5 May 17	2 Fraser, Lt. K. M.	5 May 18	20 Aug. 17	
18 June 17	1 Salter, Lt. E.	18 June 18	29 June 17	
20 June 17	2 Rogers, Lt. H.	20 June 18	. . . .	
5 Oct. 17	2 Guile, Lt. F. L.	5 Oct. 18	29 Jan. 18	I. A. R. O.
27 Oct. 17	Upton, Lt. J. A. E. *(Actg. Capt.)*	27 Oct. 18	5 Nov. 17	Attd. Adjt.
31 Jan. 18	Hargrave, 2nd-Lt. J. F. H. . .	30 Jan. 18	7 Feb. 18	Attd.
31 Aug. 18	2 Davis, 2nd-Lt. H. C. M. . .	31 Aug. 18	6 Sep. 18	Attd.
. . .	2 Skone, 2nd-Lt. H. A.	. . . .	. . . .	Attd.
19 Aug. 18	2 Brandt, Lt. H. G.	19 Aug. 18	18 Oct. 18	I. A. R. O.

3rd Queen Alexandra's Own Gurkha Rifles—*contd.*

1st Commn. or date of entering service.	Names and Rank.	Army rank.	Present appointment in Battalion.	Remarks.
	Attached.			
3 Dec. 14	*2 Taylor,* W. R. *(temp. Capt.)*	*6 Aug. 17*	*27 July 17*	*I. A. R. O. Staff Capt. A. H. Q.*
24 Dec. 14	1 Cooke, Lt. G. P.	24 Dec. 15	Mar. 15	I. A. R. O. S. & T. Corps.
11 June 15	2 Turner, Lt. R. L.	11 June 16	. . .	I. A. R. O. Adjt.
20 Nov. 14	2 O'Brien, Lt. M. C.	20 Nov. 15	. .	I. A. R. O.
5 Feb. 15	2 Robertson, Lt. R. A.	5 Feb. 16	. .	I. A. R. O.
12 Oct. 16	2 Mocatta, Lt. C. H.	12 Oct. 16	23 Jan. 18	I. A. *(on probn.)*
6 Aug. 15	1 Peake, Lt. A. J.	6 Aug. 16	13 Jan. 17	I. A. R. O.
7 Oct. 15	1 Corbett, Lt. V. W.	7 Oct. 15	11 May 17	I. A. *(on probn.)*
5 May 17	1 Mayer, 2nd-Lt. G. H.	5 May 17	22 Aug. 17	I. A. R. O.
5 May 17	1 Duke, Lt. W. F.	5 May 18	22 Aug. 17	I. A. R. O.
25 May 15	Lowe, Lt. H. A.	. . . .	14 Feb 18	I. A. R. O. *(on probn.)*
15 Jan. 18	1 Hosgood, 2nd-Lt. J. W.	15 Jan. 18	. . . .	I. A. R. O.
15 Aug. 15	1 Roe, Lt. K. T.	26 Dec. 16	11 Apl. 18	I. A. *(on probn.)*
31 May 15	Gibby, Lt. D.	31 May 16	. . . .	R. W. Fus.—I. A. *(on probn.)*
. . . .	Poingdestre, Lt. C. le G., *(Actg. Capt.)*	. . . .	. . . .	I. A. R. O.
6 July 17	1 Carstens, Lt. J. E.	6 July 18	. . . .	S. A. Infy.—I. A. *(on probn.*
19 June 15	2 Langlands, Lt. E. W. *(Actg. Capt.)*	19 June 16	. . . .	N. Staff. R.—I. A. *(on probn.)*
19 Aug. 15	2 Sheehan, Lt. A. F. C.	19 Aug. 16	. . . .	Lond R.—I. A. *(on probn.)*
24 Dec. 14	1 Cockhill, Lt. S. R.	1 Nov. 16	18 Aug. 18	I. A. *(on probn.)*
7 Nov. 16	1 Kirkpatrick, Lt. A. R.	7 May 18	3 Aug. 18	I. A. *(on probn.)*
31 Oct. 17	1 Hill, 2nd-Lt. B. B.	31 Oct. 17	6 June 18	I. A. *(on probn.)*
2 June 16	Parisotti, Lt. L.	2 June 17	. . . .	
21 Mar. 16	1 Cleveland, Lt. E. A. M.	21 Mar. 17	. . . .	I. A. *(on probn.)*
30 Jan. 16	1 Davies, Lt. E. L.	30 Jan. 17	. . .	I. A. *(on probn.)*
. . . .	2 Paget, Lt. G. H.	5 May 17	14 Oct. 18	I. A. *(on probn.)*

Subadar-Majors.

1st Commn.	Names and Rank.	Jamadar.	Subadar.	Subadar-Major.	Remarks.
8 Nov. 85	1 Gunjbir Ale *Bahadur* (52, 56)	5 May 03	1 Dec. 07	15 Feb. 11	*Order of Br. I., 2nd Cl.*
1 June 96	1 Bhim Sing Thapa, I.O.M., M. C. (52, 56).	1 June 14	24 Nov. 14	26 Sep. 15	
21 Feb. 95	3 Hari Sing Burathoki (56)	1 Apl. 12	13 Feb. 15	3 Feb. 17	

Subadars.

1st Commn.	Names and Rank.	Jamadar.	Subadar.	Remarks.
10 Feb. 83	1 Kirpa Ram Pun (52, 56)	3 May 04	1 Feb. 09	
27 Mar. 87	*1 Tul Sing Thapa* (52, 56, 72a)	*24 Dec. 07*	*18 Apl. 11*	*With 2nd Bn.*
5 Feb. 88	2 Gajbir Thapa (56)	6 Apl. 06	21 Oct. 11	
8 Dec. 87	1 Bakht Sing Rana (52, 56, 66a)	1 Feb. 09	23 Dec. 12	
17 Dec. 90	2 Dhanbir Thapa (66a)	1 Feb. 09	1 May 13	
24 Dec 91	1 Dalkesar Gurung, I.M.O.	15 Feb. 11	1 Apl. 14	
22 Feb. 91	1 Shibcharan Burathoki (56, 60)	1 Apl. 11	15 Sep. 14	
19 Dec. 01	*1 Budhibal Thapa*	*16 Sep. 14*	*16 Sep. 14*	
24 Dec. 90	1 Lachhman Gurung, I.O.M.	15 Sep. 14	5 Mar. 15	
3 Dec. 91	1 Keshbir Gurung	15 Sep. 14	5 Mar. 15	
10 Sep. 94	2 Hari Chand	1 May 13	9 Mar. 15	
18 Jan. 98	2 Arjun Thapa	10 Aug. 14	11 Mar. 15	
18 Mar. 91	2 An Sing Thapa (56, 58)	21 Oct. 11	14 Aug. 15	
1 Nov. 96	2 Damar Sing Gurung, M. C.	27 Jan. 15	27 Mar. 16	
3 Dec. 94	3 Kharak Sing Gurung	10 Feb. 15	27 Mar. 16	
11 Jan. 91	3 Durgamani Thapa	6 Apl. 15	27 Apl. 16	
17 Nov. 97	2 Bagbir Gharti	10 Aug. 14	3 Feb. 17	
10 Dec. 90	3 Karna Sing Thapa	2 Feb. 15	3 Feb. 17	
. . . .	2 Nawal Sing Rana	4 May 15	3 Feb. 17	
. . . .	3 Hari Baran Rana	10 May 15	3 Feb. 17	
11 Feb. 98	2 Balbir Thapa, I.O.M.	17 Sep. 15	3 Feb. 17	
. . . .	2 Parja Gurung	4 Oct. 15	3 Feb. 17	
28 Dec. 92	Randhoj Rana, I.O.M.	9 Aug. 15	3 Feb. 17	
13 Nov. 93	3 Chitrabir Thapa (56)	15 Sep. 14	3 Apl. 17	
1 Sep. 93	1 Chandrabir Thapa	15 Sep. 14	1 July 17	
. . . .	3 Gopal Rana *(Retd.)*	. . . .	. . . .	Re-employed 26 Mar. 15, Depôt.
2 Apl. 00	3 Bahadur Thapa, I.O.M.	4 Jan. 16	22 Nov. 17	
16 Nov. 98	2 Tula Gurung (56, 66a)	25 Sep. 15	1 Apl. 18	
20 Dec. 90	2 Tilbir Thapa	19 Jan. 16	11 Apl. 18	
18 Jan. 99	2 Dhanbir Gurung	3 Feb. 17	10 Apl. 18	
. . . .	2 Bhowan Sing Sahi *(Retd.)*	. . . .	. . . .	Re-employed 8 Aug. 18.
26 June 00	2 Man Sing Basnet	3 Feb. 17	21 June 18	

3rd Queen Alexandra's Own Gurkha Rifles—*concld.*

1st Commn. or date of entering service.	Names and Rank.	Army rank.	Remarks.
	Jemadars.		
4 Dec. 93	1 Puran Ale	16 Sep. 14	
22 Jan. 03	1 Dhanu Gurung	16 Sep. 14	Gurkha Adjt., 16 Sep. 14.
16 Nov. 94	*1 Bajiram Pun*	*5 Mar. 15*	
23 Dec. 94	*1 Dhansing Ale*	*10 Apl. 15*	
3 Mar. 96	1 Bahadur Thapa	21 May 15	
22 Nov. 94	1 Dhiraj Gurung	29 Nov. 15	
20 Nov. 95	1 Shamsher Thapa	4 Jan. 16	
28 Jan. 96	1 Shamsher Gurung	28 Jan. 16	
8 Jan. 91	2 Bhawan Sing Rana	24 Feb. 16	
7 Feb. 98	2 Kalu Gurung	27 Mar. 16	
....	2 Gamer Sing Thapa	27 Mar. 16	
6 Nov. 92	3 Dalbir Thapa	31 Mar. 16	
....	2 Asbir Gurung, I.O.M.	22 Apl. 16	
16 Mar. 96	1 Tulbar Pun (56)	20 Nov. 16	
25 Sep. 96	1 Ramchandra Ale	20 Dec. 16	
....	2 Judbir Ale	9 Dec. 16	
....	2 Churamani Thapa	15 Dec. 16	
....	2 Hari Sing Thapa	1 Jan. 17	
....	2 Parman Thapa	1 Jan. 17	
18 Jan. 99	2 Dhanbir Gurung	3 Feb. 17	
1 Nov. 00	3 Dalbir Gurung	3 Feb. 17	
18 Jan. 99	2 Dhanbir Thapa	3 Feb. 17	
18 Dec. 91	3 Damar Sing Gurung	3 Feb. 17	
....	3 Guman Sing Gurung	3 Feb. 17	
19 Jan. 97	1 Nathu Thapa	3 Apl. 17	
10 Jan. 98	1 Bhimraj Gharti (56)	1 Mar. 17	
15 Jan. 97	1 Bhanbir Gharti	1 July 17	
8 Aug. 99	1 Padam Sing Thapa (48*a*, 51)	30 July 17	
....	1 Jita Gurung	20 Nov. 17	
....	1 Tul Sing Thapa	20 Nov. 17	
...	1 Birbhan Gurung	20 Nov. 17	
....	1 Kharke Thapa	20 Nov. 17	
....	1 Sarabjit Gurung	20 Nov. 17	
...	1 Ude Sing Gurung	1 Dec. 17	
4 Mar. 03	3 Rupdhan Pun	23 Sep. 17	
8 Oct. 01	3 Nain Sing Thapa	23 Sep. 17	
20 Dec. 03	3 Lalbahadur Gurung	23 Sep. 17	
....	2 Rup Sing Thapa	24 Nov. 17	
....	3 Surbir Thapa	22 Nov. 17	
....	3 Bhagatbir Thapa	22 Nov. 17	
....	3 Narain Sing Rana	22 Nov. 17	
. a ..	3 Ajab Sing Gurung	22 Nov. 17	
....	2 Dhan Raj Gurung	21 Nov. 17	
....	2 Harak Sing Gharti, I.O.M.	21 Nov. 17	
....	2 Gane Gurang, I.O.M.	23 Nov. 17	
....	2 Harak Sing Gurung	24 Nov. 17	
....	2 Lalbahadur Bura	14 Dec. 17	
....	1 Bhabesur Gurung	1 Jan. 18	
....	1 Damar Sing Thapa	10 Jan. 18	
11 Dec. 04	1 Ude Sing Ghale	1 Mar. 18	
10 Feb. 98	1 Keser Sing Thapa	1 Apl. 18	
9 Dec. 99	1 Daudbir Thapa	1 Apl. 18	
13 Nov. 14	1 Bhagat Bir Ale	1 Apl. 18	
30 Jan. 04	1 Bishan Sing Bisht	25 May 18	
....	Lachhiman Thapa	17 Aug. 18	
22 Dec. 00	1 Shibu Gharti	1 May 18	
1 Dec. 07	1 Kirpa Sing Chand (56)	1 May 18	
....	2 Dalbahadur Nagarkoti	1 Apl. 18	
....	2 Karke Rana	21 Apl. 18	
....	2 Churamani Rana	10 Apl. 18	
....	2 Mansing Mal	11 Apl. 18	

3rd Battalion, 3rd Queen Alexandra's Own Gurkha Rifles.

Raised in Egypt on 3rd February 1917.

1st Commn. or date of entering service.	Names and Rank.	Army rank.	Present appointment in Battalion.	Remarks.
	Commandant.			
16 Jan. 95	Channer, Maj. G. K., D.S.O. (*temp. Lt.-Col.*)	16 Jan. 13	3 Feb. 17	2-3rd G. R. *Lc.*, *ex I.,m.c.*
	Company Commanders.			
20 July 98	Shaw, Maj. H. M. D.	1 Sep. 15	3 July 17	2nd-in-Comd.
25 Jan. 99	Glynton, Maj. G. M., *p.s.c.*	1 Sep. 15	3 Feb. 17	
	Company Officer.			
15 Nov. 15	Thomas, Lt. G. H.	15 Nov. 16	8 Apl. 17	3rd G. R. Qr.-Mr., 1 Sep. 17.
18 Sep. 15	Irwin, Lt. H. N., M. C.	18 Sep. 16	23 May 17	Attd.
	Attached.			
19 Aug. 14	Daniels, Lt. C. J. F. W.	1 Sep. 15	3 Feb. 17	I. A. R. O.
13 Nov. 14	Lock, Lt. H. G.	13 Nov. 15	23 May 17	I. A. R. O.
24 Feb. 15	Russell, 2nd-Lt. C.	24 Feb. 15	23 May 17	I. A. R. O.
12 Mar. 15	Storrs, Lt. G. N., M.C.	12 Mar. 16	3 Feb. 17	I. A. R. O. Adjt., 3 Feb. 17.
29 Oct. 15	Simson, Lt. G. N.	29 Oct. 16	3 Feb. 17	

		Jemadar.	Subadar.	Subadar-Major.	
	Subadar-Major.				
21 Feb. 95	Hari Sing Burathoki	1 Apl. 15	13 Feb. 13	3 Feb. 17	
	Subadars.				
3 Dec. 94	Kharak Sing Gurung	10 Feb. 15	27 Mar. 16		
11 Jan. 91	Durgamani Thapa	6 Apl. 15	27 Apl. 16		
10 Dec. 90	Karam Sing Thapa	2 Feb. 15	3 Feb. 17		
	Haribakan Rana	10 May 15	3 Feb. 17		
14 Dec. 92	Randhoj Rana	9 Sep. 15	3 Feb. 17		
	Gopal Rana (Retd.)	....	....		Re-employed 26 Mar. 15.
2 Apl. 00	Bahadur Thapa, I.O.M.	22 Nov. 17	....		Gurkha Adjt., 3 Feb. 17.
	Jemadars.				
6 Nov. 92	Dalbir Thapa	31 Mar. 16			
18 Dec. 91	Damar Sing Gurung	3 Feb. 17			
8 Nov. 97	Guman Sing Gurung	3 Feb. 17			
27 Dec. 91	Jabbarbahadur Gurung	3 Feb. 17			
9 Nov. 99	Birkhadhoj Limbu	1 Oct. 15			
...	Narain Sing Rana	22 Nov. 17			
...	Bhagatbir Thapa	22 Nov. 17			
...	Surbir Thapa	22 Nov. 17	With 4th G.R.		
...	Ajab Sing Gurung, I.D.S.M.	22 Nov. 17			

4th Battalion, 3rd Queen Alexandra's Own Gurkha Rifles.

Formed at Rawalpindi on 1st October 1916.

Composition—4 Companies of Gurkhas.

1st Commn. or date of entering service.	Names and Rank.	Army rank.	Present appointment in Battalion.	Remarks.
	Commandant.			
7 Mar. 94	Nicolay, Lt.-Col., B. U.	5 Aug. 18	3 Aug. 18	2-4th G. R.
	Company Commanders.			
17 Jan. 00	Bailey, Maj. W. B.	1 Sep. 15	1 Oct. 16	1-2nd G. R. 2nd-in-Comd.
2 July 13	Morton, Lt. T. M. (*actg. Capt.*)	5 May 16	7 July 17	I. A.
26 Feb. 15	Phillips, Lt. H. J. L. (*actg. Capt.*)	26 Feb. 16	1 Oct. 16	I. A. R. O.
22 Sep. 14	Pitcairn, Lt. D. N.	22 June 16	18 June 17	I. A.
	Company Officers.			
14 Aug. 15	Cowan, Lt. D. T., M.C. (*actg. Capt.*)	28 Aug. 17	5 Aug. 17	Attd. *Adjt.*
14 Jan. 15	Knowles, Lt. G. J.	14 Jan. 16	1 Oct. 16	I. A. R. O.
15 Aug. 14	*Barker, Lt. R. A.*	*15 May 16*	*26 Jan. 18*	*N. Lan. R.—With Nepalese Cont. I. A. (on probn.)*
24 Oct. 14	Lewis, Lt. F. H.	24 Aug. 16	4 Aug. 17	Attd. With 1-141st Infy.
2 Dec. 14	Davis, Lt. H.	2 Sep. 16	24 Aug. 17	Attd.
24 Feb. 15	Rusby, Lt. L. S.	24 Nov. 16	22 Jan. 18	I. A. (on probn.)
26 May 16	VC Barter, Lt. F., M.C.	26 May 17	18 May 17	Attd.
20 July 16	*Ferguson, Lt. W. A. M.*	*12 Feb. 17*	*3 Oct. 16*	*Attd. 1-5th G. R.*
29 June 16	French, Lt. F. H.	28 May 17	31 July 18	2-3 G. R.
19 July 16	Molesworth, Lt. R. C.	19 July 17	9 Nov. 16	Attd.
6 Oct. 16	Oliver, Lt. T. W.	6 Oct. 17	27 Nov. 16	I. A. R. O.
1 Jan. 17	Simpson, Lt. W. A.	1 Jan. 18	26 Jan. 18	R. W. Kent R., I. A. (*on probn.*)
1 Feb. 17	Harrison, Lt. K. H.	1 Feb. 17	24 Apl. 18	I. A. R. O.
1 Mar. 17	Kerr, 2nd-Lt. J. C.	1 Mar. 18	3 Mar. 18	I. A. R. O.
18 June 17	Middleton, Lt. T. B.	18 June 18	18 June 17	Attd.
6 Sep. 17	*Gordon, Lt. N. H.*	*6 Sep. 18*	*4 Apl. 18*	*I. A. R. O. Cant. Magistrate, Nowshera.*
5 Oct. 17	*James, Lt. W. R.*	*5 Oct. 17*	*24 Jan. 18*	*I. A. (T. C.)*
31 Oct. 17	Taylor, 2nd-Lt. A. H. S.	31 Oct. 17	24 Jan. 18	I. A. R. O.
31 Oct. 17	*Hall, 2nd-Lt. R. F.*	*31 Oct. 17*	*15 Feb. 18*	*I. A. R. O.*
19 Dec. 16	Logan, Lt. T.	19 Dec. 18	1 Apl. 18	I. A. R. O.
19 Dec. 16	Mercer, Lt. A. R.	19 Dec. 18	10 July 18	I. A. R. O.
21 Dec. 17	Forward, 2nd-Lt. F. C. M.	21 Dec. 17	21 Mar. 18	Attd. With 2-123 Rif.
2 Feb. 18	Jones, 2nd-Lt. O. C.	2 Feb. 18	26 Feb. 18	K. R. Rif. C. I. A. (*on probn.*)
31 Jan. 18	Greer, 2nd-Lt. U. Mc. G.	1 Feb. 18	15 Feb. 18	I. A. Qr. Mr., (*temp.*)
25 Oct. 17	Adams, 2nd-Lt. A. E. St. J.	25 Oct. 17	. . . .	R. Fus. I. A. (on probn.)
1 May 18	Whitehead, 2nd-Lt. F. W.	1 May 18	28 Nov. 18	I. A. (on probn.)
31 Aug.	Hurst, 2nd-Lt. T. R.	31 Aug. 18	21 Oct. 18	Attd.

		Jemadar.	Subadar.	Subadar-Major.
	Subadar-Major.			
1 Nov. 90	Kharaksing Rana, M.C. (58, 66*a*)	16 June 10	25 Oct. 16	
	Subadars.			
9 Dec. 01	*Budhibal Thapa*	*18 Sep. 14*	*3 Oct. 16*	
12 Nov. 94	Partiman Rana	10 Aug. 09	6 Oct. 14	2-131st U. P. R.
8 Jan. 99	Indrajit Thapa (56)	1 May 10	4 May 15	
2 Apl. 97	Narbir Thapa (56, 60*a*)	1 Jan. 14	31 Oct. 15	
12 Dec. 04	Dhanraj Pun (56)	16 Oct. 14	20 Sep. 16	
21 Nov. 96	Gopal Gurung	16 Oct. 14	12 Aug. 16	
4 Dec. 93	*Puran Ale*	*16 Sep. 14*	*1 Oct. 17*	*Attd. 3-3rd G. R.*
22 Feb. 99	Sarabjit Gurung	29 Apl. 15	22 Dec. 17	
	Jemadars.			
4 Mar. 94	Ranpati Gurung, I.D.S.M.	2 Nov. 14		
16 Nov. 94	Bajiram Pun	5 Mar. 15		
23 Dec. 94	Dhansing Ale (56, 58)	20 Apl. 15	. . . .	Attd. 3-3rd G. R.
26 Feb. 08	Kumsing Gurung	29 July 15		
13 Feb. 00	Dhanbahadur Rana (46)	25 Sep. 15		
12 Nov. 09	Shamsher Gurung	25 July 16		
22 Dec. 02	Parbir Thapa	25 July 16	. . . .	Jemdr. Adjt.
8 Dec. 98	Singbir Gurung	7 Feb. 17		
1 Nov. 02	*Sherbahadur Gurung*	*17 Feb. 17*	. . . .	*Attd. 3-3rd G. R.*
12 Nov. 07	Harkraj Gurung (56)	22 Dec. 17		
5 Dec. 98	Manbahadur Gurung	22 Dec. 17		
18 Dec. 04	Bhabesur Gurung	1 Jan. 18		
24 Nov. 07	Damarsing Thapa	10 Jan. 18		
Nov. 07	Dharam Sing Gurung	1 Apl. 18		

Regt. Centre—Bakloh.

Uniform—Dark Green.
Facings—Black.

4th Gurkha Rifles.

1st Battalion . . .
2nd Battalion . . .

. Raised at Pithoragarh, in 1857, by Lieutenant D. Macintyre, as the Extra Gurkha Regiment. Became the 19th Regiment of Bengal Native Infantry in 1861, and the 4th Gurkha Regiment later in the same year; 4th Gurkha Regiment, in 1886; 4th Gurkha (Rifle) Regiment, in 1891; present designation in 1901.

The 2nd Battalion was raised at Bakloh in 1886, by Major M. J. King-Harman.

"Ali Masjid" "Kabul, 1879" "Kandahar, 1880" "Afghanistan, 1878-80"
"Chitral" "Tirah" "Punjab Frontier" "China, 1900."

COLONEL—Lieut.-General *Sir* Arthur G. F. Browne, K.C.B., D.S.O. . . 1 January 1906

HONORARY COLONEL—*Hony.* Lieut.-General *His Excellency Maharaja Sir* Chandra Shamsher Jang, *Rana Bahadur*, G.C.B., G.C.S.I., G.C.V.O., Prime Minister of Nepal. 3 October 1906.

1st Commn. or date of entering service.	Names and Rank.	Army rank.	Present appointment in Battalion.	Remarks.
	Commandants.			
24 Dec. 90	1 Tillard, Lt.-Col. A. B., D.S.O. .	24 Dec. 16	21 Nov. 17	
7 Mar. 94	2 Nicolay, *Lt.-Col.* B. U. . .	3 Aug. 18	3 Aug. 18	With 4-3rd G. R.
	Company Commanders. (8)			
29 Feb. 95	*2 Maxwell, Maj. D. W., p.s.c.* .	*20 Feb. 13*	*10 Mar. 17*	2nd in Comd. *With 2-50th Rifles.*
20 July 98	2 Money, Maj. E. F. D., D.S.O. .	1 Sep. 15	15 Mar. 15	With 1-142nd Infy.
3 Aug 98	*1 Hogg, Maj. C. M. T.*, D.S.O., *p.s.c.*	*1 Sep. 15*	*3 Aug. 15*	*[Mhow Divn.* 2nd-in-Comd.. *G. S. O. 2nd gr.*
5 Jan. 01	1 Collins, Maj. L. P., D.S.O. .	5 Jan. 16	22 May 15	With 1-131st. R.
8 Jan. 01	*2 Massy, Maj. F. S.*	*8 Jan. 16*	*25 May 16*	*With 3-6th G. R.*
26 June 01	1 Scott, Maj. H. St. G. S., D.S.O.	6 Feb. 16	19 Aug. 16	Comdt. 2nd Bn. (*Offg.*)
8 May 01	*1 Wylie, Maj. M.*	*8 May 16*	*10 Mar. 17*	*Lv. ex I, m. c.*
18 Jan. 02	1 Roe, Maj. C. D., D.S.O. . .	18 Jan. 17	3 Aug. 18	
	Company Officers. (18)			
2 Mar. 04	*2 Lentaigne, Capt. E. C.*, D.S.O.	*2 Mar. 13*	*1 May 06*	*Lv., ex I., m.c.*
18 Jan. 05	2 Bell-Kingsley, Capt. H. E. W., D.S.O.	18 Jan. 14	9 Aug. 06	
5 Aug. 05	*1 Phayre*, Bt. Maj. *R. B*, M.C., *[L]*	*3 June 18*	*12 Dec. 06*	
24 Jan. 06	2 Gardiner, Capt. W. A. (*Actg. Maj.*)	24 Jan. 15	31 Jan. 08	
24 Jan. 06	1 Hartwell, Capt. J. R., D.S.O. .	24 Jan. 15	24 Mar. 07	With 2-1st G. R. 2nd-in-Comd.
19 Jan. 07	2 Arnott, Capt. A. M. (*Actg. Maj.*)	1 Sep. 15	20 Mar. 08	4-11th G. R.
9 Sep. 08	*2 Farwell, Capt. W. L.* . . .	*1 Sep. 15*	*25 Nov. 09*	*Lv., ex I. .m.c*
17 Jan. 11	1 Brandon, Capt. R. V. . . .	1 Sep. 15	20 Aug. 15	With 3-2nd G. R.
19 Jan. 12	2 Borrowman, Capt. C. G. . .	19 Jan. 16	27 Oct. 14	Cadet College, Quetta.
14 Jan. 14	2 Brunlees, Capt. L. H. . .	14 Jan. 18	11 Oct. 14	Adjt.
1 Oct. 14	2 Walton, Capt. L. C. J. B.	1 Oct. 18	16 Dec. 14	Attd. 3-9th G. R.
23 Dec. 14	2 Weallens, Lt. W. R. W. .	23 Dec. 15	28 Feb. 15	With Nepalese Contgt.
19 Jan. 15	2 Burgan, Lt. N.	19 Jan. 16	25 May 17	Attd. With 2-11th G. R.
11 Mar. 15	2 Harris, Lt. E. N. (*Actg. Capt.*)	11 Mar. 16	24 Feb. 18	Attd. With 2-4th G. R.
16 June 15	2 Tirebuck, Lt. C. T. . .	16 June 16	13 Feb. 16	Attd.
15 Nov. 15	1 Harrison, Lt. A. M. L., M.C. .	15 Nov. 16	18 Nov. 15	Adjt. (*Offg.*)
17 June 16	1 Sanderson, Lt. P. C. . . .	17 Mar. 18	11 June 18	Attd.
16 Aug. 16	1 Hughes, Lt. F. E. C. . . .	16 Aug. 17	11 Nov. 16	Attd. With 4-39th Garh
14 Nov. 16	1 Ryan, Lt. B. A.	14 Nov. 17	. . . .	Attd. With 3-5th G. R.
30 Jan. 17	1 Lillie, 2nd-Lt. C. J. W. .	30 Jan 17	3 Feb. 17	Attd. With 2-3rd G. R.
18 June 17	1 Barltrop, Lt. A. B. . . .	18 June 18	. . . .	With 2-3rd G. R.
1 Sep. 18	1 Etherington, 2nd-Lt. R. A. .	1 Sep. 18	1 Sep. 18	Attd. With 1-5th L. I.
. . .	1 Scott, 2nd-Lt. J. G.	. . .	. . . .	Attd. With 2-11th G. R.
20 Dec. 17	2 Weeks, 2nd-Lt. C. A. . . .	20 Dec. 17	20 Mar 18	Attd.
1 Oct. 18	1 Lentaigne, 2nd-Lt. W. D. A.	10 Oct 18	10 Oct. 18	Attd.
1 Oct. 18	2 Lowsley, 2nd-Lt. G L. .	1 Oct. 18	. . . .	Attd.
. . .	2 Albon, 2nd-Lt. R. G. E. W. .	. . . .	. . . .	Attd.
27 Oct. 17	Owens, Lt. T. D. C. . . .	27 Oct. 18	. . . .	Attd.
21 Dec. 17	Williams, Lt. S. M. . . .	21 Dec. 18	. . . .	Attd. With 2-150th Rif.
	Attached.			
24 Jan. 14	1 Dennys, Capt. C. H. M. . .	24 Jan. 18	17 Feb. 18	Conn. Rang. I. A. (*on probn.*)
4 May 15	1 Bunning, Lt. W. H., M.C. .	4 May 16	30 Mar. 18	Aust. Imp. Force. I. A. (*on probn.*)
4 May 15	1 Curtler, Lt. W. L.	4 May 16	2 Dec. 17	Worc. R. I. A. (*on probn.*)
9 July 15	1 Lawson, Lt R. A. . .	9 July 16	9 July 15	I. A. R. O. Offg. Qr.-Mr.
22 June 15	1 Read, Lt., J. E.	22 June 16	12 Apl. 18	Hamps. R. I. A. (*on probn.*)
23 Aug. 15	1 Lawrence, Lt. A. E., M.C. (52, 207*a*)	23 Aug. 15	21 Jan. 18	Shrops. L.I., I. A. (*on probn.*)
25 Nov. 15	1 Fulton, Lt. E. A. C.	25 Nov. 16	10 Jan. 18	High. L. I. I. A. (*on probn.*)
6 Oct. 15	1 McLean, Lt. K.	6 Oct. 16	25 Nov. 16	I. A. R. O.
16 Jan. 16	1 Capes, Lt. G. A.	16 Jan. 17	16 Aug. 17	Hamps. R., I. A. (*on probn.*)

4th Gurkha Rifles—*contd.*

1st Commn. or date of entering service.	Names and Rank	Army rank.	Present appointment in Battalion.	Remarks.
	Attached—*contd.*			
26 Jan. 16	1 Barron, Lt. A. S. N.	26 Jan. 17	27 Feb. 18	A. S. C., I. A. (*on probn.*)
26 Apl. 16	1 Gamble, Lt. L.	26 Apl. 17	....	E. Lan. R. I. A. (*on probn.*)
28 Nov. 16	1 Sienesi, Lt. A. R.	28 Nov. 17	11 Apl. 18	R. M. A. I. A. (*on probn.*)
13 Dec. 16	1 Bradford, Lt. F. K.	13 Dec. 17	....	North'n Fus., I. A. (*on probn.*)
28 July 17	1 Maynard, 2nd-Lt. K.	28 July 17	2 Feb. 18	I. A. (*T. C.*)
1 Oct. 17	1 Belben, Lt. H. R.	23 Aug. 18	20 Jan. 18	I. A. (*T. C.*)
27 Oct. 17	1 Madge, 2nd-Lt. H. E.	27 Oct. 17	1 Dec. 17	
27 Oct. 17	1 Tervitt, 2nd-Lt. J. F. T.	27 Oct. 15	....	
31 Jan. 18	1 Bodenham, 2nd-Lt. K. L.	31 Jan. 18	19 Feb. 18	Attd. With 1-119th Punj.
10 July 18	1 Stott, 2nd-Lt. H. W. M.	10 July 18	2 Dec. 18	I. A. (*T. C.*)
20 June 18	1 Hullock, 2nd-Lt. E. R.	20 June 18	2 Dec. 18	I. A. R. O.
22 July 18	1 Case, 2nd-Lt. C. E.	22 July 18	2 Dec. 18	I. A. R. O.
19 Nov. 14	2 Bunn, Lt. W. (*Actg. Capt.*)	19 Nov. 15	16 Sep 17	I. A. R. O.
1 Nov. 15	2 Barstow, Lt. H. L.	1 Nov. 16	....	M. G. Corps. I. A. (*on probn.*)
27 Nov. 14	2 Goodall, Lt. G. M., M.C. (*Actg. Capt.*)	27 Nov. 15	20 Jan. 17	I. A. R. O.
29 Aug. 14	2 Seymour, Lt. A. H. (*Actg. Capt.*)	29 Aug. 15	30 Nov. 17	I. A. R. O.
10 May 16	2 Walsh, Lt. L. E.	10 May 17	15 Mar. 17	I. A. R. O.
23 Apl. 15	2 Bristow, Lt. C. H.	23 Apl. 16	2 July 15	I. A. (*on probn.*) *Lv., ex I., m. c.*
25 July 15	2 Ryder, Lt. R. C.	25 July 16	17 Aug. 15	I. A. R. O.
29 July 15	2 Nye, Lt. A. R., M.C.	29 Apl. 17	8 Sep. 15	I. A. (*on probn.*)
....	2 Hamber, Lt. L. G. W. (*Actg. Capt.*)	....	....	
16 Sep. 17	Mitchell, 2nd-Lt. H. R. W.	16 Sep. 18		
21 Dec. 15	2 Partridge, Lt. E. C.	1 Dec. 16	19 Jan. 16	I. A. R. O.
31 Jan. 18	2 Fullerton, 2nd-Lt. G. A.	31 Jan. 18	31 Jan. 18	Attd.
30 Sep. 15	2 Atkinson, Lt. D. J.	30 Sep. 16	12 Jan. 18	R. Dub. Fus., I. A. (*on probn.*) S. S. O. Bakloh.
5 May 17	2 Dodd, Lt. J. R.	5 May 18	20 Aug. 17	I. A. R. O. With 2nd Bn.
10 Mar. 17	2 Thomson, Lt. W. R.	10 Mar. 18	30 Mar. 17	I. A. R. O.
20 Mar. 16	2 Moore, Lt. A. J.	20 Mar. 18	1 Apl. 17	I. A. R. O.
25 Mar. 16	2 Modget, Lt. H. R.	25 Mar. 17	8 May 17	I. A. R. O.
30 Jan. 17	2 Ennis, Lt. J. I.	30 Jan. 18	9 Feb. 17	I. A. R. O.
15 Feb. 15	2 deClermont, Lt. A. C.	15 Feb. 18	15 Feb. 15	I. A. R. O.
5 Nov. 17	2 Robertson, Lt. J. Macl.	5 Nov. 18	15 Mar. 18	I. A. R. O.
5 Nov. 17	2 Strange, 2nd-Lt. F.	5 Nov. 17	15 Mar. 18	I. A. (T. C.)
13 Jan. 15	2 Clarkson, Lt. R. M.	13 Jan. 16	6 Sep. 18	High. L. I., I. A. R. O. (*on probn.*)
1 Aug. 18	2 Perrin, 2nd-Lt. L. C.	1 Aug. 18	26 Aug. 18	I. A. R. O.

		Jemadar.	Subadar.	Subadar-Major.	
	Subadar-Majors.				
13 Apl. 86	2 Debi Chand, *Sardar Bahadur* (56)	1 Dec. 96	1 July 00	10 June 05	*Order of Br. I., 1st. Cl. Lv., m.c. to 15th Sep. 18.*
1 Aug. 90	*1 Sher Sing Rana* (56)	*8 May 03*	*3 Aug. 05*	*1 Dec. 12*	
1 Apl. 84	1 Chamu Thapa (56)	1 Apl. 05	1 Jan. 09	21 Dec. 14	
	Subadars.				
8 Mar. 86	2 Bansrai Gurung, *Bahadur*	1 Apl. 99	1 May 04	*Order of Br. I., 2nd Cl.*	
17 May 86	1 Pirthi Sing Rana (56)	1 Apl. 08	1 Jan. 12		
17 Sep. 87	2 Madan Sing Rai (56)	1 Oct. 07	1 Jan. 13		
25 Nov. 90	2 Dhan Lal Gurung	1 May 10	1 May 14		
1 Feb. 01	2 Santbir Gurung	12 Aug. 14	12 Aug. 14		
24 Sep. 87	2 Hari Sing Thapa	12 Aug. 14	12 Aug. 14		
1 Jan. 96	2 Mahabir Thakur Bahadur	12 Aug. 14	23 Oct. 14	*Order of Br. I., 2nd Cl.*	
10 Nov. 96	1 Surdal Thapa	1 Mar. 13	20 Dec. 14	With 49th Bengalis.	
8 Aug. 98	1 Ramkishen Rana (56), D.S.M.	1 Feb. 13	28 Dec. 14		
10 Nov. 98	*2 Antbir Gurung,* I.O.M. (66a)	*1 Jan. 13*	*1 Feb. 15*		
19 Nov. 00	*2 Shiam Sing Rana* (46)	*1 May 14*	*13 Mar. 15*		
7 Nov. 03	1 Gunjbir Gurung (56), I.O.M.	21 Dec. 14	14 Sep. 15		
24 Dec. 92	2 Ram Sing Sahi	23 Oct. 14	12 Dec. 15		
14 Dec. 00	2 Karan Bahadur Thapa	9 Apl. 15	22 Jan. 16		

4th Gurkha Rifles—*concld.*

Subadars—*contd.*

1st Commn. or date of entering service.	Names and Rank.	Army rank.		Remarks.
4 May 99	2 Ram Saran Gurung	21 Aug. 14	22 Apl. 16	
14 Jan. 01	2 Puran Bahadur Pun . . .	22 Apl. 15	9 July 16	
11 Nov. 07	2 Jaman Sing Gurung, I.O.M. . .	12 July 15	16 Sep. 16	
3 Jan. 01	2 Chhabu Thapa, I.D.S.M. . .	1 June 16	21 Sep. 16	
3 Dec. 97	2 Pahal Sing Thapa, ...	1 July 13	10 Feb. 17	
6 Jan. 00	1 Harikishan Gurung	8 Aug. 14	1 June. 17	
19 Sep. 99	1 Ramkishan Dura	19 Dec. 14	1 Jan. 18	
25 Dec. 98	2 Jamin Sing Gurung	21 June 15	11 Mar. 17	
	1 Nain Sing Gharti (*Retd.*) .	. . . . .	. . . .	Re-employed, 1 Mar. 15. Depôt.
	2 Lachman Gurung (*Retd.*) .	. . . . .	. . . .	Re-employed, 14 Dec. 15.
. . . .	1 Singbir Gharti (*Retd.*) . .	. . . . .	. . . .	Re-employed 1 Sep. 15.
26 June 96	2 Jogi Chand	17 Oct. 18	24 May 18	

Jemadars

1st Commn. or date of entering service.	Names and Rank.	Army rank.	Remarks.
14 Jan. 99	2 Kharkbir Ale	1 Jan. 14	
8 Dec. 96	*1 Nain Sing Gurung* . .	*8 Aug. 14*	*12 Aug. 15.*
23 Nov. 98	2 Balbir Thapa	12 Aug. 14	
4 Dec. 98	1 Pahal Sing Gurung	19 Dec. 14	
16 Nov. 00	*1 Jitbahadur Rana* . . .	*20 Dec. 14*	*With Con. Sec. No. I. Gurkhas.*
7 Nov. 96	1 Belbahadur Gurung . .	{ 28 Dec. 14 28 Apl. 15	
	2 Bhagatbir Thapa	22 Feb. 15	
17 Mar. 97	1 Narbir Thapa	5 Apl. 15	Gurkha Adjt.
19 Mar. 97	2 Budhival Thapa	5 Apl. 15	With 1st Bn.
26 Dec. 99	1 Lohasur Gurung	28 Apl. 15	
10 Jan. 10	1 Hira Sing Thapa (46) . . .	2 May 15	
25 Jan. 02	1 Chandarbir Thapa . . .	23 July 15	
16 Feb. 09	2 Dila Sing Gurung	17 Oct. 15	
22 Aug. 05	2 Rikhesor Khattri	22 Jan. 16	
15 Nov. 06	2 Hirajit Thapa	22 Jan. 16	
23 July 00	2 Chet Sing Thapa	22 Jan. 16	
20 Dec. 00	2 Chamu Thapa	22 Jan. 16	
18 Jan. 01	2 Jhup Sing Pun	22 Jan. 16	
1 Oct. 06	2 Lal Sing Thapa	13 May 16	Gurkha Adjt.
17 June 08	2 Chatur Sing Gurung . . .	10 June 16	
20 Jan. 01	2 Bahadur Thapa	9 July 16	
21 Nov. 06	2 Dilbahadur Ale	9 July 16	
23 Nov. 99	1 Ganj Sing Gurung	22 July 16	
6 Nov. 08	2 Narbir Thapa	16 Sep. 16	
13 May 03	1 Amar Sing Pun, I.D.S.M. . .	1 Feb. 17	
4 Nov. 04	1 Ragbir Thapa, I.D.S.M. . . .	1 Feb. 17	
6 Feb. 01	2 Bale Thapa (56)	1 Mar. 17	
12 Jan. 08	1 Lokhbir Ale (56), I.D.S.M. . .	1 May 17	
15 Dec. 98	2 Damar Sing Gurung . . .	1 May 17	
12 Feb. 01	1 Jaspati Gurung	1 Nov. 17	
14 Jan. 14	1 Randhoj Gurung	1 Nov. 17	
	1 Amar Sing Rana (*Retd.*) . .	. . . . .	Re-employed, 6 Mar. 15. Depôt.
	2 Santbir Thapa (*Retd.*) . . .	. . . . .	Re-employed, 10 Dec. 15.
4 Nov. 14	1 Haimard Rana.	1 Jan. 18	
19 Nov. 00	1 Jitman Gurung	1 Jan. 18	
16 Jan. 05	1 Shamsher Jang Thaker . .	1 May 18	

Regt. Centre—Abbottabad.

Uniform—Dark Green. *Facings*—Black.

5th Gurkha Rifles (Frontier Force).

1st Battalion
2nd Battalion

Raised at Abbottabad, in 1858, by Capt. H. M. F. Boisragon, as the 25th Punjab Infantry or Hazara Gurkha Battalion. Became the 7th Regt. of Infy. (or Hazara Gurkha Bn.), Punjab Irregular Force, 1861, and the 5th Gurkha Regt., or Hazara Gurkha Bn., 1861; the 5th Gurkha Regt., the Hazara Gurkha Bn., 1886; 5th Gurkha Regt., 1887; 5th Gurkha (Rifle) Regt., 1891; the 5th Gurkha Rifles, 1901: present designation, 1903.

The 2nd Battalion was raised at Abbottabad in 1886, by Major E. Molloy.

"Peiwar Kotal" "Charasiah" "Kabul, 1879" "Kandahar, 1880" "Afghanistan, 1878-80" "Punjab Frontier."

COLONEL—Maj.-Genl. Sir John M. Sym, K.C.B. . . . 30th April 1915.

1st Commn. or date of entering service.	Names and Rank.	Army rank.	Present appointment in Battalion.	Remarks.
	Commandants.			
9 Nov. 89	2 Johnson, Lt.-Col. C. R. . . .	9 Nov. 15	22 Oct. 16	
1 Mar. 90	1 Nightingale, Lt.-Col. M. R. W., C.M.G., D.S.O. [L] (*temp. Br.-Genl.*)	1 Mar. 16	1 Mar. 17	*Brig. Comdr.*
	Company Commanders. (8)			
19 Oct. 87	*1 Boddam, Lt.-Col. E. B. C., p.s.c (Q) (temp. Brig.-Genl.)* . .	*3 June 13*	*16 May 13*	*Br. Genl. i/c Administration 6th (Poona) Dn.*
5 Dec. 91	*2 Browne, Bt.-Col. H.J.P., p.s.c. (Q) (temp. Br.-Genl.)*	*3 June 16*	*22 Oct. 16*	*2nd-in-Comd. Br. Genl., Genl. Staff., S. Command.*
2 June 94	1 Villiers Stuart, Maj. W. D. (Q) (*temp. Br.-Genl.*)	2 June 12	6 May 09	Inspr. of Infantry (N).
10 Oct. 94	*2 Holdich, Bt.-Lt.-Col. H. A., D.S.O., p.s.c. (Q) (temp. Br.-Genl.)*	*3 June 15*	*1 June 09*	*Dir. of Quarterin rmy. Hdqrs.*
10 Oct. 94	2 Bateman-Champain, Bt.-Lt. Col. C. E.	20 Aug. 18	2 Feb. 08	
22 Jan. 98	1 Skipwith, Maj. F., *p.s.c.* (*Actg. Lt.-Col.*)	1 Sep. 15	22 Oct. 16	Comdt. (*temp.*)
6 Jan. 00	*1 Crowdy, Maj. J. D., D.S.O., p.s.c.*	*1 Sep. 15*	*22 Oct. 16*	*Bde. Maj.*
17 Jan. 00	1 Sanders, Maj. G. P. . . .	1 Sep. 15	1 Mar. 17	
	Company Officers. (18)			
18 Jan. 02	2 Molloy, Maj. H. T. . . .	18 Jan. 17	16 Mar. 04	With 2-6th G. R.
18 Jan 05	2 Lane, Capt. H. R. C. . .	18 Jan. 14	3 May 06	With 3-5th G. R.
5 Aug. 05	2 Duncan, Bt.-Maj. R. C. . .	3 June 18	28 Jan. 07	
17 Aug. 07	2 Neville Rolfe, Capt. A. . . .	1 Sep. 15	24 Nov. 08	
9 Sep 08	*1 Erskine, Capt. K. C. S., M.C.* .	*1 Sep. 15*	*16 Nov. 09*	*Bde. Maj.*
14 Oct. 08	2 Gouldsbury, Bt. Maj. C. A. .	1 Sep. 15	17 Mar. 14	
6 Sep. 11	2 Wellesley, Capt. H. A., M.C. .	6 Sep. 15	7 Nov. 12	
19 Jan. 12	2 Bruce, Capt. R. M., M. C. . .	19 Jan. 16	8 Oct. 14	
22 Jan. 13	2 Lemarchand, Capt. F. E. . .	22 Jan. 17	13 Mar. 14	With 4-39th Garh. Rif.
14 Jan. 14	1 Nepean, Capt. H.D.H. Y., D.S.O.	14 Jan. 18	10 Oct. 14	Qr.-Mr., 14 Jan. 16.
15 Aug. 14	*2 Loch, Capt. H. N.* . . .	*15 Aug. 18*	*3 Dec. 15*	*With R. A. F.*
1 Oct. 14	2 Spottiswoode, Capt. A. U. M.	1 Oct. 18	17 Dec. 14	
11 Nov. 14	2 Maconchy, Lt. G. A. (*Actg. Capt.*)	11 Nov. 15	26 Jan. 15	Adjt. (*temp.*)
1 Sep. 15	2 Silberbaver, Lt. H. L. (*Actg. Capt.*).	1 Sep. 16	23 Jan. 18	Attd.
15 Sep. 15	2 Marindin, Lt. J. F. . . .	15 Sep. 16	6 Dec. 15	Attd.
15 Nov. 15	2 Coates, Lt. J. C.	15 Nov. 16	19 Nov. 15	Attd. With Guides Infy.
24 Nov. 15	2 Grey, Lt. R. W. H.	24 Nov. 16	1 Feb. 16	Attd.
18 Apl. 16	1 Baker, Lt. C. S. (*Actg. Capt.*)	18 Apl. 17	25 Apl. 16	Attd.
18 Apl. 16	1 Fox, Lt. W. B. O.	18 Apl. 17	25 Apl. 16	Attd.
29 June 16	2 D'Oyley-Hugues, Lt. (*Actg. Capt.*).	29 June 17	16 Mar. 18	Attd. Adjt. (*temp.*)
29 June 16	1 Macintosh, Lt. K. J., M.C. (*Actg. Capt.*)	29 June 17	2 July 16	Attd. Adjt. (*temp.*)
29 June 16	1 Thomson, Lt. A. P. Q. . . .	29 June 17	2 July 16	Attd.
29 June 16	1 Petrie, Lt. J. F.	29 June 17	2 July 16	Attd. With 2nd Guides Infy.
14 Nov. 16	1 Fisher, Lt. F. A. B.	14 Nov. 17	24 Nov. 16	Attd. With 3-8th G. R.
14 Nov. 16	1 Walker, 2nd-Lt. J. E. . . .	14 Nov. 16	29 Nov. 16	Attd. With 9th Horse.
27 Oct. 17	2 Garvie, 2nd-Lt. K. C. . . .	27 Oct. 17	12 Nov. 17	Attd.
27 Oct. 17	2 Platts, Lt. M. A.	27 Oct. 18	5 Nov. 17	Attd.
27 Oct. 17	2 Alexander, 2nd-Lt. L. A. . .	27 Oct. 17	5 Nov. 17	Attd.
31 Aug. 18	1 Plunkett, 2nd-Lt. H. R. S. . .	31 Aug. 18	5 Sep. 18	Attd.
12 Sep. 17	1 Broadway, 2nd-Lt. P. R. . .	12 Sep. 17	. . . .	Attd.
1 Feb. 18	2 Green, 2nd-Lt. E. G. H. . .	1 Feb. 18	7 Feb. 18	Attd.
1 Oct. 18	1 Stables, 2nd-Lt. F. H. A. . .	1 Oct. 18	7 Oct. 18	Attd.
14 July 15	1 Cameron, Lt. R.T. (*Actg. Capt.*)	14 July 16	8 Dec. 17	Attd.
31 Aug. 18	2 Diack, 2nd-Lt. H. J. W. . .	31 Aug. 18	Sep. 18	Attd.

5th Gurkha Rifles (Frontier Force)—*contd.*

Attached.

1st Commn. or date of entering service.	Names and Rank.	Army rank.	Present appointment in Battalion.	Remarks.
14 Jan. 15	1 Cosby, Capt. N. R. C., M.C. .	19 Sep. 16	14 Jan. 15	I. A. R. O.
18 Dec. 14	2 Broad, Lt. R. N. D.	18 Dec. 14	5 Feb. 17	I. A. R. O.
25 July 15	2 March, Lt. J. C.	25 July 16	17 Aug. 15	I. A. R. O.
24 June 16	1 Tregale, Lt. J. W. W. . . .	24 June 17	21 Aug. 16	I. A. R. O.
10 Nov. 16	2 Mckenzie, Lt. R. D. . . .	10 Nov. 17	31 Jan. 17	I. A. R. O.
29 May 15	2 Johnstone, Lt. N. W. W. . .	29 May 16	. . . .	Gen. List.
1 Oct. 15	2 Fisher, Lt. F. B.	1 Oct. 16	5 May 17	I. A. R. O.
10 Nov. 16	1 Whitby, Lt. R. B. H.	10 Nov. 17	28 May 17	I. A. R. O.
11 Dec. 14	2 Gladstone, Lt. A. C.	31 Dec. 15	14 Mar. 16	I. A. R. O.
1 Dec. 16	2 Cousins, Lt. C. W.	1 Dec. 16	20 Jan. 16	I. A. R. O. Qr. Mr.
20 June 17	1 Simpson, Lt. C. E.	20 June 18	5 Oct. 17	I. A. R. O.
17 Mar. 15	1 Macartney, Lt. H. J. . . .	17 Mar. 16	11 Dec. 17	I. A. R. O. (*on probn.*)
20 June 17	2 Legge, 2nd-Lt. W. E. . . .	20 June 17	8 Oct. 17	I. A. R. O.
29 Mar. 18	1 Edwards, 2nd-Lt. H. B. . . .	29 Mar. 18	15 July 18	I. A. R. O.
20 Jan. 16	2 Hobbs, Lt. J. M., M.C. . . .	20 Jan. 17	26 Apl. 18	Aust. I. F. I. A. (*on probn.*)
25 June 16	1 Cormack, Lt. R. K. W. . . .	25 June 17	3 Sep. 18	I. A. (*on probn.*)
29 Nov. 15	1 Eastabrook, Lt. I.	29 Aug. 17	5 July 18	R. War. R. I. A. (*on probn.*)
19 Sep. 17	2 Paterson, Lt. G. F.	19 Sep. 18	. . . .	R. Innis. Fus. I. A. (*on probn.*)
29 Aug. 17	1 Winn, Lt. F. B.	29 Aug. 18	. . . .	I. A. R. O., 3-5th G. R.

Subadar-Majors.

1st Commn. or date of entering service.	Names and Rank.	Jemadar.	Subadar.	Subadar-Major.	Remarks.
1 Nov. 86	2 Amar Sing Thapa, *Sardar Bahadur* (56, 58, 66), M.C. . .	1 Nov. 97	1 Nov. 97	16 July 05	*Order of Br. I., 1st Cl.*
9 Dec. 90	1 Dhanraj Gurung, I.O.M. (56) .	1 Dec. 04	1 Jan. 08	1 Jan. 18	

Subadars.

1st Commn. or date of entering service.	Names and Rank.	Jemadar.	Subadar.	Remarks.
4 Dec. 86	2 Ram Sing Thapa (56, 66) .	1 Dec. 07	May 11	
2 Nov. 87	2 Bilbikram Rana (56, 66*a*) . .	17 Sep. 09	1 Jan. 12	
20 Oct. 93	2 Pahal Sing Rana (56) . . .	1 May 11	6 Apl. 14	
11 Nov. 02	1 Dhansing Gurung (46, 56) I.D.S.M.	16 Jan. 14	5 June 15	With 1-11th G. R.
3 Dec. 98	2 Arjun Thapa (56)	1 May 13	3 July 15	
7 Dec. 94	2 Puranbir Rana	1 Jan. 14	16 Dec. 15	
18 Dec. 94	1 Budhiman Gurung	4 June 15	25 Aug. 16	
12 Nov. 00	1 Sahabir Gurung	5 June 15	25 Aug. 16	
13 Nov. 02	1 Shamsher Gurung	8 Aug. 15	15 Feb. 17	
18 Nov. 03	1 Parshad Thapa	5 June 15	1 Apl. 17	
14 Nov. 01	1 Budhichand Bura	1 July 15	15 Feb. 17	
20 Dec. 02	1 Sasidhar Thapa	8 Aug. 15	29 Sep. 17	
27 Nov. 00	1 Siria Gharti	28 June 15	1 Oct. 17	
8 Dec. 98	2 Maidhar Gharti	3 July 15	22 Sep. 17	
13 May 02	2 Chint Ram Burathoki, I.D.S.M. .	9 Aug. 15	29 Sep. 17	
14 Nov. 01	2 Bhawan Sing Thapa	1 Jan. 16	29 Dec. 17	
14 Mar. 98	1 Kulbahadur Gurung	16 June 16	17 Dec. 17	
6 Nov. 99	2 Sahbir Gurung	10 Aug. 17		
17 Nov. 03	1 Udebhir, Gurung	16 June 16	24 Sep. 18	
	2 Amar Sing Thapa	15 Jan. 16	2 une 18	
27 Oct. 97	2 Narbahadur Gurung, I.D.S.M. .	1 Jan. 17	30 July 18	

5th Gurkha Rifles (Frontier Force)—*concld.*

1st Commn. or date of entering service.	Names and Rank.	Army rank.	Remarks.
		Jemadars.	
11 Nov. 94	1 Sukhlal Rana	1 May 13	
10 Nov. 96	1 Biraj Gurung	16 Oct. 14	
2 Jan. 00	2 Ghiria Thapa	17 Apl. 15	
3 Dec. 98	2 Chandra Sing Gurung . . .	3 July 15	
13 Nov. 99	1 Kabiram Gurung	16 June 16	
7 Jan. 05	1 Balwant Rana	16 June 16	
17 Feb. 04	1 Dandabir Thapa (56) . . .	16 June 16	
24 Nov. 04	1 Gorıa Pun (46)	12 Aug. 16	With 1-11th G. R.
30 Jan. 05	1 Karnabahadur Rana	25 Aug. 16	With 1-11th G. R.
7 Nov. 00	2 Ran Sing Thapa	1 Jan. 17	
30 Mar. 87	2 Padam Sing Thapa	1 Jan. 17	
8 Feb. 99	2 Kasiram Thapa	1 Jan. 17	
4 Nov. 01	2 Balbir Bura	1 Jan. 17	
16 Feb. 99	1 Kansbir Rana	15 Feb. 17	
19 Aug. 96	1 Bhadria Khattri	15 Feb. 17	
11 Dec. 04	1 Budh Sing Gurung	15 Feb. 17	
8 Feb. 02	2 Ransur Pun	1 Apl. 17	
15 Mar. 03	2 Nain Sing Thapa	1 May 17	Jemdr Adjt.
4 Feb. 03	2 Dilman Sing Thapa (56) . .	1 June 17	
20 Jan. 05	1 Kulbir Rana	19 Aug 17	
1 Feb. 04	1 Manbir Thapa	29 Sep. 17	
30 Oct. 06	1 Damar Sing Thapa, I.D.S.M. .	29 Sep. 17	
29 Jan. 05	1 Kharak Sing Gurung, I.D.S.M.	29 Sep. 17	Jemdr. Adjt.
9 Mar. 99	1 Dhanraj Gurung	29 Sep. 17	
12 Nov. 07	2 Jitbahadur Thapa	29 Sep. 17	
11 Jan. 06	1 Harkbir Gurung (56) (45 *a*) .	1 Nov. 17	
20 Jan. 05	1 Kalu Gurung	1 Nov. 17	
19 Feb. 06	1 Muktibahadur Rana (56) . .	1 Mar. 18	
27 Nov. 95	2 Nand Ram Sahi	6 Apl. 18	
31 Jan. 06	1 Lokbadahur Thapa . .	18 May 18	
17 Nov. 04	1 Kharjit Gurung . . .	6 July 18	
15 Dec. 04	2 Indrabir Khan . . .	18 May 18	
12 Nov. 03	2 Puran Sing Thapa . .	18 May 18	
10 Mar. 03	2 Chandar Sing Gurung . .	18 May 18	
26 Jan. 05	1 Bhairab Sing Gurung . . .	3 Aug. 18	
20 Nov. 02	2 Hastadal Thapa	24 June 18	
16 Feb. 04	2 Harka Sing Gurung, I.D.S.M. .	30 July 18	
. . . .	Hastabir Rana	24 Sep. 18	

3rd Battalion, 5th Gurkha Rifles (Frontier Force).

Formed at Ferozepore on 29th November 1916 by Lt.-Col. C. A. Roosmale-Cocq.

1st Commn. or date of entering service.	Names and Rank.	Army rank.	Present appointment in Battalion.	Remarks.
	Commandant.			
22 Jan. 96	Barlow, Maj. C. P. (*Actg. Lt.-Col.*)	22 Jan. 14	22 July 18	2-8th G. R.
	Company Commanders.			
30 Apl. 02	Rombulow-Pearse, Maj. A. B.	16 Apl. 16	1 Dec. 16	2-6th G. R. 2nd-in-Comd.
18 Jan. 05	Lane, Capt. H. R. C.	18 Jan. 14	16 Sep. 17	2-5th G. R.
4 Aug. 14	Barlow, Lt. A. E., M.C. (*actg. Capt.*) (12, 51, 207*a*).	4 May 16	19 Oct. 17	Attd.
9 June 15	Tudor-Craig, Lt. P. C. (*actg. Capt.*) (207*a*, 51).	9 June 16	29 Oct. 17	Attd.
	Company Officers.			
15 Aug. 14	Chenevix-Trench, Lt. H.	15 May 16	1 Mar. 18	Rif. Brig., I. A. (*on probn.*)
16 Nov. 14	White, Lt. O. J. (207*a*)	14 Aug. 16	23 June 17	I. A. (*on probn.*) With 4-11th G. R.
29 Mar. 15	Rice, Lt. A. K. (207*a*, 46)	29 Dec. 16	17 July 17	Lond. R. I. A. (*on probn.*)
31 Mar. 15	Brennan, Lt. R., M.C. (207*a*)	4 Feb. 16	22 Oct. 17	R. Ir. Fus. I. A. (*on probn.*)
17 June 15	Carling, Lt. T. E.	17 Mar. 17	Jan. 18	S. Wales Bord, I. A. (*on probn.*) With 2-50th Rif.
10 Oct. 15	Sturdy, Lt. J. R., M.C. (*actg. Capt.*) (207*a*)	10 Oct. 16	22 Oct. 17	1st Suff. R. I. A. (*on probn.*) Staff Capt.
7 Nov. 15	Watt, Lt. W. R., M.C. (207*a*)	7 Aug. 17	3 Aug. 17	With 1-5th G. R.
16 Aug. 16	Bentham, Lt. V. H. (207*a*)	16 Aug. 17	1 Mar. 18	R. E., I. A. (*on probn.*) With 1-7th G. R.
5 Aug. 16	Jones, Lt. L. W. N.	. . . . .	12 Feb. 18	I. A. R. O. (*on probn.*)
5 Sep. 16	Fairley, Lt. R. S. (207*a*)	. . . . .	27 Jan. 18	I. A. R. O. (*on probn.*)
14 Nov. 16	Ryan, Lt. B. A.	14 Nov. 17	8 Feb. 18	1-4th G. R.
14 Nov. 16	Wooly, Lt. J. H. C.	14 Nov.	6 Dec. 16	2-1st G. R. Attd. 19th Lcs.
5 Aug. 6	Pinches, Lt. A. G.	5 May 18	17 Apl. 18	I. A. (*on probn.*)
5 May 17	Mill, Lt. G. S. (*Actg. Capt.*) (51, 207*a*)	5 May 18	12 Aug. 17	I. A. R. O. Adjt.
29 Aug. 17	Winn, Lt. F. B. (207*a*)	29 Aug 18	27 Jan. 18	I. A. R. O. (*on probn.*) Qr. Mr.
5 Oct. 17	Wright, Lt. A. A. (207*a*)	5 Oct. 18	20 Jan. 18	I. A. (*T. C.*)
31 Sep. 17	Campbell, 2nd-Lt. D.	31 Sep. 18	23 Jan. 18	I. A. R. O. (*on probn.*)
5 Nov. 17	Brown, Lt. E. R.	5 Nov. 18	14 Mar. 18	I. A. (*T. C.*)
5 Nov 17	Mockridge, Lt. A. P. (207*a*)	5 Nov. 18	14 Mar. 18	I. A. (*T. C.*)
21 Dec. 17	Harrison, 2nd-Lt. T. R.	21 Dec. 17	22 Mar. 18	Attd.
31 Jan. 18	Grey, 2nd-Lt. A. R. (46, 207*a*)	31 Jan. 18	6 Feb. 18	
31 Jan. 18	Leonard, 2nd-Lt. R. G.	31 Jan. 18	6 Feb. 18	
29 May 18	Robson, 2nd-Lt. R. N.	29 May 18	22 Nov. 18	E. Lan. R., I. A. (*on probn.*).
22 July 18	Adams, 2nd-Lt. E. J.	22 July 18	. . . . .	I. A. (*T. C.*)
27 June 18	Alexander, 2nd-Lt. W.	27 June 18	. . . . .	I. A. R. O.

		Jemadar.	Subadar.	Subadar-Major.
	Subadar-Major.			
28 Dec. 91	Bhagatbir Thapa (72*a*)	1 Dec. 12	28 Dec. 14	5 Sep. 18
	Subadars.			
3 Jan. 90	Hastabir Thapa (12, 52, 56, 72*a*)	27 Sep. 08	12 Feb. 16	
1 Nov. 94	Karbir Gurung (12, 52, 56, 72*a*)	2 Apl. 13	27 Sep. 16	
1 Nov. 99	Embahadur Sahi (72*a*)	12 Mar. 15	24 Oct. 16	
8 Jan. 00	Lalitbahadur Mal, I.D.S.M., (12)	16 May 15	24 Oct. 16	
7 Jan. 05	Sarunia Gurung	28 Dec. 14	1 Jan. 17	
1 Nov. 05	Dharmdhoj Limbu	10 Aug. 15	9 Dec. 17	

3rd Battalion, 5th Gurkha Rifles (Frontier Force)—*contd.*

1st Commn. or date of entering service.	Names and Rank.	Army rank.	Present appointment in Battalion.	Remarks.
	Jemadars.			
21 Oct. 04	Kabir Thapa	27 Sep. 16		
10 Nov. 98	Kalbir Gharti	27 Sep. 16	Offg. Gurkha Adjt.	
8 Jan. 01	Birsing Khanka	24 Oct. 16		
6 Mar. 03	Lalbir Limbu	1 Jan. 17		
27 Oct. 03	Chandrabir Burathoki (56)	1 Jan. 17		
22 Oct. 03	Dharmraj Gurung (56)	1 Jan. 17		
9 Nov. 00	Bandu Gurung (60*a*)	27 Jan. 17		
6 June 00	Dalbahadur Thapa	3 Dec. 17		
11 Nov. 04	Kubir Sing Adhikar	9 Dec. 17		
15 Nov. 00	Narbir Gurung	28 Dec. 17		
26 Dec. 08	Lalbir Rai (56)	1 Apl. 18		
25 Dec. 05	Himlal Khattri	1 Apl. 18		
12 Nov. 08	Ranbahadur Limbu (56)	16 Oct. 18		

Regt. Centre—Abbottabad.

Uniform—Dark Green.
Facings—Black.

6th Gurkha Rifles.

1st Battalion . .

2nd Battalion . .

Raised at Chaubiaganj, in Cuttack, 1817, by Capt. S. Fraser, as the Cuttack Legion. Became the Rangpur Light Infantry Battalion, 1823; the 8th (or Rangpur) Local Light Infy. Bn., 1826; the 8th (or Assam) Local Lt. Infy. Bn., 1828; the 1st Assam Lt. Infy., 1844; the 46th Regt. of Bengal Native Infantry, 1861; the 42nd Regt. of B. N. I., 1861; the 42nd (Assam) Regt. of B. N. (Lt.) Infy., 1864; the 42nd (Assam) Regt. of Bengal (Lt.) Infy., 1885; the 42nd Regt., Gurkha Lt. Infy., 1886; the 42nd (Gurkha) Regt. of B. (Lt.) Infy., 1889; the 42nd Gurkha (Rifle) Regt. of Bengal Infy., 1891; the 42nd Gurkha Rifles, 1901; present designation, 1903.

The 2nd Battalion was raised at Abbottabad in 1904 by Lt.-Col. F. C. Colomb from a nucleus of half of the 1st Battalion.

"Burma, 1885-87."

1st Commn. or date of entering service.	Names and Rank.	Army rank.	Present appointment in Battalion.	Remarks.
	Commandants.			
9 Sep. 91	2 Preston, Lt.-Col. G. A. (Q) . .	9 Sep. 17	16 Apl. 18	
1 Sep. 93	1 Little, Lt.-Col. W. C.	11 June 18	17 Mar. 18	
	Company Commanders. (8)			
9 Nov. 89	*1 Lloyd, Lt.-Col. J. H. (Q) . .*	*9 Nov. 15*	*17 Mar. 14*	*2nd-in-Comd. D. A. Q. M. G. 8th (Lucknow) Divn.*
28 June 90	2 Strange, Lt.-Col. A. J. (Q) .	28 June 16	2 May 15	2nd-in-Comd.
5 Sep. 96	2 Rice, Maj. B. A. McH., D.S.O.	5 Sep. 14	19 Feb. 10	
20 Jan. 97	1 Sherlock, Maj. G. W. S., *p.s.c.*	20 Jan. 15	10 May 10	With 2-21st Punj.
23 Mar. 97	1 Allanson, Bt. Lt.-Col. C. J. L., C.I.E., D.S.O., *p.s.c* (*temp. Lt.-Col.*)	23 Mar. 15	19 Feb. 12	
20 July 98	*2 McLaughlin, Maj. H. D. . .*	*1 Sept. 15*	*19 Feb. 12*	*Malwa Bhil Corps.*
20 Jan. 00	2 Ransford, Maj. F. M. . . .	1 Sept. 15	16 Apl. 18	With 2-26th Punj.
19 May 00	1 Bredin, Maj. A. N.	1 Sept. 15	11 June 18	
	Company Officers. (18)			
30 Apl. 02	1 Jackson, Maj. D. R. H. . .	2 Mar. 16	23 Feb. 05	With 1-11th G. R.
30 Apl. 02	2 Rombulow-Pearse, Maj. A. B.	16 Apl. 16	17 Aug. 07	With 3-5th G. R.
22 Oct. 02	*1 Cornish, Maj. A. W. D.,* D.S.O., M.C.	*22 Oct. 17*	*27 Feb. 05*	*A. M. S. to G. O. C., N. Comd.*
3 Aug. 04	*2 Hackett, Capt. H. M. M.*, M.C.	*3 Aug. 13*	*1 Mar. 06*	D. A. A. G., 15th Div.
13 Aug. 04	*1 Abbott, Capt. F. B.,* D.S.O. *(Actg. Maj.)*	*13 Aug. 13*	*8 Nov. 05*	
28 Jan. 05	1 Ryan, Capt. D. G. J., D.S.O. .	28 Jan. 14	16 June 07	
27 July 05	1 Brown, Capt. E. C. . . .	27 July 14	2 Feb. 11	
24 Jan. 06	2 Blackett, Capt. C. P. (*Actg. Maj.*)	24 Jan. 15	9 Nov. 08	
17 Aug. 07	2 Bernard, Capt. R. P. St.V., M.C., D.S.O. [*I*]	1 Sept. 15	22 Nov. 08	With 8th G. R.
17 Aug. 07	1 King-Salter, Capt. N. H. . .	1 Sept. 15	7 Nov. 08	Adjt.
20 Jan. 09	2 Strahan, Capt. G. C. . .	1 Sept. 15	12 Mar. 11	Qr.-Mr., 24 Jan. 15.
29 Jan. 10	2 Barstow, Capt. T. C. E. . .	1 Sept. 15	4 May 11	Adjt.
3 Sep. 10	*1 Whittall, Capt. J. M. . .*	*1 Sept. 15*	*29 Dec. 11*	*G. S. O., 3rd gr.*
15 Aug. 14	1 Mercer, Lt. L. F.	1 Sept. 15	10 Nov. 15	With 2nd Bn.
1 Oct. 14	2 Smith, Capt. M. H. H. . .	1 Oct. 18	17 Dec. 14	With 1-50th Kumaon Rif.
14 Nov. 16	1 Collingridge, Lt. H. V. (*Actg. Capt.*)	19 May 16	22 Nov. 16	Adjt. (*Temp.*)
18 Apl. 16	2 Warhurst, Lt. A. E. . . .	18 Apl. 17	. .	Attd. With 1-2nd G. R.
14 Nov. 16	1 Castells, Lt. E.	14 Nov. 17	22 Nov. 16	
30 Jan. 17	1 Crove, Lt. G. R.	30 Jan. 18	4 Feb. 17	Attd.
. . .	Smith, 2nd-Lt. T. N.	. . . .	. . . .	Attd.
21 Dec. 17	1 Bowen, 2nd-Lt. W. O. (48*a*) .	21 Dec. 17	27 Apl. 18	Attd. With 2-131st Regt.
1 Oct. 18	2 Dobbin, 2nd-Lt. F. Le F. . .	1 Oct. 18	7 Oct. 18	Attd.
12 May 16	2 Crozier, Lt. W. H. . . .	12 May 17	15 Dec. 17	Attd.
25 Sep. 16	2 Voyle, Lt. S. G.	27 Sep. 17	11 Dec. 17	Attd.
2 Mar. 17	2 Robinson, Lt. H.	2 Mar. 18	13 Dec. 17	Attd.

6th Gurkha Rifles—*contd.*

1st Commn. or date of entering service.	Names and Rank.	Army rank.	Present appointment in Battalion.	Remarks.
	Attached.			
18 Jan. 02	2 Molloy, Maj. H. T.	18 Jan. 17	23 Mar. 18	
18 Dec. 14	2 Barton, Lt. R.M.S. (*Actg. Capt.*)	18 Dec. 15	9 Sep. 15	I. A. R. O.
13 Jan. 15	2 Bookey, Lt. J. T. B. (*Actg. Capt.*)	. . . .	6 Aug. 18	52nd Sikhs.
12 Feb. 15	1 Snodgrass, Lt. J.	12 Feb. 16	. . . .	I. A. R. O.
14 Jan. 15	1 Hearsey, Lt. N. N.	14 Jan. 16	1 May 15	I. A. R. O.
19 Mar. 15	2 Gill, Lt. G. T.	19 Mar. 16	6 July 15	I. A. R. O.
25 July 15	2 Lynes, Lt. R. H.	25 July 16	21 Aug. 15	I. A. R. O.
29 July 15	1 Hardman, Lt. P. L. . . .	29 July 16	9 Sep. 15	I. A. R. O.
5 Sep. 15	2 Faulkner, Lt. R., M.C. . . .	5 Sep. 16	4 Sep 18	R. W. Su r. R., I. A. (*on probn.*)
12 Oct. 15	2 Peattie, 2nd-Lt. G. D. . . .	12 Oct. 15	29 Nov. 17	I. A. R. O.
29 Oct. 15	1 Campbell, Lt. T. E. . .	29 Oct. 16	23 Dec. 15	I. A. R. O.
22 Nov. 16	1 Beaumont, Lt. H.	22 Nov. 17	2 Feb. 17	I. A. R. O.
8 Aug. 17	2 Thewlis, 2nd-Lt. N. E. . .	8 Aug. 17	3 Nov. 17	I. A. R. O.
26 Sep. 17	2 Hingston, Lt. F. P. R. . . .	26 Sep. 18	28 Jan. 18	I. A. R. O.
9 Oct. 17	2 Weir, Lt. A. D.	9 Oct. 18	11 Dec. 17	I. A. R. O.
8 Aug. 17	1 Rogers, 2nd-Lt. A. L. . . .	8 Aug. 17	3 Nov. 17	I. A. R. O.
27 Oct. 17	1 Ward, Lt. G. R.	27 Oct. 18	4 Nov. 17	
10 Oct. 14	1 Moffatt, Lt. J.	10 Oct. 15	10 Dec. 17	I. A. R. O.
27 Mar. 15	1 Clarabut, Lt. R. B. (*Actg. Capt.*)	27 Dec. 16	27 Apl. 18	R. W. Kent R. I. A. (*on probn.*)
6 Sep. 15	1 Wells, Lt. W. G. H.	6 Sep. 15	12 Dec. 17	A. S. C. I. A. (*on probn.*)
26 Apl. 16	1 Barker, Lt. H. A. E. . . .	26 Apl. 18	27 Jan. 18	I. A. R. O.
27 Mar. 17	1 Cook, 2nd-Lt. W. G. . . .	27 Mar. 18	27 Jan. 18	Rif. Brig. I. A. (*on probn.*)
9 Oct. 17	1 Simmonds, Lt. H.	9 Oct. 18	10 Dec. 17	I. A. R. O.
31 Jan. 18	1 Sinker, 2nd-Lt. R. C. . . .	31 Jan. 18	7 Feb. 18	
5 Nov. 17	1 Brooks, Lt. J. B.	5 Nov. 18	15 Mar. 18	I. A. R. O.
29 Mar. 18	2 Allen, 2nd-Lt. R. A. . . .	29 Mar. 18	15 July 18	I. A. R. O.
29 Mar. 18	2 Macleod, 2nd-Lt. I. N. . .	29 Mar. 18	22 July 18	I. A. R. O.
19 Feb. 18	2 Power, 2nd-Lt. C. C. . . .	19 Feb. 18	15 July 18	I. A. R. O.
20 Dec. 15	1 Wheeler, Lt. G. E. . . .	20 Dec. 16	4 Sep. 18	R. W. Surr. R. I. A. (*on probn.*)
. . . .	1 Smith, 2nd-Lt. T. A. . . .	. . . .	5 Sep. 18	I. A. (*on probn.*)
10 Feb. 16	1 Belcher, Lt. G. W., V.C. . .	10 Feb. 16	2 Nov. 18	Lond. R. I. A. (*on probn.*)
7 Oct. 15	1 Harrison, Lt. F. S.	7 Oct. 15	3 Sep. 17	I. A. R. O.

		Jemadar.	Subadar.	Subadar-Major.	
	Subadar-Majors.				
16 Nov. 91	2 Manbahadur Thapa (56) .	5 Nov. 04	12 Jan. 08	12 Jan. 11	
20 Dec. 93	1 Gambir Sing Pun, M.C., I.O.M. (66*a*)	11 Jan. 06	19 Apl. 09	19 May 15	

	Subadars.			
1 July 96	1 Deodat Thapa (58)	1 Mar. 06	1 May 10	
13 Nov. 99	2 Tula Ram Rana. . . .	13 Jan. 09	12 Jan. 11	
20 Jan. 97	2 Harku Thapa (56)	1 Mar. 09	1 Jan. 13	
28 Dec. 99	2 Santabir Rana (56)	1 Oct. 09	2 Apl. 15	
31 Aug. 97	2 Santbir Thapa (58) . . .	1 Dec. 09	19 Apl. 15	
7 Dec. 03	1 Kulbahadur Thapa, I.O.M. (56)	1 May 11	23 May 15	
12 Jan. 00	1 Satalsing Thapa, I.O.M. . .	1 Oct 12	19 May 15	
2 Jan. 99	2 Sukumbar Thapa (56) . . .	1 Jan. 10	19 June 15	
24 Jan. 05	1 Balsing Thapa, I.O.M. . . .	16 Oct. 14	21 July 15	
21 Oct. 02	1 Nandlal Gurung, I.O.M. . .	16 Oct. 04	1 Sep. 16	
14 Jan. 86	1 Jitman Gurung (*Retd.*)	. . . .	. . . .	Re-employed, 28 Apl. 15.
6 Dec. 97	1 Mandar Sing Thapa . . .	1 Dec. 13	1 May 16	
9 Dec. 04	1 Dalbahadur Thapa, I.O.M. . .	16 Oct. 14	1 Oct. 16	
13 Jan. 05	1 Kharak Bahadur Rana . .	16 Oct. 14	1 Oct. 16	1—11th G. R.
13 Apl. 95	2 Bharatsing Bhandari . . .	1 Jan. 13	1 Jan. 17	
23 Jan. 00	1 Nandbir Rana	11 May 15	6 Feb. 17	1—11th G. R.
15 Mar. 00	2 Dhanbir Gurung	19 Apl. 17	13 May 17	
28 Nov. 04	2 Seoraj Pun	28 Dec. 14	18 May 18	
1 Dec. 01	2 Brishan Bir Thapa . . .	19 Apl. 15	2 Sep. 18	
26 Feb. 08	1 Khark Singh Gurung . . .	8 July 16	14 July 18	

6th Gurkha Rifles—*concld.*

1st Commn. or date of entering service.	Names and Rank.	Army rank.	Remarks.
	Jemadars.		
28 Oct. 96	1 Meharman Thapa (66) .	1 Nov. 06	Convalst. Sec. D. Dun.
11 Mar. 99	1 Ranbahadur Gurung . . .	1 May 13	
7 Nov. 99	2 Amar Bahadur Gurung (56) (48a)	2 Apl. 15	
16 Nov. 04	1 Dhanraj Thapa, I.D.S.M. . .	23 May 15	1–11th G. R.
8 Mar. 00	1 Dhanniram Thapa . . .	5 June 15	
18 Feb. 99	2 Gamparsad Thapa	16 June 15	
8 Dec. 99	2 Ranjan Rana	19 June 15	
9 June 95	2 Hari Sing Gurung	4 July 15	Qr. Mr. Jem.
19 Dec. 04	1 Chandra Sing Gurung . . .	20 July 15	
14 Mar. 00	2 Jasbir Thapa	16 Aug. 15	Gurkha Adjt.
4 Nov. 98	2 Pahalman Gurung	17 Nov. 15	
5 Nov. 00	2 Gajbir Rana	17 Nov. 15	
8 Oct. 04	2 Bag Sing Gurung, I.D.S.M. .	3 Mar. 16	
2 Sep. 00	1 Puran Sing Gharti, I.D.S.M. .	8 July 16	
4 Mar. 05	1 Mahesur Gurung	8 July 16	1—11th G. R.
19 Dec. 04	1 Lalbir Thapa	8 July 16	
6 Dec. 02	1 Nanda Sing Gurung, I.D.S.M. .	8 July 16	
28 July 06	1 Indrabir Gurung	8 July 16	
21 Dec. 04	1 Gambir Sing Thapa	8 July 16	Gurkha Adjt.
5 Feb. 05	1 Dhanbahadur Gurung . . .	8 July 16	
6 Nov. 04	1 Sherbahadur Thapa	8 July 16	
20 Nov. 04	1 Rattan Thapa	8 July 16	
. . . .	1 Hirarup Sahi, I.S.O. . . .	22 Aug. 15	*Temp.* Assam Mily. Police.
28 Jan. 97	1 Bahadur Gharte	17 Dec. 15	
18 Dec. 06	1 Kharkabir Pun	1 Oct. 16	
15 Jan. 05	1 Dalsur Rana	1 Oct. 16	
11 Nov. 04	2 Jitbahadur Rana (56) . . .	1 Mar. 17	
17 Nov. 00	1 Rewantia Thapa (56) . . .	1 Mar. 17	
18 Jan. 00	1 Bala Sing Gurung (56) .	1 June 17	
20 Feb. 05	2 Railbahadur Gurung (48, 51) .	1 June 17	Gurkha Adjt.
18 Jan. 06	2 Jahar Sing Gurung . . .	1 June 17	
16 Mar. 08	1 Dallu Ale	1 Sep. 17	
4 Apl. 99	1 Ahmad Nur	1 Dec. 17	Qr.-Mr. Jemdr.
19 Jan. 05	2 Gobiraj Gurung	18 May 18	
24 Nov. 03	2 Manbir Gurung	18 May 18	Qr. Mr. Jem.
9 Oct. 07	2 Hast Lal Roka	24 May 18	
22 Nov. 04	2 Goria Thapa	2 Sep. 18	
22 Dec. 01	2 Narbhu Thapa	1 Oct. 18	
	Kumsing Gurung . . .	14 July 18	

3rd Battalion, 6th Gurkha Rifles.

Formed at Rawalpindi in February 1917.

Composition—2 Companies of Magars and Gurungs, 1 Coy. of Khas and 1 Coy. of Limbus and Rais.

1st Commn, or date of entering service.	Names and Rank.	Army rank.	Present appointment in Regiment.	Remarks.
	Commandant.			
5 Aug. 96	Jewett, Maj. A. E. (*Actg. Lt.-Col.*)	5 Aug. 14	28 May 18	1-8th G. R.
	Company Commanders. (4)			
8 Jan. 01	1 Massy, Maj. F. S.	8 Jan. 17	27 Jan. 17	2-4th G. R. 2nd-in-Comd.
8 May. 15	Bruce, Lt. J. G. (*actg. Capt.*)	8 May 16	13 Aug. 17	I. A. Attd.
28 Aug. 15	Jones, Lt. J. K.	28 Aug. 16	30 Aug. 17	I. A. Attd.
27 Jan. 16	Qualtrough, Lt. E. F. (*actg. Capt.*)	27 Jan. 17	16 Feb. 18	I. A. (*on probn.*)
	Company Officers.			
10 June 15	Sheperd, Lt. A. V. W. (*actg Capt.*)	10 Mar. 17	14 June 17	Attd. Qr.-Mr.
14 June 15	Shuker, Lt. H.	14 Mar. 17	10 Dec. 17	I. A. (*on probn.*)
20 Feb. 16	Littledale, Lt. A. G.	20 Feb. 17	22 Jan. 17	I. A. R. O.
7 Apl. 16	Kevill-Davies, Lt. E. F. W. H.	7 Apl. 16	1 Dec. 17	Attd.
10 May 16	Gilman, Lt. H. C. R. (*actg. Capt*)	10 May 17	29 Jan. 17	I. A. R. O. Adjt.
5 May 17	Reiner, Lt. N. A.	5 May 18	19 Aug. 17	I. A. R. O.
5 May 17	McMillan, Lt. T. B.	5 May 18	3 Oct. 17	I. A. R. O.
27 June 17	Fuller, 2nd-Lt. G.	27 June 17	15 Feb. 18	I. A R. O.
10 Aug. 17	Emmott, 2nd.-Lt. R. St. B.	10 Aug. 17	22 Jan. 18	I. A. R. O.
15 Jan. 18	Whiffin, 2nd-Lt. F. D.	15 Jan. 18	2 June 18	I. A. R. O. Qr. Mr.
16 Oct. 15	Bett, Lt. F. J. H.	16 July 17	15 Apl. 18	I. A. (*on probn.*)
11 Aug. 15	Pulling, Lt. H. D., M.C. (*actg. Capt.*)	11 Aug. 16	30 Mar. 18	I. A. (*on probn.*)
1 July 16	Pulling, Lt. G. H.	1 July 17	20 July 18	I. A. (*on probn.*)
15 July 18	Jenkins, 2nd-Lt. W. A.	15 July 18	3 Dec. 18	I. A. R. O.

		Jemadar.	Subadar.	Subadar-Major.	
	Subadar-Major.				
16 Dec. 97	Bhadrabir Thapa	1 Jan. 12	6 Feb. 17	1 Aug. 17	(tempy.).
	Subadars.				
11 Nov. 93	Ranbir Thapa	14 Feb. 13	1 Jan. 16		
25 Feb. 96	Mahantbir Rana	1 Jan. 14	12 Aug. 15		
28 Dec. 99	Motilal Limbu (56)	8 May 14	1 Jan. 17		
3 Jan. 00	Pirthibir Bura	16 Aug. 15	1 Jan. 17		
15 Nov. 98	Bombahadur Gurung	17 Mar. 15	6 Feb. 17	Lv. 6 mos.	
15 Apl. 92	Jaman Sing Karki	11 Mar. 15	1 Aug. 17		
15 Dec. 02	Birbahadur Lama	11 Aug. 15	1 Sep. 17		
	Jemadars.				
1 Nov. 99	Gagan Bahadur Sahi	1 Jan. 13			
1 Dec. 00	Mahabir Thapa	22 July 16			
18 Jan. 05	Hastabir Thapa	6 Feb. 17			
12 Nov. 02	Khimu Rana	6 Feb. 17			
24 Dec. 04	Narbahadur Gurung	1 Aug. 17			
7 Dec. 04	Kesar Ale	1 Sep. 17			
18 Nov. 06	Prithilal Gurung (46)	15 Apl. 18			
14 Mar. 12	Parta Sing Gurung	15 Apl. 18			
29 Jan. 06	Manbahadur Mal	15 Apl. 18			
14 Dec. 08	Maite Bura	15 Apl. 18			
9 Jan. 02	Moti Lal Thapa	15 Apl. 18			
25 Oct. 09	Daljit Pun	15 Apl. 18			
10 Dec. 04	Bhujaram Sahi	16 Apl. 18			

Regt. Centre—Quetta.

Uniform—Dark Green
Facings—Black.

7th Gurkha Rifles.

1st Battalion . .

2nd Battalion . .

Raised at Thayetmyo in 1902, by Major E. Vansittart as the 8th Gurkha Rifles, became the 2nd Battalion, 10th Gurkha Rifles, 1903; present designation in 1907.

The 2nd Battalion was raised at Quetta in 1907, by Major N. G. Woodyatt.

COLONEL—Major-General N. G. Woodyatt, C.B.

1st Commn. or date of entering service.	Names and Rank.	Army rank.	Present appointment in Battalion.	Remarks.
	Commandants.			
14 Sep. 87	*1 Powell, Lt.-Col. W. B.,* C.M.G.	*14 Sep. 13*	*15 Apl. 16*	
10 Nov. 88	*2 Haldane, Lt.-Col. C. L.,* C.M.G., D.S.O. [*l*]	*10 Nov. 14*	*17 Oct. 16*	
	Company Commanders. (8)			
30 Jan. 89	*1 Lee, Lt.-Col. A. W. H.* (Q) .	*30 Jan. 15*	. . .	*2nd-in-Comd.—Burma Mily. Police.*
29 Oct. 90	2 McLeod, Lt.-Col. T. J., D.S.O. (Q)	29 Oct. 16	19 Nov. 07	*Lv., ex I.*
9 Sep. 93	2 Longhurst, Bt.-Lt.-Col. A. L., *p.s.c.* (Q) [L]	3 June 16	2 Apl. 11	
6 Mar. 95	*2 Arbuthnot, Maj. A. H.*	*6 Mar. 13*	*19 Apl. 13*	*With 3-2nd G. R.*
14 Aug. 95	*1 Vaughan, Bt.-Col. L. R.,* C.B., D.S.O., *p.s.c.* (*temp. Maj.-Genl.*)	*1 Jan. 17*	*30 Mar. 13*	
20 July 98	*1 Lakin, Maj. J. H. F., p.s.c.* (*temp. Lt.-Col.*)	*1 Sep. 15*	*25 Nov. 14*	*A. Q. M. G.*
27 July 98	1 Molony, Maj. A. D. . . .	*1 Sep. 15*	30 Mar. 13	
	Company Officers. (18)			
17 Jan. 00	1 Faris, Maj. J. G.	1 Sep. 15	27 Sep. 07	2nd-in-Comd. (*temp.*)
11 Aug. 00	*2 Johnston, Maj. W.,* M.C. . .	*1 Sep. 15*	*27 Sep. 07*	
8 Jan. 01	*1 Brett, Maj. F. L. S.* . . .	*8 Jan. 16*	*30 Nov. 07*	*With 3-9th G. R.*
29 Jan. 02	2 Colenso, Maj. J. E.	29 Jan. 17	27 Sep. 07	[*Quetta.*
19 Aug. 03	*2 Harcourt, Maj. E. S.,* M.C. [*L*]	*19 Aug. 18*	*11 Jan. 08*	*G. S. O., 2nd gr., Cadet College,*
10 Oct. 03	*2 Exham, Maj. H.*	*10 Oct. 18*	*27 Sep. 07*	*Staff Capt., 34th Bde.*
9 Jan. 04	1 Anderson, Capt. M. G [L] .	9 Jan. 13	25 Dec. 07	
20 Apl 04	*2 Wilson, Bt.-Maj. N. M.,* D.S.O.	*1 Jan. 17*	*6 Feb. 10*	
3 Dec. 04	2 Whitaker, Capt S. S., M. C.	3 Dec. 13	27 Sep. 07	
29 Aug. 06	1 Robson, Capt. H. W. C. . .	29 Aug. 15	8 Nov. 07	
9 Sep. 08	1 Henstock, Capt. R. P. . . .	1 Sep. 15	12 Nov. 09	With 1-39th Garh. Rif.
8 Sep. 09	*2 Channer, Capt. G. O. DeR.* .	*1 Sep. 15*	*20 Jan. 11*	
5 Sep. 11	*1 Mackay, Capt. A. S.,* M.C. . .	*5 Sep. 15*	*24 Nov. 13*	
20 Jan. 12	1 Edwards, Capt. N. H. (*Actg. Maj.*)	20 Jan. 16	12 Mar. 13	
24 Aug. 12	*1 Johanson, Capt. J. L.* . .	*24 Aug 16*	*2 Oct. 14*	*With Nepalese Contgt.*
14 Jan. 14	1 White, Capt. B. G.	14 Jan. 18	12 Oct. 14	With 3-7th G. R.
1 Oct. 14	1 Williams, Capt. W. R. B. .	1 Oct. 18	19 Dec. 14	With 2nd Bn.
11 Nov. 14	1 Ross, Capt. A. J. H.	11 Nov. 18	20 Feb. 15	With 2nd Bn.
29 June 16	1 Clemence, Lt. S.	14 Feb. 16	29 Jan. 16	Attd.
19 Apl. 15	*2 Boileau, Lt. W. L. S.* (*Actg.-Capt.*)	*19 Apl. 16*	. . . .	*Nepalese Contgt.*
16 June 15	1 Carruthers, Lt. H. St. J. (*Actg.-Capt.*)	16 June 16	31 Oct. 15	Attd. With 2-11th G. R.
22 Dec. 15	1 Mercer, Lt. A. V. A. . .	22 Dec. 16	29 Feb. 16	Attd.
27 Oct. 17	1 Rouse, Lt. R. P. . . .	27 Oct. 18	29 Oct. 17	Attd. With 1-131st R.
27 Oct. 17	2 Blackwood, Lt. C. V. R. (207*a*).	27 Oct. 18	7 Nov. 17	
21 Dec. 17	2 Burden, 2nd-Lt. G. N. . . .	21 Dec. 17	23 May 18	Attd. With 4-11th G. R.
27 Oct. 16	1 Hill, Lt. D. R. O. (*actg. Capt.*)	27 Oct. 16	28 July 17	Attd.
17 June 15	1 Taylor, Lt. G. D.	17 June 16	5 July 17	Attd.
11 Aug. 16	1 Caruth, Lt. A. C. (*Actg. Capt.*)	11 Aug. 17	12 Apl. 17	Attd. Qr. Mr. (*temp.*)
24th Apl. 18	Sadlier, 2nd-Lt. J. W. V. . .	24 Apl. 18	29 Oct. 18	Attd.
	Attached.			
9 Sep. 08	2 Mockler, Capt. E. C.. . . .	1 Sep. 15	9 Oct. 10	2-1st G. R.
2 Jan. 09	2 Strahan, Capt. G. C. . . .	1 Sep. 15	. . . .	2-6th G. R.
27 Jan. 92	1 Cattell, Lt.-Col. G. L.. . .	27 Jan. 18	20 Apl, 17	Civil. Comdt. (*temp.*)
15 July 15	1 Williams, Lt. B. T.	. . . .	4 Aug. 17	S. Wales Bord.—I. A. R. O. (*on probn.*)

7th Gurkha Rifles—*contd.*

1st Commn. or date of entering service.	Names and Rank.	Army rank.	Present appointment in Battalion.	Remarks.
	Attached—*contd.*			
27 Nov. 14	*2 Bampton, Lt. R. W.*	*27 Nov. 15*	*28 Aug. 15*	*I. A. R. O.*
5 Feb. 15	*2 Clark, Lt. A. M.*	*5 Feb. 16*	*17 Apl. 15*	*I. A. R. O.*
30 Apl. 15	1 Lattey, Lt. H. P. T.	30 Apl. 16	1 Sep. 15	I. A. R. O.
24 Dec. 14	1 Crace, Lt. J. H.	24 Dec. 15	22 Aug. 16	I. A. R. O.
6 Oct. 16	1 Bailey, Lt. W. A.	6 Oct. 17	21 Nov. 16	I. A. R. O.
14 Jan. 15	2 Mills, Lt. A. W. F.	14 Jan. 16	14 Jan. 15	I. A. R. O.
1 Jan. 16	2 Lentaigne, 2nd-Lt. J. I. N.	1 Jan. 16	17 Aug. 16	I. A. R. O.
14 Aug. 16	2 Parry, Lt. N. E.	14 Aug. 17	26 Feb. 17	I. A. R. O.
5 Jan. 17	1 Crosby-Browne, Lt. D.	5 Jan. 18	19 Mar. 17	I. A. R. O.
27 Nov. 14	2 Fitzpatrick, Lt. G. V.	27 Nov. 15	....	I. A. R. O.
7 May 15	2 Walton, Lt. J. H.	7 May 16	....	I. A. R. O.
29 July 15	2 Pain, Lt. H. W., M.C.	29 July 16	....	I. A. R. O.
20 Nov. 14	2 Holdsworth, Lt. P. B. C.	20 Nov. 15	18 Jan. 18	I. A. R. O.
8 Aug. 17	2 Hurley, 2nd Lt. F. S.	8 Aug. 17	31 Oct. 17	I. A. R. O.
6 May 15	2 Abigail, Lt. R. P.	1 July 17	23 May 18	I. A. (*on probn.*)
1 Dec. 17	2 Burke, Lt. A. O. L.	1 Dec. 18	23 May 18	I. A. (*on probn.*)
1 Nov. 17	2 Newman, 2nd-Lt. G.	1 Nov. 17	17 Mar. 18	I. A. (T. C.)
6 June 15	1 MacDonald, Lt. P. S.	6 June 15	29 Jan. 18	R. N. B.—I. A. R. O. (*on probn.*)
12 Mar. 16	2 Macaulay, Lt. A. D. C.	12 Mar. 17	24 Oct. 17	I. A. R. O.
1 Mar. 17	1 Cowling, 2nd-Lt. C. S.	1 Mar. 17	2 Aug. 18	E. Lan. R.—I. A. (*on probn.*)
25 Apl. 15	1 Cockram, Lt. L. H.	...	18 Oct. 18	L'pool R. I. A. (*on probn.*)
26 Apl. 17	1 Hector, Lt. S. A.	...	27 Nov. 18	I. A. (*on probn.*)
1 Dec. 17	Cowling, Lt. C. S. C.	1 Dec. 18	...	

Subadar-Majors.

		Jemadar.	Subadar.	Subadar-Major.	Remarks.
10 Apl. 86	2 Kishandhoj Rai, *Sardar Bahadur*.	1 Nov. 99	10 June 05	27 Sep. 07	*Order of Br. Ind., 1st Cl.*
15 Sep. 93	*1 Narpati Rai*, I.D.S.M. (56)	*1 July 07*	*1 Nov. 07*	*7 June 16*	*With 2-11th G. R.*
14 Dec. 94	1 Hasturam Gurung (56)	27 Sep. 07	1 Dec. 11	25 May 18	

Subadars.

9 Jan. 88	2 Harkbir Ale (56)	16 May 02	29 Sep. 07	
31 Mar. 94	2 Kubir Sing Limbu, I.O.M., I.D.S.M. (56)	8 Nov. 00	1 Jan. 08	
9 Nov. 90	2 Balbahadur Limbu, *Bahadur* (56, 58, 72*a*).	1 Jan. 08	1 Oct. 10	*Order of Br. Ind., 2nd Cl.*
2 Nov. 90	1 Gopal Sing Rai (56)	1 Nov. 07	15 June 12	
24 Mar. 90	*1 Rajpal Rai, Bahadur* (56).	*8 Apl. 09*	*20 Dec. 12*	*Order of Br. Ind., 2nd Cl.*
15 Sep. 92	1 Agam Limbu (56, 66)	1 Mar. 10	11 May 14	
4 Apl. 96	2 Birihang Limbu (56)	13 Mar. 08	2 Oct. 14	
16 Dec. 91	1 Padamdhoj Gurung (56, 58)	1 Oct. 10	1 June 15	
25 Jan. 91	2 Birdhoj Limbu (56, 58)	1 Jan. 09	24 July 15	
9 Dec. 93	2 Sahabir Thapa, I.O.M. (56, 66, 72*a*)	1 Aug. 09	5 July 15	
20 Nov. 94	2 Rahardan Rai (56, 66)	28 Sep. 10	24 July 15	
28 Oct. 92	2 Gangabir Rana, M.V.O. (56, 66*a*)	1 Oct. 10	24 July 15	
3 May 95	1 Makordhoj Rai (56)	1 Jan. 12	16 Sep. 15	
21 Nov. 98	1 Mandhoj Rai (56, 66)	1 May 14	7 Oct. 15	
25 Apl. 97	*1 Gajraj Rai* (56)	*5 Feb. 15*	*16 Nov. 15*	*With 2-7th G. R.*
5 Nov. 97	1 Jitbahadur Rai (56, 58)	20 Dec. 12	7 June 16	With 2-7th G. R.
28 Apl 99	2 Dhaul Bir Rai (56)	12 Oct. 14	7 June 16	With 4-11th G. R.
29 Dec 96	1 Judhabir Thapa (56)	1 June 15	12 June 16	
1 Mar 00	1 Sing Bahadur Limbu (56)	5 Feb. 15	22 Sep. 16	With 2-7th G. R.
...	2 Mahadal Limbu	....	11 Nov. 16	Re-employed.
9 Nov. 95	2 Sasiram Thapa (66)	24 July 15	18 Jan. 17	
22 Dec. 97	1 Lahang Limbu	5 Apl. 15	18 Feb. 17	With 2-7 G. R.
...	2 Naine Lama, *Bahadur*	24 July 15	26 Feb. 17	*Order of Br. Ind., 2nd Cl.*
24 Nov. 99	1 Chehar Sing Thapa (56)	16 Nov. 15	10 Nov. 17	
...	1 Sher Bahadur Khattri (*Retd.*)	....	....	Re-employed, 4 Sep. 17.
...	1 Dhansur Rai	....	....	Re-employed, 18 Jan. 18.
...	1 Bhagiman Limbu	....	....	Re-employed, 18 June 18.
20 Dec. 02	1 Harkabahadur Limbu	27 Dec. 15	31 Dec. 17	
7 July 00	1 Bhogbir Limbu (66)	12 June 16	25 May 18	
17 Mar. 03	1 Kharka Sing Limbu (66)	1 July 16	25 May 18	
....	2 Manbir Limbu, I.O.M.	....	....	Re-employed.
14 Dec. 02	1 Surjiman Rai	19 Oct. 16	2 Sep. 18	
...	2 Dal Sing Basnet	26 Nov. 16	7 July 18	

7th Gurkha Rifles—*concld.*

1st Commn. or date of entering service.	Names and Rank.	Army rank.	Remarks.
	Jemadars.		
7 Jan. 95	2 Shyamlal Khattri (56, 66*a*, 72*a*)	12 Oct. 14	
29 Oct. 95	2 Karandhoj Ghalle	23 Mar. 15	
4 Oct. 99	*1 Lachman Limbu (56)* . . .	*17 Apl. 15*	
15 Jan. 96	2 Hastabir Limbu (66) . . .	5 July 15	
...	2 Budhbir Thapa	24 July 15	
7 Nov. 94	1 Lalbir Thapa	23 Feb. 16	With 2-7th G. R.
19 Nov. 97	1 Gristiram Rai	7 June 16	With 2-7th G. R.
25 Aug. 93	1 Purandhoj Limbu (72*a*) . .	12 June 16	
15 Jan. 03	1 Narbir Limbu (56)	12 July 16	
...	2 Narjit Gurung	22 Sep. 16	
1 Mar. 01	1 Jaspatti Limbu	14 Jan. 17	With 2-7th G. R.
...	2 Ranser Limbu	18 Jan. 17	
1 Feb. 97	1 Jasbir Lama	26 Feb. 17	With 2-7th G. R.
30 Jan. 03	1 Hastabir Limbu	10 Apl. 17	With 4-11th G. R.
3 Apl. 03	1 Padamlal Limbu	11 June 17	Gurkha Adjt. (*offg.*)
4 Jan. 98	1 Nar Sing Rai	10 Oct. 17	
1 June 93	1 Rajdhoj Rai	10 Oct. 17	
1 June 03	1 Gurul Sing Gurung	10 Oct. 17	
...	2 Hazar Sing Rai (*Retd.*) . .	. . . .	Re-employed.
...	2 Gokul Rai (*Retd.*) . . .	. . .	Re-employed.
...	2 Sirgange Rai (*Retd.*) . . .	. . .	Re-emloyed.
14 Dec. 02	1 Janaklal Lama	17 Mar. 17	With 2-7th G. R.
18 Jan. 97	2 Raghbir Limbu	11 Sep. 17	
3 May 99	2 Karbir Limbi	12 Aug. 17	With 4-11th G. R.
27 Dec. 03	1 Rupdal Rai	31 Dec. 17	
20 Jan. 03	1 Ja bahadur Limbu	22 Feb. 18	
1 Jan. 03	1 Jogbir Limbu	6 May 18	
1 Jan. 03	1 Lal Bahadur Rai	25 May 18	
23 Jan. 03	1 Hastibir Limbu	25 May 18	
4 Mar. 03	1 Ambardhoj Limbu	25 May 18	
25 May 03	1 Harkabahadur Limbu . . .	25 May 18	
...	2 Harkadhoj Thapa	24 May 18	
...	2 Mandhoj Lama	24 May 18	
3 Jan 06	1 Thaman Sing Rai	2 Sep. 18	
11 Apl. 08	1 Jusbahadur Limbu	2 Sep. 18	
...	Thamu Limbu	7 July 18	

3rd Battalion, 7th Gurkha Rifles.

Formed at Quetta on 11 June 1917.

Composition—4 companies of Gurkhas.

1st Commn. or date of entering service.	Names and Rank.	Army rank.	Present appointment in Regiment.	Remarks.
8 Oct. 90	Sealy, Lt.-Col. A. E.	24 May 16	5 Aug. 17	
	Company Commanders. (4)			
19 Jan. 07	Welman, Capt. W. H. L. J. (*Actg. Maj.*)	1 Sep. 15	9 July 17	2-3rd G. R. 2nd-in-Comd (*Offg.*)
14 Jan. 14	White, Capt. B. G. (*Actg. Capt.*)	14 Jan. 18	11 Feb. 18	1-7th G. R.
12 July 15	Burch, Lt. F. W. (*Acting Capt.*) M. C.	12 July 16	4 Oct. 17	Attd.
6 Jan. 15	Delafons, Lt. R. A.	6 Oct. 16	3 Aug. 17	Sch. of Musky., Satara.
	Company Officers.			
23 June 15	Greer, Lt. R. D. (207*a*)	23 June 16	3 Mar. 18	R. Ir. Fus., I. A. (*on probn.*)
19 July 15	Taylor, Lt. G. L.	19 July 16	...	L'pool. R. I. A. (*on probn.*)
19 Apl. 15	Young, Lt. J. W.	19 Jan. 17	15 Dec. 17	Midd'x. I. A. (*on probn.*)
21 Mar. 16	Hartnoll, Lt. F. B. (*Actg. Capt.*)	3 Aug. 17	18 June 17	I. A. R. O. Adjt.
8 Oct 15	Fielder, Lt. G. H.	1 July 17	3 Mar. 18	E. Yorks., R. I. A. (*on probn.*)
17 Aug. 16	O'Neil Ready, Lt. C. W.	17 Aug. 17	7 Aug. 17	R. F. C. I. A. R. O. (*on probn.*)
20 June 17	Shelverton, Lt. H. M	20 June 18	7 Oct. 17	I. A. R. O.
5 Oct. 17	Bethell, Lt. C. V.	5 Oct. 17	23 Jan. 18	I. A. (*T. C.*)
5 Oct. 17	Hodder, Lt. E. M.	5 Oct. 17	23 Jan. 18	I. A. (*T. C.*)
31 Oct. 17	Hickson, 2nd-Lt. J. C. A.	31 Oct. 17	17 Feb. 18	I. A. R. O.
27 June 17	Westrupp, 2nd-Lt. P.	27 Mar. 18	15 Dec. 17	Lond. R. I. A. (*on probn.*)
1 Oct. 18	Alban, 2nd-Lt. R. G. E. W.	1 Oct. 18	11 Oct. 18	
...	Clery, 2nd-Lt. J. A.	...	...	I. A. R. O.
1 May 17	Brown, 2nd-Lt. G. R.	1 May 17	13 Dec. 17	With 4-11th G. R.

		Jemadar.	Subadar.	Subadar-Major.	
	Subadar-Major.				
16 Jan. 95	Man Sing Thapa (56)	15 Jan. 11	1 June 15	Tempy. 6 July 17	
	Subadars.				
28 Aug. 99	Kulman Rai (56, 66)	15 Apl. 15	14 Jan. 17		2-7th G. R.
9 Mar. 94	Birbahadur Limbu (56)	12 Oct. 14	11 June 17		
25 Jan. 00	Sarbadhoj Rai	14 July 15	11 June 17		
11 Jan. 99	Tilokbahadur Limbu (56)	7 Oct. 15	11 June 17		
25 Dec. 02	Birbahadur Limbu	11 June 17	12 June 17		
3 May 00	Bombahadur Limbu	11 June 17	12 June 17		
4 Dec. 13	Kalu Sing Gurung (56)	11 June 17	12 June 17		
20 Apl. 05	Rapot Ram Rabha (56)	11 June 17	12 June 17		
	Jemadars.				
6 Jan. 01	Narjit Gurung	22 Sep. 16	2-7th G. R.		
16 Nov. 06	Megraj Limbu	11 June 17			
16 July 05	Bhaktabir Limbu (56)	12 June 17			
18 Sep. 07	Puranbahadur Sunwar (61)	12 June 17			
12 Nov. 07	Bahadur Gurung (46)	12 June 17			
18 May 08	Dambardhoj Limbu (56)	12 June 17			
8 Dec. 07	Gunjaman Rai	12 June 17			
24 Nov. 07	Digbahadur Rai (56)	3 Dec. 17			
1 Jan. 18	Mahaprasad Rai	1 Jan. 18			
19 Feb. 03	Ranlal Rai	2 Jan. 18			
9 Nov. 04	Dalsir Rai	2 Jan. 18			
17 Oct. 07	Malhang Limbu (61)	1 Mar. 18			
19 Sep. 02	Bhagesur Gurung (56)	1 Apl. 18			

Regt. Centre—Shillong.

Uniform—Dark Green.
Facings—Black.

8th Gurkha Rifles.

1st Battalion .
2nd Battalion .

Raised at Sylhet in 1824, by Capt. P. Dudgeon, as the 16th or Sylhet Local Battalion. Became the 11th or Sylhet Local (Light) Infantry Battalion, 1826; the 48th Regiment of Bengal Native Infantry, 1861; the 44th Regt. of Ben. N. I., 1861; the 44th (Sylhet) Regt. of Ben. Nat. (Light) Infy., 1864; the 44th (Sylhet) Regt. of Ben. (Light) Infy., 1885; the 44th Regt. Gúrkha (Light) Infy., 1886; the 44th (Gurkha) Regt. of Ben. (Light) Infy., 1889; the 44th Gúrkha (Rifle) Regt. of Ben. Infy., 1891; the 44th Gúrkha Rifles, 1901; 8th Gurkha Rifles in 1903; present designation, 1907.

The 2nd Battalion was formed at Gauhati in 1835, by Capt. W. Simonds, from *Sebundy* details already existing, and was originally styled the Assam Sebundy Corps. Became the Lower Assam Sebundy Corps, 1839; the 1st Assam Sebundy Corps, 1839; the 2nd Assam Light Infantry, 1844; the 47th Regt. of Bengal Native Infantry, 1861; the 43rd Regt. of Ben. N. I., 1861; the 43rd (Assam) Regt. of Ben. Nat. (Light) Infy., 1864; the 43rd (Assam) Regt. of Ben. (Lt.) Infy., 1885; the 43rd Regt., Gúrkha Light Infantry, 1886; the 43rd (Gúrkha) Regt. of Ben. (Lt.) Infy., 1889; the 43rd Gúrkha (Rifle) Regt. of Ben. Infy., 1891; the 43rd Gúrkha Rifles, 1901; the 7th Gurkha Rifles, 1903; present designation, 1907.

"Burma, 1885-87."

COLONEL—General *Sir* Horace M. Evans, K.C.B. 13 May 1904.

1st Commn. or date of entering service.	Names and Rank.	Army rank.	Present appointment in Battalion.	Remarks.
	Commandants.			
6 July 89	1 Roosmale-Cocq, Lt.-Col. C. A.	6 July 15	2 Sep. 17	
20 Feb. 92	2 Wall, Lt.-Col. A. C. (Q) . .	20 Feb. 18	29 July 18	
	Company Commanders. (8)			
23 Oct. 89	2 Cassels, Lt.-Col. K. S. (Q) .	23 Oct. 15	31 Mar. 14	2nd-in-Comd. With 2-3rd G. R.
13 Aug. 92	*1 Stansfeld, Maj. C. G., p.s.c.(Q) (Actg. Lt.-Col.)*	*13 Aug. 10*	*29 July 18*	*2nd-in-Comd. With 3-39th Garh. Rif.*
22 Jan. 96	2 Barlow, Maj. C. P. (*Actg. Lt.-Col.*)	22 Jan. 14	12 Feb. 18	With 3-5th G. R.
5 Aug. 96	1 Jewett, Maj. A. E. (*Actg. Lt.-Col.*)	5 Aug. 14	6 Nov. 14	With 3-6th G. R.
22 Jan. 98	2 VC Grant, Maj. J. D., *p.s.c.* (*Actg. Lt.-Col.*)	1 Sep. 15	6 Nov. 14	With 3-11th G. R.
20 July 98	*1 Coleridge, Bt.-Lt.-Col. J. F. S. D., D.S.O., p.s.c. (temp. Br.-Genl.)*	*1 Jan. 17*	*18 Feb. 17*	
18 Apl. 00	2 Giffard, Maj. D. H. R. (*Actg. Lt.-Col.*)	1 Sep. 15	2 Sep. 17	With 3-8th G. R.
8 Jan. 01	*1 Molesworth, Maj. A. L. M.* .	*8 Jan. 16*	*29 July 18*	*With Nepalese Cont.*
	Company Officers. (18)			
8 Jan. 01	2 Skene, Maj. J. G.	8 Jan. 16	28 Aug. 03	*Lv., m.c., ex Ind.*
29 Jan. 02	1 Orchard, Capt. D. S. . . .	29 Jan. 11	14 Jan. 09	With 4-39th Garh. Rifles.
19 Aug. 03	2 Buckland, Maj. G. C. B., D.S.O., M.C.	19 Aug. 18	4 July 05	*G. S. O., 2nd-Gr., Quetta.*
9 Jan. 04	2 Ross, Capt. E. J., M.C. . . .	9 Jan. 13	25 Mar. 06	
25 Jan. 08	1 Harington, Capt. H. R. (*actg. Maj.*)	1 Sep. 15	18 Apl. 09	With 2nd Bn.
8 Sep. 09	1 Kennedy, Capt. A. M. . .	1 Sep. 15	13 Nov. 10	
29 Jan. 10	1 Meade, Capt. H. R. C. .	1 Sep. 15	2 Apl. 11	With 1-50th Kumaon Rifl.
3 Sep. 10	1 Abbott, Capt. J. F., M.C. . .	1 Sep. 15	20 Nov. 11	
3 Sep. 10	2 MacLean, Capt. L. L. . .	1 Sep. 15	31 Mar. 13	R. A. F.
14 Jan. 14	2 Gordon, Capt. H. S. . . .	14 Jan. 18	13 Oct. 14	Adjt., 9 Dec. 17.
14 Dec. 14	1 Beckett, Lt. R. S. B. . . .	16 Mar. 17	19 Mar. 15	R. A. F.
18 Apl. 16	2 Hewetson, Lt. J. J.	18 Apl. 17	26 Apl. 16	With 1st Bn.
18 Apl. 16	2 Skene, Lt. S. G. S.	18 Apl. 17	26 Apl. 16	With 1st Bn.
29 June 16	2 O'Brien, Lt. B. P. T., M.C. .	29 June 17	7 July 16	
29 June 16	2 Mannock, Lt. R. C., (*actg. Capt.*)	29 June 17	7 July 16	With 2-9th G. R.
14 Nov. 16	2 Beckett, Lt. A.	14 Nov. 17	20 Nov. 16	With 4-11th G. R.
14 Nov. 16	2 Campbell, Lt. C. M. St. A. .	14 Nov. 17	22 Nov. 16	With 1-2nd G. R.
14 Nov. 16	2 Berkeley, Lt. S. S. H. . .	14 Nov. 17	20 Nov. 16	2-6th G. R.
14 Nov. 16	2 Frith, Lt. J. W.	14 Nov. 17	20 Nov. 16	With 4-11th G. R.
18 June 17	2 Hailstone, Lt. G. R.	18 June 18	27 June 18	With 3-8th G. R.
27 Oct. 17	1 Moore, Lt. C. S.	27 Oct. 18	8 Nov. 17	
29 Jan. 18	2 Hardman, 2nd-Lt. E. F. C. .	29 Jan. 18	6 Feb. 18	

8th Gurkha Rifles—*contd.*

1st Commn. or date of entering service.	Names and Rank.	Army rank.	Present appointment in Battalion.	Remarks.
	Attached.			
1 Dec. 15	1 Nicholson, Lt. E. M. F. (*Actg. Capt.*)	6 May 16	20 Jan. 16	I. A. R. O.
15 Mar. 16	1 Waddingham, Lt. C. A. . .	15 Mar. 17	30 Apl. 16	I. A. R. O.
3 June 16	2 Baird, Lt. C. J.	3 June 16	8 Dec. 17	M. G. Corps. I. A. (*on probn.*)
10 Aug. 16	2 Knights, Lt. F.	10 Aug. 17	9 Dec. 17	M. G. Corps. I. A. (*on probn.*)
8 Aug. 17	1 Irving, 2nd-Lt. M.	8 Aug. 17	8 Aug. 17	I. A. R. O.
16 Sep. 14	2 Graham, Lt. M. D. . . .	3 Feb. 15	25 Jan. 18	I. A. R. O.
11 May 15	2 Allan, Lt. R. J., M.C. . . .	11 May 16	19 Jan. 18	Sco. Rif. I. A. (*on probn.*)
2 June 15	2 Dunn, Lt. D. T.	2 June 16	24 Dec. 17	I. A. R. O.
5 Feb. 16	2 Duncan, Lt. A. W. (*Actg. Capt.*)	5 Feb. 17	9 Dec. 17	N. Z. Force. I. A. (*on probn*)
20 June 15	2 Innes, Lt. J. H., M.C. (*Actg. Capt.*)	20 June 16	20 Jan. 18	Arg. & Suth'd Highrs. I. A. (*on probn.*)
23 June 15	2 Burrell, Lt. K.	23 Mar. 17	9 Dec. 17	Ches. R. I. A. (*on probn.*)
25 July 16	2 Averill, Lt. J. E.	25 July 18	12 Feb. 18	Ches. R. I. A. (*on probn.*)
24 May 17	2 Hartshorn, Lt. E. P., D.C.M. .	24 May 17	20 Jan. 18	I. A. (*on probn.*) *Lv., ex Ind., 6 mos.*
11 Oct. 17	2 Kingett, Lt. A. E.	11 Oct. 18	10 Dec. 17	I. A. R. O.
21 Dec. 15	2 Stevenson, Lt. T. J. . . .	1 July 17	30 Mar. 18	K. R. Rif. C. I. A. (*on probn.*)
6 Jan. 16	2 Pettigrew, Lt. F. C. . . .	6 July 17	17 May 18	Lond. R. I. A. (*on probn.*)
1 Oct. 17	2 Walton, Lt. G. W.	1 Oct. 17	23 Dec. 17	I. A. (*T. C.*)
30 Oct. 17	2 Finglas, 2nd-Lt. U. C. N. . .	30 Oct. 17	29 Mar. 18	I. A. R. O. Offg. Qr.-Mr.
31 Oct. 17	2 Stutter, Lt. J. E.	31 Oct. 18	18 Feb. 18	I. A. R. O.
21 Dec. 16	2 Stewart, Lt. C.	21 Dec. 18	12 Feb. 18	I. A. R. O.
8 Apl. 15	1 Rutledge, Lt. L. H. N. . .	8 Jan. 17	22 Jan. 18	R. Innis. Fus. I. A. (*on probn.*)
5 Sep. 16	1 Petts, Lt. C. E. (*Actg. Capt.*) .	5 Sep. 17	26 Jan. 18	Notts. & Derby. R. I. A. R. O. (*on probn.*) Depot Adjt.
26 Sep. 17	1 Murphy, 2nd-Lt. F. . . .	26 Sep. 17	27 Jan. 18	I. A. R. O. (*on probn.*)
5 Nov. 17	1 Virgo, Lt. F. R.	5 Nov. 18	14 Mar. 18	I. A. R. O.
1 Aug. 17	2 Huddleston, 2nd-Lt. J. B. . .	1 May 18	30 Mar. 18	Leic. R. I. A. (*on probn.*)
17 Mar. 15	1 Irwin, Lt. N. L. C., D.S.O. (Temp. Lt.)	17 Oct. 16	20 Aug. 18	R. W. Fus. I. A. (*on probn.*)
20 Dec. 17	1 Lund, 2nd-Lt. C. E.	20 Dec. 17	21 Mar. 18	I. A. (*on probn.*)
1 Oct. 18	1 Gardner, 2rd-Lt. J. H. D. . .	1 Oct. 18	8 Oct. 18	I. A. (*on probn.*)
9 Oct. 17	Oliver, Lt. E. S.	9 Oct. 18	8 Sep. 18	I. A. R. O.
1 Dec. 15	2 Bateman, Lt. N. G. . . .	1 Dec. 16	. . . .	I. A. R. O.

		Jemadar.	Subadar.	Subadar-Major.	
	Subadar-Majors.				
1 Feb. 88	1 Rukman Sahi *Bahadur* (56,72*a*)	16 July 02	1 Apl. 06	1 June 12	*Order of Br. I. 2nd, Cl.*
4 Nov. 99	2 Rimane Thapa	9 Aug. 14	8 July 15	1 Feb. 18	

		Jemadar.	Subadar.	Remarks.
	Subadars.			
15 Nov. 89	*1 Indrabir Thapa (66a)* . . .	*23 Dec. 09*	*1 July 12*	
20 Mar. 92	1 Ransur Rana (65, 66*a*) . .	22 Aug. 10	23 Dec. 12	
20 Oct. 89	2 Bagdal Rana (66)	9 Apl. 11	30 Oct. 14 17 Jan. 15	
2 Dec. 90	1 Bagbir Thapa (66)	13 Nov. 10	1 Apl. 15	
16 Jan. 94	1 Dulsing Thapa, M.C. . .	15 Aug. 11	15 May 15	
19 Nov. 94	1 Ranbir Ghaley (56, 72*a*) . .	1 July 12	2 June 15	
15 Dec. 93	1 Dhanbir Thapa (II) (66) .	1 June 12	24 July 15	
11 Dec. 02	2 Basante Sahi (52, 56) . .	10 May 15	26 Sep. 15	
20 Nov. 00	2 Abir Thapa	1 Nov. 14	25 Feb. 16	With 2-131st. R.
28 Nov. 99	2 Dhanbir Thapa (56) . . .	19 Apl. 15	26 Feb. 16	
26 Oct. 98	1 Tulbir Gurung	4 Apl. 15	24 Apl. 16	
1 Mar. 98	1 Narbahadur Gurung . . .	15 May 15	31 May 16	
16 June 97	1 Kharak Sing Thapa, I.O.S.M.	4 Dec. 15	29 June 16	
11 Nov. 02	1 Gomu Basnet	4 Dec. 15	30 June 16	
21 Dec. 99	1 Manbir Thapa	4 Dec. 15	30 June 16	
30 Nov. 01	1 Dila Sing Gurung	20 May 16	12 Aug. 16	
26 Dec. 02	2 Jasbir Gurung (48*a*). . . .	26 Sep. 15	21 Jan. 17	
18 Nov. 01	1 Damdar Thapa.	12 June 16	18 Feb. 17	
1 Apl. 09	1 Karkbahadur Rana . . .	12 Aug. 16	18 Feb. 17	
25 Jan. 06	1 Ramnath Kachari . . .	13 Nov. 16	18 Feb. 17	With 4-11th G. R.
	1 Balbir Thapa	14 Dec. 16	15 Mar. 17	
27 Nov. 05	2 Dhanbahadur Gurung (56)	12 July 15	2 June 17	With 2-3rd GR.
28 Aug. 05	2 Ram Sing Rana (56) . .	8 July 15	17 June 17	Instr. Musky. Sch.
3 Sep. 02	2 Krishna Lal Chattri . . .	25 Sep. 15	1 Dec. 17	
12 Dec. 05	2 Laibir Pun (46)	26 Sep. 15		
7 June 04	Dalbahadur Thapa	1 Dec. 16	5 July 18	
13 Mar. 01	Rup Sing Thapa	18 Feb. 17	11 Aug. 18	
19 Jan. 03	Birban Gurung	18 Feb. 17	11 Aug. 18	

8th Gurkha Rifles—*concld.*

1st Commn. or date of entering service.	Names and Rank.	Army rank.	Remarks.
	Jemadars.		
16 Dec. 94	2 Mahadbir Thapa	24 Nov. 14	
10 Nov. 97	1 Parbir Thapa	4 Jan. 15	
11 Feb. 99	1 Hira Sing Thapa	10 May 15	
1 Mar. 98	1 Bahadur Thapa	2 June 15	
10 Nov. 97	2 Nawalbir Gurung	26 Feb. 16	
14 May 06	2 Chabi Lal Gurung (46, 58)	26 Feb. 16	
1 Aug. 10	2 Rikhi Ram Khattri	26 Feb. 16	
14 Jan. 06	2 Tekbahadur Pun (56)	26 Feb. 16	Qr. Mr. Jem.
3 Dec. 03	1 Maniraj Thapa (46, 48*a*)	30 June 16	
13 Nov. 03	2 Hastal Sing Gurung (56)	14 July 16	
8 Jan. 06	2 Balbhadra Thapa (56)	14 July 16	
8 Nov. 00	2 Tikaram Mal	15 Oct. 16	
26 Nov. 98	1 Kharakbahadur Pun	21 Jan. 17	With 4-11th G. R.
10 Nov. 99	1 Basurey Thapa	18 Feb. 17	
26 Oct. 03	1 Mandhoj Gurung	13 Feb. 17	With 4-11th G. R.
6 June 02	1 Ashu Gurung	15 Mar. 17	
4 Nov. 00	2 Narsing Thapa	6 Apl. 17	
9 Jan. 99	2 Jitman Pun	6 Apl. 17	
25 Oct. 05	2 Subhan Sing Gurung (56)	6 Apl. 17	
9 Nov. 02	2 Chandra Sing Gurung	6 Apl. 17	
10 Nov. 05	2 Narbir Pun	11 Apl. 17	
4 Jan. 99	1 Garbhey Pun	22 Apl. 17	
16 Nov. 06	1 Dalia Thapa	22 Apl. 17	
2 Mar. 02	1 Mandhoj Burathoki	7 May 17	
12 Jan. 99	2 Mani Raj Gurung	17 June 17	
14 Mar. 04	2 Bahadur Malla	18 June 17	Jem. Adjt.
6 Mar. 02	1 Budhibal Thapa M.C.	15 July 17	
23 Oct. 01	1 Lalbahadur Thapa,	2 Oct. 17	
3 Feb. 06	2 Kumraj Thapa (56)	1 Dec. 17	
22 Dec. 08	2 Dhansur Gurung (56)	1 Feb. 18	

3rd Battalion, 8th Gurkha Rifles.

Formed at Lansdowne on 11th July 1917.

Composition—4 companies Magars and Gurungs.

1st Commn. or date of entering service.	Names and Rank.	Army Rank.	Present appointment in Regiment.	Remarks.
	Commandant.			
18 Apl. 00	Giffard, Maj. D. H. R. (*Actg. Lt.-Col.*)	1 Sep. 15	31 May 18	1-8th Gurkhas.
	Company Commanders.			
21 Jan. 03	Hill, Maj. R. D. O.	21 Jan. 18	5 July 18	2nd G. R. 2nd-in-Comd.
4 Dec. 14	Field, Lt. G. L. (*Actg. Capt.*)	4 Dec. 15	13 Dec. 17	I. A. R. O.
12 June 15	O'Brien, Lt. E. C.	12 June 16	7 Dec. 17	Attd.
7 July 16	Stuart, Lt. T. C. (*Actg. Capt.*)	7 July 17	6 Dec. 17	Gord Highrs. I. A. R.O
	Company Officers.			
	Attached.			
10 Jan. 16	Webb, Lt. F. B. (*Actg. Capt.*)	10 Jan. 17	23 Feb. 18	Attd.
27 Oct. 17	Forsyth, Lt. I. M.	27 Oct. 18	7 Nov. 17	Attd.
29 June 16	Alsop, Lt. R. P. (*Actg. Capt.*)	29 June 17	3 June 18	Lond. R., I. A. (*on probn.*) Adjt.
7 July 16	Palmer, Lt. W. J.	7 July 17	7 June 18	I. A. (*on probn.*)
7 Apl. 17	Shearman, Lt. W. W.	7 July 17	28 Jan. 18	I. A. R. O.
15 Apl. 16	Jackson, Lt. L. E. W. (*Actg. Capt.*)	15 Apl. 17	20 July 18	Sea. Highrs. I. A. (*on probn.*)
20 June 17	Madge, Lt. K. A. (51)	20 June 18	5 Oct. 17	I. A. R. O.
29 Aug. 17	Daniel, 2nd-Lt. F. L.	29 Aug. 18	14 Dec. 17	I. A. R. O. (*on probn.*)
21 Nov. 17	Weir, 2nd-Lt. G. T.	21 Nov. 17	23 Nov. 18	North'd Fust,, I. A. (*on probn.*)
5 Nov. 17	Haynes, Lt. S. J.	5 Nov. 17	14 Mar. 18	I. A. (T. C.)
15 Jan. 18	Meredith, 2nd-Lt. D. F.	15 Jan. 18	12 June 18	I. A. R. O.
15 Jan. 18	Patton, 2nd-Lt. A. G.	15 Jan. 18	12 June 18	I. A. R. O. Qr. Mr.
15 July 17	Sale, 2nd-Lt. C. A.	15 Apl. 18	2 Nov. 18	R. Fus. I. A. (*on probn.*)
18 July 18	Lewis, 2nd-Lt. H. C.	18 July 18	2 Dec. 18	I. A. (T. C.)
		Jemadar.	Subadar.	Subadar-Major.
	Subadar-Major.			
2 Jan. 91	Makhan Sing Gurung, Bahadur	5 Dec. 08	1 June 12	6 Oct. 17 *Order of Br. Ind., 2 Cl.*
	Subadars.			
1 Nov. 87	Mewa Thapa, Bahadur, I.D.S.M.	5 Nov. 03	18 Mar. 07	Re-employed.
9 Feb. 06	Balbahadur Thapa	1 Dec. 16	17 June 17	
25 Nov. 98	Parbir Pun	2 Dec. 14	6 Oct. 17	
10 Dec. 98	Surbir Gurung (48a)	7 June 15	30 Oct. 17	
6 Nov. 99	Firthasing Thapa	30 May 16	1 Jan. 18	
3 Dec. 06	Parbir Rana	18 June 17	1 Jan. 18	
21 Feb. 97	Narjit Gurung	1 Jan. 18	12 Apl. 18	
16 Feb. 07	Kharakbahadur Thapa	1 Jan. 18	12 Apl. 18	
	Jemadars.			
1 Feb. 04	Karmadhoj Limbu	9 Aug. 15	....	
1 Aug. 01	Mahaser Limbu	10 Jan. 16	....	
20 June 07	Sher Sing Khattri (51)	1 Jan. 18	....	Gurkha Adjt., 1 Apl. 18.
18 Nov. 08	Abir Thapa, I.D.S.M. (61)	1 Jan. 18	....	
11 Nov. 08	Tirthe Pun (50)	1 Jan. 18	....	
23 Nov. 09	Bale Rana	1 Jan. 18	....	Qr. Mr. Jem.
26 May 03	Parte Gurung. (72a)	30 May 18	....	
7 Feb. 06	Purnabir Thapa (50)	12 Apl. 18	....	
26 Nov. 06	Dhanjit Pun (48a)	12 Apl. 18	....	
17 Nov. 07	Bhopal Pun	12 Apl. 18	....	
8 Jan. 09	Bhagidhoj Limbu	12 Apl. 18	....	
17 Oct. 10	Narbahadur Gurung (51, 61)	12 Apl. 18	....	
3 Jan. 10	Bhimbahadur Thapa (50, 61)	12 Apl. 18	....	

Regt. Centre—Dehra Dun. *Uniform*—Dark Green *Facings*—Black.

9th Gurkha Rifles.

1st Battalion . . . .
2nd Battalion . . . .

Raised at Fatehgarh in 1817, by Major C. S. Fagan, as the Fatehgarh Levy. Became the Mainpuri Levy, 1819, the 1st. Bn., 32nd Regiment of Bengal Native Infantry, 1823; the 63rd Regt. of Bengal N. I., 1824; the 9th Regiment of Bengal N. I., 1861; the 9th Regt. of Bengal Infy., 1885; the 9th (Gurkha Rifle) Regt. of Bengal Infy., 1894; 9th Gurkha Rifles, 1901; present designation, 1904.

The 2nd Battalion was raised at Dehra Dun by Major C. Y. Crommelin in 1904.

"Bhurtpore" "Sobraon" "Afghanistan, 1879-80" "Punjab Frontier."

COLONEL—

1st Commn. or date of entering service.	Names and Rank.	Army rank.	Present appointment in Battalion.	Remarks.
	Commandants.			
19 July 93	1 Bradshaw, Lt.-Col. C.R., *p.s.c.* (l)	22 June 18	22 June 18	With 2-9th G. R.
21 Sep. 89	2 Bateman-Champain, Lt.-Col. H. F., C.M.G. (*l*) (*temp. Br.-Genl.*)	21 Sep. 15	15 Mar. 17	
	Company Commanders. (8)			
2 Nov. 95	1 Bignell, Maj. B. H. (*Actg. Lt.-Col.*)	2 Nov. 13	21 May 18	2nd-in-Comd. Comdt. (*temp.*)
22 Jan. 96	*1 Marsh, Bt.-Lt.-Col. F. G.,* C.M.G., *p.s.c.* [L]	*3 June 16*	*30 Apl. 09*	*Mily. Attaché, Meshed.*
20 Feb. 97	*1 Duncan, Maj. H. C.,* D.S.O., *p.s.c.* (*temp. Lt.-Col.*) . .	*20 Feb. 15*	*7 Feb. 12*	*A. Q. M. G.*
22 Jan. 98	*2 Twiss, Bt.-Lt.-Col. W. L. O.,* M.C. *p.s.c.* [L] (*temp. Lt.-Col.*) .	*3 June 17*	*18 Nov. 12*	*G. S. O., 1st Gr., Army Hdqrs.*
4 May 98	2 Mackinnon, Maj. A. C. B.	1 Sept. 15	1 Apl. 13	With 3-9th G. R.
	Company Officers. (18)			
20 Jan. 00	2 VC Wheeler, Maj. G. C. . .	1 Sept. 15	15 Feb. 05	
5 Jan. 01	2 Collingridge, Maj. H. F. .	1 Sept. 15	14 Feb. 05	
4 Nov. 03	1 Mullaly, Maj. H. . . .	4 Nov. 18	19 Jan. 06	With 2-9th G. R.
18 Jan. 05	*1 Ostrehan, Capt. M.* . . .	*18 Jan. 14*	*20 Apl. 06*	*With Assam Mily. Police.*
22 June 05	2 Jukes, Bt.-Maj. A. H., D.S.O. (*Actg. Maj.*)	1 Jan. 17	18 Apl. 07	2nd-in-Comd.
29 Aug. 06	2 Blandy, Capt. R., M.C. . .	22 Aug. 15	12 Nov. 08	With 4-11th G. R.
25 Jan. 08	*1 Seton, Capt. H. W.* . . .	*1 Sept. 15*	*26 Apl. 09*	*Lv., ex I., 6 mos., m.c.*
20 Jan. 09	*2 Murray, Capt. R. G. H.,* M.C. .	*1 Sept. 15*	*19 Mar. 10*	*R. A. F.*
18 Sep. 09	2 Russell, Capt. R. W. . .	1 Sept. 15	19 Sep. 14	
19 Jan. 12	2 Gillan, Capt. G. V. B. . .	19 Jan. 16	7 Oct. 14	With Pol. Dept.
20 Jan. 12	1 Poynder, Capt. F. S. (*Actg. Maj.*)	20 Jan. 16	1 May 15	2nd-in-Comdt. (*temp.*)
24 Aug. 12	1 Kemp, Capt. T. N. C., M.C. .	24 Aug. 16	6 Nov. 13	With Nepalese Cont.
14 Jan. 14	2 Penfold, Capt H. DeL. . .	14 Jan. 18	13 Oct. 14	Adjt. (*Offg.*)
8 Aug. 14	2 Wilson, Capt. R. H. B. . .	8 Aug. 18	11 Mar. 15	Qr.-Mr. (*Offg.*)
11 Aug. 15	1 Pritchard, Lt. C. D. (*Actg. Capt.*)	11 Aug. 16	21 Nov. 16	Attd.
15 Aug. 15	1 Pryor, Lt. O. J.	15 Aug. 15	7 Nov. 17	Attd. With 2-9th G. R.
28 Nov. 14	1 Noad, Lt. C. K.	28 Aug. 15	2 Aug. 17	Attd.
2 Sep. 14	1 Crampton, Lt. G. P. (*Actg. Capt.*)	2 June 15	30 June 17	Attd.
18 Feb. 16	1 Pugh, Lt. L. B.	18 Feb. 17	15 June 17	Attd.
29 June 16	2 Fischer, Lt. T. G. T. . . .	29 June 17	. . . .	Attd. *Lv., ex I., 6 mos., m.c.*
29 June 16	2 Allsebrook, Lt. M. B. . . .	29 June 17	. . . .	Attd.
19 July 16	1 Wilsey, Lt. F. H. W. . . .	19 July 17	10 Nov. 16	Attd. With 1-8th G. R.
27 Oct. 16	1 Widdicombe, Lt. J.L.T. (207*a*)	27 Oct. 17	4 Jan. 17	Attd. With 2-9th G. R.
30 Jan. 17	1 Smith, Lt. E. T.	30 Jan. 18	3 Feb. 17	Attd. Qr.-Mr. (*Offg.*)
18 June 17	1 Hartland, Lt. B. S.	18 June 17	26 June 17	Attd. With 2-9th G. R.
18 June 17	1 Lovett, Lt. O. deT.	18 June 18	26 June 17	Attd. With 2-9th G. R.
27 Oct. 17	1 Smyth, Lt. H. M.	27 Oct. 18	7 Nov. 17	Attd.
31 Aug. 18	1 Roche, 2nd-Lt. I. A.	31 Aug. 18	6 Sep. 18	Attd.
21 Dec. 17	2 Bell, 2nd-Lt. J. V.	21 Dec. 17	21 Mar. 18	Attd.
21 Dec. 17	1 Connolly, 2nd-Lt. D. R. V. .	21 Dec. 17	23 May 18	Attd.
31 Aug. 18	2 Platts, 2nd-Lt. S. A.	31 Aug. 18	5 Sep. 18	Attd.
	Attached.			
19 Aug. 03	2 Beadon, Capt. V.	19 Aug. 12	. . . .	Civil, Burma.
31 Dec. 14	1 Osbourne, Lt. D. G. S. . .	11 Dec. 15	. . . .	I. A. R. O., A. R. O. for Gurkhas.
7 May 15	1 Graham, Lt. H. . . .	7 May 16	30 July 15	I. A. R. O. With Nepalese Cont.
21 May 15	2 Gladstone, Lt. S. D., M. C. (*Actg. Capt.*)	21 May 16	25 July 15	I. A. R. O.
25 July 15	2 Lawrence, Lt. N. S.	25 July 16	17 Aug. 15	I. A. R. O.
10 May 16	2 Russell, Lt. R. T., D.S.O. . .	10 May 17	10 July 16	I. A. R. O.
1 Nov. 14	2 Jones, Lt. R. S. (*Actg. Capt.*)	1 Aug. 16	24 Feb. 18	Bord. R., I. A. R. O. (*on probn.*) Adj.
14 Oct. 16	2 Scott, Lt. C. W. F.	14 Oct. 16	. . . .	I. A. With 3-11th G. R.
2 Sep. 15	2 Paterson, Lt. K.	2 June 17	8 Dec. 17	Suff. R., I. A. (*on probn.*)
1 Oct. 17	2 Chapman, Lt. E. J.	1 Oct. 17	21 Jan. 18	I. A. (*T. C.*)

9th Gurkha Rifles—*contd.*

1st Commn. or date of entering service.	Names and Rank.	Army rank.	Present appointment in Battalion.	Remarks.
	Attached—*contd.*			
5 May 17	2 Parry, Lt. J. G.	5 May 18	17 Aug. 17	I. A. R. O.
5 May 17	2 Woodman, Lt. W. F.	5 May 18	17 Aug. 17	I. A. R. O.
21 June 17	1 Parry-Jones, 2nd Lt. W. H.	27 Mar. 18	8 Mar. 18	R. Fus. I. A. (*on probn.*) Adjt.
6 Oct. 16	1 Iceton, Lt. E. A.	6 Oct. 17	31 Mar. 18	Aust. Infy. I. A. (*on probn.*)
23 Mar. 17	1 Davis, Lt. R., M.C.	23 Mar. 18	2 Mar. 18	Leic. R. I. A. (*on probn.*)
25 July 17	1 Martyr, Lt. H.	25 July 18	2 Mar. 18	E. Lan. R. I. A. (*on probn.*)
19 Dec. 17	Tungate, 2nd-Lt. W. C.	19 Dec. 18	31 Mar. 18	I. A. R. O. (*on probn.*)
10 Aug. 17	1 Clarke, 2nd-Lt. W. H.	3 Nov. 17	. . . .	I. A. R. O.
19 Apl. 15	2 Pickthall, Lt. C. M. (*Actg. Capt.*)	17 Apl 16	13 Mar. 17	I. A. R. O.
10 May 16	2 Berey, Lt C. E.	10 May 17	15 Apl. 17	I. A. R. O.
5 Oct. 17	1 Leitch, Lt. R. A.	5 Oct. 17	21 Jan. 18	I. A. R. O.
5 Oct. 17	1 Leach, Lt. A. G.	5 Oct. 18	28 Jan. 18	I. A. R. O.
5 Oct. 17	1 Brown, Lt. T. H. C.	5 Oct. 17	21 Jan. 18	I. A. (*T. C.*)
5 Oct. 17	1 Richmond, Lt. M. G.	5 Oct. 17	21 Jan. 18	I. A. (*T. C.*)
31 Oct. 17	1 Brand, 2nd-Lt. F. J.	31 Oct. 17	24 Jan. 18	I. A R. O. (*on probn.*)
12 May 16	1 Brown, Lt. A. F. W. (*Actg. Capt.*)	12 May 16	26 Jan. 18	Sco. Rif. I. A. (*on probn.*)
11 Feb. 16	2 Aron, Lt. D. J.	11 Feb. 17	1 May 18	Lond. R. I. A. (*on probn.*)
12 Aug. 15	2 Hurst, Lt. N., M.C.	12 May 17	29 Apl. 18	R. Dub. Fus. I. A. (*on probn.*)
12 Oct. 15	2 Fagan, Lt. H. A., M.C.	12 July 17	19 May 18	York R. I. A. (*on probn.*)
7 Apl. 15	2 Howell, Lt. H. R.	7 Apl. 17	18 May 18	Suff R. I. A. (*on probn.*)
26 Jan. 16	1 Fawcett, Lt. R.B., M.C. (*Actg. Capt.*)	26 Jan. 17	31 Mar. 18	R. Sco. Fus. I. A. (*on probn.*) Adjt. (*Offg.*)
29 June 16	2 Mannock, Lt. R. C. (*Actg. Capt.*)	29 June 17	7 July 16	
31 Oct. 17	Stringer, Lt. B. de. H.	31 Oct. 18	. . . .	Attd.
24 Apl. 18	Hime, 2nd-Lt. E. H.	24 Apl. 18	24 Oct. 18	I. A. (U. L.)
15 July 18	Hudson, 2nd-Lt. E.	15 July 18	1 Dec. 17	I A R. O.
17 July 18	Allen, 2nd-Lt. G. O.	17 July 18	1 Dec. 18	I. A. R. O.

Subadar-Majors.

		Jemadar.	Subadar.	Subadar-Major.	
1 June 90	2 Prem Sing Bisht, *Bahadur* (72*a*)	1 Jan. 00	1 Oct. 40	1 Feb. 13	*Order of Br. I., 1st cl.*
15 Nov. 95	1 Mehar Sing Khattri (56), I.O.M.	1 May 08	3 Nov. 14	9 Mar. 16	
	2 Bishan Sing Kathait, *Hony. Capt., Sardar Bahadur* (*Retd.*)	. . . .	. . . .	. . . .	Re-employed, 23 Apl. 15.

Subadars.

22 Oct. 92	1 Surbir Karki, *Bahadur* (58)	10 Dec. 04	22 Feb. 07	*Order of Br. I., 2nd cl.*
15 Feb. 86	2 Jhagarnand (56)	1 Nov. 04	1 Nov. 08	
1 Jan. 89	2 Jogichand (56, 66*a*)	10 Dec. 04	1 May 11	
1 Dec. 04	*2 Madho Sing Khattri (56, 66a)*	*10 Dec. 04*	*1 Feb. 13*	
6 May 87	1 Sher Sing Adhikari (66*a*)	10 Dec. 04	1 June 14	
19 Mar. 89	2 Puran Sing Khandka (56)	1 Feb. 08	1 July 14	
15 Apl. 90	2 Padam Sing Thapa, *Bahadur* (56)	1 Dec. 09	18 Sep. 14	*Order of Br. I., 2nd cl.*
8 Mar. 00	1 Shibdhoj Mal, I.O.M.	10 Aug. 14	12 Mar. 15	
1 Dec. 94	2 Rup Sing Khandka (66*a*)	1 May 11	1 May 15	
2 Feb. 96	1 Tikaram Majhi, *Bahadur*	22 Oct. 14	16 May 15	*Order of Br. I., 2nd cl.*
4 Jan. 90	2 Gumani Sahi (72*a*)	10 May 11	18 June 15	
25 Jan. 97	2 Ratan Sing Basnet (51)	1 Feb. 13	1 Aug. 15	
18 Feb. 96	1 Narbir Khattri	12 Mar. 15	14 Jan. 16	
4 Jan. 96	2 Lal Sing Basnet	31 May 15	27 Feb. 16	
18 Oct. 93	1 Bhimbahadur Bania	14 Jan. 16	2 Mar. 16	
11 Oct 95	1 Gobindram Khattri	10 Aug. 14	8 Mar. 16	
12 Nov. 99	1 Kalu Khattri	14 Jan. 16	8 Mar. 16	
. . . .	Damodhar Khattri, I.D.S.M.	3 Nov. 14	5 June 17	
5 Mar. 96	2 Kharrak Sing Khattri	15 Sep. 14	7 July 16	
25 Jan. 97	2 Ganesh Bahadur Khattri	18 Sep. 14	12 Sep. 16	
19 Nov. 95	2 Champa Sing Thapa	18 Sep. 14	22 Nov. 16	
16 Dec. 97	2 Naru Panre	18 Sep. 14	25 Feb. 17	
12 Dec. 94	2 Kesar Mal	1 Oct. 15	14 Mar. 17	
1 Mar. 97	2 Himbahadur Sahi, I.O.M.	18 June 15	24 Apl. 17	
4 Dec. 95	2 Wazirsing Mahtra	29 Mar. 15	. . . .	
	1 Haribhagat Sahi (*Retd.*)	. . . .	. . . .	Re-employed, 26 Mar. 15.
18 July 95	Bhagat Ram Khatri	1 Sep. 15	15 Jan. 18	

9th Gurkha Rifles—*concld.*

1st Commn. or date of entering service.		Names and Rank.	Army rank.	Remarks.
		Jemadars.		
15 Nov. 97	2	Himlal Karki	18 Sep. 14	
14 Jan. 99	2	*Bombahadur Khattri*	*1 May 15*	
	2	Durgadut Sahi	18 June 15	
6 Dec. 08	2	Panchbir Mal, I.O.M.	2 Mar. 16	
14 Nov. 64	1	Kulman Khattri, I.O.M., I.D.S.M.	17 Apl. 16	
4 Mar. 97	2	Lalu Khattri, I.O.M.	18 Apl. 16	
22 Mar. 01	2	Kishan Sing Bohra, I.O.M.	18 Apl. 16	
20 Nov. 00	1	Damarbahadur Sen	18 Apl. 16	
30 Nov. 04	1	Govind Adhikari	18 Apl. 16	Gurkha Adjt.
16 Mar. 03	2	Dhanbahadur Sen	18 Apl. 16	
2 Jan. 05	1	Dalbahadur Bhandari (56)	24 Apl. 16	
25 Dec. 96	2	Moti Persad Sahi	5 July 16	
25 Feb. 96	2	Lachman Khandkai	19 Oct. 16	
4 Mar. 99	2	Bahadur Thapa	19 Oct. 16	
13 Jan. 95	1	Arjun Khattri	1 Dec. 16	
1 Mar. 97	1	Chandrabir Khattri	1 Dec. 16	
18 Apl. 99	2	Harak Sing Khattri	25 Jan. 17	
14 Nov. 99	2	Balnar Sing Khattri	25 Jan. 17	
18 Nov. 00	1	Kharakbahadur Mal	9 Feb. 17	
12 Nov. 96	1	Jagbahadur Thapa	9 Feb. 17	
8 Jan. 95	2	Jaibahadur Mal	25 Feb. 17	
3 Mar. 98	2	Mandhoj Sen, I.O.M.	13 Mar. 17	
15 Dec. 99	1	Dilbahadur Khattri (56)	1 Apl. 17	With 3-3rd G. R.
7 Jan. 99	1	Dasrath Khattri (46)	1 Apl. 17	With 3-3rd G. R.
23 Feb. 04	1	Jitbahadur Khattri	1 Apl. 17	. . .
7 Dec. 03	1	Bishanbahadur Chand	5 June 17	
14 June 02	1	Baldeo Sahi	5 June 17	With 3-11th G. R.
15 Dec. 99	2	Bhotbahadur Pande	6 Dec. 17	
20 Dec. 99	2	Sassu Mal	20 Dec. 17	
6 Dec. 98	2	Kishenbahadur Khatri	15 Jan. 18	
18 Nov. 04	1	Gorakh Bahadur Mul	1 Mar. 18	
30 Nov. 04	1	Samar Bahadur Mul, I.D.S.M.	1 Mar. 18	
24 Nov. 04	1	Chandra Birkhatri	1 Mar. 18	
9 Feb. 00	2	Hari Sing Bhandari	10 Mar. 18	Qr. Mr. Jemdr.

3rd Battalion, 9th Gurkha Rifles.

Formed at Dehra Dun on 6 June 1917.

Composition— 4 companies of Gurkhas.

1st Commn. or date of entering service.	Names and Rank.	Army rank.	Present appointment in Regiment.	Remarks.
	Commandant.			
4 May 98	Mackinnon, Maj. A. C. B. (*Actg. Lt.-Col.*)	1 Sep. 15	12 June 17	2-9th G. R.
	Company Commanders. (4)			
8 Jan. 01	Brett, Maj. F. L. S.	8 Jan. 16	7 Sep. 18	1-7th G. R. 2nd-in-Comd. (*temp*).
14 Sep. 09	Bamber, Capt. C. C.	1 Sep. 15	24 Aug. 18	I. A. (*on probn.*)
1 Oct. 14	Walton, Capt. L. C. J. B.	1 Oct. 18	4 Feb. 18	2-4th G. R.
7 Sep. 15	Gribble, Lt. D. S. (*Actg. Capt.*).	7 Sep. 15	1 Feb. 18	I. A. R. O.
	Company Officers.			
5 June 15	King, Lt. R. M. E.	5 June 16	17 Apl. 18	Attd.
10 Sep. 14	Fawcett, Lt. W. L., M.C.	10 June 16	25 Feb. 18	Attd. Adjt., 1 Mar. 18.
27 July 15	Compston, Lt. P.	27 July 17	20 Oct. 17	I. A. (*on probn.*)
15 Mar. 16	Davies, Lt. V. K. N.	15 Mar. 17	28 Oct. 17	Attd.
2 Mar. 16	Dow, 2nd-Lt. D. J., M.C.	2 Mar. 17	21 Sep. 17	Sea. Highrs. I. A. (*on probn.*)
12 Aug. 15	Jones, Lt. K. C. J., M.C.	12 May 17	27 Nov. 17	Attd.
25 Oct. 16	Cartwright, Lt. T. H.	25 Oct. 17	12 Feb. 18	I. A. R. O.
5 May 17	Hollis, Lt. F. S.	5 May 18	22 Aug. 17	I. A. R. O., Qr.-Mr., 1 May 18.
27 June 17	Hartman, 2nd-Lt. V. J.	27 Mar. 18	12 Feb. 18	Dorset R. I. A. (*on probn.*)
15 Jan. 15	Meiklejohn, 2nd-Lt. W.	15 Jan. 18	7 June 18	I. A. R. O.
7 June 18	Pelly, 2nd-Lt. C. T.	7 June 18	12 Oct. 18	I. A. R. O.
22 July 18	White, 2nd-Lt. L. S.	22 July 18	28 Nov. 18	I. A. R. O.

		Jemadar.	Subadar.	Subadar-Major.
	Subadar-Major.			
15 Feb. 86	Jhagarnand	1 Nov. 04	1 Dec. 09	1 July 17
	Subadars.			
27 Dec. 90	Durga Dutt Sahi (58)	18 June 15	5 June 17	
16 Mar. 00	Tikaram Khattri, I.D.S.M.	20 June 16	5 June 17	
18 Jan. 97	Kirtiman Khattri	5 June 17	1 Oct. 17	
27 Mar. 03	Shiuprasad Thapa	5 June 17	1 Oct. 17	
1 Nov. 97	Wazir Sing Khattri, I.D.S.M.	5 June 17	1 Oct. 17	
13 May 01	Bhagwan Sing Thakur	5 June 17	1 Oct. 17	
14 Jan. 99	Bombahadur Khattri	1 May 17	1 Jan. 18	
	Jemadars.			
24 Oct. 04	Himlal Thapa (48*a*, 46)	5 June 17		
16 Nov. 02	Rambhagat Adhikari	5 June 17	Gurkha Adjt.	
90 Dec. 02	Kharak Bahadur Khattri	1 Oct. 17		
8 Mar. 05	Dalip Sing Karki	1 Oct. 17		
12 Mar. 05	Tikaram Kanwar, I.O.M. (46)	1 Oct. 17		
29 Nov. 04	Amar Sing Karki (48*a*)	1 Oct. 17		
2 Jan. 05	Kulbahadur Thapa	1 Oct. 17		
24 June 04	Balbir Thapa	3 Dec. 17		
9 Jan. 05	Lachman Sing Oli	1 Jan. 18		
24 Dec. 04	Kabiraj Khattri	1 July 18		
2 Jan. 05	Padamdhoj Mal (48*a*, 51)	1 July 18		
15 Mar. 10	Deoraj Thapa	1 July 18		

Regt. Centre 1st Battn.—Maymyo. *Uniform*—Dark Green.
Regt. Centre 2nd Battn.—Almora. *Facings*—Black.

10th Gurkha Rifles.

1st Battalion
2nd Battalion

Raised in 1890, by Lt.-Col. C. R. Macgregor, from the Kubo Valley Police Battalion, and was originally designated the 1st Regiment of Burma Infantry. Became the 10th Regt. (1st Burma Bn.) of Madras Infy., 1891, the 10th Regt. (1st Burma Rifles), Madras Infy., 1892; the 10th Regt. (1st Burma Gurkha Rifles), Madras Infy., 1895; the 10th Gurkha Rifles, 1901; present designation, 1903.

The 2nd Battalion was raised in 1908, by Lt.-Col. G. H. C. Colomb from a wing of the 1st Battalion.

1st Commn. or date of entering service.	Names and Rank.	Army rank.	Present appointment in Battalion.	Remarks.
	Commandants.			
8 Oct. 90	1 Coningham, Lt.-Col. F.E., D.S.O.	8 Oct. 16	3 Jan. 18	
28 June 93	2 Gale, Lt.-Col. E. S.	3 July 18	3 July 18	
	Company Commanders. (8)			
3 May 90	*1 Orman, Lt.-Col. F. L.*	*3 May 16*	*21 Mar. 04*	*Burma Mily. Police.*
16 Feb. 98	*2 Stewart, Maj. J. M. V.*	*1 Sep. 15*	*3 Jan. 09*	*R. O. for Gurkhas.*
6 Dec. 99	1 Bethell, Maj. L. A.	1 Sep. 15	3 Jan. 18	2nd-in-Comd.
14 Feb. 00	1 Simpson, Maj. J.	1 Sep. 15	10 Sep. 14	
18 Apl. 00	2 Gatherer, Maj. R. G. T.	1 Sep. 15	20 Jan. 15	2 11th G. R.
18 Aug. 00	*2 Weekes, Maj. H. E.*	*1 Sep. 15*	*1 Oct. 14*	*G. S. O., 3rd-gr., Ambala Bde.*
	Company Officers. (18)			
18 May 01	*2 Clarke, Maj. B. G. S. [l]*	*8 May 16*	*4 Dec. 08*	*D. A. Q.M. G., A. Hdqrs.*
8 May 01	1 Baker, Maj. W. B.	8 May 16	6 Nov. 02	With Nepalese Cont.
18 Jan. 02	1 Coningham, Maj. R.E., p.s.c.	1 Jan. 16	22 Nov. 08	
25 Feb. 03	*2 Seymour, Maj. M. H.*	*25 Feb. 18*	*25 Jan. 14*	*Staff Capt.*
16 Mar. 04	1 Williams, Capt. H. R. (*Actg. Maj.*)	16 Mar. 13	15 Mar. 09	2nd-in-Comd. (*temp.*)
24 Jan. 06	2 Merritt, Capt. B. T. B. (*Actg. Maj*)	24 Jan. 15	7 Sep. 09	2nd-in-Comd. (*temp.*)
29 Aug. 06	1 Crossfield, Capt. E. A. K.	29 Aug. 15	16 Mar. 09	
6 Oct. 06	*1 Mylne, Capt. R. H.*	*1 Sep. 15*	*4 Oct. 11*	*Staff Capt.*
17 Aug. 07	2 Davidson, Capt. H. B.	1 Sep. 15	20 Nov. 08	
9 Sep. 08	*2 Battye, Capt. T. H.*	*1 Sep. 15*	*11 June 14*	*With Nepalese Cont.*
14 Oct. 08	2 Bald, Capt. E. A.	1 Sep. 15	26 Oct. 14	With 3-11th G. R.
3 Sep. 10	2 Lemon, Capt. R. L.	1 Sep. 15	8 Jan. 16	Adjt., 10 Mar. 16.
6 Sep. 11	1 Mullaly, Capt. B. R.	6 Sep. 15	20 June 14	Adjt., 8 Feb. 17.
20 Jan. 12	1 Williams, Capt. J. H.	20 Jan. 16	21 Mar. 13	
22 Jan. 13	1 Ashmore, Capt. E. J. C., M.C.	22 Jan. 17	19 Mar. 14	With 2-3rd G. R.
15 Nov. 15	1 Laughton, Lt. J. W.	15 Nov. 16	27 Nov. 15	Attd.
29 June 16	1 Tucker, Lt. B. H. G. (*Actg. Capt.*) (48a)	29 Jan. 17	29 June 16	Attd.
26 Mar. 16	2 Loup, Lt. L. A.	26 Mar. 17	14 June 17	Attd.
7 Apl. 16	2 Corfield, Lt. V. C. W.	7 Apl. 17	15 Nov. 16	Attd. With 4-11th G. R.
18 Apl. 16	1 Hailstone, Lt. R.	18 Apl. 17	29 Apl. 16	Attd. With 4-11th G. R.
5 May 15	2 Gifford, Lt. F. R.	5 May 16	13 June 17	Attd.
29 June 16	1 Mannock, 2nd-Lt. W. A.	29 June 16	29 June 16	Attd.
29 June 16	1 Hervey, Lt. W. F	29 June 17	29 June 16	Attd.
29 June 16	1 Lindquist, Lt. W. H. H.	29 June 17	29 June 16	Attd. With 1-1st G. R.
6 July 16	2 Frost, Lt. C. C. J.	6 July 17	14 June 17	With Sig. Ser.
30 Jan. 17	2 Ennis, Lt. J. I.	30 Jan. 18	9 Feb. 17	Attd. With 1-10th G. R.
4 Oct. 17	2 Larymore, Lt. P. A. D.	4 Oct. 18	9 Feb. 17	Attd.
81 June 17	2 Carruthers, Lt. R. J.	18 June 18	5 July 17	Attd. Qr.-Mr.
18 June 17	2 Warran, Lt. W. S. E.	18 June 18	5 July 17	Attd. With No. 3 M. G. Corps.
27 Oct. 17	2 Craig, Lt. A. D.	27 Oct. 18	1 Nov. 17	Attd. With 2-11th G. R.
21 Dec. 17	1 Rivers, 2nd-Lt. W. E.	21 Dec. 17	27 Mar. 18	Attd.
21 Dec. 17	2 Christie, 2nd-Lt. G. H.	21 Dec. 17	27 Mar. 18	Attd.
31 Aug. 18	1 Alderson, 2nd-Lt. M. G.	31 Aug. 18	21 Sep. 18	Attd.
31 Aug. 18	2 Murray, 2nd-Lt. S. K.	31 Aug. 18	1 Sep. 18	Attd.
31 Jan. 18	2 Strump, 2nd-Lt. R.	31 Jan. 18	6 Feb. 18	Attd.
5 May 15	2 Davies, Lt. K. C. J.	5 May 16	6 Dec. 17	Attd.
	Attached.			
14 May 15	2 Hicks, Lt. L. J. (*a*)	14 May 16	31 Oct. 15	I. A. R. O. R. T. O., Ambala.
30 July 15	2 Dennehy, Lt. H. G.	30 July 16	15 Sep. 15	I. A. R. O.
22 Oct. 14	2 Day, Capt. E. V. G.	22 June 16	20 Jan. 18	Rif. Brig. I. A. (*on probn.*)
19 Apl. 15	2 Evans, Lt. J. L. (*Actg. Capt.*)	22 Aug. 18	19 Apl. 16	R. W. Fus. I. A. R. O. (*on probn.*)
7 Oct. 14	Thatcher, Lt. A. F. B.	7 July 16	19 Dec. 17	Som. L. I., I. A. (*on probn.*)
20 June 17	1 Nicolson, Lt. H. G.	20 June 18	1 Oct. 17	I. A. R. O.
20 June 17	1 Marlow, Lt. T.	20 June 18	10 Jan. 18	I. A. R. O.

10th Gurkha Rifles—*contd.*

1st Commn. or date of entering service.	Names and Rank.	Army rank.	Present appointment in Battalion.	Remarks.
	Attached—*contd.*			
19 Dec. 17	2 Walmisley, Lt. H. E.	19 Dec. 18	1 Apl. 18	I. A. R. O. (*on probn.*)
30 July 15	1 McGeorge, Lt. D. J. (*Actg. Capt.*)	30 July 16	15 Sep. 15	I. A. R. O. Qr. Mr.
28 Dec. 16	2 White, 2nd-Lt. F. J.	28 Dec. 17	15 Feb. 18	Lond R. I. A., (*on probn.*)
26 June 17	2 Henke, 2nd-Lt. F. E.	26 June 17	30 Jan. 18	R. Ir. Rif. I. A. R. O. (*on probn.*)
16 Nov. 15	2 MacIver, Lt. R. J.	16 Aug. 17	1 Apl. 18	R. Ir. Regt. I. A. (*on probn.*)
31 Oct. 16	1 McNaught, Lt. D.	31 Oct. 17	. . . .	I. A. R. O.

		Jemadar	Subadar.	Subadar-Major.	
	Subadar-Majors				
13 Apl. 86	2 Chittahang Limbu, I.O.M., *Sardar Bahadur* (66*a*).	1 Mar. 00	1 Mar. 06	23 Oct. 08	*Order of Br. I., 1st Cl.*
9 Mar. 87	1 Bhola Sing Kandari, *Bahadur*	27 Nov. 97	1 Dec. 00	8 May 14	*Order of Br. I., 2nd Cl.*
	Subadars.				
6 Dec. 91	1 Mandhoj Lama (56, 66*a*)	1 Oct. 03	16 Sep. 08		
21 Feb. 93	1 Chakrabadur Limbu (56, 66*a*)	12 Aug. 07	16 Sep. 08		
25 Oct. 92	1 Jitman Gurung (56)	1 June 08	16 Sep. 08		
28 Jan. 91	2 Chhetradhoj Limbu, *Bahadur* I.O.M. (56).	1 June 08	19 Sep. 08		*Order of Br. I., 2nd Cl.* With 1st Bn.
13 May 94	1 Asjit Rai (Retd.)		8 May 14		Re-employed, 1 Feb. 16. Gurkha Adjt.
21 Sep. 94	1 Singbir Rai (56, 66*a*)	16 Sep. 08	8 May 14		
13 Jan. 95	1 Harkabir Limbu (56, 66*a*)	16 Sep. 08	2 Feb. 15		
1 Nov. 93	2 Lakhdhoj Rai (56)	1 May 12	15 Apl. 15		
14 May 97	*1 Jitman Rai* (56, 66*a*)	*1 Mar. 11*	*21 June 15*		*With 2nd Bn.*
2 Feb. 98	*1 Lachhman Sunwar* (56)	*8 May 14*	*30 July 15*		
6 July 99	1 Bowraj Limbu (66*a*)	. . . .	7 Aug. 15		
1 Dec. 96	1 Surbir Limbu (56, 66*a*)	17 Sep. 11	8 Aug. 15		
8 Feb. 98	2 Bajirdhoj Rai, I.O.M. (66*a*)	1 July 13	7 Aug. 15		
24 Feb. 98	1 Santabir Rai (56) (66*a*)	8 May 14	7 Sep. 15		
2 Dec. 97	2 Sukman Rai (66*a*)	11 Sep. 14	21 Aug. 15		
24 Dec. 99	1 Phaudasing Limbu (66*a*)	5 Apl. 15	7 Sep. 15		With 3-11th G. R.
14 Nov. 02	1 Manbahadur Rai (56, 66*a*)	5 Apl. 15	7 Sep. 15		
11 Aug. 99	2 Ram Sing Burathoki, I.O.M.	29 June 15	9 Aug. 15		
4 Dec. 00	2 Lakhbahadur Limbu	10 Aug. 15	14 Sep. 15		
1 Dec. 00	1 Saharmukhi Limbu (56, 66*a*)	16 Sep. 14	3 May 17		With 3-11th G. R.
9 Jan. 02	1 Garbhe Limbu (66*a*)	13 Nov 17			
30 Apl. 04	1 Phoudasing Limbu (56)	12 Dec. 17	. . . .		With 3-11th G. R.
7 Jan. 04	1 Mangal Sing Limbu	30 July 15	26 May 18		
26 Jan. 03	1 Shiambahadur Limbu (66*a*)	26 July 18	26 July 18		
12 June 97	1 Karbir Rai (66*a*)	26 July 18	26 July 18		
27 Jan. 03	1 Bakhatbahadur Limbu	23 July 17	23 July 17		
7 Jan. 04	1 Mangalsing Limbu (58)	26 July 18	26 July 18		
7 Oct. 04	2 Mitralal Thapa	7 Aug. 15	15 Oct. 18		
28 Jan. 10	2 Sher Singh Rawat (56)	8 Aug. 15	14 July 18		

10th Gúrkha Rifles—*concld.*

1st Commn. or date of entering service.	Names and Rank.	Army rank.	Remarks.
	Jemadars.		
14 Nov. 02	1 Kalibahadur Rai (66*a*) (56)	7 Sep. 15	
23 Jan. 02	1 Sahaman Thapa (66*a*)	5 Oct. 15	
1 Aug. 08	1 Kajimn Lama (57)	5 Oct. 15	Gurkha Adjt.
19 Mar. 03	2 Chature Rai (56)	10 Jan. 16	
1 Feb. 98	2 Sanak Sing Limbu (56)	10 Jan. 16	
7 Dec. 05	2 Narbahadur Gurung	10 Jan. 16	
5 Aug. 99	2 Samardhoi Rai	10 Jan. 16	With 1st Bn.
27 Dec. 02	2 Baldhoj Limbu	21 Apl. 16	
23 Aug. 05	2 Goberdhan Gurun (56)	21 Apl. 16	Gurkha Adjt.
4 Dec. 02	1 Kajiman Limbu (56)	1 Mar. 17	
11 Dec. 05	1 Aitabahadur Limbu (56, 60*a*)	1 Mar. 17	
11 Nov. 97	2 Dhanman Rai (46)	1 Mar. 17	
10 Nov. 06	1 Singbaj Limbu	1 Apl. 17	With 3-11th G. R.
13 Nov. 05	1 Dawatsering Lama	31 Mar. 17	Depôt Qr. Mr. Jemdr.
28 Oct. 08	1 Jagatbahadur Rai	8 Apl. 17	
31 Jan. 00	1 Ujirdhoj Rai	3 May 17	
18 Mar. 03	2 Ujir Sing Limbu	1 July 17	
27 May 01	2 Harka Sing Thapa	1 July 17	
2 Feb. 06	2 Udahang Limbu	1 July 17	
8 Jan. 06	2 Bhagiman Limbu	1 July 17	
	2 Narpati Lama (*Ret.*)	. . . .	Re-employed, 4 Sep. 16.
29 Mar. 09	1 Dhanbahadur Limbu (46, 56)	1 Oct. 17	
17 Mar. 99	1 Gajbir Rana	13 Nov. 17	
19 Mar. 06	1 Bhawan Sing Chettri	11 Dec. 17	Qr. Mr. Jemdr.
6 Nov. 07	1 Lachman Limbu	1 Jan. 18	
20 Feb. 07	1 Birman Rai (56)	1 Mar. 18	
1 Jan. 18	1 Indradhoj Limbu	1 Jan. 18	
28 Feb. 03	1 Karandhoj Limbu	23 July 17	
12 Feb. 06	1 Sunambir Limbu	26 May 18	
4 Nov. 08	1 Jaidhan Rai	26 May 18	
1 Nov. 08	1 Jasbahadur Limbu	26 May 18	
28 Nov. 05	Narbahadur Limbu	24 July 18	
. . . .	Dangnath Limbu	7 Nov. 18	
. . . .	Sahabir Limbu	15 Oct. 18	

1st Battalion, 11th Gurkha Rifles.

Formed at Kut on 18 May 18.

Composition—4 Companies Gurkhas.

1st Commn. or date of entering service.	Names and Rank.	Army rank.	Present appointment in Regiment.	Remarks.
	Commandant.			
27 July 98	Johnson, Maj. A. E. (*Actg. Lt.-Col.*)	1 Sep. 15	. . . .	2-1st G. R.
	Company Commanders. (4)			
30 Apl. 02	Jackson, Maj. D. R H . . .	19 Mar. 16	3 Nov. 18	1-6th G. R.
	Company Officers.			
19 Mar. 15	Ransford, Lt. W. R.	19 Mar 16	18 May 18	I. A. R. O.
5 May 15	Templer, Lt. H. E.	5 May 16	18 Sep. 18	I. A. (*on probn.*)
1 June 15	Hartnoll, Lt. F. H.	1 June 16	7 Sep. 18	I. A. R. O. (*on probn.*)
25 July 15	Leir, Lt. J. R. C. V.	25 July 16	18 May 18	I. A. R. O.
8 Sep. 15	Swan, Lt. J. C. K.	8 Sep. 16	18 May 18	I. A. R. O. Qr.-Mr.
17 Jan. 16	Gouldsbary, Lt. C. P. . . .	27 Jan. 17	18 May 18	I. A. R. O. Adj. (Offg.)
14 Aug. 17	MacVay, 2nd-Lt. S.	14 Aug. 17	14 Sep. 18	Lancs. Fus. I. A. (*on probn.*)
7 July 17	Harder, Lt. J. R.	7 July 17	18 May 18	I. A. R. O.
18 June 17	Clarke-Bell, 2nd-Lt. D.M. . .	18 June 17	6 Sep. 18	
5 Oct. 17	Fitzpatrick, 2nd-Lt. H. J. . .	5 Oct. 17	15 Sep. 18	I. A. R. O.
27 Oct. 17	Gibbs, Lt R. K.	27 Oct. 18	9 Sep. 18	
18 Dec. 17	May, 2nd-Lt.	10 Dec. 17	11 Sep. 18	I. A. R. O.
26 Jan. 18	Kidd, 2nd-Lt. R. J.	26 Jan. 18	10 Sep. 18	
7 June 18	Baily, 2nd-Lt. J.	7 June 18	8 Oct. 18	I. A.
31 Aug. 18	Hotton, 2nd-Lt. H. O. . . .	31 Aug. 18	. . . .	Attd.
1 Sep. 18	Donnelly, 2nd-Lt. H. R. . . .	1 Sep. 18	13 Sep. 18	Attd.
		Jemadars.	Subadars.	Subadar-Majors.
	Subadar-Major.			
11 Nov. 02	Dhansing Gurung, I.D.S.M. (46, 56)	16 Jan. 18	18 May 18	I. O. M. (2nd cl.)
	Subadars.			
1 June 00	Haricharan Thapa (58) . . .	12 Jan. 11	18 May 18	
13 Jan. 05	Kharak Bahadur Rana, I.D.S.M. .	16 Oct. 11	18 May 18	
23 Jan. 00	Nandbir Rana	11 May 15	18 May 18	
16 Nov. 04	Dhanraj Thapa, I.D.S.M. . . .	23 May 15	18 May 18	
4 Mar. 03	Mahesur Gurung	8 July 16	18 May 18	
20 Jan. 05	Parasabir Rana.	12 Aug. 16	18 May 18	
6 Jan. 00	Narbir Pun	16 Dec. 15	16 Oct. 18	
	Jemadars.			
31 Oct. 00	Bhowanand Thapa	17 Apl. 15		
20 Jan. 99	Narpatti Bura	17 Apl. 15		
4 Jan. 98	Asbahadur Gurung	19 Apl. 15		
15 Mar. 99	Rabidhoj Rana (56)	15 Jan 16		
24 Nov. 04	Goria Pun (46)	12 Aug. 16		
1 June 00	Sham Sher Thapa	1 Jan. 17	Qr.-Mr. Jemdr. (*temp.*)	
15 Mar. 03	Ramparsad Gurung	1 June 17		

2nd Battalion, 11th Gurkha Rifles.

Formed at Baghdad, 24th May 1918.

Composition—4 Companies Gurkhas.

1st Commn. or date of entering service.	Names and Rank.	Army rank.	Present appointment in Regiment.	Remarks.
	Commandant.			
18 Apl. 00	Gatherer, Maj. R. G. T. (*actg. Lt. Col.*)	1 Sep. 15	7 Jan. 18	2-10th G. R.
	Company Commanders. (4)			
10 Oct. 03	Baldwin, Maj. M. C., M.C. . .	10 Oct. 18	7 Sep. 18	2nd-in-Comd.
5 Mar. 15	Jerram, Lt. M. R. K. (*actg. Capt.*) M.C.	5 Mar. 16	24 May 18	I. A. R. O.
16 June 15	Carruthers, Lt. H. St. J. (*actg. Capt.*)	16 June 16	24 May 18	1-7th G. R.
1 Oct. 15	Duncan, Lt. D. L. (*actg. Capt.*) .	1 Oct. 16	24 May 18	2-2nd G. R. Qr.-Mr.
	Company Officers.			
19 Jan 15	Burgan, Lt. N. (*actg. Capt.*) M.C.	19 Jan. 16	24 May 18	Adjt. 1-4th G. R.
10 Mar. 16	Leawood, Lt. R. R.	10 Mar. 17	12 Sep. 18	I. A. (*on probn.*)
19 July 16	Molesworth, Lt. R. C. . . .	19 July 17	24 May 18	
5 Sep. 16	Ballance, 2nd-Lt. S. H. . . .	5 June 17	3 Sep. 18	
27 Oct. 17	Craig, Lt. A. D.	27 Oct. 18	7 Sep. 18	
27 Oct. 17	Weatherdon, Lt. H. E. . . .	27 Oct. 18	3 Sep. 18	
31 Aug. 18	McGregor, 2nd-Lt. W. D. . .	31 Aug. 18	13 Sep. 18	Attd.
31 Aug. 18	Foskett, 2nd-Lt R. M. . . .	31 Aug. 18	13 Sep. 18	Attd.
31 Aug. 18	Scott, 2nd-Lt. J. G.	31 Aug. 18	10 Sep. 18	Attd.
	Attached			
1 Oct. 15	Poingdestre, Lt. C. Le G. . . .	1 Oct. 16	17 July 18	I. A. R. O.
29 Oct. 15	Davies, Lt. A. T.	29 Oct. 16	24 May 18	I. A. R. O. *Lv., ex. I., m. c., 6 m.*
29 Aug. 16	Richards, Lt. J. C.	29 Aug. 17	24 Nov. 18	Gen. List. I. A. (*on probn.*)
5 Aug. 16	Millar, Lt. W. B.	5 Feb. 18	1 Dec. 18	T. F. Attd. I. A.
5 May 17	Gray, Lt. F. W.	5 May 18	6 Sep. 18	I. A. R. O.
5 May 17	Armstrong, Lt. W. L. . . .	5 May 18	8 Aug. 18	I. A. R. O.
28 July 17	Tyndall, 2nd-Lt. C. T. A. . .	28 July 17	10 Sep. 18	I. A. R. O. (*on probn.*)
5 Oct. 17	Laurie, 2nd-Lt. G.	5 Oct. 17	. . . .	I. A. R. O.
15 June 18	Collins, 2nd-Lt. F. J. . . .	15 June 18	24 Nov. 18	Gen. List. I. A. (*on probn.*)
26 Mar. 18	Gunn, 2nd-Lt. C. F.	26 Mar. 18	1 Dec. 18	T. F. Attd. I. A.
26 Mar. 18	Inglis, 2nd-Lt. A.	26 Mar. 18	20 Nov. 18	T. F. Attd. I. A.
28 May 18	Bird, 2nd-Lt. H.	28 May 18	1 Dec. 18	T. F. Attd. I. A.
28 May 18	Cliyde, 2nd-Lt. J. C.	28 May 18	30 Nov. 18	T. F. Attd. I. A.
15 July 18	Cooper, 2nd-Lt. H. L. . . .	15 July 18	5 Dec. 18	I. A. R. O.

		Jemadars.	Subadars.	Subadar-Majors.
	Subadar-Major.			
15 Sep. 93	Narpatti Rai, I.D.S.M. . . .	1 July 07	1 Nov. 07	7 June 16
	Subadars.			
13 Dec. 95	Sirilal Thapa	16 Dec. 13	20 Dec. 14	
11 Dec. 03	Autarsing Gurung	23 Oct. 14	3 Nov. 15	
6 June 99	Jangraj Limbu	1 June 15	1 July 16	
13 May 99	Narbahadur Rai, I. O. M. . .	7 Aug. 15	10 Jan. 17	
9 Nov. 96	Deobahadur Gurung	23 Apl. 15	3 Feb. 17	
9 Nov. 00	Thamansing Thapa	10 June 15	12 Feb. 17	
	Jemadars.			
5 Dec. 97	Sukbir Limbu	6 Oct. 15		
16 Feb. 98	Santabir Bura	9 Mar. 16		
11 Dec. 00	Sirikishn Thakur	1 June 16	Adjt. Jem.	
1 Dec. 04	Parmandhoj Thapa.	1 Mar. 17		
16 Nov. 98	Rudirom Roka	14 Apl. 17		
15 Jan. 97	Bhanbir Gharti	1 July 17		
8 Dec. 98	Bhirban Gurung	20 Nov. 17	Qr. Mr. Jem., 1 Nov. 18.	
. . . .	Udesing Gurung	1 Dec. 17		
12 Sep. 02	Darme Rai	17 Apl. 18		
	Tilok Sing Limbu	16 July 18		

3rd Battalion, 11th Gurkha Rifles.

Formed at Baghdad 25 May 18.

Composition—2 companies Gurkhas, 2 companies Garhwalis.

1st Commn. or date of entering service.	Name and Rank.	Army rank.	Present appointment in Regiment.	Remarks.
	Commandant.			
22 Jan. 98	VC Grant, Maj. J. D. (*Actg. Lt.-Col.*) *p.s.c.*	1 Sep. 15	4 June 18	2-8th G. R.
	Company Commanders.			
14 Oct. 08	Bald, Capt. F. A.	1 Sep. 15	25 May 18	1-10th G. R.
10 Sep. 15	*Columbine, Lt. R. H.*	*10 Sep. 16*	*6 July 18*	*I. A. (on probn.) Adjt.*
	Company Officers.			
13 Sep. 15	Shier, Lt. M. B.	13 Sep. 16	4 Nov. 18	I. A. (*on probn.*)
14 Oct. 15	Scott, Lt. C. W. F.	14 Oct. 16	25 May 18	2-9th G. R.
7 May 16	Briggs, Lt. R. A.	7 May 17	14 Aug. 18	Notts & Derby. R., I. A. (*on probn.*)
22 Oct. 16	Jones, Lt. R. A.	22 Oct. 17	3 Sep. 18	I. A. (*on probn.*)
10 Nov. 16	Hunter, Lt. H. C. A.	10 Nov. 17	25 May 18	I. A. R. O.—Qr. Mr.
5 May 17	Parry, Lt. J. G.	5 May 18	25 May 18	I. A. R. O.
8 May 17	Connell, Lt. H. W.	8 May 18	25 May 18	I. A. R. O.
7 Aug. 17	Beazeley, 2nd-Lt. G. A.	7 Aug. 17	. . . .	Attd.
5 Oct. 17	Ross, Lt. D. A.	5 Oct. 18	3 Sep. 18	I. A. R. O.
7 Nov. 17	Grant, 2nd-Lt. A. O. (207*a*)	7 Nov. 17	7 Sep. 18	I. A. R. O.
	Shephard, Lt. G. C.	. . . .	. . . .	
28 Dec. 14	Livingstone, Lt. P. (*acting Capt.*)	28 Dec. 14	23 Nov. 18	
31 July 18	Wilson, 2nd-Lt. W.	31 July 18	23 Nov. 18	
29 Aug. 18	Morton, 2nd-Lt. E. T.	28 Aug. 18	23 Nov. 18	R. Highrs. T. F. attd. I. A.

		Jemadar.	Subadar.	Subadar-Major.
	Subadar-Major.			
24 Dec. 99	Phaudasing Limbu (66*a*)	20 Aug. 18	25 May 18	
	Subadars.			
6 July 99	Bowraj Limbu (66*a*)	7 Aug. 15	30 Oct. 18	
18 Mar. 03	Rajnande Limbu	5 Apl. 15	18 Sep. 18	
1 Dec. 00	Saharmukhi Limbu (56, 66*a*)	3 May 17	25 May 18	
30 Apl. 04	Phaudasing Limbu (II) (56)	18 Dec. 17	25 May 18	
18 July 95	Bagat Ram Khattri	15 Jan. 18	25 May 18	
	Jemadars.			
11 Mar. 01	Bahadur Rai	7 Aug. 15	. . . .	
14 Dec. 02	Tule Thapa	21 Apl. 16	. . . .	
15 Jan. 99	Amarsing Bohra	11 July 16	25 May 18	
16 Jan. 99	Hastabir Limbu (46)	1 Mar. 17	. . . .	
10 Nov. 06	Singbaj Limbu	1 Apl. 17	25 May 18	
14 June 02	Baldeo Sahi	5 June 17	. . . .	
24 Feb. 06	Asbahadur Limbu (56)	1 Oct. 17	. . . .	
20 Dec. 99	Sasumal	20 Dec. 17	25 May 18	

4th Battalion, 11th Gurkha Rifles.

Formed at

Composition—4 companies Gurkhas.

1st Commn. or date of entering service.	Name and Rank.	Army rank.	Present appointment in Regiment.	Remarks.
	Commandant.			
14 Aug. 95	Dundas, Maj. W. L. (*Actg. Lt.-Col.*)	14 Aug. 13	. . . .	2-3rd G. R.
	Company Commanders.			
29 Aug. 06	Blandy, Capt. R., M.C., (*Actg. Maj.*)	29 Aug. 15	. . . .	2-9th G. R.
19 Jan. 07	Arnott, Capt. A. M. (*Actg. Maj.*)	1 Sep. 15	. . . .	2-4th G. R.
	Company Officers.			
29 Oct. 14	Williamson, Lt. F. (*Actg. Capt.*)	5 Sep. 16	15 June 18	I. A. R. O.
18 Dec. 14	Johnson, Lt. E. P.	18 Dec. 15	24 May 18	I. A. R. O.
14 Jan. 15	Chester, Lt. A. J. B., M.C.	14 Jan. 16	24 May 18	
12 Feb. 15	Snodgrass, Lt. J.	12 Feb. 16	2 July 18	I. A. R. O. Adjt.
28 Mar. 17	Berkeley, Lt. G. F. (*Actg. Capt.*)	28 Mar. 18	16 July 18	I. A. R. O.
18 Apl. 16	Hailstone, Lt. R. (*Actg. Capt.*)	18 Apl. 17	11 July 18	1-10th G. R.
25 Jan. 15	Hurst, Lt. R. H.	25 Oct. 16	30 July 18	1st G. B., Linc. R., I. A. (*on probn.*)
10 Mar. 15	Tewson, Lt. R. G.	1 Sep. 16	10 Sep. 18	R. W. Fus., I. A. (*on probn.*)
14 Aug. 16	White, Lt. O. J.	14 Aug. 16	12 June 18	I. A. (*on probn.*) Qr. Mr.
7 Apl. 16	Corfield, Lt. V. C. W.	7 Apl. 17	24 May 18	2-10th G. R.
22 Nov. 16	West, Lt. E.	22 Aug. 18	13 Aug. 18	3-7th G. R.
4 Mar. 16	Patrick, Lt. P. J.	4 Mar. 17	17 July 18	I. A. R. O.
1 May 17	Brown, Lt. G. R.	1 May 18	1 July 18	3-7th G. R.
21 Dec. 17	Burden, 2nd-Lt. G. N.	21 Dec. 17	9 Sep. 18	2-7th G. R.
26 Mar. 18	Morris, 2nd-Lt. J. J.	26 Mar. 18	5 Oct. 18	I. A. R. O. (*on probn.*)
7 June 18	Williams, 2nd-Lt. S. J.	7 June 18	6 Oct. 18	I. A. R. O.
31 Aug. 18	Hutton, 2nd-Lt. H. O.	31 Aug. 18	6 Sep. 18	
20 May 17	Monckton, 2nd-Lt. F. E.	20 May 18	2 Sep. 18	7th Lrs.
26 May 17	Kissane, 2nd-Lt. R.	1 Sep. 18	24 May 18	3-3rd G. R.
14 Nov. 16	Beckett, Lt. A.	14 Nov. 17	7 July 18	1-8th G. R.
14 Nov. 16	Frith, Lt. J. W.	14 Nov. 17	7 July 18	2-8th G. R.
. . . .	Barr, 2nd-Lt. A.	. . . .	2 Sep. 18	

1st Commn. or date of entering service.	Name and Rank.	Jemadars.	Subadars.	Subadar-Majors.	Remarks.
	Subadar-Major.				
16 Dec. 96	Birkhasing Thapa	15 Apl. 15	18 Apl. 16	24 May 18	2-1st G. R.
	Subadars.				
3 Feb. 93	Manbir Gurung	18 Apl. 16	24 May 18	. . . .	2-1st G. R.
20 Dec. 90	Tilbir Thapa	19 Jan. 16	24 May 18	. . . .	2-3rd G. R.
	Parman Thapa	1 Jan. 17	24 May 18	. . . .	
28 Apl. 99	Dhanbir Rai	12 Oct. 14	7 June 16	. . . .	2-7th G. R.
15 Dec. 02	Anbir Pun	18 Feb. 17	24 May 18	. . . .	1-8th G. R.
25 Jan. 06	Ram Nath Cachari	13 Nov. 16	18 Feb. 17	. . . .	1-8th G. R.
15 Nov. 96	Balindhar Rai	3 Oct. 15	24 May 18	. . . .	1-7th G. R.
2 Apl. 94	Ganjaram Thapa	24 Dec. 15	26 Mar. 17	. . . .	
26 Nov. 97	Budhiman Gurung	20 Dec. 14	27 May 17	. . . .	
	Jemadars.				
	Dalbir Burathoki	24 May 18	. . . .	. . . .	
	Ganjasing Gurung	. . . .	. . . .	. . . .	
	Dhanraj Gurung, I.O.M.	24 May 18	. . . .	. . . .	
	Surbir Thapa	22 Nov. 17	. . . .	. . . .	3-3rd G. R.
3 May 99	Karbir Limbu	12 Aug. 17	. . . .	. . . .	2-7th G. R.
	Gristi Ram Rai	. . . .	. . . .	. . . .	
16 Oct. 03	Mandhaj Gurung	18 Feb. 17	. . . .	. . . .	1-8th G. R.
26 Nov. 98	Kharka Bahadur Pun	21 Jan. 17	. . . .	. . . .	1-8th G. R.
	Rajman Thapa	24 May 18	. . . .	. . . .	
	Saburam Rai	24 May 18	. . . .	. . . .	
	Tilocsing Limbu	16 July 18	. . . .	. . . .	
15 Jan. 96	Hastabir Limbu	5 July 15	. . . .	. . . .	2-7th G. R.

Uniform—Scarlet.
Facings—Blue.

Resident's Escort (Nepal).

Katmandu.

(Raised 5th April 1816.)
Commandant (*ex-officio*).—Resident in Nepal.

1st Commn or date of entering service.	Names and Rank.	Army rank.	Present appointment in Regiment.	Remarks.
		Jemadar.	Subadar.	Subadar-Major.
	Subadar.			
14 Apl. 93	Sobh Narain Sukul	23 July 09	15 May 16	
	Jemadar.			
1 Mar. 00	Kundal Sukul	15 May 16		

Establishment.—1 Subadar—1 Jemadar—5 Havildars—6 Naicks—2 Buglers—64 Sepoys—1 Sub-Assistant Surgeon.

Mounted Escort.—1 Duffadar—2 Sowars.

Uniform—Blue. *Facings*—White.

COLONEL-IN-CHIEF—Field Marshal *H. R. H.* Arthur, W. P. A., *Duke of* CONNAUGHT and STRATHEARN, K.G., K.T., K.P., G.C.B., G.C.S.I., G.C.M.G., G.C.I.E., G.C.V.O., G.B.E., A.D.C. 10 Dec. 12.

Date of 1st Comn.	Names and Rank.	First appt. to Corps.	Present appointment.	Remarks.
	Director of Supplies and Transport.			
5 Feb. 87	**Vaughan**, Brig.-Genl. R. E., C.B.	30 July 90	5 Oct. 16	A. H. Qrs.
	Deputy Directors of Supplies and Transport.			
28 Nov. 88	**Codrington**, Lt.-Col. H. de B.	12 May 93	11 Nov. 18	D. D. T., A. H. Qrs.
10 Oct. 94	**Young**, Maj. (*temp. Lt.-Col.*) H. N. (L.)	22 Mar. 99	Mar. 18	D. D. S., A. H. Qrs.
	Assistant Directors of Supplies and Transport.			
25 Nov. 85	**Bell**, Lt.-Col. R. M.	16 Mar. 91	12 Sep. 14	5th D. S. C. (*Lv., ex Ind., m.c., 12 mos. from 19 Feb. 18.*)
25 Aug. 86	*Hallowes, Lt.-Col. (temp. Col.) F. W.*, C.I.E.	*15 May 90*	*8 May 16*	*Dir., Farms, A. H. Qrs.*
16 Nov. 87	**Thomason**, Lt.-Col. A. F.	12 Sep. 91	1 Apl. 17	7 D. S. C.
11 Feb. 88	**Dickson**, Lt.-Col. J. H., C.I.E.	6 July 92	1 Apl. 17	9 D. S. C.
5 Feb. 87	**Lindsay**, Bt.-Col. H. A. P., C.M.G.	27 June 93	9 July 17	2 D. S. C.
22 Aug. 88	**Fulton**, Bt.-Col. F. C. R.	12 May 93	27 Sep. 17	6 D. S. C.
5 Feb. 87	**Roddy**, Lt.-Col. H. H., C.B.	9 Oct. 93	4 Jan. 18	1 D. S. C.
21 Dec. 89	**Swan**, Lt.-Col. J. S.	28 Jan. 95	1 Jan. 18	10 D. S. C.
24 July 86	**Burlton**, Lt.-Col. A. R.	12 May 90	31 Dec. 15	3 D. S. C.
8 July 89	**Forteath**, Lt.-Col. F. W. H.	12 May 93	28 Feb. 18	8 D. S. C.
11 Feb. 88	**Brooke**, Lt.-Col. H. M. M.	20 Oct. 93	6 Oct. 18	4 D. S. C.
	Supply and Transport Officers, 1st class.			
24 July 86	*Burlton, Lt.-Col. A. R.*	*12 May 90*	*12 May 12*	
25 Aug. 86	*Hallowes, Lt.-Col. F. W.*, C.I.E.	*15 May 90*	*15 May 12*	
25 Nov. 85	*Bell, Lt.-Col. R. M.*	*16 Mar. 91*	*16 Mar. 13*	
5 Feb. 87	Hennessy, Lt.-Col. (*temp. Col.*) J. P. C., C.M.G.	17 June 91	17 June 13	*Lv., ex I., m. c., 6 mos. from 4 Sep. 18.*
16 Nov. 87	*Thomason, Lt.-Col. A. F.*	*12 Sep. 91*	*12 Sep. 13*	
14 Sep. 87	Jones, Lt.-Col. (*temp. Col.*) H. H.	4 Feb. 92	4 Feb. 14	D. D. S. & T., N. C.
11 Feb. 88	*Dickson, Lt.-Col. J. H.*, C.I.E.	*6 July 92*	*6 July 14*	
11 Feb. 88	**Vaughan**, Lt.-Col. E. G., D.S.O.	9 Aug. 92	9 Aug. 14	O. C. "E" S. D. C.
11 Feb. 88	*Moore, Lt.-Col. C. H. G.*, C.M.G., D.S.O.	*31 Jan. 93*	*31 Jan. 15*	D. D. S. & T., Sn. Comd.
22 Aug. 88	*Fulton, Bt.-Col. F. C. R.*	*12 May 93*	*12 May 15*	
28 Nov. 88	*Codrington, Lt.-Col. H. deB.*	*12 May 93*	*12 May 15*	
8 July 89	*Forteath, Lt.-Col. F. W. H.*	*12 May 93*	*12 May 15*	
5 Feb. 87	*Lindsay, Bt.-Col. H. A. P.*, C.M.G. [*L*].	*27 June 93*	*27 June 15*	
5 Feb. 87	*Roddy, Lt.-Col. H. H.*, C.B.	*9 Oct. 93*	*9 Oct. 15*	
11 Feb. 88	*Brooke, Lt.-Col. H. M. M.*	*20 Oct. 93*	*20 Oct. 15*	
11 Feb. 88	**Annesley**, Bt.-Col. A. S. R.	25 July 94	25 July 16	6 D. S. C.
23 Mar. 89	**Geoghegan**, Lt.-Col. F. F., C.I.E.	27 Aug. 94	27 Aug. 16	"A" S. D. C.
21 Sep. 89	**Smith**, Lt.-Col. (*temp. Col.*) W. F. (*e*)	26 Sep. 94	26 Sep. 16	Deputy Director of Stocks, A. H. Qrs.
21 Sep. 89	**Dallas**, Lt.-Col. A. E., C.M.G.	13 Oct. 94	13 Oct. 16	5 D. S. C.
21 Dec. 89	*Swan, Lt.-Col. J. S.*	*28 Jan. 95*	*28 Jan. 17*	
25 July 91	**Preston**, Lt.-Col. E. E. [L].	23 Aug. 95	23 Aug. 17	"C" S. D. C.
10 Nov. 88	**Amesbury**, Lt.-Col. W. L. R.	3 Oct. 95	3 Oct. 17	S. T. C. Trng. Estabt., Rawalpindi.
13 May 89	**Byers**, Lt.-Col. C. B.	13 Apl. 96	13 Apl. 18	O. C., "H." S. D. C.
	Attached.			
10 Mar. 83	**Leonard**, Col. A. W.	13 Oct. 14	13 Oct. 14	"A" S. D. C.
12 May 83	**Palin**, Col. G. W., C.B., C.I.E.	1 Apl. 17	1 Apl. 17	"B" S. D. C.
9 Sep. 82	**Cripps**, Col. A. W., C.B.	. . . .	31 Dec. 15	"B" S. D. C.
13 Aug. 79	**Carter**, Lt.-Col. E. J.	. . . .	10 Oct. 14	"A" S. D. C.
22 Oct. 81	**Mathias**, Lt.-Col. L. J.	. . . .	16 Oct. 14	"A" S. D. C.
28 Jan. 82	**Lincoln**, Lt.-Col. S. W.	. . . .	12 Feb. 15	"A" S. D. C.
2 Aug. 82	**Nicholson**, Lt.-Col. R. W.	. . . .	14 June 15	O. C., "A" S. D. C.
5 Dec. 83	**Sanders**, Col. G. L'H.	. . . .	3 Mar. 17	No. 8 D. S. C. On deputation.
19 June 86	**Hawkins**, Col. G. A.	. . . .	6 Mar. 15	"J" S. D. C.
16 Nov. 87	**Webb**, Lt.-Col. A. E.	. . . .	18 Mar. 15	O. C., "K" S. D. C.
7 Feb. 85	**Ditmas**, Lt.-Col. A. R.	28 Oct. 90	29 Aug. 14	D. A. D. S. & T., Karachi Bde.
31 July 16	**Gwynn**, Lt. (*temp. Maj.*) C. H. (I. A. R. O.)	. . . .	1 Dec. 17	Asst. Contr. of Contracts, A. H. Qrs.

D. S. C.=Divisional Supply Company.
S.D.C.=Supply Depôt Company.

Supply and Transport Officers, 2nd class.

Date of 1st Comn.	Names and Rank.	First appt. to Corps.	Present appointment.	Remarks.
12 May 91	**Fox,** Lt.-Col. L. W.	30 Jan. 96	30 Jan. 13	D. A. D. S., 6 D. S. C.
29 Oct. 90	**Haydon,** Lt.-Col. W. P., D.S.O.	16 Jan. 97	16 Jan. 14	A. D. S. & T. Temy. 6 D. S. C.
29 Nov. 90	**Newell,** Lt.-Col. H. A. [*l*]	16 Jan. 97	16 Jan. 14	"A" S. D. C.
23 May 91	**Lushington** Lt.-Col. W. N.	20 Nov. 97	20 Nov. 14	Inspr., S. &. T., S. Comd.
9 Sep. 91	**Chamier,** Lt.-Col. W. St. G.	15 Apl. 98	15 Apl. 15	Inspr., S. & T., N. Comd.
20 Feb. 92	**Cates,** Maj. W. I.	25 May 98	25 May 15	4 D. S. C.
27 Jan. 92	**Harington,** Lt.-Col. A. W. D.	19 Oct. 98	19 Oct. 15	3 D. S. C.
31 Jan. 94	**Babington,** Maj. A. H.	3 Mar. 99	3 Mar. 16	*Lv., m.c., ex I.*
10 Oct. 94	*Young, Major (temp. Lt.-Col.) H. N. [L[illegible]*	*22 Mar. 99*	*22 Mar. 16*	D. D. S., A. H. Qrs.
30 Oct. 94	**Davies,** Major H. W.	17 Apl. 99	17 Apl. 16	4 D. S. C.
16 Jan. 95	**Smith,** Maj. J. L.	8 July 99	8 July 16	3 D. S. C.
17 Dec. 92	**Muscroft,** Bt.Lt.-Col. W. St. C.	13 July 99	13 July 16	I. D. S. C.
14 Aug. 95	**Manson,** Maj. J. A. P.	8 Aug. 99	8 Aug. 16	O. C. "L" S. D. C.
20 Feb. 95	**[illegible]avis,** Maj. G. H.	27 Sep. 99	27 Sep. 16	7 D. S. C.
28 Nov. 94	*Sharp, Major (temp. Lt.-Col.) A., C.M.G. [l].*	*4 Oct. 99*	*4 Oct. 16*	*Addl. Depy. Secy., Army Dept.*
14 Aug. 95	**Ryves,** Maj. R. H.	4 Oct. 99	4 Oct. 16	T. R. O. 7 D. S. C.
28 Nov. 94	**Synge,** Major (*temp. Lt.-Col.*) M., D.S.O. (*e*).	19 Mar. 00	19 Mar. 17	"F" S. D. C.
30 Aug. 93	*Heyland, Maj. A. K.,* D.S.O.	*1 May 00*	*1 May 17*	*Executive Commissariat Officer, Port Blair.*
20 Feb. 97	**Steele,** Maj. C. E. B.	1 May 00	1 May 17	"G" S. D. C.
6 June 96	**Shuttleworth,** Maj. A. R. B., *p.s.c.*	4 June 00	4 June 17	O. C., "B" S. D. C.
20 Jan. 97	**Sherrard,** Bt.-Lt.-Col. J. W.	4 June 00	4 June 17	1 D. S. C.
12 Dec. 94	**Thomas,** Maj. R. S.	4 Oct. 01	4 June 17	O. C., No. 2 M. T. Coy.
12 Dec. 94	**Lang,** Maj. C. F. G.	13 Feb. 03	4 June 17	O. C., No. 4 Mule Depôt, Karachi.
6 Mar. 95	**Wardell,** Maj. G. F. E.	18 Jan. 01	4 June 17	"A" S. D. C. Embkn. Bombay.
29 May 95	**Douglas,** (*temp. Lt.-Col.*) Maj. H. A.	22 Jan. 04	4 June 17	A. D. S. & T. A. H. Q.
14 Aug. 95	**Forbes,** Maj. E. E.	6 June 04	14 Aug. 17	3 D. S. C.
14 Aug. 95	**Duxbury,** Maj. C. D. (*e*)	14 Aug. 05	14 Aug. 17	*Lv., ex I., on m. c., 6 mos.*
22 July 96	**Wildes,** Maj. C. S. D.	8 Mar. 00	22 July 17	I. D. S. C.
5 Sep. 96	**Hughes-Hallett,** Bt. Col. (Bt.-Lt.-Col.) R. L.	16 June 02	5 Sep. 18	*Lv., ex I., on m. c., 6 mos.*

Supply and Transport Officers, 3rd class.

Date of 1st Comn.	Names and Rank.	First appt. to Corps.	Present appointment.	Remarks.
20 Jan. 97	**Murray,** Maj. F. S. J.	18 Oct. 05	20 Jan. 14	"F" S. D. C.
20 Jan. 97	**Innes-Lillingston,** Maj. F. F.	4 Sep. 03	20 Jan. 14	"K" S. D. C.
20 Jan. 97	**Middlemass,** Maj. R. W. H.	16 July 03	20 Jan. 14	6 D. S. C.
24 Mar. 97	**White,** Maj. H. P. M.	14 Feb. 03	24 Mar. 14	2 D. S. C.
1 Sep. 97	**Hart,** Maj. (*temp. Lt.-Col.*) E. G., D. S. O. [*l*].	26 Feb. 06	1 Sep. 14	9 D. S. C.
8 Sep. 97	**Saunders,** Maj. E. H., D.S.O.	24 Nov. 05	8 Sep. 14	2 D. S. C.
15 Sep. 97	**Syer,** Maj. H. H.	10 July 06	15 Sep. 14	2 D. S. C.
1 Dec. 97	**Watts,** Maj. H. P.	30 Jan. 03	1 Dec. 14	[illegible] D. S. C.
22 Jan. 98	**Watling,** Maj. L. D.	15 May 06	22 Jan. 15	"A" S. D. C.
22 Jan. 98	**Poole,** Maj. (*temp. Lt.-Col.*) I. M. [illegible], D.S.O.	5 July 01	22 Jan. 15	Aden.
22 Jan. 98	**Dunlop,** Bt.-Lt.-Col. (*temp. Lt.-Col.*) W. B., D.S.O.	29 Jan. 02	22 Jan. 15	2 D. S. C.
4 May 98	**Brunker** Maj. J. R.	25 June 06	4 May 15	"D" S. D. C.
7 May 98	**Morris,** Maj. C. O., D.S.O.	21 July 04	7 May 15	2 D. S. C.
20 July 98	**Watts,** Maj. A. H.	13 Sep. 04	20 July 15	2 D. S. C.
20 July 98	**Timbrell,** Maj. T., D.S.O. (*e*)	21 July 05	20 July 15	Inspector of Stocks, S. Area, Poona.
20 July 98	**Home** Maj. J. E.	6 Sep. 06	20 July 15	"I" S. D. C.
27 July 98	**Hogg,** Maj. G. E. M.	3 June 03	27 July 15	2 D. S. C.
3 Aug. 98	**Fraser,** Maj. A. C.	30 Jan. 03	3 Aug. 15	Inspector of Stocks, N. E. Area, Ambala.
3 Aug. 98	**Na[illegible]ke,** Maj. A. S.	1 July 03	3 Aug. 15	Inspector of Stocks, N. W. Area, Rawalpindi.

D. S. C. = Divisional Supply Company.
S. D. C. = Supply Depôt Company.

Supply and Transport Officers, 3rd class—*contd.*

Date of 1st Comn.	Names and Rank.	First appt. to Corps.	Present appointment.	Remarks.
25 Jan. 99	**Ball**, Maj. T. L.	21 Feb. 06	25 Jan. 16	2nd D. S. C.
25 Jan. 00	**Hume-Spry**, Maj. W. E., D.S.O.	3 Nov. 03	25 Jan. 16	D. A. D. S. and T., S. C.
25 Jan. 99	**Stewart**, Maj. A. F.	14 Sep. 04	25 Jan. 16	4th D. S. C.
11 Feb. 99	**Twiss**, Maj. (*temp. Lt.-Col.*) H. W. F. [L].	10 Feb. 06	11 Feb. 16	A. D. S., A. H. Qrs.
11 Feb. 99	**Kendall**, Maj. S. R. G.	27 Feb. 05	11 Feb. 16	3 D. S. C.
20 May 99	**Porch**, Bt.-Lt.-Col. (*temp. Lt.-Col.*) E. A., M.C.	22 Jan. 09	20 May 16	5 D. S. C.
20 May 99	**Lyster**, Maj. C. J. H., M.C.	1 Aug. 06	20 May 16	S. & T. Tr. Estabt. Rawalpindi.
2 Aug. 99	**Sweeny**, Maj. R. L. C., D.S.O., M.C.	30 Jan. 03	2 Aug. 16	No. 4, D. S. C.
12 Aug. 99	**Salmon**, Maj. W. H. B.	12 Aug. 05	18 Oct. 18	"[illegible]" S. D. C.
13 Aug. 99	**Geoghegan**, Maj. O. A. (*e*).	6 Apl. 06	12 Aug. 16	"A." S. D. C.
6 Jan. 00	**Eliot**, Maj. H. S.	1 July 06	6 Jan. 17	"B." S. D. C.
6 Jan. 00	**Walter**, Maj. E., C.I.E. [*l*].	7 May 06	6 Jan 17	Office of Contr. of Contracts, A. H. Qrs.
12 Jan. 00	**Wilkinson**, Maj. G. H., D.S.O.	26 Jan. 09	12 Jan. 17	3 D. S. C.
17 Jan. 00	**Kemble**, Maj. (*temp. Lt.-Col.*) P. B.	27 Apl. 06	17 Jan. 17	1 D. S. C.
20 Jan. 00	**Saunders**, Maj. P. C.	9 Aug. 06	20 Jan. 17	9 D. S. C.
10 Jan. 00	**Ruel**, Maj. W. G., D.S.O.	20 Dec. 04	20 Jan. 17	"B." S. D. C.
20 Jan. 00	**Corbyn**, Maj. M. McL.	14 June 09	20 Jan. 17	Asst. to D. D. Stocks, A. H. Qrs.
20 Jan. 00	**Lancaster**, Maj. E. H.	10 July 05	20 Jan. 17	8 D. S. C.
20 Jan. 00	**MacMahon**, Bt.-Lt.-Col. (*temp. Lt.-Col.*) H. F. E., M.C.	3 May 04	20 Jan. 17	2 D. S. C.
20 Jan. 00	**McEnery**, Maj. R. T. (*temp. Lt.-Col.*)	25 Aug. 09	20 Jan. 17	3 D. S. C.
4 May 01	**Steward**, Maj. E. M.	6 July 08	14 Feb. 17	D. A. D. S. & T., N. C.
4 May 01	**Price**, Maj. H. W.	25 July 08	19 Feb. 17	6 D. S. C.
3 Mar. 00	**Edward-Collins**, Maj. C. E.	14 Oct. 06	3 Mar. 17	7 D. S. C.
11 Apl. 00	**Bond**, Maj. G. W., D.S.O.	7 Oct. 06	11 Apl. 17	5 D. S. C.
18 Apl. 00	**Sargent**, Maj. A. E. E., D.S.O., M.C.	3 May 06	18 Apl. 17	3 D. S. C.
21 Apl. 00	**Carroll**, Maj. H. O.	3 July 05	21 Apl. 17	7 D. S. C.
2 May 00	**Hext**, Maj. G. T. B., D.S.O.	25 Sep. 05	2 May 17	1 D. S. C.
5 May 00	**Hext**, Maj. J. E.	21 Sep. 06	5 May 17	7 D. S. C.
5 May 00	**Rowan-Robinson**, Bt. Lt.-Col. J. R., D.S.O.	30 Sep. 06	5 May 17	3 D. S. C.
5 May 00	**Smith**, Maj. S. H. P.	4 Oct. 06	5 May 17	"J" S. D. C.
5 May 00	**Messenger**, Maj. H. T. K.	9 June 08	5 May 17	"A" S. D. C.
5 Jan. 01	**Hext**, Maj. C. W.	18 Aug. 08	17 May 17	3 D. S. C.
22 May 00	**Brown**, Bt.-Lt.-Col. W.	9 Mar. 09	22 May 17	8 D. S. C.
22 May 00	**Rennison**, Maj. A. J., D.S.O.	18 Apl. 04	22 May 17	4 D. S. C.
23 May 00	**Shine**, Maj. E. P. F.	13 Dec. 05	23 May 17	3 D. S. C.
26 May 00	**Mayo**, Maj. G. L. L.	11 June 08	26 May 17	D. A. D. T., A. H. Qrs.
25 July 00	**Ewart**, Maj., (*temp. Lt.-Col.*) R. F., D.S.O.	21 Nov. 05	25 July 17	8 D. S. C.
28 July 00	**Clarke**, Maj. G. C. C.	21 Nov. 05	28 July 17	"B" S. D. C.
28 July 00	**Pearse**, Maj. G.	21 Nov. 05	28 July 17	7 D. S. C.
28 July 00	**Hewett**, Maj. M. S., D.S.O.	26 Nov. 08	28 July 17	"G" D. S. C.
11 Aug. 00	*Alexander, Maj. (temp. Lt.-Col.) H. M., D.S.O. (e).*	*19 June 08*	*11 Aug. 17*	*I. M. B.*
18 Aug. 00	**Wace**, Maj. S. L.	10 Sep. 06	18 Aug. 17	Derajat Bde.
29 Aug. 00	**Watney**, Maj. C. W.	27 Aug. 10	29 Aug. 17	1 D. S. C.
4 May 01	**Jones**, Maj. L. A. M.	31 May 04	8 Oct. 17	"E." S. D. C.
21 Dec. 00	**Christie**, Maj. H. G. F.	4 July 05	21 Dec. 17	8 D. S. C.
5 Jan. 01	**Murray**, Maj. G. B.	25 Apl. 04	5 Jan. 18.	4 D. S. C.
5 Jan. 01	**Peyton**, Maj. (*Temp. Lt.-Col.*) A. H.	9 Aug. 08	5 Jan. 18	Bombay Bde.
5 Jan. 01	**Hall**, Maj. H. C.	30 July 10	5 Jan. 18	7 D. S. C.
8 Jan. 01	*Mortimer, Maj. P.*	*1 May 06*	*8 Jan. 18*	*Bandipur—Kashmir.*
8 Jan. 01	**Rendall**, Maj. J. C. S.	8 Sep. 04	8 Jan. 18	Bombay Bde.
8 Jan. 01	**Flint**, Maj. A. L. A.	14 June 10	8 Jan. 18	2 D. S. C.
23 Oct. 01	**Aylmer**, Maj. G., M.C.	9 Mar. 09	6 Feb. 18	2nd D. S. C.
14 Sep. 01	**Campbell**, Maj. W., D.S.O.	24 July 05	24 Mar. 18	8th D. S. C.
16 Aug. 02	**Dela Motte**, Maj. R. B., D.S.O.	29 Mar. 09	28 Mar. 18	"C." S. D. C.
24 July 01	**Stevenson-Hamilton**, Maj. O. C. J.	7 Nov. 10	15 Oct. 18	"F" D. S. C.
1 May 01	**Lloyd**, Bt. Lt.-Col. C. R., D.S.O.	22 Jan. 04	1 May 18	2 D. S. C.
4 May 01	**Kirby**, Bt. Lt.-Col. J. T., D.S.O. [*l*].	20 Dec. 04	4 May 18	8 D. S. C.
8 May 01	**Boyce**, Maj. C. J., D.S.O.	5 Dec. 05	8 May 18	1. D. S. C.
27 July 01	**Hogan**, Maj. E. M. A. J.	22 Jan. 09	27 July 18	Ambala Bde.

D. S. C.=Divisional Supply Company.
S. D. C.=Supply Depôt Company.
I. M. B.=Indian Munitions Board.

Date of 1st Comn.	Names and Rank.	First appt. to Corps.	Present appointment.	Remarks.

Attached (temporary—3rd Class).

Date of 1st Comn.	Names and Rank.	First appt. to Corps.	Present appointment.	Remarks.
5 Feb. 87	**Leicester,** Maj. B., late Ches. R.	2 Dec. 14	5 Mar. 15	4 D. S. C.
27 Sep. 89	**Chambers,** Maj. (*Hony. Lt.-Col.*) W. B., V.D., late Ind. Volrs.	19 Apl. 15	19 Apl. 15	"A" S. D. C.
17 June 91	**Stewart,** Lt.-Col. R. J. T., C.M.'s Dept.	16 Apl. 15	16 Apl. 15	4 D. S. C.
9 Apl. 92	**Hislop,** Maj. A. F., I. A. (Retd.)	18 Nov. 14	15 Feb. 15	"G." S. D. C.
23 Dec. 93	**Fenning,** Maj. W.M., C. M.'s Dept.	20 Sep. 15	20 Sept. 15	3 D. S. C.
19 Feb. 01	**Robinson,** Maj. (*temp. Lt.-Col.*) W. P., D.S.O., A. S. C.	24 Jan. 17	19 Feb. 18	A. H. Q.

Supply and Transport Officers, 4th class.

Date of 1st Comn.	Names and Rank.	First appt. to Corps.	Present appointment.	Remarks.
8 Sep. 00	**Clark,** Maj. A. S.	8 June 08	8 Sep. 13	"G" S. D. C.
19 Oct. 01	**Rice,** Maj. G. D., D.S.O. . . .	23 Nov. 05	19 Oct. 14	2 D. S C.
6 Nov. 01	**Chanter,** Maj. A. D.	17 July 08	6 Nov. 14	1 D. S. C.
4 Dec. 01	**Wheatley,** Maj. W. P. R., D.S.O.	10 July 08	4 Dec. 14	1 D. S. C.
3 Sep. 02	**Brook,** Maj. T. F. (206) . .	14 May 12	16 Dec. 14	1 D. S. C.
18 Jan. 02	**Woodhouse,** Bt.-Lt.-Col. P. St. J. R.	7 Sep. 08	28 Jan. 15	2 D. S. C.
27 Aug. 02	**Cook,** Maj. W. K.	5 Sep. 10	27 Aug. 13	6 D. S. C.
22 Oct. 02	**Western,** Maj. C. M. T. . .	5 Apl. 10	22 Oct. 15	8 D. S C.
22 Oct. 02	**Clarke,** Maj. R. C., M.C. . . .	7 June 10	22 Oct. 15	2 D. S. C.
22 Oct. 02	**Beck,** Maj. R. C.	15 May 11	22 Oct. 15	"B." S. D. C.
22 Oct. 02	**Bird,** Maj. W. F.	30 May 10	22 Oct. 15	3 D. S. C.
9 Jan. 04	**Jarrett-Kerr,** Maj. H. J. . . .	11 Apl. 11	12 Jan. 16	"H." S. D. C.
21 Jan. 03	**Stainton,** Maj. F. H. . . .	5 Oct. 10	21 Jan. 16	8 D. S C.
22 Apl. 03	**Trelawny,** Maj. L. C., M.C. .	20 Dec. 10	22 Apl. 16	3 D. S. C.
16 May 03	**Duke,** Maj. A. A. G.	5 June 12	16 May 16	Derajat Bde.
15 July 03	**Fuller,** Maj. H. F.	29 Nov. 09	15 July 16	3 D. S. C.
15 July 03	**Shaw,** Maj. H. P. J. (206) .	10 May 12	15 July 16	7 D. S. C.
29 July 03	**Shaw,** Maj. W. W.	1. Nov. 14	1 Nov. 16	"B." S. D. C. Lv., ex I., on m. c., 6 mos.
10 Oct. 03	**Langrishe,** Maj. R. B. . . .	12 Oct. 13	10 Oct. 16	"F." S. D. C.
10 Oct. 03	**Comyn,** Capt. G. V., M.C. . .	27 May 12	10 Oct. 16	8 D. S. C.
10 Oct. 03	**McCallum,** Capt. R.E., R. Scots. (206).	19 Nov. 12	10 Oct. 16	10 D. S. C.
4 Nov. 03	**Cooper,** Capt. G. B. A. . . .	10 Apl. 11	4 Nov. 16	"H." S. D. C.
4 Nov. 03	**Byrch,** Capt. A. E. B. . . .	11 May 10	4 Nov. 16	"K." S. D. C.
18 Nov. 99	**Stevenson,** Capt. H. D. M., M.C. (206).	27 Apl. 12	18 Nov. 16	1 D. S. C.
9 Jan. 04	**Christopher,** Capt. (*Bt.-Maj.*) C. de L.	13 Apl. 10	9 Jan. 17	3 D. S. C.
9 Jan. 04	**Pogson,** Capt. F. V. [*l*] . . .	16 July 10	9 Jan. 17	7 D. S. C.
9 Jan. 04	*Cobb, Capt. C. . . . .*	*8 Dec. 13*	*9 Jan. 17*	*Derajat Bde.*
4 June 04	**Bridge,** Capt. R.E. A., D.S.O. . .	12 July 10	4 June 17	8 D. S. C.
29 July 04	**Edwards,** Capt. W. M., R. G. A. (205*a*).	13 July 14	29 July 17	6 D. S. C.
29 July 04	**Pengelley,** Capt. (*temp. Maj.*) G. H.	17 Jan. 11	29 July 17	4 D. S. C.
13 Aug. 04	**Harrison,** Capt. C. H. . . .	28 June 12	13 Aug. 17	"B." S. D C.
13 Aug. 04	**Rebsch,** Capt. (*temp. Maj.*) W. K., D.S.O.	21 May 14	13 Aug. 17	Bannu Bde.
13 Aug. 04	**Bird,** Capt. A. C.	7 Nov. 10	13 Aug. 17	"K" S D C.
3 Sep. 04	**Lloyd,** Capt. C. G. (204) M.C. .	26 Oct. 12	3 Sep. 17	Derajat Bde.
3 Dec. 04	**Whittell,** Capt. H. M. . . .	9 Oct. 11	3 Dec. 17	"A." S. D. C
17 May 05	**Hogan,** Capt. E. O. (205*a*) .	27 Nov. 13	17 May 18	1 D. S C.
20 May 05	**Wright,** Capt. (*temp. Maj.*), A. St. J.	28 Aug. 11	20 May 18	A. H. Qrs.
20 May 05	**Philips,** Capt. J. W.	19 June 11	20 May 18	3 D. S. C.

D. S. C. = Divisional Supply Company.
S. D. C. = Supply Depôt Company.

Supply and Transport Officers, 4th Class—*contd.*

Date of 1st Commn.	Names and Rank.	First appt. to Corps.	Present appointment.	Remarks.
20 May 05	**Scott,** Capt. (Ty. Major) G.C., R. Ir. Rif. (206)	15 Jan. 13	20 May 18	Kohat Bde.
21 June 05	**FitzGerald,** Capt. A. (*e*)	13 Aug. 08	21 June 18	7 D. S. C.
5 Aug. 05	**Cookson,** Capt. S. M.	19 Mar. 14	5 Aug 18	"B." S. D. C.
5 Aug. 05	**Woodhouse,** Capt. F.S.	23 Mar. 14	5 Aug. 18	7 D. S. C.
5 Aug. 05	**Chesney,** Capt. E. C. [L]	11 June 14	5 Aug. 18	A. H. Qrs.
23 May 06	**Frost,** Bt.-Maj. F. D., M.C.	21 Sep. 11	11 Sep. 18	6 D. S. C.
29 Nov. 05	**Burlton-Bennet,** Capt. J. A.	21 May 18	29 Nov. 18	8 D. S. C.

Attached (temporary—4th class).

Date of 1st Commn.	Names and Rank.	First appt. to Corps.	Present appointment.	Remarks.
22 Jan. 81	**Brook,** Capt. R. W. C. (late Wilts. R.)	10 Feb. 15	10 Feb. 15	"G". S. D. C.
4 Mar. 98	**O'Connor,** Maj. M. S. S., I.A. R. O.	12 May 15	12 May 15	4 D. S. C.
13 Nov. 14	**Watson,** Capt. J., I.A.R. O.	19 Feb. 15	25 June 15	A H. Qrs.
30 Dec. 04	**Egerton,** Capt. W., I. A. R. O.	15 Dec. 14	....	8 D. S. C.
23 May 00	**Moran,** Maj. J.	....	....	9 D. S. C.
25 Oct. 15	**Dunn,** Capt. F. E., I. A. R. O. V. D.	1 Nov. 15	1 Nov. 15	Karachi Bde.
30 Oct. 14	**Evans,** Capt. R. J., I. A. R. O.	1 Nov. 14	16 Mar. 16	Office of D. D. Stocks.
11 Dec. 14	**MacDougall,** Capt. D., I.A. R. O.	23 Dec. 14	1 July 16	7 D. S. C.
22 Jan. 15	**Carter,** Capt. F. G., I.A.R. O.	1 Feb. 15	14 Nov. 16	2 D. S. C.
1 June 17	**Edwards,** Lieut. (temp. Capt.) O., I.A.R.O.	6 Nov. 16	1 June 17	"A" S. D. C.

Supply and Transport Officers, 5th class.

Date of 1st Commn.	Names and Rank.	First appt. to Corps.	Present appointment.	Remarks.
29 July 04	*Crawford, Capt. A. E., R. A. (206)*	*21 Oct. 14*	*21 Oct. 14*	3 D. S. C.
16 Aug. 05	**Mulliken,** Capt. E. M.	16 Oct. 14	16 Oct. 14	8 D. S. C.
29 Nov. 05	**Daly,** Capt. C. D.	20 Jan. 13	29 Nov. 14	Bannu Bde.
20 Dec. 05	**Barrett,** Capt. (*temp. Maj.*) B. H. H., A. S. C.	22 May 14	20 Dec. 14	Army Hd. Qrs.
24 Jan. 06	**Beattie,** Capt. J. O.	26 Oct. 14	24 Jan. 15	"D" S. D. C.
24 Jan. 06	**Jacomb,** Capt. F. B. W.	22 Oct. 14	24 Jan. 15	3 D. S. C.
24 Jan. 06	**Rae,** Capt. G. R., 7th Rajputs	3 Aug. 13	24 Jan. 15	8 D. S. C.
15 Aug. 06	**Markwick,** Capt. H. W.	23 Sep. 11	15 Aug. 15	A. H. Q.
29 Aug. 06	**Dyce,** Capt. A. G.	5 Oct. 14	29 Aug. 15	Karachi Bde.
29 Aug. 06	**Lucas,** Capt. C. A.	8 Jan. 15	29 Aug. 15	Derajat Bde.
19 Jan. 07	**Cripps,** Capt. E. J.	22 Jan. 13	19 Jan. 16	8 D. S. C.
16 Feb. 07	**Herapath,** Capt. C. B., 16th Rajputs	27 May 14	16 Feb 16	4 D. S. C.
4 May 07	**May,** Capt. C. D., 5th Infantry	21 Oct. 14	4 May. 16	2 D. S. C.
11 Dec. 07	**Gahan,** Capt. C. G. D., E. Surr. R. (206)	29 Apl. 14	11 Dec. 16	6 D. S. C.
11 Dec. 07	**Jones,** Capt. B. L.	15 May 14	11 Dec. 16	9 D. S. C.
18 Dec. 07	**Daly,** Capt. C. K., 119th Infy.	16 Oct. 14	18 Dec. 16	9 D. S. C.
25 Jan. 08	**Tennant,** Bt.-Maj. C. A. M.	15 Oct. 14	25 Jan. 17	1 D. S. C.
22 Feb. 08	**Deighton,** Capt. F. J.	26 Oct. 14	22 Feb. 17	1 D. S. C.
9 Sep. 08	**Brodrick,** Capt. W. Le C.	29 Oct. 14	9 Sep. 17	"G" S. D. C. Lv., *6* mos. ex. I.
9 Sep. 08	**Hughes,** Capt. G. V., M.C.	16 Oct. 14	9 Sep. 17	7 D. S. C.
9 Sep. 08	**Metcalf,** Capt. E. D. T., M. C.	14 Oct. 14	9 Sep. 17	2 D. S. C.
14 Oct. 08	**Gilbert,** Capt. H. H. (206)	22 May 14	14 Oct. 17	3 D. S. C.
16 Dec. 08	**Hickie,** Capt. F. C.	11 Nov. 14	16 Dec. 17	5 D. S. C.
20 Jan. 09	**Cassidy,** Capt. V. C.	2 Nov. 14	20 Jan. 18	2 D. S. C.
23 July 09	**Bryson,** Capt. A. B.	29 Oct. 14	23 July 18	2 D. S. C.
18 Sep. 09	*Williams, Capt. (temp. Maj.) S. C. (206)*	*15 May 14*	*18 Sep. 18*	*D. A. Q. M. G., Quetta.*

D. S. C. = Divisional Supply Company.
S. D. C. = Supply Depôt Company.

Date of 1st Commn.	Names and Rank.	First appt. to Corps.	Present appointment.	Remarks.
	Attached—(5th class).			
27 Nov. 14	**Burroughs**, Lt. G. H. G., I. A. R. O.	21 Dec. 14	2 Dec. 14	5 D. S. C.
4 Dec. 14	**Ralph**, Capt. S. A., I. A. R. O.	23 Jan. 15	26 Mar. 15	9 D. S. C.
20 June 13	**Wigfall**, Capt. R. H., I. A. R. O.	10 Dec. 14	7 Apl. 15	"A" S. D. C.
31 Dec. 14	**Hodges**, Lt. H. S., I. A. R. O.	13 Jan. 15	16 July 15	10 D. S. C.
12 Feb. 15	**Roger**, Capt. G. B., I. A. R. O.	22 Feb. 15	16 July 15	10 D. S. C.
13 Nov. 14	**Foreman**, Lt. C. L., I. A. R. O.	22 Jan. 15	9 Oct. 15	Bannu Bde.
20 Nov. 14	**Kennedy**, Lt. J. P., I. A. R. O.	29 Jan. 15	20 Nov. 15	1 D. S. C.
20 Nov. 14	**Hurst**, Lt. L., I. A. R. O.	26 Jan. 15	1 Dec. 15	7 D. S. C.
30 Oct. 14	**Earp**, (*temp. Capt.*) Lt. A. C., I. A. R. O.	6 Nov. 14	6 Nov. 15	M. T. School, Rawalpindi.
27 Nov. 14	**Backhouse**, Lt. P. B., I. A. R.O.	21 Jan. 15	16 Nov. 15	2 D. S. C.
24 Dec. 14	**O'Hagan**, Lt. R., I. A. R. O.	2 Jan. 15	16 Nov. 15	8 D. S. C.
24 Dec. 14	**Tyler**, Lt. B. M. C., I. A. R. O.	18 Jan. 15	18 Nov. 15	7 D. S. C.
23 Oct. 14	*Gayer, Lt. J. H. C.*	*4 Nov. 14*	*28 Jan. 16*	9 D. S. C.
13 Nov. 14	**Brotherton**, Lt. (*temp. Capt.*) J. B., I. A. R. O.	17 Nov. 14	15 Dec. 15	"A" S. D. C.
20 Nov. 14	**Davies**, Lt. A., I. A. R. O.	10 May 15	3 Feb. 16	Bombay Bde.
20 Nov. 14	**Ingram**, Lt. G. L., I. A. R. O.	26 Jan. 15	12 Jan. 16	7 D. S. C.
20 Nov. 14	**Horwood**, Lt. J. C., I. A. R. O.	21 Jan. 15	28 Jan. 16	3 D. S. C.
4 Dec. 14	**Panckridge**, Lt. ((*temp. Capt.*) H. R., I. A. R. O.	20 Jan. 15	20 Jan. 16	
27 Nov. 14	**Cumberlege**, Lt. J. H., I. A. R. O.	20 Jan. 15	28 Jan. 16	5 D. S. C.
21 Dec. 14	**Macdonald**, Lt. L. T., I. A. R. O.	6 May 15	3 Feb. 16	6 D. S. C.
24 Dec. 14	**Fawcett**, Lt. A. H., I. A. R. O.	10 May 15	3 Feb. 16	6 D. S. C. *Lv., ex I., 6 mos. on m. c. from 26th November 1918.*
31 Dec. 14	**Flashman**, Lt. H. C., I. A. R.O.	7 Jan. 15	28 Jan. 16	4 D. S. C.
31 Dec 14	**Baird**, Lt. R. V., I. A. R. O.	20 Jan. 15	28 Jan. 16	7 D. S. C.
8 Jan. 15	**Maltby**, Lt. G., I. A. R. O.	6 May 15	3 Feb. 16	4 D. S. C.
30 Oct. 14	**Crawford**, Lt. (*tempy. Capt.*) H., I. A. R. O.	9 Nov. 14	5 May 16	3 D. S. C.
30 Oct. 14	**Richards**, Lt. (*tempy. Capt.*) A., I. A. R. O.	2 Nov. 14	5 May 16	6 D. S. C.
30 Oct. 14	**Hurst**, Lt. (*Temp. Capt.*) W. F. I. A. R. O.	12 Nov. 14	5 May 16	Karachi Bde.
28 May 15	**Rodgers**, Lt. (*Temp. Capt.*) J. J., I. A. R. O.	31 May 15	3 Feb. 16	"A" S. D. C.
27 Nov. 14	**Allan**, Lt. A., I. A. Res.	22 Jan. 15	5 May 16	1 D. S. C.
23 July 15	**Chevallier**, Lt. F. T., I. A. R. O.	16 Aug. 15	12 Feb. 16	2 D. S. C.
14 May 15	**Tice**, Lt. C. A., I. A. R. O.	2 Aug. 15	12 Feb. 1[illegible]	Derajat Bde.
27 Nov. 14	**McDermott**, Lt. C. H., I.A. R.O.	23 Jan. 15	9 Apl. 16	8 D. S. C.
30 Apl. 15	**Coles**, Lt. [illegible]. E. H., I. A. R. O.	22 May 15	23 Feb. 16	8 D. S. C.
27 Nov. 14	**Pook**, Lt. J. De. C., I. A. R. O.	11 Dec. 14	6 May 16	"K" S. D. C.
14 May 15	**Corlett**, Capt. M. J. J., I. A. R. O.	4 Aug. 15	28 Feb. 16	7 D. S. C.
27 Nov. 14	**Duke**, Lt. W. S., I. A. R. O.	20 Jan. 15	9 Apl. 16	3 D. S. C.
4 Dec. 14	**Siedle**, Lt. F. J., I. A. R. O.	21 Jan. 15	9 Apl. 16	3 D. S. C.
24 Dec. 14	**Smyth**, Lt. G. C. G., I. A. R. O.	8 May 15	26 Apl. 16	4 D. S. C.
24 Dec. 14	**Nesbitt**, Lt. E. A., I. A. R. O.	11 Jan. 15	5 May 16	9 D. S. C.
8 Jan. 15	**Armitage**, Lt. C. C., I. A. R. O.	27 Apl. 15	20 May 16	"D" S. D. C.
15 Jan. 15	*Lesmond, Lt. C. A., I. A. R. O.*	*22 Jan. 15*	*9 Apl. 16*	2 D. S. C.
29 Jan. 15	**Phillips**, Lt. H. L., I. A. R. O.	1 Feb. 15	15 Jan 16	8 D. S. C.
29 Jan. 15	**Webb**, Lt. A. B., I. A. R. O.	5 June 15	17 May 16	6 D. S. C.
5 Feb. 15	**Viney**, Lt. J. E., I. A. R. O.	2 Feb. 15	22 Mar. 16	1 D. S. C.
8 Feb. 15	**Stevens**, Lt. J. R., I. A. R. O.	8 Feb. 15	9 Apl. 16	"J" S. D. C.
12 Feb. 15	**Gardiner**, Lt. R. C., I. A. R. O.	26 Feb. 15	5 May 16	2 D. S. C.
18 Feb. 15	**Leedham**, Lt. W., I. A. R. O.	25 Feb. 15	18 Apl. 16	"H" S. D. C.
18 Feb. 15	**Lutter**, Lt. W. F., I. A. R. O.	17 Apl. 15	18 Apl. 16	3 D. S. C.
18 Feb. 15	**Crealock**, Lt. W. E. W., I. A. R. O.	28 Feb. 15	10 Apl. 16	1 D. S. C.
18 Feb 15	**Spiller**, Lt. (*temp. Capt.*) T. R., I. A. R. O.	18 Feb. 15	9 Apl. 16	7 D. S. C.
5 Mar. 15	**O'Connor**, Lt. (*temp. Capt.*) G. B., I. A. R. O.	9 Mar. 15	8 May 16	3 D. S. C.
24 Dec. 14	**Lester**, Lt. R., I. A. R. O.	1 May 15	1 May 16	7 D. S. C.
2 Apl. 15	**Hardwicke**, Lt. C. B., I. A. R. O.	17 Apl. 15	5 May 16	2 D. S. C.
2 Apl. 15	**Harper**, Lt. G., I. A. R.O.	21 Apl. 15	....	2 D. S. C. *Lv., ex I., 6 mos., from 7th July 1918.*
2 Apl. 15	**Scholefield**, Lt. M. T., I.A. R. O.	19 Apl. 15	8 May 16	Derajat Bde.
2 Apl. 15	**Sherlock**, Lt. W. W., I.A. R. O.	16 Apl. 15	6 May 16	1 D. S. C.
16 Apl. 15	**Ashton**, Lt. E. M., I. A. R. O.	30 Apl. 15	6 May 16	1 D. S. C.
30 Apl. 15	**Walker**, Lt. A. W., I. A. R. O.	7 May 15	10 Apl. 16	1 D. S. C.
30 Apl. 15	**Martin**, Lt. F. B., I. A. R. O.	20 May 15	2 June 16	Derajat Bde.
11 June 15	**Sulley**, Lt. (*Temp. Capt.*) J. P., I. A. R. O.	24 June 15	18 Apl. 16	"F" S. D. C.
6 Nov. 14	**O'Dea**, Lt. M. L. B. P., I. A. R.O.	4 June 15	22 Mar. 16	2 D. S. C.

D. S. C. = Divisional Supply Company.
S. D. C. = Supply Depôt Company.

Date of 1st Commn.	Names and Rank.	Date of attachment to Corps.	Present appointment.	Remarks.
	Attached—5th Class—*contd.*			
24 Dec. 14	**McGurk,** Lt. (*temp.* *Capt.*) N. A., I. A. R. O.	5 May 15	19 Mar. 16	9 D. S. C.
25 Apl. 13	**Travers,** Capt. O. A., I. A. R. O.	20 Aug. 15	18 Apl. 16	3 D. S. C.
5 Feb. 15	**Jackson,** Lt. H., I. A. R. O.	19 Feb. 15	15 Feb. 16	4 D. S. C.
2 Apl. 15	**Nissen,** Lt. (*temp. Capt.*) T. A., I. A. R. O.	10 Apl. 15	3 Feb. 16	"A" S. D. C.
9 July 15	**Walsh,** Capt. G. G., I. A. R. O.	9 July 15 } 26 May 16 }	26 May 16	4 D. S. C.
18 June 15	**Fraser,** Lt. H. J., I. A. R. O.	24 June 15	24 June 16	5 D. S. C.
18 June 15	**Harman,** Lt. (*temp. Capt.*) O. A., M.C., I. A. R. O.	30 Aug. 15	18 Sep. 16	6 D. S. C.
2 Oct. 14	**Hornsby,** Capt. N., I. A. R. O.	28 Aug. 15	27 Nov. 16	1 D. S. C.
27 Nov. 14	**Willson,** Lt. C. B., I. A. R. O.	5 Dec. 14	29 July 16	8 D. S. C.
24 Dec. 14	**Mellish,** Lt. A., I. A. R. O.	18 May 15	15 Aug. 16	8 D. S. C.
31 Dec. 14	**Ashton,** Lt. (*temp. Capt.*) G. R., I. A. R. O.	31 July 15	5 Aug. 16	1 D. S. C.
12 Feb. 15	**Bell,** Lt. H., I. A. R. O.	28 Feb. 15	14 July 16	9 D. S. C.
5 Mar. 15	**Hickle,** Lt. J. C., I. A. R. O.	5 May 15	31 July 16	"J" S. D. C.
2 Apl. 15	**Cameron,** Lt. (*temp. Capt.*) G. S., I. A. R. O., M. C.	15 Apl. 15	24 July 16	9 D. S. C.
2 Apl. 15	**Humphrey,** Lt. D. A. R., I. A. R. O.	19 Apl. 15	7 Aug. 16	8 D. S. C.
2 Apl. 15	**Mercer,** Lt. A. W., I. A. R. O.	4 June 15	5 July 16	"F" S. D. C.
9 Apl. 15	**Gibb,** Lt. J. B., I. A. R. O.	16 Apl. 15	24 July 16	"J" S. D. C.
18 June 15	**Smith,** Lt. R. N. D., I. A. R. O.	28 June 15	24 July 16	6 D. S. C. *Lv., ex-I., m. c., from 25-9-17 to 24-2-19.*
27 Aug. 15	**George,** Lt. N. M., I. A. R. O.	3 Sep. 15	11 June 16	7 D. S. C.
16 Oct. 15	**Yeates,** Lt. R. M. M., I. A. R. O.	16 Nov. 15	16 Nov. 16	7 D. S. C.
18 June 15	**Inglis,** Lt. T. A., I. A. R. O.	15 Sep. 15	10 June 16	7 D. S. C.
27 Aug. 15	**Cove,** Lt. V. D., I. A. R. O.	1 Sep. 15	1 Sep. 16	"A" S. D. C.
15 May 16	**Judge,** Lt. (*temp. Capt.*) C. E. M., M.B.E.	15 May 16	20 Feb. 17	A. H. Qrs.
12 Oct. 15	**Whitamore,** Lt. (*temp. Capt.*) A. G., I. A. R. O.	17 Nov. 15	17 Nov. 16	5 D. S. C.
30 Oct. 14	**Dunning,** Capt. H. G., I. A. R. O.	1 Nov. 14	1 Mar. 17	- D. S. C.
22 Jan. 15	**Bowles,** Lt. R. F., I. A. R. O.	28 Apl. 15	1 Mar. 17	"C" S. D. C.
10 Feb. 16 } 23 Nov. 14 }	**Tanner,** Lt. H., I. A. R. O.	3 Mar. 16	1 Mar. 17	2 D. S. C.
3 Feb. 16	**Davis,** Lt. W., I. A. R. O.	3 Mar. 16	4 May 17	10 D. S. C.
3 Feb. 15	**Pollaco,** Lt. (*temp. Capt.*) R. H., I. A. R. O.	4 Mar. 17	9 Mar. 17	Bombay Bde.
3 Feb. 15	**Stephenson,** Capt. F. H. A., I. A. R. O.	19 Mar. 17	19 Mar. 17	8 D. S. C.
6 Aug. 15	**Howland,** Lt. C., I. A. R. O.	28 Aug. 15	28 Aug. 16	"E" S. D. C.
17 Apl. 16	**Cole,** Lt. (*temp. Capt.*) W. J.	17 Apl. 16	17 Apl. 17	2 D. S. C.
23 Dec. 15	**Baldrey,** Lt. (*temp. Capt.*) C. H., I. A. R. O.	16 Jan. 16	15 Jan. 17	3 D. S. C.
14 Mar. 16	**Davidson,** Lt. J., I. A. R. O.	3 Apl. 16	3 Apl. 17	3 D. S. C.
12 Feb. 15	**Dixon,** Lt. R. C., I. A. R. O.	18 Feb. 15	18 Feb. 17	"A" S. D. C.
19 Feb. 16	**Breslauer,** Lt. L., I. A. R. O.	25 Feb. 16	25 Feb. 17	"A" S. D. C.
25 Apl. 15	**Jenkin,** Lt. R. T., I. A. R. O.	22 Jan. 16	22 Jan. 17	3 D. S. C.
20 Feb. 16	**Austin,** Lt. J. C.	7 June 16	7 June 17	"B" S. D. C. *Lv., ex I., 6 mos., from 23-8-18.*
17 Apl. 16	**Nierses,** Lt. W.	6 May 16	6 May 17	"A" S. D. C.
23 Apl. 15	**West,** Lt. A. P. J.	1 May 16	1 May 17	4 D. S. C.
17 Apl. 16	**Bisphan,** Lt. G.	2 May 16	2 May 17	1 D. S. C.
17 Apl. 16	**Ricketts,** Lt. E. M.	15 May 16	15 May 17	1 D. S. C.
17 Apl. 16	**Fisk,** Capt. E. A.	15 May 16	15 May 17	10 D. S. C.
17 Apl. 16	**Boyd,** Lt. W. S. J.	30 May 16	30 May 17	1 D. S. C.
28 Apl. 16	**McRoberts,** Lt. (*temp. Capt.*) P. F.	1 June 16	1 June 17	Bombay Bde.
7 May 16	**Cullen,** Lt. J. G.	3 June 16	3 June 17	2 D. S. C.
18 June 15	**Wainwright,** Lt. V. P.	17 Aug. 15	17 Aug. 16	4 D. S. C.
25 Mar. 16	**Hill,** Lt. D. G., I. A. R. O.	3 June 16	3 June 17	Karachi Bde.
12 Oct. 15	**Considine,** Lt. W. J.	23 Oct. 15	23 Oct. 16	2 D. S. C.
29 May 15	**Cattell,** Lt. A. S.	28 Jan. 16	28 Jan. 17	3 D. S. C.
11 June 15	**Angelo,** Lt. M. O.	22 June 15	22 June 16	8 D. S. C.
30 Oct. 14	**Keeling,** Lt. E. P.	14 July 16	14 July 17	2 D. S. C.
21 Sep. 16	**Corlett** Lt. D. S.	. . . .	21 Sep. 17	6 D. S. C.
12 July 16	**Mileham,** Lt. C. A., I. A. R. O.	31 July 16	31 July 17	"F." S. D. C.
14 Jan. 16	**Croft,** Lt. T. de C.	27 Jan. 16	27 Jan. 17	2 D. S. C.
12 Aug. 16	**Haskew,** Lt. L. C., I. A. R. O.	15 Aug. 16	15 Aug. 17	7 D. S. C.
27 Nov. 14	**Mildmay,** Lt. A. B. C.	8 May 16	8 May 17	3 D. S. C.
12 Aug. 16	**Barnett,** Lt. A. H., I. A. R. O.	25 Aug. 16	25 Aug. 17	7 D. S. C.
18 Mar. 16	**Marshall,** Lt. D. F.	7 Apl. 16	7 Apl. 17	Derajat Bde.
12 Aug. 16	**Thornewill,** Capt. A. S., I. A. R. O.	28 Aug. 16	28 Aug. 17	3 D. S. C.

D. S. C.=Divisional Supply Company.
S. D. C.=Supply Depôt Company.

Date of 1st Commn.	Names and Rank.	Date of attachment to Corps.	Present appointment.	Remarks.
	Attached—(5th class)—*contd.*			
15 May 16	**McIntosh,** Lt. A.	20 May 16	20 May 17	Bombay Bde. On Depn.
16 Jan. 15	**Kendle,** Lt. W. E. C., I. A. R. O.	3 Sep. 16	3 Sep. 17	"A." S. D. C.
18 Dec. 14	**Rogers,** Lt. W. P.	15 Dec. 16	15 Dec. 17	5 D. S. C.
1 July 16	**Aspinwall,** Lt. W. H. E.	13 July 16	13 July 17	"A" S. D. C.
14 Sep. 16	**Byron,** Lt. F. H., I. A. R. O.	11 Oct. 16	11 Oct. 17	6 D. S. C.
20 July 16	**Stirling,** Lt. F.	24 July 16	24 July 17	"A" S. D. C.
5 Oct. 16	**Williams,** Lt. A. C., I. A. R. O.	12 Oct. 16	12 Oct. 17	8 D. S. C.
15 July 16	**Booth,** Lt. W. M.	31 July 16	31 July 17	"A" S. D. C.
7 Oct. 16	**Finlay,** Lt. J., I. A. R. O.	20 Oct. 16	20 Oct. 17	7 D. S. C.
20 Oct. 16	**Barlow,** Lt. L., I. A. R. O.	26 Oct. 16	26 Oct. 17	6 D. S. C.
15 Oct. 16	**Meares,** Lt. W. K., I. A. R. O.	27 Oct. 16	27 Oct. 17	"J." S. D. C.
18 Oct. 16	**Norton,** Lt. G. A., I. A. R. O.	1 Nov. 16	1 Nov. 17	7 D. S. C.
24 Nov. 16	**Soar,** Lt. S. F., I. A. R. O.	1 Dec. 16	1 Dec. 17	7 D. S. C.
27 Nov. 16	**Johnston,** Lt. A. H., I. A. R. O.	14 Dec. 16	14 Dec. 17	3 D. S. C.
1 Aug. 16	**Knowles,** Lt. G. B.	26 Aug. 16	26 Aug. 17	4 D. S. C.
17 Apl. 16	**MacGregor,** Lt. D. R.	17 Apl. 16	17 Apl. 17	2 D. S. C.
7 Aug. 16	**Quilter,** Lt. G. J. H.	26 Aug. 16	20 Aug. 17	10 D. S. C.
11 Aug. 16	**Routliff,** Lt. J. I.	4 Sep. 16	4 Sep. 17	10 D. S. C.
29 May 16	**Lyons,** Lt. A. J.	13 Sep. 16	13 Sep. 17	"A" S. D. C.
13 Sep. 16	**Yerbury,** Lt. S. B.	12 Oct. 16	12 Oct. 17	"A" S. D. C.
1 July 15	**Daunt,** Lt. E. W. F. T.	1 Oct. 16	1 Oct. 17	"D" S. D. C.
24 Dec. 14	**Lewis,** Lieut. C. C. C.	2 Nov. 16	2 Nov. 17	7 D. S. C.
28 Aug. 15	**Thompson,** Lt. J. S.	24 Sep. 15	24 Sep. 16	Karachi Bde.
30 July 15	**Jamieson,** Lt. G.	16 Aug. 15	16 Aug 16	2 D. S. C.
24 Sep. 15	**Brewin,** Lt. G. G.	6 Oct. 15	6 Oct. 16	Bombay Bde.
5 Mar. 15	**Keatinge,** Lt. G.	14 June 15	14 June 10	10 D. S. C.
12 Oct. 15	**Simmons,** Lt. C. E.	6 Nov. 15	6 Nov. 16	8 D. S. C.
6 Dec. 15	**McWhinnie,** Lt. W.	15 Dec. 15	15 Dec. 16	Bombay Bde.
23 Dec. 15	**Harding,** Lt. W. F.	10 Jan. 16	10 Jan. 17	3 D. S. C.
24 Jan. 16	**Kent,** Lt. (*temp. Capt.*) H. S.	7 Feb. 16	7 Feb. 17	8 D. S. C.
24 Feb. 16	**Rawlings,** Capt. G. N.	24 Feb. 16	24 Feb. 17	Bannu Bde.
1 Dec. 15	**David,** Capt. T.	29 May 16	29 May 17	8 D. S. C.
5 May 16	**Stafford,** Lt. H. C. C.	25 May 16	25 May 17	9 D. S. C.
14 Apl. 16	**Turner,** Lt. N. T. B.	18 May 16	18 May 17	8 D. S. C.
29 Apl. 16	**Marriott,** Lt. A. E.	29 May 16	29 May 17	7 D. S. C.
1 June 16	**Williams,** Lt. J. C.	1 June 16	1 June 17	1 D. S. C.
1 July 16	**Stewart,** Lt. C. W.	29 July 16	29 July 17	3 D. S. C.
19 July 16	**Harris,** Lt. C. P.	5 Aug. 16	5 Aug. 17	5 D. S. C.
8 Sept. 15	**Abrahall,** Lt. F. T. E.	19 Aug. 16	15 Aug 17	8 D. S. C.
14 Aug. 16	**Thomson,** Lt. H. H.	22 Aug. 16	22 Aug. 17	1 D. S. C.
14 Aug. 16	**Peerman,** Lt. T. C.	25 Aug. 16	25 Aug. 17	6 D. S. C.
26 Mar 15	**Harris,** Lt. G. W.	4 Sep. 16	4 Sep. 17	Kohat Bde.
14 Jan. 15	**Wreford,** Lt. R. G.	27 Sep. 16	27 Sep. 17	Asst. to S. & T. Officer, on special duty at Kashmir.
20 Sep. 16	**Goodwin,** Lt. A. E.	13 Oct. 16	13 Oct. 17	6 D. S. C.
25 Sep. 16	**Beams,** Lt. R. H.	13 Oct. 16	13 Oct. 17	"A" S. D. C.
30 Sep. 16	**Unger,** Lt. V. C. W.	11 Oct. 16	11 Oct. 17	1 D. S. C.
21 Oct. 16	**Christie,** Lt. T. M.	1 Nov. 16	1 Nov. 17	"A" S. D. C.
18 Oct. 16	**Kemp,** Lt. W. G. J.	6 Nov. 16	6 Nov. 17	1 D. S. C.
1 Nov. 16	**Bull,** Lt. C. H.	8 Nov 16	8 Nov. 17	6 D. S. C.
10 Mar. 16	**Fletcher,** Lt. J. A.	5 Dec. 16	5 Dec. 17	1 D. S. C.
10 July 16	**Lupton,** Lt. F. E.	6 Dec. 17	6 Dec. 17	"A" S. D. C.
18 Feb. 15	**Mulgrue,** Lt. J. A.	7 Dec. 16	7 Dec. 17	1 D. S. C.
19 Apl. 16	**Sherwood,** Lt. W. O.	26 Dec. 16	26 Dec. 17	1 D. S. C.
26 Sep. 14	**Kittoe,** Capt. M. C. H. S.	16 Jan. 17	16 Jan. 18	1 D. S. C.
16 Nov. 16	**Smith,** Lt. H. A.	29 Dec. 16	29 Dec. 17	1 D. S. C.
20 Oct. 16	**Mcdonald,** Lt. H. P.	29 Jan. 17	29 Jan. 18	"A" S. D. C.
16 Dec. 16	**Howard,** Lt. J. G.	30 Dec. 16	30 Dec. 17	"D" S. D. C.
18 Dec. 14	**Mathews,** Lt. (*temp. Capt.*). J. M.	10 Jan. 17	10 Jan. 18	"A." S. D. C.
20 Feb. 16	**Perrin,** Lt. F. H.	21 Aug. 16	13 Feb. 18	3 D. S. C.
12 Feb. 15	**Lloyd,** Lt. (*temp. Capt.*) A. C.	1 May 15	. . .	4 D. S. C.
30 July 15	**Zobel,** Lt. M.	5 Aug. 15	. . .	"F" S. D. C.
4 Dec. 14	**Head,** Lt. H. K.	21 Aug. 15	. . .	1 D. S. C.
27 Aug. 15	**Evans,** Lt. A. G.	16 Sep. 15	. . .	3 D. S. C.
9 Sep. 15	**Masters,** Lt. H. F.	25 Sep. 15	. . .	8 D. S. C.
3 Nov. 15	**Hearn,** Capt. E. R.	3 Dec. 15	. . .	6 D. S. C.
30 Sep. 15	**Smith,** Lt. R. E.	24 Dec. 15	. . .	2 D. S. C.
25 Jan. 16	**Page,** Lt. E. R.	30 Jan. 16	. . .	8 D. S. C.
8 Feb. 16	**Hewer,** Lt. J. A.	14 Feb. 16	. . .	9 D. S. C.
15 Feb. 16	**Thompson,** Lt. B. N.	24 Feb. 16	. . .	8 D. S. C.
17 Apl. 16	**Fortescue,** Lt. L. S.	2 May 16	. . .	3 D. S. C.
17 Apl. 16	**Gilbert,** Lt. R. S.	12 May 16	. . .	7 D. S. C.
19 May 16	**Pooley,** Lt. J. P. F.	29 May 16	. . .	6 D. S. C.
8 Mar. 16	**Boden,** Lt. J. F. W., I. A. R. O.	29 May 16	. . .	2 D. S. C.
9 July 15	**Pearson,** Lt. C. McM., I. A. R. O.	6 June 16	. . .	4 D. S. C.

D. S. C.=Divisional Supply Company.
S. D. C.=Supply Depôt Company.

Date of 1st Commn.	Names and Rank.	Date of attachment to Corps.	Present appointment.	Remarks.

Attached—(5th class)—*concld.*

1 Jan. 16	**Fleimann,** Lt. P. M. J. . . .	2 July 16	. . .	2 D. S. C.
1 June 16	**Lewis,** Lt. G. W.	19 July 16	. . .	2 D. S. C.
5 Aug. 16	**Maloy,** Lt. T. H.	22 Aug. 16	. . .	Bombay Bde.
7 Oct. 15	**Corfield,** Lt. H. R. . . .	8 Sep. 16	. . .	9 D. S. C.
1 Feb. 17	**Metcalf,** Lt. F., I. A. R. O. . .	13 Feb. 17	2 Feb. 18	"H" S. D. C.
10 Feb. 17	**Thorpe,** Lt. A. J., I. A. R. O. .	24 Feb. 17	24 Feb. 18	4 D. S. C.
21 Feb. 17	**Holland,** Lt. G. R., I. A. R. O. .	24 Mar. 17	24 Mar. 18	"A" S. D. C.
20 Mar. 17	**Gadsdon,** Lt. H. A., I. A. R. O.	2 Apl. 17	2 Apl. 18	"B" S. D. C.
28 Mar. 17	**Miller,** Lt. T., I. A. R. O. . .	7 Apl. 17	7 Apl. 18	6 D. S. C.
19 Apl. 17	**Davies,** Lt. G. P., I. A. R. O.	19 Apl. 17	19 Apl. 18	2 D. S. C.
7 Mar. 17	**Powis,** Lt. H. B., I. A. R. O.	8 May 17	8 May 18	2 D. S. C.
10 Apl. 17	**Graham,** Lt. T. C. V., I. A. R. O.	24 May 17	24 May 18	5 D. S. C.
25 June 17	**Rivett,** Lt. W. C., I. A. R. O. .	25 June 17	25 June 18	"J" S. D. C.
25 June 17	**Morris,** Lt. H. L., I. A. R. O. .	9 July 17	9 July 18	5 D. S. C.
2 July 17	**Haywood,** Lt. J. S., I. A. R. O. .	13 July 17	13 July 18	3 D. S. C.
2 July 17	**Campbell,** Lt. G. E. F., I A.R.O.	18 July 17	18 July 18	Derajat Bde. 3 D. S. C.
2 July 17	**Porritt,** Lt. A. H., I. A. R. O. .	30 July 17	30 July 18	"A" S. D. C.
7 May 17	**Norwood,** Lt. W. S., I. A. R. O.	2 Aug. 17	2 Aug. 18	1 D. S. C.
27 July 16	**Burnett** Lt. G., I. A. R. O. . .	14 Sep. 17	14 Sep. 18	1 D. S. C.
6 Sep. 16	**Fry,** Lt. W. J.	16 Sep. 16	. . .	2 D. S. C.
17 June 15	**Hayes,** Lt. T. (Yorks R.) . .	18 Sep. 16	18 Sep. 17	9 D. S. C.
6 Sep. 16	**Badcock,** Lt. W. B.	18 Sep. 16	. . .	2 D. S. C.
25 Aug. 16	**Gibson,** Lt. D. A.	21 Sep. 16	. . .	3 D. S. C.
25 Sep. 16	**Hodson,** Lt. C.	14 Oct. 16	. . .	Ambala Bde.
1 Oct. 16	**Williams,** Lt. E. P.	16 Oct. 16	. . .	3 D. S. C.
7 Oct. 16	**Baldwin,** Lt. G. W.	19 Oct. 16	. . .	Bannu Bde.
18 Oct. 16	**Rivett,** Lieut. E. G.	21 Oct. 16	. . .	Ambala Bde.
21 Oct. 16	**Recton,** Lt. T.	2 Nov. 16	. . .	3 D. S. C.
28 Oct. 16	**Bothwell,** Lt. P. J.	6 Nov. 16	. . .	3 D. S. C.
15 Oct. 16	**Boxer,** Capt. F. P.	21 Nov. 16	. . .	2 D. S. C.
25 Jan. 15	**Keatinge,** Lt. F. W. G., M. C. .	5 Dec. 16	. . .	7 D. S. C.
21 Nov. 16	**Smith,** Lt. V. S.	5 Dec. 16	. . .	1 D. S. C. .
13 May 15	**Down,** Lt. G. O.	8 Dec. 16	. . .	1 D. S. C.
5 Dec. 16	**Platts,** Lt. A.	11 Dec. 16	. . .	3 D. S. C.
27 Jan. 16	**Fitzpatrick,** Lt. E. J. . . .	10 Jan. 17	. . .	Bombay Bde.
25 Mar. 15	**Chowns,** Lt. W. P.	12 Jan. 17	. . .	4 D. S. C.
13 Mar. 15	**Hambly,** Lt. R. L.	19 Jan. 17	. . .	2 M. T. Coy.
30 Aug. 15	**Russell,** Lt. B. H.	29 Jan. 17	. . .	4 D. S. C.
4 June 16	**Britts,** Lt. M. H. G.	7 Feb. 17	. . .	"A" S. D. C.
2 Feb. 17	**Crawshaw,** Lt. H. E. J. . . .	12 Feb. 17	. . .	"A" S. D. C.
26 Feb. 17	**Craig-Adams,** Lt. R. L. . . .	3 Mar. 17	. . .	3 D. S. C.
1 Nov. 15	**Frankau,** Capt. G. N. . . .	10 Mar. 17	. . .	1 D. S. C.
1 Mar. 17	**Lewis,** Lt. F. W. H.	20 Mar. 17	. . .	4 D. S. C.
15 Mar. 17	**Damerell,** Lt. W. J.	23 Mar. 17	. . .	2 D. S. C.
10 Mar. 17	**Pell,** Lt. T. O.	25 Mar. 17	. . .	M. T. School, Rawal Pindi.
15 Mar. 17	**Wilkinson,** Lt. E. J.	5 Apl. 17	. . .	10 D. S. C.
2 Apl. 17	**Henderson,** Lt. A. N. . . .	16 Apl. 17	. . .	"C" S. D. C.
17 Apl. 17	**Phillips,** Lt. E.	23 Apl. 17	. . .	"A" S. D. C.
10 Apl. 17	**Jenkins,** Lt. S. E.	24 Apl. 17	. . .	9 D. S. C.
20 Apl. 17	**Beaton,** Lt. J. S. C.	27 Apl. 17	. . .	"A" S. D. C.
20 Apl. 17	**Harris,** Lt. L.	3 May 17	. . .	5 D. S. C.
17 Apl. 17	**Cann,** Lt. M.	7 May 17	. . .	"K" S. D. C.
17 Apl. 17	**Lancaster,** Lt. C. E. A. . . .	20 May 17	. . .	1 D. S. C.
21 Feb. 17	**Evatt,** Lt. C. A., I. A. R. O. .	7 Mar. 17	7 Mar. 18	5 D. S. C.
2 Sep. 15	**Toovey,** Lt. E. H., I. A. R. O	9 Mar. 17	9 Mar. 18	2 D. S. C.
16 Sep. 14	**Gibbs,** Capt. F. J., I. A. R. O. .	24 Mar. 17	24 Mar. 18	2 D. S. C.
22 Mar. 17	**Ball,** Lt. W. C., I. A. R. O. . .	1 Apl. 17	1 Apl. 18	Karachi Bde.
20 Mar. 17	**Jones,** Lt. F. L., I. A. R. O. . .	2 Apl. 17	2 Apl. 18	5 D. S. C.
24 Mar. 17	**Laythorpe,** Lt. F. W. H., I. A. R. O.	10 Apl. 17	10 Apl. 18	5 D. S. C.
10 Apl. 17	**Giles,** Lt. L., I. A. R. O. . . .	23 Apl. 17	23 Apl. 18	8 D. S. C.
19 Apl. 17	**Longstaffe,** Lt., J. W., I.A.R.O.	2 May 17	2 May 18	2 D. S. C.
16 May 17	**Adams,** Lt. E. F., I. A. R. O. .	30 May 17	30 May 18	6 D. S.

D. S. C.=Divisional Supply Company.
S. D. C.=Supply Depôt Company.

Date of 1st Commn.	Names and Rank.	Date of attachment to Corps.		Remarks.

Supply and Transport Officers, 6th class.

Date of 1st Commn.	Names and Rank.	Date of attachment to Corps.		Remarks.
11 Dec. 09	**Cowdell**, Capt. A. F., N. Lan. R. (206)	16 Aug. 14	16 Aug. 14	1 D. S. C.
3 Sep. 10	**Mathias**, Capt. L. W. H., 128th Prs.	21 Nov. 14	21 Nov. 14	8 D. S. C.
3 Sep. 10	**Scott**, Capt. W. R. A.	16 Sep. 14	16 Sep. 14	6 D. S. C.
3 Sep. 10	**Grey**, Capt. P.	21 Oct. 14	21 Oct. 14	2 D. S. C.
8 Jan. 11	**Brander**, Capt. R. W. . . .	14 Oct. 14	14 Oct. 14	"C" S. D. C.
22 May 09	**Gompertz**, Capt. A. W. B., 94th Infy.	3 May 16	. . . .	4 D. S. C.

Attached—(6th class).

Date of 1st Commn.	Names and Rank.	Date of attachment to Corps.	Remarks.
15 Dec. 14	**Haddick**, Capt. V. A. (Leins. R.)	8 Aug. 15	"J" S. D. C.
. . . .	**Dix**, 2nd-Lt. W., R. W. Kent. R.	20 Feb. 17	3 D. S. C.
22 Sep. 15	**Thomas**, Lt. A. R.	4 Oct. 15	6 D. S. C.
14 Jan. 16	**Ross**, Lt. T. R.	1 Feb. 16	6 D. S. C.
17 Apl. 16	**Hope**, Lt. J. W.	8 May 16	7 D. S. C.
17 Apl. 16	**Williams**, Lt. A.	9 May 16	"H" S. D. C.
17 Apl. 16	**McCorquodale**, Lt. A. R. . .	6 May 16	8 D. S. C.
15 May 16	**Miller**, Lt. A. D.	1 June 16	8 D. S. C.
18 June 15	**Hales**, Lt. W. P.	19 July 16	"G" S. D. C.
25 Mar. 16	**Holmes**, Lt. (*temp. Capt.*) A. H. R.	22 Apl. 16	7 D. S. C.
30 May 16	**Jesse**, Lt. W.	8 June 16	"F" S. D. C.
1 July 16	**Wells**, Lt. R. W.	17 July 16	8 D. S. C.
1 July 16	**Brodie**, Lt. W. S. C.	17 July 16	1 D. S. C.
10 June 16	**Angus**, Lt. H. C.	20 July 16	Karachi Bde.
6 July 16	**Smith**, Lt. K. C.	20 July 16	5 D. S. C.
1 July 16	**Daly**, Lt. D. J.	28 July 16	8 D. S. C.
15 June 16	**Smith**, Capt. C. M., I. A. R. O.	1 Aug. 16	3 D. S. C.

D. S. C. = Divisional Supply Company.
S. D. C. = Supply Depôt Company.

First Commn.	Names and Rank.	Date of attachment to Corps.	Remarks.

Attached—(6th Class)—*concld.*

28 July 16	**McNair,** Lt. G. D.	14 Aug. 16	2 D. S. C.
1 Aug. 16	**Parsons,** Lt. F. F.	18 Aug. 16	6 D. S. C.
24 Dec. 14	**Manley,** Lt. H.	2 Sep. 16	7 D. S. C. *Lv., ex Ind., m. c.*
8 Jan. 15	**Spow-rt,** Lt. W. C.	13 Sep. 16	2 D. S. C.
26 Aug. 16	**McAlpin,** Lt. D. F. C.	11 Sep. 16	7 D. S. C.
5 Sep. 16	**Blackburn,** Lt. M. E. M.	21 Sep. 16	9 D. S. C.
2 Aug. 15	**George,** Lt. A. B.	29 Sep. 16	9 D. S. C.
1 Apl. 16	**Manson,** Lt. C. E. F.	25 Oct. 16	5 D. S. C.
15 Oct. 16	**Chapman,** Lt. A. J. B.	23 Oct. 16	8 D. S. C.
17 Oct. 16	**Lucas,** Lt. C. St. J.	24 Oct. 16	8 D. S. C.
20 Oct. 16	**O'Brien,** Lt. R. W.	30 Oct. 16	2 D. S. C.
21 Oct. 16	**Morley,** Lt. M. H. A.	31 Oct. 16	5 D. S. C.
15 Oct. 16	**Biddle,** Lt. W. C.	3 Nov. 16	5 D. S. C.
20 Oct. 16	**Meares,** Lt. G. C. K.	4 Nov. 16	5 D. S. C.
28 Oct. 16	**Roberts,** Lt. C. R. J.	6 Nov. 16	5 D. S. C.
6 Nov. 16	**Bradley,** Lt. G. G.	20 Nov. 16	2 D. S. C. *Lv., ex Ind., m. c.*
20 Mar. 15	**DeBrath,** Lt. E. E.	29 Nov. 16	8 D. S. C.
10 Oct. 16	**Haycroft,** Lt. (*temp. Capt.*) J. S.	10 Oct. 16	Chinese Porter Corps.
11 Oct. 16	**Anderson,** Lt. M. L.	11 Oct. 16	Chinese Porter Corps.
11 Oct. 16	**Elkins,** Lt. W. H.	11 Oct. 16	Chinese Porter Corps.
6 Nov. 16	**Harvey,** Lt. R. H. M.	30 Nov. 16	5 D. S. C.
20 Oct. 16	**Fielding,** Lt. H	...	8 D. S. C.
7 Nov. 15	**Macgrigor,** Lt. G. N. A.	29 Dec. 16	4 D. S. C.
10 Jan. 17	**Seymour-Higgins,** Lt. A. C. P.	22 Jan. 17	4 D. S. C.
1 Feb. 17	**Vickery,** Lt. R. M.	6 Feb. 17	2 D. S. C.
11 Nov. 16	**Davidson,** Lt. G. S.	4 Dec. 16	8 D. S. C.
1 Dec. 16	**Mayles,** Lt. R. W. J.	12 Dec. 16	3 D. S. C.
1 Jan. 17	**Evans,** Lt. I. A.	19 Jan. 17	8 D. S. C.
27 Jan. 17	**Curry,** Lt. G. R.	31 Jan. 17	5 D. S. C.
25 Jan. 17	**Todman,** Lt. A. McN.	10 Feb. 17	7 D. S. C.
19 Sep. 15	**Wigfall,** Lt. C. S.	20 Feb. 17	1 D. S. C.
10 Feb. 17	**Jones,** Lt. T. J. W.	14 Feb. 17	"B" S. D. C.
4 June 16	**Britts,** 2nd-Lt. M. H. G.	7 Feb. 17	"A" S. D. C.
18 Feb. 17	**Oxenham,** 2nd-Lt. S. H.	25 Feb. 17	Ambala Bde.
21 Feb. 17	**Hindley,** Lt. C. E. H.	9 Mar. 17	"J" S. D. C.
21 Feb. 17	**Pearman,** 2nd-Lt. H. A.	8 Mar. 17	6 D. S. C.
11 Jan. 17	**Fawcett,** Lt. H. E. M.	2 Feb. 17	6 D. S. C.
3 Apl. 15	**Kendall,** 2nd-Lt. H. J.	20 Dec. 16	"H" S. D. C.
1 Mar. 17	**Jones,** Lt. J. D.	17 Mar. 17	7 D. S. C.
1 Oct. 15	**Priestly,** 2nd-Lt. E. C.	21 Jan. 17	"B" S. D. C.
15 Mar. 17	**Temple,** Lt. B.	19 Mar. 17	Karachi Bde.
24 Mar. 17	**Rendle,** Lt. J. M.	27 Mar. 17	3 D. S. C.

D. S. C. = Divisional Supply Company.
S. D. C. = Supply Depot Company.

Date of First Commn.	Names and Rank.	Date of attachment to Corps.	Remarks.

Attached—6th Class—*contd.*

19 Oct. 14	*Judkins, Capt. B. E. H.*	*12 Apl. 17*	On Mule Purchasing Duty in North China. *Medical leave in Japan.*
5 Apl. 17	**Beaty**, Lt. P. C. D.	20 Apl. 17	9 D. S. C.
16 Mar. 17	**Sampson**, Lt. R. W. K.	23 Apl. 17	9 D. S. C.
11 Apl. 17	**Jackson**, Lt. H.	27 Apl. 17	3 D. S. C.
14 Apl. 17	**Petts**, Lt. H. J.	14 May 17	9 D. S. C.
5 Apl. 17	**Wilson**, Lt. C. H.	16 May 17	2 D. S. C.
12 May 17	**Swaffer**, Lt. H. P.	26 May 17	2 D. S. C.
18 May 17	**Hoerder**, 2nd-Lt. P.E. R.	1 June 17	5 D. S. C.
23 May 17	**Miller**, 2nd-Lt. A. R. C.	2 June 17	"A" S. D. C.
6 June 17	**Hitchcock**, 2nd-Lt. F. C.	15 June 17	Derajat Bde.
1 June 17	**McDonald**, 2nd-Lt. D. W.	19 June 17	3 D. S. C.
1 June 17	**Murley**, 2nd-Lt. P. A.	21 June 17	3 D S. C.
1 June 17	**Newquist**, 2nd-Lt. H. N.	21 June 17	3 D. S C.
1 June 17	**Craydon**, 2nd-Lt. C. C.	21 June 17	7 D. S. C.
4 June 17	**Kirkman**, 2nd-Lt. W. A.	25 June 17	7 D. S. C.
	Chastel, DeBoinville, Lt. G. A.	6 July 17	3 D. S. C.
27 June 17	**Stevens**, 2nd-Lt. R. P.	7 July 17	5 D. S. C.
25 June 17	**Rogerson**, 2nd-Lt. J. F.	13 July 17	2 D. S. C.
2 July 17	**Savage**, 2nd-Lt. R.	13 July 17	6 D. S. C.
1 June 17	**Eldridge**, 2nd-Lt. F. L.	14 July 17	"F" S. D. C.
6 June 17	**Page**, 2nd-Lt. W. H.	16 July 17	3 D. S. C.
2 July 17	**Winters**, 2nd-Lt. A. E.	17 July 17	3 D. S. C.
18 Apl. 15	**Steele**, 2nd-Lt. J. T.	19 July 17	5 D. S. C.
2 July 17	**Mudie**, 2nd-Lt. P. U.	20 July 17	7 D. S. C.
4 July 17	**Woods**, 2nd-Lt. H. S.	23 July 17	3 D. S. C.
28 June 17	**Stirling**, 2nd-Lt. S. W.	27 July 17	2 D. S. C.
8 July 17	**Surrey**, 2nd-Lt. H. P.	30 July 17	8 D. S. C.
18 July 17	**Northam**, Lt. R. M.	1 Aug. 17	4 D. S. C.
	Harricks, 2nd-Lt. N.	3 Aug. 17	"A" S. D. C.
25 Oct. 15	**Cooke**, 2nd-Lt. J. C. C., (A. S. C., *On probn*).	5 Aug. 17	Bannu Bde.
5 Aug. 17	**Betteridge**, 2nd-Lt. R. C.	5 Aug. 17	"K" S. D. C.
12 Apl. 15	**Bostock-Wilson**, Lt. E. L., (A. S. C., *On probn.*)	6 Aug. 17	"B" S. D. C.
7 Aug. 17	**Coulson**, Lt. W. B.	7 Aug. 17	8 D. S. C.
26 July 17	**Vogwill**, Lt. J. W.	15 Aug. 17	6 D. S. C.
6 Aug. 17	**Marshall**, 2nd-Lt. E. C.	16 Aug. 17	2 D. S. C.
24 July 17	**Pirie**, Lt. G. W.	17 Aug. 17	6 D. S. C.
14 Aug. 17	**Stobbart**, 2nd-Lt. J.	22 Aug. 17	5 D. S. C.
14 Aug. 17	**Gravett**, 2nd-Lt. G. W.	25 Aug. 17	"J" S. D. C.
20 July 17	**Dufall**, 2nd-Lt. J. B.	27 Aug. 17	"A" S. D. C.
7 Aug. 17	**Shillitoe**, 2nd-Lt. G. R.	....	9 D. S. C.
	Jones, Lt. R. E., (A. S. C., *on probn.*)	26 Nov. 17	9 D. S. C.
11 Sep. 17	**Shove**, Lt. H. L.	9 Oct. 17	6 D. S. C.
6 Aug. 17	**Shaw**, 2nd-Lt. J. A.	....	1 D. S. C.
21 Aug. 17	**Lewis**, 2nd-Lt. C. L.	3 Sep. 17	9 D. S. C.
6 Sep. 17	**Johnson**, Lt. H. E.	....	3 D S C.
29 Aug. 17	**Davies**, 2nd-Lt. S. C.	....	4 D. S. C.
4 Sep. 17	**Clarke**, Lt. M. F. C.	12 Sep. 17	1 D. S. C.
5 Aug. 14	**Raikes**, Capt. M. H.	....	Karachi Bde.

D. S. C.=Divisional Supply Company.
S. D. C.=Supply Depôt Company.

Attached 6th Class—*contd.*

Date of First Commn.	Name and Rank.	Date of attachment to Corps.	Remarks.
14 Sep. 17	Guider, Lt. M. E.	24 Sep. 17	7 D. S. C.
13 Sep. 17	Dance, 2nd-Lt. (*temp. Lieut.*) A. J.	25 Sep. 17	8 D. S. C.
14 Sep. 17	Crozier, 2nd-Lt. D.	28 Sep. 17	3 D. S. C.
16 Sep. 17	Mitchell, 2nd-Lt H. R. W.	3 Oct. 17	4 D. S. C.
3 Oct. 17	Cutler, 2nd-Lt. G. J.	3 Oct. 17	"A" S. D. C.
3 Oct. 17	Kilgour, 2nd-Lt. R.	3 Oct. 17	5 D. S. C.
14 Sep. 17	Nicoll, Lt. W. F.	5 Oct. 17	7 D. S. C.
23 Sep. 17	Dumper, 2nd-Lt. E.	5 Oct. 17	3 D. S. C.
5 Oct. 17	Fox 2nd-Lt. C. B.	5 Oct. 17	"H" S. D. C.
16 Sep. 17	Moss, Lt. E. M.	6 Oct 17	"B" S. D. C.
3 Feb. 17	Morgan, Lt. E. J. L.	6 Oct. 17	9 D. S. C.
11 Sep. 17	Bedingfield, Lt. R. H. N.	9 Oct. 17	"C" S. D. C.
20 June 17	Henderson, 2nd-Lt. D. S.	11 Oct. 17	2 D. S. C.
4 Oct. 17	Burgess, 2nd-Lt. W. M.	16 Oct. 17	1 D. S. C.
8 Oct. 17	Dunham, 2nd-Lt. H. H. (A. S. C.)	16 Oct. 17	(*On probn.*) "A" S. D. C.
26 Sep. 17	Rouse, 2nd-Lt. T. A.	18 Oct. 17	1 D. S. C.
	Phillips, 2nd-Lt. H. W.	21 Oct. 17	3 D. S. C.
8 Aug. 17	Marten, 2nd-Lt. S. A.	23 Oct. 17	"H" S. D. C.
4 Oct. 17	Thomas, 2nd-Lt. M. L.	24 Oct. 17	8 D. S. C.
4 Oct. 17	Albrecht, 2nd-Lt. C. L.	27 Oct. 17	7 D. S. C.
4 Oct. 17	Durston, 2nd-Lt. E. R.	29 Oct. 17	9 D. S. C.
26 Sept. 17	Allen, Lt. G. M. H.	3 Nov. 17	1 D. S. C.
29 Aug 17	Hawkins, 2nd-Lt H. W.	6 Nov. 17	4 D. S. C.
31 Oct. 17	West, Lt. G. F.	8 Nov. 17	10 D. S. C.
25 Oct. 17	Tancred, Lt. G. M. W.	9 Nov. 17	"K" S. D. C.
12 Oct. 17	Hill, 2nd-Lt. F. G. A.	10 Nov. 17	5 D. S. C.
10 Oct. 17	Oldall, 2nd-Lt. C. H.	12 Nov. 17	"G" S. D. C.
25 Oct. 17	Gerard, 2nd-Lt. E. L.	13 Nov. 17	8 D. S. C.
1 Nov. 17	Riggs, 2nd-Lt. A. J.	16 Nov. 17	1 D. S. C.
20 Nov. 17	Boucher, 2nd-Lt. A. S. H.	24 Nov. 17	7 D. S. C.
20 Nov. 17	Holmes, 2nd-Lt. J. F.	28 Nov. 17	2 D. S. C.
8 Nov. 17	Staley, 2nd-Lt. G.	29 Nov. 17	"A" S. D. C.
1 Nov. 17	Thornton, 2nd-Lt. G. E.	3 Dec. 17	1 D. S. C.
25 May 16	Hamlet, Lt. R. O. (A. S. C.)	3 Dec. 17	(*On probn.*) 2 D. S. C.
12 June 17	Dempster, Lt. J. (A. S. C.)		(*On probn.*) 2 D. S. C.
13 Oct. 17	Grant-Peterkin, 2nd-Lt. M. H.		"D" S. D. C.
25 Oct 17	Williams, Lt. D.		"J" S. D. C.
21 Nov. 17	Street, 2nd-Lt. W. F.	5 Dec. 17	8 D. S. C.
10 Nov. 17	Roughton, 2nd-Lt. G. E.		"G" S. D. C.
10 Dec. 17	Perry, 2nd-Lt. P. E.	10 Dec. 17	6 D. S. C.
10 Dec. 17	Saunders, 2nd-Lt. J.	10 Dec. 17	4 D. S. C.
10 Dec. 17	Russell, 2nd-Lt. H. G.	17 Dec. 17	8 D. S. C.
14 Sep. 16	Joyce, Lt. T. H.	15 Dec. 17	7 D. S. C.
28 Dec. 17	Judah, 2nd-Lt. D. A.	10 Jan. 18	"H" S. D. C.
4 Dec. 17	Rushton, 2nd-Lt. J. P.	10 Jan. 18	2 D. S. C.
4 Jan. 18	Sullivan, 2nd-Lt. G. M. R.	21 Jan. 18	8 D. S. C.
15 Jan. 18	Moylan, 2nd-Lt. T. J.	25 Jan. 18	3 D. S. C.
7 Dec. 14	Caulfield-Giles, Lt. H. R.	26 Jan. 18	*On probn.*) "F" S. D. C.
22 Dec. 17	Syddall, 2nd-Lt. J.	28 Jan. 18	3 D. S. C.
17 Jan. 18	Pearson, 2nd-Lt. W. G.	28 Jan. 18	"D" S. D. C.
9 Jan 18	Woodgate, 2nd-Lt. H. W.	28 Jan. 18	2 D. S. C.
17 Jan. 18	Forbes, 2nd-Lt. T. C.	29 Jan. 18	5 D. S. C.
18 Jan. 18	Perry, 2nd-Lt. C. A. S.	31 Jan. 18	8 D. S. C.
21 Jan. 18	Knyvett-Hough, 2nd-Lt. C. A.	4 Feb. 18	8 D. S. C.
23 Jan. 18	Hilton, 2nd Lt. J. G. M.	5 Feb. 18	3 D. S. C.
1 Jan. 17	Boynton-Lee, Lt. F.	9 Feb. 18	5 D. S. C.
7 Nov. 15	Squire, Lt. L. G.	13 Feb. 18	7 D. S. C.
5 Sep. 16	Towndrow, 2nd-Lt. F. E.	13 Feb. 18	2 D. S. C.
6 Feb. 18	Mackaness, 2nd-Lt. S. G.	19 Feb. 18	8 D. S. C.
5 Oct. 18	Parker, 2nd-Lt. (*temp. Major*) R. N.	20 Dec. 17	Asst. Contr. of Contracts.

D. S. C.=Divisional Supply Company.
S. D. C.=Supply Depôt Company.

Attached 6th class—*contd.*

Date of First Commn.	Names and Rank.	Date of attachment to Corps.	Remarks.
15 Apl. 17	Ingram, Lt. F. E.	15 Apl. 17	Bannu Bde.
14 Dec. 17	Saunders, Lt. H. A. C.	14 Dec. 17	2 D. S. C.
11 Dec. 17	Smith, Lt. S.	28 Dec. 17	"A" S. D. C.
30 Dec. 17	Vanstone, 2nd-Lt. H. W.	30 Dec. 17	1 D. S. C.
20 Dec. 17	Stone, 2nd-Lt. J. E.	2 Jan. 18	7 D. S. C.
4 Oct. 17	Murphy, 2nd-Lt. J. J.	4 Jan. 18	"B" S. D. C.
24 Dec. 17	Edge, 2nd-Lt. G. P.	10 Jan. 18	Bombay Bde.
	Wood, Lt. E. W. M.	13 Jan. 18	"A" S. D. C.
5 Sep. 17	Startin, 2nd-Lt. H. A.	13 Jan. 18	Karoon Front, Persia.
5 Oct. 17	Hewitt, 2nd-Lt. S. H., I. A. (T. C.)	16 Jan. 18	9 D. S. C.
19 Jan. 18	Mellis, 2nd-Lt. J. S. H., I. A. (T. C.)	29 Jan. 18	7 D. S. C.
1 Feb. 18	Carrington, 2nd-Lt. S. S.	19 Feb. 18	6 D. S. C.
19 Feb. 18	Mackey, 2nd-Lt. R.	28 Feb. 18	2 D. S. C.
	Prain, 2nd-Lt. W.	26 Feb. 18	1 D. S. C.
5 Oct. 17	Kilner, 2nd-Lt. F. W., I. A. R. O.	27 Feb. 18	9 D. S. C.
15 Dec. 16	Cox, 2nd-Lt. H.	1 Mar. 18	"A" S. D. C.
	Lloyd, 2nd-Lt. J. H.	4 Mar. 18	"A" S. D. C.
19 Feb. 16	McKenna, 2nd-Lt. C.	8 Mar. 18	"F" S. D. C.
1 Nov. 17	Frost, 2nd-Lt. F. H., I. A. R. O.	16 Mar. 18	"K" S. D. C.
1 Oct. 17	Bundy, 2nd-Lt. F. C.	16 Mar. 18	"F" S. D. C.
24 June 16	Threlkeld, Lt. S., I. A. R. O.	19 Mar. 18	7 D. S. C.
	King, 2nd-Lt. H. A., I. A. (T. C.)	25 Mar. 18	"G" S. D. C.
12 Mar. 18	Lowe, 2nd-Lt. T. W., I. A. (T. C.)	27 Mar. 18	1 D. S. C.
16 Feb. 18	Woods, 2nd-Lt. C. F.	30 Mar. 18	7 D. S. C.
	Lee, 2nd-Lt. G. T. H.	22 Apl. 18	8 D. S. C.
9 Apl. 18	McIver, 2nd-Lt. A. C.	23 Apl. 18	"C" S. D. C.
20 June 17	Jennings, Lt. J. M.	24 Apl. 18	5 D. S. C.
1 May 18	Clarke, 2nd-Lt. A. A. M.	1 May 18	8 D. S. C.
4 Apl. 17	James, 2nd-Lt. F.	2 May 18	4 D. S. C.
5 Oct. 17	Chisholm, 2nd-Lt. K. M.	2 May 18	Kohat Bde.
2 May 18	Seeley, 2nd-Lt. A. C.	7 May 18	"H" S. D. C.
2 May 18	Chenery, 2nd-Lt. A. F.	7 May 18	"H" S. D. C.
	Radcliffe, Lt. H.	7 May 18	"F" S. D. C.
27 July 17	Patterson, 2nd-Lt. A. A. C.	8 May 18	8 D. S. C.
	Mackenzie, Lt. J. F.	8 May 18	"F" S. D. C.
15 Dec. 14	Mather, Lt. O. H.	8 May 18	8 D. S. C.
	Pallant, 2nd-Lt. H. C.	8 May 18	7 D. S. C.
	Lewis, 2nd-Lt. E. S.	8 May 18	"B" S. D. C.
	Knight, 2nd-Lt. W.	9 May 18	1 D. S. C.
25 Jan. 17	Barber, 2nd-Lt. A. J.	9 May 18	1 D. S. C.
	Medley, 2nd-Lt. J.	9 May 18	"K" S. D. C.
	Cude, 2nd-Lt. J.	9 May 18	"J" S. D. C.
5 Nov. 17	Ashton, 2nd-Lt. F. C.	9 May 18	"A" S. D. C.
	Auston, Lt. G. T.	10 May 18	9 D. S. C.
	Black, Lt. A. S.	10 May 18	"F" S. D. C.
27 Feb. 17	Maclarron, Lt. C. M.	10 May 18	5 D. S. C.
25 Mar. 16	White, 2nd-Lt. M. J.	11 May 18	"D" S. D. C.
5 May 17	Ross, Lt. C. G.	11 May 18	"J" S. D. C.
29 Aug. 17	Harris, Lt. A. F. R.	13 May 18	3 D. S. C.
29 Aug. 17	Middleweek, 2nd-Lt. L. W. S.	13 May 18	"K" S. D. C.
5 May 17	Gibbs, 2nd-Lt. A. R.	13 May 18	"D" S. D. C.
5 May 17	Collett, 2nd-Lt. H. S.	13 May 18	9 D. S. C.
26 Sep. 17	McGorry, 2nd Lt. T. F. E.	7 May 18	7 D. S. C.
20 Dec. 16	Myles, Lt. R. H. S.	14 May 18	5 D. S. C.
28 June 15	Whitley, Lt. L. G. M.	15 May 18	1 D. S. C.
6 May 18	Box, 2nd-Lt. E. G.	15 May 18	"H" S. D. C.
13 Apl. 15	English, Lt. B. M.	16 May 18	Ambala Bde.
	Aird, 2nd-Lt. M.	20 May 18	4 D. S. C.
	Conway, 2nd-Lt. H. D.C.M., M.M.	20 May 18	8 D. S. C.
16 May 18	Weston Foster, 2nd-Lt. A. W.	20 May 18	8 D. S. C.
5 Nov. 17	Thurston, 2nd-Lt. C. W.	21 May 18	8 D. S. C.
8 May 18	Mathalone, 2nd-Lt. M.	22 May 18	"C" S. D. C.
	Obee, 2nd-Lt. L. E.	22 May 18	8 D. S. C.
11 Dec. 14	Walpole, Lt. P. W.	22 May 18	1 D. S. C.
26 June 16	Soden, Lt. H. L. B.	22 May 18	"H" S. D. C.
2 Apl. 17	Tapsell, Lt. E. G.	29 May 18	5 D. S. C.
13 May 18	Peters, 2nd-Lt. A. S. C.	29 May 18	"F" S. D. C.
5 May 17	Houston, 2nd-Lt. C.	1 June 18	10 D. S. C.
1 May 17	Elton, 2nd-Lt. S., I. A. R. O.	7 June 18	[illegible] D. S. C.

D. S. C. = Divisional Supply Company.

S. D. C. = Supply Depôt Company.

Date of First Commn.	Names and Rank.	First apptt. to Corps.	Remarks.

Attached 6th Class—*contd.*

5 Nov. 17	Rogers, 2nd-Lt. J. S., I. A. R. O. .	5 June 18	2 D. S. C.
	Rostron, 2nd-Lt. H.	5 June 18	2 D. S. C.
1 Sep. 17	Roberts, 2nd-Lt. C. J., I. A. (T. C.)	5 June 18	2 D. S. C.
21 May 18	Hawker, 2nd-Lt. G. V.	5 June 18	"J" S. D. C.
15 Jan. 18	Diacano, 2nd-Lt. H. V., I. A. R. O.	6 June 18	"F" S. D. C.
	Tyfe, 2nd-Lt. W. S.		"C" S. D. C.
16 Jan. 16	Nicholas, Lt. W. A.	. . . .	"A" S. D. C.
31 Aug. 17	Hartley, 2nd-Lt. G. C.	10 May 18	10 D. S. C.
5 Oct. 17	Brett, 2nd-Lt. C. E.	. . . .	"A" S. D. C.
7 Jan. 18	Bowden, 2nd-Lt. A. H. M. . . .	11 July 18	7 D. S. C.
19 Jan. 18	Holyoak, 2nd-Lt. A. H. P. I. A. (T. C.)	19 Jan. 18	"K" S. D. C.
23 July 17	Crocker, Lt. A., I. A. R. O. . .	. . .	"C" S. D. C.
15 Feb. 18	Walker, Lt. E. M., I. A. R. . .	. . .	"A" S. D. C.
	Sergeant, 2nd-Lt. J. M.	. . . .	7 D. S. C.
13 Apl. 18	Heyne, 2nd-Lt. G. A. F., I. A. R. O.	. . . .	8 D. S. C.
19 Mar. 18	Hicks, 2nd-Lt. E. S., I. A. R. O. .	. . . .	2 D. S. C.
5 Aug. 14	Flemington, Capt. R. R. . .	. . . .	Ambala Bde.
28 Mar. 18	Gurton-Hobbs, 2nd-Lt. A. G., I. A. (T. C.)	6 May 18	3 D. S. C.
5 Apl. 18	Rhodes, 2nd-Lt. G. E.	. . . .	6 D. S. C.
6 Apl. 18	Carlile, 2nd-Lt. R. L. H., I. A. .	. . . .	"G" S. D. C.
6 Apl. 18	Rose, 2nd-Lt. G. H., I. A. R. O. .	. . . .	"G" S. D. C.
9 Apl. 18	Brocklebank, 2nd-Lt. H. J. . . .	. . . .	"J" S. D. C.
9 Apl. 18	Logan, 2nd-Lt. J., I. A. (T. C.) .	. . . .	6 D. S. C.
26 Sep. 16	Groombridge, Lt. K. C., I. A. R. O.	. .	"G" S. D. C.
3 May 18	Butt, 2nd-Lt. H., I. A. R. O. . .	. . . .	7 D. S. C.
9 May 18	Parsons, 2nd-Lt. J. P. D., I. A. R. O.	9 May 18	"H" S. D. C.
8 Aug. 14	Prinsep, Capt. D. W. McL (I. A.).	10 Apl. 18	Ambala Bde.
20 Nov. 14	Liddell, Capt. J. A., I. A. R. O. .	23 Apl. 18	4 D. S. C.
	Stewart Martin, 2nd-Lt. T. S., I. A. R. O.	7 Apl. 18	Derajat Bde.
18 Dec. 14	Hawkins, 2nd-Lt. R. P., I. A. R. O.	. . . .	3 D. S. C.
27 Nov. 14	Edkins, Capt. R. H., I. A. R. O. .	10 Apl. 18	Kohat Bde.
4 Dec. 14	Whitaker, Lt. B., I. A. R. O. . .	20 Apl. 18	Bannu Bde.
27 Nov. 14	Rust, F. H., I. A. R. O.	5 Apl. 18	2 D. S. C.
20 Nov. 14	Lacey, 2nd-Lt. F. A., I. A. R. O. .	6 Apl. 18	2 D. S. C.
21 July 15	Gosselin, 2nd-Lt. H. de V., I. A. R. O.	12 Apl. 18	2 D. S. C.
16 Aug. 16	Windham, 2nd-Lt. J. B. R., I. A. R. O.	9 Apl. 18	Derajat Bde.
4 Mar. 15	Morris, 2nd-Lt. H. C., I. A. R. O. .	10 Apl. 18	2 D. S. C.
9 Apl. 15	Deacon, 2nd-Lt. C. G., I. A. R. O.	7 Apl. 18	4 D. S. C.
19 Mar. 15	Swayne, 2nd-Lt. I. A. R. O. . .	14 May 18	4 D. S. C.

D. S. C.=Divisional Supply Company.

S. D. C.=Supply Depôt Company.

Attached 6th Class—*contd.*

First Commn.	Names and Rank.	Date of attachment to Corps.	Remarks.
5 Nov. 17	Barraclough, 2nd-Lt. A. E., I. A. (*T. C.*)	4 June 18	3 D. S. C.
1 Oct. 17	Roberts, 2nd-Lt. S. N., I. A. (*T.C.*)	5 June 18	3 D. S. C.
1 Oct. 17	Cullingford, 2nd-Lt. F. H., I. A. (*T. C.*)	6 June 18	3 D. S. C.
1 Oct. 17	Buttress, 2nd-Lt. E. S., I. A. R. O.	23 July 18	Bannu Bde.
	Bromley, 2nd-Lt. B. E.	13 June 18	2 D. S. C.
	Burrows, 2nd-Lt. S. J.	4 June 18	3 D. S. C.
24 June 18	Carlton, 2nd-Lt. R. F. C. L. C.	....	9 D. S. C.
12 July 18	Pugh, 2nd-Lt. J. C.	....	9 D. S. C.
5 July 18	Turner, 2nd-Lt. C. L.	....	"G" S. D. C.
26 June 18	Jarvis, 2nd-Lt. H. L.	8 July 18	9 D. S. C.
4 Sep. 12	Herbert, Capt. H. R.	....	"B" S. D. C.
10 May 18	Collings, 2nd-Lt. F. W., I.A.R.O. (*T. C.*)	....	"B" S. D. C.
26 Sep. 16	Worden, 2nd-Lt. L.	9 July 18	3 D. S. C.
5 Nov. 17	Smart, 2nd-Lt. J. N.	15 July 18	9 D. S. C.
	Gillman, 2nd-Lt. B. B.	15 July 18	9 D. S. C.
	Inglis, Lt. W. M.	....	"B" S. D. C.
20 Jan. 17	Carew, 2nd-Lt. T. P.	2 July 18	7 D. S. C.
21 Jan. 18	Blockley, 2nd-Lt. J. F. G., I. A. (*T. C.*)	1 July 18	9 D. S. C.
5 Nov. 17	Fournier, 2nd-Lt. A. E., I. A. (*T. C.*)	17 July 18	Bannu Bde.
5 Nov. 17	Sanders, 2nd-Lt. H. W., I.A.R.O.	16 July 18	"F" S. D. C.
	Firth, 2nd-Lt. H. V.	....	1 D. S. C.
5 Nov. 17	Gamble, 2nd-Lt. M. F., I.A.R.O.	18 July 18	1 D. S. C.
5 Nov. 17	Pearcy, 2nd-Lt. F. G., I.A.R.O.	....	10 D. S. C.
	Phillips, Lt. E. J.	8 July 18	2 D. S. C.
	Hamilton Smith, Lt. R.	15 July 18	9 D. S. C.
6 Dec. 16	Anderson, Lt. G., I. A. (*on probn.*)	9 July 18	2 D. S. C.
31 Jan. 18	Wells, 2nd-Lt. H. V., I. A.	5 July 18	2 D. S. C.
3 June 16	Weston, 2nd-Lt. N. A.	10 July 18	2 D. S. C.
5 Oct. 17	Spencer, 2nd-Lt. W. G., I. A. (*T. C.*)	12 July 18	Bannu Bde.
5 May 15	Ham, Lt. P. S., I. A. (*on probn.*)	12 July 18	"F" S. D. C.
	Crichton, Lt. W. M.	4 July 18	8 D. S. C.
5 Nov. 17	Pole, 2nd-Lt. G. E., I. A. (*T. C.*)	22 July 18	8 D. S. C.
4 Dec. 14	Beck, Lt. H. C.	....	"G" S. D. C.
31 Oct. 17	Cadman, 2nd-Lt. A. L.	....	"G" S. D. C.
1 Nov. 15	Short, Lt. S. E.	28 June 18	Kohat Bde.
31 Oct. 17	Harrison, 2nd-Lt. H. S.	5 July 18	9 D. S. C.
31 Oct. 17	Harris, 2nd-Lt. E. J.	9 July 18	2 D. S. C.
5 May 17	Finlayson, Lt. T.	....	"B" S. D. C.
30 May 17	Goulden, 2nd-Lt. C. F.	9 July 18	"F" S. D. C.
	Glendenning, Lt. H.	....	Derajat Bde.
28 Sep. 17	Stevenson, 2nd-Lt. J.	....	6 D. S. C.
31 Oct. 17	Miller, 2nd-Lt. R. S. C.	4 July 18	5 D. S. C.
	Hughes, 2nd-Lt. T. E.	4 July 18	3 D. S. C.
20 Aug. 14	O'Callaghan, 2nd-Lt. F. J.	2 July 18	3 D. S. C.
9 Aug. 15	Gordon, Lt. M., I. A. (*on probn.*)	5 July 18	4 D. S. C.
1 Aug. 17	Groocock, 2nd-Lt. G. H., I. A. (*on probn.*)	....	"G" S. D. C.
10 July 18	Field, 2nd-Lt. H.	15 Aug. 18	"C" S. D. C.
10 July 18	Toon, 2nd-Lt. H. E.	19 Aug. 18	"B" S. D. C.
1 Apl. 17	Beynon, 2nd-Lt. G.	25 July 18	"C" S. D. C.
11 July 18	Turner, 2nd-Lt. H. A.	9 Aug. 18	"F" S. D. C.
29 June 18	Kirk, 2nd-Lt. R. T. F.	15 July 18	"C" S. D. C.
21 Mar. 18	Jones, 2nd-Lt. C. A.	30 Aug. 18	4 D. S. C.
18 July 18	Whetham, 2nd-Lt. C. L.	....	"A" S. D. C.
1 May 18	Walker, 2nd-Lt. J., I. A. (*T. C.*)	....	"D" S. D. C.
27 Oct. 15	Tuckey, Lt. R. E. S.	20 July 18	1 D. S. C.
5 May 17	Cox, 2nd-Lt. F. B.	20 July 18	"F" S. D. C.
	Sedgley, Lt. C. H.	22 July 18	7 D. S.C.
5 Oct. 17	Clack, 2nd-Lt. G. H., I. A. (*T. C.*)	25 July 18	8 D. S. C.
	Archard, Lt. G. Q.	....	"B" S. D. C.
3 June 15	Legge, 2nd-Lt. G. E.	2 Aug. 18	3 D. S. C.
20 June 17	Fraser, 2nd-Lt. W. C.	19 Aug. 18	7 D. S. C.

D. S. C.=Divisional Supply Company.
S. D. C.=Supply Depôt Company.

First Commn.	Names and Rank.	Date of attachment to Corps.	Remarks.
	Attached 6th Class—*contd.*		
	Cohen, Lt. N. H.	17 Aug. 18	7 D. S. C.
1 Mar. 17	Robertson, Lt. R. B., I.A.R.O. . .	29 Apl. 18	9 D. S. C.
16 July 18	Anderson, 2nd-Lt.	3 Sep. 18	1 D. S. C.
29 July 18	McQuade, 2nd-Lt. J. L.	19 Aug. 18	"C" S. D. C.
	Wigner, 2nd-Lt. G. D.	5 Aug. 18	7 D. S. C.
	Bickers, 2nd-Lt. S. M.	. . . . .	"B" S. D. C.
	Cox, 2nd-Lt. A. P.	. . . . .	"A" S. D. C.
24 Jan. 16	Parsons, 2nd-Lt. L. C.	31 July 18	3 D. S. C.
24 Dec. 14	Shelverton, Lt. E. M.	. . . . .	"K" S. D. C.
1 Sep. 17	Jackson, Lt. J., I. A. R. O. . . .	22 Apl. 18	3 D. S. C.
25 Mar. 15	Handcock, 2nd-Lt. W. J.	4 May 18	7 D. S. C.
1 Jan. 15	Helmore, 2nd-Lt. G. D.	. . . . .	2 D. S. C.
5 Oct. 17	Chisholm, Lt., K. M.	. . . . .	Kohat Bde.
20 July 15	Sanders, Lt. W. G., I. A. R. O.	. . . . .	2 D. S. C.
16 July 16	Pitt, Lt. F., M. C.	28 Mar. 18	7 D. S. C.
5 May 15	Magner, Lt. J.	. . . . .	9 D. S. C.
5 Oct. 17	Brett, 2nd-Lt. C. E.	. . . . .	"A" S. D. C.
5 Oct. 17	Howitt, 2nd-Lt. G.	. . . . .	Derajat Bde.
29 Apl. 18	Stuart, 2nd-Lt. R., I. A. (*T. C.*) .	. . . . .	8 D. S. C.
20 May 18	Hill, 2nd-Lt. K. M. R.	18 June 18	7 D. S. C.
30 May 18	Maynard, 2nd-Lt. N. C., I. A. R. O.	. . . . .	Ambala Bde.
21 May 18	Roper, 2nd-Lt. G. E., I.A. R. O. .	21 May 18	"B" S. D. C.
	Forster, Lt. F. J.	. . . . .	"A" S. D. C.
23 Sep. 14	Hunt, Capt. G. E. D.	. . . . .	A. H. Q.
27 Nov. 14	Fraser, Lt. A. H. H., I. A. R. O. .	. . . . .	"A" S. D. C.
28 Oct. 15	Wallace, Lt. A. A.	24 June 18	8 D. S. C.
	Lewis, 2nd-Lt. E. S.	. . . . .	"B" S. D. C.
	Brown, 2nd-Lt.	. . . . .	"B" S. D. C.
	Dickenson, 2nd-Lt. A.	. . . . .	Bannu Bde.
	Watney, Lt. R. W.	. . . . .	Ambala Bde.
12 May 16	Bottrill, Lt. E. R.	. . . . .	7 D. S. C.
	Petty, 2nd-Lt. H. G.	. . . . .	4 D. S. C.
5 May 17	Baxter, 2nd-Lt. W. G. E., I.A.R.O.	. . . . .	73rd Bde.
18 Nov. 16	Harvey, Lt. C. W., I. A. R. O. .	13 June 18	7 D. S. C.
5 Jan. 18	Roberts, 2nd-Lt. A. C., I. A. (*T.C.*)	13 June 18	3 D. S. C.
	Phillips, Lt. E. J.	. . . . .	2 D. S. C.
12 May 16	Morrison, Lt. R. M. S., I. A. R. O.	9 May 18	2 D. S. C.
	Hartles, Lt. W. E.	9 May 18	2 D. S. C.
	Carroll, Lt. F. H.	. . . . .	Bombay Bde.
11 Dec. 17	Stevens, 2nd-Lt. H. C.	. . . . .	6 D. S. C.
5 Oct. 17	Mullarkey, 2nd-Lt. P. S. . . .	8 July 18	Bannu Bde.
7 Apl. 15	Davey, Lt. H. R.	25 May 18	2 D. S. C.
	Bartlett, 2nd-Lt. P. S.	. . . . .	Kohat Bde.
14 July 15	Williams, Lt. E. C.	9 May 18	2 D. S. C.
18 Oct. 18	Norton, Lt. G. A.	. . . . .	7 D. S. C.
	Harvey, 2nd-Lt. A. J.	6 May 18	2 D. S. C.
8 Nov. 15	Fortescue, Lt. A.	14 May 18	7 D. S. C.
24 Dec. 14	Cooke, G. P.	15 May 18	7 D. S. C.
28 Nov. 14	Manning, Lt. T. W.	18 May 18	7 D. S. C.
18 June 17	Champion, Lt. H. C.	19 May 18	4 D. S. C.
27 Nov. 16	Skinner, Lt. W. E., I. A. R. O. .	19 May 18	2 D. S. C.
15 Jan. 18	Murphy, Lt. L. A., I. A. R. O. .	10 June 18	"E" S. D. C.
15 Jan. 18	Allen, Lt. C. E., I. A. (*T.C.*) . .	20 June 18	7 D. S. C.
27 May 18	Kingham, 2nd-Lt. P., I. A. R. O. .	24 June 18	10 D. S. C.
4 June 18	Webb, 2nd-Lt. R. M.	10 June 18	"E" S. D. C.
15 May 18	Ramsay, 2nd-Lt. W. F. C., I. A. R. O.	25 May 18	"C" S. D. C.
3 June 18	Searle, 2nd-Lt. H. G., I. A. R. O. .	12 July 18	10 D. S. C.
9 Aug. 17	Bull, 2nd-Lt. A. H.	15 July 18	"B" S. D. C.
29 Aug. 17	Galliene, 2nd-Lt. J. W. H. . . .	16 July 18	"C" S. D. C.
18 June 17	Odling, Lt. R. E.	16 July 18	"C" S. D. C.
15 Aug. 17	Melvin, 2nd-Lt. J. W.	16 July 18	2 D. S. C.
5 June 18	Gillespie, 2nd-Lt. J.	27 June 18	4 D. S. C.
29 May 18	Deane, 2nd-Lt. N. G., I. A. R. O. .	24 July 18	3 D. S. C.
30 May 18	Evans, 2nd-Lt. J. D., I. A. R. O. .	27 June 18	"F" S. D. C.
	Ross, 2nd-Lt. J. B.	24 June 18	"A" S. D. C.

D. S. C.=Divisional Supply Company.
S. D. C.=Supply Depôt Company.

First Commn.	Names and Rank.	Date of attachment to Corps.	Remarks.

Attached 6th Class—*concld.*

	Owen, 2nd-Lt. S. F., I. A. (*T. C.*) .	. . . .	9 D. S. C.
1 June 16	Annely, Lt. F. R. C., I. A. R. O. .	22 July 18	10 D. S. C.
19 Jan. 18	Hordern, 2nd-Lt. R. De. L. .	. . . .	6 D. S. C.
4 Jan. 15	Gamble, 2nd-Lt. V. F.	. . . .	Ambala Bde.
31 Aug. 17	Bowden, Lt. C. C. H.	11 June 18	8 D. S. C.
5 Aug. 16	Riddock, 2nd-Lt. J. D.	13 June 18	8 D. S. C.
31 Oct. 17	Harrison, 2nd-Lt. J. S.	25 June 18	7 D. S. C.
	Segar, 2nd-Lt. W. J.	9 June 18	7 D. S. C.
	Vaughan, 2nd-Lt. E. A.	3 Sep. 18	2 D. S. C.
	Rees, 2nd-Lt. C. H.	30 Aug. 18	"C" S. D. C
29 Aug. 17	McCombe, Lt. A.	29 Aug. 18	2 D. S. C.
19 Aug. 17	Weaver, 2nd-Lt. B. H.	22 Aug. 18	3 D. S. C.
26 Nov. 17	Joshua, 2nd-Lt. W. J.	6 Oct. 18	8 D. S. C.
	Young, Lt. E. G. D.	20 Aug. 18	"A" S. D. C.
	Knowles, 2nd-Lt. L. T.	. . . .	7 D. S. C.
5 Oct. 17	Van Houten, 2nd-Lt. F., I. A. (*T. C.*)	. . . .	2 D. S. C.
	McKinnon, 2nd-Lt. J. A. . . .	. . . .	4 D. S. C.
	McIntyre, 2nd-Lt. J. P.	. . . .	Derajat Bde.
	Rose, 2nd-Lt. F. W.	. . . .	Bannu Bde.
5 May 17	Porter, 2nd-Lt. G. H. J.	. . . .	4 D. S. C.
26 Nov. 17	Darby, 2nd-Lt. G.	30 Aug. 18	7 D. S. C.
	Bishop Smith, 2nd-Lt. J.	. . . .	"B" S. D. C.
	Coleman, 2nd-Lt. P. W. . . .	. . . .	"B" S. D. C.
5 Jan. 18	Winter, 2nd-Lt. J. H., I. A. (*T.C.*)	2 Sep. 18	Derajat Bde.
	Thompson, 2nd-Lt. G. V. L. . .	. . . .	"B" S. D. C.
	Harvey, 2nd-Lt. C. W.	. . . .	"B" S. D. C.
	Farmer, 2nd-Lt. T. M.	. . . .	"B" S. D. C.
	Crabtree, 2nd-Lt. H.	. . . .	
	Watson, 2nd-Lt. J. J.	. . . .	"C" S. D. C.
1 Dec.	Lindsay, Lt. H. C. C., I. A. R. O.	4 Sep. 18	"C" S. D. C.
	Barry, 2nd-Lt. J., I. A. (*T. C.*) . .	. . . .	"F" S. D. C.
	Bunch, 2nd-Lt. J. B. S., I. A. R. O.	. . . .	"J" S. D. C.
	Cheadle, 2nd-Lt. G.	. . . .	"K" S. D. C.
	Wilkins, 2nd-Lt. W. S.	. . . .	9 D. S. C.
	Gardener, 2nd-Lt. A. O. T. . .	. . . .	3 D. S. C.
	Pescod, 2nd-Lt. R. S., I. A. R. O. .	. . . .	"F" S. D. C.
	Butchart, 2nd-Lt. A. D., I. A. R. O.	. . . .	"J" S. D. C.
	Cullen, 2nd-Lt. L.	. . . .	"B" S. D. C.
26 Sep. 17	Crockett, 2nd-Lt. N. S.	. . . .	7 D. S. C.
	Warren, 2nd-Lt. E. H.	. . . .	"A" S. D. C.
6 Sep. 18	Petch, 2nd-Lt. P. B.	. . . .	
8 Dec. 17	Ashworth, 2nd-Lt. S. L., I.A.R.O.	8 Dec. 17	
20 July 17	Carew, 2nd-Lt. T. P.	2 July 18	5 D. S. C.
3 June 15	Legge, 2nd-Lt. G. E.	2 Aug. 18	3 D. S. C.
2 June 18	Marchant, 2nd-Lt. L. B., I.A.R.O.	9 Oct. 18	"H" S. D. C.
7 June 18	Park, 2nd-Lt. W. B., I. A. R. O. .	9 Oct. 18	"H" S. D. C.
7 June 18	Brown, 2nd-Lt. H. W.	12 Oct. 18	1 D. S. C.
	Ash, 2nd-Lt. L.	20 Oct. 18	3 D. S. C.
29 Feb. 17	Benn, 2nd-Lt. F.	31 Oct. 18	1 D. S. C.
	Beard, Lt. R. H. A.	2 Nov. 18	1 D. S. C.
	Symth, 2nd-Lt. C. A.	2 Nov. 18	8 D. S. C.
1 Oct. 18	Smith, 2nd-Lt. R. W. H. . . .	5 Nov. 18	8 D. S. C.
1 Feb. 18	Skinner, 2nd Lt. J. R. R. . . .	13 Nov. 18	9 D. S. C.
	O'Meara, 2nd-Lt.	13 Nov. 18	8 D. S. C.
27 Sep. 18	Joyes, Lt. A., I. A. R. O. . . .	15 Nov. 18	5 D. S. C.
	Ridlington, Lt. H. T.	16 Nov. 18	1 D. S. C.
	Day, 2nd-Lt. I. H.	23 Nov. 18	3 D. S. C.

D. S. C. = Divisional Supply Company.
S. D. C. = Supply Depôt Company.

Attached, 6th Class.

Date of 1st Commn.	Names and Rank.	Date of attachment to Corps.	Remarks.
	Shubart, 2nd-Lt. S. D.	26 Oct. 18.	7 D. S. C.
	Holmes, 2nd-Lt. A. C.	. . . .	"D" S. D. C.
	Frisby, 2nd-Lt. W. T.	22 Nov. 18	"A" S. D. C.
	Dorrington, 2nd-Lt. G. H.	. . . .	"B" S. D. C.
	Harkess, 2nd-Lt. A. E.	22 Nov. 18	"F" S. D. C.
	Biggs, 2nd-Lt. W.	8 Nov. 18	...
	Girling, 2nd-Lt. W. H.	. . . .	2 D. S. C.
	Soar, 2nd-Lt. T.	. . . .	"A" S. D. C.
	Parker, 2nd-Lt. L.	. . . .	"E" S. D. C.
	Silvester Mayers, 2nd-Lt. H.	. . . .	"B" S. D. C.
	Staines, 2nd-Lt. F. J.	. . . .	"E" S. D. C.
	George, 2nd-Lt. J. G.	. . . .	"A" S. D. C.
	Ford, 2nd-Lt. B. H.	23 Nov. 18	2 D. S. C.
	Stewart, 2nd-Lt. J. J.	28 Nov. 18	"B" S. D. C.
	Dobson, 2nd-Lt. G. H.	. . . .	...
	Lee, 2nd-Lt. C. G.	. . . .	...
	Brook, 2nd-Lt. T. F.	. . . .	...
	Walden, 2nd-Lt. W. C.	. . . .	...
	Levick, 2nd-Lt.	. . . .	3 D. S. C.
	Morgan, 2nd-Lt. G. A.	. . . .	...
	Smith, 2nd-Lt. A. G. V.	. . . .	...
	Lowcock, 2nd-Lt. D. R.	. . . .	2 D. S. C.
	Skinner, 2nd-Lt.	. . . .	9 D. S. C.
	Purcell Gildea, Lt. E. J.	. . . .	"J" S. D. C.
	Green, 2nd-Lt. F.	. . . .	2 D. S. C.
	James, 2nd-Lt. W. H.	. . . .	7 D. S. C.
	France, Lt. (*Ty. Capt.*) E.	15 Sep. 18	...
	Paxton, Capt. W. M.	3 Oct. 18	"A" S. D. C.
	Smith, 2nd-Lt. E.	. . . .	7 D. S. C.
	Anderson, 2nd-Lt. G. K.	. . . .	"A" S. D. C.
	Crowe, 2nd-Lt. T.	. . . .	"F" S. D. C.
	Brander, 2nd-Lt. A. A.	5 Oct. 18	7 D. S. C.
	Evans, 2nd-Lt. J. W. P.	25 Oct. 18	"A" S. D. C.
	Mair, 2nd-Lt. A.	. . . .	"K" S. D. C.
	Robertson, 2nd-Lt. A. L.	. . . .	7 D. S. C.
	Treeby, 2nd-Lt. H. J.	22 Oct. 18	"F" S. D. C.
	Gain, 2nd-Lt. C. T. I	. . . .	"H" S. D. C.
	Baring Durban, 2nd-Lt. E. C.	11 Oct. 18	"A" S. D. C.
	Hyder, 2nd-Lt. H. S. (I. A., T. C.)	. . . .	"B" S. D. C.
	Knight, 2nd-Lt. W. N.	. . . .	"K" S. D. C.
	Price, 2nd-Lt. A. W.	. . . .	"B" S. D. C.
	Stannard, 2nd-Lt. G. E.	. . . .	"F" S. D. C.
	Maguire, Lt. J. J., A. S. C.	. . . .	4 D. S. C.
	Park, 2nd-Lt. H. H.	. . . .	7 D. S. C.
	Graburn, 2nd-Lt. N. E. B.	. . . .	"B" S. D. C.
	Page, 2nd-Lt. T. S.	. . . .	"H" S. D. C.
	McWhinnie, 2nd-Lt. R. S.	. . . .	"K" S. D. C.
	Cary, 2nd-Lt. W. A.	. . . .	"J" S. D. C.
	Dickson, 2nd-Lt. E. J.	11 Oct. 18	"G" S. D. C.
	Munro, 2nd-Lt. T. O. A.	15 Oct. 18	"B" S. D. C.
	Lloyd, 2nd-Lt. C. H.	18 Oct. 18	"F" S. D. C.
	Rushton, 2nd-Lt. E. H.	. . . .	"A" S. D. C.
	Day, Lt. L. T. L.	. . . .	2 D. S. C.
	Barron, 2nd-Lt. F. H.	. . . .	"B" S. D. C.
	Bannatyne, 2nd-Lt. N. E.	. . . .	2 D. S. C.
	Goode, 2nd-Lt. H. H.	. . . .	"H" S. D. C.
	Westlake, 2nd-Lt.	15 Oct. 18	"B" S. D. C.

OFFICERS OF THE A. S. C. ATTACHED TO THE S. AND T. CORPS.

Date of Commn.	Names and Rank.	First apptt. to Corps.	Remarks.
11 Oct. 15	Jones, 2nd-Lt. J. E.	7 Mar. 17	"F" S. D. C.
26 Oct. 16	Yule, 2nd-Lt. R.	8 Mar. 17	5 D. S. C.
10 Nov. 16	Sangster Green, 2nd-Lt. G. C. L.	8 Mar. 17	"J" S. D. C.
10 Nov. 16	Boshier, 2nd-Lt. L.	8 Mar. 17	2 D. S. C.
10 Nov. 16	Winn, 2nd-Lt. F. G.	8 Mar. 17	2 D. S. C.
26 Oct. 16	Wyrall, 2nd-Lt. E.	8 Mar. 17	5 D. S. C.
26 Oct. 16	Bonar, 2nd-Lt. H. H.	8 Mar. 17	2 D. S. C.
26 Oct. 16	Field, 2nd-Lt. J.	8 Mar. 17	6 D. S. C.
26 Oct. 16	Farnsworth, Lt. C.	8 Mar. 17	4 D. S. C.
10 Nov. 16	Revell, Lt. L.	8 Mar. 17	2 D. S. C.
10 Nov. 16	Goodwin, 2nd-Lt. W. E.	8 Mar. 17	8 D. S. C.
26 Oct. 16	Logan, 2nd-Lt. A. T.	8 Mar. 17	3 D. S. C.
26 Oct. 16	Leaver, 2nd-Lt. J. B.	9 Mar. 17	3 D. S. C.
10 Nov. 16	Davies, 2nd-Lt. A. F.	9 Mar. 17	7 D. S. C.
26 Oct. 16	Henderson, 2nd-Lt. H. W.	9 Mar. 17	5 D. S. C.
10 Nov. 16	Stevenson, 2nd-Lt. P. R.	10 Mar. 17	2 D. S. C.
10 Nov. 16	Hodgson, 2nd-Lt. J. F.	10 Mar. 17	8 D. S. C.
26 Oct. 16	Whyte, 2nd-Lt. L. J. de B.	14 Mar. 17	Bannu Bde.
26 Oct. 16	Stirling, Lt. C. McK.	14 Mar. 17	4 D. S. C.
26 Oct. 16	Moore, Lt. C. H.	14 Mar. 17	"G" S. D. C.
26 Oct. 16	Cottam, Lt. J. S.	14 Mar. 17	4 D. S. C.
26 Oct. 16	Plumer, 2nd-Lt. G. W. H.	14 Mar. 17	"B" S. D. C.
26 Oct. 16	Giles, Lt. M. G.	14 Mar. 17	4 D. S. C. *Lv., ex I., 6 mos. from 13 Sep. 18.*
10 Nov. 16	Pow, 2nd-Lt. J. S.	17 Mar. 17	7 D. S. C.
10 Nov. 16	King, 2nd-Lt. C. A.	17 Mar. 17	7 D. S. C.
21 Dec. 16	Woodman, 2nd-Lt. J.	12 May 17	"B" S. D. C.
6 Sep. 15	Mansfield, 2nd-Lt. L. G.	12 May 17	3 D. S. C.
15 Nov. 15	Wilcock, 2nd-Lt. C. A.	12 May 17	3 D. S. C.
15 Oct. 15	Reynolds, 2nd-Lt. A. T. M.	12 May 17	5 D. S. C.
15 Oct. 15	Rippen, 2nd-Lt. A. C.	12 May 17	2 D. S. C.
21 Dec. 16	Berthoud, 2nd-Lt. L. A.	30 May 17	3 D. S. C.
21 Dec. 16	Young, 2nd-Lt. J. C.	16 June 17	8 D. S. C.
21 Dec. 16	Moss, 2nd-Lt. S. G.	16 June 17	8 D. S. C.
21 Dec. 16	Neylan, 2nd-Lt. P. J.	18 June 17	"F" S. D. C.
21 Dec. 16	Simpson, 2nd-Lt. C. A.	18 June 17	4 D. S. C.
1 Dec. 16	Adkins, 2nd-Lt. E. W.	31 July 17	7 D. S. C.
19 Aug. 15	Simkin, Lt. F. E.	31 July 17	8 D. S. C.
11 Sep. 15	Cowper, 2nd-Lt. W. E.	31 July 17	8 D. S. C.
13 Nov. 14	Stonex, 2nd-Lt. W.	1 Aug. 17	2 D. S. C.
21 Dec. 16	Brown, 2nd-Lt. S. M.	1 Aug. 17	7 D. S. C.
31 Aug. 15	Froom, Lt. C.	1 Aug. 17	Kohat Bde.
16 May 15	Waltlake, 2nd-Lt. J. H.	1 Aug. 17	7 D. S. C.
21 Dec. 16	Ovens, 2nd-Lt. G. L.	2 Aug. 17	2 D. S. C.
1 Jan. 16	Ross, 2nd-Lt. A. L.	3 Aug. 17	7 D. S. C.
21 Dec. 16	Ramsey, 2nd-Lt. H. L.	4 Aug. 17	4 D. S. C. *Lv., ex I., 6 mos., m. c., from 20 Sep. 18.*
21 Dec. 16	Saveall, 2nd-Lt. E. W.	4 Aug. 17	Karachi Bde.
10 Nov. 16	Irvine, 2nd-Lt. H. E. G.	30 Oct. 17	2 D. S. C.

D. S. C. = Divisional Supply Company.

S. D. C. = Supply Depôt Company.

Transport Corps.

Corps.	Names and Rank.	Present appointment.	Apptt. in Corps.
1st Mule Corps (Cav. Bde.)	Hashmat Ali, Ressaidar, I.O.M.	12 May 15	
2nd Mule Corps (Cav. Bde.)	Sardar Khan, Jemadar	23 June 16	
3rd Mule Corps (Cav. Bde.)	Mohemmad Ismail, Jemadar	27 Dec. 15	
4th Mule Corps (Cav. Bde.)			
5th Mule Corps (Cav. Bde.)	Hassoo Khan, Jemadar	19 Dec. 16	
6th Mule Corps	Khuda Dad Khan, Jemdr.	17 Jan. 16	
7th Mule Corps	Alla Ditta, Jemadar	15 Oct. 16	
8th Mule Corps	Alam Sher, Jemdr.	17 Jan. 16	
9th Mule Corps	Ghulam Muhammad, Ressaidar	1 Sep. 13	
	Bhawal Din, Jemdr.	29 Dec. 16	
	Jahan Dad, Jemdr.	1 Oct. 16	
	Ghulam Rasul, Jemdr	16 Jan. 17	
10th Mule Corps	Amir Ali, Jemdr.	1 June 17	
	Allah Din, Jemadar	2 Sep. 14	
	Baisakhi Ram, Jemadar	14 Jan. 16	
11th Mule Corps			
12th Mule Corps	Sunder Singh, Jemdr.	5 Aug. 12	
13th Mule Corps			
14th Mule Corps	Mauladad Khan, Jemadar	11 Mar. 16	
15th Mule Corps	Mardan Shah, Jemadar	1 Oct. 16	
16th Mule Corps	Sham Singh, Jemadar	1 Oct. 16	
17th Mule Corps	Gulam Nabi, Jemadar	13 Jan. 16	
18th Mule Corps	Ram Ditta, Jemadar	20 Feb. 15	
19th Mule Corps	Khan Alam, Jemadar	3 May 16	
20th Mule Corps	Fatteh Khan, Jemadar	17 Jan. 16	
21st Mule Corps	Tabiat Singh, Jemadar	20 Feb. 15	
22nd Mule Corps	Kehr Singh, Jemadar	20 Feb. 15	
23rd Mule Corps	Baz Khan, Jemadar	3 Oct. 15	
24th Mule Corps	Ramzan, Jemdr.	13 Nov. 17	
25th Mule Corps	Jaffar Ali, Jemadar	23 Feb. 16	
26th Mule Corps	Gama, Jemadar (*tempy. Ressaidar*)	2 Feb. 16	
27th Mule Corps	Dwarka Singh, Subadar (*temp.*)	25 Aug. 05	
	Ram Singh, Ressaidar	2 Sep. 14	
	Ali Mardan, Ressdr.	12 Aug. 16	
	Sham Singh, Ressdr.	3 May 16	
28th Mule Corps	Akbar Khan, Jemadar	23 Sep. 15	
	Shad Mohammad, Jemadar	10 Dec. 17	
	Assa Singh, Jemdr.	1 Jan. 17	
29th Mule Corps	Ali Bahadur, Jemadar	22 Mar. 18	
30th Mule Corps	Ilam Din, Jemadar	18 May 14	
31st Mule Corps	Mahsow, Jemdr.	22 Dec. 16	
32nd Mule Corps	Samandar Khan, Jemdr.	2 Nov. 14	
33rd Mule Corps	Kaim Din, Jemadar	1 Jan. 18	
34th Mule Corps	Karam Ali, Jemadar	11 May 15	
35th Mule Corps	Jahan Dad Khan, Jemadar	18 Dec. 15	
36th Mule Corps	Ali Muhammad, Jemdr.	24 Feb. 17	
37th Mule Corps			
38th Mule Corps	Labh Singh, Jemdr.	4 May 18	
39th Mule Corps	Mardan Shah, Jemdr.	1 Oct. 18	
40th Mule Corps	Sher Ali Khan, Jemdr.	19 July 16	
41st Mule Corps	Shamas Din, Jemdr.	3 May 16	
42nd Mule Corps	Najab Khan, Jemadar	14 Feb. 17	
43rd Mule Corps	Karam, Khan Ressdr.	20 Feb. 15	
44th Mule Corps			
45th Mule Corps	Syed Ahmed Khan, Jemdr.	20 Apl. 18	
46th Mule Corps	Saif Ali Jemdr.	12 Apl. 17	
47th Mule Corps	Baz Khan, Jemdr.	11 May 15	
48th Mule Corps	Gulab Khan, Jemdr.	17 Jan. 16	
49th Mule Corps	Karam Dad, Jemdr.	24 June 18	
50th Mule Corps	Kirpal Singh, Subedar	14 July 17	Patiala State.

Transport Corps—*contd.*

Corps.	Names and Rank.	Present appointment.	Apptt. in Corps.
51st Mule Corps	Hira Singh, Subedar	14 Nov. 13	Patiala State.
52nd Mule Corps	Mohamed Ali, Jemadar	9 Dec. 15	
53rd Mule Corps	Mohamed Ali Nabi, Ressaidar	1 July 14	
54th Mule Corps	Mehar Ali Khan, Jemadar	22 Mar. 17	
55th Pack Mule Corps.	Nabi Buksh, Khan Jemdr.	15 Jan. 18	
56th Mule Corps	Ajit Singh, Risaldar (*S. & T. Reserve*)	23 Mar. 17	
57th Mule Corps	Umer Khan, Ressdr.	14 Jan. 16	
58th Mule Corps	Karim Din, Jemdr.	5 Nov. 17	
59th Mule Corps	Karam Khan, Jemdr.	. . . .	
60th Mule Corps	Alla Din, Jemdr.	2 Nov. 17	
61st Mule Corps	Amir Khan, *Bahadur*, Ressdr.	24 May 13	
62nd Mule Corps	Prem Singh, Jemdr.	13 Jan. 16	
63rd Mule Corps	Nasir Din, Ressaidar	1 Dec. 16	
64th Mule Corps	Nadar Khan, Jemdr.	22 Mar. 18	
65th Mule Corps	Mahmud Khan, Jemdr.	22 Mar. 18	
66th Mule Corps	Mehdi Khan, Ressdr.	9 Oct. 14	
67th Mule Corps	Ali Mohamed Khan, Jemdr.	13 Jan. 16	
68th Mule Corps	Jamal Din, Jemdr.	13 Jan. 16	
69th Mule Corps	Khan Bahadur, Jemdr.	11 May 18	
70th Mule Corps	Sardar Khan, Jemdr.	11 May 18	
71st Mule Corps	Ghulam Husain Khan, Jemdr.	11 May 18	
72nd Mule Corps	Hars Ram, Jemdr.	11 May 18	
No. 1 Ekka Corps	Surbuland Khan, Risldr.	1 Apl. 17	
No. 2 Ekka Corps	Dost Mohamed, Risldr.	28 Nov. 16	
No. 1 Mule Depôt	Yar Mohamed, Jemdr.	5 Nov. 17	
	Piran Ditta, Jemdr.	5 Nov. 17	
	Najif Khan, Jemdr.	. . . .	
	Abdul Rasool, Jemdr.	. . . .	
	Fatteh Din, Jemdr.	1 Oct. 16	
	Jewan Dass, Jemdr.	22 Mar. 18	
	Makhan Singh, Ressdr.	. . . .	
No. 2 Mule Depôt	Fateh Shah, Ressdr.	17 Jan. 16	
	Jagat Singh, Ressdr.	1 Sep. 13	
	Kirpal Singh, Jemdr.	5 Nov. 17	
	Mahbullah, Jemdr.	5 Nov. 17	
	Aulia Khan, Jemdr.	22 Mar. 18	
	Ghulam Mohamed, Ressdr.	1 Sep. 13	
	Lall Khan, Jemdr.	20 Nov. 18	
	Dewa Singh, Risldr. (*S. & T. Reserve*)	29 Oct. 09	
	Het Ram, Risldr. (*S. & T. Reserve*)	26 May 11	
	Amar Singh, Ressdr. (*S. & T. Reserve*)	12 Aug. 16	
	Jhanda Singh, Ressdr. (*S. & T. Reserve*)	1 Dec. 16	
	Bachitar Singh, Ressdr. (*S. & T. Reserve*)	7 Mar. 18	
	Arjan Singh, Ressdr. (*S. & T. Reserve*)	7 Mar. 18	
	Mohamed Taqi Shah, Ressdr. (*S. & T. Reserve*).	7 Mar. 18	
	Ajit Singh, Ressdr. (*S. & T. Reserve*)	12 Aug. 18	
	Janmeji Singh, Ressdr. (*S. & T. Reserve*)	7 Mar 18	
No. 3 Mule Depôt	Rang Baz Khan, Ressdr.	16 Dec. 09	
	Dewan Shah, Jemdr.	5 Nov. 17	
	Saif Ali, Jemdr.	5 Nov. 17	
	Sumunder Shah, Jemdr.	1 June 18	
	Ram Ditta, Jemdr.	21 Feb. 15	
	Benares Khan, Jemdr.	13 July 17	
	Fateh Khan, Jemdr.	30 Oct. 16	
	Ganpat Singh, Ressdr. (*S. & T. Reserve*)	3 May 07	
	Pandit Dewat Ram, Ressdr. (*S. & T. Reserve*.)	11 Aug. 05	
	Kunwar Digbijai Singh, Ressdr. (*S. & T. Reserve*).	26 Aug. 10	
	Prem Singh, Ressdr. (*S. & T. Reserve*)	13 Jan. 17	
	Abbe Ram, Ressdr. (*S. & T. Reserve*)	15 July 10	
	Syed Muzaffar Ali Shah, Ressdr. (*S. & T. Reserve*).	26 Aug. 10	

Transport Corps—*contd.*

Corps.	Names and Rank.	Present appointment.	Apptt. in Corps.
No. 4 Mule Depôt	Atta Mohammed, Subedar	. . . .	Re-employed.
No. 5 Mule Depôt	Painda Khan, Ressdr.	17 Jan. 16	
	Mohamed Mohsin, Jemdr.	11 May 15	
	Mohamad Khan, Jemdr.	2 Aug. 16	
	Fazil Din, Jemdr.	22 Mar. 18	
	Nanak Chand, Jemdr.	4 May 18	
	Shir Singh, Ressdr. (*S. & T. Reserve*)	20 Jan. 11	
	Mohd. Husain Khan, Ressdr. (*S. & T. Reserve*).	14 July 11	
	Syed Timur Ali Shah, Ressdr. (*S. & T. Reserve*).	3 Nov. 17	
	Ramalingam Pillai (*S. & T. Res.*)	26 Feb. 18	
Supply (Porter) Cooly Corps.	Hamel Sing, Jemadar	3 Oct. 15	
	Pir Khan, Jemadar	3 Oct. 15	
No. 1 Dt. Bullock Corps.	Mukram Singh, Ressaidar	6 Sep. 16	
	Nasiruddin, Ressaidar	11 Dec. 16	
	Hasso Khan, Ressaidar	13 Dec. 16	
	Moti Singh, Jemadar	6 Sep. 16	
No. 2 Dt. Bullock Corps.	Shujat Ali, Jemdr. (*Ag. Ressdr.*)	12 Aug. 16	
	Sheo Shanker Pathak, Ressaidar	7 Sep. 16	
	Mangal Singh, Ressaidar.	1 Jan. 18	
No. 3 Jat Bullock Corps.	Dewan Chand, Jemadar (*Actg. Ressdr.*)	2 Sep. 14	
	Feroze Khan, Jemadar (*Actg. Ressdr.*)	10 Apl. 17	
	Syed Issrail Shah, Ressdr.	19 Aug. 18	
	Khader Patcha, Ressdr.	10 May 17	
No. 4 Dt. Bullock Corps.	Sardar Sujan Singh, Ressdr. (*Actg. Risaldr.*)	. . .	S. & T. Res.
	Shaikh Wada, Ressdr.	. . .	
No. 5 Dt. Bullock Corps.	Bharu Singh, Ressdr.	22 Dec. 16	
	Irshad Ali, Ressdr.	June 18	
	Jhanda Singh, Ressdr.	Dec. 18	
	Subramanta Pillai, Ressaidar (*S. & T. Res.*)	16 June 17	
No. 6 Bullock Corps.	Kishnaswami Naidu, Resldr.	21 Oct. 17	
	Abdul Samad, Resldr.	23 May 18	
	A. T. Sivansankaran, Ressdr.	21 May 18	
No. 7 Bullock Corps.	Mohamed Nurul Haq, Ressdr. (*S. & T. Res.*)	. . .	
	Syed Umar, Ressdr.	24 Apl. 18	
	Mohamed Haji, Resdr.	24 Apl. 18	
	Kundan Singh, Ressdr. (*S. & T. Res.*)	. . .	
No. 8 Bullock Corps.	Budruddin, Risldr. (*S. & T. Res.*)	. . .	
	Ranishwar Singh, Resldr.	17 Apl. 15	
	Shewbulak Singh, Ressdr.	26 July 18	
	Norendro Singh, Ressdr. (*S. & T. Res.*)	16 Apl. 18	
No. 9 Bullock Corps.	Mohd. Yusuf Ali Shah, Ressdr	23 Aug. 18	
	Newal Singh, Jemdr.	17 May 15	
	Hari Singh, Jemdr.	30 Apl 18	
No. 11 Bullock Corps.	Ditta Ram, Rissaldar	15 July 10	
1st Govt. Camel Corps.	Lall Khan, Ressdr.	22 Oct. 15	
	Sikendar Shah, Ressdr.	9 June 17	
	Dina Nath, Ressdr.	21 Nov. 18	
	Aslam Khan, Ressdr.	17 June 18	
	Faiz Buksh, Ressdr.	21 June 18	
2nd Govt. Camel Corps.	Fateh Khan, Ressdr.	12 Apl. 16	
	Nanak Singh, Risldr.	29 Sep. 15	
	Naiz Ahmed Khan, Ressdr.	10 Oct. 12	
	Faiz Buksh, Ressdr.	21 June 18	
	Ghulam Hasain Khan, Ressdr	1 Dec. 13	
	Dost Muhammad Khan, Ressdr.	23 Dec. 16	Asst. to Inspr. of Camel Corps.
3rd Govt. Camel Corps.	Sharaf Ali, Risldr. (*S. & T. Res.*)	13 Feb. 16	
	Harnam Singh, Ressdr.	30 Oct. 11	
	Fazaldad Khan, Ressdr.	21 Mar. 16	
	Ram Biswas Singh, Ressdr.	21 Oct. 16	
4th Govt. Camel Corps.	Suchet Singh, Risaldar	7 May 07	On staff of Inspector of Camel Corps.
	Ghulam Hussain Khan, Ressdr.	1 Dec. 13	
	Umar Daraz Khan, Ressdr.	20 July 14	
	Dost Muhammad, Ressdr.	13 June 18	
	Ali Mahammad, Ressdr.	25 June 18	
5th Govt. Camel Corps.	Nur Din, Ressaldar (*S. & T. Res.*)	20 Mar. 14	
	Gopi Chand, Ressaidar	30 Sep. 15	
	Akbar Khan, Ressdr.	19 Jan. 16	
	Painda Khan, Ressdr.	20 Oct. 16	
	Khuda Buksh, Risaldar	17 Feb. 09	On staff of Insp., Camel Corps.

Transport Corps—*contd.*

Corps.	Names and Rank.	Present appointment.	Station.
6th Govt. Camel Corps.	Hazara Singh, Risaldar (*S. & T. Reserve*)	19 Oct. 16	
	Benni Singh, Ressaidar	18 Oct. 16	
	Wazir Beg, Risldr.	22 Aug. 14	
	Dadan Khan, Ressdr.	19 Jan. 16	
	Ram Biswas Singh, Ressaidar	21 Oct. 16	
7th Govt. Camel	Parduman Singh, Risldr. (*Actg.*) (*S. & T. Res.*)	20 Feb. 16	
	Ghulam Mohd. Khan, Ressdr.	22 Aug. 14	
	Tara Singh, Ressaidar	15 Oct. 16	
	Feroz Khan, Ressaidar	15 Oct. 16	
	Hansraj Madhok, Ressdr.	3 Aug. 17	
8th (Patiala) Camel Corps.	Hardam Singh, Risaldar	. . .	Patiala State.
	Gurbachan Singh, Risaldar	. . .	Patiala State.
	Sant Singh, Ressdr.	. . .	Patiala State.
	Narain Singh, Ressdr.	. . .	Patiala State.
9th Govt. Camel Corps.	Ghulam Haidar Khan, Ressdr	. . .	
10th Govt. Camel Corps.	Mohammad Fazil Khan, Risaldar	1 June 14	
	Rattan Singh, Risaldar	12 Dec. 16	
	Badshah Hussain, Ressdr.	28 Aug. 16	
	Gulab Singh, Ressdr.	28 Aug. 16	
50th Camel Corps	Nawab Khan, Risaldar	15 May 06	
	Ghulam Haidar Khan, Ressdr.	7 Nov. 18	
	Kala Khan, Ressdr.	28 Oct. 18	
51st Camel Corps	Rajwali Khan, Risaldar	1 Oct. 08	
	Shahwali Khan, Risaldar	20 July 14	
	Ali Akbar Shah, Ressdr.	8 Mar. 16	
	Nizam Din, Ressdr	26 Oct. 18	
52nd Camel Corps	Zaman Khan, Risaldar	1 Apl. 16	
	Jamroz Ali Khan, Ressaidar	30 Sep. 15	
	Gurdit Singh, Ressdr. (*S. & T. Res.*)	17 May 12	
	Sultan Khan, Ressaidar	4 Feb. 16	
53rd Camel Corps	Muhammad Afzal Khan, Risaldar	1 Jan. 07	
	Jai Ram, Risaldar	6 Sep. 16	
	Alam Shah, Ressdr.	4 July 18	
	Abdullah Khan, Ressaidar	6 Sep. 16	
54th Camel Corps	Mehr Ali Khan, Risaldar	1 Apl. 10	
	Rurh Singh, Ressdr. (*Actg. Risldr.*)	3 Oct. 11	
	Sher Mohamad Khan, Ressdr.	1 Oct. 16	
	Mehdi Khan, Ressdr. (*S. & T. Res.*)	18 Nov. 18	
55th Camel Corps	Mohamed Ayoob Khan, Risldr.	6 Nov. 08	
	Nasir Khan, Risaldar	1 Oct. 09	
	Sunder Singh, Ressaidar	1 Apl. 16	
	Fida Mohammed, Ressdr.	27 Feb. 17	
56th Camel Corps	Zabar Khan, Risaldar	8 June 11	
	Inder Singh, Ressaidar (*Actg. Risldr.*)	1 June 14	
	Abdul Razak Khan, Ressdr.	19 Sep. 16	
	Sher Muhammed Khan, Ressaidar	6 Sep. 16	
57th Camel Corps	Munir Khan, *Khan Sahib*, Risaldar	8 Jan. 03	
	Rahim Ullah Khan, Risaldar	6 Aug. 11	
	Kamal Shah, Ressaidar	18 Jan. 16	
	Muhammad Akbar (*Retd.*) Risaldar	. . .	
59th Grantee Camel Corps.	Dost Muhammad Khan, Ressdr.	11 Apl. 16	
	Ghulam Sarwar Khan, Ressdr.	30 Sep. 15	
	Ahmad Khan, Ressdr.	30 Sep. 15	
	Muhammad Khan, Ressdr.	2 May 17	
	Abdul Ghani, Ressdr.	8 Jan. 18	
60th Grantee Camel Corps.	Alam Sher Khan, Risaldar	1 Apl. 09	
	Hakim Khan, Risaldar	11 Mar. 16	
	Sammunder Khan, Ressaidar	29 Mar. 16	
	Ghulam Kadir Khan, Ressdr.	. . .	

Corps.	Names and Rank.	Present appointment.	Apptt. in Corps.

Transport Corps—*concld.*

61st Grantee Camel Corps.	Mohammad Nasir Khan, Risaldar	9 Nov. 14	
	Rattan Singh, Risldr. (*S. & T. Res.*)	26 May 11	
	Mohd. Aslam Khan, Ressdr.	21 Nov. 18	
	Larasil Khan, Ressdr.	21 Nov. 18	
62nd Grantee Camel Corps.	Mir Alam Khan, Risaldar	10 Mar. 14	
	Abdul Gafoor Khan, Ressdr.	27 Feb. 13	
	Mohamad Shafi, Ressdr.	20 Mar. 17	
	Ajab Khan, Ressdr.	7 Sep. 16	
65th Camel Corps	Dilbagh Singh, Risldr. (*S. & T. Res.*)	20 Mar. 14	
	Mian Mardan Ali, Ressdr (*S. & T. Res.*)	12 Aug. 10	
	Ali Asghar, Ressdr.	7 Apl. 16	
	Rahim Bakhsh, Ressdr.	7 Apl. 16	
71st Camel Corps.	Inder Singh, Risaldar (*S. & T. Res.*)	15 July 10	
	Topan Ram, Ressdr.	7 Nov. 18	
	Malik Fateh Khan, Ressdr. (*S. & T. C. Res.*)	7 Nov. 18	
	Hira Singh Ressdr. (*S. & T. C. Res.*)	28 Oct. 18	
72nd Camel Corps.	Karam Ali Khan, Risaldar	3 Feb. 16	Patiala State.
	Kehar Singh, Risldr.	4 Feb. 16	Patiala State.
	Mul Raj, Ressaidar	18 Jan. 16	
	Mohammad Khan, Ressaidar	19 Jan. 16	
No. 6th Camel Corps.	Sher Mohd. Khan, Ressdr. (*S. & T. Res.*)	15 July 10	
	Rahmat Khan, Risldr.	12 Sep. 12	
	Hakim Ali Khan	15 May 18	

TRANSPORT REGISTRATION OFFICERS.

Stewart, Lt-Col. R. J. T.	....	Dera Ismail Khan Circle.
Deighton, Capt. F. J.	1 Sep. 16	Peshawar.
Shine, Maj. E. P. F.	10 Aug. 14	Ambala.
Leicester, Maj. B., late Ches. R.	2 Dec. 14	Quetta.
Ryves, Maj. B. H.	Mar. 18	Agra.
Harvey, Lt. A. J.	....	Rawalpindi.
Craig-Adams, Lt. R. L.	19 Apl. 18	Ambala.

First Commission.	Rank and Name.	Jemadar.	Ressaidar.		District of Residence.
INDIAN OFFICERS OF THE SUPPLY AND TRANSPORT CORPS RESERVE.					
Risaldars.					
21 July 05	Indar Singh	21 July 05	. . . .	15 July 10	Jullundur.
11 Aug. 05	Rana Sher Mohamed Khan, *Rao Bahadur*	11 Aug. 05	. . . .	15 July 10	Hoshiarpur.
11 Aug. 05	Raja Singh, *Rai Sahib*	11 Aug. 05	. . . .	15 July 10	Ambala.
3 May 07	Ganpat Singh	3 May 07	. . . .	15 July 10	Gurgaon.
27 Nov. 07	Chaudhri Amar Singh	27 Nov. 07	15 July 10	26 May 11	Bulandshahr.
11 Aug. 05	Ratan Singh	11 Aug. 05	. . . .	26 May 11	Ambala.
11 Aug. 05	Dewat Ram	11 Aug. 05	15 July 10	20 Mar. 14	Rohtak.
24 Nov. 06	Nur Din	24 Nov. 06	15 July 10	20 Mar. 14	Lyallpur.
24 Nov. 06	Dilbagh Singh	24 Nov. 06	15 July 10	20 Mar. 14	Lyallpur.
26 May 11	Het Ram	. . . .	26 May 11	9 July 15	Hissar.
1 Feb. 07	Sujjan Singh	. . . .	. . . .	11 Dec. 16	Lahore.
15 July 10	Dillu Ram	. . . .	. . . .	11 Dec. 16	Hoshiarpur.
20 Sep. 12	Budruddin	. . . .	20 Sep. 12	23 Mar. 17	Agra.
25 Jan. 07	Ajit Singh	25 Jan. 07	15 July 10	23 Mar. 17	Amritsar.
29 Oct. 09	Dewa Singh	29 Oct. 09	15 July 10	23 Mar. 17	Amritsar.
17 May 12	Muhamad Amir Khan	. . . .	17 May 12	23 Mar. 17	Khanpur.
17 May 12	Farid Khan	. . . .	17 May 12	23 Mar. 17	Multan.
Ressaidars.					
11 Aug. 05	Ganda Singh	11 Aug. 05	15 July 10		Ambala
11 Aug. 05	Roda Ram	11 Aug. 05	15 July 10		Jubbulpore.
8 Feb. 07	Umar Daraz Khan	8 Feb. 07	15 July 10		Jubbulpore.
3 May 07	Hazura Singh	3 May 07	15 July 10		Ambala.
1 Nov. 07	Sadanandan Singh	1 Nov. 07	15 July 10		Sultanpur.
29 Oct. 09	Purdumman Singh	29 Oct. 09	15 July 10		Ludhiana.
12 Dec. 09	Harnam Singh	3 Dec. 09	15 July 10		Hoshiarpur.
15 July 10	Abbe Ram	. . . .	15 July 10		Rohtak.
12 Aug. 10	Quazi Mohamed Akbar Khan	. . . .	12 Aug. 10		Hazara.
12 Aug. 10	Mian Mardan Ali	. . . .	12 Aug. 10		Mianwali.
26 Aug. 10	Chaudhri Nihal Singh	. . . .	26 Aug. 10		Bijnor.
26 Aug. 10	Thakur Kundan Singh	. . . .	26 Aug. 10		Aligarh.
26 Aug. 10	Kanwar Dig Bijay Singh	. . . .	26 Aug. 10		Bulandshahr.
26 Aug. 10	Syed Muzzaffar Ali Shah	. . . .	26 Aug. 10		Meerut.
28 Oct. 10	Hoshiar Singh	. . . .	28 Oct. 10		Karnal.
20 Jan. 11	Gujar Singh	. . . .	20 Jan. 11		Amballa.
20 Jan. 11	Shib Singh	. . . .	20 Jan. 11		Gurdaspur.
27 May 10	Muhammad Nur-ul-Hak	27 May 10	26 May 11		Meerut.
27 May 10	Ahmad Khan	27 May 10	26 May 11		Peshawar.
26 May 11	Mai Sukh	. . . .	26 May 11		Hissar.
26 May 11	Md. Maksud Ali Khan	. . . .	26 May 11		Saharanpore.
7 July 11	Shamsul Haq	. . . .	7 July 11		Jhang.
14 July 11	Muhammad Hussain Khan	. . . .	14 July 11		Ferozepore.
27 May 10	Sher Ali	27 May 10	28 July 11		Gujrat.
4 Aug. 11	Mukhtiar Singh	. . . .	4 Aug. 11		Rohtak.
17 May 12	Abdul Ghani	. . . .	17 May 12		Hissar.
17 May 12	Gholam Kadar Khan	. . . .	17 May 12		Muzaffargarh.
17 May 12	Kadar Baksh	. . . .	17 May 12		Muzaffargarh.
17 May 12	Gurdit Singh	. . . .	17 May 12		Hoshiarpur.
11 July 13	Ghulam Haider Khan	. . . .	11 July 13		Rawal Pindi.
20 Mar. 14	Thakur Shiva Dhyan Sinha	. . . .	20 Mar. 14		Aligarh.
21 May 15	Fateh Mohamed	. . . .	21 May 15		Gujrat.
11 Aug. 16	Amar Singh	. . . .	11 Aug. 16		Gurdaspur.
1 Dec. 16	Jhandu Singh Man	. . . .	1 Dec. 16		Manawala.
15 Dec. 16	Hazdas Singh	. . . .	15 Dec. 16		Amritsar.
15 Dec. 16	Lakhi Singh	. . . .	15 Dec. 16		Hissar.
15 Dec. 16	Sheo Ram	. . . .	15 Dec. 16		Hissar.
15 Dec. 16	Munshi Ram	. . . .	15 Dec. 16		Hissar.
12 Jan. 17	Ghulam Sarwar Khan	. . . .	12 Jan. 17		Multan.
13 Jan. 17	Sardar Kumar Prem Singh	. . . .	13 Jan. 17		Saran.
16 June 17	T. M. Subramania Pillai	. . . .	16 June 17		Secunderabad.
3 Aug. 17	Hans Raj Madhok	. . . .	3 Aug. 17		Gujranwalla.
23 Oct. 17	Syed Timur Ali Shah	. . . .	23 Oct. 17		Meerut.
26 Feb. 18	M. Ramalingam Pillai	. . . .	26 Feb. 18		Secunderabad.
7 Mar. 18	Sardar Sahib Chowdhery . achitar Singh	. . . .	7 Mar. 18		Karnal.
7 Mar. 18	Arjan Singh	. . . .	7 Mar. 18		Ferozepore.
7 Mar. 18	Mohamed Taqi Shah	. . . .	7 Mar. 18		Multan.
7 Mar. 18	Janmeja Singh	. . . .	7 Mar. 18		Lahore.
14 Mar. 18	Sham Singh	. . . .	14 Mar. 18		Pensioner, 22nd Cav.
12 Mar. 18	Sheo Ram	. . . .	22 Mar. 18		,, 6th Cav.
16 Apl. 18	Norendro Singh	. . . .	16 Apl. 18		Hardoi.
7 May 18	Ranjit Ram	. . . .	7 May 18		Ferozepur.
7 May 18	Jas Ram	. . . .	7 May 18		Ferozepur.
7 May 18	Hiluluddin	. . . .	7 May 18		Multan.
7 May 18	Mahindar Singh	. . . .	7 May 18		Ambala.
7 June 18	Sirdar Norendra Singh	. . . .	7 June 18		Hardoi.
12 Aug. 18	Ajit Singh	. . . .	12 Aug. 18		Gurdaspur.
19 Aug. 18	Syed Israil Shah	. . . .	19 Aug. 18		Meerut.
11 Oct. 18	Tayab Ali Shah	. . . .	11 Oct. 18		Meerut.
11 Oct. 18	Hanza Ali Shah	. . . .	11 Oct. 18		Meerut.

TRANSPORT VETERINARY ASSISTANTS, 1ST GRADE.

Ranking as Jemadars.

Names.	Date of entering service.	Date of permanent rank.	Remarks.
Shaikh Ali	11 June 90	30 Apl. 13	Mandalay, 25th M. C.
Hem-raj	10 Apl. 78	30 Apl. 13	Lahore Cantt.
Muhammad Ally	7 Dec. 86	30 Apl. 13	No. 2 Mule Depôt.
Muhammad Shaffi	11 June 87	30 Apl. 13	23rd Mule Corps.
Ali Muhammad, *Khan Sahib*	25 June 97	30 Apl. 13	1st M. C.
Abdullah	3 Sep. 97	30 Apl. 13	6th M. C.
Saiyid Ismail	9 Jan. 92	31 Mar. 14	No. 5 Bullock Corps.
Abdul Rasul	11 Aug. 93	1 July 14	On Service.
Muhammad Abbas	16 Dec. 95	21 Nov. 14	57 S. C. C.
Sharaf Ali Khan	6 Aug. 97	20 Feb. 15	34th Mule Corps.
Niamat Khan	23 Sep. 97	20 Feb. 15	7th Mule Corps.
Ghulam Mohamed Imandin	18 Aug. 92	24 Dec. 17	Aden Local Transport.
Muhammad Suleman	8 May 99	1 Dec. 15	No. 2 Ekka Corps (Supernumerary).

TRANSPORT VETERINARY ASSISTANTS, 2nd GRADE.

Warrant Officers.

Names.	Date of entering service.	Date of permanent Rank.	Remarks.
Bhagwan Singh	30 June 97	1 Apl. 06	
Mangal Sen	12 July 99	1 May 06	29th M. C. With Camel Corps, Nushki-Seistan.
Mulk Shah	18 Feb. 95	17 Aug. 06	32nd M. C.
Saiyid Kamal	30 May 91	30 Apl. 09	7th Bullock Corps.
Vishnu Malhar Bandevadikar	1 Nov. 04	3 Sep. 10	No. 3 Mule Depôt,
Wilayat Ali	27 Feb. 95	11 May 11	On field service.
Muhammad Sahib	26 Jan. 90	11 Aug. 11	On field service.
Ahmad Jan	27 Jan. 00	30 Apl. 13	24th M. C.
Nur Muhammad	30 Mar. 00	30 Apl. 13	24th M. C.
Ishar Singh	7 Feb. 96	2 Nov. 14	54th Camel Corps.
Nur Muhammad Khan	21 June 00	30 Apl. 13	55th Camel Corps.
Sham Singh	17 Dec. 00	30 Apl. 13	On field service.
Sadullah Khan	8 May 01	22 May 14	4th M. C.
Kalab Hussain	19 Feb. 00	20 July 14	10th M. C.
Ali Muhammad	12 Feb. 92	2 Sep. 14	5th M. C.
Shujait Ali	16 Mar 00	20 Feb. 15	No. 1 Mule Depôt.
Ghulam Mohiuddin	17 May 01	20 Feb. 15	On field service.
Amir Ahmed	19 Feb. 00	11 May 15	8th M. C.
Golam Mohamed	3 June 01	3 Jan. 18	Bandar Abbas.

ARMY SERVICE CORPS (MECHANICAL TRANSPORT, INDIA).

Majors.

Crispin, E. H. (*Maj., 8 Jan. 15)	1 June 16

Captains.

Reyner, B. * † (Actg. Maj., 17 Feb. 17)	26 July 15
Douglas, F. W. * † (Actg. Maj., 17 Feb. 17)	16 Dec. 15
Bartlett, R. F. (Actg. Maj., 17 Feb. 17)	5 July 16
	8 July 16
Doudney, R. V. * † (Actg. Maj., 9 Apl. 18)	18 Sep. 16
May, * K. V.	1 July 17
Berridge, H. H.	29 Nov. 17

Captains—Special Reserve.

Thomas, N. S. (temp. Maj. 10 June 16)	29 Nov. 17

Lieutenants.—Special Reserve.

Thomas, N. S. (tempy. Maj. 10 June 16)	M. T.	23 Feb. 16
Ross, J. W. (on probn.)		29 Dec. 16

Lieutenants.

Mutch, C. H. * (Actg. Capt., 5 May 17)	1 Aug. 15
Eden, G. M. (T) (Actg. Capt., 2nd Oct. 16)	1 June 16
Gorman, J. (Actg. Maj. 13 Sep. 18)	1 June 16
Stevens, H. L. *	27 Sep. 16
Thompson, S. M. *	18 Oct. 16
Nasmith, S. U. (Actg. Capt., 8 Feb. 18)	23 Nov. 16
	1 June 16
Backhouse, B. S. *	1 July 17
Brock, W. *	1 July 17
Burcher, F. D. H. *	1 July 17
Symington, K. B. *	1 July 17
Cafferata, (temp. Capt., 1 Oct. 18) B. J. *	30 Sep. 17
Dalrymple, B. W. *	13 Dec. 17
Drummond, J. G. L. *	16 Apl. 18
Messenger, J. W. *	16 Apl. 18
Walford, J. N. *	16 Apl. 18

Second-Lieutenants.

Statham, W. *	20 Mar. 16
Evans, L. E. *	24 Apl. 16
Hyland, H. F. *	21 Aug. 16
Beale, R. E. C. *	16 Oct. 16
Singleton, J. *	4 Dec. 16
Hamblyn, H. H. *	14 Mar. 17
Potter, W. N. *	14 Mar. 17
Tuckey, G. * (Actg. Capt. 6 Aug. 18)	14 Mar. 17

* Temporary.
† Acting.
(T) Territorial.

OFFICERS SERVING ON GROUP H. QRS., ETC.

Locke-Elliott, Lt.-Genl. *Sir* E., K.C.B., D.S.O., Adviser, Indian Labour.

Myles (*Temp. Capt.*) A. V., I. A., Retd. List, Commanding 119th Labour Coy.

Perrott, 2nd-Lt. (*Temp. Lt.-Col.*) H. R. S., C. S. (I. A. R. O.), Commanding 80th Labour Group Headquarters.

Playfair, Lt.-Col. A. (I. A.), 27th (Lushai) Indian Labour Corps, Commanding 62nd Labour Group Headquarters.

Scott, 2nd-Lt. (*Temp. Lt.-Col.*) J. E., I. A. R. O., Commanding 36th Labour Group Headquarters.

Cole, Lt.-Col. H. W. G., C.S.I., Commanding 21st Labour Group Headquarters.

Lashkari, Temp. Lt. P. F., I. M. S., 62nd Labour Group Headquarters.

Narayan, Temp. Lt. R., I. M. S., 36th Labour Group Headquarters.

Sanyal, Lt. S. K. (I. M. S.), 68th Labour Group Headquarters.

McNair, Lt. D. C. (I. M. S.), 68th Labour Group Headquarters.

Carwithen, Lt.-Col E. T. (I. A.), Commanding 75th Labour Group Headquarters.

Corps.	Names, Rank and Corps.	Remarks.
	OFFICERS SERVING WITH COMPANIES.	
33rd (Bihar) Company	Starky, Lt. W. B. (*Temp. Capt.*), Brit. Ser. (T. F.)	
	Manners, Temp. Lt. R., Gen. List	
	Pogose, Temp. Lt. R., Brit. Ser.	
	Clarke, Temp. 2nd-Lt. N. (I. A. R. O.)	
	Anderson, Temp. 2nd-Lt., Brit. Ser.	
55th (U. P.) Company	Innes, Capt. H. D'A., Brit. Ser. (T. F.)	
	Hallahan, 2nd-Lt. R. J. (I. A. R. O.)	
	Carter, 2nd-Lt. G. T. (I. A. R. O.)	
78th (Burman) Company	Street, 2nd-Lt. (*Temp. Capt.*) W. (I. A. R. O.)	
	Head, 2nd-Lt. H. (I. A. R. O.)	
	MacFarquhar, Temp. 2nd-Lt. J., Brit. Ser.	
	Grove, Temp. 2nd-Lt. W. E. U., Brit. Ser.	
	Ashe, Temp. 2nd-Lt. A. E., Brit. Ser.	
	Macara, Temp. 2nd-Lt. J., Brit. Ser.	
79th (U. P.) Company	Rae, Temp. 2nd-Lt. A. E. P., Brit. Ser.	
	Bogue, Temp. 2nd-Lt. W., Brit. Ser.	
	Laughton, Temp. 2nd-Lt. G. G., Brit. Ser.	
	Cook, Temp. 2nd-Lt. A. W., Brit. Ser.	
	Whitten, 2nd-Lt. A. M., Brit. Ser.	
80th (Santal) Company	Gordon, 2nd-Lt. (*Actg. Capt.*) G., Brit. Ser.	
	White, Temp. 2nd-Lt. H. P. G., Brit. Ser.	
	Moore, Temp. 2nd-Lt. A. W., Brit. Ser.	
	Campbell, Temp. Lt. R. B., Brit. Ser.	
	Lusty, Temp. 2nd-Lt. G. H., Brit. Ser.	
81st (Santal) Company	Winsland, Lt. (*Temp. Capt.*) F. E., Brit. Ser.	
	Metcalf, Temp. 2nd-Lt. E. (I. A. R. O.)	
	Bay, Temp. 2nd-Lt. G. J., Brit. Ser.	

Corps.	Names, Rank and Corps.	Remarks.

OFFICERS SERVING WITH COMPANIES—*contd.*

82nd (M. P.) Company	Twigg, Capt. H. D. (Brit. Columbia Regt.), Brit. Ser.	
	Walker, Temp. 2nd-Lt., Brit. Ser.	
	Tavner, Temp. 2nd-Lt. E. J., Brit. Ser.	
83rd (H. W. F.) Company	Beattie, Lt. (*Temp. Capt.*) M. H., Brit. Ser.	
	Beresford-Barber, Temp. 2nd-Lt. H., Brit. Ser.	
84th (Garo) Company	Gilmore, Lt. (*Temp. Capt.*) J. E., Brit. Ser.	
	Burton, Temp. 2nd-Lt. W. H., Brit. Ser.	
	Lansdale-Ruthven, Temp. 2nd-Lt. H. P., Brit. Ser.	
85th (Kumaon) Company	Howard, Lt. G. B. (I. A.)	
	Conroy, Temp. 2nd-Lt. H. T., Brit. Ser.	
	Cowley-Brown, 2nd-Lt. (*Temp. Capt.*) H. T., Brit. Ser.	

OFFICERS SERVING WITH DEPÔTS.

Combined Labour Corps Depôt, Lucknow.	Graham, Capt. G. D.	Comdt.
	Pettigrew, Lt. W.	Adjt. & Qr.-Mr.
	Woods, Lt. N.	
	Deacon, Lt. H.	Acct. Offr.
	Turner, Capt. J. C. H.	
	Hitchcock, 2nd-Lt. W. H. H.	
	Eagon, 2nd-Lt. V. G.	
	Utting, 2nd-Lt. F. A. K.	
	Korb, 2nd-Lt. G. L.	
	Gregory, 2nd-Lt.	
	Wallace, 2nd-Lt.	
	Walker, Lt. E. C.	
Combined Labour Corps Depôt, Gauhati.	Mitchell, 2nd-Lt. J.	
Combined Labour Corps Depôt, Ranchi.	McGregor, 2nd-Lt. K. S.	
	Johnson-Coombs, 2nd-Lt. K. S.	
	Hayling, 2nd-Lt. A. A.	
	Terry, 2nd-Lt. E.	
	Clarke, 2nd-Lt. J. W.	
	Barber, 2nd-Lt. F. F.	
Combined Labour Corps Depôt, Bhimtal.	Wilson, 2nd-Lt. I. L. D.	
	Tulloch, 2nd-Lt. D. A.	
	Park, 2nd-Lt.	
	Huggins, Capt.	

Corps.	Names, Rank and Corps.	Remarks.

OFFICERS SERVING WITH DEPÔTS—*contd.*

Corps.	Names, Rank and Corps.	Remarks.
No. 40 France Labour Corps Depôt, Sagaing.	Littlewood, Lt. (*Actg. Lt.*) R. V., I. A. R. O.	Adjt.
	Maung Lu Oh, Jemdr.	
	Tun Zam, Jemdr.	
	Ya Shuan, Jemdr.	

OFFICERS SERVING WITH CORPS.

Corps.	Names, Rank and Corps.	Remarks.
1st Labour Corps	Wilson, Lt. (*ty. Capt*) J. M., I.A.R.O.	
	Upton, Lt. G. H. M., R. E.	
	Hayes, Lt. E. R. C., I.A.R.O.	
	Evans, 2nd-Lt. E. M	
	Green, 2nd-Lt. E. W. W.	
	Kapoor Singh, Subdr.	
	Sheikh Ahmed, Jemdr.	
	Basawa Singh, Jemdr.	
	Jewan Singh, Jemdr.	
	Basheshar Nath, Jemdr.	
	Allahdad Khan, Jemdr.	
	Harnam Singh, Jemdr.	
	Dulia Singh, Jemdr.	
2nd Labour Corps	English, Lt. A., I.A.R.O.	
	Crossle, Capt. H. C., I.A.R.O.	
	Lawrie, G., I.A.R.O.	
	Anderson, F. M.	
	Abdul Ghani, Subdr. Major, S. & M.	
	Khuda Dad, Subdr.	
	Sharim Khan, Subdr.	
	Nek Alam, Subdr.	
	Inder Singh, Subdr.	
	Gulah Khan, Jemdr. Qr.-Mr.	
	Id Buksh, Jemdr.	
	Jumma Khan, Jemdr.	
	Jagat Singh, Jemdr.	
	Narain Singh, Jemdr.	
	Bhan Singh, Jemdr.	
3rd Labour Corps		
4th Labour Corps	Shaw, Capt. F. C.	
	Mundock-Smith, 2nd-Lt., M.	
	Dawes, 2nd-Lt. F. L., I.A.R O.	
	Nehal Singh, Subdr.-Major	
	Nehal Singh, 934, Subdr.	
	Ram Singh, Subdr.	
	Meher Singh, Subdr.	
	Kishen Singh, 933, Jemdr.	
	Bhan Singh, 936, Jemdr.	
	Jhanda Singh, 938, Jemdr.	
	Hazara Singh, 1191, Jemdr.	
	Jewand Singh, Jemdr.	
	Sohan Singh, Jemdr.	
	Arjim Singh, 1194, Jemdr.	
	Lal Singh, 1195, Jemdr.	
	Mohiuddin Shariff, 1535, Jemdr.	
	Tilak Singh, 2513, Subdr.	
	Natha Singh, 1543, Jemdr.	

Corps.	Names, Rank and Corps.	Remarks.
	OFFICERS SERVING WITH CORPS—*contd.*	
5th Labour Corps	Hoff, Lt. S. R., I.A.R.O.	
	Lace, Capt. E. W., I.A.R.O.	
	Young, Lt. E. H., I.A.R.O.	
	Dickins, 2nd-Lt. E. K., I.A.R.O.	
	Beck, 2nd-Lt. J. P.	
	Tippetto Aylmer, 2nd-Lt. G. A.	
	Sabhapâthy, 1544, Subdr. Major K.	
	Mooni Rathnam, Subdr.	
	Rasundram, Subdr., A.	
	Ramdass, 11, Jemdr., V.	
	Arogyasami, 7, Jemdr., C.	
	Muthu Karappan, Jemdr.	
	Jeevaratnam, 1000, Jemdr.	
	Bagavathi, 1002, Jemdr., L.	
	Chidambaram, 28, Jemdr.	
	Parasuramen 23, Jemdr.	
	Aziz Khan, Jemdr.	
	Ramasami, Jemdr.	
6th Labour Corps	Neave, Capt. L. H.	
	Nicachi, Lt. J. C., I.A.R.O.	
	Hanrahan, Lt. M., I.A.R.O.	
	Hebberd, Lt. E. St. J., I.A.R.O.	
	Rixon, 2nd-Lt. F.	
	Ennes, 2nd-Lt. C.	
	Bahadur Shah, 1206, Subdr.	
	Hira Singh, 1041, Subdr.	
	Sunder Singh, 1039, Subdr.	
	Lakha, Jemdr.	
	Shakrulla Khan, Jemdr.	
7th Labour Corps	Maher, Lt. C. B., I.A.R.O.	
	FitzHerbert, Lt. G. A., I.A.R.O.	
	Narayan Giri, 1036, Subdr.	
	Chatter Singh, 1320, Subdr.	
	Pala Singh, 3100, Jemdr.	
	Batan Singh, 922, Jemdr.	
	Pyra Singh, 559, Subdr.	
	Keshan Singh, 1809, Subdr.	
8th Labour Corps	Cox, *Capt.* E. O., I.A.R.O.	
	Robinson, Lt. G. J., I.A.R.O.	
	Redmond, Lt. C. J., I.A.R.O.	
	Watson, 2nd-Lt. A. O.	
	Thurton, 2nd-Lt. B. W.	
	Hayes, 2nd-Lt. C. R.	
	Narain Singh, 1866, Subdr.	
	Ushnak Singh, 3635, Subdr.	
	Varyan Singh, 1867, Jemdr.	
	Pakhar Singh, 1865, Jemdr.	
9th Labour Corps	Keir, Lt. G., I.A.R.O.	
	Fraser, 2nd-Lt. D. H., I.A.R.O.	
	Rule, Lt. W. H. P., I.A.R.O.	
	Kirpa Singh, Risaldar	
	Mewa Singh, Subdr. Major	
	Faujdar Khan, 502, Jemdr.	
	Zekhraj, Jemdr.	
10th Jail Labour Corps	Knox, Lt. A. E.	
	Hughes. Lt. R. R.	
	Rodgers, 2nd-Lt. E. W. F.	
	Jones 2nd-Lt. R. B.	
	Sawan Singh, Jemdr.	
	Hardit Singh, Jemdr.	
	Bhan Sase, Jemdr.	
11th Jail Labour Corps	Neilson, Capt. E. G., I.A.R.O.	
	Moore, Lt. W. G. T	
	Carmichael, Lt. C., I.A.R.O.	
	Todd, 2nd-Lt. A. E.	
	Nana Nikam, Subdr.-Maj.	
	Narayan Rao Sindhi, 1598, Subdr.	
	Sudan Singh, 1658, Jemdr.	
	Esu Vase, 1619, Subdr.	

OFFICERS SERVING WITH CORPS—*contd.*

Corps.	Names, Rank and Corps.	Remarks.
12th Assam Labour Corps	Evanson, Lt. G. A., I. A. R. O.	
	St. Romaine, 2nd-Lt. E. C., I. A. R. O.	
	Vernieux, Lt. C. O., I. A. R. O.	
	Spencer, 2nd-Lt. B. G.	
	Mohd. Jafar, 980, Jemdr.	
	Harnam Singh 1537, Jemdr.	
	Bhawan Singh Kunwar, 1018, Jemdr.	
	Pir Mahd., Jemdr.	
	Sunder Singh (Actg. Jemdr.)	
13th Labour Corps	McGavin, Capt. A. E. C., I. A. R. O.	
	LeFeuvre, 2nd-Lt. C. O., I. A. R. O.	
	Wharton, Lt. F., I. A. R. O.	
	Corridon, 2nd-Lt. A., I. A. R. O.	
	Halsey, 2nd-Lt. G. E. C.	
14th Labour Corps	Turner, Capt. C. A. G., I. A. R. O.	
	Beart, 2nd-Lt. D. P., I. A. R. O.	
	Weston, Lt. J. S. N., I. A. R. O.	
	Lysons, 2nd-Lt. K. C., I. A. R. O.	
	Mason, 2nd-Lt. T. W.	
	Hart, 2nd-Lt. W. C.	
15th Labour Corps	Sinclair, Capt. W.	
	Fisher, Lt. S.	
	Rose, 2nd-Lt. R.	
	Davidson, 2nd-Lt. H. H.	
	Barnett, 2nd-Lt. B. C.	
	Macpherson, 2nd-Lt. D.	
16th Labour Corps	Herdon, Capt. A. H.	Comdt.
	Hopewell, 2nd-Lt. G. M.	Adjt.
	Thompson, Lt. E.	
	Wiggins, Lt. D. G.	
	Fitx Patrick, 2nd-Lt. L.	
	McAdam, Lt. J.	
17th Labour Corps	Crawford, Lt. (*temp. Capt.*) N. S., I. A. R. O.	
	Hopkins, Lt. J. A., I. A. R. O.	
	Ferguson, 2nd-Lt. W. E., I. A. R. O.	
	Mitchell, Lt. P. W., I. A. O.	
	Grant, Lt. E. R., I. A. R. O.	
18th Labour Corps	Martyn, Capt. J. K., M.C., I. A. R. O.	
	Barton, Lt. R. G.	
	Bartlett, 2nd-Lt. A. H.	
	Callaway, 2nd-Lt. F. M.	
	Russell, 2nd-Lt. E. T., I. A. R. O.	
19th Labour Corps	Woods, (*temp. Capt.*) P. W. H., I. A. R. O.	
	Collis, 2nd-Lt. W. M., I. A. R. O.	
	Dunlop, 2nd-Lt. H. L., I. A. R. O.	
20th Labour Corps	Taylor, Capt. A. W. D. C.	
	Robertson, Lt. E. A.	
	Hardaker, 2nd-Lt. H. L. W.	
	O'Leary, 2nd-Lt. R. F.	
	Moreton, 2nd-Lt. T.	
	Clifford, 2nd-Lt. G. A.	
No. 1 Military Prisoners Labour Company.	Lawson, Capt. E. F.	
	Gulab Khan, 254, Subdr.	

Corps.	Names, Rank and Corps.	Remarks.
	OFFICERS SERVING WITH DEPÔTS.	
1st Labour Corps Depôt, Berhampore.	Welchman, Lt. (*temp. Capt.*) G. H., I. A. R. O.	Comdt.
	Maybury, 2nd-Lt. M. A. K.	
	Brits, 2nd-Lt. E.	
	Cross, 2nd-Lt. P. L.	
	Mohammed Yakoob, Subdr.	
	Abdul Razak, Subdr.	
Combined Labour Corps Depôt, Jubbulpore.	Keyworth, Lt.-Col. W.	Comdt.
	Ommaney, Maj. A. H.	
	Hargreaves, Capt. H., I. A. R. O.	
	Turner, Capt. V. E.	
	Young, Capt. J.	
	Baldwin, 2nd-Lt. R. A.	
	Stark, 2nd-Lt. J.	
	Walford, 2nd-Lt. H. G.	
	Hogan, 2nd-Lt. W. E.	
	Fay, 2nd-Lt. A. T.	
	Preston, 2nd-Lt. L. C. E. W.	
	Shawali Khan, Jemdr.	Depôt Staff.
	Bikarma Singh, Jemdr.	Quarter Master.
Combined Labour Corps Depôt, Calcutta.	Halliday, Lt.-Col. H. M. (I. A.)	Comdt.
	Townsend-Smith, Capt. W. (I. A. R. O.)	Adjt. and Qr.Mr.
	Stubbs, Lt. H. H. (I. A. R. O.)	
	Baldwin, Lt. D. F. (I. A. R. O.)	
	Walsh, 2nd-Lt. J. E. (I. A. R. O.)	
	Lamb, A. R.	
	Rae, Lt. F. L. (I. A. R. O.)	
	Daspati Singh, Subdr.	
	Gujjar Singh, Jemdr.	
Combined Labour Corps Depôt, Cannanore.	Duncan, Capt. D. J., I. A. R. O.	
	Landeman, Capt.	
	Korbe, 2nd-Lt. P. A.	
	Philipze, 2nd-Lt. B. S.	
	Travers Drapes, 2nd-Lt.	
	Patmore, 2nd-Lt. S. H.	
Combined Labour Corps Depôt, Ghazipur.	Norris, Capt. C. E.	Comdt.
	Clarke, 2nd-Lt. F. H.	
	Mohamed Khan, Jemdr. (*Pensioned*)	
	Hazara Singh, Jemdr. (*Pensioned*)	
	Bhag Singh, Jemdr. (*Pensioned*)	
Combined Labour Corps Depôt, Poona.	Quilter, Capt. T. H., I. A. R. O.	Comdt.
	Wailling, Capt. W., I. A. R. O.	
	Eldridge, 2nd-Lt. G.	
	Brotherston, Lt. (*Actg. Capt.*) G. F.	
	Cowley-Brown, Capt. R.	
	Balu Mhadick, Subdr. (*Pensioned*)	
	Ganda Singh, Subdr. (*Acting*) (*Pensioned*)	
	OFFICERS SERVING WITH CORPS.	
1st Porter Corps	Ramalingam Pillai, Subdr.	
	Rule, 2nd-Lt. D.	
	Hogan, Capt. P. C.	
	Fossey, Lt. R. H.	
	Abdul Rahim, Jemdr.	
	Kama, Jemdr.	
2nd Porter Corps	Holmes, Capt. A. H. R., I. A. R. O.	Comdt.
	Fellingham, 2nd-Lt. F. W.	
	Abdul Razool, 291, Jemdr.	
	Nabi Bux, 5452, Jemdr.	
	Natha Singh, 651, Jemdr.	
3rd Porter Corps	Thomson, Capt. J. S.	
	Wells, 2nd-Lt. R. W.	
	Finnigan, 2nd-Lt. L. D.	
	Hamel Singh, Lhr.-290, Jemdr.	
	Pir Khan, Rp.-472, Jemdr.	
	Dya Ram, 674, Jemdr.	
	Sharaf Uddin, 748, Jemdr.	
4th Porter Corps	Hall, Lieut. E. S.	
	Gulam Haider, 6244, Jemdr.	
	Minbahdur Mal, 1381, Jemdr.	
	Sawan, Singh, Jemdr.	

Corps.	Names, Rank and Corps.	Remarks.

OFFICERS SERVING WITH DEPÔTS.

Corps.	Names, Rank and Corps.	Remarks.
103rd Labour Corps	Welby, Capt. W. Gotch, 2nd-Lt. C. W. Kelly, 2nd-Lt. R. J. Mathey, 2nd-Lt.	
107th Labour Corps	Maxon, Capt. R. J. Pittar, 2nd-Lt. A. H. O'Brien, 2nd-Lt. W. G.	
108th Labour Corps	Wray, Capt. G. H. (Brit. Genl. List) Farn, Lt. E. A. (I. A. R. O.) Nelson, 2nd-Lt. S. (I. A. R. O.) Mecredy, 2nd-Lt. W. H. (I. A. R. O.) Pentham, 2nd-Lt. J. (I. A. R. O.) MacKenzie, 2nd-Lt. K. S. (I. A. R. O.)	
111th Labour Corps	Gerrard, temp. Capt. E. F. Ferguson, Lt. R. H. Maclean, Lt. H. G. Willey, 2nd-Lt. V. S. Powell, 2nd-Lt. G. A.	
113th Labour Corps	Shaw, Capt. G. W. Drysdale, Lt. J. Royan, 2nd-Lt. A. McCord, 2nd-Lt. J. Murphy, 2nd-Lt. H. W.	
115th Labour Corps	McPherson, Capt. J. Williams, Lt. A. T. Pargick, 2nd-Lt. C. E. Hardinge, 2nd-Lt. G. W. P. Elliot, 2nd-Lt. J. J. Bell, 2nd-Lt. J. N.	
116th Labour Corps	Farnon, Lt. L. R. St. J. Barty, 2nd-Lt. J. E. Mortimer, 2nd-Lt. A. E.	
117th Labour Corps	Coventry, Capt. W. A. (I. A. R. O.) Wakefield, Lt. A. G. (I. A. R. O.) O'Connor, 2nd-Lt. A. F. (I. A. R. O.) Furey, 2nd-Lt. T. F. (I. A. R. O.) Morrison, 2nd-Lt. M. D. (I. A. R. O.) Tobin, 2nd-Lt. J. (I. A. R. O.)	
120th Labour Corps	Hutchison, Capt. F. H. Winter, 2nd-Lt. G. W. Blackburn, 2nd-Lt. D. A. Wilson, 2nd-Lt. F. Torry, 2nd-Lt. S. E. Collins, 2nd-Lt. J. L.	
144th Labour Corps	Burns, Capt. A. R. Owen-Smith, Lt. R. Loughten, 2nd-Lt. J. A. Cracknell, 2nd-Lt. H. W.	

OFFICERS SERVING WITH CORPS—*contd.*

Corps.	Names, Rank and Corps.	Remarks.
5th Jail Porter Corps	Dickens, Capt. W. S.	
	Bridgewater, Lt. A. G. B.	
	Fordham, Lt. H.	
	Stone, 2nd-Lt. R. C.	
	Imamuddin, 2282, Jemdr.	
	Nazir Ali, 2777, Jemdr.	
	Balbir Basnet, Jemdr.	
	Khan Singh, 453, Jemdr.	
6th Jail Porter Corps	R. de M. Michaelson-Yeates, Capt. R. M. M.	Comdt.
	Kishen Singh, 3196, Jemdr.	
	Fazal Din, 1390, Jemdr.	
	Mohd. Akram, Khan, Jemdr.	
	Diwan Ali Jemdr.	
7th Porter Corps	Miller, Temp. Capt. A. D.	Comdt.
	Duff, Lt. Baron B. H.	
	Clarke, Lt. A. B.	
	Imam Din, Jemdr.	
	Mohd. Ishaq, Jemdr.	
	Karim Bux, Jemdr.	
	Baquar Khan, Jemdr.	
8th Jail Porter Corps	Knight, 2nd-Lt. H.	
	Matthews, Capt. H.	Comdt.
	Chamarattes, 2nd-Lt. R. D.	
	Karam Khan, Jemdr.	
	Ismail Khan, Jemdr.	
	Sawan Singh, Jemdr	
9th Porter Corps		
10th Porter Corps	Falgar, Temp. Capt. R. J.	Comdt.
	Inglefield-Dall, Lt. B. M.	
	D'Cruz, 2nd-Lt. H. J.	
	Syed Abbas, Subdr.	
	Mohd. Azam, Subdr.	
	Anthony, Subdr. O.	
	Mohd. Azizuddin, Subdr.	
	Labh Singh, Jemdr.	
11th Porter Corps	Lownes, Capt. J. R.	
	Nurana Singh, Subdr.-Maj.	
	Sreepal Singh, Jemdr.	
	Kala Singh, Jemdr.	
	Bishan Singh, Jemdr.	
12th Porter Corps	McGrath, 2nd-Lt. (*temp. Capt.*) W. H., I. A. R. O.	
	Delmege, Lt. C. H.	
	Fforde, Lt. G. F.	
	Morris, 2nd-Lt. J. S., I. A. R. O.	
	Mills, 2nd-Lt. H. C., I. A. R. O.	
14th Porter Corps	Hoare, (Temp. Capt.) W.	
	Fielding, Lt. A.	
	Brewin, 2nd-Lt. A. D.	

OFFICERS SERVING WITH DEPÔTS.

Corps.	Names, Rank and Corps.	Remarks.
1st Porter Corps Depôt, Pallavaram.	Borebank, 2nd-Lt. F. W.	
	Smith, 2nd-Lt. T. S.	
	Scott, 2nd-Lt. E. S.	
	Dickins, Capt. E.	
	Blackett, Lt. W. S.	
	Abdul Razack Subadar (*Pensioned*)	
	Munnsmy Jemdr. (*Pensioned*)	
Combined Labour & Porter Corps Depôt, Lahore.	Hart, Capt. E. J. A.	Comdt.
	Marchant, 2nd-Lt. H. L. M.	
	Bruce, 2nd-Lt. R. W.	
	Hardaker, 2nd-Lt. H. L. W., I. A. R. O.	
	Boor Singh, Jemdr.	

Corps.	Names, Rank and Corps.	Remarks

OFFICERS SERVING WITH CORPS—*concld.*

Corps.	Names, Rank and Corps.	Remarks
Combined Porter Corps Depôt, Allahabad.	McGill, Capt. D. S.	Comdt.
	Saw, Lt. G. A.	Adjt.
	Critchell, 2nd-Lt. D. E. R.	Qr.-Mr.
	Barry, 2nd-Lt. F.	
	Ganga Ram, Resaldr.	
	Seikh Amer Ahmed, Jemdr.	
	Brijbhukhan Singh, Jemdr.	
Combined Porter Corps Depôt, Kamptee	Howell, Capt. H. W. H., I. A. R. O.	
	Phillips, Lt. P. B., I. A. R. O.	
	Fuller, 2nd-Lt. R. W.	
Combined Jail Labour and Porter Corps Depôt, Bareilly. (10 and 11 Combined L. C. Depôt. 5, 6, 8, 10, and 12 Combined P. C. Depôt).	Daly, Capt. D. J.	Comdt.
	Garstin, Lt. E. L.	
	Keegan, 2nd-Lt. B. A.	Adjt. & Qr.-Mr.
	Anding, 2nd-Lt. J. B.	
	Curran, 2nd-Lt. P. J. E.	
	Farrell, Lt. H. H.	
Followers' Central Depôt, Meerut.	Willoughby-Harris, 2nd-Lt. A. J.	
	Anwar Khan, Jemdr.	
	Naki Ram, Jemdr.	
	Imdad Khan, Jemdr.	
	Theokaram Singh, Jemdr.	
	Bostan Khan, Jemdr.	
	Himayat Ali Khan, Jemdr.	
Followers' Central Depôt, Kirkee.	Clabby, Lt. D. N. J.	
	Nagabhushanam, Subdr.	
	Gulab Khan, Subdr	
	Abaji Rao Shinde, Jemdr.	
Followers' Central Depôt, Lucknow.	Slater, Capt. A. C. N.	
	Chetwynd-Palmer, 2nd-Lt. R.	
	Smyth, 2nd-Lt. E. G.	
	Narayan Rao Goghle, Jemdr.	
Followers' Central Depôt, Amritsar.	McBrien, Lt. (*temp. Capt.*) C. H., I. A. R. O.	
	Dickie, 2nd-Lt. W.	
	Moudell, 2nd-Lt. E.	
	Harris, 2nd-Lt. H. F. E., I. A. R. O.	
	Barkat Ali Khan, Subdr.-Major	
	Huzura Singh, Subdr.	
	Chanda Singh, Subdr.	
	Indar Singh, Subdr.	
	Mahr Ali Khan, Jemdr.	
	Abdul Rahim Khan, Jemdr.	
Syce Corps Depôt, Sehore	Hoyle, Capt. A. L.	Comdt.
	Stoddart, 2nd-Lt. R. A.	Asst. Comdt.
Followers Central Depôt, Rawalpindi.	Kelly, Capt. P. A.	
	Mansefild, Lt. M. G.	Asst. Comdt.
	Mohammad Quadir, Jemdr.	

COMBINED LABOUR DEPÔT, DADAR BOMBAY.

Appointment.	Name and Rank.	Remarks.
Commandant	Waller, Lieut.-Col. E., I. A.	
Assistant Commandant	Molony, Maj. A. D., 7th G. R.	
Storekeeper	Goddall, 2nd-Lieut., (*Temp. Capt.*) F. E., I. A. R. O.	
Officer Commanding, Works Directorate Camp.	Kavanagh, Lieut., (*Temp. Capt.*) C., I. A. R. O.	
Officer Commanding, Wadi Bunder Camp.	Betteridge, Lieut. (*Temp. Capt.*) G. W., I. A. R. O.	
Officer Commanding, Rest Camp.	Greening, Lieut. (*Temp. Capt.*) A. H., I. A. R. O.	
Commanding "C" Camp	Burnard, Lieut. W.	
Commanding "B" Camp	Pogson, Lieut. C. A.	
Adjutant	Pain, Lieut. S. C.	
Commanding "D" Camp.	Metcalfe, Lieut. J. H.	
Commanding Reception Camp	Underdown, Lieut. A. H.	
Commanding "A" Camp	Reeks, Lieut. F.	
Adjutant & Quartermaster, Chinese Section.	Goby, Lieut. J. H.	
	MacArthur, 2nd-Lieut. A. G.	
	Over, 2nd-Lieut. J.	
	Morris, 2nd-Lieut. A. E. C.	
	Gammon, 2nd-Lieut. F. G.	
Draft Conducting Officer for Chinese.	Madden, Lieut. J. W.	
	Focken, 2nd-Lieut. F. C.	
	Brown, 2nd-Lieut. J. A. H.	
	Engley, 2nd-Lieut. E. R.	

INDIAN OFFICERS.

	Buland Khan, *Bahadur*, Subadar Major	
	Sk. Mohomed Hussain, Subadar	
	Vishnu Narvekar, Subadar	
	Mohomed Khan, Risaldar	
	Mansab Ali Khan, Risaldar	
	S. Mohomed Jaman, Risaldar	
	Kehar Singh, Subadar	
	Sk. Suleman, Subadar	
	Krishna Rao Gadke, Subadar	
	Mohomed Abas Khan, Risaldar	
	Khuda Bux Khan, Subadar	
	Yesu Rawal, Subadar	
	Arjan Bhagat, Subadar	
	Abdulla Khan, Subadar	
	Gopal Singh, Jemadar	
	Arjan Singh, Jemadar	
	Suleman Khan, Jemadar	

COMBINED LABOUR DEPÔT, KARACHI.

Commandant	Shephered, Capt. W. I.	
Assistant Commandant	Wardle, 2nd-Lieut. W. F.	

INDIAN OFFICERS.

	Kala Singh, Subadar	
	Mohammed Faizulla, Subadar	
	Ghulam Jilani, Subadar	

TECHNICAL RECRUITING OFFICE, CALCUTTA.

Technical Recruiting Officer	Kearns, Major W. I.	
Asst. Technical Recruiting Officer.	Presswell, Capt. T. G.	

TECHNICAL RECRUITING OFFICE, BOMBAY.

Act. Technical Recruiting Officer.	Bennett, Capt. H. B.	
Asst. Technical Recruiting Officer.	De-Bourbel, Lieut. G.	
Chief Instructor, Oil Engine Drivers & Ice Mechanics School, Bombay.	Wilmer, Capt. G. T.	

Appointment.	Name and Rank.	Remarks.

TECHNICAL RECRUITING OFFICE, LAHORE.

Technical Recruiting Officer	Dawson, Lieut. R. C.	

RAILWAY TRAINING CAMP, PURI.

Officer Commanding	Petrie, Capt. J. C. E.	
Asst. Commanding	O'Sullivan, 2nd-Lieut. A. G. J.	
Adjutant	Hunter, Lieut. E.	
Asst. Comdt., 128 R. C. C., (Railway Training Camp, Puri).	Humphreys, 2nd-Lieut. E. A. F.	
Asst. Comdt., 125 R. C. C. Nushki Exn. (Railway Training Camp, Puri.)	Kewley, 2nd-Lieut. J.	

RAILWAY TRAINING CAMP, JUBBULPORE.

Commandant (Temp.)	Russell, Lieut. C. S. C.	
Attached	Codd, 2nd-Lieut. H. C.	
Attached	MacArthur, Lieut. R. G.	

RAILWAY TRAINING CAMP, SAHARANPORE.

Officer Commanding	Spring, Capt. G. C.	
Asst. Commandant	Burstal, Lieut. A. C.	
Attached	Higson, 2nd-Lieut. W.	
Attached	Tucker, 2nd-Lieut. F. G.	
Accounts Officer	De-Brooy, 2nd-Lieut. H. T.	
Attached, Asst. Comdt., Nushki Exn., 97th Coy. (Railway Training Camp, Saharanpore).	Allen, Lieut. A. I. L.	
Comdt., 122 R. C. C. (Railway Training Camp, Saharanpore).	Briggs, Lieut. R. C.	
Attached Asst. Comdt., Nushki Exn. 98th Coy. (Railway Training Camp, Saharanpore).	Robinson, Lieut. C.	
Comdt., 94th Coy. (Railway Training Camp, Saharanpore).	Broome, Lieut. L. R.	
Comdt., 90th Coy. (Railway Training Camp, Saharanpore).	Bourne, 2nd-Lieut. W.	

CLERKS TRAINING SCHOOL, SAIDAPET MADRAS.

Commandant	Ormsby, Major M. H.	
Attached	Miller, 2nd-Lieut. A. G.	
Attached	Frazer, 2nd-Lieut. W. G.	

RAILWAY TRAINING CAMP, TIRUPATTUR.

Asst. Commandant	Clarke, Lieut. H.	
Offg. Commandant	Marshall, Lieut. A. D.	
Adjutant & Quartermaster	Stanford, 2nd-Lieut. W. H. C.	
Recruiting Officer, Shanghai	Manners, Major C. M.	
Attached	Farnborough, Capt. J.	
Attached	Cockell, Capt. E. L.	
Attached	Mitchell, Lieut. E. J. R.	
D. A. A. G. (Recruitment India) Singapore.	Hume, Lieut.-Col. W. P.	

MEDICAL SERVICES.

ROYAL ARMY MEDICAL CORPS.

Within a laurel wreath surmounted by a crown the rod of Æsculapius with a serpent entwined.

"In Arduis Fidelis."

Uniform—Blue. *Facings*—Dull Cherry.

First Commn.	Names.	Date of present rank.	Date of compln. Indian tour or four years as Colonel or date of age for retirement.	Date of present appointment.	Remarks.

REMOVED FROM THE CORPS AND STILL ON THE ACTIVE LIST.

Director, Medical Services in India. (1)

5 Feb. 81	*O'Donnell, T. J.*, C.B., D.S.O. *Maj.-Genl. (temp. Lt.-Genl.) (temp.)*	*18 Feb. 15*	. . . .	*13 July 16*	*Simla.*

Deputy Directors, Medical Services. (2)

2 Aug. 84	**Anderson,** L. E., C.B., Major-Genl.	4 May 12	26 Mar. 21	13 May 16	D. D. M. S., S. Comd.
30 May 85	**Cree, G.,** C.B., C.M.G. . . Major-Genl.	26 Dec. 17	3 July 22	15 Mar. 18	D. D. M. S., 8th (Lucknow) Dn.

Colonels.

28 July 86	**Davidson,** J. S., M.B. .	1 Mar. 15	28 Feb. 19	6 May 16	A. D. M. S., Jubbulpore and Jhansi Bdes.
28 July 86	**Fallon, J.**	1 Mar. 15	28 Feb. 19	5 Apl. 17	A. D. M. S., Bareilly, Garhwal and Dehra Dun Bdes.
27 July 92	**Condon,** E. H., M.D. . . .	26 Dec. 17	25 Dec. 21	1 Mar. 18	A. D. M. S., Abbottabad Bde.
30 Jan 92	**Beyts, W. G.** V. H. S.	26 Dec. 17	25 Dec. 21	11 May 16	A. D. M. S., Bombay Bde.
28 July 91	**Hardy, W. E.**	7 Apl. 18	6 Apl. 22	. . . . .	A. D. M. S., 1st (Peshawar) Divn.
27 July 92	**More, L. P.,** M.B. (161, 170, 206).	2 June 18 1 Jan. 17	1 June 22	24 Oct. 18	A. D. M. S., Bangalore Bde.

First Commn.	Names.	Date of present rank.	Date of compln. Indian tour.	Date of present appointment.	Remarks.

ROYAL ARMY MEDICAL CORPS.

Lieutenant-Colonels.

[In the case of the Senior Officers the *second* date is that of selection for increased pay under article 358 of the Royal Warrant.]

First Commn.	Names.	Date of present rank.	Date of compln. Indian tour.	Date of present appointment.	Remarks.
1 Feb. 90	**Stone**, C. A., M.D. . . .	21 May 12	. . . . .	. . . .	Comdg. Stn. Hospl., Maymyo.
29 July 90	**Mawhinny**, R. J. W., C.B. .	11 Oct. 12	18 Sep. 17	. . . .	Offg. A. D. M. S., 2nd (R'Pindi) Dn.
30 Jan. 92	**Hennessy**, J., C.B., M.B. .	7 July 14	10 Dec. 17	. . . .	Comdg. Stn. Hospl., Nowshera.
27 July 92	**McNaught**, J. G., M.D. (169, 206).	30 Oct. 14	. . . .	. . . .	Comdg. Stn. Hospl., Bangalore.
27 July 92	**O'Reilly**, H. W. H., M.B. .	1 Mar. 15	12 Nov. 18	3 Apl. 18	Offg. A. D. M. S., Southern Bde.
27 July 92	**Read**, H. W. K.	1 Mar. 15	14 Jan. 18	. . . .	Comdg. Freeman Thomas War Hospl., Bombay.
30 Jan. 93	**Mangin**, F. M.	1 Mar. 15	11 Feb 18	. . . .	
29 Jan. 94	**Clark**, E. S., M.B.	1 Mar. 15	12 Nov. 18	. . . .	Comdg. Stn. Hospl., Karachi.
29 Jan. 95	**Staddon**, H. E.	1 Mar. 15	4 Mar. 18	11 Feb. 18	Comdg. Colaba War Hospl.
29 July 93	**Killery**, St. J. B.	1 Mar. 15	18 Sep. 17	. . . .	President, Standing Medl. Board, Bombay.
28 Jan 98	**Watts**, B., D.S.O. (170, 207) .	1 Mar. 15	. . . .	. . . .	Comdg. Stn. Hospl., Ambala.
28 Jan. 98	**Macpherson**, J D. G, M.B. (207)	1 Mar. 15	. . . .	. . . .	Comdg. King George's War Hospital, Poona.
28 Jan. 98	**Ross**, N. H., M.B. . . .	1 Mar. 15	8 Oct. 18	. . . .	Comdg. Stn. Hospl., Peshawar.
28 Jan. 99	**Barrow**, H. P. W., C. M. G., D. S. O.	1 Mar. 15	. . . .	. . . .	
27 July 98	**Hall**, S. O., (95, 172*a*, 207) .	1 Mar. 15	8 Oct. 17	. . . .	Bangalore.
27 July 98	**Cowan**, J., M.B.	1 Mar. 15	28 Feb. 17	. . . .	Comdg. Stn. Hospl., Meerut.
7 July 98	**Walton**, H. B. G.	1 Mar. 15	. . . .	. . . .	Comdg. Stn. Hospl., Bareilly.
28 Jan. 99	**Winkfield**, W. B.	1 Mar. 15	6 Dec. 15	. . . .	Comdg. Stn. Hospl., Madras.
28 Jan. 99	**Lloyd**, R. H. (207) . . .	1 Mar. 15	9 Nov. 16	. . . .	Comdg. Stn. Hospl., Wellington.
28 Jan. 99	**Packer**, H. D.	1 Mar. 15	18 Dec. 18	. . .	Comdg. Stn. Hospl., Secunderabad.
28 Jan. 99	**Gill**, J. G., D.S.O.	1 Mar. 15	. . . .	. . . .	Comdg. Stn. Hospl., Belgaum.
28 Jan. 99	**Lauder**, T. C., M.B. . . .	1 Mar. 15	22 Nov. 22	. . . .	Comdg. Stn. Hospl., Kamptee.
27 July 99	**Palmer**, F. J.	1 Mar. 15	4 Jan. 16	4 Jan. 18	Comdg. Stn. Hospl., Allahabad.
27 July 99	**Prescott**, J. J. W., D.S.O. .	1 Mar. 15	. . . .	. . . .	Comdg. Cumballa War Hospl., Bombay.
28 Jan. 99	**Wroughton**, A. O. B, D.S.O.	1 Mar. 15	. . . .	. . . .	Comdg. Officer's Hospl., Nasik.
27 July 99	**Falkner**, P. H. E., F.R.C.S.I.	1 Mar. 15	. . . .	. . . .	Comdg. No. 44 Genl. Hospl., Deolali.
27 July 99	**Langstaff**, J. W., D.S.O. .	1 Mar. 15	. . . .	. . . .	A. D. M. S. A. H. Qrs., Simla.
27 July 99	**Seeds**, A. A., M.D. . . .	1 Mar. 15	13 Feb. 17	. . . .	Comdg. Stn. Hospl., Jullundur.
17 Nov. 99	**Knox**, E. B., M.D. . . .	26 Dec. 17	. . . .	. . . .	Comdg. B. G H., Aden.
4 Dec. 99	**Smith**, C. S., M.B. . . .	26 Dec. 17	6 Jan. 17	. . . .	
21 June 00	**Spiller**, W. M. H., M.B. .	26 Dec. 17	13 Dec. 18	. . . .	Comdg. Stn. Hospl., Cawnpore.
21 June 00	**Douglass**, P. C. . . .	26 Dec. 17	9 Jan. 17	. . . .	Comdg. Stn. Hospl., Jubbbulpore.

ROYAL ARMY MEDICAL CORPS.

Lieutenant-Colonels—*concld.*

First Commn.	Names.	Date of present rank.	Date of compln. Indian tour.	Date of present appointment.	Remarks.
25 Apl. 00	**Cotton**, F. W. . . .	26 Dec. 17	18 Dec. 17	. . . .	Comdg. Station Hospl., Mhow.
25 Apl. 00	**Weston**, A. F.	26 Dec. 17	. . . .	. . . .	Poona.
30 May 00	**Lauder**, F. P.	26 Dec. 17	21 June 22	. . .	Lv. in Ind., m. c., 3 mos., 7 Oct. 18.
25 Apl. 00	**Argles**, R. L.	26 Dec. 17	11 Sep. 18	. . . .	Comdg. Stn. Hospl., Rawal Pindi.
29 Nov. 00	**Kennedy**, J. C., M.D. .	3 Mar. 18	20 Nov. 18	. . . .	

Majors.

First Commn.	Names.	Date of present rank.	Date of compln. Indian tour.	Date of present appointment.	Remarks.
29 Jan. 94	**Tibbits**, W., M.B. . . .	29 Oct. 05	23 Sep. 18	. . . .	Comdg. Stn. Hospl., Sialkot.
21 June 00	**Ronayne**, C. R. L., M.B. .	21 June 12	. . . .	. .	Meerut.
21 June 00	**Popham**, R. L. . . .	21 June 12	9 Jan. 19	22 Feb. 17	L, ex India, m. c., 6 mos., Decr. 18.
2 Jan. 01	**Skinner**, R., Mc K. . .	2 Oct. 12	9 Jan. 19	. . . .	Comdg. Stn. Hospital, Risalpur.
29 Jan. 01	**Parkes**, E. E., M.B. . . .	29 Oct. 12	20 Jan. 19	. . . .	*Lv., ex Ind., m. c., 6 mths., 1 Sep. 18.*
29 Jan. 01	**Sampey**, A. W., M.B. . .	27 June 13	13 Feb. 17	. . . .	
6 July 01	**Adye-Curran**, S. M. . .	26 July 13	3 Nov. 15	. . . .	Comdg. Stn. Hospl., Dinapore.
5 Dec. 01	**Cotterill**, L., M.B. (50) . .	5 Dec. 13	10 Dec. 18	. . . .	Wellington. Comdt., Enteric Convalescent Depôt.
1 Sep. 02	**Balck**, C. A. J. A. (59, 60) .	1 Mar. 14	12 Sep. 21	. . . .	Comdg. Stn. Hospl., and Malaria Convalescent Depôt, Dagshai.
1 Sep. 02	**Rutherford**, R.	1 Mar. 14	. . . .	. . . .	Comdg. Stn. Hospl., Rangoon.
1 Sep. 02	**Mackenzie**, J., M.B., (205*a*)	1 June 14	21 Oct. 18	. . . .	D. A. D. M. S. (San.), 16th (Indian) Dn.
1 Sep. 02	**Hayes**, A. H. (206) . . .	1 June 14	20 Jan. 19	11 Aug. 15	Comdg. Stn. Hospl., Naini Tal; in charge Enteric Fever Conv. Depôt.
1 Sep. 02	**Clarke**, F. A. H. (207) . .	1 June 14	21 Jan. 18	17 Apl. 15	D. A. D. M. S. (Sany.), 5th (Mhow) Dn.
1 Sep. 02	**Browne**, W. W. (*temp Lt.-Col.*)	1 June 14	11 Feb. 18	3 July 17	A. D. M. S., A. H. Qrs.
1 Sep. 02	**Franklin**, R. J. (162, 172*a*, 206)	1 Sep. 14	22 Dec. 17	. . . .	Lv. ex. India, m. c.
31 Jan. 03	**Dawson**, F. W. W., M.B. .	31 Oct. 14	9 Nov. 16	. . . .	
31 Jan. 03	**Wilmot**, R. C.	31 Oct. 14	. . . .	. . . .	Comdg. Stn. Hospl., Agra.
31 Jan. 03	**Pennefather**, E. M. . .	31 Oct. 14	20 Nov. 18	. . . .	Comdg. Stn. Hospl., Fyzabad.
31 Jan. 03	**Carmichael**, D. G., M.B. (207)	31 Oct. 14	11 Feb. 16	. . . .	Comdg. S. H. Lahore Cantt. Splst. in Midwifery and Diseases of Women and Children.

ROYAL ARMY MEDICAL CORPS—*contd.*

Majors—*concld.*

First Commn.	Names.	Date of present rank.	Date of compln. Indian tour.	Date of present appointment.	Remarks.
31 Jan. 03	Patch, B. G.	31 Oct. 14	10 Feb. 19	. . . .	Comdg. Stn. Hospl., Kirkee. Splst., in Otology, Laryngology and Rhinology, 5th, Poona and 9th Dns.
31 Jan. 03	Stack, H. T., M.B. (207)	31 Jan. 15	26 Jan. 17	. . . .	D. A. D. M. S. (Sany.), Burma Dn.
30 Jan. 04	Lewis, R. R.	1 July 15	. . . .	. . . .	Wellington. Splst. in Dermatology, 9th Dn.
30 Jan. 04	*Johnstone, D. P.*	*1 July 15*	*2 Jan. 17*	*23 Nov. 18*	*Surgeon to H. E. the Governor of Bombay.*
30 Jan. 04	Bowie, S. C.	1 July 15	. . . .	. . . .	Comdg. Stn. Hospl., Jhansi.
30 Jan. 04	Arthur, A. S., M.B.	1 July 15	. . . .	. . . .	Secunderabad.
30 July 04	Dwyer, P., M.C., M.B. (*temp. Lt.-Col.*)	1 July 15	. . . .	. . . .	A.D.M.S., A. H. Qrs.
30 July 04	Winckworth, H. C.	1 July 15	10 Feb. 19	. . . .	Splst. in Advanced Op. Surgery, 4th (Quetta) Dn.
30 July 04	Frost, A. T., M.B.	1 July 15	11 Sep. 18	. . . .	Deolali. Splst. in Dermatology, Bombay Brigade.
30 July 04	Doig, K. A. C.	1 July 15	12 Nov. 18	. . . .	Comdg. Stn. Hospl., Nasirabad. Splst. in Prevention of Disease.
31 Jan. 05	Wallace, G. S.	15 Oct. 15	8 Oct. 18		
31 Jan. 05	Power, P., M.B.	15 Oct. 15	23 Sep. 18	. . . .	Comdg. Stn. Hospl., Poonamallee. Splst. in Dermatology, 9th (S'bad.) Dn.
31 July 05	Fawcett, C. E. W. S.	15 Oct. 15	10 Feb. 19	. . . .	D. A. D. M. S. (Mob.), 5th (Mhow) Division.
31 July 05	Rees, G. H., M.B.	15 Oct. 15	. . . .	. . . .	Comdg. Stn. Hospl., Lucknow.
31 July 05	Keane, M.	15 Oct. 15	20 Jan. 19		
3 Jan. 06	Leslie, T. C. C.	30 Jan. 18	9 Jan. 19	. . . .	Comdg. Stn. Hospl., Darjeeling.
30 Jan. 06	Gibson, L. G. (165, 206)[*l*]	30 Jan. 18	11 Feb. 15	. . . .	Comdg. Stn. Hospl., Kasauli.
30 Jan. 06	Sexton, T. W. O.	30 Jan. 18	. . . .	. . . .	Peshawar.

Captains.

First Commn.	Names.	Date of present rank.	Date of compln. Indian tour.	Date of present appointment.	Remarks.
28 Jan. 07	Boyd, J. E. M., M.C.	28 July 10	16 Feb. 15	. . . .	
28 Jan. 07	Mackenzie, D. F., M.C., M.B.	28 July 10	1 Feb. 15	. . . .	
29 July 07	Dawson, G. F., M.C., M.B. (207)	29 Jan. 11	30 Nov. 14	. . . .	Comdg. Stn. Hospl., Roorkee.
29 July 07	Petit, G., M.C.	29 Jan. 11	1 Feb. 15	. . . .	Bangalore.
29 July 07	Stuart, F. J., M.B.	29 Jan. 11	1 Feb. 15	. . . .	Poonamallee.

ROYAL ARMY MEDICAL CORPS—*contd.*

Captains—*contd.*

First Commn.	Names.	Date of present rank.	Date of compln. Indian tour.	Date of present appointment.	Remarks.
29 July 07	**Spong, W. A.**, M.B. (95) (*temp. Maj.*)	29 Jan. 11	28 Sep. 15	. . . .	D. A. D. M. S., Distribution, Karachi.
29 July 07	**Hingston, J. C. L.** (*temp. Maj.*)	29 Jan. 11	7 Sep. 15	21 Apl. 17	D. A. D. M. S. (Sany.), Meerut Dn.
4 Feb. 08	**Ellcome, J. E.**	4 Aug. 11	4 Mar. 15	. . . .	Bangalore.
4 Feb. 08	**McCombe, J. S.**, D.S.O., M.B., (*temp. Maj.*)	4 Aug. 11	10 Feb. 15	23 Apl. 17	D. A. D. M. S., Army Hdqrs., India.
1 Aug. 08	**Ryles, C.** (*temp. Maj.*) . .	1 Feb. 12	. . . .	. . . .	D. A. D. M. S. (Sany.), 2nd (Rawalpindi) Dn.
1 Aug. 08	**Jones, J. B.**, M.B.	1 Feb. 12	7 Sep. 15	. . . .	Muttra. [Jutogh.
1 Aug. 08	**O'Rorke, C. H.**, M.B. . .	1 Feb. 12	7 Feb. 16	. . . .	Comdg. Stn. Hospl.,
1 Aug. 08	**Lane, J. W.**, M.D.	1 Feb. 12	7 Feb. 16	. . . .	Lebong. Comdg. Stn. Hospl.
30 Jan. 09	**Dykes, S. S.**, M.B.	30 July 12	26 Jan. 17	. . . .	
30 Jan. 09	**Edwards, H. R.**	30 July 12	. . . .	. . . .	Deolali.
31 July 09	**Comyn, K.**, M.D.	31 Jan. 13	. . . .	. . . .	Deolali.
31 July 09	**Dickson, T. H.**, M.B. (206)	31 Jan. 13	22 Dec. 18	. . . .	Comdg. Stn. Hospl., Dum Dum.
31 July 09	**Stoney, E. C.**, M.B. . .	31 Jan. 13	2 Jan. 17		
29 July 10	**Harold, C. H. H.**, M.D. (*temp. Maj.*) (207).	29 Jan. 14	12 Nov. 17		
29 July 10	**Fyffe, E. L.**, M.B. (207) . .	29 Jan. 14	28 Oct. 17	. . . .	Wellington. Splst. Adv. operative Surgery.
29 July 10	**Bridges, R. F.**, M.B. (*temp. Maj.*)	29 Jan. 14	17 Oct. 17	13 Aug. 14	Peshawar. D. A. D. M. S. (Sanry.), 1st (Peshawar) Dn.
27 Jan. 11	**Finny, C. M.**, M.B. (206) .	27 July 14	6 Mar. 18		
27 Jan. 11	**Steven, W. S. R.**, M.B. (95).	27 July 14	4 Mar. 18	. . . .	Poona [Deolali.
27 Jan. 11	**Wilson, G.**, M.C., M.B. (206)	27 July 14	11 Feb. 18	. . . .	O. C. R. A. M. C. Depôt.
27 Jan. 11	**Frost, W. A.**, M.B. (207) . .	27 July 14	4 Mar. 18	. . . .	Comdg. Stn. Hospl., Hyderabad (Sind).
27 Jan. 11	**Reynolds, D.**, M.B. (207) (*temp. Maj.*)	27 July 14	11 Feb. 18	. . . .	D. A. D. M. S. (Mobn.). 16th (Indian) Dn.
27 Jan. 11	**Seaver, C. D. K.**	27 July 14	10 Jan. 18	. . . .	Ferozepore.
27 Jan. 11	**Levack, J. S.**, M.B. . . . . (*temp. Maj.*)	7 July 14	18 Dec. 17	25 Oct. 18	D. A. D. M. S. (Embn.) Karachi.
27 Jan. 11	**Fretz, W. L. E.**, M.B. . (*temp. Maj.*)	27 July 14	12 Nov. 17	1 Sept. 18	Addl. D. A. D. M. S., 9th (Sec'bad.) Dn.
27 Jan. 11	**Hayes, P.**, M.B.	27 July 14	21 Nov. 17	. . . .	Comdg. Stn. Hospl., Neemuch. Splst. in Dermatology, 5th (Mhow) Dn.
28 July 11	**Laird, W. B.**	28 Jan. 15	20 Nov. 18	. . . .	Comdg. Stn. Hospl., Sabathu.
28 July 11	**Davidson, R.**, M.B. (207(*a*))	28 Jan. 15	9 Jan. 19	. . . .	Cannanore.
28 July 11	**Blackmore, H. S.** . . . (*temp. Maj.*)	28 Jan. 15	10 Dec. 18	. . . .	D. A. D. M. S. (San.) 4th (Quetta) Dn.
28 July 11	**Elliott, J. M.**, M.B.	28 Jan. 15	12 Nov. 18	. . . .	Rawalpindi. Splst. in Dermatology, 2nd Dn.

ROYAL ARMY MEDICAL CORPS—*contd.*

Captains—*contd.*

First Commn.	Names.	Date of present rank.	Date of compln. Indian tour.	Date of present appointment.	Remarks.
28 July 11	**Skrimshire**, F. R. B.	28 Jan. 15	12 Nov. 18	....	Bangalore. Splst. in Ophthalmology, 9th (Sec.) Dn.
28 July 11	**Bruce**, D. W., M.B.	28 Jan. 15	12 Nov. 18	....	Mhow.
28 July 11	**Buckley**, L., M.B.	28 Jan. 15	8 Oct. 18	....	Meerut, Splst. in Dermatolgoy.
28 July 11	**Sykes**, S. P., M.B. (207)	28 Jan. 15	23 Sep. 18	....	Splst. in Ophthalmology, 2nd (Rawalpindi) Divn.
28 July 11	**Wells**, H. J. G., M.B.	28 Jan. 15	23 Sep. 18	....	Registrar-Freeman-Thomas War Hospl., Bombay.
28 July 11	**Stewart**, W.	28 Jan. 15	....	....	Comdg. Stn. Hospl., Purandhar.
28 July 11	**Heale**, A. S.	28 Jan. 15	30 Sep. 18	....	Bombay.
26 Jan. 12	**Biggam**, A. G., M.B.	30 Mar. 15	9 Jan. 19	....	Commanding Stn. Hospl., Cherat.
26 Jan. 12	**Mallam**, R. K., M.B.	30 Mar. 15	9 Jan. 19	...	
26 Jan. 12	**Large**, D. T. M., M.B. (207)	30 Mar. 15	20 Jan. 19	...	Staff Surgeon, Bombay.
25 Jan. 12	**Pratt**, W. W., M.B.	30 Mar. 15	10 Feb. 19	...	Lv., ex India, m.c., 3 mos., 9 Sep. 18
26 Jan. 12	**Hood**, A., M.B. (*temp. Maj.*)	30 Mar. 15	20 Jan. 19	...	D. A. D. M. S. (Mob.), 2nd (R'Pindi) Dn.
26 Jan. 12	**Strachan**, E. A., M.B.	30 Mar. 15	20 Nov. 18	...	Rawalpindi.
26 Jan. 12	**Blaikie**, C. J. (207)	30 Mar. 15	20 Jan. 19	...	Meerut—Splst. in Electrical Science.
26 Jan. 12	**Stevenson**, W. B., M.B. (*temp. Maj.*) (137)	30 Mar. 15	20 Jan. 19	...	D. A. D. M. S. (Sany.), Ambala Bde.
26 Jan. 12	**Ritchie**, J. L., M.B., F.R.C.S.	30 Mar. 15	20 Jan. 19	...	Rawalpindi.
26 Jan. 12	**Richardson**, D. T., M.B.	30 Mar. 15	20 Jan. 19	...	Aden.
26 Jan. 12	**Wynne**, O. W. J.	30 Mar. 15	11 Sep. 18	...	Rawalpindi. Splst. in Adv. Operative Surgery.
26 Jan. 12	**Blake**, G. A., M.B. (95)	30 Mar. 15	9 Jan. 19	...	Peshawar. Splst. in Dermatology (temp.), 1st Peshawar Dn.
26 Jan. 12	**Barry**, S. J. (*temp. Maj.*)	30 Mar. 15	18 Dec. 18	...	Poona. Officiating D. A. D. M. S. (Mobn.), Poona Division.
26 Jan. 12	**Allott**, H. W. L.	30 Ma . 15	20 Nov. 18	...	Rawalpirdi.
26 July 12	**Whitby**, E. V., M.B.	30 Mar. 15	....	...	Deccan Brit. War Hospl. Poona. Temp. Splst. in Electrical Science.
•	**Stevens**, N. W., M.B. (*Bt. Maj.*)	30 Mar. 15 3 June 17	....	......	Registrar, Colaba War Hospl.
26 Jan. 12	**Davies**, A. A. M. (206)	30 Mar. 15	20 Jan. 13	...	Comdg. Stn. Hospl., [Benares.

ROYAL ARMY MEDICAL CORPS—*contd.*

Captains—*concld.*

First Commn.	Names.	Date of present rank.	Date of compln. Indian tour.	Date of present appointment.	Remarks.
23 Jan. 13	**Sealy,** H. N.	30 Mar. 15	. . . .	. . . .	Comdg. Stn. Hospl., Amritsar.
24 Jan. 13	**Dyas,** G. E., M. C.	30 Mar. 15	. . . .	. . . .	Bangalore. Splst. in Electrical Science.
24 Jan. 13	**Bridges,** A. B. H.	30 Mar. 15	1 Oct. 16	. . . .	Surgeon to H. E. the Commander-in-Chief.
	Warburton, P. D.	30 Mar. 15	. . . .	. . . .	Deolali.
30 July 14	**Reid,** S. D., M.B.	30 Mar. 15	27 Aug. 23	. . . .	Delhi.
24 Aug. 13	**Greiller,** E. F. W.	5 Feb. 18	. . . .	. . . .	Bangalore.
	Smyth, F. G. A.	16 Feb. 18	. . . .	. . . .	Mhow.
17 Aug. 14	**Monro,** D. C., M.B.	17 Feb. 18	. . . .	. . . .	Poona.
26 May 13	**Mulholland,** H. H., M.B.	15 Mar. 18	. . . .	. . . .	Peshawar.
26 Sep. 14	**Hadden,** D. H., M.C., M.B.	26 Mar. 18	. . . .	. . . .	Poona.

Lieutenants.

First Commn.	Names.	Date of present rank.	Date of compln. Indian tour.	Date of present appointment.	Remarks.
9 Sep. 14	**Robb,** A. L., M.B. (*Tempy. Capt.*).	1 Jan. 17	. . . .	. . . .	Jubbulpore.
17 Nov. 14	**Lipscomb,** F. M. (*Tempy. Capt.*).	1 Jan. 17	. . .	. . . .	
	Macleod, D. N., M.B. (*Tempy. Capt.*).	1 Jan. 17	. . . .	. . .	Deolali.
9 May 14	**Myles,** R. B., M.B. (*Tempy. Capt.*).	1 Jan. 17	. . . .	. . . .	Rawalpindi. Splst. in Electrical Service.
6 Aug. 14	**Stowers,** R., M.C. (*Tempy. Capt.*).	1 Jan. 17	23 Sep. 23	. . . .	Kamptee.
9 Aug. 15	**Sarsfield,** T. H. (*Tempy. Capt.*)	1 Jan. 17	. . . .	. . . .	
10 Nov. 14	**Anderson,** R. A. (*Tempy. Capt.*).	1 Jan. 17	. . . .	. . . .	Ahmadnagar. Splst. in Prevention of Disease.
6 Aug. 16	**Binning,** J. A. (*Tempy. Capt.*)	1 Jan. 17	. . . .	. . . .	Aden.
1 Oct. 15	**Walker,** J. H. C.	1 Oct. 15	. . . .	. . . .	Comdg. Stn. Hospl., Attock.
22 Jan. 16	**Bowie,** T. C. (*Tempy. Capt.*).	1 Jan. 17	. . . .	. . . .	
4 May 16	**Coats,** D. H., M.B. (*Tempy. Capt.*).	1 Jan. 17	. . . .	. . . .	
13 July 16	**Pank,** P. E. D. (*Tempy. Capt.*)	1 Jan. 17	. . . .	. . . .	R. A. M. C. Depôt, Deolali.
15 July 16	**Moulson,** G.	1 Jan. 17	. . . .	. . . .	Bombay.
20 Jan. 17	**Batterham,** D. J.	16 Feb. 17	. . . .	. . . .	

Quartermasters.

First Commn.	Names.	Date of present rank.	Date of compln. Indian tour.	Date of present appointment.	Remarks.
12 Aug. 14	**Stevens,** S. *hon. Capt.*	12 Aug. 14 12 Aug. 17	. . . .	. . . .	Attached Army H' Qrs. India.

ROYAL ARMY MEDICAL CORPS—*contd.*

Captains (Special Reserve).

First Commn.	Names.	Date of present rank.	Date of compln. Indian tour.	Date of present appointment.	Remarks.
17 Dec. 99	Sandeman, C. S.	17 June 13	. . . . .	. . .	Meerut.
23 Mar. 10	Holland, E. T.	28 Sep. 13	. . . . .	. . .	
6 Feb. 11	Routh, L. M., M.D.	24 Aug. 14	. . . . .	. . .	Colaba War Hospl.
10 Jan. 10	Bignold, C. A., M.B. (*Temp. Maj.*)	5 Sep. 14	. . . . .	. . .	Offg. D. A. D. M. S. (Sany)., Poona Divn.
7 Aug. 14	Williamson, J. B., M.B.	7 Sep. 14	. . . . .	. . .	Thayetmyo. Comdg. Stn. Hospl.
7 Apl. 14	O'Sullivan, K. L.	7 Oct. 14	. . . . .	. . .	
5 Aug. 14	Dickson, R. C., M.B.	1 Apl. 15	. . . . .	. . .	Sialkot.
3 July 12	Lea, E. T. H.	1 Apl. 15	. . . . .	. . .	Karachi. Splst. in Prevention of Disease.
5 Nov. 12	D'Avray, A. D. (206, 205*a*)	1 Apl. 15	. . . . .	. . .	Karachi.
14 May 13	Fraser, A. D., M.B.	1 Apl. 15	. . . . .	. . .	Port Blair.
17 May 13	Brown, B. W., M.B.	1 Apl. 15	. . . . .	. . .	Cawnpore.
13 Mar. 12	Hayman, J. R.	1 Apl. 15	. . . . .	. . .	Peshawar. Splst. in Electrical Science.
30 Sep. 14	Jones, E. B., M.B.	1 Apl. 15	. . . . .	. . .	Bangalore.
30 Sep. 14	Minch, J. B., M.B.	1 Apl. 15	. . . . .	. . .	Risalpore.
30 Sep. 14	Campbell, W., M.B.	1 Apl. 15	. . . . .	. . .	Jubbulpore.
2 Sep. 14	Moynan, R. N. O., M.B.	1 Apl. 15	. . . . .	. . .	Colaba.
3 Sep. 14	Quinlan, W. T.	1 Apl. 15	. . . . .	. . .	Meerut. Splst. in Adv. Operative Surgery (temp.).
19 Sep. 14	Linton, E. C., M.B.	1 Apl. 15	. . . . .	. . .	Calcutta Splst. in Dermatology (temp.)
28 Aug. 14	Haworth, J. K. J., M.B.	1 Apl. 15	. . . . .	. . .	Lucknow. Splst. in Adv. Operative Surgery (temp.)
14 Aug. 14	Tandy, O. C. S., M.B.	1 Apl. 15	. . . . .	. . .	Deolali.
8 Sep. 14	Griffith, F.	1 Apl. 15	. . . . .	. . .	Aungban.
6 Sep. 14	Pyper, J. C.	1 Apl. 15	. . . . .	. . .	Deolali.
6 Sep. 14	Stallard, J. P., M.B.	1 Apl. 15	. . . . .	. . .	
8 Aug. 14	Taylor, J., M.D.	1 Apl. 15	. . . . .	. . .	Calcutta. Splst. Prevention of Disease.
2 Feb. 14	Huban, J. P., M.B.	1 Apl. 15	. . . . .	. . .	Bolarum.
7 Sep. 14	Evans, D. G., M.B.	1 Apl. 15	. . . . .	. . .	Rangoon. Splst. in Midwifery and Diseases of Women and Children.
9 Sep. 14	Greiller, B., M. C.	1 Apl. 15	. . . . .	. . .	Peshawar.
10 Sep. 14	Birch, W. S.	1 Apl. 15	. . . . .	. . .	R. Pindi.
17 Sep. 14	Foley, J. E.	6 Apl. 15	. . . . .	. . .	Peshawar.
28 Sep. 14	Heaton, T. B., M.B.	6 Apl. 15	. . . . .	. . .	Bombay.
18 Sep. 14	Smith, H.	7 Apl. 15	. . . . .	. . .	Delhi.
19 Sep. 14	Cameron, N., M.B.	7 Apl. 15	. . . . .	. . .	Wellington.
15 Oct. 14	Roddy, F. A.	18 Apl. 15	. . . . .	. . .	Rawalpindi.
17 Oct. 14	Pooley, J. S.	24 Apl. 15	. . . . .	. . .	Deolali.
20 Oct. 14	Mason, C. A.	24 Apl. 15	. . . . .	. . .	Meerut.
3 Nov. 14	Watson, G. W., M.B.	11 May 15	. . . . .	. . .	Poona.
19 Nov. 14	Hawkins, A. B.	28 May 15	. . . . .	. . .	Comdg. Stn. Hospl., Meiktila.
21 Nov. 14	Marksman, L. B. C., M.B. (207)	28 May 15	. . . . .	. . .	Belgaum. Splst. in Prevention of Disease Poona Dn. (*temp.*)
25 Nov. 14	Dickson, W. M., M.B. (207)	28 May 15	. . . . .	. . . .	
17 Dec. 14	Addison, H. L.	16 June 15	. . . . .	. . . .	Bangalore.
16 Jan. 15	Scott, G. M., M.C., M.B.	16 July 15	. . . . .	. . . .	Poona.
1 Feb. 15	Mann, E. W., M.B.	13 Aug. 15	. . . . .	. . . .	Poona.
9 May 14	Norrie, F. H. B., M.B.	10 Oct. 15	. . . . .	. . . .	Jhelum.
5 Aug. 14	Lyons, J. F.	24 Oct. 15	. . . . .	. . . .	Bangalore.
24 July 15	McClurkin, T.	24 Jan. 16	. . . . .	. . . .	
9 Mar. 14	Menzies, T., M.B.	24 Jan. 16	. . . . .	. . . .	
17 Aug. 14	Sharpe, W. W. S.	25 Apl. 16	. . . . .	. . . .	Rawalpindi.
20 Aug. 14	Graham, M. J., M.B.	25 Apl. 16	. . . . .	. . . .	Nowshera.
23 Nov. 15	Davies, T.R.	1 June 16	. . . . .	. . . .	Jhansi. Splst. in Prevention of Disease. (*temp.*).

ROYAL ARMY MEDICAL CORPS—*contd.*

Captains (Special Reserve)—*contd.*

First Comm.	Names.	Date of present rank.	Date of compln. Indian tour.	Date of present appointment.	Remarks.
23 Mar. 16	Wright, H. D. (95)	3 Oct. 16	. . . . .	. . . . .	Quetta.
18 Apl. 16	Irving, G., M.B.	19 Oct. 16	. . . . .	. . . . .	Rawalpindi.
12 May 14	Lawrence, R. D., M.B. (170, 207).	2 Nov. 16	. . . . .	. . . . .	Nowshera.
4 June 14	Cullenan, J.	1 Jan. 17	. . . . .	. . . . .	Bareilly.
1 Aug. 16	Greer, W. N.	1 Feb. 17	. . . . .	. . . . .	Shwebo.
1 Aug. 16	Fish, E. W.	1 Feb. 17	. . . . .	. . . . .	Colaba. Splst. in Adv. Op. Surgery.
4 Aug. 16	Isbister, R., M.B.	4 Feb. 17	. . . . .	. . . . .	Rawalpindi.
15 Aug. 16	Steven, A. R.	15 Feb. 17	. . . . .	. . . . .	Madras.
25 Aug. 16	Harrison, C. C.	25 Feb. 17	. . . . .	. . . . .	Calcutta.
1 Sep. 16	MacCaskie, F. B.	1 Mar. 17	. . . . .	. . . . .	Bombay.
16 July 16	Albinson, J. H.	2 Mar. 17	. . . . .	. . . . .	Delhi.
4 Sep. 16	Kitchen, J. E.	4 Mar. 17	. . . . .	. . . . .	Peshawar.
. . . .	Davin, A. L. V., M. B.	12 Apl. 17	. . . . .	. . . . .	Rawalpindi.
7 Aug. 16	Kirby, P. R. E.	8 May 17	. . . . .	. . . . .	Poona.
15 Sep. 16	Miller, R. W. H.	15 June 17	. . . . .	. . . . .	Lahore, Splst in Electrical Science.
9 Aug. 14	Macnamara, J. P., M. B.	1 July 17	. . . . .	. . . . .	Rawalpindi.
. . .	Brown, W. H.	16 Aug. 17	. . . . .	. . . . .	Lahore Cantt.
19 Aug. 14	Cullen, J. A. W.	10 Oct. 17	. . . . .	. . . . .	Rawalpindi, Splst. in Prev. of Disease.
1 Nov. 16	Fleming, N. B. B.	1 Nov. 17	. . . . .	. . . . .	Colaba, Splst. in Ophthalmology.
1 Nov. 16	Honeyman, T. J.	1 Nov. 17	. . . . .	. . . . .	Lahore Cantt.
1 Dec. 16	Clarke, R. V.	1 Dec. 17	. . . . .	. . . . .	Colaba.
18 Dec. 16	Morris, W. M., M.B.	18 Dec. 17	. . . . .	. . . . .	Bareilly.
9 Nov. 16	Simpson, R. G.	1 Jan. 18	. . . . .	. . . . .	Lahore Cantt.
16 Oct. 16	Clyde, D., M. B.	. . . .	. . . . .	. . . . .	Bareilly.
1 Nov. 16	Hickson, E B.	1 Feb. 18	. . . . .	. . . . .	Arangaon.
1 Nov. 16	Banks, J. R.	1 Feb. 18	. . . . .	. . . . .	Secunderabad.
4 Nov. 16	Riggall, P.R	1 Feb. 18	. . . . .	. . . . .	Deolali.
7 Nov. 16	Simons, G. E. L.	1 Feb. 18	. . . . .	. . . . .	Peshawar.
1 Feb. 17	Vertue, H. St. H.	1 Feb. 18	. . . . .	. . . . .	
5 Feb. 17	Blackmore, L. A.	5 Feb. 18	. . . . .	. . . . .	Quetta.
5 Feb. 17	Parker, E., M.C.	5 Feb. 18	. . . . .	. . . . .	Fyzabad.
23 Oct. 16	Agar, W.	7 Feb. 18	. . . . .	. . . . .	Deolali.
10 Feb. 17	Frith, F. C. A.	10 Feb. 18	. . . . .	. . . . .	Deolali.
20 Nov. 16	Newton, E.	20 Feb. 18	. . . . .	. . . . .	Agra.
22 Feb. 17	Woodside, R.	22 Feb. 18	. . . . .	. . . . .	Kirkee.
26 Nov. 16	Wilson, G. S.	26 Feb. 18	. . . . .	. . . . .	Lv., ex I. m. c., 6 mos.
26 Feb. 17	Fitch, A. A.	26 Feb. 18	. . . . .	. . . . .	Kirkee, Splst. in Mental Science, Poona Dn.
26 Feb. 17	Mudge, J. B.	26 Feb. 18	. . . . .	. . . . .	Colaba.
1 Dec. 16	Williams, T. P.	1 Mar. 18	. . . . .	. . . . .	Poona.
1 Mar. 17	Dykes, H. A. G.	1 Mar. 18	. . . . .	. . . . .	Deolali., Splst. in Adv. Op. Surgery.
4 Mar. 17	Mason, W. F.	4 Mar. 18	. . . . .	. . . . .	Nowshera.
1 Jan. 17	Hayman, F. K.	2 Apl. 18	. . . . .	. . . . .	
8 Apl. 17	Humphreys, R. M.	8 Apl. 18	. . . . .	. . . . .	Bombay.
1 Feb. 17	McIlroy, H. D.	1 May 18	. . . . .	. . . . .	Nowshera.
1 May 17	Wiggins, R.	1 May 18	. . . . .	. . . . .	
4 May 17	Griffin, A. J. B.	4 May 18	. . . . .	. . . . .	Bangalore.
26 Feb. 17	Tolmie, J. A. (*To take rank and precedence from 16 May 17*).	10 May 18	. . . . .	. . . . .	Wellington.
16 May 17	Hughes, L.W.	16 May 18	. . . . .	. . . . .	Poona.

Lieutenants (Special Reserve).

First Comm.	Names.	Date of present rank.	Date of compln. Indian tour.	Date of present appointment.	Remarks.
1 Mar. 17	Airth, O. F. D.	1 Mar. 17	. . . . .	. . . . .	Deolali.
17 Apl. 17	Beaumont, D. C.	17 Apl. 17	. . . . .	. . . . .	Deolali.
19 Apl. 17	Collins, J. C.	19 Apl. 17	. . . . .	. . . . .	

ROYAL ARMY MEDICAL CORPS—*contd.*

Lieutenants (Special Reserve)—*contd.*

First Commn.	Names.	Date of present rank.	Date of compln. Indian tour.	Date of present appointment.	Remarks.
21 Apl. 17	Carpenter, J. E.	21 Apl. 17	. . . . .	. . . . .	Rawalpindi.
26 Apl. 17	Powell, W. I. FitzG.	26 Apl. 17	. . . . .	. . . . .	Mhow.
26 Apl 17	Andreae, E. B.	26 Apl. 17	. . . . .	. . . . .	Dehra Dun.
27 Apl. 17	Wilcock, A. G. E.	27 Apl 17	. . . . .	. . . . .	Rawalpindi.
1 May 17	Allison, J., M.B.	1 May 17	. . . . .	. . . . .	Bareilly.
1 May 17	Pritchett, H. N.	1 May 17	. . . . .	. . . . .	Peshawar.
10 May 17	Norman, J. M.	10 May 17	. . . . .	. . . . .	Bombay.
11 May 17	Hubbard, G. R.	11 May 17	. . . . .	. . . . .	Maymyo, Splst. in Dermatology.
19 May 17	Gill, L.	19 May 17	. . . . .	. . . . .	Swebo.
1 June 17	White, W. H.	1 June 17	. . . . .	. . . . .	Kirkee.
11 June 17	Masson, K.	11 June 17	. . . . .	. . . . .	Poona.
12 June 17	Martin, J. A.	12 June 17	. . . . .	. . . . .	Quetta.
9 July 17	Bowl, W. T. G.	9 July 17	. . . . .	. . . . .	Kirkee.
19 July 17	Irwin, C. G., M.B.	19 July 17	. . . . .	. . . . .	Delhi.
1 Aug. 17	Dykes, H. B., M.B.	1 Aug. 17	. . . . .	. . . . .	Bombay.
1 Aug. 17	Shaw, T. W., M.B.	1 Aug. 17	. . . . .	. . . . .	Bombay.
2 Aug. 17	Titterton, C. M.	2 Aug. 17	. . . . .	. . . . .	Kirkee.
9 Aug. 17	Grant, P. F. A.	9 Aug. 17	. . . . .	. . . . .	Karachi.
9 Aug. 17	Fulton, G. K., M.B.	9 Aug. 17	. . . . .	. . . . .	Rangoon.
9 Aug. 17	Fleming, T., M.B.	9 Aug. 17	. . . . .	. . . . .	Bhamo.
9 Aug. 17	Hurst, G. F.	9 Aug. 17	. . . . .	. . . . .	Colaba War Hospl.
9 Aug. 17	Irving, E. P.	9 Aug. 17	. . . . .	. . . . .	Abbottabad.
9 Aug. 17	Weatherhead, W. A.	9 Aug. 17	. . . . .	. . . . .	
9 Aug. 17	Fraser, R. J. L., M.B.	9 Aug. 17	. . . . .	. . . . .	
16 Aug. 17	Brown, W. W., M.B.	16 Aug. 17	. . . . .	. . . . .	Jullundur.
16 Aug. 17	Milne, A. Y., M.B.	16 Aug. 17	. . . . .	. . . . .	Deolali.
16 Aug. 17	Aperjis, H. D.	16 Aug. 17	. . . . .	. . . . .	Risalpur.
4 Feb. 18	Carruthers, C. F. J., M.B.	4 Feb. 18	. . . . .	. . . . .	Ahmednagar.
8 Apl. 18	Neighbour, P. M.	8 Apl. 18	. . . . .	. . . . .	

Territorial Force.

Colonels.

First Commn.	Names.	Date of present rank.	Date of compln. Indian tour.	Date of present appointment.	Remarks.
. . . .	Sheen, A. W., M.D.	26 Dec. 17	. . . . .	. . . . .	Southern Command Area, Consulting Surgeon.

Majors.

First Commn.	Names.	Date of present rank.	Date of compln. Indian tour.	Date of present appointment.	Remarks.
. . . .	Walker, J.	. . . .	. . . . .	. . . . .	Lucknow.
14 Nov. 00	Kendle, F. W.	21 Feb. 13	. . . . .	. . . . .	34 G. H., Deolali.
7 Feb. 00	Halton, H.	25 Mar. 15	. . . . .	. . . . .	Comdg. Stn. Hospl., Muttra.
11 July 03	Jefferiss, F. B., F.R.C.S.E. (325).	11 July 15	. . . . .	. . . . .	Rawalpindi.

Captains.

First Commn.	Names.	Date of present rank.	Date of compln. Indian tour.	Date of present appointment.	Remarks.
10 Nov. 14	Evans, D. E., M.B. (Maj. 7 Feb. 17).	10 May 15	. . . . .	. . . . .	Jullundur.
23 Mar. 09	Webb-Johnson, C., M.B. (327).	4 Dec. 12	. . . . .	. . . . .	Calcutta.—Splst. in Midwifery and Diseases of Women and Children.

ROYAL ARMY MEDICAL CORPS—*contd.*

Territorial Force—*contd.*

Captains—*contd.*

First Commn.	Names.	Date of present rank.	Date of compln. Indian tour.	Date of present appointment.	Remarks.
12 Oct. 14	**Simpson**, H. C.	12 Oct. 14	....	....	Calcutta.
....	**Warren**, S. H.,	23 Nov. 14	....	....	In Medl. charge Enteric Con. Depôt, Wellington.
1 Apl. 15	**Phillips**, W. J., M.B.	1 Apl. 15	....	....	Colaba.
2 Jan. 14	**Van Someren**, A. G. V., M.B., F.R.C.S.E. (312 *h*).	1 Apl. 15	....	....	Splst. in Adv. Op. Surgery, Lahore Cantt.
†26 June 12 *12 Nov. 14	**Townley**, J. S. (298)	1 Apl. 15	....	....	Splst. in Diseases of Women and Children, 5th (Mhow) Dn. Comdg. Stn. Hospl., Mt. Abu.
26 Sep. 14	**Murray**, T.	1 Apl. 15	....	....	Deolali.
12 Oct. 14	**Granger**, A. P., M.B.	12 Apl. 15	....	....	Satara.
15 Nov. 14	**Gosse**, P. H. G.	16 May 15	....	....	Poona.
20 Dec. 14	**Hummell**, J. J., M.B.	20 June 15	....	....	Poona.*
1 Oct. 11	**Tough**, F. W. K., F.R.C.S. (Edin.)	16 Sep. 15	....	....	
18 Mar. 15	**Heron**, J. M., M.D.	18 Sep. 15	....	....	Bangalore.—Splst. in Midwifery and Diseases of Women and Children.
7 May 15	**Rogers**, C. D., M.B.	7 Nov. 15	....	....	Poona.
29 Oct. 14	**Pelly**, H. N.	5 Dec. 15	....	....	Poona.
4 Jan. 16	**Hibbert**, C., M.D.	4 July 16	....	....	Jubbulpore.
7 June 15	**Reed**, R. B., M.B.	7 Dec. 16	....	....	Poona.
18 Mar. 17	**Glover**, B. T. J., M.B.	18 Mar. 17	....	....	Bangalore.—Splst. in Prevention of Disease.

Temporary Lieutenant-Colonels.

First Commn.	Names.	Date of present rank.	Date of compln. Indian tour.	Date of present appointment.	Remarks.
17 Feb. 05	**Maturin**, F. H., M.B. (Lt.-Col., Hants R.)	12 July 17‡ 7 Oct. 17§ 8 Oct. 14§	....	....	O. C. 34 Genl. Hospl., Deolali.
23 July 92	**Haswell**, J. F., M.D. (Lt.-Col., 4 Bn., Bord. R.) (T.F.)	13 July 17‡	....	....	Quetta.

Temporary Majors.

First Commn.	Names.	Date of present rank.	Date of compln. Indian tour.	Date of present appointment.	Remarks.
9 Mar. 15	**Pannett**, C. A., F.R.C.S., Capt. R.A.M.C. (T.F.)	8 July 16	....	....	Splst. Adv. Operative Surgery, 9th Dn., Secunderabad.
2 Dec. 08	**Bosanquet**, W. C., M.B., Capt. R.A.M.C. (T.F.)	24 Aug. 16	....	....	Bombay.
3 Aug. 14	**Joy**, H. R. L., M.D.	1 Nov. 16	....	....	
Jan. 96	**Cosens**, F. R. S. (Maj., 4 Bn., Devon. R.) (T.F.)	6 Mar. 13§ 12 July 17‡	....	....	Quetta.

† Date of first combatant commission.
* Date of joining R. A. M. C., T. F.
§ Date of present rank in Territorial Force.
‡ Date of temporary transfer to the R. A. M. C.

ROYAL ARMY MEDICAL CORPS—*contd.*

Temporary Captains.

First Commn.	Names.	Date of present rank.	Date of compln. Indian tour.	Date of present appointment.	Remarks.
	Kisch, H. A., M.B., F.R.C.S.	17 Feb. 15	. . . .	. . .	Bangalore, Splts. in Otology, Laryngology and Rhinology.
†15 Sep. 08 *3 Sep. 14	Cane, L. B., M.D.	22 May 12 15 Apl. 15	. . . .	. . . .	Meiktila.
11 Nov. 14	McKnight, M.	11 May 15	. . . .	. . . .	Comdg. Stn. Hospl., Multan.
8 Aug. 14	Butler, P. P.	8 Aug. 15	. . . .	. . . .	Deolali.
10 Aug. 14	Burgess, J. B., M.D.	10 Aug. 15	. . . .	. . . .	Ferozepore.
12 Aug. 14	Holroyd, T. H., M.B.	12 Aug. 15	. . . .	. . . .	Meerut, Temp. Splst. in Dermatology, Meerut Dn.
15 Aug. 14	Peake, H. G.	15 Aug. 15	. . . .	. . . .	Ambala.
22 Aug. 14	Dawson, W., D.S.O., M.B.	22 Aug. 15	. . . .	. . . .	Aden.
11 Sep. 14	Roberts, R. E., M.B.	11 Sep. 15	. . . .	. . . .	Rawalpindi.
12 Sep. 14	Thomas, D. J.	12 Sep. 15	. . . .	. . . .	Splst. in Prevention of Diseases, Abbottabad.
2 Oct. 14	Howlett, L. W., M.B.	2 Oct. 15	. . . .	. . . .	Poona.
5 Oct. 14	Aikman, J. H., M.C., M.B.	5 Oct. 15	. . . .	. . . .	Nowshera. Splst. in Prevention of Disease.
10 Oct. 14	Reader, N., M.B.	10 Oct. 15	. . . .	. . . .	Lv., ex India, m. c.
10 Oct. 14	Fetherstonhaugh, W. E., M.B.	10 Oct. 15	. . . .	. . . .	Peshawar.
10 Oct. 14	Murray, M. J.	10 Oct. 15	. . . .	. . . .	Wellington.
12 Oct. 14	Costobadie, L. P.	12 Oct. 15	. . . .	. . . .	Mhow.
6 Nov. 14	Constant, C. F.	6 Nov. 15	. . . .	. . . .	Lucknow, Splst. in Electrical Science.
10 Nov. 14	Crombie, L., M.B.	10 Nov. 15	. . . .	. . . .	Manmad.
16 Nov. 14	Woolf, A. E. M., M.B., F.R.C.S.	16 Nov. 15	. . . .	. . . .	Ambala.—Splst. in Advanced Operative Surgery, Ambala Bde.
21 Nov. 14	Askey, S. G., M.B.	21 Nov. 15	. . . .	. . . .	Aden.
1 Dec. 14	Evans, W. S.	1 Dec. 15	. . . .	. . . .	Poona.
2 Dec. 14	Marjoribanks-Marcar, E.	2 Dec. 15	. . . .	. . . .	Agra.
12 Dec. 14	Von-Braun, C. R. B.	12 Dec. 15	. . . .	. . . .	Sialkot.
16 Dec. 14	Chapman, E.S., M.D., F.R.C.S.	16 Dec. 15	. . . .	. . . .	
8 Jan. 15	Harpur, H. P., M.B.	8 Jan. 16	. . . .	. . . .	Jubbulpore.
15 Jan. 15	White, C. F. Orr	15 Jan. 16	. . . .	. . . .	Poona.
18 Jan. 15	Wood, W. C. S.	18 Jan. 16	. . . .	. . . .	Rawalpindi.
21 Jan. 15	Murphy, R. W., M.D.	21 Jan. 16	. . . .	. . . .	Quetta.—In collateral charge Cadet College.
28 Jan. 15	Turner, J. R.	28 Jan. 16	. . . .	. . . .	Jullundur.
30 Jan. 15	Radley, S. B., M.B., F.R.C.S.	31 Jan. 16	. . . .	. . . .	Bellary, Splst. in Adv. Operative Surgery.
1 Feb. 15	Browning, S. H.	1 Feb. 16	. . . .	. . . .	Jubbulpore. Splst. in Prevention of disease (*temp.*).
7 Feb. 15	Herriot, T. P., M.B.	7 Feb. 16	. . . .	. . . .	Jullundur. Splst. in Prevention of Disease.
10 Feb. 15	Leach, A., M.B.	10 Feb. 16	. . . .	. . . .	Rawalpindi.
10 Feb. 16	Mclaren, D. J., M.B.	10 Feb. 16	. . . .	. . . .	Barrackpore, Comdg. Stn. Hospl.
15 Feb. 15	Anthonisz, C. F.	15 Feb. 16	. . . .	. . . .	Wellington.
15 Feb. 15	Fox, H. W.	15 Feb. 16	. . . .	. . . .	Poona.
20 Feb. 15	Beaver, H. A., M.B.	20 Feb. 16	. . . .	. . . .	Poona.
22 Feb. 15	Coffin, S. W.	22 Feb. 16	. . . .	. . . .	Malapuram.
27 Feb. 15	Laborda, F. E. R.	27 Feb. 16	. . . .	. . . .	Lucknow. Splst. in Otology, etc. (temp.)
14 Mar. 15	Metcalfe, G. C.	14 Mar. 16	. . . .	. . . .	Lucknow.
14 Mar. 15	Buchanan, J. V.	14 Mar. 16	. . . .	. . . .	Aden.
20 Mar. 15	Scholtz, C. J.	20 Mar. 16	. . . .	. . . .	Secunderabad.
29 Mar. 15	Gunn, A. R., M.B.	29 Mar. 16	. . . .	. . . .	Ambala. Splst. in Ophthalmology, Ambala Bde.
1 Apl. 15	Bryan, F., M.B.	1 Apl. 16	. . . .	. . . .	Bombay.
2 Apl. 15	Stone, E. R.	2 Apl. 16	. . . .	. . . .	Lahore. Splst. in Mental Science, N. Command.

* Date of first Temporary commission, R. A. M. C.
† Date of first commission R. A. M. C., T.F.

ROYAL ARMY MEDICAL CORPS—*contd.*

Temporary Captains—*contd.*

First Commn.	Names.	Date of present rank.	Date of compln. Indian tour.	Date of present appointment.	Remarks.
4 Apl. 15	**Carruthers**, B. M.	4 Apl. 16			Multan.
7 Apl. 15	**Adam**, G. M., M.C., M.B.	7 Apl. 16			Agra.
9 Apl. 15	**Smith**, A. H. D., M.C., M.B.	9 Apl. 16			Deolali.
13 Apl. 15	**Wood**, G. H., M.B.	13 Apl. 16			Bombay.
14 Apl. 15	**Alexander**, W. A., M.B.	14 Apl. 16			Peshawar.
14 Apl. 15	**Maguire**, P. J.	14 Apl. 16			Roorkee.
26 Apl. 15	**Crawford**, A.	26 Apl. 16			
10 May 15	**Hammond**, J. F.	10 May 16			Calcutta. Splst. in Adv. Operative Surgery.
11 May 15	**Strickland**, H. F., F.R.C.S.	11 May 16			
27 May 15	**Fettes**, D.	27 May 16			Quetta.
1 June 15	**Kerr**, F. A., M.B.	1 June 16			Agra.
2 June 15	**Cochrane**, A., M.B.	2 June 16			
7 June 15	**Harrison**, C. E.	7 June 16			Bombay.
7 June 15	**Smith**, E. D., M.B.	7 June 16			Bombay.
8 June 15	**Brown**, A., M.B.,	8 June 16			Rawalpindi.
10 June 15	**Hare**, W. F., M.B.	10 June 16			Bangalore.
1 July 15	**Wishart**, W. H.	1 July 16			Rawalpindi.
1 July 15	**Underwood**, J. E. A.	1 July 16			Jubbulpore. Splst. in Electrical Science.
10 July 15	**Whyte**, D.	10 July 16			Rawalpindi.
14 July 15	**O'Regan**, W. F., M.B.	14 July 16			Peshawar.
31 July 15	**Sproule**, C. L.	31 July 16			
4 Aug. 15	**Leiper**, J. A., M.B.	4 Aug. 16			Deolali.
9 Aug. 15	**Clark**, G. S., M.D.	9 Aug. 16			Ferozepore.
9 Aug. 15	**Cross**, F. G., F.R.C.S.	9 Aug. 16			Bangalore. Splst. in Adv. Operative Surgery.
10 Aug. 15	**Highet**, H.	10 Aug. 16			Mandalay.
12 Aug. 15	**Duffy**, J. V.	12 Aug. 16			Deolali.
18 Aug. 15	**Story**, W. G. T., M.B.	18 Aug. 16			Poona.
26 Aug. 15	**McSwiney**, S. A., M.B.	26 Aug. 16			Shillong.
25 Sep. 16	**Milsom**, E. G. D.	25 Sep. 16			Bombay.
1 Oct. 15	**Baldwin**, E.	1 Oct. 16			Peshawar.
12 Oct. 15	**Moloney**, W. J., M.B.	12 Oct. 16			Karachi.
19 Oct. 15	**Hornsby**, W., M.B.	19 Oct. 16			Deolali.
4 Nov. 15	**Keogh**, F. de C.	4 Nov. 16			Poonamallee.
7 Nov. 15	**Ryan**, P. C. H., M.D.	7 Nov. 16			Poona.
8 Nov. 15	**Rutherford**, A., M.B.	8 Nov. 16			Deolali. Splst. in Prevention of Disease.
28 Dec. 15	**Stewart**, H.	28 Dec. 16			Belgaum.
1 Jan. 16	**Briscoe**, J. R.	1 Jan. 16			Dehra Dun.
1 Jan. 16	**Smith**, W. H. R.	1 Jan. 17			Bangalore.—Splst. in Mental Science.
1 Jan. 16	**Masters**, A. M., M.D.	1 Jan. 17			Rawalpindi.—Splst. in Midwifery and Diseases of Women and Children.
1 Jan. 16	**Manson**, D., M.B.	1 Jan. 17			Anandi, Ahmednagar.
1 Jan. 16	**Sewell**, D. L., M.B.	1 Jan. 17			Colaba—Splst. in Otology, etc. (temp.).
12 Jan. 16	**Illingworth**, R. E.	12 Jan. 17			
14 Jan. 16	**Cameron**, E. H., M.B., F.R.C.S.	14 Jan. 17			Secunderabad.—Splst. in Ophthalmology (temp.).
19 Jan. 16	**Eadie**, R. B., M.B.	19 Jan. 17			Peshawar.
20 Jan. 16	**Potter**, J. W., M.B.	20 Jan. 17			Risalpur.
1 Feb. 16	**Wilson**, H. M., M.B.	1 Feb. 17			Belgaum.
4 Feb. 16	**Berg**, G. C.	4 Feb. 17			Bombay.
17 Feb. 16	**Drury**, A. N., M.B.	17 Feb. 17			Secunderabad.
17 Feb. 16	**Hannigan**, H.	17 Feb. 17			Meerut.
2 Mar. 16	**Lamont**, G. A., M.B.	2 Mar. 17			Wellington.
4 Mar. 16	**Scott**, A. A. B., M.B.	4 Mar. 17			Lahore Cant.
4 Mar. 16	**Dowdall**, J. J., M.B.	4 Mar. 17			Dehra Dun.
8 Dec. 15	**Prall**, S. R.	8 Mar. 17			Ahmednagar.
20 Mar. 16	**Wallace**, V.	20 Mar. 17			Madras.
20 Mar. 16	**Marshall**, W. H.	20 Mar. 17			Bangalore.
23 Mar. 16	**Davies**, E. I.	23 Mar. 17			Deolali.
1 Apl. 16	**Littlewood**, S.	1 Apl. 17			Lahore.
1 Apl. 16	**Harvey**, J., M.B.	1 Apl. 17			Deolali.

ROYAL ARMY MEDICAL CORPS—*contd.*

Temporary Captains—*contd.*

First Commn.	Names.	Date of present rank.	Date of compln. Indian tour.	Date of present appointment.	Remarks.
1 Apl. 16	Milne P.	1 Apl. 17			Kirkee.
1 Apl. 16	Trow, C. E. A.	1 Apl. 17			Campbellpore.
7 Apl. 16	Saunders, E. G., M.B.	7 Apl. 17			Nasik.
7 Apl. 16	Walshe, J. J.	7 Apl. 17			Karachi.
13 Apl. 16	Black, K., F.R.C.S.	13 Apl. 17			Bombay.—Splst. in Adv. Operative Surgery (temp.)
17 Apl. 16	Hughes, E. E., M.B., F.R.C.S.	17 Apl. 17			
20 Apl. 16	Fraser, G. H., M.B.	20 Apl. 17			Ambala.
20 Apl. 16	Dickson, J., M.C., M.B.	20 Apl. 17			Deolali.
24 Apl. 16	Ronaldson, G. W., M.B.	24 Apl. 17			
25 Apl. 16	Sargent, E. L. K.	25 Apl. 17			Karachi. Splst. in Adv. Op. Surgery (temp.).
25 Apl. 16	Doorly, A. R. C., M.B.	25 Apl. 17			Deolalali.
25 Apl. 16	Johnstone, J. M., M.B.	25 Apl. 17			Deolali.
25 Apl. 16	deMowbray, R. M., F.R.C.S.	25 Apl. 17			Deolali—Splst. in Adv. Op. Surgery (temp.).
26 Apl. 16	Harries, D. J., F.R.C.S.	26 Apl. 17			Deolali.—Surg. Splst.
26 Apl. 16	Brandon, G. N., M.R.C.S.	26 Apl. 17			Splst. in Ophthalmology, Deolali.
26 Apl. 16	Jones, T. L., M.B.	26 Apl. 17			Deolali.—Splst. in Otology, Laryngology and Rhinology (temp.).
26 Apl. 16	Keyms, J., M.D.	26 Apl. 17			Aden. Splst. in Ophthalmology.
26 Apl. 16	Marsden, W. G., M.B.	26 Apl. 17			Deolali.
26 Apl. 16	Evans, T. G., M.D.	25 Apl. 17			Deolali. X-Ray Splst.
26 Apl. 16	Cairns, J., M.B. (*Temp Maj.*)	26 Apl. 17			D. A. D. M. S. (Sany). 8th (Lucknow) Dn.
26 Apl. 16	Burn, S. A., M.R.C.S.	26 Apl. 17			
28 Apl. 16	Calneck, S. H.	28 Apl. 17			
1 May 16	Arthur, G., M.B.	1 May 17			Poona.
1 May 16	Farquharson, D. A., M.B.	1 May 17			Bareilly.
1 May 16	Moriarty, J. J., M.B.	1 May 17			Karachi.
1 May 16	Ferguson, H. F., M.B.	1 May 17			Colaba.
1 May 16	Miller, R. W., M.B.	1 May 17			Tank.
1 May 16	Barrow, R. H. Balfour, M.B.	1 May 17			Meerut.—Splst. in Ophthalmology (temp.).
3 May 16	Vanderwert, J. A. H.	3 May 17			Poonamallee.
8 May 16	Knobel, W. B., M.D.	8 May 17			Bangalore.
9 May 16	Watson, T.T.B.	9 May 17			Rawalpindi.
9 May 16	Michelmore, R.G.	9 May 17			Rawalpindi.
9 May 16	Haworth, A. N.	9 May 17			Deolali.
12 May 16	Dargan, P. A.	12 May 17			Abbottabad.
15 May 16	Jackson, T. H.	15 May 17			Lahore Cant.
25 May 16	Hirsch, V. R.	25 May 17			
31 May 16	Tennant, C. S., M.B.	31 May 17			Rawalpindi.
1 June 16	Hawes, A. J., M.C., M.B.	1 June 17			Poona.
4 June 16	Pearce, A. H. B.	4 June 17			Mhow.
8 June 16	O'Malley, D. V., M.B.	8 June 17			
17 June 16	Davey, C. F., M.B.	17 June 17			Dagshai.
19 June 16	Law, T. S., M.B.	19 June 17			Jhansi.
20 June 16	Mayo, H. R., M.B.	20 June 17			Meerut, Splst. in Midwifery and Diseases of Women and Children.
29 June 16	Ainsley, J. E.	29 June 17			Delhi.
6 July 16	Jones, T.	6 July 17			Ambala.
10 July 16	Crichton, B. D.	10 July 17			Cawnpore.
11 July 16	Robertson, M. K.	11 July 17			Roorkee.
14 Aug. 16	Goss, J., M.B.	12 July 17			Poona. Temp. Splst. in Electrical Science.
16 July 16	Stoddard, R. H., M.D.	16 July 17			
21 July 16	Garden, H. W., M.B.	21 July 17			Anandi, Ahmednagar.
24 July 16	Darling, T. N.	24 July 17			Poona.
2 Aug. 16	Cox, C. H.	2 Aug. 17			Bareilly, Splst. in Prevention of Disease.

ROYAL ARMY MEDICAL CORPS—*contd.*

Temporary Captains—*contd.*

First Commn.	Names.	Date of present rank.	Date of compln. Indian tour.	Date of present appointment.	Remarks.
2 Aug. 16	**Fenton**, H. F., M. B.	2 Aug. 17			Colaba.
7 Aug. 16	**Pinches**, H. I.	7 Aug. 17			Abbottabad.
7 Aug. 16	**McCaskie**, H. B.	7 Aug. 17			Rawalpindi. Staff Surgeon A.
7 Aug. 16	**Sharp**, J. E., M.B.	7 Aug. 17			Poona.
8 Aug. 16	**Pinnock**, V. St. L., M.B.	8 Aug. 17			Poona.
10 Aug. 16	**Stockley**, C. I.	10 Aug. 17			
31 Aug. 16	**Symons**, H. J. H., M.C.	31 Aug. 17			Mhow.
1 Sep. 16	**Gemmill**, W.	1 Sep. 17			Bombay.
15 Sep. 16	**Binning**, A., M.D.	15 Sep. 17			Poona, temp. Splst. in Mental Science, S. Comd.
15 Sep. 16	**Leney**, R. J. B.	15 Sep. 17			Bangalore.
16 Sep. 16	**Harley-Mason**, R. J.	16 Sep. 17			Bombay.
7 Oct. 16	**Bury**, G. W., M.D., F.R.C.S.E.	7 Oct. 17			Colaba, Specialist in Adv. Op. Surgery.
23 Oct. 16	**Smith**, T. St. C., M.B.	23 Oct. 17			Deolali.
24 Oct. 16	**Hall**, P. A.	24 Oct. 17			Lahore.
24 Oct. 16	**McDonogh**, C. L., M.B.	24 Oct. 17			Poona.
18 Dec. 16	**Challans**, F., M.D.	18 Dec. 17			Bombay.
9 Feb. 17	**Daly**, J. T.	9 Feb. 18			Bombay.
11 Feb. 17	**Simpson**, F. T., M.C., M.D.	11 Feb. 18			Bombay.
16 Feb. 17	**Bernard**, E. A.	16 Feb. 18			Ahmednagar.
1 Mar. 17	**Airth**, O. F. D., M.A.	1 Mar. 18			Deolali.
6 Apl. 17	**Reichwald**, M. B., M.B.	6 Apl. 18			Quetta.
14 Apl. 17	**Steedman**, M. T. W.	14 Apl. 18			Bombay.
15 Apl. 17	**Higgins**, D.	15 Apl. 18			Bombay.
16 Apl. 17	**Eddie**, R. W.	16 Apl. 18			Bareilly.
10 May 17	**DeMorgan**, H. A.	10 May 18			Colaba.
21 May 17	**Bush**, W. H.	21 May 18			Deolali.
21 May 17	**Milne**, C. I., M.B.	21 May 18			Peshawar.
22 May 17	**Gregory**, A.	22 May 18			Mhow.
25 May 17	**Shepherd**, G. F.	25 May 18			Kirkee.
28 May 17	**Nolan**, J. N. G., M.D.	28 May 18			Poona.
28 May 17	**Latham**, H. W.	28 May 18			Lahore.
28 May 17	**Overend**, H. F.	28 May 18			Bangalore.
30 May 17	**Henson**, P. H.	30 May 18			Bangalore.
1 June 17	**Hutchinson**, L., M.D.	1 June 18			Peshawar.
1 June 17	**Iles**, J. H.	1 June 18			Ambala.
1 June 17	**Lillco**, W., M.D.	1 June 18			Poona.
1 June 17	**Duncan**, R., M.B.	1 June 18			
4 June 17	**Pratt**, G. A.	4 June 18			Bangalore.
4 June 17	**Rotheram**, W.	4 June 18			Multan.
4 June 17	**Mearns**, J. A., M.B.	4 June 18			Quetta.
5 June 17	**Ticehurst**, C. B.	5 June 18			Karachi, Splst. in Advanced Op. Surgery, (*tempy.*)
6 June 17	**Wilson**, J. S., M.D.	6 June 18			Poona.
6 June 17	**Werden**, L. H.	6 June 18			Ferozepore.
7 June 17	**Ritchie**, D. D.	7 June 18			Peshawar.
7 June 17	**Ferguson**, J.	7 June 18			Meerut.
11 June 17	**Lawrence**, C. A.	11 June 18			
14 June 17	**Gerrard**, C. B., M.B.	14 June 18			Poona.
14 June 17	**Lyall**, C. H. G.	14 June 18			Colaba, Mental Splst., Bombay Bde.
14 June 17	**Morgan**, E.	14 June 18			Poona.
14 June 17	**Dockrell**, M.	14 June 18			Deolali.
14 June 17	**Corbett**, R. C.	14 June 18			
15 June 17	**Strachan**, R. S.	15 June 18			Hyderabad.
15 June 17	**Crooks**, F.	15 June 18			Poona, Splst. Advs. Op. Surgery.
15 June 17	**Gray**, A. H., M.D.	15 June 18			Poona.
15 June 17	**Fraser**, T.	15 June 18			Madras.
15 June 17	**Powell**, J. A.	15 June 18			Secunderabad.

ROYAL ARMY MEDICAL CORPS—*contd.*

Temporary Captains—*concld.*

First Commn.	Names.	Date of present rank.	Date of compln. Indian tour.	Date of present appointment.	Remarks.
15 June 17	**Brash,** C.	15 June 18			Wellington.
15 June 17	**Patchett,** R. T. A.	15 June 18			Deolali.
15 June 17	**Fairweather,** D. O.	15 June 18			Deolali.
16 June 17	**Farlinger,** A. C., M.D.	16 June 18			Campbellpore.
18 June 17	**Cran,** H. R.	18 June 18			Bombay.
23 June 17	**Marshall,** H. F.	23 June 18			Poona.
19 June 17	**Holroyd.** J. H. B.	19 June 18			Colaba.
19 June 17	**Parkinson,** W. H.	19 June 18			Bombay.
25 June 17	**Couldrey,** T. R.	25 June 18			Colaba.
12 July 17	**Reynolds,** R. J., M.B.	12 July 18			Karachi, Splst. in Electrical Science.
19 July 17	**Stamp,** L. D.	19 July 18			Jhansi.
19 July 17	**Nicholson,** F. D.	19 July 18			Bangalore.
23 July 17	**Haines,** R. L.	23 July 18			
26 July 17	**Ledger,** A. V., M.D.	26 July 18			Poona.
26 July 17	**Butler,** F.	26 July 18			
28 July 17	**Atkin,** E. E., M.B.	28 July 18			Rawal Pindi.
9 Aug. 17	**Barron,** S.	9 Aug. 18			Allahabad, Splst. Prevention of Disease.
9 Aug. 17	**Gordon,** G. S.	9 Aug. 18			Multan.
9 Aug. 17	**Eggar,** W. H., M.B.	9 Aug. 18			Meerut
9 Aug. 17	**Wyper,** J., M.B.	9 Aug. 18			
10 Aug. 17	**Dove,** J. H.	10 Aug. 18			Abbottabad.
16 Aug. 17	**Bernard,** E. A.	16 Aug. 18			Poona.
23 Aug. 17	**Macdonald,** G. G.	23 Aug. 18			Rangoon, Splst. in Prev. of Disease.
20 Sep. 17	**Wardrop,** D., M. B.	20 Sep. 18			Mhow.
12 Oct. 17	**Cope,** R.	12 Oct. 18			Poona.
6 Nov. 17	**Wrench,** G. T.	6 Nov. 18			Karachi.
9 Nov. 17	**Weinman,** L. O.	9 Nov. 18			Bangalore.

Temporary Lieutenants.

First Commn.	Names.	Date of present rank.	Date of compln. Indian tour.	Date of present appointment.	Remarks.
10 Sep. 18	**Turnbull,** J. N.	10 Sep. 18			Jhansi.
2 Oct. 18	**Stewart,** J.	1 Oct. 18			Bangalore.

Quartermasters.

First Commn.	Names.	Date of present rank.	Date of compln. Indian tour.	Date of present appointment.	Remarks.
25 Nov. 15	**Sumner,** Lieut. G. A.	25 Nov. 15			Bombay.
5 July 16	**Andrews,** Lieut. M.	5 July 16			Deolali.
1 Aug. 16	**Dawson,** Lieut. W.	1 Aug. 16			Deolali.
14 Feb. 18	**Wheway,** Lieut. T.	14 Feb. 18			Deolali

ROYAL ARMY MEDICAL CORPS—*contd.*

Retired Officers of the Royal Army Medical Corps temporarily re-employed.

First Commn.	Names.	Date of present rank.	Date of compln. Indian tour.	Date of present appointment.	Remarks.
			Date of Retirement.	Date of re-employment.	
4 Feb. 82	**Barratt,** Col. H. J. (207) . .	2 Oct. 11	26 Dec. 17	27 Dec. 17	A. D. M. S., Meerut Dn.
4 Aug. 83	**Tate,** Col. A. E., C.M.G. (207)	31 Dec. 12	26 Dec. 17	27 Dec. 17	A. D. M. S., 4th (Quetta) Dn.
30 May 85	**Molesworth,** Lt.-Col. R. E. (71, 206).	. . . . .	. . . . .	24 Nov. 14	Comdg. Stn. Hospl., Ahmednagar.
	Rennie, Lt.-Col. S. J. (D. I.) (71, 207).	. . . . .	. . . . .	6 Apl. 15	Comdg. Stn. Hospl., Delhi.
3 Feb. 83	**Stuart,** Lt.-Col. J. R. . . .	. . . . .	. . . . .	27 Nov. 15	Comdg. Stn. Hospl., [Pachmari.
31 Jan. 85	**Lilly,** Lt.-Col. A. T. I. . .	. . . . .	. . . . .	26 Dec. 17	Comdg. Stn. Hospl., Ranikhet.
30 May 85	**Blackwell,** Lt.-Col. C. T., M.D.	. . . . .	. . . . .	26 Dec. 17	Comdg. Stn. Hospl., Ferozepore.
1 Feb. 90	**Winter,** Lt.-Col. H. E. (207) .	. . . . .	. . . . .	26 Dec. 17	Comdg. Deccan War Hospl., Poona.
27 July 92	**Austin,** Major R. F. E. . .	. . . . .	. . . . .	28 June 15	Calcutta, Comdg. Stn. Hospl.

UNATTACHED LIST.

Local, Temporary and Honorary Rank and Temporary Commissions— Dental Surgeons.

First Commn.	Name.	Date of present rank.	Date of present appointment.	Remarks.
26 Feb. 15	Captain T. C. Cregan . . .	26 Feb. 16	. . . . .	......
27 Jan. 16	Captain F. E. Welton . . .	27 Jan. 17	. . . . .	Ferozepore. On one month's leave to England.
13 Mar. 16	Captain W. C. Holburn . . .	13 Mar. 17	. . . . .	Bangalore.
3 Apl. 16	Captain W. Y. Kingston . .	3 Apl. 17	. . . . .	Bombay.
4 Apl. 16	Captain E. E. Cunnah . . .	4 Apl. 17	. . . . .	Meerut.
10 Apl. 16	Captain D. G. Fisher . . .	10 Apl. 17	. . . . .	Delhi.
1 May 16	Captain D. Edwards . . .	1 May 17	. . . . .	Belgaum.
10 May 16	Captain A. Coupe	10 May 17	. . . . .	Karachi.
19 May 16	Captain F. E. Smith . . .	19 May 17	. . . . .	Lucknow.
24 May 16	Captain W. M. Ferguson . .	24 May 17	. . . . .	Maymyo.
12 June 16	Captain B. C. Imrie. . . .	12 June 17	. . . . .	Lahore.
12 June 16	Captain R. J. Gardner. . . .	12 June 17	. . . . .	Kirkee.
15 June 16	Captain J. Fraser	15 June 17	. . . . .	Jubbulpore.
12 July 16	Captain W. J. Marmion . .	12 July 17	. . . . .	......
11 Oct. 16	Captain J. H. Markham . .	11 Oct. 17	. . . . .	Bombay.
1 July 16	Captain C. Fox Pit	1 July 17	. . . . .	Bangalore.
10 July 16	Captain L. W. Barrett . . .	10 July 17	. . . . .	Allahabad.
28 July 16	Captain S. V. Shrimpton . .	28 July 17	. . . . .	Quetta.
1 Aug. 16	Captain L. R. Haydon . .	1 Aug 17	. . . . .	Peshawar.
21 Aug. 16	Captain H. D. Stephens . .	21 Aug. 17	. . . . .	Sialkot.
24 Aug. 16	Captain G. L. Elmitt . .	24 Aug. 17	. . . . .	Secunderabad.
1 Sep. 16	Captain H. S. Swan	1 Sep. 17	. . . . .	Bangalore.
4 Sep. 16	Captain J. H. Beckett . . .	4 Sep. 17	. . . . .	Peshawar.
14 Sep. 16	Captain A. F. Milne. . . .	14 Sep. 17	. . . . .	
17 Sep. 16	Captain E. R. Saul	17 Sep. 17	. . . . .	Ambala.
14 Sep. 16	Captain S. I. Davidson . .	14 Sep. 18	. . . . .	Quetta.
11 Dec. 16	Captain H. R. Humby . . .	11 Dec. 17	. . . . .	Lucknow.
10 Feb. 17	Captain R. Y. Richmond . .	10 Feb. 18	. . . . .	Poona.
12 Feb. 17	Captain P. K. Jackson . .	12 Feb. 18	. . . . .	Jubbulpore.
1 June 17	Captain A. B. Cross . . .	1 June 18	. . . . .	Secunderabad.
1 Aug. 17	Captain A. V. Milnes . . .	1 Aug. 18	. . . . .	Calcutta.
17 Sept. 17	Captain R. W. Markham . .	17 Sep. 18	. . . . .	Nowshera.
1 Nov. 17	Capt. P. S. Humm	1 Nov. 18	. . . . .	Ambala Bde.
17 Nov. 17	Capt. H. Thomson	17 Nov. 18	. . . . .	Poona.
	Lieut. A. B. H. Cole . . .	. . . . .	. . . . .	1st (Peshawar) Divn.

Radiographers.

First Commn.	Name.	Date of present rank.	Date of present appointment.	Remarks.
13 July 16	Capt. T. W. Barnard . .	13 Jan. 18	. . . . .	Colaba.
28 July 16	Capt. E. G. Hill	18 Jan. 18	. . . . .	Bombay.
20 July 16	Capt. L. F. Bates	20 Jan. 18	. . . . .	Secunderabad.
20 July 16	Capt. W. J. Dundas . . .	20 Jan. 18	. . . . .	Karachi.
22 July 16	Capt. W. Cameron	22 Jan. 18	. . . . .	Quetta.
9 Aug. 16	Capt. T. B. Wolverson . .	9 Feb. 18	. . . . .	Rawalpindi.
5 Oct. 16	Capt. K. G. F. Collender . .	5 Apl. 18	. . . . .	Quetta.
15 Nov. 16	Capt. E. T. Worswick .	15 May 18	. . . . .	Wellington.
24 Jan. 18	Lieut. A. W. Jolly . . .	24 Jan. 18	. . . . .	Allahabad.

QUEEN ALEXANDRA'S MILITARY NURSING SERVICE FOR INDIA.

Chief Lady Superintendents.

Names.	Date of commencement of 1st period of service.	Date of present grade.	Date of completion of present period of service.	If 1st, 2nd, 3rd or 4th period of service.	Remarks.
Waterhouse Agnes May, R. R. C.	*24 Dec. 91*	*19 Mar. 14*	...	...	Simla.
Rait, Helen Anna Macdonald, R. R. C.	• Dec. 93	18 Sept. 17	...	...	Simla.

Lady Superintendents—4.

Names.	Date of commencement of 1st period of service.	Date of present grade.	Date of completion of present period of service.	If 1st, 2nd, 3rd or 4th period of service.	Remarks.
Lloyd, Isabel Mary Annie	*25 Nov. 93*	*13 Dec. 14*	...	...	*Rawal Pindi.*
Hill, Christian Frances	18 Dec. 95	12 Dec. 15	10 Jan. 19	Fourth	Lucknow. *Lv., m. c., ex Ind.*
Dunwoodie, Lallah Bessie	25 Nov. 96	20 Mar. 16	29 Mar. 19	"	Bombay.
Truman, Mary Rose	18 Dec. 98	12 Dec. 16	22 Dec. 20	"	Bangalore.
Reynolds, Hester Jane	18 Apl. 99	18 Sep. 17	13 Feb. 19	"	Lucknow.

Senior Nursing Sisters—16.

Names.	Date of commencement of 1st period of service.	Date of present grade.	Date of completion of present period of service.	If 1st, 2nd, 3rd or 4th period of service.	Remarks.
Sykes, Ethel	20 Mar. 96	5 Oct. 07	17 Feb. 19	Fourth	Darjeeling, Calcutta.
Harris, Ada Marion, *R. R. C.*	26 Aug. 99	1 Jan. 09	17 Nov. 21	"	Poona.
Harvey, Minnie Elizabeth	11 Oct. 99	7 Nov. 09	5 Dec. 21	"	Bombay.
Gilmore, Alice Maud, R. R. C.	15 Mar. 00	14 Dec. 12	24 Apl. 22	"	Peshawar.
Crane, Blanche	17 Aug. 00	4 Feb. 13	23 Aug. 22	"	Multan. *Leave, m.c.*
Gray, Mary Emily, *R. R. C.*	26 Dec. 00	24 Nov. 13	16 Feb. 23	"	Poona, Mhow.
Warrack, Catherine Lumsden, R. R. C.	11 Dec. 01	11 Dec. 13	12 Dec. 18	Third	Mhow. *Leave, m.c., ex Ind.; 6 mths.*
Reid, Emily Josephine	3 Oct. 02	7 Nov. 13	16 July 19	"	Jhansi.
Stroughill, Mabel Anna, *R.R.C.*	3 Oct. 02	7 Mar. 14	22 Sep. 19	"	Aden.
Gilmore, Mary Gertrude *R.R.C.*	14 Oct. 02	19 Mar. 14	17 July 19	"	
McGowan, Chloe Stanley	28 Oct. 02	13 Dec. 14	24 Jan. 19	"	Ambala.
Lamb, Violet Isabel	5 Jan. 03	25 Mar. 15	12 July 19	"	Quetta.
Cunningham, Ethel Mary	2 Oct. 03	12 Dec. 15	25 Oct. 19	"	Secunderabad. *Lv., ex Ind., one year.*
Wilson, Jeanie Stewart Ramsey, *R. R. C.*	2 Oct. 03	20 Mar. 16	5 Sep. 19	"	
Knapp, Marion Domville, R.R.C.	7 Oct. 03	12 Dec. 16	17 Oct. 19	"	Secunderabad.
Goodwyn, Alice	13 Oct. 03	18 Sep. 17	15 Oct. 19	"	Jullundur.

Nursing Sisters—50.

Names.	Date of commencement of 1st period of service.	Date of present grade.	Date of completion of present period of service.	If 1st, 2nd, 3rd or 4th period of service.	Remarks.
Allman, Lyllian	2 Oct. 03	2 Oct. 03	14 Oct. 19	"	Abbottabad
Mills, Sybil Geraldine	13 Oct. 03	13 Oct. 03	20 Oct. 19	"	Jubbulpore.
White, Lucy Angela A.R.R.C.	27 Nov. 03	27 Nov. 03	19 Nov. 19	"	Colaba.
Aldridge, Winifred Mary	27 Nov. 03	27 Nov. 03	26 Jan. 20	"	Meerut.
Stonehouse, Edith Jane	28 Jan. 04	28 Jan. 04	11 Feb. 20	"	Rangoon.

Names.	Date of commencement of 1st period of service.	Date of present grade.	Date of completion of present period of service.	If 1st, 2nd, 3rd or 4th period of service.	Remarks.
Nursing Sisters—*concld.*					
Hornsby, Mary Elizabeth Tyrell A.R.R.C.	10 Feb. 04	10 Feb. 04	21 Feb. 20	Third	
Kelso, Ethel	8 Apl. 04	8 Apl. 04	12 Oct. 20	,,	*Leave ex Ind., one year.*
Rooke, Hilda	1 Oct. 04	1 Oct. 04	8 Oct. 20	,,	*Lv., ex Ind., one year, from 24 Apl. 18.*
French, Agnes Gordon .	1 Oct. 04	1 Oct. 04	29 Sep. 20	,,	
Walker, Wilhelmine . .	28 Feb. 06	28 Feb. 06	24 Apl. 22	,,	Quetta. (Cadet College Hospt.)
Tippetts, Melanie Elaine .	9 Jan. 07	9 Jan. 07	15 Jan. 23	,,	Allahabad.
Cawley, Kate	5 Feb. 07	5 Feb. 07	2 Mar. 23	,,	Kamptee.
O'Brien, Catherine . . .	16 Feb. 07	16 Feb. 07	22 Mar. 23	,,	Calcutta.
Harrison, Isabella Louisa .	20 Sep. 07	20 Sep. 07	22 Sep. 18	Second	*Leave ex Ind., one year.*
Haines, Gertrude R.R.C. .	8 Jan. 08	8 Jan. 08	7 Jan. 19	,,	
Marshall, Edith Osborne, A.R.R.C.	22 Jan. 08	22 Jan. 08	16 Jan. 19	,,	
Anderson, Clara Hawley .	16 May 08	16 May 08	7 Apl. 19	,,	Lahore.
Exshaw, Phœbe R.R.C. .	30 Sep. 08	30 Sep. 08	27 July 19	,,	
McNally, Irene Annie . .	6 Jan. 09	6 Jan. 09	5 Jan. 19	,,	*Leave m. c., ex India.*
Vernon, Olive	5 Mar. 09	5 Mar. 09	2 Aug. 19	,,	Bareilly.
Smith, Frances Amy Lawrence.	11 Mar. 10	11 Mar. 10	10 Mar. 20	,,	*Lv., m. c., ex Ind., from 20 May 18.*
Wardell, Mary . . .	7 Dec. 10	7 Dec. 10	6 Dec. 20	,,	Nowshera.
Maclean, Minnie Eda, A. R. R. C.	7 Dec. 10	7 Dec. 10	13 Dec. 20	,,	
Johnson, Constance Grace	16 Mar. 11	16 Mar. 11	14 Mar. 21	,,	Karachi.
Nevile, Rosamond Lucy .	29 Nov. 11	29 Nov. 11	28 Nov. 21	,,	Sialkot.
Veech, Annie Gertrude . .	29 Nov. 11	29 Nov. 11	28 Nov. 21	,,	*Leave ex. Ind.,* 1 year.
Wilson-Green, Margaret Alice.	18 Oct. 12	18 Oct. 12	17 Oct. 17	,,	Lucknow.
Harley, Isabel	15 Feb. 13	15 Feb. 13	14 Feb. 18	First	*Lv., m. c., ex Ind., from 17 Sep. 16.*
Green, Ethel A.R.R.C. .	15 Feb. 13	15 Feb. 13	14 Feb. 23	Second	
Hart, Alice Maud . . .	15 Feb. 13	15 Feb. 13	14 Feb. 23	,,	Quetta.
Beresford, Nora . . .	24 Sep. 13	24 Sep. 13	23 Sep. 18	First	Quetta.
Kellett, Millicent Louisa .	10 Jan. 14	10 Jan. 14	9 Jan. 19	,,	Aden.
Porter, Dorothea Annie .	6 Mar. 14	6 Mar. 14	5 Mar. 19	,,	Peshawar.
Scott, Marion Louisa . .	11 Aug. 14	11 Aug. 14	10 Aug. 19	,,	Lucknow.
MacLean, Williamina .	14 Nov. 14	14 Nov. 14	13 Nov. 19	,,	Allahabad.
Dudgeon, Cecilia Winifred	27 Feb. 15	27 Feb. 15	26 Feb. 20	,,	*Lv., m. c., ex Ind.*
West, Dorothea A.R.R.C. .	27 Mar. 15	27 Mar. 15	26 Mar. 20	,,	
Rabbidge, Mary Dorothy .	13 Nov. 15	13 Nov. 15	12 Nov. 20	,,	Multan.
Shaw, C. H.	18 Apl. 16	21 Apl. 16	17 Apl. 21	,,	Delhi.
Talbot, E. F.	4 May 16	29 Apl. 16	3 May 21	,,	Aden.
Moriarty, E.	25 Apl. 16	29 Apl. 16	24 Apl. 21	,,	Bombay.
Looke, Lilian	6 May 16	6 May 16	5 May 21	,,	Nowshera.
Bradley, K.	29 Sep. 16	29 Sep. 16	28 Sep. 21	,,	Kamptee.
Anderson, E. J. M. . .	7 Oct. 16	7 Oct. 16	6 Oct. 21	,,	Poona.
Keeble, Ida.	27 Oct. 16	27 Oct. 16	26 Oct. 21	,,	Nowshera.
Smith, Winifred May . .	12 Aug. 17	12 Aug. 17	11 Aug. 22	,,	Multan.
Homan, Eileen Maud . .	11 Aug. 17	11 Aug. 17	10 Aug. 22	,,	Aden.
Bazalgette, Mrs. Helen .	24 Sep. 17	24 Sep. 17	23 Sep. 22	,,	Quetta.
Montgomerie, Noel Rose	24 Sep. 17	24 Sep. 17	23 Sep. 22	,,	*Lv., m. c., ex Ind.*
Graham, Kathleen . . .	11 Nov. 17	11 Nov. 17	10 Nov. 22	,,	Bombay.
Clarke, C. V.	11 Jan. 18	11 Jan. 18	10 Jan. 23	,,	Meerut.
Kennedy, M. G. . . .	11 Jan. 18	11 Jan. 18	10 Jan. 23	,,	Lucknow.
Bosanquet, Pauline, Catherine, Mary.	12 Feb. 18	12 Feb. 18	11 Feb. 23	,,	Lucknow.
Webb, Constance, Isabel .	12 Feb. 18	12 Feb. 18	11 Feb. 23	,,	Lucknow.
Comeron, Elizabeth Margaret.	12 Feb. 18	12 Feb. 18	11 Feb. 23	,,	Meerut.
O'Connor, Norah Frances Mary.	23 Feb. 18	23 Feb. 18	22 Feb. 23	,,	Meerut.
Campbell, Helen, Dorothea	2 Sep. 18	2 Sep. 18	1 Sep. 23	,,	Meerut.
Scott, Margaret Dears . .	2 Sep. 18	2 Sep. 18	1 Sep. 23	,,	Meerut.
Irvin, Agnes Mary . . .	2 Sep. 18	2 Sep. 18	1 Sep. 23	,,	Meerut.
Anderson, Beatrice. . .	2 Sep. 18	2 Sep. 18	1 Sep. 23	,,	Lucknow.
Carmichael, Kathleen May Roxburgh.	2 Sep. 18	2 Sep. 18	1 Sep. 23	,,	Rawalpindi.
Kennedy, Violet . . .	2 Sep. 18	2 Sep. 18	1 Sep. 23	,,	Bombay.
Low, May Mildred Edith .	3 Sep. 18	3 Sep. 18	2 Sep. 23	,,	Lucknow.
McPherson, Emily May .	15 Nov. 18	15 Nov. 18	14 Nov. 23	,,	Quetta.

INDIAN MEDICAL SERVICE.

(Vide *Royal Warrant of 28th May 1913*.)

Note.—As all officers of the I. M. S. must qualify in Hindustani by the Lower Standard, the reference 207 denoting that qualification is omitted opposite the name of all officers above the rank of Lieutenant.

Uniform—Blue. Facings—Black Velvet.

Estabt.	Names.	Date of birth.	First Commn.	Date of rank.	Will complete tour of service.	Date of present appointment.	F. rule.	Remarks.
	Director-General. (1)							
	Major-General.							
B.	**Edwards,** *The Hon'ble* William Rice C.B., C.M.G., M.D., K.H.P. [R] (170).	17 May 62	1 Apl. 86	1 Apl. 15	7 Jan. 23	8 Jan. 18	IA	Director-General, Indian Medical Service.
	Major-Generals—(6) Civil (3) Mily (3). { Bengal 4 Madras 1 Bombay 1 }							
Bo.	**Lyons,** *The Hon'ble* Robert William Steele, M.D., K.H.P. [R].	4 Sep. 59	1 Apl. 82	11 Jan. 14	10 Jan. 19	11 Jan. 14	75	Surgn.-Genl. with the Govt. of Bombay.
B.	**Manifold,** Courtenay Clarke, C.B., C.M.G., M.B.	3 Apl. 64	31 Mar. 87	29 Dec. 17	23 June 23	24 June 18	IA	D. D. M. S., 9th (Secunderabad) Division.
B.	**Robinson,** *The Hon'ble* William Henry Banner, C.B.	7 Dec. 63	1 Apl. 86	8 Jan. 18	11 Mar. 23	12 Mar. 18	...	Surgeon-Genl. with the Govt. of Bengal.
B.	**Hehir,** Patrick, C.B., C.M.G., C.I.E., M.D. (157)	27 May 59	1 Apl. 86	13 Mar. 18	26 May 19	14 June 17	IA	D. D. M. S., Northern Command.
B.	**Hendley,** Harold, M.D., K.H.S. (170)	16 Apl. 61	31 Mar. 84	21 June 18	15 Apl. 21	6 Sep. 18	IA	D. D. M. S., 4th (Quetta) Division.
M.	**Giffard,** *The Hon'ble* Gerald Godfray, C.S.I.	19 Jan. 67	31 Mar. 90	6 July 18	5 July 23	6 July 18	...	Surgn.-Genl. with the Govt. of Madras.

Colonels (16) { Civil (6) / Mily. (10) } { Bengal 9 / Madras 4 / Bombay 3 }

Bo.	V.C., *Crimmin, John*, C.B., C.I.E., V.D., K.H.P. (Supt.) (*a*)	*19 Mar. 59*	*30 Sep. 82*	*1 Oct. 13*	*26 Oct. 18*	*22 Oct. 14*	75	*A.D.M.S., Kohat Brigade.*
Bo.	**Dyson**, Thomas Edward, M.B. (161, 170, 206)	25 Nov. 60	30 Sep. 86	10 Jan. 14	9 Jan. 19	1 July 16	IA	A. D. M. S., 16th Indian Division. *Lv.*, *m. c.*, 4 *mos.*, *29 Oct. 1918.*
B.	**Banatvala**, *The Hon'ble* Hormasjee Eduljee, C.S.I. K. H. S. (106).	20 Oct. 59	31 Mar. 84	2 Apl. 14	1 Apl. 19	2 Apl. 14	75	Inspector-General, Civil Hospitals and Prisons, Assam.
M.	**Strickland**, Percy Charles Hutchison	14 Aug. 63	1 Oct. 87	15 May 14	14 May 19	17 July 16	IA	I. G. C. H., Burma.
B.	**Mactaggart**, *The Hon'ble* Charles, C.I.E., M.B.	27 Mar. 61	1 Apl. 86	29 Mar. 15	28 Mar. 20	29 Mar. 15		Insptr.-Genl. of Civil Hospitals, & Offg. Sany. Commr., U. Provinces.
B.	**Daly**, Joseph Thomas, M.B.	3 Jan. 62	1 Apl. 86	1 Apl. 15	31 Mar. 20	1 April 15		A. D. M. S., Presidency Brigade.
B.	**Bell**, George James Hamilton, C.I.E., M.B. (105.)	28 Feb. 61	1 Apl. 86	21 June 15	20 June 20	16 Jan. 16		Inspector-General, Civil Hospitals, Bihar and Orissa. Comb. leave 6 mos., 28-6-18.
B.	**Fooks**, Henry	26 Jan. 63	1 Apl. 86	12 July 15	11 July 20	12 July 15		A. D. M. S., Burma Division.
M.	**Carr-White**, Percy, M.B. (206)	20 June 65	31 Mar. 90	7 June 16	6 June 21	29 June 16		A. D. M. S., Ambala Brigade.
Bo.	**Jennings**, William Ernest, M.D.	1 Jan. 65	1 Oct. 87	1 Oct. 16	25 Oct. 21	26 Oct. 16		A.D.M.S., Poona Division.
B.	**Green**, *The Hon'ble* Charles Robert Mortimer, M.D., F.R.C.S.	21 May 63	30 Sep. 86	11 Nov. 16	12 Apl. 23	13 Apl. 18		I. G., C. H., C. P.
M.	**Molesworth**, William, C.I.E., M.B., V. H. S.	30 Dec. 65	31 Mar. 90	22 Jan. 17	21 Jan. 22	30 Oct. 16		Dy. D. M. S. in India.
B.	**Garvie**, John, M.B. (170)	3 May 63	31 Mar. 87	25 Mar. 17	24 Mar. 22	31 Jan. 18		A. D. M. S., Allahabad & Fyzabad Brigades.
B.	**Mac Watt**, The Hon'ble Robert Charles, C.I.E., M.B., F.R.C.S., (161, 169, 206)	22 Jan. 65	1 Oct. 87	8 Jan. 18	25 Aug. 23	26 Aug. 18		Insptr.-Genl. of Civil Hospls., and Chief Plague and Malaria Medical Officer, Punjab.
Bo.	**Smith**, John Blackburn, C.B., M.B., V.H.S.	21 Jan. 65	30 Sep. 89	3 Sep. 18	19 Oct. 23	20 Oct. 18		A. D. M. S., Sialkot and Jhelum Bdes.
Bo.	**Cleveland**, Henry Francis, C.I.E. [L] (88, 169, 204)	30 Nov. 63	30 Sep. 89	1 Oct. 18	24 Oct. 23	25 Oct. 18		Offg. A. D. M. S., 16th (Indian) Division.
M	**Stodart**, Thomas, M.B. (105, 117)	13 Mar. 68	29 July 93	12 Oct. 18	11 Oct. 23	23 Feb. 18		A. D. M. S., Bannu Bde.
M	**Pridmore**, Walter George, C.M.G., M.B.	23 Sep. 64	28 July 91	4 July 18	31 Oct. 23	1 Nov. 18		A. D. M. S., Secunderabad Bde.
B	**Close**, Joseph Kinnear, M.D.	22 Dec. 64	1 Oct. 87	15 Aug. 18	7 Nov. 23	8 Nov. 18		A. D. M. S., Derajat Bde.

(*a*) Retained in service from 27 Oct. 18.

INDIAN MEDICAL SERVICE—*contd.*

Lieutenant-Colonels—(41) {Bengal 21, Madras 11, Bombay 9}

Specially selected for increased pay under Article 8 of the Royal Warrant of 28th May 1913.

[*The first date in the fifth column is the date of an officer's advancement.*]

Estabt.	Names.	Date of birth.	First Commn.	Date of rank.	Completes 30 years' pension service.	Date of present appointment.	Remarks.
	Lieutenant-Colonels. (Selected)						
M.	*Dewes, Frederick Joseph* (105) (*Supy.*) (*b*)	*3 Jan. 61*	*1 Oct 87*	*6 Dec. 10* *1 Oct. 07*	*17 Dec. 17*	. . . .	*S. M. O., I. M. S., Lahore Cant.*
Bo.	*Hojel, James Graham,* C.I.E., M.B. (*Supy.*)	*6 Nov. 63*	*31 Mar. 88*	*15 Apl. 11* *31 Mar. 08*	*23 June 18*	. . . .	*Officer Comdg. Lady Hardinge War Hospl., Bombay.*
M.	**Sutherland,** William Dunbar, M.D.	13 Sep. 66	31 Mar. 90	16 Apl. 11 31 Mar. 10	25 June 20	. . . .	Imperial Serologist, M. C., Calcutta, & Chemical Examiner to Govt.
M.	*Wright, Edmund Hasell* (*Supy.*) (*b*)	*14 Sep. 63*	*31 Mar. 90*	*7 June 11* *31 Mar. 10*	*26 June 20*	. . . .	*Civil, Coorg.*
M.	**Johnston,** Charles Arthur, C.B., D.S.O., M.B.	28 Feb. 67	31 Mar. 90	26 July 11 31 Mar. 10	26 June 20	. . . .	O. C. Indn. Stn. Hospl., Quetta.
M.	**Entrican,** James, M.D. (64, 105, 206)	2 Sep. 64	28 July 91	1 Feb. 12 28 July 11	27 July 21	. . . .	Civil Surgeon & Director, Vaccine Depot, Meiktila, Burma.
B.	*Henderson, Selby Herriot, M. B.* (*Supy.*) (*b*)	*28 Mar. 61*	*30 Sep. 86*	*25 Mar. 12* *30 Sep. 06*	*10 Nov. 16*	. . . .	*Inspector-General of Prisons, U. Provinces.*
B.	**Hall,** Edmund Alexander William, M.B. (169)	29 Nov. 64	30 Sep. 86	18 July 12 30 Sep. 06	10 Nov. 16	. . . .	Civil 24-Parganas, and Medl. Insptr. of Emigrant (Colonial Emigration), Bengal. Offg. Protector of Emigrants and Supdt. of Emigrants, Calcutta in addition.
B.	*Sunder, Charles Edward,* M.D. (*Supy.*) (*b*)	*22 Nov. 59*	*31 Mar. 87*	*5 Mar. 13* *31 Mar. 07*	*21 Sep. 17*	. . . .	*Civil, Patna.*
Bo.	*Prall, Samuel Esmond,* M.B. (170) (*Supy.*) (*b*)	*19 Apl. 62*	*29 Sep. 88*	*15 Mar. 13* *29 Sep. 08*	*19 Dec. 18*	. . . .	*O. C. Indn. Stn. Hospl., Karachi.*
M.	*Donovan, Charles,* M.D. (*Supy.*) (*b*)	*19 Sep. 63*	*28 July 91*	*30 June 13* *28 July 11*	*27 July 21*	. . . .	*Profr. of Physiology, M. C., & Surgn., 14th Dist., Madras.*
130	*Hudson, Charles Tilson,* C.M.G. (94, 206) (*Supy.*)	*27 Nov. 66*	*30 Mar. 89*	*1 Oct. 13* *30 Mar. 09*	*24 June 19*	*31 Mar. 16*	
B.	*Vost, William,* M. B. (*Supy.*) (*b*)	*23 Nov. 61*	*31 Mar. 87*	*24 Dec. 13* *31 Mar. 07*	*4 July 17*	. . . .	*P. of W. Hospital, Bellary.*

B	**Davidson**, David MacDonald	25 Mar. 65	1 Oct. 87	11 May 14 1 Oct. 07	17 Dec. 17	. . . .	Civil, Lahore, & Profr. of Forensic Medicine, Toxicology & Midwifery, K. E. Medical College, Lahore.
	Maynard, Frederic Pinsent, M.B., F.R.C.S. (*Supy.*)	*10 Mar. 64*	*1 Oct. 87*	*25 May 14* *1 Oct. 07*	*18 Mar. 18*	21 *Nov. 05*	*Profr. of Ophthalmic Surgery, Medl. Coll., Calcutta.*
H.	*Nott, Arthur Holbrook*, M.D. (*Supy.*) (*b*)	*28 Nov. 61*	*1 Oct. 87*	*1 Jan. 15* *1 Oct. 07*	*17 Dec. 17*	. . . .	*O. C. Indn. Stn. Hospl., Meerut.*
B.	*Coleman, Albert* (*Supy.*) (*b*)	*20 Dec. 62*	*1 Oct. 87*	*25 Feb. 15* *1 Oct. 07*	*4 Dec. 17*	. . . .	*Civil, Ambala.*
B.	*White, William Westropp*, C.B., M.D. (*Bt.-Col.*) *Tempy. Col.* (*Supy.*) (*b*)	*10 June 62*	*1 Oct. 87*	*29 Mar. 15* *1 Oct. 07* *3 June 15* (*a*)	*17 Dec. 17*	. . . .	——.
Bo.	**Jackson**, James, C.I.E., M.B.	23 Sep. 66	31 Mar. 90	1 Apl. 15 31 Mar. 10	25 June 20	. . .	Inspr.-Genl. of Prisons, Bombay.
B.	**Woodwright**, William Henry Edward, F.R.C.S.I. .	10 July 65	1 Oct. 87	21 June 15 1 Oct. 07	17 Dec. 17	. . . .	Civil and Lunatic Asylum, Bareilly.
B.	*Buchanan, Sir Walter James*, K.C.I.E., M.D. (*Supy.*) (*b*)	*12 Nov. 61*	*1 Oct. 87*	*12 July 15* *1 Oct. 07*	*4 Mar. 18*	. . .	*I. G. of Prisons, Bengal.*
Bo.	**Street**, Ashton, M.B., F.R.C.S. (169)	10 June 64	31 Mar. 90	25 Dec. 15 31 Mar. 10	25 June 20	. . .	Senior Surgeon and S. M. O., J. J. Hospl., and Principal and Profr. of Surgery, G. M. C., Bombay.
B.	*Roberts, Sir James Reid, Kt.*, C.I.E., M.B., F.R.C.S. . (*Supy.*) (*b*)	*24 Jan. 61*	*31 Mar. 88*	*31 Dec. 15* *31 Mar. 08*	*23 June 18*	. . . .	*Civil Surgeon, Simla* (*West*).
B.	*Gee, Frederick William*, C.I.E., M.B. (*169*) (*Supy.*) (*b*)	*11 Mar. 63*	*31 Mar. 88*	*24 Jan. 16* *31 Mar. 08*	*23 June 18*	. . . .	——.
B.	**Drake-Brockman**, Herbert Edward, F.R.C.S.E. .	11 Feb. 65	29 Sep. 88	14 Mar. 16 29 Sep. 08	16 Jan. 19	. . . .	O. C. Indn. Stn. Hospl., Peshawar.
B.	**Lane**, William Byan, C.I.E. (170)	8 Feb. 66	29 Sep. 88	1 Apl. 16 29 Sep. 08	16 Nov. 18	. . . .	——.
B.	**Lumsden**, Philip James, M.B.	2 Feb. 64	29 Sep. 88	1 Apl. 16 29 Sep. 08	19 Dec. 18	. . . .	Offg. A. D. M. S. Ahmednagar Bde.
Bo.	**Irvine**, Thomas Walter, M.B. (169)	21 Oct. 65	28 July 91	1 Apl. 16 28 July 11	27 July 21	. . .	R. S. and C. M. O. N.-W. F. P.
M	**Mitter**, Robert King, M.B. [L] (206)	23 Aug. 64	30 Jan. 92	7 June 16 30 Jan. 12	29 Jan. 22	10 May 18	O. C. I. T. War Hospital, Khandwa.
E.	**Ozzard**, Fairlie Russell (*l*) (170)	22 July 65	30 Mar. 89	23 June 16 30 Mar 09	24 June 19	. . . .	——
B.	**Calvert**, John Telfer, M.B.	4 Feb. 64	30 Mar. 89	21 Sep. 16 30 Mar. 09	1 July 19	1 Apl. 12	Principal & Profr. of Medicine, Medl. Coll., Calcutta.
Bo.	**Swinton**, Francis Edward	9 June 66	27 July 92	1 Oct. 16 27 July 12	26 July 22	. . . .	M. S. K., Madras.
B.	**Jennings**, Edgar	3 Aug. 64	30 Mar. 89	11 Nov. 16 1 Jan. 16 (*a*) 30 Mar. 09	24 June 19	. . . .	——
M.	**Bhola**, Nauth, C.I.E. (L) (170, 203)	23 Jan. 66	29 July 93	22 Jan. 13 30 Jan 17	29 July 23	. . . .	——.
B.	*Melville, Henry Bruce*, M.B. (*Supy.*) (*b*)	*4 Sep. 63*	*30 Sep. 89*	*25 Mar. 17* *30 Sep. 09*	*21 Dec. 19*	. . . .	*Residency Surgeon, Kashmir.*

(*a*) Date of promotion to Brevet-Colonel.
(*b*) Retained in the service after attaining the age of 55 years.

INDIAN MEDICAL SERVICE—*contd.*

Estabt.	Names.	Date of birth.	First Commn.	Date of rank.	Completes 30 years' pension service.	Date of present appointment.	Remarks.
	Lieutenant-Colonels—(Selected)—*contd.*						
Bo.	**Burnett,** Sidney Harvey, M.B.	31 May 68	27 July 92	1 Apl. 17 27 July 12	26 July 22	. . . .	Presidency Surgeon, 1st Dist. with attached duties and A. G. Surgeon, G. I. Hospital, Bombay.
B.	*Vaughan, Joseph Charles Stoelke,* M.B. (*b*)	*29 June 62*	*30 Sep. 89*	*1 Apl. 17* *30 Sep. 09*	*28 Nov. 19*	. . . .	*Offg. I. G. C. H., Bihar and Orissa.*
B.	**Macnab,** Allan James, C.B., F.R.C.S. (156, 169) (*Bt.-Col.*)	17 Aug. 64	31 Mar. 90	9 June 17 31 Mar. 16 14 Jan. 16 (*a*)	25 Aug. 20	. . . .	—
B.	**Wimberley,** Charles Neil Campbell, C.M.G., M.B. (64, 66*a*) (*tempy. Colonel*).	23 Sep. 67	31 Mar. 90	22 Sep 17 31 Mar 10 18 July 16 (*c*)	26 June 20	. . .	—
B.	**Crawford,** James Muir, O.B.E., M.B.	5 Aug. 66	31 Jan. 91	30 Sep. 17 31 Jan. 11	30 Jan. 21	. .	Civil and Lunatic Asylum, Benares.
B.	*Singh, Bawa Jiwar,* C.I.E. (*105, 156, 161, 169*) (*b*)	*6 May 63*	*31 Jan. 91*	*18 Oct. 17* *31 Jan. 11*	*30 Jan. 21*	. . . .	*I. G. of Prisons, Bihar and Orissa.*
M.	**Foulkes,** Thomas Howard, F.R.C.S. (195)	2 Apl. 70	29 July 93	18 Dec. 17 30 Jan. 13	29 July 23	. . . .	A. D. M. S., Aden.
B.	*James, Charles Henry,* C.I.E., F.R.C.S. (*b*)	*4 Aug. 63*	*31 Jan. 91*	*19 Dec. 17* *31 Jan. 11*	*30 Jan. 21*	. . .	*Civil, Punjab. Chief Medical Officer, & Civil Surgeon, Delhi.*
B.	**O'Kinealy,** Frederick	19 Aug. 66	31 Jan. 91	8 Jan. 18 31 Jan. 11	30 Jan. 21	. .	Surgn. Supdt., Presy. Genl. Hospl., Calcutta.
B.	**Buist,** Arthur William Treminhere, M.B.	7 Aug. 66	31 Jan. 91	23 Feb. 18 31 Jan. 11	30 Jan. 21	. . . .	Rawalpindi.
B.	**Deare,** Benjamin Hobbs	29 Apl. 67	28 July 91	5 Mar. 18 28 July 11	27 July 21	7 Apl. 12	Profr. of Clinical Medicine and M. Medica, Medl. Coll. and 2nd Physician, College Hospital, Calcutta.
Bo.	**Jackson,** Thomas, M.B.	18 Jan. 65	27 July 92	1 Apl. 18 27 July 12	26 July 21	. . . .	Surgn. Supdt., St. George's Hospl., Bombay.
B.	**Smith,** Sidney Browning, C.M.G. (170)	12 Nov. 66	28 July 91	24 June 18 28 July 11	27 July 21	. . . .	Offg. A. D. M. S., Karachi Bde.
M.	**Berry,** Alfred Eugene	21 June 69	29 Jan. 94	4 July 18 29 July 13	28 Jan. 24	. . . .	—
M.	**Pearce,** Charles Ross, M.B.	28 June 68	28 July 94	6 July 18 29 Jan. 14	27 July 24	. . . .	Director, Pasteur Institute and Bacteriologist to the Govt. of Burma and in additional charge of duties of Pathologist, Genl. Hospital, Rangoon, Burma.
Bo.	**Bennet,** Hugh, M.B., F.R.C.S.E. (170)	23 Sep. 65	29 Jan. 94	3 Sep. 18 29 July 13	28 Jan. 24	. . . .	Kohat.
Bo.	**Bennet,** Vivian Boae, M.B., F.R.C.S. (206)	20 Apl. 67	28 July 94	1 Oct. 18 29 Jan. 14	27 July 24	. . . .	—

Lieutenant-Colonels.

M.	*Castor, Richard Henderson* (105) (*b*)	*12 May 61*	*30 Sep. 86*	*30 Sep. 06*	*10 Nov. 16*	. . .	*Civil, Amherst Distt. and in charge, Moulmein Dist Jail, Burma.*
B.	*Buchanan, Andrew,* M.D. (*b*)	*17 May 61*	*31 Mar. 87*	*31 Mar. 07*	*3 July 17*	. . .	*Civil and Supdt., Lunatic Asylum and Med. School. [Nagpur*
B.	*Fischer, Lewis Gordon* (*b*)	*3 Aug. 62*	*31 Mar. 87*	*31 Mar. 07*	*4 July 17*	. . .	*Civil and Jail, Dehra Dun.*
M.	*Stewart, Thomas William* (105. 113) (*b*)	*12 Oct. 62*	*1 Oct. 87*	*1 Oct. 07*	*26 Dec. 17*	. . .	*Conv. Hospl., Camp, Karachi.*
B.	*O'Gorman, Patrick Wilkins,* C.M.G. (*b*)	*1 Mar. 60*	*31 Mar. 88*	*31 Mar. 08*	*23 June 18*	. . .	*O. C., Indn. Stn. Hospl., Nowshera.*
B.	*Smith, Henry,* C.I.E., M.D. (*b*)	*16 Aug. 62*	*31 Mar. 90*	*31 Mar. 10*	*26 June 20*	. . .	*Civil, Amritsar.*
B.	**Frost,** George Hewitt, M.B. (150, 170)	2 Apl. 67	28 July 91	28 July 11	27 July 21	. . .	O. C. Indn Stn. Hospl., Mhow.
B.	**Wood,** Harry Stotesbury, M.B. (149, 156)	21 Mar. 65	28 July 91	28 July 11	27 July 21	28 Aug. 18	Civil Sibsagar. *Pl. lv., 6 wks., 12 Nov. 18.*
M.	**Penny,** Jeremiah (150)	5 July 64	28 July 91	28 July 11	27 July 21	14 Nov. 18	O. C., Indn. Stn. Hospl., Jullundhur.
M.	**Graves,** Douglas Henry McDonell, M.B.	19 Oct. 68	28 July 91	28 July 11	27 July 21	. . .	O. C., Indn. Stn. Hospl., Mandalay.
B.	**Haig,** Patrick Balfour, C.B., M.B.	25 Aug. 66	27 July 92	27 July 12	26 July 22	. . .	Civil, Ajmere, and C.M.O., Rajputana.
B.	**Maddox,** Ralph Henry, M.B. (162, 170, 206)	27 June 64	27 July 92	27 July 12	26 July 22	. . .	—.
B.	**Hugo,** Edward Victor, C.M.G., M.D., F.R.C.S., V. H. S. (156, 167, 170). (*Ty. Col.*)	5 Jan. 65	27 July 92	27 July 12	26 July 22	. . .	—.
B.	**Smith,** Herbert Austen, C.I.E., M.B. (170)	3 May 66	27 July 92	27 July 12	26 July 22	9 Oct. 16	Surgeon to H. E. the Viceroy.
B.	**Smith,** George McIver Campbell, M.B. (156,170)	13 May 69	27 July 92	27 July 12	26 July 22	. . .	—.
M.	**Macrae,** John Lewis, M.B.	8 Aug. 68	27 July 92	27 July 12	26 July 22	. . .	O. C., Indn. Stn. Hospl. Mardan.
B.	**Milne,** Charles, M.B. (170)	4 Aug. 69	30 Jan. 93	30 Jan. 13	29 Jan. 23	. . .	—.
B.	**Young,** William, M.B., F.R.C.S.E.	6 May 71	30 Jan. 93	30 Jan. 13	29 Jan. 23	. . .	Civil & Jail, Cawnpore.
B.	**Bourke,** John Joseph, M.B. [(206)	1 Apl. 65	30 Jan. 93	30 Jan. 13	29 Jan. 23	. . .	Offg. Mint Master, Calcutta.
B	**Chatterton,** Bernard Robert, M.D., F.R.C.S.I.	14 Dec. 68	30 Jan. 93	30 Jan. 13	29 Jan. 23	. . .	O. C. No. 1, I. G. H., Karachi. Deputation duty.
B.	**Williams,** Charles Edward, M.D. (162, 170, 206)	1 Nov. 65	30 Jan. 93	30 Jan. 13	29 Jan. 23	. . .	Sanitary Commissioner, Burma.
B.	**Ogilvie,** Walter Holland, C.M.G., M.B. (170) (Tempy. Colonel).	2 Sep. 69	30 Jan. 93	{ 30 Jan. 13 { 2 Sep. 16(*c*)	29 Jan. 23	. . .	—.
B.	**Heard,** Richard, M.D. (170) V. H. S.	10 Jan. 70	30 Jan. 93	{ 29 Nov 15(*a*) { 30 Jan. 13	29 Jan. 23	. . .	Inspector of Medical Services, Army Head Quarters.
B.	**Parry,** Edgar Rowe, M.B.	15 Nov. 68	30 Jan. 93	30 Jan. 13	29 Jan. 23	. . .	O. C., Indn. Stn. Hospl., Allahabad.
B.	**More,** Paxton St. Clair, M.B.	14 June 67	30 Jan. 93	30 Jan. 13	29 Jan. 23	. . .	O. C., Indn. Stn. Hospl., Rawalpindi.
M.	**Bidie,** George, M.D., F.R.C.S.E.	14 Sep. 64	30 Jan. 93	30 Jan. 13	29 Jan. 23	28 July 18	Ag. Civil, Coonoor.
B.	**Stevens,** Cecil Robert, M.D., F.R.C.S.	14 Mar. 67	29 July 93	30 Jan. 13	29 July 23	. . .	Profr. of Surgery, Medl. College and Surgn. to the College Hospital, Calcutta.
B.	**Rogers,** *Sir* Leonard, KT., C.I.E., F.R.S., M.D., F.R.C.P., F.R.C.S.	18 Jan. 68	29 July 93	30 Jan. 13	29 July 23	17 Feb. 06	Profr. of Pathology, Medl. Coll., and Bacteriologist to Government, Calcutta.
B.	**Newman,** Earnest Alan Robert, C.I.E., M.D.	12 May 67	29 July 93	30 Jan. 13	29 July 23	. . .	Civil, Dacca. [Lucknow.
B.	**Birdwood,** Gordon Travers, M.D.	24 Jan. 67	29 July 93	30 Jan. 13	29 July 23	. . .	Civil and Profr. of Midwifery, King George's M. C.
B.	**Turner,** Reginald George, C.M.G., D.S.O., F.R.C.S. (170)	5 Feb. 70	29 July 93	30 Jan. 13	29 July 23	. . .	Poona.
R.	**Davidson,** James, D.S.O., M.D. [L] (207)	27 Nov. 65	29 July 93	30 Jan 13	29 July 23	. . .	Offg. A. D. M. S., Abbottabad Bde.
B.	**Barry,** Cecil Charles Stuart (106), C.I.E.	10 Dec. 67	29 July 93	29 July 13	28 July 23	. . .	Medl. Supdt., General Hospital, Rangoon. *Pl. lv., 6 wks. 28 Nov. 18.*
B.	**Gould,** Jay, C.B.E., M.B. (170)	20 Jan. 67	29 July 93	29 July 13	28 July 23	. . .	*Lv., m. c., 1 mo., 2. Dec. 18*
B.	**Mulvany,** John (166, 170)	9 Apl. 70	29 July 93	29 July 13	28 July 23	. . .	Supdt., Central Jail, Alipur. *Lv. ex I., 1 yr. 11 mos. 7 days 19 Sep. 18.*
B.	**Burden,** Henry, C.I.E., F.R.C.S. (*tempy. Colonel*)	22 Apl. 67	29 Jan. 93	{ 29 July 13 { 1 Apl. 18(*c*)	28 Jan. 24	. . .	—.
B.	**Bowle-Evans,** Charles Harford, C.M.G., M.B. (162 169, 206).	19 Oct. 67	29 Jan. 94	29 July 13	28 Jan. 24	. . .	—.
B.	**MacLeod,** Ewan Cameron	18 Aug. 68	29 Jan. 94	29 July 13	28 Jan. 24	. . .	Civil. Lakhimpur, Assam.

(*a*) Date of promotion to Brevet-Colonel. (*b*) Retained in service after attaining age of 55. (*c*) Date of tempy. rank and Colonel whilst holding appointment of A. D. M. S.

INDIAN MEDICAL SERVICE—*contd.*

Lieutenant-Colonels—*contd.*

Estabt.	Names.	Date of birth.	First Commn.	Date of rank.	Completes 30 years' pension service.	Date of present appointment.	Remarks.
B.	**Fisher**, John, D.S.O., M.B.	8 Jan. 67	29 Jan. 94	29 Jan. 14	28 Jan. 24		——*Sick lv. extd. to 24 Mar. 19.*
B.	**Harriss**, Stanley Arthur, M.B. (170)	4 Jan. 69	29 Jan. 94	29 Jan. 14	28 Jan. 24		O. C., Indn. Stn. Hospl., Secunderabad.
B.	**Sutherland**, David Waters, C.I.E., M.D., (169)	18 Dec. 71	28 July 94	29 Jan. 14	27 July 24		Principal & Prof. of Medicine, K. E. Medical College, [Lahore.
B.	**Granger**, Thomas Arthur, C.M.G., M.B., *V.H.S.* (169).	12 Aug. 71	28 July 94	28 July 14 1 Jan. 14(*a*)	27 July 24	1 July 18	A. D. M. S., Army Hd. Qrs.
B.	**Bamfield**, Harold John Kinnahan, D.S.O. (170)	18 Sep. 70	28 July 94	28 July 14	27 July 24		O. C., Indn. Stn. Hosptl., Lansdowne.
B.	**Grant**, John Wemyss, M.B. (170)	5 Apl. 69	28 July 94	28 July 14	27 July 24		M. S. K., Calcutta.
B.	**Hayward**, William Davey, M.B.	1 Apl. 70	28 July 94	28 July 14	27 July 24		Prof. of Midwifery, G. M. Coll., Bombay, and Physician B. M. and P. Hospl.
Bo.	**Evans**, Samuel, M.B.	17 Sep. 66	29 Jan. 95	29 July 14	28 Jan. 25		
B.	**Hamilton**, John Archibald, C.M.G., M.B., F.R.C.S.E.	4 Nov. 69	29 Jan. 95	29 July 14	27 Jan. 25		——.
Bo.	**McDonald**, James Haldane, M.B.	30 Nov. 69	29 Jan. 95	29 July 14	28 Jan. 25		O. C., Indn. Stn. Hospl., Abbottabad.
B.	**Stevens**, Algernon Francis	30 Aug. 69	29 Jan. 95	29 Jan. 15	28 Jan. 25		O. C., I. T. War Hospital, Meerut.
B.	**Bensley**, Clement Henry	27 Sep. 70	29 Jan. 95	29 Jan. 15	28 Jan. 25		Insptr.-Genl. of Prisons, C. P., *s.p.t.*
B.	**Watling**, Francis Hammond, M.B. (170)	14 Sep. 69	29 Jan. 95	29 Jan. 15	28 Jan. 25		O. C., Indn. Stn. Hospl, Barrakpore.
M.	**Wall**, Frank, C.M.G.	21 Apl. 68	29 Jan. 95	31 May 14(*a*) 29 Jan. 15	28 Jan. 25	15 Jan. 17	*Lv., ex I., m. c., 6 mos., Feb. 18. Extn. to 18 Feb. 19.*
M.	**Mathew**, Charles Montague (125*a*)	10 July 69	29 Jan. 95	29 Jan. 15	28 Jan. 25		O. C., Indn. Stn. Hospl., Kamptee.
B.	*Stephenson, John*, M.B., F.R.C.S. (L) (156, 170, 206)	6 *Feb.* 71	29 *July* 95	29 *Jan.* 15	28 July 25		*Civil, Punjab. Principal and Profr. Biology, Govt. Coll., Lahore.*
B.	**Windsor**, Frank Needham, M.B. (170, 306)	2 May 68	29 July 95	29 Jan. 15	28 July 25		O. C., Indn. Stn. Hospl., Jhansi.
B.	**Waters**, Ernest Edwin, M.D. (157)	22 July 72	29 July 95	29 Jan. 15	28 July 25		Civil, Howrah.
B.	**Leventon**, Asher, F.R.C.S.I.	29 Apl. 70	29 July 95	29 Jan. 15	28 July 25		Supdt., Campbell Medical School and Hospl., Calcutta.
Bo.	**King**, Arthur Fredrick William, F.R.C.S.E.	13 Nov. 71	29 July 95	29 Jan. 15	28 July 25		O. C., Indn. Stn. Hosp, Belgaum and ag. Civil Surgeon.
Bo.	**Standage**, Robert Frazer	5 Apl. 68	29 July 95	29 Jan. 15	28 July 25		——.
Bo.	**Gibbs**, Andrew Armstrong (*b*)	8 Mar. 71	29 July 95	29 Jan. 15	28 July 25		M. S. K., Lahore Cant.
M.	**Illington**, Edmund Moritz, F.R.C.S.E.	19 June 70	29 July 95	29 Jan. 15	28 July 25		S. M. O., Govt. General Hospl. & 2nd Surgeon with Port & Marine duties; M. O. The Penitentiary Madras. *s. p. t.*
B.	**Chapman**, Philip Francis, M.B. (170)	11 Feb. 70	29 July 95	29 July 15	28 July 25		——.
Bo.	**Hooton**, Alfred (172, 206)	15 Mar. 70	29 July 95	29 July 15	28 July 25		——.
M.	**Blenkinsop**, Frederic Linton, M.B. (206)	20 Jan. 70	29 July 95	29 July 15	28 July 25		——.
Bo.	**Knapton**, Henry Alfred Forbes	24 Sep. 68	29 July 95	29 July 15	28 July 25		O. C., Indn. Stn. Hospl. Staff Surgeon, Civil Surgeon, etc., Ahmednagar.
B.	**Cochrane**, Archer William Ross, M.B., F.R.C.S.	22 Nov. 72	29 Jan. 96	29 July 15	28 Jan. 26		Profr. of Medicine, King George's Medical College, Lucknow, *s.p.t.*
B.	**Clemesha**, William Wesley, M.D. (*temp Col.*).	19 Apl. 71	29 Jan. 96	29 July 15 1 July 17(*c*)	28 Jan. 26	9 Nov. 15	
B.	**Black**, James Alexander, M.B.	1 Mar. 70	29 Jan. 96	29 July 15	28 Jan. 26		Chemical Examiner, Punjab.
B.	**Wilson**, Roger Parker, F.R.C.S.	13 May 70	29 Jan. 96	29 July 15	28 Jan. 26		Profr. of Clinical and Operative Surgery, Medl. College, and Surgn. to the College Hospl., Calcutta.

B.	**Lindesay,** Victor Edward Hugh, M.B. (124*a*, 125*b*, 156, 170)	24 May 69	29 Jan. 96	29 July 15	28 Jan. 26	10 July 18	O. C. Troops, A. T. "Vasna."
B.	**Robertson,** James Currie, C.M.G., C.I.E., M.B.	26 Nov. 70	29 Jan. 96	29 July 15	28 Jan. 26	17 May 12	On mily. duty under the War Office.
B.	**Rainier,** Norman Robinson Jones (162)	16 Apl. 71	29 Jan. 96	29 July 15	1 Feb. 26	. . . .	O. C. Indn. Stn. Hospl., Jubbulpore.
B.	**Perry,** Edmund Ludlow, D.S.O. (170)	28 June 72	29 Jan. 96	29 July 15	28 Jan. 26	. . . .	——.
M.	**Niblock,** William James, M.B., F.R.C.S.I.	10 July 71	29 Jan. 96	29 July 15	28 Jan. 26	. . . .	Profr. of Surgery & Actg. Profr. of Biology, M. C. and 1st Surgn., Genl. Hospl., Madras.
M.	**Rost,** Ernest Reinhold (105)	22 Aug. 72	29 Jan. 96	29 July 15	28 Jan. 26	. . . .	O. C. Indn. Stn. Hospl., Calcutta.
Bo.	**Bakhle,** Chintaman Ramchandra (162, 169, 206)	12 Feb. 71	29 Jan. 96	29 Jan. 16	28 Jan. 26	. . . .	
Bo.	**Kukday,** Krishnaji Vishnoo (162, 206)	15 Sep. 70	29 Jan. 96	29 Jan. 16	28 Jan. 26	. . . .	Karachi.
B.	**Dawes,** Christopher Dering	22 Mar. 72	29 Jan. 96	29 Jan. 16	28 Jan. 26	17 July 18	O. C. Indn. Stn. Hospl., Bangalore.
M.	**Symons,** Thomas Henry, O.B.E.	17 May 72	29 Jan. 96	29 Jan. 16	28 Jan. 26	. . . .	
B.	**Walton,** Herbert James, M.D., F.R.C.S. (206)	19 Jan. 69	29 July 96	29 Jan. 16	28 July 26	30 Dec. 17	O. C. Troops, A. T. "Erinpura."
B.	**Ainsworth,** Hugh, M.B., F.R.C.S. (169, 306)	28 Sep. 71	29 July 96	29 Jan. 16	28 July 26	. . . .	O. C. Station Hospital (Nepalese Troops), Kakool.
Bo.	**Smith,** Frank Addinsell, M.D. (170)	26 Feb. 70	29 July 96	29 Jan. 16	28 July 26	. . . .	Resy. Surgn., Indore, and A. M. O., Central India, & in collateral charge of the Southern States of Central India Agency. Training School for Indian Cadets, Indore.
M.	**Cornwall,** John Wolfran, M.D.	2 Aug. 70	29 July 96	29 Jan. 16	28 July 26	. . . .	Director, Pasteur Institute, Southern India, Coonoor.
M.	**Miller,** Alfred, M.B.	29 Apl. 69	29 July 96	29 Jan. 16	28 July 26	. . . .	Profr. of Chemistry, Ag. Prof. of Medical Jurisprudence and Principal, Medl. Coll., Chemical Examiner, Madras.
M.	**Hammond,** Frederick Arthur Lucas (105, 124*a*)	21 Nov. 71	29 July 96	29 Jan. 16	29 July 26	. . . .	Civil, Burma. *Lv., ex I., m.c., 6 mos., 18 Mar. 17. Extn., m. c., one year and 6 mos.*
M.	**Brown,** Henry Robert (170)	9 June 68	29 July 96	29 Jan. 16	28 July 26	. . . .	O. C., War Hospl., Poona.
M.	**Fleming,** Archibald Nicol, D.S.O., M.B., F.R.C.S.E. (172*a*)	9 Sep. 70	29 July 96	29 Jan. 16	28 July 26	. . . .	——.
M.	**Dee,** Peter, M.B. (105)	18 Oct. 69	29 July 96	29 Jan. 16	28 July 26	. . . .	Civil, Rangoon (East and West), & Supdt., Burma Govt. Medl. School, Rangoon.
M.	**Mell,** Felix Oswald Newton, C.I.E., M.B.	9 Aug. 66	29 July 96	29 Jan. 16	28 July 26	. . . .	——. *Lv., ex I., m.c., mos.—April 18. Exten. to 23 Feb. 1919.*

GENERAL LIST.

B.	**Woolley,** Jasper Maxwell, M.D. (162, 170)	21 Dec. 72	28 Jan. 97	29 July 16	27 Jan. 27	. . . .	O. C. Indn. Stn. Hospl., Lucknow.
E.	**Hugo,** James Henry, D.S.O., M.B. (170)	16 July 70	28 Jan. 97	29 July 16	28 Jan. 27	. . . .	O. C., Indn. Stn. Hospl., Lahore Cantt.
N.	**Kenrick,** William Hamilton (170)	17 Oct. 68	28 Jan. 97	29 July 16	27 Jan. 27	. . . .	——.
W.	**Price,** Raymound Herbert, M.B., F.R.C.S.E. (170)	23 Sep. 68	28 Jan. 97	29 July 16	27 Jan. 27	. . . .	——.
Sec.	**Bryson,** Reginald, F.R.C.S.E. (162, 195)	16 June 72	28 Jan. 97	29 July 16	27 Jan. 27	. . . .	Surgeon, 1st Dist., Supt., Medl. School, Rayapuram, Madras, and Medl. Inspr. of Emigrants.
E.	**Lane,** Clayton Arbuthnot, M.D.	30 Aug. 68	28 Jan. 97	28 Jan. 17	27 Jan. 27	. . . .	Civil, Hughli, 1917, also Civil, Serampur.
E.	**Kelly,** Thomas Bernard, D.S.O., F.R.C.S.E.	11 Mar. 70	28 Jan. 97	28 Jan. 17	27 Jan. 27	. . . .	O.C., Indn. Stn. Hospl., Dehra Dun.
N.	**Watson,** Charles Henry, D.S.O.	14 Jan. 70	28 Jan. 97	28 Jan. 17	27 Jan. 27	. . . .	O. C., Station Hospital (Nepalese Troops), Dehra Dun.
W.	**Baines,** Edgar Francis Eardley	2 Feb. 71	28 Jan. 97	28 Jan. 17	27 Jan. 27	. . . .	——.
W.	**Sealy,** Geoffrey Orr Fern, F.R.C.S. (94, 100, 162, 206)	22 Apl. 73	28 Jan. 97	28 Jan. 17	. . . .	. . . .	——.
E.	**Murray,** John George Patrick, M.D., F.R.C.S.E. (170)	17 Mar. 75	28 July 97	28 Jan. 17	27 July 27	. . . .	Civil, Bihar and Orissa.
N.	**Anderson,** Samuel, M.B. (172)	22 May 74	28 July 97	1 Jan. 16(*a*) 28 Jan. 17	27 July 27	. . . .	——.
W.	**Hutchinson,** Francis Hope Grant, M.B. (170, 206)	28 Oct. 70	28 July 97	28 Jan. 17	27 July 27	. . . .	Sany. Comsr., Bombay.

(*a*) Date of promotion to Brevet-Lieutenant-Colonel.
(*b*) Qualified in a course of Instruction at the London School of Economics.
(*c*) Date of temporary rank.

INDIAN MEDICAL SERVICE—*contd.*

Estabt.	Names.	Date of birth.	First Commn.	Date of rank.	Completes 30 years' pension service.	Date of present appointment.	Remarks.
	Lieutenant-Colonels—*concld.*						
W.	**Marjoribanks,** James Leslie, M.D. (88, 132, 153, 161, 185, 206).	14 Nov. 74	28 July 97	28 Jan. 17	27 July 27	. . . .	———.
Sec.	**Fenton,** Alexander, M.B. (104, 181, 206)	29 Nov. 70	28 July 97	28 Jan. 17	. . . .	. . . .	———.
Sec.	**Knox,** Robert Welland, D.S.O., M.B., F.R.C.S. (162, 170, 206)	6 Sep. 73	28 July 97	28 Jan. 17	. . . .	. . . .	———.
E.	**Rait,** John Walter Forbes, M.B.	5 Apl. 73	28 Jan. 98	28 July 17	27 Jan. 28	. . . .	———.
N.	**O'Meara,** Eugene John, F.R.C.S.	12 July 74	28 Jan. 98	28 July 17	27 Jan. 28	. . . .	Civil Surgeon and Principal, Medl. School, Agra.
N.	**Tate,** Godfrey, M.B. (169)	5 Mar. 73	28 Jan. 98	28 July 17	27 Jan. 28	1 Oct. 16	Civil Surgeon, Simla, East.
N.	**Baird,** Roy Fearon (167, 169)	16 Mar. 72	28 Jan. 98	28 July 17	27 Jan. 28	. . . .	Peshawar. *Lv. ex* I., m. c. up to 31 Jany. 1919.
N.	**Gage,** Andrew Thomas, M.B.	14 Dec. 71	28 Jan. 98	28 July 17	27 Jan. 28	. . . .	Supdt., Royal Botanical Gardens, Calcutta.
W.	**McPherson,** George, C.I.E., M.B., F.R.C.S.E. (170)	25 Oct. 72	28 Jan. 98	28 July 17	27 Jan. 28	. . . .	
W.	**Hunt,** Spencer, M.B. (124*a*)	27 Apl. 74	28 Jan. 98	28 July 17	27 Jan. 28	. . . .	Agency Surgn., Baghelkhund.
W.	**Condon,** deVere, M.D. (162)	13 Apl. 74	28 Jan. 98	28 July 17	27 Jan. 28	. . . .	———. *Lv. m. c. 4 mos.* — *Sep. 18.*
Sec.	**Lethbridge,** William, M.B. (170, 206)	3 Mar. 72	28 Jan. 98	28 July 17	27 Jan. 28	. . . .	———. *Lv. m. c. 2 mos. 8 days. 16 Sep. 18.*
Sec.	**Kirkpatrick,** Henry, M.B.	20 Oct. 71	28 Jan. 98	28 Jan. 18	27 Jan. 28	. . . .	Profr. of Ophthalmology, Medl. College, Supdt., Govt. Ophthalmic Hospl., and M. O., Civil Orphan Asylum, Madras.
Sec.	**Fayrer,** Frederick Durand Sterling	28 Oct. 69	28 Jan. 98	28 Jan. 18	27 Jan. 28	. . . .	———. *Lv. m. c. Exten. to 24 Mar. 19.*
Sec.	**Chitale,** Padmakar Krishna (206)	13 June 71	28 Jan. 98	28 Jan. 18	27 Jan. 28	. . . .	O. C. Indn. Stn. Hospl., Nasirabad.
N.	**Hunter,** Thomas, M.D. (206)	18 Aug. 73	27 July 98	28 Jan. 18	26 July 28	. . . .	Civil, Naini Tal.
N.	**Battye,** Walter Rothney, D.S.O., M.B., F.R.C.S. (162,	20 Jan. 74	27 July 98	28 Jan. 18	26 July 28	. . . .	———.
N.	**Hutcheson,** George, M.B.	17 Dec. 74	27 July 98	28 Jan. 18	26 July 28	. . . .	Civil, Allahabad.
Sec.	**Liston,** William Glen, C.I.E., M.D.	30 July 73	27 July 98	28 Jan. 18	26 July 28	. . . .	Dir., Bact. Laby., Bombay.
N.	**Boulton,** Harold, M.B. (165, 169)	7 Sep. 72	27 July 98	28 Jan. 18 / 1 Jan. 16(*a*)	26 July 28	. . . .	———.
W.	**Anthony,** Richard William, M.B., F.R.C.S.E.	18 May 74	27 July 98	28 Jan. 18	26 July 28	. . . .	Offg. O. C. No. 1 I. G. H., Karachi.
W.	**Tucker,** Ernest Frederick Gordon, M.B.	1 Nov. 71	27 July 98	28 Jan. 18	26 July 28	. . . .	1st Physn. J. J. Hospl., and Profr. of Medicine, Clinical Medicine, and Therapeutics and Actg. Profr. of Pathology, G. M. Coll., Bombay.
W.	**Stewart** George Edward, M.B., F.R.C.S.E. (94, 169, 170).	17 Dec. 73	27 July 98	28 Jan. 18	26 July 28	. . . .	———.
W.	**Thompson,** Frank Stuart Corbitt, M.B. (72)	19 June 72	27 July 98	28 Jan. 18	26 July 28	. . . .	Supdt., Presy. Jail, Bengal.
W.	**Novis,** Thomas Shepherd, F.R.C.S. (170)	3 Dec. 74	27 July 98	28 Jan. 18	26 July 28	. . . .	Bombay.
W.	**Watson,** John William (206, 170, 172)	8 Nov. 74	27 July 98	28 Jan. 18	26 July 28	. . . .	Agency Surgeon, Bundelkhand.
E.	**Leicester,** John Cyril Holdich, M.D., F.R.C.S.	29 Mar. 72	28 Jan. 99	28 July 18	27 Jany. 29	. . . .	———.
E.	**Innes,** Hubert, M.D.	24 Aug. 73	28 Jan. 99	28 July 18	27 Jany. 29	. . . .	———.
N.	**Willmore,** William Southwick, M.D.	7 Aug. 72	28 Jan. 99	28 July 18	27 Jany. 29	. . . .	O. C., I. T. W. H., Karachi.
N.	**Walter,** Albert Elijah (124*a*, 125*b*)	18 Dec. 71	28 Jan. 99	28 July 18	27 Jany. 29	. . . .	Supdt., X-ray Institute, Dehra Dun.
N.	**Hudson,** Corrie, D.S.O., F.R.C.S.E. (71 124*a*, 162, 206).	5 Apl. 74	28 Jan. 99	28 July 18	27 Jany. 29	. . . .	O. C., School of Instruction, R'pindi.
W.	**Hutchinson,** Leslie Thomason Rose, M.D.	19 May 72	28 Jan. 99	28 July 18	27 Jany. 29	. . . .	Professor of Physiology, G. M. College, Bombay and Presidency Surgeon, 1st Dist.

N.	**Fleming**, Alexander Mathers, M.B., F.R.C.S. E. (170)	15 Sep. 73	28 Jan. 99	28 July 18	27 Jany. 29	O. C. Mdn. Stn. Hospl., Dera Ismail Khan.
N.	**Ward**, Ellacott Leamon	28 Apl. 73	28 Jan. 99	28 July 18	27 Jany. 29	I. G. of Prisons, Punjab.
N.	**Walker**, John Norman	30 Mar. 73	28 Jan. 99	28 July 18	27 Jany. 29	Civil, Mussoorie.
N.	**Roberts**, Vivian Heathcote, F.R.C.S.E. (71, 89*a*, 124*a*, 125*b*, 165, 170, 206).	20 July 74	28 Jan. 99	28 July 18	27 Jany. 29	Nowshera. Specialist in Advanced Operative Surgery.
Sec.	**Ross**, Thomas Spinks	10 Oct. 72	28 Jan. 99	28 July 18	27 Jany. 29	Health Officer, Secunderabad.
N.	**Pearson**, William MacMullen, M.B.	28 Dec. 73	28 Jan. 99	28 July 18	27 Jany. 29	———.
Sec.	**Kemp**, David Claude	8 June 71	28 Jan. 99	28 July 18	. . . .	O. C. Indn. Stn. Hospl., Ambala.
					Date of present appointment.	

Majors.

N.	**Cruddas**, Hamilton Maxwell, C.M.G. (169)	1 Aug. 74	28 Jan. 99	28 July 10	. . . .	———.
Sec.	**Groube**, George Patrick Thomson	1 July 76	28 Jan. 99	28 Jan. 11	. . . .	Jullundur.
N.	**Greig**, Edward David Wilson, C.I.E., M.D.	20 Mar. 74	27 July 99	28 Jan. 11	. . . .	Karachi.
N.	**McKechnie**, William Ernest, M.B.	11 Aug. 74	27 July 99	28 Jan. 11	. . . .	Offg. O. C. No. 37 I. G. H., Karachi.
N.	**Harvey**, William Frederick, M.B. (162, 170, 206)	28 Apl. 73	27 July 99	28 Jan. 11	. . . .	Dir., C. R. I., Kasauli.
N.	**Forster**, William Charles Hughan, M.B. (169)	15 June 74	27 July 99	28 Jan. 11	. . . .	Sany. Commissioner, Punjab, *s. p. t.*
E.	**Urwin**, John Johnson, M.B., F.R.C.S. (170, 206)	12 Mar. 71	27 July 99	28 Jan. 11	. . . .	———.
F.	**McCay**, David, M.D.	26 Feb. 73	27 July 99	28 Jan. 11	. . . .	Prof. of Physiology, Medl. Coll., Calcutta, and 2nd Physician, Med. Cobl., Hospls.
N.	**Fry**, Arthur Brownfield, D.S.O., M.D. (170, 206)	9 May 73	27 July 99	28 Jan. 11	. . . .	———.
W.	**Maddock**, Edward Cecil Gordon, M.D., F.R.C.S.E.	29 June 76	27 July 99	28 Jan. 11	. . . .	
W.	**Dickinson**, William Henry, M. B.	11 Apl. 72	27 July 99	28 Jan. 11	. . . .	Chemical Analyser to Govt. of Bombay.
W.	**Tuke**, Arthur William, F.R.C.S.I.	20 Mar. 71	27 July 99	28 Jan. 11	. . . .	Civil, Ahmednagar.
Sec.	**Stewart**, George Herbert, M.B. (105, 120, 124*a*)	8 Feb. 75	27 July 99	28 Jan. 11	. . . .	———.
N.	**Pelle**, Harry Diamond	2 Oct. 72	27 July 99	27 July 11	. . . .	———.
N.	**Cowin**, Douglas Henry Fawcett	2 June 73	27 July 99	27 July 11	. . . .	———.
Sec.	**Tucker**, William Hancock	7 June 73	27 July 99	27 July 11	. . . .	O. C. Indn. Stn. Hospl., Cannanore.
Sec.	**Chaudhuri**, Manmatha Nath, M.B.	26 July 71	27 July 99	27 July 11	. . . .	Rawalpindi.
E.	**MacGilchrist**, Archibald Currie, M.D.	24 May 72	27 Jan. 00	27 July 11	. . . .	O. C. Indn. Stn. Hospl., Agra.
E.	**Megaw**, John Wallace Dick, M.B. (124*a*)	8 Feb. 74	27 Jan. 00	27 July 11	. . . .	Professor of Pathology and temporary Principal, King Georges' Medical College, Lucknow.
E.	**Thurston**, Edward Owen, M.B., F.R.C.S.	12 Jan. 73	27 Jan. 00	27 July 11	. . . .	Resy. Surgn. and *ex-officio* Asstt. to the Resident, [Nepal, 14 Nov. 17.
N.	**Browse**, George, D.S.O., M.D. (169) (*Bt.-Lt.-Col. and temp. Col.*)	4 Apl. 74	27 Jan. 00	{ 1 Jan. 16 (*a*) / 27 July 11 }	. . . .	———.
N	**Goodbody**, Cecil Maurice, C.I.E., D.S.O., F.R.C.S.I. (*Bt. Lt.-Col.*)	12 Nov. 74	27 Jan. 00	{ 22 Dec. 16 (*a*) / 27 July 11 }	. . . .	———.
B.	**Steen**, Robert, M.D. (199)	31 Jan. 47	27 Jan. 00	27 July 11	. . . .	*Lv. m.c. e x I., 6 mos.,—Novr. 18.*
Sec.	**Elwes**, Fredrick Fenn, C.I.E., M.D. [170)]	3 May 75	27 Jan. 00	27 July 11	. . . .	Prof. of Medicine, M. C., 1st Phyn., Genl. Hospl., and Surgn., 2nd Dist Also E. the Governor, Madras.
N.	**Matthews**, Ernest Albert Churchward, D.S.O. M. B., **V. H. S.**	3 Aug. 74	27 Jan. 00	27 July 11	. . . .	———.
W.	**Stephen**, Lessel Phillip, M.B., F.R.C.S.E.	20 July 74	27 Jan. 00	27 July 11	. . . .	Karachi.
Sec.	**Gilbert**, Leonard Erskine, M.D. (105)	20 Jan. 74	27 Jan. 00	27 July 11	. . . .	———.
N.	**Stokes**, Thomas George Nesbitt, M.B.	12 Dec. 75	27 Jan. 00	27 July 11	28 Nov. 18	Sany. Commr, C. P.
N.	**Mackenzie**, Harry Malcolm, M.B. (170)	14 May 75	27 Jan. 00	27 July 11	17 Aug. 18	Prof. of Pathology, K. E. Med. Coll., and Dy. Sany. Commr., Punjab, [in addition.
N.	**Thornely**, Michael Harris, F.R.C.S.E. (99)	2 June 75	27 Jan. 00	27 July 11	. . . .	Civil, Bihar and Orissa, offg. Cuttack.
N.	**Belt**, Francis Victor Owen, M.D. (106, 120)	14 Aug. 74	27 Jan. 00	27 July 11	. . . .	———.

(*a*) Date of promotion to Bt.-Lt.-Colonel.

INDIAN MEDICAL SERVICE—*contd.*

Estabt.	Names.	Date of birth.	First Commn.	Date of rank.	Date of present appointment.	Remarks.
	Majors—*contd.*					
W.	**Murphy,** William O'Sullivan, M.B. (206)	13 Apl. 75	27 Jan. 00	27 July 11	6 Nov. 18	O. C. Troops A. T. "Varsova."
N.	**Corry,** Mathew, M.D. (170)	10 Nov. 73	27 Jan. 00	27 July 11	. . . .	*Lv. m.c., 2 mos. 10 days, 27 July 18. Exten. 12 wks. 6 Oct. 18.*
W.	**Murison,** Cecil Charles, F.R.C.S.E.	2 Jan. 74	27 Jan. 00	27 July 11	. . . .	O. C., Indn. Stn. Hospl., Cawnpore.
Sec.	**Williams,** Herbert Armstrong, D.S.O., M.B. (105)	11 Jan. 75	27 Jan. 00	27 July 11	. . . .	——.
E.	**Graham,** James Drummond, M.B. (*Bt.-Lt.-Col.*)	21 Jan. 75	28 June 00	28 Dec. 11 3 June17(*a*)	. . . .	——
E.	**Sprawson,** Cuthbert Allan, M.D. (*Ty. Lt.-Col.*)	1 Mar. 77	28 June 00	28 Dec. 11	. . . .	——.
E.	**Mackelvie,** Maxwell, M.B., F.R.C.S.E. (170) (*Bt.-Lt.-Col.*)	10 Jan. 77	28 June 00	28 Dec. 11 1 Jan. 18(*a*)	. . . .	——
E.	**Lapsley,** William, M.B. (*Act Lt.-Col.*)	28 July 75	28 June 00	28 Dec. 11	. . . .	——
W.	**Cazaly,** William Henry, M.B.	24 Sep. 73	28 June 00	28 Dec. 11	. . . .	——. *Sick lv. extend. to 27 Feb. 19.*
E.	**Coppinger,** Walter Valentine, M.D., F.R.C.S.I., D.S.O.	1 Mar. 75	28 June 00	28 Dec. 11	. . . .	Civil, Burma, Ty. Offg. Medl. Supdt., General Hospital, Rangoon.
W.	**Spitteler,** Alfred, M.B. (94, 169)	10 June 72	28 June 00	28 Dec. 11	. . . .	——.
Sec.	**Oxley,** James Charles Stewart, F.R.C.S.E. (106)	15 Nov. 74	28 June 00	28 Dec. 11	. . . .	——.
N.	**Deas,** Leonard Joseph, Montagu. M.B., F.R.C.S.E. (*Bt.-Lt.-Col.*)	12 July 73	28 June 00	28 Dec. 11 3 June17(*a*)	. . . .	——.
W.	**Houston,** William Mitchell, M.D.	2 Nov. 75	28 June 00	28 Dec. 11	. . . .	Health Officer, Port of Bombay, *s. p. t.*
W.	**Keys,** William David Acheson, M.D.	22 Feb. 73	28 June 00	28 Dec. 11	. . . .	——.
W.	**Young,** George Joseph Grafton, M.B. (94, 169, 100)	22 Apl. 72	28 June 00	28 Dec. 11	. . . .	O. C. Indn. Stn. Hospl., Kirkee.
Sec.	**Good,** James, M.B. (105, 162, 181, 206)	14 Oct. 74	28 June 00	28 Dec. 11	15 Apl. 18	O. C. Troops A. T. "Sicilia"
Sec.	**Chalmers,** Alexander, M.B., F.R.C.S.I. (162, 206)	20 Dec. 72	28 June 00	28 Dec. 11	. . . .	O. C. Indn Stn. Hospl., Dinapore.
Sec.	**Hamilton,** William Gavin	4 Oct. 74	28 June 00	28 Dec. 11	18 Nov. 18	O. C. Indn. Stn. Hospl., St. Thomas Mount, Madras.
Sec.	**Godkin,** Samuel Robert, D.S.O., F.R.C.S.I.	29 Sep. 78	28 June 00	28 Dec. 11	. . . .	*Lv., m. c., ex I., 6 mos., —Feb. 18. Exten to 18 Feb. 19.*
N.	**Melville,** Charles William Francis, M.B., F.R.C.S.E. (170) (*Actg. Lt.-Col.*)	27 Mar. 77	29 Jan. 01	29 July 12	. . . .	——.
N.	**McCarrison,** Robert, M.D., F.R.C.P. (182*b*) (*Bt.-Lt.-Col.*)	15 Mar. 78	29 Jan. 01	29 July 12 1 Jan. 18(*a*)	. . . .	Special duty. Indian Research Fund Assotn.
E.	**Masson,** James, M.B., F.R.C.S.E. (*Actg. Lt.-Col.*)	28 Apl. 75	29 Jan. 01	29 July 12	. . . .	——.
E.	**Wells,** Norman Septimus, M.B.	27 Dec. 76	29 Jan. 01	29 July 12	. . . .	——.
E.	**Anderson,** William Maurice, M.D. (169, 206)	29 Dec. 73	29 Jan. 01	29 July 12	. . . .	——.
N.	**Leonard,** William Hugh, F.R.C.S. (66, 170) (*Bt.-Lt.-Col.*)	12 Feb. 76	29 Jan. 01	20 July 12 1 Jan. 18(*a*)	. . . .	
E.	**Young,** Andrew Watson Cook, M.B. (170) (*Bt.-Lt.-Col.*)	25 Feb. 76	29 Jan. 01	29 July 12 3 June16(*a*)	4 Nov. 17	O. C. Troops, H. S. "Dongola."
N.	**Swan,** James Graham Goodenough, C.I.E., M.B.	20 Oct. 75	29 Jan. 01	29 July 12	. . . .	——.
N.	**Dalziel,** Robert McLauchlan, M.B., F.R.C.S.E.	13 Jan. 76	29 Jan. 01	29 July 12	. . . .	G. R. Depôt, Gorakhpore.
Sec.	**Robb,** James Jackson, M.B.	21 Oct. 75	29 Jan. 01	29 July 12	21 Nov. 18	O. C. Troops, H. S. "Vita."
Sec.	**Ruzzak,** Shaik Abdur	1 Nov. 73	29 Jan. 01	29 July 12	. . . .	S. M. O., Trichinopoly.
Sec.	**Foster,** Robert Basil Boothby, M.B.	14 Sep. 76	29 Jan. 01	29 July 12	. . . .	O. C. Indn. Stn. Hospl., Bakloh.
E.	**McKendrick,** Anderson Gray, M.B., F.R.S.E.	8 Sep. 76	27 June 01	27 Dec. 12	. . . .	Director, Pasteur Institute of India, Kasauli.
E.	**Moses,** Owen St. John, M.D., F.R.C.S.E. (169)	6 Oct. 73	27 June 01	27 Dec. 12	15 Dec. 17	Offg. Supdt., Central Lunatic Asylum, Berhampur, *s. p. t.* & Offg. Civil., Murshidabad.
N.	**Sumner,** Frederick William, M.B., F.R.C.S.E. (172*a*)	4 Mar. 75	27 June 01	27 Dec. 12	. . . .	O. C. Indn. Stn. Hospl., Risalpur.
N.	**Nutt,** Harold Rothery, M.D., F.R.C.S.	28 Dec. 76	27 June 01	27 Dec. 12	. . . .	On duty at Netley in England.

N.	**Ritchie,** William Duncan, M.B.	4 Apl. 77	27 June 01	27 Dec. 12	. . . .	———.
N.	**Scott,** Norman Emil Henry, C.I.E. M.B. (156)	19 Dec. 75	27 June 01	27 Dec. 12	. . . .	Special duty, Persian Gulf.
N.	**Fleming,** John Kenneth Sprot (*temp. Lt.-Col.*)	5 July 74	27 June 01	27 Dec. 12	23 Feb. 17	A. D. M. S., Army Hd. Qrs.
N.	**Hepper,** Evelyn Charles (actg. Lt.-Col.) (169)	10 Dec. 76	27 June 01	27 Dec. 12	. . . .	———.
N.	**Southon,** Charles Edward, M.B. (170)	11 Jan. 76	27 June 01	27 Dec. 12	. . . .	O. C. Indn. Stn. Hospl., Jhelum.
E.	**Fowler,** George	29 Apl. 74	27 June 01	27 Dec. 12	. . . .	———.
N.	**Husband,** James, M.B., F.R.C.S.E. (169, 206)	6 Aug. 76	27 June 01	27 Dec. 12	. . . .	O. C. Indn. Stn. Hospl., Shahjahanpur.
N.	**Foster,** Henry Bertram, M.D.	10 Oct. 74	27 June 01	27 Dec. 12	. . . .	
B.	**Kerans,** George Charles Lovell, D.S.O. (170)	10 Oct. 77	27 June 01	27 Dec. 12	8 Dec. 16	D. A. D. M. S. (Mobn.), 8th (Lucknow) Division.
W.	**McConaghy,** Christopher Birdwood, M.B.	25 Mar. 77	27 June 01	27 Dec. 12	. . . .	Agency Surgeon, Eastern States of Rajputana, Bharatpur, in visiting charge of the office of Resy. Surgn., Jaipore and Agency Surgeon Kotah and Jhalawar, in addition to his own duties.
N.	**Illius,** Henry Warwick, F.R.C.S.E.	9 Oct. 75	27 June 01	27 Dec. 12	. . . .	———.
M.	**Browne,** Edward Wemyss	7 Aug. 74	27 June 01	27 Dec. 12	. . . .	O. C. Indn. Stn. Hospl., Poona.
W.	**Christian,** John Beresford (94, 161, 169, 176, 185, 206)	2 Apl. 73	27 June 01	27 Dec. 12	. . . .	Civil Surgeon and Supdt., Medical School, Hyderabad (*s.p.t.*)
W.	**Murphy,** Andrew, M.B.	16 May 73	27 June 01	27 Dec. 12	. . . .	O. C. Indn. Stn. Hospl. and M. S., K., Rangoon.
Sec.	**Brassey,** Laurence Percival, M.B.	3 May 76	27 June 01	27 Dec. 12	. . . .	———.
Sec.	**Marr,** Colin Forbes, M.B. (106)	17 May 78	27 June 01	27 Dec. 12	. . . .	Medl. StoreKeeper, Bombay.
Sec.	**Bose,** Satis, M.B. (112)	11 June 73	27 June 01	27 Dec. 12	. . . .	Specialist in Ophthalmology, Calcutta.
Sec.	**O'Neill,** Patrick Laurence	28 Feb. 76	27 June 01	27 Dec. 12	. . . .	———.
E.	**Stanley,** Edmund Hamilton Blake	17 Mar. 75	29 Jan. 01	29 Jan. 13	. . . .	Karachi.
N.	**Thompson,** Frederick Troughton, M.B. (170)	1 May 77	27 June 01	27 June 13	. . . .	Cantt. Genl. Hospl., Jullundur.
N.	**Lister,** Alfred Ernest John, M.B., F.R.C.S.	22 May 77	29 Jan. 02	29 July 13	. . . .	Profr. of Physiology, King George's V Medl. College, Lucknow.
N.	**Williams,** Thomas Samuel Beauchamp, M.B. (*Bt. Lt.-Col.*) (162, 169)	5 May 77	29 Jan. 02	29 July 13 3 June 17 (*a*)	. . .	D. A. D. M. S. (Sanitary), Karachi.
E.	**Abbott,** Samuel Herbert Lee, M.B. (94, 170)	6 Sep. 77	29 Jan. 02	29 July 13	. . . .	———.
E.	**Lloyd,** Richard Ernest (206)	20 Sep. 75	29 Jan. 02	29 July 13	14 Dec. 13	Civil, Bengal. *Lv.. ex. I., 1 year and 2 mos, 23 Sept. 17.*
E.	**Clements,** John Edmund, M.B.	25 Feb. 75	29 Jan. 02	29 July 13	. . . .	———
N.	**Steen,** Hugh Barkley, M.D.	23 Nov. 75	29 Jan. 02	29 July 13	24 Dec. 16	Professor of Midwifery, Medl. College, & Obstetric Physician & Surgeon, Medl. College Hospl., Calcutta. (*s.p.t.*)
N.	**Bisset,** Ernest, M.B.	1 Dec. 76	29 Jan. 02	29 July 13	. . . .	O. C. Indn. Stn. Hospl., Ferozepur.
E.	**Bradley,** Richard James, M.B. (156, 170)	1 Mar. 76	29 Jan. 02	29 July 13	. . . .	———.
N.	**Overbeck-Wright,** Alexander William, M.B.	30 May 79	29 Jan. 02	29 July 13	. . . .	——— Ag. Superintendent, Central Lunatic Asylum, Yervada and Medical Officer, Central Prison.
N.	**Munro,** David, C.I.E., M.B., F.R.C.S.E., (165, 169, 206)	23 June 78	29 Jan. 02	29 July 13	. . . .	———
W.	**McCoy,** John William	6 Apl. 75	29 Jan. 02	29 July 13	. . . .	Civil, Assam.
Bo.	**Carter,** Robert Markham, C.B., F.R.C.S. (*Bt. Lt.-Col.*)	18 Oct. 75	29 Jan. 02	29 July 13 26 Apl. 16 (*a*)	. . . .	———.
Sec.	**Willcocks,** Roger Durant, M.B.	10 Aug. 76	29 Jan. 02	29 July 13	4 Feb. 18	D. M. and S. O., Kurnool., D. M. S. O., the Nilgiris, s. *p. t.*
W.	**Paterson,** Thomas George Ferguson, D.S.O., M.B. (170)	6 Dec. 76	29 Jan. 02	29 July 13	. . . .	———.
W.	**Baker,** Dodington George Richard Sherston (172)	22 July 77	29 Jan. 02	29 July 13	. . . .	——— ———.
Sec.	**Tyrrell,** Jasper Robert Joly, M.B.	14 June 76	29 Jan. 02	29 July 13	. . . .	——— O. C. Indn. Stn. Hospl., Muttra.
Sec.	**Rai,** Dewan Ganpat (L) (159, 204)	11 Nov. 75	29 Jan. 02	29 July 13	. . . .	Actg. D. M. and S. O. and Supdt., Dist. Jail, Madura, also Supdt. Medical School, Madura.
Sec.	**Scroggie,** William Reith John, C.I.E.	23 Jan. 76	29 Jan. 02	29 July 13	. . . .	———.
N.	**Gloster,** Thomas Henry, M.B. (206) (*aa*)	25 Aug. 72	29 Jan. 02	29 July 13	. . . .	———.
E.	**Knapp,** Herbert Henry George, M.D. (105, 170) (*i*)	3 Feb. 70	29 Jan. 02	29 July 13	. . . .	I. G. of Prisons, Burma.
Sec.	**Justice,** William Adolphus, M.B. (*j*)	26 May 70	29 Jan. 02	29 July 13	1 Feb. 12	Sany. Commsr., Madras.

(*a*) Date of promotion to Bt. Lt.-Colonel.

INDIAN MEDICAL SERVICE—*contd.*

Estabt.	Names.	Date of birth.	First Commn.	Date of rank.	Date of present appointment.	Remarks.
	Majors—*contd.*					
N.	**Barnardo,** Frederick Adolphus Fleming, C.I.E., M.B., F.R.C.S.E., M. R.C.P., M.A., (165, 193a) (206) (*Bt. Lt.-Col.*) (*Ty. Col.*) (K.)	4 June 74	26 July 02	26 Jan. 14; 1 Jan. 17(*); 1 May 18(*c*)	1 May 18	A. D. M. S., Embn. Staff, Bombay.
W.	**McPherson,** James, M.B., F.R.C.S.E.	16 Oct. 76	26 July 02	26 Jan. 14	. . . .	Residency Surgeon, Persian Gulf.
E.	**Franklin,** George Denne. M.B. (157, 182*b*)	28 July 77	26 July 02	26 Jan. 14	27 Aug. 17	*Lv., 1 mo. ex. India.*
E.	**Gourlay,** Charles Aikman, D.S.O., M.D.	29 Sep. 77	26 July 02	26 Jan. 14	. . . .	——.
F.	**Ross,** William Charles, M.B.	23 July 78	26 July 02	26 Jan. 14	. . . .	Sany. Commr., Bihar and Orissa.
N.	**Lloyd,** Robert Archer, D.S.O., M.D.	14 Apl. 78	26 July 02	26 Jan. 14	. . . .	——
E.	**Kunhardt,** John Conrad Gie	2 Aug. 75	26 July 02	26 Jan. 14	. . . .	Bombay.
N.	**Walker,** Ernest Alexander, M.B., F.R.C.S.E. (105, 140, 169).	20 Oct. 80	26 July 02	26 Jan. 14	. . . .	——.
E.	**Cook,** Lewis, M.B., F.R.C.S.	24 Oct. 76	26 July 02	26 Jan. 14	. . . .	8th Division.
E.	**Scott,** Leonard Bodley, M.D.	4 Mar. 77	26 July 02	26 Jan. 14	. . . .	O. C. Indn. Stn. Hospl.. Multan.
N.	**Davys,** Gerard Irvine, M.D.	11 Sep. 79	26 July 02	26 Jan. 14	. . . .	Food Analysis Laby., Kasauli.
W	**Betts,** Alfred John Vernon, M.B.	15 May 74	26 July 02	26 Jan. 14	. . . .	——.
N.	**Wilson,** Fredrick Ernest, M.B. (162, 169, 206)	21 Apl. 79	26 July 02	26 Jan. 14	. . . .	Specialist in advanced Operative Surgery, Indn. Troops, Bangalore.
W.	**Patton,** Walter Scott, M.B. (172 *a*)	7 Oct. 76	26 July 02	26 Jan. 14	. . . .	——. *Lv., ex. I. Exten to 3 Apl.18.*
N.	**Trafford,** William Leigh, M.B.	6 Dec. 77	26 July 02	26 Jan. 14	. . . .	——, *Lv., ex I., m.c., 7 Jan. 18. Exten to 17 Oct. 18.*
W.	**Paymaster,** Behramji Barjarji (206)	12 Apl. 75	26 July 02	26 Jan. 14	. . . .	O. C. Indn. Stn. Hospl., Fatehgarh.
N.	**Mackworth,** Norman Walter, M.B., F.R.C.S.E.	29 Sep. 78	22 July 02	26 Jan. 14	. . . .	O. C. Indn. Stn. Hospl., Bareilly.
N.	**Ross,** Henry, O.B.E., M.B., F.R.C.S.I. (*temp. Lt. Col.*) (170, 206)	29 June 77	66 July 02	26 Jan. 14	25 Apl. 18	Asst. Director-Genl., I. M. S. (Stores).
Sec.	**Forrest,** John, M.B. (71)	22 Oct. 76	26 July 02	26 Jan. 14	. . . .	O. C. Indn. Stn. Hospl., Roorkee.
W.	**Crossle,** Howard, M.D.	4 Dec. 79	26 July 02	26 Jan. 14	. . . .	Joint Civil Surgeon, Peshawar, I/c. Jail, Peshawar & M O., Mohmand [Militia.
Sec.	**Hirsch,** Leonard, F.R.C.S.E. (local Lt.-Col.)	14 Aug. 79	26 July 02	26 Jan. 14	. . . .	O. C., Marine Lines War Hospl., Bombay.
Sec.	**Maunsell,** Edward Charles Cecil, M.D.	15 Feb. 77	26 July 02	26 Jan. 14	. . . .	Staff Surgeon, Bangalore.
Sec.	**Illius,** John Warwick, F.R.C.S.E.	9 Oct. 75	26 July 02	26 Jan. 14	. . . .	——.
Sec.	**O'Keeffe,** Daniel Stanislaus Aloysius, M.B.	13 Nov. 77	26 July 02	26 Jan. 14	. . . .	——.
Sec.	**Cameron,** John Philip (125*b*)	22 May 79	26 July 02	26 Jan. 14	. . . .	O. C. Indn. Stn. Hospital, Delhi. *Sick lv. to 24 Feb. 18.*
E.	**Greig,** Alexander William (105)	25 Aug. 75	29 Jan. 02	29 Jan. 14	. . . .	——.
Sec.	**Harley,** Thomas William, M.B. (206)	10 Dec. 76	29 Jan. 02	29 Jan. 14	. . . .	Cantt. Genl. Hopl., Ambala.
W.	**Barron,** Robert Macpherson, D.S.O., (107, 170) *Bt. Lt.-Col.*)	24 Mar. 74	29 Jan. 02	29 Jan. 14; 1 Jan. 18(*)	. . . .	——.
N.	**Murray,** John Hanna, M.D. (*e*)	24 Mar. 73	1 Sep. 02	1 Mar. 14	. . . .	Civil and Jails, Port Blair, and Offg. S. M. O. and Civil Surgn., [Port Blair.
E.	**Mackie,** Frederic Percival, M.D., F.R.C.S.	19 Feb. 75	1 Sep. 02	1 Mar. 14	. . . .	——.
E.	**Connor,** Frank Powell, D.S.O., F.R.C.S.	16 Sep. 77	1 Sep. 02	1 Mar. 14	. . . .	Civil B. and Orissa, Temply. on Mily. duty.
E.	**Christophers,** Samuel Richard, C.I.E., M.B.	27 Nov. 73	1 Sep. 02	1 Mar. 14; *1 Jan. 11	. . . .	——.
E.	**Emslie-Smith,** Harry, M.B.	4 Dec. 76	1 Sep. 02	1 Mar. 14	. . . .	——.
E.	**Dutton,** Hugh Reginald	21 Nov. 75	1 Sep. 02	1 Mar. 14	. . . .	D. A. D. M. S. (Mobn.), 9th Secunderabad Divn.
E.	**Nesfield,** Vincent Blumhart, F.R.C.S. (92)	12 Oct. 79	1 Sep. 02	1 Mar. 14	. . . .	Jubbulpore *Lv., m. c.*
W.	**Brown,** Henry Martyn, M. B. (162, 206)	6 Mar. 75	1 Sep. 02	1 Mar. 14	. . . .	——.
N.	**Webb,** Hugh George Stiles (71, 169) (*f*)	13 Oct. 75	1 Sep. 02	1 Mar. 14; 28 Nov. 08	. . . .	O. C. Indn. Stn. Hospl., Dhond.
N.	**Jeudwine,** Wilfrid Wynne, C.M.G., M.D.	19 May 77	1 Sep. 02	1 Mar. 14	. . . .	——.
N.	**Young,** Thomas Charles McCombie, M.B. (161, 206)	26 June 80	1 Sep. 02	1 Mar. 14	. . . .	Sany. Commr., Assam.

N.	**Jolly**, George Adam, M.B., F.R.C.S.E. (*l*) (107, 139*a*, 161, 166, 169, 205(*a*) (*b*).	14 June 79	1 Sep. 02	1 Mar. 14	. . . .	——.
N.	**Brown**, Henry Coddington, C.I.E. M.B.	28 Oct. 76	1 Sep. 02	1 Mar. 14	. . . .	Asstt. Director, Central Research Institute, Kasauli.
N.	**Colinson**, Walter Julius, M.B. (206)	17 Sep. 77	1 Sep. 02	1 Mar. 14	. . . .	O. C. Labour Corps War Hospl., Dadar.
N.	**Dunn**, Cuthbert Lindsay (162, 165, 169, 206)(*g*) .	15 May 75	1 Sep. 02	1 Mar. 14	. . . .	Karachi. Specialist in the prevention of disease.
N.	**Melhuish**, Herbert Michael Henry (*Bt. Lt.-Col.*) (170).	22 Sep. 76	1 Sep. 02	{ 1 Mar. 14 / 3 June 17* }	. . . .	——.
Sec.	**Saigol**, Raghber Dayal, F.R.C.S.E. (105)	24 Oct. 76	1 Sep. 02	1 Mar. 14	. . . .	Ophthalmic Surgn., Supdt. of Jail and M.O., Govt. Plague and [Contagious Diseases Hospitals, Rangoon.
W.	**Bulteel**, Cecil Edward (94, 172*a*)	17 Aug. 77	1 Sep. 02	1 Mar. 14	. . . .	O. C., Indn. Stn. Hospl., Arangaon.
W.	**Lunham**, John Lumsden, M.B., F.R.C.S.I. (162, 172*a*, 206).	10 Mar. 79	1 Sep. 02	1 Mar. 14	. . . .	——.
Sec.	**Rogers**, Frederick Colin	1 Nov. 77	1 Sep. 02	1 Mar. 14	. . . .	M.C., 4th I. C. Section & Comdg., I. T. War Hospl., Bolarum.
N.	**Halliday**, Herbert, M.B. (170)	27 May 76	26 July 02	26 July 14	. . . .	——.
W.	**Shaw**, William Samuel Jagoe, M.D.(*a*)	Feb. 71	31 Jan. 03	31 July 14	. . . .	Supdt., Asylum & M. O., Central Prison Hospl., Yeravda, on deputation to Burma.
N.	**Broome**, Harold Holkar, M.B., F.R.C.S.	6 Mar. 75	31 Jan. 03	31 July 14	. . . .	Prof. of Operative Surgery K. E. Medl. College, and Medl. Supdt., Mayo Hospital, Lahore.
E.	**White**, Frederick Norman, C.I.E., M.D. . . .	11 June 77	31 Jan. 03	31 July 14	25 Apl. 18	Sany. Commr. with the Govt. of India, *s. p. t.*
E.	**Seymour**, Charles Gibbons (157)	9 Sep. 81	31 Jan. 03	31 July 14	. . . .	O. C., Indn. Stn. Hospl., Saugor.
N.	**Heron**, Davis, C.I.E., M.B., F.R.C.S.E. (124, 161, 206)	8 Jan. 78	31 Jan. 03	31 July 14	. . . .	Offg. M. O. & Vice-Consul, Seistan & Kain.
N.	**Keates**, Henry Crewe, M.D.	14 May 77	31 Jan. 03	31 July 14	. . . .	Karachi.
N.	**Reynolds**, Leethem, M.B.	29 Mar. 75	31 Jan. 03	31 July 14	. . . .	O. C., Indn. Stn. Hospl., Aurangabad.
Sec.	**Needham**, Richard Arthur, D.S.O., M.B. (*temp. Lt.-Col.*) (124).	31 July 77	31 Jan. 03	31 July 14	25 Oct. 18	Dy. Dir. Genl., I. M. S.
W.	**Goil**, Dwarka Prasad, M.B., F.R.C.S.E. (161, 170) .	1 Aug. 79	31 Jan. 03	31 July 14	. . . .	O. C., Indn. Stn. Hospl., Anandi.
Sec.	**Kirkwood**, James, M.B., F.R.C.S.E.	11 Dec. 79	31 Jan. 03	31 July 14	. . . .	D. M. & S. O. Anantapur, R. M. O., Genl. Hospl., Madras.
N.	**Lauddie**, Abdurrahman Khan, M.B. (94, 162) . .	18 June 77	1 Sep. 02	1 Sep. 14	. . . .	Lucknow.
W.	**Odlum**, William Henry (*c*)	2 Aug. 73	14 Nov. 00	†{ 1 Sep. 14 / *6 Dec. 07 }	. . . .	*Sick lv. to 2 Mar. 19.*
Sec.	**Ba-ket**, Maung, M.B.	4 Nov. 75	1 Sep. 02	1 Sep. 14	. . . .	Secunderabad.
Sec.	**Hingston**, Clayton Alexander Francis	6 May 77	1 Sep. 02	1 Sep. 14	. . . .	Profr. of Midwifery, Medl. Coll., & Supdt., Govt. Maternity Hospl., Madras.
N.	**Taylor**, Ernest Charles, M.B. (169) (*d*)	27 June 75	31 Jan. 03	31 Jan. 15	. . . .	War Hospl., Lahore Cant.
Sec.	**Whitmore**, Alfred, M.D. (105)	16 May 76	31 Jan. 03	31 Jan. 15	. . . .	O. C., Indn. Stn. Hospl., Tank.
E.	**Kelsall**, Robert, D.S.O., M.D. (105) (*h*) . . .	18 Oct. 75	31 Aug. 03	28 Feb. 15	. . . .	——.
E.	**Brodribb**, Charles Hildred, M.B.	30 Oct. 76	31 Aug. 03	28 Feb. 15	. . . .	Staff Surgeon "A," and M. O. i/c Civil Hospl., Secunderabad.
N.	**Macmillan**, John McCallum Anderson, M.B., F.R.C.S. (206).	14 Dec. 76	31 Aug. 03	28 Feb. 15	. . . .	——.
N.	**Gill**, Clifford Allchin (165, 170) (*Bt. Lt.-Col.*) . .	12 Aug. 78	31 Aug. 03	28 Feb. 15	. . . .	——.

(*aa*) Service as Plague Doctor counting as service for pension 8 Sep. 98 to 1 Apl. 02.
(*a*) " " " " " " 7 Oct. 99 to 5 Feb. 03.
(*b*) Overstayed leave from 16 Nov. to 20 Dec. 14.
(*c*) Service in South Africa counting as service for pension 18 Apl. to 13 Nov. 00.
(*d*) " " " " " " 10 Mar. 00 to 17 July 01.
(*e*) " " " " " " 30 Sep. 99 to 12 June 00.
(*f*) " " " " " " 23 May 01 to 31 July 02.
(*g*) " " " " " " *8 Feb. 00 to 9 Mar. 01, 11 Apl. 01 to 18 Oct. 01 and 26 Nov. 01 to 31 Aug. 02.
(*h*) " " " " " " 24 Mar. 00 to 9 Apl. 01 and 21 Mar. 02 to 2 July 03.
(*i*) Service as Plague Doctor counting as service for pension 18 Nov. 97 to 3 June 98.
(*j*) " " " " " 26 Jan. 98 to 31 Mar. 02.
(*k*) Service in South Africa as counting for pension 18 May 1900 to 15 Aug. 1900 and 4 March 01 to 22 Feb. 1902.
(*l*) Date of Ty. rank.

* Date of Brevet promotion.
† Date of rank.

INDIAN MEDICAL SERVICE—*contd.*

Estabt.	Names.	Date of birth.	First Commn.	Date of rank.	Date of present appointment.	Remarks.
	Majors—*contd.*					
W.	**Steel**, Richard Francis, M.B., F.R.C.S.E. (169, 205*a*, 206, 162).	28 Apl. 78	31 Aug. 03	28 Feb. 15	. . . .	Superintendent of Mahabelshvar.
Sec.	**Ingram**, Arthur Charles, M.D., M.R.C.P., D.P.H.	28 Mar. 77	31 Aug. 03	28 Feb. 15	. . . .	——.
Sec.	**Maconachie**, Gordon William, M.B. (206) (*Actg. Lt.-Col.*)	13 Nov. 81	31 Aug. 03	28 Feb. 15	. . . .	——.
Sec.	**Bradfield**, Ernest William Charles, O.B.E., M.B., F.R.C.S.E.	28 May 80	31 Aug. 03	28 Feb. 15	. . . .	——.
Sec.	**Brierley**, Charles Isherwood (169) (*c*)	24 Mar. 79	31 Aug. 03	28 Feb. 15	. . . .	Civil, Peshawar and Agency Surgeon, Khyber.
Sec.	**Hunter**, John Brown Dalzell, M.B., F.R.C.S.E. (94, 162, 169, 206) (*Actg. Lt.-Col.*).	27 Jan. 79	31 Aug. 03	28 Feb. 15	. . . .	——.
Sec.	**Harris**, Edward Temple, D.S.O., M.B.	27 Apl. 78	31 Aug. 03	28 Feb. 15	. . . .	Civil, Mandalay and in Addl. Charge, Central Jail, Mandalay.
N.	**Drake**, Hugh Basil	25 Aug. 79	30 Jan. 04	30 July 15	. . . .	——.
N.	**Hodgson**, Ernest Charles D.S.O.	22 Oct. 78	30 Jan. 04	30 July 15	. . . .	*Lv., 2 mos., July 18.*
N.	**McGillivray**, William Sim, M.D.	4 Jan. 82	30 Jan. 04	30 July 15	. . . .	O. C., Indn. Stn. Hospl., Baleli.
N.	**Gillitt**, William, C.I.E., M.D. (*temp. Lt.-Col.*)	26 Dec. 79	30 Jan. 04	30 July 15	. . . .	——.
N.	**Barber**, Charles Harrison, M.B. (169)	10 Mar. 77	30 Jan. 04	30 July 15	10 July 18	O. C., Troops, A. T. "Syria."
N.	**Tarr**, William, M.D., F.R.C.S.E. (94, 162, 165, 169, 206)(*a*).	21 June 76	30 Jan. 04	30 July 15	8 Oct. 18	Civil, Jubbulpore.
N.	**Watts**, Hugh, M.B.	18 Sep. 79	30 Jan. 04	30 July 15	. . . .	O. C., Indn. Stn. Hospl., Moradabad.
W.	**Jones**, Ivor Davenport, M.D. (206) (*Actg. Lt.-Col.*)	9 Sep. 79	30 Jan. 04	30 July 15	. . . .	Ahmednagar. Specialist in Advanced Operative Surgery.
W.	**Finlayson**, Walter Taylor, D.S.O.	14 July 77	30 Jan. 04	30 July 15	. . . .	Mhow.
W.	**McCowen**, William Thomas	29 Jan. 79	30 Jan. 04	{ 30 July 15 29 Aug. 12(*b*)	. . . .	——.
W.	**Stanger-Leathes**, Hugh Ellis	1 Nov. 78	30 Jan. 04	30 July 15	2 Oct. 18	D. A. D. M. S., Sn. Command.
Sec.	**Roberts**, Edmund Arthur, D.S.O. (162, 206)	15 Jan. 79	30 Jan. 04	30 July 15	. . . .	——.
Sec.	**Quirke**, Michael Joseph, M.B. (206)	5 Sep. 79	30 Jan. 04	30 July 15	. . . .	——.
Sec.	**Holmes**, John Morgan, M.B. (95)	5 Feb. 79	30 Jan. 04	30 July 15	. . . .	Health Officer, Delhi and Notified Area in addn., *s. p. t.*
Sec.	**White**, Maurice Forbes, M.B. (106, 206)	9 July 77	30 Jan. 04	30 July 15	. . . .	——.
W.	**Owens**, Terence Francis	8 Oct. 78	31 Aug. 03	31 Aug. 15	. . . .	Chemical Exr. to Govt. Burma and Police Surgn., Genl. Hospl., [Rangoon.
Sec.	**Harvey**, Alexander William Montgomery, M.B. (157) (*Bt. Lt.-Col.*).	4 Apl. 81	31 Aug. 03	{ 31 Aug. 15 3 June 17(*b*) }	. . . .	——.
N.	**Brayne**, William Frederick, M.B.	4 Mar. 76	30 Jan. 04	30 Jan. 16	. . . .	Peshawar.
W.	**Irani**, Merwan Sorab (161, 165, 206)	17 Nov. 78	30 Jan. 04	30 Jan. 16	. . . .	——.
W.	**Jones**, Seymour Whitworth	9 Sep. 79	30 Jan. 04	30 Jan. 16	. . . .	——.
W.	**Anderson**, John, M.B. (169) (*Actg. Lt.-Col.*)	17 Dec. 76	30 Jan. 04	30 Jan. 16	. . . .	——.
Sec.	**Hirst**, Geoffrey Gratrix (169)	24 Jan. 79	30 Jan. 04	30 Jan. 16	. . . .	P. A. to A. D. M. S. Meerut Divn.
E.	**Matson**, Horace Sidney, M.B. (105)	18 Oct. 79	1 Sep. 04	1 Mar. 16	. . . .	Poona. Specialist in Surgery.
E	**Stewart**, Francis Hugh, M.B.	11 Sep. 79	1 Sep. 04	1 Mar. 16	. . . .	——.
N.	**Proctor**, Alfred Henry, M.D., D.S.O.	21 Jan. 79	1 Sep. 04	1 Mar. 16	. . . .	——.
N.	**Wells**, Robert Tait, M.B. (162*a*)	24 Apl. 78	1 Sep. 04	1 Mar. 16	. . . .	——.
Sec.	**Macrae**, Ian Macpherson, M.B. (206) (*Actg. Lt.-Col.*)	2 Oct. 82	1 Sep. 04	1 Mar. 16	. . . .	——.
Sec.	**Shaw**, Charles Cecil Connock, M.D. (206)	4 June 77	1 Sep. 04	1 Mar. 16	. . . .	——.
Sec.	**Babington**, James William Herbert, M.D. (206)	9 Sep. 77	1 Sep. 04	1 Mar. 16	. . . .	——.
W.	**Peebles**, Alexander Spalding Mackie, M.D.	20 Nov. 76	1 Sep. 04	1 Mar. 16	. . . .	——.
W.	**Shettle**, Francis Broughton (172*a*)	25 Nov. 78	1 Sep. 04	1 Mar. 16	. . . .	O. C. Indn. Stn. Hospl. and Civil Surgn. Bannu.

W.	**Hamilton**, Arthur Francis, M.B., F.R.C.S. (116, 206)	13 May 80	1 Feb. 05	1 Aug. 16		Civil Surgeon and Supdt. B. J. Medical School, Poona, s.*p.t.*
K.	**McNeight**, Arthur Anderson, M.B.(L) (162, 172*a*, 206).	9 July 80	31 Jan. 05	11 July 10* / 1 Aug. 16†	13 Oct. 17	D. A. D. M. S., A. H. Qrs.
N.	**Buckley**, Hugh Clive, M.D., F.R.C.S.E. (149, 206)	31 Oct. 80	1 Feb. 05	1 Aug. 16		O/C. Memorial War Hospl., Cawnpore.
N.	**MacWatters**, Mathew Robert Cecil, M.B., F.R.C.S.	4 Jan. 78	1 Feb. 05	1 Aug. 16		Peshawar. Specialist in Advanced Operative Surgery.
N.	**Hamilton**, William Haywood, D.S.O., F.R.C.S. (161) (Bt.-Lt-Col.)	21 Dec. 80	1 Feb. 05	1 Jan. 18(*b*) / 1 Aug. 16		——.
W.	**Cunningham**, John, M.D.	9 July 81	1 Feb. 05	1 Aug. 16		Asst. Director, Central Research Institute, Kasauli, s.*p.t.*
W.	**Falk**, Herman, M.B.	18 Feb. 78	1 Feb. 05	1 Aug. 16		Lv., ex I., m. c. 6 mos.—July 18.
N.	**Cameron**, Alexander, M.B. (124, 169)	18 Nov. 79	1 Sep. 04	1 Sep. 16		——.
E.	**White**, Arthur Denham, M.B.	26 Feb. 79	1 Feb. 05	1 Feb. 17		1st Resident Surgn., Presy. Genl. Hospl., and 2nd Resdt. Surgn., [Calcutta, s.*p.t.*
N.	**Wilson**, Norman Methven	23 Feb. 81	1 Feb. 05	1 Feb. 17		Meerut. Specialist in Advanced Operative Surgery.
N.	**O'Neill**, John Stevenson, M.C., M.B. (157)	6 Dec. 77	1 Feb. 05	1 Feb. 17		——.
M.&B.	**Nealor**, William Stewart (105)	20 Apl. 80	31 Jan. 05	*31 Mar. 11 / †1 Feb. 17		Quetta. Specialist in Electrical Science and Opthalmology.
N.	**Reinhold**, Carl Henry, M.C., F.R.C.S.E. (172*a*)	4 Dec. 80	1 Sep. 05	Mar. 17		——.
N.	**Wright**, William Dundas, M.B. (206)	9 Nov. 80	1 Sep. 05	1 Mar. 17		——.
E.	**Whitamore**, Vernon Northwood	7 June 79	1 Sep. 05	1 Mar. 17		——.
W.	**Powell**, William Jackson, M.B. (172, 206)	17 Dec. 81	1 Sep. 05	1 Mar. 17		O. C., War Hospl., Ambala.
U. P.	**Pierpoint**, Harry William, F.R.C.S. (169, 206)	6 July 79	1 Feb. 06	1 Aug. 17	21 July 17	Asst. Dir. Genl., I. M. S. (Sany.)
U. P.	**Stevenson**, William David Henderson, C.I.E., M. D.	6 Dec. 78	1 Feb. 06	1 Aug. 17	14 Nov. 18	Civil and Jail, Moradabad.
U.&P.	**Cook**, Henry Patullo, M.B., F.R.C.S.E. (162, 169, 206)	13 Nov. 82	1 Feb. 06	1 Aug. 17		——.
L. P.	**Fraser**, William James, M.B., F.R.C.S.E. (162)	5 June 81	1 Feb. 06	1 Aug. 17		——.
L. P.	**Godson**, Charles Aubrey, M.C.	15 Nov. 80	1 Feb. 06	1 Aug. 17		——.
L. P.	**Lee**, Reginald Henry, M.B.	19 May 80	1 Feb. 06	1 Aug. 17		Meerut.
B. &A.	**Mearns**, William Anderson, M.B.	15 Nov. 79	1 Feb. 06	1 Aug. 17		——.
M.&B.	**Cullen**, William Barbour Alexander Kennedy M.D. (105, 162, 195, 206).	11 Mar. 81	1 Feb. 06	1 Aug. 17		
N.	**Palmer**, Cadwallader Edwards, M.B.	24 Dec. 80	1 Sep. 05	1 Sep. 17		O. C., Indn. Stn. Hospl., Dharmsala.
E.	**Newland**, Broderick Edward Middleton	19 Aug. 78	1 Sep. 05	1 Sep. 17		——.
E.	**Lack**, Lewis Albert Hodgkinson, M.B. (105)	27 Sep. 80	1 Sep. 05	1 Sep. 17		O. C., Indn. Stn. Hospl., Almora.
L.P.	**McDonald**, Edgar John Cecil	22 Sep. 80	1 Sep. 05	1 Sep. 17		O. C., Ind.n Stn. Hospl., Fyzabad.
E.	**Boyd**, John Francis	17 Sep. 81	1 Sep. 05	1 Sep. 17		Lv., m. c. 2 mos. 25 July 18. *Extn. 10 Nov. 18 to 31 Dec. 18.*
E.	**Sodhi**, Narendra Singh, M.C., (161, 204)	9 Aug. 82	1 Sep. 05	1 Sep. 17		Lv., ex I., m. c., 6 mos.—Aug. 18.
Sec.	**Gray**, William Cowan, M.B.	4 Oct. 78	1 Sep. 05	1 Sep. 17		——.
B.&A.	**Gharpurey**, Khandu Ganpatrao (162, 205(*a*), 206)	28 July 80	1 Feb. 06	1 Feb. 18		
L. P.	**Mills**, Percy Strickland, M.B. (162, 206)	28 May 83	1 Feb. 06	1 Feb. 18		Dy. M. S. K. (Bombay).
L. P.	**FitzGerald**, Desmond Charles Villiers, M.C.	24 Nov. 82	1 Feb. 06	1 Feb. 18		O. C., Indn. Stn. Hospl., Manmad.
L. P.	**Kennedy**, Robert Siggins, D.S.O., M.C., M.B. (199, 206)	8 Dec. 82	1 Feb. 06	1 Feb. 18		——.
L. P.	**Higham**, Bernard, M.B. (172*a*, 206)	14 Dec. 80	1 Feb. 06	1 Feb. 18		*Lv., ex I., m. c.*
M.&B.	**Heffernan**, Patrick, M.B. (*d*)	13 May 78	1 Feb. 06	1 Feb. 18		
U. P.	**Wernicke**, Frank Phillips, M.B., D.S.O.	19 Apl. 81	1 Feb. 06	1 Feb. 18		O. C. Marine Lines Conv. Section, Bombay.
B.&A.	**Hutchison**, Henry Stewart, M.B. (167, 169)	8 May 83	1 Feb. 06	1 Feb. 18		——.
M.&B.	**Croly**, Robert George Gibbon, M.B.	16 Sep. 81	1 Feb. 06	1 Feb. 18		——.
M.&B.	**Crump**, Stanley Trefusis	13 Nov. 78	1 Feb. 06	1 Feb. 18		——.
M.&B.	**Skinner**, James MacGregor, M.B.	14 Apl. 80	1 Feb. 06	1 Feb. 18		——.
L. P.	**O'Brien**, Charles Richard, M.B.	4 Apl. 79	1 Feb. 06	1 Mar. 18		S. M. O., Miranshah.
L. P.	**Hume**, Norman Haliburton, M.B.	8 Oct. 80	1 Feb. 06	1 Mar. 18		——.
B.&A.	**Church**, Duncan Macdonald Cochrane, M.B.	28 Dec. 80	1 Feb. 06	1 Mar. 18		

(*a*) Service in South Africa counting as service for pension 24 Mar. 00 to 27 Mar. 01.
(*c*) " " " " " 18 Nov. 01 to 8 Nov. 03.
(*d*) On tempy. non-effective list from 3 July 18.

(*b*) Date of Brevet promotion.
* Date of joining I.M.S.
† Date of rank.

INDIAN MEDICAL SERVICE—*contd.*

Comd. or area.	Names.	Date of birth.	First Commn.	Date of rank.	Date of present appointment.	Remarks.
	Majors—*concld.*					
U. P.	**Taylor**, John, D.S.O., M.D. (170)	14 Feb. 84	1 Sep. 06	1 Mar. 18	. . . .	——.
U. P.	**Stewart**, Alexander Dron, M. B., F.R.C.S.E. (157)	22 June 83	1 Sep. 06	1 Mar. 18	10 May 18	D. A. D. M. S., (Sany.) Bombay Bde.
U. P.	**Bott**, Robert Henry, M.B., F. R. C. S. (157, 172*a*)	22 Aug. 82	1 Sep. 06	1 Mar. 18	. . . .	Profr. of Surgery, K. E. Medl. College, Lahore, s. *p. t.*
L. P.	**Morison**, John, M.B. (206)	6 Nov. 79	1 Sep. 06	1 Mar. 18	. . . .	——.
U. P.	**Haughton**, Samuel George Steele, M.B. (169)	3 Sep. 83	1 Sep. 06	1 Mar. 18	. . . .	
U. P.	**Cragg**, Francis William, M.D. (172*a*, 206)	19 May 82	1 Sep. 06	1 Mar. 18	6 May 18	D. A. D. M. S. (Sany.) Ahmadnagar Bde.
L. P.	**Millar**, George McGregor, M.B. (172*a*) (205*a*, 206)	12 Sep. 80	1 Sep. 06	1 Mar. 18	. . . .	——.
B.&A.	**Thorburn**, Harold Hay, C.I.E., M.B. (169)	1 Dec. 82	1 Sep. 06	1 Mar. 18	. . . .	——.
B.&A.	**Smalley**, James	20 Sep. 78	1 Sep. 06	4 Apl. 18	. . . .	——.
L. P.	**Salisbury**, Francis Hugh, M.B.	12 Mar. 80	1 Sep. 06	4 Apl. 18	. . . .	Labour Corps War Hospl., Dadar.
M.&B.	**Berkeley-Hill**, Owen Alfred Rowland, M.D.	22 Dec. 79	2 Feb. 07	2 Aug. 18	. . . .	On deputation with Commission of Inquiry into Administration of Languno in East Africa
L. P.	**Harnett**, Walter Lidwell, M.B., F.R.C.S. (206)	7 June 79	2 Feb. 07	2 Aug. 18	. . . .	——.
L. P.	**Sandes**, John Drummond, M.B.	5 Mar. 83	2 Feb. 07	2 Aug. 18	26 Apl. 17	Surgn. to H. E. the Govr. of Bengal.
U. P.	**Napier**, Alexander Harper, M.B.	28 Sep. 80	2 Feb. 07	2 Aug. 18	. . . .	——.
U. P.	**Grisewood**, Arnold Egbert, M.B.	5 Feb. 82	2 Feb. 07	2 Aug. 18	. . . .	——.
L. P.	**Graham**, David Livingstone, M.B., F.R.C.S.E. (169)	1 Sep. 82	2 Feb. 07	2 Aug. 18	. . . .	——.
U.P.	**Nicholson**, Roger Brighouse, M.C. (172*a*)	7 Nov. 80	2 Feb. 07	2 Aug. 18	12 Mar. 18	D. A. D. M. S. (Distn.) Bombay.
M.&B.	**Cruickshank**, James Alexander, M.C., M.D.	15 Jan. 83	2 Feb. 07	2 Aug. 18	20 Aug. 16	Enteric Depôt, Parel, Specialist in Prevention of Disease.
M.&B	**Phillips**, John Alfred Steele (Bt.-Maj.) (95, 172*a*)	6 Oct. 82	2 Feb. 07	2 Aug. 18 2 Aug. 18	. . . .	
L. P.	**Rennie**, Patrick Manson, M.B.	8 Aug. 84	2 Feb. 10	2 Aug. 18	. . . .	Karachi.
B.&A.	**Inman**, Harold Mundee	9 Apl. 80	2 Feb. 07	2 Aug. 18	. . . .	——.
L. P.	**Munro**, Edmund Brodie, M.B. (15*a*, 206)	26 Oct. 82	2 Feb. 07	7 Sep. 18	. . . .	——.
B & A	**Fawcett**, Frederick O'Dowda (172*a*)	5 Dec. 82	2 Feb. 07	7 Sep. 18	. . . .	*Lv. ex. I., m. c. 6 mos.—Sep. 18.*
U. P.	**Cross**, Claude Harold (169, 206)	5 Oct. 82	1 Sep. 06	1 Sep. 18	. . . .	*Lv., m. c. 2 mos. 12 days. 18 Sep. 18 Extn. 1 m. 30 Nov. 18.*
U. P.	**Chambers**, Robert Alexander, M.B. (94, 157, 162,	1 Oct. 81	1 Sep. 06	1 Sep. 18	. . . .	——.
L. P.	**Simpson**, Norman Skinner	1 Jan. 81	1 Sep. 06	1 Sep. 18	. . . .	Aurangabad.
U. P.	**Singh**, Kanwar Shumshere	21 June 79	1 Sep. 06	1 Sep. 18	. . . .	Civil and Jail, Jullundur.
M.&B.	**Leslie**, Andrew Smith, M.B. (105, 206)	24 June 82	1 Sep. 06	1 Sep. 18	. . . .	Cawnpore.
M.&B.	**Lorimer**, Alexander Patrick Gordon, M.B.	9 Dec. 80	1 Sep. 06	1 Sep. 18	. . . .	——.
M.&B.	**Scott**, Herbert Bodley (105)	14 Dec. 82	1 Sep. 06	1 Sep. 18	. . . .	——.
M.&B.	**Fraser**, Frederick Charles	15 Feb. 80	1 Sep. 06	1 Sep. 18	. . . .	Ag. P. A. to the S.G., Bombay.
	Captains.					
L. P.	**Malcomson**, Greer Edmund, M.D.	30 Sep. 78	1 Feb. 06	1 Mar. 09*a* 22 Feb. 15*b*	1 Mar. 17	Kohat. Specialist in Prevention of Disease.
U. P.	**Williams**, William Percival Gould, M.B. (206)	30 Nov. 81	2 Feb. 07	2 Feb. 10	. . . .	O. C. Indn. Stn. Hospl., Montgomery.
M.&B.	**Mehta**, Siavax Byramjee, F.R.C.S.E. (162, 169, 204, 206)	15 Jan. 81	2 Feb. 07	2 Feb. 10	. . . .	9th (Secunderabad) Division.
U. P.	**Holroyd**, Gilbert, M.B.	17 Sep. 80	2 Feb. 07	2 Feb. 10	. . . .	5th (Mhow) Division. *Lv., m. c. 2 mos.—Nov. 18*
M.&B.	**Tarapore**, Pheroze Kharsedji (105, 161, 205*a*)	12 Nov. 79	2 Feb. 07	2 Feb. 10	. . . .	——.
B. &A.	**Kamat**, Dwarkanath Dharmaji	29 Feb. 80	2 Feb. 07	2 Feb. 10	. . . .	5th (Mhow) Division.
L. P.	**Chuckerbutty**, Sites Chunder	26 Dec. 81	2 Feb. 07	2 Feb. 10	. . . .	Offg. Civil, Assam.

L. P.	**Babonau,** Alexander Frederick, M.B. (*temp. Major*) (162, 206)	5 Oct. 82	2 Feb. 07	2 Feb. 10	16 May 18	D.A.D.M.S. (Mobn.), 4th Quetta Divn.
L. P.	**James,** Joseph Frain, M.B.	1 Nov. 82	1 Sep. 06	7 Mar. 10	. . . .	Karachi.
M.&B.	**Williams,** William Edward Rees, M.B.	27 May 83	2 Feb. 07	7 Mar. 10	9 Jan. 17	——.
B.&A.	**Bhathena,** Sorabji Jamasji (95, 172*a*)	24 Jan. 79	2 Feb. 07	7 Mar. 10	. . . .	Nasirabad.
U. P.	**Acton,** Hugh William	26 June 83	27 July 07	27 July 10	. . . .	Malana Depôt. Dagshai.
L. P.	**Green-Armytage,** Vivian Bartley, M.D.	14 Aug. 82	27 July 07	27 July 10	. . . .	R. M. O., Med. Coll. Hospl., Calcutta. s. *p. t.*
U. P.	**Dickson,** Arthur Norman, M.C., M.D.	27 Nov. 80	27 July 07	27 July 10	. . . .	——.
B.&A.	**Coullie,** Alexander Glover, M.B., F.R.C.S.E.	27 May 81	27 July 07	27 July 10	. . . .	Profr., Hygiene and Bacteriology, Medl. Coll., & R. M. O., Genl. Hospl., Madras. s. *p. t. Lv., ex I., m.c., 12 mos., 1 Nov. 17. Exten. 3 mos.*
M.&B.	**Russell,** Alexander James Hutchison, M.D. (206)	30 Aug. 82	27 July 07	27 July 10	. . . .	
M.&B.	**Wright,** Robert Ernest, M.B. (*Bt.-Maj.*)	2 Oct. 84	27 July 07	{ 29 Oct. 15*c* / 27 July 10 }	21 Aug. 18	Enteric Depôt, Parel.
U. P.	**Rai,** Dewan Hakumat, M.C., M.B. (94, 162, 169, 204)	15 May 83	27 July 07	27 July 10	. . . .	——.
M.&B.	**Riddell,** William Hunter, M.C., M.B. (127)	22 Sep. 82	27 July 07	27 July 10	. . . .	*Lv., ex I., m.c., 6 mos., —Oct. 17. Extn. to 3 Oct. 18.*
M.&B.	**Barker,** Frederic Allan, M.B.	17 Oct. 82	27 July 07	27 July 10	. . . .	——.
U. P.	**Coutts,** Duncan, M.B.	7 Nov. 82	30 July 06	{ *29 July 13 / †30 July 09 }	. . . .	Karachi.
B.&A.	**Thomas,** Arnold Newall, D.S.O., M.B.	7 Aug. 80	27 July 07	27 July 10	. . . .	——.
B.&A.	**Knowles,** Robert (206)	30 Oct. 83	1 Feb. 08	1 Feb. 11	11 Dec. 16	Director, Pasteur Institute & Research Laboratory, Shillong.
L. P.	**Brierley,** Wilfrid Edward, M.B., F.R.C.S. (206)	17 Jan. 81	1 Feb. 08	1 Feb. 11	. . . .	D. I. Khan.
L. P.	**Lapsley,** James Burne, M.B., M.C. (162, 206)	16 Aug. 84	1 Feb. 08	1 Feb. 11	. . . .	
L. P.	**Shorten,** James Alfred, M.B. (161, 206)	21 Nov. 82	1 Feb. 08	1 Feb. 11	. . . .	R. M. O., Med. Coll. Hospl., Calcutta. s. *p.*
U. P.	**Sewell,** Robert Beresford Seymour	5 Mar. 80	1 Feb. 08	1 Feb. 11	. . . .	O. C. Indn. Stn. Hospital, Bikaner.
M.&B.	**Fielding,** Charles Henry, M.B. (105)	8 Nov. 80	1 Feb. 08	1 Feb. 11	. . . .	——.
M.&B.	**Watson,** William Linton	21 June 82	1 Feb. 08	1 Feb. 11	. . . .	——.
B. &A.	**Hanafin,** John Berchmans, F.R.C.S.I.	10 Aug. 82	29 July 07	{ †29 July 10 / *15 July 14 }	. . . .	
B.&A.	**Simpson,** Wilfrid James, M.B. (162, 206)	4 Feb. 85	1 Feb. 08	1 Feb. 11	. . . .	——.
L. P.	**Stevenson,** Frederick, M.B. (149)	1 June 84	1 Feb. 08	1 Feb. 11	. . . .	——.
M.&B.	**Barnett,** James William, M.B. (95, 162)	7 May 85	1 Feb. 08	1 Feb. 11	. . . .	——.
M.&B.	**Middleton-West,** Stephen Harold, M.C., M.B.	17 May 80	1 Feb. 08	1 Feb. 11	. . . .	Cawnpore.
U. P.	**Puri,** Madan Lal (95, 162, 169, 205*a*)	13 Apl. 85	1 Feb. 08	1 Feb. 11	. . . .	——.
U. P.	**Pal,** Satya Charan (98, 205*a*)	24 Dec. 81	1 Feb. 08	1 Feb. 11	. . . .	Rawalpindi.
U. P.	**Townsend,** Reginald Stephen, M.C., M.B.	10 Nov. 82	1 Feb. 08	1 Feb. 11	4 May 18	E. M. D. Embarkation Staff, Bombay.
U. P.	**Lloyd,** Reginald Broughton, M.B. (66)	31 Aug. 81	1 Aug. 08	1 Aug. 11	. . . .	——.
L. P.	**Munro,** Archibald Campbell, M.D.	16 Jan. 86	1 Aug. 08	1 Aug. 11	. . . .	——.
U. P.	**Chopra,** Ram Nath, M.B. (95, 161, 169, 204, 206)	17 Aug. 82	1 Aug. 08	1 Aug. 11	. . . .	2nd (Rawalpindi) Division.
U. P.	**Tresidder,** Alfred Geddes, M.D.	15 Dec. 81	1 Aug. 08	*1 Aug. 11	. . . .	Staff Surgeon, Poona. Specialist in Midwifery and Diseases of [Women and Children.
B.&A.	**Jolly,** Gordon Gray, M.B. (94, 100, 172*a*)	6 Apl. 86	1 Aug. 08	1 Aug. 11	. . . .	——.
M.&B.	**Stott,** Hugh, O. B. E., M.B. (71)	18 July 84	1 Aug. 08	1 Aug. 11	. . . .	Specialist in Prevention of Disease, Civil. Medical, Tochi and M. O. [N. W. Militia.
L. P.	**McNeill,** Alister Argyll Campbell, M.B.	21 Nov. 84	1 Aug. 08	1 Aug. 11	. . . .	——.
M.&B.	**Khan,** Abdus Sattar (162, 204)	16 May 82	1 Aug. 08	1 Aug. 11	. . . .	——.
M.&B.	**Graham,** George Frederick, M.D. (172*a*, 206)	11 Jan. 85	1 Aug. 08	1 Aug. 11	. . . .	O. C. Indn. Stn. Hospl., Shillong.
U. P.	**Wadia,** Maneck Dhunjishaw (153, 155, 161*a*, 162, 165, 172*a*, 204, 206) (*l*)	5 July 83	1 Aug. 08	1 Aug. 11	. . . .	M. O. I/c. P. of War Camp, Sumerpur and O. C. Ind. Stn. Hospital, [Erinpura.

† Date of rank. * Date of joining I.M.S.

(*a*) Period from 2nd Sep. 1913 to 21st Feb. 1915 not to reckon for promotion. (*b*) Date of re-admission into the I.M.S.

(*c*) Date of Brevet promotion.

INDIAN MEDICAL SERVICE—*contd.*

Comd, or area.	Names.	Date of birth.	First Commn.	Date of rank.	Date of present appointment.	Remarks.
	Captains—*contd.*					
U. P.	**Murison,** Taylor David	7 Nov. 82	1 Aug. 08	1 Aug. 11	. . . .	Karachi, Specialist in Mental Science.
B.&A	**Vazifdar,** Sohrab Shapoorji (162, 169, 203)	1 Aug. 83	1 Aug. 08	1 Aug. 11	. . . .	——.
M.&B.	**Nelson,** John Joseph Harper, M.C., M.D., F.R.C.S.E.	12 Jan. 82	1 Aug. 08	1 Aug. 11	. . . .	——.
U. P.	**Phipson,** Edward Selby, D.S.O., M.B. (205*a*)	10 Mar. 84	1 Aug. 08	1 Aug. 11	14 Oct. 18	H. O. Simla, & *p. t.*
L. P.	**Smith,** Fleet Floyd Strother, M.D.	30 July 83	1 Aug. 08	1 Aug. 11	. . . .	Dehra Dun.
M.&B.	**Symes,** Arthur Jessop, M.B.	12 Apl. 83	1 Aug. 08	1 Aug. 11	. . . .	
B.&A.	**Little,** Gerald Lewis Colhoun, M.B., F.R.C.S. (Edn.) (*a*)	2 Oct. 83	1 Aug. 08	1 Aug. 11	18 Dec. 15	Staff Surgeon, Quetta.
L. P.	**Boyd,** Thomas Crawford, F.R.C.S.I. (172*a*)	26 Aug. 86	1 Aug. 08	1 Aug. 11	. . . .	Lahore Cantt.
M.&B	**James,** Gwilym Gregory, M.B. (Bt.-Maj.)	24 June 84	30 Jan. 09	{ 1 Jan. 18*b* 30 Jan. 12 }	. . . .	——.
L. P.	**Keyworth,** William David, M.B. (169, 206)	31 Jan. 81	30 Jan. 09	30 Jan. 12	. . . .	——.
U. P.	**Gale,** Berkeley, M.B. (169)	5 Feb. 84	30 Jan. 09	30 Jan. 12	. . . .	X-Ray Institute, Dehra Dun.
B.&A.	**Horne,** John Howard, M.B. (206)	6 Sep. 84	30 Jan. 09	30 Jan. 12	. . . .	——.
M.&B.	**King,** Harold Holmes, M.B.	17 Aug. 84	30 Jan. 09	30 Jan. 12	. . . .	——.
U. P.	**Flowerdew,** Richard Edward, M.B. (206) (*temp. Maj.*)	26 Sep. 86	30 Jan. 09	30 Jan. 12	1 Aug. 18	D. A. D. M. S., (Sany.), Frontier Brigades.
M.&B.	**Shand,** John Glendinning Bryden, M.B.	11 July 81	30 Jan. 09	30 Jan. 12	. . . .	
L. P.	**Lee,** Alfred John, M.B. (15*b*)	6 Mar. 82	30 Jan. 09	30 Jan. 12	. . . .	——.
L. P.	**Sarkar,** Sureswar (98)	12 Dec. 80	1 Aug. 08	4 June 12	. . . .	Lv., m. c., 6 mos. 13 days, 14 Jan. 18. Extn. 6 mos.
U. P.	**Smith,** Charles Harold, M.D., F.R.C.S.	19 June 83	31 July 09	31 July 12	. . . .	Specialist in Advanced Operative Surgery. O. C. Frontier War Hospital, Bannn
U. P.	**Dick,** Alan MacDonald, M.B., F.R.C.S. (206) (*Bt.-Maj.*)	30 July 84	31 July 09	31 July 12	. . . .	——.
L. P.	**Evans,** Thomas John Carey, M.C., F.R.C.S. (*Bt.-Maj.*)	26 June 84	31 July 09	31 July 12	. . . .	——.
B.&A.	**Holgate,** Maurice James, M.B. (95, 169)	15 May 84	31 July 09	31 July 12	. . . .	——.
L. P.	**Bomford,** Trevor Laurence, M.B.	8 Feb. 85	31 July 09	31 July 12	. . . .	——.
M.&B.	**Lynn,** Graham Rigby, D.S.O., M.B.	21 Jan. 85	31 July 09	31 July 12	. . . .	——.
U. P.	**Mackenzie,** Louis Hope Lovat, M.B. (*Bt.-Maj.*) (169, 174, 206).	1 Nov. 81	31 July 09	{ 31 July 12 3 June 17(*b*) }	. . . .	——. ——.
L. P.	**Eckstein,** John McDougall	23 Nov. 81	31 July 09	31 July 12	. . . .	Indian Troops War Hospital, Ambala.
B.&A.	**Jack,** William Andrew Morton, M.B.	13 July 83	31 July 09	31 July 12	. . . .	——.
L. P.	**Anderson,** Alexander Charles	12 Mar. 83	31 July 09	31 July 12	. . . .	——.
B. &A.	**Cooper,** Duncan Gordon, M.B.	24 May 84	31 July 09	31 July 12	. . . .	——.
M.&B.	**Forsyth,** William Leonard, M.B.	1 May 87	31 July 09	31 July 12	. . . .	Supdt., European Lunatic Asylum, Ranchi. (Tempy.)
L. P.	**Thakur,** Keshav Sadashiv (169)	2 Apl. 85	31 July 09	31 July 12	. . . .	Lahore Cantt.
L. P.	**Rahman,** Mohamed Abdur (162 206) (*g*) (*temp. Maj.*)	22 Oct. 81	31 July 09	31 July 12	. . . .	——.
M.&B.	**Hodge,** Edward Humfrey Vere, M.B.	28 Aug. 83	31 July 09	31 July 12	20 June 18	Attd. Medl. Branch, Army Hd. Qrs.
U. P.	**Burke,** Gerald Tyler, M.B. (206)	31 May 82	31 July 09	31 July 12	. . . .	——.
L. P.	**Gibson,** Herbert Robert Burnett, M.B.	12 June 85	31 July 09	31 July 12	. . . .	——.
M.&B.	**Nicholson,** Mark Alleyne, M.B.	10 June 85	31 July 09	31 July 12	. . . .	——.
B.&A.	**Kureishi,** Mozaffer Din Ahmed (161, 204, 206)	14 Oct. 84	30 Jan. 09	24 Oct. 12	. . . .	——.

B.&A.	**Kolapore,** Framroze Jamsetjee (204, 205*a*)	13 Oct. 82	29 Jan. 10	29 Jan. 13	. . . . .	Kohat.
U. P.	**Kennedy,** Edward Galwey, M.B.	20 June 87	29 Jan. 10	29 Jan. 13	. . . . .	——.
L. P.	**MacGregor,** Robert Forrester Douglas, M.C., M.B. (157, 199, 206)	5 Jan. 85	29 Jan. 10	29 Jan. 13	. . . . .	——.
M.&B.	**Sheppard,** Arthur Lewin, M.B.	30 Apl. 82	29 Jan. 10	29 Jan. 13	. . . . .	——.
B.&A.	**Gilroy,** Paul Knighton, M.B.	7 June 85	29 Jan. 10	29 Jan. 13	. . . . .	Special duty, Persia.
M.&B.	**Kernahan,** Joseph Arthur Alexander	11 Sep. 82	29 Jan. 10	29 Jan. 13	. . . . .	Special duty, Persia.
L. P.	**Irvine,** Maurice Lionel Corrie, M.B.	7 June 86	29 Jan. 10	29 Jan. 13	. . . . .	Bareilly.
L. P.	**Kirwan,** Ernest William O'Gorman, M.B.	18 May 87	29 Jan. 10	29 Jan. 13	. . . . .	——.
B.&A.	**Macdonald,** John Valentine, M.C., M.B. (155*a*)	14 Feb. 85	29 Jan. 10	29 Jan. 13	. . . . .	——.
U. P.	**Duncan,** George Lawrence, M.B.	6 Feb. 84	29 Jan. 10	29 Jan. 13	. . . . .	Peshawar.
U.P.	**Palit,** Anath Nath, F.R.C.S.E.	14 Oct. 83	29 Jan. 10	29 Jan. 13	. . . . .	2nd-in-Command, Indn. Stn. Hospl., Sitapur.
M.&B.	**Robson,** Hubert Alan Hirst, M.B.	17 Apl. 85	29 Jan. 10	29 Jan. 13	. . . . .	Derajat, Bde.
U. P.	**Howlett,** Cecil George, M.B.	30 May 82	29 Jan. 10	29 Jan. 13	. . . . .	——.
U. P.	**Bharucha,** Phirozshah Byramji, D.S.O., F.R.C.S.	21 Feb. 86	30 July 10	30 July 13	. . . . .	Civil, Chaman.
U. P.	**Hingston,** Richard William George, M.B.	17 Jan. 87	30 July 10	30 July 13	. . . . .	Dharamsala.
M.&B.	**Newton-Davis,** Charles, M.C., M.B.	18 Dec. 82	30 July 10	30 July 13	. . . . .	——.
U. P.	**Clifford,** Reginald Charles, D.S.O., M.C.	17 Sep. 84	30 July 10	30 July 13	. . . . .	——.
M.&B.	**Newcomb,** Clive, M.B.	3 Oct. 82	30 July 10	30 July 13	. . . . .	——.
L. P.	**Shortt,** Henry Edward, M.B.	15 Apl. 87	30 July 10	30 July 13	. . . . .	——.
U. P.	**Hughes,** Thomas Arthur, M.B. (169, 206)	6 Dec. 84	30 July 10	30 July 13	. . . . .	——.
B. & A.	**Khan,** Latafat Husain	27 Aug. 83	30 July 10	30 July 13	. . . . .	——.
L. P.	**Herrick,** Robert deStretton Berkeley, D.S.O.	23 Dec. 84	30 July 10	30 July 13	. . . . .	——.
L. P.	**Batra,** Hargobind Lal, M.C. (204, 206)	14 June 85	30 July 10	30 July 13	. . . . .	——.
M.& B.	**Purvis,** Murray, M.B.	16 Sep. 86	30 July 10	30 July 13	. . . . .	——.
M.& B.	**Taylor,** Duncan McNab, M.B.	6 Aug. 86	30 July 10	30 July 13	. . . . .	——.
L. P.	**Walker,** Walter Oliphant, M.B.	12 June 86	30 July 10	30 July 13	. . . . .	——.
L. P.	**Scott,** John, D.S.O., M.B.	28 June 86	28 Jan. 11	28 Jan. 14	. . . . .	——.
M.&B.	**Hay,** Frederick William, M.B.	15 May 86	28 Jan. 11	28 Jan. 14	. . . . .	*Lv., m. c., ex I. Extn. up to 16 Jan. 19.*
B.& A.	**Tate,** George (153)	10 Apl. 84	28 Jan. 11	28 Jan. 14	. . . . .	Madras.
						Tempy. Dy. M. S. K., Madras.
M.&B.	**Hepworth,** Sidney Milverton, M.B.	11 June 86	28 Jan. 11	28 Jan. 14	. . . . .	——.
L. P.	**Sen,** Jyoti Lal, M.C., M.B. (205*a*)	22 Oct. 83	28 Jan. 11	28 Jan. 14	. . . . .	Benares.
M.&B.	**Pandalai,** Krishen Gopinath, M.B. (195)	20 May 85	28 Jan. 11	23 Apl. 14	. . . . .	Secunderabad.
M.&B.	**Wood,** Charles Albert, M.C., M.B. (95)	16 May 86	28 Jan. 11	23 Apl. 14	. . . . .	——.
L. P.	**Cormack,** Harry Slater, M.C., M.B., F.R.C.S.E.	6 May 85	28 Jan. 11	28 July 14	. . . . .	Cawnpore.
B.& A.	**Connellan,** Percival Sandys	17 July 85	29 July 11	29 July 14	. . . . .	——.
U. P.	**Sinton,** John Alexander, M.B.	2 Dec. 84	29 July 11	29 July 14	. . . . .	——.
M.&B	**Armstrong,** Edward Randolph, M.D. (*temp. Maj.*)	28 May 85	29 July 11	29 July 14	15 June 16	D. A. D. M. S., Army Hd. Qrs.
U. P.	**Murphy,** Denis FitzGerald, M.B.	8 Nov. 83	29 July 11	29 July 14	. . . . .	—— Dardoni.
L. P.	**Stocker,** Charles James, M.C., M.B.	24 June 86	29 July 11	29 July 14	. . . . .	——.
B.&A.	**Doyle,** Eric Edward (94, 169, 206) (*temp. Maj.*)	18 Sep. 86	29 July 11	29 July 14	28 Mar. 17	D. A. D. M. S. (Mobn.) 1st (Peshawar) Divn.
B.&A.	**Plumptre,** Cyril Macdonald	27 Dec. 84	29 July 11	29 July 14	. . . . .	——.
L. P.	**Duncan,** Archibald Wallace	11 July 84	29 July 11	29 July 14	. . . . .	——.
M.&B.	**Brandenbourg,** Leon Francis, M.B. (95)	11 Dec. 84	29 July 11	29 July 14	. . . . .	——.
L. P.	**Barker,** Henry Lewis, M.D.	8 Aug. 85	29 July 11	29 July 14	. . . . .	4th Division.
L.P.	**Penny,** Edward Austen, M.B. (*temp. Maj.*)	27 Dec. 84	29 July 11	29 July 14	1 Nov. 18	D. A. D. M. S. (Mobn.) Meerut Division.

(*a*) On temporary Half-Pay List from 15th Nov. 1914 to 4th Dec. 15.
(*b*) Date of Bt. promotion.

INDIAN MEDICAL SERVICE—*contd.*

Comd. or area.	Names.	Date of birth.	First Commn.	Date of rank.	Date of present appointment.	Remarks.
	Captains—*contd.*					
L. P.	**Harland**, George Blenkhorn, M.D.	27 Nov. 83	29 July 11	29 July 14	. . . .	Meerut. Temp. Specialist in Dermatology.
L. P.	**Goss**, Edward Slade, M.C.	25 Sep. 83	29 July 11	22 Oct. 14	. . . .	——.
B. & A.	**Candy**, Ronald Herbert, M.B.	13 June 86	27 Jan. 12	27 Jan. 15	. . . .	Deesa.
M. & B.	**Veale**, Philip John, M.B.	25 Sep. 86	27 Jan. 12	27 Jan. 15	. . . .	——.
B. & A.	**Bharucha**, Jamasji Cursetji 169	9 July 87	27 Jan. 12	27 Jan. 15	. . . .	Karachi.
L. P.	**Hingston**, Henry, M.B. (206)	22 Nov. 85	27 Jan. 12	27 Jan. 15	. . . .	——.
U. P.	**Cursetjee**, Heerajee Jehangir Manockjee, M.B. (206)	14 Aug. 85	27 Jan. 12	27 Jan. 15	. . . .	——.
M. & B.	**Anderson**, Frederick Jasper, M.C., M.B.	10 Mar. 86	27 Jan. 12	27 Jan. 15	. . . .	——.
L. P.	**Gow**, Peter Fleming, D.S.O., M.B.	28 June 85	27 Jan. 12	27 Jan. 15	. . . .	——.
U. P.	**Martin**, John Simson Stuart, M.B.	18 June 88	27 Jan. 12	27 Jan. 15	. . . .	——.
U. P.	**Morrison**, Robert Victor, M.D.	1 Sep. 87	27 Jan. 12	27 Jan. 15	. . . .	Deolali.
M. & B.	**Jones**, James Walker, D.S.O., M.B.	22 Dec. 87	27 Jan. 12	27 Jan. 15	. . . .	——.
L. P.	**Hislop**, James Hall, M.C., M.B.	9 June 86	27 Jan. 12	27 Jan. 15	. . . .	——.
M. & B.	**Anderson**, Laurence Allfrey Pelham	18 Apl. 86	27 July 12	27 July 15	. . . .	——.
M. & B.	**Paton**, William Calder, M.C., M.B.	29 Jan. 86	27 July 12	27 July 15	. . . .	——.
B. & A.	**Hance**, James Bennett, M.B. (*Actg. Lt.-Col.*)	2 Apl. 87	27 July 12	27 July 15	. . . .	——.
U. P.	**Gordon**, Stephen, M.C.	4 Dec. 84	27 July 12	27 July 15	. . . .	——.
L. P.	**Rowntree**, Harold Kirkby, M.C., M.B.	27 Oct. 87	27 July 12	27 July 15	. . . .	O. C. No. 11 B. D. M. S., Bombay.
U. P.	**Thomson**, Graham, Yalden, M.B.	17 Jan. 86	27 July 12	27 July 15	. . . .	——.
B. & A.	**Eminson**, Basil Franklin, M.B.	24 Sep. 87	27 July 12	27 July 15	. . . .	O. C. Indn. Stn. Hospl., Maymyo.
L. P.	**Kennedy**, Anthony	20 Feb. 88	27 July 12	27 July 15	. . . .	——.
L. P.	**McIver**, Colin (169)	21 Oct. 84	27 July 12	27 July 15	. . . .	Loralai.
L. P.	**John**, Jordan Constantine, M.B.	22 Sep. 86	27 July 12	27 July 15	. . . .	——.
U. P.	**Porter**, Richard Reginald Maitland, M.B.	16 Oct. 88	25 Jan. 13	1 Sep. 15	. . . .	——.
U. P.	**Sweet**, Robert, D.S.O., M.B.	12 June 87	25 Jan. 13	1 Sep. 15	. . . .	——.
M. & B.	**Calvert**, Edward, M.B.	13 Aug. 87	25 Jan. 13	1 Sep. 15	. . . .	Jhansi.
M. & B.	**Webb**, John Robert Douglas (*temp. Major*)	4 Jan. 88	25 Jan. 13	1 Sep. 15	1 Aug. 17	D. A. D. M. S (Sany.), Embn. Staff, Bombay.
L. P.	**Phelan**, Francis (169)	18 Mar. 87	25 Jan. 13	1 Sep. 15	. . . .	——.
U. P.	**Macrae**, Archibald Campbell, M.B.	9 Feb. 89	25 Jan. 13	1 Sep. 15	. . . .	——
L. P.	**Kapur**, Nawin Chand	25 Aug. 86	25 Jan. 13	1 Sep. 15	. . . .	——.
B. & A.	**Hill**, Arthur Hilary Clifton	19 Mar. 88	25 Jan. 13	1 Sep. 15	. . . .	——.
L. P.	**Holmes**, Joseph Francis	14 Apl. 85	25 Jan. 13	1 Sep. 15	. . . .	——.
M. & B.	**Bal**, Narayan Krishna, M.C.	10 Sep. 85	25 Jan. 13	1 Sep. 15	. . . .	——.
B. & A.	**Haji**, Haji Sulaiman Gulam-hosseir, M.C. (207)	27 Apl. 85	25 Jan. 13	1 Sep. 15	. . . .	Rawalpindi.
U. P.	**Sokhey**, Saheb Singh, M.B.	15 Dec. 87	26 July 13	1 Sep. 15	. . . .	——.
M. & B.	**Doraisamy**, Subramania	10 Jan. 88	26 July 13	1 Sep. 15	. . . .	——.
U. P.	**Seddon**, Allan, M.B.	14 May 89	26 July 13	1 Sep. 15	. . . .	——.
M. & B.	**Findlay**, James, M.B.	19 Aug. 86	26 July 13	1 Sep. 15	. . . .	Poona Divn.
B. & A.	**Spackman**, William Collis, M.B.	23 Sep. 89	26 July 13	1 Sep. 15	. . . .	——.
L. P.	**De**, Jyotish Chandra, M.B.	23 Aug. 87	26 July 13	1 Sep. 15	. . . .	——.
M. & B.	**Mehta**, Nanulal Maganlal	31 Dec. 87	26 July 13	1 Sep. 15	. . . .	——.
L. P.	**Easton**, Robert Morrison, M.B.	1 May 87	26 July 13	1 Sep. 15	. . . .	——. *Lv., m. c., extended to 13 Mar. 19.*
L. P.	**Allen**, Charles Henry Powell (207)	5 Dec. 85	26 July 13	1 Sep. 15	. . . .	——.

B.&A.	**Martin**, Reginald Victor	20 Mar. 89	26 July 13	1 Sep. 15	. . . .	
U. P.	**Mahony**, George Henry, M.B.	31 Dec. 88	31 Jan. 14	1 Sep. 15	. . . .	——.
U. P.	**Covell**, Gordon, M.B.	20 Oct. 87	31 Jan. 14	1 Sep. 15	. . . .	*Lv., ex I., m.c. Extn to 23 Jany. 1918.*
M.&B.	**Stewart**, William Ross, M.B. (*tempy. Major*)	16 Mar. 89	31 Jan. 14	1 Sep. 15	. . . .	——.
M.&B.	**Rao**, Kotyvenkata Ramna (169)	14 Jan. 88	31 Jan. 14	1 Sep. 15	23 May 18	D. A. D. M. S. (Embn.), Bombay.
L. P.	**Moses**, John Gregory Owen, M.B.	23 Apl. 90	31 Jan. 14	1 Sep. 15	. . . .	Nowshera.
U. P.	**Chand**, Hari, M.C.	26 Aug. 86	31 Jan. 14	1 Sep. 15	. . . .	*Lv., m. c. Extn. 6 mos.*
M.&B.	**Mahadevan**, Venkatasubba	1 Feb. 88	31 Jan. 14	1 Sep. 15	. . . .	Chitral.
*U. P.	**Bilderbeck**, Alured Charles Lowther O'Shee, M.B.	22 Aug. 86	31 Jan. 14	1 Sep. 15	. . . .	——.
B.&A.	**Van Reenen**, Jacob William, M.B.	29 Sep. 89	31 Jan. 14	1 Sep. 15	. . . .	M. O., 2/5th Gurkha Rifles, Abbottabad.
U. P.	**Beatson**, Basil Fraser, (207)	7 Apl. 90	31 Jan. 14	1 Sep. 15	. . . .	2nd (Rawal Pindi.) Divn.
L. P.	**Roche**, Maurice James, M. C., M.B.	18 Mar. 89	31 Jan. 14	1 Sep. 15	. . . .	——.
U. P.	**Puri**, Neehal Das, M.B.	5 Dec. 88	31 Jan. 14	1 Sep. 15	. . . .	*Lv., ex I., m. c., from 1 Dec. 17.*
L. P.	**Roy**, Prabodh Chandra, M.B. (207)	21 Sep. 87	31 Jan. 14	1 Sep. 15	. . . .	——.
B.&A.	**Das**, Monindranath, M.C.	25 Aug. 87	31 Jan. 14	1 Sep. 15	. . . .	——.
M.&B.	**Vaidya**, Jagannath Balkrishna (207)	14 Jan. 88	31 Jan. 14	1 Sep. 15	. . . .	——.
B.&A.	**Hennessy**, Joseph Martin Reeves	7 May 86	31 Jan. 14	1 Sep. 15	. . . .	——.
L.P.	**Cowper**, Alfred Glen	18 Nov. 86	31 Jan. 14	1 Sep. 15	. . . .	——.
L.P.	**Lupton**, William Mawhood	1 Feb. 88	31 Jan. 14	1 Sep. 15	. . . .	——.
M.&B.	**Brown**, Hubert Horan	28 Mar 86	13 Jan. 14	1 Sep. 15	. . . .	Karachi.
B. P.	**Baker**, Charles Henry Neil, M.C.	18 June 87	31 Jan. 14	1 Sep. 15	. . . .	——.
M.&A.	**Pigeon**, John Walter	25 Dec. 86	1 Aug. 14	1 Sep. 15	. . . .	——.
M.&B.	**Treston**, Maurice Lawrence (172*a*)	9 Feb. 91	1 Aug. 14	1 Sep. 15	. . . .	——.
B.&A.	**Vieyra**, Peter, M.B.	19 May 88	1 Aug. 14	1 Sep. 15	. . . .	——.
L. P.	**Mitra**, Bhamini Mohan	21 Jan. 91	1 Aug. 14	1 Sep. 15	. . . .	——.
U. P.	**Savage**, Philip	3 Nov. 90	1 Aug. 14	1 Sep. 15	. . . .	Dinapore.
U. P.	**Chand**, Amir, M.B. (206)	9 Sep. 89	1 Aug. 14	1 Sep. 15	. . . .	——.
L. P.	**Lee**, Robert, M.B.	7 Oct. 89	1 Aug. 14	1 Sep. 15	. . . .	Amballa.
L. P.	**Jatar**, Nilkanth Shriram	26 May 87	1 Aug. 14	1 Sep. 15	. . . .	——.
M.&B.	**Sastry**, Tadepally Sankara, D.S.O., M.B., C.M.	Sep. 89	1 Aug. 14	1 Sep. 15	. . . .	——.
M.&B.	**Jamal-ud-din**, M.B. (204, 205*a*)	Nov. 90	1 Aug. 14	1 Sep. 15	. . . .	Cannanore.
L. P.	**Chenoy**, Ferozeshah Bapuji	26 Dec. 88	1 Aug. 14	1 Sep. 15	. . . .	——.
U. P.	**Venugopal**, Sadanala Bashiam	9 May 88	1 Aug. 14	1 Sep. 15	. . . .	——.
L. P.	**Martin**, Charles de Carteret, M.B. (172)	3 Jan. 88	1 Aug. 14	1 Sep. 15	. . . .	Derajat Bde.
M.&B	**Smith**, Joseph Henry, M.B.	27 May 89	1 Aug. 14	1 Sep. 15	. . . .	——.
	Molony, John Barré de Winton, M.B., F.R.C.S.	6 June 86	1 Aug. 14	1 Sep. 15	. . . .	——.
U.P.	**Alexander**, Harold George, F.R.C.S.	7 Feb. 88	1 Dec. 14	1 Dec. 15	. . . .	——.
B.&A.	**Wilson**, Owen, M.B.	6 Mar. 91	16 July 15	16 July 16	. . . .	——.
B.&A.	**Liston**, John Joseph, M.B.	6 Dec. 91	16 July 15	16 July 16	. . . .	——.
U.P.	**Batra**, Kanshi Ram	10 Aug. 90	16 July 15	16 July 16	. . . .	——.
L.P.	**Singh**, Bawa Harkishan	14 Dec. 89	16 July 15	16 July 16	. . . .	——.
M.&B.	**Chopra**, Pindi Dass, M.B.	1 June 88	16 July 15	16 July 16	. . . .	Santa Cruz.
L.P.	**Unger**, Oscar Ramsey	12 Sep. 89	16 July 15	16 July 16	. . . .	——.
	Grant, Ian Dingwall, M.B.		16 July 15	16 July 16	. . . .	Bannu Bde.
	†**Khan**, Mahomed Moosa	. . . . .	20 Oct. 15*	17 July 16	. . . .	——.
	†**Mukerji**, Satyendra Nath	. . . . .	8 Dec. 15*	17 July 16	. . . .	——.
	†**Murray**, Herbert Edward, M.B.	. . . . .	8 Dec. 15*	17 July 16	. . . .	——.
	†**Bose**, Ambuj Nath	21 Aug. 90	13 Mar. 16*	17 July 16	. . . .	Allahabad.

* The order of seniority of all these officers is provisional.
† Tempy. Mobilized service rendered prior to 17 July 1915 counts towards pension. Service rendered from 17 July 1915 counts towards pension and promotion.

INDIAN MEDICAL SERVICE—*contd.*

Captains—*concld.*

Names.	Date of birth.	Date of appointment.	Date of rank.	Remarks.
†**Fry**, Augustine Sargood (206)	. . . . .	13 Mar. 16*	17 July 16	——.
†**Green**, John Gerald James	. . . . .	13 Mar. 16*	17 July 16	Sick lv. extended to 8 Dec. 18.
†**Hogg**, William Peat, M.C.	. . . . .	13 Mar. 16*	17 July 16	——.
†**Vance**, Robert Lancelot	27 Aug. 90	30 May 16*	17 July 16	——.
†**White**, James Herbert Grove	23 Nov. 89	30 May 16*	17 July 16	——.
†**Collins**, Hugh Michael	23 Oct. 90	30 May 16*	30 Sep. 16	——.
†**Hanna**, George Boyle	20 Nov. 87	30 May 16*	30 Sep. 16	——.
Brown, Andrew Hunter, M.B.	. . . . .	20 Oct. 15*	20 Oct. 16	——.
†**Heppenstall**, Clement Hoyle (*a*)	. . . . .	13 Mar. 16*	2 Nov. 16	16th Indian Divn.
Cotter, Edward, M.B.	. . . . .	8 Dec. 15*	8 Dec. 16	——.
†**Bourke**, Ulick Joseph	10 June 85	30 May 16*	30 May 17	——.
†**Thomas**, David Rees	12 Apl. 89	30 May 16*	30 May 17	——.
†**Mecredy**, Cedric Cowan	1 Aug. 90	1 Aug. 16*	20 July 17	Hospital Ship "Syria."
†**Baxter**, Claude Wells Woolloton, M.C.	9 Apl. 91	1 Aug. 16*	1 Aug. 17	——.
†**Briggs**, Norman (H)	. . . .	8 Aug. 17	10 Aug. 17	——.
†**Hay**, Robert, M.B. (A)	. . . .	17 July 15	7 Sep. 17	——.
†**Ramsey**, George Arthur Stuart, M.D. (B)	. . . .	17 July 15	22 Sep. 17	——.
†**Shanks**, George, M.D. (C.)	. . . .	17 July 15	24 Sep. 17	——.
†**Ashton**, Basil Cedne, M.B. (D)	. . . .	17 July 15	23 Nov. 17	——.
†**Anand**, H. S. (J)	. . . .	9 Nov. 17	26 Nov. 17	i-c Jail, Kohat.
†**Philp**, James Leslie Ruthven, M.B. (M)	14 Aug. 89	29 Jan. 18	29 Jan. 18	174 I. F. A.
†**Ingram**, William John Sivewright, M.B., (N)	10 Sep. 89	29 Jan. 18	29 Jan. 18	Dhond.
†**Malone**, Reginald Hampstead, M.D. (L)	19 Jan. 86	20 Jan. 18	6 May 18	Central R search Institute, Kasauli
†**Sondhi**, S. D., M.C., M.B., (K)	. . . .	9 Nov. 17	20 July 18	——.
†**Kharegat**, Rustam Merwan, M.B. (E)	. . . .	11 Sep. 15	11 Sep. 18	——.

Lieutenants.

†**Shah**, Jelal Moochool (*Ty. Capt.*) (F.)		17 Dec. 15*	——.
†**Crombie**, William Maurice, M.B. (*Ty. Capt.*) (G.)		22 May 16*	*Lv., m. c., 6 mos.,* —*Aug. 18.*
†**Bird**, Joseph Godfrey, M.B. (*Ty. Capt.*)		23 Jan. 17*	——.
†**Verdon**, Philip (*Ty. Capt.*) (I.)		8 Aug. 17*	Specialist in Ophthalmology, 5th Dn.
†**Will**, William Miller, M.B.	18 Feb. 88	29 Jan. 18	4th Divn.

* The order of seniority of all these officers is provisional.

† { Temporary mobilised service rendered prior to 17th July 1915 counts towards pension.
Service rendered from 17th July 1915 counts towards pension and promotion. }

(*a*) Service rendered after 2nd November 1915 counts towards pension and promotion.

(A) Counts for pension from 7 Sep. 14.
(B) Do. 22 Sep. 14.
(C) Do. 24 Sep. 14.
(D) Do. 23 Nov. 14.
(E) Do. 11 Sep. 15.
(F) Do. 17 Dec. 15.
(G) Do. 22 May 16.
(H) Counts service for pension and promotion from 10 Aug. 14.
(I) Do. Do. 1 Dec. 15.
(J) Do. Do. 26 Nov. 14.
(K) Do. Do. 20 July 15.
(L) Do. Do. 6 May 15.
(M) Counts service for pension from 28 Sep. 14.
(N) Do. 30 Sep. 14.

INDIAN MEDICAL SERVICE—*contd.*

Names.	Date of birth.	Date of appointment.	Date of rank.	Remarks.
Temporary Lieutenant-Colonels.				
Row, R., M.D. (*a*)	. . . .	1 Mar. 18	4 Oct. 18	Hony. M. O., Lady Hardinge War Hospl.
Bhajekar, V. N., F.R.C.S.E. (*a*)	. . . .	1 Mar. 18	4 Oct. 18	Do. do.
Temporary Majors.				
Rowcroft, G. F., D.S.O.	. . . .	8 Jan. 15		
Woollatt, P. C.	. . . .	21 May 17	21 May 17	
Judah, D., M.D. (*a*)	. . . .	1 Mar. 18	. . . .	Hony. M. O. Lady Hardinge War Hospl.
Deshmukh, G. V., M.D., F.R.C.S. (*a*)	. . . .	1 Mar. 18	. . . .	Do. do.
Duggan, J. N. (*c*)	. . . .	1 Mar. 18	. . . .	Do. do.
Gilder, M. D. D., M.D., F.R.C.S. (*a*)	. . . .	1 Mar. 18	. . . .	Do. do.
Contractor, A. K., M.D. (*a*)	. . . .	1 Mar. 18	. . . .	Do. do.
Shirgaokar, S. R., M.D. (*a*)	. . . .	1 Mar. 18	. . . .	Do. do.
Fernandes, C. (*a*)	7 Nov. 66	20 Oct. 17	17 May 18	Bombay.
Engineer, S. K., M.R.C.P. (Edin.) (*a*)	5 Apl. 79	28 Dec. 15	18 May 18	——.
Temporary Captains.				
Walker, J. R. H.	. . . .	22 Dec. 14	22 Dec. 14	——.
Semon, H. C., M.D.	. . . .	7 Jan. 15	7 Jan. 15	——.
Bishop, T. H.	. . . .	18 Apl. 15	18 Apl. 15	——.
Francis, E. E.	. . . .	21 May 15	21 May 15	——.
Falkiner, J. McI.	. . . .	12 Aug. 15	18 Aug. 15	——.

Phatak, V. M.		8 Oct. 14	8 Oct. 15	——.
Oonawala, J. H.		18 Oct. 14	18 Oct. 15	——.
Postwalla, R. M.		18 Oct. 14	18 Oct. 15	——.
Paymaster, S. A.		31 Oct. 14	31 Oct. 15	Sialkot.
Shah, B. Z.		5 Nov. 14	5 Nov. 15	——.
Blaker, P. S.		5 Nov. 14	5 Nov. 15	——.
Mirajkar, V. R.		5 Nov. 14	5 Nov. 15	——.
D'Souza, A. J.		5 Nov. 14	5 Nov. 15	——.
Berry, P. N., M.B.		15 Nov. 14	15 Nov. 15	——.
Shaikh, A. H., M.B.		15 Nov. 14	15 Nov. 15	——.
Warden, S. B., M.B.	26 June 88	15 Nov. 14	15 Nov. 15	Kirkee.
Modi, L. S., F.R.C.S.I.		15 Nov. 14	15 Nov. 15	——.
Ghosh, S. M.		15 Nov. 14	15 Nov. 15	——.
Canteenvalla, J. P.		15 Nov. 14	15 Nov. 15	Dehra Dun.
Ranaday, S. G.		19 Nov. 14	19 Nov. 15	——.
Kolaporewalla, P. J.		20 Nov 14	20 Nov. 15	——.
Nariman, J. K.		23 Nov. 14	23 Nov. 15	——.
Berryman, R. C. P.		24 Nov. 14	24 Nov. 15	——.
Williams, A.		26 Nov. 14	26 Nov. 15	——.
Bharucha, K. B.		26 Nov. 14	26 Nov. 15	——.
Daboo, E. R., M.C., M.B.		26 Nov. 14	26 Nov. 15	——.
Bamboat, N. H.		26 Nov. 14	26 Nov. 15	——.
Sheorey, N. L., M.B.		26 Nov. 14	26 Nov. 15	——.
Laskari, P. F., M.B.		26 Nov. 14	26 Nov. 15	——.
Khaw, I. J. O. K., M.B.		26 Nov. 14	26 Nov. 15	——.
Mallya, G. B.	4 Apl. 90	26 Nov. 14	26 Nov. 15	Bombay.
Prasad, B., M.B.		2 Dec. 14	2 Dec. 15	——.

(a) Hony.

INDIAN MEDICAL SERVICE—*contd.*

Temporary Captains.—*contd.*

Names.	Date of birth.	Date of appointment.	Date of rank.	Remarks.
Yin, M. B., M.B.		2 Dec. 14	2 Dec. 15	Mandalay.
Banerjee, P.		2 Dec. 14	2 Dec. 15	——.
Banker, S. S.		5 Dec. 14	5 Dec. 15	——.
Bhat, S. V.		7 Dec. 14	7 Dec. 15	——.
Thwaites, W. St. C., M.B.		9 Dec. 14	9 Dec. 15	——.
Kapur, K. L.		9 Dec. 14	9 Dec. 15	——.
Rao, S. R.		9 Dec. 14	9 Dec. 15	——.
Ganapathy, C. M., M.B.		14 Dec. 14	14 Dec. 15	——.
Bhandari, M. G., M.B.		14 Dec. 14	14 Dec. 15	——.
Gai, N. J., M.B.		20 Dec. 14	20 Dec. 15	——.
Gokhale, V. B.		4 Jan. 15	4 Jan. 16	——.
Fyzee, A. A. H.		4 Jan. 15	4 Jan. 16	——.
Dadachanji, R. K., F. R. C. S. E.		4 Jan. 15	4 Jan. 16	——.
Master, F. S.	9 Oct. 83	4 Jan. 15	4 Jan. 16	Baroda.
Billimoria, S. D.		4 Jan. 15	4 Jan. 16	——.
Mehta, K. K., M.B.		4 Jan. 15	4 Jan. 16	Lansdowne.
Choksy, J. N. H., M.B.	11 Jan. 91	4 Jan. 15	4 Jan. 16	——.
Dias, B. G. V.		4 Jan. 15	4 Jan. 16	——.
Ajinkya, R. N., M.B.		12 Feb. 15	12 Feb. 16	——
Menon, T. K., M.B. (95)			19 Feb. 16	——.
Joshi, M. S.		25 Feb. 15	25 Feb. 16	——.
Parry, J. H.		9 Mar. 15	9 Mar. 16	——.
Ambler, F. B.		28 Mar. 16	28 Mar. 16	——.
Banerjee, U. N.		8 May 15	8 May 16	Mardan.
Gupta, P. K.		8 May 15	8 May 16	——.
Chatterjee, N.		8 May 15	8 May 16	——.
Ghosh, P. N.		8 May 15	8 May 16	——.
Agate, V. N., M.B.		11 May 15	11 May 16	——.
Norman, C. E. R.		14 May 15	14 May 16	——.
Malhotra, R. C., M.B.		15 May 15	15 May 16	——.
Dastur, H. S.		17 May 15	17 May 16	——.
Spencer, R. B., M.B.		17 May 15	17 May 16	——.
Chalam, B. S.		18 May 15	18 May 16	Ferozepore.
Khin, H. A., M.B.		19 May 15	19 May 16	——.
Mitra, S. L., M.B.		20 May 15	20 May 16	Fyzabad.
Goldie, E. A. M. J., M.C.,		20 May 15	20 May 16	——.
Jog, S. C.		21 May 15	21 May 16	4th Division.
Kapila, B. B., M.B.		4 June 15	4 June 16	——.

Ubhaya, N. R. R.	. . . . .	14 June 15	14 June 16	———.
Fernandez, C. J.	. . . . .	19 June 15	19 June 16	———.
Pestonji, A. B., F.R.C.S.E., D.P.H.	2 Nov. 84	21 June 15	21 June 16	Bombay. Surgical specialist, 28 Mar. 17.
Mehta, N. B.	. . . . .	21 June 15	21 June 16	Sialkot.
Singh, M. A., M.B.	. . . . .	27 June 15	27 June 16	———.
Contractor, S. C., M.B.	. . . . .	14 July 15	14 July 16	———.
Wania, H. J., M.B. (207)	. . . . .	17 July 15	17 July 16	———.
Naik, N. R.	. . . . .	23 July 15	23 July 16	Dehra Dun.
Patel, K. K.	. . . . .	29 July 15	29 July 16	———.
Dabholkar, A. Y., M.C., M.B.	. . . . .	2 Aug. 15	2 Aug. 16	———.
de Sousa, P. J.	. . . . .	6 Aug. 15	6 Aug. 16	———.
Mohan, G. V. R., M.B.	. . . . .	6 Aug. 15	6 Aug. 16	———.
Kaikini, V. M., M.B.	. . . . .	10 Aug. 15	10 Aug. 16	———.
Gupta, S. C. S., M.D., F.R.C.S.	. . . . .	10 Aug. 15	10 Aug. 16	Bombay.
Amin, K. V.	. . . . .	10 Aug. 15	10 Aug. 16	Sitapur.
Master, K. S., M.B.	. . . . .	10 Aug. 15	10 Aug. 16	———.
Primalani, C. H., M.B.	. . . . .	13 Aug. 15	13 Aug. 16	———.
Luthera, J. N., M.B.	. . . . .	13 Aug. 15	13 Aug. 16	———.
Phadke, S. K., M.B.	. . . . .	20 Aug. 15	20 Aug. 16	———.
Dadachanji, R. E., M.B.	. . . . .	21 Aug. 15	21 Aug. 16	———.
Khosla, R. N., M.B.	. . . . .	23 Aug. 15	23 Aug. 16	———.
Gwynne, J. R.	. . . . .	23 Aug. 15	23 Aug. 16	———.
Chowdhury, H. C. R.	. . . . .	25 Aug. 15	25 Aug. 16	———.
Kapadia, R. N., M.B.	. . . . .	25 Aug. 15	25 Aug. 16	———.
Rajan, H. S., M.B.	. . . . .	26 Aug. 15	26 Aug. 16	———.
Bose, S. K., M.B.	. . . . .	28 Aug. 15	28 Aug. 16	2nd-in-command Indian Station Hospital, Bankipur.
Machado, L. S., F.R.C.S.I.	. . . . .	2 Sep. 15	2 Sep. 16	———.
Dutt, A. C., M.B.	. . . . .	8 Sep. 15	8 Sep. 16	———.
Aich, N. B., M.B.	. . . . .	9 Sep. 15	9 Sep. 16	———.
Sen, S. K., M.B.	. . . . .	11 Sep. 15	11 Sep. 16	Rawalpindi.
Mukerjee, S. B., M.B.	. . . . .	13 Sep. 15	13 Sep. 16	———.
Shirgaonkar, J. V.	29 July 83	14 Sep. 15	14 Sep. 16	Rawalpindi.
Vakil, P. R., M.B.	. . . . .	4 Oct. 15	4 Oct. 16	———.
Ramsay, A. Macl., M.B.	. . . . .	17 Oct. 15	17 Oct. 16	———.
Kanal, V. M., M.B.	. . . . .	17 Oct. 15	17 Oct. 16	———.
Dhawan, M. L., M.B.	. . . . .	28 Oct. 15	28 Oct. 16	———.
Dhondy, B. S., M.B.	. . . . .	28 Oct. 15	28 Oct. 16	———.
Balkrishna, M.B.	. . . . .	1 Nov. 15	1 Nov. 16	———.
Menon, M. P. K.	. . . . .	. . . .	2 Dec. 16	———.
Godinho, P. X.	. . . . .	17 Dec. 15	17 Dec. 16	———.
Gadgil, S. B., F.R.C.S.	. . . . .	17 Dec. 15	17 Dec. 16	Bombay.
Sathe, V. L., F.R.C.S.(E.)	. . . . .	24 Dec. 15	24 Dec. 16	I. T. C. Hospl., Poona.
Karode, R. N.	. . . . .	24 Dec. 15	24 Dec. 16	———.
Panikkar, K. P., M.B.	. . . . .	1 Jan. 16	1 Jan. 17	R. M. O., Bangalore.
Subramanyam, M., M.B.	. . . . .	18 Jan. 16	18 Jan. 17	———.
Kaushash, R.	. . . . .	14 Feb. 16	14 Feb 17	Peshawar.
Chukerbuti, J. C., M.B.	. . . . .	17 Feb. 16	17 Feb. 17	———.
Naidu, B. P. B., M.B.	. . . . .	19 Feb. 16	19 Feb. 17	———.
Karaka, D. P.	. . . . .	19 Feb. 16	19 Feb. 17	———.
Panton, A. W., M.B.	. . . . .	19 Feb. 16	19 Feb. 17	———.

INDIAN MEDICAL SERVICE—*contd.*

Temporary Captains—*contd.*

Names.	Date of birth.	Date of appointment.	Date of rank.	Remarks.
Ghosh, A. M., M.B.		19 Feb. 16	19 Feb. 17	——.
Burjorjee, B. N., M.B.		20 Feb. 16	20 Feb. 17	——.
Kalsavala, N. B.		26 Feb. 16	26 Feb. 17	——.
Menon, C. V. A.		26 Feb. 16	26 Feb. 17	——.
La Frenais, A. C. L.		4 Mar. 16	4 Mar. 17	——.
Dutta, N., M.B.		10 Mar. 16	10 Mar. 17	——.
Nag, S., M.B.		10 Mar. 16	10 Mar. 17	——.
David, J. S.		11 Mar. 16	11 Mar. 17	——.
Khandwalla, M. T., M.B.		11 Mar. 16	11 Mar. 17	——.
Majmundar, P. T.		7 Apl. 16	7 Apl. 17	——.
Matthai, P. M.		17 Apl. 16	17 Apl. 17	Dargai.
Dady, M.		17 May 16	17 May 17	Dadar.
Winnington-da-Costa, A. F.		20 May 16	20 May 17	——.
Gupta, I. D.		23 May 16	23 May 17	——.
Sufi, I. A.		24 May 16	24 May 17	——.
Mogul, D. D.		27 May 16	27 May 17	——.
Bhagat, M. L., M.B.		31 May 16	31 May 17	——.
Basu, K. S.		1 June 16	1 June 17	——.
Darukhanawala, K. A. (*a*)		1 June 16	1 June 17	——.
Malhoutra, G. D.		3 June 16	3 June 17	——.
Bhargava, M. L., M.B.		5 June 16	5 June 17	——
Bhal, N. S.		6 June 16	6 June 17	Lahore Cantt.
Sewal, R. N.		6 June 16	6 June 17	——.
Ayya, Y. V.		6 June 16	6 June 17	——.
Wadia, J. R.		6 June 16	6 June 17	——.
Das, M.		7 June 16	7 June 17	——.
Chavan, N. M.		7 June 16	7 June 17	——.
Mayer, C. A. L.		9 June 16	9 June 17	Baroda
Gothaskar, S. B., M.B.		10 June 16	10 June 17	——.
Varlava, D. D., M.B.		10 June 16	10 June 17	Kohat.
Kharki, K. B. S.		10 June 16	10 June 17	Delhi.
Rodrigues, F.		10 June 16	10 June 17	I. T. War Hospl., Laho
Mitra, S. C. (*d*)		11 June 17	11 June 17	Calcutta.
Mukharji, J. G.		12 June 16	12 June 17	——.
Secluna, G.		12 June 16	12 June 17	——.
Munim, M. D.		12 June 16	12 June 17	——.
Fonseca, I.		13 June 16	13 June 17	——.
Reardon, W. A.		13 June 16	13 June 17	Jubbulpore.

Mukerji, P. C.		13 June 16	13 June 17	O. C. Ind. Stn. Hospital, Ajmere.
Mehta, G. D.		14 June 16	14 June 17	———.
Raja, R. N.		14 June 16	14 June 17	———.
Gidwani, B. (E.) B., M.B.		14 June 16	14 June 17	Meerut.
Ghosh, M. C.		14 June 16	14 June 17	Bangalore.
Mukerji, H. P.		15 June 16	15 June 17	Meerut.
Summanwar, N. R. C.		16 June 16	16 June 17	———.
Nayaran, R.		16 June 16	16 June 17	———.
Krishnaswami, P., M.B.	31 Dec. 88	17 June 16	17 June 17	———.
Morris, N. B.		17 June 16	17 June 17	Bombay.
Rao, N. M., M. B.		17 June 16	17 June 17	———.
Nangpaul, D. C., M.B. (89b)		17 June 16	17 June 17	———.
Rajaratnam, T. V. P., M.B.		18 June 16	18 June 17	———.
Krishnan, N. N., M.B.		18 June 16	18 June 17	Harnai.
Chowdhury, J.		19 June 16	19 June 17	———.
Shroff, A. D., M. B.		19 June 16	19 June 17	———.
Paonaskar, S. G., M.B.		19 June 16	19 June 17	Bareilly.
Khanna, T. R.		20 June 16	20 June 17	———.
Ram, G.		22 June 16	22 June 17	———.
Phatak, S. A.		23 June 16	23 June 17	Saugor.
Lal, S. B.		23 June 16	23 June 17	Karachi.
Dutt, S., M. C.		24 June 16	24 June 17	———.
Mukerjee, S. C.		24 June 16	24 June 17	———.
Krishnamurti, A. S., M.B.		25 June 16	25 June 17	———.
Singh, N.		25 June 16	25 June 17	———.
Chandorkar, B. R.		27 June 16	27 June 17	———.
Pestonjee, R.		5 July 16	5 July 17	———.
Naidu, C. R., M.B.		7 July 16	7 July 17	———.
Das, B., M. B.		8 July 16	8 July 17	———.
Dogra, P. N., M.B.		8 July 16	8 July 17	———.
Gupta, M. R., M.B.		11 July 16	11 July 17	I. T. H., Kirkee.
Poh, L. H.		14 July 16	14 July 17	———.
Bhushan, V., M.B.		15 July 16	15 July 17	———.
Bhal, P. L., M. B.		16 July 16	16 July 17	Egypt.
Sabawala, B. P., F.R.C.S.E. (c)	20 Mar. 85	19 July 17	19 July 17	Bombay.
Gupta, D. N.		19 July 16	19 July 17	Fyzabad.
Khanna, L. C., M.B.		20 July 15	20 July 17	Kohat.
Marsh, E. W.		25 July 16	25 July 17	———.
Uberoi, B. D., M.B.		26 July 16	26 July 17	———.
Sampson, J. C.		26 July 16	26 July 17	———.
Desai, U. L.		29 July 17	29 July 17	Bangalore.
Kuriyan, A. T.		4 Aug. 16	4 Aug. 17	———.
Chaudhuri, S. K. (b)		6 Aug. 16	6 Aug. 17	Lady Chelmsford War Hospl., Benares.
Rikhya, M. L., M.B.		8 Aug. 16	8 Aug. 17	———.
Kapur, G. D., M.B.		12 Aug. 16	12 Aug. 17	———.
Shah, S. H. U., M.B.		13 Aug. 16	13 Aug. 17	Indn. Stn. Hospital, Bannu.
Chohan, N. K.		14 Aug. 16	14 Aug. 17	———.
Thapar, K. S., M.B.		15 Aug. 16	15 Aug. 17	———.

(a) Hony. tempy. Lieut. for the period of his honorary service in the War Hospl., Datia.
(b) " " " " " " in Benares.
(c) Hony. tempy. Capt. for his honorary services in Bombay.
(d) " " " " " in Calcutta.

Names.	Date of birth.	Date of appointment.	Date of rank.	Remarks.
		Temporary Capts.—*contd.*		
Rashid, M. A.		15 Aug. 16	15 Aug. 17	———.
Singh, S.		15 Aug. 16	15 Aug. 17	Nowshera.
Oberai, G. R., M.B.		19 Aug. 16	19 Aug. 17	———.
Kumar, D. R., M.B.		21 Aug. 16	21 Aug. 17	———.
Chandra, J., M.B.		21 Aug. 16	21 Aug. 17	Peshawar.
Butt, A. H., M.B.	8 Jan. 92	21 Aug. 16	21 Aug. 17	I/c X-Ray Section, D. I. Khan.
Abdulla, M., M.B.		23 Aug. 16	23 Aug. 17	———.
Dogra, A. S., M.B.		30 Aug. 16	30 Aug. 17	———.
Das, P.		1 Sep. 16	1 Sep. 17	———.
Pillai, M. K., M.B.		4 Sep. 16	4 Sep. 17	Maymyo.
Uberoi, T. S., M.B.		5 Sep. 16	5 Sep. 17	———.
Talwar, K. C. (95)		7 Sep. 16	7 Sep. 17	———.
Tandan, P. L.		11 Sep. 16	11 Sep. 17	Peshawar.
Mehra, A. C., M.B.		16 Sep. 16	16 Sep. 17	———.
Kochhar, J. R., M.B.		16 Sep. 16	16 Sep. 17	———.
Nijhawan, G., M.B.		17 Sep. 16	17 Sep. 17	———.
Dhar, J., M.B.		18 Sep. 16	18 Sep. 17	———.
Contractor, K. H., M.B.		22 Sep. 16	22 Sep. 17	———.
Khan, G. A., M.B.		23 Sep. 16	23 Sep. 17	———.
Mhatre, S. W.		25 Sep. 16	25 Sep. 17	Kohat.
Rasul, S. G.		25 Sep. 16	25 Sep. 17	———.
Kripalani, J. K.		26 Sep. 16	26 Sep. 17	Comdg. No. 15 Sany. Sec., Karachi.
Apte, V. K., M.C.		26 Sep. 16	26 Sep. 17	———.
Subramanyam, R. B.		26 Sep. 16	26 Sep. 17	Peshawar.
Mithavala, M.A.		29 Sep. 16	29 Sep. 17	Amritsar.
Chowdhury, P. C., M.B.	15 Nov. 90	30 Sep. 16	30 Sep. 17	Bombay, Dadar.
Kacker, R. K.		1 Oct. 16	1 Oct. 17	Bareilly.
Vakil, N. H.		1 Oct. 16	1 Oct. 17	———.
Madan, K. R.		1 Oct. 16	1 Oct. 17	———.
Driver, J. C.	4 Mar. 91	4 Oct. 16	4 Oct. 17	Bombay.
Gupte, V. B.		5 Oct. 16	5 Oct. 17	———.
Bhandare, V. G.		5 Oct. 16	5 Oct. 17	Allahabad.
Bharucha, P. C., M.D. (*b*)	Nov. 82	5 Oct. 17	5 Oct. 17	Ahmednagar.
Singh, D. R. R. (*d*)		5 Oct. 17	5 Oct. 17	Bombay.
Mitra, J. M.		7 Oct. 16	7 Oct. 17	———.
Bodas, N. M.		7 Apl. 17	7 Oct. 17	———.
Hattiangadi, S. K.		13 Oct. 16	13 Oct. 17	———.
Pereira, A. X.		15 Oct. 16	15 Oct. 17	———.
Kapadia, P. R.		20 Oct. 16	20 Oct. 17	Ambala.
Basu, S. C.		21 Oct. 16	21 Oct. 17	Bareilly.
Nulkor, A. K.		31 Oct. 16	31 Oct. 17	Poona.
DeSouza, F. B.		2 Nov. 16	2 Nov. 17	———.
Mistri, A. B.	9 Oct. 83	2 Nov. 16	2 Nov. 17	Bombay.

Grandhi, M. R., M.B.		3 Nov. 16	3 Nov. 17	———.
Datta, R.		7 Nov. 16	7 Nov. 17	Meerut.
Miller, W. G.		8 Nov. 16	8 Nov. 17	———.
Parab, V. K.		8 Nov. 16	8 Nov. 17	Lansdowne.
Maitra, N.		18 Nov. 16	18 Nov. 17	Peshawar.
Bapat, V. B.		22 Nov. 16	22 Nov. 17	Poona,
Banerjee, H. S.		1 Dec. 16	1 Dec. 17	———.
Miri, G. A.		2 Dec. 16	2 Dec. 17	———.
Daruwalla, P. P.		8 Dec. 16	8 Dec. 17	———.
Varde, D. N.		8 Dec. 16	8 Dec. 17	———.
Hildreth, G. A., M. B.		11 Dec. 16	11 Dec. 17	———.
Banerji, A. N., M. B.		13 Dec. 16	13 Dec. 17	———.
Byramji, P.		15 Dec. 16	15 Dec. 17	———.
Dhairyawan, K. K. R.		18 Dec. 16	18 Dec. 17	Lucknow.
Daruwalla, J. D.		18 Dec. 16	18 Dec. 17	———.
Kapadia, F. N. (c) M.D.	5 Mar. 81	5 Aug. 18	. . . .	Bombay.

Temporary Captains.

Mitra, P. N.		8 Jan. 17	8 Jan. 18	———.
Sonpar, K. N.		8 Jan. 17	8 Jan. 18	Chitral—Civil medical charge.
Satpute, N. R.		10 Jan. 17	10 Jan. 18	Nowshera.
Quraishi, N. A. (169)		22 Jan. 17	22 Jan. 18	———.
Nagarkar, V. R.		31 Jan. 17	31 Jan. 18	———.
Pant, P.		3 Feb. 17	3 Feb. 18	Karachi.
Guilfoyle, J. M., M. B., B.CH.	12 Aug. 88	5 Feb. 17	5 Feb. 18	Bombay.
Das, A.		7 Feb. 17	7 Feb. 18	———.
Bhatia, K.		8 Feb. 17	8 Feb. 18	———.
Anand, R. R. M.		9 Feb. 17	9 Feb. 18	———.
Khan, P. M. T.		11 Feb. 17	11 Feb. 18	———.
Joachim, B. I. H.		15 Feb. 17	15 Feb. 18	Quetta.
Pillay, A. P.		18 Feb. 17	18 Feb. 18	Deolali.
Duggal, S. S.		20 Feb. 17	20 Feb. 18	Nowshera.
Dhunjibhoy, J. E.		20 Feb. 17	20 Feb. 18	———.
Patel, K. S. (a) M.B., B.S.	10 July, 87	25 Feb. 17	25 Feb. 18	Bombay.
Galvin, J. S.		1 Apl. 17	5 Mar. 18	———.
Brooks, E. C.		8 Mar. 17	8 Mar. 18	———.
Kanga, J. S.		8 Mar. 17	8 Mar. 18	———.
Lester, A. R.		12 Mar. 17	12 Mar. 18	———.
Roberts, C. L. D.		12 Mar. 17	12 Mar. 18	———.
Dewan, R. K.		16 Mar. 17	16 Mar. 18	Karachi, Ambala.
Kikani, C. M.		18 Mar. 17	18 Mar. 18	———.
Nayar, T. K.		21 Mar. 17	21 Mar. 18	———.
Desai, N. K.		1 Apl. 17	1 Apl. 18	Peshawar.
Abreu, A. J.		2 Apl. 17	2 Apl. 18	———.
Kaka, D. L.		3 Apl. 17	3 Apl. 18	———.
Menon, M. V.		5 Apl. 17	5 Apl. 18	Jubbulpore—Specialist in Dermatology. It. **War Hospital, Khandwa.**
Kapur, D. D.		7 Apl. 17	7 Apl. 18	———.
Rodrigues, J. B.		8 May 17	8 May 18	Jubbulpore.
Bisni, S. F.		10 May 17	10 May 18	———.

(a) Hony. Tempy. Lieut. for the period of his honorary service in the Marine Lines War Hospl., Bombay.
(b) „ „ Capt. „ „ „ „ „ „ „ „
(c) „ Capt.
(d) Temporary Captain for the period of his honorary services in Allahabad.

INDIAN MEDICAL SERVICE—*contd.*

Names.	Date of birth.	Date of appointment.	Date of rank.	Remarks.
		Temporary	**Captains**—*contd.*	
Valladares, W.		13 May 17	13 May 18	———.
Patnak, M. V.		17 May 17	17 May 18	———.
Campbell, A.		19 May 17	19 May 18	Karachi Brigade. Temporary specialist in Otology.
DeSouza, J. F.		24 May 17	24 May 18	———.
Dalvi, N. A.		14 June 17	14 June 18	———.
Chowdhuri, K. B.		18 June 17	18 June 18	Peshawar.
Joshi, N. N.		19 June 17	19 June 18	———.
D'Souza, J. A. A. (A.)	4 Dec. 78	20 June 17	20 June 18	Bombay.
Bhatavadekar, D. M.		23 June 17	23 June 18	———.
Arjani, N. K., B.A., L.M. & S.	22 Jan. 70	28 June 17	28 June 18	Dadar.
Tolat, K. V., M.B., B.S.	24 Aug. 90	29 June 17	29 June 88	Bombay.
Andrews, H. J., M.B.E.		30 June 17	30 June 18	Moradabad.
Maitra, M. M.		30 June 17	30 June 18	Rawalpindi.
Warma, J. D.		30 June 17	30 June 18	———.
Bannerjee, L. M.		1 July 17	1 July 18	———.
Gupta, B. L.		1 July 17	1 July 18	Patiala.
Dikshit, H. K.		2 July 17	2 July 18	Dharmsala.
Saxena, V. D.		2 July 17	2 July 18	———
Chandra, R.		3 July 17	3 July 18	Jodhpur.
Chaudhury, G. C.		3 July 17	3 July 18	———
Modak, N. W.		3 July 17	3 uly 18	Poona.
Khanna, C.		3 July 17	3 July 18	———
Mukarji, N. B.		4 July 17	4 July 18	Sitapur.
Charnalia, G. R.	23 Dec. 92	4 July 17	4 July 18	———.
Cheriyan, P. V.		5 July 17	5 July 18	———.
Limaye, V. P.		5 July 17	5 July 18	Peshawar.
Barreto, A. de G.		5 July 17	5 July 18	Karachi.
Prasad, M. (95)		5 July 17	5 July 18	Karachi.
D'Souza, W. M.		6 July 17	6 July 18	Nasirabad
Maneckshaw, H. F. (95)		6 July 17	6 July 18	R. M. O., Amritsar,
Faruki, S. M. A.		6 July 17	6 July 18	Poona Dn. Specialist in Opthalmology.
Gadgil, B. B.		6 July 17	6 July 18	Poona.
Singh, G.		7 July 17	7 July 18	Nowshera.
Agtey, R. N.		9 July 17	9 July 18	Jhelum.
Deb, M. L.		10 July 17	10 July 18	Multan.
Sharma, R. S.		11 July 17	11 July 18	———.
Ezekiel, J.		12 July 17	12 July 18	———.
Bhargava, D. P.		12 July 17	12 July 18	———.
Gupta, U. M.		12 July 17	12 July 18	Poona.
John, D.		12 July 17	12 July 18	Kohat. X-Ray specialist.
Sharma, A. N.		12 July 17	12 July 18	———.
Kishor, N.		12 July 17	12 July 18	Karachi.
Fazul-ddin, M.		13 July 17	13 July 18	———.

Maitra, G. C.	. . . . .	14 July 17	14 July 18	Karachi.
Basu, P. N. (95)	. . . . .	16 July 17	16 July 18	Karachi.
Varma, A. S.	. . . . .	16 July 17	16 July 18	R. Alipore.
Goyal, K. N. (95)	. . . . .	16 July 17	16 July 18	Karachi.
Pal, M. N.	. . . . .	16 July 17	16 July 18	Lahore Cantonment.
Ghosh, S.	. . . . .	16 July 17	16 July 18	———.
Jafri, Z. H.	. . . . .	16 July 17	16 July 18	Agra.
Alavi, H. A.	. . . . .	17 July 17	17 July 18	Lahore Cantt.
Sikdar, J. M.	. . . . .	17 July 17	17 July 18	Delhi.
Bhattacharya, S. C.	. . . . .	17 July 17	17 July 18	Poona.
Richardson, J. F.	. . . . .	17 July 17	17 July 18	Kotgarh.
Anantan, K. M.	. . . . .	18 July 17	18 July 18	Bellary.
Lal, S.	. . . . .	18 July 17	18 July 18	Amballa.
Bardhan, P. K.	. . . . .	20 July 17	20 July 18	I/c. R. T. Depôt, Saugor.
Kapur, S. N. (172*a*)	. . . . .	20 July 17	20 July 18	Delhi.
Ram, H.	. . . . .	21 July 17	21 July 18	Jullundhur.
Mukerjee, P.	. . . . .	21 July 17	21 July 18	———.
Ananthaya, U.	. . . . .	21 July 17	21 July 18	Neemuch.
Pal, N. B. (95)	. . . . .	22 July 17	22 July 18	———.
Khanna, D. D.	. . . . .	22 July 17	22 July 18	Lahore Cantt.
Shah, B. B.	Oct. 87	24 July 17	24 July 18	Miranshah.
Guha, P. C.	. . . . .	24 July 17	24 July 18	———.
Mitter, H. C.	Apl. 94	24 July 17	24 July 18	Bannu.
Basu, R. C.	. . . . .	25 July 17	25 July 18	———.
Ray, G. C.	. . . . .	25 July 17	25 July 18	Karachi.
Phani, N.	. . . . .	25 July 17	25 July 18	Poona.
Bhalla, D. R.	. . . . .	25 July 17	25 July 18	———.
Mallik, B. B.	. . . . .	27 July 17	27 July 18	Poona.
Das, L.	. . . . .	27 July 17	27 July 18	Karachi.
Sen-Gupta, A. K.	. . . . .	27 July 17	27 July 18	Deolali.
Gupta, P. R. (95)	. . . . .	28 July 17	28 July 18	Fyzabad.
Dube, K. C. (*b*)	. . . . .	28 July 17	28 July 18	Benares.
Castellino, L.	. . . . .	28 July 17	28 July 18	Karachi.
Hathiramani, H. T. (95)	. . . . .	28 July 17	28 July 18	Karachi.
Parmanand	. . . . .	28 July 17	28 July 18	Abbottabad.
Gupta, U. S.	. . . . .	29 July 17	29 July 18	Peshawar.
Rau, G. R. D.	. . . . .	30 July 17	30 July 18	———.
Gill, B. S.	. . . . .	30 July 17	30 July 18	———.
Das, S. P.	May 85	30 July 17	30 July 18	Tank.
Tarapore, B. S.	24 Apl. 80	30 July 17	30 July 18	———.
Prasad, R.	. . . . .	30 July 17	30 July 18	Almora.
Banerji, K. K. (95)	. . . . .	31 July 17	31 July 18	Karachi.
Dad, K.	. . . . .	31 July 17	31 July 18	Nepalese Contingent, Kakul.
Nunan, W., M. D. (*b*)	. . . . .	12 Sep. 18	12 Sep. 18	———.

Temporary Lieutenants.

Basu, N.	. . . . .	1 Aug. 17	———.	
Patney, S. L.	. . . . .	2 Aug. 17	———.	
Masters, N. F.	. . . . .	2 Aug. 17	Kirkee.	
Ray, J. N.	. . . . .	2 Aug. 17	———.	

(*a*) Honorary Temporary Lieutenant for the period of his honorary service in the Maine lines, War Hospital, Bombay.
(*b*) " " " " " in Benares.

INDIAN MEDICAL SERVICE—*contd.*

Names.	Date of birth.	Date of appointment.	Remarks.
Temporary Lieutenants—*contd.*			
Korke, V. T.		2 Aug, 17	Ferozepore.
Nayar, T. M.		12 Nov. 14	——.
Tha, R. R. H. O.		13 Mar. 16	——.
Paul, G. E.		12 Aug. 16	——.
Masina, P. M.		5 Jan. 17	——.
Rai, H. S.		23 Jan. 17	——.
Madgavkar, V. D.		29 Jan. 17	——.
Ullah, M. S.		5 Feb. 17	Peshawar.
Sen, R.		7 Feb. 17	——.
Nawaz, M.		8 Feb. 17	——.
da Costa, J. A.		3 Mar. 17	——.
Hesterlow, A. M. V.		14 Mar. 17	——.
Murty, P. V. R.		17 Mar. 17	——.
Nawab-ud-Din		27 Mar. 17	Nowshera.
Cursetji, H. R.		31 Mar. 17	——.
Bhatia, S. L.		8 Apl. 17	——.
Chatterji, A. C.		17 Apl. 17	——.
Chhatre, K. N.		19 Apl. 17	——
Dhunjibhoy, F.		14 May 17	——
Bhurke, G. M.		16 May 17	——
Kamakaka, B. H.		17 May 17	——
Rozdon, S. (*a*)		2 June 17	——.
Chandra, S. N.		7 June 17	——
Saldanha, M. J. (95)		7 June 17	——.
Dotivala, N. M. P.	20 Sep. 90	23 June 17	I. S. H. Kirkee.
Mandlik, G. S.		28 June 17	Poona.
Awan,, K. B., M. B.		28 June 17	——.
Mathur, H. G. D.		30 June 17	——.
Sud, R. N.		1 July 17	——.
Saldanha, A..		2 July 17	——.
Nath, V.		3 July 17	——.
Hensman, H. S.		4 July 17	——.
Aggarwal, S. N.		4 July 17	——
Sanyal, D.		5 July 17	——.
Bharucha, J. C.		7 July 17	——.
Bazaz, M. S. (95)		9 July 17	——.
Singh, S.		9 July 17	——.
Khera, R. L.		11 July 17	——.
Wacher, H. S.		11 July 17	——.
Shahane, N. R.		16 July 17	——.
Sen, H. K.		24 July 17	——.
Sondhi, R. C.		25 July 17	——.
Sen, A. K.		31 July 17	——.

Ahmad, D. U.		2 Aug. 17	——.
Khan, M. A.		2 Aug. 17	Nepalese Contingent, Kakul.
Sen, P. N.		3 Aug. 17	——.
Chatterjee, N. N., M.B. (95)	15 July 94	3 Aug. 17	Tank.
Bal, J. N.		3 Aug. 17	Nasirabad.
Saxena, B. B.		3 Aug. 17	Bareilly.
Bhaduri, D. N.		4 Aug. 17	——.
Dani, R. G.		4 Aug. 17	——.
Vania, S. D.		4 Aug. 17	——.
Sanyal, S. K.		4 Aug. 17	——.
Basu, B.		5 Aug. 17	Karachi.
Mukharji, J. M. (95)		6 Aug. 17	Calcutta.
Prabhaker, C. J.		7 Aug. 17	Derajat Brigade.
Ghosh, S. N. (95)		7 Aug. 17	Calcutta.
Chatterjee, N. N.		7 Aug. 17	Peshawar.
Khan, M. A.		7 Aug. 17	Nowshera.
Giml, H. D.		6 Aug. 17	——.
Sen, H. C.		8 Aug. 17	Karachi.
Shankar, P. U.		8 Aug. 17	Amballa.
Bhandari, S. L.		8 Aug. 17	——.
Ray, L. K.		8 Aug. 17	Peshawar.
Roy, K. Sen		8 Aug. 17	Peshawar.
Duarte, A. M.		9 Aug. 17	——.
Menon, E. K.		9 Aug. 17	——.
Mukerjee, C. P.		9 Aug. 17	Nowshera.
Shaw, S. C.		9 Aug. 17	——.
Simon, A. I.		9 Aug. 17	Bangalore.
Devasagayam, A.		9 Aug. 17	——.
Tun, A.		9 Aug. 17	——.
Banerjee, S. N.		10 Aug. 17	Mardan.
Mukerjee, S. K.		10 Aug. 17	——.
Ganguli, P. P.		10 Aug. 17	Rawalpindi.
Bhattacharya, R.		10 Aug. 17	X Ray Specialist, I. T. War Hospl., Khandw
Gupta, K. C. D.		10 Aug. 17	Peshawar.
Palit, S. K.		10 Aug. 17	Poona.
Gupta, R. S.		10 Aug. 17	Ferozepore.
Eichidana, K. H.		11 Aug. 17	Poona.
Freitas, A. M.		12 Aug. 17	Poona.
Basu, K. M. (95)		12 Aug. 17	Roorkee.
Ghosh, N. N. (95) M. B.	23 Oct, 94	12 Aug. 17	Bombay.
Mallik, P. N.		12 Aug. 17	Jullundur.
Gupta, B. S. (169)		13 Aug. 17	Peshawar.
Nandi, C. L. (95)		14 Aug. 17	Dacca.
Pal, A. C.		14 Aug. 17	Peshawar.
Lyngdoh, H.		14 Aug. 17	Sitapur.
Ray, K. M.		14 Aug. 17	Hyderabad (Sindh).
Banarji, J. K.		15 Aug. 17	——.
Mukerji, A. K. (207)		15 Aug. 17	Bakloh.

(*a*) Hony. temporary Lieutenant for the period of his hony. service in Benares.
(*b*) Hony. Ty. Captn.

INDIAN MEDICAL SERVICE—*contd.*

Names.	Date of birth.	Date of appointment.	Remarks.
		Temporary Lieutenants—*contd.*	
Mitra, K. C.		16 Aug. 17	Jhansi.
Ghosh, P.		16 Aug. 17	Sialkot.
Mitter, S. C.		17 Aug. 17	———.
Sarkar, D. L.		17 Aug. 17	Aurangabad.
Munshi, B. C.		17 Aug. 17	Calcutta.
Chaudhuri, B. N., M. B.	15 June 89	17 Aug. 17	Dadar L. C. W. Hospl., Bombay.
Ghosh, J. M.		18 Aug. 17	I/c No. 8 Sany. Sec., Kohat.
Kapur, A. N.		19 Aug. 17	Deesa.
Ghosh, N.	12 Jan. 93	20 Aug. 17	F. W. H., Tank.
Luthra, B. N.		20 Aug. 17	———.
Subramanian, N. S.		20 Aug. 17	Bannu.
Das, U. N. (95)		21 Aug. 17	Karachi.
Solomon, M. (95)		21 Aug. 17	Karachi.
Sen-Gupta, B. C.		21 Aug. 17	———.
Contractor, C. M., M. B., B. C.	2 Feb. 89	22 Aug. 17	Bombay.
Conceicao, F. D. P. H.	24 Apl. 86	22 Aug. 17	Bombay.
Ghosh, S. N. (95)		22 Aug. 17	Karachi.
Khanna, A. R.		22 Aug. 17	———.
John, K. I.		22 Aug. 17	Nowshera.
Singh, B.		22 Aug. 17	———.
Mazagonwala, E. B.		24 Aug. 17	Poona.
Ittycheria, C. C.		24 Aug. 17	Peshawar.
Chatterje, J.		25 Aug. 17	———.
Chakravarty, N.		25 Aug. 17	Derajat Bde.
Gonsalves, A.		25 Aug. 17	———.
Doshi, V. J.		25 Aug. 17	Montgomery.
Hajra, B. B. (95)		25 Aug. 17	Karachi.
Jafarey, M. A. (95)		25 Aug. 17	Peshawar.
Lahiri, P. N. (95)		27 Aug. 17	———.
Pathak, S. M.		27 Aug. 17	Saugor.
Basu, J. N.		27 Aug. 17	Comdg. No. 8 Amb. Train, Karachi.
Bose, H. N.		27 Aug. 17	Saharanpur.
Thwaites, W. St. C.		27 Aug. 17	———.
Verghese, G.		27 Aug. 17	———.
Agarwala, G. C.		27 Aug. 17	———.
Soganadan, A. D.		27 Aug. 17	———.
Mascarenhas, V. E.		28 Aug. 17	Rawalpindi.
Knight, H. W.		28 Aug. 17	———.
Nimonkar, S. A.		29 Aug. 17	Poona.
Dalal, H. P.		29 Aug. 17	Khandwa.
Basu, R. K.		29 Aug. 17	Sialkot.
Mukand, S. N.		29 Aug. 17	D'Dun.
Kapadia, M. S.		30 Aug. 17	Poona.
Mullan, N. D.		30 Aug. 17	Poona.
Chaudury, A. N. R.		30 Aug. 17	Saugor.

Bose, Y. M.		31 Aug. 17	———.
Mukherji, M.		31 Aug. 17	Allahabad.
Menezes, J. P.		31 Aug. 17	Poona.
Francis, P. V.		1 Sep. 17	Bannu.
Madhok, M. C.		1 Sep. 17	Derajat Brigade.
Maitra, S. S.		1 Sep. 17	———.
Banerjea, **H**. D., M.B.		1 Sep. 17	Bombay.
Alagappan, S. C.		3 Sep. 17	———.
Oka, M. G.	23 Oct. 84	3 Sep. 17	Poona.
Sen, N. K.		3 Sep. 17	———.
Dasgupta, K.		3 Sep. 17	———.
Iyer, R. S.		4 Sep. 17	Miranshah.
Toyle, R. C.		4 Sep. 17	Secunderabad.
Puri, S. R.		4 Sep. 17	———.
Moolky, D. M.		5 Sep. 17	———.
Masurekar, V. R.		6 Sep. 17	———.
Benjamin, G. I.		7 Sep. 17	———.
Chatterjee, B.		8 Sep. 17	Bhim Tal.
Venugopaul, K. (95)		8 Sep. 17	Karachi.
Thomas, W. C.		9 Sep. 17	Bannu.
Dhavle, D. S.		10 Sep. 17	Poona.
D'Silva, A. (95)		10 Sep. 17	Karachi.
Elijah, N.		11 Sep. 17	Belgaum.
Venkataramanan, C. S.		11 Sep. 17	Secunderabad.
Sen, G. C.		11 Sep. 17	16th Division.
Maitra, **A**. P.	4 Sep. 93	12 Sep. 17	———.
Gupta, K. C. S. (95)		12 Sep. 17	Karachi.
Advani, G. B. (95)		12 Sep. 17	Karachi.
Trivedi, S. V.		12 Sep. 17	Arangaon.
Bose, D. P.	7 Mar. 92	12 Sep. 17	Jhansi.
Menon, V. C. G.		12 Sep. 17	———.
Nandi, A. K.		14 Sep. 17	16th Division.
Venkatachalam, K.		14 Sep. 17	———.
Bhasin, D. R.		14 Sep. 17	———.
Mitra, N. C.		15 Sep. 17	Nowshera.
Koshi, E. J.		17 Sep. 17	Hangu.
Modak, Y. V.		17 Sep. 17	———.
deNetto, F.		18 Sep. 17	Erinpura.
Rau, M. S.		19 Sep. 17	———.
Mehta, N. D.		20 Sep. 17	Rawalpindi.
Kanagasabapathy, R.		21 Sep. 17	Poona.
Parasuram, A. R.		21 Sep. 17	Rangoon.
Chaudhari, B. R.		22 Sep. 17	———.
Gore, S. N. D.		24 Sep. 17	Sialkot.
Chowdhury, N. N.		25 Sep. 17	———.
Saldanha, J. L.		25 Sep. 17	Karachi.
Chatterji, R.		26 Sep. 17	Deoli.
De, S.		26 Sep. 17	Kamptee.
Chatarji, M. N.		26 Sep. 17	Karachi.
Menon, V. K. K.		26 Sep. 17	Jubbulpore.

INDIAN MEDICAL SERVIC E—*contd.*

Temporary Lieutenants—*contd.*

Name.	Date of birth.	Date of appointment.	Remarks.
Krishnan, H.		27 Sep. 17	———.
Sundaresan, V. R.	16 June 83	27 Sep. 17	Followers', Central Depôt, Kirkee.
Ram, G. S. (95)		28 Sep. 17	Karachi.
Menon, C. G.		29 Sep. 17	Maymyo.
Otto, D. S.		30 Sep. 17	———.
Dhingra, C. L.		30 Sep. 17	Bannu.
Tomas, E. K.		1 Oct. 17	———.
Rai, R. N. (95)		1 Oct. 17	Karachi.
Menon, A. K.		1 Oct. 17	———.
Garewal, A. S.		1 Oct. 17	———.
McNair, D. C.		1 Oct. 17	———.
Rai, K.		3 Oct. 17	Risalpur.
Dass, J.		4 Oct. 17	Jullundur.
Ghosh, B. C.		6 Oct. 17	———.
Mathulla, K. M.		6 Oct. 17	Karachi.
Singh, R.		6 Oct. 17	———.
Sharma, R. N.		8 Oct. 17	———.
Banerjee, L. G.		8 Oct. 17	Karachi.
Ramanathan, P. A. A.		10 Oct. 17	Bannu.
Kashy, P. K.		11 Oct. 17	———.
Bishan, G.		11 Oct. 17	Karachi.
Srinivasamurti, G.		12 Oct. 17	———.
Natesan, V. R.		15 Oct. 17	Poona.
Gupte, Y. P.		16 Oct. 17	———.
Nathan, S. K.		17 Oct. 17	1st Division.
Acharya, B. V.		18 Oct. 17	———.
Menon, K. G.		18 Oct. 17	———.
Chopra, J. M.		19 Oct. 17	Deesa.
Ayyar, R. S. H.		20 Oct. 17	Poona.
Mordaunt, L. J. P.		20 Oct. 17	168 I. F. A.
Dastoor, P. A.		21 Oct. 17	Rawalpindi.
Pothan, K. E.		21 Oct. 17	Fategarh.
Sanyal, S.		25 Oct. 17	———.
Menon, P. B.		27 Oct. 17	Multan.
Bagchi, K. P.		28 Oct. 17	Lucknow.
Malkani, B. R.	5 Nov. 93	29 Oct. 17	I. S. H., Kirkee.
Lalkaka, K. A. J.		30 Oct. 17	Karachi.
Ashraf, M. S.		31 Oct. 17	Peshawar.
Bhattacharya, K. C.		1 Nov. 17	———.
Madhavan, O. C.		3 Nov. 17	Poona.
Somayaji, S. L.		6 Nov. 17	Secunderabad.
Nath, B.		6 Nov. 17	Peshawar.
Kotwal, N. S. (95)		6 Nov. 17	Peshawar. *Lv., m. c., 3 mos., 15 Aug. 18.*

Mathew, K. M.		8 Nov. 17	——.
Bhatji, B. R.		9 Nov. 17	Secunderabad.
Menon, C. K.		13 Nov. 17	Mhow.
Parasuram, G. R. P.		19 Nov. 17	Jubbulpore.
Sabhesan, D. K. I. K.		22 Nov. 17	Allahabad.
Banerji, S. C.		23 Nov. 17	Nasirabad.
Pillai, S. K.		25 Nov. 17	Karachi.
Sharma, J. K.		26 Nov. 17	——.
Rao, B. D.		26 Nov. 17	——.
Basu, S. C.		28 Nov. 17	Rawalpindi.
Chatterjee, B. B.		29 Nov. 17	——.
Bhandarkar, P. R.		2 Dec. 17	Karachi.
Banerji, R.		3 Dec. 17	——.
Sinha, J. C.		8 Dec. 17	Roorkee.
Dalal, K. R.		8 Dec. 17	Karachi.
Sen, S. K.		8 Dec. 17	Amballa.
Mankodi, K. K. (95)		12 Dec. 17	4th (Quetta) Divn.
Sarkar, S. C.		14 Dec. 17	Lucknow.
Khanna, G. N.		16 Dec. 17	Malakand.
Joshi, K. G.		17 Dec. 17	Karachi.
Madhok, A. N. (95)		17 Dec. 17	Derajat Bde.
Trivedi, C. P. (*a*)		17 Dec. 17	Kirkee.
Rau, B. S.		19 Dec. 17	——.
Khan, W.		20 Dec. 17	Peshawar.
Padbidri, G. R.		23 Dec. 17	Aden.
Desouza, P.		24 Dec. 17	Aden.
Clarke, J. H.		28 Dec. 17	——.
Subramanyam, K. S.		29 Dec. 17	Lansdowne.
Narayanaras, D.		31 Dec. 17	Thal.
Krishnan, B. T.		2 Jan. 18	——.
Menon, K. P.		6 Jan. 18	——.
Mahiuddin, A. K. M.		8 Jan. 18	——.
Noble, A.		8 Jan. 18	——.
Sinha, J. D.		9 Jan. 18	Ferozepore.
Chakrabatti, D. C.		12 Jan. 18	Agar.
Bagchee, K.		14 Jan. 18	——.
Rau, R. V.		18 Jan. 18	Bangalore.
Muthukrishnan, A.		18 Jan. 18	Mandi.
Lele, S. C.		19 Jan. 18	Belgaum.
Gupta, J. C.		25 Jan. 18	Muttra.
Rao, C. S. R. V.		26 Jan. 18	Ambala.
Menon, E. K.		31 Jan. 18	Bhamo.
Nyan, K.		3 Feb. 18	——.
Kapur, M. K. (*b*)		5 Feb. 18	Lahore.
Colwell, H. H.		11 Feb. 18	——.
Gupta, I. N. S.		12 Feb. 18	——.
Mukerji, J. N.		14 Feb. 18	——.
Vaswani, T. P.		15 Feb. 18	Karachi.

(*a*) Hony. Ty. Lt. for the period of his hony. services at Meerut.
(*b*) ,, ,, ,, ,, employment at Lahore Cantt.

INDIAN MEDICAL SERVICE—*contd.*

Names.	Date of birth.	Date of appointment.	Remarks.
		Temporary Lieutenants—*contd.*	
Ninan, V. T.	. . . . .	16 Feb. 18	Delhi.
Ram, A.	. . . . .	18 Feb. 18	——.
Thirupod, K.	. . . . .	18 Feb. 18	Lansdowne.
Taylor, E. T. N.	. . . . .	20 Feb. 18	——.
Variava, H. E.	. . . . .	21 Feb. 18	——.
Dube, P. L.	. . . . .	21 Feb. 18	Barelily.
Tampi, G. S.	. . . . .	22 Feb. 18	Quetta.
Ayyar, G. S. A. S.	. . . . .	22 Feb. 18	——.
DeSouza, V.	. . . . .	22 Feb. 18	Jubbulpore.
Printer, M. D.	. . . . .	25 Feb. 18	——.
Limaye, G. G.	. . . . .	26 Feb. 18	——.
Rao, M. N.	. . . . .	28 Feb. 18	Ferozepore.
Gracias, C. M.	. . . . .	28 Feb. 18	Derajat Bde.
Sastri, N. S. N.	. . . . .	1 Mar. 18	Cawnpore.
Thomas, K. M.	. . . . .	1 Mar. 18	Kamptee.
Iyengar, B. R.	. . . . .	4 Mar. 18	Agra.
Oommen, C.	. . . . .	5 Mar. 18	Secunderabad.
Zacharias, P. T.	. . . . .	5 Mar. 18	Karachi.
Visvanathan, S.	. . . . .	5 Mar. 18	Cannanore.
Ramaswami, N. S.	. . . . .	7 Mar. 18	Bellary.
Dutta, J.	. . . . .	8 Mar. 18	Guna.
Ghose, K. M.	. . . . .	11 Mar. 18	Jullundur.
Nambyar, K. K.	. . . . .	12 Mar. 18	Karachi.
Bhewandiwalla, K. S.	. . . . .	15 Mar. 18	——.
Mandhle, M. B.	11 July 91	17 Mar. 18	Jhansi.
Woodruff, K. M.	. . . . .	19 Mar. 18	——.
Thomas, R. T.	. . . . .	20 Mar. 18	Patiala.
Chavan, S. G.	. . . . .	20 Mar. 18	——.
Joseph, J. P.	. . . . .	21 Mar. 18	Multan.
Dubash, M. N.	. . . . .	24 Mar. 18	——.
Nambiar, P. K. R.	. . . . .	25 Mar. 18	——.
Thapar, D. R.	. . . . .	25 Mar. 18	——.
Karve, S. D.	. . . . .	27 Mar. 18	——.
Gopalaswami, K. S.	. . . . .	9 Apl. 18	Mardan.
Sequeira, J. C.	. . . . .	17 Apl. 18	——.
Kumar, M.	. . . . .	17 Apl. 18	School of Instruction, Rawalpindi.
Majumdar, N. N.	. . . . .	18 Apl. 18	——.
Mirchandani, D. H.	. . . . .	19 Apl. 18	Karachi.
Sirker, K. K.	. . . . .	19 Apl. 18	——.
Ray, S. K.	. . . . .	19 Apl. 18	Kohat.
Chatterjee, R. K.	. . . . .	20 Apl. 18	Derajat Bde.

Bhargava, M. P.		20 Apl. 18	Jullundur.
Rodrigues, A. C.		27 Apl. 18	Karachi.
Iyer, C. K. A.		30 Apl. 18	Jhelum.
Lopes, A. V.		2 May 18	———.
Barrett, J. H.		2 May 18	———.
Dias, A. P.		4 May 18	———.
Dudani, G. L.		11 May 18	Ambala.
Mohile, K. G.		13 May 18	———.
Nalwa, I. S.		16 May 18	Aden.
Frenchman, J. E.		19 May 18	———.
Wagle, S. G.		20 May 18	———.
TrivIdi, J. G.	27 July 84	20 May 18	I. S. H., Kirkee.
Gupte, P. S.		24 May 18	Ambala.
Kulkani, N. H.		27 May 18	———.
Gnanamuttu, G. D.		28 May 18	Lahore Cantt.
Bhumbhwal, R. L.		3 June 18	———.
Mohamed, S. G.		3 June 18	Sick lv. extend to 28 Jan. 19.
deSIbra, G. P.		3 June 18	———.
Mukharji, K. D.		4 June 18	Lucknow.
Saha, N. N.		6 June 18	———.
Adhya, J. K.		8 June 18.	———.
Banerjee, S. C.		11 June 18	Barrackpore.
Bhargawa, R. S.		12 June 18	———.
Bhattacharji, S. N.		17 June 18	Jhansi.
Mody, M. S. H.		18 June 18	I. T. C. Hospl., Poona.
Kar, A. K.		21 June 18	Erinpura.
Bhansali, N. P.		21 June 18	Secunderabad.
Albuquerque, J. W. F. M.B., B.S.	14 Apl. 94	22 June 18	Bombay.
Matthew, T. C.		24 June 18	Secunderabad.
Roy, A. C.		25 June 18	Anandi.
Bhat, A. R.		25 June 18	Karachi.
Aiyar, N. S.		3 July 18	———.
Jain, B. R.		8 July 18	Campbellpore.
Bhat, R. P.		10 July 18	———.
Krishna Moorthy, Y. V.		6 July 18	———.
Row, C. K.		12 July 18	———.
Apte, G. D.		12 July 18	Arangaon.
Kutumbiah, P.		18 July 18	———.
Thambiah, S.		20 July 18	———.
Sarma, C. S. S.		20 July 18	———.
Daruwala, B. P.		22 July 18	Deolali.
Kumar, G.		23 July 18	Amballa Bde.
Pillay, C. K.		23 July 18	———.
Das, K. K.		25 July 18	Calcutta.
Bagchi, P. N.		25 July 18	Dhond.
Waghray, K.		26 July 18	Aurangabad.
Mazumdar, D. C.		26 July 18	Anandi.
Goil, J. N.		28 July 18	Belgaum.
Khan, S. A.		30 July 18	Dadar.
Rao, M. S.		30 July 18	———.

INDIAN MEDICAL SERVICE—*contd.*

Temporary Lieutenants—*concld.*

Name.	Date of birth.	Date of appointment.	Remarks.
Bose, K. L.		30 July 18	Karachi.
Nigam, K. S.		31 July 18	Delhi.
Manohar, M.		1 Aug. 18	——.
Mazumdar, L. B.		2 Aug. 18	——.
Baliga, B. P.		4 Aug. 18	——.
Kapadia, F. N.		5 Aug. 18	——.
Sethi, B. N. S.		5 Aug. 18	Ferozepore.
Ganda, R. C.		5 Aug. 18	Lahore Cantt.
Ray, G. C.		7 Aug. 18	Dinapore.
Doctor, D. B.		7 Aug. 18	Karachi.
Badekar, V. H.		8 Aug. 18	Dadar.
Husain, M. A.		9 Aug. 18	Roorkee.
Kapoor, J.		9 Aug. 18	——.
Siddique, Y. M.		9 Aug. 18	——.
Kehar, D.		9 Aug. 18	——.
Krishnaswami, C. V.		11 Aug. 18	Poona.
Bharucha, D. H.		11 Aug. 18	Meerut.
Benegal, V. N.		19 Aug. 18	——. School of Instruction, Rawal Pindi.
Pereira, J. A.		12 Aug. 18	Poona.
Deish, D.		12 Aug. 18	Dehra Dun.
Ranade, Y. B.	28 Mar. 91	16 Aug. 18	Agar.
Vaid, H. B.		16 Aug. 18	——.
Agarkar, N. P.		18 Aug. 18	Delhi.
Rajam, R. V.		19 Aug. 18	——.
Gupta, P. R. D.		21 Aug. 18	Fyzabad.
De, P.		26 Aug. 18	——.
Gavill, S. N.		27 Aug. 18	——.
Myer, B. D.		27 Aug. 18	——.
Shroff, J. D.		30 Aug. 18	——.
Bajpayee, A.		30 Aug. 18	——.
Bindra, S.		4 Sep. 18	——.
Lajmi, V. S.		5 Sep. 18	——.
Thapar, K. R.		7 Sep. 18	——.
Basu, P. L.		7 Sep. 18	——.
Gupta, A. N.		11 Sep. 18	——.
Ghosh, N. N.		11 Sep. 18	——.
doRosario, V.		11 Sep. 18	——.
Ghosh, L. M.		12 Sep. 18	——.

Ghosal, I.		12 Sep. 18	———.
Whig, P. L.		13 Sep, 18	Amballa.
Banerjea, D. N.		14 Sep. 18	———.
Majumdar, D. C.		14 Sep. 18	Agra.
Rao, H. S.		15 Sep. 18	Khandwa.
Uppal, K. C.		16 Sep. 18	———.
Gazder, J. D.		16 Sep. 18	———.
Vardya, B. K.		19 Sep. 18	———.
Garg, G. C.		19 Sep. 18	Meerut.
Kalapesi, H. D.		25 Sep. 18	———.
Prasad, R.		27 Sep. 18	———.
Gupta, P.		28 Sep. 18	———.
Velankar, S. V.		29 Sep. 18	———.
Sen, L. M.		1 Oct. 18	Delhi.
Gupta, B. N.		4 Oct. 18	Miranshah.
Chatterjee, S. N.		6 Oct. 18	———.
Ghosh, P. M.		7 Oct. 18	Bangalore.
Pal, B. N.		7 Oct. 18	Dehra Dun.
Singh, B(a)		8 Oct. 18	———.
Telang, V. D.		11 Oct. 18	———.
Sen, G. D.		11 Oct. 18	Bannu.
Gole, G. B.		11 Oct. 18	Bangalore.
Dutt, N. B.		17 Oct. 18	———.
Basu, A.		18 Oct. 18	———.
Sawhney, M. R.		18 Oct. 18	Meerut.
Bose, R. C.		23 Oct. 18	———.
Gupta, J. N. S.		30 Oct. 18	———.
Sen, A. K.		1 Nov. 18	———.
Mankad, N. P.		3 Nov. 18	———.
Banerjee, B. C.		4 Nov. 18	———.
Mukherjee, H. P.		9 Nov. 18	———.
Ghosh, N. N.		13 Nov. 18	———.
Basu, H. N.		13 Nov. 18	———.

(a) Hony.

RETIRED OFFICERS OF THE INDIAN MEDICAL SERVICE TEMPORARILY RE-EMPLOYED.

First Commn.	Name.	Date of re-employment.	Where employed.	REMARKS.
29 July 93	**Kilkelly,** Lt.-Col. P. P., M.B.	1 Jan. 15	Bombay	Resy. Surgn., Western States of Rajputana.
2 Oct. 80	**Fooks,** Lt.-Col. G. E.	29 Jan. 15	Bombay	Civil Surgeon, Dharwar.
31 Mar. 88	**Prasad,** Lt.-Col. K., M.D.	29 Jan. 15	Burma	Civil and Jail, Myaungmya.
1 Apl. 82	**Anderson,** Lt.-Col. A. V., M.B.	29 Jan. 15	Bombay	Dy. Sany. Comr., W. R. D.
29 Sep. 83	**Shore,** Lt.-Col. R., M.D.	5 Feb. 15	F. & P. Deptt.	Resy. Surgn., Mewar, Udaipur.
30 Mar. 78	**Warliker,** Lt.-Col. D. P.	6 Feb. 15	Burma	Civil and Jail, Thayetmyo.
1 Oct. 87	**Lamont,** Lt.-Col. J. C., M.B.	10 Feb. 15	Punjab	Prof. of Anatomy, K. E. Med. Coll., Lahore.
1 Oct. 72	**Ahmed,** Lt.-Col. Z. A., M.D.	26 July 15	U. P.	Civil and Jail, Gonda.
2 Oct. 80	**Khareghat,** Lt.-Col. M. P. M.	10 Dec. 14	Bombay	Supdt., Matheran and Ag. Presy. Surgn., 2nd Dist.
31 Mar. 87	**Meyer,** Lt.-Col. C. H. L.	15 Jan. 16	Bombay	Civil, Bombay.
31 Mar. 87	**Herbert,** Lt.-Col. H.	15 Jan. 16	Bombay	——.
27 July 92	**Hulbert,** Lt.-Col. J. G.	31 Dec. 15	U. P.	Civil and Jail, Fyzabad.
28 July 91	**Duer,** Lt.-Col. C., M.B., F.R.C.S.	26 Oct. 16	Burma	Civil and Jail, Maymyo.
31 Mar. 79	**Gimlette,** Lt.-Col. G. H. D., C.I.E., M.D.	11 Sep. 14	Mily.	Bangalore.
30 Jan. 93	**Langston,** Lt.-Col. T. A. O.	6 Nov. 14	Mily.	——.
28 July 91	**Palk,** Lt.-Col. (Local rank) C. H. L., M.B., F.R.C.S.E.	8 Dec. 14	Madras	I. G. of Prisons, Madras, *s. p. t.*
29 Jan. 94	**Thomson,** Major C., M.B.	19 Dec. 14	Mily.	——.
1 Apl. 86	**Sinha,** Major N. P.	30 Jan. 15	Bengal	——.
30 Mar. 78	**Smith,** Major F. C.	. . . .	Mily.	——.
27 July 92	**Green,** Lt.-Col. D. R., M.D.	29 July 18	Bengal	Offg. Superintendent, Central Jail, Alipore.

DEPARTMENTAL OFFICERS WITH HONORARY RANK.

INDIAN MEDICAL DEPARTMENT—(465).

[Note.—Assistant Surgeons seconded for Civil employment are shown in italics.]

Senior Assistant Surgeons with the honorary rank of Major (24)* { Bengal 14, Madras 4, Bombay 6 }

Date of First Commn.	Names.	Estabt.	Date of birth.	Date of present rank.	F. rule.	Remarks.
29 Nov. 00	**St. Romaine,** Eugene Alfred (170).	B.	12 Jan. 67	1 Jan. 03 m. 29 Nov. 15	N	Bangalore.
17 Dec. 04	**Corridon,** Richard . . .	B.	11 Mar. 64	14 Nov. 07 m. 21 Sep. 17	75	8th Division.
13 June 06	*Lynsdale, Albert William Jordon (207) (Supy.)*	*M.*	*7 Dec. 61*	*20 Nov. 07* m. 21 Sep. 17	N	*Asstt. to Professor of Materia Medica, Medl. College, Madras.*
8 Dec. 05	*Parkinson, John Thomas (207) (Supy.)*	*B.*	*8 Sept. 61*	*29 Mar. 08* m. 21 Sep. 17	N	*Supdt., Central Prison, Bareilly.*
17 Mar. 07	*Curran, James Francis (105, 161, 206). (Supy.).*	M.	3 Dec. 61	24 June 08 m. 21 Sep. 17	N	*(Civil Burma) Sargodha. Lv., m.c., 6 mos., 16 June 18.*
22 Aug. 06	**Paterson,** Arthur Robert, L.S.A., Lond. (Supy.) (207).	B.	25 Dec. 59	17 Apl. 09 m. 21 Sep. 17	N	Army Hd. Qrs. Estabt., Simla.
1 Mar. 06	**Gardiner,** Arthur Butler (Supy.).	Bo.	23 Mar. 61	31 Jan. 10 m. 21 Sep. 17	N	Ammunition Factory, Kirkee.
7 June 07	*Hogan, William John Alexander (Supy.)*	*B.*	*26 July 60*	*31 Jan. 10* m. 21 Sep. 17	N	*Civil and Jail, Saharanpur.*
20 Nov. 07	**Hart,** William Peter Bell (Supy.) (207).	M.	Mar. 63	30 Mar. 10 m. 21 Sep. 17	N	Bangalore.
27 Apl. 08	**Hankins,** Robert Forbes Hart (207) *(Supy.)*	M.	4 May 62	4 May 10 m. 21 Sep. 17	N	Burma Division.
20 Oct. 07	**Orman,** Charles Henry (Supy.).	B.	6 Nov. 61	23 Jan. 11 m. 21 Sep. 17	N	Dalhousie.
16 Nov. 07	*Brown, Robert (Supy.)* . .	*B.*	*7 Oct. 60*	*31 Mar. 11* m. 21 Sep. 17	N	*Civil, Sambalpur. Offg. Champaran.*
16 Nov. 07	**Fosconolo,** Henry William (Supy.)	B.	5 Nov. 61	31 Mar. 11 m. 21 Sep. 17	75	Calcutta.
14 Nov. 07	*Owen, Robert James (207) (Supy.)*	*B.*	*13 May 61*	*22 Nov. 11* m. 21 Sep. 17	N	*Civil Surgeon, Wano and M.O. Southern Waziristan Militia.*
29 Mar. 08	**Chiodetti,** Valentine Vincent (169, 207). (Supy.)	B.	7 Mar. 62	13 Mar. 12 m. 21 Sep. 17	N	Ferozepore.
16 Dec. 06	*Prescott, Henry William De Blaquire (207) (Supy.).*	Bo.	*8 Oct. 62*	*19 Apl. 12* m. 21 Sep. 17	N	*(Civil, Bombay.) Barrackpore.*
16 Dec. 06	**Pruce,** Charles Augustus (170) (Supy.)	Bo.	10 Aug. 62	19 Apl. 12 m. 21 Sep. 17	N	Madras.
26 June 08	**Lee,** Joseph (69) . .	B.	10 Mar. 70	23 Sep. 12 m. 21 Sep. 17	N	Rawalpindi.
23 July 08	*Sharples, Richard (Supy.)* .	*B.*	*8 June 60*	*23 Sep. 12* m. 21 Sep. 17	75	*Dy. Supdt., Campbell Medl. School & Hospl., Calcutta.*
23 July 08	*Courtney, Michael, I.S.O., (207) (Supy.)*	*B.*	*9 Sep. 62*	*23 Sep. 12* m. 21 Sep. 17	N	*Supdt., Central Jail, Montgomery.*
23 July 08	**Murphy,** Richard Thomas .	B.	10 Feb. 64	23 Sep. 12 m. 1 Jan. 17	N	Agra.
1 Aug. 08	**Baldry,** Thomas (Supy.) .	B.	2 Feb. 63	7 Dec. 12 m. 21 Sep. 17	N	Cordite Factory, Aruvankadu.
28 Dec. 06	**Devine,** Alfred (Supy.) . .	Bo.	21 Feb. 62	9 Aug. 13 m. 21 Sep. 17	N	Jutogh.
6 June 09	**Quadros,** Louis George (Supy.)	Bo.	25 Aug. 62	25 Nov. 13 m. 21 Sep. 17	N	Quetta.
1 Dec. 10	*Macqueen, James Ernest Barton (Supy.)*	*Bo.*	*24 Sep. 62*	*1 Dec. 13* m. 21 Sep. 17	N	*(Civil, Bombay.) Bombay.*
1 Dec. 10	**Alphonso,** Albino Graciano (Supy.).	Bo.	24 Aug. 63	1 Dec. 13 m. 21 Sep. 17	N	Poona.
2 Aug. 08	*Bonnar, Thomas Henry (Supy.)*	*B.*	*6 June 63*	*8 Dec. 13* m. 21 Sep. 17	N	*Civil, Kohima, Naga Hills (Assam).*
23 Sep. 08	*Farmer, Christopher Alfred (Supy.)*	*B.*	*25 Apl. 63*	*8 Dec. 13* m. 21 Sep. 17	75	*(O. & R. Ry.) Lucknow.*
23 Sep. 08	*Robertson, John (Supy.) (97)*	*B.*	*8 Jan. 63*	*8 Dec. 13* m. 21 Sep. 17	N	*H. E. the Viceroy's Dispy.*
23 Sep. 08	*Neal, William Daniel (Supy.)*	*B.*	*12 June 62*	*8 Dec. 13* m. 21 Sep. 17	N	*Asst. to Supdt., Medl. Coll. Hospls., Calcutta.*
23 Sep. 08	**Archer,** Arthur John (Supy.)	B.	20 Mar. 63	8 Dec. 13 m. 21 Sep. 17	N	Cawnpore.

* In ranks of Major and Captain—excluding supernumeraries and those seconded.

DEPARTMENTAL OFFICERS WITH HONORARY RANK.

INDIAN MEDICAL DEPARTMENT—(465)—*concld.*

Senior Assistant Surgeons with the honorary rank of Major (24)* { **Bengal 14**, **Madras 4**, **Bombay 6** }
—*concld.*

Date of First Commn.	Names.	Estabt.	Date of birth.	Date of present rank.	F. rule.	Remarks.
5 Feb. 11	**Cabral,** Lewis Mathew (95, 207).	Bo.	21 Sep. 65	1 Jan. 14 *m.* 21 Sep. 17	N	Poona.
12 Jan. 09	**DeLanty,** Harry William (170). (Supy.)	B.	25 June 63	28 Jan. 14 *m.* 21 Sep. 17	N	Subathu.
17 Apl. 09	**Weston,** George Oswald (Supy.).	B.	27 Apl. 63	1 Feb. 14 *m.* 21 Sep. 17	N	Egypt.
23 Sep. 09	*Galvin, Michael*	*B.*	*21 Mar. 64*	*12 Aug. 14* *m.* 21 Sep. 17	N	*Supdt. and Medl. Offr., Juvenile Jail, Alipore.*
23 Sep. 09	**Greenwood,** Alfred (Supy.)	B.	5 Sep. 63	12 Aug. 14 *m.* 21 Sep. 17	N	Jubbulpore.
o Sep. 09	**Newing,** Francis James Edmund L'Estrange (Supy.).	B.	16 Sep. 62	13 Dec. 14 *m.* 13 Dec. 17	N	Nowshera.
7 Sep. 10	*Sadleir, Arthur Willoughby Woodward (170, 207).*	*B.*	*1 Mar. 67*	*25 Dec. 14* *m.* 25 Dec. 17	N	*(B. B., & C. I. Ry.) 5th Division, Mhow.*
7 Sep. 10	*Harvey, Noble Spear* . . .	*B.*	*20 Sep. 67*	*25 Dec. 14* *m.* 25 Dec. 17	N	*Supdt., Central Prison, Benares, U. P.* s. *p. t.*
7 Sep. 10	*Bedell, Elliot Clive (162, 206).*	*B.*	*9 Sep. 67*	*25 Dec. 14* *m.* 25 Dec. 17	N	*(O. & R. Ry.) Fyzabad.*
26 Dec. 10	*McCarthy, Patrick (105)* . .	*B.*	*19 Feb. 68*	*25 Dec. 14* *m.* 25 Dec. 17	N	*Civil and Jail, Akyab.*
26 Dec. 10	**Durham,** William Robert (170, 207).	B.	10 June 68	25 Dec. 14 *m.* 25 Dec. 17	75	Amballa.
23 Jan. 11	*Kerr, Charles William Ernest (95, 162, 165, 169, 206).*	*B.*	*15 Oct. 69*	*10 Mar. 15* *m. 10 Mar. 18*	N	*(N. W. Ry.) Lahore Cant.*
15 June 11	*Fox, Charles James (162, 169, 206).*	*B.*	*14 Dec. 69*	*19 Mar. 15* *m. 19 Mar. 18*	N	*Asstt. Dir., Pasteur Institute of India, Kasauli.*
22 Nov. 11	*Greson, Edward James (170, 270).*	*B.*	*5 May 69*	*19 Mar. 15* *m. 19 Mar. 18*	75	*(Civil, Burma.) Calcutta.*
24 June 08	*deSilva, Godfrey Francis (207).*	*M*	*17 Dec. 65*	*19 Mar. 15* *m. 19 Mar. 18*	N	*(Civil, Madras.) Poona.*
24 June 08	*Gibson, Joseph Francis (192, 206),*	*M.*	*11 Oct. 70*	*19 Mar. 15* *m. 19 Mar. 18*	N	*(Civil, Madras.) Madras.*
24 June 08	*Dunn, Fridolin Joseph* . .	*M*	*6 Mar. 70*	*19 Mar. 15* *m. 19 Mar, 18*	N	*(Civil, Madras.) Ahmednagar.*
24 June 08	**McKie,** Robert	M.	7 June 70	*19 Mar. 15* *m. 19 Mar. 18*	N	Wellington.
24 June 08	*D'Sylva, William Robert (207)*	*M.*	*26 Sep. 68*	*19 Mar. 15* *m. 19 Mar. 18*	N	*(Civil, Madras.) Secunderabad.*
11 Sep. 09	*Coshan, Francis William Anthony,*	*M.*	*23 May 69*	*19 Mar. 15* *m. 19 Mar. 18*	N	*(Civil, Madras.) Tempy. mily. duty. Bangalore.*
11 Sep. 09	*Pereira, Robert Bernard (95, 192),*	*M.*	*7 June 70*	*19 Mar. 15* *m. 19 Mar. 18*	N	*(Civil, Madras.) Poona.*
11 Sep. 09	*Gonsalves, Daniel Randolph, F. R.C.S. (Edin.)*	*M.*	*2 Oct. 72*	*19 Mar. 15* *m. 19 Mar. 18*	N	*(Civil, Madras.) Bellary.*
11 Sep. 09	*deCruz, Felix George (105, 147)*	*M.*	*7 Feb. 73*	*19 Mar. 15* *m. 19 Mar. 18*	N	*(Civil, Madras.) Tempy. mily. duty. Belgaum.*
11 Sep. 09	*Rodrigues, Lambert Kenneth,* F.R.C.S.E. *(105, 121, 147, 193, 206). [(106).*	*M.*	*19 Nov. 74*	*19 Mar. 15* *m. 19 Mar. 18*	N	*(Civil, Burma.) Meerut. Lv., m.c., 2 mos. 18 Nov. 18.*
11 Sep. 09	**Pereira,** Edward Julian Christopher.	M.	28 Nov. 73	*19 Mar. 15* *m. 19 Mar. 18*	N	Ahmednagar.

*In ranks of Major and Captain—excluding supernumeraries and those seconded.

DEPARTMENTAL OFFICERS WITH HONORARY RANK.

INDIAN MEDICAL DEPARTMENT.

[NOTE.—Assistant Surgeons seconded for Civil employment are shown in italics.]

Senior Assistant Surgeons with the honorary rank of Captain.

Date of First Commn.	Names.	Estabt.	Date of birth.	Date of present rank.	F. rule.	Remarks.
19 Apl. 12	**Raymond**, Celestine (Supy.)	Bo.	7 Mar. 62	23 Mar. 16	N	Dadar.
11 Mar. 11	**Apcar**, Martyrose Mackertich Owen John.	B.	1 Mar. 67	30 Apl. 16	75	Multan.
15 Feb. 10	**Macqueen**, Kenneth George Sydney (f) (95).	M.	28 Oct. 71	4 Nov. 16	N	6th Division.
12 Mar. 12	*Bœrmel, Reginald Alexand r (99)*	B.	*19 Oct. 68*	*5 Nov. 16*	75	*(Civil, Bengal.) Nasirabad.*
28 May 12	*Hefferman, William St. Michael, D. T. M. (170).*	B.	*29 Sep. 67*	*5 Nov. 16*	N	*Civil and Jail, Bassein, Burma.*
23 Sep. 12	**Griffiths**, Ernest Arnim Carapiet (99, 169).	B.	5 Aug. 67	5 Nov. 16	N	Dinapore.
7 Dec. 12	*Fleming, James Francis (169).*	B.	*20 Nov. 65*	*6 Nov. 16*	N	*Supdt., Punjab Lunatic Asylum, Lahore.*
7 Dec. 12	*Murphy, Edwin Joseph (106, 120, 162, 170, 206)*	B.	*31 May 68*	*6 Nov. 16*	N	*(Civil, Burma.) Tempy. mily. duty.*
7 Dec. 12	*Babonau, Robert Gunn*	B.	*4 Aug. 71*	*6 Nov. 16*	N	*Naini Tal.*
7 Dec. 12	*Mills, Harry George Charles.*	B.	*7 June 69*	*6 Nov. 16*	N	*(Civil, Bihar and Orissa.) 2nd Division.*
7 Dec. 12	**Prins**, Edward Gerald Alfred	B.	24 Dec. 69	6 Nov. 16		Secunderabad.
23 Dec. 12	**DeSilva**, Louis Xavier	Bo.	3 Feb. 64	21 Feb. 17	N	Mount Abu.
9 Aug. 13	**Rodrigues**, Peter Hyacinth (94, 169, 172a) (Supy.) [162, 206).	Bo.	2 Nov. 62	7 Mar. 17	N	Dockyard Dispy., Bombay.
15 Feb. 10	*Thomas, Arthur William, (107,*	M.	*1 Oct. 75*	*4 May 17*	N	*(Civil, Madras.)*——
25 Nov. 13	**DeSouza**, Peter Nicholas (169, 207)	Bo.	17 Feb. 64	9 Aug. 17	N	——
1 Dec. 13	**DeCunha**, Alfred (04, 169, 207)	Bo.	1 Jan. 66	10 Aug. 17		Karachi.
10 Mar. 13	*Byers, George Francis (165), (169)*	B.	*3 Apl. 68*	*16 Sep. 17*	75	*(N. W. Ry.)* Bareilly.
10 Mar. 13	*Bancroft, Charles Reginald Washington.*	B.	*11 July 68*	*16 Sep. 17*		*Civil, Goalpara, Assam.*
8 Dec. 13	**White**, William Thomas	B.	5 July 69	16 Sep. 17		Freeman-Thomas War Hospl., Bombay.
31 Dec. 13	*Cox, Herbert William Valentine (170).*	B.	*14 Feb. 71*	*2 Feb. 18*		*(Civil Punjab). Tempy. Mily. duty, A. T. "Erinpura."*
31 Dec. 13	**Gallaway**, Herbert James (170)	B.	14 Dec. 69	2 Feb. 18		Fort Sitabuldi.
1 Jan. 14	**Wale**, Christopher Charles Augustus (supy.)	Bo.	20 Sep. 63	2 *Nov.* 17		Kamptee.
19 Mar. 15	*Almeida, Anthony Edwin (207) (supy.)*	Bo.	*27 May 64*	*26 Feb. 18*		*(Civil, Bombay.) Kuldana.*
19 Mar. 15	**Beale**, Alfred (95, 162, 170) (supy.)	Bo.	5 Sep. 64	26 Feb 18		Medical Store Depôt, Bombay.
1 Jan. 14	*Charters, William Christopher Montague.* [(94, 169, 207).	B.	*12 Feb. 70*	*20 Mar. 18*		*(Civil, Punjab.) Tempy. mily. duty, Wellington.*
1 Jan. 14	**Holmes**, George Cornelius Francis	B.	24 Aug. 69	20 Mar. 18		Peshawar.
28 Jan. 14	*Mills, Percival Beatson (170, 207)*	B.	*5 Nov. 71*	*27 Apl. 18*		*(Civil, U. P.) Khandwa.*
28 Jan. 14	*Brachio, John Jeremiah Alexander*	B.	*11 July 71*	*27 Apl. 18*		*(Civil, Bengal.) Mily. duty, Lucknow.*
28 Jan. 14	*O'Brien, George Patrick (106, 170, 196, 207)*	B.	*22 Aug. 70*	*27 Apl. 18*		*(Civil, Burma.) Wellington.*
28 Jan. 14	*Thompson, Christopher George (170)*	B.	*25 Apl. 70*	*27 Apl. 18*		*(Civil, U. P.) Karachi.*
28 Jan. 14	*Mansfield, Henry (170)*	B.	*22 Dec. 70*	*27 Apl. 18*		*(Civil, Bengal.) Tempy. mily. duty, Rawalpindi.*
28 Jan. 14	**Archer**, Edward Joseph '. (170, 207)	B.	12 Apl. 70	27 Apl. 18		Sargodha.
1 Feb. 14	**Stone**, William Joseph Key (170)	B.	26 Feb. 72	25 June 18		Supdt., Mily. Pupil class, Medl. College, Calcutta.
23 Mar. 16	**Nunes**, Joseph Mathias (supy.)	Bo.	5 Sep. 63	24 Aug. 18		Bombay.
13 Dec. 14	*Deeks, William Charles Lobo (170, 206)*	B.	*26 Mar. 73*	*5 Sep. 18*		*(Civil, Punjab.) Tempy. mily. duty. H. S. "Varela."*
13 Dec. 14	*Harvey, James Adolphus Fanshaw (170)*	B.	*29 June 71*	*5 Sep. 18*		*(Civil, C. P.) Tempy. Mily. duty.*
13 Dec. 14	*Rodgers, Rivers Thomas (169)*	B.	*13 Sep. 72*	*5 Sep. 18*		*Supdt., Centl. Jail, Jubbulpore.*
13 Dec. 14	*Turner, John Nelson (165, 207)*	B.	*3 Apl. 71*	*5 Sep. 18*		*Civil, Gangtok, Sikkim.*
13 Dec. 14	*Mann, Henry Valentine (170)*	B.	*3 Feb. 72*	*5 Sep. 18*		*(Civil, Bengal.) Calcutta.*
13 Dec. 14	**Dewey**, Harold Vernon (95, 170, 207)	B.	24 Apl. 72	*5 Sep. 18*		Rawalpindi.
21 Feb. 17	*Chamarett, George Robert (94, 169, 207).*	Bo.	*5 Aug. 64*	*5 Sep. 18*		*(Civil, Punjab.) Secunderabad.*
21 Feb. 17	**D'Souza**, Louis (170, 188)	Bo.	8 Jan. 65	5 Sep. 18		G. C. Factory, Jubbulpore.

DEPARTMENTAL OFFICERS WITH HONORARY RANK.

INDIAN MEDICAL DEPARTMENT.

Senior Assistant Surgeons with the honorary rank of Lieutenant (23).* { Bengal 14, Madras 2, Bombay 7 }

Date of First Commn.	Names.	F. rule.	Date of birth.	Date of present rank.	Remarks.
24 June 08	**Cartman,** Stephen Ansell (105, 207) (*Supy.*)	M.	2 May 62	24 June 08	Wellington.
30 Mar. 10	*Craggs, Henry Clement (161, 170, 207)*	M.	13 Jan. 75	30 Mar. 10	*(Civil, Madras.) Tempy. Mily. duty ——. Lv., m.c., 2 mos. 14 days, 28 Sep. 18.*
30 Mar. 10	*Carreck, Frederick, Frank Rath (104)*	M.	5 Dec. 76	30 Mar. 10	*(Civil, Madras.) Tympy. Mily. duty. ——*
30 Mar. 10	*Truter, Albert Wilhelm (105, 207)*	M.	24 Aug. 76	30 Mar. 10	*(Civil, Burma.) Saugor.*
4 May 10	*Hamlin, Alfred Edward (105, 207) L. R. C. P. (Lond.) M. R. C. S. (Eng.) (121, 181).*	M.	8 Mar. 75	4 May 10	*(Civil, Burma.) Sialkot.*
1 Apl. 13	**Brown,** Arthur George (105, 206)	M.	28 Feb. 72	1 Apl. 13	Medical Store Depôt, Calcutta.
25 Dec. 14	**Masterson.** William John (170)	B.	20 Apl. 72	25 Dec. 14	Medl. Store Depôt, Lahore.
1 Mar. 15	*Malir, Joseph*	B.	15 Sep. 69	1 Mar. 15	*(N.-W. Ry.)*
1 Mar. 15	**Holmes,** Joseph Alexander Hercules (170).	B.	15 Mar. 71	1 Mar. 15	
10 Mar. 15	*Gillson, William John*	B.	3 June 70	10 Mar. 15	*(Civil, Bengal.) Ishapore.*
10 Mar. 15	*Deeks, Frederick George Handel*	B.	11 Mar. 72	10 Mar. 15	*(Plague duty, Allahabad.)*
10 Mar. 15	**Bamford,** William (170)	B.	2 Sep. 72	10 Mar. 15	——
19 Mar. 15	**Mersh,** William George (207)	B.	8 Feb. 71	19 Mar. 15	——
10 Mar. 13	**O'Donoghue,** Patrick Edward (Supy.).	B.	14 June 70	21 Mar. 15	8th Division.
3 June 15	**McDonald,** James Joseph (*Supy.*) (*f*), I.	B.	20 June 74	3 June 15	16th (Indian) Division. *Lv., m.c., 1 yr., 1 Sep. 17. Extn. 6 mos.*
3 June 15	**Brown,** William Henry (*Supy.*) (*f*) II.	B.	9 Jan. 75	3 June 15	Deolali.
1 Dec. 15	**Duckworth,** Edwin Victor (*f*) (*Supy.*) III.	B.	11 May 71	1 Dec. 15	Bangalore.
30 Apl. 16	**Montgomery,** Joseph Pierre	B.	9 Jan. 68	30 Apl. 16	Cossipore.
30 Apl. 16	*Brookes, William Lindfield (104, 140, 149, 162, 169, 206).*	B.	22 Jan. 72	30 Apl. 16	*(Civil, Burma.) Pr. of War Camp, Thayetmyo.*
20 Sep. 16	**Stewart,** Hugh Norman (95) (*tempy.*)	B.	16 Aug. 72	20 Sep. 16	
5 Nov. 16	*Kidby, William Thomas*	B.	21 Dec. 72	5 Nov. 16	*(Civil, Punjab.) Tempy. mily. duty. ——*
5 Nov. 16	*Matthews, Ernest George Albert (207).*	B.	26 Feb. 74	5 Nov. 16	*(Civil, U. P.) Tempy. Mily. duty. ——*
5 Nov. 16	*Baptist, Arthur Ambrose Emmanuel (162, 165, 169, 206) L.S.A. Lond. III.*	B.	1 Nov. 73	5 Nov. 16	*(Civil, Bihar and Orissa.) Mily. duty, Cawnpore.*
6 Nov. 16	*Perdriau, Alexander Douglas Cunningham (Supy.). IV (f) (170, 207).*	B.	22 Sep. 72	6 Nov. 16	*(Civil, C. P.) Muttra.*
6 Nov. 16	*Matthews, John Gustavus Thomas*	B.	21 Nov. 71	6 Nov. 16	
6 Nov. 16	*Peachy, Edwin Augustus (170)*	B.	6 May 73	6 Nov. 16	*(Civil, Burma.) Lebong.*
6 Nov. 16	*Martin, William Jeremiah Phillips*	B.	15 July 70	6 Nov. 16	*(Civil, Punjab.) Meerut.*
6 Nov. 16	**Meneaud,** Charles John Augustus	B.	15 Mar. 70	6 Nov. 16	——
7 Mar. 17	*Deane, George Archibald (207)*	Bo.	1 Aug. 67	7 Mar. 17	*(Civil, Bombay.) Poona.*
7 Mar. 17	*Pollock, Claude Reginald (207).*	Bo.	12 July 67	7 Mar. 17	*(Civil, Burma.) Cawnpore.*
7 Mar. 17	*Whittenbury, Joseph Henry (161, 170 206).*	Bo.	7 Oct. 66	7 Mar. 17	*(Civil, Bombay.) Campbellpore.*
7 Mar. 17	*Netscher, Francis Hubert Dean*	Bo.	18 July 66	7 Mar. 17	*(Civil, Bombay.) Tempy. mily. duty, Bellary.*
7 Mar. 17	**DeSouza,** Louis Joseph	Bo.	13 Mar. 67	7 Mar. 17	6th Division.
4 May 17	*Passanha, Waldemar Herman*	M.	18 Sept. 76	4 May 17	*No. 138 C. F. A., Sect. "A," Aden.*
4 May 17	*Trutwein, Percy St. Clair*	M.	8 Oct. 76	4 May 17	*X-Ray Dept., Ripon Haspl., Simla.*

* Excluding supernumeraries and those seconded.

(*f*) Specially promoted under the provisions of para. 470, A. R., I., Vol. II.

INDIAN MEDICAL DEPARTMENT.

Senior Assistant Surgeons with the Honorary rank of Lieutenant—*contd.*

Date of First Commn.	Names.	Estabt.	Date of birth.	Date of present rank.	Remarks.
4 May 17	**Stracey,** Patrick St. Clair (111, 193).	M.	11 Aug. 77	4 May 17	2nd (Rawal Pindi) Dn.
9 Aug. 17	**Noronha,** Dioginho . .	Bo.	12 Sep. 65	9 Aug. 17	Jhansi.
9 Aug. 17	*MacGregor, Alexander Hepburn (207).*	*Bo.*	*9 Nov. 66*	*9 Aug. 17*	*(Civil, Bombay.) Calicut.*
9 Aug. 17	**Paes,** Anthony (170) . .	Bo.	21 Dec. 65	9 Aug. 17	———.
9 Aug. 17	*Vincent, George Walter .*	*Bo.*	*8 Jan. 68*	*9 Aug. 17*	*(Civil, Burma.) Malapuram.*
9 Aug. 17	*Ives, Robert George (95, 170)*	*Bo.*	*20 Apl. 75*	*9 Aug. 17*	*(Civil, Bombay.) Tempy. Mily. duty—*
9 Aug. 17	*Kelly, Claude Cyril (94, 170, 207).*	*Bo.*	*1 May 72*	*9 Aug. 17*	*Ty. Mily. duty.*
9 Aug. 17	*Kirkpatrick, William Ernest (185.)*	*Bo.*	*7 Sep. 74*	*9 Aug. 17*	*(Civil, Bombay.) 5th Division. Lv., m. c., 20 mos. 14 dys., 11 Oct. 16. Exten. 2 mos.*
9 Aug. 17	*DeSousa, Alfred Baldwin .*	*Bo.*	*26 Feb. 75*	*9 Aug. 17*	*(Civil Bombay.)*
9 Aug. 17	**Lawrence,** John Letwitch (94, 178, 207).	Bo.	20 Jan. 73	9 Aug. 17	Bombay Brigade.
10 Aug. 17	**James,** Arthur	Bo.	16 Apl. 74	10 Aug. 17	Medical Store Depôt, Rangoon.
2 Aug. 17	**Thoy,** Guildford (170) .	B.	30 Jan. 74	2 Aug. 17	4th Divn.
4 Aug. 17	**Benson,** Herbert Edward IV.	B.	27 Aug. 70	4 Aug. 17	No. 3 Co., A. B. C., Lahore Cantt.
25 Oct. 17	**Hardaker,** Alexander John (162, 169).	B.	4 June 71	25 Oct. 17	———.
25 Decr. 17	*Marshall, William John .*	*B.*	*9 Dec. 72*	*25 Decr. 17*	*(Civil, Bengal.) 8th Division.*
2 Feb. 18	*Culpeper, Arthur George .*	*Bo.*	*15 Mar. 73*	*2 Feb. 18*	*(Civil, Burma.) Fyzabad.*
24 Feb. 18	*Pettigrew, Reginald William*	*Bo.*	*31 Oct. 73*	*24 Feb. 18*	*(Civil, Bombay.) Jhansi.*
24 Feb. 18	*Smith, Frederick Russell .*	*B.*	*13 Sep. 73*	*24 Feb. 18*	*(Quarantine Medl. Offr.) Mily. duty. ——*
26 Feb. 18	**Hussey,** William George St. John.	B.	23 Dec. 71	26 Feb. 18	Poonamallee.
26 Feb. 18	**Netscher,** Walter Arthur Clifford (Supy.)	Bo.	2 Jan. 71	26 Feb. 18	S. A. Venereal Laby., Poona.
5 Mar. 18	*Shemain, Nicholas Philip (188a, 193).*	*B.*	*15 Feb. 77*	*5 Mar. 18*	*(Civil, Bombay.) Kirkee.*
5 Mar. 18	**Kenyon,** Louis Percival (95, 161, 170, 206).	B.	29 Jan. 73	5 Mar. 18	M. M. S., Mhow.
20 Mar. 18	*Mullins, Charles (165, 170) .*	*B.*	*29 Aug. 73*	*20 Mar. 18*	*(Offg. Civil, Assam.) Tempy. mily. duty.*
20 Mar. 18	*Stewart, Donald Day (104, 140, 170).*	*B.*	*1 Jan. 72*	*20 Mar. 18*	*(Civil, Burma.) Tempy. mily. duty.—*
20 Mar. 18	*Cearns, George William I.*	*B.*	*17 Feb. 74*	*20 Mar. 18*	*(Civil, U. P.) ——.*
27 Apl. 18	**Rebeiro,** Walter Reginald	B.	25 Sep. 73	27 Apl. 18	H. S. "Loyalty."
25 June 18	*Knight, Frank Ernest . .*	*B.*	*28 Jan. 75*	*25 June 18*	*(Civil, Bengal.) ——*
25 June 18	*Thomas, John Douglas (170)*	*B.*	*4 Sep. 72*	*25 June 18*	*(Civil, Bengal.) Calcutta.*
25 June 18	*Langhorne, Sydney Graham (170).*	*B.*	*9 Feb. 74*	*25 June 18*	*(N. W. Ry.) No. 3, A. B. C., Lahore Cant.*
25 June 18	*Stuart, William Willoughby, L.R.C.P. & S.I.*	*B.*	*15 Feb. 72*	*25 June 18*	*(Civil, C. P.) Tempy. mily. duty. Lv., m. c., 2 mos., 12 Apl. 18, ex. 12 weeks.*
25 June 18	**Smith,** Benjamin Lazarus (170).	B.	4 Dec. 73	25 June 18	———.
2 July 18	*Thompson, Herbert Charles (170).*	*B.*	*3 Aug. 73*	*2 July. 18*	*(Civil, U. P.) Tempy. mily. duty, Rawalpindi.*
2 July 18	*Bates, Adolphus Vincent (Supy.)*	*B.*	*1 Aug. 74*	*2 July. 18*	*Asstt. to Supdt., Presy. Genl. Hospl., Calcutta.*
2 July 18	**Hanvey,** Gilbert Reuben Joseph (Supy.)	B.	27 Apl. 74	2 July 18	Medl. Mobn. Stores, Lucknow.
24 Aug. 18	*Scott, Lionel . . . . .*	*Bo.*	*7 Dec. 75*	*24 Aug. 18*	*(Civil, Bombay.) Tempy. mily. duty, Rawalpindi.*
24 Aug. 18	*Andeen, George Fredrick, L. A. H. (Dub.) (94, 162, 169).*	*Bo.*	*5 Apl. 77*	*24 Aug. 18*	*(Civil, Bombay.) Lady Willingdon War Hospl., Poona.*
24 Aug. 18	**Thipthorp,** Ernest Ebenezer (207).	Bo.	19 Mar. 70	24 Aug. 18	Madras.

Date of first column.	Names.	Estabt.	Date of birth.	Date of present rank.	Remarks.

INDIAN MEDICAL DEPARTMENT.

Senior Assistant Surgeons with the Honorary rank of Lieutenant—*concld.*

5 Sep. 18	*Coxe, Charles Robert . . .*	B.	*18 Aug. 72*	*5 Sep 18*	*(Plague duty) Neemuch.*
5 Sep. 18	*Stone, Richard Julian Key (170).*	*B.*	*23 Mar. 75*	*5 Sep. 18*	*(Salt Deptt.)* ——
5 Sep. 18	*Doyle, James (170) . . .*	*B.*	*1 Jan. 73*	*5 Sep 18*	*(Civil, C. P.) Quetta Division.*
5 Sep. 18	*Byrne, Henry Reynold (162)*	*Bo.*	*14 Oct. 77*	*5 Sep. 18*	*(Civil, Bombay.) Fort Lahore.*
5 Sep. 18	*Hendricks, St. John Edward*	*Bo.*	*27 Apl. 76*	*5 Sep. 18*	*(Civil, C. P.) Tempy. mily duty.*
5 Sep. 18	*Donnelland, Joseph Henry Arnold (111, 188a) . . .*	*Bo.*	*14 Dec. 78*	*5 Sep. 18*	*(Civil, Bombay.) Tempy. Mily. duty, Bombay.*
5 Sep. 18	**Thompson,** Charles . .	Bo.	28 Aug. 76	5 Sep. 18	Mhow.
20 Sep. 18	**Greene,** Henry Alexander (95, 162, 165, 169, 207). (Supy.)	B.	23 Mar. 74	20 Sep 18	——.

MEDICAL WARRANT OFFICERS—(418).

INDIAN MEDICAL DEPARTMENT.

Assistant Surgeons, 1st Class.

Ranking as Conductors.

Names.	Estabt.	Date of birth.	Date of warrant rank.	Date of appointment to present class.	Remarks.
Menezes, John Francis (159) . .	*Bo.*	*1 July 67*	*14 Sep. 87*	*14 Sep. 06*	*(Civil, Bombay.) Karachi.*
Pereira, Dominic Frank (207) (Supy.)	Bo.	30 Mar. 62	20 Sep. 86	1 Jan. 07	Lucknow.
Williams, Clarence John (106, 206)	B.	22 Feb. 73	5 Feb. 95	15 Feb. 14	Meiktila.
Turner, William Warnett (169) . .	*B.*	*18 Oct. 72*	*4 June 94*	*15 Apl. 14*	*(S. P. Ry.)* ——.
Mathews, Frederick William, L.R.C.P. & S.I., L. M., DUB.	*B.*	*5 Mar. 75*	*5 Feb. 95*	*15 Apl. 14*	*(Civil, U. P.) Quetta.*
Davies, Edward Aubrey II	*B.*	*27 July 74*	*5 Feb. 95*	*15 Apl. 14*	*(Civil, Burma.) Darjeeling.*
Cutler, Frederick George	*B.*	*3 Dec. 72*	*5 Feb. 95*	*15 Apl. 14*	*(Civil, C. P.) Jubbulpore.*
Johnson, Edgar Osborne (170) .	B.	20 Mar. 73	19 Feb. 94	15 Apl. 14	Multan.
Parrott. William FitzAllan (95, 170, 207)	B.	25 Feb. 73	17 Sep. 94	27 Apl. 14	Peshawar.
Ellis, Samuel Neville	B.	28 May 72	5 June 95	5 June 14	Lucknow.
Richardson Hector Alfred (172a) .	B.	22 Oct. 75	18 Sep. 95	18 Sep. 14	Sub-charge Medl. Mobn. Stores, Lahore Cantt.
Callaghan, George Maurice (*f*) (Supy.) (193) VII.	M.	20 May 78	31 Mar. 99	11 Nov. 14	Kirkee.
Twells, Thomas William . . .	*B.*	*26 Aug. 75*	*20 Dec. 95*	*11 Nov. 14*	*(Civil, U. P.) Roorkee.*
Whitley, Arthur Henry	*B.*	*18 Jan. 76*	*10 Feb. 96*	*11 Nov. 14*	*(Civil, U. P.) Bareilly.*
O'Connor, William (95, 165, 169, 172a).	*B.*	*21 Oct. 74*	*10 Feb. 96*	*11 Nov. 14*	*Certifying Surgeon of Factories, Bengal.*
Cornelius, Alfred Bertram (170) .	*B.*	*19 Mar. 75*	*10 Feb. 96*	*11 Nov. 14*	*(Civil, Bengal.)* ——

(*f*) Specially promoted under the provisions of para. 470, A. R., I., Vol. I.

INDIAN MEDICAL DEPARTMENT.

Assistant Surgeons, 1st Class—*contd.*

Ranking as Conductors.

Names	Estabt.	Date of birth.	Date of warrant rank.	Date of appointment to present class.	Remarks.
Shave, Edward Desmond (165, 169)	B.	1 Sep. 75	10 Feb. 96	11 Nov. 14	(Civil, Punjab.) 9th Division.
Brooks, Thomas Henderson (105, 125a, 140, 148b, 169, 196)	B.	27 Aug. 72	17 Sep. 94	11 Nov 14	(Civil, Burma.) Khanspur.
Howatson, George Archibald Simpson (170)	B.	6 Oct. 76	10 Feb. 96	11 Nov. 14	(Civil, U. P.) Tempy mily. duty. ——
Clarke, Albert Edward (172a)	B.	7 June 75	16 May 96	11 Nov. 14	(Civil, Punjab.) Tempy. mily. duty, Delhi.
White, Albert Alexander Wellington	B.	30 Aug. 74	16 May 96	11 Nov. 14	Allahabad.
O'Leary, Francis Henry	B.	5 Sep. 73	10 Feb. 96	11 Nov. 14	(Civil, Burma.) 8th Division.
Lopes, Archibald Percival (193)	B.	7 Mar. 76	16 Feb. 97	11 Nov. 14	(Civil, Bengal.) Military duty. ——Lv, m.c. 4 mos. 27 days, 17 June 18. Exten. 1 mo.
Fox, Ernest Clayton Ross (170, 207)	B.	24 Aug. 76	16 Feb 97	11 Nov. 14	Pasteur Institute, Kasauli. Temporary military duty, Bombay.
Gleeson, Frank Herbert, L.R.C.P.&S.I. L.M. (169)	B.	6 Mar. 76	16 Feb. 97	11 Nov. 14	(Civil, Bengal.) Mily. duty.
Perkins, Gilbert Henry (95, 162, 165, 169, 206)	B.	4 Jan. 76	16 Feb. 97	11 Nov. 14	I. M. Stores, 2nd (R. Pindi) Dn.
Milburn, John Cecil (170, 207)	B.	30 Mar. 75	16 Feb. 97	11 Nov. 14	(Civil, Burma.) Dalhousie.
Foy, James Robert (106, 162, 170, 206)	B.	21 Apl. 76	16 Feb. 97	11 Nov. 14	(N. W. Ry.) Meerut.
Jones, George Towers	B.	7 July 74	16 Feb. 97	11 Nov. 14	1st Division, Risalpur.
Thipthorp, William Henry	B.	9 May 74	16 Feb. 97	11 Nov. 14	R. B. S. Ry. Mily. duty, Rawalpindi.
Wredden, Joseph Longman (170, 207)	B.	3 Mar. 75	1 Apl. 97	11 Nov. 14	Dockyard Dispy., Kidderpore.
Dyer, Augustus Wilhelm (165, 169)	B.	25 Apl. 76	1 Apl. 97	11 Nov. 14	——
Mendis, Francis Xavier	B.	30 Dec. 74	10 Feb. 96	11 Nov. 14	(Civil, Bengal.) Attock.
Duckworth, Arthur Robbins	B.	4 Mar. 74	16 Feb. 97	11 Nov. 14	(Civil, Bengal.) Quetta.
Duckworth, Henry (Supy.) VII	M	24 July 77	1 Apl. 97	11 Nov. 14	(Civil, Burma.) Tempy. mily. duty, Mandalay.
Jennings, George Samuel (94, 162, 169, 206)	B.	25 Dec. 75	1 Oct. 97	11 Nov. 14	(Civil, U. P.) Amballa.
Fleming, Henry Leonard Osborne (95, 170)	B.	10 Aug. 75	10 Feb. 96	7 Dec. 14	(Civil, Bengal.) 1st Division, Peshawar.
McCurtis, Lawrence	B.	20 June 76	12 Mar. 98	12 Mar. 15	(Civil, Bengal.) Agra.
Rebeiro, Eustace Francis (94, 169)	B.	22 July 75	12 Mar. 98	12 Mar. 15	Burma Division.
Holehouse, Alfred (170) VII	B.	29 July 76	12 Mar. 98	12 Mar. 15	(Civil, Bengal.) Agra.
D'Roche, Edmund	B.	7 Apl. 74	5 Feb. 95	18 Apl. 15	1-7 Hants, Aden.
Wiseham, Willoughby Richard Kelly (95, 162, 170, 206)	B.	28 May 75	13 June 98	13 June 15	Mily. duty. ——
Garrod, Herbert John Joseph (169)	B.	1 Mar. 78	22 Feb. 99	22 Feb. 16	(Civil, U. P.) Tempy. mily duty. ——
Hurst, Francis Gerald (170)	B.	12 June 77	22 Feb. 99	22 Feb. 16	(Civil, Bengal.) Tempy. mily. duty. ——
Fox, Sydney John Vere (170)	B.	17 Aug. 77	22 Feb. 99	22 Feb. 16	(Civil, Bengal.) Tempy. mily. duty.
Hottinger, Edward Francis (170)	B.	21 Nov. 76	22 Feb. 99	22 Feb. 16	(Civil, Punjab.) Hosur.

Names.	Estabt.	Date of birth.	Date of warrant rank.	Date of appointment to present class.	Remarks.

INDIAN MEDICAL DEPARTMENT.

Assistant Surgeons, 1st Class—*concld.*

Ranking as Conductors.

Names.	Estabt.	Date of birth.	Date of warrant rank.	Date of appointment to present class.	Remarks.
Crow, Charles George (105, 170)	B.	4 Sep. 76	22 Feb. 99	22 Feb. 16	(Civil, Burma.) Gharial.
Chalke, James Charlemagne	B.	8 June 76	22 Feb. 99	22 Feb. 16	On H. E. the Govr. of Bengal's Staff.
deGruyther, Cecil Guy	B.	16 Oct. 76	22 Feb.* 99	22 Feb. 16	(Civil, Bengal.) Lv., m.c., 2 mos., 28 Nov. 18.
Chambers, Lofty Elgin	B.	31 Mar. 76	22 Feb. 99	22 Feb. 16	H. S. "Varsova."
Rebeiro, Cyril Wilfred (170)	B.	18 Feb. 78	22 Feb. 99	22 Feb. 16	(Civil, Burma.) Jhansi.
Duckworth, George Edwin (105)	M.	28 Feb. 79	31 Mar. 99	31 Mar. 16	(Civil, Burma.) Thamakan, Burma.
Duckworth, Percival Thomas (106, 193, 207). VI	M.	13 June 79	31 Mar. 99	31 Mar. 16	Medl. Store Depôt, Madras.
Gernon, Anthony Walter	M.	9 Feb. 78	31 Mar. 99	31 Mar. 16	Dehra Dun.
Gormanly, Patrick James (170)	B.	30 Dec. 75	16 Feb. 97	17 Apl. 16	———
Mahoney, Ebenezer Robert (172a, 193)	M.	12 Apl. 76	30 June 98	17 Apl. 16	(Civil, Madras.) 1st Division.
Gibbon, Douglas Arnold VIII (*f*)	B.	24 Aug. 79	17 Feb. 00	17 Aug. 16	Barian.
Murray, Robert Edmund (94, 169, 170, 270) VIII	B.	16 Mar. 78	17 Feb. 00	17 Feb. 17	(Civil, Punjab.) 1st Division.
Shillong, James Henry	M.	13 June 78	31 Mar. 00	31 Mar. 17	(Civil, Madras.) Poonamallee.
Sandways, Percy Neil	M.	10 Dec. 77	31 Mar. 00	31 Mar. 17	(Civil, Burma.) Mily duty.—
Hart, Maurice Roberts Wilson, M.B.E. (106)	M.	16 Nov. 79	31 Mar. 00	31 Mar. 17	(Civil, Madras.) Tempy. mily. duty, Nowshera.
Hart, James Alexander (193)	M.	16 Nov. 76	31 Mar. 00	31 Mar. 17	(Civil, Madras.) Aden.
Lynsdale, Henry Herbert	M.	2 Jan. 77	31 Mar. 00	31 Mar. 17	(Civil, Burma.)
Moody, William Ernest (94, 207, 169) IX (*f*).	Bo.	19 Feb. 80	1 Oct. 00	1 Apl. 17	Henjam.
Harvey, Ashley Edwin Dunbar (170)	B.	18 May 73	16 Feb. 97	15 Apl. 17	Asst. to the Senior Medl. Offr., Port Blair, and in m./c. Ross Hospl.
King, George Harold (207)	M.	27 Feb. 75	1 Apl. 97	15 Apl. 17	Bellary.
Pinto, Maurice Camillus (106)	B.	15 July 76	13 July 98	15 Apl. 17	6th Division. ———
Fitzpatrick, Robert John	B.	19 Oct. 78	17 Feb. 00	15 Apl. 17	Fort Govindgarh.
Browne, Robert Fredric (193)	M.	7 Mar. 79	31 Mar. 00	21 Apl. 17	Jubbulpore.
Scanlon, Patrick Augustus (106)	M.	6 Sep. 77	1 July 00	1 July 17	(Civil, Burma.) Tempy. mily. duty, Bombay. Sub-charge Colaba War Hospl.
Stevenage, Emanuel Antony (94, 162, 169, 207) (Supy.) X. IX	M.	20 Dec. 78	31 Mar. 01	1 Oct. 17	No. 9 Coy., A. B. C., Secunderabad.
Williams, John George	Bo.	21 Apl. 79	1 Oct. 00	1 Oct. 17	(Civil, Bombay.) 1st Division.
Lopez, Vincent Joseph (95, 193, 206) XI (*f*).	M.	18 Nov. 78	31 Mar. 01	1 Oct. 17	Rawalpindi.
Thompson, Harry Victor James (206) XII (*f*).	M.	24 May 78	31 Mar. 01	1 Oct. 17	No. 10 Coy., A. B. C.
Munrowd, Henry George Hemmings	B.	23 Nov. 77	4 Mar. 01	4 Mar. 18	(Offg. Civil, Assam.) Meerut.
Bergin, Francis Aloysius	M.	28 Apl. 79	31 Mar. 01	31 Mar. 18	Rangoon.
McGrath, Patrick John (162, 206) X XI XII	M.	18 Mar. 79	31 Mar. 03	31 Mar. 18	(Civil, Madras.) ——
Rehling, Geoffrey Carl (95, 162, 169, 172a, 206) XIII (*f*)	Bo.	26 Nov. 79	1 Nov. 99	20 Apl. 18	Aden.
Torpy, Arthur Thomas	Bo.	20 Nov. 78	1 Oct. 00	21 Apl. 18	Medl. Store Depôt, Bombay.
DePenning, Hubert Felix (94, 100, 162, 169, 207).	Bo.	17 Jan. 79	27 Sep. 01	27 Sep. 18	(Plague duty, Baluchistan.) 2nd Division.
Smyth, Stanislaus George (172a, 207)	Bo.	31 Oct. 79	27 Sep. 01	27 Sep. 18	Asst. to Civil Surgeon, Delhi.
Easdon, Lionel Vivian O'Brien (95)	Bo.	15 Nov. 80	27 Sep. 01	27 Sep. 18	(Civil, Bombay.) Peshawar.
Maher, William Hugh (206)	Bo.	15 Jan. 79	27 Sep. 01	27 Sep. 18	Malapuram.
Otto, Hermann Frank	Bo.	12 Oct. 79	27 Sep. 01	27 Sep. 18	(Civil, Bombay.) Wellington.

Assistant Surgeons, 2nd Class.

Ranking as Conductors.

Names.	Estabt.	Date of birth.	Date of warrant rank.	Date of appointment to present class.	Remarks.
Blinkworth, Kenneth William (170)	B.	30 Sep. 81	13 Mar. 02	13 Mar. 14	(Offg. Civil, Assam.) Tempy. mily. duty. ——

(*f*) Specially promoted under the provisions of para. 470, A. R., I., Vol. II.

Names.	Estabt.	Date of birth.	Date of warrant rank.	Date of appointment to present class.	Remarks.

INDIAN MEDICAL DEPARTMENT.

Assistant Surgeons, 2nd Class—*contd.*

Ranking as Conductors.

Taylor, Clement George	*M.*	*6 Jan. 79*	*31 Mar. 02*	*31 Mar. 14*	*(Civil, Madras.) Tempy. mily. duty. A. T. "Syria."*
Montgomery, Charles William Taylor.	B.	5 Aug. 76	1 Nov. 99	20 Apl. 14	Bombay.
Lawrence, Alfred George XIII	Bo.	14 Apl. 70	1 Nov. 99	20 Apl. 14	Kirkee.
Boyne, Archibald Raymond Hastings (94, 170, 207)	Bo.	23 May 80	27 Sep. 01	*20 Apl. 14*	——.
Bennett, William Robert (95, 106)	M.	18 Oct. 79	31 Mar. 01	21 Apl. 14	——.
White, Donald Edward Everard (§)	*Bo.*	*8 Apl. 81*	*1 Oct. 02*	*1 Oct. 14*	*(Civil, Bombay.) Dehli.*
Kelly, Hubert Ronald	Bo.	21 June 79	1 Oct. 00	28 Dec. 14	H. M. S. "Minto."
Jaensch, Louis Victor (207)	*B.*	*30 June 82*	*2 Mar. 03*	*2 Mar. 15*	*(Civil, U. P.) ——.*
Holmes, Frank Walton (169)	*B.*	*19 Oct. 82*	*2 Mar. 03*	*2 Mar. 15*	*(Civil, U. P.) Dehra Dun.*
McCurtis, Ambrose (95, 162, 165, 170, 185, 206)	*B.*	*11 Mar. 82*	*2 Mar. 03*	*2 Mar. 15*	*Supdt., Central Prison, Naini, U. P. s. p. t.*
Salt, James Samuel (169, 206)	*B.*	*3 Oct. 81*	*2 Mar. 03*	*2 Mar. 15*	*Resy. Dispy., Bushire, Persian Gulf.*
Bell, Philip	B.	29 Oct. 81	2 Mar. 03	2 Mar. 15	Lahore Cantt.
Fordham, Henry James John (170)	*B.*	*25 Aug. 82*	*2 Mar. 03*	*2 Mar. 15*	*(Civil, Punjab.) Agra.*
Gibson, Thomas Joseph, L.M.S. *(Lond.),* L.R.F.P. *&* S. *(Glas.) &* L. M. *(Dub.)* (193)	*M.*	*9 Feb. 81*	*31 Mar. 03*	*31 Mar. 15*	*(Civil Hospl., Mercara) Bangalore. Lv. m. c., 42 days from 25 Sep. 18.*
Huffton, Joseph Hillier Spencer (106, 193)	M.	2 Sep. 82	31 Mar. 03	31 Mar. 15	Secunderabad.
Gernon, Lawrence Peter	M.	7 Aug. 81	31 Mar. 03	31 Mar. 15	Wellington.
Delaney, Clement Davidson	M.	26 Mar. 80	31 Mar. 03	31 Mar. 15	Bangalore.
Gueiselar, John Alexander (Supy.)(94, 100, 147, 161, 169, 193, 206), XIV.	*Bo.*	*1 June 80*	*31 Mar. 03*	*31 Mar. 15*	*Civil Surgeon, Sibi.*
Rosair, Harry Basil, L.R.C.P. *&* S.I., L.A.H. *&* L.M. *(Dub.)* D. T.M.	*M.*	*1 May 82*	*31 Mar. 03*	*19 Apl. 15*	*(Civil, U. P.) 9th Division.*
Vincent, Henry XV(*f*)	Bo.	12 June 82	26 Sep. 04	3 June 15	Poona.
Flanagan, Harold James Clarke XVI(*f*) (193).	B.	20 Jan. 82	31 Mar. 05	3 June 15	——.
Warner, Charles Mark Ernest XVII (*f*) (162, 172*a*, 206)	B.	19 Sep. 85	7 June 05	3 June 15	No. 6 Coy., A. B. C., Poona.
Raphael, Stephen Charles XVIII. (*f*)	B.	2 Dec. 85	5 Mar. 06	3 June 15	——.
Cotton, Eugene Avron XIX (*f*)	Bo.	10 Nov. 84	30 June 06	3 June 15	M. S. Depôt, Bombay.
Gillson, Edward Henry XX (*f*) (95)	B.	11 Oct. 85	1 June 07	3 June 15	
Steinhoff, Frederick George (162, 206)	*Bo*	*19 June 83*	*16 Sep. 03*	*16 Sep. 15*	*Tempy. Mily. duty, H. S. "Vita."*
Browne, Percy Walter Roland	Bo.	26 Sep. 82	16 Sep. 03	16 Sep. 15	Poona.
White, Eric Douglas	Bo.	27 Nov. 82	16 Sep. 03	16 Sep. 15	6th Division.
Peters, Joseph Hugh Charles(172*a*)	Bo.	31 Oct. 81	16 Sep. 03	16 Sep. 15	Karachi.
Francis, Leonard Arnodd (207)	Bo.	15 Dec. 81	16 Sep. 03	16 Sep. 15	——.
Mathews, Valentine George, L.R.C. S.I., L.A. H. *(Dub.)*	*B.*	*14 Feb. 81*	*24 Sep. 03*	*24 Sep. 15*	*(Civil, U. P.) Delhi.*
Hughes, Reginald Hastings (95, 105, 206)	M.	17 Feb. 84	31 Mar. 04	1 Oct. 15	Medl. (Mobn.) Stores, 9th (Secunderabad) Division.
Rodrigues, James Michael, M.C. XXI(*f*).	B.	13 Jan. 83	7 June 05	1 Jan. 16	Meerut.
Dudley, Gerald Walter XXII (*f*)	Bo.	31 July 85	31 Oct. 05	1 Jan. 16	Embn. Staff, Bombay.
Moody, Cyrus Murray (94, 162, 170, 207)	Bo.	27 Dec. 82	4 Jan. 04	4 Jan. 16	M. S. D., Bombay.
Keelan, Robert Samuel (162, 165, 169)	B.	5 Nov. 81	25 Feb. 04	25 Feb. 16	*(Civil, U. P.) Tempy. mily. duty. Deolali.*
Foy, Frank Herbert (165, 170, 207)	B.	24 [illegible] 82	25 Feb. 04	25 Feb. 16	Meerut.
Holmes, Leslie Stuart (169)	B.	20 Aug. 84	25 Feb. 04	25 Feb. 16	Meerut. *Lv., m. c., 6 mos. in I. Exten. 60 days.*
Butcher, Thomas Bertram (170)	*B.*	*13 Oct. 83*	*25 Feb. 04*	*25 Feb. 16*	*(Civil U. P.) Tempy. mily.*
King, Henry Anderson	B.	4 Oct. 82	25 Feb. 04	25 Feb. 16	Bareilly. *[duty.*
Bayley-deCastro, Alexander	B.	1 Dec. 82	25 Feb. 04	25 Feb. 16	Cownpore.
Thompson, Ralph Hope Gorman	B.	21 Apl. 80	25 Feb. 04	25 Feb. 16	Khandwa.
Mackay, Donald Lionel (95, 172*a*)	B.	1 Apl. 83	25 Feb. 04	25 Feb. 16	Barian. *[duty.—*
Mason, Colin Hope (172a)	*B.*	*15 Nov. 82*	*25 Feb. 04*	*25 Feb. 16*	*(Civil, C. P.) Tempy. mily.*
Bartels, George Keyes Ashworth(147)	B.	18 Aug. 83	25 Feb. 04	25 Feb. 16	Medl. Store Depôt, Madras.
Deatker, Frederick Tucker (66, 169, 207)	*B.*	*7 Dec. 82*	*25 Feb. 04*	*25 Feb. 16*	*(N. W. R.) No. 5 Coy., A. B. C., Mhow.*

(*f*) Specially promoted under the provisions of para. 470, A. R., I., Vol. II.

INDIAN MEDICAL DEPARTMENT.

Assistant Surgeons, 2nd Class—*contd.*

Names.	Estab.	Date of birth.	Date of warrant rank.	Date of apptt. to present class.	Remarks.
		Ranking as Conductors.			
Elloy, Arthur Leonard (172*a*)	B.	2 Mar. 82	25 Feb. 04	25 Feb. 16	Sub-Ch., X-Ray apparatus, Bareilly.
Phillips, Henry Cecil (172*a*)	B.	14 Nov. 82	25 Feb. 04	25 Feb. 16	(Civil, Punjab.) Bombay.
Simmons, Bertram Hope (170)	B.	29 Dec. 83	25 Feb. 04	25 Feb. 16	(Civil, Bengal.) Rawalpindi.
deGruyther, Alexander Nicoll (172*a*)	B.	10 June 82	25 Feb. 04	25 Feb. 16	Meerut. 7 Coy. A.B.C.
Stuart, Thomas Joseph	B.	24 Aug. 81	25 Feb. 04	25 Feb. 16	Poona.
Lincoln, Cecil Henning (170)	B.	21 Nov. 83	25 Feb. 04	25 Feb. 16	Quarantine Medical Officer, Mohammerah.
Gibbon, Cecil Bertram Benjamin (207)	B.	30 Dec. 83	25 Feb. 04	25 Feb. 16	Moradabad.
Duckworth, George Francis (105, 207)	M.	25 Aug. 82	31 Mar. 04	31 Mar. 16	(Civil, Madras.) 9th Division. Wellington.
Stewart, Charles Netscher	M.	1 Mar. 82	31 Mar. 04	31 Mar. 16	9th Division, Lv., m.c., 6 mos., 6-6-17. Extn., m. c., 12 mos. Exten. 3 mos. without pay from 6 Sep. 18.
Claudius, Clarence John (105)	M.	25 Oct. 82	31 Mar. 04	31 Mar. 16	(Civil, Burma.) Bangalore.
O'Rose, Francis Joseph (193)	M.	24 Sep. 82	31 Mar. 04	31 Mar. 16	(Civil, Burma.) Agra.
Copcutt, Ernest Edward (95)	M	23 Oct. 82	31 Mar. 04	31 Mar. 16	Karachi.
Gomez, Albert Edward (193, 207)	M	8 Feb. 82	31 Mar. 04	31 Mar. 16	——.
Shunker, Cecil Percival Voncoutre (162, 206)	M.	14 Nov. 83	31 Mar. 04	31 Mar. 16	Sub-Charge, Govt. House Dispy., Madras.
Perreira, William Archibald Clement (*f*) XXIII.	B.	16 Oct. 81	31 Mar. 05	31 Mar. 16	——.
Murray, Charles William (95, 172*a*).	Bo.	2 Nov. 81	29 Apl. 04	29 Apl. 16	——.
Dalton, Garnett Michael	B.	16 Feb. 83	6 June 04	6 June 16	(N. W. Ry.) Peshawar.
Warner, William George Henry (95), (207).	B.	14 Jan. 83	6 June 04	6 June 16	(Civil, Bengal.) 10 B. S. Hospl., Aden.
Eates, Elkins Austin XXIV (*f*)	B.	18 Aug. 83	31 Oct. 05	1 July 16	Delhi.
Aitkins, George Vernon	B.	25 May 83	24 Sep. 04	24 Sep. 16	——.
Young, Harold Augustus	B.	27 Feb. 82	26 Sep. 04	26 Sep. 16	Civil, Bengal, Kirkee.
Aitkins, Grant Routh	B.	6 Feb. 84	26 Sep. 04	26 Sep. 16	——.
Wells, Clarence Archibald (169)	B.	3 May 83	26 Sep. 04	26 Sep. 16	R.I.M.S. "Mayo."
O'Brien, Charles Patrick	B.	27 Sep. 81	26 Sep. 04	26 Sep. 16	(Civil, C. P.) ——.
Hasle, Arthur William (172*a*)	B.	26 Oct. 82	26 Sep. 04	26 Sep. 16	Offg. Civil, Assam. Tempy Mily. duty.—
Aitkins, Louis Duckworth Routh (50, 172*a*).	B.	27 Nov. 82	26 Sep. 04	26 Sep. 16	Rawalpindi.
DePenning, Herbert Clement (94, 100, 162, 169).	Bo.	7 Dec. 82	26 Sep. 04	26 Sep. 16	Supdt., Mily. Pupils class, G. M. College, Bombay.
Cody, William Ernest	Bo.	12 May 83	26 Sep. 04	26 Sep. 16	Poona.
Braganza, John Benedict Valentine (95, 170)	Bo.	26 Jan. 82	26 Sep. 04	26 Sep. 16	(Civil, Bombay.) J. J. Hospl., Bombay.
D'Souza, James Berchmans	Bo.	3 Sep. 81	26 Sep. 04	26 Sep. 16	H.M.S. "Hardinge."
Hudson, Alexander Leslie	Bo	31 Jan. 84	26 Sep. 04	26 Sep. 16	Quarantine Medical Officer, Bushire, Persian Gulf.
XIV					
XV					
Dunn, Richard Luke XXV (*f*)	M.	9 Aug. 82	31 Mar. 05	1 Oct. 16	HinduRao's Hospital, Delhi.
Ross, Alfred Newton XXVI (*f*) (94, 98, 111, 147, 162, 165, 169, 198, 206).	M.	19 Apl. 84	31 Mar. 05	1 Oct. 16	Medl. Mobn. Stores, Peshawar. Attd. A. H. Qrs., Medl. Branch.
Beer, William Arthur XXVII (*f*) (95, 172 (*a*), 207).	M.	5 Dec. 82	31 Mar. 05	1 Oct. 16	Mhow.
Boilard, Ernest Hugh (95)	B.	13 July 80	3 Nov. 04	3 Nov. 16	(Civil, C.P.) ——.
Braganza, Francis Bruno Anthony, M.C.	B.	6 Oct. 82	27 Feb. 05	27 Feb. 17	No. 7 Coy. A.B.C. Meerut.
Lobo, Clement	B.	13 Dec. 83	27 Feb. 05	27 Feb. 17	——.
Fordham, Walter Charles William (94, 95, 12*a*).	B.	12 Sep. 84	27 Feb. 05	27 Feb. 17	——.
Creais, Edward John (172*a*)	B.	22 Apl. 85	27 Feb. 05	27 Feb. 17	(Civil, Burma.) Tempy. Mily. duty.—
Traynor, Thomas Willick (169)	B.	4 Jan. 85	27 Feb. 05	27 Feb. 17	Asstt. to Civil Surgeon, Lahore.
Clarkson, Ruxton (95)	B.	25 Apl. 82	27 Feb. 05	27 Feb. 17	(Civil, Bengal.) Tempy. Mily. duty.—
Feegrade, Egbert Stanley	B.	21 Jan. 84	27 Feb. 05	27 Feb. 17	(Civil, Burma.) Mily duty.—
Blaker, Henry Beauchamp, D. T. M.	B.	30 May 83	27 Feb. 05	27 Feb. 17	(Offg. Civil, Assam.) Tempy. Mily. duty.—
Lobo, Edward Demosthenes	B.	11 Jan. 83	27 Feb. 05	27 Feb. 17	Kamptee.
Browne, Arthur Elijah (95, 172*a*), D. C. M.	B.	4 Jan. 85	27 Feb. 05	27 Feb. 17	——.
Campbell, Arthur Duncan (94, 100, 162, 169).	B.	19 Jan. 85	27 Feb. 05	27 Feb. 17	(Civil, Bengal.) 4th Division.
Johnstone, James Glansey (95)	B.	26 Mar. 83	27 Feb. 05	27 Feb. 17	Quarantine Medical Officer, Lingah. and offg. Vice Consul, Lingah.
Crunden, Edward George (169)	B.	24 July 83	27 Feb. 05	27 Feb. 17	(Civil, Assam.) Tempy. mily. duty.—

(*f*) Specially promoted under the provisions of para. 470, A. R., I., Vol. II.

INDIAN MEDICAL DEPARTMENT.

Assistant Surgeons, 2nd Class—*contd.*

Ranking as Conductors.

Names.	Estabt.	Date of birth.	Date of warrant rank.	Date of apptt. to present class.	Remarks.
Ireland, Richmond Clement (169, 207)	B.	10 Oct. 82	27 Feb. 05	27 Feb. 17	Chakrata.
Thorpe, Percy	B.	23 Apl. 83	27 Feb. 05	27 Feb. 17	Peshawar.
Doyle, Gerald William (95)	B.	27 Nov. 84	27 Feb. 05	27 Feb. 17	Abbottabad.
Dewey, Joseph Owen (169)	*B.*	*4 May 84*	*27 Feb. 05*	*27 Feb. 17*	(*Civil, Punjab.*) *Malaria Convalescent Depôt, Dagshai.*
FitzGerald, Loftus Austen Wattell (95, 172*a*).	B.	1 Nov. 82	27 Feb. 05	27 Feb. 17	Cawnpore.
Frost, Dallas Melville (*f*) 172*a*, XXVIII	B.	20 Apl. 85	5 Mar. 06	5 Mar. 17	Nowshera.
Blaker, Gerald Hugh (*f*) (95, 172*a*) XXIX	B.	22 Jan. 85	5 Mar. 06	5 Mar. 17	16th Indian Divn.
Anderson, Charles Lionel (95, 111, 153, 172*a*, 207).	M.	6 Dec. 81	31 Mar. 05	31 Mar. 17	Sub-charge, M. M. Stores. Quetta.
Berlie, Herbert Clay (193)	*M.*	*22 June 84*	*31 Mar. 05*	*31 Mar. 17*	*Quarantine Medical Officer, Bunder Abbas, Persian Gulf.*
L'Fleur, Joseph Francis (172*a*, 207)	M.	10 June 84	31 Mar. 05	31 Mar. 17	——.
Pereira, John Paul (105, 193).	B.	26 June 82	31 Mar. 05	31 Mar. 17	Maymyo.
Lawrance, George Hope (95, 169)	M.	11 July 84	31 Mar. 05	31 Mar. 17	——.
Pereira, John Michael (94, 147)	M.	29 May 81	31 Mar. 05	31 Mar. 17	——.
Rieley, Stanley Desmond (207)	*M.*	*27 July 82*	*31 Mar. 05*	*31 Mar. 17*	(*Agri. Research Inst.*) *Tempy. mily. duty.* ——.
Westrap, Alexander Edmond (162, 206). XXIII XVI	M.	21 Apl. 83	31 Mar. 05	31 Mar. 17	Secunderabad.
Gomez, Aubin Joseph (94, 100, 162, 169, 193).	M.	1 May 82	31 Mar. 05	31 Mar. 17	1st Division.
Dunlop, Charles William Doig (95)	M.	1 Apl. 83	31 Mar. 05	31 Mar. 17	Peshawar.
Gomez, Stephen Clarence XXV XXVI	B.	26 Dec. 81	31 Mar. 05	31 Mar. 17	St. Thomas' Mount.
Cabral, Joseph Reginald Herbert	B.	9 May 84	31 Mar. 05	31 Mar. 17	Sub-charge, X-Ray Apparatus, Maymyo.
Ariss, Vincent Herbert	B.	11 July 82	31 Mar. 05	31 Mar. 17	——.
Hart, Wilfred Ernest Frend XXVII	*M.*	*18 Apl. 83*	*31 Mar. 05*	*31 Mar. 17*	(*Civil, Burma.*) *Secunderabad.*
Coombes, Melville George	M.	17 Jan. 83	31 Mar. 05	31 Mar. 17	——.
Raphael, Alfred Joseph (Supy) (172*a*) (*f*) XXX XVII XXI	B.	5 Apl. 84	5 Mar. 06	5 Sep. 17	Hyderabad (Sind.)
Francis, Leopold Vasconcellos DaConceicao.	Bo	29 Feb. 84	27 Sep. 05	27 Sep. 17	——.
Fratel, Eugene (95, 162, 206,) (172*a*)	Bo.	25 Jan. 85	27 Sep. 05	27 Sep. 17	No. 2 Coy., A.B.C., Rawal Pindi.
Aquino, Frank (95, 172*a*)	Bo.	20 Mar. 83	27 Sep. 05	27 Sep. 17	——.
O'Connor, William Walter	Bo.	20 Mar. 83	27 Sep. 05	27 Sep. 17	Delhi.
Marchant, Cecil Harcourt, (*f*) XXXI (95, 161, 172*a*, 206)	M.	30 Mar. 86	31 Mar. 06	Oct. 17	Divl. Laby., Wellington.
Nicholas, Dudley Hope Joseph (f) XXXII	*B.*	*18 July 85*	*31 Mar. 06*	*1 Oct. 17*	*M. O. Junior Sectt. Estabt., Simla.*
Gillespie, Harry Rupert Sylvester (95, 169)	B.	31 Dec. 82	31 Oct. 05	31 Oct. 17	——.
Courtney, Frederick Ignatius (207) XXII	B.	12 Feb. 85	31 Oct. 05	31 Oct. 17	——.
Woodsell, John Wilson, M.C. (95, (94, 172*a*)	Bo.	8 July 83	31 Oct. 05	31 Oct. 17	——.
Johnstone, Thomas Sleman XXIV	B.	22 Aug. 84	31 Oct. 05	31 Oct. 17	No. 2, A.B.C., Rawal Pindi.
Harvey, George James	B.	23 May 85	31 Oct. 05	31 Oct. 17	——.
Oakley, George Urbain	Bo.	2 Jan. 84	31 Dec. 05	31 Dec. 17	——.
Bouche, Benjamin Jasper (96, 165, 169, 172*a*) (*f*) XXXIII	...	27 Feb. 88	4 Mar. 08	4 Mar. 18	16th Indian Divn. Lv. m.c.
Coombes, William Kenneth (95, 172*a*) (*f*) XXXIV	...	29 Mar. 87	4 Mar. 08	4 Mar. 18	Sub-Ch., X'Ray Apparatus, Rawal Pindi.
Dunn, James Joseph Francis XXX	B.	7 Jan. 86	5 Mar. 06	5 Mar. 18	——.
Frederick, Charles William Enoch (172*a*)	B.	28 June 84	5 Mar. 06	5 Mar. 18	Bombay.

(*f*) Specially promoted under the provisions of para. 470, A. R., I., Vol. II.

Names	Estabt.	Date of birth.	Date of warrant rank.	Date of apptt. to present class.	Remarks.
INDIAN MEDICAL DEPARTMENT.					
Assistant Surgeons, 2nd Class—*concld.*					
Ranking as Conductors.					
Peters, Cecil William Edwin (95 172*a*)	B.	13 Jan. 85	5 Mar. 06	5 Mar. 18	Rawalpindi.
Rodgers, Maundy Charles Rodgers (95, 172*a*)	B.	4 Dec. 84	5 Mar. 06	5 Mar. 18	Rawalpindi.
XXVIII					
Mathews, Arthur Ernest	*B.*	*12 Jan. 85*	*5 Mar. 06*	*5 Mar. 18*	(*Civil, Assam*) *Tempy.* [*mily. duty.* ——.
Martin, John Henry Addison. (172*a*, 207).	B.	23 Dec. 82	5 Mar. 06	5 Mar. 18	——.
Thomas, William Herbert George (207).	B.	17 Aug. 84	5 Mar. 06	5 Mar. 18	Secunderabad.
Browne, Anthony Mark Francis (95)	*B.*	*7 Oct. 84*	*5 Mar. 06*	*5 Mar. 18*	(*Civil, U. P.*) *Quetta.*
Mullins, Horace (169) . .	B.	22 Sep. 83	5 Mar. 06	5 Mar. 18	Bangalore.
Cade, Osborne Halliburton (94) .	B.	21 Feb. 85	5 Mar. 06	5 Mar. 18	No. 4 Coy., A.B.C., Quetta.
XVIII					
XXIX					
Whitbread, Frederick George (95) .	M	21 May 86	31 Mar. 06	31 Mar. 18	Ahmednagar.
Blazey, William Henry (94, 172*a*)	B.	12 Dec. 82	31 Mar. 06	31 Mar. 18	Dalhousie.
XXXI					
Sherard, Wilfred George . .	B.	7 Oct. 82	31 Mar. 06	31 Mar. 18	——.
Jackson, Septimus George (94) .	M.	16 Sep. 82	31 Mar. 06	31 Mar. 18	Mobn Stores, 4th (Quetta) Divn.
XXXII					
Lopez, James William Clement, D. C. M. (147, 193)	M.	27 Sep. 83	31 Mar. 06	31 Mar. 18	——.
Walker, Eldon Bertram . . .	B.	11 Aug. 83	31 Mar. 06	31 Mar. 18	
D'Costa, Joseph Aloysius . .	B.	16 Sep. 83	31 Mar. 06	31 Mar. 18	——.
Ashe, John Willoughby Kenneth (193)	M.	26 July 83	31 Mar. 06	31 Mar. 18	Secunderabad.
Sandways, William George, (94, 169)	M.	1 July 84	31 Mar. 06	31 Mar. 18	——.
Gomes, Arthur Louis (95) (*f*) . .	Bo	25 Dec. 85	31 Mar. 07	31 Mar. 18	R. I. M. S. "Palinurus."
XXXV					
Gibbs, John Walker (94, 172*a*) (*f*) .	*M.*	*5 Aug. 84*	*31 Mar. 07*	*31 Mar. 18*	(*Civil Madras*) *Bangalore.*
XXXVI					
O'Reilly, Cecil Charles Gillespie .	B.	24 Mar. 86	5 Mar. 06	1 Apl. 18	——.
Shaw, Ebenezer Lionel Hugh .	B.	10 Nov. 84	1 June 06	1 June 18	
Newbold, John Walter, (Supy.) 6)	Bo.	23 Mar. 87	10 Aug. 07	10 Aug. 18	——.
XXXVII.					
Nery, Percival Edward (94, 172*a*)	M.	18 May 82	31 Mar. 06	31 Aug. 18	——.
Sinclair, Arthur Carr (206) .	Bo.	3 Jan. 84	1 Oct. 06	1 Oct. 18	No. 8 Coy., A. B. C., [Lucknow
Richardson, Joseph Murray (162, 172*a*, 206)	*Bo.*	*13 Mar. 86*	*1 Oct. 06*	*1 Oct. 18*	(*Civil, C. P.*) *Mhow.*
Howard, James Eyre	*Bo.*	*24 May 86*	*1 Oct. 06*	*1 Oct. 18*	(*Civil, Bombay.*) *A. T.* [*"Ellora."*
Cuzen, Edward McLelland (95, 169)	Bo.	20 Aug. 84	1 Oct. 06	1 Oct. 18	
Vanderhide, Rumold Francis . .	*Bo.*	*1 July 86*	*1 Oct. 06*	*1 Oct. 18*	(*Civil, Burma.*) *Nasirabad.*
Dyer, John Charles	*Bo.*	*29 Oct. 83*	*1 Oct. 06*	*1 Oct. 18*	*Rawalpindi.*
Emile, Vincent Wallace Ashley .	Bo.	13 Dec. 85	1 Oct. 06	1 Oct. 18	Peshawar.
Pinto, John Anthony (169) . .	Bo.	3 Feb. 84	1 Oct. 06	1 Oct. 18	——.
Ne'Roy, William Cyril (95, 169) (§)	Bo.	27 Dec. 85	1 Oct. 06	1 Oct. 18	Karachi.
Freeman, Owen Richard Cowley	Bo.	14 June 85	1 Oct. 06	1 Oct. 18	Poona.
Lincoln, Christopher Joseph (95, 172*a*)	Bo.	20 June 85	1 Oct. 06	1 Oct. 18	——.
Martin, William Stanislaus (94, 147, 172*a*, 193) (*f*) *XXXVIII.*	*M.*	*10 Apl. 85*	*31 Mar. 07*	*1 Oct. 18*	(*Civil, Madras.*) *Tempy. mily. duty. Meerut.*
Ottmann, Gerald Walter (94, 95, 172*a*)	B.	20 Oct. 81	8 Oct. 06	8 Oct. 18	——.
Fido, George Russell (95, 172*a*) .	Bo.	22 Nov. 84	31 Oct 05	31 Oct. 18	Hospitalship "Sarsavati." [17-8-17.
Assistant Surgeons, 3rd Class.					
Ranking as Sub-Conductors.					
XIX					
Richardson, George Alexander (88, 162, 206)	*Bo.*	*7 Sep. 75*	*18 Sep. 95*	*18 Sep. 02*	*Vice-Consul, Hodeida.*
Martin, Abel (111)	Bo.	22 Apl. 84	29 Dec. 06	29 Dec. 13	M. S. Depôt, Madras.
Wale, Claude Willoughby . .	Bo.	13 Feb. 85	29 Dec. 06	29 Dec. 13	Pathancote.
D'Silva, Edward	Bo.	7 Aug. 84	29 Dec. 06	29 Dec. 13	Bombay.
Grant, Ernest Ralph	Bo.	8 Apl. 86	29 Dec. 06	29 Dec. 13	Bombay.
Emile, Cecil Allan Ashley . .	Bo.	1 Nov. 84	29 Dec. 06	29 Dec. 13	Quetta.
Gomes, Dominic Vincent (172*a*) .	Bo.	19 July 84	31 Dec. 06	31 Dec. 13	Jubbulpore.
White, Kenneth William Everard (95)	Bo.	11 July 85	31 Dec. 06	31 Dec. 13	Neemuch.
Bedell, Percival Stanley (95, 162, 165, 69, 206)	*B.*	*21 Sep. 84*	*1 Mar. 07*	*1 Mar. 14*	(*Civil, Bengal*). *Mily. duty.*
Elloy, Kenneth Percival (95, 172*a*)	B.	16 July 84	1 Mar. 07	1 Mar. 14	3rd Division.
Vardon, Arnold Carapiet (94, 169).	B.	14 Nov. 85	4 Mar. 07	4 Mar. 14	Bombay.

(§) Qualified for promotion.

(*f*) Specially promoted under the provisions of para. 470, A. R. I., Vol. II.

INDIAN MEDICAL DEPARTMENT.

Assistant Surgeons, 3rd Class—*contd.*

Ranking as Sub-Conductors.

Names.	Estabt.	Date of birth.	Date of warrant rank	Date of apptt. to present class.	Remarks.
Bedell, Hubert Charles . . .	B.	8 Dec. 86	4 Mar. 07	4 Mar. 14	Multan.
McGuire, Gilbert (165) . . .	*B.*	*15 May 86*	*4 Mar. 07*	*4 Mar. 14*	*(Civil, Punjab.) Jullundur.*
Gloria, Julius Septimus. . . .	B.	12 July 84	4 Mar. 07	4 Mar. 14	———.
Holmes, Ralph (95, 172a) . . .	*B.*	*15 July 86*	*4 Mar. 07*	*4 Mar. 14*	*Civil Hospl., Nairobi, Br. East Africa.*
Burns, Herbert Percy . . .	B.	9 Mar. 87	4 Mar. 07	4 Mar. 14	Divl. Liby., Lucknow.
Dolphin, Maurice James Xavier (172a).	*B.*	*18 Nov. 86*	*4 Mar. 07*	*4 Mar. 14*	*(Lower Ganges Bridge, Paksey.) Tempy. mily. duty, Burhan.*
Thoy, Albert Robinson (95, 169, 172a)	B.	6 Oct. 86	4 Mar. 07	4 Mar. 14	Divl. Laby., Peshawar.
Michael, Donald Fairbairn (§) .	B.	8 Feb. 87	7 Mar. 07	4 Mar. 14	Karachi.
Murphy, Stanley Beresford William (172a).	B.	28 Oct. 86	4 Mar. 07	4 Mar. 14	———.
Mackey, George (172a)	*B.*	*15 Sep. 86*	*4 Mar. 07*	*4 Mar. 14*	*Asstt. to the Director, C.R. Inst., Kasauli.*
Marchant, Thomas James (172a)	B.	17 Apl. 85	4 Mar. 07	4 Mar. 14	———.
Fordham, Frederick Percival Gilson (94, 169).	B.	26 Oct. 86	4 Mar. 07	4 Mar. 14	———.
Pell, Edward James	B.	29 Dec. 85	4 Mar. 07	4 Mar. 14	Meerut.
Mitchell, William Patrick Smythe	*B.*	*27 Feb. 86*	*4 Mar. 07*	4 Mar. 14	*(M. O., Bastar State.)*——Ambala.
Victor, Albert Charles (172a) (§).	B.	17 Oct. 86	4 Mar. 07	4 Mar. 14	Enteric Fever Convalescent Depôt, Naini Tal.
Dingavan, George Stanley . .	B.	28 Nov. 85	4 Mar. 07	4 Mar. 14	Lucknow.
Cordeiro, Edward (94, 169) . .	Bo.	13 Sep. 86	20 Mar. 07	20 Mar. 14	Karachi.
Hardaker, Frederick George (172a) (94).	Bo.	15 June 85	20 Mar. 07	20 Mar. 14	Sialkot.
Smith, Robert George (95) (§) .	Bo.	24 Feb. 85	31 Mar. 07	31 Mar. 14	Rawalpindi.
Kiddle, John St. Alban (95) . .	Bo.	22 July 85	31 Mar. 07	31 Mar. 14	Ramandroog.
Brown, John Michael Mary (95).	Bo.	8 Oct. 83	31 Mar. 07	31 Mar. 14	Lahore Cantt.
XXXV					
XXXVI					
XXVIII					
Davies, Charles Henry Lawrance (94, 162, 169, 206)	*M.*	*28 July 85*	*31 Mar. 07*	*31 Mar. 14*	*(Civil, Persian Gulf.) Tempy. mily. duty, Colaba.*
Lafrenais, Hugh Maurice . . .	*M.*	*12 July 87*	*31 Mar. 07*	*31 Mar. 14*	*(Civil, Madras.) Tempy. mily. duty, Bombay.*
Rebeiro, Francis Victor deVilhena .	M.	24 Feb. 84	31 Mar. 07	31 Mar. 14	Jhansi.
Rodrigues, Reginald Hugh Sausman	B.	30 Sep. 85	31 Mar. 07	31 Mar. 14	Rangoon.
XX					
Allan, Leslie Kenneth	B.	*23 Aug. 84*	*11 June 07*	11 June 14	*(Civil, Punjab.) Bangalore.*
Smith, Charles Percival	B.	30 Jan. 86	11 June 07	11 June 14	Bombay.
Cox, Cecil George	B.	27 Feb. 85	11 June 07	11 June 14	Agra.
Muller, Christian Gerald (95, 147, 172a, 193, 207) .	B.	6 Nov. 84	30 June 07	30 June 14	———.
Stracey, Stanislaus Russel (94, 111, 147, 161, 172a, 185, 193, 206).	*M.*	*2 July 86*	*30 June 07*	*30 June 14*	*X'Ray Instt., D. Dun.*
LaBouchardiere, Basil Richard (105, 106, 172a).	M.	18 July 85	30 June 07	30 June 14	———.
Conquest, Cuthbert Hugh Claude	M.	18 Sep. 83	30 June 07	30 June 14	Bangalore.
Pereira, John George (94, 95, 147, 169, 172a, 192, 193)	B.	11 Sep. 83	30 June 07	30 June 14	———.
Owers, Frederick Fleetwood. . .	Bo.	24 June 85	10 Aug. 07	10 Aug. 14	Bombay.
XXXVII					
Fratel, William	Bo.	28 Jan. 87	10 Aug. 07	10 Aug. 14	———.
XXXIII					
Quick, Christopher Francis Henry (95, 172a)	...	*28 July 87*	*4 Mar. 08*	4 Mar. 15	*Dalhousie.*
Parkinson, Richard Henry Francis, D. C. M. (95, 172a)	...	23 Nov. 86	4 Mar. 08	4 Mar. 15	———.
Shephard, Charles William (94, 172a)	...	26 Oct. 87	4 Mar. 08	4 Mar. 15	H. M. S. "Dalhousie."
Barboza, Aloysius Herbert Vivian (172a)	...	28 July 86	4 Mar. 08	4 Mar. 15	Poona. *Lv., m.c., 3 mos., 18 Nov. 18.*
XXXIV					
Taylor, Arthur Octavius	.	5 May 85	4 Mar. 08	4 Mar. 15	Wellington.
Jacob, Lewis Alexander (207) . .	.	6 May 83	4 Mar. 08	4 Mar. 15	Peshawar.
Fraser, Albert Glen Leslie (94, 169) .	.	10 Jan. 87	4 Mar. 08	4 Mar. 15	Risalpur.
Toussaint, Hugh Wiltshire (95, 172a).	.	5 Nov. 86	4 Mar. 08	4 Mar. 15	Nowshera.
Carroll, Ferdinand (172a) . . .	.	8 Sep. 85	4 Mar. 08	4 Mar. 15	Peshawar.

(§) Qualified for promotion.

Name.	Estabt.	Date of birth.	Date of warrant rank.	Date of apptt. to present class.	Remarks.

INDIAN MEDICAL DEPARTMENT.

Assistant Surgeons, 3rd Class—*contd.*

Ranking as Sub-Conductors.

Name.	Estabt.	Date of birth.	Date of warrant rank.	Date of apptt. to present class.	Remarks.
Casling, Paul Vernon (95, 172*a*, 207)		3 Jan. 87	4 Mar. 08	4 Mar. 15	No. 1 A. B. C., Peshawar.
McLeavy, Henry Evelyn Beaumont		18 Jan. 85	4 Mar. 08	4 Mar. 15	Bombay.
Hogan, Herbert Frederick		17 Nov. 86	4 Mar. 08	4 Mar. 15	———.
Baron, Frederick Charles		10 July 84	5 Mar. 06	23 Mar. 15	———.
Hill, Eric Rudolph (*f*) XXXIX		5 Sep. 89	18 Apl. 10	18 Apl. 15	4th Division.
Honey, Bryan St. Clair		8 Apl. 87	6 May 08	6 May 15	———.
Murphy, Heyward Norman (*f*) XL		6 Mar. 90	15 Apl. 12	3 June 15	———.
Doyle, Henry		2 June 87	1 Sep. 08	1 Sep. 15	Bombay.
Mason, George Herbert		15 Nov. 86	1 Sep. 08	1 Sep. 15	Bombay.
Rogers, John Auty (172*a*, 207)		3 Dec. 86	1 Sep. 08	1 Sep. 15	———.
Pownes, Herbert Clarence		*10 Aug. 86*	*1 Sep. 08*	*1 Sep. 15*	*(Temporary, Civil, Bengal.) Tempy. mily. duty——.*
Heppotette, John Emmanuel Reid		10 July 85	8 Sep. 08	*1 Sep. 15*	*(Lower Ganges Bridge.)* Rawalpindi.
Claudius, Albert Henry Thomas		*5 Jan. 88*	*1 Mar. 09*	*1 Mar. 16*	*(Civil, U. P.) Tempy. mily. duty——.*
Shunker, William Harold Douglas		27 Sep. 87	1 Mar. 09	1 Mar. 16	———.
Dees, John Sylvester (95)		15 Sep. 88	1 Mar. 09	1 Mar. 16	Belgaum.
Pope, William Victor		15 Mar. 85	1 Mar. 09	1 Mar. 16	Meerut, Sub-Ch., Medl. Mobn. Stores.
Jackson, Decimus Stephen (105)		21 Feb. 88	1 Mar. 09	1 Mar. 16	———.
Gomez, Archibald Frederick		26 July 87	1 Mar. 09	1 Mar. 16	Divisional Laby., Maymyo.
Peacock, Gerald Eustace (172*a*)		29 Dec. 88	1 Mar. 09	1 Mar. 16	Peshawar.
Kelly, Robert William		13 Jan. 88	1 Mar. 09	1 Mar. 16	———.
Theodore, James Herbert		19 July 87	1 Mar. 09	1 Mar. 16	———.
Duckworth, Eugene		6 Aug. 87	1 Mar. 09	1 Mar. 16	———.
Greenway, Allan Leslie (172*a*)		19 Nov. 88	1 Mar. 09	1 Mar. 16	Mily. duty.
Rowley, Francis Hope Jarvis		22 Sep. 87	1 Mar. 09	1 Mar. 16	———.
Burke, Charles Henry		29 Jan. 88	1 Mar. 09	1 Mar. 16	Rawalpindi.
Durning, John George		16 Jan. 89	1 Mar. 09	1 Mar. 16	Dalhousie.
Dudley, Andrew Ernest (95)		31 May 89	1 Mar. 09	1 Mar. 16	Jullundur.
Hanson, Albert Donohue (172*a*)		29 May 88	1 Mar. 09	1 Mar. 16	Peshawar.
Leopold, Richard Thornhill		10 Sep. 88	1 Mar. 09	1 Mar. 6	Lahore Cantt.
Asquith, Frank Herbert		*11 Sep 88*	*1 Mar. 09*	*1 Mar. 16*	———.
Cahill, John Joseph		28 June 88	1 Mar. 09	1 Mar. 16	Bombay.
Cowper, Glen Joseph (94)		22 Feb. 88	1 Mar. 09	1 Mar. 16	Peshawar.
Bell, Alan Rothwell (172*a*, 207)		19 Sep. 89	1 Mar. 09	1 Mar. 16	———.
Elkins, Frank Cuthbert Ambridge		3 Mar. 88	1 Mar. 09	1 Mar. 16	———.
Perkins, John William (*f*) XLI		15 Sep. 89	17 Apl. 11	17 Apl. 16	Peshawar.
Wrafter, George Tresham (*f*) XLII		9 June 93	20 Apl. 14	1 July 16	Lahore Cantt.
D'Arcy, Arthur Frederick Joseph		27 Oct. 87	1 Sep. 09	1 Sep. 16	———.
Blaker, Harold Bartholomew (95)		5 Aug. 88	1 Sep. 09	1 Sep. 16	Rawalpindi.
Hendricks, William St. Alban (172*a*)		29 Oct. 87	1 Sep. 09	1 Sep. 16	H. M. S. "Dufferin."
Lewis, Arthur Elloy		1 Sep. 87	1 Sep. 09	1 Sep. 16	Rawalpindi.
Caynor, Samuel Stanley (172*a*) XXXIX		7 Nov. 88	1 Sep. 09	1 Sep. 16	———.
Sargent, Henry Lawrence (*f*) XLIII		*9 Aug. 88*	*18 Apl 10*	*18 Oct. 16*	*Turkhana, Brit. East Africa.*
D'Abreu, Aubyn Reginald		*7 Nov. 88*	*18 Apl. 10*	*18 Apl. 17*	*(Offg. Civil, Punjab.) Tempy. mily. duty——.*
D'Silva, Harold Aloysius Henry (94) (95) (147) (170) (193)		16 Apl. 88	18 Apl. 10	18 Apl. 17	Ahmednagar.
Mackenzie, Raymond Frank (95)		*14 Dec. 89*	*18 Apl. 10*	*18 Apl 17*	*R. I. M. S. "Lawrence."*
Browne, Walter Albert (95)		6 Aug. 89	18 Apl. 10	8 Apl. 17	Karachi.
DeCunha, John Henry (205*a*)		4 Apl. 88	18 Apl. 10	18 Apl. 17	———.
Quick, Albert Norman (95, 172*a*)		16 July 89	18 Apl. 10	18 Apl. 17	Cherat.
Hardy, Guy William		21 Aug. 88	18 Apl. 10	18 Apl. 17	———.
Beveridge, Robert Laird Wilson		24 Mar. 88	18 Apl. 10	18 Apl. 17	Kirkee.
Menezes, Jeremias Simeon (95, 172*a*)		20 Oct. 88	18 Apl. 10	18 Apl. 17	Karachi.
Hallums, William Blaney (172*a*) XLIII		22 Sep. 89	18 Apl. 10	18 Apl. 17	Rawalpindi.
Furey, Henry George		23 May 88	18 Apl. 10	18 Apl. 17	Calcutta.
Mann, Alfred Charles Stephens (95, 172*a*)		9 Feb. 89	18 Apl. 10	18 Apl. 17	———.
Mascarenhas, Colin John		10 June 88	18 Apl. 10	18 Apl. 17	———.
Emmett, Percival William		24 May 90	18 Apl. 10	18 Apl. 17	Rawalpindi.
Pacheco, Julian Norman Joseph (*f*) XLIV		15 Apl. 90	26 June 11	26 June 17	Hospitalship "Madras".
Tobin, Patrick William		28 May 85	30 June 07	1 Aug. 17	———.
Lewis, Roderick Paul (*f*) XLV		6 Feb. 90	15 Apl. 12	15 Oct. 17	———.
Morton, Frederick (*f*) XLVI		21 Jan. 91	17 Apl. 11	17 Oct. 17	———.
Fanaken, Patrick Frank		28 Jan. 90	18 Oct. 10	18 Oct. 17	Meerut.
Holt, Edwin Brook (*f*) XLVII		8 July 90	15 Apl. 12	15 Apl. 18	———.

(*f*) Specially promoted under the provisions of para. 470, A. R., I., Vol. II.

Names.	Date of birth.	Date of warrant rank.	Date of apptt. to present class.	Remarks.

INDIAN MEDICAL DEPARTMENT.

Assistant Surgeons, 3rd Class—*concld.*

Ranking as Sub-Conductors.

Names.	Date of birth.	Date of warrant rank.	Date of apptt. to present class.	Remarks.
Carleton, Alan William	6 Jan. 90	17 Apl. 11	17 Apl. 18	Khandwa.
Bamford, Albert Victor Wallace	15 Nov. 89	17 Apl 11	17 Apl. 18	———.
McNally, Edward Walker	17 Dec. 88	17 Apl. 11	17 Apl. 18	Kuldana.
Rozario, George Stanislaus	22 Dec. 89	17 Apl. 11	17 Apl. 18	Bangalore.
Rozario, Christopher Joseph (172*a*)	22 Dec. 89	17 Apl. 11	17 Apl. 18	———.
Day, Arthur Henry	25 Feb. 91	17 Apl. 11	17 Apl. 18	Aden.
Vincent, Julius Francis deSouza	11 Aug 89	17 Apl. 11	17 Apl. 18	I. G. Hospital, Belgaum.
Bartley, Allen St. Clair (207)	19 Mar. 90	17 Apl. 11	17 Apl. 18	Ferozepore.
King, Graydon Procter	15 May 88	17 Apl. 11	17 Apl. 18	Meerut Divl. Laboratory.
Fernandez, Manuel Tom. XLI XLVI	21 Nov. 89	17 Apl. 11	17 Apl. 18	———.
Nyss, Alcysius Neville	*2 Oct. 90*	*17 Apl. 11*	*17 Apl. 18*	*Br. E. Africa—Serenli. Lv. m c. 3 mos., 5 July 18.*
de Monte Arthur, Norman (89*a*, 172*a*), M.C.	26 July 90	17 Apl. 11	17 Apl. 18	———.
Falloon, Terence St. Aubyn (95, 172*a*)	18 Aug. 90	17 Apl. 11	17 Apl. 18	———.
Windsor, Henry Carlyle William	9 Oct. 91	17 Apl. 11	17 Apl. 18	Ahmednagar.
Hayter, Robert Thomas Mansfield (95)	25 Aug. 89	17 Apl. 11	17 Apl. 18	Peshawar.
Lobo, Mark Vincent Jeremias	25 Apl. 89	17 Apl. 11	17 Apl. 18	R. I. M. S. "Nearchus."
Osbourne, Arthur Benjamin	21 Apl. 89	17 Apl. 11	17 Apl. 18	Aden.
deGama, Ciriaco Elias	16 Mar. 89	17 Apl. 11	17 Apl. 18	Roorkee.
Johnson, Francis Herbert	*3 Dec. 90*	*23 June 11*	*23 June 18*	*I/c X-Ray Deptt., Simla.*
Mendis, John Cecil	18 Apl. 91	23 June 11	23 June 18	Attock.
Donowan, Algernon Benjamin XLIV	6 July 89	23 June 11	23 June 18	Bombay.
Nicholas, Malcolm Joseph (*f*) XLVIII	5 Apl. 89	15 Apl. 12	15 Oct. 18	Wellington.

Assistant Surgeons, 4th Class.

GENERAL LIST.

Ranking as Sub-Conductors.

Names.	Date of birth.	Date of warrant rank.	Remarks.
Deatker, Francis Gilbert Frederick	25 Nov. 90	15 Dec. 11	Bombay.
Menezes, Luis Damian Chrysostom (95)	1 Dec. 91	15 Apl. 12	Bombay.
Corner, Cecil Gerald Sedgeley (95, 169)	24 Aug. 91	15 Apl. 12	Station Hospital, Amballa.
Martin, Francis Gregory	7 Dec. 90	15 Apl. 12	———.
Bartley, James Hartigan	7 Dec. 90	15 Apl. 12	Cawnpore.
D'Silva, Mervyan Joseph (95, 172*a*)	13 Jan. 92	15 Apl. 12	Rawalpindi.
Wells, Harold Arthur Thompson	19 May 90	15 Apl. 12	———.
Lupprian, Ernest Victor (95)	20 June 91	15 Apl. 12	Bombay.
Rodgers, Wilfred Edwin XLVIII	9 Dec. 89	15 Apl. 12	*M*———.
Quanbrough, Frederick Kinchant	16 May 91	15 Apl. 12	Jubbulpore.
Blakeman	3 Mar. 91	15 Apl. 12	———.
Brodie, Anthony Herbert (95, 172*a*)	5 Jan. 91	15 Apl. 12	———.
Woutersz, Eustace Weville Colvin	25 Jan. 90	15 Apl. 12	———.
D'Silveira, John Luke Isidore	18 Oct. 89	15 Apl. 12	———.
Pereira, Joseph Celestin	22 June 90	15 Apl. 12	Wellington.
Thyne, William Harold	11 Apl. 91	15 Apl. 12	———.
Giblin, Hubert Thomas	30 Dec. 90	15 Apl. 12	Rawalpindi.
Sargent, Walter Nicholas (95)	*28 Aug. 91*	*15 Apl. 12*	*Bura, British East Africa.*
Cordeiro, Paul Edwin	5 Dec. 90	15 Apl. 12	———.
Campbell, Henry Gordon Montgomery	13 Dec. 90	15 Apl. 12	4th Division.
Bailie, William Herbert. XLVII	21 Apl. 90	15 Apl. 12	Hyderabad.
Vincent, Bosworth (95)	12 Aug. 90	15 Apl. 12	———.
Rose-Meyer, Joseph Francis Clarence	*8 Oct. 90*	*15 Apl. 12*	*(Civil, U. P.) 1st Division. Lv. m.c., 2 mos.,—Aug. 18.*
DeCunha, Charles Michael	6 July 90	15 Apl. 12	———.
Devine, Charles (172*a*)	27 Aug. 90	15 Apl. 12	———.
Donoghue, Patrick James	23 Mar. 91	15 Apl. 12	———.
Mason, Robert Douglas XL	22 July 90	15 Apl. 12	———.
Johnson, William	10 Apl. 91	15 Apl. 12	———.
Cooper, Louis Bernard (95)	22 Oct. 90	15 Apl. 12	———.
Michael, Leslie Hope Fairbairn XLV	13 July 91	15 Apl. 12	H.M.S. "Northbrook."
McGuire, John Parnell	13 Nov. 90	15 Apl. 12	———.
Fruvall, Harold Andrew	19 Nov. 90	15 Apl. 12	Secunderabad.
Martin, Albert Edward	20 Aug. 90	15 Apl. 12	———.
Lawrie, Lawrence John	20 July 90	15 Apl. 12	———.

INDIAN MEDICAL DEPARTMENT—*contd.*

Assistant Surgeons, 4th Class—*contd.*

Ranking as Sub-Conductors—contd.

Names.	Date of birth.	Date of warrant rank.	Remarks.
Smith, Charles Joseph (95)	7 Sep. 90	15 Apl. 12	Nowshera.
Forbes, Ronald Binnie	15 Feb. 91	15 Apl. 12	——.
Gale, Richard Campbell	9 Jan. 92	15 Apl. 12	——.
Sweeney, James Edward	1 Aug. 90	15 Apl. 12	——.
Rego, Anao Saturnino	12 Jan. 89	15 Apl. 12	Bombay.
Garthwaite, Thomas Edmund	9 June 90	15 Apl. 12	——.
Phillips, Irenaeus Pinto	15 Dec. 89	15 Apl. 12	——.
Doran, Patrick Joseph (95)	3 Oct. 90	15 Apl. 12	Peshawar.
Hixon, Arthur James (172*a*)	1 June 91	15 Apl. 12	——.
Cartner, James Roland (95, 172*a*)	12 June 90	15 Apl. 12	Bangalore.
Major, Dennis Donald James (172*a*)	26 Dec. 90	15 Apl. 12	Secunderabad.
Easdon, Horace Verney (95)	20 June 91	24 June 12	Bombay.
Dique, John Stephen	29 Oct. 89	27 June 12	Port Blair.
Lamond, John Dougald Malcom, M.C.	15 May 91	16 Dec. 12	Bombay. Exten. lv. 2 mos.
Rowe, William Joseph	21 Dec. 89	16 Dec. 12	——.
Barnard, James Stephen Edward	12 Nov. 90	16 Dec. 12	——.
Holmes, Benjamin Oliver Samuel Herbert	18 Dec. 89	16 Dec. 12	——.
Smith, Robert Henry	18 Apl. 91	15 Mar. 13	——.
Rishworth, Henry Richard	8 Feb. 91	21 Apl. 13	——.
Martin, Charles Andrew (172*a*)	27 Apl. 92	21 Apl. 13	Mily. duty.
Steele, Mortimer Lawrence Axen (83*b*)	15 June 91	21 Apl. 13	Aden.
Phaure, Albert Edgard Alphonse	21 Aug. 90	21 Apl. 13	——.
Vieyra, Cyril Francis	29 Jan. 91	21 Apl. 13	Amballa.
Smith, Richard Oliver Ashley	6 July 91	21 Apl. 13	——.
Lynsdale, Bruce St. Clair Fitzgerald	14 May 91	21 Apl. 13	——.
Boodrie, Eric Hugh Bower (95, 172*a*)	5 Jan. 91	21 Apl. 13	——.
Stuart, Herbert Reginald (95)	10 Dec. 89	21 Apl. 13	——.
Edge, Victor Alexander (172*a*)	3 Mar. 93	21 Apl. 13	No. 8 Coy., A. B. C., Lucknow.
Firth, Henry Barclay	14 Sep. 92	21 Apl. 13	Burhan.
Allen, Alexander Louis Grant	10 Apl. 91	21 Apl. 13	1st Division.
Hinder, Bertram Anderson Parker	3 June 90	21 Apl. 13	——.
Doyle, Henry Herbert	7 Nov. 92	21 Apl. 13	Aden.
Feegrade, Alexander Archibald (95)	8 Sep. 90	21 Apl. 13	——.
VanCuylenberg, Victor Altendorf	21 Sep. 90	21 Apl. 13	——.
DeSouza, Sebastian Alex	28 Apl. 91	21 Apl. 13	——.
Luxa, Henry Joseph	21 June 91	21 Apl. 13	Ahmednagar.
Hartley, George McGuire (172*a*)	13 Apl. 92	1 July 13	Mhow.
Coutts, James Stephen Percival	23 Dec. 91	2 July 13	——.
Hyde, John Aurelian	22 Apl. 91	2 July 13	Egypt.
Ince, Henry Thomas	23 Aug. 90	6 Oct. 13	Jubbulpore.
Mason, Harold St. John	23 Apl. 92	15 Dec. 13	——.
Agate, Arthur Richard	27 June 91	15 Dec. 13	——. [Sept 18.
Bamford, Charles Walker	23 Feb. 93	20 Apl. 14	Lahore Cantt. Lv. *m. c.* 2 mos. 23
Clifford, John Hudson	28 Mar. 93	20 Apl. 14	——.
Sterling, John McNish	24 Jan. 92	20 Apl. 14	——.
Holt, Douglas Gooch	27 May 93	20 Apl. 14	Belgaum.
Aling, George Edwin	4 Jan. 94	20 Apl. 14	——.
XLII			
Graham, Frederick Mantgue	8 Mar. 92	20 Apl. 14	——.
Vernal, Eric Easton Leigh	26 Sep. 92	20 Apl. 14	Jullundur. [Corps.
Barker, Albert Victor	30 July 92	20 Apl. 14	4th Division. No. 4. Coy., A. B.
Thomas, Risdale Charles Perriman	21 Dec. 90	20 Apl. 14	——.
Robertson, Kenneth Ernest Robert	1 Sep. 93	20 Apl. 14	Gharial.
LeFeuvre, Eugene Eustache	1 Apl. 92	20 Apl. 14	——.
Taylor, Robert Marshall	8 Sep. 92	20 Apl. 14	——.
Ivatts, Edwin	14 Dec. 92	20 Apl. 14	Bangalore.
Eate, Samual Nathan (95)	28 Apl. 91	20 Apl. 14	——.
Duckworth, Herbert Cecil	4 Dec. 91	20 Apl. 14	——.
L'Fleur, Mary Joseph Percival	14 Feb. 93	20 Apl. 14	——.
Goodman, James Garnett, M. C. (No. 95)	21 July 93	20 Apl. 14	——.
Matthews, Vivian Matthew	23 Mar. 91	20 Apl. 14	2nd Division.
Hendrick, Uriel St. John	24 Aug. 92	20 Apl. 14	Sialkot.
Bampton, William John De Beaufort	6 Nov. 92	20 Apl. 14	——.
McGuire, Charles William	4 June 92	20 Apl. 14	Peshawar.
Rodrigues, George David	1 Mar. 92	20 Apl. 14	——.

INDIAN MEDICAL DEPARTMENT—*contd.*

Assistant Surgeons, 4th Class—*concld.*

Ranking as Sub-Conductors—concld.

Names.	Date of birth.	Date of warrant rank.	Remarks.
Fonseca, Francis Ignatius Alexis	1 July 92	20 Apl. 14	Poona.
Evans, Percy Stanbrook	20 May 94	20 July 14	———.
Turner, Albert Wilfred	16 Oct. 92	20 July 14	———.
McDonald, Andrew Joseph Patrick	24 July 92	20 July 14	———.
Thompson, Charles Louis	7 Feb. 92	20 July 14	4th Division.
Evans, Harry Stanbrook (172*a*)	25 Aug. 92	20 July 14	H. S. "Erinpura."
Pigott, George Borkman	7 Aug 92	5 Oct. 14	Rawalpindi.
Greenway, Lovell Austin Noel	26 Dec. 92	5 Oct. 14	8th Division, Allahabad.
Boucher, George John	12 Apl. 93	5 Oct. 14	S. Hospl., Aden.
Michael, Norman Alfred	24 Oct. 93	5 Oct. 14	———.
Mascarenhas, Robert	24 Dec. 91	5 Oct. 14	———.
Vanderwart, Stanley Christian	4 Aug. 92	5 Oct. 14	———.
Croning, Henry Lionel	19 Jan. 93	5 Oct. 14	———.
Boswell, Trevelyan Kenneth Clarke	8 Oct. 92	5 Oct. 14	———.
Mahoney, Cyril Clare	28 Sep. 93	22 Feb. 15	R. I. M. S. "Lhasa."
Haige, Henry Clarence	24 June 94	22 Feb. 15	———.
Lazaro, Thomas	27 Dec. 92	22 Feb. 15	———.
D'Netto, Cecil Innocent (95)	13 July 93	22 Feb. 15	———.
Bussell, Edward Warwick	11 Oct. 93	22 Feb. 15	———.
Aitkins, Stephen Reginald	9 Aug. 93	22 Feb. 15	Peshawar.
Todd, Norman Charles (95)	19 Feb. 94	22 Feb. 15	———.
Upshon, Cyril Oscar	8 Sep. 92	22 Feb. 15	2nd Division.
Marks, Eric Blessington Murray FitzHenry	16 Oct. 94	22 Feb. 15	———.
Monisse, William Thomas Joseph	21 Dec. 92	22 Feb. 15	———.
Oliver, William Bertram	6 July 93	22 Feb. 15	———.
Battye, Albert	29 Oct. 91	22 Feb. 15	Poonamallee.
Moul, Samuel William Arney	25 June 93	22 Feb. 15	———.
Elkins, Temple Ivor Harris	7 May 92	22 Feb. 15	———
Meyers, Joseph Willoughby	10 Mar. 92	22 Feb. 15	Rawalpindi.
Eyles, Clive Mark Egerton	12 May 94	22 Feb. 15	Mhow.
Weekin, Henry Kenneth	30 Dec. 92	22 Feb. 15	———.
Rowe, John Benjamin (95)	8 May 93	22 Feb. 15	———.
Mackay, Donald Murray	30 June 94	22 Feb. 15	———.
Burghall, Clarence James Davis (95)	4 Aug. 92	22 Feb. 15	Peshawar.
Pateman, Ernest Albert Townsend	11 June 94	22 Feb. 15	———.
Symonds, Cyril	4 Sep. 93	22 Feb. 15	Rawalpindi.
Rose, George Xavier Archibald (95, 169)	25 Dec. 93	22 Feb. 15	Peshawar.
Blofeld, Victor Arthur George	25 Mar. 93	22 Feb. 15	———.
Hall, Ellis James	14 Sep. 94	22 Feb. 15	———.
James, Ernest Robert	2 Dec. 93	22 Feb. 15	———.
DeRavara, Reginald Walker	13 May 92	22 Feb. 15	———.
Johnson, Robert Bertram	10 Jan. 94	22 Feb. 15	———.
Armstrong, Valentine Oral	15 Aug. 93	4 Oct. 15	———.
Crean, Leo Emmanuel	3 Jan. 92	4 Oct. 15	———.
Houston, James Joshua	2 Oct. 95	6 Mar. 16	———.
La Frenais, Lambert Norman	1 Feb. 94	6 Mar. 16	Mhow.
De Noronha, Manuel Xavier	26 Sep. 94	6 Mar. 16	———.
Forrest, Frederick Mervin Faulkland	18 Nov. 94	6 Mar. 16	Meerut.
Watts, Alfred Dunstan	11 Sep. 95	6 Mar. 16	———.
Hamilton, William Westland	25 Mar. 94	6 Mar. 16	Rawalpindi.
Ogden, John Lionel (95)	8 Dec. 94	6 Mar. 16	Cherat.
McGuire, James	14 July 93	6 Mar. 16	———.
Wylde, Clarence James	16 Nov. 95	6 Mar. 16	———.
Godwin, Wilton James Ross	5 Sep. 94	6 Mar. 16	Rawalpindi.
Sweeney, Edward Reginald Stanislos	11 Mar. 95	6 Mar. 16	"Kaipun."
O'Callaghan, Clarence	3 Dec. 94	6 Mar. 16	Allahabad.
Hynes, Cyril Edward	26 May 94	20 Oct. 16	Tank.
Campbell, Robert James Foy	13 Nov. 96	20 Oct. 16	———.
Gibbs, Robert Victor	24 Aug. 94	20 Oct. 16	———.
Netscher, Charles Stewart	18 May 96	22 Apl. 18	Bangalore.
Amor, James Anthony	17 May 97	22 Apl. 18	Poona.
Arkile, Richard Neville	1 May 97	22 Apl. 18	Multan.
Ross, Archibald Duncan	16 Aug. 95	22 Apl. 18	———.
Murphy, Hugh Archer	23 May 97	22 Apl. 18	Quetta.

INDIAN MEDICAL DEPARTMENT—*concld.*

Temporary Assistant Surgeons, 4th Class.

The order of Seniority of all these Subordinates is provisional.

Names.	Date of birth.	Date of apptt.	Remarks.
Ferguson, Egbert George	15 July 94	15 Nov. 16	———.
Norton, Wilfred John Mary	25 Feb. 95	15 Nov. 16	———.
Backhurst, Leonard George	30 Apl. 95	16 Nov. 16	———.
Thyer, William Ernest	20 Jan. 96	16 Nov. 16	———.
Acton, Harold Leonard	16 Dec. 95	20 Nov. 16	———.
Bryan, George Michael	4 Sep. 92	20 Nov. 16	———.
Hewson, George Charles	4 Oct. 95	20 Nov. 16	———.
McCue, William Thomas	17 May 95	20 Nov. 16	Gharial.
Ross, Sidney Reginald	27 Jan. 95	20 Nov. 16	———.
Stideston, Dudley Dare	13 Dec. 96	20 Nov. 16	———.
Symonds, Horace	17 Nov. 97	20 Nov. 16	———.
Towning, Walter Ernest	20 June 96	20 Nov. 16	Durat.
Wright, Cecil	28 Oct. 93	20 Nov. 16	———.
Alexander, Raymond James	6 Apl. 94	23 Nov. 16	Nowshera.
Johnson, Charles Potter	28 Nov. 96	23 Nov. 16	Peshawar.
Matthews, Arthur George Teasdale	29 July 97	24 Nov. 16	———.
Morris, Percy George	6 Sep. 96	24 Nov. 16	———.
Reid, Allan Maxwell	9 Feb. 94	24 Nov. 16	Amballa.
Nugent, Lionel St. Dominic Hugh Vere (172*a*)	20 Feb. 96	29 Nov. 16	Quetta.
Fisher, Mervin	13 June 96	6 Dec. 16	Wellington.
McDermott, Reuben Manley	24 Nov. 94	9 Dec. 16	———
Connor, James Michael	16 Mar. 97	22 Mar. 17	Rawalpindi.
Dessa, Ainslie Clifford William	20 Feb. 17	22 Mar. 17	———.
Easey, Robert	30 Mar. 96	22 Mar. 17	Aden.
Francis, Henrix Victor	15 July 96	22 Mar. 17	Amballa.
Freeman, John St. Quentin	12 Apl. 97	22 Mar. 17	———
James, Henry Kerr	10 Dec. 95	22 Mar. 17	———.
Rose, Leopold William	11 July 96	22 Mar. 17	Bombay.
Sealy, Donald Charles	5 May 97	22 Mar. 17	Aden.
D'Sylva, George Hermengeld	13 Apl. 96	23 May 17	Nowshera.
Hopper, Reginald McNally	3 Dec. 95	23 May 17	Rawalpindi.
Bellgard, Sydney James	1 Jan. 98	29 Apl. 18	Sailkot. Exten lv. m. c. 1 mo. 11 Nov. 18.
Brooks, Arthur Gordon	10 Oct. 97	29 Apl. 18	Lahore Cantt.
D'Silva, Norman Augustine Tessiere	6 Sep. 95	29 Apl. 18	———.
D'Souza, Joseph Fermino Aloysius	21 June 97	29 Apl. 18	Peshawar.
D'Souza, Louis John Maximian	30 Oct. 98	29 Apl. 18	Aden.
George, Arthur Whitby	28 June 97	29 Apl. 18	Ferozepore.
Nolan, Allan Albert Morris	13 Dec. 98	29 Apl. 18	Peshawar.
Pierce, Percy Alexander	2 Nov. 95	29 Apl. 18	Rawalpindi.
Prins, Gerald Eugene Arthur	1 Apl. 95	29 Apl. 18	Secunderabad.
Sexton, Joseph Anthony	24 Nov. 97	29 Apl. 18	Bombay.
Williams, Ethelred Oliver	20 Aug. 97	29 Apl. 18	Multan.
Burnett, Feridun Melik-Beglar	24 Feb. 99	2 July 18	4 (Quetta) Division.
Coombs, Stanley Hugh Jamieson	5 Dec. 97	2 July 18	Meerut.
Daniell, Maximillian Ivan Van Greicken Quinlivan	22 Feb. 98	2 July 18	Jubbulpore.
Rappa, Arthur Alexander	19 July 97	2 July 18	Lahore Cantt.

INDIAN MEDICAL DEPARTMENT (BENGAL).

SUB-ASSISTANT SURGEONS.

Senior Sub-Assistant Surgeons, 1st Class. (28)*

Ranking as Subadar.

Names.	Date of birth.	Date of entering service.	Date of present class.	Remarks.
Muhammad Khan (E.), *Khan Sahib, Bahadur* (*Order of Br. Ind., 2nd class*). (Hony. Asst. Surgeon, 18-1-18) (Supy.)	17 Mar. 63	8 Apl. 84	29 Nov. 00	24th Punjabis. Hyderabad (Sind).
Sher Muhammad (E.) *Khan Sahib, Bahadur* (*Order of Br. Ind., 2nd class*). (Hony. Asstt. Surgn., 30-11-17.)	Aug. 65	6 Apl. 87	1 Jan. 03	Govr.-Genl.'s Body-Guard.
Saiyid Ali Naki (E.), *Khan Sahib, Bahadur.* (Hony. Asst. Surgeon, 8-3-18.)	*29 June 68*	*4 Apl. 88*	*21 Mar. 04*	Lucknow.
Tilok Singh (E.) *Bahadur* (Hony. Asst. Surgn., 8 Nov. 18).	Apl. 70	15 Apl. 90	21 Mar. 04	War Hospital, Ambala.
Nurul Hasan (E.) *Khan Sahib* (Supy.) (*f*) I	Mar. 73	2 Apl. 92	15 Oct. 04	1st Division.
Shah Mir Khan (E.) (Supy.)	23 May 61	10 Apl. 85	25 Apl. 06	Sitapore.
Niaz Ali Khan (E.), *Bahadur*	31 May 64	13 Apl. 87	25 Apl. 06	Jubbulpore.
Kehar Singh, Sardar Sahib (E.)	*20 Feb. 68*	*7 May 91*	*1 Oct. 09*	Tank.
Shaikh Hussain Ali (E.) *Bahadur* (*Order of Br. India, 2nd class*) (Supy.)	27 Dec. 61	29 Mar. 81	10 Jan. 10	———.
Ajaib Singh *Rai Sahib* (E.)	25 Feb. 67	19 Apl. 87	15 Apl. 10	———.
Muhammad Mihrdin (E.), *Khan Sahib.* (Supy.)	21 Feb. 62	1 Apl. 81	15 Apl. 10	1st Sappers & Miners, Roorkee.
Azhar Husain (E.) (Supy.) (*f*) II	9 Feb. 71	7 May 91	13 Mar. 11	Meerut.
Gurditt Singh (E.), *Rai Sahib* (Supy.)	15 May 61	6 Apl. 83	28 Dec. 11	I. S. Trans. Corps, Bharatpur.
Karam-chand (E.), *Bahadur*	June 70	13 Apl. 91	16 Jan. 12	———.
Niranjan-das (E.)	Mar. 66	4 Apl. 88	23 May 12	32nd Sikh Pioneers.
Thakur-das (E.) *Rai Sahib*	15 Aug. 65	19 Apl. 87	1 Feb. 13	———.
Saiyid Turáb Ali (E.) (Supy.)	1 Feb. 63	17 Aug. 86	15 Oct. 13	Calcutta.
Sital Ram Varma (E.) *Bahadur.* (*Order of Br. Ind., 2nd class.*)	Nov. 64	19 Apl. 87	10 Feb. 14	1-144th Infy., Bharatpur.
Alah-banda (E.), *Bahadur*	2 Jan. 68	7 Apl. 88	10 Feb. 14	Lahore Cantt. *Lv. m.c., 2 mos., 4 Oct. 18.*
Faizullah (E.)	8 Mar. 67	8 Feb. 89	12 Mar. 14	Karachi.
Muhammad Zamán Khán (E.)	5 Oct. 67	25 Mar. 89	1 Apl. 14	8th Division.
Raghu-nàth (E.)	Jan. 69	4 Apl. 89	10 Apl. 14	Ferozepore.
Kishan-chand (E.) *Bahadur*	June 68	4 Apl. 89	15 May 14	16th Divison. *Lv. m.c., 6 weeks, 15 Nov. 18.*
Pandit Shankar-dás (E.) I. O. M.	Sep. 64	4 Apl. 88	28 May 14	33rd Ps., Fatehgarh.
Muzzammil Husain (E.)	13 Nov. 69	16 Aug. 89	16 Sep. 15	3-6th G. R.
Mahadeo-pershad (E.) I. D. S. M., Supy. (*f*) III.	1 Jan. 76	1 May 99	1 Nov. 15	16th Division. *Lv. m.c., 6 mos., — Feb. 18.* Extn. 6 weeks. Extn. 6 weeks, 24 Oct. 18.
Pir-bakhsh, *Khan Sahib, Bahadur* (E.)	July 69	15 Apl. 90	1 Nov. 15	37th Lancers.
Ajudhya-Parshad Misir (E) (*f*) (Supy.) IV.	1 June 78	1 May 98	1 Dec. 15	I. T. War Hospl., Khandwa.
Haribans Lal (E)	Nov. 69	15 Apl. 90	1 Feb. 16	26th Light Cavalry, Aden.
Barkat Ram, *Bahadur* (E.) (*f*) (Supy.) V.	Mar. 73	12 Apl. 93	15 May 16	———.
Bhagwan Singh (E.)	15 June 66	15 Apl. 90	15 May 16	I. S. Cavalry, Alwar.
Rafiud-din Khan (E.), *Bahadur* (*Order of Br. India, 2nd Cl.*)	Dec. 70	15 Apl. 90	23 May 16	2-5th G. Rifles.
Ihsan Ali (E.)	18 Apl. 71	5 May 90	27 Dec. 16	———.
Nabi Ahmad Sadiki (E.) *Bahadur* (*Order of Br. India, 2nd class*).	June 72	13 Apl. 91	1 Jan. 17	———.
Abdul Ghani (E.) (*Supy.*) (*f*) VI.	*28 Apl. 76*	*1 May 96*	*21 Feb. 17*	*Allahabad.*
Saifud-din, *Khan Sahib* (E.) I. D. S. M. *	Apl. 71	13 Apl. 91	21 Feb. 17	———.
Shankar-das (E)	Feb. 71	13 Apl. 91	1 Feb. 18	S.-C. S. M. O., I.M.S. Office. Quetta.
Pandit Natthu Ram (E.)	Mar. 68	4 Apl. 88	16 Feb. 18	———.
Kapúr Chand (E.)	Mar. 72	3 Apl. 91	17 Mar. 18	Bannu.
Mumtaz Ali Khan Bhatti (E.) (Supy.) (*f*) VII.	Dec. 77	13 July 99	7 July 18	Seistan.

(*f*) Specially promoted under the provisions of para. 470, A. R., I., Vol. II.
(E) Passed in English.
* Excluding seconded.

INDIAN MEDICAL DEPARTMENT (BENGAL)—*contd.*

Senior Sub-Assistant Surgeons, 1st Class—*contd.*

Names.	Date of birth.	Date of entering service.	Date of present rank.	Remarks.
Rivazuddin (E.) I.O.M.	16 Jan. 69	5 May 90	7 July 18	27th Punjabis.
Bir Singh (E.) I.D.S.M.	*8 Nov. 70*	*7 May 91*	*1 Nov. 18*	*Sikh Congt. and K. A. Rifles, Fort Mangocha. Lv., m. c., 6 mos., 15-4-18.*
Mahendu Lál Garg (E.)	4 Apl. 70	7 May 91	1 Nov. 18	Muttra.

Senior Sub-Assistant Surgeons, 2nd Class. (28)*

Ranking as Jemadar.

Names.	Date of birth.	Date of entering service.	Date of present rank.	Remarks.
Asad Ali (E.)	7 Aug. 71	4 May 91	10 Apl. 14	Lucknow.
Khuda-bakhsh (E.)	Sep. 71	14 July 91	15 May 14	25th Cavalry, Meerut.
Debi-dayal (E.)	Aug. 69	14 July 91	28 May 14	Attd. S. and T. Corps, Peshawar.
Dharm Singh (E.)	Feb. 72	2 Apl. 92	8 Feb. 15	———.
Har-parshád (E.), I.D.S.M., *Rai Sahib* I	May 71	2 Apl. 92	23 Feb. 15	———.
Saiyid Nazim Ali (E)	4 Apl. 72	1 May 92	1 Aug. 15	Rawalpindi.
Narayan-das (E) II	12 May 72	1 May 92	1 Sep. 15	Amballa.
Abdul Aziz Khan (E.) (*f*) (Supy.) VIII	11 June 74	26 Mar. 94	1 Dec. 15	Benares.
Mohan Lal (E.) (Supy.) (*f*) IX	5 June 73	1 May 95	1 Dec. 15	Derajat Brigade.
Sundar Singh (E.) (Supy.) (*f*) X	July 74	23 Oct. 95	1 Dec. 15	———.
Abdul Ghafur (E.) (Supy.) (*f*)	18 Sep. 73	1 May 96	1 Dec. 15	8th Division.
I.D.S.M. XI	12 Feb. 78	1 May 99	1 Dec. 15	8th Division.
Sundar Singh (E) (Supy.) XII	May 69	14 July 91	15 May 16	———.
Sultan Ali (E.)	28 July 71	1 May 92	15 May 16	———.
Abdus Shakúr Khán (E.)	Nov. 72	12 Apl. 93	23 May 16	Minto Hospital, Malakand.
Rustam Ali (E.)	15 Mar. 72	12 Apl. 93	26 June 16	Burg Hari Singh Hospital, Peshawar Distt.
Nizamud-din (E.) V				
Dare Khan (E)	*July 70*	*12 Apl. 93*	*27 Dec. 16*	*Mombassa, B. E. Africa.*
Amir Hamzá (E.)	June 71	12 Apl. 93	27 Dec. 10	Sialkot.
Pohlo Ram (E.) (106) (Supy.) (*f*) I.D.S.M. XIII	June 73	12 Apl. 93	1 Jan. 17	Dehra Dun.
Hira Singh, *Rai Bahadur* (E.) (*f*) (Supy.) XIV.	Jan. 75	9 July 96	1 Jan. 17	2-35 Sikhs. Lahore Cantt.
Bhagwan Singh (E.) I.O.M. (Supy.) (*f*) XV.	12 Oct. 73	1 May 95	1 Jan. 17	14th Sikhs.
Abdulah (E.) (Supy.) (*f*) XVI	Jan. 76	23 Oct. 95	1 Jan. 17	46th Punjabis.
Ghaus Muhammad (E.) I.O.M. (Supy.) (*f*) XVII.	Apl. 75	10 Apl. 96	1 Jan. 17	———.
Saiyid Abdul Basit (E.) (Supy.) (*f*) XVIII. (*a*)	10 May 77	14 Apl. 97	1 Jan. 17	X-Ray Institute, Dehra Dun.
Bal-mukand (E.) (Supy.) (*f*) XIX	27 Apl. 79	30 Mar. 00	1 Jan. 17	———.
Mithu Lall (E.) (Supy.) (*f*) XX	25 Sep. 79	24 Apl. 01	1 Jan. 17	Depôt 74th Punjabis, Attock.
Dayal Singh (E.) (Supy.) (*f*) XXI	28 Feb 81	27 Mar. 02	1 Jan. 17	Peshawar.
Kishan Singh (E.) (Supy.) (*f*) XXII XIII	28 Dec 81	27 Mar. 02	27 Mar. 17	1st Division.
Bulta Mal (E)	July 71	12 Apl. 93	5 May 17	———.
Wali-ullah (E)	20 Sep. 68	9 Oct. 91	1 June 17	48th Pioneers.
Abdur Razzak (E) (Supy.) (*f*) XXIII	15 Mar. 82	10 July 02	10 July 17	———.
Muhammad Husain (E) (Supy.) (*f*) XXIV.	30 Dec. 82	10 July 02	10 July 17	———
Ghulam Dastgir Khan (E.)	25 Dec. 70	1 May 93	1 July 17	Bareilly.
Hira Lal (E.)	2 Aug. 71	1 May 93	1 July 17	Moradabad.
Sri Ram (E) VIII	18 Nov. 70	7 May 91	20 Oct. 17	Jubbulpore. Br. F. Africa.
Dula Ram (E.) (*Supy.*) (*f*) *XXV*	Mar. 74	17 Apl. 95	1 Jan. 18	Br. E. Africa.
Daulat Singh (E.) I.O.M., (*Supy.*) (*f*) XXVI.	11 Nov. 74	1 May 96	1 Jan. 18	21st Mountain Battery.
Narayan-parshad Sukul (E.), I.D.S.M. *Supy.* (*f*) XXVII	25 Nov. 79	19 Apl. 00	1 Jan. 18	30th Laneers.
Nagina Singh (E.) (*Supy.*) (*f*) XXVIII	5 Apl. 79	19 Apl. 00	1 Jan. 18	7th Division.
Brindaban (E.)	Jan. 73	10 Apl. 94	1 Feb. 18	9 Hodson's Horse.

(*f*) Specially promoted under the provisions of para. 470, A. R., I., Vol. II.
(E.) Passed in English.
* Excluding seconded.

INDIAN MEDICAL DEPARTMENT (BENGAL)—*contd.*

Senior Sub-Assistant Surgeons, 2nd Class (28).*

Ranking as Jemadar.

Names.	Departmental No.	Date of birth.	Date of entering service.	Date of present rank.	Remarks.
Muhammad Raza Khan (E.) (Supy.) (*f*). XXIX		15 Nov. 75	1 May 96	15 Mar. 18	
Zorawar Singh (E.) (Supy.) (*f*) XXX		1 Oct. 82	2 Apl. 03	2 Apl. 18	Br. East Africa Prot.
Madho-parshad (E.) (Supy.) (*f*) XXXI		1 Apl. 83	2 Apl. 03	2 Apl. 18	127th Baluchis.
Ram-ditta Mal (E.) (Supy.) (*f*) XXXII		20 Mar. 83	2 Apl. 03	2 Apl. 18	112th Infy.
Barkatullah (E.) (Supy.) (*f*) XXXIII		16 Aug. 81	2 Apl. 03	2 Apl. 18	37th Dogras.
Gurditt Singh (E.) (Supy.) (*f*) XXXIV		12 June 81	1 May 03	1 May 18	
Hakim Singh (E.), (Supy.) (*f*) XXXV		20 Apl. 81	1 July 02	1 July 18	23rd Pioneers.
Rawail Singh (E.), I.D.S.M.		Oct. 73	10 Apl. 94	16 Feb. 18	16th Indian Divn. *Lv., m.c., 3 mos., —Aug. 18. Exten. 2 mos.*
Muhammad Faiyaz Khan (E.) . .		30 June 72	May 94	17 Mar. 18	———.
Hardyal Singh (E.)		30 July 72	1 May 94	7 July 18	Civil Dispy.
Sirájuddin (E.)		3 May 71	7 May 91	5 Sep. 18	———.
Fazlur Rahman (E.)		25 Jan. 70	1 May 92	15 Sep. 18	29th Mt. Battery
Jagat Singh (E.)		21 Feb. 71	1 May 93	20 Oct. 18	27th Mt. Battery.
Karam Husain (E.)		Apl. 74	17 Apl. 95	1 Nov. 18	2-54th Sikhs.

Sub-Assistant Surgeons, 1st Class.

Warrant Officers.

Names.	Departmental No.	Date of birth.	Date of entering service.	Date of present rank.	Remarks.
Abbas Ali (E.) (Supy.) (b) XXXVI .	*873*	*June 74*	*23 Oct. 95*	*15 Sep. 99*	———.
Mohan Lal (E.) (Supy.) (*a*) XXXVII	891	Apl. 74	10 Apl. 96	10 Apl. 01	Quetta.
Miyan Jáni (E.)	778	27 Dec. 70	1 May 93	1 May 03	
Abdul Hamid Khan (E.) . . .	741	21 Apl. 69	1 May 02	1 Oct. 03	Bikaner.
Ram Narain (E.)	807	21 July 72	1 May 94	1 May 04	Ahmedabad.
Mirza Muhammad Yûsuf Beg (E.) .	631	9 Nov. 66	20 Apl. 88	16 Oct. 04	Fatehgarh.
Muhammad Amir (E.) (Supy.) (*f*) XXXVIII.	859	29 Dec. 72	1 May 95	17 Dec. 04	
Tika Ram (E.) (Supy.) (*a*) (*f*) XXXIX	917	10 Jan. 73	1 May 96	17 Dec. 04	1-3rd G. R.
Maiku Lal (E.)	825	23 Mar. 73	17 Apl. 95	17 Apl. 05	Bakloh.
Lachhman-das (E.) XXV	829	Sep. 74	17 Apl. 95	17 Apl. 05	———.
Kaka Ram (E.)	834	Jan. 76	17 Apl. 95	17 Apl. 05	9th Division.
Gokalchand (E.)	838	May 74	17 Apl. 95	17 Apl. 05	Ambala.
Narain-das (E.)	840	Jan. 73	17 Apl. 95	17 Apl. 05	Dehra Dun.

* Excluding seconded.

(*a*) Promoted for Field Service. (*b*) Promoted for service in Seistan.

(E.)—Passed in English.

(*f*) Specially promoted under the provisions of para. 470, A. R., I., Vol. II.

INDIAN MEDICAL DEPARTMENT (BENGAL)—*contd.*

Sub-Assistant Surgeons, 1st Class—*contd.*

Names.	Departmental No.	Date of birth.	Date of entering service.	Date of present class.	Remarks.
Lachhman Singh (E.)	842	Jan. 75	17 Apl. 95	17 Apl. 05	Miranshah.
XV					
Wahidyar Khan (E.), I.D.S.M.	852	4 Aug. 74	1 May 95	1 May 05	Bareilly.
IX					
Gaya-din Sharma (E.)	857	15 May 73	1 May 95	1 May 05	69th Punjabis, Agra.
Hira Lal (E.)	858	25 June 73	1 May 95	1 May 05	Meerut.
XXXVIII					
Ram Singh (E.), I.O.M.	862	25 June 74	1 May 95	1 May 05	Simla.
Jiwan Mal (E.)	867	Mar. 75	13 July 95	13 July 05	23rd Mt. Battery, Abbottabad.
X					
Badri-parshad (E.)	748	15 Jan. 69	1 May 92	19 Oct. 05	Attached 3-5th Gurkha Rifles, Peshawar. *Lv., m. c., 3 mos., from 15 Oct. 18.*
Sundar-das (E.)	870	Jan. 76	23 Oct. 95	23 Oct. 05	Derajat Brigade. *Lv., m.c., 3 mos., — Oct. 18.*
XVI					
XXXVI					
Ram Singh (E.)	805	1 Aug. 73	1 May 94	23 Oct. 05	——.
Maharaj Kishan (E.)	874	Aug. 73	23 Oct. 95	23 Oct. 05	Peshawar.
Tarachand (E.)	876	Oct. 75	10 Apl. 96	10 Apl. 06	D. I. Khan.
XVII					
Jwala Singh (E.)	*884*	*July 75*	*10 Apl. 96*	*10 Apl. 06*	
Gulab-din (E.)	885	Mar. 75	10 Apl. 96	10 Apl. 06	D. I. Khan.
XXXVII					
Fazl-din (E.)	892	Aug. 73	10 Apl. 96	10 Apl. 06	56th Camel Corps.
Kirpa Ram (E.)	894	June 72	10 Apl. 96	10 Apl. 06	2-4th G. Rifles.
Ghulam Muhammad (E.)	897	Mar. 74	10 Apl. 96	10 Apl. 06	124th Baluchis.
Hira Singh (E.)	899	1 Sep. 74	1 May 96	1 May 06	D. I. Khan.
Varyam Singh (E.)	*900*	*1 Sep. 74*	*1 May 96*	*1 May 06*	*Nyasaland Protectorate, Zomba.*
Kedar Nath Atrish (E.)	901	10 Sep. 75	1 May 96	1 May 06	Dehra Dun.
XXIX					
Tula Ram Joshi (E.)	904	17 Mar. 75	1 May 96	17 Aug. 06	Barrackpore.
Sham Singh (E.)	911	5 Apl. 73	1 May 96	13 Oct. 06	——.
Nur Muhammad (E.)	908	1 Nov. 72	1 May 96	15 Oct. 06	37th Dogras.
Mota Singh (E.)	912	15 Apl. 74	1 May 96	17 Oct. 06	
Shiu-parshad (E.)	905	23 Sep. 75	1 May 96	21 Oct. 06	——.
Arjan-das Gosain (E.), I.O.M.	914	10 Sep. 74	1 May 96	3 Dec. 06	M. S. Depôt, Lahore.
XXVI					
XLI					
Kesar Singh (E.), I.D.S.M. (Supy.) (*f*)	926	Oct. 75	9 July 96	13 Jan. 07	I. S. T. Corps, Jaipur.
XL VI.					
Muhammad Ishak (E.), I.D.S.M.	*924*	*Feb. 76*	*9 July 96*	*22 Apl. 07*	*Hongkong Co., R. A.*
XL					
Karm Singh (E.)	729	June 72	2 Apl. 92	8 June 07	Nowshera.
Gurditt Singh Saini (E.)	927	Mar. 73	9 July 96	26 June 07	2-6th G. Rifles.
XIV					
Chuni Lal (E.)	932	Mar. 73	26 Oct. 96	3 Sep. 07	
Salagram (E.)	933	Nov. 73	26 Oct. 96	14 Sep. 07	
Gopal-das (E.)	934	Nov. 76	6 Apl. 97	17 Oct. 07	117th Mahrattas.
Dhyan Singh (E.)	909	6 May 75	1 May 96	20 Oct. 07	
XI					
Shaik Nur-Allah, *Khan Sahib* (E.)	913	6 Apl. 75	1 May 96	20 Oct. 07	——.

(*f*) Specially promoted under the provisions of para. 470, A. R., I., Vol. II. (E.) – Passed in English.

INDIAN MEDICAL DEPARTMENT (BENGAL)—*contd.*

Sub-Assistant Surgeons, 1st Class—*contd.*

Names.	Departmental No.	Date of birth.	Date of entering service.	Date of present class.	Remarks.
Bichha Ram (E.)	831	17 Aug. 74	17 Apl. 95	21 Oct. 07	Karachi.
Muhammad Umar (E.), I.D.S.M.	937	May 76	5 Apl. 97	15 Dec. 07	Bhowali.
XVIII					
Altaf Husain (E.)	944	20 Aug. 75	14 Apl. 97	1 Apl. 08	——.
Awadh Bihari Lal (E).	785	15 Aug. 71	29 Sep. 93	15 Apl. 08	Karachi.
Ghulam Ahmad (E.) (144*a*)	952	22 Feb. 76	7 Apl. 98	27 July 08	Peshawar.
Hira Singh (E)	953	13 Aug. 73	1 May 98	6 Aug. 08	Bannu Brigade.
IV					
Kundan Lall (E.)	947	9 Nov. 74	14 Apl. 97	19 Oct. 08	Dehra Dun.
Khurshaid Ali Khan (E.) (105)	853	Feb. 75	1 May 95	1 Mar. 09	——.
Kashi Ram (E.)	963	Feb. 77	11 Apl. 99	11 Apl. 09	Kohat.
Muhammud Husain (E.) (144*a*)	965	June 77	11 Apl. 99	11 Apl. 09	15th Lancers.
Basant Singh (E.)	*966*	*Oct. 76*	*11 Apl. 99*	*11 Apl. 09*	*Br. East Africa.*
Fazl-i-Ilaihi (E.)	967	Dec. 77	11 Apl. 99	11 Apl. 09	Kashgar.
Muhammad Sharif (E.)	970	12 May 80	1 May 99	1 May 09	Derajat Brigade.
Kishor Singh (E)	971	12 June 76	1 May 99	1 May 09	4th Rajputs.
XII					
Brindaban (E.)	973	10 Apl. 77	1 May 99	1 May 09	7th Dn.
Dhani Ram (E.)	975	4 Aug. 77	1 May 99	1 May 09	
Mulchand Sharma (E.)	976	Dec. 76	1 May 99	1 May 09	Br. East Africa.
III					
Ghulam Haidar (E.), I.D.S.M.	980	Oct. 78	13 July 99	13 July 09	1-5th G. R.
IX					
Kishan Singh (E.)	984	17 Jan. 77	20 Dec. 99	20 Dec. 09	Lahore Cantt.
Khilanda Ram Diwan (E.)	*985*	*Mar. 78*	*30 Mar. 00*	*30 Mar. 10*	*Jullundhur.*
Mahbub Ali Khan (E.) *Khan Sahib.*	987	May 79	30 Mar. 00	30 Mar. 10	With Brit. Agent, Kabul.
Tekchand (E.), I. O. M.	988	11 Nov. 79	30 Mar. 00	30 Mar. 10	23rd Pioneers.
XIX					
Zahirud-din Khan (E.) Khan Sahib (Supy.) (f) XLI	*1075*	*20 Mar. 82*	*2 Apl. 03*	*1 Apl. 10*	*Khorasan Agency Dispy. Meshed. Exten. of lv. 3 mos.*
Miraj-ud-din (E.)	968	Nov. 75	1 Apl. 99	11 Apl. 10	Lahore Cantt. *Lv., m. c. 8 mos. 9 days, 5 Sep. 18.*
XXVII					
XXVIII					
Muhammad Sakhawatullah Khan (E.)	1000	10 June 78	19 Apl. 00	19 Apl. 10	Karachi.
Santokh Singh (E.)	1003	8 Apl. 80	19 Apl. 00	19 Apl. 10	Imperial S. Infantry, Alwar.
Ata Muhammad (E.)	1012	Aug. 78	2 July 00	2 July 10	
Lakshman-parshad (E.)	1013	26 Nov. 78	2 July 00	2 July 10	Meerut.
Bhaiya Lal (E.) (Supy.) (*f*) XLII	863	10 Dec. 75	1 May 95	27 Jan. 11	Lucknow. *Lv., m. c. 8 mos. Nov. 18.*
Ram Lal (E.)	940	7 May 75	14 Apl. 97	31 Mar. 11	
Ghulam Husain (E.)	*1025*	*Dec. 79*	*31 Mar. 01*	*31 Mar. 11*	——.
Saiyid Saadat Ali (E.)	1026	20 Oct. 80	31 Mar. 01	31 Mar. 11	19th Lancers.
Kehar Singh (E.)	1014	4 May 79	2 July 00	17 Apl. 11	——.
Kaka Ram (E.)	1021	Feb. 78	31 Mar. 01	17 Apl. 11	1st Division.
Harnam Singh (E.)	1005	7 Dec. 80	19 Apl. 00	19 Apl. 11	Peshawar.
XX					
Muhammad Hafiz Khan (E)	1033	11 Mar. 80	22 Apl. 01	22 Apl. 11	
Jodh Singh (E.)	1035	15 Aug. 80	22 Apl. 01	22 Apl. 11	Meerut.

(E.)—Passed in English.
(*f*) Specially promoted under the provisions of para. 470, A. R., I., Vol. II.

INDIAN MEDICAL DEPARTMENT (BENGAL)—*contd.*

Sub-Assistant Surgeons, 1st Class—*contd.*

Names.	Departmental No.	Date of birth.	Date of entering service.	Date of present class.	Remarks.
Mahabir-parshad (E.) (196)	1036	2 Sep. 78	22 Apl. 01	22 Apl. 11	——.
XLII					
Saiyid Muhammad Ejaz (E.) I.D.S.M.	1006	18 Aug. 75	19 Apl. 00	15 Oct. 11	4th Division. *Lv. m. c. 2 mos.—Sep. 18.*
Uttam Singh (E.)	1031	31 May 79	22 Apl. 01	22 Oct. 11	7th Division.
XXII					
Arjun Singh (E.)	1042	3 Dec. 80	27 Mar. 02	27 Mar. 12	Impl. Service Infy., Alwar.
Ram Singh (E.)	1044	18 Oct. 82	27 Mar. 02	27 Mar. 12	Mhow. Alwar.
Daya Singh (E.)	1029	10 Aug. 80	22 Apl. 01	15 Apl. 12	Sub-charge, X-Ray apparatus, 1st (Peshawar) Dn.
Naranjan-das (E.)	1027	25 Mar. 79	31 Mar. 01	15 Apl. 12	Nowshera.
Amin-chand (E.)	1043	3 Mar. 83	27 Mar. 02	15 Apl. 12	Bannu Brigade.
Parmanand Misra (E.), I.D.S.M. (106)	1048	5 July 79	1 May 02	1 May 12	
Ali Husain Khan (E.)	1049	20 Sep. 81	1 May 02	1 May 12	Depôt 23rd Cavalry, Bareilly.
Natha Singh (E.)	1052	12 Mar. 81	1 May 02	1 May 12	Kohat Brigade, 37th Lancers.
Ishwari-datt (E.)	1054	5 July 80	1 May 02	1 May 12	Lansdowne.
Amar-chand (E.)	1055	18 June 80	1 May 02	1 May 12	35th Sikhs.
Nazir Khan (E.)	1059	1 July 79	1 May 02	1 May 12	Agra.
XXIII					
Daulat Ram (E.)	1066	22 Dec. 81	10 July 02	10 July 12	Lansdowne.
Kanhaiya Lal (E.)	1067	15 Sep. 81	10 July 02	10 July 12	Bakloh.
Kishan Singh (E.)	*1070*	*7 Sep. 81*	*10 July 02*	*10 July 12*	*Nyasaland Protectorate, Mungoche.*
Saiyid Ahmad (E.)	1072	16 Apl. 81	1 Oct. 02	1 Oct. 12	3rd S. H., Rawalpindi.
Mihrab Khan (E.)	1047	14 Aug. 80	1 May 02	21 Oct. 12	Meerut.
Kartar Singh (E.)	1051	17 Oct. 81	1 May 02	21 Oct. 12	Riverain Dett., Survey Deptt.
Khuda-bakhsh (E.)	1053	1 July 80	1 May 02	21 Oct. 12	M. S. Depôt, Calcutta.
Jaswant Singh (E.)	1074	16 Apl. 80	19 Jan. 03	19 Jan. 13	——.
XLI					
Diwan-chand (E.)	*1076*	*11 Apl 83*	*2 Apl. 03*	*2 Apl. 13*	*Uganda Protectorate, Kisumu.*
XXX					
Harnam Singh (E.), I.O.M.	1078	1 Oct 82	2 Apl. 03	2 Apl. 13	34th Pioneers.
Anant Ram Ungle (E.)	1079	15 Apl. 82	2 Apl. 03	2 Apl. 13	21st Cavalry.
Abdul Majid (E.), I.D.S.M.	1083	1 Apl. 82	2 Apl. 03	2 Apl. 13	——.
Jai Gopal (E.)	1086	2 Oct. 82	2 Apl. 03	2 Apl. 13	Dehra Dun.
Sawa Singh (E.)	*1088*	*Jan. 80*	*2 Apl. 03*	*2 Apl. 13*	Peshawar.
XXXI					
Ratan-chand Marwah (E.)	1091	16 Mar. 79	2 Apl. 03	2 Apl. 13	Folls. Hospl., Sunny Bank.
Indar-Singh (E.)	1092	1 Aug. 80	2 Apl. 03	2 Apl. 13	Rawalpindi.
XXXII					
Karim-bakhsh (E.)	1096	5 May 80	2 Apl. 03	2 Apl. 13	16th Division.
Barkat Ram (E.)	1097	13 June 79	2 Apl. 03	2 Apl. 13	Impl. S. Cavy., Alwar State.
Bishan Singh (E.)	1082	14 Aug. 77	2 Apl. 03	21 Apl. 13	——.
Niyaz Muhammad Khan (E.)	1060	5 May 80	1 May 02	1 May 13	Peshawar. *Lv. m. c. 3 mos.—Oct. 18.*
XXXIV					——.
Kesar Chand (E.)	*1100*	*10 Nov. 82*	*1 May 03*	*1 May 13*	*Br. East Africa Protectorate, Eldoret Dispy.*
Bu-ali Khan (E.)	1102	12 Feb. 82	1 May 03	1 May 13	Meerut.
Abdul Hakim Khan (E.)	1103	23 Dec. 79	1 May 03	1 May 13	Q. V. O. Corps of Guides.
Mathura-das (E.)	1104	15 Nov. 82	1 May 03	1 May 13	——.
Kishan Lal (E.)	1106	25 Jan. 80	1 May 03	1 May 13	Dehra Dun.
Sukh Ram-das (E.)	*1107*	*11 Oct. 80*	*1 May 03*	*1 May 13*	Lahore.
Bhagwan Singh (E.)	1109	20 May 81	1 July 03	1 July 13	Jullundhur.
Hukam Singh (E.), I.O.M.	*1111*	*25 Mar. 81*	*1 July 03*	*1 July 13*	*Uganda Protectorate.*
XXIV					
Ishar Singh (E.), I.O.M.	1115	21 Mar. 80	1 Oct. 03	1 Oct. 13	——.

(E.)—Passed in English.

Names.	Departmental No.	Date of birth.	Date of entering service.	Date of present class.	Remarks.

INDIAN MEDICAL DEPARTMENT (BENGAL)—*contd.*

Sub-Assistant Surgeons, 1st Class—*contd.*

Names.	Departmental No.	Date of birth.	Date of entering service.	Date of present class.	Remarks.
XXXV					
Hidayatullah (E.)	1050	5 June 79	1 Mar. 02	21 Oct. 13	——. [*Africa.*
Lal Singh Sandhu (E.)	*1030*	*12 Apl. 80*	*1 Oct. 05*	*12 Mar. 14*	*Somali Scouts, B. East.*
Nagindar Singh (E.), I.O.M.	1116	1 Apl. 82	23 Mar. 04	23 Mar. 14	1st Division
Hira Singh (E.)	1117	Apl. 82	23 Mar. 04	23 Mar. 14	Phillour.
Kaka Ram (E.)	1119	Feb. 81	23 Mar. 04	23 Mar. 14	Peshawar.
Badri Parshad (E.)	1114	3 Sep. 82	1 Oct. 03	19 Apl. 14	13th Rajputs.
Abdul Hakim (E.)	1108	1 Aug. 77	1 May 03	20 Apl. 14	1st Division.
XXXIII					
Muhammad Zaki (E.)	1121	20 Apl. 79	1 May 04	1 May 14	S. M. O., I. M. S. Office & Stores, Mhow. Lv., m.c. 3 mos.,—June 18.
Sham Singh (E.)	1122	10 Mar. 83	1 May 04	1 May 14	Amballa.
Maula-bakhsh (E.), I.O.M.	1124	13 Oct. 76	1 May 04	1 May 14	Resy. Hospl., Bushire.
Parma Nand Sharma (E.)	1126	26 Sep. 79	7 July 04	7 July 14	Jullundur. *Lv. m. c., 3 mos., Oct. 18.*
Ishwari-parshad Sharma (E.)	1129	3 May 80	1 Oct. 04	19 Oct. 14	——.
Karta Ram (E.)	1105	5 Nov. 80	1 May 03	20 Oct. 14	——.
Rala Ram (E)	743	15 June 71	1 May 92	21 Jan. 15	——.
Ganesh Dás (E.)	1132	Oct. 84	5 Apl. 05	5 Apl. 15	——.
Panna Lal (E.)	1133	18 Jan. 84	5 Apl. 05	5 Apl. 15	Karachi.
Dilbag Rae (E.)	1134	Aug. 85	5 Apl. 05	5 Apl. 15	S.-C. Frontier War Hospl., Bannu.
Harkishan Dass (E.)	*1136*	*Jan. 82*	*5 Apl. 05*	*5 Apl. 15*	*Br. East Africa Protectorate, Embu Dispy.*
Abdul Said (E.)	1137	Aug. 84	5 Apl. 05	5 Apl. 15	25th Mountain Batty.
Abdul Hakim Faruqi (E.)	1138	Aug. 77	5 Apl. 05	5 Apl. 15	Peshawar.
Moti Ram (E.)	1139	Aug. 82	5 Apl. 05	5 Apl. 15	76th Punjabis.
Mahdi Hasan (E.)	1140	20 Mar. 83	5 Apl. 05	5 Apl. 15	——.
Fasl Karim (E.)	*1141*	*Apl. 82*	*5 Apl. 05*	*5 Apl. 15*	*Kohat.*
Shankar Lal (E.)	1128	10 Dec. 83	1 Oct. 04	18 Apl. 15	1-5th G. Rifles.
Acharmal (E.)	1142	May 84	5 Apl. 05	18 Apl. 15	Military duty.
Inayatullah (E.)	1015	20 Apl. 77	2 July 00	24 June 15	——.
Jit Singh (E.)	1150	June 83	11 July 05	11 July 15	——.
Muhammad Yasuf Bhatti (E.)	1155	20 Dec. 80	11 July 05	11 July 15	Peshawar.
Parmanand (E.)	1156	May 82	11 July 05	11 July 15	Station Hospital.
Narain Singh (E)	1149	1 Nov. 80	1 May 05	21 Nov. 15	Impl. Service Camel Corps, Bikaner.
Bhup Chand (E.)	1159	Jan. 85	6 Apl. 06	6 Apl. 16	Chitral.
Lachhman Singh (E.)	1160	10 Apl. 84	6 Apl. 06	6 Apl. 16	H. S. Vasna.
Jaswant Singh (E.)	*1161*	*1 June 84*	*6 Apl. 06*	*6 Apl. 16*	*Br. E. Africa Protectorate.*
Abdul Wahid (E.)	1162	18 Mar. 84	6 Apl. 06	6 Apl. 16	*Civil, Chitral.*
Abdul Karim (E.)	1163	24 Nov. 85	6 Apl. 06	6 Apl. 16	——.
Chuhar Singh (E.)	1164	10 Apl. 83	6 Apl. 06	6 Apl. 16	59th Rifles.
Waryam Singh (E.)	1165	Sep. 85	6 Apl. 06	6 Apl. 16	9th Division. *Lv. m.c., 2 mos., 22 days, 13 Oct. 18.*
Ram Lal Abrole (E.) I.D.S.M.	1167	28 Mar. 84	6 Apl. 06	6 Apl. 16	2-6th Gurkhas.
Girdhari Lal (E.)	1168	Aug. 81	6 Apl. 06	6 Apl. 16	Chitral.
Shiv Ramdas (E.)	1170	30 Mar. 83	6 Apl. 06	6 Apl. 16	54th Sikhs F. H.
Zafar Husain (E.), I.O.M.	1171	20 Sep. 84	6 Apl. 06	6 Apl. 16	——.
Deo-raj (E) *Rai Sahib* (Supy.) (*f*)	1169	Jan. 85	29 Apl. 07	29 Apl. 16	——.
XLIII					
Jwala Singh (E.)	1172	15 July 84	1 May 06	1 May 16	——.
Achhru Ram (E.)	*1173*	*7 Jan. 85*	*1 May 06*	*1 May 16*	*Uganda Protectorate.*
Ziladar Khan (E.)	1178	18 July 83	1 May 06	1 May 16	72nd Punjabis.
Siddiq Ahmad (E.)	1179	11 Aug. 86	1 May 06	1 May 16	8th Division.
Sawan Singh (E.)	1183	5 Aug. 86	1 May 06	1 May 16	16th Indian Division.
Kirpa Singh (E.)	1184	9 July 83	1 May 06	1 May 16	2-34th Sikh Pioneers.
Dewan Singh (E.)	1186	15 Apl. 84	1 May 06	1 May 16	129th Baluchis.
Kehar Singh Chandail (E.), I.O.M.	1189	11 Apl. 84	1 Oct. 06	1 Oct. 16	——.
Harnam Singh Gill (E.)	1185	10 Aug. 83	1 May 06	16 Oct. 16	——.
Abdus Sattar Khan (E.)	986	29 Oct. 77	30 Mar. 00	17 Oct. 16	——.
Aswini Kumar Sarkar (Supy.) (*f*)	1291	17 May 88	3 Apl. 10	3 Apl. 17	——.
XLIV					
Atar Singh (E.)	1190	1 July 84	1 Oct. 06	15 Apl. 17	——.

(E). Passed in English.

(*f*) Specially promoted under the provisions of para. 470, A.R., I., Vol. II.

INDIAN MEDICAL DEPARTMENT (BENGAL)—*contd.*

Sub-Assistant Surgeons, 1st Class—*contd.*

Names.	Departmental No.	Date of birth.	Date of entering service.	Date of present class.	Remarks.
Shiv Ram (E.)	1191	28 Nov. 83	29 Apl. 07	29 Apl. 17	——.
Muhammad Din (E.)	1192	16 Mar. 85	29 Apl. 07	29 Apl. 17	Fyzabad.
Himmat Singh (E.)	1193	15 Aug. 85	29 Apl. 07	29 Apl. 17	1st Divn.
Lal Singh Bedi (E.) I.O.M. XLIII.	1194	17 Oct. 84	29 Apl. 07	29 Apl. 17	——.
Mulk-raj Singh (E.)	1197	15 Mar. 85	29 Apl. 07	29 Apl. 17	Karachi.
Mul Singh (E.), I.O.M.	1198	15 July 84	29 Apl. 07	29 Apl. 17	37th Lancers.
Guranditta (E.)	1199	1 Apl. 84	29 Apl. 07	29 Apl. 17	87th Punjabis.
Nand Lal (E.) (105)	1200	3 June 85	29 Apl. 07	29 Apl. 17	Bangalore.
Bishun-das Sawhney (E.)	1201	25 Feb. 87	29 Apl. 07	29 Apl. 17	Folls.' Hospl., Campbellpore.
Milkhi Ram (E.)	*1203*	*1 Aug. 83*	*29 Apl. 07*	*29 Apl. 17*	*Br. E. Africa.*
Muhammad Barkat-ullah (E.)	1204	30 Apl. 85	29 Apl. 07	29 Apl. 17	Lahore Cantt. *Exten. lv. 1 mo., 14 June 18.*
Ram Saran Mehra (E.)	1207	15 May 84	29 Apl. 07	29 Apl. 17	Depôt, 2-5th G. R.
Bashi Ram (E.)	1208	28 Mar. 87	29 Apl. 07	29 Apl. 17	Aden.
Diwan-chand (E.)	1209	27 Mar. 86	29 Apl. 07	29 Apl. 17	Peshawar.
Fazl-Shah (E.)	1210	24 Dec. 86	29 Apl. 07	29 Apl. 17	——.
Ahmad-bakhsh (E.)	1212	17 May 85	29 Apl. 07	29 Apl. 17	Peshawar.
Raghunath Singh (E)	1177	1 Sep. 86	1 May 06	1 May 17	Fort Shabkadar.
Bal-mukand (E.)	1215	9 Feb. 87	1 May 07	1 May 17	3rd Horse.
Kanhaiya Lal Sukul (E.)	1216	20 Mar. 85	1 May 07	1 May 17	——.
Indar Singh (E.)	1217	15 June 87	1 May 07	1 May 17	——.
Kundan Lal (E.), I.O.M.	1219	28 May 86	1 May 07	1 May 17	56th Rifles, F. F.
S. Saadat Ali (E.)	1220	1 May 85	1 May 07	1 May 17	Military duty.
Atar Singh Ilamwala (E.)	1221	5 May 80	1 May 07	1 May 17	154 I. F. A.
Chuhar Khan (E.)	*1222*	*1 Oct. 85*	*1 May 07*	*1 May 17*	*Br. East Africa Protectorate.*
Muhammad Hussain (E.)	1224	25 Apl. 85	1 May 07	1 May 17	112th Infy.
Mit Singh (E.)	1225	12 Oct. 84	1 May 07	1 May 17	S. M. O's. Office, Bannu.
Narindar Singh (E.)	1226	19 June 87	1 May 07	1 May 17	1-6th G. R.
Indar Singh (E.)	1228	15 Mar. 85	1 May 07	1 May 17	Derajat Brigade.
Sant Singh (E.)	1231	20 Feb. 85	1 May 07	1 May 17	35th Sikhs, Peshawar.
Gur-dayal Kashop (E.)	1232	29 Jan. 83	1 May 07	1 May 17	Bombay.
Sampuran Singh (E.)	1233	15 Apl. 87	1 May 07	1 May 17	I. S. T. Corps, Bharatpur.
Shamsher Jang (E.)	1236	28 Oct. 86	1 Oct. 07	1 Oct. 17	——.
Bhagwan-das (E.)	1213	28 Mar. 79	29 Apl. 07	21 Oct. 17	Tank.
Daya Ram (E.)	1227	1 July 85	1 May 07	21 Oct. 17	24th Punjabis.
Kehar Singh (E.)	1238	5 Aug. 84	1 Oct. 07	21 Oct. 17	
Narayan-das (E)	990	Dec. 78	25 Jan. 16	22 Oct. 17	S. M. C. 27th Mountain Battery. *Exten. lv.m.c. 1 m.*
Rahim-bakhsh (E.)	1214	12 July 79	1 May 07	21 Apl. 18	99th Infy.
Naurang Singh (E.) (§)	1223	18 June 82	1 May 07	21 Apl. 18	Ambala.
Mahammad Abdullah (E.)	1240	1 Apl. 86	1 May 08	1 May 18	1st Division.
Gopal-datt Joshi (E.)	1243	5 July 86	1 May 08	1 May 18	——.
Kudrat Ali (E.)	1244	15 July 86	1 May 08	1 May 18	Kohat.
Rahim Bakhsh (E.)	1245	10 June 86	1 May 08	1 May 18	——.
Kishan Singh (E.) I.D.S.M.	1246	15 July 87	1 May 08	1 May 18	11th Rajputs, Chitral.
Mula Singh (E.), I.O.M.	1247	5 June 86	1 May 08	1 May 18	Jullundur.
Waryam Singh (E.)	*1248*	*4 July 86*	*1 May 08*	*1 May 18*	*Br. E. Africa.*
Sawan Singh (E.)	1249	3 July 86	1 May 08	1 May 18	——.
Lakha Singh (E.)	1250	14 Feb. 84	1 May 08	1 May 18	37th C.F.A.
Partab Singh (E.) (§)	1251	15 June 86	1 May 08	1 May 18	——.
Bachint Singh (E.)	1252	7 June 87	1 May 08	1 May 18	Khyaragali.
Lajpat Rae (E.)	1254	3 Aug. 86	1 May 08	1 May 18	Ahmedabad.
Ram Chabila Singh (E.)	1255	10 July 81	1 May 08	1 May 18	——.
Basant Singh (E.)	1257	12 Aug. 84	1 May 08	1 May 18	——.
Karam-chand Kapur (E.)	1258	1 Mar. 86	1 May 08	1 May 18	Bannu Brigade.
Dhanraj (E.)	1260	12 Apl. 87	1 May 08	1 May 18	57th Rifles.
Wadhawa Singh (E.)	1261	1 Nov. 85	1 May 08	1 May 18	8th Divn.
Nur Hasan (E.)	1262	22 Feb. 87	1 May 08	1 May 18	——.
Ganpat Rae (E.)	1265	8 Apl. 86	1 May 08	1 May 18	——.
Chhujju Singh (E.)	1268	7 Apl. 87	1 May 08	1 May 18	13th Lancers.
Amar Nath Puri (E.)	1270	16 June 86	1 May 08	1 May 18	5th Cavalry.
Sudama Ram (E.)	1271	8 Mar. 85	1 May 08	1 May 18	16th Indian Divn. *Lv. m.c., 3 mos., 26 Sep. 18.*
Labhu Ram (E.)	1272	1 Apl. 85	1 May 08	1 May 18	34th Sikh Pioneers.
Diwan Chand (E.)	*1273*	*24 Apl. 84*	*1 May 08*	*1 May 18*	*Uganda Protectorate.*
Hukam Singh (E.) I.O.M.	1274	27 June 87	1 May 08	1 May 18	——.
Shambhu Nath (E.)	1275	10 Apl. 85	1 May 08	1 May 18	——.

(E)—Passed in English. (§)—Qualified for promotion.

NDIAN MEDICAL DEPARTMENT (BENGAL)—*contd.*

Sub-Assistant Surgeons, 1st Class—*concld.*

Names.	Departmental No.	Date of birth.	Date of entering service.	Date of present class.	Remarks.
Gian-chand (E.)	1277	7 Oct. 85	1 May 08	1 May 18	29th Battery.
Bashi Ram Sharma (E.)	1278	2 June 86	1 May 08	1 May 18	Lahore Cantt.
Prem Nath (E.)	1282	5 Oct. 86	1 May 08	1 May 18	H. S. Vasna.
Kishan Singh (E.)	1269	7 Feb. 87	1 May 08	1 June 18	
Brij Lal (E).	1283	24 Nov. 86	20 July 08	20 July 18	4th Division.
Raghunath Sahai (E.)	1284	15 Aug. 86	20 July 08	20 July 18	69th Punjabis, Aden.
Amar Singh (E.)	1286	29 Oct. 86	20 July 08	20 July 18	———.
Gur-dayal Singh (E.)	1288	11 July 87	1 Oct. 08	1 Oct. 18	———. Jhansi.
Chiranji Lal (E.)	1256	15 Oct. 85	1 May 08	1 Nov. 18	Dehli.

Sub-Assistant Surgeons, 2nd Class.

Names.	Departmental No.	Date of birth.	Date of entering service.	Date of present class.	Remarks.
Rala Singh (E.)	1002	4 Apl. 79	2 Apl. 03	21 Apl. 08	1-144th Infy., Bharatpur.
Gokul Chand (E.)	1120	Oct. 79	23 Mar. 04	23 Mar. 09	Risalpur.
XLIV					
Salig Ram (E.)	1235	11 Apl. 84	6 July 07	19 Oct. 14	———. Lahore Cantt.
Muhammad Zainul Abedin	1292	5 Mar. 89	3 Apl. 10	3 Apl. 15	
Durga Parshad Misra	1294	3 Aug. 86	1 May 10	1 May 15	
Kali Mohan Chanda	1295	6 Nov. 87	1 May 10	1 May 15	Peshawar., *Lv. m.c., 2 mos., Novr. 18.*
Safi Muhammad Faruqee	1296	3 Jan. 88	1 May 10	1 May 15	———.
Basant Singh	1297	4 Dec. 87	1 May 10	1 May 15	D. I. Khan.
Fasl Ilahi, Khan Sahib	*1300*	*19 Mar. 88*	*13 June 10*	13 June 15	*Consulate Dispy., Kashgar, Seistan.*
Arjun Singh	1301	19 Sep. 87	13 June 10	13 June 15	1st Divn.
Wilayat Shah	*1303*	*26 Dec. 86*	*13 June 10*	*13 June 15*	*Br. E. Africa Pro.*
Bhagat Ram	1304	20 Dec. 87	13 June 10	*13 June 15*	I. S. Troops, Bikaner.
Ganga-parshad	1306	15 Mar. 82	1 Oct. 10	1 Oct. 15	
Bhupendra-nath Lahiri	1307	21 Jan. 90	8 Apl. 11	8 Apl. 16	———.
Jogendra-nath Khan	1308	18 Apl. 88	8 Apl. 11	8 Apl. 16	———.
Zul-fikar Hyder	1309	27 Sep. 90	18 Apl. 11	18 Apl. 16	
Satyaranjan Bhowmick	1310	9 July 87	18 Apl. 11	18 Apl. 16	———.
Abdul Karim	*1313*	*3 Mar. 89*	*1 May 11*	*1 May 16*	*British East Africa.*
Sansar-Chand	1314	1 July 88	1 May 11	1 May 16	X Ray, Nowshera.
Gopi-Nath Agarwal, M.C.	1315	1 Jan. 91	1 May 11	1 May 16	Dehra Dun.
Sahib Singh	1316	15 Mar. 91	1 May 11	1 May 16	———.
Chhattar Singh	1317	1 Mar. 87	1 May 11	1 May 16	7th Dn.
Ganga Singh	1318	15 May 86	1 May 11	1 May 16	7th Dn.
Murar Rao Khare	1319	20 Oct. 87	13 June 11	13 June 16	I. T. H., Ahmednagar.
Bhagwan Singh	1320	27 Feb. 89	13 June 11	13 June 16	Cantt. Dispy., Campbellpore.
Mewa Ram Vashisht	1322	1 July 88	13 June 11	13 June 16	7th Dn.
Muhammad-din	1323	21 Feb. 87	13 June 11	13 June 16	Zhob Militia, Fort Sandeman.
Laddha Ram	1324	4 June 89	13 June 11	13 June 16	1st Divn.
Ata Muhammad	1325	14 Nov. 88	13 June 11	13 June 16	35th Sikhs. Turco-Persian Frontier.
Ram-datta Kumar	1326	15 May 88	13 June 11	13 June 16	Q. V. O. Corps of Guides.
Sardar Khan	1327	23 Feb. 88	13 June 11	13 June 16	Lansdowne.
Bhag Singh	1328	3 Oct. 89	13 June 11	13 June 16	2-5th G. R., Abbottabad.
Harnam Singh	1329	15 Aug. 84	13 June 11	13 June 16	Dehra Dun.
Amar Nath	1330	27 Jan. 83	13 June 11	13 June 16	———.
Gujjarmal Varma	1331	16 Apl. 89	13 June 11	13 June 16	8th Dn.
Upendra Kumar Gangoli	1333	11 Oct. 89	22 July 11	22 July 16	20th Coy., 3rd S. & M.
Karim Bakhsh	1334	1 Aug. 82	22 Jan. 12	22 Jan. 17	Risalpur.
Shaikh Ali Jan	1335	20 Apl. 87	22 Jan. 12	22 Jan. 17	7th Lancers, Ferozepore.
Gokul Singh	1336	1 Mar. 82	22 Jan. 12	22 Jan. 17	Ambala.
Shaikh Farid	1337	6 Feb. 82	22 Jan 12	22 Jan. 17	1st Divn.
Rajendra Kishore Sen Gupta	1339	17 Jan. 87	1 Apl. 12	1 Apl. 17	
Dinesh Chandra Sen	1340	28 May 90	1 Apl. 12	1 Apl. 17	
Dhirendra Nath Das Gupta	1341	26 Sep. 90	1 Apl. 12	1 Apl. 17	
Bidhu Bhusan Bose	1342	30 June 89	1 Apl. 12	1 Apl. 17	
Ananda Gopal Bandyopadhyay	1344	25 June 90	1 Apl. 12	1 Apl. 17	4th Divn.
Sharafat Ali Khan	1345	21 Apl. 89	1 Apl. 12	1 Apl. 17	Derajat Bgde.
Nagendra Nath Das Gupta	1346	25 June 89	1 Apl. 12	1 Apl. 17	8th Division.
Soroshi Ranjan Sen	1347	23 Mar. 88	1 Apl. 12	1 Apl. 17	138-C. F. A., Aden.
Pratul Chandra Chakravarti	1349	14 Jan. 92	12 Apl. 12	12 Apl. 17	Kamptee. Lv. m.c., 2 mos., —Apl. 18.

(E)—Passed in English. (§)—Qualified for promotion.

INDIAN MEDICAL DEPARTMENT (BENGAL)—*contd.*

Sub-Assistant Surgeons, 2nd Class—*concld.*

Names.	Departmental No.	Date of birth.	Date of entering class.	Date of present class.	Remarks.
Khageswar Bhuyan	1350	8 July 91	12 Apl. 12	12 Apl. 17	
Shaikh Muhammad Abdul Mattalib	1352	11 June 90	12 Apl. 12	12 Apl. 17	A. T. "Erinpura."
Lal Singh	1354	1 Apl. 90	16 Apl. 12	16 Apl. 17	1st Divn.
Balram Das	1356	1 May 88	16 Apl. 12	16 Apl. 17	2-39th G. R.
Yaqub Ali	1357	15 May 89	16 Apl. 12	16 Apl. 17	Zhob Militia Corps, Fort Sandeman.
Brahma-Swarup Bhatnagar	1358	15 Dec. 89	16 Apl. 12	16 Apl. 17	Bareilly.
Birendra Nath Das	1359	16 Nov. 88	16 Apl. 12	16 Apl. 17	Bareilly.
Mathura-parshad Sarswit, I.D.S.M.	1360	19 Nov. 88	16 Apl. 12	16 Apl. 17	——.
Waryam Singh	1361	28 Oct. 90	16 Apl. 12	16 Apl. 17	——.
Raj Singh	1362	24 Dec. 84	12 June 12	12 June 17	A. T. "Erinpura."
Sri Ram	1363	15 Mar. 88	12 June 12	12 June 17	41st Dogras, Dargai.
Roshan Lal	1364	6 Oct. 90	12 June 12	12 June 17	Derajat Brigade.
Muhammad Husain	1366	16 Mar. 83	12 June 12	12 June 17	Lahore Cantt.
Muhammad Said	1367	5 Mar. 90	12 June 12	12 June 17	84th Punjabis.
Jatindra Nath Neogi	1369	18 Aug. 91	4 July 12	4 July 17	
Saiyid Jafar Husain	1370	10 June 88	4 July 12	4 July 17	
Diwan Singh (E.)	1196	3 Apl. 85	29 Apl. 17	12 Sep. 17	35th Sikhs.
Chamman Lal (E.) (Supy.) (*f*) XLV	1263	15 June 83	1 May 08	1 Jan. 18	Derajat Brigade.
Alah-ditta (E.) (Supy.) (*f*) XLVI	1281	6 June 85	1 May 08	1 Jan. 18	Bannu Brigade.
Mubarak Shah Khan, I.D.S.M.	1372	1 Mar. 90	20 Jan. 13	20 Jan. 18	
Muhammad Salim	1373	15 Feb. 86	20 Jan. 13	20 Jan. 18	——.
Sundar Singh	1374	1 June 85	20 Jan. 13	20 Jan. 18	32nd Sikh Pioneers.
Dinesh Chandra Sarkar	1375	19 Sep. 89	10 Apl. 13	10 Apl. 18	8th Divn.
Saiyad Muhamad Ali	1377	18 Nov. 88	10 Apl. 13	10 Apl. 18	Meerut.
Harendra Kumar Guha	1379	7 Nov. 88	22 Apl. 13	22 Apl. 18	No. 3 Coy., S. & M.
Biraj Mohan Ghosh	1380	10 July 88	22 Apl. 13	22 Apl. 18	——.
Mahman Singh	1381	15 May 89	12 May 13	12 May 18	2-6th G. R. On leave till 14 July 18.
Surayan Singh	1382	1 Oct. 87	12 May 13	12 May 18	——.
Ishar Singh	1383	20 Nov. 90	12 May 13	12 May 18	Bannu.
Rati-lal Keshav-lal Modi	1384	15 July 92	12 May 13	12 May 18	
Shamsud-din	1385	2 July 92	12 May 13	12 May 18	——.
Muhammad Mohsin	1386	1 May 89	12 May 13	12 May 18	Peshawar. *Lv. m.c., 2 mos., Nov. 18.*
Ata Muhammad Khan, I.O.M.	1388	8 Sep. 90	12 May 13	12 May 18	——.
Tulsi Ram	1389	26 Feb. 89	18 July 13	18 July 18	——. Miranshah.
Balu Ram Suchadeva	1390	29 Jan. 90	18 July 13	18 July 18	
Hasan-din Dar	1391	1 May 90	18 July 13	18 July 18	No. 6 Coy., S. & M.
Azam Ali Khan	1392	16 June 91	18 July 13	18 July 18	——.
Hari Chand	1393	1 Nov. 90	18 July 13	18 July 18	Bannu Brigade.
Jawand Singh	1394	1 Aug. 88	18 July 13	18 July 18	1-6th G. Rifles.
Mulk-raj Puri	1395	5 May 90	18 July 13	18 July 18	2/35th Sikhs.
Das Ram	1396	29 Aug. 91	18 July 13	18 July 18	Tempy. duty, Peshawar Division.
Chaudhri Ram	1397	2 May 90	18 July 13	18 July 18	1-1st G. R., Dharamsala.
Harditt Singh	1398	14 Nov. 86	18 July 13	18 July 18	Amritsar.
Muhammad Rafi	1399	1 July 85	18 July 13	18 July 18	Dagshai.
Ganga Singh	1365	4 Jan. 84	12 June 12	12 June 18	Tempy. duty, Peshawar Divn.
Bhagh Ali	1400	10 Oct 85	14 Nov. 13	14 Nov. 18	2nd Divn. *Lv, m. c., 1 mo.*
Vishnu Sakha Ram Panjre	1401	16 Oct. 86	14 Nov. 13	14 Nov. 18	9th Divn. *Lv., m. c., 1 year, 24-10-17.*

Sub-Assistant Surgeons, 3rd Class.

Names.	Departmental No.	Date of birth.	Date of entering class.	Remarks.
Rameshwar-dayal	998	15 Mar. 78	19 Apl. 00	Cant. Hospital, Pachmarhi.
Barkat Ali (E.)	1218	15 Oct. 86	1 May 07	——.
XLV				
XLVI				
Rasul Shah (E.)	1285	7 Jan. 86	20 July 08	——.
Saiyid Asghar Ali (E.)	*1289*	*10 Mar. 80*	*1 Oct. 08*	*Brit. East Africa.*
Balbhadar-parshad Misra	1368	18 Mar. 84	12 Dec. 13	Jubbulpore.
Harnam Singh	1402	1 Apl. 85	28 Jan. 14	2nd Reserve Batn., Gur. Rifles.
Surja Narayan Ghosh	1376	12 Jan. 90	10 Apl. 13	8th Divn.—Lv. m. c., 2 mos., 10-9-18.
Sudhiriswar Sen Gupta, I.D.S.M.	1402½	25 June 90	29 Apl. 14	——.
Mukand Ram	1403	25 July 85	1 May 14	127 Baluchis, East Africa.
Muhammad Abdul Rahman	1405	13 Apl. 89	1 May 14	7th Division.
Surajpal, I.D.S.M.	1406	7 July 91	1 May 14	7th Division.
Naurata Ram	1407	5 Apl. 91	1 May 14	59th Scinde Rifles.
Jugal Kishor	1409	1 July 91	1 May 14	
Saiyid Abdul Wahab	1411	23 Oct. 83	27 June 14	3rd Division.
Hazari Mal	1412	15 Dec. 90	27 June 14	Military duty.

(*f*) Specially promoted under the provisions of para. 470, A. R., I., Vol. II.

INDIAN MEDICAL DEPARTMENT (BENGAL)—*contd.*

Sub-Assistant Surgeons. 3rd Class—*contd.*

Names.	Departmental No.	Date of birth.	Date of entering class.	Remarks.
Shiam Singh	1413	31 Dec. 92	1 Oct. 14	10th Jats.
Sewa Singh	1414	28 May 90	1 Oct. 14	
Ganda Ram	1415	1 May 91	30 Oct. 14	4th Division.
Ujagar Singh	1417	13 June 83	30 Oct. 14	——.
Buta Mal	1418	15 July 92	30 Oct. 14	4th Division.
Ahmad Husain	1419	9 Jan. 85	30 Oct. 14	——.
Muhammad-din	1420	1 Dec. 86	30 Oct. 14	——.
Abdul Raoof Khan	1421	20 Aug. 89	17 Apl. 15	1st Division.
Hazara Singh	1422	1 Feb. 89	17 Apl. 15	25th Cavalry.
Sucha Singh	1423	31 Dec. 91	17 Apl. 15	——.
Prem Narain Misra	1424	1 Oct. 91	17 Apl. 15	1st (Peshawar) Division.
Bakhshi Ram	1425	10 July 93	17 Apl. 15	1st (Peshawar) Division. Lv., m.c., 3 mos., 14-7-18. *Exten. 3 mos.*
Sunder Singh	1426	14 Apl. 91	17 Apl. 15	87th Panjabis.
Atar Singh Chatwal	1427	5 Oct. 92	17 Apl. 15	1st (Peshawar) Division.
Vishwa Nath Shukla	1428	15 July 91	17 Apl. 15	1st (Peshawar) Division.
Balwant Singh	1429	1 Aug. 90	17 Apl. 15	
Muhammad Ishaq Siddiqi	1430	29 Aug. 91	12 Apl. 16	——
Muhammad Ismail	1431	30 July 94	12 Apl. 16	——. Bannu.
Partap Singh	1432	10 Feb. 93	12 Apl. 16	——.
Prem Singh	1433	2 Apl. 93	12 Apl. 16	
Hari Singh	1434	2 Nov. 92	12 Apl. 16	——.
Sita Ram Sabarwal	1435	27 Sep. 92	12 Apl. 16	
Shambhu Nath	1436	17 Jan. 92	12 Apl. 16	9th Division.
Bishan Singh	1437	1 Sep. 95	9 June 16	Hospital ship "Ellora."
Brij Lal Ghai	1438	1 Oct. 93	9 June 16	Bombay Bde.
Hashim Ali	1439	21 Jan. 96	9 June 16	2nd Division.
Mubarak Ali	1441	1 May 91	9 June 16	2nd Division.
Fazal Hasan	1416	3 Oct. 90	30 Oct. 16	——.
Daulat Ram	1442	17 Jan. 95	9 Apl. 17	4th Division.
Harnam Sinhg	1443	2 Jan. 94	9 Apl. 17	——.
Muhammad Ayyub	1444	May 96	9 Apl. 17	——.
Muhammad Shafiq	1445	1 July 94	9 Apl. 17	——.
Raghubir Singh	1446	5 Mar. 92	9 Apl. 17	
Santa Singh	1447	9 Oct. 93	9 Apl. 17	——.
Jiwan Singh	1448	13 Mar. 95	9 Apl. 17	——.
Kulwant Singh	1449	25 Aug. 94	9 Apl. 17	
Bansidhar Tripatti	1450	3 June 95	9 Apl. 17	Chitral.
Jodh Singh	1451	12 Dec. 94	9 Apl. 17	——.
Bhagwant Lal Medan	1452	1 June 94	9 Apl. 17	——.
Muhammad Yakub	1453	Oct. 94	9 Apl. 17	——.
Arjan Singh	1454	5 Aug. 93	16 July 17	——.
Arjun Singh Sekhon	1455	4 Sep. 93	28 July 17	——.
Saiyid Irshad Husain Alavi	1456	5 Feb. 94	5 Apl. 18	
Shib Charan Das	1457	7 Apl. 98	5 Apl. 18	
Jiwan Singh	1458	24 Feb. 96	5 Apl. 18	
Abdul Aziz Khan Durrani	1459	1 Oct. 96	5 Apl. 18	Bareilly
Brij Lal Fakay	1460	12 Nov. 95	5 Apl. 18	
Baijnath Lal	1461	1 Aug. 94	5 Apl. 18	
Rajwant Singh	1462	1 Apl. 96	5 Apl. 18	
Dasaundha Singh	1463	2 Jan. 96	5 Apl. 18	
Indra Datta Dikshit	1464	15 May 96	5 Apl. 18	
Shamsud-din	1465	30 Oct. 95	5 Apl. 18	
Awtar Singh Grewal	1466	15 Nov. 94	5 Apl. 18	
Har Chand Singh	1467	9 Feb. 95	5 Apl. 18	
Abdul Ghafur Buxie	1468	1 Jan. 98	5 Apl. 18	
Sapuran Singh	1469	17 Aug. 94	5 Apl. 18	
Sowa Singh Bajwa	1470	5 June 92	5 Apl. 18	
Trimbakla Hargovind Trivedi	1471	7 Apl. 97	5 Apl. 18	
Amar Nath Chopra	1472	15 Mar. 97	5 Apl. 18	
Amar Singh Marati	1473	28 Aug. 96	5 Apl. 18	Ambala.
Indar Singh	1474	1 Oct. 95	5 Apl. 18	
Bishandas Sharma	1475	16 Dec. 95	5 Apl. 18	
Balwant Singh	1476	2 Jan. 99	5 Apl. 18	
Manoharlal Ghei	1477	18 Aug. 97	5 Apl. 18	
Fateh Chand Prashara	1478	1 Apl. 95	5 Apl. 18	
Ganda Ram Chauhan	1479	2 Dec. 96	5 Apl. 18	
Brindaban Bakshi	1480	19 Feb. 97	5 Apl. 18	
Narsingh Das Khurana	1481	1 Jan. 96	5 Apl. 18	
Bachan Singh Uppal	1482	25 Oct. 94	5 Apl. 18	Kohat.
Balmukand Nehraw	1483	30 July 96	5 Apl. 18	
Jasmer Singh Gill	1484	23 Mar. 95	5 Apl. 18	
Tuljashanker Hargovind Trivedi	1485	2 July 98	5 Apl. 18	
Bakshi Ram Rai	1486	2 Dec. 96	5 Apl. 18	

INDIAN MEDICAL DEPARTMENT (BENGAL)—*contd.*

Sub-Assistant Surgeons, 3rd Class—*contd.*

Names.	Departmental No.	Date of birth.	Date of entering class.	Remarks.
Pyare Jan Siddiqi	1487	5 July 96	5 Apl. 18	
Ganga Singh Punn	1488	1 Mar. 97	5 Apl. 18	
Gopal Chandra	1489	10 Nov. 97	5 Apl. 18	
Balmukund Vaid	1490	1 June 97	5 Apl. 18	Derajat Bde.
Bansi Ram Sharma	1491	15 Mar. 95	5 Apl. 18	Derajat Bde.
Mithanlal Vaish	1492	30 Nov. 95	5 Apl. 18	Derajat Bde.
Diwan Singh Bharara	1493	12 June 97	5 Apl. 18	Derajat Bde.
Jagdish Chandra Mohan	1494	29 Sep. 98	5 Apl. 18	Peshawar.
Kishan Singh Sumra	1495	15 Jan. 99	5 Apl. 18	Derajat Bde.
Bhai Kartar Singh Gaga	1496	13 Sep. 97	5 Apl. 18	Derajat Bde.
Jaswant Singh Sethi	1497	14 Oct. 93	5 Apl. 18	Derajat Bde.
Kartar Singa Bedi	1498	2 Jan. 97	5 Apl. 18	Amballa.
Bala Parshad Srivastav	1499	1 Jan. 98	5 Apl. 18	Derajat Bde.
Balwant Singh Bindra	1500	Mar. 94	5 Apl. 18	Derajat Bde.
Jai-mal Singh Bachher	1501	1 June 96	5 Apl. 18	———
Atma Ram Paul	1502	12 Oct. 98	5 Apl. 18	Derajat Bde.
Gitaram Paras Ram Shyngle	1503	8 May 95	5 Apl. 18	Derajat Bde.
Brindaban Sharma	1504	22 Dec. 96	5 Apl. 18	Derajat Bde.
Anil Chandra Sen	1505	10 July 96	5 Apl. 18	Derajat Bde.
Kehar Singh Gill	1506	21 Apl. 97	5 Apl. 18	Derajat Bde.
Dina Nath Vasudeva	1507	20 Apl. 94	5 Apl. 18	———
Sirdar Hamid Hasan	1508	20 Mar. 95	5 Apl. 18	———
Eadwine Franklin Das	1509	14 June 95	5 Apl. 18	*Lv., m.c., 3 mos., Oct. 18.*
Kartar Singh Grewal	1510	25 May 96	5 Apl. 18	Bannu.
Fazl-ul Rahman	1511	4 Feb. 96	5 Apl. 18	Bannu.
Kartar Singh Jauhal	1512	20 Oct. 97	5 Apl. 18	Bannu.
Gurbachan Singh Grewal	1513	5 Sep. 96	5 Apl. 18	Miranshah F. War Hospl.
Devindar Singh	1514	28 Dec. 98	5 Apl. 18	Bannu.
Ranjit Singh	1515	1 June 96	5 Apl. 18	Miranshah F. War Hospl.
Narayan Singh Dhariwal	1516	10 Sep. 97	5 Apl. 18	Kohat.
Srish Chandra Chakravarti	1517	15 Aug. 96	5 Apl. 18	Kohat.
Basharat Ali Khan	1518	3 Dec. 95	5 Apl. 18	
Daulat Singh	1519	1 Mar. 97	5 Apl. 18	
Hari Ram Parasar	1520	1 Apl. 95	5 Apl. 18	
Partap Singh	1521	12 Apl. 96	5 Apl. 18	
Sohan Singh Sidhu	1522	7 Sep. 97	5 Apl. 18	
Sher Singh Bhullar	1523	5 Dec. 95	5 Apl. 18	
Yogendrapal Singh	1524	2 Jan. 96	5 Apl. 18	
Durga Das Asal	1525	1 Apl. 96	5 Apl. 18	
Ganga Singh	1526	25 May 97	5 Apl. 18	4th Division.
Bagla Parshad Banerjee	1527	2 Feb. 95	5 Apl. 18	8th Division.
Hazura Singh Grewal	1528	1 May 98	5 Apl. 18	
Haveli Ram Ghei	1529	2 Jan 98	5 Apl. 18	
Ghulam Asghar Khan Khalidie	1530	21 Mar. 97	5 Apl. 18	
Sultan Singh Ahluwalia	1531	5 Mar. 96	5 Apl. 18	
Pradyumn Singh Bedi	1532	6 May 96	5 Apl. 18	
Salig Ram Beni Madho Vyas	1533	20 Jan. 95	5 Apl. 18	
Dhari Ram Bhatia	1534	9 Jan. 96	5 Apl. 18	
Sohan Lal Chopra	1535	12 Jan. 98	5 Apl. 18	8th Division.
Brij Raj Singh	1536	7 Mar. 98	5 Apl. 18	8th Division.
Pritam Singh	1537	15 Oct. 98	5 Apl. 18	
Udham Singh	1538	22 Aug. 97	5 Apl. 18	8th Division.
Kazi Jalilud-din Ahmad	1539	15 June 98	5 Apl. 18	
Abdur Rahman Khan	1540	19 Sep. 95	5 Apl. 18	
Bankim Chandra Mukerjee	1541	3 Jan. 97	5 Apl. 18	8th Division.
Dharmatma Singh	1542	4 July 99	5 Apl. 18	
Ram Singh Dharmwal	1543	10 July 94	5 Apl. 18	8th Division.
Sri Krishna Vashishtha	1544	29 Oct. 96	5 Apl. 18	Dehra Dun.
Tela Singh Bajwa	1545	1 Mar. 95	5 Apl. 18	
Albert Victor Dalip Singh	1546	2 Oct. 96	5 Apl. 18	
Chiranji Lal Shah	1547	26 Nov. 98	5 Apl. 18	
Hari Krishan Gupta	1548	5 Oct. 99	5 Apl. 18	
Sant Singh	1549	5 Apl. 98	5 Apl. 18	8th Division.
Ignatius Wellington	1550	27 Sep. 95	5 Apl 18	8th Division.
Saiyid Qamer Ali	1551	6 Jan. 96	5 Apl. 18	
Mehta Abinasi Ram	1552	4 Mar. 97	5 Apl. 18	
Kundan Lal Jaini	1553	13 Feb. 96	5 Apl. 18	
Krishn Chandra Varma	1554	28 Oct. 98	5 Apl. 18	
Jeshanker Manishanker Raval	1555	20 Aug. 97	5 Apl. 18	
Lal Chandra Meara	1556	12 Dec. 96	5 Apl. 18	
Faiz Muhammad Khan Chaudhry	1557	1 Apl. 97	5 Apl. 18	
Arjun Singh Deol	1558	18 Aug. 96	5 Apl. 18	1-30th Punjabis.

INDIAN MEDICAL DEPARTMENT (BENGAL)—*contd.*

Sub-Assistant Surgeons, 3rd Class—*contd.*

Names.	Depart-mental No.	Date of birth.	Date of entering class.	Remarks.
Mit Singh Bhasauria	1559	15 Apl. 98	5 Apl. 18	
Jani Mirza Ghazanfar Beg	1560	23 Mar. 93	5 Apl. 18	Meerut.
Mirza Shamshad Beg	1561	24 Jan. 95	5 Apl. 18	
Lal Singh Gill	1562	22 June 97	5 Apl. 18	
Milkha Singh Dhindsa	1563	18 Nov. 95	5 Apl. 18	
Harnam Singh Gill	1564	3 Apl. 99	5 Apl. 18	Bareilly.
Peshaurilal Mehta	1565	18 Jan. 98	5 Apl. 18	
Gursaran Singh Grewal	1566	1 May 98	5 Apl. 18	
Kesar Ram Minocha	1567	20 Nov. 96	5 Apl. 18	Peshawar.
Abdur Rahim	1568	25 July 97	5 Apl. 18	
Crawford Thoburn Wilson	1569	28 Nov. 95	5 Apl. 18	Roorkee.
Bansi Dhar Tewari	1570	13 Jan. 94	5 Apl. 18	Roorkee.
Chanan Singh Deol	1571	12 Aug. 97	5 Apl. 18	Delhi.
Bala Datt Joshi	1572	1 Jan. 95	5 Apl. 18	Delhi.
Tika Ram Joshi	1573	15 Jan. 95	5 Apl. 18	Dehra Dun.
Barkat Ram	1574	5 May 96	5 Apl. 18	Peshawar.
Chandra Pall Gupta	1575	27 Mar. 97	5 Apl. 18	
Sham Lal Bawa	1576	18 June 96	5 Apl. 18	
Bansi Lal Kochhar	1577	1 Apl. 97	5 Apl. 18	
Samuel Peter Joseph	1578	11 Apl. 95	5 Apl. 18	
Ganesh Das Bhandhari	1579	22 May 96	5 Apl. 18	———
Prem Chandra Singh	1580	29 Mar. 96	5 Apl. 18	I. T. Hospl., Ahmednagar.
Panchbir Singh Gorkha	1581	5 Jan. 97	5 Apl. 18	
Charles Alexander	1582	20 Aug. 97	5 Apl. 18	
Banarsidas Verma	1583	11 Feb. 98	5 Apl. 18	I. T. Hospl., Ahmednagar.
Govind Lal	1584	18 Dec. 95	5 Apl. 18	
Alfred Edgar Dial	1585	8 Sep. 97	5 Apl. 18	
Kanhaiya Lal	1586	20 Mar. 96	5 Apl. 18	
Sher Muhammad Aulie	1587	31 Oct. 95	5 Apl. 18	
Sardar Singh Gill	1588	10 Dec. 97	5 Apl. 18	
Saiyid Faiyaz Hussain Rizvi	1590	Jan. 98	5 Apl. 18	*Lv., m c., 2 mos.—Augt. 18.*
Mirza Muhammad Azim Barlas	1591	4 Mar. 96	5 Apl. 18	
Abdul Majid Khan Haphysi	1592	3 Apl. 97	5 Apl. 18	Bareilly.
Khazan Singh	1593	28 July 92	5 Apl. 18	I. T. Hospl., Ahmednagar.
Milkhi Ram	1594	30 Oct. 95	5 Apl. 18	
Saiyid Abbas Haidar	1595	18 Jan. 97	5 Apl. 18	
Chiranji Lal Johar	1596	1 Jan. 97	5 Apl. 18	
Bir Bal Malhautra	1597	22 May 97	5 Apl. 18	Mhow.
Lakhpat Singh	1598	22 Sep. 95	5 Apl. 18	Jhansi.
Autar Singh Bhangu	1599	23 Feb. 98	5 Apl. 18	D. I. Khan.
Kapur Singh Grewal	1600	7 July 98	5 Apl. 18	
Sada Ram	1601	3 Jan. 96	5 Apl. 18	Jhansi.
Baba Harcharan Singh	1602	3 Apl. 96	5 Apl. 18	Jhansi.
Sardara Singh Kokari	1603	22 July 98	5 Apl. 18	Jhansi.
Mathura Parshad Soti	1604	22 Dec. 95	5 Apl. 18	Jhansi.
Vijaya Narayan Gupta	1605	21 Aug. 97	5 Apl. 18	Jhansi.
Purusottam Saran Vaish	1606	25 Sep. 95	5 Apl. 18	Jhansi.
Puran Singh	1607	10 Nov. 95	5 Apl. 18	
Jia Lal	1608	25 Aug. 96	5 Apl. 18	Mhow.
Jiwan Lal	1609	2 Jan. 97	5 Apl. 18	
Bishan Das Minocha	1610	2 Nov. 98	5 Apl. 18	Somaliland.
Jagir Singh Grewal	1611	8 Mar. 99	5 Apl. 18	Rawalpindi.
Saiyid Muhammad Ahmad Naqvi	1612	7 Dec. 95	5 Apl. 18	
Bhavaneshwar Singh Rajput	1613	5 Oct. 96	5 Apl. 18	
Baldeo Parsad Gupta	1614	3 Jan. 98	5 Apl. 18	
Sant Ram Varma	1615	16 June 96	5 Apl. 18	9th Division.
Nand Lal Gupta	1616	23 July 98	5 Apl. 18	
Raghunath Das Vadara	1617	2 Oct. 98	5 Apl. 18	
Babu Lal Sharma	1618	15 Nov. 97	5 Apl. 18	Agra.
Moinuddin Bilgrami	1619	4 Jan. 96	5 Apl. 18	
Dharmanand Pande	1620	12 June 96	5 Apl. 18	4th Division.
Mukand Lal Varma	1621	18 Dec. 97	5 Apl. 18	
Sher Singh Liddar	1622	13 Aug. 97	5 Apl. 18	
Albon Joseph Ransom	1623	23 Feb. 96	5 Apl. 18	
Gopal Singh Dhaliwal	1624	1 Sep. 96	5 Apl. 18	
Sukh-ram Das	1625	30 Sep. 95	5 Apl. 18	Ferozepore.
John Lloyd George	1626	30 June 95	5 Apl. 18	Multan.
Gain Ishmael Nazarene	1627	9 Apl. 98	5 Apl. 18	Multan. *Lv., m. c., 3 mos., 15-7-18.*
Chunni-Lal Srivastava	1628	13 Mar 96	5 Apl. 18	E. Persia.
Pindi Das Bali	1629	16 Nov. 97	5 Apl. 18	Jullundur.
Nand Kumar Tiwari	1630	12 Apl. 95	5 Apl. 18	Agra.

INDIAN MEDICAL DEPARTMENT (BENGAL)—*concld.*

Sub-Assistant Surgeons, 3rd Class—*concld.*

Names.	Departmental No.	Date of birth.	Date of entering class.	REMARKS.
Daya Ram	1631	2 Jan. 96	5 Apl. 18	
Madhu Sudanpal	1632	12 Apl. 95	5 Apl. 18	Lahore Cantt.
Sohan Singh Pannun	1634	25 Dec. 94	5 Apl. 18	Amritsar.
Muhammad Khurshaid Jan	1635	1 Aug. 96	5 Apl. 18	Lahore Cant.
Kartar Singh Bewal	1636	1 Dec. 95	5 Apl. 18	Lahore Cant.
Durga Parsad Khatri	1638	8 Mar. 98	5 Apl. 18	Ambala.
Saiyid Ishtiyaq Ali	1639	15 Jan. 96	5 Apl. 18	
Pyare-Lal Gahlot	1640	21 Aug. 97	5 Apl. 18	Ambala.
Raghu-nandan Parsad Geol	1641	25 July 96	5 Apl. 18	Jhansi.
Nur Alam	1642	4 July 95	5 Apl. 18	Ambala.
Nibaran Chandra Roy	1644	7 Mar. 97	5 Apl. 18	Quetta.
Mul Chandra	1645	5 July 95	5 Apl. 18	Persia.
Bhag Chandra	1646	5 Jan. 98	5 Apl. 18	Bakloh, Amritsar.
Jogesh Chandra Chakravarti	1647	25 Oct. 96	5 Apl. 18	
Rama Nand	1648	22 Jan. 95	5 Apl. 18	
Kirpashanker Girdharilal Dave	1649	5 June 98	5 Apl. 18	
Raghbir Singh Bedi	1650	20 Mar. 96	5 Apl. 18	
Santan Singh Bisht	1651	28 Jan. 96	5 Apl. 18	
Saiyid Akhtar Husain Naqvi	1652	22 Nov. 96	5 Apl. 18	Multan.
Babu Parsad Bhargava	1653	10 Mar. 96	5 Apl. 18	
Banwari-Lal Sharma	1654	3 Jan. 95	5 Apl. 18	Jhelum.
Ahmad Husain Khan	1655	1 Dec. 98	5 Apl. 18	
Shaikh Muhammad Shafaat Husain Siddeqi	1656	25 Apl. 99	5 Apl. 18	
Pyare Singh	1657	6 June 96	5 Apl. 18	
Abdul Majid Khan Shirwani	1658	20 Jan. 96	5 Apl. 18	
Walaiti Ram Nehraw	1659	28 Feb. 95	5 Apl. 18	
Muhammad Ahmad Shamshi	1660	2 July 96	5 Apl. 18	
Chunni-Lal Gandhi	1661	July 98	5 Apl. 18	
Arjun Singh Dhillon	1662	1 Jan. 98	5 Apl. 18	
Bhawani Datt Misra	1663	2 June 98	5 Apl. 18	
Hamid Husain	1664	10 Jan. 97	5 Apl. 18	
Janki Parsad Bhatnagur	1665	3 Nov. 95	5 Apl. 18	
Muhammad Sadik Ali Quraishi	1666	2 Feb. 97	5 Apl. 18	Peshawar.
Prem Narayan Singh	1667	2 Mar. 98	5 Apl. 18	Peshawar.
Mata Din Singh Qanaujaya	1668	30 Dec. 95	5 Apl. 18	Peshawar.
Kazi Muhammad Nazir Hasan	1669	18 Oct. 95	5 Apl. 18	Peshawar.
Muhammad Yaqub Khan	1670	10 Aug. 98	5 Apl. 18	
Muhammad Wasil Khan	1671	4 Mar. 97	5 Apl. 18	Peshawar.
Sulaiman Khan	1672	1 July 96	5 Apl. 18	Peshawar.
Saiyid Muhammad Ewaz Zaidi	1673	27 Jan. 96	5 Apl. 18	Peshawar.
Premendu Mukerji	1674	25 Nov. 96	5 Apl. 18	Peshawar.
Hari Ram Mehta	1675	25 May 97	5 Apl. 18	Peshawar.
Sudhanya Kumar Ghosh	1676	24 Oct. 94	5 Apl. 18	Peshawar.
Fazl Husain Chohan	1678	22 Sep. 95	5 Apl. 18	Lahore Cantt.
Raiji Ranchodji Baria	1679	5 June 95	5 Apl. 18	Peshawar.
Gajraj Singh Gorkha	1680	20 Sep. 95	5 Apl. 18	Peshawar.
Ram Saran Sharma	1681	14 Dec. 95	5 Apl. 18	Peshawar.
Prayag Das Dvivedi	1682	21 Sep. 95	5 Apl. 18	Nasirabad.
Moazzam Ali Talukdar	1683	23 Dec. 96	5 Apl. 18	Peshawar.
Panna-Lal Sur	1684	18 May 96	5 Apl. 18	Peshawar.
Sukhbasi Lal Saksena	1685	5 Apl. 96	5 Apl. 18	Peshawar.
Said-ud-din Ahmad	1686	17 Feb. 98	5 Apl. 18	Delhi.
Mohan Lal Fulji Thakur	1687	14 Aug 98	5 Apl. 18	Quetta.
Muhammad Tasawar Khan Afridi	1688	3 July 96	5 Apl. 18	Salonika.
Badri Parshad Sharma	1689	2 Jan. 96	5 Apl. 18	Multan.
Tek Chand	1690	7 Oct. 94	3 May 18	Ambala Bde.
Maharaj Singh	1691	. . . .	4 Oct. 18	Dehli.
Gaya Parsad Saksena	1692	3 Nov. 95	4 Oct. 18	Dehli.

INDIAN MEDICAL DEPARTMENT (MADRAS).

SUB-ASSISTANT SURGEONS.

Senior Sub-Assistant Surgeons, 1st Class (7).*

Ranking as Subadars.

Names.	Date of birth.	Date of entering service.	Date of present class.	Remarks.
J. Sundaram Pillai (104)	11 May 66	30 June 88	21 Apl. 07	——.
T. Ramanujulu Nayudu	20 July 67	30 June 88	11 Dec. 08	73rd Carnatic Infy.
P. Sivaprakasa Mudaliar	13 June 67	30 June 88	15 May 09	Central Stores, Bangalore.
S. Solomon	13 Jan. 68	29 June 89	15 Sep. 11	116th Mahrattas.
Abdur Rahim Khan	3 June 69	29 June 89	1 Mar. 17	81st Pioneers.
N. Sadasiva Pillai, *Rai Sahib* (Supy.) (*f*) (I).	21 May 68	30 June 90	1 Nov. 18	Central Stores, Secunderabad.
P. Ratnam Pillai (106)	21 July 69	29 June 89	1 Nov. 18	I. T. W. H., Bolarum.

Senior Sub-Assistant Surgeons, 2nd Class, excluding Supy. and Seconded (7).

Ranking as Jemadars.

Names.	Date of birth.	Date of entering service.	Date of present class.	Remarks.
I				
T. A. Muttusāmi	15 Jan. 69	30 June 90	15 Feb. 16	8th Division.
M. A. Thomas Pillai (Supy.) (*f*) II .	12 Dec. 69	30 June 91	1 Jan. 17	Rangoon.
Robert Sahu (105, 106) (Supy.) (*f*) III.	9 Jan. 72	30 June 93	1 Jan. 17	Ajmere.
M. Kadir Sharif (106)(Supy.) (*f*) IV .	22 Aug. 73	30 June 93	1 Jan. 17	——.
P. Hira Lal (106) (Supy.) (*f*) V . . .	27 July 71	30 June 93	1 Jan. 17	Depôt 121st Pioneers, Meerut.
M. Hira Singh (Supy.) (*f*) VI . . .	5 Feb. 74	1 July 94	1 Jan. 17	——.
Narikuti Ramunni (Supy.) (*f*) VII . .	6 Mar. 74	1 July 94	1 Jan. 17	——.
S. Jasudasan, I. D.S.M. (Supy.) (*f*) VIII	25 Dec. 72	30 June 95	1 Jan. 17	8th Division.
V. Virasami Nāyudu (104) (Supy.)(*f*) IX	1 Apl. 75	31 Mar. 98	1 Jan. 17	——.
Samuel Manikkam (Supy.) (*f*) X . .	30 Apl. 79	31 Mar. 01	1 Jan. 17	33rd Cavalry.
M. Amirtanadan Pillai	18 Mar. 65	30 June 90	1 Mar. 17	Rangoon.
Muhammad Fakhr-ud-din Sahib . . .	23 Apl. 69	30 June 91	1 Mar. 17	N'bad.
Muhammad Riazuddeen (Supy.) (*f*) XI .	30 June 75	30 June 95	1 Jan. 18	9th Division.
II				
N. Vijyarāghava Mudali (106) . . .	13 July 70	30 June 93	1 Mar. 18	——.
A. J. Schaffter (Supy.) (*f*) XII . . .	16 Nov. 82	31 Mar. 03	31 Mar. 18	16th Indian Division.
III				

Sub-Assistant Surgeons, 1st Class.

Warrant Officers.

Names.	Deptl. No.	Date of birth.	Date of entering service.	Date of present class.	Remarks.
IV					
V					
M. Kelu	1257	23 May 72	30 June 93	1 July 03	Mandalay.
D. Bhagwant Rao	1223	26 Mar. 69	7 Aug. 93	1 Aug. 03	——.
P. K. Kelappan	1250	10 Jan. 71	30 June 93	17 Apl. 04	Cannanore.
VI					
VII					
Muhammad Abdul Ghafur (106)	1267	30 June 70	1 July 94	2 July 04	8th Division.
T. Sankaran	1270	16 June 72	1 July 94	2 July 04	Hospitalship "Varela."
M. Appalasami Nayudu (106) .	1271	15 Nov. 74	1 July 94	2 July 04	——.
C. K. Gopalan	1272	[illegible] June 74	1 Oct. 94	2 Oct. 04	Trichinopoly.
P. S. Kozundaivēlu Mudaliar .	127[illegible]	2 Oct. 69	1 Oct. 94	2 Oct. 04	Mhow. *Lv.*, *m.c.*, *3 mos.—Aug 18.*
K. Adinarayana	1264	22 June 69	1 July 94	17 Oct. 04	G. R. Depôt, Gorakhpore.

* Excluding seconded.

(*f*) Specially promoted under the provisions of para. 470, A. R., I., Vol. II.

INDIAN MEDICAL DEPARTMENT (MADRAS)—*contd.*

Sub-Assistant Surgeons, 1st Class—*concld.*

Names.	Departmental No.	Date of birth.	Date of entering service.	Date of present class.	Remarks.
XI					Bangalore.
P. Antnunatham Pillai (105)	1285	14 Apl. 72	30 June 95	1 July 05	Cannanore.
Kunniyil Raman	1289	15 Apl. 74	30 June 95	1 July 05	
VIII					——
J. Adinārāyanasāmi Nāyudu	1287	15 Jan. 71	30 June 95	16 Oct. 05	M. S. Depôt, Calcutta.
O. Kunhi Raman	1279	30 June 75	30 June 95	16 Oct. 05	Jatta.
Muhammad Ataullah Khan	1280	20 Mar. 74	30 June 95	16 Oct. 05	——.
S. C. Dorasami Pillai	1306	28 June 75	1 July 96	21 Apl. 07	
G. Joyce	1311	15 Oct. 76	30 June 97	30 June 07	
Saiyid Abdul Latif	1312	16 Aug. 75	30 June 97	30 June 07	Maymyo.
Odan Govindan	1315	17 Dec. 75	30 June 97	20 Mar. 08	——.
E. M. Ayyulu Nāyudu	1314	Feb. 75	30 June 97	2 Nov. 08	Medl. Store Depôt, Rangoon
M. Kuppu Sāmi Pillai	1321	17 Nov. 73	17 July 97	13 Jan. 09	
IX					——.
Muhammad Abdul Azim Khan	1327	3 Dec. 75	30 June 98	1 Nov. 09	Secunderabad.
P. A. Chengalvarāyan	1328	23 Mar. 75	31 Mar. 99	24 Nov. 09	1st (Peshawar) Divn.
C. Parthasaratny Nayagar (105)	1333	1 July 78	31 Mar. 00	1 Apl. 10	Bangalore.
B. V. Sārangapani Pillai	1341	Aug. 77	31 Mar. 00	1 Apl. 10	——. Bannu.
M. Abdul Rahim, I.O.M.	1344	3 Oct. 80	31 Mar. 01	31 Mar. 11	
Muhammad Bahāud-din	1345	May 78	31 Mar. 01	31 Mar. 11	9th Division.
Gudalore Ranganayakulu Nayudu, *Rao Sahib*	1350	3 Feb. 82	31 Mar 02	31 Mar. 12	Bolarum.
Esarapu Venkatasami Nayudu	1349	Oct. 80	31 Mar. 02	15 Apl. 12	
Lal Muhammad	1339	20 Oct. 81	10 Mar. 02	21 Oct. 12	Karachi.
K. A. Rajagopal Achari (106)	1354	16 Mar. 81	31 Mar. 03	31 Mar. 13	——.
B Thavamony Joel	1355	1 July 81	31 Mar. 03	31 Mar. 13	
XII					(Civil, Port Blair). Karachi.
T. R. Govindasami Pillai, I.O.M.	1359	25 Nov. 83	31 Mar. 03	31 Mar. 13	——.
M. Subbayya Reddiar (14)	1360	16 Nov. 81	31 Mar. 03	31 Mar. 13	——.
C. K. Kesavalu Nayudu	1361	3 July 80	31 Mar. 03	31 Mar. 13	——.
X					——.
Chingleput Manikkam	1356	16 Apl. 84	31 Mar. 03	21 Apl. 13	St. Thomas Mount. *Lv., m. c,*
G. R. Samuel	1362	13 Nov. 80	31 Mar. 03	21 Apl. 13	*3 mos., 15 Nov. 18.*
					Karachi.
D. Gnanasiromani	1365	3 Jan. 80	31 Mar. 03	20 Oct. 13	(Civil, Port Blair).
B. H. Gurudayal Singh	1369	1 July 81	31 Mar. 04	31 Mar. 14	(Civil, Port Blair) *9th Division.*
S. Muhammad Husain	1371	16 July 83	31 Mar. 04	31 Mar. 14	1st Division.
N. Kadir Khan	1372	16 July 83	31 Mar. 04	31 Mar. 14	Karachi.
N. Asirvadam	1370	30 June 80	31 Mar. 04	19 Apl. 14	16th Indian Divn.
J. B. D'Cruz	1378	8 Jan. 81	4 July 04	4 July 14	——.
P. P. V. Ramunni Nayar	1383	16 July 82	31 Mar. 05	31 Mar. 15	Followers' Hospital, Bangalore
T. Kunhi Raman	1385	16 Nov. 83	31 Mar. 05	31 Mar. 15	——.
Muhammad Abdur Rauf	1387	1 July 83	31 Mar 05	17 Oct. 15	
Culpankam Virasami Rajogopal Pillai	1390	5 Sep. 84	31 Mar. 06	31 Mar. 16	73rd C. Infy.
M. Tandavarayan Pillai	1396	7 July 86	15 Oct. 06	15 Oct. 16	
V. K. Nasiruddin (Supy.) XIII					Karachi.
(*f*)	1411	4 July 86	12 Mar. 08	12 Mar. 17	——.
Moses Solomon Vincent	1394	18 Mar. 84	11 Oct. 06	31 Mar. 17	9th Division.
Dumpoor Penchaloo Nayudu	1397	23 Nov. 85	31 Mar. 07	31 Mar. 17	——.
Puthussery Kunhi Raman	1399	18 Jan. 86	1 Apl. 07	1 Apl. 17	H. S. "Guilford Castle."
Daniel P. Joseph	1400	July 84	1 Apl. 07	1 Apl. 17	127 I. F. A.
Cornelius Sundararaj Gopal, I.D.	1401	14 Aug. 85	1 Apl. 07	1 Apl. 17	
S.M.					Colombo.
Arangil Chandu	1404	6 Nov. 86	1 Apl. 07	1 Apl. 17	Rangoon.
N. S. Daniel	1405	23 Nov. 86	1 Apl. 07	1 Apl. 17	——.
Ignatius Daniel Joseph	1403	8 Jan. 85	1 Apl. 07	15 Apl. 17	5th L. Infy.
Joseph Chinnapen Aseervathun	1406	6 June 83	17 May 07	21 Oct 17	Hospitalship "Letitia."
Ramisetty Subbayya	1409	28 Dec. 85	12 Mar. 08	12 Mar. 18	——.
Vyasam Venkataramayya	1410	15 Aug. 85	12 Mar. 08	12 Mar. 18	
XIII					——.
Chathoth Kannan	1414	4 Mar. 85	12 Mar. 08	12 Mar. 18	Hospitalship "Letitia."
M. K. Yusuff Khan	1415	16 Sept. 87	12 Mar. 08	12 Mar. 18	——.
C. Celestian	1417	? Dec. 84	15 June 08	15 June 18	

(*f*) Specially promoted under the provisions of para. 470, A. R., I., Vol. II.

INDIAN MEDICAL DEPARTMENT (MADRAS)—*concld.*

Sub-Assistant Surgeons, 2nd Class.

Names.	Depart-mental No.	Date of birth.	Date of entering service.	Date of present class.	Remarks.
Mekala Appalasami	1419	3 Oct. 86	14 Mar. 09	14 Mar. 14	——.
T. V. Damodoram	1421	13 Dec. 87	12 Mar. 10	12 Mar. 15	Meiktila.
C. V. Ekambaram Mudali . .	1422	July 86	12 Mar. 10	12 Mar. 15	H. S. "Varsova."
V. Swaminathan	1423	June 87	12 Mar. 10	12 Mar. 15	M. S. Depôt, Rangoon.
Daniel Israel Njanapragasam, I.O.M.	1425	7 Oct. 87	14 Mar. 11	14 Mar. 16	——.
Pottekat Kazhungil Kesava Menon	1426	2 Aug. 87	14 Mar. 11	14 Mar. 16	——, Chitral.
M. A. Narayanasami Nayudu .	1427	15 July 89	7 Mar. 12	7 Mar. 17	Aden.
T. G. Krishnasami Pillai . .	1431	27 Sep. 91	7 Mar. 12	7 Mar. 17	——.
M. Sesha Reddi	1434	12 June 91	7 Mar. 12	7 Mar. 17	——.
P. T. Varadarajulu Nayakar .	1437	6 Sep. 90	7 Mar. 12	7 Mar. 17	Secunderabad.
Muhammad Usman	1440	14 Dec. 88	7 Mar. 12	7 Mar. 17	Ahmednagar.
Muhammad Abdul Rahim . .	1441	12 Dec. 88	22 July 12	22 July 17	D. 145, I. F. A.
M. Venkatasami Nayudu .	1442	10 Sep. 89	22 July 12	22 July 17	——.
T. B. Karumbayram (Supy.) (*f*) XIV	1456	15 Sep. 92	5 Mar. 15	17 Aug. 17	——.
Syed Budruddin	1443	12 May 89	13 Mar. 13	13 Mar. 18	——.
S. R. Antonymuthu	1444	18 Mar. 90	13 Mar. 13	13 Mar. 18	——.
C. M. Tholasy	1445	28 Aug. 91	13 Mar. 13	13 Mar. 18	——.

Sub-Assistant Surgeons, 3rd Class.

Names.	Depart-mental No.	Date of birth.	Date of entering service.	Remarks.
F. Jesudasan	1447	13 Dec. 87	6 Mar. 14	——.
Abdur Rauf Khan.	1448	5 Nov. 91	8 July 14	——.
T. C. Narasimhan	1449	2 Feb. 91	8 July 14	Attock.
A. Abdul Aleem	1450	17 Mar. 93	8 July 14	4th Rajputs.
M. Narayanan Nair	1451	15 Aug. 89	8 July 14	Palavaram.
P. P. Kuppa	1452	19 Feb. 89	5 Oct. 14	——.
Matayi Govindan	1453	19 Feb. 92	5 Mar. 15	——.
R. Jacob	1454	24 June 92	5 Mar. 15	Aden.
Muzhar Husain	1455	15 Sep. 92	5 Mar. 15	——.
XIV				
T. M. Sundaram Pillai . . .	1458	6 Oct. 91	7 July 15	"D" Sec., 138 C. F. A., Aden.
G. J. Balraj Nadar	1459	25 Feb. 92	7 July 15	1st (Peshawar) Divn.
K. V. Kunhi Raman	1460	30 June 92	7 July 15	1st (Peshawar) Divn.
Ponnuswmny Dinadayalan . .	1461	26 June 89	30 Aug. 15	——.
N. Gopalswami	1462	12 July 92	6 Mar. 16	1-16 Pioneers. *Lv., m. c., 3 mos.—29 Sep. 18.*
Clewent Ram Singh	1464	1 Aug. 91	6 Mar. 16	12 Combd. F., Amb.
M. S. Mariadas	1465	7 Aug. 92	29 June 16	Derajat Brigade.
T. V. Narayanan Nair . . .	1465	13 Dec. 94	4 Mar. 18	
Vellore Gangadara Ranganatham	1467	20 Mar. 95	4 Mar. 18	
P. V. Viswanatha Iyer . . .	1468	21 July 93	4 Mar. 18	
S. Ranganadham Pillai . . .	1469	22 Sep. 96	4 Mar. 18	Bangalore.
T. S. Hiruswamy Rao . . .	1470	15 Sep. 92	9 Apl. 18	
Tasadduq Husain	1471	15 June 97	22 June 18	

(*f*) Specially promoted under the provisions of para. 470, A. R. I., Vol. II.

Names.	Date of birth.	Date of entering service.	Date of present class.	Remarks.

INDIAN MEDICAL DEPARTMENT (BOMBAY).

SUB-ASSISTANT SURGEONS.

Senior Sub-Assistant Surgeons, 1st Class—(8).*

Ranking as Subadars.

Names.	Date of birth.	Date of entering service.	Date of present class.	Remarks.
Ram Rao Gangadhar	11 Mar. 69	16 Sep. 91	13 Aug. 04	H. S. "Varsova."
Atma Ram Bapuji (Supy.)	Jan. 61	5 Sep. 82	1 Jan. 05	Belgaum.
Itu Chawan, *Rao Sahib* (Supy.) . . .	10 Sep. 61	5 Sep. 82	13 Feb. 12	Followers' Hospital, Poona.
Mirza Muhamed Beg, *Khan Bahadur* (Supy.) (*f*) I	3 Apl. 72	14 Sep. 96	16 Jan. 16	Consulate Dispy., Birjand.
Dhondi Nagooji Vizrae, *Rao Sahib* . .	May 66	16 Sep. 87	21 Mar. 16	———.
D. P. DeSouza	24 May 66	16 Sep. 88	15 Apl. 15	34 C. C. H.
Balwant Rao Gore	10 July 66	16 Sep. 88	5 May 16	Karachi. *Lv. on m. c., 3 mos.—Sep. 18.*
Shaikh Miran Shaikh Rahim, *Khan Sahib*	5 Jan. 70	16 Sep. 88	10 Sep. 16	Poona.
Ganga Ram Hariba (Supy.) II	26 Dec. 71	16 Sep. 94	17 Jan. 17	I. T. H., Kirkee.
Narayan Moreshvar	6 Feb. 69	16 Sep. 91	17 Jan. 17	Peshawar.
Sadik Aaron	20 May 71	16 Sep. 92	15 Apl. 17	Mhow.
Subhanji Ellappa Bhorgay	10 Apl. 73	16 Sep. 93	17 Sep. 17	102nd Grenadiers, Bushire.

Senior Sub-Assistant Surgeons, 2nd Class, excluding Supy. and Seconded (9).

Ranking as Jemadars.

Names.	Date of birth.	Date of entering service.	Date of present class.	Remarks.
Elijah Abraham, *Bahadur* (*Order of Br. Ind., 2nd class*)	20 May 72	16 Sep 93	15 Apl. 16	———.
Ram-chandra Ganu Naidoo	31 Mar. 74	16 Sep. 93	5 May 16	———.
Dhondo Mahadeo	19 Feb. 71	16 Sep. 94	30 June 16	Poona.
Lakshman Prasad Jurawan Singh . . .	3 Nov. 72	16 Sep. 93	10 Sep. 16	———.
Munisami Naidu, I.O.M.	19 June 70	16 Sep. 94	15 Sep. 16	Thal.
II				
Benjamin David	15 Feb. 73	28 Aug. 96	1 Jan. 17	———.
Amba Shankar Morarji (Supy.) (*f*) III .	24 June 76	9 Aug. 97	1 Jan. 17	Kirkee.
Shankar Vishnu Gole (Supy.) (*f*) IV. .	5 Mar. 79	16 Sep. 01	1 Jan. 17	S. C. Frontier War Hospl., Miranshah.
Saiyid Abdul Kadur, Moulvi	10 May 73	11 Mar. 96	15 Apl. 17	Kirkee.

Sub-Assistant Surgeons, 1st Class.

Warrant Officers.

Names.		Date of birth.	Date of entering service.	Date of present class.	Remarks.
Kali-das Chhote-Lal	223	2 Nov. 70	16 Sep. 95	17 Oct. 17	Hyderabad (Sindh).
III					
Sulaiman Gulab	237	10 June 76	9 Aug. 97	9 Aug. 07	8th Division.
Raoji Daji Gholap	238	15 Dec. 74	9 Aug. 97	20 Oct. 07	Karachi.
Benjamin Reuben	217	28 Sep. 72	16 Sep. 95	18 Oct. 09	
Kali-charan Shiu-ratan Misra . .	250	13 July 78	3 Sep. 00	3 Sep. 10	Mhow.
Sahedeo Ram-chandra Khanolkar .	251	7 July 77	3 Sep. 00	3 Sep. 10	———.
Shrirang Ram-chandra	253	2 Mar. 77	3 Sep. 00	16 Apl. 11	———.
IV					
Mohan-Lal Nagesar Shukla . . .	263	25 Apl. 78	16 Sep. 01	15 Apl. 12	Nasirabad. *Lv., m.c., 3 mos —26 Oct. 18.*
Dattatraya Bal-krishna Maliker .	260	19 June 79	16 Sep. 01	15 Apl. 12	Peshawar.
Raghu-nath Bal-krishna Alate . .	272	16 Mar. 80	16 Sep. 02	16 Sep. 12	109th Infy.
Waman Hari Supekar	274	1 Aug. 81	16 Sep. 02	16 Sep. 12	Mhow.
Amrat-Lal Jas-Karan Kothari . .	269	12 Mar. 79	16 Sep. 02	21 Oct. 12	———.
Shiv Ram Babaji Patel	271	20 June 81	16 Sep. 02	21 Apl. 13	116 Mahrattas.

* Excluding seconded.

(*f*) Specially promoted under the provisions of para. 470, A. R. I., Vol. II.

INDIAN MEDICAL DEPARTMENT (BOMBAY)—*contd.*

Sub-Assistant Surgeons, 1st Class—*concld.*

Names.	Departmental No.	Date of birth.	Date of entering service.	Date of present class.	Remarks.
Vroj-Lal Umed Ram Pandit, I.O.M.	282	5 Nov. 80	16 Sep. 03	16 Sep. 13	——.
Dali Shankar Mulji Trivedi	285	5 Feb. 82	16 Sep. 03	16 Sep. 13	Poona.
Udhav Sakha Ram Shindge	286	21 Nov. 81	16 Sep. 03	16 Sep. 13	——.
Edward James	252	2 Nov. 78	3 Sep. 00	20 Oct. 13	——.
Kashi-nath Ram-chandra Kalamkar	288	15 Feb. 82	11 Apl. 04	11 Apl. 14	Roorkee.
Binden Pandhari-nath Godbale	293	28 Feb. 82	17 Apl. 05	18 Apl. 15	46th Punjabis.
Pran-Shankar Purusottam Raval	294	23 Nov. 83	15 Sep. 05	15 Sep. 15	Aden.
Jata Shankar Jeshta Ram Bhatt	295	8 Nov. 83	15 Sep. 05	15 Sep. 15	Brit. East Africa.
Magan-Lal Dhanji Udani	296	17 Aug. 82	15 Sep. 05	15 Sep. 15	——.
Munshi Shaikh Ahmad Shaikh Lal	297	20 Dec. 82	15 Sep. 05	15 Sep. 15	Risalpur.
Ramkrishna Ganpat Shinde, M.C.	298	15 July 82	15 Sep. 05	15 Sep. 15	6th Division.
Arjun Atmaji Redkar	304	20 July 83	15 Sep. 05	16 Oct. 15	Poona.
Bell, Robert Samuel	305	16 Aug. 83	4 Apl. 06	4 Apl. 16	5th Light Infy.
Solomon D'Rozario	303	22 Dec. 81	15 Sep. 05	17 Apl. 16	43rd Erinpura Regt.
Shaikh Lal Abdul Karim	307	3 Apl. 84	4 Apl. 06	17 Apl. 16	2nd Lancers.
Wadilal Mohan-lal	308	19 May 82	17 Apl. 06	17 Apl. 16	7th Dn.
Aziz-ul Hakk	310	1 Aug. 84	17 July 06	17 July 16	
Ferris, George Julian, I.D.S.M.	*312*	*10 Feb. 86*	*15 Sep. 06*	*15 Sep. 16*	*Mn. Batty., R. G. A., Egypt.*
Dattatraya, Ganga-dhar Kale	313	10 Apl. 85	15 Sep. 06	15 Sep. 16	122nd Infy.
Hormasji Manekji Limbuwalla	314	26 Nov. 83	15 Sep. 06	15 Sep. 16	——.
Shaikh Azimuddin Shaikh Ismail	318	18 Dec. 85	15 Sep. 06	15 Sep. 16	——.
Ram-krishna Anant Parab	321	23 Feb. 85	15 Sep. 06	15 Sep. 16	——.
Fazal Ahmad (Supy.) (*f*) V	316	6 Feb. 83	15 Sep. 06	1 Jan. 17	Dettd. Native Hospl., Chuppti.
Atma Ram Bhukhen-das Talati	326	19 Nov. 87	16 Sep. 07	16 Sep. 17	No. 26 I. G. Hospl. and X-Ray Asstt., Aden.
Shiu-parshad Krishna-lal Vyas	328	29 Nov. 86	16 Sep. 07	16 Sep. 17	——.
Chhote Lal Mathurbhai Amin	329	8 Oct. 86	16 Sep. 07	16 Sep. 17	Sub-charge, Charbar.
Benjamin Sudamji Powar	330	1 July 86	16 Sep. 07	16 Sep. 17	Lahore Cantt.
Shaikh Mahiboob Muhammad Sahib	331	5 May 84	16 Sep. 07	16 Sep. 17	17th Cavy.
Shaikh Amarbhai Daudbhai	323	3 Mar. 85	15 Sep. 06	21 Oct. 17	Mhow.
Karmeswar-das	324	15 May 82	5 Jan. 07	21 Oct. 17	
Ram Rao Anand Rao Doctor	325	15 June 85	16 Sep. 07	21 Oct. 17	——
John David Baily	334	17 Jan. 85	6 Jan. 08	6 Jan. 18	9th Division.
Shaikh Usmanmia Husainmia	336	11 Sep. 86	3 Feb. 08	3 Feb. 18	——.
Ganesh Raghunath Shirodker	337	16 Aug. 87	16 Sep. 08	16 Sep. 18	X-Ray Sect.
Gordhandas Jethabhai Desai	338	6 July 87	16 Sep. 08	16 Sep. 18	——.
Yeshvant Shiveramji Masoji	339	20 Jan. 87	16 Sep. 08	16 Sep. 18	——.
Balwant Narayan Masurker	341	16 July 88	16 Sep. 08	16 Sep. 18	Jodhpur.
Joseph Appaji Karvekar	*342*	*31 Oct. 86*	*16 Sep. 08*	*16 Sep. 18*	*Br. East Africa Protectorate, Makindu.*
Waman Ramkrishna Rajpathak	343	19 Dec. 87	16 Sep. 08	16 Sep. 18	——.
Maganlal Kanji	345	27 Dec. 85	16 Sep. 08	16 Sep. 18	——.
Vithal Manadeo Satpute	347	1 Jan. 87	16 Sep. 08	16 Sep. 18	——.
Bhalchandra Sakharam Shede	349	11 Oct. 87	16 Sep. 08	16 Sep. 18	——.
Aaron Ezekeil Ziradker	350	27 Sep. 84	16 Sep. 08	16 Sep. 18	112th Infy.
Shanker Laxman Matker	351	14 Feb. 87	16 Sep. 08	16 Sep. 18	——.
Yeshvant Ramchandra Jadhav	353	15 Dec. 87	16 Sep. 08	16 Sep. 18	Jhansi.
Kazi Saiyd Abdul Kayum	309	7 Aug. 85	17 Apl. 06	20 Oct. 18	——.
Shaikh Abdulla Shaikh Rahim	335	14 Dec. 84	3 Feb. 08	20 Oct. 18	Poona.
Krishna Ladoo Matker	340	11 Aug. 87	16 Sep. 08	20 Oct. 18	In ch. X-Ray, Mhow.
Rajaram Gangadhar Khanolker	344	23 Sep. 88	16 Sep. 08	20 Oct. 18	Office of S. M. O., I. M. S., Ahmednagar and Staff Surgeon, Ahmednagar.

Sub-Assistant Surgeons, 2nd Class.

Names.	Departmental No.	Date of birth.	Date of entering service.	Date of present class.	Remarks.
John Peter Andrews	*354*	*7 Nov. 87*	*7 Jan. 09*	*7 Jan. 14*	*Br. East Africa Protectorate, Moench.*
Govind Ramchandra Jejuraker	355	11 Nov. 88	20 Apl. 09	20 Apl. 14	Karachi.
Balwantrao Dalsukhram Vyas	359	29 Sep. 87	14 Sep. 09	14 Sep. 14	Cantt. Hospl., Kamptee.
Hasham Khan Hayat Khan Pathan	357	30 July 86	14 Sep. 09	19 Oct. 14	Sehore.
Haris-chandra Govindrao Shirke	361	25 May 89	26 Jan. 10	26 Jan. 15	——.
Bapalal Himmatlal Yagnik	362	15 Mar. 86	26 Jan. 10	26 Jan. 15	——.
Vishwanath Bhagawantrao Mohite	360	18 Dec. 85	14 Sep. 81	14 Sep. 15	——.
Raghunath Sitaram Patki	363	26 Dec. 87	14 Sep. 10	14 Sep. 15	38th Dogras.

(*f*) Specially promoted under the provisions of para. 470, A. R. I., Vol. II.

INDIAN MEDICAL DEPARTMENT (BOMBAY)—*contd.*

Sub-Assistant Surgeons, 2nd Class—*concld.*

Names.	Departmental No.	Date of birth.	Date of entering service.	Date of present class.	Remarks.
Keshavlal Harjiwan Bhatt	*364*	*28 Aug. 88*	*14 Sep. 10*	*14 Sep. 15*	*Br. East Africa.*
Ramchandra Bhaurao Shinde	365	5 Sep. 88	14 Sep. 10	14 Sep. 15	Neemuch.
Kaikobad Hormusji Dumree	366	24 Sep. 90	14 Sep. 10	14 Sep. 15	———.
Kala Chaku Vaghela	367	18 Mar. 87	14 Sep. 10	14 Sep. 15	———.
Purushottam Damodar Mahatekar	368	5 Sep. 89	14 Sep. 10	14 Sep. 15	39th C. I. Horse.
Nilkhant Ramchandra Mauskar	371	10 Jan. 86	14 Sep. 10	14 Sep. 15	———.
Chintaman Gopal Gogate, I.D.S.M.	374	18 June 86	14 Sep. 10	14 Sep. 15	Cantt. Hospl., Nowshera.
Govind Ramchandra Upalekar	375	15 Jan. 88	14 Sep. 10	14 Sep. 15	Divl. Laby., R. Pindi.
Shripad Hari Sathe	376	18 Dec. 86	14 Sep. 10	14 Sep. 15	1st Division.
Vasant Ganesh Anjarlekar	378	16 Jan. 87	14 Sep. 10	14 Sep. 15	Derajat Brigade.
Ambalal Devshanker Bhatt	377	14 Dec. 88	14 Sep. 10	14 Sep. 16	———.
Ganpat Kanoji Rao Rane, I.O.M.	379	19 July 89	14 Sep. 11	14 Sep. 16	21st Coy., 3rd S. & M.
Ramchandra Bal-Krishna Kusurkar	380	15 May 88	14 Sep. 11	14 Sep. 16	Mhow.
Kalathil Gopalan	382	17 Mar. 91	14 Sep. 11	14 Sep. 16	7th Division.
Vasudev Ram-Chandra Madheker	383	16 Oct. 89	14 Sep. 11	14 Sep. 16	———.
Shiv Ram Vithal Petekar	384	11 July 89	14 Sep. 11	14 Sep. 16	Lv., M.C., 3 mos.—July 18.
Sakharam Sonu Parab	385	17 Oct. 88	14 Sep. 11	14 Sep. 16	127 I. F. A. *Lv., m.c., 3 mos., 18 Aug. 18.*
Shankar Krishna Gole	386	19 Aug. 89	14 Sep. 11	14 Sep. 16	———.
Dattatraya Mahaniraja Makashir	387	1 Oct. 89	14 Sep. 11	14 Sep. 16	Kirkee.
Chunilal Himattram Bhatt	388	1 Nov. 87	14 Sep. 11	14 Sep. 16	———.
Gangaram Ramkrishna Nabar	389	16 Oct. 88	14 Sep. 11	14 Sep. 16	8th Division.
Hirjeebhai Narbhoram Dave	391	10 Jan. 89	14 Sep. 11	14 Sep. 16	Karachi.
Mahbubali Badulla Shaikh	392	3 Sep. 88	14 Sep. 11	14 Sep. 16	Agra.
Raghu-Nath Balwant Limaye	393	28 Jan. 88	14 Sep. 11	14 Sep. 16	Barian.
V					
Dhondo Narayan Kadhe	394	25 Feb. 90	1 Apl. 12	1 Apl. 17	———.
Arjunlal Gagabhai Jotania (Supy.) (*f*) VI	435	28 July 92	20 Sep. 15	17 Aug. 17	———.
Sitaram Vishnu Sathe (Supy.) (*f*) VII	440	3 Mar. 91	20 Sep. 15	17 Aug. 17	
Shankar Gopal Patwardhan	395	28 Feb. 89	18 Sep. 12	18 Sep. 17	Poona.
Pandurang Ramchandra Dhavle	396	3 Mar. 89	18 Sep. 12	18 Sep. 17	Satara.
Dhirajlal Virjibhai Bhatt	398	15 Nov. 89	18 Sep. 12	18 Sep. 17	
Chiman Rao Anand Rao Doctor	401	8 July 89	18 Sep. 12	18 Sep. 17	———.
Shaikh Muhammad Dadasahib, I.O.M.	402	4 Jan. 88	18 Sep. 12	18 Sep. 17	126 I. F. A.
Balkrishna Trimbak Lele	404	21 Sep. 88	18 Sep. 12	18 Sep. 17	2-4th Gurkha Rifles.
Raghuel Isudas Harry	405	24 Aug. 88	18 Sep. 12	18 Sep. 17	No. 7 M.B., Egypt.
Dinanath Babu Rao Sirsat	406	26 Oct. 90	18 Sep. 12	18 Sep. 17	———.
Vasant Sakharam Gaikawad	407	17 July 88	18 Sep. 12	18 Sep. 17	5th Light Infy.
Chhotelal Krishnaram Bhatt	409	5 Feb. 91	18 Sep. 12	18 Sep. 17	———.
Ramchandra Gopal Thitte	411	15 May 89	18 Sep. 12	18 Sep. 17	1st Brahmins.
Ramkrishna Mahadeo Haldankar	412	5 July 89	18 Sep. 12	18 Sep. 17	1st Division.
Vaman Balwant Unde	413	18 Oct. 86	18 Sep. 12	18 Sep. 17	
Vishvanath Narayan Divshikar	414	4 Feb. 89	3 Apl. 13	3 Apl. 18	Bombay.
Narhar Balwant Salunke	415	9 July 90	3 Apl. 13	3 Apl. 18	Karachi.
R. G. Joglekar	416	7 Oct. 89	15 Sep. 18	15 Sep. 18	
Chhotelal Duryaw Singh Verma	417	27 Nov. 92	15 Sep. 18	15 Sep. 18	Peshawar.
Manikal Motilal Talati	418	12 Sep. 90	15 Sep. 18	15 Sep. 18	———.
Nabi Jan Husain Ali Shaikh	419	29 May 90	15 Sep. 18	15 Sep. 18	———.
Kakoobhai Jayantilal Trivedi	420	6 July 91	15 Sep. 18	15 Sep. 18	———.

Sub-Assistant Surgeons, 3rd Class.

Names.	Departmental No.	Date of birth.	Date of entering service.	Remarks.
Shaik Ghulam Mohamed Lal Sahib	421	4 Apl. 89	5 Jan. 14	127 I. F. A.
Amrit Balwant Rajurkar	422	11 July 89	6 Apl. 14	F. War Hospl., Bannu.
Anantprasad Shivprasad Mehta	423	30 Oct. 91	25 Aug. 14	———.
Philip Somabhai Kantharia	424	26 Aug. 92	25 Aug. 14	———.
Edmund Paul Lotlikar	425	10 Dec. 91	25 Aug. 14	127 I. F. A.
Keshav Waman Khuperkar, I.O.M.	426	21 Nov. 88	25 Aug. 14	———.
Moreshawar Balwant Bhagwat	427	7 Oct. 91	25 Aug. 14	7th Rajputs, Aden.
Venktesh Anant Chandragiri	428	6 May 89	25 Aug. 14	152nd C. C. F. A.

(*f*) Specially promoted under the provisions of para. 470, A. R., I., Vol. II.

INDIAN MEDICAL DEPARTMENT (BOMBAY)—*concld.*

Sub-Assistant Surgeons, 3rd Class—*concld.*

Names.	Departmental No.	Date of birth.	Date of entering service.	Remarks.
Manibhai Rambhai Desai	429	10 May 94	20 Sep. 15	Rawalpindi.
Homi Kawasji Kanga	431	21 July 93	20 Sep. 15	1st (Peshawar) Division.
Ganesh Hari Patange	432	27 Nov. 92	20 Sep. 15	
Maruti Raoji Vipat	433	9 Dec. 91	20 Sep. 15	—.
Chhotubhai Bhikubhai Joshi	434	15 June 94	20 Sep. 15	—.
VI				
Damodar Janardan Prabhavalkar	436	1 May 93	20 Sep. 15	—.
Ranchhodlal Popatlal Parikh	438	4 Mar. 93	20 Sep. 15	Karachi.
Shambhuial Jethalal Shah	439	4 Feb. 94	20 Sep. 15	Karachi.
VII				
Bhailal Motilal Shah	441	16 Dec. 91	20 Sep. 15	Bombay.
Motabhai Pirozshah Bharucha	443	15 Apl. 92	20 Sep. 15	Jhelum.
Somnath Mahadeo Rawal	444	4 Jan. 92	20 Sep. 15	—.
Dattatraya Keshaverao Chitnis	445	18 Dec. 93	17 July 16	—.
Ganpath Ramaya Guru	446	23 Dec. 92	17 July 16	—.
Shamaldas Dullabhji Desai	447	10 Aug. 93	17 July 16	Nowshera.
Fulchand Dalsukhram Mehta	448	5 Jan. 94	17 July 16	—.
Amritrao Chimanrao Malawde	449	15 Dec. 96	17 July 16	—.
Vinayek Krishna Altekar	450	14 June 93	17 July 16	—.
Abdulmia Ahmedmia Bukhari	451	12 Sep. 93	17 July 16	—.
Erach Cursetji Dadabhai	452	18 May 92	17 July 16	—.
Hiralal Ranchhodlal James	453	18 Dec. 92	17 July 16	Nowshera.
Ramniklal Karsanlal Joshipura	454	14 Apl. 92	17 July 16	—.
Michael Simon Judas DeSouza	455	28 Dec. 92	17 July 16	31st D. C. O. Lancers.
Shivailal Trambaklal Oza	456	27 Jan. 92	17 July 16	4th Division.
Verji Punja Thaker	458	15 Nov. 89	17 July 16	—.
Kuppaswami Panchanadham Pillai	459	6 Aug. 92	17 July 16	Nowshera.
Vishwanath Vasudeo Datye	460	24 Nov. 92	17 July 16	—.
Shriniwas Balasudram Mudiar	461	7 Dec. 94	11 July 16	8th Division.
Hanmant Dattatrya Bedeskar	462	21 Mar. 93	26 July 17	Ahmednagar.
Vinayak Narayan Deuskar	463	4 Nov. 95	24 Sep. 17	—.
Jamsetji Pestonji Patel	464	9 Oct. 94	24 Sep. 17	Derajat Brigade.
Manishanker Karunashanker Shukla	465	11 July 96	24 Sep. 17	—.
Harilal Narsinh Pancholi	466	26 Dec. 93	24 Sep. 17	—.
Keshavlal Dulabhji Udani	467	25 Sep. 94	24 Sep. 17	Nowshera.
Laxuman Narayan Gogate	468	27 Nov. 94	24 Sep. 17	—.
Mohamed Eusuf Ali Khan Tabib	469	6 July 96	24 Sep. 17	—.
Prem Roy Parsuram Pandya	470	8 Jan. 95	24 Sep. 17	—.
Ganesh Vasudeo Sakriker	471	5 May 95	24 Sep. 17	—.
Shambhu Chidamber Padki	472	5 Sep. 94	24 Sep. 17	—.
Bhailal Dahyabhai Amin	473	7 Sep. 94	24 Sep. 17	—.
Jivabhai Lallubhai Amin	474	21 Oct. 94	24 Sep. 17	—.
Hiralal Parshotamdash Sheth	475	5 July 94	24 Sep. 17	—.
Sundar Vishnu Vedak	476	3 Oct. 93	24 Sep. 17	—.
Sunderlal Tribhuvan Bhatt	477	20 Aug. 96	24 Sep. 17	Kemptee.
Dattatraya Govind Mahajan	478	10 Nov. 94	24 Sep. 17	—.
Ramchandra Bhogilal Bhatt	479	24 Apl. 94	24 Sep. 17	—.
Prabhushanker Dayashanker Dave	480	16 June 91	24 Sep. 17	4th Division.
Narhar Vishwanath Bal	481	26 June 96	1 Oct. 17	—.
Gopal Hari Behere	482	24 Dec. 94	1 Oct. 17	—.
Hanmant Shrinivas Dandi	483	17 May 95	1 Oct. 17	—.
Shankar Ramchandra Abhyankar	484	12 Jan. 94	1 Oct. 17	—. Tank.
Vishwanath Mahadeo Paranjpe	485	16 Apl. 95	1 Oct. 17	Miranshah Hospl.

INDIAN MEDICAL DEPARTMENT.

Temporary Sub-Assistant Surgeons.

3rd Class.

The order of seniority of all these subordinates is provisional.

Names.	Date of appointment.	Remarks.
Chanan Das	20 Nov. 16	—.
Chajju Ram	20 Nov. 16	—.
Jagat Ram	20 Nov. 16	—.
Kanshi Ram	20 Nov. 16	—.
Devi Dayal	22 Nov. 16	—.
Behari Lal	22 Nov. 16	—.
Sadashiv Waman Gupte	24 Nov. 16	—.
Ramchandra Gangadhar Jathar	24 Nov. 16	—.
Dattatraya Shripad Mahabal	24 Nov. 16	—.
Kesheo Abajee Pradhan	24 Nov. 16	—.
Pattanath Sankaran	24 Nov. 16	1-3rd Gurkha Rifles.
Muhammad Husain	11 Dec. 16	—.
A. Balaji Singh	11 Dec. 16	—.
Muhammad Sattar Sharif	11 Dec. 16	Derajat Brigade. *Lv. m. c. 3 mos. 13 Sep. 18.*
Karam Chand	15 June 17	33rd Cavalry.
Balwant Rai Datt	15 June 17	Kohat Brigade.
Mohammad Abdul Latif	15 June 17	Tank.
Nur Mohammad	15 June 17	Depôt 127th Baluchis.
Dayal Singh	15 June 17	—.
Gobind Ram	18 July 17	—.
Abdul Karim	4 Sep. 17	2-12th Pioneers.
Abdul Aziz Khan	23 April 18	Rawalpindi Division.
Ali Ahmad	23 April 18	Peshawar.
Babu Singh	23 April 18	Lahore Division.
Bishn Das	23 April 18	" "
Brij Lal	23 April 18	Rawalpindi Division.
Dalip Singh	23 April 18	Peshawar Division. *Lv., m. c., 3 mos., 5 Aug. 18.*
Fateh Muhammad Khan	23 April 18	Lahore Division.
Gauri Parsad Varma	23 April 18	" "
Ghulam Muhammad Khan	23 April 18	Rawalpindi Division.
Gian Singh	23 April 18	Quetta Division.
Gokal Singh	23 April 18	Jhelum.
Gulab Rai	23 April 18	Rawalpindi Division.
Gurmukh Shah	23 April 18	Peshawar Division.
Habibullah	23 April 18	—.
Harbans Lal	23 April 18	Rawalpindi Division.
Jagdish Chandra Madan	23 April 18	—.
Jai Singh	23 April 18	Quetta Division.
Kartar Singh	23 April 18	Peshawar Division.
Kedar Nath	23 April 18	Lahore Division.
Mangal Singh	23 April 18	Quetta Division.
Manghu Lal	23 April 18	Rawalpindi Division.
Matlab Khan	23 April 18	Peshawar.
Mehar Singh Chowla	23 April 18	—.
Mihr Ali	23 April 18	—.
Muhammad Afzal Khan	23 April 18	Peshawar Division.
Muhammad Ismail	23 April 18	Peshawar.
Muhammad Munir	23 April 18	5th (Mhow) Division.
Nirmal Singh	23 April 18	Quetta Division.
Prem Singh	23 April 18	Peshawar Division.
Pritam Singh	23 April 18	Quetta Division.
Sewa Ram	23 April 18	Peshawar.
Sohan Singh	23 April 18	Peshawar Division.
Sundar Singh	23 April 18	—.
Udham Singh	23 April 18	Lahore Division.
Umrao Singh	23 April 18	Ambala.
Uttam Chand Malhotra	23 April 18	Lahore Division.
Rishi Lal Vaid	25 April 18	—.
Autar Singh	28 April 18	Jhelum.

SUB-ASSISTANT SURGEONS WHO SERVED WITH THE LATE HYDERABAD CONTINGENT AND HAVE NOT ENTERED THE INDIAN MEDICAL DEPARTMENT.

(Vide *Letter from Department of Military Supply, No. 4315-G., dated 19th July 1907.*)

No.	Name.	Date of birth.	Date of entering service.	Date of present rank.	REMARKS.
	Senior Sub-Assistant Surgeons, 1st Class.				
	Nil.				
	Senior Sub-Assistant Surgeons, 2nd Class.				
17	Muhammad Abdul Karim	Apl. 63	15 June 85	4 Mar. 05	Central Medical Stores, Bolarum.
	Sub-Assistant Surgeons, 1st Class.				

ARMY HOSPITAL CORPS.

Commanding Companies.

D. OR A. D., M. S. OF DIVNS. AND ADEN BDE.

No. 1 Company, 1st (Peshawar) Division.
No. 2 „ 2nd (Rawal Pindi) „
No. 3 „ 3rd (Lahore) „
No. 4 „ 4th (Quetta) „
No. 5 „ 5th (Mhow) „
No. 6 Company, Poona Division.
No. 7 „ Meerut „
No. 8 „ 8th (Lucknow) „
No. 9 „ 9th (Secunderabad) „
No. 10 „ Burma „

No. 11 Company, Aden Brigade.

ARMY BEARER CORPS.

Commanding Companies,

Nos. 1 to 9—D. A. D., M. S. (Mobn.) of Division.
Nos. 10 and 11—A. D. M. S. of Division and Aden Bde.

No. 1 Company, 1st (Peshawar) Division.
No. 2 „ 2nd (Rawal Pindi) „
No. 3 „ 3rd (Lahore) „
No. 4 „ 4th (Quetta) „
No. 5 „ 5th (Mhow) „
No. 6 „ 6th (Poona) „
No 7 Company, Meerut Division.
No. 8 „ 8th (Lucknow) „
No. 9 „ 9th (Secunderabad) „
No. 10 „ Burma „
No. 11 „ Aden Brigade.

VETERINARY SERVICES.

ARMY VETERINARY SERVICE.

UNIFORM—*Blue*. FACINGS—*Maroon cloth*.

First Commn.	Names.	Present rank.	Date of completion of Indian Tour.	Remarks.
	Removed from the Corps as Colonels and still on the Active List.			
30 July 84	**Nuthall, C. E.,** C. B., C. M. G. (Tempy. Brig.-Genl.)	4 June 10	. . . .	Dir. of Vety. Services in India.
	Larnder, E. W.	26 July 16	. . . .	D. D. V. S., S. Comd.
12 June 89	**Newsom, A. C.,** C.M.G. . . .	23 Sep. 17	10 Dec. 1917	
11 Sep. 89	**Eassie, F.,** C.M.G., D.S.O. (Bt.-Col.)	1 Jan. 18	1 Feb. 1915	
	ARMY VETERINARY CORPS.			
	Lieutenant-Colonels.			
	Sawyer, H. T., D.S.O.	25 Apl. 14	. . . .	D. D. V. S., Army Hd. Qrs.
3 Feb. 92	*Hunt, F. D.*	*15 Dec. 15*	. . . .	*A. R. D., Ahmednagar.*
12 June 95	**Edwards, W. B.**	23 Sep. 17	. . . .	D. D. V. S., N. Comd.
	Majors.			
22 Jan. 98	**McGowan, J. A. B.** (*actg. Lt.-Col., 16 June 17*).	22 Jan. 13	4 Mar. 1915	
6 July 98	**Williams, A. J.,** D.S.O. F.R.C.V.S. (*actg. Lt.-Col., 5 Nov. 17*) . .	6 July 13	9 Jan. 1917	
13 Apl. 01	**Deacon, A. F.**	10 July 15	. . . .	D. A. D. V. S., 5th (Mhow) Dn.
13 Apl. 01	**Welch, H. C.**	10 July 15	. . . .	*Lv. ex. 1., 6 mths., m. c.* [Dn.
2 Nov. 01	**Jackson, G. T. T.**	10 July 15	. . .	D. A. D. V. S., 9th (Secunderabad)
2 Nov. 01	**Steel, C. E.** (*temp. Lt.-Col.*) . .	10 July 15	30 Nov. 1914	
2 Nov. 01	**Gillett, E. S.,** C.I.E. (206). (*Temp. Lt.-Col.*)	10 July 15 / 25 Feb. 18	. . . .	Dy. Dir. Remt. Directorate, A. Hd.-Qrs.
2 Nov. 01	**Beatty, P. V.**	10 July 15	23 Sep. 1918	
15 Apl. 02	**Glasse, M. St. G.** (**temp. Lt.-Col.*)	10 July 15	28 Sep. 1914	
15 Apl. 02	**Nicol, W. H.** (207) (**temp. Lt.-Col.*)	10 July 15	10 Dec. 1914	
15 Apl. 02	**Roberts, N. d'E.** (*Bt.-Lt.-Col.*) .	10 July 15	. . . .	A. R. D., Hosur.
16 May 03	**Ludgate, W.,** D.S.O.	10 July 15	2 Jan. 1917	D. A. D. V. S., Meerut Dn.
16 May 03	**Jelbart, W. A.**	10 July 15	2 Jan. 1917	
16 May 03	**Nicholas, J.**	10 July 15	28 Feb. 1917	
16 May 03	**Allen, H.**	10 July 15	. . . .	A. R. D., Sargodha.
16 May 03	**Matthews, R. C.,** D.S.O. . . .	10 July 15	2 Jan. 1917	
16 May 03	**Leaning, A.,** D.S.O.	10 July 15	26 Jan. 1917	
6 Feb. 04	**Macdonald, D.**	10 July 15	13 Sep. 1916	
6 Feb. 04	**Holness, H. J.,** D.S.O. (207) . .	10 July 15	26 Sep. 1916	
6 Feb. 04	**Russell, E. C.,** F.R.C.V.S. .	10 July 15	. . . .	A. R. D., Babugarh.
13 Apl. 01	**Dalgleish, A. J.**	13 July 16	. . . .	D. A. D. V. S., 4th (Quetta) Dn.
2 Nov. 01	**Porteous, R.**	2 Nov. 16	2 Jan. 1917	
	Captains.			
3 Feb. 06	**Horner, A. L.** M.C., (*actg. Maj., 29 Aug. 16*)	3 Feb. 11	18 Dec. 1918	
13 Feb. 07	**Stewart, H. C.** (*temp Maj., 25 Feb. 18*)	13 Feb. 12	. . . .	D. A. D. V. S., Poona Dn.
9 Sep. 08	**Hilliard, J. J.** (*Actg. Maj., 1 Feb. 17*)	9 Sep. 13	28 Apl. 1915	
9 Sep. 08	**Gooderidge, R. A.** (*Actg. Maj., 5 Feb. 17*)	9 Sep. 13	4 Nov. 1915	
7 Oct. 08	**Murray, C. A.** (*Actg. Maj., 12 Aug. 17*) (*172a*).	7 Oct. 13	4 Nov. 1915	
9 Sep. 08	**Bowhay, A. B.** (*Actg. Maj., 5 July 17*)	20 Oct. 13	2 Mar. 1916	
3 Sep. 10	**Doyle, E. C.,** D. S. O. (*Tempy. Maj.*)	3 Sep. 15 / 4 Aug. 16	11 Sep. 1917	
3 Sep. 10	**Nicholl, E. Mc. K.** (*Actg Maj., 4 May 17*)	3 Sep. 15	14 Jan. 1918	
3 Sep. 10	**Thwaytes, R. C. G.**	3 Sep. 15 / 8 Oct. 17	11 Sep. 1917	[D. I. Khan.
15 Feb. 11	**Kelly, G. A.** (*temp. Maj., 25 Feb. 18*)	3 Sep. 15	20 Nov. 1918	D. A. D. V. S., Independent Bde., [Dn.
16 Sep. 11	**Hayes, F. B.** (*temp. Maj., 25 Feb. 18.*)	3 Sep. 15	10 Apl. 1918	D. A. D. V. S., 2nd (Rawalpindi)
16 Sep. 11	**Hogg, F.**	3 Sep. 15	18 Dec. 1918	
17 Feb. 12	**Lewis, H. D.** (*temp. Maj. 25 Feb. 18.*)	3 Sep. 15	12 Nov. 1918	D. A. D. V. S., 8th (Lucknow) Dn.
17 Feb. 12	**Beck, D'A. S.** (*temp. Maj.*). .	3 Sep. 15	10 Dec. 1918	
17 Feb. 12	**Stephenson, H.,** M. C. . .	3 Sep. 15	18 Dec. 1918	
17 Feb. 12	**Shea, T. L.**	3 Sep. 15	10 Feb. 1919	
21 Sep. 12	**Davis, T. J.** (*temp. Maj.*) *2 Nov. 18*)	3 Sep. 15	11 Sep. 1923	D. A. D. V. S., 1st (Peshawar) Dn.
7 Sep. 14	**Ward, J. A.** (*temp. Maj.*) . .	7 Sep. 15	18 Nov. 1917	
18 Jan. 15	**Hay, W.**	18 Nov. 17	11 Sep. 1923	Poona.

* Lt.-Col. while employed as A. D. V. S. 16 June 1917.

ARMY VETERINARY CORPS (TEMPORARY).

Captains.

First Commn.	Names.	Present rank.	Date of completion of Indian tour.	Remarks.
13 Oct. 14	**Simpson**, R. W., M.C. (*Actg. Maj., 25 Jan. 17*).	13 Oct. 14	. . . .	
31 Aug. 14	**Isherwood**, R.	31 Aug. 15	. . . .	A. R. D., Sargoda.
4 Sep. 14	**Lang**, W. W.	4 Sep. 15	. . . .	D. A. D. V. S., 16th (Indian) Dn
8 Oct. 14	**Sikes**, F. G. (*temp. Maj., 25 Feb. 18.*)	8 Oct. 15	. . . .	D. A. D. V. S., Army Hd.-qrs.
24 Oct. 14	**Blackburn**, J.	24 Oct. 15	. . . .	Quetta.
6 Nov. 14	**Hunter**, T. S.	6 Nov. 15	. . . .	O. C. No. 18, I. M. V. S., Sibi.
11 Nov. 14	**Barry**, G.	11 Nov. 15	. . . .	A. R. D., Saharanpur.
14 Jan. 15	**Hobbs**, E. V.	14 Jan. 16	. . . .	Mhow.
6 Feb. 15	**Jerrom**, J. H. G.	6 Feb. 16	. . . .	Special duty.
10 Mar. 15	**Rose**, G. A.	10 Mar. 16	. . . .	Lahore.
10 Mar. 15	**Childs**, T.	10 Mar. 16	. . . .	*Lv., ex I.*
11 Mar. 15	**Cunningham**, J.R.	11 Mar. 16	. . . .	Nowshera.
15 July 15	**Murison**, R. E.	15 July 16	. . . .	Muttra.
30 Aug. 15	**McDougall**, W. F.	31 Aug. 16	. . . .	Special duty.
20 Jan. 16	**O'Neill**, A.	17 Jan. 17	. . . .	Meerut.
25 Mar. 16	**Wright**, N.	25 Mar. 17	25 Mar. 16	*Lv., m.c., Ex. I., 6 mths.*
29 Mar. 16	**Flanders**, C. M.	29 Mar. 17	. . . .	Agra. (*Spl. Duty.*)
18 Apl. 16	**Shannon**, J. A.	18 Apl. 17	. . . .	
24 July 16	**Prichard**, L. E.	24 July 17	. . . .	
24 July 16	**Bannatyne**, T.	24 July 17	. . . .	Nowshera.
24 July 16	**Rouse**, A.	24 July 17	. . . .	Risalpur.
25 July 16	**Power**, J. A.	25 July 17	. . . .	A. R. D., Quetta.
31 July 16	**Coutts**, J. D.	31 July 17	. . . .	*Lv., m c., ex I., 6 mths.*
1 Aug. 16	**Howie**, G.	1 Aug. 17	. . . .	A. R. D., Aurangabad.
4 Aug. 16	**Brinsmead**, H.	4 Aug. 17	. . . .	Bangalore.
19 Sep. 16	**McFarlane**, A. McP.	19 Sep. 17	. . . .	Risalpur.
9 Oct. 16	**Burndred**, W. G.	9 Oct. 17	. . . .	*Lv., m.c., ex I., 6 mths.*
13 Nov. 16	**Croken**, I. E.	13 Nov. 17	. . . .	Special Duty.
3 Jan. 17	**Calder**, C. K.	3 Jan. 18	. . . .	Delhi.
15 Jan. 17	**Mackie**, C. (*temp. Maj. 1 Oct. 18*)	15 Jan. 18	. . . .	D. A. D. V. S. (Embkn.), Karachi
27 Jan. 17	**Robertson**, J.	27 Jan. 18	. . . .	Meerut.
21 Feb. 17	**Pethick**, R. S.	21 Feb. 18	. . . .	Mhow.
	O'Gogarty, M. G.	11 Mar. 18	. . . .	
23 Mar 17	**Moncrieff**, J. I. S.	23 Mar. 18	. . . .	Karachi.
11 Apl. 17	**Bentham**, W. G.	11 Apl. 18	. . . .	Special Duty.
1 May 17	**Bosworth**, T. J.	1 May 18	. . . .	
8 May 17	**McDougall**, A. H.	8 May 18	. . . .	Rawalpindi.
15 May 17	**Dunn**, H. S. A.	15 May 18	. . . .	Bareilly.
15 May 17	**Hayes**, J. W.	15 May 18	. . . .	O. C. No. 17, I. M. V. S.
11 June 17	**Tay**, B.	11 June 18	. . . .	A. R. D., Ahmednagar.
23 July. 17	**Donne**, F.	23 July 18	. . . .	Special Duty.
1 Aug. 17	**McElligot**,	1 Aug. 18	. . . .	Special Duty.
4 Aug. 17	**Arnold**, T. F.	6 Aug. 18	. . . .	A. R. D., Mona.
13 Aug. 17	**Pearce**, E. A.	13 Aug. 18	. . . .	Mhow.
13 Aug. 17	**Christopher**, F.	13 Aug. 18	. . . .	Dalband.
13 Aug. 17	**Shallcross**, E. P.	13 Aug. 18	. . . .	
15 Aug. 17	**McAlister**, N. H.	15 Aug. 18	. . . .	Special duty.
18 Aug. 17	**Barry**, J. J. M.	18 Aug. 18	. . . .	
26 Aug. 17	**Mackenzie**, F. McD.	26 Aug. 18	. . . .	

Lieutenants.

First Commn.	Names.	Present rank.	Date of completion of Indian tour.	Remarks.
17 Sept. 17	**Dier**, H. V.	17 Sept. 17	. . . .	
25 Oct. 17	**Melck**, G. H.	25 Oct. 17	. . . .	Cambellpore.
12 Nov. 17	**Rice**, J. P.	12 Nov. 17	. . . .	Rawalpindi.
12 Nov. 17	**Petrie**, W. S.	12 Nov. 17	. . . .	Special duty.
14 Dec. 17	**Lawrence**, G. C.	14 Dec. 17	. . . .	
4 Jan. 18	**Blampied**, T. LeQ.	4 Jan. 18	. . . .	Lucknow.
8 Jan. 18	**Conder**, C. S.	8 Jan. 18	. . . .	
10 Jan. 18	**Litt**, J.	10 Jan. 18	. . . .	
10 Jan. 18	**Shaw**, T. A.	10 Jan. 18	. . . .	
10 Jan. 18	**Metcalfe**, A. N.	10 Jan. 18	. . . .	
10 Jan. 18	**Beattie**, R.	10 Jan. 18	. . . .	
15 Jan. 18	**Horris**, J. P. A.	15 Jan. 18	. . . .	
26 Jan. 18	**Weidenbach**, L. P.	26 Jan. 18	. . . .	Saugor.
1 Feb. 18	**Barry**, W. E.	1 Feb. 18	. . . .	
8 Feb. 18	**Wishart**, D. G.	8 Feb. 18	. . . .	Ambala.
10 Feb. 18	**Banham**, G. F.	10 Feb. 18	. . . .	Special duty.
18 May 18	**McGuirk**, O.	18 May 18	. . . .	Rawalpindi.
1 June 18	**Sanderson**, F. H.	1 June 18	. . . .	Lahore
3 June 18	**Urquhart**, W.	3 June 18	. . . .	Agra.
15 July 18	**Smith**, E. K.	15 July 18	. . . .	
23 July 18	**Simpson**, K. S.	23 July 18	. . . .	Quetta.
24 July 18	**Macrae**, D. E.	24 July 18	. . . .	Belgaum.
24 July 18	**Williams**, J. B.	24 July 18	. . . .	Karachi.
2 Aug. 18	**Richardson**, S. G. P. L.	2 Aug. 18	. . . .	Rawalpindi.
12 Aug. 18	**John**, W. D.	12 Aug. 18	. . . .	Meerut.

ARMY VETERINARY CORPS (SPECIAL RESERVE).

Captains.

First Commn.	Names.	Present rank.	Date of completion of Indian tour.	Remarks.
5 Aug. 14	**Daubney**, R.	5 Aug. 15		Sedentary Duty, Poona.
Sep. 14	**White**, J. H. M.	4 Aug. 17		Lahore.
..	**Wallace**, E.	4 Aug. 17		Sialkot.
3 Jan. 17	**Woodland**, P. F.	3 Jan. 18		Kohat.
3 Jan. 17	**Bowes**, E. C.	3 Jan. 18		Ferozepore.
1 Feb. 17	**Farelly**, M.	1 Feb. 18		Derajat Bde.

Lieutenants.

First Commn.	Names.	Present rank.	Date of completion of Indian tour.	Remarks.
18 Sep. 17	**Beaumont**, J. W.	18 Sept. 17		
16 Jan. 18	**Driver**, H. C.	16 Jan. 18		
26 Jan. 18	**Haywood**, J. D.	16 Jan. 18		Peshawar.

ARMY VETERINARY CORPS (TERRITORIAL FORCE).

Captains.

First Commn.	Names.	Present rank.	Date of completion of Indian tour.	Remarks.
July 08	**Lloyd**, T. W.	3 July 13		Special Duty.
Feb. 00	Batt, A. E. G.	30 Sept. 14		Special Duty.
	Dawson, L. A. F.	26 Aug. 15		Secunderabad.

INDIAN ARMY RESERVE OF OFFICERS (VETERINARY BRANCH).

Captains.

First Commn.	Names.	Present rank.	Date of completion of Indian tour.	Remarks.
1 Nov. 14	**Weathrall**, C. H. G.	1 Nov. 15		Jubbulpore.
1 Dec. 14	**Stevenson**, W. (*temp. Maj., 25 Feb. 18*)	4 Dec. 15		D. A. D. V. S., Burma Dn., Rangoon.
1 July 15	**Kerr**, P. J. (*temp. Maj., 13 May 18*)	1 July 16		D. A. D. V. S., Ambala Bde.
10 Feb. 16	**MacGregor**, A. D.	10 Feb. 17		Dera Ismail Khan.
27 Feb. 17	**Assheton Smith**, T. (*temp. Maj., 13 May 18*)	27 Feb. 18		Bangalore.
28 Feb. 17	**Pool**, W. A.	28 Feb. 18		Sohawa.
19 Apl. 17	**Meadows**, D. (*Tempy. Maj.*)	19 Apl. 18 1 Jan. 18		D. A. D. V. S.

Lieutenants.

First Commn.	Names.	Present rank.	Date of completion of Indian tour.	Remarks.
9 July 15	*Farbrother, E. S.*	*9 July 15*		
13 Dec. 17	**Hickey**, S. G. M.	13 Dec. 17		Special Duty.

OFFICERS OF THE ARMY VETERINARY DEPARTMENT TRANSFERRED TO THE INDIAN CIVIL VETERINARY DEPARTMENT.

First Commn.	Names.	Present rank.	Date of arrival in India.	Present apptt.	Remarks.
	Colonels.				
27 Aug. 84	**Pease**, H. T., C.I.E. (162, 206). (Supy.)	3 Sep. 10	31 Mar. 85	1 Apl. 12	Principal, Punjab Vety. College, Lahore.
27 Aug. 84	**Evans**, G. H., C.I.E., A.D.C., (105, 206).	10 Feb. 13	31 Mar. 85	25 Oct. 11	Supdt., C. V. D., Burma, Rangoon.
21 June 93	**Farmer**, J., C.I.E., F.R.C.V.S. (162, 165, 170, 206).	20 July 17	11 Oct. 93	11 June 12	Chief Supdt., C. V. D., Punjab, Lahore.
	Lieutenant-Colonels.				
21 June 93	**Smith**, A., F.R.C.V.S. (206).	29 July 14	11 Oct. 93	4 June 12	Principal, Bengal Vety. College, Calcutta and Supdt., Vety. Deptt., Bengal.
25 July 94	**Walker**, G. K., C.I.E., O.B.E., F.R.C.V.S. (170, 206).	20 July 17	2 Jan. 96	12 June 12	Supdt. C. V. D.. Bombay. *Lv., C. L., 6 mths., 3 Apl. 18.*

DEPARTMENTAL OFFICERS WITH HONORARY RANK.

First Commn.	Names.	Date of birth.	Date of grade or rank.	F. rule.	Remarks.
	ORDNANCE DEPARTMENT, NORTHERN ARMY.				
	Commissaries with the honorary rank of Major—8.				
20 May 98	Thorne, Theophilus (b) (207).	18 May 60	25 May 06 m. 16 Feb. 07	75	Karachi.
18 Mar. 05	Todd, Thomas Robert (207).	30 Nov. 66	17 June 08 m. 21 Sep. 17	N	Ferozepore.
25 May 06	Hobbs, William Christian (207).	16 Feb. 65	13 Sep. 11 m. 21 Sep. 17	N	Allahabad.
23 Jan. 07	Bowder, William Edmund (207).	5 July 64	11 Nov. 12 m. 21 Sept. 17	N	Rawalpindi.
29 Dec. 08	Sharpe, Robert (207) (b) .	28 Sep. 63	5 Aug. 13 m. 21 Sep. 17	N	
11 Nov. 12	Allwright, Samuel Robert (170, 207 (b).	1 Sep. 61	3 Aug. 14 m. 21 Sep. 17	N	Kirkee.
5 Aug. 13	Maher, Francis (166, 206).	19 June 66	19 Sep. 14 m. 21 Sep. 17	N	
17 Jan. 14	Miller, Frank Edwin (170).	24 June 66	15 Jan. 15 m. 21 Sep. 17	N	Ferozepore.
9 Jan. 14	Harding, George (170, 207) (b).	8 Dec. 62	18 May 15 m. 21 Sep. 17	N	Bombay.
14 Jan. 07	O'Donald, Edward (170) (c) (IX).	12 Nov. 64	1 Apl. 16 m. 21 Sep. 17	N	Special duty Viceroy's Camp, Delhi.
22 Jan. 11	Villiers, Arthur . . .	31 May 65	1 Sep. 16 m. 2 Sep. 16	N	Army Hdqrs., India (Ord. Bch.), Simla.
6 Feb. 14	Hardaker, Christopher Bennett (95, 207).	9 Mar. 66	1 Sept. 16 m. 21 Sep. 17	N	Allahabad.
8 May 14	Utting, Frederick James (170, 207).	21 Aug. 67	8 Dec. 17	N	Rawalpindi.
3 Aug. 14	Robertson, Alexander	6 Sep. 64	28 Sep. 18	N	
	Deputy Commissary with the honorary rank of Captain—1.				
18 May 15	Sydenham, John Philip (c) (III). [(c) (II).	30 Oct. 65	3 June 15	N	Kirkee—Asst. Inspr. of Ammn. No. 2 charge. [I. S.
18 May 15	Ramsay, William John	1 Dec. 68	1 Jan. 16	N	Dum Dum—Ch. Examnr. (Mily.
18 May 15	Digby, Charles (c) (V) . [(IV).	27 June 66	1 Jan. 16	N	Calcutta.—Ch. Clerk, Office of Dir. of Ord. Factories. [Sec.
18 May 15	Allen, Joseph Henry (c)	10 Oct. 69	1 July 16	N	Aruvankadu.—Hd. Ovsr., Cordite
18 May 15	Chisholme, Andrew Charles (207) (b) (I).	31 July 61	1 Jan. 17	N	Kirkee. Facty.—Hd. Ovsr. Laby.
19 Sep. 14	Cahill, Michael (207) (b) .	22 Aug. 61	28 Sep. 18	N	Rifle Factory, Ishapore.
15 Jan. 15	Cuerdon, George Henry Edward.	24 Jan. 70	28 Sep. 18	N	Kirkee Fy.—Ovsr.—Examn. Secn.
15 Jan. 15	Petri, Arthur Theodore	18 Oct. 64	28 Sep. 18	N	Dum-Dum.
	Assistant Commissaries with the honorary rank of Lieutenant—5				
1 Sep. 06	Handley, Frederick (b) (VII).	23 Mar. 60	1 Sep. 06 12 July 17	N	Cawnpore.
23 May 12	Williams, Friend Isaac (VI).	14 June 69	23 May 12 1 Sep. 16	N N	Allahabad.
	(I)				
	(II)				
	(III)				
	(IV)				
	(V)				
18 May 15	Dickinson, Joseph Jonathan (95).	2 Mar. 66	18 May 15	N	Rawalpindi.
3 June 15	Smith, Robert (c) (IX) .	4 Dec. 71	3 June 15	N	Ind. Munitions Board.
3 June 15	Stanley, James Charles 51, 95 (c) (XIII).	3 Dec. 64	3 June 15	N	
1 July 16	Gardner, George (c) (95, 170) (XII).	19 Dec. 67	1 July 16	N	Aruvankadu, Cordite Factory.
1 July 16	Todd, Edward James (c) (94) (XIV).	14 Jan. 76	1 July 16	N	Jubbulpore, G.C. Factory.
	(VI)				
1 Jan. 17	Finch, Frederick George (c) (X).	29 Sep. 65	1 Jan. 17	N	
1 July 17	Hicks, William (95, 170) (c) (VIII).	9 Sep. 67	1 July 17 8 Dec. 17	N	Karachi.
	(VII)				
1 July 17	Rix, Walter Henry Burton (95, 170, 207).	3 July 69	12 July 17	N	Fort William.
8 Dec. 17	Wallingford, Williams (170)	16 July 66	8 Dec. 17		Dum Dum. Ovsr. Laby.
	(VIII)				
1 July 18	Harrison, Edward John (c)	11 Nov. 70	1 July 18		Kirkee.
1 July 18	Hughes, Robert William, (c). [170.).	11 Aug. 77	1 July 18		
28 Sep. 18	Balmforth, Jonathan. (95,	20 Dec. 68	28 Sep. 18	N	

(b) Supy. On attaining 55 years of age.
(c) Supy. Specially promoted under the provisions of para. 470, A. R. I., Vol. II.

ORDNANCE DEPARTMENT, SOUTHERN ARMY.

Commissaries with the honorary rank of Major—7.

First Commn.	Names.	Date of birth.	Date of grade or rank.	F. rule.	Remarks.
12 Aug. 02	*Smith, Benjamin Edwin*	31 Jan. 65	28 June 05 m. 14 Jan. 11	75	*Army Hdqrs., India, Q. M. G's Br.*
20 Sep. 02	*Collinson, John Carson (b).*	11 Jan. 61	18 Aug. 07 m. 20 Sept. 17	N	*Fort William.*
16 Feb. 04	**Hartley**, George Richard (106, 192, 207).	13 Mar. 64	29 Sep. 08 m. 21 Sep. 17	N	Madras.
1 Nov. 04	*Rush, John Shipman (207) (b).*	5 Apl. 62	21 May 10 m. 4 Apl. 17	N	*Bombay. Inspr. of Explosives.*
29 Nov. 00	*Walmsley, Robert, (94, 95, 169, 207). (b)*	6 Dec. 62	21 May 10 m. 29 Nov. 15	N	*Quetta.*
28 June 05	*Craven, Albert (172a, 207) (b) (Tempy. Capt. in Army).*	13 Sep. 62	2 May 13 m. 1 Jan. 17	N	
8 Sep. 08	**Duggan**, John (207)	10 Feb. 64	15 Apl. 15 m. 21 Sep. 17	N	Madras.
29 Sep. 08	*Blackwood, Albert (207) (b)*	26 Mar. 63	11 Jan. 16 m. 21 Sep. 17	N	*Rawalpindi.*
6 Oct. 08	*Howard, Charles (207)*	4 June 65	10 May 16 m. 21 Sep. 17	N	*Cawnpore—Exmr., Genl. Stores, I. S.*
2 Dec. 11	*Coleburt, James Garrett (95) (b).*	30 May 62	10 May 16 m. 21 Sep. 17	N	*Ferozepore.*
1 Jan. 11	**Hayhurst**, John Henry Joseph.	8 May 64	12 June 16 m. 21 Sep. 17	N	Agra.
15 Jan. 12	*Power, William (207) (b)*	13 Apl. 63	30 May 17 m. 21 Sep. 17	N	*Kirkee Fact ry.*
4 Mar. 12	**Ross**, Rees (95, 169)	23 Feb. 64	13 Sep. 17 m. 21 Sep. 17	N	Rangoon.
4 July 12	**Jones**, Edwin	13 Jan. 66	6 Dec. 17	N	
4 Sep. 13	*Green, William Edward*	10 Sep. 64	26 Mar. 18	N	*Dum-Dum.—Hd. Ovsr., Laby.*
4 Sep. 13	**Staveley**, William (95)	8 Mar. 67	26 Mar. 18	N	Jubbulpur Depôt.
12 July 14	**Kendall**, Walter John (105, 193, 207).	18 Jan. 72	13 Apl. 18	N	

Deputy Commissaries with the honorary rank of Captain—4.

First Commn.	Names.	Date of birth.	Date of grade or rank.	F. rule.	Remarks.
12 June 16	*Wilson, Andrew (a) (XVI) (b).*	9 Mar. 62	1 July 16	N	*Bombay.*
3 June 15	**Walters**, Richard (169) (XV).	30 Mar. 67	1 Jan. 17 13 Apl. 18		
15 Apl. 15	**Hewes**, Ernest Edgar	17 Sep. 68	13 Sep. 17	N	Cossipore—Ishapore Br.
15 Aug. 15	**Kaye**, Frederick William	14 Dec. 63	6 Dec. 17	N	Cossipore.
11 Jan. 16	**Holden**, Henry (95, 207) (XV).	25 Apl. 65	26 Mar. 18	N	

Assistant Commissaries with the honorary rank of Lieutenant—7.

First Commn.	Names.	Date of birth.	Date of grade or rank.	F. rule.	Remarks.
3 June 15	*Chambers William John (XVII).*	5 Oct. 72	3 June 15 13 Apl. 18	N	*Quetta.*
3 June 15	*Johnston, Charles (XIX) (XVI). (a)*	18 June 74	3 June 15	N	*Cossipore—Asst. Controller of War Accounts.*
1 Jan. 17	*Goodall, Alfred (a) (XVIII)*	24 Nov. 67	1 Jan. 17	N	*Cossipore—Chief Exr. No. 2 charge.*
1 Jan. 17	*Wood, Joseph Henry (95, 170) (a) (XX).*	18 Nov. 69	1 Jan. 17	N	*Fort William.*
9 Mar. 17	*McManus, John*	15 Sept. 70	9 Mar. 17	N	*Army Hdqrs., Ind. (Q. M. G.'s Br.*
9 Mar. 17	**Bray**, Thomas	20 Mar. 67	9 Mar. 17	N	
30 May 17	*Ward, Thomas Edward (95) (b).*	14 Oct. 62	30 May 17	N	*Karachi.*
1 July 17	*Stephens, Harry Frederick (a) (XXI).*	10 May 70	1 July 17	N	*Dum-Dum—Ov sr., Laby. Fuze.*
13 Sep. 17	**Burke**, Robert Henry (94, 161, 169, 204).	18 Oct. 65	13 Sep. 17	N	Q etta.
14 Oct. 17	**Bell**, Henry Gerald	19 Jan. 66	14 Oct. 17	N	Ferozepore.
6 Dec. 17	**Southgate** Ernest James (66a)	28 Apl. 68	6 Dec. 17	N	Madras.
26 Mar. 18	**Murphy**, William Robert. (XVII)	21 Jan. 72	26 Mar. 18	N	
16 Apl. 18	*Hanlon, George James Patrick.*	13 Nov. 65	16 Apl. 18		

(*a*) Supy. Specially promoted under the provisions of para. 470, A. R., I., Vol. II.
(*b*) Supy. on attaining the age of 55 years.

SUPPLY AND TRANSPORT CORPS, BENGAL.

Commissaries with the honorary rank of Major—4.

First Commn.	Names.	Date of birth.	Date of grade or rank.	F. rule.	Remarks.
17 Dec. 04	*McWalters, William (64, 207) (b).*	10 June 61	8 July 12 m. 21 Sept. 17	N	"K." S. D. Coy.
29 Dec. 05	*Doyle, Michael Joseph (b)*	13 Oct. 60	8 Aug. 13 m. 21 Sep. 17	N	(Supy.) "J." S. D. Coy.
18 Jan. 07	**Grindley,** William Henry (207).	12 Jan. 65	23 Nov. 13 m. 21 Sep. 17	N	"B" S. D. Coy.
10 Oct. 08	**Ryan,** Edward John (71, 165, 169, 206).	17 Oct. 64	6 Dec. 14 m. 21 Sep. 17	N	"B" S. D. Coy.
4 Apl. 09	*Hewitt, James (64, 170) (b).*	17 May 61	13 Oct. 15 m. 21 Sept. 17	N	(Supy.) Amritsar.
27 Feb. 13	*Duthie, Thomas (170, 207) (b).*	1 Aug. 61	1 Jan. 16 m. 27 Sep. 17	N	(Supy.) "L" S. D. Coy.
21 Dec. 10	*Howell, Richard (65, 206) (b).*	3 July 60	10 June 16 m. 21 Sep. 17	N	(Supy.) "A" S. D. Coy.
1 Sep. 14	**Whittaker,** Thomas (207).	13 Aug. 67	3 Sep. 16 28 May 18 m. 21 Sep. 17	N	No. 2 D. S. Coy.
30 Oct. 12	*Lyttle, William John (170, 207) (a).*	3 Sep. 70	1 Jan. 17 m. 21 Sep. 17	N	No. 8 D. S. Coy.
6 Feb. 13	*Ricketts, Frederick Rodney Ross (169, 207) (b).*	23 June 62	27 Sep. 17	N	"A" S. D. Coy.
5 Apl. 13	*White, George Ernest (170, 207) (b).*	10 Mar. 63	27 Sep. 17	N	No. 7 D. S. Coy.
8 Aug. 13	*Williams, William (65, 207) (b).*	3 Apl. 63	27 Sep. 17	N	No. 3 D. S. Coy.
23 Nov. 13	*Gilbertson, George (65, 71, 94, 136, 161, 165, 169, 185, 204) (b).*	14 Oct. 60	3 Apl. 18	N	(Supy.) No. 8 D. S. Coy.
27 Mar. 12	**Carrigan,** Peter . . .	1 Oct. 65	3 Apl. 18	N	No. 3 D. S. Coy.

Deputy Commissaries with the honorary rank of Captain—5.

First Commn.	Names.	Date of birth.	Date of grade or rank.	F. rule.	Remarks.
3 June 15	**Tibbs,** James (71, 207)	2 Jan. 65	3 June 15 23 June 17		No. 2 D. S. Coy.
3 June 15	**Holmes,** Richard John	5 Dec. 71	3 June 15 27 Sep. 17	N	Rawalpindi, M. T. School.
26 June 08	*Lloyd, Thomas William, (165, 169, 207)*	8 July 65	1 Jan. 16 28 May 18	N	P. A. to Q. M. G. in Ind.
3 Sep. 16	*Kemp, William John (a).*	3 Sep. 68	10 Sep. 16	N	Ambala Bde.
17 May 16	*Bristow, Robert Lancelot (170, 207) (a).*	1 Aug. 67	17 May 17	N	Karachi Bde.
3 June 15	*Leitch, George (172a), (a).*	21 Oct. 72	1 July 17	N	No. 2 D. S. Coy.
13 Oct. 15	*Lang, Alfred (207) (b).*	5 Oct. 61	27 Sep. 17	N	No. 3 D. S. Coy.
26 June 08	**Rumbold,** William (71, 170, 207).	12 Nov. 68	3 Apl. 18	N	No. 3 D. S. Coy.
30 Apl. 16	**Salew,** George Augustus (64).	21 Dec. 70	28 May 18		No. 3 D. S. Coy.

(a) Supy. Specially promoted under the provisions of para. 470, A.R., I., Vol. II.
(b) Supy. On attaining the age of 55 years.

SUPPLY AND TRANSPORT CORPS, BENGAL—concld.

Assistant Commissaries with the honorary rank of Lieutenant—9.

First Commn.	Names.	Date of birth.	Date of grade or rank.	R. rule.	Remarks.
23 May 12	Wright, Ralph (a) (165, 207)	15 Oct. 70	23 May 12	N	No. 7 D. S. Coy.
23 May 12	Hayman, William Charles (a).	21 Sep. 72	23 May 12	N	Karachi Bde.
3 June 15	Edgerley, William (169)	10 Oct. 71	3 June 15 10 Sep. 16		No. 7 D. S. Coy.
3 June 15	Bosworth, James Albert (a)	21 Jan. 73	3 June 15	N	No. 8 D. S. Coy.
3 June 15	Jolley, Henry John (71, 170, 207) (a).	10 Dec. 74	3 June 15		Bannu Bde.
1 Jan. 16	McGinn, James Joseph, (207) (a).	3 Jan. 72	1 Jan. 16 27 Sep. 17		"A" S. D. Coy.
1 Jan. 16	Payne, Charles (65)	14 Apl. 69	1 Jan. 16 3 Apl. 18		Ambala Bde.
1 Jan. 16	Kee, William Johnston (a).	9 Jan. 70	1 Jan. 16 28 May 18		"F" S. D. Coy. (M. T. School, Meerut.)
1 Jan. 16	Ballin, Julius (a)	4 Feb. 76	1 Jan. 16		No. 8 D. S. Coy.
1 Jan. 16	Sly, Frederick Charles (a)	22 Sep. 72	1 Jan. 16		No. 3 D. S. Coy.
1 Jan. 16	Gordon, William Duncan (170, 207) (a).	10 Nov. 71	1 Jan. 16		"J" S. D. Coy.
1 Jan. 16	Watson, Percy Hamilton (65, 170, 207) (a).	19 July 76	1 Jan. 16		No. 3 D. S. Coy.
30 Apl. 16	Powell, Frank James	6 Dec. 65	30 Apl. 16		No. 3 D. S. Coy.
1 July 16	King, William Gilbert (a)	6 Dec. 75	1 July 16		A. Hd. Qrs.
10 Sep. 16	Nash, Albert Edward (64, 71) (a).	3 May 74	10 Sep. 16	N	No. 2 D. S. Coy.
29 Nov. 16	Spinks, Charles William Frederick (207).	12 Mar. 71	29 Nov. 16	N	"A" S. D. Coy.
1 Jan. 17	Cheeseman, Edward James (170, 207) (a).	27 May 75	1 Jan. 17	N	No. 8 D. S. Coy.
1 Jan. 17	Grimley, Henry Barton (a)	11 Jan. 74	1 Jan. 17	N	No. 2 D. S. Coy.
7 Feb. 17	Clements, Isaac (a)	11 Nov. 73	7 Feb. 17	N	No. 2 D. S. Coy.
23 June 17	Rose, Alfred Mortimer (95, 165, 169, 206) (a).	12 Oct. 67	23 June 17	N	"F" S. D. Coy.
1 July 17	Emmerson, Mathew (a)	26 Aug. 72	1 July 17	N	"H" S. D. Coy.

SUPPLY AND TRANSPORT CORPS, 9th (Secunderabad) and Burma Divisions.

Commissary with the honorary rank of Major—1.

First Commn.	Names.	Date of birth.	Date of grade or rank.	R. rule.	Remarks.
8 May 07	Goodhall, Edward James (207) (b).	19 July 63	11 June 14 m. 3 June 15	N	(Supy.) No. 9 D. S. Coy.
1 Nov. 07	King, James Alfred (207)	21 May 65	19 July 18	N	"A" S. D. Coy.

Deputy Commissaries with the honorary rank of Captain—2.

First Commn.	Names.	Date of birth.	Date of grade or rank.	R. rule.	Remarks.
19 Aug. 12	Woledge, Henry Walter (b)	31 Oct. 62	24 Mar. 18	N	Bombay Bde.
20 Sep. 13	Bloomfield, James Aldous (b).	24 Mar. 61	24 Mar. 18	N	(Supy.) "A" S. D. Coy.
11 June 14	Lewis, Charles (b)	28 May 62	24 Mar. 18	N	No. 9 D. S. Coy.
24 Mar 16	Henry, Robert (207)	28 Feb. 67	24 Mar. 18		"F" S. D. Coy.
31 Oct. 17	Hoare, William Isaac (a)	14 Dec. 65	24 Mar. 18	N	No. 9 D. S. Coy.
1 Jan. 16	Penn, William Henry	10 May 65	19 July 18	N	No. 9 D. S. Coy.

Assistant Commissaries with the honorary rank of Lieutenant—3.

First Commn.	Names.	Date of birth.	Date of grade or rank.	R. rule.	Remarks.
3 June 15	Morris, John Arthur (95, 169, 207) (b)	2 July 63	3 June 15 24 Mar. 18		"A" S. D. Coy.
1 Jan. 16	Courtney, John (65)	25 Oct. 68	1 Jan. 16 19 July 18	N	No. 9 D. S. Coy.
1 Jan. 16	Bartram, Arthur (207) (a)	10 Sep. 70	1 Jan. 16		"E" S. D. Coy.
19 July 18	Cooper, James Edwin	10 Jan. 68	2 July 18		No. 9 D. S. Coy.

(a) Supy. Specially promoted under the provisions of para. 470, A. R. I., Vol. II.
(b) Supy. On attaining the age of 55 years.

First Commn.	Names.	Date of birth.	Date of grade or rank.	F.rule.	Remarks.

SUPPLY AND TRANSPORT CORPS, BOMBAY.

Commissaries with the honorary rank of Major or Lieut.-Colonel—2.

27 July 96	Orr, Richard, (94. 169, 207) (Hony. Lt.-Col., 1 Jan. 17). b)	5 July 61	11 Mar. 05	N	(Supy.) No. 6 D. S. Coy.
27 July 96	Smith, James Muir (94, 207). (b)	18 Dec. 59	22 Apl. 08 m. 27 July 11	75	(Supy.) No. 6 D. S. Coy.
7 Sep. 04	**Skinner,** Frederick John (95, 170, 207).	1 Apl. 68	18 Dec. 14 m. 1 Jan. 17	N	Karachi.
6 Apl. 12	Graveson, Henry (207) (b)	2 Apl. 62	2 Oct. 16 m. 21 Sep. 17	N	(Supy.) "G" S. D. Coy.
18 Dec. 14	Ruddock, William Bernard (b)	13 Nov 62	2 Apl. 17 m. 21 Sep. 17	N	(Supy.) "G" S. D. Coy.
21 Sep. 13	Bray, Edwin Morris (95, 124a, 206) (b)	6 Feb. 62	13 Nov. 17	N	(Supy.) "G" S. D. Coy.
5 July 16	Sayers, Charles (65) (b) .	27 Feb. 63	13 Nov. 17		(Supy) "A" S. D. Coy.
1 Jan. 16	**Morris,** James (95, 169, 207).	22 July 70	27 Feb. 18		"G" S. D. Coy.

Deputy Commissaries with the honorary rank of Captain—1

6 Feb. 17	Webb, George (b) (65, 207)	10 Sep. 63	27 Feb. 18		(Supy.) "A" S. D. Coy.
2 Apl. 17	**Hall,** Edmund Septemus (65).	4 Sep. 68	10 Sep. 18		No. 5 D. S. Coy.

Assistant Commissaries with the honorary rank of Lieutenant– 3.

3 June 15	**Douglas,** Alfred Ernest (65, 170, 207).	4 July 71	3 June 15 27 Feb. 18	N	No. 6 D. S. Coy.
1 July 16	**Bingham,** Ernest George (65), M.C.	22 Apl. 73	1 July 16 13 Nov. 17		No. 5 D. S. Coy.
6 Feb. 17	Page, Albert Frederick Arthur (64) (a).	1 June 74	6 Feb. 17		"A" S. D. Coy.
10 Sep. 18	Nagle, Edmund Henry Baillie.	22 June 75	10 Sep. 18		"C." S. D. Coy.
10 Sep. 18	Fowles, Samuel (65, 66, 207) (a).	9 Feb. 72	10 Sep. 18		No. 6 D. S. Coy.

SUPPLY AND TRANSPORT CORPS—AMALGAMATED LIST.

Assistant Commissary with the honorary rank of Lieutenant.

1 Jan. 16	Darby, William Alfred (a)	20 Jan. 74	1 Jan. 16	N	"K" S. D. Coy.
22 Mar. 18	**Humphreys,** Alfred .	1 June 75	22 Mar. 18		Army Hdqrs.

DEPARTMENTAL OFFICERS WITH HONORARY RANK TEMPORARILY RE-EMPLOYED FROM THE RETIRED LIST.

Supply and Transport Corps.

Cubbon, Deputy Commissary and Honorary Captain, R. "G" S. D. Coy.
Ellis, Commissary and Honorary Major, T. J. "L" S. D. Coy.
Emmett, Deputy Commissary and Honorary Captain, W. H. "H" S. D Coy.
Green, Commissary and Honorary Major. W. W. "K" S. D. Coy.
Hutchins, Commissary and Honorary Captain, E. E. "L" S. D. Coy.
Phillips, Deputy Commissary and Honorary Captain, P. "A" S. D. Coy.
Wilsey, Commissary and Honorary Major, H. W. "F" S. D. Coy.

(a) Supy. Specially promoted under the provisions of para. 470, A. R. I., Vol. II.
(b) Supy. On attaining the age of 55 years.

First Commn.	Names.	Date of birth.	Date of grade or rank.	F. rule.	Remarks.
MILITARY WORKS SERVICES AND PUBLIC WORKS DEPARTMENT, INDIA.					
Commissaries with the honorary rank of Major—2.					
8 May 10	**Farrington,** George (104, 180, 196).	13 June 64	14 June 13 / *m.* 21 Sep. 17	N	Asst. Engr., Meiktila Dn., Meiktila, Chindwin Cir., Burma.
1 Nov. 11	**Collins,** Cecil . . .	21 Aug. 64	22 Dec. 17	N	Asst. Engr., 2nd gr., M. W., Deolali.
Deputy Commissaries with the honorary rank of Captain—3.					
7 Nov. 11	**Curley,** Thomas . . .	10 Aug. 64	3 Jan. 18	N	Asst. Engr., Chindwin Dn.' Monywa, Chindwin Circle, Burma.
Assistant Commissaries with the honorary rank of Lieutenant—6.					
2 May 12	**Sykes,** John	21 May 65	2 May 12	N	Asst. Engr., U. P. *Ser.* (*temply. with M. W. S., Allahabad.*)
12 May 12	**Gell,** William Thomas (169).	20 Dec. 66	12 May 12	N	Asst. Engr., 2nd gr., M. W., Nowshera.
14 June 13	*Hart, Frederick William* .	*13 Sep. 67*	*14 June 13*	N	*Instr. in Applied Science in Photography, Thomason College, Rurki.*
5 July 13	**Sheridan,** Richard John	15 Nov. 67	5 July 13	N	Asst. Engr., 2nd gr., M. W., [Bombay.
25 May 15	*Thomas, Frederick,* (*a*) .	*9 July 60*	*25 May 15*	N	*Supy. Sub-Engr., 2nd gr., M. W., Lucknow.*
26 Aug. 15	*Cannell, John McArthur* (*a*).	*27 Aug. 60*	*26 Aug. 15*	N	*Supy. Asst. Engr., 2nd gr., M. W., Lucknow.*
27 Aug. 15	**Tayler,** Henry John Dore	12 Oct. 69	27 Aug. 15	N	Burma, P. W. D. *Ser. temply. with M. W. S., Nowshera.*
1 Jan. 16	*McDonough, Albert Dorsey* (*169*) (*b*)	*8 Apl. 71*	*1 Jany. 16*		*Asst. Engr., M. W., Cawnpore.*
1 Jan. 16	*O'Connor, Charles Edward Butler* (*b*)	*20 Mar. 73*	*1 Jany. 16*	N	*Asst. Engr., 3rd gr., M. W., Agra.*
1 July 17	*Hall, Thomas Charlie* (*b*)	*10 Aug. 72*	*1 July 17*		*Asst. Engr., 2nd gr., M. W., Rawalpindi.*
15 Oct. 17	*Evans, James John* (*b*) .	*3 June 66*	*15 Oct. 17*		*Asst. Engr., M. W., Dalhousie.*
22 Dec. 17	**Hamilton,** William . .	18 June 65	20 Dec. 17		Asst. Engr., 2nd gr., M. W., Jhelum.
3 Jan. 18	*Tivey, William Henry* (*a*)	*6 Oct. 60*	*3 Jan. 18*		*Sub-Engr., 1st gr., U. P.*
3 Jan. 18	**Stuart,** Henry Hewitt	6 Oct. 65	3 Jan. 18		Asst. Engrs., 2nd gr., M. W., Bangalore.
26 May 18	*Cuerdon, Thomas* (*b*) . .	*28 Feb. 76*	*26 May 18*		*Asst. Engr., Bengal P. W. D. (temply. with M. W. S., Fort William.)*

DEPARTMENTAL OFFICER WITH HONORARY RANK TEMPORARILY RE-EMPLOYED FROM THE RETIRED LIST.

Military Works Services.

Beck, Comsy. & Hony. Maj. T. (M. W. S.)

PUBLIC WORKS DEPARTMENT, MADRAS.

First Commn.	Names.	Date of birth.	Date of grade or rank.	F. rule.	Remarks.
Commissaries with the honorary rank of Major—1.					
14 May 07	*Gasson, Henry Thomas* .	*30 Oct. 61*	*2 Oct. 11* / *m. 21 Sep. 17*	N	*2nd S. and M.* (*Supy.*)
2 Oct. 11	**Briant,** Charles Henry .	23 Mar. 67	7 Aug. 18	N	Sub-Engr., 2nd gr. and Offg. Ex. Asst. Engr., Koraput Dn., 1st Circle.
Deputy Commissary with the honorary rank of Captain—1.					
5 Jan. 13	**Phelan,** James O'Connor (111, 197, 207). . . .	12 Feb. 69	7 Aug. 18	N	Sub-Engr., 1st gr., and Asst. Engr. Supy. (*temply. with M. W. S., Ahmednagar.*)
Assistant Commissaries with the honorary rank of Lieutenant—2.					
22 Nov. 15	**Cuthbert,** Robert . .	2 June 65	22 Nov. 15 / 1 Jan. 17	N	*Sub-Engr.,* Supy. *1st gr.,* (*Prov. P.*) *and Asst. Engr.* (*Lent to Marine Dept.*)
7 Aug. 18	**Power,** Micheal . .	14 Oct. 67	7 Aug. 18		Sub-Engr., 2nd gr.. (*Prov. P.*) Hd-qrs. Sub-divn., Kistna Central Dn. II Cir.

(*a*) Supy. on attaining the age of 55 years.
(*b*) Supy. specially promoted under the provisions of para. 470, A. R., I., Vol. II.

First Commn.	Names.	Date of birth.	Date of grade or rank.	F. rule	Remarks.

PUBLIC WORKS DEPARTMENT, BOMBAY.

Commissary with the honorary rank of Major—1.

28 Jan. 04	Hudson, John Robert (58, 67a, 207).	22 Nov. 65	19 Sep. 05 m. 21 Sep. 17	N	(Supy.) 3rd Sappers and Miners.

BARRACK DEPARTMENT, PUNJAB-BENGAL.

Commissary with the honorary rank of Major—1.

17 May 08	Rennick, Denis (a) . .	24 Nov. 59	24 Feb. 11 m. 21 Sep. 17	N	(Supy.) B. M., M. W., Ambala.
17 May 08	Harvey, William (a) . .	20 Sep. 60	24 Nov. 14 m. 21 Sep. 17	N	(Supy.) B. M., M.W., Poona.
3 July 12	Clay, Francis James . .	30 Dec. 62	20 Sep. 15 m. 21 Sep. 17	N	B. M., M. W., Lucknow.
23 Oct. 13	Chisholm, Alexander (95, 169) (c)	4 Aug. 63	30 Dec. 17	N	B. M., M. W., Meerut.
22 Nov. 14	Hilton, Frank Leonard .	11 Jan. 67	4 Aug. 18	N	B. M., M. W., Bombay.

Deputy Commissary with the honorary rank of Captain—1.

20 Sep. 15	Kemp, John (a) . .	8 Feb. 61	4 Aug. 18	N	(Supy.) B. M., M. W., Secunderabad.
8 Feb. 16	Courtenay, Robert Benjamin.	11 Nov. 66	4 Aug. 18		B. M., M. W., Allahabad.

Assistant Commissaries with the honorary rank of Lieutenant—3.

24 Oct. 16	Armitage, John . .	28 July 65	24 Oct. 16		B. M., M. W., Bombay.
1 Jan. 17	Greaves, William Henry (a)	1 Sep. 62	1 Jan. 17 30 Dec. 17		B. M., M. W., Quetta.
1 Jan. 17	Slade, Frank, (a) . . .	22 May 62	1 Jan. 17 30 Dec. 17		(Supy.) B. M., M. W., Nowshera.
1 May 12	O'Connor, Patrick . .	28 May 65	30 Dec. 17		(Supy.) B. M., M. W., Cawnpore.
4 Aug. 18	Mooney, Henry Austin .	12 Feb. 71	4 Aug. 18		B. M., M. W., Rawalpindi.

BARRACK DEPARTMENT, MADRAS.

Commissaries with the honorary rank of Major—2.

24 May 13	Reuben, William . .	6 Feb. 66	8 July 16 m. 21 Sep. 17	N	B. M., 1st cl., M. W., Rangoon Dn., Burma.

ARMY CLOTHING DEPARTMENT.

Under the Control of Indian Munitions Board.

Commissary with the honorary rank of Major.—1.

27 July 05	Phillips, George Henry	31 Jan. 71	21 Sep. 17	N	Army Clothing Fy., Madras.

Deputy Commissary with the honorary rank of Captain.—1.

15 Jan. 14	McEntee, John . .	23 Nov. 71	15 Jan. 18	N	Army Clothing Fy., Bombay.

Assistant Commissary with the honorary rank of Lieutenant.

3 June 15	Barson, Albert Henry (a).	7 Feb. 75	3 June 15		Army Clothing Fy., Alipore.
1 Jan. 17	Johnson, Ernest (169) (a)	21 Apl. 74	1 Jan. 17		Asst. Controller (Clothing), I. M B., Simla.

POSTS AND TELEGRAPHS.

Assistant Commissary with the honorary rank of Lieutenant.

2 Dec. 16	Packwood, Isaac	24 May 65	2 Dec. 16		Telegraph Master. (On deptn. to Army Dept.)

(a) Supy. Specially promoted under the provisions of para. 470, A. R., I., Vol. II.
(b) Supy. under the terms of A. D. letter No. 52146—1-A. G. 10, dated 6th August 1917.
(c) Supy. on attaining the age of 55 years.

First Commn.	Name.	Date of birth.	Date of grade or rank.	F. rule	Remarks.
	INDIA MISCELLANEOUS LIST.				
	Commissaries with the honorary rank of Major—4.				
11 Jan. 05	**Arbery**, James, M.B.E.	24 May 64	5 Nov. 11 m. 14 Aug. 12	N	Survey of India Dept.
1 Mar. 07	**Taylor**, Thomas . . .	16 Mar. 65	1 Feb. 13 m. 1 July 16	N	I. A. R. O. (*T. C.*)
7 Sep. 04	**Pepper**, Henry (51, 52, 207).	28 Nov. 67	19 Jan. 15 m. 21 Sept. 17	N	Army Hdqrs. (Q.-M.-G.'s Bch.).
1 May 10	*Duffy, Thomas Augustine (a) (temp. Capt. 22 Apl. 16.)*	*2 Dec. 71*	*1 Jan. 16 m. 21 Sept. 17*	N	*Office of the Financial Adviser, Military Finance.*
5 Nov. 11	*Davis, William (170) (a)*	*5 Oct. 69*	*1 Jan. 16 m. 21 Sept. 17*	N	*Divl. Staff Office, 1st (Peshawar) Dn.*
3 Feb. 12	*Bartlett, Edward Lawrence (a).*	*18 Oct. 71*	*1 Jan. 16 m. 21 Sept. 17*		Cavy. School, Saugor.
15 Mar. 09	**Buegg**, Alfred James .	17 Apl. 65	15 Aug. 16 m. 21 Sept. 17	N	I. A. R. O. (*T. C.*)
	Deputy Commissaries with the honorary rank of Captain—4.				
1 Dec. 11	**Salmon**, Thomas Francis	13 June 65	19 Jan. 15	N	Mily. Prison and Det. Bk. Staff, Poona.
1 Mar. 13	*Crosbie, Christopher Stephen (a).*	*14 Dec. 67*	*3 June 15*	N	*Army Hdqrs. (G. S. Bch.).*
17 Mar. 12	*Clarke, James Edward (a)*	*20 Apl. 74*	*1 Jan. 16*	N	*Army Hdqrs. (Mily. Secy.'s Bch.).*
24 Oct. 13	*Swinnerton, John Christopher (a).*	*23 Dec. 68*	*1 Jan. 16*	N	*Divl. Staff Office, Burma Divn.*
2 May 10	*Martin, Edgar Henry (a)*	*18 Aug. 70*	*1 July 16*	N	S. Comd. Office.
17 Mar. 12	*Dawson, Frank Donald (207).*	*25 Oct. 67*	*1 Jan. 17*	N	*Army Hdqrs. (A. G.'s Bch.).*
17 Mar. 12	*Drysdale, David (207) (a)*	*23 May 70*	*1 Jan. 17*	N	*Army Hdqrs. (Mily. Works Bch.).*
1 Apl. 13	*Reynolds, Harry Lewis .*	*28 Feb. 66*	*1 July 17*	N	*Lv. in I., 1 yr., 16 Jan. 18 (pending Ret.).*
	Assistant Commissaries with the honorary rank of Lieutenant—8.				
1 Apl. 13	*Bridge, John Henry (a) .*	*13 Dec. 67*	*1 Apl. 13*	N	*Asst. R. O., Muttra. (Tempy.) S. S. O. & Cantt. Magte., Muttra.*
8 Feb. 14	**Elliott**, Joseph . .	3 Feb. 66	8 Feb. 14	N	Army Remount Dept.
15 June 14	**Rushton**, Samuel Kelsall.	19 Oct. 72	15 June 14	N	Army Hdqrs. (A. G.'s Bch.).
3 June 15	*Richardson, Francis Lionel (a).*	*12 Nov. 76*	*3 June 15*	N	I. A. R. O. (*T. C.*)
3 June 15	*Joyner, Harry (a) (95, 170).*	*15 Sep. 78*	*3 June 15*	N	*Divl. Staff Office, 1st (Peshawar) Dn.*
3 June 15	*Godfrey, Richard, (170) (a)*	*3 Apl. 67*	*3 June 15*	N	*Divl. Staff Office, 5th (Mhow) Dn.*
3 June 15	*Davis, Harry John Alfred (a)*	*13 Jan. 76*	*3 June 15*	N	*Divl. Staff Office, Meerut Dn.*
3 June 15	*Cheers, Joseph McGregor (a)*	*26 Apl. 74*	*3 June 15*	N	I. A. R. O. (*T. C.*)
1 Jan. 16	*Graham, Ronald (a) . .*	*3 May 71*	*1 Jan. 16*	N	*Survey Dept.*
1 Jan. 16	*Vousley, Frederick (a) .*	*25 Mar. 69*	*1 Jan. 16*	N	I. A. R. O. (*T. C.*)
1 Jan. 16	*Cameron, Angus James (temp. Capt.) (a).*	*18 May 75*	*1 Jan. 16*	N	*Army Hdqrs. (G. S. Bch.).*
1 Jan. 16	*Roffey, Henry Richard,* D. C. M. *(a)*	*2 June 79*	*1 Jan. 16*	N	*Army Hdqrs. (G. S. Bch.).*
1 July 16	*Somerville, George Aytoun (a) (95).*	*28 July 64*	*1 July 16*	N	*Staff College, Quetta.*
1 July 16	*Edwards, William Henry (162, 170, 206) (a).*	*23 Mar. 73*	*1 July 16*	N	I. A. R. O (*T. C.*)
1 Jan. 17	*Ricks, George (a) . .*	*29 Oct. 61*	*1 Jan. 17*		*Prisoners of War Camp, Thayetmyo.*
1 Jan. 17	*Parks, Hedley Charles (a)*	*2 Jan. 76*	*1 Jan. 17*		I. A. R. O. (*T. C.*)
1 July 17	*Fressanges, George Francis (a).*	*22 Sept. 73*	*1 July 17*		*Army Hdqrs. (G. S. Bch.).*
6 Aug. 17	*Marketis, William Philip (b)*	*19 Sep. 76*	*6 Aug. 17*	N	*Embkn. Staff, Bombay.*
6 Aug. 17	*Freeman, Thomas Henry (b)*	*20 Apl. 80*	*6 Aug. 17*	N	*Embkn. Staff, Bombay.*
27 Sep. 18	*Murray, Charles John (c)*	*2 June 71*	*27 Sep. 18*		I. A. R. O. (*T. C.*)
27 Sep. 18	*Prior, Charles James (c)*	*10 Apl. 75*	*27 Sep. 18*		I. A. R. O. (*T. C.*)
27 Sep. 18	*Norris, William Albert (c)*	*8 Oct. 78*	*27 Sep. 18*		I. A. R. O. (*T. C.*)
27 Sep. 18	*Jackson, Charles Gowan (c)*	*9 Nov. 77*	*27 Sep. 18*		I. A. R. O. (*T. C.*)
27 Sep. 18	*Forde, Andrew (c) . .*	*5 Oct. 80*	*27 Sep. 18*		I. A. R. O. (*T. C.*)
27 Sep. 18	*Morrow, Charles Thomas (c)*	*27 Nov. 76*	*27 Sep. 18*		I. A. R. O. (*T. C.*)
27 Sep. 18	*Creak, Evan Victor (c) .*	*15 June 79*	*27 Sep. 18*		I. A. R. O. (*T. C.*)
27 Sep. 18	**Hill**, Alfred John (*c*) . .	29 Apl. 73	27 Sep. 18		I. A. R. O. (*T. C.*)
27 Sep. 18	**Marks**, Frederick Charles (*c*)	26 Dec. 70	27 Sep. 18		I. A. R. O. (*T. C.*)
27 Sep. 18	**Danskin**, John William (*c*)	13 May 75	27 Sep. 18		I. A. R. O. (*T. C.*)
27 Sep. 18	**Marshall**, Percy Harold (*c*)	16 June 78	27 Sep. 18		I. A. R. O. (*T. C.*)
1 Nov. 18	**Ginns**, Lewis Bertrand (*c*)	2 June 81	1 Nov. 18		I. A. R. O. (*T. C.*)
20 Nov. 18	**Colter**, Parr (*c*) . . .	22 Aug. 73	20 Nov. 18		I. A. R. O. (*T. C.*)
26 Nov. 18	**Coward**, Arthur John (*c*)	11 Jan. 72	27 Sep. 18		I. A. R. O. (*T. C.*)
26 Nov. 18	**Wilson**, Thomas George (*c*)	16 Dec. 73	27 Sep. 18		I. A. R. O. (*T. C.*)

(*a*) Supy. Specially promoted under the provisions of para. 470, A. R. I., Vol. II.
(*b*) Supy. under the terms of A. D. letter No. 52146—1-A. G. 10, dated 6th August 1917.
(*c*) Supy. under the terms of Army Instructions (India) No. 1112, dated 24th September 1918.

DEPARTMENTAL OFFICERS WITH HONY. RANK.

MISCELLANEOUS LIST, MADRAS.

Commissaries with the honorary rank of Major.

First Commn.	Names.	Date of birth.	Date of grade or rank.	F. rule.	Remarks.
11 Sep. 12	**Coulter,** Robert	30 Oct. 63	14 Sept. 16 *m.* 21 Sept. 17	N	Supdt., Det. Bk., Dagshai. *Temply. at Ahmednagar.*
1 Jan. 13	**Hanly,** John	4 Nov. 61	5 Jan. 17 *m.* 21 Sept. 17	N	Supdt., Det. Bk., Lucknow.

Assistant Commissary with the honorary rank of Lieutenant.

First Commn.	Names.	Date of birth.	Date of grade or rank.	F. rule.	Remarks.
1 Jan. 16	**Twiddy,** George William	24 Feb. 72	1 Jan. 16 24 June 18	N	Divl. Office, 8th Lucknow Divn.
17 June 18	**Kenny,** Robert	19 Mar. 75	17 June 18	"	
22 Sep. 18	**Baldrey,** Frederick John	25 Nov. 77	22 Sep. 18		

DEPARTMENTAL WARRANT OFFICERS.

Note.—All Departmental Warrant Officers are serving under the new (1889) Furlough Rules.

ORDNANCE DEPARTMENT, NORTHERN ARMY.

NOTE.—The names shown in italics are those of Warrant Officers of the Ordnance Dept. seconded for employment in manufacturing establishments, or in offices, stated opposite their names or who are supernumerary or promotion for field services.

Conductors—38.

Names.	Date of birth.	Date of rank.	Remarks.
Henderson, James (95, 170, 207)	*15 Aug. 65*	*18 Mar. 05*	*Jubbulpore—Hd. Ovsr., Timber Sec.*
(X)			
Sullivan, William Charles	29 May 67	11 Jan. 07	Ferozepore.
Norbury, Albert (207)	*2 Nov. 70*	*10 Sep. 07*	*Calcutta—Office of Dir. of Ord. Factories.*
(XI)			
Shortridge, John James Francis (88, 94, 153, 161, 165, 169, 185, 204, 207).	10 Oct. 74	28 Jan. 09	Allahabad.
Misson, William (95)	5 Dec. 67	26 Mar. 09	
McInnes, Peter (170)	*28 June 68*	*4 Mar. 10*	*Kirkee Facty.—Ovsr., Laby.*
Bennett, John (170).	*21 Oct. 66*	*4 Mar. 10*	*Aruvankadu.—Ovsr., Cordite Sec.*
(XII)			
Snook, Frederick (170)	30 Apl. 72	12 Aug. 10	Allahabad.
Beedan, Albert William (58, 95, 170)	29 Mar. 71	10 Jan. 11	
Proctor, William David (30, 58, 207)	3 Nov. 72	22 Jan. 11	
Davis, Enoch (52, 66, 170, 207)	*27 Apl. 71*	*26 Mar. 11*	*Aruvankadu—Ovsr., Cannon Cartridge.*
Grout, Thomas (58)	30 July 69	18 May 11	
Johnson, Harry George	*4 Feb. 71*	*6 June 11*	*Cawnpore.—Ovsr., Stores.*
(XIII)			
Bailey, William Alfred.	*11 Sep. 73*	*21 June 11*	*Jubbulpore.—Ovsr., Timber Sec.*
Kybett, Charles Henry (95)	4 July 71	21 June 11	
Ives, William Percival	*23 Mar. 73*	*1 Oct. 11*	*Dum Dum.—Ovsr., Laby.*
Rose, John (95, 207)	16 June 74	1 Oct. 11	*Lv., m. c., ex I., up to 19 Dec. 18.*
Bunnett, Frederick (95, 170)	7 Oct. 75	1 Oct. 11	Rawalpindi.
(XIV).			
Godbold, Oswald Miles (170, 207)	*10 Apl. 76*	*23 June 12*	*Dum-Dum—Examnr.*
Andrews, Frederick John	7 Oct. 69	16 Sep. 13	
Walsh, Patrick	*2 Oct. 72*	*11 Oct. 13*	*Kirkee.—Ovsr., Laby.*
Warmington, William Thomas	11 Nov. 70	11 Oct. 13	
Thorne, George	2 May 67	{11 Oct. 13 {30 Apl. 16	
Rush, William	11 May 72	11 Oct. 13	Quetta.
Spooner, Alfred (94, 172)	11 Apl. 71	7 Jan. 14	Quetta.
Ancrum, Herbert Henry	*26 Feb. 72*	*9 Jan. 14*	*Simla—Army Hdqrs. (Ord. Bch.).*
Gordon, Lovell Charles	5 June 73	9 Jan. 14	Karachi.
Sanderson, Edward Francis	*4 July 75*	*1 Feb. 14*	*Balasore—(Offg.) Chief Exmr., Proof and Expl. Estt.*
Smith, Peter (170)	9 Aug. 70	6 Feb. 14	Quetta.
Boughtflowers, John (94, 170)	3 Mar. 72	18 Feb. 14	Ferozepore.
Aydon, Walter William (95, 169)	21 June 73	11 Mar. 15	Mhow. Clerk to D. A. D. O. S.
Geraghty, Patrick	30 Jan. 70	23 Mar. 15	
Bunce, Arthur.	*22 Aug. 74*	*26 Mar. 15*	*Dum-Dum—Hd. Ovsr. Stores.*
Hyde, Charles Bernard	26 Jan. 78	26 Mar. 15	
Warwick, Thomas	25 Mar. 74	18 May 15	Allahabad.
Nicholas, Thomas (95, 205)	18 Dec. 71	{16 June 15 {11 May 17	
Fitter, James Henry (153).	*14 Aug. 74*	*16 June 15*	*Kirkee—Hd. Ovsr.—Stores.*
Bedwell, Percy	*16 Apl. 75*	*16 June 15*	*Army Hdqrs. (Ord. Bch.), Simla.*
Grayson, George Robert	*25 July 71*	*16 June 15*	*Dum-Dum—Asst. Ovsr., Laby.*
Ward, Harry (95)	*18 Aug. 76*	*16 June 15*	*Aruvankadu—Hd. Ovsr., Stores.*
Archer, James Ernest (95)	20 Aug. 75	17 July 15	Madras.
Maley, John (165, 170).	13 Jan. 73	18 July 15	Rawalpindi.
Shaw, Arthur (95)	*3 Mar. 72*	*12 July 16*	*Jubbulpore—Ovsr., Examn.*
Pemberton, Henry	17 Oct. 75	22 Aug. 16	
Wood, Alban Hubert	3 Nov. 74	1 Sep. 16	
Boswell, James Edwin	11 Nov. 70	27 Sep. 16	Ferozepore.
Nicholl, Henry (94, 170)	*2 Oct. 74*	*26 May 17*	*Cossipore—Ovsr., Stores.*
Robinson, John Robert (172a)	*26 Nov. 77*	*26 May 17*	*Aruvankadu—Ovsr., Stores.*
Clements, Henry Stanley (95, 170).	13 Apl. 75	26 May 17	Ferozepore.
McNamara, John Daniel (170)	14 Oct. 80	8 Dec. 17	

ORDNANCE DEPARTMENT, SOUTHERN ARMY.

Conductors—46.

Names.	Date of birth.	Date of rank.	Remarks.
Coates, William	15 Oct. 69	18 Aug. 07	
Humphreys, Edward (8, 12) (XVIII) * * *	3 Jan. 72	18 Aug. 07	Madras.
Payne, Charles (169, 207)	16 Jan. 70	27 Dec. 08	*Lv., m.c., in I., up to 22 Dec. 18.*
Jeffrey, John Euphrates	10 Mar. 70	13 Feb. 09	
Evans, Alfred Harry	12 Mar. 65	18 Feb. 09	Agra.
Mitchell, Benjamin Martin	18 July 72	6 Mar. 09	Rangoon.
Goodair, Alfred George (XIX) * * *	24 Sep. 69	19 Oct. 09	Kirkee.
Thompson, Henry (XX)	*8 Nov. 72*	*22 Jan. 10*	*Cossipore.—Hd. Ovsr., Stores.*
Irwin, Joseph (XXI) * * *	*25 Sep. 68*	*18 Oct. 10*	*Asst. Controller of War Accounts.*
Jones, Thomas Powel	15 Jan. 73	18 Oct. 10	
Ashfield, Edwin	*12 Oct. 73*	*4 Jan. 11*	*Army Hdqrs. (Ord Bch.), Simla.*
Greenford, Charles	2 Feb. 71	4 Jan. 11	Jubbulpore Depôt.
Fitzpatrick, Charles (95, 169, 170)	*27 Oct. 70*	*4 Jan. 11*	*Ishapore—Hd. Ovsr., Stores.*
White, Arthur Ernest (94, 169, 207)	2 Oct. 69	4 Jan. 11	Kirkee.
Ford, Harry (95, 206)	7 Sep. 73	23 Jan. 11	Rawalpindi.
Smith, William Mogford (95, 170)	21 July 70	11 Oct. 11	Quetta.
Parke, Robert (94, 104, 161, 169, 185, 193, 206)	*29 May 74*	*26 Oct. 11*	*Jubbulpore.—Hd. Ovsr., Stores.*
Hynes, Charles Frederick John	2 Jan. 71	2 Dec. 11	
Malcolm, John	30 Nov. 71	2 Dec. 11	
Freed, Ernest (58)	11 Sep. 69	6 May 12	Allahabad.
Selby, Herbert Richard (95, 172*a*, 193)	21 Mar. 71	5 June 12	
Breach, James Benjamin	3 June 74	4 July 12	Madras.
Benns, Albert (206)	3 Nov. 72	29 May 13	
Hinkley, Alfred Michael	12 Dec. 72	4 Sep. 13	
Walker, Benjamin Nettleton	5 Mar. 73	19 Dec. 13	Ferozepore.
Norkett, Harry William (94, 95, 106, 162, 165, 169, 185, 206)	18 Jan. 72	4 Apl. 14	
Graham, George	1 Nov. 71	12 July 14	Ferozepore.
Sansom, Arthur (94, 95, 169)	2 Dec. 74	22 Nov. 14	
Villiers, Alfred (94, 169, 185)	29 Sep. 79	15 Apl. 15	Kirkee.
Young, Frederick William, (95, 170, 185), (XXII)	*17 Mar. 77*	*3 June 15* / *10 May 16*	
Bardens, Arthur Thomas (94, 162, 169, 185, 206)(*a*) (XXIII)	19 Nov. 73	3 June 15 / 30 July 16	Kirkee.
Tye, James	*7 July 72*	*15 Aug. 15*	*Kirkee—Examr., Insp. Sec., No. 2 charge.*
Tomlinson, James	13 Aug. 74	15 Aug. 15 / 1 Mar. 16	*Lv., in & ex I., m. c., up to 18 Oct. 18.*
Pendery, George Bellville	7 Dec. 67	15 Aug. 15	
Phillips, George Stanley (94, 170)	9 Mar. 76	11 Jan 16	Rawalpindi.
Kent, William	*7 Dec. 71*	*17 Mar. 16*	*Cawnpore.—Ovsr., Central Sec.*
Davis, John (XXII)	5 May 72	17 Mar. 16	Jubbulpore Depôt.
Sharpe, William (95, 172*a*, 193)	4 Nov. 72	10 May 16	Kirkee.
Standen, William Henry (95, 172*a*)	29 Feb. 71	12 June 16	Allahabad.
James, Benjamin	20 June 70	17 June 16	Ferozepore.
Williams, Frank (94) (a) (XXIV)	*10 May 76*	*1 July 16*	
Fox, Ernest (XXIII) * * *	*27 Dec. 69*	*30 July 16*	*Dum Dum.—Ovsr., Stores.*
Johnson, Ernest Victor (95) D.C.M.	15 July 76	30 May 17	Allahabad.
Grant, Frederick William (170)	*8 July 75*	*1 July 17*	*Jubbulpore.—Ovsr., Stores.*
Eldridge, James Henry (51, 95, 172*a*)	28 May 70	1 July 17	
Purkis, Alfred Edwin (94, 193)	31 Oct. 76	12 July 17	
Cormack, William (95, 106)	30 July 74	6 Dec. 17	
Foster, Tom	14 Apl. 71	26 Mar. 18	
Chiesman, Albert	6 June 73	13 Apl. 18	
Wilson, Thomas (94, 95)	5 Nov. 75	16 Apl. 18	Rawalpindi.
Sanderson, Richard James (193, 207)	*26 Feb. 76*	*1 Aug 18*	*Kirkee.—Asst. Ovsr., Laby., Ammn. Facty.*
Tomlin, William	15 June 72	1 Aug. 18	
Denton, Thomas	3 Feb. 74	23 Sep. 18	Madras.

Sub-Conductors.

Names.	Date of birth.	Date of rank.	Remarks.
Marner, Thomas (58, 95, 170)	9 Sep. 72	8 Oct. 11	Madras.
Green, Herbert (95, 172*a*)	9 Mar. 72	11 Oct. 11	
Dunbar, Goodwyn Alexander (XXIV) * * *	*1 July 77*	*26 Oct. 11*	*Cossipore.—Ovsr. Examn.*

(*a*) Supy. Specially promoted under the provisions of para. 470, A. R., I., Vol. II.

ORDNANCE DEPARTMENT—GENERAL LIST.

Conductors.

Names.	Date of birth.	Date of rank.	Remarks.
Hill, Heber Murdock (153, 165, 169, 170, 207) (b) (XXVIII).	*1 July 76*	*1 Jan. 16*	*Army Hdqrs. Q. M. G.'s Br., Simla.*
Reed, Thomas (95) (b) (XXVII)	*1 July 76*	*1 Jan. 17*	*Fort William.*
Biggs, John Burles (94, 169, 193) (b) (XXVI)	*13 Jan. 79*	*30 May 17*	*Rangoon—Ovsr. Army Clothing*
McDougall, John Leigh Needham, M.C. (b) (XXV).	*8 July 77*	*1 July 17*	*[Depot.*
Posth, Charles (b) (XXXI) *	*26 Nov. 77*	*28 June 18*	*P. W. D., Delhi.*

Sub-Conductors—78.

Names.	Date of birth.	Date of rank.	Remarks.
Robinson, Arthur James (XXIX) (a)	*1 June 75*	*25 May 09*	*Ishapore—Asst. Controller, War*
Moore, Patrick	25 Oct. 74	30 Oct. 11	*[Accounts.*
Wade, John	6 Feb. 73	31 Oct. 11	
Heapy, George (95, 172a, 193, 207)	5 Aug. 80	1 Nov. 11	Kirkee. S. Comd. Office.
Pelly, Alfred (94, 95, 169, 170)	26 Apl. 74	28 Nov. 11	Karachi.
Langrish, Frank (95)	5 Oct. 76	2 Dec. 11	
Bertram, George Frederick (95, 172a)	*28 Aug. 75*	*15 Jan. 12*	*Cossipore.—Ovsr., Stores.*
Dudman, Charles Henry (94)	21 Aug. 76	6 Mar. 12	
Roberts, Henry John (94)	18 Apl. 73	11 Mar. 12	Quetta.
(XXV)			
(XXVI) * * * *			
Warner, Frederick Charles (95)	24 Nov. 76	24 Apl. 12	*[and Saddlery Facty.*
Parsons, Charles Henry (193)	*30 Mar. 73*	*26 Apl. 12*	*Cawnpore.—Asst. Ovsr., Harness.*
McCarthy, Edwin Alfred Reeves (95)	10 Apl. 77	26 Apl. 12	Rawalpindi.
Bolam, Arthur Roland (95, 172a)	9 Sep. 73	6 May 12	*[raty.*
Bowles, Frederick Charles	*11 Aug. 74*	*15 May 12*	*Dum Dum.—Asst. Ovsr., Labo-*
Welton, William (170)	23 Sep. 77	1 June 12	*[Cannon Cartridge.*
Jefferies, Henry	*20 Jan. 71*	*5 June 12*	*Aruvankadu—Asst. Ovsr.*
Stigings, John Charles (95)	24 July 74	23 June 12	
Tate, Frederick Richard (95, 169)	9 June 74	4 July 12	
Whitehead, Walter (95)	17 May 74	12 Oct. 12	Allahabad.
Wood Thomas	12 Sep. 74	11 Nov. 12	Fort William.
(XXVII)			
(XXVIII)			
Fannon, John Thomas (172a)	26 Dec. 78	23 Mar. 13	
Murray, George Alexander	15 Oct. 78	23 Mar 13	Kirkee.
Perriton, William Lewis (95, 106, 172a)	9 July 75	15 Apl. 13	Allahabad.
Nuttall, Robert William	5 Sep. 77	2 May 13	
Docking, Alfred Charles	3 Dec. 78	19 May 13	Ft. William.
Jackson, Joseph Edward	13 Feb. 74	29 July 13	
Harrison, Herbert	28 Sep. 73	26 Aug. 13	Bombay.
Hall, George Charles (50, 172a)	7 Jan. 75	4 Sep. 13	Rawalpindi.
Byrne, Cecil (95, 172a)	1 Sep. 76	14 Sep. 13	
(XXIX) * * *			
Harris, Henry George (170)	8 Feb. 76	16 Sep. 13	*[ratory.*
Snowdon, John Henry (95, 169, 206)	*10 Sep. 76*	*11 Oct. 13*	*Dum Dum—Asst. Ovsr., Labo-*
Welch, George Arthur James	29 July 80	1 Nov. 13	Rangoon.
Dyer, Walter Lawrence Edmund (170)	*29 Dec. 76*	*17 Nov. 13*	*Cawnpore—Asst. Ovsr., Stores.*
Jones, Richard Walker (94, 95, 169, 172a)	5 Feb. 75	17 Nov. 13	Rangoon.
Ryall, Thomas	17 Nov. 75	19 Dec. 13	
Pilkington, John (60a, 94, 162, 169, 185, 206)	31 Dec. 73	9 Jan. 14	
Robinson, Joseph Henry (12, 71, 95, 170, 207)	17 July 74	4 Feb. 14	Rangoon. *[Central Sec.*
Squires, Joseph Henry	*26 Dec. 72*	*6 Feb. 14*	*Aruvankadu.—Asst. Ovsr.,*
Hinton, Thomas Alfred (12, 95)	28 Jan. 78	6 Feb. 14	
FitzGerald, Leo (169, 170)	*21 July 79*	*18 Feb. 14*	*Calcutta.—Office of D. O. F.*
Mansell, James	26 June 75	18 Feb. 14	
Colquhoun, Thomas James	26 Sep. 77	8 Mar. 14	Ferozepore.
Turner, Oliver (94, 95, 172a)	9 June 76	10 Mar. 14	Kirkee.
Smith, Thomas	*12 Jan. 77*	*17 Mar. 14*	*Dum Dum—Asst. Ovsr., Stores*
Shepherd, Rowley James	*4 Aug. 75*	*12 May 14*	*Dum Dum—Asst. Ovsr., Central Sec.*

(a) Supy. Specially promoted under the provisions of para. 470, A. R., I., Vol. II.
(b) Specially promoted, vide A. D. Notfn. No 1459, dated 28 June 18.
* Temporary Conductor from 15 Mar. 17 to 28 June 18.

Names.	Date of birth.	Date of rank.	Remarks.

ORDNANCE DEPARTMENT—GENERAL LIST—*contd.*

Sub-Conductors—*contd.*

Names.	Date of birth.	Date of rank.	Remarks.
Behenna, William Henry	3 May 78	12 May 14	
Pegg, Thomas (95, 172*a*)	28 Sep. 79	28 June 14	
Tutton, Charles (95)	5 June 79	11 July 14	
Pearce, Thomas	11 Nov. 76	12 July 14	Quetta.
Dickinson, John (95, 172*a*)	28 July 80	30 Aug. 14	Rawalpindi.
Rayner, Arthur	*1 Aug. 79*	*22 Nov. 14*	*Indian Munitions Board.*
Rice, Thomas (72*b*)	16 Sept. 77	22 Nov. 14	Madras.
Pearce, Egbert	31 Mar. 79	2 Jan. 15	Allahabad.
Mackenzie, Alexander	*14 Mar. 80*	*19 Mar. 15*	*Kirkee—Asst. Controller of War Accounts.*
Chambers, James Thomas (12)	26 Nov. 80	19 Mar. 15	Kirkee.
Collingwood, Charles Frederick (95)	30 Jan. 76	3 Mar. 15	Rawalpindi.
Blofeld, George Stanislaus (71, 72*a*, 207)	23 June 75	26 Mar. 15	Madras.
Potts, Thomas (94, 95, 169, 172*a*, 193)	25 Oct. 75	15 Apl. 15	*Lv. m c. in I., up to 5 Feb. 19.*
Davies, Charles Alexander	4 Mar. 77	18 May 15	Allahabad.
Matthews, James (XXX)	9 Oct. 81	3 June 15 1 Jan. 18	Rawalpindi.
White, John Henry (a) (XXXII)	*12 June 78*	*3 June 15* *1 Jan. 18*	
Ogbourne, Thomas (a) (XXXIII)	*12 July 78*	*3 June 15* *1 Jan. 18*	*Rangoon.*
Overy, Alexander Earnest Richard (a) (XXXV)	*19 July 79*	*3 June 15* *1 Jan. 18*	Lucknow. Clerk to D.A.D.O.S.
Thorpe, Thomas William (95)	3 Oct. 76	16 June 15	
Sanders, Arthur	27 Apl. 77	17 July 15	Ferozepore.
Styles, Albert Francis	19 Jan. 77	14 Aug. 15	Rangoon.
Scragg, Samuel (95)	11 Feb. 78	15 Aug. 15	Agra.
Calcott, Thomas (a) (XXXVI)	*20 Aug. 81*	*1 Jan. 16* *1 Jan. 18*	
Richardson, Ralph	25 Sep. 81	11 Jan. 16	Rawalpindi.
Freeston, Walter Lewis	3 Nov. 77	17 Mar. 16	Ferozepore.
Cumming, William Thomas	2 June 76	10 May 16	
Webb, Albert	24 Dec. 78	12 June 16	
Hull, Albert	14 June 79	17 June 16	Kirkee.
Barrows, John Cecil Russell (a)	*24 Jan. 83*	*1 July 16*	
Carney, Martin	21 May 76	12 July 16	Ferozepore.
Jones, William John	11 June 76	30 July 16	Mussoorie,—clerk to D. A. D. O. S.
Cook, Albert	27 Mar. 74	22 Aug. 16	
Standley, Richard	*13 June 80*	*1 Sep. 16*	*Army Hdqrs. (Ord. Bch.)*
Hennessy, Edmund Patrick	17 Feb. 80	1 Sep. 16	
McCrea, Thomas	10 Dec. 76	27 Sep. 16	
Eagle, Herbert (172a) (a) (XXXIV)	*29 Aug. 82*	*1 Jan. 17* *1 Jan. 18*	Madras.
Gypson, James Carr (94, 161, 165, 172a, 185, 206) (XXXVIII) (a).	*27 Feb. 83*	*1 Jan. 17* *1 Apl. 18*	
Bridson, William Earnest (a)	*24 Dec. 84*	*1 Jan. 17*	*Lahore Clothing Depot.*
Robinson, George Alexander	10 Oct. 77	12 Jan. 17	
Roper, Charles Russell	9 Apl. 75	1 Feb. 17	
Mayo, Archie Robert (95)	5 Nov. 76	26 May 17	N. Comd. Office.
March, Alexander Maitland (95)	20 Aug. 77	30 May 17	Quetta.
Sharpe, John Charles	*19 Dec. 75*	*1 July 17*	*Dum Dum—Asst. Examr. Inspn. Sec. No. 2 Charge.*
Langton, John	25 Mar. 76	1 July 17	Madras.
Sanderson, Thomas (a) (XXXVII)	*21 Jan. 81*	*1 July 17* *1 Apl. 18*	*Army Hdqrs. (Q. M. G.'s Br.)*
McCully, Alfred James (a) XXXIX.	*2 Oct. 81*	*1 July 17* *1 Oct. 18*	*Army Hdqrs. Aeronautics Directorate.*
Bolton, John Stacey (a)	*28 Aug. 83*	*1 July 17*	*Army Hdqrs. (Q. M. G.'s Br.)*
James, William Richard, (169, 172*a*, 207)	21 Feb. 78	12 July 17	
Stone, Walter	*27 Mar. 79*	*6 Dec. 17*	*Army Hdqrs. (Ord. Bch.)*
Minns, Ernest Walter (95) (XXX)	9 Nov. 76	6 Dec. 17	Rawalpindi.
Gradwell, Joseph	23 May 78	1 Jan. 18	
Roberts, Charles William	14 Oct. 76	1 Jan. 18	Ferozepore.
Wymer, Ernest (94, 95, 169, 172(*a*), 207)	13 Apl. 80	1 Jan. 18	Rawalpindi.
Penlington, Alfred Thomas	*14 Dec. 80*	*1 Jan. 18*	*Jubbulpore—Asst. Ovsr., Stores.*

(*a*) Supy. Specially promoted under the provisions of para. 470, A. R., I., Vol. II.

ORDNANCE DEPARTMENT—GENERAL LIST—*concld.*

Sub-Conductors—*concld.*

Names.	Date of birth.	Date of rank.	Remarks.
James, William	*24 Aug. 77*	*1 Jan. 18*	*Jubbulpore—Overseer.*
(XXXI)			
Grogan, Thomas	11 Nov. 77	1 Jan. 18	
(XXXII)			
Gibson, Archibald (95)	29 July 75	1 Jan. 18	
(XXXIII)			
Davis, Edward	*23 Nov. 79*	*1 Jan. 18*	*Cosssipore— Asst. Ovsr., Central Sec.*
Cheetham, Henry Joseph	*6 Nov. 77*	*1 Jan. 18*	*Army Hdqrs. (Q. M. G.'s Br.)*
Richardson, George (132, 153, 169, 170)	*4 Sep. 78*	*1 Jan. 18*	*Kirkee Fy.—Asst. Ovsr., Stores.*
Newberry, John (95)	27 June 77	1 Jan. 18	Quetta.
Chambers, Robert (95)	9 July 74	1 Jan. 18	Quetta.
Smith, Herbert Frank (95, 172*a*)	1 Apl. 79	1 Jan. 18	
Rolfe, Alfred Thomas (106)	22 Apl. 77	1 Jan. 18	
Hartnoll, Philip Edward (95)	23 Nov. 81	1 Jan. 18	
Watmore, Archibald	8 Sep 74	1 Jan. 18	
Notley, Thomas (169, 172*a*)	22 Feb. 82	1 Jan. 18	Bangalore—Clerk to D.A. D.O.S
Cullinan, John	14 Mar. 80	1 Jan. 18	Army Hdqrs. (Ord. Bch.)
Leplar, Walter	10 Dec. 77	1 Jan. 18	Jubbulpore Depôt.
Wiles, Fitzharding (95)	29 Oct. 76	1 Jan. 18	
Green, Edward Frederick (95)	24 June 77	1 Jan. 18	
(XXXIV)			
Senior, Andrew Thomas	9 Nov. 79	1 Jan. 18	
O'Neill, Edward Martin Marie	24 June 83	1 Jan. 18	Madras.
Adams, Charles William	*25 July 79*	*1 Jan. 18*	*Calcutta—Office of D. O. F.*
Kelly, Joseph Aloysius	21 June 75	1 Jan. 18	Allahabad Govt. Tannery.
Moroney, William Magna (95, 169, 172*a*)	21 Aug. 77	1 Jan. 18	
Cordwell, Edwin George (95)	1 Jan. 78	1 Jan. 18	
Ford, James (172*a*)	10 May 77	1 Jan. 18	
Atkinson, John Thomas (94, 95, 172*a*)	11 Sep 79	1 Jan. 18	Ferozepore.
Hall, Ernest Arthur Main	*3 July 80*	*1 Jan. 18*	*Dum Dum—Asst. Exmr.*
Marsh, Henry William (94, 95)	16 Dec. 76	1 Jan. 18	
Brazier, Herbert Owen	16 May 78	1 Jan. 18	Army Hdqrs. (Ord. Bch.)
Keniston, Lionel Robert Hill	13 Dec. 79	1 Jan. 18	Bombay Clothing Depôt.
Ockelford, Arthur	14 Nov. 77	1 Jan. 18	
(XXXV)			
Mahoney, Thomas (170, 207)	4 Nov. 77	1 Jan. 18	
Allen, George (95)	4 Jan. 79	1 Jan. 18	
Francis, John Lowndes	8 May 78	1 Jan. 18	Rawal Pindi.
(XXXVI)			
Connolly, Edward Thomas	8 Dec. 79	1 Jan. 18	Keamari Clothing Depôt.
Roberts, Shadrack Frank	6 Sep. 81	1 Jan. 18	Allahabad.
Braham, Arthur Morris Frederick	28 Mar. 80	1 Jan. 18	
Batt, Bertram John	17 Dec. 80	1 Jan. 18	
Chapman, Arthur	16 Apl. 79	1 Apl. 18	Kirkee.
Reynolds, Frederick Joseph	*13 Feb. 79*	*1 Apl. 18*	*Calcutta,—Office of D. O. F.*
Gibson, Richard	7 Nov. 80	1 Apl. 18	
Berry, Arthur Cecil	17 June 80	1 Apl. 18	
(XXXVII)			
Eade, James	26 Oct. 80	1 Apl. 18	
Sealy, Samuel Charles	9 Aug. 80	1 Apl. 18	Kirkee, A. O. Examn.
Lewis, Smith	13 Sep. 81	1 Apl. 18	Kirkee, A. O. Stores.
(XXXVIII)			
Malone, Michael	15 Dec. 80	1 July 18	Quetta.
Spanier, Leon (172*a*)	1 May 80	1 July 18	
Nance, Frederick	28 Feb. 80	1 July 18	
Morley, Henry	8 Jan. 82	1 July 18	
Coleman, James Williams (95, 172*a*)	21 July 79	1 July 18	
Collins, Albert Edward (94, 95, 185)	29 Apl. 80	1 Oct. 18	Quetta.
Marchant, Joseph Daniel	3 Dec. 80	1 Oct. 18	
Pritchard, Edward James	*8 July 80*	*1 Oct. 18*	*Army Hdqrs. (Aeronautics Directorate.)*
Sheppard, Frederick (95, 170)	27 May 79	1 Oct. 18	
(XXXIX)			
Rollins, Charles (95, 169, 172*a*)	8 Apl. 81	1 Oct. 18	
Skinner, William (95, 162, 172*a*, 206, 207)	9 July 82	1 Oct. 18	
Charles, Barker	14 May 81	1 Oct. 18	Jubbulpore, A. O. Central Sec.

SUPPLY AND TRANSPORT CORPS, BENGAL.

Conductors.

Names.	Date of birth.	Date of rank.	Remarks.
Secombe, Arthur (207)	10 Apl. 73	29 July 11	No. 8 D. S. Coy.
Alexander, Hugh Mansfield	23 Apl. 68	30 Apl. 12	No. 2 D. S. Coy.
Adams, Edward Hamilton (95, 170, 207)	28 June 72	26 Feb 13	No. 8 D. S. Coy.
Tebbs, Harry	2 Apl. 74	1 Feb. 14	No. 3 D. S. Coy.
Tyler, Arthur (170, 207)	6 May 76	6 Dec. 14	No. 7 D. S. Coy.
Franklin, Charles	16 July 75	27 Apl. 15	"K." S. D. Coy.
Stevenson, William (71, 207)	17 May 77	3 June 15 3 Apl. 18	"B." S. D. Coy.
Belchem, Arthur Frederick	11 Apl. 68	3 July 15	No. 2 D. S. Coy.
Byers, Robert (65)	9 June 71	22 July 15	No. 8 D. S. Coy.
Dinwoodie, John	*8 May 72*	*13 Oct. 15*	*Army Hdqrs. (Q. M. G.'s Bch.).*
Poole, Frederick Abner	27 Oct. 71	14 Oct. 15	No. 8 D. S. Coy.
Creagh, Charles James, M.M. (65)	1 Apl. 73	14 Oct. 15 1 Aug. 16	"J." S. D. Coy.
Morris, Edward	1 Apl. 74	2 Dec. 15	No. 1 D. S. Coy.
Pratt, Alfred John (66, 71)	24 May 75	27 Dec. 15	"J." S. D. Coy.
Hughes, William Benjamin (64, 71, 206)	22 May 75	1 Jan. 16 3 Sep. 16	No. 2 D. S. Coy.
Rossiter, Bertram Edward (*a*)	17 July 76	1 Jan. 16 23 June 17	No. 7 D. S. Coy.
Bolland, John Frederick (65, 207)	13 Aug. 71	20 Mar. 16	No. 8 D. S. Coy.
Ryan, Patrick Charles (94, 170, 207)	13 Jan. 75	30 Apl. 16	No. 5 D. S. Coy.
Power, Thomas Vivian	16 Oct. 76	30 Apl. 16	"L." S. D. Coy.
Ewing, James	3 July 73	17 May 16	No. 1 D. S. Coy.
Mason, Patrick John	15 Nov. 79	10 June 16	"B." S. D. Coy.
Stoneham, Thomas Ellison	28 Jan. 75	1 Aug. 16	No. 1 D. S. Coy.
Cotton, Albert	15 June 75	10 Sep. 16	No. 1 D. S. Coy.
Smith, Ernest Llewellyn (170, 207)	15 June 77	29 Nov. 16	No. 3 D. S. Coy.
Curtis, Henry Charles (95, 99, 170, 207)	26 Aug. 72	7 Feb. 17	No. 2 D. S. Coy.
Millwright, Charles	21 May 75	27 Sep. 17	No. 8 D. S. Coy.
Hayes, William (172*a*)	5 Nov. 77	28 May 18	No. 8 D. S. Coy.
Firbank, Thomas	1 Mar. 75	26 Oct. 18	No. 7 D. S. Coy.

Sub-Conductors.

Names.	Date of birth.	Date of rank.	Remarks.
Tapp, William	6 Jan. 74	10 Feb. 10	No. 1 D. S. Coy.
Teather, Robert Thomas (95, 162 165, 168, 206)	19 Dec. 77	1 Mar. 10	"J." S. D. Coy.
Spiller, William (65)	18 Jan. 76	9 Apl. 10	Kohat Brigade.
Sadller, Richard William Clark Travers	27 Apl. 74	21 Dec. 10	"L." S. D. Coy.
Holmes, Walter	18 Oct. 75	13 Feb. 11	"C." S. D. Coy. Attd. from No. 2 D. S. Coy.
Rogers, Edward (64, 165, 172*a*)	15 Aug. 78	11 Mar. 11	Ambala Bde.

SUPPLY AND TRANSPORT CORPS, 9th (Secunderabad) and Burma Divisions.

Conductors.

Names.	Date of birth.	Date of rank.	Remarks.
Hunt, Edward Clarke (207)	4 Nov. 67	11 Oct. 09	"A." S. D. Coy., Bombay.
Skehan, William James	19 July 69	12 July 10	"E." S. D. Coy.
Poulter, John	24 Jan. 65	26 Sep. 13	"E." S. D. Coy.
Hargreaves, Frederick William (207)	27 Aug. 64	20 July 14	No. "A." S. D. Coy.
Meynell, Robert	15 Apl. 66	9 May 15	No. 9 D. S. Coy.
Thomas, William (*a*)	*25 July 75*	*3 June 15*	*No. 9 D. S. Coy.*
Maule, John Horace, D.C.M. (207)	28 Feb. 67	24 Mar. 16	No. 9 D. S. Coy.
Terry, William	29 Nov. 70	25 May 17	No. 10 D. S. Coy.
Fitzgerald, Patrick	3 Dec. 67	31 Oct. 17	No. 9 D. S. Coy.
Turner, David	1 Apl. 69	24 Mar. 18	No. 10 D. S. Coy.
Waters, Albert	14 Apl. 73	1 May 18	S. Commd. office.
Huskisson, Alfred Edward	6 Aug. 74	2 July 18	"E." S. D. Coy.
Costello, Charles Patrick	19 May 71	19 July 18	No. 9 D. S. Coy.

(*a*) Supy. Specially promoted under the provisions of para. 470, A. R., I., Vol. II.

Names.	Date of birth.	Date of rank.	Remarks.

SUPPLY AND TRANSPORT CORPS, 9th (Secunderabad) and Burma Divisions—*contd.*

Sub-Conductors.

Tobin, Walter James	5 Apl. 70	1 Sep. 08	"A." S. D. Coy.

SUPPLY AND TRANSPORT CORPS, BOMBAY.

Conductors.

Williams, Charles Herbert (65, 207)	13 July 75	10 Feb. 10	No. 6 D. S. Coy.
Simon, Peter Patrick (66, 94, 169, 206)	25 Aug. 76	19 July 10	No. 5 D. S. Coy.
West, John (12, 58, 65, 66, 71)	27 Mar. 72	10 Oct. 14	"A." S. D. Coy.
Mortlock, Joseph (65, 66)	30 Mar. 74	18 Dec. 14	No. 5 D. S. Coy.
Kelly, Thomas (65, 95, 207)	28 Oct. 74	14 Feb. 15	"C." S. D. Coy.
Cole, Robert Frank	9 Sep. 77	1 Jan. 16 2 Oct. 16	No. 4 D. S. Coy.
Johnson, George	*12 Oct. 74*	*1 Oct. 16*	*No. 6 D. S. Coy.*
Davidson, Albert Willie Eugene (95)	*21 Mar. 75*	*1 Oct. 16*	*On Spl. Ser., China.*
Kettlewell, James (66)	2 May 77	1 Oct. 16	Karachi Brigade.

Sub-Conductors—23.

SUPPLY AND TRANSPORT CORPS—AMALGAMATED LIST.

Conductors.

Names.	Date of birth.	Date of rank.	Remarks.
Carey, Frederick (a)	*20 May 76*	*3 June 15*	No. 6 D. S. Coy.
Holdaway, Frank (64, 71, 170)	27 Aug. 73	1 Jan. 16 2 Oct. 16	No. 4 D. S. Coy.
Eagles, James (170) (a)	*9 July 78*	*1 Jan. 16*	No. 4 D. S. Coy.
Bennett, Ernest (a)	*19 Dec. 78*	*1 Jan. 16*	No. 6 D. S. Coy.
Dunne, John, M.C. *(a)*	*5 Sep. 77*	*1 July 16*	No. 6 D. S. Coy.
Tattersall, Oliver (a)	*14 Feb. 79*	*1 Jan 17*	No. 2 D. S. Coy.
Gates, Thomas (a)	*1 Sep. 77*	*1 Jan. 17*	No. 2 D. S. Coy.
Lyons, Patrick (95, 170, 207)	25 Feb. 80	6 Feb. 17	No. 1 D. S. Coy.
Williams, Henry Charles (169, 207)(a)	*28 Nov. 77*	*6 Feb. 17*	No. 1 D. S. Coy.
Clark, Alfred (207)	19 Mar. 76	6 June 17	No. 7 D. S. Coy.
Edgerley, Henry Frank (a)	*9 Dec. 78*	*31 Oct. 17*	No. 7 D. S. Coy.
Olley, Arthur Henry	15 Apl. 77	13 Nov. 17	No. 1 D. S. Coy.
Elliott, Herbert James (185) (a)	*23 Jan. 78*	*13 Nov. 17*	"A." S. D. Coy.
Green, William	4 Mar. 79	18 Dec. 17	No. 3 D. S. Coy.
Dodd, William (169, 172a)	1 June 72	27 Feb. 18	No. 1 D. S. Coy.
Curtis, James Ernest (a)	*31 July 76*	*1 May 18*	*"C." S. D. Coy.*
Lambourne, Albert Edward (a)	*25 July 80*	*28 May 18*	*Aden.*
Fitch, Alfred Percy (a)	*13 Oct. 76*	*2 July 18*	*"A" S. D. Coy., Bombay.*
Gregory, Albert (95)	13 May 75	10 Sep. 18	"C." S. D. Coy.

Sub-Conductors.

Names.	Date of birth.	Date of rank.	Remarks.
Wilson, Arthur Frederick	2 June 75	19 July 10	No. 5 D. S. Coy.
Tynan, Patrick (a)	*15 July 82*	*15 July 11*	No. 6 D. S. Coy.
Stephens, George Richard Tomlin	26 May 71	7 Aug. 11	No. 2 D. S. Coy.
Neale, Thomas	28 June 75	23 Sep. 11 1 Sep. 14	No. 6 D. S. Coy.
Cullen, Roland Richard Berthold	23 May 77	22 Feb. 12	No. 2 D. S. Coy.
Gale, James Frank	13 Jan. 75	6 Apl. 12	No. 7 D. S. Coy.
Perry, Frederick (95)	*13 Sep. 78*	*6 Apl. 12*	No. 5 D. S. Coy.
Taylor, William	15 July 75	1 July 12	No. 7 D. S. Coy.
Crowley, Charles (95)	29 June 73	19 Aug. 12	No. 4 D. S. Coy.
Swain, John Henry (95, 165, 169,	27 Sep. 76	1 Sept. 12	"L." S. D. Coy.
Dempster, Thomas 170, 206, 207).	24 Sep. 78	10 Oct. 12	"G." S. D. Coy.
Knight, Warburton Richard (65, 95)	17 June 81	8 Feb. 13	No. 4 D. S. Coy.
Bould, Cecil Henry	30 Apl. 75	26 Feb. 13	Bannu Bde.
Whitley, Charles	5 Sep. 71	20 Mar. 13	No. 6 D. S. Coy.
Snelling, William Thomas	27 Sep. 78	5 Apl. 13	No. 5 D. S. Coy.
Bantock, Alvin Ernest	31 Dec. 77	20 May 13	No. 5 D. S. Coy.
Ingram, James	14 May 72	8 Aug. 13	No. 7 D. S. Coy.
Cubitt, Arthur	4 Mar. 77	18 Sept. 13	No. 7 D. S. Coy.
Howe, Alfred John. (71)	23 Nov. 76	26 Sept. 13	No. 3 D. S. Coy.
Morris, Thomas (207)	13 July 73	10 Oct. 13	Ambala Bde.
Pye, James	11 June 79	1 Dec. 13	No. 2 D. S. Coy.
Pain, Henry Bervill (64, 95, 170)	8 Mar. 80	1 Dec. 13	Bannu Bde.
Dunlop, Albert	4 May 77	24 Dec. 13	No. 4 D. S. Coy.
Wright, Fredrick Lawrence Hall	30 Aug. 78	30 Dec. 13	No. 5 D. S. Coy.
Sadler, Frank (95)	5 Apl. 75	6 Jan. 14	No. 1 D. S. Coy.
Foort, Frederick	25 Jan. 80	23 Jan. 14	No. 1 D. S. Coy.
Morgan, Edward	12 July 76	1 Feb. 14	No. 4 D. S. Coy.
Jones, Henry	28 Oct. 77	7 Mar. 14	No. 5 D. S. Coy.
Appleby, Thomas	13 Jan. 73	2 June 14	"J." S. D. Coy.
Atkinson, Errol Russell	1 Apl. 79	20 July 14	No. 4 D. S. Coy.
Howard, Charles	6 Nov. 78	10 Oct. 14	No. 8 D. S. Coy.
Ward, John	7 Dec. 75	6 Dec. 14	No. 2 D. S. Coy.
Martin, Mark	13 Feb. 75	1 Feb. 15	No. 7 D. S. Coy.
Morris, George	25 Nov. 74	4 Feb. 15	No. 5 D. S. Coy.
Biggerstaff, Thomas James	27 July 80	6 Feb. 15	No. 6 D. S. Coy.
King, George	8 Apl. 71	19 Feb. 15	"L." S. D. Coy.
Webb, Tom Harris	1 June 82	3 Mar. 15	No. 7 D. S. Coy.
Samways, James	9 Mar. 78	8 May 15	No. 8 D. S. Coy.
Waugh, Charles Robert	19 Aug. 78	9 May 15	No. 3 D. S. Coy.
Levings, Frederick George, D.C.M.	. . . .	3 June 15 7 Feb. 17	No. 3 D. S. Coy.
Halloran, John	. . . .	3 June 15 3 Apl. 18	No. 2 D. S. Coy.
Price, Frederick William John (a)	. . . .	3 June 15	"H" S. D. Coy.
Mitchell, John (a)	13 Nov. 80	3 June 15	No. 7 D. S. Coy.
Addison, Lancelot John (a)	. . . .	*3 June 15*	No. 2 D. S. Coy.
Woodcroft, James	2 Dec. 74	1 July 15	No. 2 D. S. Coy.
Cullen, Douglas Egerton	5 Mar. 83	3 July 15	No. 8 D. S. Coy.

(*a*) Supy. Specially promoted under the provisions of para. 470, A. R., I., Vol. II.

Names.	Date of birth.	Date of rank.	Remarks.

SUPPLY AND TRANSPORT CORPS—AMALGAMATED LIST—*concld.*

Sub-Conductors—*concld.*

Names.	Date of birth.	Date of rank.	Remarks.
Bruce, Herbert	. . . .	22 July 15	No. 7 D. S. Coy.
Purslow, Frederick George	. . . .	13 Oct. 15	"A" S. D. Coy.
Tovey, William	2 Nov. 81	2 Dec. 15	No. 7 D. S. Coy.
Napier, John Francis	15 Jan. 77	2 Dec. 15	No. 9 D. S. Coy.
Ogden, Arthur William	13 Sep. 79	27 Dec. 15	"A" S. D. Coy.
Clarke, Joseph	13 May 81	1 Jan. 16 30 Apl. 16	No. 1 D. S. Coy.
Terry, Charles Edward (*a*)	19 Oct. 80	1 Jan. 16 18 Dec. 17	No. 9 D. S. Coy.
Jennings, Ralph	3 Apl. 81	1 Jan. 16 28 Oct. 18	No. 4 D. S. Coy.
Senior, Ernest Wilford (*a*)	5 Nov. 81	*1 Jan. 16*	No. 9 D. S. Coy.
Grainge, Booth (*a*)	18 Dec. 78	*1 Jan. 16*	No. 4 D. S. Coy.
Woledge, Arnold Robinson (*a*)	6 Dec. 86	*1 Jan. 16*	No. 3 D. S. Coy.
McNamara, Maurice (*a*)	6 Feb. 83	*1 Jan. 16*	No. 3 D. S. Coy.
Garrad, Albert Timothy	. . . .	16 Feb. 16	Bombay Bde.
Mitton, Frederick	20 Feb. 79	15 Mar. 16	No. 8 D. S. Coy.
Jennings, Joseph William	20 Nov. 79	20 Mar. 16	No. 3 D. S. Coy.
Adlam, Ernest Thomas	16 Oct. 78	24 Mar. 16	No. 7 D. S. Coy.
Tingey, John Montague	4 Mar. 77	30 Apl. 16	No. 7 D. S. Coy.
Roderick, Cecil Beresford	26 Nov. 78	30 Apl. 16	No. 1 D. S. Coy.
Booth, Arthur	. . . .	30 Apl. 16	No. 2 D. S. Coy.
Mackey, Frank	16 Nov. 81	30 Apl. 16	No. 3 D. S. Coy.
Hewson, James William	18 July 82	30 Apl. 16	No. 7 D. S. Coy.
Mackey, Charles Edward	16 Nov. 82	30 Apl. 16	"K" S. D. Coy.
Milner, Edwin James	23 Nov. 81	30 Apl. 16	No. 4 D. S. Coy.
Mathews, Francis	23 June 81	10 June 16	No. 1 D. S. Coy.
Balch, Harry George	21 Dec. 81	15 June 16	No. 4 D. S. Coy.
Clarke, Harry Alfred	. . . .	1 July 16 2 Oct. 16	No. 2 D. S. Coy.
Walsh, Edward Thomas (*a*)	. . . .	*1 July 16*	No. 6 D. S. Coy.
Helliwell, Seth (*a*)	. . . .	*1 July 16*	No. 3 D. S. Coy.
Reid, Donald (*a*)	. . . .	*1 July 16*	No. 8 D. S. Coy.
Stickland, Herbert	. . . .	5 July 16	No. 2 D. S. Coy.
Searle, Robert Percival	. . . .	1 Aug. 16	No. 1 D. S. Coy.
Glover, Albert Claude	2 Oct. 80	1 Aug. 16	No. 9 D. S. Coy.
Lea, William Henry	. . . .	10 Sept. 16	No. 2 D. S. Coy.
Bagg, Percy Robert	. . . .	1 Oct. 16	No. 6 D. S. Coy.
Wilks, Cecil Stanley	. . . .	22 Oct. 16	No. 2 D. S. Coy.
Smith, John William	31 July 80	29 Nov. 16	No. 9 D. S. Coy.
Letts, Frederick Charles (*a*)	. . . .	*1 Jan. 17*	No. 8 D. S. Coy.
Dudding, William, D.C.M. (*a*)	. . . .	*1 Jan. 17*	No. 1 D. S. Coy.
Haywood, William Theodore (*a*)	. . . .	*1 Jan. 17*	No. 5 D. S. Coy.
Henshaw, Albert Harold	23 Jan. 80	2 Jan. 17	No. 4 D. S. Coy.
Hempsall, Edgar	. . . .	6 Feb. 17	No. 6 D. S. Coy.
Brown, Walter Hugh	. . . .	13 Feb. 17	No. 3 D. S. Coy.
Hurley, Henry Arthur James	. . . .	24 Feb. 17 30 Sep. 17	Kohat Bde.
Morel, James Arthur Courtnay	. . . .	2 Apl. 17	No. 3 D. S. Coy.
Newman, Eugene William	. . . .	28 May 17	"F" S. D. Coy.
Lawson, William Wilfred	. . . .	6 June 17	"A" S. D. Coy.
Develin, Arthur Archibald	. . . .	23 June 17	"C" S. D. Coy.
Burnett Henry (*a*)	. . . .	*1 July 17*	
Kinghorn, Peter	. . . .	21 July 17	"A" S. D. Coy.
Carter, James (*a*)	. . . .	*30 July 17*	"C" S. D. Coy.
Downes, Ernest Frederick Fussell	. . . .	27 Sep. 17	No. 3 D. S. Coy.
Rogers, Herbert Oswald	. . . .	28 Oct. 17	No. 2 D. S. Coy.
Bennett, Walter	. . . .	31 Oct. 17	"L" S. D. Coy.
Dixie, William James	. . . .	13 Nov. 17	No. 3 D. S. Coy.
Rogers, William Stanford (*a*)	. . . .	*18 Dec. 17*	No. 1 D. S. Coy.
Fullford, Harry	. . . .	27 Feb. 18	No. 6 D. S. Coy.
Pepper, John William	. . . .	24 Mar. 18	No. 7 D. S. Coy.
Simpson, Charles Alexander Ladwick	. . . .	1 Apl. 18	No. 3 D. S. Coy.
Mugford, Percy William	. . . .	1 May 18	No. 3 D. S. Coy.
Williams, Alfred Edward	. . . .	28 May 18	No. 10 D. S. Coy.
Taylor, Alfred Duncan (*a*)	. . . .	*1 July 18*	*No. 6 D. S. Coy.*
Davidson, William Leith ***Lingerton*** (*a*)	. . . .	*1 July 18*	*No. 5 D. S. Coy.*
Currie, Charles James Richard Whatcott	. . . .	2 July 18	No. 2 D. S. Coy.
Davies, Edward John	. . . .	19 July 18	No. 6 D. S. Coy.
Crowther, Charles	. . . .	10 Sep. 18	No. 1 D. S. Coy.
Tanner, Peter (*a*)	. . . .	*10 Sep. 18*	*Ambala Bde.*
Shepherd Sydney (*a*)	. . . .	*23 Oct. 18*	*No. 8 D. S. Coy.*
Cooper, Albert Thomas	. . . .	26 Oct. 18	No. 1 D. S. Coy.
Leicester, Oswald Howard	. . . .	7 Nov. 18	No. 2 D. S. Coy.

(*a*) **Supy. Specially promoted under the provisions of para. 470, A. R., I., Vol. II.**

Names.	Date of birth.	Date of rank.	Remarks.

DEPARTMENTAL WARRANT OFFICERS TEMPORARILY EMPLOYED FROM THE RETIRED LIST.

Supply and Transport Corps.

Acott, Sub-Conductor H. T. . .	. . . . .	. . . . .	"F" S. D. Coy.
D'Auvergne, Sub-Conductor V. .	. . . . .	. . . . .	
Hunter, Conductor T. J. . . .	. . . . .	. . . . .	No. 7 D. S. Coy.

MILITARY WORKS SERVICES AND PUBLIC WORKS DEPARTMENT, INDIA.

Conductors—24.

⋮	⋮	⋮	⋮
Bayliss, Charles Edward	15 Oct. 66	11 Oct. 10	Sub-Engr., 2nd gr., P. W. D. (temply. with M. W. S., Peshawar).
Sanderson, George Wm. . . .	1 Feb. 65	1 June 11	Sub-Engr., 2nd gr., M. W., Kirkee.
Sullivan, Thomas Edward . . .	8 May 70	7 Nov. 11	Sub-Engr., 3rd gr., M. W.
Ryan, Patrick	6 Mar. 70	16 Nov. 11	Sub-Engr., 2nd gr., M. W.—Offg. Gn. Engr., Ferozepore.
Mitchell, Arthur James (172*a*, 207) .	23 May 69	12 May 12	Sub-Engr., 2nd gr., M. W., Mhow.
Sheath, Frederick George	*5 Jan. 64*	*23 Sep. 12*	*Press Room Foreman, Govt. Press, Allahabad.*
Mackenzie, Thomas	16 Jan. 67	23 Sep. 12	Sub-Engr., 2nd gr., M. W., Jubbulpore.
Smith, James Norman	10 Feb. 73	14 June 13	Sub-Engr., 2nd gr., M. W., Quetta.
Clapton, Oliver	29 Aug. 67	1 Apl. 14	Sub-Engr., 3rd gr., M. W., Ahmednagar.
King, William Henry	*1 June 30*	*20 Aug. 14*	*2nd S. & M.*
Rich, Harry William Wallace . .	14 Sep. 70	25 May 15	Sub-Engr., 1st gr., P.W.D., Bombay (*temply. with M. W. S., Risalpur*).
Bailey, William	27 July 66	9 July 15	Sub-Engr., 3rd gr., M. W., Chakrata.
Cunningham, Charles James . .	1 Jan. 65	26 Aug. 15	Sub-Engr., 3rd gr., P. W. D., Bombay, Dharwar Dist.
Innes, Henry Walter	8 Aug. 72	27 Aug. 15	Sub-Engr., 3rd gr., M. W., Secunderabad.
Mathew, John Drysdale	3 Feb. 67	6 Oct. 15	Sub.-Engr., 3rd gr., U. P., P. W. D., (*temp. with M. W. S.*)
Quigley, Joseph	21 Dec. 72	1 Jan. 16	Sub-Engr., 3rd gr., M. W.
Brown, William James	30 Apl. 74	30 Apl. 16	Sub-Engr., 3rd gr., (*s. p. t.* 2nd gr.), P. W. D., Bombay. (*supy.*) (*Temply. with M. W. S.*)
Inward, Arthur Albert	18 Oct. 71	27 June 16	Sub.-Engr., 3rd gr., M. W., Aden.
Everard, Richard (a)	*25 Mar. 74*	*1 July 16*	*Supvr., 2nd gr., M. W.*
Harvey, Albert	27 Feb. 71	28 Oct 16	Sub-Engr., 3rd gr., M. W., Quetta.
Stone, John Philip (a)	*26 Dec. 72*	*1 July 17*	*Sub Engr., 3rd gr., M. W., Poona.*
Allaway, George Arthur	20 Sep. 70	22 Dec. 17	Sub-Engr., 3rd gr., Peshawar.
Rees, Richard	21 May 64	3 Jan. 18	Sub-Engr., 3rd gr., M.W., Jhansi.
Harton, John Gower	3 July 70	28 Feb. 18	Supvr., 1st gr., Bengal P. W. D. (*temply. with M. W. S., Cawnpore*).
Eltham, James Frederick	. . . .	*28 May 18*	*1st S. & M.*

Sub-Conductors—41.

Hewson, Fredrick Munford . . .	5 Dec. 75	18 July 10	Sub-Engr., 3rd gr., M. W., Agra.
Forsyth, Douglas George . . .	26 Dec. 73	16 July 11	Supvr., 1st gr., M. W.
Jameson, Alexander Neville . .	12 May 79	1 Nov. 11	Supvr., 1st gr., M. W.
Cuerden, Charles	1 Apl. 78	6 Nov. 11	Supvr., 1st gr., M. W., Quetta.
Lansley, George Edward	*12 Oct. 69*	*7 Nov. 11*	*P. A. to the Principal, Thomason College, [Rurki.*
O'Connor, Gerald Butler . . .	16 June 75	16 Nov. 11	Supvr., 1st gr., M. W., Meerut.
Briant, Robert Joseph	1 July 76	16 July 12	Supvr., 1st gr., M. W., Bangalore.
Casey, Cornlius O'Connor . . .	31 May 75	16 July 12	Supvr., 1st gr., M. W., Meerut.
Lock, John	20 Mar. 71	17 July 12	Supvr., 2nd gr., M. W., Nowshera.
Garrity, William Henry	10 Apl. 71	23 Sep. 12	Supvr., 1st gr., M. W., Jhansi.
Holmes, George Henry	9 Mar. 79	14 June 13	Supvr., 1st gr., M. W.
Woodward, Martin	14 Jan. 75	5 July 13	Supvr., 2nd gr., M. W., Aden. No. 6 E. F. [Park.
Tripp, William Herbert	*1 Jan. 76*	*16 July 13*	*2nd Asst. Master, Thomason College, Rurki.*
Cuerden, Herbert	11 Mar. 80	16 July 13	Supvr., 1st gr., M. W., Dehra Dun.
Rooks, Reuben	7 Aug. 78	13 Mar. 14	Sub-Engr., 1st gr. (*Bombay P. W. D.*) (*Temply. with M.W.S. Ahmedabad.*)
Brown, George Joseph Flowerdew .	16 July 79	1 Apr. 14	Supvr., 2nd gr., M. W.
Wright, Frederick	8 Dec. 72	13 May 14	Supvr., 1st gr., M. W., Quetta.
Burns, Owen Harcourt	11 Oct. 75	1 Sep. 14	Supvr., 1st gr., M. W., Ranikhet.
Foley, Maurice	17 May 77	4 Jan. 15	P. W. Inspr., N. W. Ry.
Barwise, Albert Harold Osborne (a) .	*11 May 71*	*3 June 15*	Ovsr., 1st gr., M. W., Peshawar.
Tovell, Albert Frederick	22 Mar. 76	11 June 15	Supvr., 1st gr., (*tempy.*) M. W. Jullundur.

NOTE.—The names shown in italics are those of warrant officers supernumerary to the establishment of the P. W. D. for extra-departmental positions, as stated opposite their names.

(*a*) Supy. Specially promoted under the provisions of para. 470, A. R., I., Vol. II.

Names.	Date of birth.	Date of rank.	Remarks.

MILITARY WORKS SERVICES AND PUBLIC WORKS DEPARTMENT, INDIA—*concld.*

Sub-Conductors—*concld.*

Names.	Date of birth.	Date of rank.	Remarks.
Wilkins, Chas. Edward William (b)	*16 Jan. 73*	*9 July 15*	*Offg. 1st Asst. Master, Thomason College, Rurki. Lv. up to 15 Apl. 18.*
Wilson, Robert	2 Dec. 68	26 Aug. 15	Supvr., 2nd gr., M. W.
Wilkinson, Herbert Eli	13 Mar. 79	6 Oct. 15	Sub-Engr., 2nd gr. *Bom. P. W. D.*, Poona.
McNevin, Frederick Lionel	9 Nov. 80	20 Nov. 15	Sub-Engr., 2nd gr., Bombay. P. W. D. (*Tempy. with M. W. S., Shahjehanpur.*)
Stenson, William	13 Sep. 79	20 Nov. 15	Supvr., 1st gr., Central Provinces. (*Temply. with M. W. S., Jubbulpur.*)
Jackson, George	6 Aug. 77	20 Nov. 15	Sub-Engr., Punjab. (*Temply. with M. W. S., Aden.*)
Baxter, George Frederick	18 Dec. 79	1 Jan. 16	Supvr., 2nd gr., M. W., Peshawar.
Kearsley, William (a)	*1 Sep. 79*	*1 Jan. 16*	*Ovsr., 1st gr., M. W. S.*
Arnall, Francis Thomas	19 July 74	11 Feb. 16	Permanent Way Inspr., N. W. Ry.
Ball, Francis Charles	7 Oct. 79	30 Apl. 16	P. W. D. Bombay and Sub-Engr., 3rd gr., (*2nd spt.*), Karachi.
Haywood, William George	23 July 79	27 June 16	Supvr., 2nd gr., M. W., Cawnpore.
Young, Robert Vernon	1 June 82	16 July 16	Supvr., 2nd gr., M. W., Quetta.
Beckett, Daniell	7 Sep 73	16 July 16	Supvr., 2nd gr., M. W., Ranikhet.
Kilby, Hugh James	27 June 78	9 Nov. 16	Supvr., 2nd gr., M. W., Ferozepore.
Lys, Arthur James (a)	*29 Apl. 87*	*1 July 17*	*Ovsr., 1st gr., M. W., Jhelum.*
Dewberry, Albert	26 June 79	15 Dec. 17	P. W. Inspr., N. W. Ry.
Lettington, Arthur William	26 Jan. 82	22 Dec. 17	Supvr., 2nd gr., M. W.
Leonard, Percy Darling	24 July 69	3 Jan. 18	Supvr., 2nd gr., M. W., Meerut.
White, Alfred Louis Percy	27 Feb. 81	28 Feb. 18	Supvr., 2nd gr., P. W. D. Bengal (*temply. with M. W. S., Secunderabad*).
Chapman, Frank Edward	28 Dec. 81	18 Mar. 18	Supvr., 2nd gr., P. W. D., Bengal. *Lv., m. c., in I., 6 mths, 26 Aug. 18.*
Holden, Sidney	15 Jan. 79	17 May 18	Supvr., 2nd gr., M. W.
Goode, Robert Adolphus	1 July 18		
Cullen, James	1 July 18		

Temporary Sub-Conductor.

Names.	Date of birth.	Date of rank.	Remarks.
Gurney, Stratton	26 June 76	. . .	Lecturer in Civil Engineering, Thomason College, Rurki.

DEPARTMENTAL WARRANT OFFICER TEMPORARILY RE-EMPLOYED FROM THE RETIRED LIST. MILITARY WORKS SERVICES.

Perry, Condr. J. (Ordnance Dept.)

PUBLIC WORKS DEPARTMENT, MADRAS.

Conductors—7.

Names.	Date of birth.	Date of rank.	Remarks.
Mahony, Joseph Stanislaus	18 Aug. 69	14 May 07	Sub-Engr., 2nd gr., *provl.* (*temply. with M. W. S.*)
Furnell, Henry Adam	26 Nov. 66	10 Sep. 08	Sub-Engr., 5th gr., (*Prov. P*) Kistna Eastern Dn., II Circle.
Sheridan, Robert Stanislaus	1 Mar. 70	8 July 11	Supvr., 1st gr., Sub-Engr., 6th gr., T R. Godavary Northern Divn., I. Circle.
Whittaker, James	29 Oct. 71	2 Oct. 11	Sub-Engr. 4th gr., *Seconded* (*temply. with M. W. S.*)
England, Philip Henton	5 Oct. 68	15 Jan. 13	Sub-Engr., 3rd gr. Edn. Dept.
Lisles, Richard Albert	26 Jan. 74	1 Jan. 17	Sub-Engr., 3rd gr., *tempy Seconded* (*temply. with M. W. S.*)
Thomas, John Henry	5 Apl. 72	7 Aug. 18	Sub-Engr., VI gr., (*Provl. P.*) *Lv. C.L., 3 mths, 18 Oct. 18.*

Sub-Conductors.

Names.	Date of birth.	Date of rank.	Remarks.
Gwynn, Frank Eykyn	10 Apl. 73	14 May 07	Sub-Engr., 5th gr. *Lv., m. c., 2 yrs., 14 Apl. 17.*
Wilson, John William	20 Oct. 72	10 Sep. 08	Supr., 2nd gr., and tempy. Sub-Engr., Vizigapatam Dn., I Circle.
Baker, Alfred Farmer	28 Dec. 78	10 Sep. 08	Sub-Engr., 4th gr., tempy. (*temply. with M. W.S.*)
Ray, Charles Basil	6 Mar. 75	8 July 11	Sub-Engr., 5th gr., *Provl.* (*temply. with M. W. S.*)
Grant, Thomas	28 Sep. 79	15 Jan. 13	Sub-Engr., 5th gr. (*Prov. P*) *Seconded* (*temply with M W. S.*)
Taylor, Henry Skeuce	20 Aug. 82	27 Sep. 13	Sub-Engr., 5th gr., *provl. Seconded* (*temply. with M. W. S.*).

(*a*) Supy. Specially promoted under the provisions of para. 470, A. R., I., Vol. II.
(*b*) Supy. *vide* A. D. Notification No. 963, dated 1st October 1915.

Names.	Date of birth.	Date of rank.	Remarks.

PUBLIC WORKS DEPARTMENT, BOMBAY.

Conductors.

Names.	Date of birth.	Date of rank.	Remarks.
Naughton, Thomas Henry	31 Aug. 75	16 July 15	Sub-Engr., 1st gr. *(Temply. with M. W. S. Kohat.)*
Dobson, William Ewart	4 Dec. 74	16 July 17	Sub-Engr., 1st gr., Lake Whiting Dist.

BARRACK DEPARTMENT, PUNJAB—BENGAL.

Conductors—12.

Names.	Date of birth.	Date of rank.	Remarks.
Cant, Henry Douglas	27 Oct. 65	23 June 13	B. M., M.W., Wellington.
Fisher, George	19 Sep. 67	29 Nov. 13	B. M., M. W., Bangalore.
Busby, Thomas (c)	*29 May 62*	*2 Nov. 14*	*(Supy.) B. M., M. W., Mhow.*
Flight, Edgar	2 Nov. 68	13 Nov. 14	B. M., M. W., Peshawar.
Tobin, Alfred Edward (105)	9 Nov. 68	24 Nov. 14	B. M., M. W., Lahore Cant.
Parr, Thomas William	4 Dec. 65	20 Sep. 15	B. M., M. W., Jhansi.
Bristow, Henry James	7 June 68	8 Feb. 16	B. M., M. W., Fort William.
Fox, George Corbett	17 Oct. 72	24 Oct. 16	A. B. M., M. W., Jubbulpore.
Holehouse, Walter	20 Mar. 73	22 May 17	A. B. M., M. W., Barrackpore.
Stanfield, Thomas Edward (106, 207)	29 Aug. 66	29 May 17	A. B. M., M. W., Sialkot.
Liddy, John	16 Apl. 68	1 Sep. 17	A. B. M., M. W., Murree.
Willis, William	14 Feb. 64	30 Dec. 17	A. B. M., M. W., Fyzabad.
Tilley, Frank Lewis	3 Oct. 67	4 Aug. 18	B. M., M. W., Jullundur.

Sub-Conductors—21.

Names.	Date of birth.	Date of rank.	Remarks.
Oliver, Frederick	22 Aug. 66	1 Mar. 07	A. B. M., M. W., Meerut.
Gregory, Thomas Moore	4 Dec. 66	23 Dec. 11	A. B. M., M. W., Peshawar.
Nunn, Charles Arthur (b)	*6 June 78*	*17 Mar. 12*	*Supy. A. B. M., M. W., Khyra Gali.*
Knight, Frank Percival	3 Apl. 75	19 Mar. 12	A. B. M., M. W., Multan Cant.
Marsh, Thomas	8 May 66	3 July 12	A. B. M., M. W., Aden.
Shevlin, William	27 Jan. 73	16 Sep. 12	A. B. M., M. W., Ambala.
Montrose, Ernest	7 May 70	23 June 13	A. B. M., M. W., *Lv., m.c., 6 mths. ex. I., 29 May 18.*
Tytler, Samuel	17 Nov. 75	23 Oct. 13	A. B. M., M. W., Karachi.
Tyrrell, Robert Christopher	23 Feb. 80	8 Feb. 14	A. B. M., M. W., Madras.
Carroll, Thomas	25 July 72	24 Nov. 14	A. B. M., M. W., Deolali.
Paull, John Booden (b)	*1 July 77*	*3 June 15*	*(Supy.) A. B. M., M. W.,* Ahmednagar.
Runcorn, John Edwin	24 Feb. 78	20 Sep. 15	A. B. M., M. W., Chakrata.
Netherwood, Robert George	13 Feb. 74	15 Oct. 15	A. B. M., M. W., Secunderabad.
Harvey, Albert Frederick	6 Oct. 85	16 Dec. 15	A. B. M., M. W., Poona.
Orridge, Frederick Francis	16 Nov. 76	20 Dec. 15	A. B. M., M. W., Bombay.
Watson, Sidney John	19 July 74	8 Feb. 16	A. B. M., M. W., Delhi.
Watson, William George	4 Feb. 78	28 June 16	A. B. M., M. W., Jalapahar.
Bloor, John (b)	*30 Oct. 83*	*1 July 16*	*(Supy.) A. B. M., M. W., Karachi.*
Whitman, Harry	26 Nov. 74	24 Oct. 16	A. B. M., M. W., Poona.
Rogers, James (b)	*27 July 78*	*1 Jan. 17*	*A. B. M., M. W.*
Borrett, James Bryant (b)	*13 Dec. 80*	*1 Jan. 17*	*A. B. M., M. W., Lucknow.*
Walke, William Francis	8 Oct. 75	22 May 17	A. B. M., M. W., Lahore Cant.
Webb, Charles Edward	5 Jan. 77	29 May 17	A. B. M., M. W.
Stobie, David (b)	*2 July 78*	*1 July 17*	*A. B. M., M. W.,* Bangalore.
McKew, Percy John (b)	*23 Jan. 76*	*1 July 17*	*A. B. M., M. W., Bombay.*
Gingell, George Henry	25 Jan. 71	1 Sep. 17	A. B. M., M. W., Dalhousie.
Kite, Edward Frank	15 June 76	30 Dec. 17	Burma. P. W. D.
Corner, Arthur Hugh Sedgely	21 June 71	4 Aug. 18	A. B. M., M. W., Secunderabad.

(*b*) Supy. Specially promoted under the provisions of para. 470, A. R., I., Vol. II.
(*c*) Supy. on attaining the age of 55 years.

Names.	Date of birth.	Date of rank.	Remarks.

ARMY CLOTHING DEPARTMENT.

Under the control of the Indian Munitions Board.

Conductor—1.

Ventham, William George	17 Sep. 72	15 Jan. 14	Boot Depôt, Cawnpore.

Sub-Conductors—3.

Daglish, George	1 Jan. 74	5 Mar. 12	Clothing Contracts Depôt, Delhi.
Alderton, John Robert	22 Feb. 79	2 Aug. 13	Army Clothing Facty., Shahjahanpur.
Mackay, Robert	8 Oct. 75	15 Jan. 14	Army Clothing Fy., Karachi.

POSTS AND TELEGRAPHS.

Conductors.

Newington, Francis Ernest	7 Dec. 64	15 Nov. 12	Tel. Master.
Rollins, William	28 June 64	1 June 13	Tel. Master.
Bailey, William	1 Aug. 66	19 Apl. 18	Telegraphist.
Hughes, Thomas	31 Mar. 70	23 Sep. 18	Telegraphist.

Sub-Conductors.

Coady, John William	22 Oct. 71	26 Jan. 15	Telegraphist.

INDO-EUROPEAN TELEGRAPH DEPARTMENT, PERSIA.

Conductor.

Stephens, G. H.	13 May 77	1 Aug. 14	

Sub-Conductor.

Peters, N. M.	19 Jan. 75	1 Aug. 14	*Lv., 8 mths., 22 Dec. 17.*

INDIA MISCELLANEOUS LIST.

Conductors—35.

Peel, Arthur William	8 Jan. 70	1 Jan. 08	Army Hdqrs. (G. S. Bch.).
Hopper, George	14 Jan. 67	19 Nov. 08	Mily. Farms Dept.
Wright, John	20 Oct. 68	31 Jan. 11	Mily. Prison and Det. Bk. Staff, Sialkot.
Tucker, Charles	18 Aug. 64	1 May 12	Army Remount Dept.
Cloy, Colin	13 Oct. 72	4 May 12	Army Remount Dept.
Fox, James	28 Oct. 78	1 Apl. 13	Army Hdqrs. (Q. M. G.'s Bch.).
Stratton, John George	5 Oct. 72	2 Nov. 13	Army Hdqrs. (Q. M. G.'s Bch.).
Newton, Lewis	20 Nov. 71	7 Mar. 14	Army Rem't. Deptt.
Kenaelly, William	20 Jan. 76	7 Nov. 14	Indian Munitions Board (*Temply.*)
Farrell, Robert (a)	*16 Dec. 76*	*3 June 15*	
Kenny, James Leonard	14 Apl. 79	3 June 15 31 Oct. 16	Army Staff Office, S. Army.
Hurst, William	30 Sep. 76	3 June 15 29 Oct. 18	Divl. Staff Office, 6th (Poona) Dn.
Bryce, John (a)	*30 Sep. 74*	*3 June 15*	Bombay Bde. *Staff Office, 6th* (*Poona*) *Dn.*
Rosser, John Campbell	22 May 77	1 Nov. 15	Army Hdqrs. (Q. M. G.'s Bch.). *Lv., ex I, m.c., 6 mths.*
Hill, Thomas William (a)	*22 Mar. 73*	*1 Jan. 16*	*Divl. Staff Office, 5th* (*Mhow*) *Dn.*
Waugh, Robert Travers (a)	*24 Jan. 80*	*1 Jan. 16*	*Army Hdqrs.* (*A. G.'s Bch.*)
Erswell, Harry Walter (a)	*1 Mar. 80*	*1 Jan. 16*	*Army Hdqrs.* (*Ord. Bch.*).

(*a*) Supy. Specially promoted under the provisions of para. 470, A. R., I., Vol. II.

INDIA MISCELLANEOUS LIST—*contd.*

Conductors—35—*contd.*

Names.	Date of birth.	Date of rank.	Remarks.
Coombs, Mosdels Heriot (a)	*1 July 80*	*1 Mar. 16*	*A. G.'s Bch., Record Section, Bombay.*
Smith, Walter (a)	*27 Feb. 79*	*1 July 16*	*Survey of Deptt.*
Mackie, Robert (a)	*21 Aug. 76*	*1 July 16*	*Army Staff Office, S. Army.*
Evans, George (a)	*1 Dec. 81*	*1 July 16*	*Army Hdqrs. (Mily. Secy.'s Bch.).*
Chapman, Arthur (a)	*8 July 74*	*1 July 16*	*Temply. empld. with Nepalese contingent.*
Chandler, Frank	15 Oct. 74	15 Aug. 16	Mily. Farms Depôt.
Baiey, George William Bryant (a)	*14 Dec. 74*	*1 Jan. 17*	*Dvl. Staff Office, 9th Dn.*
Robertson, William (a)	*18 June 66*	*1 Jan. 17*	*Sch. of Musky, Pachmari.*
Hill, William	18 Oct. 68	15 July 17	Mily. Prison & Det. Bk. Staff.
Flowerdew, John Lionel (a)	*17 Oct. 72*	*18 Feb. 18*	*Mily. Farms Dept.*
Harrison, William	21 Apl. 75	8 Feb. 18	Army Hdqrs. (G. S. Bch.).
Atkinson, Stephen	28 July 78	1 July 18	*Northern Commd. Office.*
Richmond, Reginald Hamilton	20 Mar. 75	23 Oct. 18	Army Hdqrs. (Q. M. G.'s Bch.).

Sub-Conductors—62.

Names.	Date of birth.	Date of rank.	Remarks.
Lovick, Ronald Douglas	13 July 81	1 May 10	Mily. Farms Dept.
Lynch, Cyril Mark	15 Sep. 75	1 May 10	Mily. Farms Dept.
Weston, William	20 Jan. 75	1 May 10	Army Hdqrs. (Q. M. G.'s Bch.).
Weeks, Alfred Henry	28 Nov. 74	1 May 10	Army Hdqrs. (Mily. Secy.'s Bch.).
Smith, Henry William	1 Apl. 79	18 Oct. 10	Army Rem't. Deptt.
FitzGerald, Gerald	6 June 78	20 Oct 10	Army Hdqrs. (Q. M. G.'s Bch.).
Bartrop, Charles Gordon	15 July 73	31 Jan. 11	Army Hdqrs. (G. S. Bch.).
Pluck, Harry	29 Dec. 73	6 Sep. 11	Army Hdqrs. (Q. M. G.'s Bch.).
Goodwin, Thomas	12 Feb. 77	6 Oct. 11	Army Rem't. Deptt.
Hartnett, George Washington (95, 170, 193).	10 Apl. 78	5 Nov. 11	Divl. Staff Office, 2nd (Rawalpindi) Dn.
Walker, Herbert	4 Nov. 76	17 Mar. 12 18 Sep. 13	*Southern Commd. Office.*
Greensmith, Joe	29 Jan. 79	1 Apl. 12	Army Hdqrs. (Medl. Bch.).
Baxter, William	15 Mar. 65	22 July 12	Poona Mily. Prison.
Strudwick, George (94, 169, 247)	25 Apl. 77	30 Sep. 12	Mily. Farms Dept.
Draper, Francis Hastings	14 Feb. 71	9 Dec. 12	Mily. Farms Dept.
Trewhella, Matthew	5 June 72	1 Feb. 13	Mily. Farms Dept.
Hogg, Francis Arthur	12 Dec 80	1 Mar. 14	Army Hdqrs. (G.S. Bch.).
Wilson, George	2 Jan. 79	7 Mar. 14	Indian Cadet College, Indore.
Nelson, Wilfrid Clarance	10 Jan. 77	7 Mar. 14	Army Rem't. Deptt.
Day, Percival Edward	16 Dec. 71	7 Mar. 14	Mily. Prison and Det. Bk. Staff.
Wells, Horatio Joseph	18 July 77	17 Mar. 14	Mily. Farms Dept. *Lv., ex I., 12 mos., m. c.*
Byrne, Frederick	25 Sep. 76	1 Apl. 14	Mily. Farms Dept.
Hill, Arthur	28 May 78	1 Apl. 14	Mily. Farms Dept.
Pownall, John Fletcher	17 Jan. 74	11 Apl. 14	Mily. Farms Dept.
Green, William Joseph	29 June 77	16 Apl. 14	Mily. Farms Dept.
Sidney, Charles Stephen	13 Mar. 75	18 Apl. 14	Prisoners' War Camp, Thayetmyo.
Lewis, William Thomas	26 Jan. 73	1 July 14	Army Hdqrs. (A. G.'s Bch.).
Pe ley, John	22 Feb. 82	5 Oct. 14	Army Rem't. Deptt.
Powell, Henry Douglas, D.C.M.	11 May 72	16 Nov. 14	Mily. Farms Dept.
Hirtes, Albert Ernest (64) (a)	*17 Apl. 73*	*3 June 15*	*Indian Sub. Vety. Corps.*

(a) Supy. Specially promoted under the provisions of para. 470, A. R., I., Vol. II.

INDIA MISCELLANEOUS LIST—*concld.*

Sub-Conductors—*concld.*

Names.	Date of birth.	Date of rank.	Remarks.
Richards, William John (95, 170) (a)	*1 June 74*	*3 June 15*	*Embkn. Staff, Madras.*
Smith, Edward Whitworth (a)	*3 Nov. 77*	*3 June 15*	*Army Hdqrs. (Medl. Br.)*
Whittick, Ernest George (a)	*12 Feb. 79*	*3 June 15*	*Mily. Farms Dept.*
Macdonald, Hector Ian (a)	*9 Aug. 85*	*3 June 15*	*Army Dept., Secretariat.*
Gardner, John William Bertrand (a)	*13 Jan. 86*	*3 June 15*	*Army Hdqrs. (Q. M. G.'s Bch.)*
Clift, Edgar (a)	*23 Dec. 81*	*3 June 15*	*Army Hdqrs. (A. G.'s Bch.)*
Reilly, James Henry Joseph (a)	*24 Sep. 78*	*3 June 15*	*Divl. Staff Office (1st Peshawar Dn.)*
Wetton, John Lewis (a)	*29 July 80*	*3 June 15*	*Army Hdqrs. (M. S. Br.)*
Pay, Edward (a)	*26 Sep. 82*	*3 June 15*	*Army Hdqrs. (A. G.'s Br.)*
Morris, Wilfred (a)	*2 Dec. 79*	*3 June 15*	*Divl. Staff Office, 2nd Rawalpindi Dn.*
Harris, George William (a)	*7 June 78*	*3 June 15*	*Poona Bde. Office.*
Fry, Fredrick Arthur (a)	*17 Feb. 83*	*3 June 15*	*Army Hdqrs. (Medl. Bch.)*
Gilling, Henry Tertius	20 Jan. 72	1 Nov. 15	Mily. Farms Dept.
Wardell, Charles (a)	*19 Oct. 78*	*1 Jan. 16*	*Ambala Bde. Staff Office.*
Court, Edgar Maurice	10 Nov. 76	1 Jan. 16	Mily. Farms Dept.
Jackson, Saville Britian (a)	*7 Nov. 81*	*1 Jan. 16*	*Army Hdqrs., A. G.'s Bch.*
Sisley, James Lewis (a)	*19 Sep. 83*	*1 Jan. 16*	*Army Hdqrs., Medl. Bch.*
Hewitt, William Robert (a)	*30 Jan. 88*	*1 Jan. 16*	*Army Hdqrs., Q. M. G.'s Bch.*
Terry, Thomas (a)	*4 Sep. 78*	*1 Jan. 16*	*Army Hdqrs., Q. M. G.'s Bch.*
O'Neill, George Philip (a)	*14 Feb. 81*	*1 Jan. 16*	*Divl. Staff Office, 9th (Sec'bad) Dn.*
McDermott, Brian James (a)	*10 Feb. 91*	*1 Jan. 16*	*Army Hdqrs, G. S. Bch.*
Hubert, John (95, 172a) (a)	*16 Dec. 79*	*1 Jan. 16*	*Divl. Staff Office, 4th Quetta Dn.*
Keiger, Frederick Gustave	3 Oct. 76	1 Feb. 16	Mily. Farms Dept.
Smyth, Richard (95, 162, 169, 206, 165). (a)	*9 Dec. 80*	*1 July 16*	*Dera Ismail Khan.*
Walsh, James Edward (a)	*18 June 86*	*1 July 16*	*Army Hdqrs., M. W. Bch.*
Watson, James Horton (95, 172a) (a)	*23 June 75*	*1 July 16*	*Karachi Bde.*
McCarthy, William Thomas (170)	16 Mar. 78	15 Aug. 16	Army Hdqrs., A. G.'s Bch.
Wyatt, Alexander	1 Sep. 77	31 Oct. 16	Army Hdqrs., A. G.'s Bch.
Maude, Harold (a)	*30 May 79*	*1 Jan. 17*	*Embkn. Staff, Karachi.*
Redding, James	14 Jan. 77	1 Jan. 17	Army Rem't. Deptt.
Somerfield, George James (a)	*21 Mar. 82*	*1 Jan. 17*	*Divl. Staff Office, Meerut Dn.*
Howey, John Robe (a)	*28 Feb. 86*	*1 Jan. 17*	*Army Hdqrs. (Q. M. G.'s Bch.)*
Boyd, Robert William (a)	*12 Mar. 75*	*1 Jan. 17*	*Meerut Cavy. Bde. Office.*
Freeman, William Elgy (a)	*8 Dec. 80*	*1 Jan. 17*	*Army Hdqrs. (Q. M. G.'s Bch.)*
Walsh, Hector (a)	*9 Apl. 84*	*1 Jan. 17*	*Army Hdqrs. (Ord. Bch.)*
Spires, James (a)	*20 Oct. 82*	*1 July 17*	*Army Hdqrs. (A. G.'s Bch.)*
Roberts, Charles (a)	*3 Feb. 72*	*1 July 17*	*I. S. Vety. Corps.*
Allen, Samuel William Joseph	25 July 76	15 July 17	Army Rem't. Deptt.
Walsh, Thomas Edward	20 Nov. 76	14 Jan. 18	Army Rem't. Deptt.
Ambrose, Arthur	22 Dec. 83	28 Jan. 18	Army Hdqrs.
Addinall, Frederick William	. . . .	1 July 18	
Crane, Walter James (a)	. . . .	*1 July 18*	
Ellis, Alfred Edward	20 Jan. 78	23 Oct. 18	Army Hdqrs. (G. S. Bch.)
Cook, Henry Charles	. . . .	30 Oct. 18	Army Rem't. Deptt.

(a) **Supy.** Specially promoted under the provisions of para. 470, A. R., I., Vol. II.

NON-DEPARTMENTAL WARRANT OFFICERS.

Names.	Date of birth.	Date of rank.	Rank and appointment.
Field, Frederick	24 Aug. 64	29 July 03	Sergt.-Maj., Army Vety. Sch., Poona.
Fitzgerald, John	30 Mar. 67	5 Nov. 03	Sergt.-Maj., Sch. of Cookery, Poona.
Baird, Frederick	1 Feb. 75	22 June 10	Sergt.-Maj., Sch. of Musky., Changla Gali.
Hill, John	27 May 80	6 Oct. 13	Sergt.-Maj., Cavy. Sch., Saugor.
Johnstone, John	15 Dec. 82	19 Mar. 15	Garrison Sergt.-Maj., Fort William.
Jeffs, William	9 Feb. 76	11 Oct. 15	Transport Sergt.-Maj., Jhansi. (*Temply. doing duty at Pachmari.*)
Ford, Richard	30 Dec. 70	1 Dec. 15	Farrier Sergt.-Maj., I. S. V. C.
Collins, George	17 June 83	29 Feb. 16	Sergt.-Maj., Sch. of Musky., Satara.
Rans, Charles Alexander	26 Nov. 71	23 Oct. 16	Attd. 3rd Gn. Bn. Bedf. R.
Sutcliffe, Wilson	6 June 81	16 Dec. 16	Sergt.-Maj., Cadet Coll., Quetta.
Kaye, George Fennell	1 Feb 73	19 Oct. 17	Transport Sergt.-Maj., Bombay.
Clarke, Frederick Richard	24 May 80	20 Dec. 17	Transport Sergt.-Maj., Bombay.
Ball, Frederick George	3 Sep. 77	15 Aug. 18	Transport Sergt.-Maj., Fort William.
Barker, James William	15 Jan. 83	13 Nov. 18	Garrison Sergt.-Maj., Allahabad.
Palmer, Albert Victor	....	1 July 18	Farrier Qrmr. Sergt.

OFFICERS IN CIVIL EMPLOYMENT.

Names, Rank and Corps.	Date of first appointment to Civil Dept.	Remarks.
SURVEY OF INDIA.		
Surveyor General of India.		
Burrard, Col. *Sir* S. G., K.C.S.I., F.R.S., Brit. Ser.	2 Sep. 84	
Superintendents.		
Renny-Tailyour, Col. T. F. B., C.S.I., Brit Ser.	1 Apl. 88	Supdt., Southern Circle.
Lenox-Conyngham, Col. G. P., F.R.S., Brit. Ser.	16 July 89	Supdt., Trigonometrical Survey.
Crichton, Lt.-Col. R. T., C.I.E. (Supy. List.)	16 Dec. 89	Supdt., Northern Circle.
Ryder, Col. C. H. D., C.I.E., D.S.O., Brit. Ser. [*l*]	29 Nov. 91	Dy. Dir. of Surveys.
Coldstream, Lt.-Col. W. M., R.E.	25 Nov. 94	Temply. reverted to Mily. duty.
Deputy Superintendents.		
Robertson. Lt.-Col. C. L., C.M.G., R.E.	3 May 94	Offg. Supdt., Eastern Circle.
Pirrie, Bt.-Col. F. W., C.M.G., C.I.E (Supy. List)	11 June 95	*Lv., ex I., 10 mos. 12 dys., 1 May 18.*
Turner, Lt.-Col. H. H., R.E.	22 Apl. 97	Offg. Supdt., Map Publication.
Beazeley, Lt.-Col. G. A., D.S.O., R.E.	12 Oct. 97	
Crosthwait, Lt.-Col. H. L., R.E.	22 Nov. 97	
Tandy, Maj. E. A., R.E.	31 Jan. 98	Offg. Supdt.
Mears, Lt.-Col. A., [L] (Supy. List)	21 Nov. 97	In chge. Surveyor General's Office, Calcutta.
Wood, Bt. Lt.-Col. H., R.E.	16 Nov. 98	Temply. reverted to Mily duty.
Cowie, Maj. H. McC., R.E.	12 Nov. 98	In chge. Nos. 13, 14 Parties, and Computing Office, Dehra-Dun.
Gunter, Bt. Lt-.Col., C. P., R.E.	10 July 99	
Rich, Maj. E. T., C.I.E., R.E.	3 July 99	
McHarg, Bt. Lt.-Col. A. A., D.S.O., R.E. [*l*]	9 Jan. 00	In chge. No. 19 Party.
Tandy, Maj. M. O'C., D.S.O., R.E.	17 Jan. 00	Temply. reverted to Mily. duty.
Thomas, Maj. R. H., R.E.	17 July 01	Temply. reverted to Mily. duty.
Thuillier, Maj. L. C. (Supy. List.)	12 Sep. 01	
Browne, Bt. Lt.-Col. (*Ag. Lt.-Col.*) C. M., C.M.G., D.S.O., R.E.	8 Jan. 03	Temply. reverted to Mily. duty.
Phillimore, R. H., D.S.O., R.E.	28 June 03	Temply. reverted to Mily. duty.
MacLeod, Maj. M. N., D.S.O., M.C., R.E.	25 Oct. 04	Temply. reverted to Mily. duty.
Hunter, Maj. (*temp. Lt.-Col.*) F. F., D.S.O.	17 Aug. 05	Temply. reverted to Mily. duty.
Crosthwait, Maj. L. G.	13 Oct. 05	
Assistant Superintendents.		
Couchman, Maj. (*Ag. Lt.-Col.*) H. J., M.C., R.E.	30 Jan. 06	(*a*) Temply. reverted to Mily. duty.
Trenchard, Maj. O. H. B., R.E.	16 Jan. 06	(*a*) In charge No. 1 Drawing Office, Calcutta.
Hamilton, Maj. S. W. S., D.S.O., R.E.	14 Jan. 06	(*a*) Temply. reverted to Mily. duty.
Cotter, Maj. V. R.	30 Oct. 05	(*a*) Temply. reverted to Mily. duty.
Campbell, Maj. J. D., R.E.	5 Dec. 06	(*a*) Temply. reverted to Mily. duty.
Pye, Maj. K. W., D.S.O., R.E.	17 Mar. 06	(*a*) Temply. reverted to Mily. duty.
Foster, Maj. R.	12 Dec. 06	(*a*) Temply. reverted to Mily. duty.
Morshead, Maj. H. T., D.S.O., R.E.	3 Dec. 06	(*a*) Temply. reverted to Mily. duty.
Gwyn, Maj. A. H.	17 Feb. 08	(*a*)
King, Capt. F. J. M., R.E.	27 Mar. 07	(*a*) In chge. Photo-Litho Office, Calcutta.
Lewis, Capt. C. G., R.E.	7 Nov. 07	(*a*)
Thompson, Maj. C. M.	1 Nov. 07	Temply. reverted to Mily. duty.
Mason, Bt.-Maj. K., M.C., R.E.	29 Aug. 09	Temply. reverted to Mily. duty.
Scott, Capt. F. B.	12 Apl. 10	Temply. reverted to Mily. duty.
Perry, Capt. W. E., M.C., R.E.	19 Apl. 10	(*a*) In chge. No. 1 Party.
Wauchope, Capt. R. S.	4 July 10	Attd. M. W. S.
Nosworthy, Bt.-Maj. F. P., D.S.O., M.C., R. E.	1 Nov. 13	(*On probn.*)—Temply. reverted to Mily. duty.
Roome, Capt. H. E., M.C., R.E.	26 Feb. 13	(*On probn.*)—Temply. reverted to Mily. duty.
Jackson, Capt. L. H.	7 Sep. 14	(*On probn.*)—Temply. reverted to Mily. duty.

(*a*) Officiating Dy. Supdt.

Names, Rank and Corps.	Date of first appointment to Civil Dept.	Remarks.

PUBLIC WORKS AND RAILWAY DEPARTMENTS.

RAILWAY REVENUE ESTABLISHMENT.

North-Western Railway.

Waghorn, Bt.-Col. (*temp. Br-Genl.*) W. D., R.E.	Nov. 89	Temply. reverted to Mily. duty.
Anderson, Bt. Lt.-Col. C. F., R. E. D.S.O. [L]	19 Jan. 95	Temply. reverted to Mily. duty.
Cameron, Lt.-Col. H. A., C.I.E., R.E. .	20 Mar. 97	Controller of Traffic, Ry. Dept. (Ry. Bd.), Simla.
Freeland, Bt. Lt.-Col. (*temp. Maj.-Genl.*) H. F. E., C.B., D.S.O., M.V.O., R.E.	8 June 98	Dir. Genl. of Transportation, Ry. Dept. (Ry. Board), Simla.
Walton, Capt. G., R.E.	8 Apl. 12	Temply. reverted to Mily. duty.
King, Capt. C. J. S., R.E.	31 July 14	Temply. reverted to Mily. duty.
Phipps, Capt. C. C., M.C., R.E. . . .	1 Aug. 14	Temply. reverted to Mily. duty.

Oudh and Rohilkhand Railway.

Budden, Bt.-Maj. F. H., M.C., R.E.. .	24 Jan. 10	Temply. reverted to Mily. duty.

Eastern Bengal Railway.

Murray, Col. V., C.B., C.M.G., R. E. .	7 Jan. 89	Temply. reverted to Mily. duty.
Watson, Maj. C. S. M. C., D.S.O., R.E..	31 May 05	Temply. reverted to Mily. duty.
Manton, Capt. (*temp. Lt.-Col.*) L., D.S.O., R.E. [*l*]	1 Sep. 11	Temply. reverted to Mily. duty.

Traffic Officers unattached.

Magniac, Bt. Lt.-Col. (*temp. Col.*) *C. L.*, C.M.G., R. E.	7 Feb. 97	(*D.T.M. Supy.*)—Temply. reverted to Mily. duty.

ENGINEER ESTABLISHMENT.

Chief Engineer.

First Class.

Scudamore, Col. W. V., R.E. . . .	17 Nov. 89	P. W. D.—*Lv., 2 yrs., 21 Jan. 18.*

Second Class.

Blakeway, Col. J. P., C.M.G., R.E. . .	31 May 89	Cen. Provinces (P. W. D.)—Chief Engr., Buildings and Roads Branch, P. W. D., Cen. Provs., & also Chief Engr., Irrigation (in addn.).
Rushton, Lt.-Col. H. W., D.S.O., R. E.	12 Sep. 91	Assam (P.W.D.).—Temply. reverted to Mily. duty.

Superintending Engineers.

First Class.

Rivett-Carnac, Lt.-Col. S. G., R.E. .	16 Jan. 97	Cen. Provs.—*Lv. 6 mos., 19 June 18.*
Smyth, Lt.-Col. W. C., R.E.	5 Oct. 90	Chief Engr., 2nd Cl. (s. *p.t. Supy.*)—Agent & Chief Engr., H. H. the Nizam's State Ry.
Westropp, Lt.-Col. F. M., D.S.O., R.E. .	27 June 91	Bombay (P. W. D.).—Temply. reverted to Mily. duty.
Wilkinson, Lt.-Col. C. W., C.M.G., D.S.O., R.E.	6 July 01	Rly. Dept.—Temply. reverted to Mily. duty.

Second Class.

Garrett, Maj. A. ff., R. E.	8 Feb. 97	Cent. Prov., P. W. D.—Temply. reverted to Mily. duty.
Muter, Maj. R. S., R.E.	29 June 91	(*1st. cl.*, s. *p. t—Supy.*)—Rly. Dept.—Temply. reverted to Mily. duty.

Third Class.

Holme, Maj. A. S., R.E. [L]. . . .	16 Feb. 97	Punjab (P. W. D.)—Temply. reverted to Mily. duty.
Hearn, Bt. Lt.-Col. G. R., R.E. . . .	19 June 94	(*S. Engr.*, 2nd cl., s. *p. t.—Supy.*) Ry. Dept.—Temply reverted to Mily. duty.

Names, Rank and Corps.	Date of first appointment to Civil Dept.	Remarks.
PUBLIC WORKS AND RAILWAY DEPARTMENTS—*contd.*		
ENGINEER ESTABLISHMENT—*contd.*		
Executive Engineers.		
Anderson Maj. E. P., D.S.O., R.E.	15 Nov. 03	Rly. Dept.—Temply. reverted to Mily. duty.
Barnardiston, Maj. (*Ag. Lt.-Col.*) E., D.S.O., R.E.	5 July 93	(*S. Engr.*, 2nd cl., (*s.p.t.*)—*Supy.*) Rly. Dept.—Temply. reverted to Mily. duty.
Barron, Maj. W. E., R.E.	31 Oct. 04	Rly. Dept.—O. & R. Ry.
Battye, Maj. B. C., D.S.O., R.E.	16 May 14	Punjab (P. W. D.)—Temply. reverted to Mily. duty.
Birney, Bt.-Lt.-Col. (*temp. Col.*) C. F., D.S.O., R.E.	1 Nov. 01	Rly. Dept.—Temply. reverted to Mily. duty.
Brough, Bt. Lt.-Col. (*temp. Brig.-Genl.*) A., D.S.O., R.E.	16 July 97	Rly. Dept.—Temply. reverted to Mily. duty.
Burn, Maj. R. N., R.E.	5 July 04	Rly. Dept.—Temply. reverted to Mily. duty.
Chesney, Maj. C. H. R., D.S.O., R.E.	24 Dec. 10	Punjab (P. W. D.)—Templ., reverted to Mily. duty.
Clarke, Maj. (*Ag. Lt.-Col.*) C. J., R.E.	1 June 96	(*Supdg. Engr.*, *3rd cl.* *s.p.t.* *Supy.*) Rly. Dept.—Temply. reverted to Mily. duty.
Cowie, Maj. (*temp. Lt.-Col.*) H. E. C., D.S.O., R.E.	13 June 95	(*S. Engr.*, *3rd cl. s. p. t. & Supy.*) Rly. Dept.—Temply. reverted to Mily. duty.
Crookshank, Bt.-Col. (*temp. Brig.-Genl.*) S. D'A., C.B., C.I.E., D.S.O., M.V.O., R.E.	8 Jan. '92	U.P. (P.W.D.)—Temply. reverted to Mily. duty.
Eustace, Maj. (*temp. Lt.-Col.*) F. R. H., D.S.O., R.E.	14 Feb. 98	Ry. Dept.—Temply. reverted to Mily. duty.
Everett, Bt.-Maj. M., D.S.O., R.E.	17 Sep. 11	Assam P. W. D.—Temply. reverted to Mily. duty.
Garforth, Maj. W., D.S.O., M.C., R.E.	11 Oct. 08	Punjab (P. W. D.)—Temply. reverted to Mily. duty.
Giles, Lt.-Col. V, D.S.O., R.E.	19 Nov. 10	United Provs. (P. W. D.)—Temply. reverted to Mily. duty.
Graeme, Maj. (*Act'g. Lt.-Col.*) J. A., D.S.O., R.E.	23 Oct. 08	United Provs. (P. W. D.)—Temply. reverted to Mily. duty.
Haig, Maj. W. de H., D.S.O., R.E.	31 Mar. 09	United Provs. (P. W. D.)—Temply. reverted to Mily. duty.
Hogg, Bt.-Lt.-Col. C. C. H., R.E.	20 Oct. 10	U. P. (P. W. D.)—Temply. reverted to Mily. duty.
Hogg, Maj. P. G. H., D.S.O., R.E.	16 June 09	U. P. (P. W. D.)—Temply. reverted to Mily. duty.
Hopkins, Maj. L. E., D.S.O., R.E.	15 Jan. 95	(*Supdg. Engr.*, *2nd cl. temp.*) Rly. Dept.
Izat, Maj. (*Act'g. Lt.-Col.*) W. R., D.S.O., R.E.	17 Apl. 91	Rly. Dept.—Temply. reverted to Mily. duty.
Johnston, Capt. D. S., R.E.	21 Sep. 09	Bombay (P. W. D.)—Temply. reverted to Mily. duty.
Kent, Maj. L. M., R.E.	5 Mar. 09	Rly. Dept.—Temply. reverted to Mily. duty.
Lynch, Maj. C. St. J., D.S.O., R.E.	12 Mar. 06	Rly. Dept.—Temply. reverted to Mily. duty.
MacGeorge, Maj. J. B., R.E.	7 Dec. 08	United Provs. (P. W. D.)—Temply. reverted to Mily. duty.
Macrae, Maj. W., R.E.	13 Feb. 07	Rly. Dept.—Temply. reverted to Mily. duty.
Manley, Maj. E. N., D.S.O., R. E.	15 June 98	(*Supdg. Engr. 3rd cl.* (*temply. Supy.*) Temply. reverted to Mily. duty.
Marryat, Capt. J. R., R.E.	18 Nov. 08	Rly. Dept.—Temply. reverted to Mily. duty.
Mathews, Maj. W. F., R.E.	13 May 99	*Offg.* Dy. Agent, O. & R. Ry.
Mount, Maj. (*temp. Lt.-Col.*) A. H. L., R.E.	1 Sep. 06	Rly. Dept.—Temply. reverted to Mily. duty.
Osborne, Maj. G. F. F., D.S.O., R.E.	6 Jan. 04	Rly. Dept.—Temply. reverted to Mily. duty. *Lv., ex Ind., m. c., 1 yr., 4 Jan. 18.*
Pollard-Lowsley, Bt. Lt.-Col. (*Actg. Lt.-Col.*) H. deL., C.I.E., D.S.O., R.E.	16 Apl. 99	Cen. Provs, P. W. D.—Temply. reverted to Mily. duty.
Pollock, Maj. H. C., R.E.	15 Mar. 09	United Provs. (P. W. D.)—Temply. reverted to Mily. duty.
Porter, Maj. M. T., M.C., R.E.	12 Oct. 08	Ry. Dept.—Temply. reverted to Mily. duty.
Rhodes, Bt.-Maj. (*temp. Lt.-Col.*) G. D., D.S.O., R.E.	4 Aug. 12	Rly. Dept.—Temply. reverted to Mily. duty.
Roberts, Maj. W. H., D.S.O., M.C., R.E.	28 Feb. 13	Punjab (P. W. D.)—Temply. reverted to Mily. duty.
Russell, Maj. (*temp. Col.*) W. K., D.S.O., R.E.	2 Apl. 96	*Supdtg. Engr.*, *3rd cl.* *s.p.t.* *Supy.* Rly. Dept.—Temply. reverted to Mily. duty.
Smyth, Maj. (*temp. Lt.-Col.*) G. J. W., D.S.O., R.E.	17 June 96	*Supdtg. Engr.*, *3rd cl.* *s.p.t.* *Supy.* Rly. Dept.—Temply. reverted to Mily. duty.
Sopwith, Maj. G. E., M.C., R.E.	21 June 09	Punjab (P. W. D.)—Temply. reverted to Mily. duty.
Stallard, Capt. (*temp. Lt.-Col.*) R. H., R.E.	22 July 10	Rly. Dept.—Temply. reverted to Mily. duty.
Townshend, Capt. F. H. E., M.C., R.E.	11 Oct. 09	Bombay (P. W. D.)—Temply. reverted to Mily. duty.
Walker, Maj. A. D., R.E. [L]	1 May 05	Rly. Dept.—Temply. reverted to Mily. duty
Walton, Bt.-Lt.-Col., C., D.S.O., R.E.	10 May 01	Rly. Dept.—Temply. reverted to Mily. duty.
Woodhouse, Capt. H. L., M.C., R.E.	21 Mar. 11	Rly. Dept.—Temply. reverted to Mily. duty.
Assistant Engineers.		
Carson, Capt. C. F., M.C., R.E.	22 July 13	Rly. Dept.—Temply. reverted to Mily. duty.
Fitzpatrick, Bt.-Maj. N. T., D.S.O., M.C., R.E.	5 May 14	Bombay (P. W. D.)—Temply. reverted to Mily. duty.
Gosset, Capt. H. H. E., M.C., R.E.	15 July 13	Rly. Dept.—Temply. reverted to Mily. duty.
Roberts, Capt. J. R., M.C., R.E.	. . . .	Rly. Dept.—Temply. reverted to Mily. duty.

CANTONMENT MAGISTRATES' DEPARTMENT.

Names, Rank and Corps.	Date of substantive appointment to Department.	Date of appointment to present grade.	Remarks.
1st Grade Magistrates. (3)			
Leslie, Lt.-Col. W. C. C., Supy. List	*1 Oct. 92*	*1 Jan. 10*	*Dehra Dun, Supy.*
Waller, Lt.-Col. E., Supy. List	2 Apl. 95	7 Mar. 12	Temp. Mily. duty.
Field, Lt.-Col. C. W., Supy. List	*13 Nov. 96* / *8 Feb. 05*	*7 Feb. 16*	*Secunderabad. Supy.*
Vaughan, Lt.-Col. R. R., Supy. List	18 May 97	13 Mar. 16	Abbottabad.
Herbert, Lt.-Col. C., D.S.O., Supy. List	*1 Jan. 00*	*30 Jan. 18*	*Ranikhet, Supy.*
Harvest, Lt.-Col. H. de V., Supy. List	3 May 99	30 Jan. 18	Quetta.
2nd Grade Magistrates. (6)			
Walker, Lt.-Col. F. S., Supy. List	4 Feb. 01	13 Feb. 15	Jhansi.
Prince, Lt.-Col. W. C. S., Supy. List [L]	11 May 03	19 Feb. 16	Lucknow.
Wimberley, Lt.-Col. F. G. A., Supy. List	11 May 03	13 Mar. 16	Peshawar.
Cowie, Lt.-Col. W. A. L., Supy. List	*11 May 03*	*20 Oct. 17*	*Chakrata. Supy.*
Lawrenson, Lt.-Col. T. G. P., Supy. List	19 May 03	20 Oct. 17	Cawnpore.
Barry, Maj. W. T., Supy. List	25 Jan. 04	30 Jan. 18	Temp. Mily. duty.
3rd Grade Magistrates. (31)			
Hodgson, Lt.-Col. W. G., Supy List	17 July 04	1 Nov. 05	Kasauli.
Knowles, Maj. J. K., Supy. List	6 Oct. 04	29 Oct. 06	Jullundur.
Kirkwood, Lt.-Col. A. T., Supy. List	21 Oct. 04	30 Dec. 06	Rangoon.
Burn, Lt.-Col. G. C., Supy. List	13 Nov. 04	1 Mar. 07	Mhow.
Warren-Codrington, Lt.-Col. H. O., Supy. List	15 May 05	31 Mar. 07	Temp. Mily. duty.
Cumberlege, Lt.-Col. C. J., Supy. List	22 Aug. 05	21 May 07	Temp. Mily. duty.
Palin, Maj. R. H., Supy. List	29 Oct. 06	1 Jan. 08	Temp. Mily. duty.
Browning, Lt.-Col. T. C., Supy. List	30 Dec. 06	2 Feb. 08	Rawalpindi.
Lyon, Lt.-Col. J. W. H., Supy. List	1 Mar. 07	6 Feb. 09	Temp. Mily. duty.
Peck, Lt.-Col. J. H., Supy. List	19 May 07	1 July 09	Poona.
Walker, Maj. A. M. E. H. T., Supy. List	21 May 07	1 Jan. 10	Temp. Mily. duty.
Burton, Lt.-Col. R. W., Supy. List	6 Sep. 07	30 Jan. 10	Temp. Mily. duty.
Hunt, Lt.-Col. V. deV.	1 Jan. 08	19 Mar. 10	Cadet College, Wellington.
Anderson, Lt.-Col. W. C., Supy. List	2 Feb. 08	26 Apl. 10	Temp. Mily. duty.
Trotter, Maj. E. G. S., Supy. List. [L]	6 Feb. 09	30 Aug. 10	Temp. Mily. duty.
Anderson, Maj. K. E.	7 Apl. 09	13 Dec. 10	Temp. Mily. duty.
Carwithen, Lt.-Col. E. T.	1 July 09	6 Sep. 11	C. P.
Leslie, Lt.-Col. P. N., Supy. List	1 Jan. 10	30 Oct. 11	*Lv. ex Ind., m.c., 6 mos., from 28 Mar. 18.*
Scott, Maj. R., Supy. List	30 Jan. 10	7 Mar. 12	*Lv., in India, m.c., 1 yr. 14 days from 16 Oct. 17.*
Cattell, Lt.-Col. G. L.	30 Aug. 10	12 Sep. 12	Temp. Mily. duty.
Woodward, Maj. C. P.	13 Dec. 10	3 Jan. 13	Army Hdqrs.—Q. M. G.'s Bch.
Paterson, Maj. H. F. W.	30 Oct. 11	22 Jan. 13	Temp. Mily. duty.
Beyts, Maj. R. F. S.	7 Mar. 12	19 Aug. 13	Temp. Mily. duty.
Trafford, Maj. E. A.	26 Jan. 12	13 Feb. 15	Ferozepore.
Fenning, Maj. W. M., Supy. List	*25 July 10* / 7 Mar. 12	20 Dec. 15	Attd. S. and T. Corps.
Lindesay, Maj. (*temp. Lt.-Col.*) F. S.	8 Aug. 12	7 Feb. 16	Temp. Mily. duty.
Biscoe, Maj. J. V. M.	6 Sep. 12	19 Feb. 16	Temp. Mily. duty.
Reid, Maj. W. D.	3 Jan. 13	13 Mar. 16	Bareilly.
Stewart, Lt.-Col. R. J. T.	22 Jan. 13	19 May 17	Attd. S. and T. Corps.
Moran, Maj. J.	19 Aug. 13	20 Oct. 17	Attd. S. and T. Corps.
Thorne, Maj. A. C. D.	6 Jan. 15	30 Jan. 18	Temp. Mily. duty.
Peart, Maj. D. G.	. . . . .	19 Dec. 17	Meerut (*re-employed*).
Assistant Magistrates. (8)			
Molony, Maj. O. H. C.	13 Feb. 15	13 Feb. 15	Kirkee (*on deputation Deolali.*)
Pott, Maj. H. P., 5th Infy.	20 Dec. 15	20 Dec. 15	Temp. Mily. duty.
Tucker, Maj. H. B.	7 Feb. 16	7 Feb. 16	Temp. Mily. duty.
Extra Assistant Magistrates. (3)			
Sardar Khan, Subdr.-Maj., *Khan Bahadur*, C.I.E.	18 Aug. 10	18 Aug. 10	Lucknow.
Sardar Hira Singh, Risldr.	5 Sep. 10	5 Sep. 10	Lv. in I., m.c., 5 mths., 1 Aug. 18.
Rannu Thapa, Subdr.-Maj., *Rai Sahib*	1 Oct. 10	1 Oct. 10	Meerut.
Attached.			
Irving, Lt. J. H. R., I. A. R. O.	. . . . .	16 May 18	Nasirabad.
Davies, Lt. L. P., I. A. R. O.	. . . . .	19 June 18	Belgaum.
Gordon, Lt. N. H., I. A. R. O.	. . . . .	15 July 18	Nowshera.
Flashman, Capt. T. C., I. A. R. O.	. . . . .	23 Aug. 18	Rawalpindi.
Hamilton, Capt. A. H. C., I A. R. O.	. . . . .	2 Sep. 18	Allahabad.
Deeks, 2nd-Lt. T. B., I. A. R. O.	. . . . .	1 Aug. 18	Multan.
Chatfield, Lt. H. S., I. A. R. O.	. . . . .	15 Sep. 18	Aurangabad.
Lincoln, Lt. E. H., I. A. R. O.	. . . . .	16 Aug. 18	Sialkot.

INDIAN MUNITIONS BOARD.

Ordnance Factories.

Names.	Date of appointment.	Designation.
Babington, Colonel D. M., C.I.E.	1 Jan. 18	Superintendent, Cordite Factory, Aruvankadu.
Blank, Lt. A. L. (I. A. R.)	16 Apl. 18	Assistant Superintendent, Harness and Saddlery Factory, Cawnpore.
Brown, Major B. H., R. A.	1 Jan. 18	Assistant Superintendent, Ammunition Factory, Dum Dum.
Cowie, Major D. G., R. A.	13 Apl. 18	Superintendent, Ammunition Factory, Dum Dum.
Fraser, Major T C., R. A.	8 Apl. 18	Assistant Director, Ordnance Factories, Calcutta.
Hare, Maj. G. A., R. A.	3 Aug. 18	Assistant Superintendent, Ammunition Factory, Kirkee
Ingram, Lt. G., R. E.	2 Aug. 18	Asst. to Consltg. Elec. Engr., Cossipore.
Jacques, 2nd-Lt. F.	7 Aug. 18	Asst. Supt., Gun Carr. Factory, Jubbulpore.
Kinloch, Capt. R. I. H., R. A.	1 Jan. 18	Assistant Superintendent, Rifle Factory, Ishapore.
Lane, Lt.-Col. F. C., R. A.	1 Jan. 18	Superintentdent, Harness and Saddlery Factory, Cawnpore.
Larmour, Maj. L. C., R. A.	7 Nov. 18	Assistant Superintendent, Harness and Saddlery Factory, Cawnpore.
Lenfestev, Major L. d'E., R. A.	1 Jan. 18	Assistant Superintendent, Gun and Shell Factory, Cossipore.
Mair, Captain R.	4 Mar. 18	Executive Engineer, Ordnance Factories Extension Scheme, Kirkee.
Morris, Lt. H. McG.	22 Jan. 18	Assistant Engineer, Ordnance Factories Extension Scheme, Cossipore.
Newman, 2nd-Lt. B. Leigh	25 Feb. 18	Assistant Engineer, Ordnance Factories Extension Scheme, Ishapore.
Parbury, Major E., R. A.	1 Jan. 18	Superintendent, Gun and Shell Factory, Cossipore.
Rhodes, Major G. E.	1 Apl. 18	Ex. Engineer, Ordnance Factories Works Department, Calcutta.
Sitwell, Major N. S. H., R. A.	1 Jan. 18	Superintendent, Ammunition Factory, Kirkee.
Sturrock, Lt.-Col. G. C., R. A.	1 Jan. 18	Deputy Director of Ordnance Factories, Calcutta.
Tomkins, Lt.-Col. E. L., R. A.	1 Jan. 18	Superintendent, Gun Carriage Factory, Jubbulpore.
Tute, Major C. S., R. A.	1 Jan. 18	Assistant Superintendent, Gun Carriage Factory Jubbulpore.
Urquhart, Major A. M., R. A.	1 Jan. 18	Assistant Superintendent, Cordite Factory, Aruvankadu.
Vines, Major C. E., R. A.	20 Feb. 18	Assistant Superintendent, Gun and Shell Factory Cossipore, (Ishapore Branch.)
Wheeler, Lt. P. J., R. F. A.	1 Jan. 18	Assistant Superintendent, Gun and Shell Factory Cossipore.
Wiles, Lt. H. H., I. A. R. O.	1 Jan. 18	Danger Building Officer, Cordite Factory, Aruvankadu.,
Young, Brigadier-General H. A., C.I.E., R. A.	1 Jan. 18	Director of Ordnance Factories, Calcutta.

Army Clothing Factories, and Munitions (Textiles) Depot.

Names.	Date of appointment.	Designation.
Aitken, 2nd-Lt. W. G.	19 Aug. 18	Asst. Inspr. of Army Boots, Cawnpore.
Anderson, 2nd-Lt. J. H.	18 June 18	Asst. Supt., A. C. Factory, Bombay.
Anderson, Capt. R. W. A.	29 June 18	Clothing Contracts Depot, Lucknow.
Archer, 2nd-Lt. A. S.	24 Sep. 18	Assistant Inspector of Clothing, Sialkot.
Belcher, 2nd-Lt. A. R.	24 Sep. 18	Assistant Snperintendent, Army Clothing Factory Jullundur.
Bell, 2nd-Lt. T. W.	12 July 18	Asst. Supt., A. C. Factory, Jullundur.
Beasant, Lt. C. A., I. A. R. O.	20 Aug. 17	Assistant Superintendent, Army Clothing Factory, Shahjahanpur.
Botterill, 2nd-Lt. P. W.	21 June 18	Asst. Supt., A. C. Factory, Madras.
Bowden, Major W. H., I. A.	20 Aug. 17	Superintendent, Army Clothing Factory, Madras.
Bowler, Lt. H. M., I. A. R. O.	20 Aug. 17	Inspector of Clothing, Jullundur.
Bryer, 2nd-Lt. P. J., I. A. R. O.	9 May 18	Inspector of Clothing, Agra.
Buck, 2nd-Lt. A. G.	20 Aug. 17	Assistant Controller (Textile Stores), Munitions Depot, Lahore.
Colquhoun, Major A. F. R., Brit. Ser.	20 Aug. 17	Superintendent, Army Clothing Factory, Sialkot.
Cooper, Lt. R. E., I. A. R. O.	20 Aug. 17	Assistant Superintendent, Army Clothing Factory, Shahjahanpur.
Davidson, Lt. D. J., I. A. R. O.	20 Aug. 17	Supt., A. C. Factory, Shahjahanpore.
Denison, 2nd-Lt. O. G.	28 June 18	Asst. Controller (Textile Stores), Mnnitions Depot; Bombay.
Edwards, 2nd-Lt D. J., I. A. (*temp*)	20 Sep. 18	Asst. Supdt. Army Clothing Factory, Shahjahanpur.
Galvin, Lt. O. H., I. A. R. O.	20 Aug. 17	Inspector of Clothing, Madras.
Garson, Lt. I. G. M.	20 Aug. 17	Assistant Superintendent, Army Clothing Factory, Jullundur.
Gray, 2nd-Lt. T.	23 Sep. 18	Assistant Superintendent, Army Clothing Factory, Agra.
Greenway, Major C. D. K., Res. of Offrs., E. Kent R.	20 Aug. 17	Superintendent, Army Clothing Factory, Agra.
Harvey, 2nd-Lt. H.	22 Aug. 18	Asst. Inspr. of Army Boots, Cawnpore.
Hirst, 2nd-Lt. J.	. . . .	Assistant Superintendent, Army Clothing Factory, Alipore
Holt, Lt. E., I. A. R. O.	20 Aug. 17	Superintendent, Army Clothing Factory, Alipore.
Jacobs, 2nd-Lt. E. J.	3 Aug. 18	Asst. Inspr. of Army Boots, Calcutta.
Lane, Lt.-Col. F. C., R. A.	20 *Aug.* 17	*Government Inspector of Army Boots, Cawnpore.* (Vide *also Ordnance Factories Officers* ante.)

Names.	Date of appointment.	Designation.
		INDIAN MUNITIONS BOARD—*contd.*
Lemon, Lieut. W. S., I.A.R.O.	20 Aug. 17	Assistant Superintendent, Army Clothing Factory, Jullundur.
Levy, Lieut. M. P.	22 July 18	Assistant Superintendent, Army Clothing Factory, Alipore.
Ling, 2nd-Lieut. F. G., I.A.R.O.	20 Aug. 17	Assistant Superintendent, Army Clothing Factory, Shahjahanpur.
Loewe, 2nd-Lieut. H. M. J.	20 Aug. 17	Inspector of Clothing, Alipore.
Mackenzie, 2nd-Lt. D.	18 Sep. 18	Assistant, Inspector of Clothing, Madras.
McDonald, Lieut. S. H., I.A.R.O.	20 Aug. 17	Clothing Contracts Depot. Ahmadabad.
Mckee, 2nd-Lieut. R.	2 Sep. 18	Assistant Inspector of Army Boots, Madras.
MacMillan, 2nd-Lieut. H. A.	27 June 18	Assistant Superintendent, Army Clothing Factory, Madras.
Manduell, Lieut. J. S., I.A.R.O.	20 Aug. 17	Clothing Contracts Depot, Bangalore.
Murphy, Lt. D. W.	19 Sep. 18	Asst. Superintendent, Army Clothing Factory, Sialkot.
Meese, Lt. W. A. B.	23 Sep. 18	Assistant Superintendent, Army Clothing Factory Jullundur.
Miller, 2nd-Lt. T. S.	30 Sep. 18	Assistant Inspector of Clothing, Bombay.
Nichols, Lieut. J. E., I.A.R.O.	20 Aug. 17	Assistant Superintendent, Army Clothing Factory, Sialkot.
Nickels, Lieut. T., I.A.R.O.	20 Aug. 17	Assistant Superintendent, Army Clothing Factory, Alipore.
Paige, Lieut. R. H., I.A.R.O.	20 Aug. 17	Assistant Superintendent Army Clothing Factory, Alipore.
Paterson, 2nd-Lt. A. H.	27 Sep. 18	Assistant Superintendent, Army Clothing Factory, Alipore.
Petters, 2nd-Lieut. H. D., I. A. R. O.	10 July 18	Clothing Contracts Depot, Lahore.
Phillips, 2nd-Lieut. P. F., I.A.R.O.	29 Aug. 18	Assistant Superintendent, Army Clothing Factory, Madras.
Reynolds, 2nd-Lt. N.	19 Sep. 18	Assistant Superintendent, Army Clothing Factory, Sialkot.
Richards, 2nd-Lt. C. G.	23 Sep. 18	Assistant Superintendent, Army Clothing Factory, Agra.
Riley, Lieut. (*tempy. Capt.*) D., I.A.R.O.	20 Aug. 17	Superintendent, Army Clothing Factory, Bombay.
Rockley, Lieut. F. M., I.A.R.O.	20 Aug. 17	Assistant Superintendent, A. C. Factory, Bombay.
Rose, 2nd-Lieut. A. S., I.A.R.O.	2 Oct. 18	Assistant Superintendent, Army Clothing Factory, Alipore.
Savage, 2nd-Lieut. C. T.	24 June 18	Assistant Inspector of Clothing Factory, Agra.
Sharman, 2nd-Lt. E. P.	19 Sep. 18	Assistant Superintendent, Army Clothing Factory, Shahjahanpur.
Spragge-White, Lieut. T. E., I.A.R.O.	20 Aug. 17	Inspector of Clothing, Bombay.
Stuart-Prince, Lieut. (*tempy. Capt.*) D., I.A.R.O.	20 Aug. 17	Clothing Contracts Depot, Lahore.
Vessey, Lieut. (*temp. Capt.*) S. J., I.A.R.O.	20 Aug. 17	Superintendent, Army Clothing Factory, Jullundur.
Walsh, 2nd-Lieut. D. T.	18 June 18	Clothing Contracts Depot, Patna.
Watkins, Lieut. F. J., I.A.R.O.	20 Aug. 17	Assistant Superintendent, Army Clothing Factory, Sialkot.
Webb, 2nd-Lieut. H. G.	20 Aug. 17	Assistant Superintendent, Army Clothing Factory, Alipore.
Wemyss, Capt. A. B., I. A. R. O.	....	Assistant Superintendent, Army Clothing Factory, Alipore.
Williams, Lt. C. A. L., I.A.R.O.	20 Aug. 17	Clothing Contracts Depot, Delhi.
Wilson, 2nd-Lt. A. T.	....	Assistant Controller (Textile Stores), Munitions (Textiles Depot), Cawnpore.
Woods, 2nd-Lt. L. H. W.	30 Oct. 18	Asst. Superintendent, Army Clothing Factory, Alipore.
		Miscellaneous.
Alexander, Lieut.-Col. H. M., D.S.O.	15 Aug. 18	Controller (Clothing) Indian Munitions Board.
Bailey, Lt. H. E., R. F. A.	....	Assistant Controller (Oils and Paints), Calcutta.
Bally, Maj. E. M.	1 Feb. 18	Superintendent, Government Tannery, Allahabad.
Bishop, Lieut. H. C., I.A.R.O.	24 May 18	Assistant Controller (Electrical and Mechanical), Calcutta.
Brown, Lieut. W. Bremner, R.G.A.	....	Assistant to the War Office Representative (Hides.) Madras.
Clarke, Lieut. W. D. M., I.A.R.O.	4 Aug. 17	Assistant Controller (Hides), Bombay.
Churchill, Lt.-Col. A.	1 Oct. 18	Deputy Controller (Munitions Manufacture), Madras.
Collinge, 2nd-Lt. J.	1 Oct. 18	Assistant Controller (Clothing Contracts), Indian Munitions Board.
Cothay, Captain H. W.	8 Oct. 18	Assistant Controller (Clothiug Contracts) Indian Munitions Board.
Craig, Lieut. W. J., I.A.R.O.	16 May 18	Deputy Controller (Munitions), United Provinces, Cawnpore.
Eliot, Maj. (*temp. Lt.-Col.*) M., I. A.	20 Aug. 17	Dir of A. C. Factories, Calcutta.
Frith, Lieut. W. H.	18 Jan. 18	Assistant Controller (Textiles Manufacture), Indian Munitions Board.

INDIAN MUNITIONS BOARD—*concld.*

Miscellaneous—*contd*

Names.	Date of appointment.	Designation.
Gabbett, Maj. K. J., I. A.	20 Aug. 17	Deputy Controller (Clothing), Indian Munition Board.
Gibbons, 2nd-Lieut. A. J., I.A.R.O.	15 Dec. 17	Assistant Controller (Hardware, Implements & Metals), Bombay.
Greswell, Capt. E. A., I.A.R.O.	28 Apl. 17	Deputy Controller (Timber Supplies), Punjab.
Griffiths, Capt. W. S., I.A.R.O.	. . .	Deputy Controller, (Hardware, Implements & Metals), Bombay.
Guthrie, Capt. A., I. A.	6 June 17	Deputy Controller (Hides), Madras.
Hemsworth, Lieut. D. D., R.E.	6 Sep. 18	Assistant Controller (Munitions Manufacture), Calcutta.
Holt, Lieut. E., I.A.R.O.	20 Aug. 17	Asststant Director, Army Clothing Factories, Calcutta.
Horton-Bennett, Capt. A. G. R., I.A.R.O.	20 Aug. 17	Assistant Controller, Clothing, Indian Munitions Board.
Kay, Lieut. T. A. I.A.R.O.	10 Feb. 18	Assistant Controller (Jute Manufactures), Calcutta.
Keyte, Lieut. P. E. I.A.R.O.	1 Aug. 17	Assistant Controller (Stores), Bombay.
Langer, Lieut. W. C. I.A.R.O.	22 Feb. 18	Deputy Controller (Hardware, Implements & Metals.), Calcutta.
Lee, 2nd-Lieut. I.A.R.O.	9 July 18	Assistant Controller (Electrical & Mechanical), Calcutta.
MacGregor, Lieut.-Col. J., R. E.	10 Nov. 17	Controller (Shipbuilding), Indian Munitions Board, Calcutta.
Markwick, Lieut. W. L.	5 Nov. 17	Chemical Assistant, Esociet Tannin Research Factory, Maihar.
McKerrow, Major H. B.	3 Sep. 17	Deputy Controller (Textiles), Indian Munitions Board.
Mellor, Capt. R. W., R. F. A.	9 Mar. 18	Deputy Controller (Hides), Calcutta.
Morrison, Maj. J., R.A.	1 Jan. 18	Second Military Assistant (Ordnance Factories), Indian Munitions Board.
Noble, Capt. T. L., I.A.R.O.	18 Sep. 18	Assistant Director of Army Clothing Factories, Calcutta.
Palmer, Lt.-Col. C. C., R. A.	1 Jan. 18	Military Assistant (Ordnance Factories), Indian Munitions Board.
Percy, Capt. C. A., I.A.R.O.	4 May 18	Controller of Munitions, Karachi.
Phillips, 2nd-Lieut. R. G. T., I.A.R.O.	12 Sep. 18	Assistant Controller of Munitions, Karachi.
Savage, Lt. W. C., I.A.R.O.	2 May 18	Assistant Controller (Munitions Manufacfure), Calcutta.
Sellers, Capt. W. L	20 Aug. 17	Deputy Director of Army Clothing Factories, Calcutta.
Sherrard, Capt. G. C., I.A.R.O.	1 July 18	Deputy Controller, Agr. Requirements, Mesopotamia—Simla.
Simpson, Lt. A.	17 Jan. 18	Assistant Controller (Textiles), Indian Munitions Board.
Simpson, Lt. L. C., I.A.R.O.	5 Dec. 17	Assistant Controller (Hardware, Implements & Metals), Bombay.
Stuart, Maj.-Genl. Sir R. C. O., K.C.S.I.	16 Feb. 17	Member, Indian Munitions Board.
Sutcliff, Lt. L. B.	16 Feb. 18	Assistant Controller (Hides), Bombay.
Townshend, Lt. C. A.	10 Nov. 17	Assistant Controller (Shipbuilding Design), Indian Munitions Board, Calcutta.
Turnbull, Lt. W. J., R. E.	28 Mar. 18	Assistant Mettalurgist, Sakchi.
Wallis, 2nd-Lt. S. A., R. E.	8 May 17	Assistant Controller (Timber Supplies), Bombay.
Webb, Lt. C. E.	17 May 18	Assistant Controller (Hides), Bombay.
White, Lt. D. G.	28 Sep. 18	Assistant Controller (Clothing), Indian Munitions Board.
Wilson, Lt.-Col. R., B.I.A.R.O.	28 Jan. 18	Controller (Jute Manufactures), Calcutta.
Yulll, 2nd-Lt. A. F., I.A.R.O.	2 Dec. 18	Assistant Controller (Oils and Paint), Bombay.

GOVERNMENT OF INDIA.

Names, Rank and Corps.	Date of first appointment to Civil Dept.	Remarks.
Ames, Maj., H. L. R.	4 May 17	Asst. Poll. Resdt.
Anderson, Maj. G. H.	5 Dec. 10	Poll. Offr.—*Lv., c. l., 1 yr., 30 Oct. 17.*
Atkinson, Col. (*Ty. Maj.-Genl.*) E. H. deV., C.B., C.M.G., C.I.E., R.E.	1 Nov. 01	Principal, Thomason Civil Engr. Coll., Rurki.
Bannerman, Lt.-Col., A. D'A. G., C.V.O., C.I.E. (Supy. List).	8 Nov. 92	Resdt., cl. II.—Resdt., Kashmir.
Bayley, Lt.-Col. S. F.	25 Mar. 89	Poll. Offr.—(*Temply. reverted to Mily. duty.*)
Beadon, Lt.-Col. H. C. (Supy. List)	23 Nov. 95	Deputy Commr., Delhi.
Bender, Lt. W. E. G.	1 Jan. 17	
Bourke, Lt.-Col. J. J., M.B., I.M.S.	4 July 97	(*offg.*) Mint Master, Calcutta.
Burden, Lt.-Col. H., C.I.E., F.R.C.S., I.M.S. (*actg. Col.*)	25 Nov. 95	Agcy. Surgn., 2nd cl.—(*Temply. reverted to Mily. duty.*)
Cardew, Maj. A. M., R. E.	1 July 16	Addl. Asst. Secy.—Foreign and Poll. Dept.
Coldstream, Maj. J. C. (Supy. List)	. . .	Secy. Central Employment and Labour Board.
Colvin, Capt. E. J. D. (temp. Maj.)	4 Feb. 12	(*Temply. reverted to mily. duty.*)
Corry, Maj. M., M.D., I. M. S.	26 May 02	(Offg.) C. M. O. and Civil Surgn., Delhi. (*Temply. reverted to Mily. duty.*)
Cox, Hony. Maj.-Genl. *Sir* P. Z., G.C.I.E., K.C.S.I., K.C.I.E. (Supy. List).	23 Dec. 90	Resdt., 1st cl.—Persian Gulf.
Deas, Maj. (Bt. Lt.-Col.) L. J. M., M.B., F.R.C.S.E., I.M.S.	27 May 05	Residcy. Surgn., 2nd cl.—(*Temply. reverted to Mily. duty.*)
Douglas, Lt.-Col. M. W., C.I.E. (Supy. List).	18 Sep. 90	Supdt., P. B. and Ch. Commr., Andaman and Nicobar Islands.
Drake, Maj. H. B., I.M.S.	25 July 11	Dy. Assay Master, Bombay. (*Temply. reverted to Mily. duty.*)
Edwards, *The Hon'ble* Maj.-Genl. W. R., C.B., C.M.G., M.D., K. H. P. [R]	8 Jan. 18	D. G., Ind. Med. Service.
Evans-Gordon, Capt. K. A. G., 39 Horse.	12 May 13	(*Temply. reverted to Mily. duty.*)
Field, Maj. D. M.	16 Jan. 17	Poll. Offr.—Under Secy. to the Govt. of India, Foreign and Poll. Dept.
Flowerdew, Capt. R. E., M.B., I.M.S. (*temp. Maj.*)	19 Mar. 14	Second Medl. Offr., Port Blair. (*Temply. reverted to Mily. duty.*)
Fowle, Capt. T. C. W., 40 Pathans	6 June 14	Poll. Offr.—(*Temply. reverted to Mily. duty.*)
Franklin, Maj. G. D., M.B., I.M.S.	1 Feb. 07	Agy. Surgn., 2nd cl. (*Temply. reverted to Mily. duty.*
Gabriel, Maj. C. H. (Supy. List).	19 Sep. 04	Poll. Offr.—First Asst. to Resdt. in Kashmir. (First) (*b*).
Goodbody, (*Bt.-Lt.-Col.*) C. M., C.I.E., D.S.O., F.R.C.S.I., I.M.S.	10 Mar. 05	Residcy. Surgn., 2nd cl.—(*Temply. reverted to Mily. duty.*)
Gough, Lt.-Col. H. A. K., C.I.E. (Supy. List.)	24 Apl. 98	Poll. Offr.—*on leave.*
Grey, Lt.-Col. W. G. (L) (Supy. List)	1 May 92	H. B. M.'s Consul General at Khorasan, Meshed.
Hamilton, Lt.-Col. R. E. A., C.I.E. (Supy. List).	28 June 98	Poll. Offr.—Ajmer-Merwara.
Haig, Lt.-Col. P. B., C.B., M.B., I.M.S.	16 June 03	*Offg.* Agcy. Surgn., 1st cl. Civil, Ajmere and C. M. O., Rajputana.
Haig, Lt.-Col. T. W., C.M.G. (L)	7 July 92	Poll. Offr.—H. B. M.'s Consul-General at Ispahan.
Haworth, Maj. L. B. H. (L) (Supy. List)	2 May 01	Poll. Agt. and H. B. M.'s Consul at Maskat.
Heyland, Maj., A. K., D.S.O.	11 Jan. 18	Executive Commissariat Officer, P. B.
Heron, Maj. D., C.I.E., M.B., F.R.C.S.E., I.M.S.	10 Dec. 12	*Offg.* Medl. Offr., Seistan Consulate, and *e.-o.* H. B. M.'s Vice-Consul, Seistan and Kain.
Holmes, Maj. J. M., M. B., I. M. S.	17 Oct. 17	Hlth. Offr., Delhi.
Hughes, Lt.-Col. F. T. C. (Supy. List)	11 Dec. 05	Dy. Assay Master, Calcutta.
Hugo, Lt.-Col. J. H., D.S.O., I.M.S.	19 Mar. 02	*Tempy. on Mily. duty.*
Hunt, Lt.-Col. S., M.B., I.M.S.	3 Apl. 07	Agcy. Surgn., 2nd cl.—Baghelkhand.
Hunter, Maj. J. B. D., M.B., F.R.C.S.E., I.M.S. (*actg. Lt.-Col.*)	9 Oct. 07	Agcy. Surgn., 2nd cl.—(*Temply. reverted to mily. duty.*)
James, Lt.-Col. C. H., C.I.E., F.R.C.S., I. M. S.	1 Oct. 17	Chief Medl. Offr. Civil Surgn., Delhi.
Jones, Capt. (*Staff Capt.*) R. T.	16 Apl. 18	Spl. duty, For. and Poll. Dept.
Kennion, Lt.-Col. R. L., C.I.E. (Supy. List).	8 Apl. 93	Poll. Offr.—H. B. M.'s Consul at Kermanshah.
Knox, Lt.-Col. S. G., C.S.I., C.I.E. (Supy. List).	22 Nov. 94	Poll. Offr.—Spl. duty, Basra.
Lawrence, Maj. H. R.	28 Apl. 04	Poll. Offr.—(*Temply. reverted to mily. duty.*)
Lorimer, Maj. D. L. R., C.I.E. (Supy. List).	9 Mar. 03	Poll. Offr.—*Lv., c. l., 2 yrs., 24 Nov. 17.*
Luard, Lt.-Col. C. E. (Supy. List)	9 Dec. 97	Poll. Offr.—Poll. Agent, Southern State of Central India, Malwa.
Lyall, Maj. R. A., D.S.O. (Supy. List.)	10 Nov. 01	Poll. Offr.—(*Temply. reverted to mily. duty.*)

Names, Rank and Corps.	Date of first appointment to Civil Dept.	Remarks.
GOVERNMENT OF INDIA—*continued.*		
McCarrison, Maj. (*Bt.-Lt.-Col.*) R., M.D., F.R.C.P., I.M.S.	9 Nov. 04	Agcy. Surgn., 2nd cl.—On Mily. duty.
McConaghy, Maj. C. B., M.B., I.M.S.	25 Apl. 04	Agcy. Surgn., 2nd cl.—Agcy. Surgn., E. R. States. *Offg.* Resdency. Surgn., Jaipur & Agcy. Surgn., Kotah & Jhalawar.
McCormick, Lt.-Col. A. L. C., R.E.	7 June 07	Mint Master, Calcutta.
Mackenzie, Maj. C. F., C.I.E. (L)	15 Feb. 05	Poll. Offr.—Spl. duty, Basra.
Macnab, Bt.-Col. A. J., C.B., F.R.C.S., I.M.S.	10 Dec. 05	Agcy. Surgn., 2nd cl.—(*Temply. reverted to Mily. duty.*)
Macnabb, Maj., R. J.	14 Aug. 09	Poll. Offr.—(*Temply. reverted to mily. duty.*)
Macpherson, Lt.-Col. A. D. (Supy. List)	25 June 98	Poll. Offr. Resdt., W. R. States.
Macrae, Lt.-Col. J. L., M.B., I.M.S.	4 Aug. 06	Civil Surgn. (*Temply. reverted to Mily. duty.*)
Melville, Lt.-Col., H. B., M.B., I.M.S.	12 Nov.	Agcy. Surgn., 2nd Class & Resy. Surgn., Kashmir.
Murray, Maj. J. H., M.D., I.M.S.	26 Apl. 05	Sr. Medl. Offr., Civil Surgn., & Asst. Supdt. Jail, Port Blair.
Muspratt-Williams, Lt.-Col. C. A., C.I.E., R.A.	9 Sep. 98	Chief Inspr. of Explosives.
Neale, Maj. W. G. (Supy. List)	25 July 06	Poll. Offr.—Deputy Secy., Foreign & Poll. Dept.
Needham, Maj. (*temp. Lt.-Col.*) R. A., D.S.O., M.B., I.M.S.	22 May 11	Dy. Dir. Genl.—I.M.S.
O'Connor, Lt.-Col. W. F. T., C.I.E. (Supy. List)	26 Aug. 99	Poll. Offr.—Resident, Nepal.
Ogilvie, *Maj.* G. D. (Supy. List).	17 Feb. 05	Asst. Secy., Army Dept. (s.*p.t.*)
Oliver, Maj. D. R. G. (L)	17 July 06	Poll. Offr.— Asst., Resdt., Kashmir for Leh.
Peart, Maj. C. L. (L), 106 Pio.	28 Apl. 12	Secy., Board of Examrs.—(*Temply. reverted to Mily. duty.*)
Pierpoint, Maj. H. W., F.R.C.S., I.M.S.	6 Apl. 12	Offg. Agcy. Surgn., 2nd cl. (*Temply. reverted to Mily. duty.*)
Pollard-Lowsley, Maj. (*Bt. Lt.-Col.*) H. deL., C.I.E., D.S.O., R. E.	1 Apl. 99	Under Secy. to the Govt. of India, P. W. D., Simla.—(*Temply. reverted to Mily. duty.*)
Prideaux, Lt.-Col. F. B., C.S.I., C.I.E. (Supy. List)	1 July 95	Poll. Offr.—*Lv. C. L., 1 yr., 21 Nov. 18.*
Roberts, Lt.-Col. *Sir* J. R., Kt., C.I.E., M.B., F.R.C.S., I. M. S.	26 June 97	Civil Surgn., Simla (West).
Roberts, Maj. W. H., D.S.O., M.C., R.E.	29 Apl. 12	Spl. duty, under Chief Commr., Delhi.—(*Temply. reverted to Mily. duty.*)
Robertson, Lt.-Col. J. C., C. M. G., C.I.E., M.B., I.M.S.	17 May 12	On Mily. duty under the War Office.
Ross, Maj. (*temp. Lt.-Col.*) H., O.B.E., M.B., F.R.C.S.I., I.M.S.	18 Oct. 16	Asst. Dir. Genl., I. M. S. (Stores).
Rowlandson, Maj. H. W., 82nd Punj.	11 June 14	Comdt. (*Services with Army Dept.*)
St. John, Maj. H. B., C.I.E. (Supy. List)	27 May 99	Poll. Agt., Quetta.
Scott, Maj. N. E. H., C.I.E., M.B., I.M.S.	5 Dec. 06	Agcy. Surgn., 2nd cl. *Spl. duty, Persian Gulf.*
Smith, Lt.-Col. H. A., C.I.E., M.B., I. M. S.	12 Oct. 16	Surgn. to H. E. the Viceroy.
Smith, *the Hon'ble* Lt.-Col. J. M., C.V.O., C.I.E. (Supy. L.)	28 June 87	Resdt., 1st cl.—Agt. to the G. G. of Rajputana & C. C., Ajmer-Merwara.
Stanton, Lt., C.A., I.A.R.O.	20 Aug. 17	Asst. Commdt. and Asst. D.S.P., Port Blair.
Stevens, Capt. L. C.	1 June 18	Secy., Central Publicity Board.
Stevenson, Maj. W. D. H., C.I.E., M.A., M.D., D.Ph., I.M.S.	14 Nov. 18	Asst. D. G., I.M.S. (*Sany.*)
Stewart, Maj. H., C.I.E., (Supy. List)	2 Nov. 99	Poll. Offr.—*Lv., C. L. to 24 Sep. 18.*
Sutherand, Lt.-Col. W. D., M.D., I.M.S.	5 Jan. 93	Imperial Serologist and Chemical Examiner to Govt. Calcutta.
Sykes, Lt.-Col. *Sir* P. M., K.C.I.E., C.M.G., (Supy. List) (*tempy.* Bd.-Genl.)	22 Nov. 02	Poll. Offr.—On Spl. duty, South Persian Rifles, Persian Gulf.
Tate, Lt.-Col. G., M.B., I.M.S.	1 Oct. 16	Civil Surgn., Simla (East).
Thorburn, Maj. H. H., C.I.E., M.B., I.M.S.	14 Apl. 11	Offg. Agcy. Surgn., 2nd cl. Medl. Offr., South Persian Rifles.
Trevor, Lt.-Col. A.P., C.S.I. C.I.E., [*l*] (Supy. List)	14 June 98	Poll. Offr. *Lv., C. L., 1 yr., 13 Dec. 17.*
Tyrrell, Maj. J. R. J., M.B., I.M.S.	1 Jan. 06	Agcy. Surgn.—(*Temply. reverted to Mily. duty.*)
Wallis, Maj. H. J. K., 20 Horse	12 Nov. 10	Dy. Assay Master, s.*p.t.*, Bombay Mint. (*Temply. reverted to Mily. duty.*)
Walter, Lt.-Col. A. E., O.B.E., I.M.S.	2 July 05	Supdt., X-Ray Institute, Dehra Dun.
Watson, Lt.-Col. J. W., I.M.S.	12 Oct. 03	Agcy. Surgn.—(*Temply. reverted to Mily. duty.*)
Weir, Maj. J. L. R. (*l*) (Supy. List)	29 Jan. 08	Poll. Offr.—(*On spl. duty in Persian Gulf.*)
White, Maj. F. N., C.I.E., M.B., I.M.S.	24 Feb. 13	Sany. Commr. with Govt. of India, (s.*p.t.*)
Wickham, Capt. E. T. R.	10 Sep. 17	Offr. on Spl. Duty under Foreign & Poll. Dept.
Williams, Capt. S., 130th Baluchis	13 July 12	Poll. Offr. Asst. Poll. Agt., Chilas.
Willis, Maj. G. H., C.I.E., M.V.O., R.E.	22 June 07	Mint Master, Bombay.
Wilson, Bt.-Lt.-Col. A. T., C.M.G., C.I.E., D.S.O. [*l*]	25 July 09	Poll. Offr.—(*On special duty in Persian Gulf.*)
Wright, Lt.-Col. E. H. I.M.S.	27 Mar. 03	*Offg.* Civil Surgn., Coorg.
Young, Maj. (*Bt.Lt.-Col.*) A. W. C., M.B., I.M.S.	1 Aug. 12	Hlth. Offr., Delhi.—(*Temply. reverted to Mily. duty.*)

Names, Rank and Corps.	Date of first appointment to Civil Dept.	Remarks.

MADRAS.

Governor.

His Excellency the Right Hon'ble JOHN SINCLAIR *Baron* PENTLAND OF LYTH, P.C., G.C.S.I., G.C.I.E. . . . 30 Oct. 12.

Military Secretary.

Money, Capt., W. S. E., 22nd (Sam Browne's) Cavalry . . . . 5 Oct. 17.

Aides-de-Camp.

Capt. L. M. **Peet,** 6th Jat Infy. . . . . 19 Oct. 16.
Capt. R. G. S. **Hale,** 6th Bn. Worc. R. . . . . 17 Nov. 17.
Capt. A. S. **Balfour,** R.I.M. (Hony.) . . . . 1 Jan. 17.
Hony. Capt. V. S. **Alexander,** *Sirdar Bahadur* (*retd.*) . . . . 5 Apl. 18.

Commandant, H. E. the Governor's Body Guard.

Lt. (*Actg. Capt.*) **Smith,** D. E., I.A.R.O. . . . . 14 Feb. 18.

Surgeon.

Maj. F. F. **Elwes,** C.I.E., I.M.S. (Actg.) . . . . 14 Nov. 14.

TRAVANCORE.

Faunce, Lieut.-Col. R. DeL., 26th Cavy.	2 Nov. 16	Comdt., Nayar Bde.
Antram, Capt. E. W., 1st G. R. .	25 June 17	Comdt., 1st Bn., Nayar Bde.

BOMBAY.

Governor.

His Excellency Sir George Ambrose Lloyd, G.C.I.E., D.S.O. . . . . 16 Dec. 18.

Military Secretary.

Capt. A. K. **MacEwan,** 227th Bde., R.F.A. (*Tempy.*) . . . . 18 July 16.

Aides-de-Camp.

Capt. G. A. **Laverton,** 2/4 Wilts. (*Tempy.*) . . . . 14 Nov. 16.
The Hon'ble Lt.-Col. P. R. **Cadell,** C.I.E., I.C.S., Bo. Bn., I. D. F. (Hony.) . 15 Jan. 15.
Capt. E. W. **Huddleston,** C.I.E., R.I.M. (Hony.) . . . . 1 Nov. 13.
Capt. N. F. **Paton,** Bo. Bn., I.D.F. (Hony.) . . . . 22 June 17.
Capt. L. A. L. *Lord* **Mountbatten,** G.C.V.O., K. R. Rif. C. . . . . 23 July 18.
Capt. A. W. **Mackintosh,** R. Def. Corps . . . . 24 July 18.
Hony. Capt. *Meherban* **Abdul Majid Khan,** Nawab of **Savanur** (Hony.) 1 May 13.
Hony. Capt. *Maharawal Shri* **Ranjitsinhji Mansinhji,** Raja of **Baria** (Hony.) . . . .
Hony. Capt. *Meherban Sir* **Parashramrav Ramchandrarav,** K.C.I.E., Chief of **Jamkhandi** (Hony.) . . . . 1 Sep. 16.
Subdr.-Maj. **Balkrishna Rao,** *Bahadur,* 110th Mah. L. I. . . . . . . 1 Oct. 16.

Surgeon.

Capt. C. M. **Rigby,** R.A.M.C. . . . . 5 Jan. 15.

Barrett, Maj. C. C. J., C.I.E. . . .	29 Oct. 07	Asst. Poll. Agt.—Spl. duty, Persian Gulf. *Lv., C.L., 6 mths. 1 June 18.*
Beale, Lt.-Col. W. (Supy. List) .	9 Jan. 95	Poll. Agt., Mahi Kantha.
Berthon, Lt.-Col. H. W. (Supy. List)	14 Mar. 99	Poll. Agt., Kathiawar, Halar Prant.
Burke, Lt.-Col. R. J. C. (Supy. List)	30 Oct. 00	Asst. Poll. Agt.—(*Temply. reverted to Mily. duty.*)
Carter, Lt.-Col. J. R. B. G. (Supy. List).	20 May 96	Poll. Agt., Kathiawar. Gohilwar Prant. Lv. m.c., 6 mths. ex I.
Coghill, Lt.-Col. N. S. (Supy. List) .	30 May 96	Poll. Agt., Palanpur.
Condon, Lt.-Col. J. K. (Supy. List) .	1 Aug. 98	Poll. Agt., Kathiawar.
Dixon, Lt. M. J., I. A. Res. . . .	22 Apl. 17	A.-D.-C. to Poll. Resdt., Aden.
Gordon, Capt. (*Tempy. Maj.*) J. de la H., M. C., 67 Punj.	26 Dec. 11	Asst. Poll. Agt.—(*Temply. reverted to Mily. duty.*)

Names, Rank and Corps.	Date of first appointment to Civil Dept.	Remarks.
	BOMBAY—*concld.*	
Harold, Lt.-Col. C. F. (Supy. List)	30 Aug. 96	Asst. Poll. Agt.—Southern Mahratta Country States.
Harrison, Maj. H. (Supy. List.)	21 Dec. 96	Poll. Agt.—Kathiawar, Gohilwad Prant.
Hume, Maj. S. G. W., 7th Raj.	. . . .	Asst. Resdt., Perim.
Jacob, Lt.-Col. H. F., C.S.I. (Supy. List).	22 Nov. 92	First Asst. Poll. Resdt., Aden.—Liaison Officer, Cairo.
Lang, Capt. L. E., M.C.	27 Mar. 09	Asst. Poll. Agt.—(*Temply. reverted to Mily. duty.*)
Meek, Maj. A. S.	16 Oct. 07	Asst. Poll. Agt.—Spl. duty Persian Gulf.
Merewether, Lt.-Col. J. W. B. C.I.E. (Supy. List.)	7 Aug. 93	Poll. Agt.—*Lv., ex Ind., 18 mos., 14 days, 24th July 14.*
Mosse, Maj. A. H. E. (Supy. List)	18 Dec. 01	Poll. Agt.—Kathiawar Jhalawar Prant.
Murphy, Capt. G. P., 82 Punj.	23 Apl. 13	Asst. Poll. Agt.—(*Temply. reverted to Mily. duty.*)
O'Brien, Lt.-Col. E. (Supy. List)	19 Sep. 97	Poll. Agt., Administrator, Porbandar State.
Oldfield, Lt.-Col. T. A. F. R (Supy. List).	15 May 98	Poll. Agt.—Sorath Prant, Kathiawar.
Pottinger, Lt.-Col. R. S. (Supy. List).	11 Aug. 93	Poll. Agt.—Cutch & *Actg.* Agt. to the Governor in Kathiawar.
Reilly, Maj. B. R. (Supy. List)	17 Apl. 08	Asst. Poll. Resdt., Aden.
Strong, Maj. H. S. (Supy. List)	23 Oct. 99	Asst. Poll. Resdt., Aden.
Tate, Capt. J. C.	15 Feb. 09	Asst. Poll. Agt.—(*Temply. reverted to Mily. duty.*)
Walker, Lt.-Col. G. K., C.I.E., O.B.E., F.R.C.V.S., C.V.D.	25 July 94	Supdt., C. V. D., Bombay. *Lv., m. c., ex. I., 6 mths., 3 Apl. 18.*
Webb, Lt. (tempy. Capt). A. W. T.	9 Nov. 11	Asst. Poll. Agt. (*Temply. reverted to Mily. duty.*)
Webber, Maj. M. R. H.	. . . .	Employed for Education of H. H. the Gaekwar of Baroda's Grand-children.
Wightwick, Capt. H. M.	30 Sep. 09	Asst. Poll. Resdt., Aden.
Wilberforce-Bell Capt. H. (*Temp. Maj.*)	5 Jan. 10	Asst. Poll. Agt.—(*Temply. reverted to Mily. duty.*
Wodehouse, Lt.-Col. F. W., C.I.E. (Supy. L.)	9 Feb. 91	Sr. Poll. Agt.—S. M. Country States and Resdt., Kolhapur
Wood, Lt.-Col. W. M. P. C.I.E. (Supy. List).	16 June 99	1st Asst. Poll. Resdt., Aden.

BENGAL.

Governor.

His Excellency the Right Hon'ble LAWRENCE JOHN LUMLEY DUNDAS, *Earl of* RONALDSHAY, G.C.I.E., Mar. 17.

Military Secretary.

Capt. H. G. **Vaux**, Brit. Ser. . . . . 27 Mar. 17.

Surgeon.

Maj. J. D. **Sandes**, I.M.S. . . . . 26 Apl. 17.

Aides-de-Camp.

Capt. R. W. **Hyde**, 1st Gn. Bn. Linc. R.	5 Mar. 18.
Lieut. D. **Balfour**, Lothians and Border Horse	12 Jan. 18.
2nd-Lt. C. B. **Leon**, 21st (Empress of India's) Lrs. (Extra)	4 Apl. 18.
Capt. E. A. **Hackett-Smith**, I. A. (Extra)	9 Nov. 18
Col. C. H. **Shepherd**, D.S.O., Comdt., 1st Gn. Bn. Linc. R. (Hony.)	21 Aug. 18.
Col. C. M. **Pearce**, 7th E, Indian Ry. Bn. V.D. (Hony.)	27 Mar. 17.
Lt.-Col. (Hony. Col.) D. A. **Tyrie**, 1st (Cossipore) Bde., Mobile Arty., V.D. (Hony.)	27 Mar. 17.
Lt.-Col. (Hony Col.) R. S. **Hawkins**, 42nd Assam-Bengal Ry. Bn., V.D. (Hony.)	27 Mar. 17.
Lt.-Col. R. **Glen**, 41st Eastern-Bengal Coy., V.D. (Hony.)	27 Mar. 17.
Capt. D. F. **Vines**, O.B.E., R. I. M. (Hony.)	27 Mar. 17.
Risaldar **Faiz Muhammad Khan**, I.O.M., 1st Lrs.	27 Mar. 17.

Commandant, H. E. the Governor's Body Guard.

Capt. P. V. **Douetil**, I. A. R. O. . . . . 15 Feb. 18.

Hirst, Maj. F. C. (Supy. List)	27 Jan. 00	Dir. of Surveys, Bengal & Assam. (*On Mily. duty.*)
Macpherson, Maj. D. I., 10 Jats.	6 May 12	Comdt., Mily. Police.—Dacca Bn., Dacca.
Pritchard, Maj. P. A. R.	8 Apl. 08	Supdt. of Police, 3rd gr.—(*Temply. reverted to Mily. duty.*)
Smith, Lt.-Col. A., F.R.C.V.S., C. V. D.	21 June 93	Principal, Bengal Vety. College.
Ravenhill, Lt. (*temp. Capt.*) M. H. (I. A. R. O.).	18 Dec. 17	Asst. Comdt., Mily. Police, Dacca Bn., Dacca.

Names, Rank and Corps.	Date of first appointment to Civil Dept.	Remarks.

BIHAR AND ORISSA.

Lieutenant-Governor.

The Hon'ble Sir EDWARD ALBERT GAIT, K.C.S.I., C.I.E., I.C.S. . . . . 19 Nov. 15.

Private Secretary.

Aides-de-Camp.

Lt. E. D. T. **Rowley**, 1st Bn. K. O. Sco. Bord. . . . . 23 Nov. 18.
Maj. A. T. **Peppe**, V. D., Chota Nag. L. H. (*Hony.*) . . . . . . 20 Nov. 18.
Maj. J. A. M. **Wilson**, V.D., Bihar L. H. (*Hony*) . . . . . . 20 Sep. 18.
Subadar-Major **Sita Ram Singh**, late 16th Rajputs (*Hony.*) . . 15 Oct. 14.
Hony. Capt. Sardar Bahadur **Hira Singh**, late 16th Rajputs (*Hony.*). 16 Oct. 12.

Bell, *The Hon'ble* Col. G. J. H., C.I.E., M.B., I.M.S.	6 Jan. 92	*Lv., C. L. 6 mths., 28 June 18.*
Connell-McDowell, Maj. A. E. J. .	25 Mar. 08	Supdt. of Police, 3rd gr. (*Temply. reverted to Mily. duty.*)
Hodgson, Maj. W. H. (supy.) . .	2 Jan. 08	Supdt., Police, 4th gr., Monghyr.
Vaughan, Lt.-Col. J. C. S., M.B., I.M.S.	5 June 96	I. G., Civil Hospls. (*Temply*).

BURMA.

Lieutenant-Governor.

The Hon'ble Sir REGINALD HENRY CRADDOCK, K.C.S.I., I.C.S. . . . . 15 Feb. 18.

Private Secretary.

Meade, Maj. W. L., I. A., 95th Infy. . . . . 5 Jan. 18.

Aides-de-Camp.

Lt.-Col. J. L. W. **ffrench-Mullen**, C.I.E. (*Hony.*) . . . . 28 Aug. 17.
Lt. L. C. **Graves**, 18th Rif. Bde. . . . . 12 July 18.
Hony. Capt. **Muzaffar Khan**, *Sardar Bahadur, late* 72 Punj. . . . . 31 July 08.
Hony. Capt. *Sardar Bahadur* **Amar Singh**, *Rai Bahadur*, I.O.M., Burma Mily. Police . . . . 12 Sep. 08.
Subadar (*Offg. Sub-.Maj.*) **Maung Aung Bwin**, K.S.M. Burma Mily. Police 7 Nov. 16.

Abbey, Lt-Col. W. B. T. (Supy. List.)	8 May 97	Dy. Commr., 2nd gr., Comdt., Chin Hills Bn., Falam.
Alves, Maj. J. W. J.	8 Mar. 13	Asst. Commr., 2nd gr., *prov. sub.*—(*temply. reverted to*
Aplin, The Hon'ble Lt.-Col. S. L., C.S.I. (Supy. List).	16 Apl. 91	Commr., Mandalay Dn., Addl. Member, Imp. Legislative Council.
Barker, Capt. R. W.	7 Feb. 12	Asst. Commr., 1st gr., *s.p.t.* (*Temply. reverted to Mily. duty.*)
Batten, Maj. H. H.	29 Nov. 05	Dy. Commr., 3rd gr., *s.p.t.* (*Temply. reverted to Mily. duty.*)
Bayett, Lt. H. M., I. A. R.O. . . .	. . . .	Asst. Comdt., Mily. Police, Chindwin Bn., Monywa.
Bazett, Maj. S.	15 Mar. 05	Dy. Commr., 3rd gr.—(*Temply. reverted to Mily. duty.*)
Beadon, Maj. V. (L.)	11 Feb. 10	Dy. Commr., 3rd gr., *s.p.t.* (*Temply. reverted to Mily. duty.*)
Bigg Wither, Maj. F. (L), (Supy. List).	22 Sept. 04	Dy. Commr., 2nd gr. *s.p.t.*, Yamethin.
Bowen, Lt.-Col. C. E. (Supy. List) .	25 Feb. 97	Dy. Commr., 2nd gr., Monywa.
Bowley, 2nd-Lt. G. W., *I. A.* (*T. C.*).	. . . .	Asst. Comdt., Mily. Police, Chindwin Bn., Monywa.
Broome, Capt. W. G. K., 89 Punj. .	6 Nov. 13	Asst. Comdt., Mily. Police, Spl. Duty, Monywa.
Browne, Maj. H. F. M.	14 July 08	Asst. Commr., 1st gr., *Prov. sub.*—(*Temply. reverted to Mily. duty.*)
Burne, Maj. L. E. L. (Supy. List) .	27 July 02	Dy. Commr., 3rd gde., *prov. sub.*—Comdt., Mily. Pol. Chindwin Bn., Monywa.
Butterfield, Maj. E., D.S.O. (Supy. List).	1 May 06	Dy. Commr., 3rd gr., *prov. sub.*—Comdt., Mily. Police, N., Shan States Bn., Lashio.
Candy, Capt. H. R., 73 Carn. Infy. .	3 May 18	Asst. Comdt., Mily. Police, Myitkyina Bn., Htawgaw.
Carter, Maj. J. F. C.	3 Sep. 05	D. S. P., 3rd gr., *prov. sub.*—(*Seconded.*)
Cooper, Lt. R. F., 41 Dogras . . .	28 July 17	Asst. Comdt., Mily. Police, Myitkyina Bn., Ft. Harrison.

Names, Rank and Corps.	Date of first appointment to Civil Dept.	Remarks.
	BURMA—*contd.*	
Copplestone, Lt. F. E., I. A. (*on probn.*).	. . .	Asst. Comdt., Mily. Police, Chindwin Bn., Monywa.
Corbett, Maj. B. A., 13th Rajputs	. . .	(*Temply. reverted to Mily. duty.*)
DesVœux, Lieut.-Col. H. (Supy. List)	21 Dec. 90	Commr. (*Seconded*) I. G., Police, Burma.
Duff, Lt. J. M., 18th Bn. Rif. Brig.	. . .	Asst. Comdt., Mily. Police, Chindwin Bn., Monywa.
Dyne, 2nd-Lt. C. F., 3rd Gn. Bn., Bedf. R.	. . .	Asst. Comdt., Mily. Police, Chindwin Bn., Monywa.
Entrican, Lt.-Col. J., M.D., I.M.S.	16 Nov. 17	Civil Surgn., Meiktila.
Ewing, Capt. R. R., 61 Pio.	19 June 16	Comdt., Mily. Police, Spl. Duty, Nomalin.
French, Lt. F. G., 36 Sikhs	28 July 17	Asst. Comdt., Mily. Police, Chin Hills Bn., Kanpet et.
ffrench-Mullen, Lt.-Col. J. L. W., C.I.E.	17 Mar. 08	D. I. G., Mily. Police, on spl. duty, Maymyo.
Finnie, Maj. A. McL.	2 Jan. 08	D. S. P., 3rd gr., *prov. sub.*—(*Temply. reverted to Mily. duty.*)
Gould, Capt. S. C.	28 May 10	D. S. P., 3rd gr., *prov. sub.*—(*Temply. reverted to Mily. duty.*)
Grey, Capt. W. A. S.	12 Nov. 08	Dy. Commr., 3rd gr., *s. p. t.*—(*Temply. reverted to Mily. duty.*)
George, 2nd-Lt. A. G., 3rd Gn. Bn., Bedf. R.	. . .	Asst Comdt., Mily. Police, Chindwin Bn., Monywa.
Green, Capt. J. E. S., 18th Bn., Rif. Brig.	. . .	Asst. Comdt., Mily. Police, Chindwin Bn., Monywa.
Hackett, Maj. T. D. H.	12 Mar. 10	Dy. Commr., 3rd gr., *s.p.t.*, Comdt., Mily. Police, Spl. Duty Monywa.
Hickie, Maj. C. C.	14 Nov. 08	D. S. P., 3rd gr., *prov. sub.*—(*Temply. reverted to Mily. duty.*)
Hollamby, Lt. W. C., I. A. (*T.C.*)	. . .	Asst. Comdt., Mily. Police, Chindwin Bn., Monywa.
Jackson, Lt. H. S., 1 Lancers	28 July 17	Asst. Comdt., Mily. Police, Myitkyina Bn., Myitkyina.
Key-Mowat, Lt. C. G.	. . .	Asst. Comdt., Mily. Police, Chindwin Bn., Monywa.
Kirkwood, Lt.-Col. A. T.	8 Apl. 15	Cant. Magte., Rangoon.
Knapp, Major H. H. G., M.A., M.D., D.T.M. & H., I.M.S.	11 Jan. 16	I. G. of Prisons, Burma.
Langtry, Lt.-Col. H. V. M. (Supy. List)	30 July 01	Dy. Commr., 2nd gr., *s.p.t.* Adj., Mily. Police, Toungoo Bn., Toungoo.
Lee, Lt.-Col. A. W. H. 1-7 G. Rif.	27 Aug. 14	Comdt., Mily. Police.—Myitkyina Bn., Myitkyina.
Lethbridge, Maj. A., 54 Sikhs	26 Apl. 14	Asst. Comdt., Mily. Police. *Lv., c. l., ex I., 1 yr., 1 Jan. 18.*
Lord, Maj. H. P. M. [L] (Supy. List)	9 Dec. 04	Asst. Commr., 1st gr., *prov. sub.*—Headqrs., Asst. and Addl. Dist. Magte., Moulmein.
McCann, Maj. H. H., 69 Punj.	. . .	Comdt. Mily. Police, Putao Bn., Ft. Hertz.
Macdonald, Maj. R. W., D.S.O. (Supy. List).	23 Dec. 04	D. S. P., 3rd gr.,—(*Temply. reverted to Mily. duty.*)
Macnabb, Lt.-Col. D. J. C., C.S.I. (Supy List).	14 Apl. 87	Commr.—Magwe Dn.
Malet, Capt. A. W., (125*a*) 38 Dogras	27 Apl. 14	Asst. Comdt., Mily. Police—(*Temply. reverted to Mily. duty.*)
Massy, Maj. W. I. J., 12 Prs.	26 Dec. 11	(*Tempy. reverted to Mily. duty.*)
Mercer, Lt. W. D. F., 1st Gn. Bn., R. I. R., Regt.	. . .	Asst. Comdt., Mily. Police, Chindwin Bn., Monywa.
Nairne, Capt. R. G., 61 Pio.	26 Jan. 15	Asst. Comdt., Mily. Police, S. Shan States Bn., Loi Mwe.
Neil, Lt. J. A., 94th Infy.	15 Nov. 17	Asst. Comdt., Mily. Police—Spl. Duty, Mawtack.
Nethersole, Lt.-Col. F.R. (Supy. List)	27 June 98	Dy. Commr., 2nd gr. (*offg. 1st gr.*), Tharrawaddy.
Newcombe, Maj. G. H. [L] (Supy. List).	22 Apl. 07	Dy. Commr., 3rd gr., *prov. sub.*, Comdt., Mily. Police.—Mandalay Bn., Mandalay.
Obbard, *The Hon'ble* Lt.-Col. O. J. (Supy. List)	21 May 90	Commr., Pegu Dn.
Orman, Lt.-Col. F. L., 10 G. R.	5 July 05	Comdt., Mily. Police —D. I. G. Mily. Pol., Rangoon., *s.p.t.*
Ormiston, Lt.-Col. T. L. (Supy. List)	28 Apl. 98	Dy. Commr., 1st gr. *s. p. t.* Excise Commr., Rangoon.
Owens, Lt.-Col. F. C. (Supy. List)	8 June 97	Dy. Commr., 2nd gr. Comdt., Mily. Police, Bhamo Bn., Bhamo.
Patrick, Maj. R. M. F., M. C., 42nd Deoli Regt.	31 July 16	Asst. Comdt., Mily. Police.—*Offg.* Comdt. S. S. S. Bn., Taunggyi.
Pelly, Capt. H. R.	25 Jan. 14	Asst. Commr., 2nd gr.—*prov. sub.* (*Temply. reverted to Mily. duty.*)
Percy-Smith, Bt. Lt.-Col. E. S.(Supy. List).	4 July 03	D. S. P., 3rd gr.—(*Temply. reverted to Mily. duty.*)
Prescott, Maj. H. C.	10 Jan. 08	D.S.P., 3rd gr., *prov. sub.*—(*Temply. reverted to Mily. duty.*)
Prothero, Capt. J. E. S.,	. . .	Dy. Commr., 3rd gr., *prov. sub.* (*Temply. reverted to Mily. duty.*)
Richardson, Maj. G. G., 105th Infy.	17 Apl. 14	(*Temply. reverted to Mily. duty.*)
Roberts, Maj. A. S. B. (Supy. List)	11 Aug. 01	Dy. Commr., 2nd gr., *s.p.t.* Comdt., Mily. Police. *Lv., 6 mths,*
Sandeman, Maj. A. P. (Supy. List)	9 Apl. 04	Dy. Commr., 3rd gr., *prov. sub.*—(*Temply. reverted to Mily. duty.*)
Slacke, Maj. G. C.	17 Feb. 12	Asst. Commr., 1st gr., *s.p.t.*—(*Temply. reverted to Mily. duty.*)
Slater, Maj. A. F. M.	16 Aug. 06	D. S. P., 3rd gr., *prov. sub.*—(*Temply. reverted to Mily. duty.*)
Stedman, Lt. H., M. C., 59 Scinde Rifles.	. . .	Asst. Comdt., Mily. Police, Chin Hills Bn., Tiddim.

Names, Rank and Corps.	Date of first appointment to Civil Dept.	Remarks.
		BURMA—*concld.*
Stone, Lt.-Col. W. R. (Supy. List)	24 Mar. 95	Dy. Commr., 1st gr.—*Lv., ex I., 23 mos. 23 dys., 11 April 17*
Strickland, Col. P. C. H.	17 July 16	I. G., Civil Hospls., Burma.
Stuart, Maj. D. N.	14 May 06	Dy. Commr., 3rd gr., *prov. sub.*—(*Temply. reverted to Mily. duty.*)
Thomson, Maj. A. J. D.	24 Sep. 09	D. S. P., 3rd gr., *prov. sub.*—(*Temply. reverted to Mily. duty.*)
Todd, 2nd-Lt. H. C., 18th Bn. Rif. Brig.	. . .	Asst. Comdt., Mily. Police, Chindwin Bn., Monywa.
Turner, Capt. J., 114th Mahrattas	25 Dec. 16	Asst. Comdt., Mily. Police—Bhamo Bn., Bhamo.
Vaughan, Capt. H. R., 41st Dogras	22 Nov. 17	Asst. Comdt., Mily. Police, *Offg.* Comdr., Shwebo Bn., Shwebo.
Ward, Lt. L. H., I. G. Rifles	. . .	Asst. Comdt., Mily. Police.—Myitkyina Bn., Myitkyina.
Wellborne, Maj. C. deM.	16 Aug. 09	D. S. P., 3rd gr., *prov. sub.*—(*Temply. reverted to Mily. duty.*)
Williams, Maj. G. R. K. (Supy. List)	20 June 02	Dy. Commr., 3rd gr.—*Offg.* Adjt., Mily. Police.—Rangoon Bn., Rangoon.
Wright, Maj. G. D., 98 Infy.	25 Jan. 15	Asst. Comdt., Mily. Police—Chindwin Bn., Monywa.
Yorston, Capt. C., 3rd Gn. Bn., Bedf. R.	. . .	Asst. Comdt., Mily. Police, Chindwin Bn., Monywa.

PUNJAB.

Lieutenant-Governor.

The Hon'ble Sir MICHAEL FRANCIS O'DWYER, G.C.I.E., K.C.S.I. . . . . 26 May 13.

Private Secretary.

Lt.-Col. E. C. **Bayley**, C.I.E. . . . . 6 Mar. 07.

Aides-de-Camp.

Hony. Capt. **Ghulam Muhammad Khan**, *Sardar Bahadur, late* 29th Mn. Battery (*Hony.*) . . . . 1 Jan. 16.
Hony. Capt. **Surja**, *Sardar Bahadur*, late 48th Pioneers (*Hony.*) . . . . . . 1 Jan. 16.
Hony. Capt. **Bishen Singh**, *Sardar Bahadur, late* 57th Wilde's Rifles (*Hony.*) . . 1 Jan. 16.

Barton, Lt.-Col. A. E. (Supy. List)	14 Dec. 91	Dy. Commr., 1st gr.—Jhelum.
Beadon, Lt.-Col. H. C. (Supy. List)	23 Nov. 95	Dy. Commr., 2nd gr.—Delhi.
Buck, Lt.-Col. C. H. (Supy. List)	29 Oct. 96	Dy. Commr., 2nd gr.—Kangra.
Burlton, Lt.-Col. P.S.M. (Supy. List)	12 Oct. 91	*Offg.* Commr., Jullunder.
Coldstream, Maj. J. C. (Supy. List)	14 Apl. 00	Dy. Commr., 3rd gr. (*Temply. reverted to Mily. duty.*)
Dennys, Lt.-Col. H. T., C.I.E.	26 Nov. 88	I.-G. of Police, Punjab.
Douglas, Lt.-Col., M. W., C.I.E. (Supy. List).	18 Sep. 90	Dy. Commr., 1st gr. (*Serving under Govt. of India.*)
Elliott, Lt.-Col. A.C., C.B.E. (Supy. List)	8 Mar. 93	Dy. Commr., 1st gr.—Gurgaon.
Ferrar, Maj. M. L. (L) (Supy. List.)	5 Sep. 01	Asst. Commr., 1st gr.—Jullunder.
Frizelle, Lt.-Col. J. (Supy. List)	18 Sep. 97	Dist. & Sessions Judge, 3rd gr. (*Temply. reverted to Mily. duty.*)
Irvine, Lt.-Col. A. A. (Supy. List)	13 Dec. 95	Dist. & Sessions Judge, 3rd gr., Ambala.
Knollys, Maj. R. W. E. (Supy. List)	17 Jan. 01	Dist. & Sessions Judge, 4 gr., Delhi.
MacWatt, Col. R. C., C.I.E., M. B., F.R.C.S.	16 Mar. 93	Inspr.-Genl., Civil Hospls.
Nicolas, Maj. F. C. (Supy. List)	23 Oct. 02	Asst. Commr., 1st gr. (*Temply. reverted to Mily. duty.*)
O'Brien, Lt.-Col. A. J., C.I.E. (L) (Supy. List).	24 Oct. 96	Dy. Commr., 2nd gr.—Gujranwala.
Pease, Col. H. T., C.I.E., C.V.D.	31 Mar. 85	Principal, Punjab Vety. College, Lahore.
Roe, Lt.-Col. B. O. (Supy. List)	30 Oct. 94	Dist. and Sessions Judge, 3rd gr.—Campbellpore.
Sanford, Lt.-Col. G. B. (Supy. List)	12 Mar. 06	Asst. Commr., 1st gr.—(*Temply. reverted to Mily. duty.*)
Thompson, Lt.-Col. C. P. (Supy. List)	10 Sept. 89	Commr., Multan Dn.
Young, Lt.-Col. *Sir* F. P., K.B.E., C.I.E. (Supy. List).	24 Dec. 88	Commr., Rawalpindi. Dn.

UNITED PROVINCES OF AGRA AND OUDH.

Lieutenant-Governor.

The Hon'ble Sir SPENCER HARCOURT BUTLER, K.C.S.I., C.I.E., I.C.S. . . 15 Feb. 18.

Private Secretary.

Mr. T. **Lister**, I.C.S. . . . . 15 Feb. 18.

Aides-de-Camp.

(*Services temply. with C.-in-C.*)

Lt.-Col. E. A. **Neville**, *late* B. & N. W. Ry. V. Rif., V.D. (*Hony.*)
Col. J. H. E. **Beer**, C.I.E., V.D. (*Hony.*) . . . . 25 Oct. 12.
Lt.-Col. J. **Walker**, V. D. (*Hony.*) . . . . 14 Dec. 12.
Risaldar-Maj. **Qudrat Khan**, *Bahadur* (*Hony.*) . . . . *Re-appointed* 1 Jan. 16.
Hony. Capt. **Chamu Sing Burathoki**, I.O.M. (*Hony.*) . . . . . . 4 Jan. 17.

Names, Rank and Corps.	Date of first appointment to Civil Dept.	Remarks.

UNITED PROVINCES OF AGRA AND OUDH—*concld.*

Loch, Capt. P. G., 97 Infy.	18 Mar. 13	Poll. Dept.—Poll. Agent, Koweit.
Mactaggart, Col. C., C.I.E., I.M.S.	14 Sep. 90	Inspr. Genl. of Civil Hospls. and in charge Sany. Commr.'s Office, U. P.
Noel, Capt. E. W. C., C.I.E.	7 Dec. 12	Poll. Dept.—H. B. M.'s Vice-Consul & *ex-officio* Asst. to the Pol. Resdt. in the Persian Gulf, Ahwaz.
Sandes, Capt. E. W. C., R.E., M.C.	15 Oct. 10	Asst. Principal.—(*Temply. reverted to Mily. duty.*)
Thomson-Glover, Capt. J. W., 35 Sikhs.	11 Apl. 14	Asst. Collr., 2nd cl.—(*Temply. reverted to Mily. duty.*)

ASSAM.

Chief Commissioner.

The Hon'ble Mr. N. D. Beatson-Bell, C.S.I., C.I.E. . . . . 1 Apl. 18.

(*Shillong.*)

Banatvala, *Hon'ble* Col. H. E., C.S.I., K.H.P., I.M.S.	12 Mar. 93	Inspr. Genl. of Civil Hospls. and Prisons, Assam.
Carter, Lt. W. H., I.A.R.O., 2-6th G. R.	1 Feb. 18	Asst. Commdt., 4th Bn., Assam Rifles, Imphal.
Cloete, Maj., H. D. M.C., 92 Punj.	1 Sep. 17	(*Offg.*) Comdt., 2nd (Lakhimpur) Bn., Assam Rifles, Sadiya.
Cole, Lt.-Col. H. W. G., C.S.I. (Supy. List.)	12 Apl. 90	Dy. Commr., 1st gr.—Comdt., 22 Manipur Labour Corps. (*Temply. reverted to Mily. Duty.*)
Coote, Capt. M. C., 107 Prs.	13 Mar. 14	Asst. Comdt., 4th Bn., Assam Rifles, Imphal.
Copeland, Lt. J. H., I. A. R. O., 8th G. R.	17 Aug. 17	Asst. Comdt., 2nd (Lakhimpur) Bn., Assam Rifles, Sadiya.
Davies, Capt. G. B., 48th Prs.	11 Mar. 17	Asst. Comdt., 1st (Lushai Hills) Bn., Assam Rifles, Aijal.
Falkland, Capt. H. L. F., 13 Raj.	7 Mar. 14	Offg. Comdt., 1st (Lushai Hills) Bn., Assam Rifles, Aijal.
Gurdon, *The Hon.* Lt.-Col. P. R. T., C.S.I. (Supy. List).	2 June 86	Commr.—Assam Valley Dists., Gauhati.
Halliday, Lt.-Col. H. M. (Supy. List)	7 Dec. 92	Dy. Commr., 1st gr.—*Ser. under Army Dept.*
Halliday, Lt. H. I., I. A. R. O.	26 Nov. 17	Asst. Comdt., 1st (Lushai Hills) Bn., Assam Rifles, Lungleh.
Hebbert, Capt. P. B. 44th Infy.	3 Oct. 17	Asst. Comdt., 2nd (Lakhimpur) Bn., Assam Rifles, Sadiya.
Herbert, The Hon'ble Lt.-Col. D. (Supy. List).	21 Mar. 91	Dy. Commr., 1st gr., Inspr. Genl. of Police, Assam.
Hooper, Lt. E. J., *I.A.R.O.*	28 May 17	Asst. Comdt., 4th Bn., Assam Rifles, Imphal, Manipur.
Kennedy, Lt.-Col. W. M., C.I.E. (Supy. List).	20 Feb. 94	Dy. Commr., 2nd gr.—(Chairman of the Assam Labour Board).
Lowry-Corry, Capt. W., 23rd Cav.	14 May 17	P. A. to the Chief Commr.
Montefiore, Capt. C. E., 110 Infy.	11 Mar. 14	Asst. Comdt., 3rd (Naga Hills) Bn., Assam Rifles, Kohima.
Ostrehan, Capt. M., 1-9th G. R.	12 Feb. 18	Asst. Comdt., 4th Bn., Assam Rifles, Imphal.
Parry, Lt. N. E., I.A.R. O., 2-7th G. R.	3 Feb. 18	Asst. Comdt., 2nd (Lakhimpur), Bn. Assam Rifles, Sadiya.
Playfair, Lt.-Col. A. (Supy. List)	19 Sep. 94	Dy. Commr., 2nd gr. (*offg. 1st gr.*), and Comdt., 27th Lushai Labour Corps. (*Temply. reverted to Mily. duty.*)
Prior, Lt. H. C., I.A.R.O., 2-1st G. R.	4 Feb. 18	Asst. Comdt., 3rd (Naga Hills), Bn. Assam Rifles, Kohima.
Row, Lt.-Col. G. R., I. A. (Retd.)	15 May 17	Offg. Comdt., 4th Bn., Assam Rifles, Imphal, Manipur.
Sanderson, Lt. J. F. B., 4th Oxford and Bucks. L.I.	22 Sept. 17	Asst. Comdt., 3rd (Naga Hills) Bn., Assam Rifles, Kohima.
Shakespear, Col. L. W., C.B., I. A.	15 Nov. 17	Depty. Inspr. Genl., Assam Rifles (*temply.*), Shillong.
Tuker, Capt. F. I. S., 1-2nd G. R.	23 Feb. 18	Asst. Comdt., 4th Bn., Assam Rifles.
Vickers, Maj. A., 48th Prs.	30 Apl. 17	Offg. Comdt., 3rd (Naga Hills) Bn., Assam Rifles, Kohima.
Walker, Lt. G. D., I.A.R.O., 2-8th G. R.	28 Jan. 18	Asst. Comdt., 4th Bn., Assam Rifles, Imphal.

BALUCHISTAN.

Agent to the Governor-General and Chief Commissioner.

The Hon'ble Mr. Henry Robert C. Dobbs, C.S.I., C.I.E., I.C.S. . . . 5 Dec. 17.

Biscoe, Maj. H. V. [L] (Supy. List)	23 Apl. 06	Poll. Dept. *Com. Lv., 1 yr., 18 Apl. 17.*
Bruce, Maj. C. E. (Supy. List)	15 Nov. 01	Poll. Agt.—Loralai.
Chenevix-Trench, Maj. R.H. (Supy. List)	12 Nov. 01	Poll. Agt & Dy. Comsr., Sibi.
Cohen, Maj. A. M. [L], 1-4th R. W. Kent R.	29 May 18	Personal Asst. to A. G. G. & C. C.
Daukes, Maj. C. T. (Supy. List)	16 Dec. 03	Asst. Poll. Agt.—Zhob. (*Temply. reverted to mily duty.*)
Dew, Lt.-Col. A. B., C.S.I., C.I.E. (Supy List)	21 Dec. 97	Resdt., 2nd class, Revenue and Judicial Commr.
Finnis, Capt. H. C.	12 Jan. 12	Poll. Offr.—(*Temply. reverted to Mily. duty.*)
Harris, Maj. T. G. M. (Supy. List)	5 Mar. 07	First Asst. to the Hon'ble A.G.G. & C. C. in Baluchistan.

BALUCHISTAN—*concld.*

Names, Rank and Corps.	Date of first appointment to Civil Dept.	Remarks.
Hutchinson, Maj. W. G., O.B.E. (Supy List)	6 Apl. 03	Poll. Agt.—Chagai.
Jacob, Lt.-Col. A. L., C.I.E., O.B.E., (Supy. List)	20 May 95	Poll. Agt.—Zhob.
McConaghey, Lt.-Col. A., C.I.E., O.B.E. (Supy. List).	25 May 92	*C. L., 23 m. 2d., 3 Apl. 17.*
Maclagan, Col. R. S., C.B., C.S.I., Br. Ser.	Aug. 83	Secy. (P. W. D.) to A. G. G.
Pritchard, Maj. H. R. N.	24 June 04	Asst. Poll. Agt., & Asst. Commr., Quetta & Pishin.
Ramsay, Lt.-Col. A. D. G., C.I.E., (Supy. List).	16 July 98	Poll. Agt., Kelat, and Poll. Agt. in charge of the Bolan Pass.
St. John, Maj. H. B., C.I.E., (Supy. List.)	27 May 99	Poll. Agt. and D. C., Quetta-Peshin.
Walker, Capt. G. B., M.C. (169)	1 Apl. 14	2nd Asst. to the Hon'ble Agt. & C. C. in Baluchistan.

CENTRAL INDIA.

Agent to the Governor-General.

The Hon'ble Mr. OSWALD VIVIAN BOSANQUET, C.S.I., C.I.E., I.C.S. . . . 15 Nov. 1913.

(*Indore.*)

Names, Rank and Corps.	Date of first appointment to Civil Dept.	Remarks.
Beville, Lt.-Col. F. G., C.I.E.	10 Jan. 90	Offr. on spl. duty in C. I.
Colvin, Capt. E. J. D. (*Temp. Maj*).	4 Feb. 13	Poll. Agt., Baghelkhand.
Hunt, Lt.-Col. S., I.M.S.	18 Jan. 09	Agcy. Surgn., 2nd cl.—Baghelkhand, Sutna.
Luard, Lt.-Col. C. E.	9 Dec. 97	Poll. Agt., S. States of C. I., Manpur.
McLaughlin, Maj. H. D., 6 G. R.	7 June 09	Comdt., Malwa Bhil Corps. Indore.
Minchin, Lt.-Col. A. B., C.I.E.	9 Nov. 96	Poll. Agt., Bundelkhand, Nowgong.
Smith, Lt.-Col. F. A., M.D., I.M.S.	3 May 00	Agcy. Surgn., 1st cl.—Resdy. Surgn. and A. M. O., Indore.
Spence, Lt.-Col. P. T. A. (Supy. List).	31 Dec. 89	Poll. Agt., Malwa.

CENTRAL PROVINCES.

Chief Commissioner.

The Hon'ble Sir BENJAMIN ROBERTSON, K.C.S.I., K.C.M.G., C.I.E., I.C.S. . 3 Aug. 1912.

(*Nagpur.*)

Names, Rank and Corps.	Date of first appointment to Civil Dept.	Remarks.
Champion, Lt. P. H., W. Kent Yeo.	19 Aug. 18	P. A. to the Chief Commr.
Morris, Lt.-Col. D. O. (Supy. List)	29 May 93	Dist. and Sessns. Judge, 1st class (*provl.*), Saugor.
Plowden, Lt.-Col. T. C. (Supy. List)	26 Nov. 94	Dy. Commr., 1st cl., Betul.
Green, The Hon. Col. C. R. M., M.D., F.R.C.S., I.M.S.	13 Apl. 18	Inspr.-Genl., Civil Hospls.
Scrafton, Maj. J. E. B.	25 July 10	Dist. Supdt. of Police, 4th cl. (*offg. 3rd cl.*) (*Reverted temply. to Mily. duty.*)

Names, Rank and Corps.	Date of first appointment to Civil Dept.	Remarks.

HYDERABAD.

Resident.

The Hon'ble Sir STUART FRASER, K.C.S.I., C.I.E. . . . . July 1

Brodribb, Major C. H., I.M.S. . .	10 Sep. 16	Offg. Resdy. Surgn., Hyderabad.
Whicher, Capt. J. D'A.	9 May 18	P. A. to the Resdt.

MYSORE AND COORG.

Resident and Chief Commissioner.

The Hon'ble MR. H. V. COBB, C.S.I., C.I.E. . . . . 7 Apl. 16.

(*Bangalore.*)

James, Maj., E. H. S. (Supy.) . . .	May 01	First Asst. to the Resdt.

NORTH-WEST FRONTIER PROVINCE.

Chief Commissioner and Agent to the Governor-General.

The Hon'ble LIEUT.-COL *Sir* GEORGE OLAF ROOS-KEPPEL, G.C.I.E., K.C.S.I., [L], 7th July 1908.

(*Peshawar.*)

Anson, Maj. G. F. W. (Supy. List). .	8 Mar. 07	Poll. Dept.—Asst. Commr. Thal.
Blakeway, Lt.-Col, D. B., C.I.E. (Supy. List).	23 Nov. 94	Poll. Dept.—Resdt., 2nd cl.—Rev. Commr., N.-W. F. Prov.
Brett, Maj. J. A. (Supy. List). . .	27 Aug. 06	Poll. Dept.—Asst. Commr., Mardan.
Brierley, Maj. C. I., I.M.S. . .	19 Apl. 07	Agcy. Surgn., 2nd cl.—Civil Surgn., Peshawar & Agcy. Surgn., Khyber.
Broke-Smith, Maj. P. W. L., D.S.O., R.E.	14 Aug. 18	P. A. to the Secy., Chief Commr., P. W. D.
Crossle, Maj. H., M.D., I.M.S. . .	20 May 06	Agcy. Surgn., 2nd cl.—Joint Civil Surgn., Peshawar.
Crosthwaite, Maj. C. G. (Supy. List).	16 Jan. 03	Poll. Dept.—Dy. Commr., Bannu.
Crosthwaite, Maj. J. S. (Supy. List). .	24 May 08	Poll. Dept.—Asst. Commr. & Sub-divl. Offr., Tank.
Fraser, Col. H. A. D., R. E. . . .	Jan. 18	Secy. (P. W. D.) to C. C. & A. G. G.
Garratt, Maj. R. (Supy. List) . . .	26 July 02	Poll. Dept.—Poll. Agt.—Tochi.

Names, Rank and Corps.	Date of first appointment to Civil Dept.	Remarks.

NORTH-WEST FRONTIER PROVINCE—*concld.*

Garstin, Maj. W. A. M. (Supy. List)	18 Feb. 08	Poll. Dept.—Asst. Secy. to Chief Commr., N. W. F. P.
Griffith, Maj. R. E. H., C.I.E.	19 Nov. 08	Poll. Dept.—Asst. Commr., Charsadda.
Heale, Maj. R. J. W., O.B.E. (Supy. List.)	9 Jan. 03	Poll. Dept.—Poll. Agt., Kurram.
Humphrys, Maj. F. H. (Supy. List).	6 Dec. 03	Poll. Dept.—*Lv., c. l., 2 yrs., 14 Nov. 17.*
Irvine, Lt.-Col. T. W., F.R.C.S.E., I.M.S.	1 Nov. 02	Agcy. Surgn., 1st cl.—C. M. O., N.-W. F. Provce.
Keen, Lt.-Col. W. J., C.I.E. (Supy. List)	14 Oct. 98	Poll. Agt.—Dir, Swat, and Chitral.
Minchin, Lt.-Col. C. F., D.S.O. (Supy. L.)	15 July 91	Poll. Dept.—Divl. and Sessions Judge, Derajat.
Prescott, Maj. C. W. (Supy. List)	26 Feb. 07	Poll. Dept.—Asst. Commr., Nowshera.
Reilly, Maj. N. E.	1 June 10	Poll. Dept.—Asst. Pol. Agt., Chitral.
Wilson, Maj. D. G. (Supy. List).	9 July 06	Poll. Dept.—Dist. Judge, Peshawar.
Yule, Maj. R. A. (Supy. List).	3 Jan. 05	Poll. Offr.—(*Addl.*) Divl. and Sessions Judge, Peshawar.

RAJPUTANA.

Agent to the Governor-General and Chief Commissioner of Ajmer-Merwara.

The Hon'ble LT.-COL. J. MANNERS SMITH, V.C. C.V.O., C.I.E., 4 Jan. 1918.

(*Mount Abu.*)

Benn, Lt.-Col. R. A. E., C.I.E. (Supy. List)	21 June 95	Resdt., Jaipur.
Dawson, Capt. A. A. F. C. H., 20 Inf.	1 Sep. 12	Asst. Comdt., Mewar Bhil Corps and Asst. Pol. Supdt., Hilly Tracts, Mewar.
Dawson, Bt.-Col. C. H. (Retd.)	14 Feb. 15	Comdt., Mewar Bhil Corps & Poll. Supdt., Hilly Tracts, Mewar.
Durand, Capt. E. P. M., I. A.	19 Mar. 18	Asst. Poll. Agt., E. States.
Hamilton, Lt.-Col. R. E. A., I. A.	. . . .	First Asst. to the Agt. to the Governor-Genl., Rajputana.
Hanson, Maj. G. J. G., 69th Punj.	15 May 15	Guardian to the Maharaj Kunwar, Bikaner.
Haig, Lt.-Col. P. B., C.B., M.B., I.M.S.	16 June 03	Agcy. Surgn., 1st cl., Civil Surgn., Ajmer and C. M.O. in Rajputana.
Fisher, Lt.-Col. J., D.S.O., I.M.S.	9 May 02	(*Tempy. reverted to Mily. duty.*)
Grey, Maj. A. J. H. (Supy. List)	20 Oct. 06	Asst. Resdt., Mewar. *Lv., c.l., 1 yr. 6 mths., 12 Apl. 18.*
Kilkelly, Lt.-Col. P. P., M. B. (retd.)	1 Jan. 15	Resdy. Surgn., W. States of Rajputana.
Macpherson, Lt.-Col. A. D., I. A.	25 June 98	Poll. Agt., W. States of Rajputana.
McConaghy, Maj. C. B., M. B.	25 Apl. 04	Agcy. Surgn., E. States.
Morris, Lieut. J. C. T., I. A. R. O.	. . . .	Adjt., Mewar Bhil Corps.
Patterson, Lt.-Col. S. B., I. A.	. . . .	Commr., Ajmer-Merwara.
Peacock, Lt.-Col. H. B. (Supy. List)	15 Dec. 89	Pol. Agt., *Lv., C. L., 2yrs. 2 Aug. 18.*
Shore, Lt.-Col. R. (Retd.), M.D., I.M.S.	5 Feb. 15	Resdy. Surgn., Mewar, Udaipur, and Agcy. Surgn., Mewar Bhil Corps.
Windham, Lt.-Col. C. J., C.I.E. (Supy. List.)	30 July 93	Poll. Agt., Kotah & Jhalawar.
Williams, Maj. (Bt. Lt.-Col.) T. S. B., I.M.S.	10 Dec. 08	(*Temply. reverted to Mily. duty.*)
Wilson, Maj. F. E., M.B., I.M.S.	16 Dec. 07	(*Tempy. reverted to Mily. duty.*)

ECCLESIASTICAL ESTABLISHMENT, BENGAL.

SEE OF CALCUTTA.

Lord Bishop of Calcutta and Metropolitan in India and the Island of Ceylon.

The Most Revd. George Alfred Lefroy, D.D., D.C.L. . . . . . 19 Feb. 13

Archdeacon and Commissary.

The Ven'ble Walter Kelly Firminger, M.A., B.D., B. Lit. . . . . . 1 Jan. 14

Domestic Chaplain.

The Revd. J. Godber, M. A. . . . . . 1 Dec. 18

Registrar of the Archdeaconry and Secretary to the Most Revd. the Lord Bishop of Calcutta and Metropolitan in India and the Island of Ceylon, and Registrar of the Dioceses of Calcutta, Lahore, Lucknow and Nagpur.

H. G. Pearson, Esq. (*Bar.-at-Law*) (on leave) . . . . . 1 Nov. 06
J. W. Langford James, Esq., M. A., (*Bar.-at-Law*) Offg. . . . . .

Bishop of Lahore.

The Right Revd. Henry Bickersteth Durrant, D.D. . . . . . 6 Aug. 13

Archdeacon and Commissary.

The Ven'ble James Greensill Skottowe Syme, M.A. . . . . . 15 Nov. 16

Domestic Chaplain.

The Revd. Henry Stanley Stephenson, B.A. . . . . . 19 July 18

Bishop of Lucknow.

The Right Revd. George Herbert Westcott, D.D.† . . . . . 6 Nov. 10

Archdeacon and Commissary.

The Ven'ble Percy Hugh Chapman, M.A., LL.D. . . . . . 17 Aug. 12

Bishop of Nagpur.

The Right Revd. Eyre Chatterton,* D.D. . . . . . 3 Apl. 03

Archdeacon and Commissary

The Ven'ble Cyril Price, M.A. (207) . . . . . 22 Aug. 10

Apptd.	Names.	Date of rank.	Remarks.
	Senior Chaplains.		
1893	**Chapman,** the Ven'ble Percy Hugh, M.A., LL.D.	14 Dec. 06	Archd. Lucknow and Chapln., Nain Tal.
1893	**Collins,** The Revd. James Henry.	24 July 06	Dagshai.
1894	**Syme,** the Ven'ble James Greensill Skottowe, M.A.	24 Jan. 07	*Lv., c. l., 1 yr. 2 mths., ex I., 28 Nov. 17.*
1894	**Brookes,** the Revd. Canon Joshua Alfred Rowland, M.A.	1 Apl. 07	*Lv., c. l., 14 mths., ex I., 31 Jan. 18.*
1898	**Stuart,** the Revd. Canon Robert William Hall, B.A.	10 May 08	*Lv., ex. I., yr. 6 mths., 1 Jan. 18.*
1895	**Stanley,** the Revd. Albert Edward, M.A. . . .	14 May 08	(*Lv., ex I., 2 yrs., 28 Feb. 17.*)
1896	**Menzies,** the Revd. Canon Henry, M.A. (207) .	17 Apl. 09	Lucknow (Civil).
1899	**Canney,** the Revd. Duncan Arnold	12 May 09	Chapln., Cawanpore.
1897	**Price,** the Ven'ble Cyril, M.A. (207)	14 Apl. 10	*Lv., ex. I., 2 yrs., 20 Aug. 18.*
1900	**Muspratt,** the Revd. Walter, M.A.	18 Apl. 10	Risalpur.
1897	**Smith,** the Revd. Canon Joseph Frank, B.D., A.K.C.	9 June 10	Chapln., St. James's Church, Calcutta.
1900	**Hickox,** the Revd. Sidney Ernest, M.A. . .	8 Aug. 10	Sukkur, Sindh.
1898	**Anstey,** the Revd. Canon Harry Christopher Scott, M.A.	11 Mar. 11	Pachmarhi.
1898	**Firminger,** the *Ven'ble* Walter Kelly, M.A., B.D., B. Lit.	24 May 11	Archdn. and Chapln., St. John's Church, Calcutta.
1898	**Stewart,** the Revd. Charles, B.A.	22 Nov. 11	Campbelpore.
1899	**Stokoe,** the Revd. Cecil George, M.A. . . .	26 June 12	Chapln., St. Paul's Cathedral, Calcutta.
1900	**Clarke,** the Revd. Walrond Lugard, B.A. . .	13 Dec. 13	Chapln., Kamptee.
1901	**Keeling,** the Revd. Ernest William Philipps, B.A.	31 Aug. 16	Sabathu.
1901	**Ninis,** the Revd. Canon Richard Duncan, M.A. .	12 Mar. 14	Meerut.
1901	**Cole,** the Revd. Alex. Barnet Farquharson, M.A. (206).	28 Mar. 14	Hazara.
1904	**Markby,** the Revd. Frederick Edward, M.A. . .	29 July 14	Jullunder.
1901	**Smith,** the Revd. Henry Tregarthen Percy, M.A.	19 Aug. 14	*On Leave.*
1901	**Bell,** the Revd. William Lachlan, M.A. . .	3 Sep. 14	*Placed under Army Dept.*
1904	**Thomas,** the Revd. Edward Swayne, B.A. . .	1 Dec. 14	Ferozepore.
1901	**Kitching,** the Revd. Willie Lorimer Walton, M.A.	19 Dec. 14	Lucknow Cant.
1902	**Wheeler,** the Revd. Canon Hugh Trevor, M.A. .	21 Dec. 14	*Offg.* Archdeacon, Simla.
1902	**Drawbridge,** the Revd. William Hamilton, M.A.	4 Mar. 15	Cuttack.
1904	**Molony,** the Revd. Percival John, M.A. . . .	9 Oct. 15	Pachmari.
1902	**Parker,** the Revd. William Almair Hedley, M.A.	30 Nov. 15	Fort William, Calcutta.
1903	**Martin,** the Revd. Frederick Walter, M.A. . .	24 Dec. 15	Offg. Archdeacon & Chapln., Nagpur.
	Rintoul, the Revd. Charles Randolph, M. A. . .	6 Oct. 16	Nowshera.
1903	**Clough,** the Revd. Edmund Richard . . .	22 Nov. 16	Garrison Chaplain, Jubbulpore.

* Acts as Comsy. to the Lord Bishop of Calcutta in the Saugor and Narbada Divisions.
† Acts as Comsy. to the Lord Bishop of Calcutta in part of the Province of Agra.

Apptd.	Names.	Date of rank.	Remarks.
1903	**Fagan**, the Revd. Hugh William Farquharson, B.A.	25 Nov. 16	Dalhousie.
1903	**Crozier**, the Revd. Philip Horsfall	5 Dec. 16	Darjeeling.
1903	**King**, the Revd. John Blakeney, B.A.	20 Dec. 16	Multan.
1904	**Buckwell**, the Revd. Frederick Charles	11 Jan. 17	Delhi.
1904	**Castle**, the Revd. Willie Whichello, B.A.	29 Mar. 17	Hyderabad, Sind.
1906	**Stephenson**, the Revd. Henry Stanley, B.A.	9 June 17	Bishop's Chaplain.
1904	**Penley**, the Revd. Horace Octavius, M.A.	27 Oct. 17	Chapln. St. Thomas' Church, Calcutta.
1907	**Selwyn**, the Revd. Arthur Lewis Henry, B.A.	21 Dec. 17	*Placed under Army Dept.*
1904	**Campbell**, the Revd. Rowland William, M.A.	25 Dec. 17	Peshawar.
1908	**Irwin**, the Revd. Benjamin Christopher Bulteel, B.A.	26 Apl. 18	Allahabad Cant.
1907	**Proby**, the Revd. Rudolph Simon, Bennertz, B.A.	11 May 18	Quetta.
1905	**Maunsell**, the Revd. Arthur Persse Gabbet, B.D. (204)	25 May 18	Lahore.
1905	**Ridsdale**, the Revd. Arthur Cyril, M.A.	12 Sep. 18	Lebong.
1906	**Wardell**, the Revd. Arthur Francis Gregson	6 Nov. 18	Raipur.
1905	**Godber**, the Revd. John, M.A.	1 Dec. 18	Bishop's Chapln.

Junior Chaplains.

Apptd.	Names.	Date of rank.	Remarks.
1906	**Williams**, the Revd. Jas. Ernest Harris, M.A.	16 Mar. 09	Sialkote.
1906	**Henry**, the Revd. William Ernest Charles, M.A.	4 Aug. 09	Derajat (Kohat).
1909	**Carter**, the Revd. Bryan Bernard, B.A.	27 Oct. 09	Chapln., Sagour.
1907	**Dixon**, the Revd. Thomas Harold, M.A. (169, 204, 206)	21 Mar. 10	Rawalpindi.
1907	**O'Conor**, the Revd. William Owen, M.A.	5 May 10	*Placed under Army Dept.*
1910	**Ledgard**, the Revd. Ralph Gilbert, B.A.	6 May 10	Ranikhet.
1909	**Dyer**, the Revd. Basil Saunders, B.A.	31 Aug. 10	Barrackpur.
1910	*Barne, the Revd. George Dunsford, M.A.*	*11 May 11*	*Principal, Military Asylum, Sanawar.*
1909	**Padfield**, the Revd. George Augustus Selwyn (206)	1 Aug. 11	*On leave.*
1908	**Meyler**, the Revd. Edward Mowbray, M.A.	26 Nov. 11	Allahabad. (civil).
1909	**Crabtree**, the Revd. Harold Gaythorne, M.A. (172*a*)	24 Feb. 12	*Lv., ex I., 9 mths.*
1909	**Horwood**, the Revd. Kenric Cosens, B.A.	18 Mar. 12	Amraoti.
1912	**England**, the Revd. Herbert George, M.A. (206)	4 Nov. 12	Ambala.
1909	**Jones**, the Revd. John Strand, B.A.	21 Nov. 12	Delhi (Camp).
1912	**Clarke**, the Revd. Richard Charles Bolingbroke, M.A.	24 Nov. 12	Jubbulpore.
1909	**Irwin**, the Revd. Ronald John Beresford, D.S.O., M.C., M.A.	25 Nov. 12	*Placed under Army Dept.*
1912	**Hemming**, the Revd. Charles Henry	21 Dec. 12	Karachi.
1910	**Cotton**, the Revd. Ben, M.A.	15 Feb. 13	Bareilly.
1910	**Smith**, the Revd. Francis Herbert, B.A.	25 Mar. 13	(*On furlough.*)
1913	**Roberts**, the Revd. Arthur Betton	13 May 13	Nasirabad.
1913	**Bennet**, the Revd. Norman, M.A.	9 Aug. 13	Benares Cant.
1910	**Bridges**, the Revd. Francis Llewellyn, M.A.	21 Oct. 13	Neemuch.
1912	**Cohu**, the Revd. Clifford John, M.A.	26 Feb. 14	Jhansi.
1911	**Spooner**, the Revd. Harold, M.C.	6 Apl. 14	*Placed under Army Dept.*
1911	**Birch**, the Revd. Ormonde Winstanley, M.C.	4 Nov. 14	*Placed under Army Dept.*
1911	**Bill**, the Revd. Sydney Alfred, M.A.	15 Dec. 14	Mussoorie.
1913	**Thomson**, the Revd. Thomas Albert	5 Jan. 15	Addl. Chapln., St. John's Church, Calcutta.
1913	**Williams**, the Revd. Henry Frank-Fulford, M.A.	20 Feb. 16	*Lv. P. L., 3 mths., 1 Dec. 18.*
1913	**Carden**, the Revd. Henry Craven, M.A.	23 Feb. 16	*Placed under Army Dept.*
1913	**McKelvie**, the Revd. Robert Fitz Stanley, M.A.	2 Nov. 16	*Placed under Army Dept.*
1913	**Wilkinson**, the Revd. Ernest Roland, M.A.	4 Nov. 16	Jr. Chapln., St. Paul's Cathedral, Calcutta.
1914	**Talbot**, the Revd. Alfred Dixon	18 Oct. 17	Lucknow Cant.
1914	**Dunlop**, the Revd. Douglas Lyall Chandlee, M.A.	18 Oct. 17	Placed under Army Dept.
1914	**Maynard**, the Revd. Bertram Martin	19 Oct. 17	Muttra.
1915	**Grimes**, the Revd. Cecil John	26 Jan. 18	Placed under Army Dept.
1915	**Tambling**, the Revd. Francis Gordon Hill	11 Apl. 18	Kidderpore.
1916	**Noake**, the Revd. Arthur Raynor, M.A.	23 Jan. 18	*Placed under Army Dept.*
1916	**Young**, the Revd. Ernest Joseph, B.A.	6 Apl. 18	Chapln., St. Thomas Church, Howrah.
1917	**Newton**, the Revd. Richard Palmer, M.A.	5 Feb. 17	Dinapore.
1917	**Marshall**, the Revd. Norman Edwyn, M.A.	2 June 17	Placed under Army Dept.

Apptd.	Names.	Date of rank.	Remarks.
	Probationary Chaplains.		
1916	**Hosbons,** the Revd. Percy Rowland	10 Apl. 18	West Ridge, Pindi.
1916	**Weller,** the Revd. John Reginald, B.A. . . .	4 Oct. 16	*Placed under Army Dept.*
1916	**Gillespy,** the Revd. Francis Roebuck, B.A. . .	14 Nov. 16	Quetta., Asst. Chapln.
1917	**Saunders,** the Revd. Charles John Godfrey (137, 206)	24 Feb. 17	Chapln., Cawnpore.
1917	**Rigg,** the Revd. Arthur Cecil Peitroni, M.A. . .	16 Apl. 17	Chapln., Agra.
1917	**Broughton,** the Revd. Arthur Hardwicke, M.A. .	16 Apl. 17	Chapln., Meerut.
1917	**Kettlewell,** the Revd. Harold Herbert, B.A. .	8 Jan. 18	Lahore Cant.
1918	**Martin,** the Revd. Frederick William . . .	10 Apl. 18	Mhow, 2nd Chapln.
1918	**Hare,** the Revd. Arthur Neville, B.A.	12 Apl. 18	Fyzabad.
	Tempy. Chaplains.		
1917	**Petter,** the Revd. Lionel St. George	4 Aug. 17	Sialkot.
1918	**Howard,** the Revd. D. H., M.A.	12 Aug. 18	Lahore.

CHURCH OF SCOTLAND.

Presidency Senior Chaplain.

1897	**Thomson,** the Revd. William, M.A.	22 Feb. 10	Presy. Sr. Chapln., Church of Scotland, Bengal, and Sr. Chapln., St. Andrew's Church, Calcutta.

Senior Chaplains.

1902	**McNeill,** the Revd. John Henry Horton, M.A., B.D. (169, 206).	30 Oct. 13	*Lv., 18 mths. to 20 Jany. 18.*
1903	**Nelson,** the Revd. Allan Manson	24 Mar. 14	Lahore.
1905	**Wright,** the Revd. James Johnstone	14 Apl. 16	*Lv., 3 yrs. m. c., to 13 Apl. 19.*
1903	**Jamieson,** the Revd. Robert George, M.A. (205*a*)	26 Oct. 16	Allahabad.

Junior Chaplains.

1910	**Macpherson,** the Revd. George Cook, M.A., B.D.	5 Sep. 16	Rawalpindi.
1912	**Dodd,** the Revd. George Edward, M.A., B.D. .	21 Jan. 14	Attd. 1 Gn. Bn. Sco. Rif.
1914	**McCaul,** the Revd. Matthew Wilson, B.A. . .	30 Sep. 17	R. Sc. Fus.

Temporary Chaplains.

1916	**Deans,** the Revd. William	1 Dec. 16	Cawnpore.
1916	**Silver,** the Revd. Alexander, M.A.	1 Dec. 16	*Placed under Army Dept.*
1917	**Roche,** the Revd. G. R. M.	12 Jan. 17	Peshawar.
1917	**McLean,** the Revd. L., B.D.	16 Jan. 17	*Placed under Army Dept.*
1917	**Fairlie,** the Revd. John A., M.A.	7 Mar. 17	2nd Chapln., St. Andrew's Church, Calcutta.
1918	**MacKinnon,** the Revd. N.D., B.D.	5 Feb. 18	*Placed under Army Dept.*

ECCLESIASTICAL ESTABLISHMENT, MADRAS.

Bishop of Madras.

The Right Revd. Henry Whitehead, D.D. . . . . 29 June 99.

Archdeacon of Madras, Bishop's Commissary and Sr. Joint Chapln. of the Cathedral.

The Ven'ble Lionel Edgar Cox, M.A. . . . . 13 Mar. 10.

Actg. Domestic Chaplain to the Bishop.

The Revd. George Augustus Arthur Wright, M.A. . . . . 25 July 18.

Registrar and Secretary.

Frederick Rowlandson, Esq., B.A., LL.B. [Ootacamund] . . . . 1 Dec. 75.

Apptd.	Names.	Date of rank.	Remarks.
	Senior Chaplains.		
1897	**Bull,** the Revd. Edmund	12 Nov. 07	St. Thomas' Mount with Pallavaram.
1806	**Giles,** the Revd. Clement Douglas, M.A.	1 Mar. 09	Chapln. of Fort St. George, Madras.
1898	**Heycock,** the Revd. Francis Wheaton, M.A.	21 Mar. 11	Mercara and Mysore.
1900	**Nuttall,** the Revd. Frank, M.A.	29 Oct. 13	Ootacamund.
1900	**Creak,** the Revd. Ettrick Havelock, M.A.	16 Dec. 13	*Lv., m. c., 1 yr., 11 July 18.*
1900	**Morton,** the Revd. Bertram Mitford, M.A., A.Sc.	29 Dec. 13	Junior Joint Chapln. of St. George's Cathedral, Madras.
1901	**Flynn,** the Revd. Hugh Hamilton, A.K.C.	24 Mar. 14	Secunderabad, Deccan.
1901	**Stone,** the Revd. Henry Cecil Brough, M.A.	24 Mar. 14	*Ser. placed under Army Dept.*
1901	**Lowsley,** the Revd. Warin Ashbel, M.A.	20 Sep. 14	Coimbatore.
1901	**Jervis,** the Revd. Ernest Oswald, M.A.	8 Oct. 14	*Ser. placed under Army Dept.*
1901	**Piers,** the Revd. Samuel Octavius, M.A.	1 Dec. 14	Chapln., Bellary.
1903	**Bridge,** the Revd. Henry Noel, M.A.	10 Feb. 16	Chapln., Trichinopoly.
1903	**Proctor,** the Revd. Francis Owen, M.A.	11 Jan. 17	Chapln., Coonoor.
1904	**Wright,** the Revd. George Augustus Arthur, M.A.	1 Nov. 17	Actg. Domestic Chapln.
1905	**Sell,** the Revd. Charles Edward, B.A.	13 Oct. 08	Chapln., Cocanada.
	Junior Chaplains.		
1905	**Smith,** the Revd. George Cecil Augustus, M.A.	26 Dec. 08	Chapln., Wellington.
1906	**Brownrigg,** the Revd. Ernest Graham, M.A.	29 Sep. 09	*Ser. placed under Army Dept.*
1906	**Borlase,** the Revd. John Jennings Dingle, B.A., LL.B., LL.D.	11 Dec. 09	Chapln., St. Mark's, Bangalore.
1908	**Loasby,** the Revd. Harry Clement, M.A.	2 Feb. 10	Chapln., Vizagapatam.
1910	**Hacking,** the Revd. Henry, M.A.	18 Mar. 10	*Ser. placed under Army Dept.*
1908	**Careless,** the Revd. William Edward, M.A.	5 Feb. 11	*Lv., m. c., 1 yr., 15 Dec. 18.*
1908	**Beeley,** the Revd. Ben Darcey	23 Dec. 11	Second Chapln., Wellington.
1909	**Bull,** the Revd. Francis Faulkner	11 Oct. 12	Chapln., Bolarum, Deccan.
1910	**Jones,** the Revd. Hugh, M.A.	23 Apl. 13	Calicut and Cannanore.
1915	**Langdale-Smith,** the Revd. Richard Marmaduke, B.A.	1 Apl. 18	Hebbal Camp, Bangalore.
	Probationary Chaplains.		
1917	**Chrichton,** Walter Richard, M.A.	31 Oct. 17	Chapln., Holy Trinity Church, Bangalore
1917	**Price,** Julian Sydney	15 Apl. 17	Chapln., St. John's Church, Bangalore.
1918	**Price,** Arthur, B.A.	14 Jan. 18	Chapln., Poonamallee.
	Temporary Chaplains.		
1917	**Edmonds,** the Revd. Herbert James, M.A.	1 Sept. 17	Sr. Joint Chapln., St. Matthias' Church, Vepery, Madras.
1918	**James,** the Revd. R.	1 July 18	Chapln., Trimulgherry, Deccan.

CHURCH OF SCOTLAND.

Presidency Senior Chaplain.

1897	**Gillan,** the Revd. David Hedley, M.A., B.D.	21 Oct. 10	Presy. Sr. Chapln., and Chapln., St. Andrew's Church, Egmore, Madras.

Senior Chaplain.

1901	**Philip,** the Revd. James Gibson, M.A.	22 Oct. 11	Bangalore.

Junior Chaplains.

1907	**Mitchell,** the Revd. James Donald, M.A., B.D.	2 Feb. 09	Secunderabad.

ECCLESIASTICAL ESTABLISHMENT, BOMBAY.

Bishop of Bombay.

The Right Revd. Edwin James Palmer, M.A., D.D. (*on leave*) . . . . 28 May 08.

Archdeacon and Commissary.

The Ven'ble C. M. Barham, M.A. (*on leave*). . . . . 17 Sep. 13.

Domestic Chaplain.

The Revd. P. H. Loyd . . . .

Registrar.

John Cuthbert Greenside Bowen, Esq. . . . . 17 Nov. 94.

Apptd.	Names.	Date of rank.	Remarks.
	Senior Chaplains.		
1895	**Barham**, the *Ven'ble*, C. M., M.A.	24 Sep. 08	*Lv. C. L., 2 yrs., 20 Sep. 18.*
1899	**Kennelly**, the Revd. W. J. McK., M.A., (206), (162)	28 Sep. 09	St. Paul's Church, Poona.
1900	**Hatchell**, the Revd. Christopher Frederic Wellesley M.A.	14 Apl. 10	Chapln., Colaba and Actg. Archdeacon of Bombay.
1900	**Mould**, the Revd. Horace, B.A.	16 Nov. 10	No. 5 Training Bn. Camp, Kirkee.
1900	**Foote**, the Revd. Harold	30 June 13	*Lv., m. c. 3 mths., 14 Nov. 18.*
1902	**D'Alessio**, the Revd. Edward Samuel John, B.A..	4 Oct. 15	Chapln., Ahmedabad.
1903	**Tibbs**, the Revd. Philip Graydon	7 Apl. 16	Chapln., Kirkee.
1904	**Arnould**, the Revd. Henry Lloyd, M.A. .	11 Jan. 07	Chapln., St. Mary's, Poona.
	Junior Chaplains.		
1906	**Hill**, the Revd. Edward Eustace, M.A. . . .	11 May 09	Chapln., Spl. Duty, Poona.
1908	**Martin**, Revd. Herbert Craven, M.A. . . .	20 Jan. 10	*Lv., p. l., 3 mths., 5 Nov. 18.*
1907	**Hamerton**, the Revd. Frederick William Mountgarrett, M.A.	3 Nov. 10	Snr. Presy. Chapln., Bombay.
1908	**Collier**, the Revd. Charles Bernard Grey, M.A.	12 Feb. 11	Chapln., Ahmednagar.
1908	**Hewitt**, the Revd. George	27 June 11	*Ser. placed under Army Dept.*
1909	**Harvey**, the Revd. George Frederick, B.A. .	23 Apl. 12	Belgaum.
1909	**Sawtell**, the Revd. William Arthur . . .	17 Dec. 12	Chapln., Steamer Pt., Aden.
1910	**Ryall**, the Revd. Charles Richard, B.A. . .	29 Apl. 13	*Lv., m. c.*
1910	**Mason**, the Revd. Charles Douglas Thomas, M.A.	16 Dec. 13	Chapln., Deolali.
1911	**Dart**, the Revd. John Lovering Campbell, B.A.	30 Sep. 14	Garrn., Chapln., Bombay.
1915	**Martindale**, the Revd. H.	10 May 16	*Ser. placed under Army Dept.*
	Probationary Chaplains.		
1912	**Hall**, the Revd. John Chappé, M.A.		Chapln., Mt. Abu.
1914	**Wormald**, the Revd. Robert Leonard, M.A. .	31 Oct. 14	*Ser. placed under Army Dept.*
1917	**Ashley-Brown**, Rev. W.	8 Dec. 17	With A. Trans. Erinpura for 6 mths. from 24 Sep. 18.
1918	**Gaul**, the Revd. A. C.		Asst. Chapl., Deolali.
	Temporary Chaplains.		
	Alsop, the Revd. W. W. N., M.A.	. . . . .	Deccan War Hospital.
	Garrad, the Revd. W. R.	. . . . .	Cumballa War Hospital.
1917	**Kerridge**, the Revd. H. J.	4 Apl. 17	Pasan Camp, Kirkee, Poona.
	McFarlane, the Revd. W. E.	. . . . .	War Hosptl., Colaba.
	Scott, the Revd. F. H., M.A.	. . . . .	Chapln., Camp, Aden.
	Streatfeild, the Revd. S. F., B.A. . . .	. . . . .	Chapln, Ghorepuri.
	Bruce-Kerr, The Revd. G. H., M.A. . . .	. . . . .	Parel Hospital, Bombay.
	Twentyman, the Revd. W. P.	. . . . .	Chapln., Byculla.
	Perry, the Revd. T. T.	. . . . .	Asst. Chapl., Deolali.

CHURCH OF SCOTLAND.

Apptd.	Names.	Date of rank.	Remarks.
	Presidency Senior Chaplain.		
1897	**Matthew**, the Revd. John Crombie, B.D. . .	22 May 14	Presidcy. Sr. Chapln., Poona.
	Senior Chaplains.		
1899	**Cant**, the Revd. Alan, B.SC., B.D.	16 Dec. 10	Karachi.
1901	**Gordon**, the Revd. James Drummond, B.SC., B.D.	22 Mar. 14	Sr. Chapln., St. Andrew's Church, Bombay.
1903	**Macfarlane**, the Revd. Andrew, B.D. (206)	27 Nov. 15	Ser. placed under Army Dept.
	Junior Chaplain.		
1913	**Lee**, the Revd. Robert Ewing, B.D.	23 Oct. 16	Ser. placed under Army Dept.
1916	**MacOmish**, R.	. . . .	Temp. Chapln., Quetta.

ECCLESIASTICAL ESTABLISHMENT, RANGOON.

Bishop of Rangoon.

The Right Revd. Rollestone Sterritt Fyffe, D.D. (*Lv., ex I., 8 mos. m.c.*) . . . . 21 Jan. 10.

Registrar of the Diocese and Archdeaconry.

Clareace Hamlyn, Esq. (*Bar.-at-Law*) . . . . 20 June 02

Apptd.	Names.	Date of rank.	Remarks.
	Senior Chaplains.		
1894	**Seeley**, the Revd. George Henry	27 Dec. 08	Chapln., Maymyo.
1901	**Ellaby**, the Revd. George Alfred, M.A.	22 Jan. 14	Chapln., Rangoon Cant.
1904	**Price**, the Revd. Howel Evans	29 Aug. 17	Chapln., Shwebo.
	Junior Chaplains.		
1909	**Johnson**, the Revd. Wilfrid Harry Cowper, M.A. (*Hony.*)	23 Feb. 12	Cathedral, Rangoon.
1912	**Park**, the Revd. William Robert, B.A., C.I.E. . . .	26 Mar. 12	*Ser. placed under Army Dept.*
	Anderson, the Revd. Nicol Keith, M.A.	9 Oct. 14	Mandalay.
	Thursfield, the Revd. Gerald Arthur Richard, M.A.	18 Jan. 17	*Ser. placed under Army Dept.*
	Lister, the Revd. John George, M.A.	1 Jan. 18	Chapln., Meiktila.

CHAPLAINS OF THE WESLEYAN CHURCH.

Date of Appointment.	Names.	Remarks.
	Bengal.	
1 Jan. 96	**Buckley**, the Revd. Abner William	*Offg.* Supdtg. Wesleyan Chapln. in India, Simla.
1 Dec. 91	**Knott**, the Revd. Alfred Ernest, D.S.O.	Services placed under Army Department.
31 Jan. 95	**Walker**, the Revd. George Christopher, B.A. . .	Lucknow.
1 Mar. 95	**Thomas**, the Revd. William Wallace	Services placed under Army Department.
1 Oct. 97	**Brown**, the Revd. Alexander Dowie	Services placed under Army Department.
1 Dec. 00	**Harris**, the Revd. Theophilus, M.C.	Mhow.
1 Feb. 02	**Revnell**, the Revd. Arthur Jesse	Services placed under Army Department.
1 Feb. 02	**Atkins**, the Revd. Frederick Richard	Quetta.
1 Mar. 09	**Porri**, the Revd. Louis Peter	Ambala.
23 Nov. 13	**Hart**, the Revd. Frank	Rawalpindi.
1 Jan. 15	**Jenkin**, the Revd. Harry Justhan	Lahore.
15 Feb. 16	**James**, the Revd. Percival Walter	Services placed under Army Department.
24 June 16	**Brunt**, the Revd. Percy C., B.A.	Services placed under Army Dept.
7 Aug. 16	**Jones**, the Revd. Albert	Services placed under Army Dept.
24 Dec. 16	**Young**, the Revd. J. Mervyn	Services placed under Army Dept.
1 Apl. 17	**Gentle**, the Revd. Charles Edward	Peshawar.
1 Nov. 17	**James**, the Revd. Fred.	Calcutta.
	Madras.	
1 Nov. 16	**Vanes**, the Revd. J. Alfred, B.A.	Bangalore.
1 Feb. 18	**Wenyon**, the Revd. Francis A.	Secunderabad.
1 July 18	**Newham**, the Revd. H. Holliday	Temp. Chapln., Willington.
20 July 18	**Nightingale**, the Revd. Albert E.	Temp. Chapln., Bangalore.
	Bombay.	
1 Dec. 10	**Mawe**, the Revd. Montague Denoon, B.A. . . .	Temp. Major, R. F. A.
1 Feb. 13	*Wright, the Revd. Arthur Yeomans*	
1 Jan. 11	**Garland**, the Revd. Alfred	Bombay.
15 Nov. 16	**Sheriff**, the Revd. Thomas Henry	Temp. Chapln., Bombay.
7 Aug. 16	**Bolam**, the Revd. James Ernest	Poona.
15 May 17	**Allwood**, the Revd. Eric Sanders	Temp. Chapln., Poona.

TEMPORARY HALF-PAY LIST.

No.	Former Corps.	Names.	Date of birth.	Date of first commission.	Date of appt. to Indian Army.	Date of rank substantive.	When placed on half pay.
		INDIAN ARMY.					
		Lieutenant-Colonels.					
	R. War. R.	**Macpherson,** Roderick George (50, 71, 170)	5 Jan. 69	12 Sep. 88	7 Apl. 91	12 Sep. 14	12 Sep. 18
	Hamps. R.	**Carleton,** Henry Anthony, D.S.O. (50, 53, 66a, 71, 105, 167)	12 Sep. 68	24 Apl. 89	22 July 91	24 Apl. 15	12 Sep. 18
		Majors.					
	Unattd. List	**Crookshank,** Claude Kennedy (51, 65, 66a, 167) [*l*]	24 Apl. 80	17 Jan. 00	9 Apl. 01	15 Sep. 15	8 Apl. 17
	Conn. Rang.	**Buller,** James Henry George, D.S.O. (162, 169)	16 Aug. 81	8 Jan. 01	16 Nov. 03	8 Jan. 16	10 Aug. 18
	Ches. R.	**Cramer-Roberts,** Marmaduke Torin, D.S.O. (51, 52, 66a, 157) *Service for pension—Embd. Mila. and attd. Reglrs. from 4 Sep. 00 to 29 Apl. 02; for promotion, from 1 Mar. to 18 July 01 and from 19 Aug. 01.*	14 Dec. 80	30 Apl. 02	12 June 05	1 Apl. 16	20 Dec. 16
	Unattd. List	**Plowden,** Chas. Terence Chichele (66, 137, 162, 169, 205a)	6 Feb. 83	27 Aug. 02	13 Jan. 04	27 Aug. 17	8 Mar. 18
5	Unattd. List	**Hadow,** Hubert Rowley (50, 53, 66a, 66, 95, 167, 172a)	12 Feb. 84	21 Jan. 03	8 Apl. 04	21 Jan. 18	9 May 16
		Captains.					
	N. Staff. R.	**Harkness,** Henry Law (51, 71)	18 May 79	23 May 00	11 Oct. 02	23 May 09	12 Oct. 13
	R. A.	**MacIvor,** Ivar (50, 71)	9 Jan. 80	6 Jan. 00	24 May 03	6 Jan. 09	9 Sep. 14
	Unattd. List	**Mackenzie,** John Montolieu Hay (169)	17 Aug. 85	5 Aug. 05	8 Dec. 06	5 Aug. 14	1 Feb. 15
	S. Staff. R.	**Allen,** Guy Francis Seymour (51, 65, 71, 72b, 137) *Service for pension—In Ranks from 30 Jan. 00 to 24 June 01 and Embd. Mila., from 21 Dec. 01 to 10 Oct. 02 for promotion from 21 Dec. 01 to 10 Oct. 02, forfeited from 9 Apl. 11 to 6 Mar. 12 for promotion.*	14 Mar. 83	28 Jan. 03	12 Sep. 07	7 Mar. 12	1 Mar. 15
5	York. R.	**Bone,** Archibald Frederick (51, 162, 167, 174) [L] *Service for pension—In Ranks—Impl. Yeo. from 12 Feb. 00 to 19 Aug. 01 (half to count) and Embd. Mila. from 4 Mar. to 16 May 02; for promotion from 11 Mar. 02.*	31 May 82	17 May 02	1 Dec. 03	11 Mar. 11	18 Mar. 15
	Unattd. L.	**Joynson,** Kenneth Bright (51, 71, 167, 170)	4 Mar. 84	9 Jan. 04	9 Mar. 05	9 Jan. 15	3 Apl. 15
	Unattd. List	**Melville,** Edward Patrick Alexander (66a, 71, 167, 172a) *Forfeited from 14 Feb. to 3 Mar. 09; for promotion.*	6 Dec. 80	14 Feb. 00	10 May 01	4 Mar. 09	13 June 15
	E. Kent, R.	**Herbert-Lynch,** Edward (51, 52, 144, 156)	6 Oct. 80	21 Feb. 00	16 Sep. 03	21 Feb. 09	19 Aug. 15
	Essex R.	**Hynes,** Harry Geo. Claude (51, 52, 66, 157, 175) [*l*]	12 Jan. 83	25 Mar. 03	16 Dec. 05	25 Mar. 12	8 Oct. 16
10	R. Fus.	**Grey,** William Archibald Swinton (104)	17 July 83	4 June 04	1 Feb. 06	4 June 13	21 Sep. 16
	S. W. Bord	**Growse,** Hugh Edward, (51, 52, 66a, 95, 167)	3 Oct. 85	27 Sep. 05	23 June 10	27 Sep. 14	29 Nov. 16
	Unattd. List	**MacIntyre,** Donald (51, 52, 157)	23 June 83	19 Aug. 03	18 Dec. 04	19 Aug. 12	1 Dec. 16
	Unattd. List	* **Morton,** Nathaniel (207) (*Service for pension, from 24 Aug. 10.*)	28 Dec. 86	8 Sep. 08	26 Apl. 13	1 Sep. 15	3 June 17
14	Unattd. List	**Tranchell,** Henry George (59, 71, 137)	21 Feb. 88	25 Jan. 08	13 Mar. 09	1 Sep. 15	8 Mar. 18
		Lieutenants.					
	W. Rid. R.	**Thomas,** Tudor Ivor Griffiths (57, 71, 172l (*Ty. Capt.*)	18 Feb. 86	23 May 06	13 Sep. 03	23 Aug. 08	11 Mar. 14
2	Unattd. List.	**Dickinson,** Wm. Michael Kington (51, 52, 57, 71) (*Tempy. Capt., 14 June 16*) *Service for pension from 10 Feb. 09.*	2 May 85	22 Feb. 08	1 Apl. 10	22 May 10	25 May 15
		INDIAN MEDICAL SERVICE.					
		Captain.					
1		**Rundall,** Laurence, M.B.	1 Oct. 78	26 July 02	...	26 July 05	17 Sep. 06

* Has not passed the final Examination.

SUPERNUMERARY LIST.

OFFICERS OF THE INDIAN ARMY IN PERMANENT CIVIL EMPLOY TRANSFERRED TO THE SUPERNUMERARY LIST.

(*Vide para.* 305, *A. R., I., Vol. II.*)

Lieutenant-Colonels.

Names.	Date of transfer to Supernumerary List.	Remarks.
Stratton, W. C. R., C.I.E. (161) . . . . . .(a)	9 May 95	Poll. Employ.
Macdonald, F. W. P. (50, 94, 161, 165, 169) [L](a)	13 Apl. 98	Poll. Employ.
Minchin, C. F., D.S.O. (50, 94, 161, 165, 168, 175) .	30 Mar. 04	Poll. Employ.
Leslie, W. C. C. (161)	18 May 00	Cant. Magte.'s Dept.
Gurdon, *Hon.* P. R. T., C.S.I. (162) . . .(a)	2 June 96	Civil Employ.
VC Smith, J. M., C.V.O., C.I.E. (50) . . . . (a)	27 June 97	Poll. Employ.
Field, C. W. (74, 169)	8 Feb. 15	Cant. Magte.'s Dept.
Aplin, S. L., C.S.I. (65, 104)	15 Apl. 01	Civil Employ.
Haig, T. W., C.M.G. (106, 153, 170, 203) [L] . .	7 July 02	Poll. Employ.
Cox, *Sir* P. Z., G.C.I.E., K.C.S.I. (88, 153, 161) (*Hon. Maj.-Gen.*)	23 Dec. 00	Poll. Employ.
Douglas, M. W., C.I.E. (162, 165)	18 Sep. 00	Civil Employ.
Bayley, S. F. (161)	24 Mar. 99	Poll. Employ.
McConaghey, A., C.I.E. (57, 94, 161, 169) . . .	25 May 02	Poll. Employ.
Crichton, R. T., C.I.E. (50, 162).	16 Dec. 99	Survey Dept.
Macnabb. D. J. C., C.S.I. (104)	13 Apl. 97	Civil Employ.
Young, *Sir* F. P., K.B.E., C.I.E.	24 Dec. 98	Civil Employ.
Herbert, C., D.S.O. (156)	16 Mar. 09	Cant. Magte.'s Dept.
Waller, E. (159, 170)	14 May 03	Cant. Magte.'s Dept.
Burlton, P. S. M. (50, 170)	13 Oct. 01	Civil Employ.
Hughes, F. T. C. (50, 161)	11 Dec. 15	Poll. Employ.
Cowie, W. A. L. (107)	20 Aug. 12	Cant. Magte.'s Dept.
Bannerman, A. D'A. G., C.V.O., C.I.E. (161) .	8 Nov. 02	Poll. Employ.
Spence, P. T. A. (161, 169)	1 Jan. 00	Poll. Employ.
Thompson, C. P. (94, 169).	19 Sep. 99	Civil Employ.
Barton, A. E.	15 Dec. 01	Civil Employ.
Obbard, O. J. (104)	20 May 00	Civil Employ.
Cole, H. W. G., C.S.I. (57)	11 Apl. 00	Civil Employ.
Herbert, D.	21 Mar. 01	Civil Employ.
Beville, F. G., C.I.E. (161)	9 Jan. 00	Poll. Employ.
Halliday, H. M. (57, 90, 98)	27 Dec. 02	Civil Employ.
Wodehouse, F. W., C.I.E. (89, 133, 153) . . .	28 Jan. 01	Poll. Employ.
Roos-Keppel, *Hon. Sir* G. O., G.C.I.E., K.C.S.I., (106, 162, 199) [L]	19 Sep. 07	Chief Commr. & A. G. G., N.-W. F. Province.
Hodgson, W. G. (50, 57, 195)	3 Mar. 14	Cant. Magte.'s Dept
Merewether, J. W. B., C.I.E. (50, 88, 106, 154) .	7 Aug. 03	Poll. Employ.
Grey, W. G. (169, 204) [L].	1 May 12	Poll. Employ.
Windham, C. J., C.I.E.	30 July 03	Poll. Employ.
Peacock, H. B. (161)	15 Dec. 99	Poll. Employ.
Carter, J. R. B. G. (50, 133)	20 May 06	Poll. Employ.
Kennion, R. L., C.I.E., (161, 169)	7 Apl. 03	Poll. Employ.
Vaughan, R. R. (170)	18 May 07	Cant. Magte.'s Dept.
Benn, R. A. E., C.I.E. (94, 169)	21 June 05	Poll. Employ.
Jacob, H. F., C.S.I. (162, 175, 204)	22 Nov. 02	Poll. Employ.
Harvest, H. de V. (104, 153, 162, 204)	20 Apl. 08	Cant. Magte.'s Dept.
Pirrie, F. W., C.M.G., C.I.E. (*Bt.-Col.*) (170) . .	12 June 05	Survey Dept.
DesVoeux, H. (104).	21 Dec. 00	Civil Employ.

(*a*) Under Mily. Fur. Rules, 1875.

Lieutenant-Colonels—*concld.*

Names.	Date of transfer to Supernumerary List.	Remarks.
Stone, W. R. (104)	24 Mar. 05	Civil Employ.
Bowen, C. E. (50, 104)	25 Feb. 07	Civil Employ.
Ware, F. C. W., C.I.E. (57, 94, 170)	25 Aug. 06	Poll. Employ.
Prince, W. C. S. (50, 106, 154, 162, 169, 195, 202, 204) [L]	2 Apl. 11	Cant. Magte.'s Dept.
Morris, D. O.	17 May 03	Civil Employ.
Sykes, *Sir* P. M., K.C.I.E., C.M.G.	22 Nov. 12	Poll. Employ.
Beale, W. (89, 132)	9 Jan. 05	Poll. Employ.
Playfair, A. (50, 66, 145, 161)	19 Sep. 04	Civil Employ.
Blakeway, *Hon.* D. B., C.I.E. (161, 165, 169)	24 Nov. 04	Poll. Employ.
Condon, J. K. (104, 133, 161, 204)	1 Aug. 08	Poll. Employ.
Knox, S. G., C.S.I., C.I.E. (50, 88, 94, 100, 127, 162, 169)	22 Nov. 04	Poll. Employ.
Dew, A. B., C.S.I., C.I.E. (50, 94, 100, 156, 161, 169)	21 Dec. 07	Poll. Employ.
Mears, A. (50, 162) [L]	19 Nov. 04	Survey Dept.
Nethersole, F. R.	21 Nov. 07	Civil Employ.
Oldfield, T. A. F. R. (153)	27 June 08	Poll. Employ.
Plowden, T. C. (50, 170)	15 May 08	Civil Employ.
Burton, R. W.	18 July 16	Cant. Magte.'s Dept.
Peck, J. H. (66*a*, 94, 107, 170)	1 May 16	Cant. Magte.'s Dept.
Lyon, J. W. H. (50, 60, 148*a*, 195)	20 June 15	Cant. Magte.'s Dept.
Pottinger, R. S. (107, 162)	11 Aug. 03	Poll. Employ.
Harold, C. F. (50, 132, 154, 170)	30 Aug. 06	Poll. Employ.
Cumberlege, C. J. (50, 61, 71, 153, 169)	7 Nov. 14	Cant. Magte.'s Dept.
Roe, B. O.	31 Oct. 04	Civil Employ.
Lawrenson, T. G. P. (65, 71, 105, 169)	19 May 13	Cant. Magte.'s Dept.
O'Connor, W. F. T., C.I.E. (65, 125, 156, 162, 169, 174, 199)	26 Aug. 09	Poll. Employ.
Beadon, H. C. (169)	24 Nov. 05	Civil Employ.
Walker, F. S. (57, 169)	2 May 09	Cant. Magte.'s Dept.
Browning, T. C. (50, 59)	16 June 15	Cant. Magte.'s Dept.
Elliott, A. C. (169)	9 Mar. 03	Civil Employ.
Prideaux, F. B., C.S.I., C.I.E. (162, 170)	1 July 05	Poll. Employ.
Gough, H. A. K., C.I.E. (94, 156, 169)	24 Apl. 08	Poll. Employ.
Kennedy, *Hon.* W. M., C.I.E. (50)	20 Feb. 04	Civil Employ.
Drummond, A. B. (94, 156, 162, 169)	7 Apl. 05	Poll. Employ.
Sanford, G. B. (50, 57, 106, 165)	12 Mar. 10	Civil Employ.
Ormiston, T. L. (50, 104, 162)	28 Apl. 08	Civil Employ.
Berthon, H. W. (133)	14 Mar. 09	Poll. Employ.
Hamilton, R. E. A., C.I.E. (57, 157, 162, 169)	28 June 08	Poll. Employ.
Leslie, P. N. (50, 66*a*)	3 Aug. 17	Cantt. Magte.'s Dept.
Coghill, N. S. (89, 132)	30 May 06	Poll. Employ
Langtry, H. V. M. (164)	30 July 11	Civil Employ.
Jacob, A. L., C.I.E., O.B.E. (50, 93, 161, 169)	20 May 05	Poll. Employ
Minchin, A. B., C.I.E. (94, 162, 165, 169)	9 Nov. 06	Poll. Employ.
McConaghey, F. (94, 170)	21 June 08	Poll. Employ.
Buck, C. H.	29 Oct. 06	Civil Employ.
Irvine, A. A. (162)	14 Dec. 05	Civil Employ.
Trevor, A. P., C.S.I., C.I.E. (161, 169) [*l*]	14 June 08	Poll. Employ
Burn, G. C. (105)	22 Aug. 14	Cant. Magte.'s Dept.
O'Brien, A. J., C.I.E. (161, 204) [L]	24 Oct. 06	Civil Employ.
Wimberley, F. G. A.	24 Aug. 11	Cantt. Magte.'s Dept.
Kirkwood, A. T. (104, 167)	10 June 14	Cantt. Magte.'s Dept.
Wood, W. M. P., C.I.E. (154, 172)	16 June 09	Poll. Employ.
Macpherson, A. D. (162, 169)	25 June 08	Poll. Employ.
Smith, C. A. (144*a*, 162, 169)	28 June 09	Poll. Employ.
O'Brien, E. (88, 111, 132, 153, 162, 169)	19 Sep. 07	Poll. Employ.
Warren-Codrington, H. O. (50, 66, 153, 162)	25 July 14	Cantt. Magte.'s Dept.
Luard, C. E. (71)	9 Dec. 07	Poll. Employ.
Harrison, H. (50, 65, 133, 162, 165)	21 Dec. 06	Poll. Employ.
Ramsay, A. D. G., C.I.E., O.B.E. (88, 161, 196)	16 July 08	Poll. Employ.
Patterson, S. B. A. (66, 169)	24 June 08	Poll. Employ.
Abbey, W. B. T. (104, 117, 125*a*, 140, 162, 181, 202)	8 May 07	Civil Employ.
Keen, W. J., C.I.E. (169)	14 Oct. 08	Poll. Employ.

Majors.

Names.	Date of transfer to Supernumerary List.	Remarks.
Owens, F. C. (94, 103, 113, 162)	8 June 07	Civil Employ.
Anderson, W. C. (51, 170, 195)	31 Dec. 16	Civil Employ.
Frizelle, J. (162, 170)	18 Sep. 07	Civil Employ.
Bigg-Wither, F. (88, 100, 103, 127, 129, 169, 180, 185, 204) [L]	22 Sep. 14	Civil Employ.

Names.	Date of transfer to Supernumerary List.	Remarks.
Majors—*concld.*		
Palin, R. H. (57, 94, 169)	10 June 15	Cant. Magte.'s Dept.
Knollys, R. W. E. (50, 144*a*, 162, 169)	17 Jan. 11	Civil Employ.
Stewart, H., C.I.E. (50, 66, 161, 169)	15 July 11	Poll. Employ.
Coldstream, J. C. (94, 166, 169)	14 Apl. 10	Civil Employ.
Nicolas, F. C. (57, 65, 94, 165)	23 Oct. 12	Civil Employ.
St. John, H. B., C.I.E. (94, 162, 170)	27 May 09	Poll. Employ.
Fenning, W. M. (162)	3 Aug. 11	Cant. Magte.'s Dept.
Haworth, L. B. H. (94, 159, 165, 169) [L]	2 May 11	Poll. Employ.
Strong, H. S. (132, 153)	23 Oct. 09	Poll. Employ.
Barry, W. T. (50, 66, 71, 137)	25 Jan. 14	Cant. Magte.'s Dept.
Trotter, E. G. S., (51, 71, 125*e*, 127, 137) [L]	2 Mar. 17	Civil Employ.
James, E. H. S. (50, 57, 94, 162, 169)	3 May 11	Poll. Employ.
Hirst, F. C. (170)	27 Jan. 10	Dir. of Surveys, Bengal and Assam.
Garratt, R. (51, 157, 161, 169)	26 July 12	Poll. Employ.
Williams, G. R. K. (59, 104, 170)	20 June 12	Civil Employ.
Chenevix-Trench, R. H. (127, 162, 169)	15 Nov. 11	Poll. Employ.
Ferrar, M. L. (57, 94, 165, 169, 203) [L]	5 Sep. 11	Civil Employ.
Lorimer, D. L. R., C.I.E., (161, 169)	15 Nov. 11	Poll. Employ.
Anderson, G. H. (100, 162, 169)	9 Mar. 13	Poll. Employ.
Hutchinson, W. G., O.B.E. (51, 162, 167, 170)	5 Dec. 11	Poll. Employ.
Lyall, R. A., D.S.O. (162, 169)	6 Apl. 13	Poll. Employ.
Bruce, C. E. (72, 162, 169)	10 Nov. 11	Poll. Employ.
Keyes, T. H., C.I.E., (94, 162, 165, 167, 170) [L] (**temp. Lt. Col. 5 Jan. 18*).	2 Feb. 13	Poll. Employ.
Thuillier, L. C.	12 Sep. 11	Survey Dept.
Crosthwaite, C. G. (162, 169)	16 Jan. 13	Poll. Employ.
Mosse, A. H. E. (89, 132)	18 Dec. 11	Poll. Employ.
Burne, L. E. L. (104, 120, 140, 167, 170)	27 July 12	Civil Employ.
Burke, R. J. C. (132, 153)	30 Oct. 10	Poll. Employ.
Lord, H. P. M. (104, 137, 170) [L]	9 Dec. 14	Civil Employ.
Knowles, J. K.	16 Mar. 14	Cant. Magte.'s Dept.
Roberts, A. S. B. (71, 104)	11 Aug. 11	Civil Employ.
Scott, R. (58, 71, 169)	17 Sep. 17	Cant. Magte.'s Dept.
Grey, A.J.H. (94, 124, 125*b*, 161, 172, 175)	20 Oct. 16	Poll. Employ.
Sandeman, A.P. (104, 113), (*temp. Lt.-Col., 17 Sep. 18*)	9 Apl. 14	Civil Employ.
Yule, R. A. (51, 137, 162, 169)	3 Jan. 15	Poll. Employ.
Campbell, W. L., C.I.E. (125*b*, 125*c*, 169, 199) (L)	19 Nov. 15	Poll. Employ.
Newcombe, G. H. (104, 157, 162, 170) [L]	22 Apl. 17	Civil Employ.
Pritchard, H. R. N. (71, 169, 188*a*)	24 June 14	Poll. Employ.
Mackenzie, C. F. (137, 161) [L]	15 Feb. 15	Poll. Employ.
Walker, A. M. E. H. T. (204)	4 May 16	Cantt. Magte.'s Dept.
Daukes, C. T. (162, 169)	16 Dec. 13	Poll. Employ.
Wilson, D. G. (51, 162)	9 July 16	Poll. Employ.
Percy-Smith, E. S. (*Bt. Lt.-Col.*) (104, 125*a*, 140, 181)	4 July 13	Civil Employ.
Rae, M. E. (51, 169)	9 Mar. 17	Poll. Employ.
Oliver, D. R. G. (167, 169)	18 July 16	Poll. Employ.
Biscoe, H. V. (71, 88, 94, 100, 161, 169)	23 Apl. 16	Poll. Employ.
Prescott, C. W. (162, 172)	26 Feb. 17	Poll. Employ.
Butterfield, E., D.S.O. (51, 59, 104, 167)	1 May 16	Civil Employ.
Lawrence, H. R. (169)	28 Apl. 14	Poll. Employ.
Humphrys, F. H. (161, 166, 169)	6 Dec. 13	Poll. Employ.
Heale, R. J. W., O.B.E. (162, 165, 169, 204)	9 Jan. 13	Poll. Employ.
Brett, J. A. (12, 66*a*, 94, 16, 169)	27 Aug. 16	Poll. Employ.
Anson, G. F. W. (51, 127, 161)	8 Mar. 17	Poll. Employ.
MacDonald, R. W., D.S.O. (51, 104, 120, 120*a*, 146, 166, 169).	23 Dec. 14	Civil Employ.
Ogilvie, G. D. (161, 169)	17 Feb. 15	Poll. Employ.
Field, D. M. (66*a*, 161, 165, 169, 205*a*)	11 Mar. 17	Poll. Employ.
Bailey, F. M. C.I.E. (72*b*, 162, 199) (*temp. Lt.-Col. 1 Apl. 18*)	24 Dec. 15	Poll. Employ.
Weir, J. L. R. (71, 94, 162, 170, 199) [*l*]	21 May 18	Poll. Employ.
Crosthwaite, J. S. (94, 162, 169)	24 May 18	Poll. Employ.
Harris, T. G. M. (51, 169)	5 Mar. 17	Poll. Employ.
Hodgson, W. H. (50, 57, 98, 137)	2 Jan. 18	Civil Employ.
Gabriel, C. H. (162, 169)	19 Sep. 14	Poll. Employ.
Reilly, B. R. (66, 88, 137, 153)	17 Apl. 18	Poll. Employ.
Neale, W. G. (66*a*, 88, 161, 169)	25 July 16	Poll. Employ.
Barrett, C. C. J. (66, 66*a*, 88, 94, 132, 153, 170)	29 Oct. 17	Poll. Employ.
Meek, A. S. (57, 132, 195)	16 Oct. 17	Poll. Employ.

* Without pay and allowances of the rank.

Names, Rank, and Corps.	Date of Sub. Lt.-Col.	Date of succession to Colonel's allowance.

OFFICERS ATTAINING COLONEL'S ALLOWANCE.

AMALGAMATED LIST OF OFFICERS OF THE GENERAL LISTS IN THE THREE PRESIDENCIES SHOWING ORDER OF THEIR SUCCESSION TO THE COLONEL'S ALLOWANCE UNDER G. G. O. 473 of 1883.

ESTABLISHMENT—13.

Beckett, Col. William Henry, B.I. *u.s.l*	*20 Feb.* 85	*20 Feb.* 97
Hutchins, Col. Alfred George, M.I. *u.s.l.*	*20 Feb.* 85	*20 Feb.* 97
Quin, Maj.-Genl. Thomas James, B.I. *u.s.l.*	*4 Mar.* 85	*4 Mar.* 97
Begbie, Maj.-Genl. Elphinstone Waters, C.B., D.S.O. *u.s.l.*	*4 Apl.* 85	*4 Apl.* 97
Jennings, Genl. Sir *Robert Melvill,* K.C.B., B.C. *u.s.l.*	*20 May* 85	*20 May* 97
Marsh, Maj.-Genl. Frank Hale Berwick, B.I. *u.s.l.*	*20 May* 85	*20 May* 97
Tregear, Maj.-Genl. Sir *Vincent William,* K. C.B., B.I. *u.s.l.*	*11 June* 85	*11 June* 97
Marriott, Col. Edwin Metcalfe Leslie, B.I. *u.s.l.*	*9 July* 85	*8 Oct.* 07
Weston, Col. George Edward, M.I. *u.s.l.*	*4 Aug.* 85	*16 Apl.* 10
Chatterton, Col. Frank William, C.I.E., B.I. *u.s.l.*	*20 Sep.* 85	*19 Aug.* 10
Mainwaring, Col. Edward Phillipson, B.I. *u.s.l.*	*20 Dec.* 85	*30 Mar.* 12
Carruthers, Col. Joseph George Thompson, B.I. u.s.l.	*8 June* 86	*18 Oct.* 15
Bruce, Col. Andrew McCrae, C.B., B.I. ® *u.s.l.*	*4 Nov.* 86	*5 Nov.* 15
Stevens, Lt.-Col. Malcolm Wilkinson, Bo. I. *u.s.l.*	27 Dec. 87	

LIST OF OFFICERS ADMITTED TO THE INDIAN STAFF CORPS AFTER THE 12th SEPTEMBER 1866, AND BEFORE 1st JULY 1881, SHOWING THEIR ORDER OF SUCCESSION TO THE COLONEL'S ALLOWANCE UNDER G. G. O. No. 473 of 1883.

(The names in *italic* are those of officers who have already succeeded to the Colonel's allowance.)

LATE BENGAL—(20).

Names and rank.	Date of Sub. Lt.-Col.	Date of succession to Colonel's allowance.
M *Sconce, Col. J., u.s.l.*	*9 Dec. 79*	*9 Dec. 91*
M *Mallock, Col. H. A., u.s.l.*	*9 Dec. 79*	*9 Dec. 91*
Evans, Genl. Sir *H. M.*, K.C.B. *u. s. l.* ®	*20 June 84*	*20 June 96*
Willis, Col. J. L. N., u. s. l.	*13 Nov. 84*	*13 Nov. 96*
Wilmer, Col. J. R., u. s. l.	*22 June 85*	*22 June 97*
Pitcher, Col. D. G., u. s. l.	*3 Sep. 85*	*3 Sep. 97*
Marshall, Col. C. H. T. u.s.l.	*12 Oct. 85*	*12 Oct. 97*
Campbell, Col. J. E., u. s. l.	*10 Dec. 85*	*10 Dec. 97*
Rivaz, Col. V., C.B., *u. s. l.*	*8 June 86*	*8 June 98*
Money, Col. E. A., C.B., *u. s. l.*	*16 July 86*	*16 July 98*
King-Harman, Col. M. J., u. s. l.	*2 Oct. 86*	*2 Oct. 98*
Temple, Col. J. A., u. s. l.	*19 Dec. 86*	*19 Dec. 98*
Carey, Lt.-Col. H. R. Le M., u. s. l.	*15 Jan. 87*	*13 Feb. 00*
Turner, Col. A. H., C.B., *u. s. l.* ®	*24 May 87*	*17 Nov. 00*
VC *Hammond, Col.* Sir *A. G.*, K.C.B., D.S.O., *u.s.l.* ®	*7 June 87*	*8 Sep. 08*
Sandeman, Col. J. E., u. s. l.	*7 June 87*	*28 Aug. 09*
Brownlow, Col. C. C., C.B., *u. s. l.* ®	*8 June 87*	*27 June 10*
VC *Biddulph, Col. J., u. s. l.*	*22 Oct. 87*	*26 June 14*
Little, Lt.-Col. J. A., u. s. l.	*18 Mar. 88*	*22 Mar. 15*

Kelly, Col. J. G., C.B., *u. s. l.*
Rennick, Lt.-Col. A. deC., *u. s. l.*
Martelli, Lt.-Col. N. C., *u. s. l.*
Cuninghame, Col. D. S., *u. s. l.*
Salmon, Col. W. H., *u. s. l.*
Butler, Lt.-Col. J., *u. s. l.*
Money-Simons, Lt.-Col. J. J., *u. s. l.*
Talbot, Lt.-Col. *Sir* A. C., K.C.I.E., *u. s. l.*
Aberigh-Mackay, Lt.-Col. J. L., *u. s. l.*
Gordon, Col. R., *u. s. l*
Abbott, Col. (*Hony. Brig.-Genl.*) H. A., C.B., *u. s. l.*
Maxwell, Lt.-Col. H. St. P., C.S.I., *u. s. l.*
Clibborn, Lt.-Col. J., C.I.E., *u. s. l.*
Garbett, Col. C. H. V., *u. s. l.*
Wheatley, Col. H. S., C.B., *u. s. l.*
Wingate, Col. G., C.I.E., *u. s. l.*
Macmullen, Lt.-Col. W. H. F., *u. s. l.*
Barrow, Genl. *Sir* E. G., G.C.B. [R]
Hewett, Lt.-Col. W. S., *u. s. l.*
Rundall, Col. F. M., C.B., D.S.O., *u. s. l.*
Pears, Lt.-Col. T. C., *u. s. l.*
Lean, Lt.-Col. W. W., *u. s. l.*
Money, Col. G. A., *u. s. l.*
Read, Col. (*Hony. Brig.-Genl.*) H., C.B., *u. s. l.*
Gartside-Tipping, Lt.-Col. R. F., C.B., *u. s. l.*
Worlledge, Col. J. F. [L] *u. s. l.*
Jones, Lt.-Col. A. E., *p.s.c.* [L] *u. s. l.*
Hawkins, Bt.-Col. F., *u. s. l.*
Barrett, Genl. *Sir* A. A., K.C.B., K.C.V.O. [R]
Cook, Lt.-Col. W., *u. s. l.*
Lumsden, Lt.-Col. H. R. W., *u. s. l.*
Onslow, Bt.-Col. R. C. *u. s. l.*
Balfour, Bt.-Col. J. H.
Whistler, Bt.-Col. A. E.
Bunbury, Maj.-Genl. W. E., C.B., *p.s.c.*
Hodgson, Bt.-Col. G. B.
Thomson, Col. W. D.
Young, Maj.-Genl. C. F. G.
Younghusband, Maj.-Genl. *Sir* G. J., K.C.I.E., C.B., *p.s.c.*
Hutchinson, Col. F. P., C.B., [L]
Templer, Bt.-Col. H.
Justice, Bt.-Col. C. LeG.
Ostrehan, Lt.-Col. F. G. R., *u. s. l.*
Hancock, Bt.-Col. F. H.
Twigg, Col. R. H., C.B.

LATE MADRAS—(6).

	Date of Sub. Lt.-Col.	Date of succession to Colonel's allowances.
Bredin, Col. A., u. s. l.	*11 Mar. 85*	*11 Mar. 97*
Swinton-Skinner, Col. E., u. s. l.	*4 Jan. 86*	*7 May 09*
Cooke, Lt.-Col. C. B., u. s. l.	*8 Mar. 90*	*8 Dec. 14*
Newill, Lt.-Col. J. H. u. s. l.	*4 July 91*	*7 Dec. 16*
Robertson, Lt.-Col. Sir D., K.C.S.I., *u. s. l.*	*8 Aug. 91*	*28 Apl. 17*
Rolland, Col. (hon. Brig.-Genl.) S. E., u. s. l.	*9 Nov. 92*	*8 May 18*

Pryce, Lt.-Col. D. D., *u. s. l.*
Hawkes, Col. G., *u. s. l.*
Lawford, Col. E. E., M. *u. s. l.*
Cadell, Lt.-Col. H. F., *u. s. l.*
Hastings, Maj.-Genl. E. S., C.B., D.S.O.
Juxon-Jones, Lt.Col. J. H. *u. s.l.*
Preston, Lt.-Col. J. E., C.B., D.S.O., *u. s. l.*
Presgrave, Col. E. R. J., C.B., D.S.O., *u. s. l.*
Jones, Lt.-Col. R. G., C.I.E., *u. s. l.*
Aspinall, Lt-Col. H. H. H., *u. s. l.*
Keary, Lt.-Genl. *Sir* H. D'U., K.C.B., K.C.I.E., D.S.O.
Stevens, Lt.-Col. G. B., *u. s. l.*
Purdon, Lt.-Col. D. W., C.M.G., *u. s. l.*
King, Col. H. T., *u. s. l.*
Yate, Col. F. H.
Plumer, Lt.-Col. T. H.

LATE BOMBAY—(7).

Humfrey, Col. F. T., u. s. l.	*20 Sep. 82*	*26 Sep. 94*
Tandy, Col. H. S., u. s. l.	*12 June 87*	*12 June 99*
Hunter, Lt.-Col. J. M., C.S.I., *u. s. l.*	*12 Apl. 90*	*12 Apl. 02*
Leckie, Col. F. W. V., u. s. l.	*5 July 90*	*5 July 02*
Jones, Lt.-Col. W. H. D. u. s. l.	*20 July 90*	*7 May 03*
Forjett, Lt.-Col. F. H., u. s. l.	*10 June 91*	*31 May 16*
Melliss, Col. Sir H., K.C.S.I. Ⓡ *u. s. l.*	*10 June 91*	*23 Nov. 16*

Grant, Col. J., *u. s. l.*
Kirkwood, Col. J. N.S., *u. s. l.*
Wetherall, Col. W. A., *u. s. l.*
Macbay, Lt.-Col. W. G. W.
Walker, Col. J. N., *u. s. l.*
Yate, Col. C. E., C.S.I., C.M.G., *u. s. l.*
Thomas, Col H. R. D., *u. s. l.*
Ferris, Lt.-Col. W. B., *u. s. l.*
Aslett, Lt.-Col. W. C., *u. s. l.*
Beckham, Lt.-Col. L. G., *u.s.l.*
Peirse, Bt.-Col. C. E., *u. s. l.*
Gott, Col. G. A.
Forbes, Bt.-Col. A. W., C.B.
Aplin, Bt.-Col. P. J. H., D.S.O. [L]
Fryer, Lt.-Col. G. W. S., *u. s. l.*

RETIRED OFFICERS OF THE INDIAN ARMY LIABLE TO BE RECALLED TO ACTIVE SERVICE (A. R., I., VOL. I, Para. 700). (*Clause 99, I. A. C. of '93.*)

Names and rank.	Date of admission to I. A.	Date of birth.	
Anderson, Maj. Charles Kirkpatrick	30 Jan. 94	31 Jan. 69	1919
Badcock, Maj. George Henry	28 Apl. 94	15 June 71	1921
Barnes, Maj. Ernest	22 Apl. 92	16 Aug. 69	1919
Beaman, Maj. George Philip Roberts	25 Oct. 94	11 Mar. 69	1919
Beville, Maj. Harry Gilbert Peyton	15 Apl. 92	2 Oct. 69	1919
Brownlow, Lt. Celadon Charles Southerland	12 Mar. 07	23 Dec. 85	1930
Boys, Capt. Herbert Charles	20 June 04	20 Sep. 82	1927
Campbell, Maj. Ian. Hamilton	24 Nov. 92	9 Dec. 69	1919
Carey, Capt. William Reginald	3 July 02	3 Jan. 79	1924
Casserly, Maj. James Henry	5 June 93	25 July 69	1919
Craik, Maj. James	18 July 96	24 Nov. 71	1921
Creagh, Lt.-Col. Edward Cottingham	15 Oct. 92	8 July 69	1924
Davidson, Maj. Arthur Noel	1 Feb. 95	24 Apl. 71	1921
Delme-Radcliffe, Maj. Seymour Arthur	14 Aug. 97	26 Aug. 73	1923
Dunscombe, Maj. Nicholas Blake	22 May 94	22 July 69	1919
Fellowes, Maj. Halford Le Mesurier	13 Apl. 95	17 Aug. 74	1924
Fellows, Capt. Bertram Charles	22 Nov. 98	1 Aug. 77	1922
Fiddes, Capt. John Montgomery	7 Aug. 07	4 June 84	1929
Gabbet, Maj. Alexander Cecil	1 June 93	10 Aug. 71	1921
Gib, Maj. Howard Alaric	30 Nov. 94	31 May 71	1921
Goldthorp, Maj. Francis Herbert	4 Mar. 92	14 Aug. 70	1920
Gordon, Capt. Robert Napier Steuart	3 Oct. 99	16 Oct. 79	1924
†**Greig,** Maj. William Best	16 Mar. 92	24 Mar. 70	1920
Henderson, Capt. Joseph Hedley	15 Oct. 03	1 Apl. 79	1924
Heyland, Lt. William Frank (*a*)	1 Apl. 10	27 Oct. 89	1933
Hislop, Maj. Arthur Fowler	31 July 94	5 May 71	1921
Holland, Maj. Edward Frederick	19 Nov. 97	24 July 76	1926
Home, Capt. William Elton	17 Apl. 01	8 Feb. 81	1926
Hornby, Maj. Montague Leyland, D.S.O.	27 Mar. 92	23 July 70	1920
Hughes, Capt. Arthur Pitcairn	13 Oct. 99	5 July 78	1923
Hughes, Maj. John Edward	20 Dec. 93	22 Nov. 71	1921
Hunter-Gray, Capt. George Thornton	25 Feb. 03	17 Sep. 79	1924
Hutchinson, Maj. George Stuart Monteth	15 Mar. 03	12 Nov. 83	1933
Iles, Maj. Francis William	23 Oct. 97	27 Dec. 76	1926
Jones, Maj. Thomas William Hathway	29 Oct. 97	11 June 76	1926
Knox-Niven, Maj. Hugh William	31 Mar. 97	11 July 76	1926
Lane, Maj. Frank Macdonald	1 Mar. 92	1 Feb. 70	1920
Lash, Maj. Augustus Oliver	31 July 95	10 Feb. 70	1920
Leman, Lt. John Downton	10 Mar. 11	4 Aug. 90	1935
Lindesay, Maj. George William Guy	25 Nov. 93	5 Nov. 70	1920
Miller, Capt. Hugh	13 Apl. 97	11 Feb. 75	1920
Moore, Maj. George Dashwood Maxwell	18 Mar. 92	25 July 69	1919
Morgan, Capt. John Henry Woodside	5 Oct. 05	2 Feb. 77	1922
Myles, Lt. Andrew Vere	1 Nov. 10	25 Apl. 90	1935
Norton, Lt. William Richmond Tattenham	8 Mar. 04	1 Nov. 83	1928
Nuttall, Maj. John Rattray	26 Feb. 94	27 Nov. 69	1919
Owen, Lt. Owen	22 July 05	27 Apl. 84	1929
Paddon, Capt. Stanley Somerset Wreford	24 Mar. 05	28 Mar. 81	1926
Pritchard, Capt. Douglas James	29 Aug. 00	10 Mar. 78	1923
Reid, Lt. Alexander Daniel	11 Oct. 01	2 Feb. 82	1927
Rice, Maj. Harry Arthur Harrington	22 Apl. 93	13 Aug. 68	1918
Riley, Maj. Charles Brook	22 Aug. 98	18 Feb. 77	1927
Rimington, Lt. Michael Ballantyne	21 Sep. 05	18 May 81	1926
Rogers, Lt. Archibald Clement Campbell	25 Apl. 04	29 July 84	1929
Ross, Lt. William Robert	27 Apl. 97	9 Feb. 76	1921
Rouse, Maj. Archibald Henry Tyldem	30 Oct. 92	29 May 69	1919
Simson, Capt. Rupert	10 Mar. 06	21 Aug. 84	1929
Strong, Maj. Norman Baskervyle Glegg	30 Nov. 93	13 Oct. 71	1921
Tisdall, Lt. Archibald James	13 Mar. 14	9 Sep. 93	1938
Tristram, Maj. Charles Edward	16 Oct. 92	19 Nov. 68	1918
Turton, Maj. Hugh Mitchell	2 Feb. 94	29 Dec. 75	1920
Urmston, Capt. Herbert Neville	5 Nov. 08	24 Mar. 85	1930
Wace, Lt. Denyss Chamberlaine	30 Nov. 07	13 Jan. 86	1931
Ward, Maj. George Louis Stuart	16 Apl. 92	2 Apl. 69	1919
White, Capt. Oliver Wardlaw	11 Oct. 99	9 Dec. 78	1923
Withers, Capt. Alfred Ross	5 Nov. 01	4 Aug. 70	1923

(*a*) With Machine Gun Corps. † Recalled to Army Service from 21st September 1915.

VICTORIA CROSS, ORDERS OF KNIGHTHOOD, ETC.

OFFICERS OF THE INDIAN SERVICES AND ROYAL (LATE INDIAN) ARTILLERY AND ENGINEERS.

A LIST OF MILITARY RECIPIENTS OF THE "VICTORIA CROSS."

Adams, Maj.-Genl *Sir* R. B., K.C.B. [L] . 9 Nov. 97
Barter, Lt. F., M.C., 3rd G. R. . . . . .
Boisragon, Bt.-Col. G. H., 5th G. R. . . 12 July 92
𝕸 Cadell, Col. T., C.B. . . . . 29 Apl. 62
Chatta Singh, Sepoy, 9th Infy. . . . . 21 June 16
Cobbe, Maj.-Genl. *Sir* A. S., K.C.B., C.S.I., D.S.O., *p.s.c.* . . . . 6 Oct. 02
Costello, Maj. E. W., C.M.G., 22nd Punj. *p.s.c.* 9 Nov. 97
Creagh, Genl. *Sir* O'M., G.C.B., G.C.S.I. . 17 Nov. 79
Crimmin, Col. J., C.B., C.I.E., V.D., 𝕶. 𝕳 P. I.M.S. . . . . 17 Sep. 89
Darwan Sing Negi, Jemadar, 39th Garh Rif. 23 Nov. 14
Grant, Bt.-Col. C. J. W. (retd.) . . . . 26 May 90
Grant, Maj. J. D., 8th G. R., *p. s. c.* . . 6 July 04
Hammond, Col. *Sir* A. G., K.C.B., D.S.O. . 15 Oct. 81
Khudadad, Naik, 129th Baluchis. . . . 31 Oct. 14
Kulbir Thapa, Rifleman, 2129, 3rd G. R. . 18 Nov. 15
Lala, Lance-Naik, 501, 41st Dogras. . . 13 May 16
𝕸 Lyster, Lt.-Genl. H. H., C.B. . . . . 2 Sept. 59
Martin-Leake, Surgn.-Maj. A., F.R.C.S., V.D., 1—36th Bengal Nagpur Railway B. (*a*) (*tempy. Capt. R. A. M. C.*) . . . 13 May 02
Melliss, Maj.-Gen. *Sir* C. J., K.C.B., K.C. M.G. [L] . . . . 15 Jan. 01
Mir Dast, I.O.M., Subadar, 57th Rif. 29 June 15
𝕸 Probyn, Genl. *Rt. Hon. Sir* D. M., G.C.B., G.C.S.I. G.C.V.O., I.S.O. . . . . 18 June 58
Ridgeway, Col. R. K., C. B. . . . . . 8 May 80
Shahamad Khan, Naik, 1605, Punj . . 26 Sep. 16
Sinton, Capt., J. A., I. M. S. . . . . . 21 June 16
Smith, Lt.-Col. J. M., C.V.O., C.I.E., Supy. List. . . . . 12 July 92
Smyth, Capt, J. G., 15th Sikhs . . . 29 June 15
Walker, Maj.-Genl. W. G., C.B. . . . 22 Apl. 03
𝕸 Watson, Genl. *Sir* J., G.C.B. . . . . 18 June 58
Wheeler, Maj. George Campbell, 9th G. R. . . . . 8 June 17
Wylly, Maj. G. G. E., Guides, *p.s.c.* . . 23 Nov. 00

(*a*) Also awarded Clasp to Victoria Cross.

OFFICERS OF THE ROYAL (LATE INDIAN) ARTILLERY AND ENGINEERS.

Hill-Johnes, Lt.-Genl. *Sir* J., G.C.B. Arty. Apl. 58
𝕸 Thackeray, Col. *Sir* E. T., K.C.B., Engrs. 29 Apl. 62

KNIGHTS GRAND CROSS, KNIGHTS COMMANDERS AND COMPANIONS OF THE MOST HONOURABLE ORDER OF THE BATH.

Military Knights Grand Cross (G.C.B.).

Barrett, Genl. *Sir* A. A., K.C.S.I., K.C.V.O. 𝕬. 𝕯. 𝕮. . . . . 3 Jan. 18
Barrow, Genl. *Sir* E. G., G.C.S.I. . . . 25 June 06
𝖁𝕮 Creagh, Genl. *Sir* O'M., G.C.S.I. . 25 June 09
Egerton, Fd. Marshal. *Sir* C. C., D.S.O. . 24 June 04
𝕸 Lumsden, Genl. *Sir* P. S., C.S.I. . . . 3 July 85
Pratap Singh, *His Highness Maharaja Bahadur, Sir* G.C.S.I., G.C.V.O., A. D. C., of Jodhpur, (Hony. Lieut.-Genl.), (Hony. Col., 34th Horse). . . . . 1 Jan. 18
𝖁 𝕮. 𝕸 Probyn, Genl. *Rt. Hon. Sir* D.M., G.C.B. (Civil) G.C.S.I., G.C.V.O., I.S.O. (*Extra Eq.*) . . . . 2 June 10
Scallon, Genl. *Sir* R.I., K.C.I.E., D.S.O., . 1 Jan. 18
𝖁𝕮𝕸 Watson, Genl. *Sir* J. . . . . 26 June 02

Civil Knight Grand Cross (G.C.B.).

Nepal, Hony. Lieut.-Genl. *His Excellency Maharaja Sir* Chandra Shumsher Jung, *Rana Bahadur*, G.C.S.I., G.C.V.O. (Hony. Col., 4th G. R.), *Prime Minister and Marshal* of—(*Hony.*) . . . . . 9 July 08
𝖁𝕮𝕸 Probyn, Genl. *Rt. Hon. Sir* D. M., G.C.B. (Mily.), G.C.S.I., G.C.V.O., I.S.O. (*Extra Eq.*) . . . . 11 July 02
Ridgeway, Col. *Rt. Hon. Sir* J. W., G.C.M.G., K.C.S.I. . . . . 9 Nov. 06

Military Knights Commanders (K.C.B.).

𝖁𝕮 Adams, Maj.-Genl. *Sir* R. B. [L] . . 14 June 11
Bayly, Maj.-Genl. *Sir* A. W. L., K.C.M.G., C.S.I., D.S.O. *p.s.c.* . . . . . . 10 June 11
Bikaner, Hony. Maj.-Genl., *His Highness, Maharaja Sir* Ganga Singh, *Bahadur*, G.C.S.I., G.C.I.E., A.D.C., *Maharaja of*—2nd Lrs. . . . . 1 Jan. 18
Birdwood, Genl. *Sir* W. R., K.C.S.I., K.C.M.G., C.I.E., D.S.O. 𝕬. 𝕯. 𝕮. . . . 4 June 17
Bower, Maj.-Genl. *Sir* H. . . . . . 23 Aug. 12
Brooking, Maj. Genl., *Sir* H. T., K.C.M.G. 23 Jun. 16
Browne, Lt.-Genl. *Sir* A. G. F., D.S.O. . 25 June 09
Campbell, Lt.-Genl. *Sir* F., D.S.O. . . . 3 June 16
𝖁𝕮 Cobbe, Maj.-Genl. (*tempy. Lt.-Genl.*) *Sir* A. S., C.S.I., D.S.O. . . . . 10 Mar. 17
Cox, Lt.-Genl. *Sir* H. V., K.C.M.G., C.S.I. 3 June 18
De Brath, Lt.-Genl. *Sir* E., C.I.E., *p.s.c.* . 22 June 14
Egerton, Lt.-Genl. *Sir* R. G., K.C.I.E. . 3 June 16
Elliot, Lt.-Genl. *Sir* E. L., D.S.O. . . 26 June 02
Evans, Genl. *Sir* H. M. . . . . . . 20 June 11
Freyer, Surgn. Lt.-Col. *Sir* P.J., I.M.S. . 4 June 17
𝕸 Gatacre, Maj.-Genl. *Sir* J. . . . . 28 June 07
Grover, Genl. *Sir* M. H. S., K.C.I.E. [L] . 22 June 14
Hamilton, Surgn.-Genl. *Sir* H., M.D. . . 3 June 13
𝖁𝕮 Hammond, Col. *Sir* A. G., D.S.O. . . 1 Jan. 03
Hawkes, Maj.-Genl. *Sir* H. M. P., C.S.I. . 20 June 11
Hudson, Lt.-Genl. *Sir* H., C.I.E. . . . 3 June 18
Jacob, Lt.-Genl. *Sir* C. W. . . . . . 1 Jan. 17
Jennings, Genl. *Sir* R. M. . . . . . 25 June 09
Keary, Lt.-Genl. *Sir* H.D'U., K.C.I.E., D.S.O. 3 June 16
𝕸 Mackenzie, Col. *Sir* A. R. D. . . . 28 June 07
McLeod, Genl. *Sir* D. J. S., K.C.I.E., D.S.O. . . . . 3 June 12
Mansfield, Maj.-Genl. *Sir* H. . . . . 1 Jan. 14
Martin, Lt.-Genl. *Sir* A. R. . . . . . 24 June 10
𝖁 𝕮 Melliss, Maj.-Genl. *Sir* C. J., K.C.M.G. [L] . . . . 29 Oct. 15
Nixon, Genl. *Sir* J. E. [L] [R.] . . . . 20 June 11

KNIGHTS GRAND CROSS, KNIGHTS COMMANDERS AND COMPANIONS OF THE MOST HONOURABLE ORDER OF THE BATH—*contd.*

Military Knights Commanders (K. C. B.)—*concld.*

Pearson, Genl. *Sir* A. A. . . . . 20 June 11
Phayre, Lt.-Genl. *Sir* A. . . . . 1 Jan. 16
Powell, Maj. Genl. *Sir* C. H. [L] . . . . 1 Jan. 17
Ramsay, Maj.-Genl. *Sir* J. G. . . . . . 20 June 11
Richardson, Lt.-Genl. *Sir* G. L. R., C.S.I., C.I.E. . . . . 25 June 00
Sym, Maj.-Genl. *Sir* J. M. . . . . . . 26 June 08
𝔐 Thornton, Dy. Surgn.-Genl. *Sir* J. H., M.B. . . . . 24 June 04
Toker, Maj.-Genl. *Sir* A. C. . . . . 29 June 06
Tregear, Maj.-Genl. *Sir* V. W. . . . . 25 June 09
Watkis, Lt.-Genl. *Sir* H. B. B., *p.s.c.* . . 3 June 15
𝔐 Westmacott, Maj.-Genl. *Sir* R., D.S.O. 20 May 98
Woon, Lt.-Genl. *Sir* J. B. . . . . . . 20 June 11

Civil Knights Commanders (K.C.B.).

Chamberlain, Col. *Sir* N.F. FitzG., K.C.V.O., C.B. (*Mily.*) . . . . 11 Aug. 03
Ross, Maj. *Sir* R., K.C.M.G., F.R.S., F.R.C.S., I.M.S. . . . . 19 June 11

Military Companions (C.B.).

Abbott, Hony. Brig.-Gen. H. A. . . . . 20 May 98
𝔐 Aitchison, Genl. C. T. . . . . 29 May 75
𝔐 Anderson, Genl., H. C. . . . . 28 June 07
Atkinson, Hony. Brig.-Genl. F. G. . . . 26 June 08
Barnes, Col. O. . . . . 3 June 93
Barratt, Maj. Genl. W. C., D.S.O. . . . . 20 June 11
Barrow, Maj.-Genl. *Sir* G. deS., K.C.M.G., *p.s.c.* [*l*] . . . . 3 June 15
Beatson, Col. C. H., I.M.S . . . . 28 June 07
𝔐 Beckett, Col. S. . . . . 25 Aug. 85
Begbie, Maj.-Genl. E. W., D.S.O. . . . 20 May 96
Beville, Lt.-Col. C. H. . . . . 3 June 16
Beynon, Maj Genl. *Sir* W. G. L., K.C.I.E., D.S.O. . . . . 3 June 15
Bingley, Maj-Genl. *Sir* A. H., K.C.I.E. . . 3 June 15
Biscoe, Lt.-Genl. W. W. . . . . 20 May 96
Bookey, Col. J. T. B., I.M.S. . . . . . 29 Nov. 00
Boome, Bt.-Col., E. A. . . . . 3 June 18
Bradshaw, Maj.-Genl. L. J. E. . . . . . 28 June 07
Brander, Bt.-Col. H. R., *p.s.c.* . . . . . 16 Dec. 04
Bromhead, Col. *Sir* B. P., *Bart.* . . . . 9 Apl. 89
Brown, Lt.-Col. W. H. . . . . 3 June 16
Brownlow, Bt.-Col. C. C. . . . . . . 23 May 00
Bruce, Col. A. McC. . . . . 23 May 00
Bruce, Bt.-Col. The Hon. C. G., M.V.O . 1 Jan. 18
Buchanan, Col. (*temp. Brig.-Genl.*) K. J. . 14 Aug. 08
Bunbury, Maj.-Genl. W. E., *p.s.c.* . . . . 20 June 11
Burton, Col. E. B. . . . . 20 June 11
V.C. 𝔐 Cadell, Col. T. . . . . 28 June 07
Campbell, Col. *Sir* R. N., K.C.M.G., C.I.E., M.B. . . . . 24 June 10
Carnegy, Maj.-Genl. P. M. . . . . . . 20 June 11
Carruthers, Col. (*temp. Brig.-Genl.*) R. A., C.M.G. . . . . 14 Aug. 08
Carter, Bt. Lt.-Col. R. M., I.M.S., ... 1 Jan. 18
Cartwright, Col. C. M., C.M.G. . . . . . 3 June 16
Cassels, Bt.-Col. (*temp. Brig.-Genl.*) R. A., D.S.O., I. A. . . . . 26 Mar. 18
Chamberlain, Col. *Sir* N. F. Fitz G., K.C.B. (*Civil*), K.C.V.O . . . . 29 Nov. 00
𝔐 Chamier, Maj.-Genl. F. E. A., C.I.E. . . 28 June 07
Christopher. Maj.-Genl. L. W. . . . . . 20 May 98
Clery, Brig.-Genl. C. B. L. . . . . . . 18 Aug. 16
𝔐 Clifford, Lt.-Genl. R. C. R. . . . . . 30 May 91
Climo, Maj.-Genl. S. H., D.S.O. . . . . 3 June 16
Coats, Bt.-Col. G. H. B. . . . . . . 25 June 09
Cole, Col. E. H. C.M.G. . . . . 1 Jan. 16
Cook, Maj.-Genl. J. . . . . 3 June 99
Cookson, Maj.-Genl. G. A., C.M.G. . . . 20 June 11
Cowper, Maj.-Genl. M., C.I.E. . . . . . 3 Jun. 16
Cox, Col. F. W. H. . . . . 3 June 16
V.C. Crimmin, Col. J., C.I.E., M.D., I.M.S. 3 June 3
Cripps, Col. A. W., S. & T. Corps . . . 1 Jan. 16
Crocker, Col. (*temp. Br.-Genl.*) S. F. . . 3 June 16
Davison, Maj.-Genl. K. S. . . . . . . 26 June 08
Dawson, Col. H. L., C.V.O. . . . . . . 29 Nov. 00
Deane, Bt.-Col. G. W. . . . . 30 June 05
Delamain, Maj.-Genl. *Sir* W. S., K.C.M.G., D.S.O. . . . . 1 Jan. 14
Dick, Col. A. R. [L]. . . . . 3 June 13
Dobbie, Col. (*temp. Brig.-Genl.*) W. H. . 14 June 12
Drummond, Maj.-Genl. *Sir* F. H. R., K.C.I.E. . . . . 26 June 08
Dunsterville, Col. K. S. (retd.) . . . . 1 Jan. 18
Dunsterville, Maj.-Genl. L. C. . . . . 3 June 16
Durand, Col. A. G. A., C.I.E., *p.s.c.* . . 12 July 92
Dyce, Col. G. H. C. . . . . 21 Jan. 96
Dyer, Col. (*temp. Brig.-Genl.*), R. E. H. . 4 June 17
Eardley-Wilmot, Maj.-Genl. R. . . . . . 20 May 96
Edwardes, Brig.-Genl. I. S. M., C.M.G., D.S.O. . . . . 3 June 18
Edwards, Col. (*temp. Brig.-Genl.*) Fitz J. M., C.M.G., D.S.O., A. D. C . . . . 1 Jan. 18
Edwards, *The Hon'ble* Maj.-Genl. W. R., C.M.G., M.D., I.M.S. . . . . 22 June 14
Eustace, Maj. Genl. A. H., D.S.O. . . . 1 Jan. 16
Even, Bt.-Col. G. E. . . . . 25 June 09
Ewart, Maj.-Genl. *Sir* R. H., K.C.M.G., C.I.E., D.S.O. . . . . 3 June 15
Eyre, Bt.-Col. E. H. . . . . 26 May 94
Fairbrother, Bt.-Col. W. T. . . . . . . 25 June 09
Fane, Maj. Genl. *Sir* V. B., K.C.I.E. . . . 1 Jan. 14
Fasken, Maj.-Genl. C. G. M. . . . . . 6 Sep. 04
Fasken, Col. W. H. . . . . 1 Jan. 16
Fenton, Hony. Brig.-Genl. A. B. . . . 28 June 07
Fisher, Col. J. [L] . . . . 23 Aug. 12
Fitz-Hugh, Hony. Maj.-Genl. A. . . . . 22 Feb. 81
Fowler, Maj.-Genl. C. A., D.S.O. . . . . 1 Jan. 16
Fowler, Maj.-Genl. F. J., D.S.O. . . . . 1 Jan. 16
Fry, Maj.-Genl. C. I. . . . . 3 June 15
Gartside-Tipping, Lt.-Col. R. F. . . . . 29 Nov. 00
Geoghegan, Lt.-Col. S., *p.s.c.* [L] . . . 29 Oct. 15
Gordon, Col. J. H., D.S.O. . . . . . . 3 June 93
Gordon, Lt.-Col. J. L. R. . . . . . . 3 June 16
Gordon, Bt.-Col. L. A. . . . . 25 June 09
Grainger, Surgn.-Genl. T., M.D., I.M.S. 20 June 11
Gray, Maj.-Genl. W. duG. . . . . . . 24 June 04
Gregory, Bt. Col., C. L. . . . . 3 June 18
Gwatkin, Col. F. S. . . . . 26 June 02
Haig, Col. P. B., M.B., I. M. S. . . . . 1 Jan. 17
Hall, Hony. Brig.-Genl. L. M. M. . . . 28 June 07
Handcock, Lt.-Genl. A. G. . . . . . . 3 June 93
Hastings, Maj.-Genl. E. S. D.S.O. . . . 29 June 06
Heath, Bt.-Col. H. H. R. . . . . . . 30 June 05
Hehir, Maj.-Genl. P., C.M.G., C.I.E., M.D., F.R.C.S., I.M.S. . . . . 29 Oct. 15
Hendley, Maj.-Genl. C. E. . . . . . . 1 Jan. 16
Herbert, Maj. Genl. L., C.V.O. . . . . . 14 June 12
Herbert, Col L. N. . . . . 1 Jan. 17
Hill, Maj.-Genl. J., D.S.O. . . . . . . 1 Jan. 18
Hobday, Hony. Maj.-Gen. T. F. . . . . 22 June 97
Hogg, Maj.-Genl. G. C. . . . . 20 May 96
Holland, Maj.-Genl. H. W. . . . . . . 2 June 69
Holman, Bt.-Col., H. C., C.M.G., D.S.O. . 1 Jan. 18
Huggins, Lt.-Col. P. G., D.S.O. [L] . . . 25 June 09
Hughes, Maj.-Genl. C. F. . . . . . . 22 June 97
Hunter, Maj.-Genl. J. G. . . . . . . 3 June 13
Hutchinson, Col. F. P. [L] . . . . . . 14 June 12
Irvine, Lt.-Col. G. B., I.M.S. (retd.) . . 3 June 16
James, Col. H. . . . . 20 June 11
James, Maj.-Genl. W. B., C.I.E., M.V.O. . 1 Jan. 18
Johnston, Lt.-Col. C. A., D.S.O., I. M. S. 4 June 17

ORDER OF THE BATH—*concld.*

Military Companions (C. B.)—*concld.*

Jopp, Hony. Brig.-Genl. J.	26 May 94
Kelly, Bt.-Col. J. G.	16 July 95
Kennedy, Col. H.	3 June 16
Kerr, Bt.-Col. M. A.	16 Dec. 04
Knight, Maj. Genl., W. C., C.S.I., D.S.O., *p.s.c.* [L.]	1 Jan. 16
Knox, Bt.-Col., A. W. F.	1 Jan. 18
Lucas, Bt.-Col., F. G., D.S.O.	1 Jan. 18
Lyons-Montgomery, Col. H. F.	29 June 06
VC M Lyster, Lt.-Genl. H. H.	22 Feb. 81
McKay, Col. H. K., C.I.E., I.M.S.	29 June 06
M McInroy, Col. C.	26 May 94
Macandrew, Maj.-Genl., H. T. M., D.S.O.	1 Jan. 18
MacGeorge, Lt.-Col. H. K.	18 Aug. 16
Macintyre, Maj.-Gen. D. C. F.	23 Aug. 11
Mackenzie-Kennedy, Maj.-Genl. E. C. W.	24 June 10
M Maclean, Maj.-Genl. C.S., C.I.E.	22 Feb. 81
MacLeod, Bt.-Col. R. W.	28 June 07
Macnab, Bt.-Col. A. J., I.M.S.	3 June 18
Maconchy, Col. (Hony. Brig.-Genl.) E. W. S. K., C.M.G., C.I.E., D.S.O.	20 June 11
Maitland, Maj.-Genl. P. J.	20 May 96
M Malcolmson, Maj.-Genl. J. H. P.	19 July 79
Malleson, Maj.-Genl. W., C.I.E.	18 Aug. 16
Manifold, Lt.-Gl. C.C., C.M.G., M.B., I.M.S.	22 June 14
Manning, Hon. Brig.-Gen. *Sir* W. H., K.C.M.G.	7 Aug. 03
Massy, Col. H. S.	26 June 03
Masters, Col. A.	24 June 04
Mayne, Col. R. C. G.	4 Oct 98
Miles, Col. P. J.	3 June 18
Molesworth, Hony. Brig.-Genl. E. H.	28 June 07
Money, Col. E. A.	22 June 97
Monteith, Bt.-Col. J.	29 June 06
Muir, Col. C. W., C.I.E.	26 June 02
Murray, Lt.-Col. C. F. T., M.V.O., S. & T. Corps	3 June 16
Muspratt, Col., F. C., C.M.G.	1 Jan. 18
Norie, Col. C. E. de M., C.M.G., D.S.O.	3 Jan. 16
O'Connor, Lt.-Col. P. F., I.M.S.	26 June 02
O'Malley, Bt.-Col. W. A. D'O.	25 June 09
Palin, Col. G. W., C.I.E.	22 June 14
Palin, Maj.-Genl. P. C., C.M.G.	3 June 16
Patch, Col. R.	20 May 96
Peterson, Col. F. H., D.S.O. [*l*]	1 Jan. 16
Pirie, Maj.-Genl. C. P. W.	1 Jan. 16
Pollock, Maj.-Genl. J. A. H.	26 June 03
M Pratt, Col. H. M.	10 Apl. 89
Prendergast, Col. C.G.	1 Jan. 18
Presgrave, Col. E. R. J., D.S.O.	26 June 08
Preston, Lt.-Col. J. E., D.S.O.	25 June 09
Price, Col. C. H. U., D.S.O.	22 June 14
Pulley, Bt.-Col. C.	24 June 04
Radford, Bt.-Col. O. C., C.I.E.	30 June 05
Rainey-Robinson, Col. R. M., C.M.G.	14 June 12
Rattray, Bt.-Col. C.	4 June 17
Read, Hony. Brig.-Genl. H.	16 Dec. 04
VC Ridgeway, Col. R. K.	30 June 05
Ridgeway, Bt.-Col. R. T. I.	3 June 16
Rivaz, Col. V.	23 May 00
Roberts, Col. (*temp. Brig.-Genl.*), H. L., M.V.O., (*l*)	4 June 17
Robinson, Lt.-Col. G. W.	3 June 16
Robinson, Maj.-Genl. W. H. B., I.M.S.	29 Oct. 15
Roche, Col. H. J.	18 Feb. 15
Roddy, Lt.-Col. H. H.	3 June 16
Rodwell, Col. E. H., *p.s.c.* [L].	20 June 11
Rundall, Bt.-Col. F. M., D.S.O.	28 June 07
Schalch, Col. V. A.	24 Jan. 04
Scott, Maj.-Genl. T. E., C.I.E., D.S.O.	4 June 17
Selwyn, Col. C. H.	1 Jan. 17
Seton, Bt.-Col. *Sir* B. G., *Bart,* I.M.S.	4 June 17
Shakespear, Col., L. W.	3 June 18
Shea, Maj.-Genl J. S. M., C.M.G., D.S.O., *p.s.c., q.s.*	18 Feb. 15
Shore, Col. O. B. S. F., C.I.E., D.S.O., *p.s.c.* [L]	1 Jan. 14
Smith, Lt.-Col., J. B., M.B., I.M.S.	3 June 18
Smyth, Bt.-Col. E. W.	21 Jan. 96
Somerset, Col. (*temp. Brig.-Genl.*), C. W., C.M.G., M.V.O.	1 Jan. 18
Steele, Col. St. G. L. [L].	20 June 11
Stevenson, Bt.-Col. F.	29 June 05
M Stewart, Hon. Maj.-Genl. G.	21 June 87
M Stewart, Maj.-Genl. J. C.	30 May 91
Stewart, Maj.-Genl. J. M.	20 June 11
Stewart, Maj.-Genl. *Sir* N. R., *Bart.*	29 Nov. 00
Sullivan, Col. E. L., C.M.G.	3 June 18
Swann, Maj.-Genl. J. C.	6 Sep. 04
Swayne, Bt.-Col. *Sir* E. J. E., K.C.M.G.	6 Sep. 04
Thackwell, Col. C. G. R., D.S.O.	14 June 12
Thomas, Col. F. H. S., S. & T. Corps	1 Jan. 10
Tighe, Maj.-Genl. *Sir* M. J., K.C.M.G., C.I.E., D.S.O.	20 June 11
Tulloch, Maj.-Genl. J. W. G.	16 Mar. 06
Turner, Col. A. H.	22 June 97
Turner, Maj.-Genl. J. G. [L].	20 June 11
Twigg, Col. R. H.	3 June 13
M Upperton, Maj.-Genl. J.	18 Nov. 82
Vaughan, Bt.-Col., (*temp. Brig.-Genl.*), R.E.	1 Jan. 18
Vaughan, Bt.-Col., L.R., D.S.O.	3 June 18
Waddell, Lt.-Col. L. A., C.I.E., M.B.	16 Dec. 04
Wadeson, Maj.-Genl. F. W. G.	3 June 15
VC Walker, Maj.-Genl. W. G.	22 June 14
Wallace, Maj.-Genl. A.	29 June 08
Walton, Col. W. C., C.M.G.	4 June 17
Wapshare, Lt.-Genl. R., C.S.I.	3 June 15
Watson, Col. H.D., C.M.G., C.I.E., M.V.O.	3 June 18
Watson, Maj.-Genl. W. A., C.M.G., C.I.E., *p.s.c.*	1 Jan. 14
Welchman, Bt.-Col. E. W. St. G.	28 June 07
Westropp, Col. G. R. C.	20 June 11
Wheatley, Col. H. S.	30 June 05
White, Bt.-Col. W. W., M.D., I.M.S.	1 Jan. 16
Wickham, Col. W. J. R.	22 Aug. 02
Widdicombe, Lt.-Col. G. T.	3 June 15
Wigram, Bt.-Col. K., D.S.O.	3 June 18
Willoughby, Col. I.M.E., C.M.G.	3 June 18
Woodyatt, Maj. Genl. N. G.	3 June 16
Yeilding, Col. W. R., C.I.E., D.S.O.	26 June 08
Young, Hony. Brig.-Gen. G. F.	22 June 97
Younghusband, Maj.-Genl. *Sir* G. J., K.C.M.G., K.C.I.E., *p.s.c.*	29 Nov. 00
Younghusband, Col., (*temp. Brig.-Genl.*) L. N., C.M.G.	4 June 17

Civil Companions (C. B.).

Cordeaux, Maj. H. E. S., C.M.G.	9 Nov. 04
Durand, Lt.-Col. *Sir* E.L., *Bart.*	13 Feb. 87
Forbes, Brig.-Genl. A. W.	1 Jan. 17
Harrington, Hony Lt.-Col. *Sir* J. L., K.C.M.G., K.C.V.O.	11 July 17
Hayes-Sadler, Lt.-Col. *Sir* J., K.C.M.G.	11 Jul 17
Wigram, Bt.-Col. C., C.S.I., C.V.O. *p.s.c.*, 18th Lrs. *Eq.*	3 June 18

OFFICERS OF THE ROYAL (LATE INDIAN) ARTILLERY AND ENGINEERS.

Military Knights Grand Cross (G. C. B.)

VC M Hills-Johnes, Lt.-Genl. *Sir* J., Arty.	3 June 93

Military Knights Commanders (K. C. B.)

Chapman, Genl. *Sir* E. F., Arty., *p.s.c.*	29 June 06
Graham, Maj.-Genl. *Sir* T., Arty.	26 June 08

OFFICERS OF THE ROYAL (LATE INDIAN) ARTILLERY AND ENGINEERS—*concld.*

Civil Knight Commander (K. C. B.)

𝔙ℭ 𝔐 Thackeray, Col. *Sir* E. T., C.B., (Mil.) Engrs. . . . . 22 June 94

Military Companions (C. B.)

𝔐 Bonham, Hony. Col. J., Arty. . . . . . 29 June 06
Borradaile, Col. G. W., Arty. . . . . . 18 Nov. 82
Caldecott, Maj.-Genl. F. J., Arty. . . . 26 May 94
Lovett, Maj.-Genl. B., C.S.I., Engrs. . . 10 Apl. 89
Swinley, Maj.-Genl. G., Arty. . . . . . 3 June 93
𝔙ℭ., 𝔐. Thackeray, Col. *Sir* E.T., K.C.B. (Civil), Engrs. . . . . 22 June 94
Tillard, Maj.-Genl. J. A., Arty. . . . . 1 July 87
𝔐 Ward, Maj.-Genl. F. W., Arty. *p.s.c.* . 26 May 94
Wauhope, Bt.-Col. R. A., C.M.G., C.I.E., Engrs. . . . . 30 June 05

Civil Companions (C. B.)

Trotter, Lt.-Col. *Sir*, H. K.C.M.G., Engrs. . 20 Apl. 80
Tulloch, Maj. H., Engrs. . . . . 8 Jan. 95

KNIGHTS GRAND COMMANDERS, KNIGHTS COMMANDERS AND COMPANIONS OF THE MOST EXALTED ORDER OF THE STAR OF INDIA.

Knights Grand Commander (G.C.S.I.).

Barrow, Genl. *Sir* E. G., G.C.B. . . . . 18 Feb. 16
Bikaner, Hony. Maj.-Genl. *His Highness* Maharaja *Sir* Ganga Singh, *Bahadur,* G.C.I.E., K.C.B., A. D. C., *Maharaja* of —2nd Lrs. . . . . 12 Dec. 11
𝔙ℭ Creagh, Gen. *Sir* O'M., G.C.B. . . 12 Dec. 11
Gwalior, Hony. Lieut.-Genl. *His Highness Sir* Madho Rao Scindia, *Bahadur,* G.C.V.O., G.B.E., A.D.C., *Maharaja* of—(Hony.-Col., 1st Lrs) . . . . . . 25 May 95
Hewett, *Hon. Sir* J. P., C.I.E., late Hon. Col., Ind. Vols. . . . . 12 Dec. 11
Hyderabad, Hony. Liet.-Genl. *His Exalted Highness Nawab Mir Sir* Usman Ali Khan *Bahadur, Fateh Jang,* G.B.E., *Nizam* of—(Hony. Col., 20th H.) . . 12 Dec. 11
Jaipur, Hony. Maj.-Genl. *His Highness Maharajadhiraja Sir* Sawai Madho Singh, *Bahadur,* G.C.I.E., G.C.V.O.,G.B.E., *Maharaja* of—(Hony.-Col., 13th Rajputs) 1 Jan. 88
Jammu and Kashmir, Hony. Lt.-Genl. *His Highness Maharaja Sir* Partab Singh, *Bahadur,* G.C.I.E., G.B.E., *Maharaja* of —(Hony.-Col., 37th Dogras) . . . . 25 May 92
Kapurthala, Hony. Lieut.-Col. *His Highness Maharaja Sir* Jagatjit Singh, *Bahadur, Maharaja of*— . . . . 12 Dec. 11
Kolahpur, Hony. Col., *His Highness Sir* Shahu Chhatrapati *Maharaj,* G.C.I.E., G.C.V.O., L.L.D., *Maharaja* of — (Hony. Col., 103rd Inf.) . . . . 1 Jan. 95
Kotah, Hony. Lt.-Col. *His Highness Maharao Sir* Umed Singh, *Bahadur,* G.C.I.E., G.B.E., *Maharao* of—(42nd R.) . . 12 Dec. 11
Mac Donnell, *Rt. Hon.* A.P. Lord, K.C.V.O., late Hon. Col., Ind. Vols. . . . . . 22 June 97
Mysore, Hony. Col. *His Highness Maharaja* Sri Krishnaraja Wadiyar *Bahadur,* G.B.E., *Maharaja* of—(Hony. Col., 26th Lt. Cav.) . . . . 1 Jan. 07
Nepal, Hony. Lieut.-Genl. *His Excellency Maharaja Sir* Chandra Shamsher Jang, *Bahadur, Rana,* G.C.B., G.C.V.O. (Hony. Col., 4th G.R.), *Prime Minister and Marshal* of— . . . . 2 Jan. 05
Pentland, *Rt. Hon.* J., Baron — of Lyth, G.C.I.E., Hony. Comdt. 1st Mad. Gds., I. D. F. . . . . 3 June 18
Pratap Singh, *His Highness Maharaja Bahadur, Sir,* G.C.B., G.C.V.O., A.D.C., of Jodhpur (Hony. Lt.-Genl.), (Hony. Col., 34th Horse) . . . . 22 June 97
𝔙ℭ 𝔐. Probyn, Genl. *Rt. Hon. Sir* D. M., G.C.B., G.C.V.O., I.S.O. . . . . . . 20 June 11
Rewa, Hony. Lt.-Col. *His Highness Maharaja Sir* Venkat Raman Singh, *Bahadur,* Maharaja of — . . . . . . 22 June 97
Willingdon, *Rt. Hon.* F., Baron — of Ratton, G.C.I.E., G.B.E., Hony. Comdt., 4th Bo. L. Horse, I. D. F. . . . . 3 June 18

Knights Commanders (K.C.S.I.).

Alwar, Hony. Lt.-Col. *His Highness Maharaja Sir* Jey Singh, *Bahadur,* K.C.I.E., *Maharaja* of — . . . . 1 Jan. 09
Baillie, Maj. D. C., late Ind. Vols. . . . 1 Jan. 14
Barnes, *Sir* H. S., K.C.V.O., late Hon. Col., Ind. Vols. . . . . 1 Jan. 03
Barr, Lt.-Col. *Sir* D. W. K. . . . . 26 June 02
Barrett, Genl. *Sir* A. A., G.C.B., K.C.V.O., 𝔄. 𝔇. ℭ. . . . . 29 Oct. 15
Bhavnagar, Hony. Lieut.-Col. *His Highness Maharaja Sir* Bhavsinhji Takhatsinhji, *Maharaja of*— . . . . . . 24 June 04
Bhim Shumshere Jung Bahadur, Rana, K.C.V.O., *Genl. Sir C.-in.C., in Nepal* (Hony) . . . . 1 Jan. 17
Birdwood, Genl. *Sir* W.R., K.C.B., K.C.M.G., C.I.E., D.S.O., *q.s.,* 𝔄. 𝔇. ℭ. . . . . 1 Jan. 15
Butler, *Hon'ble* Hony. Co *Sir* S. H., C.I.E., Rangoon Rif. I. D. F. . . . . .
Colvin, *Hony. Col.* Hon. *Sir* E. G., late Ind. Vols. . . . . 12 Dec. 11
Cox, Lt.-Col. *Sir* P. Z., G.C.I.E. . . . . 29 Oct. 15
Daly, Lt.-Col. *Sir* H., K.C.I.E. . . . . 3 June 16
Dunlop-Smith, Lt.-Col. *Sir* J. R., K.C.V.O., C.I.E. . . . . 1 Jan. 10
Gait, *Hon'ble* Lt.-Col. *Sir* E. A., C.I.E., 1st Bihar L. Horse, I. D. F. . . . . 3 June 15
Hooper, Hony. Col. *Sir* W. R., I.M.S., 𝔎.𝔋.𝔖. . . . . 1 Jan. 03
Jind, Hony. Lieut.-Col. *His Highness Maharaja Sir* Ranbir Singh *Rajendra Bahadur,* G.C.I.E., *Maharaja of*— . . . . 1 Jan. 09
Jitendra Narayan Bhup Bahadur, *H. H. Maharaja—of Cooch Behar,* Hony. Lt., N. B. Mtd. Rif. . . . . 23 Dec. 16
Kishengarh, Hon. Lt.-Col. *His Highness Maharajadhiraj Sir* Madan Singh, *Bahadur,* K.C.I.E., *Maharaja* of— . . . 12 Dec. 11
Lethbridge, Brigade-Surgn.-Lieut.-Col. *Sir* A. S., M.D. . . . . 20 May 96
Lovett, *Hon'ble* Lt.-Col. *Sir* H. V., Late Ind. Vols. . . . . 1 Jan. 16
Maler Kotla, Hony.-Maj. *His Highness, Nawab Sir* Ahmed Ali Khan *Bahadur, Nawab* of—
Melliss, Col. *Sir* H. . . . . 22 June 97
Navanagar, Hony. Lt.-Col., *His Highness Maharaja Shri Sir* Ranjit Sinhji Vibhaji, *Maharaja of*— . . . . 24 June 04
𝔐. Newmarch, Maj.-Genl. *Sir* O. R. . . 26 May 94
O'Dwyer, *Hon'ble* Hony. Col. *Sir* M. F., G.C.I.E., Punj. L. Horse, I. D. F. . . . 3 June 13

KNIGHTS GRAND COMMANDERS, KNIGHTS COMMANDERS AND COMPANIONS OF THE MOST EXALTED ORDER OF THE STAR OF INDIA—*contd.*

Knights Commanders (K.C.S.I.)—*contd.*

Ratlam, Hony. Col. *His Highness Sir* Sajjan Singh, *Raja* of—(39th H.) . . 25 June 09
Ridgeway, Col. *Rt. Hon. Sir* J. W., G.C.B., G.C.M.G. . . . . 20 July 85
Robertson, Lt.-Col. *Sir* D. . . . . . . 1 Jan. 03
Robertson, *Hon'ble* Hony. Col. *Sir* B., K.C.M.G., C.I.E., 1st Madras Guards . . 12 Dec. 11
Roos-Keppel, *Hon'ble* Lt.-Col. *Sir* G. O., G.C.I.E. [L]. . . . . 3 June 13
Sirmur, Hony. Lieut.-Col. *His Highness Maharaja Sir* Amar Parkash *Bahadur, Maharaja of*— . . . . 3 June 15
Younghusband, Lt.-Col. *Sir* F. E., K.C.I.E. 1 Jan. 17

Companions (C.S.I.).

Anderson, Maj. A. J, 3rd Gp. Garr. Arty. I.D.F. . . . . 23 Aug. 18
Aplin, Lt.-Col. S. L. . . . . 3 June 16
Archer, Lt.-Col. C., C.I.E. . . . . 12 Dec. 11
Banatvala, *Hon'ble* Col. H. E., K.H.P. . 4 June 17
Bannerman, Surgn.-Gen. W. B., M.D., I.M.S. 12 Dec. 11
Bayly, Maj.-Gen. *Sir* A. W. L., K.C.B., K.C.M.G., *p.s.c.* . . . . 1 Jan. 09
Beatson-Bell, *Hon'ble* Maj. N. D., C.I.E., 44th Cal. Sco. . . . . 3 June 15
Buckley, the Hon'ble L. E., Lt.-Col., S.P. M. Rif. . . . . 3 June 18
Carey, Lt.-Col. B. S., C.I.E., V.D., U. Bur. Rif. . . . . 22 June 14
Cleghorn, Surgn.-G J. M.D., K.H.S. . . 22 June 97
V.C. Cobbe, Maj.-Genl. (*temp. Lt.-Genl.*) *Sir* A. S., K.C.B., D.S.O. . . . . . . 1 Jan. 18
Cole, Lt.-Col. H. W. G. . . . . . . 1 Jan. 13
Colvin, *Hon'ble* Hony. Col. *Sir* E. G., late Ind. Vols. . . . . 29 June 06
Cox, Lt.-Genl. *Sir* H. V., K.C.B, K.C.M.G. 12 Dec. 11
Curtis, Capt. G. S., Poona Rif. I. D. F. . . 12 Dec. 11
Dallas, Lt.-Col. C. M. . . . . 12 Dec. 11
Dew, Lt.-Col. A. B., C.I.E. . . . . . . 3 June 15
M Dickinson, Lt.-Col. W. . . . . . 2 June 69
Eliott, Lt.-Col. F. H. . . . . 3 June 15
Gabriel, Maj. E. V., C.M.G., C.V.O., 1-17th C. I. Ry. Bn. I. D. F. . . . .
Giffard, Lt.-Col. G. G., I.M.S. . . . . 3 June 13
Goad, Col. H. . . . . 28 June 07
M Grey, Col. L. J. H. . . . . 1 Jan. 77
Gurdon, *Hon'ble* Lt.-Col. P. R. T. . . . 12 Dec. 11
Hailey, *Hon'ble* Lt.-Col. W. M., C.I.E., 3rd Punj. Rif. . . . . 3 June 15
Haji Hafiz Obeidullah Khan, A.D.C., *Nawabzada*, Hony. Maj. (Brig. Genl. & C.-in-C.) Bhopal State Forces) . . . . 14 June 12
Harris, Surgn.-Genl. G. F. A., M.D., F.R.C.P., I.M.S. . . . . 12 Dec. 11
Harrison, Col. T. A. . . . . 3 June 18
Hawkes, Maj.-Genl. *Sir* H. M. P., K.C.B. . 25 June 09
Herbert, Lt.-Col. C. . . . . 1 Jan. 09
Hopwood, Bt.-Col. (*temp. Maj.-Genl.*) H. R., I. A. . . . . 1 Jan. 18
Hose, Lt.-Col. J. W., Agra Rif. . . . . 12 Dec. 11
Hunter, Lt.-Col. J. M. . . . . 3 June 99
Hutchinson, Lt.-Genl. H. D. . . . . . 1 Jan. 03
Hutchinson, Lt.-Col. J. B. . . . . . . 24 June 04
Impey, Lt.-Col. L., C.I.E. . . . . . 4 June 17
Jacob, Lt.-Col. H. F. . . . . . . 4 June 17
Kaye, Lt.-Col. C., C.I.E. . . . . 1 Jan. 17
Kennedy, Lt.-Col. W. P. . . . . . . 29 June 06
Knight, Maj.-Genl. W. C., C.B., D.S.O. 1 Jan. 17
Knox, Lt.-Col. S. G., C.I.E. . . . . . . 4 June 17
Lambagraon, Hony. Lt.-Col. *Raja Sir* Jai Chand, K.C.I.E. *Raja* of— . . . . . 1 Jan. 13
Lawrence, the Hon'ble H. S., Hony. Comdt., Sind. Rif. I. D. F. . . . . 3 June 18
LeMesurier, Lt.-Col. H., C.I.E., E. Bengal Rif. . . . . 24 June 10
M Lumsden, Genl. *Sir* P. S., G.C.B. . 25 Mar. 70
Macnabb, Lt.-Col. D. J. C. . . . . . 1 Jan. 13
Marlow, Col. B. W., C.I.E. . . . . . 4 June 17
Maxwell, Lt.-Col. H. St. P. . . . . . 25 May 92
McMahon, Lt.-Col. *Sir* A. H., G.C.M.G., G.C.V.O., K.C.I.E. . . . . 22 June 97
Miley, Col. J. A. . . . . 22 June 97
Montgomery, Lt.-Col. J. A. L. . . . . 1 Jan. 03
Prideaux, Lt.-Col. F. B., C.I.E. . . . . 4 June 17
Ramsay, Lt.-Col. *Sir* J., K.C.I.E. . . . 12 Dec. 11
Reid, Col. L. H. . . . . 14 June 12
Richardson, Maj.-Genl. C. W. G. . . . 1 Jan. 18
Richardson, Lt.-Genl. *Sir* G. L. R., K.C.B., C.I.E. . . . . 24 July 01
Saunders, *Hon'ble* Lt.-Col. A. L., V.D., 10th Naini Tal Coy., I. D. F. . . . . 3 June 13
Sinclair, Surgn. Genl. D., M.B. . . . 31 Dec. 98
Sirmur, Hony. Lt.-Col. *Kumar* Bir Bikram Singh, of—, C.I.E. 1st S. & M. . . . 22 June 14
Stevenson, Surgn.-Genl. H. W., I.M.S. . 14 June 12
Taylor, Lt.-Col. A. W. N. . . . . . . 23 Aug. 18
Trevor, Maj. A. P., C.I.E., Poll. Resdt. . 1 Jan. 18
Trevor, Col. G. H. . . . . 30 May 91
Walsh, *the Hon'ble* Capt. E. H. C., 1st Bihar L. H. . . . . 12 Dec. 11
Wapshare, Lt.-Genl. R., C.B. . . . .
Wigram, Bt.-Lt.-Col. C., C.B., C.V.O., 18th Lrs. *p.s.c.*, Eq. . . . . 12 Dec. 11
Willoughby, Lt.-Genl. M. W. . . . . 6 June 85
Woods, Lt.-Col. A. E., Supy. List. . . 14 June 12
Wylie, Maj.-Genl. H. . . . . 22 Feb. 81
Yate, Col. C. E., C.M.G. . . . . . . 15 Feb. 87
Younghusband, *Hony.* Col. A. D., late Ind. Vols. . . . . 28 June 07

OFFICERS OF THE ROYAL (LATE INDIAN) ARTILLERY AND ENGINEERS.

Companions (C.S.I.).

Home, Col. F. J., Engrs. . . . 25 May 92
Lovett, Maj.-Genl. B., C.B., Engrs. . . 6 Feb. 73

KNIGHTS GRAND CROSS, KNIGHTS COMMANDERS AND COMPANIONS OF THE MOST DISTINGUISHED ORDER OF SAINT MICHAEL AND SAINT GEORGE.

Knight Grand Cross (G.C.M.G.).

Ridgeway, Col. *Rt. Hon.* Sir J. W., G.C.B., K.C.S.I. . . . . 1 Jan. 00
McMahon, Lt.-Col. *Sir* A. H., G.C.V.O., K.C.I.E., C.S.I. . . . . 14 Nov. 16

KNIGHTS GRAND CROSS, KNIGHTS COMMANDERS AND COMPANIONS OF THE MOST DISTINGUISHED ORDER OF SAINT MICHAEL AND SAINT GEORGE —*contd.*

Knights Commanders (K.C.M.G.).

Barrow, Maj.-Genl. *Sir* G. de S., C.B. . . . 1 Jan. 18
Bayly, Maj.-Gen. *Sir* A. W. L., K.C.B., C.S.I., D.S.O., *p.s.c.* . . . 24 Jan. 17
Birdwood, Genl. *Sir* W. R., K.C.B., K.C.S.I., C.I.E., D.S.O., *q.s.*, A.D.C. . . . 3 Jan. 15
Brooking, Maj.-Genl. *Sir* H. T., K.C.B. . . 19 Oct. 17
Campbell, Col. *Sir* R. N., C.B., C.I.E., I.M.S. 4 June 17
Carmichael, Hony. Col. *Rt. Hon.* T. D. *Baron*—of Skirling, G.C.I.E., late Ind. Vol. 3 Nov. 11
Cox, Lt.-Genl. *Sir* H. V., K.C.B., C.S.I . . 29 Oct. 15
Delamain, Maj. Genl. *Sir* W. S., C.B., D.S.O. 23 Aug. 18
Ewart, Maj.-Genl. *Sir*, R. H., C.B., C.I.E., D.S.O. . . . 1 Jan. 18
Harrington, Hony. Lt.-Col. *Sir* J. L., K.C.V.O., C.B. . . . 25 June 09
Hayes-Sadler, Lt.-Col. *Sir* J., C.B. . . . 9 Nov. 07
Manning, Hony. Brig.-Gen. *Sir* W. H., C.B. 6 Sep. 04
V.C. **Melliss**, Maj. Genl. *Sir* C. J., K.C.B. [L] . . . 23 Aug. 18
Robertson, Hony. Col. *Hon. Sir* B., K.C.S.I., C.I.E., 1st Madras Guards. . 22 June 14
Ross, Maj. *Sir* R., K.C.B., F.R.S., F.R.C.S., I.M.S. . . . ...
Swayne, Bt.-Col. *Sir* E. J. E., C.B. . . 24 June 10
Tighe, Maj-Genl. *Sir* M. J., C.B., C.I.E., D.S.O. . . . 28 Mar. 16
Treherne, Surg.-Genl. *Sir* F. H. . . . 4 June 17
Younghusband, Maj.-Genl *Sir* G. J., K.C.I.E., C.B., *p.s.c.* . . . 29 Oct. 15

Companions (C.M.G.).

Abbott, Bt.-Lt.-Col. L. H. 11th Rajputs . . 3 June 16
Anderson, Lt.-Col. W. C., 1st G. R. . . 1 Jan. 16
Andrew, Lt.-Col. (*temp. Brig.-Genl.*) A. W., 114th Mah. . . . 26 Mar. 18
Bainbridge, Lt.-Col. W. F., D.S.O., *p.s.c.*, 54th Sikhs . . . 29 Oct. 15
Baird, Bt. Lt.-Col., H. B. D., D.S.O. . . 1 Jan. 18
Bald, Maj. (*tempy. Lt.-Col.*) J. A. . . 1 Feb. 16
Banbury, Lt.-Col., W. E. . . .
Bateman-Champain, Lt.-Col. H. F., 9th G. R. 3 June 8
Boome, Bt.-Col. E. H., C.B. . . . 4 June 17
Bowle-Evans, Lt.-Col. C. H., M.B., I. M. S. . 3 June 15
Bradley, Lt.-Col. F. G. (Retd.) . . . 3 June 16
Brownlow, Lt.-Col. (*temp. Brig.-Genl.*), d'A. C, C.I.E. . . . 4 June 17
Campbell, Lt.-Col. (*temp. Brig.-Gen.*) L. W. Y. . . . 3 June 16
Campbell, Maj. (*temp. Lt.-Col.*), W. N., D.S.O., (retd.) . . . 4 June 17
Carey, Lt.-Col. & Bt.-Col. (*temp. Brig.-Genl.*), O. W. . . . 4 June 17
Carruthers, Col. (*temp. Br.-Genl.*) R.A., C.B. 1 Jan. 16
Cartwright, Col., C. M., C.B. . . . 1 Jan. 18
Chamney, Maj. H., V.D., late Ind. Vols. . 29 Nov. 00
Chaytor-White, Lt.-Col. J., M.D., (retd.) I. M. S. . . . 4 June 17
Cobbe, Bt. Col. H. H., D.S.O. . . . 1 Jan. 17
Cole, Col. E. H., C.B. . . . 1 Jan. 18
Colomb, Col. (*temp. Br.-Genl.*). G. H. C. 4 June 17
Coleridge, Bt. Lt.-Col., J. F. S. D., D.S.O., 8th G. R. . . . 3 June 18
Cookson, Maj.-Genl. G. A., C.B. . . 3 June 16
Cordeaux, Maj. H. E. S., C.B. . . . 26 June 02
V.C. **Costello**, Bt. Lt.-Col., E. W., D.S.O. . 1 Jan. 18
Creagh, Col. A. H. D., M.V.O. . . . 3 Jun. 16
Cruddas, Maj. H. M., I. M. S. . . . 1 Jan. 16
Cullen, Lt.-Col. E. H. S., D.S.O., M.V.O., 34 Prs. . . . 1 Jan. 16
Dale, Lt.-Col. G. A., 19th Punj. . . . 4 June 17
Dallas, Col. A. E, S. & T. Corps . . 15 Feb. 17
Davies, Lt.-Col. & Bt.-Col. (*temp Brig.-Genl.*), C. H., D.S.O. . . . 4 June 17
DeLisle. Col. George de Sausmarez, (retd.) 4 June 17
Davidson, Bt.-Col., (*temp. Brig.-Genl*)., S. R. . . . 4 June 17
Douglas, Col. (*temp. Big.-Genl.*) J. A. C.I.E., [L] . . . 3 June 13
Drake-Brockman, Lt.-Col. D. H. . . . 3 June 15
Edwardes, Brig.-Genl. S. M., C.B., D.S.O. . ...
Edwards, Col. (*temp. Br.-Genl.*) Fitz J. M., C.B., D.S.O., A.D.C. . . . 1 Jan. 17
Edwards, Lt.-Col. R. M. (Retd.) . . . 3 June 16
Edwards, *The Hon'ble* Maj. Genl. W. R. C.B., M.D. . . . 29 Nov. 00
Elsmie, Lt.-Col.(*temp. Br.-Genl.*) A. M. S., *p.s.c.*, [*l*] 56th Rif. . . . 29 Oct. 15
Fagan, Lt.-Col. E.A., D.S.O. . . . 1 Jan. 18
Frazer, Col. G. S., I. A. . . . 3 Jun. 16
Gabriel, Maj. E. V., C.S.I., C.V.O., 1-17th Bn. B. B., & C. I. Ry. Bn., I.D.F. . . 3 June 18
Gaussen, Lt.-Col. J. R., D.S.O. 3rd H. . 3 June 16
Glasfurd, Lt.-Col. A. I. R., D.S.O. . . 4 June 17
Granger, Lt. Col. T. A., M.B., I.M.S. . . 3 June 18
Gray, Lt.-Col. F. W. B., D.S.O., 57th Rif. 18 Feb. 15
Gregson, Capt. E. G., C.I.E., I.A.R.O. 18 Dec. 17
Grimston, Col. S. B. . . . 3 June 16
Gwyn-Thomas, Lt.-Col. G. . . . 3 June 18
Haig, Lt.-Col. T. W., Supy. List [L]. . 14 June 12
Haldane, Lt.-Col. C. L., [*l*] 7 G. R., D.S.O. 29 Oct. 15
Hamilton, Lt.-Col. J. A., M.B., F.R.C.S., I. M. S. . . . 1 Jan. 16
Harding, Lt.-Col. M. F. . . . 3 June 18
Hehir, Maj. Genl. P., I. M. S., C.B., C.I.E. . . . 15 Feb. 17
Hennessy, Lt.-Col. J. P. C. S. & T. Corps 3 June 15
Holman, Bt.-Col. (*temp. Br.-Genl.*) H. C., C.B., D.S.O., 16th Cav., *p.s.c.* [L]. . 18 Feb. 15
Holland-Pryor, Bt.-Col., (*temp. Br.-Genl.*) P., D.S.O., M.V.O. . . . 1 Jan. 18
Hornby, Bt.-Col. M. L., (retd.) D.S.O. . 4 June 17
Huddlestone, Comdr. W. B., R.I.M. . 22 Dec. 16
Hudson, Lt.-Col., C. T., I.M.S. . . . 1 Jan. 18
Hugo, Lt.-Col. E.V., M.D., F.R.C.S., I.M.S. 4 June 17
Jacob, Lt.-Col., A. Le. G., C.I.E., D.S.O. 1 Jan. 18
Jeudwine, Maj. W. W., I.M.S. . . . 1 Jan. 16
Johnson, Col. (*temp. Brig.-Genl.*) F. E., D.S.O. . . . 18 Feb. 14
Jones, Lt.-Col. (*temp. Maj.-Genl.*) L.C., M.V.O., 7th Lrs. . . . 3 June 16
Keily, Lt.-Col. F. P. C., D.S.O. . . . 4 June 17
Knatchbull, Lt.-Col. (*temp. Brig.-Genl.*) G. W. C. . . . 1 Feb. 16
Leeds, Lt.-Col. T. L., D.S.O., 59th Rif. . 3 June 18
Lees, Lt.-Col., C. H. B., 53rd Sikhs. . . 4 June 17
Lindsay, Bt.-Col. H. A. P., S. & T. Corps 3 June 15
McCrea, Lt.-Col. A. C. . . . 3 Jan. 16
Maclachlan, Bt.-Col. T. R., 92nd Punj. . 29 Oct. 15
MacLeod, Lt.-Col. J. N., C.I.E., M.B., F.R.C.S., I. M. S., (retd.) . . . 4 June 17
Maconchy, Col. E. W. S. K., C.B., C.I.E., D.S.O. . . . 24 Jan. 17
Manifold, Maj. Genl. C. C., C.B., M.B., I. M. S. . . . 4 June 17
Marsh, Bt. Lt.-Col., F. G. . . . 1 Jan. 18
Maxwell, Bt. Col. L. . . . 1 Jan. 17
Moore, Lt.-Col. C. H. G., D.S.O., S. & T. Corps . . . 3 June 15
Muspratt, Col. F. C., C. B. . . . 1 Jan. 16
Napier, Lt.-Col. *Hon.* H. D. [L] . . . 28 June 07
Nepean, Lt.-Col. H. E., C.B., . . . 4 June 17
Norie, Col. (*temp. Brig.- Genl.*). C. E. de M., C.B., D.S.O. . . . 4 June 17
Nightingale, Lt.-Col., M. R. W., D.S.O. 1 Jan. 18
Ogilvie, Lt-Col. (*temp. Col.*), W. H., M.B., I.M.S. . . . 4 June 17
O'Gorman, Lt.-Col. P. W., I.M.S. . . 3 June 18
Owen, Lt.-Col. C. W., C.I.E., I.M.S. (retd.) 15 Feb. 87
Palin, Maj.-Genl. P. C., C.B. . . .
Palmer, Lt.-Col. H. I. E. . . . 3 June 16
Peck, Bt.-Col. A. W. . . . 1 Jan. 18

KNIGHTS GRAND CROSS, KNIGHTS COMMANDERS AND COMPANIONS OF THE MOST DISTINGUISHED ORDER OF SAINT MICHAEL AND SAINT GEORGE—*contd.*

Companions (C.M.G.)—*contd.*

Perry, Lt.-Col. F. F., F.R.C.S.I., (retd.) I.M.S. . 4 June 17
Pirrie, Bt. Col. F. W., C.I.E. (Supy. List) . 3 June 18
Pitcher, Bt. Lt.-Col., D. Le. G. . . . . 1 Jan. 18
Powell, Lt.-Col. W. B., 7th G. R. . . . 19 Oct. 16
Prain, Lt.-Col. *Sir* D., Kt., C.I.E., F.R.S., M.B. 1 Jan. 12
Price, Bt.-Col. (Ty. Br.-Genl.) C. U. . 3 June 16
Pridmore, Lt.-Col. W. G., M.B., I.M.S. . 29 Oct. 15
Pritchard, Lt.-Col. A. G. . . . . 3 June 16
Pryce, Bt.-Lt.-Col. H. E. *Cap.* R., *p.s.c.* 3 June 15
Purdon, Col. D. W. . . . . 1 Jan. 17
Rainey Robinson, Col. R. M., C.B., (retd.) 1 Jan. 17
Rind, Bt.-Col. A. T. S. A. . . . . . 15 Feb. 17
Robertson, Lt.-Col. J. C., C.I.E., M.B., I. M. S. . . . . 8 Aug. 17
Savage, Col. W. H. . . . . 1 Jan. 17
Shairp, Maj. (*temp. Lt.-Col.*) A., S. & T. Corps [*l*] . . . . 29 Oct. 15
Shakespear, Lt.-Col. J. C.I.E., D.S.O., Supy. List . . . . 1 Jan. 17
Shea, Maj. Genl. J. S. M., C.B.; D.S.O. . 1 Jan. 18
Shepparu, Lt.-Col. (*temp. Col.*) G. S. . 3 June 15
Shirley, Lt.-Col. W., I. A. (Retd.) . . 24 Jan. 17
Skeen, Maj.-Genl. A. . . . . . . 3 June 16
Smith, Lt.-Col., S. B., I. M. S. . . . 4 June 17
Soady, Col. (*temp. Brig.-Genl.*), G. J. F. 4 June 17
Somerset, Col. C. W., C. B., M.V.O. . . 1 Jan. 17
Souter, Lt.-Col. H. M. W., D.S.O. . . 4 June 17
Southey, Bt.-Col. W. M. . . . . . 18 Feb. 15
Stack, Lt.-Col. C. S. . . . . . . 18 Aug. 16
Stainforth, Lt.-Col. H. G. . . . . . 3 June 15
Stevens, Lt.-Col. N. M. C. [L] . . . 3 Jun. 15
Sweeny, Lt.-Col., T. H., F.R.C.S.I., (retd.) I.M.S. . . . . 4 June 17
Sullivan, Col. E. L. C.B. . . . . . 18 Feb. 15
Sykes, Lt.-Col. (*temp. Br.-Genl.*) *Sir* P. M., K.C.I.E. . . . . 26 June 02
Taylor, Bt. Lt.-Col. R. O'B., C.I.E., 19th Lrs. . . . .
Templer, Lt.-Col. C. F. . . . . . 1 Jan. 16
Thompson, Col. W. D. . . . . . 3 June 18
Thornton, Lt.-Col. C. E. . . . . . 4 June 17
Tomkins, Bt.-Lt.-Col. H. L., D.S.O. . . 3 Jan. 16
Turner, Lt.-Col. R. W., D.S.O., F.R.C.S., I.M.S. . . . . 3 June 18
Wall, Lt.-Col. E. W. . . . . . . 4 June 17
Wall, Bt.-Col. F., I.M.S. . . . . . 3 June 15
Walton, Col. (*temp. Br.-Genl.*) W.C., C.B. . . . . 3 June 16
Watson, Col. H. D., C.I.E., M.V.O., Eq. . 29 Oct. 15
Watson, Maj.-Genl. W. A., C.B., C.I.E. 4 June 17
Willougby, Col. M.E., C.B., *p.s.c.* [L] . 1 Jan. 14
Wilson, Bt. Lt.-Col. A. T., C.I.E., D.S.O. [1] . . . . 14 June 12
Wilson, Capt. N.F.J., C.B.E., R. I. M.
Wimberley, Lt.-Col. (*temp. Col.*), C. N.C., M.B., I.M.S. . . . . 4 June 17
Wingate, Col., A. W. S., I. A. (Retd.) . 3 June 16
Winter, Bt.-Col. C. B., D.S.O. . . . 4 June 17
Yate, Col. C. E., C.S.I. . . . . . . 2 June 88
Young, Col. F. DeB. . . . . 23 Nov. 16
Younghusband, Col. (*temp. Brig.-Genl.*) L. N., C.B. . . . .

OFFICERS OF THE ROYAL (LATE INDIAN) ARTILLERY AND ENGINEERS.

Knight Commanders (K.C.M.G.).

Sanderson, Lt. *Sir* P., Arty. . . . . . 3 June 96
Trotter, Lt.-Col. *Sir* H., C.B., Engrs. . . 29 June 06

Companions (C.M.G.).

Wauhope, Bt.-Col. R.A., C.B., C.I.E., Engrs. 30 June 05
Western, Lt.-Col. J. H., Engrs. . . . 2 June 88

KNIGHTS GRAND COMMANDERS, KNIGHT COMMANDERS AND COMPANIONS OF THE MOST EMINENT ORDER OF THE INDIAN EMPIRE.

Knights Grand Commanders (G.C.I.E.).

Bikaner, Hony. Maj.-Genl. *His Highness Maharaja Sir* Ganga Singh, *Bahadur*, G.C.S.I., K.C.B., A.D.C., *Maharaja* of—2nd Lrs. . . . . 1 Jan. 07
Carmichael, Hony. Col. *Rt. Hon.* T. D. *Baron*—of Skirling, K.C.M.G., late Ind. Vols. . . . . 3 Nov. 11
Cox, Lt.-Col. *Sir* P. Z., K.C.S.I. . . . . 25 Aug. 17
Jaipur, Hony. Maj.-Genl. *His Highness Maharajadhiraja Sir* Sawai Madho Singh, *Bahadur*, G.C.S.I., G.C.V.O., G. B.E., *Maharaja* of—(Hony. Col., 13th Rajputs) . . . . 23 May 00
Jammu and Kashmir, Hony. Lt.-Genl. *His Highness Maharaja Sir* Partab Singh, *Bahadur*, G.C.S.I., G.B.E., *Maharaja* of—(Hony. Col., 37th Dogras) . . 12 Dec. 11
Jind, Hony. Lieut.-Col., *His Highness Maharaja Sir* Ranbir Singh *Rajendra Bahadur*, K.C.S.I., *Maharaja* of— . . 1 Jan. 16
Khairpur, Hony. Lieut.-Col., *His Highness Mir Sir* Imam Baksh Khan *Talpur*, *Mir* of— . . . . 12 Dec. 11
Kolahpur, Hony. Col. *His Highness Sir* Shahu Chhatrapati Maharaj, G.C.S.I., G.C.V.O., LL.D., *Maharaja* of—(Hony. Col. 103rd Infy.) . . . . 12 Dec. 11
Kotah, Hony. Lt.-Col. *His Highness Maharao Sir* Umed Singh *Bahadur*. G.C.S.I., G.B.E., *Maharao* of —(42nd Regt.) . . . . 28 June 07
O'Dwyer, *Hon'ble* Hony. Col. *Sir* M. F., K.C.S.I., Punj. L. H. . . . . . 4 June 17
Patiala, Hony. Maj. Genl., *His Highness Maharajadhiraja Sir* Bhupindar Singh Mahindar *Bahadur*, G.B.E., *Maharaja* of— . . . . 12 Dec. 11
Pentland, *Hon.* J. *Baron*—of Lyth—Hony. Comdt. 1st Mad. Gds., G.C.S.I. . 4 Sep. 12
Rampur, Hony. Col. *His Highness Nawab Sir* Muhammad Hamid Ali Khan, *Bahadur*, G.C.V.O., A.D.C. *Nawab of*—(9th Horse) . . . . . 1 Jan. 08
Roos-Keppel, *Hon'ble* Lt.-Col. *Sir* G.O., K.C.S.I. . . . . 4 June 17
Willingdon, *Rt. Hon.* F. *Baron*—of Ratton—G.C.S.I., G.B.E., Hony. Comdt. 4th Bo. L. Horse, I.D.F. . . . . . . 12 Mar. 13

ORDER OF THE INDIAN EMPIRE—*contd.*

Knights Commanders (K.C.I.E.).

Alwar, Hony. Lt.-Col. *His Highness Maharaja Sir* Jey Singh, *Bahadur, Maharaja* of— . . . . 12 Dec. 11
Barwani, Capt. *His Highness, Rana, Sir* Ranjit Singh, *Rana* of— . . . . . 1 Jan. 18
Beynon, Maj.-Genl. W. G. L., C.B., D.S.O. 17 Sept. 17
Bingley, Maj.-Genl. *Sir* A. H., C.B. . 19 Dec. 18
Buchanan, Lt.-Col. *Sir* W. J., M.D. . . 3 June 18
Burt, *Hon'ble* Col. Comdt. *Sir* H. P., V.D., 2-24th N.-W. Ry. Bn. . . . . . . 12 Dec. 11
Daly, Lt.-Cl. *Sir* H., K.C.S.I. [L] . 12 Dec. 11
Drummond, Maj.-Gen. *Sir* F. H. R., C.B. 14 June 12
Earle, *Hon'ble*, Hony. Col. *Sir* A., Surma V. L. H. . . . . 12 Dec. 11
Egerton, Maj.-Genl. (*Ty. Lt.-Genl.*) *Sir* R. G., K. C. B. . . . . 25 Aug. 17
Fane, Maj.-Genl. *Sir* V. B., C.B. . . . 1 Jan. 18
Grover, Genl. *Sir* M. H. S., K.C.B.[L] . 12 Dec. 11
Jamkhandi, Hony. Capt. *Meherban Sir* Parashramrav. Ramchandrarav (Bhausaheb Patwardhan) Chief of— . 1 Jan. 13
Jaora, Hony. Maj. *His Highness Nawab Sir* Muhammad Iftikar Ali Khan *Bahadur Nawab* of — . . . . 12 Dec. 11
Judha Shumshere Jung Bahadur, Rana, *Genl., Comdg. Southern Divn. in Nepal* (Hony.) . . . . 23 Dec. 16
Keary, Maj.-Genl, *Sir* H. D'U., K. C. B., D. S. O. . . . . 25 Aug. 17
Kishengarh, Hony. Lt.-Col. *His Highness* Maharajadhiraj *Sir* Madan Singh, *Bahadur*, K.C.S.I., *Maharaja* of— . . . . 1 Jan. 09
Lambagraon, *Raja Sir* Jai Chand, C.S.I. *Raja of*— . . . . 3 June 18
McLeod, Genl. *Sir* D. J. S., K.C.B., D.S.O. . 26 June 08
McMahon, Lt.-Col. *Sir* A. H., G.C.M.G., G.C.V.O., C.S.I. . . . . 1 Jan. 06
Hon'ble Sir Malik Umar Hayat Khan, M.V.O., Hony. Maj., *Tiwana* of Shahpur (18th Lrs.) . . . . 3 June 16
Hari Singh, Raja, A. D. C., Commander-in-Chief of the Kashmir Army . . . . 1 Jan. 18
Muhammad Ali Beg, Hony. Lt.-Col. *Nawab, Afsar-i-Jang, Afsar ud-Daula*, M.V.O., *Bahadur*, 20th Horse. . . . 26 June 08
Ramsay, Lt.-Col. *Sir* J., C.S.I. . . . 3 June 15
Scallon, Lt.-Genl. *Sir* R. I., G.C.B., D.S.O. . . . . 24 June 10
Simpson, Surgn.-Genl. *Sir* B., M.D . . 15 Feb. 87
Sykes, Lt.-Col. (*tempy. Br.-Genl.*) *Sir* P. M., C.M.G. . . . . 3 June 15
Talbot, Lt.-Col *Sir* A. C. . . . . . 12 Aug. 95
Thornhill, Lt.-Col. *Sir* H. B. . . . . 12 Dec. 11
Vankaner, Hony. Capt. *Rai Saheb Sir* Amarsinhji Banesinhji, *Raj Saheb* of— 12 Dec. 11
Younghusband, Lt.-Col. *Sir* F. E., K.C.S.I. 10 Dec. 04
Youghusband, Maj.-Gen. *Sir* G. J., K.C.M.G., C.B., *p.s.c.* . . . . 1 Jan. 13

Companions (C.I.E.).

Adamson, Lt.-Col. C. H. E. . . . . 24 May 88
Alcock, Lt.-Col. A. W., M.B. . . . . 1 Jan. 03
Alderman, Capt. R. E., I.A.R.O. . . . 1 Jan. 18
Allanson, Bt. Lt.-Col. C. J. L., D.S.O., *p.s.c.*, 6th G. R. . . . . 1 Jan. 16
Anderson, Lt.-Col. J., M.B., I.M.S. . . 4 June 17
Archer, Lt.-Col. C., C.S.I., Supy. List. . 1 Jan. 06
Bagshawe, 2nd-Lt. E. L., 4th Bomb. Gr. Garrn. Arty. . . . . 18 Aug. 16
Bagshawe, Lt.-Col. F. W. . . . . . 19 Dec. 18
Bailey, Maj., F. M. (Supy. List) . . . 1 Jan. 15
Bannerman, Lt.-Col. A. D'A. G., C.V.O., Supy. List . . . . 26 June 03
Barnado, Bt. Lt.-Col. F. A. F., I.M.S. . 3 June 18
Barrett, Maj. C. C. J., I. A. . . . . 1 Jan. 18
Barron, Capt. C. A., 5th Punj. L. H. . . 12 Dec. 11
Barry, Lt.-Col. C. C. S., I. M. S. . . . 3 June 15
Bate, Col. T. E. L., I.M.S. . . . . 26 June 02
Bayley, Lt.-Col. E. C., 15th Lrs. . . . 12 Dec. 11
Bayley, Lt. V., 4th Simla Rif. . . . . 1 Jan. 17
Beatson-Bell, *Hon'ble* Maj. N. D., C.S.I., 44th Cal. Sco. . . . . 1 Jan. 06
Beer, Hony. Col. J. H. E., V.D., Muss. Rif. 25 June 09
Bell, Col. G. J. H., M.B., I.M.S. . . . 22 June 14
Benn, Lt.-Col. R. A. E., Supy. List . . 1 Jan. 04
Berkeley, Capt. A. M., I. A. R. O. . . 3 June 18
Berkeley, Maj.-Genl. J. C. . . . . 24 May 88
Beville, Lt.-Col. F. G., Supy. List . . 26 June 08
Bickford, Lt.-Col. W. W., 107th Prs. . . 1 Jan. 13
Birdwood, Genl. *Sir* W. R., K.C.B., K.C.S.I., K.C.M.G., D.S.O., *q.s.*, A.D.C. . Jan. 08
Blakeway, Lt.-Col. D. B. . . . . . 14 Aug. 08
Boxwell, Maj. A. . . . . 3 June 16
Brown, Maj. H. C., M.B., I.M. S. . . 1 Jan. 18
Browne, Col. S. H., M.D. . . . . . 1 Jan. 96
Browne, Surgn.-Genl. W. R., M.D. . . . 1 Jan. 06
Browning, Lt.-Col. W. B., I.M.S. . . . 31 Dec. 90
Brownlow, Lt.-Col. d'A. C., C. M. G. . . 18 Aug. 16
Burden, Lt.-Col. H., F.R.C.S., I.M.S. . . 12 Dec. 11
Butler, *Hon'ble* Hony. Col. *Sir* S. H., C.I.E., late Ind. Vols. . . . . . .
Cadell, Lt.-Col. P. R., 15th Bom. Btn. . 3 June 13
Campbell, Maj. A. Y. G., 10th S. P. Mtd. Rif. . . . . 2 Jan. 11
Campbell, Lt.-Col. C. F., *Eq.* . . . . 19 Mar. 06
Campbell, Lt.-Col. R. H., 6th Bangalore Coorg and Mysore Btn. . . . . . 14 June 12
Campbell, Col. *Sir* R. N., K.C.M.G., C.B., M.B. 25 June 09
Campbell, Maj. W. L. [*l*] (Supy. List.) . 3 June 13
Carey, Lt.-Col. B. S., C.S.I., V.D., U. Bur. V. Rif. . . . . 2 June 93
Carr-Hall Maj. (*temp. Lt.-Col.*) R. E. . 4 June 17
Carter, Lt.-Col. G. L., 106th Prs. . . . 4 June 17
Carter, Capt. *the Hon'ble* F. W. C.B.E. Cal. Scots . . . . 3 June 18
Caruana, Col. A. F. . . . . 3 June 18
M Chamier, Maj.-Genl. F. E. A., C.B. . . 26 June 02
Chatterton, Col. F. W. . . . . . . 6 May 94
Chenevix-Trench, Lt.-Col. G. F. . . . 2 Jan. 11
Christophers, Maj. S. R., M.B., I.M.S. . 1 Jan. 15
Church, Maj.-Genl. T. R. . . . . . 24 May 83
Cleveland, Bt.-Col. H. F., V.H.S., I.M.S. 1 Jan. 18
Clibborn, Lt.-Col, J. . . . . 9 Nov. 01
Cooper, Hony. Col. *Sir* W. E., *Kt.*, 16th Cawnpore Rif. . . . . 1 Jan. 97
Cowper, Maj.-Genl. M., C.B. . . . . 14 June 12
Crichton, Lt.-Col. R. T., Supy. List . . 1 Jan. 09
VC., Crimmin, Col. J., C.B., V.D., K.H.P., I.M.S. . . . . 1 Jan. 01
Crookshank, Maj. S. D'A., M.V.O., Punj. L.H. . . . . 14 June 12
Crosthwaite, Lt. H. R., 1st Madras Gds. . 4 June 17
Cunliffe, Maj. J. B., 2nd (Madras) Group Garr. Arty. . . . . 1 Jan. 16
Darley, Lt.-Col. J. R., D.S.O., 119th Infy. 1 Jan. 18
Davern, Lt. F. B., 3rd Gp. Garrn. Arty. I.D.F. . . . . 23 Aug. 18
DeBrath, Lt.-Genl. *Sir* E., K.C.B., *p. s. c.* 1 Jan. 03
Dempster, Maj. F. E., late Ind. Vols. . 1 Jan. 96
Dennys, *The Hon'ble* Col. G. W. P., I.M.S. . . . . 1 Jan. 15
Dennys, Lt.-Col. H. T. . . . . . . 3 June 15
Dew, Lt.-Col., A. B., C.S.I., Supy. List . 12 Dec. 11
Dickson, Maj. H. R. P., 29th Lrs. . . 25 Aug. 17
Dickson, Lt.-Col. J. H., S. & T. Corps . 1 Jan. 17
Douglas, Col. J. A., C. M. G. . . . . 25 Aug. 17
Douglas, Comdr. W. R. B., R. I. M. . . 1 Jan. 17
Douglas, Lt.-Col. M. W., Supy. List . . 26 June 03
Drew, Col. A. B. H., I. A. . . . . . 3 June 18
Dunlop, Hony. Col. R. W. L., V.D., Bo. V. Rif. . . . . 1 Jan. 13

ORDER OF THE INDIAN EMPIRE—*contd.*

Companions (C.I.E.)—*contd.*

Dunlop Smith, Lt.-Col. *Sir* J. R., K.C.S.I., K.C.V.O.	1 Jan. 01
Durand, Col. A. G. A., C.B., *p.s.c.*	22 June 97
Elwes, Maj. F. F., M.B., I.M.S.	12 Dec. 11
Evans, Col. G. H., C.V.D.	12 Dec. 11
Ewart, Maj.-Genl. *Sir* R. H., K.C.M.G., C.B., D.S.O., A.D.C.	14 June 12
Farmer, Lt.-Col. J., I. C. V. D.	1 Jan. 17
Ferard, Capt. H. C., U. P. Horse	1 Jan. 98
ffrench-Mullen, Lt.-Col. J. L. W.	14 June 12
Filgate, Lt.-Col. T. R., V.D., 1st Bihar L. H.	1 Jan. 13
Flynn, Lt. J. D., E. I. 7th E. I. R. Bn.	1 Jan. 17
Gait, *Hon'ble* Lt.-Col. *Sir* E. A., K.C.S.I., 1st Bihar L. H.	1 Jan. 07
Gee, Lt.-Col. F. W., I.M.S.	3 June 18
Geoghegan, Lt.-Col. F. E., S. & T. Corps	19 Dec. 18
Gilchrist, Maj. W. F. C., 52nd Sikhs.	23 Aug. 18
Gillitt, Maj. W., M.D., I. M. S.	3 June 16
Gimlette, Lt.-Col. G. H. D., M.D.	9 Nov. 01
Girdhar Singh, Lt.-Col., Bharatpur I. S. Troops	11 Aug. 17
Gleadowe-Newcomen, Hony. Col. A. H., V.D., late U. P. Horse (S. Regt.)	1 Jan. 07
Godfrey, Lt.-Col. S. H., Supy. List [L]	14 Aug. 08
Goodbody, Bt. Lt.-Col. C. M., D.S.O., F.R.C.S.I., I. M. S.	1 Jan. 18
Goodwin, Col. F., V.D., B., B. and C. I. Ry. Rif.	2 Jan. 11
Gordon, Hony. Col. G. G., V.D., late N. B. Mtd. Rif.	2 Jan. 11
Gordon, Bt.-Col. J. C. F.	31 July 97
Gordon, Lt.-Col. R. F. C., 15th Lrs.	12 Dec. 11
✠ Gordon, Genl. W.	1 Jan. 78
Gough, Lt.-Col. H. A. K., Poll. Dept.	1 Jan. 18
Greig, Maj. E. D. W., M.B., I.M.S.	22 June 14
Greig, Maj. J. G., 121st Prs.	12 Dec. 11
Grey, Lt.-Col. A., V.D., late Punj. L. H.	24 June 10
Grice, Lt.-Col. (Hony Col.) W. T., V.D.	3 June 15
Griffith, Capt. R. F. H.	1 Jan. 17
Gurdon, Lt.-Col. B. E. M., D.S.O., (retd.)	23 May 00
Haidar Ali Khan, Lt.-Col., Kashmir I. S. Troops	11 Aug. 17
Hailey, *Hon'ble* Lt.-Col. W. M., C.S.I., 1st Punj. Rif.	12 Dec. 11
Halliday, *Hon.* Maj. Sir F. L., Kt., M.V.O., late 2nd C. V. R.	1 Jan. 09
Hallowes, Lt.-Col., F.W., S. & T. Corps	1 Jan. 18
Hamilton, Lt.-Col. R. E. A.	3 June 18
Hathi, Lt.-Col., E., G. I. S. Troops	25 Aug. 17
Hay, Lt.-Col. W. N., D.S.O.	1 Jan. 17
Hehir, Maj.-Genl. P., C.B., C.M.G., I. M. S.	3 June 18
Heron, Maj. D., M B., F.R.C.S.E.	3 June 18
Hickley, Lt.-Col. V. N., V.D., 1st Bihar L.H.	3 June 13
Hignell, 2nd-Lt. S. R., N. 8th U. P. Horse	4 June 17
Hill, Hony. Maj. T. H., I.S.M.D.	9 Nov. 01
Hogg, Maj. and Bt. Lt.-Col. R. E. T., 38th Horse	12 Dec. 11
Hojel, Lt.-Col. J. G., I.M.S.	1 Jan. 17
Holdsworth, Lt.-Col. J. J., V.D., late Gorakhpur L. H.	31 Dec. 98
Holland, Maj. R. E., 4th Simla Rif.	1 Jan. 17
Holloway, Col. (*temp. Maj.-Genl.*) B., *p.s.c.* [L]	3 June 15
Huddleston, Hony. Lt.-Col. G., V.D., late E. I. R. Vol. Rif. Corps	26 June 03
Hudson, Lt.-Genl. *Sir* H., K.C.B.	26 June 03
Hughes-Buller, Capt. R. B., late Assam V. L. H.	26 June 08
Impey, Lt.-Col. L., C.S.I., Supy. List	12 Dec. 11
Izat, Lt. (*temp. Maj.*) J., I. A. R. O.	1 Jan. 18
Jackson, Lt.-Col. J., M.B., I.M.S.	3 June 16
Jacob, Lt.-Col. A. L.	22 June 14
Jacob, Lt.-Col. A. Le G., C.M.G., D.S.O., 106th Prs.	1 Jan. 07
James, Lt.-Col. C. H., I.M.S.	14 June 12
James, Maj.-Genl. W. B., C.B., M.V.O.	14 June 12
✠ Jarrett, Col. H. S.	1 Jan. 95
Jeffery, Major W. H., 73rd Inf.	22 June 14
Jones, Lt.-Col. R. G.	12 Dec. 11
Kaye, Lt.-Col. C., C.S.I., 20th Inf.	1 Jan. 13
Keen, Maj. W. J., (Supy. List)	1 Jan. 16
Kemball, Lt.-Col. C. A., Supy. List [L].	1 Jan. 03
Kennedy, Maj. W. M. (Supy. List)	1 Jan. 16
Kennion, Lt.-Col. R. L., Pol. Dept	1 Jan. 18
Keyes, Bt.-Lt.-Col. T. H.	4 June 17
King, Col. W. G., M.B.	3 June 98
Khan Muhammad Akbar Khan, Capt.	11 Aug. 17
Knowles, Hony. Col. J. G., V.D., Surma V. L. H.	1 Jan. 13
Knox, Lt.-Col. S. G., C.S.I., Supy. List	25 June 09
Lane, Col. C. T.	25 May 95
Lane, Lt.-Col. W. B., I.M.S.	1 Jan. 18
Leigh, Lt.-Col. H. P. P.	1 Jan. 92
LeMesurier, Lt.-Col. H., C.S.I., E. B. Rif. I.D.F.	1 Jan. 03
Levett-Yeats, Lt. G. A., V.D., U. P. Horse	4 June 17
Liston, Maj. W. G., M.D., I.M.S.	1 Jan. 13
Lloyd, Lt.-Col. Fitz W., V.D., Assam V. L. H.	22 June 14
Loch, Lt.-Col. G. H.	1 Jan. 01
Lorimer, Lt.-Col. D. L. R.	4 June 17
Lowis, Lt. F. C., I. A. R.	22 June 14
Lyall, *Hon'ble* 2nd-Lt. F. F., 1st Bihar L. Horse I.D.F.	1 Jan. 16
McPherson, Maj. G., I. M.S.	3 June 18
MacDermott, Lt. (*temp. Capt.*) J. W., I. A. R. O.	1 Jan. 18
Mackenzie, Maj. C. F., I. A.	1 Jan. 18
Mackenzie, Maj. J., 35th Sikhs	12 Dec. 11
MacLaughlin, Lt.-Col. A. J. M., V.D., late Surma V. L. H.	28 Jan. 07
✠ Maclean, Maj.-Genl. C. S., C.B.	20 Mar. 88
MacLeod, Lt.-Col. J. N., C. M. G., M.B., F.R.C.S., Edin, I.M.S. Retd.	1 Jan. 08
Maconchy, Col. E. W., S. K., C.B., C.M.G., D.S.O.	1 Jan. 09
Macrae, Hony. Col. A. W., V.D., Malabar Rif.	12 Dec. 11
Mactaggart, Col. C., M.B., I.M.S.	12 Dec. 11
MacWatt, Col. R. C., I.M.S.	1 Jan. 16
Mahon, Maj. (*temp. Col.*) H. J., I.D.F.	19 Dec. 18
Malleson, Maj.-Genl. W., C.B.	1 Jan. 10
Marlow, Col. B. W., C.S.I.	12 Dec. 11
Masson, Hony. Col. *Sir* D. P., Kt., V.D., late 1st P. V. R.	1 Jan. 98
McConaghey, Lt.-Col. A., Supy. List	12 Dec. 11
McKay, Col. H. K., C.B., I.M.S.	31 Dec. 98
Meade, Lt.-Col. M. J.	1 Jan. 09
Mell, Lt.-Col., F. O. N., I.M.S.	25 Aug. 17
Melville, Lt.-Col. H. G., I.M.S.	3 June 18
Merewether, Lt.-Col. J. W. B., Pol. Dept.	3 June 16
Miller, Capt. H., Assam V. L. H.	1 Jan. 17
Minchin, Lt.-Col. A. B., Supy. List	21 May 98
Mir Ausaf Ali Khan, *Sardar*, Comdt., Maler Kotla I. S. Sappers	24 July 01
Molesworth, Col. W., C.B.E., M.B., V.H.S., I.M.S.	22 June 14
Mollison, Maj. E. J., 125 Rif.	3 June 16
Money, Lt.-Col. E. D., D.S.O., 1st G. R.	12 Dec. 11
Moore, Lt. P. L., S. Prov. M. Rif.	14 June 12
Muhi-ud-din Khan, Risaldar-Maj., *Sardar Bahadur* 31st Lrs, A. D. C.	11 Aug. 17
Muir, Col. C. W., C.B.	15 July 17
Munro, Maj. D., I. M. S.	25 Aug. 17
Murray, Capt. S. G. C.	3 June 18
Natha Singh, Maj.-Genl., I. S. Troops	11 Aug. 17
Nauth, Lt.-Col., B., I.M.S.	25 Aug. 17
Newman, Lt.-Col. E. A. R., M.D.	3 June 18
Noel, Capt. E. W. C., Poll. Dept.	1 Jan. 18
O'Brien, Maj. A. J., Supy. List [L].	1 Jan. 06
O'Connor, Maj W. F. T., Supy. List	16 Dec. 04
O'Meara, Lt.-Col. C. A. E.	4 June 17
Owen Lt.-Col. C. W., C.M.G., I.M.S.	24 May 81
Paige, Capt. C. P., 109th Inf.	17 Oct. 17
Palin, Col. G. W., C.B., S. and T. Corps	12 Dec. 11
Park, Revd. W. R.	25 Aug. 17
Parkin, Lt.-Col. H.	25 June 09
Patel, Hony. Lt. B. D., Bal. Vol. Rif.	24 June 04
Peacocke, Lt.-Col. T.G., I.A.	3 June 16
Pearson, Lt. J. R., 7th (S) U. P. H.	3 June 13
Pease, Col. H. T., Punj. L. H.	1 Jan. 07
Peile, Bt.-Col. S. C. F.	1 Jan. 01
Perry, Lt.-Col. F. F., F.R.C.S.	26 June 08

ORDER OF THE INDIAN EMPIRE—*concld.*

Companions (C.I.E.)—*concld.*

Pigott, Hony. Col. R. E. P., V.D., 1-17th B., B. and C. I. Ry. Bn. . . . . . 12 Dec. 11
Pirie, Bt.-Col., F. W., C.M.G. . . . . 25 Aug. 17
Pooran Singh, Kapurthala I. S. Troops 11 Aug. 17
M Porteous, Col. C. A. . . . . . . 25 May 95
Prain, Lt.-Col. *Sir* D., *Kt.*, C.M.G., F.R.S., M.B. . . . . 29 June 06
Prideaux, Maj. F. B., C.S.I., Supy. List . 12 Dec. 11
Pye, Lt.-Col. W. E. . . . . 1 Jan. 17
Radford, Bt.-Col. O. C., C.B. . . . . 24 July 01
Ramsay, Maj. A. D. G., Supy. List . . 12 Dec. 11
Rawlins, Lt.-Col, A. K., D.S.O., 24th Punj. . . . . 25 Aug. 16
Rawlinson, Lt.-Col. C. B., Supy. List . 2 Jan. 05
Redl, Lt.-Col. E. A. F. . . . . . . 4 June 17
Rice, Maj. S. M. . . . . 1 Jan. 17
Richardson, Lt.-Genl. *Sir.* G. L. R., K.C.B., C.S.I. . . . . 28 July 93
Rivett-Carnac, Hony. Col. J. H., V.D., 22nd B. & N. W. Ry. 13th . . . . 1 Jan. 78
Roberts, Lt.-Col. *Sir* J. R., *Kt.*, M.B., F.R.C.S., I.M.S. . . . . 12 Dec. 11
Robertson, *Hon'ble*, Hony. Col. *Sir* B., K.C.S.I., K.C.M.G., 1st Madras Gds. . 1 Jan. 98
Robertson, *Hon'ble* Lt.-Col. J.C., C.M.G., M.B., I.M.S. . . . . 1 Jan. 14
Rogers, Lt.-Col. *Sir* L., *Kt.*, M.D., F.R.C.P., F.R.C.S. I.M.S. . . . . . 12 Dec. 11
Rose, Brig.-Genl. J. L., I. S. Troops . . 1 Jan. 18
Royle, Lt. J. R. E. J. . . . . 28 June 86
St. John, Maj. H. B. . . . . . . 3 June 13
Sardar Khan, Subdr. Maj., *Khan Bahadur, late* 69th Punj. . . . . . 24 July 01
Scott, Maj. N. E. H., I.M.S. . . . . 3 June 18
Scott, Maj.-Genl. T. E., C.B., D.S.O. . . 24 July 01
Screggfe, Maj. W. R. J., I.M.S. . . . 3 June 18
Shakespear, Lt.-Col. J., C.M.G., D.S.O. Supy. List. . . . . 20 May 96
Shore, Col. (*tempy. Br.-Genl.*) O. B. S. F., C.B., D.S.O., *p.s.c.* [L] . . . 1 Jan. 16
Singh, Lt.-Col. the Hon'ble Bawa Jiwan, I.M.S. . . . . 1 Jan. 18
Simpson, Lt.-Col. J. H., U. P. Horse. . 1 Jan. 13
Sirmur, Hony. Lt.-Col. Bir Bikram Singh, *Raj Kumar* of—, C.S.I. (1 S. and M.). . 21 May 98
Smith, Lt.-Col. H., M.D., I.M.S. . . . 1 Jan. 18
Smith, Lt.-Col. H. A., I.M.S.. . . . 3 June 18
VC Smith, Lt.-Col. J. M., C.V.O., Supy. List . . . . 1 Jan. 94
Spence, Lt.-Col. (Ty. Col.) A. H. O. . 4 June 17
Stanyon, Hony.Col. H. J., V.D., Nagpur Rif. A.D.C. . . . . 25 May 95
Steel, Bt.-Lt. Col. R. A., 17th Cav. . 3 June 16
Stevenson, Maj. W. D. H., M.D., I.M.S. 1 Jan. 18
Stewart, Capt. F. H., 4th Cal. Sco. . 12 Dec. 11
Stewart, Maj. H. . . . . 1 Jan. 13
Stokes, Bt.-Lt.-Col. C. B. . . . . . 1 Jan. 17
Stratton, Lt.-Col. W. C. R., Supy. List. 12 Dec. 11
Streatfeild, Lt. H. C., Late Bihar L. H. 1 Jan. 13
Sutherland, Lt.-Col. D. W., M.D., M.R.C.P., F.R.S., I.M.S. . . . . 4 June 17
Swan, Maj. (*temp. Lt.-Col.*) J. G. G., M.B., I. M. S. . . . . 1 Jan. 18
Taylor, Bt.-Lt.-Col., R.O'B., C.M.G., 19th Lrs. *p.s.c.* [L] . . . . . . 14 June 12
Tegart, Lt. C. A., 2nd C. V. R. . . . 1 Jan. 17
Temple, Lieut.-Col. *Sir* R. C., *Bart.* . 1 Jan. 94
Thomson, Lt.-Col. S. J., I.M.S. . . 1 Jan. 98
Thorburn, Capt. H. H. . . . . . 4 June 17
Thubron, 2nd-Lt., J. B. S., 45th Aden Rif., I.D.F. . . . . 4 Sep. 18
Tighe, *Sir* Maj.-Genl. M. J., K.C.M.G. C.B., D.S.O. . . . . 14 June 12
Trevor, Maj. A. P., C.S.I., [*l*] Supy. List 12 Dec 11
Tucker, Maj.-Genl. L. H. E. . . . 1 Jan. 91
Vale Capt. S. D., R.I.M. . . . . 3 June 18
Waddell, Lt.-Col. L. A., C.B., M.B. . 24 July 01
Waddington, Maj. C. W., M.V.O., 1-17th B., B. & C. I. Ry. Bn. . . . 24 June 04
Wagstaff, Maj. L. C., 2nd Rajput L. I. 17 Oct. 17
Walker, Maj. G. K., F.R.C.V.S., I. C. V.D. . . . . 3 June 13
Walker, Lt.-Col. *Sir* J. L., Kt., V.D., late 2nd P. V. R. . . . . . 24 May 88
Walter, Maj. E., S. & T. Corps . . 1 Jan. 17
Ware, Lt.-Col. F. C. W., Supy. List. . 9 Nov. 01
Watson, Col. H. D., C.B., C.M.G., M.V.O., *Eq.* . . . . 1 Jan. 06
Watson, Maj.-Genl. W.A., C.B., C.M.G., *p.s.c.* . . . . 19 Mar. 06
White, Maj. F. N., M.D., I.M.S. . . 3 June 16
Wilson, Bt.-Lt.-Col. A.T., C.M.G., D.S.O. 25 Aug. 17
Wilson, Capt. D. W., 4th Bombay L. H 1 Jan. 17
Windham, Lt.-Col. C. J., Supy. List. . 3 June 15
Wingate, Col. G. . . . . 9 Nov. 01
Withers, Lt. E. C., R.I.M. . . . . 1 Jan. 17
Wodehouse, Lt.-Col. F. W., Supy. List 14 June 12
Wood, Lt.-Col., W. M. P. . . . . . 4 Sep. 18
Yates, Maj. J. A., D.S.O. 103 Inf. . . . 25 Aug. 17
Yasin Khan, *Hony. Capt. Sardar Bahadur*, I.O.M., *late* 24th Inf. . . 24 July 01
Yeilding, Col. W R., C.B., D.S.O. . . 25 May 95
Young, Lt.-Col. Sir F. P., K.B.E. Supy. List . . . . 3 June 98

OFFICERS OF THE ROYAL (LATE INDIAN) ARTILLERY AND ENGINEERS.

Knight Commander (K.C.I.E.).

Thuillier, Col. *Sir* H. R., R.E.

Companions (C.I.E.).

Marshall, Maj.-Genl. G. F. L., Engrs. . . 2 Jan. 93
Merriman, Col. W., Engrs. . . . . . 1 Jan. 90
M Tennant, Hony. Lt.-Genl. J. F., Engrs. . . . . 1 Jan. 79
Tomkins, Genl. W. P., Engrs. . . . 1 Jan. 90

KNIGHTS GRAND CROSS, KNIGHTS COMMANDERS, COMMANDERS, AND MEMBERS OF THE 4TH AND 5TH CLASSES OF THE ROYAL VICTORIAN ORDER.

Knights Grand Cross (G.C.V.O.).

Charles, Lt.-Col. (temp. Surgn.-Genl.) *Sir* R. H., M.D., F.R.C.S.I., G.B.E. . . 4 Feb. 12
Gwalior, Hony. Lieut.-Genl. *H. H. Sir* Madho Rao Scindia, *Bahadur*, G.C.S.I., G.B.E., A.D.C., *Maharaja* of (Hony. Col., 1 Lrs.) . . . . 31 Mar. 03
Jaipur, Hony. Maj.-Genl., *His Highness Sir* Sawai Madho Singh, *Bahadur*, G.C.S.I., G.C.I.E., G.B.E., *Maharaja* of (Hony. Col., 13 Rajputs) . . . . 31 Mar. 03
Kolhapur, Hony. Col. *His Highness Sir* Shahu Chhatrapati, Maharaj, G.C.S.I., G.C.I.E., LL.D., (Hony. Col., 103rd Inf.) Maharaja of . . . .
McMahon, Lt.-Col. *Sir* A. H., G.C.M.G., K.C.I.E., C.S.I. . . . . 12 Dec. 11
Nepal, Hony. Lieut.-Genl. *His Excellency Maharaja Sir* Chandra Shumsher Jung, *Rana Bahadur*, G.C.B., G.C.S.I. (Hony. Col., 4 G. R.) *Prime Minister and Marshal* of.
Pratap Singh, *His Highness Maharaja Bahadur, Sir*, G.C.B., G.C.S.I., A.D.C. of Jodhpur (Hony. Lt.-Genl.), (Hony. Col., 34th Horse) . . . .
VC M Probyn, Genl. *Rt. Hon. Sir* D. M., G.C.B., G.C.S.I., I.S.O, *Eq.* . . .
Rampur, Hony. Col., *His Highness Nawab Sir* Muhammad Hamid Ali Khan, *Bahadur*, G.C.I.E., A.D.C., *Nawab* of—

KNIGHT'S GRAND CROSS, KNIGHT'S COMMANDERS, COMMANDERS, AND MEMBERS OF THE 4TH AND 5TH CLASSES OF THE ROYAL VICTORIAN ORDER—*concld.*

Knight's Commanders (K.C.V.O.).

Barrett, Genl. *Sir* A. A., G.C.B., K.C.S.I., A. D. C. . . . . 4 Jan. 12
Bell, Maj.-Genl. *Sir* J. A. . . . . 27 Nov. 11
Bhim Shumshere Jung Bahadur, Rana, K.C.S.I., *Genl. Sir*, C.-in-C. in Nepal . ...
Chamberlain, Col. *Sir* N. F. Fitz. G., K.C.B. 12 July 11
Dunlop-Smith, Lt.-Col. *Sir* J. R., K.C.S.I., C.I.E. . . . . 1 Jan. 16
Harrington, Hony. Lt.-Col. *Sir* J.L., K.C.M.G., C.B. . . . . 9 Nov. 08

Commanders (C.V.O.).

Bannerman, Lt.-Col. A. D.'A. G., C.I.E., Supy. List . . . . 12 Dec. 11
Dawson, Col. H. L., C.B. . . . . 22 Aug. 02
Fletcher, Lt.-Col. H. A. . . . . 17 May 09
Gabriel, Maj. E. V., C.S.I., C.M.G., 1-17th B., B. & C. I. Ry. Btn., I. D. F. . . . 5 Jan. 09
Hennell, Col. *Sir* R., *Kt.*, D.S.O. . . . 2 July 10
Herbert, Maj.-Genl. L., C.B. . . . 31 Mar. 03
Smith, Lt.-Col. J. M., C.I.E., Supy. List . . . . 12 Dec. 11
Wigram, Bt.-Lt.-Col. C., C.B., C.S.I., *p. s. c.*, 18 Lrs. *Eq.* . . . . 31 Mar. 03

Members, 4th class (M.V.O.).

Bamber, Col. C. J., I.M.S. Retd. . . . 12 Dec. 11
Bliss, Maj. J. A., D.S.O., 64th Prs. . . . 24 July 13
Bruce, Bt.-Col. *Hon.* C. G., C.B. . . . 21 July 03
Buist, Lt.-Col. A. H., Guides . . . . 10 Jan. 12
Campbell, Maj. H., D.S.O., Guides. . . . 24 July 06
Carnegy, Lt.-Col. C. G. . . . . 12 Dec. 11
Crookshank, Maj. S. D'A., C.I.E., Punj. L. Horse, I. D. F. . . . . 12 Dec. 11
Creagh, Col. A. H. D., C.M.G. . . . 12 Dec. 11
Cullen, Lt.-Col. E. H. S., C.M.G., D.S.O., 34th Prs. . . . . 12 Dec. 11
Freeth, Hony. Lt.-Col. W. . . . . 14 Oct. 02
Hawes, Lt.-Col. C. H., D.S.O., 23rd Cav. . 26 July 13
Holland, Lt.-Col. G. L. . . . . 12 Dec. 11
Holland-Pryor, Bt.-Col. P. (*tempy. Br.-Genl.*) C.M.G., D.S.O. . . . . 9 July 09
James, Maj.-Genl. W. B., C.B., C.I.E. . . 12 Dec. 11
Jones, Bt.-Col. (*tempy. Maj.-Genl.*) L. C., C.M.G., 7th Lrs., *p.s.c.* [*l*] . . . 6 July 11
Keighley, Lt.-Col. V. A. S., 18th Lrs. . . 4 Jan. 11
Malcolm, Maj. P., D.S.O. . . . . 24 Apl. 23
Hon'ble Sir Malik Umar Hayat Khan, K.C.I.E., Hony. Maj., *Tiwana* of Shahpur 18th Lrs. . . . . 12 Dec. 11
McBarnet, Maj. A. E., D.S.O. . . . 12 Aug. 04
Mears, Bt.-Lt.-Col. C. D., 16th Cav. . 12 Dec. 11
Muhammad Ali Beg, Hony. Lt.-Col. *Nawab Afsar-i-Jang, Afsar-ud-daula, Bahadur, Sir*, K.C.I.E., 20th Horse. . . . . 15 Feb. 06
Murray, Lt.-Col. C. F. T., C.B., S. & T. Corps . . . . 12 Dec. 11
Norman, Bt.-Col. C. L., D.S.O., Guides, *p.s.c.* . . . . 13 July 08
Pennington, Lt.-Col. A. W. . . . . 25 July 07
Ricketts, Lt.-Col. P. E., D.S.O., 18th Lrs.
Roberts, Col. (*tempy Br.-Genl.*) H. L., C.B. [*l*] . . . . 19 Mar. 06
Somerset, Col. C. W., C.B., C.M.G. . 12 Dec. 11
Strachey, Lt.-Col. J. . . . . 1 July 06
Waddington, Maj. C. W., C.I.E., 1-17th B., B. & C. I. Ry. Btn., I. D. F. . . 24 July 11
Waterfield, Lt.-Col. A. C. M., *p. s. c.* [*l*]. 9 Nov. 02
Watson, Col. H. D., C.B., C.M.G., C.I.E., *Eq.* . . . . 27 Dec. 11

Members, 5th class (M.V.O.).

Btn. Mirza Karim Beg, Hony Capt., *Sardar Bahadur, late* 20th Horse.
Mirza Umrao Beg, Hony. Capt., *Sardar Bahadur, late* 31st Lrs. . . . 1 May 03
Tegart, Lt. C. A., C.I.E., Cal. V. Rif. (2nd Bn.), I. D. F. . . . . 4 Jan. 12

KNIGHT'S GRAND CROSS, KNIGHT'S COMMANDERS, COMMANDERS OFFICERS, AND MEMBERS OF THE MOST EXCELLENT ORDER OF BRITISH EMPIRE.

Knight's Grand Cross.

Gwalior, Hony. Lt.-Genl., *His Highness Sir*, Madho Rao Scindia, *Bahadur*, G.C.S.I., G.C.V.O., A.D.C., *Maharaja of*—(Hony. Col., 1st Lrs.) . . . . 4 Dec. 17
Hyderabad, Lt.-Genl., *His Exalted Highness Nawab Mir, Sir* Usman Ali Khan *Bahadur, Fateh Jang, Nizam* of—(Hony. Col., 20th Horse) . . . . 4 Dec. 17
Jaipur, Hony. Maj.-Genl., *His Highness Maharajadhiraja Sir* Sawai Madho Singh, *Bahadur*, G.C.S.I., G.C.I.E., G.C.V.O., LL.D., *Maharaja of*— . . 1 Jan. 18
Jammu and Kashmir, Hony. Lt.-Genl., *His Highness Maharaja Sir* Partab Singh, *Bahadur*, G.C.S.I., C.I.E., *Maharaja* of—(Hony. Col., 37th Dogras). 1 Jan. 18
Kotah, Hony. Lt.-Col., *His Highness Maharao Sir* Umed Singh, *Bahadur*, G.C.S.I., G.C.I.E., *Maharao of*— . . 1 Jan. 18
Mysore, Hony. Col., *His Highness Maharaja* Sri *Sir* Krishnaraja Wadiyar *Bahadur, Maharaja of*—(Hony. Col., 26th Lt.Cav.) . . . . 4 Dec. 17
Patiala, Hony. Maj.-Genl., *His Highness Maharajadhiraja Sir* Bhupindar Singh Mahindar *Bahadur*, G.C.I.E., *Maharaja of*— . . . . 1 Jan. 18
Willingdon, *Rt. Hon.* F. *Baron*—of Ratton, G.C.S.I., G.C.I.E., Hony. Comdt., Bo. L. Horse, I. D. F. . . . . 4 Dec. 17

KNIGHT'S GRAND CROSS, KNIGHT'S COMMANDERS, COMMANDERS, OFFICERS, AND MEMBERS OF THE MOST EXCELLENT ORDER OF BRITISH EMPIRE—*contd.*

Knight's Commanders.

Young, Lt. Col. *Sir* F. P., C.I.E. . . . 3 June 18 1

Commanders.

Carter, The Hon'ble F. W., C.I.E. Capt. 44th Cal. Sco., I. D. F. . . . . 3 June 18
Elliott, Lt.-Col. A. C. . . . . . 3 June 18
Gould, Lt.-Col., J., M.B., I. M. S. . . . 4 June 17
Molesworth, Col. W., C.I.E., M.B., V.H.S., I.M.S. . . . . 19 Dec. 18
Wilson, Capt. N. F. J., C.M.G., R.I.M. . 3 June 18
Wilson-Johnston, J., I.C.S., Lt., 5th Punj. Lt. Horse, I. D. F. . . . . 3 June 18

Officers.

Armstrong, 2nd-Lt. J. W., 3rd Punj. Rif., I. D. F. . . . . 3 June 18
Ashford, Capt. J., 3rd Punj. Rif., I. D. F. 3 June 18
Bonner, Lt.-Col. T. W., 1-13th G. I. P. Ry. Bn., I. D. F. . . . . 3 June 18
Bradfield, Maj. E. W. C., I. M. S. . . 3 June 18
Burkitt, 2nd-Lt. F. H., 3rd Punj. Rif., I. D. F. . . . . 3 June 18
Cochran, Capt. A., Col. Gp. Gn. Ary., I. D. F. . . . . 3 June 18
Copeland, Capt. D. P., Att. 8th G. R. . . 3 June 18
Crawford, Lt.-Col J. M., M.B., I.M.S. . . 3 June 18
Crum, Maj. W. E., 3rd Cal. Lt. Horse, I. D. F. . . . . 3 June 18
Fraser, W. S. Capt., 2-17th B., B. & C.I. Ry. Bn. I. D. F. . . . . . . 3 June 18
Holland, Lt. (*temp. Maj.*) G. H. R., I.A.R.O. 19 Dec. 18
Heale, Maj. R. J. W., Poll. Dept. . . . 4 Dec. 17
Hotson, Capt. J.E.B., I.C.S., I. A. R. O. 3 June 18
Howard, W. H. K. (2nd-Lt.) Temp. Lt.-Col., I. A. R. O. . . . . 3 June 18
Hutchinson, Maj. W. G. . . . . . . 3 June 18
Jacob, Lt.-Col. A. L., C.I.E., Poll. Dept. 7 Jan. 18
Puech, Capt. A. G., V. O., I. A. R. O. . 3 June 18
Pugh, Col.A. J., Comdt., Cal. L. H. . . 3 June 18
Ramsay, Maj , A D.G , C.I.E., Poll. Dept. 7 Jan. 18
Ross, Maj. H., I.M.S. . . . . . . 3 June 18
Stott, Capt. H., I. M. S. . . . . . . 3 June 18
Symons, Lt.-Col., T. H., I.M.S. . . 4 Dec. 17
Travers, W. L. Capt., N. B. M. Rif. . . 3 June 18
Vines, Capt. O. F , R.I.M. . . . . 3 June 18
Vivian, V. P. T., 2nd-Lt. I A. R. O. . . 3 June 18
Wadderburn, D. St. C., Maj. 7th E. I. Ry. Bn. . . . . 3 June 18
Walker, Lt.-Col. G. K., C.I.E., F.R.C.V.S., I.C.V.D. . . . . 19 Dec. 18
Walter, Lt.-Col. A. E., I.M.S. . . . . 19 Dec. 18

Members.

Ahmad Nawaz Khan, *Sadhozai Nawab* of Dera, Hony. Capt . . . . . . 3 June 18
Andrews, A. J. (Temp. Lt.) I. M. S. . . 3 June 18
Arbery, J., Comy. & *Hony.* Maj., I. M. L. 3 June 18
Bishan-Singh, Kathait Hony. Capt. *Sardar Bahadur* . . . . 3 June 18
Devasahayam, Hony. Capt. *Sardar Bahadur* 3 June 18
Green, T. G. Maj., V.D., 2nd Nagpur Rif. . 3 June 18
Hanwant Singh, Resaldar-Maj. and Hony. Capt., *Sardar Bahadur* (retd.) . . . 4 Dec. 17
Tharle-Hughes, Lt. R., I.A.R.O. . . . . 4 Dec. 17
Judge, Lt. C.E.M. . . . . 7 Jan. 18
Malik, Muhammad, Mubaraz Khan, Hony. Capt. . . . . 3 June 18

COMPANIONS OF THE "DISTINGUISHED SERVICE ORDER."

Companions (D.S.O.).

Abbott, Capt. F. B., 6th G. R. . . . . 25 Aug. 15
Adams, Lt.-Col., F., 20th Horse . . . 3 June 18
Alexander, Bt.-Col. A. deV. . . . . 26 June 92
Alexander, Bt. Lt.-Col.E. C., 55th Rif. . 2 Sep. 02
Alexander, Maj. H. M., S. & T. Corps. . 8 Nov. 15
Alexander, Lt.-Col. H. S., 44th Infy. . . 3 June 18
Allanson, Bt.-Lt.-Col. C. J. L., C.I.E., 6th G. R., *p.s.c.* . . . . 4 Nov. 15
Allen, Lt.-Col. H. M. . . . . 10 July 08
Anderson, Maj. B. E. . . . . 3 June 16
Anderson, Bt.-Lt.-Col. J. F. H. . . . 3 June 16
Andrew, Brig.-Genl. F. A. . . . . . . 3 June 17
Aplin, Col. P. J. H. [L] . . . . . 6 Sep. 03
Apthorp, Maj. S. E., 96th Infy. . . . 15 Mar. 16
Armstrong, Col. O. C. . . . . . . 26 June 02
Arthur, Maj. L. F., 26th Cav. *p.s.c.* [L] . 1 Jan. 17
Ashburner, Maj. H. W. . . . . . . 3 June 16
Auchinleck, Capt. C. J. E. . . . . . 3 June 17
Badcock, Lt.-Col. F. F., 6th G. R. . . 12 July 92
Bagshawe, 2nd-Lt. E. L., C.I.E., 4th G. Garr. Arty. . . . . 1 Jan. 18
Bainbridge, Lt.-Col. W. F., C.M.G., 54th Sikhs, *p.s.c.* . . . . 29 Nov. 00
Baird, Brig.-Genl. H. B. D., C.M.G., 12th Cav. *p.s.c.* . . . . 3 June 15
Baker, Capt. W. H. G., 31st Lrs. . . . 3 June 17
Baldwin, Brig.-Genl., G. M., 25th Cav. . 21 Jan. 90
Ball, Maj. L. P. 5th Infy. . . . . . 3 June 18
Bamfield, Lt.-Col , H. J. K., I. M. S., . 3 June 17
Barge, Capt., K., M.C. . . . . . . 1 Jan. 18
Barratt, Maj.-Gen. W. C., C.B. . . . 11 Oct. 97
Barron, Maj. R. M., I.M.S. . . . . . 1 Jan. 18
Batten, Col. F. G., I. A. . . . . . 1 Jan. 17
Battye, Maj. I. U. Guides . . . . . 3 June 16
Battye, Lt.-Col. W. R., M.B., F.R.C.S., I.M.S. . . . . 1 Feb. 16
Bayly, Maj.-Genl. *Sir* A. W. L., K.C.B., K.C.M.G., C.S.I. . . . . 1 July 87
Beaman, Capt. A. A. H., 1st Lrs. . . 3 June 16
Beatty, Lt.-Col., G. A. H., 9th Horse . . 3 June 17*
Becher, Lt.-Col. G. A., 8th Cav. . . . 3 June 18
Begbie, Maj.-Genl. E. W., C.B. . . . 30 May 91
Bell. Lt.-Col. R. C. (retd.) . . . . . 1 Jan. 17
Bell-Kingsley, Capt., H. E. W. . . . 3 June 17
Bernard, Capt. R. P. St. V., M.C., 6th G. R. . . . . 1 Jan. 18
Beynon, Maj.-Genl. *Sir* W. G. L., K.C.I.E., C.B. . . . . 16 July 93
Bhuracha, Capt. P. B., F.R.C.S., I.M.S. . 1 Jan. 16
Bignell, Maj. R. L., I. A. . . . . . 1 Jan. 18
Birdwood, Genl. *Sir* W. R., K.C.B., K.C.S.I., K.C.M.G., C.I.E., *q.s.*, A.D.C. . . . 16 July 08
Bliss, Lt.-Col. J. A., M.V.O., 64th Prs. . 1 Jan. 18
Bond, Maj. G. W. . . . . 3 June 16
Borradaile, Col. H. B. [L] . . . . . 16 July 95
Bridge, Capt. R. E. A. . . . . . . 3 June 17
Bridges, Lt.-Col. A. H., 116th Mah. . . 3 June 17
Bromilow, Maj. D. G. . . . . 3 June 17
Browne, Lt.-Genl. *Sir* A.G.F., K.C.B. . . 28 July 90

* Bar to D. S. O.

COMPANION OF THE "DISTINGUISHED SERVICE ORDER."—*contd.*

Companion (D. S. O.)—*contd.*

Browne, Lt.-Col. A. P. . . . . 3 June 18
Browse, Lt.-Col., G., M.D., I. M. S. . . 3 June 17
Bruce, Lt.-Col. G. D., 61st Prs. . . . 1 Jan. 16
Buckland, Capt. G. C. B, 8th G. R. . 1 Jan. 15
Buller, Maj. J. H. G., 57th Rifl. . . . 26 July 17
Burnett, Capt. A. H . . . . 3 June 17
Butterfield, Maj. E. . . . . 1 Jan. 16
Cadell, Maj. J. G., 45th Sikhs . . . . 3 June 17
Cahusac, Capt. C.F., 36th Horse . . . 17 Sep. 17
Campbell, Lt.-Col. A. J. . . . . 3 June 17
Campbell, Col. C. C. . . . . . . 1 July 87
Campbell, Lt.-Genl. *Sir* F., K.C.B. . . 23 May 98
Campbell, Maj. H., M.V.O. . . . . 1 Jan. 18
Campbell, Maj. W., S. & T. Corps . . 23 Nov. 16
Campbell, Maj. (*tempy. Brig.-Genl.*) W. N., C.M.G. (retd.) . . . . 3 June 16
Capper, Lt.-Col., A. S. . . . . . 3 June 17
Carleton, Lt.-Col. H A. . . . . . . 3 June 17
Carnegy, Lt.-Col. G. P. O. . . . . . 3 June 16
Cassels, Bt.-Col. R. A., C.B., 32nd Cav. . 1 Jan. 18
Casson, Bt.-Lt.-Col. W. F. S., 27th Cav. 1 Jan. 16
Catty, Bt. Lt.-Col. T. C. . . . . . . 3 June 17
Caulfeild, Bt.-Col. G. N. . . . . . 29 Dec 93
Chambers, Maj. P. R., 10th Horse . . . 18 Aug. 16
Channer, Maj. G. K., I. A. . . . . . 26 Mar. 18
Chamier, Maj. J. A., 33rd Punj. . . . . 1 Jan. 17
Chatterton, Lt.-Col. G. D. L., 66th Punj. . 25 Nov. 80
Chope, Capt. A. J. H., 2nd G. R. . . . 29 Oct. 15
Clifford, Capt. R. C., I.M.S. . . . . 19 Oct. 16
Climo, Maj.-Genl. S. H., C.B. . . . 20 May 98
VC Cobbe, Lt.-Genl. *Sir* A. S., K.C.B., C.S.I., *p.s.c.* . . . . 29 Nov. 00
Cobbe, Bt. Col. H. H., C.M.G., 13th Lrs. 29 Nov. 00
Cobham, Maj. H. W. (Retired) . . . .
Cochran, Bt. Lt.-Col. G. W., 81st Prs. . 18 Aug. 16
Colan, Capt. H. N., 67th Punj. . . . . 1 Jan. 16
Colan, Lt.-Col. W. R. B. . . . . . 3 June 17
Coleridge, Bt.-Lt.-Col. J. F. S. D., C.M.G., *p.s.c.*, 8 G. R. . . . . 3 June 16
Collins, Maj. L. P., 4th G. R. . . 18 Mar. 15
Collum, Capt. J. A., 26th Cav. . . . 3 June 17
Connor, Maj. F. P., F.R.C.S., I.M.S. . . 1 Jan. 18
Cook, Lt.-Col. C. C., 7th Lrs. . . . 18 Aug. 16
Cooke, Brig.-Genl. H. F. Prs. . . . . 1 Jan. 17
Coppinger, Maj. W. V., M.D., F.R.C.S.I., I. M. S. . . . . 3 June 18
Coringham, Lt.-Col. F. E., 10th G. R. . 1 Jan. 18
Cornish, Capt. A. W. D., M.C. . . . 3 June 18
VC Costello, Lt.-Col. E. W., C.M.G. . . 3 June 17
Craik, Maj. J. . . . . 1 Jan. 18
Cramer Roberts, Maj. M. T., ... 3 June 15
Craster, Maj. G. . . . . 1 Jan. 18
Crawford, Capt. J. D., M.C., 89th Punj. . 17 Sep. 17
Crowdy, Maj. J. D., 5th G. R. . . . . 3 June 18
Cullen, Lt.-Col. E. H. S., C.M.G., M.V.O., 34 Prs. . . . . 23 Aug. 12
Cuming, Bt. Lt.-Col. R. J., 48th Prs. . . 3 June 18
Daniell, Maj. J. A. S. 14th Sikhs . . . 3 June 15
Darley, Lt.-Col. J. R., C.I.E., 119th Infy. 18 Aug. 16
Davidson, Col. J., M.D., I.M.S. [L]. . . 23 Aug. 12
Davies, Lt.-Col. C. H., C.M.G. 53rd Sikhs 3 July 05
Davis, Lt.-Col. G. J., 22nd Punj. . . . 3 June 16
Dayrell, Maj. W. S., 72nd Punj. . . . 3 June 18
Deane, Bt. Lt.-Col. D., 12th Cav. . . . 3 June 18
de Burgh, Maj. E., *p.s.c.*, 9th Horse . . 3 June 16
De Butts, Capt. F. C., M.C. 55th Rifl. . 1 Jan. 17
de Courcy, Maj., the Hon. M. W. R. . . 3 June 17
Delamain, Maj.-Genl. Sir W. S., K.C.M.G., C.B. . . . . 14 Apl. 06
De la Motte, Maj. R. B. S. & T. Corps . 3 June 16
Dent, Maj. W., 103rd Infy. . . . . . 1 Jan. 16
Dobbs, Lt. Col. C. F., 94th Infy. . . . 1 Jan. 17
Dopping-Hepenstal, Lt.-Col. M. E., 1st G.R. . . . . 3 June 16
Doyle, Maj. I. P., I.M.S. . . . . . 23 Aug. 89
Drysdale, Capt. A. E., M.C. . . . . 3 June 16
Duncan, Maj. H. C., 9th G. R., *p.s.c.* . 29 Oct. 15
Dundas, Maj. P. H., 6th Jats. . . . . 18 Feb. 15
Dunlop, Lt.-Col. W. B., S. & T. Corps . 3 June 15
Dunsford, Brig.-Genl. F. P. S. . . . . 3 June 17
Dyer, Maj. H. R., 26th Horse . . . . 26 June 18
Dyke, Lt.-Col. P. H. . . . . 1 Jan. 18
Eadie, Maj. J. I. . . . . 3 June 17
Edwardes, Brig.-Genl. S M., C.B., C.M.S. 16 July 95
Edwards, Col. Fitz. J. M, C.B., C.M.G., A.D.C. . . . . 23 May 98
Edwards, Col. J. B. . . . . . . 21 Jan. 96
Egerton, Fd. Marshal *Sir* C. C., G.C.B. 30 May 91
Elliot, Lt.-Genl. *Sir* E. L., K.C.B. . . 1 July 87
Elliot, Lt.-Col. W. H. W., M.B. . 26 June 02
Ellis, Maj, S. G. V. . . . . 3 June 17
Etheridge, Maj. F., 7th Rajputs . . . 3 June 18
Eustace, Maj. Gen. A. H., C.B. . . . . 6 Sep 94
Evatt, Hony. Brig.-Genl. J. I. . . . . 9 Jan. 09
Ewart, Maj. R. F., S. & T.C. . . . . 26 May 18
Ewart, Maj.-Gen. *Sir* R. H., K.C.M.G., C.B., C.I.E. . . . . 27 Aug. 95
Fagan, Lt.-Col. (*tempy. Br.-Genl.*) E. A., C.M.G., 36th Horse . . . . . . 1 Jan. 17*
Fagan, Maj. B. J. . . . . 1 Jan. 18
Farran, Maj. G. L., M.C., 4th Cav. . . June 17
Fetherstonhaugh, Lt.-Col. W. A., 8th Cav. 1 Jan. 17
Fleming, Lt.-Col. A. N. . . . . . 1 Jan. 18
Finlay, Maj. R. F. 58th Rif. . . . . 3 June 16
Finlayson, Maj. W. T. . . . . . . 1 Jan. 18
Firth, Lt.-Col. R. A., 2-10th G. R. . . 1 Feb. 16
Fisher, Lt.-Col. J., M.B, I.M.S. . . 20 May 98
FitzGerald, Maj. C.R.L., 126th Infy. . 3 June 16
Foster, Maj. W. M. A. . . . . . 3 June 17
Fowler Col. F. J., C.B. . . . . . 23 Aug. 89
Fowler, Col. C. A., C.B. . . . . . 16 July 08
Franklin, Maj. (*temp. Brig.-Genl.*) H. S. E., 15th Sikhs . . . . 1 Jan. 16
Fry, Maj. A. B., M.B., I. M. S. . . . 3 June 17
Galbraith, Maj., E. D., 55th Rif. . . 1 Jan. 18
Gaussen, Lt.-Col. J. R., C.M.G., 3rd Horse 29 Nov. 00
Gilbert, Lt.-Col. G. E. L. . . . . . 16 July 08
Giles, Bt. Lt.-Col. E. D., C.M.G., *p.s.c.*, 35th Horse . . . . 3 June 16
Glasfurd, Lt.-Col. A. I. R., *p.s.c.*, 46th Punj. 3 June 16
Godkin, Maj. S. R., F.R.C.S.I., I. M. S. 3 June 17
Godwin, Bt.-Col. C. A. C. . . . . . 1 Jan. 18
Goodbody, Bt.-Lt-Col. C. M., C.I.E. . . ...
Gordon Bt. Col. S. V. . . . . . . 21 Jan. 96
Gordon, Capt. D. H., 34th Sikhs . . . 17 Mar. 17
Gordon, Col. J. H., C.B. . . . . . 1 July 87
Gold, Maj. G. . . . . 3 June 18
Gourlay, Maj. C. A., M.D., I.M.S. . . . 1 Jan. 18
Gow, Capt. P. F., I.M.S. . . . .
Grant, Maj. G. P., 106th Prs. *p.s.c.* . . 2 Sep. 92
Gray, Lt.-Col. F. W. B. Col., C.M.G., 57th Rif. . . . . 16 July 98
Green, Maj. W. G. K. . . . . . . 1 Jan. 18
Griffin, Lt.-Col. C. P. G. . . . . . 29 Nov. 00
Cunning. Maj. G. H., 21st Cav. . . . 1 Jan. 18
Gurdon, Lt.-Col. B. E. M., C.I.E., (retd.). 10 July 95
Gwyn-Thomas, Lt.-Col. G., C.M.G. . .
Hadew, Bt.-Col. R. C. . . . . . . 21 Jan. 90
Haldane, Lt.-Col. C. L., C.M.G., 7th G. R. 1 Jan. 18
Hamer, Maj. M. A., M.C. . . . . . 3 June 16
Hamilton, Bt. Lt.-Col. W. H., I.M.S. . 3 June 16
VC Hammond, Col. *Sir* A. G., K.C.B . 7 Dec. 88
Hance, Capt. H. M., M.C., I. A. R. O. 22 Sep. 16
Harbord, Maj. (*temp. Brig.-Genl.*) C. R., 30th Lrs. . . . . 1 Jan. 18
Harley, Maj. A. B. . . . . 3 June 17
Harold, Comdr. A. E., R.I.M. . . . . 3 June 16
Harris, Maj. E. T., M.B. I.M.S. . . . 3 June 17
Hart. Maj. E. G. S. & T. C. . . . . 1 Jan. 18
Hartley, Maj. A. F., 11 . . . . . . 3 June 17
Hartwell, Capt. J. R. . . . . . 23 Nov. 16
Hastings, Maj.-Gen. E. S., C.B. . . . 1 July 87
Hastings, Bt.-Maj. W. H., 92nd Punj. . 1 Jan. 17
Haughton, Maj. E. J. H., 105th Mah. . . 1 Jan. 18
Hawes, Lt.-Col. C. H., M.V.O. . . . . 3 June 16
Hawkes, Maj. W. C. W., 106th Prs. . . 1 Jan. 17

* Bar to D. S. O.

COMPANIONS OF THE "DISTINGUISHED SERVICE ORDER"—*contd.*

Companions (D. S. O.)—*contd.*

Hay, Lt.-Col. C. J. B., Corps of Guides . 1 Jan. 16
Hay, Lt.-Col. W. N., C.I.E., 129th Baluchis 3 June 18
Haydon, Lt-Col. W. P. . . . . . . 3 June 17
Headlam, Comdr. E. J., R.I.M. . . . 1 Jan. 16
Hemsley, Maj. C., 64th Prs. . . . . 3 June 18
Henderson, Maj. K., *p.s.c.*, 39th Garh. Rif. 3 June 16
Henderson, Maj. R. W. . . . . . . 3 June 18
Henegan, Lt.-Col. J. . . . . . . 29 Dec. 93
Hennell, Col. *Sir* R., *Kt*, C.V.O. . . 1 July 78
Herbert, Lt.-Col. C., Supy. List . . . 27 Aug. 95
Herrick, Capt. R. de S. B., I.M.S. . . . 3 June 16
Hesketh, Maj. W., 12th Cav. . . . . 3 June 18
Hewett, Maj. G., 48th Prs. . . . . 1 Jan. 16
Hewett, Maj. M. S., S. & T. C. . . . 1 Jan. 18
Hext, Maj. G. T. B., S. & T. Corps . . 1 Jan. 17
Heyland, Maj. A. K., S. & T. Corps . . 3 June 15
Hickey, Lt. P. F. B., I. A. R. O. . . . 1 Jan. 18
Hill, Lt.-Col. H. C., 110th Mah. . . . 19 Oct. 16
Hill, Maj.-Genl. J., C.B. . . . . . 16 May 09
Hodgson, Maj. E. C., I.M.S. . . . . 3 June 18
Hodgson, Maj. G.C., 32nd Prr. . . . 16 Dec. 04
Hogg, Maj. C. M. T., 4th G. R. *p.s.c.* . . 1 Jan. 17
Holdich, Brig.-Genl. H. A. . . . . 3 June 16
Holland-Pryor, Col. (*tempy. Brig.-Genl*) P., C.M.G., M.V.O. . . . . . . 3 June 16
Holman, Maj.-Genl. H. C., C.B., C.M.G., *p.s.c.* [L]. . . . . 29 Nov. 00
Hornby, Bt-Col. M. L., C.M.G. . . . 27 Aug. 95
Hoskyn, Maj. J. C. M., 44th Infy. . . 26 Mar. 18
Hudson, Maj. C., I.M.S. . . . . 6 Sep. 04
Huggins, Lt.-Col. P. G., C.B. [L]. . . 1 July 87
Hugo, Lt.-Col. J. H., M.B., I.M.S. . . 20 May 98
Hulseberg, Lt.-Col. H. 127th Infy. . . 1 Jan. 16*
Hume-Spry, Maj. W. E. . . . . . 3 June 16
Hunt, Maj. F. E., 29th Lrs. . . . . 18 Mar. 01
Hunt, Maj. H. R. A., 25th Punj. . . . 3 June 16
Hunter, Maj F. F., Civil Employ . . 3 June 17
Hutchinson, Lt.-Col. C. A. K., 41st Dogras . . . . 1 Jan. 16
Inkson, Maj. N. L., I. A. R. O. . . . . 17 Mar. 17
Inskip, Capt. R. D. . . . . 3 June 16
Irwin, Revd. R. J. B., M.C., M.A. . . . 3 June 16
Jackson, Lt.-Col. H. W. . . . . . 1 Jan. 18
Jackson, Maj. F. A., 21st Cav. . . . . 1 Jan. 18
Jacob, Bt.-Col. A. Le G., C.M.G., C.I.E., 106th Prs. . . . . 4 Oct. 98
Jatar, Lt. N S., I.M.S. . . . . . . 3 June 17
Johnston, Lt.-Col. C. A., C.B., I.M.S. . 1 Jan. 17
Johnston, Maj. P. D. C., 8th Rajputs . . 3 June 18
Johnston, Maj. R. . . . . 3 June 16
Jones, Maj. A. N. G., 36th Horse, . . 1 Jan. 18
Jones, Lt.-Col. H. J. . . . . . . 16 July 95
Jones, Capt. J. W., I.M.S. . . . . 22 Sept. 16
Jukes, Bt.-Maj. A. H., 9th G. R. . . 15 Mar. 16
Keary, Lt.-Genl. *Sir* H. D'U., K.C.B., K.C.I.E. . . . . 18 Mar. 92
Keen, Maj. F. S., 45th Sikhs . . . . 1 Jan. 16
Keily, Lt.-Col. F. P. C., C.M.G., 125th Rif. 15 Apl. 16
Kelly, Lt.-Col. T. B. F. R. C. S. I. M. S. 1 Jan. 18
Kelsall, Maj. R., I.M.S. . . . . . 3 June 16
Kemmis, Maj A. W. M. 10th Lr. . . . 1 Jan. 18
Kennedy, Maj. R. S. M.C., M.B., I.M.S. 3 June 15
Kennedy-Craufurd-Stuart, Maj.C., 127th Infy. . . . . 7 Nov. 15
Keogh, Lt.-Col. J. B., 32 Lrs. . . . 3 Mar. 17
Kerans, Maj. G. C. L., I. M. S. . . . 1 Jan. 16
Kinch, Lt. A. S., R., I.M.. . . . . 3 June 16
Kirby, Bt. Lt.-Col. J. T., S. & T. Corps . Mar. 17
Kirkpatrick, Maj. W . . . . . . 2 Sep. 02
Koebel, Maj. F. E., 51st Sikhs . . . 16 Aug. 17
Knight, Maj. Gen. W. C, C.B., C.S.I., *p.s.c.* [L] . . . . 22 Aug. 02
Knott, Rev. A. E. . . . . 25 Aug. 17
Knowles, Maj. G., 2nd Lrs. . . . . 12 Sep. 02*
Knox, Lt.-Col. E. F. . . . . . . 3 June 16
Knox, Lt.-Col. R. W., M.B., F.R.C.S., I. M. S. . . . . 29 Oct. 15
Lambert, Maj. W. J., 29th Lrs. . . . 1 Jan. 17
Leeds, Lt.-Col. T. L., C.M.G., 59th Rif. . 23 Nov. 17
Lentaigne, Capt. E. C. . . . . . . 3 June 17
Leslie, Lt.-Col. W.S. . . . . 3 June 17
Lewis, Capt. H. V., M.C., 129th Baluchis. 3 June 17
Lind, Maj. (*temp. Lt.-Col.*) A. G., 58th Rif. 1 Jan. 18
Lloyd, Bt. Lt.-Col. C. R. . . . . . 3 June 17
Lloyd, Maj. R. A. M.D., I.M.S. . . . 17 Sep. 7
Lock, Brig.-Genl. F. R. E. . . . . . 3 June 17
Low, Lt.-Col. R. B., 9th Horse . . . 21 Jan. 96
Lucas, Brig.-Genl. F. G., C.B., 5th G. R. . 20 May 98
Lugard, Maj. E. J. . . . . 28 July 90
Lukin, Lt.-Col. R. C. W. . . . . . 3 June 15
Lunn, Capt. A. J., 82nd Punj. . . . . 26 Apl. 17
Lyall, Maj. R. A., I. A.. . . . . . . 1 Jan. 17
Lynck, Capt. J. B. . . . . 1 Jan. 18
Lynn, Capt. G. R., M.B., I.M.S. . . . 3 June 17
McBarnet, Maj. A. E., M. V. O. . . . . 2 Sep. 02
McConaghy, Major J. G., 25th Cav. . 3 June 16
McCudden, Maj. J. H., M. C., 21st Cav. . . . . 1 Jan. 18
McKenna, Maj. J. C., 16th Rajputs . . 19 Oct. 16
McLeod, Genl. *Sir* D J.S., K.C.B., K.C.I.E. 1 July 87
McLeod, Maj. D. K., Guides . . . . 1 Jan. 17
McLeod, Maj. T. J., 2-7th G. R. . . . 1 Jan. 16
McRae, Maj. H. St. G. M., 45th Sikhs . 1 Jan. 17
McWatters, Bt. Lt.-Col. H. C. . . . 3 June 16
McVean, Lt.-Col. D. A. D., 45th Sikhs . 2 Sep. 02
Macandrew, Maj.-Genl. H. J. M., C.B., *q.s.* . . . . 29 Nov. 02
Macdonald, Bt. Maj. H., 11th Lrs. . . 3 June 18
Macdonald, Maj. R. W. . . . . . 3 June 16
Mackenzie, Lt.-Col., C. . . . . . 3 June 17
Mackenzie, Maj. H. J., 51st Sikhs . . 1 Jan. 17
Macleod, Maj. N. . . . . 3 June 17
Macleod, Maj. R. D. C. . . . . . 3 June 18
Macmullen, Maj. (*temp. Brig.-Genl.*) C. N., *p.s.c.*, 15th Sikhs . . . . 29 Oct. 15
Maconchy, Col. E. W. S. K, C.B., C.M.G., C.I.E. . . . . 19 Nov. 91
Macpherson, Maj. J., 124th Infy. . . 29 Nov. 00
Macquoid, Lt.-Col. C.E.E.F.K., 4th Cav. 20 May 93
Macrae, Maj. J. C., 19th Punj. . . . 18 Aug. 16
Mahon, Maj. A. E., 55th Rif. . . . 3 June 18
Malcolm, Maj. P., M.V.O. . . . . . 2 Sep. 02
Marsh, Lt. B. C., R.I.M. . . . . . 3 June 16
Marsh, Maj. C. H., 16th Lrs. . . . 29 Nov. 00
Marshall, A. Maj. 28th Cav. . . . . 3 July 15
Marshall, Bt. Maj. J. S., 35th Sikhs . . 1 Jan. 17
Massy, Maj. S. D., 29th Punj. . . . . 3 June 16
Mathias, Capt. W. H., 128th Prs. . . 23 Aug. 18
Matthews, Maj. E. A. C., I.M.S. . . . 3 June 18
Maude, Maj. E. A., 26th Punj . . . . 3 June 17
Mayne, Capt. A. G. O. M. . . . . . 3 June 17
Medlicott, Maj. H. E. . . . . . . 3 June 18
Mills, Maj. A. M., 18th Lrs. . . . . 1 Jan. 17*
Milward, Bt. Lt.-Col. C. A, 53rd Sikhs . 3 June 17
Minchin, Lt-Col.C. F., Supy. List. . 29 Nov. 00
Mitford, Maj. J. P, 98th Infy. . . . 1 Jan. 17
Moberly, Bt.-Lt-Col B. R., 56th Rif., *p.s.c.* . . . . 29 Oct. 15
Moberly, Lt.-Col. (*Ty. Brig. Genl.*) F. J., 25th Punj, *p.s.c.* . . . . 10 Aug. 13
Moens, Bt.-Lt.-Col. A. W. H. M. . . . 3 June 17
Money, Lt.-Col. E. D., C. I. E. . . . 3 June 18
Money, Maj. E. F. D., 4th G. R. . . . 1 Jan. 17
Moore, Lt-Col. C. H. G., S. & T. Corps. 16 Dec. 04
Moore, Maj. H. D., 130th Baluchis. . . 3 June 17
More, Maj. J. C., 51st Sikhs . . . . 3 June 18
Morgan, Maj., B. E., ... 3 June 17
Morgan, Maj. M. H. L., 62nd Punj. [L] 29 Oct. 15
Morin, Maj. A. H., I. A. R. O. . . . . 3 June 17
Morris, Maj. C. O., S. & T. Corps. . . 3 June 16
Morris, Bt.-Col. G. M., 62nd Punj. . . 17 Mar. 17
Muirhead, Maj. J. A., 1st Lrs. . . . . 3 June 17
Murphy, Lt.-Col. W. R., I.M.S. . . . 28 July 91
Murray, Maj. K. D. B., 59th Rif. *p.s.c.* . 1 Jan. 17
Muspratt, Lt.-Col. S. F., 12th Cav. . . 3 June 16
Needham, Maj. R. A., M.B., I. M. S. . . 1 Jan. 10
Nopean, Lt.- D. H. Y., 5th G. R. . . .
Newnham, Lt.-Col. C. C. . . . . . 3 June 18

* Bar to D. S. O.

COMPANIONS OF THE "DISTINGUISHED SERVICE ORDER"—*contd.*

Companions (D.S.O.)—*concld.*

Nicholson, Capt. F. L., M.C., 37th Dogras 3 June 18
Nightingale, Brig.-Genl. M. R. W., C.M.G., 5 G. R. [L] 25 Aug. 15
Nisbet, Bt. Lt-Col. T., 28th Cav. 17 Mar. 17
Nixon, Capt. C. H. F. 3 June 16
Norie, Col. C. E. de M., C.B., C.M.G., 2nd G. R. *p.s.c.*[*l*] 26 June 02
Norie, Maj. F. H. 18 Feb. 15
Norman, Brig-Genl. C. L., M.V.O. 3 June 17
Norman, Bt.-Col. W. H. 3 June 17
Nugent, Capt. J. F. H. 3 June 16
O'Donnell, Maj.-Gen. H., C.B. 23 Aug. 89
O'Rorke, Capt. J. M. W. 3 June 18
Ogg, Maj. A. C., 7th Rajputs 18 Aug. 16
Ogle, Bt. Lt.-Col. N. 3 June 17
Orr, Bt.-Col. G.M. 1 Jan. 18
Parkin, Maj. J. F., 113th Infy. 3 June 18
Paterson, Maj. T. G. F., M.B., I.M.S. 3 June 17
Peel, Maj. B. G., 81st Prs. 19 Oct. 16
Pepys, Major. G. L., 57th Rif. *p.s.c.* [L] 29 Oct. 15
Perry, Lt.-Col., E. L., I.M.S. 3 June 17
Peterson, Brig.-Genl. F. H., C.B., [*l*] 27 Aug. 95
Philipps, *Hony.* Maj.-Genl. 29 Nov. 00
Phipson, Capt. E. S., M.B., I. M. S. 1 Feb. 61
Pitman, Capt. C. R. S., M.C., 27th Punj. 8 Mar. 17
Pocock, Lt.-Col. P. F., 120th Infy. 23 Aug. 18
Ponsonby, Capt., D. G. 3 June 17
Poole, Maj. I. M. C. 3 June 16
Pope, Maj. S. B., *p.s.c.*, 58th Rif. 3 June 16
Popham, Bt. Lt.-Col. E. L., 26th Cav. 3 June 17
Pott, Maj. E. H., 1st Lrs. 3 June 17
Pott, Capt. D., M.C. 3 June 17
Pratt, Maj. H. R. E., 35th Sikhs 20 May 98
Presgrave, Col. E. R. J., C.B., *u.s.l.* 12 Feb. 95
Preston, Lt.-Col. J. E., C.B., *u.s.l.* 25 Nov. 86
Preston, Maj. W. V. P. 3 June 16
Price, Col. C. H. U., C.B. 7 July 99
Proctor, Maj. A. H., M.D., I.M.S. 3 June 18
Pryce, Lt.-Col. H. E. R., 18th Infy. 3 June 17
Pryor, Capt. T., I. A. R. O. 1 Jan. 18
Ralph, Lt.-Col., A. C., 4th Rajputs 3 June 18
Ralston, Maj. W. H., M.C. 3 June 17
Rawlins, Lt.-Col. A. K., 24th Punj. 20 May 98
Raymond, Maj. E. D., M.C., 30th Lrs. 17 Sep. 17
Rebsch, Capt. W. K., 92nd Punj. 3 June 16
Reid, Tempy.-Maj. A. D., (Lt. retd., I. A.), Innis. Fus. 3 June 17
Rennison, Maj. A. J., S. & T. Corps 23 Nov. 16
Reyne, Capt. G. von R., 76th Punj. 19 Oct. 16
Rice, Maj. B. A. Mc.H., 6th G. R. 3 June 18
Rice, Maj. G. D., S. & T. Corps 3 June 18
Ricketts, Lt.-Col. P. E., M.V.O., 18th Lrs. 1 Jan. 17
Riddell, Maj. A. 3 June 17
Risley, Maj. C. G., 18th Lrs. 1 Jan. 17
Roberts, Maj. E.A., I. M. S., 1 Jan. 18
Roberts, Maj. W. B. 101st Grs. 1 Jan. 17
Robinson, Maj. D. G., 46th Punj. 1 Jan. 16
Rodgerson, Capt. A. P. 18 Jan. 18
Roe, Maj. C. D., 1-4th G. R. 23 July 15
Ross, Maj. A. C., 20th Horse 18 Feb. 15
Ross, Maj. G. W. 3 Jun. 16
Ross, Bt. Lt.-Col. R. C., 6th Infy. 1 Jan. 16
Rowan-Robinson, Bt. Lt.-Col. J. R., S. & T. Corps 23 Nov. 16
Rowcroft, Maj. G. F. 16 May 99
Rowlandson, Bt. Lt.-Col. M. G. D., 38th Dogras 3 June 18
Ruel, Maj. W. G., S. & T. Corps 1 Jan. 17
Rundall, Col. F. M., C.B. 19 Nov. 91
Russell, Lt. R. T., I. A. R. O. 30 Mar. 17
Ryan, Capt. D. G. J., 6th G. Rif. 25 Aug. 15
Rybot, Maj. N. U. L., 76th Punj. 1 Jan. 16
St. John, Bt. Lt.-Col. (*temp. Brig.-Genl*) R.S. 3 June 17
Salkeld, Maj. H. Y. 22nd Lrs. 4 Mar. 18
Sangster, Maj. P. B., 2nd Lrs. 3 June 16
Sargent, Maj. A. E. E., M.C., S. & T. Corps 1 Jan. 17
Saunders, Maj. E. H., S. & T. Corps 29 Nov. 00
Saunders, Maj. M., 36th Sikhs 3 June 17
Scale, Maj. J. D., I. A. 5 Mar. 17
Scallon, Lt.-Genl. *Sir* R. I., G.C.B., K.C.I.E. 2 Aug. 88
Scoones, Capt. G. A. P., M.C. 3 June 17
Scott, Lt. C. A., R.I.M. 3 June 15
Scott, Maj. G. B., 27th Punj 29 Oct. 15
Scott, Maj. H. L., M.C. 3 June 17
Scott, Maj. H. St. G.S., 4th G. R. 29 Oct. 15*
Scott, Capt. I., I. M. S. 25 Mar. 18
Scott, Maj.-Genl. T. E., C.B., C.I.E. 11 Oct. 97
Selby, Lt.-Col. W., F.R.C.S., I.M.S. 20 May 98
Senior, Col. H. W. R. 3 June 18
Seton-Browne, Lt.-Col., C. L. S., 55th Rif. 10 June 97
Shakespear, Bt.-Maj. G. F. C., M.C. 1 Jan. 15
Shakespear, Lt.-Col. J., C.M.G., C.I.E., Supy List 28 July 99
Shea, Maj., A. G. 3 June 17
Shea, Maj.-Gen. J.S.M., C.B., C.M.G., *p.s.c., q.s.* 10 Sep. 01
Sheepshanks, Capt. R. H., 12th Cav. 29 Oct. 15
Shepherd, Maj. C. I., 53rd Sikhs 27 Mar. 17
Sheringham, Capt. A. T., 121st Prs. 3 June 16
Sherston, Capt. J. R. V., M.C. 1 Jan. 18
Shore, Col. O. B. S.-F., C.B., C.I.E., *p.s.c.* [L] 29 Nov. 09
Sidgwick, Hony. Maj. C.K.D., V.D. 1-17th B., B. & C. I. Ry. Btn., I. D. F. 21 Aug. 02
Skinner, Maj. A. B. 1 Jan. 18
Skinner, Col. G. J. 28 July 90
Skinner, Maj. H. T., 29th Punj. 3 June 16
Smith, Maj. C. W. J., 29th Punj. 1 Jan. 17
Smith, Bt. Lt.-Col. H. H. 3 June 16
Souter, Lt.-Col. H. M. W., C.M.G. 3 June 17
Stable, Capt. R. H., 122nd Infy. 16 Aug. 17
Stewart, Bt.-Col. J. H. K., 39th Garh. Rif., *p.s.c.* 3 June 15
Strong, Bt. Lt.-Col. A. D., 10th Lrs. 1 Jan. 16
Sutherland, Maj. R. O. 3 June 17
Sweeny, Maj. R.L.C., M.C., S.& T. Corps 1 Jan. 17
Sweet, Lt.-Col., E. H. 3 June 17
Sweet, Capt. R., I.M.S. 3 June 16
Sykes, Lt.-Col. W. A., M.B., I.M.S. 1 July 87
Synge, Maj. M., S. & T. Corps 1 Jan. 17
Tarver, Lt.-Col. A. L., 124th Infy. *p.s.c.* 16 July 08
Taylor, Capt. J., M.B., I.M.S. 3 June 15
Taylor, Lt.-Col. H. N. 29 Dec. 93
Taylor, Maj. N. C. 3 June 16
Teversham, Bt.-Col. R. K. 1 July 87
Thackwell, Col. C. G. R., C.B. 20 May 98
Thomas, Capt. A. N., M.B., I.M.S. 3 June 17
Thomas, Lt.-Col. G., 2nd Lrs. 14 Nov. 16
Thomas, Lt.-Col. H. St. G., R. Suss. 3 June 17
Thompson, Bt.-Col. W. O. 27 Aug. 95
Thomson, Bt. Lt.-Col. A. G., 58th Rif., *p.s.c.* 3 June 15
Thomson, Lt.-Col. A. H. G., 56th Punj. 3 June 17
Thornhill, Bt. Lt.-Col. C. J. M. 1 Jan. 18
Thornhill, Capt. C. M., M.C. 3 June 17
Tighe, Maj.-Genl. *Sir* M. J., K.C.M.G., C.B., C.I.E. 23 Aug. 89
Tillard, Lt.-Col. A. B., 1-3rd G. R. 20 May 98
Timbrell, Maj. T., S. and T. Corps. 17 Sep. 17
Tobin, Maj. H. W., 128th Prs. 18 Aug. 16
Tomkins, Lt.-Col. H. L., C.M.G., 28th Punj. 20 May 98
Toogood, Lt. C. G., I. A. 30 Mar. 15
Torrie, Maj. C. J., 30th Punj. 10 Mar. 17
Townsend, Capt. F. O., I. A. R. O. 17 Sep. 16
Trench, Capt. C. F. 7th Lrs. 25 Aug. 16
Turnbull, Maj. G. O., 26th Punj. 1 Jan. 17
Turner, Lt.-Col. A. G. B. 3 June 17
Turner, Lt.-Col. R. G., C.M.G., I.M.S. 1 Jan. 17
Twiss, Maj. E. K., 10th Jats. 1 Jan. 17
Twiss, Maj. H. W. F. 3 June 18
Tyndall, Maj. H. S., 40th Pathans 1 Jan. 17
Tytler, Col. H. C. 3 June 18
Unwin, Bt.-Col. G. B. 16 Jul. 06
Vaughan, Lt.-Col. E. G., S. & T. Corps. 23 Aug. 12

* Bar to D. S. O.

COMPANIONS OF THE "DISTINGUISHED SERVICE ORDER"—*concld.*

Companions (D.S.O.)—*concld.*

Vaughan, Bt.-Col. L. R., C.B., 7th G. R., *p.s.c.* . . . 3 June 15
Venning, Maj. F. E. W. . . . 3 June 17
Villiers Stuart, Maj. J. P., 55th Rif., *p.s.c.* 3 June 15
Walker, Capt. C. W. G., 37th Dogras . 1 Jan. 18
Wallace, Maj. E. C. L., 30th Punj. . 3 June 18
Ward, Lt.-Col. A., 30th Punj. . . . 3 June 17
Watson, Maj. C. H., I.M.S. . . . 3 June 17
Watt, Lt.-Col. D. M., 2nd G. R., *p.s.c.* 10 Mar. 15*
Webb, Maj. W. F. R. . . . . 25 Nov. 16
Wernicke, Maj. F. P., B.M., I.M.S. . . 3 June 18
Wheatley, Maj. W. P. R. . . . . 3 June 17
Westlake, Bt.-Col. A. P. . . . . 1 July 87
M. Westmacott, Maj.-Genl. *Sir* R., K.C.B. 28 Jul. 90
Whitaker, Capt. S. S., M.C., 7th G. R. . 18 Oct. 17
Whitehead, Bt.-Lt.-Col. J., *p.s.c.*, 1st Brah. . . . . 3 June 16
Wigram, Bt.-Lt.-Col. K., C.B., 2nd G. R., *p.s.c.* . . . . 1 Jan. 17
Wilford, Maj E. E., 30th Lrs. . . . 1 Jan. 17
Wilkinson, Maj. G. H., S. and T. Corps 3 June 18
Willans, Maj. T. J., 57th Rif. . . . 3 June 15
Williams, Maj. A. F. C., 31st Lrs. . . 29 Nov. 00
Williams, Maj. H. A., M.B., I.M.S. . . 29 Nov. 00
Williams, Maj. L. G., 5th Cav. . . . 18 Aug. 16
Willis, Maj. E. S. C., 58th Rif. . . . 22 Aug. 02
Willis, Maj. G. C. W., 31st Lrs. . . . 29 Oct. 15
Willoughby, Maj. D. V. . . . . . 3 June 17
Wilson, Bt. Lt.-Col. A. T., C.M.G., C.I.E. 1 Jan. 16
Wilson, Lt.-Col. J. A., 8th G. R. . . . 23 Aug. 12
Wilson, Bt.-Maj. N. M., 7th G. R. . . 1 Jan. 16
Wilson, Bt. Lt.-Col. R. C. M.C., 114th Mahrattas . . . . . 3 June 18
Wilson-Johnson, Bt. Lt.-Col. W. E. . . 3 June 19
Winter, Bt.-Col. C. B., C.M.G. . . . 3 June 16
Wood, Lt.-Col. E. J. M. . . . . 3 June 17
Worgan, Maj. R. B., 20th Horse . . . 1 Jan. 17
Wright, Lt.-Col. F. W., M.B., I.M.S. . 1 July 87
Wyatt, Maj. E. R. C., 106th Prs., *p.s.c.* . 1 Jan. 17
Wylie, Maj. G. G. E. . . . . . 1 Jan. 18
Wynter, Capt. J. R., 52nd Sikhs . . 31 May 16
Yates, Maj. J. A., C.I.E., 103rd Infy. . . 3 June 18
Yates, Maj. R. J. B. . . . . . 1 Jan. 18
Yeilding, Col. W. R., C.B., C.I.E. . . 30 May 91
Young, Lt.-Col. H. G. . . . . . 3 June 17

COMPANIONS OF THE IMPERIAL SERVICE ORDER.

Companions (I.S.O.).

Anscomb, Maj. A. M., V.D., 27th Bal. Coy.
Bean, Lt.-Col. L. C. D., V.D., late N. W. Ry. V. Rif.
Bird Lt.-Col., (*Hony.-Col.*), F. D., V.D., 1st Madras Gds.
Courtney, S. A. S., Hony. Maj. M., I. M. D. D. . . . . 3 June 18
Girard, Maj. R. G., V.D., Cal. V. Rif.
Goulding, Col. H. R., V.D., late 1st Punj. V. Rif.
Hawkins, Hon. Maj. W., V.D., 3rd Punj. Rif.
Luker, Maj. F., V.D., Alld. V. Rif.
Maflin, Hony. Maj. G. H., V.D., *late* Cal. V. Rif.

V.C. M. Probyn, Genl *Rt. Hon. Sir* D. M., G.C.B., G.C.S.I., G.C.V.O., Eq.

THE MILITARY CROSS (M. C.).

Abbot, Capt. J. F., 8th G. R. . . . 3 June 17
Abbott, Capt. R. S 38th Horse . . . 1 Jan. 17
Acworth, Capt. D. H., 55th Rif. . . . 1 Jan. 15
Adlard, Lt. F. A., I. A. R. O. . . . 26 Mar. 17
Aitken, 2nd-Lt. (T. Capt.), A. B. . . 3 June 17
Aitkenhead, Lt. J. D., I. A. R. O. . . 26 Mar. 17
Alexander, Maj. C., 81st Prs. . . . 3 June 17
Alexander, Maj. R. G., 11th Lrs. . . 1 Jan. 16
Amar Sing Thapa, Sub-Maj. . . . 18 Jan. 18
Ambrose, Lt. R. D., 123rd Rif. . . . 26 Mar. 18
Anderson, Capt. F. J., M.B., I.M.S. . . 3 June 17
Anderson, Capt. J. McK. 109th Infy. . 3 June 17
Andrewes, Capt. C. L. . . . . 3 June 18
Andrews, Capt. H. A. . . . . 3 June 16
Anthony, Subdr., I.D.S.M., 2nd S. and Minners . . . . . 3 June 18
Apte, Temp. Lt. V. K., I. M. S. . . 26 Mar. 18
Arrindell, Lt. C. D. S. M., I. A. R. O. . 1 Jan. 18
Arsla Khan, I.O.M., Subdr.-Maj., *Bahadur*, 57th Rif. . . . . 3 June 15
Arthur, Capt. D., I.M.S. . . . . 19 Oct. 16
Atkins, Maj. G. W., 25th Punj. . . . 1 Feb. 16
Aylmer, Maj. G., S. & T. Corps. . . 3 June 16
Aylward, Lt. D., I.A.R.O. . . . . 3 June 18
Baddeley, Maj. R. J. H., 15th Lrs. . . 3 June 16
Bailey, Capt. P. F. H. . . . . 3 June 16
Baines, 2nd Lt. R. H. . . . . 3 June 17
Bainbridge, Capt. D., 57th Rif. . . . 1 Jan. 16
Baker, Capt. C. H. N., I. M. S. . . 3 June 17
Baker, Lt. W. S., I. A. R. O. . . . 30 Mar. 17
Bakht Bahadur Adhikari, Subdr., 9th G. R. . . . . . . 1 Jan. 16
Bal, Lt. N. K., I. M. S. . . . . 29 Oct. 15
Baldwin, Capt. M.C. . . . . . 3 June 16
Bapty, Lt. P. S., I. A. R. O. . . . 1 Jan. 16
Barge, Capt. K., D.S.O., 17th Cav. . . 1 Jan. 17
Barron, Capt. A. H. D. . . . . 1 Jan. 18
Barlow, 2nd-Lt., W. N. C. . . . . 3 June 17
Barstow, Capt. A. E., 15th Sikhs . . 13 Jun. 15
V.C., Barter, Lt. F., 3rd G. R. . . . 26 July 18
Batra, Capt. H. L., I. M. S. . . . . 1 Jan. 18
Baxter, Lt. C. W. W., I. M. S. . . . 26 Mar. 18
Bayly, 2nd-Lt. P. F. B., I. A. R. O. . . 31 May 16
Beazley, Capt. W. E., 54th Sikhs . . 1 Jan. 17
Benton, Maj. W. B., 105th Infy. . . 29 Oct. 17*
Bennett, Capt. A. D., 19th Punj. . . . 3 June 17
Bennett, Lt. W. R. M., I. A. R. O. . . 3 June 17
Beresford, Capt. G. de la P. . . . 1 Jan. 18
Bernard, Capt. R. P. St. V., D.S.O., 6th G. R. . . . . 25 Feb. 17
Bhim Sing Thapa, Subadar-Major, I.O.M., 3rd G. R. . . . . . . 3 June 18
Bignell, Capt. G. N., 29th Punj. . . 13 Feb. 17
Bingham, Asst. Comm. & Hon. Lt., E. G. 1 Jan. 18
Bishan Sing Rawat, Jemdr., 39th Garh. Rif. 3 June 18
Bir Singh, Subdr., 15th Sikhs . . . 18 Feb. 15
Bland, 2nd-Lt G. H. . . . . . 3 June 17
Blandford, Lt. A. W., I. A. R. O. . . 3 June 16
Blanford, Capt. P. T., 84th Punj. . . 3 June 16
Blandy, Capt. R. . . . . . . 3 June 17
Bowers, Lt., P. L., I. A. R. O. . . . 3 June 17
Boyd, Capt. O. T., 5th Cav. . . . 19 Aug. 16
Braganza, F. B. A., Asst. Surgeon, (*3rd Class*) I.S.M.D.
Bray, Capt. N. N. E., 18th Lrs. . . . 4 Feb. 18
Bremner, Capt C. E. U., 29th Punj. . 13 Feb. 17
Brewer, Capt. J. C. . . . . . 3 June 16
Brook-Hunt, Lt. G. L. . . . . . 1 Jan. 18
Brock, Capt. B. DL., 126th Infy. . . 3 June 15
Brooke, Capt. A. S., 18th Infy. . . . 1 Jan. 18
Bruce, Maj. G. E. . . . . . . 3 June 16
Bruce, Capt. R. M., 5th G. R. . . . 3 June 18
Brunskill, Capt. G. S., 47th Sikhs . . 18 Feb. 15
Buckland, Capt., G. C. B., D.S.O. . . 3 June 17
Budhipal Thapa, Subdr., 3rd G. R. . . 26 July 18
Burdett, Capt. E. W., 48th Prs. . . . 19 Oct. 16
Burne, Capt. F. O. M., 64 Prs. . . . 3 June 18
Burne, Lt. R. H., I. A. R. O. . . . 26 Mar. 18
Burnett, Maj. R. F. D., 42nd Regt. . . 2 Oct. 15
Bur Singh, Ressdr., 16th Cav. . . . 22 May 18

* Bar to M. C.

THE MILITARY CROSS (M. C.)—*contd.*

Butler, Capt. R. B.	23 Nov.	16
Campbell, Lt. A., I.A.R.O.	3 June	18
Cameron, Lt. G. S., I.A.R.O.	3 June	18
Cameron, Lt. D. A.	18 July	17
Campbell, Capt. F. C. G., 40th Pathans	1 Jan.	16
Campbell, Maj. J.	3 June	16
Cardew, Capt. H. S., 39th Garh. Rif.	19 Oct.	16
Carruthers, Maj. J. H. De W., 1-39th Garh. Rif.	1 Jan.	16
Chambers, Capt. G. H. I. A.	1 Feb.	16
Champion de-Crespigny, Capt., H.	3 June	17
Chester, Lt. A. J. B., I. A. R. O.	3 June	16
Christie, Capt., R. C.	3 June	17
Clarke, Capt. A. C. K. S., 15th Lrs.	1 Jan.	16
Clarke, Maj. R. C., S. & T. Corps.	3 June	17
Clifford, Capt. R. C., I.M.S.	1 Jan.	16
Cloete, Maj. H. D.	3 June	16
Coates, Maj. S. B.	3 June	16
Combe, Maj. S. B., 47th Sikhs	1 Jan.	16
Comyn, Capt. G. V., S. & T. Corps	3 June	18
Cooper, Capt. G. C.	3 June	17
Cooper, Lt. H. G.	1 Jan.	18
Cooper, Capt. K. E., 110th Infy.	29 Oct.	15
Corbett, Capt., T. W.	3 June	18
Cormack, Capt. H. S., M.B., F. R. C.S., I.M.S.	1 Jan.	16
Cornish, Maj. A. W. D., D. S. O., 6th G. R.	1 Feb.	16
Corse-Scott, Lt. E. J., 2nd G. R.	3 June	15
Cosby, Capt. N. R. C., I. A. R. O.	3 June	16
Courtney, Hony. Maj. S. A. S., I. S. M. D.	3 June	18
Coxe, Lt. K. H., I A R O,	8 Sep.	18
Crawford, Capt. J. D., D. S. O.	3 June	16
Cruickshank, Capt. V. A., I. M. S.,	18 Aug.	16
Cubbon, Capt. F. R.	18 July	17*
Cullen, 2nd-Lt , A B., I.A.R.O.	3 June	17
Curtis, 2nd-Lt. A. C., U. L.	3 July	15
Dabholkar, Tempy. Capt. A. Y., I. M. S.	26 Mar.	17
Daboo, Lt. E. R., I.M.S.	31 Mar.	17
Damar Sing Gurung, Subadar, 3rd G. R.	3 June	18
Dane, Capt. R.	3 June	17
Dann, Lt. C. F., 28th Punj.	1 Jan.	18
Das, Capt. M., M. B., I. M. S.	1 Jan.	18
Dattajirao Khanvilkar, Jemdr., 110th Infy.	29 Oct.	15
Daunt, Capt. C. O. B., 38th Horse	4 Feb.	18
Davidson, Lt. J. R., I. A. R. O.	1 Jan.	18
Davies, Capt. C. N., I. M. S.	26 Mar.	18
De Butts Capt. F. C., D.S.O., 55th Rif.	3 June	15
Deedes, Capt. R. B., 31st Punj.	1 Jan.	16
De la Motte, Capt. H. T., (retd.)	13 Feb.	17
Dening, Capt. R.	3 June	18
De Rosse, Lt. J. A., I.A.R.O.	3 June	18
Dhanlal Gurung, Jemdr., 5th G. R.	3 June	16
Dhan Sing Negi, Subadar, 39th Garh. Rif.	18 Feb.	15
Dickson, Capt. A. N., M. D., I. M. S.	3 June	17
Dods, Lt. E. R. S., 102nd Grs.	17 Mar.	17
Dowland, Lt G. R., I. A. R. O.	26 Mar.	18
Drysdale, Capt. A. E., 47th Sikhs, D.S.O.	3 Jul.	15
Dunlop, Maj. C. A. M.	3 June	16
Dunne, Sub-Condr. J., S. & T. Corps	19 Oct.	16
Dunning, 2nd-Lt. A. G. A., I. A. R. O.	3 June	17
Durand, Lt. P. F., I. A. R. O.	3 June	16
Dutt, Lt. S., I. M. S.	4 Feb.	15
Dyce, Maj. H. L., 9th Horse	1 Jan.	16
Eales, Lt. C. H. H., Guides	1 Jan.	16
Eastmond, 2nd-Lt , A., I. A. R. O.	3 June	17
Eccles, Capt. M., 119th Infy.	29 Oct.	15
Edelston, Lt. E., I. A. R. O.	3 June	15
Edward-Collins, Capt. G.	3 June	16
Eiloart, Lt. B. H., I. A. R. O.	3 June	17
Elkington, Capt. E. H. O.	3 Jun.	16
Erskine, Capt. K. C. S., 5th G. R.	1 Jan.	16
Evans, Capt., D., McN.	18 June	17
Evans, Bt. Maj. T. J. C., F.R.C.S., I. M. S.	3 Jun.	16
Ewart, 2nd-Lt. J. A., I. A. R. O.	3 June	16
Fane, Lt. A. G. C., I. A. R. O.	1 Jan.	18
Farman Ali, Sub.-Maj.	3 June	16
Farran, Maj. G. L., D. S. O., 4th Cav.	3 June	16
Farquhar, Capt. F. R., 36th Horse	26 May	17
Farquharson, Lt. F. A.	1 Jan.	18
Faunthorpe, Lt.-Col. J. C., M. C. U. P. Horse	1 Jan.	17
Feltham, Lt. W. I. P., 52nd Sikhs	1 Jan.	18
Ferguson, Lt. D. H., I. A. R. O.	3 June	16
Fieldsend, 2nd-Lt., C. E., I. A. R. O.	3 June	16
Finnis, Capt. H., 53rd Sikhs	31 May	16
Fitz Gerald, Capt. D. C. V., I.M.S.	3 June	16
FitzGerald, Capt. G. M., 19th	14 Nov.	16
Fleming, Maj. W. E., 41st Dogras	27 Mar.	15
Ford, Capt. J. M. R.	1 Jan.	18
Fowler, Lt. E. K., 57th Rif.	3 Jun.	15
Fox, Capt. F. N., 39th Garh. Rifle	3 June	18
Fox Male, Maj. W. C. E. T.	3 Jun.	16
Francis, Capt. R. F., 47th Sikhs	1 Jan.	16
Fraser, Capt. W. A. K., 39th Horse	1 Jan.	16
Fraser-Tytler, Capt. W. K., 25th Cav.	3 June	18
Frost, Bt. Maj. F. D.	3 Jun.	16
Fulton, Capt. J. D.	3 June	17
Galbraith, 2nd-Lt., I. W. Ls.	16 Aug.	17
Gall, Capt. L., 25th Cav.	3 Jun.	16
Gambir Sing Pun, Sub.-Maj., 6 G. R.	29 Oct.	15
Garrett, Capt. K. A., 51st Sikhs	20 Apl.	17
Garstin, Capt. H. A. 21st Cav.	26 Sept.	17
Gatherer, Capt. A. H. N. 46th Punjs.	19 Oct.	16
Gauri Shankar Dube, Subadar, *Bahadur*, 1st S. and M.	18 Feb.	15
Geary, Lt. H. V., I. A. R. O.	3 Jun.	16
Gibson, Capt. J. G., 14th Lrs.	1 June	18
Gilroy, Capt. P. K., M.B.	1 Jan.	18
Gladstone, 2nd-Lt. S. D., I. A. R. O.	30 Mar.	17
Godfree, 2nd-Lt., A. S. Horse	16 Aug.	17
Goad, Capt., C. E. G. B.	8 Jun.	17
Goldfrap, Capt. H. W., 103rd Infy.	15 Mar	16
Goldie, Ty. Lt. E. A. M. J., I. M. S.	3 June	18
Goldsmith, Bt Maj. H. A., 95th Infy.	18 Aug.	16
Goodall, Lt. G. M., I. A. R. O.	1 Jan.	18
Gopinath Agarwal, Sub-Asst. Surgn. (*2nd Class*)	3 Jun.	16
Gordon, Capt. J. de la H., 6th Infy.	1 Jan.	16
Gordon, Maj. J. G. B.	3 Jun.	16
Gordon, Lt., C. A.. I. A. R. O.	3 June	17
Gordon, Capt. S., I.MS.	...	
Goss, Bt. Maj. E.S., I.M.S.	3 Jun.	16
Gough, Lt. W. G. H., 2nd G. R.	24 Jun.	16
Goût, Maj. P. J., 94th Infy.	3 Jun.	16
Grace, Capt. H. G., 21st Cav.	20 Apl.	17
Gracey, Lt. D. D., 1st G. R.	17 Mar.	17
Graham, Capt. L. C. T., 9th Horse	4 Feb.	18
Grant, Capt. R. A. P., 112th Infy.	23 Aug.	18
Gray, Maj. E St. C., 34th Horse	1 Jan.	16
Gray, 2nd-Lt. J. W., 28th Punj.	31 Mar.	17
Gregory, Lt. M. S., I. A. R. O.	3 June	18
Griffith, Maj. W. R. C.	3 Jun.	16
Griffith-Griffin, Bt.-Maj. F. M., 129th Baluchis	3 Jun.	15
Grover, Capt. A. C.	1 Jan.	18
Guise, Lt. H. R. C., I.A.R. O.	3 June	18
Gul Akbar, Subadar—Bahadur	3 Jun.	16
Gulam Ali, Subadar	25 Aug.	17
Guthrie, Capt., I. D.	1 Jan.	18
Gwatkin, Capt. F.	3 June.	18
Hackett, Capt. H. M. M., 6th G. R.	3 June	18
Haji, Capt., H. S. G., I. M. S.	1 Jan.	18
Hall, Capt., W. D., 5th Jats.	26 Sept.	17
Halliley, Lt. W.S., I. A. R. O.	19 Oct.	16
Hamer, Maj. M. A., D. S. O., 129th Baluchis.	3 Jun.	15
Hamilton, Capt., A. F.	3 June	17
Hance, Capt. H. M., D. S. O., I. A. R. O.	23 Jul.	15
Harbord, Capt. L. B., 44th Infy.	1 Jan.	17
Harcourt, Capt. E.S., 7th G. R.	1 Jan.	16
Hari Chand, Capt., I. M. S.	3 June	17
Haridhoj Kattri, Subadar, 9th G. R.	18 Feb.	15
Harman, Lt. O. A., I. A. R. O.	3 June	18
Harnath Singh, Bahadur, I.O.M.	11 Apl.	18
Harris. Lt. (*Tempy. Capt.*) A. E. C.	3 June	17
Harris, Lt. F. O., I. A. R. O.	27 Mar.	17
Harris, Rev. T.,	14 Jan.	14
Harrison, Lt. A. M. L., 4th G. R.	24 Feb.	17
Harvey, Capt. C. O.	3 June	18
Hasan Shah, Jemdr.	3 June	18
Hawinda, Jemdr., 58th Rifl.	1 Jan.	16
Hawkes, Lt. G. E. Sikhs	26 Mar.	18

*Bar to M. C.

THE MILITARY CROSS (M. C.)—*contd.*

Haycraft, Capt. W. C. S.	3	Jun.	16
Hayes, Lt. L., I. A. R. O.	1	Jan.	18
Heath, Capt. L. M.	3	Jun.	16
Henderson, Maj. K. D. M., 17th Infy.	3	June	17
Hendry, Lt. F. C., I. A. R. O.	1	Jan.	16
Herbert, Lt. Y. A., I. A. R. O.	3	June	18
Heron-Jones, Lt. C. V., I. A. R. O.	1	Jan.	16
Hertz-Smith, Capt., G. V.	3	June	17
Hewer, J. A., I. A. R. O.	3	June	18
Hickman, Capt., H. T. D.	18	June	17
Higginson, 2nd-Lt., N., I. A. R. O.	19	Nov.	17
Hill, Lt. A. L., I. A. R. O.	1	Jan.	16
Hinds, Capt. W. A. G., 103rd Infy.	15	Mar.	16
Hingston, Capt. R. W. G., I.M.S.	25	Nov.	16
Hislop, Capt. J.H.	3	Jun.	16
Hogg, Lt. W. P., I. M. S.	1	Jan.	18
Howson, Maj. G., 4th Cav.	3	Jun.	15
Hornor, Capt. F. H. F., 107th Prs.	1	Jan.	16
Hughes, Maj. C. F. W., 15th Sikhs	3	Jun.	16
Hughes, Capt. G V., S. & T. C.	1	Jan.	18
Hunt, Capt. C. E., 34th Prs.	3	Jun.	15
Ibbotson, 2nd-Lt. A. W., I. A. R. O.	1	Jan.	17
Iman Din Khan, Jemadar (Actg. Subr.) R. G. A.	26	Apl.	17
Inchha Ram, Subadar, 6th Jat L. I.	1	Jan.	15
Indar Singh, Subadar, 58th Rif.	18	Feb.	15
Inglis, Lt. H. G., I. A. R. O.	31	May	16
Inskip, Capt. R. D., 59th Rif.	1	Jan.	16
Irwin, Rev. R. J. B., D.S.O.	14	Jan.	16*
Ives, Capt. G., I. A. R. O.	3	Jun.	1
Jackson, Capt. M. H., 29th Lrs.	1	Jan.	1
Jacob, Capt. R. M., 130th Baluchis	1	Jan.	1
Jai Singh, Ressaider, 6th Cav.	3	Mar.	17
James, Capt. F. H, 104th Rif.	*31	May	16
James, Capt. W. R., 89th Punj.	24	Jun.	16
Jarvis, Capt. C., 26th Horse	1	Jan.	17
Jerram, Lt., M. R. K., I. A R. O.	3	June	17
Johnston, Maj. W., 7th G. R.	1	Jan.	16
Johnson, Capt. G. S., 61st Prs.	3	June	18
Jones, Capt., B. I.	3	June	17
Joyce 2nd-Lt. A. H. B., Punj.	26	Mar.	18
Keatinge, Lt. F. W. G., I. A. R. O.	3	Jun.	16
Kellie, Capt. G. J. D., 30th Punj.	3	June	18
Kemp, Capt. T. N. C., 9th G. R.	24	Jan.	16
Kennedy, Capt. E. E.	1	Jan.	18
Kennedy, Lt. K. H. H., I. A. R. O.	3	June	17
Kennedy, Capt. R. S., D. S. O., I.M.S.	1	Jan.	17
Kensington, Maj. E. C.	3	Jun.	16
Kerr, Maj. A. M. 30th Dogras	15	Feb.	17
Kharak Sing Rana, Subadar, 2nd G. R.	18	Feb.	15
Kiddle, Lt. J. A. C., I. A. R. O.	24	Jun.	16
Laing, Capt. S. V. B., 76th Punj.	1	Jan.	16
Lakhi Ram, Subadar, 6th Jat L. I.	18	Feb.	15
Lamond, Asst. Surgn. J. D. M.	3	Jun.	16
Lane, Capt. C. R. C.	3	June	18
Lane, Maj. J. T. H., 39th Garh. Rif.	18	Feb.	15
Lang Lt. J. H., Rajputs	26	Mar	18
Lang, Capt. L. E., Poll. Employ.	3	June	17
Lapsley, Capt. J. B., M.B., F.R.C.S., I. M. S.	3	June	18
Layng, Capt. T. M. I. A.	1	Jan.	16
LeCornu, Capt. C. J. S.	3	Jun.	16
LePatourel, Lt. E. D.S.O., C., I. A. R. O.	19	Oct.	16
Leigh-Clare, Lt, H. J. L., I. A. R. O.	25	Aug.	17
Leith-Ross, Capt. W., 55th Rif.	3	June	18
Lewis, Capt. H. V., D.S.O., 129th Baluchis	3	Jun.	15*
Lindop, 2nd-Lt. A. H. E., I. A. R. O.	3	June	17
Lloyd, Capt. C. G. S., T. Corps	3	June	18
Lloyd-Evans, Lt. H. J., I. A. R. O.	3	June	15
Lockhart, Capt. R. McG. M., 51st Sikhs	8	Mar.	18
Logan, 2nd-Lt., A. F., U. L.	4	Nov.	15
Lucas, Lt. A. G., Baluchis	26	Mar.	18
Lumb, Capt. F. G. E., 39th Garh. Rif.	18	Feb.	15
Lunn, Capt. A. J., 82nd Punj.	17	Mar.	17
Lyster, Maj. C. J. H.	3	Jun.	16
Macartney, Capt. F. A., 53rd Sikh.	27	Mar.	17
MacBrayne, Maj. R. J., 15th Sikhs	1	Jan.	16
Macdonald, Capt., D. T.	3	June	17
Macdonald, 2nd-Lt., L. J., I. A. R. O.	3	June	17
Macdonald, Capt. J. V., M. B., I.M.S.	13	Feb.	17
MacGregor, Capt. R. F. D., I.M.S.	3	Jun.	16
Mackay, Capt. A. S.	3	June	18
MacKenzie, Lt. J. MacK., I. A. R. O.	3	June	17
Macintosh, Lt. K. J.	18	Jan.	18
MacMahon, Bt. Lt.-Col. H. F. E.	3	Jun.	16
Mair, Lt. R., I. A. R. O.	3	Jun.	16
Malla Singh, Subadar, 3rd S. and M.	18	Feb.	15
Maltby, Capt. C. M., 95th Infy.	3	Jun.	16
Mangin, Capt. E. B., 107th Prs.	3	Jun.	15
Marsh, Capt. H. F. F., 2nd G. R.	31	May	16
Marshall, Capt. H. H., 48th Prs.	3	June	18
Marshall, Lt. J., I. A. R. O.	1	Jan.	18
Marshall Lt. J. G., I. A. R. O.	1	Jan.	18
Martin, Capt. G. D., 9th Infy.	18	Feb.	15
Martin, Maj. J.	3	Jun.	16
Martyn, 2nd-Lt., J. K., I. A. R. O.	3	June	17
Maynard, Maj. F. H.	3	Jun.	16
McCleverty Capt., G M. 2nd G. R.	3	June	17
McConnell, 2nd-Lt. H. L. C., I. A. R. O.	1	Jan.	18
McCudden, Maj. I. H., D. S. O., 21st Cav.	16	May	16
McLean, Lt. A. M., I. A. R. O.	3	June	18
McPherson, Capt. A. B., 6th Infy	1	Jan.	16
McSwiney, Lt. H. F. C., 3rd G. R.	18	Feb.	15
Mein, Capt. D. B., 55th Rif.	3	Jul.	15
Mellis, Capt. A. R., I. A. R. O.			
Menzies, Lt. F. A., I. A. R. O.	24	Jun.	16
Meredith, Capt. A. E., 114th Mah.	3	June	18
Messinier, E. B., Asst. Surgeon, I.S.M.D.	3	Jul.	15
Metcalfe Capt. E. D., 3rd Horse	3	June	18
Metcalfe, Capt. E. D. T., S. and T. Corps	3	June	17
Middleton-West, Capt., S. H, M.B., I M. S.	3	Jun.	17
Moody, Capt. H. H., 13th Lrs	1	Jan.	16
Moore, Lt A. N. F., I. A R. O.	1	Jan.	18
Moore, Lt. C. F. F., 123rd Rif.	3	Jun.	15
Moore, Capt. E. C. F., I. A. R. D.	3	June	18
Moriarty, 2nd-Lt. J., I. A. R. O.	8	Mar.	17
Morice, Maj. R. C. L., 37th Dogras	15	Apl.	16
Morris 2nd-Lt. H. S., I. A. R. O.	3	Jun.	17
Morse, Capt. L. H.	3	Jun.	16
Mosse, Capt. C. O. H., 120th Infy.	1	Jan.	16
Moss-Blundell, Capt. R. S, 27th Lt. Cav.	3	Jun.	17
Mukand Singh, Risaldar *Bahadur*, 2nd Lrs.	4	Feb.	18
Mulloy, Capt. N. F. C., Lrs.	16	Aug.	17
Munn, Capt. F. L. R., 46th Punj.	1	Jan.	16
Murray, Capt. R. G. H., 9th G. R.	10	Mar.	15
Murray, Capt. E. M., of Guides	15	Apl.	16
Nain Sing Chinwarh, Subadar, *Bahadur*, 39th Garh. Rif.	1	Jan.	15
Nethersole, Capt. J., 25th Cav.	3	Jun.	15*
Nicholson, Capt. F. H., D.O.S.	3	Jun	16
North, Capt. K. A., att. 34th Prs.	1	Jan.	18
Northey, Capt. W., 1st G. R.	1	Jan.	18
Noyes, Capt. C. D.	3	Jun.	16
O'Brien, Lt. B. P. T., 8th G. R.	16	Aug.	17
Oddie, Lt., W. P., I. A. R. O.	16	Aug.	17
Ogley, Lt. E. F.	18	Jun.	17
O'Neill, Capt. H. J. D., 127th Infy.	3	Jun.	16
O'Neill, Capt. J. S., M.B., I.M.S.	3	Jun.	15
O'Neill, Capt. W. H. L., Sikhs	8	Mar.	18
Ozanne, Capt. G. D., 99th Infy.	1	Jan.	17
Page, Lt. W. W. K.	4	Feb.	18
Pain, Lt. H. W., I. A. R. O.	1	Jan.	18
Palin, Lt. C. W., 129th Baluchis.	17	Apl.	17
Pancham Sing Mahar, Jemadar, 39th Garh. Rif	10	Mar.	15
Parbat Chand, Subadar-Major, 59th Rif.	15	Apl.	15
Parker, tempy. Lt. A. H., 3rd Punj. Rif.	1	Jan.	16
Paterson, Maj. F. W. J., 66th Punj.	1	Jan.	16
Paterson, Capt. G. F. J., 34th Prs.	1	Jan.	16
Paterson, 2nd-Lt, L. M.	3	June	17
Paton, Capt. W. C., M.B., I.M.S.	1	Jan.	16
Patrick, Capt. R. M. F., 42nd Regt.	1	Jan.	16
Peck, Lt. L. J.	18	Jan.	18
Pender, Lt. G. T.	1	Jan.	18
Pelly, Ty. Capt. E. G.	1	Jan.	18
Phayre, Bt. Maj. R. B.	3	Jun.	16
Pim, Lt. H. M., 24th Punj.	1	Jan.	16
Pitman, Capt. C. R. S., D.S.O.	3	Jun.	16
Pogson, Capt. C. A., 117th Mah.	29	Oct.	15
Pola Khan, Jemdr., 101 Grs.	18	June	18
Porch, Bt.-Col. E. A., S. & T. Corps	3	June	16*
Ponting, Capt T. J.	3	June	17
Pott, Capt. D., 13th Lrs.	1	Jan.	16
Pottinger, Capt. J. A., 30th Punj.	1	Jan.	16
Powell, Lt. A. E., I. A. R. O.	3	June	18

*Bar to M. C.

THE MILITARY CROSS (M. C.)—*contd.*

Powell, Bt. Lt.-Col., D. H., 105th Infy. . 1 Jan. 17
Powell, Lt. W. N., 16th Cav. . . . . . 3 June 18
Preston, Lt. E. B. C. . . . . 1 Jan. 18
Rackstraw, Lt. G. I. A. R. O. . . . . . 3 June 18
Radcliffe-Smith, Maj. N. M. R. . . . 3 June 17
Radley, Capt. H. P . . . . 3 June 17
Rai, Capt. D. H., I.M.S. . . . . . 3 Jun. 16
Ralston, Capt. W. H., 47th Sikhs . . . 1 Jan. 16
Ramkrishna Ganpat Shinde, 2nd Class Sub.-Asst. Surgn. No. 298 . . . 3 Jun. 15
Rana Jodha Jang Bhadhur, 2nd-Lt., I. L. Forces . . . . 4 Nov. 15
Raymond, Maj. E. D., 30th Lrs. . . . 1 Jan. 16
Raymond, Capt. M. C. . . . . 3 June 18
Raynor, Capt. C. A., 48th Prs. . . . 19 Oct. 16
Reed, Lt. G., I. A. R. O. . . . . . 1 Jan. 18
Reed, Maj. H. R. B., 39th Garh. Rif. . 1 Jan. 16
Reed, Lt. L. N., I. A. R. O. . . . . 1 Jan. 18
Reinhold, Capt. C. H., F.R.C.S., I. M.S. . 3 June 17
Reinhold, Maj. H. E. . . . . . . 3 Jun. 16
Renny, Maj. H. B. . . . . 3 Jun. 16
Rice, Lt. R. A. . . . . 1 Jan. 18
Richardson, Lt. R., 55th Infy. . . . . 1 Jan. 18
Riddell, Capt. W. H., M.B., I.M.S. . . 24 Jun. 16
Ridge, Lt. K. J. T., 99th Infy. . . . . 3 June 17
Roberts, Lt. F. C., I. A. R. O. . . . . 1 Jan. 18
Robert, 2nd-Lt. V. G., I. A. R. O., 129th Baluchis . . . . 3 June 17
Robertson, Lt. A. McL, I. A. R. O. . . 3 June 18
Robinson, Lt. D. C., 36th Sikhs . . . 17 Mar. 17
Rodrigues, Asst. Surgn. (*2nd Class*) J. M. 3 June 16
Rooth, Lt. A. V., I. A. R. O. . . . . 26 Mar. 18
Ross, Capt. E. J., 8th G. R. . . . . 3 June 18
Rowntree, Capt. H. K., M.B., I.M.S. . . 3 June 16
Rundle, Lt., C. A. G. . . . . . . 3 June 17
Salusbury-Trelawny, Lt. J. M., I. A. R. O. . . . . 3 June 17
Sangram Sing "Negi," Subadar, 39th Garh. Rif. . . . . 3 Jun. 16
Sant Singh, Subadar, 34th Prs. . . . . 1 Jan. 15
Sarabjit Gurung Bahadur, Subadar, 2nd G. R. . . . . 17 Mar. 17
Sargent, Maj. A. E. E., D.S.O., S. & T. Corps . . . . 1 Jan. 16
Savory, Capt. R. A. 14th Sikhs . . . 3 Jun. 16
Saxton, Capt. P. D. . . . . 3 June 17
Scoones, Capt. G. A. P., D.S.O., 2nd G. R. . . . . 1 Jan. 16
Scott, Capt. J. B. . . . . 1 Jan. 18
Scott, Maj. H. L., 1st G. R. . . . . 1 Jan. 16
Scott, Capt. K. R., 31st Punj. . . . . 3 June 18
Scratton, Capt. G. W. B., 19th Lrs. . . 3 June 17
Searle, Lt. C. S., 97th Infy. . . . . 23 Jul. 15
Sen, Capt. J. L., I.M.S. . . . . . 3 June 16
Shakespear, Capt. G. F. C., 88 Infy. . 3 Jun. 16
Shaw, Lt. D. S., I. A. R. O. . . . . . 8 Mar. 18
Shearer, Lt. E. J., 31st Lrs. . . . . 2 Apl. 17
Shearer, Capt., J. E. . . . . . . 3 June 17
Sher Afzal, Subadar . . . . . . 3 June 17
Sherston, Capt. J. R. V., 11th Lrs. . . 18 Feb. 15
Shirley, Capt. S. R., 54th Sikhs . . . 1 Jan. 16
Sitaram Sellar, Jemadar, 117th Mah. . . 29 Oct. 15
Smith, Capt. A. D., 2nd G. R. . . . 3 Jun. 15
Smith, Lt. V. W., I. A. R. O. . . . 4 Feb. 18
Sneyd, 2nd-Lt., R. I. S., I. A. R. O. . . 3 June 17
Sodhi, Capt, N. S., I. M. S. . . . . 3 June 17
Sohan Singh, Jemadar, 24th Punj. . . . 29 Oct. 15
Sondhi, Ty. Lt. S. D., I.M.S. . . . . . 3 June 16
Spooner, Revd. H. . . . . 19 Oct. 16
Spurgin, Capt., E. W., Lrs. . . . . . 16 Aug. 17
Staples Capt. E. C., 11th Rajputs . . . 15 Mar. 16
Stead, Lt. E A., 35th Sikhs . . . . 1 Jan. 18
Stedman, Lt. H., I. A. R. O. . . . . 1 Jan. 17
Steel, Capt. J. G., I. A. R. O. . . . . 17 Apl. 17
Steuart, Lt. F. G., I. A. R. O. . . . 3 June 18
Stevens, Capt., C. F. L. . . . . . . 25 Aug. 17
Stevenson, Capt. H. D. M. . . . . 3 June 17
Stewart, Maj. A. E., 124th Infy. . . . . 1 Jan. 17
Stewart, 2nd-Lt., J. A., I. A. R. O. . . . 3 June 17
Stocker, Lt. C. J., M.B., I. M. S. . . . 3 Jun. 16
Stockley, Maj. C. H., 66th Punj. . . . 19 Oct. 16
Stone, Capt. A. G., 3rd G. R. . . . . 1 Jan. 17
Storrs, Lt. G. N., I. A. R O. . . . . 26 Mar. 18
Story, Capt. J. A., 61st Prs. . . . . . 1 Jan. 17
Sweeny, Maj. R. L. C., D S.O., S. and T. Corps. . . . . 3 Jun. 16
Tate, 2nd-Lt. J., I.A.R.O. . . . . . 26 Mar. 17
Tedman, Ty. Lt. C., I.A.R.O. . . . 24 Jun. 16
Thackwell, Capt. A. C. A., 81st Prs. . . 1 Jan. 17
Thakur Singh, Subadar, 47th Sikhs. . . 1 Jan. 15
Thomson, Capt. A. R., 7th Raj. . . . . 29 Oct. 15
Thorburn, Lt. R. . . . . 1 Jan. 18
Thorn, 2nd-Lt. T. C., I.A.R.O. . . . 5 Mar. 17
Thornhill, Capt. C. M. 24th Punj. . . . 1 Jan. 16
Thornton, 2nd-Lt. R. T., I.A.R.O. . . . 13 Feb. 17
Tinley, Lt. F. B. N., 20th Horse . . . 3 Jun. 15
Tomkins, Lt. G. L., 128th Prs. . . . . 3 June 18
Townsend, Capt. R. S., M.B., I.M.S. . . 3 Jun. 16
Trail, Capt. W. S , 57th Rif. . . . . 3 Jun. 16
Trelawny, Capt. L. C. . . . . . . 3 June 18
Tuite-Dalton, Capt. E. G. T., 3rd G. R. . 26 Mar. 17
Twiss, Maj. W. L. O., 9th G R. . . . 1 Jan. 16
Venables, Capt. A. V., I.A.R.O. . . . 3 Jun. 16
Vernon, Maj. H. B. . . . . . . 3 Jun. 16
Vigors, Capt. M. D. . . . . 25 June 17
Walker, Capt. B. W. G., 23rd Cav. . . 3 June 17
Walker, Capt. G. B., 31st Punj. . . . 18 Aug. 16
Wallis, Capt. B. H., 107th Prs. . . . . 1 Jan. 16
Watson, Bt. Maj. T. N. . . . . . . 3 June 17
Watkis, Capt. H. L., 37th Lrs. . . . . 3 Jun. 15
Wellesley, Capt. H. A. . . . . . 3 June 17
Wells, Maj. F. W. A. . . . . . . 3 Jun. 16
Wells, Capt. T. R., 33rd Punj. . . . . 1 Jan. 16
Wemyss. Lt. A. F., I.A.R.O. . . . . 3 Jun. 16
Whigham-Teasdale, 2nd-Lt. D. G., I.A.R.O. . . . . 26 May 17
Whitakar, Capt. S. S., 7th G. R. . . . 18 Oct. 17
White, Lt. C. R., I. A. R. O. . . . . 3 Jnne 18
Whitworth, Capt. D. E., 2nd Lrs. . . . 4 Feb. 18
Wigley, Capt. P. J. R., 90th Punj. . . . 1 Jan. 16
Willoughby, Capt. M. G. P. . . . . . 3 Jun. 16
Wilson, Maj. D. D., 17th Cav. . . . . 3 Jun. 16
Wilson, R. Bt. Lt.-Col. C., D. S. O. . . 3 June 16
Woakes, 2nd-Lt. R. B., I. A. R. O. . . 3 June 17
Wolseley-Bourne, Temp. Capt. J. F. . . 2 Sept. 18
Wood, Capt. C. A., M.B., I.M.S. . . . 1 Jan. 16
Woodseil, Sub-Condr., T. W., I. S. M. D. 16 Aug. 17
Woodward, Lt. R. G , 52nd Sikhs . . . 26 Mar. 18
Young, Lt. (*tempy. Capt.*) K. deL., 8th Cav. 1 Jan. 17
Zaman Khan, Subadar *Bahadur*, 129th Baluchis . . . . 1 Jan. 15
Zorawar Singh, Capt., Indian Land Forces 1 Jan. 16

THE ROYAL RED CROSS.

1st class.

Barker, Miss M. E., Retd. Q. A. M. N.S. I. 21 Apl. 99
Clay, Mrs. I. M., Retd. Q. A. M. N. S. I. 28 Nov. 05
Cusins, Miss C. L., Q.A.M.N.S.I. . . . 1 Jan 17
Exshaw, Miss P., Q. A. M. N S. I. . . 25 Feb. 18
Gilmore, Miss A.M., Q.A.M.N.S.I. . . . 1 Jan. 17
Gilmore, Miss. M. G., Q. A. M. N. S. I. 3 June 18
Gray, Miss M. E., Q. A. M. N. S. I. . . 10 Nov. 08
Haines, Miss G., Q. A. M. N. S. I. . . 25 Feb. 18
James, Miss J. M., Retd. Q. A. M. N. S.I. . . . . 11 June 07
Kelly, Miss E. S., Retd. Q. A. M. N. S. I. . . . . 19 Feb. 14
Warrack, Miss C. L., Q. A. M. N. S. I. . 3 Jun. 16
Waterhouse, Miss A. M., Q. A. M. N. S. I. . . . . 10 Dec. 01
Watt, Miss P. F., Q.A. M. N. S. I. . . 3 Jun. 15
Weighell, Miss A. J., Q. A. M. N. S. I.
Wildman, Miss E. A., Retd., Q. A. M. N. S. I. . . . . 9 Jul. 08

2nd class.

Green, Miss E., Q. A. M. N. S. I. . . 25 Feb. 18
Knapp, Miss M. D., Q. A. M. N. S. I. . 1 Jan. 16
Lowe, Miss. A. R. I., Q. A. M. N S. I. 3 Jun. 16
Marshall, Miss, E. Q., Q. A. M. N. S. I. 3 June 1
MaLean, Miss M. E., Q. A. M. N. S. I. . 8 June 13
Stuart, Miss M. G., Q. A. M. N. S. I. (*temp.*) . . . . 25 Feb. 18
West, Miss D., Q. A. M. N. S. I. . . . 25 Feb. 18

KAISAR-I-HIND MEDAL.

1st class.

Barnes, Maj. E. . . . 2 Jan. 05
Bell, Lt.-Col. C.T., R.A. . . . .
Bikaner, Hony. Col. *H. H. Sir* Ganga Singh, *Bahadur*, G.C.S.I., G.C.I.E., A.D.C., *Maharaja* of—2nd Lrs. 23 May 00
Bingley, Maj.-Genl. A. H., C.B., C.I.E. . 23 May 00
Bott, Capt. R. H., M. B., F.R.C.S., I.M.S. 4 June 17
Campbell, Col. R. N., C.B., C.I.E., M.B. . 23 May 00
Dawson, Bt.-Col. C. H. . . . 1 Jan. 03
Dyson, Col. T. E., M.B. . . . 9 Nov. 01
Gwalior, Hony. Maj.-Genl. *H. H. Sir* Madho Rao Scindia, *Bahadur*, G.C.S.I., G.C.V.O., A.D.C. *Maharaja* of —(Hony. Col., 1st Lrs.). . . . 23 May 00
Gwyther, Lt.-Col. A., M.B., F.R.C.S., I.M.S. 12 Dec. 11
Haig, Col. P. B., C.B., M.B., I.M.S. . . 3 Jun. 13
Hamilton, Maj. R. E. A. . . . 2 Jan. 05
Harvest, Lt.-Col. H. de V., Supy. List . 25 Jun. 09
Hogan, Sr. Asst. Surgn. Hy. Capt. W. J. A., I.S.M.D. . . . 22 Jun. 14
Husband, Maj. J. H., M.B., F.R.C.S.E.*, I.M.S. . . . 1 Jan. 15
Hutchinson, Maj. W. G., Supy. List . . 24 Jun. 10
Irvine, Lt.-Col. T.W., M.B., F.R.C.S.E., I.M.S. . . . 2 Jan. 11
Jacob, Col. *Sir* S. S., K.C.I.E., C.V.O. . 9 Nov. 01
James, Lt.-Col., C. H., C.I.E., I.M.S. . . 23 May 00
Lyons, *the Hon'ble* Surgn.-Genl. R.W.S., M.D., K.H.P. . . . 1 Jan. 03
Carrison, Maj. R., M.D. . . . 24 Jun. 10

McCloghry, Col. J., I.M.S. . . . 9 Nov. 01
MacWatt, Lt.-Col. R.C., M.B. . . . 1 Jan. 08
Morrison, Senr. Asst. Surgn., Hony. Capt. J. 1 Jan. 08
O'Meara, Maj. E. J., F.R.C.S. . . . 1 Jan. 09
Pedley, Surgn.-Lieut.-Col., T. F., M.D., V.D., Rangoon Port Def. Vols. . . .
Pitcher, Col. D. G. J. . . . 23 May 00
Poynder, Lt.-Col. J. L., I.M.S . . 26 Jun. 02
Prasad, Lt.-Col. K., I.M.S. . . . 4 June 17
Pritchard, Lt.-Col. C. H. . . . 1 Jan. 15
Rost, Maj. E. R., I.M.S. . . . 1 Jan. 13
Shore, Lt.-Col. R., M.D. . . . 1 Jan. 00
Shoubridge, Maj. C.A. G., 112th Infy. . 1 Jan. 09
Smith, Lt.-Col. H., M. D., I.M.S. . . 2 Jan. 11
Southon, Maj. C. E., M.B., I. M. S. . 22 Jun. 14
Stokes, Surgn. Maj. W., M.B., 29th Sth. Indn. Ry. Btn., I.D.F. . . . 1 Jan. 14
Tucker, Maj. W. H., I.M.S. . . . 12 Dec. 11
Tyrrell, Maj. J. R. J., M.B., I.M.S. . 12 Dec. 11
Vaughan, Lt.-Col. J.C.S., M.B. . . . 1 Jan. 10
Wake, Lt.-Col. E. St. A. . . . 1 Jan. 04
Walter, Maj. A. E., I.M.S. . . . 12 Dec. 11
Ward, Maj. E. L., I.M.S. . . . 1 Jan. 13
Waterhouse, Miss A.M., R.R.C., Q.A.M. N.S.I. . . . 1 Jan. 17
Westcott, The Revd. F. . . . 1 Jan. 18
Wilkins, Col. J. S., D.S.O., I.M.S. . . 1 Jan. 01
Wilkinson, Lt.-Col. E., I. M. S. . . 9 Nov. 01
Younghusband, Lt.-Col. *Sir* F. E., K.C.I.E. . . . 1 Jan. 01

2nd class.

Abdur Razzak Khan, Sr. Sub-Asst. Surgn. Subdr., I.S.M.D. . . . 12 Dec. 11
Anscomb, Maj. A. M., I.S.O., V.D., 27th Bal. Coy., I. D. F. . . . 2 Jan. 05
Baker, Dy. Comsy. Hy. Maj. T., Ord. Dept. . . . 23 May 00
Banks, Surgn.-Maj. C., Cal. Rif., I. D. F. 2 Jan. 11
Bayley, Lt.-Col. E. C., C.I.E., 15th Lrs. . . . 24 Jun. 12
Beatson-Bell, *Hon'ble* Maj. N. D., C.S.I., C.I.E., 44th Cal. Sco., I.D.F. . . . 23 May 00
Beville, Lt.-Col. F. G., C.I.E. . . . 23 May 00
Bolster, Miss A., Retd. Q. A. M. N. S. I. . . . 14 Jun. 13
Cattell, Maj. G. L. . . . 1 Jan. 12
Clerke, Sr. Asst. Surgn. Hony. Maj. L. A. H., I.S.M.D. . . . 3 Jun. 13
Clutterbuck, Lt.-Col. P. H., V.D., 7th (S) U. P. Horse . . . 12 Dec. 11
Crow, Asst. Surgn. C. G., I.S.M.D. . . 3 Jun. 16
Deane, Asst. Surgn. G. A., I.S.M.D. . 22 June 1
Dewes, Lt.-Col. F. J., I.M.S. . . . 1 Jan. 09
Ffrench, Lt.-Col. T. . . . 26 Jun. 02
Fleming, *Hon.* Lt. J. F., I.S.M.D. . . 3 Jun. 16
Fraser, Capt. R. T., 2nd Surma Valley L. H. . . . .
Grant, Lt.-Col. J. W., M.B. . . . 23 May 00
Hanwant Singh, Risldr. Maj. (Hony. Capt.) *Sirdar Bahadur, late* 31st. Horse . . . 22 June 14
Haworth, Maj. L. B. H. . . . 23 May 00
Hayes, Miss M. L., Q. A. M. N. S. I. . 12 Dec. 11
Holbrooke, Maj. B.F.R., 124th Bal. Infy. . . . 1 Jan. 01
Holden, Maj. H. N., 5th Cav. . . 23 May 00
Hunter, Depy. Comy. Hony. Capt. J. . . 26 Jun. 02
Ihsan Ali, Sub-Asst. Surgn., I.S.M.D. . 12 Dec. 11
Kelly, Asst. Surgn., C. C., I.S.M.D. . . 3 Jun. 16
Kelly, Miss E. S., R.R.C., (Retd.) Q. A.M. N. S. I. . . . 12 Dec. 11
Knollys, Maj. R. W. E. . . . 1 Jan. 01
Knox, Maj. R. W., I.M.S. . . . 1 Jan. 09
Kreyer, Lt.-Col. F. A. C. . . . 2 Jan. 11
Lawrence, Capt. H. R. . . . 1 Jan. 15

McDonald, Hon. Lt. J. J., I.S.M.D. . . 1 Jan. 14
Macdonald, Capt. C. R., Bihar L. Horse . 22 Jun. 14
Macleod, Lt.-Col. J. N. C.I.E., M.B., F.R.C.S. . . . 9 Nov. 01
Maddox, Lt.-Col. R. H., I.M.S. . . . 26 Jun. 02
Makkhan Singh, Sub-Asst. Surgn., I. S. M. D. . . . 1 Jan. 13
Mohammed Khan, Sr. Sub-Asst. Surgn., I.S.M.D. . . . 1 Jan. 14
Moore, Nursg. Sister D. L. T., Q. A. M. N. S. I. . . . 1 Jan. 08
Morris, Maj. R.L. . . . 23 May 00
Murphy, Hony. Lt., E. J., I.S.M.D.
Niranjan Das, Sr. Sub-Asst. Surgn., Jemdr. I.S.M.D. . . . 12 Dec. 11
O'Brien, Maj. E., Supy. List . . . 9 Nov. 01
Orman, Sr. Asst. Surgn. Hy. Capt. C. H., I.S.M.D. . . . 12 Dec. 11
Owen, Sr. Asst. Surgn., Hony. Capt. R. J. . . . 29 Jun. 06
Palin, Maj. R. H. . . . 1 Jan. 15
Peters, Lt.-Col. C. T., I.M.S. . . . 3 June 13
Pinney, Maj. J. C. D. . . . 23 May 00
Plowden, Maj. T. C., Supy. List. . . 9 Nov. 01
Powell, Maj. J., V.D., Punj. L. Horse . . 1 Jan. 14
Rait, Miss H. A. M., Q. A. M. N.S.I. . . 12 Dec. 11
Robarts, Capt. C. S. H., 75th Infy. . 1 Jan. 09
Robinson, Lt.-Col. W.H.B. . . . 1 Jan. 01
Rocke, Capt. C. E. A. S., S. & T. Corps . 1 Jan. 16
Sadlier, Hon. Capt. A. W. B., I. S.M.D.
Salkield, Capt. T., 3rd Punj. Rif., I. D. F. . . . 14 June 12
Scotland, Lt-Col. D. W., M.B. . . . 23 May 00
Shaikh Ali Shabash, Sr. Sub-Surgn., I.S.M.D. . . . 12 Dec. 11
Sher Singh, Subdr., *late* 1st S. & M. . . . 3 Jun. 13
Stevenson, Surgn.-Genl. H. W., I.M.S. . 23 May 00
Stewart, Maj. H., C.I.E. . . . 23 May 00
Weighell, Miss A. J. . . . 3 Jun. 15
Wildman, Miss E. A., R.R.C., Q. A. M. N. S. I. Retd. . . . 12 Dec. 11
Wiseman, Depy. Comsy. Hony. Capt. C. S. . . . 3 Jun. 13

* Bar to M. C.

GOOD SERVICE PENSIONS.

OFFICERS WHO ARE RECEIVING GOOD SERVICE PENSIONS AS REWARD FOR DISTINGUISHED OR MERITORIOUS SERVICE.

1 V.C. M. Hills-Johnes, Lt.-Genl. *Sir* J., G.C.B., Art. Retd. . . . . . . 15 Dec. 81

Dickinson, Lt.-Col. W., C.S.I., Retd. . 9 June 87

Tillard, Maj.-Genl. J. A., C.B., Art. Retd. . . . . 19 June 93

Atkinson, Maj.-Genl. J. R. B., Retd. . 9 Jan. 95

5 Arnott, Lt.-Col. J., M.D., I.M.S., Retd. . 2 Mar. 96

Willoughby, Maj.-Genl. J. F., Retd. . 27 Oct. 96

Biscoe, Lt.-Genl. W. W., C.B., Retd. . 2 July 97

Maitland, Major-Genl. P. J., Retd. . . 2 July 97

Eardley-Wilmot, Maj.-Genl. R., C.B., Retd. . . . . 4 Sep. 97

10 Cook, Maj.-Genl. J., C. B., Retd. . . 4 May 98

Stewart, Maj.-Genl. *Sir* N. R., *Bart.*, C.B., Retd. . . . . 16 Apl. 99

Melliss, Col. *Sir* H., K C.S.I., *m. s. l.* . 12 June 96

McLeod, Genl. *Sir* D. J. S., K.C.B., K.C.I.E., D.S.O., Retd. . . . . 25 July 99

Patch, Col. R., C.B., Retd. . . . . . 5 July 02

15 Egerton, Fd. Marshal *Sir* C. C., G.C.B., D.S.O. . . . . 11 July 03

V.C. Creagh, Genl. *Sir* O'M., G.C.B., G.C.S.I., Retd. . . . . 28 Nov. 03

Barrow, Genl. *Sir* E. G., G.C.B., G.C.S.I. 11 Jan. 04

Christopher, Maj.-Genl. L. W., C.B., Retd. 8 Aug. 04

Elliot, Lt.-Genl. *Sir* E. L., K.C.B., D.S.O., Retd. . . . . 27 July 05

20 Richardson, Lieut.-Genl. *Sir* G. L. R., K.C.B., C.S.I., C.I.E., Retd. . . . . 16 Mar. 06

Hutchinson, Lt.-Genl. H. D., C.S.I., Retd. 24 Mar. 07

Browne, Lt.-Genl. *Sir* A. G. F., K.C.B., D.S.O., Retd. . . . . 13 Feb. 08

Pearson, Genl. *Sir* A. A., K.C.B. . . . 19 June 08

Scott, Maj. Genl. *Sir* C. H., K.C.B., R.A., Retd. . . . . 7 July 08

25 Martin, Lt.-Genl. *Sir* A. R., K.C.B., Retd. 4 Nov. 08

DeBrath, Lt.-Genl. *Sir* E., K.C.B., C.I.E., Retd. . . . . 12 Nov. 08

Nixon, Genl. *Sir* J. E., K.C.B. . . [L] 2 May 09

Hawkes, Maj.-Genl. *Sir* H. M. P., K.C.B., C.S.I., Retd. . . . . 3 Sep. 09

Mansfield, Maj.-Genl. *Sir* H., K.C.B., Retd. . . . . 7 Sep. 09

30 Woon, Genl. *Sir* J. B., K.C.B. . . . 10 June 10

V.C. Adams, Maj.-Genl. *Sir* R. B., K.C.B., Retd. . . . . 15 June 10

Pollock, Maj.-Genl. J. A. H., C.B., Retd. 27 June 10

Bayly, Maj.-Genl. *Sir*, A. W. L., K.C.B., K.C.M.G., C.S.I., D.S.O., Retd. 17 Feb. 11

Gray, Maj.-Genl. W. duG., C.B., Retd. . 11 July 11

35 Scallon, Genl. *Sir* R. I., G.C.B., K.C.I.E., D.S.O. . . . . 23 Oct. 11

Bannerman, Maj.-Genl. W. B., C.S.I., M.D., I.M.S., K.H.P. Retd. . 26 July 11

Macdonald, Maj.-Genl. *Sir* J. R. L., K.C.I.E., C.B., R.E., Retd. . . . 26 Feb. 12

Barrett, Genl. *Sir* A. A., G.C.B., K.C.S.I., K.C.V.O., A. D. C. . . . 25 Oct. 12

Grover, Genl. *Sir* M. H. S., K.C.B., K.C.I.E. . . . . 5 Sep. 13

40 Lyons, *the Hon'ble* Maj.-Genl. R. W. S., M.D., I.M.S., K.H.P. . . 10 Jan. 14

Grainger, Genl. T., C.B., M.D., I.M.S., K.H.S. Retd. . . . . . 2 Apl. 14

V.C. Aylmer, Lt.-Genl. *Sir* F. J., K.C.B., R.E. . . . . 20 Apl. 14

Bradshaw, Maj.-Genl. L. J. E., C.B., Retd. . . . . 22 May 14

Watkis, Genl. *Sir* H. B. B., K.C.B. . 18 Oct. 14

45 Edwards, Maj. Genl. W. R., C.B., C.M.G., M.B., K.H.P., I.M.S. . . . 1 Apl. 15

46 Powell, Maj.-Genl. C. H., C.B. . . 5 Nov. 15

*
*
*
*

Number.	Names, Rank, Corps or late Corps.	Date of promotion in the Order.
	ORDER OF BRITISH INDIA.	
	NOTE.—The names shown in *italics* are those of Officers on the Pension Establishment.	
	First Class, with the title of "Sardar Bahadur"	
	Permanent.	
	Abdul Aziz, Subadar-Maj. (Hony. Capt.), 25th Madras Infy.	3 June 93
	Abdul Aziz, Risaldar-Maj. (Hony. Capt.), 5th Cav., A.-D.-C.	6 May 03
	Abdul Karim Khan, Risaldar-Maj., Govr.-Genl.'s Body Guard	21 Mar. 04
	Abdul Karim Khan, Risaldar-Maj. (Hony. Capt.), 30th Lrs., A.-D.-C.	12 Dec. 11
	Abdullah Khan, Risaldar (Hony. Lt. & Risaldar-Maj.), Aden Troop	16 Feb. 92
	Abdullah Khan, Subadar, I.O.M., *106th Prs.*	13 Jan. 16
	Abiram Gurung, Subadar-Maj., (Hony. Capt.), 8th G. R.	4 Jan. 07
	Adhar Singh, Subadar-Maj. (Hony. Capt.), 7th D. C. O. Rajputs	15 June 12
	Ajab Khan, Hon'ble, Subdar-Major (Hony. Capt.) I. O. M., 76th Punj.	4 Aug. 16
	Akbar Ali, Subadar-Maj. (Hony. Capt.), 24th Hazara Mountain Battery	17 Sep. 00
	Alahdin, Subadar Maj. (Hony. Capt.), 130th K. G. O. Baluchis	20 Mar. 11
	Alam Khan, Subadar-Major, I.O.M., *Queen Victoria's Corps of Guides (Infy.)*	19 Jan. 16
	Alexander, Subadar Maj. (Hony. Capt.), 2nd Q. O. S. and M.	12 Dec. 11
	Ali Gauhar Khan, Risaldar-Maj. (Hony. Capt.), 27th Lt. Cav.	7 Nov. 06
	Ali Haidar Khan, Risaldar-Maj. (Hony. Capt.), 38th Horse	13 Dec. 11
	Ali Muhammad Khan, Risaldar-Maj. (Hony. Capt.), 2nd Lrs.	31 May 01
	Amar Singh, Subadar-Maj. (Hony. Capt.), 19th Punj.	12 Dec. 11
	Amar Singh, Thapa, Subadar-Maj., 5th. G R. (F. F.)	12 Dec. 11
	Arbela, Subadar-Maj., (Hony. Capt.), 20th Punj. Infy.	26 Apl. 01
	Arbel Singh, Subadar-Maj. (Hony. Capt.), 38th Dogras	26 Oct. 12
	Arsia Khan, Subadar, (Hony. Lieut.), 20th Punj. Infy.	16 Feb. 87
	Asadulla Khan, Subadar-Maj., 22nd Punj.	3 June 18
	Atar Singh, Subadar-Maj. (Hony. Capt.), 76th Punj.	30 Aug. 03
	Atar Singh, Sub.-Maj., I.O.M., *47th Sikhs*	13 Jan. 14
	Attru, Subadar, (Hony. Lieut.), 56th Rifles (F. F.)	12 Dec. 11
	Badal Singh, Sub.-Maj., 89th Punj.	22 June 14
	Bahadur Singh, Risaldar-Maj., I.O.M., Corps of Guides	15 June 12
	Bahadur Ali Khan, Subadar-Maj. (Hony. Capt.), I.O.M., *51 Sikhs*	26 Dec. 06
	Bakshi Jaswant Singh, 39th Horse	3 June 16
	Balak Singh, Subadar-Maj., 8th Rajputs	16 Mar. 16
	Balbir Rana, Subadar (Hony. Lieut.), 3rd G. R.	8 July 07
	Beli Ram, SubadarMaj. (Hony. Capt.), I.O.M., *53rd Sikhs*	2 July 12
	Bhagat Singh, Subadar-Maj., (Hony. Capt.) 45th Sikhs	3 June 13
	Bhagwan Singh, Sub.-Maj. (Hony. Capt.), I.O.M., *14th Sikhs*	12 Dec. 11
	Bhai Khan, Risaldar, 54th Camel Corps	22 June 14
	Bhairo Singh (Hony. Capt.) Subadar-Maj., I.O.M., *119th Infy.*	3 June 15
	Bhanga Singh, Subadar-Maj. (Hony. Capt.), Derajat Mtn. Batty.	21 June 00
	Bhupal Thapa, Subadar-Maj. (Hony. Capt.), 39th Garh. Rif.	22 Sep. 92
	Bijai Singh, Subadar, 91st Punj.	4 June 17
	Binda Singh, Sub.-Maj. (Hony. Capt.), 4th P. A. V. Rajputs	12 Dec. 11
	Birbal Nagarkoti, Subadar-Maj. (Hony. Capt.), 7th G. R.	17 May 01
	Bisesar Tiwari, Sub.-Maj. (Hony. Capt.), I.O.M., *1st Brahmans.*	12 Aug. 13
	Bishan Singh, Subadar-Maj. (Hony. Capt.) 57th Rif.	24 Oct. 12
	Bishn Singh, Subadar-Major, 32nd Sikh Prs.	11 Nov. 15
	Bishn Singh Kathait, Subadar-Maj. (Hony. Capt.), 9th G. R.	22 June 12
	Bishnu Thapa, Subadar-Maj. (Hony. Capt.), 5th G. R.	3 Aug. 06
	Chamu Singh Burathoki, Subadar-Major (Hony. Capt.), 3rd G. R.	8 Apl. 03
	Chandi Singh, Subadar-Maj. (Hony. Capt.), 1st Brahmans.	24 Oct. 99
	Chaudhri Khan, Subadar-Maj. (Hony. Capt.), 92nd Punj.	31 Mar. 01
	Chinnasami, Subadar-Maj. (Hony. Capt.), 2nd S. and M.	14 Jan. 08
	Chittahang Limbu, Subadar-Maj., I.O.M., 2nd Bn., 10th G. R.	3 June 18
	Diwan Singh, Subadar-Maj., 24th Punj.	4 June 17
	Dalman Ale, Subadar-Maj. (Hony. Capt.), I.O.M., *2nd G. R.*	23 May 12
	Dasaundha Singh, Subadar, 22nd Punj.	12 June 17
	Daud Khan, Subadar-Maj., (Hony. Capt.), 15th Madras Infy.	17 May 01
	Dayal Singh, Sub-Maj. (Hony. Capt.), 54th Sikhs F. F.	12 Dec. 11
	Dayal Singh, Subadar-Major, 102nd Grs.	7 Apl. 16
	Debi Chand, Subadar-Maj., 4th G. R.	22 Feb. 17
	Devasahayam, Subadar-Maj. (Hony. Capt.), 2nd S. and M. (Hony. A.-D.-C. to H. E. the C.-in-C.)	5 May 12
	Dost Muhammad, Subadar-Maj., 106th Prs.	14 June 12
	Faiz Khan, Risaldar-Maj. (Hony. Capt.), 36th Horse	12 Dec. 11
	Faiz Talab Khan, Risaldar-Maj. (Hony. Capt.), Corps of Guides	23 Oct. 03
	Faizulla Khan, Subadar (Hony. Lieut.), Bengal S. and M.	27 Aug. 97
	Fakir Muhammad, Subadar (Hony. Lt.), 29th Punj.	21 Sept. 11
	Fazl Khan, Subadar-Maj. (Hony. Capt.), 82nd Punj.	10 Oct. 05
	Fateh Din, Subadar, 23rd Peshawar Mtn. Batty.	3 June 15
	Fateh Khan, Subadar-Major, 19th Punj.	1 Jan. 17
	Fazal Muhammad, Subadar, 29th Punj.	4 June 17
	Ganga Sewak Pande, Subadar-Maj., (Hony. Capt.), 23rd Infy.	20 Dec. 97
	Garkha, Subadar-Maj. (Hony. Capt.), 38th Dogras	12 Dec. 11
	Ghan Singh, Subadar, 76th Punj.	21 Nov. 17
	Ghulam Muhammad, Subadar-Maj. (Hony. Capt.), 29th Mtn. Batty.	13 Aug. 01

Number.	Names, Rank, Corps or late Corps.	Date of promotion in the Order.
	ORDER OF BRITISH INDIA—*contd.*	
	First Class, with the title of "Sardar Bahadur"—*contd.*	
	Ghulam Muhammad, Subadar-Maj. (Hony. Capt.), 27th Punj.	13 Dec. 11
	Ghulam Muhammad Khan, Sub.-Maj., (Hony. Capt.), 22nd Punj.	12 Dec. 11
	Girdhari, Subadar-Maj. (Hony. Capt.), 42nd Deoli Regt.	4 Jan. 18
	Gopal Ram Lama, Subadar-Maj. (Hony. Capt.), 2nd G. R.	24 June 13
	Gujan, Sub.-Maj. (Hony. Capt.), 6th Jat L. I.	21 Oct. 14
	Gurdatt Singh, Subadar-Maj. (Hony. Capt.), 15th Sikhs	28 June 07
	Gurdatt Singh, Risldr.-Maj. (Hony. Capt.), 12th Cav.	19 Sep. 10
	Gurdit Singh, Risaldar-Major, 6th Cav.	9 Jan. 14
	Hakim Khan, Subadar, 72nd Punj.	2 May 13
	Hakim Singh, Ressaidar, 8th Mule Corps	15 June 12
	Hamzullah Khan, Risaldar-Maj. (Hony. Capt.), 17th Cav.	2 May 11
	Hanooman Singh, *Risaldar-Maj, 20th Deccan Horse*	3 June 18
	Hanwan Singh, Rilsadar-Maj. (Hony. Capt.), 3rd S. Horse	15 Jan. 08
	Hara Singh, Subdr.-Maj. (Hony. Capt.), I.O.M., 91st Punj.	17 July 09
	Harditt Singh, Sub.-Maj. (Hony. Capt.), 21st Kohat Mtn. Batty. (F. F.)	12 Dec. 11
	Hazara Singh, Subadar-Maj. (Hony. Capt.), 35th Sikhs	16 Apl. 06
	Hazura Singh, Subadar-Maj. (Hony. Capt.), 90th Punj.	17 Feb. 13
	Hussain Khan, Subadar-Maj. (Hony. Capt.), 15th Madras Infy.	8 May 80
	Jagan Nath Pande, Subadar-Maj. (Hony. Capt.), 25th Bombay Infy.	5 Dec. 89
	Jagat Singh, Risaldar-Maj., (Hony. Capt.), 16th Lrs.	21 Sep. 84
	Jahangir Khan, Risaldar-Maj. (Hony. Capt.), 31st D. C. O. Lrs.	30 Mar. 99
	Jai Singh, Subadar-Maj. (Hony. Capt.), Corps of Guides	11 May 88
	Jiwand Singh, Subadar-Maj. (Hony. Capt.), 45th Sikhs	20 Mar. 04
	Judhbir Thapa, Subadar-Maj. (Hony. Capt.), 2nd G. R. (Hony. A.-D.-C. to H. E. the C.-in-C.)	4 Jan. 94
	Jwala Singh, Subadar-Maj. (Hony. Capt.), 45th Sikhs	17 May 01
	Jwala Singh, Risaldar (Hony. Lieut.) Corps of Guides.	2 Oct. 10
	Jwala Singh, Sub.-Maj. (Hony. Capt.), 35th Sikhs	3 Nov. 13
	Jwala Singh, Subadar, 32nd Sikh Prs.	14 June 12
	Jwala Singh, Sub.-Major, 34th Sikh Prs.	7 Aug. 14
	Kadam Khan, Risaldar-Maj. (Hony. Capt.), 4th Cav.	15 July 98
	Kaku Singh, Subadar-Maj. (Hony. Capt.), 35th Sikhs	24 Jan. 02
	Kala Singh, Subadar-Major (Hony. Capt.), 56th R.	30 May 06
	Kalu Thapa, Subadar-Maj. (Hony. Capt.), 44th G. R.	2 May 94
	Kamal-ud-din, Ressaidar (Hony. Lieut.), Govr.-Genl.'s Body Guard	12 July 06
	Kamardin, Sub.-Maj. (Hony. Capt.), I.O.M., 126th Balu. Infy.	6 Sep. 13
	Kandhai Singh, Subadar-Maj. (Hony. Capt.), 6th Infy., Hyderabad Contingent	1 May 97
	Kanchan Singh, Risaldar (Hony. Lieut.), 2nd Bombay Lrs.	16 Apl. 89
	Kanhai Parshad Dube, Subadar-Maj. (Hony. Capt.), 3rd Brahmans	30 Mar. 96
	Karmdad Khan, Sub.-Major, 124th Infy.	25 Apl. 14
	Khushlal Singh, Subadar-Maj. (Hony. Capt.), 27th Punj.	26 Nov. 10
	Khuslal Singh, Subadar-Maj., 72nd Punj.	3 June 15
	Kirpa Ram Thapa, Subadar-Maj. (Hony. Capt.), I.O.M., 2nd G. R.	7 Oct. 02
	Kishandhoj Rai, Sub.-Maj., 2nd Bn., 7th G. R.	22 June 14
	Kishn Singh, Sub.-Maj., 36th Sikhs	1 Oct. 15
	Kushal Singh Burathoki, Subdr.-Maj. (Hony. Capt.), 3rd G. R.	25 Apl. 10
	Lal Khan, Risaldar-Maj. (Hony. Capt.), 33rd Light Cav.	23 Aug. 98
	Lal Singh, Risaldar-Maj. (Hony. Capt.), 14th Bengal Lrs.	2 May 94
	Lehna Singh, Subadar-Maj., (Hony. Capt.), 54th Sikhs	3 Feb. 07
	Madaman, Subadar, 57th R.	17 Feb. 17
	Madho Sing Rana, Subadar-Maj. (Hony. Capt.), 4th G. R., A.-D.-C.	9 Feb. 99
	Magar Singh, Subadar-Maj. (Hony. Capt.), 26th Punj.	29 Mar. 07
	Mahatab Singh, Subadar-Major (Hony. Capt.), 48th Prs.	2 Sep. 02
	Mahrab Ali Khan, Risaldar-Maj., (Hony. Capt.), 3rd Lrs, Hyderabad Contingent	1 May 98
	Malapa, Subadar, 40th Pathans	3 June 16
	Malik Ahmad'yar Khan, Risaldar-Maj. (Hony. Capt.), 18th Lrs.	13 July 06
	Malik Ghulam Muhammad Khan, Ris.-Maj., *Khan Bahadur*, 39 K. G. O. C. I. H.	12 Dec. 11
	Malik Sher Bahadur Khan, Ris.-Maj., 26th Cav.	18 Jan. 15
	Mana Khan, Subadar-Major (Hony Capt.), 33rd Punj.	1 Jan. 03
	Mangal Singh, Risaldar-Maj. (Hony. Capt.), 3rd Bengal Cav.	10 Oct. 99
	Man Singh, Subadar-Maj. (Hony. Capt.), 14th Sikhs	13 Dec. 98
	Mansur Khan, Subadar-Maj., 55th Coke's Rif.	3 June 13
	Mathura Singh, Risaldar-Maj. (Hony. Capt.), 6th Bengal Cav.	23 Feb. 92
	Mihan Singh, Subadar-Maj. (Hony. Capt.) 28th Punj.	12 Dec. 11
	Mihr Din, Subadar-Maj., 1st S. & M.	11 Oct. 17
	Mihr Khan, Subadar-Maj. (Hony. Capt.), 29th Punj.	12 Dec. 11
	Mir Alam, Subadar-Maj. (Hony. Capt.), 28 Punj. Infy.	3 Jan. 91
	Mir Alam, Sub.-Maj., 58th Vaughan's Rif. (F. F.)	16 June 13
	Mir Haidar Shah Khan, Risaldar-Maj. (Hony. Capt.), 37th Lrs.	3 June 13

Number.	Names, Rank, Corps or late Corps.	Date of promotion in the Order.

ORDER OF BRITISH INDIA—*contd.*

First Class, with the title of "Sardar Bahadur"—*contd.*

	Mir Jafar Khan, Risaldar-Major, I.O.M., *9th Horse (A. D. C. to H. E. the C.-in-C.*	*21 Nov. 13*
	Mirza Umrao Beg., Ris.-Maj. (Hony. Capt.), M. V. O., *31st D. C. O. Lrs.*	*20 June 11*
	Mit Singh, Subadar-Maj., (Hony. Capt.) 53rd Sikhs (F. F.). (Hony. A. D. C., to H. E. the Viceroy)	*16 July 09*
	Mobin Khan, Subadar-Major (Hony. Capt.), 21st Punj.	*3 June 16*
	Moti Negi, Subadar-Maj. (Hony. Capt.), 39th Garh. Rif.	*8 Jan. 94*
	Muhammad Akbar Ali Khan, Subadar-Major, 7th Lrs.	1 Jan. 97
	Muhammad Ali Beg, Risaldar-Maj. (Hony. Capt.), 9th Horse	*30 Sep. 14*
	Muhammad Amin Khan, Risaldar-Maj. (Hony. Capt.), 23rd Cav.	*19 Dec. 90*
	Muhammad Amin Khan, Ris.-Maj., 15th Lrs.	23 Apl. 13
	Muhammad Aslam Khan, Risaldar, 11th Lrs.	3 June 16
	Muhammad Ibrahim, Subadar-Maj., (Hony. Capt.), 10th Bombay Infy.	*9 Oct. 93*
	Muhammad Ismail, Subadar-Maj., I.O.M., 32nd Mn. Batty.	4 Dec. 16
	Muhammad Khan, Subadar-Maj. (Hony. Capt.), 28th Bombay Prs.	*20 May 99*
	Muhammad Saiyid Khan, Risaldar-Maj. (Hony. Capt.), 1st Lrs.	*24 Nov. 10*
	Muhammad Shihabuddin, Subadar (Hony. Lieut.), 95th Infy.	*17 May 01*
	Muhi-Ud-Din Khan, Ris.-Maj., 31st Lrs.	28 Aug. 14
	Mukhmad Jan, Subadar, 59 Scinde Rif. (F. F.)	3 June 15
	Mul Singh, Subadar-Major, 126th Bal. Infy.	1 Jan. 17
	Musaffar Khan, Subadar-Maj. (Hony. Capt.), 72 Punj. (A.-D.-C. to H. H. the L.-G. of Burma)	*28 Mar. 01*
	Nakchhed Singh. Subadar-Maj. (Hony. Capt.), 16th Rajputs (The Lucknow Regiment)	*20 Mar. 96*
	Nand Singh, Subadar-Maj. (Hony. Capt.), 23rd Punj. Prs.	*14 Feb. 96*
	Narayan Singh, Subadar-Maj. (Hony. Capt.), 34th Sikh Prs.	*26 June 08*
	Narayan Singh, Subadar-Maj., (Hony Capt), I.O.M., *56th Punj. R. (Frontier Force)*	*12 Dec. 11*
	Narbahadur Gurung, Sub-Maj. (Hony. Capt.), 2nd G. R.	*9 Oct. 11*
	Nath Singh, Risaldar-Maj., 37th Lrs.	*2 Jan. 17*
	Nathu Singh Sunar, Subadar Maj. (Hony. Capt.), 39th Garh. Rif.	*14 June 12*
	Narayan Singh, Subadar-Maj., (Hony. Capt.), 51st Sikhs (F. F.)	*5 Oct. 05*
	Nawab Khan, Subdr-Maj., (Hony. Capt.) I.O.M., *30th Mtn. Batty.*	*11 Sep. 09*
	Nawal Sing Rana, Sub-Maj. (Hony. Capt.), I.O.M., *8th G. R.*	*23 May 12*
	Niyaz Khan, Subadar, (Hony. Lieut.), 57th Rif. (F. F.)	*12 Dec. 11*
	Nur Khan, Subadar-Maj. (Hony. Capt.), 61st Prs.	*3 June 13*
	Prakas Rae, Subadar-Maj. (Hony. Capt.), 91st Punj.	*28 May 02*
	Phula Singh, Sub-Maj (Hony. Capt.), 12th Prs	*12 Dec. 11*
	Prem Singh, Subadar-Maj. (Hony. Capt.), I.O.M. *32nd Sikh Prs. (Hony. A.-D.-C. to H. E. the C.-in-C.)*	*5 Nov. 08*
	Rahim Khan, Subdr.-Maj. (Hony. Capt.) 107th Prs.	*7 Jan. 10*
	Ramchandar Rao Mahadik, Risaldar-Major (Hony. Capt.), 31st D. C. O. Lrs.	*17 May 01*
	Ram Singh, Subadar, 32nd Sikh Prs.	1 Jan. 17
	Ramzan Khan, I.O.M., Subadar, 53rd Sikhs	*12 Mar. 03*
	Rekha Ram, Subadar-Maj. (Hony. Capt.), 6th Jat. L. Infy.	*12 Dec. 11*
	Rur Singh, Subadar-Maj. (Hony. Capt.), 29th Madras Infy.	*17 May 01*
	Rustam Khan, Subdr.-Maj. (Hony. Capt.), 17th Infy.	*17 July 09*
	Saiyid Ali, Subadar Maj. (Hony. Capt.) 53rd Sikhs (F. F.)	*18 Mar. 11*
	Saiyid Husain, Subadar-Maj. (Hony. Capt.), 61st Prs.	*30 Aug. 05*
	Sandhya-das, Subdr.-Maj. (Hony. Capt.), 25th Punj.	*19 Feb. 10*
	Santbir Gurung, Subadar-Maj. (Hony. Capt.), I.O.M. *2nd G. R.* (Hony. A.-D.-C. to H. E. the C.-in-C.)	*10 Sep. 02*
	Sarabjit Thapa, Subadar-Maj. (Hony. Capt.), 2nd 1st G. R.	*3 Feb. 13*
	Sardar Janmeja Singh, Risaldar-Maj. (Hony. Capt.), 21st Cav.	*5 Feb. 12*
	Sarfaraz Khan, Subadar-Maj. (Hony. Capt.), Corps of Guides	*21 June 00*
	Sawan Singh, Subadar (Hony. Lieut.), 107th Prs.	*15 June 03*
	Sayyid Amir, Subadar-Maj., (Hony. Capt.), 1st Madras Lrs.	*18 Apl. 89*
	Sayyid Jalal, Subadar-Maj., (Hony. Capt.), 21st Madras Prs.	*1 May 98*
	Sayyid Yakub, Subadar-Maj., (Hony. Capt.), 27th Madras Infy.	*16 Apl. 90*
	Shah Muhammad, Subadar-Maj., (Hony Capt.), 25th Punj. Infy.	*27 Sep. 01*
	Shaikh Ahmad, Subadar-Maj., (Hony. Capt.), 1st Madras Lrs.	*26 Oct. 92*
	Shaikh Amir, Subadar-Maj., (Hony. Capt.), 25th Madras Infy.	*1 Jan. 92*
	Shaikh Babu, Subdr.-Major, 116th Mah.	3 June 16
	Shaikh Husain, Subadar-Maj. (Hony. Capt.), 101st Grs.	*4 Nov. 99*
	Shaikh Ibrahim, Subadar (Hony. Lieut.), 116th Mah.	*14 June 12*
	Shaikh Imam, *Khan Sahib*, Subadar-Major, 103rd Mah. L. Infy.	4 Jany. 16
	Shaikh Ismail, Subadar-Maj. (Hony. Capt.), 7th Bombay Infy.	*23 Apl. 90*
	Shaikh Mahbub, Subadar-Maj. (Hony. Capt.), 23rd Bombay Infy.	*17 May 01*
	Shaikh Mahtabdin, Subadar-Maj. (Hony. Capt.), 31st Madras Infy.	*2 June 93*
	Shaikh Muhammad, Subadar-Maj. (Hony. Capt.), 28th Bombay Prs.	*7 Jan. 99*
	Shaikh Wali Ahmad, Subadar-Maj. (Hony. Capt.), 1st Infy., Hyderabad Contingent	*26 June 95*

Number.	Names, Rank, Corps or late Corps.	Date of promotion in the Order.

ORDER OF BRITISH INDIA—*contd.*

First Class, with the title of "Sardar Bahadur"—*concld.*

	Sher Bas, Subadar-Maj. (Hony. Capt.), 52nd Sikhs	3 Dec. 98
	Sher Khan, Risaldar-Major, (Hony. Capt.) 1st Lrs.	7 Oct. 05
	Sher Singh, Subadar-Maj. (Hony. Capt.), 28th Mtn. Batty.	6 May 01
	Sher Singh, Risaldar-Maj. (Hony. Capt.), 13th D. C. O. Lrs. (Hony. A.-D.-C. to H. E. the C.-in-C.)	17 May 01
	Sikandar Khan, Risaldar-Maj. Khan Sahib (Hony. Capt.), 11th Lrs.	4 Aug. 07
	Singbir Ghale, Subdr.-Maj. (Hony. Capt.), 13th G. R.	25 Aug. 10
	Sirput Pudwal, Subadar-Maj. (Hony. Capt.), 128th Prs.	26 Apl. 02
	Sital Sing Lama, Subadar (Hony. Lieut.), 2nd G. R.	3 Apl. 12
	Siwa Moduck, Subadar-Maj. (Hony. Capt.), 101st Grs.	22 June 06
	Sudh Singh, Subadar, (Hony. Lieut.), Bengal S. and M	29 Aug. 93
	Suhel Singh, Subadar, 56th Punj. Rif.	31 Mar. 15
	Sundar Singh, Subadar-Maj. (Hony. Capt.), 36th Sikhs	29 Sep. 03
	Sundar Singh, Subadar-Maj., (Hony. Capt.), 84th Punj.	12 Dec. 11
	Surja, Subdr.-Maj. (Hony. Capt.), 48th Prs.	12 Oct. 16
	Suraj Mal, Risaldar, 14th Murray's Jat Lrs.	22 Mar. 17
	Tesu Khanka, Subadar-Maj., (Hony. Capt.), 1st G. R.	27 Feb. 95
	Thakur Singh, Risaldar, I. O. M., 21st Cav. F. F.	17 Nov. 13
	Tirath Ram, Risaldar-Maj. (Hony. Capt.), Corps of Guides	20 July 07
	Umdah Singh, Risaldar-Maj. (Hony. Capt.), 22nd F. F.	21 Mar. 04
	Umrao Singh Rawat, Subadar-Maj. (Hony. Capt.), 2nd Bn., 39th Gar. Rif.	3 June 13
	Waryam Singh, Subadar-Maj. (Hony. Capt.), 47th Sikhs	7 Mar. 02
	Waryam Singh, Subadar-Maj., 21st Punj	2 Feb. 17
	Wassawa Singh, Subadar, I.O.M., 32nd Sikh Prs.	12 Oct. 16
	Yar Muhammad, Subdr.-Maj., (Hony. Capt.), 127th Baluch L. I. (Hony. A.-D.-C. to H. E. the C.-in-C.)	4 Oct. 09
	Yasin Khan, Subadar-Maj. (Hony. Capt.), C.I.E., 24th Punj. Infy.	15 Apl. 98

Senior Sub-Assistant Surgeons.

	Abdur Rahman, 1st class, Bengal Estabt.	4 June 17
	Ahmadulla Khan, Khan Bahadur, 1st class, Bengal Establishment	4 Oct. 07
	Bhuraya Julaya, Rai Bahadur, 1st Class, Bombay Establishment	12 Dec. 11
	Suhail Singh, 1st Class, Bengal Establishment	2 Jan. 15
	Shaikh Amir-bakhsh, Khan Bahadur, 1st class, Bengal Establishment	11 Apl. 02

Supernumerary.

	Bakhan Sing Thakur, Subadar-Maj. (Hony. Capt.), 44th G. R.	13 Aug. 97
	Baldeo Singh, Risaldar-Maj. (Hony. Capt.), 14th Murray's Jat Lrs.	1 Jan. 03
	Ghulam Muhammad, Subadar-Maj. (Hony. Capt.), 22nd Derajat Mtn. Batty.	1 Jan. 03
	Gul Mawaz Khan, Risaldar-Major, *Bahadur*, 18th Lrs.	30 Nov. 17
	Gulzada, Subadar, 126th Infy.	15 Oct. 15
	Habibullah Khan, Subadar, 1st S. and M.	16 Dec. 17
	Hashmat Ali, Ressaider, I.O.M., 1st Mule Corps-S. and M.	17 Aug. 17
	Hira Singh, Subadar-Major (Hony. Capt.), S. and M.	1 Jan. 03
	Hira Singh, Subadar-Maj. (Hony. Capt.), 1st Sikh Infy.	19 July 01
	Jag Singh, Subadar (Hony. Lieut.), Bengal S. and M	1 Jan. 03
	Kasim Ali, Subadar, *Bahadur*, 31st Div. Sig. Coy., S. and M.	19 July 18
	Khitab Gul, Sub.-Maj., 120th Infy.	4 Aug. 15
	Khushal Singh, Subadar, *Bahadur*, 27th Punj.	19 July 18
	Khwaja Muhammad Khan, Risaldar, Q. V. O. Corps of Guides (F. F.)	1 Jan. 15
	Lachman, Singh Subadar, *Bahadur*, 32nd Sikh Prs.	19 July 18
	Lall Mast, Subadar Hony. Lieut., 29th Bombay Infy.	13 Aug. 97
	Mewa Singh, Subadar-Major. Bahadur, 48th Prs.	2 Aug. 18
	Mohibulla Khan, Subadar, I.O.M., S. Waz. Militia.	30 Mar. 17
	Muhammad Beg, Risaldar-Maj. (Hony. Capt.), 26th Lt. Cav.	1 Jan. 03
	Muissuddin Khan, Risaldar-Maj. (Hony. Capt.), 15th Lrs.	1 Jan. 03
	Natha Singh, Subadar-Maj., I.O.M., 34th Sikhs, Prs.	17 Aug. 17
	Nasir Khan, Subadar-Major. 59th Scinde Rif.	4 Aug. 15
	Nur Alam, Subadar-Major., 28th Mn. Batty.	4 Aug. 15
	Prem Sing Bisht, Subadar-Major, *Bahadur*, 9th G. R.	19 July 18
	Ram Singh, Subadar-Maj. (Hony. Capt.), 28th Bombay Prs.	13 Aug. 97
	Sakharam Rao Jagtap, Subdr. 117th Mah.	4 Aug. 15
	Sakt Chand, Subadar Major, 24th Punj.	4 Aug. 15
	Shaikh Farid, Risaldar-Maj. (Hony. Capt.), 26th Lt. Cav.	1 Jan. 03

ABSTRACT.

No. on roll.	Permanent Establishment.	Vacancies.	Supernumeraries.	REMARKS.
Combatant Ranks.	...	...	...	...
Sub-Assistant Surgeons	...	...	...	...

Number.	Names, Rank, Corps or late Corps.	Date of admission to the Order.
	ORDER OF BRITISH INDIA—*contd.*	
	Second Class, with the title of "Bahadur."	
	Permanent.	
	Abdul Amir, Subadar, 130th Baluchis	4 June 17
	Abdul Aziz, Subadar-Maj., 81st Prs.	22 June 14
	Abdul Ghani, Subadar, 4th Punj. Infy.	15 Apl. 98
	Abdul Hak, Subadar, 91st Punj.	2 Jan. 02
	Abdul Hak, Subadar, 46th Punj.	12 Dec. 11
	Abdul Karim, Subadar-Maj., 75th Carnatic Infy.	13 May 12
	Abdulla Khan, Subadar, 27th Punj.	22 Dec. 02
	Abdulla, Jemadar, 20th Infy.	8 July 07
	Abdur Rahman, Subadar-Maj., 3rd Madras Infy.	22 July 94
	Abdur Rahim Khan, Subadar, Bengal S. and M.	9 Oct. 96
	Abdur Razack, Subadar-Major, 73rd Carnatic Infy.	19 Jan. 18
	Abdur Razzak, Subadar-Major, 63rd Infy.	3 June 16
	Abdus Sattar, Sub.-Maj., 63rd Palamcottah L. Infy.	2 Aug. 11
	Ahmad Khan, Risaldar-Maj., 29th Lrs.	14 June 12
	Ahmed Khan, Ressaidar and Woordie-Major, Khan Sahib, 11 Lrs.	6 July 93
	Ahmed Khan, Risaldar-Maj., 36th Horse	22 Feb. 17
	Akbar Ali, Subadar-Maj., 1st S. and M.	3 June 15
	Ali Bakhsh, Subadar-Maj., Bengal S. and M.	21 June 87
	Ali Hussain, Risaldar-Maj., 1st Punj. Cav.	15 Apl. 98
	Amir Hussain, Subadar, 1st Madras Lrs.	18 Sept. 88
	Amir Khan, Subadar, 55th Coke's Rif.	25 Nov. 14
	Amir Khan, Risaldar, 61st Grantee Camel Corps	4 Sept. 12
	Amir Muhammad Khan, Risaldar-Maj., 20th Horse	12 Oct 16
	Arif Khan, Subadar, 17th Madras Infy.	28 Jan. 97
	Arjun Singh, Risl.-Major, 23rd Cav.	11 Nov. 15
	Arsla Khan, Subadar, I.O.M., 53rd Sikhs	12 Oct. 16
	Ashraf Ali Khan, Risaldar-Major, 3rd Horse	13 Jan. 14
	Attar Khan, Subadar, 33rd Punj.	3 June 15
	Azhagarsami, Sub.-Maj., 73rd Carnatic Infy.	12 Dec. 11
	Azad Khan, Subadar, Frontier Constabulary	
	Bachitar Singh, Sub.-Maj., 31st Punj.	14 June 12
	Badan Singh, Subadar, 36th Sikhs	3 Feb. 13
	Bainta, Subadar, 37th Dogras	2 Nov. 87
	Bakhshish Singh, Subadar-Maj., 3rd Sikh Infy.	13 Dec. 98
	Balkrishna Rao, Subadar-Maj., 110th Mah. L. Infy.	4 June 17
	Ballia, Subadar, No. 3 Field Battery, Hyderabad Contingent	1 May 97
	Balwant Singh Mahat, Subadar-Maj., 1st G. R.	13 Feb. 01
	Bansraj Gurung, Subadar, 4th G. R.	10 Aug. 14
	Bapurao Gaekwad, Subadar-Maj., 103rd Mah. L. Infy.	4 June 17
	Baryam Singh, Subadar, 47th Sikhs	1 May 07
	Basawa Singh, Sub.-Maj., 92nd Punj.	3 Apl. 12
	Bas Khan, Subadar-Major, 59th Scinde Rif. (F. F.)	31 Dec 05
	Bedhawa Singh, Subadar, 22nd Punj.	10 Feb. 05
	Bela Singh, Subadar, I.O.M., 55th Coke's Rif.	1 Jan. 17
	Beli Singh, Subadar-Maj., 38th Dogras	3 June 15
	Bhag Singh, Subadar-Maj., 37th Dogras	7 Dec. 06
	Bhagatbir Gurung, Subadar-Major, 1st G. R.	3 June 16
	Bhagat Singh, Subadar-Maj., 62nd Punj.	12 Dec. 11
	Bhagwan Giri, Sub.-Maj., 10th G. R.	12 Dec. 11
	Bhola Sing Kandari, Subadar-Maj., 10th G. R.	3 June 15
	Birbal, Subadar-Maj, 74th Punj.	31 Oct. 17
	Birbhan Thapa, Subdr.-Major, 28th G. R.	3 June 16
	Bir Sing Bohra, Subadar-Maj., 9th G. R.	24 Oct. 00
	Bisesar Singh, Subdr.-Maj., Rajput L. I.	11 Apl. 10
	Bishn Singh, Sub.-Maj., 59th Rif. (F. F.)	12 Dec. 11
	Bishn Singh, Risaldar-Maj., 12th Cav.	5 Sep. 03
	Bostan Khan, Subadar, 53rd Sikhs	31 Mar. 15
	Buland Khan, Sub.-Maj., 105th Mah. L. Infy.	12 Dec. 11
	Bunyad Ali, Subadar-Maj., 1st S. and M.	31 July 06
	Buta Singh, Subadar-Maj., 10th Bengal Infy.	17 May 01
	Chaman Khan, Risaldar-Maj., 35th Horse	17 Oct. 12
	Chanda Singh, Subadar, 27th Punj.	6 Sep. 13
	Chanda Singh, Subadar, Queen Victoria's Own Corps of Guides	16 June 13
	Chanda Singh, Subadar-Maj., 33rd Burma Infy.	17 May 01
	Chandan Singh, Subadar, 95th Infy.	17 May 01
	Chandarchand, Subadar, 28th Madras Infy.	23 Nov. 83
	Chanderdip Singh, Subadar, 11th Rajputs	17 May 01
	Chendriah, Subadar-Maj., 21st Madras Prs.	21 July 87
	Darshanu, Subdr.-Maj., 67th Punj.	24 Jan. 10
	Dayal Singh, Subadar, 20th D. C. O. Infy.	15 June 12
	Daya Ram, Subadar-Major, 2-67th Punj.	3 June 16
	Dhaku Kubre, Subadar, 25th Bombay Infy.	5 Dec. 89
	Durga Singh, Subadar, 21st Punj.	2 July 12
	Farzand Ali Khan, Risaldar, 6th K. E. O. Cav.	16 Mar. 16
	Fateh Singh, Subadar-Maj., 32nd Prs.	17 May 01
	Fazil Khan, Subadar-Maj., 7th Bombay Infy.	12 Dec. 95
	Fazldad Khan, Khan Sahib, Risl.-Major, 12th Cav.	4 Jan. 16
	Fazl-dad Khan, Subadar-Maj., 30th Punj.	2 Dec. 01

Number.	Names, Rank, Corps or late Corps.	Date of admission to the Order.
	ORDER OF BRITISH INDIA—*contd.*	
	Second Class, with the title of "Bahadur"—*contd.*	
	Ganda Singh, Risaldar, 22nd Cav.	3 June 15
	Ganga Datt, Risaldar-Maj., 2nd Lrs	24 Sept. 17
	Ganpati Kenjle, Subadar, 105th Mah. L. I.	9 Mar. 17
	Ghaus Muhammad Khan, Risaldar-Maj., 21st Cav.	26 Nov. 14
	Ghulam Ali, Subadar-Maj., 30th Punj.	*29 Mar. 07*
	Ghulam Muhiuddin, Subadar, 27th Punj.	*12 Dec. 11*
	Ghulam Muhiuddin Khan, Risaldar-Maj., 17th Cav.	*14 June 12*
	Ghulam Rasul Khan, Subadar-Maj., 5th Lt. Infy.	*6 Apl. 02*
	Gobardhan Tiwari, Subadar, 5th Bengal Infy.	*21 June 87*
	Gopal Singh, Subadar-Maj., 12th Madras Infy.	*29 July 91*
	Gopal Singh, Risaldar-Maj., 19th Lrs.	*15 June 12*
	Gopal Singh, Risaldar, 7th Lancers	3 June 18
	Gopi, Sub.-Maj., 30th Punj.	6 Aug. 15
	Govind Rao, Risaldar, 30th Lrs.	1 Jan. 17
	Govindaraju, Sub-Maj., 2nd S. & M.	*14 Jan. 12*
	Gula Jan, Sub.-Maj., 1st S. & M.	*12 Dec. 11*
	Gulab Singh, Subadar, 14th Sikhs (Hy. A.D.C. to L. G., Punj.)	*12 Jan. 11*
	Gulmir, Sub.-Maj., 28th Punj.	*26 Apl. 15*
	Gunjbir Ale, Subadar-Maj., 3rd G. R.	21 Nov. 17
	Gurdatt Singh, Sub.-Maj., 14th Mah.	*12 Dec. 11*
	Gurdit Singh, Subadar, 57th Rif.	*3 June 13*
	Gyan Singh, Subadar, 72nd Punj.	*28 May 02*
	Gyan Singh, Subadar, 72nd Punj.	3 June 16
	Habibullah Khan, Subadar-Maj., 46th Punj.	4 June 17
	Hans Ram Sing Negi, Subadar, 39th Garh. Rif.	22 June 14
	Hanwat Singh, Subadar-Maj., 10th Jats.	*1 Jan. 03*
	Harjalu, Subadar-Maj., 52nd Sikhs	*14 June 12*
	Harkabir Thapa, Subdr.-Major, I.O.M., 5th G. R.	3 June 16
	Hasan Khan, Subadar, 3rd S. & M.	*28 Oct. 06*
	Hasan Muhammad, Subadar-Maj., 104th Rif.	11 Oct. 17
	Haskelji Israil, Subadar, 28th Bombay Prs.	*2 Mar. 96*
	Hayat Muhammad Khan, Risaldar, 52nd Camel Corps	21 Oct. 14
	Hayatullah, Subdr.-Maj., 54th Sikhs	*11 Sep. 09*
	Hukum Singh, Ris.-Maj., 33rd Lt. Cav.	*3 Oct. 10*
	Hukam Singh, Act.-Subadar-Major, 2-32nd Sikh Prs.	3 June 18
	Ibrahim Ap Khan, Risaldar-Major, 28th Cav.	31 Mar. 15
	Ilahi Khan, Risaldar-Maj., 33rd Lt. Cav.	*17 May 01*
	Illahi Bakhsh, Subadar, 121st Prs.	*9 Mar. 17*
	Isar Singh, Sub.-Maj., 93rd Burma Infy.	*12 Dec. 11*
	Isar Singh, Risaldar-Maj., 22nd Cav. (F. F.)	*22 June 14*
	Ismail Khan, Risaldar, 1st Lrs.	*12 Dec. 11*
	Jafir Khan, Risaldar-Maj., 1st Bom. Lrs.	*28 July 95*
	Jagat Singh, Subadar, 32nd Burma Infy.	*4 Nov. 97*
	Jalal Khan, Subadar, 19th Punj.	3 June 15
	Jawand Singh, Risaldar, 38th Horse	*12 Dec. 11*
	Jiwan Singh, Subadar, 23rd Sikh Prs.	*1 Apl. 05*
	Jokaran Rai, Subadar, 6th G. R.	*15 Apl. 05*
	Joseph, Subadar-Maj., 81st Prs.	*11 Dec. 02*
	Jowand Singh, Subadar, Jullundur Mn. By.	*30 Dec. 91*
	Kala Singh, Risaldar, 2nd Lrs.	*17 May 01*
	Kalandar Khan, Risaldar-Maj., 7th Lrs.	*30 Aug. 08*
	Kalandar Khan, Subadar-Major, 91st Punj.	1 Jan. 17
	Kalka Singh, Subadar, 97th Infy.	3 June 15
	Kamal Khan, Subadar, I.O.M., 54th Sikhs	*5 May 12*
	Kanha, Subadar, 37th Dogras.	*13 Dec. 11*
	Karam Sher Khan, Subadar-Major, 18th Infy.	3 June 15
	Karam Singh, Risaldar-Major, 13th Lrs. (Hony. A.D.C. to H.E. the Viceroy)	*9 Mar. 17*
	Karim Dad Khan, Risaldar-Maj., 20th Horse	*17 Dec. 98*
	Karim Khan, Subadar-Maj., 76th Punj.	*17 May 01*
	Kartar Singh, Risaldar-Major, 12th Cav.	4 Dec. 16
	Kehar Sing, Subadar, 14th Sikhs	*3 June 15*
	Kesar Singh, Subadar-Maj., I.O.M., 30th Punj.	*3 June 13*
	Keshri Pershad, Subadar-Maj., Mewar Bhil Corps	22 June 14
	Kesu Sawant, Subadar, 128th Prs.	*19 Nov.*
	Khan Bahadur, Subadar-Maj., 22nd Punj.	*10 Feb. 02*
	Khem Singh, Subadar, 31st Mtn Batty.	18 Nov. 10
	Khubi Ram, Resaldar-Maj., 4th Lrs.	*14 Jan. 08*
	Khudadad Khan, Subadar, Burma Military Police.	...
	Kishen Singh, Subadar, I.O.M., 22nd Derajat Mtn. Batty. (F. F.)	*12 Dec. 11*
	Krishna Appaji, Subadar, 3rd S. & M.	*15 Sep. 08*
	Krishna Singh, Subadar, Maj., 63rd Lt. Infy.	*12 Aug. 13*
	Kusahar Singh, Subadar-Maj., 2nd Bengal Infy.	*5 July 94*
	Labh Singh, Risaldar-Maj., 29th Lrs.	22 June 14
	Lachman Dube, Subr.-Maj., 3rd Brahmans	*10 Sep. 09*
	Laik Singh, Subadar-Maj., 41st Dogras	*12 Dec. 11*
	Lalbihari Singh, Subadar-Maj., 8th Rajputs	*17 May 01*
	Lal Sing Gurung, Subadar, 5th G. R.	3 June 16
	Mahabal Tiwari, Sub.-Maj., 110 Mah. Lt. Infy.	*12 Dec. 11*
	Mahadeorao Sondkar, 116th Mah.	3 June 15

Number.	Names, Rank, Corps or late Corps.	Date of admission to the Order.
	ORDER OF BRITISH INDIA—*contd.*	
	Second Class, with the title of "Bahadur"—*contd.*	
	Mahtab Khan, Subadar, Quetta Mtn. Batty.	*1 Dec. 95*
	Makhan Sing Gurung, Subadar, 8th G. R.	23 May 12
	Mal Singh, Subadar-Maj., 34th Prs.	*17 Nov. 01*
	Malik Dost Muhammad Khan, Risl.-Major, 17th Cav.	7 Apl. 16
	Malik Muhammad Hayat Khan, Ris.-Maj., 13th Lrs. (Hy. A.D.C. to L.G., Punj.)	*12 Dec. 11*
	ManjitGharti, Subadar, 2nd G. R.	3 Jnne 18
	Mall Singh, Subadar, 45th Sikhs	3 June 18
	Maluk Singh, Subdr.-Major, 33rd Punj.	19 Jan. 16
	Mansaram Pun, Subadar, 5th G. R.	4 June 17
	Man Singh, Subadar, 58th Rif.	*2 Sep. 02*
	Man Singh, Subadar, 15th Sikhs	*15 June 03*
	Man Singh, Subadar-Maj., 23rd Sikh Prs.	2 Jan. 17
	Mangal, Singh, Risaldar, 16th Cav.	*12 Dec. 11*
	Mangal Singh, Subadar-Major, 35th Sikhs	1 Jan. 17
	Maria Cruz, Subadar-Major, 2nd S. & M.	3 June 18
	Matapalt Singh, Subadar, 11th Rajputs	4 June 17
	Megchand Gharti, Subadar-Maj., 2nd G. R.	*18 Nov. 08*
	Mewa Thapa, Subadar, 8th G. R.	*9 Apl. 14*
	Mihr Khan, Subadar-Maj., 4th Punj. Infy.	*15 Feb. 96*
	Mir Abbas, Subadar-Maj., 64h Prs.	*23 May 01*
	Mir Akbar, Subadar-Major, 27th Punj.	12 Oct. 16
	Mir Baz Khan, Subadar, 51st Sikhs (F. F.)	4 Jan. 18
	Mir Dad Khan, Risaldar 9th Horse	1 Jan. 17
V. C.	**Mir Dast**, Subadar, I.O.M., 57th Rif.	17 Dec. 15
	MirMumtaz Huslan, Risaldar, 30th Lrs.	3 June 18
	Misri Khan, Risaldar-Maj. (Hony. Lieut.), 18th K. G. O. Lrs.	*29 Nov. 01*
	Mitt Singh, Subadar-Maj., 14th Sikhs	*19 Feb. 03*
	Mirza Kutb-ud-din, Subadar, Bengal S. & M.	*26 Jan. 95*
	Mohan, Subadar, 37th Dogras	*26 Apl. 11*
	Mohan Singh, Subadar-Maj., 8th Rajputs	*17 June 05*
	Molar Singh, Subadar, I. O. M., 53rd Sikhs	31 Jan. 16
	Moti, Subadar, 28th D. C. O. Infy.	*3 Nov. 13*
	Moti Gurung, Subdr., 3rd G. R.	*23 Apl. 10*
	Motilal Lama, Subadar-Maj., 2nd G. R.	22 June 14
	Muhammad Akbar, Sub.-Maj., 26th Punj.	12 Dec. 11
	Muhammad Akram Khan, Risaldar-Maj., 1st Lrs.	14 June 17
	Muhammad Bakhsh, Subadar-Maj., 93rd Burma Infy.	22 June 14
	Muhammad Hanif, Risaldar, 19th Lrs.	3 June 15
	Muhammad Zaman, Subdr-Major, 62nd Punj.	3 June 16
	Muhammad Hayat, Subadar-Maj., 73rd Carnatic Infy.	*9 Oct. 09*
	Muhammad Jafar, Subadar, 81st Prs.	*3 June 13*
	Muhammad Raza Khan, Ressaidar, 2nd Lrs.	*21 Nov. 82*
	Mul Raj, Subadar, 69th Punj.	*3 June 16*
	Mula Singh, Subadar-Major, 25th Punj.	*1 Jan. 17*
	Mumtaz Ali Khan, Rasaldar, 39th Horse	13 Jan. 16
	Munisami, Subadar-Maj., 6th Madras Infy.	*14 Aug. 99*
	Musall Khan, Risaldar, 22nd Cav. (F. F.)	*21 Oct. 04*
	Naganna, Subadar, 8th Madras Infy.	*23 June 98*
	Nain Singh Gurung, Sub.-Major, 7 G. R.	10 July 15
	Nanak Sing, Subadar, 56th Rif. (F. F.)	3 June 15
	Nand Singh, Subadar-Maj., 1st S. & M.	*3 June 13*
	Naojadik Singh, Subadar-Maj., 11th Rajputs	*4 Jan. 07*
	Narasimhulu, Subadar, 80th Carnatic Infy.	*12 Dec. 11*
	Narayan Singh, Risaldar, 2nd Lrs.	*21 Nov. 82*
	Narayan Singh, Subadar-Maj., 47th Sikhs	*8 Apl. 02*
	Narayan Singh, Risaldar, 5th Cav.	*23 Apl. 13*
	Natha Singh, Risaldar, 36th Horse	*17 May 01*
	Nawab Shah, Subadar, The Frontier Garrison Arty.	*29 July 06*
	Nasir Ali Khan, Ris.-Maj., 34th Horse	*29 Mar. 11*
	Padam Sing Barathoki, Subadar-Maj., 1st G. R.	22 June 14
	Pahlwan Khan, Rishaldar-Maj., 12th Cav.	*7 Mar. 14*
	Panjab Singh, Subadar-Maj., 2nd Punj. Infy.	*21 June 87*
	Partab Chand, Subadar, 4th G. R.	*12 Dec. 11*
	Poshoti Mari, Subadar, 3rd S. and M.	4 June 17
	Prem Singh, Subadar-Maj., 51st Sikhs	*28 Apl. 02*
	Pudai Singh, Jemadar, 28th Mule Corps	*21 Nov. 13*
	Punjab Singh, Risaldar-Maj., 6th Bom. Cav.	*1 Aug. 94*
	Purusottam, Sub.-Maj., 74th Punj.	*16 Apl. 11*
	Ralla Singh, Subadar, 1st Bn. 19th Punj.	3 June 18
	Ram Das, Subadar-Maj., 61st Prs.	22 June 14
	Ram Singh, Subadar-Maj., 36th Sikhs	*10 Dec. 01*
	Ram Singh, Subadar, I. O. M., 82nd Punj.	4 June 17
	Ramabalamb Singh, Sub.-Maj., 13th Rajputs	*11 Nov. 10*
	Ramasami, Subadar-Maj., Madras S. and M.	*17 May 01*
	Ramasing, Subadar, 2nd S. and M.	*12 Dec. 11*
	Rambaksh Singh, Sub.-Major, 16th Rajputs	9 Sep. 14
	Ranjit Singh, Risldr.-Maj., 32nd Lrs.	*4 Oct. 09*
	Ratan Singh, Risaldar, 3rd Horse	*12 Dec. 11*

Number.	Names, Rank, Corps or late Corps.	Date of admission to the Order.
	ORDER OF BRITISH INDIA—*contd.*	
	Second Class, with the title of "Bahadur"—*contd.*	
	Rudrnarain Nagarkoti, Subadar-Maj., 1st G. R.	4 June 17
	Rukman Sahl, Subadar-Maj., 8th G. R.	1 Jan. 17
	Rup Chand, Risaldar-Major, 33rd Cav.	25 Apl. 14
	Rur Singh, Subadar, 1st Bn. 32nd Sikh Prs.	24 Nov. 17
	Sahaj Ram, Risaldar, 7th Lrs.	*14 June 12*
	Saiyid Abbas, Subadar-Maj., 64th Prs.	*12 Dec. 11*
	Sajawat Khan, Risaldar, 58th Silladar Camel Corps	*1 Jan. 06*
	Salih Khan, Subadar, 20th Infy.	15 June 12
	Sangat Singh, Subadar, 45th Sikhs	*16 Dec. 04*
	Sankabir Thapa, Subadar, 5th G. R.	*3 June 13*
	Sanman Nagar Koti, Subadar, 3rd G. R.	*3 June 13*
	Sant Singh, Ris.-Maj., 5th Cav.	*12 Dec. 11*
	Santa Singh, Risaldar-Maj., I. O. M., 33rd Lt. Cav.	17 Feb. 17
	Santbir Gurung, Subadar-Maj., 4th G. R.	*12 Dec. 11*
	Sapuran Singh, Sub.-Maj., 15th Sikhs	*12 Dec. 11*
	Sardar Khan, Risaldar-Maj., 10th Lrs.	*11 Sep. 12*
	Sardara Singh, Subadar-Major, 54th Sikhs	*3 June 16*
	Sarfaraz Ali Khan, Ris.-Maj., 12th Cav.	*21 Sep. 11*
	Satal Singh Bohora, Subadar, 44th G. R.	*9 Oct. 96*
	Sayyid Abbas, Subadar, 2nd Madras Lrs.	*18 Apl. 88*
	Shah Gul, Subadar, *late* 58th Rif.	3 June 13
	Shahzab Mir, Ressaidar, 11th Lrs.	*15 June 12*
	Shahibullah, Sub.-Maj., 40th Pathans	12 Dec. 11
	Shaikh Byleem, Subadar-Maj., 23rd Madras Infy.	*21 June 87*
	Shaikh Chaud, Subadar-Maj., 3rd S. and M.	*30 Aug. 12*
	Shaikh Hamid, Subadar, 2nd Madras Infy.	*7 Dec. 82*
	Shaikh Hussain Ali, 1st Cl. Senior Asst. Surgeon, I.S.M.D.	11 May 17
	Shaikh Husain-bakhsh, Subadar-Maj., 17th Infy.	*17 May 03*
	Shaik Hussain, Risaldar, 33rd Cav.	*3 June 11*
	Shaikh Jamal, Risaldar-Maj., 3rd Sind Horse	*21 Oct. 81*
	Shaikh Muhammad Afzal Husain, Ris.-Maj., 29th Lrs.	*16 Oct. 10*
	Shaikh Sillamon, Subadar, 17th Madras Infy.	*21 June 87*
	Shakir Khan, Subadar, 19th Punj.	*12 Dec. 11*
	Shamsuddin, Subadar-Maj., 21st Madras Prs.	*5 Jan. 99*
	Sharif Khan, Subadar, 21st Punj.	*12 Dec. 11*
	Sher Baz Khan, Risaldar-Maj., 10th Lrs.	*8 Feb. 04*
	Sher Muhammad Khan, Risaldar, 35th Horse	*1 Jan. 06*
	Sher Singh, Subadar, 34th Sikh Prs.	4 June 17
	Shiu Lal, Risaldar, 16th Cav.	*15 Dec. 05*
	Shiu Narayan Singh, Sub.-Maj., 123rd R.	18 Nov. 10
	Shiumbar Singh, Subadar-Major, I.O.M., 7th Rajputs	3 June 11
	Sikandar Khan, Subadar-Maj., 105th Mah. Lt. Infy.	*17 May 08*
	Singbir Gharti, Subadar, 4th G. R.	*29 July 04*
	Sohan Singh, Subadar, 72nd Punj.	4 June 17
	Sohan Singh, Ressaidar, 2nd Lrs.	3 June 18
	Somayya, Subadar, 81st. Prs.	*3 June 15*
	Suchait Singh, Risalda, 30th Lrs.	*18 Mar. 11*
	Sudama, Subaa, 25*th* Punj.	22 Mar. 17
	Sultan Jan Kh n, Risaldar-Maj., Army Remount Dept.	2 Feb. 17
	Sultan Sharif, Subadar, 81st Prs.	3 June 16
	Sundar, Subadar, Corps of Guides	*20 Sep. 01*
	Sundar Singh, Subadar-Maj., I.O.M, 45th Sikhs	22 Nov. 17
	Sundar Singh, Subadar, I. M. O. 14th Sikhs	*10 Mar. 11*
	Sundar Singh, Ris.-Maj., 31st Lrs.	*1 Mar. 11*
	Sundar Singh, Sub.-Maj., I.O.M., 82nd Punj.	*12 Dec. 11*
	Sundar Singh, Subadar, 32nd Sikh Prs.	*3 June 13*
	Sur Singh, Subadar-Major, 1 S. and M.	*2 June 05*
	Tagat Sing Rawat, Subadar-Maj., I.O.M., 1-39th Garh. Rif.	1 Jan. 15
	Taj Muhamamad, Subadar-Maj., 26 Bombay Infy.	*4 Nov. 99*
	Tejbir Thapa, Subadar-Maj. (Hony. Capt.), 2 G. R.	*28 July 99*
	Thaman Sing Mahat, Subadar, 5th G. R.	*12 Aug. 01*
	Thamman Singh, Subadar, 15th Sikhs	*20 June 07*
	Tilok Singh, Risaldar-Maj., 31st Lrs.	*17 May 01*
	Tiroovengadum, Subadar, 4th Madras Prs.	*29 May 87*
	Tota Singh, Subadar-Major, 1st Bn., 124th Infy.	3 June 18
	Tulja Ram, Jemadar, 2nd Horse	*24 July 06*
	Turra Baz Khan, Subadar-Major 51st Sikhs F. F.	3 June 18
	Tursam khan, Ressaidar, I.O.M., 27th Mule Corps	3 June 16
	Ujagar Singh, Subadar, 2nd Punj.	*7 Mar. 02*
	Umar Bakhsh, Subadar, 27th Bombay Infy.	*9 Nov. 90*
	Unkar Singh, Subadar, 20th Infy.	22 June 14
	Uttam Singh Subadar-Maj., 105 Mah. L. Infy.	*30 Aug. 05*
	Uttam Singh-Risaldar, Major, 10th Lrs.	3 June 18
	Venkatanarasayya, Sub.-Maj., 80th Carnatic Infy.	12 Dec. 11
	Wali Dad, Subadar, 84th Punj.	*3 Aug. 08*
	Wazir Khan, Sub.-Maj., 99th Infy.	13 Dec. 11
	Zard Ali, Subadar, 58th Rif.	*30 Aug. 08*

Number.	Names, Ranks, Corps, or late Corps.	Date of admission to the Order.
	ORDER OF BRITISH INDIA—*contd.*	
	Second Class, with the title of "Bahadur"—*contd.*	
	Supernumerary.	
	Abdul Ali, Subadar, 58th Rif.	1 Jan. 15
	Abdul Aziz Khan, Risaldar-Maj., 19th Lrs.	18 Aug. 16
	Abdul Aziz Khan, Subr., 1st S. and M.	18 Nov. 16
	Abdul Ghafur Khan, Risaldar-Maj., 1st Lrs.	*19 July 01*
	Abdul Shakur Khan, Subadar, 119th Infy.	19 July 18
	Afzal Khan, Subadar-Major, 82nd Punj.	4 Aug. 15
	Agdi Singh, Subr., 102nd Grs.	18 Nov. 16
	Agia Ram, Risaldar.-Maj., 16th Cav.	18 Nov. 16
	Ahmad Kutti, Subadar, 2nd Moplah Rif.	*1 Jan. 03*
	Allah Din, Jemadar, S. & T. Corps	
	Alam Khan, Ressaidar, S. and T. Corps	19 July 18
	Alladitta Khan, Subadar-Major, 109th Infy.	16 Dec. 16
	Ali Dost, Subadar, 106th Prs.	1 Jan. 18
	Arsla Khan, Subadar, 57th Rif.	1 Jan. 15
	Ashik Ali Khan, Subadar-Major, 17th Infy.	1 Jan. 03
	Amar Singh, Risaldar-Maj., 38th Horse	18 Aug. 16
	Amir Khan, Ressaidar, 2nd Mule Corps	4 Aug. 15
	Amir Muhammad Khan, Ressaldar, 25th Cav.	1 Jan. 18
	Awal Khan, Resaldar-Major, 4th Cav.	4 Aug. 15
	Bagga Singh, Ris-Maj, 36th Horse	19 Apl. 17
	Balaji Mule, Sabadar, 101 Grs.	*1 Jan. 03*
	Balbahadur Khattri, Subadar, 9th G. R.	1 Jan. 15
	Balbahadur Limbu, Subadar, 7th G. R.	24 July 15
	Baldeo Singh, Subadar, 8 Rajputs	*1 Jan. 03*
	Balwant Singh, Risaldar-Maj., 3rd Horse	18 Aug. 16
	Baryam Singh, I.O.M., Subadar, 3rd S. and M.	24 July 15
	Bhag Singh, Subadar-Major, I.O.M., 37th Dogras	19 July 18
	Bhagat Singh, Subabar, 51st Sikhs	3 June 16
	Balwant Rao Sawant, I.O.M., 117th Mah.	4 Aug. 15
	Bihari Singh, Subadar-Maj., 95th Infy.	*1 Jan. 03*
	Bije Sing Kandari, Subadar, 1-39th Garh. Rif.	*4 Aug. 15*
	Brahmdeo Singh, Subadar-Major, 99th Infy.	19 July 18
	Casim Ali, Subadar, 31st (Divl. Sig.) Coy	17 Sep. 15
	Chager Singh, Sub.-Maj., 37th Dogras	18 Nov. 16
	Chanda, Subadar-Major, 42nd Deoli Regt.	19 July 18
	Chanda Singh, Risaldar (*Actg.-Ris-Maj.*) 29th Lrs.	19 Apl. 17
	Chanda Singh, Subadar, 1st S. & M.	1 Jan. 18
	Chandrabir Thapa, Subadar, 9th G. R.	4 Aug. 15
	Chhetradhoj Limbu, Subadar, I.O.M., 10th G. R.	19 July 18
	Dalpat Singh, Subadar-Major, I.O.M., 6th Jat. Lt. Infy.	19 July 18
	Dayal Singh, Risaldar, Cav.	8 Mar. 18
	Dewa Singh, Risaldar, 9th Horse	*1 Jan. 03*
	Dhuman Khan, Jemadar, 53rd Sikhs	8 Aug. 15
	Dilawar Khan, Risaldar, 38th Horse	19 Apl. 17
	Dost Muhammad Khan, 19th Lrs.	27 July 17
	Diwan Singh, Subadar, 30th Punj.	24 Aug. 15
	Dula Singh, Subadar, Major, 48th Prs.	4 Aug. 15
	Faqirullah, Subadar-Major, 54th Sikhs	19 July 18
	Farman Ali, Subadar-Major, 92nd Punj.	4 Aug. 15
	Fateh Ali Khan, Subadar, Khan Sahib, 1st S. and M.	*4 Aug. 15*
	Fateh Sing Newar, Subadar-Maj., 2nd G. R.	1 Jan. 15
	Fateh Singh, Ris-Maj, 6th Cav.	19 Apl. 17
	Fateh Singh, Subadar, I.O.M., 24th Punj.	19 July 18
	Fazal Din, Subadar, Sappers and Miners, Ry. Bn.	3 June 18
	Fazal Hussain, Subadar, 93rd Infy.	3 June 16
	Fazl Shah, Subr., 1st S. and M.	18 Nov. 16
	Firoz Ali, Subadar-Maj., I.O.M., 1st Labour Corps	17 Aug. 17
	Firoz Khan, Subadar-Major, 56th Punj. Rif.	4 Aug. 15
	Gajraj Singh, Subadar-Maj., 7th Rajputs	*19 July 01*
	Gama Khan, Jamadar, S. and T. Corps.	19 July 18
	Gambhir Singh Gurung, Subadar-Major, 2-3rd G. R.	4 Aug. 15
	Ganpat Mahadeo, Subadar, 3rd S. and M.	4 Aug. 13
	Ganga Dat, Ris.-Maj., 2nd Lrs.	19 Apl. 17
	Gauri Charan, Subadar, I.O.M., 59th Rif.	17 Aug. 17
	Gauri Sankar Dube, Subadar, 1st S. and M.	4 Aug. 15
	Gauri Shankar, I.O.M., 1st Cl. Asst. Surgn., I.S.M.D.	1 Jan. 18
	Ghansar Singh, Risaldar 12th Cav.	1 Jan. 18
	Ghulam Muhammad, Ressaidar, 9th Mule Corps, S. & T.	17 Aug. 17
	Ghulam Dastaghir Khan, 29th Lrs.	19 Apl. 17
	Gopal Singh, Risaldar-Maj., 10 Lrs.	*1 Jan. 03*
	Govind Singh, Subr., 9th Bhopal Infy.	18 Nov. 16
	Gul Akbar, M. C., Subadar, 24th Punj.	4 Aug. 15
	Gunga Sahi, Risaldar, 1st Bombay Lrs.	*13 Aug. 97*
	Gurdayal Singh, Subadar-Maj., 23rd Punj. Pns.	*13 Aug. 97*
	Gurdit Singh, Subadar, Mountain Battery	3 June 18
	Gyan Singh, Subadar, 12th Prs.	17 Aug. 17
	Hamir Singh, Risaldar, 34th Horse	1 Jan. 17
	Hari Singh, Subr., 107th Prs.	18 Nov. 16

Number.	Names, Ranks, Corps, or late Corps.	Date of admission to the Order.
	ORDER OF BRITISH INDIA—*contd.*	
	Second Class, with the title of "Bahadur"—*contd.*	
	Supernumerary—*contd.*	
	Harnam Singh, Subadar, 56th Rif.	3 June 16
	Hashmat Ali, Ressaidar, I.O.M., 1st Mule Corps	30 June 16
	Hasan Khan, Subadar-Major, 8th Prs.	3 June 16
	Hashmat Dad Khan, Sub dar, 107th Prs.	4 Aug. 15
	Hayat Ali Beg, I. O. M., Risaldar 29th Lrs.	19 July 18
	Hira Singh, Risaldar-Maj., 30th Lrs.	18 Aug. 16
	Hira Singh, Ris.-Maj., 19th Lcs.	19 Apl. 17
	Husain Baksh Khan, Risaldar Major, 34th Horse	1 Jan. 17
	Husamuddin Khan, Risaldar-Maj., 23rd Cav.	17 Aug. 17
	Hussain Khan, Risaldar-Maj., 2nd Ls., Hyderabad Contingent	*13 Aug. 97*
	Imam Ali, Subadar, 57th Rif.	4 Aug. 15
	Imam Din, Subadar, 90th Punj.	19 July 18
	Inchha Ram, Subadar, 6th Jat Lt. Infy.	4 Aug. 15
	Ishar Singh, Risaldar, Corps of Guides	*1 Jan. 03*
	Ismail Khan, Subadar, 3rd S. and M.	4 Aug. 15
	Jaganath Singh, Subadar, 13th Madras Infy.	*13 Aug. 91*
	Jagandar Singh, Subadar-Major, 69th Punj.	4 Aug. 15
	Jagat Sing Rawat, I.O.M., Subadar, 39th Garh. R.	*1 Jan. 15*
	Jaimal Singh, Subr., 36th Sikhs	18 Nov. 16
	Jai Ram, Risaldar, Hodson's Horse	20 Apl. 17
	Jan Muhammad, Subadar, 69th Punj.	*24 Jan. 02*
	Jamit Singh, Subadar-Maj., 25th Punj. Infy.	*13 Aug. 97*
	Joseph, Subadar, 2nd S. and M.	19 July 18
	Jwala Singh, I.O.M., Subadar, 27th Jacob's Mtn. Batty.	4 Aug. 15
	Kaberay Karki, Subadar-Major, 8th G. R.	*1 Jan. 03*
	Kallaperumal, Subadar 61st Prs.	3 June 18
	Kalu Gurung, Subadar, I. O. M., 1st G. R.	19 July 18
	Kalu Khan, Subadar, 62nd Punj.	4 Aug. 15
	Kandasami, Subadar, 61st Prs.	*1 Jan. 03*
	Kapur Singh, Subadar, 1st Labour Corps	17 Aug. 17
	Karak Bahadur Thakur, I.O.M., Subadar, 3rd G. R.	4 Aug. 15
	Karim Khan, Subadar, 3rd S. & M.	1 Jan. 18
	Karm Singh, Subadar, 128th Prs.	19 July 18
	Karnabir Thapa, Subadar-Maj., 7th G. R.	*1 Jan 03*
	Kesar Singh, Subadar-Major, 29th Punj.	4 Aug. 15
	Kharak Singh, Subabar, I. O. M., 32nd Sikh Prs.	19 July 18
	Khwaja Muhammad, Jemadar, 19th Lrs.	20 Apl. 17
	Khem Singh, Subadar, 36th Sikhs	4 Aug. 15
	Khusal Khan, Risaldar, Lrs.	27 July 17
	Kishan Chand, 1st Cl. Asst. Surgn., I.S.M.D.	1 Jan. 18
	Konsal Singh, Risaldar, Horse	18 Mar. 18
	Konsal Singh, Risaldar, 20th Horse	19 July 18
	Krishna Bhosle, Subada -Major, 128th Prs.	19 Aug. 18
	Krishnasami, Subadar-Maj., 64th Prs.	17 July 17
	Kudrat Khan, Risaldar-Maj., 33rd Lt. Cav.	*24 Jan. 02*
	Labh Singh, Subadar-Major, 107th Prs.	4 Aug. 15
	Labh Singh, Subadar, 51st Sikhs	17 Aug. 17
	Labh Singh, Subadar-Major, 24th Punj.	19 July 18
	Lakha Singh, Risaldar, Lrs.	20 Apl. 17
	Lal Singh, Subadar, Punjabis	3 June 18
	Lehna Singh Subadar, 48th Prs.	24 July 15
	Lehra, Subadar, 32nd Div. Sig. Coy., S. and M.	19 July 19
	Mahabir Thakur, Subadar, 4th G. R	19 July 18
	Mahbub Khan, Risaldar, 21st Cav.	17 Aug. 17
	Maingha Singh, Subadar, I. O. M., 12th Prs.	19 July 18
	Makbul Khan, Risaldar-Maj., 8th Cav.	*13 Aug. 97*
	Makhar Singh, Kawar, 2-39th Garh. Rif.	4 Aug. 15
	Makhmad Azam, I.O.M., Subadar, 129th Baluchis	4 Aug. 15
	Malik Ahmadyar Khan, Risaldar, 18th Lrs.	1 Jan. 17
	Malikdad Khan, Risaldar, *Khan Sahib*, Aden Troop	15 Oct. 15
	Malik Khan Muhammad, Risaldar-Maj., 9th Horse	18 Aug 16
	Maluk Khan, Ressaidar, 4 Mule Corps, S. & T.	17 Aug. 17
	Mangal Singh, Subadar-Major, 53rd Sikhs	3 June 16
	Mardan Khan, Risaldar, Lrs.	27 July 17
	Mazhar Ali Khan, Ressaidar, 1st Lrs.	*1 Jan. 03*
	Mauladad Khan, Subadar, 126th Baluchistan Infy.	*1 Jan. 03*
	Mehar Bakash, Subadar Divisional Signal Company	3 June 18
	Mehar Singh, Subadar, 41st Dogras	4 Aug. 15
	Mewa Singh, Subadar-Maj., Prs.	1 Jan. 18
	Michael, Subadar, 2nd S. & M.	3 June 16
	Mir Alam Khan, 22nd Cavalry	30 Nov. 17
	Mota Singh, Subada -Major, I. D. S. M., 2nd Battalion 23rd Sikh Pioneers	
	Mir Muhiyuddin, Regimentdar, Com dt. Mysore Lrs. I.S.T.	3 June 18
	Muhammad Feroz Khan, Subabar, 31st Punj.	3 June 18
	Muhammad Abdul Hafiz, Subadar Prs.	3 June 18
	Muhammad Azim, I.O.M. Subadar, 2nd Punj.	1 Jan. 18
	Muhammad Hasan, Subadar, 106th Prs.	1 Jan. 18
	Mahammad Saiyid, Subabar, 82nd Punj.	19 July 18
	Nabi Ahmed Sadiki, Sr. Sub-Asst. Surgn 1st class., I.S.M.D.	8 Mar. 18

Number.	Names, Ranks, Corps or late Corps.	Date of admission to the Order.
	ORDER OF BRITISH INDIA—*contd.*	
	Second Class with the title of "Bahadur"—*contd.*	
	Supernumerary—*contd.*	
	Nain Singh Gurung, 1st Bn., 4th G. R.	3 June 16
	Nain Singh Chinwarh, Subadar, 2-39th Garh. Rifl.	4 Aug. 15
	Naine Lama, Subadar, 7th G. R.	19 July 18
	Nand Ram, Subadar-Major, 10th Jats	3 June 17
	Nanjappa, Subadar, 64th Prs.	19 July 18
	Narayan Singh, Subadar, Labour Corps	1 Jan. 18
	Narayan Singh, Ressaidar, 21st Cav.	19 July 18
	Natha Singh (II), I.O.M., Subadar, 34th Sikh Prs.	4 Aug. 15
	Nathe Khan, Ressaidar, 1st Lrs.	19 July 18
	Net Ram, Risaldar-Maj., 7th Lrs.	*13 Aug. 95*
	Niamat Ullah, 1st S. & M.	1 Jan. 18
	Nigahia Ram, Risaldar Major, 20th Horse	1 Jan. 17
	Munir Khan, *Khan Sahib*, Risaldar, C. Corps	15 Mar. 18
	Niyaz Muhammad Khan, Risaldar-Maj., 15th Ls.	*1 Jan. 03*
	Nur Muhammad Khan, Major, Rampur I.S.I.	3 June 18
	Padam Singh Thapa, Subadar, G. R.	1 Jan. 18
	Painda Khan, Jemadar (Ty. Ressaidar), S. & T. Corps	17 Aug. 17
	Pande Khan, Subadar-Maj., Ry. Bn.	1 Jan. 18
	Parbhudan Singh, Subadar, 13th Rajputs.	3 June 16
	Parmod Singh, Subadar, 52nd Sikhs	*3 June 15*
	Partab Singh, Risaldar, 13th Lrs.	17 July 17
	Pertab Singh, Subadar, 36th Sikhs	19 July 18
	Pir Baksh, 1st-Cl. Assistant Surgeon, *Khan Saheb*, I.S.M.D.	1 Jan. 18
	Pratab Singh, Risaldar, 30th Lrs.	27 July 17
	Prem Singh, Risaldar, 20th Horse	18 Aug. 16
	Qasim Ali, Subadar, 31st (Divl. Sig.) Coy.	17 Sep. 15
	Rab Nawaz, Ressaidar, 11th Lrs.	21 Aug. 15
	Rafi-ud-din Khan, 2nd-Cl. Senr. Sub-Assistant Surgn.	18 Aug. 16
	Rajpal Rai, Subadar-Maj., 7th G. R.	1 Jan. 18
	Raj Taleb, I.O.M., Subadar, 58th Rif.	4 Aug. 15
	Ramasami, Subadar, 2nd S. and M.	19 July 18
	Ram Kishan Bisht, Subadar-Maj., 2-39th Garh. Rif.	*1 Jan. 03*
	Ram Rup Singh, Subadar, I.O.M., 1st S. and M.	19 July 18
	Ram Singh, Subadar-Maj., 67th Punj.	*1 Jan. 03*
	Ram Singh, Risaldar, 9th Horse	*20 Apl. 17*
	Sada Singh, Jemdr., 1st S. and M.	4 Aug. 15
	Saddha Singh, Risldr., 4th Cav.	4 Aug. 15
	Sadik Muhammad Khan, I.O.M., Risaldar, 36th Horse.	19 July 18
	Sajjut Khan, Ressaidar, 33rd Light Cav.	*1 Jan. 03*
	Santuji, Bapuji, Capt. Indrose Transport Corps I.S.T.	3 June 18
	Sardar Khan, Subadar, 3rd S. and M.	19 July 18
	Saudagar Singh, Subadar, 47th Sikhs	4 Aug. 15
	Sawan Singh, I.O.M., Subadar, 24th Punj.	14 July 15
	Senbir Gurung, I.O.M., Subadar, 4th G. R.	4 Aug. 15
	Santo Singh, Subadar, 22th Mtn. Btty.	1 Jan. 18
	Sarabjit Gurung, Subadar, 2nd G. R.	4 Aug. 15
	Shadim Khan, Subadar, 30th Mtn. Batty.	3 June 16
	Shah Singh, Subadar-Maj., 14th Sikhs	17 Aug. 17
	Shaikh Budruddin, Ressaidar, S. & T. Reserve	7 Apl. 17
	Shaikh Ismail, Subadar, 94 Infy.	*1 Jan. 03*
	Shaikh Yasin, Subadar, 116th Mar. Lt. Infy.	4 Aug. 15
	Shamsher Gurung, Subadar, 8th G. R.	*4 Aug. 15*
	Sher Baz, Subadar, 27th Mtn. Batty.	25 Nov. 16
	Sher Baz, Subadar, S. & M.	1 Jan. 18
	Sher Gul, Subadar, 51st Sikhs	19 July 18
	Sher Shah, Subadar, 27th Punj.	19 July 18
	Shib Lal, Subadar, 6th Jat Lt. Infy.	4 Aug. 15
	Shikari, Subadar-Maj.	1 Jan. 18
	Shiuratan Singh, Subadar-Major, 2nd Rajput Lt. Infy.	4 Aug. 15
	Sri Bhadhur Singh, Subadar, 120th Rajputana Infy.	4 Aug. 15
	Subha Ram, Ressaidar, Ls.	8 Mar. 18
	Sundar, Subr.-Maj., 28th Punj.	18 Nov. 16
	Sundar Singh, Subadar-Major, 89th Punj.	4 Aug. 15
	Sundar Singh, Subadar-Majar, I.O.M., 45th Sikhs	19 July 18
	Sundar Singh, Subadar, 41st Dogras	4 Aug. 15
	Sundar Singh, Subadar, 1st S. and M.	4 Aug. 15
	Sundar Singh, Subadar, 29th Mtn. Batty.	3 June 15
	Sundar Singh, Subadar, 62nd Punj.	19 July 18
	Sundar Singh, Risaldar, 18th Lrs.	1 Jan. 17
	Suraj Singh, I.O.M., Risaldar, 2nd Lrs.	19 July 18
	Surbir Karki, 9 G. R.	4 Aug. 15
	Tagat Singh Rawat, Subadar-Maj., I.O.M.	
	Tawakil Khan, Ressaidar, 18th Ls.	30 Dec. 16
	Tega, Subadar, 37th Dogras	3 June 16
	Thakur Moti Singh, I.D.S.M., Lieut-Col., Bikaner Camel Corps, I. S. Troops	

Number.	Names, Ranks, Corps or late Corps.	Date of admission to the Order.

ORDER OF BRITISH INDIA—*concld.*

Second Class, with the title of "Bahadur"—*concld.*

Supernumerary—*concld.*

Number.	Names, Ranks, Corps or late Corps.	Date of admission to the Order.
	Thakur Jor Singh, Risaldar, 32nd Lrs.	19 July 18
	Thakur Kishen Singh, I.O.M., Major Bikanir Camel Corps I.S.T.	3 June 18
	Thakur Mul Singh, Risaldar Major, 26th Lt. Cav.	16 Dec. 16
	Thakur Singh, Subadar-Major, M.C., 47th Sikhs	19 July 18
	Thakur Singh, Subadar, 91st Punj.	19 July 18
	Tikaram Majhi, Subadar, 1st Bn., 9th G. R.	3 June 16
	Tulsi Ram Gharti, Subadar-Maj., 2nd G.	17 Aug. 17
	Uma Rawat, Subadar, 119th Infy.	4 Aug. 15
	Umar Din, Subadar-Major, 125th Rif.	4 Aug. 15
	Umar Khan, Ressaidar, (Ty. Subadar) S. & T. Corps	17 Aug. 17
	Veerappen, Subadar, 61st Prs.	1 Jan. 18
	Viraraghavulu, Subadar, 64th Prs.	1 Jan. 18
	Wazir Khan, Subadar-Maj., 5th Punj. Infy.	*13 Aug. 97*
	Wazir Singh, Subadar, 27th M. Btry.	1 Jan. 18
	Zaman Khan, Subadar, 129th Baluchis	4 Aug. 15
	Zaman Ali, Jemadar, S. & T. Corps	1 Jan. 18

Number.	Names, Ranks, Corps or late Corps.	Date of admission to the Order.
	ORDER OF BRITISH INDIA.	
	Senior Sub-Assistant Surgeons.	
	Permanent.	
	Bishun Nath, 1st class, Bengal Estabt.	*11 Apl. 02*
	Fasl Ilahi, 1st class, Bengal Estabt.	*7 Aug. 10*
	Muhammad Khan, *Khan Sahib,* 1st class, Bengal Estabt.	2 Jan. 15
	Nasiruddin Khan, Khan Sahib, 1st class, Bengal Estabt.	*12 Dec. 11*
	Ram Singh, 1st class, Bengal Estabt.	22 June 14
	Saiyid Ali Naki, Khan Sahib, 1st class, I.S.M.D.	2 Apl. 14
	Shaikh Kabirud-din, Khan Bahadur, 1st class, Bengal Estabt.	*1 Apl. 02*
	Sher Muhammad, *Khan Sahib,* 1st class, Bengal Estabt.	4 June 17
	Supernumerary.	
	Alah-banda, 1st class, Bengal Estab.	3 June 16
	Barkat Ram, 1st class, I. S. M. D., Bengal Estabt.	19 July 18
	Debi-ditta Saithi, 1st class, Bengal Estabt.	*1 Jan. 05*
	Karam Chand, 1st class, I. S. M. D., Bengal Estabt.	19 July 18
	Niaz Ali Khan, 1st class, I. S. M. D., Bengal Estabt.	19 July 18
	Pati Ram, Rai Bahadur, 1st class, Bengal Estabt.	*1 Jan. 03*
	Tilok Singh, 48, 1st class, I. S. M. D.	4 Aug. 15
	Williamson, J., 1st class, I. S. M. D., Madras	1 Jan. 17

ABSTRACT.

No. on roll.	Permanent Establishment.	Vacancies.	Supernumeraries.	Remarks.
Combatant Ranks	...	...		
Sub-Assistant Surgeons	...	...		

Number.	Names and Rank and Corps.	Date of admission to or promotion in the Order.
	ORDER OF BRITISH INDIA.	
	IMPERIAL SERVICE TROOPS.	
	First Class, with the title of "Sardar Bahadur."- 16.	
	Abdul Ghani, Inspr.-General, Gwalior State Forces	*2 Jan. 11*
	Daud Khan, Comdt., Alwar Lrs.	*1 Jan. 04*
	Farman Ali Khan, Maj.-Genl., Adj. Genl., Jammu and Kashmir Army	12 Dec. 11
	Girdhar Singh, Lieut.-Colonel, Commandant, Bharatpur I. S. Infy.	12 Dec. 11
	Gurnam Singh, Comdr.-in-Chief, Jhind State	1 Jan. 03
	Hardas Singh, Commander-in-Chief, Naba State	*12 Dec. 11*
	Hashim Nawaz Jang, Comdt., 2nd Lrs.	1 Jan. 04
	Khajur Singh, Maj.-Genl., Genl. Officer Comdg. 2nd Brigade Kashmir Troops, Satwari	12 Dec. 11
	Mirza Karim Beg, M. V. O. (Hony. Capt.) Comdt., Bhopal Lrs.	1 Jan. 03
	Nand Singh, Comdt., Patiala Lrs.	1 Jan. 03
	Natha Singh, *Sardar Bahadur,* Comdt., Alwar Infy..	1 Jan. 03
	Nehal Singh, Maj.-Genl., late Comdg. 2nd Bde., Kashmir I. S. Troops	*2 Jan. 11*
	Sundar Singh, *Sardar,* Comdt., Patiala Infy.	1 Jan. 03
	Sunit Singh, Lieut.-Colonel, Kashmir Lrs.	*28 Mar. 08*
	Suraj Pershad, Comdt., Gwalior Transport Corps	*13 Aug. 97*
	FRONTIER MILITIA CORPS AND MILITARY POLICE.	
	First class, with the title of "Sardar Bahadur."	
	Amar Singh, (Hony. Capt.) Subadar-Major, I.O.M., *Rai Bahadur, Burma Military Police (A. D. C. to H. H. the L.-G., Burma)*	*22 June 14*
	Azad Khan, Subadar-Maj., *Khan Sahib,* Frontier Constabulary	26 Oct. 17
	Harnam Singh, (Hony. Capt.) Subadar-Major, Burma Military Police	*3 June 17*
	Jaswas Khan, Subadar-Maj., Khyber Rif.	3 June 15

Number.	Names and Ranks and Corps.	Date of admission to or promotion in the Order.

ORDER OF BRITISH INDIA.

FRONTIER MILITIA CORPS AND MILITARY POLICE—*concld.*

First class with the title of "Sardar Bahadur"—*concld.*

	Mir Ahmad, Subadar, Khyber Rif. Rif.	12 Dec. 11
	Mir Hussan, Sub.-Maj., Southern Waziristan Militia	*6 Jan. 01*
	Muhammad Ali Khan, Subadar-Major, I.O.M., *Khan Sahib,* Khyber Rif.	12 Dec. 11
	Muhammad Husain, Subadar, Kurram Militia	12 Dec. 11
	Saran Singh, Subadar-Major, Burma Military Police	14 Nov. 15
	Sohawa Singh (Hony. Capt.) Subadar-Major, I.O.M., *Rai Bahadur, Burma Military Police*	*3 June 13*
	Tor Khan, I.O.M., Subadar-Major, N. W. Militia	14 June 21

Second class, with the title of "Bahadur."—22.

	Abhiram Thapa, Subadar-Major, Assam Military Police	30 Aug. 93
	Ahmad Mir, *Khan Sahib,* Zhob Militia, Risaldar-Major	3 June 16
	Anokh Singh, Subadar-Major, Burma Military Police	3 June 11
	Arjun Rai, Subadar-Major, Assam Military Police	*12 Dec. 11*
	Bolkhan, Subadar-Major, Frontier Constabulary	3 June 16
	Daniel, E., Subadar, Burma Military Police	3 June 16
	Fateh Mahomed Khan, Subadar-Major, Khan Bahadur, Burma Military Police	*3 June 13*
	Haidar Khan, Subadar, Khyber Rif.	4 June 17
	Hans Ram, Subadar, Maj., *Rai Bahadur,* Burma Mily. Police	26 Oct. 17
	Jamaluddin, Sub-Maj., Assam Mily. Police	14 Nov. 15
	Jangsher Gurung, Subadar, *Rai Bahadur,* Burma Military Police	22 June 14
	Jasbahadur Thapa, Subadar-Major, I.O.M., *Rai Bahadur,* Burma Military Police	14 June 12
	Jawas Khan, Subadar-Major, I.O.M., Khyber Rif.	12 Dec. 11
	Jhuttan Singh, Subadar-Major, Rai Bahadur, Burma Military Police	*3 June 13*
	Kochi Khan, Subadar, Khyber Rif,	14 June 12
	Khudadad Khan, Subadar, Burma Military Police	3 June 15
	Nand Lall Thapa, Subadar, I.O.M., *Burma Military Police*	*12 Dec. 11*
	Pall Singh, *Rai Bahadur,* Sub-Maj., Burma Mily. Police	21 Jan. 15
	Pir Muhammad Khan, Subadar-Major, Burma Military Police	3 June 15
	Ram Narain, Subadar-Major, Burma Military Police	3 June 15
	Ranbahadur Suba, *Rai Bahadur,* Subadar-Major, Burma Military Police	12 Dec. 11
	Sadiq Khan, Subadar, Khan Bahadur, Burma Military Police	*22 June 14*
	Sanjai Suba, Subadar. Assam Military Police	*3 June 14*
	Sardara Sing, Subadar-Major, Burma Military Police	*12 Dec. 11*
	Sher Singh, Subadar, Burma Military Police	22 June 14

Supernumerary.

	Bagbir Yakha, Subadar, Burma Military Police	3 June 16

IMPERIAL SERVICE TROOPS.

First class, with the title of "Sardar Bahadur."

Supernumerary.

	Bakshish Singh, Maj.-Gen., Mily. Secy. to H. H. the Maharajah of Patiala	14 June 12
	Bilkere Chamaraj Urs, Comdt., Mysore I. S. Lrs.	3 June 16
	Daud Khan, Colonel, Comdt., Alwar I. S. Infy.	3 June 16
	Dharm Singh, Bahadur, I.O.M., Maj., Comdt. Kashmir Mtn. Batty.	27 July 17
	Din Dayal, Lieut.-Colonel, Comdt., I. S. Infy., Bikaner	3 June 13
	Ganesh Lal, *Major, Bahadur,* Bharatpur Infy.	4 Aug. 15
	Haiderali Khan, Lt.-Col., *Bahadur,* Commdg. 2nd Kashmir I. S. R.	28 Apl. 17
	Kanwar Jeoraj Singh, Bahadur, Lt.-Col., Comdt. Bikaner Camel Corps	27 July 17
	Kishen Singh, Lt.-Col., *Bahadur*	14 Apl. 17
	Lutf Ali Khan, Maj., *Bahadur*	14 Apl. 17
	Mahomed, Baksh, Khan, *Bahadur,* Maler Kotla Sappers	1 Jan. 18
	Muhammad Afzal Khan, Lieut.-Col., Comdt., Bhawalpur Camel Corps	4 June 17
	Muhammad Ramzan Khan, Col., Comt., 2nd Patiala I. S. Infy.	4 June 17
	Natha Singh, *Major-General Bahadur,* Commandant, Jhind Infy.	4 Aug. 15
	Thakur Pratap Singh, Bahadur, Comdt. Jodhpur Lrs.	27 July 17

Second Class, with the title of "Bahadur."

	Abdul Majid Khan, Jemadar, Bahawalpur Lrs	*13 Aug. 97*
	Abdul Qayum Khan, Maj., Comdt., Bhopal I. S. Lrs.	4 June 11
	Balwan Singh, Maj., Superintendent, Mily. Stores, Jammu	12 Dec. 13
	Bhagwan Singh, Maj.-Genl. G.O.C., Kashmir State Troops	*1 Jan. 01*
	Bhur Singh, Colonel, *Sardar Bahadur,* Comdt., Alwar I. S. Lrs.	12 Dec. 14
	Dharm Singh, Maj., I.O.M., Commdt., No. 1st Kashmir I. S. Arty.	22 June 17
	Didar Singh, Risaldar-Maj., Jhind Lrs.	*13 Aug. 97*
	Durga Sing, Lt.-Col., I.O.M., 3rd Kashmir I. S. Rif.	12 Dec. 11

Number.	Names and Ranks and Corps.	Date of admission to or promotion in the Order.

IMPERIAL SERVICE TROOPS—*concld.*

Second class with title of "Bahadur"—*concld.*

	Ganeshi Lal, Major, Bharatpur I. S. Infy.	14 June 12
	Gurbaksh Singh, Asst. Comdt., Bikaner Camel Corps	27 Mar. 08
	Harnam Singh, Sardar Comdt., Faridkot I. S. Sappers	12 Dec. 11
	Kala Singh, Subadar, 1st Patiala Infy.	28 July 16
	Kanhai Parshad Dube, Asst. Comdt., Sirmur State I. S. Sappers	14 June 12
	Lachman Singh, Subadar, Sirmur Sappers	1 Jan. 04
	Lutf Ali Khan, Major, Comdt., Indore State I. S. Transport Corps	14 June 14
	Mahomed Bakhsh Khan, Asstt. Comdt., Maler-Kotla Sappers	1 Jan. 03
	Mehr Muhammad Khan, Sardar, Maler-Kotla Sappers	*20 May 98*
	Quadar Baksh Khan, Comdt., Maler Kotla I. S. Sappers	12 Dec. 11
	Rawal Sri Dhan Singhji, Comdt., Bhavnagar Lrs.	*13 Aug. 97*
	Samandar Khan, Maj.-Genl. G. O. C., 1st Brigade Kashmir, I. S. Troops, Gilgit	28 May 09
	Sardar Kishen Singh, Comdt., Bharatpur I. S. Transport Corps	12 Dec. 11
	Sardar Natha Sing, Comdt., Jind I. S. Infy.	2 Jan. 11
	Sardar Pooran Singh, Commander-in-Chief, Kapurthala State I. S. Troops	14 June 12
	Sher Singh, Sardar, Nabha Infy.	*10 May 98*
	Shripatrao Ingley, Lt.-Col., Comdt., 4th Gwalior I. S. Infy.	2 Jan. 11
	Thakur Dalip Singh, Commandant, Tehri I. S. Sappers	1 Jan. 14
	Thakur Dip Singh, Rao Bahadur, Adjt., Impl. Cadet Corps	*13 Aug. 97*
	Thakur Partap Singh, Comdt., 1st Sirdar Risala, Jodhpur	12 Dec. 11

Supernumerary.

	Achhroo Singh, Colonel in the Patiala I. S. Troops	1 Jan. 19
	Abdul Karim Khan, Maj.	28 Apl. 17
	Agar Singh (I), Risaldar, I. S. Troops	10 Apl. 17
	Alif Khan, Subadar in No. 1st Kashmir Mtn. Btry. I. S. Troops	1 Jan. 18
	Aman Singh, Sqdn. Comdr., I. S. Troops	10 Apl. 17
	Anop Singh, Squadron Comdr., Jodhpur Lrs.	27 July 17
	Baldev Singh, Lt.-Col.	28 Apl. 17
	Barkat Khan, Maj., 1st Company, Maler Kotla I. S. Sappers	17 Aug. 17
	Bhawani Singh, Maj., Indore Mounted Escort	17 Aug. 17
	Bhagwan Singh, Major in the Tehri I. S. Troops	1 Jan. 18
	Bishan Singh, Major in the Faridkot I. S. Troops	1 Jan. 18
	Bulwant Singh, Major in the Patiala I. S. Lancers	19 July 18
	Durga Datt Mahey, Maj.	28 Apl. 17
	Dattaji Rao Rakda, Major in the 3rd Gwalior I. S. Troops	1 Jan. 18
	Elijah Abrahim, 2nd Cl. Sub.-Asst. Surgn., I.S.M.D.	22 Feb. 18
	Ganpat Rao, Nimbalkar, Lt.-Col., Comdt. Gwalior Infy.	27 July 17
	Girdhari Sing, Commandant, 4th Gwalior Infy	3 June 16
	Hunumunth Rao Powar, Risaldar, Mysore State Lrs.	27 July 17
	Isa Khan, Subadar, Asst. Comdt., Bahawalpur Mtd. Rif.	3 June 16
	Kallan Singh, Lt.	28 Apl. 17
	Kanwar Jeoraj Singh, Major, Bikaner Camel Corps	3 June 16
	Mahomed Azmatullah, Capt., 2nd Hyderabad Lrs.	3 June 16
	Mehr Bakhsh, Subadar, 1st Company, Maler Kotla, I. S. Sappers	17 Aug. 17
	Mit Singh, Lt.-Colonel in the Maler Kotla I. S Troops	1 Jan. 18
	Muhammad Din, Sub-Asst. Surgn. (Jemadar), Kashmir I. S. Rif.	4 June 17
	Narain Singh, *Sardar Bahadur,* Captain in the Bharatpur I. S. Troops	1 Jan. 18
	Nihal Singh, Maj., *Sardar*	28 Apl. 17
	Paney Singh, Captain in the Jodhpur I. S. Lancers	19 July 18
	Sardar Gurbuksh Singh, Comdt., Patiala Infy.	3 June 16
	Sher Singh, Maj., Tehri Sappers	17 Aug. 17
	Thakur Hukam Singh, Major in the Jaipur I. S. Trans. Corps	19 July 18
	Thakur Sheonath Singh, Maj., Bikaner Camel Corps	27 July 17

Note.—The names shown in *italics* are those of Indian Officers who are not actually on Imperial Service rolls.

THE INDIAN ORDER OF MERIT (I.O.M.).

Note.—The names in *italics* are those of members on the pension establishment.

1st Class.

Corps.	Names and Rank.	In what rank admitted.	Date of admission to the Order, and subsequent promotion.	Authority. G. G. O. or A.D. Notfn.
26th Horse	**Wazir Singh**, Jemadar	Jemadar	18 Sept. 17	No. 115, 19 Jan 18
C. of Guides	**Alam Khan**, Subadar-Major, *Bahadur*	Subadar-Major	8 Oct. 15	No. 102, 28 Jan. 16
3rd S. and M.	**Gurmukh Singh**, Naik, 3108	Naik	2 Mar. 16	No. 1387, 18 Nov. 16
9th Infy.	**Jai Lal**, Jemadar	Jemadar	21 Apl. 17	No. 8, 1 Jan. 18
4th Sikhs	*Bhagwan Singh, Subadar-Major (Hony. Capt.), Sardar Bahadur*	*Jemadar*	*11 Dec. 97*	*No. 100, 27 Jan. 96*
	Jallal, Jemadar I. O. M.			
24th Punj.	*Yasin Khan, Subadar-Maj.*, C.I.E., *Sardar Bahadur, Hony. Capt.*	*Subadar-Maj.*	*27 July 97*	*No. 276, 9 Mar. 00*
32nd Sikh Prs.	**Wassawa Singh**, Subadar, *Sardar Bahadur*	Subadar	6 May 04	No. 691, 15 July 04
5th Sikh Prs.	**Bir Singh**, *Havildar*, 2103	Havildar	16 June 17	No. 945, 16 June 16
47th Sikhs	**Sucha Singh**, Subadar	Subadar	23 July 15 17 Aug. 17	No. 1359, 17 Aug. 17
51st Sikhs.	*Hussain Shah, Jemadar*	*Jemadar*	*10 June 97*	*Nos. 865, 6 Aug. 97 and 1029, 10 Sep. 97*
53rd Sikhs	*Mit Singh, Subadar-Maj. (Hony. Capt.), Sardar Bahadur (Hony. A. D. C. to H. E. the Viceroy)*	Subadar	18 Nov. 02	No. 381, 24 Apl. 03
56th Punj.	**Ayub Khan**, Jemadar	Jemadar	4 June 17	L.G. P. 5493, 4 June 17
58th Rif.	**Suhel Singh**, Subadar	Subadar	4 May 15	No. 563, 11 June 15
58th Rif.	**Muhammad Arbi**, Subadar	Subadar	19 Nov. 17	No. 1178, 3 June 18
59th Rif.	*Kamal Khan, Subadar*	*Jemadar*	*18 Nov. 02*	*No. 381, 24 Apl. 03*
117th Mah.	**Balwant Rao Sawant**, Subadar-Maj., *Bahadur*	. . . . .	17 Aug. 17	No. 1357, 17 Aug. 17
122nd Infy.	**Ksha Rawat**, Naick, 22	Do.	14 May 12	No. 839, 30 Aug. 12
127th Baluchis	**Sarbuland**, Subadar (*Actg. Sub-Maj.*)	Subadar	5 Aug. 17	No. 8, 1 Jan. 18
6th G. R.	*Singiah Lama, Havildar, 2754*	*Naick*	*14 Mar. 91*	*No. 862, 4 Sep. 91*
8th G. R.	*Kaberaj Kharki, Subadar-Major, Bahadur*	*Subdr.-Maj.*	*6 July 04*	*No. 957, 7 Oct. 04*
Do.	*Nawal Sing Rana, Subadar-Maj., Sardar Banauur Hony. Capt.*	*Subadar*	*6 July 04*	*No. 957, 7 Oct. 04*
I. S. M. D.	**Pargan Singh**, I.O.M., Sub-Asst. Surgn., 1230	. . . .	8 Jan. 16	No. 524, 12 May 16
3rd Kashmir Rif.	**Durga Singh**, Lieut.-Col., *Bahadur*	Lieut.-Col.	10 Sept. 15	No. 874, 10 Sept. 15
Burma Military Police.	*Amar Singh*, Subadar-Major, *Sirdar Bahadur ; Rai Bahadur. (Hony. Capt.)* (A.-D.-C. to His Honour the Lieutenant-Governor of Burma)	Do.	2 Feb. 91	No. 1180, 18 Dec. 01
3rd G.	**Bahadur Thapa**, Subadar, I.O.M.	Subadar	11 Apl. 18	No. 2536, 26 Oct. 18

2nd Class.

Corps.	Names and Rank.	In what rank admitted.	Date of admission to the Order, and subsequent promotion.	Authority.
2nd Lrs.	*Mazhar Ali Khan, Ressaidar*	*Sowar*	*11 Dec. 79*	*No. 432, 10 June 87*
Do.	*Imam Ali Khan, Jemadar*	*Lance Duffadar*	*12 Aug. 00*	*No. 115, 8 Feb. 01*
Do.	**Faiz Muhammad Khan**, Risaldar	Risaldar	27 Aug. 15	No. 819, 27 Aug. 15
Do.	**Muhammad Umar Faruk Khan**, Jemadar	Jemadar	17 Aug. 17	No. 1359, 17 Aug. 17
Do.	**Shahzad Khan**, Sowar, 2283	Sowar	17 June 18	No. 2536, 28 Oct. 18
3rd Lrs.	**Suraj Singh**, Risaldar	Risaldar	3 June 16	No. 849, 29 July 16
3rd Skinner's Horse.	*Diwan Singh, Duffadar, 231*	*Sowar*	*1 Oct. 97*	*No. 1313, 26 Nov. 97*
Do.	*Sukhdeo Singh, Sowar, 345*	*Do.*	*1 Oct. 97*	*No. 1313, 26 Nov. 97*
Do.	*Shaikh Jamal, Risaldar-Major*	*Ressaidar*	*28 July 80*	*No. 60, 4 Feb. 81*
4th Cav.	**Puran**, Acting Lance-Dafadar, 2205	Actg. Lnce.-Daffadar.	6 Jan. 16	No. 524, 12 May 16
Do.	**Khazan Singh**, Lce.-Dafadar, 2055	Lce.-Dafadar	24 May 16	No. 1435, 25 Nov. 16
6th Cav.	**Karam Singh**, Lance-Dafadar, 2386	Lance-Dafadar	4 Feb. 18	No. 2185, 20 Sept. 18
7th Lrs.	*Kalandar Khan Risaldar-Maj., Bahadur*	*Jemadar*	*9 May 87*	*No. 755, 23 Sept. 87*
Do	*Gurdutt Singh, Ressaidar*	*Sowar*	. . .	*No. 49, 14 Feb. 98*
Do.	**Lal Singh**, Kot-Dafadar, 2762	Kot Duffadar	10 Dec. 15	No. 1244, 10 Dec. 15
Do.	**Kehar Singh**, Sowar, 3365	Sowar	17 Aug. 17	No. 1357, 17 Aug. 17
Do.	**Muhammad Akbar Khan**, Risaldar-Major, *Sardar Bahadur*	Risldr.-Maj.	22 Nov. 15	No. 1631, 19 July 18
9th Hodson's Horse.	*Sumandar Khan, Duffadar, 1453*	*Lce.-Duffadar*	*20 Mar. 85*	*No. 298, 29 May 85*
Do.	*Kesur Singh, Trumpeter, 1139*	*Trumpeter*	*20 Mar. 85*	*No. 298, 29 May 85*
Do.	*Poorun Singh, Duffadar, 1133*	*Lce.-Duffadar*	*20 Mar. 85*	*No. 298, 29 May 85*
Do.	*Abdulla Khan, Sowar*	*Sowar*	. . .	*No. 298, 29 May 85*
Do.	*Rahmat Khan, Sowar, 2067*	Do.	*12 Sep. 97*	*No. 935, 19 Aug. 98*
Do.	*Mir Jaffar Khan, Risaldar-Major, Sardar Bahadur*	*Kot-Duffadar*	*18 Oct. 97*	*No. 1423, 24 Dec. 97*
Do.	**Abdullah Khan**, Sowar, 3456	Sowar	25 Nov. 16	Lon. Gaz., 25 Nov. 16
Do.	**Amir Muhammad**, Duffadar, 2782	Duffadar	26 Apl. 17	Lon. Gaz., 26 Apl. 17

Corps.	Names and Rank.	In what rank admitted.	Date of admission to the Order.	Authority.
	THE INDIAN ORDER OF MERIT—*contd.* 2nd Class—*contd.*			G. G. O. or A. D. Notfn.
9th Hodson's Horse.	**Hakim Singh**, Duffadar, 2764	Duffadar	26 Apl. 17	Lon. Gaz., 26 Apl. 17
Do.	**Sarfaraz Khan**, Duffadar, 2967	Duffadar	26 Apl. 17	Lon. Gaz., 26 Apl. 17
10th Lrs.	**Hayat Muhammad**, Sowar, 2793	Sowar	25 Nov. 16	Lon. Gaz., 25 Nov. 16
11th Lrs.	*Dayal Singh, Duffadar*	*Duffadar*	*27 Apl. 59*	*No. 976, 5 July 59*
Do.	*Jawala Singh, Kot-Duffadar, 2034*	*Sowar*	*22 Aug. 95*	*No. 184, 16 Feb. 96*
Do.	**Prem Singh**, Kot-Duffadar, 2417	Do.	27 July 97	No. 49, 14 Jan. 98
11th Lrs.	*Arur Singh, Sowar, 1920*	*Sowar*	*27 July 97*	*No. 49, 14 Jan 98*
Do.	**Bhagwan Singh**, Jemadar	Jemadar	23 May 18	No. 2536, 26 Oct. 18
Do.	*Pratab Singh, Lce.-Daffadar, 1842*	*Lce.-Duffadar*	*1 Aug. 97*	*No. 49, 14 Jan. 98*
Do.	*Phangan Singh, Sowar, 2349*	*Sowar*	*2 Aug. 97*	*No. 49, 14 Jan. 98*
Do.	*Lal Budin, Lance-Duffadar, 2070*	*Do.*	*2 Aug. 97*	*No. 49, 14 Jan. 98*
Do	*Indar Singh, Duffadar, 2472*	*Do.*	*16 Sep. 97*	*No. 232, 25 Feb. 98*
Do.	*Sant Singh, Duffadar, 2171*	*Lce.-Duffadar*	*17 Sep. 97*	*No. 232, 25 Feb. 98*
Do.	**Jit Singh**, Lance-Dafadar, 2889	Do.	29 Oct. 15	Lon. Gaz., 29 Oct. 15
Do.	**Ganga Singh**, Actg. Lance-Dafadar, 2614.	Actg. Lance-Duffadar.	29 Oct. 15	Lon. Gaz., 29 Oct. 15
12th Cav.	**Narain Singh**, Acting Lnce.-Dafadar, 650.	Do.	7 Feb. 16	No. 583, 26 May 16
Do.	**Partab Singh**, Dafadar, 455	Daffadar	7 Feb. 16	No. 583, 26 May 16
13th Lrs.	*Sewa Singh, Duffadar, 171*	*Do.*	*9 Aug. 97*	*No. 394, 15 Apl. 98*
Do.	**Fazal Hussain**, Risaldar-Maj.	Do.	9 Aug. 97	No. 394, 15 Apl. 98
Do.	*Ghulam Muhiyuddin, Lce.-Duffadar, 1103*	*Lce.-Duffadar*	*9 Aug. 97*	*No. 394, 15 Apl. 98*
Do.	*Hira Singh, Sowar, 698*	*Sowar*	*9 Aug. 97*	*No. 394, 15 Apl. 98*
14th Do.	*Kunhaiya Lall, Ressaidar*	*Duffadar*	*11 Dec. 79*	*No. 131, 27 Feb. 80*
Do.	**Rati Ram**, Risaldar	Jemadar	5 Sep. 15	No. 102, 28 Jan. 16
Do.	**Dhan Singh**, Sowar, 2664	Sowar	5 Sep. 15	No. 102, 28 Jan. 16
Do.	Hargyan Singh, Risaldar	Risaldar	2 Apl. 17	Force "D" Order No. 68, 2 Apl. 17.
Do.	**Pahlad Singh**, Lance-Dafadar, 2587	Lance-dafadar	17 Aug. 17	No. 1359, 17 Aug. 17
15th Do.	*Rabb Nawaz Khan, Jemadar, Khan Bahadur*	*Jemadar*	*3 Mar. 95*	*No. 1065, 11 Oct. 95*
Do.	**Malik Mihr Khan**, Jemadar	Do.	11 Jan. 16	No. 187, 18 Feb. 16
16th Cav.	**Amrik Singh**, Sowar, 2018	Sowar	17 Aug. 17	No. 1357, 17 Aug. 17
Do.	**Sher Singh**, Sowar, 1887	Do.	17 Aug. 17	No. 1357, 17 Aug. 17
17th Do.	**Wazir Khan**, Ressaidar	Jemadar	6 Feb. 16	No. 1268, 20 Oct. 16
Do.	**Dost Muhammad Khan**, Ressaidar (A. D. C. to H. E. the C.-in-C.)	Do.	30 Mar. 00	No. 875, 27 Sep. 01
Do.	Dharam Singh, Actg. Lance Dafadar, 3004	Lance-Dafadar.	29 Dec. 16	No. 676, 5 May 17
18th Lrs.	**Mohara Khan**, Lance-Duffadar, 1995	Lance-Duffadar	16 Mar. 17	Force "A" Order No. 40, 29 May 17.
Do.	**Adalat Khan**, Jemadar	Jemadar	1 Dec. 17	France 5th Cav. Divn. Order No. 1687, 18 Dec. 17.
Do.	**Mohamed Khan**, Jemadar	Do.	1 Dec. 17	France 5th Cav. Divn. Order No. 1687, 18 Dec. 17.
Do.	**Mohammad Khan**, Duffadar	Duffadar	14 June 17	Force "A" Order No. 55, 24 July 17.
19th Do.	*Partab Singh, Duffadar*	*Do.*	...	*No. 553, 13 Aug. 61*
Do.	*Rahmatali Khan, Ressaidar*	*Ressaidar*	...	*13 Mar. 64*
Do.	*Kishun Singh, Duffadar*	*Do.*	*5 Jan. 79*	*No. 366, 2 May 79*
Do.	*Muhammad Ishak, Duffadar, 1130*	*Duffadar*	*19 Apl. 80*	*No. 563, 27 Sep. 80*
Do.	*Ram Singh, Sowar, 1047*	*Sowar*	*19 Apl. 80*	*No. 563, 27 Sep. 80*
Do.	*Gulab Singh, Duffadar, 1071*	*Do.*	*19 Apl. 80*	*No. 563, 27 Sep. 80*
Do.	*Mehdi Khan, Kot-Duffadar, 1985*	*Duffadar*	*23 Jan. 09*	*No. 3, 1 Jan. 10*
Do.	*Mehr Khan, Lance-Duffadar, 2236*	*Lce.-Duffadar*	*23 Jan. 09*	*No. 3, 1 Jan. 10*
Do.	**Yusuf Ali Khan**, Sowar, 2981	Sowar	23 Jan. 09	No. 3, 1 Jan. 10
Do.	**Hashim Khan**, Kot-Dafadar, 2939	Kot-Dafadar	25 Nov. 16	Lon. Gaz., 25 Nov. 16
Do.	**Muhamad Hussein**, Sowar, 2205	Sowar	25 Nov. 16	Lon. Gaz., 25 Nov. 16
Do.	**Sarbuland Khan**, Dafadar, 3379	Dafadar	25 Nov. 16	Lon. Gaz., 25 Nov. 16
Do.	Bishan Singh, Jemadar	Jamadar	18 Jan. 17	No. 676, 5 May 17
Do.	**Mehar Singh**, Dafadar, 2810	Dafadar	4 Feb. 18	No. 2185, 20 Sep. 18
20th Deccan Horse.	*Ghafur Khan, Duffadar, 835*	*Trooper*	*19 June 87*	*No. 371, 4 May 88*
Do.	*Hussain Ali Khan, Duffadar, 572*	*Do.*	*9 Aug. 87*	*No. 911, 18 Nov. 87*
Do.	*Mir Murad Ali, Duffadar, 402*	*Sowar*	*30 Oct. 71*	*No. 614, 24 June 92*
Do.	*Kifayat Khan, Duffadar, 881*	*Do.*	*30 Oct. 91*	*No. 614, 24 June 92*
Do.	*Shah Nur Khan, Duffadar, 338*	*Do.*	*14 Apl. 99*	*No. 294 of 1901*
Do.	*Mustafa Khan, Duffadar, 333*	*Do.*	*14 Apl. 99*	*No. 294 of 1901*
Do.	**Abdur Rahman Khan**, Duffadar	Do.	14 Apl. 99	No. 294 of 1901
Do.	**Sardar Singh**, Duffadar, 295	Duffadar	10 Mar. 15	No. 356, 16 Apl. 15
Do.	**Shankar Rao**, Duffadar, 255	Do.	10 Mar. 15	No. 356, 16 Apl. 15
21st Cav.	*Buta Singh, Sowar*	*Do.*	...	*No. 533, 17 Sept. 80*
Do.	*Mowaz Khan, K. Dafadar*	*Do.*	...	*No. 533, 17 Sept. 80*
Do.	*Thakur Singh, Risaldar Sardar Bahadur*	*Do.*	*3 Nov. 94*	*No. 41, 11 Jan. 95*
Do.	*Dildar Khan, Risaldar*	*Lance-Duffadar*	*3 Nov. 94*	*No. 41, 11 Jan. 95*
Do.	**Man Singh**, Jemadar	Sowar	3 Nov. 94	No. 41, 11 Jan. 95

Corps.	Names and Rank.	In what rank admitted.	Date of admission to the Order.	Authority.
	THE INDIAN ORDER OF MERIT—*contd.*			
	2nd Class—*contd.*			G. G. O. or A. D Notfn.
21st Cav.	**Kunda Singh**, Risaldar	Do.	3 Nov. 94	No. 41, 11 Jan. 95
Do.	**Jehangir Beg**, Jemadar	Jemdr.	24 May 08	No. 526, 26 June 08
Do.	Mathar Singh, Lance-Daffadar, 3192	Lance-Dafadar	1 Jan. 17	No. 676, 4 May 17
Do.	**Bakshish Singh**, Lance-Duffadar, 3653	Do.	30 Apl. 17	No. 1531, 14. Sep. 17
Do.	**Channan Singh**, Sowar, 3719	Sowar,	27 Apl. 18	No 1571, 13 July 18
25th Do.	*Matva Singh, Duffadar, 1678*	*Do.*	*10 Dec. 79*	*No. 252, 23 Apl. 80*
Do.	*Zeenoodeen, Duffadar, 1863*	*Do.*	*10 Dec. 79*	*No. 252, 23 Apl. 80*
Do.	*Wali Mahomed, Jemadar*	...	...	
26th Do.	**Sher-Baz Khan** Risaldar	Sowar	9 Aug. 97	No. 394, 15 Apl. 98
Do.	**Zaman Khan**, Risaldar	Risaldar	16 Dec. 16	No. 1510, 16 Dec. 16
Do.	**Muhammad Amin**, Ressaidar	Ressaidar	6 July 17	No. 1362, 17 Aug. 17
29th Lrs.	**Hayat Ali Beg**, Ressaidar	Jemadar	18 Feb. 15	No. 313, 2 Apl. 15
31st Lrs.	*Ramchandar-rao Mahadik, Risaldar-Maj. Sardar Bahadur, Hony. Capt.*	*Naick*	*12 Dec. 86*	*No. 954, 2 Dec. 87*
Do.	*Kasim Ali Khan, Jemadar*	*Do.*	*30 Dec. 86*	*No. 954, 2 Dec. 87*
32nd Do.	**Abdul Rahman Khan**, Lance-Duffader 3445.	Lance-Duffader	28 June 17	No. 8, 1 Jan. 18
33rd Cav.	**Arjan Singh**, Lce.-Duffadar, 2571	Do.	9 Feb. 15	No. 284, 19 Mar. 15
33rd Cav.	**Santa Singh**, Risaldar Maj., *Bahadur*	Risaldar	3 Mar. 15	No. 630, 3 July 15
Do.	**Bishan Singh**, Dafadar, 2492.	Dafadar	3 Mar. 15	No. 630, 3 July 15
Do.	**Buda Singh**, Sowar, 2829	Sowar	3 Mar. 15	No. 630, 3 July 15
Do.	**Mazhar Khan**, Sowar. 2858	Sowar	8 Jan. 16	No. 524, 12 May 16
34th Poona Horse.	*Gul Sher Khan, Lance-Naick*	*Lce.-Naick*	*28 July 80*	*No. 60, 4 Feb. 81*
Do.	*Bapu Rao Khanvelker, Sowar*	*Sowar*	*28 July 80*	*No. 60, 4 Feb. 81*
Do.	*Ameer Khan, Duffadar*	*Do.*	*28 July 80*	*No. 60, 4 Feb. 81*
Do.	**Madho Singh**, Sowar, 2743	Sowar	2 Nov. 14	No. 111, 29 Jan. 15
36th Horse	**Muhamad Nur Khan**, Ressaldar	Ressaldar	25 Nov. 16	Lon. Gaz., 25 Nov. 16
Do.	Hajee Ahmed, Dafadar, 2786	Dafadar	29 Dec. 16	No. 676, 5 May 17
Do.	Hardit Singh, Dafadar, 3036	Dafadar	1 Jan. 17	No. 676, 5 May 17
38th Horse	Dalip Singh, Sowar 2870	Sowar	30 Aug 17	No. 115, 19 Jan. 18
C. of Guides	*Jai Singh, Subdr.-Maj., Sardar Bahadur*	*Naick*	*1 Aug. 57*	*No. 517, 30 Mar. 58*
Do.	*Konder Singh, Duffadar*	*Sowar*	*2 Apl. 79*	*No. 533, 14 June 79*
Do.	*Yukub, Duffadar*	*Do.*	*2 Apl. 79*	*No. 533, 14 June 79*
Do.	*Guldast, Havildar*	*Naick*	...	*No. 1228, 29 Nov. 95*
Do.	*Suba Singh, Duffadar, 1070*	*Lce.-Duffadar*	*4 Apl. 95*	*No. 1228, 29 Nov. 95*
Do.	*Makhmadin, Subadar*	*Sepoy*	*13 Apl. 95*	*No. 1228, 29 Nov. 95*
Do.	*Sundar, Subadar, Bahadur*	*Havildar*	*13 Apl. 95*	*No. 1228, 29 Nov. 95*
Do.	*Senu, Havildar, 2376*	*Lce.-Naick*	*13 Apl. 95*	*No. 1228, 29 Nov. 95*
Do.	*Tirath Ram, Risaldar-Maj., Sardar Bahadur, Hony. Capt.*	*Ressaidar*	*1 Aug. 97*	*No. 49, 14 Jan. 98*
Do.	*Sirandas, Sowar, 1473*	*Sowar*	*1 Aug. 97*	*No. 49, 14 Jan. 98*
Do.	*Nasir, Sowar, 1408*	*Do.*	*1 Aug. 97*	*No. 49, 14 Jan. 98*
Do.	*Panjaba, Sowar, 1297*	*Do.*	*1 Aug. 97*	*No. 49, 14 Jan. 98*
Do.	*Pir Mahomed, Daffadar, 1274*	*Do.*	*1 Aug. 97*	*No. 49 14 Jan. 98*
Do.	*Sadr-Ud-Din, Kot-Duffadar, 840*	*Kot-Duffadar*	*1 Aug. 97*	*No. 49, 14 Jan. 98*
Do.	*Sarfaraz Khan, Subadar-Maj., Sardar Bahadur, Hony. Capt*	*Subadar-Major*	*2 Aug. 97*	*No. 49, 14 Jan. 86*
Do.	*Dar Jan, Sepoy, 1998*	*Sepoy*	*2 Aug. 97*	*No. 49, 14 Jan. 98*
Do.	*Jowahir Singh, Sepoy, 3446*	*Do.*	*2 Aug. 97*	*No. 49, 14 Jan 98*
Do.	**Bahadur Singh**, Risaldar-Maj., *Sardar Bahadur.*	Jemadar	17 Aug. 97	No. 49, 14 Jan. 98
Do.	*Mahomed Khan, Duffadar, 1455*	*Sowar*	*17 Aug. 97*	*No. 49, 14 Jan. 98*
Do.	*Alaf Khan, Sowar, 1219½*	*Do.*	*17 Aug. 97*	*No. 49, 14 Jan. 98*
Do.	*Bura Khan, Duffadar, 806½*	*Duffadar*	*17 Aug. 97*	*No. 49, 14 Jan. 98*
Do.	*Sikandar Khan, Subadar*	*Jemadar*	*16 Sep. 97*	*No. 232, 25 Feb. 98*
Do.	*Khawaja Nur Naick, 390*	*Do.*	*16 Sep. 97*	*No. 232, 25 Feb. 98*
Do.	**Mahomed Afzal**, Havildar, 3858 (*a*)	Do.	16 Sep. 97	No. 232, 2 Feb. 98
Do.	*Dewa Singh, Sepoy, 3912*	*Do.*	*16 Sep. 97*	*No. 232, 25 Feb. 98*
Do.	*Mohammad Nasim. Havildar, 4525 (b)*	*Sepoy*	...	*G.G.O. 908, 13 Oct. 05*
Do.	*Manga, Havildar, 3595*	*Do.*	*29 May 08*	*No. 526, 26 June 08*
Do.	*Nur Zaman, Sepoy, 5186 (c)*	*Sepoy*	*29 May 08*	*No. 526, 26 June 08*
Do.	**Kishan Singh**, Havildar, 4744	Havildar	17 Sep. 15	No. 907, 17 Sep. 15
Do.	**Man Bir Thapra**, Subadar, Major	Subadar Major	8 June 18	No. 2536, 26 Oct. 18
3rd L. H. C.	*Mahomed Khan, Dafadar*	*Trooper*	*21 June 58*	*No. 319, 11 Mar. 59*
Do.	*Mahomed Khan, Duffadar 14*	*Do.*	*15 Nov. 59*	*No. 16, 15 Nov. 59*
Frontier Garn. 21st Mtn. Batty.	*Karam Singh, Lance-Naik, 424*	*Lance-Naik*	*23 Sep. 15*	*No. 906, 17 Sep. 15*
Do.	**Jan Muhammad**, Gunner-Havildar	Gunner-Naik	23 Sep. 15	No. 906, 17 Sep. 15
Do.	**Mohamed Baksh**, Pay-Havildar, 546,	Pay-Havildar	...	No. 849, 29 July 16
Do.	**Chanda**, Subadar	Subadar	2 Jan. 17	No. 676, 5 May 17
33rd Mtn. Batty	**Habib Khan**, Havildar, 22	Havildar	...	No. 630, 3 July 15
Do.	**Havildar Latif**, Havildar, 30	Do.	4 Aug. 15	No. 727, 4 Aug. 15

(*a*) Transfd. to Br. Cen. Africa, 16 Sep. 1898. (*b*) Discharged at his own request, 1 April 1910.
(*c*) Transferred to Reserve.

Corps.	Names and Rank,	In what rank admitted.	Date of admission to the Order.	Authority.
	THE INDIAN ORDER OF MERIT—*contd.* **2nd Class**—*contd.*			
				G.G.O. or A. D. Notfn.
26th Mtn. Batty.	*Nihal Singh, Havildar, 328*	*Havildar*	*10 June 97*	*No. 865, 6 Aug. 97*
Do.	*Rur Singh, Driver Havildar, 473*	*Driver Havldr.*	*10 June 97*	*No. 865, 6 Aug. 97*
Do.	**Jwala Singh**, Subadar	Gunner	10 June 97	No. 865, 6 Aug. 97
Do.	*Diwan Singh*	*Do.*		*G. G. O., 6 Aug. 97*
Do.	*Utam Chand, Naick, 2357*	*Naick*	*10 June 97*	*No. 1211, 29 Oct. 97*
Do.	*Dulla Khan, Jamadar*	*Gunner*	*10 June 97*	*No. 1211 29 Oct. 97*
Do.	**Fazal Ilahi**, Gunner, 2811	Do.	23 Sep. 15	No. 906, 17 Sep. 15
Do.	**Gurdit Singh**, Gunner-Havildar, 1096	Havildar	29 Aug. 15	No. 1081, 29 Oct. 15
Do.	**Bir Singh**, Gunner-Naik, 2275	Naik	2 Jan. 17	No. 676, 5 May 17
Do.	**Indar Singh**, Driver-Havildar, 1125	Driver-Havildar	2 Jan. 17	No. 676, 5 May 17
Do.	**Muhammad Bakhsh**, Jemadar	Jemadar	2 Jan. 17	No. 676, 5 May 17
27th Mtn. Batty.	*Thakoor Dass, Gunner, 89*	*Gunner*	*3 Dec. 87*	*No. 391, 11 May 88*
Do.	**Gopala**, Naik, 665	Naik	1 Jan. 16	No. 187, 18 Feb. 16
Do.	**Bhan Singh**, Gunner-Havildar, 704	Havildar	21 Nov. 17	No. 676, 5 May 17
Do.	**Sher Baz**, Subadar	Subadar	21 Nov. 17	No. 676, 5 May 17
Do.	**Sirdar Khan** Gunner 1063	Gunner	18 Oct. 17	No. 1179, 3rd June 18
28th Mtn. Batty.	*Alam-Sher, Subadar*	*Naick*	*10 Dec. 79*	*No. 252, 23 Apl. 98*
Do.	*Hossein Mahomed, Naick, 78*	*Do.*	*27 July 97*	*No. 49, 14 Jan. 98*
Do.	*Jiwan, Trumpeter, 253*	*Trumpeter*	*16 Sep. 97*	*No. 232, 25 Feb. 98*
Do.	*Nihala, Driver, 299*	*Driver*	*16 Sep. 97*	*No. 232, 25 Feb. 98*
Do.	*Ishar Singh, Jemadar*	*Jemadar*	*16 Sep. 97*	*No. 232, 25 Feb. 98*
30th Mtn. Batty.	*Nawab Khan, Subadar-Maj., Sardar Bahadur Hony. Capt.*	*Do.*	*16 Sep. 97*	*No. 232, 25 Feb. 98*
Do.	**Baryam Singh**, Gunner-Havildar, 342	Gunner-Havildar	7 Feb. 16	No. 583, 26 May 16
Do.	**Gajja Singh**, Driver Havildar, 322	Driver Havildar	7 Feb. 16	No. 583, 26 May 16
Do.	**Maya Singh**, Driver, 642	Driver	7 Feb. 16	No. 583, 26 May 16
32nd Mtn. Batty.	**Muhammad Ismail**, Subadar Major *Sardar Bahadur*	Havildar-Major	10 June 97	No. 785, 16 July 97
Attd. R. H. A.	Feteh Jang, Indian Driver 1394	Driver	25 Nov. 16	Lon. Gaz., 25 Nov. 16
No. 1 Kashmir Mtn. Batty.	**Dharm Singh**, Maj.	Havildar	13 Apl. 95	No. 745, 10 July 95
Do.	**Balwan Singh**, Lt.-Col.	Do.	13 Apl. 95	No. 745, 16 July 95
1st S. and M.	*Hazara Singh, Jemadar, A. Coy.*	*Sapper*	*1 Dec. 91*	*No. 179, 19 Feb. 92*
Do.	*NabiBakhsh, Naick*	*Do.*	*10 Mar. 95*	*No. 743, 16 July 95*
Do.	*Chanda Singh, Naick, 4148*	*Do.*	*10 Mar. 05*	*No. 743, 16 July 95*
Do.	*Natha Singh, Jemadar, B. Coy.*	*Naick*	*16 Sep. 97*	*No. 232, 25 Feb. 98*
Do.	*Gaj Raj Tewari, Havildar, 4310*	*Sapper*	*18 Nov. 02*	*No. 381, 24 Apl. 03*
Do.	*Sat Narain, Naick, 4880*	*Do.*	*18 Nov. 02*	*No. 381, 24 Apl. 03*
Do.	*Badri Pande, Havildar, 4362*	*Lance-Naick*	*18 Nov. 02*	*No. 381, 24 Apl. 03*
Do.	**Chagatta**, Jemadar, 279, 3 Co.	Sapper	26 May 04	No. 957, 7 Oct. 04
Do.	*Usman Khan, Havildar, 3921*	*Naick*	*6 July 04*	*No. 1169, 10 Dec 04*
Do.	**Ram Rup Singh**, Jemadar	Jemadar	12 Nov. 14	No. 111, 29 Jan. 15
Do.	**Sheik Abdul Rahman**, Sapper, 2192	Sapper	4 May 15	No. 563, 12 June 15
Do.	**Jiwa Khan**, Sapper, No. 1694	Do.	3 June 15	Lon. Gaz., 23 June 15
Do.	**Sawan Singh**, Naik, 764	Naik	16 June 17	No. 945, 16 June 17
Do.	**Bhullan Khan**, Naik, 785	Do.	16 June 17	No. 945, 16 June 17
Do.	**Hardit Singh**, Havildar, 1364	Havildar	17 Aug. 17	No. 1359, 17 Aug. 17
Do.	**Alam Khan**, Jemadar	Jemadar	17 Aug. 17	No. 1359, 17 Aug. 17
2nd S. and M.	*Ramasami, Subdr.-Maj., Bahadur*	*Subadar*	*26 July 97*	*No. 49, 14 Jan. 98*
Do.	**Burland**, Subadar	Do.	17 Aug. 17	No. 1359, 17 Aug. 17
Do.	**Loganadun**, Havildar, 4170	Havildar	17 Aug. 17	No. 1359, 17 Aug. 17
3rd S. and M.	*Asha Giri, Havildar, 1450*	*Naick*	*26 July 97*	*No. 49, 14 Jan. 98*
Do.	**Baryam Singh**, Subadar, *Bahadur*	Do.	20 Dec. 01	No. 264, 28 Mar. 02
Do.	**Dalip Singh**, Naik, 2855	Do.	17 Nov. 14	No. 19, 19 Feb. 15
Do.	**Dalip Singh**, Sapper, 3795	Sapper	28 Oct. 14	No. 111, 29 Jan. 15
Do.	**Firoz Ali**, Subadar	Jemadar	17 Nov. 14	No. 1168, 24 Dec. 14
Do.	**Ghulam Nabi**, Jemadar	Havildar	8 Dec. 14	No. 284, 19 Mar. 15
Do.	**Nur Dad**, Lce-Naik, 3743	Lce-Naik	8 Dec. 14	No. 284, 19 Mar. 15
Do.	**Ghulam Haidar**, Sapper, 3998	Sapper	8 Dec. 14	No. 284, 19 Mar. 15
Do.	**Mulla Singh**, Subadar	Subadar	18 Feb. 15	No. 313, 2 Apl. 15
Do.	**Muhammad Din**, Subadar	Jemadar	4 Aug. 15	No. 727, 4 Aug. 15
Do.	**Hari Tingri**, Lance-Naik, 4193	Lance-Naik	19 Nov. 15	No. 1159, 19 Nov. 15
Do.	**Son Singh**, Sapper, 2878	Sapper	10 Dec. 15	No. 1244, 10 Dec. 15
Do.	**Uttam Singh**, Jemadar	Jemadar	1 Jan. 16	No. 187, 18 Feb. 16
Do.	**Liyakat Ali**, Jemadar	Do.	1 Jan. 16	No. 187, 18 Feb. 16
Do.	**Muhamad Baksh**, Driver Havildar, 3144.	Driver Havildar	1 Jan. 16	No. 187, 18 Feb. 16
Do.	**Bir Singh**, Subadar	Subadar	17 Apl, 16	No. 1385, 18 Nov. 16
Do.	**Mahtab Khan**, Sapper, 4116	Sapper	23 Mar. 16	No. 1387, 18 Nov. 16
Do.	**Dhondibarao Barge**, Subadar	Subadar.	17 Aug. 17	No. 1359, 17 Aug. 17
Do.	**Lalu Rajanna**, Havildar, 2423	Havildar	17 Aug. 17	No. 1359, 17 Aug. 17
Do.	**Abdul Aziz**, Sapper, 4587	Sapper	17 Aug. 17	No. 1357, 17 Aug. 17
Do.	**Inder Singh**, Sapper, 3075	Do.	17 Aug. 17	No. 1357, 17 Aug. 17

THE INDIAN ORDER OF MERIT—*contd.*

2nd Class—*contd.*

Corps.	Names and Rank.	In what rank admitted.	Date of admission to the Order.	Authority. G. G. O. or A. D. Notfn.
3rd S. and M.	**Saiyid Hussain**, Naik, 3955	Naik	17 Aug. 17	No. 1357, 17 Aug. 17
32nd Divl. Sig. Company.	**Bhag Singh**, Sapper, 42	Do.	17 Apl. 16	No. 1385, 18 Nov. 16
Lahore Divl. Sig. Coy.	**Alah Rakkha Khan**, Jemadar	Jemadar	1 Jan. 16	No. 187, 18 Feb. 16
1st Brahmans	*Bissesar Tewari, Subadar-Maj., Sardar Bahadur, Hony. Capt.*	*Naick*	*12 May 86*	*No. 167, 14 Feb. 90*
Do.	*Jwala Parshad Awasthi, Jemadar*	*Sepoy*	*1 Aug. 86*	*No. 654, 10 Aug. 88*
2nd Rajput L I.	*Harparsad Misser, Havildar, 2382*	*Havildar*	*12 June 86*	*No. 561, 6 July 88*
Do.	*Bisesar Singh, Subadar-Maj.*	*Sepoy*	*19 June 86*	*No. 654, 10 Aug. 88*
Do.	**Loknath Singh**, Subadar-Maj.	Subadar	8 Mar. 16	No. 1385, 18 Nov. 16
Do.	**Multan Singh**, Subadar	Do.	18 Apl. 16	No. 1385, 18 Nov. 16
Do.	**Multan Singh**, Sepoy, 1782	Sepoy	8 Mar. 16	No. 1385, 18 Nov. 16
5th Infy.	**Mahmud Ali Khan**, Sepoy, 3266	Do.	12 Apl. 16	No. 1385, 18 Nov. 16
Do.	**Wahid Ali Khan**-Subadar-Major	Subadar-Major	30 June 17	No. 8, 1 Jan. 18
Do.	**Abdul Khan**, Sepoy 3291	Sepoy	19 Aug. 17	No. 1179, 3 June 18
6th Infy.	**Jai Lal**, 1821	Havildar	10 Mar. 16	No. 356, 16 Apl. 15
Do.	**Dalpat Singh**, Subadar-Major	Subadar-Major.	21 Jan. 16	No. 524, 12 May 16
Do.	**Juglal**, Pay Havildar, 1906	Pay Havildar	21 Jan. 16	No. 524, 12 May 16
Do.	**Chandgi**, Havildar, 2798	Havildar	21 Jan. 16	No. 524, 12 May 16
Do.	**Matu**, Lance-Naik, 3171	Lance-Naik	21 Jan. 16	No. 524, 12 May 16
Do.	**Hardwari**, Naik, 2822	Naik	21 Jan. 16	No. 524, 12 May 16
Do.	**Harman**, Sepoy, 3375	Sepoy	21 Jan. 16	No. 524, 12 May 16
Do.	**Chellaperumaul**, Naik, 3968	Naik	26 Mar. 17	
Do.	**SayidAbdul Wahib**, Naik, 3641	Naik	26 Mar. 17	
Do.	**Sadi**, Sepoy 2472	Sepoy	22 Apl. 16	No. 1385, 18 Nov. 16
Do.	**Kalu Ram**, Sepoy, 2495	Do.	22 Apl. 16	No. 1385, 18 Nov. 16
7th Rajputs	**Dwarka Singh**, Sepoy, 1586	Do.	3 July 15	No. 630, 3 July 15
Do.	**Gokaran Singh**, Sepoy, 1311	Do.	3 July 15	No. 630, 3 July 16
Do.	**Jhandu Singh**, Sepoy, 3390	Do.	8 July 17	No. 1362, 17 Aug. 17
9th Infy.	**Mugil Khan**, Sepoy, 3943	Do.	12 Apl. 16	No. 1385, 18 Nov. 16
11th Rajputs	**Shiuambar Singh**, Subadar-Maj.	Subadar-Maj	17 Aug. 17	No. 1357, 17 Aug. 17
Do.	**Ramkishor Singh**, Sepoy, 1829	Jemadar	15 Oct. 15	No. 1023, 15 Oct. 15
Do.	**Shiu Prosad Singh**, Naik 1634	Naik	10 May 17	No. 1170, 20 July 17
Do.	**Ram Nath Singh**, Lance-Naik 2189	Lance Naik	10 May 17	No. 1170, 20 July 17
12th Prs.	**Maingha Singh**, Jemadar	Jemadar	1 Jan. 16	No. 187, 18 Feb. 16
13th Rajputs	*Drigpal Singh, Sepoy, 3785*	*Do.*	*28 Oct. 86*	*No. 395, 27 May 87*
Do.	**Bakhtawar Singh**, Subadar	Subadar	3 June 15	No. 526, 3 June 15
14th Sikhs	*Nehal Singh, Havildar*	*Do.*	*13 Nov 63*	*No. 78, 29 Jan. 64*
Do.	*Budh Singh, Subadar*	*Do.*	*21 Nov. 78*	*No. 164, 21 Feb. 79*
Do.	*Man Singh, Subadar.-Maj., Sardar Bahadur, Hony. Capt.*	*Havildar*	*21 Nov. 78*	*No. 164, 21 Feb. 79*
Do.	*Sundar Sing, Subadar*	*Do.*	*10 Mar. 95*	*No. 742, 16 July 95*
Do.	*Matab Sing, Jemadar*	*Do.*	*10 Mar. 95*	*No. 742, 16 July 95*
Do.	*Burh Sing, Havildar, 1402*	*Lance-Naick*	*10 Mar. 95*	*No. 742, 16 July 95*
Do.	*Jaimal Sing, Sepoy, 1357*	*Sepoy*	*10 Mar. 95*	*No. 742, 16 July 95*
Do.	*Sham Sing, Naick, 1577*	*Do.*	*10 Mar. 95*	*No. 742, 16 July 95*
Do.	*Saddah Sing, Havildar, 1629*	*Do.*	*10 Mar. 95*	*No. 742, 16 July 95*
Do.	*Bhag Sing, Havildar, 1693*	*Do.*	*10 Mar. 95*	*No. 742, 16 July 95*
Do.	*Prem Sing, Subadar*	*Do*	*10 Mar. 95*	*No. 742, 16 July 95*
Do.	*Mal Sing, Sepoy, 1725*	*Do.*	*10 Mar. 95*	*No. 742, 16 July 95*
Do.	*Badan Sing, Naick, 1419*	*Do.*	*10 Mar. 95*	*No. 742, 16 July 95*
Do.	*Ganga Sing, Sepoy, 1922*	*Do.*	*10 Mar 95*	*No. 742, 16 July 95*
Do.	*Dayal Sing, Sepoy, Reserve, 1674*	*Do.*	*10 Mar. 95*	*No. 742 16 July 95*
Do.	*Atar Singh, Subadar-Major*	*Jemadar*	*17 Apl. 95*	*No. 744, 16 July 95*
Do.	*Partab Singh, Havildar, 1252*	*Sepoy*	*17 Apl. 95*	*No. 744, 16 July 95*
Do.	*Gurja Singh, Havildar, 772*	*Lce.-Naick*	*17 Apl. 95*	*No. 744, 16 July 95*
Do.	*Atar Singh, Naick, 1822*	*Sepoy*	*17 Apl. 95*	*No. 744, 16 July 99*
Do.	*Kaka Singh; Jemadar*	*Do.*	*19 Oct. 97*	*No. 99, 27 Jan. 95*
Do.	*Bogga Singh, Havildar, 1752*	*Do.*	*19 Oct. 97*	*No. 99, 27 Jan. 96*
Do.	*Jodh Singh, Naick*	*Do.*	...	*No. 742, 16 July 95*
Do.	**Hazara Singh**, Lance-Naik, 3760	Lance-Naik	3 June 16	No. 849, 29 July 16
15th Sikhs.	*Waryam Singh, Jemadar*	*Jemadar*	*18 Oct. 97*	*No. 430, 22 Apl. 59*
Do.	*Kishn Singh, Havildar, 3315*	*Naick*	*18 Oct. 97*	*No. 430, 22 Apl. 98*
Do.	*Bela Singh, Havildar, 2669*	*Lance-Naick*	*18 Oct. 97*	*No. 430, 22 Apl. 98*
Do.	*Bhola Singh, Sepoy, 2847*	*Sepoy*	*18 Oct. 97*	*No. 430, 22 Apl. 98*
Do.	*Hazara Singh, Naick, 2683*	*Lce.-Naick*	*6 Nov. 97*	*No. 430, 22 Apl. 98*
Do.	*Jiwan Singh, Sepoy, 3304*	*Sepoy*	*6 Nov. 97*	*No. 430, 22 Apl. 98*
Do.	*Amar Singh, Sepoy, 3379*	*Do.*	*6 Nov. 97*	*No. 430, 22 Apl. 98*
Do.	*Hira Singh, Subadar*	*Havildar*	*16 Nov. 97*	*No. 12, 6 Jan. 96*
Do.	*Sahib Singh, Sepoy, 3036*	*Sepoy*	*11 Dec. 97*	*No. 99, 27 Jan. 96*
Do.	**Phuman Singh**, Subadar	*Do.*	*11 Dec. 97*	No. 99, 27 Jan. 99
Do	*Golab Singh, Sepoy, 3184*	*Do.*	*11 Dec. 97*	*No. 99, 27 Jan 99*
Do.	*Karpal Singh, Naick, 3385*	*Do.*	*11 Dec. 97*	*No. 99, 27 Jan 99*
Do.	**Bishn Singh**, Havildar, 4011	Havildar	3 June 15	Lon. Gaz., 23 June 15
Do.	**Bakshi Singh**, Lance-Naick, 698	Sepoy	3 June 15	Lon. Gaz., 23 June 15

THE INDIAN ORDER OF MERIT—*contd.*

2nd Class—*contd.*

Corps.	Names and Rank.	In what rank admitted.	Date of admission to the Order.	Authority. G. G. O. or A. D. Notfn.
15th Sikhs	**Mangal Singh**, Naik, 1311	Lance-Naik	23 July 15	No. 815, 27 Aug. 15
Do.	**Mahan Singh**, Havildar, 4727	Havildar	1 Jan. 16	No. 187, 18 Feb. 16
17th Infy.	**Ali Bahadur Khan**, Subadar,	Subadar	4 June 17	L. G. 4 June 17
Do.	**Muhammad Sialamn Khan**, Subadar Actg. Subdr.-Maj.	Do.	3 July 17	No. 8, 1 Jan. 18
18th Infy.	*Ramcharitr Singh, Jemadar*	*Havildar*	*28 Oct. 86*	*No. 395, 27 May 87*
19th Punj.	**Lal Khan**, Lance Naik, 1194	Lance Naik	8 Mar. 16	No. 1385, 18 Nov. 16
Do.	**Abbaz Khan**, Sepoy, No. 2712	Sepoy	17 Aug. 17	No. 1359, 17 Aug. 17
20th Infy.	*Abdullah Khan, Jemadar*	*Havildar*	*23 Feb. 90*	*No. 515, 6 June 90*
Do.	*Hasan Khan, Subadar*	*Subadar*	*25 Sep. 97*	*No. 232, 25 Feb. 98*
Do.	*Lali, Havildar, 4607*	*Sepoy*	*25 Sep. 97*	*No. 232, 25 Feb. 98*
Do.	**Surain Singh**, Bugler, 822	Bugler	13 Nov. 14	No. 1168, 24 Dec. 14
Do.	**Masin Khan**, Acting Subdr.-Maj.	Actg. Subdr.-Maj	10 Dec. 15	No. 1244, 10 Dec. 15
Do.	**Radhu**, Lance-Naik, 1058	Lance-Naik	10 Dec. 15	No. 1244, 10 Dec. 15
Do.	**Mewa Singh**, Havildar, 204	Havildar	10 Dec. 15	Lon. Gaz., 23 June 15
Do.	*Said Ali, Sepoy*	*Bugler*	...	*No. 232, 25 Feb. 98*
Do.	*Sadu Singh, Havildar, 3531*	*Havildar*	*16 May 08*	*No. 526, 26 June 08*
Do.	**Ram Singh**, Havildar, 4780	Naick	16 May 08	No. 526, 26 June 08
Do.	**Churu**, Sepoy, 2158	Sepoy	17 Aug. 17	No. 1359, 17 Aug. 17
22nd Punj.	**Bawa Singh**, Sepoy, 845	Do.	2 June 17	No. 868, 2 June 17
Do.	**Lal Singh**, Havildar, 4717	Havildar	17 Aug. 17	No. 1357, 17 Aug. 17
23rd Prs.	*Chunda Singh, Sepoy*	*Sepoy*	...	*G. G. O., 27 May 58*
Do.	*Boolaka Singh, Sepoy*	*Do.*	...	*G. O. C. C., 5 Oct. 64*
Do.	*Futteh Singh, Havildar*	*Havildar*	...	*No. 782, 22 July 69*
Do.	**Mangal Singh**, Havildar, 3985	Do.	21 Sept. 17	No. 115, 19 Jan. 18
Do.	**Phagga Singh**, Naik, 4670	Naik	9 Sept. 17	No. 115, 19 Jan. 18
Do.	*Jhunda Singh, Naick*	*Do.*	...	*No. 1260 of 1879.*
Do.	*Goordial Singh, Subadar-Major*	*Havildar*	*6 Oct. 79*	*No. 1260, 26 Dec. 79*
Do.	*Bachan Singh, Havildar, 3017*	*Do.*	*30 Oct. 91*	*No. 614, 24 June 92*
Do.	*Prem Singh, Sepoy, 3039*	*Do.*	*21 Nov. 91*	*No. 614, 24 June 92*
Do.	*Isar Singh, Jemadar*	*Naick*	*15 Dec. 91*	*No. 614, 24 June 92*
Do.	*Nand Singh, Havildar, 2558*	*Havildar*	*15 Dec. 91*	*No. 614, 24 June 92*
Do.	*Sundar Singh, Havildar, 3322*	*Sepoy*	*26 Nov. 01*	*No. 403, 9 May 02*
Do.	*Jhanda Singh, Naick, 3259*	*Do.*	*26 Nov. 01*	*No. 403, 9 May 02*
Do.	**Jowalla Singh**, Sepoy, 4129	Do.	6 May 04	No. 957, 7 Oct. 04
Do.	*Tilok Singh, Sepoy, 3752*	*Do.*	*19 May 04*	*No. 957, 7 Oct. 04*
Do.	**Rolla Singh**, Sepoy, 3324	Do.	26 June 04	No. 957, 7 Oct. 04
Do.	**Teja Singh**, Havildar, 4584	Havildar	...	No. 630, 3 July 15
Do.	**Binda Singh**, Jemadar	Jemadar	17 Aug. 17	No. 1359, 17 Aug. 17
Do.	**Phagga Singh**, Naik, 4670	Naik	9 Sep. 17	No. 115, 19 Jan. 18
Do.	**Mangul Singh**, Havildar, 3985	Havildar	21 Sep. 17	No. 115, 19 Jan. 18
24th Punj.	*Maroof Shah, Sepoy, 1765*	*Do.*	*1 Sep. 80*	*No. 639, 19 Nov. 80*
Do.	*Wadhawa Singh, Havildar, 3580*	*Sepoy*	*26 July 97*	*No. 49, 14 Jan. 98*
Do.	**Futteh Singh**, Jemadar	Do.	26 July 97	No. 49, 14 Jan. 98
Do.	**Mir Gul**, Sepoy, 3457 (*a*)	Do.	27 July 97	No. 49, 14 Jan. 98
Do.	*Said Hassan, Havildar, 3345*	*Do.*	*27 July 97*	*No. 49, 14 Jan. 98*
Do.	**Umar Khan**, Subadar	Do.	27 July 97	No. 49, 14 Jan. 98
Do.	*Ali Gul, Havildar, 2549*	*Havildar*	*27 July 97*	*No. 49, 14 Jan. 98*
Do	*Mir Badshah, Sepoy, 3451*	*Sepoy*	*27 July 97*	*No. 49, 14 Jan. 98*
Do.	*Mira Baz, Havildar, 2507*	*Havildar*	*27 July 97*	*No. 49, 14 Jan. 98*
Do.	**Yarak**, Sepoy, 318	Sepoy	4 Aug. 15	No. 727, 4 Aug. 15
Do.	**Lal Singh**, Lance-Naik, 4943	Lance-Naick	4 Aug. 15	No. 727, 4 Aug. 15
Do.	**Gossain**, Lance-Naik, 4358	Do.	19 Nov. 15	No. 1159, 19 Nov. 15
Do.	**Parmodh Singh**, Sepoy, 4812	Sepoy	19 Nov. 15	No. 1159, 19 Nov. 15
Do.	**Mangal Singh**, Sepoy, 4022	Do.	19 Nov. 15	No. 1159, 19 Nov. 15
Do.	**Lachman Singh**, Sepoy, 242	Do.	19 Nov. 15	No. 115, 9 Jan. 18
26th Punj.	*Mirbaz Shah, Havildar, 2643*	*Do.*	*8 May 86*	*No. 631, 17 Sep. 86*
Do.	*Umar Jan, Subadar*	*Lance-Havildar*	*8 May 86*	*No. 631, 17 Sep. 86*
Do.	**Jan Gul**, Subadar,	Subadar	2 June 17	No. 868, 2 June 17
Do.	**Gulan Khan**, Naik, 759	Naik	8 Feb. 17	No. 868, 2 June 17
27th Punj.	*Asa Sing, Subadar, Bahadur*	*Sepoy*	*21 Nov. 78*	*No. 230, 13 Mar. 79*
Do.	*Goordit Singh, Sepoy*	*Do.*	*21 Nov. 78*	*No. 230, 13 Mar. 77*
Do.	*Mangal Singh, Havildar*	*Havildar*	*14 Dec. 86*	*No. 519, 8 July 89*
Do.	**Bhagat Singh**, Subadar	Subadar	17 Apl. 16	No 1385, 17 Nov. 17
Do.	**Pala**, Jemadar	Jemadar	2 June 17	No. 868, 2 June 17
Do.	**Rama**, Subadar	Subadar	5 Apl. 77	No. 8, 1 Jan. 18
28th Punj.	*Mir Alam, Subdr.-Maj., Sardar Bahadur*	*Naick*	*6 Nov. 63*	*No. 78, 29 Jan. 64*
Do.	*Akbar Khan, Sepoy, 1669*	*Sepoy*	*19 Dec. 79*	*No. 252, 23 Apl. 80*
Do.	*Hari Singh, Havildar, 2992*	*Havildar*	*21 Feb. 08*	*No. 392, 15 May 08*
Do.	**Munshi**, Havildar, 4178	Sepoy	21 Feb. 08	No. 392, 15 May 08
Do.	*Gulmir, Subadar-Major*	*Subadar-Major*	*7 Jan. 16*	*No. 637, 31 June 16*
Do.	**Adam Khan**, Naik, 465	Naik	6 Apl. 16	No. 1385, 18 Nov. 16
Do.	**Jan Baz**, Lance Naik, 4573	Lance Naik	6 Apl. 16	No. 1385, 18 Nov. 16
Do.	**Rajwali**, Lance-Naik, 1057	Do.	21 Apl. 17	No. 9, 1 Jan. 18
Do.	**Gulab Din**, Jemadar	Jemadar	5 Nov. 17	No. 1885, 16 Aug. 18
29th Punj.	*Dewan Singh, Sepoy, 2899*	*Sepoy*	*4 Apl. 91*	*No. 522, 12 May 91*
Do.	*Haidar Hussain, Sepoy, 2603*	*Do.*	*3 Nov. 94*	*No. 41, 11 Jan. 95*

Corps.	Names and Rank.	In what rank admitted.	Date of admission to the Order.	Authority.
	THE INDIAN ORDER OF MERIT—*contd.*			
	2nd Class—*contd.*			G. G. O. or A. D. Notfn.
29th Punj.	*Gul Muhammad, Naik, 4050*	*Naik*	*6 Sep. 14*	*No. 1141, 18 Dec. 14*
30th Punj.	*Kesar Singh, Subadar-Maj.* Bahadur	*Lance-Havildar*	*16 Jan. 98*	*No. 430, 22 April 98*
Do.	**Labh Singh**, Subadar	Subadar	4 Oct. 18	No. 2342, 4 Oct. 18
Do.	**Sundar Singh**, Jemadar	Jemadar	3 Aug. 17	No. 8, 1 Jan. 18
31st Punj.	*Ram Singh, Jemadar*	*Naick*	*27 July 97*	*No. 49, 14 Jan. 98*
Do.	*Khiwa Khan, Sepoy, 2145*	*Sepoy*	*27 July 97*	*No. 49, 14 Jan. 98*
Do.	*Rulia Singh, Sepoy, 2445*	*Do.*	*27 July 97*	*No. 49, 14 Jan. 98*
Do.	*Futteh Khan, Sepoy, 1983*	*Do.*	*27 July 97*	*No. 49, 14 Jan. 98*
Do.	*Walidad Khan, L. Havildar, 730*	*L. Havildar*	*27 July 97*	*No. 49, 14 Jan. 98*
Do.	*Atma Singh, Havildar, 1335*	*Naick*	*27 July 97*	*No. 49, 14 Jan. 98*
Do.	*Kala Singh, Sepoy, 1897*	*Sepoy*	*27 July 97*	*No. 49, 14 Jan. 98*
Do.	*Fazl Khan, Drummer, 1226*	*Bugler*	*27 July 97*	*No. 49, 14 Jan. 98*
Do.	*Wazir Ali Khan, Lance-Naick, 266*	*Lance-Naick*	*26 Apl. 98*	*No. 99, 27 Jan. 99*
32nd Prs.	*Lal Singh, Sepoy, 2667*	*Sepoy*	*21 Nov. 91*	*No. 614, 24 June 92*
Do.	*Jhanda Singh, Naick, 1179*	*Naick*	*15 Dec. 91*	*No. 614, 24 June 92*
Do.	*Sundar Singh, Jemadar*	*Sepoy*	*7 Feb. 95*	*No. 657, 21 June 95*
Do	*Prem Singh, Subdr.-Maj., Sardar Bahadur. Hony. Capt.* (Hony. A.-D.-C. to H. E. the C.-in-C.)	*Subadar*	*13 Apl. 95*	*No. 745, 15 July 95*
Do.	*Wadhawa Singh, Havildar, 1699*	*Havildar*	*13 Apl. 95*	*No. 745, 16 July 95*
Do.	*Esur Singh, Sepoy, 2835*	*Do.*	*13 Apl. 95*	*No. 970, 13 Sep. 95*
Do.	**Bhagwan Singh**, Sepoy, 3405	Sepoy	6 May 04	No. 1143, 2 Dec. 04
Do.	**Jhanda Singh**, Subadar,	Naick	6 May 04	No. 17, 23 Mar. 06
Do.	*Kesar Singh, Subadar*	*Subadar*	*26 May 04*	*No. 957, 7 Oct. 04*
Do.	*Labh Singh, Havildar, 2657*	*Havildar*	*26 May 04*	*No. 957, 7 Oct. 04*
Do.	**Mastan Singh**, Naick	Sepoy	. . . .	No. 356, 16 Apl. 15
Do.	**Kharak Singh**, Jemadar	Jemadar	1 Jan. 16	No. 187, 18 Feb. 16
Do.	**Mangal Singh**, Havildar, 3623	Havildar	1 Jan. 16	No. 187, 18 Feb. 16
33rd Punj.	*Mana Khan, Subadar-Major, Sardar Bahadur, Hony. Capt.*	*Jemadar*	*30 Jan. 90*	*No. 572, 20 June 90*
Do.	*Narain Singh, Naick*	*Private*	*7 Feb. 92*	*No. 709, 22 July 92*
Do.	**Bahadur Khan**, Sub-Maj.	Sub-Maj	3 June 16	No. 849, 29 July 16
Do.	**Maluk Singh**, Sub-Maj.	Do.	3 June 16	No 849, 29 July 16
Do.	**Attar Khan**, Subadar-Maj.	Subadar.	15 Sep. 15	No. 219 of 1917.
34th Prs.	**Gopal Singh**, Naick, 1561	Bugler	13 May 08	No. 526, 26 June 08
Do.	**Pahara Singh**, Naick, 1694	Sepoy	18 May 08	No. 526, 26 June 05
Do.	**Pala Sing**, Havildar, 1148	Havildar	18 Feb. 15	No. 313, 3 April 15
Do.	**Nikka Singh**, Havildar, 1871	Do.	23 Nov. 14	No. 111, 29 Jan. 15
Do.	**Mastan Sing**, 4563	Sepoy	10 Mar. 14	No. 365, 16 April 15
Do.	**Tota Sing**, Lce.-Naick, 2775	Naick	18 Feb. 15	No. 313, 3 April 15
Do.	**Bir Sing**, Naick, 2071	Do.	18 Feb. 15	No. 313, 3 April 15
Do.	**Natha Singh**, Subadar	Subadar	4 May 15	No. 563, 12 June 15
Do.	**Jwala Singh**, Subadar-Major, *Sardar Bahadur.*	Subadar-Major	1 Jan. 16	No. 187, 18 Feb. 16
Do.	**Gujа Singh**, Naik, 1907	Naik	1 Jan. 16	No. 187, 18 Feb. 16
Do.	**Mangal Singh**, Subadar	Subadar	17 Aug. 17	No. 1359, 17 Aug. 17
Do.	**Maghar Singh**, Subadar	Do.	5 Mar. 17	No. 8, 1 Jan. 18
35th Sikhs.	*Kurm Singh, Havildar*	*Havildar*	*16 Sep. 97*	*No. 332, 25 Feb. 98*
Do.	*Waryam Singh, Jamadar*	*Sepoy*	*16 Sep. 97*	*No. 332, 25 Feb. 98*
Do.	**Bir Singh**, Havildar, 2103	Havildar	29 Oct. 15	Lon. Gaz., 29 Oct. 15
Do.	**Lal Singh**, Lance-Havildar, 2479	Lance-Havildar	29 Oct. 15	Lon. Gaz. 29 Oct. 15
Do	**Bir Singh**, Sepoy, 3420	Sepoy	5 Nov. 17	No. 1885, 16 Aug. 18
36th Sikhs.	*Sundar Singh, Havildar, 470*	*Do.*	*27 Oct. 95*	*No. 578, 15 May 96*
Do	*Sundar Singh, Naick, 330*	*Do.*	*3 Sep. 97*	*No. 133, 4 Feb. 98*
Do.	**Gurmukh Singh**, Subadar	Do.	3 Sep. 97	No. 133, 4 Feb. 98
Do.	*Sobha Singh, Jemadar*	*Do.*	*3 Sep. 97*	*No. 133, 4 Feb. 98*
Do.	*Jiwan Singh, Sepoy, 1354 (a)*	*Do.*	*3 Sep. 97*	*No. 133, 4 Feb. 98*
Do.	*Ghulla Singh, Sepoy, 1146*	*Do.*	*13 Sep. 97*	*No. 133, 4 Feb. 98*
Do	*Attar Singh, Naick, 1078*	*Havildar*	*13 Sep 97*	*No. 133, 4 Feb. 98*
Do.	*Hira Singh, Naick, 1183*	*Sepoy*	*13 Sep. 97*	*No. 133, 4 Feb. 98*
Do	*Sujan Singh, Naick, 1046*	*Do.*	*13 Sep. 97*	*No 133, 4 Feb. 98*
Do.	*Sher Singh, Havildar, 368*	*Do.*	*13 Sep. 97*	*No. 133, 4 Feb. 98*
Do.	*Mehmapgh, Sin Seoy, 823*	*Do.*	*13 Sep. 97*	*No 133, 4 Feb. 98*
Do.	**Chajja Singh**, Jemadar	Do.	13 Sep. 97	No. 133, 4 Feb. 98
Do.	**Badan Singh**, Havildar, 1369	Do.	13 Sep. 97	No. 133, 4 Feb. 98
Do.	*Phuman Singh, Havildar, 1597*	*Do.*	*13 Sep. 97*	*No. 133, 4 Feb. 98*
Do.	*Ghuna Singh, Naik, 1600*	*Do.*	*13 Sep. 97*	*No. 133, 4 Feb. 98*
Do.	*Bela Singh, Sepoy, 1295*	*Do.*	*13 Sep. 97*	*No. 133, 4 Feb. 98*
Do.	*Rur Singh, Sepoy, 180*	*Do.*	*13 Sep. 97*	*No. 133, 4 Feb. 98*
Do.	*Harnam Singh, Naick, 1589*	*Do.*	*13 Sep. 97*	*No. 133, 4 Feb. 98*
Do.	*Harnam Singh, Naick, 817*	*Lance-Naick*	*13 Sep. 97*	*No. 133, 4 Feb. 98*
Do.	*Natha Singh, Sepoy, 1539*	*Sepoy*	*13 Sep. 97*	*No. 133, 4 Feb. 98*
Do.	*Bhagwan Singh, Sepoy, 1588*	*Do.*	*13 Sep. 97*	*No. 133, 4 Feb. 98*
Do.	*Sada Singh, Lance-Naik, 807*	*Lance-Naick*	*13 Sep. 97*	*No. 133, 4 Feb. 98*
Do.	*Thaman Singh, Sepoy, 1741*	*Sepoy*	*13 Sep. 97*	*No. 133, 4 Feb. 98*
Do.	*Bhagwan Singh, Sepoy, 2277*	*Do.*	*3 June 15*	*Lon. Gaz., 23 June 15*
Do.	**Sundar Singh**, Subadar	Subadar	12 Apl 16	No. 1385, 18 Nov 16
Do.	**Massa Singh**, Jemadar	Jemadar	12 Apl. 16	No. 1385, 18 Nov. 16

Corps.	Names and Rank.	In what rank admitted.	Date of admission to the Order.	Authority.

THE INDIAN ORDER OF MERIT—*contd.*

2nd Class—*contd.*

Corps.	Names and Rank.	In what rank admitted.	Date of admission to the Order.	G. G. O. or A. D. Notfn.
36th Sikhs	**Pritam Singh,** Naik, 3013	Naik	17 Apl. 16	No. 1385, 18 Nov. 16
Do.	**Prem Singh,** Lance Naik, 3648	Lance-Naik	12 Apl. 16	No. 1385, 18 Nov. 16
Do.	**Hari Singh,** Naik, 2948	Naik	15 Apl. 16	No. 1385, 18 Nov. 16
Do.	**Kartar Singh,** Naik, 3539	Naik	16 Feb. 17	No. 115, 19 Jan. 18
Do.	**Fakir Singh,** Naik, 3126	Naik	17 Aug. 17	No. 1359, 17 Aug. 17
Do.	**Sewa Singh,** Jemadar	Jemadar	17 Aug. 17	No. 1359, 17 Aug. 17
37th Dogras	*Hakum, Subadar*	*Naick*	*31 Jan. 79*	*No. 736, 8 Aug. 79*
Do.	**Surjan,** Sepoy, 1441½	Sepoy	7 Jan. 16	No. 524, 12 May 16
Do.	**Bir Singh,** Sepoy, 2370	Do.	7 Jan. 16	No. 524, 12 May 16
Do.	**Devi Singh,** Sepoy, 2124	Do.	7 Jan. 16	No. 524, 12 May 16
Do.	**Munshi,** Jemadar	Jemadar	21 Jan. 16	No. 524, 12 May 16
Do.	**Sadhu,** Havildar, 1678	Havildar	21 Jan. 16	No. 524, 12 May 16
Do.	**Bhag Singh,** Subadar-Maj.	Subadar-Maj	17 Aug. 17	No. 1359, 17 Aug. 17
Do.	**Bijho,** Havildar, 1610	Havildar	17 Aug. 17	No. 1359, 17 Aug. 17
38th Dogras	**Giyar Singh,** Jemadar	Do.	4 May 15	No. 563, 11 June 15
Do.	**Munshi,** Sepoy, 2163	Sepoy	13 Jan. 16	No 524, 12 May 16
39th Garh. Rif.	*Jagat Singh Rawat, Subadar*	*Sepoy*	*2 Apl. 91*	*No. 728, 31 July 91*
Do.	**Alam Singh Negi,** Havildar, 1810	Havildar	23 Nov. 14	No. 117, 29 Jan. 15
Do.	**Ganesh Singh Sajwan,** Rifleman, 1342	Rifleman	9 Nov. 14	No. 111, 29 Jan. 15
Do.	**Partab Rana,** Rifleman, 2417	Do.	18 Feb. 15	No. 313, 3 April 15
Do.	**Jawahiru Negi,** Rifleman, 2285	Do.	4 May 15	No. 563, 12 June 15
Do.	**Banchu Negi,** Rifleman, 2480	Do.	4 May 15	No. 563, 12 June 15
Do.	**Butha Sing Negi,** Jemadar	Do.	4 May 15	No. 563, 12 June 15
Do.	**Jaman Sing Bisht,** Naik, 1283	Do.	4 May 15	No. 563, 12 June 15
40th Pathans	*Abdul Ahmad, Havildar, 3561*	*Naick*	*23 Mar. 91*	*No. 952, 2 Oct. 91*
40th Pathans	**Gul Zaman,** Subadar	Subadar	19 July 17	No. 8, 1 Jan. 18
Do	*Mahomed Khan, Sepoy, 4080 (a)*	*Sepoy*	*16 Sep. 97*	*No. 232, 25 Feb. 98*
Do.	*Sharif, Havildar, 1352*	*Havildar*	*26 June 04*	*No. 1143, 2 Dec. 04*
Do.	*Payo, Lance-Naick*	*Lance-Naick*	*26 June 04*	*No. 1143, 2 Dec. 04*
Do.	*Akbar Shah, Sepoy, 3650*	*Sepoy*	*6 July 04*	*No. 1143, 2 Dec. 04*
Do.	**Lehna Singh,** Jemadar	Jemadar	3 June 15	Lon. Gaz., 23 June 15
Do.	**Abas Khan,** Havildar, 4551	Havildar	23 Dec 15	Lon. Gaz., 23 Dec. 15
40th Pathans	**Jehandad Khan,** Subadar *Bahadur*	Subadar	1 Jan. 16	No. 187, 18 Feb. 16
Do.	**Punna Khan,** Naik, 4799	Naik	2 June 17	No. 868, 2 June 17
Do.	**Gul Zaman,** Subadar	Subadar	19 July 17	No. 8, 1 Jan. 18
41st Dogras	*Gopi, Sepoy, 2096*	*Sepoy*	*20 Oct. 97*	*No. 430, 22 Apl. 98*
Do.	**Gujar Singh,** Jemadar	Havildar	4 May 15	No. 563, 12 June 15
Do.	**Tara Chand,** *Subadar*	Subadar	21 Jan. 16	No. 1302, 8 Aug. 17
Do	**Hoshnack Singh,** Sepoy, 340	Sepoy	9 Jan 16	No. 524, 12 May 16
Do.	**Surjan,** Naik, 547	Naik	9 Jan. 16	No. 524, 12 May 16
Do	**Rao,** Naik, 510	Do.	13 Jan. 16	No 524, 12 May 16
Do.	**Fauju,** Sepoy, 1259	Sepoy	13 Jan. 16	No. 524, 12 May 16
Do.	**Munshi,** Sepoy, 2136	Do.	13 Jan. 16	No. 524, 12 May 16
42nd Regt.	**Modha,** Lance-Naik, 3081	Lance-Naick	17 Dec. 15	No. 1284, 17 Dec. 15
Do.	**Chattar Budja,** Sepoy, 3184	Sepoy	17 Dec. 15	No. 1284, 17 Dec. 15
Do.	**Lachmi Narain,** Sepoy, 3020	Do.	17 Dec. 15	No. 1284, 17 Dec. 15
45th Sikhs	*Lal Singh, Sepoy*	*Havildar*	*7 Mar. 79*	*No. 605, 4 July 79*
Do.	*Mahtab Singh, Havildar, 3005*	*Sepoy*	*27 Oct. 95*	*No. 578, 15 May 99*
Do.	*Uttam Singh, Jemadar*	*Jemadar*	*26 July 95*	*No. 49, 14 Jan. 98*
Do.	*Bhola Singh, Sepoy*	*Sepoy*	*2 Aug. 97*	*No. 49, 14 Jan. 98*
Do.	**Natha Singh,** Subadar	Lance-Naick	26 July 97	No. 49, 14 Jan. 98
Do.	*Jeewa Singh, Naick, 3038*	*Sepoy*	*27 July 97*	*No. 49, 14 Jan. 98*
Do.	*Vir Singh, Havildar, 2729*	*Lance-Naick*	*29 July 97*	*No. 49, 14 Jan 98*
Do.	*Prem Singh, Jemadar*	*Sepoy*	*30 July 97*	*No. 49, 14 Jan 98*
Do.	*Nand Singh, Havildar, 3021*	*Do.*	*2 Aug. 97*	*No. 49, 14 Jan. 98*
Do.	*Wassan Singh, Jemadar*	*Havildar*	*5 Dec. 01*	*No. 403, 9 May 02*
Do.	**Roda Singh,** Havildar, 3792	Sepoy	5 Dec. 01	No. 403, 9 May 02
Do.	**Channan Singh,** M.S.M., Lance-Naik, 525	Lance-Naik	16 Feb. 17	No. 115, 19 Jan. 18
Do.	**Channan Singh,** Lance-Naik 525	Lance-Naik	2 June 17	No. 868, 2 June 17
Do.	**Sundar Singh,** Subadar Maj.	Subadar Maj.	17 Aug 17	No. 1359, 17 Aug. 17
Do.	**Narain Singh,** Subadar	Subador	17 Aug. 17	No. 1359, 17 Aug. 17
Do.	**Bagga Singh,** Naik, 77	Naik	17 Aug. 17	No. 1359, 17 Aug. 17
Do.	**Battan Singh,** Sepoy. 1197	Sepoy	17 Aug. 17	No. 1359, 17 Aug. 17
Do.	**Mall Singh,** Lance-Naik, 615	Lance-Naik	17 Aug. 17	No. 1359, 17 Aug. 17
46th Punj	*Said Baz, Havildar, 275*	*Sepoy*	*8 May 86*	*No. 631, 17 Sep 86*
Do.	**Hagat Muhammad,** Jemadar,	Jemadar	17 Aug. 17	No. 1359, 17 Aug. 17
47th Sikhs	*Wurriam Singh, Subdr.-Maj., Sardar Bahadur, Hony. Capt.*	*Do.*	*10 Dec. 79*	*No. 252, 23 Apl. 80*
Do.	*Sham Singh, Havildar, 2680*	*Do.*	*27 Oct. 95*	*No. 578, 15 May 96*
Do.	*Pertab Singh, Havildar, 259*	*Do.*	*27 Oct. 95*	*No. 578, 15 May 96*
Do.	*Atar Singh, Subadar-Maj., Sardar Bahadur, Hony. Capt.*	*Havildar*	*24 Feb. 98*	*No. 99, 27 Jan. 99*
Do.	**Bakshi Singh**	Subadar	10 Mar. 15	No. 356, 16 Apl. 15
Do.	**Harnam Singh**	Do.	10 Mar. 15	No. 356, 16 Apl. 15

(*a*) Transferred to Reserve.

THE INDIAN ORDER OF MERIT—*contd.*

2nd Class—*contd.*

Corps.	Names and Rank.	In what rank admitted.	Date of admission to the Order.	Authority. G. G. O. or A. D. Notfn.
47th Sikhs	**Narain Singh**, Jemadar	Havildar	10 Mar. 15	No. 356, 16 Apl. 15
Do.	**Gujjan Singh**, Havildar, 514	Do.	4 May 15	No. 563, 12 June 15
Do.	**Rur Singh**, Sepoy, 2270	Sepoy	4 May 15	No. 563, 12 June 15
Do.	**Harnam Singh**, Havildar, 1078	Havildar	23 Mar. 17	No. 8, 1 Jan. 18
Do.	**Natha Singh**, Sepoy, 4639	Sepoy	23 Mar. 17	No. 8, 1 Jan. 17
Do.	**Mastan Singh**, Havildar, 728	Havildar	5 Nov. 17	No. 1885, 16 Aug. 18
48th Ps.	**Dayal Singh**, Havildar, 943	Do.	17 Aug. 17	No. 1357, 17 Aug. 17
51st Sikhs	*Narayan Singh, Subadar*	*Do.*	*3 May 81*	*No. 351, 24 June 81*
Do.	*Narayan Singh, Subadar-Maj., Sardar Bahadur (Hony. Capt.)*	*Subadar*	*10 June 97*	*No. 785, 16 July 97*
Do.	*Sheo Singh, Jemadar*	*Sepoy*	*10 June 97*	*No. 865, 6 Aug. 97*
Do.	*Maha Singh, Havildar, 3363*	*Havildar*	*10 June 97*	*No. 865, 6 Aug. 97*
Do.	*Muhan Singh, C. Havildar*	*Do.*	*10 June 97*	*No. 865, 6 Aug. 97*
Do.	*Habibullah, Sepoy, 4666*	*Sepoy*	*10 June 97*	*No. 865, 6 Aug. 97*
Do.	*Shah Sowar, Havildar, 3900*	*Lance-Naick*	*10 June 97*	*No. 865, 6 Aug. 97*
Do.	*Jalandhar, Lance-Naick, 4287*	*Do.*	*10 June 97*	*No. 865, 6 Aug. 97*
Do.	*Mahomed Baksh, Havildar, 4031*	*Havildar*	*10 June 97*	*No. 865, 6 Aug. 97*
Do.	**Prem Singh**, Naick, 1310	Naick	6 Jan. 16	No. 524, 12 May 16
Do.	**Jawand Singh**, Lnce. Naick, 1342	Lnce.-Naick	7 Jan. 16	No. 524, 12 May 16
Do.	**Khan Zaman**, Sepoy, 1116	Sepoy	7 Jan. 16	No. 524, 12 May 16.
Do.	**Chanda Singh**, Jemadar	Jemadar	21 Jan. 16	No. 524, 12 May 16
Do.	**Makmad Din**, Jemadar	Do.	6 Apl. 16	No. 1385, 18 Nov. 16
Do.	**Mangal Singh**, Havildar, 840	Havildar	6 Apl. 16	No. 1385, 18 Nov. 16
Do.	**Suhel Singh**, Havildar, 1193	Do.	6 Jan. 16	No. 1335, 18 Nov. 16
Do.	**Khankhel**, Naik, 1189	Naik	8 Mar. 16	No. 1387, 18 Nov. 16
Do.	**Tika Khan**, Subada	Subadar	17 Aug. 17	No. 1359, 17 Aug. 17
Do.	**Arjan Singh**, Subadar	Do.	22 Apl. 17	No. 8, 1 Jan. 18
Do.	**Kalya**, Havildar, 993	Havildar	22 Apl. 17	No. 8, 1 Jan. 18
Do.	**Nikka Singh**, Naik, 1031	Naik	22 Apl. 17	No. 8, 1 Jan. 18
52nd Sikhs	*Jai Singh, Naick*	*Do.*	...	*No. 639, 19 Nov. 80*
Do.	*Partab Singh, Sepoy*	*Sepoy*	...	*No. 639, 19 Nov. 80*
Do.	**Ghulam Hassan**, Naik, 2665	Naik	8 Mar. 16	No. 1387, 18 Nov. 16
53rd Sikhs	*Debi Singh, Sepoy, 2196*	*Do.*	*20 Oct. 97*	*No. 430, 22 Apl. 98*
Do.	*Bell Ram, Subadar-Major, Sardar Bahadur (Hony. Capt.)*	*Jamadar*	*20 Oct. 97*	*No. 430, 22 Apl. 98*
Do.	**Maggar Singh**, Naick, 2019	Sepoy	20 Oct. 97	No. 430, 22 Apl. 98
Do.	*Jai Singh, Lance.-Naick, 1427*	*Lance-Naick*	*6 Nov. 97*	*No. 430, 22 Apl. 98*
Do.	*Ramzan Khan, Subadar, Sardar Bahadur*	*Subadar*	*2 Dec. 97*	*No. 430, 22 Apl. 98*
Do.	**Ali Sher**, Havildar, 2182	Havildar	20 May 08	No. 526, 26 June 08
Do.	**Hira Singh**, Sepoy, 2797	Sepoy	20 May 08	No. 526, 26 June 08
Do.	**Arsla Khan**, Subadar *Bahadur*	Jemadar	24 May 08	No. 526, 26 June 08
Do.	**Faiz Talab**, Jemadar	Do.	3 Dec. 15	No. 1213, 3 Dec. 15
Do.	**Bahadur Shah**, Naik, 3543	Naik	3 Dec. 15	No. 1213, 3 Dec. 15
Do.	*Allah Khan, Sepoy, 3218*	*Sepoy*	*3 Dec. 15*	*No. 1213, 3 Dec. 15*
Do.	**Chanan Singh**, Subadar	Jemadar	7 Jan. 16	No. 524, 12 May 16
Do.	**Shinghar Singh**, Havildar, 3112	Havildar	13 Jan. 16	No. 524, 12 May 16
Do.	**Fateh Muhammad**, Havildar, 2973	Do.	7 Jan. 16	No. 524, 12 May 16
Do.	**Kehar Singh**, Naik, 3573	Naik	6 Jan. 16	No. 524, 12 May 16
Do.	**Shingar Singh**, Sepoy, 194	Sepoy	8 Mar. 16	No. 1385, 18 Nov. 16
54th Sikhs	*Wasira, Havildar, 3474*	*Havildar*	*19 Mar. 91*	*No. 354, 10 Apl. 91*
Do.	*Ganesha Singh, Naick, 3585*	*Naick*	*19 Mar. 91*	*No. 354, 10 Apl. 91*
Do.	*Tulsi Ram, Subadar*	*Havildar*	*3 Apl. 95*	*No. 1228, 29 Nov. 95*
Do.	*Bhola Singh, Lance-Naick, 629*	*Lance-Naick*	*18 Nov. 02*	*No. 381, 24 Apl. 03*
Do.	**Bishen Singh**, Havildar, 1877	Sepoy	18 Feb. 08	No. 392, 15 May 08
Do.	**Buta Khan**, Havildar, 2095	Havildar	2 June 17	No. 868, 2 June 17
Do.	**Ali Bahadur**, Jemadar, 2437	Do.	20 June 17	No. 558, 16 Mar. 18
Do.	**Qasim Ali**, Jemadar	Jemadar	24 June 17	No. 1887, 2 Nov. 17
55th Rif.	*Kaisroo, Havildar*	*Havildar*	...	*G. O. C. C., 17 Jan. 59*
Do.	**Bela Singh**, Subadar	Bugler	10 June 97	No. 865, 6 Aug. 97
Do.	*Ishar Singh, Havildar, 878*	*Lance-Naick*	*10 June 97*	*No. 865, 6 Aug. 97*
Do.	*Nurdad, Sepoy, 774*	*Sepoy*	*10 June 97*	*No. 865, 6 Aug. 97*
Do.	*Daud Shah, Subadar*	*Havildar*	*25 June 04*	*No. 749, 29 July 04*
Do.	*Kalandar, Sepoy, 1780*	*Sepoy*	*18 May 08*	*No. 526, 26 June 08*
Do.	**Teja Singh**, Jemadar,	Naick	24 May 08	No. 526, 26 June 08
Do.	*Basant Singh, Sepoy, 1360*	*Sepoy*	*24 May 08*	*No. 526, 26 June 08*
Do.	**Usman Khan**, Sepoy, 1695	Do.	24 Oct. 14	No. 111, 20 Jan. 15
Do.	**Khoda Bakhsh**, Havildar, 1267	Havildar	21 Nov. 16	No. 676, 5 May 17
56th Rifles	**Khwaja Muhammad**, Subadar	Jemadar	13 Jan. 16	No. 524, 12 May 16
Do.	**Allah Ditta**, Naik, 1600	Naik	7 Jan. 16	No. 524, 12 May 16
Do.	**Jetha Singh**, Sepoy, 2184	Sepoy	7 Jan. 16	No. 524, 12 May 16
Do.	**Jagat Singh**, Havildar, 1016	Havildar	13 Jan. 16	No. 524, 12 May 16
Do.	**Sapuran Singh**, Havildar, 1078	Do.	12 Jan. 16	No. 524, 12 May 16
Do.	**Harnam Singh**, Subr.-Major, *Bahadur*	Subadar-Major	8 Mar. 16	No. 1385, 18 Nov. 16
Do.	*Jodh Khan, Subadar*	*Subadar*	*17 Aug. 17*	*No 1359, 17 Aug. 17*

Corps.	Names and Rank.	In what rank admitted.	Date of admission to the order.	Authority.
	THE INDIAN ORDER OF MERIT—*contd.*			
	2nd Class—*contd.*			G. G. O. or A. D. Notfn.
56th Rif.	**Bishen Singh,** Lance-Naik, 2414	Lance-Naik	22 Apl. 17	No. 8, 1 Jan. 18
57th Rif.	**Lehna Singh,** Lance-Naick, 1503 (*a*)	Do.	20 Apl. 01	No. 683, 26 July 01
Do.	**Mir Dast,** Jemadar, 836	Do.	18 May 08	No. 526, 26 June 08
Do.	**Arsla Khan,** Subadar-Major	Subadar	20 May 08	No. 526, 26 June 08
Do.	**Dadan,** Sepoy, 2820	Sepoy	20 May 08	No. 526, 26 June 08
Do.	**Imam Ali,** Sepoy, 2913	Do.	20 May 08	No. 526, 26 June 08
Do.	**Said Akbar,** Lance-Naik, 2718	Lance-Naik	23 Nov. 14	No. 111, 29 Jan. 15
Do.	*Daulat Khan, Sepoy, 3576*	*Sepoy*	*23 Nov. 14*	*No. 111, 29 Jan. 15*
Do.	**Lalak,** Lance-Naik, 2544	Lance-Naik	18 Feb. 15	No. 313, 2 Apl. 15
Do.	**Mangal Singh,** Jemadar	Jemadar	3 June 15	Lon. Gaz., 23 June 15
Do.	**Atma Singh, Naik,** 2609	Naik	3 June 15	Lon. Gaz., 23 June 15
Do.	**Gagna, Havildar,** 2630	Havildar	23 July 15	No. 815, 27 Aug. 15
Do.	**Palla Ram,** Sepoy, 2632	Sepoy	1 Jan. 16	No. 187, 18 Feb. 10
Do.	**Diwan Singh,** Sepoy, 3402	Do.	3 June 16	No. 849, 29 July 16
Do.	**Salim Khan,** Havildar, 2172	Havildar	11 Apl. 17	No. 1644, 28 Sep. 17
58th Rif.	*Man Sing, Subadar, Bahadur*	*Do.*	*10 Dec. 79*	*No. 252, 23 Apl. 80*
Do.	**Umar Khan,** 3680	Do.	10 Dec. 79	No. 252, 23 Apl. 80
Do.	**Karam Singh,** Havildar, 1811	Do.	31 Oct. 14	No. 111, 29 Jan. 15
Do.	**Harchand Singh,** Subadar	Do.	10 Mar. 15	No. 356, 16 Apl. 15
Do.	**Roshan Khan,** Havildar, 1838	Havildar	10 Mar. 15	No. 256, 16 Apl. 15
Do.	**Saidak,** Havildar, 3572	Do.	10 Mar. 15	No. 356, 16 Apl. 15
Do.	**Lal Badshah,** Naik, 3032	Naik	10 Mar. 15	No. 336, 16 Apl. 15
Do.	**Sher Khan,** Lance-Naik, 2834	Do.	10 Mar. 15	No. 356, 16 Apl. 15
Do.	**Isar Sing,** Sepoy 2742	Sepoy	10 Mar. 15	No. 356, 16 Apl. 15
Do.	**Sihel Singh,** Jemadar	Jemadar	18 Feb. 15	No. 313, 2 Apl. 15
Do.	**Santa Singh,** Havildar, 1925	Havildar	1 Jan. 16	No. 187, 18 Feb. 16
Do.	**Kashmir Singh,** Naik, 2830	Naik	1 Jan. 16	No. 187, 18 Feb. 16
Do.	**Phangan Singh,** Lance-Naik, 3131	Lance-Naik	1 Jan. 16	No. 187, 18 Feb. 16
Do.	**Lal Khan,** Subadar	Subadar	19 Nov. 17	No. 1178, 3 June 18
59th Rif.	*Amar Sing, Sepoy, 4230*	*Do.*	*24 Apl. 08*	*No. 526, June 08*
Do.	**Buta Singh** Lance-Naik, 3705	Lance-Naik	4 May 15	No. 563, 12 June 15
Do.	**Abdul Wahab,** Havildar, 3063	Havildar	10 Mar. 15	No. 356, 16 Apl. 15
Do.	*Dost Mohamed, Havildar, 3191*	*Do.*	*10 Mar. 15*	*No. 356, 16 Apl. 15*
Do.	*Mohamed Jan, Havildar, 3638*	*Do.*	*10 Mar. 15*	*No. 356, 16 Apl. 15*
Do.	**Muzafar Khan,** Havildar, 3663	Do.	10 Mar. 15	No. 356, 16 Apl. 15
Do.	*Biaz Gul, Lance-Naik 3902*	*Lance-Naik*	*12 Dec. 14*	*No. 111, 29 Jan. 15*
Do.	**Nasir Khan,** Subadar-Maj., *Sardar Bahadur.*	Subadar-Major	1 Jan. 16	No. 187, 18 Feb. 16
Do.	**Khan Gul,** Subadar	Subadar	8 Mar. 16	No. 1385, 18 Nov. 16
Do.	**Gauri Charan,** Subadar	Do.	18 Apl. 16	No. 1385, 18 Nov. 16
Do.	**Sunit Chand,** Jamadar	Havildar	8 Mar. 16	No. 1385, 18 Nov. 16
Do.	**Kalandar Khan,** Jemadar	Do.	18 Apl. 16	No. 1385, 18 Nov. 16
Do.	**Tola Singh,** Subadar	Subadar	8 Mar. 16	No. 1387, 18 Nov. 16
Do.	**Ali Faqir,** Lance Naik, 4551	Lance-Naik	8 Mar. 16	No. 1387, 18 Nov. 16
Do.	**Sahib-i-Haq,** Subadar	Subadar,	5 Nov. 17	No. 1885, 16 Aug. 18
62nd Punj.	*Karm Singh, Subadar*	*Lance-Naik.*	*7 Feb 95*	*No. 657, 21 June 95*
Do.	*Sher Khan, Sepoy, 1878*	*Sepoy*	*3 Feb. 15*	*No. 525, 3 June 15*
Do.	**Nihal Singh,** Sepoy, 1434	Sepoy	3 June 16	No. 849, 29 July 16
Do.	**Muhammad Khan,** Havildar, 1618	Havildar	8 Feb. 17	No. 115, 19 Jan. 18
66th Punj.	*Canda Singh, Subadar*	*Naik*	*26 July 97*	*No. 59, 14 Jan. 98*
69th Punj.	*Baryam Singh, Subadar*	*Havildar*	*16 Nov. 97*	*No. 430, 22 Apl. 98*
Do.	**Jagandar Singh,** Sub-Maj. Bahadur	Sub-Maj.	3 June 16	No. 849, 29 July 16
Do.	**Muhammad Khan (I),** Subadar	Subadar	3 June 16	No. 849, 29 July 16
Do.	**Sardar Khan,** Jemadar	Jemadar	3 June 16	No. 849, 29 July 16
Do.	**Nidham Singh,** Lnce-Naik, 440	Lance-Naik	3 June 16	No. 849, 29 July 16
Do.	**Kirpa Ram,** Sepoy, 649	Sepoy	3 June 16	No. 849, 29 July 16
Do.	**Sher Muhammad,** Lance Naik, 2081	Lance-Naik	11 Sep. 17	No. 1938, 9 Nov. 17
71st Punj.	*Hira Singh, Havildar*	*Sepoy*	. . .	*G. O. C. C. 9 Apl. 58*
72nd Punj.	*Gama, Sepoy, 638*	*Havildar*	*28 Mar. 91*	*No. 647, 3 July 51*
Do.	**Fateh-Dad,** Sepoy, 646 (*b*)	Do.	28 Mar. 91	No. 647, 3 July 91
Do.	*Karam-dad, Havildar, 242*	*Do.*	*28 Mar. 91*	*No. 647, 3 July 91*
Do.	*Gulab Khah, Havildar, 380*	*Do.*	*28 Mar. 91*	*No. 647, 3 July 91*
Do.	*Din Muhammad, Sepoy, 395*	*Do.*	*28 Mar. 91*	*No. 647, 3 July 91*
Do.	*Fazl-din, Sepoy, 401 (b)*	*Do.*	*28 Mar. 91*	*No. 647, 3 July 91*
Do.	*Fakir Muhammad, Sepoy, 402 (b)*	*Do.*	*28 Mar. 91*	*No. 647, 3 July 91*
Do.	*Muhammad Alam, Sepoy, 407 (b)*	*Do.*	*28 Mar. 91*	*No. 647, 3 July 91*
Do.	*Shams-din, Sepoy, 411 (b)*	*Do.*	*28 Mar. 91*	*No. 647, 3 July 91*
Do.	*Kashu, Sepoy, 415 (b)*	*Do.*	*28 Mar. 91*	*No. 647, 3 July 91*
Do.	*Karam-Ilahi, Sepoy, 426*	*Do.*	*28 Mar. 91*	*No. 647, 3 July 91*
Do.	*Miran-bakhsh, Sepoy, 439 (b)*	*Do.*	*28 Mar. 91*	*No. 647, 3 July 91*
Do.	*Ghulam Muhammad, Sepoy, 449 (b)*	*Do.*	*28 Mar. 91*	*No. 647, 3 July 91*
Do.	*Saif Ali, Sepoy, 454*	*Do.*	*28 Mar. 91*	*No. 647, 3 July 91*
Do.	*Mahtab-din, Sepoy, 460*	*Do.*	*28 Mar. 91*	*No. 647, 3 July 91*
Do.	*Mihr-din, Sepoy, 259*	*Do.*	*28 Mar. 91*	*No. 647, 3 July 91*
Do.	*Mangu, Sepoy, 277 (b)*	*Do.*	*28 Mar. 91*	*No. 647, 3 July 91*
Do.	*Chaugatta, Jemadar*	*Do.*	*28 Mar. 91*	*No. 647, 3 July 91*

(*a*) Transferred to Reserve. (*b*) Transferred to Active Reserve.

THE INDIAN ORDER OF MERIT—*contd.*

2nd Class—*contd.*

Corps.	Names and Rank.	In what rank admitted.	Date of admission to the Order.	Authority. G. G. O. or A. D. Notfn.
72nd Punj.	Rangoo Rest, Sepoy	Sepoy	28 Mar. 91	No. 647, 3 July 91
Do.	Gopal Singh, Subadar-Major, Bahadur	Subdr.-Major	25 Apl. 91	No. 862, 4 Sep. 91
Do.	Hira Singh, Naick, 212	Naik	6 Aug. 00	No. 1086, 5 Dec. 92
Do.	**Pakhar Singh,** Sepoy, 2632	Sepoy	22 Aug. 13	No. 709, 22 Aug. 13
73rd Infy.	Mahomed Hayath, Subadar-Maj., Bahadur	Naik	8 Mar. 87	No. 590, 29 July 87
74th Punj.	Narayan Singh, Havildar, 2815	Sepoy	22 Aug. 95	No. 578, 15 May 96
Do.	**Dost Muhammad Khan,** Subadar-Major	Subadar	3 June 16	No. 849, 29 July 16
75th Infy.	Ramkrishnamma, Havildar, 1490	Private	23 Aug. 87	No. 542, 29 June 88
76th Punj.	Ajab Khan, Hon'ble Sub.-Maj., (Hony. Capt. Sardar Bahadur)	Sub.-Maj.	3 Sep. 15	No. 849, 3 Sep. 15
Do.	Mehdi Khan, Subdr.	Jemdr.	3 Sep. 15	No. 849, 3 Sep. 15
Do.	Kesar Singh, Havildar, 994	Havildar	3 Sep. 15	No. 849, 3 Sep. 15
Do.	Karam Dad, Naik, 1029	Lance-Naik	3 Sep. 15	No. 849, 3 Sep. 15
Do.	**Shiv Ram,** Sepoy, 1549	Sepoy	3 Sep. 15	No. 849, 3 Sep. 15
Do.	Bagga Khan, Sepoy, 398	Do.	19 Nov. 15	No. 1159, 19 Nov. 15
Do.	**Hukam Singh,** Lance-Naik, 533	Lance-Naik	19 Nov. 15	No. 1159, 19 Nov. 15
Do.	**Harnam Singh,** Havildar, 1528	Havildar	. . .	No. 1203, 3 Dec. 15
Do.	**Muhammad Sadik,** Naik, 2352	Naik		No. 1203, 3 Dec. 15
Do.	**Hussaijn Shah,** Subadar-Maj.	Subadar-Maj.	17 Aug. 17	No. 1357, 17 Aug. 17
82nd Punj.	Sundar Singh, Subadar-Maj.	Do.	13 Sep. 97	No. 133, 4 Feb. 98
Do.	**Ram Singh,** Subadar, *Bahadur*	Sepoy	28 Oct. 97	No. 430, 22 Apl. 98
Do.	**Kesar Singh,** Sepoy, 1443	Sepoy	8 Mar. 16	No. 1387, 18 Nov. 16
Do.	Nawab Khan, Lance-Naik, 2565	Lance-Naik	8 Feb. 17	No. 115, 19 Jan. 18
Do.	**Ali Muhammad,** Sepoy, 3237	Sepoy	17 Aug. 17	No. 1359, 17 Aug. 17
Do.	**Kalu Ram,** Havildar, 1053	Havildar	9 Feb. 17	No. 1767, 2 Aug. 18
83rd L. Infy.	Dailiah, Private, 1955	Private	7 Feb. 92	No. 709, 22 July 92
84th Punj.	**Muhammad Zman,** Sepoy, 2423	Sepoy	17 Aug. 17	No. 1359, 17 Aug. 17
87th Punj.	Sayyid Yakub, Subadar-Major, Sardar-Bahadur.	Subadar	8 Mar. 87	No. 590, 29 July 87
Do.	Abdul Rusack, Private, 885	Private	Mar. 87	No. 590, 29 July 87
Do.	Permaul, Private, 1305	Do.	8 Mar. 87	No. 590, 29 July 78
89th Punj.	Mahe Khan, Sepoy, 1058 (c)	Sepoys, Burma Mily. Police	28 Mar. 91	No. 647, 3 July 91
Do.	Uttam Singh, Havildar, 237	Sepoys, Burma Mily. Police	6 Jan. 93	No. 825, 25 Aug. 93
Do.	Ganda Singh, Havildar, 241	Sepoys, Burma Mily. Police	6 Jan. 93	No. 825, 25 Aug. 93
89th Punj.	**Dasunda Singh,** Sepoy, 2131	Sepoy	3 June 15	No. 524, 3 June 15
Do.	**Indar Singh,** Sepoy, 2316	Do.	29 Oct. 15	Lon. Gaz., 29 Oct. 15
Do.	**Hira Tiwari,** Colour-Havildar, 1088	Colour-Havildar	1 Jan. 16	No. 187, 18 Feb. 16
Do.	**Mahtab Ali,** Naik, 1841	Naik	11 Mar. 16	No. 1387, 18 Nov. 16
Do.	**Zaman Shah,** Subadar	Subadar	13 Apl. 16	No. 1385, 18 Nov. 16
Do.	**Ram Persan Tiwari,** Havildar, 1749	Havildar	13 Apl. 16	No. 1385, 18 Nov. 16
90th Punj.	Karam-din, Sepoy, 406	Naik	28 Mar. 91	No. 647, 3 July 91
Do.	**Partab Singh,** Sepoy, 2478	Do.	19 Nov. 15	No. 1159, 19 Nov. 15
Do.	**Balkaran Misir,** Naik, 1592	Do.	17 Aug. 17	No. 1359, 17 Aug. 17
Do.	**Jhanda Singh,** Naik, 1481	Do.	17 Aug. 17	No. 1359, 17 Aug. 17
91st Punj.	Prakas Rae, Subadar-Maj., Sardar Bahadur. (Hony. Capt.)	Subadar	19 Feb. 91	No. 1180, 18 Dec. 91
Do.	Hara Singh, Subadar-Maj., Sardar Bahadur. (Hony. Capt.)	Jemadar	9 Feb. 92	No. 740, 29 July 92
Do.	Bahadur Khan, Subadar	Do.	26 Dec. 92	No. 826, 25 Aug. 93
Do.	**Suleman Khan,** Sepoy, 3275	Sepoy	27 Sep. 15	No. 1203, 3 Dec. 15
Do.	**Ata Muhammad,** Sepoy, 3729	Sepoy	15 Apl. 17	No. 8, 1 Jan. 18
92nd Punj.	Amir, Sepoy, 456 (b)	Sepoy	28 Mar. 91	No. 647, 3 July 01
Do.	**Muhammad Azim,** Havildar, 1753	Havildar	3 June 15	No. 524, 3 June 15
Do.	**Mihan Singh,** Naik, 1775	Naik	3 June 15	No. 524, 3 June 15
Do.	**Farman Ali,** Subadar-Major, *Bahadur*	Subadar-Maj.	6 Jan. 16	No. 1385, 18 Nov. 16
Do.	Hazara Singh, Jemadar	Jemadar	20 Jan. 16	No. 524, 12 May 16
Do.	**Partab Singh,** Jemadar	Subadar-Maj.	7 Apl. 16	No. 1387, 18 Nov. 16
Do.	**Sher Singh,** Subadar	Subadar	17 Aug. 17	No. 1359, 17 Aug. 17
93rd Burma Infy.	Jawand Singh, Subadar	Private	7 Feb. 92	No. 709, 22 July 92
Do.	**Muhammad Bakhsh,** Subr-Major, *Bahadur*	Subadar	18 Apl. 16	No. 1385, 18 Nov. 16
Do.	**Indar Singh,** Subadar	Jemadar	18 Apl. 16	No. 1385, 18 Nov. 16
Do.	**Karam Dad,** Subadar	Subadar	2 June 17	No. 868, 2 June 17
Do.	**Channan Singh,** Havildar, 2569	Havildar	2 June 17	No. 868, 2 June 17
Do.	**Lachman Singh,** Subadar	Subadar	17 Aug. 17	No. 1359, 17 Aug. 17
33rd Regt. (3rd B. Bn.) M.I., now 93rd Burma Infy.	Badawa Singh, late Private	Do.	20 Feb. 92	Madras C.O., No. 421, 2 Aug. 92.
96th Berar Infy.	Mehar Singh, Sepoy, 2331	Sepoy	15 Oct. 15	No. 1023, 15 Oct. 15
Do.	**Surjan,** Sepoy, 1356 1668.	Do.	29 Oct. 15	No. 1080, 29 Oct. 15

(*b*) Transferred to Active Reserve.
(*c*) Transferred to Reserve on 29th Feb. 1904.

THE INDIAN ORDER OF MERIT—*contd.*

2nd Class—*contd.*

Corps.	Names and Rank.	In what rank admitted.	Date of admission to the Order.	Authority. G. G. O. or A. D. Notfn.
97th Infy.	**Bishnath Singh,** Sepoy, 3330	Sepoy	7 Jan. 16	No. 524, 12 May 16
Do.	**Rachpal Singh,** Havildar, 2159	Havildar	21 Jan. 16	No. 524, 12 May 16
Do.	**Ramkumar Singh,** Subadar	Subadar	22 Jan. 16	No. 524, 12 May 16
101st Grs.	*Bhagwan Singh, Private, 1057*	*Private*	*27 July 80*	*No. 222, 15 Apl. 81*
Do.	**Fazil Khan,** Sepoy, 1222	Sepoy	3 June 15	No. 524, 3 June 15
Do.	**Rahim Khan,** Subadar	Subadar	30 May 18	No. 2536, 26 Oct. 18
Do.	**Sewaz Khan,** Private, 811	Private	15 May 15	No. 1245, 15 May 15
102nd Grs.	**Jafar Ali,** Subadar	Subadar	11 Jan. 15	No. 284, 19 Mar. 15
Do.	**Muhammad Ali,** Subadar, 335	Subadar	12 Feb. 17	No. 607, 23 Apl. 18
Do.	**Ganga Ram,** Jemadar	Jemadar	20 Feb. 17	No. 115, 19 Jan. 17
Do.	**Ganga Ram Singh,** Subadar	Subadar	17 Aug. 17	No. 1359, 17 Aug. 11
Do.	*Muhammad Ali, Subadar*	*Subadar*	*17 Aug. 17*	*No. 1359, 17 Aug. 17*
103 M. L. I.	**Dhoni Dhanaude,** Sepoy, 3173	Sepoy	10 Dec. 15	No. 1244, 10 Dec. 15
Do.	**Atmaram Chatge,** Colour-Havildar, 3060	Colour-Havildar	10 Dec. 15	No. 1244, 10 Dec 15
Do.	**Gopalrao Palande,** Jemadar	Do.	10 Dec. 15	No. 1244, 10 Dec. 15
104th Rif.	*Butta Singh, Havildar, 756*	*Naik*	*22 June 98*	*No. 137, 3 Feb. 99*
Do.	**Subal Singh,** Subadar	Subadar	17 Nov. 14	No. 1168, 24 Dec. 15
Do.	**Kheta Ram,** Rifleman, 2866	Rifleman	4 Dec. 14	No. 284, 19 Mar. 15
Do.	**Dhanna Ram,** Rifleman, 2578	Do.	4 Dec. 14	No. 284, 19 Mar. 15
Do.	**Maula Dad,** Rifleman, 2090	Do.	4 Dec. 14	No. 284, 19 Mar. 15
Do.	**Narsu Singh,** Naik, 2263	Naik	8 Dec. 14	No. 284, 19 Mar. 15
Do.	**Mangej Singh,** Rifleman, 2694	Rifleman	10 Dec. 15	No. 1244, 10 Dec. 15
Do.	**Mukh Ram,** Subadar	Subadar	17 Aug. 17	No. 1357, 17 Aug. 17
Do.	**Ganpat Singh,** Color-Havildar, 2241	Colour-Havildar	17 Aug. 17	No. 1357, 17 Aug. 17
105th Mah.L.I.	**Chimaji Gaurd,** Subadar	Subadar	9 June 17	No. 894, 9 June 17
106th Prs.	**Abdullah Khan,** Subadar	Sapper	2 Dec. 91	No. 179, 19 Feb. 92
Do.	*Muhammad Akbar, Jemadar*	*Private*	*1 Dec. 00*	*No. 355 of 1901.*
Do.	**Gulam Hussain,** Sepoy, 1822	Sepoy	17 Feb. 17	No. 236, 17 Feb. 17
Do.	**Kurban,** Sepoy, 2796	Sepoy	17 Feb. 17	No. 236, 17 Feb. 17
107th Prs.	**Khan Zaman,** Subadar	Subadar	10 Mar. 15	No. 356, 16 Apl. 15
Do.	**Nathu Singh,** Bugler, 4050	Bugler	4 May. 15	No. 563, 12 June 15
110th Infy.	**Shah Nawaz Khan,** Jemadar	Havildar	15 Oct. 15	No. 1022, 15 Oct. 15
Do.	**Mhadu Dalvi,** Havildar, 1966	Havildar	3 Mar. 17	No. 319 of 1917.
11th Infy.	*Hyder Beg, Lce.-Naik, 1148*	*Lce.-Naik*	*7 Dec. 14*	*No. 284, 19 Mar. 15*
Do.	**Sakaram More,** Lance-Naik, 1498	Do.	4 Aug. 15	No. 727, 4 Aug. 15
Do.	**Laxuman Joth,** Sepoy (Ward Orderly),	Sepoy	10 Dec. 15	No. 1244, 10 Dec. 15
116th Mah.	**Raja Ram Bwege,** Private, 1650	Private	7 Jan. 16	No. 524, 12 May 16
Do.	**Dianu Shinde,** Private, 1793	Do.	7 Jan. 16	No. 524, 12 May 16
Do.	**Murari Sinde,** Subadar	Jemadar	7 Jan. 16	No. 1387, 8 Nov. 16
117th Mah.	*Jaimal Sing, Cr.-Havildar*	*Cr.-Havildar*	*30 Jan. 90*	*No. 674, 18 July 90*
Do.	**Jairam Dalvi,** Havildar, 2437	Havildar	4 Aug. 15	No. 727, 4 Aug. 15
Do.	**Vithu Kadam,** Havildar, 2474	Havildar	4 Aug. 15	No. 727, 4 Aug. 15
Do.	**Mahadeorao Nalaude,** Subadar	Subadar	10 Dec. 15	No. 1244, 10 Dec. 15
119th Infy.	**Bhairo Sing,** Subadar-Major, *Bahadur* (*Hony. Capt.*)	Subadar	19 Dec. 03	No. 748, 29 July 04
Do.	**Natha Singh,** Naik, 3887	Naik	4 Aug. 15	No. 727, 4 Aug. 15
Do.	**Uma Rawat,** Havildar, 3639	Havildar.	4 Aug. 15	No. 727, 4 Aug. 15
Do.	**Ramlal Singh,** Private, 3486	Private	4 Aug. 15	No. 727, 4 Aug. 15
Do.	**Unad Singh,** Subadar	Subadar	*10 Dec. 15*	*No. 1244, 10 Dec. 15*
Do.	*Sarfaraj Khan Subadar*	*Subadar*	10 Dec. 15	No. 1244, 10 Dec. 15
Do.	**Gul Muhammad,** Havildar	Havildar	10 Dec. 15	No. 1244, 10 Dec. 15
Do.	**Sarfaraz Khan,** Subadar	Subadar	10 Dec. 15	No. 1244, 10 Dec. 15
120th Infy.	*Ali Sher Khan, Trumpeter, 1220*	*Trumpeter*	*11 Sep. 00*	*No. 46, 18 Jan. 01*
Do.	**Durga Rawat,** Subadar	Subadar	7 Dec. 14	No. 284, 19 Mar. 15
Do.	**Hukam Singh,** Sepoy, 1566	Sepoy	7 Dec. 14	No. 284, 19 Mar. 15
Do.	**Ganesh Ram,** Colour-Havildar, 873	Colour-Havildar	4 Aug. 15	No. 727, 4 Aug. 15
Do.	**Lachman Singh,** Sepoy, 1712	Sepoy	4 Aug. 15	No. 727, 4 Aug. 15
Do.	**Kum Singh,** Sepoy, 1513	Sepoy	4 Aug. 15	No. 727, 4 Aug. 15
Do.	**Khitab Gul,** Subadar-Maj., *Sardar Bahadur.*	Subadar-Maj.	17 Aug. 17	No. 1357, 17 Aug. 17
Do.	**Girwar,** Havildar, 1009	Havildar	17 Aug. 17	No. 1357, 17 Aug. 17
123rd Rif.	*Ramjirao Dalvi, Naick, 237*	*Naick*	*3 Sep. 86*	*No. 733, 7 Sep. 88*
Do.	*Rajbarao Palanda, Sepoy, 331*	*Do.*	*3 Sep. 86*	*No. 733, 7 Sep. 88*
1-123rd Out Ram's Rif.	**Bhim Singh,** Jemadar	Jemadar.	21-22 N. 17	No. 1178, 3 June 18
Do.	**Bhura Ram,** Subadar	Subadar	21 22 N. 17	No. 1178, 3 June 18
Do.	**Hans Ram,** Jemadar	Jemadar	22 Nov. 17	No. 1178, 3 June 18
Do.	**Arjun Ram,** Jemadar	Jemadar	21-22 N. 17	No. 1178, 3 June 18
Do.	**Nanig Ram,** Jemadar	Jemadar	11 Apl. 18	No. 2536, 26 Oct. 18
124th Infy.	*Maya Singh, 2376*	Naick	*22 Apl. 03*	*No. 1013, 9 Oct. 03*
Do.	**Ayub Khan,** Jemadar	Jemadar	29 Oct. 15	Lon. Gaz., 29 Oct. 15
Do.	**Inayat Ali Shah,** Sepoy, 606	Sepoy	. . .	No. 1512, 16 Dec. 16
Do.	**Abdul Karim,** Sepoy, 573	Do.	. . .	No. 1512, 16 Dec. 16
Do.	**Muhammad Afzal,** Subadar	Subadar.	9 June 17	No. 894, 9 June 17
Do.	**Darweza Khan,** Subadar	Subadar	5 Nov. 17	No. 1629, 19 July 18
123rd Rif.	**Har Nath Singh,** Subadar	Jemadar	21-22 Nov. 17.	No. 1178, 3 June 18

THE INDIAN ORDER OF MERIT—*contd.*

2nd Class—*contd.*

Corps.	Names and Rank.	In what rank admitted.	Date of admission to the Order.	Authority. G. G. O. or A. D. Notfn.
124th Infy.	Bhagwan Singh, Colour-Havildar, 2468.	Colour-Havildar	17 Aug. 17	No. 1359, 17 Aug. 17
Do.	Sarwar Khan, Sepoy, 702	Sepoy	5 Nov. 17	No. 1885, 16 Aug. 18
125th Infy.	Balsing, Rifleman, 3100	Rifleman	7 Jan. 16	No. 524, 12 May 16
Do.	Sher Khan, Havildar, 1771	Havildar	22 Apl. 16	No. 1387, 18 Nov. 16
Do.	Goru Ram, Rifleman, 2436	Rifleman	22 Apl. 16	No. 1387, 18 Nov. 16
Do.	Feroz Khan, Lance-Naik, 2394	Lance Naik	21 Apl. 17	No. 8, 1 Jan. 18
126th Inf.	*Bulaka Singh, Subadar*	*Havildar*	*6 Jan. 94*	*No. 728 of 1894.*
Do.	*Kamar Din, Subadar-Major, Sardar Bahadur, Hony. Capt.*	*Subadar*	*12 Dec. 00*	*No. 355 of 1901.*
Do.	*Dost Muhammad, Private, 2029*	*Private*	*12 Dec. 00*	*No. 355, 26 Apl. 01*
Do.	Labh Singh, Subadar	Subadar	17 Aug 17	No. 1359, 17 Aug. 17
Do.	Mukhmad Shah, Sepoy, 5564	Sepoy	5 Nov. 17	No. 1885, 16 Aug. 18
127th Bal. L.I.	*Nur Muhammad, Private, 2737*	*Do.*	*9 Oct. 98*	*No. 277, 9 Mar. 00*
Do.	*Sharif Khan, Havildar, 2627*	*Do.*	*9 Oct. 98*	*No. 277, 9 Mar. 00*
Do.	Nur Dad, Subadar	Do.	9 Oct. 98	No. 277, 9 Mar. 00
Do.	*Shah Zad Shah, Havildar, 153*	*Do.*	*9 Oct. 98*	*No. 277, 9 Mar. 00*
Do.	*Khuda Bux Khan, Private, 188*	*Do.*	*9 Oct. 98*	*No. 277, 9 Mar. 00*
Do.	*Fazal Khan, Private, 304*	*Do.*	*9 Oct. 98*	*No. 277, 9 Mar 00*
Do.	*Karam Dad, Private, 162*	*Do.*	*9 Oct. 98*	*No. 277, 9 Mar. 00*
Do.	*Nur Muhammad, Private, 31*	*Do.*	*9 Oct. 98*	*No. 277, 9 Mar. 00*
Do.	Subey Khan, Private, 767 (a)	Do.	9 Oct. 98	No. 277, 9 Mar. 00
Do.	*Shere Baz, Private, 295*	*Do.*	*9 Oct. 98*	*No. 277, 9 Mar. 00*
Do.	*Shahzada Khan, Private, 1361*	*Do.*	*9 Oct. 98*	*No. 277, 9 Mar. 00*
Do.	*Hamid Khan, Subadar*	*Subadar*	*20 Dec. 01*	*No. 264, 28 Mar. 02*
Do.	Harak Bahadur Thapa	Jemadar	8 Feb. 15	No. 313, 3 Apl. 15
Do.	Zergun Shah, Subadar	Subadar	3 June 16	No. 849, 29 July 16
Do.	Alim Khan, Lce.Naik, 32	Lce-Naik	3 June 16	No. 849, 29 July 16
Do.	Gulam Muhammad, Havildar, 4071	Havildar	9 June 17	No. 894, 9 June 17
128th Prs.	Indar Singh, Jemadar	Jemadar	3 June 15	No. 524, 3 June 15
Do.	Ganda Singh, Havildar, 3397	Havildar	3 June 15	No. 524, 3 June 15
Do.	Abdul Latif Khan, Jemadar	Jemadar	7 Jan. 16	No. 524, 12 May 16
129th Baluchis	Sar Amir, Naik, 4280	Naik	31 Oct. 14	No. 111, 29 Jan. 15
Do.	Mir Badshah, Jemadar	Jemadar	10 Mar. 14	No. 356, 16 Apl. 15
Do.	Shaib Jan, Sepoy, 453	Sepoy	10 Mar. 14	No. 356, 16 Apl. 15
Do.	*Hobab Gul, Naik, 3814*	*L. Naik*	*18 Feb. 15*	*No. 313, 2 Apl. 15*
Do.	*Redi Gul, Sepoy, 4333*	*Sepoy*	*18 Feb. 15*	*No. 313, 2 Apl. 15*
Do.	Makhmad Azam, Subadar	Subadar	15 Apl. 15	No. 473, 21 May 15
Do.	Nekamal, Lance-Naik, 118	Lce-Naik	15 Apl. 15	No. 473, 21 May 15
Do.	Saiday Khan, Sepoy, 250	Sepoy	15 Apl. 15	No. 473, 21 May 15
Do.	Raji Khan, Sepoy, 471	Sepoy	3 Jan. 15	Lon. Gaz., 23 Jan. 15
Do.	Fateh Haider, Jemadar	Havildar	1 Jan. 16	No. 187, 18 Feb. 16
Do.	Mansabdar, Sepoy, 639	Sepoy	2 June 17	No. 808, 2 June 17
Do.	Mirjan, Sepoy, 436	Sepoy	9 June 17	No. 894. 9 June 17
Do.	Mirza Khan, Havildar, 3151	Havildar	9 June 17	No. 894, 9 June 17
130th Baluchis	*Hyder Khan, Subadar-Major*	*Subdr.-Maj.*	*27 July 80*	*No. 356, 23 June 82*
Do.	Fazal Ilahi, Lance-Naik 4291	Lance-Naik	9 June 17	No. 894, 9 June 17
Do.	Karam Ilahi, Private, 820	Private	9 June 17	No. 894, 9 June 17
Do.	Malik Aman, Subadar	Subadar	9 June 17	No. 894, 9 June 17
1st G. R.	*Balbhadar Gurung, Havildar, 368*	*Do.*	*3 Nov. 94*	*No. 41, 11 Jan. 95*
Do.	*Dhan Singh Bhandari, Naick*	*Bugler*	*3 Nov. 94*	*No. 41, 11 Jan. 95*
Do.	*Lachman Thapa, Naick, 29*	*Naick*	*3 Nov. 94*	*No. 41, 11 Jan. 95*
Do.	*Narsingh Gharti, Sepoy, 48*	*Sepoy*	*3 Nov. 94*	*No. 41, 11 Jan. 95*
Do.	*Bulber Ghurtie, Subadar*	*Do.*	*20 Dec. 75*	*No. 454, 17 Apl. 76*
Do.	*Basu Gurung, Havildar, 1041*	*Do.*	*3 Nov. 94*	*No. 41, 11 Jan. 95*
Do.	Sasidhar Thapa, Jemadar	Jemadar	10 Mar. 15	No. 356, 16 Apl. 15
Do.	Malik Aman, Jemadar	Jemadar	9 June 17	No. 894, 9 June 17
Do.	Abdul Amir, Subadar	Subadar	4 June 17	No. 872, 4 June 17
Do.	Jitman Thapa, Rifleman, 2719	Jemadar	10 Mar. 15	No. 356, 16 Apl. 15
Do.	Ramkishen Thapa, Naik, 1843	Naik	23 July 15	No. 815, 27 Aug. 15
Do.	Khamba Sing Gurung, Rifleman, 2734	Rifleman	25 June 15	No. 707, 30 June 15
Do.	Pahalwan Gurung, 2397, Rifleman	Do.	25 June 15	No. 707, 30 June 15
Do.	Kalu Gurung, Subadar	Subadar	8 Mar. 16	No. 1385, 18 Nov. 16
Do.	Dirjia Thapa, Subadar	Do.	17 Apl. 16	No. 1385, 18 Nov. 16
Do.	*Dhanbir Sahai, Jemadar*	*Do.*	*11 Sep. 97*	*No. 1314, 26 Nov. 97*
Do.	*Tika Ram Lama, Havildar, 1279*	*Naick*	*20 Oct. 97*	*No. 430, 22 Apl. 98*
Do.	*Kirpa Ram Thapa, Subadar-Maj., Sardar Bahadur, Hony. Capt.*	*Subadar*	*20 Oct. 97*	*No. 430, 22 Apl. 98*
Do.	Bhagatbir Thapa, Subadar	Subadar	17 Aug. 17	No. 1359, 17 Aug. 17
Do.	Jagat Chand, Lance-Naik, 2578	Lance-Naik	17 Aug. 17	No. 1359, 17 Aug. 15
Do.	Judbir Thapa, Rifleman, 3158	Rifleman	17 Aug. 17	No. 1359, 17 Aug. 15
2nd G. R.	*Dalman Ale, Subadar-Major, Sardar Bahadur, Hony. Capt.*	*Havildar*	*16 Nov. 97*	*No. 430, 22 Apl. 98*
Do.	Bahanbir Rana, Subadar	Lce.-Havildar	3 Jan. 02	No. 403, 9 May 02

(a) Transferred to Reserve.

THE INDIAN ORDER OF MERIT—*contd.*

2nd Class—*contd.*

Corps.	Names and Rank.	In what rank admitted.	Date of admission to the Order.	Authority. G. G. O. or A. D Notfn.
2nd G. R.	**Dan Sing Gurung**, Subadar	Rifleman	20 Oct. 97	No. 430, 22 Apl. 98
Do.	*Santbir Gurung, Subadar-Maj., Sardar Bahadur, Hony. Capt.*	*Subadar*	*16 Dec. 97*	*No. 430, 22 Apl. 98*
Do.	**Deosar Thapa**, Rifleman, 4144	Rifleman	3 Dec. 11	No. 436, 10 May 12
Do.	**Padamdhoj Gurung**, Naik, 1618	Naik	2 Nov. 14	No. 111, 29 Jan. 15
Do.	**Arjun Rana**, Subadar, 1959	Subadar	10 Mar. 14	No. 356, 16 Apl. 15
Do.	**Bhagat Bahadur Gurung**, Rifleman, 1959.	Rifleman	10 Mar. 15	No. 356, 16 Apl. 15
Do.	**Tula Gurung**, Rifleman, 2265	Do.	10 Mar. 14	No. 356, 16 Apl. 15
Do.	**Dan Sing Lama**, Subadar	Subadar	4 May 15	No. 568, 12 June 15
Do.	**Hastobir Roka**, Rifleman, 2515	Rifleman	4 May 15	No. 563, 12 June 15
Do.	**Jagtea Pun**, Rifleman, 3266	Do.	4 May 15	No. 563, 12 June 15
Do.	**Manjit Gurung**, Rifleman, 4024	Do.	4 May 15	No. 563, 12 June 15
Do.	**Partiman Gurung**, Rifleman 2477	Do.	4 May 15	No. 563, 12 June 15
Do.	**Ujir Sing Gurung**, Rifleman, 1599	Do.	4 May 15	No. 563, 12 June 15
Do.	**Bhim Bahadur Gurung**, Naik 1957	Naik	1 Jan. 16	No. 187, 18 Feb. 16
Do.	**Anarupe Rana**, Rifleman 2693	Rifleman	1 Jan. 16	No. 187, 18 Feb. 16
Do.	**Dhanbir Sonwar**, Jemadar	Jemadar	16 Feb. 17	No. 115, 19 Jan. 18
Do.	**Padan Singh Thapa**, Rifleman, 4991	Rifleman	16 Feb. 17	No. 115, 19 Jan. 18
Do.	**Gorea Gurung**, Naik, 3843	Naik	17 Aug. 17	No. 1359, 17 Aug. 17
Do.	**Jagea Rana**, Rifleman, 2896	Rifleman	30 Mar. 17	No. 1192, 21 July 17
Do.	**Lachman Gurung**, Subadar	Rifleman	11 Sep. 97	No. 1314, 26 Nov. 97
Do.	*Lalbir Thapa, Naik, 2227*	*Do.*	*20 Oct. 97*	*No. 430, 22 Apl. 98*
Do.	*Pahalman Sahai, Rifleman, 2201*	*Do.*	*20 Oct. 97*	*No. 430, 22 Apl. 98*
Do.	*Chamu Singh Barathoki, Subadar-Major, Sardar Bahadur, Hony. Capt.*	*Subdr.-Major*	*13 Dec. 97*	*No. 430, 22 Apl. 98*
Do.	*Gopal Gurung, Drum-Maj., 2252*	*Bugler*	*13 Dec. 97*	*No. 430, 22 Apl. 98*
Do.	*Deb Singh Bisht, Havildar, 267*	*Havildar*	*13 Dec. 97*	*No. 430, 22 Apl. 98*
Do.	**Jaman Sing Thapa**, Lance-Naik, 1757	Lance-Naik	6 Nov. 14	No. 111, 29 Jan. 15
Do.	**Jaman Sing Khattri**, Lance-Naik, 2016.	Do.	13 Nov. 14	No. 111, 29 Jan. 15
Do.	**Bahadur Thapa**, Havildar, 1569	*Havildar*	*4 May 15*	*No. 353, 12 Jan. 15*
Do.	**Harak Singh**, Gharti Lance-Naik, 1805	*Lance-Naik*	*4 May 15*	*No. 353, 12 Jan. 15*
Do.	**Gane Gurung**, Rifleman, 1749	*Do.*	*4 May 15*	*No. 353, 12 Jan. 15*
Do.	**Kharak Bahadur Ale**, Bugler, 2252	Bugler	10 Mar. 15	No. 356, 16 Apl. 15
Do.	*Ghambhir Sing Gurung, Sub-Major*	*Sub-Major*	...	*No 946, 9 July 15*
Do.	**Bhim Sing Thapa**, Subadar	Subadar	...	No. 646, 9 July 15
Do.	**Jit Sing Gurung**, Subadar	Do.	1 Jan. 16	No. 187, 18 Feb. 16
3rd G. R.	**Asbir Gurung**, Jemadar	Jemadar	11 Dec. 17	No. 1178, 3 June 18
Do.	**Abdul Rahim**, Sub-Asstt. Surgeon	Sub-Asstt. Surgn.	9 Apl. 18	No. 2536, 26 Oct. 18
Do.	**Balbir Thapa**, Subadar	3rd Q. A. O. Gurkha Rifles.	12 Nov. 17	No. 1178, 3 June 18
Do.	**Dhanraj Gurung**, Jemadar	Jemadar	10 Apl. 18	No. 2536, 26 Oct. 18
Do.	**Man Singh Thapa**, Rifleman, 3741	Rifleman	10 Apl. 18	No. 2536, 26 Oct. 18
Do.	**Randhoj Rana**, Subadar	Subadar	21 22 Nov. 17.	No. 1178, 3 June 15
Do.	**Lal Bahadur Gurung**, Jemadar	Jemadar	21 Nov. 17	No. 1178, 3 June 18
4th G. R.	*Manbir Limbu, Subadar*	*Havildar*	*25 Apl. 91*	*No. 841, 26 Aug. 92*
Do.	*Surbir Gurung, Naick, 1151*	*Rifleman*	*28 Oct. 97*	*No. 430, 22 Apl. 98*
Do.	*Har Singh Bisht, Jemadar*	*Havildar*	*Dec. 97*	*No. 430, 22 Apl. 89*
Do.	*Setia Thapa, Rifleman, 1467*	*Rifleman*	*11 Dec. 97*	*No. 430, 22 Apl. 98*
Do.	**Diwan Singh Gurung**, Naik, 4252	Naick	10 Mar. 15	No. 356, 16 Apl. 15
Do.	**Parbir Gurgung**, Rifleman, 4509	Rifleman	10 Mar. 15	No. 356, 16 Apl. 15
Do.	**Gungbir Gurung**, Subadar	*Jamadar*	*4 May 15*	*No. 563, 12 June 15*
Do.	**Khamba Sing Gurung**, Rifleman, 4015	*Rifleman*	*4 May 15*	*No. 563, 12 June 15*
Do.	**Wazir Sing Burathoki**, Rifleman, 4578	*Do.*	*4 May 15*	*No. 563, 12 June 15*
Do.	**Bhandoj Rai**, Rifleman, 2417	*Do.*	3 June 15	Lon. Gaz., 23 June 15
Do.	**Bhakat Singh Rana**, Havildar, 1840	Havildar	3 June 15	Lon. Gaz., 23 June 15
Do.	**Senbir Gurung**, Subadar, Bahadur	Subadar	23 July 15	No. 815, 27 Aug. 15
Do.	**Dhanjit Gharti**, Subadar	Do.	27 Sep. 15	No. 906, 17 Sep. 15
Do.	**Dhanraj Gurung**, Subadar	Do.	23 Sep. 15	No. 906, 17 Sep. 15
Do.	**Antbir Thapa**, Subadar	Subadar	1 Jan. 16	No. 187, 18 Feb. 16
Do.	**Kharakbir Thapa**, Jemadar	Jemadar	1 Jan. 16	No. 187, 18 Feb. 16
Do.	**Mandhoj Thapa**, Rifleman, 776	Rifleman	10 May 17	No. 1170, 20 July 17
Do.	**Jaman Singh Gurang**, Jamadar	Jemadar	17 Aug. 17	No. 1359, 17 Aug. 17
5th G. R.	*Moti Ram Thapa, Subadar*	*Sepoy*	*8 June 88*	*No. 692, 24 Aug. 88*
Do.	*Bhagatbir Thapa, Subadar*	*Lce.-Naick*	*12 Dec. 91*	*No. 287, 18 Mar. 92*
Do.	*Nandia Thapa, Bugler, 75*	*Bugler*	*2 Dec. 91*	*No. 179, 19 Feb. 92*
Do.	*Bandho Pun, Sepoy, 2211*	*Sepoy*	*2 Dec. 91*	*No. 659, 8 July 92*
Do.	**Harakbir Thapa**, Subadar-Maj.	Do.	20 Dec. 91	No. 683, 15 July 92
Do.	*Karbir Burathoki, Subadar*	*Lance-Naick*	*26 Nov. 97*	*No. 430, 22 Apl. 98*
Do.	*Goria Rana, Subadar*	*Do.*	*13 Nov. 97*	*No. 430, 22 Apl. 98*
Do.	*Maniram Pun, Jemadar*	*Rifleman*	*23 Nov. 97*	*No. 430, 22 Apl. 98*
Do.	*Birkhdhoj Khattri, Subadar*	*Do.*	*21 Dec. 97*	*No. 12 6 Jan. 99*

THE INDIAN ORDER OF MERIT—*contd.*
2nd Class—*contd.*

Corps.	Names and Rank.	In what rank admitted.	Date of admission to the Order.	Authority. G. G. O. or A. D. Notfn.
5th G. R.	*Motiram Thapa, Jemadar*	*Rifleman*	*8 Oct. 97*	*No. 430, 22 Apl. 98*
Do.	*Kharu Gurung, Rifleman, 1327*	*Do.*	*2 Dec. 97*	*No. 430, 22 Apl. 98*
Do.	**Dhanjit Gharti**, Subadar	Subadar	. . .	No. 906, 17 Sept. 15
Do.	**Dhanraj Gurung**, Subadar	Do.	. . .	No. 906, 17 Sept. 15
Do.	**Tejbir Thapa**, Jemadar	Jemadar	1 Jan. 16	No. 680, 17 June 16
Do.	**Naharsing Gurung**, 562	Lce.-Naik	3 June 16	No. 849, 29 July 16
Do.	**Dal Bahadur Thapa**, Rifleman, 2455	Rifleman	18 Nov. 16	No. 1391, 18 Nov. 16
6th G. R.	*Narbahadur Thapa, Jemadar*	*Sepoy*	*22 Nov. 79*	*No. 1000, 26 Oct. 94*
Do.	*Judhbir Gurung, Subadar-Maj.*	*Naick*	*19 June 86*	*No. 306, 29 Apl. 87*
Do.	*Bhim Singh Thapa, Havildar, 2533*	*Lance-Naick*	*6 Apl. 89*	*No. 655, 19 July 89*
Do.	*Sahabir Thapa, Subadar*	*Sepoy*	*4 May 89*	*No. 796, 6 Sep. 89*
Do.	*Kaman Singh Thapa, Subadar*	*Naick*	*20 Oct. 97*	*No. 430, 22 Apl. 98*
Do.	*Jaimani Thapa, Lance-Naick, 2694*	*Sepoy*	*24 Mar. 91*	*No. 862, 4 Sep. 91*
Do.	**Gambir Sing Pun**, Subadar-Major	Sub-Maj	17 Sep. 15	No. 906, 17 Sep. 15
Do.	**Sahabir Thapa**, Subadar 1	Subadar	17 Sep. 15	No. 906, 17 Sept. 15
Do.	**Dalbahadur Thapa**, Jemadar	Jemadar	17 Sep. 15	No. 906, 17 Sept. 15
Do.	**Bal Singh Thapa**, Jemadar	Do.	29 Aug. 15	No. 1080, 29 Oct. 15
Do.	**Budhiram Gurung**, Lance-Naik, 2661	Lance-Naik	29 Aug. 15	No. 1681, 29 Oct. 15
Do.	**Hembahadur Rana**, Lance-Naik, 343	Do.	29 Aug. 15	No. 1081, 29 Oct. 15
Do.	**Chandra Singh Gurung**, Rifleman, 29	Rifleman	29 Aug. 15	No. 1081, 29 Oct. 15
Do.	**Harka Gurung**, Rifleman, 1088	Do.	29 Aug. 15	No. 1081, 29 Oct. 15
Do.	**Santabir Gurung**, Jemadar	Havildar	29 Aug. 15	No. 1081, 29 Oct. 15
Do.	**Nandial Gurung**, Jemadar	Do.	21 Aug. 15	No. 849, 28 July 16
Do.	**Kulbahadur Thapa**, Subdr.	Subadar	3 June 16	No. 849, 29 July 16
Do.	**Sataising Thapa**, Subadar	Do.	1 Jany. 16	No. 680, 17 June 16
7th G. R.	**Harkraj Rai**, Havildar, 858	Havildar	24 July 15	No. 1159, 19 Nov. 15
Do.	**Narman Rai**, Rifleman, 1085	Rifleman	24 July 15	No. 1159, 19 Nov. 15
Do.	**Budhiman Limbu**, Havildar, 664	Havildar	26 Dec. 16	No. 676, 5 May 17
Do.	**Ranbahadur Thapa**, Rifleman, 2029	Rifleman	17 Aug. 17	No. 1357, 17 Aug. 17
Do.	**Dhojman Rai**, Rifleman, 2053	Do.	17 Aug. 17	No. 1357, 17 Aug. 17
Do.	**Sivinanda Limbu**, Rifleman, 2332	Do.	17 Aug. 17	No. 1357, 17 Aug. 17
Do.	**Birbahadur Rai**, Rifleman, 2440	Do.	17 Aug. 17	N . 1357, 17 Aug. 17
Do.	**Nar Bahadur Limbu**, Rifleman, 2475	Do.	17 Aug. 17	No. 1357, 17 Aug. 17
Do.	**Raidhan Rai**, Rifleman, 2639	Do.	17 Aug. 17	No. 1357, 17 Aug. 17
Do.	**Kubir Singh Limbu**, Subadar	Subadar	17 Aug. 17	No. 1357, 17 Aug. 17
Do.	**Sahabir Thapa**, Subadar	Do.	17 Aug. 17	No. 1357, 17 Aug. 17
Do.	**Kharak Bahadur Sunwar**, Rifleman	Rifleman	17 Aug. 17	No. 1359, 17 Aug. 17
8th G. R.	*Dhup Chand Thakur, Subadar*	*Havildar*	*24 Mar. 91*	*No. 862, 4 Sep. 91*
Do.	*Gomo Basnet, Subadar*	*Do.*	*20 Jan. 94*	*No. 666, 6 July 94*
Do.	**Jangbir Rana**, Subadar	Havildar	6 July 04	No. 957, 7 Oct. 04
Do.	*Ramu Gurung, Naick, 2239*	*Naick*	*6 July 04*	*No. 957, 7 Oct. 04*
Do.	**Karagbir Gurung**, Rifleman, 21	Rifleman	6 July 04	No. 957, 7 Oct. 04
Do.	*Prahlad Gurung, Jemadar*	*Sepoy*	*28 Mar. 91*	*No. 647, 3 July 91*
Do.	*Jitbahadur Gurung, Sepoy, 2353*	*Do.*	*28 Mar. 91*	*No. 647, 3 July 91*
Do.	*Kabiram, Gurung Subadar*	*Do.*	*28 Mar. 91*	*No. 647, 3 July 91*
Do.	*Ranjit Thapa, Havildar, 3590*	*Do.*	*28 Mar 91*	*No. 647, 3 July 91*
Do.	*Balu Gurung, Sepoy, 2703*	*Do.*	*28 Mar. 91*	*No. 647, 3 July 91*
Do.	*Tilaram Thapa, Sepoy, 2716*	*Do.*	*28 Mar. 91*	*No. 647, 3 July 91*
Do.	**Bhim Sing Rana**, Subadar	Do.	28 Mar. 91	No. 647, 3 July 91
Do.	*Gambhir Rai, Lance-Havildar, 1611*	*Lance-Havr.*	*28 Mar. 91*	*No. 647, 3 July 91*
Do.	*Jati Ram, Sepoy, 1435*	*Sepoy*	*28 Mar. 91*	*No. 647, 3 July 91*
Do.	*Maita Sing Lama, Sepoy, 1079*	*Do.*	*28 Mar. 91*	*No. 647, 3 July 91*
Do.	**Jaichand Thakur**, Subadar	Subadar	19 Nov. 11	No. 69, 29 Jan. 12
Do.	**Hari Persad Thapa**, Havildar, 3110	Havildar	23 Nov. 14	No. 111, 29 Jan. 15
Do.	**Kumbha Sing Gurung**, Lce-Naik-4232	Lance-Naik	3 June 16	No. 849, 29 July 16
Do.	**Bajbaran Gurung**, Rifleman, 3609	Rifleman	3 June 16	No. 849, 29 July 16
Do.	**Chhetra dhoj Limbu**, Subadar	Subadar	3 June 16	No. 849, 29 July 16
Do.	**Pharsaidhoj Rai**, Subdr.	Do.	3 June 16	No. 849, 29 July 16
Do.	**Narbahadur Rai**, Jemadar	Do.	3 June 16	No. 849, 29 July 16
Do.	**Ram Sing Burathoki**, Jemader	Jemadar	3 June 16	No. 849, 29 July 16
Do.	**Kharak Bahadur Chettri**, Rifleman, 1480.	Rifleman	17 Aug. 17	No. 1359, 17 Aug. 17
9th G. R.	**Mehar Singh Khattri**, Subadar	Subadar	4 May 15	No. 563, 12 June 15
Do.	**Panchbir Mal**, Rifleman, 2721	Rifleman	18 Feb. 15	No. 313, 3 Apl. 15
Do.	**Shibdhoj Mal**, Jemadar	Jemadar	4 May 15	No. 563, 12 June 15
Do.	**Tika Ram Kunwar**, Rifleman, 2589	Rifleman	3 June 15	Lon. Gaz., 23 June 15
Do.	**Karanbahadur Khattri**, Havildar, 1179.	Havildar	19 Jan. 16	No. 524, 12 May 16
Do.	**Randhoj Karki**, Rifleman, 2820	Rifleman	17 Apl. 16	No. 1385, 18 Nov. 16
Do.	**Himbahadur Sahi**, Jemadar	Jemadar	17 Aug. 17	No. 1359, 17 Aug. 17
Do.	**Lalu Khattri**, Jemadar	Do.	17 Aug. 17	No. 1359, 17 Aug. 17
10th G.R.	**Bajirdhoj Rai**, Subadar	Do.	17 Sep. 15	No. 906, 17 Sep. 15
Do.	**Chittahang Limbu**, Subdr.-Maj. Bahadur.	Subdr.-Maj.	1 Jany. 16	No. 680, 17 June 16
1st Mule Corps	**Hashmat Ali**, Jemadar	Jemadar	29 Aug. 15	No. 1081, 29 Oct. 15

THE INDIAN ORDER OF MERIT—*concld.*

2nd Class—*contd.*

Corps.	Names and Rank.	In what rank admitted.	Date of admission to the Order.	Authority.
				G. G. O. or A. D. Notfn.
9th Mule Corps	**Muhammad Fazil,** Kot-Dafadar	Kot-Dafadar	27 Aug. 15	No. 819 27 Aug. 15
Do.	**Bahadur Shah,** Lance-Naik, 1106	Lance-Naik	29 Aug. 15	No. 1081, 29 Oct. 15
Do.	**Bir Singh,** Driver, 1350	Driver	29 Aug. 15	No. 1081, 29 Oct. 15
Do.	**Bhiva Jayaram**	Do.	11 Mar. 04	No. 4396B., 21 Dec. 04
13th Mule Corps	**Hashan,** Driver, 97	Do.	17 Nov. 14	No. 20, 20 Jan. 15
Do.	*Sher Baz, Driver, 940*	*Do.*	*17 Nov. 14*	*No. 313, 3 Apl. 15*
19th Mule Corps	**Ali Akbar,** Naik, 780	Naik	25 Mar. 17	No. 8, 1 Jan. 18
21st Mule Corps	**Rodha,** Driver, 722	Do.	4 Aug. 15	No. 727, 4 Aug. 15
Do.	**Suchet,** Driver, 796	Do.	4 Aug. 15	No. 727, 4 Aug. 15
Do.	**Kwaja,** Driver, 632	Do	4 Aug. 15	No. 727, 4 Aug. 15
Do.	*Sher Baz,* Lance-Naik, 176	Lance-Naik	4 Aug. 15	No 727, 4 Aug. 15
27th Mule Corps	*Toorsum Khan, Ressaidar*	*Daffadar*	*1 Aug. 97*	*No. 49, 14 Jan. 98*
30th Mule Corps	**Afridi,** Driver, 1181	Driver	10 Dec. 15	No. 124 11 Dec. 15
Do.	**Allah Ditta,** Driver, 1158	Driver	10 Dec. 15	No. 1244, 10 Dec. 15
31st Mule Corps	**Sayyid Abdulla,** Driver, 1357	Do.	1 Apl. 15	No. 426, 7 May 15
I. S. M. D.	*Shunkur Dass, Hospl. Asst.*	*2nd Cl. H. A.*	*13 Dec. 78*	*No. 365, 2 May 79*
Do.	*Ahmadullah Khan, Senior H. A.*	*Senior H. A.*	*19 Mar. 91*	*No. 951, 2 Oct. 91*
Do.	*Ganuram, 2nd Class Senior Hospl., Asst.*	*2nd Cl. H. A.*	*3 Nov. 94*	*No. 41, 11 Jan. 95*
Do.	*Kashi Ram, 1st grade Hospl. Asst., 595*	*1st Gr. H. A.*	*27 July 97*	*No. 49, 14 Jan. 98*
Do.	**Kehar Singh Chandall,** Hospl. Asst. 1189.	3rd Cl. H. A.	. . .	No. 526, 26 June 08
Do.	**Maula Baksh,** Sub-Asst. Surgn., 1st class, 1124.	Sub-Asst. Surgn.	4 May 15	No. 563, 12 June 15
Do.	**Tek Chand,** Sub-Asstt. Surgn., 1st class, 988.	Do.	4 May 15	No. 563, 12 June 15
Do.	**Rane, G. K. R.,** Sub-Asstt. Surgn., 3rd class, 370.	Do.	4 May 15	No. 563, 12 June 15
Do.	**Pandit Shanker Dass,** Sr. Sub-Asst. Surgn.	Sr. Sub-Asst. Surgn.	12 Dec. 14	No. 111, 29 Jan. 15
Do.	**Harnam Singh**	Sub-Asst. Surgn.	23 Nov. 14	No. 111, 29 Jan. 15
Do.	**Vroj-Lal Umed Ram Pandit**	1st class, Sub-Asst. Surgn.	7 Dec. 14	No. 284, 19 Mar. 15
Do.	**Nagindar Singh,** Sub-Asst. Surgn., 1st class 1116.	Sub-Asst. Surgn.	10 Mar. 15	No. 365, 16 Apl. 15
Do.	**Munasami Ramasami,** Sub-Asst. Surgn., 1st class.	Do.	17 Sep. 15	No. 907, 17 Sep. 15
Do.	**Keshav Waman Khuperkar,** Sub-Asst.-Surgn., 3rd class.	Do.	17 Sep. 15	No. 907, 17 Sep. 15
Do.	**Shaikh Muhammad Dadasahib,** Sub-Asst.-Surgn., 3rd class.	Do.	17 Sep. 15	No. 907, 17 Sep. 15
Do.	**RamSingh,** Sub-Asst. Surgn., 1st class 862	Do.	1 Jan. 16	No. 187, 18 Feb. 16
Do.	**Pargan Singh,** Sub-Asst. Surgn., 2nd class, 1230.	Do.	1 Jan. 16	No. 187, 18 Feb. 16
Do.	**Ghaus Muhammad,** Sub-Asst. Surgn., 1st class, 880.	Do.	1 Jan. 16	No. 680, 17 June 16
Do.	**Bhagwan Singh,** Sub-Asst. Surgn., 1st class 851	Do.	3 June 16	No. 849, 29 July 16
Do.	**Ishar Singh,** Sub-Asst. Surgn., 1st class	Do.	3 June 16	Nos. 8 &9, 29 July 16
Do.	**Daniel, Israel Nuanapragasam,** 2nd class, Sub-Asst. Surgn.	2nd class, Sub-Asst. Surgn.	25 Nov. 16	Lon. Gaz. 25 Nov. 16
Do.	**Hukam Singh,** Sub-Asst. Surgn., 1st Cl. 1111.	Sub Asst. Surgn.	1 Feb. 17	Lon. Gaz 1 Feb. 17
Do.	**Mula Singh,** Sub-Asst. Surgn., 2nd Cl. 1247.	Sub. Asst. Surgn.	1 Feb. 17	Lon. Gaz. 1 Feb. 17
Do.	**Daulat Singh,** Sub-Asst. Surgn., 1st Cl.	Do.	2 Jan. 17	No. 676, 5 May 17
Do.	**Tangore Ramaswami Govindaswami Pillai,** 1359.	Sub. Asst. Surgn. 1st class.	4 June 17	No. 1151, 21 July 17
Do.	**Mul Singh,** Sub-Asst. Surgn., 2nd class, 1198.	Do.	7 Aug. 17	No. 1359, 17 Aug. 17
Do.	**Riyazud-Din,** Sub-Asst. Surgn. 2nd class	Sub. Asst. Surgn.	17 Apl. 16	No. 1385, 17 Nov. 16
Do.	**Arfan Dass Gosain,** Sub-Asst. Surgn. 1st class, 914.	Sub. Asst. Surgn. II. Cl.	. . .	No. 1151, 20 July 17
Govt. Mayo Salt Mines.	*Ghulam Hyder Khan, Hospl. Asst.*	*Hospl. Asst.*	. . .	*No. 443, 30 July 08*
Border Mil. Pol	*Muhamed Zafar Khan, Jemadar*	*Jemadar*	*13 Mar. 96*	*No. 259, 21 Mar. 02*
Fron. Pol. Bn.	**Sangram Singh,** Havildar-Maj.	Havildar-Maj.	10 Feb. 91	No. 482, 8 May 91
Do.	**Chandrabir Thappa,** Sepoy	Sepoy	10 Feb. 91	No. 482, 8 May 91
Do.	**Boi Khan,** Subadar-Maj., *Khan Sahib, Bahadur.*	Subadar-Maj.	26 May 17	No. 1887, 2 Nov. 17
Khyber Rifl.	**Muhammad Ali Khan,** Subadar-Maj., *Sardar Bahadur, Khan Sahib.*	Jemadar	. . .	No. 123. 8 Feb. 89
Do.	**Jawas,** Subadar-Maj., *Bahadur*	Subadar	25 Aug. 97	No. 639, 3 June 98
Do.	**Mohibulla,** Subadar	Do.	30 June 05	No. 838, 29 Sep. 05
Kurram Militia	**Yar Baz,** Jemadar	Havildar	. . .	No. 392, 15 May 08
Do.	**Sawar Ali,** Sowar, 4256	Sowar	4 Oct. 14	No. 301, 26 Mar. 15
N.-W. Militia.	**Tor Khan,** Subadar-Maj., *Sardar Bahadur.*	Do.	1 Sep. 97	No. 547, 13 June 02

Corps.	Names and Ranks.	In what rank admitted.	Date of admission to the Order.	Authority.

THE INDIAN ORDER OF MERIT—*contd.*

2nd Class—*contd.*

Corps.	Names and Ranks.	In what rank admitted.	Date of admission to the Order.	G. G. O. or A. D. Notfn.
N. W. Militia	*Kamar Din, Sepoy*	*Sepoy*	*22 Sep. 05*	*No. 221, 15 June 06*
Do.	*Abbas Khan, Sepoy*	*Do.*	*22 Sep. 05*	*No. 221, 15 June 06*
Do.	**Darim**, 1st Gde. Dafadar, 262	Daffadar	7 Jan. 15	No. 301, 26 Mar. 15
Do.	**Pat Khan**, Jemadar	Jemadar	3 June 15	No. 526, 3 June 15
Do.	**Gul Rakhim**, Jemadar	Do.	24 Aug. 16	No. 1513, 16 Dec. 16
So. Waziristan Militia.	**Turra Baz**, Subadar	Subadar	16 June 17	No. 945, 16 June 17
Zhob Militia	**Hayat Khan**, Jemadar	Havildar	26 Apl. 05	No. 1037, 1 Dec. 05
Do.	**Ali Jan**, Jemadar	Sepoy	26 Apl. 05	No. 1037, 1 Dec. 05
Do.	**Sultan Khan**, Kot-Havildar	Do.	26 Apl. 05	No. 1037, 1 Dec. 05
Do.	**Amin Khan**, Naick	Naick	4 Dec. 05	No. 189, 1 June 06
Do.	**Majid**, Sepoy	Sepoy	14 Dec. 05	No. 189, 1 June 06
Do.	**Syed Rasul**, Lance-Dafadar	Sowar	8 Nov. 12	No. 1053, 8 Nov. 12
Alwar Infy.	**Khan Muhammad Khan**, Subadar	Subadar	4-5 Nov. 17	No. 1179, 3 June 18
Bharatpur Infy.	**Balbir Singh**, Subadar	Subadar	3 June 16	No. 849, 29 July 16
Do.	**Ganeshi Lal**, Major (*Acting Lt.-Col.*)	Sardar Bahadur	4 Oct. 18	No. 2342, 4 Oct. 18
Bikanir Camel Corps.	**Kishen Singh**, Subadar	Do.	22 Apl. 03	No. 877, 9 Sep. 04
Do.	**Ali Khan**, Sepoy, 1534	Sepoy	20 Nov. 14	No. 1168, 24 Dec. 14
Do.	**Chur Singh**, Jemadar, No. 1 Company, Jemadar.	Jemadar	1 Apl. 16	No. 849, 29 July 16
Do.	**Riyazuddin**, Senior Sub-Asst. Surgn., 2nd class.	Sr. Sub-Asst. Surgn.	18 Nov. 16	No. 1385, 18 Nov. 16
Do.	**Kundar Lal**, Sub-Asst. Surgn., 2nd class, 1219.	Sub-Asst. Surgn.	18 Nov. 16	No. 1385, 18 Nov. 16
Do.	**Lal Singh**, Sub-Asst. Surgn.	Sub-Asst. Surgn.	18 Nov. 16	No. 1385, 18 Nov. 16
Burma Mtd. Rifl.	**Uttam Singh**, Sowar, 68	Sowar	27 Jan. 18	No. 1703, 26 July 10
Faridkot Impl. S. & M.	**Nand Singh**, Sub Maj. (*Hony. Capt.*) Sardar Bahadur (*retd.*).	Sub Maj.	1 Feb. 17	Lon. Gaz. 1 Feb. 17
Gwalior Ls.	**Ibrahim**, Sowar, 1120	Sowar	17 Feb. 17	No. 236, 17 Feb. 17
Jind Infy. I. S. Troops	**Harnam Singh**, Subadar	Subadar	3 June 15	No. 526, 3 June 15
Do.	**Sadda Singh**, Sepoy, 2238	Sepoy	2 June 17	No. 868, 2 June 17
Jind Infy.	**Bhagwan Singh**, Subadar	Subadar	21 Nov. 16	No. 676, 5 May 17
Jodhpur I. S. L.	**Dal Singh**, Duffadar	Duffadar	12 Jan. 01	No. 419, 10 May 01
Do.	**Gulab Singh**, Sowar, 1392	Sowar	1 Jan. 17	No. 676, 5 May 17
Do.	**Dhoukal Singh**, Sowar, 1296	Sowar	23 Sept. 17	No. 115, 19 Jan. 18
Jodhpur Lancers.	**Moti Singh**, Sowar, 1740	Do.	. . .	No. 849, 25 July 16
Tehri I. S. Sappers.	***Hasara Singh***	*Sapper*	*2 Dec. 91*	*No. 179, 19 Feb. 92*
Kashmir Rifl.	**Danya**, Havildar, 58	Havildar	3 Mar. 95	No. 744, 16 July 95
Do.	*Johar, Sepoy, 182*	*Do.*	*3 Mar. 95*	*No. 744, 16 July 95*
Do.	*Lala, Sepoy, 953*	*Do.*	*3 Mar. 95*	*No. 744, 16 July 95*
Do.	**Labha**, Naick	Do.	7 Mar. 95	No. 743, 16 July 95
Do.	*Awi Singh, Sepoy, 247*	*Do.*	*7 Apl. 95*	*No. 744, 16 July 95*
Do.	*Bhagwan Singh, Maj.-Genl.*	*Major*	*17 Apl. 95*	*No. 744, 16 July 95*
Do.	*Gandib Singh, Major*	*Subadar*	*17 Apl. 95*	*No. 744, 16 July 95*
Do.	*Jitman, Subadar*	*Jemadar*	17 Apl. 95	No. 744, 16 July 95
Do.	**Niku**, Sepoy, 578	Sepoy	17 Apl. 95	No. 744, 16 July 95
Do	*Sundar Khattri, Sepoy, 123*	*Do.*	*20 Dec. 91*	*No. 683, 15 July 92*
Do.	*Nagda, Subadar*	*Do.*	*20 Dec. 91*	*No. 683, 15 July 92*
Do.	*Bahadur Gurung, Havildar*	*Do.*	*5 Mar. 93*	*No. 509, 26 May 93*
Do.	*Randir Singh Subadar*	*Subadar*	*3 June 15*	*No. 526, 3 June 15*
Do.	**Bal Bahadur Chetti**, Sepoy, 310	Sepoy	3 June 15	No. 526, 3 June 15
Do.	**Dal Bahadur Thapa**, Sepoy 1275,	Do.	3 June 15	No. 526, 3 June 15
Do.	**Atta Ullah**, Havildar, 841	Havildar	21 Nov. 16	No. 676, 5 May 17
Do.	**Hafiz Ali**, Sepoy, 1157	Sepoy	21 Nov. 16	No. 676, 5 May 17
Do.	**Rahim Ali**, Sepoy, 1467	Sepoy	21 Nov. 16	No. 676, 5 May 17
I. S. Rifl.	**Devi Singh**, Lance-Naik, 1151	Lance-Naik	21 Nov. 16	No. 676, 5 May 17
Do.	**Man Bir**, Lance-Havildar, 129	Lance-Havildar	27 Oct. 16	No. 676, 5 May 17
Kashmir I. S. Mtn. Batty.	**Sarup Singh**, Jemadar	Jemadar	25 Sep. 17	No. 1178, 3 June 18
Do.	**Khushal Khan**, Jemadar	Do	25 Sep. 17	No. 1178, 3 June 18
Mysore Lrs.	**A. Lingaraj Urs**, Jemadar	Do.	23 Nov. 16	No. 637, 3 June 16
Assam M. P.	**Mansur Rai**, Subadar	Subadar	3 June 16	No. 849, 29 July 16
Do.	*Buta Singh, Subadar*	*Havildar*	*20 Jan. 90*	*No. 666, 6 July 04*
Do.	**Bakshi Ram Dogra**, Subadar	Do.	2 Mar. 94	No. 666, 6 July 94
Do.	**Narbahadur Gurung**, Rifleman, 1632	Rifleman	1 Jan. 16	No. 187, 18 Feb. 16
Do.	**Budhiman Thapa**, Rifleman, 1348	Do.	1 Jan. 16	No. 187, 18 Feb. 16
Burma M. P	*Sowan Singh, Subadar*	*Jemadar*	*12 Dec. 92*	*No. 1022, 2 Nov. 94*
Do.	*Jia Khan, Naick*	*Naick*	*5 Feb. 93*	*No. 59, 19 Jan. 94*
Do.	*Mir Zaman, Sepoy*	*Sepoy*	*7 Feb. 93*	*No. 59, 19 Jan. 94*

THE INDIAN ORDER OF MERIT—*concld.*

2nd Class—*concld.*

Corps.	Names and Ranks.	In what rank admitted.	Date of admission to the Order.	Authority, C. G. O. or A. D. Notfn.
Burma M. P.	*Sohawa Singh (Hony. Capt.) Subadar-Maj., Sardar Bahadur, Rai Bahadur.*	*Subadar*	*15 Jan. 94*	*No. 793, 17 Aug. 94*
Do.	*Shaikh Rahim Bahhsh, Subadar-Maj., Khan Bahadur.*	*Jemadar*	*19 Aug. 86*	*No. 950, 17 Oct. 90*
Do.	**Man Bir Thapa,** Subadar, *Rai Bahadur*	Havildar	13 Dec. 92	No. 1022, 2 Nov. 94
Do.	**Anbahadur Thapa,** Havildar	Sepoy	13 Dec. 92	No. 1022, 2 Nov. 94
Do.	**Jasbahadur Thapa,** Subadar-Maj. *Bahadur, Rai Bahadur.*	Subadar	19 Oct. 00	No. 1054, 19 Oct. 00
Do.	**Nand Lall Thapa,** Subadar, *Bahadur*	Do.	28 Dec. 00	No. 1239, 28 Dec. 00
Do.	**Tumpao Tu,** Naik, 4225	Naik	27 Feb. 14	No. 207, 27 Feb. 14
Do.	**Latong La,** Lance-Naik, 4690	Lance-Naik	27 Feb. 14	No. 207, 27 Feb. 14
Do.	**Kala Singh,** Sowar, 2519	Sowar	24 Feb. 15	No. 1435, 25 Nov. 16
Do.	**Saif Ali,** Lance-Naik, 831	Lance-Naik	10 Aug. 16	No. 676, 5 May 17
Do.	**Chandri Chand,** Subadar	Subadar	20 Feb. 17	No. 115, 19 Jun. 18
Do.	**Khan Singh,** Sepoy, 2549	Sepoy	2 June 17	No. 868, 2 June 17
Do.	**Bhawani Datt,** Colour-Havildar, 1429	Colr.-Hevr.	5 Dec. 17	No. 1885, 16 Aug. 18
Indian Field Ambulance Corps.	**Jaganaut,** Stretcher Bearer, 2421	Stretcher Bearer	18 Feb. 15	No. 313, 3 Apl. 15
Army Bearer Corps.	**Behari,** Havildar, 8001	Havildar	18 Feb. 15	No. 20, 20 Jan. 15
Do.	**Hussain Khan,** Dooly Bearer, 6694	Dooly Bearer	29 Oct. 15	No. 1080, 29 Oct. 15
Patiala F. S. Infy.	**Kala Singh,** Subadar	Subadar	3 June 16	No. 849, 29 July 16
Malay States Guides.	**Kehar Singh,** Havildar, 2639	Havildar	5 Aug. 16	No. 886, 5 Aug. 16
Do.	**Sawan Singh,** Naik, 1631	Naik	5 Aug. 16	No. 886, 5 Aug. 16
Do.	**Sarwan Singh,** Sepoy, 2512	Sepoy	5 Aug. 16	No. 886, 5 Aug. 16

THE INDIAN ORDER OF MERIT (CIVIL DIVISION).

3rd Class.

Corps.	Names and Rank.	In what rank admitted.	Date of admission to the Order and subsequent promotion.	Authority.
				Home Deptt. Notfn. (Public).
2nd Punj.	*Sultan Ahmad*	*Sepoy*	...	*G.G.O. No. 647, dated 3 July 91.*
Do.	*Booti*	*Do.*	...	*G. G. O. No. 647, 3 July 91*
Do.	*Khairoo*	*Do.*	...	*G. G. O. No. 647, 3 July 91.*
Do.	*Ghulam Nabi*	*Do.*	...	*G. G. O. No. 647, 3 July 91.*
Do.	*Mira Bux*	*Do.*	...	*G. G. O. No. 647, 3 July 91.*
Do.	*Nawab Shah*	*Do.*	...	*G. G. O. No. 647, 3 July 91.*
Do.	*Chandoo*	*Do.*	...	*G. G. O. No. 647, 3 July 91.*
Do.	*Jooma*	*Do.*	...	*G. G O. No. 647, 3 July 91.*
55th Rif.	*Badan Singh*	*Do.*	...	*G. G. O. No. 403, dated 9 May 02.*
13th Lrs.	**Niaz Ali**	Recruit	6 Mar. 06	No. 562, 6 Mar. 06.
34th Horse	**Sultan Muhammad Khan**, Subadar	Ressaidar	21 Mar. 12	No. (Political) 358, 21 Mar. 12.
21st Mn. Batty.	**Chaughatta**, Driver-Naick	Driver-Naik	6 Mar. 06	No. 563, 6 Mar. 06.
22nd Mn. Batty	*Kishen Singh, Subadar*	*Havildar*	*6 Nov. 02*	*No. 3441, 6 Nov. 02.*
57th Rif (*late 4th Punj. Infy.*)	**Jagat Singh**, Jemadar	Lance-Naik	25 Feb. 03	No. 553, 25 Feb. 03.
93rd Burma Infy.	**Haidar Khan**, Sepoy	Sepoy	23 Jan. 05	No. 160, 23 Jan. 05.
117th Mah.	*Shaikh Rahiman, Private*	*Private*	*1 June 05*	*No. 1195, 1 June 05.*
Zhob Militia	**Malak**, Lance-Daffadar	Lce.-Duffadar	20 June 07	No. 1451, 20 June 07.
N. Waziristan Militia.	**Ghani Shah**, Subadar	Havildar	18 Mar. 08	No. 741, 18 Mar. 08.
So. Waziristan Militia.	**Kawan Khan**, Subadar	Subadar	27 Aug. 09	No. 3266, 27 Aug. 09.
Late Kohat Border Mily. Police.	**Muhammad Shah**, Sowar	Sowar	13 Aug. 08	No. 1845, 13 Aug. 08.
Do.	**Nur Shah**, Jemadar	Jemadar	25 June 09	No. 2605, 25 June 09.
Mekran Levy Corps.	**Abdul Hakim**, Havildar	Havildar	12 Aug. 13	No. (Political) 446, 12 Aug. 13.
Do.	**Gulzar Khan**, Naik	Naik	12 Aug. 13	No. (Political) 446, 12 Aug. 13.
Frontier Constabulary	**Mauladad Khan**, Jemadar	Jemadar	18 Sep. 14	No. (Political) 1450, 18 Sep. 14.
	Jahangir, Naik.	Naik	18 Sep. 14	No. (Political) 1450, 18 Sep. 14.
Do.	**Karim Khan**, Sepoy	Sepoy	18 Sep. 14	No. (Political) 1450, 18 Sep. 14.
Burma Mil. Police	**Mitung Tang**, Havildar	No. 600 C. (Home Dept.)	30 Dec. 16	
Kalat Levies	**Ghulam Nabi**, *Khan Sahib*, Risaldar.	Risaldar	31 May 17	No. 267 (Public), 31 May 17.

Corps.	Names and Ranks.	Date.	Authority.
	INDIAN DISTINGUISHED SERVICE MEDALLISTS.		
	NOTE.—The names in *italics* are those of members on the pension establishment.		
			A. D. Notfn.
1st Lrs.	Nishan Ali, Sowar, 1735	27 Aug. 15	No. 820, 27 Aug. 15.
Do.	Ruknuddin, Jemadar	28 Jan. 16	No. 103, 28 Jan. 16.
Do.	Abdul Majid Khan, Trumpeter, 961	28 Jan. 16	No. 103, 28 Jan. 16.
2nd Lrs.	Abdul Latif Khan, Resaidar	1 Jan. 17	No. 676, 5 May 17.
Do.	Lahora Singh, Sowar, 1917	19 July 18	No. 1629, 19 July 18.
Do.	Nihalu, Dafadar, 1584	26 Oct. 18	No. 2536, 26 Oct. 18.
4th Cav.	Karam Singh, Dafadar, 333	1 Apl. 15	No. 426, 8 May 15.
Do.	Abdul Majid Beg, Lance-Dafadar	12 May 16	No. 525, 12 May 16.
Do.	Ghulam Muhammad Khan, Sowar, 1877	12 May 16	No. 525, 12 May 16.
Do.	Shoo Ram, Dafadar, 2100	3 Jun. 16	No. 849, 29 July 16.
6th Do.	Amir Singh, Jemadar	1 Jan. 17*	No. 219, 17 Feb. 17.
Do.	Arjan Singh, Jemadar	1 Jan. 17	No. 219, 17 Feb. 17.
Do.	Bachittar Singh, Jemadar	1 Jan. 17	No. 219, 17 Feb. 17.
Do.	Uday Chand Dafadar, 829	1 Jan. 17	No. 219, 17 Feb. 17.
Do.	Jai Singh, Risldr.	19 July 18	No. 1629, 19 July 18.
Do.	Gurmukh Singh, Lance-Dafadar, 2339	19 July 18	No. 1629, 19 July 18.
Do.	Mul Singh, Lance-Dafadar, 2621	20 Sep. 18	No. 2185, 20 Sep. 18.
Do.	Raja Singh, Sowar, 2943	20 Sep. 18	No. 2185, 20 Sep. 18.
7th Lrs.	Ran Singh, Dafadar, 3574	4 Aug. 15	No. 728, 4 Aug. 15.
Do.	Bakhtawar Singh, Sowar 3548	4 Aug. 15	No. 728, 4 Aug. 15.
Do.	Nanha Singh, Jemadar	10 Dec. 15	No. 1248, 10 Dec. 15.
Do.	Ranjit Singh, Lance-Dafadar, 3074	10 Dec. 15	No. 1248, 10 Dec. 15.
Do.	Chet Ram, Ressaidar	18 Nov. 16	No. 1388, 18 Nov. 16.
Do.	Thakur Singh, Jemadar	18 Nov. 16	No. 1388, 18 Nov. 16.
Do.	Tulsi Ram, Dafadar, 3590	18 Nov. 16	No. 1388, 18 Nov. 16.
Do.	Saran Singh, Lance-Dafadar, 2838	17 Aug. 17	No. 1357, 17 Aug. 17.
Do.	Genda Singh, Actg. Lance-Dafadar, 3684	17 Aug. 17	No. 1357, 17 Aug. 17.
Do.	Nanha Singh, Actg. Lance-Dafadar, 3237	17 Aug. 17	No. 1357, 17 Aug. 17.
Do.	Kehr Singh, Actg. Lance-Dafadar, 3429	17 Aug. 17	No. 1357, 17 Aug. 17.
Do.	Abdul Shakur Khan, Sowar, 3320	17 Aug. 17	No. 1357, 17 Aug. 17.
Do.	Ghulam Mohammad Khan III, Sowar, 3654	17 Aug. 17	No. 1357, 17 Aug. 17.
Do.	Ali Muhammad, Sowar, 3886	17 Aug. 17	No. 1357, 17 Aug. 17.
Do.	Jug Lal, Sowar, 3465	17 Aug. 17	No. 1357, 17 Aug. 17.
Do.	Bujan Singh, Sowar, 3536	17 Aug. 17	No. 1357, 17 Ang. 17.
Do.	Parmeshri Singh, Sowar, 3934	17 Aug. 17	No. 1357, 17 Aug. 17.
Do.	Magar Singh, Sowar, 3717	17 Aug. 17	No. 1357, 17 Aug. 17.
Do.	Subah Singh, Sowar, 3735	17 Aug. 17	No. 1357, 17 Aug. 17.
Do.	Sultan Singh, Sowar, 3464	17 Aug. 17	No. 1357, 17 Aug. 17.
Do.	Hem Singh, Kot-Dafadar, 2951	19 July 18	No. 1631, 19 July 18.
Do.	Jaimal Singh, Sowar, 3947	19 July 18	No. 1631, 19 July 18.
9th Horse	Mir Hussain, Sowar, 2870	18 Nov. 16	No. 1388, 18 Nov. 16.
Do.	Pretem Singh, A. Lce. Daf., 3150	23 June 17	No. 972, 23 June 17.
Do.	Shamsuddin Khan, Sowar, 3341	23 Jude 17	No. 972, 23 June 17.
Do.	Sujjan Singh, Sowar, 3090	23 June 17	No. 972, 23 June 17.
Do.	Sikandar Khan, Dafadar, 2845	19 July 18	No. 1629, 19 July 18.
10th Lrs.	Rattan Singh, Lance-Dafadar, 2916	19 July 18	No. 1629, 19 July 18.
Do.	*Uttam Singh, Risaldar-Major*	*19 July 18*	*No. 1629, 19 July 18.*
11th Lrs.	*Shahzad Mir, Ressaidar*	*28 June 07*	*No. 585, 28 June 07.*
Do.	Ram Singh, Lance-Naik, 163	21 Nov. 16	No. 676, 5 May 17.
Do.	Lashkar Singh, Lance-Naick, 368	12 June 17	No. 868, 2 June 17.
Do.	Ram Singh, Ressaidar, 11 Lancers	22 Nov. 17	No 1887, 2 Nov. 17.
Do.	Ram Suran, Sowar, 302	19 July 18	No. 1629, 19 July 18.
Do.	Ram Singh, Dafadar, 2889	26 Oct. 18	No. 2536 26 Oct. 18.
Do.	Dalip Singh, Dafadar, 114	26 Oct. 18	No. 2536, 26 Oct. 18.
12th Cav.	Ashaq Khan, Ressaidar	17 Aug. 17	No. 1360, 17 Aug. 17.
Do.	Bhan Singh, Kot-Dafadar 371	17 Aug. 17	No. 1360, 17 Aug. 17.
Do.	Waryam Singh, Dafadar, 608	13 July 18	No. 1571, 13 July 18.
Do.	Mehr Singh, Lance-Dafadar, 660	13 July 18	No. 1571, 13 July 18.
13th Do.	Karm Singh, Risaldar-Major	28 June 07	No. 585, 28 June 97.
Do.	Kaim Din, Dufadar, 372	5 Jan. 17	No. 676, 5 May 17.
Do.	Sard Chand, Dafadar, 380	2 June 17	No. 868, 2 June 17.
Do.	Lall Man, Lance-Dafadar, 112	2 June 17	No. 868, 2 June 17.
Do.	Dost Muhammad, Actg. Lance-Dafadar, 2085	2 June 17	No. 868, 2 June 17.
Do.	Puran Singh, Sowar, 2531	2 June 17	No. 868, 2 June 17.
Do.	Ujagar Singh, Sowar, 2582	2 June 17	No. 868, 2 Jnne 17.
Do.	Jagpal Singh, Jemadar	2 Aug. 18	No. 1767, 2 Aug. 18.
14th Lrs.	Amir Singh, Jemadar	20 Apl. 17	No. 597, 20 Apl. 17.
Do.	Deepchand, Sowar, 2978	17 Aug. 17	No. 1360, 17 Aug. 17.
Do.	Luttas Singh, Sowar, 3111	17 Aug. 17	No. 1360, 17 Aug. 17.
Do.	Amir Singh, Jemadar	17 Aug. 17	No. 1360, 17 Aug. 17.
Do.	Hargyan Singh, Risaldar.	17 Aug. 17	No. 1360, 17 Aug. 17.
Do.	Mansa Ram, Risaldar	17 Aug. 17	No. 1360, 17 Aug. 17.
Do.	Mehun Singh, Lance-Dafadar, 3105	17 Aug. 17	No. 1360, 17 Aug. 17.
Do.	Nandlal, Sowar, 3155	17 Aug. 17	No. 1360, 17 Aug. 17.

*Bar to I. D. S. M.

INDIAN DISTINGUISHED SERVICE MEDALLISTS—*contd.*

NOTE.—The names in *italics* are those of members on the pension establishment.

Corps.	Names and Ranks.	Date.	Authority. A. D. Notfn.
14th Lrs.	Bhani Singh, Jemadar	1 Jan. 18	No. 9, 1 Jan. 18.
Do.	Chandgi Ram, Lance-dafadar	1 Jan. 18	No. 9, 1 Jan. 18.
Do.	Abdul Wahas Khan, Ward Orderley, 2671	1 Jan. 17	No. 219, 17 Feb. 17.
15th Do.	Bakhtawar Singh, Dafadar, 387	4 Aug. 15	No. 728, 4 Aug. 17.
Do.	Ibrahim Khan, Dafadar, 346	1 Jan. 16	No. 187, 18 Feb. 16.
Do.	Allahad Khan, Lce. Dafadar, 2176	1 Jan. 16	No. 187, 18 Feb. 16.
Do.	Rahmatullah Khan, Actg. Lce-Dafadar, 1741	27 Apl. 16	No. 737, 30 June 16.
Do.	Dost Muhammad Khan, Sowar, 2626	27 Apl. 16	No. 737, 30 June 16.
Do.	Tulsi Ram, Dafadar, 379	17 Nov. 16	No. 1388, 17 Nov. 16.
16th Cav.	Phuman Singh, Lance-Dafadar, 1527	. . . .	No. 630, 3 July 15.
Do.	Phuman Singh, Sowar, 996	4 Aug. 15	No. 728, 4 Aug. 15.
Do.	Nikka Singh, Sowar, 1289	4 Aug. 15	No. 728, 4 Aug. 15.
Do.	Kirpa Singh, Sowar, 1725	19 Nov. 15	No. 1160, 19 Nov. 15.
Do.	Atma Singh, Sowar, 2003	19 Nov. 15	No. 1160, 19 Nov. 15.
Do.	Dhula Sing, Sowar, 2053	12 May 16	No. 525, 12 May 16.
Do.	Ismail Khan, Sowar, 2855	12 May 16	No. 525, 12 May 16.
Do.	Alaudin Khan, Ressaidar	12 May 16	No 525, 12 May 16.
Do.	Kala Singh, Kot-Dafadar, 1103	19 July 18	No. 1631, 19 July 18.
Do.	Gujar Singh, Kot-Dafadar, 1390	26 Oct. 18	No. 2536, 26 Oct 18.
17th Cav.	Khan Sahib, Sowar, 1980	20 Oct. 16	No. 1269, 20 Oct. 16.
18th Do.	*Ali Muhammad Khan, Duffadar*	*28 June 07*	*No. 728, 4 Aug. 15.*
Do.	Dost Muhammad Khan, Ressaidar	18 Nov. 10	No. 933, 18 Nov. 10.
Do.	Murad Khan, Naik, 1979	18 Nov. 10	No. 933, 18 Nov. 10.
Do.	Muhammad Khan, Sowar, 2009	18 Nov. 10	No. 933, 18 Nov. 10.
Do.	Tarja Khan, No. 4155	1 Jan. 17	No. 676, 5 May 17.
Do.	Bhagwan Singh, Dafadar, 1941	1 Jan. 17	No. 219, 17 Feb. 17.
Do.	Lal Singh, Sowar, 1595	1 Jan. 17	No. 219, 17 Feb. 17.
Do.	Hadayat Khan, Sowar, 2466	1 Jan. 17	No 219, 17 Feb. 17.
Do.	Mohammad Sharif Khan, Sowar, 2686	1 Jan. 17	No. 219, 17 Feb. 17.
Do.	Allah Ditta Khan, Dafadar, 193	18 Dec. 17	France 5th Cav. Div. Or. No. 1687, 18 Dec. 17.
Do.	Fahid Khan, Dafadar, 2147	18 Dec. 17	France 5th Cav. Dn. Or. No. 1687, 18 Dec. 17.
Do.	Ghulam Mahummad Khan, Lance-Dafadar, 2293	18 Dec. 17	France 5th Cav. Dn. Or. No. 1687, 18 Dec. 17.
Do.	Hayat Khan, Lance-Dafadar, 2202	18 Dec. 17	France 5th Cav. Dn. Or. No. 1687, 18 Dec. 17.
Do.	Makhan Khan, Sowar, 2400	18 Dec. 17	France 5th Cav. Dn. Or. No. 1687, 18 Dec. 17.
Do.	Talib Hussain Khan, Sowar, 2533	18 Dec. 17	France 5th Cav. Dn. Or. No. 1687, 18 Dec. 17.
Do.	Khuda Vaksh Khan, Jemadar	23 Jan. 18	France 5th Cav. Dn. Or. No. 1720, 23 Jan. 18.
Do.	Adalat Khan, Jemadar	23 Jan. 18	France 5th Cav. Dn. Or. No. 1720, 23 Jan. 18.
Do.	Muhammed Jan Khan, Sowar, 2411	23 Jan. 18	France 5th Cav. Dn. Or. No. 1720, 23 Jan. 18.
Do.	Khuda Baksh Khan, Kot-Dafadar, 1575	1 Jan. 18	No. 219, 17 Feb. 17.
Do.	Malik Sultan Khan, Risldr.	19 July 18	No. 1629, 19 July 18.
Do.	Sultan Khan, Sowar, 2385	19 July 18	No. 1629, 19 July 18.
Do.	Fatch Khan, Driver, 13196	19 July 18	No. 1629, 19 July 18.
19th Do.	Abdul Aziz Khan, Risaldar-Major	1 Jan. 10	No. 4, 1 Jan. 10.
Do.	Muhammad Juma Khan, Kot. Dafadar, 2383	1 Jan. 10	No. 4, 1 Jan. 10.
Do.	Amir Khan, Daffadar, 2631	1 Jan. 10	No. 4, 1 Jan. 10.
Do.	Gulbar Khan, Dafadar, 2947	1 Jan. 18	No. 5.
Do.	Hira Singh, *Bahadur*, Risldr.-Maj.	19 July 18	No. 1629, 19 July 18.
Do.	Moti Singh, Risldr.	19 July 18	No. 1629, 19 July 18.
Do.	Bishan Singh, Sower, 2077	20 Sep. 18	No. 2185, 20 Sep. 18.
Do.	Nadar Khan, Sowar, 3761	20 Sep. 18	No. 2185, 20 Sep. 18.
Do.	Sundar Singh, Sowar, 3485	20 Sep. 18	No. 2185, 20 Sep. 18.
Do.	Sarbaland Khan, Jemadar, I. O. M.	26 Oct. 18	No. 2536, 26 Oct. 18.
Do.	Ali Muhammad, Dafadar, 3659	26 Oct. 18	No. 2536, 26 Oct. 18.
20th Horse	Gokal Singh, Sowar, 830	10 Mar. 15	No. 356, 16 Apl. 15.
Do.	Ali Mohammed, Sowar	25 Nov. 16	Lon. Gaz., 25 Nov. 16.
Do.	Sayed Ghulam Mahbub, Salutri Major, 340	1 Jan. 17	No. 676, 5 May 17.
Do.	Shaikh Faiz-ud-Din Ressaidar	19 July 18	No. 1629, 19 July 18.
21st Cav.	*Thakur Singh, Risaldar*	*1 Jan. 08*	*No. 1, 1 Jan. 08.*
Do.	Dewa Singh, Dafadar, 2589	26 June 08	No. 527, 26 June 08.
Do.	Ali Sher Khan, Resaidar	17 Feb. 17	Lon. Gaz. 1 Jan. 17.
Do.	Santokh Singh, Dafadar 3090	16 June 17	Lon. Gaz. 16th June 17.
Do.	Bahadur Sher Kot Dafadar 3071	10 Nov. 17	No. 1908, 10 Nov. 17.
Do.	Rasul Khair, Ressaldar	16 Nov. 17	No.
Do.	Farzand Ali, Ressaidar	1 Jan. 18	No. 9, 1 Jan. 18.
Do.	Kunda Singh, Resaldar	1 Jan. 18	No. 9, 1 Jan. 18.
Do.	Rasul Khan, Kot-Dafadar, 3971	16 Aug. 18	No. 1885, 16 Aug. 18.

INDIAN DISTINGUISHED SERVICE MEDALLISTS

NOTE.—The names in *italics* are those of members on the pension establishment.

Corps.	Names and Ranks.	Date.	Authority. A. D. Notfn.
22nd Cav.	Azam Ali, Risaldar	17 Feb. 17	No. 219, 17 Feb. 17.
Do.	Wagid Ali, Dafadar, 705,	17 Aug 17	No 1360, 17 Aug. 17.
Do.	Abdulla Khan, Kot-Dafadar, 1902.	13 July 18	No. 1571, 13 July 18.
Do.	Suraj Din, Sowar, 2159	13 July 18	No 1571, 13 July 18.
23rd Do.	Hasamuddin Khan, Risaldar-Major	26 June 08	No. 527, 26 June 08.
Do.	Ismail Jan Dafadar	2 Nov. 17	No. 1887, 2 Nov. 17.
Do.	Bahadur Singh, Lance-Naik, 1466	1 Jan. 18	No. 9, 1 Jan. 18.
25th Do.	Kale Khan, Risaldar	15 Aug. 13	No. 691, 15 Aug. 13.
Do.	Ali Shan Khan, Dafadar, 2966	15 Aug. 13	No. 691, 15 Aug. 13.
Do.	Diwan Khan, Lance-Daffadar, 3284	15 Aug. 13	No. 691, 15 Aug. 13.
Do.	Budh Singh, Sowar, 3992	22 June 14	No. 578, 22 June 14.
Do.	Sher Ali Khan, Jemadar	3 June 15	No. 529, 3 June 15.
Do.	Sundar Singh, Sowar, 3843	3 June 15	No. 529, 3 June 15.
Do.	Ibrahim Khan, Dafadar, 3236	3 June 18	No. 1180, 3 June 18.
Do.	Janak Singh, Sowar, 4122	3 June 18	No. 1180, 3 June 18.
Do.	Palwan Khan, Actg. Lance-Dafadar, 3612	3 June 18	No. 1180, 3 June 18.
Do.	Sehju Ram, Dafadar, 3616	3 June 18	No. 1180, 3 June 18.
Do.	Sohan Singh, Lance-Dafadar, 3936	3 June 18	No. 1180, 3 June 18.
Do.	Subha Singh, Kot-Dafadar, 3204	3 June 18	No. 1180, 3 June 18.
26th Do.	Aidan Singh, Dafadar, 1228	16 Dec. 16	No. 1511, 16 Dec. 16.
Do.	Ramnath Singh, Lce.-Dafadar, 1226	16 Dec. 16	No. 1511, 16 Dec. 16.
Do.	Fateh Khan, Lce.-Dafadar, 1009	16 Dec. 16	No. 1511, 16 Dec. 16.
Do.	Muhammad Amin, Ressaidar	3 Aug. 17	No. 1265, 3 Aug. 17.
Do.	Feroze Khan, Ressaidar	17 Aug. 17	No. 1363, 17 Aug. 17.
Do.	Dario Singh, Dafadar, 905	17 Aug. 17	No. 1 63, 17 Aug. 17.
27th Do.	Ram Lal, Actg. Lance-Dafadar, 1016	19 Nov. 16	No. 1160, 19 Nov. 15.
Do.	Khazan Singh, Sowar, 1612	19 Nov. 15	No. 1160, 19 Nov. 15.
Do.	Dalpat Singh, Dafadar, 851	18 Nov. 16	No. 1388, 18 Nov. 16.
29th Lrs.	Darayo Singh, Dafadar, 2334	29 Dec. 16	No. 676, 5 May 17.
Do.	Hira Singh, Sowar, 3094	19 Nov. 17	No. 115, 18 Jan 18.
Do.	Mohamed Hafiz, Lance-Dafadar, 2673	29 Dec. 16	No. 676, 5 May 17.
Do.	Kahul Singh, Risaldar	4 June 17	Lond Gaz. 4 June 17.
Do.	Natha Singh, Jemadar	19 July 18	No. 1629, 19 July 18.
Do.	Ranchander, Sowar, 2328	19 July 18	No. 1629, 19 July 18.
30th Do.	Kapur Singh, Kot-Dafadar, 2708	3 June 16	No. 849, 29 July 16.
31st Lrs.	Muhi-ud-din Khan, Risaldar-Major	28 June 07	No. 691, 15 Aug. 13.
Do.	Ajun Khan, Jemadar	25 Nov. 16	No. 41, 13 Jan. 17.
32ndDo.	Bahadur Singh, Sowar, 3935	2 June 17	No. 068, 2 June 17.
Do.	Ashgar Ali Khan, Sowar, 3633	17 Aug. 17	No. 1360, 17 Aug. 17.
Do.	Fateh Mohammad Khan, Dafadar, 1821	17 Aug. 17	No 1360, 17 Aug. 17
Do.	Fazar Ali Khan, Jemadar	1 Jan. 18	No. 9, 1 Jan. 18.
Do.	Sundar Singh, Ressaidar	1 Jan. 18	No. 9, 1 Jan. 18.
Do.	Jiwan Singh, Dafadar, 2786	16 Aug. 18	No. 1885, 16 Aug. 18.
33rd Do.	Mansaram, Dufadar, 2292	19 Mar. 15	No. 285, 19 Mar. 15.
Do	Buta Singh, Sowar, 3080	19 Mar. 15	No. 285, 19 Mar. 15.
Do.	Mangal Singh, Sowar, 3357	19 Mar. 15	No. 285, 19 Mar. 15.
Do.	Hidayat Ali Khan, Ressaidar	2 July 15	No. 630, 3 July 15.
Do.	Abdul Sattar Khan, Ressaidar	2 July 15	No. 630, 3 July 15.
Do.	Saadulla Khan, Sowar, 2696	2 July 15	No. 630, 3 July 15.
Do.	Mangal Singh, Dafadar, 2952	1 Jan. 16	No. 187, 18 Feb. 16.
Do.	Kharak Singh, Actg. Lance-Dafadar	12 May 16	No. 524, 13 May 16.
Do.	Alam Ali Khan, Dafadar, 2731	12 May 16	No. 524, 13 May 16.
Do.	Abdullah Khan, Sowar, 3027	. . . .	No. 313, 2 Apl. 15.
Do.	Firman Shah, Sowar, 3154	. . .	No. 313, 2 Apl. 15.
Do.	Darbara Singh, Lance-Dafadar, 3382	18 Nov. 16	No. 1386, 18 Nov. 16.
Do.	Yusaf Khan, Dafadar, 2994	19 July 18	No. 1631, 19 July 18.
Do.	Mazhar Ali Khan, Dafadar, 2869	19 July 18	No. 1631, 19 July 18.
34th Horse	Badan Singh, Rissaidar	10 Mar. 15	No. 356, 16 Apl. 15.
Do.	Bulwant Singh, Ressaidar		4 March 18.
Do.	Sattar Shah, Risaldar	11 Jan. 14	
Do.	Rawat Singh, Risaldar	4 June 17	Lond. Gaz., 4 June 17.
Do.	Amir Khan, Ressaidar	19 July 18	No. 1629, 19 July 18.
Do.	Nadir Ali Khan, Dafadar, 3216	19 July 18	No. 1629, 19 July 18.
35th Horse	Dalip Singh, Dafadar, 3562	20 Sep. 18	No. 2185, 20 Sep. 18.
Do	Khuda Baksh Khan, Jemadar	1 Jan. 17	No. 676, 5 May 17.
36th Horse	Sahib Singh, Kote Dafadar, 615	1 Jan. 18	4 March 18.
Do.	Lalla Jan, Sowar, 3150	1 Jan. 17	No. 676, 5 May 17.
37th Lrs.	Muhammad Khan, Dafadar, 3036 (retd)	19 July 18	No. 1629, 19 July 18.
Do.	Dan Sing, Actg. Lance-Dafadar, 3115	7 Aug. 17	No. 1357, 17 Aug. 17.
38th Horse	Pertab Singh, Dafadar 2562	19 July 18	No. 1629, 19 July 18.
Do.	Shev Singh, Dafadar 2183	19 Nov. 17	No. 115, 18 Jan. 18.
Do.	Natha Singh, Risaldar	19 Nov. 17	No. 115, 18 Jan. 18.
Do.	Bostan Khan, Ressaidar	19 July 18	No. 1629, 19 July 18.
Do.	Tara Singh, Actg. Lance-Dafadar, 2788	19 July 18	No. 1629, 19 July 18.
39th Horse	Abdul Rahman Khan, Ressaidar	19 July 18	No. 1629, 19 July 18.
		2 Aug. 12	No. 759, 2 Aug. 12.

INDIAN DISTINGUISHED SERVICE MEDALLISTS—*contd.*

NOTE.—The names in *italics* are those of members on the pension establishment.

Corps.	Names and Ranks.	Date.	Authority. A. D. Notfn.
39th Horse	Harbant Singh, Dafadar, 2421	2 Aug. 18	No. 1767, 2 Aug. 18.
Do.	Ali Muhammad Khan, Sowar, 2588	26 Oct. 18	No. 2536, 26 Oct. 18.
Aden Troop	Aoliya Khan, Dafadar, 405	28 June 07	No. 586, 28 June 07.
Guides	*Moghal Baz Khan, Ressaidar*	*28 June 07*	No. 586, 28 June 07.
Do.	Khwaja Muhammad Khan, Risaldar, *Bahadur*	26 June 08	No. 527, 26 June 08.
Do.	Sharbat Ali, Jemadar	24 June 09	No. 600, 24 June 09.
Do.	Sher Muhammad, Dafadar, 1844	17 Oct. 13	No. 864, 17 Oct. 13.
Do.	Ghafur Shah, Havildar, 4363	21 Nov. 13	No. 958 21 Nov. 13.
Do.	Bhan Singh, Sepoy, 5510	23 July 15	No. 815, 27 Aug. 15.
Do.	Sher Muhammad, Sowar, 2466	27 Aug. 15	No. 820, 27 Aug. 15.
Do.	Taj Muhammad, Havildar, 4442	28 Jan. 16	No. 103, 28 Jan. 16.
Do.	Jagtu, Sepoy, 5312	28 Jan. 16	No. 103, 28 June 16.
Do.	Arsala Khan, Havildar, 4830	18 Nov. 16	No. 1386, 18 Nov. 16.
Do.	Ahmad Khan, Naik, 5555	17 Aug. 17	No. 1360, 17 Aug. 17.
Do.	Bishan Singh, Subadar	3 June 18	No. 1180, 3 June 18.
Do.	Mian Singh, Naik, 5303	3 June 18	No. 1180, 3 June 18.
Do.	Saida Khan, Subadar	26 Oct. 18	No. 2536, 26 Oct. 18.
Do.	Allah Khan, Naik, 234	26 Oct. 18	No. 2536, 26 Oct. 18.
Do.	Kabiraj, Sepoy, 5672	26 Oct. 18	No. 2536, 26 Oct. 18.
Do.	Manbahadur, Lance-Naik	26 Oct. 18	No. 2536, 26 Oct. 18.
1st S. & M.	*Bunyad Ali, Subadar-Major, Bahadur*	*28 June 07*	*No. 586, 28 June 07.*
Do.	Ramautar Patak, Subadar	26 June 08	No. 527, 26 June 08.
Do.	Sultan, Jemadar	9 Aug. 12	No. 780, 9 Aug. 12.
Do.	Diwan Ali, Jemadar	9 Aug. 12	No. 780, 9 Aug. 12.
Do.	Nawab Khan, Jemadar	15 Nov. 12	No. 1077, 15 Nov. 12.
Do.	Suba Singh, Sapper	. . . .	No. 313, 2 Apl. 15.
Do.	Chagatta, I. O. M., Colour-Havildar	. . . .	No. 313, 2 Apl. 15.
Do.	Abdul Aziz, Jemdr.	. . . .	No. 313, 2 Apl. 15.
Do.	Sunder Singh, Subadar	10 Mar. 15	No. 356, 16 Apl. 15.
Do.	Sucha Singh, Havildar, 372	10 Mar. 15	No. 356, 16 Apl. 15.
Do.	Basant Singh, 708, Sapper	10 Mar. 15	No. 356, 16 Apl. 15.
Do.	Bhawani Singh, Sapper, 91	10 Mar. 15	No. 356, 16 Apl. 15.
Do.	Mehar Baksh, Jemadar	4 May 15	No. 563, 12 June 15.
Do.	Niamatullah, Jemadar	4 May 15	No. 563, 12 June 15.
Do.	Basant Singh, Havildar, 312	29 Oct. 15	Lon. Gaz., 29 Oct. 15.
Do.	Hari Singh, Lance-Naik, 1910	29 Oct. 15	Lon. Gaz., 29 Oct. 15.
Do.	Fazl Shah, Subadar	1 Jan. 16	No. 187, 18 Feb. 16.
Do.	Magh Singh, Naik, 859	1 Jan. 16	No. 187, 18 Feb. 16.
Do.	Bhawani Datt, Lce. Naik, 1029	1 Jan. 16	No. 187, 18 Feb. 16.
Do.	Rambharose Misr, Havildar, 107	29 July 16	No. 849, 29 July 16.
Do.	Bholan Khan, Naik, 785	3 June 16	No. 849, 29 July 16.
Do.	Nawab Khan, Naik, 24	3 June 16	No. 849, 29 July 16.
Do.	Sawan Singh, Naik, 764	3 June 16	No. 849, 29 July 16.
Do.	Khairullah Khan, Lce-Naik	3 June 16	No. 849, 29 July 16.
Do.	Bostan, Sapper, 33	3 June 16	No. 849, 29 July 16.
Do.	Nur Hussain, 32	3 June 16	No. 849, 29 July 16.
Do.	Ram Bharose Misr, Havildar, 107	20 Nov. 16	No. 1386, 20 Nov. 16.
Do.	Harnam Singh, I Jemadar	18 Nov. 16	No. 1388, 18 Nov. 16.
Do.	Alam Kkan, Jemadar	18 Nov. 16	No. 1388, 18 Nov. 16.
Do.	Ramayan Singh, Havildar 185	18 Nov. 16	No. 1388, 18, Nov. 16.
Do.	Sharaf Din, Naik, 939	18 Nov. 16	No. 1388, 18 Nov. 16.
Do.	Kirpal Singh, Naik, 1178	18 Nov. 16	No. 1388, 18 Nov. 16.
Do.	Jiwa Singh, Colour Havildar, 565	16 Dec. 16	No. 1511, 16 Dec. 16.
Do.	Bishen Singh, Naik, 507	2 June 17	No. 868, 2 Jan. 17.
Do.	Pala Singh, Sapper, 1339	2 June 17	No 868, 2 Jan. 17.
Do.	Allah Rakha, Jemadar	17 Aug. 17	No. 1360, 17 Aug. 17.
Do.	Allah Rakha, Lance-Naik, 3164	17 Aug 17	No. 1360, 17 Aug. 17.
Do.	Amir Ali, Tempy. Naik, 2986	17 Aug. 17	No. 1360, 17 Aug. 17.
Do.	Feroze Khan, Havildar, 467	17 Aug. 17	No. 1360, 17 Aug. 17.
Do.	Haidar Zaman, Sapper, 2607	17 Aug. 17	No. 1360, 17 Aug. 17.
Do.	Hari Singh, Lance-Naik, 2259	17 Aug. 17	No. 1360, 17 Aug. 17.
Do.	Ishar Singh, Colour-Havildar, 339	17 Aug. 17	No. 1360, 17 Aug. 17.
Do.	Jamit Singh, Lance-Naik, 2446	17 Aug. 17	No. 1360, 17 Aug. 17.
Do.	Kala Singh, Naik, 1139	17 Aug. 17	No. 1360, 17 Aug. 17.
Do.	Khair Din, Naik, 1973	17 Aug. 17	No. 1360, 17 Aug. 17.
Do.	Nur Akhmad, Havildar, 1066	17 Aug. 17	No. 1360, 17 Aug. 17.
Do.	Rala Singh, Sapper, 2803	17 Aug. 17	No, 1360, 17 Aug. 17.
Do.	Sarfaroz, Lance-Naik, 1900	17 Aug. 17	No. 1360, 17 Aug. 17.
Do.	Yakhudin, Jemadar, 4 Coy.	17 Aug. 17	No. 1360, 17 Aug. 17.
Do.	Mihr Din, Subadar-Maj.	1 Jan. 18	No. 5, 1 Jan. 18.
Do.	Mihr Din, Subadar-Major, *Bahadur*	1 Jan. 18	No. 9, 1 Jan. 18.
Do.	Ghulam Akbar, Naik, 2173	1 Jan. 18	No. 9, 1 Jan. 18.
Do.	Nowsher Khan, Havildar, 1007	1 Jan. 18	No. 9, 1 Jan. 18.
Do.	Harnam Singh, Jemadar	19 July 18	No. 1629, 19 July 18.
Do.	Gujar Singh, Jemadar	19 July 18	No. 1629, 19 July 18.

INDIAN DISTINGUISHED SERVICE MEDALLISTS—*contd.*

NOTE—The names in *italics* are those of members on the pension establishment.

Corps.	Names and Ranks.	Date.	Authority. A. D. Notfn.
2nd S. & M.	Lakshamaya, Havildar, 2650	19 Nov. 15	No. 1160, 10 Nov. 15.
Do.	Bochan, Jemadar	19 Nov. 17	No. 115, 18 Jan. 18.
Do.	Raman, Naik, 3324	19 Nov. 15	No. 1160, 19 Nov. 15.
Do.	Chinnasawmi, Sapper, 2408	19 Nov. 15	No. 1160, 19 Nov. 15.
Do.	Munisawmi, Sapper, 3292	19 Nov. 15	No. 1160, 19 Nov. 15.
Do.	Said Ghafar, Sapper, 3641	19 Nov. 15	No. 1160, 19 Nov. 15.
Do.	Joseph, Jemadar	3 June 16	No. 849, 29 July 16.
Do.	R. Kothandapani, Jemadar	3 June 16	No. 849, 29 July 16.
Do.	Hanamantu, Subadar	18 Nov. 16	No. 1388, 18 Nov. 16.
Do.	David, Naik, 4427	18 Nov. 16	No. 1388, 18 Nov. 16.
Do.	Krishnasami, Lce. Naik, 3604	18 Nov. 16	No. 1388, 18 Nov. 16.
Do.	Antoni, Jemadar	2 June 17	No. 868, 2 June 17.
Do.	Bo Chan, Jemadar	2 June 17	No. 868, 2 June 17.
Do.	Ayyum Permaul, Lance-Naik, 3746	17 Aug. 17	No. 1360, 17 Aug. 17.
Do.	Govindu, Sapper, 2620	17 Aug. 17	No. 1360, 17 Aug. 17.
Do.	Ponnusami, Jemadar	17 Aug. 17	No. 1360, 17 Aug. 17.
Do.	Subramani, Havildar, 3047	17 Aug. 17	No. 1360, 17 Aug. 17.
Do.	Gopalsami, Sapper 2722	1 Jan. 18	No 9, 1 Jan. 18.
Do.	Bo Thaw, Jemadar	19 July 18	No. 1629, 19 July 18.
3rd S. & M.	*Ali Bahadur, Jemadar*	*4 May 8*	*No. 563, 12 June 15.*
Do.	Fateh Khan, Havildar, 2268	19 Mar. 15	No. 285, 19 Mar. 15.
Do.	Data Din Badai, Jemadar	4 May 15	No. 563, 12 June 15.
Do.	Nur Alam, Jemadar	3 June 15	Lon. Gaz., 23 June 15.
Do.	Mahamed Baksh, Driver Havildar, 3144	3 June 15	Lon. Gaz. 23 June 15.
Do.	Indar Singh, Sapper, 4052	3 June 15	Lon. Gaz. 23 June 15.
Do.	Ismail Khan, Subadar, *Bahadur*	. . . .	No. 313, 2 Apl. 15.
Do.	Gangaji Rao Khanvalkar, Havildar, 1930	4 Aug. 15	No. 728, 4 Aug. 15.
Do.	Narayan Halatkar, Naick, 2899	4 Aug. 15	No 728, 4 Aug. 15.
Do.	Said Hussain, Lance-Naick, 3955	4 Aug. 15	No. 728, 4 Aug. 15.
Do.	Krishna Bonsle, Havildar, 2151	19 Nov. 15	No. 1160, 19 Nov. 15.
Do.	Bagga Singh, Lance-Naik, 4012	19 Nov. 15	No. 1160, 19 No. 15.
Do.	Maroti Jadhao, Havildar, 3348	1 Jan. 16	No. 187, 18 Feb. 16.
Do.	Harnam Singh, Naik, 2720	1 Jan. 16	No. 187, 18 Feb. 16.
Do.	Shaikh Ramzan, Lance-Naik, 2869	1 Jan. 16	No. 187, 18 Feb. 16.
Do.	Saleh Muhamad, Sapper, 3450	1 Jan. 16	No. 187, 18 Feb. 16.
Do.	Gulam Kadar, Havildar, 4049	3 June 16	No. 849, 29 July 16.
Do.	Chanda Singh, Havildar, 3167	18 Nov. 16	No. 1388, 18 Nov. 16.
Do.	Sahib Singh, Naik, 3120	18 Nov. 16	No. 1386, 18 Nov. 16.
Do.	Genda Singh, Lce-Naik 4171	18 Nov. 16	No. 1386, 18 Nov. 16.
Do.	Jhanda Singh, Sapper, 4074	18 Nov. 16	No. 1386, 18 Nov. 16.
Do.	Darlal Singh, Havildar, 2390	1 Feb 17	Lon. Gaz. 1 Feb. 17.
Do.	Ganpat Kharade, Havildar, 2941	2 June 17	No. 868, 2 June 17.
Do.	Muhammad Juma, Naik, 3194	2 June 17	No. 868, 2 June 17.
Do.	Taj Din, Naik, 3904	17 Aug. 17	No. 1360, 17 Aug. 17.
Do.	Samundar Khan, Havildar, 3376	17 Aug. 17	No 1357, 17 Aug. 17.
Do.	Gajraj Singh, Sapper, 3111	17 Aug. 17	No. 1357, 17 Aug. 17.
Ry. Bn. S. & M.	Alif Khan, Subadar	3 June 18	No. 1180, 3 June 18.
Do.	Mangal Singh, Jemadar	3 June 18	No. 1180, 3 June 18.
Do.	Fateh Khan, Subadar	3 June 18	No. 1180, 3 June 18.
Do.	Deodatt Singh, Sapper, 3728	1 Jan. 18	No. 9, 1 Jan. 18.
Do.	Chanan Singh, Havildar, 2914	1 Jan. 18	No. 9, 1 Jan. 18.
Do.	Habibullah Khan, Jemadar	1 Jan. 18	No. 9, 1 Jan. 18.
Do.	Jiwan Singh, Naik, 4310	1 Jan. 18	No. 9, 1 Jan. 18.
Do.	Rahmat Khan, Naik, 3683	1 Jan. 18	No. 9, 1 Jan. 18.
Do.	Amir Zaman Khan, Subadar	19 Nov. 17	No. 115, 18 Jan. 18.
Do.	Fazil Din, Subadar	19 Nov. 17	No. 115, 18 Jan. 18.
Do.	Shev Baz, Subadar	19 Nov. 17	No. 115, 18 Jan. 18.
12th Div. Sig. Coy.	Asir Ram, Jemadar	19 Nov. 15	No. 187, 18 Feb. 16.
Do.	Shamanagal Singh, Sepoy, 2057	12 May 16	No. 525, 12 May 16.
25th Ry. Coy.	Faju, Sapper, 367	21 Nov. 17	No. 676, 5 May 17.
Do.	Mohamed Sadio, Lance-Naik, 382	21 Nov. 17	No. 676, 5 May 17.
26th Ry. Coy.	Pala Singh, Havildar, 4969	10 Dec. 15	No. 1249, 10 Dec. 15.
3rd Divl Sig. Coy.	Muhammad Fazal, Naik, 50	19 July 18	No. 1629, 19 July 18.
31st Div. Sig. Co. S. & M.	Dharm Singh, Lance-Naick, 1560	9 Aug. 12	No. 780, 9 Aug. 12.
Do.	Mawaz Din, Havildar, 28	3 June 16	No. 849, 29 July 16.
Do.	Wali Dad, Sapper, 53	3 June 16	No. 849, 29 July 16.
Do.	Puran Singh, Havildar, 10	18 Nov. 16	No. 1388, 18 Nov. 16.
Do.	Ismail Khan, Colour-Havildar, 26	17 Aug. 17	No. 1360, 17 Aug. 17.
Do.	Sant Singh, Havildar, 47	17 Aug. 17	No. 1360, 17 Aug. 17.

Corps.	Names and Ranks.	Date.	Authority.

INDIAN DISTINGUISHED SERVICE MEDALLISTS—*contd.*

NOTE.—The names in *italics* are those of members on the pension establishment.

Corps.	Names and Ranks.	Date.	A. D. Notfn.
31st. Div. Sig Co. S. & M.	Chamel Singh, Naik, 24	19 July 18	No. 1629, 19 July 18.
Divl. Sig. Coy.	Maluk Singh, Havildar, 24	1 Jan. 18	No. 9, 1 Jan. 18.
32nd Divl. Sig. Coy.	Allah Dad Khan, Lce-Naik, 57	18 Nov. 16	No. 1388, 18 Nov. 16.
Do.	Sowar Khan, Sowar, 73	18 Nov. 16	No. 1388, 18 Nov. 16.
33rd Divl. Sig. Coy.	Chinsawmy, Sapper, 20	12 May 16	No. 525, 12 May 16.
Do.	Appasawmy, Sapper, 99	12 May 16	No. 525, 12 May 16.
34th Divl. Sig. Coy.	Murugesam, Jemadar	4 Aug. 15	No. 728, 4 Aug. 15.
Do.	Rangasami, Sapper, 45	4 Aug. 15	No. 728, 4 Aug. 15.
Do.	Abdul Jabbar, Sapper, 20	4 Aug. 15	No. 728, 4 Aug. 15.
Do.	Syed Abdul Rahim, Sapper, 81	10 Dec. 15	No. 728, 4 Aug. 15.
Do.	Subramani, Sapper, 80	17 Aug. 17	No. 1357, 17 Aug. 17.
Do.	Baramuddin, Sapper, 4071	17 Aug. 17	No. 1357, 17 Aug. 17.
35th Divl. Sig. Coy.	Wariam Singh, Sepoy, 2314	12 May 16	No. 525, 12 May 16.
Do.	Saif Ali, Lce-Naik, 19	18 Nov. 16	No. 1388, 18 Nov. 16.
41st Wire Sig.	Pailhwan Khan, Havildar, 5	17 Aug. 17	No. 1360, 17 Aug. 17.
Lahore Sig. Coy.	Khan Zaman, Naik, 44	4 May 15	No. 563, 12 June 15.
Do.	Mushtak Hussain, Lance-Naik, 23	4 May 15	No. 563, 12 June 13.
Do.	Ghulam Ayud Din, Sapper, 45	4 May 15	No. 563, 12 June 13.
Do.	Changa, Sapper, 4204	3 June 15	Lon. Gaz., 23 June 15.
Do.	Gujar Singh, Sapper, 7	3 June 15	Lon. Gaz., 23 June 15.
12th Brigade Sig. Section.	Angamathu, Lance-Naik, 1759	19 Nov. 15	No. 1160, 19 Nov. 15.
30th Brigade Sig. Section.	Khan Bahadur, Sapper, 105	10 Nov. 15	No. 1160, 19 Nov. 15.
Do.	Muhammad Khan, Naik, 39	19 Nov. 15	No. 1160, 19 Nov. 15.
Do.	Gobindu., Sapper, 94	19 Nov. 15	No. 1160, 19 Nov. 15.
Indian Tel Coy.	Hari Persad, Sub-Inspr. (Grades as Subadar)	21 Nov. 16	No 676, 5 May 17.
Indian Sig. Coy.	Jai Singh, Sapper, 42	3 June 15	Lon. Gaz., 23 June 15.
Ind. Postl. Service.	Gossain, Subadar	1 Jan. 18	No. 9, 1 Jan. 18.
21st Mtn. By.	Bhola Singh, Driver, 571	2 Jan. 17	No. 676, 5 May 17.
Do.	Gazan Singh, Gunner-Havildar, 376	2 Jan. 17	No. 676, 5 May 17.
Do.	Narain Singh, Driver, 667	2 Jan. 17	No. 676, 5 May 17.
Do.	Rajah, Jemadar	2 Jan. 17	No. 676, 5 May 17.
Do.	Sarwar Khan, Driver, 726	2 Jan. 17	No. 676, 5 May 17.
Do.	Nar Singh, Gunner, 1023	1 Jan. 18	No. 9, 1 Jan. 18.
22nd Mtn. By.	*Mangal Singh, Driver, 766*	*26 June 08*	*No. 527, 26 June 08.*
Do.	Ghazan Khan, Havildar, 773	1 Jan. 18	No. 9, 1 Jan. 18.
Do.	Aurungzeb, Gunner, 917 (Lance-Naik)	3 June 18	No. 1180, 3 June 18.
Do.	Santokh Singh, Driver, 1153	26 Aug. 18	No.
23rd Mtn. Batty.	Chet Singh, Gunner, 200	. . . .	No. 630, 3 July 15.
Do.	Ganayah Singh, Lance-Naick, 40	. . . .	No. 630, 3 July 15.
Do.	Kishen Singh, Gunner, 46	. . . .	No. 630, 3 July 15.
Do.	Sher Khan, Driver Naick, 226	4 Aug. 15	No. 728, 4 Aug. 15.
Do.	Harnam Singh, Gunner, 535	4 Aug. 15	No. 728, 4 Aug. 15.
Do.	Jwala Singh, Gunner, 52	12 May 16	No. 525, 12 May 16.
Do.	Ali Akbar, Naik, 103	18 Nov. 16	No. 1388, 18 Nov. 16.
Do.	Mehr Singh, Gunner, 34	18 Nov. 16	No. 1388, 18 Nov. 16.
24th Mtn. By.	Muhammad Afsar, Driver, 791	4 Oct. 18	No. 2342, 4 Oct. 18.
25th Mtn. By.	Haidar Khan, Jemadar	17 Aug. 17	No. 1356, 17 Aug. 17.
26th Mtn. By.	Nikka Singh, Naik, 1239	29 Oct. 15	No. 1081, 29 Oct. 15.
Do.	Gaiian Singh, Gunner-Naik	2 Jan. 17	No. 676, 5 May 17.
Do.	Ghulam Mohamed, Havildar-Major	2 Jan. 17	No. 676, 5 May 17.
Do.	Hem Singh, Subadar	2 Jan. 17	No. 676, 5 May 17.
Do.	Liakat Ali, Qr.-Master-Havildar, 1354	2 Jan. 17	No. 676, 5 May 17.
Do.	Hassan Ghulam, Jemadar	17 Aug. 17	No. 1360, 17 Aug. 17.
27th. Do.	Bachan Singh, Naick, 727	3 June 15	No. 530, 3 June 15.
Do.	Sundar Singh, Lance Naik, 1141	3 June 16	No. 849, 29 July 16.
Do.	Bhan Singh, Havildar, 704	3 June 16	No. 849, 29 July 16.
Do.	Imam Din, Lance-Naik, 970	21 Nov. 16	No. 676, 5 May 17.
Do.	Jahan Dad, Driver-Havildar, 247	21 Nov. 16	No. 676, 5 May 17.
Do.	Narayan Singh, Havildar, 783	1 Jan. 18	No. 9, 1 Jan. 18.
28th Mnt. By.	Mehr Khan, Gunner, 1211	3 June 15	No. 530, 3 June 15.
Do.	Rajwali, Gunner, 1138	21 Nov. 16	No. 676, 5 May 17.
Do.	Fateh Ali, Lance-Naik, 561	9 Dec. 16	No. 676, May 17.
Do.	Narain Singh, Gunner, 412	9 Dec. 16	No. 676, May 17.
Do.	Sant Singh, Havildar, 471	9 Dec. 16	No. 676, May 17.

INDIAN DISTINGUISHED SERVICE MEDALLISTS—*contd.*

NOTE.—The names in *italics* are those of members on the pension establishment.

Corps.	Names and Ranks.	Date.	Authority. A. D. Notfn.
30th Mtn. Batt.	Daya Singh, Jemadar	24 Sep. 15	No. 942, 24 Sept. 15.
Do.	Kishan Singh, Jemadar	24 Sep. 15	No. 942, 24 Sept. 15.
Do.	Phuman Singh, Gunner, 592	19 Nov. 15	No. 1160, 19 Nov. 15.
Do.	Mota Singh, Gunner, 480	19 Nov. 15	No. 1160, 19 Nov. 15.
Do.	Sardar Khan, Gunner, 141	19 Nov. 15	No. 1160, 19 Nov. 15.
Do.	Harim Khan, Gunner, 752	19 Nov. 15	No. 1160, 19 Nov. 15.
Do.	Lal Beg, Naik, 425	19 Nov. 15	No. 1169, 19 Nov. 15.
Do.	Imam Ali, Naik, 471	10 May 17	No. 1170, 20 July 17.
Royal Art.	Muhammad Alam, Naik, 9195	17 Aug. 17	No. 1360, 17 Aug. 17.
Do.	Manowar Khan, Havildar-Maj., 2287	17 Aug. 17	No. 1360, 17 Aug. 17.
Do.	Sultan Ali, Jemadar	17 Aug. 17	No. 1360, 17 Aug. 17.
Do.	Arjan Singh, Lance-Naik, 6627	19 July 18	No. 1629, 19 July 18.
Do.	Colunda, Indian Driver, 3088	19 July 18	No. 1629, 19 July 18.
Do.	Durra Khan, Naik, 1630	19 July 18	No. 1629, 19 July 18.
Do.	Khazan Singh, Driver, 9558	20 Sep. 18	No. 2185, 20 Sep. 18.
Indian Coast Arty.	Langar Khan, Havildar, 23228	15 Oct. 15	No. 1024, 15 Oct. 15.
Do.	Said Ahmad, Havildar, 117	10 Dec. 15	No. 1248, 10 Dec. 15.
Do.	Bostan Khan, Lance-Naik, 131	10 Dec. 15	No. 1248, 10 Dec. 15.
Do.	Mila Singh, Sepoy, 465	10 Dec. 15	No. 1248, 10 Dec. 15.
Do.	Gulsher, Havildar, 4440	10 Dec. 15	No. 1248, 10 Dec. 15.
Do.	Rhoda Singh, Havildar, 4517	10 Dec. 15	No. 1248, 15 Oct. 15.
Ind. Divl. Ammunition Col.	Multani Ram, Jemadar	17 Aug. 17	No. 1360, 17 Aug. 17.
Do.	Fazal Dad Havildar-Major, 4133	19 July 18	No. 1629, 19 July 18.
Do.	Abdul Qayum, Driver, 27652	20 Sep. 18	No. 2185 20 Sep. 18.
Do.	Bashir Ahmed, Driver, 21316	20 Sep. 18	No. 2185, 20 Sep. 18.
Do.	Nathoo Khan, Driver, 32810	20 Sep. 18	No. 2185, 20 Sep. 18.
Do.	Tara Datt, Dirver, 18115	20 Sep. 18	No. 2185, 20 Sep 18.
2nd Rajputs.	Badan Singh, Lance Naik, 2393	3 June 15	No. 528, 3 June 15.
Do.	Ramlal Singh, Havildar, 911	4 Aug. 15	No. 728, 4 Aug. 15.
Do.	Indar Singh, Subadar	15 Oct. 15	No. 1026, 15 Oct. 15.
Do.	Raghubir Singh, Naik, 2119	15 Oct. 15	No. 1026, 15 Oct. 15.
Do.	Indar Singh, Subadar	....	No. 1023, 15 Oct. 15.
Do.	Raghubir Singh, Jemadar	....	No. 1023, 15 Oct. 15.
Do.	Bakhtawar Singh, Havildar, 1889	18 Nov. 16	No. 1386, 18 Nov. 16.
Do.	Munshi Singh, Sepoy, 1875	18 Nov. 16	No. 1386, 18 Nov. 16.
Do.	Pahlad Singh, Jemadar	18 Nov. 16	No. 1388, 18 Nov. 16.
Do.	Desraj Singh, Havildar, 1424	18 Nov. 16	No. 1388, 18 Nov. 16
Do.	Mukund Singh, Havildar, 1859	18 Nov. 16	No. 1388, 18 Nov. 16.
Do.	Ramsarup Singh, Sepoy, 1860	18 Nov. 16	No. 1388, 18 Nov. 16.
Do.	Kanhaiya Singh, Sepoy, 2241	18 Nov. 16	No. 1388, 18 Nov. 16.
Do.	Chunni Singh, Sepoy, 2304	18 Nov. 16	No. 1388.
3rd Brahmans	*Lachman Dube, Subadar-Maj.*	*1 Jan. 08*	*No. 1, 1 Jan. 08.*
4th Rajputs	Faqira Singh, Subadar	....	No. 630, 3 July 15.
5th Infy.	Rahamat Khan, Jemadar	15 July 16	No. 793, 15 July 16.
Do.	Ajmeri, Colour-Havildar, 1973	15 July 16	No. 793, 15 July 16.
Do.	Munsab Khan, Sepoy, 1915	15 July 16	No. 793, 15 July 16.
Do.	Faiz Muhammad Khan, 2126	15 July 16	No. 793, 15 July 16.
Do.	Mazhar Khan, Sepoy, 2728	15 July 16	No. 793, 15 July 16.
Do.	Azizuddin Khan, Subadar	21 Nov. 16	No. 676, 5 May 17.
Do.	Rahimdad Khan, Sepoy, 2579	21 Nov. 16	No. 676, 5 May 17.
Do.	Karam Ilahi Khan, Ty. Lance-Naik, 3094	1 Jan. 18	No. 91, Jan. 18,
Do.	Ghulam Nabi-Khan, Havildar, 2506	1 Jan. 18	No. 91, Jan. 11.
Do.	Maula Bakhsh, Havildar, 2517	1 Jan. 18	No. 91, Jan. 11.
Do.	Sikhdar, Jemadar	3 June 18	No. 1110, 3 June 18.
Do.	Maru Khan, Naik, 1985	3 June 18	No. 1180, 3 June 18.
6th Infy.	Rambhagat, Sepoy, 2960	4 May 15	No. 563, 12 June 15.
Do.	Ratna, Subadar	1 Jan. 16	No. 187, 18 Feb. 16.
Do.	Har Lal, Havildar, 1767	1 Jan. 16	No. 187, 18 Feb. 16.
Do.	Hardwari, Naik, 2822	1 Jan. 16	No. 187, 18 Feb. 16.
Do.	Bhup Singh, Bugler, 3106	1 Jan. 16	No. 187, 18 Feb. 16.
Do.	Lehri, Sepoy, 3415	1 Jan. 16	No. 187, 18 Feb. 16.
Do.	Lukhi Ram, Jemadar	....	No. 313, 2 Apl. 15.
Do.	Badlu, Havildar, 1548	....	No. 313, 2 Apl. 15.
Do.	Risal, Sepoy, 3426	....	No. 313, 2 Apl. 15.
Do.	Harpul, Havildar, 1696	....	No. 313, 2 Apl. 15.
Do.	Maru, Lce. Naik, 2685	....	No. 313, 2 Apl. 15.
Do.	Bagmal, Sepoy, 1004	....	No. 313, 2 Apl. 15.
Do.	Mula, Havildar, 1574	....	No. 313, 2 Apl. 15.
Do.	Nobat, Sepoy, 3852	....	No. 313, 2 Apl. 15.
Do.	Shiulal, Subadar	12 May 16	No. 525, 12 May 16.
Do.	Jai Chand	12 May 16	No. 525, 12 May 16.
Do.	Mangli, Lance-Naik, 4301	12 May 16	No. 525, 12 May 16.
Do.	Sukhram, Naik, 3619	18 Nov. 16	No. 1386, 18 Nov. 16
		18 Nov. 16	No. 1388, 18 Nov. 16.

INDIAN DISTINGUISHED SERVICE MEDALLISTS—*contd.*

NOTE.—The names in *italics* are those of members on the pension establishment.

Corps.	Names and Ranks.	Date.	Authority.
			A. D. Notfn.
6th Infy.	Kanaya, Sepoy, 3574	18 Nov. 16	No. 1388, 18 Nov. 16.
Do.	Desram, Sepoy, 3751	18 Nov 16	No. 1388, 18 Nov. 16
Do.	Ram Singh, Sepoy, 4075	18 Nov. 16	No. 1388, 18 Nov. 16.
Do.	Gurdhala, Sepoy, 3934	3 June 18	No. 1178, 3 June 18
7th Rajputs	Bhawani Pershud Singh, Havildar, 1176	20 Mar. 13	No. 260, 20 Mar. 13.
Do.	Brijmohan Singh, Subadar	19 Mar. 15	No. 285, 19 Mar. 15.
Do.	Ramdhar Singh, Naick, 1567	. . . .	No. 630, 3 July 15.
Do.	Matan Singh, Sepoy, 2165	. . . .	No. 630, 3 July 15.
Do.	Gurdatt Singh, Havildar, 1192	. . . .	No. 630, 3 July 15.
Do.	Sirdar Singh, Subadar	10 Dec. 15	No. 1248, 10 Dec. 15.
Do.	Umrao Singh, Colour-Havildar, 1479	17 Aug. 17	No. 1357, 17 Aug. 17.
Do.	Bhola Singh, Sepoy, 2134	17 Aug. 17	No. 1357, 17 Aug. 17.
Do.	Chandika Singh, Sepoy, 2890	6 Sep. 18	No. 2060, 6 Sep. 18.
8th Rajputs	Daulat Singh, Subadar	17 Aug. 17	No. 1357, 17 Aug. 17.
Do.	Ashar Mutca Singh, 1880	17 Aug. 17	No. 1357, 17 Aug. 17.
9th Infy.	Abdul Latif, Sepoy, 2069	3 June 15	Lon. Gaz., 23 June 15.
Do.	Angad Pande, Naik, 2867	1 Jan. 16	No. 187, 1 Feb. 16.
Do.	Bhure Singh, Subadar-Major	. . . .	No. 313, 2 Apl. 15.
Do.	Amar Singh, Havildar, 2203	. . . .	No. 313, 2 Apl. 15.
Do.	Ishar Singh, Havildar, 4132	12 May 16	No. 525, 12 May 16.
Do.	Ramzan Khan, 2708, Havildar	12 May 16	No. 525, 12 May 16.
Do.	Sher Muhammad Khan, Jemadar	17 Aug. 17	No. 1360, 17 Aug. 17.
Do.	Sheo Chand, Lance-Naik, 347	17 Aug. 17	No. 1360, 17 Aug. 17.
Do.	Maima Singh, Lance-Naik (Tempy. Naik), 8433	17 Aug. 17	No. 1360, 17 Aug. 17.
10th Jats	Pirdhan Sepoy, 2299	3 June 16	No. 529, 3 June 15.
Do.	Bhola, Sepoy, 3363	18 Nov. 15	No. 1388, 18 Nov. 16.
11th Rajputs	Narain Singh, Subadar	25 July 18	No. 1704, 26 July 18.
13th Rajputs	Daulat Singh, Sepoy, 1566	3 June 15	No. 530, 3 June 15.
Do.	Girdhari Singh, Sepoy, 1870	. . . .	No. 530, 3 June 15.
14th Sikhs	Bhagwan Singh, Sepoy, 4475	17 Sep. 15	No. 906, 17 Sept. 15.
Do.	Bir Singh, Naik, 3694	17 Sep. 15	No. 960, 17 Sept. 15.
Do.	Gurditt Singh, Ward-Orderly, 3550	17 Sep. 15	No. 960, 17 Sept. 15.
Do.	Sham Singh, Subadar-Maj.	3 June 16	No. 849, 29 July 16.
Do.	Sucha Singh, Lce-Naik, 3658	3 June 16	No. 849, 29 July 16.
Do.	Amar Singh, Lce.-Naik, 3243	3 June 16	No. 849, 29 July 16.
Do.	Jagat Singh, Sepoy, 4118	3 June 16	No. 849, 21 July 16.
Do.	Kala Singh. Sepoy, 2648	3 June 16	No. 849, 29 July 16.
Do.	Chanan Singh, Sepoy, 4506	3 June 16	No. 849, 29 July 16.
Do.	Jiwan Singh, Sepoy, 2248	3 June 16	No. 849, 29 July 15.
Do.	Jagndiar Singh, Sepoy, 4330	3 June 16	No. 849, 29 July 16.
Do.	Gujjan Sepoy, 4156	3 June 16	No. 849, 29 July 16.
Do.	Lal Singh, Sepoy, 4353	3 June 16	No. 829, 29 July 16.
Do.	Bhog Singh, Sepoy, 4857	3 June 16	No. 849, 29 July 16.
Do.	Chanan Singh, Sepoy, 4393	3 June 16	No. 849, 29 July 16.
Do.	Sundar Singh, Sepoy, 3171	3 June 16	No. 849, 29 July 16.
Do.	Lal Singh, Sepoy, 4151	3 June 16	No. 849, 29 July 16.
Do.	Bharpur Singh, Sepoy, 4386	3 June 16	No. 849, 29 July 16.
Do.	Chanda Singh, Drummer, 4144	3 June 16	No. 849, 29 July 16.
Do.	Indar Singh, Drummer, 3246	3 June 16	No. 849, 29 July 16.
Do.	Kala Singh, Sepoy, 4361	3 June 16	No. 849, 29 July 16.
Do.	Jiwan Singh, Sepoy, 4505	3 June 16	No. 849, 29 July 16.
Do.	Sundar Singh, Sepoy, 948	3 June 16	No. 849, 29 July 16.
Do.	Bhag Singh, Lance-Naik, 4538	3 June 16	No. 849, 29 July 16.
Do.	Harnam Singh, Lance-Naik, 3738	3 June 16	No. 849, 29 July 16.
Do.	Phuman Singh, Sepoy, 4147	3 June 16	No. 849, 29 July 16.
Do.	Sarmukh Singh, Sepoy, 4292	3 June 16	No. 849, 26 July 16.
Do.	Likal Singh, Sepoy, 4019	3 June 16	No. 849, 26 July 16.
Do.	Bachan Singh, Sepoy, 4295	3 June 16	No. 849, 29 July 16.
Do.	Mangal Singh, Sepoy, 4540	3 June 16	No. 849, 29 July 16.
Do.	Indar Singh, Sepoy, 4146	3 June 16	No. 849, 29 July 16.
Do.	Gujar Singh, Sepoy, 2860	3 June 16	No. 849, 29 July 16.
Do.	Chanan Singh, Sepoy, 4232	3 June 16	No. 849, 29 July 16.
Do.	Ram Singh, Lce.-Naik, 3657	3 June 16	No. 849, 29 July 16.
Do.	Gajjan Singh, Sepoy, 4573	3 June 16	No. 849, 29 July 16.
Do.	Khem Singh, Sepoy, 3903	3 June 16	No. 849, 29 July 16.
Do.	Munshi Singh, Sepoy, 4252	3 June 16	No. 849, 29 July 16.
Do.	Bir Singh, Drummer, 3100	3 June 16	No. 849, 29 July 16.
Do.	Narayan Singh, Subadar	31 Aug. 17	No. 1441, 31 Aug. 17.
Do.	Jiwan Singh, Subadar	3 June 18	No. 1178, 3 June 18.
Do.	Jaswant Singh, Jemadar	19 July 18	No. 1629, 19 July 18.
15th Sikhs	Tilok Singh, Sepoy, 529	4 May 15	No. 563, 12 June 15.
Do.	Lal Singh, Sepoy, 702	23 July 15	No. 815, 27 Aug. 15.
Do.	Sucha Singh, Sepoy, 638	23 July 15	No. 815, 27 Aug. 15.
Do.	Sapuran Singh, Sepoy, 962	23 July 15	No. 815, 27 Aug. 15.
Do.	Bhagwan Singh, Lce.-Naik, 53	1 Jan. 16	No. 815, 27 Aug. 15.
Do.	Diwan Singh, Sepoy, 219	1 Jan. 16	No. 815, 27 Aug. 15
Do.	Wazir Singh, Jemdr.	. . . .	No. 313, 2 Apl. 15.

INDIAN DISTINGUISHED SERVICE MEDALLISTS—*contd.*

NOTE—The names in *italics* are those of members on the pension establishment.

Corps.	Names and Ranks.	Date.	Authority. A. D. Notfn.
15th Sikhs	Bir Singh, Jemadar	....	No. 313, 2 Apl. 15.
Do.	Bishn Singh, Naik, 4576	....	No. 313, 2 Apl. 15.
Do.	Sarain Singh, Sepoy, 1255	23 July 15	No. 815, 27 Aug. 15.
Do.	Fateh Singh, Sepoy, 1036	23 July 15	No. 815, 27 Aug. 15.
Do.	Ujagar Singh, Sepoy, 1001	23 July 15	No. 815, 27 Aug. 15.
Do.	Sundar Singh, Sepoy, 1249	23 July 15	No. 815, 27 Aug. 15.
Do.	Ganda Singh, Sepoy, 1339	23 July 15	No. 815, 27 Aug. 15.
Do.	Harnam Singh, Sepoy, 1360	23 July 15	No. 815, 27 Aug. 15.
Do.	Phuman Singh, Subadar	3 June 16	No. 849, 29 July 16.
Do.	Bhagwan Singh, Havildar, 4616	3 June 16	No. 849, 29 July 19.
Do.	Bishn Singh, Havildar, 3989	3 June 16	No. 849, 29 July 16.
Do.	Bachan Singh, Havildar, 4572	3 June 16	No. 849, 29 July 16.
Do.	Budh Singh, Havildar, 4429	3 June 16	No. 849, 29 July 19
17th Infy.	Abdul Ghafoor, Lce.-Naik, 2251	8 Nov. 17	No. 1388, 18 Nov. 16.
Do.	Ali Bahadur Khan, Subadar	4 June 17	Lond. Gaz., 4 June 17.
Do.	Niamat Khan, Sepoy, 2472	1 Jan. 18	No. 9, 1 Jan. 18.
Do.	Babu Khan, Sepoy, 3119	1 Jan. 18	No. 9, 1 Jan. 18.
Do.	Muhammad Ali Khan, Sepoy	1 Jan. 18	No. 9, 1 Jan. 18.
Do.	Hayat Khan, Subadar	3 June 18	No. 1180, 3 June 18.
Do.	Munshi Khan, Sepoy, 3166	3 June 18	No. 1180, 3 June 18.
Do.	Qadir Baksh, Sepoy, 2797	3 June 18	No. 1180, 3 June 18.
Do.	Abdul Rahman Khan, Havildar, 1736	3 June 18	No. 1180, 3 June 18.
Do.	Allah Nur Khan, Sepoy, 3117	3 June 18	No. 1180, 3 June 18.
Do.	Allah Ditta Khan, Sepoy, 3052	3 June 18	No. 1180, 3 June 18.
Do.	Ghulam Maula Khan, Naik, 2663	3 June 18	No. 1180, 3 June 18.
Do.	Ibrahim Khan, Naik, 2958	3 June 18	No. 1180, 3 June 18.
Do.	Karim Baksh, Sepoy, (Actg. Naik), 2715	3 June 18	No. 1180, 3 June 18.
Do.	Qasim Ali Khan, Sepoy, (Acting Naik), 1846	3 June 18	No. 1180, 3 June 18.
Do.	Muhammad Ali Khan, Naik, 2556	3 June 18	No. 1180, 3 June 18.
19th Punj.	Jalal Khan, Subadar	26 June 08	No. 527, 26 June 08.
Do.	Sundar Singh, Sepoy, 1249	23 July 15	No. 527, 26 June 08.
Do.	*Ganda Singh, Sepoy, 1339*	*23 July 15*	*No. 527, 26 June 08.*
Do.	Ralla Singh, Subadar	17 Aug. 17	No. 1357, 17 Aug. 17.
Do.	Ghulam Muhammad, Subadar	17 Aug. 17	No. 1357, 17 Aug. 17.
Do.	Sher Ahmed, Havildar, 372	17 Aug. 17	No. 1357, 17 Aug. 17.
Do.	Kishen Singh, Bugler, 1455	17 Aug. 17	No. 1357, 17 Aug. 17.
20th Punj.	Sher Ali, Naik, 370	4 Aug. 15	No. 728, 4 Aug. 15.
Do.	Ghulam Haidar, Lance-Naik, 1099	10 Dec. 15	No. 1248, 10 Dec. 15.
Do.	Kirpa, Lance-Naik, 1483	10 Dec. 15	No. 1248, 10 Dec. 15.
Do.	Bhima, Sepoy, 1937	10 Dec. 15	No. 1248, 10 Dec. 15.
Do.	Bachittar, Sepoy No. 1576	17 Aug. 17	No. 1360, 17 Aug. 17.
21st Punj.	Fateh Ali, Havildar, 4287	1 Jan. 17	No. 676, 5 May 17.
Do.	Habib Khan, Subadar	8 Aug. 17	No. 1305, 8 Aug. 17.
22nd Punj.	*Badr Din, Havildar, 3802*	*26 June 08*	*No. 1248, 10 Dec. 15.*
Do.	Jehandad Khan, Jemadar	26 June 08	No. 1248, 10 Dec. 15.
Do.	*Partab Singh, Sepoy, 3997*	*26 June 08*	*No. 1248, 10 Dec. 17.*
Do.	Jawala Singh, Sepoy, 4838	26 June 08	No. 1248, 10 Dec. 15.
Do.	Said Ahmad, Havildar, 117	....	No. 1248, 10 Dec. 15.
Do.	Bostan Khan, Lce.-Naik, 131	....	No. 1248, 10 Dec. 15.
Do.	Mila Singh, Sepoy, 465	....	No. 1248, 10 Dec. 15.
Do.	Gulshar, Havildar, 4440	....	No. 1248, 10 Dec. 15.
Do.	Rhoda Singh, Havildar, 4517	....	No. 1248, 10 Dec. 15.
Do.	Hakim, Khan, Jemadar,	17 Aug. 17	No. 1357, 17 Aug. 17.
Do.	Ganga Singh, Havildar, 4272½	17 Aug. 17	No. 1357, 17 Aug. 17.
Do.	Dalep Singh, Sepoy 222½	17 Aug. 17	No. 1057, 17 Aug. 17.
23rd Prs.	*Ram Singh, Havildar*	*28 June 07*	*No. 586, 28 June 07.*
2/23 Prs.	Natha Singh, Lance Naik, 4143	19 Nov. 17	No. 115, 18 Jan. 18.
Do.	Phumman Singh, Havildar 4059	8 Nov. 16	No. 1386 18 Nov. 16.
Do.	Lachman Singh, Sepoy 145	17 Aug. 17	No. 1360, 17 Aug. 17.
24th Punj.	Gul Akbar, M.C. Subadar	28 June 07	No. 586, 28 June 07.
Do.	Fazal Din, Naik, 3900	12 Dec. 11	No. 101
Do.	Lachman Sing, Havildar, 2540	3 June 15	Lon. Gaz., 23 Jan. 15.
Do.	Pal Singh, Sepy, 514	4 Aug. 15	No. 728, 4 Aug. 15.
Do.	Vir Singh, Sepy, 405	4 Aug. 15	No. 728, 4 Aug. 15.
Do.	Labh Singh, Sepy, 4909	4 Aug. 15	No. 728, 4 Aug. 15.
Do.	Sardar Khan, Naiki 4889	4 Aug. 15	No. 728, 4 Aug. 15.
Do.	Lal Khan, Lance-Naik, 141	4 Aug. 15	No. 728, 4 Aug. 15.
Do.	Ualasmir, Naik, 4668	19 Nov. 15	No. 1169, 19 Nov. 15.
Do.	Labh Singh, Naik, 4755	19 Nov. 15	No. 1169, 19 Nov. 15.
Do.	Kharak Singh, Naik, 4510	19 Nov. 15	No. 1169, 19 Nov. 15.
Do.	Khajan Singh, Naik, 474	19 Nov. 15	No. 1169, 19 Nov. 15.
Do.	Gunda Singh, Lance-Naik, 4810	19 Nov. 15	No. 1169, 19 Nov. 15.
Do.	Haidar Khan, Naik, 4650	19 Nov. 15	No. 1169, 19 Nov. 15.
Do.	Sohnu, Sepoy, 709	19 Nov. 15	No. 1169, 19 Nov. 15.
Do.	Lala, Sepy, 30	19 Nov. 15	No. 1169, 19 Nov. 15.
Do.	Ram Singh, Sepoy, 131	19 Nov. 15	No. 1169, 19 Nov. 15.

Corps.	Names and Ranks.	Date.	Authority.
	INDIAN DISTINGUISHED SERVICE MEDALLISTS *contd.*		
	NOTE.—The names in *italics* are those of members on the pension establishment.		
			A. D. Notfn.
24th Punj.	Gindu, Naik, 4487	19 Nov. 15	No. 1169, 19 Nov. 15.
Do.	Siama, Sepoy, 4950	19 Nov. 15	No. 1169, 19 Nov. 15.
Do.	Jiwan Singh, Sepoy, 4956	19 Nov. 15	Ne. 1169, 19 Nov. 15.
Do.	Yar Akhmad, Sepoy, 544	19 Nov. 15	No. 1169, 19 Nov. 15.
Do.	Nawab Ali, Sepoy, 663	19 Nov. 15	No. 1169, 19 Nov. 15.
Do.	Thakur Singh, Sepoy, 640	18 Nov. 16	No. 1386, 18 Nov. 16.
Do.	Bhagwan Singh, Havildar, 4301	17 Aug. 17	No. 1357, 17 Aug. 17.
Do.	Kishen Singh, Sepoy, 441	17 Aug. 17	No. 1357, 17 Aug. 17.
25th Punj.	Sadhu Singh, I., Jemadar	28 June 07	No. 586, 28 June 07.
Do.	Indar Singu, Subadar	4 Apl. 16	No. 638, 3 June 16.
Do.	Mir Afzul, Havildar, 2811	18 Nov. 16	No. 1386, 18 Nov. 16.
Do.	Hakam Khan, Subadar	Do.	No. 1386, 18 Nov. 16
Do.	Kajir Khan, Jemadar, 2361	Do.	No. 1386, 18 Nov. 16.
Do.	Arjan Singh, Lance-Naik, No. 2832	17 Aug. 17	No. 1360, 17 Aug. 17.
Do.	Sawan Sing, Havildar, 2370	1 Jan. 18	No. 9, 1 Jan. 18.
Do.	Sher Bahadur, Jemadar	16 Aug. 18	No. 1885, 16 Aug. 18.
26th Punj.	Sher Singh, Sepoy, 4711	Do.	No. 1388, 17 Nov. 16.
Do.	Devi Dyal, Ward Orderly, 4484	1 Jan. 17	No. 676, 5 May 17.
Do.	Channan Singh, Lance-Naik, 1199	17 Aug. 17	No. 1360, 17 Aug. 17.
Do.	Ghulam Haidar, Lance-Naik No. 1144	17 Aug. 17	No. 1360, 17 Aug. 17.
Do.	Hazara Singh, Sepoy, 1755	17 Aug. 17	No. 1360, 17 Aug. 17.
27th Punj.	Fateh Khan, Subadar, 4277	17 Nov. 16	No. 1386, 17 Nov. 16.
Do.	Natha Singh, Lance-Naik, 4316	Do.	No. 1386, 17 Nov. 16.
Do.	Mir Akbar, Subadar Major	Do.	No. 1388, 18 Nov. 16.
Do.	*Basawa Singh, Jemadar*	*Do.*	*No. 1583, 17 Nov 16.*
Do.	*Hira Singh, Havildar, 2244.*	*Do.*	*No. 1388, 17 Nov 16*
Do.	Makhmad Din, Lance-Naik, 4642	Do.	No. 1388, 17 Nov. 16.
Do.	Santa Singh, Lance-Naik, 4851	Do.	No. 1388, 17 Nov. 16.
Do.	Bela Singh, Colour-Havildar, 4339	2 June 17	No. 868, 2 June 17.
Do.	Zain Din, Havildar, 4211	2 June 17	No. 868, 1 June 17.
Do.	Dhulla Khan, Sepoy, 4880	2 June 17	No. 868, 1 June 17.
Do.	Aman, Gul, Subadar,	17 Aug. 17	No. 1360, 17 Aug. 77.
Do.	Bhagat Singh, Havildar No. 4546	17 Aug. 17	No. 1360, 17 Aug. 17.
Do.	Chandu, Sepoy, 422	1 Jan. 18	No. 9, 1 Jan. 18.
28th Punj.	Umraz Khan, Subadar	26 June 08	No. 527, 26 June 08.
Do.	Wadhawa Singh, Havildar, 3475	12 May 16	No. 525, 12 May 16.
Do.	Isa Khan, Sepoy, 913	18 Nov. 16	No. 1386, 18 Nov. 16.
Do.	Amir Khan, Sepoy, 717	Do.	No. 1386, 18 Nov. 16.
Do.	Mohommed Amin, Sepoy, 929	Do.	No. 1386, 18 Nov. 16.
Do.	Multan, Havildar, 4186	Do.	No. 1388, 18 Nov. 16.
Do.	Gul Baz, Lance-Naik, 810	Do.	No. 1388, 18 Nov. 16.
Do.	Raja Khan, Jemadar	17 Aug. 17	No. 1360, 17 Aug. 17.
Do.	Nikab Gul, Havildar, 120	17 Aug. 17	No. 1360, 18 Nov. 16.
Do.	Ketu Ram, Naik, 4520	1 Jan. 18	No. 9, 1 Jan. 18.
Do.	Karsa Singh, Sepoy, 219	1 Jan. 18	No. 9, 1 Jan. 18.
Do.	Rajwali, Naik, 1057	16 Aug. 18	No. 1885, 16 Aug. 18.
29th Punj.	Bhagwan Singh, Havildar, 3587	3 June 15	No. 530, 3 June 15.
Do.	Kanwar Udham Singh	19 Nov. 17	No. 115, 18 Jan. 18.
Do.	Pala Singh, Havildar, 3591	10 Dec. 15	No. 1249, 10 Dec. 15.
Do.	Albel Singh, Sepoy, 4902	1 Jan. 16	No. 563, 12 June 16.
Do.	Gulaba, Lance-Naik, 4060	21 Nov. 16	No. 676, 5 May 17.
Do.	*Mangtu, Sepoy, 73*	*Do.*	*No. 676, 5 May 17.*
Do.	*Puran Sing, Sepoy, 1675*	*Do.*	*No. 676, 5 May 17.*
Do.	Baz Khan, Havildar, 3123	3 June 18	No. 1180, 3 June 18.
30th Punj.	Sher Khan, Havildar, 4391	28 Jan. 16	No. 103, 28 Jan. 16.
Do.	Gurdit Singh, Havildar, 3870	4 Oct. 18	No. 2342, 4 Oct. 18.
Do.	Hakam Singh, Jemadar	17 Apl. 17	No. 894, 9 June 17.
Do.	Ismail Khan, Lance-Naik, 4260	9 June 17	No. 894, 9 June 17.
Do.	Lall Singh, Havildar, 3911	9 June 17	No. 849, 9 June 17.
Do.	Said Muhammad, Lance Naik, 4479	4 Oct. 18	No. 2342, 4 Oct. 18.
Do.	Thakur Singh, Subadar	9 June 17	No. 849, 9 June 17.
Do.	Ali Ahmed, Havildar, 3933	1 Jan. 18	No. 9, 1 Jan. 18.
Do.	Fateh Ali, Sepoy 4678	3 June 18	No. 1180, 3 June 18.
Do.	Prabh Dial, Jemadar	4 Oct. 18	No. 2342, 4 Oct. 18.
31st Punj.	Pakhar Singh, Lance-Naik, 4489	18 Nov. 16	No. 1386, 18 Nov. 16.
	Gajjan Singh Lance-Naik, 4213		No. 1360, 7 Aug. 17.
32nd Prs.	*Rur Singh, Subadar*	*28 June 07*	*No. 586, 28 June 07.*
Do.	Sant Singh, Subadar	28 June 07	No. 586, 28 June 07.
Do.	*Chattar Singh, Jemadar*	*1 Jan. 08*	*No. 1, 1 Jan. 08.*
Do.	*Sundar Singh, Subadar,* Bahadur	*9 Aug. 12*	*No. 780, 9 Aug. 12.*
Do.	Budha Singh, Subadar, 2969	9 Aug. 12	No. 780, 9 Aug. 12.
Do.	Wasawa Singh, Subadar	23 July 15	No. 815, 27 Aug. 15.
Do.	Hira Singh, Jemadar	26 Apl. 17	Lon. Gaz., 26 Apl. 17.
Do.	Ishar Singh, Havildar, 3813	17 Aug. 17	No. 1360, 17 Aug. 17.
Do.	Mundar Singh, Naik, 4102	1 Jan. 18	No. 9, 1 Jan. 18.
	Sham Singh, Jemadar, No.		No. 388, 17 Nov. 16.
33rd Punj	Muzarab Shah, Subadar	3 June 16	No. 849, 29 July 16.

Corps.	Names and Ranks.	Date.	Authority.
	INDIAN DISTINGUISHED SERVICE MEDALLISTS—*contd.*		
	NOTE.—The names in *italics* are those of members on the pension establishment.		
			A. D. Notfn.
33rd Punj.	Farman Ali, Havildar, 1635	3 June 16	No. 849, 29 July 16.
Do.	Lachman Singh, Sepoy, 2209	3 June 16	No. 849, 29 July 16.
Do.	Mowaz, Sepoy, 2260	3 June 16	No. 849, 29 July 16.
Do.	Mehr Singh, Havildar, 1187	1 Jan. 18	No. 9, 1 Jan. 18.
	Chaudry Khan Naik, 1029	17 Feb. 17	No. 219, 17 Feb. 17.
Do.	Abbas, Naik, 1877	3 June 18	No. 1180, 3 June 18.
Do.	Ali Akbar, Naik, 1861	3 June 18	No. 1180, 3 June 18.
Do.	Lal Khan, Naik, 2329	3 June 18	No. 1180, 3 June 18.
Do.	Mohammad Roshan, Sepoy, 2733	3 June 18	No. 1180, 3 June 18.
34th Prs.	Natha Singh, Subadar	26 June 08	No. 527, 26 June 08.
Do.	Chanda Singh, Naik, 857	26 June 08	No. 527, 26 June 08.
Do.	*Bir Singh, Sepoy, 1414*	*26 June 08*	*No. 527, 26 June 08.*
Do.	Prem Singh, Havildar, 1576	1 Jan. 16	No. 187, 18 Feb. 16.
Do.	Sher Singh, Subadar	. . .	No. 313, 2 Apl. 15.
Do.	Narayan Singh, Havildar, 1452	. . .	No. 313, 2 Apl. 15.
Do.	Ishar Singh, Sepoy, 2578	. . .	No. 313, 2 Apl. 15.
Do.	Tota, Lance-Naick, 2775	. . .	No. 313, 2 Apl. 15.
Do.	Sant Singh, Sepoy, 3063	. . .	No. 313, 2 Apl. 15.
Do.	Pala, Havildar, 1148	. . .	No. 313, 2 Apl. 15.
Do.	Gopal Singh, Sepoy, 3013	. . .	No. 313, 2 Apl. 15.
Do.	Bir, Naick, 2078	. . .	No. 313, 2 Apl. 15.
Do.	Katha Singh, Sepoy, 2398	.	No. 313, 2 Apl, 15.
Do.	Maghar Singh, Subadar	3 June 16	No. 849, 29 July 16.
Do.	Sundar Singh, Havildar, 1862	3 June 16	No. 849, 29 July 16,
Do.	Ishar Singh, Sepoy, 2709	3 June 16	No. 849, 29 July 16.
Do.	Sadhu Singh, Sepoy, 1264	3 June 16	No, 845, 29 July 16.
Do.	Sham Singh, Havildar, 1560	18 Nov. 16	No. 1388, 18 Nov. 16.
Do.	Mangal Singh, Havildar, 1740	Do.	No. 1388, 18 Nov. 16.
Do.	Mahtab Singh, Sepoy, 4307	17 Aug. 17	No. 1360, 17 Aug. 17.
Do.	Mal Singh, Sepoy, 3870	1 Jan. 18	No. 9, 1 Jan. 18.
35th Sikhs	Mangal Singh (I), Subadar-Major	28 June 07	No. 586, 28 June 07.
Do.	Asa Singh, Sepoy, 3201	15 Apl. 15	No. 473, 21 May 15.
Do.	Mula Singh, Havildar, 2882	1 Jan. 16	No. 873, 21 May 15.
Do.	Fauja Singh, Havildar, 2242	3 June 16	No. 849, 29 July 16.
Do.	Budh Singh, Sepoy, 3788	18 Nov. 16	No. 1386, 18 Nov. 16.
36th Sikhs	*Badan Singh, Subadar*, Bahadur	*28 June 07*	*No. 586, 28 June 07.*
Do.	Nand Singh, Havildar, 2187	19 Nov. 15	No. 1160, 19 Nov. 15.
Do.	Sundar Singh, Havildar, 2123	18 Nov. 16	No. 1386, 18 Nov. 16.
Do.	Dall Singh, Havildar, 2240	Do.	No. 1386, 18 Nov. 16.
Do.	Natha Singh, Havildar, 2320	Do.	No. 1386, 18 Nov. 16.
Do.	Indar Singh, Lance-Naik, 2324	Do.	No. 1386, 18 Nov. 16.
Do.	Raulla Singh, Sepoy, 2042	Do.	No. 1386, 18 Nov. 16.
Do.	Bishen Singh, Sepoy, 3672	Do.	No. 1386, 18 Nov. 16.
Do.	Nahar Singh, Sepoy, 3402	Do.	No. 1388, 18 Nov. 16.
Do.	Bakshish Singh, Sepoy, 4157	Do.	No. 1388, 18 Nov. 16.
Do.	Atma Singh, Sepoy, 4439	Do.	No. 1388, 18 Nov. 16.
Do.	Narain Singh, Sepoy, 4391	2 June 17	No. 868, 2 June 17.
Do.	Badhawa Singh, Sepoy, 3006	17 Aug. 17	No. 1360, 17 Aug. 17.
Do.	Bant Singh, Sepoy, 3981	17 Aug. 17	No. 1360, 17 Aug. 17.
Do.	Kata Singh, Naik, 2254	17 Aug. 17	No. 1360, 17 Aug. 17.
Do.	Kishen Singh, Sepoy, 3796	17 Aug. 17	No. 1360, 17 Aug. 17.
Do.	Ram Singh, Sepoy, Naik, 2396	17 Aug. 17	No. 1360, 17 Aug. 17.
Do.	Thakar Sing, Lance-Naik, 1799	17 Aug. 17	No. 1360, 17 Aug. 17.
37th Dogras	Daya Ram, Havildar, 1682	12 May 16	No. 525, 12 May 16.
Do.	Bijho, Naik, 1620	Do.	No. 525, 12 May 16.
Do.	Ram Dayal, Lnce-Naick, 2089	Do.	No. 525, 12 May 16.
Do.	Birbul, Lance-Naik, 2017	Do.	No. 525, 12 May 16.
Do.	Gunga Ram, Sepoy, 2207	Do.	No. 525, 12 May 16.
Do.	Kirpa, Sepoy, 2353	Do.	No. 525, 12 May 16.
Do.	Mahant Naik, 1941	17 Aug. 17	No. 1360, 17 Aug. 17.
Do.	Gujar Singh, Jemadar	1 Jan. 16	No. 187, 18 Feb. 16.
Do.	Thakur Dya, Sepoy, 2061	1 Jan. 16	No. 187, 18 Feb. 16.
Do.	Subhkaran Sepoy, 2409	12 May 16	No. 525, 10 May 16.
Do.	Malari Sepoy, 2267	12 May 16	No. 525, 10 May 16.
38th Dogras	Lachman, Sepoy, 2357	17 Aug. 17	No. 1360, 17 Aug. 17.
Do.	Rup. Sing, Sepoy, 2102	17 Aug. 17	No. 1360, 17 Aug. 17.
39th Garh. Rif.	Kedar Sing Rawat, Subadar	4 May 15	No. 563, 12 June 15.
Do.	Goman Singh Negi, Jemadar	4 May 15	No. 563, 12 June 15.
Do.	Dangwa Ramola, Lance-Naik, 1321	4 May 15	No. 563, 12 June 15.
Do.	Chandar Singh Negi, Rifleman, 1598	4 May 15	No. 563, 12 June 15.
Do.	Gopal Sing Pharswan, Rifleman, 1465	4 May 15	No. 563, 12 June 15.
Do.	Dan Sing Negi, Subadar	. . .	No. 313, 2 Apl. 15.
Do.	Raichand Negi, Rifleman, 1085	. . .	No. 313, 2 Apl. 15.
Do.	Bhola Bisht, Bugler, 2697	. . .	No. 313, 2 Apl. 15.
Do.	Keshi Bisht, Rifleman, 2854	. . .	No. 313, 2 Apl. 15.
Do.	Kutalu Bisht, Rifleman, 1760	. . .	No. 313, 2 Apl. 15.

Corps.	Names and Ranks.	Date.	Authority.
	INDIAN DISTINGUISHED SERVICE MEDALLISTS—*contd.*		
	NOTE.—The names in *italics* are those of members on the pension establishment.		
			A. D. Notfn.
397h Gar. Rif.	Lachman Singh Rawat, Jemadar	. . . .	No. 313, 2 Apl. 15.
Do.	Bir Sing Danu, Havildar, 617	. . . .	No. 313, 2 Apl. 15.
Do.	Kedar Sing Mahar, Naik, 1480	. . . .	No. 313, 2 Apl. 15.
Do.	Keshar Sing Rana, Rifleman, 289	. . . .	No. 313, 9 Apl. 15.
Do.	Nain Sing Rawat, Rifleman, 870	. . . .	No. 313, 2 Apl. 15.
19th Garh. Rif.	Ranjir Sing Pandir, Havildar	. . . .	No. 313, 2 Apl. 15.
Do.	Diwan Singh Padhujar, Havildar	. . . .	No. 313, 2 Apl. 15.
Do.	Mohan Sing Negi, Havildar, 195	3 June 16	No. 849, 29 July 16.
Do.	Deb Sing Aswal, Lance-Naik, 324	3 June 16	No. 849, 29 July 16.
40th Pathans	Sahibullah, Subadar-Major	28 June 07	No. 586, 28 June 07.
Do.	*Ambar, Naik, 660*	*1 Jan. 08*	*No. 1, 1 Jan. 08.*
Do.	Haidar Ali, Sepoy, 3435	3 June 15	Lon. Gaz., 23 June 15.
Do.	Shiraz, Colour-Havildar, 3372	1 Jan. 16	Lon. Gaz., 23 June 15.
Do.	Chamel Singh, Sepoy, 4079	21 Nov. 16	No. 676, 5 May 17.
Do.	Mohamed Jafar Khan, Sepoy, 2718	Do.	No. 676, 5 May 17.
Do.	Mukhtiara, Havildar, 3893	1 Jan. 18	Do.
Do.	Sahib Shah, Naik, 3952	1 Jan. 18	Do.
41st Dogras	Jamit Singh, Lance-Naik, 1090	4 May 15	No. 563, 12 June 15.
Do.	Sidhu, Sepoy, 306	10 Mar. 15	No. 356, 16 Apl. 15.
Do.	Sodagar Singh, Havildar, 409	4 May 15	No. 563, 12 June 15.
Do.	Mehar Singh, Subadar, *Bahadur*	1 Jan. 16	No. 563, 12 June 15.
Do.	Gujar Singh, I.O.M., Jemadar	1 Jan. 16	No. 563, 12 June 15.
Do.	Surjan Singh, Naik, 547	1 Jan. 16	No. 563, 12 Jnne 15.
Do.	Dhiyan Singh, Lce.-Naik, 1389	1 Jan. 16	No. 563, 12 June 15.
Do.	Thakur Dyal, Sepoy, 2061	1 Jan. 16	No. 563, 12 June 15.
Do.	Subha Singh, Subadar	12 May 16	No. 525, 12 May 16.
Do.	Subh Karan, Sepoy, 2409	Do.	No. 525, 12 May 16.
Do.	Ram Singh, Lance.-Naik, 338	Do.	No. 525, 12 May 16.
Do.	Malari, Sepoy, 2267	12 May 16	No. 525, 12 May 16.
Do.	Jodha, Havildar, 253	Do.	No. 525, 12 May 16.
Do.	Thakur Singh, Sepoy, 1501	Do.	No. 525, 12 May 16.
Do.	Bhoda, Lance-Naik, 665	Do.	No. 525, 12 May 16.
43rd Regt.	Mema, Lance-Naik, 3496	2 June 17	No. 868, 2 June 17.
Do.	Raja, Sepoy, 3003	17 Aug. 17	No. 1360, 17 Aug. 17.
45th Sikhs	Fateh Singh, Sepoy, 1036	23 July 15	No. 1360, 17 Aug. 17.
Do.	Ujagar Singh, Sepoy, 1001	23 July 15	No. 1360, 17 Aug. 17.
Do.	Mehar Singh, Jemadar	25 Feb. 16	No. 228, 25 Feb. 16.
Do.	Puran Singh, Sepoy, 1121	25 Feb. 16	No. 228, 25 Feb. 16.
Do.	Sundar Singh, Sepoy, 948	. . . .	No. 736, 1 July 16.
Do.	Bhag Singh, Lance.-Naik, 4538	. . . .	No. 736, 1 July 16.
Do.	Sarup Singh, Havildar, 180	17 Aug. 17	No. 1360, 17 Aug. 17.
Do.	Kaka Singh, Havildar, 325	2 Aug. 18	No. 1767, 2 Aug. 18.
46th Punj.	Abdul Ghafur Khan	28 Jan. 16	No. 103, 28 Jan. 16.
Do.	Badi-ul-Zaman, Subadar	28 Jan. 16	No. 103, 28 Jan. 16.
Do.	Habibullah Khan, Subadar-Major	28 Jan. 16	No. 130, 28 Jan. 16.
Do.	Firoz Khan, Subadar	19 July 18	No. 1629, 19 July 18.
47th Sikhs	Wariam Singh, Jemadar	28 June 07	No. 586, 28 June 07.
Do.	Mit Sing, Naik, 1189	4 May 15	No. 563, 12 June 15.
Do.	Kehr Singh, Sepoy, 1308	4 May 15	No. 563, 12 June 15.
Do.	Mota Singh, Naik, 1336	15 Apl. 15	No. 473, 21 May 15.
Do.	Waryam Singh, Sepoy 1791	15 Apl. 15	No. 473, 21 May 15.
Do.	Kishen Singh, Naik, 2116	1 Jan. 16	No. 187, 18 Feb. 16.
Do.	Bhola Singh, Havildar, 228	. . . .	No. 313, 2 Apl. 15.
Do.	Jagat Singh, Naik, 355	. . . .	No. 313, 2 Apl. 15.
Do.	Kesar Singh, Sepoy, 1410	. . . .	No. 313, 2 Apl. 15.
Do.	Bhuta Singh, Sepoy, 2266	. . . .	No. 313, 2 Apl. 15.
Do.	Bhagat Singh, Havildar, 482	. . . .	No. 313, 2 Apl. 15.
Do.	Mit Singh, Subadar	18 Nov. 16	No. 1386, 18 Nov. 16.
Do.	Mal Singh, Havildar, 790	Do.	No. 1386, 18 Nov. 16.
Do.	Amar Singh, Lnce-Naik, 1835	18 Nov 16	No. 1386, 18 Nov. 16.
Do.	Kehr Singh, Lnce-Naik, 1672	Do.	No. 1386, 18 Nov. 16.
Do.	Nikka Singh, Sepoy, 1742	Do.	No. 1386, 18 Nov. 16.
Do.	Dewa Singh, Havildar, 1090	Do.	No. 1388, 18 Nov. 16.
Do.	Mastan Singh, Naik, 728	Do.	No. 1388, 18 Nov. 16.
Do.	Dial Singh, Sepoy, 2362	Do.	No. 1388, 18 Nov. 16.
Do.	Jowala Singh, Sepoy, 3319	Do.	No. 1388, 18 Nov. 16.
Do.	Bakshi Singh, Havildar, 1716	2 June 17	No 868, 2 June 17.
Do.	Kishan Singh, Naik, 1384	Do.	No. 868, 2 June 17.
Do.	Baghat Singh, Naik, 1429	Do.	No. 868, 2 June 17.
Do.	Sunjar Singh, Lance-Naik, 2370	Do.	No. 868, 2 June 17.
Do.	Arjan Singh, Lance-Naik, 2039	17 Aug. 17	No. 1360, 17 Aug. 17.
Do.	Gajjan Singh, Lance-Naik, 4213	Do.	No. 1360, 17 Aug. 17.
Do.	Dalip Singh, Naik, 1824	16 Aug. 18	No. 1885, 16 Aug. 18.
48th Sikhs	*Ganga Singh, Subadar*	*4 Aug. 15*	*No. 728, 4 Aug. 15.*
Do.	Hukmi, Naik, 365	4 Aug. 15	No. 728, 4 Aug. 15.
Do.	Harchand, Havildar, 345	4 Aug. 15	No. 728, 4 Aug. 15.

INDIAN DISTINGUISHED SERVICE MEDALLISTS—*contd.*

NOTE.—The names in *italics* are those of members on the pension establishment.

Corps.	Names and Ranks.	Date.	Authority. A. D. Notfn.
48th Sikhs	Nand Singh, Naik, 613	19 Nov. 15	No. 1160, 19 Nov. 15.
Do.	Girdhara Singh, Jemadar	19 Nov. 15	No. 1160, 19 Nov. 15.
Do.	Dial Singh, Havildar, 994	19 Nov. 15	No. 1160, 19 Nov. 15.
Do.	Dewa Singh, Havildar, 279	19 Nov. 15	No. 1160, 19 Nov. 15.
Do.	Bhola Singh, Sepoy, 884	12 May 16	No. 525, 12 May 16.
Do.	Katha Singh, Naik, 583	17 Aug. 17	No. 1357, 17 Aug. 16.
Do.	Halphul, Naik, 1367	Do.	No. 1357, 17 Aug. 16.
Do	Dewak Ram, Naik, 1439	Do.	No. 1357, 17 Aug. 16.
Do.	Binda Singh, Sepoy, 713	17 Aug. 17	No. 1357, 17 Aug. 16.
51st Sikhs	Khial Zada, Sepoy, 1940	Do.	No. 1357, 17 Aug. 16.
Do.	Khan Khel, Lnce-Naik, 1189	Do.	No. 1357, 17 Aug. 16.
Do.	Jalal Khan, Naik, 1056	Do.	No. 1357, 17 Aug. 16.
Do.	Amir Khan, Havildar, 1832	Do.	No. 1357, 17 Aug. 16.
Do.	Karram Singh, Lnce-Naik, 436	18 Nov. 16	No. 1386, 18 Nov. 16.
Do.	Bhagwan Singh, Subadar	Do.	No. 1388, 18 Nov. 16.
Do.	Sher Gul, Subadar	Do.	No. 1388, 18 Nov. 16.
Do.	Labh Singh, Subadar	Do.	No. 1388, 18 Nov. 16.
Do.	Akbar Khan, Subadar	Do.	No. 1388, 18 Nov. 16.
Do.	Sundar Singh, Havildar, 225	Do.	No. 1388, 18 Nov. 16.
Do.	Prem Singh, Havildar, 1288	Do.	No. 1388, 18 Nov. 16.
Do.	Barkat Shah, Havildar, 1446	Do.	No. 1388, 18 Nov. 16.
Do.	Habib Shah, Sepoy, 2042	Do.	No. 1388, 18 Nov. 16.
Do.	Sundar Singh, Sepoy, 2843	Do.	No. 1388, 18 Nov. 16.
Do.	*Arjan Sing, Havildar*	*28 June 07*	*No. 563, 12 June 15.*
Do.	Dalip Singh, Sepoy, 1775	17 Aug. 17	No. 1360, 17 Aug. 17.
Do.	Gurdit Singh, Lance-Naik, 928	Do.	No. 1360, 17 Aug. 17.
Do.	Kishan Singh, Sepoy, 2084	Do.	No. 1360, 17 Aug. 17.
Do.	Tail Singh, Naik, 1337	Do.	No. 1360, 17 Aug. 17.
Do.	Hira Singh, Havildar, 1072	1 Jan. 18	No. 9, 1 Jan. 18.
Do.	Gul Badshah, Naik, 950	1 Jan. 18	No. 9, 1 Jan. 18.
Do.	Labh Singh, Lance-Naik, 1673	1 Jan. 18	No. 9, 1 Jan. 18.
Do.	Mangal Singh, Sepoy, 2782	1 Jan. 18	No. 9, 1 Jan. 18.
Do.	Ujagar Singh, Subadar	1 Jan. 18	No. 9, 1 Jan. 18.
Do.	Arjan Singh, Subadar	16 Aug. 18	No. 1885, 16 Aug. 18.
Do.	Akbar Khan, Subadar, I.D.S-M.	26 Oct. 18	No. 2536, 26 Oct. 18.
Do.	Jowala Singh, Lance-Naik, 1960	26 Oct. 18	No. 2536, 26 Oct. 18.
Do.	Isar Singh, Sepoy, 2053	26 Oct. 18	No. 2536, 26 Oct. 18.
Do.	Mangal Singh, Havildar, 1502	26 Oct. 18	No. 2536, 26 Oct. 18.
52nd Sikhs	Ata Khan, Subadar	26 June 08	No. 527, 26 June 08.
Do.	Chur Khan, Lance-Naik, 2520	4 May 15	No. 527, 26 June 08.
Do.	Karam Khan, Subadar	3 June 15	No. 529, 3 June 15.
Do.	Kapur Chand, Col. Havildar, 2055	18 Nov. 16	No. 1386, 18 Nov. 16.
Do.	Sohnu, Havildar, 1955	Do.	No. 1386, 18 Nov. 16.
Do.	Ranja, Lance-Naik, 3026	Do.	No. 1386, 18 Nov. 16.
Do.	Indar Singh, Havildar, 2266	Do.	No. 1388, 18 Nov. 16.
Do.	Nuruddin, Sepoy	28 June 07	No. 586, 28 Jan. 07.
Do.	Sundar Singh, Naik, 2949	2 June 17	No. 868, 2 June 17.
Do.	Harsukh, Naik, 2303	Do.	No. 868, 2 June 17.
Do.	Dhani Ram, Naik, 2696	16 Aug. 18	No. 1885, 16 Aug. 18.
53rd Sikhs	*Sayyid Ali, Subadar, Sardar Bahadur*	*26 June 08*	*No. 527, 26 June 08.*
Do.	Kishn Singh, Havildar, 1956	26 June 08	No. 527, 26 June 08.
Do.	Chanan Singh, Havildar, 2235	26 June 08	No. 527, 26 June 08.
Do.	Kehr Singh, Sepoy, 3573	26 June 08	No. 527, 26 June 08.
Do.	Jamal Din, Havildar, 2584	12 May 16	No. 525, 12 May 16.
Do.	Bhola Singh Lance-Naik, 3625	12 May 16	No. 525, 12 May 16.
Do.	Ram Singh, Lance-Naik, 3360	12 May 16	No. 525, 12 May 16.
Do.	Jawahir Singh, Sepoy, 3586	12 May 16	No. 525, 12 May 16.
Do.	Farman Ali, Sepoy, 103	12 May 16	No. 525, 12 May 16.
Do.	Allah Dad, Sepoy, 3736	12 May 16	No. 525, 12 May 16.
Do.	Mehr Singh, Sepoy, 3910	12 May 16	No. 525, 12 May 16.
Do.	Harmel Singh, Havildar, 2747	12 May 16	No. 525, 12 May 16.
Do.	Painda Khan, Sepoy, 3576	12 May 16	No. 625, 12 May 16.
Do.	Fateh Mohammed, Bugle Major, 2534	18 Nov. 16	No. 1386, 18 Nov. 16.
Do.	Arjan Singh, Havildar, 2710	Do.	No. 1388, 18 Nov. 16.
Do.	Bazm Ali, Sepoy 376	Do.	No. 1388, 18 Nov. 16.
Do.	Jiwan Singh, Sepoy, 549	2 June 17	No. 868, 2 June 17.
Do.	Hari Singh, Havildar, 2723	17 Aug. 17	No 1360, 17 Aug. 17.
Do.	Hazara Singh, Jemadar	Do.	No. 1360, 17 Aug. 17.
Do.	Gurdial Singh, Jemadar	26 Oct. 18	No. 2536, 26 Oct. 18.
Do.	Bahadur Singh, Lance-Naik, 4203	26 Oct. 18	No. 2536, 26 Oct. 18.
Do.	Mangal Singh, Havildar, 3887	26 Oct. 18	No. 2536, 26 Oct. 18.
54th Sikhs	Golodu, Subadar	26 June 08	No. 1360, 17 Aug. 17.
Do.	*Hashmat Ali, Havildar, 1312*	*26 June 08*	*No. 1360, 17 Aug. 17.*
Do.	Mian Khan, Havildar, 2545	26 June 08	No. 1360, 17 Aug. 17.
Do.	Delavara Singh, Sepoy, 3577	29 July 16	No. 849, 29 July 16.
Do.	Udam Singh Kanwar, Subadar	17 Aug. 16	No. 115, 19 Jan. 18.

INDIAN DISTINGUISHED SERVICE MEDALLISTS—*contd.*

NOTE.—The names in *italics* are those of members on the pension establishment.

Corps.	Names and Ranks.	Date.	Authority.
			A. D. Notfn.
54th Sikhs	Jhanda Singh, Sepoy, 28	18 Nov. 16	No. 1388, 18 Nov. 16.
Do.	Ragha, Jemadar	..	No. 772, 19 May 17.
Do.	Sultan Baksh, Havildar, 3320		No. 559, 16 March 18.
Do.	Mir Zamin, Naik, 2463		No. 559, 16 March 18.
Do.	Thenu Ram, Sepoy 2821	17 Aug. 17	No. 1360, 17 Aug. 17.
Do.	Muhammad Sher, Jemadar	Do.	No. 1360, 17 Aug. 17.
Do.	Kesar Singh, Subadar	2 Nov. 17	No. 1887, 2 Nov. 17.
Do.	Bachant Singh, Havildar, 2635	2 Nov. 17	No. 1887, 2 Nov. 17.
Do.	Lakhmir Khan, Sepoy, 559	26 Oct. 18	No. 2536, 26 Oct. 18.
55th Rifl.	Nazar Din, Sepoy, 1713	28 June 07	No. 586, 28 Jan. 07.
Do.	Mansur Khan, Subadar-Maj., Sardar-Bahadur	26 June 08	No. 527, 26 June 08.
Do.	*Daud Shah, Subadar*	*26 June 08*	*No. 527, 26 June 08.*
Do.	Fauja Singh, Jemadar	26 June 08	No. 527, 26 June 08.
Do.	Baghi Shah, Naik, 1414	26 June 08	No. 527, 26 June 08.
Do.	Summar Gul, Sepoy, 2118	26 June 08	No. 527, 26 June 08.
Do.	Mehr Khan, Sepoy, 2625	15 Apl. 15	No. 473, 21 May 15.
Do.	Turai, Naik, 1195	21 Nov. 16	No. 676, 5 May 17.
Do.	Channan Singh, Havildar, 1614	2 Nov. 17	No 1887, 2 Nov. 17.
Do.	Bagga Singh, Sepoy, 1881	2 Nov. 17	No. 1887, 2 Nov. 17.
Do.	Hardit Singh, Subadar	3 June 18	No. 1180, 3 June 18.
Do.	Ibrahim Khan, Subadar	3 June 18	No. 1180, 3 June 18.
Do.	Ghulam Nabi, Lance-Naik, 2406	3 June 18	No. 1180, 3 June 18.
Do.	Hoshnaki, Havildar, 1567	3 June 18	No. 1180, 3 June 18.
Do.	Mowiz Khan, Sepoy, 3403	3 June 18	No. 1180, 3 June 18.
56th Do.	*Sarfraz Khan, Subadar*	*26 June 08*	*No. 527, 26 June 08.*
Do.	Firoz Khan, Subadar-Major, Bahadur	...	No. 630, 3 July 15.
Do.	Bakshish Sing, Havildar, 867	12 May 16	No. 525, 12 May 16.
Do.	Indar Singh, Havildar, 1008	12 May 16	No. 525, 12 May 16.
Do.	Nur Ahmad, Sepoy, 2696	12 May 16	No. 525, 12 May 16.
Do.	Ditta Khan, Sepoy, 2448	12 May 16	No. 525, 12 May 16.
Do.	Budh Singh, Lnce-Naik, 662	12 May 16	No. 525, 12 May 16.
Do.	Attar Sing, Sepoy, 1656	12 May 16	No. 525, 12 May 16.
Do.	Gulab Khan, Sepoy, 2142	12 May 16	No. 525, 12 May 16.
Do.	Baghel Singh, Havildar, 679	12 May 16	No. 525, 12 May 16.
Do.	Madar Khan, LceNaik, 1723	3 June 16	No. 849, 29 July 16.
Do.	Wazir Singh, Col-Havr., 449	28 Nov. 16	No. 1386, 18 Nov. 16.
Do.	Mangal Singh, Havildar 1303	Do.	No. 1386, 18 Nov. 16.
Do.	Sujawal Khan, Naik, 1794	Do.	No. 1386, 18 Nov. 16.
Do.	Bishan Singh, Sepoy, 2414	Do.	No. 1386, 18 Nov. 16.
Do.	Suba Singh, Sepoy, 2498	Do.	No. 1388, 18 Nov. 16.
Do.	Dheru Khan, Sepoy, 2694	Do.	No. 1388, 18 Nov. 16.
Do.	Bakhshi Chand, Subadar-Major	31 Mar. 17	Force D. Order No. 66, 31 Mar. 17.
Do.	Bagha Sigh, Sepoy, 2353	2 June 17	No. 868, 2 June 17.
Do.	Dhuman, Sepoy, 2640	17 Aug. 17	No. 1360, 17 Aug. 17.
Do.	Samundar Khan, Naik, 2255	Do.	No. 1360, 17 Aug. 17.
Do.	Hakim Khan, Naik, 1946	1 Jan. 18	No. 9, 1 Jan. 18.
57th Do.	Bishn Singh, Subadar-Maj., *Bahadur*	1 Jan. 08	No. 1, 1 Jan. 08.
Do.	Bhagat Singh, Subadar	26 June 08	No. 527, 26 June 08.
Do.	*Munir, Jemadar*	*26 June 08*	
Do.	Jagat Singh, Subadar	26 June 08	No. 527, 26 June 08.
Do.	*Bahadur, Sepoy, 1506*	*26 June 08*	*No. 527, 26 June 08.*
Do.	*Basid Khan, Sepoy, 2534*	*26 June 08*	*No. 527, 26 June 08.*
Do.	Lal Mir, Sepoy, 2558	26 June 08	No. 527, 26 June 08.
Do.	Gyan Singh, Sepoy, 2587	26 June 08	No. 527, 26 June 08.
Do.	*Nihal Singh, Sepoy, 1898*	*1 Jan. 09*	*No. 5, 1 Jan. 09.*
Do.	*Gulsada, Havildar, 1302*	*1 Jan. 10*	*No. 4, 1 Jan. 10.*
Do.	*Hira, Lance-Naik, 1809*	*1 Jan. 10*	*No. 4, 1 Jan. 10.*
Do.	Bur Singh, Jemadar	15 Apl. 15	No. 473, 21 May 15.
Do.	Karim Khan, Havildar, 2595	15 Apl. 15	No. 473, 21 May 15.
Do.	Narayan, Naik, 2583	15 Apl. 15	No. 473, 21 May 15.
Do.	Sahib Sher, Naik, 1760	15 Apl. 15	No. 473, 21 May 15.
Do.	Ram Saran, Sepoy, 2702	15 Apl. 15	No. 473, 21 May 15.
Do.	Sar Mast, Havildar, 2540	3 June 15	Lon. Gaz., 23 June 15.
Do.	Alvas Khan, Sepoy, 3484		No. 313, 2 Apl. 15.
Do.	Mir Badshah, Sepoy, 2108		No. 313, 2 Apl. 15.
Do.	Mir Badshah, Sepoy, 3119		No. 313, 2 Apl. 15.
Do.	Mir Baz, Sepoy, 3223		No. 313, 2 Apl. 15.
Do.	Umbras Khan, Naik, 1597	21 Nov. 16	No. 676, 5 May 17.
Do.	Mangtu, Havildar, 2624	1 Jan. 18	No. 9, 1 Jan. 18.
Do.	Rupa Singh, Subadar	1 Jan. 18	No. 9, 1 Jan. 18.
Do.	Hira Singh, Jemadar	3 June 18	No. 1180, 3 June 18.
Do.	Kala Khan, Naik, 2594	3 June 18	No. 1180, 3 June 18.
Do.	Kapur Singh, Havildar, 2614	3 June 18	No. 1180, 3 June 18.
Do.	Sher Muhammad, Sepoy, 3591	3 June 18	No. 1180, 3 June 18.
58th Do.	Fazl Dad, Havildar, 2198	4 May 15	No. 563, 12 June 15.

Corps.	Name and Ranks.	Date.	Authority.
	INDIAN DISTINGUISHED SERVICE MEDALLISTS—*contd.*		
	NOTE.—The names in *italics* are those of members on the pension establishment.		A. D. Notfn.
58th Rif.	Raj Talab, Subadar	10 Mar. 15	No. 356, 16 Apl. 15.
Do.	Sardar, Naik, 3066	4 May 15	No. 563, 12 June 15.
Do.	Ajun, Havildar, 2763	10 Mar. 15	No. 356, 16 Apl. 15.
Do.	Sarfaraz, Havildar, 3136	10 Mar. 15	No. 356, 16 Apl. 15.
Do.	Dewa Singh, Naick, 2758	10 Mar. 15	No. 356, 16 Apl. 15.
Do.	Sergun Shah Naick, 2634	10 Mar. 15	No. 356, 16 Apl. 15.
Do.	Hawinda, Jemadar	1 Apl. 15	No. 425, 8 May 15.
Do.	Nar Naikaz, 3080	1 Apl. 15	No. 328, 8 May 15.
Do.	Phuman Singh, Subadar	1 Apl. 15	No. 426, 7 May 15.
Do.	Zar Baz, Naick, 3080	1 Apl. 15	No. 426, 7 May 15.
Do.	Said Azghar, Lce-Naik, 3567	1 Apl. 15	No. 426, 7 May 15.
Do.	Dewa Singh, Sepoy, 3374	1 Apl. 15*	No. 426, 7 May 15.
Do.	Maluk Singh, Sepoy, 3133	1 Apl. 15	No. 426, 7 May 15.
Do.	Hamid Khan, Jemadar		No. 313, 2 Apl. 15.
Do.	Indar Singh, Jemadar		No. 313, 2 Apl. 15.
Do.	Baidullah, Naik, 3404		No. 313, 2 Apl. 15.
Do.	Sundar Singh, Havildar, 2164		No. 313, 2 Apl. 15
Do.	Lashkar, Havildar, 3212		No. 313, 2 Apl. 15.
Do.	Muhammad Amin, Lce-Naik, 2934	3 June 16	No. 849, 29 July 16.
59thDo.	Mukhmad Jan, Subadar	26 June 08	No. 527, 26 June 08.
Do.	*Jahan Dad Khan, Jemadar*	*26 June 08*	*No. 527, 26 June 08.*
58th Rif.	*Sansar Singh, Jemadar*	*26 June 08*	*No. 527, 26 June 08.*
Do.	*Mir Nabbi Hussain, Havildar, 2623*	*26 June 08*	*No. 527, 26 June 08.*
Do.	*Ralla Singh, Havildar, 2894*	*26 June 08*	*No. 527, 26 June 08.*
Do.	Mobin Khan, Subadar, 2917	26 June 08	*No. 527, 26 June 08.*
Do.	*Attar Singh, Sepoy, 3556*	*26 June 08*	*No. 527, 26 June 08.*
Do.	*Kaka Singh, Sepoy, 4442*	*26 June 08*	*No. 527, 26 June 08.*
Do.	Amir Ali, Naik, 3529	4 May 15	No. 563, 12 June 15.
Do.	Lal Khan, Sepoy, 4731	10 Mar. 15	No. 356, 16 Apl. 15.
Do.	Chur Khan, Lce.-Naik, 2520		No. 563, 12 June 15.
Do.	Akbar Khan, Sepoy, 4845	4 May 15	No. 563, 12 June 15.
Do.	Fazal Dad, Havildar, 3233	3 June 18	No. 1178, 3 June 18.
Do.	Rahim Ali, Lance-Naik, 3885	3 June 18	No. 1178, 3 June 18.
Do.	Bagh Singh, Lance-Naik, 2927	3 June 18	No. 1178, 3 June 18.
Do.	Yar Khan, Havildar-Major, 2446	3 June 18	No. 1180, 3 June 18.
59th Rif.	Bishn Singh, Subadar	1 Jan. 16	No. 187, 18 Feb. 16.
Do.	Niaz Gul, Havildar, 4264		No. 313, 2 Apl. 15.
Do.	Nur Ali, Havildar, 4016	3 June 16	No. 849, 29 July 16.
Do.	Nathu, Havildar, 4111	8 Nov. 16	No. 1386, 18 Nov. 16.
Do.	Phumman Singh, Lance-Naik 4770	18 Nov. 16	No. 1386, 18 Nov. 16.
Do.	Saida Khan, Lance-Naik, 3601	18 Nov. 16	No. 1388, 18 Nov. 16.
Do.	Abuzar Khan, Havildar, 4116	17 Aug. 17	No. 1360, 17 Aug. 17.
Do.	Hazrat Khan, Havildar, 4183	17 Aug. 17	No. 1360, 17 Aug. 17.
Do.	Devi Dyal, Havildar, 3452	16 Aug. 18	No. 1885, 16 Aug. 18.
Do.	Niaz Gul, Havildar, 4342	16 Aug. 18	No. 1885, 16 Aug. 18.
61st Prs.	Muhammad Abdul Hafiz, Subadar	1 Feb. 17	Lon. Gaz. 1 Feb. 17.
Do.	Ayyacanoo, Colour Havildar, 2696	1 Feb. 17	Lon. Gaz. 1 Feb. 17.
Do.	Abdur Rahim, Colour Havildar, 1641	1 Feb. 178	Lon. Gaz. 1 Feb. 17.
Do.	Tambusami, Jemadar	1 Jan. 18	No. 9, 1 Jan. 18.
Do.	Kannayaram, Havildar, 2956	3 June 18	No. 1180, 3 June 18.
Do.	Laul Khan, Sabadar	3 June 15	No. 1180, 3 June 18.
62nd Punj.	Aman Shah, Naik, 1209	4 May 15	No. 187, 18 Feb. 16.
Do.	Bari Sher, Naik, 13	10 Mar. 1	No. 356, 16 Apl. 15.
Do.	*Bishan Singh, Havildar, 437*	*3 Dec. 15*	*No. 1214, 3 Dec. 15.*
Do.	Mahl Sigh, Havildar, 558	18 Nov. 16	No. 1388, 18 Nov. 16.
Do.	Channan Singh, Sepoy, 1451	17 Aug. 17	No. 1360, 17 Aug. 17.
Do.	Sher Zaman Khan, Subadar	17 Aug. 17	No. 1360, 17 Aug. 17.
Do.	Sunder Singh, Havildar 55	17 Aug. 17	No. 1360, 17 Aug. 17.
63rd Inf.	Rajagopal, Havildar, 2644	9 June 17	No. 894, 9 June 17.
64th Prs.	Rahman Sharif, Havildar-Major, 3661	15 Oct. 15	No. 1025, 15 Oct. 15.
Do.	Qadir Beg, Havildar, 4030	15 Oct. 15	No. 1025, 15 Oct. 15.
Do.	Barmayya, Naik, 4977	17 Aug. 17	No. 1360, 17 Aug. 17.
Do.	P. Narayanasami, Jemadar	17 Aug. 17	No. 1360, 17 Aug. 17.
Do.	Gopalsami, Subadar	3 June 18	No. 1180, 3 June 18.
Do.	Madurai, Private, 1029	1 Jan. 18	No. 9, 1 Jan. 18.
66th Punj.	Dhir Singh, Sepoy, 1931	10 Dec. 15	No. 1248, 10 Dec. 15.
Do.	Santa Singh, Sepoy, 1903	18 Nov. 16	No. 1388, 18 Nov. 16.
Do.	Dharm Singh, Jemadar	17 Aug. 17	No. 1360, 17 Aug. 17.
Do.	Muhammad Khan, Subadar	17 Aug. 17	No. 1357, 17 Aug. 17.
Do.	Sunder Singh, Jemadar	17 Aug. 17	No. 1357, 17 Aug. 17.
Do.	Muzffar Khan, Havildar, 600	17 Aug. 17	No. 1357, 17 Aug. 17.
Do.	Mit Singh, Naik, 318	17 Aug. 17	No. 1357, 17 Aug. 17.
Do.	Channan Singh, Naik, 1181	17 Aug. 17	No. 1357, 17 Aug. 17.
67thDo.	Gauhar Ali, Naik, 1164	19 Nov. 15	No. 1160, 19 Nov. 15.
Do.	Sharif Khan, Havildar, 163	19 Nov. 15	No. 1160, 19 Nov. 15.
Do.	Indar Singh, Jemadar	19 Nov. 15	No. 1160, 19 Nov. 15.
Do.	Lachman Singh, Subadar	19 Nov. 15	No. 1160, 19 Nov. 15.
Do.	*Khem Singh, Jemadar*	*19 Nov. 15*	*No. 1160, 19 Nov. 15.*

* Bar to Indian Distinguished Service Medal.

Corps.	Names and Ranks.	Date.	Authority.
	INDIAN DISTINGUISHED SERVICE MEDALLISTS—*contd.*		
	NOTE.—The names in *italics* are those of members on the pension establishment.		
			A. D. Notfn.
67th Punj.	Khansi Ram, Havildar, 580	19 Nov. 15	No. 1166, 19 Nov. 15.
Do.	Mahan Singh, Havildar, 132	19 Nov. 15	No. 1160, 19 Nov. 15.
Do.	Chartu, Lance-Naik, 898	19 Nov. 15	No. 1160, 19 Nov. 15.
Do.	Allah Ditta, Sepoy, 90	19 Nov. 15	No. 1160, 19 Nov. 15.
Do.	Khoja Khan, Lance-Naik, 1397	17 Aug. 17	No. 1360, 17 Aug. 17.
Do.	Mansifdar Khan, Naik, 1005	17 Aug. 17	No. 1360, 17 Aug. 17.
Do.	Tiloka Singh, Havildar, 99	17 Aug. 17	No 1360, 17 Aug. 17.
Do.	Sera, Havildar, 4142	17 Aug. 17	No. 1357, 17 Aug. 17.
Do.	Nur Khan, Sepoy, 2332	17 Aug. 17	No. 1359, 17 Aug. 17.
69th Do.	Labh Singh, Naik, 465	26 Feb. 15	No. 208, 26 Feb. 15.
Do.	Gheba Khan, Jemadar	3 June 16	No. 849, 29 July 16.
Do.	Ghulam Ali, Naik, 862	3 June 16	No. 849, 29 July 16.
Do	Gurdit Singh, Naik, 466	3 June 16	No. 849, 29 July 16.
Do.	Punna Singh, Naik, 433	3 June 16	No. 849, 29 July 16.
Do.	Bhulla Singh, Sepoy, 1734	3 June 16	No. 849, 29 July 16.
Do.	Mahommed Fazal, Sepoy 1209	18 Nov. 16	No. 1388, 18 Nov. 16.
Do.	Jalal Din, Lance-Naik, 2450	9 Nov. 17	No. 1938, 9 Nov. 17.
Do.	Parbhu Singh, Lance-Naik, 1444	9 Nov. 17	No 1938, 9 Nov. 17.
Do.	Magar Singh, Lance-Naik, 1302	9 Nov. 17	No. 1938, 9 Nov. 17.
Do.	Muhammad Din, Lance-Naik, 1877	9 Nov. 17	No. 1938, 9 Nov. 17.
Do.	Zaman Ali, Lance-Naik, 1669	16 Aug. 18	No. 1885, 16 Aug. 18.
72nd Punj.	Bishan Singh, Naick, 2064	22 Aug. 13	No. 710, 22 Aug. 13.
Do.	Bazkhan, Sepoy, 3146	18 Nov. 16	No. 1386, 18 Nov. 16.
Do.	Feroze Khan, Naik, 2438	17 Aug. 17	No 1360, 17 Aug. 17.
74th Punj.	Wariam Singh, Subadar	17 Aug. 17	No. 1360, 17 Aug. 17.
75th Inf.	Shaikh Farid, Lance-Naik, 3254	....	No. 1248, 10 Dec. 15.
76th Punj.	Shan Khan, Naik, 793	....	No. 630, 3 July 15.
Do.	Ghajja Singh, Lance-Naik, 709	3 Sep. 15	No. 850, 3 Sep. 15.
Do.	Karim Khan, Sepoy, 1815	3 Sep. 15	No. 850, 3 Sep. 15.
Do.	Gul Sher, Bugler, 198	3 Sep. 15	No. 850, 3 Sep. 15.
Do.	Lal Khan, Lance-Naik, 172	3 Sep. 15	No. 850, 3 Sep. 15.
Do.	Bari Sher, Lance-Naik, 991	3 Sep. 15	No. 850, 3 Sep. 15.
Do.	Badhan, Sepoy, 1588	3 Sep 15	No. 850, 3 Sep. 15.
Do.	Sujawal Khan, Sepoy, 1761	3 Sep. 15	No. 850, 3 Sep. 15.
Do.	Sher Dil, Sepoy, 1495	3 Sep. 15	No. 850, 3 Sep. 15.
Do.	Firoz Khan, Sepoy, 688	3 Sep. 15	No. 850, 3 Sep. 15.
Do.	Madar Ali, Sepoy, 1866	3 Sep. 15	No. 850, 3 Sep. 15.
Do.	Bahadur Shah, Subadar	19 Nov. 15	No. 1060, 19 Nov. 15.
Do.	Ganga Singh, Lance-Naik, 915	19 Nov. 15	No. 1060, 19 Nov. 15.
Do.	Dharam Singh, Sepoy, 1546	19 Nov. 15	No. 1060, 19 Nov. 15.
Do.	Thakur Singh, Sepoy, 1365	19 Nov. 15	No. 1060, 19 Nov. 15.
Do.	Badlu, Lance-Naik, 473	19 Nov. 15	No. 1060, 19 Nov. 15.
Do.	Mathra, Lance-Naik, 280	19 Nov. 15	No. 1060, 19 Nov. 15.
Do.	Isar Singh, Sepoy, 529	19 Nov. 15	No. 1060, 19 Nov. 15.
Do.	Hussain Shah, Sub-Maj.	19 Nov. 15	No. 1060, 19 Nov. 15.
Do.	Gulzara Singh, Jemadar	19 Nov. 15	No. 1060, 19 Nov. 15.
Do.	Hukam Dad, Havildar, 762	19 Nov. 15	No. 1260, 19 Nov. 15.
Do.	Zaman Khan, Lance-Naik, 883	19 Nov. 15	No. 1060, 19 Nov. 15.
Do.	Firoz Khan, Naik, 1218	19 Nov. 15	No. 1060, 19 Nov. 15.
Do.	Sattar Muhammad, Sepoy, 1053	19 Nov. 15	No. 1060, 19 Nov. 15.
Do.	Said Zaman, Havildar, 259	10 Dec. 15	No. 1250, 10 Dec. 15.
Do.	Hamidullah, Naik, 1302	10 Dec. 15	No. 1250, 10 Dec. 15.
Do.	Afsar Ali, Sepoy, 999	10 Dec. 15	No. 1250, 10 Dec. 15.
Do.	Bostan Khan, Sepoy, 996	10 Dec. 15	No. 1250, 10 Dec. 15.
Do.	Gobardhan, Sepoy, 1638	10 Dec. 15	No. 1250, 10 Dec. 15.
Do.	Harphul, Sepoy, 1797	10 Dec. 15	No. 1250, 10 Dec. 15.
Do.	Bela Singh, Sepoy, 1610	10 Dec. 15	No. 1250, 10 Dec. 15.
Do.	Ghulam Muhammad Khan, Subadar	17 Aug. 17	No. 1357, 17 Aug. 17.
Do.	Narani Singh, Havildar, 414	17 Aug. 1	No. 1357, 17 Aug. 17.
Do.	Gul Sher Khan, Havildar, 73	17 Aug. 17	No. 1357, 17 Aug. 17.
Do.	Ram Nath, Sepoy, 1941	17 Aug. 17	No. 1357, 17 Aug. 17.
79th Inf.	Syed Madar, Havildar, 1551	12 Dec. 11	No. 1016, 12 Dec. 11.
Do.	Angamuttu, Lance-Naik, 1759	19 Nov. 15	No. 1160, 19 Nov. 15.
81st Prs.	Chinnatambi, Havildar, 3149	17 Aug. 17	No. 1360, 17 Aug. 17.
Do.	Anthony David, Jemadar	19 July 18	No. 1629, 19 July 18.
82nd Punj.	Hira Sigh, Subadar	18 Nov. 16	No. 1388, 18 Nov. 16.
Do.	Sadda Ram, Sepoy, 2000	18 Nov. 16	No. 1388, 18 Nov. 16.
Do.	Abdul Khan, Havildar No. 1395	17 Aug. 17	No. 1360, 17 Aug. 17.
Do.	Dalip Singh, Subadar	17 Aug. 17	No. 1360, 17 Aug. 17.
Do.	Ghulam Muhiyuddfn, Subadar	17 Aug. 17	No. 1360, 17 Aug. 17.
Do.	Mam Chand, Havildar No. 773	17 Aug. 17	No. 1360, 17 Aug. 17.
84th Punj.	Mansha Ram, Lance-Naik, 1340	21 Nov. 16	No. 676, 5 May 17.
Do.	Surat Singh, Lance-Naik, 1303	21 Nov. 16	No. 676, 5 May 17.
87th Punj.	Sarban Singh, Subadar	26 June 08	No. 1016, 12 Dec. 11.
Do.	*Sewa Singh, Jemadar*	*28 June 07*	*No. 586, 28 June 07.*
Do.	Sohan Singh, Sepoy, 3212	2 June 15	No. 868, 2 June 17.
Do.	Nadir Ali, Sepoy, 2589	2 June 15	No. 868, 2 June 17.

INDIAN DISTINGUISHED SERVICE MEDALLISTS—*contd.*

NOTE.—The names in *italics* are those of members on the pension establishment.

Corps.	Names and Ranks.	Date.	Authority. A. D. Notfn.
88th Inf.	Mohamed Hussain, Havildar, 2529	21 Nov. 16	No. 676, 5 May 17.
89th Inf.	Bur Singh, Sepoy, 290	26 Feb. 15	No. 208, 26 Feb. 15.
Do.	Buta Singh, Sepoy, 2028	26 Feb. 15	No. 208, 26 Feb. 15.
Do.	Dasundar Singh, Sepoy, 2131	26 Feb. 15	No. 208, 26 Feb. 15.
Do.	Harnam Singh, Havildar, 1528	29 Oct. 15	Lon. Gaz., 29 Oct. 15.
Do.	Muhammad Sadik, Naik, 2352	29 Oct. 15	Lon. Gaz., 29 Oct. 15.
Do.	Indar Singh, Naik, 1793	29 Oct. 15	No. 1081, 29 Oct. 15.
Do.	Ramji Misr, Naik 2029	1 Jan. 16	No. 187, 18 Feb. 16.
Do.	Karam Dad, Naik 1485	1 Jan 16	No. 187, 18 Feb. 16.
Do.	Muhammad Khan, Sepoy, 2102	1 Jan 16	No. 187, 18 Feb. 16.
Do.	Bagga Singh, Jemadar	3 June 16	No. 849, 29 July 16.
Do.	Baldeo Singh, Jemadar	3 June 16	No. 849, 29 July 16.
Do.	Narayan Singh, Naik, 1998	3 June 16	No. 849, 29 July 16.
Do.	Bur Singh, Havildar, 1476	18 Nov. 16	No. 1386, 18 Nov. 16.
Do.	Santa Singh, Lance-Naik, 2278	18 Nov. 16	No. 1386, 18 Nov. 16.
Do.	Amir Dad, Lance-Naik 2370	18 Nov. 16	No. 1386, 18 Nov. 16.
Do.	Ram Diwan Tewari Sepoy, 1647	18 Nov. 16	No. 1386, 18 Nov. 16.
Do.	Mit Singh, Sepoy, 1773	18 Nov. 16	No. 1386, 18 Nov. 16.
Do.	Hazura Singh, Sepoy, 2588	18 Nov. 16	No. 1386, 18 Nov. 16.
Do.	Kishen Singh, Sepoy, 1991	18 Nov. 16	No. 1386, 18 Nov. 16.
Do.	Ghulam Mahammad, (Now Jemadar) Col. Havr., 1334	18 Nov. 16	No. 1388, 18 Nov. 16.
Do.	Gurdit Singh, Havildar, 1653	18 Nov. 16	No. 1388, 18 Nov. 16.
89th Inf.	Jageo Dube, Havildar, 1776	18 Nov. 16	No. 1388, 18 Nov. 16.
Do.	Harnam Singh, Sepoy, 2495	18 Nov. 16	No. 1388, 18 Nov. 16.
Do.	Sher Muhammad, Sepoy, 2578	18 Nov. 16	No. 1388, 18 Nov. 16.
Do.	Mehr Singh, Sepoy, 2123	1 Jan. 18	No. 9, 1 Jan. 18.
90th Punj.	Bishan Singh, Havildar, 1155	19 Nov. 15	No. 1160, 19 Nov. 15.
Do.	Sobha Singh, Naik, 1282	19 Nov. 15	No. 1160, 19 Nov. 15.
Do.	Mangal Singh, Sepoy, 2215	19 Nov. 15	No. 1160, 19 Nov. 15.
Do.	Gurdit Singh, Sepoy, 2211	19 Nov. 15	No. 1160, 19 Nov. 15.
Do.	Bhagat Singh, Sepoy, 1776	1 Jan. 16	No. 1160, 19 Nov. 15.
Do.	Man Singh, Sepoy, 2936	18 Nov. 16	No. 1392, 18 Nov. 16.
Do.	Harnam Singh, Subadar	17 Aug. 17	No. 1360, 17 Aug. 17.
Do.	Narayan Singh, Subadar	17 Aug. 17	No. 1360, 17 Aug. 17.
Do.	Cakhar Khan, Lance-Naik, 2754	1 Jan. 18	No. 9, 1 Jan. 18.
91st Punj.	Narinjan Singh, Sepoy, 3059	26 Dec. 16	No. 676, 5 May 17.
Do.	Kishen Singh, Sepoy, 3143	26 Dec. 16	No. 676, 5 May 17.
Do.	Dunnia, Sepoy, 4160	1 Jan. 18	No. 9, 1 Jan. 18.
92nd Punj.	Farman Ali, Subadar-Major	3 June 15	No. 528, 3 June 15.
Do.	Sucha Singh, Sepoy, 2636	1 Jan 16	No. 528, 3 June 15.
Do.	Hari Singh, Sepoy, 2119	12 May 16	No. 525, 12 May 16.
Do.	Fateh Mahammud, Sepoy, 1461	12 May 16	No. 525, 12 May 16.
Do.	Painda, Khan, Sepoy, 2704	12 May 16	No. 525, 12 May 16.
Do.	Basant Singh, Sepoy, 2005	12 May 16	No. 525, 12 May 16.
Do.	Rur Singh, Naik, 2117	12 May 16	No. 525, 12 May 16.
Do.	Dalip Singh, Lance-Naik, 2638	18 Nov. 16	No. 1386, 18 Nov. 16.
Do.	Tek Singh, Naik, 2052	18 Nov. 16	No. 1388, 18 Nov. 16.
Do.	Harnam Singh, Lance-Naik, 2353	18 Nov. 16	No. 1388, 18 Nov. 16.
Do.	Manga Khan, Lance-Naik, 1585	18 Nov. 16	No. 1388, 18 Nov. 16.
Do.	Channan Singh, Lance-Naik, 2907	18 Nov. 16	No. 1388, 18 Nov. 16.
Do.	Mast Khan, Sepoy, 2470	18 Nov. 16	No. 1388, 18 Nov. 16.
Do.	Sampooran Singh, Sepoy, 2484	18 Nov. 16	No. 1388, 18 Nov. 16.
Do.	Sohan Singh, Sepoy, 2022	18 Nov. 16	No. 1388, 18 Nov. 16.
Do.	Khan Zaman, Subadar	17 Aug. 17	No. 1360, 17 Aug. 17.
Do.	Harnam Singh, Havildar, 2488	17 Aug. 17	No. 1360, 17 Aug. 17.
Do	Kishan Singh, Naik, 2478	17 Aug. 17	No. 1360, 17 Aug. 17.
Do.	Sapooran Singh, Havildar, 1993	17 Aug 17	No. 1360, 17 Aug. 17.
Do.	Wasakha Singh, Havildar, 2361	17 Aug. 17	No. 1360, 17 Aug. 17.
Do.	Feroze Khan, Lance-Naik, 2613	1 Jan. 18	No. 9, 1 Jan. 18.
Do.	Khushi Muhammad, Sepoy, 3847	1 Jan. 18	No. 9, 1 Jan. 18.
93rd Infy.	Mihan Singh, Sepoy, 2616	4 May 15	No. 563, 12 June 15.
Do.	Bhan Singh, Sepoy, 2828	12 May 16	No. 525, 12 May 16.
Do.	Alam Khan, Sepoy, 3060	18 Nov. 16	No. 1386, 18 Nov. 16.
Do.	Fateh Khan, Sepoy, 3133	18 Nov. 16	No. 1386, 18 Nov. 16.
Do.	Bhan Singh, Havildar, 1994	17 Aug. 17	No. 1360, 17 Aug. 17.
Do.	Sahib Singh, Havildar, 2019	17 Aug. 17	No. 1360, 17 Aug. 17.
Do.	Shaikh Abdulla, Jemadar	1 Jan. 18	No. 9, 1 Jan. 18.
95th Infy.	Masnad Ali, Havildar, 4195	19 Mar. 15	No. 285, 19 Mar. 15.
Do.	Rampershad Singh, Jemdr.	4 Aug. 15	No. 728, 4 Aug. 15.
Do.	Babu Singh, Lance-Naik, 4324	4 Aug. 15	No. 728, 4 Aug. 15.
Do.	Bhairom Singh, Sepoy, 4584	4 Aug. 15	No. 728, 4 Aug. 15.
Do.	Bhagwant Singh, Sepoy, 4836	4 Aug. 15	No. 728, 4 Aug. 15.
Do.	Sewa Singh, Sepoy, 4823	4 Aug. 15	No. 728, 4 Aug. 15.
96th Infy.	Jadunandan Singh, Subadar	19 July 18	No. 1629, 19 July 18.
97th Infy.	Kanhaiya, Subadar-Major, 3733	12 May 16	No. 525, 12 May 16.
Do.	Mamchand, Sepoy, 3733	12 May 16	No. 525, 12 May 16.

INDIAN DISTINGUISHED SERVICE MEDALLISTS—*contd.*

NOTE.—The names in *italics* are those of members on the pension establishment.

Corps.	Names and Ranks.	Date.	Authority. A. D. Notfn.
97th Infy.	Shiblal, Havildar, 3261	12 May 16	No 525, 12 May 16.
Do.	Shaikh Faqir Muhammad, Havildar, 2188	12 May 16	No. 525, 12 May 16.
Do.	Nathu Ram, Havildar, 2553	12 May 16	No. 525, 12 May 16.
Do.	Ussanah, Bugler, 2858	12 May 16	No. 525, 12 May 16.
Do.	Jagat Singh, Sepoy, 2680	12 May 16	No. 525, 12 May 16.
Do.	Parbhu, Sepoy, 2646	12 May 16	No. 525, 12 May 16.
Do.	Raghunath Singh, Sepoy, 3259	12 May 16	No. 525, 12 May 16.
Do.	Hazari Singh, Havildar, 2648	12 May 16	No. 525, 12 May 16.
Do.	Shibaram Singh, Sepoy, 3548	12 May 16	No. 525, 12 May 16.
Do.	Ganeth Singh, Lance-Naik, 3416	12 May 16	No. 525, 12 May 16.
Do.	Ram Lal, Sepoy, 3736	12 May 16	No. 525, 12 May 16.
Do.	Shiu Sahai, Sepoy, 3788	12 May 16	No. 525, 12 May 16.
99th Infy.	Ami Chand, Lance-Naik, 4391	21 Nov. 16	No. 676, 5 May 17.
101st Grs.	Sabdal Khan, Sepoy, 905	3 June 15	No. 530, 3 June 15.
Do.	Umar Khan, Sepoy, 1226	3 June 18	No. 1178, 3 June 18.
Do.	Dilbar Shah, Sepoy, 1968	3 June 18	No. 1178, 3 June 18.
Do.	Abdul Satar Khan, Lance-Naik, 2819	26 Oct. 18	No. 2536, 26 Oct. 18.
102nd Do.	Ganga Ram, Havildar, 2289	19 Mar. 15	No. 530, 3 June 15.
Do.	Sultan Ahmad, Sepoy, 2936	19 Mar. 15	No. 530, 3 June 15.
Do.	Tula Ram, Havildar, 3072	12 May 16	No. 525, 12 May 16.
Do.	Mansar Ali, Havildar, 3139	12 May 16	No. 525, 12 May 16.
Do.	Sheoram, Private, 2924	12 May 16	No. 525, 12 May 16.
Do.	Sanwal Ram, Havildar, 2461	12 May 16	No. 525, 12 May 16.
Do.	Shedu Ram, Private, 2890	12 May 16	No. 525, 12 May 16.
102nd Grs.	Sirdara Ram, Private, 2294	12 May 16	No. 525, 12 May 16.
Do.	Surja Ram, Private, 3245	12 May 16	No. 525, 12 May 16.
Do.	Khan Muhammad, Subadar	18 Nov. 16	No. 1388, 18 Nov. 16.
Do.	Ram Dayal Singh, Havildar, 1967	2 June 17	No. 868, 2 June 17.
Do.	Feroz Khan, Sepoy, 3124	2 June 17	No. 868, 2 June 17.
Do.	Ahmed Khan, Sepoy No. 3206	17 Aug. 17	No. 1360, 17 Aug. 17.
Do.	Girdhari Ram, Sepoy, No. 3220	17 Aug. 17	No. 1360, 17 Aug. 17.
Do.	Tulsi Ram, Sepoy, No. 3791	17 Aug. 17	No. 1360, 17 Aug. 17.
103rd Infy.	Yeshwant Rao Bhosle, Subadar	24 Sep. 15	No. 942, 24 Sep. 15.
Do.	Baburao Gaekwad, Subadar	10 Dec. 15	No. 1248, 10 Dec. 15.
Do.	Shaikh Maqtum, Jemadar	10 Dec. 15	No. 1248, 10 Dec. 15.
Do.	Hamid Khan, Havildar, 2873	10 Dec. 15	No. 1248, 10 Dec. 15.
Do.	Sayyid Lal, Bugle-Major, 2469	10 Dec. 15	No. 1248, 10 Dec. 15.
Do.	Dinu Jadhow, Color-Havildar, 2343	17 Aug. 17	No. 1357, 17 Aug. 17.
Do.	Atmaran Rane, Havildar, 3111	17 Aug. 17	No. 1357, 17 Aug. 17.
Do.	Kashabha Suijore, Sepoy, 3645	17 Aug. 17	No. 1357, 17 Aug. 17.
Do.	Gopal Rao Palande, Jemadar, I.O.M.	17 Aug. 17	No. 1357, 17 Aug. 17.
104th Rif.	Mewa Ram, Havildar, 2346	12 Dec. 11	No. 1016, 12 Dec. 11.
Do.	Mohru Ram, Havildar, 2435	19 Mar. 15	No. 285, 19 Mar. 15.
Do.	Teja Ram, Lance-Naik, 2898	4 Aug. 15	No. 728, 4 Aug. 15.
Do.	Rahim Baksh, Subadar	10 Dec. 15	No. 1248, 10 Dec. 15.
Do.	Khiu Singh, Lance-Naik, 2245	10 Dec. 15	No. 1248, 10 Dec. 15.
Do.	Chandgi Ram, Rifleman, 3205	10 Dec. 15	No. 1248, 10 Dec. 15.
Do.	Chokha Ram, Rifleman, 2907	10 Dec. 15	No. 1248, 10 Dec. 15.
Do.	Uda Ram, Rifleman, 1628	10 Dec. 15	No. 1248, 10 Dec. 15.
Do.	Ramlal Ram, Naik, 3087	10 Dec. 15	No. 1248, 10 Dec. 15.
Do.	Maula Bakhsh, Lance-Naik, 1625	10 Dec. 15	No. 1248, 10 Dec. 15.
Do.	Hari Singh, Lance-Havildar, 2538	17 Aug 17	No. 1360, 17 Aug. 17.
Do.	Rahimdad Khan, Havildar, 1912	17 Aug. 17	No. 1357, 17 Aug. 17.
Do.	Sharaf Khan, Colour-Havildar, 1170	17 Aug. 17	No. 1357, 17 Aug. 17.
Do.	Ganesha Ram, Naik, 3069	17 Aug. 17	No. 1357, 17 Aug. 17.
Do.	Jhonta Singh, Lance-Naik, 1615	17 Aug. 17	No. 1357, 17 Aug. 17.
Do.	Parsa Ram, Rifleman, 3766	1 Jan. 18	No. 9, 1 Jan. 18.
105th Infy.	Daulat Rao Bhosle, Subadar	18 Nov. 16	No. 1386, 18 Nov. 16.
Do.	Bala More, Havildar, 2680	17 Aug. 17	No. 1360, 17 Aug. 17.
Do.	Nama Thange, Sepoy, 2271	17 Aug. 17	No. 1360, 17 Aug. 17.
Do.	Vitohaba Jagtap, Sepoy, 3241	17 Aug. 17	No. 1360, 17 Aug. 17.
Do.	Ramchandar Shinde, Lance-Naik, 2511	17 Aug. 17	No. 1360, 17 Aug. 17.
Do.	Maroti Bodke, Subadar	17 Aug. 17	No. 1360, 17 Aug. 17.
Do.	Saiyid Suliman, Subadar	1 Jan. 18	No. 9, 1 Jan. 18.
106th Prs.	Dost Muhammad, Subadar-Major, *Sardar Bahadur*	28 June 07	No. 285, 19 Mar. 15.
Do.	Ali Dost, Subadar	1 Jan. 10*	No. 4, 1 Jan. 10.
Do.	Haidar Ali, Col. Havildar, 829	18 Nov. 16	No. 1338, 18 Nov. 16.
Do.	Ram Ali, Sepoy, 1239	18 Nov. 16	No 1388, 18 Nov. 16.
Do.	Muhammad Hasan, Subadar	17 Feb. 17	No. 237, 17 Feb. 17.
Do.	Wali Muhammad, Subadar	17 Feb. 17	No. 237, 17 Feb. 17.
Do.	Kalbi Hussain, Havildar, 43	17 Feb. 17	No. 237, 17 Feb. 17.
Do.	Abdul Karim, Naik, 1437	17 Feb. 17	No. 237, 17 Feb. 17.
Do.	Abdul Hakim, Sepoy, 1465	17 Feb. 17	No. 237, 17 Feb. 17.
Do.	Haji Muhammad, Sepoy, 2091	17 Feb. 17	No. 237, 17 Feb. 17.

* Bar to Indian Distinguished Service Medal.

INDIAN DISTINGUISHED SERVICE MEDALLISTS—*contd.*

NOTE.—The names in *italics* are those of members on the pension establishment.

Corps.	Names and Ranks.	Date.	Authority. A. D. Notfn.
106th Prs.	Ali Juma, Jamadar (now Subadar)	17 Aug. 17	No. 1357, 17 Aug. 17.
Do.	Ali Nazar, Naik, 1533	17 Aug. 17	No. 1357, 17 Aug. 17.
Do.	Haider, Havildar, 202	1 Jan. 18	No. 9, 1 Jan. 18.
Do.	Seyed Muhammad, Lance-Naik, 1267	1 Jan. 18	No. 9, 1 Jan. 18.
107th Do.	Achhar Sing, Naik, 2980	4 May 15	No. 563, 12 June 15.
Do.	Phaga Sing, Sepoy, 3991	4 May 15	No. 503, 12 June 15.
Do.	Fateh Muhammad Khan, Subadar	1 Jan. 16	No. 187, 18 Feb. 16.
Do.	Jetta Singh, Sepoy, 3475	1 Jan. 16	No. 187, 18 Feb. 16.
Do.	Labh Singh, Subadar	....	No. 313, 2 Apl. 15.
Do.	Hashmat Dad Khan, Subadar	....	No. 313, 2 Apl. 15.
Do.	Baghat Singh, Havildar, 3417	....	No. 313, 2 Apl. 15.
Do.	Nasrullah Khan, Havildar, 3214	12 May 16	No. 525, 12 May 16.
Do.	Abdul Rahman, Jemadar	18 Nov. 16	No. 1386, 18 Nov. 16.
Do.	Habib Khan, Jemadar	18 Nov. 16	No. 1386, 18 Nov. 16.
Do.	Krishna Chorpade, Jemadar	18 Nov. 16	No. 1386, 18 Nov. 16.
Do.	Basawa Singh, Havildar, 2753	18 Nov. 16	No. 1386, 18 Nov. 16.
Do.	Sabat, Khan, Lance-Naik, 4573	18 Nov. 16	No. 1386, 18 Nov. 16.
Do.	Ganiya Singh, Lance-Naik, 3757	18 Nov. 16	No. 1386, 18 Nov. 16.
Do.	Nizam Khan, Naik, 3293	18 Nov. 16	No. 1388, 18 Nov. 16.
109th Infy.	*Anant Bhogle, Lance-Naik, 2663*	*28 Oct. 16*	*No. 1295, 28 Oct. 16.*
Do.	Dost Muhammad, Sepoy, 2867	16 Dec. 16	No. 1511, 16 Dec. 16.
110th Infy.	Hari Sawant, Subadar	19 Mar. 15	No. 5, 2 Jan. 11.
Do.	Vishnu Ghone, Jemadar	19 Mar. 15	
Do.	Shankar Fatak, Sepoy, 2113	17 Aug. 17	No. 1357, 17 Aug. 17.
Do.	Laxuman Ghorpade, Sepoy, 2140	17 Aug. 17	No. 1357, 17 Aug. 17.
Do.	Shaikh Yasin, Subadar-Maj., *Bahadur*	17 Aug. 17	No. 1357, 17 Aug. 17.
112th Infy.	Mehr Khan, Sepoy, 3243	18 Nov. 16	No. 1386, 18 Nov. 16.
113 Do.	*Parsadi Sing, Subadar*	*4 May 15*	*No. 563, 12 June 15.*
Do.	Mehr Din, Sepoy, 933	17 Aug. 17	No. 1357, 17 Aug. 17.
116th Mah.	Rama Pokle, Lance-Naik 2149	18 Nov. 16	No. 1386, 18 Nov. 16.
Do.	Vithal Rao Kathar, Naik, 826	18 Nov. 16	No. 1388, 18 Nov. 16.
117th Mah.	Vishnu Aire, Havildar, 1883	4 Aug. 15	No. 728, 4 Aug. 15.
Do.	Dadu Aggave, Havildar, 2742	4 Aug. 15	No. 728, 4 Aug. 15.
Do.	Abdul Gaffar, Private, 2915	4 Aug. 15	No 728, 4 Aug. 15.
Do.	Sakharam Sawant, Sepoy, 2074	10 Dec. 15	No. 1248, 10 Dec. 15.
Do.	Ganpat Khedkar, Sepoy, 1741	10 Dec. 15	No. 1248, 10 Dec. 15.
Do.	Maruti Savant, Lance-Naik, 2271	10 Dec. 15	No. 1248, 10 Dec. 15.
Do.	Abajirao Kadam, Sepoy, 2432	10 Dec. 15	No. 1248, 10 Dec. 15.
Do.	Anandrao Kadam, Sepoy, 2790	10 Dec. 15	No. 1243, 10 Dec. 15.
Do.	Ganpat Rane, Sepoy, 1776	10 Dec. 15	No. 1248, 10 Dec. 15.
Do.	Luximan, Panaskar, Sepoy, 2262	10 Dec. 15	No. 1248, 10 Dec. 15.
117th Mah.	Narsingrao Sindhi, Sepoy, 3094	17 Aug. 17	No. 1357, 17 Aug. 17.
Do.	Sadashiw Gaode, Sepoy, 3329	17 Aug. 17	No. 1357, 17 Aug 17.
119th Infy.	Dewa Rawat, Lance-Naik, 4025	4 Aug. 15	No. 728, 4 Aug. 15.
Do.	Gokal. Private, 4166	4 Aug. 15	No. 728, 4 Aug. 15.
Do.	Dhanpa Rawat, Private, 4112	4 Aug. 15	No. 728, 4 Aug. 15.
Do.	Makna Rawat, Lance-Naik, 4229	4 Aug. 15	No. 728, 4 Aug. 15.
Do.	Nabi Bux, Ward Orderly, 2939	4 Aug. 15	No. 728, 4 Aug. 15.
Do.	Muhammad Safi Khan, Havildar, 3606	10 Dec. 15	No. 1248, 10 Dec. 15.
Do.	Hukma, Rese vist, 3318	17 Aug. 17	No. 1357, 17 Aug. 17.
Do.	Baloo Singh, Sepoy, 4526	17 Aug. 17	No 1357, 17 Aug. 17.
Do.	Gulla Ram, Jemadar	17 Aug. 17	No. 1360, 17 Aug. 17.
Do.	Shikari, Subadar Tempy. Sub.-Maj.	17 Aug. 17	No. 1360, 17 Aug. 17.
Do.	Radmal Singh, Sepoy, 4036	17 Aug. 17	No. 1357, 17 Aug. 17.
120th Do.	Ganesh, Havildar, 978	19 Mar. 15	No. 285, 19 Mar. 15.
Do.	Gangadhar, Sepoy, 1511	4 Aug. 15	No. 728, 4 Aug. 15.
Do.	Bhima Rawat, Lance-Naik, 1089	4 Aug. 15	No 728, 4 Aug. 15.
Do.	Rawat Singh, Sepoy, 1202	4 Aug. 15	No. 728, 4 Aug. 15.
Do.	Pitha Rawat, Subadar	10 Dec. 15	No. 1248, 10 Dec. 15.
Do.	Juma Bakhsh, Sepoy, 1719	10 Dec. 15	No. 1248, 10 Dec. 15.
Do.	Ashgar Ali, Lce.-Naik, 912	18 Nov. 16	No. 1388, 18 Nov. 16.
Do.	Chuna Ram Havildar 1644	17 Aug. 17	No. 1360, 17 Aug. 17.
Do.	Lachman Singh, Havildar 1186	17 Aug. 17	No. 1357, 17 Aug. 17.
Do.	Deburam, Havildar 1046	17 Aug. 17	No. 1357, 17 Aug. 17.
Do.	Ghufur Khan, Sepoy, 1869	17 Aug. 17	No. 1357, 17 Aug. 17.
Do.	Yusuf Ali, Sepoy, 2009.	17 Aug. 17	No. 1357, 17 Aug. 17.
Do.	Zaidi Gul, Sepoy, 1915	17 Aug. 17	No. 1357, 17 Aug. 17.
Do.	Dhan Singh, Sepoy, 1246	17 Aug. 17	No. 1357, 17 Aug. 17.
121st Prs.	Dal Khan, Havildar, 1027	1 Jan. 16	No. 187, 18 Feb. 16.
Do.	Ude Ram, Sepoy, 2456	3 June 16	No. 849, 29 July 16.
Do.	Kana Ram, Sepoy, 2488	18 Nov. 16	No. 1386, 18 Nov. 16.
Do.	Phul Khan, Havildar, 2145	2 June 17	No. 868, 2 June 17.
Do.	Makhu Khan, Lance-Naik, 2580	2 June 17	No. 868, 2 June 17.
Do.	Chote Khan, Sepoy, 3812	17 Aug. 17	No 1360, 17 Aug. 17.
123rd Rifl.	Bije Singh, Subadar	28 June 07	No. 586, 28 June 07.
Do.	Ghulam Muhammad, Subadar	18 Nov. 10	No. 933, 18 Nov. 10.

INDIAN DISTINGUISHED SERVICE MEDALLISTS—*contd.*

NOTE.—The names in *italics* are those of members on the pension establishment.

Corps.	Names and Ranks.	Date.	Authority. A. D. Notfn.
123rd Rifl.	Rura Ram, Subadar	12 Dec. 11	No. 1016, 12 Dec. 11.
Do.	Ghulam Mohammed, Subadar	18 Nov. 16	No. 993, 18 Nov. 10.
Do.	Murbarik Ali, Jemadar	3 June 16	No. 849, 29 July 16.
Do.	Samrat Singh, Subadar	17 Aug. 17	No. 1360, 17 Aug. 17.
Do.	Bega Ram, Lance-Naik, 658	17 Aug. 17	No. 1357, 17 Aug. 17.
Do.	Ganpat Sing, Naik, 576	3 June 18	No. 1178, 3 June 18.
Do.	Ramkanwar Singh, Lance-Naik, 1326	3 June 18	No. 1178, 3 June 18.
Do.	Samwat Ram, Naik, 716	3 June 18	No. 1178, 3 June 18.
Do.	Sheoji Ram, Rifleman, 927	3 June 18	No. 1178, 3 June 18.
Do.	Momraj Ram, Rifleman, 1443	3 June 18	No. 1178, 3 June 18.
Do.	Sheochand Ram, Rifleman, 2003	3 June 18	No. 1178, 3 June 18.
Do.	Ganpat Singh, Rifleman, 2414	3 June 18	No. 1178, 3 June 18.
Do.	Ladu Khan, Rifleman, 2507	3 June 18	No. 1178, 3 June 18.
124th Infy.	Ghulam Hussein, Sepoy, 4321	3 June 15	Lon. Gaz., 23 Jan. 15.
Do.	Juma Khan, Naik, 4225	4 Oct. 18	No. 2342, 4 Oct. 18.
Do.	Narayan Singh, Sepoy, 4435	18 Nov. 16	No. 1386, 18 Nov. 16.
Do.	Bal Din, Jemadar	17 Aug. 17	No. 1356, 17 Aug. 17.
Do.	Sher Singh, Lance-Naik, 3277	17 Aug. 17	No. 1360, 17 Aug. 17.
Do.	Boota Khan, Havildar, 3138	1 Jan. 18	No. 9, 1 Jan. 18.
Do.	Abdullah Khan, I. O. M., Sepoy, 4565	3 June 18	No. 1180, 3 June 18.
125th Rifl.	Gul Sher, Sepoy, 2602	1 Jan. 16	No. 1386, 18 Nov. 16.
Do.	Tirka Ram, Havildar, 2044	12 May 16	No. 525, 12 May 16.
Do.	Munshi Singh, Subadar	18 Nov. 16	No. 1386, 18 Nov. 16.
Do.	Hakim Khan, Havildar, 1922	18 Nov. 16	No. 1386, 18 Nov. 16.
Do.	Seochand Ram, Lce.-Naik, 2370	18 Nov. 16	No. 1388, 18 Nov. 16.
Do.	Makanda Ram, Lce.-Naik, 2327	18 Nov. 16	No. 1388, 18 Nov. 16.
Do.	Jeysat Ram, Rifleman, 2398	18 Nov. 16	No. 1388, 18 Nov. 16.
Do.	Bachan Singh, Havildar, 1744	17 Aug. 17	No. 1360, 17 Aug. 17.
Do.	Hukma Ram, Havildar, 2020	1 Jan. 18	No. 9, 1 Jan. 18.
Do.	Nanoo Singh, Rifleman, 3482	1 Jan. 18	No. 9, 1 Jan. 18.
Do.	Ragunath Singh, Havildar, 2261	16 Aug. 18	No. 1885, 16 Aug. 18.
126th Infy.	Bud Singh, Jemadar	1 Jan. 10	No. 4, 1 Jan. 10.
Do.	Guldar Shah, Jemadar	15 Oct. 15	No. 1024, 15 Oct. 15.
Do.	Ahmad Khan, Havildar	15 Oct. 15	No. 1024, 15 Oct. 15.
Do.	Wazir Khan, Havildar	15 Oct. 15	No. 1024, 15 Oct. 15.
Do.	Kampoo Khan, Lance-Naik, 5425	16 Aug. 18	No. 1885, 16 Aug. 18.
127th Bal. L. I.	Raj Wali, Subadar	28 June 07	No. 586, 28 June 07.
Do.	Reshmin, Havildar, 290	4 Oct. 18	No. 2342, 4 Oct. 18.
Do.	Fateh Muhammad, Sepoy, 672	18 Nov. 16	No. 1388, 18 Nov. 16.
Do.	Dad Khan, Sepoy, 246	1 Jan. 16	Lon. Gaz., 14 Dec. 16.
Do.	Tarsum, Sepoy, 447	1 Jan. 18	No. 9, 1 Jan. 18.
Do.	Amalmir, Sepoy, 416	3 June 18	No. 1180, 3 June 18.
Do.	Hayat Muhammad, Naik, 489	3 June 18	No. 1180, 3 June 18.
Do.	Jemadar, Sepoy, 759	3 June 18	No. 1180, 3 June 18.
Do.	Mirsher, Havildar, 3394	4 Oct. 18	No. 2342, 4 Oct. 18.
Do.	Muhammad Sadiq, Lance-Naik, 4104	3 June 18	No. 1180, 3 June 18.
Do.	Karam Bakhsh, Subadar	3 June 18	No. 1180, 3 June 18.
128th Prs.	Mallekhan VI, Havildar, 3368	12 May 16	No. 525, 12 May 16.
Do.	Labh Singh, Havildar, 2862	12 May 16	No. 525, 12 May 16.
Do.	Sardar Khan, Private, 215	12 May 16	No. 525, 12 May 16.
Do.	Ghosi Khan, Private, 207	12 May 16	No. 525, 12 May 16.
Do.	Rakhmat Shah, Lance-Naik, 3500	12 May 16	No. 525, 12 May 16.
Do.	Jiwan Singh, Private, 2965	12 May 16	No. 525, 12 May 16.
Do.	Krishna Bhosle, Subadar	18 Nov. 16	No. 1388, 18 Nov. 16.
Do.	Karm Singh, Subadar	18 Nov. 16	No 1388, 18 Nov. 16.
Do.	Sheikh Hubdar Ali, Sepoy, 4062	1 Jan. 17	Lon. Gaz., 1 Feb. 17.
Do.	Dhunde Khan, Naik, 3369	17 Aug. 17	No. 1360, 17 Aug. 17.
Do.	Dula Singh, Lance-Naik, 4444	17 Aug. 17	No. 1360, 17 Aug. 17.
Do.	Teja Singh, Subadar	17 Aug. 17	No. 1360, 17 Aug. 17.
129th Baluchis	Nur Khan, Jemadar	4 May 15	No. 563, 12 June 15.
Do.	Aulyia Khan, Sepoy, 2012	4 May 15	No. 563, 12 June 15.
Do.	Mehrab Gul, Sepoy, 4267	4 May 15	No. 563, 12 June 15.
Do.	Mustkim, Sepoy, 4305	1 Jan. 16	No. 563, 12 June 15.
Do.	Shak Uran, Lance-Naik, 294	3 June 18	No. 1180, 3 June 18.
Do.	Kassib, Sepoy, 105	. . .	No. 313, 2 Apl. 15.
Do.	Lal Sher, Sepoy, 2813	. . .	No. 313, 2 Apl. 15.
Do.	Jafar Khan, Sepoy, 3600	. . .	No. 13, 2 Apl. 15.
Do.	Said Ahmed, Sepoy, 4182	. . .	No. 313, 2 Apl. 15.
Do.	Ghulam Jilani, Jemadar	3 June 16	No. 849, 29 July 16.
Do.	Burhan Ali, Sepoy, 1172	3 June 16	No. 849, 29 July 16.
Do.	Boota Khan, Sepoy, 1342	21 Nov. 16	No. 676, 5 May 17.
Do.	Taj Muhammad, Havildar, (Retd.) 1894	17 Aug. 17	No. 1357, 17 Aug. 17.
Do.	Fateh Khan, Naik,	2 Nov. 17	No. 1887, 2 Nov. 17.
Do.	Nur Khan, Subadar	1 Jan. 18	No. 9, 1 Jan. 18.
Do.	Durani, Subadar	4 Oct. 18	No. 2342, 4 Oct. 18.
130th Baluchis	Muhammad Ali, Havildar, 3380	25 Feb. 16	No. 220, 25 Feb. 16.

INDIAN DISTINGUISHED SERVICE MEDALLISTS—*contd.*

NOTE.—The names in *italics* are those of members on the pension establishment.

Corps.	Names and Ranks.	Date.	Authority. A. D. Notfn.
130th Baluchis	Lal Khan, Sepoy, 337	21 Nov. 16	No. 676, 5 May 17.
Do.	Mirza Khan, Sepoy, 799	21 Nov. 16	No. 676, 5 May 17.
Do.	Sultan Bakhsh, Sepoy, 4297	21 Nov. 16	No. 676, 5 May 17.
Do.	Chaman Khan, Private, 69	9 June 17	No. 894, 9 June 17.
Do.	Fateh Mohammad, Lance-Naik, 304	9 June 17	No. 894, 9 June 17.
Do.	Muhammad Hussain, Naik, 3183	9 June 17	No. 894, 9 June 17.
Do.	Sultan Khan, Havildar, 4006	9 June 17	No. 894, 9 June 17.
Do.	Shah Baz Khan, Private, 4996	9 June 17	No. 894, 9 June 17.
Do.	Sultan Khan, Subadar	1 Jan. 18	No. 9, 1 Jan. 18.
Do.	Nowsher, Subadar	1 Jan. 18	No. 9, 1 Jan. 18.
Do.	Rajwali, Naik, 3946	3 June 18	No. 1180, 3 June 18.
Do.	Waris Khan, Lance-Naik, 1863	3 June 18	No. 1180, 3 June 18.
1st G. R.	Padam Thapa, Naick	28 June 07	No. 586, 28 June 07.
Do.	Baliram Thapa, Rifleman, 2814	10 Mar. 15	No. 356, 16 Apl. 15.
Do.	Kansi, Gurung, Rifleman, 2515	10 Mar. 15	No. 356, 16 Apl. 15.
Do.	Khambasing Gurung, Rifleman, 2734	3 June 15	Lon. Gaz., 23 June 15.
Do.	Phalman Gurung, Rifleman, 2397	3 June 15	Lon. Gaz., 23 June 15.
Do.	Jagat Sher Gurung, Havildar, 1912	28 Jan. 16	No. 103, 28 Jan. 16.
Do.	Lalbir Thapa, Rifleman, 1946	1 Jan. 16	No. 103, 28 Jan. 16.
Do.	Chitabir Thapa, Rifleman, 2650	1 Jan. 16	No. 103, 28 Jan. 16.
Do.	Nundu Gurung, Subadar	*3 June 16	No. 849, 29 July 16.
Do.	Jangia Rana, Jemadar	3 June 16	No. 849, 29 July 16.
Do.	Akbar Ali, Havildar, 1955	3 June 16	No. 849, 29 July 16.
Do.	Ramia Gharti, Lnce.-Naik	3 June 16	No. 849, 29 July 16.
Do.	Bishan Singh, Thapa, Rifleman, 2723	3 June 16	No. 849, 29 July 16.
Do.	Kalu Gurung, Rifleman, 1842	3 June 16	No. 849, 29 July 16.
Do.	Harkbir Limbu, Subadar	18 Nov. 16	No. 1386, 18 Nov. 16.
Do.	Jit Bahadur Thapa, Subadar	18 Nov. 16	No. 1386, 18 Nov. 16.
Do.	Kharak Singh Thapa, Havildar, 1963, I. D. S. M.	18 Nov. 16	No. 1386, 18 Nov. 16.
Do.	Kharak, Bahadur Gurang, Lce.-Naik, 2958	18 Nov. 16	No. 1386, 18 Nov. 16.
Do.	Tilbikram Rana, Havildar, 2022	2 June 17	No. 868, 2 June 17.
Do.	Guman Sing Gurung, Jemadar	17 Aug. 17	No. 1360, 17 Aug. 17.
Do.	Lal Sing Rana, Jemadar	17 Aug. 17	No. 1360, 17 Aug. 17.
Do.	Narain Mal, Naik, 3551	17 Aug. 17	No. 1360, 17 Aug. 17.
Do.	Paras Bahadur Thapa, Rifleman, 4383	17 Aug. 17	No. 1360, 17 Aug. 17.
Do.	Padan Sing Thapa, Rifleman, 122	17 Aug. 17	No. 1360, 17 Aug. 17.
Do.	Kalu Ram Thapa, Subadar	8 Aug. 17	No. 1305, 8 Aug. 17.
Do.	Bhairap Singh Gurung, Rifleman, No. 2599	17 Aug. 17	No. 1360, 17 Aug. 17.
Do.	Dhan Sing Gurung, Bugle-Maj., 1883	2 Nov. 17	No. 1887, 2 Nov. 17.
Do.	Balbir Gharti, Rifleman, 2840	2 Nov. 17	No. 1887, 2 Nov. 17.
Do.	Hemlal Thapa, Rifleman, 3640	2 Nov. 17	No. 1887, 2 Nov. 17.
Do.	Maniraj Gurung, Rifleman, 3974	2 Nov. 17	No. 1887, 2 Nov. 17.
Do.	Rubrabir Thapa, Rifleman, 2954	1 Jan. 18	No. 9, 1 Jan. 18.
1-2 Do.	Gurung, Havildar Bahadur, 3719	19 Nov. 17	No. 115, 18 Jan. 18.
2nd G. R.	Aslal Rai, Naik	8 June 07	No. 586, 28 June 07.
Do.	Ahiman Rana, Subadar	8 Sep. 18	No. 117, 8 Sep. 18.
Do.	Setea Gurung, Naik	28 June 07	No. 586, 28 June 07.
Do.	Bhagatbir Thapa, Naik, 1776	21 Jan. 08	No. 1, 1 Jan. 08.
Do.	Kharak Bahadur Thakur, Subadar	8 June 07	No. 586, 28 June 07.
Do.	Bag Sing Thapa, Havildar, 1628	20 Mar. 15	No. 356, 16 Apl. 15.
Do.	Bhimbahadur Thapa, Naik, 1517	10 Mar. 15	No. 356, 16 Apl. 15.
Do.	Amar Sing Pun, Lance-Naik, 23[illegible]	10 Mar. 15	No. 356, 16 Apl. 15.
Do.	Ram Pershad Thapa, Naik, [illegible]	2 Apl. 15	No. 313, 2 Apl. 15.
Do.	Gamer Sing Bura, Naik 167[illegible]	2 Apl. 15	Do.
Do.	Sher Sing Ghale, Lce.-Naik, [illegible]	2 Apl. 15	Do.
Do.	Ran Patti Gurung, Havildar, [illegible]	2 Apl. 15	Do.
Do.	Judhia Sarki, Havildar, 1280	2 Apl. 15	Do.
Do.	Janglai Gurung, Havildar 1432	2 Apl. 15	Do.
Do.	Patiram Thapa, Naik, 1593	2 Apl. 15	Do.
Do.	Kalu Gurung, Rifleman, 2518	2 Apl. 15	Do.
Do.	Pertab Sahi, Jemadar	18 Nov. 16	No. 1386, 18 Nov. 16.
Do.	Khampa Sing Gurung, Rifleman, 4487	18 Nov. 16	No. 1388, 18 Nov. 16.
Do.	Keshar Singh Rau, Havildar, 3699	30 Mar. 17	No. 1191, 27 July 17.
Do	Bahadur Gurung, Havildar, 3719	2 June 17	No. 868, 2 June 17.
Do.	Puran Singh Gurung, Lance-Naik, 4212	2 June 17	No. 868, 2 June 17.
Do.	Randhoj Limbu, Lance-Naik, 4814	2 June 17	No. 868, 2 June 17.
Do.	Bhagtia Gurung, Rifleman, 4439	2 June 17	No. 868, 2 June 17.
Do.	Gunan Singh Gurung, Jemadar	17 Aug. 17	No. 1360, 17 Aug. 17.
Do.	Lal Singh Rana	17 Aug. 17	No. 1360, 17 Aug. 17.
Do.	Narain Mal, Naik, 3551	17 Aug. 17	No. 1360, 17 Aug. 17.
Do.	Padam Singh Thapa, Rifleman, 122	17 Aug. 17	No. 1360, 17 Aug. 17.
Do.	Parsar Bahadur Thapa, Rifleman, 4383	17 Aug. 17	No. 1360, 17 Aug. 17.
Do.	Budhi Bal Thapa, Lance-Naik, 4466	1 Jan. 18	No. 9, 1 Jan. 18.
Do.	Gangasing Gharti, Rifleman, 547	1 Jan. 18	No. 9, 1 Jan. 18.

* Bar to Indian Distinguished Service Medal.

INDIAN DISTINGUISHED SERVICE MEDALLISTS—*contd.*

NOTE.—The names in *italics* are those of members on the pension establishment.

Corps.	Names and Ranks.	Date.	Authority. A. D. Notfn.
2nd G. R.	Makardhoj Rai, Lance-Naik, 1695	1 Jan. 18	No. 9, 1 Jan. 18.
Do.	Narbir Bura, Rifleman, 2933	1 Jan. 18	No. 9, 1 Jan. 18.
3rd G. R.	Barn Sing Thapa, Subadar	4 May 15	No. 563, 12 June 15.
Do.	Puran Sing Thapa, Jemadar	4 May 15	No. 563, 12 June 15.
Do.	Motilal Gurung, Havildar, 1313	4 May 15	No. 563, 12 June 17.
Do.	Ganpati Thapa, Rifleman, 2118	. .	No. 313, 2 Apl. 15.
Do.	Khark Sing Gurung, Rifleman, 2533	3 June 16	No. 849, 29 July 16.
Do.	Sirdhoj Rai, Rifleman, 1108	3 June 16	No. 849, 29 July 16.
Do.	Kharke Pun, Havildar, 1985	*15 Feb. 18	Lon. Gaz., 18 Feb. 18.
Do.	Dalip Sing Rana, Rifleman, 2726	3 June 18	No. 1178, 3 June 18.
Do.	Mang Sing Thapa, Rifleman, 3704	3 June 18	No. 1178, 3 June 18.
Do.	Pothiman Limbu, Rifleman, 2829	3 June 18	No. 1178, 3 June 18.
Do.	Amar Sing Gurung, Rifleman, 2651	3 June 18	No. 1178, 3 June 18.
Do.	Dharam Sing Rana, Havildar, 1793	3 June 18	No. 1178, 3 June 18.
Do.	Ajab Sing Gurung, Havildar, 1733	3 June 18	No. 1178, 3 June 18.
Do.	Narain Sing Rana, Havildar, 2827	3 June 18	No. 1178, 3 June 18.
Do	Dame Thapa, Piper, 2789	3 June 18	No. 1178, 3 June 18.
Do.	Amar Sing Thapa, Lance-Naik, 2803	3 June 18	No. 1178, 3 June 18.
Do.	Randhoj Gurung Havildar, 1879	3 June 18	No. 1178, 3 June 18.
Do.	Man Sing Mal, Havildar, 2582	3 June 18	No. 1178, 3 June 18.
Do.	Kushal Singh Bisht, Rifleman, 3060	26 Oct. 18	No. 2538, 26 Oct. 18.
Do.	Chandra Sing Gurung, Naik, 3182	3 June 18	No. 1178, 3 June 18.
Do.	Setia Pun, Rifleman, 3548	26 Oct. 18	No. 2536, 26 Oct. 18.
Do.	Ajudhe Rana, Lance-Naik, 3399	3 June 18	No. 1178, 3 June 18.
Do.	Pairam Pun, Lance-Naik, 3322	26 Oct. 18	No. 2536, 26 Oct. 18.
Do.	Manbirshing Gurung, Havaildar, 1878	26 Oct. 18	No. 2536, 26 Oct. 18.
Do.	Harke Gharti, Rifleman, 3627	26 Oct. 18	No. 2536, 26 Oct. 18.
Do.	Kishen Bahadurchettri, Rifleman, 3286	26 Oct 18	No. 2536, 26 Oct. 18.
Do.	Bhagat Bir Thapa, Rifleman, 3477	26 Oct. 18	No. 2536, 26 Oct. 18.
4th Do.	Deotinarain Newar, Rifleman, 4203	3 June 15	Lon. Gaz., 23 June 15.
Do.	Patiram Kunwar, Rifleman, 3994	3 June 15	Lon. Gaz., 23 June 15.
Do.	Asbir Rana, Lance-Naik, 798	23 July 15	No. 815, 27 Aug. 15.
Do.	Lachman Gurung, Lance-Naik, 4945	23 July 15	No. 815, 27 Aug. 15.
Do.	Garbha Sing Gurung, Rifleman, 1074	23 July 15	No. 815, 27 Aug. 15.
Do.	Lokbir Ale, Havildar, 4620	3 June 16	No. 849, 29 July 16.
Do.	Amar Sing Pun, Havildar, 4098	3 June 16	No. 849, 29 July 16.
Do.	Chhabu Thapa, Jemadar	2 Jun. 17	No. 868, 2 June 17.
Do.	Gamir Singh Thapa, Rifleman, 3388	17 Aug. 17	No. 1360, 17 Aug. 17.
Do.	Udepun, Rifleman, 3850	17 Aug. 17	No. 1360, 17 Aug. 17.
Do.	Karna Sing Gurung, Havildar, 2537	17 Aug. 17	No. 1360, 17 Aug. 17.
Do.	Ram Kishen Rana, Subadar	8 Aug. 17	No. 1305, 8 Aug. 17.
Do.	Ragbur Thapa, Jemadar	8 Aug. 17	No. 1305, 8 Aug. 17.
Do.	Nande Gharti, Rifleman, 263	10 May 17	No. 1170, 20 July 17.
Do.	Kharak Sing Pun, Havildar, 4290	2 Nov. 17	No. 1887, 2 Nov. 17.
Do.	Balbahadur Gurung, Naik, 131	2 Nov. 17	No. 1887, 2 Nov. 17.
Do.	Gagan Sing Rana, Lance-Naik, 4512	2 Nov. 17	No. 1887, 2 Nov. 17.
Do.	Dhanbahadur Gurung, Rifleman, 4588	2 Nov. 17	No. 1887, 2 Nov. 17.
Do.	Taune Gurung, Rifleman, 910	2 Nov. 17	No. 1887, 2 Nov. 17.
Do.	Balbir Pun, Rifleman, 672	2 Nov. 17	No. 1887, 2 Nov. 17.
Do.	Hari Kishen Thapa, Rifleman, 2529	1 Jan. 18	No. 9, 1 Jan. 18.
Do.	Puranbahadur Thapa, Naik, 3299	1 Jan. 18	No. 9, 1 Jan. 18.
5th G. R.	Ratanbir Thapa, Lance-Naik, 4544	17 Sep. 15	No. 906, 17 Sep. 15.
Do.	Chintaram Bura, Havildar, 7824	1 Jan. 16	No. 680, 17 June 16.
Do.	Narbahadur Gurung, Naik, 1489	1 Jan. 16	No. 680, 17 June 16.
Do.	Dalbir Chand, Rifleman, 4766	1 Jan 16	No. 680, 17 June 16.
Do.	Mansa Ram Pun, Subdr.	3 June 16	No. 849, 29 July 16.
Do.	Dhansing Gurung, Subdr.	3 June 16	No. 849, 29 July 16.
Do.	Gopia Thapa, Rifleman, 2578	3 June 16	No. 849, 29 July 16.
Do.	Dharm Sing Gurung, Lce Naik, 2298	18 Nov. 16	No. 1388, 18 Nov. 16.
Do.	Balbir Rana, Rifleman, 2068	18 Nov. 16	No. 1388, 18 Nov. 16.
Do.	Gambak Sing Mal, Rifleman, 2627	18 Nov. 16	No. 1388, 18 Nov. 16.
Do.	Damar Sing Gurnug, Havildar, 4367	16 Aug. 18	No. 1885, 16 Aug. 18.
6th G. R.	Balbir Rana, Havildar, 42	17 Sep. 15	No. 1388, 18 Nov. 16.
Do.	Narbir Thapa, Havildar, 285	17 Sep. 15	No. 1388, 18 Nov. 16.
Do.	Lal Sing Thapa, Rifleman, 56	17 Sep. 15	No. 1388, 18 Nov. 16.
Do.	Danbir Thapa, Rifleman, 1108	17 Sep. 15	No. 1388, 18 Nov. 16.
Do.	Dhaoraj Thapa, Jemadar	1 Jan. 16	No. 1388, 18 Nov. 16.
Do.	Kulbahadur Thapa, Subadar	. . .	No. 736, 1 July 16.
Do.	Indrajit Gurung, Naik, 724	3 June 16	No. 849, 29 July 16.
Do.	Puransing Gharti, Havildar, 4388	3 June 16	No. 849, 29 July 16.
Do.	Kesar Rana, Rifleman, 604	3 June 16	No. 849, 29 July 16.
Do.	Nabirai Gurung, Rifleman, 706	3 June 16	No. 849, 29 July 16.
Do.	Nanda Sng Gurung, Havildar.	3 June 16	No. 849, 29 July 16.
Do.	Anrup Bura, Rifleman, 1972	17 Aug. 17	No. 1360, 17 Aug. 17.

* Bar to Indian Distinguished Service Medal.

Corps.	Names and Ranks.	Date.	Authority.
	INDIAN DISTINGUISHED SERVICE MEDALLISTS—*concld.*		
	NOTE.—The names in *italics* are those of members on the pension establishment.		A. D. Notf n.
7th G. R.	Budhiman Rai, Rifleman, 664	3 Sep. 15	No. 830, 3 Sep. 15.
Do.	Kamrup Negi, Lance-Naik, 1961	19 Nov. 15	No. 1160, 19 Nov. 15.
Do.	Narman Rai, Lance-Naik, 1037	19 Nov. 15	No. 1160, 19 Nov. 15.
Do.	Harakbahadur Gurney, Rifleman, 1542	19 Nov. 15	No. 1160, 19 Nov. 10.
Do.	Ranbahadur, Lance-Naik, 1546	19 Nov. 15	No. 1160, 19 Nov. 15.
Do.	Chunahang, Jemadar	19 Nov. 15	No. 1160, 19 Nov. 15.
Do.	Jagit Rai, Lance-Naik, 1428	19 Nov. 15	No. 1160, 19 Nov. 15.
Do.	Lalbahadur Limbu, Lance-Naik, 1169	19 Nov. 15	No. 1160, 19 Nov. 15.
Do.	Talab Limbu, Rifleman, 1379	19 Nov. 15	No. 1160, 19 Nov. 15.
Do.	Harakdho, Rai, Lance-Naik, 2005	19 Nov. 15	No. 1160, 19 Nov. 15.
Do.	Ramdal Rai, Naik, 778	19 Nov. 15	No. 1160, 19 Nov. 15.
Do.	Randhoj Rai, Rifleman, 1984	26 Dec. 16	No. 676, 5 May 17.
Do.	Narpati Rai, Subadar-Maj.	17 Aug. 17	No. 1360, 17 Aug. 17.
Do.	Padan Sing Rai, Lance-Naik, 1952	17 Aug. 17	No. 1360, 17 Aug. 17.
Do.	Partiman Lama, Naik, 761	17 Aug. 17	No. 1357, 17 Aug. 17.
Do.	Panchdhoj Rai, Naik, 793	17 Aug. 17	No. 1357, 17 Aug. 17.
2—7 G. R.	Randhoj Limbo, Lance Naik, 4814	19 Nov. 17	No. 115, 18 Jan. 18.
Do.	Karan Sing Rai, Rifleman, 2339	17 Aug. 17	No. 1357, 17 Aug. 17.
8th G. R.	Mewa Thapa, Subadar	28 June 07	No. 586, 28 June 07.
Do.	*Jit Bahadur Gurung, Rifleman, 2941*	*28 June 07*	No. 586, 28 June 07.
Do.	*Kaku Thapa, Havildar, 1750*	*1 Jan. 08*	*No. 1, 1 Jan. 08.*
Do.	Narbahadur Gurung, Jemadar	9 Aug. 12	No. 780, 9 Aug. 12.
Do.	Bhairab Sahi, Havildar, 2446	9 Aug. 12	No. 780, 9 Aug. 12.
Do.	Kalia Pun, Lance-Naik, 3202	9 Apl. 14	No. 347, 9 Apl. 14.
Do.	Bhawansing Gurung, Rifleman, 3454	9 Apl. 14	No. 347, 9 Apl. 14.
Do.	Chandre Thapa, Lance-Naik, 3206	3 June 16	No. 849, 29 July 16.
Do.	Abir Thapa, Rifleman,	3 June 16	No. 849, 29 July 16.
Do.	Kishne Burathoki, Rifleman 4529	3 June 16	No. 849, 29 July 16.
Do.	Kharak Sing Thapa, Subadar	17 Aug. 17	No. 1360, 17 Aug. 17.
Do.	Khiman Sing Rana, Naik, 4290	17 Aug. 17	No. 1360, 17 Aug. 17.
Do.	Khairaj Gharti, Rifleman, 4666	1 Jan. 18	No. 9, 1 Jan. 18.
Do.	Lal Singh Gurung, Naik, 3490	1 Jan. 18	No. 9, 1 Jan. 18.
Do.	Damodahar Khattri, Jemadar	10 Mar. 15	No. 849, 29 July 16.
Do.	Lalit Bahadur Mal, Havildar, 1515	10 Mar. 15	No. 849, 29 July 16.
Do.	Man Singh Rana, Lance-Naik, 4138	26 Oct. 18	No. 2536, 26 Oct. 18.
Do.	Harihoj Khattri, Subadar	. . .	No. 313, 2 Apl. 15.
Do.	Kharak Bahadur Rana, Subadar	19 July 18	No. 1629, 19 July 18.
9th G. R.	Jaman Singh Khattri, Lce.-Naik, 2016	. . .	No. 313, 2 Apl. 15.
Do.	Gajbir Bisht, Rifleman, 2492	. . .	No. 313, 2 Apl. 15.
Do.	Ran Bahadur Sahi, Rifleman, 3055	. . .	No. 313, 2 Apl. 15.
Do.	Jowar Sing Khattri, Lance Naik, 2096	10 Mar. 15	No. 356, 16 Apl. 15.
Do.	Sirikisan Khattri, Lance Naick, 2314	10 Mar. 15	No. 356, 16 Apl. 15.
Do.	Manbahadur Sahi, Rifleman, 2722	10 Mar. 15	No. 356, 16 Apl. 15.
Do.	Kharak Bahadur Bhandari, Rifleman, 2799	10 Mar. 15	No. 356, 16 Apl. 15.
Do.	Dhan Bahadur Khattri, Rifleman, 2588	10 Mar. 15	No 356, 16 Apl. 15.
Do.	Gambhir Sing Bohra, Havildar, 1922	3 June 15	No. 646, 9 July 15.
Do.	Kulman Khattri, Naik, 2027	3 June 15	No. 646, 9 July 15.
Do.	Samar Bahadur Mal, Lance-Naik, 2118	3 June 15	No. 646, 9 July 15.
Do.	Balbahadur Khattri, Rifleman, 2885	3 June 15	No. 646, 9 July 15.
Do.	Dinbahadur Kunwar, Rifleman, 3075	3 June 15	No. 646, 9 July 15.
Do.	Ratanbir Khattri, Subadar	3 June 16	No. 849, 29 July 16.
Do.	Dalbahadur Sen, Rifleman, 2868	3 June 16	No. 849, 29 July 16.
Do.	Newal Sing Khattra, Rifleman, 3276	16 Sep. 16	No. 1096, 16 Sep. 16.
Do.	Tikaram Khattri, Jemadar	18 Nov. 16	No. 1386, 18 Nov. 16.
Do.	Bahadur Sahi, Naik, 2219	18 Nov. 16	No. 1386, 18 Nov. 16.
Do.	Puran Bahadur Karki, Naik, 2456	18 Nov. 16	No. 1386, 18 Nov. 16.
Do.	Kubir Sing Thapa, Rifleman, 2771	18 Nov. 16	No. 1386, 18 Nov. 16.
Do.	Astal Rai, Rifleman, 2856	18 Nov. 16	No. 1386, 18 Nov. 16.
Do.	Bahadur Sahi, Naik, 2642	17 Aug. 17	No. 1360, 17 Aug. 17.
Do.	Wazir Singh, Havildar, 1326	17 Aug. 17	No. 1360, 17 Aug. 17.
Do.	Mohanlal Ale, Rifleman, 3151	1 Jan. 18	No. 9, 1 Jan, 18.
Do.	Teklbahadur Khattri, Lance-Naik, 3557	1 Jan. 18	No. 9, 1 Jan. 18.
10th G. Rif.	Ambir Gurung, Rifleman, 256	17 Sep. 15	No. 906, 17 Sep. 15.
Do.	Sirinian Rai, Rifleman, 898	17 Sep. 15	No. 906, 17 Sep. 15.
Do.	Saimdhoj Rai, Rifleman, 723	1 Jan. 16	No. 680, 17 June 16.
Do.	Ramsingh, Burathoki, Subadar	. . .	No. 736, 1 July 16.
Do.	Dilsingh Limbu, Rifleman, 43	. . .	No. 736, 1 July 16.
Do.	Subidat Rai, Havildar, 2217	. . .	No. 736, 1 July 16.
Do.	Dil Sing Limbu, Rifleman, 43	3 June 16	No. 849, 29 July 16.
Do.	Subidat Rai, Havildar, 2217	3 June 16	No. 849, 29 July 16.
Do.	Harsing Thapa, Rifleman, 356	3 June 16	No. 849, 29 July 16.
Do.	Budbir Limbu, Rifleman, 362	3 June 16	No. 849, 29 July 16.
Do.	Jamansing Sunwar, Rifleman, 611	3 June 16	No. 849, 29 July 16.
Do.	Panche Rai, Rifleman, 179	3 June 16	No. 849, 29 July 16.

INDIAN DISTINGUISHED SERVICE MEDALLISTS—*contd.*

NOTE.—The names in *italics* are those of members on the pension establishment.

Corps.	Names and Rank.	Date.	Authority. A. D. Notfn.
Labour Corps.			
48th Coy.	Litap Shah, Havildar, 1355	20 Sep. 18	No. 2185, 20 Sep. 18.
49th Do.	Khawas Khan, Havildar, 1227	20 Sep. 18	No. 2185, 20 Sep. 18.
M. G. Corps.	Amar Singh, Jemadar	19 July 16	No. 1629, 19 July 18.
2nd M. Corps	Abdullah Shah, Driver, 1406	1 Jan. 16	No. 187, 18 Feb. 16.
Do.	Mirzaman, Naik, 533	18 Nov. 16	No. 1388, 18 Nov. 16.
4th M. Corps	Ghulam Haidar, Lce-Naik, 1006	18 Nov. 16	No. 1386, 18 Nov. 16.
5th M. Corps	Muhammad Shah, Ressaidar	12 May 16	No. 525, 12 May 16.
7th M. Corps.	Ibrahim, Naik, 988	1 Jan. 18	No. 9, 1 Jan. 18.
9th M. Corps	Fateh Khan, Kot Duffadar, 182	1 Jan. 16	No. 525, 12 May 16.
Do.	Bahawul Din, Kot Duffadar, 82	1 Jan. 16	No. 525, 12 May 16.
Do.	Khan Gul, Naik, 694	1 Jan. 16	No. 187, 18 Feb. 16.
Do.	Mahomed Khan, Naik, 843	1 Jan. 16	No. 187, 18 Feb. 16.
11th Do.	Painda Khan, Naik, 831	1 Apl. 15	No. 426, 8 May 15.
Do.	Rafi-ud-Din, Lance-Naik, 1526	1 Apl. 15	No. 426, 8 May 15.
Do.	Chedu Beg, Driver, 605	1 Apl. 15	No. 426, 8 May 15.
Do.	Faqir Mahomed, Driver, 1315	1 Apl. 15	No. 426, 8 May 15.
Do.	Shubrati, Driver, 986	1 Apl. 15	No. 426, 8 May 15.
Do.	Hiram Baksh, Naik	14 May 15	No. 563, 12 June 15.
Do.	Ibrahim Khan, Kot-Dafadar, 604	7 Aug. 17	No. 1360, 17 Aug. 17.
13th M. Corps	Kashi Ram, Naik, 7112	11 Jan. 18	No. 9, 1 Jan. 18.
14th Do.	Kaim Din, Driver, 1127	17 Aug. 17	No. 1360, 17 Aug. 17.
Do.	Mullah Singh, Naik, 644	19 July 18	No. 1629, 19 July 18.
Do.	Gul Zaman, Lance-Naik, 198	6 Sep. 18	No. 2060, 6 Sep. 18.
19th Do.	Ali Akbar, Naik, 780	17 Aug. 17	No. 1360, 17 Aug. 17.
20th Do.	Sahib Din, Naik, 371	17 Aug. 17	No. 1357, 17 Aug. 17.
22nd Do.	Ishar Singh, Quartermaster-Dafadar, 1309	19 July 18	No. 1629, 19 July 18.
23rd Do.	Khair Din, Driver, 931	17 Aug. 17	No. 1360, 17 Aug. 17.
26th Do.	Amroo, Driver, 501	17 Aug. 17	No. 1357, 17 Aug. 17.
27th Do.	Rup Singh, Driver, 862	1 Jan. 16	No. 563, 12 June 15.
Do.	Muhammad Husain, Driver, 1126	12 May 16	No. 525, 12 May 16.
30th Do.	Wilayat Khan, Driver, 1299	10 Dec. 15	No. 1248, 10 Dec. 15.
32nd Do.	Mir Said, Lance-Naik, 1198	17 Aug. 17	No. 1360, 17 Aug. 17.
Do.	Samundar Khan, Jemadar	17 Aug. 17	No. 1360, 17 Aug. 17.
Do.	Bagh Hussain, Driver, 1693	1 Jan. 18	No. 9, 1 Jan. 18.
33rd Do.	Abdulla, Driver, 2062	3 June 15	No. 528, 3 June 15.
35th Do.	Jahan Khan, Jemadar	19 July 18	No. 1629, 19 July 18.
38th Do.	Sundar Singh, Lce.-Naik, 319	1 Jan. 18	No. 528, 3 June 15.
50 Camel Corps	*Nasir Mahommed, Risaldar*	*26 June 08*	*No. 527, 26 June 08*
Divl. Train, 33rd Transport Corps.	Sahib Din, Naik, 980	29 Oct. 15	No. 1081, 29 Oct. 15.
I. M. S. D.	Abdus Sattar Khan, 2nd gr. Hospl. Asst., 986	1 Jan. 10	No. 4, 1 Jan. 10.
Do.	Nabi Ahmad Sidiki, 1st cl. Sub. Asst. Surgn., 695	9 Apl. 14	No. 347, 9 Apl. 14.
Do.	Gaurie Shankar, Senior Sub-Asst. Surgn., 1st Class	4 May 15	No. 563, 12 June 15.
Do.	Zafar Husain, Sub-Asst. Surgn., 1171, 3rd Class	4 May 15	No. 563, 12 June 15.
Do.	Wadhawa, Naik, 4007 (112 Indian Field Ambulance)	4 May 15	No. 563, 12 June 15.
Do.	Mahadeo Parshad, Sr. Sub-Asst. Surgn., 2nd cl.	3 June 15	Lon. Gaz., 23 June 15.
Do.	Narayanparshad Sukul, Sub-Asst. Surgn., 1st cl.	3 June 15	Lon. Gaz., 23 June 15.
Do.	Mohan Lal, Sub-Assistant Surgeon, 1st cl., 854	19 Nov. 15	No. 1160, 19 Nov. 15.
Do.	Barkatullah, Sub-Assistant Surgeon, 1st cl., 1080	19 Nov. 15	No. 1160, 19 Nov. 15.
Do.	Fazal Ahmad, Sub-Assistant Surgeon, 3rd cl., 316	19 Nov. 15	No. 1160, 19 Nov. 15.
Do.	T. Narayanasami Saminathan Nayadu, Sr. Sub-Assistant Surgeon, 1st cl.	17 Dec. 15	No. 1285, 17 Dec. 15.
Do.	Pohlo Ram, Sub-Asst. Surgn., 1st cl., 772	1 Jan. 16	No. 187, 18 Feb. 16.
Do.	Kishan Singh, Sub-Asst Surgn., 2nd cl., 1246	1 Jan. 16	No. 187, 18 Feb. 16.
Do.	Mathura-parshad Sarswit, Sub-Asst. Surgn., 3rd cl., 1360	1 Jan. 16	No. 187, 18 Feb. 16.
Do.	Bhagwan Das, Sub-Asst. Surgn., 2nd cl.	12 May 16	No. 525, 12 May 16.
Do.	Bishan Singh, Sub-Asst. Surgn., 1st cl.	12 May 16	No. 525, 12 May 16.
Do.	Mulchand Sharma, Sub-Asst. Surgn., 1st cl.	12 May 16	No. 525, 12 May 16.
Do.	Muhammad Umar, Sub-Asst. Surgn., 1st cl., 937	. . .	No. 313, 2 Apl. 15.
Do.	Tika Ram, Sub-Asst. Surgn., 1st cl., 917	3 June 16	No. 849, 29 July 16.
Do.	Jaswant Singh, Sub-Asst. Surgn., 1st cl., 1074	18 Nov. 16	No. 1388, 18 Nov. 16.
Do.	Karta Ram, Sub-Asst. Surgn., 1st cl., 1105	18 Nov. 16	No. 1388, 18 Nov. 16.
Do.	Sudama Ram, Sub-Asst. Surgn., 2nd cl., 1271	18 Nov. 16	No. 1388, 18 Nov. 16.
Do.	Chuhar Singh, Sub-Asst. Surgn., 2nd cl., 1164	18 Nov. 16	No. 1388, 18 Nov. 16.
Do.	Harnam Singh, Sub-Asst. Surgn., 3rd cl., 1329	18 Nov. 16	No. 1388, 18 Nov. 16.
Do.	Abdul Ghafur, Sub-Asst. Surgn., 1st cl, 910	1 Feb. 17	Lon. Gaz., 1 Feb. 06.
Do.	Saivid Muhammad Ejaz, Sub-Asst. Surgn., 1st cl., 1006	1 Feb. 17	Lon. Gaz., 1 Feb. 06.
Do.	Sudhiriswar Sen Gupta, 3rd cl. Sub-Asst. Surgeon, 1402½	17 Aug. 17	No. 1360, 17 Aug. 17.
Do.	Mubarak Shah Khan, 3rd cl. Sub-Asst. Surgeon, 1372	17 Aug. 17	No. 1360, 17 Aug. 17.
Do.	Rawail Singh 1st cl. Sub-Asst. Surgeon, 798	17 Aug. 17	No. 1360, 17 Aug. 17.
Do.	Suraj Pal, 3rd cl. Sub-Asst. Surgeon, 1406	17 Aug. 17	No. 1360, 17 Aug. 17.
Do.	Reuben, Benjamin, 1st class Sub-Asst. Surgn.	17 Aug. 17	No. 1357, 17 Aug. 17.
Do.	Saiyid Abdul Basit, 1st class Sub-Asst. Surgn.	17 Aug. 17	No. 1357, 17 Aug. 17.

INDIAN DISTINGUISHED SERVICE MEDALLISTS—*contd.*

NOTES.—The names in italibs are *italics* of members on the pension establishment.

Corps.	Names and Rank.	Date.	Authority. A. D. Notfn.
I. S. M. D.	Cornelius Sundararaj Gopal, 2nd class Sub-Asst. Surgeon	17 Aug. 17	No. 1357, 17 Aug. 17.
Do.	Ghulam Haidar, 1st cl. Sub-Asst. Surgn., 980	1 Jan. 18	No. 9, 1 Jan. 18.
Do.	Abdul Majid, 1st cl. Sub-Asst. Surgn., 1083	1 Jan. 18	No. 9, 1 Jan. 18.
Do.	Chintaman Gopal Gogate, 2nd cl. Sub-Asst. Surgn., 374	1 Jan. 18	No. 9, 1 Jan. 18.
Do.	Wahidyar Khan, 1st cl. Sub-Asst. Surgn., I.S.M.D.	1 Jan. 18	No. 9, 1 Jan. 18.
*	Ali-al-Mansur, Arab Scout Officer, Mesopotamia	20 July 17	No. 1171, 21 July 17.
Do.	Mehdi Hassan Khan, 1st cl. Civil Sub-Asst. Surgn., 352	12 July 18	No. 1572, 12 July 18.
Do.	Bir Singh, 2nd cl. Senior Sub-Asst. Surgn.	12 July 18	No. 1572, 12 July 18.
Do.	Muhammad Ishak, Isl cl. Sub-Asst. Surgn. 924	19 July 18	No. 1629, 19 July 18.
Do.	Selvadori Jasudasan, 2nd cl. Senior Sub.-Asst. Surgn.	19 July 18	No. 1629, 19 July 18.
Do.	Saif-ud-Din, 1st cl. Senior Sub-Asst. Surgn., *Khan Sahib*	20 Sep. 18	No. 2185, 20 Sep. 18.
Do.	Har Parshad, 2nd cl. Senior Sub-Asst. Surgn., *Rai Sahib*	20 Sep. 18	No. 2185, 20 Sep. 18.
Army Bearer Corps.	Madan Singh, Bearer, 4349	1 Jan. 16	No. 187, 18 Feb. 16.
Do.	Mangli, Lance-Naik, 8037	3 June 15	No. 187, 18 Feb. 16.
Do.	Khushali, Naik, 7017, Army Bearer Corps	4 May 15	No. 187, 18 Feb. 16.
Do.	Surjoo, Lance-Naik, 11018, Army Bearer Corps	4 May 15	No. 187, 18 Feb. 16.
Do.	Maston Singh, Bearer, 4397	1 Jan. 16	No. 187, 18 Feb. 16.
Do.	Ram Charan, Lce.-Naik, 7074	1 Jan. 16	No. 187, 18 Feb. 16.
Do.	Beni, Bearer, 7204	1 Jan. 16	No. 187, 18 Feb. 16.
Do.	Chabi, Bearer, 7259	1 Jan. 16	No. 187, 18 Feb. 16.
Do.	Ram Lal, Bearer, 7357	12 May 16	No. 525, 12 May 16.
Do.	Surjoo, Lce.-Naik, 7032	. . .	No. 313, 2 Apl. 15.
Do.	Ram Sabatu, Bearer, 8316	. . .	No. 313, 2 Apl. 15.
Do.	Achroo, Naik, 3039	. . .	No. 313, 2 Apl. 15.
Do.	Shivcharan, Bearer, 7208	12 May 16	No. 525, 12 May 16.
Do.	Budi Ram, Stretcher Bearer, 4539	. . .	No. 736, 1 July 16.
Do.	Fateh Muhammad, Naik, 4567	. . .	No. 736, 1 July 16.
Do.	Habbu, Lce-Havildar, 7010	3 June 16	No. 849, 29 July 16.
Do.	Fateh Mahomed, Naik, 4567	3 June 16	No. 849, 29 July 16.
Do.	Dhami, Lce-Naik, 4263	3 June 16	No. 849, 29 July 16.
Do.	Jaisi Ram, Bearer, 3260	3 June 16	No. 849, 29 July 16.
Do.	Budi Ram, Bearer, 4539	3 June 16	No. 849, 29 July 16.
Do.	Beni, Bearer, 7198	3 June 16	No. 849, 29 July 16.
Do.	Budba, Bearer, 7314	3 June 16	No. 849, 29 July 16.
Do.	Fakir Chand, Bearer, 3519	3 June 16	No. 849, 29 July 16.
Do.	M. ti, Bearer, 7257	3 June 16	No. 849, 29 July 16.
Do.	Sitar Mohammad, Bearer, 2408	3 June 16	No. 849, 29 July 16.
Do.	Gunga Singh, Bearer, 4614	3 June 16	No. 849, 29 July 16.
Do.	Beni, Lce-Naik, 8032	18 Nov. 16	No. 1388, 18 Nov. 16.
Do.	Ram Singh, Bearer, 3513	18 Nov. 16	No. 1388, 18 No. 16.
Do.	Fatha, Bearer, 3369	18 Nov. 16	No. 1388, 18 No. 16.
Do.	Mahomed Khan, Bearer, 4304	18 Nov. 16	No. 1388, 18 No. 16.
Do.	Pitam, Berar, 7405	18 Nov. 16	No. 1388, 18 No. 16.
Do.	Komil, Bearer, 7159	18 Nov. 16	No. 1388, 18 No. 16.
Do.	Sher Ahmed, Naik 1495	12 June 17	No. 868, 2 June 17.
Do.	Bagannah (II) Bearer, 9216	9 June 17	No. 894, 2 June 17.
Do.	Latchman (IV), Bearer, 9236	9 June 17	No. 894, 9 June 17.
Do.	Muhammed Ismail, Bearer, 9239	9 June 17	No. 894, 9 June 17.
Do.	Muhamm d Kashim, Bearer, 9240	9 June 17	No. 894, 9 June 17.
Do.	Nannu, Lance-Havildar, 7007	17 Aug. 17	No. 1360, 17 Aug. 17.
Do.	Misri, Naik, 7054	17 Aug. 17	No. 1360, 17 Aug. 17.
Do.	Mahabir, Bearer, 8192	17 Aug. 17	No. 1360, 17 Aug. 17.
Do.	Hansraj, Bearer, 7747	17 Aug. 17	No. 1360, 17 Aug. 17.
Do.	Karim Bakhsh, Lance-Naik (Acting Naik)	26 Oct. 18	No. 2536, 26 Oct. 18.
Do.	Kanjan, Lance-Naik, 7161	26 Oct. 18	No. 2536, 26 Oct. 18.
Do.	Ramboo Lal, Bearer, 18171	17 Aug. 17	No. 1360, 17 Aug. 17.
Do.	Sheo Lal, Bearer, 8111	17 Aug. 17	No. 1360, 17 Aug 17.
Do.	Tulsi, Lance-Naik, 7043	17 Aug. 17	No. 1360, 17 Aug. 17.
Do.	Udai Ram, Bearer, 18528	17 Aug. 17	No. 1360, 17 Aug. 17.
Do.	Besakhi, Lance-Havildar, 3042	1 Jan. 18	No. 9, 1 Jan. 18.
Do.	Data Din, Bearer, 7294	1 Jan. 18	No. 9, 1 Jan. 16.
Do.	Mangta, Naik, 7018	1 Jan. 18	No. 9, 1 Jan. 18.
Do.	Ram Lal, Bearer, 17642	1 Jan. 18	No. 9, 1 Jan. 18.
Do.	Sunnoo, Bearer, 213	1 Jan. 18	No. 9, 1 Jan. 18.
Do.	Venketasawmy, Bearer, 9249	19 July 18	No. 1629, 19 July 18.
Do.	Angiah, Bearer, 9257	19 July 18	No. 1629, 19 July 18.
Do.	Rathan, Bearer, 9179	19 July 18	No. 1629, 19 July 18.
Do.	Agumi Mistri, Lance-Havilder, 8785	19 July 18	No. 1629, 19 July 18.
Army Hospital Corps.	Bansi, 2nd Grade Ward Servant, 2157	19 July 18	No. 1629, 19 July 18.
Postal Corps	Muzaffar Ahmad, Subadar	3 June 18	No. 1180, 3 June 18.
Do.	Bishan Dass, Subadar	3 June 18	No. 1180, 3 June 18.
Do.	Abdul Karim	25 Oct. 18	No. 2538, 25 Oct. 18.

* Hony. Reward.

Corps.	Names of Rank.	Date.	Authority.
	INDIAN DISTINGUISHED SERVICE MEDALISTS—*contd.*		
	NOTE.—The names in *italics* are those of members on the pension establishment.		
			A. D. Notfn.
Postal Corps	Bindraban	25 Oct. 18	No. 2538, 25 Oct. 18.
Do.	Shah Ghufar Khan	25 Oct. 18	No. 2538, 25 Oct. 18.
Chauffeur	Sailendra Nath Basu, Naick	. . .	No. 348, 31 Mar. 16.
S. A. Ind. Bearer Corps.	Lal Bahadur Jugmohan Singh, Pte., 74	21 Nov. 16	No. 676, 5 May 17.
	Imperial Service Troops.		
Alwar Infantry	Ghulham Hasan Khan, Havildar, 13	3 June 18	No. 1180, 3 June 18.
Do.	Malang Khan, Havildar, 89	3 June 18	No. 1180, 3 June 18.
Alwar Lrs.	Kesri Singh, Jemadar, 777	1 Jan. 17	No. 676, 5 May 17.
Do.	Muhammad Ali Khan, Risaldar	17 Aug. 17	No. 1357, 17 Aug. 17.
Alwar	Zulfeqar Ali, Subadar	3 June 18	No. 1178, 3 June 18.
Bikaner Camel Corps.	Bakhtawar Singh, Commandant	28 June 07	No. 586, 28 June 07.
Do.	Faiz Ali Khan, Sepoy, 115	24 Dec. 14	No. 1169, 24 Dec. 14.
Do.	Dulwant Singh, Sowar	13 Apl. 16	No. 849, 29 July 16.
Do.	Thakur Moti Singh, Asst. Comdt.	3 June 16	No. 639, 3 June 16.
Do.	Balwant Singh, Sowar, 900	3 June 16	No. 640, 3 June 16.
Do.	Sadul Singh, Jemadar	3 June 16	No. 641, 3 June 16.
Do.	Sugan Singh, Naick, 1326	3 June 16	No. 641, 3 June 16.
Do.	Sadul Singh, Jemadar	3 June 16	No. 849, 29 July 16.
Do.	Sugan Singh, Naik, 1326	3 June 16	No. 849, 29 July 16.
Do.	Thakur Balu Singh, Capt.	26 Apl. 17	Lon. Gaz., 26 Apl. 17.
Do.	Bhur Singh, Jemadar	3 June 18	No. 1180, 3 June 18.
Bharatpur Trans. Corps	Roda Khan, Jemadar	28 June 07	No. 586, 28 June 07.
Do.	Tek Chand, Jemadar	19 July 18	No. 1629, 19 July 18.
Bharatpur Infy.	Sanwaliya Singh, 2089	3 June 16	No. 849, 29 July 16.
Do.	Harmukh Singh, Jemadar	1 Feb. 17	Lon. Gaz., 1 Feb. 17.
Do.	Hira Naik, 2247	4 Oct. 18	No. 2342, 4 Ocr. 18.
Do.	Chatar Sen, Havildar, 1753	1 Feb. 17	Lon. Gaz., 1 Feb. 17.
	Nathi, Naik, 2020	4 Oct. 18	No. 2342, 4 Oct. 18.
Do.	Peary Lal, Havildar, 1926	1 Jan. 18	No. 9 Jan. 18.
Do.	Ram Chand, Naik, 1691	4 Oct. 18	No. 2342, 4 Oct. 18.
Do.	Zormal, Lance-Naik, 1817	4 Oct. 18	No. 2342, 4 Oct. 18.
1st Patiala Infantry.	Partap Singh, Jemadar	3 June 18	No. 1180, 3 June 18.
Do.	Peary Lal, Havildar, 1926	1 Jan. 18	No. 9, 1 Jan. 18.
Do.	Rattan Singh, Naik Tempy. Havildar, 344	26 Oct. 18	No. 2536, 26 Oct. 18.
Do.	Sewa Singh, Sepoy, 725	26 Oct. 18	No. 2536, 26 Oct. 18.
Do.	Bhagwan Singh, Capt.	26 Oct. 18	No. 2536, 26 Oct. 18.
Do.	Dharm Singh, Subadar	26 Oct. 18	No. 2536, 26 Oct. 18.
Bhavnagar Lancers.	Mohbat Sinhji, Risaldar	3 June 16	No. 639, 3 June 16.
Do.	Bapoo Bhai, Risildar	17 Aug. 17	No. 1360, 17 Aug. 17.
Burma Mtd. Infy.	Chanan Singh, Dafadar, 26	1 Jan. 18	No. 9, 1 Jan. 18.
Do.	Partab Singh, Jemadar	1 Jan. 18*	No. 9, 1 Jan. 18.
Do.	Gulzar Singh, Risaldar	26 July 18	No. 1703, 26 July 18.
Do.	Saif Ali, Dafadar, 15	26 July 18	No. 1703, 26 July 18.
Do.	Wariam Singh, Atg. Lance-Dafadar, 41	26 July 18	No. 1703, 26 July 18.
Do.	Savdeqar Singh, Sowar, 147	26 July 18	No. 1703, 26 July 18.
Do	Man Singh, Sowar, 80	26 July 18	No. 1703, 26 July 18.
Mysore I. S. Lrs.	Subbaraya Urs, Risaldar	3 June 15	No. 528, 3 June 15.
Do.	Syid Ahmed, Risaldar	3 June 16	No. 639, 3 June 16.
Do.	Kasi Rao Sindhay, Jemadar	3 June 16	No. 639, 3 June 16.
Do.	Venkoji Rao Powar, Jemadar	3 June 16	No. 639, 3 June 16.
Mysore Lancers.	Limbaji Power, Jemadar	3 June 18	No. 1180, 3 June 18.
Do.	Abdul Aziz Khan, Lance-Dafadar, 590	3 June 18	No. 1180, 3 June 18.
Do.	Madar Singh, Lance-Dafadar, 1123	3 June 18	No. 1180, 3 June 18.
Do.	Nahar Singh, Sowar 1892	3 June 18	No. 1180, 3 June 18.
1st Kashmir Mountain By.	Kummel Singh, Naik, 86	3 June 18	No. 1180, 3 June 18.
2nd Kashmir Rifl.	Mardan Ali, Subadar	3 June 15	No. 530, 3 June 15.
Do.	Billu, Sepoy, 1091	3 June 15	No. 530, 3 June 15.
Do.	Saif Ali, 6	3 June 15	No. 530, 3 June 15.
Do.	Madhu, Havildar, 2	3 June 15	No. 530, 3 June 15.
Do.	Bhagwan Singh, Sepoy, 9	3 June 15	No. 530, 3 June 15.

* Bar to Indian Distinguished Service Medal.

Corps.	Names of Rank.	Date.	Authority.
	INDIAN DISTINGUISHED SERVICE MEDALISTS—*contd.*		
	NOTE.—The names in italics are those of members on the pension establishment.		A. D. Notfn.
	Imperial Service Troops—*contd.*		
3rd Kashmir Rifl.	Ganga Ram, Sepoy, 2476	3 June 15	No. 530, 3 June 15.
Do.	Devi Singh, Sepoy, 1131	3 June 15	No. 530, 3 June 15.
Do.	Bambahadur, Jemadar	21 Nov. 16	No. 676, 5 May 17.
Do.	Harku, Havildar, 82	21 Nov. 16	No. 676, 5 May 17.
Do.	Indru, Lance Naik, 694	21 Nov. 16	No. 676, 5 May 17.
3rd Gwalior I. S. Infy.	Raghojirao Khandolkar, Lt.	1 Jan. 18	No. 9, 1 Jan. 18.
Do.	Wamanrao Sawant, Capt.	3 June 18	No. 1180, 3 June 18.
4th Gwalior Infy.	Bhaskar Rao Deodhar, Capt.	3 June 18	No. 1180, 3 June 18.
Do.	Niyaz Ali, Naik, 1956	3 June 18	No. 1180, 3 June 18.
3rd Gwalior Infy.	Waman Gopal Gokhale, Lt.	3 June 18	No. 1180, 3 June 18.
Jaipur Trans. Corps.	Hukam Singh, Major	19 July 18	No. 1631, 19 July 18.
Do.	Kesar Singh, 1st Grade Military Sub-Asst. Surgn.	19 July 18	No. 1631, 19 July 18.
Do.	Ram Dewa, Lance-Dafadar, 3936	19 July 18	No. 1631, 19 July 18.
Do.	Khema, Driver, 4448	19 July 18	No. 1631, 19 July 18.
Do.	Nawab Khan, Driver, 6339	19 July 18	No. 1631, 19 July 18.
Do.	Fakir Muhammad, Jemadar	19 July 18	No. 1629, 19 July 18.
Do.	Jawana, Lance Dafadar, 3810	19 July 18	No. 1629, 19 July 18.
Jind Infy.	Gugar Singh, Havildar, 1367	3 June 15	No. 530, 3 June 15.
Do.	Sadhu Singh, Sepoy, 2276	3 June 15	No. 530, 3 June 15.
Do.	Lakha Singh, Sepoy, 2287	3 June 15	No. 530, 3 June 15.
Do.	Bihara Singh, Capt.	3 June 16	No. 849, 29 July 16.
Do.	Jagta Singh, Sepoy, 2030	3 June 16	No. 849, 29 July 16.
Do.	Ghulam Haidar, Havildar, 1927	21 Nov. 16	No. 676, 5 May 17.
Do.	Bahal Singh, Subadar	3 June 18	No. 1180, 3 June 18.
Do.	Bishen Singh, Subadar	3 June 18	No. 1180, 3 June 18.
Jodhpur Lrs.	Chiman Singh, Dafadar, 151	3 June 16	No. 849, 29 July 16.
Do.	Bahadur Singh, Sowar, 1086	19 July 18	No. 1629, 19 July 18.
Do.	Aidan Singh, Dafadar, 1186	19 July 18	No. 1629, 19 July 18.
Rampur Infy.	Muhammad Dastgir Khan, Havildar, 75	3 June 18	No. 1180, 3 June 18.
Do.	Abbas Ali Khan, Subadar	3 June 18	No. 1180, 3 June 18.
Do.	Dilwar Husain Khan, Subadar	3 June 18	No. 1180, 3 June 18.
Do.	Naiz Ahmad Khan, Havildar, 110	3 June 18	No. 1180, 3 June 18.
Indore State Transport.	Ganpat Rao, Compounder, 226	29 Oct. 15	No. 1081, 29 Oct. 15.
Indore I. S. Transport.	Dasharat Singh, Driver, 128	29 Oct. 15	No. 1081, 29 Oct. 15.
Kotla Malar, I. S. Sappers.	Chagi Singh, Havildar, Art. 268	17 Aug. 17	No. 1360, 17 Aug. 17.
Tehri I. S. Troops.	Sant Ram, Artcr, Havildar	1 Jan. 18	No. 9, 1 Jan. 18
Faridkot I. S. Sappers and Miners.	Chait Singh, Jemadar	3 June 18	No. 1180, 3 June 18.
Do.	Moti Singh, Jemadar	3 June 18	No. 1180, 3 June 18.
Faridkot S. and M.	Thana Singh, Naik, 654	3 June 18	No. 1180, 3 June 18.
1st Hyderabad Lancers.	Muhammad Fakruddin, Jemadar	26 Oct. 18	No. 2536, 26 Oct. 18.
Do.	Abdul Basita Khan, Dafadar, 557	26 Oct. 18	No. 2536, 26 Oct. 18.
	Militia and Levies.		
Khyber Rif.	Mohibulla, Subadar, I.O.M.	28 June 07	No. 586, 28 June 07.
Do.	Muhammad Ali Khan, Subadar-Major, *Khan Sahib*, *Sardar Bahadur*, I.O.M.	28 June 07	No. 1081, 29 Oct. 15.
Do.	Jawas, Subadar-Major, *Bahadur*, I.O.M.	26 June 08	No. 527, 26 June 08.
Do.	Gul Zir, Jemadar	9 Apl. 15	No. 339, 9 Apl. 15.
Do.	Taus Khan, Jemadar	16 Dec. 16	No. 1515, 16 Dec. 16.
Do.	Shamal, Havildar, 1888	16 Dec. 16	No. 1515, 16 Dec. 16.
Kurram Militia	Muhammad Hussain, Subadar, *Sardar Bahadur*	26 June 08	No. 527, 26 June 08.
Do.	Akbar Ali, Subadar	2 Jan. 11	No. 5, 2 Jan. 11.
Do.	Said Asghar, Subadar	25 July 14	No. 670, 25 July 14.
Do.	Faqir Hassan, Jemadar	5 Feb. 15	No. 143, 5 Feb. 15.
N. W. Militia	Khusal Khan, Naik, 2268	1 Jan. 08	No. 1, 1 Jan. 08.
Do.	Malham Khan, Subadar	23 June 10	No. 522, 23 June 10.
Do.	*Zar Khan, Subadar*	*23 June 10*	*No. 522, 23 June 10.*
Do.	Zarim, Lce Dafadar, 368	20 Sep. 12	No. 911, 20 Sep. 12.
Do.	Sardar Khan, Subadar	22 Aug. 13	No. 710, 22 Aug. 13.
Do.	Nawab Khan, Havildar, 1916	22 Aug. 13	No. 710, 22 Aug. 13.
Do.	Zalim, Jemadar	26 Mar. 15	No. 302, 26 Mar. 15.
Do.	Zarif Khan, Jemadar	26 Mar. 15	No. 302, 26 Mar. 15.

INDIAN DISTINGUISHED SERVICE MEDALISTS—*contd.*

NOTE.—The names in *italics* are those of members on the pension establishment.

Militia and Levies—*contd.*

Corps.	Names and Rank.	Date.	Authority. A. D. Notfn.
N. W. Militia	Khajai, Naick, 3025	26 Mar. 15	No. 302, 26 Mar. 15.
Do.	Tor Khan, Sepoy, 5042	26 Mar. 15	No. 302, 26 Mar. 15.
Do.	Amir Khan, Sepoy, 5058	26 Mar. 15	No. 302, 6 Mar. 15.
Do.	Zamir Ullah, Sepoy, 4509	26 Mar. 15	No. 302, 26 Mar. 15.
Do.	Pat Khan, Jemadar	9 Apl. 15	No. 339, 9 Apl. 15.
Do.	Khial Din, Cr. Havildar, 4789	9 Apl. 15	No. 339, 9 Apl. 15.
Do.	Khan Dais, Naick, 4223	9 Apl. 15	No. 339, 9 Apl. 15.
Do.	Tawahir Din, Sepoy, 4503	9 Apl. 15	No. 339, 9 Apl. 15.
Do.	Ain-ud-Din, Sepoy 5523	9 Apl. 15	No. 339, 9 Apl. 15.
Do.	Mir Shah Jan, Sepoy, 5538	9 Apl. 15	No. 339, 9 Apl. 15.
Do.	Mohi Khan, Sepoy, 2135	9 Apl. 15	No. 339 9 Apl. 15.
Do.	Zar Khan, Lance-Naick, 2284	3 June 15	No. 529, 3 June 15.
Do.	Syad Akhmad, Drill Naick, 2203	3 June 15	No. 529, 3 June 15.
Do.	Malang Khan, Havildar, 167	24 Aug. 16	No. 1514, 16 Dec. 16.
S. W. Militia	Kawan Khan, Subadar, I.O.M.	1 Jan. 09	No. 5, 1 Jan. 09.
Do.	Gul Anar, late Havildar, 1931	1 Jan. 09	No. 5, 1 Jan. 09.
Do.	Muhammad Ayub Khan, late Jemadar	1 Jan. 10	No. 4, 1 Jan. 10.
Do.	Gulaki, Havildar, 1553	15 Aug. 14	No. 731, 15 Aug. 14.
Do.	Ghulam, Naick, 2412	15 Aug. 14	No. 731, 15 Aug. 14.
Do.	Tik Khan, Jemadar	16 Dec. 16	No. 1515, 16 Dec. 16.
Do.	Zari Gul, Jemadar	16 Dec. 16	No. 1515, 16 Dec. 16.
Do.	Maqam Khan, Subadar	9 Mar. 17	No. 358, 10 Mar. 17.
Zhob Militia	Muhammad Ismail Khan, Risaldar-Major	28 June 07	No. 586, 28 June 07.
Do.	Sham Beg, Sepoy	28 June 07	No. 586, 28 June 07.
Do.	Hunar, Sepoy	1 Jan. 08	No. 1, 1 Jan. 08.
Do.	Mitha Khan, Risaldar	1 Jan. 09	No. 5, 1 Jan. 09.
Do.	Ahmad Mir, Risaldar-Major	28 June 07	
Mekran Levy Corps.	Sayyid Muhammad, Jemadar	16 Sep. 16	No. 1095, 16 Sep. 15.
Do.	Gwaram Khan, Jemadar	16 Sep. 16	No. 1095, 16 Sep. 15.
Do.	Nabi Bakhsh, Dafadar, 251	16 Sep. 16	No. 1095, 16 Sep. 15.
Do.	Dad Khuda, Lnce-Naik, 257	16 Sep. 16	No. 1095, 16 Sep. 15.
Chagai Levy Corps.	Radho, Jemadar	17 Aug. 17	No. 1356, 17 Aug. 17.
Sarhadi Levy Corps.	Edookhan, Risaldar	17 Aug. 17	No. 1356, 17 Aug. 17.
Sistan Levy Corps.	Barat Ali, Jemadar	17 Aug. 17	No. 1356, 17 Aug. 17.
Do.	Juma Jalal, Jemadar	17 Aug. 17	No. 1356, 17 Aug. 17.

Military Police.

Corps.	Names and Rank.	Date.	Authority.
Naga Hills	Arjun Rai, Subadar	28 June 07	No. 586, 28 June 07.
Do.	Harksing Rai, Jemadar	9 Apl. 14	No. 347, 9 Apl. 14.
Lakhimpur Mily. Police Battn.	Jangbir Lama, Jemadar	6 Sep. 12	No. 854, 6 Sep. 12.
Do.	Dalbahadur Thapa, Havildar, 1120	6 Sep. 12	No. 854, 6 Sep. 12.
Lushai Hills	Jamansing Rai, Havildar, 920	15 Nov. 12	No. 1077, 15 Nov. 12.
Mily. Police Battn.	Nainsing Sahi, Havildar, 970	15 Nov. 12	No. 1077, 15 Nov. 12.
Burma Mily. Police.	Jangia Lama, Subadar	27 Feb. 14	No. 208, 27 Feb. 14.
Do.	Narjit Limbu, Naick, 1303	27 Feb. 14	No. 208, 27 Feb. 14.
Do.	Jangdip Limbu, Lance Naick, 2112	27 Feb. 14	No. 208, 27 Feb. 14.
Do.	Parbal Thapa, Lance Naick, 2598	27 Feb. 14	No. 208, 27 Feb. 14.
Do.	Alla Sa, Lance Naick, 4771	27 Feb. 14	No. 208, 27 Feb. 14.
Do.	Shamdhoj Rai, Sepoy, 4072	27 Feb. 14	No. 208, 27 Feb 14.
Do.	Maitraj Rai, Rifleman, 2175	3 June 16	No. 849, 29 July 16.
Do.	Khan Ahmed, Jemadar	21 Nov. 16	No. 676, 5 May 17.
Do.	Bhim Singh, Sepoy, 2536	17 Aug. 17	No. 1[illegible]60, 17 Aug. 17.
Do.	Muhammad Khan, Lance-Naik, 4816	3 June 18	No. 1180, 3 June 18.
Do.	Nar Singh, Sepoy, 2684	16 Aug. 18	No. 1885, 16 Aug. 18.
Malay State Guides.	Bogh Singh, Jemadar	5 Aug. 16	No. 887, 5 Aug. 16.
Do.	Santa Singh, Naik, 1632	Do.	No. 887, 5 Aug. 16.
Aden Troop	Diwan Ali Khan, Jemadar	5 Aug. 16	No. 887, 5 Aug. 16.
Do.	Najaf Khan, Lance-dafadar, 297	2 Nov. 17	No. 1887, 2 Nov. 17.
Do. (Chin Hills Bn.)	Budhiman Thapa, Rifleman, 1348	1 Jan. 16	No. 563, 12 Jun. 16.
Assam Mily. Police.	Gajman Rai, Jemadar	18 Nov. 16	No. 1386, 18 Nov. 16.
Front. Constabulary.	Zaffar Khan, Subadar	2 Nov. 17	No. 1887, 2 Nov. 17.
Do.	Mad Ashgar, Jemadar	2 Nov. 17	No. 1887, 2 Nov. 17.

OFFICERS HOLDING RANK IN THE ARMY, ETC., AUTHORISED TO WEAR FOREIGN ORDERS ON ALL OCCASIONS.

Adams, Lt.-Col. F., D.S.O., 20th Horse.	Ordre de la Couronne Officier (*Belgium*). Croix de Guerre (*Belgium*).
Adye, Bt.-Col. D. R., 98th Infy.	St. Anne, 3rd class, with swords (*Russia*).
Ali Dost, Subdr., 106th Prs. . .	Croix de Guerre (*France*).
Allanson, Bt. Lt.-Col. C. J. L., C.I.E., D.S.O., 6th G. R.	Croix de Guerre (*France*).
Anderson, Bt.-Col. W. C., C.M.G., 1st G. R.	Order of St. Anne, 3rd class, with swords (*Russia*).
Andrew, Lt.-Col. A. W., C.M.G.	Order of St. Stanislas, 3rd class, with swords (*Russia*).
Andrew, Lt.-Col. F. A., D.S.O., 128th Prs.	Croix de Guerre (*France*).
Arrindell, Lt., C.D.S.M., M.C., I.A.R.O.	Croix de Guerre (*France*).
Auchinleck, Maj. C. J. E., D.S.O., 62nd Punj.	Croix de Guerre (*France*).
Aylmer, Maj. G., M. C., S. & T. Corps.	Legion of Honour Croix de Guerre (*France*).
Badcock, Lt.-Col. F. F., D.S.O.	Order of St. Stanislas, 3rd class, with swords (*Russia*).
Bainbridge, Lt.-Col. W. F., C.M.G., D.S.O., I.A.	Nile, 4th class, (*Egypt*).
Baird, Bt. Lt. Col. H.B.D., C.M.G. D.S.O., 12th Cav.	Croix de Guerre (*France*).
Bakhat Bahadur Adhikari, Subdr., 1-9th G. R.	Cross of St. George, 4th class (*Russia*).
Ball, Maj. L P., 5th Infy. . .	Croix de Guerre (*France*).
Barratt, Maj.-Genl. W. C., C.B., D.S.O.	Brilliant Star of Zanzibar, 2nd class.
Barrow, Genl. *Sir* E. G., G.C.B., G.C.S.I.	Medjidieh, 5th class.
Barrow, Maj.-Genl. *Sir* G. deS., K.C.M.G., C.B.	Legion of Honour Croix de Commandeur (*France*). Order of the Nile, 2nd class (*Egypt*).
Battine, Lt.-Col. R. St. C., D.S.O, 21st Cav.	St. Stanislas, 2nd class (*Russia*).
Battye, Lt.-Col. W. R., D.S.O., M.B., F.R.C.S., I.M.S.	Legion of Honour Croix de Chevalier (*France*).
Bax, Col. W. I. (retd.) . . .	Bahaduri-i-Afghania, 2nd. class (*Afghanistan*).
Bennett, Maj. E.E.S., 66th Punj	Croix de Guerre (*France*).
Bhim Sing Thapa, M.C., I.O.M., Sub.-Maj., 3rd G. R.	White Eagle, 5th class, with swords (*Serbia*).
Bignell, Bt.-Maj. R. L., D.S.O., 41st Dogras.	Order of St. Stanislas, 3rd class, with swords (*Russia*).
Birdwood, Lt.-Genl. Sir W. R., K.C.B., K.C.S.I., K.C.M.G., C.I.E., D.S.O., 𝔄.𝔇.ℭ., I.A.	Legion of Honour (Grand Officier) (*France*). Legion of Honour Croix de Guerre (*France*). Grand Officier de l'Ordre de la Couronne (*Belgium*). Croix de Guerre (*Belgium*). Order of the Nile (2nd class) (*Egypt*).
Bliss, Lt.-Col. J. A., D.S.O., M.V.O., 64th Prs.	White Eagle, 4th class, with swords (*Serbia*).
VC Boisragan, Col. G. H. . . I.A.	Nile, 4th class (*Egypt*).
Boome, Bt.-Col. E. H., C.B., C.M.G.	Order of St. Anne, 3rd class, with sword (*Russia*).
Bradley, Bt.-Col. R. J., M.B., I.M.S.	White Eagle, 4th class, with swords (*Serbia*).
Bridges Lt.-Col. A. H., D.S.O., 116th Mah.	Croix de Guerre, (*Belgium*).
Brooking, Maj.-Genl., *Sir* H. T., K.C.B., K.C.M.G.	Karageorge, 3rd class with Sword (*Serbia*).
Brown, Bt. Lt.-Col. W., S. & T. Corps.	Karageorge, 4th class, with swords (*Serbia*).
Browne, Lt.-Col. A. P. . . .	St. Anne, 3rd class, with swords (*Russia*).
Browne, Bt.-Col. H. J. P., 5th G.R.	White Eagle, 3rd class, with swords (*Serbia*).
Brownlow, Lt.-Col. d'A. C., C.M.G., C.I.E.	Order of St. Stanislas, 2nd class, with swords (*Russia*).
Buckland, Maj. G. C. B., D.S.O., M.C., 8th, G. R.	Croix de Guerre (*France*).
Burrows, Capt. H. M., 51st Sikhs.	White Eagle, 5th class, with swords (*Serbia*).
Buswell, Lt. G., I.A.R.O. . .	Croix de Guerre, (*Belgium*).
Butler, Bt. Maj. R. B., M.C., 30th Lrs.	White Eagle, 5th class, with swords (*Serbia*).
Cahusac, Capt. C. F., D.S.O., 36th Horse.	White Eagle, 5th class, with swords (*Serbia*).
Cameron, *Hony.* Lt. A. J. . .	Medaille Militaire (*France*).

OFFICERS HOLDING RANK IN THE ARMY, ETC., AUTHORISED TO WEAR FOREIGN ORDERS ON ALL OCCASIONS—*contd.*

Campbell, Col. L. W. Y., C.M.G.	White Eagle, 3rd class, with swords (*Serbia*). Legion of Honour Comdr. (*France*.)
Campbell, Maj. W. N., C.M.G., D.S.O., I. A. (retd.).	Order of St. Vladimir, 4th class, with swords (*Russia*).
Carey, Maj. T. de B., 119th Infy.	White Eagle, 4th class, with swords (*Serbia*).
Carey, Col. O. W., C.M.G., I A.	Order of St. Anne, 3rd class, with swords (*Russia*).
Carleton, Lt.-Col. H. A., D.S.O., 90th Punj.	White Eagle, 4th class, with swords (*Serbia*).
Carnegy, Bt.-Col. C. G., M.V.O.	Order of St. Stanislas, 2nd class.
Carruthers, Col. R. A., C.B., C.M.G.	Legion of Honour, Officer (*France*).
Cassels, Lt.-Col. G. R., 123rd Rif.	Order of the Nile, 3rd class (*Egypt*).
Cassells, Col. R. A., C.B., D.S.O., 32nd Lrs.	White Eagle, 4th class, with swords (*Serbia*).
Chagatta, Jemadar, I.O.M., 1st S. & M.	Cross of St. George, 4th cl. (*Russia*).
Chauncy, Maj. C. H. K., 124th Infy.	Order of St. Stanislas, 3rd class, with swords (*Russia*).
Chenevix-Trench, Lt.-Col. G. F., C.I.E. (retd).	White Eagle, 4th class (*Serbia*). Order of the Crown Commendatori (*Italy*).
Chope, Maj. A. J. H., D.S.O., 2nd G. R.	Nile, 4th class (*Egypt*).
Christie, Capt. R. C., M.C., 102nd Grs.	White Eagle, 5th class, with swords (*Serbia*).
Churchill, Maj. W. M. (retd.)	Order of St. Anne, 3rd class (*Russia*).
Climo, Maj.-Genl. I. S. H., C.B., D.S.O., I. A.	Nile, 3rd class (*Egypt*).
Cobban, Maj. C. L. 43rd Regt.	Order of St. Stanislas, 2nd class.
V.C. Cobbe, Maj.-Genl. *Sir*, A. S., K.C.B., C.S.I., D.S.O.	Order of St. Maurice and St. Lazarus, Commander.
Colan, Lt.-Col. W. R. B., D.S.O., 67th Punj.	Order of St. Stanislas, 3rd class, with swords (*Russia*).
Combe, Bt., Lt.-Col. S. B., D.S.O., M.C., 47th Sikhs.	Legion d' Honneur Croix de Chevalier (*France*).
Compton, Maj. C. W. McG. 69th Punj.	White Eagle, 5th class, with swords (*Serbia*).
Cook, Lt.-Col. C. C., D.S.O.	White Eagle, 4th class, with swords (*Serbia*).
Cooper, Maj. G. C., M.C., 121st Prs.	Croix de Guerre (*France*).
Cooper, Lt.-Col. W. G. 34th Horse.	Order of St. Maurice and St. Lazarus, Cavalier.
V.C. Costello, Lt.-Col. E.W., C.M.G., D.S.O., 22nd Punj.	Croix de Guerre (*France*).
Cotton, Lt.-Col., W. L., 5th Infy.	Legion d'Honneur Croix d'Officier (*France*).
Cowper, Maj.-Gen. M., C.B., C.I.E.	White Eagle, 2nd class, with swords (*Serbia*).
Cox, Lt.-Genl. *Sir* H. V., K.C.B., K.C.M.G., C.S.I.	White Eagle, 2nd class, with swords (*Serbia*).
Cox, Lt. O., I.A.R.O.	Croix de Guerre, (*France*).
Crawford, Bt. Maj. J. D., D.S.O., M.C., 89th Punj.	White Eagle, 5th class, with swords (*Serbia*).
Creagh, Col. A. H. D., C.M.G., M.V.O.	White Eagle, 4th class, with swords (*Serbia*).
Crocker, Col. S. F., C.B.	Order of St. Stanislas, 2nd class, with swords (*Russia*).
Cullen, Lt.-Col. E. H. S., C.M.G., D.S.O., M.V.O., 34th Prs.	Order of St. Stanislas, 3rd class, with swords (*Russia*).
Cuming, Lt.-Col. R. J., D.S.O., 48th Prs.	White Eagle, 4th class, with swords (*Serbia*).
Cursetjee, Capt. H. J. M., M.B., I.M.S.	White Eagle, 5th class, with swords (*Serbia*).
Dale, Lt.-Col. G. A., C.M.G., 19th Punj.	Croix de Guerre (*France*).
Dalkesar Gurung, Subadar, I.O.M., 3rd G. R.	Cross of St. George, 4th class (*Russia*).
Davidson, Bt.-Col. S. R., C.M.G., 47th Sikhs.	St. Anne, 2nd class, with swords (*Russia*).
Davies, Col. C. H., C.M.G., D.S.O.	Order of St. Stanislas, 2nd class, with swords (*Russia*).
Daya Singh, Subadar, I.D.S.M., 30th Mn. Batty.	White Eagle, 5th class, with swords (*Serbia*).
Deas, Bt. Lt.-Col. L. J. M., M.B., F.R.C.S.E., I.M.S.	Croix de Guerre (*France*).
DeCourcy, Maj. *Hon.* M. W.R. D.S.O., 32nd Prs.	White Eagle, 5th class, with swords (*Serbia*).
De la Motte, Maj., R. B., D.S.O., S. & T. Corps.	White Eagle, 5th class, with swords (*Serbia*).
Delme-Radclife, Maj. S. A., (retd.).	Order of the Crown of Italy Cavalier.
Dent, Bt. Lt.-Col. W., D.S.O., 103rd L. Infy.	White Eagle, 5th class, with swords (*Serbia*).

OFFICERS HOLDING RANK IN THE ARMY, ETC., AUTHORISED TO WEAR FOREIGN ORDERS ON ALL OCCASIONS—*contd.*

Dick, Col. A. R., C.B.	Legion d'honneur Croix d'Officier.
Dobbs, Lt.-Col. C. F., D.S.O., 94th Infy.	St. Anne, 3rd class, with swords (*Russia*).
Donald, Capt. C. D., I. A. R. O.	Silver medal for Mily. valour (*Italy*).
Dopping-Hepenstal, Lt.-Col. Maj., M.E., D.S.O., 1st G. R.	Croix de Guerre (*France*).
Douglas, Col. J. A., C.M.G., C.I.E.	White Eagle, 3rd class, with swords (*Serbia*).
Douglas, Capt. S. R. (late I.M.S.) Retd.	Ordre de la Couronne Officier (*Belgium*).
Drummond, Lt.-Col. A. B.	Izzat-i-Afghania (*Afghanistan*).
Drummond, Maj.-Genl. *Sir* F. H. R., K.C.I.E., C.B.	Izzat-i-Afghania (*Afghanistan*).
Duckworth, Hony. Lt. E. V., I. M. Dept.	Croix de Guerre (*France*).
Dundas, Maj. W. L., 3rd G. R.	Order of the Nile, 4th class (*Egypt*).
Dunsford, Lt.-Col. F.P.S., C.I.E., 2nd. L. Infy.	Croix de Guerre (*France*).
Dyke, Lt.-Col. P. H., D.S.O., 130th Bal.	St. Stanislea, 3rd class, with swords (*Russia*).
Eadie, Maj. J. I., D.S.O., 97th Infy.	White Eagle, 5th class, with swords (*Serbia*).
Eastmead, Lt.-Col. C. S., 3rd G. R. [L].	Order of the Nile, 3rd class (*Egypt*).
Edwardes, Col. S. M., C.B., C.M.G., D.S.O., I. A.	Legion d'Honneur Croix d'Officier, (*France*).
Edwards, Lt.-Col. A. C., 102nd Grs.	White Eagle, 4th class, with swords (*Serbia*).
Edwards, Col. Fitz. J. M., C.B., C.M.G., D.S.O., A. D. C.	Order of the Nile, 3rd class (*Egypt*).
Egerton, Maj., J. B., 23rd Cav.	Croix de Guerre (*France*).
Egerton, Lt.-Genl. Sir R. G., K.C.B., K.C.I.E.	Karageorge, 2nd class, with swords (*Serbia*).
Elliot, Lt.-Genl. *Sir* E. L., K.C.B., D.S.O. (retd).	Legion d'Honneur Croix deCommandeur (*France*).
Erskine, Capt. K. C. S., M.C., 5th G R.	Legion d'Honneur Croix deChevalier (*France*).
Ewart, Maj.-Genl. *Sir* R. H., K.C.M.G., C.B., C.I.E., D.S.O.	St. Stanislas, 2nd class, with swords (*Russia*). Legion d'Honneur, Croix de Commandeur (*France*).
Fagan, Maj. B. J., D.S.O., 17th Infy.	Croix de Guerre (*France*).
Fane, Maj.-Genl. *Sir* V. B., K.C.I.E., C.B., I. A.	Croix de Guerre (*France*).
Falk, Maj. H., M.B., I.M.S.	White Eagle, 5th class, with swords (*Serbia*).
Ferguson, Lt. D. H., M.C., I.A. R. O.	Silver medal for Mily. Valour (*Italy*).
Fetherstonhaugh, Bt. Lt.-Col. W. A., D.S.O., 8th Cav.	Legion of Honour Croix de Chevalier (*France*).
Field, Maj. D. M. (Supy.) List	Khidmat (*Afghanistan*).
Filgate, Lt.-Col. T. R., C.I.E., V.D., I. D. F.	Croix de Chevalier (*France*).
Finlay, Maj. R. F., D.S.O., 58 Rif.	Croix de Guerre (*France*).
Fowler, Maj.-Genl. F. J., C.B., D.S.O.	Karageorge, 3rd class, with swords (*Serbia*).
Franks, Capt. K. F., 117th Mah.	White Eagle, 5th class, with swords (*Serbia*).
Fry, Maj. A.B., D.S.O. M.D., I.M.S.	Legion d'Honneur Croix de Chevalier (*France*).
Fry, Maj.-Genl. C. I., C.B.	Brilliant Star of Zanzibar, 4th class.
Fulton, Col. F. C. R., S. and T. Corps.	White Eagle, 3rd class, with swords (*Serbia*).
Gabbett, Hony. Lt.-Col. A. C., (Retd.)	White Eagle, 4th class, with swords (*Serbia*).
Gabriel, Maj. E.V. C.S.I., C.V.O., I.D.F.	Order of the Crown of Italy (Officer).
Geoghegan, Lt.-Col. S., C.B., I.A.	Nile, 3rd class (*Egypt*).
Galbraith, Maj. E. D., D.S.O., 55th Rif.	Legion d'Honneur Croix de Chevalier.
Garden, Lt.-Col. A.R.H., (retd.)	Order of St. Stanislas, 2nd class.
Godson, Maj. C. A., I.M.S.	Silver Medal for Military Valour. (*Italy*).
Gordon, Bt.-Col. S. D. Retd.	Medjidieh, 5th class.
Goss, Capt. E. S., M.C., I.M.S.	White Eagle, 5th class, with swords (*Serbia*).
Gouldsbury, Bt. Maj. C. A., 5th G. R.	Croix de Guerre (*France*).

OFFICERS HOLDING RANK IN THE ARMY, ETC., AUTHORISED TO WEAR FOREIGN ORDERS ON ALL OCCASIONS—*contd.*

Green-Armytage, Capt. V. B., M.D., I.M.S.	White Eagle, 5th class, with swords (*Serbia*). Legion d'Honneur Croix de Chevalier (*France*).
Grimston, Col. S.B., C.M.G.	Legion d'Honneur Croix de Chevalier (*France*).
Guthrie, Lt. A., Madras Arty. Vols.	White Eagle, 5th class, with swords (*Serbia*).
Gwatkin, Capt. F., M.C., 18th Ls.	Croix de Guerre (*Belgium*).
Gyan Singh, Subadar Major, I. A.	Croix de Guerre (*France.*)
Hains, Capt. J. B., 10th Jats.	Order of St Stanislas, 3rd Class (*Russia*).
Haldane, Lt.-Col. C. L., C.M.G., D.S.O., 7th G. R.	Nile, 4th class (*Egypt*).
Hall, Maj. E. G. 117th Mah.	Karageorge, 4th class, with swords (*Serbia*).
Hamilton, Bt. Lt.-Col., W. H., D.S.O., F.R.C.S., I.M.S.	Order of the Crown of Italy, Cavalier.
Hammick, Capt. E., 17th Infy.	Croix de Chevalier d l'Ordre de la Couronne (*Belgium.*)
Harrington, Maj. *Sir* J. L., K.C.M.G., K.C.V.O., C.B., Retd.	Star of Ethiopia, 3rd class (*Abyssinia*).
Hay, Capt., F. W., M.B., I.M.S.	Croix de Guerre (*France*).
Hawes, Maj. C. M., 20th Infy.	White Eagle, 5th class, with swords (*Serbia*).
Hawkes, Maj. W. C.W., D.S.O., 106th Prs.	Legion d'Honneur Croix de Chevalier (*France*).
Herbert, Col. L. N., C.B.	White Eagle, 3rd class, with swords (*Serbia*).
Hill, Maj. H. W. D., 16th Cav.	White Eagle, 3rd class with swords (*Serbia*).
Hill, Maj.-Genl. J., D.S.O.	White Eagle, 3rd class, with swords (*Serbia*). Order of the Nile, 2nd class (*Egypt*).
Holman, Col. H. C., C.B., C.M.G., D.S.O., 16th Cav., *p.s.c.* [L].	St. Stanislas, 2nd class, with sword (*Russia*). Legion d'Honneur Croix d'Officier (*France*).
Holmes, Maj. C. W. A., 116th Mah.	Croix de Guerre (*France*).
Holmes, Lt.-Col. G. V.	Order of St. Anne, 3rd class, with swords (*Russia*).
Home, Col. J. M., *p.s.c.* [L]	St. Anne, 2nd class, with swords (*Russia*).
Hopwood, Col. H. R., C.S.I., 33rd L. Cav.	White Eagle, 4th class, with swords (*Serbia*).
Hoskyn, Maj. J. C. M., D.S.O. 44th Infy.	Order of the Crown of Italy, Cavalier.
Houston, Lt.-Col A. M., 69th Punj.	Order of St. Stanislas, 3rd class, with swords (*Russia*).
Hugo, Lt.-Col. J. H., D.S.O., M.B., I.M.S.	Croix de Chevalier (*France*).
Hunt, Maj. H. R. A., D.S.O., 25th Punj.	Legion d'Honneur Croix de Chevalier (*France*).
Hutcheson Maj. W. C., 3rd Brahman	Croix de Guerre (*France*).
Ismail Khan, Subadar, I.D.S.M., 3rd S. & M.	Cross of St. George, 4th class (*Russia*).
Jacob, Lt.-Genl. *Sir* C. W., K.C.B.	St Vladimir, 4th class, with swords (*Russia*). Grand Officier Croix de Guerre (*Belgium*) Ordre de la Couronne, (*Belgium*).
Jai Singh, Subadar, 37th Dogras.	Cross of St. George, 3rd class (*Russia*).
Jatar, Capt. N. S., I.M.S., D.S.O.	White Eagle, 5th class, with swords (*Serbia*).
Jennings, Genl. *Sir* R. M., K.C.B.	Osmanieh, 4th class.
Jollie, Lt. Col. F.	Medjidieh, 4th class.
Jones, Col. L. C., C.M.G., M.V.O., 7th Lrs.	Order of the Crown of Italy Officer.
Kapadia, empy. Lt. R. N., M.B., I.M.S.	Croix de Guerre (*France*).
Keary, Lt.-Genl. *Sir* H. d'u., K.C.B., K.C.I.E., D.S.O.	Karageorge, 2nd class, with swords (*Serbia*).
Keefer, Surgn.-Maj. W. N.	Osmanieh, 4th class.

OFFICERS HOLDING RANK IN THE ARMY, ETC., AUTHORISED TO WEAR FOREIGN ORDERS ON ALL OCCASIONS—*contd.*

Keily, Lt.-Col. F. P. C., C.M.G., D.S.O., 125th Rif.	Order of St. Stanislas, 3rd class, with swords (*Russia*).
Kilkelly, Lt.-Col. P. F., M.B.	Brilliant Star of Zanzibar, 4th class.
Knox, Col. A. W. F., C.B., 58th Rif.	St. Vladimir, 3rd class (*Russia*). St. Anne, 2nd class (*Russia*). Legion d'Honneur, Croix d'Officier, (*France*).
Knox, Lt.-Col. R. W., D.S.O., M.B., F.R.C.S.E., I.M.S.	White Eagle, 4th class, with swords (*Serbia*).
Knox-Niven, Maj. H. W., Retd.	Croix de Guerre (*France*).
Landon, Maj. C. R. H. P., 35th Horse.	Croix de Guerre (*France*).
Lane, Bt.-Maj. H. P. 16th Cav.	Order of St. Anne, 2nd class, with swords (*Russia*).
Lang, Bt. Maj. W. H. 16th Cav.	Croix de Guerre (*France*).
Lapsley, Capt. J. B., M.B., F.R.C.S., I.M.S.	Croix de Guerre (*France*).
Leeds, Lt.-Col. T. L. C.M.G., D.S.O., 59th Rif.	Order of St. Anne, 3rd class, with swords (*Russia*).
Lees, Maj. C. H. B., C.M.G., 51st Sikhs.	Croix de Guerre (*France*).
Leslie, Bt.-Col. W. S., D.S.O., 31st Punj.	White Eagle, 3rd class, with swords (*Serbia*).
Lewis, Capt. H. V., D.S.O., M.C., 129th Bal.	Silver Medal for Military Valour (*Italy*).
Lewis, Capt. R. F., 21st Cav.	White Eagle, 5th class, with swords (*Serbia*).
Lind, Maj. A. G., D.S.O., 58th Rif.	Croix de Guerre (*France*).
Lucas, Col. F. G., C.B., D.S.O., I. A.	Legion d' Honneur Croix d' Officer (*France*).
McConaghy, Maj. J. G., D.S.O., 25th Cav.	White Eagle, 4th class, with swords (*Serbia*).
McMahon, Lt.-Col. *Sir* A. H., G.C.V.O., K.C.I.E., C.S.I., Retd.	Izzat-i-Afghania (*Afghanistan*). Sadari, 1st class (*Afghanistan*).
Macdonald, Bt.-Mai. H., 11th Lrs.	Croix de Guerre (*France*).
Mackinnon, Maj. A. C. B., 9th G. R.	White Eagle, 4th class, with swords (*Serbia*).

Macausland, Capt. R. A., 45th Sikhs.	Croix de Guerre (*France*).
Maclachlan, Col. T. R., C.M.G., 92nd Punj.	Order of St. Stanislas, 3rd class, with swords (*Russia*).
MacMullen, Col. C. N., D.S.O., 15th Sikhs.	Legion d'Honneur Croix d'Officer (*France.*)
Mahon, Maj. H. J., C.I.E., I.D.F.	Croix de Chevalier (*France*).
Maitland, Maj. G. R., D.S.O., 14th Lrs.	Croix de Guerre, (*Belgium*).
Maitland, Maj.-Genl. P. J., C.B., Retd.	Izzat-i-Afghania (*Afghanistan*).
Malcolm, Lt.-Col. W. L.	Order of St. Stanislas, 3rd class, with swords (*Russia*).
Malleson, Maj.-Genl.) W., C.B., C.I.E.	Hurmat (*Afghanistan*).
Malyon, Capt. F. H., [L], 21st Punj.	Medaille d'Honneur (*France*).
Manifold, Maj.-Genl. C. G., C.B., C.M.G., M.B., I.M.S.	Croix de Guerre (*Belgium*).
Man Singh, Subadar-Maj., 23rd Prs.	Croix de Guerre (*France*).
Marsh, Bt. Lt.-Col. F. G., C.M.G., 9th G. R.	Order of St. Vladmir, 4th class, with swords, and bow (*Russia*). Order of St. Stanislas 2nd class, with sword (*Russia*). Order of St. Vladmir, 3rd class (with sword).
Martin, Maj. G. D., M.C., 9th Infty.	White Eagle, 5th class, with swords (*Serbia*).
Mears, Lt.-Col. C. D., M.V.O., 16th Cav.	Order of St. Stanislas 3rd class, with swords (*Russia*).
Medlicott, Maj. H. E., D.S.O., 3rd Horse.	Croix de Guerre, (*Belgium*).
V.C. Melliss, Maj.-Gen. *Sir* C. J., K.C.B. K.C.M.G., [L].	Brilliant Star of Zanzibar, 2nd class.
Melliss, Col. *Sir* H., K.C.S.I.	Medjidieh, 4th class.
Mercer, Lt. L. F., 6th G. R.	Silver Medal for Military Valour (*Italy*).
Meynell, Maj. H., 29th Lrs.	Legion of Honour Croix de Chevalier (*France*).
Michael, Subadar, *Bahadur*, 2nd S. and M.	White Eagle, 5th class, with swords (*Serbia*).
V.C. Mir Dast, Subadar, I.O.M., 57th Rifl.	Cross of St. George, 3rd class (*Russia*).

OFFICERS HOLDING RANK IN THE ARMY, ETC., AUTHORISED TO WEAR FOREIGN ORDERS ON ALL OCCASIONS—*contd.*

Mockler, Lt.-Col. G. H. G., [L]. — St. Anne, 2nd class, with swords (*Russia*).

Morse, Capt. L. H, M.C. 62nd Punj. — White Eagle, 5th class, with swords (*Serbia*).

Moti Singh, Jemadar . . . — Croix de Guerre (*France*).

Muhammed Din, Subadar, I.O.M. — Croix de Guerre (*France*).

Murray, Lt.-Col. E. R. B., 90th Punj. — St. Stanislas, 3rd class, with swords (*Russia*).

Nash, Hony. Lt. A. E., S. & T. Corps. — Medaille Militaire (*France*).

Nasir Khan, Sub.-Maj., I.O.M. 59th Rif. — Medal of St. George, 1st class (*Russia*).

Newall, Bt.-Maj. C. L. N., 2nd G. R. — Legion d'Honneur, Croix d'Officier (*France*). Order of the Crown of Italy—Officer—(*Italy*).

Nisbet, Bt. Lt.-Col. T., D.S.O., 28th L. Cav. — Croix de Guerre (*France*).

Norie, Col. C. E. de M., C.B., C.M.G., D.S.O. — White Eagle, 3rd class, with swords (*Serbia*).

Norman, Bt.-Col. W. W., Retd. — St. Anne, 3rd class with swords (*Russia*). Izzat-i-Afghania (*Afghanistan*).

O'Grady, Bt. Lt.-Col., H. de C., 52nd Sikhs. — Legion de Honneur Croix d'Officer (*France*).

O'Rorke, Capt. J. M. W., D.S.O., 25th Cav. — Legion of Honour Croix de Chevalier (*France*).

Ogle, Bt. Lt.-Col. N., D.S.O., 67th Punj. — White Eagle, 5th class, with swords (*Serbia*).

Orr., Bt.-Col., G. M., D.S.O., 11th Lrs. — Order of St. Maurice and St. Lazarus Cavalier (*Italy*).

Orton, Bt. Lt. Col. S. B., 39th Garh. Rif. — Croix de Guerre (*France*). Order of St. Stanishlas 2nd class with swords (*Russia.*)

Ottley, Lt.-Col. W. J., 23rd Prs. — White Eagle, 4th class, with swords (*Serbia*).

Owen, Lt.-Col. C. W., C.M.G., C.I.E., I.M.S. — Hurmat (*Afghanistan*).

Paddon, Bt.-Maj. S S. W., (retd.) — Order of St. Anne. 3rd class.

Palin, Maj. Genl. P. C., C.B. — Karageorge, 3rd class, with swords (*Serbia*). Order of the Nile, 2nd class (*Egypt*).

Patton, Maj. W. S., I. M.S., M.B. — White Eagle, 4th class, with swords (*Serbia*).

Pearson, Genl. *Sir* A. A., K.C.B, 124th Infy. — Brilliant Star of Zanzibar, 1st class.

ennefather-Evans, Col. G., 19th Punj. — Order of St. Stanislas, 3rd class, with swords (*Russia*).

ilips, Maj. J. C., 28th Punj. . — Order of the Crown of Italy Cavalier.

Picot, Lt.-Col. H. P., I.A. (*Ret.*). — Legion of Honour (*France*).

Pitcher, Lt.-Col. D. LeG. C.M.G., 39th C. I. Horse. — Order of St. Maurice and St. Lazarus Officer.

Ponting, Capt. T. J., M.C., 92nd Punj. — Croix de Guerre (*France*).

Pope, Maj. S. B., D.S.O., 58th Rif. — Legion d'Honneur Croix de Chevalier.

Powell, Lt. D., 30th Punj.. . . — Croix de Guerre (*France*).

Pratap Singh, Hony. Lt.-Genl., *His Highness Maharaja Bahadur Sir*, G.C.B., G.C.S.I., G.C.V.O., A.D.C. (of Jodhpur) (Hon. Col. 34th Horse.) — Legion of Honour, Grand Officer (*France*).

Prendergast, Col. C. G., C.B. . — Legion d'Honneur, Croix d'Officier (*France*).

Price, Col. C. U., C.M.G., 130th Bal. — St. Anne, 3rd class, with swords (*Russia*).

Prideaux, Lieut. G. V. A., I. A. R. O. — White Eagle, 5th class, with swords (*Serbia*).

Pryce, Lt.-Col. H. E. ap R., C.M.G., D.S.O., 18th Infy. — Croix de Guerre (*Belgium*).

Puransing Gurung, Subadar, 1st G. R. — Cross of St. George, 4th class (*Russia*).

Puran Singh, Jemadar I. A.. . — Croix de Guerre (*France*).

Ramsay, Maj. A. D. G., C.I.E. O.B.E. — Izzat-i-Afghania (*Afghanistan*).

Ratlam, Hony. Col. *H. H. Raja Sir* Sajjan Singh, K.C.S.I., *Raja* of —. — Legion of Honour, Croiz d'Officier (*France*.)

Rattray, Col. (*tempy. Br.-Genl.*) C., C.B. — Order of St. Stanislas, 2nd class, with swords (*Russia*).

OFFICERS HOLDING RANK IN THE ARMY, ETC., AUTHORISED TO WEAR FOREIGN ORDERS ON ALL OCCASIONS—*contd.*

Ridgeway, Col. *Rt. Hon. Sir* J. W., G.C.B., G.C.M.G., K.C.S.I.	Bahaduri-i-Afghania, 1st class (*Afghanistan*).
Rivett-Carnac, Col. E. H. . . .	Medjidieh, 5th class.
Robertson, Maj. D. E, 11th Lrs.	Legion d'Honneur Croix de Chevalier (*France*).
Ross, Maj. D. B., 19th Punj. .	Croix de Chevalier, (*France*).
Rowlandson, Col. M. A. . . .	Osmanieh, 4th class.
Russell, Maj. G. H., 126th Infy.	Croix de Guerre (*France*).
Saiyed Abdul Ghaffur, Subadar	Croix de Guerre (*France*).
Scale, Maj. J. D. D.S.O., 87th Punj.	Order of St. Vladimir 4th class with swords and Bow (*Russia*) Order of Star of Roumania (Knight) (*Roumania*). Order of St. Anne, 3rd class (*Russia*) Legion d'Honneur, Croix de Chevalier, (*France*).
Scott, Maj. Genl. T. E, C.B., C.I.E., D.S.O.	Brilliant Star of Zanzibar, 2nd class.
Shakespear, Col. L. W., C.B. .	Order of St. Stanislas, 2nd class, with swords (*Russia*).
Shallow, Lt. G., I. A. R. O. .	White Eagle, 5th class, with swords (*Serbia*).
Sham Singh, Sub.-Maj., D.S.M. 14th Sikhs.	Medaille Militaire (*France*).
Shea, Maj.-Genl. J. S. M., C.B., C.MG., D.S.O.	Legion of Honour Croix de Commander (*France*). Order of the Nile, 2nd class (*Egypt*).
Shearer, Capt. J. E., M.C., 26th Punj.	Croix de Guerre (*France*).
Shore, Col. O. B. S. F., C.B., C.I.E., D.S.O.	White Eagle, 3rd class, with swords (*Serbia*).
Shuttleworth, Lt.-Col. D. I., 3rd G. R.	Croix de Guerre (*France*).
V.C. Sinton, Capt. J. A., M.B., I.M.S.	Order of St. George, 4th class (*Russia*).
Skeen, Maj.-Genl. A., C.M.G. .	Croix de Guerre (*France*).
Smith, Capt. F. F. S., I.M.S.	Order of St. Sava, 5th class (*Serbia*).
V.C. Smith, Lt.-Col. J. M., C.V.O., C.I.E.	Izzat-i-Afghania (*Afghanistan*).
Smithers, Maj. L. S. H., 17th Infy.	White Eagle, 5th class with sword (*Serbia*).
V.C. Smyth, Capt., J. G., 15th Sikhs.	St. George, 4th class (*Russia*).
Spankie, Capt. C. G., 42nd Infy.	White Eagle, 5th class, with swords (*Serbia*).
Stack, Lt.-Col. C. S., C.M.G. 33rd L. Cav.	White Eagle, 4th class, with swords (*Serbia*).
Stainton, Maj. T. H., S. & T. Corps.	White Eagle, 5th class, with swords (*Serbia*).
Steel, Col. R. A., C.I.E., 17th Cav.	Order of St. Stanislas, 2nd class (*Russia*) Ordre de le Couronne Officier (*Belgium*). White Eagle, 3rd class, with swords, (*Serbia*). Order of the Rising Sun, 3rd class, (*Japan*).
Stevens, Lt.-Col. N. M. C., C.M.G.	Order of St. Anne, 3rd class, with swords (*Russia*).
St. John, Lt.-Col. R. S., D.S.O., 20th Infy.	Karageorge, 4th class, with swords (*Serbia*).
Stewart, Lt.-Col. F. T. . .	Order of St. Stanislas, 3rd class, with swords (*Russia*).
Stewart, Col. J. H. K., D.S.O., 39th Garh. Rif.	Croix de Guerre (*France*).
Storr, Major C. L., 54th Sikhs	Legion of Honour Croix de Chevalier (*France*).
Sultan, Subadar, I.D.S.M. 1st S. & M.	Karageorge, 4th class, with swords (*Serbia*).
Tanner, Bt.-Col. C. O. O., 127th Infy.	Brilliant Star of Zanzibar, 4th class.
Tarver, Col. A. L., D.S.O., A.D.C., 124th Infy.	Croix de Guerre (*France*).
Thornhill, Bt. Lt.-Col. C. J. M., D.S.O., 6th Infy.	Legion d'Honneur, Croix de Chevalier, (*France*).
Thornhill, Capt. C. M., M.C., D.S.O., 24th Punj.	Croix de Guerre (*France*).
Tighe, Maj.-Genl. *Sir* M. J., K.C.M.G., C.B., C.I.E., D.S.O.	Brilliant Star of Zanzibar, 2nd class.

OFFICERS HOLDING RANK IN THE ARMY, ETC., AUTHORISED TO WEAR FOREIGN ORDERS ON ALL OCCASIONS—*concld.*

Tika Khan, Subadar-Maj. 58th Rif. — White Eagle, 5th class, with swords (*Serbia*).

Todd, Capt. R., 27th Punj. — Silver Medal for Military Valour (*Italy*).

Toker, Maj.-Genl. *Sir* A. C., K.C.B., — Osmanieh, 4th class.

Townsend, Capt. F. O., D.S.O., I. A. R. O. — White Eagle, 5th class, with swords (*Serbia*).

Tree, Capt. H. B. LeD., 112th Infy. — Nile, 4th class (*Egypt*).

Tucker, Lt.-Col. W. H., I.M.S. — Legion d'Honneur Croix de Chevalier (*France*).

Tulloch, Maj.-Gen. J. W. G., C.B. Ret. — Brilliant Star of Zanzibar, 2nd class. Rising Sun, 3rd class (*Japan*).

Turnbull, Maj. B., 23rd Prs. — White Eagle, 5th class, with swords (*Serbia*).

Turner, Capt. C. J. R., 15th Sikhs. — Nile, 4th class (*Egypt*).

Twiss, Lt.-Col. W. L. O., M.C., 9th G. R. — Legion of Honour Croix de Chevalier (*France*).

Tyndall, Maj. H. S, D.S.O., 40th Pathans. — Croix de Guerre (*France*).

𝕸 Upperton, Maj.-Genl. J., C.B. — Medjidieh, 3rd class.

Vicary, Bt. Maj. E. S., 16th Cav. — Croix de Guerre (*France*).

Villiers-Stuart, Lt.-Col. J. P., D.S.O., 55th Rif. — Legion d'Honneur Croix de Chevalier.

Wallace, Maj.-Genl. A., C.B. — Nile, 2nd class (*Egypt*).

Ward, Lt.-Col. H. C. S., 2nd Lrs. — White Eagle, 4th class, with swords (*Serbia*).

Warner, Maj. W. W., I.A. (*Ret.*) — Order of St. Anne, 3rd class (*Russia*).

Watson, Col. H. D., C.M.G., C.I.E., M.V.O., I.A. — Nile, 3rd class (*Egypt*).

Watson, Maj.-Genl. W. A., C.B., C.M.G., C.I.E., I.A. — Nile, 2nd class (*Egypt*).

White, Maj. M.F., M.B., I.M.S. — Croix de Guerre (*France*).

Whitby, Maj. H. F. 19th Lrs. — Croix de Guerre.

Whitehead, Lt.-Col. J., D.S.O., 1st Brahmans. — Legion of Honour Croix de Officer (*France*).

Wigram, Col. K., C.B., D.S.O., 1-2nd G. R. — Legion of Honour Croix de Chevalier (*France*). Croix de Guerre (*Belgium*).

Wilkinson, Maj. G. H, S. & T. Corps. — Order of Crown of Italy Cavalier.

Williams, Maj. A. F. C., D.S.O., 31st Lrs. — St. Stanislas, 3rd class, with swords (*Russia*).

Williams, Lt.-Col. L. G., D.S.O., 5th Cav. — Legion d'Honneur Croix de Chevalier (*France*).

Willoughby, Maj. D. V., D.S.O., 1st Brahmans. — Legion d'Honneur Croix de Chevalier (*France*).

Willoughby, Col. M. E., C.B., C.M.G. — Order of St. Anne, 2nd class, with swords (*Russia*).

Wilson, Lt.-Col., R. C., M.C., 114th Mah. — Croix de Guerre (*France*).

Wilson-Johnston, Bt. Lt.-Col. W. E., D.S.O., 36th Sikhs — White Eagle, 4th class, with swords (*Serbia*).

Winter, Bt.-Col. C. B., C.M.G., D.S.O., 112th Infy. — White Eagle, 4th class, with swords (*Serbia*).

Wintle, Lt.-Col. C. E. H., 114th Mah. — Croix de Guerre (*France*).

Wood, Lt.-Col. E.J. M., D.S.O. — Order of St. Anne 3rd class, with swords (*Russia*).

Wood, Maj. H. O. B., 8th Rajputs. — White Eagle, 4th class, with swords (*Serbia*).

Worsley, Major, C. F. M., 21st Cav. — Croix de Guerre (*France*).

Yakhudin, Subadar, I.D.S.M., 1st S. & M. — White Eagle, 5th class, with swords (*Serbia*).

Yate, Col. C. E., C.S.I., C.M.G. — Hurmat (*Afghanistan*).

Young, Lt.-Col. H. G., D.S.O., 22nd Cav. — Croix de Guerre (*France*).

Younghusband, Col. L. N., C.B., C.M.G. — White Eagle, 4th class, with swords (*Serbia*).

OFFICERS OF THE ROYAL (LATE INDIAN) ARTILLERY AND ENGINEERS.

Borradaile, Col. G. W., C.B., Arty. — Medjidieh, 3rd class.

Free, Col. J. F., Arty. — Medjidieh, 3rd class.

Hebbert, Col. W. S., Arty. — Medjidieh, 3rd class.

𝕸 Moncrieff, Hon. Col. *Sir* C. C. S., K.C.S.I., K.C.M.G., Engrs. — Medjidieh, 1st class.

Taylor, Col. P. T. H., Arty. — Medjidieh, 3rd class.

Western, Hony. Lt.-Col. J. H., C.M.G., Engrs. — Medjidieh, 1st class.

UNITED SERVICE INSTITUTION OF INDIA.

Prize Essay Gold Medallists.

Year.	Names and Rank.
1873	Colquhoun, Capt. J. A. S., R.A.
1874	Colquhoun, Capt. J. A. S., R.A.
1879	St. John, Maj. O. B. C., R.E.
1880	Barrow, Lieut. E. G., S.C.
1882	Mason, Lieut. A. H., R.E.
1883	Collen, Maj. E. H. H., S.C.
1884	Barrow, Capt. E. G., S.C.
1887	Yate, Lieut. A. C., S.C.
1888	Maude, Capt. F. N., R.E.
1888	Young, Maj. G. F., S.C. (specially awarded a silver medal).
1889	Duff, Capt. B., S.C.
1890	Maguire, Capt. C. M., S.C.
1891	Cardew, Lieut. F. G., S.C.
1893	Bullock, Maj. G. M., Devon. R.
1894	Carter, Capt. F. C., North'd. Fus.
1895	Neville, Lt.-Col. J. P. C., S.C.
1896	Bingley, Capt. A. H., S.C.
1897	Napier, Capt. G. S. F., Oxf. L. I.
1898	Mullaly, Maj. H., R.E.
1898	Clay, Capt. C. H., S.C. (specially awarded a silver medal).
1899	Neville, Col. J. P. C., S.C.
1900	Thuillier, Capt. H. F., R.E.
1900	Lubbock, Capt. G., R. E. (specially awarded a silver medal).
1901	Ranken, Lt.-Col. G. P., S.C.
1902	Turner, Capt. H. H. F., S.C.
1903	Hamilton, Maj. W. G., D.S.O., Norf. R.
1903	Bond, Capt R. F. G., R.E. (specially awarded a silver medal).
1904	MacMunn, Maj. G. F., D.S.O., R.F.A.
1905	Cockerill, Maj. G. K., R. War. R.
1907	Wood, Maj. E. J. M., 99th Deccan I.
1908	Jeudwine, Maj. H. S., R. A.
1909	Molyneux, Maj. E. M. J., D.S.O., 12th Cav.
1909	Elsmie, Lt.-Col. A. M. S., C.M.G., *p.s.c.* (*l*). 56th Rif. (specially awarded a silver medal).
1911	Petrie, D., Esq., M.A., Punj. Police.
1912	Carter, Maj. B. C. M., L'pool Regt.
1913	Thomson, Maj A. G., 58th Rif.
1914	Bainbridge, Lt.-Col. W. F., D.S.O., 51st Sikhs.
1914	Norman, Maj. C. L., M.V.O., Guides (specially awarded a silver medal).
1915	No award.
1916	Crum, Maj. W. E., Cal. L. H.
1917	Blaker, Maj. W. F., R. F. A.
1918	Gompertz, Capt. A. V., M.C., R. E.

MACGREGOR MEMORIAL SILVER MEDALLISTS.

Instituted in the year 1887.

Officers.

Year.	Names and Rank.
1889	VC Bell, Col. M. S., C.B., R.E. (specially awarded a gold medal).
1890	Younghusband, Capt. F. E., King's Dgn. Grds.
1891	Sawyer, Major H. A., S.C.
1892	Vaughan, Capt. H. B., S.C.
1893	Bower, Capt. H., 17th B.C. (specially awarded a gold medal).
1894	O'Sullivan, Major G. H. W., R.E.
1895	Davies, Capt. H. R., Oxf. L.I.
1896	Cockerill, Lieut. G. K., 28th P.I.
1897	Swayne, Capt. E. J. E., 16th B.I.
1898	Walker, Capt. H. B., D.C. L.I.
1899	Douglas, Capt. J. A., 2nd B.L.
1900	Wingate, Capt. A. W. S., 14th B.L.
1901	Burton, Major E.B., 17th B.L.
1902	Ray, Capt. M. R. E., S.C.
1903	Manifold, Lt.-Col. C. C., I.M.S.
1904	Fraser, Capt. L. D., R.G.A.
1905	Rennick, Maj. F., 40th Pathans (specially awarded a gold medal).
1906	Shahzada Ahmad Mir, Risaldar, 36th Horse.
1907	Nangle, Capt. M. C., 92nd Punj.
1908	Gibbon, Capt. C. M., R. Ir. Fus.
1910	Sykes, Maj P. M., C.M.G., late 2nd D. G. (specially awarded a gold medal)
1910	Turner, Capt. F. G., R.E.
1911	Leachman, Capt. G. E., R. Suss. R.
1911	Gurmukh Singh, Jemdr., 93rd Burma Infy.
1912	Pritchard, Capt. B. E. A., 83rd L. I. (specially awarded a gold medal).
1912	Wilson, Lieut. A. T., C.M.G., 32nd Prs.
1913	Abbay, Capt. B. N., 27th Cav.
1914	Bailey, Capt. F. M., I. A.
1914	Morshead, Capt. H. T., R. E.
1915	Waterfield, Capt. F. C., 45th Sikhs.
1916	No award.
1917	No award.
1918	Nowel, Capt. E.. W. C., Political Dept.

Non-Comd. Officers and men.

Year.	Names and Rank.
1891	Ramzan Khan, Havildar, 3rd Sikhs.
1892	Jaggat Singh, Havildar, 19th P. I.
1893	Fazaldad Khan, Dafadar, 17th B.C.
1894	Mul Singh, Sowar, 6th B.C.
1895	Gunga Dyal Singh, Sepoy, 2nd B.I.
1896	Ghulam Nabi, Sepoy, Guides.
1897	Shahzad Mir, Dafadar, 11th B.L.
1898	Adam Khan, Havildar, Guides.
1899	Mihr Din, Naick, B. S. and M.
1900	Gurdit Singh, Havildar, 45th Sikhs.
1901	Sundar Singh, Colour Havildar, 31st Bur. L. I.
1902	Tilbir Bhandari, Havildar, 9th G.R.
1903	Ghulam Hussain, Lance Dafadar, Guides.
1904	Moghal Baz, Dafadar, Guides.
1905	Madho Ram, Havildar, 8th G.R.
1906	Ghafur Shah, Lance Naik, Guides
1907	Sheikh Usman, Havildar, 103rd Mah. L. I.
1908	Malang, Havildar, 56th Rif.
1909	Muhammad Raza, Havildar, 106th Pioneers.
1910	Sher Jung, *Khan Bahadur*, Survey of India.
1912	Mohibullah, Lance Dafadar, Guides.
1913	Sirdar Khan, Sowar, 39th Horse.
1913	Waratong, Havildar, Burma Mily. Police (specially awarded a silver medal).
1914	Haidar Ali, Naick, 106th Hazara Prs.
1915	Ali Juma, Havildar, 106th Hazara Prs.
1916	Abdur Rahman, Naik, 21st Punj.
1916	Zarghun Shah, Havildar, 58th Rif. (specially awarded a silver medal).
1917	Main Afraz Gul, Sepoy, Khyber Rif.
1918	No ward.

LIST OF BRITISH SERVICE OFFICERS ON THE ACTIVE LIST (WHO ARE NOT GENERAL OFFICERS) QUALIFIED IN HINDUSTANI BY THE HIGHER STANDARD.

NOTE.—The names on this List are not indexed as they are arranged alphabetically.

Name, Rank and Corps.

Adam, Maj. R. F., R.A.
Agnew, Maj. H. C., R.E.
Ainslie, Lt.-Col. C., R. E.
Ainsworth, Bt.-Lt.-Col. W. J., D.S.O., Durh. L. I.
Aitkein, Capt. A. B., M.C., R. E.
Alexander, Capt. P. S., 8 Hussrs.
Alexander, Maj. W. N. S., D.S.O., Conn. Rang.
Allardyce, Maj. J. G. B., D.S.O., R.A.
Allason, Maj. W., D.S.O., Bedf. R.
Allen, Maj. H., D.S.O., R. A.
Allsup, Maj. E. S., D.S.O., R.A.
Anderson, Lt.-Col. A. T., R.A.
Anderson, Bt.-Lt.-Col. C. F., D.S.O., R.E.
Anderson, Maj. E. P., D.S.O., R.E.
Anderson, Bt.-Lt.-Col. N. G., C.M.G., D.S.O., *p.s.c.*, A. S. Corps.
Andrewes, Maj. F. E., D.S.O., R.A.
Applin, Maj. R. V. K., D.S.O., 14th Hussars.
Armitage, Maj. F. A. W., D.S.O., W. York. R.
Armstrong, Lt.-Col. W. H., E. York. R.
Ashburner, Bt.-Lt.-Col. L. F., D.S.O., M.V.O., *p.s.c.*, R. Fus.
Ashby, Capt. H. D., R.A.
Aspinall, Bt.-Lt.-Col. C. F., C.M.G., D.S.O., *p.s.c.*, R. Muns. Fus.
Atkinson, Col. E. H. de V., C.B., C.M.G., C.I.E.
Austin, Bt.-Col. H. H., C.M.G., D.S.O., *p.s.c.*, R.E.
Babington, Col. D.M., C.I.E., R.A.
Baddeley, Col. C. E., C.M.G.
Bagnall, Maj. H. G., R.A.
*Baily, Capt. R. H., R. Warwick R.
Bainbridge, Col. P.A., C.M.G.
Baird, Bt.-Lt.-Col. A. W.F., C.M.G., D.S.O., *p.s.c.*, Gord. Highrs.
Ball, Maj. K. M., R.A.
Ballard, Col. C. R., C.B., *p.s.c.*, Norf. R.
Bamber, Capt. C. C., R. War. R.
Baring, Maj. T. E., Rif Brig.
Barker, Maj J. S., M.V.O., R.F.
Barlow, Maj. N. W., Hamps. R.
Barnes, Capt. L. E., *p.s.c.*, R.E.
Barrington, Maj. J. F., R.A.
Barron, Maj. A. F. N., R. A.
Barrow, Lt.-Col.H. P. W., C. M. G., R.A.M.C.
Bartholomew, Lt.-Col. H. J., D.S.O., Worc. R.
Bartholomew, Bt.-Lt.-Col. W. H., C.M.G., D.S.O., *p.s.c.*, R.A.
Barton, Lt.Col. H. J., R.E.
Barton, Lt.-Col. P., D.S.O., R.A.
Bate, Bt.-Lt.-Col. T. R. F., R.A.
Bateson, Maj. J. H., D.S.O., R.A.
Bayly, Maj. A. R., R.A.
Bayley, Bt.-Lt.-Col. L. S., R.A.
Beach, Bt.-Lt.-Col. W. H., C.M.G., D.S.O., R.E.
Beamish, Maj. W. S., R.A.
Beasley, Maj.J. H. M., D.S.O., R.
Beeman, Maj. S. W., L'pool R.
Bell, Maj. A. H., D.S.O., R.E.
*Bell, Capt. M. D., R.A.
Bent, Bt.-Col. A. M., C.M.G., R. Muns. Fus.
Berthon, Bt.-Lt.-Col. C. P., E. York. R.
Bertie-Clay, Lt.-Col. N. S., R. A.
Bethell, Bt.-Lt.-Col. H. K., C.M.G., D.S.O., *p.s.c.*, 7 Husrs.
Beynon, Lt.Col. H. L. N., C.M.G., *p.s.c.*, R.A.
Biddulph, Bt.-Lt.-Col. H., R.E.
Biddulph, Lt.-Col. M., Army Pay Dept.
Bignell, Capt. R. G., R. A.
Binham, Capt. C. H., R.A.
Bingham, Maj. C. H. M., D.S.O., *p.s.c.*, A. S. Corps.
Binney, Capt. E. V., D.S.O., R.E.
Birch, Capt. P. Y., R.A.
Bird, Bt.-Lt.-Col. A. J. G., D.S.O., R.E.
Bird, Bt.-Maj. C. A., R. E.
Birkett, Maj. R. M., D.S.O., R. Suss. R.
Birney, Maj. C. F., D.S.O., R.E.
Black, Bt.-Maj. C. H. G., D.S.O.
Blackham, Col. R. J., C.I.E., D.S.O., R.A.M.C.
Blair, Maj. J. M., D.S.O., Gord. Highrs.
Bland, Bt. Col. W. St. C., R.A.
Bliss, Lt.-Col. T. G. C., Army Pay Dept.
Blomfield, Capt. H. M., 1-5th E. Kent R.
Blore, Lt.-Col. H. R., D.S.O., *p.s.c.*, K. R. Rif. C.
Boase, Maj. G. O., R. A.
Boger, Lt.-Col. D. C., Ches. R.
Boileau, Bt.-Lt.-Col. G. H., D.S.O., R.E.
Bond, Maj. E. L., D.S.O., R.A.
Bond, Bt.-Maj. L. V., R. E.
Bond, Lt.-Col. R. F. G., R.E.
Bonham-Carter, Bt.-Lt.-Col. C., D.S.O., *p.s.c.*, R. W. Kent R.
Booth, Bt.-Maj. F., R. Lanc. R.
Booth, Bt.-Lt.-Col. T. M., D.S.O., Gord. Highrs.
Borrett, Bt.-Lt.-Col. O. C., D.S.O., R. Lanc. R.
Boutflower, Capt. E. C., W. Rid. R.
Bowen, Lt.-Col. H. W., R.A.
Bowes, Col., W. H., C.B., *p.s.c.*
Boyce, Lt.-Col. H. A., D.S.O., *p.s.c.*, R.A.
Boyce, Capt. W. W., R.A.M.C.
Boyd, Maj. S., D.S.O., R.E.
Boyd, Lt.-Col. S. O., R.A.
Bradney, Bt-Maj. E., R. E.
Braine, Bt.-Lt.-Col. H. E. R. R., D.S.O., *p.s.c.*, R. Muns. Fus.
Braithwaite, Lt.-Col. W. G., C.M.G., D.S.O., *p.s.c.*, R. W. Fus.
Brake, Bt.-Col. H. E. J., C.B., D.S.O., R.A.
Brander, Maj. G. G. S., Suff. R.
Brancker, Bt.-Lt.-Col. W. S., *p.s.c.*, R.A.
†Bray, Col. C. A., C.B., C.M.G., Army Pay Dept.
Brett, Col. H. G., R.A.
Brews. Lt.-Col. R. F., R.A.
Bright, Lt.-Col. R. A., R.A.
Brind, Bt.-Lt.-Col. J. E. S., D.S.O., R. A.
Broke-Smith, Maj. P. W. L., D.S.O., R. E. A.
Brooke, Capt. A. F., R.A.
Brooke, Lt. J. A., 6th Dgns.
*Brown, Maj. B. H., *p.a.c.*, R.A.
Brownlow, Capt. C. A. L., R.A.
Brownlow, Col. C. W., R.A.
Brunker, Col. C. M., D.S.O.
Bryant, Maj. J. M., R.A.
Buchanan, Maj. F. E., R. Sc. Fus.
*Buckland, Capt. G. N., R.A.
Burdett, Maj. A.B., York & Lancs. R.
Burke, Maj. H. F., D.S.O., R.A.
Burn, Maj. R. N., R.E.
Burne, Capt. A. H., D.S.O., R.A.
*Burne, Capt. S. F., R.A.
Burney, Capt. A. E. C., M.C., R.A.
Bushell, Capt. C. W., R.E.
Bussell, Capt. S. F. B., R.A.
Butler, Maj. B. A. B., R.A.
Butler, Col. H. H.
Byrne, Maj. J. D., R. A.
Cameron, A. H., R.A., M.C.
Campbell, Maj. A., D.S.O., R.E.
Campbell, Capt. D. D. H., M.C., R.A.
Campbell, Col. H. M., C.M.G.
Campbell, Capt., J. D., R.E.
Campbell, Maj. K. G., D.S.O., R.A.
Campbell, Col. M.S.C., C.I.E., R.A.
Campbell, Lt.-Col. W. McL., M.V.O., R. Highrs.
Campion, Capt. D. J. M., R.A.
Carew, Lt. R. L. O., R.A.
Carey, Bt.-Lt.-Col. A. B., C.M.G., R. E.
Carey, Lt.-Col. H. E., R.A.
Carey, Maj. W. L. J., R.A.
Caroe, Capt. O. K., 1-4 R. W. Surr. R.
Carr, Maj. H. G., R.A.
Carson, Capt. C. F., R.E.
Carter, Lt.-Col. B. C. M., C.B., C.M.G., L'pool R.
Carter, Lt.-Col. E. P., R.A.
Cartwright, Maj. A.L., Dur. L. I.
Cartwright, Lt.-Col. G. N., D.S.O., R.A.
Castle, Maj. R.W., R.A.
Castle, Maj. S. M., Notts. & Derb. R.
Cave-Browne, Capt. W., R.E.
Cavendish, Capt. F. G., Leins. R.
Cavendish, Bt.-Lt.-Col. F.W.L.S.H. D.S.O., *p.s.c.*, 9th Lcrs.
Cayley, Col. W. de S., C.B. C.M.G.
Chadwick, Capt. E. W., R. A.
Chancellor, Maj. *Sir* J. R., K.C.M.G., D.S.O., *p.s.c.*, R.E.
Chapman, Capt. A. C., R.A.
Chapman, Lt.-Col. L. J., C.M.G., R.A.
Charles, Bt.-Col. J. R. E., D.S.O., *p.s.c.*, R. E.
Charlton, Lt.-Col. C. E. C. G., R.A.
Charteris, Bt.-Lt.-Col. J., D.S.O., *p.s.c.*, R.E.
Chater, Capt. A. F., R.E.

* Passed Proficiency test. † Hony. Maj.-Genl.

LIST OF BRITISH SERVICE OFFICERS ON THE ACTIVE LIST (WHO ARE NOT GENERAL OFFICERS) QUALIFIED IN HINDUSTANI BY THE HIGHER STANDARD.

Name, Rank and Corps.

Church, Lt.-Col. G. R. M., C.M.G., R.A.
Clarke, Bt.-Maj. B. F., Leic. R.
Clarke, Maj. C. J., R.E.
Clarke, Capt. E. H., R.E.
Clarke, Maj. H., R.A.
Clarke, Maj. H. B., R.E.
Clarke, Maj. M. O., R. Fus.
Clarkson, Lt.-Col. B. St. J., C.M.G., Dorset R.
Clayton, Capt. I. N., R.A.
Clifford, Maj. W., D.S.O., North'd. Fus.
Climo, Maj. P. H., R. A.
Clinton, Capt. W. L., K. R. Rif. C.
Cobbe, Maj. I. S., R. A.
Coffin, Lt.-Col. Campbell, C.M.G., *p.s.c.*, R.E.
Coffin, Lt.-Col. Clifford, D.S.O., *p.s.c.*, R. E.
Colbeck, Maj. B. B., D.S.O., R. A.
Colchester-Wemyss, Maj. J. M., R. Scots.
Cole, Maj. A. F., R. A.
Collins, Capt. H. S., D.S.O., Shrop. L. I.
Colvile, Maj. A. M., R. A.
V. C., Colvin, Maj. J.M.C., *p.s.c.*, R. E.
Cook, Lt.-Col. H. R., R. A.
Cooper, Lt.-Col. P.T., R. A.
Copeland, Col., R. J., M.B., R. A. M. C.
Corbett, Maj. E. F., R. A.
Cornish-Bowden, Maj. J. H. T., D.S.O., D.C.L.I.
Corrie, Maj. G. G. W., R. A.
Corrie, Lt.-Col. W.F.T., R. A.
Cotter, Maj. E. B., R. A.
Cotter, Capt. G. E. S., R. A.
Cotter, Maj. H. J., D.S.O., R. A.
Cotter, Capt. R.K., R. A.
Courtice, Maj. J. G., D.S.O., R. A.
Cowan, Maj. J. deB., R. A.
Cowdell, Capt. A. F., N. Lan. R.
Cowie, Maj. D. G., R. A.
Cowper, Maj. L. I., R. Lanc. R.
Cox, Maj. C. H. F., R. A.
Cox, Capt. M. N., M.C., W. Rid. R.
Cox, Maj. R. F., 13th Hussars.
Craster, Col. S. L.
Craven, Bt.-Lt.-Col. A. J., R. E.
Craven, Maj. W. S. D., R. A.
Crawford, Capt. A. E., R. A.
Creswell, Maj. E .F., D.S.O., R. A.
Cripps, Maj. C. W., R. A.
Crocker, Maj. B. E., Welsh R.
Crofton, Maj. R. B., R.A.
Crole-Wyndham, Col. W. G., C.B., *p.s.c.*
Crooke, Lt.-Col. C. D. P., C.M.G., Suff. R.
Crookenden, Maj. A., *p.s.c.*, Ches. R.
Crookshank, Lt.-Col. A. A., R. E.
Crowdy, Maj. C. R., R. A.
Crowe, Col. J. H. V., C.B., *p.s.c.*
Cruickshank, Lt. A. J., R.E.
Cruickshank, Maj. H. St. J., R. A.
Cruickshank, Maj. P. H., R. A.
Cumberlege, Bt.-Lt.-Col. A. F., R.E.

Name, Rank and Corps.

Cumming, Lt.-Col. H. R., *p.s.c.*, Dur. L. I.
*Cunningham, Capt. A. G., M.C., R. A.
Cunningham, Maj. A. H., R. E.
Curtis, Capt. A. D., R. A.
Dallas, Col. A. G., C.B., C.M.G., *p.s.c.*
Dallas, Col. J.
Daubeney, Col. E. K., D.S.O.
Davie, Lt.-Col. K. M., *p.s.c.*, Glouc. R.
*Davies, Capt. A., R. A. M. C.
Davies, Maj. L. M., R. A.
Davies, Capt. O. H., R. A.
Davies, Lt.-Col. W. P. L., D.S.O., *p.s.c.*, R. A.
Dawson, Capt., A.G., Middx. R.
Dawson, Capt. A., M.B., R. A. M. C.
Day, Capt. C. L., R.A.
Dean, Capt. H. H., Dorset. R.
Deane, Capt. J. A. L., R. A.
De Brett, Lt.-Col. H. S., D.S.O., *p.s.c.*, R.A.
Deed, Capt. L. C. B., R. E.
Delaforce, Lt.-Col. E.F. C.M.G., R.A.
Dennistoun, Lt.-Col. J. G., R. A.
De Pree, Lt.-Col. H. D., C.M.G., *p.s.c.*, R. A.
Deverell, Bt.-Lt.-Col. C.J., *p.s.c.*, W. York. R.
De-Sausmarez, Lt.-Col. C., D.S.O., R. A.
Dickins, Maj. F., R. A.
Dickson, Maj. W. E. R., R. E.
Dickson, Capt. T. H., R.A.M.C.
Digby, Maj. A. K., D.S.O., R. A.
Diggle, Capt. P. G. W., Bord. R.
Dobbyn, Maj. G. H. W., R. A.
Donovan, Maj. C. C., R. A.
Doran, Col. W. R. B., C.B., D.S.O., *p.s.c.*
Dorman, Maj. L. C., Worc. R.
Dorrien-Smith, Maj. E. P., D.S.O., Shrops. L. I.
Douglas, Lt.-Col. A. P., C.M.G., R. A.
*Douie, Capt. F. McC., D.S.O., M.C., R.E.
Dowding, Maj. H. C. T., *p.s.c.*, R. A.
Dowell, Col. G. C., C.M.G.
Down, Capt. C. B., R. A.
Drake-Brockman, Capt. V., R.A.
Drew, Maj. J. S., Cam'n. Highrs.
Dreyer, Maj. G. V., R. A.
Dudgeon, Col., F. A., C.B., *p.s.c.*
Duke, Maj. B. L., R. A.
Duncan, Lt.-Col. J., C.M.G., D.S.O., *p.s.c.*, R. Sc. Fus.
Dundas, Maj. R. C., R. Scots.
Dunn, Maj. E. G., *p.s.c.*, R. I. Rif.
Dunsterville, Lt.-Col. E. L., R. E.
Durell, Col. A. J. V., Army Pay Dept.
Dutton, Capt. R. M. L., M.C., R.A.
Dyson, Lt.-Col. H. B., D.S.O., R. Scots.
Earle, Maj. R. G., D.S.O., R. E.
East, Lt.-Col. L. W. P., C.M.G., D.S.O., *p.s.c.*, R. A.
Easton, Maj. F. A., R. A.
Eckersley, Col. E., M.B., R. A. M. C.

Name, Rank and Corps.

Eddowes, Maj. W. B., Manch. R.
Edmond, Maj. J. H. R., A.
*Edwards, Capt. W. M., R. A.
Elgood, Capt. C. L., 10 Bn., Midd'x R.
Elkington, Lt.-Col. R. J. G., C.M.G., R. A.
Elles, Maj. H. J., D.S.O., R. E.
Ellington, Bt.-Lt.-Col. E. L., C.M.G., *p.s.c.*, R. A.
Elliott, Capt. B. H., R. A.
Elton, Lt-Col. F. A. G. Y., R. A.
Evans, Maj. A. S., R. E.
Evans, Capt. B. L., 1-4th R. W. Surr. R.
Evans, Maj. L. P., D.S.O., R. Highrs.
Evans, Col. U. W., C.B., R. E.
Evans, Lt.-Col. W., D.S.O., R. A.
Everard, Maj. C. J., R. A.
Everett, Col. H. J., C.B., *p.s.c.*
Everett, Capt. M., D.S.O., R. E.
Ewbank, Col. W., *p.s.c.*, R. E.
Fairholme, Col. W E., C.M.G., M.V.O., *p.s.c.*
Fanshawe, Maj. L. A., D.S.O., R. A.
Farfan, Maj. A. J. T., R. A.
Farmar, Lt.-Col. W. C. R., R. A.
Farquhar, Maj. J., D.S.O., *p.s.c.*, R. A.
Fergusson, Bt.-Lt.-Col. A. C., R. A.
Fergusson, Maj. V. M., D.S.O., *p.s.c.*, R. A.
Field, Maj. K. D., R. A.
Finnis, Maj. F. A., R. A.
Finny, Capt. C. M., R. A. M. C.
*Firth, Capt. D., W. Rid. R.
Fisher, Bt-Lt.-Col. B. D., D.S.O., *p.s.c.*, 17th Lcrs.
Fooks, Lt. O.J. F., 14th Hussrs.
Foote, Lt.-Col. H. B., R. A.
Forbes, Capt. R. M. N., R. A.
Ford, Lt.-Col. C. H., C.M.G., R. A.
Forster, Maj. D., D.S.O. *p.s.c.*, R.E.
Fowle, Lt.-Col. T. E., *p.s.c.*, *late of* Bedf. R.
Fowler, Maj. E. G., R. A.
Fox, Maj. B. H., R. E.
Franklin, Maj. R. J., R. A. M. C.
Franks, Bt.-Col. G. McK., *p.s.c.*, R. A.
Fraser, Lt.-Col. L. D., C.M.G., R.A.
Fraser, Lt.-Col. T., C. B., C.M.G., *p.s.c.*, R. E.
Fraser, Maj. T. C., R.A.
Freeland, Bt.-Lt.-Col. H. F. E., C.B., D.S.O., M.V.O., R. E.
Freeland, Capt. K. F., R. A.
Freeth, Maj. C. J. D., R. A.
Frith, Col. H. C., C.B., *late of* Som. L. I.
Fryer, Col. F. A. B., *late of* 6th Dgns.
Fuller, Col. R. W., D.S.O.
Gahan, Capt. C. G. D., E. Surr. R.
Gale, Capt. H. J. G., R.A.
Game, Lt.-Col. P. W., D.S.O. *p.s.c.*, R. A.
Gardiner, Maj. H. W., R. A.
Garratt, Maj. C. H., Army Pay Dept.
*Garrett, Capt. R. C., R. A.
Garstin, Lt.-Col. H. E., R. A.
Gatt, Maj. J. E. H., M.D., R. A. M. C.

* Passed Proficiency test.

LIST OF BRITISH SERVICE OFFICERS ON THE ACTIVE LIST (WHO ARE NOT GENERAL OFFICERS) QUALIFIED IN HINDUSTANI BY THE HIGHER STANDARD.

Name, Rank and Corps.

Gauntlett, Capt. H. G., Linc. R.
Geddes, Lt.-Col. G. H., C.B., R. A.
Geddes, Col. J. G., C.B.
Gee, Maj. E. D. F., R. A.
Gibbon, Bt. Lt.-Col. C. M., *p.s.c.*, R. Ir. Fus.
Gibson, Maj. A. W., R.A.M.C.
Gibson, Capt. L. G., R.A.M.C.
Gifford, Maj. F. R., Midd'x. R.
Gillespie, Maj. R. St. J., R. E.
Gillett, Maj. E. S., A. V. Corps.
Girdwood, Capt. A.C., D.S.O., North'd. Fus.
Glanville, Col. F., D.S.O.
Glasgow, Maj. A. E., D.S.O., R. Sus. R.
Gloster, Col. G. M., C.M.G.
Gogarty, Lt.-Col. H. E., *p.s.c.*, Worc. R.
Gordon, Capt. A. D., M.C. R. Berk. R.
Gordon, Lt.-Col. A. F., C.M.G.,D.S.O., *p.s.c.*, Gord. Highrs.
Gordon, Lt.-Col. H., D.S.O. Leic. R.
Gordon, Col. L. G. F., C.B., D.S.O.,
Goring-Jones, Lt.-Col. M. D.,C.M.G., Dur. L.I.
Gorton, Lt.-Col. R. StG., *p.s.c.*, R.A.
Gosset, Lt.-Col. F. W., D.S.O., *p.s.c.*, R. A.
Græm, Capt. J. A., R. E.
Graham, Maj. C., North'n. R.
Graham, Maj. C. A. L., D.S.O., R. A.
Graham, Capt. C. P., D.S.O., Welsh. R.
Graham, Col. M. D., C.B.,
Graham, Maj. R. C. D., R. A.
Grant, Maj. H. F. L., *p.s.c.*, R. A.
Grant, Bt.-Col. P. G., C.M.G., R. E.
Gray, Maj. R. A., R. Ir. Fus.
Greathed, Lt.-Col. R. N., *p.s.c.*, R. A.
Green, Lt.-Col. S. F. St. D., M. D., R. A.M.C.
Greene, Maj. A. C. R., *p.s.c.* R. A.
Greenhill-Gardyne, Maj. A. D., Gord. Highrs.
Greenshields, Capt. D. J., M.C. R., A.
Greenstreet, Bt. Lt.-Col. C. B. L., R.E.
Gregory, Maj. A. J. R., R. A.
Gribbon, Maj. W. H., R. Lanc. R.
Grier, Col. H. D., C.B., R. A.
Grinlinton, Maj. J. L., D.S.O., R. A.
Grose, Maj. J., R. A.
Grove, Maj. P. L., Chesh. R.
Gunn, Lt.-Col. G., Cam'n. Highrs.
Gunter, Maj. A. C., R. A.
Gunter, Maj. C. P., R. E.
Hacker, Capt. E. S., A. S. Corps.
Haig, Maj. A. G., D.S.O., R. A.
Haig, Maj. C. H., Leic. R.
Haig, Capt. W. de H., R. E.
Haines, Lt-Col. H. A., M.D., R. A. M. C.
Hambro, Bt. Lt.-Col., P. O., *p.s.c.*, 15th Hussrs.
Hamilton, Col. A. B., C.B., *p.s.c.*
Hamilton, Capt. E. G., D.S.O., M.C., Con. Rang.

Name, Rank and Corps.

Hamilton, Capt. H. W., 5th Drgn. Grds.
Hamilton, Col. L. A. H., *p.s.c.*
Hamilton, Capt. R., R.E.
Hamilton, Col. W. G., C.B., D.S.O., *p.s.c.*
Hamilton-Johnston, Capt. D. C., R. Highrs.
Hanbury, Maj. P. L., D.S.O., Shrop. L. I.
Hancock, Maj. M. P., R. Fus.
Hardcastle, Maj. E. L., R.A.
Harding Newman, Maj. E., D.S.O., R. A.
Hare, Maj. G. A., R. A.
Hare, Lt.-Col. R. H., C.M.G., D.S.O., M.V.O., *p.s.c.*, R. A.
Harris, Capt. C. D., Shrop. L. I.
Hart-Synnot, Maj. A. H. S., D.S.O., *p.s.c.*, E. Sur. R.
Harvey, Maj. C. G. S., R. A.
Haskard, Capt. D. D., R. A.
Haslam, Maj. B. J., *p.s.c.*, R. E.
Haslam, Maj. G. W. P., R. Ir. R.
Hassells-Yates, Maj. G. A., R. A.
Hawthorn, Maj. G. M. P., D.S.O., L'pool. R.
Hayes, Maj. A. H. R. A. M. C.
Hayes-Sadler, Maj. J., R.A.
Haymes, Maj. R. L., D.S.O., R. A.
Haywood, Maj. A. H. W., D.S.O., R. A.
Headlam, Maj. T. A., E. York. R.
Heales, Capt. A. G., E. Surr. R.
Heard, Col. E. S., *p.s.c.*
Hearn, Maj. G. R., R.E.
Heath, Col. G. M., C.B., D.S.O., *p.s.c.*
Heath, Maj. M. G., R. W. Surr. R.
Heaton-Ellis, Capt. J.S., R. A.
Hebbert Lt. G. K. P., R. A.
Hebert, Capt. A., R. A.
Hemming, Lt.-Col., N. M., C.B., R. E.
Henderson, Maj. E. G., *p.s.c.*, R. E.
Henderson, Maj. H. E., R. A.
Henderson, Capt. N. G. B., R. Highrs.
Heneker, Maj.-Genl. W. C. G., C.B., D.S.O.
Henry, Lt.-Col. J., R. A.
*Hepper, Lt.-Col. L. L., R.A.
Heycock, Lt.-Col. C. H., R.E.
Heywood, Capt. T. G. G., R. A.
Hildreth, Maj. H. C., F.R.C.S., R.A.M.C.
Hill, Capt. A. F. S., *late* R.E.
Hill, Maj. E. F. J., M.C., R.E.
Hill, Maj. G. N., R.A.
Hill, Maj. R. R., *p.s.c.*, R. A.
Hingley, Maj, S. H. Midd'x R.
Hobart, Capt. P. C. S., D.S.O., M.C., R.E.
Hodgson, Maj. P. E., R.E.
Hodson, Maj. H. V. R., N. Staff. R.
*Hogan, Capt. E. O., R.A.
Hogg, Maj. C. C. H.,C.M.G., R.E.
Hogg, Capt. D. McA., R. E.
Hogg, Maj. P. G. H., D.S.O., R. E.
Holdich, Capt. G. W. V., D.S.O., *p.s.c.*, R. A.
Holland, Lt. G. W., 2nd Liverpool Regt.

Name, Rank and Corps.

Holland, Maj. R. T., M.C., R.A.
Hollins, Maj. C. E., Linc. R.
Holme, Maj. A. S., R.E.
Holt, Col. M. P. C., C.B., D.S.O., *late* R.A.M.C.
Home, Lt.-Col., R.E., D.S.O., *p.s.c.* R.A.
Hope, Maj. J. W., D.S.O., *p.s.c.*, R.A.
Horn, Maj. R., M.C., Sea. Highrs.
Horsfall, Maj. A. G., W. Rid. R.
Howard, Bt.-Lt.-Col., T. N. S. M., W. York. R.
Hughes, Maj. C. G. E., Ches. R.
Hughes, Maj. E. L., D.S.O., North'n. R.
Hume-Spry, Maj. L., D.S.O., *p.s.c.*, W. York. R.
Humphrys, Col. C. V.
Hunt, Maj. W. M., M.C., R.A.
Iggulden, Col. H. A.
Iles, Maj. F. A., R.E.
Iles, Lt.-Col. H. W., R. A.
Incledon-Webber, Maj. A. B., D.S.O., R. Ir. Fus.
Ingham, Maj. R. J. Fitz G., D.S.O., *p.s.c.*, R. A.
Ingoldby, Capt. C. M., R.A.M.C.
Innes, Capt. G. Vd'A., R Sco. Fus.
Irvine, Maj. A. E., D.S.O. Dur. L. I.
Irvine-Fortescue, Capt. A., M.B., R. A.M.C.
Irwin, Capt. A. P. B., D.S.O., E. Surr. R.
Jackson, Col. S. C. F., D.S.O.
Jarrett, Capt. A. F. V., M.C., R. A.
Jefferd, Capt. W. W., Middx. R.
Jenkin, Maj. F. C., R.A.
Jenkins, Capt. W. M., R.A.
Jennings, Col. H. A. K., C.I.E., R.A.
Jennings, Lt.-Col. W., R. A.
Jennings-Bramly, Lt-Col. A. W., D.S.O.
*Jervis, Capt. H. S., *p.s.c.*, R. Munster Fus.
Johnson, Col. E. P., R.E.
Johnson, Col. F. E., C.M.G., D.S.O.
Johnson, Lt.-Col. R. M., *p.s.c.*, R.A.
Johnson, Lt.-Col. T. P., A. S. Corps.
*Johnston, Maj. C. E. L., R.A.
Johnston, Maj. R. H., D.S.O., Linc. R.
Jones, Capt. A. E. B., M.D. R.A.M.C.
Julius, Maj. S. deV. A., *p.s.c.*, R. Suss. R.
Haye, Col. R.A., R.A.
Kelly, Bt.-Lt.-Col. C. R., R.A.
Kelly, Capt. E. H., D.S.O., M.C., R.E.
Kelly, Lt.-Col. H. E. T., C.M.G., R.A.
Kemble, Lt.-Col. F. A., R.A.
Kemp, Lt.-Col. G. C., C.B., R.E.
Kennedy, Col. A. A., C.M.G., *p.s.c.*, 3rd Hussrs.
Kennedy, Maj. A. C., R.A.
Kennedy, Capt. G., R. Muns. F.
Kenyon, Capt. H. E., R.A.
Keogh, Maj. J. H., R.A.

* Passed Proficiency test. †Passed Interpreter's test.

LIST OF BRITISH SERVICE OFFICERS ON THE ACTIVE LIST (WHO ARE NOT GENERAL OFFICERS) QUALIFIED IN HINDUSTANI BY THE HIGHER STANDARD—*contd.*

Name, Rank and Corps.

Kettlewell, Maj. H. W., Shrops. L. I.
Kiddle, Lt.-Col. F., C.M.G., M.B., R.A.M.C.
King, Capt. C. J. S., R. E.
King, Maj. R. G., R. E.
Kinsman, Maj. G. R. V., D.S.O., R.A.
Kirby, Capt. H. A., M.C., R.A.
Kirke, Bt. Lt.-Col. W. M. St. G., D.S.O., *p.s.c.*, R.A.
Kitchin Capt. E. G., R. E.
*Knapp, Lt.-Col. K.K., C.M.G., R. A.
Knight, Capt. J. H., R.A.
Knox, Bt. Lt.-Col. H. H. S., *p.s.c.*, North'n. R.
Lambert, Lt.-Col. T. S., *p.s.c.*, E. Lan. R.
*Landon, Capt. R. P., R.A.
Lane, Lt.-Col. F. C., R. A.
Lang, Lt.-Col. G. G., D.S.O., W. York. R.
Langhorne, Capt. A. P. Y., D.S.O., R.A.
Langhorne, Maj. J. A. D., D.S.O., R.A.
Lascelles, Lt.-Col. A. E., *p.s.c.*,
Latham, Capt. E., R. F. A.
Lauder, Maj. F. P., R.A.M.C.
*Law, Maj. C. A., Wilt. R.
Law, Maj. J. P., Devon. R.
Lawrence, Col. H. D., *p.s.c.*,
Lawrence-Archer, Maj. J. H., R.A.
Lawson, Capt. C. G., R.A.
Layard, Capt. C. P. J., M.C., R.A.
Leachman, Bt. Lt.-Col. G. E., C.I.E., R. Suss. R.
Leader, Col. H. P., C.B.
Leah, Maj. T. C., R. A.
Lean, Col. K. E., C.B.
LeBreton, Maj. E. P., R. E.
Lecky, Col. R. St. C., C.B.
Lee, Lt. D. J. N., R. G. A. (S. R.)
LeFanu, Capt. R., Leic. R.
Legard, Lt.-Col. D'A., *p.s.c.*, 17 Lcrs.
Lenfestey, Maj. L. d'E., R.A.
Leslie, Lt.-Col. G. A. J., C.M.G., R. E.
Leslie-Smith, Capt. G. L., R. A.
Lesslie, Lt.-Col. W. B., C.M.G., R.E.
Lethbridge, Lt.-Col. S., R.A.
Lewin, Maj. E. O., D.S.O., R.A.
Lewis, Capt. C. G., R.E.
Lewis, Maj. H. A., R.A.
Lewis, Capt. L. A., R.A.M.C.
Lewis, 2nd-Lt. N. C., 1-4 Border R.
Lewis, Capt. W. H., M.C., R. A.
Liddell, Col. W. A., C. B.
Lilley, Capt. C. C., Worc. R.
Lindsay, Maj. W. D., R.A.
Lindsell, Capt. W. G., M.C., R. A.
Ling, Capt. C. G., M.C., R.E.
Lipsett, Maj. L. J., C.M.G., *p.s.c.*, R. Ir. R.
Lister, Maj. F. H., D.S.O., R.A.
Little, Col. C. B., C.M.G.
Little, Capt. E. M., R.A.
Loch, Maj. G. C. B., R. E.
Loch, Maj. S. G., D.S.O., *p.s.c.*, R.E.
Long, Capt. W. E. L., R.A.
Loring, Capt. E. J., M.C., R.E.
Loveday, Maj. F. W. *p.s.c.*, R.A.
Lowis, Lt.-Col. P. S., R.A.
Lubbock, Maj. G., D.S.O., R. E.
Luke, Bt.-Lt.-Col. T. M., D.S.O., O.B.E., R.A.
Lyall, Lt. W. J., R. E.
Lyle, Lt.-Col. G. S. B., R.A.
Lynch, Capt. C. St. J., R.E.
Lyon-Campbell, Lt.-Col. C. H. D., R. Fus.
Lyon, Lt.-Col., F., C.M.G., D.S.O., *p.s.c.*, R.A.
Lyon, Capt. N. H. B., Middx. R.
McCallum, Capt. R. E., R. Scots.
McClintock, Maj. R. L., D.S.O., R. E.
McCulloch, Bt. Maj. A. J., *p.s.c.*, 7th Dgn. Gds.
McEwen, Capt. O. R., R.A.M.C.
McGowan, Maj. W. H., R.A.
McHarg, Maj. A. A., R. E.
McLean, Capt. G. N., Gord. Highrs.
McLellan, Lt. J. S., 24 Bn., Rif. Bde.
McLeod, Col. W. K.
McNaught, Lt.-Col. J. G., M.D., R.A. M. C.
McVittie, Bt. Lt.-Col. R. H., A.O.D.
MacCarthy, Maj. I. A. O., R.A.M.C.
MacClellan, Maj. G. P., R.A.
MacFarlan, Maj. J. B., R.A.
Mac Farlane, Col. D. A., C.B., D.S.O.
Mackenzie, Lt.-Col. F. W., R.A.
Mackenzie, Capt. H.H., R.A.
*Mackenzie, Maj. J., M.B., R. A. M. C.
Mackenzie, Maj. J. H., *p.s.c.*, R. Scots.
Mackintosh, Lt.-Col. J. B., R. A.
Maclagan, Col. R. S., C.B., C.S.I.
MacMunn, Bt.-Col. G. F., C.B., D.S.O., *p.s.c.*, R.A.
Magniac, Maj. C. L., C.M.G., R.E.
Magniac, Maj. M., D.S.O., Lanc. Fus.
Malcolm, Bt. Col. N., D.S.O., *p.s.c.*, A. & S. Highrs.
Marden, Bt. Col. T. O., C.M.G., *p.s.c.*, Welsh R.
Marindin, Bt. Lt.-Col. A. H., *p.s.c.*, R. Highrs.
Marindin, Maj. C. C., R.A., *p.s.c.*
Marrable, Col. A. G., C.B., *p.s.c.*
Marsh, Maj. A. C. E., R.A.
Marshall, Bt. Lt.-Col. F. J., *p.s.c.*, Sea. Highrs.
Marshall, Maj. H. S., D.S.O., R.A.
Marshall, Lt.-Col. T. E., R. A.
Martin, Capt. E. C. DeR., M.C. York. L.I.
Mascall, Maj. M. E., R.A.
Massie, Lt.-Col. R. H., R.A.
Mason, Capt. G. K. M., D.S.O., 14th Hussrs.
Mathew-Lannowe, Maj. E. B., *p.s.c.*, R. W. Sur. R.
Matthews, Maj. J., R. A. M. C.
Mawhinny, Lt.-Col. R. J. W., C.B., R.A.M.C.
May, Bt. Lt.-Col. R. S., D.S.O., *p.s.c.*, R. Fus.
Maynard, Bt.-Lt.-Col. C. C. M., D.S.O., *p.s.c.*, Devon. R.
Mecredy, Maj. H., R.A.
Medill, Maj. P. M., R. A.
*Meikle, L. & R., 2-7 Hamps R.
Melvill, Maj. T. P., 17th Lcrs.
Metcalfe, Lt.-Col., F. H., R.A.
Meynell, Maj. G., *p.s.c.*, Shrops. L. I.
Milman, Maj. L. C. P., R.A.
Milman, Maj. O. R. E., R.A.
Milne-Home, Capt. A. C., North'd. Fus.
Milner, Maj. E. F., R. Ir. R.
Mitchell, Capt. T., R. Suss. R.
Moberly, Maj. A. H., D.S.O., R.A.
Mocatta, Capt. V. E., 14 Hussrs.
Moffitt, Maj. F. W., D.S.O., *p.s.c.*, Essex R.
Molesworth, Maj. Capt. F. C., R.E.
Molesworth, Capt. G. N., Som. L.I.
Molesworth, Bt.-Col. R. P., R.A.
Montgomery, Capt. W. A. M., S. Lan. Regt.
Moore, Capt. H. T., R. A.
More, Lt.-Col. L.P., M.B., R.A. M.C.
Morgan, Maj. F. A. S., R.A.
Morris, Maj. R. C., R.A.
Morrison, Maj. J., R.A.
Morshead, Capt. H. T., R. E.
Morton, Col. W. R.
Moutray, Capt. A. G., Conn. Rang.
Mudge, Bt.-Lt.-Col., A., W. Surr. R.
Murray, Maj. A. D., R.A.
Murray, Maj. H. F. F., R. Highrs.
Murray, Lt.-Col. J. A. S., A.O.D.
Murray, Lt.-Col. J. H. S., R.E.
Murray, Lt. K.M., 1st S. Lancs.
Murray, Maj. T. D., Leins. R.
Murray, Lt.-Col. V., C.M.G., R. E.
Murray-Lyon, Capt. D. M., M.C., High L. I.
Muter, Maj. R. S., R.E.
Muspratt-Williams, Lt.-Col. C. A., C.I.E., R.A.
Nash, Maj. H. E. P., R. Scots.
Nation, Capt. C. F., R.E.
*Nation, Capt. R. F., R. Fus.
Nevinson, Lt.-Col. T. St. A. B-L. R.A.
Newington, Capt. C., Bedf., R.
Newnham, Capt. L. A., Midd'x. R.
Newton, Capt. P. I., R.A.
Newton, Maj. T. C., R.A.
Nicholson, Col. C. L., C.M.G., *p.s.c.*, E. Lan. R.
Nicholson, Maj. E. H., R. Fus.
Nicolls, Col. E. G., C.B.
Nixon, Capt. E. J., M. C., R.A.
Norman, Lt.-Col. H. H., M.B., R.A.M.C.
Normand, Maj. S. R., R. A.
*North, Capt. H. N., R.E.
Norton, Lt.-Col. C. E. G., 7th Hussars.
Norton, Maj. E. F., M.C., R.A.
Nunn, Capt. J. H., R. A.
O'Donnell, Col. H., *p.s.c.*
O'Dowda, Bt.-Col. J. W., C.M.G., *p.s.c.*, R. Dub. F.
Officer, Capt. W. G., W. Rid. R.
Ogg, Lt.-Col. G. S., C.I.E. R.A.
Ogg, Maj. W. M., D.S.O., *p.s.c.*, R.A.
O'Leary, Col. T. E., C.B., C.M.G.
O'Malley, Capt. T. F., R. Muns. Fus.

* Passed Proficiency test.

LIST OF BRITISH SERVICE OFFICERS ON THE ACTIVE LIST (WHO ARE NOT GENERAL OFFICERS) QUALIFIED IN HINDUSTANI BY THE HIGHER STANDARD—*contd.*

Name, Rank and Corps.

Oldman, Bt. Lt.-Col. R. D. F., C.M.G., D.S.O., Norfolk R.
Ollivant, Bt. Lt.-Col. A. H., C.M.G., *p.s.c.*, R.A.
Onslow, Lt.-Col. C. C., C.M.G., Bedf. R.
Orr, Maj. H. M. C., Linc. R.
Osborn, Maj. W. L., D.S.O., R. Suss. R.
Osborne, Maj. H.P., D.S.O., *p.s.c.*, Midd'x. R.
Ouseley, Lt.-Col. R. G., C.B., C.M.G., D.S.O., R. A.
Ovens, Lt.-Col. R. M., C.M.G., S. Staff. R.
Paine, Maj. J. H., D.S.O., R.A.
Pakenham, Maj. G. de la P. B., D.S.O., *p.s.c.*, Bord. R.
Palmer, Bt. Lt.-Col. C. C., *p.a.c.*, R. A.
Palmer, Lt.-Col. C. E., D.S.O., R. A.
Palmer, Lt.-Col. H. R., R.A.
Pargiter, Capt. R. B., R.A.
Parbury, Maj. E., R.A.
Parker, Col. F. J.
Parker, Col. J. L., C.M.G., R.A.
Parker, Maj. W. M., C.M.G., A. S. Corps.
Parkin, Capt. H. D., A. S. Corps.
Patch, Lt-Col. F. R., R. A.
Patterson, Capt. J., R. A.
Payn, Maj. W. A., *p.s.c.*, Shrops., L. I.
Payne, Capt. D. W., R.A.
*Peake, Capt. F. G., W. Rid. R.
Pedley, Col. S. H., C.B., R. W. Kent R.
Peel, Capt. L., York. R.
Peile, Lt.-Col. A. J., R.A.
Peiniger, Maj. R. F., R.A.
Pemberton, Bt.-Maj. S., M.C., R.E.
Perceval, Lt.-Col. C. J., C.M.G., D.S.O., *p.s.c.*, R.A.
Percey-Smith, Maj. D. C., D.S.O., Middx. R.
Percival, Capt. D., R.A.
Perkins, Lt.-Col. A. E. J., R.A.
Perram, Capt. G. T. C., R.A.
Perreau, Lt.-Col. A. M., C.M.G., R.A.
Perry, Maj. S., R.A.
Perry, Capt. W. E., R.E.
Phelips, Capt. R. M., R. Lanc. R.
Phillimore, Capt. P. G., A. S. Corps.
Phillips, Maj F. R., D.S.O., R.A.
Phillips, Col. H. deT., C.M.G., R.A.
Phillips, Lt.-Col. T. R., C.M.G., R.A.
Phipps, Lt.-Col. C. F., D.S.O., R.A.
Pigott, Lt.-Col. V. R.
Pike, Maj. W. N., E. York. R.
Pilcher, Lt.-Col. A. J., R.E.
Pile, Capt. F. A., R.A.
Pinwill, Bt. Lt.-Col. W. R., *p.s.c.*, L'pool R.
Platt, Lt.-Col. E. H. R. C. R., Som. L. I.
Plowden, Capt. G. F., M.C., Ox. & Bucks. L. I.
Plummer, Lt.-Col. E. W., D.S.O., R.A.
Plunkett, Maj. E. A., *p.s.c.*, Linc. R.
Pollard, Lt.-Col. A. E. St. V., Bord. R.
Pollard, Maj. C. J. K., R.A.
Pollard, Bt. Col J. H. W., C.M.G., *p.s.c.*, R. Scots Fus.
Pollock, Lt.-Col. C. E., R.A.M.C.
Ponsonby, Capt. H. C., M.C., K. R. Rif. C.
Poole, Bt.-Col. A. J., R. War. R.
Poole, Maj. F. G., D.S.O., Middx. R.
*Potter, Maj. C. F., D.S.O., *p.s.c.*, R.A.
Powell, Maj. A. E., R.E.
Powell, Col. S. H., C.B.
Power, Maj. R. E., E. Kent. R.
Pownall, Capt. H. R., R.A.
Prendergast, Col. D. G.
Preston, Capt. C. O'D., R.A.
Price, Capt. J., R.A.
Prince, Maj. P. E., R.E.
Pritchard, Bt. Lt.-Col. H. L., D.S.O., R.E.
Purcell, Maj. V. E., R.E.
Purser, Maj. L. M., R.A.M.C.
Queripel, Maj. L. H., D.S.O., R.A.
Quinton, Lt.-Col. F. W. D., R.A.
Radcliffe, Bt. Lt.-Col. F. W., C.I.E., *p.s.c.*, Dorset. R.
Radcliffe, Lt.-Col. W. C. A., R.A.
Raikes, Maj. L. T., D.S.O., R.A.
Ramsay, Maj. H. A., D.S.O., *p.s.c.*, R.A.
Ramsden, Maj. J. V., D.S.O., R.A.
Rasch, Maj. *Sir* F. C., 6th Dgn. Gds.
Rawlence, Capt. M., R.E.
Rawson, Capt. C. D., D.S.O., R.A.
Ready, Maj. B. T., Welsh R.
Redfern, Maj. A. F., A. S. Corps.
Reed, Maj. C., D.S.O., R.A.
Reeves, Maj. S. R., R.A.
Renny, Col. S. M., C.I.E.
Reynolds, Col. A. R.
Riach, Lt.-Col. A. H. D., R. E.
Riall, Maj. C. P. B., E. York. R.
Riccard, Maj. J. S., R.A.
Rich, Maj. E. T., R.E.
Richey, Maj. F. W., R.A.
Rimington, Maj.-Genl. J. C., C.B.
Roberts, Maj. W. H., MC., R.E.
Robertson, Maj. D. S., *p.s.c.*, R. Sc. Fus.
Robertson, Maj. W., *p.s.c.*, R.E.
*Robertson-Glasgow, Maj. N., R.A.
Robinson, Lieut., E., R.A.
Robinson, Lt.-Col. P. M., C.M.G., R. W. Kent R.
Robinson, Lt.-Col. S. W., D.S.O., R.A.
Roe, Col. C. H., C.I.E.
Rolland, Maj. A., D.S.O., R.E.
Rolland, Maj. C. E. T., R.A.
Rooke, Maj. E. H., R.E.
Roome, Capt. H. E., M.C., R.E.
Rose, Maj. D'A. L., 1-9, Bn., Hamps. R.
Ross, Maj. A. J., D.S.O., R.E.
Ross, Bt.-Col. R. J., *p.s.c.*
Rothwell, Maj. R. S., R.A.
Rountree, Lt. A. N., A.S.C.
Rouse, Col. H., C.B., D.S.O., R.A.
*Routh, Maj. G. M., D.S.O., R.A.
Row, Capt. A. W. W., Midd'x. R.
Rowley, Lt.-Col. F. G. M., C.M.G., Midd'x. R.
Rowan-Robinson, Maj. H., D.S.O., *p.s.c.*, R.A.
Rudkin, Lt.-Col. W. C. E., D.S.O., R.A.
Russell, Capt. A. G., M.C., R.A.
Russell, Maj., W. K., R.E.
*Ryan, Lt. C. E., M.C., R. A.
Ryder, Lt.-Col. C. H. D., C.I.E., D.S.O., R.E.
Sadler, Maj. H., 6th Dgn. Gds.
St. John, Col. G. F. W., C.B.
Salt, Maj. H. F., R.A.
Sanders, Bt.-Lt.-Col A. R. C., D.S.O., *p.s.c.*, R.E.
Sanders, Lt.-Col. G. A. F., *p.s.c.*, R.E.
Sanders, Maj. W. O. S., R.A.
Sandes, Capt. E. W. C., M.C., R.E.
Sandford, Maj. D. A., D.S.O., R.A.
Sandford, Capt. H. R.
Sandilands, Maj. H. R., North'd. Fus.
Sandys, Lt.-Col. W. B. R., C.M.G., R.A.
Sargeaunt, Maj. H. G., R.A.
Sarson, Maj. E. V., R.A.
Saulez, Capt. A. T., R.A.
Saunders, Lt.-Col. F. W., R.E.
Schrottky, Maj. C. E. G., R.A.
Scafe, Maj. W. E., D.S.O., Devon R.
Scarlett, Maj. J. A., D.S.O., R.A.
Scott, Maj. A. C., R.E.
Scott, Bt.-Col. E. W. W., A. S. Corps.
Scott, Capt. G. C., R. Ir. Rif.
Scott, Capt. M. A. H., R.E.
Scovil, Capt. F. H., R.A.
Scrase-Dickins, Col. S. W., C.B.
Scudamore, Col. W. V., R.E.
Seabroke, Maj. G. T., E. Lan. R.
Seagrim, Maj. A. H., Leins. R.
Seal, Lt. J. R., I. G. B., E. York R.
Seymour, Maj. B. K. R. Rif. C.
Shaw, Lt. G. C., R.A.
Shaw, Maj. G. D. A., R.A.
Shekleton, Col. H. P., C.B., C.M.G. *p.s.c.*
Sheppard, Bt.-Col. S. H., D.S.O. *p.s.c.*, R.E.
Sherer, Lt.-Col. J. D., R.A.
Short, Lt.-Col. W. A., C.M.G., R.A.
Simon, Maj. H. S. P., D. of Corn. L. I.
Sitwell, Maj. N. S. H., R.A.
Sitwell, Col. W. H., C.B., D.S.O., *p.s.c.*
Skinner, Col. F. St. D., *p.s.c.*
Skinner, Maj. R. B., R.E.
Sloan, Bt.-Col. J. M., D.S.O., M.B. R. A. M. C.
Smallwood, Lt.-Col. F. G., C.V.O., R.A.
Smart, Maj. G. E., R.A.
Smith, Lt.-Col. E. O., C.M.G
Smith, Col. G. B., C.B.
Smith, Capt. R. C., Dur. L. I.
Smith, Bt.-Col. W. E. B., C.M.G., Linc. Regt.
Smith, Col. W. D., C.B., R. Sc. Fus.
Smyth, Lt.-Col. G. A., M.C., R.A.
Smyth, Maj. G. J. W., R.E.
Smyth, Capt. G. O. M.C., S., R.A.
Smyth, Col. R. N., D.S.O.
Smyth-Osbourne, Maj. G. N. T. *p.s.c.*, Devon. R.
Sneyd, Capt. D. G. T., M.C., R.A.
Snowdon, Maj. H. S. K., R.A.

* Passed Proficiency test.

LIST OF BRITISH SERVICE OFFICERS ON THE ACTIVE LIST (WHO ARE NOT GENERAL OFFICERS) QUALIFIED IN HINDUSTANI BY THE HIGHER STANDARD—*contd.*

Name, Rank and Corps.

Somerville, Capt. J. A. H. B., R.A.
Somerville, Maj. W. A.T.B., D.S.O., R. Lanc. R.
Sorsbie, Col. R. F.,
Spaight, Maj. T. H. L., R.E.
Spalding, Maj. W. B., R.A.
Span, Maj. H. J. B., Welsh Regt.
Spencer, Maj. F. E., M.C., R.A.
Speranza, Lt.-Col. W. S., R.E.
Sprague, Maj. L. C., *p.s.c.*, R. Ir. Rif.
Squires, Capt. E. K., D.S.O., M.C., R.E.
Stack, Maj. G. H., D.S.O., R.E.
Stagg, Capt. M., R.E.
Stanbrough, Maj. L. K., D.S.O. R.A.
Standbridge, Maj. W. J., R. A.
Statham, Lt.-Col. J. C. B., C.M.G., R.A.M.C.
Stebbing, Capt. H. T., R.A.
Steel, Maj. E. A., D.S.O., R.A.
Stewart, Lt.-Col. C. G., C.M.G., D.S.O., *p.s.c.*, R.A.
Stewart, Lt.-Col. D. B., D.S.O., R.A.
Stewart, Lt.-Col. J. A., A.O.L.
Stewart, Maj. P. A. V., D.S.O., *p.s.c.*, K. O. Sco. Bord.
Stewart, Maj. R. S., *p.s.c.*, L'pool R.
Stewart, Maj. W. M., *p.s.c.*, Cam'n Highrs.
Stirling, Maj. *Sir* G. M. H., Bt., D.S.O., Essex Regt.
Stockdale, Lt.-Col. H. E., C.M.G., D.S.O., R.A.
Stockwell, Col. G. C. I., D.S.O., *p.s.c.*
Stockley, Maj. E. N., D.S.O., R.E.
Stockley, Lt.-Col. H. R., C.I.E., R.E.
Stoehr, Capt. C. F., R.E.
Stokes, Col. A., C.B., C.M.G., D.S.O.
Stokes-Roberts, Col. E. R. B.
Stoney, Capt. H. H., N. Staff. R.
Strange. Col. R. G.
Strick, Maj. J. A., D.S.O., Shrops. L.I.
Strover, Maj. M. R., R.A.
Stuart, Lt.-Col. D. M.
Sturrock, Maj. G. C., R.A.
Swayne, Maj. O. R., D.S.O., R.A.
Sweetman, Lt.-Col. M. J., E. York. R.
Swettenham, Maj. W. A. W., R.A.
Swinburne, Capt. T. A. S., D.S.O., R.E.
Sykes, Maj. F. B., D.S.O., R.A.
Sykes, Bt. Lt.-Col. F. H., C.M.G., *p.s.c.*, 15th Hussrs.
Tagart, Col. H. A. L., C.B., D.S.O., *p.s.c.*
Tancock, Lt.-Col. O. K., C.M.G., R.A.
Tanner, Maj. F. C., D.S.O.R. Scots.
Tarver, Lt.-Col. W. K., A. S. Corps.
Taylor, Col. E. T., *p.s.c.*
Taylor, Maj. G. J. S., R.A.
Temple, Maj. T., R. A.
Templer, Lt.-Col. W. F., Army Pay Dept.
Tennent, Capt. C. B., R.A.
Thom, Capt. J. H., D.S.O., R.A.
Thomas, Maj. C. S. W., R.A.
Thomas, Lt.-Col. H. M., R.A.

Thompson, Maj. A. J., A. V. Corps.
Thorp, Maj. A., R.A.
Thorpe, Bt. Lt.-Col., E. I. de S., Bedf. R.
Thuillier, Lt.-Col. H. F., C.B., C.M.G., R.E.
Tidswell, Col. E. C., D.S.O.
Tighe, Lt.-Col. F. A., R.A.
Todd, Capt. H. C., M.B. R. A. M. C.
Todd, Capt. V. O., R. Lan. R.
Toke, Bt. Lt.-Col. R. T., Welsh Regt.
Tollemache, Capt. L. deO., Linc.R.
Tomkins, Lt.-Col. E. L., R.A.
Toogood, Maj. A. S., R. War. R.
Toogood, Maj. C., D.S.O., Linc. R.
Toppin, Maj. S. M., M.C., R.A.
Tovey, Maj. G. S., D.S.O., R.A.
Townshend, Capt. P. A. K., R. Berk. R.
Traill, Capt. T. B., D.S.O., R. SC. Fus.
Traill, Maj. W. H., *p.s.c.*, E. Lanc. R.
Traill, Maj. W. S., D.S.O., R.E.
Trench, Maj. D. leP., M.C., R.A.
*Trenchard, Capt. G. B. B., R.A.
Trent, Lt.-Col. G. A., C.M.G., North'n. R.
Triscott, Col. C. P., C.B., D.S.O.
Troup, Capt F. C. A., R.A
Tudor, Col. E. A. T., R.E.
Tulloch, Col., J. A. S., C.B., *p.s.c.*, R.E.
Tudor, Lt.-Col. H. H., C.M.G., R.A.
Turner, Lt.-Col. F.C., C.M.G., *p.s.c.*, North'd. Fus.
Turner, Col. M. N., C.B.
Turner, Maj. W. M. FitzH., R.A.
Tuson, Col. H. D., C.M.G., *p.s.c.*,
*Tute, Capt. C.S. R.A.
Tweedie, Maj. H. C., D.S.O., N., Staff. R.
Tweedie, Maj. J. L. F., Glouc. R.
Twining, Maj.-Genl. P. G., C.B., C.M.G., M.V.O., R.E.
Twiss, Maj. F. A., M.V.O., R.A.
Tyler, Maj. R. M., *p.s.c.*, Dur. L. I.
Tyrrell, Maj. J. F., R.A.
Tytler, Maj. R. A. N., Gord. Highrs.
Underhill, N., R. G. A.
Urquhart, Maj. A. M., R.A.
Venables, Maj. J. D., R. W. Fus.
Ventris, Capt. E. F. V., R.A.
Villiers, Maj. E. F., C.M.G., D.S.O., R. Suss. R.
Vincent, Maj. B., C.M.G., *p.s.c.*, 6th Drgns.
Vines, Maj. C. E., R.A.
Wace, Maj. E. G., D.S.O., *p.s.c.*, R.E.
Wagstaff, Maj. C. M., C.I.E., D.S.O., *p.s.c.*, R.E.
Walden, Tempy. 2nd-Lt., A. E., R. G. A.
Walker, Capt. E. C., R.E.
Walker, Maj. E. W. M., R.A.
Walker, Maj. H. A., D.S.O., *p.s.c.*, R. Fus.
Walker, Maj. J. B., D.S.O., R.A.

Walker, Lt.-Col. M., R.A.
Walker, Maj. W. B., R.A.
Wallace, Capt. R. F. H., *p.s.c.*, R. Highrs.
Waller, Maj. R. S. DeW., R.A.
Walshe, Lt Col. F. W. H., *p.s.c.*, R.A.
Walter, Col. J. Mac N., C.B., D.S.O., *p.s.c.*
Wanliss, Lt.-Col. C.
Warburton, Maj. W. M., R.A.
Ward, Maj. B. S., R.A.
Wardrop, Lt.-Col. A. E., C.M.G., R.A.
Warrington-Morris, Maj. W. D., R.A.
Watling, Lt.-Col. F. W., *p.s.c.*, R.E.
V.C. Watson, Lt.-Col. T. C., R.E.
Watson, Capt. V. C., R.A.
Watts, Lt.-Col. C. D. R., C.M.G., A. O. D.
Wavell, Maj. A. P., M.C., *p.s.c.*, R. Highrs.
Way, Lt.-Col. L., R. A. M. C.
Webb, Maj. E. C., F.R.C.V.S., A. V. Corps.
Webb, Maj. R. E., York and Lanc. R.
Weddell, J. M., Capt., R. A. M. C.
Weeding, Maj. T., R. W. Surr. R.
Weedon, Lt.-Col. F. F., R. E.
Wellesley, Lt.-Col. R. A. C., C.M.G., R.A.
Wells, 2nd-Lt., P. G., 2nd L'pool Regt.
West, Maj. A. H. D., R. A.
Weston, Maj. R. S., C.M.G., Manch. R.
White, Bt. Lt.-Col., J. R., R. E.
White, Maj. R. F., Essex R.
Whitting, Maj. E. Le G., M.C., R.A.
Wickham, Capt. J. C., D.S.O., R.E.
Wilkes, Maj. G. S., R.A.
Wilkins, Capt. D. A., E. Kent, R.
Wilkins, Maj. G. H. C., R.A.
Wilkins, Lt.-Col. H. St. C., *p.s.c.*, R. W. Surr. R.
Wilkinson, Maj. A. C., D.S.O., R.A.
Wilkinson, Maj. H. B. des. V., *p.s.c.*, Dur. L. I.
*Williams, Maj. H. J., 1st Drgn. Guards.
Williams, Bt., Lt.-Col. O. de L. D.S.O. R. W.Fus.
Williams, Maj. R. C., D.S.O., R.A.
Williams, Bt.-Col. W. de L., D.S.O., *p.s.c.*, Hamp. R.
Williamson Oswald, Col. O. C., R.A.
V.C., Willis, Maj. R. R., Lan. Fus.
Wilson, Capt. E. R. C., R.A.
Wilson, Capt. G., R. A. M. C.
Wilson, Maj. J. H. B., W. Rid. R.
*Wilson, Capt. R. E., D.S.O., R.A.
Winsloe, Lt.-Col. A. R., D.S.O., R.E.
Winsloe, Bt. Lt.-Col. H. E., D.S.O., R.E.
Winter, Lt.-Col. O. de l'E., C.M.G., D.S.O., R. A.

* Passed Proficiency test.

LIST OF BRITISH SERVICE OFFICERS ON THE ACTIVE LIST (WHO ARE NOT GENERAL OFFICERS) QUALIFIED IN HINDUSTANI BY THE HIGHER STANDARD—*concld.*

Name, Rank and Corps.	Name, Rank and Corps.	Name, Rank and Corps.
Wintour, Col. F., C.B., *p.s.c.*	Woodhouse, Capt. J. D. F., 14th Hussrs.	Yarde.-Buller, Col. *Hon'ble* H., D.S.O., M.V.O., *p.s.c.*
Wise, Capt. P.K., D.S.O., R.War. R.	Woods, Lt.-Col., G. G., R.A.	Yatman, Bt.-Col. C., D.S.O., North'd Fus.
Witts, Capt. F.V.B., D.S.O., M.C., R.E.	Wright, Capt. S. H., R.A.	Yeatman, Capt. G. D., Dorset. R.
Wolley-Dod, Col. O. C., C.B., D.S.O., *p.s.c.*	Wyatt, Lt.-Col. F. O., M. V. O., R.A.	Young, Col. C. A., C.M.G., R.A.M.C.
Wood, Maj. H., R.E.	Wylde, Lt.-Col. R. D., R.A.	*Young, Bt. Lt.-Col. E. de L., R.E.
Wood, Bt.-Col. T. B., C.M.G., *p.s.c.*, R.A.	Wynne, Lt.-Col. H.E.S., D.S.O., R.A.	Young, Lt.-Col. H. A., R.A.
Woodhouse, Capt. H. L., M.C., R.E.	Wynter, Lt.-Col. F. A., D.S.O., R.A.	Young, Maj. H. G., D.S.O., R.A.

* Passed Proficiency test.

NON-EFFECTIVE LIST.

OFFICERS OF THE INDIAN ARMY ON THE UNEMPLOYED SUPERNUMERARY LIST.

Note.—The names on this list are not indexed as they are arranged alphabetically.

Generals.

Names, Rank and Corps.	Date of Birth.	Date of 1st Commn.	Date of Rank.	Date of Transfer to U. S. List.
M. Aitchison, C. T., C.B.	28 May 23	10 June 42	22 Jan. 89	8 Jan. 87
M. Anderson, H. C., C.B. (*Col., 4 Rajputs*)	28 May 26	2 Sep. 42	1 Apl. 94	2 Sep. 87
M. Baillie, J. C. P., B. I.	8 Sep. 35	20 Dec. 52	1 Apl. 94	1 Jan. 84
Burrows, G. R. S., Bo. I	18 Feb. 27	1 Jan. 44	17 Aug. 90	1 Juy 86
M. Cockburn, H. A., B. I.	10 Nov. 31	13 June 51	1 Apl. 94	6 Mar. 87
M. Edwardes, *Sir* S. de B., K.C.B., Bo. I. (*Col., 102 Grs.*)	22 Mar. 40	20 Feb. 57	9 Nov. 96	1 Apl. 94
Evans, *Sir* H. M., K.C.B., ® (*Col., 8 G. R.*)	8 Dec. 41	20 June 58	9 Nov. 01	9 Dec. 03
M. Gordon, *Sir* T. E., K.C.B., K.C.I.E., C.S.I., ®	12 Jan. 32	21 Aug. 49	1 Apl. 94	29 Jan. 93
M. Gordon, W., C.I.E.	10 Feb. 24	11 June 41	1 Apl. 94	10 Feb. 86
Halliday, J. G., M. I.	9 May 22	9 Sep. 38	1 Dec. 88	8 Jan. 83
Jennings, *Sir* R. M., K.C.B., B.G.L.C. (*Col., 6 K. E. O. Cavy.*).	20 Nov. 41	20 May 59	28 Jan. 05	1 Apl. 06
M. Lumsden, *Sir* P. S., G.C.B., C.S.I., ®	9 Nov. 29	10 Dec. 47	17 Aug. 90	9 Nov. 96
M. Miles, J.	28 Nov. 27	3 Feb. 45	1 Apl. 94	28 Nov. 85
V.C. M. Probyn, *Rt. Hon. Sir* D. M., G.C.B., G.C.S.I., G.C.V.O., I.S.O., B. C., ®. (*Col., 11 K. E. O. Lancers*).	21 Jan. 33	20 Oct. 49	1 Dec. 88	1 July 82
M. Thomson, *Sir* M., K.C.I.E., B. I.	1 Apl. 32	20 Dec. 53	1 Apl. 94	2 July 85
Tomkins, W. P., C.I.E.	11 June 41	8 June 60	1 Apl. 98	12 June 01
Walker, G. W.	6 Apl. 23	9 Dec. 42	28 Nov. 85	1 Feb. 82
V.C. M. Watson, *Sir* J., G.C.B. (*Col., 13 D. C. Lancs.*)	4 Sep. 29	3 Feb. 48	1 Mar. 91	1 July 86
Young, J. N.	4 Feb. 24	21 Aug. 40	1 Mar. 91	4 Feb. 86

Lieutenant-Generals.

Names, Rank and Corps.	Date of Birth.	Date of 1st Commn.	Date of Rank.	Date of Transfer to U. S. List.
M. Clifford, R. C. R., C.B. ® (*Col., 22 Cavy. F.F.*)	30 June 39	13 Dec. 56	7 Feb. 99	12 Sep. 00
Halliday, G. T., B. C. (*Col., 16 Cavalry*)	4 Nov. 41	20 Dec. 58	4 Nov. 00	9 Nov. 01
Handcock, A. G., C.B., ® (*Col., 6 Jat L. I.*)	24 Feb. 40	11 Dec. 58	21 Feb. 00	20 Jan. 01
Hankin, E.L.	22 Nov. 31	10 Dec. 47	23 Sep. 92	10 Dec. 92
M. Iredell, F. S., Bo. I. (*Col., 114 Mahrattas*)	. . . .	10 Dec. 54	1 Jan. 93	19 Dec. 91
Jones, G. T., B. I.	5 Feb. 39	26 Aug. 58	1 Jan. 93	22 Apl. 90
V.C. M. Lyster, H. H., C.B., ®	24 Dec. 30	20 Sep. 48	1 Sep. 91	1 July 92
M. McNair, E. J., B. I.	9 July 38	26 July 56	20 Dec. 91	24 Dec. 92

OFFICERS OF THE INDIAN ARMY ON THE UNEMPLOYED SUPERNUMERARY LIST—*contd.*

Lieutenant-Generals—*contd.*

Names, Rank and Corps.	Date of Birth.	Date of 1st Commn.	Date of Rank.	Date of Transfer to U. S. List.
Morton, B. W. D.	24 Aug. 29	9 June 48	10 Dec. 92	24 Aug. 87
𝔐. Phelps, A., Bo. I.	24 Oct. 37	20 Dec. 53	24 Dec. 92	1 Jan. 93
𝔐. Smith, O. L.	30 July 28	12 June 47	1 July 92	30 July 86
Stewart, J. M., B. I.	19 June 41	20 Oct. 58	10 Feb. 94	1 Apl. 94
𝔐. Story, P., B. I.	27 Jan. 40	11 Dec. 57	1 Jan. 93	1 Jan. 93
Strahan, C.	6 May 43	8 June 60	28 Apl. 01	1 Oct. 04
Tyrrell, F. H., M. I. (*Col., 74 Punjabis*)	5 Nov. 40	5 May 57	1 Aug. 94	1 Nov. 94
Willoughby, M. W., C.S.I., Ⓡ (*Col. 104 Rifles*)	18 Mar. 33	9 June 49	9 Dec. 93	9 June 94

Major-Generals.

Names, Rank and Corps.	Date of Birth.	Date of 1st Commn.	Date of Rank.	Date of Transfer to U. S. List.
𝔐. Bartleman, W. F., B.I.	26 Apl. 40	16 Sep. 57	5 Oct. 93	12 Feb. 95
Begbie, E. W., C.B., D.S.O., Ⓡ, M. I.	15 June 42	4 Apl. 59	6 June 98	21 Dec. 02
Berkeley, J. C., C.I.E., M. I.	23 Jan. 39	10 Oct. 57	1 Jan. 93	1 Jan. 93
𝔐. Boyd, J. M.	4 Oct. 37	20 Feb. 54	9 June 94	4 Oct. 96
𝔐. Campbell, A. E.	18 Mar. 34	20 Jan. 51	1 July 92	18 Mar. 92
𝔐. Chambers, R. M., Bo. I.	7 June 33	21 July 58	1 July 91	7 June 91
𝔐. Chamier, F. E. A., C.B., C.I.E.	13 May 33	20 Feb. 50	8 Oct. 90	8 Oct. 90
𝔐. Chapman, H. (*Col., 8 Cavalry*)	11 Oct. 35	15 July 56	9 Dec. 93	12 Sep. 95
Church, T. R., C.I.E.	5 Jan. 31	14 June 50	13 June 91	5 Jan. 89
Coape-Smith, H.	24 Sep. 29	20 Dec. 48	31 May 90	24 Sep. 87
Cuningham, C. A. (*Col., 127 Bal. L. I.*)	29 Jan. 42	20 June 58	12 Jan. 99	29 Jan. 04
Currie, F., B. Cav.	24 Nov. 41	20 Nov. 58	6 June 98	6 June 98
𝔐. Dalmahoy, P. C., B. I.	7 Sep. 40	13 Dec. 56	20 Aug. 90	20 Aug. 90
Dixon, E. G., M. I.	15 Dec. 37	20 Feb. 57	3 Jan. 91	5 Apl. 93
𝔐. Elliot, H. R., M. I.	11 Dec. 36	20 Aug. 55	13 June 91	5 Oct. 93
Elliott, E. D.	30 Sep. 38	11 Dec. 57	1 Apl. 95	15 May 00
Elton, H. S., Ⓡ (*Col., 76 Punjabis*)	30 Dec. 41	20 Oct. 58	10 June 98	9 Jan. 01
𝔐. Evans, L. E., B. I.	17 Mar. 37	13 Dec. 56	1 July 91	22 Mar. 93
Fagan, J. L.	4 Feb. 43	11 June 59	6 Sep. 99	4 Feb. 02
𝔐. Fisher, J. F. L.	10 Dec. 32	20 Dec. 48	31 May 90	31 May 90
𝔐. Franks, W. A., B. I.	14 June 38	4 Oct. 55	1 Apl. 94	1 Apl. 94
𝔐. Galloway, J. M. C., M. C., Ⓡ (*Col., 27 Cav.*)	19 Apl. 40	1 Sep. 57	25 May 94	19 Apl. 99
𝔐. Gatacre, *Sir* J., K.C.B. (*Col., 123 Rifles*)	14 Feb. 41	13 June 57	10 Dec. 97	10 Dec. 02
𝔐. Gurdon, E. P.	4 June 33	20 Jan. 52	10 Dec. 92	20 Jan. 90
Henderson, P. D., C.S.I., M. C.	19 Aug. 40	20 Feb. 57	16 Sep. 95	16 Sep. 95
Hogg, G. C., C.B., Ⓡ (*Col., 34 Poona Horse*).	13 Sep. 42	27 Oct. 58	20 Sep. 97	13 Sep. 04
𝔐. Howey, W., B. I	7 Aug. 38	4 Feb. 58	1 Apl. 94	1 Apl. 94
Hughes, C. F., C.B. Ⓡ	19 Mar. 44	7 June 61	21 Feb. 00	19 Mar. 03
Hunt, R. A. C., Bo. I. (*Col., 105 Infy.*)	23 Aug. 38	4 Oct. 57	6 Feb. 91	30 July 93
𝔐. Jacob, W.	14 Dec. 37	20 Jan. 55	22 Oct. 94	2 May 97
Lawrence, W. A. (*Col., 17 Cavalry*)	5 July 43	4 Feb. 61	8 Nov. 97	5 July 02
𝔐. LeGeyt, P. H.	22 Dec. 34	20 Feb. 51	23 Sep. 92	23 Sep. 92
𝔐. Maclean, C. S., C.B., C.I.E., Ⓡ (*Col., 21 Cavalry*)	15 Dec. 36	14 Dec. 53	2 Sep. 93	1 Feb. 96
Magrath, B. H. W., Ⓡ	4 Feb. 32	20 Dec. 49	31 May 90	4 Feb. 90
Marsh, F. H. B., B. I.	26 July 41	20 May 59	12 Aug. 97	12 Aug. 97
𝔐. Moore, C. A., Bo. C., Ⓡ	12 Nov. 39	4 Nov. 56	25 Apl. 95	25 Apl. 09
𝔐. Prendergast, G. A. (*Col., 15 Lancers*)	11 Mar. 34	20 Mar. 50	3 Jan. 91	19 Nov. 93
Quin, T. J., B. I.	1 May 24	4 Mar. 50	6 June 98	6 June 98

OFFICERS OF THE INDIAN ARMY ON THE UNEMPLOYED SUPERNUMERARY LIST—*contd.*

Major-Generals—*contd.*

Names, Rank and Corps.	Date of Birth.	Date of 1st Commn.	Date of Rank.	Date of Transfer to U. S. List.
M. Rowcroft, G. C.	17 Sep. 31	10 Aug. 49	17 Aug. 90	17 Sep. 89
Rutherfurd, T. W.	4 Sep. 32	14 June 51	23 Sep. 92	4 Sep. 90
Saward, M. H.	22 Dec. 40	9 Dec. 59	23 June 96	1 Apl. 03
Searle, A. T.	29 Dec. 30	12 Dec. 49	8 Oct. 90	29 Dec. 88
Shakespear, G. R. J., ®	14 Oct. 42	7 Sep. 60	27 Sep. 99	25 Oct. 02
Smith, J. B.	12 June 39	26 June 56	3 Mar. 95	12 June 98
M. Stansfeld, H. H., B. I.	10 Oct. 39	13 Dec. 56	17 Aug. 90	23 Sep. 90
M. Stewart, J. C., C.B. (*Col., 7 Hariana Lancers*)	20 June 40	4 Sep. 57	29 Oct. 95	20 June 99
Swiney, J.	9 Feb. 32	12 Apl. 49	31 May 90	9 Feb. 90
Sym, *Sir* J. M., K.C.B., ®	15 Feb. 39	11 Dec. 58	4 Oct. 96	27 Sep. 99
M. Szczepanski, H. C. A.	21 Jan. 41	13 June 57	10 Dec. 97	10 Dec. 97
Toker, *Sir* A. C., K.C.B., ® (*Col., 18 Infantry*)	10 Dec. 43	4 May 60	2 May 97	2 May 02
Tregear, *Sir* V. W., K.C.B., B. I.	25 June 42	11 June 59	12 Aug. 97	1 Oct. 02
M. Upperton, J., C.B.	10 June 38	10 June 54	20 Sep. 94	20 Sep. 94
M. Walker, J. G. D., M. C.	4 Sep. 42	4 Oct. 58	12 Sep. 97	6 June 98
M. Ward, F. W., C.B.	6 Apl. 40	11 Dec. 57	25 May 95	25 May 00
Waterhouse, J.	24 July 42	10 June 59	27 June 99	10 June 04
M. Westmacott, *Sir* R., K.C.B., D.S.O., ® (*Col., 128 Pioneers*).	16 Mar. 41	19 Mar. 59	20 June 99	16 Mar. 03

Colonels (with honorary rank of Brigadier-General).

Names, Rank and Corps.	Date of Birth.	Date of 1st Commn.	Date of Rank.	Date of Transfer to U. S. List.
Abbott, H. A., C.B. (*Col., 15 Sikhs*)	22 Jan. 49	15 Feb. 68	15 Feb. 98	22 Jan. 06
M. Adam, F. J. S.	22 Sep. 36	20 Dec. 56	20 Dec. 86	15 Dec. 94
Gilchrist, R. A.	26 Nov. 45	4 Aug. 65	4 Aug. 95	26 Nov. 02
Jopp, J., C.B.	13 Mar. 40	20 Oct. 56	20 Oct. 86	13 Mar. 97
Read, H., C.B.	19 Aug. 52	15 Jan. 73	15 Jan. 03	19 Aug. 09
Rolland, S. E. (Col., 86 Carnatic Infantry).	7 Nov. 46	9 Nov. 66	29 Aug. 93	7 Nov. 09

Colonels.

Names, Rank and Corps.	Date of Birth.	Date of 1st Commn.	Date of Rank.	Date of Transfer to U. S. List.
M. Anderson, C. J., ®	2 Sep. 35	20 May 53	20 May 83	2 Sep. 92
Baillie, R.	28 Jan. 57	27 Jan. 76	27 Jan. 05	28 Jan. 14
M. Barnes, O., C.B.	23 Dec. 34	4 Mar. 55	4 Mar. 85	4 Mar. 93
Beckett, W. H., B. I.	18 Feb. 37	20 Feb. 59	20 Feb. 89	18 Feb. 94
Beddy, E.	11 Jan. 37	20 Feb. 57	20 Feb. 87	20 Feb 95
M. Biddulph, J.	25 July 40	20 Jan. 58	24 Feb. 88	25 July 97
Bishop, E. B.	28 Sep. 43	2 Dec. 62	29 Aug. 93	2 Dec. 00
M. Boyle, *Hon.* R. E.	3 June 37	10 Dec. 54	10 Dec. 84	3 June 94
Bedin, A.	10 Oct. 37	11 Mar. 59	11 Mar. 89	10 Oct. 94
Bromhead, *Sir* B. P., *Bt.*, C.B.	24 Oct. 38	18 Jan. 59	18 Jan. 89	18 Jan. 97
Brownlow, C. C., C.B. [R]	20 Nov. 43	8 June 61	1 Sep. 91	20 Nov. 00
Bruce, A. McC., C.B., B. I., ®	5 Jan. 42	4 Nov. 60	4 Nov. 90	5 Jan. 99
Bruce, E. A., ®	9 Aug. 40	6 Jan. 59	1 July 87	9 Aug. 97
Byng, T. R.	9 Dec. 39	11 June 59	11 June 89	9 Dec. 96
V.C. M. Cadell, T., C.B.	5 Sep. 35	17 Apl. 54	17 Apl. 84	5 Sep. 92
Campbell, J. E.	3 Sep. 42	10 Dec. 59	10 Dec. 89	3 Sep. 99

OFFICERS OF THE INDIAN ARMY ON THE UNEMPLOYED SUPERNUMERARY LIST—*contd.*

Colonels—*contd.*

Names, Rank and Corps.	Date of Birth.	Date of 1st Commn.	Date of Rank.	Date of Transfer to U. S. List.
𝔐. Cantor, C. H.	21 Nov. 36	13 Dec. 56	13 Dec. 86	13 Dec. 94
Carruthers, J. G. T., B. I.	11 Nov. 40	8 June 60	8 June 90	11 Nov. 97
𝔐. Chapman, A. R.	26 Oct. 36	4 Jan. 56	4 Jan. 86	4 Jan. 94
Chatterton, F. W., C.I.E., B. I.	1 June 39	20 Sep. 59	20 Sep. 89	1 June 96
𝔐. Chrystie, G.	7 Sep. 41	12 Dec. 57	12 Dec. 87	7 Sep. 98
Clarke, T. G.	10 June 35	26 Jan. 53	26 Jan. 83	10 June 92
Codrington, G. H. F.	29 May 37	13 June 57	13 June 87	13 June 95
Cooke-Collis, M. C.	8 Oct. 50	30 Dec. 71	30 Dec. 01	8 Oct. 07
Cowie, T. R.	12 Feb. 43	4 July 60	4 July 90	12 Feb. 00
Cuninghame, D. S.	5 June 47	1 Dec. 65	29 Oct. 98	25 June 04
Currie, A., Bo. C.	9 Dec. 40	27 Oct. 58	27 Oct. 88	9 Dec. 97
deKantzow, C. A.	27 June 35	14 Sep. 53	14 Sep. 83	27 June 93
Ewing, J.	13 Mar. 36	7 Dec. 55	7 Dec. 85	7 Dec. 93
Eyre, E. H., C.B., Ⓡ	26 Nov. 38	4 Mar. 58	1 July 87	4 Mar. 96
Fancourt, St. J. M., C.B.	30 Sep. 47	1 Dec. 69	26 Mar. 95	30 Sep. 94
𝔐. Furlong, M.	11 Aug. 39	4 Jan. 57	4 Jan. 87	11 Aug. 96
Garbett, C. H. V.	19 Mar. 49	17 Nov. 69	11 July 01	19 Mar. 06
Georges, H. W. E.	1 Sep. 56	26 Feb. 78	26 Feb. 07	1 Sep. 13
Goldie, J. O.	8 May 41	4 Sep. 58	4 Sep. 88	8 May 98
Gordon, H. W.	21 May 37	6 May 54	6 May 84	21 May 94
Gordon, J. H., C.B., D.S.O.	25 Jan. 39	20 Apl. 57	20 Apl. 87	25 Jan. 96
Gordon, L. A., C.B.	28 May 57	14 June 77	23 Oct. 06	28 May 14
Gordon, R.	18 Jan. 50	8 Jan. 68	8 Jan. 98	18 Jan. 07
Grant, J.	17 Sep. 47	30 June 65	30 June 95	17 Sep. 04
𝔐. Grey, L. J. H., C.S.I.	1 July 40	13 Dec. 56	13 Dec. 86	1 July 97
Grierson, J.	24 Aug. 39	12 Dec. 57	12 Dec. 87	24 Aug. 96
Griffith, R.	1 Feb. 33	13 June 52	13 June 82	13 June 90
VC Hammond, *Sir* A. G., K.C.B., D.S.O., Ⓡ	28 Sep. 43	7 June 61	12 Feb. 90	28 Sep. 00
Hancock, G. E.	28 Jan. 41	10 June 59	10 June 89	28 Jan. 98
Hatchell, D. T.	2 Apl. 39	12 June 58	12 June 88	12 June 96
Hawkes, G.	10 Oct. 51	19 Feb. 70	19 Feb. 00	10 Oct. 08
Hawkins, F.	18 Sep. 56	21 Sep. 74	10 Feb. 04	18 Sep. 13
Hesketh, R. W.	8 Sep. 37	20 Mar. 57	20 Mar. 87	20 Mar. 95
Hogg, T. W.	30 July 41	4 Dec. 57	4 Dec. 87	30 July 98
Humfrey, F. T., Ⓡ	25 Jan. 36	26 Sep. 56	22 Nov. 83	25 Jan. 93
Hutchins, A. G., M.I.	9 Jan. 41	20 Feb. 59	28 Nov. 88	9 Jan. 98
Jacob, G. A.	21 Aug. 40	13 Dec. 56	13 Dec. 86	21 Aug. 87
𝔐. Jacob, H. B.	30 July 41	20 Dec. 57	20 Dec. 87	30 July 98
Jacob, *Sir* S. S., K.C.I.E., C.V.O.	14 Jan. 41	10 Dec. 58	10 Dec. 88	14 Jan. 98
𝔐. Jarrett, H. S., C.I.E.	17 June 39	13 Dec. 56	13 Dec. 86	17 June 96
Kelly, J. G., C.B.	28 Nov. 43	11 Sep. 63	10 July 95	28 Nov. 00
Killgour, F.	9 Apl. 41	11 July 57	11 July 87	9 Apl. 98
King, H. T.	20 Feb. 56	1 May 78	23 May 07	20 Feb. 13
King-Harman, M. J.	1 Mar. 43	2 Oct. 60	2 Oct. 90	1 Mar. 00
Kirkwood, J. N. S.	4 Jan. 46	1 Sep. 65	1 Sep. 95	4 Jan. 03
Lane, C. T., C.I.E.	20 Feb. 42	11 Dec. 58	1 Dec. 88	20 Feb. 99
Lawford, E. A. M.	11 Mar. 56	9 Aug. 73	9 Aug. 02	11 Mar. 13
Leckie, F. W. V., Ⓡ	4 Apl. 47	5 July 64	26 July 94	4 Apl. 04
Lecky, G.	20 July 41	12 Nov. 60	12 Nov. 90	20 July 98
Lugard, H. I.	8 Jan. 37	14 Dec. 53	14 Dec. 83	8 Jan. 94
𝔐. Luxmoore, C. T. P.	16 Aug. 39	4 Aug. 57	4 Aug. 87	16 Aug. 96
𝔐. McInroy, C., C.B.	3 Mar. 38	17 Mar. 55	17 Mar. 85	3 Mar. 95
McMullin, J.	11 Aug. 34	16 Nov. 55	16 Nov. 85	11 Aug. 89
McNair, A. L.	12 Oct. 43	27 Aug. 61	29 Aug. 93	12 Oct. 00

OFFICERS OF THE INDIAN ARMY ON THE UNEMPLOYED SUPERNUMERARY LIST—*contd.*

Colonels—*contd.*

Names, Rank and Corps.	Date of Birth.	Date of 1st Commn.	Date of Rank.	Date of Transfer to U. S. List.
M. McNeill, D.	25 June 38	8 Dec. 55	8 Dec. 85	25 Jan. 95
Macdonald, D.	10 Dec. 35	20 Oct. 53	20 Oct. 83	10 Dec. 92
M. Mackenzie, *Sir* A. R. D., K.C.B., B.C., ®	30 Sep. 35	30 Dec. 54	23 July 82	17 Jan. 93
M. Macnagkten, A. W., Bo. C.	19 May 35	4 Mar. 54	4 Mar. 84	29 May 92
M. Macpherson, J. D.	16 May 41	19 Dec. 57	19 Dec. 87	16 May 98
Mainwaring, E. P., B. I.	13 Oct. 41	20 Dec. 59	20 Dec. 89	13 Oct. 98
Major, F. W.	12 Aug. 40	10 Dec. 58	10 Dec. 88	12 Aug. 97
M. Mallock, H. A.	9 Apl. 34	9 Dec. 53	9 Dec. 83	9 Apl. 89
Marriot, E. M. L., B. I.	2 July 40	9 July 59	9 July 89	2 July 97
Marshall, C. H. T.	8 July 41	12 Oct. 59	12 Oct. 89	8 July 98
M. Martin, C., C.B., B. C.	3 Feb. 34	20 Nov. 51	1 Oct. 81	20 Nov. 89
Matthews, C. R.	11 Nov. 38	20 Mar. 55	20 Mar. 85	11 Nov. 95
Melliss, *Sir* H., K.C.S.I., [R]	2 Apl. 47	30 June 65	30 June 95	2 Apl. 04
M. Menzies, O.	30 Aug. 37	26 Aug. 54	26 Aug. 84	30 Aug. 94
Michell, J. W. A.	22 Nov. 40	4 Apl. 58	4 Apl. 88	22 Nov. 97
Middleton, H. J. J.	31 Mar. 56	28 Oct. 76	28 Oct. 05	31 Mar. 13
Money, E. A., C.B.	16 Aug. 43	16 July 60	16 July 90	16 Aug. 00
Money, G. A.	24 Apl. 53	11 Sep. 72	20 Nov. 02	24 Apl. 10
M. Ommanney, E. L., C.S.I.	24 Aug. 34	9 June 55	9 June 85	9 June 93
Onslow, R. C.	26 July 57	11 Sep. 76	11 Sep. 05	26 July 14
Pearse, C. J.	20 July 34	20 June 53	20 June 83	20 July 91
Peart, G. R.	17 July 40	12 Aug. 59	12 Aug. 89	12 Aug. 97
Peirse, C. E.	6 June 54	13 Nov. 75	7 Mar. 05	6 June 11
Philips, J.	4 Dec. 37	27 Nov. 57	27 Nov. 87	27 Nov. 95
Pitcher, D. G.	2 Nov. 39	3 Sep. 59	3 Sep. 89	2 Nov. 96
M. Plowden, C. H.	22 Feb. 37	14 Feb. 54	14 Feb. 84	22 Feb. 94
M. Porteous, C. A., C.I.E.	27 July 39	13 Dec. 56	13 Dec. 86	27 July 95
Pratt, H. M., C.B., ®	24 Oct. 38	13 Dec. 56	2 Mar. 85	24 Oct. 12
Presgrave, E. R. J., C.B., D.S.O.	29 June 55	11 Feb. 75	11 Feb. 04	29 June 93
Preston, B. H.	13 Feb. 36	10 June 54	10 June 84	13 Feb. 98
Prideaux, W. F., C.S.I.	30 Apl. 40	12 Jan. 60	12 Jan. 90	12 Jan. 90
Rivaz, V., C.B.	26 June 42	8 June 60	8 June 90	26 June 99
M. Ross, W. H.	23 Mar. 35	20 Dec. 52	20 Dec. 82	23 Mar. 92
Rowlandson, M. A.	12 Nov. 41	20 Aug. 58	20 Aug. 88	12 Nov. 98
Rundall, F. M., C.B., D.S.O.	18 May 51	9 Mar. 72	8 June 02	18 May 08
Salmon, W. H.	2 Feb. 48	12 Jan. 66	12 Jan. 96	3 Feb. 05
Sandeman, J. E.	3 Sep. 45	7 June 61	29 Aug. 93	3 Sep. 00
M. Sconce, J.	2 Apl. 36	9 Dec. 53	9 Dec. 83	2 Apl. 93
Scott, T. A.	29 Aug. 39	13 June 57	13 June 87	29 Aug. 96
Scott, W.	3 Jan. 43	12 Mar. 60	12 Mar. 90	3 Jan. 00
Seton, W. J.	10 Oct. 36	20 July 55	20 July 85	10 Oct. 93
Shaw, E.	28 Sep. 41	12 Dec. 57	12 Dec. 87	28 Sep. 98
Shepherd, C. E.	22 Mar. 43	19 Dec. 60	19 Dec. 90	22 Mar. 00
Skinner, G. J., D.S.O.	16 June 41	16 Sep. 59	16 Sep. 89	16 June 98
Smith, J. MacD	9 Aug. 37	20 Apl. 55	20 Apl. 85	9 Aug. 94
M. Stack, C. E., Bo. C.	2 Apl. 36	20 Feb. 53	31 Oct. 82	2 Apl. 93
M. St. Maur-Wynch, H. St. M.	23 Dec. 33	12 June 52	12 June 82	12 June 90
Swinhoe, C.	29 Aug. 38	27 July 55	27 July 85	29 Aug. 95
Swinton-Skinner, E.	29 Jan. 43	4 Jan. 60	4 Jan. 90	29 Jan. 00
M. Tabuteau, T. R.	7 Apl. 38	11 Dec. 57	11 Dec. 87	11 Dec. 95
Tandye, H. S., ®	30 June 43	12 June 61	29 Aug. 93	30 June 00
Temple, J. A.	4 Apl. 42	19 Dec. 60	19 Dec. 90	4 Apl. 99
Teversham, R. K., D.S.O.	6 Sep. 56	10 Sep. 75	10 Sep. 04	6 Sep. 13
Thomas, H. R. D.	9 Dec. 48	10 Feb. 69	10 Feb. 99	9 Dec. 05
Thomas, R. M. B.	14 Sep. 40	6 Jan. 59	6 Jan. 89	14 Sep. 97
Trevor, G. H., C.S.I.	29 Jan. 40	12 June 58	12 June 88	29 Jan. 97

OFFICERS OF THE INDIAN ARMY ON THE UNEMPLOYED SUPERNUMERARY LIST—*contd.*

Colonels—*concld.*

Names, Rank and Corps.	Date of Birth.	Date of 1st Commn.	Date of Rank.	Date of Transfer to U. S. List.
Turner, A. H., C.B.	12 Oct. 42	24 May 61	1 Sep. 91	12 Oct. 99
M. Tweedie, M.	18 Mar. 38	20 Sep. 56	20 Sep. 86	18 Mar. 95
Vallings, A.	27 July 38	20 Sep. 58	20 Sep. 88	20 Sep. 96
Vansittart, E.	19 Apl. 56	10 Sep. 75	10 Sep. 06	19 Apl. 13
M. Vibart, E. D. H.	20 Aug. 37	4 Sep. 54	4 Sep. 84	20 Aug. 94
Walker, J. N.	22 Aug. 47	20 July 67	20 July 97	22 Aug. 04
Weldon, F.	30 July 36	26 Sep. 55	26 Sep. 85	26 Sep. 93
Weston, G. E., M. I.	25 Aug. 41	4 Aug. 59	4 Aug. 89	25 Aug. 98
Wetherall, W. A.	8 Apl. 47	30 Jan. 66	30 Jan. 96	8 Apl. 04
Wheatley, H. S., C.B.	16 May 51	3 Aug. 70	24 Oct. 00	16 May 08
White, F. P. L.	15 May 52	24 July 72	10 Feb. 04	15 May 09
Wilkins, W. H.	6 Oct. 41	12 Dec. 57	12 Dec. 87	6 Oct. 98
Willis, J. L. N.	29 Apl. 41	13 Nov. 58	13 Nov. 88	29 Apl. 98
Wilmer, J. R.	20 Jan. 41	22 June 59	22 June 89	20 Jan. 98
Wilson, E. H., C.B.	17 Sept. 45	25 July 65	25 July 95	17 Sep. 02
M. Wilson, W. H.	13 Sep. 39	20 Dec. 56	20 Dec. 86	13 Sep. 96
Wingate, G., C.I.E.	21 Nov. 52	23 Sep. 71	23 Sep. 01	21 Nov. 09
Woodhouse, H., B. I.	15 May 38	11 Dec. 58	16 July 88	15 May 96
Worlledge, J. F.	16 Apl. 56	28 Feb. 74	10 Feb. 04	16 Apl. 13
M. Wright, H. C.	30 May 35	4 Apl. 55	4 Apl. 85	4 Apl. 93
Yate, C.E., C.S.I., C.M.G.	28 Aug. 49	9 Nov. 67	13 Jan. 01	28 Aug. 06
Young, G.	12 Aug. 38	20 Aug. 58	20 Aug. 88	20 Aug. 96
Young, G. G.	5 Jan. 37	13 June 56	13 June 86	13 June 94

Lieutenant-Colonels.

Names, Rank and Corps.	Date of Birth.	Date of 1st Commn.	Date of Rank.	Date of Transfer to U. S. List.
Aberigh-Mackay, J. L.	6 June 44	13 Feb. 67	13 Feb. 93	6 June 99
Aslett, W. C.	12 Jan. 56	13 June 74	13 June 00	12 Jan. [illegible]
Aspinall, H. H. H.	24 Dec. 56	22 Mar. 77	22 Mar. 03	24 Dec. 11
Barr, *Sir* D. W. K., K.C.S.I.	29 Nov. 46	20 Sep. 64	20 Sep. 90	20 Nov. 01
Beckham, L. G.	14 Jan. 56	13 June 74	13 June 00	[illegible] Jan. 11
Butler, J.	14 Mar. 46	20 Feb. 66	20 Feb. [illegible]	14 Mar. 01
Cadell, H. F.	24 Nov. 52	12 Feb. 73	12 Feb. 00	24 Nov. 07
Carey, H. R. LeM.	28 June 42	15 Jan. 61	15 Jan. 87	28 June 97
Clibborn, J., C.I.E.	8 Dec. 47	14 July 69	[illegible] July 95	8 Dec. 02
Cook, W.	19 Nov. 57	10 Sep. [illegible]	10 Sep. 01	19 Nov. 12
Cooke, C. B.	17 Mar. 44	8 Mar 64	8 Mar. 90	17 Mar. 99
Ferris, W. B	29 Oct. 51	18 Aug. 69	18 Aug. 95	29 Oct. 06
Forjett, F. H.	28 Jan. 42	[illegible] June 65	10 June 91	28 Jan. 97
Fryer, G. W. S.	16 Sep. [illegible]	21 Aug. 78	21 Aug. 04	16 Sep. 13
Gärtside-Tipping, R. F., C.B.	17 Aug. 52	3 Aug. 72	3 Aug. 99	17 Aug. 07
Hewett, W. S.	30 Dec. 51	30 Dec. 71	30 Dec. 97	30 Dec. 06
Hunter, J. M., C.S.I.	[illegible] May 44	12 Apl. 64	12 Apl. 90	31 May 99
Hutchinson, J. B., C.S.I.	22 Mar. 44	17 July 63	17 July 89	17 July 01
Jones, A. E.	6 Sep. 55	28 Feb. 74	28 Feb. 00	6 Sep. 10
Jones, F. L.	5 Aug. 56	10 Sep. 75	10 Sep. 03	5 Aug. 11
Jones, R. G., C.I.E.	26 May 55	13 June 74	13 June 01	26 May 10
Jones, W. H. D.	4 Nov. 44	26 July 64	26 July 90	4 Nov. 99
Juxon-Jones, J. H.	1 Feb. 55	10 June 74	10 June 00	1 Feb. 10
Lawson, H.	31 Jan. 55	12 Aug. 74	12 Aug. 01	31 Jan. 11
Lean, W. W.	19 June 53	11 Sep. 72	11 Sep. 98	19 June 08
Little, J. A.	7 Feb. 40	18 Mar. 62	18 Mar. 88	7 Feb. 95
Lumsden, H. W.	9 Sep. 53	6 Sep. 76	6 Sep. 02	9 Sep. 08

OF ICERS OF THE INDIAN ARMY ON THE UNEMPLOYED SUPERNUMERARY LIST—*concld.*

Lieutenant-Colonels—*concld.*

Names, Rank and Corps.	Date of Birth.	Date of 1st Commn.	Date of Rank.	Date of Transfer to U. S. List.
Macbay, W. G. W.	23 Feb. 45	23 Mar. 67	23 Mar. 93	23 Feb. 00
Macmullen, W. H. F.	13 June 48	28 Oct. 71	28 Oct. 97	13 June 03
Maisey, F. C.	7 July 51	23 Sep. 71	23 Sep. 97	7 July 06
Martelli, N. C.	13 May 43	11 Oct. 64	11 Oct. 90	13 May 98
Maxwell, H. St. P., C.S.I.	17 Mar. 50	6 Feb. 69	6 Feb. 95	17 Mar. 05
Meade, J. W. B.	30 Dec. 56	25 Jan. 77	25 Jan. 03	30 Dec. 11
Monck-Mason, H. E.	11 Feb. 56	10 Sep. 75	10 Sep. 02	11 Feb. 11
Money-Simons, J. J.	26 Sep. 44	11 Jan. 67	11 Jan. 93	29 Sep. 99
Newill, J. H.	12 Sep. 44	4 July 65	4 July 91	12 Sep. 99
Ostrehan, F. G. R.	20 Aug. 59	11 May 78	11 May 04	20 Aug. 14
Pears, T. C.	27 June 51	29 May 72	29 May 98	27 June 06
Piers, W. B.	4 Apl. 59	1 May 78	1 May 04	4 Apl. 14
Preston, F. G.	18 Nov. 55	28 Feb. 74	28 Feb. 00	18 Nov. 10
Preston, J. E., C.B., D.S.O.	30 Nov. 55	21 Sep. 74	21 Sep. 00	30 Nov. 10
Pryce, D. D.	23 May 48	8 July 68	8 July 94	23 May 03
Purdon, D. W.	17 July 53	1 May 78	1 May 04	17 July 08
Rennick, A. deC.	2 May 44	6 July 64	6 July 90	2 May 99
Robertson, *Sir* D., K.C.S.I.	24 June 47	8 Aug. 65	8 Aug. 91	24 June 02
Stevens, G. B. (tempy. Col.)	5 Feb. 59	19 Sep. 77	19 Sep. 03	5 Feb. 14
Stevens, M. W., Bo. I.	24 July 44	27 Dec. 61	27 Dec. 87	24 July 99
Talbot, *Sir* A. C., K.C.I.E.	3 June 45	15 Jan. 67	15 Jan. 93	3 June 00
Wright, E. L.	28 Oct. 58	11 May 78	11 May 04	28 Oct. 13
Yate, A. C.	28 Feb. 53	22 Sep. 76	22 Sep. 02	28 Feb. 08

RETIRED PAY.

Note :—The names on this list are not indexed as they are arranged alphabetically.

Names, Rank and Corps.	Date of Birth.	Date of 1st Commn.	Date of Rank.	Date of Retirement.
Abbott, Lt.-Col. George Masson	22 Aug. 41	31 Mar. 63	31 Mar. 89	22 Apl. 01
Abbott, Bt.-Col. Herbert Edward Stacey, D.S.O., Engrs.	6 Apl. 55	17 Aug. 74	1 Oct. 04	1 Oct. 06
Acton, Lt.-Col. Henry Lowry Barnwell	3 Aug. 61	11 Aug 80	11 Aug. 09	11 Aug. 12
𝔙𝔆 Adams, Maj.-Genl. *Sir* Robert Bellew, K.C.B. [R]	26 July 56	11 Sep. 76	19 Mar. 06	1 Dec. 11
Adamson, Lt.-Col. Charles Henry Ellison, C.I.E.	16 Sep. 46	8 Jan. 68	8 Jan. 94	25 Oct. 06
Adamson, Hony. Col. David, B.	29 Aug. 40	5 Jan. 59	10 May 87	10 May 87
Adutt, *Hony. Capt. Dy. Comsy.* Henry, Ord., So.	3 Sep. 50	16 Sep. 94	2 Jan. 04	1 Nov. 04
Adye, Lt.-Col. Arthur	6 June 55	13 June 74	13 June 00	6 July 08
Alban, Maj. Clifton Frederick Samuel	14 July 54	13 June 74	13 June 95	16 Aug 96
Albert, *Hy. Lt. Asst. Comsy.* William, P.W.D.	11 July 49	19 Jan. 04	19 Jan. 04	11 July 04
Alexander, Bt.-Col. Aubrey deVere, D.S.O.	28 May 49	8 Dec. 69	8 Dec. 99	15 Feb. 08
Alexander, Maj. Frederick Gordon	21 July 50	7 Dec. 70	7 Dec. 90	1 May 96
Alexander, Lieut. James	. . . .	. . . .	23 Nov. 41	26 Apl. 45
Alexander, Maj. Robert Gerald, M.C.	25 Apl. 80	20 May 99	1 Sep. 15	14 July 17
Alexander, Maj.-Genl. Robert, M. Arty.	13 Aug. 40	9 Dec. 59	1 Apl. 97	23 Apl. 79
Allen, Lt.-Col. Bedford Morant	6 Oct. 52	24 Apl. 72	24 Apl. 98	25 July 98
Allen, Maj. Hugh Morris, D.S.O., M.	6 Oct. 67	14 Sep. 87	14 Sep. 05	4 Oct. 13
Allum, *Hy. Maj. Dy. Comsy.* Edwin William, Ord., B.	. . . .	29 May 79	11 Aug. 82	4 Jan. 84
Alves, Lt.-Col. Malcolm Arbuthnot, B. Engrs.	16 June 42	8 June 60	16 June 87	16 June 87
Alves, *Hy. Capt. Dy. Comsy.* William, S. & T. Corps, B.	11 May 60	25 Mar. 00	17 Oct. 03	1 Oct. 05
Anderson, *Hy. Capt. Dy. Comsy.* Alexander, P. W. D., I.	19 Sep. 50	1 Dec. 01	27 Feb. 04	19 Sep. 05
Anderson, Maj.-Genl. Arthur William Leslie, M. Inf. (*Col. 63 L. I.*)	6 June 42	4 Dec. 59	8 Dec. 98	19 Dec. 98
Anderson, Lt.-Col. Clinton Cortlandt	29 Nov. 60	18 Feb. 80	18 Feb. 06	1 Mar. 10
Anderson, Maj. Charles Kirkpatrick, B.	31 Jan. 69	23 Mar. 89	23 Mar 07	24 Mar. 14
Anderson, Col. Edward Brooke	18 July 49	19 Feb. 67	19 Feb. 97	21 Mar. 05
𝔐 Anderson, Col. John Watson Macclesfield, Bo.	28 Dec. 37	20 Jan. 55	20 Jan. 85	23 Mar. 88
Anderson, Col. Rolland Frederick Hart	28 Apl. 59	13 Aug. 79	6 June 07	13 Aug. 11
Anderson, Maj. Wm. Paul.	16 Jan. 58	6 Aug. 79	6 Aug. 99	1 Mar. 05
Andrews-Speed, Col. Henry Speed, Engrs.	6 June 52	12 Sep. 72	6 Aug. 03	7 June 06
Angelo, Maj.-Genl. Fredk. William Pakenham (*Col. 9 Hodson's Horse*).	15 Sep. 59	30 Jan. 78	17 Feb. 11	6 May 11
Angelo, Lt.-Col. John Cecil Cartlandt	30 July 69	29 Nov. 90	29 Nov 16	12 Dec. 16
𝔐 Angelo, Hon. Col. Richard Fisher, B.	3 Sep. 35	4 Jan. 55	1 Aug. 83	1 Aug. 83
Anwyl-Passingham, Maj. Robert Townshend	16 Oct. 67	30 Jan. 86	30 Jan. 04	13 Feb. 07
Appleton, Bt.-Col. Henry, Engrs.	23 Nov. 55	17 Aug. 74	13 Aug. 04	4 Apl. 08
Arbuthnot, Maj. Archibald Ernest, M. Cav.	. . . .	20 Feb. 57	2 Dec. 74	5 Aug. 78
Arbuthnot, Lt.-Col. George Herbert, M.	16 Dec. 62	9 Sep. 82	9 Sep. 08	1 Jan. 11
Archer, Lt.-Col. Charles, C.S.I., C.I.E.	18 Aug. 63	9 Sep. 82	9 Sep. 08	19 Aug. 16
Argles, Maj. Owen Charles	6 July 66	25 Aug. 86	25 Aug. 04	20 Nov. 08
Armstrong, Maj. (Ty. Lt.-Col.) Arthur Knox	3 June 64	28 Apl. 86	28 Apl. 04	29 Apl. 04
Armstrong, Hony. Lt.-Col. John Arthur, B. Engrs.	14 Nov. 40	8 June 60	28 Mar. 87	28 Mar. 87
Armstrong, Col. Oliver Carleton, D. S. O.	16 Oct. 59	11 May 78	11 May 07	11 May 10
Arnott, Col. Napoleon, Engrs.	9 Mar. 44	22 Dec. 65	19 Oct 95	9 Mar. 99
Arthur, *Hy. Maj. Comsy.* Harry Robert, I.O.D.	16 Mar. 61	10 Aug. 06	23 June 12	3 July 14
Arthur, Hony. Capt. John Richard, S. & T. Corps, B.	15 June 64	5 July 09	5 July 16	2 Oct. 16
Ashby, Lt.-Col. John Shuckburgh	29 Jan. 58	17 Jan. 77	22 Mar. 02	23 Mar. 04
Atkinson, *Hy. Capt. Dy. Comsy.* Charles, M. W. S., B.	18 Nov. 35	21 July 82	21 July 82	16 Mar. 88
Atkinson, *Hony.* Br. Gl. Francis Garnett, C.B., I.A.	10 Nov. 57	1 Sep. 76	1 Sep. 14	1 Sep. 14
Atkinson, Lt.-Col. George Charles	17 Apl. 58	28 Oct. 77	28 Oct. 03	28 Mar. 09
Atkinson, *Hy. Capt. Dy. Comsy.* Harry Crowther	4 Dec. 65	1 Mar. 08	29 Nov. 12	1 Feb. 13
Atkinson, Maj.-Genl. John Richard Breeks, B. Inf. [R]	8 Apl. 44	7 June 61	15 Sep. 94	16 Sep. 95
Atkinson, Col. Stephen Edward	22 Sep. 41	20 Sep. 58	10 Sep. 88	10 Feb. 91
Audain, Maj. & Bt.-Lt.-Col. Guy Mortimer	10 Oct. 64	23 Aug. 84	23 Aug. 02	15 Nov. 06
Austin, *Hy. Capt. Dy. Comsy.* William Campbell, Bk. M.	12 Oct. 46	24 Oct. 93	29 Nov. 95	8 May 96
Ayerst, Lt.-Col. William	30 Aug. 60	6 Feb. 84	6 Feb. 10	7 Feb. 10
𝔐 Aytoun, Lt.-Col. Andrew, Bo. Arty.	10 Jan. 25	8 June 43	18 May 63	27 May 64
Babington, Col. Charles Walter, B. Inf.	30 Mar. 41	4 May 60	4 May 90	13 June 92
Babington, Lt.-Col. Stephen, Bo.	9 Nov. 44	27 Apl. 61	27 Apl. 87	24 June 89

RETIRED PAY—*contd.*

Names, Rank and Corps.	Date of Birth.	Date of 1st Commn.	Date of Rank.	Date of Retirement.
Badcock, Bt.-Lt.-Col. George Henry, P.	15 June 71	14 Sep. 92	1 Jan. 17	16 Oct. 13
Baddeley, Bt.-Col. William Lewis Clinton, R.E.	16 Oct. 50	15 Dec. 71	19 Dec. 01	16 Oct. 07
⚔ Badgley, Col. William Francis, B.	10 Nov. 38	4 Apl. 57	4 Apl. 87	28 June 89
Bagley, Maj. Richard George	27 Mar. 65	9 May 88	9 May 06	9 May 13
Bailey, *Hy. Capt. Dy. Comsy.* George, Ord. B.	1 Dec. 39	11 June 88	15 July 94	1 Dec. 94
Bailey, Lt.-Col. Thomas James	3 Feb. 44	5 Aug. 64	5 Aug. 90	26 Sep. 92
Bailey, Maj. William Augustus	19 Aug. 66	31 Dec. 87	31 Dec. 05	26 Apl. 16
Bairnsfather, Capt. Hugh William, M.	....	20 Nov. 59	13 Apl. 73	23 Sep. 74
Bairnsfather, Lt.-Col. Peter Robt.	3 Feb. 50	2 Nov. 72	2 Nov. 98	2 Nov. 04
Bairnsfather, Maj. Thomas Henry	27 Nov. 59	13 Aug. 79	13 Aug. 99	14 Aug. 99
Baker, *Hy. Capt. Dy. Comsy.* Alfred, P.W.D., B.	21 Feb. 47	20 Aug. 90	11 May 96	3 Apl. 01
Baker, *Hy. Capt. Comsy.* Joseph, S. & T. Corps, Ben.	6 Dec. 52	1 July 02	29 Dec. 05	23 July 07
Baker, Lt.-Col. Louis Samuel	15 Apl. 58	28 Oct. 77	28 Oct. 03	30 Oct. 05
Baker, Hony. Maj.-Genl. Robert John, M.	....	11 Feb. 45	17 Jan. 77	27 Oct. 76
Baker, *Hy. Maj. Dy. Comsy.* Thomas, Ord. M.	31 July 50	6 Oct. 99	18 Oct. 02	1 Apl. 96
Baker, Col. Walter Way, R.E.	23 May 59	18 Dec. 78	25 Sep. 07	9 Apl. 11
Balfe, Col. Edmund, *p.s.c.*	10 June 46	30 June 69	30 June 99	10 June 03
Bamford, *Hy. Capt. Dy. Comsy.* Francis, Misc., Bo.	31 May 68	20 May 05	22 Nov. 06	6 Dec. 06
Bannerman, Lt.-Col. Arthur Johnston, Bo.	26 Apl. 39	11 June 58	28 Oct. 82	28 Oct. 84
Barclay, Col. Ernest Armond	14 Aug. 50	6 July 70	6 July 00	13 July 08
Barnes, *Comsy. Hy. Capt.*, Charles	18 June 63	15 Feb. 05	1 Nov. 08	1 Mar. 13
Barnes, Maj. Ernest	16 Aug. 69	15 Feb. 89	15 Feb. 07	28 July 09
Barnet, Bt. Col. Horace Hutton, R. E.	6 Mar. 56	28 Jan. 75	1 Oct. 04	6 Mar. 09
Barnett, Capt. Robert Percy Sanders.	13 Nov. 59	13 Aug. 79	13 Aug. 90	26 Sep. 95
Barton, Hony. Maj.-Genl. Howard James, B.	10 July 36	26 Oct. 58	17 Apl. 85	17 Apl. 85
Barton, Bt.-Col. Maurice Charles, D.S.O., Engrs. (I.P.)	13 Apl. 52	2 May 72	14 Mar. 03	14 Jan. 07
Barton, Maj. Thomas Steward	4 July 67	5 Feb. 87	5 Feb. 05	1 Mar. 05
Barwell, Capt. William Blunt, B.	....	4 Feb. 55	4 Feb. 67	15 Aug. 72
Basevi, Hon. Col. Charles Edward, Bo. Arty.	24 Sep. 36	8 June 55	21 Dec. 81	21 Dec. 81
Batterbury, *Hy. Capt. Comsy.* Charles, S. & T. Corps, B.	9 Aug. 51	29 Apl. 01	11 Feb. 05	4 Oct. 05
⚔ Battye, Capt. Charles Forbes, B.	....	....	2 Sep. 67	25 Dec. 72
Baugh, Maj. Guilum Scott	29 June 02	12 Nov. 70	12 Nov. 90	31 Dec. 95
Baugh, Maj. Mackenzie Walcot	27 Oct. 62	10 Mar. 83	10 July 01	10 Mar. 07
Bax, Col. William Ironside, B. Cav.	12 May 43	4 Sep. 60	16 Feb. 87	11 Oct. 92
Bayly, Maj.-Genl. *Sir* Alfred William Lambert, K.C.B., C.S.I., D.S.O. [R] (*Col. 126 Infy.*)	18 Feb. 56	13 June 74	19 Mar. 06	19 Mar. 12
Beale, Bt. Col. Anthony	15 Apl. 61	13 Aug. 79	6 June 07	13 Aug. 11
Beaman, Maj. George Philip Roberts, Bo.	11 Mar. 69	3 May 90	3 May 08	20 Dec. 13
Beames, Maj. David	25 July 62	9 Sep. 82	10 July 01	16 Dec. 02
Beard, *Hy. Lieut. Asst. Comsy.* William Misc. List, I	9 Nov. 50	23 Mar. 03	23 Mar. 03	9 Nov. 05
Beatley, *Hy. Maj. Dy. Comsy.* Herbert Henry, Misc., B.	24 Apl. 52	23 Dec. 89	1 July 17	20 Mar. 01
Beaufort, *Hy. Maj. Comsy.* Frederick, P. W. D., India.	20 Sep. 54	5 July 07	1 Mar. 09	20 Sep. 09
Beavan, Col. Reginald	20 Sep. 43	4 May 60	4 May 90	11 June 92
Becher, Col. Arthur William Reddie, B. Cav.	6 Dec. 42	20 Dec. 59	20 Dec. 89	2 Feb. 92
Beck, *Hy. Capt. Comsy.* Thomas, M. W. S. and P. W. D., I.	4 July 55	30 Sep. 07	8 May 10	4 July 10
Beckett, Hy. Capt. Dy Asst. Comsy. George Stevens, P. W. D., B.	....	26 Jan. 81	26 Jan. 81	15 Sep. 82
⚔ Beckett, Col. Stephen, C.B., B.	26 Apl. 40	13 June 57	13 June 87	1 Nov. 89
Beckett, *Hy. Capt. Dy. Asst. Comsy.* William, Ord., B.	5 Nov. 56	12 Aug. 02	3 July 17	18 Nov. 02
Beddek, Maj. Edward Escott	2 Jan. 68	14 Sep. 87	14 Sep. 05	15 Sep. 07
Begbie, Lt.-Col. Alfred Glynn, R. E.	25 Sep. 42	1 July 61	16 Dec. 88	10 July 93
Begbie, Lt.-Col. Arundel Sinclair, B.	3 Sep. 65	25 Aug. 56	12 Oct. 11	1 Sep. 13
Begbie, Bt.-Col. Francis Richard	2 Apl. 47	20 July 66	22 July 96	14 Sep. 09
Bell, Col. Alex. Wm. Croker	26 Mar. 49	7 July 69	7 July 99	7 July 07
Bell, Lt.-Col. Charles Thornhill	1 Dec. 62	1 Oct. 82	1 Apl. 11	7 Jan. 17
Bell, Lt.-Col. Henry Phillips	12 Aug. 66	4 May 87	4 May 13	20 Aug. 14
Bell-Clothier, Col. Robert Freebairn	26 Feb. 53	12 Aug. 74	12 Aug. 04	2 Sep. 16
Bannett, *Hy. Capt. Asst. Comsy.* James, Ordn., So. C.	28 Jan. 58	13 Mar. 05	13 Mar. 05	8 Sep. 08
Bentinck, Capt. Reginald Joseph	5 Sep. 66	5 Feb. 87	5 Feb. 98	25 Apl. 01
Beresford, Hony. Maj.-Genl. George De la Poer, M.	16 Aug. 30	10 June 48	29 Dec. 80	29 Dec. 80
Beresford, Hon. Lt.-Col. George William, B.	1 June 44	26 Oct. 61	23 June 87	23 June 87
Benwell, Maj. Reginald Limour	6 Oct. 70	10 Oct. 91	10 Oct. 09	28 Apl. 18

RETIRED PAY—*contd.*

Names, Rank and Corps.	Date of Birth.	Date of 1st Commn.	Date of Rank.	Date of Retirement.
Beresford, Maj. Jas. Hugh Brownlow de la Poer, B. Inf.	30 Mar. 62	23 May 85	29 Nov. 00	24 May 05
Berkeley, Lt.-Col. Robert Bruce	8 May 64	7 Feb. 85	7 Feb. 11	16 Feb. 14
Bernard, Lt.-Col. Edward Henry	5 Oct. 62	22 Oct. 81	13 Nov. 07	14 July 09
Berry, *Hy. Capt. Dy. Comsy.* Edwin, Ord. B.	22 Oct. 47	28 June 95	13 Apl. 01	22 Oct. 02
Beville, Maj. Harry Gilbert Peyton	2 Oct. 69	9 Nov. 89	9 Nov. 07	9 Nov. 09
Bicknell, Lt.-Col. Maldion Byron, R.A.	4 Apl. 62	22 Feb. 82	1 Feb. 10	22 Feb. 14
Biddulph, Bt.-Col. Stephen Francis	4 Feb. 58	5 Sep. 77	17 Jan. 07	4 Feb. 10
Biggs, Col. Henry Vero, D.S.O., R.E.	9 May 60	30 July 79	27 May 10	29 Nov. 12
Billing, *Hy. Maj. Comsy.* Walter, Ordce., So.	14 June 56	16 Feb. 04	1 July 17	1 Nov. 09
Birch, Col. William James Alexander	14 Aug. 46	1 Dec. 65	1 Dec. 95	1 Dec. 97
Bird, Col. Wm. John Butterworth	27 Nov. 52	15 Dec. 71	15 Dec. 01	16 Sep. 04
Birdwood, Lt.-Col. George Christopher McDowall	28 Mar. 63	9 Sep. 82	9 Sep. 08	11 July 09
Birdwood, Col. William Spiller	2 Oct. 54	21 Sep. 74	10 Feb. 04	19 Mar. 09
Birkett, Lieut. Draycot De La Grade	....	....	30 Apl. 58	17 June 60
Biscoe, Lt.-Genl. Wm. Walter, C.B., B. C. [R]. (*Col. 19 Lancs.*)	12 Aug. 41	4 Feb. 60	19 Mar. 00	1 June 01
Bishop, *Hy. Lt. Asst. Comsy.* James, S. and T. Corps, Sec. and Bur.	12 July 55	1 June 07	1 June 07	12 July 10
Bishop, Lt.-Col. Leslie Trevor	31 Jan. 45	2 Mar. 66	2 Mar. 92	2 Mar. 04
Black, Hony. Maj.-Genl. James, Bo.	....	29 Jan. 43	25 Nov. 74	5 Oct. 74
Black, Maj.-Genl. William Campbell (*Col., 101st Grendrs.*)	2 June 46	12 Jan. 66	25 Oct. 02	12 Jan. 04
Blain, *Hony. Lieut., Asst. Comsy.* William, P. W. D., India	14 May 53	29 Nov. 00	29 Nov. 00	1 May 05
Blake, *Hy. Lt., Asst. Comsy.* Henry Albert Randolph, Ord., So. C.	11 May 65	14 May 04	14 May 04	8 Sep. 08
Blake, *Hy. Capt. Dy. Comsy.* John, Comst. Trans., B.	12 Aug. 40	29 Apl. 91	13 Apl. 95	13 Aug. 95
Blake, *Hon. Lieut. Asstt. Comsy.* Thomas George, P. W. D., M.	1 Mar. 50	18 Jan. 05	18 Jan. 05	1 Mar. 05
Blanshard, Lt.-Col. Frederick Hardy	15 June 44	16 Feb. 64	16 Feb. 90	25 Mar. 92
M. Blathwayt, Hon. Lt.-Col. Linley, B.	....	11 Aug. 57	12 Jan. 80	12 Jan. 80
Blazey, *Hy. Capt. Dy. Comsy.* George, Ord., M.	21 Dec. 37	31 July 85	16 Mar. 88	21 Dec. 92
Bleakin, Capt. Charles Woolley	....	....	20 Jan. 72	27 Dec. 72
Blood, Genl. *Sir* Bindon, G.C.B., Engrs. (R)	7 Nov. 42	19 Dec. 60	13 Dec. 06	7 Nov. 07
Blunt, Col. Ernest, C.B., Engrs.	4 Nov. 51	6 Jan. 72	22 Oct. 03	2 Apl. 08
Boileau, Lt.-Col. Herbert Edward	14 Dec. 61	23 Apl. 81	23 Apl. 07	23 Apl. 09
Boileau, Maj. Lewis Maltby	14 Aug. 49	22 July 68	22 July 88	1 Aug. 92
Bolton, Maj. Harry John	4 Sep. 50	17 July 72	17 July 92	5 Feb. 96
Bond, Bt. Lt.-Col. Chetwynd Rokeby Alfred	2 June 63	10 May 82	10 July 01	10 May 02
Bond, Hony. Maj.-Genl. Francis George, C.B., C.M.G., R.E.	10 Aug. 56	2 Feb. 76	18 Aug. 16	10 Aug. 13
M. Bonham, Hy. Col. John, C.B., B. Arty.	....	9 Dec. 52	1 July 74	1 July 74
Bonham-Carter, Lt.-Col. Herman, R. E.	20 Apl. 63	28 July 83	1 Aug. 08	30 Jan. 10
Bonner, *Hy. Lt. Dy. Asst. Comsy.* Edwin, Ord., M.	....	20 Nov. 80	20 Nov. 80	25 Apl. 81
Bonnor, Lieut. George Hawkins	....	....	14 Mar. 58	3 June 61
M. Bonus, Maj.-Genl. Joseph, Bo. Engrs.	14 Sep. 36	8 June 55	25 Nov. 86	30 Nov. 88
Booth, *Hy. Lieut. Asst. Comsy.* Henry Gilbert	16 Aug. 64	26 June 08	26 June 08	15 Mar. 12
Borradaile, Col. George William, C.B., Bo. Arty.	13 June 38	11 Dec. 57	14 July 86	12 July 94
Borradaile, Col. Harry Benn, D.S.O., B.	4 Oct. 60	14 Jan. 80	8 June 07	30 Dec. 12
Borthistle, *Hy. Capt. Dy. Comsy.* John William, Ord. M.	17 Aug. 39	15 Mar. 89	28 Jan. 93	12 July 93
Boswell, Maj. William Legh	26 Apl. 59	30 Jan. 78	30 Jan. 98	15 Mar. 02
Boughey, Col. George Fletcher Ottley, C.S.I., R. E.	23 Jan. 44	25 June 62	9 Apl. 95	12 Oct. 97
Boulderson, Lt.-Col. Henry Allan Balfour, M.	17 Dec. 54	13 June 75	13 June 01	12 July 07
Bourchier, Maj. Arthur Charles Francis, B.	15 Apl. 64	25 Aug. 83	25 Aug. 01	25 Aug. 07
Bower, Maj.-Genl. *Sir* Hamilton, K.C.B., B.	1 Sep. 58	2 Oct. 80	15 Feb. 09	30 Jan. 14
Bowring, Lt.-Col. George	3 June 58	11 Sep. 76	11 Sep. 02	18 May 06
Bowyer, Lt.-Col. Wentworth Grenville, R. E.	10 Nov. 50	15 Dec. 71	18 Oct. 02	1 Mar. 96
Boyd, *Hy. Capt. Comsy.* Henry Augustin, M. W. S. and P. W. D., India	26 May 55	6 Sep. 07	17 Feb. 10	8 May 10
Boyd, Hony. Maj.-Genl. Mossom, Bo.	25 Jan. 31	3 Apl. 47	25 Nov. 81	25 Nov. 81
Boys, Capt. Herbert Charles	20 Sep. 82	18 Jan. 02	18 Jan. 11	14 June 12
Bradley, Lt.-Col. Frederick Gardner, C.M.G., M.	2 Mar. 60	31 Jan. 80	1 June 04	31 Jan. 11
Bradshaw, Major Frederick Ewart, B.	7 Feb. 65	6 Feb. 84	6 Feb. 02	15 Dec. 04
Bradshaw, Maj.-Genl. Laurence Julius Elliott, C.B. [R], B.	2 Dec. 57	10 Sep. 75	7 Oct. 07	5 Feb. 16
Bragg, *Hy. Lt. Dy. Asst. Comsy.* Robert Albert, M.	5 Sep. 44	4 Oct. 93	4 Oct. 93	1 Mar. 98
Braid, Maj Arthur Reade, R.A.	23 Sep. 61	26 July 81	4 Dec. 99	10 Aug. 07
Brander, Col. Herbert Ralph, C.B.	8 Oct. 61	14 Jan. 80	1 June 04	14 Jan. 12
Branson, Col. Charles, Edward Douglas, B. Inf.	2 Jan. 42	8 Sep. 59	8 Sep. 89	12 Mar. 91
Bray, *Hy. Lieut., Asst. Comsy.*, Walter James, Ordce., So. C.	1 Oct. 60	28 June 05	28 June 05	9 Aug. 08
Breakey, *Hy. Maj., Comsy.*, William, Ordce., N. C.	17 Feb. 55	1 Apl. 02	5 Dec. 08	23 Oct 09

Names, Rank and Corps.	Date of Birth.	Date of 1st Commn.	Date of Rank.	Date of Retirement.
RETIRED PAY—*contd.*				
Bredin, *Hy. Capt., Comsy.*, Edgar, Ord., N. C.	25 July 51	21 Jan. 95	19 Dec. 97	25 July 06
Bremner, Col. Harry John	27 Nov. 59	13 Aug. 79	10 Oct. 04	1 Oct. 11
Brereton, Col. Charles William, M. Arty.	28 Sep. 39	11 Dec. 58	13 May 89	18 Dec. 93
Briggs, Lt.-Col. Robert Graham, M. Inf.	. . . .	9 Oct. 59	10 Mar. 80	10 Mar. 80
Broadbent, Col. John Edward, C.B., Engrs.	14 Apl. 45	17 July 66	22 Jan. 98	14 Apl. 02
Broadfoot, Maj. William, B. Engrs.	15 Oct. 41	8 June 60	30 June 81	30 June 81
Brockman, Lt.-Col. Percy William Drake	3 Nov. 63	25 July 82	6 Mar. 07	25 July 14
Brookes, *Hy. Lieut., Asst. Comsy.*, James, So.	25 Sep. 59	1 May 09	1 May 09	1 May 09
Broome, Col. Arthur Plantagenet, B. Inf.	30 Jan. 41	12 Oct. 59	12 Oct. 89	20 Nov. 91
Broome, Bt.-Col. George Simpson	10 Dec. 62	10 May 82	16 Jan. 10	10 May 14
Broome, Maj. Robert Dennis	28 Dec. 57	8 Mar. 79	8 Mar. 99	8 Apl. 01
Broome, Bt.-Col. William Arthur	6 Apl. 50	30 Sep. 68	30 Sep. 98	18 Dec. 06
Brotherston, *Hy. Capt., Dy. Comsy.* William, M.	3 Mar. 37	8 May 63	28 Jan. 91	3 Mar. 02
Brough, Col. William Richard Charles, M. Arty.	12 July 41	11 Dec. 58	20 May 85	20 May 85
Brown, Bt. Col. Charles Allen, Bo.	23 Sep. 58	10 Feb. 81	29 Mar. 10	31 Mar. 13
Brown, Lt.-Col. Harry Troup	3 Apl. 61	22 Oct. 81	22 Oct. 07	11 May 08
Brown, Hony. Capt. Dy. Comsy, Simon	24 Mar. 63	12 July 10	11 June 14	24 Mar. 18
Browne, *Hony. Maj. Comsy.* E. A.	6 Dec. 59	8 Apl. 05	1 Jan. 16	24 Nov. 17
Browne, Lt.-Genl. *Sir* Arthur Geo. Fred., K.C.B., D.S.O. [R] (*Col. 4 G. R.*)	21 June 51	28 Oct. 71	21 Mar. 08	15 Feb. 09
Browne, Maj., Henry Earnest (half-pay List)	19 June 74	10 Oct. 94	10 Oct. 12	10 Oct. 17
Browne, Capt. William	4 July 61	14 Jan. 80	14 Jan. 91	14 Jan. 00
Browne, Lt.-Col. William Henry	16 May 45	20 Oct. 61	20 Oct. 87	10 May 95
Browne, Lt.-Col. William Henry	19 Mar. 44	30 May 64	3 May 87	8 Sep. 03
Browning, Lt.-Col. Herbert Arrott	8 Feb. 61	22 Jan. 81	22 Jan. 07	7 Feb. 15
Brownlow, Col. Cecil Barry, B.	24 Dec. 54	10 Sep. 75	10 Sep. 04	10 Sep. 13
Brownlow, Lieut. Celadon Charles Sutherland	23 Dec. 85	24 Jan. 06	24 Apl. 08	1 Sep. 14
Bruce, Lt.-Col. William Alexander Macdonald	10 July 66	14 Sep. 87	14 Sept. 13	3 Oct. 15
Brumby, *Hy. Lt., Asst. Comsy.*, Thomas William	10 Feb. 39	6 June 87	27 July 88	10 Feb. 94
Buchanan, Lieut. George Alexander Leslie	4 Oct. 64	16 Dec. 85	16 Dec. 85	30 Oct. 95
Buckland, Lt.-Col. Philip Arnold	30 July 51	24 Apl. 72	24 Apl. 98	24 Apl. 04
Buckle, Hony. Maj.-Genl. Charles Randolph, M. Arty.	29 Oct. 35	7 Dec. 55	21 July 86	21 July 86
Budd, Hony. Maj.-Genl. Richard Carruthers, M.	21 May 34	20 July 54	15 Dec. 87	15 Dec. 87
Buist, Lt.-Col. David, Simson	8 Mar. 64	25 Nov. 85	25 Nov. 03	1 Dec. 11
Bullock, Col. Robert	14 June 42	11 Dec. 58	11 Dec. 88	19 Apl. 01
Burchell, *Hy. Lt., Asst. Comsy.*, John	. . . .	. . . .	29 July 82	31 Mar. 88
Burgess, *Hy. Maj., Asst. Comsy.*, Alexander James, Ord. N.	24 Oct. 58	7 Sep. 04	7 Sep. 04	1 May 08
Burgess, Lt.-Col. Forsyth Francis Robert	9 Oct. 50	8 July 68	8 July 94	18 Sep. 06
Burke, *Hony. Capt. Comsy.* Peter Joseph, M.	26 Sep. 58	5 Aug. 06	1 June 07	26 Sep. 13
Burn, Lt.-Col. Arthur George	10 June 56	10 Mar. 75	10 Mar. 01	10 Mar. 13
Burne, Col. Knightley Poytnz, M.	10 Nov. 58	23 Oct. 80	11 June 07	23 Oct. 12
Burne, Lt.-Col. Newdigate Addington Knightley	23 Nov. 55	11 Sep. 76	11 Sep. 02	6 Dec. 06
Burrard, Maj. Charles	15 Feb. 67	24 July 86	24 July 04	24 July 11
Burton, Lt.-Col. Charles William Westbrooke	28 June 58	11 Sep. 76	11 Sep. 03	20 Nov. 14
Burton, Col. Edmund Boteler, C.B.	22 Apl. 61	13 Aug. 79	6 June 07	31 May 13
Burton, Col. Francis Charles	14 Feb. 44	11 July 65	20 Feb. 97	26 Feb. 04
Burton, Maj. Fredric Nuthall	18 July 66	14 Sep. 87	14 Sep. 05	10 Oct. 11
Butchart, Capt. John, B. Arty.	26 Sep. 40	10 June 59	23 Nov. 72	11 Mar. 74
Buttanshaw, Hony. Col. Thomas, B.	. . . .	12 Dec. 49	1 Aug. 79	1 Aug. 79
Buttanshaw, Lt. William Henry, B. Infy.	19 Aug. 35	4 Sep. 54	11 June 56	9 May 63
Cahusac, Col. William Fremantle, C.M.G.	26 Aug. 57	30 Jan. 78	30 Jan. 07	20 June 11
Caldecott, Maj.-Genl. Francis James, C.B., Bo. Arty.	29 Apl. 42	8 June 60	19 Mar. 94	1 Oct. 07
Caldecourt, *Hony. Capt. Dy. Comsy.* George Frederick John	3 Jan. 63	4 July 10	2 May 12	3 Jan. 18
Calli, *Hy. Capt., Comsy.* Gabriel, Misc., India	19 June 50	20 Mar. 01	10 Nov. 04	15 Mar. 05
Callow, *Hy. Capt. Comsy.* Joseph, Ordce., N.	17 Feb. 54	22 Oct. 02	14 Nov. 04	23 Mar. 07
Cameron, Maj. Donald Hay	21 Aug. 67	23 July 87	23 July 05	23 July 11
Camilleri, Lt.-Col. John Mary	25 June 65	14 Jan. 85	13 Nov. 10	20 Sept. 12
Campbell, Maj.-Genl. Alexander Henry Edward, M. Cav.	11 Sep. 35	20 Mar. 54	1 May 86	1 May 86
Campbell, Lt.-Col. Alexander William Dennistown	8 Sep. 48	30 Dec. 71	30 Dec. 97	1 Feb. 00
Campbell, Col. Colin Charles, D.S.O.	18 June 42	22 Dec. 60	22 Dec. 90	10 Aug. 93
Campbell, Lt.-Col. Charles Ferguson, C.I.E., *Eq.*	23 Nov. 63	9 Sep. 82	9 Sep. 08	27 Oct. 08
Campbell, Lt.-Col. Colin Powys	28 May 59	13 Aug. 79	13 Aug. 05	1 Sep. 05
Campbell, Lt.-Col. Charles Walter, B. Cav.	4 Apl. 36	12 Oct. 59	24 Jan. 84	24 Jan. 84
Campbell, Maj. Ian Hamilton, B.	9 Dec. 69	3 Sep. 90	8 Sep. 08	3 Sep. 13
Campbell, Col. William Mussen, R. Art.	11 Aug. 48	13 Jan. 69	1 Feb. 97	11 Aug. 05
Campbell, Maj. William Nevile, C.M.G., D.S.O.	27 Dec. 63	6 Feb. 84	6 Feb. 02	17 June 03
Candy, Lt.-Col. James Molesworth	16 Mar. 52	2 Nov. 73	2 Nov. 99	2 Nov. 10
Cann, *Hy. Capt., Comsy.* Charles, Bk. Dept., M.	10 Sep. 52	5 Mar. 04	29 Dec. 05	10 Sep. 07
Canterbury, *Hony. Maj. Comsy.*, John James, S. & T. Corps, B.	29 Dec. 50	31 Aug. 01	1 July 17	29 Dec. 05
Cardew, Maj. Francis Gordon	1 June 62	10 Mar. 83	10 July 01	10 Mar. 07
Carey, William Reginald	3 Jan. 79	21 Apl. 00	23 Mar. 10	14 July 16
Cargill, Lt.-Col. Spencer, B. Arty.	30 July 38	11 June 58	25 May 85	25 May 85
Carleton, Bt.-Col. Henry Augustus	6 May 58	9 Oct. 77	9 Oct. 06	9 Oct. 09

RETIRED PAY—*contd.*

Names, Rank and Corps.	Date of Birth.	Date of 1st Commn.	Date of Rank.	Date of Retirement.
Carpendale, Lt.-Col., John Maxwell	15 Apl. 58	11 Sep. 77	11 Sep. 03	31 Jan. 09
Carpendale, Bt.-Col. Percy Maxwell	16 Apl. 59	11 May 78	11 May 07	12 May 10
Carpendale, Lt.-Col. William Maxwell, B.	14 May 62	22 Oct. 81	22 Oct. 07	2 Nov. 13
Carter, Lt.-Col. Edward James	9 Nov. 58	13 Aug. 79	13 Aug. 05	13 Aug. 11
Carter, Hony. Lt.-Col. Stanley, Bo.	18 Oct. 41	19 Dec. 60	15 Aug. 86	15 Aug. 86
Carthew-Yorstoun, Lt.-Col. Morden Ewart	8 Oct. 58	26 Mar. 79	26 Mar. 05	23 May 08
Casserly, Maj. James Henry	25 July 69	10 Oct. 91	10 Oct. 09	11 Oct. 16
Cates, Lt.-Col. George Edward Hyde	7 Apl. 57	11 Sep. 76	11 Sep. 02	7 Apl. 12
Caulfeild, *Hony.* Brig.-Genl. Francis William John	7 July 59	14 Jan. 80	11 Apl. 17	14 Jan. 12
Caulfeild, Bt.-Col. Gordon Napier, D.S.O.	27 Jan. 62	22 Jan. 81	11 June 07	18 Feb. 13
Cavenagh, Hony. Maj.-Genl. Gordon, B.	. . . .	18 Jan. 45	28 Oct. 76	1 Aug. 76
Cawood, Lt.-Col. George Crane	1 Mar. 61	11 Aug. 80	11 Aug. 06	11 Aug. 11
Chaldecott, Lt.-Col. Algernon Winn, B.	27 Jan. 64	28 July 83	28 July 09	1 Aug. 13
Chalmers, Capt. Robert Clarence Hilliard	16 Feb. 65	25 Aug. 86	25 Aug. 97	23 Sep. 03
Chaloner, *Hy. Maj., Comsy.* William, P.W.D., India	13 June 57	29 Aug. 06	1 July 17	1 Mar. 09
Chalwin, *Hony. Capt., Asst. Comsy.* George Walter, Ord., So.	4 May 59	18 Aug. 07	18 Aug. 07	4 Mar. 08
Chamberlain, Col. *Sir* Neville Francis FitzGerald, K.C.B., K.C.V.O.	13 Jan. 56	9 Aug. 73	6 Jan. 94	1 Nov. 01
Chapman, Maj. Francis Robert Henry	22 Oct. 58	12 Feb. 76	18 Oct. 02	14 Mar. 97
Chapman, *Hy. Capt., Dy. Comsy.* Thomas, P. W. D., B.	2 Jan. 44	12 June 84	11 May 96	7 Jan. 94
Charlesworth, Comsy. & Hony. Maj. Thomas William	7 Apl. 61	2 May 13	21 Sep. 17	15 Feb. 18
Charters, *Hy. Lt., Asst. Comsy.* Christopher, Comst. Trans., B.	31 Oct. 39	21 Nov. 90	21 Nov. 90	1 Nov. 94
Cheape, Capt. Charles Henry Morris	6 Aug. 55	9 Aug. 73	9 Aug. 85	5 Dec. 94
Chenevix-Trench, Lt.-Col. George Frederick, C.I.E.	23 June 59	22 Jan. 79	22 Jan. 05	10 Apl. 12
St. Cherry, Col. Francis Sawbridge, M. Cav.	1 Dec. 40	4 May 57	3 May 84	4 May 90
Chesney, Col. Harold Frank, R.E.	7 May 59	31 Jan. 78	18 May 07	7 May 11
Chesney, Lt.-Col. Kellow, B.	17 Nov. 61	22 Jan. 81	22 Jan 07	1 Feb. 11
Chooet, *Hy. Capt., Dy. Comsy.* George, Comst. M.	15 Mar. 36	22 Mar. 83	17 May 86	15 Mar. 91
Christie, Lt.-Col. James Harry	19 July 59	1 May 78	1 May 04	30 Oct. 11
Christopher, Maj.-Genl. Leonard William, C.B. [R]	22 Jan. 48	9 Mar. 67	1 Apl. 02	7 Nov. 08
Chrystie, Col. William, M. Engrs.	16 Aug. 34	11 June 53	11 July 81	8 Jan. 83
Churchill, Capt. Wandril Maurice	29 Oct. 82	28 Jan. 03	28 Jan. 12	29 Jan. 18
Clark-Kennedy, Col. Arthur Harry	12 Dec. 57	11 Nov. 77	11 Nov. 06	27 Mar. 09
Clarke, Maj. George Lecot Philip	3 June 68	14 Sep. 87	14 Sep. 05	1 Dec. 07
Clarke, Bt.-Col. Henry, Engrs.	17 Sep. 48	3 Jan. 69	1 Jan. 98	17 Sep. 05
Clarke, Col. Henry Stephenson, B. Arty.	28 Nov. 39	12 Dec. 57	10 May 85	10 May 85
Clarke, Col. Sydenham Cubbon, M. Engrs.	3 Mar. 41	9 Dec. 59	12 Aug. 90	8 Mar. 91
Clementi, Col. Montagu, B.	25 Apl. 39	11 June 58	11 June 88	11 June 90
Clerk, Maj. Edward, M. Cav.	. . . .	16 May 44	26 Apl. 59	10 Apl. 63
Clerk, Lt.-Col. Robert Mildmay, M. Inf.	7 Dec. 44	4 Mar. 61	4 Mar. 87	5 Feb. 90
Cliff, *Hy. Capt. Dy. Comsy.* William, S. and T. Corps, B.	10 Mar. 49	29 Oct. 91	2 Jan. 04	11 Mar. 04
Clifford, *Hy. Capt. Comsy.* William Julian, Misc., Bo.	9 Apl. 57	29 Apl. 02	5 June 06	1 May 08
Close, Capt. Arthur John	15 Nov. 72	24 July 91	24 July 02	5 Nov. 11
Coats, Bt.-Col. George Henry B., C.B.	30 Sep. 52	28 Oct. 71	28 Oct. 01	21 Jan. 10
Cobham, Maj. (Tempy. Brig.-Genl.) Horace Walter, D.S.O.	1 July 66	9 May 88	9 May 06	27 Oct. 10
Cochrane, Lt. William Joseph	. . . .	20 Feb. 58	18 Sep. 60	1 Oct. 64
Cockerill, Capt. Robert Charles	31 July 65	29 Aug. 85	29 Aug. 96	17 May 98
Cocksedge, *Dy. Comsy. and Hony. Capt.* Samuel	16 Nov. 56	20 Sep. 09	1 Nov. 11	6 Nov. 11
Cole, Lt.-Col. Frederick Temple, R. A.	11 Aug. 61	23 Feb. 81	18 May 09	25 Apl. 13
Cole, Lt.-Col. Henry Hardy, R. E.	10 Mar. 43	17 Dec. 62	17 Dec. 89	17 Dec. 94
Cole, Bt.-Col. Robert Arthur	12 June 57	30 Jan. 78	30 Jan. 07	1 Mar. 10
Cole, *Hy. Capt. Dy. Comsy.* Thomas George, S. & T., B.	28 Aug. 49	9 Oct. 99	2 Jan. 04	28 Aug. 4
Coleman, *Hy. Capt. Dy. Comsy.* James Turner, I. Mis. List.	9 Nov. 62	1 June 07	1 June 11	15 June 14
Coles, Maj. Colin Hennessey Read, B. Inf.	6 Jan. 66	7 Feb. 85	7 Feb. 03	14 May 05
Collier, Hony. Col. Clarence Augustus, Bo.	. . . .	3 Aug. 46	14 June 76	14 June 76
Collins, *Hy. Capt. Comsy.* George, Ord., N.	20 July 55	1 Apl. 03	4 Jan. 06	18 Mar. 09
Collins, Lt.-Col. George Atkins	15 June 55	13 June 74	13 June 00	2 Jan. 04
Collins, Hony. Capt. Dy. Comst. Williams James	3 June 63	11 Dec. 19	12 May 1	4 June 13
Collyer, *Hy. Capt. Dy. Comsy.* George, Comst. Trans., Bo.	19 Aug. 41	18 Feb. 84	1 Oct. 91	19 Aug. 96
Colomb, Col. Francis Cracroft, B	4 Mar. 61	22 Jan. 81	5 Nov. 07	8 Jan. 14
Colomb, Lt.-Col. Richard Pasley, Bo.	23 June 58	23 July 79	23 July 05	23 June 13
Colquhoun, Maj. Herbert William Coxworthy	29 Apl. 64	23 May 85	23 May 03	3 July 11
Colquhoun, Lt.-Col. James Andrew Sutherland, B. Arty.	30 Nov. 39	10 Dec. 58	16 Jan. 87	10 Dec. 92
Colvile, Capt. Augustus Henry Asgill	. . . .	. . . .	20 Sep. 61	15 Dec. 70
Colvill, *Hy. Capt. Comsy.* Alfred Benjamin, Bk., M.	17 June 51	30 Mar. 03	14 Feb. 06	28 Mar. 01

RETIRED PAY—*contd.*

Names, Rank and Corps.	Date of Birth.	Date of 1st Commn.	Date of Rank.	Date of Retirement.
Colvin, Lt.-Col. John Russell Colquhoun	9 Sep. 58	13 Aug. 79	13 Aug. 05	7 Apl. 08
Congreve, Maj. Claude Vyvian	18 Oct. 62	10 May 82	10 July 01	10 May 06
Congreve, Maj. Stewart Melvill	21 Nov. 57	12 Feb. 76	12 Feb. 97	1 Mar. 97
Connor, *Hy. Maj. Comsy.* Peter	8 June 60	11 Nov. 11	1 July 17	8 May 14
Cook, Maj.-Genl. James, C.B. [R]. (*Col. 36 Sikhs*)	22 Mar. 44	8 June 61	12 Sep. 00	7 Oct. 00
Cooke, Col. William	13 Jan. 47	7 Nov. 65	29 Aug. 93	7 Nov. 97
Cooke, Capt. William Nevison Macdonald	6 Aug. 61	10 May 82	10 May 93	31 Jan. 00
Cooper, Maj. Edmund Saffrey	9 July 62	19 Dec. 83	19 Dec. 01	19 Mar. 05
Copus, *Hy. Capt. Dy. Comsy.* David, Ordce. So.	18 Dec. 58	1 Nov. 04	18 Aug. 07	6 Oct. 08
Corbett, Col. Frank Vincent, R. E.	15 Dec. 43	15 Jan. 64	12 Aug. 95	19 Dec. 98
Corbett, Maj. Henry George Lyon	9 June 70	9 Apl. 92	9 Apl. 10	9 Apl. 15
Cordeaux, Maj. Harry Edward Spiller, C.B., C.M.G.	15 Nov. 70	2 June 94	3 June 12	3 June 12
Cordue, Lt.-Col. William George Ranger, R.E.	22 Aug. 63	25 July 82	L. G. 9 Nov. 15 1 June 07	4 Oct. 15
Cormac, *Hy. Lt. Dy. Asst. Comsy.* James, P. W. D., M.	3 Sep. 38	15 Mar. 93	15 Mar. 93	3 Sep. 93
Corse-Scott, Capt. Alexander James, B.	. . . .	26 Jan. 66	26 Jan. 78	4 Mar. 86
Cosgrove, *Hy. Capt. Dy. Comsy.* Bernard, Ordce., So. C.	7 Sep. 56	1 Nov. 04	18 Aug. 07	19 Aug. 10
Cotterill, Maj. Hugh Edward	11 Aug. 66	21 Sep. 89	1 Sep. 07	27 Sep. 09
Couper, Lt.-Col. Edward Edmonstone	15 Dec. 60	11 Aug. 80	11 Aug. 06	11 Aug. 12
Cousins, *Hy. Lt. Asst. Comsy.* Frederick Sydney	1 Mar. 59	23 Nov. 13	23 Nov. 13	1 Mar. 14
𝔐. Coussmaker, Lt.-Col. George, Bo.	. . . .	4 Jan. 58	1 Apl. 83	1 Apl. 83
Coussmaker, Lt.-Col. Macclesfield Forbes, Bo.	25 May 43	19 Dec. 60	19 Dec. 86	10 May 98
Cowley, Bt. Col. James William	9 Nov. 60	Jan. 80	9 June 07	14 Jan. 12
Cox, *Hony. Maj. Comsy.* George, Ord., N.	21 Feb. 53	14 Apl. 02	1 July 17	4 Jan. 06
Craig, *Hy. Capt. Dy. Comsy.* Samuel, S. and T. Corps, B.	6 Jan. 49	13 Aug. 95	28 Dec. 00	1 Apl. 02
Craik, Maj. James, D.S.O.	24 Nov. 71	7 Nov. 91	7 Nov. 09	7 Nov. 11
Craster, Lt.-Col. James Cecil Balfour	14 July 54	20 Nov. 75	20 Nov. 01	13 June 02
𝔐. Craster, Col. William Robert, B. Arty.	26 Sep. 36	13 June 56	2 Mar. 85	27 Nov. 90
Crawford, *Hy. Capt., Asst. Comsy.,* William, Ord.B.	. . . .	16 Sep. 82	4 Jan. 84	14 Mar. 84
Crawley, Lt.-Col. George Burridge	18 Mar. 58	30 Jan. 78	30 Jan. 04	30 Jan. 10
VC Creagh, Genl. *Sir* O'Moore, G.C.B., G.C.S.I., [R] (*Col., 129th Baluchis*)	2 Apl. 48	2 Oct. 66	11 Dec. 07	4 May 14
Creak, Col. Henry Charles, B. Cav.	2 Nov. 37	4 Mar. 59	10 May 87	10 May 87
Cresswell, Lt.-Col. Charles Arthur	14 Aug. 44	12 Jan. 66	12 Jan. 92	23 June 98
Crommelin, Bt. Col. Clarence Yule, B.	25 Dec. 60	22 Oct. 81	10 Dec. 07	22 Oct. 13
Cronin, John Joseph	25 Sep. 60	11 Aug. 80	11 Aug. 06	25 Sep. 15
Crouch, *Hy. Capt. Dy. Comsy.* Tom, Bk., M.	28 Feb. 53	14 Nov. 93	6 Dec. 95	5 Jan. 97
𝔐. Crowe, Hony. Maj.-Genl. Thomas Carlisle, Bo. Arty.	. . . .	10 Dec. 47	31 Dec. 78	31 Dec. 78
Crowther, Bt.-Col. Robert Theodore	17 June 60	14 Jan. 80	8 June 07	14 Jan. 12
Cubbon, *Hy. Capt. Dy. Comsy.* Richard, S. and T. Corps, Bo.	22 Aug. 53	10 Sep. 04	13 Oct. 05	22 Aug. 08
Cubitt, Maj. William Martin	13 June 64	23 Aug. 84	23 Aug. 02	6 Dec. 09
Cullen, *Hy. Capt. Comsy.* William, S. & T. Corps, B.	12 Feb. 50	28 Dec. 00	28 Mar. 03	11 Feb. 05
Cumberland, Maj.-Genl. William Bentinck, B. Arty.	21 July 33	12 June 52	17 Nov. 85	17 Nov. 85
𝔐. Cumming, Capt. George Peter	. . . .	12 Dec. 57	12 Dec. 69	10 Jan. 71
Cunningham, Lt.-Col. Allan Joseph Champneys, B. Engrs.	29 Oct. 42	8 June 60	9 Jan. 87	5 Aug. 91
Currie, Lt.-Col. John William	21 Jan. 56	12 Feb. 76	12 Feb. 02	13 Feb. 02
Curtis, Capt. Ernest Henry	. . . .	20 Oct. 59	20 Oct. 71	8 Nov. 72
Custance, Maj. Hugh Lionel	7 Nov. 58	17 May 79	17 May 99	30 Mar. 04
Cutler, *Hy. Lt. Dy. Asst. Comsy.* Joseph George, Ord., B.	7 Oct. 38	11 June 88	11 June 88	9 Oct. 91
Dale, Lt.-Col. Henry Montague, M. Inf.	25 July 45	20 Jan. 62	20 Jan. 88	28 Feb. 94
Dallas, Lt.-Col. Charles Mowbray, C.S.I.	30 Aug. 61	22 Jan. 81	22 Jan. 07	30 Aug. 16
Daly, Lt. Edward Dermat Hamilton, B.	7 Nov. 53	9 Aug. 73	9 Aug. 74	26 Mar. 76
Daly, Sir Hugh, Lt-Col., K.C.S.I., K.C.I.E.	29 May 60	22 Oct. 81	22 Oct. 07	7 Apl. 16
Daly, *Hy. Capt. Dy. Comsy.* John, Bk., M.	16 Nov. 34	22 Feb. 87	3 May 89	16 Nov. 89
Daniell, Maj. Frank William	16 Mar. 68	11 Feb. 88	11 Feb. 06	15 Nov. 09
𝔐. Daniell, Lt.-Col. Henry Smith, Bo. Cav.	24 Feb. 39	4 Nov. 56	11 May 82	11 May 82
Daniell, Maj. Robert Holford	7 Jan. 47	8 Jan. 68	8 Jan. 88	15 Oct. 92
Davey, *Hy. Capt. Dy. Asst. Comsy.* Thomas, Ord., Bo.	16 Sep. 51	11 Feb. 00	11 Feb. 00	22 Dec. 00

Names, Rank and Corps.	Date of Birth.	Date of 1st Commn.	Date of Rank.	Date of Retirement.
RETIRED PAY—*contd.*				
Davidson, Maj. Arthur Noel.	24 Apl. 71	13 Feb. 91	13 Feb. 09	1 Mar. 11
Davidson, Lt.-Col. Charles	20 Feb. 59	14 Jan. 80	14 Jan. 06	14 Jan. 08
Davidson, Maj. George, R. E.	18 May 51	6 Jan. 72	15 Jan. 91	18 May 01
Davies, *Hy. Lt. Dy. Asst. Comsy.* Charles, Misc. M.	20 Aug. 46	12 Nov. 00	12 Nov. 00	20 Aug. 01
Davies, Lt.-Col. Henry Samuel Price	31 Oct. 56	1 May 78	1 May 04	31 Oct. 11
Davies, Lt.-Col. John	12 June 57	16 Feb. 78	16 Feb. 04	9 Nov. 11
Davis, Lt.-Col. Alfred Thornton, B. Inf.	28 Aug. 42	20 Oct. 59	20 Oct. 85	10 Feb. 83
Davis, Col. Cecil	4 Mar. 61	23 Apl. 81	19 Mar. 12	1 Apl. 14
Davis, Col. Richard Percival, B.	23 July 40	22 Aug. 57	22 Oct. 83	22 Oct. 83
Davis, Col. Thomas Arnoll, B. Arty.	21 Feb. 41	12 Dec. 57	13 Dec. 88	3 Mar. 93
Davis, *Hony.-Lieut. Asst. Comsy.* William, P. & T.	15 Oct. 62	15 Nov. 12	15 Nov. 12	28 Dec. 16
Dawes, Bt.-Col. William Myers	26 Sept. 58	19 Feb. 81	14 Aug. 08	29 Mar. 13
Dawson, Lt.-Genl. Alexander Henry, M. Arty.	15 Oct. 27	12 Dec. 45	1 Oct. 82	19 Jan. 83
Dawson, Bt.-Col. Charles Hutton, Bo.	9 Nov. 59	23 Oct. 80	18 Mar. 08	23 Oct. 12
Dawson, Lt.-Col. Francis William, Supy. List.	27 May 58	23 July 79	23 July 05	23 July 11
Dawson, Col. Harry Leonard, C.V.O., C.B.	5 Dec. 54	12 Nov. 73	12 Nov. 99	21 Nov. 07
Day, Bt.-Col. John George, Engrs.	29 July 54	11 Sep. 73	10 Feb. 04	15 Nov. 05
Deane, Lt-Col. Frederick Bernard, M.	1 Aug. 57	10 Sep. 76	10 Sep. 02	16 July 05
Deane, Col. George Williams, C.B.	1 Nov. 50	13 Jan. 69	30 Mar. 01	18 Jan. 07
DeBrath, Lt.-Genl. Ernest, C.B., C.I.E. [R]	12 Dec. 58	11 Sep. 76	17 Feb. 11	3 May 11
deKantzow, Lieut. Alfred.	. . . .	. . . .	23 Nov. 56	9 Feb. 55
Delamain, Bt.-Col. Frank Gun	6 Apl. 58	12 Feb. 76	20 Apl. 05	17 Jan. 10
dela Motte, Capt. Hubert Thomas, M.C.	12 July 95	4 Aug. 14	8 Aug. 18	17 Aug 18
deLautour, Col. Edward Joseph, B. Arty.	10 Mar. 42	8 June 60	1 Sep. 91	28 June 14
DeLisle, *Hony.* Brig.-Genl. George deSausmarez, C.M.G.	12 May 62	10 May 82	25 Nov. 10	10 May 14
Delme-Racliffe, Maj. Seymour Arthur.	26 Aug. 73	17 Dec. 92	15 Jan. 11	16 Jan. 11
Dempster, Col. Cathcart	11 July 48	9 July 68	22 Jan. 96	9 July 00
Dempster, *Hy. Capt. Dy. Comsy.* John, I.O.D.	22 Mar. 63	29 Nov. 00	29 Sep. 03	15 Aug. 15
Denis-de-Vitre, Maj.-Genl. William, Bo. Arty.	18 Mar. 40	11 Dec. 59	4 Feb. 85	4 Feb. 85
Dennys, Lt.-Col. Charles John.	29 May 52	26 Mar. 74	26 Mar. 00	21 Mar. 07
Denton, *Hy. Capt. and Comsy.* Amos, S. and T. Corps, B.	23 Mar. 49	1 Jan. 01	2 Jan. 04	19 Mar. 04
dePalezieux-Falconnet, Maj.-Genl. Grenville Pulteney, M. Engrs.	23 May 34	12 June 52	4 Mar. 86	4 Mar. 86
Devine, *Hy. Capt. and Dy. Comsy.* Joseph	27 Feb. 49	16 Oct. 96	1 Dec. 01	27 Feb. 04
deWilton, Maj. Albert Walter	3 Feb. 62	10 Mar. 83	10 July 01	10 Mar. 03
Dewing, Lt.-Col. Robert Henry, M.	6 Aug. 63	23 Aug. 84	26 Oct. 08	1 Oct. 12
Dikie, Hony. Maj.-Genl. John Elford, C.B., R. E.	26 Apl. 56	17 Aug. 74	13 Aug. 04	26 Apl. 13
Dickinson, Lt.-Col. William, C.S.I., Bo. [R.]	7 July 31	8 Jan. 48	19 Dec. 76	1 July 82
Dickson, *Hy. Capt. Dy. Comsy.* John Aaron, Comst. Trans. M.	10 Mar. 45	26 Aug. 87	3 Nov. 93	10 Mar. 00
Dickson, *Hy. Capt. Dy. Comsy.* Richard, Ord., B.	28 Sep. 54	19 Dec. 97	15 Nov. 01	9 Oct. 02
Dixon, Col. Peter Eden	2 July 61	23 Feb. 81	15 Nov. 08	27 Feb. 13
Dobbin, Lt.-Col. George Miller, B. Arty.	. . . .	9 Dec. 50	4 Feb. 74	4 Feb. 74
Dobbin, Lt-Col. William James Knowles	1 June 56	13 Oct. 77	3 Oct. 03	13 Oct. 05
Dobbs, Lt.-Col. Alexander Hugh	9 Sep. 59	1 May 78	1 May 04	1 May 10
Dobbs, Col. George Cadell	3 July 49	7 July 69	7 July 99	7 July 07
Domenichetti, Lt.-Col. Francis Henry, M.	17 Nov. 66	30 Jan. 86	30 Jan. 12	15 Feb. 14
Donahgue, *Hy. Capt. Dy. Comsy.* Lawrance, M.	8 July 56	12 July 06	10 Sep. 08	8 July 11
Donaldson, *Hy. Capt. Dy. Comsy.* Charles Edward, Misc. I.	8 May 63	29 Nov. 00	17 Apl. 06	15 May 06
Donlea, *Hy. Capt. Dy. Comsy.* Thomas, Comst. Trans. B.	29 Apl. 36	2 Nov. 84	2 Jan. 89	29 Apl. 91
Donnan, William	21 Sep. 67	11 Feb. 88	11 Feb. 14	9 Mar. 16
Donovan, *Hy. Capt. Dy. Comsy.* James, P.W.D., B.	4 June 51	1 Nov. 95	7 Jan. 99	1 June 00
Dorkins, *Dy. Comsy. H Capt.*, Alfred James, I. Ord. Dept.	9 Jan. 59	29 Nov. 00	5 Aug. 13	9 Jan. 14
Dorsey, *Hy. Capt. Dy. Comsy.* George Mathias, Ord. M.	7 Jan. 35	31 July 85	16 Mar. 88	7 Jan. 90
Doughty, *Hy. Lt. Dy. Asst. Comsy.* William, Comst. Trans., M.	20 May 47	3 Nov. 93	3 Nov. 93	5 May 99
Doveton, Capt. Albert Mongles	3 Aug. 67	22 Aug. 88	22 Aug. 99	13 June 08
Doveton, Lt.-Col. Henry, Bo. Engrs.	21 May 39	10 Dec. 58	31 Dec. 85	31 Dec. 85
Dowden, Col. Thomas Freeman, Bo. Engrs.	26 June 37	12 Dec. 56	21 June 85	27 May 93
Dowling, *Hy. Capt. Dy. Comsy.* Michael J.	18 May 38	2 Mar. 82	3 Apl. 92	21 May 93
Downing, Lt.-Col. John George	27 Nov. 53	20 Nov. 75	20 Nov. 01	15 June 09
Doyle, *Hy. Lt. Asst. Comsy.* Henry James, Ord., S. A.	15 Jan. 57	4 Mar. 08	4 Mar. 15	Jan. 12
Drage, Capt. Godfrey, D.S.O. (*Maj. Res. of Offr., Br. Se.*)	19 June 68	23 Mar. 89	23 Mar. 00	9 June 07
Drake, *Hy. Capt. Comsy.* Thomas Oakley, Ord., N. C.	6 Apl. 59	22 Oct. 02	18 Mar. 05	28 June 05
Drake-Brockman, Col. Ralph Rhenius Evans, M. Engrs.	30 Sep. 41	19 Dec. 60	22 Apl. 93	1 Oct. 96

RETIRED PAY—*contd.*

Names, Rank and Corps.	Date of Birth.	Date of 1st Commn.	Date of Rank.	Date of Retirement.
Dressner, Capt. Charles John Bernhard Hough	30 Sep. 55	11 Sep. 76	11 Sep. 87	11 Sep. 90
Drew, *Hy Capt.. Comsy.* George, Bk., M.	22 Nov. 51	20 Mar. 00	3 Apl. 04	1 Mar. 05
Driesen, *Hy. Capt., Comsy.* William Frederick, S. and T., B.	18 Mar. 60	11 Feb. 05	4 Apl. 09	4 Oct. 10
Drury, Lt.-Col. Francis Macdonald	29 Nov. 53	26 Mar. 73	26 Mar. 00	1 Nov. 01
Ducat-Hamersley, Maj. Hugh Charles Claude, B.	12 June 66	30 Jan. 86	30 Jan. 04	9 Mar. 11
Ducat, Col. Charles Merewether, Bo.	5 Sep. 60	22 Jan. 81	21 July 08	21 Feb. 13
Ducate, Lt.-Col. Claude Tulloch	29 Jan. 64	16 Mar. 87	16 Mar. 13	26 Apl. 17
Dun, Hony Maj.-Genl. Charles William, M.	. . . .	10 June 42	28 Oct. 76	31 July 76
Dunlop, *Hy. Capt. Dy. Comsy.* Henry William, Ord. M.	4 Jan. 33	16 Sep. 82	24 Sep. 85	4 Jan. 88
Dunlop-Smith, Lt.-Col. *Sir* James Robert, K.C.S.I., K.C.V.O., C.I.E.	24 Aug. 54	13 Aug. 79	13 Aug. 05	13 Aug. 11
Dunn, *Hy. Capt., Dy. Comsy.* George, M.W.S., Bo.	2 Aug. 43	17 Aug. 91	23 Dec. 92	2 Dec. 94
Dunscombe, Maj. Nicholas Blake	22 July 69	3 Oct. 90	8 Oct. 08	14 Nov. 14
Dunsterville, Col. Knightley Stalker, C.B., *Arty.*	25 Jan. 57	19 Aug. 75	10 Jan. 05	11 Oct. 08
Duperior, Maj.-Genl Henry William	9 Sep. 51	4 Jan. 71	24 Mar. 06	18 Apl. 11
Durand, Col. Algernon George Arnold, C.B., C.I.E.	31 Mar. 54	21 Dec. 72	27 Apl. 96	6 Jan. 01
M. Durand, Hony. Col. Charles James, B.	23 Sep. 40	4 Aug. 57	17 Aug. 86	17 Aug. 86
Durand, Lt.-Col. *Sir* Edward Law, *Bart.*, C.B.	5 June 45	10 Oct. 65	10 Oct. 91	10 Oct. 93
Dyce, Col. George Hugh Cole, C.B.	19 Feb. 46	30 Dec. 64	30 Dec. 94	19 Feb. 03
Dyke, Col. Edward Hart, B. Arty.	11 Nov. 37	12 June 57	1 July 85	6 Dec. 90
Dyson, *Hy. Capt. Comsy.* James Beverley, B.	23 Dec. 58	11 July 04	4 Jan. 08	24 Dec. 13
Eales, Lt. Col. Cecil Moncrieff	28 Feb. 61	22 Jan. 81	22 Jan. 07	23 Jan. 07
Eardley-Wilmot, Maj. Henry	3 Mar. 54	12 Feb. 76	12 Feb. 97	30 May 02
Eardley-Wilmot, Col. Irton	8 Apl. 59	1 May 78	1 May 07	1 May 11
Eardley-Wilmot, Maj.-Genl. Revell, C.B., [R.], B. Inf.	29 Aug. 42	4 Mar. 60	12 Feb. 95	7 Sep. 98
Easter, *Hy. Lt. Dy. Asst. Comsy.* George Obadiah, M.	7 Jan. 49	4 Feb. 03	4 Feb. 03	7 Jan. 04
Eckford, Maj. John James Haldane Black	14 June 64	25 Aug 83	25 Aug. 01	1 Mar. 05
Eden, Col. William Thomas, Bo.	21 Apl. 38	20 Dec. 55	9 Mar. 82	9 Mar. 82
Edgell, Lt.-Col. Edward Arnold	13 Oct. 65	9 Dec. 84	25 Sep. 09	15 Mar. 13
Edwardes, Maj. James Graham, Bo. Arty.	17 July 35	12 June 58	5 Nov. 84	8 Jan. 70
Edwards, *Hy. Capt., Comsy.* Arthur, Misc. India	23 Sep. 55	20 May 98	17 Apl. 06	1 Mar. 07
Edwards, Hon. Br.-Genl. John Burnard, D.S.O., B.	6 May 57	30 Jan. 78	12 Feb. 07	31 Jan. 13
Edwards, Lt.-Col. Roderick Mackenzie, C.M.G.	26 July 64	7 Feb. 85	7 Feb. 11	15 Mar. 11
Egerton, Genl. *Sir* Charles Comyn, G.C.B., D.S.O. [R.] (*Col., 23 Cavy.*)	10 Nov. 48	8 June 67	28 Oct. 06	5 Feb. 07
Egerton, Lt.-Col. Charles Philip	7 Mar. 63	6 Feb. 84	6 Feb. 10	10 Dec. 17
Eliot, Col. Geoffrey Lawrence	4 Oct. 48	8 Jan. 68	8 Jan. 98	8 Jan. 00
Eliott, Lieut.-Col. Supy. List, Francis Hardinge, C.S.I.	1 Apl. 62	22 Oct. 81	3 Aug. 85	1 Apl. 17
Elliot, Lt.-Genl. *Sir* Edward Locke, K.C.B., D.S.O. [R]	28 Jan. 50	22 Jan. 68	30 June 06	3 Oct. 11
Elliott, Bt.-Col. George Hambley	10 Jan. 48	13 July 67	16 Aug. 99	30 Sep. 99
Ellis, Col. Charles Conyngham C.B., Engrs.	2 Mar. 52	2 May 72	1 Oct. 03	2 Mar. 09
Ellis, *Hy. Maj. Comsy.* Thomas James, S. & T. Corps, B.	14 May 53	31 Jan. 90	31 Jan. 05	14 May 08
Elliston, Lt.-Col. Edmund Charles	23 Feb. 43	22 Mar. 64	22 Mar. 90	1 Oct. 92
Enriquez, Lt.-Col. Albert Dallas	17 Mar. 54	14 Nov. 74	14 Nov. 00	1 Jan. 10
Emmett, *Hony. Capt. Dy. Comsy.* William Henry, S. & T. Corps, B.	. . .	29 Sep. 39	5 Dec. 61	5 Dec. 61
Evatt, Major Francis Willie	20 Sept. 63	6 May 85	6 May 03	6 May 09
Evatt, *Hony* Br.-Genl. John Thorold, D.S.O. (*Col. 39 Garh. Rif.*)	1 Aug. 61	23 Apl. 81	23 Apl. 13	23 Apl. 13
Even, Bt.-Col. George Eusebe, C.B., Bo.	12 Aug. 55	21 Sep. 75	21 Sep. 04	19 Feb. 13
Eyre, Lt.-Col. Gervas Selwyn	25 July 51	12 June 69	12 June 95	18 Feb. 02
Eyre, Maj. Thomas Houghton	31 July 48	30 Dec. 71	30 Dec. 91	2 Jan. 93
Eyre, Lt.-Col. Vincent George Lawrence	31 July 48	22 Jan. 67	22 Jan. 93	22 Jan. 95
Ezechiel, *Hy. Maj. Dy. Comsy.* James, Comst.-Trans., Bo.	23 July 50	22 Sep. 81	18 Oct. 02	1 Nov. 99
Fagan, Lt.-Col. Christopher George Forbes	15 Mar. 56	11 Feb. 75	11 Feb. 01	15 Feb. 07
Fagan, Maj. Henry Horace Frederick	9 Apl. 58	11 May 78	11 May 98	15 May 08
Fagan, Maj. Hugh Rollo	4 Aug. 65	29 Aug. 85	29 Aug. 03	15 Sep. 06
Fairbrother, Bt.-Col. William Tomes, C.B.	19 July 56	11 Feb. 75	11 Feb. 04	1 July 14
Faithfull, Lt.-Col. Henry Turner	5 Oct. 53	12 Feb. 74	12 Feb. 02	12 Feb. 02
Falkland, *Hy. Capt. Comsy.* Lancelot, S. & T. Corps, B.	4 July 54	20 May 98	1 June 05	1 Oct. 08
Falcon, Lt.-Col. Robert Worgan, B.	18 Feb. 64	10 Mar. 83	10 Mar. 09	10 Mar. 13
Farrant, Lt.-Col. Harland FitzLyddon, M.	28 Oct. 57	1 Oct. 75	1 Oct. 01	1 Oct. 13

Names, Rank and Corps.	Date of Birth.	Date of 1st Commn.	Date of Rank.	Date of Retirement.
RETIRED PAY—*contd.*				
Fasken, Maj.-Gen. Charles Grant Mansell, C.B., B. (*Col., 52nd Sikhs*)	24 May 55	13 June 74	11 Dec. 07	2 Jan. 14
Faulds, *Hy. Maj. Dy. Comsy.* John Rankine, Misc., Bo.	27 Mar. 46	15 Mar. 87	3 Oct. 00	27 Mar. 08
Fayrer, Lt.-Col. Jas. Outram Spens	20 Feb. 61	22 Jan. 81	22 Jan. 07	23 Jan. 07
Fellows, Bt.-Maj. Bertram Charles, M.	1 Aug. 77	4 Aug. 97	1 Jan. 17	28 Apl. 13
Fellowes, Maj. Halford LeMesurier	17 Aug. 74	31 Jan. 94	31 Jan. 12	1 Mar. 15
Fenton, Lt.-Col. Layard Livingston, Bo.	14 Oct. 49	8 Jan. 70	8 Jan. 96	9 Jan. 02
Fenton, Hony. Br.-Genl. Alexander Bulstrode, C.B.	4 Jan. 56	12 Nov. 73	20 Oct. 76	12 Nov. 11
Fenwick, Col. George Carew	1 June 48	21 Nov. 65	27 June 96	10 May 91
Fenwick, Lt.-Col. John Bourchier, Bo.	....	10 Dec. 54	8 Sep. 79	8 Sep. 79
Ferris, Col. James Ludlow	25 Aug. 39	4 Mar. 59	4 Mar. 89	4 Nov. 96
Ffrench, Lt.-Col. Thomas	23 Dec. 62	14 May 84	14 May 10	1 Aug. 10
Fiddes, Capt. John Montgomery	4 June 84	4 July 03	20 Apl. 14	18 Jan. 15
Field, Lt.-Col. William Caldwell Faure	18 Aug. 52	8 May 72	8 May 98	8 May 00
Finch, Maj.-Genl. Alexander Urquahart Hamilton, Bo. Engrs.	....	13 June 51	31 Dec. 78	31 Dec. 78
Finnimore, *Hy. Capt. Dy. Comsy.* James Payne, Misc., M.	4 Aug. 58	27 June 06	24 May 08	25 May 08
Firth, Major Henry Vivian	9 Aug. 69	22 June 92	6 Oct. 10	22 June 12
FitzGerald, Lt.-Col. Charles Mordaunt	4 July 52	30 Dec. 71	30 Dec. 97	1 July 04
Fitz Hugh, Hony. Maj.-Genl. Alfred, C.B., B.	21 Jan. 37	7 Dec. 55	14 June 86	14 June 86
Fixott, Col. Jas. Laing, Arty.	6 Feb. 53	9 Jan. 73	1 Feb. 02	13 Feb. 07
Flannery, *Hy. Capt. Comsy.* Michael Joseph, M. W. S., Bo.	20 June 59	18 July 05	23 Aug. 07	15 Mar. 08
Fleming, Maj. Henry Lawrence	6 Nov. 64	10 Nov. 86	10 Nov. 04	26 Dec. 06
Fleming, Lt.-Col. John Murchison	12 Aug. 58	13 Aug. 79	13 Aug. 05	25 Aug. 11
Fletcher, Lt.-Col. Henry Arthur, C.V.O., B. Cav.	20 Apl. 43	4 Sep. 60	4 Sep. 86	3 Mar. 89
Floate, *Hy. Capt. Asst. Comsy.* Charles, Ord., M.	....	24 Sep. 85	25 Sep. 87	25 Sep. 87
Floord, Maj. Edward Russell	7 Sep. 67	11 Feb. 88	11 Feb. 06	11 Apl. 09
𝔐. Foord, Hony. Maj.-Genl. William Octavius, M.	21 Oct. 38	20 Feb. 57	16 July 87	16 July 87
𝔐. Foote, Hony. Col. Frank Bernard, B.	....	11 Dec. 49	2 Feb. 78	11 Dec. 77
Forbes, Capt. Edward Esmé	1 Sep. 55	11 Feb. 75	11 Feb. 86	11 Feb. 95
Forbes, Maj. Lindsay Anstruther	21 Oct. 65	10 Nov. 86	10 Nov. 04	24 Feb. 07
Ford, Maj. Charles Annesley Wilbraham	31 July 66	14 Sep. 87	14 Sep. 05	26 Mar. 11
Ford, Capt. Charles Wilbraham, P.W.D., B.	....	24 Jan. 39	24 Jan. 54	10 Sep. 58
Fordyce, Maj. Alexander Dingwall	21 Mar. 60	22 Jan. 79	22 Jan. 99	23 Jan. 99
Fordyce, Bt.-Col. John Fraser Dingwall	6 Nov. 50	4 Feb. 69	3 Apl. 02	10 Aug. 02
Formby, Lt.-Col. Reginald Frederick Robert, M.	18 Jan. 59	19 Feb. 81	19 Feb. 07	31 Mar. 13
Forster, Lt.-Col. William David, M. Arty.	....	14 June 50	1 Aug. 72	1 Aug. 72
Forth, Maj. Clarence Thomas Willoughby	12 Apl. 63	6 May 85	6 May 03	1 Nov. 09
Foss, Lt.-Col. Kenneth Mackenzie	14 Mar. 56	14 June 77	2 July 03	10 June 09
Foster, *Hy. Capt. Dy. Comsy.* Henry, Ord., M.	28 Jan. 38	16 Mar. 88	4 Feb. 91	28 Jan. 93
Foster, Lt.-Col. Louis Maurice	6 Feb. 64	6 Feb. 84	3 Mar. 07	29 Apl. 10
Foxton, *Hy. Capt., Asst. Comsy.* Thomas, Ord., B.	....	18 Nov. 80	11 Aug. 82	16 Sep. 82
Foy, *Comsy. and Hony. Maj.* John, S. & T., B.	29 Mar. 60	10 Dec. 08	26 Feb. 13	5 Apl. 13
Francis, Col. George Frankland	6 Mar. 51	5 Nov. 70	5 Nov. 00	5 Nov. 08
Fraser, Lt.-Col. Edward Alexander	8 Oct. 46	13 Sep. 64	13 Sep. 90	13 Sep. 96
Fraser, Lt.-Col. James, B. Inf.	3 Apl. 42	7 June 61	7 June 87	1 June 89
Fraser, Col. Thomas Lyons, Bo.	26 Mar. 37	10 Dec. 53	10 Dec. 81	10 Dec. 81
Fraser, Capt. William Forbes Mackenzie Ian	22 Mar. 58	2 July 79	2 July 90	5 July 99
Fraser, Maj. Walter Simon	21 May 64	Mar. 87	23 Mar. 05	2 Feb. 10
Free, Col. John Frederick, B. Arty.	31 Aug. 41	12 June 58	1 Nov. 86	21 July 92
Freeman, Lt.-Col. Francis Peere Williams, B. Arty.	29 Sep. 36	7 Dec. 55	9 Dec. 82	9 Dec. 82
Freeman, *Asst. Comsy. Hony. Capt.* Thomas, I.O.D.	7 Jan. 59	23 June 12	23 June 12	7 Jan. 14
Freeth, Lt.-Col. William, M.V.O., M.	31 Mar. 41	11 Dec. 58	20 Mar. 83	20 Mar. 83
Fryer, Bt.-Col. Leslie Charles	7 Mar. 61	23 Apl. 81	11 June 07	25 Apl. 09
Fulton, Bt.-Col. Robert	12 July 52	17 Feb. 72	17 Feb. 02	5 Apl. 10
Gabbett, Maj. Alexander Cecil, M.	10 Aug. 71	8 Oct. 90	8 Oct. 08	22 Mar. 13
Gaisford, Capt. James	15 Mar. 68	22 Aug. 88	25 Aug. 99	25 Nov. 04
Gale, *Dy. Comsy. Hony. Maj.* Arthur, I.M. List.	7 Feb. 59	2 Dec. 02	10 May 08	24 Oct. 13
Gale, Bt.-Col. Walter Andrew, late R. E.	9 Nov. 55	28 Jan. 75	31 Dec. 04	6 May 11
Garden, Lt.-Col. Archibald Ross Hervey, Bo.	21 Nov. 67	11 Feb. 88	11 Feb. 14	12 Feb. 14
Garnett, *Comsy. and Hony.* Maj. William	10 Feb. 58	10 Sep. 04	1 Oct. 08	31 Jan. 13
Garrett, Lt.-Col. Annesley John	4 Jan. 46	29 May 66	29 May 92	15 Mar. 99
𝔐 Garrett, Capt. Newson Dunnell, B. Arty.	7 May 39	12 Dec. 57	10 Dec. 67	24 Feb. 72
Garrett, Lt.-Col. Robert Vernon, B.	25 Nov. 55	11 Feb. 75	11 Feb. 02	15 Nov. 04
Garsia, Capt. Christopher, B.	....	23 Mar. 58	23 Mar. 70	1 July 82
Garstin, Col. George Lindsay	10 June 51	3 Sep. 70	3 Sep. 00	6 Nov. 08
Gataker, Capt. Melmoth William, B.	....	4 Sep. 58	4 Sep. 70	19 May 71
Gayer, Col. Edward Alexander	5 May 64	10 Mar. 83	23 Nov. 13	31 Aug. 17
Gerrard, Maj. Halford Dumergue	5 July 55	29 Apl. 74	29 Apl. 94	30 Apl. 94
Gib, Maj. Howard Alaric	31 May 71	14 Sep. 92	14 Sep. 10	25 Oct. 16
Gibbs, Col. George Reynell, B.	11 Jan. 55	25 June 58	25 June 88	25 June 90

RETIRED PAY—*contd.*

Names, Rank and Corps.	Date of Birth.	Date of 1st Commn.	Date of Rank.	Date of Retirement.
Gibbons, *Dy. Comsy. & Hony. Capt.*, James Henry	12 June 65	21 Mar. 11	14 June 13	11 Oct. 18
Gibbs, Lt. Col. Mossom Innis	1 Jan. 41	30 Dec. 71	30 Dec. 97	9 Feb. 00
Gilbert, Lt. Col. Gerald Edwin Lloyd, D.S.O.	23 Jan. 64	9 May 85	9 May 11	6 Sep. 12
Giles, Lt.-Col. William	16 Apl. 59	11 May 78	5 Apl. 05	11 May 06
Gill, *Hy. Capt. Dy. Asst. Comsy.* Thomas, Comst., B.	. . . .	21 Oct. 78	28 July 82	28 July 82
Giltrap, *Hy. Capt. Dy. Asst. Comsy.* Robert, Ord., B.	13 Sep. 37	22 Feb. 91	22 Feb. 91	13 Sep. 92
Glennie, Col. Edward, Engrs.	13 Mar. 51	2 Aug. 71	31 Dec. 00	1 Aug. 07
Goad, Col. Howard, C.S.I.	15 Sep. 57	10 Sep. 75	10 Sep. 04	1 Apl. 08
Goater, *Dy. Comsy. Hy. Capt.* Cecil, I.M.L.	11 Apl. 66	31 Jan. 11	31 Jan. 11	8 Feb. 14
Godfrey, Lt.-Col. Stuart Hill, C.I.E.	2 June 61	22 Jan. 81	22 Jan. 07	
Godwin-Austen, Lt.-Col. Henry Homersham, B.	. . . .	26 Dec. 51	2 Mar. 78	12 June 77
Goldie, Col. Barre John, B. Engrs.	6 Apl. 40	10 Dec. 58	16 Sep. 90	8 June 92
Goldingham, Capt. George Agnew	. . . .	. . . .	21 May 63	1 Dec. 63
Goldney, Maj. Frederick Charles Napier, B.	9 Sep. 49	10 Nov. 69	10 Nov. 89	11 Nov. 89
Goldsmid, Maj. George Stuart	3 Jan. 56	11 Feb. 75	11 Feb. 96	12 Feb. 96
Goldthorp, Maj. Francis Herbert	14 Aug. 70	28 June 90	28 June 08	1 July 14
Gompertz, Lt.-Col. Bowes Thorpe Montague	1 Nov. 47	11 Jan. 67	11 Jan. 93	11 Jan. 95
Goodman, *Hy. Maj. Comsy.* William, Ord., N. C.	24 July 57	22 Oct. 01	1 July 17	31 Dec. 05
Goodwin, *Hy. Capt. Comsy.* Thomas Arthur, P. W. D., Bo.	13 July 50	12 Oct. 03	28 Jan. 04	13 July 05
Goodwyn, Maj. Henry Edward, D.S.O., Engrs.	11 Oct. 55	28 Jan. 75	12 Dec. 94	1 Feb. 02
Gordon, Lt.-Col. Alex. Evans	31 July 45	18 Apl. 65	18 Apl. 91	28 June 06
Gordon, Lt.-Col. John Adrias Davidson, B. Inf.	18 Dec. 44	20 June 61	20 June 87	1 May 94
Gordon, Col. John Charles Frederick, C.I.E.	4 May 49	10 Nov. 69	10 Nov. 99	2 Mar. 09
Gordon, Maj. Peter Lawrence, M. Cav.	. . . .	4 Oct. 56	4 Oct. 76	17 Apl. 74
Gordon, Lt.-Col. Philip James, B.	2 June 59	22 Jan. 81	22 Jan. 07	23 Sep. 13
Gordon, Capt. Robert Napier Steuart (*a*)	16 Oct. 79	20 July 98	20 July 07	8 July 10
Gordon, Bt. Col., Stewart Douglas, B.	8 Sep. 55	21 Sep. 74	12 Oct. 04	21 Sep. 12
Gordon, Bt.-Col. Stannus Verner, D.S.O.	29 Aug. 46	9 Aug. 64	23 Jan. 97	11 Aug. 02
Goschen, Lt. Henry	. . . .	13 Dec. 56	22 Oct. 58	29 Aug. 66
Gowan, Lt.-Col. Walter Edward, B. Inf.	23 Dec. 44	8 June 61	8 June 87	6 Feb. 92
Grace, Maj. Charles Hallyburton Campbell	2 May 66	16 Nov. 87	16 Nov. 05	16 Nov. 14
Gracey, Col. Thos., C.S.I., Engrs.	21 Jan. 43	17 Dec. 62	12 Aug. 95	21 Jan. 00
Graham, Maj.-Genl. Archibald William, Bo.	. . . .	14 June 45	2 Mar. 78	2 Mar. 78
Graham, Hony. Maj.-Genl. George Fergus, B.	16 May 36	14 Jan. 54	10 Apl. 86	10 Apl. 86
Kt. Graham, Hony. Maj.-Genl. George Farquhan Irving, B.	3 Feb. 40	20 Dec. 56	1 Aug. 87	1 Aug. 87
Graham, Maj.-Genl. *Sir* Thomas, K.C.B., B. Arty.	20 Jan. 42	12 June 58	5 Apl. 01	1 Dec. 94
VC Grant, Bt.-Col. Charles James William, M.	14 Oct. 61	10 May 82	10 Feb. 10	5 Nov. 12
Grant, Bt.-Col. Francis Charles, B.	17 May 58	11 Aug. 80	18 Nov. 09	11 Aug. 12
Kt. Grant, Col. Francis William, B.	4 Nov. 39	20 Apl. 57	20 Apl. 87	6 July 89
Grant, Maj. Ian Hope	30 Nov. 66	25 Aug. 86	25 Aug. 04	21 May 07
Grant, Maj. John, Bo.	28 Nov. 47	8 Jan. 68	8 Jan. 88	10 May 88
Kt. Grant, Col. John Pyne, Bo.	16 Apl. 38	4 Jan. 56	4 Jan. 86	1 Jan. 90
Grant, Col. Suene, R.E.	23 Mar. 51	2 Aug. 71	8 Aug. 02	23 Mar. 08
Grant, *Hy. Capt. Dy. Comsy.* Thomas, Misc. M.	7 Sep 49	14 Nov. 93	27 Nov. 95	10 July 96
Grantham, Bt.-Col. Charles Fulford	1 Aug. 57	13 Oct. 77	13 Oct. 06	7 Aug. 11
Graves, Col. Somerset Henry Paul	20 Nov. 46	2 Sep. 68	2 Sep. 98	1 Jan. 10
Gray, *Hy. Lt., Dy. Asst. Comsy.* Thomas, M.W.S., B.	3 May 39	24 Mar. 91	23 Mar. 91	3 May 94
Gray, Maj.-Genl. William duGard, C.B., [R] M.	9 Sep. 56	10 Sep. 75	1 Dec. 06	12 Oct. 12
Green, *Hy. Capt. Dy. Comsy.* George Harry, So.	26 Sep. 63	11 Juy 10	2 May 13	12 July 14
Green, *Comsy. & Hony. Capt.* Walter William, S. & T., N.	5 July 58	3 May 05	21 Dec. 10	6 Feb. 13
Greig, Col. Philip Henry, Bo. Arty.	27 Sep. 41	12 Dec. 57	1 Jan. 88	1 June 94
Greig, Maj. William Best	24 Mar. 70	29 Oct. 90	4 Mar. 09	3 Dec. 10
Grevelink, *Hy. Capt. Dy. Comsy.* George, S. & T., B.	17 Aug. 50	26 Aug. 87	11 Jan. 99	24 June 01
Grey, Col. Edward, S. and T. Corps	12 May 60	11 Aug. 80	11 Aug. 09	12 Jan. 11
Grey, Lt.-Col. Wm. Francis Hungerford	18 Feb. 49	3 June 68	3 June 94	24 Mar. 01
Griffin, Lt.-Col. Cecil Pender Griffiths, D.S.O.	22 Jan. 64	7 Feb. 85	7 Feb. 11	1 Mar. 12
Griffiths, Hony. Lt.-Col. Charles	19 Oct. 62	10 May 82	10 May 08	11 Nov. 11
Griffiths, *Hy. Capt. Comsy.* David, S. and T., B.	4 Jan. 53	19 Mar. 04	9 Dec. 07	4 Jan. 08
Grove, Lt. Henry Montgomery	27 Feb. 67	5 Feb. 87	1 Nov. 88	17 Oct. 03

(*a*) Tempy. Lieut.-Col.

RETIRED PAY—*contd.*

Names, Rank and Corps.	Date of Birth.	Date of 1st Commn.	Date of Rank.	Date of Retirement.
Gully, Lt.-Col. William, B. Arty.	. . . .	7 Dec. 55	10 Feb. 75	10 Feb. 75
Gurr, *Hy. Maj., Comsy.*, George William, Ord.	23 June 57	20 May 98	29 Dec. 08	23 June 12
Gwatkin, Col. Fred. Stapleton, C.B.	30 Jan. 49	27 Mar. 72	22 Jan. 99	1 Nov. 04
Hadow, Bt.-Col. Reginald Campbell, D.S.O.	6 July 51	27 Apl. 70	15 Feb. 00	27 Apl. 02
Haines, Lt.-Col. Robert Lewis	20 May 55	28 Jan. 75	21 Dec. 01	28 Jan. 05
Hall, Maj. George James Cadell	12 Dec. 61	10 May 82	10 July 01	15 Sep. 06
Hall, Col. Herbert Sidney George	6 Nov. 59	11 May 78	11 May 07	1 July 10
Hall, *Hony.* Br.-Gl. Lewis Montgomery, Murray, C.B.	10 Sep. 55	1 Oct. 73	1 Oct. 03	7 Mar. 10
Halliday, *Hy. Lt. Dy. Asst. Comsy.* John, Ord., B.	. . . .	24 Nov. 81	24 Nov. 81	8 Dec. 86
Hamilton, Bt.-Col. Alexander, B.	27 Feb. 62	11 Aug. 80	11 June 07	11 Aug. 12
Hamilton, Maj. Claude, B.	2 Sep. 64	25 Aug. 83	25 Aug. 01	26 Aug. 07
Hamilton, Lt.-Col. Claude Leslie	16 Apl. 63	10 May 82	10 May 08	10 May 14
Hamilton, Lt. Edward Chetwood, B.	. . . .	13 Feb. 66	10 Nov. 69	1 July 82
Hamilton, Lt.-Col. George Fred. Dashwood	18 May 60	31 Jan. 80	31 Jan. 06	5 Oct. 10
Hammond, Col. Henry Albany, M.	10 Mar. 42	11 Dec. 58	11 Dec. 88	1 Jan. 91
Hamond, Col. Robert Thomas, M. Arty.	12 Dec. 39	12 Dec. 57	10 Dec. 84	10 Dec. 84
Hankin, Lt.-Col. Frederick George, M.	. . . .	11 Dec. 49	27 June 74	24 June 74
Harcourt, Maj.-Genl. Philip Hammer, Bo. Arty.	30 Jan. 36	9 Dec. 53	3 June 84	3 June 84
Hardaker, *Hy. Lt. Asst. Comsy.* Richard, Comst.-Trans., B.	2 Oct. 36	17 Nov. 86	17 Nov. 86	2 Oct. 91
Harding, Capt. Edward Frank	28 Feb. 69	16 July 90	19 July 01	1 Jan. 10
Hardy, *Hony. Capt. Comsy.* Charles Richard, Misc., M.	27 Feb. 52	6 May 98	1 June 02	27 Feb. 07
Harrington, Maj. *Sir* Jn. Lane, K.C.M.G., K.C.V.O., C.B.	6 Dec. 64	6 June 88	6 June 06	9 Mar. 09
Harris, *Hony. Capt. Comsy.* Alfred, M.W.S., Bo.	23 Aug. 52	28 Sep. 03	18 July 05	23 Aug. 07
Harris, Arthur Philip Desborough	17 Dec. 62	22 Oct. 81	11 June 07	22 Oct. 13
Harris, Lt. Francis Anderson, B.	26 Oct. 57	10 Sep. 75	10 Sep. 75	6 Nov. 86
Harris, Capt., George Edward	. . . .	. . . .	30 Dec. 69	18 Apl. 72
Harris, Col. Thomas Marshall, Bo. Arty.	8 Nov. 28	11 June 47	1 Oct. 77	1 Oct. 82
Harrison, Col. Charles William Ingleby, B. Engrs.	28 June 39	11 June 58	12 Aug. 90	17 Nov. 94
Harrison, Capt. Douglas Charles William	23 Mar. 55	11 Feb. 75	11 Feb. 86	11 Feb. 95
Harrison, *Hy. Capt. Dy. Comsy.* Joseph, Ord., B.	22 July 50	11 Feb. 95	1 Apl. 99	5 Dec. 99
Harrison, Maj. William Christopher Warde, M.	6 Feb. 68	28 June 90	28 June 08	6 Dec. 12
Hart, *Hony. Capt. Comsy.* John, P.W.D., Ind.	4 Aug. 52	11 July 01	19 Jan. 04	4 Aug. 07
Hartigan, Lt.-Col. Arthur Edwin Stewart	1 July 62	10 Mar. 83	10 Mar. 09	11 Mar. 09
Hartigan, Bt.-Col. Edward Ross	2 Apl. 60	1 Feb. 80	23 June 09	1 Feb. 12
Hastings, Lt.-Col. Warren, B.	19 July 52	28 Feb. 74	28 Feb. 00	27 Oct. 12
Hatch, Lt.-Col. Alfred Edmund	28 Apl. 63	25 Aug. 83	25 Aug. 09	28 Aug. 09
Hatch, Col. Arthur Vincent	10 Sep. 60	14 Jan. 80	8 June 07	14 Jan. 12
Hatherell, Lieut.-Col. William George, Bo.	18 Feb. 61	22 Oct. 81	31 May 07	22 Oct. 13
Haughton, Col. Thomas Hutchinson	8 May 54	11 Feb. 77	11 Feb. 06	1 Jan. 10
Hawkes, Maj.-Genl., *Sir* Henry Montague Pakington, K.C.B., C.S.I. [R]	25 Aug. 55	23 Nov. 72	1 May 05	29 Aug. 11
Hawks, Maj. Frederick Welman	20 Oct. 70	21 Sep. 89	4 Aug. 91	15 Nov. 11
Hawks, Lt.-Col. Thomas Bate	31 Oct. 61	9 Sep. 82	4 June 67	9 Sep. 14
Hay, Maj. Arthur	27 July 67	9 May 88	21 Oct. 14	16 May 08
Hay, Lt.-Col. Henry Thomas Horatio	7 Dec. 63	23 Aug. 84	30 Mar. 09	24 Aug. 11
Hayes, Bt.-Col. Clarence Henry	3 Oct. 55	1 Oct. 74	21 Mar. 04	6 Jan. 12
Hayes-Sadler, Lt.-Col. *Sir* James, K.C.M.G., C.B.	11 Oct. 51	6 July 70	6 July 96	11 Mar. 97
Hayne, Maj. Arthur Gorham-Howard	1 Feb. 52	3 Feb. 72	3 Feb. 92	21 Mar. 96
Haywood, *Hy. Lt. Asst. Comsy.* William, P.W.D., B.	11 July 46	16 Oct. 96	1 June 00	11 July 01
Hazelgrove, Bt. Col. Harry Seymour, Bo.	25 Dec. 60	22 Jan. 81	3 Oct. 09	22 Jan. 13
Headwards, *Hy. Capt. Comsy.* John James, Ordce., N.C.	10 June 61	22 Oct. 02	18 Mar. 05	20 Jan. 08
Heath, Col. Harry Heptinstall Rose, C.B.	24 June 50	24 July 72	22 Jan. 00	10 Sep. 10
Heathcote, *Hy. Capt. Comsy.* Otho, D. Ordce., N.	17 June 53	21 Jan. 95	19 Dec. 97	17 June 08
Hebbert, Col. William Stanley, M. Arty.	24 Aug. 41	10 June 59	2 Jan. 86	8 July 93
Heenan, *Hy. Lt. Asst. Comsy.* Michael, P.W.D., B.	29 Nov. 43	7 Nov. 94	11 May 96	29 Nov. 98
Heming, Lt.-Col. Dempster, M. Inf.	2 Dec. 43	16 Nov. 61	16 Nov. 87	4 Feb. 94
Hemsworth, *Hy. Lt. Asst. Comsy.* Daniell, Ord., M.	6 Oct. 35	25 Sep. 87	25 Sep. 87	6 Oct. 90
Henchy, Lt.-Col. Robert Cameron, M. Arty.	. . . .	8 June 49	1 Aug. 72	1 Aug. 72
Henderson, *Hy. Capt. Dy. Comsy.* Fulton, Ord., M.	16 Mar. 33	1 Nov. 83	10 Jan. 87	16 Mar. 88

Names, Rank and Corps.	Date of Birth.	Date of 1st Commn.	Date of Rank.	Date of Retirement.
RETIRED PAY—*contd.*				
Henderson, *Hony. Capt. Comsy.* John, P. W. D., Bo.	10 Nov. 53	28 Nov. 83	13 July 05	19 Sep. 05
Henderson, Capt. Joseph Hedley	1 Apl. 79	4 Apl. 00	4 Apl. 09	25 Aug. 11
Henry, *Hony. Capt. Comsy.* Thomas, M. W. S., Bo.	15 Aug. 50	1 Sep. 97	18 Sep. 03	18 July 05
Henry, Lt.-Genl. George, C. B., Engrs. ® . . .	23 Aug. 46	8 Jan. 68	26 Dec. 07	4 Nov. 08
Hennel, Lt.-Col. *Sir* Reginald, *Kt.*, D.S.O., Bo. Inf.	11 June 44	25 July 61	25 July 87	19 Nov. 89
Hennessy, *Hy. Capt. Dy. Comsy.* Thomas, B. .	23 Oct. 58	17 May 08	24 Feb. 11	23 Oct. 13
Hepper, Maj. Harry Albert Lawless, R.E. . . .	30 Jan. 70	14 Feb. 90	30 Jan. 10	31 Mar. 12
Herbert, Lt.-Col. Charles, C.S.I., B.	16 May 54	13 June 74	13 June 00	13 Sep. 12
Herbert, Maj.-Genl. Lionel, C.V.O., C.B. . . .	30 Nov. 60	22 Jan. 81	2 Jan. 14	11 Mar. 14
Heron, the Revd. John, M.A., M.B., Presy. Sr. Chapln.	. . . .	. . . .	. . . .	17 Mar. 18 11 Mar. 14
Herschel, Col. John, B. Engrs.	29 Oct. 37	12 Dec. 56	10 Mar. 86	
Hewson, *Hy. Lt. Asst. Comsy.* James Thomas, Ord., B.	. . . .	17 Oct. 77	4 Dec. 78	9 Apl. 80
Heyland, Lieut. William Frank, S.	27 Oct. 89	20 Jan. 09	20 Apl. 11	12 Oct. 13
Heyman, Maj. Charles Henry Cockburn . . .	12 June 60	13 Aug. 79	13 Aug. 99	22 Apl. 00
Heywood, Col. John Magee, B. Engrs. . . .	2 July 37	13 June 56	3 May 86	10 Dec. 89
Higginson, Lt.-Col. Henry Shakespear, B. Arty. .	15 Dec. 42	8 June 60	1 Jan. 89	8 June 89
Hight, Maj. Edward Lancelot.	5 Mar. 58	11 May 78	11 May 98	13 May 98
Highway, *Hy. Capt. Dy. Comsy.* Evon, P.W.D., B.	21 Jan. 47	11 May 96	1 June 00	1 Dec. 01
Hildebrand, Lt.-Col. William Hatton	6 July 65	29 Aug. 85	29 Aug. 11	17 Feb. 12
Hill, Lt.-Col. Frank Barton	13 Feb. 67	6 June 88	6 June 14	6 June 16
Hill, Col. Rowland Henry Thomas, M. Inf. . .	7 July 45	2 Jan. 62	17 May 90	6 Mar. 91
VC M Hills-Johnes, Lt.-Genl. *Sir* James, G.C.B., B. Arty. [R]	20 Aug. 33	11 June 53	31 Dec. 83	30 June 88
Hindle, *Hy. Capt. Dy. Comsy.* Thomas, Ord., B.	22 July 43	27 Feb. 92	1 Dec. 94	21 Jan. 95
Hislop, Maj. Arthur Fowler	5 May 71	9 Apl. 92	9 Apl. 10	10 Apl. 12
Hitchins, Bt.-Col. Chas. Henry Macintire C.M.G.	30 Jan. 60	31 Jan. 80	21 Feb. 09	31 Jan. 12
Hobbs, *Hy. Capt. Dy. Comsy.* William, M. W. S., Bo.	18 Sep. 48	1 June 96	16 Aug. 03	18 Sep. 03
Hobday, Col. James Ramsay	16 July 50	13 Jan. 69	13 Jan. 99	15 Apl. 07
Hobday, Hony. Maj.-Genl. Thomas Francis, C.B.	4 Jan. 47	23 July 65	24 July 95	25 July 03
Hocking, *Hy. Maj. Comsy.* Alfred W., Ordce., N. C.	4 Aug. 59	7 Apl. 03	23 Jan. 07	29 Dec. 08
Hodding, Maj. James Sweet	1 Jan. 67	11 Feb. 88	11 Feb. 06	1 Mar. 08
Hodgson, Col. Charles Nesbit, B. Inf.	15 Jan. 42	20 Apl. 59	1 Jan. 88	29 May 91
Hogg, Maj. Hardinge Cornwallis, Bo.	3 Dec. 47	12 Oct. 67	12 Oct. 87	3 Aug. 89
Hogg, 2nd-Lt. Reginald William Threlfall, U. L..	9 Dec. 87	29 Aug. 06	29 Aug. 06	4 Jan. 09
Hogg, Lt.-Col. Theodore Chas. Mackenzie Trower, B..	13 May 67	25 Aug. 86	25 Aug. 12	20 Oct. 13
Hogge, Bt.-Col. George Sapte	10 Feb. 61	14 Jan. 80	8 June 07	14 Jan. 12
Holdich, Col. *Sir* Thos. Hungerford, K.C.M.G., K.C.I.E., C.B., Engrs.	13 Feb. 43	17 Dec. 62	1 Sep. 91	13 Feb. 00
Hole, Lt.-Col. Frederick, M.	15 May 39	4 Jan. 60	4 Jan. 86	8 Feb. 88
Holland, Maj. Edward Frederick	24 July 76	5 Aug. 96	5 Aug. 14	12 Apl. 17
Holland, Hony. Maj.-Genl. Henry William, C.B., Bo.	. . . .	25 July 41	30 June 77	20 Apl. 77
Holland, *Hy. Capt. Comsy.* John Arthur, Ordce. Dept.	18 Feb. 61	20 May 98	10 Apl. 08	2 May 13
Holloway, Bt.-Col. Edward Leigh	20 Nov. 59	25 Feb. 80	5 Nov. 08	25 Feb. 12
Holton, *Hy. Capt. Comsy.* William Henry, Misc. List, M..	30 Sep. 62	1 Oct. 03	4 Oct. 07	1 July 09
Home, Col. Frederick Jervis, C.S.I., B. Engrs. . .	22 Oct. 39	11 June 58	12 Aug. 90	26 May 95
Home, Capt. William Elton, P.	8 Feb. 81	14 Feb. 00	8 Apl. 10	19 Mar. 14
Hood, Lt.-Col. Earnest Frederick, B. . . .	24 Sep. 67	21 Dec. 87	21 Dec. 13	1 Jan. 14
Horlick, *Hy. Lt. Dy. Asst. Comsy.* Cyrus George, Bk., M.	2 July 54	30 Dec. 00	30 Dec. 00	1 Apl. 03
Hornby, Bt.-Lt.-Col. Montague Leyland, C.M.G., D.S.O.	23 July 70	30 Jan. 89	3 June 16	29 Apl. 09
Horniman, *Hy. Capt. Dy. Comsy.* Richard, Ord., B.	27 Mar. 43	22 Feb. 91	1 Dec. 94	19 Aug. 96
Hornsby-Drake, Lt.-Col. Alfred Western Hatchell, M.	30 Nov. 45	26 July 64	29 Sep. 87	29 Sep. 87
Houison-Craufurd, Maj. (*Hony. Lt.-Col.*) John Archibald	10 May 62	22 Oct. 81	10 July 01	25 Oct. 05
oHw, Lt. Co Arthur Pemberton	15 Apl 68	11 Feb. 88	13 July 14	11 Feb. 17
Howard, Lt.-Col. Thomas, B. Engrs.	27 Feb. 39	22 June 59	27 Feb. 87	1 May 90
Howe, Maj. Rupert Bohun Blunt	8 Nov. 65	8 Dec. 86	8 Dec. 04	8 Dec. 10
Howell, Lt.-Col. Auberon Arthur, B.	2 Nov. 65	29 Aug. 85	29 Aug. 11	12 Nov. 13
Howlett, Col. Arthur	27 Feb. 50	22 Aug. 68	22 Aug. 98	1 Jan. 02
Hudleston, Col. John, M.	16 Jan. 42	4 Aug. 58	4 Aug. 88	9 Sep. 90
Hudson, Capt. Arthur Keith	26 Feb. 69	9 Nov. 89	9 Nov. 00	14 Feb. 10
Hudson, Maj. Wemyss.	1 Dec. 63	25 Aug. 83	25 Aug. 01	13 Feb. 06
Huggins, Lt.-Col. Ponsonby Glenn, C.B., D.S.O. .	21 Jan. 57	9 Apl. 76	19 Apl. 02	19 Apl. 14
Hughes, Maj. Arthur Pitcairne	5 July 78	20 July 98	16 Jan. 18	28 Sep. 10
Hughes, Maj. John Edward, M.	22 Nov. 71	3 Sep. 92	3 Sep. 10	12 Dec. 12
Hughes, Col. William Gwynne, M.	28 Mar. 41	12 June 58	12 June 88	12 June 89
Hume, Lt.-Col. Walter Francis, Bo. Inf. . . .	21 June 43	8 June 61	8 June 87	24 Aug. 89

RETIRED PAY—*contd.*

Names, Rank and Corps.	Date of Birth.	Date of 1st Commn.	Date of Rank.	Date of Retirement.
Hunter, *Hy. Maj. Comsy.* James, M. W. S., and P. W. D., I.	18 Aug. 57	4 Aug. 07	1 July 17	17 Feb. 10
Hunter, Lt.-Col. John, B. Arty.	. . . .	14 June 45	1 Aug. 72	1 Aug. 72
Hunter, Lt., Malcolm Russell	21 Jan. 66	29 Aug. 85	29 Aug. 85	11 Oct. 95
Hutchins, *Hy. Capt. Comsy.* Ethelred Elder, S. & T. Corps, B	3 April 54	22 Nov. '00	1 July 02	4 Apl. 09
Hutchins, Col. Herbert Leonard	13 Aug. 55	10 Sep. 75	10 Sep. 04	15 Mar. 08
Hutchinson, Maj. Geo. Stuart Menteth	12 Nov. 83	18 Jan. 02	18 Jan. 11	30 Apl. 17
Hutchinson, Lt.-Genl. Henry Doveton, C.S.I. [R] (*Col. 3 G. R.*)	13 Sep. 47	6 July 67	1 June 07	21 Mar. 08
Hutchinson, Lt.-Col. Jas. Wm. Caldwell	20 Jan. 60	22 Jan. 79	24 April 05	22 Jan. 07
Hutchinson, *Hy. Capt. Asst. Comsy.* Nicholas, M.W.S., B.	. . . .	1 Dec. 82	17 May 84	24 Apl. 85
Huyshe, Lt.-Col. Dunbar Fraser, B. Arty.	3 Dec. 42	9 Dec. 59	1 Jan. 89	14 Apl. 89
Hynds, *Hy. Lt. Asst. Comsy.* James, Ord. N. A.	11 Nov. 56	10 Aug. 06	10 Aug. 06	11 Nov. 11
Iles, Maj. Francis William	27 Dec. 76	5 Aug. 96	5 Aug. 14	6 Aug. 14
Iredell, Col. James Shrubb, Bo.	7 Oct. 39	13 Dec. 56	29 Oct. 86	29 Oct. 86
Impey, Lt.-Col. Lawrence, C.I.E.	17 June 62	10 Mar. 83	10 Mar. 09	17 June 17
Jackson, Maj. Cecil.	2 June 61	11 Aug. 80	11 Aug. 00	12 Aug. 00
Jackson, Lt.-Col. Eric Colvin Sutherland, B. Inf.	25 May 42	20 Sep. 60	20 Sep. 86	1 Mar. 89
Jackson, Maj.-Genl. George D'Aguilar, B. Cav.	20 July 43	20 Nov. 59	24 Nov. 98	8 Dec. 98
Jackson, Lt.-Col. John	8 Nov. 61	11 Aug. 80	1 June 04	11 Aug. 12
Jackson-Jellie, *Hy. Capt. Dy. Comsy.* Robert, Ordce., So.	19 Oct. 58	1 Apl. 03	28 June 05	29 June 10
Jacob, Maj. William Swinton*	23 Aug. 65	8 Dec. 86	8 Dec. 04	11 Jan. 08
Jacob, Col. William Vesey FitzGerald, B.	9 May 37	8 Dec. 55	8 Dec. 85	22 Jan. 88
James, Capt. Francis Polglase	10 June 68	18 Feb. 91	10 July 01	20 May 09
James, Lt.-Col. Montague	8 Apl. 42	11 Jan. 67	11 Jan. 93	15 Sep. 09
James, Maj. Montague Gifford, B.	14 June 68	14 Sep. 87	14 Sep. 05	16 Sep. 07
James, Col. Herbert, C.B.	12 Nov. 59	11 May 78	11 May 04	12 Nov. 11
Jamieson, Bt.-Col. Alister Wm., B.	8 Mar. 47	9 Nov. 66	9 Nov. 96	9 Nov. 04
Jeffcoat, Maj. Fred. Hugh. Hiddingh	24 July 67	16 Nov. 87	16 Nov. 05	16 Nov. 10
Jennings, Bt.-Col. Robert Henry, C.S.I., R.E.	13 Oct. 52	12 Sep. 72	21 May 03	13 Oct. 09
Jephson, *Hy. Lt. Asst. Comsy.* Henry, Ord., Bo.	31 Oct. 45	10 Feb. 94	10 Feb. 94	11 Feb. 00
Jermyn, Col. Turenne, B.	12 Apl. 62	22 Oct. 81	11 June 07	22 Oct. 13
Johnson, *Hy. Maj. Dy. Comsy.* Charles, Ord., M.	28 July 39	17 May 86	6 Oct. 85	28 July 94
Johnson, Maj. Clarence Arnold Keatinge, B.	6 Apl. 70	21 Sep. 89	21 Sep. 07	1 Nov. 12
Johnson, *Hy. Capt. Dy. Comsy.* George Henry, P. W. D., Bo.	27 June 40	14 Dec. 92	1 Oct. 94	9 Apl. 95
Johnson, Lt.-Col. Gilbert Ward	31 Dec. 66	14 Sep. 87	14 Sep. 13	14 Sep. 14
Johnstone, Maj. Beresford Assheton	28 Mar. 61	23 Apl. 81	23 Apl. 01	23 Apl. 05
St. Johnstone, Col. James William Hope, B.	29 Jan. 36	20 Dec. 52	20 Dec. 82	24 Feb. 87
Johnstone Maj. Robert FitzRoy MacLean	30 Sep. 59	11 May 78	11 May 98	14 Oct. 98
Johston-Saint, Capt., Percy,	27 June 86	17 Aug. 07	1 Sep. 15	1 July 18
Jones, *Hy. Lt. Asst. Comsy.* Frank Sainty, I. M. L.	1 Jan. 72	17 Mar. 12	17 Mar. 12	1 Apl. 14
Jones, Maj. Jenkin	5 June 61	22 Jan. 81	22 Jan. 01	1 June 02
Jones, Maj. Thomas William Hathway	11 June 76	29 Feb. 96	1 Mar. 14	12 Aug. 14
Jopp, Col. Keith Alexander, Bo. Engrs.	20 July 38	12 June 57	1 July 85	23 Dec. 91
Keenan, *Hy. Lt., Dy. Asst. Comsy.* James, M. W. S., B.	18 June 31	12 June 84	12 June 84	30 July 86
Kellie, Maj. Archibald Henry	7 Mar. 61	1 Feb. 79	1 Feb. 99	15 Oct. 02
Kellie, Lt.-Col. Edward Chamier	24 Nov. 49	21 Aug. 69	21 Aug. 95	1 Dec. 07
Kelly, Col. James, Unatd., M.	22 Dec. 30	28 Sep. 58	28 Sep. 88	1 Sep. 92
Kelly, *Hy. Lt. Asst. Comsy.* Jno., P. W. D., M.	7 Feb. 45	3 Sep. 93	15 Feb. 94	4 Apl. 99
Kelly, *Hy. Capt. Dy. Comsy.* Patrick, P.W.D., Ind.	30 Sep. 52	5 May 06	2 Jan. 07	30 Sep. 07
Kelly *Hy. Capt. Dy. Comsy.* Patrick, Ord. S. C.	4 Mar. 57	28 June 05	21 May 10	4 Mar. 12
Kemball, Lt.-Col. Charles Arnold, C.I.E.	27 June 60	22 Jan. 81	22 Jan. 07	26 Apl. 14
Kendall, Maj. John	23 June 66	9 May 85	9 May 03	9 May 09
Kennedy, Lt.-Col. Willoughby Pitcairn, C.S.I.	10 Dec. 50	8 Feb. 70	8 Feb. 96	31 Mar. 06
Kenny, Col. Henry Torrens, Bo.	3 Sep. 58	23 Oct. 80	11 June 07	31 Dec. 12
Keown, Maj. John Charrier	6 May 64	Feb. 84	1 Feb 02	27 Dec. 05
Kerr, *Hy. Lt., Dy. Asst. Comsy.* James, Bo.	11 Jan. 40	31 Mar. 88	31 Mar. 88	1 Sep. 88
Kerr, Bt.-Col. Mark Ancrum. C.B., B.	2 May 59	13 Aug. 79	6 June 07	13 Aug. 11
Kerrich, Col. Leonard Wilkinson Cleveland, M.	25 Aug. 59	22 Jan. 79	2 June 07	22 Jan. 11
Keswick, *Hony. Maj. Comsy.*, Robert	26 Sep. 61	12 Aug. 02	1 July 17	10 May 16
Ketchen, Maj.-Genl. Isaac, M. Arty.	3 Feb. 39	12 June 57	29 Sep. 90	1 Dec. 89
Kettle, *Hy. Capt. Asst. Comsy.* Henry D., S. & T. Corps, Bo.	5 July 54	11 Mar. 05	1 July 17	5 July 09
Kettlewell, Bt.-Col. Edward Alexander	19 Jan. 60	13 Aug. 79	6 June 07	13 Aug. 11

RETIRED PAY—*contd.*

Names, Rank and Corps.	Date of Birth.	Date of 1st Commn.	Date of Rank.	Date of Retirement.
Key, *Hy. Lt. Asst. Comsy.* John, Ord., B.	11 June 33	12 June 84	12 June 84	22 June 88
Key, *Hy. Lt. Asst. Comsy.* John Alfred, M. W.S., M.	28 June 54	11 Aug. 01	11 Aug. 01	10 Mar. 04
Keyworth, *Hony.* Lt.-Col. Walter, M.	30 Apl. 68	23 Mar. 89	23 Mar. 07	27 Mar. 13
Kingston, Maj. Frederick Anderson, B.	4 Mar. 65	30 Jan. 86	30 Jan. 04	9 July 12
Kinsman, Col. Harold John, B. Arty.	14 Aug. 38	12 June 58	9 Apl. 90	15 Apl. 91
Kirkpatrick, *Hy. Capt. Comsy.* Charles Henry, Bo.	1 Apl. 72	5 Dec. 09	12 Mar. 14	15 Mar. 14
Kirkpatrick, Maj. William, D.S.O., M. . . .	2 Feb. 63	25 Aug. 83	25 Aug. 01	1 Jan. 05
M. Kitson, Capt. Edward	. . . . .	12 Dec. 57	12 Dec. 69	11 Feb. 70
Knight, *Hy. Capt. Comsy.* Charles Herbert, Bo. M.L.	12 Aug. 64	17 Feb. 06	27 Mar. 10	12 June 14
Knight, *Hy. Lt. Dy. Asst. Comsy.* William Henry, M. W. S., B.	28 Oct. 38	22 June 92	22 June 92	28 Oct. 93
Kreyer, Lt.-Col. Frederick August Christian, Bo. .	8 Apl. 59	1 May 78	1 May 04	13 Dec. 10
Lacey, *Hy. Capt. Comsy.* James Ord. Bo. . . .	28 Nov. 50	1 Sep. 92	6 Oct. 99	21 Mar. 05
Lalor, *Hy. Capt. Dy. Comsy.* John, P. W. D., M. .	2 Nov. 46	15 Apl. 90	15 Feb. 94	2 Nov. 01
Lamb, Bt.-Col. Henry Campbell	9 Nov. 48	5 May 69	5 May 99	28 June 07
Lambert, *Hy. Lt. Asst. Comsy.*, Vaughan Edwin, P. W. D., I.	11 Dec. 55	17 Feb. 10	17 Feb. 10	11 Dec. 10
Land, *Hy. Capt. Comsy.* Joseph John, Ord., N. C.	7 Sep. 52	20 May 98	1 Apl. 03	1 Apl. 05
Landon Col. Aislabie, B. Inf.	29 July 42	8 June 60	8 June 90	6 Aug. 91
Lane, Maj. Frank Macdonald	1 Feb. 70	1 Mar. 90	1 Mar. 08	25 Apl. 15
Lane, Capt. Thomas Edward Moore	19 Jan. 74	31 Jan. 94	31 Jan. 03	4 Feb. 12
Lang, Lt.-Col. Arthur George Boileau	14 July 61	10 May 82	30 Sept. 07	20 May 14
Lang, Col. Richard Tickell Montgomery, B. Cav.	12 Jan. 42	12 Oct. 59	12 Oct. 89	23 Mar. 95
Lash, Lt. Henry Andrew.	13 Mar. 68	9 May 88	10 Mar. 90	25 May 98
Laskey, *Hy. Maj. Comsy.* Walter William, M. .	8 Apl. 60	4 Aug. 08	1 July 17	23 Feb. 14
Lassalle, Lt.-Col. James Hector Charles Greenhill, B. Inf.	. . . .	20 Feb. 60	28 Apl. 82	28 Apl. 82
Lauderdale, Lt.-Col. Frederick Henry *Earl of* B. .	16 Dec. 40	19 June 60	1 Jan. 86	1 Jan. 89
Laurence, Col. R.T.R., R.E.	30 June 64	15 Feb. 84	2 June 13	15 Sept. 16
Lawrence, Lt.-Col. Henry John, B. Inf. . . .	16 Sep. 45	4 Jan. 62	4 Jan. 88	28 Mar. 95
Lawrence, *Hy. Capt. Comsy.* John, M. . . .	11 June 59	1 Feb. 07	26 Sept. 13	11 June 14
Lawson, *Hy. Capt. Dy. Comsy.* Alexander, P. W. D., Bo.	1 Feb. 52	25 June 94	1 Oct. 96	1 July 97
Lawson-Tancred, Maj. *Sir* T. S., Bart. . . .	14 May 70	11 Sep. 89	11 Sep. 07	14 Oct. 11
Leach, Br. Genl. Harold Pemberton, C.B., D.S.O., Engrs.	14 Apl. 51	2 Aug. 71	6 Jan. 00	14 Apl. 08
Leader, Maj. Thomas Alexander Fred., M. . .	30 June 47	11 Jan. 67	11 Jan. 87	11 Jan. 91
Leahy, *Hy. Capt. Asst. Comsy.* Patrick, Comst., B.	. . . .	16 Sep. 82	11 May 85	11 May 85
Lean, *Hy. Lt. Asst. Comsy.* Edwin, P. W. D., B. .	5 Dec. 47	4 Aug. 99	4 Aug. 99	5 Dec. 02
Leason, *Hy. Capt. Asst. Comsy.* Frederick William, B.	29 Nov. 58	17 May 08	17 May 08	29 Nov. 13
LeBailly, Capt. Alexander Cartland	7 Dec. 63	10 May 82	10 May 93	1 Jan. 96
LeBreton, Maj. William Inglis, Bo. . . .	16 May 45	23 Dec. 64	23 Dec. 84	5 June 88
Lee, Maj.-Genl. Henry Herbert, Bo. Engrs. . .	4 Oct. 38	11 Dec. 57	4 Feb. 86	4 Feb. 86
Lee, *Hony. Maj., Comsy.*, Lennie	8 Sep. 63	10 Sep. 05	10 July 12	22 Dec. 17
Lee, *Hy. Capt. Dy. Asst. Comsy.* Thomas, Comst., B.	6 Nov. 32	7 Dec. 83	17 Mar. 87	8 July 87
Legat, *Hy. Lt. Dy. Asst. Comsy.* Thomas, Ord., Bo.	2 June 54	18 Nov. 96	28 Nov. 96	15 Sep. 02
M. Leggett, Col. Joseph Beauchamp, M. . . .	22 Sep. 39	16 Aug. 56	16 Aug. 86	1 Dec. 90
Leigh, Lt.-Col. Henry Percy Poingdestre, C.I.E. .	28 Dec. 51	4 Jan. 71	4 Jan. 97	4 Jan. 09
Leishman, Maj. John Thomas, Bo. Arty. . . .	16 Dec. 35	8 June 55	1 Dec. 85	25 July 70
M. Leith, Maj. Thomas, Bo.	. . . .	3 Feb. 48	3 Feb. 68	3 Feb. 68
Leman, Lieut. John Downton	4 Aug. 90	29 Jan. 10	29 Apl. 12	1 Jan. 15
Lester, Hony. Maj.-Genl. William Charles, Bo. .	15 Oct. 29	13 June 46	15 Sep. 81	15 Sep. 81
Leveson-Gower, Maj. Chas. Cameron . . .	30 June 66	28 Apl. 86	28 Apl. 04	28 Apl. 06
Lewin, Lt.-Col. Wilfred Hale	18 May 56	11 Aug. 80	1 June 04	18 May 11
Lewis, Lt.-Col. Charles Henry	22 Oct. 61	30 July 84	30 July 10	26 Apl. 12
Lewis, Bt.-Col. David Silvanus	23 Dec. 58	11 May 78	11 May 07	11 May 10
Light, Lt.-Col. Reginald Hough	30 Aug. 59	13 Aug. 79	1 June 04	13 Aug. 11
Light, William Alfrey	19 Sept. 72	28 Jan. 93	28 Jan. 11	16 Apl. 18
Lightfoot, Maj. Thomas William	11 Nov. 65	22 Aug. 88	22 Aug. 06	23 Aug. 12
Lilly, Lt.-Col. George William	7 Jan. 63	14 May 84	14 May 10	5 June 10
Lincoln, Lt.-Col. Sydney Woodward, Bo. . . .	8 Mar. 60	28 Jan. 82	28 Jan. 08	28 Jan. 14
Lindesay, Maj. George William Guy	5 Nov. 70	29 Oct. 90	29 Oct. 08	27 Apl. 10
Lindsell, Maj. Philip Barber.	27 Nov. 56	11 June 80	11 Sep. 08	11 Sep. 00
Link, *Hy. Maj. Comsy.* Willie Cresswell . . .	6 Aug. 59	1 Apl. 03	10 Aug. 06	3 Aug. 14
Little, Maj. Ignatius Mary	15 Aug. 76	1 Sep. 95	1 Mar. 15	8 Mar. 18
Lloyd, Maj. Arthur Malcolm	2 Sept. 60	11 May 78	11 May 98	1 Nov. 02
Lloyd, Major Charles Edward	12 June 62	1 Sep. 81	10 July 01	10 Dec. 03
Lloyd, Lt.-Col. Edward	24 Sept. 48	16 Oct. 65	16 Oct. 92	16 Oct. 94
Lloyd, Maj. John Hayes, Bo.	12 Jan. 41	4 April 54	4 Apl. 74	1 July 82
M. Lloyd, Col. Richard Mercer, Bo.	27 July 39	4 Mar. 56	4 Mar. 86	24 Nov. 82
Loch, Lt.-Col. Granville Henry, C.I.E., B. . . .	26 Jan. 59	21 Aug. 78	21 Aug. 04	26 Jan. 14

RETIRED PAY—*contd.*

Names, Rank and Corps.	Date of Birth.	Date of 1st Commn.	Date of Rank.	Date of Retirement.
Locke, *Hony. Capt. Comsy.* Charles Robert, B.D., M.	4 Jan. 57	1 Feb. 04	14 Feb. 07	1 April 11
Lockwood, Capt. Henry Boileau	. . . .	20 Feb. 8	13 Nov. 74	12 Apl. 75
Lodge, Lt.-Col. Frank, Bo. Arty.	28 June 41	8 June 60	24 Nov. 82	12 June 78
Lofts, *Asst. Comsy. Hony. Lt.* Henry Edward, B.D., B.	10 Oct. 57	24 Feb. 11	24 Feb. 11	3 July 12
Logan-Home, Lt.-Col. Ferdinand Cospatric	21 Sep. 61	23 Apl. 81	23 Apl. 07	24 Apl. 13
Lonergan, *Hy. Maj. Comsy.* James	4 July 57	11 Feb. 00	1 July 17	4 July 12
Long, Maj. William Christopher	22 Mar. 77	27 Jan. 00	27 Jan. 12	27 Jan. 16
Longe, Col. Francis Bacon, C.B., R. E.	31 Oct. 56	2 Feb. 76	1 Oct. 05	31 Oct. 11
Loudon, Maj. Francis Arthur	23 Feb. 63	10 May 82	10 July 01	22 May 06
Lovell, Maj. Arthur Neville	25 Mar. 67	21 Dec. 89	21 Dec. 07	23 Dec. 09
Lovett, Maj.-Genl. Beresford, C.B., C.S.I., B. Engr.	16 Feb. 39	11 June 58	1 Nov. 92	17 Nov. 94
ℳ. Low, Capt. John Alves, B. Arty.	11 June 40	12 Dec. 57	19 Aug. 67	30 Sep. 69
Lowndes, Hony. Maj.-Genl. Thomas, M.	5 Sep. 35	10 Dec. 54	1 Aug. 87	1 Aug. 87
Lowry, Maj. Frederick James Sharples	10 Sep. 59	1 May 78	1 May 98	7 July 00
Lowry, Lt.-Col. Henry Ward, M.	26 Jan. 59	19 Feb. 81	19 Feb. 07	19 Feb. 13
Lowry, Col. William Henry, M.	2 Dec. 55	29 Nov. 76	29 Nov. 02	2 Dec. 12
Lucas, Col. Hugh Claude Edward	27 May 49	21 Aug. 69	21 Aug. 99	25 Jan. 09
Lugard, Maj. Edward James, D.S.O.	23 Mar. 65	10 Nov. 86	10 Nov. 04	10 Nov. 06
Lumsden, *Hy. Lt. Asst. Comsy.* Robert, S. and T. Corps, B.	21 June 50	11 Mar. 04	11 Mar. 04	21 June 05
Lushington, Lt.-Col. Edward Charles Mortimer	15 Dec. 55	27 June 75	27 June 01	27 June 07
Lynch, Col. Jas. Beverly, D.S.O.	1 July 48	8 Jan. 68	8 Jan. 98	1 Feb. 00
Lyne, Maj. Charles Virgil Nunez	13 Mar. 67	11 Feb. 88	11 Feb. 06	11 Feb. 08
Lyon, *Hy. Capt. Dy. Comsy.* William Frederick, P.W.D., Bo.	11 Oct. 48	2 Apl. 95	1 July 99	12 Oct. 03
Lyons-Montgomery, Col. Hugh Frederick, C.B.	4 Jan. 56	13 June 75	20 May 02	1 May 08
Lys, Col. Arthur Mariott, M.	14 Sept. 37	4 Mar. 56	4 Mar. 86	15 Apl. 88
Lyster, Col. William Henry	5 Aug. 46	16 Oct. 66	16 Oct. 96	22 Oct. 04
Lyttle, *Hy. Lt. Asst. Comsy.* Andrew, Comst. Trans., B. [M.	12 Dec. 39	4 June 86	4 June 86	14 Dec. 94
McArdle, Patrick	15 Apl. 65	29 July 11	6 Dec. 14	3 Sept. 16
McBarnet, Maj. Alexander Edward, D.S.O., M.V.O., B.	26 Sep. 65	3 July 83	3 July 07	26 Oct. 13
McCausland, Lt.-Col. William Henry, M.	22 July 45	12 Jan. 66	31 Dec. 87	31 Dec. 87
McDermott, *Hy. Maj. Comsy.* John, Misc. List, India	17 July 52	19 Oct. 92	3 Jan. 03	17 Apl. 06
McIntyre, Brig.-Genl. Hugh David, M.	4 Dec. 60	14 Jan. 80	3 Apl. 17	15 Apl. 12
McKay, Lt.-Col. Thompson	12 Oct. 61	9 Sep. 82	1 Dec. 05	25 Oct. 08
McKenzie, Francis, *Comsy. & Hony. Capt.*	18 Apl. 62	5 Aug. 13	19 Sep. 14	5 Jan. 16
McLaurin, Asst. Comsy. and Hony. Lt. George Charter, I A. O.	22 June 74	1 July 17	1 July 17	12 Sep. 18
McLeod, Genl. *Sir* Donald James Sim, K.C.B., K.C.I.E., D.S.O., M. G. L. C. [R]. (*Col., 28 Lt. Cavy.*)	22 Feb. 45	20 July 61	7 May 06	8 May 06
McMahon, Lt.-Col. Sir Arthur Henry, G.C.M.G., G.C.V.O., K.C.I.E., C.S.I.	28 Nov. 62	10 Mar. 83	10 Mar. 09	10 Mar. 15
McMahon, *Hy. Lt. Dy. Asst. Comsy.*, Robert., Misc., M.	1 Nov. 47	21 Jan. 02	21 Jan. 02	5 Nov. 02
McNamara, *Hy. Maj. Comsy.* Thomas Joseph Ord. No.	18 May 56	27 June 02	1 July 17	18 May 11
McNeile, Lt.-Col. Charles, B.	22 Aug. 41	11 Dec. 58	11 Dec. 84	9 Mar. 88
ℳ. McRae, Col., John Graham	16 Mar. 41	2 Feb. 58	2 Feb. 88	25 May 91
MacAlister, *Hy. Capt. Dy. Comsy.* Robert Walter P.W.D., M.	31 Jan. 49	1 Sep. 98	2 Nov. 01	31 Jan. 04
Macartney, Lt.-Col. Henry Frederick Tucker	11 June 56	22 Jan. 79	17 Oct. 06	28 Jan. 07
ℳ. Macaulay, Hony. Col., Charles Edward, B.	26 May 39	12 Dec. 57	5 May 84	5 May 84
Macausland, Lt.-Col. Redmond Conyngham Samuel, B.	29 Dec. 52	13 Nov. 72	13 Nov. 08	13 Nov. 04
Macdonald, Lt.-Col. Charles Edward Wylde	24 Sep. 54	28 Feb. 74	28 Feb. 00	21 June 06
Macdonald, Col. Clarence Herbert, Bo.	9 Sep. 61	1 Sep. 80	11 June 07	1 Sep. 12
Macdonald, Maj.-Genl. *Sir* James Ronald Leslie, K.C.I.E., C.B. [R]	8 Feb. 62	22 Feb. 82	11 Apl. 08	22 Jan. 13
Macey, *Hony. Capt. Comsy.* Thomas Yarnold	5 Nov. 56	16 Nov. 02	6 Nov. 07	5 Nov. 11
Machell, Maj. Edward John, B. Arty.	6 Sep. 39	11 Dec. 58	24 Mar. 80	24 Mar. 08
Mackean, Lt. Ralph Ashenhurst	7 July 88	9 Sep. 08	9 Dec. 10	19 Sep. 17
Mackenzie, Col. James Stuart Fraser	6 Jan. 43	8 June 60	8 June 90	7 Oct. 91
Mackenzie, Lt.-Col. Thomas Harding	6 Apl. 49	3 Feb. 69	3 Feb. 95	6 Apl. 05
Maclaverty, Col. Archibald Iver, M. Arty.	31 Aug. 42	8 June 60	27 July 90	8 June 91
Maclean, *Hy. Capt. Comsy.* Malcolm McPhie, Ord., B.	14 Nov. 50	28 July 94	12 Aug. 02	16 May 05
Maclean, Capt. Frederick Allan	9 Nov. 78	14 Feb. 00	14 Feb. 09	15 Feb. 18
MacLeod, Col. Roderick William, C.B.	16 Dec. 51	19 Oct. 72	19 Nov. 02	23 Dec. 10

RETIRED PAY—*contd.*

Names, Rank and Corps.	Date of Birth.	Date of 1st Commn.	Date of Rank.	Date of Retirement.
Maconchy, Col. Ernest Wm. Stuart King, C.B., C.I.E., D.S.O., M. (*Hony.* Brig.-Genl. 19 May 17)	18 June 60	28 Jan. 82	11 June 07	28 Jan. 14
Macpherson, Lt.-Col. Duncan Alexander Allen, B.	18 Feb. 50	7 July 69	7 July 95	3 Apl. 98
Mahon, Maj. Genl. Reginald Henry, C.B., C.S.I., R. A.	1 May 59	31 Jan. 78	19 Mar. 11	1 May 11
Mainwaring, Lt.-Col. Francis George Lawrence.	8 Nov 51	30 Dec. 71	30 Dec. 97	10 Dec. 99
Maitland, Maj.-Genl. Pelham James, C.B. [R] Bo.	27 June 47	2 May 66	1 Apl. 02	28 June 04
Malcolm, (*Hony.* Lt.-Col.) Pulteney, D.S.O., M.V.O.	16 Aug. 61	11 Aug. 80	27 Mar. 17	11 Aug. 04
𝔐. Malcolmson, Hony. Maj.-Genl. John Henry Porter, C.B., Bo.	20 Oct. 32	8 June 49	1 Oct. 82	1 Oct. 82
Malet, Maj. Guilbert Edward Wyndham, B. Arty.	12 July 39	11 Dec. 58	2 Aug. 79	1 Aug. 79
Malpass, *Hy. Capt. Dy. Comsy.* Arthur, P.W.D., Ind.	2 July 54	29 Aug. 04	4 Aug. 07	6 Sep. 07
Maltby, Lt.-Col. Edward Pennycuick, M. Inf.	13 Aug. 44	4 Jan. 61	1 Jan. 87	1 Jan. 87
Maltby, Lt.-Col. Francis Crichton	26 Oct. 47	16 Oct. 66	16 Oct. 92	16 Oct. 04
Mander, Col. Alfred Thomas, Bo. Engrs.	21 Feb. 36	7 Dec. 55	1 July 85	23 Sep. 89
Mander, Maj.-Genl. Frederick Day, Bo. Inf. (*Col. 103 L. I.*)	4 Mar. 42	2 Aug. 59	7 Sep. 98	10 Sep. 98
Manera, Lt.-Col. Joseph Scott Gowrie	6 July 57	22 May 76	22 May 02	12 May 08
Manley, *Hy. Capt. Dy. Comsy.* Clifford, Misc. B.	1 Oct. 52	21 June 93	1 Jan. 02	9 July 03
Manning, Hony. Brig.-Genl. *Sir* William Henry, K.C.M.G., K.B.E., C.B.	19 July 3	25 Aug. 86	10 Feb. 04	6 Dec. 10
Mansel, Lt.-Col. Charles Grenville	20 Dec. 50	8 July 68	8 July 94	8 July 00
Mansel, Col. William Grenville	21 Sep. 52	30 Dec. 71	10 Feb. 04	30 Dec. 09
Mansfield, *Sir* Maj.-Genl. Herbert K.C.B. [R]	30 Aug. 55	20 Nov. 75	1 May 05	1 Sep. 12
Mantz, *Hy. Capt. Dy. Asst. Comsy.* James Thomas, Ord., B.	22 Sep. 52	27 May 95	18 Oct. 02	24 Sep. 99
Marchant, *Hy. Capt. and Comsy.* Edward, B. M.	11 Nov. 49	21 Dec. 98	3 Apl. 04	11 Nov. 04
Marr, *Hy. Capt. Dy. Comsy.* William, P.W.D., M.	8 Nov. 40	4 Dec. 85	21 Apl. 91	8 Nov. 95
Marriott, Lt.-Col. Leslie Hawthorne	12 Jan. 67	9 May 88	May 14	16 May 14
Marshall, Maj.-Genl. George Frederic Leycaster, C.I.E., R.E., B.	27 Mar. 43	8 June 60	1 Apl. 97	9 Nov. 97
Marshall, *Hy. Maj. Dy. Comsy.* William, Ord., M.	. . . .	15 July 82	10 Jan. 87	10 Jan. 87
Marshall, Bt.-Col. William Simpson	4 Mar. 56	21 Sep. 74	10 Feb. 04	18 Apl. 07
Martin, Lt.-Genl. *Sir* Alfred Robert, K.C.B. [R]	30 Mar. 53	2 Dec. 74	27 July 09	21 Sep. 12
Martin, Lt.-Col. Gerald Ward.	19 May 47	22 June 67	22 June 93	20 May 00
Martindale, Major Cecil Sydney de Butts.	28 May 61	11 Aug. 80	11 Aug. 00	1 Nov. 09
Mason, Maj.-Genl. Henry Macan (*Col., 30 Lancers*)	9 Aug. 50	8 Feb. 70	29 June 04	14 Nov. 06
Mason, *Hy. Capt. Dy. Comsy.* John, M.W.S., Bo.	3 Apl. 49	2 Dec. 94	11 Sep. 97	16 Aug. 03
Massy, Lt.-Col. Charles Francis	10 Dec. 45	26 May 65	26 May 91	27 Apl. 00
Massy, Col. Harry Stanley, C.B.	12 July 55	28 Feb. 74	10 Feb. 04	28 Feb. 06
Masters, Col. Alexander, C.B.	7 Mar. 48	30 Nov. 70	15 Feb. 01	30 Nov. 08
Mathew-Lannowe, Maj.-Genl. Brownlow Hugh, Bo. Engrs.	24 Dec. 35	8 June 55	9 Apl. 86	9 Apl. 86
Mathias, Lt.-Col. Leonard John, Bo.	26 Nov. 60	22 Oct. 81	22 Oct. 07	22 Oct. 13
Mathias, Lt.-Col. Vincent Lewis, M.	5 Apl. 43	22 Mar. 64	25 June 86	25 June 86
Mauditt, Lt.-Col. Frank Rowley Metcalfe Chas. De. Rittich, M	12 Mar. 62	21 Mar. 83	21 Mar. 09	29 Aug. 13
Maxwell, Maj. Alexander Gordon	8 Mar. 67	11 Sep. 89	11 Sep. 07	3 Aug. 09
Maxwell, Lt.-Col. Harrie Malcolm, F.R.C.V.S., I.C.V.D.	29 July 59	11 June. 84	7 Dec. 09	29 July 14
Maynard, Capt. Charles Walsham, B. Arty.	10 Oct. 29	8 June 49	27 Aug. 58	20 Feb. 67
Mayne, Col. Richard Charles Graham, C.B.	27 Aug. 52	21 Dec. 72	10 Nov. 00	1 Oct. 05
Meade, Lt.-Col. John deCourcy Dashwood	23 Dec. 54	9 Aug. 73	9 Aug. 99	9 Aug. 11
Meade, Lt.-Col. Malcolm John, C.I.E.	4 Feb. 54	9 Aug. 73	9 Aug. 99	9 Aug. 11
Meagher, Maj. Dennis John	3 Dec. 60	2 Nov. 01	2 Aug. 03	29 Oct. 09
Medley, Lt.-Col. Ernest James	10 Aug. 59	11 May 78	11 May 04	11 May 06
Meiklejohn, Lt.-Col. John Forbes, B. Arty.	5 Dec. 39	10 Dec. 58	3 July 85	3 July 85
Mein, Lt.-Col. Alexander Bowes	9 Feb. 53	30 Dec. 71	30 Dec. 97	3 Mar. 01
Mein, Bt.-Col. Alexander Lechmere, R.E.	15 July 54	29 Apl. 73	24 Jan. 04	15 July 11
Mein, Col. John Edmund	16 Sep. 51	6 July 70	6 July 00	1 Mar. 10
Mellor, *Hy. Capt Comsy.* Daniel, Alfred, Bk. Dept. M.	24 May 58	30 Dec. 00	6 June 04	5 May 10
Melvill, Lt.-Col. Philip James	11 Apl. 58	1 May 78	1 May 04	19 Nov. 08
Melville, Bt.-Col., John Swinton	3 Aug. 60	11 Ang. 80	19 Aug. 07	11 Aug. 12
Mercer, Col. Charles Archibald, B.	7 July 47	8 June 67	8 June 97	1 July 05

RETIRED PAY—*contd.*

Names, Rank and Corps.	Date of Birth.	Date of 1st Commn.	Date of Rank.	Date of Retirement.
Mercer, *Hy. Maj. Comsy.* Stephen Minchin, S. & T. Corps, B.	30 July 53	31 May 95	1 July 17	1 June 05
Merewether, Lt.-Col. Herbert Duncan	8 Oct. 61	9 Sep. 82	9 Sep. 08	9 Sep. 14
Merriman, Col. William, C.I.E., Bo. Engrs.	2 Apl. 38	12 Dec. 56	1 July 85	27 May 93
Michell, Maj. Henry, B. Inf.	. . . .	. . . .	22 May 63	22 May 63
Middlecoat, Col. Francis, M.	20 June 39	4 Jan. 58	4 Jan. 88	17 Jan. 91
Miley, Col. James Aloysius, C.S.I.	22 Oct. 45	18 July 65	18 July 95	8 Dec. 03
Miller, Capt. Hugh	11 Feb. 75	22 Jan. 96	1 Mar. 06	1 Feb. 08
𝔐. Miller, Hony. Maj.-Genl. James, B.	26 Oct. 35	8 June 55	4 Nov. 86	4 Nov. 86
Miller, *Hy. Capt. Dy. Comsy.* Thomas, Misc. India	13 Sep. 58	7 Feb. 03	10 Nov. 04	15 May 05
Milne, Lt. Robert	. . . .	12 Dec. 57	20 July 60	27 July 67
Mitchell, Col. George Warder, Bo.	29 Nov. 58	11 May 78	11 May 07	30 Sep. 13
𝔐. Mitford, Hony. Maj.-Genl. Reginald Colville William, B.	23 Feb. 39	25 July 55	25 July 86	25 July 86
Mitton, *Hy. Capt. Dy. Comsy.* William Hamilton, S. and T. Corps, B.	25 Aug. 51	17 Dec. 03	1 June 05	25 Aug. 06
Mocatta, Lt.-Col. Daniel Edward	30 Dec. 61	10 May 82	23 Nov. 07	10 May 14
Mockler, Maj.-Genl. Edward, Bo. Inf.	18 Sep. 42	2 Aug. 59	7 Sep. 98	20 Nov. 98
Molesworth, Hony. Br.-Gl., Edward Hogarth, C.B.	2 May 54	4 Feb. 73	14 Sep. 12	5 Nov. 11
Montanaro, Col. Alfred	4 Oct. 53	14 Oct. 71	14 Oct. 97	1 Nov. 11
Montefiore, Lt.-Col. Emanuel, Bo. Arty.	1 Mar. 42	8 June 60	20 June 81	20 June 81
Monteith, Maj. Arthur Mackworth	18 Sep. 53	9 Aug. 73	9 Aug. 93	9 Aug. 97
Monteith, Col. John, C.B.	24 June 52	13 Nov. 72	13 Nov. 02	14 Nov. 10
Montgomery, Bt.-Col. Charles Arnulph Shrewsbury, Bo.	19 Nov. 60	25 Feb. 80	10 June 07	29 Mar. 12
Montgomery, Lt.-Col. Jas. Alexr. Lawrence, C.S.I.	10 Oct. 49	14 Sep. 67	14 Sep. 93	14 Sep. 05
Montgomery, Maj. Thomas Rodger Arundel Gayer.	28 Dec. 57	5 Oct. 78	5 Oct. 98	5 Oct. 02
Moon, *Hy. Maj. Dy. Comsy.* James	. . . .	16 Sep. 68	15 Jan. 84	15 Jan. 81
Mooney, *Hy. Capt. Dy. Comsy.* John, M. W. S. Bo.	17 May 47	11 Dec. 91	2 Dec. 94	1 June 96
Moore, Maj. Frederic Thornton Trevor	20 May 70	8 June 89	8 June 07	11 Aug. 11
Moore, *Hy.* Lt.-Col. George Dashwood, Maxwell	25 July 69	3 May 90	17 July 16	3 May 11
Moore, *Hy. Capt. Dy. Asst. Comsy.* George F.	. . . .	2 Mar. 82	31 Dec. 87	31 Dec. 87
Moore, Lt.-Col. George Henry John	13 Jan. 59	22 Jan. 79	1 June 04	22 Jan. 07
Moore, Lt.-Col. *Sir* George Montgomerie John, Kt., C.I.E., M. Arty.	22 Nov. 44	18 July 65	19 Jan. 87	19 Jan. 87
Moore, Maj. Martin James, B. Cavy.	21 Apl. 43	20 Oct. 59	20 Oct. 79	25 Nov. 80
Moorhead, *Hy. Maj. Dy. Comsy.* James, Mics., B.	25 Nov. 38	20 Sep. 85	8 May 88	25 Nov. 93
Mootham, *Hy. Capt. Dy. Comsy.* Delme G., Comst.-Trans., M.	12 Apl. 43	17 Feb. 8	1 July 92	4 Nov. 93
Morgan, Capt. John Henry Woodside	2 Feb. 77	8 Sep. 97	8 Sep. 06	19 Nov. 14
Morris, Maj. John George	22 Nov. 56	28 Feb. 74	28 Feb. 94	15 May 94
Morris, *Hy. Capt. Comy.*, Richard, S. & T. Corps	19 June 59	4 Jan 08	24 Dec. 13	28 June 14
Morris, Maj. Robert Lee, B.	12 Sep. 68	30 Jan. 89	30 Jan. 07	3 Jan. 13
Morrison, *Hony. Capt. & Comsy.* Henry, Misc. India	10 Nov. 49	21 Feb. 97	9 July 03	10 Nov. 04
Morrison, *Hy. Capt. Dy. Comsy.* John William, P. W. D., I.	5 May 51	5 Dec. 02	15 Mar. 04	5 May 06
Morrison, 2nd-Lt. William Hay, I. A. Res	. . . .	26 July 15	26 July 15	24 Nov. 16
Morton, *Hy. Lt. Asst. Comsy.* Arthur William, S. and T. Corps, B.	19 Jan. 68	3 May 05	3 May 05	19 July 10
Morton, Lt.-Col. Edward Ross, B.	25 Sep. 64	25 Nov. 85	25 Nov. 11	27 Nov. 13
Mottet, Lt.-Col. Arthur Charles, M. Inf.	6 Nov. 41	4 Sep. 59	29 Mar. 82	29 Mar. 82
Mottet, Col. de la Fontaine Henry Edward, M.	. . . .	20 Dec. 50	24 Dec. 79	24 Dec. 79
Mudget, *Hy. Capt. Dy. Comsy.* Charles, Comst. Trans., B.	9 Apl. 40	20 Mar. 89	18 Nov. 91	12 Apl. 95
Muir, Col. Charles Wemyss, C.B., C.I.E.	12 Apl. 50	7 July 69	7 July 95	7 July 07
Muir, Col. William James Wemyss	29 Oct. 40	11 Dec. 58	11 Dec. 88	21 June 92
Mullins, Bt.-Col. Willoughby Brooking	19 Aug. 59	14 Jan. 80	9 June 07	14 Jan. 12
Mumford, *Hony. Capt. Comsy.* Robert, Ord. N.	6 Apl. 56	27 June 02	7 Apl. 03	6 Apl. 11
𝔐. Murray, Hony. Maj.-Genl. Alan, B.	28 Feb. 40	20 Dec. 56	4 Mar. 87	4 Mar. 87
Murray, Lt.-Col. George	18 Feb. 61	14 Jan. 80	30 Oct. 05	1 May 08
Murray, Maj. Henry Brooke	29 Mar. 62	14 May 84	14 May 02	1 May 05
Nurrison, *Hy. Capt. Dy. Comsy.* William Alexander, Comst.-Trans., B.	13 Apl. 40	2 Nov. 84	8 Feb. 90	14 Apl. 95
Myles, Lieut., Andrew Vere	25 Apl. 90	8 Sep. 09	8 Dec. 11	8 Sep. 15
Napier, *Hy. Capt. Dy. Comsy.* Charles, S. & T. Corps, Bur.	8 May 52	15 Apl. 06	17 Aug. 06	8 May 07

RETIRED PAY—*contd.*

Names, Rank and Corps.	Date of birth.	Date of 1st Commn.	Date of Rank.	Date of Retirement.
Napier of Magdala, Lt.-Col. Robert Wm., *Lord*, B.	11 Feb. 45	20 Jan. 62	5 Apl. 83	5 Apl. 83
Napier, Lt.-Col. *Hon.* Henry Dundas, C.M.G.	16 Feb. 64	23 Aug. 84	23 Aug. 10	25 Aug. 12
Naylor, *Hy. Capt. Comsy.* Francis	8 July 61	1 Apl. 03	10 Aug. 06	4 Apl. 08
Neal, *Hy. Capt. Dy. Comsy.* James, Ord., P.	25 Aug. 37	21 Jan. 81	16 Sep. 82	25 Aug. 92
Nelson, *Hy. Maj. Comsy.* George Henry, S. & T. Corps, Bo.	11 Mar. 50	22 Sep. 81	13 Apl. 04	11 Mar. 05
𝔐. Nepean, Col. Herbert Augustus Tierney, M.	17 Apl. 39	26 Sep. 55	26 Sep. 85	1 May 90
Neville, Hony. Br.-Gl., Josiah Philip Crampton	23 Nov. 46	11 Dec. 66	15 Mar. 97	11 Jan. 05
Newell, *Hy. Lt., Dy. Asst. Comsy.* Joseph Edward, Ord., Bo.	16 May 50	1 Dec. 94	1 Dec. 94	1 Aug. 99
Newmarch, Lt.-Col. Lindsay Sherwood	23 Mar. 57	13 Aug. 79	13 Aug. 05	23 Mar. 12
𝔐. Newmarch, Hony. Maj.-Genl. *Sir* Oliver Richardson, K.C.S.I., B.	30 Oct. 34	17 Mar. 55	1 Nov. 87	1 Nov. 87
Newnham, Lt.-Col. Arthur Tristram Herbert, Bo.	17 Jan. 61	22 Jan. 81	27 Jan. 07	22 Jan. 13
Newnham, Maj. Walter Fawcett	17 Jan. 67	14 Sep. 87	14 Sep. 05	19 Mar. 12
Newnham-Smith, Lt.-Col. Edward Deedes	3 Oct. 49	27 Feb 69	27 Feb. 95	15 Mar. 01
Newton, *Comsy. and Hy. Maj.*, George, S. & T.	28 May 63	24 July 11	21 Sep. 17	19 Aug. 18
Nicholls, Lt.-Col. Thomas, B. Inf.	26 Mar. 43	8 June 61	8 June 87	2 Sep. 89
Nicholson, *Hy. Capt., Dy. Comsy.* John, P. W. D., M.	15 Feb. 39	18 May 88	15 Mar. 93	15 Feb. 94
Nicholson, Maj. Thomas Maccallum	4 Mar. 77	4 Aug. 97	4 Aug. 15	19 Sep. 18
Norie, Maj. Frank Hay, D.S.O., B.	8 Mar. 70	4 Mar. 91	4 Mar. 09	4 Mar. 12
Norman, Col. William Wylie	11 Mar. 60	23 Oct. 80	21 Dec. 08	23 Oct. 12
Nortong, Lieut. Wm., Richmond, Tottenham	1 Nov. 83	18 Jan. 02	18 Apl. 04	16 Sep. 08
𝔐. Noverre, Col. William Lynd, B.	10 Dec. 40	20 Nov. 57	1 Jan. 85	1 Jan. 85
Nurse, Lt.-Col. Charles George	22 Mar. 62	22 Jan. 81	22 Jan. 07	23 Jan. 09
Nurse, Maj. Henry Harvey	2 Nov. 63	5 Feb. 87	5 Feb. 05	13 Feb. 08
Nuttall, Maj. John Rattray	19 Nov. 50	10 Feb. 69	10 Feb. 95	10 Feb. 01
Nutting, Maj. Harold Bligh,	28 June 82	30 Apl. 02	30 Apl. 17	11 Sept. 18
𝔐. Obbard, Hony. Maj.-Genl. Thomas, M.	28 Jan. 35	10 Dec. 53	30 Sep. 86	30 Sep. 86
O'Brien, *Hy. Maj. Dy. Comsy.* Terence, Comst. B.	17 July 32	7 Aug. 85	20 Mar. 87	2 Sep. 87
O'Brien, *Comsy. Hy. Capt.* William	2 Jan. 66	5 June 06	25 June 14	16 Apl. 17
O'Donnell, *Comsy. Hy. Capt.* Charles, Ord. N.A.	15 Jan. 60	7 Feb. 12	8 May 14	15 Jan. 15
O'Donnell, Lt.-Col. George Boodrie	26 Oct. 52	19 Jan. 76	19 Jan. 02	1 Jan. 10
O'Donnell, *Hy. Capt. Dy. Comsy.* Patrick, M. W. S., Bo.	10 Sep. 42	23 Dec. 92	1 June 96	11 Sep. 97
O'Donoghue, Bt.-Col. Montague Ernest	16 Feb. 59	11 May 78	11 May 07	11 May 10
O'Gorman, *Hy. Maj. Dy. Comsy.* Cornelius, Comst., B.	. . . .	16 Sep. 82	2 Nov. 84	9 Jan. 85
O'Grady, *Hy. Lt. Asst. Comsy.* Stephen, Ordce. Dept., So. C.	11 July 55	6 Oct. 08	6 Oct. 08	11 July 10
O'Loughlen, *Hy. Lt. Asst. Comsy.* Patrick Michael, Ord., B.	26 Jan. 38	3 Apl. 87	3 Apl. 87	26 Jan. 93
O'Malley, Col. William Arthur D'Oyly, C.B., B.	6 Jan. 53	12 Feb. 73	25 June 03	8 Nov. 13
O'Neill, *Hy. Capt., Dy. Comsy.* John, P. W. D., I.	11 May 51	24 Feb. 02	28 Feb. 03	11 May 06
O'Sullivan, Col. Gerald Hope Wildig, Engrs.	9 Mar. 53	6 Jan. 72	29 Nov. 00	17 Feb. 06
O'Sullivan, *Hy. Lt. Dy. Asst. Comsy.* John., Bo.	20 Dec. 34	1 Jan. 88	1 Jan. 88	20 Dec. 89
Ogilvie, Col. Francis Dashwood, B. Cav.	. . . .	20 Apl. 46	23 June 75	1 May 75
Ogley, *Hy. Lt. Dy. Asst. Comsy.* Lewis, S. & T., B	18 Dec. 48	18 Mar. 87	18 Mar. 87	17 Dec. 03
Oldfield, Lt.-Col. Christopher George, R. G. A.	28 Feb. 63	14 Feb. 83	17 July 11	1 Mar. 14
Oldham, Lt.-Col. George Wingate, Bo. Engrs.	8 Mar. 40	11 June 58	30 Nov. 82	30 Nov. 82
Ommanney, Col. Montagu Willes, B. Arty.	27 Mar. 40	11 Dec. 57	19 July 87	1 Apl. 92
Onslow, Lt.-Col. Francis Montgomery, M. Cav.	9 Nov. 43	4 Sep. 60	7 May 86	7 May 86
Oppenheim, *Hy.* Maj.. *Comsy.* Arthur Edwin, Misc., Bo.	23 Apl. 59	5 Feb. 95	3 Dec. 98	12 Dec. 05
Orchard, Capt. John Waller	12 Sep. 64	10 Nov. 86	10 Nov. 97	25 May 05
Ormsby, Capt. Stuart Frederick	11 Aug. 92	6 Sep. 11	21 Nov. 16	7 May 15
Orr, Maj. John Lock, Bo.	3 May 66	16 May 88	16 May 06	29 May 13
Orr, Capt. Sutherland Alexander Mackay	4 May 69	10 Oct. 91	10 July 01	10 Oct. 10
Orr, Bt.-Col. William John	13 Mar 49	7 July 69	7 July 99	31 Dec. 07
Ottley, Col. *Sir* John Walter, K.C.I.E., Engrs.	22 July 41	15 Jan. 64	12 Aug. 95	22 July 98
Ouchterlony, Lt.-Col. Thomas Heathcote, Bo. Arty.	. . . .	12 Dec. 57	9 Sep. 80	9 Sep. 80
Owen, Lt.-Col. Conrad Agnew, Bo. Cav.	20 May 43	27 Oct. 61	1 Oct. 86	1 Oct. 86
Owen, Lieut. Owen	27 Apl. 84	22 Apl. 03	22 July 05	31 Oct. 07
Oxley, Lt.-Col. Charles Rawson, M. Inf.	12 June 38	4 Apl. 59	4 Apl. 85	24 July 88
Ozzard, Maj. Hudson Henry	17 Mar. 57	13 June 74	13 June 94	4 Apl. 99
Paddon, Maj. Stanley Somerset Wreford, W.	28 Mar. 81	15 Aug. 00	15 Aug. 09	23 Feb. 14

RETIRED PAY—*contd.*

Names, Rank and Corps.	Date of Birth.	Date of 1st Commn.	Date of Rank.	Date of Retirement.
Pakenham, Lt.-Col. William Wingfield Verner	14 Nov. 57	30 Jan. 78	30 Jan. 04	28 Apl. 06
Paleologus, Maj. William Constantine	9 July 70	22 Nov. 91	23 Mar. 07	24 Mar. 14
Palmer, Col. Clarence Henry	4 Aug. 40	12 June 58	12 June 88	15 Sep. 91
Parfitt, *Hy. Lt., Dy. Asst. Comsy.* William Christopher, Ord. B.	7 Aug. 52	20 May 98	20 May 98	7 Mar. 99
Park, *Comsy. Hy. Capt.* James, I. O. D.	6 Feb. 59	25 May 06	7 Feb. 12	6 Feb. 14
Parker, Maj. Harry Prevost England	20 Dec. 64	29 Aug. 85	29 Aug. 03	29 Aug. 10
Parker, Maj. James	14 Dec. 66	30 Jan. 86	30 Jan. 04	27 Feb. 10
Parker, Maj. George Marcus Godfrey	2 Oct. 71	3 Sep. 92	3 Sep. 10	13 Nov. 18
Parker, Lt.-Col. Neville Thornton	14 June 64	25 Nov. 85	25 Nov. 11	29 Nov. 18
Parkin, Lt.-Col. Henry, C.I.E., Bo.	10 Dec. 58	1 May 78	1 May 04	10 Dec. 13
Parrott, Lt.-Col. Benjamin Alexander Napier	11 Oct. 50	8 July 68	8 July 94	1 May 98
Parsons, Col. Arthur Dewar, M. Cav.	27 Sep. 34	20 Jan. 56	18 Oct. 83	20 Oct. 89
Patch, Col. Robert, C.B. [R]	25 July 42	12 June 63	7 Dec. 91	9 Nov. 04
Paterson, *Hy. Lt. Asst. Comsy.* Alexander, Comst., B.	19 Apl. 38	28 June 90	28 June 90	20 Apl. 93
Patton, *Hy. Capt. Comsy.* John Edward, P. W. D., M.	18 Jan. 50	4 Apl. 99	31 Jan. 04	18 Jan. 05
Paul, Bt.-Col. Ernest Theodore, B.	14 Apl. 58	1 Jan. 76	22 Jan. 06	1 Jan. 14
Paxton, Maj. Archibald Francis Pinkney	6 Dec. 62	28 Feb. 85	28 Feb. 03	23 Mar. 09
Payne, Col. Charles Drummond Passmore	20 July 41	12 Jan. 59	12 Jan. 86	7 Aug. 91
Pearse, Maj. Sydney Arthur, M. D.S.O.	14 Oct. 62	14 May 84	14 May 02	19 Jan. 05
Pearson, Col. George Falconer, M.	. . . .	14 Jan. 46	28 Oct. 76	1 Aug. 76
Peart, Maj. Denis George	13 June 66	25 Aug. 86	25 Aug. 04	6 Sep. 11
Peile, Bt.-Col. Solomon Charles Frederick, C.I.E.	11 Dec. 55	9 Aug. 74	9 Aug. 00	11 Dec. 10
Pendleton, *Hy. Capt. Comsy.* George, Bk. Dept., M.	28 May 62	15 Oct. 04	29 Oct. 07	22 June 08
Penrose-Thackwell, Bt.-Col. Edward Rawdon	8 Sep. 56	12 Feb. 76	12 Feb. 05	1 Jan. 90
Penstone, *Hy. Capt. Dy. Comsy.* Giles	15 Sep. 39	20 July 82	8 Oct. 90	8 Oct. 90
Penton, Maj.-Genl. Herbert Edward (*Col., 107 Pio.*)	5 Apl. 50	8 Jan. 68	10 Dec. 02	9 June 08
Perkins, Maj. Norman Chichester	15 Sep. 60	13 Aug. 79	13 Aug. 99	9 Apl. 00
Perks, *Hy. Lt. Dy. Comsy.* Arthur, P. W. D., B.	30 Apl. 45	29 Nov. 98	29 Nov. 98	30 Apl. 00
Perry, Maj. George Edgar John	12 May 69	3 May 90	3 May 08	3 May 10
Persse, Col. Edward, M.	2 Mar. 38	4 Mar. 86	4 Mar. 86	4 Mar. 86
Peterson, Maj. Cecil Herbert	21 Oct. 67	16 Nov. 87	16 Nov. 05	17 Nov. 11
Peyton, Col. Algernon George, B.	3 Oct. 59	11 Aug. 80	6 Apl. 10	19 Aug. 12
Peyton, Col. Lumely Scobell, B.	3 June 54	13 June 74	8 Apl. 04	13 June 12
Phillips, Col. Alfred Lucian, B.	24 Mar. 61	22 Jan. 81	16 Oct. 08	24 Mar. 13
Philipps, *Hon.* Maj.-Genl. Ivor, D.S.O., *p.s.c.*	9 Sep. 61	12 May 83	10 July 16	20 July 03
Phillips, *Hy. Capt., Dy. Comsy.*, Phillip	6 Apl. 57	20 May 98	22 Apl. 08	6 Apl. 12
Phillipps, Lt.-Col. Charles Reginald, Bo.	7 Feb. 58	10 Sep. 75	10 Sep. 01	28 Sep. 13
Phillott, Bt.-Col. Douglas Craven	28 June 60	14 Jan. 80	14 Jan. 06	28 Apl. 12
Picot, Lt.-Col. Henry Phillip	29 Apl. 57	21 Sep. 74	21 Sep. 00	4 Nov. 03
Pierson, Col. James Frederick, M. Arty.	6 Nov. 40	12 June 58	12 June 82	12 June 82
Pigou, Maj. Frederick Hugo	31 Oct. 66	22 Aug. 88	22 Aug. 06	5 Aug. 09
Pike, *Hy. Maj. Comsy.* Jacob	28 June 59	27 June 05	4 July 12	28 June 14
Pinney, Maj. John Charles Digby	5 July 67	8 June 89	8 June 07	23 Apl. 12
Pipe, *Hy. Maj. Comsy.* Edward Alfred, Ord. N. A.	23 Mar. 57	4 Jan. 06	1 July 17	31 Mar. 11
Plowden, Lt.-Col. Walter Francis Courtenay Chichele, B.	6 Sep. 52	19 Oct. 72	19 Oct. 98	6 Sep. 07
Pluckrose, *Hy. Capt. Comsy.* Edward, Bk., Madras	29 Mar. 52	18 Apl. 03	30 Mar. 06	21 Sep. 06
Plummer, *Hy. Maj. Dy. Comsy.* Samuel Charles, P. W. D., M.	. . . .	20 Jan. 78	5 Dec. 87	5 Dec. 87
Pollock, Maj.-Genl. John Archibald Henry, C.B. [R] (*Col., 51 Sikhs*)	1 Dec. 56	13 June 74	19 Mar. 06	29 June 11
Pollock, Col. John Graham, M. Arty.	15 Nov. 41	10 June 59	13 May 89	10 June 90
Pond, Lt.-Col. Alexander Donald Charters	5 Sep. 58	1 Nov. 79	1 Feb. 07	1 Nov. 07
Poole, Col. Arthur, Bo.	8 Apl. 41	11 Nov. 58	28 Feb. 87	28 Feb. 87
Poole, Maj. William Thomas Conway	1 Apl. 69	1 Mar. 90	1 Mar. 08	28 Oct. 14
Porteous, Col. James Edward	21 Dec. 46	22 Sep. 65	22 Sep. 95	26 Jan. 99
Porter, Col. Atwell Robert	25 Aug. 49	21 Aug. 69	15 Oct. 99	13 Mar. 05
Porter, Maj. Herbert Edward	6 Dec. 57	11 Nov. 77	11 Nov. 97	4 Nov. 07
Porter, *Hy. Lt. Asst. Comsy.* William	. . . .	15 Aug. 67	15 Aug. 67	18 Aug. 70
Powell, *Hy. Capt. Dy. Comsy.* William, Comst., B.	18 Apl. 34	1 July 84	17 July 87	20 Mar. 89
Powis, *Hy. Lt. Asst. Comsy.* Charles, Misc., B.	17 Nov. 52	25 Nov. 93	17 Nov. 52	21 Feb. 97
Prendergast, Maj.-Genl. Charles Lewis, B. Inf. (*Col. 28 Punjabis*)	15 Oct. 43	4 Mar. 60	31 Jan. 99	11 Feb. 99
Pressey, Lt.-Col. Arthur	13 Feb. 60	28 Jan. 82	3 June 07	28 Jan. 11
Price, Comsy. *Hony.* Maj.. Ernest	21 July 61	10 Sep. 08	1 Apl. 18	1 Jan. 17

Names, Rank and Corps.	Date of Birth.	Date of 1st Commn.	Date of Rank.	Date of Retirement.

RETIRED PAY—*contd.*

Names, Rank and Corps.	Date of Birth.	Date of 1st Commn.	Date of Rank.	Date of Retirement.
Price, *Hony. Lt., Asst. Commsy.* Edwin John	12 Feb. 62	6 Feb. 14	6 Feb. 14	12 Feb. 17
Priestley, Lt.-Col. Fredk. Jos. Blakiston	18 Apl. 58	30 Jan. 78	1 Feb. 05	30 Jan. 06
Priestley, Lt.-Col. George William	26 Dec. 62	9 Sep. 82	30 Dec. 10	31 Dec. 10
Pringle, Lt.-Col. Alexander	27 Oct. 50	30 Dec. 72	30 Dec. 98	23 Aug. 01
Pringle, *Hy. Maj. Dy. Comsy.* Robert, Misc., Bo.	19 Nov. 47	8 Nov. 86	20 July 98	7 May 99
Prior, Col. William, B.	13 Oct. 59	13 Aug. 79	28 Sep. 08	13 Nov. 12
Pritchard, Maj. Alfred Bassett	5 Dec. 57	11 Nov. 78	11 Nov. 98	10 May 03
Pritchard, Capt. Douglas James, B.	10 Mar. 78	16 Feb. 98	16 Feb. 07	16 Feb. 14
Proudfoot, Col. Andrew William	30 Nov. 48	24 Nov. 69	24 July 99	15 Mar. 09
Pulford, Col. Russell Richard, C.I.E., Engrs.	20 Dec. 45	24 Mar. 65	27 Nov. 96	1 Oct. 02
Pulley, Bt.-Col. Charles, C.B.	21 Oct. 51	28 Oct. 71	20 May 98	10 Aug. 06
Pulleyn, *Comsy., and Hy. Maj.*, George Herbert, M.C.	27 Sep. 62	4 Oct. 10	1 Jan. 17	20 Oct. 10
Quilter, *Hy. Lt. Asst. Comsy.* John Howlett, M. W. S., B.	23 Mar. 36	14 June 83	14 June 83	23 Mar. 91
Quin, Lt.-Col. Edward	25 Mar. 42	12 May 63	12 May 89	10 June 96
Quin, Maj. Thomas, M.	16 Nov. 61	11 Aug. 80	11 Aug. 00	28 Oct. 04
Radcliff, Bt.-Col. Samuel Garnett	16 Oct. 59	19 Feb. 81	10 Feb. 08	30 Apl. 09
Radcliffe, Lt.-Col. Arthur Wilbraham Twining	11 Feb. 54	9 Aug. 73	9 Aug. 99	1 Sep. 02
Radford, *Hy. Lt. Asst. Comsy.* James Thomas, S. & T. Corps	24 Sep. 59	4 Oct. 04	4 Oct. 04	1 Nov. 07
Radford, Col. Oswald Claude, C. B., C.I.E.	4 Nov. 50	26 Sep. 71	26 Sep. 01	1 Jan. 10
Raikes, Hony. Maj.-Genl. Charles Lewis, M.	20 July 37	8 June 55	8 June 87	8 June 87
Raikes, Col. Percy Bury, B. Arty.	28 June 39	11 Dec. 57	13 Dec. 84	13 Dec. 84
Rainey-Robinson, Col. Robert Maximilian, C.B., C.M.G.	28 Apl. 61	22 Jan. 81	11 June 07	14 Feb. 13
Rainford, *Hy. Capt. Dy. Comsy.* William, Ord., M.	20 Nov. 39	28 Jan. 93	28 July 94	20 Nov. 94
Raitt, Lt.-Col. Henry George Bowen	11 Aug. 62	10 Mar. 83	10 Mar	20 Dec. 15
Ralph, Maj. Arthur Jeffreys	23 Aug. 69	8 June 89	8 June 07	1 July 09
Ramsay, Lt.-Col. Henry Lushington	15 Sep. 54	28 Feb. 74	23 Feb. 00	19 Mar. 01
Ramsay, Capt. James Alexander	. . . .	20 Feb. 50	16 July 64	31 Dec. 64
Ramsay, Maj.-Genl. *Sir* John George, K.C.B. (*Col. 24 Punjabis*)	5 Nov. 56	11 Feb. 75	17 May 10	1 May 12
Ramsay, The Hon. *Sir* John, K.C.I.E., C.S.I.	5 Dec. 62	9 Sept. 82	9 Sep. 08	5 Dec. 17
Ramsden, Col. Herbert Frecheville Smyth	6 Mar. 56	10 Sep. 75	21 Apl. 05	13 Apl. 08
Ramsden, Lt.-Col. William Charles, B. Inf.	10 Feb. 44	8 June 61	8 June 87	5 Sep. 89
Ranken, Bt.-Col. George Patrick	13 Oct. 59	13 Aug. 79	6 June 07	16 Aug. 11
Ransom, Bt.-Col. James Mann, Bo.	16 July 61	1 Feb. 81	26 Sep. 09	1 Feb. 13
Rattray, Bt.-Col. Rullion Hare	1 June 59	30 Jan. 78	30 Jan. 07	30 Jan. 10
Raven, Maj. Alexander William Nicholas	9 Mar. 70	21 Dec. 89	21 Dec. 07	22 Dec. 13
Ravenshaw, Lt.-Col. Charles Withers	29 Sep. 51	8 May 72	8 May 98	25 Mar. 06
Ravenshaw, Maj. Harold Alexander	24 Dec. 56	24 Mar. 76	24 Mar. 96	4 Apl. 00
Raverty, Maj. Henry George, Bo. Inf.	. . . .	28 June 43	12 Aug. 62	12 Aug. 62
Rawlings, *Hony. Maj. Comsy.* Frederick Gilbert, Ord., S.C.	14 June 57	12 Aug. 02	1 July 17	13 Sep. 06
Rawlinson, Lt.-Col. Charles Brook, C.I.E.	10 Apl. 66	9 May 85	9 May 11	12 May 17
Read, *Hy. Capt. Comsy.* Robert, P. W. D.	15 Mar. 49	28 June 94	16 Oct. 96	15 Mar. 04
Reader, *Hy. Capt. Dy. Comsy.* William Cory, P. W. D., M.	2 Apl. 51	1 Mar. 98	2 Nov. 01	26 Mar. 03
Reeves, *Hony. Maj. Comsy.* William, S. & T. Corps	27 Feb. 58	9 Dec. 07	6 Feb. 13	26 Feb. 13
Reid, Maj. Charles Clements	30 Aug. 61	22 Oct. 81	10 July 01	23 Oct. 05
Reid, Col. Charles Edward, M. Arty.	23 Sep. 40	12 June 58	27 July 86	27 July 86
Reid, Col. Lestock Hamilton, C.S.I.	22 July 57	1 May 78	1 May 07	1 May 13
Reilly, Lt.-Col. Rawdon Edward Dennys	15 Nov. 52	31 Aug. 72	31 Aug. 98	6 Sep. 10
Reilly, *Hy. Capt., Dy. Comsy.* Thomas, P. W. D.	29 May 51	19 Jan. 04	28 Feb. 06	29 May 06
Rennick, Col. Robert Henry Francis, B.	2 Apl. 42	18 Dec. 61	5 Oct. 92	1 Mar. 05
Renny, Bt.-Col. Alexander MacWhirter	27 Oct. 54	11 Sep. 73	11 Sep. 99	11 Sep. 11
Renny, Bt.-Col. George Blakiston	24 Sep. 51	27 Mar. 72	10 Feb. 04	27 Mar. 10
Renton, Lt.-Col. Robert Ross, B.	21 May 62	30 Jan. 84	30 Jan. 10	30 Jan. 13
Repton, Maj. Frederick William	2 Nov. 60	13 Aug. 79	13 Aug. 99	7 May 02
Retallick, Lt.-Col. John Mark Anthony	13 Nov. 57	10 Sep. 75	10 Sep. 01	28 Oct. 04
Rhubottom, *Hony. Capt., Comsy.* Robert, I. O. D.	11 Nov. 57	22 Oct. 02	18 Mar. 05	11 Nov. 12
Rice, Hony. Maj.-Genl. Harry Chippindale Plunkett, B.	. . . .	20 Dec. 53	1 Feb. 86	1 Feb. 86
Richards, Lt.-Col. Edward Willoughby	27 June 65	5 July 84	5 July 10	20 Aug. 14
Richards, *Hony. Capt., Comsy.*, Frederick, Ord., So	14 Nov. 49	3 June 97	29 July 04	14 Nov. 04
Richardson, Lt.-Genl. *Sir* George Lloyd Reilly, K.C.B., C.S.I. C.I.E. [R] (*Col., 18 Lancers*)	20 Sep. 47	23 Mar. 66	28 Oct. 06	7 Apl. 08
Ricketts, Maj. Lancelot Hamilton	23 Sep. 67	8 Dec. 88	8 Dec. 06	8 Dec. 11
Rideout, Col. Fred. Charles Wood	4 Dec. 59	22 Jan. 79	4 June 07	24 Mar. 09
Ridgeway, Col. *the Rt. Hon. Sir* Joseph West, G.C.M.G., G.C.B., K.C.S.I., B. Inf.	16 May 44	4 Dec. 60	2 Mar. 85	Aug. 89
VC Ridgeway, Col. Richard Kirby, C.B.	18 Aug. 48	8 Jan. 68	8 Jan. 98	9 Jan. 06
Riley, Maj. Charles Brook	18 Feb. 77	2 Jan. 97	2 Jan. 15	2 Aug. 16
Rimington, Lieut. Michael Ballantyne	18 May 81	11 May 00	21 Apl. 02	1 Jan. 11
Rind, Bt.-Col Alexander Thomas Seton Abercromby, C.M.G.	28 Mar. 47	2 Oct. 66	2 Oct. 96	27 Feb. 99

RETIRED PAY—*contd.*

Names, Rank and Corps.	Date of Birth.	Date of 1st Commn.	Date of Rank.	Date of Retirement.
Rippon, Lt.-Col. George, M.	18 Nov. 61	11 Aug. 80	1 June 04	1 Dec. 12
Rishworth, *Hy. Lt. Asst. Comsy.* John Richard	21 Sep. 57	16 Oct. 08	16 Oct. 08	9 Aug. 12
𝔐. Ritchie, Hon. Maj.-Genl. John, Bo. Arty.	31 Jan. 34	11 June 53	22 Nov. 86	22 Nov. 86
Rivett-Carnac, Col Ernest Henry, B.	30 June 57	12 Feb. 76	29 May 05	12 Feb. 14
Roberton, Lt.-Col. Peter Archibald, B. Inf.	. . . .	7 July 38	31 Dec. 61	31 Dec. 61
Roberts, Lt.-Col. Allan Scott, B. Inf.	29 Aug. 45	20 Feb. 62	20 Feb. 88	18 Dec. 94
Roberts, Hony. Maj.-Genl. John, B.	. . . .	27 Sep. 52	28 Sep. 83	20 Jan. 13
Roberts, Lt.-Col. Montgomery Browne, B.	22 Oct. 63	23 May 85	23 May 11	31 Dec. 87
Robinson, *Hy. Capt. Dy. Comsy.* Henry, Ord., B.	15 June 46	1 Feb. 95	20 May 98	15 June 01
Robinson, Lt.-Col. Charles Walter, S. & T. Corps.	23 Mar. 76	5 Aug. 96	5 Aug. 14	4 Apl. 18
Rodgers, *Hy. Lt. Dy. Asst. Comsy.* James, Misc., B.	17 Nov. 31	27 Apl. 85	27 Apl. 85	17 Dec. 86
Rodney Brown, Maj. Walter Medlicott, B.	29 July 68	21 Dec. 89	21 Dec. 07	23 Dec. 12
Roffey, *Hy. Capt. Comsy.* Lewis Jas. Bk. M.	6 Feb. 52	5 Apl. 95	3 Jan. 97	6 Feb. 07
Rogers, Lieut. Archibald Clement Campbell	29 July 84	21 Jan. 03	21 Apl. 05	1 Dec 09
Rogers, *Hy. Lt. Dy. Asst. Comsy.* Thomas Edwin, S. & T., Bo.	19 Nov. 46	1 Nov. 99	1 Nov. 99	6 Nov. 01
Rooke, Lt.-Col. Alfred Shipton	7 Dec. 59	19 Feb. 81	19 Feb. 07	7 Mar. 12
Rooke, Major Benjamin Proctor Simpson	12 Jan. 66	28 Apl. 86	28 Apl. 04	28 Apl. 06
Rose, Col. Hugh	30 Nov. 60	11 May 78	11 May 07	1 Jan. 12
Ross, Capt. Charles Edmund	9 Sept. 63	6 Feb. 84	6 Feb. 95	17 June 98
Ross, Lt. John William	18 July 66	5 Feb. 87	29 Jan. 88	27 Feb. 97
Ross, *Hy. Capt. Comsy.* Peter, P. W. D.	23 Mar. 53	1 June 00	27 Feb. 04	21 Mar. 08
Ross, Lieut. Wm. Robert	9 Feb. 76	22 Jan. 96	22 Apl. 98	28 Nov. 04
Rouse, Major Archibald Henry Tylden	29 May 69	28 June 90	28 June 08	1 June 16
Routleff, *Hy. Capt. Dy. Comsy.* William John, Comst., B.	8 Feb. 35	11 May 85	14 Aug. 89	8 Feb. 90
Row, Lt.-Col. George Russell	10 June 62	22 Jan. 81	22 Jan. 07	19 Apl. 13
Rowcroft, Bt.-Col. (tempy. Maj. in I.M.S.) George Francis, D.S.O.	4 May 61	13 Aug. 79	1 June 04	20 Nov. 07
Royle, Lt. Joseph Ralph Edward John, C.I.E.	. . . .	7 June 61	15 June 62	30 Jan. 67
Ruck, *Hony. Capt. Comsy.* James, P. W. D.	15 Jan. 58	14 May 07	2 Oct. 11	15 Jan 13
𝔐. Ruggles, Hony. Maj.-Genl. John, B. (*Col. 19 Punjabis*)	. . . .	13 June 45	11 Nov. 76	1 Sep. 76
Rule, *Hony. Capt. Comsy.* William	2 Oct. 56	1 Mar. 05	10 Sep. 08	2 Oct. 11
Russell, Col. Bruce Bremner, R. E.	25 Aug. 59	18 Dec. 78	1 Oct. 09	23 Oct. 12
Russell, Col. George Algernon, B. Arty.	3 June 39	11 Dec. 57	1 Apl. 87	12 Dec. 88
Rutherford, Col. James, Bo. Inf.	9 Jan. 42	4 Aug. 59	4 Aug. 89	10 Sep. 98
Ryall, Lt.-Col. Edward Charles	1 Apl. 63	9 Sep. 82	1 Jan. 09	17 Feb. 09
Ryan, Lt.-Col. Edward Henry, B. Arty.	. . . .	12 June 57	30 Dec. 78	30 Dec. 78
Ryan, *Hy. Capt. Dy. Comsy.* Philip Justin, P.W.D., B.	20 Apl. 36	10 Jan. 83	1 Aug. 89	20 Apl. 91
Rynd, Lt.-Col. Frederick Cecil, M.	24 Oct. 61	22 Jan. 81	22 Jan. 07	7 Dec. 13
Sage, Bt. Col. Charles Arthur Ross	22 Jan. 51	8 July 68	8 July 98	8 July 00
Salkeld, Col. Charles Edward, B.	13 June 41	8 June 60	8 June 90	1 Jan. 91
Samuells, Col. William Leycester, B.	14 July 41	20 Nov. 57	20 Nov. 87	3 Apl. 90
Sanderson, Lt. *Sir* Percy, K.C.M.G., M. Arty.	. . . .	9 Dec. 59	9 Dec. 59	8 Feb. 70
Sandilands, *Hy. Lt., Dy. Asst. Comsy.*, Alexander, Ord., Bo.	24 June 52	1 Aug. 99	1 Aug. 99	16 Mar. 00
Sandwith, Lt.-Col., John Robert	24 July 53	28 Apl. 76	28 Apl. 02	28 Apl. 08
Saulez, Maj. Edmund, Bo.	21 Feb. 67	14 Sep. 87	14 Sep. 05	4 Oct. 12
Saunders, *Hony. Capt. Comsy.* Frederick	7 Aug. 62	19 May 04	20 Sept. 07	11 Oct. 11
Saw, *Hy. Capt., Dy. Comsy.*, Henry, Ord., B.	29 Aug. 47	13 Sep. 92	21 Jan. 95	12 Nov. 98
Sawyer, Bt.-Col. Herbert Anthony	16 Sep. 47	2 Oct. 66	2 Oct. 96	16 May 99
Schalch Col. Vernon Ansdell, C.B.	7 Sep. 49	29 May 69	29 May 99	12 June 07
Schneider, Maj. Reginald Percy Congreve	1 Feb. 66	25 Aug. 86	25 Aug. 04	1 Sep. 07
Scott, Bt.-Col. *Sir* Buchanan, K.C.I.E., Engrs.	14 Jan. 56	2 Aug. 71	31 Dec. 00	14 Jan. 05
Scott, Col. Courtenay Harvey Saltern, B.	. . . .	17 Aug. 52	17 Aug. 80	17 Aug. 80
Scott, *Hony. Capt. Dy. Comsy.* John, P. W. D.	21 Mar. 56	23 Sep. 08	4 July 10	21 Mar. 11
Scott, *Hony. Maj. Comsy.* Thomas John, Bo.	5 Oct. 53	6 Oct. 99	1 July 17	1 Nov. 05
Seagrim, Lt.-Col. Albert, B.	. . . .	12 Mar. 52	12 Mar. 78	12 Mar. 78
Sealy, Lt.-Col. Charles William Henry, Bo.	16 Sep. 46	17 July 66	17 July 92	17 Apl. 12
Searle, Bt.-Col. Alfred Ernest Stuart, Bo.	16 Mar. 60	13 Aug. 79	7 Mar. 08	13 Aug. 11
Searle, Lt.-Col. Percy Cormack	17 Jan. 69	11 Feb. 88	11 Feb. 14	14 Mar. 14
Sellens, *Hy. Capt. Dy. Comsy.* Sidney Hubert, P.W.D.	23 May 57	1 Mar. 09	21 Mar. 11	12 May 12
Selman, Hony. Lt. Asst. Comsy. Benjamin, M.W.S.	7 Nov. 56	6 Nov 11	6 Nov. 11	7 Nov. 11
Sexton, Maj. John Joseph O'Brien	30 Dec. 66	25 Aug. 86	25 Aug. 04	1 Jan. 07

RETIRED PAY—*contd.*

Names, Rank and Corps.	Date of Birth.	Date of 1st Commn.	Date of Rank.	Date of Retiremen
Seymore, Capt. Isaac William, Ord.	23 Mar. 58	19 Nov. 06	4 Mar. 12	23 Mar. 13
Shakespear, Lt.-Col. William Frederic, B.	21 Mar. 61	9 Aug. 82	9 Aug. 08	16 Nov. 12
Sharpe, *Hy. Maj., Dy. Comsy.*, John Henry, Comst., B.	. . . . .	11 July 77	26 Feb. 84	20 Nov. 85
Shaw, Col. Edward William, M.	24 July 37	12 Dec. 56	26 Apl. 83	26 Apl. 83
Shaw, Col. George Jocelyn	23 Jan. 57	1 Apl. 76	1 Apl. 07	25 Sep. 11
Shaw, Maj. Herbert Stuart, B.	2 Jan. 67	10 Nov. 88	10 Nov. 06	11 May 13
Shelley, Maj. Archibald Douglas Graham, Engrs.	28 Apl. 62	26 July 81	9 June 99	10 Dec. 05
Shepherd, Lt.-Col. William, B. Engrs.	16 Feb. 42	9 Dec. 59	10 Jan. 89	12 Feb. 91
Sheppard, Lt.-Col. Charles Henry, M. Inf.	5 May 45	1 Mar. 67	1 Mar. 88	28 June 94
Sherard, Col. Ralph Woodchurch, Bo.	11 Nov. 60	22 Jan. 79	5 June 07	22 Jan. 11
Shewen, Col. (Hony. Bdg.-Genl.) Mansel Travers, Bo.	10 May 62	1 Feb. 80	10 June 07	20 Jan. 14
S uirley, Lt.-Col. William, C.M.G., B.	11 Dec. 66	5 Feb. 87	5 Feb. 13	6 Feb. 13
Shortland, Col. Daniel Vincent, M. Arty.	14 Oct. 41	11 Dec. 58	10 Dec. 88	11 Dec. 93
Gl. Sibthorpe, Hony. Maj. Genl. Lester Horatio, Bo.	25 June 33	20 Jan. 51	2 Feb. 85	2 Feb. 85
Sirvester, *Hy. Capt. Comsy.*, George Frederick, M.W.S.	16 Sep. 60	21 Mar. 08	11 Nov. 11	14 June 13
Simpson, *Hy. Capt., Comsy.* Benjamin, Misc., India	24 Dec. 55	11 Feb. 03	1 Mar. 07	1 Mar. 08
Simpson, Col. Robert James Brown, M.	20 Feb. 40	20 Feb. 58	20 Feb. 88	21 Feb. 88
Simson, Capt. Rupert	20 Aug. 84	18 Jan. 05	18 Jan. 14	11 June 14
Sinclair, *Hy. Maj., Dy. Comsy.*, Andrews, S., Ord., B.	14 May 32	11 Aug. 82	13 May 87	2 Apl. 87
Singleton, *Hony. Capt., Comsy.*, William Joseph, P. W. D., India	23 Sep. 53	22 Jan. 05	2 Feb. 08	23 Sep. 08
Skeaf, *Hy. Capt., Dy. Comsy.*, Walter Henry	24 July 56	8 Aug. 04	1 Oct. 08	24 July 11
Slaney, *Hy. Lieut., Asst. Comsy.* Henry, P.W.D., M.	12 July 51	31 Jan. 04	31 Jan. 04	12 July 06
Smart, Col. Alexr. William, Arty.	1 Oct. 48	15 July 68	10 July 97	19 Oct. 05
Smart, Maj. Edward de Sansmarez	8 Aug. 59	23 Oct. 80	23 Oct. 00	24 Oct. 00
Smiles, *Hy. Capt., Dy. Comsy.* John Oswald, S. & T., M.	4 Oct. 49	15 Mar. 91	2 Jan. 04	4 Oct. 04
Smith, Maj. Allan	2 Dec. 51	30 Dec. 71	30 Dec. 91	1 Aug. 96
Smith, *Hy. Capt., Comsy.* Charles	14 May 52	22 Feb. 00	31 Jan. 04	14 May 07
Smith, Maj. George Rainier deHeriez, B.	13 Aug. 67	10 Nov. 87	16 Nov. 05	18 Nov. 12
Smith, *Hy. Capt., Dy. Comsy.*, Henry, M. W. S., B.	. . . . .	2 June 69	16 Sep. 82	16 Sep. 82
Smith, *Hy. Capt., Dy. Comsy.*, James, S. & T. Corps	15 Apl. 51	9 Mar. 00	12 May 05	15 Apl. 06
Smith, Col. John Graham	30 Mar. 56	21 Sep. 75	21 Sep. 04	1 Feb. 08
Smith, *Hy. Capt., Dy. Comsy.*, James Henry, Ord., B.	[illegible] May 44	20 Nov. 84	27 Feb. 92	6 Mar. 96
Smith, Maj. Owen Annesly	8 Mar. 67	24 July 86	24 July 04	19 Apl. 12
Smith, *Hy. Capt., Dy. Comsy.*, Samuel, Ord., B.	9 May 46	19 Feb. 86	25 June 94	9 May 01
Smith, Col. Thomas Hugh	8 Nov. 61	31 Jan. 80	9 June 07	23 Feb. 12
Smith, *Hy. Capt., Comsy.*, William, Ord., So.	21 May 55	11 Feb. 00	1 Nov. 04	21 May 10
Smith, Col. William Graham, B. Inf.	2 Jan. 39	11 Aug. 59	11 Aug. 89	11 Jan. 92
Smith, *Hy. Maj. Comsy.*, Wm. Henry, Ord. M.	26 May 52	9 May 01	1 July 17	18 Mar. 05
Smyth, Col. Etwall Walter, C.B., B. Inf.	9 Nov. 43	7 June 61	28 Dec. 93	5 Dec. 98
Gl. Smyth, Col. Ralph Graham, B. Engrs.	14 Mar. 36	7 Dec. 55	1 July 85	7 Dec. 89
Snee, *Hy. Lt., Asst. Comsy.*, Michael, P. W. D.	29 Aug. 40	19 Jan. 04	19 Jan. 04	29 Aug. 04
Snell, Lt.-Col. Francies William	2 Aug. 48	23 Jan. 69	23 Jan. 95	1 July 11
Solbe, Lt.-Col. Charles Edward DeLisle, M.	21 Jan. 65	16 Nov. 87	16 Nov. 13	17 Nov. 13
Southam, *Hy. Capt. Dy. Comsy.* Thomas, Ord., M.	4 Sep. 47	6 Jan. 94	8 Nov. 01	12 Aug. 08
Southey, Lt.-Col. Henry Herbert	28 Mar. 67	5 Feb. 87	5 Nov. 13	2 Apl. 14
Sparkes, *Hy. Capt., Asst. Comsy.*, Thomas George, Misc., India	18 Jan. 60	15 May 05	15 May 05	1 Jan. 07
Sparling, Lt.-Col. James Philip	24 Feb. 54	28 Feb. 74	28 Feb. 00	1 Apl. 08
Spence, *Hy. Cat., Dy. Comsy.*, James, Ord., M.	31 Jan. 40	13 Aug. 84	16 Mar. 88	31 Jan. 95
Spencer, Col. Tomas Edward	14 Nov. 47	22 July 68	22 July 98	22 July 06
Spens, Col. Argndel Thomas, Bo.	16 May 35	20 Mar. 53	20 Sep. 81	20 Sep. 81
Splane, *Hony. Capt., Comsy.*, George Gibson, Ord. Sc.	19 Nov. 51	21 Feb. 93	20 Nov. 94	19 Nov. 06
Spratt-Bowring, Col. Fredk. Thomas Nelson, C.B., Engrs.	21 Oct. 47	13 Jan. 69	18 May 98	21 Oct. 04
Stainforth, Bt.-Col. Lesley Charles Hamilton, B.	23 June 60	22 Jan. 81	24 Aug. 07	22 Jan. 13
Stainforth, Col. William, M. Inf.	7 Oct. 43	4 Nov. 61	29 Oct. 91	1 Feb. 94
Stanton, Col. Edward Charles, R. E.	4 June 57	14 Aug. 76	17 Jan. 06	4 Jan. 14
Stanton, *Hy. Capt. Comsy.*, Martin Joseph, S. & T. Corps	10 Feb. 53	27 July 03	11 Feb. 05	9 Dec. 07
Steel, Col. John Philip, B. Engrs.	7 July 36	7 Dec. 55	1 July 85	30 Mar. 92
Steele, *Hy. Capt. Comsy.*, William Henry, Ord. B.	14 Sep. 51	22 Mar. 95	1 Apl. 03	4 May 05
Stephen, *Hy. Lt. Asst. Comsy.*, James, Comst.-Trans., B.	19 Mar. 40	14 Aug. 89	14 Aug. 89	20 Mar. 95
Stevens, Maj. Charles	25 May 57	11 Nov. 76	11 Nov. 96	13 Nov. 97
Stevens, Col. Charles Frederick	13 June 57	19 Apl. 76	19 Apl. 05	19 Apl. 14
Stevens, Lt.-Colonel Morton (Supy. List)	19 Feb. 61	5 Dec. 83	14 June 85	1 Mar. 17

RETIRED PAY—*contd.*

Names, Rank and Corps.	Date of Birth.	Date of 1st Commn.	Date of Rank.	Date of Retirement.
Stevenson, Col. Francis, C.B.	8 Mar. 51	8 Feb. 70	8 Feb. 00	8 Feb. 08
Stevenson, Col. Kenlis Fergus, M.	20 Sep. 39	4 Aug. 57	4 Aug. 87	1 May 90
M. Stewart, Hony. Maj.-Gl. Geo., C.B., B. (*Col. Guides*)	20 July 39	20 Sep. 56	31 Oct. 87	31 Oct. 87
Stewart, Maj.-Genl. *Sir* Norman Robert, *Bart.*, C.B. [R]	27 Sep. 51	30 Dec. 71	30 Sep. 00	26 Aug. 04
Stewart, Maj. William	8 Oct. 59	21 June 79	21 June 99	4 Dec. 99
Stock, Col. Henry John, Bo.	15 Dec. 40	4 Jan. 58	15 May 87	15 May 87
Stockley, Col. Henry Watts, Bo. Arty.	23 July 36	7 Dec. 55	1 July 85	3 June 89
Stoddart, Bt.-Col. Charles Henry	17 July 43	11 Sep. 63	25 May 94	30 Dec. 01
Stokes, *Hy. Capt. Comsy.*, George James, Ord., N.C.	19 Sep. 55	1 Apl. 02	1 Apl. 03	19 Sep. 10
Stokes, Maj. William Allen, R.E.	29 Jan. 72	25 July 90	25 July 10	11 Oct. 12
Stooks, Maj. Charles Sumner, D.S.O.	24 Nov. 75	28 Sep. 95	28 Sep. 13	5 July 18
Stopford, Col. William Henry James, Bo. Cav.	26 Aug. 39	11 Feb. 58	11 Feb. 88	15 Mar. 90
Stotherd, Lt.-Col. Edward Augustus Wood, B.	6 Mar. 64	5 Feb. 87	5 Feb. 13	5 July 13
Strachey, Capt. Bertram	29 Sep. 65	30 Jan. 86	30 Jan. 97	7 Dec. 03
Strachey, Lt.-Col. John, M.V.O.	5 May 57	2 July 79	2 July 05	1 Oct. 07
Straker, Col. Charles Doyle Anderson, M. Arty.	12 Mar. 39	10 Dec. 58	7 May 88	31 Aug. 93
Strettell, Lt.-Col. Arnold Dashwood.	15 Mar. 45	4 Mar. 62	4 Mar. 88	16 Apl. 94
Strong, Maj. Norman Baskervyle Glegg	13 Oct. 71	18 June 92	18 June 10	18 June 12
Stuart, Lt.-Col. Alex. Percy Dunbar Carmatt, Bo.	1 July 65	8 Dec. 86	8 Dec. 04	8 Dec. 13
Stuart, Lt.-Col. Charles John Lewis	8 July 54	27 June 79	27 June 00	31 July 02
Studdy, Col. Thomas James Charles Aylmer, B. Arty.	11 Dec. 39	12 Dec. 57	11 May 88	25 Jan. 96
Sutton, Bt.-Col. Henry George	26 May 60	13 Aug. 79	6 Aug. 08	13 Aug. 11
Swann, Maj.-Genl. John Christopher, C.B.	22 Apl. 56	10 Sep. 75	21 Mar. 08	20 Mar. 14
Swayne, Bt.-Col. *Sir* Eric John Eagles, K.C.M.G., C.B.	14 May 63	10 Mar. 83	18 Aug. 05	10 Mar. 07
Swayne, Col. Harold George Carlos, R.E.	14 Feb. 60	18 Feb. 80	31 Dec. 10	23 Feb. 14
Sweeney, Hy. Capt. Comsy., Joseph Donald	16 Oct. 59	14 July 09	5 Aug. 12	15 Sep. 14
Sweeny, *Hy. Lt. Asst. Comsy.*, Patrick, Ord., M.	27 Jan. 52	20 May 98	20 May 98	1 Nov. 99
Swete, Lt.-Col. Charles Denroche	1 Mar. 43	25 June 61	25 June 87	5 June 92
Swiney, Lt.-Col. Geo. Walter Brandon	10 Nov. 55	26 Aug. 76	26 Aug. 02	26 Aug. 04
Swinley, Maj.-Genl. George, C.B., B. Arty.	24 Apl. 42	6 Nov. 60	17 Nov. 95	1 Apl. 97
Swinton, Lt.-Col. Arthur, B. Arty.	11 Jan. 41	9 Dec. 59	1 Jan. 89	1 Jan. 94
Tacchi, *Hony. Maj. Comsy.*, George, S. & T. Corps, B.	20 Mar. 58	10 Sep. 04	1 Oct. 08	16 Dec. 10
Tayler, Maj. William Fothergill Cooke	1 Mar. 68	30 Jan. 89	30 Jan. 07	6 Feb. 09
Taylor, Lt.-Col. Albert FitzWilliam, B.	14 Nov. 42	12 Nov. 59	12 Nov. 85	1 Mar. 87
Taylor, Lt.-Col. Douglas James Orr	21 Nov. 59	1 May 78	22 June 05	22 June 06
Taylor, *Hy. Lt. Dy. Asst. Comsy.*, Frederick, Ord., B	1 Jan. 48	3 June 97	3 June 97	1 Jan. 03
Taylor, Bt.-Maj. Francis Heale	14 July 66	5 Feb. 87	5 Feb. 98	21 Apl. 10
Taylor, Maj. George Arthur Campbell, M.	22 Jan. 69	21 Sep. 89	21 Sep. 07	1 Dec. 12
Taylor, Lt.-Col. Hugh Neufville, D.S.O., M.	20 Dec. 59	28 Jan. 82	1 June 04	1 Feb. 14
M. Taylor, Col. James Best, M.	17 Apl. 40	13 Dec. 56	13 Dec. 86	29 Dec. 88
Taylor, Col. Pierce Thomas Henry, B. Arty.	3 Dec. 40	12 June 58	9 Apl. 86	9 Apl. 86
Taylor, Lt.-Col. Robert Eyre Sullivan	19 Aug. 56	23 Jan. 78	23 Jan. 04	30 Aug. 10
Taylor, Col. Robert Francis, M. Inf.	28 Aug. 41	9 Dec. 59	9 Dec. 89	25 July 90
Taylor, *Comsy. and Hony. Capt.* George	1 Sep. 59	16 Dec. 10	10 June 16	23 Nov. 16
Temple, Col. Edward, B.	28 Nov. 38	15 May 56	15 May 86	29 Aug. 88
Temple, Lt.-Col. *Sir* Richard Carnac, *Bart.*, C.I.E.	15 Oct. 50	4 Jan. 71	4 Jan. 97	25 Jan. 04
Templer, Maj. Chas. Bertram	9 Apl. 60	11 Aug. 80	11 Aug. 00	14 Jan. 05
Tennant, Col. Claude Cambridge	2 Dec. 59	26 Nov. 79	19 Dec. 08	2 Dec. 11
Ternan, Lt.-Col. Henry Breffney	16 Oct. 47	22 Feb. 68	22 Feb. 94	26 Feb. 07
V.C. M. Thackeray, Col. *Sir* Edward Talbot, K.C.B., B. Engrs.	19 Oct. 36	9 Dec. 54	25 Nov. 84	9 Apl. 88
Thackwell, *Hony.* Brig.-Genl., Osbert Montague Roche, R. E.	7 Sep. 60	19 May 80	13 Sep. 17	13 Sep. 17
Thomas, Maj.-Genl. Charles Frederick, B. Inf.	12 Sep. 43	8 June 61	3 July 99	9 July 99
Thomas, Col. Edgar Hastings, M.	1 Dec. 37	16 Aug. 56	16 Aug. 89	13 June 89
Thomas, *Hy. Maj. Comsy.* William, Ord., N. C.	3 June 58	1 Apl. 02	1 Apl. 03 / *m.* 21 Sep. 17	3 June 13
Thompson, Bt.-Col. David Montgomery	5 June 61	14 Jan. 80	14 Jan. 09	1 Feb. 12
Thompson, Lt.-Col. Harry Arthur Hale, B.	25 Apl. 66	5 Feb. 87	5 Feb. 13	6 Feb. 13
Thompson, Col. Ross, M. Engrs.	7 Aug. 38	11 June 58	2 Mar. 85	1 Dec. 94
Thompson, Col. William Anson	28 May 59	22 Jan. 79	3 June 07	13 Mar. 13
Thompson, Capt. William Thomas Fetherstonhaugh	23 Apl. 78	22 Jan. 98	22 Jan. 07	4 July 16
Thomson, Maj. David Birdwood	5 Mar. 63	10 May 82	10 July 01	17 May 04
Thomson, Lt.-Col. Mowbray Townshend	10 Sep. 60	11 May 78	1 Feb. 05	11 May 06
Thornhill, Lt.-Col. *Sir* Henry Beaufoy, K.C.I.E., C.M.G.	7 Jan. 54	23 Apl. 73	23 Apl. 99	7 Jan. 12
Thuillier, Col. *Sir* Henry Ravenshaw, K.C.I.E., B. Engrs.	26 Mar. 38	12 June 57	7 Oct. 87	26 Mar. 95
Thuillier, Col. (Hony. Brig.-Genl.) Willoughby	22 Feb. 60	13 Aug. 79	1 June 04	25 Oct. 11

RETIRED PAY—*contd.*

Names, Rank and Corps.	Date of Birth.		Date of 1st Commn.		Date of Rank.		Date of Retirement.	
Thurburn, Col. Jas. White, C.S.I., Engrs.	3 Dec.	48	7 July	69	20 May	98	3 Dec.	05
Tighe, Lt.-Col. Michael Augustus	27 Apl.	61	23 Apl.	81	23 Apl.	07	30 July	07
Tighe, Lt.-Col. Stanislans Michael. B.	15 June	67	29 Aug.	85	29 Aug.	11	29 Aug.	13
Tillard, Maj.-Genl. John Arthur, C.B., B. Arty.[R.]	13 Sep.	37	12 Dec.	57	28 Nov.	90	1 Apl.	95
Tilley, Maj. William F., R.E.	22 Apl.	63	30 June	85	19 Apl.	03	15 July	11
Tisdall, Lt Archibald James	9 Sep.	93	22 Jan.	13	13 Mar.	14	8 Dec.	16
Todd, *Hy. Capt. Dy. Comsy.* Thomas, Ord., B.	28 June	40	1 Jan.	94	27 May	95	28 June	95
Tomlin Money Shewan, Maj Robert Ernest, R. Engrs.	19 May	66	16 Sep.	85	21 Feb.	03	26 June	13
Tottenham, Lt.-Col. Robert Loftus	29 Oct.	65	9 May	85	9 May	11	13 May	11
Townsend, Lt.-Col, the *Hon.* Edward Copleston	11 Mar.	64	25 Aug.	83	25 Aug.	09	26 Sep.	18
Trevelyan, Lt.-Col. Walter Raleigh, Bo.	....	.	4 Oct.	57	20 May	83	20 May	83
Trevor, Col. Francis Randolph, M.	18 May	37	10 May	57	9 Oct.	85	9 Oct.	85
Trevor, Maj. Harry, B. Inf.	7 July	61	22 Jan.	81	22 Jan.	01	1 Feb.	05
Trevor, Col. Salsbury Thomas, B. Engrs.	18 Sep.	34	29 Dec.	52	1 Oct.	81	30 Mar.	89
Tristram. Maj. Charles Edward, M.	19 Nov.	68	25 Mar.	91	25 Mar.	09	19 Apl.	14
Trotter, Lt.-Col. *Sir* Henry, C.B., K.C.M.G., B. Engrs	30 Aug.	41	8 June	60	29 July	87	5 Dec.	90
Trotter, Lt.-Col. John Mowbray, B.	6 Sep.	42	19 Dec.	60	8 July	84	8 July	84
Trotter, Lt.-Col. Robert Francis	1 Nov.	49	8 Dec.	69	8 Dec.	95	12 Feb.	10
Troup, Maj. William Bedford, B. Arty.	....	.	12 Dec.	57	16 Nov.	78	16 Nov.	78
Tucker, Maj.-Genl. Louis Henry Emile, C.I.E., B. Infy.	18 Oct.	43	19 Dec.	60	11 Feb.	99	27 Apl.	99
Tucker, Capt. William Fane	....	.	20 Dec.	59	5 July	69	25 Jan.	72
Tuite, Bt.-Col. Mark Antony	2 May	57	1 May	78	1 May	07	1 May	10
Tulloch, Maj. Hector, C.B., M. Engrs.	16 Apl.	35	9 Dec.	53	5 July	72	12 Mar.	73
Tulloch, Maj.-Genl. John Walter Graham, C.B.	2 Nov.	61	11 Aug.	80	1 May	08	1 Apl.	12
Turberville, Col. John Picton, M.	6 Nov.	37	10 Dec.	54	25 Feb.	83	25 Feb.	83
Turkington, *Hy. Capt. Asst. Comsy.*, Wm. Horatio, Ord., N. Cir.	18 Apl.	64	20 May	98	20 May	98	16 May	04
Turner, Col. Archer Lloyd Marischal, R.A.	14 Mar.	60	30 July	79	31 July	05	1 Aug.	11
Turner, Hony. Br.-Genl. George Harvey, Bo.	26 Aug.	61	23 Apl.	81	11 June	07	23 Apl.	13
Turner, *Hony. Lt. Asst. Comsy.* George James, I.	31 Jan.	56	4 Mar.	07	4 Mar.	07	23 Apl.	13
Turner, *Hony. Maj. Comsy.* James Wilson	9 Nov.	57	15 Sep.	01	29 Dec.	08	31 Jan.	11
Turton, Maj. Hugh Mitchell, B.	29 Dec.	70	18 June	92	18 June	10	9 Nov.	12
M. Tweddell, Col. Francis, B. Infy.	26 Sep.	40	27 Aug.	59	2 Mar.	85	18 June	13
Tweddell, Maj. Francis	11 Sep.	63	23 Aug.	84	23 Aug.	02	15 May	90
Twemlow, Maj.-Genl. Edward D'Oyley, Bo. Engrs.	20 Nov.	40	9 Dec.	59	1 Apl.	95	20 Jan.	05
Tydeman, *Hy. Maj. Comsy.* Edwin James	24 Nov.	56	2 Jan.	07	23 Sep.	08	21 May	96
Tyers, *Hy. Lt. Asst. Comsy.* Thomas, P. W. D., B.	23 Oct.	39	1 May	90	1 May	90	1 Nov.	11
Ulyett, *Hy. Capt., Comsy.* James Henry, P. W. D. M.	10 Sep.	53	2 Nov.	01	14 May	07	23 Oct.	94
Underwood, *Hony. Capt. Comsy.* Andrew William Ord. Br.	7 Dec.	62	27 June	05	4 July	12	10 Sep.	08
Unwin, Col. William Heathcote, B.	20 May	40	20 Dec.	57	1 Aug.	84	12 June	16
Urmston, Capt. Herbert Neville	24 Mar.	85	23 May	06	23 May	16	1 Aug.	84
Vallance, Maj. John Dunscombe, M.	....	.	12 Dec.	49	12 Dec.	69	25 Novr.	16
Vanrenen, Lt.-Col. Adrian John Hebron	28 Feb.	64	10 Mar.	83	10 Mar.	09	1 Apl.	73
Vans-Agnew, Lt.-Col. John	23 Aug.	59	13 Aug.	79	13 Aug.	05	31 Mar.	09
Vaughan, Bt.-Col. Henry Bathrust	27 Feb.	58	17 Apl.	80	25 Oct.	05	23 Aug.	05
Vaugan, Maj. Percy Balderston	13 July	59	13 Aug.	79	13 Aug.	99	17 Apl.	12
Vaughan-Arbuckle, Lt.-Col. Benjamin, B. Arty.	10 Sep.	39	12 June	58	21 July	87	14 Aug.	99
Vautin, *Hy. Lt. Asst. Comsy.* Henry Claude	13 Dec.	37	20 Dec.	89	28 Dec.	89	1 Sep.	93
Vera-O'Sullivan, *Hy. Capt. Dy. Comsy.*, Timothy Murthogh Comst., B.	18 Nov.	35	2 Jan.	89	20 Nov.	90	14 Dec.	92
Vickers, Capt. Cuthbert	28 Sep.	70	12 Mar.	92	10 July	01	21 Nov.	90
Vidal, Maj. Leonard Hugh	13 Oct.	56	1 May	78	31 Mar.	86	12 Sep.	08
Vivian, Bt.-Col. Frederick Guy	23 Dec.	52	4 Jan.	71	1 Apl.	99	17 Nov.	99
Voyle, Lt.-Col. Francis Ranken Cave, B.	6 Apl.	45	9 Oct.	63	1 Mar.	85	4 Jan.	03
Vyvyan, Col. Beville Granville, B. Infy.	13 May	42	8 June	60	31 Dec.	87	1 Mar.	85
Vyvyan, Col. Richard Octavius, B. Infy.	21 June	40	4 Feb.	60	4 Feb.	90	31 Dec.	87
Wace, Lieut. Denyss Chamberlaine	13 Jan.	86	5 Aug.	05	5 Nov.	07	11 Mar.	92
Waddington, Hony. Maj.-Genl. Thomas, Bo.	26 Apl.	27	21 June	47	31 Dec.	81	3 Apl.	12
M. Wake, Col. Alfred James, B. Arty.	21 June	34	9 Dec.	54	8 Dec.	83	31 Dec.	81
Wake, Lt.-Col. Edward St. Aubyn	20 July	62	6 Feb.	84	6 Feb.	10	18 Apl.	85
Walcott, Col. Edmund Scopoli, C.B., Bo. Infy.	15 Apl.	42	9 Mar.	60	2 Mar.	85	13 Feb.	15
Walden, *Hy. Capt. Comsy.* George Honry, P. W.D., M.	13 Aug.	61	18 May	04	8 Sep.	07	1 Apl.	89
Waldron, Bt.-Col. Henry Francis Kelly	21 Jan.	59	1 May	78	5 Dec.	07	23 June	13
Walker, *Hy. Lt. Asst. Comsy.* James, P. W. D., B.	9 Aug.	48	1 July	97	1 July	97	1 Nov.	10
							9 Aug.	03

RETIRED PAY—*contd.*

Names, Rank and Corps.	Date of Birth.	Date of 1st Commn.	Date of Rank.	Date of Retirement.
Walker, Lt.-Col. John Bridges, Bo. Arty.	12 Mar. 40	12 June 58	18 Oct. 82	18 Oct. 82
Walker, Bt.-Col. Percy Gerald	25 Sep. 57	9 July 79	5 June 07	29 July 11
Wallace, Dunlop, Maj. Arthur	18 Jan. 66	30 Jan. 86	30 Jan. 04	15 Mar. 10
Waller, Bt.-Col. Edmund Augustus, Engrs.	15 Dec. 53	29 Apl. 73	10 Feb. 04	22 May 07
Waller, *Hy. Capt. Dy. Comsy.* Frederick, Ord., Bo.	29 Mar. 47	21 Jan. 95	19 Dec. 97	3 Aug. 98
Waller, Maj.-Genl., John Edmund, B. Infy.	6 Oct. 41	4 July 59	1 Aug. 94	11 Aug. 97
Waller, Maj., Robert Jocelyn, B.	3 May 44	22 Mar. 64	22 Mar. 84	28 Aug. 88
Wallis, Maj. Hugh Ryves	30 June 63	30 Mar. 87	3 Oct. 06	16 Feb. 10
Walters, *Hy. Maj. Comsy.* Lewis, S. & T. Corps, B.	22 Dec. 55	29 Nov. 00	1 July 17	21 Dec. 10
Wapshare, Lt.-Col. Arthur	6 Dec. 51	28 Oct. 71	28 Oct. 97	5 Nov. 03
Ward, Lt.-Col. Arthur Edward	25 Dec. 46	21 Nov. 65	21 Nov. 91	1 Mar. 94
Ward, Maj. George Louis Stuart, B.	2 Apl. 69	3 May 90	3 May 08	3 May 13
Ward, Col. John, M.	19 Aug. 40	4 Nov. 57	30 Mar. 87	30 Mar. 87
Ward, Col. William, Bo. Arty.	25 Nov. 38	12 June 58	18 June 85	18 June 85
Warde, Lt.-Col. Hugh North	21 Feb. 63	6 Feb. 84	6 Feb. 10	10 Feb. 10
Warden, Lt.-Col. Alfred Walter	13 July 65	6 Feb. 84	6 Feb. 10	7 Feb. 10
Warner, Bt. Lt.-Col. William Ward	14 Mar. 67	4 May 87	4 May 05	21 Dec. 07
Warren, Maj. Richard Pennefather	5 Dec. 61	22 Oct. 81	10 July 01	10 Nov. 05
Waterfield, Maj. Bertram Clarke	17 Nov. 68	19 Sep. 88	19 Sep. 06	5 Oct. 09
Watson, Maj. George Frederick	27 July 57	28 Oct. 76	28 Oct. 96	16 June 01
Watson, Lt.-Col. George Herbert	30 Mar. 57	11 Sep. 76	11 Sep. 02	1 Jan. 10
Watson, *Hy. Lt. Asst. Comsy.* John, P. W. D., M.	22 Dec. 48	7 Jan. 99	7 Jan. 99	22 Dec. 03
Watt, *Hy. Maj., Comsy.*, Alexander, P. W. D., M.	4 Nov. 50	27 July 96	1 July 17	10 Apl. 05
Watts, *Hy. Capt. Comsy.* Harry	27 Nov. 63	26 Nov. 01	29 Nov. 12	13 Jan. 15
Watts, Col. John Briscoe	5 Aug. 45	2 Oct. 63	20 June 96	17 Dec. 96
Wauhope, Col. Robert Alexander, C.B., C.M.G., C.I.E., Engrs.	14 Apl. 55	29 Apl. 73	28 Aug. 99	31 Aug. 05
Way, Lt. Henry George, M.	20 May 56	13 June 74	13 June 74	20 Nov. 84
Webb, Maj. Henry North	4 Mar. 49	8 July 69	8 July 89	24 Nov. 93
Webster, Lt.-Col. Thomas	21 Oct. 62	5 Dec. 83	14 Oct. 08	5 Dec. 09
Welchman, Col. Arthur John Tregonwell, B. Infy.	10 Nov. 43	20 Dec. 59	7 July 90	16 Aug. 92
Welchman, Edmund Walter St. George, C.B.	13 Feb. 57	11 July 74	11 July 00	11 July 12
Welchman, Maj. Frederick Deffield	7 June 52	30 Nov. 70	30 Nov. 90	30 Nov. 94
Weller, Maj. Alexander Thomas	12 June 50	12 Nov. 70	12 Nov. 90	31 May 96
Weller, Col. George Herbert	27 Jan. 60	13 Aug. 79	6 June 07	18 Apl. 08
Welman, Maj. Harvey	31 Jan. 61	14 Jan. 80	14 Jan. 00	31 Dec. 00
West, Capt. Alexander Arthur	19 Mar. 69	8 June 89	8 June 00	20 May 08
Western, Lt.-Col. James Halifax, C.M.G., B. Engrs.	17 May 42	8 June 60	27 Jan. 87	7 Jan. 87
Western, Col. John Sutton Edward, *p.s.c.*	26 Mar. 57	12 Feb. 78	12 Feb. 07	15 May 09
Westmorland, Col. Isaac Peat, B. Engrs.	12 Nov. 38	12 Dec. 56	3 Feb. 84	43 Feb. 84
Westropp, Col. George Ralph Collier, C.B.	21 June 59	14 Sep. 78	1 June 07	21 June 11
Westwood, *Hy. Capt. Comsy.* Thomas, P. W. D., India	29 Aug. 55	28 Feb. 06	21 Mar. 08	1 June 08
Wheeler, *Hy. Lt. Asst. Comsy.* Arthur, P.W.D., India	6 Mar. 53	4 Aug. 07	4 Aug. 07	5 Mar. 08
Wheeler, *Hony. Capt. Comsy.* George, S. & T.	7 Nov. 52	25 May 99	4 Oct. 04	1 June 07
M. Wheeler, Hony. Maj.-Genl. Robert, B.	22 Feb. 39	8 Dec. 55	31 Dec. 87	31 Dec. 82
Wheler, Col. Charles Stuart	24 Aug. 51	30 Dec. 71	20 Nov. 03	1 June 11
Whitbread, *Hy. Lt. Dy. Asst. Comsy.*, Arthur, Misc., B.	25 Nov. 51	7 Mar. 01	7 Mar. 01	7 Feb. 03
White, Capt. Oliver Wardlaw	9 Dec. 78	27 July 98	27 July 07	27 Jan. 09
Whittall, Bt.-Col. Francis Vaughan	23 Mar. 61	22 Jan. 81	15 Aug. 07	13 Feb. 13
Whyte, Bt.-Col. Charles William Frederick	1 May 62	10 May 82	12 June 09	10 May 14
Whyte, Lt.-Col. John Frederick	7 June 65	1 Sep. 88	1 Sep. 10	1 Sep. 16
Wickham, *Hy. Maj. Comsy.* Arthur Frederick, Ordce., So.	10 Apl. 53	22 Mar. 95	9 Apl. 08	10 Apl. 08
Wickham, Col. William James Richard, C.B.	22 Jan. 60	13 Aug. 79	10 Feb. 04	23 Nov. 13
Wiggens, Lt.-Col. Francis Edwin, B. Inf.	21 Sep. 42	10 Dec. 59	10 Dec. 85	1 Mar. 88
Wilbond, *Hy. Capt. Comsy.* Joseph Henry, Ord., So	23 Jan. 52	22 Aug. 94	20 Sep. 02	23 Jan. 07
Wild, *Hy. Maj. Dy. Comsy.* William, Bo.	....	16 Sep. 68	18 Feb. 84	18 Feb. 84
Wildey, *Hy. Capt. Dy. Comsy.* Richard Samuel, Ord. B.	2 Oct. 43	19 Jan. 92	1 Dec. 94	8 Apl. 95
Wilkieson, Maj. Claude William	24 Sep. 62	10 Mar. 83	10 July 01	15 Jan. 04
Wilkinson, Lt.-Col. Alfred Reginald, B. Inf.	26 Mar. 43	4 Jan. 60	4 Jan. 86	20 Mar. 89
Wilkinson, *Hy. Capt. Dy. Comsy.* James Elworthy	30 Nov. 61	1 Nov. 61	1 Feb. 13	1 Apl. 13
Willes, Maj. George Frederick	29 Mar. 57	24 June 77	24 June 97	25 June 01
Williams, *Hy. Capt., Comsy.* Albert Henry, Ord. N.	28 Dec. 58	1 Apl. 03	4 Jan. 06	22 Apl. 10
Williams, Maj. Cecil Howard	21 Jan. 65	10 Nov. 86	10 Nov. 04	19 Jan. 09
Williams, Maj. Chas. Stanley	1 Apl. 64	6 Feb. 84	6 Feb. 02	28 Mar. 04

RETIRED PAY—*concld.*

Names, Rank and Corps.	Date of Birth.	Date of 1st Commn.	Date of Rank.	Date of Retirement.
Williams, Bt.-Col. George Augustus, B.	12 Nov. 56	10 Sep. 75	31 Mar. 06	10 Sep. 13
Williamson, Col. Cyril Venn Wilton	30 July 58	27 Jan. 76	27 Jan. 05	30 July 10
Williamson, Lt.-Col. Michael	4 Apl. 65	4 May 87	8 May 13	15 May 13
Willmott, *Hy. Capt. Dy. Comsy.* Edward Hortan-Battinson, S. & T. Corps, Sec.	21 Sep. 60	20 May 98	15 Apl. 06	16 Oct. 08
Willoughby, Maj.-Genl. James Fortnom Bo. Inf. [R] (*Col. 33rd Light Cavy.*)	28 June 44	27 Jan. 61	19 Apl. 99	20 Apl. 99
Wilmer, Lt.-Col. Algernon Henry	11 May 57	6 Oct. 76	6 Oct. 02	4 Apl. 05
Wilmer, Bt.-Capt. William. B.	. . . .	3 July 60	3 July 72	1 July 82
Wilmot, Lt. Edward Revel Eardley, B. Arty.	. . . .	20 Nov. 30	31 Jan. 35	5 Feb. 38
Wilsey, *Comsy. Hony. Capt.* Harry Willasey, B.	23 Nov. 58	10 Sep. 04	10 Dec. 08	23 Nov. 13
Wilson, Bt.-Col. Charles Henry Luttrell Fahie, R.A., I.O.D.	24 Aug. 58	9 Oct. 77	14 July 09	24 May 12
Wilson, Col. George Frederick, R.E.	6 Jan. 51	2 Aug. 71	18 Jan. 01	6 Jan. 08
Wilson, Lt.-Col. John Francis	29 July 56	10 Sep. 75	10 Sep. 01	10 Sep. 07
Wilson, *Hy. Maj. Comsy.* Thomas, Ord. B.	30 Nov. 58	27 June 02	1 July 17	13 Jan. 05
Wilson, *Hy. Capt. Comsy.* William, S. & T. Corps, Bengal	10 Dec. 53	22 Nov. 00	1 Apl. 02	10 Dec. 08
Wilson, Col. William Bernard	23 Apl. 47	24 Apl. 69	20 Mar. 01	8 Oct. 07
Wiltshire, *Hy. Lt. Asst. Comsy.* Charles, S. & T., B.	1 Sop. 46	28 Dec. 00	28 Dec. 00	4 Sep. 01
Wimberley, Lt.-Col. Charles Irvine	16 Sep. 61	29 July 82	29 July 08	14 Jan. 09
Wingate, Col. Alfred Woodrow Stanley, B., C.M.G.	3 May 61	22 Oct. 81	24 Aug. 08	19 Dec. 13
Wise, *Hy. Capt. Dy. Comsy.* Charles Driver	6 Dec. 35	10 Apl. 85	7 Dec. 90	7 Dec. 90
Wiseman, *Hy. Capt. Dy. Comsy.* Charles Sheriffe	30 May 37	27 July 88	29 Aug. 91	30 May 92
Wiseman, *Hy. Lt. Asst. Comsy.* Frederick, P. W. D.	5 July 52	30 June 06	30 June 06	5 July 07
Withers, Capt. Alfred Ross, P.	4 Aug. 78	21 Apl. 00	21 Apl. 09	4 Mar. 13
Woods, Lt.-Col. Albert Edward, C.S.I.	29 May 62	22 Oct. 81	22 Oct. 07	29 May 17
Wood, Maj. Cyril Edward	17 Mar. 66	25 Aug. 86	25 Aug. 04	25 Aug. 10
Wood, Lt.-Col. Edward Jas. Fandon, B.	4 Feb. 50	1 Jan. 73	1 Jan. 99	1 Jan. 05
Wood, Capt. Francis Ludlow	9 Aug. 70	3 May 90	3 May 01	8 Sep. 05
Wood, Lieut. Ralph Bulkely Manners	23 July 80	21 Feb. 00	23 Oct. 00	8 May 13
Woodcock, Maj.-Genl. Charles Samuel, B. Arty.	. . . .	10 Dec. 41	1 Aug. 72	1 Aug. 72
Woodcock, Col. John, M.	. . . .	1 Mar. 47	27 Feb. 75	6 Jan. 75
Woodcock, Lt.-Col. William Charles Marmaduke, B.	26 Sep. 65	6 May 85	1 Sep. 09	12 Nov. 12
Woodhouse, Bt.-Col. Henry Opie	1 May 47	10 Apl. 66	15 Sep. 97	1 Aug. 99
M. Woodward, Col. Walter William, Bo. Arty.	6 May 34	9 Dec. 52	1 Oct. 81	17 May 89
Wray, Lt.-Col. John Willoughby	9 Sep. 51	12 Nov. 70	12 Nov. 96	10 Mar. 02
Wylie, Maj.-Genl. Henry, C.S.I., B. Inf.	24 July 44	4 Oct. 61	10 Oct. 99	15 Apl. 00
Wyllie, Lt.-Col. Frederick	2 July 57	22 Feb. 79	22 Feb. 05	9 Mar. 07
Wyllie, Lt.-Col. John Alfred	7 July 55	10 Sep. 76	10 Sep. 02	7 Apl. 09
Wyllie, Lt.-Col. Robert John Humphrey	3 May 45	4 Aug. 63	4 Aug. 89	1 Apl. 92
Yate, Col. William Gordon	1 Oct. 54	12 Nov. 73	10 Feb. 04	12 Nov. 05
Yeilding, Col. William Richard, C.B., C.I.E., D.S.O.	13 Jan. 56	29 Nov. 76	7 Sep. 04	25 Feb. 10
Young, Maj. Arthur Thomas	24 Nov. 60	22 Jan. 81	22 Jan. 01	23 Jan. 01
Young, Capt. Charles Wilson	26 Aug. 54	28 Feb. 74	28 Feb. 85	17 Feb. 93
Young, Col. Edward Archibald	6 June 48	15 Dec. 69	15 Dec. 99	7 Jan. 10
Young, Hony Br.-Genl. George Frederick, C.B.	5 May 46	30 June 65	30 June 91	26 Nov. 03
Young, Lt.-Col. Harry Howlett	28 Jan. 51	27 Aug. 70	27 Aug. 96	27 Aug. 02
Young, Maj. Henry Pottinger	2 Jan. 50	27 Feb. 69	27 Feb. 89	21 July 91
Young, *Hy. Capt., Dy. Comsy.* John Wallis Lombard, S. & T., B.	29 July 56	23 July 07	24 July 11	29 July 11
Young, Lt.-Col. Julian Henry	24 Dec. 57	1 May 78	1 May 04	21 Dec. 06
Young, *Hy. Lt., Dy. Asst. Comsy.* John Joseph, P. W. D., B.	4 Aug. 44	23 Oct. 94	23 Oct. 94	4 Aug. 99
Young, Col. Thomas Henry Binny	19 Nov. 41	25 June 58	25 June 88	1 Feb. 99
Young, Maj. Thomas Simpson	23 Aug. 62	23 May 85	23 May 03	22 Apl. 06
Young, Lt.-Col. William Hope	29 July 51	29 May 73	29 May 99	30 May 99
Younghusband, Lt.-Col. *Sir* Francis Edward, K.C.S.I., K.C.I.E.	31 May 63	10 May 82	10 May 08	10 May 10

RETIRED PAY.

(Medical Services.)

Note.—The names on this list are not indexed as they are arranged alphabetically.

Names and Rank.	Date of Rank.	Date of Retirement.
Adams, Lt.-Col. Charles, M.B., M.	2 Apl. 01	8 July 01
Adey, Surgn.-Maj. Arthur Wellesley George, Bo.	1 July 73	1 May 84
Ahmed, Lt.-Col. Zalnoor Allee, M.D., B.	1 Oct. 92	19 July 03
Alcock, Lt.-Col. Alfred William, C.I.E.	1 Oct. 05	29 Dec. 07
Alderton, *Asst. Surgn.* Charles Henry Arthur, L.R.C.P. & S.E., L.F.P.S., Glas.	31 Mar. 15	5 Feb. 15
Allan, *Sr. Asst. Surgn., Hony. Capt.*, George Robert	5 June 07	2 June 09
Allison, *Sr. Asst. Surgn., Hony. Capt.* Alfred Albert	14 Nov. 07	23 July 11
Allison, Lt.-Col. Hazlett, M.D., M.	30 Sep. 03	23 Dec. 03
Alpin, Lt.-Col. William George Patrick, M.D.	1 Apl. 04	6 Jan. 10
Anderson, Lt.-Col. Adam Rivers Steele, M.B.	30 Mar. 09	11 Sep. 18
Anderson, Lt.-Col. Alexander Vass, M.B.	1 Apl. 02	19 July 07
Anderson, Maj. Dugald Nairne, M.B., Sec.	27 July 11	1 Sep. 12
Anderson, *Sr. Apothy. Hy. Sgn. Lt.* James, M.	9 Dec. 91	4 Mar. 93
Anderson, Lt.-Col. John, C.I.E., M.B.	30 Sep. 98	1 Apl. 10
Anderson, Lt.-Col. Joseph William Townsend, F.R.C.S.	2 Oct. 00	15 Apl. 06
Armstrong, Lt.-Col. Henry	30 Apl. 98	30 June 08
Armstrong, Lt.-Col. James, B.	31 Mar. 94	17 Oct. 99
Armstrong, Maj. and Bt. Lt.-Col. Wilfred Ernest Arbuthnot	30 Jan. 04	12 Mar. 07
Arnim, Lt.-Col. Henry Charles Leffler	31 Mar. 08	13 June 13
Arnott, Bde.-Sgn. Lt.-Col. James, M.D., Bo. [R]	30 Apl. 91	1 Apl. 98
Ashworth, *Sr. Apothy. Hy. Sgn.-Lt.* James George, M.	7 Nov. 87	9 July 91
Avetoom, Lt.-Col. Sarkies Thaddeus, Bo.	2 Apl. 01	1 Oct. 01
Bailey, *Sr. Asst. Sgn. Hy. Lt.* James Albert, B.	15 Apl. 01	16 Aug. 01
Bailey, *Sr. Asst. Sgn. Hy. Capt.* John Charles	31 May 04	1 Jan. 06
Bain, Lt.-Col. David Stuart Erskine	31 Mar. 99	5 Apl. 10
Bainbridge, Sgn.-Gl. George, M.D., Bo.	10 Aug. 97	30 Oct. 02
Baker, *Sr. Asst. Sgn. Hy. Capt.*, Daniel Thomas, B.	24 May 99	25 July 91
Baker, Lt.-Col. George Henry, M.B., B.	1 Oct. 05	24 Dec. 13
Baker, Surgn. Lt.-Col. Oswald, B.	1 Apl. 93	6 July 98
Baker, Lt.-Col. Richard John, M.D.	2 Apl. 01	12 Aug. 09
Bamber, Col. Charles James, M.V.O.	12 July 10	12 July 15
Banks, Col. Samuel O'Brien, Bo.	18 Sep. 95	13 Nov. 00
Bannerman, Maj.-Genl. *The Hon'ble* William Burney, C.S.I., M.D. [R]	26 July 11	6 July 18
Barretto, *Sr. Asst. Sgn. and Sgn. Lt.* Peter, Bo.	18 Apl. 94	18 July 86
Barry, Lt.-Col. Thomas David Collis	31 Mar. 07	18 Nov. 09
Barter, Bde.-Surgn. Jos. Francis, M.	26 Mar. 86	21 Apl. 90
Basu, Maj. Baman Das	31 Jan. 03	1 May 08
Basu, Lt.-Col. Dharamadas	31 Mar. 97	7 July 02
Bate, Col. Thomas Elwood Lindesay, C.I.E.	12 July 05	12 July 10
Batty, Bde.-Surgn. Richard Hamilton, Bo.	24 June 87	24 June 87
Bay, *Sr. Asst. Sgn. Hony. Maj.* Thomas Archibald	6 Mar. 03 *m.* 2 Apl. 28	26 Sep. 06
Beale, *Sr. Apothy. Loc. Sgn.* Robert William, B.	1 May 70	12 Dec. 77
Beatson, Col. Charles Henry, C.B., B.	16 June 05	27 Mar. 11
Beatty, Sgn.-Gl. Thomas Barkley, M.D., Bo.	1 Apl. 80	31 Mar. 85
Bedford, Lt.-Col. *Sir* Charles Henry, *Kt.*, M.D.	30 Sep. 09	18 Dec. 11
Beech, Surgn.-Lt.-Col. Lionel, M.	1 Oct. 92	1 Mar. 93
Bennett, Lt.-Col. Charles Henry, M.D.	1 Oct. 97	11 Feb. 08
Bensley, Lt.-Col. Charles Norman	30 Sept. 06	12 Nov. 11
Bensley, Bde.-Surgn. Edwin Clement, B.	15 Nov. 85	15 Nov. 85
K, H. S. Benson, Surgn.-Genl. Percy Hugh, M. B. Ⓡ.	1 Apl. 08	26 July 11
Bigger, Lt.-Col. Samuel Ferguson, M.B., B.	30 Mar. 98	1 Sep. 03
Blaker, *Sr. Asst. Sgn. Hy. Capt.* Richard Michael, B.	17 Feb. 98	7 Mar. 00
Blood, Surgn.-Maj. Joseph, B.	31 Mar. 89	21 Dec. 92
Boileau, *Sr. Asst. Surgn. Hy. Capt.* George Thomas	23 Oct. 05	8 Mar. 06
Bond, *Sr. Apothy, Hy. Surgn. Capt.* Samuel Patterson, B.	16 July 88	22 Oct. 02
Bookey, Col. John Thomas Brownrigg, C.B. Ⓡ B.	2 May 00	16 June 05
Boucher, *Sr. Asst. Surgn. Hy. Maj.* William Alfred, B.	14 Nov. 07 *m.* 2 Apl. 18	12 Mar. 12
C. M. Boustead, Surgn.-Maj. Robinson, Bo.	1 July 73	5 Feb. 89
Bowder, *Sr. Asst. Surgn. Hy. Capt.* Arthur George, B.	16 Nov. 07	28 Jan. 14
Bown, Lt.-Col. Arthur Thomas	1 Oct. 04	14 Dec. 09
Boyd, Surgn. Lt.-Col. Herbert, B.	1 Oct. 89	15 Jan. 95
Boyd, Lt.-Col. Henry Walker Butler, F.R.C.S.I.	31 Mar. 96	29 June 06
Bradley, *Sr. Asst. Sgn. Hy. Capt.* Francis	22 Nov. 04	14 Feb. 08
Brandon, *Sr. Asst. Sgn., Hony. Maj.* Joseph, B.	20 Jan. 07	7 Dec. 12
Briscoe, *Sr. Apothy. Hony Surgn. Capt.* Charles, B.	16 July 88	19 Aug. 91

Names and Rank.	Date of Rank.		Date of Retirement.	

RETIRED PAY.

Medical Services—*contd.*

Brown, Lt.-Col. Edwin Harold, M.D., F.R.C.S.E.	30 Sept.	06	10 Nov.	09
Brown, *Sr. Asst. Surgn., Hony. Capt.* Joseph	20 Apl.	05	25 Aug.	06
Brown, *Sr. Asst. Surgn., Hony. Maj.* Julius	27 Apl.	00	22 Aug.	06
K. H. S. Browne, Col. Samuel Haslett, C.I.E., M.D.	10 Apl.	03	29 Apl.	06
Browne, *Apothy. Hony. Surgn.* Wm. Francis, B.	5 June	68	15 June	77
K. H. S. Browne, Surgn.-Genl. William Richard, C.I.E., M D. ®	11 Oct.	04	1 Apl.	08
Browning, Lt.-Col. Winthropp Benjamin, C.I.E.	31 Mar.	00	17 May	10
Bull, Lt.-Col. George Henry, M.D.	30 Mar.	98	1 July	08
Burke, Lt.-Col. William Henry, M.B.	30 Sep.	02	1 Jan.	10
Burn, *Apothy. and Hony. Asst. Surgn.* Benjamin, Bo.	12 Oct.	70	1 May	75
Burnet, *Sr. Asst. Surgn. Hony. Capt.* Isaac, B.	9 Apl.	00	15 Nov.	02
Burns, *Sr. Asst. Surgn. Hony. Capt.* Isaac, B.	26 Sep.	99	6 May	02
Burton, Lt.-Col. John Adolphus	2 Oct.	00	6 Dec.	10
Busher, *Sr. Asst. Surgn., Hy. Lieut.* William David	17 Dec.	04	26 Mar.	08
Butler, Sgn. Maj. Ebenezer Robert, M.D., Bo.	4 Oct.	66	1 Apl.	77
Butler, *Surgn.* Wm. John, M.	30 Mar.	72	25 Feb.	80
Cabral, *Sr. Asst. Surgn. Hon. Capt.* Anthony Lawrence, M.	15 Apl.	01	20 Apl.	05
Cadell, Lt.-Col. John Macfarlane, M.B., B.	1 Oct.	05	5 Mar.	13
Cadge, Lt.-Col William Hotson, B.	31 Mar.	96	15 Nov.	00
Caldeira, *Sr. Asst. Surgn. Hony. Capt.* Mansoel Santa Anna, Bo.	9 Mar.	99	16 Mar.	01
Caldwell, Surgn. Maj. Wm. Smith, M.D., B.	29 Jan.	69	4 July	81
Cama, Lt.-Col. Rustom Hormesji	31 Mar.	00	13 July	10
Cameron, Surgn. Lt.-Col. Charles, B.	27 July	79	21 Sep.	92
Cameron, *Sr. Apothy. Hony. Sgn.-Lt.* John, B.	7 Nov.	85	25 Sep.	91
Cameron, Surgn. Maj. Lewis, M.D., B.	31 Mar.	78	9 Aug.	91
Campbell, *Sr. Asst. Surgn. Hy. Lieut.* Albert	10 Aug.	02	27 Mar.	06
Campbell, *Sr. Asst. Surgn. Hony. Capt.* Geo. Henry, B.	8 July	95	26 Sep.	98
Campbell, Col. *Sir* Robert Neil, K.C.M.G., C.B., C.I.E., M.B.	2 Apl.	09	2 Apl.	14
Carleton, *Sr. Asst. Surgn. Hy. Capt.* William George	17 Dec.	04	29 Mar.	08
Carlos, *Sr. Asst. Surgn. Hony. Lt.* Norbert Henry, M.	4 May	10	1 Jan.	13
Caroll, *Sr. Asst. Surgn. Hony. Capt.* Charles	7 Feb.	04	7 June	07
Carr-Calthrop, Col. Christopher Wm., B.	2 Apl.	99	2 Apl.	04
Carrapiett, *Sr. Asst. Surgn. Hony. Surgn. Capt.* Samuel Edward, M.	30 Aug.	95	7 Dec.	96
Carroll Lt.-Col Edward Richard William Charles, B.	1 Apl.	05	25 Aug.	12
Carruthers, Col. Herbert St. Clare, M.	30 June	08	30 June	13
Carson, Lt.-Col. Walter Peter, M.B., Bo.	30 Sep.	98	28 Jan.	04
Cazalett, *Sr. Asst. Surgn., Hony. Lieut.* Alexander Grant	9 Dec.	07	6 June	09
Chalke, *Sr. Asst. Surgn. Hony. Surgn. Capt.* Stephen, M.	12 Mar.	94	29 Apl.	97
Channer, Lt. Col. Osborne Henry, M.B.	30 Sep.	96	15 Apl.	07
Charles, Lt.-Col. (*Tempy. Surgn. Genl.*) *Sir* Richard Havelock, K.C.V.O., M.D., F.R.C.S.I. (*Sergt.-Surgn. to the King*)	1 Apl.	02	20 Mar.	08
Chaves, *Sr. Asst. Surgn. Hony. Capt.* Ignatius	1 Jan.	06	16 June	07
Chaytor-White, Joshua, C.M.G., M. D.	30 Sep.	09	22 Dec.	12
M. Christison, Surgn. Genl. *Sir* Alexander *Bart.*, M.D., B.	24 Nov.	82	24 Nov.	82
Clarkson, Lt.-Col. Frank Cecil	30 Sep.	06	1 Mar.	12
K. H. S. Cleghorn, Surgn. Genl. James, C.S.I., M.D., ®, B.	29 Mar.	95	25 Oct.	98
Clement, *Sr. Asst. Surgn. Hony. Maj.* Edward Patrick	30 June	04	29 Oct.	07
Clerke, *Sr. Asst. Surgn., Hony. Major,* Louis Arthur Henry	7 June	07	15 July	16
Coates, Lt.-Col. William, M. D.	31 Mar.	97	4 July	07
Cobb, Lieut.-Col. Robert	31 Mar.	96	10 Oct.	06
Collie, Lt.-Col. Mackintosh Alexander Thomas, M.B., Bo.	31 Mar.	03	30 June	13
Collins, *Sr. Asst. Surgn. Hony. Lieut.* Robert, M.	8 June	03	26 Jan.	05
Colson, Surgn. Maj. Edward, Bo.	1 Apl.	81	24 June	89
Compigne, Surgn.-Maj. Horatio David Steel, M.D., B.	1 Oct.	78	13 Nov.	84
Connor, *Sr. Apthy. Hony. Surgn.* James, B.	11 Nov.	76	4 Oct.	34
Connor, *Sr. Asst. Surgn., Hony. Capt.* Thomas Augustus Samuel, M.	1 Apl.	03	18 Aug.	04
Constant, Bde.-Surgn. Fredk. Geo., B.	8 Jan.	82	17 Jan.	80
Conway, *Apthy. Hony. Surgn.* William, Bo.	27 May	68	15 Nov.	81
Cook, Surgn.-Genl. Henry, M.D., Bo.	21 Nov.	86	21 Nov.	86
Cookson, Surgn.-Maj. Henry. B.	20 Jan.	72	20 May	81
Cooper, Dy. Surgn. Gl. Clarence, M.D., M.	27 Dec.	81	27 Dec.	82
Cooper, *Sr. Asst. Surgn. Hony. Capt.* William, B.	17 Mar.	00	1 Jan.	03
Cooper, *Sr. Asst. Surgn. Hony. Capt.* Wm. Henry	30 Sep.	98	4 Apl.	07
Cordell, *Sr. Asst. Surgn. Hony. Maj.* Charles, B.	20 July	98	1 Jan.	94
Cotton, *Sr. Asst. Surgn. Hony. Maj.* Alexander Adolphus, B.	31 Jan.	10	13 Dec.	14
Cotton, *Sr. Asst. Surgn. Hony. Maj.* Arthur David, B.	2 Nov.	08	8 Dec.	13
Courtney, *Sr. Apothy. Hony. Capt.* James Mathew, B.	22 Sep.	86	7 Jan.	92
Courtney, Surgn. Lt.-Col. W. Michael, B.	1 Oct.	89	7 Dec.	95
Cowell, Surgn. Andrew Richard, M.D., Bo.	1 Oct.	66	7 Mar.	75
Crawford, Lt.-Col. Dirom Grey, M.B.	1 Oct.	01	5 Mar.	11
Crocker, Dy. Surgn. Genl. Henry, M.D., M.	1 Mar.	83	1 Dec.	83
Cullen, Bde. Surgn. Lt.-Col. Peter, M.D., B.	18 Apl.	88	1 June	92

RETIRED PAY.

Medical Services—*contd.*

Names and Rank.	Date of Rank.	Date of Retirement.
Dalal, Lt.-Col. Kharshedji Ardeshir, M.B.	31 Mar. 96	3 Jan. 06
Dale, *Sr. Asst. Surgn. Hony. Capt.* James, M.	5 May 99	21 May 03
Damania, Lt.-Col. Pirozshah Jamsetjee, Bo.	30 Mar. 98	28 Oct. 98
Dantra, Lt.-Col. Sorabshaw Hormasji, B.	31 Mar. 96	10 July 03
Dawson, Lt.-Col. Arthur Willan, M.D.	24 June 11	19 July 17
Davies, *Sr. Asst. Surgn., Hy. Capt.* Daniel Robert, B.	23 Jan. 11	20 Dec. 13
Davis, *Sr. Asst. Sgn. Hy. Maj.* Charles Edward	1 Jan. 06	23 Sep. 08
Davis, *Sr. Asst. Surgn. Hy. Capt.* George William, B.	23 Jan. 11	1 Feb. 14
Davis, Lt.-Col. Robert Evans Stuart, M.B., M.	30 Sep. 02	29 Oct. 10
Davis, *Sr. Asst. Surgn. Hony. Capt.* Thomas, B.	15 June 95	11 Apl. 04
Day, *Sr. Asst. Surgn. Hony. Maj.* Harry, B.	20 Jan. 07	3 Jan. 11
D'Cruz, *Sr. Asst. Surgn. Hony. Maj.* Arthur	22 Aug. 06	17 Apl. 09
Deane, Lt.-Col. Andrew, M.D., B.	4 Dec. 95	1 Apl. 00
Delany, Lt.-Col. Thomas Henry, M.B., F.R.C.S.I.	28 July 17	28 Jan. 18
DeLima, *Sr. Asst. Surgn. Hony. Surgn. Capt.* Diego Santa Anna, Bo.	31 Aug. 90	18 Apl. 94
Dennys, Col. the Hon'ble George William Patrick	16 June 10	10 Jan. 17
DeRessurriecao, *Sr. Asst. Surgn. Hony. Capt.* John Augustus *	8 Mar. 06	16 Nov. 07
DeSouza, *Sr. Asst. Surgn. Hony. Capt* Cajetan Marie, Bo.	16 June 07	19 Apl. 12
D'Santos, *Sr. Asst. Surgn. Hony. Lt.* Francis Xavier, M.	7 Aug. 08	15 June 11
Dey, Jogesh Chandra, M.B.	27 Jan. 15	25 July 16
Dias *Sr. Asst. Surgn., Hony. Capt.* Augustus Xavier, Bo.	17 Mar. 99	14 July 00
Dick, Maj. Maxwell, M.B., M.	29 Jan. 08	13 Feb. 14
Dimmock, Lt.-Col. Henry Peers, M.D., Bo.	31 Mar. 00	15 Apl. 11
Divecha, Surgn.-Maj. Framji Ruttonji, M.	31 Mar. 92	1 Nov. 97
D'Lacy, *Sr. Asst. Surgn., Hony. Capt.* Harry, B.	27 Apl. 00	17 Dec. 04
Dobson, Col. Andrew Francis, M.B.	14 May 01	28 May 06
Dobson, Lt.-Col. Edwin Francis Horatio, M.B., B.	2 Oct. 00	27 Nov. 10
Douglas, Capt. Stewart Ranken	28 Jan. 01	15 Sep. 07
Doyle, Surgn. Lt.-Col. Bernerd, B.	31 Mar. 96	30 July 02
Doyle, Lt.-Col. Francis Joseph, M.	2 Oct. 00	7 Apl. 01
Doyle, Maj. Ignatius Purcell, D.S.O., M.	30 Sep. 98	31 Dec. 03
Drake-Brockman, Bde. Surgn. Lt.-Col. Edward Forster, M.	21 Apl. 90	4 Apl. 94
Drake-Brockman, Lt.-Col. Vivian Godfrey	30 Jan. 13	13 Feb. 15
D'Ravara, *Sr. Asst. Surgn. Hy. Lt.,* Arthur John	3 Dec. 06	30 Sep. 09
Duckering, *Surgn.* S., Bo.	31 Mar. 65	21 Mar. 74
Duckworth, *Sr. Asst. Surgn. Hony. Surgn. Capt.* Geo. John, M.	12 Mar. 94	22 Apl. 96
Duer, Lt.-Col. Charles, M.B., F.R.C.S.	28 July 11	28 Nov. 13
Duke, Lt.-Col. Joshua	30 Mar. 92	1 Nov. 02
Duncan, Surgn.-Maj., Wm., M.B., B.	2 Oct. 77	4 Dec. 90
Dutt, Lt.-Col. Russick Lall, B.	1 Apl. 98	30 Oct. 99
Dymott, Surgn. Maj. Donald Frederick, M.	30 Mar. 92	14 May 97
Earle, Lieut.-Col. Hubert Malins, B.	27 July 04	27 July 12
Easden, *Sr. Asst. Surgn. Hony. Capt.* Charles, M.	1 May 97	15 Oct. 00
Eate, *Sr. Asst. Surgn. Hony. Capt.* George William, M.	21 Sep. 97	20 Nov. 00
Eates, *Sr. Asst. Surgn. Hony. Maj.* William	10 Aug. 02	20 Jan. 07
Elcum, Maj. Donald, M.D., M.	31 Mar. 89	10 Mar. 95
Elkins, *Sr. Asst. Surgn. Hy. Maj.* David Arthur	3 May 03	14 Nov. 07
Elliot, Lt.-Col. Robert Henry, M.D., F.R.C.S.	30 Jan. 12	19 Apl. 15
Elliot, Lt.-Col. William Henry Wilson, M.B., D.S.O.	31 Mar. 07	15 Nov. 08
Ellis, *Sr. Apothy. Hony. Surgn.* James Charles, B.	31 Oct. 77	13 Dec. 91
Emerson, Lt.-Col. George Augustus, M.B.	1 Oct. 97	2 Apl. 08
Evans, Col. Arthur Owen	21 Nov. 16 / 22 Jan. 12	22 Jan. 17
Evans, Lt.-Col. James Welham, M.	31 Mar. 00	9 May 00
Everard, *Sr. Apothy Hony. Lt.* Thomas, M.	9 Dec. 88	8 Dec. 91
Eyre, Lt.-Col. Montague Stokes, M.B.	30 Sep. 96	7 Apl. 07
... Fairweather, Dy. Surgn.-Genl. James, M.D., B.	19 Oct. 86	19 Oct. 86
Falconer, *Sr. Asst. Surgn. Hy. Capt.* Samuel Ebenezer, M.	15 Apl. 01	24 May 04
Farmcombe, Surgn. Maj. Thomas Beard, B.	3 Jan. 66	18 Feb. 75
Farrell, *Sr. Asst. Surgn. Hony. Lieut.* Joseph Benson, Bo.	29 Nov. 00	21 Sep. 04
Fawcett, Lt.-Col. Edward, M.	4 Apl. 94	28 Dec. 99
Fearnside, Lt.-Col. Clarence Forbes, M.B., M.	31 Mar. 10	10 Sep. 12
Fenwick, *Sr. Asst. Surgn. Hony. Lieut.* John, Bo.	4 Feb. 07	15 Sep. 10
Fernandez, *Sr. Asst. Surgn. Hony. Lt.* Anthony Philip, Bo.	6 Mar. 94	13 May 99
Ferris, Surgn. Lt.-Col. John Edward Charnock, B.	30 Mar. 92	4 Aug. 92
ffrench-Mullen, Col. Douglas, M.D., B.	25 Mar. 07	25 Mar. 12
Ffrench-Mullen, Lt.-Col. Jarlath, M.D.	30 Mar. 98	5 May 08
Fink, Major George Herbert, B.	31 Mar. 96	13 July 01
Finnamore, *Sr. Asst. Surgn. Hy. Capt.* Henry Issac, B.	3 Dec. 93	11 Oct. 99
Fisher, *Sr. Asst. Surgn. Hy. Lieut.* John	8 June 03	27 Mar. 06
Fitzpatrick, *Sr. Asst. Surgn. Hy. Surgn. Capt.* John, B.	4 Sep. 91	11 Nov. 96
Fleming, *Sr. Asst. Surgn., Hony. Maj.* James George	31 May 04	18 Mar. 09
Fonseca, *Sr. Asst. Surgn. Hy. Capt.* Dominic Bernard, B.	15 Sep. 00	1 Apl. 02
Fooks, Lt.-Col. George Ernest, Bo.	2 Oct. 00	6 Dec. 10

* Temply. re-employed.

RETIRED PAY.

Medical Services—*contd.*

Names and Rank.	Date of Rank.	Date of Retirement.
Forrest, *Sr. Asst. Surgn. Hy. Lt.* George Oliver, B.	27 Apl. 00	29 May 00
Forrester, *Sr. Asst. Surgn. Hy. Maj.* William, B.	14 Nov. 07	3 Mar. 14
Fox, *Sr. Apthy. Hy. Surgn.* Charles Leolin, B.	13 Nov. 75	21 Aug. 85
Fox, *Sr. Asst. Surgn. Hy. Capt.* Frederick George, B.	31 Jan. 10	16 Feb. 13
Fox, *Sr. Asst. Surgn. Hy. Capt.* Geo. Francis, B.	17 Oct. 98	1 Jan. 05
Fraser, Maj. Herbert St. John	29 Jan. 06	18 June 09
Fraser, *Sr. Asst. Surgn. Hy. Capt.* James, B.	23 Sep. 09	23 Sep. 12
Freyer, Surgn. Lieut.-Col. (*Temp. Hon. Col*) Peter Johnston, C.B., M.D., B.	30 Sep. 95	3 May 95
Fullerton, Lt.-Col. John Campbell, M.B., B.	31 Mar. 94	25 June 04
Galvin, *Sr. Asst. Surgn. Hy. Capt.* John B.	17 Oct. 99	9 Apl. 00
Gamack, Surgn. Maj. Alexander, M.D., M.	4 Aug. 76	1 Aug. 79
Gamlen, Capt. Robert Long, M.D.	1 Aug. 11	28 Aug. 14
Garstin, *Sr. Asst. Surgn. Hony. Capt.* George, B.	15 Apl. 01	7 Feb. 04
Gaudoin, *Sr. Asst. Surgn. Hy. Capt.* George Robert Baptist	20 Nov. 07	18 Sep. 09
Ghose, *Surgn.* Fokeer Chand, M.B., B.	1 Oct. 69	16 May 70
Gibb, *Sr. Asst. Surgn., Hy. Capt.* John	31 May 04	20 Feb. 10
Gibbons, Lt.-Col. James Barry	1 Oct. 01	17 Feb. 08
Gidney, Henry Albert John, F.R.C.S.E.	28 July 17	26 July 18
Gilbert, Lt.-Col. Clarence Edwin Llyod, B.	31 Mar. 07	21 Sep. 12
Giles, Lieut.-Col. George Michael James, M.B., F.R.C.S., B.	2 Oct. 00	15 Jan. 01
Gill, *Sr. Asst. Surgn. Hy. Maj.* George	31 May 04	24 Sep. 06
Gill, Lieut. John Henry, M.B.	26 July 01	27 Mar. 07
Gillespie, *Sr. Asst. Surgn. Hy. Capt.* Telesphorus Daniel William, Bo.	1 Apl. 02	1 Aug. 05
Gillson, *Sr. Asst. Surgn. Hy. Surgn. Capt.* Edward, M.	12 Mar. 94	6 Oct. 94
Gimlette, Lt.-Col. George Hart Desmond, C.I.E., M.D.	31 Mar. 99	1 Apl. 10
Goldsmith, *Sr. Asst. Surgn. Hy. Capt.* John, B.	31 Jan 10	8 Nov. 13
Goldsmith, Surgn.-Maj. Septimus Jesse, Bo.	1 Apl. 85	4 Mar. 91
Grainger (Maj.-Genl.) Surgn.-Genl. Thomas, C.B., M.D., K.H.S.	24 May 14	29 Apl. 18
Grant, Col. D. St. J. D., M.B., B.	1 Jan. 11	14 Mar. 17
Gray, Bde. Surgn. Robert, M.B., B.	3 June 87	3 Apl. 91
Gray, Bde. Surgn. Lieut.-Col. Wellington, M.	1 Sep. 88	12 Apl. 92
Greahy, *Sr. Asst. Surgn. Hy. Lieut.* John William, Bo.	31 Jan. 10	1 Dec. 10
Greany, Surgn.-Genl. John Philip, M.D., ®	20 June 05	1 Oct. 08
Greany, Lt.-Col. Hugh, M.D., M.	1 Oct. 01	27 Apl. 12
Green, Lt.-Col. Douglas Richard, M.D., B.	27 July 12	1 Aug. 12
Green, *Sr. Asst. Surgn. Hy. Surgn. Capt.* Francis, M.	16 Nov. 95	9 Sep. 97
Greenhow, Surgn.-Maj. Henry Martineau, B.	1 Jan. 66	20 Aug. 76
Gregg, *Sr. Asst. Surgn. Hy. Maj.* Charles Nathaniel, B.	29 Mar. 08	31 Dec. 13
Griffith, Bde. Surgn. Lieut.-Col. Griffith, B.	6 July 90	1 May 93
Griffiths, Lt.-Col. Gilbert Sunders, B.	30 Sep. 96	1 June 04
Gwyther, Lt.-Col. Arthur, M.B., F.R.C.S.E.	29 July 14	29 Jan. 18
Haig, Lieut.-Col. Percy deHaga, B.	30 Sep. 95	15 Feb. 91
Hall, Col. Geoffrey Craythorne, B.	2 Dec. 01	1 Nov. 92
Hall, Surgn. Capt. Gilbert Capel, M.	30 Sep. 86	12 Mar. 98
Hall, *Sr. Apothy., Hy. Surgn. Lt.*, Thomas Kelly, M.	9 Dec. 88	5 Jan. 92
Hamilton, Surg.-Genl. *Sir* Henry, K.C.B., M.D., R.B.	25 Mar. 07	7 Apl. 11
Hamilton, *Sr. Asst. Surgn. Hy. Surgn. Capt.* John, B.	29 Aug. 91	15 June 95
Hare, Lt.-Col. Edward Christian	30 Sep. 06	3 Aug. 17
Harding, *Sr. Asst. Surgn. Hy. Capt.* Walter Henry, B.	1 July 01	27 Nov. 03
Hardy, *Sr. Asst. Surgn. Hy. Surgn. Lieut.* Mark Arthur, M.	6 Nov. 95	6 Aug. 96
Harington, Lt.-Col. Hastings Norman Victor, M.	2 Oct. 00	28 Nov. 10
Harkness, Maj. George Francis Innes	31 Aug. 15	19 May 17
Harris, Surgn.-Genl. George Francis Angelo, C.S.I., M.D., F.R.C.P.R.	1 Apl. 12	1 Apl. 15
Harrison, Lt.-Col. Clarence Barrymore	29 July 15	10 Feb. 16
Harvey, Surgn. William, B.	1 Apl. 68	6 Apl. 79
Hassan, Major Syed, B.	2 Oct. 92	1 Jan. 99
Hatch, Lieut.-Col. William Keith, Bo.	1 Mar. 97	15 Nov. 03
Hawkes, *Sr. Asst. Surgn. Hy. Surgn.-Capt.* Henry, M.	22 Apl. 96	1 Jan. 97
Hawkins, Lieut.-Col. Frederic Davy Cæsar, B.	2 Apl. 01	26 Oct. 01
Hayden, Capt. Arthur Falconer, M.B., F.R.C.S.	1 Sep. 08	23 Jan. 12
Hazlett, Bde.-Surgn. Lieut.-Col. Henry James, M.	7 Mar. 91	6 July 97
Heard, Surgn. Maj. Samuel Thomas, M.D., M.	29 Jan. 69	1 Dec. 74
Heathcock, *Sr. Asst. Surgn., Hy. Maj.* William, B.	22 Nov. 04	3 Nov. 10
Hellein, *Sr. Asst. Surgn. Hy. Capt.* Cæsar Augustus, M.	25 June 98	21 July 00
Henderson, Surgn.-Maj. George, M.D., M.	1 July 73	20 Nov. 89
Henderson, Dy. Surgn.-Genl. John, M.D., M.	16 May 85	1 July 88
Hendley, Lt.-Col. Arthur Gervase	30 Mar. 09	25 June 09
Hendricks, *Sr. Asst. Surgn. Hy. Capt.* Wm. Edward, M.	21 Mar. 02	8 June 03
Heppolette, *Sr. Asst. Surgn. Hy. Capt.* John Edward	May 02	8 Dec. 05
Heppolette, *Sr. Asst. Surgn. Hy. Maj.* William Alexander	11 Apl. 04	2 Nov. 08
Herbert, Lt.-Col. Herbert	31 Mar. 07	20 Oct. 07
Hesterlow, *Sr. Asst. Surgn. Hy. Capt.* Joseph Nathaniel, M.	20 Nov. 00	6 Mar. 03
Hickie, *Sr. Asst. Surgn. Hy. Capt.* John, B.	9 Apl. 01	14 July 02

Names and Rank.	Date of Rank.	Date of Retirement.
RETIRED PAY.		
Medical Services—*contd.*		
Hill, *Sr. Asst. Surgn. Hy. Maj.* Thos. Henry, C.I.E., B.	10 Mar. 92	1 Nov. 00
Hodgkins, *Sr. Asst. Surgn. Hy. Maj.* Henry Charles, B.	12 June 95	28 Dec. 98
Hogan, *Sr. Asst. Surgn. Hony. Capt.* James William	1 Jan. 03	3 Aug. 06
Hollingsworth, *Sr. Asst. Surgn. Hy. Capt.* Robert, M.	1 Jan. 97	1 Apl. 00
Holmes, *Sr. Asst. Surgn. Hy. Lieut.* Henry, B.	24 July 96	10 July 00
Holmes, *Sr. Asst. Surgn. Hy. Surgn. Capt.* Joseph, B.	12 Mar. 94	17 Feb. 98
𝔊. 𝔅. 𝔖., Hooper, Hy. Col. *Sir* William Roe, K.C.S.I., B.	6 Aug. 04	12 Jan. 95
Hore, Lt.-Col. Ernest Wickham, M.B.	31 Mar. 10	7 Dec. 11
Hottinger, *Sr. Asst. Surgn. Hy. Surgn. Capt.* George Alexander, B.	12 Mar. 94	8 July 95
Hughes, Surgn. Col. David Erskine, M.D., F.R.C.S.E., Bo.	15 Sep. 92	15 Sep. 97
Hulbert, Lt.-Col. Joseph George, M.B., B.	27 July 12	27 July 14
Hunter, Lieut.-Col. Christian Bernard, B.	31 Mar. 99	15 Apl. 00
Hunter, Lt.-Col. George Yeates Cobb, B.	17 Nov. 13	25 Oct. 14
Hunter, Surgn.-Maj. Joseph, H. M.	31 Mar. 88	30 Sep. 94
Hutcheson, Col. George, M.D., B.	1 Oct. 97	1 Oct. 02
Hyde, *Sr. Asst. Surgn. Hy. Capt.* William, B.	5 Sep. 98	1 Feb. 04
Hynes, *Sr. Asst. Surgn. Hy. Maj.* George, B.	8 Dec. 05	11 Jan. 13
Irvine, Lt.-Col. Gerard Beatty, C.B., B.	31 Mar. 07	12 Oct. 18
Jackson, *Sr. Asst. Surgn. Hy. Capt.* Septimus George	20 Nov. 07	15 Feb. 10
Jackson, Surgn.-Maj. Warwick, B.	1 Apl. 80	8 Apl. 87
James, Lt.-Col. Richard, M.B.	31 Mar. 99	29 Apl. 10
James, Lt.-Col. Sydney Price, M.D.	29 Jan. 16	15 Jan. 18
Jameson, Lt.-Col. John Bland, M.B., Bo.	31 Mar. 10	30 June 13
Jameson, Surgn.-Maj. George William, B.	20 Jan. 72	4 Apl. 82
Jervis, Surgn.-Lieut.-Col. Henry Pruce, Bo.	1 Oct. 97	27 Nov. 02
Jenney, Lt.-Col. George William, M.B.	30 June 13	4 Oct. 18
Johnson, Bde.-Surgn. Lieut.-Col. Edward Reginald, B.	1 Oct 93	2 Jan. 94
Johnson, *Sr. Asst. Surgn. Hy. Maj.* Robert, B.	20 July 98	12 Mar. 99
Johnstone, *Sr. Asst. Surgn. Hy. Capt.* James, B.	8 Dec. 05	22 Nov. 11
Jones, Lt.-Col. John Lloyd Thomas, M.D.	17 Mar. 11	28 Apl. 18
Jordan, Lt.-Col. John Gregory, M.B.	10 Nov. 16	18 Apl. 18
Joubert dela Ferté, Col. Chas. Henry, M.B., F.R.C.S., B.	29 Mar. 00	29 Mar. 05
Kanga, Lieut.-Col. Jamshedji Kharshedji, M.	2 Apl. 01	18 Dec. 01
Karney, *Hy. Surgn.* William, M.	12 Feb. 79	21 Nov. 89
Keefer, Surgn.-Maj. William Napier, B.	1 Oct. 81	11 Dec. 89
Keegan, Bde.-Surgn. Lieut.-Col. Denis Francis, M.D., F.R.C.S., B.	11 Aug. 91	1 Apl. 94
Kelawala, Lt.-Col. Manekshaw Jamshedji, B.	1 Apl. 02	16 Apl. 11
Keith, Bde. Surgn.-Lieut.-Col. Joseph Forbes, M.D., Bo.	15 Sep. 92	7 May 94
Kellie, Col. George Jerome	16 Oct. 05	15 Oct. 08
Kelly, *Sr. Asst. Surgn. Hy. Capt.* James, B.	3 Mar. 99	19 Dec. 01
Kelsey, *Asst. Surgn.* A., Bo.	9 Nov. 66	12 Apl. 63
Kharegat, Lieut.-Col. Mervanji Pestanji, M.	2 Oct. 00	8 Aug. 01
Kidd, *Sr. Apthy. Hy. Surgn.*, Wm. Arnold, B.	12 June 84	11 Dec. 85
Kiddle, *Sr. Asst. Surgn. Hony. Capt.* Thomas	26 Sep. 06	21 Dec. 06
Kilkelly, Lieut.-Col. Patrick Percy, M.B., Bo.	30 Jan. 13	1 Mar. 14
King, *Sr. Asst. Surgn., Hy. Capt.* Henry William George	5 June 07	30 Mar. 10
King, Col. Walter Gawen, C.I.E., M.B.	29 Apl. 06	25 May 10
Kirtikar, Lt.-Col. Konaba Ranchoddas, Bo.	31 Mar. 07	24 May 04
Knight, *Sr. Apthy. Hy. Surgn. Lt.* William Isaac, B.	3 June 90	25 July 92
Laffrey, *Sr. Asst. Surgn. Hony. Surgn. Capt.* Jonas Isaac, M.	12 Mar. 94	1 Jan. 95
Lafond, *Sr. Asst. Surgn. Hy. Capt.* Hugh Alfred, Bo.	19 Apl. 12	23 May 15
Laing, Surgn. Lieut.-Col. James Anderson, M.D., M.	1 Apl. 90	9 Sep. 95
Lalor, Lt.-Col. Nicholas Purcell O'Gorman, M.B.	29 July 15	25 July 18
Lamont, Lt.-Col. John Charles, M.B.	1 Oct. 07	10 Mar. 08
Lane, Lt.-Col. Daniel Thomas, M.D.	1 Apl. 15	29 Mar. 18
Lancaster, Lt.-Col. John, M.B., M.	31 Mar. 94	1 Apl. 04
Langston, Lt.-Col. Thomas Alfred Ollivant	25 Dec. 15	12 Dec. 10
Lawrence, *Sr. Asst. Surgn. Hy. Capt.* Thomas Cuthbert	8 June 03	20 Nov. 07
Leahy, Lieut.-Col. Albert William Denis, M.D., B.	30 Sep. 02	5 June 03
Leapingwell, Lieut.-Col. Arthur Henry, M.	1 Apl. 93	31 Mar. 01
Lee, Lt.-Col. William Alexander	30 Sep. 96	7 Apl. 08
Lethbridge, Bde.-Surgn. Lieut.-Col. *Sir* Alfred Swaine, K.C.S.I., M.D.	9 Apl. 92	1 Apl. 98
Lewtas, Lt.-Col. John, M.D., B.	30 Sep. 95	21 Nov. 05
Lewty, *Sr. Asst. Surgn. Hy. Capt.* William Henry, B.	15 Feb. 97	6 Mar. 89
Little, Col. Charles Colhoun, M.D., M.	12 Feb. 00	12 Feb. 05
Little, Lieut.-Col. Stephen, M.D.	31 Mar. 99	22 June 09
𝔊𝔱. Loch, Dy. Surgn.-Genl. Jas. Henry, M.D., B.	1 Jan. 84	1 Jan. 89
Long, Major William Christopher	27 Jan. 12	27 Jan. 16
Lowdell, Lieut.-Col. Charles George Walter, Bo.	30 Sep. 98	30 Apl. 99
Lowson, Maj. Charles Stewart, M.B.	27 July 11	27 July 17
Luard, Capt. Hugh Bixley, M.B.	31 Mar. 90	15 Mar. 07
Luke, *Sr. Asst. Surgn. Hy. Lt.* Alexander, B.	1 Feb. 04	11 Feb. 04
McArdle, *Sr. Asst. Surgn. Hy. Capt.* Wm. Richard, B.	27 Apl. 00	31 May 04
McCalman, Lt.-Col. Hugh, M.D., Bo.	30 Sep. 96	7 Apl. 01
McCartie, Lt.-Col. Chas. Joseph, M.D., B.	30 Sep. 94	25 June 00
McCloghry, Col. James, F.R.C.S.I.	20 June 05	13 Jan. 09
McDermott, *Sr. Apothy. Hy. Surgn. Lt.* J.M.	21 Sep. 93	24 Feb. 94

RETIRED PAY.

Medical Services—*contd.*

Names and Rank.	Date of Rank.	Date of Retirement.
McDonnell, Surgn. Lt.-Col. Jas. O'Malley, M.D., F.R.C.S., B.	1 Apl. 90	22 July 95
McGann, Col. Terence Joseph, M.	1 Apl. 95	5 May 01
McKay, Col. Henry Kellock, C.B., C.I.E.®	3 Dec. 04	3 Dec. 09
McKey, *Sr. Asst. Surgn. Hy. Capt.* Patrick	1 Feb. 04	8 Dec. 07
M. V. O., McLeod, *Hony.* Col. Kenneth, M.D., B.	2 May 06	16 Apl. 92
McMullen, *Sr. Asst. Surgn. Hy. Capt.* George Churchill	1 Apl. 02	4 Oct. 07
McNaught, *Sr. Asst. Surgn. Hy. Capt.* James, B.	1 Dec. 94	17 Oct. 98
McVean, Capt., Norman Niel George Cowan	1 Sep. 09	19 Jan. 17
Macdonald, Lt.-Col. Thos. Rankin, B.	31 Mar. 00	20 June 00
MacGregor, Capt. Roderick Dear, M.B.	1 Sep. 07	28 Oct. 12
Macgregor, Surgn. Lt.-Col. John, M.D., Bo.	31 Mar. 96	29 Oct. 96
Mackenzie, Surgn. Frederic Morell, B.	3 Aug. 77	3 Aug. 77
Mackenzie, Surgn.-Major. John Thos. Denison, Bo.	23 July 70	8 May 78
Mackey, *Sr. Asst. Surgn. Hony. Capt.* James, B.	20 Jan. 01	23 Oct. 05
MacLaren, Bde.-Surgn. Lieut.-Col. Geo. Gilbert, M.D., B.	1 May 93	27 June 93
Maclaren, Lt.-Col. James Farquharson, M.B., B.	2 Apl. 01	24 June 11
MacLeod, Lt.-Col. John Norman, C.M.G., C.I.E., M.B., F.R.C.S.E.	30 Jan. 13	23 Aug. 16
Macnamara, Lt.-Col. John William Unthank, M.D.	31 Oct. 99	12 July 08
Macnamara, Surgn.-Maj. Nottidge Chas., B.	4 Nov. 66	15 Apl. 76
Macnamara, Col. Robert Joseph, M.D.	14 Mar. 16	11 Nov. 16
Macrae, Col. Roderick, C.I.E., M.B.®	12 Feb. 05	1 Mar. 10
Mahony, *Sr. Asst. Surgn. Hy. Capt.*, Ebenezer, M.	15 Apl. 01	15 Apl. 04
Mair, Lieut.-Col. Edward, M.B., B.	31 Mar. 94	3 July 02
Mallins, Lieut.-Col. Clement, M.D., M.	30 Sep. 98	1 May 99
Maloney, *Sr. Asst. Surgn. Hy. Capt.* John Barry, Bo.	13 Aug. 91	9 Mar. 99
Mantell, Surgn.-Major Alfred. Adams, M.D., B.	24 Jan. 67	5 Jan. 77
Manuk, Capt. Mack Walter, M.B., Bo.	27 July 02	1 Aug. 02
Marchant, *Sr. Asst. Surgn. Hy. Lieut.*, William, B.	15 Apl. 01	6 Jan. 05
Marks, Lt.-Col. Robert John	27 Dec. 12	26 Feb. 18
Marsden, Lt.-Col. James Cort	31 Oct. 99	10 Feb. 10
Marshall, Maj. Daniel Grove, M.B., B.	31 Mar. 00	24 June 05
Martin, Lieut.-Col. Daniel Nicholas, M.D., B.	28 Feb. 98	1 May 99
Masani, Surgn. Lieut.-Col. Harmasji Dadabhai, Bo.	31 Mar. 97	16 Jan. 96
Mason, *Sr. Asst. Surgn. Hy. Capt.* William, B.	7 Feb. 04	8 Jan. 05
Mawson, Lt.-Col. William Arthur	31 Mar. 96	14 July 06
Meneaud, *Sr. Asst. Surgn. Hy. Surgn. Lt.* Charles Robert, M.	12 Jan. 95	10 Oct. 98
Mercado, *Sr. Asst. Surgn., Hy. Capt.*, Andrew Lyon	16 Sep. 03	16 Dec. 06
Meyer, Lt.-Col. Charles Hardwick Louw, M.D., Bo.	31 Mar. 07	15 Mar. 13
Milchem, *Sr. Asst. Surgn., Hy. Capt.* George Thomas	22 Aug. 06	25 Sep. 10
Miller, *Sr. Asst. Surgn. Hy. Surgn. Capt.* Henry James, B.	3 Sep. 91	1 Dec. 94
Milne, Lt.-Col. Alexander, M.B., B.	2 Apl. 01	25 July 11
Minty, *Sr. Asst. Surgn. Hony. Maj.* Thomas William	31 Jan. 10	26 Mar. 15
Mistri, Lt.-Col. Kavasji Hormasji	2 Oct. 00	4 Mar. 08
Mitchell, *Sr. Asst. Surgn. Hy. Surgn.-Maj.* William Marcus, B.	17 Aug. 98	22 July 98
Moore, *Sr. Asst. Surgn., Hy. Capt.* John.	21 Mar. 02	17 Mar. 07
Moore, Major Thos. Chas. M.	31 Mar. 99	23 May 02
Moorhead, Surgn. Lieut.-Col. James, M.D., B.	30 Sep. 96	29 May 98
Moran, Lt.-Col. James Joseph, M.D., M.	30 Mar. 98	10 Oct. 98
Moran, Maj. Edgar John, M.B.	29 Jan. 07	1 Oct. 15
Moriarty, Col. Mathew Denis, M.D., F.R.C.S.I., B.	25 Oct. 02	26 Oct. 05
Morris, *Hy. Surgn.* Easter Alfred, M.	12 Jan. 95	4 Dec. 88
Morrison, *Sr. Asst. Surgn. Hy. Capt.*, James	3 May 01	. . . .
Morwood, Lt.-Col. James, M. D.	30 Sep. 06	11 Nov. 18
Mukerji, Lieut.-Col. Upendra Nath, M.B., B.	1 Oct. 04	17 Feb. 05
Mullane, Lieut.-Col. Patrick, M.D., B.	2 Apl. 01	1 Nov. 08
Mullen, *Sr. Asst. Surgn. Hy. Surgn. Capt.* Archibald, Bo.	12 Mar. 94	22 June 97
Mulroney, Lieut.-Col. Thomas Richard, M.D., F.R.C.S.	31 Mar. 00	13 Aug. 09
Mulvany, Surgn. Lieut.-Col. Edward, B.	30 Mar. 92	11 July 92
Murphy, *Sr. Asst. Surgn. Hy. Capt.* Daniel O'Connell, B.	12 Mar. 12	5 June 14
Murphy, *Sr. Asst. Surgn. Hy. Capt.* George	18 Feb. 01	14 Sep. 06
Murphy, *Sr. Asst. Surgn. Hy. Capt.* Michael	1 Jan. 03	31 Oct. 09
Murphy, Surgn. Lieut.-Col. Patrick, M.D., Bo.	1 Oct. 89	17 Jan. 95
Murphy, Lieut.-Col. Wm. Reid, D.S.O., B.	30 Mar. 92	15 July 99
Murray, Col. Robert Davidson, M.B.	29 Mar. 05	29 Mar. 10
Murray, Lieut.-Col. Wm. Flood, B.	2 Apl. 96	18 Dec. 99
Nailer, *Sr. Asst. Surgn. Hy. Capt.* Robert Ball	5 June 07	7 Mar. 08
Nandi, Lt.-Col. Shambhu Chandra, M.B., B.	2 Oct. 00	6 Dec. 05

RETIRED PAY.

Medical Services—*contd.*

Names and Rank.	Date of Rank.	Date of Retirement.
Nesbitt, *Sr. Apthy. Hy. Surgn. Lieut.* George, B.	3 Sep. 88	29 Aug. 93
Newland, Major Arthur, Geo. Edward, M.	2 Apl. 93	1 Aug. 99
Newman, Col. John Henry, M.D., B.	9 Mar. 95	14 July 00
Newton, Surgn. Major Isaac, B.	6 July 77	6 July 72
Newton, *Sr. Asst. Surgn. Hy. Maj.* Isaac	1 Apl. 05	23 July 08
Nicholas, *Sr. Asst. Surgn. Hy. Capt.* Cyprian Terence Montrose	5 June 07	16 Sep. 08
Nicholl, *Asst. Surgn.* Robert	20 Jan. 47	7 Nov. 55
Nicholson, Lieut.-Col. George Fredk, M.D., F.R.C.S.I., B.	3 Mar. 98	23 Feb. 99
Nolan, *Hony. Maj., Sr. Asstt. Surgn.* Arthur Herbert, B.	4 Jan. 11	13 June 12
Norman, *Sr. Apthy. Hy. Surgn. Capt.* John, M.	11 Jan. 93	17 Jan. 94
Norman, Lt. Vincent Philip	30 July 10	12 Nov. 16
O'Brien, Lieut.-Col. Joseph, M.D., B.	9 May 96	19 Aug. 98
O'Connor, Lieut.-Col. Patrick Fenelon, C.B., B.	31 Mar. 95	13 July 05
O'Connor, *Sr. Asst. Surgn. Hy. Maj.* William Frank	27 Apl. 00	31 Jan. 10
O'Dwyer, Lieut.-Col. Malachi, M.B., B.	30 Sep. 98	30 Nov. 98
O'Hara, Col. William, M.	28 May 06	7 June 11
O'Leary, *Sr. Asst. Surgn. Hy. Surgn. Capt.* Daniel, M.	12 Mar. 95	21 Sep. 97
O'Mealy, *Sr. Asst. Surgn. Hy. Capt.*, Michael, B.	14 July 02	30 June 04
O'Neal, *Sr. Asst. Surgn., Hy. Capt.* George Samuel	8 Dec. 05	4 July 10
O'Reilly, *Sr. Asst. Surgn. Hy. Capt.* Lawrence John	20 Jan. 07	12 Feb. 10
Ollenbach, *Sr. Asst. Surgn. Hy. Maj.* Donald Smith, B.	31 Jan. 10	22 Nov. 13
Owen, *Sr. Asst. Surgeon, Hy. Capt.*, Charles Arthur, F.R.C.S.E.	17 Apl. 09	1 June 09
Owen, Lieut.-Col. Chas. Wm., C.M.G., C.I.E., B.	30 Sep. 96	15 Mar. 02
Pais, *Sr. Asst. Surgn., Hy. Capt.* Simon, Bo.	16 Mar. 01	1 Apl. 05
Pals, *Sr. Asst. Surgn. Hy. Maj.* Survulo Joseph, Bo.	28 Dec. 06	23 Dec. 12
Palk, Maj. Charles Henry Leet, M.B., F.R.C.S.E.	28 July 03	1 May 09
Pank, Lieut.-Col. Philip Durrell	31 Mar. 00	17 May 10
Parakh, Surgn.-Lieut.-Col. Danjisha Narroji, Bo.	30 Sep. 94	14 Mar. 98
Parker, Lieut.-Col. Joseph, M.D., Bo.	30 Sep. 95	18 May 01
Paske, Depy. Sugn.-Genl. Chas. Thos., B.	1 Jan. 79	1 Jan. 79
Payne, Lieut. Ernest LeFevre, B.	29 Jan. 96	3 Nov. 00
Peck, Lt.-Col. Edward Surman, M.B.	12 Dec. 14	6 Dec. 13
Pedroza, Surgn.-Maj. Francis Hoskins, M.	30 Sep. 95	6 Jan. 96
Pemberton, Lt.-Col. Robert	30 Mar. 98	1 July 08
Pereira, *Sr. Asst. Surgn., Hony. Capt.* Joseph. Anthony	1 Apl. 02	1 Mar. 06
P. H. S. Perry, Lieut.-Col. Francis Frederick, C.M.G., C.I.E., F.R.C.S.	31 Mar. 99	14 June 09
Peters, Lieut.-Col. Chas. Thos., Bo.	26 Feb. 93	24 June 99
Pinch, Capt. Albert Edwin Hayward, B.	29 July 99	17 Dec. 01
Pinto, Maj. Joseph Orphine	1 Oct. 99	31 Mar. 06
Pigani, Lt.-Col. Lionel John, F.R.C.S., B.	1 Apl. 06	10 June 11
Poole, Surgn.-Maj. George Kenneth, M.D., B.	14 Mar. 67	7 May 76
Pope, Lieut.-Col. Thos. Henry, M.D., M.	30 Mar. 98	1 Mar. 04
Pope, *Sr. Asst. Surgn, Hony. Capt.* Joseph Agnew Reynaad	15 Feb. 10	15 June 16
Power, Bde.-Surgn. Richard, B.	1 Jan. 90	1 Jan. 86
Poynder, Lieut.-Col. John Leopold	31 Oct. 96	12 Dec. 09
Prain, Lt.-Col. *Sir* David, *Kt.*, C.M.G., C.I.E., F.R.S., M.B.	1 Oct. 04	31 July 06
Prasad, Lt.-Col. Kanta, M.B.	31 Mar. 08	3 June 13
Pratt, Lt.-Col. (Bt.-Col.) James John, B.	29 Sept. 03	27 Dec. 12
Prentie, *Sr. Asst. Surgn. Hy. Capt.* Joseph, B.	1 Apl. 56	1 Apl. 11
Price, Bde.-Surgn. Lieut.-Col. William, M.D., M.	1 Mar. 03	7 Apl. 96
Price, *Sr. Apthy. Hy. Surgn.* William, B.	19 May 80	24 July 85
Prichard, *Sr. Asst. Surgn., Hy. Maj.* James William	6 Mar. 03	13 Oct. 06
Prince, *Sr. Asst Surgn., Hy. Capt.*, Richard Cunning Debeaux	26 Sep. 06	16 Nov. 06
Quayle, Col. William Adair, M.D., M.	1 Apl. 08	3 Apl. 11
Ranking, Lt.-Col. Geo. Spiers Alexr. M.D., B.	31 Mar. 95	30 May 05
Ratton, Surgn.-Maj. Jas Joseph Louis, M.	30 Sep. 79	26 Jan. 88
Raye, Col. Daniell O'Connell, M.D., B.	2 Apl. 94	2 Apl. 99
Rebeiro, *Sr. Asst. Surgn. Hy. Capt.* Sebastian, Bo.	12 Feb. 97	17 Mar. 99
Reid, *Surgn.* John, B.	27 July 71	27 July 71
Reid, *Sr. Asst. Surgn. Hy. Capt.* John, B.	26 Sep. 98	18 Feb. 01
Reilly, Lieut-Col. Edwin William	1 Apl. 02	18 July 07
Reynolds, Surgn. J., Bo.	29 Apl. 97	31 Dec. 62

RETIRED PAY.

Medical Services—*contd.*

Names and Rank.	Date of Rank.	Date of Retirement.
Richards, Major, Walton Guyon, M.B.	29 July 08	15 Jan. 16
Ricketts, Lt.-Col. William Symonds Percival, M.B.	30 Mar. 09	10 June 10
Roach, *Sr. Asst. Surgn. Hy. Capt.* Henry Richard	4 Feb. 07	5 Feb. 11
Robb, Surgn.-Maj. John, M.D., Bo.	2 Nov. 89	2 Nov. 89
Roberts, Lt.-Col. Alfred Ernest, M.B.	1 Oct. 07	21 Apl. 08
Roberts, *Sr. Asst. Surgn. Hy. Capt.* Chas. Geo., M.	1 Apl. 00	21 Nov. 01
Roberts, *Sr. Asst. Surgn. Hy. Capt.* Henry, M.	29 Apl. 97	24 May 99
Roberts, Dy. Surgn.-Genl. Wm. Howland, M.D., M.	1 July 88	1 July 93
Robertson, *Sr. Asst. Surgn., Hy. Capt.* David	5 June 07	1 Dec. 07
Robertson, Col. Robert, M.B.	30 June 15	4 July 18
Robinson, *Surgn.* Ernest Lawrn, B.	31 Mar. 77	19 Jan. 87
Robinson, *Sr. Asst. Surgn. Hy. Capt.* William Henry	22 Nov. 04	12 Jan. 09
Rodrigues, *Sr. Asst. Surgn. Hy. Capt.* Anthony Philip	1 Apl. 05	9 Dec. 07
Rodrigues, *Sr. Asst. Surgn., Hy. Maj.* Albert Robert Goddard	5 June 07	4 May 10
Roe, Col. Robert Bradley, M.	5 May 16	15 May 14
Rogers-Harrison, Lieut.-Col. Augustus Napoleon, M.	7 Apl. 96	6 Jan. 02
Ross, Surgn.-Col. Geo. Cumberland, B.	1 Jan. 94	1 Oct. 97
Ross, Maj. *Sir* Ronald, K.C.B., K.C.M.G., M.	2 Apl. 93	31 July 99
Russell, Lieut.-Col. Allan Rupert Postance	30 Sep. 06	11 Oct. 07
Russell, Lieut.-Col. Edgar Geo., B.	30 Mar. 92	27 Aug. 00
Russell, *Sr. Apothy. Hony. Capt.* Hugh, B.	16 July 83	4 Oct. 89
Salaman, Bde.-Surgn. Lieut.-Col. Selim Myer, M.D., Bo.	7 May 94	31 Mar. 97
Sampson, *Sr. Asst. Sgn. Hy., Capt.* Joseph	21 Mar. 00	28 Dec. 06
Sanders, Surgn-Maj. Edwin, B.	1 Apl. 81	1 Jan. 91
Sanjana, Lt.-Col. Kavasji Cursetji	2 Oct. 00	22 Oct. 08
Sarkies, Lieut.-Col. Carapiet John, M.B., Bo.	29 Sep. 03	6 Dec. 12
Sarkies, Lt.-Col. Sarkies Carrapiet	31 Mar. 99	6 May 10
Saunders, *Sr. Apothy. Hy. Surgn. Capt.* Francis William, B.	18 Jan. 90	24 Feb. 94
Scotland, Lt.-Col. David Wilson, M. B.	30 Sep. 06	26 Mar. 07
Scott, Lieut.-Col. James, M.B.	1 Apl. 02	23 Oct. 07
Scott-Moncrieff, Lieut.-Col. William Elmsley, M.D., F.R.C.S.E.	28 July 14	29 July 14
Scroggie, *Sr. Asst. Surgn. Hy. Capt.* Wm. Reith, Bo.	14 Sep. 92	21 Mar. 00
Seaman, Bde.-Surgn. Lieut.-Col. Albert Baird, B.	22 Oct. 94	1 Oct. 96
Seton, Bt.-Col. *Sir* Bruce Gordon, *Bart.*, C.B., I.M.S.	30 Jan. 12	27 May 17
Sharman, Lt.-Col. Eric Harding, M.	30 Jan. 13	1 Aug 13
Sherman, *Sr. Asst. Surgn. Hy. Capt.* Henry Murray	2 Dec. 06	27 Apl. 08
Shore, Lt.-Col. Robert, M.D.	29 Sep. 03	25 Dec. 11
Shunker, *Sr. Asst. Surgn. Hy. Lieut.* Edwin Luke	18 Aug. 04	14 May 07
Silcock, Lt.-Col. Alexander, M.D.	2 Apl. 01	25 Mar. 09
Shillito, Surgn.-Maj. William, F.R.C.S., B.	13 Jan. 60	23 July 63
Simon, *Sr. Asst. Surgn. Hy. Capt.* James Robert	15 Apl. 04	2 Dec. 06
Simmonds, Lieut.-Col. Wm. Allason, B.	31 Mar. 95	9 Aug. 01
Simpson, Surgn.-Genl. Sir Benjamin, K.C.I.E., M.D., ®. B.	29 Mar. 85	29 Mar. 90
Sinclair, Surgn.-Genl. David, C.S.I., M.	11 Oct. 91	15 Nov. 04
Sinclair, *Apthy. Hony. Surgn.* George, Bo.	14 June 73	20 Apl. 83
Sinha, Major Narenda Prasanna, B.	1 Apl. 98	1 June 04
Smith, Surgn.-Maj Ferdinand Clarence, F.R.C.S., M.	30 Mar. 90	1 July 95
Smith, Bde.-Surgn Lieut.-Col. James, M.	12 Oct. 90	1 Apl. 95
Smith, Lt.-Col. Julian Carter Carington, M.B.	31 Mar. 99	7 Aug. 09
St. Smith-Wynne, Depty. Surgn.-Genl. William Arnold, M.	29 Jan. 77	1 Mar. 78
Smyth, Bde.-Surgn. Lieut.-Col. Fredk. Augustus, B.	1 Oct. 89	27 July 98
Smyth, Col., J., M.D.	7 June 11	31 Oct. 18
Spencer, Lt.-Col. Dhanjibhai Barjorji	31 Mar. 00	17 May 10
Spencer, Surgn.-Lieut.-Col. Thos. Chas. Howell	1 Oct. 89	31 Dec. 94
Staggs, *Sr. Asst. Surgn. Hy. Capt.* Valentine James	15 Apl. 01	1 Apl. 06
Stephen, Col. Arthur, M.B., B.	19 May 94	19 May 99
Stephens, Surgn. Lieut.-Col. Augustus Edward Richard, B.	31 Mar. 96	24 May 97
Stevenson, Surgn.-Genl. Henry Wickham, C.S.I., ®. Bo.	11 Jan. 09	10 Jan. 14
Stewart, Surgn. Lieut.-Col. Alexander Kemette, Bo.	31 Mar. 96	31 Mar. 97
Stoker, Lieut.-Col. Richard Nugent, B.	30 Sep. 94	2 Apl. 00
Sturmer, Lt.-Col. Arthur James, M.	31 Mar. 95	12 Apl. 05
Summers, *Sr. Asst. Surgn. Hy. Capt.* Joseph Seymour, Bo.	16 Dec. 06	1 Dec. 13
Sutherland, Depy. Surgn. Genl. Philip Warren, B.	18 Dec. 83	5 Feb. 89
Swaine, Lieut.-Col. Charles Lethbridge, M.D.	31 Mar. 96	23 July 07
Swaine, Lieut.-Col. Fred. Robert, M.B., B.	30 Sep. 93	1 Nov. 04
Swainson, *Sr. Asst. Surgn. Hy. Surgn. Lieut.* Edward, B.	29 Aug. 93	1 June 96

RETIRED PAY.

Medical Services—*concld.*

Names and Rank.	Date of Rank.	Date of Retirement.
Sweeny, Lt.-Col. Terence Humphrys, F.R.C.S.I., C.M.G.	30 Sept. 98	1 Mar. 09
Sykes, Lt.-Col. Joseph	31 Oct. 99	14 Dec. 09
Sykes, Lt.-Col. William Ainley, D.S.O., M.B., B.	2 Apl. 09	18 July 12
Talbot, Richard Francis Chetwynd, M.D., F.R.C.S.I.	1 Sep. 05	24 July 13
Tallent, *Sr. Asst. Surgn. Hy. Surgn. Capt.* Henry, M.	7 Dec. 96	25 June 98
Thomas, Lieut.-Col. George Tucker, M.	31 Mar. 95	26 June 05
Thomas, Lt.-Col. William Frederick, M.D.	31 Mar. 99	6 Jan. 08
Thompson, *Sr. Asst. Surgn. Hony. Capt.* William, B.	15 Apl. 01	10 Aug. 02
Thomson, Maj. Charles, M.B., B.	29 Jan. 06	30 Jan. 11
Thomson, Lt.-Col. George Sloane, M.B.	31 Mar. 08	25 Dec. 15
Thomson, Lt.-Col. Henry, M.D.	29 Sep. 03	1 Feb. 12
Thomson, Lt.-Col. & Bt.-Col. Samuel John, C.I.E.	1 Oct. 97	17 Jan. 08
Thornhill, Lt.-Col. William Henry, M.D., M.	1 Oct. 97	9 Apl. 99
K. Thornton, Dy. Surgn.-Genl. *Sir* Jas. Howard, K.C.B., M.B., B. (R)	30 June 86	13 Aug. 01
Thorold, Maj. William Grant, B.	1 Apl. 98	15 Apl. 03
Todd, *Sr. Apthy. Hy. Surgn. Lieut.* Theophilus, M.	11 Mar. 91	21 Sep. 93
Tomes, Surgn. Lieut.-Col., Arthur, M.D., B.	31 Mar. 96	13 Oct. 97
Toomey, *Sr. Asst. Surgn. Hy. Maj.* Patrick	31 May 04	17 Apl. 09
Tunnicliffe, *Sr. Asst. Surgn. Hony. Capt.* Charles, B.	7 June 07	31 Mar. 11
Tresham, *Sr. Apthy., Hy. Surgn. Capt.* Daniel John, B.	1 Sep. 85	8 Mar. 92
Trutwein, *Sr. Asst. Surgn. Hy. Surgn. Capt.* Edmund, M.	12 Mar. 94	16 Nov. 95
K. H. S. Turnbull, Surgn.-Maj.-Genl. Peter Stephenson, M.D., (R). Bo.	26 Feb. 93	2 Mar. 96
Twigg, Maj. Herbert Joseph Richard, M.B.	27 July 10	13 Jan. 13
Vaid, Lt.-Col. Cooverji Cawasjee, B.	31 Mar. 00	18 May 05
Van Geyzel, Lt.-Col. John Lawrence, M.B.	31 Mar. 00	9 July 11
Vaz, *Sr. Asst. Surgn. Hy. Capt.* Andrew John Baptist	1 Mar. 06	4 Feb. 07
Viegas, *Sr. Asst. Surgn. Hony. Capt.* John Matthew	15 Oct. 00	1 Jan. 06
Waddell, Lt.-Col. Laurence Austin, C.B., C.I.E., M.B.	31 Mar. 00	21 Oct. 06
Wade, *Sr. Asst. Surgn. Hy. Surgn. Capt.* William, B.	18 June 89	15 Feb. 97
Walker, *Sr. Asst. Surgn. Hy. Capt.* Henry Dumrie, Bo.	13 July 00	16 Sep. 03
Wall, *Sr. Asst. Surgn. Hy. Capt.* Sydney Alfred	1 Feb. 04	1 Aug. 07
Waller, *Sr. Asst. Surgn. Hy. Lt.* David, B.	3 May 03	16 July 11
Walsh, Lieut.-Col. John Henry Tull, B.	1 Apl. 04	21 July 05
Warliker, Lt.-Col. Damodar Purshotum	30 Mar. 98	1 Apl. 09
Waters, Lt.-Col. George, M.	31 Mar. 96	23 Apl. 01
Watson, Maj. Thomas Edgar, M.B., M.	29 Jan. 07	15 Mar. 13
Watts, *Sr. Asst. Surgn. Hy. Surgn. Capt.* Geo. Albert, M.	12 Jan. 95	12 Mar. 95
Webster, Charles George, Lt.-Col., F.R.C.S.E.	29 Jan. 15	29 Jan. 16
Weir, Col. Patrick Alexander, M.B.	26 Oct. 05	26 Oct. 10
Weir, Lt.-Col. Richard Rose, M.B.	30 Sep. 02	21 Feb. 10
Wells, *Sr. Asst. Surgn. Hy. Surgn. Lt.* Henry, M.	4 Mar. 93	17 Aug. 95
Weston, *Sr. Asst. Surgn., Hy. Capt.* Edwin	5 June 07	23 Dec. 07
Weston, *Sr. Asst. Surgn.. Hony. Capt.* Joseph Theophilus, B.	1 Jan. 03	1 Sep. 05
Weston, *Sr. Asst. Surgn. Hony. Capt.* William Archibald, B.	1 Nov. 00	22 Nov. 04
Whitcombe, Lt.-Col. Ernest Gerald Robert	31 Jan. 11	27 Oct. 11
White, *Sr. Asst. Surgn. Hy. Capt.* Samuel Charles, M.	12 Mar. 94	24 May 02
Whitwell, Lieut.-Col. Robert Richard Harvey, M.B., B.	31 Mar. 00	17 Sep. 05
Wilkie, Col. David, M.B. R.	2 Apl. 04	2 Apl. 09
Wilkins, Col. Thomas James Hackett	30 June 03	30 June 08
Wilkinson, Lt.-Col. Edmund, F.R.C.S., B.	28 July 11	13 Nov. 14
Willcocks, Lt.-Col. Alexr. John, M.D., B.	30 Sep. 93	1 Apl. 01
Williams, Surgn. Lt.-Col. Alfred Henry, M.B., B.	30 Mar. 92	4 July 97
Williams, Major, Charles Louis, M.D.	30 Sept. 01	19 Mar. 07
Williams, *Sr. Asst. Surgn. Hy. Capt.* William Archelaus, B.	2 June 00	11 Sep. 04
Williamson, Dy. Surgn.-Genl. Benjamin, M.B., M.	27 Jan. 84	27 Jan. 84
Williamson, *Sr. Asst. Surgeon Hy. Lt.* Joseph Hiscox	5 June 07	11 Sep. 09
Wilson, *Sr. Asst. Surgn. Hony. Capt.* George, B.	15 June 00	1 July 01
Windross, *Sr. Asst. Surgn. Hy. Maj.* Mark, B.	1 Apl. 05	6 Sep. 10
Wood, Surgn. Lt.-Col. Julius John, B.	1 Oct. 86	15 Feb. 92
Woolbert, Lt.-Col. Henry Robert, M.B., F.R.C.S., B	1 Oct. 05	30 June 13
Wortabet, Lt.-Col. Henry George Luther, M.D.	31 Mar. 99	22 June 09
Wright, Lieut.-Col. Frederick William, M.B., D.S.O., B.	1 Apl. 93	7 Dec. 05
K. Wyndowe, Dy. Surgn.-Genl. Samuel Jardine, M.D., M.	15 June 81	15 June 81
Wynne, *Sr. Asst. Surgn. Hy. Capt.* Francis Alexander	20 Mar. 05	1 Dec. 06
Younan, Lt.-Col. Arthur Charles, M.B.	1 Oct. 05	1 Jan. 15
Young, Surgn. Lt.-Col. Edward Wm., Bo.	30 Sep. 96	1 May 97
Young, Lt.-Col. James, M.B., B.	24 Oct. 97	31 Mar. 02
Youngerman, Lt.-Col. Edward Pettingall, M.B.	1 Oct. 07	2 Mar. 07

RETIRED INDIAN OFFICERS HOLDING HONORARY BRITISH RANK.

Honorary Captains.

Names, rank and late corps.	Date of rank.	Remarks.
Bhupal Thapa, Subadar-Major, *Sardar Bahadur*, 39th Garhwal Rifles	26 June 95	
Kalu Thapa, Subadar-Major, *Sardar Bahadur*, 44th Gurkha Rifles	16 Dec. 96	
Moti Negi, Subadar-Major, *Sardar Bahadur*, 39th Garhwalis	26 May 97	
Zaman Khan, Subadar-Major, *Sardar Bahadur*, 20th Infantry	20 July 98	
Judbir Thapa, Subadar-Major, *Sardar Bahadur*, 2nd Gurkha Rifles	20 Sep. 99	Hy. A.-D.-C. to H. E. the C.-in-C.
Bakhan Sing Thakur, I.O.M., Subadar-Major, *Sardar Bahadur*, 44th Gurkha [Rifles.	5 Feb. 00	
Abdullah Khan, I.O.M., Risaldar-Major, *Sardar Bahadur*, 7th Bombay Lancers	18 Apl. 00	
Tejbir Thapa, Subadar-Major, *Bahadur*, 1-2nd Gurkha Rifles	1 May 01	
Wali Muhammad, Risldr-Maj., *Sardar Bahadur*, G.G.'s Body Guard, A.-D.-C.	1 June 01	
Bhanga Singh, Subadar-Major, *Sardar Bahadur*, Derajat Mn. Batty.	24 Mar. 02	
Muhammad Khan, Subadar-Major, *Sardar Bahadur*, 28th Bo. Pioneers	10 May 02	
Yasin Khan, I.O.M., Subadar-Major, C.I.E., *Sardar Bahadur*, 24th Punjab Infantry	17 Dec. 02	
Kanhai Parshad Dube, I.O.M., Subadar-Major, *Sardar Bahadur*, 3rd Brahmans	1 Jan. 03	Hy. A.-D.-C. to H. H. the L.-G., United Provinces.
Man Singh, I.O.M., Subadar-Major, *Sardar Bahadur*, 14th Sikhs	16 Jan. 03	
Shaikh Mahbub, Subadar-Major, *Sardar Bahadur*, 23rd Bo. Infantry	31 Jan. 03	
Kaku Singh, Subadar-Major, *Sardar Bahadur*, 35th Sikhs	1 Feb. 03	
Mangal Singh, Risaldar-Major, *Sardar Bahadur*, 3rd Bengal Cavalry	1 Apl. 03	
Akbar Ali, Subadar-Major, *Sardar Bahadur*, 24th Mountain Battery	1 Nov. 03	
Mana Khan, I.O.M., Subadar-Major, *Sardar Bahadur*, 33rd Punjabis	15 Nov. 03	
Chandi Singh, Subadar-Major, *Sardar Bahadur*, 1st Brahmans	1 Jan. 04	
Sirput Pudwal, Subadar-Major, *Sardar Bahadur*, 128th Pioneers	15 Feb. 04	
Hira Singh, Subadar-Major, *Sardar Bahadur*, 1st S. and M.	1 Apl. 04	
Muizzud-din Khan, Risaldar-Major, *Sardar Bahadur*, 15th Lancers	8 May 04	
Jiwand Singh, Subadar-Major, *Sardar Bahadur*, 45th Rattray's Sikhs	1 Aug. 04	
Sher Singh, Risaldar-Major, *Sardar Bahadur*, 13th Lancers	20 Oct. 04	Hy. A.-D.-C. to H. E. the C.-in-C.
Sher Baz, Subadar-Major, *Sardar Bahadur*, 52nd Sikhs	1 Nov. 04	
Muzaffar Khan, Subadar-Major, *Sardar Bahadur*, 72nd Punjabis	1 Nov. 04	A.-D.-C. to H. H. the L.-G., Burma.
Shaikh Husain, Subadar-Major, *Sardar Bahadur*, 101st Grenadiers	3 Dec. 04	
Muhammad Amin Khan, Risaldar-Major, *Sardar Bahadur*, 23rd Cavalry	5 Jan. 05	
Jwala Singh, Subadar-Major, *Sardar Bahadur*, 45th Sikhs	1 Feb. 05	
Muhammad Beg, Risaldar-Major, *Sardar Bahadur*, 26th Light Cavalry	1 Feb. 05	
Jahangir Khan, Risaldar-Major, *Sardar Bahadur*, 31st Lancers	31 Mar. 05	
Waryam Singh, Subadar-Major, *Sardar Bahadur*, 47th Sikhs	18 Apl. 05	
Lal Khan, Risaldar-Major, *Sardar Bahadur*, 33rd Lt. Cavy.	9 July 05	
Baldeo Singh, Risaldar-Major, *Sardar Bahadur*, 14th Lancers	1 Aug. 05	
Faiz Talab Khan, Risaldar-Major, *Sardar Bahadur*, Guides	1 Nov. 05	
Sundar Singh, Subadar-Major, *Sardar Bahadur*, 36th Sikhs	1 Nov. 05	
Umdah Singh, Risaldar-Major, *Sardar Bahadur*, 22nd Cavalry	31 Dec. 05	
Mitt Singh, Subadar-Major, *Sardar Bahadur*, 29th Punjabis	16 Jan. 06	
Shaikh Ismail, Subadar-Major, *Sardar Bahadur*, 2nd Q. O. S. and M.	27 Mar. 06	
Ramchandar Rao Mahadik, I.O.M., Risaldar-Major, *Sardar Bahadur*, 31st Lancers	31 Mar. 06	
Saiyid Husain, Subadar-Major, *Sardar Bahadur*, 61st Pioneers	30 June 06	
Lehna Singh, Subadar-Major, *Sardar Bahadur*, 53rd Sikhs	1 July 06	Hy. A.-D.-C. to H. E. the C.-in-C.
Mahatab Singh, Subadar-Major, *Sardar Bahadur*, 48th Pioneers	1 July 06	
Siwa Moduck, Subadar-Major, *Sardar Bahadur*, 101st Grenadiers	1 Sep. 06	
Gurdatt Singh, Subadar-Major, *Sardar Bahadur*, 15th Sikhs	15 Dec. 06	
Magar Singh, Subadar-Major, *Sardar Bahadur*, 26th Punjabis	29 Mar. 07	
Malik Ahmadyár Khan, Risaldar Major, *Sardar Bahadur*, 18th Lancers	1 July 07	
Muhammad Ali Beg, Risaldar-Major, *Sardar Bahadur*, 9th Horse	15 Oct. 07	
Bishnu Thapa, Subadar-Major, *Sardar Bahadur*, 1-5th G.R.	1 Dec. 07	
Chamu Singh Burathoki, I.O.M., Subadar-Major, *Sardar Bahadur*, 3rd G. R.	1 Dec. 07	
Atar Singh, Subadar-Maj., *Sardar Bahadur*, 76th Punjabis	1 Mar. 08	
Sher Singh, Subadar-Maj., *Sardar Bahadur*, 28th Mtn. Battery	31 Mar. 08	
Shaikh Farid, Risaldar-Major, *Sardar Bahadur*, 26th Lt. Cavalry	10 Apl. 08	
Hazara Singh, Subadar-Maj., *Sardar Bahadur*, 35th Sikhs	16 May 08	

Names, rank and late corps.	Date of rank.	Remarks.

RETIRED INDIAN OFFICERS HOLDING HONORARY BRITISH RANK—*contd.*

Honorary Captains—*contd.*

Names, rank and late corps.	Date of rank.	Remarks.
Ali Gauhar Khan, Risaldar-Maj., *Sardar Bahadur*, 27th Light Cavalry	15 June 08	
Sikandar Khan, Risaldar-Major, *Khan Sahib, Sardar Bahadur*, 11th K. E. O., Lancers	1 Jan. 09	
Fazl Khan Subadar-Major, *Sardar Bahadur*, 82nd Panjabis	1 Jan. 09	
Tirath Ram, I.O.M., Risaldar-Major, *Sardar Bahadur*, Guides	1 Feb. 09	
Prakas Rae, I.O.M., Subadar-Major, *Sardar Bahadur*, 91st Punjabis	1 Feb. 09	
Chinna Sami, Subadar-Major, *Sardar Bahadur*, 2nd Q. O. S. and M.	7 Feb. 09	
Narayan Singh, I.O.M., Subadar-Major, *Sardar Bahadur*, 51st Sikhs	1 Mar. 09	
Gurdatt Singh, Subadar-Major, *Sardar Bahadur*, 15th Sikhs	1 Apl. 09	
Bidhi Chand, Subadar-Major, *Sardar Bahadur*, 38th Dogras	1 Apl. 09	Hy. A.-D.-C. to H. E. the C.-in-C.
Kirpa Ram Thapa, I.O.M., Subadar-Major, *Sardar Bahadur*, 2nd G. R.	1 Apl. 09	
Rae Singh, Subadar-Major, *Sardar Bahadur*, 41st Dogras	8 Apl. 09	
Lehna Singh, Subadar-Major, *Sardar Bahadur*, 54th Sikhs	15 Apl. 09	
Abdul Aziz, Risaldar-Major, *Sardar Bahadur*, 5th Cavalry, A.-D.-C.	19 Apl. 09	
Nakchhed Singh, Subadar-Major, *Sardar Bahadur*, 16th Rajputs	1 Sep. 09	
Sarfaraz Khan, I.O.M., Subadar-Major, *Sardar Bahadur*, Guides	15 Sep. 09	
Ghulam Muhammad, Subadar-Major, *Sardar Bahadur*, 29th Mountain Battery	16 Sep. 09	
Abiram Gurung, Subadar-Major, *Sardar Bahadur*, 8th G. R.	1 Jan. 10	
Ghulam Muhammad, Subadar-Major, *Sardar Bahadur*, 22nd Derajat Mountain Battery	1 Feb. 10	
Ali Muhammad Khan, Risaldar-Major, *Sardar Bahadur*, 2nd Lancers	1 June 10	
Madho Sing Rana, Subadar-Major, *Sardar Bahadur*, 4th G. R., A.-D.-C.	10 June 10	
Jagat Singh, Risaldar-Major, *Sardar Bahadur*, 16th Bengal Lancers	24 June 10	
Ghulam Hussain, Subadar-Major, *Sardar Bahadur*, 28th Punjab Infantry	24 June 10	
Hussain Khan, Subadar-Major, *Sardar Bahadur*, 15th Madras Infantry	24 June 10	
Jai Singh, I.O.M., Subadar-Major, *Sardar Bahadur*, Q. O. Corps of Guides	24 June 10	
Wajid Ali Risaldar-Major, *Sardar Bahadur*, 3rd Bengal Cavalry	24 June 10	
Sayyid Amir, Subadar-Major, *Sardar Bahadur*, 1st Madras Lancers	24 June 10	
Muhammad Bakhsh, Risaldar-Major, *Sardar Bahadur*, 3rd Bombay Cavalry	24 June 10	
Jagan Nath Pande, Subadar-Major, *Sardar Bahadur*, 25th Bombay Infantry	24 June 10	
Lehna Singh, Subadar-Major, *Sardar Bahadur*, 45th Sikhs	24 June 10	
Sayyid Yakub, Subadar-Major, *Sardar Bahadur*, 27th Madras Infantry	24 June 10	
Shaikh Ismail, Subadar-Major, *Sardar Bahadur*, 7th Bombay Infantry	24 June 10	
Matbar Singh Thapa, Subadar-Major, *Sardar Bahadur*, 1st G. R.	24 June 10	
Mir Alam, I.O.M., Subadar-Major, *Sardar Bahadur*, 28th Punjab Infantry	24 June 10	
Shaikh Amir, Subadar-Major, *Sardar Bahadur*, 25th Madras Infantry	24 June 10	
Mathura Singh, Risaldar-Major, *Sardar Bahadur*, 6th Bengal Cavalry	24 June 10	
Shaikh Ahmad, Subadar-Major, *Sardar Bahadur*, 5th Infantry, H. C.	24 June 10	
Shaikh Ahmad, Subadar-Major, *Sardar Bahadur*, 1st Madras Lancers	24 June 10	
Shaikh Mahtabdin, Subadar-Major, *Sardar Bahadur*, 31st Madras Infantry.	24 June 10	
Abdul Aziz, Subadar-Major, *Sardar Bahadur*, 25th Madras Infantry	24 June 10	
Ratan Singh, Subadar-Major, *Sardar Bahadur*, Hazara Mountain Battery	24 June 10	
Muhammad Ibrahim, Subadar-Major, *Sardar Bahadur*, 10th Bombay Infantry.	24 June 10	
Lal Singh, Risaldar-Major, *Sardar Bahadur*, 14th Bengal Lancers	24 June 10	
Tezu Khanka, Subadar-Major, *Sardar Bahadur*, 1st Gurkha Rifles	24 June 10	
Wazir Singh, Subadar-Major, *Sardar Bahadur*, 15th Sikhs	24 June 10	
Shaikh Wali Ahmad, Subadar-Major, *Sardar Bahadur*, 1st Infantry, H. C.	24 June 10	
Nand Singh, Subadar-Major, *Sardar Bahadur*, 23rd Punjab Pioneers	24 June 10	
Kandhai Singh, Subadar-Major, *Sardar Bahadur*, 6th Infantry, H. C.	24 June 10	
Ram Singh, Subadar-Major, *Sardar Bahadur*, 28th Bombay Pioneers	24 June 10	
Ganga Sewak Pande, Subadar-Major, *Sardar Bahadur*, 23rd Bombay Infantry.	24 June 10	
Narayan Singh, I.O.M., Subadar-Major, *Sardar Bahadur*, 2nd Punjab Infantry.	24 June 10	
Sayyid Jalal, Subadar-Major, *Sardar Bahadur*, 21st Madras Pioneers	24 June 10	
Mahrab Ali Khan, Risaldar-Major, *Sardar Bahadur*, 3rd Lancers, H. C.	24 June 10	
Shaikh Muhammad, Subadar-Major, *Sardar Bahadur*, 28th Bombay Pioneers.	24 June 10	
Mirza Ibrahim Beg, Subadar-Major, *Sardar Bahadur*, 3rd Bombay Infantry	24 June 10	
Arbiela, Subadar-Major, *Sardar Bahadur*, 20th Punjab Infantry	24 June 10	

RETIRED INDIAN OFFICERS HOLDING HONORARY BRITISH RANK—*contd.*

Honorary Captains—*contd.*

Names, rank and late corps.	Date of rank.	Remarks.
Rur Singh, Subadar-Major, *Sardar Bahadur,* 29th Madras Infantry	24 June 10	
Muhammad Saib, Subadar-Major, *Sardar Bahadur,* 77th Moplah Rifles	24 June 10	
Daud Khan, Subadar-Major, *Sardar Bahadur,* 15th Madras Infantry	24 June 10	
Birbal Nagarkoti, Subadar-Major, *Sardar Bahadur,* 7th G. R.	24 June 10	
Hira Singh, Subadar-Major, *Sardar Bahadur,* 1st Sikh Infantry	24 June 10	
Shah Muhammad, Subadar-Major, *Sardar Bahadur,* 25th Punjab Infantry	24 June 10	
Gopal Ram Lama, Subadar-Major, *Sardar Bahadur,* 2nd G. R.	24 June 10	
Kala Singh, Subadar-Major, *Sardar Bahadur,* 56th Rifles	24 June 10	
Santbir Gurung, I.O.M., Subadar-Major, *Sardar Bahadur,* 2nd G. R.	1 Oct. 10	Hy. A.-D.-C. to H. E. the C.-in-C.
Sher Khan, Risaldar-Major, *Sardar Bahadur,* 1st Lancers	6 Oct. 10	
Hanwant Singh, M.B.E., Risaldar-Major, *Sardar Bahadur,* 3rd Skinner's Horse.	16 Oct. 10	
Karan Sing Gurung, Subadar-Major, *Sardar Bahadur,* 1st G. R.	31 Dec. 10	
Gurdatt Singh, Risaldar-Major, *Sardar Bahadur,* 12th Cavalry	1 Feb. 11	
Kushal Singh, Burathoki, Subadar-Major, *Sardar Bahadur,* 3rd G. R.	15 Feb. 11	
Narayan Singh, Subadar-Major, *Sardar Bahadur,* 34th Sikh Pioneers	1 Mar. 11	
Mirza Umrao Beg, M.V.O., Subadar-Major, *Sardar Bahadur,* 31st Lancers	31 Mar. 11	
Dur Muhammad Khan, Risaldar, *Bahadur,* 15th Lancers	8 Apl. 11	
Rustam Khan, Subadar-Major, *Sardar Bahadur,* 17th Infantry	24 June 11	
Chaudhri Khan, Subadar-Major, *Sardar Bahadur,* 92nd Punjabis	20 July 11	
Muhammad Saiyid Khan, Risaldar, Major, *Sardar Bahadur,* 1st Lancers	16 Aug. 11	
Hamzullah Khan, Risaldar-Major, *Sardar Bahadur,* 17th Cavalry	1 Jan. 12	
Bahadur Ali Khan, Subadar-Major, *Sardar Bahadur,* I.O.M., 51st Sikhs	1 Jan. 12	
Faiz Khan, Risaldar-Major, *Sardar Bahadur,* 36th Horse	15 Jan. 12	
Narbahadur Gurung, Subadar-Major, *Sardar Bahadur,* 2nd G. R.	29 Feb. 12	
Rekha Ram, Subadar-Major, *Sardar Bahadur,* 6th Jat Light Infantry	1 Mar. 12	
Mihr Khan, Subadar-Major, *Sardar Bahadur,* 29th Punjabis	11 Mar. 12	
Hira Singh, Subadar-Major, *Sardar Bahadur,* 16th Rajputs	4 May 12	
Harditt Singh, Subadar-Major, *Sardar Bahadur,* 21st Mountain Battery	16 May 12	
Nawal Sing Rana, Subadar-Major, *Sardar Bahadur,* I.O.M., 1st Battn., 8th Gurkha Rifles.	1 June 12	
Prem Singh, Subadar-Major, *Sardar Bahadur,* I.O.M., 32nd Sikh Pioneers	1 July 12	Hy. A.-D.-C. to H. E. the C.-in-C.
Hara Singh, Subadar-Major, *Sardar Bahadur,* I.O.M., 91st Punjabis	15 July 12	
Gopala, I.O.M., Subadar-Major, *Sardar Bahadur,* 41st Dogras	1 Nov. 12	Hy. A.-D.-C. to H. H. the L.-G., Punjab.
Khushial Singh, Subadar-Major, *Sardar Bahadur,* 27th Punjabis	1 Nov. 12	
Sardar Janmeja Singh, Risaldar-Major, *Sardar Bahadur,* 21st Cavalry	17 Nov. 12	
Garkha, Subadar Major, *Sardar Bahadur,* 38th Dogras	1 Jan. 13	
Sandhya Das, Subadar-Major, *Sardar Bahadur,* 25th Punjabis	16 Jan. 13	
Bishn Singh Kathait, Subadar-Major, *Sardar Bahadur,* 2nd Bn., 9th Gurkha Rifles.	31 Jan. 13	
Dalman Ale, Subadar-Major, *Sardar Bahadur,* I.O.M., 1st Battn., 2nd Gurkha Rifles.	1 Feb. 13	
Rahim Khan, Subadar-Major, *Sardar Bahadur,* 107th Pioneers	28 Feb. 13	
Dayal Singh, Subadar-Major, *Sardar Bahadur,* 54th Sikhs	16 Mar. 13	
Devasahayam, Subadar-Major, *Sardar Bahadur,* 2nd Sappers and Miners	9 Apl. 13	Hy. A.-D.-C. to H. E. the C.-in-C.
Nur Khan, Subadar-Major, *Sardar Bahadur,* 61st Pioneers	19 Apl. 13	
Abdul Karim Khan, Risaldar-Major, *Sardar Bahadur,* Govr. Genl.'s Body-guard, A.D.C.	19 Apl. 14	
Singbir Ghale, Subadar-Major, *Sardar Bahadur,* 2nd Battn., 3rd Gurkha Rifles.	31 May 13	
Amar Singh, Subadar-Major, *Sardar Bahadur,* 19th Punjabis	1 June 13	
Sarabjit Thapa, Subadar-Major, *Sardar Bahadur,* 1st Gurkha Rifles	15 June 13	
Mihan Singh, Subadar-Major, *Sardar Bahadur,* 28th Punjabis	4 July 13	
Adhar Singh, Subadar-Major, *Sardar Bahadur,* 7th Rajputs	1 Aug. 13	
Arbel Singh, Subadar-Major, *Sardar Bahadur,* 38th Dogras	1 Aug. 13	
Yar Muhammad, Subadar-Major, *Sardar Bahadur,* 127th Infantry	15 Sep. 13	Hy. A.-D.-C. to H. E. the C.-in-C.
Alexander, Subadar-Major, *Sardar Bahadur,* 2nd Sappers and Miners	9 Oct. 13	Hy. A.-D.-C. to H.E. the Govnr. of Madras.
Mir Haider Shah Khan, Risaldar-Major, *Sardar Bahadur,* 37th Lancers	31 Dec. 13	
Phula Singh, Subadar-Major, *Sardar Bahadur,* 12th Pioneers	1 Mar. 14	
Beli Ram, I.O.M., Subadar-Major, *Sardar Bahadur,* 53rd Sikhs	1 Apl. 14	
Umrao Sing Rawat, Subadar-Major, *Sardar Bahadur,* 2nd Bn., 39th Garhwal Rifles.	4 May 14	
Gulam Muhammad, Subadar-Major, *Sardar Bahadur,* 27th Punjabis	1 July 14	
Aluhdin, Subadar-Major, *Sardar Bahadur,* 130th Baluchis	1 July 14	

RETIRED INDIAN OFFICERS HOLDING HONORARY BRITISH RANK—*concld.*

Honorary Captains—*concld.*

Names, rank and late corps.	Date of rank.	Remarks.
Saiyid Ali, Subadar-Major, *Sardar Bahadur*, 53rd Sikhs	1 July 14	
Nathu Singh Sunar, Subadar-Major, *Sardar Bahadur*, 1st Battn., 39th Garhwal Rifles.	1 July 14	
Jwala Singh, Subadar-Major, *Sardar Bahadur*, 35th Sikhs	1 July 14	
Gulam Muhammad Khan, Sabadar-Major, *Sardar Bahadur*, 72nd Punjabis.	1 July 14	
Ali Haidar Khan, Risaldar-Major, *Sardar Bahadur*, 38th Central India Horse.	1 July 14	
Nawab Khan, I.O.M., Subadar-Major, *Sardar Bahadur*, 30th Mountain Battery.	1 July 14	
Binda Singh, Subadar-Major, *Sardar Bahadur*, 4th Rajputs	1 July 14	
Abdul Karim Khan, Risaldar-Major, *Sardar Bahadur*, 30th Lancers (Gordon's Horse).	1 July 14	
Bisesar Tewari, I.O.M., Subadar-Major, *Sardar Bahadur*, 1st Brahmans	4 July 14	
Bagwan Singh, I.O.M., Subadar-Major, *Sardar Bahadur*, 14th Sikhs	4 July 14	
Bzura Singh, Subadar-Major, *Sardar Bahadur*, 90th Punjabis	9 July 14	
Kamardin, I.O.M., Subadar-Major, 126th Infantry	13 July 14	
Badal Singh, Subadar-Maj , *Sardar Bahadur*, 89th Punjabis	15 July 14	
Narayan Singh, Subadar-Major, *Sardar Bahadur*, 56th Rifles	26 Oct. 14	
Asa Singh Subadar-Major, *Sardar Bahadur*, 24th Punjabis	1 Nov. 14	
Sohawa Singh, Subadar-Major, *Sardar Bahadur*, 14 Sikhs and Bur. M. P.	8 Jan. 15	
Bishan Singh, Subadar-Major, *Sardar Bahadur*, 57th Rifles	1 June 15	*Hony.* A. D. C. to H. E. the Viceroy.
Mit Singh, Subadar-Major, *Sardar Bahadur*, 53rd Sikhs	10 June 15	
Bhagat Singh, Subadar-Major, *Sardar Bahadur*, 45th Sikhs	1 July 15	
Girdhari, Subadar-Major, *Sardar Bahadur, Rai. Bahadur*, 42nd Deoli Regt.	4 Aug. 15	
Surja, Subadar-Major, *Bahadur*, 48th Pioneers	4 Aug. 15	
Kurban Ali Khan, *Khan Bahadur*, 14th Sikhs	4 Aug. 15	
Ram Sing, Risaldar-Major, *Sardar-Bahadur*, 9th Hodson's Horse	22 Feb. 16	
Gugan, Subadar-Major, *Sardar-Bahadur*, 6 Jat Light Infy.	15 June 16	
Harnam Singh, Subadar-Major, *Sardar Bahadur*, Burma Mily. Police	30 June 16	
Mir Jafar Khan, I.O.M., Risaldar-Major, *Sardar Bahadur*, 9th Hudson's Horse.	19 July 16	*Hony.* A. D. C. to H.E. the C.-in-C.
Ajab Khan, I. O. M., Hon'ble Subadar-Major, *Sardar-Bahadur*, 76th Punjabis	15 Aug. 16	
Amar Singh, I.O.M., Subadar-Major, *Sardar Bahadur, Rai Bahadur*, Burma Mily. Police.	31 Aug. 16	
Mulik Ghulam Muhammad, Subadar-Maj. Khan, *Sardar Bahadur*, 39th King George's Own Central India Horse.	11 Oct. 16	
Mobin Khan, Subadar-Major, Sardar-Bahadur, 21st Punjabis	1 Nov. 16	
Bhairo Singh, *I. O. M.*, Subadar-Major, *Sardar Bahadur*, 119th Infantry	18 Dec. 16	
Shaikh Babu, Subadar-Major, *Sardar Bahadur*, 116th Marhattas	8 Jan. 17	
Nur Alam, Subadar-Maj., *Sardar Bahadur*, 28th Mountain Battery	1 June 17	
Bakshi Jaswant Singh, Risaldar-Maj., 39th Horse	16 July 17	
Mansur Khan, Subadar-Major, *Sardar Bahadur*, 55th Coke's Rifles	1 Sep 17	
Alam Khan' Subadar-Maj., *Sardar Bahadur*, Corps of Guides	1 Nov. 17	
Natha Singh, Risaldar-Major, *Sardar Bahadur*, 17th Lancers	1 Jan. 18	
Balak Singh, Subadar-Major, *Sardar Bahadur*, 8th Rajputs	1 Jan. 18	
Khusial Singh, Subadar-Major, *Sardar Bahadur*, 72nd Punjabis	25 Jan. 18	
Sardar **Gopal Singh**, *the Hon'ble* Risaldar, 11th Lancers	26 July 18	
Fateh Din, *Sardar Bahadur*, I.D.S.M. 23rd Min. Btty.	1 Sep. 18	

Honorary Lieutenants.

Names, rank and late corps.	Date of rank.	Remarks.
Jag Singh, Subadar, *Sardar Bahadur*, Ben. Sappers and Miners	15 May 03	
Muhammad Shiabuddin, Subadar, *Sardar Bahadur*, 2nd Infy., Hyd. Contgt.	4 Aug. 03	
Misri Khan, Risaldar-Major, *Bahadur*, 18th Lancers	1 Dec. 03	
Sawan Singh, Subadar, *Sardar Bahadur*, 107th Pioneers	28 Feb. 06	
Balbir Rana, Subadar, *Sardar Bahadur*, 3rd G. R.	1 Feb. 09	
Kamalud-din, Risaldar, *Sardar Bahadur*, G. G.'s Body Guard	7 Feb. 09	
Arsla Khan, Subadar, *Sardar Bahadur*, 20th Punjab Infantry	24 June 10	
Kanchan Singh, Risaldar, *Sardar Bahadur*, 2nd Bombay Lancers	24 June 10	
Abdullah Khan, Hony. Risaldar-Major, *Sardar Bahadur*, Aden Troop	24 June 10	
Sudh Singh, Subadar, *Sardar Bahadur*, Bengal S. and M.	24 June 10	
Lall Mast, Subadar, *Sardar Bahadur*, 29th Bombay Infantry	24 June 10	
Faizulla Khan, Subadar, *Sardar Bahadur*, Bengal S. and M.	24 June 10	
Fakir Muhammad, Subadar, *Sardar Bahadur*, 29th Punjabis	1 Jan. 12	
Jwala Singh, Risaldar, *Sardar, Bahadur* Corps of Guides	1 Feb. 12	
Sital Sing Lama, Subadar, *Sardar Bahadur*, 2nd G. R.	1 Apl. 12	
Alahdad Khan, Subadar, *Sardar Bahadur*, I.O.M., 127th Infantry	15 June 12	
Niyaz Khan, Subadar, *Sardar Bahadur*, 57th Rifles	1 July 12	
Shaikh Ibrahim, Subadar, *Sardar Bahadur*, 116th Maharattas	2 Sep. 12	
Attru, Subadar, *Sardar Bahadur*, 56th Punjabi Rifles	5 Dec. 12	
Shah Wali Khan, Subadar-Major, 108th Infy.	4 Aug. 15	
Sitaram Ghogle, Subadar-Major, 117th Mahrattas	4 Aug. 15	
Hira Singh, Subadar-Major, 14th Sikhs	4 Aug. 15	
Thakur Singh, Risaldar, *Sardar Bahadur*, I. O. M., 21st Cavalry	6 Dec. 15	
Hakim Khan, Subadar, *Sardar-Bahadur*, 72nd Punjabis	12 Jan. 15	
Thakur Bhagwant Singh, Risaldar-Maj., Light Cavalry	6 Apl. 17	
Suraj Mal, Risaldar, *Sardar Bahadur*, 14th Lancers	1 May 18	

DEPARTMENTAL WARRANT OFFICERS.

Note.—The names on this list are not indexed as they are arranged alphabetically.

Names.	Rank.	When pensioned.	Remarks.
Abbott, R. Y., Bo.	Condr.	1 Jan. 87	Ireland.
Acott, Henry Thomas, S. & T. C.	Sub-Condr.	31 Mar. 12	India.
Adams, George, B.	Sub-Condr.	11 Dec 91	Queensland.
Aldkin, Edward Russell, Bo.	Condr.	22 Dec. 96	England.
Ailnley, Samuel, M.	Condr.	23 Sep. 01	Pallavaram.
Altken, Alfred, I.	Sub-Condr.	10 June 10	India.
Alderton, William, Ordce., N.	Condr.	11 Jan. 07	India.
Allen, G. B., Bo.	Sub-Condr.	22 Jan. 86	England.
Allen, James, Bo.	Condr.	9 Feb. 86	England.
Alexander, Thomas, I. O. D.	Sub-Condr.	12 Jan. 17	England.
Anderson, Alexander, B.	Condr.	18 Apl. 84	England.
Anderson, Andrew, Bo.	Sub-Condr.	25 Feb. 97	Quetta.
Anderson, Douglas, M.	Sub-Condr.	28 Dec. 01	Madras.
Anderson, George David McEller	Condr.	1 Apl. 13	England.
Andrew, Arthur Miller, I.M.L.	Sub-Condr.	9 Dec. 12	England.
Archbold, John Arthur, B.	Condr.	5 Nov. 95	England.
Armstong, William Adolphus, B.	Sub.-Condr.	12 July 01	India.
Artlett, George John, I.M.L.	Sub.-Condr.	18 Apl. 14	England.
Asbury, George Henry, I. O. D.	Condr.	25 Jan. 18	India.
Ashby, George Henry, M.	Condr.	1 Apl. 95	Vizagapatam.
Astwick, George, Bo.	Sub-Condr.	20 Apl. 00	England.
Atherton, William, Bo.	Condr.	1 Apl. 92	England.
Bagg, George, I. M. L.	Condr.	28 July 14	England.
Baker, Fredrick Ossian Albert, B.	Condr.	8 Mar. 95	England.
Baker, William, N.	Condr.	11 Oct. 13	Out of India.
Baker, William, B.	Sub-Condr.	22 Oct. 94	India.
Baker, Wm., M.	Condr.	1 Dec. 85	Madras.
Bale, Robert, B.	Sub-Condr.	28 Apl. 93	Punjab.
Balfour, James, M.	Sub-Condr.	30 Aug. 88	Madras.
Barraclough, Joseph, M.	Condr.	1 Jan. 76	Colombo.
Barry, Charles, B.	Sub-Condr.	16 Mar. 00	Bengal.
Bartells, Frederick, M.	Condr.	30 Oct. 04	Bangalore.
Basen, Charles, Bo.	Condr.	5 Sep. 04	England.
Bay, James Charles, M.	Condr.	20 Feb. 98	Colombo.
Beardsley, Thomas, Bo.	Sub-Condr.	1 July 10	Quetta.
Beart, Frank, M.	Sub-Condr.	8 Sep. 95	Bangalore.
Beasley, L. A.	Condr.	12 Dec. 09	Persia.
Beatty, John, B.	Condr.	30 Sep. 82	Ireland.
Beaumont, Charles, Bo.	Sub-Condr.	31 Oct. 05	England.
Beck, David.	Sub-Condr.	10 May 06	England.
Beckett, Harry Garside, So.	Condr.	22 Dec. 06	England.
Beevers, William, B.	Sub-Condr.	29 June 83	Bombay.
Bell, John, B.	Condr.	18 Feb. 99	England.
Benford, Charles, B.	Sub-Condr.	1 Aug. 00	England.
Bennett, Edwin John, Ordce. S.	Condr.	5 Oct. 11	England.
Bentall, Arthur Mede, I.	Sub-Condr.	1 Oct. 09	Meerut.
Benton, William, M.	Sub-Condr.	10 Jan. 92	Madras.
Benyon, James, B.	Sub-Condr.	22 Mar. 92	India.
Bettesworth, Charles, B.	Condr.	27 July 01	India.
Bickley, William, N.	Condr.	1 Nov. 11	England.
Bidwell, James, B.	Condr.	17 Jan. 02	England.
Birks, William George, Bo.	Sub-Condr.	11 Oct. 00	India.
Blazey, William, M.	Condr.	14 Sep. 94	Madras.
Bloss, William, Bo.	Condr.	15 Aug. 99	England.
Bodfish, John James	Condr.	28 Nov. 11	England.
Bolton, Charles, M. W.	Condr.	19 Oct. 10	Rurki.
Bonner, Edwin, M.	Condr.	1 Apl. 96	Port Blair.
Booth, Alfred Stephen, So.	Condr.	1 July 10	India.
Booth, James, Ordce., So.	Sub-Condr.	24 May 06	England.
Borrett, Frederick Augustus, Bo.	Condr.	19 Mar. 09	Karachi.
Bowcher, Frank, B.	Sub-Condr.	13 June 02	India.
Bowyer, Charles, Bo.	Condr.	17 Oct. 02	India.
Boyce, Albert, Bo.	Condr.	26 Oct. 91	England.
Bradford, Thomas, B.	Sub-Condr.	12 June 91	England.
Brand, Arthur, B.	Sub-Condr.	23 Aug. 09	England.
Brennan, Martin, I. O. D.	Sub-Condr.	2 Jan. 15	England.
Brett, Henry	Sub-Condr.	21 Nov. 12	England.
Brew, George, Bo.	Condr.	9 July 87	England.
Brining, Andrew Franklin, M. W.	Sub-Condr.	15 Feb. 15	Out of India.
Brown, John Watt	Condr.	1 Nov. 06	England.
Brown, Richard James, B.	Condr.	14 Dec. 97	England.
Browne, Thomas Henry, B.	Sub-Condr.	31 May 98	India.
Bryant George Henry, N.	Condr.	26 Mar. 11	England.
Buchanan, James, Bo.	Condr.	12 Aug. 87	Ireland.
Bull, Bryan, Ordce. N.	Condr.	5 June 12	England.

DEPARTMENTAL WARRANT OFFICERS—*contd.*

Names.	Rank.	When pensioned.	Remarks.
Burgham, Russel Joseph, Bo.	Sub-Condr.	20 Sep. 01	India.
Butler, Robert Meade, M.	Sub-Condr.	1 Aug. 06	Bellary.
Cairncross, Arthur	Sub-Condr.	8 Jan. 10	England.
Calvert, Albert, N.	Condr.	18 Feb. 14	Ferozepore.
Caavert, Chas. Augustus, B.	Condr.	4 Nov. 99	England.
Cllvert, John, B.	Condr.	28 Oct. 01	England.
Calvesbert, Harry Octavius, B.	Condr.	31 July 04	India.
Cameron, Keith Ross, B.	Condr.	30 June 93	India.
Cane, A.	Sub-Condr.	20 Mar. 14	England.
Carew, William, B.	Condr.	4 July 84	Bengal.
Carter, George, B.	Condr.	25 June 98	Bombay.
Carter, James Joseph, B.	Sub-Condr.	5 Dec. 84	England.
Carter, William, M.	Sub-Condr.	21 Dec. 00	India.
Chambers, Joseph, B.	Condr.	9 May 96	England.
Chapman, Edward Henry, I. M. L.	Condr.	18 Sep. 18	England.
Charlton, Joseph, I. O. D.	Condr.	6 May 12	England.
Chase, Edmund Thomas, B.	Condr.	5 Mar. 97	England.
Chatham, James, M.	Condr.	2 Feb. 82	India.
Chisholm, Colin, B.	Sub-Condr.	18 Sep. 13	England.
Christmas, C. E.	Condr.	14 Apl. 09	England.
Church, John, Bo.	Sub-Condr.	3 Feb. 90	England.
Clarke, Charles James, B.	Condr.	4 Aug. 93	India.
Clarke, George Edward	Condr.	17 June 16	
Clarke, Henry, B.	Condr.	20 Feb. 03	England.
Clarke, John, B.	Condr.	13 Jan. 95	India.
Clarke, Peter, Bo.	Condr.	1 July 90	England.
Clarke, Thomas, So	Condr.	5 Sept. 10	England.
Clarke, William, N.	Sub-Condr.	16 Nov. 09	Cossipore.
Clements, David, Bo.	Condr.	1 Apl. 85	England.
Cogan, John, Bo.	Sub-Condr.	21 Nov. 83	India.
Coghlan, Arthur, Bo.	Condr.	27 May 92	England.
Coleman, Sydney James, B.	Condr.	23 Sep. 91	India.
Coleman, William James, B.	Sub-Condr.	1 Oct. 05	England.
Colhoun, William George, Bo.	Condr.	14 Sep. 00	India.
Collard, Samuel, M.	Condr.	23 Apl. 96	New Zealand
Collier, Patrick, B.	Condr.	15 Oct. 86	England.
Combe, James, B.	Condr.	15 Feb. 89	England.
Cook, Richard, B.	Condr.	5 Apl. 89	England.
Cooke, Alfred, B.	Condr.	11 Mar. 98	Punjab.
Cooke, John, B.	Condr.	1 Mar. 89	England.
Coombs, William Francis, Misc.	Condr.	1 May 12	England.
Cooper, John, B.	Condr.	24 Mar. 92	India.
Cooper, Joseph, B.	Sub-Condr.	20 Nov. 89	Bombay Cir.
Cooper, Walter Herbert Aitkin	Condr.	1 Jan. 16	
Cope, Ernest, M.	Condr.	13 Sep. 03	Arni.
Copeman, Nelson, M. W. S.	Condr.	1 June 11	Quetta.
Copley, William Vanster, M.	Condr.	10 Nov. 08	Madras.
Court, Frederick Thomas, S.	Condr.	7 Mar. 12	England
Cox, James, M.	Sub-Condr.	21 Mar. 95	Madras.
Cresswell, John, I. M. L.	Condr	29 Oct. 18	
Cronolly, John, Bo.	Condr.	28 Apl. 76	Poona.
Crowhurst, Donald	Condr.	10 Mar. 13	India.
Crumley, Thomas James	Sub-Condr.	16 Feb. 16	
Crummey, Bernard Smith, B.	Condr.	5 Mar. 94	Punjab.
Cullen, John Blake, B.	Sub-Condr.	10 Sep. 12	Secunderabad
Curtis, Benjamin, So.	Condr.	30 May 13	England.
Curtis, William, I. O. D.	Condr.	5 Aug. 14	England.
Dadds, Henry	Condr.	12 June 16	
Dally, Ernest, So.	Condr.	29 May 13	England.
Daly, Stephen, S. & T.	Condr.	1 Mar. 12	England.
D'Auvergne, Verner, Bo.	Sub-Condr.	19 Aug. 04	Bengal.
Davies, David, Bo.	Sub-Condr.	20 Apl. 98	England.
Davies, William, So.	Condr.	17 Nov. 07	England.
Davis, John Thomas, B.	Sub-Condr.	5 May 99	Punjab.
Davis, Ronald Archibald, I.	Condr.	14 Dec. 09	India.
Davison, William, N.	Condr.	3 Oct. 10	India.
Dawkins, Frederick Charles, M.W.S.	Sub-Condr.	17 May 18	England.
Day, George	Sub-Condr.	6 Oct. 11	England.
Day, James, B.	Sub-Condr.	16 Apl. 99	England.
Deacon, Arthur Henry	Condr.	5 Mar. 13	England.
Deane, Francis, Bo.	Condr.	17 Aug. 87	England.
Deeble, George	Condr.	7 Feb. 15	England.
Dennet, Roger John, Bo.	Condr.	31 Oct. 84	Bombay.
Dennis, John William, So.	Condr.	7 Feb. 07	England.
Devine, William, B.	Condr.	17 Oct. 95	England.
Digby, Robert	Condr.	25 July 08	England.

Names.	Rank.	When pensioned.	Remarks

DEPARTMENTAL WARRANT OFFICERS—*contd.*

Names.	Rank.	When pensioned.	Remarks
Dillon, Thomas Patrick, M.	Sub-Condr.	1 May 06	India.
Dillow, Frederick, Bo.	Condr.	21 June 05	England.
Dimmock, Robert Buckingham, P.	Condr.	1 Aug. 11	England.
Donaghy, James, Bo.	Condr.	10 Mar. 87	England.
Doolan, Williams, Bo.	Condr.	11 Mar. 05	England.
Dorsey, Arthur, M.	Condr.	6 Feb. 03	Poonamali.
Dowley, George, Bo	Sub-Condr.	1 Mar. 89	England.
Doyle, John	Condr.	12 Mar. 11	India.
Drapar, Thomas, Bo.	Condr.	20 Mar. 72	Bombay.
Duley, William, M.	Condr.	1 Apl. 96	England.
Dunbar, William, M.	Condr.	26 Apl. 12	Madras.
Dungey, Robert, Bo.	Condr.	28 Feb. 97	England.
Dunk, Fred. Stephen, M.	Condr.	11 Feb. 96	London.
Dwyer, Thomas, Misc.	Condr.	1 Jan. 08	India.
Eacott, John Henry West, I.	Condr.	27 Mar. 11	Fyzabad.
Eaglesome, George, Bo.	Schoolmaster, 1st class	8 Apl. 91	England.
Echlin, Frank, M.	Sub-Condr.	1 Sep. 12	England.
Edmunds, Thomas, I. M. List.	Sub-Condr.	2 June 13	
Edwards, Frederick, M.	Condr.	19 Oct. 88	Bangalore
Edwards, Moses, Bo.	Condr.	3 Feb. 00	India.
Edwards, Robert Richard, B.	Sub-Condr.	18 Mar. 08	England.
Eganore, Frederick John, Sec.	Sub-Condr.	16 Nov. 08	Maymyo.
Elliott, Charles M.	Condr.	1 Feb. 05	England.
Ellis, Albert Samuel William, So.	Condr.	18 Oct. 10	India.
Ellison, Frank William, I.	Sub-Condr.	8 Nov. 16	
Emery, William Lee, M.	Sub-Condr.	16 Oct. 05	Bombay.
Etches, Ernest Robert	Condr.	14 Oct. 16	
Evans, Pryce, Bo.	Condr.	18 Mar. 04	India.
Facey, Ernest George, Bo.	Sub-Condr.	3 Jan. 98	England.
Fahie, Thomas, M.	Sub-Condr.	25 Oct. 86	England.
Ferguson, Henry, B.	Condr.	16 Aug. 05	Bengal.
Ferris, Frederic, B.	Sub-Condr.	5 Mar. 03	India.
Ferris, Henry, N.	Condr.	4 Mar. 10	England.
Finniston, William John, B.	Sub-Condr.	19 Jan. 00	India.
Fisher, Thomas, B.	Sub-Condr.	15 Dec. 92	England.
Fletcher, Henry, Bo.	Condr.	26 May 01	India.
Flood, James, So.	Condr.	26 Nov. 13	Out of India.
Ford, Frederick, Bo.	Condr.	9 Oct. 96	India.
Forde, H., B.	Condr.	18 Aug. 82	India.
Foster, Henry, Bo.	Sub-Condr.	11 Sep. 06	India.
Fox, James, Bo.	Condr.	22 Oct. 84	England.
Francke, William Henry, M.	Sub-Condr.	17 Aug. 97	Rangoon.
Franks, William, M.	Sub-Condr.	1 Aug. 86	Madras.
Freeman, George, M.	Sub-Condr.	15 Aug. 65	England.
Freemantle, James Swan, M.	Condr.	7 Jan. 05	Rajhamund.
Froggat, John, Bo.	Condr.	19 Jan. 98	England.
Fryer, Frederick William, N.	Condr.	25 Feb. 08	Fort William.
Galloway, George Jas. E.	Condr.	1 Nov. 04	India.
Gardiner, George, Bo.	Condr.	10 Feb. 10	Hyderabad, Sind.
Gargrave, J., Bo.	Sub-Condr.	1 Feb. 79	England.
Garland, Joseph, B.	Condr.	25 Feb. 08	England.
Garner, John, Bo.	Condr.	29 Oct. 93	England.
Gately, Patrick, B.	Sub-Condr.	16 Sep. 82	Bengal.
Gellard, William, B.	Condr.	28 Aug. 99	Bengal.
George, Joseph, B.	Condr.	15 June 07	India.
Gerard, Mervyn, M.	Condr.	31 Aug. 84	St. Thomas' Mount.
Gibson, Charles, B.	Condr.	28 Apl. 07	India.
Gibson, John	Condr.	14 Feb. 09	England.
Giddings, Thomas, B.	Condr.	12 May 69	Europe.
Girling, George, B.	Condr.	16 Sep. 04	Bengal.
Goodburn, Frederick	Condr.	26 May 17	
Gordon, Edwin Boothe, B.	Sub-Condr.	15 Feb. 01	India.
Gorman, John, I.	Sub-Condr.	8 June 10	India.
Graham, Noah, M.	Sub-Condr.	21 Oct. 93	England.
Graham, Robert, S. & T. Corps	Sub-Condr.	30 Sep. 17	
Grant, Thomas, Bo.	Condr.	24 Mar. 97	England.
Graves, Thomas	Condr.	1 Mar. 16	
Green, Edward, B.	Sub-Condr.	17 Aug. 00	Bengal.
Gregory, John, B.	Condr.	25 April. 02	India.
Griffiths, John, B.	Condr.	15 Jan. 91	England.
Griffiths, Richards James Hasell	Condr.	1 Aug. 18	Bengal.
Griffiths, William, N.	Sub-Condr.	16 Aug. 09	Rawal Pindi.
Grimshaw, John, B.	Sub-Condr.	26 Aug. 18	Europe.
Haigh, Godfrey, M.	Condr.	12 Jan. 00	St. Thomas' Mount

DEPARTMENTAL WARRANT OFFICERS—*contd.*

Names.	Rank.	When pensioned.	Remarks.
Hale, James William, N.	Condr.	23 June 06	England.
Hall, Charles, B.	Sub-Condr.	10 Oct. 96	England.
Hall, James, N.	Condr.	24 Apl. 12	England.
Hall, James B.	Condr.	15 Nov. 07	England.
Hallas, Frederick, M.	Sub-Condr.	1 Dec. 13	Out of India.
Hammond, William, So.	Condr.	4 Jan. 11	England.
Handy, Alexander, So.	Sub-Condr.	6 Feb. 08	England.
Handy, Cebert, So.	Sub-Condr.	24 Jan. 10	England.
Hanron, Michael, B.	Sub-Condr	14 Nov. 90	New Zealand.
Harding, Benjamin Thomas, Bo.	Condr.	10 Oct. 13	Quetta.
Harding, George Harry, N.	Condr	29 Mar. 10	India.
Hardy, John, W. B.	Condr.	15 Dec. 82	Bangalore.
Harker, John James, Bom.	Sub.-Condr.	9 Feb. 17	
Harman, William, M.	Sub-Condr.	1 Mar. 79	Bombay.
Harrington, John Joseph, N.	Sub-Condr.	11 Mar. 06	England.
Harris, Henry, M.	Condr.	15 Nov. 04	Madras.
Harris, Jesse, N.	Condr.	11 Mar. 11	England.
Hart, Richard, B.	Condr.	19 Jan. 00	England.
Harwood, Albert, P. W. D.	Condr.	1 Mar. 08	England.
Hattersley, John, M.	Sub-Condr.	29 May 97	Bowringpet.
Hay, Peter, B.	Condr.	3 Nov. 00	India.
Haycock, Henry Beckford, B.	Sub-Condr.	1 June 02	India.
Hayes, Michael, So.	Condr.	19 Oct. 09	England.
Hayes, William, P.-B.	Sub-Condr.	25 Mar. 10	England.
Hayes, William John, M. W. S.	Condr.	26 June 14	India.
Hayler, Stephen	Condr.	10 Dec. 06	England.
Head, James, Bo.	Condr.	1 Mar. 92	England.
Healy, Richard Thomas, M.	Condr.	17 Mar. 06	Rangoon.
Heap, William, B.	Sub-Condr.	1 Sep. 08	St. Thomas' Mount.
Heapy, Henry, B.	Sub-Condr.	6 Apl. 12	England.
Heaton, Robert, So.	Condr.	7 Jan. 11	England.
Hedges, Ernest William, Bo.	Sub-Condr.	1 Dec. 04	England.
Hein, Henry Gerard, B.	Sub-Condr.	6 July 83	Punjab.
Hewitt, William Thomas, N.	Condr.	10 Sep. 07	England.
Hewson, James, B.	Condr.	16 Feb. 00	Bengal.
Hickey, John, N.	Condr.	1 Feb. 14	Kirkee.
Hierons George, I.	Sub-Condr.	16 Sep. 09	England.
Higgins, Joseph, B.	Condr.	22 Dec. 01	England.
Hiley, Samuel, N.	Sub-Condr.	3 Sep. 09	England.
Hill, Thomas	Sub-Condr.	5 July 12	India.
Hilton, John, So.	Condr.	28 Jan. 11	England.
Hoare, George William, B.	Condr.	14 Aug. 03	India.
Hobbs, Edward	Sub-Condr.	20 Feb. 13	England.
Hockney, Thomas William, Ordn. Dept.	Condr.	1 July 17	
Hodge, John, M.	Condr.	6 Jan. 14	Secunderabad
Hodgson, Henry.	Condr.	15 Oct. 15	India.
Holborn, John, Bo.	Condr.	21 Dec. 00	England.
Holmes, Alfred, Bo.	Condr.	1 May 91	England.
Holmes, James, B.	Condr.	14 Aug. 03	Bombay.
Horsley, Lee, B.	Sub-Condr.	2 Nov. 00	Bengal.
Howe, Edward James, I. O. D.	Condr.	21 June 12	England.
Howes, William, B.	Condr.	18 Apl. 84	England.
Hoy, John, Bo.	Sub-Condr.	11 Sep. 09	Mhow.
Hunter, James Robert, Bo.	Condr.	25 Mar. 06	India.
Hunter, Thomas John	Condr.	1 Feb. 15	India.
Hyde, William, M.	Sub-Condr.	1 June 83	Colombo.
Ingledew, George Henry, N.	Condr.	8 May 10	England.
Inglis, William, N.	Sub-Condr.	1 Nov. 06	India.
Jackson, Edwin Charles, B.	Sub-Condr.	17 June 04	Punjab.
Jackson, Thomas, Bo	Condr.	2 Mar. 03	England.
Jacobs, Maurice, B.	Condr.	15 Sep. 08	England.
James, Henry, Bo.	Sub-Condr.	19 Jan. 00	Bengal.
Jameson, Charles James, B.	Condr.	9 Sep. 03	England.
Jessiman, Robert, B.	Condr.	8 Oct. 04	India.
Johnson, William, B.	Condr.	16 Apl. 80	Punjab.
Johnston, Adrian, B.	Sub-Condr.	17 Nov. 99	India.
Johnston, James, B.	Condr.	15 May 63	New Zealand.
Johnston, Thomas	Condr.	21 Apl. 93	Nova Scotia.
Johnstone, Eustace Martin, B.	Sub-Condr.	7 May 97	Punjab.
Johnstone, John, B.	Condr.	2 Nov. 00	Bengal.
Jolly William, B.	Condr.	7 June 90	England.
Jones, James, B.	Sub-Condr.	21 Apl. 94	India.
Jones, Philip, B.	Condr.	11 Oct 01	India.
Joseph, Patrick, B.	Sub-Condr.	9 Apl. 80	Punjab.
Joyce, James Samuel, S.	Condr.	11 Dec. 01	Coconada.
Judge, Edwin Justice, M.	Sub-Condr.	1 Sep. 11	India.

Names.	Rank.	When pensioned.	Remarks.

DEPARTMENTAL WARRANT OFFICERS—*contd.*

Names.	Rank.	When pensioned.	Remarks.
Kalkhoven, Frederick, Bo.	Condr.	22 May 09	Quetta.
Kay, George, B.	Sub-Condr.	20 Apl. 88	New Zealand.
Kay, William James, N.	Condr.	1 Mar. 10	England.
Kean, Thomas, B.	Sub-Condr.	10 May 95	Punjab.
Keene, Frank Robert	Sub-Condr.	27 July 16	
Kelly, James McNamara	Condr.	6 June 17	England.
Kelly, John, M.	Condr.	16 Oct. 73	Bangalore.
Kelly, John Morris, Ord. S.	Condr.	25 June 11	England.
Kelly, Joseph, M.	Condr.	1 Aug. 97	England.
Kelly, Joseph, B.	Condr.	22 Jan. 04	India.
Kelly, William Patrick Courtenay	Sub-Condr.	....	
Kennedy, George, M.	Condr.	20 Nov. 91	Bangalore.
Kenny, James	Condr.	8 Feb. 13	Lucknow.
Kent, George, M.	Condr.	4 Aug. 02	St. Thomas' Mount.
Keogh William Ernscliffe, B.	Condr.	7 Apl. 02	Bombay.
Kernon George Edward, N.	Sub-Condr.	2 Oct. 07	Bombay.
Kerwick, Frederick James, M. W. S.	Condr.	1 Sep. 14	India.
Key, Eugene, So.	Condr.	27 May 13	New Zealand.
Kerulf, Egbert Alexis, So.	Condr.	10 Apl. 07	India.
King, Leonard Burrows, So.	Condr.	10 Mar. 14	England.
King, William Henry	Sub-Condr.	6 Sep. 11	England.
Kirkpatrich, Edwin, B.	Sub-Condr.	11 May 88	England.
Knapp, Edward George, Misc.	Condr.	1 Apl. 12	England.
Knight, William, N.	Condr.	31 Jan. 10	Allahabad.
Knox, Andrews, B.	Condr.	12 Apl. 05	England.
Lacey Charles, Bo.	Condr.	30 Mar. 87	England.
Laing, James, N.	Condr.	26 Mar. 09	Fort William.
Lalor, James, N.	Condr.	10 Jan. 11	England.
Langford, Edwin Clarence, B.	Condr.	1 July 03	England.
Leaff, Geoffrey, So.	Condr.	9 Oct. 08	England.
Lee, Edmund Hallock, M.	Condr.	17 May 98	Mandalay.
Leedam, C. W., Bo.	Condr.	23 Jan. 14	Out of India.
Lenon, James, Bo.	Sub-Condr.	13 Aug. 91	Neemuch.
Levason, William, N.	Condr.	15 June 10	England.
Lewis, Charles B.	Sub-Condr.	17 Jan. 09	England.
Lewis, Richard, N.	Condr.	10 Mar. 14	England.
Leyden, Philip Lavalle, B.	Sub-Condr.	11 Nov. 87	Bengal.
Lendsay, Eustace Herbert Chilvas	Condr.	14 Aug. 17	India.
Lilburn, Robert	Sub-Condr.	8 Nov. 09	England.
Loane, Hope, B.	Condr.	15 Sep. 08	Ambala.
Longmate, James	Condr.	20 Mar. 16	
Lucas, Herbert	Condr.	21 Nov. 12	Simla.
Lucking, Alfred, B.	Condr.	8 Jan. 10	England.
Lungley, George	Condr.	11 Oct. 11	England.
Lynch, Arthur Barrington, B.	Condr.	1 Oct. 07	Bangalore.
Lynch, Michael Thomas, M.	Condr.	19 Mar. 04	Pallavaram.
Lynch, Thomas	Condr.	26 Mar. 15	Out of India.
McAleese, William John, M.	Sub-Condr.	1 Aug. 83	India.
McCann, Charles B., B.	Condr.	14 Dec. 88	Bangalore.
McClung, Francis David, N.	Condr.	15 May 12	England.
McCollum, Robert, M.	Condr.	6 Dec. 96	Coimbatore.
McConachie, Henry, B.	Condr.	25 Nov. 03	India.
McConnachie, Joseph, Bo.	Condr.	1 May 95	England.
McCormack, John, B.	Sub-Condr.	25 Feb. 99	Bengal.
McCulloch, Ernest, M. W. S.	Condr.	1 May 12	England.
McDonald, George, M.	Condr.	10 Oct. 82	Bangalore.
McDowell, William, So.	Condr.	4 Nov. 08	England.
McHale, John Edward, Bo.	Condr.	19 July 10	Karachi.
McIntyre, Peter (Misc.)	Condr.	28 Jan. 18	India.
McKay, George William, N.	Condr.	2 Nov. 08	England.
McKie, Joseph Wason, B.	Sub-Condr.	2 Mar. 95	Canada.
McLeod, Hector, B.	Condr.	9 Dec. 95	England.
McStocker, Samuel Robert	Condr.	7 Aug. 11	England.
MacAlister, Walter Henry, M.	Sub-Condr.	27 Sept. 13	India.
Mack, William, B.	Sub-Condr.	15 Oct. 01	England.
Mackintosh, Duncan, M.	Sub-Condr.	24 Dec. 92	Arcot.
Mackintosh, Simon, N.	Sub-Condr.	22 Jan. 10	England.
Madely, Walter, M.	Sub-Condr.	19 Jan. 91	England.
Maggs, Charles, Bo.	Sub-Condr.	18 Aug. 68	Abyssinia.
Maher, Michael, B.	Condr.	10 Feb. 10	Lucknow.
Mandy, Henry Althans, N.	Sub-Condr.	15 Dec. 04	India.
Mansfield, Edgar Hall, B.	Sub-Condr.	25 Nov. 05	Bombay.
Markham, William, M.	Sub-Condr.	1 Oct. 05	Bangalore.
Marlow, George	Condr.	31 Oct. 16	
Marshall, Francis James, N.	Condr.	22 Jan. 11	England.
Martingale, Jas. John William	Condr.	2 Sep. 05	England.
Mason, Alfred, Bo.	Sub-Condr.	24 Oct. 01	India.

DEPARTMENTAL WARRANT OFFICERS—*contd.*

Names.	Rank.	When pensioned.	Remarks.
Mason, Henry George, N.	Sub-Condr.	10 Sept. 10	England.
Massey, Thomas, B.	Sub-Condr.	5 Jan. 13	India.
Massingham, Robert Martin, N.	Condr.	11 Dec. 11	England.
Mathews, William, B.	Condr.	24 Oct. 79	Bengal.
May, William, Ord. So.	Condr.	19 Mar. 15	...
Meriton, G. H.	Condr.	10 Jan. 11	England.
Merrell, Frederick, N.	Condr.	16 Sep. 13	England.
Mills, George Taylor	Condr.	7 Nov. 14	England.
Mithan, Thomas, B.	Condr.	10 Dec. 93	Punjab.
Montgomery, Richard John, M.	Condr.	1 Sep. 04	England.
Moore, James, B.	Sub-Condr.	8 Mar. 05	England.
Moore, Walter James	Condr.	18 July 11	England.
Moore, William James	Sub-Condr.	11 Apl. 06	England.
Morris, Frederick James, I.	Sub-Condr.	13 Sep. 09	England.
Morrison, Hugh, S.	Sub-Condr.	1 Dec. 07	Coimbato
Mortimer, John.	Condr.	13 May 08	England.
Myatt, Leonard George, I. M. List.	Condr.	6 Oct. 13	England.
Naughton, Thomas, B.	Condr.	19 Sep. 02	Rangoon
Neild, George, M.	Sub-Condr.	1 June 09	England.
Newbold, Gustavus Sheridan, B.	Condr.	1 Nov. 94	England.
Newby, Ernest, Bo.	Condr.	17 Nov. 95	Bombay.
Newey, Colin	Condr.	15 Mar. 05	England.
Nicholl, William	Condr.	17 Jan. 07	England.
Nicholson, William Robert, Bo.	Condr.	1 Oct. 97	India.
Noble, James Fritz, M.	Condr.	14 June 01	Malabar.
Norris, Richard, Bo.	Condr.	29 Mar. 92	England.
Norris, Robert A., M.	Condr.	15 Feb. 82	Bangalore.
Norton, Arthur Adrian, B.	Sub-Condr.	22 Feb. 12	England.
O'Brien, David, Bo.	Sub-Condr.	13 Sep. 73	Mhow.
O'Brien, James, B.	Condr.	25 Feb. 05	India.
O'Brien, John, B.	Sub-Condr.	5 Dec. 70	Punjab.
O'Farrell, Eugene.	Condr.	28 Feb. 18	
O Hare, J., M.	Condr.	10 Mar. 88	Bangalore.
O'Neill, Patrick Joseph, M.	Condr.	1 July 00	St. Thomas Mount.
Packer, James, B.	Condr.	30 June 91	England.
Painter, Frederick George, I.	Sub-Condr.	30 Oct. 09	England.
Pannell, William, B.	Condr.	14 Dec. 96	England.
Papworth, Richard, B.	Condr.	16 Sep. 04	Bengal.
Parkin, George, So.	Condr.	6 Jan. 08	England.
Parkinson, Job, So.	Condr.	6 Jan. 07	India.
Parks, Frederick William	Sub-Condr.	1 Mar. 07	Allahabad.
Pavia, Arnold, Ordce., S.	Condr.	8 June 11	India.
Pearson, William, M.	Sub-Condr.	1 Dec. 84	Madras.
Pearston, Andrew, So.	Condr.	4 Apl. 14	England.
Pembroke, Charles Herbert Slock, B.	Condr.	1 June 13	Bareilly.
Pennington, George, N.	Condr.	26 Aug. 13	England.
Penson, Henry James, B.	Sub-Condr.	20 Jan. 01	England.
Penticoss, Stephen, M.	Sub-Condr.	9 Oct. 90	Trichinopoly.
Perkins, William, M.	Condr.	24 Jan. 05	Pallavaram.
Perry, Charles, B.	Condr.	5 Mar. 97	England.
Perry, John, N.	Condr.	21 June 11	India.
Perry, Randolph Edward Joseph, Bo.	Sub-Condr.	4 Nov. 98	India.
Peterson, John	Sub-Condr.	31 Oct. 11	England.
Phelps, William Edmondson, B.	Sub-Condr.	17 Jan. 84	Calalcutt
Phillips, Joseph, M.	Condr.	1 Aug. 95	England.
Pike, Henry, B.	Condr.	20 Mar. 85	England.
Pilcher, Victor George, I.	Sub-Condr.	15 May 09	England.
Pinkney, William, So.	Sub-Condr.	24 Aug. 08	England.
Pinkstone, George, B.	Condr.	26 Feb. 82	England.
Pinson, Charles	Sub-Condr.	5 Jan. 13	Lahore.
Pitchford, Samuel, M.	Condr.	15 Oct. 09	England.
Pitts, Henry, So.	Condr.	22 Jan. 10	England.
Plumb, Harry, S. & T. Corps.	Sub-Condr.	18 Dec. 17	Out of India.
Pope, Alfred	Sub-Condr.	1 Feb. 13	India.
Preist, Henry, B.	Condr.	5 Mar. 97	England.
Preston, William Henry	Condr.	15 Nov. 17	
Quin, H., M.	Condr.	19 Jan. 90	Europe.
Quinn, John, S.	Condr.	6 Aug. 11	England.
Quinn, Roland Martin, S. & T. C.	Condr.	1 July 12	England.
Raisin, Wm. Henry	Sub-Condr.	31 Dec. 05	England.

DEPARTMENTAL WARRANT OFFICERS—*contd.*

Names.	Rank.	When pensioned.	Remarks.
Ralph, James Bernard, B.	Condr.	17 Nov. 97	India.
Ramsbottam, Joseph Ormerod, B.	Condr.	16 Feb. 00	Punjab.
Ratcliffe, Joseph, Bo.	Sub-Condr.	12 Aug. 88	England.
Raymond, Richard, M.	Sub-Condr.	9 Aug. 94	Secunderabad.
Reeks, Ernest Joseph, N.	Condr.	14 Sep. 13	England.
Rees, William M., Bo.	Condr.	27 June 09	England.
Reynolds, George Albert, B.	Condr.	24 July 85	Melbourne.
Richards, William Henry, So.	Condr.	8 Mar. 14	England.
Richardson, Isaac Wilson	Sub-Condr.	1 Mar. 12	England.
Riddle, James, B.	Condr.	8 Mar. 89	England.
Riley, James, N.	Condr.	25 Nov. 04	India.
Riordan, Denis, M.	Condr.	23 Sep. 70	Europe.
Roberts, Harry William, B.	Sub-Condr.	29 May 03	India.
Robertson, Arthur, Sec.	Condr.	11 Oct. 09	England.
Robertson, Frank William, B.	Condr.	24 Oct. 03	Ferozepore.
Robertson, George Davidson	Condr.	22 Sep. 09	India.
Robinson, George, Bo.	Condr.	9 Jan. 93	England.
Robinson, John, B.	Condr.	3 Oct. 79	Ferozepore.
Rogers, Albert Edward, So.	Condr.	10 Mar. 14	Out of India.
Rogers, James, Ord. N.C.	Condr.	7 Jan. 11	England.
Ross, Arthur Thomas, B.	Condr.	16 Sep. 04	Bengal.
Rossiter, Robert James, B.	Condr.	12 Feb. 04	India.
Rowland, Michael Henry	Sub-Condr.	15 Apl. 13	England.
Russell, Charles	Condr.	29 Mar. 11	Karachi.
Saunders, Edwin Robert, M.	Condr.	10 Apl. 07	England.
Scott, Arthur William	Condr.	29 Oct. 11	India.
Searle, Archibald Theodore	Condr.	16 Jan. 15	...
Sears, Harry, M.	Condr.	23 Nov. 95	England.
Sheath, Frank A.	Sub-Condr.	1 July 08	England.
Sheppard, William	Sub-Condr.	1 Apl. 14	Out of India.
Sheret Robert, I.	Condr.	1 Mar. 10	England.
Shubart, David Henry James, N.	Condr.	3 July 13	Allahabad.
Simons, Arthur, Ord. N.C.	Sub-Condr.	7 Dec. 10	England.
Sinclair, William, B.	Condr.	16 Nov. 03	India.
Sinnott, Martin, So.	Condr.	12 Oct. 12	England.
Slade, Fredrick William, I.	Sub-Condr.	25 Jan. 11	England.
Smith, Arthur Henry, Bo.	Condr.	27 Aug. 07	India.
Smith, Arthur William	Condr.	5 Mar. 10	England.
Smith, Cubitt John	Sub-Condr.	1 Feb. 04	India.
Smith, Joseph, M.	Sub-Condr.	16 Aug. 85	India.
Smith, John Richard, M.	Condr.	1 Oct. 04	India.
Smith, Noah, M.	Sub-Condr.	21 Oct. 93	England
Smith, Walter Albert, I. M. L.	Condr.	14 Aug. 15	England.
Smith, William, B.	Sub-Condr.	30 Dec. 13	Meerut.
Soper, Emanuel G. B.	Condr.	11 Nov. 90	England.
Southall, Henry, M.	Condr.	1 Oct. 04	England.
Southam, Edward, Bo.	Condr.	13 Apl. 06	England.
Sowman, James	Condr.	6 Oct. 11	England.
Spark, Oliver, B.	Condr.	19 July 01	Bengal.
Spooner, Thomas, B.	Condr.	15 Feb. 00	India.
Stafford, Frederick Sydney, Bo.	Condr.	2 Mar. 06	India.
Stapleton, John, S. & T. C.	Sub-Condr.	1 Mar. 13	England.
States, Charles, Bo.	Sub-Condr.	1 Dec. 13	Poona.
Steele, William, B.	Condr.	5 Dec. 95	England.
Stephens, Ernest Barrow, So.	Condr.	15 Oct. 07	England.
Stephens, Francis, B.	Condr.	16 Jan. 14	Meerut.
Stevens, George P.	Sub-Condr.	6 May 04	India.
Stewart, Edward, B.	Condr.	8 Dec. 99	Tasmania.
Stone, Robert, So.	Condr.	13 Feb. 09	India.
Suggate, Alfred, B.	Condr.	23 Jan. 94	India.
Sullivan, Daniel	Sub-Condr.	6 Mar. 14	Madras.
Symonds, William, N.	Condr.	19 June 30	India.
Tannyane, Thomas	Condr.	25 May 14	Kirkee.
Tavner, Ernest John	Sub-Condr.	24 Oct. 14	India.
Tayler, George William	Condr.	20 July 14	India.
Taylor, Bertram Charles, B.	Condr.	24 July 03	India.
Taylor, William James, B.	Condr.	1 Feb. 14	Out of India.
Thatcher, Sidney Winter	Condr.	27 Dec. 05	England.
Thompson, Gilbert Lionel, B	Sub-Condr.	31 Aug. 04	India.
Thorburn, John, I.	Condr.	1 Apl. 11	Out of India.
Thornton, Roland Everad	Sub-Condr.	6 May 16	England.
Tierney, Thomas, So.	Sub-Condr.	1 May 07	Ireland.
Tinkler, Alfred Charles	Condr.	18 Nov. 17	Out of India.

DEPARTMENTAL WARRANT OFFICERS—*concld.*

Names.	Rank.	When pensioned.	Remarks.
Trutwein, Isaac, M.	Sub-Condr.	17 Mar. 94	Madras.
Tucker, Henry, M.	Condr.	7 Nov. 96	England.
Turner, Henry, Bo.	Condr.	3 May 97	England.
Turner, Themhs, M.	Condr.	1 Apl. 03	England.
Twyman, George Edward Walter	Sub-Condr	12 Mar. 09	England.
Tyler, Mark William, Bo.	Condr.	15 Feb. 09	Quetta.
Urquhart, Charles	Sub-Condr.	20 Dec. 15	
Vahey, George Frederick, B.	Sub-Condr.	13 Feb. 11	England.
Vaughan, Richard, M.	Condr.	27 Sep. 88	St. Thomas' Mount.
Vicars, William	Condr.	17 Mar. 16	England.
Vines, Alfred, I.	Sub-Condr.	18 July 10	Quetta.
Voller, William, So.	Sub-Condr.	4 June 13	Bombay.
Wake, Ernest, B.	Sub-Condr.	8 Feb. 14	England.
Wakeman, Alfred, B.	Condr.	Dec. 93	England.
Walcroft, William, N.	Condr.	16 Nov. 08	England.
Walker, Alexander, M.	Condr.	13 Oct. 98	England.
Walker, Richard Buckley, So.	Condr.	12 Jan. 10	England.
Walker, William, Bo.	Condr.	25 Mar. 86	England.
Walmsley, Edwin, B.	Condr.	14 Dec. 00	Ferozepore.
Walters, William, M.	Sub-Condr.	1 July 12	Bangalore.
Ward, Russell William, M.	Condr.	12 Feb. 94	Vellore.
Warner, Charles, B.	Sub-Condr.	15 June 96	Bengal.
Warner, William, I. O. D.	Sub-Condr.	1 June 12	England.
Watson, John Thomas, Bo.	Condr.	15 Aug. 94	Ahmednagar.
Waugh, James Nichol, B.	Sub-Condr.	10 June 13	England.
Weaver, Robert, M.	Sub-Condr.	1 July 88	Madras.
Webster, James William, So.	Condr.	23 Jan. 11	England.
Weir, Geo. William, B.	Condr.	19 June 03	England.
Weldon, William, B.	Sub-Condr.	24 Mar. 99	Fort William.
Weller, G. H., P. W. D., India	Condr.	20 Nov. 11	India.
Wells, Sidney Herbert	Condr.	8 Mar. 15	Out of India.
Welsh, Robert T., N.	Condr.	1 Mar. 06	England.
Weston, Herbert, Laidlay	Condr.	2 Feb. 13	England.
Wheeler, William Frederick, N.	Sub-Condr	21 Oct. 12	England.
White, Charles Bernard, M.	Sub-Condr.	3 June 06	Poonamallee
Whiten, Adolphus, Bo.	Condr.	16 Sep. 96	England.
Whitley, William, Bur.	Condr.	27 Aug. 04	England.
Whitmore, Richard, B.	Condr.	14 Mar. 90	England.
Wickham, William, Bo.	Sub-Condr. (*Hony. Capt.*)	28 Feb. 90	England.
Wigley, Charles, Bo.	Sub-Condr.	28 Sep. 96	England.
Wigmore, Harry N.	Sub-Condr.	7 May 13	Out of India.
Wilden, John	Sub.-Condr.	11 June 06	India.
Williams, Alfred, So.	Condr.	25 Apl. 09	England.
Williams, George	Condr.	9 Feb. 13	England.
Williams, Henry, B.	Condr.	14 Feb. 02	India.
Williams, John, So.	Condr	20 Oct. 05	England.
Williams, Joseph, M.	Condr.	20 Mar. 75	Europe.
Williams, John, P.-B.	Condr.	5 May 10	England.
Williams, Joseph Henry, N.	Condr.	22 Mar. 09	India.
Williams, Walter, N.	Sub-Condr.	26 Sep. 10	England.
Willis, James Archibald, B.	Condr.	11 Mar. 11	England.
Willous, William, M.	Condr.	30 Aug. 04	England.
Winter, James	Condr.	23 Mar. 13	England.
Witt, Charles, M.	Condr.	9 Feb. 07	Poonamallee.
Woodard, Henry, M.	Condr.	1 Feb. 98	England.
Woodbridge, Charles Frederick	Condr.	22 July 15	...
Woodhouse, Joseph, M	Condr.	1 Aug. 76	Bangalore.
Woodthorpe, William James, So.	Condr.	27 Dec. 08	England.
Worcester, Edward	Condr.	15 Oct. 14	India.
Wright, Walter John, B.	Condr.	1 Mar. 08	England.
Youles, Arthur Thomas, B.	Condr.	12 Feb. 04	India.

MEDICAL WARRANT OFFICERS.

Note.—The names on this list are not indexed as they are arranged alphabetically.

Names.	Rank.	When pensioned.	Remarks.
Barker, Frederick William, Bo.	Asst. Surgn., 1st cl.	26 Sep. 02	India.
Baron, Thomas, B.	Apothy., 1st gr.	25 Jan. 93	Bengal.
Bickett, Richard, B.	Asst. Surg., 2nd cl.	12 Aug. 96	Bengal.
Blake, Francis, M.	Asst. Surgn., 1st cl.	30 Jan. 99	Bowringpet.
Bradshaw, Charles E., B.	Apothy.	13 Aug. 80	Bengal.
Brown, Francis, M.	Apothy., 2nd gr.	16 Feb. 81	Madras.
Butler, Richard, M.	Asst. Surgn., 2nd cl.	4 Mar. 95	Bellary.
Campbell, Frank, M.	Asst. Surgn., 1st cl.	19 July 97	India.
Carvalho, Salvadore, Bo.	Asst. Surgn., 1st cl.	25 Jan. 01	India.
Clayton, Henry, B.	Apothy., 1st gr.	9 Sep. 92	Bengal.
Coshan, John Anthony Joseph, M.	Asst. Surgn. 1st cl.	29 Jan. 00	Madras.
Crabbe, Joseph, B.	Asst. Surgn., 1st cl.	14 Aug. 03	India.
Curry, Michael Arthur, Bo.	Asst. Surgn., 3rd cl.	20 May 01	England.
Davis, Jas. Henry Edwin	Asst. Surgn., 4th cl.	1 July 07	England.
Debnam, Albert, M.	Asst. Surgn., 1st cl.	15 June 94	Bangalore.
DesBrosses, Edwin, B.	Apothy.	19 Apl. 89	Bengal.
Dias, P. A., Bo.	Asst. Apothy.	26 July 85	Bombay.
D'Miranda, M. F., Bo.	Asst. Apothy.	16 June 85	Bombay.
D'Souza, John William Napoleon, Bo.	Asst. Surgn., 1st cl.	20 Apl. 00	India.
D'Sylva, Francis, M.	Asst. Apothy.	1 Sep. 73	Madras.
Duffy, Dillon, B.	Apothy.	19 May 86	England.
Dullard, Mathew, B.	Apothy.	10 July 91	Bengal.
Dwyer, John, B.	Sr. Apothy.	19 Jan. 75	
Eliot, James Alexander, B.	Apothy.	20 Jan. 88	Punjab.
Falconer, Price James, M.	Apothy., 1st gr.	26 Feb. 93	Madras.
Fernandes, M. F., Bo.	Apothy.	6 Apl. 88	Bandora.
Fonseca, George Basil, M.	Asst. Surgn., 1st cl.	21 Dec. 95	Madras.
Francis, Dominic	Asst. Surgn., 1st cl.	20 Nov. 05	India.
Freeland, Robert, B.	Sr. Apothy.	30 Jan. 82	Ranikhet.
Fullam, William Arnold, B.	Asst. Surgn., 1st cl.	13 July 00	Bengal.
Gates, Alexander, B.	Asst. Surgn., 1st cl.	2 Apl. 98	Madras.
Gill, George, M.	Asst. Apothy.	20 Mar. 90	Belgaum.
Gooroosawmy, Pillay T., M.	Asst. Surgn., 1st cl.	27 Oct. 95	Madras.
Govindsingh, T. H., M.	Asst. Surgn., 1st cl.	25 Aug. 95	Coimbatore.
Green, Cecil Beaumont	Asst. Surgn., 3rd cl.	15 May 05	Madras.
Griffiths, Dudley James Hayfield	Asst. Surgn., 4th cl.	26 Apl. 07	India.
Grose, Ebenezer Donald, B.	Apothy.	20 June 91	Bengal.
Haiaes, George Richard	Asst. Surgn., 1st. cl.	1 July 09	India.
Hall, John Robert Charles, B.	Apothy.	8 Aug. 79	Bengal.
Haller, William Henry, M.	Asst. Apothy.	19 Aug. 87	Madras.
Hart, Reginald Henry Willick, B.	Asst. Surgn., 1st cl.	1 Nov. 13	Out of India.
Hart, Samuel Stuckey, B.	Apothy.	21 May 80	Bengal.
Hesterlow, Henry, M.	Asst. Surgn., 1st cl.	2 Feb. 98	Madras.
Hewins, John Edwin	Asst. Surgn., 2nd cl.	1 Dec. 05	Poona.
Jackson, Herbert Charles, M.	Asst. Surgn., 3rd cl.	1 Aug. 11	Madras.
Jenkins, George Robert, B.	Apothy.	24 May 83	Punjab
Johnson, Henry William B.	Apothy.	2 Feb. 83	Bengal.
Keefe, Timothy B.	Apothy.	21 July 74	England.
Killoway, Samuel	Asst. Surgn., 1st cl.	2 Mar. 08	Cuddalore.
Kirby, George B.	Asst. Surgn., 1st cl.	21 May 98	Bengal.
Legge, Robert Henry, M.	Sr. Apothy.	14 July 88	Bangalore.
Lobo, M. F., Bo.	Sr. Apothy.	18 Dec. 85	India.
Lowman, William Henry, B.	Apothy.	11 Mar. 92	Bengal.
Lynsdale, Henry Herbert Dominick, M.	Apothy., 2nd gr.	13 June 90	Kolar.

MEDICAL WARRANT OFFICERS—*concld.*

Names.	Rank.	When pensioned.	Remarks.
McDiarmid, John, M.	Apothy, 2nd gr.	29 Apl. 81	Bangalore.
McGrath, James, B.	Apothy,	11 June 73	Agra.
Macedo, J. B. S., Bo.	Apothy.	8 May 89	Karachi.
Mann, William, B.	Asst. Surgn., 1st cl.	8 Dec. 99	Bengal.
Martin, Nicholas, Bo.	Asst. Surgn., 2nd cl.	1 May 02	India.
Mascarenhas, J., Bo.	Asst. Apothy.	4 June 90	Ratnagiri.
Peacock, Edward M.	Asst. Surgn., 1st cl.	22 June 04	Mandalay.
Peters, Joseph Thomas, B.	Asst. Surgn., 1st cl.	12 Aug. 04	Bengal.
Prem, John, Bo.	Apothy.	20 July 87	Ajmir.
Raddick, John Charles, B.	Asst. Apothy.	9 Apl. 94	Vellore.
Rath, James, M.	Apothy., 2nd gr.	15 Nov. 88	Coimbatore.
Robinson, B., B.	Sr. Apothy.	1 Feb. 72	
Rose, Albert	Asst. Surgn., 1st cl.	20 June 07	Poona.
Ross, Daniel	Asst. Surgn., 1st cl.	15 Jan. 13	India.
St. Francis, J. M., Bo.	Apothy.	21 Dec. 64	Vingeria.
Saxon, Samuel, B.	Apothy.	29 Jan. 86	England.
Selvey, Richard, M.	Apothy., 2nd gr.	6 Apl. 93	Bangalore.
Sharman, William, B.	Sr. Apothy.	Dec. 78	
Shaw, George Edward	Asst, Surgn., 1st cl.	17 Oct. 12	India.
Smith, Andrew Frederick, M.	Asst. Surgn., 1st cl.	26 Apl. 00	Bellary.
Summerhayes, Henry, M.	Asst. Surgn.	4 July 71	England.
Thomas, John Knox, Bo.	Asst. Surgn., 2nd cl.	16 Feb. 00	India.
Tilbury, A. S., Bo.	Asst. Apothy.	27 Jan. 82	Bombay.
Tranfield, Thomas, B.	Apothy.	12 Apl. 89	Pallavaram.
Turner, Charles Austin William	Asst. Surgn., 3rd cl.	14 Sep. 12	Bareilly.
Verghese, Mathew, M.	Asst. Surgn., 1st cl.	9 Oct. 98	Quilon.
Wachsel, Charles Horace, B.	Asst. Surgn., 1st cl.	1 Jan. 12	Mhow.
Wale, Reginald Riddle, M.	Asst. Surgn., 1st cl.	1 Jan. 99	Bangalore.
Walker, John, M.	Sr. Apothy.	10 Sep. 87	Bangalore.
Walmsley, Henry, B.	Apothy.	20 Jan. 87	Bombay.
Watson, Henry, B.	Apothy.	14 Dec. 88	Bengal.
Winckler, Peter Arthur, M.	Asst. Surgn., 1st cl.	10 Mar. 07	Bangalore.
Wooten, Donald	Apothy., 2nd gr.	4 Aug. 92	Bangalore.

Warrant of Precedence in India.

No. 328.

HOME DEPARTMENT.

Calcutta, the 10th February 1899.

NOTIFICATION.

In supersession of Home Department Notification No. 2032, dated 1st November 1877, the following new Warrant of Precedence for India which has been approved by Her Majesty the Queen-Empress of India, and which has received Her Royal Sign Manual, as also the revised graded list of civil offices not reserved for Members of the Indian Civil Service, supplementary to the Warrant of Precedence, which has been prepared under the orders of the Governor-General in Council, are published for general information :—

I.—Warrant of Precedence approved by the Queen-Empress of India.

VICTORIA, by the Grace of God of the United Kingdom of Great Britain and Ireland, Queen, Defender of the Faith, Empress of India,—

To all to whom these Presents shall come:

Whereas it hath been represented unto Us that it is advisable that the rank and precedence of persons holding appointments in the East Indies as regulated by Our Royal Warrant, dated the 18th day of October 1876, should be altered. We do therefore hereby declare that it is Our will and pleasure that in lieu of the table laid down in Our said recited Warrant, the following table be henceforth observed with respect to the rank and precedence of the persons hereinafter named, *viz.*:—

1. Governor-General and Viceroy of India.
*2. Governors of Madras and Bombay.
3. President of the Council of the Governor-General.
4. Lieutenant-Governor when in his own territories.
5. Commander-in-Chief in India.
6. Lieutenant-Governor.
7. Chief Justice of Bengal.
8. Bishop of Calcutta, Metropolitan of India.
9. Ordinary Members of the Council of the Governor-General.
10. Commander-in-Chief of Her Majesty's Naval Forces in the East Indies.
11. Chief Justice of a High Court other than that of Bengal.
12. Bishops of Madras and Bombay.
†13. Ordinary Members of Council in Madras and Bombay.
‡14. Lieutenant-Generals Commanding the Forces, Punjab, Bengal, Madras and Bombay.
‡15. Chief Commissioners of the Central Provinces and Assam; Residents at Hyderabad and in Mysore; and Agents to the Governor-General in Rajputana, Central India and Baluchistan.
16. Puisne Judges of a High Court.
17. Chief Judge of a Chief Court.
18. Military Officers above the rank of Major-General.
§19. Comptroller and Auditor-General.
20. Additional Members of the Council of the Governor-General for making Laws and Regulations.
21. Bishops of Lahore, Rangoon and Lucknow.
22. Secretaries to the Government of India.
23. Commissioner in Sind.
24. Judges of a Chief Court; Recorder of Rangoon; and Judicial Commissioner, Burma.
‖25. Chief Secretaries to the Governments of Madras and Bombay.
26. Major-Generals; Members of a Board of Revenue; Commissioners of Revenue and Customs, Bombay; and Financial Commissioners, Punjab and Burma.
27. Judicial Commissioners, including Additional Judicial Commissioners of Oudh, the Central Provinces and Sind.
28. Additional Members of the Councils of the Governors of Madras and Bombay for making Laws and Regulations; and Members of the Legislative Council of a Lieutenant-Governor.
29. Vice-Chancellors of Indian Universities.

* The Governor of Bengal ranks in Article 2 of the table (Home Department Notification No. 2013, dated the 1st October 1913).

† The Ordinary Members of Council in Bengal rank in Article 13 of the table (Home Department Notification No. 2013, dated the 1st October 1913).

‡ Members of the Executive Council in Bihar and Orissa will rank in Article 15 of the table, but when within their own jurisdiction they shall take precedence of the officers included in Article 14 of the Warrant (Home Department Notification No. 2013, dated the 1st October 1913).

§ Royal Warrant dated 6th November (Home Department Notification No. 232-C., dated the 15th January 1915).

‖ The Chief Secretary to the Government of Bengal ranks provisionally in Article 25 of the table (Home Department letter No. 860, dated the 15th October 1916).

FIRST CLASS.

30. Members of the Indian Civil Service of 30 years' standing.

31. Advocate-General, Calcutta.

32. Commissioners of Divisions; the Superintendent of Port Blair; and Residents, Political Agents, and Superintendents drawing Rs. 2,000 a month and upwards (not being Collectors or Deputy Commissioners of British Districts), within their respective charges.

33. Chief Secretaries to Local Governments other than those of Madras and Bombay.

34. Surveyor-General of India; Directors-General of the Post Office and Telegraphs in India and of Railways; Chief Engineer, first class; Accountants-General, Military and Public Works Departments; Director, Royal Indian Marine; and Manager, North-Western Railway.

35. Bishops (not territorial) under license from the Crowr

36. Archdeacons of Calcutta, Madras and Bombay.

37. Brigadier-Generals.

38. Commissioners of Divisions.

*39. Commissioner of Northern India Salt Revenue; and Opium Agents, Benares and Bihar.

40. Secretaries and Joint Secretaries to Local Governments; and Private Secretary to the Viceroy.

SECOND CLASS.

41. Members of the Indian Civil Service of 23 years' standing and Colonels.

42. Military Secretary to the Viceroy.

43. Judicial Commissioners of Hyderabad Assigned Districts and Baluchistan; the Superintendent of Port Blair; and Residents, Political Agents, and Superintendents drawing Rs. 2,000 a month and upwards (not being Collectors or Deputy Commissioners of British Districts).

44. Inspector-General of Forests in India and Director of the Geological Survey.

45. Standing Counsel to the Government of India.

46. Directors of Public Instruction and Inspectors-General of Police and Prisons under Local Governments; and Accountants-General.

47. Survey Commissioner and Director of Land Records and Agriculture, Bombay; Commissioners of Settlements; and Controllers of Military Accounts.

48. Chief or Senior Civil Secretary to a Local Administration.

49. Chief Engineers, second and third classes; Deputy Surveyor-General; Deputy Director-General of Telegraphs in India; and Director-in-Chief, Indo-European Telegraph Department.

The Chief of the General Staff ranks in Article 14 of the table (Home Department Notification No. 727, dated 20th July 1906). The General Officers Commanding the Northern and Southern Armies also rank in Article 14 of the table (Home Department Notification No. 2445, dated the 13th September 1907). But the Chief Commissioners of the Central Provinces and the North-West Frontier Province, the Residents at Hyderabad and in Mysore and the Agents to the Governor-General in Rajputana, Central India, Baluchistan, shall, when within their own jurisdiction, take precedence of the General Officers Commanding the Northern and Southern Armies and also of the Chief of the General Staff (Home Department letter No. 620-621, dated the 3rd March 1908).

The Agent to the Governor-General and Chief Commissioner of the North-West Frontier Province ranks in Article 15 of the table (Home Department Notification No. 5923, dated the 1st April 1901).

The Chairman of the Railway Board ranks in Article 20 of the table (Home Department Notifications No. 699, dated the 13th March 1905, and No. 232-C., dated 15th January 1915).

The Bishop of Nagpur ranks in Article 21 of the table (Home Department Notifications No. 2191, dated the 27th May 1903, and No. 232-C., dated 15th January 1915).

The Joint Secretary to the Government of India in the Public Works Department and the Members of the Railway Board rank in Article 22 of the table (Home Department Notifications No. 1259, dated the 15th June 1899, No. 699, dated the 13th March 1905, and No. 232-C., dated 15th January 1915).

The Joint Secretaries to the Government of India rank in Article 22 (Home Department Notifications No. 2013, dated 1st October 1913, and No 232-C., dated 15th January 1915).

The Financial Adviser, Military Finance, rank in Article 22 (Home Department Notification, No. 328, dated the 10th February 1899).

The Inspector-General of Irrigation ranks in Article 26 of the table (Home Department Notifications No. 1479, dated the 20th June 1906, and No. 232-C., dated 15th January 1915).

The Chief Commissioner of Delhi ranks in Article 25 of the table (Home Department Notifications No. 2320, dated 16th October 1912, and 232-C., dated 15th January 1915).

The Financial Commissioner, Central Provinces, ranks in Article 27 (Home Department Notifications No. 2013, dated 1st October 1913, and No. 232-C., dated 15th January 1915).

The Advocates General, Madras and Bombay, rank immediately below Article 31 (Home Department Notifications No. 2013, dated 1st October 1913, and No. 232-C., dated 15th January 1915).

Members of the Legislative Council of a Chief Commissioner rank in Article 40 (Home Department Notification No. 2013, dated the 1st October 1913).

Director of the Indian Institute of Science rank in Article 46 (Home Department Notification No. 2013, dated the 1st October 1913).

The Additional Members of the Council of the Governor of Bengal for making Laws and Regulations rank in Article 28 of the table (Home Department Notification No. 2013, dated 1st October 1913, and No. 232-C., dated 15th January 1915).

* The Educational Commissioner with the Government of India will take rank in Article 39 of the table (Home Department Notification No. 1065, dated 12th July 1915).

The Director-General, Indian Medical Service, ordinarily takes rank in Article 26 (Home Department Notification Nos. 2297, dated 18th May 1909, and No. 232-C., dated 15th January 1917), but when holding the rank of Lieutenant-General, he ranks in Article 18 of the Warrant (Home Department Notification No. 70-C., dated 5th January 1918).

50. Divisional and District and Sessions Judges; Collectors and Magistrates of Districts, Deputy Commissioners of Districts; Deputy Superintendent of Port Blair; and the Chief Officer of each Presidency Municipality, within their respective charges.

51. Archdeacons of Lahore, Lucknow, and Rangoon.

*52. Deputy Secretaries to the Government of India.

53. The Senior Chaplains of the Church of Scotland in Bengal, Madras, and Bombay.

54. Remembrancers of Legal Affairs and Government Advocates under Local Governments.

55. Officers in the First Class Graded List of Civil Officers not reserved for Members of the Indian Civil Service and Director, Zoological Survey of India.

THIRD CLASS.

56. Members of the Indian Civil Service of 18 years' standing and Lieutenant-Colonels.

57. The Deputy Director, Royal Indian Marine.

58. The Assistant Director, Royal Indian Marine.

59. Commanders and Inspectors of Machinery, Royal Indian Marine.

60. Political Agents and Superintendents drawing less than Rs. 2,000 a month (not being Collectors or Deputy Commissioners of British Districts), within their own charges.

61. Secretaries to Local Administrations other than those already specified.

62. Consulting Engineers to the Government of India for Railways.

63. Private Secretaries to Governors.

64. Military Secretaries to Governors.

65. Administrators-General.

66. Sanitary Commissioners under Local Governments, Postmasters-General, the Comptroller, Post Office, and Conservators of Forests, first grade; Deputy Director-General of Telegraph Traffic, and Directors, Telegraph Engineering.

67. Directors of Public Instruction and Inspectors-General of Police and Prisons under Local Administrations, and Comptrollers and Deputy Auditors-General.

68. Managers of State Railways other than the North-Western Railway, Chairman of the Port Trust, Bombay, and Chairman of the Port Trust, Calcutta.

69. Vice-Chairman of the Port Trust, Calcutta; Directors of Traffic and Construction, Indian Telegraph Department; Examiners of Accounts, Public Works Department, first class; Officers of the Superior Revenue Establishment of State Railways, first class, first grade; Superintending Engineers, Public Works Department, first class; and Superintendents of the Survey of India Department, first grade.

70. Inspectors-General of Registration and Directors of Land Records and Agriculture under Local Governments.

71. Senior Chaplains other than those already specified.

72. Sheriffs within their own charges.

73. Officers in the Second Class Graded List of Civil Officers not reserved for Members of the Indian Civil Service and Emigration Agent at Madras.

FOURTH CLASS.

74. Members of the Indian Civil Service of 12 years' standing and Majors.

75. Lieutenants of over 8 years' standing; and Chief Engineers of the Royal Indian Marine.

76. Government Solicitors.

77. Inspectors-General of Registration; Sanitary Commissioners; and Directors of Land Records and Agriculture under Local Administrations.

78. Officers in the Third Class Graded List of Civil Officers not reserved for Members of the Indian Civil Service and Officers of the Postal and Telegraph Department drawing Rs. 800 a month and upwards.

The Controller of Currency will take rank in Article 32 of the table (Home Department Notification No. 1840, dated the 8th October 1914).

The Directors of Railway Construction and Railway Traffic rank in Article 34 of the table (Home Department Notification No. 1259, dated the 15th June 1899). Consuls-General rank as No. 37 (*a*) below Brigadier-Generals, and Consuls as No. 41 (*a*) below Members of the Indian Civil Service of 23 years' standing and Colonels (Foreign Department Circular letter No. 257-G., dated the 22nd February 1899).

The Revenue and Judicial Commissioner in Baluchistan ranks in Article 38 of the table when in Kalat or Las Bela or elsewhere outside the limits of his charge (Foreign Department letter No. 1284-G., dated the 1st August 1901). In British Baluchistan and the Agency territories he ranks in Article 32 (Foreign Department letter No. 613-G., dated the 11th April 1901).

The Director, Central Criminal Intelligence Department, ranks in Article 39 of the table (Home Department Notification No. 1230, dated the 10th June 1904), but Sir Charles Cleveland, K.C.I.E., K.B.E., will take rank in Article 26 while holding the above appointment. (Home Department Notification No. 1636-C., dated the 21st March 1918.)

The members of the Legislative Council of a Chief Commissioner rank in Article 40 (Home Department Notification No. 2013, dated 1st October 1913).

The Director-General of Education in India ranks in Article 44 of the table (Home Department Notification No. 1988, dated the 27th June 1902). Also Sanitary Commissioner with the Government of India (Home Department Notification No. 3975, dated the 15th December 1905).

The Director of the Indian Institute of Science ranks in Article 46 (Home Department Notification No. 2013, dated 1st October 1913).

Military Deputy Auditors General and the Senior Controller of Military Supply Accounts rank in Article 47 of the table (Home Department Notification No. 882, dated 4th June 1914).

Officers of the Public Works Engineer establishment of 19 years' standing who hold the rank of Executive Engineer rank in Article 73 (Home Department Notification No. 666, dated 11th August 1016).

* The Deputy Financial Adviser, Military Finance, ranks in Article 52 (Home Department Notification No. 1328, dated 3rd September 1918).

The entries in the above table apply exclusively to the persons entered therein, and, while regulating their relative precedence with each other, do not apply to the non-official community resident in India, the members of which shall take their place according to usage.

Officers in the above table will take precedence in order of the numbers of the entries. Those included in one number will take precedence *inter se* according to the date of entry into that number.

When an officer holds more than one position in the table, he will be entitled to the highest position accorded to him.

Officers who are temporarily officiating in any number in the table will rank in that number below permanent incumbents.

All officers not mentioned in the above table, whose rank is regulated by comparison with rank in the Army, to have the same rank with reference to civil servants as is enjoyed by military officers of equal grades.

All other persons who may not be mentioned in this table to take rank according to general usage, which is to be explained and determined by the Governor-General in Council in case any question shall arise.

Nothing in the foregoing rules to disturb the existing practice relating to precedence at Native Courts, or on occasions of intercourse with Natives, and the Governor-General in Council to be empowered to make rules for such occasions in case any dispute shall arise.

All ladies to take place according to the rank herein assigned to their respective husbands with the exception of wives of Peers and of ladies having precedence in England independently of their husbands, and who are not in rank below the daughters of Barons; such ladies to take place according to their several ranks, with reference to such precedence in England, immediately after the wives of Members of the Council of the Governor-General.

Given at Our Court at Windsor this tenth day of December, in the year of Our Lord one thousand eight hundred and ninety-eight and in the sixty-second year of Our reign.

By Her Majesty's Command,

(Signed) GEORGE HAMILTON.

The Secretary to the Railway Board ranks in Article 49 of the table (Home Department Notification No. 699, dated 13th March 1905).

Officers in charge of the Zhob, Quetta, Pishin and Thal-Chotiali districts rank in Article 50 of the table throughout their respective charges, whether British or Agency territory (Foreign Department letter No. 613-G., dated the 11th April 1901).

The Judicial Commissioner, Chota Nagpur, Chairmen of the Boards of Trustees for the improvement of the Cities of Bombay and Calcutta, and President, Rangoon Municipal Committee, within their respective charges, rank in Article 50 (Home Department Notification No. 2013, dated 1st October 1913).

The Archdeacon of Nagpur ranks in Article 51 of the table (Home Department Notification No. 3389, dated the 29th December 1903).

The Director-General of Commercial Intelligence ranks in Article 52 of the table (Home Department Notification No. 699, dated the 13th March 1905).

The Chief Conservators of Forests rank in Article 54 of the table (Home Department Notification No. 2771, dated the 21st October 1907).

The President, Forest Research Institute and College, Dehra Dun, ranks in Article 54 (Home Department Notification No. 2013, dated 1st October 1913).

The Deputy Military Accountant General and the Junior Controller of Military Supply Accounts rank in Article 55 of the table (Home Department Notification No. 882, dated 4th June 1914).

The Political Agent in Kalat ranks in Article 60 of the table (Foreign Department letter No. 613-G., dated the 11th April 1901). District Judges in Lower Burma and the Judge of the Small Cause Court, Rangoon, will take rank in Article 60 of the table when within and Article 74 when without their respective charges (Home Department Notification No. 574, dated the 9th March 1906).

The First Assistant to the Agent to the Governor-General in Baluchistan ranks in Article 61 of the table (Foreign Department letter No. 613-G., dated the 11th April 1901). This order has also been extended to the First Assistant to the Residents at Hyderabad and in Mysore, and to the Agents to the Governor-General in Rajputana and Central India (Foreign Department letter No. 1389-G., dated the 2nd September 1903).

The Chief Inspector of Explosives in India ranks in Article 62 of the table (Home Department Notification No. 1988, dated the 27th June 1902). The Consulting Architect to the Government of India also ranks in Article 62 (Home Department Notification No. 1479, dated the 20th June 1906).

The Collectors of Customs at Calcutta, Bombay, Madras, Rangoon and Karachi rank in Article 66 (Home Department Notification No. 2013, dated 1st October 1913).

The Deputy Director, Central Criminal Intelligence Department, ranks in Article 67 of the table (Home Department Notification No. 1230, dated the 10th June 1904).

The Chairman of the Port Trust, Rangoon, ranks in Article 68 of the table (Home Department Notification No. 5006, dated the 29th August 1901).

The Chairman of the Port Trust, Madras, ranks in Article 68 (Home Department Notification No. 2013, dated 1st October 1913).

The Directors of Telegraphs, 1st class, take rank in Article 69 of the table (Home Department letter No. 377, dated the 18th February 1911).

The Electrical Engineer-in-Chief, Telegraph Department, ranks in Article 69 (Home Department Notification No. 2013, dated 1st October 1913).

The Chief Examiners of Accounts, North Western Railway, Eastern Bengal State Railway and Oudh and Rohilkhand Railway, rank in Article 69 of the table (Home Department Notification (Public) No. 180-C., dated 9th January 1914).

The Mining Engineer and Metallurgist, State Railways, rank in Article 69 (Home Department Notification No. 1775, dated the 1st October 1914).

The Registrar of Joint Stock Companies, Bengal, and the Registrar of Companies, Bombay, rank in Article 78 (Home Department Notification No. 891-C., dated 22nd February 1917).

The Assistant Financial Adviser, Military Finance, takes rank in Article 78 (Home Department Notification No. 1328, dated 3rd September 1918).

The Examiner of Questioned Documents takes rank in Article 78 (Home Department Notification No. 1343, dated 5th September 1918).

I.—Supplementary graded list of Civil offices not reserved for Members of the Indian Civil Service prepared under the orders of the Governor-General in Council.

* FIRST CLASS—(No. 55 of the Warrant).

Assay Master of the Mint, Calcutta and Bombay.
Chief Judges of Presidency Courts of Small Causes.
† Commissioners of Police, Calcutta, Madras, Bombay and Rangoon.
Controller of Printing and Stationery. (*a*)
Deputy Comptroller-General.
‡ Director-General of Archæology.
Director of the Botanical Survey of India.
Director, Zoological Survey of India. (*e*)
‡ Inspector-General of Agriculture in India.
Masters of the Mint, Calcutta and Bombay.
Meteorological Reporter to the Government of India.
Superintendent of Revenue Surveys, Madras.
Superintendent, Trigonometrical Surveys.

* SECOND CLASS—(No. 73 of the Warrant).

Actuary to the Government of India. (*d*)
Adviser on Chinese Affairs in Burma.
§ Agent-General in India for the British Protectorates in Africa under the Administration of the Foreign Office.
Chief Collector of Customs, Burma.
Chief Constructor of the Royal Indian Marine Dockyard at Bombay. (*b*)
‡ Chief Inspector of Mines in India.
Chief Presidency Magistrates.

The Directors of Land Records, Directors of Agriculture and Excise Commissioners under Local Governments rank in Article 70 of the table (Home Department Notification No. 2771, dated the 21st October 1907).

Registrars of Co-operative Credit Societies under Local Governments and Comptroller of Patents rank in Article 70 (Home Department Notification No. 2013, dated 1st October 1913).

The Political Agents and Superintendents in Baluchistan drawing less than Rs. 2,000 when outside their respective charges rank in Article 73, unless their Army rank or standing in the Civil Service gives them a higher place (Foreign Department letter No. 1284-G., dated the 1st August 1901). The same rule is applicable to the Political Agent in Haraoti and Tonk (Foreign Department letter No. 938-G., dated the 21st May 1903).

Principals of Government Colleges, Officers of the Indian Agricultural Service or of the General List of the Indian Finance Department, or of the Public Works Engineer Establishment, or of the Superior Revenue Establishment of State Railways, or of the Civil Veterinary, Forest, Survey, Mines, Postal and Telegraph, Customs, or Scientific Departments, or Officers of the Indian Institute of Science, or Sanitary Engineers, not being Superintending Engineers, or Consulting Architects to Local Governments, drawing Rs. 1,250 a month and upwards, Officers of the Police Department drawing Rs. 1,200 a month and upwards, Divisional Controllers of Military Accounts, Director of Survey, Madras, Electrical Adviser to the Government of India, Judicial Commissioner, Chota Nagpur, and Secretary to the Bengal Legislative Council and Assistant Secretary to the Government of Bengal in the Legislative Department, rank in Article 73 (Home Department Notification No. 2013, dated 1st October 1913).

The Director of Statistics will take rank in Article 73 (Home Department Notification No. 964, dated 11th June 1914).

The Consulting Surveyor to the Government of Bombay will take rank in Article 73 (Home Department Notification No. 1188, dated 13th July 1914).

Vice-Consuls rank as No. 74 (*a*) below Members of the Indian Civil Service of 12 years' standing and Majors (Foreign Department Circular letter No. 257-G., dated the 22nd February 1899).

Consular office s de carrières will in their respective grades take precedence of consular officers who are not de carrière (Foreign Department Circular No. 1454-G., dated 2nd July 1912).

The Chief Accountant and the Deputy Accountant in the office of the Director of Ordnance Factories take rank respectively in Articles 75 and 78, and the Civil Engineer Adviser to the Director of Ordnance Factories in Article 77 (Home Department Notification No. 4071, dated the 12th October 1911).

Registrars of Co-operative Credit Societies and Excise Commissioners under local Administrations, rank in Article 77 (Home Department Notification No. 2013, dated 1st October 1913).

Principals of Government Colleges, officers of the Indian Educational Service and the graded Educational Service, or of the Indian Agricultural Service, or Sanitary Engineers, or Consulting Architects, or Electrical Inspectors, or Electrical Engineers drawing Rs. 1,000 a month and upwards and officers of the General List of the Indian Finance Department, or officers of the higher branch of the Military Accounts Department, or of the Public Works Engineer Establishment, or of the Superior Revenue establishment of State Railways, or of the Civil Veterinary, Forest, Survey, Police, Mines, Customs, or Scientific Departments drawing Rs. 900 a month and upwards, Assistant Commissioners, Northern India Salt Revenue, drawing Rs. 800 a month and upwards, Chemical Examiners for Customs and Excise, Collector of Income Tax, Calcutta, Director, Vaccine Institute, Belgaum, First Assistant Superintendent, Port Blair, Legal Assistant in the Legislative Department of the Government of India, officers of the Archæological Department drawing Rs. 800 a month and upwards, officers of the Excise and Salt Departments drawing Rs. 800 a month and upwards, officers of the Provincial Services of not less than 18 years' standing drawing Rs. 600 a month and upwards, Registrar to the Chief Court, Lower Burma, and Secretary to the Board of Revenue in the Department of Revenue Settlement, Survey, Land Records and Agriculture, Madras, when a member of the Provincial Civil Service, rank in Article 78 (Home Department Notification No. 2013, dated 1st October 1913).

* The entries in each class are arranged in alphabetical order.
† Home Department Notification No. 2002, dated the 8th September 1899.
‡ Home Department Notification No. 1988, dated the 27th June 1902.
§ Home Department Notification No. 3119, dated the 11th September 1903.
(*a*) Home Department Notification No. 266, dated the 1st February 1906.
(*b*) Home Department Notification No. 3802, dated the 21st September 1911.
(*d*) Home Department Notification No. 1736-C., dated 13th March 1913.
(*e*) Home Department Notification No. 532, dated the 1st July 1916.

Superintendent of the Indian Museum.

Superintendent of Stamps and Stationery.

Superintendents, Telegraph Department, 1st and 2nd grades.

Under the orders of Her Majesty's Secretary of State for India, the following table showing the relative rank of officers in the Army, Royal Navy, and Royal Indian Marine is attached † to the Warrant of Precedence for India, published with Home Department Notification No. 328, dated the 10th February 1899:—

Lieutenant-Colonels .	Captains under 3 years' and officers of corresponding rank, Royal Navy.	
	Deputy Director, Royal Indian Marine . .	But junior to all Lieutenant-Colonels.
	Assistant " " " " . .	
	Commanders, Royal Navy	
	Commanders, Royal Indian Marine	
	Staff Commanders and officers of corresponding rank, Royal Navy.	
	Inspectors of Machinery, Royal Indian Marine .	
	Chief Engineers, Royal Indian Marine . .	
Majors . . .	Lieutenants of 8 years' seniority and officers of corresponding rank, Royal Navy.	
	Lieutenants, Royal Indian Marine, over 8 years' seniority.	But junior to all Majors.
	Engineers, Royal Indian Marine, of and over 9 years' seniority.	
Captains . .	Lieutenants under 8 years' and officers of corresponding rank, Royal Navy.	
	Lieutenants, Royal Indian Marine, under 8 years' seniority.	But junior to all Army Captains.
	Engineers, Royal Indian Marine, under 9 years' seniority.	
Lieutenants . .	Sub-Lieutenants and officers of corresponding rank, Royal Navy.	
	Sub-Lieutenants, Royal Indian Marine . .	But junior to all Lieutenants.
	Assistant Engineers, Royal Indian Marine .	

§" NOTE.—Officers of the Royal Naval Reserve rank with, but after, officers of the Royal Navy and Royal Indian Marine of their corresponding rank."

† Home Department Notifications Nos. 227, dated the 27th January 1900, and 1236, dated the 26th May 1911.
§ Home Department Notification No. 1014, dated the 9th July 1915.

Reported between the 1st October and 31st December 1918.

Names.	Rank.	Corps or Department.	When.	Remarks.
		Deaths.		
Adamson, John Thomas Graves.	Maj.	I. A.	25 Oct. 18	Deesa.
Allen, James	Asst. Commy. & Hony. Lt.	I. Misc. List	22 Oct. 18	Simla.
Allen, W.	2nd-Lt.	6th Hamps. R.	12 Oct. 18	......
Alves, John William Jerome	Maj.	I. A.	29 Oct. 18	Delhi.
Ashton, Edward Kitton Venning.	Lt.	I. A.	4 Dec. 18	......
Babington, Thomas Zachary Dodson.	Lt.	I. A. R. O.	16 Oct. 18	Karachi.
Bailey, *Hon.* J. L.	Lt.	1st Brecknock Bn. S. W. Bord.	26 Oct. 18	Mhow.
Baines, Hubert	Lt.	75th Infy.	22 Oct. 18	Died of wounds received in action.
Baker, Richard Stanley	2nd-Lt.	I. A. R. O. & 34th U. Burma Bn. I. D. F.	13 Nov. 18	Rawalpindi.
Bakewell, John	Sub-Condr.	I. A. Deptts. (retd.)	17 Sep. 17	......
Batcheler, Edward Beevor	Condr.	I. A. Deptts. (retd.)	6 Jan. 17	......
Battman, James	Dy. Commy. & Hony. Capt.	P. W. D. (retd.)	21 Dec. 17	......
Bedell, Frederick Francis	Senr-Asst. Surgn. Hony. Maj.	I. M. D.	22 Oct. 18	Dehra Dun.
Bedell, Herbert John	Asst. Surgn.	I. M. D.	17 Oct. 18	Peshawar.
Beveridge, Douglas Lionel	2nd-Lt.	I. A. R. O.	7 Nov. 18	Nasirabad.
Bhargawa, R S.	Tempy. Lt.	I. M. S.	29 Oct. 18	Rawalpindi.
Blenkinsop, Frederick Henry	Lt-Col.	I. M. S. (retd.)	25 Sep. 18	......
Blois Johnson, Thomas Gordon, C.M.G., C.I.E.	Lt.-Col.	I. A.	6 Nov. 18	Murree.
Bonsey, Francis Henry William.	2nd-Lt.	I. A. R. O.	20 Dec. 18	Deolali.
Borland, Sydney Douglas	2nd-Lt.	I. A. R. O.	6 Nov. 18	Tank.
Brewer, John Giblett	Lt.	I. A. R. O.	2 Nov. 18	......
Brock, George Selby, M.B.	Capt.	I. M. S.	7 Oct. 18	Rawalpindi.
Brown, Charles King Valentine.	Maj.	102nd Grs.	25 Oct. 18	Mhow.
Brown, Edward James	2nd-Lt.	I. A. R. O.	14 Nov. 18	Baghdad.
Brown, Eric Russell Wilkie	2nd-Lt.	I. A.	19 Nov. 18	......
Brown, George	Condr.	S. & T. Corps.	25 Oct. 18	Abbottabad.
Brownson, R. D. M.B.	Capt.	R. A. M. C. (Spl. Res.)	21 Oct. 18	Peshawar.
Butt, Joseph Werner	Asst. Surgn.	I. M. D.	26 Oct. 18	Deolali.
Byrne, John Cahill	Lt.	I. A. R. O.	11 Oct. 18	Madras.
Cadogan, Thomas James	Sub-Condr.	S. & T. Corps	22 Oct. 18	Rawalpindi.
Campbell, Frederick William	2nd-Lt.	I. A. R. O.	11 Dec. 18	Chausan Kach.
Cannell, Hugh Featherstone Cameron.	Lt.	I. A.	31 Oct. 18	Died of wounds received in action.
Carbonaro, Ernest	Maj.	I. A. (retd).	30 Dec 17	......
Carpendale, Maxwell Montague.	Maj.	36th Horse	14 Oct. 18	......
Carroll, George Thomas	Sr. Asst.-Surgn & Hony. Capt.	I. M. D. (retd.)	7 Sep. 18	......
Chapman, Ernest Herbert Stuart.	Capt.	31st Lrs.	1 Nov. 18	Shiraz.
Chavasse, Pere	2nd-Lt.	I. A. R. O.	8 Oct. 18	Killed in action.
Christian, Aviet Thadeus	Lt.	I. A. R. O.	6 Nov. 18	......
Clelland Robert Marshall	2nd-Lt.	I. A. R. O.	22 Dec, 18	......
Cockburn, R. C. R.	Lt.	Wilts. R. I. A. (*onprobn.*)	25 Nov. 18	Pachmarhi.
Cogan, Lionel Gatrell	Lt.	153rd Intfy.	7 Oct. 18	Ahmednagar.
Colborne, Reginald Elgar	2nd-Lt.	I. A. R. O.	6 Nov. 18	Subathu.
Collins, Arthur Michael Austen.	Lt.	38th Dogras	28 Oct. 18	Fatehgarh.
Connell, William Patrick	2nd-Lt.	R. A.	24 Nov. 18	Accidentally killed, Mesopotamia.
Conroy, H. T.	Lt.	I. A. R. O.	28 Oct. 18	......
Cracroft-Wilson, C. W.	Lt.	4th S. Wales Bord.	12 Nov. 18	Poona.
Crocker, P. J. W.	2nd-Lt.	R. W. Kent R.	16 Nov. 18	......
Crow, H. P.	Capt.	R. A. M. C. (Spl. Res.)	9 Nov. 18	Jhansi.

Reported between the 1st October and 31st December 1918—contd.

Names.	Rank.	Corps or Department.	When.	Remarks.
		Deaths—*contd.*		
Cumming, Edward John	Capt.	120th Infy.	15 Nov. 18	Persia.
Dall, J.	2nd-Lt. (temp. Capt.)	High. L. I.	12 Oct. 18	Secunderabad.
Darley, William Hastings La Touche.	Maj.	12th Prs.	11 Oct. 18	Rouen.
Davidson, Alexander Henry	Maj.-Genl.	I. A. (retd.)	10 Aug. 18	......
Davies, Arthur	2nd-Lt.	I. A. (*T. C.*)	19 Oct. 18	Rawalpindi.
Davies, W. B.	2nd-Lt.	1st G. B. Som. L. I.	3 Nov. 18	Amritsar.
Denwood, F. W.	2nd-Lt.	Attd. 2-6th R. Suss. R.	22 Oct. 18	Lahore Cantonment.
de Rosse, Joseph Albert	Lt.	I. A. R. O.	2 Dec. 18	......
Dow, John	Capt.	I. M. S.	5 Nov. 18	Shiraz.
Dunn, James	Lt.	I. A. R. O.	Not known.	Died at Sea.
Egerton, James Boswell	Maj.	23rd Cavy.	27 Oct. 18	Killed in action.
Eyre, Frederick Vincent	Col.	B. Arty. (retd.)	2 Sep. 17	......
Fairgrieve, Alexander Oman	Lt.	I. A. (*on probn.*)	25 Dec. 18	Bannu.
Ferguson, James	Capt.	R. A. M. C. (Dental Surgn.)	9 Oct. 18	Quetta.
Fieldhouse, W. B.	2nd-Lt.	L'pool R.	1 Nov. 18	Quetta.
Fisher, Eric Whitmore	Lt. (*temp. Capt*).	I. A. R. O.	14 Oct. 18	Bombay.
Follitt, John Fraser	Lt.	I. A. R. O.	21 Oct. 18	......
Forsyth, William	Condr.	I. Misc. List	28 Oct. 18	Rawalpindi.
Ghosh, L. P.	Temp. Lt.	I. M. S.	10 Oct. 18	Ferozepore.
Gillman, Herbert Francis Webb, C.S.I.	Lt.-Col.	1st Madras Gds., I. D. F.	23 Nov. 18	Simla.
Golding, Thomas	Sub-Condr.	I. A. Deptts. (retd.)	23 Oct. 17	......
Gorton, Arthur Llewellyn	Lt.	I. A. R. O.	29 Oct. 18	Killed in action.
Graves-Knyfton, R.	Maj.	1-4th Som. L. I.	29 Oct. 18	Trimulgherry.
Greenaway, Thomas	Lt.-Col.	I. A. (retd.)	27 Sep. 18	......
Grove, Henry Leslie	Hony. Maj. Genl.	I. A. (retd.)	23 Sep. 18	......
Hales, Walter Percy	Lt.	I. A. R. O.	1 Nov. 18	Shiraz.
Hall, H. S. H. H.	Capt.	3rd R. W. Kent. R.	24 Oct. 18	Calcutta.
Hallowes, John Chaworth	Lt.	7th Hrs.	28 Oct. 18	Killed in action.
Hamilton, A. L.	Lt.	3rd High. L. I.	25-26 Oct. 18	Killed in action.
Hammick, Eustace	Capt.	17th Infy.	8 Oct. 18	Died of wounds received in action.
Hancock, Frank Herbert	Bt.-Col.	I. A.	13 Oct. 18	......
Harding, Homer	2nd-Lt.	I. A. (*T. C.*)	3 Oct. 18	Bangalore.
Hardman, Basil Brocas	Lt.	I. A. R. O.	29 Oct. 18	Killed in action.
Hardman, K.	Capt.	3rd High. L. I.	25-26 Oct. 18	Killed in action.
Harvey, John William	Lt.	I. A. R. O.	26 Dec. 18	Calcutta.
Haslam, Bernard John, D.S.O.	Maj.	R. E.	26 Aug. 18	Killed in action.
Hawkins, Henry	Condr.	I. A. Depts. (retd.)	23 Mar. 18	......
Heckroodt, B.	Capt.	2-6th R. Suss. R.	6 Oct. 18	......
Henderson, Philip Durham, C.S.I.	Major-Genl.	I. A. (U. S. L.)	19 Apl. 18	......
Henry, Andrew Ogilvie	Condr.	I. A. Deptts. (retd.)	3 July 18	......
Hepworth, G. P.	Capt.	R. F. A. (T. F.)	27 Oct. 18	Neemuch.
Hewson, Falkiner Melton, F.R.C.S.I.	Capt.	R. A. M. C.	22 Oct. 18	Mhow.
Hill, George	Sub-Condr.	M. W. S.	26 Nov. 18	......
Hobbs, Arthur Edward Singleton.	2nd-Lt. (*temp. Capt.*)	I. A. R. O.	16 Oct. 18	Mussoorie.
Holden, Hyla Napier	Lt.-Col.	5th Cavy.	26 Oct. 18	Killed in action.
Holland, Vivian Ernest	Lt. (Acg. Capt.)	7th Hrs.	8 Nov. 18	Died of wounds received in action.
Houghton, John	Sr. Apothy.	I. M. D. (retd.)	27 Dec. 17	......
Howard, Frank William	Condr.	I. A. Deptt. (retd.)	16 Feb. 18	......
Hoyne, Fox, Leslie Vincent.	Capt.	120th Infy.	....	Yazgad, Date not reported.
Irving, Alfred	2nd-Lt.	I. A.	20 Oct. 18	Killed in action.
Jackson, Wilfrid Flower	Lt.	I. A. R. O.	28 Oct. 18	Killed in action.
Jeffries, P.	Lt.	M. G. Corps (Motors).	3 Dec. 18	Peshawar.
Jukes, Andrew Monro, M.D.	Capt.	I. M. S.	18 Oct. 18	......
Kennedy, Miles Arthur Claude, D.S.O.	Capt.	8th G. R.	2 Nov. 18	Rawalpindi.
Kerigan, Henry	Asst. Commy. and Hony. Lt.	I. A. Dept. (retd.)	25 Mar. 18	......

Reported between the 1st October and 31st December 1918—contd.

Names.	Rank.	Corps or Department.	When.	Remarks.
		Deaths—*contd.*		
King, Henry Alfred	2nd-Lt.	I. A. (*T. C.*)	4 Nov. 18	......
Kitchener, Cecil	2nd-Lt.	R. G. A.	30 Oct. 18	Rawalpindi.
Law, John, M.D.	Bde. Surgn.	I. M. S. (retd.)	6 Feb. 18	......
Lawrie, F. A.	Capt.	Sco. Rif.	25 Sep. 18	Bangalore.
Lay, John Frederick	Lt.	I. A. R. O.	28 Oct. 18	Rawalpindi.
Learoyd, Geoffrey Ernest Douglas.	Capt.	21st Lrs.	29 Oct. 18	Risalpur.
Lee, George Thomas Hutteman	2nd-Lt.	I. A. R. O.	26 Oct. 18	Lucknow.
Lewis, G. K.	2nd-Lt.	13th High. L. I.	1 Nov. 18	Died of wounds received in action.
Lunn, G. A.	Lt.	2-6th R. Suss. R.	15 Oct. 18	Lahore.
Lynch, Michael John	2nd-Lt.	I. A. R. O.	15 Oct. 18	Quetta.
Lynch-Staunton, Reginald Kirkpatrick, D.S.O.	Bt.-Lt.-Col.	R. A.	7 Nov. 18	Died of wounds received in action.
McCurrich, Laurence Oliphant.	2nd-Lt. (*Temp. Capt.*).	I. A. R. O.	25 Oct. 18	Jalapahar.
McIntyre, R. J.	Temp. 2nd-Lt.	Attd. 1st High. L. I.	25-26 Oct. 18	Killed in action.
McLean, Richardson Pelman Archibald.	Lt.	I. A. R. O.	19 Nov. 18	......
Macdonald, Donald	Col.	I. A. (U. S. L.)	27 Mar. 18	......
Mander, Thomas James	2nd-Lt.	I. A. R. O.	9 Nov. 18	Mesopotamia.
Mein, Dudley Gerald	Lt.	I. A.	26 Oct. 18	Killed in action.
Melville, Harry George	Lt.-Col.	I. M. S.	7 Dec. 18	Mesopotamia.
Meredith, Maurice Neville	Lt.	I. A. R. O.	26 Oct. 18	Killed in action.
Miller, Clement Francis Patrick.	2nd-Lt.	I. A. R. O.	4 Oct. 18	Poona.
Mills, James	Sub-Condr.	I. A. Deptts. (retd.)	17 Sep. 17	......
Mohan, N. C.	Temp. Lt.	I. M. S.	10 Oct. 18	Bombay.
Morris, Wilfrid Cyril	2nd-Lt.	I. A. (*T. C.*)	15 Dec. 18	Ahmednagar.
Mosse, William Oliver Matless.	Col.	I. A. (retd.)	10 Oct. 18	......
Nevill, Edward James	Lt.	I. A. R. O.	22 Dec. 18	Bareilly.
Newman, Reuben McCarthy	Lt.	Otago, R., N. Z. Forces, I. A. R. O. (*on probn.*)	23 Nov. 18	Abbottabad.
Northey, M. A.	Lt.	3rd R. W. Kent R.	29 Oct. 18	Killed in action.
O'Brien, Thomas A.	Temp. Lt.	A. V. Corps	6 Oct. 18	Ajmere.
Owens, John	Condr.	I. A. Deptts. (retd)	11 May 18	......
Parkinson, R. F., M.C.	Capt.	R. E. (S. R.)	7 Nov. 18	Mesopotamia.
Peacock, Frank	2nd-Lt.	I. A. (*T. C.*)	3 Dec. 18	Cawnpore.
Peddar, Digby Cotes	Hony. Lt. Col.	I A. (retd.)	8 Sep. 18	......
Phillips, Percy Charles	Sub-Condr.	I. A. Deptts. (retd.)	14 June 18	......
Philson, Samuel Cowell	Col.	A. M. S. (retd.)	4 Nov. 18	Chakrata.
Pinfold, William Herbert	2nd-Lt.	I. A. R. O.	9 Oct. 18	Bombay.
Pitts, William Henry	Dy. Commy. & Hony. Capt.	P. W. D. (retd.)	8 May 18	......
St. Planck, Charles	Surgn. Genl.	I. M. S. (retd.)	23 Aug. 18	......
Ponter, W. C.	Lt.	23rd Rif. Brig.	22 Nov. 18	Bareilly.
Price, W. H.	Qr. Mr. & Hony. Lieut.	Genl. List attached 6th Bangalore, Coorg & Mysore Bn., I. D. F.	16 Oct. 18	Bangalore.
Prothero, Rowland John	Lt.	7th Hrs.	8 Nov. 18	Died of wounds received in action.
Quigley, William Grainger	2nd-Lt.	I. A. (*T. C.*)	2 Nov. 18	Subathu.
Rao, M. S. R. K., M.B.	Temp. Capt.	I. M. S.	6 Nov. 18	Aden.
Raymond, Wynne Dudley	Lt.	I. A. (*on probn.*)	26 Oct. 18	Killed in action.
Reed, Henry Gerrard	Lt.	I. A.	29 Oct. 18	Killed in action.
Reeve, W. J.	Lt.	6th Devon. R.	29 Nov. 18	Cawnpore.
Richards, William Earnest	Condr.	I. O. D.	14 Sep. 18	......
Riddoch, J. K.	Lt. (*Actg. Capt.*)	1-6th Hamps. R.	17 Oct. 18	......
Roche-Kelly, James	2nd-Lt.	2-32nd M. & S. M. Ry. Rif., I. D. F.	21 Nov. 18	Madras.

Constructors of the Royal Indian Marine Dockyards at Bombay and Kidderpore.§

Deputy Administrator-General, Bengal. (*a*)

Deputy Collector of Salt Revenue, Bombay.

Deputy Commissioner of Northern India Salt Revenue.

Deputy Commissioners of Police, Calcutta and Bombay.

Deputy Commissioner of Salt, Abkari and Customs Department, Madras.

Deputy Conservators of Forests drawing Rs. 800 a month and upwards.

Deputy Directors of Land Records and Agriculture, Madras and Burma. (*b*)

Deputy Director of the Imperial Forest School, Dehra Dun.

Deputy Directors of Revenue Settlements and Deputy Superintendents of Revenue Surveys, Madras.

Deputy Postmasters-General of the 1st, 2nd, † and 3rd‡ grades.

Deputy Superintendents, Geological Survey of India.

Deputy Superintendents, Survey of India Department.

District Superintendents of Police drawing Rs. 800 a month and upwards.

Engineer and Electrician of the Persian Gulf Section of the Indo-European Telegraph Department.

Examiners of Accounts, Public Works Department, 4th class, 1st and 2nd grades.

Executive Engineers, Public Works Department, 1st and 2nd grades.

Inspector-General of Railway Mail Service.

Judge of the City Civil Courts, Madras.

Judges of Presidency Courts of Small Causes ; and First Judge of the Small Cause Court, Rangoon.

Manager of the Cordite Factory, Aruvankadu. (*a*)

Officers of the Indian Educational Service and of the Graded Educational Service drawing less than Rs. 1,250 a month but more than Rs. 1,000 a month.

Officers of the Public Works Engineer Establishment of 12 years' standing who hold the rank of Executive Engineer. (*d*)

Officers of the Superior Revenue Establishments of the State Railways, 2nd class, 1st and 2nd grades.

Palæontologist, Geological Survey of India.

Presidency Magistrates.

Principals of Government Colleges, Officers of the Indian Educational Service and the graded Educational Service, or of the Indian Agricultural Service, or Sanitary Engineers, or Consulting Architects, or Electric Inspectors or Electrical Engineers, drawing Rs. 1,000 a month and upwards; and officers of the General List of the Indian Finance Department, or officers of the higher branch of the Military Accounts Department, or of the Superior Revenue Establishment of State Railways or of the Civil Veterinary, Forest, Survey, Police, Mines, Customs or Scientific Departments, drawing Rs. 900 a month and upwards, Assistant Commissioners, Northern India Salt Revenue, drawing Rs. 800 a month and upwards, Chemical Examiner for Customs and Excise, Collector of Income-Tax, Calcutta, Director, Vaccine Institute, Belgaum, First Assistant Superintendent, Port Blair, Legal Assistant in the Legislative Department of the Government of India, officers of the Archæological Department drawing Rs. 800 a month and upwards, officers of the Excise and Salt Departments drawing Rs. 800 a month and upwards, (*e*) officers of the Provincial Services of not less than 18 years' standing drawing Rs. 600 a month and upwards, Registrar to the Chief Court, Lower Burma, and Secretary to the Board of Revenue in the Department of Revenue Settlement, Survey, Land Records and Agriculture, Madras, when a member of the Provincial Civil Service. (*b*)

Protector of Emigrants and Superintendent of Emigration, Calcutta.

Public Prosecutor in Sind.

Registrars to the High Courts and to the Chief Court, Punjab.

Registrar of Joint Stock Companies, Bengal, (*c*) Registrar of Companies, Bombay. (*c*)

Sub-Deputy Opium Agents drawing Rs. 800 a month and upwards.

* Superintendent of Land Records and Agriculture in Sind.

---

† The entries in each class are arranged in alphabetical order.
‡ Home Department Notification No. 2292, dated the 24th July 1902.
(*a*) Home Department Notification No. 1155, dated the 10th May 1906.
(*b*) Home Department Notification No. 574, dated the 9th March 1906.
* Home Department Notification No. 1988, dated the 27th June 1902.
(*a*) Home Department Notification No. 2875, dated the 31st October 1907.
(*b*) Home Department Notification No. 2013, dated the 1st October 1913.
§ Home Department Notification No. 3802, dated the 21st September 1911.
(*c*) Home Department Notification No. 891-C., dated the 22nd February 1917.
(*d*) Home Department Notification No. 666, dated the 11th August 1917.
(*e*) Home Department Notification No. 143-C., dated the 9th November 1917.

Chief Superintendents of the Telegraph Department.
Collector of Customs and Salt Revenue, Sind.
Collectors and Magistrates of Districts ; and Deputy Commissioners of Districts, and of Settlements.
Conservators of Forests, 2nd and 3rd grades.
Consulting Surveyor to the Government of Bombay. (h)
Deputy Accountants-General under Local Governments.
Deputy Directors of Telegraphs. (g)
Deputy Inspectors-General of Police.
Deputy Superintendent of Port Blair.
Directors of the Persian Gulf Section, and of the Persian Section of the Indo-European Telegraph Department.
Director of Statistics. (i)
Directors of Telegraphs of the 2nd, 3rd and 4th classes. (e)
Divisional and District and Sessions Judges.
Emigration Agent at Madras.
Examiners of Accounts, Public Works Department, 2nd and 3rd classes.
Government Astronomer, Madras.
Government Emigration Agent at Calcutta for British Guinea and Natal, and for Trinidad, Fiji, Jamaica and Mauritius. (d)
Imperial Bacteriologist.
Inspector of Mines to the Government of India.
* Librarian, Imperial Library.
Officer in charge of the Records of the Government of India.
Officers of the Indian Educational Service, and of the Graded Educational Service, drawing Rs. 1,250 a month and upwards.
Officers of the Superior Revenue Establishment of State Railways, 1st class, 2nd and 3rd grades.
Principal of the Mayo College at Ajmer.
Principal of the Rajkumar College at Rajkot.
Principals of Government Colleges, Officers of the Indian Agricultural Service, or of the General List of the Indian Finance Department, or of the Superior Revenue establishment of State Railways, or of the Civil Veterinary, Forest Survey, Mines, Postal and Telegraph, Customs, or Scientific Departments, or officers of the Indian Institute of Science, or Sanitary Engineers, not being Superintending Engineers, or Consulting Architects to local Governments, drawing Rs. 1,250 a month and upwards, officers of the Police Department drawing Rs. 1,200 a month and upwards, Divisional (j) Controllers of Military Accounts, Director of Survey, Madras, Electrical Adviser to the Government of India, Judicial Commissioner, Chota Nagpur, and Secretary to the Bengal Legislative Council and Assistant Secretary to the Government of Bengal in the Legislative Department (k) and Secretary to the Legislative Council of the Governor of Bombay and Assistant Remembrancer of Legal Affairs, Bombay. (l)
Reporter on Economic Products.
Superintendent of the Royal Botanical Gardens, Calcutta.
Superintendents, Geological Survey of India.
Superintendents of Revenue Survey and Assessment, Bombay.
Superintendents of the Survey of India Department, 2nd grade.
Superintending Engineers, Public Works Department, 2nd and 3rd classes.
Under-Secretaries to the Government of India.

† Third Class—(No. 78 of the Warrant).

Agricultural Chemist.
Assistant Director of Dairy Farms. (f)
Assistant Inspector-General of Forests.
Assistant Secretaries to the Government of India.
Chief Chemical Examiner, Central Chemical Laboratory, Nainital. (c)
Collector of Stamp Revenue, Superintendent of Excise Revenue, and Deputy Collector of Land Revenue, Calcutta.
Commander of the steamer employed in the Persian Gulf Section of the Indo-European Telegraph Department.

---

* Home Department Notification No. 1988, dated the 27th June 1902.
† The entries in each class are arranged in alphabetical order.
(c) Home Department Notification No. 1741, dated the 27th May 1910.
(d) Home Department Notification No. 3888, dated the 27th September 1911.
(e) Home Department Notification No. 377, dated the 18th February 1911.
(f) Home Department Notification No. 1772, dated the 16th August 1912.
(g) Home Department Notification No. 3839, dated the 30th November 1905.
(h) Home Department Notification No. 1188, dated the 15th July 1914.
(i) Home Department Notification No. 964, dated the 11th June 1914.
(j) Home Department Notification No. 882, dated the 4th June 1914.
(k) Home Department Notification No. 2013, dated the 1st October 1913.
(l) Home Department Notification No. 1403, dated the 6th August 1914.

Chief Superintendents of the Telegraph Department.
Collector of Customs and Salt Revenue, Sind.
Collectors and Magistrates of Districts ; and Deputy Commissioners of Districts, and of Settlements.
Conservators of Forests, 2nd and 3rd grades.
Consulting Surveyor to the Government of Bombay. (h)
Deputy Accountants-General under Local Governments.
Deputy Directors of Telegraphs. (g)
Deputy Inspectors-General of Police.
Deputy Superintendent of Port Blair.
Directors of the Persian Gulf Section, and of the Persian Section of the Indo-European Telegraph Department.
Director of Statistics. (i)
Directors of Telegraphs of the 2nd, 3rd and 4th classes. (e)
Divisional and District and Sessions Judges.
Emigration Agent at Madras.
Examiners of Accounts, Public Works Department, 2nd and 3rd classes.
Government Astronomer, Madras.
Government Emigration Agent at Calcutta for British Guinea and Natal, and for Trinidad, Fiji, Jamaica and Mauritius. (d)
Imperial Bacteriologist.
Inspector of Mines to the Government of India.
* Librarian, Imperial Library.
Officer in charge of the Records of the Government of India.
Officers of the Indian Educational Service, and of the Graded Educational Service, drawing Rs. 1,250 a month and upwards.
Officers of the Superior Revenue Establishment of State Railways, 1st class, 2nd and 3rd grades.
Principal of the Mayo College at Ajmer.
Principal of the Rajkumar College at Rajkot.
Principals of Government Colleges, Officers of the Indian Agricultural Service, or of the General List of the Indian Finance Department, or of the Superior Revenue establishment of State Railways, or of the Civil Veterinary, Forest Survey, Mines, Postal and Telegraph, Customs, or Scientific Departments, or officers of the Indian Institute of Science, or Sanitary Engineers, not being Superintending Engineers, or Consulting Architects to local Governments, drawing Rs. 1,250 a month and upwards, officers of the Police Department drawing Rs. 1,200 a month and upwards, Divisional (j) Controllers of Military Accounts, Director of Survey, Madras, Electrical Adviser to the Government of India, Judicial Commissioner, Chota Nagpur, and Secretary to the Bengal Legislative Council and Assistant Secretary to the Government of Bengal in the Legislative Department (k) and Secretary to the Legislative Council of the Governor of Bombay and Assistant Remembrancer of Legal Affairs, Bombay. (l)
Reporter on Economic Products.
Superintendent of the Royal Botanical Gardens, Calcutta.
Superintendents, Geological Survey of India.
Superintendents of Revenue Survey and Assessment, Bombay.
Superintendents of the Survey of India Department, 2nd grade.
Superintending Engineers, Public Works Department, 2nd and 3rd classes.
Under-Secretaries to the Government of India.

† Third Class—(No. 78 of the Warrant).

Agricultural Chemist.
Assistant Director of Dairy Farms. (f)
Assistant Inspector-General of Forests.
Assistant Secretaries to the Government of India.
Chief Chemical Examiner, Central Chemical Laboratory, Nainital. (c)
Collector of Stamp Revenue, Superintendent of Excise Revenue, and Deputy Collector of Land Revenue, Calcutta.
Commander of the steamer employed in the Persian Gulf Section of the Indo-European Telegraph Department.

---

* Home Department Notification No. 1988, dated the 27th June 1902.
† The entries in each class are arranged in alphabetical order.
(c) Home Department Notification No. 1741, dated the 27th May 1910.
(d) Home Department Notification No. 3888, dated the 27th September 1911.
(e) Home Department Notification No. 377, dated the 18th February 1911.
(f) Home Department Notification No. 1772, dated the 16th August 1912.
(g) Home Department Notification No. 3839, dated the 30th November 1905.
(h) Home Department Notification No. 1188, dated the 15th July 1914.
(i) Home Department Notification No. 964, dated the 11th June 1914.
(j) Home Department Notification No. 882, dated the 4th June 1914.
(k) Home Department Notification No. 2013, dated the 1st October 1913.
(l) Home Department Notification No. 1403, dated the 6th August 1914.

Constructors of the Royal Indian Marine Dockyards at Bombay and Kidderpore.§
Deputy Administrator-General, Bengal. (*a*)
Deputy Collector of Salt Revenue, Bombay.
Deputy Commissioner of Northern India Salt Revenue.
Deputy Commissioners of Police, Calcutta and Bombay.
Deputy Commissioner of Salt, Abkari and Customs Department, Madras.
Deputy Conservators of Forests drawing Rs. 800 a month and upwards.
Deputy Directors of Land Records and Agriculture, Madras and Burma. (*b*)
Deputy Director of the Imperial Forest School, Dehra Dun.
Deputy Directors of Revenue Settlements and Deputy Superintendents of Revenue Surveys, Madras.
Deputy Postmasters-General of the 1st, 2nd, † and 3rd‡ grades.

Deputy Superintendents, Geological Survey of India.

Deputy Superintendents, Survey of India Department.

District Superintendents of Police drawing Rs. 800 a month and upwards.
Engineer and Electrician of the Persian Gulf Section of the Indo-European Telegraph Department.
Examiners of Accounts, Public Works Department, 4th class, 1st and 2nd grades.

Executive Engineers, Public Works Department, 1st and 2nd grades.

Inspector-General of Railway Mail Service.

Judge of the City Civil Courts, Madras.

Judges of Presidency Courts of Small Causes; and First Judge of the Small Cause Court, Rangoon.

Manager of the Cordite Factory, Aruvankadu. (*a*)

Officers of the Indian Educational Service and of the Graded Educational Service drawing less than Rs. 1,250 a month but more than Rs. 1,000 a month.

Officers of the Public Works Engineer Establishment of 12 years' standing who hold the rank of Executive Engineer. (*d*)

Officers of the Superior Revenue Establishments of the State Railways, 2nd class, 1st and 2nd grades.

Palæontologist, Geological Survey of India.

Presidency Magistrates.

Principals of Government Colleges, Officers of the Indian Educational Service and the graded Educational Service, or of the Indian Agricultural Service, or Sanitary Engineers, or Consulting Architects, or Electric Inspectors or Electrical Engineers, drawing Rs. 1,000 a month and upwards; and officers of the General List of the Indian Finance Department, or officers of the higher branch of the Military Accounts Department, or of the Superior Revenue Establishment of State Railways or of the Civil Veterinary, Forest, Survey, Police, Mines, Customs or Scientific Departments, drawing Rs. 900 a month and upwards, Assistant Commissioners, Northern India Salt Revenue, drawing Rs. 800 a month and upwards, Chemical Examiner for Customs and Excise, Collector of Income-Tax, Calcutta, Director, Vaccine Institute, Belgaum, First Assistant Superintendent, Port Blair, Legal Assistant in the Legislative Department of the Government of India, officers of the Archæological Department drawing Rs. 800 a month and upwards, officers of the Excise and Salt Departments drawing Rs. 800 a month and upwards, (*e*) officers of the Provincial Services of not less than 18 years' standing drawing Rs. 600 a month and upwards, Registrar to the Chief Court, Lower Burma, and Secretary to the Board of Revenue in the Department of Revenue Settlement, Survey, Land Records and Agriculture, Madras, when a member of the Provincial Civil Service. (*b*)

Protector of Emigrants and Superintendent of Emigration, Calcutta.

Public Prosecutor in Sind.

Registrars to the High Courts and to the Chief Court, Punjab.

Registrar of Joint Stock Companies, Bengal, (*c*) Registrar of Companies, Bombay. (*c*)

Sub-Deputy Opium Agents drawing Rs. 800 a month and upwards.

* Superintendent of Land Records and Agriculture in Sind.

---

† The entries in each class are arranged in alphabetical order.
‡ Home Department Notification No. 2292, dated the 24th July 1902.
(*a*) Home Department Notification No. 1155, dated the 10th May 1906.
(*b*) Home Department Notification No. 574, dated the 9th March 1906.
* Home Department Notification No. 1988, dated the 27th June 1902.
(*a*) Home Department Notification No. 2875, dated the 31st October 1907.
(*b*) Home Department Notification No. 2013, dated the 1st October 1913.
§ Home Department Notification No. 3802, dated the 21st September 1911.
(*c*) Home Department Notification No. 891-C., dated the 22nd February 1917.
(*d*) Home Department Notification No. 666, dated the 11th August 1917.
(*e*) Home Department Notification No. 143-C., dated the 9th November 1917.

Superintendent of the Indian Museum.

Superintendent of Stamps and Stationery.

Superintendents, Telegraph Department, 1st and 2nd grades.

Under the orders of Her Majesty's Secretary of State for India, the following table showing the relative rank of officers in the Army, Royal Navy, and Royal Indian Marine is attached † to the Warrant of Precedence for India, published with Home Department Notification No. 328, dated the 10th February 1899:—

Lieutenant-Colonels	Captains under 3 years' and officers of corresponding rank, Royal Navy. Deputy Director, Royal Indian Marine Assistant " " " Commanders, Royal Navy Commanders, Royal Indian Marine Staff Commanders and officers of corresponding rank, Royal Navy. Inspectors of Machinery, Royal Indian Marine Chief Engineers, Royal Indian Marine	But junior to all Lieutenant-Colonels.
Majors	Lieutenants of 8 years' seniority and officers of corresponding rank, Royal Navy. Lieutenants, Royal Indian Marine, over 8 years' seniority. Engineers, Royal Indian Marine, of and over 9 years' seniority.	But junior to all Majors.
Captains	Lieutenants under 8 years' and officers of corresponding rank, Royal Navy. Lieutenants, Royal Indian Marine, under 8 years' seniority. Engineers, Royal Indian Marine, under 9 years' seniority.	But junior to all Army Captains.
Lieutenants	Sub-Lieutenants and officers of corresponding rank, Royal Navy. Sub-Lieutenants, Royal Indian Marine Assistant Engineers, Royal Indian Marine	But junior to all Lieutenants.

§" NOTE.—Officers of the Royal Naval Reserve rank with, but after, officers of the Royal Navy and Royal Indian Marine of their corresponding rank."

† Home Department Notifications Nos. 227, dated the 27th January 1900, and 1236, dated the 26th May 1911.
§ Home Department Notification No. 1014, dated the 9th July 1915.

Reported between the 1st October and 31st December 1918.

Names.	Rank.	Corps or Department.	When.	Remarks.
Deaths.				
Adamson, John Thomas Graves.	Maj.	I. A.	25 Oct. 18	
Allen, James	Asst. Commy. & Hony. Lt.	I. Misc. List	22 Oct. 18	
Allen, W.	2nd-Lt.	6th Hamps. R.	12 Oct. 18	
Alves, John William Jerome	Maj.	I. A.	29 Oct. 18	
Ashton, Edward Kitton Venning.	Lt.	I. A.	4 Dec.	
Babington, Thomas Zachary Dodson.	Lt.	I. A. R. O.	16 Oct.	
Bailey, *Hon.* J. L.	Lt.	1st Brecknock Bn. S. W. Bord.	26 Oct	
Baines, Hubert	Lt.	75th Infy.	22 O	
Baker, Richard Stanley	2nd-Lt.	I. A. R. O. & 34th U. Burma Bn. I. D. F.	13	
Bakewell, John	Sub-Condr.	I. A. Deptts. (retd.)		
Batcheler, Edward Beevor	Condr.	I. A. Deptts. (retd.)		
Battman, James	Dy. Commy. & Hony. Capt.	P. W. D. (retd		
Bedell, Frederick Francis	Senr-Asst. Surgn. Hony. Maj.	I. M. D.		
Bedell, Herbert John	Asst. Surgn.	I. M. D.		
Beveridge, Douglas Lionel	2nd-Lt.	I. A. R. O.		
Bhargawa, R S.	Tempy. Lt.	I. M. S.		
Blenkinsop, Frederick Henry	Lt-Col.	I. M. S. (		
Blois Johnson, Thomas Gordon, C.M.G., C.I.E.	Lt.-Col.	I. A.		
Bonsey, Francis Henry William.	2nd-Lt.	I. A. R		
Borland, Sydney Douglas	2nd-Lt.	I. A.		
Brewer, John Giblett	Lt.	I. A. R.		
Brock, George Selby, M.B.	Capt.	I. M. S.		
Brown, Charles King Valentine.	Maj.	102nd Grs.		
Brown, Edward James	2nd-Lt.	I. A. R. O.	14 Nov. 18	
Brown, Eric Russell Wilkie	2nd-Lt.	I. A.	19 Nov. 18	
Brown, George	Condr.	S. & T. Corps	25 Oct. 18	Abbotta
Brownson, R. D. M.B.	Capt.	R. A. M. C. (Spl. Res.)	21 Oct. 18	Peshawar.
Butt, Joseph Werner	Asst. Surgn.	I. M. D.	26 Oct. 18	Deolali.
Byrne, John Cahill	Lt.	I. A. R. O.	11 Oct. 18	Madras.
Cadogan, Thomas James	Sub-Condr.	S. & T. Corps	22 Oct. 18	Rawalpindi.
Campbell, Frederick William	2nd-Lt.	I. A. R. O.	11 Dec. 18	Chausan Kach.
Cannell, Hugh Featherstone Cameron.	Lt.	I. A.	31 Oct. 18	Died of wounds received in action.
Carbonaro, Ernest	Maj.	I. A. (retd).	30 Dec 17	......
Carpendale, Maxwell Montague.	Maj.	36th Horse	14 Oct. 18	......
Carroll, George Thomas	Sr. Asst.-Surgn & Hony. Capt.	I. M. D. (retd.)	7 Sep. 18	......
Chapman, Ernest Herbert Stuart.	Capt.	31st Lrs.	1 Nov. 18	Shiraz.
Chavasse, Pere	2nd-Lt.	I. A. R. O.	8 Oct. 18	Killed in action.
Christian, Aviet Thadeus	Lt.	I. A. R. O.	6 Nov. 18	......
Clelland Robert Marshall	2nd-Lt.	I. A. R. O.	22 Dec. 18	......
Cockburn, R. C. R.	Lt.	Wilts. R. I. A. (*on probn.*)	25 Nov. 18	Pachmarhi.
Cogan, Lionel Gatrell	Lt.	153rd Infy.	7 Oct. 18	Ahmednagar.
Colborne, Reginald Elgar	2nd-Lt.	I. A. R. O.	6 Nov. 18	Subathu.
Collins, Arthur Michael Austen.	Lt.	38th Dogras	28 Oct. 18	Fatehgarh.
Connell, William Patrick	2nd-Lt.	R. A.	24 Nov. 18	Accidentally killed, Mesopotamia.
Conroy, H. T.	Lt.	I. A. R. O.	28 Oct. 18	......
Cracroft-Wilson, C. W.	Lt.	4th S. Wales Bord.	12 Nov. 18	Poona.
Crocker, P. J. W.	2nd-Lt.	R. W. Kent R.	16 Nov. 18	......
Crow, H. P.	Capt.	R. A. M. C. (Spl. Res.)	9 Nov. 18	Jhansi.

Reported between the 1st October and 31st December 1918—contd.

Names. | Rank. | Corps or Department. | When. | Remarks.

Deaths—*contd.*

Names: **King,** Henry Alfred; **Kitchener,** Cecil; **Law,** John, M.D.; **Lawrie,** F. A.; **Lay,** John Frederick; **Learoyd,** Geoffrey Douglas.; **Lee,** George Thomas Hutteman Ernest; **Lewis,** G. K.; **Lunn,** G. A.; **Lynch,** Michael John; **Lynch-Staunton,** Reginald Kirkpatrick, D.S.O.; **McCurrich,** Laurence Oliphant.; **McIntyre,** R. J.; **McLean,** Richardson Pelman Archibald.; **Macdonald,** Donald; **Mander,** Thomas James; **Mein,** Dudley Gerald; **Melville,** Harry George; **Meredith,** Maurice Neville; **Miller,** Clement Francis Patrick.; **Mills,** James; **Mohan,** N. C.; **Morris,** Wilfrid Cyril; **Mosse,** William Oliver Matless.; **Nevill,** Edward James; **Newman,** Reuben McCarthy; **Northey,** M. A.; **O'Brien,** Thomas A.; **Owens,** John; **Parkinson,** R. F., M.C.; **Peacock,** Frank; **Peddar,** Digby Cotes; **Philipps,** Percy Charles; **Philson,** Samuel Cowell; **Pinfold,** William Herbert; **Pitts,** William Henry; **St. Planck,** Charles; **Ponter,** W. C.; **Price,** W. H.; **Prothero,** Rowland John; **Quigley,** William Grainger; **Rao,** M. S. R. K., M.B.; **Raymond,** Henry Gerrard; **Reed,** Wynne Dudley; **Reeve,** W. J.; **Richards,** William Earnest; **Riddoch,** J. K.; **Roche-Kelly,** James.

Rank: 2nd-Lt.; 2nd-Lt.; Bde. Surgn.; Capt.; Lt.; Capt.; 2nd-Lt.; 2nd-Lt.; Lt.; 2nd-Lt.; Bt.-Lt.-Col.; 2nd-Lt. (*Temp. Capt.*); Temp. 2nd-Lt.; Lt.; Col.; 2nd-Lt.; Lt.-Col.; Lt.-Col.; 2nd-Lt.; Sub-Condr.; Temp. Lt.; 2nd-Lt.; Col.; Lt.; Lt.; Lt.; Temp. Lt.; Condr.; Capt.; 2nd-Lt.; Hony. Lt. Col.; Sub-Condr.; Col.; 2nd-Lt.; Dy. Commy. & Hony. Capt.; Surgn. Genl.; Lt. M. & Hony. Qr. Mr. & Hony. Lieut.; Lt.; 2nd-Lt.; Temp. Capt.; Lt.; Lt.; Condr.; Lt. (*Actg. Capt.*); 2nd-Lt.

Corps or Department: I. A. (T. C.); R. G. A.; I. M. S. (retd.); Sco. Rif.; I. A. R. O.; 31st Lrs.; I. A. R. O.; 13th High. L. I.; 2-6th R. Suss. R.; I. A. R. O.; R. A.; I. A. R. O.; Attd. 1st High. L. I.; I. A. R. O.; I. A. (U. S. L.); I. A. R. O.; I. M. S.; I. A. R. O.; I. A. R. O.; I. A. Deptts. (retd.); I. M. S.; I. A. (T. C.); I. A. (retd.); I. A. R. O.; Otago R., N. Z. Forces, I. A. R. O. (*on probn.*); 3rd R. W. Kent R.; A. V. Corps; I. A. Deptts. (retd); R. E. (S. R.); I. A. (T. C.); I. A. (retd.); I. A. Deptts.; A. M. S. (retd.); I. A. R. O.; P. W. D. (retd.); I. M. S. (retd.); 23rd Rif. Brig.; Genl. List attached 6th Bangalore; Coorg & Mysore Bn. I. D. F.; 7th Hrs.; I. A. (T. C.); I. M. S.; I. A. (*on probn.*); 6th Devon. R.; I. O. D.; 1-6th Hamps. R.; 2-32nd M. & S. M.; Ry. Rif., I. D. F.

When: 4 Nov. 18; 30 Oct. 18; 6 Feb. 18; 25 Sep. 18; 28 Oct. 18; 29 Oct. 18; 26 Oct. 18; 1 Nov. 18; 15 Oct. 18; 7 Nov. 18; 25 Oct. 18; 25-26 Oct. 18; 19 Nov. 18; 27 Mar. 18; 9 Nov. 18; 26 Oct. 18; 7 Dec. 18; 26 Oct. 18; 4 Oct. 18; 17 Sep. 17; 10 Oct. 18; 15 Dec. 18; 10 Oct. 18; 22 Dec. 18; 23 Nov. 18; 29 Oct. 18; 6 Oct. 18; 11 May 18; 7 Nov. 18; 3 Dec. 18; 8 Sep. 18; 14 June 18; 4 Nov. 18; 9 Oct. 18; 8 May 18; 23 Aug. 18; 22 Nov. 18; 16 Oct. 18; 8 Nov. 18; 2 Nov. 18; 6 Nov. 18; 26 Oct. 18; 29 Oct. 18; 29 Nov. 18; 14 Sep. 18; 17 Oct. 18; 21 Nov. 18.

Remarks: Rawalpindi. ......; Bangalore. ......; Rawalpindi.; Risalpur.; Lucknow.; Died of wounds received in action.; Lahore.; Quetta.; Died of wounds received in action.; Jalapahar.; Killed in action.; ......; Mesopotamia. ......; Killed in action.; Mesopotamia.; Killed in action.; Poona.; ......; Bombay.; Ahmednagar. ......; Bareilly.; Abbottabad.; Killed in action.; Ajmere.; ......; Mesopotamia.; Cawnpore. ......; Chakrata.; Bombay. ......; Bareilly.; Bangalore.; Died of wounds received in action.; Subathu.; Aden.; Killed in action.; Killed in action.; Cawnpore.; ......; Madras.

Reported between the 1st October and 31st December 1918—contd.

Names.	Rank.	Corps or Department.	When.	Remarks.
		Deaths—*contd.*		
Rogers, Leslie Alexander Gerrard.	Capt.	20th Nilgiri Malabar Bn., I. D. F.	10 Oct. 18	Coonor.
Rogers, Thomas Jerome	2nd-Lt.	I. A. (*T. C.*)	30 Oct. 18	Rawalpindi.
Rolfe, John Shirley	2nd-Lt.	I. A. R. O.	4 Oct. 18	Ambala.
Rutherfoord, Thomas Corrie, M.D.	Maj.	I. M. S.	18 Oct. 18	......
Sanderson, John Barnes	*Hony.* Lieut.	P. W. Dept. (retd.)	2 Sep. 17	W. Australia.
Savage, Wilfred Clyde Richmond.	Maj.	I. A. (*retd.*)	5 May 18	......
Scarr, Geoffrey Campbell	Lt.	I. A. R. O.	18 Nov. 18	Killed. Aeroplane accident—England.
Schofield, A. T.	Capt.	1st Kent Bn.	10 Nov. 18	Died of wounds received in action.
Sercombe, George Henry	Condr.	I. U. List, (Posts & Telegraphs).	25 Oct. 18	Ambala.
Shaw, Ernest Vernon James.	2nd-Lt.	I. A. R. O.	10 Nov. 18	Deolali.
Smith, Thomas.	Dy. Comsy. & Hony. Maj.	I. A. Deptts. (*retd.*)	4 Mar. 18	......
Smith, T.	2nd-Lt.	Notts. Yeo. attd. 7th Hrs.	23 Oct. 18	Killed in action.
Sparkes, Richard Abbot	Maj.	54th Sikhs	2 Nov. 18	Abbottabad.
Stephen, James Eliot	Lt.	19th Punj.	14 Oct. 18	Killed in action.
Stewart, Robert Locke	2nd-Lt.	I. A. (T. C.)	18 Oct. 18	Agra.
Stockley, Henry Watts	Col.	I. A. (retd.)	3 Oct. 18	......
Sutcliffe, Roland Pruett*	Lt.	I. A. (*on probn.*)	20 Oct. 18	Lahore Cantonment.
Swinhoe, Edmund Arthur	Maj.	S. & T. Corps	27 Oct. 18	Sialkot.
Sweet, Roy Thornhill, D.S.O.	Capt.	7th G. R.	....	Yazgad. Date not reported.
Swynnerton, Frederick	Lt.	I. A. R. O.	18 Dec. 18	Bombay.
Synnott, Walter Pierre	2nd-Lt.	M. G. Corps	11 Oct. 18	Mhow.
Tanner, Charles Cyril Pentin	Capt.	R. A.	5 Oct. 18	......
Tanner, John Arthur, C.P., D.S.O.	Col.	R. E. (retd.)	23 July 17	......
Tate, James Samuel	Condr.	I. O. D. (retd.)	26 Oct. 18	......
Taylor, John Frederick Oddin, B.A.	Revd.	Eccl. Est., Bengal.	....	......
Thomas, J. V.	Lt.	Attd. 3rd S. & M.	3 Dec. 18	At Sea.
Thomson, Francis Wyville, M.B.	Lt.-Col.	I. M. S. (retd.).	27 May 18	......
Thoyts, Newman Burfoot	Col.	I. A. (retd.)	20 Sep. 18	......
Tonge, H.	Lt.	14th Dur. L. I. (attd. 2nd G. B., Northld. Fusrs.)	25 Oct. 18	Lahore Cantonment.
Torkildson, Wilford	Sub-Condr.	S. & T. Corps	6 Nov. 18	Nowshera.
Townley, Bertram Russell	Lt.	I. A. R. O.	7 Nov. 18	Shiraz.
Trueman, William Ernest	2nd-Lt.	I. A. R. O.	30 Oct. 18	Subathu.
Tudor, P. B.	2nd-Lt.	R. F. A. (Spl. Res.)	1 Nov. 18	......
Turner, Duncan	2nd-Lt.	I. A. (*T. C.*)	16 Oct. 18	Trichinopoly.
Turner, W. M.	2nd-Lt.	Gen. List.	26 Oct. 18	England.
Vallings, Adolphus	Col.	I. A. (U. S. L.)	19 Apl. 18	......
Vickers, Hugh Gordon Muschamp.	Lt.	I. A. R. O.	30 Oct. 18	Killed in action.
Ward, Charles Albert John	Lt.	I. A. R. O.	15 Dec. 18	......
Warner, Clive Wynyard	2nd-Lt.	I. A. attd. 3rd Horse.	14 Oct. 18	Quetta.
Watkin, H.	Tempy. Lt.	W. Rid. R.	21 Oct. 18	Lahore Cantonment.
Weekes, S. C.	2nd-Lt.	I. A. R. O. (*on probn.*)	28 Nov. 18	Kasauli.
Weston Ronald John	Lt. (*Ty. Maj.*)	I. A. R. O.	27 Oct. 18	Rawalpindi.
Wheler, Henry Isham	Col.	I. A. (retd.)	28 July 18	......
Whitaker, Gilbert	Capt.	34th U. Bur. Bn., I. D. F.	18 Dec. 18	Namtu.
Wilkinson, George	Commy. & Hony. Capt.	I. O. D. (retd.)	28 Aug. 18	......
Williams, Robert Lukyn	Lt.	I. A. (attd. 23rd Cavy.)	27 Oct. 18	Killed in action.
Willis, Charles Fancourt, C.B.	Col.	I. M. S. (retd.)	28 Apl. 18	England.
Wilmot, Sacheverel Darwin	Capt.	R. A.	14 Oct. 18	Karachi.

Reported between the 1st October and 31st December 1918—concld.

Names.	Rank.	Corps or Department.	When.	Remarks.
Deaths—*concld.*				
Wilson, E. C.	Temp. Lt.	R. W. Kent, R.	29 Oct. 18	Killed in action.
Wilson, John Dawson, M.C.	Lt.	I. A. (*on probn.*)	13 Oct. 18	......
Wilson, John Newbald	Maj.-Genl.	I. A. (retd).	1 May 18	Tasmania.
Wood, Julian	Condr.	I. A. Deptts. (retd.)	4 Feb. 18	......
Wynne, Erick Ralph Lovatt	Lt.	I. A.	26 Oct. 18	Killed in action.
Yeld, Cyril de Clare	2nd-Lt.	I. A. R. O.	24 Oct. 18	Quetta.
Yule, George Udny, D.S.O.	Maj. (*temp. Lt.-Col.*)	R. E.	22 Dec. 18	Bombay.
The following casualty is believed to have occurred.				
Jones, Lewis Jeremy	Maj.	9th Infy.	28-29 Oct. 14	Death on or since 28-29 Oct. 1914 accepted for official purposes.
Resignations.				
Aspandiar Rustomji, Irani	Temp. Hony. 2nd-Lt.	Ind. Land Forces.	24 Sep. 18	......
Boultbee-Whall, George Henry Brooke.	Lt.	I. A. R. O.	1 Nov. 18	On transfer to British Service.
Clutterbuck, Peter Henry, C.I.E., V.D.	Lt.-Col.	8th U. P. Horse, (Supy. List). I. D. F.	30 Sep. 18	......
Datta, Manindra Chandra	Tem. Lt.	I. M. S.	23 Nov. 18	......
James, Alfred Ernest	2nd-Lt.	I. A.	15 Dec. 18	......
Lawson-Brown, Eleanor	Nursing Sister	Q. A. M. N. S. I.	...	......
Lockie, Hugh Reid	Capt.	I. A. R. O.	31 Mar. 17	......
Murphy, Terence St. Kevin	2nd-Lt.	I. A. R. O.	2 Oct. 18	......
Murray, Alexander Radcliffe Hepburn.	2nd-Lt.	I. A. R. O.	15 Aug. 18	......
Noronha, F. H.	Temp. Capt.	I. M. S.	Mar. 18	......
Norton, Harry Baker	Lt. (*Temp. Capt.*)	I. A. R. O.	30 Sep. 18	......
Pearce, Joseph Chaning	2nd-Lt.	20th Nilgiri Malabar Bn., I. D. F.	22 Aug. 18	......
Saunders, Claude Douglas	Lt.	I. A. R. O.	22 Oct. 18	Reverting to Br. Ser.
Stoddart, George Frederick, V.D.	Lt.-Col.	9th N. Bengal Mtd. Rif., I. D. F.	16 Oct. 18	......
Tyler-Cove, V. R.	Nursing Sister	Q. A. M. N. S. I.	1 Oct. 18	......
Temporary Half Pay List.				
Carleton, Henry Anthony, D.S.O.	Lt.-Col.	90th Punj.	12 Sep. 18	......
Macpherson, Roderick George.	Lt.-Col.	41st Dogras	12 Sep. 18	......
Retirements.				
Bell, Richard Carmichael, D.S.O.	Lt.-Col.	38th Horse	21 June 18	......
Blomfield, Patrick Valentine.	Capt.	2nd Lrs.	11 Nov. 18	......
Collins, William James	Dy. Commy. & *Hony.* Capt.	P. W. D.	4 June 18	......
Cresswell, John	Condr.	I. Misc. List	29 Oct. 18	......
De La Motte, Hubert Thomas, M.C.	Capt.	29th Punj.	17 Aug. 18	......
Down, Frederick Edward	Lt. (*Temp. Capt.*)	I. A. R. O.	3 Sep. 18	......
Gibbons, James Henry	Dy. Commy. & *Hony.* Capt.	M. W. S.	11 Oct. 18	......
Griffiths, Charles Robert Williamson.	Lt.	I. A. R. O.	28 Sep. 18	......

Retirements—contd.

Names.	Rank.	Corps or Department.	When.	Remarks.
Gurdon, Bertrand Evelyn Mellish, C.I.E., D.S.O.	Lt.-Col.	I. A. (Supy. List.	1 Dec. 18	......
Harrop, Edward Arthur	2nd-Lt.	I. A. (*T. C.*)	21 Aug. 18	......
Hodges, Robert William	Maj.	I. A. R. O.	9 May. 17	......
Irvine, Gerard Beatty, C.B.	Lt.-Col.	I. M. S.	12 Oct. 18	......
Jenney, George William, M.B.	Lt.-Col.	I. M. S.	4 Oct. 18	......
Kerr, Howard Low	Lt.	I. A. R. O.	24 Oct. 18	......
Lalor, Nicholas Purcell O'Gorman, M.B.	Lt.-Col.	I. M. S.	25 July 18	......
McLaurin, George Charter	Asst. Commy. & Hony. Lt.	I. O. D.	12 Sep. 18	......
Magnus, Basil James	Lt.	I. A. R. O.	12 Oct. 18	......
Morwood, James, M.D.	Lt.-Col.	I. M. S.	11 Nov. 18	......
Nicholson, Thomas Maccallum.	Maj.	S. & T. Corps	19 Sep. 18	......
Norman, F. J.	Ty. 2nd-Lt.	Gen. List	2 July 18	......
Nutting, Harold Bligh	Maj.	5th Cavy.	11 Sep. 18	......
Parker, George Marcus Godfrey.	Maj.	S. & T. Corps	13 Nov. 18	......
Parker, Neville Thornton	Lt.-Col.	I. A. (Supy. List).	29 Nov. 18	......
Pulleyn, George Herbert, M.C.	Commy. Hony. Maj.	S. & T. Corps	28 Oct. 18	......
Smith, Frederick Vincent	Lt.-Col.	4th Cav.	11 Sep. 18	......
Smyth, John, M.D.	Col.	I. M. S.	31 Oct. 18	......
Stooks, Charles Sumner, D.S.O	Maj.	5th L. Infty.	5 July 18	......
Thatcher, William Sutherland, M.C.	Lt.	I. A. R. O.	1 Sep. 18	......
Townsend, *the Hon'ble* Edward Copleston.	Lt.-Col.	I. A. (Supy. List).	26 Sep. 18	......
Vickers, Hilton	Lt.-Col.	I. A.	1 Nov. 18	......

Dismissals, Removals and Discharges.

Names.	Rank.	Corps or Department.	When.	Remarks.
Buckley, Henry Albert	2nd-Lt.	I. A. (*T. C.*)	23 Sep. 18	Dismissed the service.
Budd, Robert Whittaker	2nd-Lt.	A. S.C (M.T.C.)	...	Dismissed the service.
Gosling, John Evans	Lt.	I. A. R. O.	9 Oct. 18	Services dispensed with.
Hosking, Frederick Mervyn	Lt.	I. A. R. O.	20 Dec. 18	Services dispensed with.
Hughes, George Frederick	Lt.	I. A. R. G.	15 Nov. 18	Services dispensed with.
Jackson, John Gildart	Capt.	1st G. B., Linc. R.	...	Dismissed the service.
Jones, Albert	2nd-Lt.	I. A. (*T. C.*)	20 Sep. 18	Dismissed the service.
Pereira, Joseph Augustine	Tempy. Lt.	I. M. S.	12 Oct. 18	Services dispensed with.
Proctor, Norman Calderbank.	Lt.	I. A. R. O.	12 Oct. 18	Dismissed the service.
Scott, Walter	Lt.	I. A. R O.	15 July 18	Cashiered.
Sirkar, N. K.	Temp. Lt.	I. M. S.	19 Sep. 18	......

MEMORANDA.

Retirements :—The date of retirement of Lt.-Col. A. R. S. Anderson, I. M. S., should read "11th September 1918" and not "1st August 1918," as notified on page 1928 of the Indian Army List for October 1918.

The date of retirement of Col. R. Robertson, M.B., I. M. S., should read "4th July 1918" and not "30th June 1918," as notified on page 1928 of the Indian Army List for October 1918.

www.ingramcontent.com/pod-product-compliance
Ingram Content Group UK Ltd.
Pitfield, Milton Keynes, MK11 3LW, UK
UKHW010157180926
475500UK00003B/66

9 781843 422952

www.ingramcontent.com/pod-product-compliance
Ingram Content Group UK Ltd.
Pitfield, Milton Keynes, MK11 3LW, UK
UKHW010157180926
475500UK00003B/65

9 781843 422945